Native American Ethnobotany

Native American Ethnobotany

Daniel E. Moerman

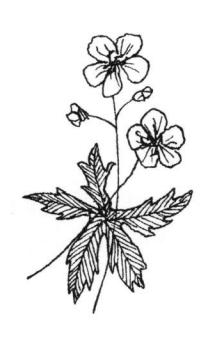

Timber Press
Portland, Oregon

*For Claudine, Jennifer, and Allison—
three wonderful women from three
generations—with love*

Copyright © 1998 by Timber Press, Inc.
All rights reserved.

Timber Press, Inc.
The Haseltine Building
133 S.W. Second Avenue, Suite 450
Portland, Oregon 97204, U.S.A.

Printed in Hong Kong

Library of Congress Cataloging-in-Publication Data

Moerman, Daniel E.
    Native American ethnobotany / Daniel E. Moerman.
        p. cm.
    Includes bibliographical references and indexes.
    ISBN 0-88192-453-9
       1. Indians of North America—Ethnobotany. 2. Ethnobotany—
North America. I. Title.
E98.B7M66 1998
581.6'097—dc21   97-32877
                                                                                CIP

Contents

Preface 7
Acknowledgments 9

Plant Use by Native Americans 11
    Plants Used as Drugs 12
    Plants Used as Foods 15
    Plants Used as Fibers and Dyes 16
    Plants with Other Uses 17
    Conclusions on Usages 18
    Sources of Information on Plant Usages 18
    Plant Usage Categories 19
    Native Americans 23

Organization of the Information in *Native American Ethnobotany* 29
    Scientific Plant Names 29
    Common Plant Names 32
    Ethnobotanical Information 32

Catalog of Plants 33

Notes 615

Bibliography 619

Plant Usage Indexes
    Index of Tribes 625
    Index of Usages 765

Plant Name Indexes
    Index of Synonyms 873
    Index of Common Names 891

Preface

In 1970, I was in the Sea Islands of South Carolina doing anthropological fieldwork for my Ph.D. dissertation in a rural black community. I was surprised to learn that sometimes when people were ill, they gathered various wild plants as medicines. Although I had not intended to do so, I began to inquire into the matter. In time, I had identified about 35 species of plants, from *Allium* to *Zanthoxylum*, which were so used. I then asked what seemed at the time to be a few simple questions: How had these people learned of these plants? Did it do any good to take them? Had anyone else ever used them, and if so, for what? *Native American Ethnobotany* is the most complete of my several efforts that have resulted from asking those naive questions. The book is intended to help anyone interested in the use of plants understand which ones have been used by native North American peoples to treat disease, for food, fiber, and dye, and for many other purposes as well.

As I began this work, I found I was in the company of some of the great American students of ethnobotany of the past—Huron H. Smith, Francis Densmore, Melvin Randolf Gilmore, Matilda Coxe Stevenson, Gladys Tantaquidgeon, and many others. Their fascinating works, too often overlooked, were found with the help of reference librarians in the recesses of one library or another across the continent. Most of the works have long been out of print or were published in specialist journals and were very hard to find.

Once found, many of the works were difficult to use. Most were not indexed, or at least not very well indexed. To answer a seemingly simple question such as, Who used Allium and for what?, took hours of digging. I decided to try to coordinate the information. I bought a few hundred index cards, the kind with sorting holes around the edges, and started filling them in. Soon I had more cards than would fit on the sorting pin. I had to start new boxes of subcategories to store the cards. Once I dropped a stack of cards on the floor. It was very discouraging.

One day in 1974 I had a chance conversation with Ralph Nichols, then the registrar at the University of Michigan—Dearborn. Among other things he was responsible for putting out the schedule of classes each term—course number, name, department, room, time, professor, and so on. He was experimenting with some new system that worked on a computer; it was used for manipulating what he called data banks. I told him what I was doing and he said I might want to give his system a try. I had played around with a computer a few years earlier and had thought it kind of fun, so I decided I would look into it. Soon I was working with a new kind of card, rectangular ones with a notch on one corner! I transferred the material from one kind of card to the other and started feeding them into the hopper at the computing center. The material was then read into a database using the TAXIR (Taxonomic Information Retrieval) system available on the university computer.

In 1975, with a summer stipend from the National Endowment for the Humanities, I filled out thousands of code sheets with material from the works by Smith, Tantaquidgeon, and others. With funds provided by the university, I hired keypunch operators to produce the thousands of cards needed for computing. The first version of the database was published in 1977 as *American Medical Ethnobotany: A Reference Dictionary* (New York: Garland Publishing). That book included 1288 plant species from 531 genera from 118 families used in 4869 different ways in 48 different societies. The book was produced from line printer copy, all in upper case. I typed the page numbers on little labels and stuck them onto the bottoms of the pages by hand, one at a time. It was crude, but it had advantages. There was no index, but each of the 4869 items appeared in the book four times—one list organized by genus, one by family, one by use, and one by tribe.

Even though that book had more than 500 pages, it was far from a comprehensive list of Native American medicinal plant use. I had not had time nor resources to code several other available sources (for example, Francis Elmore on the Navaho, Erna Gunther on the Northwest Coast peoples). And at the time there were no sources available on two major groups, the Cherokee and Iroquois. With the help of several reference librarians, I continued to add to my collection of ethnobotanical works, published and unpublished. Paul Hamel and Mary Chiltoskey's *Cherokee Plants and Their Uses* and James Herrick's monumental *Iroquois Medical Botany* appeared later in the 1970s (those two sources alone had more items in them, 4875, than all the rest of my original database).

And the database technology was changing. Personal computers began to show up on desks, supplementing mainframes, and the weakest link in mainframe computer technology—printing—was revolutionized by the introduction of laser printers. In 1984 the National Endowment for the Humanities again agreed to support my project with a grant. I established a laboratory at the University of Michigan—Dearborn, and students were hired to do the coding. Within 2 years a new database was constructed on the foundation of the original one. It was published in 1986 as *Medicinal Plants of Native America* (University of Michigan Museum of Anthropology, Technical Reports, Number 19), in two volumes with 910 pages, and included 2147 plant species from 760 genera from 142 families used in 17,634 different ways in 123 different societies. The listings were much more readable than in the earlier version because laser printing allowed the use of a variety of fonts.

As I continued to work, however, I realized that my database was incomplete in another way. Native American peoples used many species of plants as medicines but they used many plants for other purposes as well, in particular, for food. Other scholars had begun to recognize that medicinal plants were not completely distinct from food plants—indeed, there was often a substantial overlap in species that were used for both food and medicine. So I decided to broaden my net.

I applied to the National Science Foundation for funds to cover the costs of building a database of American Indian food plants to supplement the existing medicinal plant collection. As I started accumulating new food plant material, I checked again all the original material coded in *Medicinal Plants of Native America*. In many of them I found detailed discussions of food plants that we had passed by in earlier projects. But in addition to the food plants there was ethnobotanical information on dyes, fibers, and many other uses. The notion of going back over the same material for a third or fourth time seemed silly so I decided to code everything this time. The result is this book, which includes 4029 kinds of plants (species, subspecies, varieties, and so on) from 1200 genera from 242 families used in 44,691 different ways in 291 different societies. Of the 44,691 usages, 24,945 are medicinal, 11,078 are as food, 2567 as fibers, 607 as dyes, and 5494 as other kinds.

On a more personal level, my wife Claudine Farrand is a true partner in life. She provided context for this work by making life fun. My daughter Jennifer Moerman did the illustrations seen throughout the book. She is a wonderful artist and a great friend.

But our deepest debt is to those predecessors of ours on the North American continent who, through glacial cold in a world populated by mammoths and saber-toothed tigers, seriously, deliberately, and thoughtfully studied the flora of a new world, learned its secrets, and encouraged the next generations to study closer and to learn more. Their diligence and energy, their insight and creativity, these are the marks of true scientists, dedicated to gaining meaningful and useful knowledge from a complex and confusing world. That I cannot list them individually by name in no way diminishes my sense of obligation to them.

Acknowledgments

Any long and complicated project such as the work that has gone into *Native American Ethnobotany* is due to the contributions of many scholars, specialists, organizations, and friends. The National Endowment for the Humanities twice provided support. An NEH summer stipend in 1977 was a small investment that had a great importance in encouraging me to continue; it was instrumental in the production of the first published version of my database. A much more substantial Research Tools Award (NEH-RT-20408-04) in 1984 from the Endowment led to the second, larger version of the work. Subsequently, two grants from the National Science Foundation materially advanced the work. In 1987, the NSF supported (BNS-8704103) a substantial analysis of the previous database. In 1993, the Foundation provided major funding (SBR-9200674) to add all of the food plants to the database and to support a complete analysis of the flora utilized by Native American peoples. Regardless of will or wish, this project could never have been even imagined without that support.

The University of Michigan—Dearborn, my professional home for more than two decades, has supported the work in untold ways, large and small, material and personal. Various chairs and deans have been especially helpful at one time or another, among them Jim Foster, Don Levin, John Presley, Robert Simpson, Rick Straub, and Victor Wong. Several sponsored-research administrators at the university's Dearborn and Ann Arbor campuses have helped with the arcana of grant management, including Drew Buchanan, Paul Cunningham, Jim Gruber, Lee Katz, and Marty Tobin. Many of my University of Michigan colleagues have provided support, among them Katie Anderson-Levitt, Barry Bogin, Dick Ford, and Larry Radine.

Professional colleagues from around the country provided a broad range of help. John Kartesz of the University of North Carolina Herbarium provided a computerized version of his authoritative listing of North American plant synonyms; it is unclear how the project could have been completed without it. John also identified a number of odd plant names outside the range even of his comprehensive listing. Stanwyn Shetler, formerly of the Smithsonian Institution, provided a computerized version of the preliminary checklist for *Flora North America* that was invaluable in my understanding the dimensions of the North American flora. Tony Resnicek of the University of Michigan Herbarium enlisted the aid of several of his colleagues (Howard Crum, Bob Fogel, and Mike Wynne) to help with the identification of a number of nonvascular plants. Other botanists or ethnobotanists who have helped in one way or another over the years include Jim Duke and Steve Beckstrom-Sternberg of the U.S. Department of Agriculture Agricultural Research Service, Scott Peterson of the U.S. Department of Agriculture National Plant Data Collection Center, Nina Etkin and Paul Ross of the University of Hawaii, Jim Menhart of Texas A&M University, Sara Gage of the University of Washington, Brian Compton of the University of British Columbia, Jim McCormick of Mt. Royal College, Calgary, and Jim Herrick of Chittenango, New York.

This project could not have been carried out without the use of computers. A number of individuals have been particularly helpful. Bob Brill, inventor of TAXIR, one of the first database management systems, taught me the rudiments of programming. Bob Parnes provided persistent support for the notion that computers are devices to enhance communication. Betsy Alexander at Page Ahead Software and Joseph Lovett at Step2 Software provided much needed support with InfoPublisher. A dozen people at Fox Software and then at Microsoft helped with FoxPro and later with Windows.

Librarians, too, were essential. Shirley Smith and then Margaret Kruszewski, of the Mardigian Library at the University of Michigan—Dearborn, helped organize hundreds of interlibrary loan requests with patience and good humor.

A number of students helped with the coding over the years. This intensely tedious work requires a subtle blend of decisiveness and caution and is not easy. I deeply appreciate the contributions of all the people who have helped in the nitty-gritty of scientific work, among them Mary Johnson, Essam Khraizat, Sarita Kini, Grace Kazaleh, Karen Duhart, Maria Madias, and Shannon Becker. Essam and Sarita also helped with the analysis of the fiber and dye plants; we gratefully acknowledge the support of an NSF Research Experience for Undergraduates grant to support that work.

Special mention must be made of Ann Bruder. Ann worked on the coding for the work published in 1986 and was the research manager for this more recent project. She coded thousands of items and managed the work of all the other coders. Her clear sense of organization, her no-nonsense willingness to get some work done, and not least, her willingness to put up with my ambiguities and foibles, all with a genial good humor, allowed this project to be completed. I cannot imagine how it would have happened without her.

Anthropologists are often and rightly criticized for taking from the lives of other people and not giving very much back. In a small effort to address that criticism, I decided some time ago to attempt to find funds sufficient to purchase a copy of this book to give to each of the 1100 or so registered American and Canadian Indian tribes. With the help of the Development Office at the University of Michigan—Dearborn, and, in particular, the imaginative work of Susan Skramstad and Annette Ketner, we managed to raise a portion of the money for that purpose. Contributions came from both individuals and companies. We will give as many copies as we can, starting with those groups represented in these pages. We will happily accept further contributions from others and will continue to distribute books as long as we receive funds. I am deeply grateful to these contributors for their generous support of this effort: Cal M. Bewicke; Lou Johnson of the Trout Lake Farm Company; Katy Moran; William Thomson, Professor of Psychology at the University of Michigan—Dearborn; Ginger Webb; Capsugel Division of the Warner-Lambert Company; Flora, Inc.; Indena USA, Inc.; McZand Herbal, Inc.; MW International, Inc.; Nature's Herbs; and Shaman Pharmaceuticals, Inc.

Plant Use by Native Americans

Native American peoples had a remarkable amount of knowledge of the world in which they lived. In particular, they knew a great deal about plants. There are in North America 31,566 kinds (species, subspecies, varieties, and so on) of vascular plants: seed plants, including the flowering plants (angiosperms) and conifers (gymnosperms), and spore-bearing plants, including the ferns, club mosses, spike mosses, and horsetails (pteridophytes). North America is defined here as North America north of Mexico, and Hawaii and Greenland. American Indians used 2874 of these species as medicines, 1886 as foods, 230 as dyes, and 492 as fibers (for weaving, baskets, building materials, and so on). They used 1190 species for a broad range of other purposes as well, classified in this book as Other. All told, they found useful purpose for 3923 kinds of vascular plants. Native American Ethnobotany also contains information on 106 kinds of nonvascular plants (algae, fungi, lichens, liverworts, and mosses). The data for nonvascular plants are much less complete than those for vascular plants, however.

Native American Ethnobotany includes information on plant use by Native American people. Most of the plants used are native to North America, but some are not. Some are plants that were introduced into North America—some perhaps in pre-Columbian times and some certainly thereafter—and that became naturalized, growing spontaneously. Other plants are introductions that were kept in cultivation. The information in *Native American Ethnobotany* documents plant usage no doubt dating back to very early times and passed down through generations as traditional knowledge, as well as innovations in response to much more recent plant introductions.

Many species had multiple uses. Native American people used 55 different species for at least one purpose in each of the five categories: Drug, Food, Fiber, Dye, and Other. Table 1 lists the 10 champion plants with many uses in all categories, and Table 2 lists the 10 with the greatest total number of uses, regardless of category.

Table 1. The 10 plants with the greatest number of uses, and with uses in all five categories, by Native Americans

PLANT	DRUG	FOOD	FIBER	DYE	OTHER	TOTAL
Thuja plicata, western red cedar	52	6	188	1	121	368
Prunus virginiana, common chokecherry	132	163	4	2	36	337
Urtica dioica, stinging nettle	114	20	36	1	51	222
Yucca baccata, banana yucca	9	126	47	1	39	222
Cornus sericea, red osier dogwood	97	21	9	6	58	191
Heracleum maximum, common cowparsnip	112	57	2	1	17	189
Rhus trilobata, skunkbush sumac	38	69	29	11	34	181
Pseudotsuga menziesii, Douglas fir	67	18	18	1	72	176
Betula papyrifera, paper birch	28	9	59	3	76	175
Populus balsamifera, balsam poplar	103	16	18	1	35	173

Table 2. The 10 plants with the greatest number of uses, regardless of category, by Native Americans

PLANT	DRUG	FOOD	FIBER	DYE	OTHER	TOTAL
Thuja plicata, western red cedar	52	6	188	1	121	368
Achillea millefolium, common yarrow	355	3	0	1	7	366
Prunus virginiana, common chokecherry	132	163	4	2	36	337
Typha latifolia, broadleaf cattail	50	71	105	0	28	254
Acorus calamus, calamus	219	4	0	1	5	229
Urtica dioica, stinging nettle	114	20	36	1	51	222
Yucca baccata, banana yucca	9	126	47	1	39	222
Artemisia tridentata, big sagebrush	166	5	11	0	34	216
Amelanchier alnifolia, Saskatoon serviceberry	30	117	7	0	38	192
Cornus sericea, red osier dogwood	97	21	9	6	58	191

Plants Used as Drugs

There are 2582 species included in *Native American Ethnobotany* that were used medicinally by Native Americans. The listing of Plant Usage Categories toward the end of this chapter defines the many sorts of drug use that appear in the Catalog of Plants. The 10 plants with the greatest number of uses as drugs by Native Americans are

Achillea millefolium, common yarrow	355
Acorus calamus, calamus	219
Artemisia tridentata, big sagebrush	166
Lomatium dissectum, fernleaf biscuitroot	139
Prunus virginiana, common chokecherry	132
Artemisia ludoviciana, Louisiana sagewort	128
Oplopanax horridus, devil's club	128
Juniperus communis, common juniper	117
Mentha canadensis, Canadian mint	115
Urtica dioica, stinging nettle	114

The first thing people usually ask about American Indian medicinal plants is, Do they work? This it turns out is a tricky question. The short answer is, Yes. The longer answer is more interesting. What does it mean to say that a medicine "works"? Essentially it means that the medicine has the effect that we want it to have, that it meets our expectations. This means that a drug that meets one person's expectations may not meet another's, and people may therefore disagree over whether the drug works. Such disagreements usually hinge on different conceptions of health or healing. This is to say that definitions of health and well-being are often cultural matters; they are rarely simple matters of "fact."

Consider a case not from American Indian medicine but from the history of European medicine. In European tradition as recently as the early twentieth century (and very often in many other medical traditions as well) people understood that illness was in part the result of certain kinds of imbalance in the body's "humors." Certain humors (blood, bile, or phlegm, for example) accumulated at the expense of others and the healer's goal was to attempt to reestablish the proper balance. One way to achieve this was often to purge the sick individual, to cause him to vomit.

Consider also these similar cases. The Iroquois used *Lobelia inflata* to induce vomiting; they soaked the whole plant in cold water, then drank it. *Lobelia* has several English common names, among them emetic herb, vomit wort, and pukeweed. It contains a series of alkaloids but the one probably responsible for its major actions on the human body is lobeline. Besides causing emesis, some species of *Lobelia* have been used as an expectorant in cough syrups. So, does it work? Well, a physician in the European tradition and an Iroquois healer could probably agree that *Lobelia* is an effective emetic—the drug works. But why do the Iroquois want to induce vomiting? They use this emetic to cure a sufferer of "tobacco or whiskey addiction." It is at this point that the Euro-American and the Iroquois may find themselves at odds, the latter saying it does work, the former saying it does not work. Euro-American medical theory does not to my knowledge contain any theory allowing that vomiting will cure people of addiction to whiskey or cigarettes. Similarly, the Cherokee, another Native American group, are reported to use *L. inflata* as an emetic (now we have triple agreement) to cure asthma (and now a triple disagreement). Whereas the Euro-American, the Iroquois, and the Cherokee are likely to agree on one dimension of these treatments (emesis), they may disagree about others. Asking about the effectiveness of a drug, then, is not a simple biological or medical issue but also a complex problem of culture and meaning.

Issues of culture and meaning come in many different

forms. For example, in all the thousands of listings in this book one will only find 55 items categorized as Cancer Treatment, under Drug, in the Catalog of Plants. Yet cancer is the disease against which U.S. president Richard Nixon "declared war." Cancer is evidently a much more important disease in modern America than it was in native America. Why? The primary reason seems to be that Native American peoples did not suffer from cancer nearly as often as modern Americans. This may still be true although that is controversial. (For a fascinating review of the situation, see Thomas J. Csordas: The Sore That Does Not Heal: Cause and Concept in the Navajo Experience of Cancer. Journal of Anthropological Research 45:457–486. 1989.) It is less controversial for the historical and prehistoric past. Cancer is a relatively recent disease that is to some degree dependent on carcinogens, substances that "cause" or accelerate cancer. Most carcinogens are industrially manufactured artifacts like food colors, radioactive materials, and x-ray machines. An interesting exception is tobacco. Native Americans regularly used tobacco as a ceremonial smoke. A dozen or so people would occasionally share a bowl of tobacco in a pattern very different from the addicted smoking of Europeans and Euro-Americans (modern cigarettes are, of course, industrially manufactured products). The other significant carcinogen that is not an industrial product is sunlight. The differences in susceptibility to cancer in Native America and among modern Americans are generally matters of culture, and they seem to me to account for the general lack of cancer among Native American people, at least in the past, and therefore, the rarity of cancer treatments documented here.

In contrast to cancer treatments, one will find 582 treatments classified as Eye Medicine, under Drug, in the Catalog of Plants. I believe that the explanation for this is probably to be found in a similar cultural comparison: American Indians often lived in smoky houses. As the conditions that required treatment varied, so did the treatments available.

This cultural analysis does not mean that American Indian medicines are not valuable for modern conditions. Indians used *Podophyllum peltatum*, mayapple, for a broad range of things: as a cathartic, an insecticide, for rheumatism, and so on. Even though it has been so reported, it seems very unlikely that they used it as a cancer remedy. The plant has for a long time had an interesting role in Western medicine. A resin from the roots known as podophyllin is a common and moderately effective treatment for condylomata acuminata or venereal warts. Since some forms of venereal warts (caused by a virus) can be a precursor to cancer, this botanical remedy can be said to be a cancer cure. It is also the basis for the production of etoposide, a semisynthetic derivative of podophyllotoxin, a chemical found in mayapple. Etoposide is widely used in the treatment of several forms of cancer.

Another interesting example of cross-cultural drugs is taxol. Taxol is found in the leaves and bark of several species of yew, of the genus *Taxus*. Native Americans used three species of *Taxus* for a broad range of things: antirheumatic, cold remedy, lung medicine, and so on. The Tsimshian Indians of British Columbia have been reported to use the plant "for internal ailments and cancer." *Taxus* has not traditionally been an important Western medicine but, as a result of screening of plant material at the National Cancer Institute in Washington, D.C., researchers have shown that taxol is an interesting and potentially very useful cancer drug. It seems likely that this Native American drug, which was used by one tribe to treat cancer, may become an important modern cancer drug.

Of course, just because taxol "works" for ovarian cancer does not mean it "works" for rheumatism, colds, or for lungs, at least not in any very absolute sense. We must always consider the expectations and wishes of the people taking medicines. Why do we take medicines at all? Usually, we take medicines when we "do not feel well." The goal is to "feel better." For example, say I am tired, sore, have a low fever, a sore throat, and a runny nose. Occasionally, a chill runs down my spine. I have not slept well for 2 days. My wife goes into the kitchen and makes a big pot of fresh chicken soup. It is rich and flavorful; the whole house smells warm and sweet. I eat a small bowl of the soup and eat a few chunks of chicken. Along with it I eat a crusty piece of bread. I feel much better. An hour later I go to sleep and get some real rest. Are we to class chicken soup as an effective drug? Perhaps, but clearly this is different from, say, the matter of taxol and ovarian cancer. In the matter of chicken soup, much of the "effectiveness" of the treatment probably comes from the context within which the "cure" takes place. But the soup is an important part of that context and probably essential to it.

An interesting question is *why* plants have medicinal value. What does *Taxus brevifolia* care about cancer? In short the answer is, Nothing. But plants do produce a broad range of chemicals that serve a variety of purposes for the plants, some of which can be used by people to serve their own often different purposes. Phytochemists have classified plant chemicals (or phytochemicals) as either primary or secondary: primary chemicals are those involved in the basic biochemistry of plant life, particularly photosynthesis; secondary chemicals are all the rest. Drugs are generally found to belong to the class of secondary chemicals. Typically, in the past these secondary chemicals were thought not to be essential; they were considered to be the byproducts of primary processes, or to be just random. More recently, as scientists have taken a more ecological approach to plant chemistry, the view of the function of secondary chemicals has changed. Plants seem to produce many chemicals that are biologically active, in other organisms as well as in themselves, to enhance their own survival. They may produce herbicides to inhibit the growth of competing plants. For example, salicylic acid (a naturally occurring chemical from which aspirin is made) is a water-soluble phytotoxin (plant poison) that washes off the leaves of willows (*Salix*) and other plants to the ground below, inhibiting the growth of competing plants. Juglone, produced by black walnut trees (*Juglans nigra*), does the same thing.

Plants also produce substances that deter browsing by insects and other herbivores. A classic case involves the relationship between grasses and grazing mammals. Grasses rely on physical defenses (sharp leaves, high silica content) and vegetative reproduction rather than chemicals for defense: essentially, grasses grow back once they are browsed or mowed. More interesting for our present purpose, however, are plants that produce toxic or repellent chemicals. Some familiar botanical insecticides are nicotine from tobacco (*Nicotiana*) and the pyrethrin found in chrysanthemums *(Dendranthema, Leucanthemum,* and other genera formerly included in a more inclusive genus *Chrysanthemum).* Pyrethrin is the active ingredient in Raid and other common commercial insecticides. In certain cases, herbivores have adapted to these defenses. A well-known case is that of the monarch butterfly (*Danaus plexippus*). The monarch larvae feed on milkweeds (*Asclepias*), which are very toxic to other animals and insects. The larvae ingest and sequester various cardiac glycosides throughout their bodies, thereby deterring birds from feeding on them. Humans, too, can use the cardiac glycosides of milkweed when they want to cause vomiting for whatever reason (recall the discussion above about pukeweed).

Similarly, pines (*Pinus*) are generally protected against a variety of both insects and fungi by the pitch they exude from bark and leaves. Protection against insects is both physical (the sticky pitch drowns them) and chemical (the monoterpenes in pitch are toxic to many insects). Several kinds of bark beetles (*Dendroctonus* species, for example), however, can detoxify these substances and even use them as pheromones that serve to attract a sufficiently large number of beetles to overcome the physical resistance of the tree—the tree's chemical defense is turned against itself. Similarly, people have used pine pitch as a disinfectant and for a broad range of other purposes as well. Both people and other creatures can occasionally find their own uses for the secondary chemicals that plants produce.

Some cases of animal use of plant secondary chemicals are straightforward. For example, it is reasonable enough to use pyrethrin to kill insects; that is probably what the chrysanthemums make them for in the first place. Similarly, it seems reasonable to expect that substances produced by plants to protect themselves against worms that eat their roots might turn out to be useful for treating intestinal parasites. Other cases are less obvious. For example, several different plants—birches (*Betula*), willows (*Salix*), wintergreen (*Gaultheria*), *Spiraea*—produce various salicylates, among them salicin and methyl salicylate, which are precursors for aspirin. Aspirin, or acetylsalicylic acid, is a semisynthetic drug, a modification of the natural precursors. The natural products do more or less what aspirin does but aspirin has the great benefit of being much less toxic than salicin. The natural salicylates have the advantage of being readily absorbed through the skin and for that reason they are often used as the active ingredients in "sports creams" for stopping muscle pain. As noted above, the natural salicylates are probably herbicides that serve to reduce competition for growing space for the plants that produce them. But why should an herbicide stop headaches? This is a question without an obvious answer. In part it is because no one is quite sure how aspirin or the other salicylates work in the first place. They apparently inhibit the production of prostaglandins. Prostaglandins are important chemicals involved in the processes of temperature regulation, inflammation, and pain; they also help maintain the mucous layer in the gut, preventing the stomach from digesting itself. This is apparently why aspirin can upset the stomach when one takes it to relieve a headache. But it is not at all clear to me why an herbicide should be an effective inhibitor of prostaglandin production. There are many other similar cases, all of which indicate how much we have yet to learn about human and plant physiology.

Problems and Paradoxes

It is usually not too difficult to "make sense" of American Indian medicinal plant use with analyses such as those described above. Usually one can find parallels with other medicinal plant use elsewhere; often the drugs involve a secondary chemical that performs understandable functions for the plant. Sometimes it is not clear why such substances might do the things they do (as in the case of the herbicide that stops headaches) although often it is clear. There are, however, limits to such rational interpretations. It is not difficult to find cases in which different peoples—sometimes even the same peoples—are reported to use the same plant, often prepared in the same or very similar ways, for quite opposite purposes. Consider these paradoxical cases:

> The Woodlands Cree are reported to mix the fruit of *Arctostaphylos uva-ursi*, kinnikinnick, with grease and give it to

children as a treatment for diarrhea, whereas the Upper Tanana are said to eat raw kinnikinnick berries as a laxative.

The Hualapai are said to use a decoction of the leaves of *Eriodictyon angustifolium*, yerba santa, as a laxative, whereas the Paiute are said to use the same formulation as a treatment for diarrhea.

The Iroquois reportedly use a decoction of the roots of *Silphium perfoliatum*, cup plant, as an emetic (to cause vomiting), whereas the Meskwaki use the root of the same plant to alleviate the vomiting and nausea that often accompany pregnancy.

The Cherokee are said to use an infusion of the bark of *Hydrangea arborescens*, wild hydrangea, as an antiemetic for children but are also said to use the same thing to induce vomiting to "throw off disordered bile."

Similar cases can be found in which the same plant is used as both a stimulant and a sedative. There are even a few cases in which a plant is considered poisonous by one group whereas for another it is used as an antidote for poisoning.

How are we to account for such apparent contradictions? I believe there is much in medicine, any form of medicine, that resists logic, rationality, and explanation. For example, modern Western biomedicine has a long history of treatments once deemed essential that are now considered nonsense. Historians seem generally to agree, for example, that George Washington was bled to death by his physicians. It is likely that some of the treatments we use today will be ridiculed in a decade (as will some of their replacements in another decade). In addition, some physicians are better than others. The most highly trained physicians can make mistakes, as seen in many malpractice suits in our courts. There is no reason to believe that Native American physicians had a monopoly on accuracy. They doubtless made mistakes, too, and some of these mistakes are probably reported in the material in this book.

Some treatments that seem paradoxical may be homeopathic in some sense. The bulk of Western medicine is said to be allopathic. In allopathy, disease is fought by using its opposite, or something that acts against the disease, such as an antibiotic. Homeopathy treats disease by prescribing drugs, usually in very small doses, that produce symptoms resembling the disease being treated. The logic of homeopathy suggests that you should take some of the hair of the dog that bit you. It also underlies the practice of vaccination for smallpox or measles in which a controlled and attenuated case of an illness is induced to develop an immunity to a real infection. Such logic may make sense of the notion that what one group recognized as a poison, another group may have recognized as a homeopathic treatment for poisoning.

Some of the seemingly paradoxical cases may be the result of confusion on the part of the Native Americans consulted, errors in understanding by the investigator, or both. Some errors in understanding may be due to mistranslation. It is unlikely that such errors are restricted to paradoxical cases—the potential of such errors in any case urges us to be careful with *all* the information gathered here.

These reservations do not undercut the real value of the information gathered in *Native American Ethnobotany*. There is no doubt that Native Americans had a huge reservoir of real knowledge about the medicinal values of plants. But they were human beings who also made mistakes, who sometimes understood things wrongly, or differently. That there may be some errors in the information gathered here does not make everything wrong; it does mean that just because something has been reported does not make it true.

Plants Used as Foods

There are 1649 species included in *Native American Ethnobotany* that were used for food by Native Americans. They are an extremely diverse assortment of species. The listing of Plant Usage Categories toward the end of this chapter defines the many sorts of food use that appear in the Catalog of Plants. The 10 plants with the greatest number of uses as foods by Native Americans are

Prunus virginiana, common chokecherry	163
Yucca baccata, banana yucca	126
Zea mays, corn	121
Amelanchier alnifolia, Saskatoon serviceberry	117
Prosopis glandulosa, honey mesquite	79
Rubus idaeus, American red raspberry	74
Carnegia gigantea, saguaro	72
Rubus spectabilis, salmonberry	72
Rubus parviflorus, thimbleberry	71
Typha latifolia, broadleaf cattail	71

Is there a relationship between the food plants and the medicinal plants? We have noted that many medicines are plant substances that are toxic; they are often herbicides or insecticides or the like. It is not at all uncommon for medicines to be poisonous. In most societies, medicines are considered dangerous, or powerful, or are associated with strong powers in nature. They are often under the control of people with elaborate and esoteric training, for example, shamans and physicians. In Western culture, medicines are often only available by prescription, presumably because they are dangerous, or poisonous, or addicting. In any event, it seems a reasonable prediction that foods and drugs would be quite different, that these would be discrete categories.

Quite surprisingly, food and drug plants are not so discrete. There is a remarkable overlap of the two categories. [Much of the interpretation of such overlap that follows is based on a mathematical and statistical argument, details of which are complex and do not require full explanation here. Aficionados can read more about the background of this argument by consulting a paper of mine: An Analysis of the Food Plants and Drug Plants of Native North America. Journal of Ethnopharmacology 52: 1–22. 1996.]

One way to see overlap between food and drug plants is to notice that there are a number of plant families that have disproportionate shares of both kinds of plants, among which are the parsley family (Apiaceae), honeysuckle family (Caprifoliaceae), heath family (Ericaceae), oak family (Fagaceae), pine family (Pinaceae), and rose family (Rosaceae). These plants may produce nutritious and edible parts (often fruits or seeds) that attract animals, probably to facilitate plant dispersal. At the same time, however, they may produce noxious and poisonous substances to protect other aspects of their integrity.

The Rosaceae, an excellent example of this proposition, includes apples (*Pyrus*) and almonds, cherries, peaches, and pears (all *Prunus*) yet is also characterized by what we may think of as the "poisoned apple syndrome." Many members of the rose family produce nutritious and attractive fruits; they also produce quite toxic substances that occur in the leaves, bark, and pits of the fruits of many species. Among these chemicals are a class of cyanogenic glucosides (about 24 have been identified). Their toxicity occurs because they release cyanide gas when combined with certain enzymes; usually this occurs if the appropriate plant parts are crushed or chewed. Amygdalin is one of several cyanogenic glucosides produced by various species of Rosaceae that "can give rise to cyanide poisoning. Usually this is only moderate, with distress, but occasionally more serious poisoning gives rise to loss of consciousness, and serious respiratory trouble. Apnoeia and fatal collapse are exceptional, but have occurred" (F. Bodin and C. F. Cheinisse, page 162 in Poisons. New York: McGraw-Hill. 1970).

People have died eating apple pits; "poisoned apple" is redundant. In this manner the plants can attract various browsers to them to aid in dispersion of the seeds but simultaneously protect the seeds from being destroyed. At the same time, in moderation, people have made medicinal use of these chemicals, as the rose family is an important sources of plant medicines. Of 836 North American species of Rosaceae, 56 (7%) have been used medicinally by Native Americans, typically for treatment of dermatological, gastrointestinal, and gynecological problems of many sorts, 23 (3%) have been used as food, and a significantly larger number, 77 (9%), as both food and medicine.

Other good examples of food–drug overlap are found in the honeysuckle family, Caprifoliaceae. Of the North American plants in the family, 7 of the 11 genera and 30 of the 88 species were used medicinally by Native Americans, 5 of the genera and 17 of the species were used as foods, and 16 of the species used as foods were also used as medicines.

Members of the Caprifoliaceae, like those of Rosaceae, produce significant quantities of cyanogenic glycosides. One genus, *Sambucus* (elderberry), has been shown to produce a whole series of them, including holocalin, prunasin, sambunigrin, and zierin. *Sambucus* also produces a series of alkaloids, most of which probably act to inhibit excessive browsing by birds. It is the most heavily used of the seven medicinal genera in the family and is particularly interesting. Some species of *Sambucus* provide edible fruits. The earliest recipe of which I am aware that calls for elder is from Apicius's cookbook, written during the reign of Tiberius, A.D. 14–37. He recommended a sort of omelet of eggs, elderberries, pepper, wine, and *liquamen* (a sauce made of fish and salt, of the order of Worcestershire). Elderberries are generally cooked, dried, or fermented before they are eaten to moderate the effects of several emetic alkaloids. These substances are probably responsible for the common use of the plants as emetics, cathartics, or laxatives (comprising 55 of the 241 drug uses). Although most of the food uses of elder reported in *Native American Ethnobotany* involve such cooking or drying, in a number of cases it is reported that the berries were eaten raw!

So, on the one hand, there is a substantial overlap of food plants and medicinal plants. There are a number of other families with overlaps like the ones described. Overall, in the data collected here, there are 3244 species used for *either* food or drug, 987 of which are used for *both* food and drug. But on the other hand there are 662 species used for food and not medicine, and 1595 used for medicine and not food. Moreover, it is often the case that for species used for both food and medicine, different plant parts are used. For those cases in which the plant part used can be identified, it is far more likely for a root to be used for a medicine (35%) than for a food (9%). Likewise, it is far more likely for a fruit to be used for a food (47%) than for a medicine (12%). So the food–drug situation is complex, and simple assertions ("foods and medicines are different" or "foods and medicines are the same") just will not do.

Plants Used as Fibers and Dyes

There are far fewer fiber and dye plants than medicinal or food plants—442 fiber plants and 217 dye plants are documented in *Native American Ethnobotany*. In part that is because I did not seek out all possible sources of information on fiber plant usage as intensively as I did those for drugs and foods. Most of the items included here (see Plant Usage Categories toward the end of this chapter for listings of the uses that appear in the Catalog of Plants) are from comprehensive ethnobotanical works that included fiber or dye plants along with drugs, foods, and the rest. There are, however, several sources explicitly devoted to fibers that have been checked, including Barrett's *Pomo Indian Basketry*, Kelly's *Yuki Basketry*, and Kirk's *Panamint Basketry*.

The plants most widely used for fibers are a diverse lot. The 10 plants with the greatest number of uses as fibers by Native Americans are

Thuja plicata, western red cedar	188
Typha latifolia, broadleaf cattail	105
Betula papyrifera, paper birch	59
Yucca baccata, banana yucca	47
Urtica dioica, stinging nettle	36
Picea glauca, white spruce	35
Tilia americana, American basswood	35
Yucca glauca, small soapweed	35
Chamaecyparis nootkatensis, Alaska cedar	34
Apocynum cannabinum, Indian hemp	33

Most widely utilized is *Thuja plicata*. It has been used to make baskets, to make thread for sewing boxes together, and its bark has been used to make everything from mats to cords to canoes to cradles to sanitary napkins. *Typha latifolia* comes in second with similarly diverse uses as a fiber. Its leaves have been used for weaving baskets and mats, and the fuzzy down has been used to stuff mattresses and for making quilts and baby diapers. Cattail is also an important food plant—the fleshy stems and roots are eaten by many Native American groups.

The 10 plants with the greatest number of uses as dyes by Native Americans are

Alnus incana, mountain alder	53
Alnus rubra, red alder	21
Sanguinaria canadensis, bloodroot	20
Chrysothamnus nauseosus, rubber rabbitbrush	16
Rhus glabra, smooth sumac	16
Rumex hymenosepalus, canaigre dock	14
Populus deltoides, eastern cottonwood	13
Juglans nigra, black walnut	12
Rhus trilobata, skunkbush sumac	11
Juglans cinerea, butternut	9

The taxonomic distribution of dye plants is quite interesting. The following analysis is based in part on the work of my student, Essam Khraizat, who spent several months studying the dye plants in the database, and we express our appreciation to the National Science Foundation for a Research Experience for Undergraduates Award, which allowed that work to occur. Many are members of the sunflower family (Asteraceae), oak family (Fagaceae), pine family (Pinaceae), buttercup family (Ranunculaceae), or rose family (Rosaceae). The interpretation of why some plants produce important dyestuffs and others do not is similar to the analysis of medicinal plants. Oaks are a good example. The substances in the various oaks that are responsible for their ability to dye fibers are flavonoids. Depending on the plant, flavonoids may function in attracting or repelling pollinating birds and insects, regulating growth and photosynthesis, conferring antimicrobial activity, or responding to infection and injury. Although flavonoids occur in most plants, the amount varies greatly from species to species; oaks are very rich in flavonoids. There are several kinds of flavonoids, one of which is flavones. Quercitin, a flavone that occurs in high concentrations in the bark of oaks (genus *Quercus*), is a particularly strong dyestuff. One problem with modern synthetic dyes is that many are often quite toxic; some synthetic organic dyes are highly carcinogenic. Quercitin, too, is carcinogenic.

Plants with Other Uses

Many American Indian plant uses do not fit comfortably into the categories of drugs, foods, fibers, or dyes. Classified here as Other, 1074 species are documented as having such uses in *Native American Ethnobotany*. This very broad category includes all sorts of things, including hunting and fishing supplies (rods, lines, lures, traps, bows, arrows, spears, etc.), incense and fragrances, fuels, tools, and many more uses. The kinds of uses, which appear in the Catalog of Plants, are listed under Plant Usage Categories toward the end of this chapter. Among such uses are more than 750 we have classified under Ceremonial Items. Ceremonial items include a very wide range of plants or plant parts used for making talismans for various chants and religious ceremonies, for making shamans' rattles, for use in various activities at funerals, and the like. The 10 plants with the greatest number of Other uses by Native Americans are

Thuja plicata, western red cedar	121
Taxus brevifolia, Pacific yew	84
Betula papyrifera, paper birch	76
Pseudotsuga menziesii, Douglas fir	72
Pinus edulis, twoneedle pinyon	61
Cornus sericea, red osier dogwood	58
Tsuga heterophylla, western hemlock	56
Arctostaphylos uva-ursi, kinnikinnick	51
Urtica dioica, stinging nettle	51
Holodiscus discolor, ocean spray	49

Perhaps the most interesting thing about this list of 10 is that the first 7 are trees. This is somewhat surprising since only about 9% of the plants in North America are trees, yet they are disproportionately represented in this Other category. A few particular items also stand out. *Taxus brevifolia* is a relatively inconspicuous understory tree in Pacific forests made up primarily of *Pseudotsuga menziesii, Sequoia sempervirens* (redwood), or other trees. Yet 30 different western tribes are reported to have used it to make bows, several saying it made the very best bows. As discussed above under Plants Used as Drugs, the tree is also interesting as it is the source of a chemical known as taxol, which has been approved by the U.S. Food and Drug Administration as a treatment for ovarian cancer. There is much to learn by browsing through the many items listed under what otherwise appears to be a less than interesting category, Other!

Among the uses classified as Other are a few dozen cleaning agents. In particular, there are a number of plants regularly used on the Pacific Northwest coast to clean up after handling fish. Also, plant materials are used as containers, cooking tools, decorations, designs, fasteners, fuels, and weapons.

One particularly interesting category is that of Smoke Plant. There are 100 different species used in various smoking mixtures although a half dozen account for most of the 308 uses reported here. In first place is *Arctostaphylos uva-ursi*, kinnikinnick. The word kinnikinnick may be spelled five or six different ways; it comes from a Delaware Indian language and means something like "item for mixing in." It was not uncommon for American Indian people to mix together several different things to make their smoking materials. Often, in addition to kinnikinnick, they would use one of several kinds of tobacco, *Nicotiana*, as well as the leaves or bark of silky cornel, *Cornus sericea*, a plant closely related to the flowering dogwood. Other plants widely used for smoking were the dwarf sumac, *Rhus glabra*, and the prairie rose, *Rosa arkansana*.

Smoke plants often had medicinal uses as well. Often one smoked a pipe of various plant leaves and barks as a sort of prayer for better health. The medical, ceremonial, and religious elements of life were twined together, and I sort them out here at my peril.

Conclusions on Usages

There is an enormous amount of real human knowledge contained in *Native American Ethnobotany*. The earliest evidence we have of human beings using plants for medicine comes from the Middle Paleolithic site of Shanidar in northern Iraq, dated about 60,000 years ago. People have been experimenting with nature since then (and perhaps before), learning what could be eaten, what would stop bleeding or relieve pain, what would make good baskets or colors. People first came to the Americas about 15,000 years ago and have been studying the plants of the two continents ever since. Given that the floras of North America and China are remarkably alike, it is possible that the earliest Asian immigrants to North America saw recognizable plants when they got here.

Much of this accumulated knowledge of useful plants, slowly wrung from nature over millennia, has in a few centuries been lost, at least lost as a part of normal human life. There are specialists—anthropologists, ethnobotanists, phytochemists, pharmacognosists—who are aware of some portions of what this book contains. But in past times this was to a large degree the knowledge of ordinary people. Surely there were specialists, people who were more interested than others in these matters, and they may even have developed esoteric knowledge that they kept to themselves for personal profit. But generally this was normal human knowledge, part and parcel of everyday life. People walked in the world and saw plants they knew to be useful for various purposes. Their children learned of these matters as naturally as our children learn the names of baseball teams or athletic shoes or rock bands. In a world where one may buy a bottle of aspirin tablets at a grocery store for little more than the price of the bottle there is not much need for people to be able to recognize willow, black birch, spiraea, or wintergreen as naturally occurring painkillers.

And I do not believe there is a need for anyone to give up on aspirin tablets and rely on willow twigs. One need not eat medicinal plants in order to appreciate them, any more than a bird-watcher needs to eat a curlew to enjoy it. But when one knows something of the uses people have made of thousands of the wild plants around us, the plants take on a new meaning, a new value greater than their beauty or their cooling shade or their pleasant scent.

Sources of Information on Plant Usages

Native American Ethnobotany is based on the research of hundreds of scholars. I have accumulated the material included here over a period of more than 25 years. In that period, any time I saw an item containing useful information, I made a note of it. In addition, in 1993 I did an intensive search of the literature using traditional techniques such as reading bibliographies as well as using computerized search techniques. I examined approximately 500 possible sources. The information gathered here is based on material from 206 of those sources, enumerated in the Bibliography.

The criteria for selecting a source to be included were fairly simple. First, the material had to be primary, that is, based on original work with Native Americans who used the plants. I excluded all secondary material, that is, material based on prior published work. If I found an interesting secondary source I examined its bibliography to identify original sources that had yet to be consulted. In a few cases the primary source was not written by the individual who actually did the fieldwork. For example, source 59, by Catherine Fowler, is based on primary research done many years earlier by Willard Z. Park with the Northern Paiute, work that was not published at the time. As a second criterion, the source had to have reasonably clear scientific plant names, the plant identifications preferably having been made by professional botanists. Third, at least some of the information had to come from Native Americans living north of the Rio Grande. Fourth, information was coded only once even if it was published several times. In such cases I tried to use the earliest publication, but sometimes I used a later one if it had better plant identifications. Whereas the average number of plant usages coming from each source is 217.5, the range is from 1 to 2947. The oldest source was published in 1840, the most recent in 1993.

Concerning the second criterion, on the accuracy of plant identification, although sources were used in which care had been taken in this matter, no doubt some plants were misidentified and thus misnamed. Or, in some cases, if the plant were reidentified now, another identification might be made because, for example, what may have been recognized as one species at the time the plant was originally identified may now be recognized as two similar species. Which of those two (or both) species was used cannot be determined without a reidentification. Such reidentification is possible when a sample of the plant was dried, pressed, and preserved in an herbarium for reexamination. Such specimens are called vouchers. Although vouchers exist in various herbaria for some of the sources, unfortunately some sources' identifications do not rest on vouchered collections. "Suspect" identifications must be reverified by recourse to vouchers, when they exist.

And there is further work to do on the plants themselves, as documented in *Native American Ethnobotany*. For example, sometimes plants were originally identified in the sources only to the level of genus; these include the "sp." entries under various genera in the Catalog of Plants. The particular species (one or more than one species) may yet be identified, if vouchers exist. Apart from the matter of the accuracy of identification, the precision of identification in *Native American Ethnobotany* depends on that in the original sources. For example, in some species different subspecies or varieties may be recognized. If one source reported a plant named only to the level of species, and another source reported a plant to the level of subspecies or variety (or used a name that can be attributed to a particular subspecies or variety), then the usages will be listed in the Catalog of Plants under the species, or under the subspecies or variety, respectively. Thus the usages of *Rhus trilobata* by the Kiowa were reported in two sources, and because the sources used plant names that can be determined to different levels of precision, some uses are reported under the entry for *R. trilobata*, and others under that for *R. trilobata* var. *trilobata*.

Plant Usage Categories

There are 186 different plant usage categories in *Native American Ethnobotany*. They are all defined in the following list under the five major categories used in the Catalog of Plants: Drug, Food, Fiber, Dye, Other. Some care needs to be exercised in the interpretation of certain categories. Under Drug, for example, there are a number of uses described for particular diseases: Cancer Treatment, Tuberculosis Remedy, etc. Most of these disease categories have a minimum of 50 to 60 items (particular bits of information about plants used in those ways) listed under them, and some are very large with hundreds or more. Some treatments for specific diseases without a category of their own are classified under Miscellaneous Disease Remedy. A few examples of such are treatments for diabetes, flu, and measles. But listed under one of the other major usage categories, Other, are usages not for a particular disease (such as flu) but for plants said to be used for treatments such as antibiotics, or for sunstroke, or for "strengthening veins."

A subcategory that may appear under any of the major categories (Drug, Food, Fiber, Dye, Other) is Unspecified. For example, a source may state that a plant is used as a medicine but reports nothing about the particular use. "The plant is used medicinally" would be a typical report classified under Drug, Unspecified. Or under Food, for example, it is common for a source to report that a particular plant is eaten but not specify in any more detail how, what part, in what form, or when. Such usages are classified under Food, Unspecified. "Unspecified" is a very large category, including 567 reports under Drug, 2756 under Food, 8 under Fiber ("plant used for its fibers"), and 95 under Dye ("plant used as a dye"). Finally, there are 30 usages classified under Other, Unspecified. In all such cases the source reported the name of the plant but no uses for it were specified.

Drug
Abortifacient. Drug designed to eliminate a pregnancy, to induce an abortion, or to "bring on a delayed period" (an emmenagogue). See also Contraceptive.

{Drug}

Adjuvant. An item that is a subordinate element in medicines that helps them work or taste better.

Alterative. An alterative is something that changes the character of one's system. This category is used only if the term alterative actually appears in the source.

Analgesic. Drug that relieves pain.

Anesthetic. Drug that reduces the sense of touch or pain.

Anthelmintic. Drug used for the treatment of intestinal parasites.

Anticonvulsive. Drug that stops or prevents convulsions or fits.

Antidiarrheal. Drug used to stop diarrhea.

Antidote. Drug that negates the effects of a poison. See also Poison.

Antiemetic. Drug that inhibits vomiting.

Antihemorrhagic. Drug that stops hemorrhage, especially internal bleeding. See also Hemostat.

Antirheumatic (External). Drug used for rheumatism or arthritis and that is applied externally, for example, a liniment.

Antirheumatic (Internal). Drug used for rheumatism or arthritis and that is taken internally, for example, aspirin.

Basket Medicine. Drug that makes people buy baskets. If a prospective buyer picks up a basket, he will not be able to let go and will pay the price asked.

Blood Medicine. Medicine designed to purify or influence the blood.

Breast Treatment. Drug used to treat breasts.

Burn Dressing. All types of dressings applied externally to burns.

Cancer Treatment. Drug used to treat cancer or tumors. This category is used only if the word cancer explicitly appears in the source.

Carminative. Drug that relieves flatulence or "gas." Simethicone is a carminative.

Cathartic. Drug that causes evacuation of the bowels, a strong laxative; physic and purgative are synonyms of cathartic. See also Laxative.

Ceremonial Medicine. Medicine used as a part of ceremonies.

Cold Remedy. Medicine used for the relief or cure of colds.

Contraceptive. Drug used to prevent pregnancy. See also Abortifacient.

Cough Medicine. Medicine used for the relief or cure of coughs.

Dermatological Aid. Drug used to treat any conditions of the skin or hair: acne, dandruff, itching, etc.

Diaphoretic. Drug that causes sweating.

Dietary Aid. Drug that affects the diet or hunger in a situation involving illness, usually used to increase the appetite of a sick person with no appetite or to decrease it (as with "diet pills"). See also Dietary Aid, under Food, below.

Disinfectant. Drug used to eliminate "infection," literal or otherwise.

Diuretic. Drug that causes urination. See also Urinary Aid.

Ear Medicine. Drug used for earaches, deafness, or any other afflictions of the ear.

Emetic. Drug that causes vomiting.

Expectorant. Drug that promotes the ejection of mucus from the lungs, usually by spitting. See also Pulmonary Aid, and Respiratory Aid.

Eye Medicine. Drug used for any afflictions of the eye.

Febrifuge. Drug used to reduce fevers.

Gastrointestinal Aid. Drug used to treat distress of the digestive tract.

Gland Medicine. Drug used for the treatment of irregular glands: swollen, suppurating, etc.

Gynecological Aid. Drug used to treat problems surrounding pregnancy and childbirth and other problems specific to women.

Hallucinogen. Substance that induces hallucinations.

Heart Medicine. Drug used for the treatment of heart problems.

Hemorrhoid Remedy. Drug used for hemorrhoids or "piles."

Hemostat. Drug used to stop external bleeding. This category is also used for topical treatments for nosebleed (as in "leaves applied to nose for nosebleed"). See also Antihemorrhagic.

Herbal Steam. Herbs used medicinally in a steam bath for various ailments. Plants for smoking are classified as Smoke Plant, under Other, below.

Hunting Medicine. Medicine used to help a hunter find or capture prey.

Hypotensive. Drug used to reduce blood pressure.

Internal Medicine. Medicine used for various internal ailments. These are frequently very vaguely described, for example, "decoction of flowers taken for internal disorders."

Kidney Aid. Drug used for kidney troubles and the treatment of "dropsy," an anachronistic term for edema that appears often in older ethnobotanical literature.

Laxative. A mild treatment for constipation. See also Cathartic.

Liver Aid. Drug used for the treatment of various liver disorders.

Love Medicine. Drug used to procure love from another.

Miscellaneous Disease Remedy. Drug used for a particular disease, one not categorized elsewhere, such as ague, grippe, or rabies.

Narcotic. Drug that produces sleep or stupor, a strong sedative. See also Sedative.

Nose Medicine. Drug used for the treatment of various nose ailments. See also Hemostat.

Oral Aid. Drug used for the treatment of various mouth disorders.

Orthopedic Aid. Drug used for afflictions of the muscles or bones.

Other. Drug used for various conditions and ailments that are not diseases and not categorized elsewhere, such as heat prostration, meanness, and sunstroke.

Panacea. Drug used as a cure-all, a drug that will help any condition.

Pediatric Aid. Drug specifically mentioned as a treatment for children.

Poison. Substance that usually kills, injures, or impairs an organism. See also Antidote.

Poultice. Drug held against the skin.

Preventive Medicine. Drug that prevents various ailments.

Psychological Aid. Drug directed toward the mind and specifically toward issues of will or desire.

Pulmonary Aid. Drug used for lung conditions. See also Expectorant, Respiratory Aid, and Tuberculosis Remedy.

Reproductive Aid. Drug used by males or females to facilitate successful reproduction.

Respiratory Aid. Drug used to help breathing. See also Expectorant, and Pulmonary Aid.

Sedative. Drug that reduces excitement or upset, a "downer." See also Narcotic.

Snakebite Remedy. Drug used for the treatment of snakebites.

Sports Medicine. Drug used by athletes for various complaints or to enhance performance.

Stimulant. Drug that stimulates or wakes a person up, an "upper."

Strengthener. Drug used to increase strength.

Throat Aid. Drug used for afflictions of the throat.

Tonic. Drug used as a tonic for various ailments. This generic category is only used when the term tonic is specifically mentioned in the source.

Toothache Remedy. Drug used for toothaches as well as any other dental medicines.

Tuberculosis Remedy. Drug used for the treatment of tuberculosis. This category is used only when the source specifically mentions consumption, scrofula, or tuberculosis. Scrofula is a form of tuberculosis that affects the lymph nodes, especially of the neck. See also Pulmonary Aid.

Unspecified. Drug used as an unspecified medicine or to treat unspecified illnesses. This category is used if the source states something such as "plant used as a medicine" but with no more specific information.

Urinary Aid. Drug used for problems of the urinary tract and by men for sexual organ problems. See also Diuretic.

Venereal Aid. Drug used for any venereal disease: gonorrhea, syphilis, etc.

Vertigo Medicine. Drug for treating dizziness.

Veterinary Aid. Drug used to treat diseases and injuries of animals.

Witchcraft Medicine. Drug used for sorcery or magic.

Food

Appetizer. Food or drink that stimulates the appetite and is usually served before a meal.

Baby Food. Food prepared especially for babies or children.

Beverage. Beverages include beer, cider, coffee, intoxicants, tea, wine, and other drinks.

Bread & Cake. All biscuits, breads, cakes, cookies, doughnuts, dumplings, muffins, pancakes, stuffings, pones, rolls, and tortillas.

Breakfast Food. This category is used when the source specifically mentions the term breakfast.

Candy. Candies, chewing gums, confections, and taffy.

Cooking Agent. Cooking oils, curdling agents, flavorings, food colorings, food thickeners, pectin, vinegars, and yeast.

Dessert. This category is used when the source specifically mentions the term dessert.

Dietary Aid. Plant parts added to food to make it more digestible, to benefit the diet, to improve the (normal) appetite, or as a good source of vitamins. Items to help the appetite of a sick person are classed as Dietary Aid, under Drug, above.

Dried Food. Food sun or fire dried and usually stored for future use.

Fodder. Food fed to animals by people.

Forage. Food sought out by animals, usually by grazing or browsing.

Frozen Food. Food frozen and stored for future use.

Fruit. Fruit used fresh with no significant preparation.

Ice Cream. This category is used only when the source specifically mentions ice cream.

Pie & Pudding. Savory or sweet pies and puddings. Basically, a pie is a pudding with a crust.

Porridge. A soft food made by boiling grain or legume meal in a liquid until thick, including atole, boiled hot cereal, gruel, hominy, mucilaginous mass, mush, and paste.

Preservative. Food additive used to protect against decay, discoloration, or spoilage.

Preserves. Conserves, fruit butters, jams, jellies, marmalades, nut butters, pastes, preserves.

Sauce & Relish. Condiments, dips, gravies, ice cream toppings, relishes, sauces, sweet pickles, and syrups.

Snack Food. This category is used when the source specifically states the food was used for nibbles, refreshments, snacks, or tidbits. It is also used for popcorn.

Soup. Broths, chowders, pottages, soups, and stews.

Sour. Foods said to have acidic, bitter, sour, or tart tastes.

Special Food. Ceremonial foods, delicacies, rations, and food used for ritual meals and feasts.

Spice. Used to flavor, marinate, or season foods.

Staple. Foods used regularly, including cereals, flours, meals, pinoles, sugars, hominies, and rice. This category is also used when the source states that the item was used as a dependable, important, main, principal, or staple food.

Starvation Food. Food said to be used in times of emergency, famine, food shortage, or need but generally avoided in ordinary times.

Substitution Food. Food used to take the place of another food when the latter was unavailable.

Sweetener. Nectars, sugars, sweetening flavors, and syrups.

Unspecified. A very large category with 2756 items. In many sources it is common to see something such as "the root of this plant was eaten" or "the inner bark was used for food." When the food use was not specified any more clearly it has been classified as Unspecified.

Vegetable. Used for all foods described in the source as being greens, vegetables, or the like, for example, "the flower stalks were eaten as greens."

Winter Use Food. Stored foods used during the winter months.

Fiber

Basketry. Used to make all types of baskets.

Brushes & Brooms. Used to make any type of brush or broom.

Building Material. Used to make bath tubs, caches, corrals, doors, fences, floors, granaries, hearths, lodges, lumber, ramadas, roofs, sand breaks, shelters, sweat houses, thatch, wind breaks, etc.

Canoe Material. Used to make boats, canoes, dugouts, kayaks, rafts, and paddles.

Caulking Material. Used to stop up and make watertight the seams of various articles: boats, water bottles, etc.

Clothing. Used to make cloth and all types of clothing, including aprons, baby clothes, belts, capes, dresses, costumes, diapers, hats, overcoats, robes, sandals, shoes, skirts, and socks.

Cordage. Fiber made into string and ropes and used for many different applications.

Furniture. Used to make beds, chairs, cradles, etc.

Mats, Rugs & Bedding. Used to make mats for bedding, beds, blankets, cradle mats, curtains, cushions, head pads, flooring, kneeling pads, mattresses, pillows, place mats, quilts, rugs, sleeping bags, tarpaulins, towels, etc.

Other. Fiber used for various applications: as an absorbent, sanitary napkins, structures in weaving, for carving, etc.

Scouring Material. Plant parts used to polish, scour, sharpen, or smooth items.

Sewing Material. Plant parts used as embroidery stretchers, needles, thread, or for sewing things together.

Snow Gear. Used to make snow gear: skis, sleds, snowshoes, toboggans, etc.

Sporting Equipment. Used to make sporting equipment: balls, bats, hitting sticks, pucks, etc.

Unspecified. This category is used when plants were used as a source of fibrous material and fiber but a specific use is not given in the original source.

Dye

Dyes are used to impart new and often permanent colors to various articles. The colors listed are from the sources and are sorted into the following categories: Black, Blue, Blue-Black, Brown, Gray, Green, Orange, Orange-Red, Orange-Yellow, Pink, Purple, Red, Red-Brown, Red-Yellow, White, Yellow, and Yellow-Brown. When the source does not state a color, the usage is classified as Unspecified. A material used to set the dye, to attempt to make it more permanent, is classified as a Mordant.

Other

Cash Crop. Plant parts grown or gathered for sale or trade.

Ceremonial Items. Any item made from plant parts used for ceremonial purposes.

Cleaning Agent. Plant parts used to wipe or clean another object.

Containers. This category includes all types of containers: bags, boxes, buckets, carrying nets, coffins, hearths, pottery, reels, trays, and quiver cases as well as the covers, liners, and stoppers for containers.

Cooking Tools. Any item used in the preparation and eating of food.

Decorations. Used to make bouquets, carvings, decorations, hair ornaments, makeup, masks, tattoos, and wreaths. This category is also used for house plants.

Designs. Used to make designs for baskets, beadwork, carvings, and in other decorative arts.

Fasteners. Many different types of fasteners, including buttons, closure pins, glues, lashings, and tepee pegs.

Fertilizer. Plants used to increase plant production.

Fuel. Plant material used for any sort of fuel, including charcoal, firewood, matches, and tinder.

Good Luck Charm. Plant parts used in various ways to bring good luck.

Hide Preparation. Plant parts used to cure, smoke, or tan animal hides.

Hunting & Fishing Item. Used to make hunting and fishing equipment: animal calls, arrows, bows, darts, fish nets and traps, fish poison, harpoons, hooks, snares, spears, etc.

Incense & Fragrance. All types of scents, including air fresheners, deodorants, perfumes, and sachets.

Insecticide. Used to kill, remove, or repel insects.

Jewelry. Used to make anklets, beads, bracelets, earrings, finger rings, labrets, necklaces, and wristlets.

Lighting. Used to make candles, oil lamp wicks, and torches.

Lubricant. Used to lubricate working parts.

Malicious Charm. Plant parts used to curse opponents in games or horse races, to do away with evil spirits after a death, or to cause other people bad luck.

Malicious Magic. This category is used for plants that brought unwanted weather or when the source specifically mentions "malicious magic."

Musical Instrument. Plant parts used to make any types of instruments: drums, drumsticks, flutes, musical rasps, rattles, trumpets, and whistles.

Paint. Used to make and inks, paints, stains, varnishes, and face paints. Paints are different from dyes in that they are applied on the surface whereas dyes change the color of the material itself.

Paper. Used to make cards, paper, pulpwood, and toilet paper.

Plant Indicator. The presence of this plant indicated the presence of another plant.

Planting Seeds. Seeds gathered from the plant and stored for future planting.

Preservative. Plant parts used for fishhooks, cedar bark bailers, tanning paddles, and to protect articles from decay.

Protection. Plant parts used in various ways as protection from animals, athletic opponents, evil spirits, ghosts, supernatural powers, warfare, weather, and witches.

Sacred Items. Plants or plant parts dedicated or set apart for the service or worship of deity.

Season Indicator. The appearance of some plant parts indicated fishing, hunting, planting, or harvesting times, the new year, seasons, or the types of weather to come, for example, "flowers indicate it is time to plant beans."

Smoke Plant. Plant parts used for smoking materials. See also Smoking Tools, and Snuff.

Smoking Tools. Plant parts used to make pipes, pipe bowls, pipe stems, tamps, and cigarette papers. See also Smoke Plant, and Snuff.

Snuff. Used for snuff and chewing tobacco. See also Smoke Plant, and Smoking Tools.

Soap. Plant parts used as soap to clean various items, including the body.

Stable Gear. Plant parts used to make stable gear: bridle ropes, dog sled harnesses, ox yokes, saddle blankets, saddles, and stirrups.

Tools. Plant parts used to make various types of tools: awls, digging sticks, fire drills, planting sticks, seed beaters, tool handles, wedges, etc.

Toys & Games. Used to make several different kinds of toys and games: balls, children's rattles and whistles, gambling dice and sticks, popguns, slingshots, targets, etc.

Unspecified. This category is used for a few items when a source reported that a particular plant was named but apparently not used for anything.

Walking Sticks. Used to make canes, staffs, and walking sticks.

Water Indicator. Used as an indication of water location, quality of the water, or as a charm to make it rain.

Waterproofing Agent. Plant parts used to waterproof various items: baskets, bottles, boxes, hides, mittens, moccasins, water bottles, etc.

Weapon. Plant parts used to make weapons: armor, clubs, lances, spears, and war bows and arrows.

Native Americans

The names of Native American or American Indian groups is a complicated matter. Even the phrase American Indian is problematic. Christopher Columbus and his followers were naive about the location of "India." And so, many people now seem to prefer the term Native American. But it is also the case that indigenous peoples were here long before the continents of the New World were named after the Italian navigator, Amerigo Vespucci. In Canada, the generic term usually used is Native People. Yet the fact remains that many people have been quite happy with the term Indian. One of the radical movements of the 1960s, for example, was known as AIM, the American Indian Movement. Since these names often have political significance it is for all practical purposes impossible to refer generically to the indigenous population of the Americas without offending someone, but such is not the intention here.

There are similar problems with tribal names. A classic case is the name Eskimo. The word is, apparently, an English mispronunciation of a French mispronunciation of a Montagnais or Micmac word *ayashkimew*, which seems to have meant something like "eaters of raw meat," intended as a nasty insult. The Eskimo people generally call themselves Inuit or Innuit, meaning "people." I am unaware of the Inuit name for the Micmac but my guess is that it was equally insulting. Such derived names are not always insults, however. The name Navaho or Navajo is a Spanish variation on a Tewa word meaning "large arroyo with cultivated fields," a place name. The Navajo name for themselves is Dene, meaning "people," but many Navajo also call themselves Navajo. Again, the names used to refer to particular Native American tribes can occasion political debate. Referring to one group by a term that means "people" implicitly asserts that other groups are something other than people, for example. Names used in *Native American Ethnobotany* are not intended in any way to offend anyone. I have elected to use the names for peoples reported in the original sources. This means that material for "the same people" is sometimes listed under different designations, but rarely is it for the same people at the same time. There are 291 groups mentioned in *Native American Ethnobotany*, as listed below. The reference numbers below correspond to the sources enumerated in the Bibliography, from which information on each of the peoples was obtained:

Abnaki. Saint François du Lac, about 100 miles (160 km) northwest of Montréal, Quebec (144).

Acoma. A pueblo in western central New Mexico (32, 200). See also Keresan, and Keres, Western.

Alabama. Located on a state reservation on the Trinity River in Polk County, Texas (172, 177).

Alaska Native. Alaska (85).

Aleut. Seventeen Aleutian Islands lie between the Alaska Peninsula and Attu, Alaska (7, 8, 149, 189).

Algonquin. New England (22).

Algonquin, Quebec. Western Quebec (18).

Algonquin, Tête-de-Boule. Manouan, about 260 miles (420 km) north of Montréal, Quebec (132).

Anticosti. On the Ile d'Anticosti in the Saint Lawrence River, north of the Gaspé Peninsula, Quebec (143).

Apache. Various Apache groups live in Arizona and New Mexico (17, 32, 47, 121, 138).

Apache, Chiricahua & Mescalero. Chiricahua Apache ranged through western New Mexico, southeastern Arizona, and southward into Mexico; their eastern boundary began at the Rio Grande. The Mescalero Apache were principally located in

New Mexico; the Rio Grande was their western boundary (33).

Apache, Mescalero. Southern New Mexico, western Texas, and northern Chihuahua, Mexico (14, 17).

Apache, San Carlos. The San Carlos Reservation is in eastern Arizona (95).

Apache, Western. Fort Apache Reservation, Arizona (26).

Apache, White Mountain. White Mountain, Arizona (136).

Apalachee. In the sixteenth and seventeenth centuries the Apalachee lived in Spanish Florida, what is now northern Florida and southern Georgia (81).

Arapaho. Between the Platte and Arkansas Rivers in eastern Colorado and southeastern Wyoming (19, 118, 121).

Arikara. Missouri River region, especially in South Dakota (70).

Atsugewi. North of the Sierra Nevada and south of the Pit River in northeastern California (61).

Bannock. Idaho (118).

Bellabella. The village of Bella Bella, British Columbia, is about 250 miles (400 km) northwest of Vancouver (184).

Bella Coola. The village of Bella Coola, British Columbia, is near the mouth of the Bella Coola River, about 250 miles (400 km) north northwest of Vancouver (74, 150, 184). See also Kimsquit.

Blackfoot. The Blackfoot hunted over the region of Montana, Alberta, and Saskatchewan. Blackfoot is a common spelling in Canada whereas Blackfeet is more common in the United States (82, 86, 97, 114, 118, 121).

Cahuilla. Southern California (15).

California Indian. California (19, 118).

Canadian Indian. British Columbia, Alberta, and Saskatchewan (97).

Carrier. Near Fort St. James and Anahim Lake in central and northern British Columbia (31, 89, 184).

Carrier, Northern. Near Hagwelget, in northwestern British Columbia (150).

Carrier, Southern. Near Ulkatcho, in northwestern British Columbia (43, 150).

Catawba. Along the Catawba River in the Carolinas (157, 177).

Chehalis. Western central coast of Washington state (79).

Cherokee. The Cherokees are found throughout much of western North Carolina (particularly in Graham and Cherokee Counties) and in northwestern Georgia. There are also many Cherokees in Oklahoma (80, 126, 169, 177, 203, 204).

Cheyenne. Montana and Oklahoma (19, 75, 76, 82, 83, 97).

Cheyenne, Northern. Montana (82).

Chickasaw. Information in source 28 was collected by Gideon Lincecum during the years 1800–1835 in Mississippi. Most Chickasaws are now in Oklahoma (28, 177).

Chinook. Southwestern coast of Washington state (79).

Chinook, Lower. Southwestern coast of Washington state (79).

Chippewa. Also known as the Ojibwa, Chippewas are located in the upper Midwest and southern Ontario. Source 51 describes the Ojibwa of northern Minnesota, primarily the Leech Lake, Red Lake, and White Earth reservations but also including a consultant from the Bois Fort reservation. Source 53 describes the Ojibwa of northern central Minnesota and the Manitou Rapids Reserve in Ontario. Source 71 describes people from Pinconning and Lapeer, Michigan, and Sarnia, Ontario. See also Ojibwa.

Choctaw. Saint Tammany Parish, Louisiana, on the northern shore of Lake Pontchartrain, is a center of the Choctaw region. Information in source 28 was collected by Gideon Lincecum during the years 1800–1835 in Mississippi (25, 28, 158, 177).

Chumash. Southern California (15).

Clallam. Olympic Peninsula, Washington (57).

Coahuilla. Southern California coast (13).

Cochiti. Pueblo Indians of New Mexico (32). See also Keresan.

Cocopa. Southwestern Arizona, and Baja California and Sonora, Mexico, generally along the lower Colorado River (37, 64).

Cocopa & Yuma. Lower Colorado River in Arizona (37).

Cocopa, Maricopa, Mohave & Yuma. The Cocopa, Mohave, and Yuma are located on the lower Colorado River in Arizona, and the Maricopa are located on the Gila River in Arizona (37).

Coeur d'Alene. The Coeur d'Alene Reservation is in northern Idaho (89, 178).

Comanche. The Comanche Indian Reservation is near Indiahoma, Comanche County, Oklahoma (29, 32, 99).

Concow. The Round Valley Indian Reservation is in Mendocino County, California, and stretches as a band about 60 miles (100 km) broad for 84 miles (135 km) along the coast, about midway between San Francisco and the Oregon border to the north (41). See also Numlaki.

Costanoan. The territory once occupied by the Costanoan Indians is in the Coast Ranges of central California. Their territory extended from San Francisco south to Big Sur and from the Pacific coast inland to the Diablo Range foothills (21).

Costanoan (Olhonean). Central California (117).

Cowichan. Southeastern coast of Vancouver Island, British Columbia (182, 186).

Cowlitz. Southern central Washington state (79).

Cree. The Cree range from the Northwest Territories (16) through central Alberta and southwestern Saskatchewan (97) to Montana (82).

Cree, Alberta. Northern Alberta (149).

Cree, Hudson Bay. Hudson Bay region, Canada (92).

Cree, Plains. Montana and North Dakota (113).

Cree, Woodlands. Saskatchewan (109).

Creek. Information from source 28 was collected by Gideon Lincecum during the years 1800–1835 in the original Creek homeland in Georgia. Most Creeks are today in Oklahoma (28, 169, 172, 177, 203).

Crow. A Siouan tribe living in southwestern Montana and northern Wyoming (19, 82).

Dakota. The Dakota, also known as the Lakota or Sioux, live on a number of reservations in Montana, the Dakotas, Nebraska, and Minnesota (69, 70, 97, 121). See also Lakota, and Sioux.

Delaware. The Delaware originally lived on the East Coast. Today, some live in Ontario but most are in Oklahoma (176).

Delaware, Oklahoma. Oklahoma (175).

Delaware, Ontario. Ontario (175).

Diegueño. The Diegueño live throughout southernmost California, notably on the Santa Ysabel Indian Reservation in San Diego County (15, 37, 84, 88).

Eskimo. There are many communities of Inuit or Eskimo people across the Arctic regions of Alaska, Canada, and Greenland. Some sources describe individual villages whereas others are more general in their coverage. Source 149 describes some Alaskan Inuit and source 181 describes people living at Hudson Bay.

Eskimo, Alaska. Nelson Island, on the Bering Sea coast of the Yukon-Kuskokwim Delta, western Alaska (1), villages along the northern Bering Sea and Arctic Alaska (4), western and southwestern Alaska (128), and Kodiak, Alaska (149).

Eskimo, Arctic. Central Canadian Arctic and sub-Arctic (127, 128).

Eskimo, Chugach. These southern Alaskan Inuit are located east of the Kenai Peninsula along the coast to Controller Bay, Alaska (149).

Eskimo, Greenland. Greenland (128).

Eskimo, Inuktitut. Alaska, Canada, and Greenland (202).

Eskimo, Inupiat. Kotzebue in northwestern Alaska (98).

Eskimo, Kuskokwagmiut. Napaskiak, western Alaska (122).

Eskimo, Nunivak. Nunivak Island, Alaska (149).

Eskimo, Western. Lower Kuskokwim, Nunivak, and Nelson Islands on the western coast of Alaska (108).

Flathead. An Interior Salish tribe located in western Montana and Idaho, north of the Gallatin River, between the Rocky Mountains in the west and the Little Belt Range in the east (19, 82, 89, 97).

Gabrielino. Southern California (15).

Gitksan. The Gitksan live along the northern coast of British Columbia and along the Skeena River (43, 73, 74, 150).

Gosiute. In desert territory bordering the Great Salt Lake in Utah on the south and extending westward into eastern Nevada (39).

Great Basin Indian. Uintah-Ouray Reservation, Fort Duchesne, Utah (121).

Green River Group. South of Seattle, Washington (79).

Gros Ventre. An Algonquian tribe living in the Milk River area of northern Montana (19, 82, 118).

Hahwunkwut. Northern California (117).

Haida. The Haida live along the northern coast of British Columbia into southern Alaska as well as on the Queen Charlotte Islands off the coast of British Columbia (43, 79).

Haihais. Central coast of British Columbia (43).

Haisla. Central coast of British Columbia (43, 73, 74).

Haisla & Hanaksiala. Central coast of British Columbia (43).

Hanaksiala. Central coast of British Columbia (43).

Havasupai. Cataract Canyon, a side branch of the Grand Canyon in northwestern Arizona (17, 162, 197, 200).

Hawaiian. Hawaii (2, 112).

Heiltzuk. Central coast of British Columbia (43).

Hesquiat. Coast of British Columbia and Hesquiat Harbor, Vancouver Island, British Columbia (43, 185, 186).

Hidatsa. Missouri River region, South Dakota (70).

Hoh. Hoh Village is near the mouth of the Hoh River on the western side of the Olympic Peninsula, Washington (137).

Hopi. The Hopi live in several villages on the Hopi Reservation in northeastern Arizona. The Hopi Reservation is surrounded by the Navajo Reservation (17, 32, 34, 36, 42, 56, 101, 120, 138, 190, 200, 205).

Houma. Louisiana (158).

Hualapai. Northwestern Arizona (195).

Hupa. Northern California (117, 118, 148).

Huron. The Hurons lived in southern Ontario and Michigan until the early nineteenth century. Now they live in Wisconsin (3).

Iroquois. The Iroquois live throughout upstate New York and in southern Quebec (87, 124, 141, 142, 196).

Isleta. The Isleta pueblo is located on the western bank of the Rio Grande, 12 miles (19 km) south of Albuquerque, New Mexico (32, 100).

Jemez. The Jemez pueblo is located on the Jemez river about 45 miles (72 km) northwest of Albuquerque, New Mexico (17, 32, 44, 100, 200).

Kamia. Imperial Valley in southeastern California (62).

Kansa. The Kansa, who give their name to Kansas, are now located in Oklahoma (68).

Karok. The Karok are located along the Klamath River in an area paralleling the California coast from above Bluff Creek in Humboldt County to Happy Camp in Siskiyou County (6, 117, 148).

Kawaiisu. East of Bakersfield in southeastern California (206).

Keresan. There are seven pueblos near Albuquerque, New Mexico, where the languages are classified as Keresan. The Eastern Keresan pueblos are Cochiti, San Felipe, Santa Ana, Santo Domingo, and Sia. The Western Keresan pueblos are Acoma and Laguna. All of these but Santa Ana and Santo Domingo are treated separately in this book (198). See also Acoma; Cochiti; Keres, Western; Laguna; San Felipe; Sia.

Keres, Western. The Acoma and Laguna pueblos (171). See also Keresan.

Kiliwa. Lower Colorado River valley (37).

Kimsquit. Bella Coola, British Columbia (184). See also Bella Coola.

Kiowa. Southern plains near the Arkansas and Red Rivers (19, 192).

Kitasoo. Central coast of British Columbia (43).

Kitkatla. Northern coast of British Columbia (43).

Klallam. Southern shore of Vancouver Island, British Columbia, and the northern central Olympic Peninsula, Washington (78, 79).

Klamath. Southern central Oregon (45, 46, 117, 118, 163).
Koasati. Southeastern United States (177).
Konomeho. Northern California (117).
Koyukon. Huslia and Hughes, Alaska (119).
Kuper Island Indian. Kuper Island, southwest of Nanaimo, Vancouver Island, British Columbia (182).
Kutenai. Montana (82).
Kwakiutl. Northern Vancouver Island and north of Vancouver Island on the mainland coast of British Columbia (20, 182, 183).
Kwakiutl, Southern. Central western coast of British Columbia and northeastern coast of Vancouver Island, British Columbia (183, 184).
Kwakwaka'wakw. Central coast of British Columbia (43).
Laguna. The Laguna pueblo is in New Mexico (32). See also Keresan, and Keres, Western.
Lakota. Standing Rock Reservation is located on the central border of North Dakota and South Dakota (106); Rosebud Reservation is in Todd County, South Dakota (106, 139). The Lakota are also known as Dakota and Sioux. See also Dakota, and Sioux.
Lillooet. British Columbia (89).
Lolahnkok. Northern California (117).
Luiseño. Southern California near San Juan Capistrano (15, 155).
Lummi. Northwestern border of Washington, near British Columbia (79).
Lummi, Washington. Northwestern Washington state (182).
Mahuna. Southwestern California (140).
Maidu. Sacramento, California (173).
Makah. Northwestern tip of the Olympic Peninsula, Washington (43, 67, 79, 186).
Malecite. New Brunswick, Canada (116, 160).
Mandan. North Dakota along the Missouri River near the mouth of the Heart River (19, 70).
Maricopa. Southern central Arizona (32, 37, 47, 95).
Maricopa & Mohave. The Maricopa are located on the Gila River in Arizona; the Mohave are located on the lower Colorado River in Arizona (37).
Mendocino Indian. Mendocino County, California, halfway between San Francisco and the Oregon border (41).
Menominee. Wisconsin (54, 151).
Meskwaki. Tama, Iowa (152).
Mewuk. Central California (117).
Micmac. Nova Scotia, Prince Edward Island, New Brunswick east of the St. John River, and part of the Gaspé Peninsula, Quebec (40, 116, 145, 156, 159, 194).
Midoo. Central California (117).
Missouri River Indian. Montana (82).
Miwok. The Sierra Nevada together with the western foothills and a relatively small portion of the adjacent Sacramento–San Joaquin Valley, California (12).
Modesse. Northern California (117).
Modoc. Northern California and southern Oregon along the Lost River, as well as Tule, Klamath, and Clear Lakes (45).
Mohave. Lower Colorado River valley (32, 37, 168).
Mohegan. Connecticut (30, 174, 176).
Montagnais. Montagnais live throughout eastern Quebec, Lac-St.-Jean (24) and the northern coast of the Gulf of St. Lawrence and the lower St. Lawrence River (156), and Labrador (174).
Montana Indian. Montana (19, 82).
Montauk. Eastern end of Long Island, New York (30).
Nanaimo. Vancouver Island, British Columbia (182).
Nanticoke. Delaware (174, 175).
Narraganset. Rhode Island (30).
Natchez. Lower Mississippi River (172, 177).
Navajo. Arizona, New Mexico, and Utah (17, 23, 32, 47, 55, 56, 90, 97, 100, 110, 136, 138, 165, 200).
Navajo, Kayenta. Northeastern Arizona, near Monument Valley (205).
Navajo, Ramah. Western New Mexico (191).
Neeshenam. Bear River, Placer County, California (129).
Nevada Indian. Nevada (118, 121).
Nez Perce. Along the lower Snake River and its tributaries in western Idaho, northeastern Oregon, and southwestern Washington (19, 82, 97).
Nimpkish. North of Vancouver Island on the mainland of British Columbia (183).
Nisga. Northern coast of British Columbia (43).
Nisqually. South of Puget Sound, Washington (79).
Nitinaht. Southwestern coast of Vancouver Island, British Columbia, from near Jordan River to Pachena Point, extending inland along Nitinat Lake (67, 185, 186).
Nootka. Southwestern coast of Vancouver Island, British Columbia (170, 185, 186).
Nootka, Manhousat. Vancouver Island, British Columbia (186).
Northwest Indian. Oregon and Washington (19, 118).
Numlaki. Round Valley Indian Reservation is in Mendocino County, California, midway between San Francisco and the Oregon border (41). See also Concow.
Nuxalkmc. Valley of the Bella Coola River, South Bentinck Arm, Tallio Inlet, and Kimsquit in western central British Columbia (43).
Oglala. South Dakota (70).
Ojibwa. Also known as the Chippewa, Ojibwas are located in the upper Midwest and southern Ontario. Source 5 describes the Ojibwa north of lakes Superior and Huron in Ontario. Source 96 describes people on Parry Island, located on the eastern central side of Georgian Bay, Ontario. The Fort Bois Ojibwa described in source 135 live on a reservation 140 miles (225 km) northwest of Duluth, Minnesota. People described in source 153 live in northern Wisconsin. See also Chippewa.
Ojibwa, South. Red Lake and Leech Lake, Minnesota (91).
Okanagan-Colville. Okanagan is the Canadian spelling. The Okanagan occupied the Okanagan and Similkameen River val-

leys and the shores of Lake Okanagan on both sides of the U.S-Canadian border. The Colville are located on the Colville Reservation in northeastern Washington state (74, 188). See also Okanagon.

Okanagon. Okanagon is the American rather than the Canadian spelling. The Okanagon are found on the Colville Reservation in Washington state and on various reserves in British Columbia (125, 164, 178). See also Okanagan-Colville.

Omaha. Nebraska (58, 68, 70).

Oregon Indian. Oregon (118).

Oregon Indian, Warm Springs. Warm Springs, Oregon (104, 118).

Oto. Eastern Nebraska along the Platte River (70).

Oweekeno. Western central coast of British Columbia (43).

Paipai. Southwestern United States (37).

Paiute. Paiutes generally live in the Great Basin region although some live in California and Oregon. People described in sources 15 and 104 are from Surprise Valley, California. Source 111 describes the Warm Springs Reservation, Oregon. The remaining sources (32, 118, 121, 167, 180) generally describe people living in Nevada.

Paiute, Nevada. Nevada (121).

Paiute, Northern. Western Nevada (59, 60).

Paiute, Southern. Northern Arizona, southern Utah, and southern Nevada and adjacent areas of California (15).

Panamint. Death Valley, California (105).

Papago. In desert regions south of the Gila River of Arizona and extending into Sonora, Mexico (17, 32, 34, 36, 47, 95, 118, 146).

Papago & Pima. Arizona (35).

Pawnee. Missouri River region (70).

Pend d'Oreille, Upper. Montana (82).

Penobscot. Northern New England and the maritime provinces of Canada. There is a Penobscot Reservation in Maine (156, 174).

Pima. Gila and Salt River valleys of southern Arizona (17, 32, 36, 47, 95, 146).

Pima, Desert. Southwestern Arizona (133).

Pima, Gila River. North of Jalisco, Mexico, to Phoenix, Arizona (133).

Pima, Lehi. Lehi, Arizona (47).

Pit River. Northern California (41, 117).

Plains Indian. Montana (82, 97).

Poliklah. Northern California (117).

Pomo. Sonoma, Mendocino, and Lake Counties, California (9, 11, 41, 66, 117, 118).

Pomo, Calpella. Northern California (41).

Pomo, Kashaya. Western coast of Sonoma County, California (72).

Pomo, Little Lakes. Northern California (41).

Pomo, Potter Valley. Northern California (41).

Ponca. The Northern Ponca are located in Nebraska and South Dakota whereas the Southern band is located in Oklahoma (70, 94).

Potawatomi. Wisconsin (154).

Pueblo. New Mexico and Arizona (32, 34).

Puyallup. Southeastern side of Puget Sound, Washington (79).

Quileute. Western coast of the Olympic Peninsula, Washington (67, 79, 137).

Quinault. Southwestern coast of the Olympic Peninsula, Washington (79, 201).

Rappahannock. Virginia (30, 161).

Ree. Montana (19).

Rocky Boy. Rocky Boy Reservation, Montana (118).

Round Valley Indian. Round Valley, 200 miles (320 km) north of San Francisco in northern California (41).

Saanich. Southeastern side of Vancouver Island, British Columbia (182).

Salish. North of Vancouver Island on the mainland coast of British Columbia (178, 182, 183, 185).

Salish, Coast. Northern central side of Vancouver Island and north of Vancouver Island on the mainland of British Columbia (182, 183).

Salish, Cowichan. Southeastern side of Vancouver Island, British Columbia (182, 186).

Salish, Halkomelem. Vancouver Island, British Columbia (186).

Salish, Straits. Vancouver Island, British Columbia (186).

Samish. Northwestern coast of Washington, near the British Columbia border (79).

Sanel. Mendocino County, California (41).

San Felipe. The San Felipe pueblo is in New Mexico (32). See also Keresan.

San Ildefonso. The San Ildefonso pueblo is located about 20 miles (32 km) north of Santa Fe, New Mexico, on the eastern bank of the Rio Grande (138).

Sanpoil. South of the Columbia River in northeastern Washington (89, 131, 188).

Sanpoil & Nespelem. South of the Columbia River in northeastern Washington (131, 188).

Santa Clara. The Santa Clara pueblo is about 25 miles (40 km) north of Santa Fe, New Mexico, on the western bank of the Rio Grande (138).

Seminole. Southern Florida (169).

Seri. Isla Tiburon in the Gulf of California, Mexico (50).

Shasta. Along the Klamath River in northern California near the Oregon border (93, 117).

Shinnecock. Long Island, New York (30).

Shoshoni. Most Shoshoni are in Nevada but they extend into Montana and central California (82, 97, 117, 118, 121, 180).

Shuswap. Southern interior plateau of British Columbia (89, 123, 178).

Sia. Sia (sometimes Zia) is a Keresan pueblo near Albuquerque, New Mexico (199). See also Keresan.

Sierra. California (41).

Sikani. Headwaters of Peace River in British Columbia (150).

Similkameen. Various reserves in British Columbia (164).

Sioux. The Sioux, also known as the Dakota or Lakota, hunted buffalo across the Great Plains from Minnesota to Wyoming and Montana (19, 82, 118, 153). See also Dakota, and Lakota.

Sioux, Fort Peck. The Fort Peck Reservation is in northern Montana (19).

Sioux, Teton. Wyoming and Montana (52).

Skagit. Northwestern Washington state (79).

Skagit, Upper. Northern Cascade Range, Washington (179).

Skokomish. Western central side of Puget Sound, Washington (79).

Snake. Montana (19).

Snohomish. Northeastern side of Puget Sound, Washington (79).

Snuqualmie. Central Washington (79).

Songish. Southeastern side of Vancouver Island, British Columbia (182).

Southwest Indians. Southwestern United States (17).

Spanish American. Southwestern United States (32).

Spokan. The Spokan Reservation is in eastern Washington, the Colville Reservation in Washington, and the Coeur d'Alene Reservation in Idaho (178).

Squaxin. South of Puget Sound, Washington (79).

Stony Indian. Montana (82).

Swinomish. Northwestern coast of Washington state (79).

Tanaina. Near Anchorage, Alaska (149).

Tanana, Upper. Alaska between Anchorage and Fairbanks (77, 102, 115).

Tarahumara. Arizona and northern Mexico (17).

Tewa. Near Santa Fe, New Mexico (17, 32, 42, 138).

Tewa of Hano. Near Santa Fe, New Mexico (138).

Thompson. Southwestern British Columbia (89, 125, 164, 178, 187).

Thompson, Lower. On the lower Fraser River in British Columbia (89).

Thompson, Upper (Fraser Band). Along the Fraser River above Lytton, British Columbia (164).

Thompson, Upper (Lytton Band). Lytton, British Columbia (164).

Thompson, Upper (Nicola Band). Along the Nicola River from a few miles above Spences Bridge to above Nicola Lake in British Columbia (164).

Thompson, Upper (Spences Bridge). Along the Thompson River from Lytton to Ashcroft, British Columbia (164).

Tlingit. Southeastern coastline of the Alaska panhandle from Yakutat Bay to Cape Fox (79, 107, 149).

Tolowa. Northwestern California (6).

Tsimshian. Northwestern coast of British Columbia into the southeastern portion of the Alaska panhandle (43, 73, 74, 79).

Tsimshian, Coast. Western central coast of British Columbia (43).

Tubatulabal. California (193).

Ukiah. Ukiah, Mendocino County, northern California (41).

Umatilla. Along the Umatilla and Columbia Rivers in Oregon (45, 89).

Ute. Western Colorado and eastern Utah (32, 34, 38, 45, 118).

Wailaki. Round Valley Reservation in Mendocino County, northern California (41, 118).

Walapai. Cataract Canyon, northwestern Arizona (17, 162, 197).

Warihio. Lower Colorado River valley (37).

Washo. Near Lake Tahoe on the California-Nevada border (10, 118, 121, 180).

West Coast Indian. West Coast of the United States (134).

Wet'suwet'en. East of the western central coast of British Columbia (73).

Weyot. Northern California (117).

Winnebago. Originally living near Green Bay, Wisconsin, the Winnebago now occupy reservations in Wisconsin and Nebraska (70, 130).

Wintoon. Central California (117).

Yana. Northern California (147).

Yaqui. Guadalupe, Arizona (47

Yavapai. Western Arizona (17, 63, 65).

Yokia. Ukiah, Mendocino County, northern California (41).

Yokut. Central California (117).

Yuki. Round Valley, Mendocino County, northern California (9, 41, 48, 49, 103, 118).

Yuma. Lower Colorado River valley in Arizona (32, 34, 37).

Yurok. Northwestern California (6).

Yurok, South Coast (Nererner). Northwestern California (117).

Zuni. The Zuni pueblo is about 40 miles (64 km) southwest of Gallup, New Mexico (17, 27, 32, 34, 136, 166).

Organization of the Information in *Native American Ethnobotany*

Scientific Plant Names

The first principle underlying the organization of *Native American Ethnobotany* is that the book is to be used for finding all the known uses of a given plant by Native Americans. The information is, therefore, organized in the Catalog of Plants alphabetically by the scientific names of the plants.

Genera and Species

Scientific plant names have a very particular form, for example, for balsam fir, a member of the plant family Pinaceae, *Abies balsamea* (L.) P. Mill. The first two words in the name, *Abies balsamea*, are, respectively, the genus (plural, genera) and the species (plural, species), in combination called a binomial. One may think of them rather like the surname and first name of a person, "Doe, John," in the order he would be listed in a telephone directory, last name first. The names are in a special form of Latin agreed on by botanists and are usually printed in italics or underlined. In this book, seven different species of the genus *Abies* are listed together in the first pages of the Catalog of Plants. Following the species name is another term, one of importance in distinguishing instances in which the same name may have been used more than once, sometimes to apply to different plants. In our example the term is "(L.) P. Mill.," known as the author or authority for the binomial, and it represents the names of the botanists who named the species. The "L." refers to Carl Linnaeus, the inventor of binomial nomenclature for organisms, who first described the species scientifically in his book, *Species Plantarum*, published in Stockholm in 1753. But Linnaeus originally placed it in a different genus, as *Pinus balsamea* L. At the time most conifers were lumped together in the genus *Pinus*, now restricted to the true pines. The "P. Mill." refers to Philip Miller, who reassigned the species to a separate genus for firs, *Abies*, in the eighth edition of his book, *The Gardener's Dictionary*, published in London in 1768.

Some plant names are more complex than the example just given. In some species, botanists recognize a number of subspecies or naturally occurring varieties. Usually the rank of subspecies or variety is recognized in a particular species, but sometimes both ranks are used and then a subspecies may comprise more than one variety. For the purpose of completeness the subspecies (abbreviated here as ssp.) and variety (var.) names are both given in this book, for example, for the full name of Pacific red elder, a member of the Caprifoliaceae and a plant used in a dozen or so ways by a number of West Coast tribes, *Sambucus racemosa* ssp. *pubens* var. *arborescens* (Torr. & Gray) Gray.

When reference is made to an unknown species of a particular genus, the abbreviation "sp.," for species, is used. A source may have referred to one unknown species of a genus, or more than one unknown species; "sp." is used here for both kinds of references. For example, uses of firs that cannot be identified to species level are collected together under *Abies* sp.

Species that are known to result from hybridization have a multiplication sign added to their names, for example, *Apocynum* ×*floribundum* Greene, formed by the hybridization *Apocynum androsaemifolium* L. × *A. cannabinum* L.

Families

In a formal designation the names of the authorities are followed by the family names: Pinaceae, the scientific name for the pine family, which includes *Abies*, *Pinus*, and other genera, and Caprifoliaceae, the name for the elder family, which includes *Sambucus* and *Viburnum* among other genera. There are between 375 and 400 families of vascular plants worldwide, of which 286 occur in North America north of the Rio Grande. To see which genera of which families are included in *Native American Ethnobotany*, please look up the family name in the alphabetically organized Catalog of Plants. The family names of all plants, vascular and nonvascular, that are included in *Native American Ethnobotany* are listed below.

Family names for nonflowering plants are perhaps less familiar. Here are cross-references to the family names for the nonflowering plants, including nonvascular plants (algae, fungi, lichens, liverworts, and mosses), the vascular cryptogams (club mosses, spike mosses, ferns, and horsetails), and among seed plants, the gymnosperms (conifers):

Algae
 Alariaceae
 Bangiaceae
 Fucaceae
 Laminariaceae
 Leathesiaceae
 Lessoniaceae
 Palmariaceae
 Rhodymeniaceae
 Trentepohliaceae
 Ulvaceae
 Zygnemataceae
Fungi
 Agaricaceae
 Amanitaceae
 Boletaceae
 Cantharellaceae
 Exobasidiaceae
 Helvellaceae
 Hydnaceae
 Lycoperdaceae
 Pezizaceae
 Polyporaceae
 Tricholomataceae
 Ustilaginaceae
Lichens
 Alectoriaceae
 Cladoniaceae
 Lecanoraceae
 Parmeliaceae
 Peltigeraceae
 Stereocaulaceae
 Stictaceae
 Teloschistaceae
 Usneaceae
Liverworts
 Conocephalaceae
 Jungermanniaceae
Mosses
 Aulacomniaceae
 Bryaceae
 Dicranaceae
 Hylocomiaceae
 Leucobryaceae
 Leucodontaceae
 Mniaceae
 Polytrichaceae
 Pottiaceae
 Rhytidiaceae
 Sphagnaceae
Club Mosses
 Lycopodiaceae
Spike Mosses
 Selaginellaceae
Ferns
 Blechnaceae
 Cyatheaceae
 Dennstaedtiaceae
 Dryopteridaceae
 Marattiaceae
 Ophioglossaceae
 Osmundaceae
 Polypodiaceae
 Pteridaceae
 Thelypteridaceae
 Vittariaceae
Horsetails
 Equisetaceae
Conifers
 Cupressaceae
 Ephedraceae
 Pinaceae
 Taxaceae
 Taxodiaceae
 Zamiaceae

Some families of angiosperms or flowering plants have alternative names (Asteraceae or Compositae for the sunflower family, for example) and the alternative name is given in parentheses below:

Flowering Plants
 Acanthaceae
 Aceraceae
 Acoraceae
 Agavaceae
 Aizoaceae
 Alismataceae
 Amaranthaceae
 Anacardiaceae
 Annonaceae
 Apiaceae (Umbelliferae)
 Apocynaceae
 Aquifoliaceae
 Araceae
 Araliaceae
 Arecaceae (Palmae)
 Aristolochiaceae
 Asclepiadaceae
 Asteraceae (Compositae)
 Balsaminaceae
 Berberidaceae
 Betulaceae
 Bignoniaceae
 Bixaceae
 Boraginaceae
 Brassicaceae (Cruciferae)
 Bromeliaceae
 Burseraceae
 Butomaceae
 Cactaceae
 Calycanthaceae
 Campanulaceae
 Cannabaceae
 Cannaceae
 Capparaceae
 Caprifoliaceae
 Caricaceae
 Caryophyllaceae
 Celastraceae
 Chenopodiaceae
 Chrysobalanaceae
 Cistaceae
 Clethraceae
 Clusiaceae (Guttiferae)
 Combretaceae
 Commelinaceae
 Convolvulaceae
 Cornaceae
 Crassulaceae
 Cucurbitaceae
 Cuscutaceae
 Cyperaceae
 Datiscaceae
 Diapensiaceae
 Dioscoreaceae
 Dipsacaceae
 Droseraceae
 Ebenaceae
 Elaeagnaceae
 Empetraceae
 Epacridaceae
 Ericaceae
 Euphorbiaceae
 Fabaceae (Leguminosae)
 Fagaceae
 Fouquieriaceae
 Garryaceae
 Gentianaceae
 Geraniaceae
 Gesneriaceae
 Goodeniaceae
 Grossulariaceae
 Haemodoraceae
 Haloragaceae
 Hamamelidaceae
 Hippocastanaceae
 Hippuridaceae
 Hydrangeaceae
 Hydrocharitaceae
 Hydrophyllaceae
 Iridaceae
 Juglandaceae
 Juncaceae
 Juncaginaceae
 Krameriaceae
 Lamiaceae (Labiatae)
 Lauraceae
 Lemnaceae
 Lennoaceae
 Lentibulariaceae
 Liliaceae
 Linaceae
 Loasaceae
 Loganiaceae
 Lythraceae
 Magnoliaceae
 Malvaceae
 Marantaceae
 Melastomataceae
 Meliaceae
 Menispermaceae
 Menyanthaceae
 Moraceae
 Musaceae
 Myricaceae
 Myrsinaceae
 Myrtaceae
 Nelumbonaceae
 Nyctaginaceae
 Nymphaeaceae
 Olacaceae
 Oleaceae

Onagraceae	Polygalaceae	Sarraceniaceae	Thymelaeaceae
Orchidaceae	Polygonaceae	Saururaceae	Tiliaceae
Orobanchaceae	Pontederiaceae	Saxifragaceae	Typhaceae
Oxalidaceae	Portulacaceae	Scrophulariaceae	Ulmaceae
Paeoniaceae	Potamogetonaceae	Simaroubaceae	Urticaceae
Pandanaceae	Primulaceae	Simmondsiaceae	Valerianaceae
Papaveraceae	Punicaceae	Smilacaceae	Verbenaceae
Passifloraceae	Ranunculaceae	Solanaceae	Violaceae
Pedaliaceae	Rhamnaceae	Sparganiaceae	Viscaceae
Phytolaccaceae	Rhizophoraceae	Staphyleaceae	Vitaceae
Piperaceae	Rosaceae	Sterculiaceae	Xyridaceae
Pittosporaceae	Rubiaceae	Styracaceae	Zingiberaceae
Plantaginaceae	Rutaceae	Symplocaceae	Zosteraceae
Platanaceae	Salicaceae	Taccaceae	Zygophyllaceae
Plumbaginaceae	Santalaceae	Tamaricaceae	
Poaceae (Gramineae)	Sapindaceae	Theaceae	
Polemoniaceae	Sapotaceae	Theophrastaceae	

Synonyms

Some plants have in the course of time been named more than once scientifically. What was originally thought to have been two separate species, for example, may now be considered to be one variable species. The goal in this book is to provide a useful reference by which anyone can easily find all the known uses of particular plants by Native American peoples. But that would be very difficult if the "same plant" were to show up here under as many as five or ten different scientific names. And there are such cases! In the material gathered for *Native American Ethnobotany*, the champion is *Arctostaphylos uva-ursi*, kinnikinnick or bearberry, with 19; *Gutierrezia sarothrae*, broom snakeweed, had 17. *Oplopanax horridus*, devil's club, had only 15 but they are particularly interesting because four different generic names, three different species names, plus a batch of different plant name authors were found in the ethnobotanical literature:

Echinopanax horridum
Echinopanax horridum (J. E. Sm.) Dene. & Pl.
Echinopanax horridum (Smith)
Fatsia horrida B. & H.
Fatsia horrida (Sm) B. and H.
Fatsia horrida Sm. (B. and H.)
Oplopanax horridum
Oplopanax horridum (J. E. Smith) Miq.
Oplopanax horridum (Smith) Miq.
Oplopanax horridum (Sm.) Miquel.
Oplopanax horridus
Oplopanax horridus (J. E. Smith) Miq.
Oplopanax horridus (J.E. Smith) Miq.
Oplopanax horridus (Smith) Miq.
Panax horridum

One must look closely to see the differences sometimes, but they are there. Variant names for plants are known to botanists as synonyms. Some of the differences reflect differences in how the plant was classified at different times or by different people; some are the result of different conventions (in giving authors' names, for example), and some are errors of omission or other mistakes. The most "correct," current name, using the conventions adopted here, is different yet: *Oplopanax horridus* (Sm.) Torr. & Gray ex Miq. (The "ex" means that although John Torrey and Asa Gray made the taxonomic decision to place a previously described species in the genus *Oplopanax*, F. A. W. Miquel was the person who first published the binomial *Oplopanax horridus* scientifically, adopting the name that Torrey and Gray had proposed.) If a uniform plant nomenclature were not adopted for use in this book, the 157 uses of devil's club would be spread throughout the Catalog of Plants under the different names used in the various sources from which the information was drawn.

The primary source I have used for choosing the most correct, currently used plant name is John Kartesz's *A Synonymized Checklist of the Vascular Flora of the United States, Canada, and Greenland, Second Edition* (Timber Press, 1994). The author kindly sent me a computerized version of the work prior to publication that allowed me to resolve synonyms electronically. Cross-references from synonyms to the accepted scientific name are listed in the Index of Synonyms. It should be clear that I am not taking exception to the botanical classifications of the various ethnobotanists but am relying on a single modern authority to maximize the value of *Native American Ethnobotany*.

Common Plant Names

Although a correctly identified plant can be assigned one, unique scientific name, common names for the plant may be several to many. Unlike scientific names, there are no rules of nomenclature governing common names, and different people in the same place, or people in different places, have bestowed their own names on plants. Not all names so applied have been adopted by others and have been forgotten, and some plants do not have truly commonly used names at all. The common name included in the Catalog of Plants is not necessarily the one used by ethnobotanists reporting on usage. And because various ethnobotanists may have used different common names for the various usages of the plant, usually more than one common name appears in the original sources. When one is available, what has generally been chosen is, instead, a relatively standard common name established by the U.S. Department of Agriculture Natural Resources Conservation Service. The common names used by the various ethnobotanists are cross-referenced in the Index of Common Names, however. Unfortunately, with few exceptions the names applied to plants by Native Americans, their common names for the plants, have not been recorded in the sources and are not indexed. With so many Native American languages, it would require a separate, large book to treat the American Indian names of plants as comprehensively as their uses are cataloged here.

Ethnobotanical Information

Under each plant name, usages are categorized first by the five main categories—Drug, Food, Fiber, Dye, Other—in that order. Each of the five main categories is next divided alphabetically by tribe, all the names of which are listed in the previous chapter under Native Americans. Following the tribe name are all the uses of the plant, listed alphabetically according to the categories defined in the previous chapter under Plant Usage Categories.

In a few hundred instances, usages cited in the Catalog of Plants have numbered superscripts, for example, Pitch taken for any type of bad disease.[1] These numbers refer to more detailed information that would not fit in a short note, and this information is enumerated in the Notes, which precede the Bibliography. The same note may be referred to more than once in the Catalog of Plants.

The statement or statements on usage are followed by an abbreviated reference, for example, (184:197). The reference applies to all preceding statements, even if the reference is separated from the particular statement, for example, by the italicized name of a different usage category or the boldfaced name of a different tribe. The number preceding the colon refers to the source from which the information came, and sources are enumerated in the Bibliography. The number following the colon is the page number in that original source. When the original source used a plant name differs from that adopted in the Catalog of Plants, that is, the source cited a synonymous name, the plant name used in the source is also given within the parentheses, for example, within the entry for *Acer negundo* var. *interius*, (as *Negundo interius* 138:38). When differences are minor, however, such as insignificant differences in author citation, these other names are not given. Cross-references from the synonyms to the names used as the main entries in the Catalog of Plants are listed alphabetically in the Index of Synonyms.

For the reasons described above, under Common Plant Names, only one common name has been included for each plant in the Catalog of Plants. However, an extensive Index of Common Names is provided so that information on Native American plant use may be found even if one does not initially know the scientific name of the plant.

Two comprehensive Plant Usage Indexes are provided so that the information collected in the Catalog of Plants can be found in ways other than by plant name:

Index of Tribes. Categories of plant usage are arranged by names of Native American groups, as listed in the previous chapter under Native Americans. Categories of plant usage are listed first under one of the five main categories—Drug, Food, Fiber, Dye, Other—and then alphabetically under the particular usage. Plants are identified to the level of species in the Index of Tribes. If subspecies or varieties appear in the Catalog of Plants, check under those names, too, for all usages indexed. For example, one may look up Comanche and under Drug, Cold Remedy, see that *Rhus trilobata* was used. The specific ethnobotanical information and the source from which the information was obtained may be found by turning to *Rhus trilobata*, *R. trilobata* var. *pilosissima*, and *R. trilobata* var. *trilobata* in the Catalog of Plants.

Index of Usages. Plant genera are arranged alphabetically under the categories of plant usage, first under one of the five main categories—Drug, Food, Fiber, Dye, Other—and then alphabetically under the particular usage, as defined in the previous chapter under Plant Usage Categories. For example, one may look up Drug, Cold Remedy, and see that *Rhus* was used by the Cahuilla, Cheyenne, Chippewa, Comanche, and Iroquois. The specific ethnobotanical information and the sources from which the information was obtained may be found by turning to *Rhus* in the Catalog of Plants and examining the Drug, Cold Remedy, entries listed for *Rhus*.

Catalog of Plants

***Abies*, Pinaceae, conifer**

Abies amabilis (Dougl. ex Loud.) Dougl. ex Forbes, Pacific Silver Fir

- **Drug**—**Bella Coola** *Eye Medicine* Liquid pitch mixed with mountain goat tallow and used for infected eyes. *Gastrointestinal Aid* Infusion of bark taken for stomach ailments. *Throat Aid* Liquid pitch mixed with mountain goat tallow and taken for sore throat. *Tuberculosis Remedy* Infusion of bark taken for tuberculosis. (184:197) **Haisla** *Tonic* Bark and other plants used as a tonic. *Unspecified* Bark and other plants used for "sickness." (73:152) **Hanaksiala** *Gastrointestinal Aid* Infusion of bark taken for stomach ulcers. *Hemorrhoid Remedy* Infusion of bark taken for hemorrhoids. (43:173) **Kitasoo** *Unspecified* Decoction of bark used medicinally. (43:316) **Nitinaht** *Internal Medicine* Infusion of crushed bark, red alder and hemlock barks taken for internal injuries. *Preventive Medicine* Boughs placed in fire and smoke inhaled to prevent sickness. (186:71) **Oweekeno** *Cold Remedy* Pitch boiled with grease or pitch and sugar and taken for colds. (43:68) **Thompson** *Cold Remedy* Pitch taken for colds.[1] *Panacea* Pitch taken for any type of bad disease.[1] *Tuberculosis Remedy* Poultice of pitch and buttercup roots used for tuberculosis. Pitch taken for tuberculosis.[1] Decoction of boughs and/or bark taken for tuberculosis. *Unspecified* Decoction of branches taken as medicine. (187:97)
- **Food**—**Haisla** *Unspecified* Cambium used for food. (73:151) **Kitasoo** *Unspecified* Inner bark used for food. (43:316) **Nitinaht** *Candy* Hardened pitch chewed for pleasure. (186:71)
- **Fiber**—**Haisla & Hanaksiala** *Snow Gear* Boughs used as a "bush sleigh" to pull cargo across deep snow. (43:173)
- **Other**—**Haisla** *Ceremonial Items* Pitch applied to the face of mourners. **Hanaksiala** *Containers* Boughs used to line oolichan (candlefish) ripening pits. (43:173) **Nitinaht** *Hunting & Fishing Item* Long, hard knots used to make halibut hooks. *Incense & Fragrance* Boughs bundled up and used as home air fresheners. (186:71) **Thompson** *Incense & Fragrance* Boiled boughs mixed with decoctions of other plants and deer grease and used to perfume the hair.[2] (187:97)

Abies balsamea (L.) P. Mill., Balsam Fir

- **Drug**—**Abnaki** *Dermatological Aid* Gum used to make various ointments. Gum used for "slight" itches. (144:164) *Disinfectant* Used as an antiseptic. *Panacea* Leaves made into pillows and used as a panacea. (144:155) *Unspecified* Needles and wood stuffed into pillows and used for good health. (144:163, 164) **Algonquin, Quebec** *Dermatological Aid* Poultice of gum applied to open sores, insect bites, boils, and infections. *Gynecological Aid* Needles used in a sudatory for women after childbirth and for other purposes. *Heart Medicine* Roots used for heart disease. *Laxative* Needles used to make a laxative tea. *Poultice* Needles used for making poultices. *Unspecified* Needles used in a sudatory for women after childbirth and for other purposes. (18:124) **Algonquin, Tête-de-Boule** *Cold Remedy* Sap chewed for colds. (132:118) **Anticosti** *Kidney Aid* Decoction of bark and bark from another plant taken for kidney troubles. Gum eaten for kidney pains. *Throat Aid* Infusion of sap used for sore throats. (143:64) **Chippewa** *Analgesic* Gum melted on warm stone and fumes inhaled for headache. (53:338) *Antirheumatic (External)* Decoction of root used as herbal steam for rheumatic joints. (53:362) *Dermatological Aid* Gum of plant with bear grease used as an ointment for the hair. (53:350) *Herbal Steam* Gum of plant melted on warm stone as herbal steam for headache. (53:338) Decoction of root sprinkled on hot stones and used as herbal steam for rheumatism. (53:362) **Cree, Woodlands** *Abortifacient* Pitch used for menstrual irregularity. *Cold Remedy* Infusion of bark and sometimes wood taken for colds. *Cough Medicine* Infusion of bark and sometimes wood taken for coughs. *Dermatological Aid* Pitch and grease used as an ointment for scabies and boils. Poultice of pitch applied to cuts. *Tuberculosis Remedy* Infusion of bark taken for tuberculosis. Decoction of pitch and sturgeon oil used for tuberculosis. (109:21) **Iroquois** *Antirheumatic (External)* Steam from decoction of branches used as a bath for rheumatism. *Antirheumatic (Internal)* Compound decoction taken for rheumatism. (87:269) *Cancer Treatment* Poultice of gum and dried beaver kidneys applied for cancer. (141:37) *Cold Remedy* Compound decoction taken for colds. (87:269) Infusion of gum and hot milk taken as an antiseptic for colds. (141:37) *Cough Medicine* Decoction taken straight or diluted with alcohol for coughs. *Dermatological Aid* Compound decoction applied to cuts, bruises, sprains, or sores. *Gynecological Aid* Steam from decoction of branches used as a bath for parturition. *Orthopedic Aid* Decoction used as wash and poultice applied to cuts, bruises, sprains, and sores. (87:269) *Tuberculosis Remedy* Compound decoction taken during early stages of consumption. *Urinary Aid* Used for bed-wetting. *Venereal Aid* Used for gonorrhea.

34 *Abies balsamea* (Pinaceae)

(87:270) **Malecite** *Laxative* Juice used as a laxative. (116:244) *Unspecified* Pitch used in medicines. (160:6) *Venereal Aid* Infusion of bark used for gonorrhea. Infusion of bark, spruce bark, and tamarack bark used for gonorrhea. (116:257) **Menominee** *Adjuvant* Inner bark used as a seasoner for medicines. *Analgesic* Infusion of inner bark taken for chest pain. *Cold Remedy* Liquid balsam pressed from trunk used for colds. (151:45) *Dermatological Aid* Gum from plant blisters applied to sores. (54:132) *Pulmonary Aid* Liquid balsam pressed from trunk used for pulmonary troubles. *Unspecified* Poultice of fresh inner bark used for unspecified illnesses. (151:45) **Micmac** *Antidiarrheal* Buds, cones, and inner bark used for diarrhea. *Burn Dressing* Gum used for burns. *Cold Remedy* Gum used for colds. *Dermatological Aid* Gum used for bruises, sores, and wounds. *Gastrointestinal Aid* Cones used for colic. *Laxative* Buds used as a laxative. *Orthopedic Aid* Gum used for fractures. *Venereal Aid* Bark used for gonorrhea and buds used as a laxative. (40:53) **Montagnais** *Dietary Aid* Inner bark grated and eaten to benefit the diet. (156:313) **Ojibwa** *Ceremonial Medicine* Needle-like leaves used as part of the ceremony involving the sweat bath. *Cold Remedy* Balsam gum used for colds and leaf smoke inhaled for colds. (153:378) *Cough Medicine* Plant used as a cough medicine. (135:244) *Dermatological Aid* Balsam gum used for sores and compound containing leaves used as a wash. *Diaphoretic* Needle-like leaves used as part of the medicine for the sweat bath. *Eye Medicine* Liquid balsam from bark blister used for sore eyes. *Stimulant* Leaves used as a reviver and used in compound as a wash. (153:378) **Ojibwa, South** *Cold Remedy* Bark gum taken for chest soreness from colds. *Dermatological Aid* Bark gum applied to cuts and sores. *Diaphoretic* Decoction of bark used to induce sweating. *Venereal Aid* Bark gum taken for gonorrhea. (91:198) **Penobscot** *Burn Dressing* Sap smeared over burns, sores, and cuts. *Dermatological Aid* Sap used as a salve for burns, sores, and cuts. (156:309) **Potawatomi** *Cold Remedy* Needles used to make pillows, believing that the aroma kept one from having a cold. (154:121) Fresh balsam gum swallowed for colds. *Dermatological Aid* Balsam gum used as a salve for sores. *Tuberculosis Remedy* Infusion of bark taken for "consumption and other internal affections." (154:68, 69)
- **Food**—**Micmac** *Beverage* Bark used to make a beverage. (159:258)
- **Fiber**—**Algonquin, Tête-de-Boule** *Mats, Rugs & Bedding* Boughs used as mats on the tent floor. (132:118) **Cree, Woodlands** *Building Material* Boughs used to make a brush shelter. *Canoe Material* Wood used to make paddles. (109:21) **Malecite** *Mats, Rugs & Bedding* Needles and branches used for pillows and bedding. *Sewing Material* Roots used for thread. (160:6) **Micmac** *Mats, Rugs & Bedding* Boughs used to make beds. (159:258) **Ojibwa** *Canoe Material* Resin boiled twice and added to suet or fat to make a canoe pitch. (153:420) **Potawatomi** *Mats, Rugs & Bedding* Needles used to make pillows, believing that the aroma kept one from having a cold. (154:121)
- **Other**—**Malecite** *Waterproofing Agent* Pitch used for waterproofing seams in canoes. (160:6) **Micmac** *Fuel* Wood used for kindling and fuel. (159:258)

Abies concolor (Gord. & Glend.) Lindl. ex Hildebr., White Fir
- **Drug**—**Keres, Western** *Antirheumatic (External)* Infusion of foliage used as a bath for rheumatism. *Antirheumatic (Internal)* Infusion of foliage taken for rheumatism. (171:24) **Paiute** *Dermatological Aid* Poultice of fresh pitch applied to cuts. *Pulmonary Aid* Decoction of needles and bark resin taken for pulmonary troubles. *Tuberculosis Remedy* Soft bark resin eaten or infusion of bark taken for tuberculosis. *Venereal Aid* Compound decoction of resin taken for venereal disease. **Shoshoni** *Dermatological Aid* Simple or compound poultice of warm pitch applied to sores or boils. *Pulmonary Aid* Decoction of needles and bark resin taken for pulmonary troubles. (180:30) **Tewa** *Dermatological Aid* Resinous sap from main stem and larger branches used for cuts. (138:38) **Washo** *Tuberculosis Remedy* Soft bark resin eaten or infusion of bark taken for tuberculosis. (180:30)
- **Fiber**—**Mewuk** *Building Material* Branches used to line acorn caches. (117:346)
- **Dye**—**Klamath** *Brown* Bark used as a tan dye for buckskin. (45:88)
- **Other**—**Keres, Western** *Decorations* Plant used for decoration at dances and in the house. (171:24) **Klamath** *Hide Preparation* Bark used to tan buckskin. (45:88) **Tewa** *Smoking Tools* Twigs used for making pipestems. (138:38)

Abies fraseri (Pursh) Poir., Fraser's Fir
- **Drug**—**Cherokee** *Dermatological Aid* Used for wounds and ulcers. *Gastrointestinal Aid* Taken "to loosen bowels and cleanse and heal internal ulcers." *Gynecological Aid* Used for "falling of the womb," the "whites," and weak backs in females. *Kidney Aid* Burst blister, take ooze alone or with turpentine for "kidney trouble." *Laxative* Taken "to loosen bowels and cleanse and heal internal ulcers." *Pulmonary Aid* "Balsam for breast and lung complaints with pain, soreness or cough." *Urinary Aid* Used for urinary diseases. *Venereal Aid* Used for venereal diseases. (80:34)

Abies grandis (Dougl. ex D. Don) Lindl., Grand Fir
- **Drug**—**Bella Coola** *Eye Medicine* Compound of gum drawn on a hair across sore eyes. (150:50) Liquid pitch mixed with mountain goat tallow and used for infected eyes. (184:197) *Gastrointestinal Aid* Decoction of root bark or stem taken every day for stomach trouble. (150:50) Infusion of bark taken for stomach ailments. (184:197) *Throat Aid* Compound of gum from bark blisters warmed and taken for sore throat. (150:50) Liquid pitch mixed with mountain goat tallow and taken for sore throat. (184:197) *Tuberculosis Remedy* Decoction of root bark or stem taken every day for tuberculosis. (150:50) Infusion of bark taken for tuberculosis. (184:197) **Carrier, Southern** *Unspecified* Tree used as medicine. (150:50) **Chehalis** *Cold Remedy* Decoction of needles taken for colds. (79:19) **Gitksan** *Antirheumatic (External)* Poultice of compound containing bark applied for rheumatism. *Dermatological Aid* Poultice of compound containing bark applied to boils or ulcers. *Pulmonary Aid* Poultice of compound containing bark used as a chest plaster for lung hemorrhage. (150:50) **Green River Group** *Cold Remedy* Decoction of needles taken for colds. (79:19) **Hesquiat** *Dermatological Aid* Blister pitch mixed with oil rubbed on the hair and scalp because it smelled nice. Blister pitch mixed with oil rubbed on the scalp to prevent the hair from falling out. (185:41) **Karok** *Tonic* Infusion of needles taken as a tonic. (148:379) **Kwakiutl** *Ceremonial Medicine* Branches and pollen used in purification rites and ceremonies. *Cough Medicine* Decoction of pitch taken for coughs and tuberculosis. *Dermatological Aid* Pitch and grease eaten or rubbed on sores and boils. *Laxative* Decoction of pitch taken as a tonic and laxative. Pitch and catfish oil taken for constipation. *Oral*

Aid Root held in the mouth for gum boils and canker sores. *Tonic* Infusion of bark taken as a tonic to stay young and strong. *Tuberculosis Remedy* Decoction of pitch taken or pitch rubbed on chest and back for tuberculosis. (183:268) **Nitinaht** *Internal Medicine* Infusion of crushed bark, red alder and hemlock barks taken for internal injuries. *Unspecified* Boughs placed in fire and smoke inhaled to prevent sickness. (186:71) **Okanagan-Colville** *Cough Medicine* Decoction of bark taken for bad coughs. *Dermatological Aid* Needles dried, powdered, mixed with marrow, and used to scent the hair and keep from going bald. Bark dried, powdered, and rubbed on the neck and under the arms as a deodorant. *Dietary Aid* Pitch taken for a loss of appetite. *Gastrointestinal Aid* Pitch taken for ulcers. Decoction of bark taken for "bad stomachs" with loss of appetite and loss of weight. *Gland Medicine* Pitch mixed with deer marrow and applied externally each evening for goiter. *Other* Branch tips chewed for allergies caused by water hemlock. *Strengthener* Pitch taken for a general feeling of weakness. *Tuberculosis Remedy* Pitch taken for consumption. (188:23) **Okanogon** *Cathartic* Decoction of bark and gum taken as a physic. *Eye Medicine* Gum used for sore eyes. (125:41) **Saanich** *Dermatological Aid* Pitch mixed with venison suet and used for psoriasis and other skin diseases. Pitch made into a salve and used for cuts and bruises. (182:69) **Salish, Coast** *Dermatological Aid* Infusion of pounded root bark used for falling hair and dandruff. (182:69) **Shuswap** *Dermatological Aid* Poultice of soft pitch applied to sores. *Panacea* Decoction of bark taken for tuberculosis and other sickness. *Toothache Remedy* Hard pitch chewed to clean the teeth. *Tuberculosis Remedy* Decoction of bark taken for tuberculosis. (123:50) **Thompson** *Cathartic* Decoction of bark and gum taken as a physic. (164:462) *Cold Remedy* Pitch taken for colds.[1] (187:97) *Eye Medicine* Gum used for sore eyes. (125:41) Decoction of bark used as a wash for sore eyes and gum used in corners of eyes. (164:462) *Panacea* Infusion of boughs taken for any illness. Pitch taken for any type of bad disease.[1] (187:97) *Pediatric Aid* Branches thought to be of help to young girl under "magical spell." (164:509) *Tuberculosis Remedy* Poultice of pitch and buttercup roots used for tuberculosis. Pitch taken for tuberculosis.[1] *Unspecified* Decoction of branches taken as medicine. (187:97) *Venereal Aid* Very strong decoction of various plant parts taken for gonorrhea. (164:462) *Witchcraft Medicine* Branches thought to be of help to young girl under "magical spell." (164:509)

- *Food*—**Nitinaht** *Candy* Hardened pitch chewed for pleasure. (186:71) **Shuswap** *Beverage* Gum from inside the bark, next to the trunk, made into a drink. *Candy* Gum from inside the bark, next to the trunk, chewed. (123:50) **Thompson** *Beverage* Branch tips sometimes steeped to make a tea-like beverage. (187:97)

- *Fiber*—**Hesquiat** *Clothing* Boughs used by wolf dancers as decorative clothing. (185:41) **Okanagan-Colville** *Mats, Rugs & Bedding* Boughs used as a bedding base in the sweat house. (188:23) **Paiute** *Mats, Rugs & Bedding* Boughs used on the floor of sweat houses and for beds. (111:44) **Thompson** *Building Material* Bark used for covering lodges. Branches used to make temporary lodge flooring. *Canoe Material* Bark used to make canoes. *Mats, Rugs & Bedding* Branches used for bedding. (164:496) Boughs used as bedding and temporary floor coverings and changed every 2 to 3 days. (187:97)

- *Other*—**Chehalis** *Fuel* Wood used for fuel. (79:19) **Hesquiat** *Incense & Fragrance* Fragrant boughs placed under bedding as an incense. (185:41) **Nitinaht** *Hunting & Fishing Item* Long, hard knots used to make halibut hooks. *Incense & Fragrance* Boughs bundled up and used as home air fresheners. (186:71) **Thompson** *Incense & Fragrance* Boiled boughs mixed with decoctions of other plants and deer grease and used to perfume the hair.[2] *Protection* Boughs used by young girls to scrub the face and clothes so that they would not be bothered by bears. (187:97)

Abies lasiocarpa (Hook.) Nutt., Subalpine Fir
- *Drug*—**Blackfoot** *Analgesic* Needle smudge smoke inhaled for headaches. (86:79) *Ceremonial Medicine* Plant burned as ceremonial incense. (114:273) *Cold Remedy* Poultice of leaves applied for chest colds. (97:17) Poultice of plant applied for chest colds. (114:273) *Dermatological Aid* Needles used as a deodorant. (86:107) Infusion of needles mixed with grease and applied as a hair tonic. Needles packed into moccasins as a foot deodorant. (86:123) Leaves mixed with grease and used as hairdressing. Gummy secretions used on wounds as an antiseptic. (97:17) *Emetic* Infusion of resin taken as an emetic to clean the insides. (86:65) *Febrifuge* Poultice of leaves applied for fevers. (97:17) Poultice of plant applied for fevers. (114:273) *Oral Aid* Resin chewed for bad breath and pleasure. (86:123) *Pulmonary Aid* Gummy secretions taken for lung troubles. (97:17) *Stimulant* Needle smudge smoke inhaled for fainting. (86:79) *Tuberculosis Remedy* Infusion of needles taken for coughing up blood, a sign of tuberculosis. Needle smudge used to fumigate the patient with tuberculosis. (86:70) *Venereal Aid* Needle smudge used to fumigate those faces that were swollen from a form of venereal disease. (86:69) *Veterinary Aid* Needle smudge used to fumigate sick horses. Little bags of needles tied on a belt and hung around the horse's neck as a perfume. Infusion of bark given to horses for diarrhea. (86:87) Ground needles used in horse medicine bundles. (97:17) **Cheyenne** *Ceremonial Medicine* Needles burned as incense in ceremonies by persons afraid of thunder. *Stimulant* Plant used to revive a dying person's spirit. *Witchcraft Medicine* Burning needle smoke and aroma used to chase away bad influences (illness). (83:5) **Crow** *Ceremonial Medicine* Young twigs and leaves burned as incense in certain ceremonies. (19:5) *Cold Remedy* Infusion of crushed needles taken for colds. *Cough Medicine* Infusion of crushed needles taken for coughs. *Laxative* Infusion of crushed needles used for constipation. **Flathead** *Dermatological Aid* Needles dried, pounded, mixed with deer grease, and used as a hair tonic. Needles, lovage roots, buckbrush leaves, and pinedrops boiled and used to make hair grow longer. Needles pounded and used alone or mixed with grease or marrow for skin diseases. Needles pulverized into baby powder and used for rashes from excessive urination. *Oral Aid* Needles pounded, mixed with lard, and used for bleeding gums. *Pediatric Aid* Needles pulverized into baby powder and used for rashes from excessive urination. (82:2) **Gitksan** *Cold Remedy* Decoction of bark or inner bark used for colds. *Cough Medicine* Decoction of bark or inner bark used for coughs. *Misc. Disease Remedy* Decoction of bark or inner bark used for flu. *Tonic* Decoction of bark or inner bark used as a tonic. (73:152) **Kutenai** *Dermatological Aid* Gummy bark secretions used for cuts and bruises. **Montana Indian** *Cold Remedy* Poultice of needles used for colds. Infusion of needles and resinous blisters used for colds. (82:2) *Dermatological Aid* Gummy secretion from the bark

used as an antiseptic for wounds and ulcers. Plant applied to corns for easy removal. (19:5) Gummy bark secretions used as an antiseptic for wounds. *Febrifuge* Poultice of needles used for chest fevers. (82:2) *Pulmonary Aid* Gummy secretion from the bark taken for lung troubles. (19:5) **Okanagan-Colville** *Cough Medicine* Decoction of bark taken for bad coughs. *Dermatological Aid* Bark dried, powdered, and rubbed on the neck and under the arms as a deodorant. Needles dried, powdered, mixed with marrow, and used to scent the hair and keep from going bald. *Dietary Aid* Pitch taken for a loss of appetite. *Gastrointestinal Aid* Pitch taken for ulcers. Decoction of bark taken for "bad stomachs" with loss of appetite and loss of weight. *Gland Medicine* Pitch mixed with deer marrow and applied externally each evening for goiter. *Other* Branch tips chewed for allergies caused by water hemlock. *Strengthener* Pitch taken for a general feeling of weakness. *Tuberculosis Remedy* Pitch taken for consumption. (188:23) **Shoshoni** *Cold Remedy* Infusion of needles taken for colds. Infusion of resinous blisters taken for colds. (118:37) **Thompson** *Cold Remedy* Pitch taken for colds.[1] *Cough Medicine* Decoction of bark taken over a period of time for bad coughs. *Dermatological Aid* Poultice of pitch used alone for cuts or with Vaseline for sores. The pitch was smeared over injuries and covered with a bandage. In earlier times, animal fat was probably used in place of Vaseline Decoction of bark taken over a period of time for bruises. *Orthopedic Aid* Decoction of bark taken over a period of time for sprains. *Panacea* Pitch taken for any type of bad disease.[1] *Tuberculosis Remedy* Inner bark eaten as a medicine for "shadow on the chest," the beginning of tuberculosis. It made the informant very sick with aching, flu-like symptoms, but after that, she did not develop tuberculosis. Poultice of pitch and buttercup roots used for tuberculosis. Pitch taken for tuberculosis.[1] Decoction of boughs and/or bark taken for tuberculosis. *Unspecified* Decoction of branches taken as medicine. (187:97) **Wet'suwet'en** *Cold Remedy* Decoction of bark or inner bark used for colds. *Cough Medicine* Decoction of bark or inner bark used for coughs. *Misc. Disease Remedy* Decoction of bark or inner bark used for flu. *Tonic* Decoction of bark or inner bark used as a tonic. (73:152)

- **Food**—**Blackfoot** *Candy* Cones pulverized into a fine powder, mixed with back fat and marrow, and eaten as a confection. The fragments of the cones were left by squirrels and chipmunks and gathered by the Indians to make the confection. The confection was distributed during social gatherings and meetings. It was an aid to digestion as well as a delicacy. (86:100) Resin chewed for bad breath and pleasure. (86:123) Cones pulverized into a fine powder, mixed with back fat and marrow, and eaten as a confection. Cones pulverized into a fine powder, mixed with back fat and marrow, and eaten as a confection. The confection was an aid to digestion as well as a delicacy. (86:100) **Shuswap** *Unspecified* Seeds used for food. (123:50) **Thompson** *Unspecified* Inner bark used for food. (187:97)
- **Fiber**—**Hanaksiala** *Furniture* Wood used to make chairs and insect-proof storage boxes for dancing regalia. (43:174) **Okanagan-Colville** *Mats, Rugs & Bedding* Boughs used as a bedding base in the sweat house. (188:23) **Shuswap** *Mats, Rugs & Bedding* Heated branches used to make a warm bed. Branches used as floor of sweat house, after swimming to keep feet clean, and when butchering a deer. (123:50) **Thompson** *Mats, Rugs & Bedding* Boughs valued as bedding and temporary floor coverings and changed every 2 to 3 days. (187:97)
- **Other**—**Blackfoot** *Ceremonial Items* Chewed needles sprayed over the Horn ceremonial containers to purify them. Needle incense used for transfer ceremonies of Black Spring tepee design and Bear Medicine Hat bundle. (86:36) *Incense & Fragrance* Needles packed with stored items, saddle pads or burned in a household smudge. (86:107) Used for incense, primarily in association with the Medicine Pipe bundles. (86:36) Leaves used as perfume. (97:17) Leaves used in buckskin bags for delightful odor and mixed with grease to add fragrance to hair oil. (114:278) *Paint* Needles mixed with dry paint to make it smell better. *Protection* Needle smudge used for safety's sake during severe thunderstorms. (86:36) **Cheyenne** *Ceremonial Items* Used to drive away ill spirits or to revive spirits of the dying. Used by Sundancers for confidence and protection from thunder and for purification. **Flathead** *Incense & Fragrance* Needles placed on stoves as incense or pulverized and used as body and garment scents. Boughs used as incense. (82:2) **Hanaksiala** *Containers* Wood used to make chairs and insect-proof storage boxes for dancing regalia. *Designs* Tree used as a design on wooden drums. (43:174) **Nez Perce** *Ceremonial Items* Branches used to drive away ghosts and bad spirits. *Incense & Fragrance* Boughs burned and used as incense in sweat houses. (82:2) **Shuswap** *Cash Crop* Cones sold for money. *Fuel* Wood used for a long-lasting fire. (123:50) **Thompson** *Incense & Fragrance* Boiled boughs mixed with decoctions of other plants and deer grease and used to perfume the hair.[2] *Soap* Sweet smelling bough used by "old people" to wash their skin, to give their bodies a nice scent. (187:97)

Abies procera Rehd., Noble Fir

- **Drug**—**Paiute** *Cold Remedy* Dried branches stored for use as a cold remedy. Crumbled leaves smoked for colds. Mashed leaves sewn into a sack placed around the child's neck "for colds." *Cough Medicine* Decoction of leaves taken as cough medicine. *Dermatological Aid* Dried branches stored for use as a deodorant. (as *A. nobilis* 111:45)

***Abies* sp.**, Balsam Fir

- **Drug**—**Carrier** *Dermatological Aid* Poultice of chewed nodules applied to wounds. (31:71) *Eye Medicine* Poultice of pitch applied to injured eyes. *Tuberculosis Remedy* Decoction of tree warts and pitch taken for tuberculosis. (31:70) **Carrier, Northern** *Burn Dressing* Gum used as an ointment on wounds, especially on burns. *Cathartic* Decoction of bark taken as a purgative. *Dermatological Aid* Gum used as an ointment on wounds, especially on burns. (150:51) **Cheyenne** *Adjuvant* Leaves and fungus burned on coals, with or without sweet grass, and used to strengthen medicines. *Ceremonial Medicine* Leaves with fungus burned on coals, with or without sweet grass, and used ceremonially to purify sickness. (76:169) **Gitksan** *Cathartic* Gum or mashed cones used as purgative and diuretic for consumption and gonorrhea. *Dermatological Aid* Gum or mashed cones applied to cuts and sores, especially gonorrheal sores. *Diuretic* Gum or mashed cones used as purgative and diuretic for consumption and gonorrhea. *Laxative* Juicy inner bark taken for constipation. *Other* Gum or mashed cones taken for many serious ailments. *Tuberculosis Remedy* Gum or mashed cones used as purgative and diuretic for consumption. *Venereal Aid* Gum or mashed cones used as purgative and diuretic for gonorrhea. Gum or mashed cones taken for gonorrhea and applied to gonorrheal sores. **Sikani** *Dermatological Aid*

Gum applied to wounds. (150:51) **Thompson** *Gastrointestinal Aid* Decoction of shoots and bark taken for stomach trouble. *Tonic* Decoction of shoots and bark taken as a tonic. (164:462)
- *Food*—**Thompson** *Unspecified* Gum chewed and swallowed. (164:484)
- *Fiber*—**Carrier** *Building Material* Wood used to make shingles for roofs. (31:70)
- *Other*—**Carrier** *Hide Preparation* Rotten wood used to smoke and tan skins. (31:70)

Abronia, Nyctaginaceae

Abronia elliptica A. Nels., Fragrant White Sand Verbena
- *Drug*—**Hopi** *Pediatric Aid* Plant placed on child's head to induce sleep. (200:75) *Sedative* Plant placed on child's head to induce sleep. (200:36, 75)

Abronia fragrans Nutt. ex Hook., Snowball Sand Verbena
- *Drug*—**Keres, Western** *Dietary Aid* Roots ground, mixed with corn flour and eaten to give one a good appetite and to make one fat. *Psychological Aid* Roots ground, mixed with corn flour, and eaten to keep one from becoming greedy. (171:24) **Navajo** *Dermatological Aid* Plant used for boils. (90:158) *Gastrointestinal Aid* Plant taken to "remove the effects of swallowing a spider." (55:46) **Navajo, Kayenta** *Cathartic* Plant used as a cathartic. *Dermatological Aid* Plant used for insect bites. *Diaphoretic* Plant used as a sudorific. *Emetic* Plant used as an emetic. *Gastrointestinal Aid* Plant used for stomach cramps. *Panacea* Plant used as a life medicine. (205:21) **Navajo, Ramah** *Dermatological Aid* Cold infusion used as lotion for sores or sore mouth and to bathe perspiring feet. *Oral Aid* Cold infusion used as lotion for sores or sore mouth. (191:26) **Ute** *Gastrointestinal Aid* Roots and flowers used for stomach and bowel troubles. (38:32) **Zuni** *Gastrointestinal Aid* Fresh flowers eaten for stomachaches. (27:377)
- *Food*—**Acoma** *Unspecified* Roots ground, mixed with cornmeal, and eaten. **Laguna** *Unspecified* Roots ground, mixed with cornmeal, and eaten. (32:39)
- *Other*—**Keres, Western** *Ceremonial Items* Flowers made into ceremonial necklaces. (171:24)

Abronia latifolia Eschsch., Coastal Sand Verbena
- *Food*—**Clallam** *Unspecified* Roots used for food. (57:201) **Klallam** *Unspecified* Roots used for food. **Makah** *Unspecified* Roots eaten in the fall. (79:29)

Abronia turbinata Torr. ex S. Wats., Transmontane Sand Verbena
- *Drug*—**Shoshoni** *Dermatological Aid* Poultice of mashed leaves applied to swellings. (180:30)

Abronia villosa S. Wats., Desert Sand Verbena
- *Drug*—**Paiute** *Diuretic* Used as an urinary inducer. (118:41) **Shoshoni** *Burn Dressing* Poultice of mashed roots applied to burns. (180:30)

Abutilon, Malvaceae

Abutilon incanum (Link) Sweet, Pelotazo
- *Drug*—**Hawaiian** *Gastrointestinal Aid* Dried flowers eaten for gripping stomachaches. Flowers, root bark, and other plants pounded, resulting liquid heated and taken for stomachaches. (2:69)

Acacia, Fabaceae

Acacia greggii Gray, Catclaw Acacia
- *Food*—**Cahuilla** *Porridge* Dried pods ground into flour and used to make mush or cakes. *Vegetable* Pods eaten fresh. (15:29) **Diegueño** *Fodder* Used to feed domesticated animals. (88:218) **Havasupai** *Bread & Cake* Seeds stored, roasted, ground, and made into bread. (197:225) **Pima** *Unspecified* Beans formerly used for food. (146:76) **Pima, Gila River** *Starvation Food* Seeds used as "starvation food." (133:7) **Seri** *Porridge* Beans ground into a meal, mixed with water or sea lion oil, and eaten. (50:136)
- *Fiber*—**Cahuilla** *Building Material* Considered an outstanding construction material and a fine firewood. (15:29) **Havasupai** *Basketry* Split twigs used as basket material. *Brushes & Brooms* Twigs made into a brush and used to brush off metates. (197:225) **Papago** *Basketry* Thorns removed, twigs split in half lengthwise, and used to make serviceable baskets. (34:57) *Other* Twigs used for curved structures in wrapped weaving. (34:53) **Pima** *Building Material* Bushes dried, piled high, and used as brush fences. *Furniture* Branches used to make cradle frames. (47:90)
- *Other*—**Cahuilla** *Fuel* Considered an outstanding construction material and a fine firewood. (15:29) **Papago** *Hunting & Fishing Item* Fitted around deer hunters' heads and used in sizing deer head disguises. (34:40) *Incense & Fragrance* Buds and blossoms dried and used by women as perfume sachets. (34:52) *Tools* Short transverse sticks affixed to poles and used to dislodge saguaro fruits from the shafts. (34:20) Curved rods used for fleshing and dehairing animal skins. Stems peeled of bark and thorns and used to beat sheep hides to make them more pliable. (34:69) **Pima** *Fuel* Bushes dried and used for firewood. *Hunting & Fishing Item* Wood used to make bows. Used to make bows. (47:90)

Acacia koa Gray, Koa
- *Drug*—**Hawaiian** *Diaphoretic* Leaves spread out on the bed to cause the patient lying on them to sweat. *Pediatric Aid* Ashes of this and other plants applied to the mouth interior of infants for physical weakness. *Strengthener* Ashes of this and other plants applied to the mouth interior of infants for physical weakness. (2:46)
- *Fiber*—**Hawaiian** *Canoe Material* Wood used to make canoes. (2:46)

Acacia sp., Cat's Claw
- *Fiber*—**Hualapai** *Basketry* Limbs split and used to coil around the edges of baskets. *Furniture* Limbs used for cradleboard "spudi." Roots used to make the cradleboard frame. (195:14) **Walapai** *Furniture* Used for cradle frames. (17:49)

Acacia willardiana, Palo Blanco
- *Fiber*—**Seri** *Basketry* Splints used to make the woof for basketry. (50:138)
- *Other*—**Seri** *Hunting & Fishing Item* Used to make bows. (50:138)

Acalypha, Euphorbiaceae

Acalypha virginica L., Virginia Threeseed Mercury
- *Drug*—**Cherokee** *Kidney Aid* Root used for "dropsy." *Misc. Disease Remedy* Root used for pox. *Urinary Aid* Root used for "gravel." (80:61)

Acamptopappus, Asteraceae

Acamptopappus sphaerocephalus var. *hirtellus* Blake, Rayless Goldenhead

- **Drug**—Kawaiisu *Analgesic* Mashed plant used as a salve for pain. (206:9)

Acanthaceae, see *Justicia, Stenandrium*

Aceraceae, see *Acer*

Acer, Aceraceae

Acer alba, White Maple
- **Drug**—Micmac *Cough Medicine* Bark used as a cough remedy. (40:53)

Acer circinatum Pursh, Vine Maple
- **Drug**—Karok *Love Medicine* Branches used by women as a love medicine. (148:385) **Thompson** *Antidiarrheal* Wood burned to charcoal, mixed with water and brown sugar, and taken for dysentery.[3] *Misc. Disease Remedy* Wood burned to charcoal, mixed with water and brown sugar, and taken for polio.[3] (187:145)
- **Food**—Clallam *Dried Food* Sap eaten dried. *Unspecified* Sap eaten fresh. (57:197)
- **Fiber**—Makah *Basketry* Used to make baskets. (67:285) **Modesse** *Snow Gear* Branches used for the frames of snowshoes. (117:223) **Quinault** *Basketry* Straight shoots used to make openwork baskets for general household utilities. *Building Material* Poles used to hold down roof planks on houses. (79:40) **Shuswap** *Snow Gear* Used to make snowshoes. (123:56) **Skagit** *Furniture* Saplings used as swings for baby cradles. (79:40) **Thompson** *Furniture* Wood used in making baby basket frames. (187:145) *Snow Gear* Wood used to make snowshoes. (164:498) Wood used in making snowshoe frames. (187:145)
- **Other**—Chehalis *Hunting & Fishing Item* Used to make the wattleworks of fish traps. (79:40) **Karok** *Cooking Tools* Wood used to made acorn paddles. (6:15) **Lummi** *Hunting & Fishing Item* Used to make the wattleworks of fish traps. (79:40) **Nitinaht** *Containers* Wood used to make small boxes and oil containers. *Cooking Tools* Wood used to make bowls and drinking containers. (186:90) *Hunting & Fishing Item* Used to make bows. (186:76) Wood used to make bows. (186:90) **Quileute** *Hunting & Fishing Item* Used to make the wattleworks of fish traps. **Quinault** *Paint* Charcoal mixed with oil and used as black paint. **Skagit** *Tools* Saplings used to make salmon tongs. (79:40) **Thompson** *Hunting & Fishing Item* Wood gathered while green, the heart removed and used in making bows, arrows, and dip net frames. *Tools* Wood used in making implement handles. (187:145)

Acer glabrum Torr., Rocky Mountain Maple
- **Drug**—Blackfoot *Cathartic* Infusion of bark taken in the morning as a cathartic. (86:65) **Navajo, Ramah** *Panacea* Infusion of branches used for swellings, a "life medicine." (191:36) **Okanagan-Colville** *Hunting Medicine* Branch tied in a knot and placed over the bear's tracks while hunting to stop the wounded bear. (188:59) **Thompson** *Antiemetic* Decoction of wood and bark taken for nausea caused by smelling a corpse. (164:475) *Gynecological Aid* Decoction of sticks and saskatoon sticks taken to heal women's insides and stimulate lactation. The decoction was made either with two sticks each of saskatoon and rocky mountain maple, or, for a stronger medicine, four sticks each and used after childbirth to heal women's insides and to stimulate the flow of milk for nursing. *Snakebite Remedy* Decoction of four straight, young sticks used as a wash or taken for snakebites. The informant could not recall whether the decoction was taken internally or used as a wash. (187:146)
- **Food**—Blackfoot *Spice* Dried, crushed leaves used to spice stored meat. (86:100)
- **Fiber**—Okanagan-Colville *Building Material* Wood used to make the frame of a woman's sweat house. *Snow Gear* Wood used to make snowshoes. (188:59) **Thompson** *Cordage* Fibrous inner bark used to make twine. *Furniture* Wood used for cradle frames. (187:146) *Snow Gear* Wood used for making snowshoes. (164:499) Twigs used to make snowshoe frames. (164:500) Branches used whenever obtainable to make snowshoe frames. (187:146)
- **Other**—Bella Coola *Cooking Tools* Bark used to make spoons. *Tools* Straight sticks tied around a tree base, ignited, and burned until the tree fell. (184:200) **Blackfoot** *Containers* Bark used to make paint containers. (86:107) **Isleta** *Hunting & Fishing Item* Straight, young tree trunks used to make bows. (100:20) **Montana Indian** *Hunting & Fishing Item* Young twigs used for making fish net hoops. (19:5) **Okanagan-Colville** *Cooking Tools* Wood used to make screens for smoke drying racks. Wood used to make tongs for removing food from boiling water. *Fasteners* Wood used to make tepee pegs. *Fuel* Wood used for fuel. *Musical Instrument* Wood used to make drum hoops. *Tools* Wood used to make spear handles. *Toys & Games* Wood used to make a child's first bow. (188:59) **Thompson** *Hunting & Fishing Item* Wood used for making bows. (164:499) **Yurok** *Toys & Games* Fruit used to play with. (6:15)

Acer glabrum var. *douglasii* (Hook.) Dippel, Douglas Maple
- **Fiber**—Gitksan *Basketry* Inner bark used to make baskets. *Mats, Rugs & Bedding* Inner bark used to make mats. (73:153) **Haisla & Hanaksiala** *Snow Gear* Wood used to make snowshoes. (43:209)
- **Other**—Haisla *Ceremonial Items* Wood made into rattles and used by shamans. **Haisla & Hanaksiala** *Ceremonial Items* Wood used to make frontispieces for chief's masks. *Cooking Tools* Wood used to make spoons. *Tools* Wood used to make ax handles. **Hanaksiala** *Ceremonial Items* Wood used to make the skulls for the shamanistic costumes. (43:209) **Oweekeno** *Cooking Tools* Wood used to make spoons. (43:82)

Acer glabrum var. *neomexicanum* (Greene) Kearney & Peebles, New Mexico Maple
- **Food**—Apache, Chiricahua & Mescalero *Sweetener* Sap collected and boiled to obtain syrup and sugar. (as *A. neomexicana* 33:44)

Acer macrophyllum Pursh, Bigleaf Maple
- **Drug**—Klallam *Tuberculosis Remedy* Infusion of bark taken for tuberculosis. (79:39) **Kwakiutl** *Dermatological Aid* Sticky gum from the spring buds mixed with oil and used as a hair tonic. (183:275) **Thompson** *Tonic* Raw sap used as a tonic in the olden days. (187:147)
- **Food**—Clallam *Dried Food* Sap eaten dried. *Unspecified* Sap eaten fresh. (57:197) **Costanoan** *Unspecified* Seeds used for food. (21:248) **Cowichan** *Spice* Leaves used in steaming pits to flavor deer, seal, or porpoise meat. **Saanich** *Spice* Leaves used in steaming pits to flavor deer, seal, or porpoise meat. **Salish, Coast** *Unspecified* Cambium eaten in small quantities with oil. (182:77) **Thompson** *Sauce & Relish* Sap boiled to make a type of maple syrup. *Unspecified* Raw shoots used for food. *Vegetable* Sprouted seeds boiled and eaten as green vegetables. The sprouted seeds were generally bitter, but the young

shoots were considered to be quite sweet and juicy. (187:147)
- *Fiber*—**Cahuilla** *Building Material* Limbs used for house construction and considered good firewood. (15:29) **Clallam** *Canoe Material* Wood used to make canoe paddles. (57:197) **Concow** *Basketry* Inner bark used in spring to make baskets. *Clothing* Bark used to make crude dresses. (41:365) **Cowlitz** *Cordage* Bark used to make rope and tumplines. (79:39) **Karok** *Mats, Rugs & Bedding* Leaves made into mats and used to cover the layers of dried salmon stored for the winter in baskets. (148:385) **Klallam** *Canoe Material* Wood used to make canoe paddles. **Lummi** *Furniture* Wood used to make cradleboards. (79:39) **Maidu** *Basketry* Withes used as coarse twine warp and weft in the manufacture of baskets. *Sewing Material* Withes used as coiling thread. (173:71) **Makah** *Basketry* Used to make baskets. (67:285) **Nisqually** *Building Material* Boughs used to cover temporary housing. (79:39) **Nitinaht** *Canoe Material* Hard, lightweight wood used to make paddles. (186:91) **Skagit** *Canoe Material* Wood used to make canoe paddles. **Snohomish** *Canoe Material* Wood used to make canoe paddles. **Swinomish** *Furniture* Wood used to make cradleboards. (79:39) **Thompson** *Basketry* Inner bark used to make scouring pads, temporary baskets, and sometimes for weaving bags. *Scouring Material* Inner bark used to make scouring pads and sometimes for weaving bags and for temporary baskets. (187:147) **Tolowa** *Clothing* Bark fibers used to make women's skirts. (6:15)
- *Other*—**Cahuilla** *Fuel* Limbs used for house construction and considered good firewood. (15:29) **Chehalis** *Cooking Tools* Dead wood used for smoking salmon. (79:39) **Cowichan** *Cooking Tools* Large leaves used for lining baskets, wrapping fish, and placing on berry drying racks. (182:77) **Haida** *Decorations* Wood used to make wood carvings. (79:39) **Hesquiat** *Toys & Games* White wood used to make rattles. (185:60) **Karok** *Containers* Leaves placed under and between layers of the bulbs while cooking in the earth oven. (148:385) *Cooking Tools* Wood used to made acorn paddles. (6:15) Wood made into the paddle used for stirring the food in the cooking baskets. (148:385) **Kwakiutl, Southern** *Ceremonial Items* Wood used to carve masks. *Cooking Tools* Wood used to carve dishes and spoons. *Musical Instrument* Wood used to carve rattles. (183:296) **Lummi** *Cooking Tools* Leaves used to cover food cooking in pits. Wood used to make dishes and spoons. (79:39) **Nitinaht** *Ceremonial Items* Hard, lightweight wood used to make masks and ceremonial rattles. *Cooking Tools* Hard, lightweight wood used to make bowls. (186:91) **Pomo** *Toys & Games* Wood used to make dice for a gambling game. (66:14) **Pomo, Kashaya** *Toys & Games* Branches used to make staves for a dice-type gambling game. (72:70) **Quinault** *Fuel* Dead wood used for smoking salmon. (79:39) **Salish, Coast** *Fuel* Wood used as an excellent fuel. (182:77) **Skagit** *Cooking Tools* Leaves used to cover food cooking in pits. (79:39) **Skagit, Upper** *Containers* Leaves used to cover food cooking in pits. (179:42) **Snohomish** *Cooking Tools* Leaves used to cover food cooking in pits. **Squaxin** *Containers* Leaves used to lay fish on while cleaning. **Swinomish** *Cooking Tools* Dead wood used for smoking salmon. Wood used to make dishes and spoons. (79:39) **Thompson** *Containers* Inner bark used to make scouring pads, temporary baskets, and sometimes for weaving bags. *Cooking Tools* Inner bark used to make soapberry whippers. Children sometimes made miniature whisks which they used to whip the juice that was left after the dried soapberries were soaked. They made the juice with their whisks and then drank it. Leaves used to line the containers used in making ripened salmon eggs. The maple leaves were used to line the basket and were placed in layers between the eggs. The eggs were generally prepared in a birch bark basket, placed in a hole in the ground lined with birch bark, and left there until springtime when they were considered cooked. Leaves used in pit cooking, to line the pit and interspersed between the layers of food. The leaves were also used between layers of fish in fish caches. Wood used to make soapberry eating paddles. (187:147) **Tlingit** *Decorations* Wood used to make wood carvings. **Tsimshian** *Decorations* Wood used to make wood carvings. (79:39) **Wailaki** *Hunting & Fishing Item* Bark cut into 1-inch bands, fastened together into a roll, and used to catch deer. (41:365)

Acer negundo L., Box Elder

- *Drug*—**Cheyenne** *Ceremonial Medicine* Wood burned as incense for making spiritual medicines. (82:4) **Meskwaki** *Emetic* Decoction of inner bark taken as an emetic. (152:200) **Ojibwa** *Emetic* Infusion of inner bark taken as an emetic. (153:353)
- *Food*—**Apache, Chiricahua & Mescalero** *Dried Food* Inner bark scrapings dried and kept for winter use. *Sweetener* Inner bark boiled until sugar crystallizes out of it. (33:44) **Cheyenne** *Candy* Sap mixed with shavings from inner sides of animal hides and eaten as candy. (82:4) Sap boiled, added to animal hide shavings, and eaten as a relished candy. (83:13) **Dakota** *Sweetener* Sap used to make sugar. (69:366) **Montana Indian** *Sauce & Relish* Sap boiled or frozen and used as a sweet syrup. (82:4) **Ojibwa** *Beverage* Sap mixed with the sap of the sugar maple and used as a beverage. (153:394) **Omaha** *Sweetener* Sap boiled to make sugar and syrup. (68:329) Sap used to make sugar. **Pawnee** *Sweetener* Sap used to make sugar. **Ponca** *Sweetener* Sap used to make sugar. **Winnebago** *Sweetener* Sap used to make sugar. (70:101)
- *Other*—**Cheyenne** *Ceremonial Items* Wood burned during Sundance ceremonies. (82:4) *Cooking Tools* Wood used to make bowls. (83:46) *Fuel* Wood burned and used for cooking meat. (82:4) Wood used as firewood for cooking meat. (83:13) **Dakota** *Ceremonial Items* Wood made into charcoal and used for ceremonial painting and tattooing. (70:101) *Decorations* Wood used to obtain charcoal for tattooing. (69:366) **Keres, Western** *Ceremonial Items* Twigs made into prayer sticks. (171:24) **Kiowa** *Ceremonial Items* Wood burned in the altar fire of the peyote ceremony. (192:40) **Montana Indian** *Cooking Tools* Large trunk burls or knots used to make bowls and dishes. *Musical Instrument* Large trunk burls or knots used to make drums. *Smoking Tools* Large trunk burls or knots used to make pipestems. (82:4) **Navajo** *Tools* Wood used to make tubes for bellows. (55:62) **Omaha** *Ceremonial Items* Wood made into charcoal and used for ceremonial painting and tattooing. (70:101) *Decorations* Plant made into charcoal and used for tribal tattooing of girls. (68:336) **Sioux** *Ceremonial Items* Wood made into charcoal and used for ceremonial painting and tattooing. (82:4)

Acer negundo var. *interius* (Britt.) Sarg., Box Elder

- *Food*—**Cree** *Sweetener* Sap used to make sugar. (97:44)
- *Other*—**Tewa** *Smoking Tools* Twigs used for making pipestems. (as *Negundo interius* 138:38)

Acer negundo L. **var. *negundo***, Box Elder
- *Food*—**Sioux** *Staple* Sap boiled down in the spring and made into sugar. (as *Negundo aceroides* 19:16)
- *Other*—**Kiowa** *Ceremonial Items* Wood burned in the altar fire of the peyote ceremony. Wood burned in the altar fire of the peyote ceremony. (as *Negundo aceroides* 192:40)

Acer nigrum Michx. f., Black Maple
- *Drug*—**Ojibwa, South** *Antidiarrheal* Decoction of inner bark used for diarrhea. "Arbor liquore abundans, ex quo liquor tanquam urina vehementer projicitur [A tree full of sap, which shoots out forcefully, just like urine]." (91:199)
- *Food*—**Ojibwa** *Sweetener* Sap used to make sugar. (as *A. saccharinum* var. *nigrum* 135:234)
- *Other*—**Ojibwa** *Hunting & Fishing Item* Wood used to make arrows. *Toys & Games* Root used to make the bowl for the dice bowl game. (as *A. saccharinum* var. *nigrum* 135:234)

Acer pensylvanicum L., Striped Maple
- *Drug*—**Abnaki** *Respiratory Aid* Used for bronchial troubles. (144:154) **Algonquin, Quebec** *Unspecified* Infusion of plant used as a medicinal tea. *Veterinary Aid* Plant eaten by a moose with a broken bone to aid its healing. (18:196) **Iroquois** *Emetic* Decoction of bark taken as an emetic. *Laxative* Compound decoction of bark taken as a laxative. *Orthopedic Aid* Decoction of bark applied as poultice for paralysis. (87:378) **Micmac** *Antihemorrhagic* Wood used for spitting blood. *Cold Remedy* Bark used for colds. *Cough Medicine* Bark used for coughs. *Kidney Aid* Wood used for kidney trouble. *Misc. Disease Remedy* Bark used for "grippe." *Orthopedic Aid* Unspecified plant parts used for "trouble with the limbs." *Venereal Aid* Wood used for gonorrhea. (40:53) **Ojibwa, South** *Emetic* Decoction of inner bark taken as an emetic. (91:200) **Penobscot** *Antihemorrhagic* Compound infusion of plant taken for "spitting up blood." (156:311) *Dermatological Aid* Poultice of steeped bark applied to swollen limbs. (156:310) *Kidney Aid* Compound infusion of plant taken for kidney trouble. *Tonic* Compound infusion of plant taken as a tonic. *Venereal Aid* Compound infusion of plant taken for gonorrhea. (156:311)
- *Food*—**Micmac** *Beverage* Bark used to make a beverage. (159:258)
- *Other*—**Cherokee** *Fuel* Wood used for firewood. (80:44) **Ojibwa** *Hunting & Fishing Item* Wood used to make arrows. *Toys & Games* Root used to make the bowl for the dice bowl game. (135:234)

Acer rubrum L., Red Maple
- *Drug*—**Cherokee** *Analgesic* Infusion of bark taken for cramps. *Antidiarrheal* Infusion taken for dysentery. *Dermatological Aid* Infusion taken for hives. *Eye Medicine* Inner bark boiled and used with water as wash for sore eyes. (80:44) Decoction of inner bark boiled to a syrup and used as a wash for sore eyes. (203:73) *Gynecological Aid* Compound infusion of bark taken for "female trouble" and cramps. *Misc. Disease Remedy* Hot infusion of bark given for measles. (80:44) **Iroquois** *Blood Medicine* Complex compound taken as a blood purifier. *Eye Medicine* Infusion of bark used as drops for sore eyes and cataracts. *Hunting Medicine* Decoction of plants used as a wash for traps, a "trapping medicine." (87:378) **Ojibwa** *Eye Medicine* Decoction of bark used as wash for sore eyes. (153:353) **Potawatomi** *Eye Medicine* Decoction of inner bark used as an eyewash. (154:37) **Seminole** *Dermatological Aid, Hemorrhoid Remedy,* and *Orthopedic Aid* Decoction of bark used for ballgame sickness: sores, back or limb pains, and hemorrhoids. (169:269)
- *Food*—**Abnaki** *Sweetener* Used as a sweetener. (144:152) Sap used to make sugar. (144:170) **Algonquin, Quebec** *Sauce & Relish* Sap used to make syrup. *Sweetener* Sap used to make sugar. (18:99) **Iroquois** *Bread & Cake* Bark dried, pounded, sifted, and made into bread. (as *A. rubra* 196:119)
- *Fiber*—**Cherokee** *Basketry* Used to make baskets. *Building Material* Wood used for lumber. *Furniture* Wood used to make furniture. (80:44) **Malecite** *Basketry* Used to make basket splints. (160:6) **Micmac** *Basketry* Used to make basket ware. (159:258)
- *Other*—**Cherokee** *Decorations* Wood used to carve. (80:44) **Iroquois** *Cooking Tools* Wood used to make bowls. (141:53) **Ojibwa** *Designs* Leaf frequently used in the Ojibwe beadwork designs. Many leaves, flowers, and fruits furnish designs. Since the plants are sacred to their midewiwin or medicine lodge, it is common for them to use especially valuable remedies in their designs. (153:412) **Potawatomi** *Hunting & Fishing Item* Traps boiled in water with bark to deodorize the scent of the previous animal trapped. (154:116) **Seminole** *Cooking Tools* Plant used to make spoons. *Hunting & Fishing Item* Plant used to make arrowheads. *Stable Gear* Plant used to make ox yokes. (169:472)

Acer rubrum var. *drummondii* (Hook. & Arn. ex Nutt.) Sarg., Drummond's Maple
- *Drug*—**Koasati** *Dermatological Aid* Infusion of bark taken and used as a wash for gun wounds. (177:39)

Acer saccharinum L., Silver Maple
- *Drug*—**Cherokee** *Analgesic* Infusion of bark taken for cramps. *Antidiarrheal* Infusion taken for dysentery. *Dermatological Aid* Infusion taken for hives. *Eye Medicine* Inner bark boiled and used with water as wash for sore eyes. *Gynecological Aid* Compound infusion of bark taken for "female trouble" and cramps. *Misc. Disease Remedy* Hot infusion of bark given for measles. (80:44) **Chippewa** *Dermatological Aid* Bark boiled and used as a wash for old, stubborn, running sores. (71:136) **Iroquois** *Unspecified* Sap, thimbleberries, and water used to make a medicine. (196:142) **Mohegan** *Cough Medicine* Infusion of bark, removed from south side of tree, taken for cough. (174:269) **Ojibwa** *Venereal Aid* Infusion of root bark taken for gonorrhea. (135:232) **Ojibwa, South** *Antidiarrheal* Decoction of inner bark used for diarrhea. (91:198) *Diuretic* Compound decoction of inner bark taken as a diuretic. (91:199)
- *Food*—**Chippewa** *Sweetener* Sap used to make sugar. (71:136) **Dakota** *Sweetener* Sap used to make sugar. (70:100) **Iroquois** *Beverage* Sap, thimbleberries, and water used to make a drink for home consumption and longhouse ceremonies. (196:142) Sap fermented and used as an intoxicant. (196:146) *Bread & Cake* Bark dried, pounded, sifted, and made into bread. (196:119) *Sweetener* Sap used to make sugar. (196:142) **Ojibwa** *Sweetener* Sap used to make sugar. (135:234) **Omaha** *Sweetener* Sap boiled to make sugar and syrup. (68:328) Sap used to make sugar. **Ponca** *Sweetener* Sap used to make sugar. **Winnebago** *Sweetener* Sap used to make sugar. (70:100)
- *Fiber*—**Cherokee** *Basketry* Used to make baskets. *Building Material* Wood used for lumber. *Furniture* Wood used to make furniture. (80:44)

- *Dye*—**Omaha** *Black* Twigs and bark made into a black dye and used in tanning leather. Twigs, bark, and an iron stained clay mixed with tallow, roasted in a pot, and made into a black dye used in tanning leather. (68:324) Twigs and bark made into a black dye and used to color tanned hides.[4] **Winnebago** *Black* Twigs and bark made into a black dye and used to color tanned hides.[4] (70:100)
- *Other*—**Cherokee** *Decorations* Wood used to carve. (80:44) **Chippewa** *Cleaning Agent* Bark, hemlock, and swamp oak bark boiled together to make a wash to remove rust from steel or iron. The barks were boiled together and used to remove rust from steel or iron and to prevent further rusting. (71:136) **Ojibwa** *Hunting & Fishing Item* Wood used to make arrows. *Toys & Games* Root used to make the bowl for the dice bowl game. (135:234)

Acer saccharum Marsh., Sugar Maple

- *Drug*—**Iroquois** *Blood Medicine* Complex compound used as a blood purifier. *Dermatological Aid* Compound decoction of leaves used as a wash on parts affected by "Italian itch." *Eye Medicine* Compound infusion of bark used as drops for blindness. (87:378) Sap used for sore eyes. (196:142) *Pulmonary Aid* Infusion of bark with another whole plant taken by forest runners for shortness of breath. (as *A. saccharophorum* 141:52) *Unspecified* Sap, thimbleberries, and water used to make a medicine. (196:142) **Mohegan** *Cough Medicine* Inner bark used as a cough remedy. (176:69, 128) **Potawatomi** *Expectorant* Inner bark used as an expectorant. (154:37)
- *Food*—**Algonquin, Quebec** *Sauce & Relish* Sap used to make syrup. *Sweetener* Sap used to make sugar. (18:98) **Cherokee** *Sweetener* Juice used to make sugar. (80:44) Sap used to make sugar. (126:32) **Dakota** *Sweetener* Sap formerly used to make sugar. (70:100) **Iroquois** *Beverage* Sap made into sugar and used to make beer. (as *A. saccharophorum* 141:52) Sap, thimbleberries, and water used to make a drink for home consumption and longhouse ceremonies. (196:142) Sap fermented and used as an intoxicant. (196:146) *Bread & Cake* Bark dried, pounded, sifted, and made into bread. (196:119) *Sweetener* Sap used to make sugar. (as *A. saccharophorum* 141:52) Sap used to make sugar. (196:142) **Malecite** *Sauce & Relish* Used to make maple syrup. *Sweetener* Used to make maple syrup and sugar. (160:6) **Menominee** *Sweetener* Boiled sap made into maple sugar and used in almost every combination of cookery. (151:61) **Meskwaki** *Sweetener* Maple sugar used instead of salt as seasoning in cooking. (152:255) **Micmac** *Beverage* Bark used to make a beverage. *Sauce & Relish* Sap used to make maple syrup and maple sugar. (159:258) **Mohegan** *Sweetener* Sap used as a sweetening agent and to make maple syrup. (176:69) **Ojibwa** *Beverage* Sap saved to drink as it comes from the tree, alone or mixed with box elder or birch sap. *Sour* Sap allowed to sour to make vinegar and mixed with maple sugar to cook sweet and sour meat. *Sweetener* Maple sugar used to season all kinds of meats, replaced now with salt. Smith describes in detail the process by which the Ojibwe make maple syrup. Although now (1932) they use iron kettles, originally the sap and storage vessels were "made of birch bark, sewed with boiled basswood fiber or the core of the jack pine root." The vessels are rendered waterproof by the application of pitch secured by boiling jack pine cones. (153:394) **Potawatomi** *Beverage* Maple sap, as it came from the tree, drunk by children. *Candy* Children made taffy by cooling the maple sap in the snow. *Sour* Maple sap not only furnished the sugar for seasoning material but also furnished the vinegar. Sap that was allowed to become sour made a vinegar to be used in cooking venison which was afterwards sweetened with maple sugar. This corresponds somewhat to the German "sweet and sour" style of cooking. *Sweetener* Maple sugar used, instead of salt, to season all cooking. The sugar maple and the black sugar maple are found all over Wisconsin and were considered to be the most valuable trees in the forest because they furnished them their seasoning material. While they do use salt today, it is an acquired ingredient and most of the old people would prefer to have sugar for their seasoning. (154:92)
- *Fiber*—**Cherokee** *Building Material* Wood used for lumber. *Furniture* Wood used to make furniture. (80:44) **Malecite** *Canoe Material* Wood used to make paddles and oars. (160:6)
- *Other*—**Cherokee** *Decorations* Wood used to carve. (80:44) **Chippewa** *Cash Crop* Sap made into sugar and used as a commodity of intertribal commerce. Maple sugar was among the great staples in the domestic economy and was a commodity of intertribal commerce being traded to people of tribes in areas not possessing this tree. (71:135) *Cooking Tools* Used to make paddles for stirring maple sap. (53:377) **Malecite** *Lighting* Wood used to make torch handles. (160:6) **Meskwaki** *Designs* Leaf used in beadwork designs. (152:266) **Micmac** *Hunting & Fishing Item* Used to make bows and arrows. (159:258) **Ojibwa** *Cooking Tools* Wood used to make paddles for stirring maple sugar or wild rice while scorching or parching it. Wood used to make bowls and many other objects of utility. (153:413)

Acer sp., Maple

- *Other*—**Paiute** *Hunting & Fishing Item* Wood used to make bows. (111:88)

Acer spicatum Lam., Mountain Maple

- *Drug*—**Algonquin, Tête-de-Boule** *Dermatological Aid* Poultice of boiled root chips applied to wounds and abscesses. (132:118) **Iroquois** *Antihemorrhagic* Compound decoction of roots and bark taken for internal hemorrhage. (87:377) *Gastrointestinal Aid* Plant used for intestinal diseases. (142:94) **Malecite** *Eye Medicine* Infusion of outside bark used for sore eyes. Poultice of outside bark used for sore eyes. (116:248) **Micmac** *Eye Medicine* Bark used for sore eyes. (40:53) **Ojibwa** *Eye Medicine* Infusion of pith used as a wash for sore eyes and pith used to remove foreign matter. (153:353) **Potawatomi** *Cough Medicine* Compound containing inner bark used as cough syrup. (154:37)
- *Other*—**Menominee** *Designs* Leaves used as design for bead work and appliqué work. (151:73) **Ojibwa** *Designs* Three-lobed leaf was a great favorite with Ojibwe women for design work for beading. (153:413) *Hunting & Fishing Item* Wood used to make arrows. *Toys & Games* Root used to make the bowl for the dice bowl game. (135:234) **Potawatomi** *Designs* Leaves used as a pattern for bead and appliqué work. In making up a design for art work or bead work, a woman would burn deer antlers until they turned to charcoal and use this to rub on the backs of leaves. This surface was placed down upon a piece of white birch bark and rubbed until the shape and venation of the leaves were transferred to the birch bark. Then arranging with other leaves, a design would be formed which would be the pattern for the bead work. Oftentimes, this would be placed directly under

the loom so that the form and outline of the finished bead work would be a true representation of the natural object. (154:110)

Achillea, Asteraceae
Achillea millefolium L., Common Yarrow
- *Drug*—**Abnaki** *Cold Remedy* Infusion of whole plant given to children for colds. (144:174) *Febrifuge* Used for fevers. *Misc. Disease Remedy* Used for grippe. (144:154) *Pediatric Aid* Infusion of whole plant given to children for colds. (144:174) **Algonquin, Quebec** *Analgesic* Crushed leaves used as a snuff for headaches. *Cold Remedy* Used for colds. *Poultice* Leaves used for poultices. *Respiratory Aid* Used for respiratory disorders. (18:240) **Algonquin, Tête-de-Boule** *Analgesic* Decoction of leaves and flowers used for headaches. (132:118) **Bella Coola** *Breast Treatment* Leaves pounded, heated, and used for breast abscesses. (184:201) *Burn Dressing* Poultice of chewed leaves applied to burns. (150:65) Leaves pounded, heated, and used for burns. *Dermatological Aid* Leaves pounded, heated, and used for boils. *Pediatric Aid* and *Respiratory Aid* Poultice of leaves and eulachon (candlefish) grease applied to the chest and back of children for bronchitis. (184:201) **Blackfoot** *Analgesic* Infusion of plant taken or rubbed on the body to soothe the pain of gastroenteritis. (86:65) *Antirheumatic (External)* Poultice of chewed flowers applied to swollen parts. Infusion of plant applied to swellings. *Dermatological Aid* Infusion of plant applied to sores. (86:74) *Diuretic* Infusion of plant taken as a diuretic to pass the sickness with the urine. (86:69) *Gynecological Aid* Infusion of leaves taken when labor pains started and to ease the delivery. Infusion of leaves taken to expel the afterbirth. (86:60) *Liver Aid* Infusion of plant taken or rubbed on the body for liver troubles. (86:65) *Panacea* Infusion of plant rubbed on the body part affected by sickness. (86:69) *Throat Aid* Infusion of plant taken for sore throats. (86:70) *Veterinary Aid* Infusion of plant used as an eyewash for horses. (86:87) **Carrier, Southern** *Cold Remedy* Decoction of entire plant, except roots, taken for colds. *Dermatological Aid* Poultice of chewed leaves applied to swellings. *Orthopedic Aid* Poultice of chewed leaves applied to sprains. (150:65) **Chehalis** *Antidiarrheal* Decoction of leaves taken for the passage of blood with diarrhea. (79:49) **Cherokee** *Antihemorrhagic* Used for hemorrhages and spitting blood. *Dermatological Aid* Astringent leaves used for hemorrhages and bowel complaints. *Febrifuge* Infusion taken for fever. *Gastrointestinal Aid* Used for bowel complaints. *Gynecological Aid* Used for flooding. *Hemorrhoid Remedy* Used for bloody piles. *Respiratory Aid* Dried leaves smoked for catarrh. *Sedative* Infusion taken for restful sleep. *Urinary Aid* Used for bloody urine. (80:62) **Cheyenne** *Analgesic* Infusion of leaves and flowers taken for chest pains. *Antiemetic* Infusion of fresh or dried plant taken for nausea. (83:17) *Cold Remedy* Infusion of leaves used for colds. (82:6) Infusion of fresh or dried plant taken for colds. *Cough Medicine* Infusion of fresh or dried plant taken for coughs. *Diaphoretic* Infusion of fresh or dried plant taken to cause perspiring. (83:17) *Febrifuge* Infusion of leaves used for fevers. (82:6) *Heart Medicine* Infusion of leaves and flowers taken for heart troubles and chest pains. *Hemostat* Crushed leaves placed in the nose for nosebleeds. *Respiratory Aid* Infusion of plant taken or leaves rubbed on body for respiratory diseases. *Throat Aid* Infusion of fresh or dried plant taken for tickling of the throat. *Tuberculosis Remedy* Infusion of plant taken or leaves rubbed on body for tuberculosis. (83:17) **Chippewa** *Analgesic* Decoction of leaves steamed and inhaled for headache. (53:336) *Dermatological Aid* Decoction of root applied to skin "eruptions." (53:350) *Herbal Steam* Decoction of leaves sprinkled on hot stones as herbal steam for headache. (53:336) *Stimulant* Dried chewed root spit onto limbs as a stimulant. (53:364) *Veterinary Aid* Decoction of leaves and stalk applied to horses as a stimulant. (53:366) **Clallam** *Cold Remedy* Infusion of leaves used for colds. *Gynecological Aid* Infusion of leaves used during childbirth. (57:199) **Cowlitz** *Dermatological Aid* Infusion of leaves used as a hair wash. *Gastrointestinal Aid* Decoction of roots taken for stomach troubles. (79:49) **Cree, Woodlands** *Analgesic* Infusion of plant heads used to make a compress for headaches. *Antihemorrhagic* Leaves chewed for bleeding. *Febrifuge* Infusion of plant heads used to make a compress for fevers. Decoction of roots taken for fevers. *Toothache Remedy* Decoction of roots taken or roots chewed for toothaches. (109:23) **Creek** *Toothache Remedy* Plant used as toothache medicine. (172:663) **Crow** *Burn Dressing* Poultice of plant used for burns. *Dermatological Aid* Poultice of plant used for boils and open sores. (82:6) **Delaware** *Kidney Aid* Infusion of plant used for kidney disorders. *Liver Aid* Infusion of plant used for liver disorders. (176:35) **Delaware, Oklahoma** *Kidney Aid* Infusion of whole plant taken for kidney disorders. *Liver Aid* Infusion of whole plant taken for liver disorders. (175:29, 74) **Flathead** *Antirheumatic (External)* Leaves boiled and used for aching backs and legs. *Cold Remedy* Infusion of leaves used for colds. *Dermatological Aid* Leaves crushed and used for wounds. *Disinfectant* Herb used as a disinfectant. *Febrifuge* Infusion of leaves used for fevers. (82:6) **Gitksan** *Throat Aid* Decoction of young plant or root gargled for sore throat. (150:65) **Gosiute** *Analgesic* Infusion of plant used for headaches. (39:360) *Antirheumatic (External)* Poultice of plant applied to joints affected by rheumatism. (39:350) Poultice of plant applied for rheumatism. *Dermatological Aid* Poultice of plant applied to bruises. *Gastrointestinal Aid* Infusion of plant used for biliousness. (39:360) **Haisla & Hanaksiala** *Unspecified* Plants placed on heated rocks and rising vapors used for unspecified illness. (43:220) **Hesquiat** *Analgesic* Leaves chewed and the juice swallowed for any kind of internal pain. *Cough Medicine* Leaves chewed and the juice swallowed for prolonged cough. *Gastrointestinal Aid* Leaves chewed and the juice swallowed for the stomach. *Internal Medicine* Leaves chewed and the juice swallowed for internal organs. (185:61) **Iroquois** *Analgesic* Infusion of roots or leaves used internally or externally for headaches. (87:469) Plant chewed and poultice of leaves applied for neuralgia. *Anthelmintic* Infusion of leaves given to children with worms. *Anticonvulsive* Decoction of plants given and used as wash for babies with convulsions. *Antidiarrheal* Compound decoction of plants, bark, and roots taken for diarrhea. (87:470) Infusion of smashed plants taken for diarrhea. (87:471) Infusion of plant and seeds from another plant used for diarrhea. (141:64) *Antiemetic* Decoction or infusion of plants, bark, and roots taken for vomiting. (87:470) Decoction of leaves, branches, and another plant taken for vomiting and nausea. (141:64) *Antirheumatic (Internal)* Infusion of plants taken when "sore through the joints." *Blood Medicine* Compound decoction of plants, roots, and bark taken as a blood purifier. (87:470) *Emetic* Infusion of smashed plants taken as an emetic for sunstroke. (87:471) *Febrifuge* Infusion of leaves given to babies with any kind of

fever. (87:469) Infusion of smashed plants taken for fever caused by sunstroke. (87:471) Poultice of plant applied and infusion of plant used for fevers. (142:103) *Gastrointestinal Aid* Plant used for "summer complaint" and decoction of plants taken for cramps. (87:470) Decoction of plant fragments taken for digestive cramps. (141:64) *Misc. Disease Remedy* Plant used as a miscellaneous disease remedy. (87:470) *Panacea* Infusion of leaves given to babies with any kind of sickness. *Pediatric Aid* Infusion of leaves given to babies with any kind of sickness or fever. (87:469) Compound decoction of stems given to children with diarrhea. Decoction of plants given and used as wash for babies with convulsions. *Stimulant* Cold infusion given and used as wash on unconscious person who had fallen. *Venereal Aid* Compound decoction of plants, roots, and bark taken for venereal disease. (87:470) Decoction of plant tops used as a wash on parts affected by gonorrhea. (87:471) **Karok** *Dermatological Aid* Poultice of soaked stalks and leaves applied to wounds. (148:390) **Klallam** *Cold Remedy* Decoction of leaves taken for colds. *Dermatological Aid* Poultice of chewed leaves applied to sores. *Gynecological Aid* Decoction of leaves taken during childbirth. (79:49) **Kutenai** *Dermatological Aid* Leaves crushed and used for wounds. Decoction used for washing sores and other skin problems. *Disinfectant* Herb used as a disinfectant. (82:6) **Kwakiutl** *Antirheumatic (External)* Leaves used in a steam bath for rheumatism. *Cold Remedy* Poultice of leaves applied to the chest for colds. (183:278) *Dermatological Aid* Poultice of chewed leaves applied or compound rubbed on sores and swellings. (183:266) *Gynecological Aid* Poultice of leaves applied to chest for hardened breasts after childbirth. *Herbal Steam* Leaves used in a steam bath for rheumatism or general sickness. *Panacea* Leaves used in a steam bath for general sickness. (183:278) **Lakota** *Dermatological Aid* Poultice of dried and chewed plants applied to wounds and sores. (106:46) **Lummi** *Analgesic* Decoction of flowers taken for body aches. *Diaphoretic* Decoction of flowers taken to produce sweating. *Misc. Disease Remedy* Decoction of flowers taken to prevent mumps. (79:49) **Mahuna** *Toothache Remedy* Rolled leaves inserted into cavity of painful tooth. (140:24) **Makah** *Blood Medicine* Leaves chewed as a blood purifier. (67:322) Decoction of leaves taken to purify the blood. (79:49) *Cathartic* Leaves chewed "to clean one out." (67:322) *Diaphoretic* Raw leaves chewed by women to produce sweating at childbirth. (79:49) *Gynecological Aid* Plant taken by an expectant mother close to the time of birth for an easy delivery. Plant taken at the start of labor "to hurry the baby." (67:322) Decoction of leaves taken to heal the uterus after birth. Raw leaves chewed by women to produce sweating at childbirth. (79:49) *Other* Plant used like an antibiotic. *Throat Aid* Leaves chewed and the juice swallowed for sore throats. (67:322) **Malecite** *Dermatological Aid* Used as a liniment for bruises. *Orthopedic Aid* Used as a liniment for sprains. (116:244) **Mendocino Indian** *Analgesic* Infusion of leaves and flowers taken for headaches. *Dermatological Aid* Infusion of leaves and flowers used as a wash for bruises. *Eye Medicine* Infusion of leaves and flowers used as a wash for sore eyes. *Gastrointestinal Aid* Infusion of leaves and flowers taken for stomachaches. *Orthopedic Aid* Infusion of leaves and flowers used as a wash for sprains. *Tuberculosis Remedy* Infusion of leaves and flowers taken for consumption. (41:391) **Menominee** *Dermatological Aid* Poultice of dried, powdered leaves applied to swellings and sores. (54:132) Poultice of leaves used on children's rash and fresh tops used to rub on eczema. *Febrifuge* Infusion of leaves used for fevers. *Pediatric Aid* Poultice of leaves used for "the rash of children." (151:28, 29) **Micmac** *Antirheumatic (External)* Dried, powdered bark or green leaves rubbed over swellings. (194:25) *Cold Remedy* Herb used for colds. (40:53) *Dermatological Aid* Dried, powdered bark or green leaves rubbed over bruises. *Diaphoretic* Decoction of plant taken with milk to cause a sweat for colds. (194:25) *Orthopedic Aid* Herb used for swelling, bruises, and sprains. (40:53) Dried, powdered bark or green leaves rubbed over sprains. (194:25) **Miwok** *Analgesic* Dried or green mashed leaves used for pain and used during influenza epidemic. Dried or green mashed leaves used for pain. *Cold Remedy* Infusion of leaves and flowers taken for bad colds. *Misc. Disease Remedy* Infusion of leaves and flowers used externally for influenza. (12:166) **Mohegan** *Dietary Aid* Cold, compound infusion taken as an appetizer. *Gastrointestinal Aid* Cold, compound infusion taken for the stomach. (174:266) Compound infusion of leaves taken as a stomach aid and to improve the appetite. (176:75, 128) *Kidney Aid* Infusion of plant taken for the kidneys. (174:269) Simple or compound infusion of leaves taken for kidney disorders. (176:69, 128) *Liver Aid* Infusion of plant taken for the liver. (174:269) Simple or compound infusion of leaves taken for liver disorders. (176:69, 128) **Montagnais** *Febrifuge* Infusion of plant used for fever. (156:315) **Nitinaht** *Cold Remedy* Decoction of plants taken for colds. *Panacea* Plants chewed and swallowed as "medicine for everything." (186:96) *Throat Aid* Leaves chewed and the juice swallowed for sore throats. (67:322) **Ojibwa** *Ceremonial Medicine* Florets smoked for ceremonial purposes. *Febrifuge* Florets placed on coals and smoke inhaled to break a fever. (153:362) **Okanagan-Colville** *Analgesic* Infusion of roots taken for headaches. *Antidiarrheal* Infusion of roots taken for diarrhea. *Antirheumatic (External)* Decoction of whole plant used as a bath for arthritic or rheumatic pains. *Cathartic* Decoction or roots and scarlet gilia leaves taken as a physic. *Cold Remedy* Infusion of roots taken for colds. *Dermatological Aid* Leaves and stems mixed with white clematis and witches'-broom branches to make a shampoo. *Gastrointestinal Aid* Infusion of roots taken for stomachaches. *Laxative* Decoction or roots and scarlet gilia leaves taken as a laxative. *Toothache Remedy* Roots mashed and applied to the tooth for toothaches. (188:74) **Okanagon** *Dermatological Aid* Decoction used as wash for chapped hands, pimples, rashes, and insect bites. *Eye Medicine* Decoction of whole plant used as a wash for sore eyes. *Snakebite Remedy* Decoction of plant used as a wash for insect or snake bites. *Tonic* Decoction of whole plant taken as a tonic. (as *Achilles millefolium* 125:40) **Paiute** *Analgesic* Green plants smelled by old men for headaches. *Cold Remedy* Infusion of plant taken or green plants smelled for colds. (167:317) *Dermatological Aid* Poultice of crushed leaves applied to swellings. (104:196) Decoction of leaves and stems used as a liniment for skin sores. (167:317) *Eye Medicine* Cold infusion of leaves used as a wash for sore eyes. (104:197) *Orthopedic Aid* Poultice of crushed leaves applied to sprains. (104:196) *Toothache Remedy* Leaves chewed for toothache. (104:197) **Paiute, Northern** *Cold Remedy* Decoction of roots taken for chest cold. Roots dried and chewed raw for colds. *Cough Medicine* Leaves soaked and used for coughs. *Dermatological Aid* Poultice of pulverized roots applied to cuts and sores. *Diaphoretic* Leaves soaked and sprinkled on the hot rocks in the sweat bath. *Kidney Aid* Decoction of roots taken for kidney troubles. *Misc. Disease Remedy* Decoction of roots taken for flu. *Throat Aid* Root chewed and the saliva allowed to flow down the throat for sore throats. (59:128) **Potawatomi** *Stimulant* Flowers smudged on live coals to revive comatose patient. *Witchcraft Medicine* Flowers smudged on live coals to repel evil spirits. (154:47, 48) **Quileute** *Antirheumatic (External)* Poultice of boiled leaves applied to rheumatic limbs. *Febrifuge* Poultice of boiled leaves applied to rheumatic limbs for the fever. *Panacea* Decoction of leaves used as an aromatic bath for sick infants. *Pediatric Aid* Decoction of leaves used as an aromatic bath for sick infants. **Quinault** *Eye Medicine* Decoction of roots used as an eyewash. *Tonic* Decoction of roots taken as a general tonic. *Tuberculosis Remedy* Decoction of roots taken for tuberculosis. (79:49) **Saanich** *Cold Remedy* Young leaves chewed and juice swallowed for colds. *Throat Aid* Young leaves chewed and juice swallowed for sore throats. *Toothache Remedy* Poultice of leaves held in the mouth for toothaches. (182:80) **Salish** *Eye Medicine* Decoction of plants used for sore eyes. (178:293) **Shuswap** *Blood Medicine* Decoction of flowers and roots taken as a blood purifier. (123:58) *Dermatological Aid* Infusion of leaves taken for poison ivy. (123:56) **Skagit** *Antidiarrheal* Decoction of leaves taken for diarrhea. **Snohomish** *Antidiarrheal* Decoction of leaves taken for diarrhea. **Squaxin** *Dermatological Aid* Poultice of chewed leaves applied to sores. *Gastrointestinal Aid* Decoction of roots taken for stomach troubles. **Swinomish** *Other* Plant used as a bath for invalids. (79:49) **Thompson** *Antidiarrheal* Infusion of leaves given to children for diarrhea. Leaves chewed or infusion of leaves taken for dysentery. Infusion of roots or whole plant taken for diarrhea. *Antirheumatic (External)* Decoction of leaves and roots used for bathing arthritic limbs. Poultice of pounded roots used on the skin for sciatica. *Cold Remedy* Infusion of flowers taken in small quantities for colds. Infusion of roots or whole plant taken for colds. Leaves chewed or decoction of leaves taken for colds. Roots chewed or decoction of roots taken for colds. (187:166) *Dermatological Aid* Decoction of plant used as a wash for chapped hands, pimples, rashes, and insect bites. (as *Achilles millefolium* 125:40) Infusion of plant used as wash or powdered stem and leaf applied for skin problems. (164:460) Leaves and roots rubbed on sores. Poultice of mashed basal leaves used for cuts. (187:166) *Eye Medicine* Decoction of whole plant used as a wash for sore eyes. (164:460) *Gastrointestinal Aid* Infusion of roots or whole plant taken for bad stomach cramps. *Misc. Disease Remedy* Infusion of flowers taken in small quantities for influenza. *Orthopedic Aid* Leaves and roots rubbed on broken bones. *Panacea* Decoction of whole plant taken for any sickness. Decoction of plant used as a wash for any kind of sickness. *Pediatric Aid* Infusion of leaves given to children for diarrhea. (187:166) *Snakebite Remedy* Decoction of plant used as a wash for insect or snake bites. (as *Achilles millefolium* 125:40) Decoction of whole plant used as a wash for snakebites. (164:460) *Tonic* Decoction of whole plant taken as a tonic. (as *Achilles millefolium* 125:40) Decoction of whole plant taken as a tonic "for slight indisposition." (164:460) *Toothache Remedy* Mashed root placed over a tooth for toothache. *Unspecified* Roots and stems considered "a good medicine." *Urinary Aid* Infusion of flowers taken in small quantities for bladder trouble. *Venereal Aid* Root used for venereal disease. (187:166) **Ute** *Dermatological Aid* Poultice of plant applied externally to bruises. *Panacea* Infusion of plant taken for cases of sickness. (38:32)

Winnebago *Dermatological Aid* Infusion of herb used as a wash for swellings. *Ear Medicine* Wad of leaves or infusion put into ear for earache. (70:134) **Yuki** *Cold Remedy* Infusion of leaves and flowers taken for cold in the chest. *Respiratory Aid* Infusion of leaves and flowers taken for cold in the chest. (49:47)

- **Food**—**Blackfoot** *Beverage* Leaves and flowers used to make a pleasant tea. (86:100) **Haisla & Hanaksiala** *Forage* Plant eaten by bears. (43:220) **Klamath** *Preservative* Stem, leaf, and flower placed inside fish cavity as a preservative. (45:105)
- **Other**—**Clallam** *Incense & Fragrance* Seeds used as house fragrances. (57:199) **Cree, Woodlands** *Hunting & Fishing Item* Dried flowers used for lynx bait. (109:23) **Kutenai** *Incense & Fragrance* Leaves formerly used for cologne, perfume, and bath powder. (82:6) **Okanagan-Colville** *Insecticide* Leaves and stems used in a smudge to keep mosquitoes away. (188:74) **Potawatomi** *Protection* Seed heads placed on a pan of live coals to produce smoke to keep the witches away. (154:117)

Achillea millefolium var. *arenicola* (Heller) Nobs, Common Yarrow

- **Drug**—**Pomo, Kashaya** *Dermatological Aid* Mashed leaf juice used as a salve on sores. (as *A. borealis* ssp. *arenicola* 72:120)

Achillea millefolium var. *borealis* (Bong.) Farw., Boreal Yarrow

- **Drug**—**Aleut** *Analgesic* Infusion of leaves taken for pains in stomach, throat, chest, and muscles. *Cold Remedy* Infusion of leaves taken for colds, stomach pains, and throat pains. *Gastrointestinal Aid* Infusion of leaves taken for stomach pains and throat pains. *Hemostat* Leaves used as a coagulant for cuts and stuffed into nostrils for nosebleeds. *Throat Aid* Infusion of leaves taken for stomach pains, throat pains, and colds. *Tuberculosis Remedy* Infusion of leaves taken for consumption in post-Russian era. (as *A. borealis* 8:426) **Costanoan** *Dermatological Aid* Decoction of plant used as a wash for sores. Poultice of heated leaves applied to wounds to prevent swelling. *Gastrointestinal Aid* Decoction of plant taken for stomachaches. *Toothache Remedy* Heated leaves held in the mouth for toothaches. (as *A. borealis* 21:25) **Eskimo, Alaska** *Unspecified* Infusion of dried plants used for medicinal purposes. (as *A. borealis* 4:716) **Eskimo, Nunivak** *Unspecified* Infusion of dried plants used for its medicinal qualities. (as *A. borealis* 149:325) **Kwakiutl** *Dermatological Aid* Poultice of chewed or soaked and heated plant applied to swellings and sores. (as *A. borealis* 20:381)

Achillea millefolium var. *californica* (Pollard) Jepson, California Yarrow

- **Drug**—**Yurok** *Eye Medicine* Used to wash or steam aching, sore eyes. (6:15)

Achillea millefolium var. *occidentalis* DC., Western Yarrow

- **Drug**—**Carrier** *Antirheumatic (External)* Decoction of leaves and stems used as a bath for rheumatism. *Toothache Remedy* Crushed roots placed in the tooth for toothaches. (as *A. lanulosa* 31:85) **Cheyenne** *Antiemetic* Infusion of green or dried leaves taken for slight nausea. *Cold Remedy* Infusion of green or dried leaves taken for colds. *Cough Medicine* Infusion of dried, pounded plant taken for coughs. *Throat Aid* Infusion of dried, pounded plant taken for tickling in the throat. (as *A. lanulosa* 76:189) **Cree, Woodlands** *Pediatric*

Aid Flowers and wild mint flowers wrapped in a cloth, dipped in water, and used to remove teething gum pus. *Toothache Remedy* Decoction of roots and other herbs taken for teething-related sickness. Flowers and wild mint flowers wrapped in a cloth, dipped in water, and used to remove teething gum pus. (as *A. lanulosa* 109:22) **Great Basin Indian** *Dermatological Aid* Poultice of crushed, fresh plant applied to sores. *Laxative* Infusion of plant taken as a mild laxative. (as *A. lanulosa* 121:50) **Kawaiisu** *Snakebite Remedy* Dried, crushed, and powdered leaves applied to snakebite wounds. (as *A. lanulosa* 206:9) **Meskwaki** *Dermatological Aid* Decoction of stem and leaves used as a wash for "place on the body that is ailing." *Febrifuge* Infusion of leaves and blossoms taken for fever. *Misc. Disease Remedy* Infusion of leaves and blossoms taken for ague. (as *A. lanulosa* 52:210) **Montana Indian** *Cathartic* Infusion of herb used as a cathartic. (as *A. lanulosa* 19:5) **Navajo** *Dermatological Aid* Infusion of plant used as a wash for cuts and saddle sores. *Stimulant* Plant used in a "life medicine for impaired vitality." *Tonic* Plant used in a tonic. (55:79) **Navajo, Kayenta** *Analgesic* Plant used for headaches caused by weak or sore eyes. *Eye Medicine* Plant used in lotion for sore eyes caused from wearing ceremonial masks. *Febrifuge* Plant used as a fever medicine. (as *A. lanulosa* 205:44) **Navajo, Ramah** *Ceremonial Medicine* Plant used as a ceremonial emetic. *Emetic* Plant used as a ceremonial emetic. (as *A. lanulosa* 191:47) **Ojibwa** *Ceremonial Medicine* Compound containing flowering heads smoked for ceremonial purposes. *Dermatological Aid* Poultice of leaves applied to spider bite. (as *A. lanulosa* 153:362) **Paiute** *Analgesic* Poultice of fresh, mashed and boiled leaves applied to sprained ankle pains. Poultice of fresh, mashed leaves dampened with water applied with a cloth to tired, aching feet. (as *A. lanulosa* 111:118) Crushed green plant smelled for headaches. Decoction of leaves taken for headaches. Decoction of root taken for gas pains. Poultice of boiled, whole plant applied to pains or sores. Poultice of mashed leaves applied as a compress for headaches. *Antirheumatic (External)* Decoction of plant used as a liniment or wash for sores or rashes. *Blood Medicine* Decoction of plant taken as a blood tonic after childbirth. *Cold Remedy* Root chewed for colds. (as *A. lanulosa* 180:31–33) *Cough Medicine* Infusion of leaves taken as a cough medicine. (as *A. lanulosa* 111:118) *Dermatological Aid* Decoction of plant used as a liniment or wash for sores or rashes. Poultice of boiled, whole plant applied to sores. Poultice of mashed, green plant applied to swellings. Poultice of mashed leaves applied to swellings or sores. (as *A. lanulosa* 180:31–33) *Emetic* Infusion of yarrow taken as an emetic for tuberculosis and other respiratory diseases. (as *A. lanulosa* 111:118) *Eye Medicine* Strained decoction of leaves used as drops for sore eyes. *Febrifuge* Decoction of leaves used as a wash for fevers. *Gastrointestinal Aid* Decoction of root taken for gas pains. *Gynecological Aid* Decoction of plant taken as a blood tonic after childbirth. *Kidney Aid* Decoction of root believed to be good for the kidneys. (as *A. lanulosa* 180:31–33) *Respiratory Aid* Infusion of yarrow taken as an emetic for respiratory diseases. *Toothache Remedy* Poultice of fresh, mashed roots packed around an infected tooth for the pain. (as *A. lanulosa* 111:118) Green leaves or roots used in various ways for toothaches. (as *A. lanulosa* 180:31–33) *Tuberculosis Remedy* Infusion of yarrow taken as an emetic for tuberculosis. (as *A. lanulosa* 111:118) *Urinary Aid* Decoction of plant taken for bladder ailments. *Venereal Aid* Compound decoction of plant taken for gonorrhea. *Veterinary Aid* Decoction of plant used to disinfect cuts and saddle sores on horses. Poultice of boiled leaves applied to collar sores on horses. (as *A. lanulosa* 180:31–33) **Sanpoil** *Abortifacient* Decoction of stems and leaves used to cause abortion. *Cold Remedy* Decoction of root boiled until dark in color and taken while warm for colds. (131:218) **Shoshoni** *Analgesic* Crushed green plant smelled for headaches. Decoction of flower taken for stomachaches and used as a liniment for muscular pains. Decoction of leaves taken for headaches. Poultice of boiled, whole plant applied to pains or sores. *Anesthetic* Poultice of fresh roots applied to deaden pain so wound could be opened. Poultice of mashed, fresh roots applied as an anesthetic to painful wounds. *Antidiarrheal* Decoction of plant taken for diarrhea. *Antiemetic* Decoction of plant taken for upset stomach. *Antirheumatic (External)* Decoction of plant used as a liniment or wash for sores or rashes. (as *A. lanulosa* 180:31–33) *Carminative* Infusion of roots taken for gas pains. (as *A. lanulosa* 118:45) *Cold Remedy* Decoction of plant taken for colds. (as *A. lanulosa* 180:31–33) *Dermatological Aid* Poultice of whole plant applied for felon. (as *A. lanulosa* 118:43) Decoction of flower used as a wash for itching. Decoction of plant used as a liniment or wash for sores or rashes. Decoction of root used as a preliminary soak to help extract splinters. Poultice of boiled, whole plant applied to sores. Poultice of fresh roots applied to deaden pain so wound could be opened. Poultice of mashed leaves applied to swellings or sores. *Gastrointestinal Aid* Decoction of flowers taken for stomachaches or indigestion. Decoction of leaves taken for colic or dyspepsia. *Orthopedic Aid* Decoction of flowers used as a liniment for muscular pains. *Toothache Remedy* Green leaves or roots used in various ways for toothaches. **Washo** *Dermatological Aid* Poultice of mashed leaves applied to swellings or sores. (as *A. lanulosa* 180:31–33) **Zuni** *Burn Dressing* Blossoms and root chewed and juice applied before fire-eating or -walking. Poultice of pulverized plant mixed with water applied to burns. (as *A. lanulosa* 166:42)

- *Dye*—**Great Basin Indian** *Green* Leaves used to make a green dye. (as *A. lanulosa* 121:50)
- *Other*—**Ojibwa** *Ceremonial Items* and *Smoke Plant* Flower heads used in the "kinnikinnick" mixture smoked in medicine lodge ceremonies. (as *A. lanulosa* 153:417)

Achillea sibirica Ledeb., Siberian Yarrow
- *Drug*—**Cree, Woodlands** *Oral Aid* Poultice of chewed roots applied to gum sores. *Pediatric Aid* Decoction of roots and other herbs taken for teething-related sickness. *Toothache Remedy* Decoction of roots and other herbs taken for teething-related sickness. (109:23)

Achillea sp., Yarrow
- *Drug*—**Cree, Woodlands** *Analgesic* Fresh, crushed flower inserted into the nostril for headaches. *Burn Dressing* Poultice of a flower applied to burn pains. *Toothache Remedy* Poultice of chewed roots applied to toothaches. (109:23) **Mewuk** *Dermatological Aid* Poultice of bruised leaves applied to cuts and wounds. *Gastrointestinal Aid* Infusion of leaves used for stomach distress. *Pulmonary Aid* Infusion of leaves used for lung distress. (117:366)
- *Food*—**Aleut** *Unspecified* Species used for food. (7:29)
- *Other*—**Cree, Woodlands** *Hunting & Fishing Item* Plant used to make lures for traps. (109:23)

Achlys, Berberidaceae

Achlys triphylla (Sm.) DC., Sweet After Death
- **Drug**—**Cowlitz** *Tuberculosis Remedy* Infusion of leaves taken for tuberculosis. **Lummi** *Dermatological Aid* Decoction of leaves used as a hair wash. *Emetic* Infusion of smashed plants taken as an emetic. (79:31) **Paiute** *Eye Medicine* Strained infusion of dried, shredded roots used as a wash for cataracts. (111:73) **Skagit** *Dermatological Aid* Decoction of leaves used as a hair wash. *Tuberculosis Remedy* Infusion of leaves taken for tuberculosis. (79:31) **Thompson** *Veterinary Aid* Decoction of roots used as a delousing wash for sheep. (187:186)
- **Other**—**Saanich** *Insecticide* Leaves dried and hung in houses to keep flies and mosquitoes away. (182:79) **Thompson** *Insecticide* Decoction of plant used as a furniture and floor wash for lice, bedbugs, and other household pests. (187:186)

Aconitum, Ranunculaceae

Aconitum columbianum Nutt., Columbian Monkshood
- **Drug**—**Okanagan-Colville** *Poison* Plant considered highly poisonous. *Witchcraft Medicine* Used for witchcraft. (188:117)

Aconitum delphiniifolium DC., Larkspurleaf Monkshood
- **Drug**—**Eskimo, Inupiat** *Poison* Roots considered poisonous. (98:140) **Salish** *Unspecified* Plant used as a medicine. (178:294)

Aconitum fischeri, Fischer Monkshood
- **Drug**—**Gosiute** *Poison* Plant considered poisonous. (39:360)

Aconitum heterophyllum
- **Drug**—**Cree, Hudson Bay** *Poison* Plant considered poisonous. (92:303)

Aconitum maximum Pallas ex DC., Kamchatka Aconite
- **Drug**—**Aleut** *Poison* Plant possibly used at one time as a poison. (8:428)

Aconitum sp., Monkshood
- **Drug**—**Aleut** Used for fish and whale poison. (7:29) **Blackfoot** *Febrifuge* Roots used for fevers. *Poison* Plant considered poisonous. *Respiratory Aid* Roots used for acute respiratory infections. *Throat Aid* Roots used for acute throat infections. (97:34)

Acoraceae, see *Acorus*

Acorus, Acoraceae

Acorus calamus L., Calamus
- **Drug**—**Abnaki** *Carminative* Used for stomach gases. (144:154) Decoction of roots taken for stomach gas. (144:175) **Algonquin, Quebec** *Cold Remedy* Infusion of ground roots taken for colds. *Cough Medicine* Infusion of ground roots and chokecherry taken for coughs. *Gynecological Aid* Infusion of ground roots taken after childbirth and for symptoms of menopause. *Heart Medicine* Infusion of ground roots and pepperroot taken for heart disease. *Preventive Medicine* Carried on the person in order to avoid contracting a disease. (18:135) **Blackfoot** *Analgesic* Rootstock ground, mixed with tobacco, and smoked inhaled for headaches. *Gastrointestinal Aid* Poultice of crushed rootstocks and hot water applied for cramps. *Pulmonary Aid* Poultice of crushed rootstocks and hot water applied to sore chests. *Throat Aid* Poultice of crushed rootstocks and hot water applied to sore throats. *Toothache Remedy* Poultice of crushed rootstocks and hot water applied to toothaches. (97:23) **Cherokee** *Analgesic* Root chewed for headache. *Anthelmintic* Used for worms. *Anticonvulsive* Infusion given to "prevent recurrent spasms." *Antidiarrheal* Root chewed and juice swallowed for diarrhea. *Carminative* Used for flatulent colic. *Cold Remedy* Root variously chewed or used in infusion for colds. *Dermatological Aid* Used for "white swelling." *Diaphoretic* Used as a diaphoretic. *Diuretic* Used as a diuretic. *Gastrointestinal Aid* "Possesses stimulant and stomachic virtues" and used for "gravel." *Kidney Aid* Used for yellowish urine and "dropsy." *Stimulant* "Possesses stimulant and stomachic virtues" and used for "gravel." *Throat Aid* Root chewed for sore throat. *Urinary Aid* Used for flatulent colic, "white swelling," worms, yellowish urine, and "gravel." (80:28) **Cheyenne** *Analgesic* Decoction of root taken for bowel pain. (75:42) Infusion of root taken for bowel pain. (76:171) Plant smoked for headaches. *Ceremonial Medicine* Plant used in a sweat lodge ceremony. *Cold Remedy* Plant smoked or infusion of roots taken for colds. *Diuretic* Infusion of roots taken as a diuretic. (83:7) *Gastrointestinal Aid* Decoction of root taken for bowel pain. (75:42) *Laxative* Infusion of roots taken as a laxative. (83:7) *Panacea* Chewed root rubbed on skin for any illness. (75:42) Root chewed and rubbed on the skin for any illness. (76:171) *Pediatric Aid* Bit of root tied to child's clothing to keep the night spirits away. (75:42) Root tied to child's dress or blanket to keep away the night spirits. (76:171) *Witchcraft Medicine* Bit of root tied to clothes to keep night spirits away from children. (75:42) Root tied to child's dress or blanket to keep away the night spirits. (76:171) Plant used to ward off ghosts. (83:7) **Chippewa** *Cathartic* Warm infusion of root taken as a physic by children and adults. (53:344) *Cold Remedy* Decoction of root taken or snuff of pulverized root used for colds. (53:340) Infusion of plants taken for colds. *Cough Medicine* Infusion of plants taken for coughs. (71:124) *Hunting Medicine* Decoction of roots used on fish nets as a charm. (53:376) *Pediatric Aid* Decoction of root taken and dried root chewed by children for toothache. Decoction of root used by children as a gargle for sore throat. (53:342) Infusion of root taken by children and adults as a physic. (53:344) *Respiratory Aid* Infusion of plants taken for bronchial troubles. (71:124) *Throat Aid* Decoction of root gargled by children and root chewed by adults for sore throat. *Toothache Remedy* Decoction of root taken or root chewed, especially by children, for toothache. (53:342) **Cree** *Gastrointestinal Aid* Rootstock ground, mixed with water, and taken for an upset stomach. *Throat Aid* Rootstock peeled, chewed, and liquid swallowed for sore throats. (97:23) **Cree, Alberta** *Hallucinogen* Root chewed for the hallucinogenic effects. *Stimulant* Root chewed for the stimulant effects. *Unspecified* Root chewed for the medicinal effects. (149:331) **Cree, Woodlands** *Adjuvant* Roots added to any decoction to improve medicinal action. *Analgesic* Poultice of powdered roots and yellow pond lily roots or cow parsnip roots applied for headaches. *Antihemorrhagic* Decoction of rootstocks used for coughing up blood. *Antirheumatic (External)* Poultice of powdered roots and yellow pond lily roots or cow parsnip roots applied to painful joints, applied for muscle pain, and applied for rheumatism. Rootstocks used for sore muscles and rheumatic pains. *Cold Remedy* Rootstock chewed to prevent getting a cold after sweating during the winter. Roots smoked in a pipe for colds. Dried rootstock chewed for colds. *Cough Medicine*

Rootstock chewed for coughs from colds. Dried rootstock chewed for coughs. *Dermatological Aid* Decoction of rootstocks used for rash from touching nettles or other irritating plants. Poultice of powdered roots and yellow pond lily roots or cow parsnip roots applied to flesh worms. Poultice of chewed rootstock applied to cuts. *Ear Medicine* Poultice of water softened rootstock applied to the ear for earaches. *Febrifuge* Rootstocks used for severe chill. *Gastrointestinal Aid* Grated rootstocks in water taken for stomachaches. Rootstocks used for upset stomachs. *Hemostat* Poultice of chewed rootstock applied as a styptic. *Orthopedic Aid* Decoction of rootstocks used for lower back pains. Poultice of powdered roots and yellow pond lily roots or cow parsnip roots applied to limb swellings. Rootstock used for facial paralysis. *Panacea* Grated rootstocks used as an ingredient in a many herb remedy for various ailments. *Pediatric Aid* Decoction of rootstocks used for sickness related to teething. *Pulmonary Aid* Decoction of rootstocks used for whooping cough. Decoction of rootstocks used for stabbing pains in the chest. *Throat Aid* Decoction of rootstocks used for sore throats. Dried rootstock chewed for sore throats. *Toothache Remedy* Decoction of rootstocks used for sickness related to teething. Poultice of chewed rootstock applied to aching teeth. *Venereal Aid* Decoction of rootstocks used for venereal disease. (109:24) **Dakota** *Carminative* Dried roots taken as a carminative. (69:359) Plant used as a carminative and decoction taken for fever. *Ceremonial Medicine* Blades of grass used as garlands in mystery ceremonies. *Cold Remedy* Rootstock chewed, decoction taken, or smoke treatment used for colds. *Cough Medicine* Rootstock chewed as a cough remedy. *Febrifuge* Decoction of plant taken for fever. *Gastrointestinal Aid* Infusion of pounded rootstock taken for colic. *Panacea* Rootstock regarded as a panacea. *Psychological Aid* Paste of rootstock rubbed on warrior's face to prevent excitement and fear. *Toothache Remedy* Rootstock chewed for toothache. (70:69, 70) **Delaware** *Abortifacient* Infusion of roots used for suppressed menses. *Cold Remedy* Infusion of roots used for colds. *Cough Medicine* Infusion of roots used for coughs. *Gastrointestinal Aid* Roots used with sassafras roots for intestinal pains. (176:37) **Delaware, Oklahoma** *Abortifacient* Compound containing root taken for suppressed menses. *Analgesic* Compound containing root used for stomachache and intestinal pains. *Cold Remedy* Compound containing root taken for colds. (175:31, 74) *Cough Medicine* Infusion of root taken for coughs, colds, and suppressed menses. (175:31) **Delaware, Ontario** *Cold Remedy* Infusion of scraped root taken for colds. (175:31) **Iroquois** *Anthelmintic* Compound infusion of roots taken for tapeworms. *Blood Medicine* Compound infusion of roots taken as a blood remedy. (87:279) Infusion of plant and another plant given to children with poor blood circulation. (141:70) *Cold Remedy* Used for colds and sore throats from colds or singing. (87:279) *Dermatological Aid* Compound decoction taken for "boils around the abdomen of children." (87:278) *Ear Medicine* Decoction of roots used as drops in ear for earache. (87:279) *Emetic* Compound decoction of plant taken by women as an emetic for epilepsy. (87:278) *Gastrointestinal Aid* Powdered roots and cold water taken when feeling bad after eating meals. *Misc. Disease Remedy* Infusion of powdered roots taken for grippe with chills. (141:70) *Pediatric Aid* Compound decoction taken for "boils around the abdomen of children." (87:278) Infusion of plant and another plant given to children with poor blood circulation. Infusion of roots

and another plant given to children who scream during the night. *Respiratory Aid* Infusion of roots and roots from another plant used for hard respiration from lower chest pains. (141:70) *Throat Aid* Decoction of roots used as gargle for sore throat. (87:279) *Toothache Remedy* Root packed into hole of aching tooth to break up the tooth. Roots smoked and the smoke sucked into hollow tooth for toothache. *Witchcraft Medicine* Used for the detection of bewitchment. (87:278) **Lakota** *Cough Medicine* Roots chewed for coughs. *Hypotensive* Infusion of roots taken for high blood pressure. *Misc. Disease Remedy* Infusion of roots taken for diabetes. (106:48) *Orthopedic Aid* Infusion of pulverized roots and gun powder taken for arm and leg cramps. (139:26) *Throat Aid* Roots chewed for sore throat. *Toothache Remedy* Roots chewed for toothache. (106:48) **Malecite** *Cold Remedy* Infusion of one root used for colds. *Preventive Medicine* Roots chewed to prevent disease. (116:249) *Unspecified* Used for medicines. (160:6) **Menominee** *Abortifacient* Compound decoction of root used for irregular periods. (54:133) *Analgesic* Root, a very powerful remedy, used for stomach cramps. *Cathartic* Root used as a "good physic for the whole system, clearing the bile and all." (151:22, 23) *Cold Remedy* Root chewed or decoction of root used as cold remedy. (54:130) *Gastrointestinal Aid* Root used for stomach cramps. (151:22, 23) **Meskwaki** *Analgesic* Decoction of root taken for "a cramp expected in the stomach." (152:202) *Burn Dressing* Compound used for burns. *Cathartic* Plant used as a physic. (152:201, 202) *Cough Medicine* Decoction of root taken for cough. *Gastrointestinal Aid* Decoction of root taken for "a cramp expected in the stomach." *Tuberculosis Remedy* Decoction of root taken for tuberculosis. (152:202) **Micmac** *Cold Remedy* Root used for colds. *Cough Medicine* Root used for coughs. *Misc. Disease Remedy* Root used for cholera, smallpox, and other epidemics. *Panacea* Root and herb used for the prevention of disease in general and root used for disease in general. (40:53, 54) Plant used as a panacea. (156:316) *Pulmonary Aid* Root used for lung ailments, pneumonia, and pleurisy. (40:53, 54) *Unspecified* Roots chewed for medicinal use. (159:258) **Mohegan** *Abortifacient* Infusion of root taken for suppressed menses. *Analgesic* Infusion of root taken for stomach pains. (176:69, 128) *Antirheumatic (Internal)* Small pieces of root used for rheumatism. (176:128) *Cold Remedy* Infusion of root taken for colds. *Panacea* Root chewed to insure good health and root carried to "ward off sickness." (176:69, 128) *Tonic* Complex compound infusion including sweetflag root taken as spring tonic. (174:266) **Nanticoke** *Cold Remedy* Infusion of root given to infants for colds. (175:55) *Gastrointestinal Aid* Root used as colic medicine. (175:55, 84) *Pediatric Aid* Infusion of root given to infants for colds. (175:55) **Ojibwa** *Analgesic* Root used for stomach cramps. *Cathartic* Root used as a quick acting physic. *Cold Remedy* Root used for cold in the throat. *Gastrointestinal Aid* Root used for stomach cramps. (153:355) *Heart Medicine* Used as a heart stimulant. (5:2247) *Hunting Medicine* Root and sarsaparilla root made into tea and used on gill nets to bring a fine catch of whitefish. (153:428) *Throat Aid* Root chewed for sore throat. Used to make a throat tonic for singers. (5:2247) Roots chewed for sore throat. (5:2309) Root used for "a cold in the throat." (153:355) **Omaha** *Carminative* Plant used as a carminative. (68:334) Plant used as a carminative and decoction taken for fever. *Ceremonial Medicine* Blades of grass used as garlands in mystery ceremonies. *Cold Remedy* Rootstock chewed, decoction taken, or smoke treat-

ment used for colds. *Cough Medicine* Rootstock chewed as a cough remedy. *Febrifuge* Decoction of plant taken for fever. (70:69, 70) *Gastrointestinal Aid* Root chewed for stomach disorders. (58:584) Infusion of pounded rootstock taken for colic. *Panacea* Rootstock regarded as a panacea. (70:69, 70) *Tonic* Rootstock chewed as a tonic. (68:334) *Toothache Remedy* Rootstock chewed for toothache. (70:69, 70) *Veterinary Aid* Plant put into the feed of ailing horses. (58:584) **Pawnee** *Carminative* Plant used as a carminative and decoction taken for fever. *Ceremonial Medicine* Blades of grass used as garlands in mystery ceremonies. *Cold Remedy* Rootstock chewed, decoction taken, or smoke treatment used for colds. *Cough Medicine* Rootstock chewed as a cough remedy. *Febrifuge* Decoction of plant taken for fever. *Gastrointestinal Aid* Infusion of pounded rootstock taken for colic. *Panacea* Rootstock regarded as a panacea. *Toothache Remedy* Rootstock chewed for toothache. **Ponca** *Carminative* Plant used as a carminative. *Ceremonial Medicine* Blades of grass used as garlands in mystery ceremonies. *Cold Remedy* Rootstock chewed, decoction taken, or smoke treatment used for colds. *Cough Medicine* Rootstock chewed as a cough remedy. *Febrifuge* Decoction taken for fever. *Gastrointestinal Aid* Infusion of pounded rootstock taken for colic. *Panacea* Rootstock regarded as a panacea. *Toothache Remedy* Rootstock chewed for toothache. (70:69, 70) **Potawatomi** *Antihemorrhagic* Compound decoction of small amount of root taken for hemorrhage. *Respiratory Aid* Powdered root snuffed up nose for catarrh. (154:39, 40) **Rappahannock** *Gastrointestinal Aid* Chewed plant juice taken by older people for the stomach. (161:29) Infusion given to children and babies for pains and stomach cramps. *Pediatric Aid* Infusion given to children and babies for fretfulness, pains, and stomach cramps. *Sedative* Infusion given to children and babies for fretfulness. (161:30) *Tonic* Chewed plant juice taken by older people as a tonic. (161:29) **Shinnecock** *Blood Medicine* Root nibbled "to dry your blood." *Oral Aid* Root dried, cooked in sugar, and eaten for the breath. (30:118) **Sioux, Fort Peck** *Abortifacient* Used to cause abortion. *Panacea* Root chewed and swallowed as a "cure-all." (19:5) **Winnebago** *Carminative* Plant used as a carminative and decoction taken for fever. *Ceremonial Medicine* Blades of grass used as garlands in mystery ceremonies. *Cold Remedy* Rootstock chewed, decoction taken, or smoke treatment used for colds. *Cough Medicine* Rootstock chewed as a cough remedy. *Febrifuge* Decoction of plant taken for fever. *Gastrointestinal Aid* Infusion of pounded rootstock taken for colic. *Panacea* Rootstock regarded as a panacea. (70:69, 70) *Tonic* Complex compound injected via bird wing bone for general health. (130:265) *Toothache Remedy* Rootstock chewed for toothache. (70:69, 70)
- **Food**—**Abnaki** *Unspecified* Roots used for food. (144:175) **Dakota** *Unspecified* Dried root chewed for the agreeable taste. (69:359) **Lakota** *Unspecified* Leaves and stalks used for food. (139:26) **Micmac** *Beverage* Used to make a beverage. (159:258)
- **Dye**—**Chippewa** *Mordant* Plant used with bloodroot as a mordant in dyeing. (71:131)
- **Other**—**Blackfoot** *Cash Crop* Plant used for barter. (97:23) **Cheyenne** *Protection* Root tied to a child's necklet, dress, or blanket to keep away the night spirits. *Smoke Plant* Pulverized root and red willow bark used for smoking. (76:171) **Chippewa** *Protection* Decoction of roots used as a charm to "rattle snakes away." (53:376) **Omaha** *Incense & Fragrance* Leaves made into wreaths and worn around the neck or head for the pleasant odor. (58:584)

Acourtia, Asteraceae
Acourtia microcephala DC., Sacapellote
- **Drug**—**Coahuilla** *Cathartic* Decoction of plant taken to produce "a very quick passage of the bowels." (as *Perezia microcephala* 13:78)

Acourtia wrightii (Gray) Reveal & King, Brownfoot
- **Drug**—**Hualapai** *Dermatological Aid* Poultice of woolly "cotton" applied to open, bleeding wounds. (as *Perezia wrightii* 195:49) **Navajo, Kayenta** *Gynecological Aid* Plant used for difficult labor, a postpartum medicine. (as *Perezia wrightii* 205:49) **Pima** *Hemostat* Plant used as a styptic. (as *Perezia wrightii* 146:80)

Acrostichum, Pteridaceae, fern
Acrostichum danaeifolium Langsd. & Fisch., Inland Leatherfern
- **Drug**—**Seminole** *Febrifuge* Infusion of plant taken and rubbed on the body for high fevers. (169:202)
- **Other**—**Seminole** *Cash Crop* Plant sold to greenhouses. (169:506)

Actaea, Ranunculaceae
Actaea pachypoda Ell., White Baneberry
- **Drug**—**Blackfoot** *Cold Remedy* Decoction of root used for colds and coughs. *Cough Medicine* Decoction of root used for coughs and colds. (as *A. eburnea* 114:275) **Cherokee** *Dermatological Aid* Infusion of root used for itch. *Stimulant* Infusion given "to relieve and rally a patient at point of death." *Throat Aid* Infusion of root used as a gargle. *Toothache Remedy* "Will kill teeth of young people if not careful with it." (80:55) **Chippewa** *Anticonvulsive* and *Pediatric Aid* Decoction of roots taken by children and adults for convulsions. (as *A. alba* 71:130) **Iroquois** *Urinary Aid* Decoction of roots taken when "a man urinates blood." (as *A. alba* 87:321) **Meskwaki** *Analgesic* Decoction of root taken for childbirth pain. *Gynecological Aid* Decoction of root taken for childbirth pain. *Stimulant* Root used to revive and rally a patient at the point of death. *Urinary Aid* Used as a genitourinary remedy for men and women. (as *A. alba* 152:237, 238)

Actaea rubra (Ait.) Willd., Red Baneberry
- **Drug**—**Alaska Native** *Poison* Berries considered poisonous. (85:149) **Algonquin** *Analgesic* Used for stomach pains, in some seasons for males, other seasons for females. (22:142) **Blackfoot** *Cold Remedy* Decoction of roots taken for colds. *Cough Medicine* Decoction of roots taken for coughs. *Veterinary Aid* Decoction of roots used to treat horses. (97:34) **Cheyenne** *Ceremonial Medicine* Roots used in ceremonies. *Dermatological Aid* Roots used for sores. *Dietary Aid* Decoction of roots taken to improve the appetite. (83:33) *Gynecological Aid* Infusion of root pieces used by women after childbirth for increased milk flow. (82:8) Infusion of stems taken by pregnant and nursing mothers to increase milk flow. (83:22) **Chippewa** *Gynecological Aid* Decoction of root taken for excessive flowing. (53:358) **Cree, Hudson Bay** *Cathartic* Plant used as a purgative. (as *A. spicata* 92:303) **Cree, Woodlands** *Gynecological Aid* Infusion of small piece of root taken to slow heavy menstrual flow. (109:25) **Eskimo, Arctic** *Poison* Fruits considered poisonous. (128:17) **Ojibwa** *Gastrointestinal Aid* Root eaten by men for stomach troubles. *Gynecological Aid* Infusion

of root taken after childbirth "to clear up the system." (153:382) **Ojibwa, South** *Analgesic* and *Gastrointestinal Aid* Decoction of root taken for stomach pain caused by having "swallowed hair." (91:201) **Potawatomi** *Gynecological Aid* Infusion of root given "to purge the patient of afterbirth." (154:74) **Thompson** *Antirheumatic (Internal)* Decoction of root taken in a 1-teaspoon dose for arthritis. *Poison* Red and white berried plant considered extremely poisonous. *Pulmonary Aid* Decoction of plant taken for bronchial or lung trouble. (187:245)

Actaea rubra ssp. *arguta* (Nutt.) Hultén, Red Baneberry
- **Drug**—**Blackfoot** *Cold Remedy* Decoction of root used for colds and coughs. *Cough Medicine* Decoction of root used for coughs and colds. (as *A. arguta* 114:275) **Cheyenne** *Blood Medicine* Infusion of dried, pounded roots and stems used as a blood medicine. (as *A. arguta* 76:174) *Gynecological Aid* Simple or compound decoction of plant taken to increase maternal milk flow. (as *A. arguta* 75:41) Infusion of dried, pounded roots and stems taken after childbirth to make first milk pass off quickly. (as *A. arguta* 76:174) **Okanagon** *Antirheumatic (Internal)* Decoction of roots taken for rheumatism. *Dietary Aid* Decoction of roots taken for emaciation. (as *A. arguta* 125:41) **Quileute** *Dermatological Aid* Poultice of chewed leaves applied to boils. **Quinault** *Dermatological Aid* Poultice of chewed leaves applied to wounds. (as *A. arguta* 79:30) **Thompson** *Antirheumatic (Internal)* Decoction of roots taken for rheumatism. (as *A. arguta* 125:41) Decoction of root taken for rheumatism. (as *A. arguta* 164:463) *Dietary Aid* Decoction of roots taken for emaciation. (as *A. arguta* 125:41) Decoction of roots taken for emaciation. (as *A. arguta* & *A. eburnea* 164:463) *Poison* Decoction of roots considered poisonous if taken in large quantities. (as *A. arguta* & *A. eburnea* 164:512) *Venereal Aid* Decoction of root taken for syphilis. (as *A. arguta* & *A. eburnea* 164:463)

Actaea rubra (Ait.) Willd. ssp. *rubra*, Red Baneberry
- **Drug**—**Iroquois** *Antihemorrhagic* Compound decoction with roots taken for internal hemorrhage. *Antirheumatic (External)* Infusion of roots used as a wash for rheumatism. *Psychological Aid* Taken and sprinkled on head to give "young men the right sense." *Veterinary Aid* Infusion of roots given to dogs "when the dog won't hunt anymore." (as *A. spicata* ssp. *rubra* 87:321)

Actaea sp.
- **Drug**—**Micmac** *Poison* Plant considered poisonous. (145:56)

Adenocaulon, Asteraceae

Adenocaulon bicolor Hook., American Trailplant
- **Drug**—**Cowlitz** *Dermatological Aid* Poultice of leaves applied to boils. **Squaxin** *Dermatological Aid* Poultice of leaves applied to scrofula sores. *Tuberculosis Remedy* Poultice of leaves applied to scrofula sores. (79:48)

Adenostoma, Rosaceae

Adenostoma fasciculatum Hook. & Arn., Common Chamise
- **Drug**—**Cahuilla** *Antirheumatic (External)* and *Disinfectant* Decoction of leaves and branches used to bathe infected, sore, or swollen areas of the body. (15:29) **Coahuilla** *Veterinary Aid* Plant used to make a drink given to sick cows. (13:79)
- **Fiber**—**Cahuilla** *Building Material* Branches used to build ramadas and fences. (15:29) **Costanoan** *Basketry* Wood used for basketry. (21:249)
- **Other**—**Cahuilla** *Fuel* Large roots used for firewood. Wood coals used as a favorite source for roasting. *Hunting & Fishing Item* Branches used to make arrows. Branches used to make bows. *Lighting* Branches bound together and used for torches. (15:29) **Costanoan** *Hunting & Fishing Item* Wood used for arrow foreshafts. (21:249) **Diegueño** *Fuel* Used for firewood. (88:217) **Luiseño** *Fasteners* Gum used to secure stone points to arrows. *Hunting & Fishing Item* Plant used to make arrow foreshafts. (155:205)

Adenostoma sparsifolium Torr., Redshank
- **Drug**—**Cahuilla** *Antirheumatic (External)* Plant used for arthritis. *Cold Remedy* Leaves used to make a beverage for colds. *Emetic* Infusion of dried leaves taken for stomach ailments by inducing bowel movements or vomiting. *Gastrointestinal Aid* Leaves used to make a beverage for ulcers. Infusion of dried leaves taken for stomach ailments by inducing bowel movements or vomiting. *Laxative* Infusion of dried leaves taken for stomach ailments by inducing bowel movements or vomiting. *Pulmonary Aid* Leaves used to make a beverage for chest ailments. *Veterinary Aid* Poultice of plant and bacon fat applied to saddle sores on horses. (15:30) **Coahuilla** *Analgesic* Infusion of twigs taken for stomach and intestinal pain. *Cathartic* Infusion of twigs used "to produce vomit and bowel relief." *Dermatological Aid* Pulverized twigs mixed with grease and used as a salve. *Emetic* Infusion of twigs used "to produce vomit and bowel relief." *Gastrointestinal Aid* Infusion of twigs taken for stomach and intestinal pain. (13:77, 78) **Diegueño** *Gastrointestinal Aid* Infusion of plant taken for colic. *Toothache Remedy* Infusion of plant used as a mouthwash for toothaches. (88:217)
- **Food**—**Cahuilla** *Unspecified* Seeds used for food. (15:30) **Coahuilla** *Unspecified* Seeds used for food. (13:77)
- **Fiber**—**Cahuilla** *Building Material* Wood used for building material and fence posts. *Clothing* Stripped bark used as a fibrous material for women's skirts. (15:30) **Coahuilla** *Building Material* Used as building material. (13:77)
- **Other**—**Cahuilla** *Fuel* Used for firewood. Limbs used as a favorite firewood for roasting, giving a high-intensity heat. *Hunting & Fishing Item* Wood used to make arrowheads. (15:30) **Coahuilla** *Fuel* Used for fuel. (13:77) **Diegueño** *Fuel* Wood and roots used for firewood. (88:217)

Adiantum, Pteridaceae, fern

Adiantum aleuticum (Rupr.) Paris, Aleutian Maidenhair
- **Drug**—**Lummi** *Dermatological Aid* Infusion of leaves used as a hair wash. **Makah** *Antihemorrhagic* Leaves chewed for internal hemorrhages from wounds. *Dermatological Aid* Infusion of leaves used as a hair wash. *Gastrointestinal Aid* Leaves chewed for sore chest and stomach troubles. **Skokomish** *Dermatological Aid* Infusion of leaves used as a hair wash. (as *A. pedatum* var. *aleuticum* 79:14)
- **Fiber**—**Karok** *Basketry* Stems used for the designs in baskets. (as *A. pedatum* var. *aleuticum* 6:15) **Makah** *Basketry* Midribs used for the designs in basketry. **Quinault** *Basketry* Midribs used for the designs in basketry. (as *A. pedatum* var. *aleuticum* 79:14) **Tolowa** *Basketry* Dried, stored stems soaked in water and used for the designs in bas-

kets. **Yurok** *Basketry* Stems used for the designs in baskets. (as *A. pedatum* var. *aleuticum* 6:15)
- **Other**—**Pomo, Kashaya** *Jewelry* Stem inserted into ear lobe to keep hole from closing, either alone or with feathers; an earring. (as *A. pedatum* var. *aleuticum* 72:45)

Adiantum capillus-veneris L., Common Maidenhair
- **Drug**—**Mahuna** *Antirheumatic (Internal)* Plant used for rheumatism. (140:60) **Navajo, Kayenta** *Dermatological Aid* Infusion of plant used as a lotion for bumblebee or centipede stings. *Psychological Aid* Plant smoked or infusion of plant used for insanity. (205:14)

Adiantum jordanii C. Muell., California Maidenhair
- **Drug**—**Costanoan** *Analgesic* Decoction of plant used for "pain below the shoulders." *Blood Medicine* Decoction of plant used to purify the blood. *Gastrointestinal Aid* Decoction of plant used for stomach troubles. *Gynecological Aid* Decoction of plant used to expel the afterbirth and for postparturition. (21:4)
- **Fiber**—**Pomo, Kashaya** *Basketry* Dried, split stems used as a material for basket design. (72:46)
- **Other**—**Pomo, Kashaya** *Jewelry* Stem inserted into ear lobe to keep hole from closing, either alone or with feathers; an earring. (72:46)

Adiantum pedatum L., Northern Maidenhair
- **Drug**—**Cherokee** *Antirheumatic (External)* Compound decoction of root applied with warm hands for rheumatism. (80:8) Decoction of roots rubbed on area affected by rheumatism. (177:3) *Antirheumatic (Internal)* Infusion taken for rheumatism. *Emetic* Infusion of whole plant given as an emetic "in case of ague and fever." (80:34) Decoction of whole plant used as an emetic in cases of ague and fever. (203:74) *Febrifuge* Infusion of whole plant blown over head and chest of patient for fever. (80:34) Decoction of whole plant used as an emetic in cases of fever. (203:74) *Heart Medicine* Powdered leaves smoked for heart trouble. *Misc. Disease Remedy* Infusion of whole plant given as an emetic "in case of ague and fever." (80:34) Decoction of whole plant used as an emetic in cases of ague. (203:74) *Other* Given for "sudden paralytic attacks as in bad pneumonia of children." *Respiratory Aid* Powdered plant "snuffed" and smoked for asthma. (80:34) **Costanoan** *Blood Medicine* Decoction of plant used to purify the blood. *Gastrointestinal Aid* Decoction of plant used for stomach troubles. (21:5) **Hesquiat** *Respiratory Aid* Infusion of dried fronds burned to ashes, mixed with unknown, and taken for shortness of breath. Green fronds chewed for shortness of breath. *Strengthener* Infusion of dried fronds burned to ashes, mixed with unknown, and taken for strength and endurance. This infusion used especially by dancers in winter. Hesquiat dancers would take nothing but this medicine on day when they were dancing; it made them "light on their feet" and helped them continue dancing for a long time without tiring. Green fronds chewed by dancers in winter for strength and endurance. (185:29) **Makah** *Gastrointestinal Aid* Fronds chewed or eaten for "weak stomach." (67:217) **Menominee** *Antidiarrheal* Compound decoction of root used for dysentery. (54:131) *Gynecological Aid* Blades, stem, and root used for "female maladies." (151:47) **Meskwaki** *Pediatric Aid* Compound containing root and stems used for children who "turn black." (152:237) **Micmac** *Other* Herb used for fits and taken as an "agreeable decoction." (40:54) **Nitinaht** *Ceremonial Medicine* Used by dancers to make them light-footed. (186:61) **Potawatomi** *Gynecological Aid* Infusion of root taken by nursing mothers for caked breast. (154:73)
- **Fiber**—**Karok** *Basketry* Softened stems dried and used for the black designs in basket caps and other baskets. (148:377) **Makah** *Basketry* Dark petioles split in two, worked until soft, and used for black in basketry. (67:217)
- **Other**—**Karok** *Decorations* Stems used as decoration on clothing, especially on the Jump Dance dress. (148:377) **Kwakiutl, Southern** *Containers* Used to line baskets. *Cooking Tools* Used to cover berry drying racks. (183:264) **Maidu** *Decorations* Stalks used as decorative overlay twine in the manufacture of baskets. (173:71) **Pomo** *Tools* Stem inserted in a pierced ear lobe to keep the wound from closing. (66:11) **Potawatomi** *Good Luck Charm* Black stems used as hunting charms to bring good luck. (154:122)

Adiantum pedatum L. ssp. *pedatum*, Northern Maidenhair
- **Drug**—**Iroquois** *Analgesic* Decoction of plant used by children for cramps. *Antirheumatic (External)* Compound decoction of green roots used as foot soak for rheumatism. *Antirheumatic (Internal)* Compound decoction of green roots taken for rheumatism. *Diuretic* Decoction of roots taken for the cessation of urine due to gall. *Emetic* Infusion of plant induced vomiting as a remedy for love medicine. *Gynecological Aid* Compound decoction or infusion of roots taken for excessive menstruation. Decoction of roots used by "ladies to get period, cleans out" or for abortions. Plant used for abortion pains and pain when about to deliver. *Liver Aid* Decoction of roots taken for the cessation of urine due to gall. (87:258) *Orthopedic Aid* and *Pediatric Aid* Poultice of smashed plant applied to sore back of babies. (87:257) Decoction of plant used by children for cramps. *Snakebite Remedy* Poultice of wet, smashed fronds bound to snakebites. *Venereal Aid* Decoction of plant used as a wash for gonorrhea. Decoction of root taken for venereal disease and used as a wash for sores. (87:258)

Adiantum sp., Maidenhair Fern
- **Drug**—**Apalachee** *Unspecified* Plant water used for medicinal purposes. (81:98)
- **Fiber**—**Hahwunkwut** *Basketry* Plant used to make cooking bowls, mush baskets, and other small baskets. (117:183) **Poliklah** *Basketry* Stems used to make the designs on baskets. (117:170) **Wintoon** *Basketry* Used to make designs on baskets. (117:264)
- **Other**—**Hahwunkwut** *Cooking Tools* Plant used to make cooking bowls, mush baskets, and other small baskets. (117:183)

Aesculus, Hippocastanaceae
Aesculus californica (Spach) Nutt., California Buckeye
- **Drug**—**Costanoan** *Hemorrhoid Remedy* Smashed fruit applied as a salve for hemorrhoids. *Poison* Fruit used as a fish poison. *Toothache Remedy* Decoction of bark used for toothaches and loose teeth. (21:23) **Kawaiisu** *Hemorrhoid Remedy* Broken seeds used as suppositories for piles. *Poison* Raw seeds considered poisonous if eaten. (206:10) **Mendocino Indian** *Poison* Fresh fruit considered poisonous. *Toothache Remedy* Bark placed in cavity of tooth for toothaches. *Veterinary Aid* Fruit given to horses for bot worms and apt to cause an abortion in cows. (41:366) **Pomo** *Poison* Nuts used as poison. (66:14)

- *Food*—**Costanoan** *Fruit* Fruit used for food. (21:252) **Kawaiisu** *Bread & Cake* Seeds pounded, leached, boiled into a mush, made into a cake, and eaten with meat. (206:10) **Mendocino Indian** *Forage* Fruits eaten by squirrels as forage. *Fruit* Fruits roasted and eaten cold without salt. (41:366) **Miwok** *Soup* Roasted, peeled nuts ground into a meal and used to make soup. *Winter Use Food* Nuts stored for long periods and resorted to only when the acorn crop failed. (12:148) **Modesse** *Starvation Food* Nuts eaten in times of need. (117:223) **Pomo, Kashaya** *Unspecified* Boiled nuts eaten with baked kelp, meat, and seafood. Nuts were put into boiling water to loosen the husk. After the husk was removed, the nutmeat was returned to boiling water and cooked until it was soft like cooked potatoes. The nutmeat was then mashed with a mortar stone. The grounds could be strained at this stage or strained after soaking. The grounds would be soaked and leached a long time to remove the poisonous tannin. An older method was to peel the nuts and roast them in ashes until they were soft. They were then crushed and the meal was put in a sandy leaching basin beside a stream. For about 5 hours, the meal was leached with water from the stream. When the bitterness disappeared it was ready to eat without further cooking. (72:27) **Tubatulabal** *Unspecified* Nuts used for food. (193:15) **Yana** *Staple* Nuts ground into a fine meal and eaten. (147:251) **Yuki** *Unspecified* Nutmeats mashed and used for food. (49:85)
- *Other*—**Kawaiisu** *Cooking Tools* Wood sections hollowed out by burning and carved into bowls. (206:10) **Mendocino Indian** *Tools* Wood used as twirling sticks for making fire by friction. (41:366) **Pomo, Kashaya** *Hunting & Fishing Item* Wood used to make bows. Ground nuts sprinkled into pools to kill fish, a fishing method. *Tools* Wood used to make a drill stick and block for making fires. (72:27)

Aesculus flava Ait., Yellow Buckeye
- *Fiber*—**Cherokee** *Building Material* Wood used for lumber. *Furniture* Wood used to make baby cradles. (as *A. octandra* 80:27)
- *Other*—**Cherokee** *Ceremonial Items* Wood used to make masks. *Containers* Wood used to make dough trays. *Decorations* Wood used to carve. *Paper* Wood used for pulpwood. (as *A. octandra* 80:27)

Aesculus glabra Willd., Ohio Buckeye
- *Drug*—**Delaware** *Antirheumatic (External)* Nuts carried in the pocket for rheumatism. *Ear Medicine* Infusion of ground nuts mixed with sweet oil or mutton tallow and applied for earache. *Poison* Nuts ground and used as fish poison in streams. (176:30) **Delaware, Oklahoma** *Ear Medicine* Poultice of pulverized nuts with sweet oil applied for earache. *Poison* Pulverized nuts used as fish poison called "fish peyote," made the fish dizzy. (175:25, 74) **Mohegan** *Antirheumatic (External)* Carried in the pocket for rheumatism pain. (176:78)

Aesculus glabra var. *arguta* (Buckl.) B. L. Robins., Ohio Buckeye
- *Drug*—**Kiowa** *Emetic* Infusion of the inside of fruit taken as an emetic. (as *A. arguta* 192:41)

Aesculus hippocastanum L., Horse Chestnut
- *Drug*—**Iroquois** *Analgesic* Compound of powdered roots used for chest pains. *Pulmonary Aid* Compound of powdered roots used for chest pains. (87:379) **Mohegan** *Antirheumatic (External)* Horse chestnut carried in the pocket for rheumatism. **Shinnecock** *Antirheumatic (External)* Horse chestnut carried in the pocket for rheumatism. (30:121)

Aesculus pavia L., Red Buckeye
- *Drug*—**Cherokee** *Antirheumatic (External)* Nut carried in pocket for rheumatism and good luck. *Cancer Treatment* Poultice of pounded nuts used for tumors and infections. *Dermatological Aid* Poultice of pounded nuts used for tumors and infections and as a salve for sores. *Gastrointestinal Aid* Nuts used in various ways for dyspepsia and colic. (80:27) Infusion of roots taken and used as a bath for dyspepsia. (177:39) *Gynecological Aid* Cold, compound infusion given to stop bleeding after delivery. Infusion of bark and cold compound infusion of bark used in delivery. *Hemorrhoid Remedy* Nut carried in pocket for piles. *Orthopedic Aid* Poultice of pounded nuts used for "white swelling" and sprains. *Stimulant* Infusion of ground nutmeat taken to prevent fainting. (80:27)
- *Other*—**Cherokee** *Good Luck Charm* Nut carried in the pocket for good luck. (80:27)

Aesculus sp., Buckeye
- *Drug*—**Cherokee** *Poultice* Pounded nuts used in poultices. (203:74) **Creek** *Tuberculosis Remedy* Roots, a very strong medicine, used in cases of "pulmonary consumption." (172:658) Roots used for pulmonary consumption. **Koasati** *Throat Aid* Poultice of heated beans applied to the throat for tonsil troubles. (177:39)
- *Other*—**Cherokee** *Hunting & Fishing Item* Pounded roots strewed on water to "intoxicate fishes." (203:75) **Mewuk** *Tools* Wood used to make the fire drill. (117:325)

Agalinis, Scrophulariaceae

Agalinis tenuifolia (Vahl) Raf. var. *tenuifolia*, Slenderleaf False Foxglove
- *Drug*—**Meskwaki** *Antidiarrheal* Infusion used for diarrhea. (as *Gerardia tenuifolia* 152:246, 247)

Agaricaceae, see *Agaricus*

Agaricus, Agaricaceae, fungus

Agaricus campestris
- *Food*—**Delaware** *Unspecified* Salted, boiled, or fried in fat and used for food. (176:60) **Pomo, Kashaya** *Vegetable* Baked on hot rocks or in the oven or fried. (72:130)

Agaricus silvicola, Deer Mushroom
- *Food*—**Pomo, Kashaya** Plant top cooked on a flat hot rock and eaten. (72:129)

Agaricus sp.
- *Food*—**Thompson** *Dried Food* Plant strung, dried, cut, peeled, and eaten raw or roasted. (164:483)

Agastache, Lamiaceae

Agastache foeniculum (Pursh) Kuntze, Blue Giant Hyssop
- *Drug*—**Cheyenne** *Analgesic* Cold infusion of leaves taken for chest pains caused by coughing. (as *A. anethiodora* 76:186) *Cold Remedy* Infusion of leaves taken as a cold medicine. *Diaphoretic* Leaves used in a steam bath to induce sweating. *Febrifuge* Powdered leaves rubbed on the body for high fevers. (83:27) *Heart Medicine* Infusion of leaves used for a weak heart. (as *A. anethiodora* 76:186) Infusion of leaves taken to correct dispirited heart. *Herbal Steam* Leaves used in

a steam bath to induce sweating. (83:27) *Pulmonary Aid* Cold infusion of leaves taken for chest pains caused by coughing. (as *A. anethiodora* 76:186) Cold infusion of leaves taken for chest pain. (as *A. anethiodora* 75:42) *Unspecified* Infusion of leaves taken for its medicinal qualities. (83:27) **Chippewa** *Analgesic* Infusion of root taken for cold and chest pain. (as *A. anethiodora* 53:340) *Burn Dressing* Simple or compound poultice of leaves or stalk applied to burns. (as *A. anethiodora* 53:352) *Cough Medicine* Infusion of root taken for cough of "an internal cold." (as *A. anethiodora* 53:340) **Cree** *Ceremonial Medicine* Flowers frequently included in medicine bundles. (97:51) **Cree, Woodlands** *Antihemorrhagic* Infusion of stem, leaves, and other plants taken for coughing up blood. (109:26)
- *Food*—**Cheyenne** *Beverage* Leaves used to make tea. (as *A. anethiodora* 76:186) **Cree, Woodlands** *Beverage* Leaves added to store bought tea to improve the flavor. (109:26) **Dakota** *Beverage* Leaves used to make a hot, tea-like beverage taken with meals. *Sweetener* Plant used as a sweetening flavor in cooking. (as *A. anethiodora* 70:113) **Lakota** *Beverage* Used to make tea. (139:49) **Omaha** *Beverage* Leaves used to make a hot, tea-like beverage taken with meals. *Sweetener* Plant used as a sweetening flavor in cooking. **Pawnee** *Beverage* Leaves used to make a hot, tea-like beverage taken with meals. *Sweetener* Plant used as a sweetening flavor in cooking. **Ponca** *Beverage* Leaves used to make a hot, tea-like beverage taken with meals. *Sweetener* Plant used as a sweetening flavor in cooking. **Winnebago** *Beverage* Leaves used to make a hot, tea-like beverage taken with meals. *Sweetener* Plant used as a sweetening flavor in cooking. (as *A. anethiodora* 70:113)
- *Other*—**Chippewa** *Protection* Plant used as a charm for protection. (as *A. anethiodora* 53:376)

Agastache nepetoides (L.) Kuntze, Yellow Giant Hyssop
- *Drug*—**Iroquois** *Dermatological Aid* Compound infusion of plants used as wash for poison ivy and itch. (87:422)

Agastache pallidiflora ssp. *neomexicana* var. *neomexicana* (Briq.) R. W. Sanders, New Mexico Giant Hyssop
- *Drug*—**Navajo, Ramah** *Ceremonial Medicine* Plant used in ceremonial chant lotion. *Cough Medicine* Plant used for bad coughs. *Dermatological Aid* Dried, pulverized root used as dusting powder for sores or cankers. *Disinfectant* Plant used as fumigant for "deer infection." *Febrifuge* Plant used as a fever medicine. *Witchcraft Medicine* Plant used to protect from witches. (as *A. neomexicana* 191:41)
- *Food*—**Acoma** *Spice* Leaves used for flavoring. (as *A. neo-mexicana* 32:34) **Apache** *Staple* Used as one of the most important foods. **Comanche** *Staple* Used as one of the most important foods. (as *A. neo-mexicana* 32:10) **Keres, Western** *Spice* Leaves mixed with meat for seasoning. (as *Agastrache neo-mexicana* 171:24) **Laguna** *Spice* Leaves used for flavoring. (as *Agastache neo-mexicana* 32:34) **Mohave** *Staple* Used as one of the most important foods. **Paiute** *Staple* Used as one of the most important foods. **Papago** *Staple* Used as one of the most important foods. **Ute** *Staple* Used as one of the most important foods. **Yuma** *Staple* Used as one of the most important foods. (as *A. neo-mexicana* 32:10)
- *Other*—**Pima** *Cash Crop* Obtained by barter from the Papago Indians. (as *A. neo-mexicana* 32:10)

Agastache scrophulariifolia (Willd.) Kuntze, Purple Giant Hyssop
- *Drug*—**Meskwaki** *Diuretic* Infusion of root used as a diuretic. *Unspecified* Compound of plant heads used medicinally. (152:225)

Agastache urticifolia (Benth.) Kuntze, Nettleleaf Giant Hyssop
- *Drug*—**Miwok** *Antirheumatic (Internal)* Decoction of leaves taken for rheumatism. *Misc. Disease Remedy* Decoction taken for measles. (12:166) **Okanagan-Colville** *Cold Remedy* Infusion of leaves taken as a cold medicine. *Febrifuge* and *Pediatric Aid* Leaves placed in babies' blankets for fevers. (188:109) **Paiute** *Analgesic* Cold infusion of leaves used for stomach pains. *Cold Remedy* Decoction of plant taken for colds. *Dermatological Aid* Poultice of mashed leaves applied to swellings. *Gastrointestinal Aid* Cold infusion of leaves used for indigestion and stomach pains. **Shoshoni** *Cathartic* Decoction of plant taken as a physic. (180:33)
- *Food*—**Gosiute** *Unspecified* Seeds formerly used for food. (as *Lophanthus urticifolius* 39:374)

Agavaceae, see *Agave, Cordyline, Dasylirion, Manfreda, Nolina, Pleomele, Yucca*

Agave, Agavaceae

Agave americana L., American Century Plant
- *Food*—**Apache** *Dried Food* Heads and young leaves roasted, sun dried, and used immediately or stored. *Staple* Used as one of the most important foods. (32:10) **Apache, White Mountain** *Beverage* "Hearts" and roots pit baked, crushed, and fermented into an intoxicating beverage. (136:145) *Unspecified* Tubers pit baked and eaten. (136:155) *Winter Use Food* Tubers pit baked and stored for future use. (136:145) **Comanche** *Staple* Used as one of the most important foods. **Mohave** *Staple* Used as one of the most important foods. **Paiute** *Staple* Used as one of the most important foods. **Papago** *Staple* Used as one of the most important foods. (32:10) *Unspecified* Pit baked and extensively used for food. (34:16) Pit baked and used for food. (36:61) *Vegetable* Crowns with leaves removed eaten as greens in winter. Central flowering stalks eaten as greens in spring before they emerged. (34:14) Flower stalks eaten as greens. (34:16) Pit baked and used as greens. Flower stalks roasted in ashes and eaten as greens. (34:46) **Pima** *Dried Food* Fruit heads roasted, centers sun dried and used for food. *Sauce & Relish* Juice boiled and used as a syrup. *Starvation Food* Used for food in times of famine. (146:70) **Ute** *Staple* Used as one of the most important foods. **Yuma** *Staple* Used as one of the most important foods. (32:10)
- *Fiber*—**Papago** *Brushes & Brooms* Bundles of fibers used as combination brushes and combs for hair. (34:51) *Building Material* Leaves split and used for the weft of wrapped weaving in house frames. (34:53)
- *Other*—**Papago** *Cash Crop* Pit baked, kept in jars, and traded as a delicacy. (34:16) *Containers* Two-ply fiber formerly used to make carrying nets. (34:54) **Pima** *Cash Crop* Obtained by barter from the Papago Indians. (32:10)

Agave decipiens Baker, False Sisal
- *Food*—**Apache, White Mountain** *Beverage* Heart and tubers used to make a fermented drink. *Unspecified* Tubers pit baked and eaten. (136:155)

Agave deserti Engelm., Desert Agave
- *Food*—**Cahuilla** *Dried Food* Flowers parboiled to release the bitter-

ness and dried for future use. Baked leaves dried and stored for future use. Roasted, pounded stalks and leaves made into cakes and sun dried. *Unspecified* Flowers parboiled to release the bitterness and eaten. Baked leaves eaten. Roasted stalks used for food. (15:31) **Cocopa** *Unspecified* Crowns gathered and pit baked. (37:202) **Diegueño** *Unspecified* Roots and stalks baked overnight in a pit oven and used for food. (84:13) **Papago** *Unspecified* Pit baked and used for food. (36:61) **Pima** *Candy* Heads baked, sliced, dried, and eaten like candy. *Unspecified* Heads pit baked and eaten with pinole. (47:48) **Pima, Gila River** *Candy* Plant dried and used as sweets. (133:6) *Dried Food* Hearts dried and stored indefinitely. (133:4) *Staple* Hearts pit roasted and used as a staple food. (133:7)
- *Fiber*—**Cahuilla** *Basketry* Pounded leaves dried and made into cactus bags. *Clothing* Pounded leaves dried and made into shoes, sandals, and women's skirts. *Cordage* Pounded leaves dried and made into nets, slings, and cordage. Pounded leaves dried and made into nets used for baby cradles. *Scouring Material* Pounded leaves dried and made into cleaning brushes for cooking water. (15:31) **Pima** *Cordage* Dead leaves cut, beaten, and fibers twined into cords or rope. (47:48)
- *Dye*—**Cahuilla** *Unspecified* Burned stalk ash used as a dye for tattoos. (15:31)
- *Other*—**Cahuilla** *Ceremonial Items* Pounded leaves dried and made into netting used for ceremonial costumes. *Fuel* Dried stalks used for firewood. *Hunting & Fishing Item* Pounded leaves dried and made into bowstrings and snares. *Stable Gear* Pounded leaves dried and made into saddle blankets. *Tools* Thorns used as awls in basket making. Thorns used as tool for tattooing. (15:31) **Cocopa** *Cash Crop* Baked crowns obtained from Paipai and Diegueño in trade for agricultural products. (37:202)

Agave lechuguilla Torr., Lechuguilla
- *Fiber*—**Papago** *Cordage* Leaves used to make rough cordage. (34:61) **Pima** *Brushes & Brooms* Fiber used to make hairbrushes. (17:50)
- *Other*—**Papago** *Cash Crop* Fibers and the cordage made from them bartered with the Pimas for blankets and cotton. (34:61)

Agave palmeri Engelm., Palmer's Century Plant
- *Food*—**Apache, Western** *Beverage* Juice fermented into a drink. Crowns cooked, fermented in a vessel, ground, boiled, and the liquor again fermented. Juice strained and mixed with "tiswin water," a liquor of fermented maize. Flower stalk baked and chewed for juice. *Candy* Heart of the crown eaten by children as candy. *Dried Food* Plant eaten dried. *Substitution Food* Used in absence of other foods. *Unspecified* Crowns used for food. (26:169) **Papago** *Unspecified* Pit baked and used for food. (36:61)
- *Fiber*—**Apache, Western** *Sewing Material* Thorn used as needle and thread. (26:169)
- *Other*—**Apache, Western** *Decorations* Juice used by young girls to daub on their cheeks. *Paint* Juice covering pit stones after baking used to paint stripes on buckskin. *Tools* Stalk fashioned into hoe handles. *Weapon* Stalk used for a lance shaft. (26:169)

Agave parryi Engelm., Parry's Agave
- *Food*—**Apache** *Dried Food* Heads and young leaves roasted, sun dried, and used immediately or stored. Heads and young leaves roasted, sun dried, and used immediately or stored. *Staple* Used as one of the most important foods. (32:10) *Unspecified* Roots baked and eaten. (32:13) **Apache, Chiricahua & Mescalero** *Unspecified* Bulbous crowns baked in pits, pulpy centers released, pounded into thin sheets, and eaten. The Mescalero Apache were named for the food they made from mescal. In the pits where the crowns were baked, the largest rock was placed in the center and a cross made on it from black ashes. While the mescal baked, the women were supposed to stay away from their husbands, and if the crown was not completely roasted when removed from the pit, they were believed to have disobeyed. (33:35) Stalks roasted, boiled, or eaten raw. *Vegetable* Stalks boiled, dried, and stored to be used as vegetables. (33:38) **Apache, Mescalero** *Bread & Cake* Leaf bases pit cooked, made into cakes, dried, and used for food. (14:30) **Apache, Western** *Beverage* Juice fermented into a drink. Crowns cooked, fermented in a vessel, ground, boiled, and the liquor again fermented. Juice strained and mixed with "tiswin water," a liquor of fermented maize. Flower stalk baked and chewed for juice. *Candy* Heart of the crown eaten by children as candy. *Dried Food* Plant eaten dried. Plant eaten dried. *Substitution Food* Used in absence of other foods. Used in absence of other foods. *Unspecified* Crowns used for food. (26:169) **Comanche** *Staple* Used as one of the most important foods. **Mohave** *Staple* Used as one of the most important foods. **Paiute** *Staple* Used as one of the most important foods. **Papago** *Staple* Used as one of the most important foods. **Ute** *Staple* Used as one of the most important foods. **Yuma** *Staple* Used as one of the most important foods. (32:10)
- *Fiber*—**Apache, Western** *Sewing Material* Thorn used as needle and thread. (26:169)
- *Other*—**Apache, Western** *Decorations* Juice used by young girls to daub on their cheeks. Juice used by young girls to daub on their cheeks. *Paint* Juice covering pit stones after baking used to paint stripes on buckskin. *Tools* Stalk fashioned into hoe handles. *Weapon* Stalk used for a lance shaft. (26:169) **Pima** *Cash Crop* Obtained by barter from the Papago Indians. (32:10)

Agave schottii Engelm., Schott's Century Plant
- *Food*—**Papago** *Unspecified* Pit baked and used for food. (36:61)

Agave sp., Mescal Agave
- *Drug*—**Hualapai** *Dermatological Aid* Used as a facial cream. (195:55)
- *Food*—**Apache, San Carlos** *Unspecified* Pit baked and used for food. (95:257) **Havasupai** *Beverage* Leaves and young buds baked, soaked in water, and used as a drink. (200:71) **Hualapai** *Staple* Plant considered a main staple. *Sweetener* Stems, before blooming, eaten like sugar cane. *Winter Use Food* Plant stored for winter use. (195:55) **Navajo** *Beverage* Juice squeezed from baked fibers and drunk. *Dried Food* Heads baked or boiled, pounded into flat sheets, sun dried, and stored for future use. *Porridge* Dried, baked heads boiled and made into a "paste." *Soup* Dried, baked heads boiled and made into soup. *Unspecified* Heads baked and eaten. Leaves boiled and eaten. Young and tender flowering stalks and shoots roasted and eaten. (23:94) **Yavapai** *Beverage* Leaf stubs and heads pounded to express juice and used as a drink. (65:259) *Dried Food* Pounded, cooked, dried meaty centers of leaves stored in houses for later use. (65:260) *Unspecified* Flower stalk baked and soft, inner part used for food. (65:259)
- *Fiber*—**Hualapai** *Clothing* Cut, split leaves used to make sandals.

Cordage Cut, split leaves used to make rope. *Mats, Rugs & Bedding* Cut, split leaves used to make cradle mats. (195:55) **Navajo** *Cordage* Plant fibers used to make rope. (23:94) **Yavapai** *Brushes & Brooms* Grass stem brush used to transfer excess mescal juice from dish to slab. (65:259)
- **Other**—**Havasupai** *Cash Crop* Leaves and young buds baked and traded with the Hopi. *Ceremonial Items* Stalk and fiber used to make ceremonial equipment. (200:71) **Hualapai** *Containers* Crushed fibers used as an ingredient in pottery making. (195:55) **Navajo** *Cooking Tools* Leaves used to line the baking pits. (23:91) *Tools* Sharp pointed leaf tips used to make basketry awls. (23:94)

Agave utahensis Engelm., Utah Agave
- **Food**—**Havasupai** *Beverage* Plant used to make a drink. (197:66)
- **Fiber**—**Havasupai** *Brushes & Brooms* Used to make brushes for the hair and for cleaning grinding stones. To make the brushes, the dried matter of a dead and rotten leaf was knocked free from the fibers, which were then bent in two. The upper end of this brush was wrapped with a cord and the bent portion was covered with buckskin or cloth. The loose fibers were cut to the right length and hardened by burning the ends. (197:212) **Navajo** *Mats, Rugs & Bedding* Fibers used to make blankets. (55:37)
- **Other**—**Havasupai** *Cooking Tools* Used to make spoons for thin drinks. (197:212)

Ageratina, Asteraceae
Ageratina altissima (L.) King & H. E. Robins. **var. *altissima***, White Snakeroot
- **Drug**—**Cherokee** *Antidiarrheal* Taken for diarrhea, gravel, and urinary diseases. *Diuretic* Root used as a diuretic. *Febrifuge* Taken for fever. *Misc. Disease Remedy* Taken for ague. *Stimulant* Root used as a stimulant. *Tonic* Root used as a tonic. *Urinary Aid* Taken for gravel and urinary diseases. (as *Eupatorium rugosum* 80:56) **Iroquois** *Blood Medicine* Decoction of roots taken to separate venereal disease from the blood. (as *Eupatorium rugosum* 87:459) *Cathartic* Decoction of whole plant and roots taken as a physic. (as *Eupatorium rugosum* 87:458) *Diaphoretic* Decoction of roots taken and used as a sweat bath to keep patient cooled. *Gynecological Aid* Decoction or infusion of roots taken for a fallen or inflamed womb. *Panacea* Plant used for anything. *Venereal Aid* Decoction of roots taken to separate venereal disease from the blood. *Veterinary Aid* Infusion of plants given to horses to stop sweating. (as *Eupatorium rugosum* 87:459) *Witchcraft Medicine* Decoction of stems used as a witchcraft medicine. (as *Eupatorium rugosum* 87:458)

Ageratina altissima **var. *roanensis*** (Small) Clewell & Woot., White Snakeroot
- **Drug**—**Chickasaw** *Toothache Remedy* Roots chewed and held in mouth for toothache. **Choctaw** *Stimulant* and *Tonic* Used as a "warming stimulant and tonic." *Toothache Remedy* Roots chewed and held in mouth for toothache. (as *Eupatorium ageratoides* 28:288) **Meskwaki** *Diaphoretic* Used as a steaming agent in sweat bath. *Stimulant* Smudged and used to revive an unconscious patient. (as *Eupatorium urticaefolium* 152:214)

Ageratina herbacea (Gray) King & H. E. Robins., Fragrant Snakeroot
- **Drug**—**Navajo, Ramah** *Analgesic* Cold infusion taken and used as lotion for headache. *Febrifuge* Cold infusion taken and used as lotion for fever. (as *Eupatorium herbaceum* 191:51)

Ageratina occidentalis (Hook.) King & H.E. Robins., Western Snakeroot
- **Drug**—**Zuni** *Antirheumatic (External)* Ingredient of "schumaakwe cakes" and used externally for rheumatism. *Dermatological Aid* Ingredient of "schumaakwe cakes" and used externally for swelling. (as *Eupatorium occidentale* 166:50)

Agoseris, Asteraceae
Agoseris aurantiaca (Hook.) Greene, Orange Agoseris
- **Drug**—**Navajo, Ramah** *Ceremonial Medicine* Plant used as a ceremonial emetic. *Dermatological Aid* Cold infusion taken and used as lotion for arrow or bullet wounds. *Disinfectant* Cold infusion taken and used as lotion for "deer infection." *Emetic* Plant used as a ceremonial emetic. *Orthopedic Aid* Wet leaves rubbed on swollen arms, wrists, or ankles. *Panacea* Root used as a "life medicine." *Witchcraft Medicine* Cold infusion taken and used as lotion for protection from witches. (191:47)
- **Food**—**Karok** *Candy* Root juice used for chewing gum. (as *A. gracilens* 148:389)

Agoseris aurantiaca (Hook.) Greene **var. *aurantiaca***, Orange Agoseris
- **Food**—**Gosiute** *Unspecified* Leaves used for food. (as *Troximon aurantiacum* 39:383)

Agoseris aurantiaca **var. *purpurea*** (Gray) Cronq., Orange Agoseris
- **Other**—**Keres, Western** *Unspecified* Plant known and named but no use was specified. (as *A. purpurea* 171:25)

Agoseris glauca (Pursh) Raf., Pale Agoseris
- **Food**—**Thompson** *Candy* Milky juice chewed as gum. (as *A. villosa* 164:493)

Agoseris glauca **var. *dasycephala*** (Torr. & Gray) Jepson, Pale Agoseris
- **Drug**—**Okanagan-Colville** *Dermatological Aid* Infusion of entire plant used to wash sores and rashes. Poultice of latex applied to sores. *Laxative* Infusion of roots taken as a laxative. (188:74) **Thompson** *Dermatological Aid* Milky latex used to remove warts. (187:167)
- **Food**—**Okanagan-Colville** *Candy* Latex dried and used as chewing gum. (188:74) **Thompson** *Candy* Milky latex used as chewing gum. (187:167)

Agoseris retrorsa (Benth.) Greene, Spearleaf Agoseris
- **Food**—**Kawaiisu** *Vegetable* Green leaves boiled and eaten. Whole plant above the ground boiled, washed in cold water to remove bitterness, and fried in grease. (206:10)

Agoseris sp.
- **Food**—**Ute** *Unspecified* Leaves formerly used as food. (as *Troximon* 38:36)

Agrimonia, Rosaceae
Agrimonia gryposepala Wallr., Tall Hairy Agrimony

- **Drug**—**Cherokee** *Antidiarrheal* Infusion of burs taken to "check bowels." *Blood Medicine* Infusion of root taken to build up blood. *Dermatological Aid* Powdered root compound used for pox. *Dietary Aid* Infusion of root given to satisfy children's hunger. *Febrifuge* Infusion of burs taken for fever. *Gastrointestinal Aid* Cold infusion of pulverized root taken for bowels. *Gynecological Aid* Infusion of burs taken to "check discharge." *Pediatric Aid* Infusion of root given to satisfy children's hunger. (80:22) **Iroquois** *Antidiarrheal* Infusion or decoction used by children for diarrhea, "summer complaint," or vomiting. *Antiemetic* Infusion given to children for diarrhea, "summer complaint," and vomiting. (87:357) *Basket Medicine* Infusion of roots and flowers used on anything to sell, a "basket medicine." *Emetic* Decoction of plants taken for diarrhea and as emetic for "summer complaint." (87:358) *Other* and *Pediatric Aid* Infusion given to children for diarrhea, "summer complaint," and vomiting. (87:357) **Meskwaki** *Hemostat* Root used as a styptic for nosebleeds. (152:241) **Ojibwa** *Urinary Aid* Compound containing root used as a medicine for urinary troubles. (153:383, 384) **Potawatomi** *Hemostat* Plant used as styptic and infusion snuffed for nosebleed by Prairie Potawatomi. (154:76)

Agrimonia parviflora Ait., Harvestlice
- **Drug**—**Cherokee** *Antidiarrheal* Infusion of burs taken to "check bowels." *Blood Medicine* Infusion of root taken to build up blood. *Dermatological Aid* Powdered root compound used for pox. *Dietary Aid* Infusion of root given to satisfy children's hunger. *Febrifuge* Infusion of burs taken for fever. *Gastrointestinal Aid* Cold infusion of pulverized root taken for bowels. *Gynecological Aid* Infusion of burs taken to "check discharge." *Pediatric Aid* Infusion of root given to satisfy children's hunger. (80:22)

Agrimonia sp.
- **Drug**—**Iroquois** *Liver Aid* Compound decoction of roots taken for too much gall. (87:357)

Agropyron, Poaceae
Agropyron sp., Wheat Grass
- **Food**—**Paiute** *Unspecified* Species used for food. (167:243)

Agrostis, Poaceae
Agrostis perennans (Walt.) Tuckerman, Upland Bentgrass
- **Food**—**Klamath** *Unspecified* Seeds used for food. (45:91)

Aizoaceae, see *Carpobrotus, Trianthema*

Alariaceae, see *Alaria, Egregia*

Alaria, Alariaceae, alga
Alaria marginata Postels & Ruprecht, Short Kelp
- **Food**—**Hesquiat** *Dried Food* Stipes and fronds with attached herring eggs dried for later use.[6] (185:24)
- **Fiber**—**Hesquiat** *Sporting Equipment* Dried stipes use as "pucks" and hitting sticks.[5] (185:24)

Alcea, Malvaceae
Alcea rosea L., Hollyhock
- **Drug**—**Shinnecock** *Dermatological Aid* Leaves used to apply infusion of flowers to inflamed areas. (as *Althaea rosea* 30:120)

Alectoriaceae, see *Alectoria, Bryoria, Cornicularia, Pseudephebe*

Alectoria, Alectoriaceae, lichen
Alectoria fremontii Tuckerm., Black Moss
- **Food**—**Montana Indian** *Starvation Food* Long, black, hair-like lichen used as a famine food. (19:5)

Alectoria jubata
- **Food**—**Coeur d'Alene** *Unspecified* Formerly used for food. (black tree moss 178:91) **Spokan** *Unspecified* Species used for food. (black tree moss 178:344) **Thompson** *Unspecified* Plant cooked and eaten. (black moss 164:482)

Alectoria nigricans (Ach.) Nyl., Caribou Moss
- **Food**—**Eskimo, Inuktitut** *Fodder* Plant given to fawns to try to get them to eat from their hands. (202:191)
- **Other**—**Eskimo, Inuktitut** *Fuel* Dried plant used for tinder. (202:191)

Alectoria nitidula (Th. Fr.) Vain, Caribou Moss
- **Food**—**Eskimo, Inuktitut** *Fodder* Plant given to fawns to try to get them to eat from their hands. (202:191)
- **Other**—**Eskimo, Inuktitut** *Fuel* Dried plant used for tinder. (202:191)

Alectoria ochroleuca (Hoffm.) Massal., Caribou Moss
- **Food**—**Eskimo, Inuktitut** *Fodder* Plant given to fawns to try to get them to eat from their hands. (202:191)
- **Other**—**Eskimo, Inuktitut** *Fuel* Dried plant used for tinder. (202:191)

Alectoria sarmentosa
- **Drug**—**Nitinaht** *Dermatological Aid* Used for wound dressing material and as bandages. (maidenhair moss 186:55)
- **Fiber**—**Hanaksiala** *Mats, Rugs & Bedding* Plant used as mattresses at seasonal camps. (common witch's hair 43:144) **Nitinaht** *Clothing* Used for baby diapers and female sanitary napkins. (maidenhair moss 186:55)
- **Other**—**Bella Coola** *Decorations* Formerly used to decorate dance masks. (184:195) **Nitinaht** *Cooking Tools* Used for wiping salmon. (maidenhair moss 186:55)

Alectoria sp., Tree Lichen
- **Food**—**Hesquiat** *Forage* Plant browsed by deer. (185:17)

Aletes, Apiaceae
Aletes acaulis (Torr.) Coult. & Rose, Stemless Indian Parsley
- **Drug**—**Keres, Western** *Cathartic* Plant used as a cathartic. *Emetic* Plant used as an emetic. (171:25)

Aletes anisatus (Gray) Theobald & Tseng, Rocky Mountain Indian Parsley
- **Food**—**Isleta** *Sauce & Relish* Leaves eaten fresh as a relish. (as *Pseudocymopterus aletifolius* 32:47) Raw leaves eaten as a relish. (as *Pseudocymopterus aletifolius* 100:40) *Vegetable* Leaves cooked and used as greens. (as *Pseudocymopterus aletifolius* 32:47) Cooked leaves eaten as greens. (as *Pseudocymopterus aletifolius* 100:40)

Aletris, Liliaceae
Aletris farinosa L., White Colicroot
- **Drug**—**Catawba** *Antidiarrheal* Infusion of leaves taken for dysentery. (157:188) Infusion of leaves taken for bloody dysentery. (177:7) *Gas-*

trointestinal Aid Infusion of leaves taken for colic and stomach disorders. (157:188) Cold infusion of leaves taken for colic and stomach disorders. (177:7) **Cherokee** *Antirheumatic (Internal)* Taken for rheumatism. *Carminative* Taken for flatulent colic. *Cough Medicine* Taken for coughs. *Febrifuge* Tonic used for child bed fever. *Gynecological Aid* Tonic used to strengthen womb and root prevented abortion. *Liver Aid* Taken for jaundice. *Pulmonary Aid* Taken for lung diseases. *Tonic* Tonic used for child bed fever and to strengthen womb. *Tuberculosis Remedy* Taken for consumption. *Urinary Aid* Taken for "strangury," a slow, painful urination. (80:57) **Micmac** *Abortifacient* Root used as an emmenagogue. *Gastrointestinal Aid* Root used as a stomachic. *Tonic* Root used as a tonic. (40:54) **Rappahannock** *Gynecological Aid* Infusion of plant given to women and girls for "female troubles." (161:34)

Aleurites, Euphorbiaceae

Aleurites moluccana (L.) Willd., Indian Walnut
- **Drug**—**Hawaiian** *Abortifacient* Nutshells and gourds burned and the resulting smoke or fumes entered the vagina for swollen wombs. *Dermatological Aid* Baked nutmeats, other plants, and breadfruit milk applied to scrofulous sores, ulcers, and bad sores. *Gastrointestinal Aid* Flowers and other plants pounded and resulting liquid given to infants for stomach or bowel disorders. *Laxative* Nut oil used to make a very strong laxative. *Pediatric Aid* Flowers and other plants pounded and resulting liquid given to infants for stomach or bowel disorders. *Respiratory Aid* Bark and other plants pounded, resulting liquid heated and taken for asthma. *Strengthener* Nutmeats baked, ground, mixed with other plants, and eaten to build up the body. *Tuberculosis Remedy* Baked nutmeats, other plants, and breadfruit milk applied to scrofulous sores. (2:56)
- **Other**—**Hawaiian** *Fuel* Wood used for fuel. *Lighting* Nuts dried, strung together, and burned like candles. (2:56)

Alismataceae, see *Alisma, Sagittaria*

Alisma, Alismataceae

Alisma plantago-aquatica L., American Water Plantain
- **Drug**—**Cree, Woodlands** *Gastrointestinal Aid* Dried stem base eaten or grated and taken in water for heart "troubles," including heartburn. Stem base taken for stomachaches, cramps, and stomach flu. *Heart Medicine* Dried stem base eaten or grated and taken in water for heart "troubles," including heartburn. *Laxative* Stem base taken for constipation. *Misc. Disease Remedy* Stem base taken for stomach flu. *Panacea* Powdered stem base and many other herbs used for various ailments. *Stimulant* Stem base given to prevent fainting during childbirth. (109:26) **Iroquois** *Gynecological Aid* Infusion of plant used for "womb troubles." *Kidney Aid* Split roots used for lame back or kidneys. *Orthopedic Aid* Split root used for lame back or kidneys and leaf infusion used as a runner's liniment. *Other* Raw root chewed to strengthen veins. *Tuberculosis Remedy* Decoction of plant or roots or infusion of roots taken for consumption. (87:272)
- **Food**—**Iroquois** *Beverage* Plant made into a tea and used by forest runners. (141:65)

Alisma subcordatum Raf., American Water Plantain
- **Drug**—**Cherokee** *Dermatological Aid* Used as a poultice on old sores, wounds, bruises, swellings, and ulcers. *Gastrointestinal Aid* Root used for bowel complaints, sores, wounds, and bruises. (80:61)

Allenrolfea, Chenopodiaceae

Allenrolfea occidentalis (S. Wats.) Kuntze, Iodine Bush
- **Food**—**Cahuilla** *Beverage* Ground seed flour and water made into a drink. *Bread & Cake* Ground seed flour dampened, shaped, dried, and eaten as a cookie. *Porridge* Ground seed flour and water made into a mush. (15:36) **Maricopa** *Staple* Seeds harvested, winnowed, parched, ground, and the meal eaten. **Mohave** *Staple* Seeds harvested, winnowed, parched, ground, and the meal eaten. (37:187) **Pima** *Unspecified* Ripe seeds winnowed, roasted, ground, water added, cooked, and used for food. (47:69) **Pima, Gila River** *Starvation Food* Seeds used as "starvation food." (133:6) *Unspecified* Seeds used for food. (133:7) **Yuma** *Staple* Seeds harvested, winnowed, parched, ground, and the meal eaten. (37:187)

Allionia, Nyctaginaceae

Allionia incarnata L., Trailing Windmills
- **Drug**—**Navajo, Ramah** *Dermatological Aid* Cold infusion of root used as a lotion for swellings. (191:26)

Allium, Liliaceae

Allium acuminatum Hook., Tapertip Onion
- **Food**—**Gosiute** *Unspecified* Bulbs eaten in spring and early summer. (39:360) **Hoh** *Unspecified* Bulbs pit baked and used for food. (137:59) **Karok** *Unspecified* Bulbs relished by only old men and old women. (148:380) **Paiute** *Sauce & Relish* Leaves eaten as a relish. *Unspecified* Bulbs roasted and used for food. Seeded heads placed in hot ashes for a few minutes, seeds extracted and eaten. (104:102) *Vegetable* Onions eaten raw, boiled, or baked in a pit. (111:55) **Quileute** *Unspecified* Bulbs pit baked and used for food. (137:59) **Salish, Coast** *Unspecified* Strongly flavored bulbs eaten with other foods. (182:74) **Thompson** *Unspecified* Thick coated, spherical bulbs eaten. (164:482) Bulbs dug in the spring and used for food. (187:117) **Ute** *Unspecified* Bulbs and leaves used for food. (38:32)
- **Other**—**Salish, Coast** *Insecticide* Bulbs rubbed on the skin to repel insects. (182:74)

Allium anceps Kellogg, Twinleaf Onion
- **Food**—**Paiute, Northern** *Bread & Cake* Bulbs cooked on hot rocks, squeezed into cakes, and eaten. *Unspecified* Bulbs roasted in the sand and eaten. (59:44)

Allium bisceptrum S. Wats., Twincrest Onion
- **Drug**—**Mahuna** *Dietary Aid* Plant juice used as an appetite restorer. (140:62)
- **Food**—**Apache, White Mountain** *Unspecified* Bulbs eaten raw and cooked. (136:155) **Gosiute** *Unspecified* Bulbs eaten in spring and early summer. (39:360) **Paiute** *Sauce & Relish* Leaves eaten as a relish. *Unspecified* Seeded heads placed in hot ashes for a few minutes, seeds extracted and eaten. Bulbs roasted and used for food. (104:102) **Ute** *Unspecified* Bulbs and leaves used for food. (38:32)

Allium bisceptrum var. *palmeri* (S. Wats.) Cronq., Aspen Onion
- **Food**—**Havasupai** *Starvation Food* Bulbs eaten only when very hungry. (as *A. palmeri* 197:211)

Allium bolanderi S. Wats., Bolander's Onion
- **Food**—**Karok** *Unspecified* Bulbs relished by only old men and old women. (148:380) **Mendocino Indian** *Unspecified* Corms used for food. (41:322)

Allium brevistylum S. Wats., Shortstyle Onion
- **Drug**—**Cheyenne** *Dermatological Aid* Poultice of ground roots and stems applied and infusion used as a wash for carbuncles. (76:171)

Allium canadense L., Meadow Garlic
- **Drug**—**Cherokee** *Carminative* Used as a carminative. *Cathartic* Used as a mild cathartic. *Diuretic* Used as a diuretic. *Ear Medicine* Used "to remove deafness." *Expectorant* Used as an expectorant. *Kidney Aid* Used for "dropsy." *Misc. Disease Remedy* Used for scurvy. *Pediatric Aid* and *Pulmonary Aid* Tincture used to prevent worms and colic in children and used as a croup remedy. *Respiratory Aid* Used for asthma. *Stimulant* Used as a stimulant. (80:35) **Mahuna** *Dermatological Aid* Plant rubbed on body for protection from insect bites. Plant rubbed on body for protection from lizard, scorpion, and tarantula bites. *Snakebite Remedy* Plant rubbed on body for protection from poisonous snakebites. (140:63)
- **Food**—**Cherokee** *Vegetable* Boiled bulbs fried with grease and greens. (126:46) **Iroquois** *Vegetable* Cooked and seasoned with salt, pepper, or butter. Bulb, consisting of the fleshy bases of the leaves, eaten raw. (196:118) **Menominee** *Unspecified* Small, wild onion used for food. (151:69) **Meskwaki** *Spice* Dried bulb used for seasoning. *Winter Use Food* Dried bulb used for winter cookery. (152:262) **Potawatomi** *Soup* Very strong flavor of this plant, a valuable wild food, used in soup. (154:104)

Allium canadense* var. *mobilense (Regel) Ownbey, Meadow Garlic
- **Food**—**Dakota** *Sauce & Relish* Fresh, raw bulbs used as a relish. *Spice* Bulbs used as a flavor for meat and soup. *Unspecified* Fried bulbs used for food. **Omaha** *Sauce & Relish* Fresh, raw bulbs used as a relish. *Spice* Bulbs used as a flavor for meat and soup. *Unspecified* Fried bulbs used for food. **Pawnee** *Sauce & Relish* Fresh, raw bulbs used as a relish. *Spice* Bulbs used as a flavor for meat and soup. *Unspecified* Fried bulbs used for food. **Ponca** *Sauce & Relish* Fresh, raw bulbs used as a relish. *Spice* Bulbs used as a flavor for meats and soups. *Unspecified* Fried bulbs used for food. **Winnebago** *Sauce & Relish* Fresh, raw bulbs used as a relish. *Spice* Bulbs used as a flavor for meat and soup. *Unspecified* Fried bulbs used for food. (as *A. mutabile* 70:71)

Allium cepa L., Garden Onion
- **Drug**—**Mohegan** *Cold Remedy* Syrup of chopped onions taken for colds. **Shinnecock** *Cold Remedy* Syrup of chopped onions taken for colds. *Disinfectant* Used to destroy germs because of a volatile oil in roots. *Ear Medicine* Heart of onion placed in ear for earache. *Febrifuge* Onion placed in a sick room to draw fever out. *Misc. Disease Remedy* Onion placed in a sick room to draw out flu. (30:120)
- **Food**—**Haisla & Hanaksiala** *Vegetable* Bulbs used for food. (43:194) **Havasupai** *Unspecified* Bulbs used for food. (197:212) **Navajo** *Unspecified* Onions singed, to remove the strong taste, and eaten immediately. *Winter Use Food* Onions singed, to remove the strong taste, dried, and stored for winter use. (55:31) **Navajo, Ramah** *Unspecified* Species used for food. (191:20) **Neeshenam** *Unspecified* Eaten raw, roasted, or boiled. (129:377) **Oweekeno** *Unspecified* Bulbs used for food. (43:77) **Seminole** *Unspecified* Plant used for food. (169:505)
- **Dye**—**Iroquois** *Green* Bulb peelings used as a green dye for wool. *Yellow* Bulb peelings used as a yellow dye for wool. (142:104)

Allium cernuum Roth, Nodding Onion
- **Drug**—**Cherokee** *Cold Remedy* Juice taken for colds. *Dermatological Aid* Juice given to children for hives. *Febrifuge* Used as poultice for feet in "nervous fever." *Gastrointestinal Aid* Infusion taken for colic. *Kidney Aid* Juice taken after "horsemint tea" for "gravel and dropsy." *Liver Aid* Juice taken for "liver complaints." *Pediatric Aid* Juice given to children for hives and croup. *Pulmonary Aid* Poultice of fried plant put on chest for croup. Juice given to children for croup. *Respiratory Aid* Juice taken for "phthisic." *Throat Aid* Juice taken for sore throat. *Urinary Aid* Juice taken after "horsemint tea" for "gravel and dropsy." (80:47) **Isleta** *Dermatological Aid* Poultice of onions applied externally for infections. *Throat Aid* Poultice of warm onions applied externally to throat for sore throat. (100:20) **Kwakiutl** *Dermatological Aid* Poultice of soaked bulbs applied to sores and swellings. (183:272) **Makah** *Analgesic* Poultice of chewed plants applied to the chest for pleurisy pains. *Pulmonary Aid* Poultice of chewed plants applied to the chest for pleurisy pains. **Quinault** *Analgesic* Poultice of chewed plants applied to the chest for pleurisy pains. *Pulmonary Aid* Poultice of chewed plants applied to the chest for pleurisy pains. (79:24)
- **Food**—**Apache, Chiricahua & Mescalero** *Spice* Onions used to flavor soups and gravies. *Vegetable* Onions occasionally eaten raw. (33:47) **Bella Coola** *Unspecified* Bulbs eaten fresh. (184:199) **Blackfoot** *Spice* Bulbs and leaves used as flavoring. *Vegetable* Bulbs and leaves eaten raw. (97:23) **Cherokee** *Unspecified* Bulbs used for food. (80:47) **Clallam** *Unspecified* Bulbs eaten raw, cooked in pits, or fried with meat. (57:196) **Cree** *Vegetable* Species used for food. (16:485) **Flathead** *Sauce & Relish* Bulbs used as condiments. *Staple* Bulbs used as a staple food. (82:10) **Haisla & Hanaksiala** *Vegetable* Bulbs cooked and eaten and the tops eaten fresh with meat. (43:193) **Hoh** *Unspecified* Bulbs pit baked and used for food. (137:59) **Hopi** *Spice* Used for flavoring before the introduction of the cultivated onion. (200:70) *Unspecified* Eaten raw with cornmeal dumplings or fresh piki bread. (120:20) **Isleta** *Vegetable* Bulbs eaten fresh, uncooked, or boiled. *Winter Use Food* Bulbs stored for future use. (100:20) **Keres, Western** *Vegetable* Bulbs used for food. (171:25) **Klallam** *Unspecified* Bulbs used for food. (79:24) **Kutenai** *Sauce & Relish* Bulbs used as condiments. *Staple* Bulbs used as a staple food. (82:10) **Kwakiutl, Southern** *Unspecified* Bulbs cooked and used for food. (183:272) **Makah** *Unspecified* Bulbs used for food. (67:338) Bulbs eaten sparingly. (79:24) **Navajo** *Sauce & Relish* Bulbs used to make gravies. *Soup* Bulbs used to make soup. *Spice* Leaves finely chopped and used like chives in salads or sauces. (110:29) *Unspecified* Onions singed, to remove the strong taste, and eaten immediately. (55:31) *Vegetable* Bulbs cooked with other vegetables. Roasted bulbs eaten with salt and pepper. (110:29) *Winter Use Food* Onions singed, to remove the strong taste, dried, and stored for winter use. (55:31) **Navajo, Ramah** *Dried Food* Bulbs, never the tops, dried for the winter. *Unspecified* Bulbs, never the tops, eaten raw, with fried or boiled meat. (191:20) **Nitinaht** *Unspecified* Bulbs used for food. (67:338) **Ojibwa** *Vegetable* Used in the spring as an article of food, the small wild onion was sweet. (153:406)

Okanagan-Colville *Dried Food* Bulbs dried and stored for winter use. *Vegetable* Bulbs pit cooked and eaten. (188:38) **Okanagon** *Staple* Roots used as a principal food. (178:238) *Unspecified* Bulbs and leaves used for food. (125:37) **Oweekeno** *Unspecified* Bulbs used for food. (43:76) **Quileute** *Unspecified* Bulbs pit baked and used for food. (137:59) **Quinault** *Unspecified* Bulbs used for food. (79:24) **Salish, Coast** *Unspecified* Strongly flavored bulbs eaten with other foods. (182:74) **Shuswap** *Forage* Bulbs eaten by sheep and cattle. *Spice* Bulbs used to flavor dried salmon heated with dried bread on an open fire. (123:54) **Thompson** *Dried Food* Bulbs tied in bundles, partially dried, pit cooked, and used for food.[7] Bulbs dried for winter storage. The dried bulbs were sprinkled with water and became just like fresh bulbs or they were soaked overnight in water. *Special Food* Cooked bulbs considered a delicacy.[7] (187:117) *Unspecified* Bulbs and leaves used for food. (125:37) Thick bulbs cooked and eaten. (164:481)
- **Other**—Salish, Coast *Insecticide* Bulbs rubbed on the skin to repel insects. (182:74)

Allium cernuum var. *obtusum* Cockerell ex J. F. Macbr., Nodding Onion
- **Food**—Acoma *Unspecified* Bulbs used for food. (as *A. recurvatum* 32:15) **Blackfoot** *Spice* Bulbs used for flavoring. *Vegetable* Bulbs eaten raw. (as *A. recurvatum* 114:278) **Hopi** *Unspecified* Dipped in water with broken wafer bread and eaten raw. (as *A. recurvatum* 32:15) Bulbs washed and eaten raw with broken wafer bread dipped in water. (as *A. recurvatum* 138:53) **Isleta** *Unspecified* Bulbs eaten raw or boiled. **Laguna** *Unspecified* Bulbs used for food. **Tewa** *Unspecified* Dipped in water with broken wafer bread and eaten raw. (as *A. recurvatum* 32:15) Bulbs washed and eaten raw with broken wafer bread dipped in water. (as *A. recurvatum* 138:53)

Allium dichlamydeum Greene, Coastal Onion
- **Food**—Pomo, Kashaya *Vegetable* Greens and bulb eaten raw or cooked with potatoes or meats for flavoring. (72:86)

Allium douglasii Hook., Douglas's Onion
- **Food**—Okanagan-Colville *Dried Food* and *Vegetable* Bulbs dried, pit cooked, and eaten. (188:38)

Allium drummondii Regel, Drummond's Onion
- **Food**—Cheyenne *Spice* Boiled with meat, when salt scarce, to flavor the food. (as *A. nuttallii* 76:171) *Unspecified* Bulbs formerly boiled with meat and used for food. (as *A. nuttalii* 83:12) Species used for food. (as *A. nuttallii* 83:45) **Lakota** *Unspecified* Species used for food. (139:27) **Navajo, Ramah** *Unspecified* Bulbs boiled with meat. (191:20)

Allium falcifolium Hook. & Arn., Scytheleaf Onion
- **Food**—Shoshoni *Spice* Bulbs used for seasoning. (118:14)

Allium geyeri S. Wats., Geyer's Onion
- **Food**—Apache *Unspecified* Bulbs used for food. (32:15) **Apache, Chiricahua & Mescalero** *Spice* Onions used to flavor soups and gravies. *Vegetable* Onions occasionally eaten raw. (33:47) **Hopi** *Spice* Used for flavoring before the introduction of the cultivated onion. (200:70) *Unspecified* Eaten raw with cornmeal dumplings or fresh piki bread. (120:20) **Okanagan-Colville** *Dried Food* and *Vegetable* Bulbs dried, pit cooked, and eaten. (188:38)

Allium geyeri var. *tenerum* M. E. Jones, Bulbil Onion
- **Food**—Keres, Western *Spice* Bulbs used largely for seasoning. (as *A. sabulicola* 171:25) **Pueblo** *Spice* Bulbs used for seasoning. (as *A. sabulicola* 32:15)

Allium hyalinum Curran, Glassy Onion
- **Food**—Tubatulabal *Unspecified* Leaves, stalks, and heads used for food. (193:12)

Allium lacunosum S. Wats., Pitted Onion
- **Food**—Tubatulabal *Unspecified* Leaves, stalks, and heads used for food. (193:12)

Allium macropetalum Rydb., Largeflower Wild Onion
- **Food**—Navajo *Dried Food* Bulbs rubbed in hot ashes, dried, and stored for winter use. *Unspecified* Bulbs rubbed in hot ashes and eaten. (as *A. deserticolum* 32:15) Onions singed, to remove the strong taste, and eaten immediately. (as *A. deserticola* 55:31) *Vegetable* Entire plant eaten raw or cooked with meat. (as *A. deserticola* 165:221) *Winter Use Food* Onions singed, to remove the strong taste, dried, and stored for winter use. (as *A. deserticola* 55:31)

Allium nevadense S. Wats., Nevada Onion
- **Food**—Paiute, Northern *Unspecified* Whole plant eaten raw. (59:44)

Allium parvum Kellogg, Small Onion
- **Food**—Paiute, Northern *Dried Food* Bulbs dried and eaten. *Soup* Bulbs dried, ground, and cooked in soup. (59:44)

Allium peninsulare J. G. Lemmon ex Greene, Mexicali Onion
- **Food**—Tubatulabal *Unspecified* Leaves, stalks, and heads used for food. (193:12)

Allium platycaule S. Wats., Broadstemmed Onion
- **Food**—Paiute *Sauce & Relish* Leaves eaten as a relish. *Unspecified* Bulbs roasted and used for food. Seeded heads placed in hot ashes for a few minutes, seeds extracted and eaten. (104:102)

Allium pleianthum S. Wats., Manyflower Onion
- **Food**—Paiute *Sauce & Relish* Green leaves eaten as a relish. (104:102)

Allium sativum L., Cultivated Garlic
- **Drug**—Cherokee *Carminative* Used as a carminative. *Cathartic* Used as a mild cathartic. *Diuretic* Used as a diuretic. *Ear Medicine* Used "to remove deafness." *Expectorant* Used as an expectorant. *Kidney Aid* Used for "dropsy." *Misc. Disease Remedy* Used for scurvy. *Pediatric Aid* and *Pulmonary Aid* Tincture used to prevent worms and colic in children and used as a croup remedy. *Respiratory Aid* Used for asthma. *Stimulant* Used as a stimulant. (80:35)
- **Food**—Algonquin, Tête-de-Boule *Spice* Bulbs mixed with food and eaten. (132:118)

Allium schoenoprasum L., Wild Chives
- **Food**—Alaska Native *Unspecified* Bulbs used sparingly. (85:113) **Cree, Woodlands** *Spice* Leaves added to boiled fish for flavor. *Unspecified* Fresh leaves used for food. (109:26) **Eskimo, Inuktitut** *Spice* Used as a soup condiment. (202:182) **Eskimo, Inupiat** *Soup* Bulbs and leaves used to make soup. *Vegetable* Leaves eaten cooked or raw with seal oil, meat, and fish. Leaves used like raw green onions or garlic in a salad. Leaves fried with meat, fat, other greens, vinegar, salt, and

pepper and eaten as a hot salad. (98:28) **Koyukon** *Unspecified* Plant eaten raw, alone or with fish. (119:56) **Tanana, Upper** *Frozen Food* Stems and bulbs frozen for future use. *Unspecified* Stems and bulbs eaten raw, fried, or boiled. (102:15)

Allium schoenoprasum **var.** *sibiricum* (L.) Hartman, Wild Chives
- *Food*—**Anticosti** *Soup* Leaves salted and added to soup. (143:69) **Cheyenne** *Spice* Boiled with meat, when salt scarce, to flavor the food. (as *A. sibiricum* 76:171) *Unspecified* Species used for food. (as *A. sibiricum* 83:45) Bulbs formerly boiled with meat and used for food. (as *A. sibiricum* L. 83:12) **Great Basin Indian** *Unspecified* Bulbs used for food. (as *A. sibiricum* 121:46)
- *Dye*—**Great Basin Indian** *Brown* Bulb skin used as a golden brown dye. (as *A. sibiricum* 121:46)

Allium **sp.**, Onion
- *Drug*—**Blackfoot** *Analgesic* Plant smudge smoke inhaled for headaches. (86:79) *Antiemetic* Infusion of bulbs taken for vomiting and allowed the retention of food. (86:65) *Antirheumatic (External)* Infusion of bulbs, sometimes combined with *Monarda*, applied to swellings. (86:75) *Cold Remedy* Bulb smudge used to fumigate the patient for a cold. *Cough Medicine* Infusion of bulbs taken for persistent coughs. (86:70) *Dermatological Aid* Infusion of bulbs, sometimes combined with *Monarda*, applied to sores. (86:75) *Ear Medicine* Infusion of bulbs used for ear infections. (86:80) *Emetic* Infusion of plant and another plant taken and used as a steam to serve as an emetic. (86:65) *Eye Medicine* Infusion of bulbs used as an eyewash. (86:80) *Misc. Disease Remedy* Infusion of bulbs taken for a disease which caused a swollen penis and severe constipation. (86:69) *Pediatric Aid* Infusion of bulb taken by nursing mother to pass medicinal properties to the child through the milk. (86:65) *Respiratory Aid* Dried bulb snuff used to open the sinuses. Plant smudge smoke inhaled for sinus troubles. (86:70, 79) *Veterinary Aid* Bulb smudge used to fumigate horses with sinus congestion. Infusion of plant pieces and *Monarda* applied to saddle sores. (86:87) **Iroquois** *Pediatric Aid* and *Sedative* Compound infusion used for a "baby who starts suddenly, especially in sleep." (87:320) **Rappahannock** *Febrifuge* Poultice of mashed, raw onions applied for fever from inflamed injury. (161:32)
- *Food*—**Blackfoot** *Spice* Bulbs used to spice soup made of wheat and marrow. (86:100) *Vegetable* Bulbs boiled with meat. *Winter Use Food* Bulbs preserved for later use. (97:23) **Coeur d'Alene** *Vegetable* Roots used as a principal vegetable food. (178:89) **Comanche** *Unspecified* Roasted bulbs used for food. (29:520) **Costanoan** *Winter Use Food* Bulbs gathered in winter and used for food. (21:255) **Hualapai** *Vegetable* Bulbs eaten fresh. *Winter Use Food* Bulbs stored for winter use. (195:19) **Kawaiisu** *Vegetable* Tops and roots eaten raw and fresh. (206:10) **Malecite** *Unspecified* Species used for food. (160:6) **Montana Indian** *Vegetable* Onions cooked and eaten. (19:6) **Omaha** *Unspecified* Bulbs and tops eaten both raw and cooked. (68:325) **Paiute, Northern** *Unspecified* Stems rolled into a ball and eaten. *Vegetable* Bulbs eaten raw. (59:44) **Sanpoil & Nespelem** *Unspecified* Bulbous roots cooked in pits and used for food. (131:100) **Spokan** *Unspecified* Roots used for food. (178:343)
- *Dye*—**Navajo** *Green* Used for a green dye. (55:32)
- *Other*—**Blackfoot** *Incense & Fragrance* Bulbs rubbed on quivers as a deodorant. *Soap* Bulbs and water used to shine arrows. (86:107)

Allium stellatum Nutt. ex Ker-Gawl., Autumn Onion
- *Drug*—**Chippewa** *Cold Remedy* and *Pediatric Aid* Sweetened decoction of root taken, especially by children, for colds. (53:340)
- *Other*—**Chippewa** *Toys & Games* Used as toys. (53:377)

Allium textile A. Nels. & J. F. Macbr., Textile Onion
- *Food*—**Lakota** *Soup* Bulbs cooked in stews. *Unspecified* Bulbs eaten fresh or stored for future use. (106:50)

Allium tricoccum Ait., Wild Leek
- *Drug*—**Cherokee** *Antihemorrhagic* Plant eaten as a spring tonic. *Cold Remedy* Plant eaten for colds. *Ear Medicine* Warm juice used for earache. *Pulmonary Aid* Plant eaten for croup. (80:52) **Chippewa** *Emetic* Decoction of root taken as a quick-acting emetic. (53:346) **Iroquois** *Anthelmintic* and *Pediatric Aid* Decoction of plant given to children for worms. *Tonic* Decoction of plant taken as a spring tonic and "cleans you out." (87:281)
- *Food*—**Cherokee** *Unspecified* Species used for food. (80:52) Young plants boiled, fried, and eaten. (204:251) *Vegetable* Bulbs and leaves cooked like poke, with or without eggs. (126:47) **Iroquois** *Vegetable* Cooked and seasoned with salt, pepper, or butter. Bulb, consisting of the fleshy bases of the leaves, eaten raw. (196:118) **Menominee** *Winter Use Food* Large, wild onion dried for winter use. (151:69) **Ojibwa** *Dried Food* Large, bitter, wild leek gathered in spring and dried for future use. (153:406) **Potawatomi** *Vegetable* Large, wild onion used for food. (154:104)

Allium unifolium Kellogg, Oneleaf Onion
- *Drug*—**Mendocino Indian** *Poison* Plant considered poisonous. (41:323)
- *Food*—**Mendocino Indian** *Unspecified* Bulbs and leaf bases fried and eaten. (41:323) **Papago** *Unspecified* Bulbs used for food. (32:15) **Pomo** *Spice* Bulbs cooked with other bulbs as a seasoning. *Unspecified* Bulbs eaten raw or baked. (11:89) **Yuki** *Unspecified* Bulbs eaten raw or fried. (49:86)
- *Other*—**Papago** *Cash Crop* Bulbs traded for baskets, skins, or pottery. (34:18)

Allium validum S. Wats., Pacific Onion
- *Food*—**Cahuilla** *Spice* Bulbs used as a flavoring ingredient for other foods. *Vegetable* Bulbs eaten raw. (15:37)

Allium vineale L., Wild Garlic
- *Drug*—**Cherokee** *Carminative* Used as a carminative. *Cathartic* Used as a mild cathartic. *Diuretic* Used as a diuretic. *Ear Medicine* Used "to remove deafness." *Expectorant* Used as an expectorant. *Kidney Aid* Used for "dropsy." *Misc. Disease Remedy* Used for scurvy. *Pediatric Aid* and *Pulmonary Aid* Tincture used to prevent worms and colic in children and used as a croup remedy. *Respiratory Aid* Used for asthma. *Stimulant* Used as a stimulant. (80:35) **Mahuna** *Dermatological Aid* Plant rubbed on body for protection from insect bites. Plant rubbed on body for protection from lizard, scorpion, and tarantula bites. *Snakebite Remedy* Plant rubbed on body for protection from poisonous snakebites. (140:63) **Rappahannock** *Hypotensive* Raw root bulbs chewed for high blood pressure. *Pulmonary Aid* Raw root bulbs chewed for shortness of breath. (161:34)

- *Food*—**Hopi** *Unspecified* Bulb used for food. (190:159)

Alnus, Betulaceae

Alnus glutinosa (L.) Gaertn., European Alder

- *Drug*—**Rappahannock** *Panacea* Infusion of bark used according to diagnosis. (161:31)

Alnus incana (L.) Moench, Mountain Alder

- *Drug*—**Bella Coola** *Unspecified* Cones used for medicine. (184:202) **Blackfoot** *Tuberculosis Remedy* Infusion of bark taken for scrofula. (82:5) **Chippewa** *Blood Medicine* Infusion of bark taken for anemia. (71:128) *Emetic* Compound decoction of scraped inner bark taken as an emetic. (53:346) *Eye Medicine* Compound decoction of root used as a wash or compress for sore eyes. (53:360) *Gynecological Aid* Decoction of root with powdered bumblebees taken for difficult labor. (53:358) **Cree, Woodlands** *Eye Medicine* Decoction of inner bark used as a wash for sore eyes. *Laxative* Bark removed by scraping downwards used as a laxative. (109:27) **Iroquois** *Analgesic* Infusion of young plant taken for pain. *Urinary Aid* Decoction of stems and couch grass rhizomes used for thick urine. (141:38) **Kutenai** *Abortifacient* Infusion of bark taken for menstrual regulation. (82:5) **Malecite** *Oral Aid* Bark chewed and used for ulcerated mouths. (116:245) **Menominee** *Cold Remedy* Infusion of root bark taken to congest loose mucus during a cold. *Dermatological Aid* Infusion of root bark used as an astringent, healing wash for sores. Poultice of inner bark applied to swellings. *Veterinary Aid* Infusion of root bark used as a wash for horses with saddle gall. (151:26) **Meskwaki** *Antihemorrhagic* and *Pediatric Aid* Decoction of root given to children who pass blood in their stools. (152:206) **Micmac** *Oral Aid* Bark used for ulcerated mouth. (40:54) **Mohegan** *Analgesic* Infusion of twigs used as a liniment for pain of sprains, bruises, backache, and headache. (176: 69, 128) *Orthopedic Aid* Infusion of twigs used as a liniment for sprain and backache pains. (176:69, 70) **Ojibwa** *Gastrointestinal Aid* Decoction of root taken as astringent and coagulant after bloody stools. (153:358) **Potawatomi** *Antidiarrheal* Infusion of bark taken for flux. *Dermatological Aid* Juice of inner bark used as a wash for the itch. *Gynecological Aid* Infusion of bark used for "flushing the vagina." *Hemorrhoid Remedy* Infusion of bark injected rectally for piles. (154: 43) *Veterinary Aid* Powdered bark used as an astringent for horse galls. (154:116) Powdered inner bark sprinkled on galled spots on ponies. (154:43) **Shuswap** *Dermatological Aid* Decoction of bark used as a wash for sores. *Diaphoretic* Decoction of bark taken to "sweat everything out." *Unspecified* Decoction of bark taken for the body. (123:59)
- *Fiber*—**Cree, Woodlands** *Caulking Material* Wood charcoal mixed with pitch and used for sealing canoe seams. *Snow Gear* Decoction of bark applied to toboggan boards to soften them for bending. (109:27)
- *Dye*—**Algonquin, Tête-de-Boule** *Yellow* Inner bark used to make yellow dye. (132:119) **Blackfoot** *Orange* Bark boiled and used as an orange dye. *Red-Brown* Bark boiled and used as a reddish brown dye. (82:5) **Chippewa** *Black* Used with grindstone dust or black earth to make a black dye. (53:372) *Red* Inner bark boiled with other inter barks and bloodroot and used to make a red dye. (53:371) Bark boiled to make a bright red dye. (71:128) *Yellow* Inner bark pounded, steeped, and boiled to make a yellow dye. (53:373) **Cree, Woodlands** *Brown* Decoction or infusion of inner bark used to wash and restore the brown color of old moccasins. *Orange-Red* Decoction of inner bark used as a reddish orange dye for quills. *Red-Brown* Infusion of inner bark used as a reddish brown dye for hides. *Unspecified* Infusion of bark applied to darken birch bark used to make baskets. Decoction of inner bark used to darken hides. *Yellow* Decoction of catkins used as a yellow dye for quills. (109:27) **Flathead** *Orange* Bark boiled and used as an orange dye. *Red* Bark used to make a flaming red hair dye. *Red-Brown* Bark boiled and used as a reddish brown dye. **Kutenai** *Orange* Bark boiled and used as an orange dye. *Red-Brown* Bark boiled and used as a reddish brown dye. (82:5) **Menominee** *Red-Brown* Bark boiled and cloth or material immersed in boiling liquid as a reddish brown dye. (151:78) **Navajo** *Red* Powdered bark used as a reddish dye.[8] (55:39) **Nez Perce** *Orange* Bark boiled and used as an orange dye. *Red-Brown* Bark boiled and used as a reddish brown dye. (82:5) **Ojibwa** *Yellow* Inner bark used for dyeing light yellow or with other ingredients for red, red-brown, or black. (153:425) **Potawatomi** *Brown* Bark used to obtain a brown dye. *Red* Bark used to obtain a red dye. (154:116) **Shuswap** *Unspecified* Used for buckskin dye. (123:59)

Alnus incana ssp. *rugosa* (Du Roi) Clausen, Speckled Alder

- *Drug*—**Abnaki** *Dermatological Aid* Used for "slight" itches. (as *A. rugosa* 144:155) Decoction of plant, two other plants, and Vaseline used as an ointment for "slight" itches. (as *A. rugosa* var. *americana* 144:165) **Algonquin, Quebec** *Emetic* and *Laxative* Infusion of inner bark taken as an emetic and laxative. *Toothache Remedy* Root bark mixed with molasses and used for toothaches. (as *A. rugosa* 18:153) **Cherokee** *Cathartic* Infusion of roots taken as a cathartic by women during menses. *Emetic* Decoction of inner bark taken to induce vomiting when unable to retain food. Infusion of roots taken as an emetic by women during menses. *Eye Medicine* Infusion of bark rubbed into the eye for eye troubles. *Gastrointestinal Aid* Decoction of inner bark taken to induce vomiting when unable to retain food. *Gynecological Aid* Infusion of roots taken as an emetic and cathartic by women during menses. (as *A. rugosa* 177:14) **Cree, Woodlands** *Eye Medicine* Decoction of inner bark used as a wash for sore eyes. *Laxative* Bark removed by scraping downwards used as a laxative. (as *A. rugosa* 109:27) **Iroquois** *Antihemorrhagic* Compound decoction with twigs taken for internal hemorrhage. *Cathartic* Decoction of young shoot bark taken as a physic. *Emetic* Decoction of young shoot bark taken as a spring emetic. *Urinary Aid* Infusion of bark or decoction of plant taken for urinating problems. *Venereal Aid* Decoctions used internally or externally for venereal chancres or sores. *Witchcraft Medicine* Decoction used to paint a trap or bow and arrow as a charm to get game. (87:301) **Menominee** *Alterative* Infusion of inner bark used as an alterative. (as *A. rugosa* 151:26)
- *Fiber*—**Cree, Woodlands** *Caulking Material* Wood charcoal mixed with pitch and used for sealing canoe seams. *Snow Gear* Decoction of bark applied to toboggan boards to soften them for bending. (as *A. rugosa* 109:27)
- *Dye*—**Cree, Woodlands** *Brown* Decoction or infusion of inner bark used to wash and restore the brown color of old moccasins. *Orange-Red* Decoction of inner bark used as a reddish orange dye for quills.

Red-Brown Infusion of inner bark used as a reddish brown dye for hides. *Unspecified* Infusion of bark applied to darken birch bark used to make baskets. Decoction of inner bark used to darken hides. *Yellow* Decoction of catkins used as a yellow dye for quills. (as *A. rugosa* 109:27)

Alnus incana ssp. tenuifolia (Nutt.) Breitung, Thinleaf Alder
- **Drug**—**Bella Coola** *Analgesic* Poultice of compound containing buds applied for lung or hip pains. *Antirheumatic (External)* Poultice of compound containing buds applied to lung or hip pain. *Pulmonary Aid* Poultice of compound containing buds applied for lung pains. *Unspecified* Cones used for an "unspecified complaint." (as *A. tenuifolia* 150:55) **Blackfoot** *Tuberculosis Remedy* Infusion of bark taken for scrofula. (as *A. tenuifolia* 97:32) Hot drink made from bark taken for scrofula. (as *A. tenuifolia* 114:275) **Cree, Woodlands** *Eye Medicine* Decoction of inner bark used as a wash for sore eyes. *Laxative* Bark removed by scraping downwards used as a laxative. (as *A. tenuifolia* 109:27) **Gitksan** *Diuretic* Catkins and shavings eaten raw or decoction taken as a diuretic for gonorrhea. *Laxative* Crushed pistillate catkins eaten raw as a laxative. (as *A. tenuifolia* 150:55) *Unspecified* Bark and other plants used to make a salve. (73:152) *Venereal Aid* Catkins and shavings eaten raw or decoction taken as a diuretic for gonorrhea. (as *A. tenuifolia* 150:55) **Keres, Western** *Dermatological Aid* Bark ground into a powder and used on open sores. (as *A. tenuifolia* 171:25) **Okanagan-Colville** *Dietary Aid* Infusion of plant tops given to children with poor appetites. *Gynecological Aid* Decoction of plant tops and leaves taken after childbirth to "clean out." *Pediatric Aid* Infusion of plant tops given to children with poor appetites. *Toothache Remedy* Burnt ashes used to clean the teeth. (as *A. tenuifolia* 188:87) **Sanpoil** *Dermatological Aid* Decoction of bark used as a wash for sores and powder of sapwood used on sores. (as *A. tenuifolia* 131:220)
- **Fiber**—**Cree, Woodlands** *Caulking Material* Wood charcoal mixed with pitch and used for sealing canoe seams. *Snow Gear* Decoction of bark applied to toboggan boards to soften them for bending. (as *A. tenuifolia* 109:27) **Thompson** *Snow Gear* Wood used to make snowshoes. (as *A. tenuifolia* 187:188)
- **Dye**—**Apache, White Mountain** *Red-Brown* Bark used to dye deerskin and other skins a reddish brown. (as *A. tenuifolia* 136:155) **Blackfoot** *Orange* Decoction of inner bark used to make an orange dye. *Red-Brown* Decoction of inner bark used to make a red-brown dye. (as *A. tenuifolia* 97:32) **Cree, Woodlands** *Brown* Decoction or infusion of inner bark used to wash and restore the brown color of old moccasins. *Orange-Red* Decoction of inner bark used as a reddish orange dye for quills. *Red-Brown* Infusion of inner bark used as a reddish brown dye for hides. *Unspecified* Infusion of bark applied to darken birch bark used to make baskets. Decoction of inner bark used to darken hides. *Yellow* Decoction of catkins used as a yellow dye for quills. (as *A. tenuifolia* 109:27) **Isleta** *Red* Root bark, mountain mahogany root bark, and wild plum root bark used to make a red dye for buckskin. (as *A. tenuifolia* 100:21) **Jemez** *Red* Bark, mountain mahogany bark, and birch bark boiled together and used as red dye to paint moccasins. (as *A. tenuifolia* 44:20) **Keres, Western** *Red* Soaked bark rubbed on buckskin as a red dye. (as *A. tenuifolia* 171:25) **Klamath** *Orange* Fresh or dried bark boiled and used as an orange dye for coloring horse hair ropes and cinches. (as *A. tenuifolia* 45:94) **Montana Indian** *Orange* Decoction of inner bark used as an orange dye. (as *A. tenuifolia* 19:6) **Navajo** *Brown* Powdered bark used as a tan dye.[8] Bark and twigs used as a brownish dye. *Red* Powdered bark used as a reddish dye.[8] (as *A. tenuifolia* 55:39) **Navajo, Ramah** *Red-Brown* Bark used to dye buckskin a reddish brown color. (as *A. tenuifolia* 191:30) **Okanagan-Colville** *Red-Brown* Bark and wood used to make red and brown dyes. (as *A. tenuifolia* 188:87) **Tewa** *Red* Bark dried, finely ground, boiled, cooled, and used as a red dye for deerskin. (as *A. tenuifolia* 138:38) **Zuni** *Red-Brown* Bark used to dye deerskin reddish brown. (as *A. tenuifolia* 166:80)
- **Other**—**Blackfoot** *Stable Gear* Bark split, covered with raw hide, and used to make stirrups. (as *A. tenuifolia* 114:275) **Montana Indian** *Hide Preparation* Bark used for tanning. (as *A. tenuifolia* 19:6) **Thompson** *Hunting & Fishing Item* Wood used to make bows. (as *A. tenuifolia* 187:188)

Alnus rhombifolia Nutt., White Alder
- **Drug**—**Kawaiisu** *Unspecified* Plant used as medicine. (206:10) **Mendocino Indian** *Antidiarrheal* Decoction of dried bark taken for diarrhea. *Antihemorrhagic* Decoction of dried bark taken to check hemorrhages for consumption. *Blood Medicine* Decoction of dried bark taken as a blood purifier. *Burn Dressing* Poultice of dried wood applied to burns. *Diaphoretic* Decoction of dried bark taken to perspire. *Emetic* Decoction of dried bark taken as an emetic. *Gastrointestinal Aid* Decoction of dried bark taken for stomachaches. *Gynecological Aid* Decoction of dried bark taken to facilitate childbirth. *Tuberculosis Remedy* Decoction of dried bark taken to check hemorrhages for consumption. (41:332) **Pomo** *Dermatological Aid* and *Pediatric Aid* Decoction of bark used as a wash for babies with skin disease. (66:12) **Pomo, Kashaya** *Dermatological Aid* Decoction of bark used as wash for skin diseases: sores, diaper rash, peeling or itching skin. (72:19)
- **Food**—**Costanoan** *Unspecified* Inner bark used for food. (21:248) **Karok** *Preservative* Wood used to smoke salmon, eels, and deer meat. (148:382)
- **Fiber**—**Karok** *Basketry* Roots used to make baskets. (148:382)
- **Dye**—**Costanoan** *Unspecified* Juice used as a dye. (21:248) **Karok** *Unspecified* Inner bark dried, ground, mixed with flour and water, and used as a dye. (148:382) **Mendocino Indian** *Unspecified* Fresh bark used as a dye to color basket material and deerskins. **Wailaki** *Red* Fresh bark formerly chewed and used as a red dye to color fishermen's bodies for successful fishing. (41:332)
- **Other**—**Kawaiisu** *Hide Preparation* Bark smoke used to tan white buckskin yellow. (206:10) **Mendocino Indian** *Fuel* Wood used for tinder. *Hunting & Fishing Item* Young shoots used to make arrows. (41:332) **Thompson** *Incense & Fragrance* Stems sometimes used as a scent. (164:503)

Alnus rubra Bong., Red Alder
- **Drug**—**Bella Coola** *Cathartic* Decoction of bark taken as a purgative. **Carrier, Northern** *Gastrointestinal Aid* Infusion of ground inner bark injected for biliousness. **Carrier, Southern** *Dermatological Aid* Sap applied to cuts and decoction of bark taken as a purgative. (150:55) **Clallam** *Dermatological Aid* Staminate aments chewed and used for sores. *Gastrointestinal Aid* Pistillate aments chewed and used for the

stomach. *Pulmonary Aid* Pistillate aments chewed and used for the lungs. (57:198) **Cowlitz** *Analgesic* and *Orthopedic Aid* Rotten wood rubbed on the body to ease "aching bones." (as *A. oregona* 79:27) **Gitksan** *Analgesic* Infusion of stem bark used as an emetic and purgative for headache and other maladies. (150:55) *Cathartic* Bark used as a purgative. (73:152) Infusion of stem bark used as an emetic and purgative for headache and other maladies. *Cough Medicine* Decoction of bark and root taken in the morning for a cough. *Emetic* Infusion of stem bark used as an emetic and purgative for headache and other maladies. *Unspecified* Infusion of stem bark, not from root, taken for many maladies. (150:55) **Haisla** *Dermatological Aid* Bark used to make a wound dressing and wash. *Tonic* Bark used as a tonic. (73:152) **Hesquiat** *Misc. Disease Remedy* Decoction of bark used to make a medicine for internal ailments. *Tuberculosis Remedy* Decoction of bark used to make a medicine for tuberculosis. (185:62) **Hoh** *Unspecified* Infusion of bark used for medicine. (as *A. oregona* 137:61) **Klallam** *Antidiarrheal* Catkins chewed for diarrhea. (as *A. oregona* 79:27) **Kwakiutl** *Analgesic* Poultice of bark applied to sores and aches. *Antihemorrhagic* Bark held in women's mouth for blood-spitting. *Dermatological Aid* Poultice of bark applied or infusion of bark rubbed on sores, aches, and eczema. *Respiratory Aid* Infusion of bark taken for tuberculosis and asthma. *Tuberculosis Remedy* Bark held in women's mouth for tuberculosis. Infusion of bark taken or bark held in women's mouth for tuberculosis. (183:279) **Kwakiutl, Southern** *Analgesic* Poultice of bark, fresh sea wrack, and black twinberry applied for aches and pains. (183:260) **Nitinaht** *Dermatological Aid* Infusion of crushed bark, western hemlock, and grand fir barks taken for bruises. *Internal Medicine* Infusion of bark, western hemlock, and grand fir barks taken for undiagnosed internal injuries. *Orthopedic Aid* Infusion of crushed bark, western hemlock, and grand fir barks taken for broken bones and ribs. *Pulmonary Aid* Infusion of crushed bark, western hemlock, and grand fir barks taken for lung ailments. *Tuberculosis Remedy* Infusion of crushed bark, western hemlock, and grand fir barks taken for tuberculosis. (186:98) *Unspecified* Bark used for medicine. (67:243) **Pomo, Kashaya** *Dermatological Aid* Decoction of bark used as wash for skin diseases: sores, diaper rash, peeling or itching skin. (as *A. oregona* 72:19) **Quileute** *Antidiarrheal* Raw cones eaten for dysentery. (as *A. oregona* 79:27) *Unspecified* Infusion of bark used for medicine. (as *A. oregona* 137:61) **Saanich** *Tonic* Sap used as a tonic. (182:79) **Swinomish** *Cold Remedy* Decoction of bark taken for colds. *Dermatological Aid* Decoction of bark taken for scrofula sores. *Gastrointestinal Aid* Decoction of bark taken for stomach troubles. *Tuberculosis Remedy* Decoction of bark taken for scrofula sores. (as *A. oregona* 79:27) **Thompson** *Dermatological Aid* Infusion of bark used as a wash for scabby skin, eczema, and skin sores. One informant used a concentrated decoction of the bark as a wash for her uncle who had a severe allergic reaction to hops. *Toothache Remedy* Poultice of immature catkins applied to the tooth for toothache. (187:188)

- **Food**—**Clallam** *Sweetener* Sap mixed with soapberry whip as a sweetener. (57:198) **Haisla & Hanaksiala** *Preservative* Wood used to smoke fish and meat. (43:224) **Salish, Coast** *Unspecified* Cambium eaten fresh with oil in spring. (182:79) **Skagit, Upper** *Unspecified* Sap used for food. (179:42) **Swinomish** *Unspecified* Sap taken from the inside of the bark only with the incoming tide and used as food. (as *A. oregona* 79:27)

- **Fiber**—**Hesquiat** *Canoe Material* Wood used for carved dishes and canoe bailers. (185:62) **Karok** *Basketry* Stems used in basketry. (as *A. oregana* 6:16) Roots used to make baskets. (148:382) **Makah** *Furniture* Wood used to make baby cradles. (as *A. oregona* 79:27) **Nitinaht** *Canoe Material* Wood used to make canoe bailers. (186:98) **Oweekeno** *Canoe Material* Wood used to make tool handles, canoe bailers, masks, and rattles. (43:86) **Quileute** *Canoe Material* Green wood seasoned and used to make canoe paddles. (as *A. oregona* 79:27) **Tolowa** *Basketry* Stems used in basketry. **Yurok** *Basketry* Stems used in basketry. (as *A. oregana* 6:16)

- **Dye**—**Bella Coola** *Red* Bark used to make a red dye for cedar bark. (184:202) **Hesquiat** *Red* Boiled, steeped bark used as a red dye for cedar bark and other items. (185:62) **Kwakiutl, Southern** *Black* Bark used to make a black dye. *Brown* Bark used to make a brown dye. *Orange* Bark used to make an orange dye. *Red* Bark used to make a red dye. (183:296) **Lummi** *Red-Brown* Bark made into a red to brown dye and used to make fish nets invisible to fish. (as *A. oregona* 79:27) **Nitinaht** *Brown* Bark used as a brown dye for baskets. (67:243) *Red* Formerly used to make red basket dyes. (186:79) Infusion of crushed bark used to make different shades of red dye. (186:98) **Oweekeno** *Red* Bark used to make a red dye. (43:86) **Quileute** *Red-Brown* Bark made into a red to brown dye and used to make fish nets invisible to fish. **Quinault** *Red-Brown* Bark made into a red to brown dye and used to make fish nets invisible to fish. (as *A. oregona* 79:27) **Salish, Coast** *Red-Brown* Bark boiled to make a reddish brown dye and used to color fish nets, baskets, canoes, and head rings. (182:79) **Snohomish** *Red-Brown* Bark made into a red to brown dye and used to make fish nets invisible to fish. (as *A. oregona* 79:27) **Thompson** *Brown* Bark boiled in water to make a brown dye and used for mountain goat wool, cloth, and other items. (187:188) *Red* Bark used as a red dye. (as *A. oregona* 164:501) Bark boiled in water to make a red dye and used for mountain goat wool, cloth, and other items. The dye was used to color mountain goat wool and other cloth and to deepen the color of basket materials such as bitter cherry bark. Skins were tanned and dyed simultaneously by soaking them in a cooled solution of the bark. (187:188) **Tolowa** *Unspecified* Bark used to dye fibers. (as *A. oregana* 6:16) **Wintoon** *Red* Inner bark chewed and used as a dull red dye. (as *A. oregona* 117:264) **Yurok** *Unspecified* Bark used to dye fibers. (as *A. oregana* 6:16)

- **Other**—**Bella Coola** *Fuel* Wood used as fuel for smoking fish. (184:202) **Clallam** *Cooking Tools* Wood used for dishes and utensils. *Fuel* Wood used for firewood. (57:198) **Haisla** *Fuel* Wood burned for boiling oolichan (candlefish) grease. **Haisla & Hanaksiala** *Ceremonial Items* Bark dyed red and used for ritual applications. *Cooking Tools* Wood used to make dishes to serve oolichan (candlefish) grease. (43:224) **Hesquiat** *Cooking Tools* Wood used for carved dishes and canoe bailers. *Fuel* Wood considered a good fuel for smoking fish. (185:62) **Karok** *Tools* Water-soaked stems used for lashing in the sweat house. (as *A. oregana* 6:16) **Kitasoo** *Fuel* Wood used for drying and smoking salmon both as a fuel and as a flavoring agent. *Preservative* Wood used for drying and smoking salmon both as a fuel and as a flavoring agent. (43:328) **Kwakiutl, Southern** *Ceremonial Items* Wood used to carve masks. *Cooking Tools* Wood used to carve dishes and spoons. (183:296) Wood used to make bowls and large tubs for tribal feasts. (183:279) *Fuel* Wood used for fire when drying salal berry cakes. (183:

282) *Musical Instrument* Wood used to carve rattles. (183:296) **Makah** *Containers* Cones used to store elderberries in the creeks. *Cooking Tools* Leaves used to cook halibut heads and salmonberry sprouts. Wood used to make bowls and dishes. *Preservative* Wood used for smoking and drying fish. *Smoke Plant* Leaves formerly smoked. (67:243) **Nitinaht** *Ceremonial Items* Wood used for making masks, ceremonial rattles, and model canoes. *Cooking Tools* Leaves and branches placed over and beneath food in steaming pits and kettles. Wood used to make bowls. *Fuel* Wood used as a fuel for drying and smoking fish and meats. (186:98) *Preservative* Wood used for smoking and drying fish. *Water Indicator* "If you see a creek without alder along its banks, the water isn't good to drink." (67:243) **Oweekeno** *Ceremonial Items* Wood used to make masks and rattles. *Fuel* Wood used as fuel for smoking fish. *Musical Instrument* Wood used to make rattles. *Tools* Wood used to make tool handles and canoe bailers. (43:86) **Quinault** *Containers* Bark used to line pots for storing elderberries. (as *A. oregona* 79:27) **Salish, Coast** *Cooking Tools* Wood used to make dishes and spoons. *Hunting & Fishing Item* Wood used to make arrow points. (182:79) **Skagit, Upper** *Cooking Tools* Wood used to make canned food dishes, spoons, and platters. *Fuel* Wood used as fuel for smoking salmon. (179:42) **Swinomish** *Containers* Bark used to line pots for storing elderberries. (as *A. oregona* 79:27) **Thompson** *Fuel* Wood used as a fuel for smoking meat. (187:188)

Alnus serrulata (Ait.) Willd., Hazel Alder

- **Drug**—**Cherokee** *Analgesic* Used for childbirth pain and infusion of bark used for various pains. *Blood Medicine* Infusion of bark taken to purify blood and compound infusion used as a blood tonic. *Cathartic* Used as an "emetic and purgative." *Cough Medicine* Infusion of bark taken for cough. *Dermatological Aid* Used for skin eruptions and infusion used to bathe hives. *Emetic* Used as an "emetic and purgative." *Eye Medicine* Infusion of bark "rubbed and blown in eyes for drooping." (80:22) Infusion of bark rubbed into the eye for eye troubles. (177:15) *Febrifuge* Hot infusion of berries taken for fever. *Gastrointestinal Aid* Compound used in steam bath for indigestion, biliousness, and jaundice. *Gynecological Aid* Used for childbirth pain and compound infusion taken for menstrual period. *Heart Medicine* Infusion taken for heart trouble. *Hemorrhoid Remedy* Compound infusion of root taken and used as a bath for piles. *Hypotensive* Cold infusion of bark taken to purify blood or lower blood pressure. *Kidney Aid* Infusion of scraped bark made kidneys act. *Oral Aid* Infusion of bark given to babies for "thrash," a mouth soreness. *Orthopedic Aid* Used for swellings and sprains. *Pediatric Aid* Infusion of bark given to babies for "thrash," a mouth soreness. *Toothache Remedy* Compound infusion of bark held in mouth for toothache. *Urinary Aid* Compound infusion taken to "clear milky urine." (80:22)

Alnus sp., Alder

- **Drug**—**Carrier** *Antihemorrhagic* Decoction of bark taken for bleeding. *Heart Medicine* Decoction of bark taken "whenever their heart moved." (31:72) **Malecite** *Unspecified* Used to make medicines. (160:6) **Micmac** *Analgesic* Bark used for cramps. *Antiemetic* Bark used for retching. *Antirheumatic (Internal)* Bark used for rheumatism. *Cathartic* Bark used as a physic. *Dermatological Aid* Bark and leaves used for festers and bark used for wounds. *Febrifuge* Bark and leaves used for fevers and festers. *Hemostat* Bark used for bleeding. *Misc. Disease Remedy* Bark used for diphtheria. *Orthopedic Aid* Bark used for dislocations and fractures. *Pulmonary Aid* Bark used for hemorrhaging of the lungs. (40:54) **Mohegan** *Analgesic* and *Orthopedic Aid* Infusion used as wash for sprains, bruises, headaches, and backache. (174:265) **Montagnais** *Blood Medicine* Decoction of twigs taken as a blood purifier. (156:315) **Penobscot** *Analgesic* Decoction of bark taken for cramps. *Antiemetic* Decoction of bark taken for retching. *Gastrointestinal Aid* Decoction of bark taken for cramps and retching. (156:309)
- **Food**—**Paiute** *Spice* Wood used to smoke deer meat when a smoke flavor was not wanted. *Unspecified* Cambium layer and sap used for food. (111:64)
- **Dye**—**Carrier** *Black* Used as a black dye for fish nets. (31:72) **Great Basin Indian** *Orange* Bark used to make an orange dye. (121:47) **Micmac** *Unspecified* Bark used to make a dye. (159:258) **Paiute** *Orange* Bark steeped in water for an orange dye to color moccasins and to decorate knife handles. (111:64)
- **Other**—**Carrier** *Hunting & Fishing Item* Used to make fish nets. (31:72)

Alnus viridis (Vill.) Lam. & DC., Sitka Alder

- **Drug**—**Cree, Hudson Bay** *Dermatological Aid* Bark used for the astringent qualities. *Kidney Aid* Bark used for dropsy. (92:303)

Alnus viridis ssp. *crispa* (Ait.) Turrill, American Green Alder

- **Drug**—**Cree, Woodlands** *Abortifacient* Decoction of plant used in a steam treatment to bring about menstruation. (as *A. crispa* 109:27) **Eskimo, Alaska** *Dermatological Aid* Poultice of leaves used in the past for infected wounds or sores. The poultice was left in place over the wound until the leaves stuck to it and was then pulled off, removing the "poison" with it. (as *A. crispa* 1:35) **Eskimo, Inuktitut** *Antirheumatic (Internal)* Bark burned as an inhalant for "rheumatism." (as *A. crispa* 202:188) **Okanagan-Colville** *Dietary Aid* Infusion of plant tops given to children with poor appetites. *Gynecological Aid* Decoction of plant tops and leaves taken after childbirth to "clean out." *Pediatric Aid* Infusion of plant tops given to children with poor appetites. *Toothache Remedy* Burnt ashes used to clean the teeth. (as *A. crispa* 188:87) **Tanana, Upper** *Carminative* Decoction of inner bark taken for stomach gas. *Febrifuge* Decoction of inner bark taken for high fevers. *Stimulant* Branches with leaves used for steam bath switches and as a floor covering in the steam bath. (as *A. crispa* 102:5) **Thompson** *Toothache Remedy* Poultice of immature catkins applied to the tooth for toothache. (as *A. crispa* 187:188)
- **Food**—**Tanana, Upper** *Preservative* Wood used to smoke fish. (as *A. crispa* 102:5)
- **Fiber**—**Tanana, Upper** *Mats, Rugs & Bedding* Branches with leaves used for steam bath switches and as a floor covering in the steam bath. (as *A. crispa* 102:5)
- **Dye**—**Eskimo, Alaska** *Orange* Bark soaked in water to make a rusty orange dye used to color tanned skins. (as *A. crispa* 1:35) **Eskimo, Inuktitut** *Red-Brown* Bark used to make a red-tan dye. (as *A. crispa* 202:188) **Iroquois** *Brown* Used as a brown dye for wool. Used as a brown dye for wool. (as *A. mollis* 142:84) **Okanagan-Colville** *Red-Brown* Bark and wood used to make red and brown dyes. (as *A. crispa* 188:87) **Tanana, Upper** *Unspecified* Inner bark boiled and liquid used as a dye or soaked bark rubbed directly onto article to be dyed. (as *A. crispa* 102:5)

- *Other*—**Eskimo, Alaska** *Fuel* Wood used as firewood. *Smoke Plant* Wood burned to make ashes added to tobacco. (as *A. crispa* 1:35) **Eskimo, Inuktitut** *Insecticide* Bark burned to repel mosquitoes. *Preservative* Bark burned to smoke fish. (as *A. crispa* 202:188) **Tanana, Upper** *Fuel* Wood used for firewood. *Hunting & Fishing Item* Wood used to make bows. (as *A. crispa* 102:5) **Thompson** *Water Indicator* Plants used as water indicators. (as *A. crispa* 187:188)

Alnus viridis ssp. *sinuata* (Regel) Á. & D. Löve, Sitka Alder
- *Drug*—**Bella Coola** *Unspecified* Cones used for an "unspecified complaint." (as *A. sitchensis* 150:55) Cones used for medicine. (as *A. sinuata* 184:202) **Gitksan** *Antihemorrhagic* Pistillate catkins eaten for "throwing blood out." *Cathartic* Pistillate catkins crushed and eaten raw as a physic. (43:225) *Tonic* Bark and other plants used as a tonic. (as *A. crispa* ssp. *sinuata* 73:152) *Venereal Aid* Decoction of pistillate catkins taken for gonorrhea. (43:225)
- *Dye*—**Eskimo, Alaska** *Unspecified* Bark used for dying reindeer skins. (as *A. frutica* 4:715)
- *Other*—**Haisla & Hanaksiala** *Cooking Tools* Wood used to make spoons. (43:225)

Alocasia, Araceae
Alocasia macrorrhizos (L.) G. Don, Giant Taro
- *Drug*—**Hawaiian** *Analgesic* Plant, other plants, and water taken as a laxative and an appetizer for acute pain in stomach or bowels. *Burn Dressing* Plant made into a salve and used on burns. *Dietary Aid* and *Laxative* Plant, other plants, and water taken as a laxative and an appetizer for acute pain in stomach or bowels. *Love Medicine* Plant used as a stimulant, effecting a constant reminder to the one desired of his or her presence. (2:17)

Alopecurus, Poaceae
Alopecurus aequalis var. *aequalis*, Shortawn Foxtail
- *Other*—**Apache, Chiricahua & Mescalero** *Containers* Moist grass laid onto hot stones to prevent steam from escaping. (as *A. aristulatus* 33:36)

Aloysia, Verbenaceae
Aloysia wrightii Heller ex Abrams, Wright's Beebrush
- *Drug*—**Havasupai** *Analgesic* Plant boiled and taken for headaches. *Antirheumatic (Internal)* Plant boiled and taken for rheumatism. *Psychological Aid* Plant boiled and taken for slight distempers. **Walapai** *Venereal Aid* Plant used for gonorrhea. (as *Lippia wrightii* 162:285)
- *Food*—**Havasupai** *Beverage* Leaves boiled into tea. (197:238) Twigs boiled to make tea. (197:66)

Alyxia, Apocynaceae
Alyxia oliviformis Gaud., Maile
- *Drug*—**Hawaiian** *Dermatological Aid* Infusion of pounded plant and other plants used in a sweat bath for yellow blotches on the skin. (as *A. clivaeformis* 2:69)

Amanitaceae, see *Amanita*

Amanita, Amanitaceae, fungus
Amanita muscaria, Amanita

- *Drug*—**Pomo, Kashaya** *Poison* Plant considered poisonous. (72:128)

Amaranthaceae, see *Amaranthus*, *Iresine*

Amaranthus, Amaranthaceae
Amaranthus acanthochiton Sauer, Greenstripe
- *Food*—**Hopi** *Starvation Food* Used numerous times to ward off famines. (as *Acanthochiton wrightii* 200:74) *Vegetable* Cooked as greens. (as *Acanthochiton wrightii* 32:10) Cooked with meat and eaten as greens. (as *Acanthochiton wrightii* 200:74)

Amaranthus albus L., Prostrate Pigweed
- *Food*—**Apache, Chiricahua & Mescalero** *Bread & Cake* Seeds winnowed, ground into flour and used to make bread. (as *A. graecizans* 33:48) *Unspecified* Eaten without preparation or cooked with green chile and meat or animal bones. (as *A. graecizans* 33:46) **Apache, White Mountain** *Unspecified* Seeds used for food. (136:155) **Cochiti** *Vegetable* Young plants eaten as greens. (as *A. graecizans* 32:16) **Navajo, Ramah** *Staple* Threshed seeds ground into flour. (as *A. graecizans* 191:25)
- *Other*—**Navajo** *Ceremonial Items* Used with many different plants to smoke for lewdness, which was performed at the Coyote Chant. (as *A. graecizans* 55:45)

Amaranthus arenicola I. M. Johnston, Sandhill Amaranth
- *Food*—**Hopi** *Unspecified* Leaves boiled and eaten with meat. (as *A. torreyi* 56:18) Boiled with meat. (as *Amblogyne torreyi* 190:162)

Amaranthus blitoides S. Wats., Mat Amaranth
- *Food*—**Acoma** *Dried Food* Young plants boiled and dried for winter use. (32:15) *Staple* Seeds ground into meal. (32:22) *Vegetable* Young plants boiled and eaten as greens. (32:15) **Apache, White Mountain** *Unspecified* Seeds used for food. (136:155) **Hopi** *Porridge* Ground seeds used to make mush. (190:162) *Unspecified* Seeds used as food. (32:22) Seeds formerly prized as a food. (56:18) Seeds eaten for food. *Vegetable* Cooked and eaten as greens. (200:74) **Klamath** *Unspecified* Seeds used for food. (45:96) **Laguna** *Dried Food* Young plants boiled and dried for winter use. (32:15) *Staple* Seeds ground into meal. (32:22) *Vegetable* Young plants boiled and eaten as greens. (32:15) **Montana Indian** *Unspecified* Seeds formerly used as articles of the diet. *Vegetable* Used as a potherb. (19:6) **Navajo** *Forage* Plant used as sheep forage. *Porridge* Seeds ground into meal and made into stiff porridge or mixed with goat's milk and made into gruel. *Staple* Seeds ground into a meal and used for food. (55:45) *Vegetable* Boiled and eaten like spinach, boiled and fried in lard, or canned. **Pueblo** *Vegetable* Boiled and eaten like spinach, boiled and fried in lard, or canned. **Spanish American** *Vegetable* Boiled and eaten like spinach, boiled and fried in lard, or canned. (32:15) **Tewa** *Unspecified* Boiled or fried and used for food. (138:53) **Zuni** *Bread & Cake* Seeds originally eaten raw, but later ground with black cornmeal, made into balls, and eaten. (166:65)
- *Other*—**Kiowa** *Fasteners* Used to make a glue. (192:26)

Amaranthus caudatus L., Love-lies-bleeding
- *Food*—**Cocopa** *Unspecified* Fresh plants baked and eaten. *Vegetable* Plants cooked and eaten as greens. *Winter Use Food* Plants cooked, rolled into a ball, baked, and stored. **Mohave** *Unspecified* Fresh plants baked and eaten. *Vegetable* Plants cooked and eaten as greens.

Winter Use Food Plants cooked, rolled into a ball, baked, and stored for future use. (37:200)

Amaranthus cruentus L., Red Amaranth

- **Food**—**Hopi** *Cooking Agent* Plant used as a red coloring for paper bread distributed at kachina exhibitions. (as *A. paniculatus* 56:18) Heads dried and used as a brilliant pink dye for wafer bread. (200:74) **Keresan** *Vegetable* Leaves eaten as greens. (as *A. paniculatus* 198:558) **Navajo, Ramah** *Staple* Threshed seeds ground into flour. (191:25) **Sia** *Unspecified* Seeds used for food. *Vegetable* Leaves used as greens. (*Amaranthus paniculatus* 199:107) **Zuni** *Cooking Agent* Feathery part of plant ground into a fine meal and used to color ceremonial bread red.[9] (as *A. hybridus paniculatus* 166:87)
- **Dye**—**Hopi** *Red* Flowers used to color bread red for certain dances. Flowers used to color bread red for certain dances. (as *A. paniculatus* 190:162) *Unspecified* Flowers used to color piki. (42:283)
- **Other**—**Apache, White Mountain** *Paint* Flowers used as face paint. (as *A. hybridus paniculatus* 136:155) **Keresan** *Hunting & Fishing Item* Seeds used to bait snares. (*Amaranthus paniculatus* 198:558) **Zuni** *Ceremonial Items* Feathery part of plant ground into a fine meal and used to color ceremonial bread red.[9] (as *A. hybridus paniculatus* 166:87) *Paint* Crushed leaves and blossoms moistened with spittle or water and rubbed on cheeks as rouge. (as *A. hybridus paniculatus* 166:83)

Amaranthus fimbriatus (Torr.) Benth. ex S. Wats., Fringed Amaranth

- **Food**—**Cahuilla** *Porridge* Parched seeds ground into a flour and used to make mush. *Vegetable* Boiled leaves eaten as greens or used as potherbs. (15:37)

Amaranthus hybridus L., Slim Amaranth

- **Drug**—**Cherokee** *Ceremonial Medicine* Used as an ingredient in a green corn medicine. *Dermatological Aid* Astringent leaves used for profuse menstruation. *Gynecological Aid* Leaves used to "relieve profuse menstruation." (80:23) **Keres, Western** *Gastrointestinal Aid* Infusion of plant used for the stomach. (171:26)
- **Food**—**Acoma** *Dried Food* Young plants boiled and dried for winter use. *Vegetable* Young plants boiled and eaten as greens. (32:16) **Havasupai** *Bread & Cake* Seeds parched, ground fine, boiled, thickened, made into balls, and eaten as dumplings. (197:66) *Porridge* Seeds parched, ground, and used to make mush. *Soup* Seeds parched, ground, and used to make soup. (197:67) Leaves and squash flowers boiled, ground, and fresh or dried corn and water added to make soup. (197:74) *Unspecified* Seeds used for food. *Vegetable* Leaves of young plants cooked like spinach. (197:218) Young, fresh, tender leaves boiled, drained, balled into individual portions, and served. (197:66) **Keres, Western** *Unspecified* Seeds collected and ground with meal for food. *Vegetable* Young, tender plants used for greens like spinach. *Winter Use Food* Plant used as winter food by boiling and drying for winter storage. (171:26) **Laguna** *Dried Food* Young plants boiled and dried for winter use. *Vegetable* Young plants boiled and eaten as greens. (32:16)

Amaranthus palmeri S. Wats., Careless Weed

- **Food**—**Cocopa** *Unspecified* Fresh plants baked and eaten. *Vegetable* Plants cooked and eaten as greens. *Winter Use Food* Plants cooked, rolled into a ball, baked, and stored. **Mohave** *Unspecified* Fresh plants baked and eaten. *Vegetable* Plants cooked and eaten as greens. *Winter Use Food* Plants cooked, rolled into a ball, baked, and stored. (37:200) **Navajo** *Staple* Seeds ground into a meal and used for food. *Sweetener* Parched, ground seeds chewed to obtain sugar. (55:46) **Papago** *Dried Food* Seeds basket winnowed, parched, sun dried, cooked, stored, and used for food. (34:24) *Staple* Seeds ground and used as food. (32:23) *Unspecified* Seeds used for food. (36:62) *Vegetable* Leaves eaten as greens in midsummer. (34:14) Boiled and used for greens. (34:46) Greens used for food. (36:61) **Pima** *Unspecified* Leaves boiled and eaten with pinole. *Vegetable* Young, tender leaves cooked and eaten as greens. (47:47) **Pima, Gila River** *Dried Food* Leaves dried and stored for year-round use. (133:5) *Unspecified* Leaves boiled and eaten. (133:7) *Vegetable* Leaves used as greens. (133:5) **Yuma** *Staple* Seeds parched and ground into meal. (37:189) *Unspecified* Fresh plants baked and eaten. *Vegetable* Plants cooked and eaten as greens. *Winter Use Food* Plants cooked, rolled into a ball, baked, and stored for future use. (37:200)

Amaranthus powellii S. Wats., Powell's Amaranth

- **Food**—**Hopi** *Unspecified* Seeds used for food. *Vegetable* Leaves used as greens. (42:283)

Amaranthus retroflexus L., Redroot Amaranth

- **Drug**—**Cherokee** *Ceremonial Medicine* Used as an ingredient in a green corn medicine. *Dermatological Aid* Astringent leaves used for profuse menstruation. *Gynecological Aid* Leaves used to "relieve profuse menstruation." (80:23) **Iroquois** *Witchcraft Medicine* Decoction and doll used to "make a person break out like cancer." (87:316) **Keres, Western** *Gastrointestinal Aid* Infusion of plant used for the stomach. (171:26) **Mohegan** *Throat Aid* Infusion of leaves taken for hoarseness. (176:70, 128) **Navajo, Ramah** *Antidote* Stem, 3 inches long, made into snake figurine for snake infection. (191:26)
- **Food**—**Acoma** *Dried Food* Young plants boiled and dried for winter use. *Vegetable* Young plants boiled and eaten as greens. (32:15) **Apache, Chiricahua & Mescalero** *Bread & Cake* Seeds winnowed, ground into flour, and used to make bread. (33:48) *Unspecified* Leaves eaten without preparation or cooked with green chile and meat or animal bones. (33:46) **Cochiti** *Vegetable* Young plants eaten as greens. (32:16) **Iroquois** *Vegetable* Cooked and seasoned with salt, pepper, or butter. (196:117) **Isleta** *Vegetable* Fresh, tender, young leaves eaten as greens. (100:21) **Jemez** *Unspecified* Young plant used for food many generations ago. (44:20) **Keres, Western** *Unspecified* Seeds collected and ground with meal for food. *Vegetable* Young, tender plants used for greens like spinach. *Winter Use Food* Plant boiled and dried for winter storage. (171:26) **Laguna** *Dried Food* Young plants boiled and dried for winter use. *Vegetable* Young plants boiled and eaten as greens. (32:15) **Mendocino Indian** *Staple* Small, shiny black seeds used to make pinole. (41:346) **Mohegan** *Vegetable* Combined with mustard, plantain, dock, and nettle and used as mixed greens. (176:83) **Navajo** *Bread & Cake* Seeds ground, boiled, mixed with corn flour, and made into dumplings. *Porridge* Seeds ground, boiled, and mixed with corn flour into a gruel. (165:222) *Unspecified* Seeds used for food. Leaves and seeds mixed with grease and eaten. (55:46) *Vegetable* Boiled and eaten like spinach, boiled and fried in lard, or canned. (32:15) Leaves boiled and eaten like

spinach. *Winter Use Food* Leaves boiled and canned. (55:46) **Navajo, Ramah** *Bread & Cake* Seeds winnowed, ground with maize, made into bread, and used as a ceremonial food in Nightway. *Special Food* Seeds winnowed, ground with maize, made into bread, and used as a ceremonial food in Nightway. *Vegetable* Leaves used as spring greens, boiled with meat, boiled alone, or boiled and fried with meat or fat. *Winter Use Food* Seeds stored for winter use. (191:26) **Pueblo** *Vegetable* Boiled and eaten like spinach, boiled and fried in lard, or canned. **Spanish American** *Vegetable* Boiled and eaten like spinach, boiled and fried in lard, or canned. (32:15) **Tewa** *Unspecified* Boiled or fried and used for food. (138:53)

Amaranthus sp., Amaranth
- *Food*—**Gosiute** *Unspecified* Seeds formerly eaten and constituted and important source of food. (39:361) **Navajo** *Staple* Seeds ground into meal and used as food. *Sweetener* Seeds ground into meal and chewed by the handful to obtain sugar. (32:23) **Yavapai** *Vegetable* Leaves boiled for greens and sometimes mixed with dried mescal. (65:256)

Amaranthus spinosus L., Spiny Amaranth
- *Drug*—**Cherokee** *Ceremonial Medicine* Used as an ingredient in a green corn medicine. *Dermatological Aid* Astringent leaves used for profuse menstruation. *Gynecological Aid* Leaves used to "relieve profuse menstruation." (80:23)

Ambrosia, Asteraceae
Ambrosia acanthicarpa Hook., Flatspine Burr Ragweed
- *Drug*—**Zuni** *Abortifacient* Infusion of whole plant taken and used as wash for "obstructed menstruation." *Toothache Remedy* Ground root placed in tooth for toothache. (as *Gaertneria acanthicarpa* 166:51, 52)
- *Other*—**Navajo, Ramah** *Ceremonial Items* Leaf ash used as Evilway blackening. (as *Franseria acanthicarpa* 191:51)

Ambrosia ambrosioides (Cav. ex Rydb.) Payne, Ambrosia Leaf Burr Ragweed
- *Drug*—**Pima** *Analgesic* and *Antihemorrhagic* Decoction of crushed roots taken by women for pains and menstrual hemorrhage. *Cough Medicine* Poultice of warmed leaves applied to the chest to loosen a cough. *Gynecological Aid* Decoction of crushed roots taken by women for pains and menstrual hemorrhage. (as *Franseria ambrosioides* 47:103)

Ambrosia artemisiifolia L., Annual Ragweed
- *Drug*—**Cherokee** *Ceremonial Medicine* Used as an ingredient in green corn medicine. *Dermatological Aid* Crushed leaves rubbed on insect sting and infusion of leaf rubbed on hives. *Disinfectant* Juice of wilted leaves applied to infected toes. *Febrifuge* Infusion of leaf taken for fever. *Pulmonary Aid* Infusion taken for pneumonia. (80:52) **Dakota** *Antidiarrheal* Infusion of leaves and plant tops taken for bloody flux. *Antiemetic* Infusion of leaves and plant tops taken for vomiting. (69:369) **Delaware** *Blood Medicine* Poultice of plant used to prevent blood poisoning. (176:35) **Delaware, Oklahoma** *Blood Medicine* Poultice of plant applied to prevent "blood poison." (175:29) *Dermatological Aid* Poultice of plant applied to prevent "blood poisoning." (175:29, 74) **Houma** *Gynecological Aid* Decoction of root taken for menstrual troubles. (158:65) **Iroquois** *Antidiarrheal* Compound decoction of plants taken for diarrhea with bleeding. (87:468) *Heart Medicine* Infusion of roots taken for stroke. (87:469) *Orthopedic Aid* Decoction of plants taken for cramps from picking berries. (87:468) **Lakota** *Antirheumatic (External)* Infusion of leaves applied to swellings. (139:35) **Luiseño** *Emetic* Plant used as an emetic. (155:228) **Mahuna** *Dermatological Aid* Infusion of plant used as a wash for minor skin eruptions and scalp diseases. (140:13)
- *Other*—**Lakota** *Paper* Plant used for toilet paper. (139:35)

Ambrosia artemisiifolia var. *elatior* (L.) Descourtils, Annual Ragweed
- *Drug*—**Oto** *Antiemetic* Bruised leaves laid on scarified abdomen for nausea. (as *A. elatior* 70:132)

Ambrosia chamissonis (Less.) Greene, Silver Burr Ragweed
- *Drug*—**Makah** *Other* Plant used as medicine for healing. *Strengthener* Plant used as medicine for strength. (67:323)
- *Other*—**Hesquiat** *Toys & Games* Children played with stems that exuded a blood-colored juice; looked like they had been injured. (185:62)

Ambrosia psilostachya DC., Cuman Ragweed
- *Drug*—**Cheyenne** *Analgesic* Infusion of leaves and stems taken for bowel pains and bloody stools. (76:188) Infusion of leaves and stem taken for cramps in the bowels. *Antidiarrheal* Infusion of leaves and stem taken for bloody stools. (75:39) *Antihemorrhagic* Infusion of ground leaves and stems taken for bloody stools. (76:188) *Cold Remedy* Infusion of ground leaves and stems taken for colds. *Gastrointestinal Aid* Infusion of leaves and stems taken for bowel pains and bloody stools. (76:188) Infusion of leaves and stem taken for bowel cramps. (75:39) Infusion of ground leaves and stems taken for bowel cramps. (76:188) *Laxative* Infusion of ground leaves and stems taken for constipation. (83:18) **Costanoan** *Orthopedic Aid* Poultice of heated leaves applied to aching joints. (21:25) **Diegueño** *Dermatological Aid* Decoction of stems and leaves used after a hair wash as a rinse for dandruff. (84:13) *Gastrointestinal Aid* Infusion of leaves taken for stomach pains. (88:219) **Gosiute** *Eye Medicine* Poultice of steeped leaves applied to sore eyes. (39:361) **Keres, Western** *Gynecological Aid* Infusion of plant given to women during difficult labor. (171:26) **Kiowa** *Dermatological Aid* Decoction of plant used as a wash for sores. *Veterinary Aid* Decoction of plant used as a wash for sores on horses. (192:55)
- *Other*—**Kiowa** *Fuel* Used rolled up, with various sages, in the sweat houses. (192:55)

Ambrosia tenuifolia Spreng., Slimleaf Burr Ragweed
- *Drug*—**Navajo** *Gynecological Aid* Plant used to facilitate delivery of the placenta after childbirth. (as *Franseria tenuifolia* 90:151)
- *Food*—**Papago** *Dried Food* Surplus of roots sun dried on roofs and used for food. *Staple* Roots used as a staple crop. (as *Franseria tenuifolia* 34:17) *Unspecified* Roots used for food. (*Franseria tenuifolia* 36:60) *Vegetable* Stalks eaten as greens in the summer. (as *Franseria tenuifolia* 34:14)
- *Other*—**Navajo** *Smoke Plant* Herb mixed with tobacco. (as *Franseria tenuifolia* 90:151)

Ambrosia trifida L., Great Ragweed
- *Drug*—**Cherokee** *Ceremonial Medicine* Used as an ingredient in

green corn medicine. *Dermatological Aid* Crushed leaves rubbed on insect sting and infusion of leaf rubbed on hives. *Disinfectant* Juice of wilted leaves applied to infected toes. *Febrifuge* Infusion of leaf taken for fever. *Pulmonary Aid* Infusion taken for pneumonia. (80:52) **Iroquois** *Antidiarrheal* Compound decoction of plants taken for diarrhea with bleeding. *Blood Medicine* Plant used in a blood medicine. (87:468) **Lakota** *Unspecified* Seeds used medicinally. (139:35) **Meskwaki** *Psychological Aid* Root chewed to drive away fear at night. (152:210)

Amelanchier, Rosaceae

Amelanchier alnifolia (Nutt.) Nutt. ex M. Roemer, Saskatoon Serviceberry

- **Drug—Blackfoot** *Cathartic* Infusion of plant and chokecherry cambium taken as a purge. (86:68) *Ear Medicine* Decoction of berry juice used for eardrops. *Eye Medicine* Decoction of dried berries or berry juice dripped into the eye and covered with a soft hide piece. (86:80) *Gastrointestinal Aid* Berry juice taken for an upset stomach. *Laxative* Berry juice taken as a mild laxative. (86:65) *Pediatric Aid* Infusion of plant and chokecherry cambium taken by nursing mothers to pass medicinal values to baby. (86:68) **Cheyenne** *Dietary Aid* and *Pediatric Aid* Smashed fruits used to improve loss of appetite in children. (83:34) *Unspecified* Infusion of leaves used for healing. (76:176) Smashed fruits used as an ingredient for medicinal mixtures. (83:34) **Cree, Woodlands** *Cold Remedy* Decoction of sticks taken for bad colds. *Cough Medicine* Decoction of roots taken for coughs. *Diaphoretic* Decoction of stems and snowberry stems taken to cause sweating. *Febrifuge* Decoction of stems and snowberry stems taken for fevers. *Misc. Disease Remedy* Decoction of sticks taken for flu. *Pediatric Aid* Decoction of roots taken for teething sickness. *Pulmonary Aid* Decoction of roots taken for chest pains and lung infections. *Toothache Remedy* Decoction of roots taken for teething sickness. (109:28) **Flathead** *Veterinary Aid* Sharpened wood used to drain blood and other liquids from horses' swollen ankles. (82:9) **Okanagan-Colville** *Cold Remedy* Decoction of branches taken for colds. *Contraceptive* Decoction of branch ashes and pine branch or bud ashes taken to prevent having children. *Tonic* Decoction of branches taken as a general tonic. (188:120) **Pomo** *Gynecological Aid* Decoction of roots taken for too frequent menstruation. (66:13) **Thompson** *Contraceptive* Decoction of plant and bitter cherry taken as birth control. (187:253) *Gastrointestinal Aid* Decoction of bark taken for stomach troubles. (164:462) *Gynecological Aid* Warm decoction taken and used as a wash after childbirth. (164:471) Warm decoction of stems and twigs taken by women or used as a bath after childbirth. Strong decoction of bark taken by women after childbirth to hasten the dropping of the afterbirth. The decoction was taken immediately after childbirth. The medicine made from the tall variety of saskatoon was said to clean her out and help heal her insides. It was also said to stop her menstrual periods after the baby was born and hence act as a form of birth control. (187:253) *Tonic* Decoction of fresh bark taken as a tonic. (164:471)
- **Food—Atsugewi** *Dried Food* Dried, stored berries soaked in water and eaten. *Porridge* Ripe, mashed fruit added to water to form a paste and eaten without cooking. (61:139) **Bella Coola** *Fruit* Berries used for food. (184:208) **Blackfoot** *Dessert* Berries and buffalo fat used to make a soup eaten as a dessert at feasts. (97:37) *Dried Food* Berries dried and stored, some with back fat, for future use. (86:100) Berries dried for future use. (97:37) *Fruit* Crushed berries, animal fat, and dried meat used to make pemmican. Berries and fat stuffed into an intestine, boiled, and eaten like a sausage. Dried berries used to make sausages. (86:100) *Preserves* Berries used to make preserves. (86:26) *Snack Food* Berries and red osier dogwood berries used as a favorite snack reserved for men. (86:100) Berries used to make tasty snacks. (86:26) *Soup* Dried berries used to make soups. Crushed leaves mixed with blood, dried, and used to make a rich broth in winter. (86:100) *Special Food* Berries used in ritual meals. Berry soup used for most ceremonial events.[10] (86:26) *Staple* Berries used as a staple food. *Winter Use Food* Crushed berries mixed with flour for winter storage. (86:100) **Cheyenne** *Beverage* Leaves used to make tea. (76:176) Leaves used to make a red beverage tea. *Pie & Pudding* Fruits boiled, sugar and flour added, and eaten as a pudding. (83:34) *Special Food* Berries stewed for feasts. *Winter Use Food* Berries dried for winter use. (76:176) **Cree, Plains** *Dried Food* Berries crushed, dried, and stored for future use. (113:202) **Cree, Woodlands** *Dried Food* Sun dried fruit eaten cooked in water or raw as a sweet snack. Sun dried fruit eaten boiled or pounded into a pemmican. *Fruit* Fruit eaten fresh. Sun dried fruit eaten boiled or pounded into a pemmican. *Preservative* Barked split sticks, 4 inches long, boiled in sturgeon oil to keep the oil fresh during storage. *Snack Food* Sun dried fruit eaten raw as a sweet snack. (109:28) **Dakota** *Fruit* Prized berries used for food. (70:87) **Flathead** *Pie & Pudding* Dried berries mixed with flour, sugar, and water and eaten as a sweet pudding. (82:9) **Gosiute** *Dried Food* Berries mashed and dried in large quantities for winter use. *Fruit* Berries used in season. (39:361) **Great Basin Indian** *Dried Food* Berries eaten dried. *Fruit* Berries eaten fresh or added to elk or deer meat to make pemmican. (121:48) **Hesquiat** *Forage* Berries eaten by bears. *Fruit* Berries used for food. (185:72) **Karok** *Dried Food* Berries dried and stored in big baskets. *Fruit* Berries eaten fresh. (148:385) **Kitasoo** *Fruit* Fruit used for food. (43:341) **Klamath** *Fruit* Fresh berries used for food. *Unspecified* Seeds chewed for pleasure. *Winter Use Food* Dried berries stored for winter use. (45:97) **Kwakiutl, Southern** *Fruit* Berries used for food. (183:288) **Lakota** *Beverage* Petals, leaves, and small stems used to make a drink. *Fruit* Berries eaten fresh. (106:36) Fruits eaten for food. (139:56) *Starvation Food* Berries dried and eaten during famines. (106:36) **Mendocino Indian** *Fruit* Black, glaucous berries eaten fresh. (41:355) **Modesse** *Fruit* Berries used for food. (117:223) **Montana Indian** *Beverage* Berries used to make wine. (19:6) Fruits used to make wine. *Bread & Cake* Fruits sun dried, pounded, formed into patties, and stored for winter use. *Forage* Berries eaten by bears and grouse. Young stems and leaves eaten by elk, deer, moose, and mountain sheep. (82:9) *Fruit* Berries spiced and eaten. *Pie & Pudding* Berries used to make pies. (19:6) Fruits made into pies and eaten. (82:9) *Preserves* Berries used to make jam. (19:6) Fruits made into jams and jellies. *Soup* Fruits sun dried and eaten in meat stews. (82:9) *Winter Use Food* Large quantities of berries gathered and dried for winter use. (19:6) **Navajo** *Fruit* Fruits eaten for food. (55:52) **Okanagan-Colville** *Dried Food* Berries dried for future use. *Frozen Food* Berries frozen for future use. *Fruit* Berries eaten fresh, with sugar or cooked. *Pie & Pudding* Berries used to make pies and puddings. *Sweetener* Dried berries used to sweeten "Indian ice

cream." *Winter Use Food* Berries canned for future use. (188:120) **Okanagon** *Bread & Cake* Berries pressed into cakes and used for food. (125:38) *Staple* Berries used as a principal food. (178:238) **Omaha** *Fruit* Prized berries used for food. **Ponca** *Fruit* Prized berries used for food. (70:87) **Saanich** *Fruit* Berries eaten in late summer. **Salish, Coast** *Fruit* Berries eaten in late summer. (182:86) **Shuswap** *Fruit* Berries used for food. (123:65) **Thompson** *Beverage* Twigs used to make a tea-like beverage. (187:253) *Bread & Cake* Berries pressed into cakes and used for food. (125:38) Berries dried into cakes. *Dried Food* Berries dried loose like raisins. *Frozen Food* Berries frozen for future use. (187:253) *Fruit* Fruits eaten fresh in large quantities. (164:489) Berries eaten fresh or boiled. *Pie & Pudding* Dried berries and many other ingredients used to make a special pudding. The dried berries with bitterroot, flour, butter, cream, sugar, and sometimes tiger lily bulbs, avalanche lily corms, deer fat, black tree lichen, and salmon eggs were used to make a special pudding. *Preserves* Berries jammed. *Spice* Berry juice used to marinate other foods. *Sweetener* Dried berry cakes used as a sweetener for other foods. (187:253) *Unspecified* Drupes eaten wherever found. (164:487) *Winter Use Food* Fruits preserved for future use. (164:489) Berries frozen or canned for future use. (187:253) **Ute** *Dried Food* Berries dried for winter use. *Fruit* Berries used in season. (38:32) **Winnebago** *Fruit* Prized berries used for food. (70:87)

- *Fiber*—**Cree, Woodlands** *Basketry* Stems used to make rims for birch bark baskets. (109:28) **Gosiute** *Basketry* Used for basketry. *Furniture* Used for cradle frameworks. (39:361) **Karok** *Basketry* Twigs and stems used to reinforce the rims of basket hoppers for pounding acorns. Wood used as stiffening for baskets or for making handles. (148:385) **Maidu** *Basketry* Withes used to make basket rims. (173:71) **Okanagan-Colville** *Cordage* Young branches twisted into rope. (188:120)
- *Other*—**Blackfoot** *Cash Crop* Dried berries traded for tobacco. (97:37) *Ceremonial Items* Berries, elk manure, and tobacco seed planted in small prairie plot in the Tobacco Planting ceremony. (86:26) Forked sticks used in religious rituals. *Hunting & Fishing Item* Shoots used to make arrows. (97:37) *Toys & Games* Berries used in an harvesting game. Favors were asked while presenting a gift of four of the berries. The receiver was obliged to return the goodwill. Girls played a game while harvesting the berries. After some berries had been gathered the girls would sit together and hold their breath while another called out "tops, tops, tops" at a regular beat. Each girl put a berry in her bag for every call and the one who held her breath the longest won all the other girls' berries. (86:107) **Crow** *Fasteners* Wood used for tepee stakes and closure pins. (82:9) **Dakota** *Hunting & Fishing Item* Wood used for arrow shafts. (70:87) *Toys & Games* Plant used to make popgun pistons. (70:116) **Flathead** *Hunting & Fishing Item* Hard, flexible stems used for arrow shafts. (82:9) **Gosiute** *Hunting & Fishing Item* Used to make arrows. (39:361) **Karok** *Hunting & Fishing Item* Twigs used as points on arrow shafts. Wood used to make the foreshafts of salmon harpoons. (148:385) **Lakota** *Hunting & Fishing Item* Stems used to make arrows. *Toys & Games* Stems made into hoops with leather covers to use in a game. (139:56) **Mendocino Indian** *Hunting & Fishing Item* Wood used to make arrows. (41:355) **Montana Indian** *Weapon* Very hard and tough wood used for making arrows and ramrods. (19:6) **Okanagan-Colville** *Hunting & Fishing Item* Wood used to make arrows and spears. *Tools* Wood used to make digging sticks and seed beaters. (188:120) **Okanagon** *Cash Crop* Traded with the Coast Indians. (125:38) *Tools* Wood used to make root diggers and other tools. (125:39) **Omaha** *Hunting & Fishing Item* Wood used for arrow shafts. (70:87) *Toys & Games* Plant used to make popgun pistons. **Pawnee** *Toys & Games* Plant used to make popgun pistons. (70:116) **Ponca** *Hunting & Fishing Item* Wood used for arrow shafts. (70:87) *Toys & Games* Plant used to make popgun pistons. (70:116) **Salish, Coast** *Hunting & Fishing Item* Wood occasionally used to make arrows. (182:86) **Shuswap** *Hunting & Fishing Item* Stems of the young plant used for arrows. (123:65) **Thompson** *Cash Crop* Traded with the Coast Indians. (125:38) *Cooking Tools* Wood used to make salmon spreaders. *Hunting & Fishing Item* Wood used to make arrows. Wood used as reinforcement for dip net hoops. (187:253) *Tools* Wood used to make root diggers and other tools. (125:39) Wood used to make root diggers, handles, and other tools. (164:496) **Winnebago** *Hunting & Fishing Item* Wood used for arrow shafts. (70:87) *Toys & Games* Plant used to make popgun pistons. (70:116)

Amelanchier alnifolia (Nutt.) Nutt. ex M. Roemer var. *alnifolia*, Saskatoon Serviceberry

- *Food*—**Haisla** *Fruit* Berries combined with other fruits and eaten. **Hanaksiala** *Dried Food* Berries dried and eaten. (43:263) **Oweekeno** *Fruit* Berries used for food. (43:107)

Amelanchier alnifolia var. *cusickii* (Fern.) C. L. Hitchc., Cusick's Serviceberry

- *Food*—**Paiute** *Candy* Mashed berries formed into cakes, sun dried, and eaten as candy. *Dried Food* Berries eaten dried. *Fruit* Berries eaten fresh. *Winter Use Food* Mashed berries formed into cakes, sun dried for winter use, boiled, and eaten. (as *A. cusickii* 111:83)

Amelanchier alnifolia var. *semiintegrifolia* (Hook.) C. L. Hitchc., Pacific Serviceberry

- *Drug*—**Bella Coola** *Venereal Aid* Compound decoction taken for gonorrhea. (as *A. florida* 150:60)
- *Food*—**Alaska Native** *Bread & Cake* Berries used to make muffins. *Dried Food* Berries dried and used in place of raisins or currants. *Fruit* Berries eaten raw. *Pie & Pudding* Berries used to make puddings and pies. (as *A. florida* 85:75) **Chehalis** *Fruit* Fruits eaten fresh. *Spice* Fruits dried and used as seasoning in soup or with meats. (as *A. florida* 79:38) **Hoh** *Fruit* Fruits eaten for food. (as *A. florida* 137:64) **Lummi** *Dried Food* Berries dried, boiled with dog salmon, and eaten at feasts. (as *A. florida* 79:38) **Quileute** *Fruit* Fruits eaten for food. (as *A. florida* 137:64) **Sanpoil & Nespelem** *Bread & Cake* Berries dried whole or mashed, formed into cakes, and dried. *Fruit* Berries eaten raw or cooked with salmon. (as *A. florida* 131:101) **Skagit** *Fruit* Berries eaten fresh. (as *A. florida* 79:38) **Skagit, Upper** *Dried Food* Berries mashed and dried for winter use. *Fruit* Berries eaten fresh. (as *A. florida* 179:38) **Swinomish** *Dried Food* Fruits dried and eaten during the winter. *Fruit* Fruits eaten fresh. (as *A. florida* 79:38)
- *Other*—**Samish** *Hunting & Fishing Item* Wood used as the spreader in rigging halibut line. **Snohomish** *Toys & Games* Wood used to make disks for gambling games. **Swinomish** *Hunting & Fishing Item* Wood used as the spreader in rigging halibut line. (as *A. florida* 79:38)

Amelanchier arborea (Michx. f.) Fern., Common Serviceberry

- *Drug*—**Cherokee** *Anthelmintic* Compound infusion taken for worms.

Antidiarrheal Compound infusion taken for diarrhea. *Tonic* Compound infusion taken as a spring tonic. (80:54) **Iroquois** *Venereal Aid* Infusion of bark used for gonorrhea. (87:351)
- *Food*—**Cherokee** *Fruit* Berries used for food. (80:54)

Amelanchier arborea (Michx. f.) Fern. var. *arborea*, Common Serviceberry

- *Drug*—**Iroquois** *Blood Medicine* Fruits formerly used as a blood remedy. *Gynecological Aid* Fruits, infusion of small branches given to mothers after childbirth for afterpains and hemorrhages. (as *A. oblongifolia* 124:96)
- *Food*—**Blackfoot** *Dried Food* Berries dried for winter use. *Soup* Berries used with stews and soups. *Unspecified* Berries used with meats. (as *A. oblongifolia* 114:277)

Amelanchier canadensis (L.) Medik., Canadian Serviceberry

- *Drug*—**Cherokee** *Anthelmintic* Infusion of bark used as a bath and given to children with worms. *Pediatric Aid* Infusion of bark used as a bath and given to children with worms. (177:27) **Chippewa** *Antidiarrheal* Compound decoction of root taken for dysentery. (53:344) *Disinfectant* Compound decoction of inner bark used as a disinfectant wash. (53:366) *Gynecological Aid* Compound decoction of bark taken for "female weakness." (53:356) Infusion of root taken to prevent miscarriage after an injury. (53:358) **Iroquois** *Blood Medicine* Fruits formerly used as a blood remedy. *Gynecological Aid* Fruits, infusion of small branches given to mothers after childbirth for afterpains and hemorrhages. (124:96)
- *Food*—**Chippewa** *Dried Food* Fruit dried for winter use. *Fruit* Fruit eaten fresh. (71:132) **Iroquois** *Bread & Cake* Fruit mashed, made into small cakes, and dried for future use. *Dried Food* Raw or cooked fruit sun or fire dried and stored for future use. *Fruit* Dried fruit taken as a hunting food. *Sauce & Relish* Dried fruit cakes soaked in warm water and cooked as a sauce or mixed with corn bread. (196:128)
- *Other*—**Iroquois** *Season Indicator* Blossoms used as a reliable method of when to plant the corn. (196:21)

Amelanchier canadensis ssp. *laevis*, Canadian Serviceberry

- *Food*—**Menominee** *Winter Use Food* Berries dried for winter use. (151:70)

Amelanchier canadensis var. *oblongifolda* Torr. & Gray, Canadian Serviceberry

- *Food*—**Ojibwa** *Fruit* Fruit used for food. (135:236)

Amelanchier laevis Wieg., Allegheny Serviceberry

- *Drug*—**Ojibwa** *Gynecological Aid* Infusion of bark taken by expectant mothers. (153:384)
- *Food*—**Cherokee** *Fruit* Fruit used for food. (80:21) Fresh fruit used for food. (126:55) **Menominee** *Winter Use Food* Berries dried for winter use. (151:70) **Ojibwa** *Dried Food* Berries used for food and dried for winter use, the Indians preferred them to blueberries. (153:408)

Amelanchier pallida Greene, Pale Serviceberry

- *Drug*—**Pomo, Kashaya** *Gynecological Aid* Decoction of boiled roots taken to check too frequent menstruation. (72:104)
- *Food*—**Cahuilla** *Dried Food* Berries dried for future use. *Fruit* Berries eaten fresh. (15:38) **Costanoan** *Fruit* Raw fruits used for food. (21:249) **Kawaiisu** *Fruit* Fruit eaten sparingly while fresh. (206:11)
- *Fiber*—**Pomo, Kashaya** *Building Material* Stems and foliage used to thatch inland houses. (72:104)
- *Other*—**Hopi** *Hunting & Fishing Item* Used to make bows and arrows. (200:79) **Kawaiisu** *Ceremonial Items* Sticks used in a Kawaiisu tale. Coyote sharpens the ends of the sticks (some versions refer to the roots) and plants them, points upward, on one side of his house. Grizzly bear, chasing coyote over the house, jumps on the points and is killed. *Hunting & Fishing Item* Stems used to make arrows and gun cleaners. (206:11)

Amelanchier sanguinea (Pursh) DC. var. *sanguinea*, Roundleaf Serviceberry

- *Food*—**Menominee** *Winter Use Food* Berries dried for winter use. (as *A. huronensis* 151:70)

Amelanchier sp., Serviceberry

- *Drug*—**Navajo** *Emetic* Plant used as an emetic. (90:148) **Shoshoni** *Eye Medicine* Decoction of inner bark, sometimes with roots, used as drops for snow-blindness. (180:33)
- *Food*—**Abnaki** *Fruit* Fruit used for food. (144:152) Fruits eaten for food. (144:168) **Algonquin, Quebec** *Fruit* Fruit used for food. (18:90) **Carrier** *Dried Food* Berries dried for winter use. (31:75) **Coeur d'Alene** *Fruit* Berries eaten fresh, boiled and eaten, or mashed and eaten. (178:89) **Iroquois** *Fruit* Fruit used for food. (142:90) **Okanagon** *Staple* Berries used as a principal food. (178:238) **Spokan** *Fruit* Berries used for food. (178:343) **Thompson** *Preserves* Berries collected in large quantities and cured. (178:237) **Wintoon** *Fruit* Berries used for food. (117:264)
- *Other*—**Coeur d'Alene** *Tools* Wood used to make root diggers. (178:91) **Pit River** *Protection* Wood made into a heavy robe or overcoat and corset armor and used for fighting. (117:222) **Shasta** *Hunting & Fishing Item* Young shoots used to make arrows. (117:217)

Amelanchier stolonifera Wieg., Running Serviceberry

- *Drug*—**Potawatomi** *Tonic* Root bark used to make a tonic. (as *A. spicata* 154:76)
- *Food*—**Potawatomi** *Dried Food* Berries dried for winter use. *Fruit* Berries relished as a fresh food. *Winter Use Food* Berries dried and canned for winter use. (as *A. spicata* 154:107)

Amelanchier utahensis Koehne, Utah Serviceberry

- *Drug*—**Navajo** *Gynecological Aid* Plant used during labor and delivery. (90:148)
- *Food*—**Havasupai** *Forage* Fruit eaten by deer. (197:222) **Isleta** *Fruit* Fruit formerly used for food. (as *A. prunifolia* 100:21) **Navajo** *Dried Food* Berries dried for winter use. (90:148) *Fruit* Fruits eaten fresh. (as *A. prunifolia* 55:52) Berries eaten fresh. (90:148) *Winter Use Food* Fruits dried and preserved for winter use. (as *A. prunifolia* 55:52) **Paiute** *Dried Food* Berries crushed, dried, and used for food. *Fruit* Berries eaten fresh. (as *A. venulosa* 104:100)
- *Fiber*—**Havasupai** *Basketry* Wood used to make basket rims. *Furniture* Wood used to make cradleboards. (197:222)
- *Other*—**Havasupai** *Cooking Tools* Wood used to make flat parching trays. *Hunting & Fishing Item* Stems made into arrow shafts and used for hunting. *Tools* Wood used to make the spindle of the fire drill. (197:222) **Hopi** *Ceremonial Items* Plant used to make pahos (prayer sticks). *Hunting & Fishing Item* Plant used to make bows and arrows. (42:284)

Amelanchier utahensis Koehne **ssp. *utahensis***, Utah Serviceberry
- *Drug*—**Navajo, Ramah** *Ceremonial Medicine* Leaves used as emetics in various ceremonies. *Emetic* Leaves used as a ceremonial emetic. *Panacea* Dried fruit used as a "life medicine." (as *A. mormonica* 191:30)
- *Food*—**Navajo, Ramah** *Fruit* Berries eaten raw or sometimes cooked. (as *A. mormonica* 191:30)
- *Other*—**Navajo, Ramah** *Ceremonial Items* Stem used to make Evilway hoop. (as *A. mormonica* 191:30)

Amianthium, Liliaceae
Amianthium muscitoxicum (Walt.) Gray, Fly Poison
- *Drug*—**Cherokee** *Dermatological Aid* Used for itch. (80:34) Root used as a sure, but severe, cure for itch. (203:74) *Poison* Used to poison crows. (80:34) Root used as a crow poison. (203:74)

Ammannia, Lythraceae
Ammannia coccinea Rottb., Valley Redstem
- *Food*—**Mohave** *Unspecified* Seeds gathered and prepared as food. **Yuma** *Unspecified* Seeds gathered and prepared as food. (37:187)

Amoreuxia, Bixaceae
Amoreuxia palmatifida Moc. & Sessé ex DC., Mexican Yellowshow
- *Food*—**Pima, Gila River** *Unspecified* Roots used for food. (133:7)
- *Other*—**Pima, Desert** *Cash Crop* Roots used for trade. (133:6)

Amorpha, Fabaceae
Amorpha canescens Pursh, Lead Plant
- *Drug*—**Meskwaki** *Anthelmintic* Infusion of leaves used to kill pinworms or any intestinal worms. *Dermatological Aid* Infusion of leaves used for eczema. (152:227) **Ojibwa, South** *Analgesic* Decoction of root taken for stomach pain. *Gastrointestinal Aid* Decoction of root taken for stomach pain. (91:200) **Omaha** *Analgesic* Moxa of twigs applied for neuralgia. (68:334) Moxa of stems used in cases of neuralgia. (70:93) *Antirheumatic (External)* Moxa of twigs applied for rheumatism. (68:334) Moxa of stems used in cases of rheumatism. (70:93) *Dermatological Aid* Powdered, dried leaves blown into cuts and open wounds. (68:334)
- *Food*—**Oglala** *Beverage* Leaves used to make a hot tea. (70:93)
- *Other*—**Oglala** *Smoke Plant* Dried leaves crushed fine, mixed with buffalo fat, and used as a smoking material. (70:93)

Amorpha fruticosa L., Desert Indigobush
- *Fiber*—**Kiowa** *Mats, Rugs & Bedding* Long stems used as a foundation for bedding material. (192:31)
- *Other*—**Lakota** *Hunting & Fishing Item* Stems used to make arrows. (139:45) **Pawnee** *Cooking Tools* Shrub used on the ground to receive meat while butchering, to keep the meat clean. (70:93)

Amorpha nana Nutt., Dwarf Indigobush
- *Drug*—**Navajo** *Respiratory Aid* Plant used as a snuff for catarrh. (as *A. microphyllus* 55:55)

Ampelopsis, Vitaceae
Ampelopsis cordata Michx., Heartleaf Peppervine
- *Drug*—**Cherokee** *Urinary Aid* Infusion of bark taken for urinary troubles. (177:41)

Amphiachyris, Asteraceae
Amphiachyris dracunculoides (DC.) Nutt., Prairie Broomweed
- *Drug*—**Comanche** *Dermatological Aid* Poultice of boiled flowers used for eczema and skin rashes. (as *Guitierrezia dracunculoides* 99:5)

Amphicarpaea, Fabaceae
Amphicarpaea bracteata (L.) Fern., American Hogpeanut
- *Drug*—**Cherokee** *Antidiarrheal* Infusion of root taken for diarrhea. *Snakebite Remedy* Infusion of root blown on snakebite wound. (80:38) **Chippewa** *Cathartic* Compound decoction of root taken as a physic. (as *Falcata comosa* 53:346) **Iroquois** *Gastrointestinal Aid* and *Tuberculosis Remedy* Compound decoction of plants taken for a bad stomach caused by consumption. (87:365) **Lakota** *Antirheumatic (External)* Poultice of pulverized leaves applied with any salve to swellings. (139:45)
- *Food*—**Cherokee** *Bread & Cake* Underground fruit used to make bean bread. (126:45) *Unspecified* Roots used for food. (80:38) *Vegetable* Underground fruit cooked like pinto beans or added to cornmeal and hot water. (126:45) **Chippewa** *Fruit* Fruit used for food. (as *Amphicarpa monoica* 71:134) *Unspecified* Roots boiled and used for food. (as *Falcata comosa* 53:320) **Dakota** *Unspecified* Beans used for the agreeable taste and nutritive value. (as *Falcata comosa* 70:95) **Meskwaki** *Unspecified* Nuts gathered and stored in heaps by the mice, taken by the Meskwaki and used. (as *Amphicarpa monoica* 152:259) **Ojibwa** *Unspecified* Roots cooked, although really too small to be considered of much importance. *Vegetable* Beans cooked, unusual flavor imparted and eaten. (as *Amphicarpa pitcheri* 153:405) **Omaha** *Unspecified* Roots peeled, boiled, and eaten. (as *Amphicarpaea monoica* 58:341) Beans used for the agreeable taste and nutritive value. (as *Falcata comosa* 70:95) *Winter Use Food* Roots gathered from the storehouses of field mice and stored in skin bags during the winter. (as *Amphicarpaea monoica* 58:341) **Pawnee** *Unspecified* Beans used for the agreeable taste and nutritive value. **Ponca** *Unspecified* Beans used for the agreeable taste and nutritive value. **Winnebago** *Unspecified* Beans used for the agreeable taste and nutritive value. (as *Falcata comosa* 70:95)

Amsinckia, Boraginaceae
Amsinckia douglasiana A. DC., Douglas's Fiddleneck
- *Drug*—**Costanoan** *Unspecified* Plant used for medicinal purposes. (21:13)

Amsinckia lycopsoides Lehm., Tarweed Fiddleneck
- *Food*—**Atsugewi** *Bread & Cake* Parched, ground seeds made into cakes and eaten without cooking. (as *Amsinkia parviflora* 61:139) **Mendocino Indian** *Unspecified* Fresh, juicy shoots formerly used for food. (41:382)

***Amsinckia* sp.**, Fiddleneck
- *Food*—**Pima, Gila River** *Vegetable* Leaves boiled or boiled, strained, refried, and eaten as greens.[11] (133:5)

Amsinckia spectabilis Fisch. & C. A. Mey., Woolly Breeches

- *Food*—Pima *Unspecified* Young leaves rolled into balls and eaten raw. (95:264)

Amsinckia tessellata Gray, Bristly Fiddleneck
- *Food*—Gosiute *Unspecified* Seeds formerly used for food. (39:361) Kawaiisu *Vegetable* Leaves bruised by rubbing between the hands and eaten with salt. (206:11) Pima *Unspecified* Leaves eaten raw. (95:264)

Amsonia, Apocynaceae
Amsonia tomentosa Torr & Frém. **var. *tomentosa***, Woolly Bluestar
- *Drug*—Zuni *Snakebite Remedy* Compound poultice of root applied with much ceremony to rattlesnake bite. (as *A. brevifolia* 66:53)

Anacardiaceae, see *Mangifera, Rhus, Toxicodendron*

Anagallis, Primulaceae
Anagallis sp., Pimpernel
- *Drug*—Mahuna *Venereal Aid* Infusion of plant taken for gonorrhea when the bladder and urinal tract fail. (as *Centunculus* 140:70)

Ananas, Bromeliaceae
Ananas comosus (L.) Merr., Pineapple
- *Food*—Seminole *Unspecified* Plant used for food. (169:500)

Anaphalis, Asteraceae
Anaphalis margaritacea (L.) Benth. & Hook. f., Western Pearly-everlasting
- *Drug*—Algonquin, Tête-de-Boule *Burn Dressing* Poultice of boiled leaves applied to burns. (132:119) Bella Coola *Tuberculosis Remedy* Plants formerly used for tuberculosis. (184:201) Cherokee *Analgesic* Infusion steamed and inhaled for headache. *Cold Remedy* Warm infusion taken for cold and leaves smoked or chewed for colds. *Cough Medicine* Leaves and stems smoked for bronchial cough. *Eye Medicine* Infusion steamed and inhaled for blindness caused by the sun. *Respiratory Aid* Dried leaves smoked for catarrh. *Throat Aid* Used for throat infection. (80:48) Cheyenne *Ceremonial Medicine* Powdered flowers chewed and rubbed on body to protect and strengthen warrior. *Disinfectant* Smoke used to purify gift made to the spirits. (75:42) *Unspecified* Plant used as a strong medicine. (75:187) *Veterinary Aid* Plant used in various ways to make horses long-winded. (75:42) Powdered flowers used on the sole of each horse hoof to make it enduring and untiring. (76:187) Powdered flowers put on each hoof and blown between the ears for long-windedness, spirit, and endurance. (97:56) Chippewa *Antirheumatic (External)* Compound decoction of flowers used as herbal steam for rheumatism and paralysis. *Herbal Steam* and *Orthopedic Aid* Infusion of flower used as herbal steam for rheumatism and paralysis. (53:362) Delaware, Oklahoma *Tonic* Compound containing root used as a tonic. (175:74) Iroquois *Antidiarrheal* Roots and stalks used for diarrhea and dysentery. *Eye Medicine* Infusion of plants used as wash for sore eyes. *Gastrointestinal Aid* Compound decoction of roots and flowers taken for bruise on back of stomach. (87:465) *Respiratory Aid* Infusion of flowers and roots from another plant used for asthma. (141:63) Kwakiutl *Dermatological Aid* Poultice of flowers applied to sores and swellings. *Internal Medicine* Decoction of flowers taken for internal disorders. (183:278) Mahuna *Dermatological Aid* Flowers used for skin ulcers and foot sores. (as *Antennaria margaritacea* 140:11) Mohegan *Cold Remedy* Infusion of plant taken for colds. (174:265) Infusion of leaves taken as a cold medicine. (176:70, 128) Montagnais *Cough Medicine* Decoction of plant taken for cough. *Tuberculosis Remedy* Decoction of plant taken for consumption. (156:314) Nitinaht *Other* Plants rubbed on the hands to soften them for handling or touching sick people. (186:97) Ojibwa *Stimulant* Powdered flowers sprinkled on coals and smoke inhaled to revive stroke victim. (153:362, 365) Okanagan-Colville *Gastrointestinal Aid* Cooled infusion of roots and shoots taken as a laxative and emetic for a "poison stomach." (188:75) Potawatomi *Witchcraft Medicine* Flowers smoked in a pipe or smudged on coals to repel evil spirits. (154:49) Quileute *Antirheumatic (Internal)* Whole plant used as a steam bath for rheumatism. (79:48) Thompson *Misc. Disease Remedy* Decoction of dried flowers taken for rheumatic fever. (187:167)
- *Food*—Anticosti *Beverage* Flowers used to scent alcohol. (143:68)
- *Other*—Anticosti *Ceremonial Items* Infusion of plant used to "force the blood for sacrifices." (143:68) Cherokee *Smoke Plant* Dried leaves used as a substitute for chewing tobacco. (80:48) Cheyenne *Ceremonial Items* Leaves burned as incense and used to purify gifts offered to the sun or the spirits. (83:18) *Protection* Dried flowers carried or chewed and rubbed on the body as protection from danger before battle. (76:187) Okanagan-Colville *Incense & Fragrance* Leaves, stems, and flowers placed in baby cradles, pillows, or stored clothes for the good smell. (188:75) Paiute *Containers* Branches used to cover baskets filled with berries. (111:116) Potawatomi *Protection* Dried tops placed on a pan of live coals to hurt the eyes of the evil spirits and keep them away. (154:117)

Andromeda, Ericaceae
Andromeda polifolia L., Bog Rosemary
- *Drug*—Mahuna *Respiratory Aid* Plant used for catarrh. (140:24)
- *Food*—Tanana, Upper *Beverage* Leaves used to make tea. (102:8)

Andromeda polifolia var. *glaucophylla* (Link) DC., Bog Rosemary
- *Food*—Ojibwa *Beverage* Fresh or dried leaves and tips boiled for a beverage tea. (as *A. glaucophylla* 153:400)

Andropogon, Poaceae
Andropogon floridanus Scribn., Florida Bluestem
- *Drug*—Seminole *Analgesic* Roots used with a song or spell as an analgesic. (169:167) Infusion of plant taken for wolf sickness: vomiting, stomach pain, diarrhea, and frequent urination. (169:227) Infusion of roots used for moving sickness: moving pain in the waist region. (169:285) *Antidiarrheal* and *Antiemetic* Infusion of plant taken for wolf sickness: vomiting, stomach pain, diarrhea, and frequent urination. (169:227) *Cough Medicine* Infusion of plant taken and used as bath for gopher-tortoise sickness: cough, dry throat, noisy chest. (169:236) *Gastrointestinal Aid* Infusion of plant taken for wolf sickness: vomiting, stomach pain, diarrhea, and frequent urination. (169:227) *Pulmonary Aid* and *Throat Aid* Infusion of plant taken and used as bath for gopher-tortoise sickness: cough, dry throat, noisy

chest. (169:236) *Urinary Aid* Infusion of plant taken for wolf sickness: vomiting, stomach pain, diarrhea, and frequent urination. (169:227)

Andropogon gerardii Vitman, Big Bluestem
- **Drug**—**Chippewa** *Analgesic* Decoction of root taken for stomach pain. (as *A. furcatus* 53:342) *Diuretic* Simple or compound decoction of root taken as a diuretic. (as *A. furcatus* 53:348) *Gastrointestinal Aid* Decoction of root taken for stomach pain. (as *A. furcatus* 53:342) **Omaha** *Febrifuge* Decoction of blades of grass used as a wash for fevers. *Stimulant* Decoction of blades of grass taken for "general debility and languor." (as *A. furcatus* 70:68, 69)
- **Fiber**—**Omaha** *Building Material* Grass used on poles to support earth coverings of lodges. **Ponca** *Building Material* Grass used on poles to support earth coverings of lodges. (as *A. furcatus* 70:68)
- **Other**—**Apache, Chiricahua & Mescalero** *Containers* Moist grass laid onto hot stones to prevent steam from escaping. (as *A. furcatus* 33:36) Used to cover fruit and allow ripening. (as *A. furcatus* 33:39) Grass used under fruit when drying. (as *A. furcatus* 33:40) **Arikara** *Hunting & Fishing Item* Stiff, jointed stems used by little boys to make arrows.[12] **Hidatsa** *Hunting & Fishing Item* Stiff, jointed stems used by little boys to make arrows.[12] **Mandan** *Hunting & Fishing Item* Stiff, jointed stems used by little boys to make arrows.[12] **Omaha** *Toys & Games* Stiff, jointed stems used by little boys to make arrows for toy bows. **Ponca** *Toys & Games* Stiff, jointed stems used by little boys to make arrows for toy bows. (as *A. furcatus* 70:68)

Andropogon glomeratus (Walt.) B.S.P., Bushy Bluestem
- **Drug**—**Catawba** *Analgesic* Roots used for backaches. *Orthopedic Aid* Roots used for backaches. (177:5) **Rappahannock** *Dermatological Aid* Infusion of roots taken for the itch and applied to ivy poisoning. *Hemorrhoid Remedy* Compound poultice with roots applied as salve for piles. (161:28)

Andropogon sp., Broom Grass
- **Drug**—**Houma** *Gynecological Aid* and *Pediatric Aid* Decoction of plant taken by pregnant women to strengthen both mother and child. (158:65)

Andropogon virginicus L., Broomsedge Bluestem
- **Drug**—**Cherokee** *Antidiarrheal* Infusion taken to "check bowels." *Ceremonial Medicine* Used as an ingredient in green corn medicine. *Dermatological Aid* Infusion used for frostbite and sores and ooze used to bathe itch. (80:27)
- **Dye**—**Cherokee** *Yellow* Stems, alone or with onion peels, used to make a yellow dye. (80:27)

Androsace, Primulaceae
Androsace occidentalis Pursh, Western Rockjasmine
- **Drug**—**Navajo, Ramah** *Gynecological Aid* Compound decoction of whole plant used for postpartum hemorrhage. *Pediatric Aid* Compound decoction of whole plant used for birth injury. (191:38)

Androsace septentrionalis L., Pygmyflower Rockjasmine
- **Drug**—**Navajo, Ramah** *Analgesic* Cold infusion taken for internal pain. *Panacea* Plant used as "life medicine." *Venereal Aid* Compound decoction of plant taken before sweat bath for venereal disease. (191:38, 39) *Witchcraft Medicine* Plant used as a lotion to give protection from witches. (191:38)

Androsace septentrionalis ssp. *subulifera* (Gray) G. T. Robbins, Pygmyflower Rockjasmine
- **Drug**—**Navajo, Kayenta** *Analgesic* and *Witchcraft Medicine* Plant used for bewitchment and pain from witches' arrows. (205:35)

Androsace sp., Pinetorum
- **Food**—**Isleta** *Beverage* Leaves steeped in water to make a beverage. (100:22)

Anemone, Ranunculaceae
Anemone canadensis L., Canadian Anemone
- **Drug**—**Chippewa** *Dermatological Aid* Poultice of roots applied to wounds and infusion of root used as wash for sores. *Hemostat* Leaves used for nasal hemorrhages, bleeding sores, and wounds. (71:130) **Iroquois** *Anthelmintic* Decoction of roots taken for worms. *Witchcraft Medicine* Compound infusion of plants and liquor used to counteract witch medicine. (87:328) **Meskwaki** *Eye Medicine* Infusion of root used as a wash for crossed eyes, eye twitch, and eye poisoning. (152:238) **Ojibwa** *Ceremonial Medicine* Root eaten to clear throat so one can sing well in medicine lodge ceremony. *Throat Aid* Root eaten to clear throat so one can sing well in ceremonies. (153:382, 383) **Ojibwa, South** *Analgesic* and *Orthopedic Aid* Decoction of root used for pain in the lumbar region. (as *A. pennsylvanicum* 91:201) **Omaha** *Panacea* Highly esteemed medicine taken and applied externally for many illnesses. **Ponca** *Panacea* Highly esteemed medicine taken and applied externally for many illnesses. (70:82)

Anemone cylindrica Gray, Candle Anemone
- **Drug**—**Meskwaki** *Analgesic* Infusion of root taken for headache and dizzy spells. *Burn Dressing* Poultice of leaves applied to bad burns. *Eye Medicine* Decoction of stem and fruit used as a wash for sore eyes. *Psychological Aid* Used as a medicine for "crazy people." *Stimulant* Infusion of root used for headache and dizzy spells. (152:238) **Ojibwa** *Pulmonary Aid* Infusion of root taken for lung congestion and tuberculosis. *Tuberculosis Remedy* Infusion of root used for lung congestion and tuberculosis. (153:383)
- **Other**—**Ponca** *Good Luck Charm* Woolly fruits used as good luck charms when playing cards. They rubbed their hands in the smoke that resulted from burning some of the woolly fruits for good luck. Some of the chewed fruit would work as well. (70:82)

Anemone multifida Poir., Pacific Anemone
- **Drug**—**Blackfoot** *Abortifacient* Plant used to cause abortions. (86:60) *Analgesic* Ripe seed head "cotton" burned on hot coals and the smoke inhaled for headaches. (97:35) **Carrier, Southern** *Cold Remedy* Aroma of crushed leaves inhaled for head or lung colds. *Panacea* Decoction of plant, without roots, taken for any sickness. **Gitksan** *Antirheumatic (Internal)* Plant eaten or decoction of plant taken in sweat bath for rheumatism. *Diaphoretic* Eaten or decoction taken in sweat bath for rheumatism. (150:57) **Okanagon** *Hemostat* Leaves applied to nose for nosebleeds. **Thompson** *Hemostat* Leaves applied to nose for nosebleeds. (125:42) Fresh leaves used to plug nostrils and as an inhalant for nosebleed. (164:474) Wool from seed heads mixed with pitch and used inside the nostril for nosebleeds. *Poison* Plant considered very poisonous. (187:246)
- **Other**—**Thompson** *Insecticide* Strong decoction of whole plant used to kill lice and fleas. (164:513)

Anemone multifida var. *globosa* Torr. & Gray, Hudson's Anemone
- **Drug**—**Blackfoot** *Analgesic* Cottony flower burned on hot coals for headache. (as *A. globosa* 114:274, 275)

Anemone narcissiflora L., Narcissus Anemone
- **Food**—**Alaska Native** *Unspecified* Upper root ends used for food. (85:151) **Eskimo, Alaska** *Ice Cream* Leaves, other salad greens, and oil beaten to a creamy consistency and frozen into "ice cream." *Unspecified* Leaves prepared in oil together with other salad greens and beaten to a creamy consistency. (4:715)

Anemone narcissiflora ssp. *villosissima* (DC.) Hultén, Narcissus Anemone
- **Drug**—**Aleut** *Antihemorrhagic* Decoction of root taken for unspecified hemorrhage. (8:428)

Anemone sp., Buttercup
- **Drug**—**Cowlitz** *Tuberculosis Remedy* Infusion of plants taken for tuberculosis. (79:29) **Ojibwa, South** *Analgesic* Snuff of powdered, dry leaves used as an errhine for headache. (91:200) **Thompson** *Unspecified* Plant used medicinally for unspecified purpose. (164:469)

Anemone virginiana L., Tall Thimbleweed
- **Drug**—**Cherokee** *Pulmonary Aid* Infusion of root taken for whooping cough. (80:58) **Iroquois** *Antidiarrheal* Cold decoction of roots taken for diarrhea. (87:328) *Emetic* Decoction or infusion of smashed roots or plants taken as an emetic. (87:327) *Love Medicine* Decoction of roots taken as an emetic and used as a wash to cure a love medicine. Infusion of stems and roots used as a love medicine for either sex. *Tuberculosis Remedy* Decoction of roots taken for tuberculosis. (87:328) *Witchcraft Medicine* Compound infusion of smashed plants taken as an emetic to remove bewitchment. Roots placed under the pillow to dream the truth about wife's crookedness. (87:327) Root used as revenge to "kill man who played a trick on man's son." (87:328) **Menominee** *Dermatological Aid* Poultice of root applied to boils. (151:48) **Meskwaki** *Respiratory Aid* Smoke of seeds inhaled for catarrh. *Stimulant* Smoke of seedpod directed up nostril to revive sick and unconscious patient. (152:238)

Anemopsis, Saururaceae

Anemopsis californica (Nutt.) Hook. & Arn., Yerba Mansa
- **Drug**—**Cahuilla** *Cold Remedy* Infusion of plant used for colds. *Dermatological Aid* Infusion of bark used as a wash for open sores. *Gastrointestinal Aid* Infusion of plant used for stomach ulcers. Decoction of bark taken for ulcers. *Pulmonary Aid* Decoction of peeled, cut, and squeezed roots taken for pleurisy. *Respiratory Aid* Infusion of plant used for chest congestion. *Veterinary Aid* Infusion of plant used for open sores on cattle. (15:38) **Costanoan** *Analgesic* Decoction of roots used as a general pain remedy. *Dermatological Aid* Dried, powdered plant applied as a disinfectant to wounds. Infusion of plant used as a wash for sores. *Disinfectant* Dried, powdered plant applied as a disinfectant to wounds. *Gynecological Aid* Decoction of roots used for menstrual cramps. (21:8) **Diegueño** *Unspecified* Plant used as medicine. (84:15) **Isleta** *Blood Medicine* Infusion of leaves taken as a blood medicine. *Dermatological Aid* Poultice of damp leaves used on open wounds. *Disinfectant* Infusion of leaves used as a disinfectant on open wounds. *Pulmonary Aid* Infusion of leaves taken for lung hemorrhages. (100:22) **Kawaiisu** *Cold Remedy* Decoction of broken roots taken for colds. *Cough Medicine* Decoction of broken roots taken for coughs. *Dermatological Aid* Leaves used as a salve for cuts and wounds. *Misc. Disease Remedy* Decoction of broken roots taken for diabetes. *Veterinary Aid* Leaves used as a salve for livestock with cuts and wounds. (206:11) **Keres, Western** *Burn Dressing* Poultice of green, chewed leaves applied to burns. *Dermatological Aid* Dried leaves ground into a powder and used on open sores. (171:26) **Mahuna** *Dermatological Aid* Powdered plants used as a disinfectant for knife wounds. *Disinfectant* Powdered plants used as a disinfectant for knife wounds. (140:15) **Paiute** *Laxative* Infusion of roots taken as a laxative. (167:317) *Orthopedic Aid* Decoction of leaves used as a bath for muscular pains and sore feet. (180:33, 34) *Venereal Aid* Infusion of roots taken for gonorrhea. (167:317) **Papago** *Emetic* Decoction of leaves taken as an emetic. (34:65) **Pima** *Cold Remedy* Infusion of dried roots or plant taken for colds. *Cough Medicine* Roots chewed and swallowed or decoction of roots taken for coughs. *Dermatological Aid* Decoction of plant used as a wash and poultice of leaves applied to wounds. *Diaphoretic* Infusion of plant taken for colds and to cause sweating. (47:78) *Emetic* Decoction of crushed root taken as an emetic. (as *Houttuynia californica* 146:80) *Gastrointestinal Aid* Poultice of wet, powdered roots applied for stomachaches. *Other* Infusion of roots taken and used as a wash for "bad disease." *Throat Aid* Dry root held in the mouth for sore throats and infusion taken for itchy throat. (47:78) *Tuberculosis Remedy* Decoction of crushed root taken for consumption. (as *Houttuynia californica* 146:80) *Venereal Aid* Infusion of roots used as a wash for syphilis. (47:78) **Shoshoni** *Anticonvulsive* Infusion of whole plant taken for fits. (118:46) *Cold Remedy* Decoction of roots taken as a tonic for general debility following colds. *Dermatological Aid* Poultice of boiled, mashed roots applied to swellings. *Disinfectant* Decoction of roots used as an antiseptic wash. *Gastrointestinal Aid* Decoction of roots taken for stomachaches. *Tonic* Decoction of roots taken as a tonic for general debility following colds. *Venereal Aid* Decoction of plant taken for gonorrhea. (180:33, 34) **Tubatulabal** *Cold Remedy* Decoction of plant taken for heavy colds. (193:59)
- **Food**—**Kamia** *Bread & Cake* Pulverized seeds used for bread. *Porridge* Pulverized seeds cooked as mush. (62:24)

Angadenia, Apocynaceae

Angadenia berteroi (A. DC.) Miers, Pineland Golden Trumpet
- **Drug**—**Seminole** *Dermatological Aid* Decoction of roots taken and used as a wash for sores. (as *Rhabdadenia corallicola* 169:271) *Other* Decoction of roots used by men for menstruation sickness: painful abdominal swelling and impotence.[13] (as *Rhabdadenia corallicola* 169:248) Decoction of roots taken and used as a wash for chronic sickness. (as *Rhabdadenia corallicola* 169:271)

Angelica, Apiaceae

Angelica archangelica L., Norwegian Angelica
- **Food**—**Eskimo, Greenland** *Vegetable* Tender, young leaf stalks and peeled, young flowering stems eaten raw. (128:28)

Angelica arguta Nutt., Lyall's Angelica
- **Food**—**Shuswap** *Spice* Young stems eaten and used to flavor salmon

heated with dried bread over an open fire. *Vegetable* Young stems, with a celery flavor, eaten in May. (123:56)

Angelica atropurpurea L., Purplestem Angelica

- **Drug**—**Cherokee** *Abortifacient* "Root tonic" taken for obstructed menses. *Carminative* "Root tonic" taken for flatulent colics. *Cold Remedy* "Root tonic" taken for colds. *Febrifuge* "Root tonic" taken for fever. *Misc. Disease Remedy* "Root tonic" taken for ague. *Oral Aid* "Root tonic" used as gargle for sore mouth. *Sedative* "Root tonic" taken by weakly and nervous females. *Throat Aid* "Root tonic" used as gargle for sore throat. (80:23) **Delaware** *Gastrointestinal Aid* Roots used for stomach disorders. (176:33) **Delaware, Oklahoma** *Gastrointestinal Aid* Root used for stomach disorders. (175:28, 74) **Iroquois** *Analgesic* Compound infusion of plants used as steam bath to sweat out headaches. *Antirheumatic (External)* Plant or root used internally, externally, or in steam bath for rheumatism. (87:400) *Blood Medicine* Dried roots used as blood purifier. *Cold Remedy* Decoction of smashed roots taken for colds. (87:401) *Diaphoretic* Infusion of plants used as steam bath to sweat out rheumatism and headaches. (87:400) *Febrifuge* Decoction of dried roots taken for fevers and chills. *Gynecological Aid* Decoction of roots taken by women for weakness. *Misc. Disease Remedy* Decoction of smashed roots taken for the flu. *Orthopedic Aid* Poultice of roots applied to broken bones. (87:401) *Other* Decoction of roots used as steam bath for frostbite and exposure. (87:400) *Poison* Plant used as poison. *Pulmonary Aid* Infusion of roots used for pneumonia. *Witchcraft Medicine* Plant used to punish evil persons. (87:401) **Menominee** *Analgesic* Poultice of cooked, pounded root applied to painful areas. *Dermatological Aid* Poultice of cooked, pounded root applied to swellings. (151:55)
- **Other**—**Delaware, Oklahoma** *Smoke Plant* Seeds sometimes mixed with tobacco and used for smoking. (175:28) **Iroquois** *Protection* Infusion of smashed roots used as wash to remove ghosts from the house. (87:401)

Angelica breweri Gray, Brewer's Angelica

- **Drug**—**Miwok** *Analgesic* Root chewed for headaches. *Cold Remedy* Root chewed for colds. (12:166) **Paiute** *Cold Remedy* Decoction of roots used for colds or chest ailments. Hot decoction of roots taken for colds. *Cough Medicine* Dried root chewed for sore throats or coughs. *Dermatological Aid* Salve of mashed roots applied to cuts and sores. *Kidney Aid* Decoction of roots taken, instead of drinking water, for kidney ailments. *Throat Aid* Dried root chewed for sore throat and coughs. **Shoshoni** *Adjuvant* Plant used as an adjuvant to improve flavor or amplify effect of medicines. *Analgesic* and *Antirheumatic (External)* Poultice of pulped root applied to rheumatic pains or swellings. *Cold Remedy* Decoction of roots used for colds or chest ailments. Dried, shaved roots smoked in cigarettes for head colds. Hot decoction of roots and whisky taken for heavy chest colds. *Pediatric Aid* Decoction of split root given to children for whooping cough. *Pulmonary Aid* Decoction of split root in whisky given to children for whooping cough. Poultice of pulped roots applied for pneumonia. *Tonic* Decoction of roots taken in small doses as a tonic. *Tuberculosis Remedy* Decoction of roots taken for tuberculosis. *Venereal Aid* Decoction of roots taken and used as a wash for venereal diseases. *Veterinary Aid* Smoke from root compound inhaled by horses for distemper. **Washo** *Adjuvant* Plant used as an adjuvant to improve flavor or amplify effect of medicines. *Cough Medicine* Dried root chewed for sore throats or coughs. *Misc. Disease Remedy* Infusion of scraped, dried root taken for influenza. *Respiratory Aid* Infusion of scraped, dried root taken for bronchitis. *Throat Aid* Dried root chewed for sore throat or coughs. (180:34, 35)
- **Other**—**Miwok** *Protection* Chewed, rubbed on body or decoction taken to ward off snakes. (12:166)

Angelica dawsonii S. Wats., Dawson's Angelica

- **Drug**—**Blackfoot** *Antihemorrhagic* Infusion of roots taken for coughing up blood. (86:70) *Antirheumatic (External)* Poultice of chewed roots applied to swellings. (86:75) *Ceremonial Medicine* Roots used as a religious power medicine. (86:40) *Dermatological Aid* Poultice of chewed roots applied to rashes, eczema, and athlete's foot. (86:75) *Dietary Aid* Infusion of roots given to children with malnutrition. *Gastrointestinal Aid* Infusion of roots taken for intestinal ailments. (86:65) *Misc. Disease Remedy* Infusion of roots applied to mumps swellings. *Other* Infusion of roots applied for a disorder characterized by sore groins and underarms. (86:75) *Pediatric Aid* Infusion of roots given to children with malnutrition. (86:65) *Unspecified* Roots used medicinally for unspecified purpose. (86:40) *Veterinary Aid* Root smudge used to fumigate horses with nasal gleet. Infusion of roots given to horses with nasal gleet. Infusion of roots used as a wash for hoof frogs and infections. (86:87)
- **Other**—**Blackfoot** *Ceremonial Items* Roots used for individual power by ceremonialists and diviners. Roots used for collective power in designated societies. Roots used by ceremonialists to bless others with long life and good luck. Root used in the rites of most age-graded societies and in the Natoas Beaver and Medicine Pipe bundles. Root used by the Horn Society in the initiation ceremonies. The root was found in the nontransferable initiation bundles. Having been distributed by the owner of the Spear Staff bundle to the others, down to the owner of the Marten, the root was kept in the mouth and used to bless the initiates. After the ritual, the root was tied to the paint application sticks, to fortify and replenish the supernatural power inherent in the paint. The same was done with a wooden scratching pin, which was either worn in the hair or attached to one's garments. This was done in preparation for the next opening ceremony. Root pieces attached to the binding of the Spear Staff (with bundle), the head staff. The Spear Staff was used during the secret ceremony of the Horns, in which the chief ceremonialist participated in a fertility rite with the initiate's wife. Root held in the mouth during Horn Society curse ceremony. Curses were indulged in by the Horn Society as a whole. In this case a special sweat lodge was constructed. A human figure representing the condemned man was drawn in the bottom of a pit intended to contain hot stones. The stones were brought in at a designated time and placed on the drawing to represent the victim's suffering. Root used during the ritual of body painting and tied to headdresses of all members of Pigeon Society. *Good Luck Charm* Roots used for luck in games of chance. Roots tied in small bundle and attached to the tail base or bridle for luck in horse racing. Roots used by ceremonialists to bless others with long life and good luck. *Malicious Charm* Root held in the rider's mouth to cast a spell so that other horses could not pass. Juice used to spray on quirt (a rod) to cause other horses to fall behind

during a race. A piece of plant was placed in the mouth, and the rider sprayed the quirt with the juice. This quirt was not used to whip the horse. When the race began, the rider would try to get on the right side of the other horses; he carried the magic quirt but whipped his horse with a regular one. At an opportune time the rider threw back the sprayed quirt, causing the other horses to fall behind. Root held in the mouth to make the other players lazy during the hand game. (86:40)

Angelica genuflexa Nutt., Kneeling Angelica
- **Drug**—**Bella Coola** *Cathartic* Decoction of root or raw root taken as a purgative. **Gitksan** *Analgesic* Compound decoction of root taken for headache. *Eye Medicine* Compound decoction of root taken for weak eyes. (150:61)
- **Food**—**Hanaksiala** *Unspecified* Leaves and stems used for food. (43:211)
- **Other**—**Bella Coola** *Cooking Tools* Hollow stems used to make drinking straws. *Tools* Hollow stems used to make breathing tubes for hiding under water when in danger. (184:200) **Haisla & Hanaksiala** *Containers* Hollow stems used to collect liquid Sitka spruce pitch. *Incense & Fragrance* Leaves chewed and juice rubbed on the body to mask the human smell. *Soap* Leaves used with devil's club to wash the human smell from one's body. **Hanaksiala** *Toys & Games* Hollow stems used to make whistles. (43:211) **Nitinaht** *Toys & Games* Leafstalks used in children's games. (186:91)

Angelica lineariloba Gray, Poison Angelica
- **Drug**—**Paiute, Northern** *Antihemorrhagic* Decoction of dried, scraped roots taken for spitting up blood. *Pulmonary Aid* Decoction of dried, scraped roots taken for pneumonia. (59:126)

Angelica lucida L., Wild Celery
- **Drug**—**Aleut** *Analgesic* Poultice of leaves applied for internal or external pain. *Cold Remedy* Leaves used to make a tonic for colds. *Throat Aid* Leaves used to make a soothing drink for sore throats. *Tonic* Leaves used to make a tonic for colds. (8:427) **Eskimo** *Panacea* Plant used for most illnesses. *Preventive Medicine* Root eaten as a preventative medicine. *Psychological Aid* Plant used for the feeling of malaise. (149:325) **Eskimo, Inuktitut** *Unspecified* Young stems used medicinally. (202:184) **Eskimo, Kuskokwagmiut** *Disinfectant* Burning stems shaken inside and outside the house for purification. (122:31) **Kwakiutl** *Analgesic* Plant used to prepare sweat bath for localized pains. *Herbal Steam* Plant used to prepare sweat bath for general weakness and localized pains. (as *Coelapleurum gmelini* 20:376) *Hunting Medicine* Plant tied on halibut hooks as a good luck charm. (183:276) *Stimulant* Plant used to prepare sweat bath for general weakness. *Unspecified* Used on heated stones in the steam bath to dry up the patient's disease. (as *Coelapleurum gmelini* 20:376)
- **Food**—**Alaska Native** *Unspecified* Young stems and tender stalks of young leaves peeled and the juicy inside eaten raw. *Vegetable* Leaves cooked as a green vegetable or boiled with fish. (85:11) **Bella Coola** *Unspecified* Formerly used for food. (184:201) **Eskimo, Alaska** *Unspecified* Stalks, with the outer sheet peeled off, eaten raw by children and adults. Only young plants were considered good to eat because older plant became fibrous and strong tasting. Young leaves eaten with seal oil. (1:37) *Vegetable* Used like celery. (as *Coelapleurum gmelini* 4:715) **Eskimo, Inuktitut** *Unspecified* Young stems used for food. (202:184) **Eskimo, Inupiat** *Unspecified* Peeled stems and young leaves stored in seal oil for future use. (98:16) **Makah** *Unspecified* Peeled petioles used for food. (67:292)
- **Other**—**Eskimo, Alaska** *Ceremonial Items* Plant formerly used during the seal bladder festival. *Smoking Tools* Dried hollow stems formerly used as pipestems. (1:37)

Angelica pinnata S. Wats., Small-leaf Angelica
- **Drug**—**Gosiute** *Unspecified* Root used as medicine. (39:361)

Angelica sp., Angelica
- **Drug**—**Costanoan** *Analgesic* Root burned and inhaled for headaches. *Dermatological Aid* Heated leaf juice rubbed on sores. *Gastrointestinal Aid* Root chewed for stomachaches. *Orthopedic Aid* Burned twigs used to beat aching joints. (21:23) **Creek** *Analgesic* Roots used for "a dry belly-ache" and back pains. *Anthelmintic* Roots given to children as a vermifuge. *Carminative* Roots used for "friendly carminative qualities." *Gastrointestinal Aid* Roots used for all the stomach and intestinal disorders. *Orthopedic Aid* Root taken for back pain. *Pediatric Aid* Roots given to children as a vermifuge. *Sedative* Roots used for hysterics. (172:657) **Lolahnkok** *Love Medicine* Plant rubbed on hands and girl's neck to make her give in. (117:191) **Mendocino Indian** *Analgesic* Poultice of roots applied to head and ears for headaches. *Cold Remedy* Roots chewed and swallowed or smoked for colds. *Eye Medicine* Root juice used for sore eyes. *Febrifuge* Roots chewed and swallowed for fevers. *Gastrointestinal Aid* Roots chewed and swallowed for colic. *Psychological Aid* Poultice of roots applied to head and ears for nightmares. *Respiratory Aid* Root smoked for catarrh. *Snakebite Remedy* Root rubbed on legs to prevent rattlesnake bites. (41:371) **Paiute** *Dermatological Aid* and *Venereal Aid* Poultice of boiled roots applied to sores and swellings, especially venereal. (167:317) **Tewa** *Analgesic* Pulverized root taken with water for stomachache. *Antidiarrheal* Root chewed or decoction taken for diarrhea. *Antiemetic* Pulverized root taken with water for vomiting. *Gastrointestinal Aid* Root chewed or decoction taken for stomach disorders. *Gynecological Aid* Decoction of root not recommended for pregnant women because of astringency. (138:71)
- **Food**—**Aleut** *Unspecified* Species used for food. (7:29) **Mendocino Indian** *Unspecified* Fresh sprouts eaten raw. (41:370) **Neeshenam** *Vegetable* Leaves eaten as greens. (129:377)
- **Other**—**Mendocino Indian** *Good Luck Charm* Carried about the person for good luck in gambling or hunting. (41:370)

Angelica sylvestris L. Woodland Angelica
- **Drug**—**Micmac** *Cold Remedy* Root used for head cold. (40:54) Infusion of roots and spikenard roots used for head colds. (116:259) *Cough Medicine* Root used for cough. (40:54) Infusion of roots and spikenard roots used for coughs. (116:259) *Throat Aid* Root used for sore throat. (40:54) Infusion of roots and spikenard roots used for sore throats. (116:259)

Angelica tomentosa S. Wats., Woolly Angelica
- **Drug**—**Pomo, Kashaya** *Cold Remedy* Decoction of root used for colds. *Dermatological Aid* Decoction of root used as a strong wash for bathing sores. *Gastrointestinal Aid* Decoction of root used for stomachaches. *Gynecological Aid* Decoction of root used to regulate

menses and ease menstrual cramps and discomforts of menopause. *Oral Aid* Root chewed or held in the mouth to prevent bad breath. *Other* Root shavings smoked by the shaman when doctoring. *Throat Aid* Root chewed or held in the mouth to prevent sore throat. Root held in singer's mouth to prevent hoarseness and rawness of throat. (72:20) **Yana** *Analgesic* Decoction of roots taken or poultice of roots applied for headaches. *Antidiarrheal* Decoction of roots taken for diarrhea. *Cold Remedy* Decoction of roots taken for colds. *Panacea* Decoction of roots taken for colds, diarrhea, headaches, and other ailments. (147:253)
- *Food*—**Karok** *Vegetable* Leaves eaten raw as greens. (148:387) **Pomo, Kashaya** *Unspecified* Young, green shoots eaten raw. (72:20) **Yana** *Unspecified* Peeled stems eaten raw. (147:251)
- *Other*—**Karok** *Ceremonial Items* Roots used as a purification after a funeral. (148:387) **Pomo, Kashaya** *Protection* Root carried and/or hung in homes for protection. (72:20)

Angelica tomentosa var. *hendersonii* (Coult. & Rose) Di Tomaso, Henderson's Angelica
- *Drug*—**Mewuk** *Antidote* Infusion of plant taken for mussel poisoning. (as *A. hendersoni* 117:366)
- *Food*—**Mewuk** *Unspecified* Young stems eaten raw. (as *A. hendersoni* 117:366)

Angelica venenosa (Greenway) Fern., Hairy Angelica
- *Drug*—**Iroquois** *Orthopedic Aid* Poultice of plant applied to sprained muscles and twisted joints. *Poison* Roots eaten to commit suicide. (87:400)

Annonaceae, see *Annona, Asimina*

Annona, Annonaceae
Annona glabra L., Pond Apple
- *Food*—**Seminole** *Unspecified* Plant used for food. (169:509)
- *Other*—**Seminole** *Cleaning Agent* Plant used to make lye. *Cooking Tools* Plant used to make spoons. (169:509)

Annona reticulata L., Custard Apple
- *Drug*—**Seminole** *Kidney Aid* Infusion of flowers taken for kidney disorders. (169:274)
- *Food*—**Seminole** *Unspecified* Plant used for food. (169:495)

Antennaria, Asteraceae
Antennaria anaphaloides Rydb., Pearly Pussytoes
- *Drug*—**Paiute** *Dermatological Aid* Plant served as a perfume and placed with clothing or handkerchiefs. (111:116)

Antennaria dioica (L.) Gaertn., Stoloniferous Pussytoes
- *Drug*—**Gosiute** *Eye Medicine* Poultice of steeped plant applied to the eyes for snow-blindness. (39:361)

Antennaria howellii Greene, Howell's Pussytoes
- *Drug*—**Bella Coola** *Analgesic* Decoction of leaves taken for body pain, but not pain in the limbs. (150:65)

Antennaria howellii ssp. *neodioica* (Greene) Bayer, Field Pussytoes
- *Drug*—**Ojibwa** *Gynecological Aid* Infusion of herb taken after childbirth to purge afterbirth and to heal. (as *A. neodioica* 153:363)

Antennaria parvifolia Nutt., Small-leaf Pussytoes
- *Drug*—**Lakota** *Antirheumatic (External)* Used for swellings. (139:35) **Navajo, Kayenta** *Blood Medicine* Plant chewed with deer or sheep tallow as a blood purifier. (as *A. aprica* 205:44) **Navajo, Ramah** *Ceremonial Medicine* Plant used ceremonially for mad coyote bite. *Witchcraft Medicine* Cold infusion of root taken for protection from witches. (as *A. aprica* 191:47)
- *Food*—**Navajo, Kayenta** *Vegetable* Used for greens in foods. (as *A. aprica* 205:44)

Antennaria plantaginifolia (L.) Richards., Woman's Tobacco
- *Drug*—**Cherokee** *Gastrointestinal Aid* Infusion of entire plant given, especially to children, for "bowel complaint." *Gynecological Aid* Infusion taken for excessive discharge in monthly period. *Pediatric Aid* Infusion of entire plant given, especially to children, for "bowel complaint." (80:50) **Iroquois** *Gynecological Aid* Infusion of roots taken for leukorrhea. *Toothache Remedy* Decoction of plant used as a mouthwash for toothaches. (87:464) **Meskwaki** *Gynecological Aid* Infusion of leaves taken after childbirth to prevent sickness. (152:210)

Antennaria rosea Greene, Rosy Pussytoes
- *Drug*—**Okanagan-Colville** *Ceremonial Medicine* Roots dried, powdered, put into hot coals at winter dance, and smoke used to drive away bad spirits and to revive passed-out dancers. *Reproductive Aid* Leaves chewed and swallowed to increase male virility. (188:75)
- *Food*—**Blackfoot** *Candy* Leaves chewed by children for the flavor. (97:56)
- *Other*—**Blackfoot** *Smoke Plant* Leaves sometimes used in the tobacco mixture. (97:56) **Great Basin Indian** *Smoke Plant* Tiny, dried leaves used as an element of "kinnikinnick" smoking mixture. (121:50)

Antennaria rosulata Rydb., Kaibab Pussytoes
- *Drug*—**Navajo, Ramah** *Hunting Medicine* Plants from where deer have slept or browsed used for good luck in hunting. *Pediatric Aid* Compound decoction taken for birth injury. *Witchcraft Medicine* Cold infusion of root taken for protection from witches. (191:47)

Antennaria sp., Pussytoes
- *Drug*—**Natchez** *Cold Remedy* Warm infusion of tops and roots taken for colds. (172:668) Infusion of roots and tops taken for colds. (177:60) *Cough Medicine* Warm infusion of tops and roots taken for coughs. (172:668) Infusion of roots and tops taken for coughs. (177:60) **Thompson** *Cold Remedy* and *Cough Medicine* Some parts or whole plant chewed and swallowed for coughs. (164:458)

Anthemis, Asteraceae
Anthemis cotula L., Stinking Chamomile
- *Drug*—**Cherokee** *Analgesic* Used as a "sudorific and anodyne for colds." *Anticonvulsive* Used for epilepsy. *Antirheumatic (Internal)* Used for rheumatism. *Dermatological Aid* Bruised herb applied externally to "draw blister." *Diaphoretic* Used as a "sudorific and anodyne for colds." *Emetic* Used as an emetic. *Febrifuge* Used for fevers. *Kidney Aid* Used for "dropsy." *Respiratory Aid* Used for asthma. *Sedative* Used for hysterics. *Tonic* Used as a tonic. (80:32) **Iroquois** *Antidiarrheal* Compound decoction of plants, bark, and roots taken for diarrhea. *Antiemetic* Decoction or cold infusion of plants, bark, and roots taken for vomiting. (87:471) *Blood Medicine* Compound decoc-

tion of bark, plants, and roots taken as blood purifier. *Emetic* Cold infusion of stalks taken as an emetic when not feeling well and for spring fever. *Febrifuge* Cold infusion of dried roots and stems taken as an emetic for spring fever. (87:472) *Gastrointestinal Aid* Compound decoction of plants taken for too much gall and biliousness. Compound decoction of plants, bark, and roots taken for stomach cramps. (87:471) Cold infusion of stalks taken for ptomaine poisoning. Decoction of plants taken for biliousness. (87:472) *Other* Decoction of plants given to children with "red spots." (87:471) Decoction of plants given to children and adults for "summer complaint." (87:472) *Pediatric Aid* Decoction of plants given to children with "red spots." (87:471) Decoction of plant given to children for "summer complaint" and stomach cramps. (87:472) *Pulmonary Aid* Compound decoction of plants taken for shortness of breath. (87:471) *Sedative* Cold infusion of dried roots and stems taken as a sedative. (87:472) *Toothache Remedy* Root chewed for toothaches. (87:471) *Venereal Aid* Compound decoction of bark, plants, and roots taken for venereal disease. (87:472) **Karok** *Gynecological Aid* Plant used by pregnant women. (148:390) **Mendocino Indian** *Antirheumatic (External)* Infusion of plants used as a wash for rheumatism. *Cold Remedy* Infusion of plants used as a wash for severe colds. *Eye Medicine* Plant juice used as an eyewash. (41:392) **Mohegan** *Febrifuge* Cold infusion of plant taken for fever. (174:264) Cold infusion of leaves taken for fever. (176:70, 128) *Panacea* Cold infusion of leaves thought to "benefit the entire body." (176:70) **Yuki** *Poison* Plant considered poisonous. (49:94)

Anthemis sp.
- **Drug**—**Cree** *Kidney Aid* Plant and tarweed used for kidney pains. (camomile 16:494)
- **Food**—**Navajo** *Beverage* Fresh or dried plant used to make tea. (as *Cota*, Navajo tea 110:20)

Anthoxanthum, Poaceae
Anthoxanthum odoratum L., Sweet Vernalgrass
- **Food**—**Hesquiat** *Forage* Cattle used this plant for forage. (185:56)
- **Fiber**—**Abnaki** *Basketry* Used to make baskets. (144:175) **Ojibwa** *Basketry* Grass used to make baskets. (153:419) **Potawatomi** *Basketry* Used to make baskets. *Sewing Material* Used to sew buckskin when making moccasins and articles of clothing. (154:120)
- **Other**—**Abnaki** *Containers* Used to make containers. (144:156) **Ojibwa** *Ceremonial Items* In the olden times, used ceremonially because of its persistent sweet scent. (153:419)

Antidesma, Euphorbiaceae
Antidesma pulvinatum Hbd., Hame
- **Drug**—**Hawaiian** *Antiemetic* Leaves chewed and swallowed for vomiting spells. *Dermatological Aid* Infusion of pounded bark and other plants used as a wash for ulcers and scrofulous sores. *Tuberculosis Remedy* Infusion of pounded bark and other plants used as a wash for scrofulous sores. (2:39)

Antitrichia, Leucodontaceae, moss
Antitrichia curtipendula (Hedw.) Brid., Hanging Moss
- **Other**—**Hanaksiala** *Cooking Tools* Plant used in earth ovens. (43:145)

Apiaceae, see *Aletes, Angelica, Apiastrum, Apium, Berula, Carum, Chaerophyllum, Cicuta, Conioselinum, Conium, Coriandrum, Cuminum, Cymopterus, Daucus, Erigenia, Eryngium, Ferula, Foeniculum, Heracleum, Hydrocotyle, Ligusticum, Lomatium, Musineon, Oenanthe, Oreoxis, Osmorhiza, Oxypolis, Pastinaca, Perideridia, Petroselinum, Peucedanum, Pimpinella, Polytaenia, Pseudocymopterus, Pteryxia, Sanicula, Sium, Sphenosciadium, Taenidia, Tauschia, Thaspium, Zizia*

Apiaceae sp.
- **Food**—**Yana** *Unspecified* Roots roasted and eaten. (as Umbelliferae 147:251)

Apiastrum, Apiaceae
Apiastrum angustifolium Nutt., Mock Parsley
- **Food**—**Cahuilla** *Unspecified* Hair-like plant provided a small seasonal food source in wet years. (15:39)

Apios, Fabaceae
Apios americana Medik., Groundnut
- **Food**—**Cherokee** *Substitution Food* Uncooked seeds substituted for pinto beans in bean bread. (as *Glycine apios* 126:46) *Vegetable* Beans used for food. (80:24) Roots cooked like potatoes. (as *Glycine apios* 126:46) **Chippewa** *Vegetable* Tubers eaten. (as *A. tuberosa* 71:133) **Dakota** *Unspecified* Roasted or boiled tubers used for food. (as *Glycine apios* 70:94) **Delaware** *Bread & Cake* Roots dried, ground into flour, and made into bread. *Unspecified* Roots boiled and eaten as the cultivated potato. *Winter Use Food* Tuberous roots used as winter food. (as *A. tuberosa* 176:59) **Huron** *Starvation Food* Roots used with acorns during famine. (as *A. tuberosa* 3:63) **Iroquois** *Unspecified* Tubers eaten. (as *A. uberosa* 196:120) **Menominee** *Vegetable* Roots cooked with maple sugar and superior to candied yams. *Winter Use Food* Peeled, parboiled, sliced roots dried for winter use. (as *A. tuberosa* 151:68) **Meskwaki** *Vegetable* Root stocks eaten raw. *Winter Use Food* Root stocks peeled, parboiled, sliced, and dried for winter use. (as *A. tuberosa* 152:259) **Mohegan** *Cooking Agent* Dried roots ground into a flour and used for thickening stews. *Unspecified* Fresh or dried roots cooked and used for food. (as *A. tuberosa* 176:83) **Omaha** *Unspecified* Thickened root boiled until the skin came off and used for food. (as *A. apios* 68:325) Roasted or boiled tubers used for food. (as *Glycine apios* 70:94) *Vegetable* Nuts boiled, peeled, and eaten as a vegetable. (as *A. tuberosa* 58:341) **Pawnee** *Unspecified* Roasted or boiled tubers used for food. **Ponca** *Unspecified* Roasted or boiled tubers used for food. (as *Glycine apios* 70:94) **Potawatomi** *Vegetable* Wild potato was appreciated. (as *A. tuberosa* 54:103) **Seminole** *Unspecified* Plant used for food. (169:492) **Winnebago** *Unspecified* Roasted or boiled tubers used for food. (as *Glycine apios* 70:94)

Apios tuberosum, Potato Bean
- **Food**—**Cheyenne** *Unspecified* Species used for food. (as *Alycine apios* 83:45) *Vegetable* Roots used for food. (as *Alycine apios* 76:179)

Apium, Apiaceae
Apium graveolens L., Wild Celery
- **Drug**—**Houma** *Tuberculosis Remedy* Compound decoction of plant with whisky taken for tuberculosis. (158:64)

- *Food*—**Cahuilla** *Vegetable* Used as a potherb. (15:39) **Luiseño** *Vegetable* Plant used for greens. (155:230)

Apium sp.
- *Drug*—**Mahuna** *Kidney Aid* Infusion of plant taken for chronic diseases of the kidneys. (American parsley 140:68)
- *Food*—**Shoshoni** *Spice* Steeped seeds added to dishes for flavoring. (wild celery 118:29)

Aplectrum, Orchidaceae
Aplectrum hyemale (Muhl. ex Willd.) Torr., Adam and Eve
- *Drug*—**Catawba** *Analgesic* Pounded, powdered, boiled roots used for head pains. *Dermatological Aid* Pounded, powdered, boiled roots used for boils. (157:188) Poultice of beaten roots applied to boils. (177:10) **Cherokee** *Dietary Aid* Given to endow children with the gift of eloquence and to make them fat. *Pediatric Aid* Given to endow children with the gift of eloquence and to make them fat. (80:51)
- *Food*—**Cherokee** *Fodder* Roots added to the slop to make hogs fat. (80:51)

Apocynaceae, see *Alyxia, Amsonia, Angadenia, Apocynum, Catharanthus, Ochrosia*

Apocynum, Apocynaceae
Apocynum androsaemifolium L., Spreading Dogbane
- *Drug*—**Cherokee** *Veterinary Aid* Used to bathe dogs for mange. (80:32) **Chippewa** *Analgesic* Root used as snuff, herbal steam, poultice or in decoction for headache. *Anticonvulsive* Compound decoction of root taken or sprinkled on chest for convulsions. (53:336) *Cold Remedy* Weak decoction of root given only to infants for colds. (53:340) *Ear Medicine* Decoction of root poured into ear for soreness. (53:360) *Heart Medicine* Decoction of root taken for heart palpitations. (53:338) *Hemostat* Decoction of root on cotton or mashed root used as a plug in nostril for nosebleed. (53:356) *Pediatric Aid* Weak decoction of root given only to infants for colds. (53:340) *Psychological Aid* Dried, pulverized root used in various ways for insanity. *Vertigo Medicine* Dried, pulverized root used in various ways for dizziness. (53:336) **Cree, Woodlands** *Eye Medicine* Plant used for sore eyes. *Gynecological Aid* Decoction of plant used to increase lactation. (109:28) **Iroquois** *Anthelmintic* Compound infusion of roots taken for worms. *Dermatological Aid* Milk used for warts. *Gastrointestinal Aid* Compound infusion of roots taken for stomach cramps. *Gynecological Aid* Compound infusion of roots taken for evacuation of the placenta. *Liver Aid* Decoction of roots taken as a liver medicine. *Veterinary Aid* Decoction of roots mixed with feed and given to horses with worms. (87:415) **Meskwaki** *Gynecological Aid* Compound containing rind used by a woman with "an injured womb." *Kidney Aid* Root used for dropsy. (152:201) **Montana Indian** *Cathartic* Root, poisonous in large doses, used as a cathartic. *Febrifuge* Root, poisonous in large doses, used as a febrifuge. *Poison* Root poisonous in large doses and poisonous to cattle feeding on it. *Tonic* Root, poisonous in large doses, used as a tonic. (19:6) **Ojibwa** *Analgesic* Root smoke inhaled for headache. *Ceremonial Medicine* Root, considered sacred, eaten during the medicine lodge ceremony. (153:354, 355) Roots eaten during the medicine lodge ceremony. The roots are also chewed to keep the other witch doctors from affecting one with an evil charm. (153:428) *Diuretic* Infusion of root taken as a diuretic during pregnancy. *Gynecological Aid* Infusion of root taken as a diuretic during pregnancy. *Oral Aid* Root used for coated tongue and headache. *Throat Aid* Root eaten for throat trouble. (153:354, 355) **Okanagan-Colville** *Love Medicine* Leaves chewed and the juice and pulp swallowed or dried leaves smoked as an aphrodisiac. (188:72) **Potawatomi** *Diuretic* Root used as a diuretic. *Heart Medicine* Decoction of green berries used as a heart medicine. *Kidney Aid* Decoction of green berries used as a kidney medicine. *Urinary Aid* Root used as a diuretic and urinary medicine. (154:38) **Salish** *Unspecified* Plant used as a medicine. (178:294) **Sanpoil** *Gynecological Aid* Infusion of roots taken about once a week as a contraceptive. (131:219)
- *Fiber*—**Bella Coola** *Cordage* Stems dried, pounded, and used to make twine. (184:201) **Great Basin Indian** *Cordage* Root and branch outer fiber used to make nets, cordage, and thread. (121:49) **Menominee** *Cordage* Three strands of outer bark plaited into a very strong cord and cord plaited into heavier ropes. *Sewing Material* Outer bark or rind used as the finest thread material. (151:73) **Meskwaki** *Sewing Material* Outer rind or bark used for thread. (152:267) **Montana Indian** *Cordage* Bark used as a chief source for cordage. (19:6) **Ojibwa** *Sewing Material* Outer rind used for fine sewing. In the fall, when mature, this plant makes one of the strongest native fibers, stronger even than the cultivated hemp to which it is related. (153:413) **Okanagan-Colville** *Cordage* Stems used to make fiber, as a substitute for Indian hemp. (188:72) **Potawatomi** *Sewing Material* Fine divisions of bark were very strong and used as a thread for sewing on the fine beadwork. (154:111) **Thompson** *Cordage* Inner bark fiber used as thread and twine for binding and tying. (164:497)
- *Other*—**Bella Coola** *Hunting & Fishing Item* Stems dried, pounded, and used to make eulachon (candlefish) nets. (184:201) **Chippewa** *Protection* Root chewed to counteract evil charms. (53:376) **Menominee** *Hunting & Fishing Item* Three strands of outer bark plaited into a very strong cord and used for bow strings. (151:73) Plant stalk sucked by hunters to imitate fawn wanting its mother, a doe magnet. (151:79) **Okanagon** *Fasteners* Fiber used as thread or twine for binding or tying. **Thompson** *Fasteners* Fiber used as thread or twine for binding or tying. (125:39)

Apocynum cannabinum L., Indian Hemp
- *Drug*—**Blackfoot** *Dermatological Aid* Decoction of root used as a wash "to prevent hair falling out." *Laxative* Decoction of root taken as a laxative. (114:276) **Cahuilla** *Unspecified* Used for the medicinal properties and as a fibrous material. (15:39) **Cherokee** *Abortifacient* Root used for pox and "uterine obstructions." *Antirheumatic (Internal)* Used for rheumatism. *Cough Medicine* Used for coughs. *Dermatological Aid* Root used for pox. *Kidney Aid* Infusion of root taken for "dropsy" and Bright's disease. *Pulmonary Aid* Used for whooping cough. *Respiratory Aid* Used for asthma. (80:38) **Cree, Hudson Bay** *Cathartic* Decoction of chewed leaves and bark taken as a purgative. *Dermatological Aid* Poultice of chewed leaves and bark applied to wounds. *Emetic* Decoction of chewed leaves and bark taken as an emetic. *Unspecified* Plant used as medicine. (as *A. hypericifolium* 92:303) **Iroquois** *Antidiarrheal* Infusion of roots used as a wash for children with diarrhea. *Blood Medicine* Roots used as blood purifier. (87:415) *Emetic* Infusion of roots taken as a spring or summer emet-

ic. *Eye Medicine* Infusion of roots taken to clear up yellow eyes. (87: 416) *Gastrointestinal Aid* Roots used for biliousness. *Laxative* Roots used as a laxative. *Pediatric Aid* Infusion of roots used as a wash for children with diarrhea. (87:415) **Keres, Western** *Gynecological Aid* Crushed leaves rubbed on mothers' breasts to produce more and richer milk. Infusion of plant used by mothers to produce more and richer milk. (as *A. viride* 171:27) **Kutenai** *Veterinary Aid* Tops chewed and used for horses with eyes. (82:12) **Menominee** *Anthelmintic* Decoction of root taken for worms. (as *A. pubescens* 54:131) **Meskwaki** *Kidney Aid*, *Misc. Disease Remedy*, and *Panacea* Root used as a universal remedy for many things, especially dropsy and ague. (152:201) **Micmac** *Anthelmintic* Root used as a vermifuge. (40:54) **Navajo, Kayenta** *Ceremonial Medicine* Plant used as a Waterway emetic. *Emetic* Plant used as a Waterway emetic. *Other* Plant used for immersion in cold water. (as *A. suksdorfii* var. *angustifolium* 205:36) **Navajo, Ramah** *Analgesic* Decoction of plant taken for persistent stomachache. *Ceremonial Medicine* Leaves used as a ceremonial emetic and cold infusion of leaves used as a ceremonial lotion. *Emetic* Leaves used as a ceremonial emetic. *Gastrointestinal Aid* Decoction of plant taken for persistent stomachache. (as *A. sibiricum* var. *salignum* 191: 39) **Okanagan-Colville** *Contraceptive* Decoction of roots taken during monthly periods to become permanently sterile. (188:72) **Penobscot** *Anthelmintic* Infusion of root taken to expel worms. (156:310) **Thompson** *Unspecified* Decoction of root used medicinally. (164:470) *Venereal Aid* Milky stem latex used for venereal disease. (187:159)

- *Food*—**Isleta** *Candy* Gummy latex mixed with clean clay and used as chewing gum. (as *A. angustifolium* 32:31) Gum mixed with clean clay and used for chewing gum. (as *A. angustifolium* 100:22) **Karok** *Unspecified* Seeds eaten raw. (as *A. androsaemifolium* var. *nevadense* 148:388) **Kiowa** *Candy* Milky latex used as chewing gum. After the latex was squeezed from the plant, it was allowed to stand overnight, whereupon it hardened into a "white gum." Two kinds of gum were recognized: that which was left overnight, and that which was chewed only a few hours after it had been extracted from the plant. (192:47)
- *Fiber*—**Cahuilla** *Unspecified* Used for the medicinal properties and as a fibrous material. (15:39) **California Indian** *Cordage* Outer and inner bark used to make string. (118:60) **Cherokee** *Clothing* Fibers used to weave grave cloth material. *Cordage* Used to make cords. (80:38) **Chippewa** *Cordage* Fiber considered the best available for making fine cordage. (71:140) **Kutenai** *Cordage* Used to make twine and rope. *Sewing Material* Stalks split, dried, and used as thread for sewing tepee covers. (82:12) **Luiseño** *Clothing* Plant used to make aprons. (155:201) Bark fiber made into twine and used to make front aprons worn by women. *Cordage* Inner bark fiber soaked in boiling water and used to make twine. (155:202) **Mendocino Indian** *Clothing* Inner bark formerly used to make garments. *Cordage* Inner bark fiber used to make rope. *Sewing Material* Inner bark used for making thread. (41:378) **Meskwaki** *Cordage* Outer bast fiber plaited into heavy cord and two-ply cord. *Sewing Material* Outer bast fiber used as thread. (152:267) **Modoc** *Unspecified* Used as a fiber. (45:103) **Nez Perce** *Cordage* Stalks dried, split into fibers, and used to make rope. (82:12) **Okanagan-Colville** *Cordage* Stems twisted and rolled into twine. (188:72) **Okanagon** *Clothing* Inner bark used for making garments. *Cordage* Inner bark used for making rope and twine. *Sewing Material* Inner bark used for making thread. (125:39) **Paiute, Northern** *Cordage* Stem fibers twisted and plied into cordage and nets. (60: 75) *Mats, Rugs & Bedding* Used as twined weft rows for mats. (60:89) **Shuswap** *Cordage* Stems used to make bridle ropes, bowstrings, and thread for sewing baskets and buckskin. (123:57) **Thompson** *Clothing* Inner bark used for making garments. (125:39) Inner bark used for making garments. (164:498) *Cordage* Inner bark used for making rope and twine. (125:39) Inner bark used for making rope and twine. (164:498) Stems used to make string. The stems were cut in the fall, usually in October, soaked and sometimes split in half. The fibrous outer skin was peeled off and the brittle inner stem discarded. The fibrous part was then dried for indefinite storage and used to make string. Plant made into rope and used to make fish nets. (187:159) *Sewing Material* Inner bark used for making thread. (125:39) Inner bark used for making thread. (164:498) Plant made into thread and used for sewing. (187:159)
- *Other*—**Luiseño** *Containers* Bark fibers made into twine and used to make large-meshed nets for carrying bulky or heavy articles. Bark fiber made into twine and used to make net sacks for carrying acorns and other small seeds. *Hunting & Fishing Item* Inner bark fiber made into twine and used to make bowstrings. Bark fiber made into twine and occasionally used to make long nets and draw nets for catching rabbits. Bark fiber made into twine and used to make fishing nets.[14] Bark fiber made into twine and used to make slings. (155:202) Inner bark fibers made into twine and used to make bowstrings. (155:203) Plant fiber used to make bowstrings. (155:206) **Mendocino Indian** *Hunting & Fishing Item* Inner bark used to make nets. (41:378) **Nez Perce** *Containers* Used to make bags. (82:12) **Okanagon** *Cash Crop* Inner bark traded in large quantities to the Spences Bridge band. (164:498) *Hunting & Fishing Item* Inner bark used for making nets and snares. (125:39) **Paiute, Northern** *Hunting & Fishing Item* Stem fibers twisted and plied into cordage and nets. (60:75) **Shuswap** *Stable Gear* Stems used to make bridle ropes, bowstrings, and thread for sewing baskets and buckskin. (123:57) **Songish** *Hunting & Fishing Item* Used to make purse nets. (182:78) **Thompson** *Containers* Plant fiber made into sacks and used for storing potatoes, oats, or onions. (187:159) *Hunting & Fishing Item* Inner bark used for making nets and snares. (125:39) Inner bark used for making nets and snares. (164: 498) Used to make nets for catching deer. (178:246) Plants made into rope or thread and used for sewing and to make fish nets. The plants were dried, beaten with a stick to soften and loosen the fiber, and then rolled and twisted on a piece of buckskin covering the upper leg. The resulting rope or thread was used for sewing, for rope, and for fish nets. (187:159) **Yuki** *Hunting & Fishing Item* Dried, crushed stem fibers used to make fish nets and snares for deer, bears, and small game. (49:90)

Apocynum ×*floribundum* Greene, Intermediate Dogbane

- *Drug*—**Navajo, Ramah** *Ceremonial Medicine* Leaves used as a ceremonial emetic. *Disinfectant* Plant placed on hot rocks and applied to patient's head for "deer infection." *Emetic* Leaves used as a ceremonial emetic. (as *A. medium* 191:39)
- *Fiber*—**Havasupai** *Building Material* Stems used for thatch on houses. *Clothing* Bark strips braided and worn as a belt. *Cordage* Vines, with leaves removed, twisted into a rope and used by children in play. (as *A. medium* 197:236)

- *Other*—Havasupai *Toys & Games* Milky substance used by children to rub on playmates' faces and stick their eyelids together. Vines, with leaves removed, twisted into a rope and used by children in play. (as *A. medium* 197:236)

Apocynum sp., Dogbane
- *Drug*—Chippewa *Analgesic* and *Other* Snuff of dried, pulverized root used to cause sneezing for "relieving the head." (53:340)
- *Fiber*—Abnaki *Basketry* Used to make baskets. (144:156) **Shoshoni** *Cordage* Plant used to make string. (118:52) **Wintoon** *Cordage* Used to make the best string and thread. (117:276)

Aquifoliaceae, see *Ilex, Nemopanthus*

Aquilegia, Ranunculaceae
Aquilegia canadensis L., Red Columbine
- *Drug*—Cherokee *Gynecological Aid* Cold infusion used for "flux." *Heart Medicine* Infusion taken for heart trouble. (80:30) **Iroquois** *Dermatological Aid* Compound infusion of plants taken and used as a wash for poison ivy and itch. *Kidney Aid* Infusion of roots taken before meals for the kidneys. *Witchcraft Medicine* Compound used to detect bewitchment. (87:320) **Meskwaki** *Antidiarrheal* Decoction of root and leaves taken for diarrhea. *Ceremonial Medicine* Decoction of root and leaf used as a "power of persuasion at trade or council." *Gastrointestinal Aid* Root chewed for stomach and bowel troubles. *Love Medicine* Seeds used with love medicine and for smoking. *Urinary Aid* Compound containing root taken "when the contents of the bladder are thick." (152:238, 239) **Ojibwa** *Gastrointestinal Aid* Root considered a good medicine for stomach trouble. (153:383) **Omaha** *Analgesic* Infusion of crushed seeds taken for headache. *Febrifuge* Infusion of crushed seeds taken for fever. *Love Medicine* Pulverized seeds used as a love charm. **Pawnee** *Analgesic* Infusion of crushed seeds taken for headache. *Febrifuge* Infusion of crushed seeds taken for fever. *Love Medicine* Seeds used as a love charm. **Ponca** *Analgesic* Infusion of crushed seeds taken for headache. *Febrifuge* Infusion of crushed seeds taken for fever. *Love Medicine* Pulverized seeds used as a love charm. (70:82, 83)
- *Other*—Meskwaki *Incense & Fragrance* Ripe seeds used to perfume smoking tobacco. (152:273) **Omaha** *Incense & Fragrance* Chewed seed paste used among blankets or other effects by young men as perfume. (68:323) Chewed seed paste spread among clothing and used as perfume, especially by bachelors. **Pawnee** *Incense & Fragrance* Seeds used as perfume. **Ponca** *Incense & Fragrance* Crushed seeds spread among clothing and used as perfume, especially by bachelors. (70:82)

Aquilegia coerulea James, Colorado Blue Columbine
- *Drug*—Gosiute *Analgesic* and *Gastrointestinal Aid* Seed chewed or infusion of roots used for abdominal pains. *Heart Medicine* Plant used as a medicine for the heart. *Panacea* Seed chewed or infusion of roots used when "sick all over." (39:362)

Aquilegia elegantula Greene, Western Red Columbine
- *Drug*—Keres, Western *Blood Medicine* Infusion of plant used as a blood purifier. (171:27)

Aquilegia eximia Van Houtte ex Planch., Van Houtte's Columbine
- *Other*—Pomo, Kashaya *Ceremonial Items* Flowers used in dance wreaths at the Strawberry Festival. (72:39)

Aquilegia formosa Fisch. ex DC., Western Columbine
- *Drug*—Paiute *Analgesic* and *Antirheumatic (External)* Mashed fresh roots rubbed briskly on aching rheumatic joints. (180:37) *Cold Remedy* Decoction of leaves taken for colds. *Cough Medicine* Leaves chewed for coughs. (104:197) Decoction of roots taken as a cough remedy. (180:37) *Dermatological Aid* Poultice of chewed roots or leaves applied to bee stings. (104:196) Chewed seeds rubbed on body and clothing for perfume, and seeds used in a sachet and stored with clothing. (111:71) *Gastrointestinal Aid* Seeds chewed for stomachaches. (104:197) *Panacea* Plant used for a variety of maladies. (104:155) *Throat Aid* Leaves chewed for sore throats. (104:197) **Quileute** *Dermatological Aid* Poultice of chewed leaves or milky pulp from scraped roots applied to sores. (79:30) **Shoshoni** *Analgesic* Decoction of roots taken for stomachaches. *Antidiarrheal* Decoction of roots taken for diarrhea. *Dermatological Aid* Mashed ripe seeds rubbed into hair "to discourage head lice." *Emetic* Compound decoction of root taken to induce vomiting. *Gastrointestinal Aid* Decoction of roots and leaves taken for dizziness or biliousness. Decoction of roots taken for stomachaches. *Stimulant* Decoction of roots and leaves taken to counteract dizziness. *Venereal Aid* Decoction of whole plant taken for venereal diseases. (180:37) **Thompson** *Dermatological Aid* Decoction of whole plant used as a wash for the hair and scalp. (164:475) *Love Medicine* Plant used as a charm by women "to gain the affection of men." (164:507) *Strengthener* Root smeared on people's legs to increase stamina before a race. *Veterinary Aid* Root smeared on horse's legs to increase stamina before a race. (187:247)
- *Food*—Hanaksiala *Candy* Flowers sucked by children for the sweet nectar. (43:262) **Yurok** *Unspecified* Sweet nectaries inside the sepal spurs bitten off and savored mostly by the younger people. (6:17)
- *Other*—Okanagan-Colville *Good Luck Charm* Flower used as a good luck charm. (188:117) **Thompson** *Good Luck Charm* Plant used as a charm by both sexes "to retain wealth and possessions." Plant used as a charm for good luck in gambling. (164:507) Whole plant kept by people as a good luck charm. (187:247)

Aquilegia formosa Fisch. ex DC. **var. *formosa***, Crimson Columbine
- *Food*—Miwok *Vegetable* Early spring greens boiled and eaten. (as *A. truncata* 12:159) **Thompson** *Forage* Flowers used as sources of nectar by hummingbirds. (as *A. truncata* 164:516)

Aquilegia micrantha Eastw., Mancos Columbine
- *Drug*—Navajo, Kayenta *Gynecological Aid* Plant used to deliver placenta. *Hemostat* Plant used as a hemostatic. (205:22)

Aquilegia **sp.**, Garden Columbine
- *Drug*—Thompson *Strengthener* Root smeared on people's legs to increase stamina before a race. *Veterinary Aid* Root smeared on horse's legs to increase stamina before a race. (187:247)

Aquilegia triternata Payson, Chiricahua Mountain Columbine
- *Drug*—Navajo, Kayenta *Analgesic* and *Ceremonial Medicine* Plant used as a ceremonial fumigant for headaches or other severe pain. (205:22)

Arabis, Brassicaceae

Arabis drummondii Gray, Drummond's Rockcress
- *Drug*—**Okanagon** *Analgesic* Decoction of whole plant taken for pains in the lumbar region. *Kidney Aid* Decoction of whole plant taken for kidney troubles. *Orthopedic Aid* Decoction of whole plant taken for pains in the lumbar region. *Urinary Aid* Decoction of whole plant taken for bladder troubles. (125:41) **Salish** *Venereal Aid* Decoction of plants taken for gonorrhea. (178:294) **Thompson** *Analgesic* Decoction of whole plant taken for pains in the lumbar region. (125:41) Decoction of whole plant taken as a diuretic and for lower back pains. *Dermatological Aid* Poultice of fresh or dried plant applied to sores. *Diuretic* Decoction of whole plant taken as a diuretic. (164:464) *Kidney Aid* Decoction of whole plant taken for kidney troubles. *Orthopedic Aid* Decoction of whole plant taken for pains in the lumbar region. *Urinary Aid* Decoction of whole plant taken for bladder troubles. (125:41) Decoction of whole plant taken as a kidney and bladder medicine. *Venereal Aid* Strong decoction of plant taken for gonorrhea. (164:464)

Arabis fendleri (S. Wats.) Greene, Fendler's Rockcress
- *Drug*—**Keres, Western** *Gastrointestinal Aid* Infusion of plant used as a stomach medicine. (171:27) **Navajo, Ramah** *Panacea* Whole plant used as "life medicine." (191:28)

Arabis glabra (L.) Bernh., Tower Rockcress
- *Drug*—**Cheyenne** *Cold Remedy* Plant used for colds. Infusion of plant taken to check a cold when it first appears. *Panacea* Infusion of plant taken as a general preventative for sickness. *Pediatric Aid* Infusion of plant given to children as a general preventative for sickness, when sickness is about. (76:174)
- *Food*—**Cheyenne** *Beverage* Infusion of plant used as a beverage. (76:174)

Arabis holboellii Hornem., Holboell's Rockcress
- *Drug*—**Thompson** *Toothache Remedy* Leaves chewed as a strong medicine for toothache. (187:193)
- *Other*—**Navajo** *Ceremonial Items* Plant used in the Night Chant Ceremony. (55:48)

Arabis lyrata L., Lyrate Rockcress
- *Food*—**Alaska Native** *Vegetable* Rosettes of lobed leaves added to tossed salads or cooked and served as a green vegetable. (85:13)

Arabis perennans S. Wats., Perennial Rockcress
- *Drug*—**Navajo, Kayenta** *Anticonvulsive* Plant used for hiccups caused by dry throat. *Psychological Aid* Plant used for effects of a bad dream. (205:23) **Navajo, Ramah** *Analgesic* Cold infusion taken and used as lotion for general body pain. (191:28)

Arabis puberula Nutt., Silver Rockcress
- *Drug*—**Shoshoni** *Antirheumatic (External)* Crushed plant used as a liniment or mustard plaster. (180:37)

Arabis sparsiflora Nutt., Sicklepod Rockcress
- *Drug*—**Okanagan-Colville** *Antidiarrheal* Roots chewed and juice swallowed for diarrhea. *Contraceptive* Plant used for birth control. *Eye Medicine* Infusion of whole plant used as an eyewash for sore eyes. *Gastrointestinal Aid* Roots chewed and juice swallowed for heartburn. (188:91)

Araceae, see *Alocasia, Arisaema, Calla, Colocasia, Lysichiton, Peltandra, Symplocarpus, Xanthosoma*

Arachis, Fabaceae

Arachis hypogaea L., Peanut
- *Food*—**Huron** *Starvation Food* Roots used with acorns during famine. (3:63) **Seminole** *Unspecified* Plant used for food. (169:483)

Araliaceae, see *Aralia, Cheirodendron, Oplopanax, Panax*

Aralia, Araliaceae

Aralia californica S. Wats., California Spikenard
- *Drug*—**Karok** *Antirheumatic (External)* Decoction of roots used as a soak for arthritis. (6:17) **Mendocino Indian** *Cold Remedy* Decoction of dried roots taken for colds. *Febrifuge* Decoction of dried roots taken for fevers. *Gastrointestinal Aid* Decoction of dried roots taken for stomach diseases. *Pulmonary Aid* Decoction of dried roots taken for lung diseases. *Tuberculosis Remedy* Decoction of dried roots taken for consumption. (41:371) **Pomo** *Dermatological Aid* Decoction of roots used as a wash for sores and itching sores. *Panacea* Plant used as a medicine for various ailments. (66:14) **Pomo, Kashaya** *Dermatological Aid* Decoction of root applied externally for open sores and itching. (72:21)

Aralia hispida Vent., Bristly Sarsaparilla
- *Drug*—**Algonquin, Quebec** *Heart Medicine* Infusion of roots taken for heart disease. *Unspecified* Infusion of roots used as a medicinal tea. (18:205) **Potawatomi** *Alterative* Root used as an alterative. *Tonic* Root used as a tonic. (154:40)

Aralia nudicaulis L., Wild Sarsaparilla
- *Drug*—**Abnaki** *Blood Medicine* Used as a tonic to strengthen the blood. (144:154) Used to strengthen the blood. (144:170) **Algonquin, Quebec** *Kidney Aid* Infusion of roots given to children for kidney disorders. *Pediatric Aid* Infusion of roots given to children for kidney disorders. (18:205) **Algonquin, Tête-de-Boule** *Ear Medicine* Poultice of chewed roots applied to "sick" ears. (132:119) **Bella Coola** *Analgesic* Decoction of root taken for stomach pain. *Gastrointestinal Aid* Decoction of root taken for stomach pain. (150:61) Decoction of roots taken for stomach pains. (184:201) **Cherokee** *Blood Medicine* Infusion of root taken as a blood tonic. (80:53) **Chippewa** *Abortifacient* Simple or compound decoction of root taken for "stoppage of periods." (53:358) *Blood Medicine* Decoction of root taken for "humor in the blood." (53:340) Mashed root taken as a "remedy for the blood." *Dermatological Aid* Poultice of mashed, fresh root applied to sores. (53:350) *Hemostat* Chewed, fresh root, or dried, powdered root used for nosebleed. (53:356) *Veterinary Aid* Compound infusion of root applied to chest and legs of horse as a stimulant. (53:366) **Cree, Woodlands** *Dermatological Aid* Poultice of chewed roots applied to wounds to draw out the infection. *Gynecological Aid* Decoction of fruiting stalk used to stimulate lactation. *Oral Aid* Decoction of roots used to wash teething child's infected gums to prevent spread of infection. *Panacea* Powdered roots and many other herbs used for various ailments. *Pediatric Aid* Decoction of roots and other plants taken for teething sickness. Decoction of roots used to wash teething child's infected gums to prevent spread of infection. Decoction of

plant, excluding the fruit, used for childhood pneumonia. *Pulmonary Aid* Decoction of plant, excluding the fruit, used for childhood pneumonia. *Toothache Remedy* Decoction of roots and other plants taken for teething sickness. (109:29) **Delaware, Oklahoma** *Tonic* Compound containing root used as a tonic. (175:74) **Iroquois** *Blood Medicine* Infusion of roots taken as a blood medicine and blood purifier. (87:393) *Cancer Treatment* Plant used for cancer. *Cold Remedy* Plant used for colds. (87:394) *Cough Medicine* Compound infusion of roots taken as a consumption cough medicine. *Dermatological Aid* Compound decoction applied as poultice to cuts, sores, and ulcers on legs. Compound infusion of roots applied as salve on venereal disease skin cracks. (87:393) Compound infusion of roots taken for fever sores. Powdered root applied to split skin between the toes. (87:394) *Eye Medicine* Decoction of roots used as a wash for sore eyes. *Gastrointestinal Aid* Infusion of roots taken as a blood medicine or for upset stomach. (87:393) *Misc. Disease Remedy* Plant used for sugar diabetes. *Throat Aid* Compound applied as poultice for sore throat. (87:394) *Tuberculosis Remedy* Compound infusion of roots taken as a consumption cough medicine. *Venereal Aid* Compound infusion of roots applied as salve on venereal disease skin cracks. (87:393) **Kwakiutl** *Antihemorrhagic* Roasted, beaten, broken roots and grease taken for blood-spitting. *Cough Medicine* Roasted, beaten, broken roots and grease taken for coughing. (183:277) **Menominee** *Dermatological Aid* Compound decoction of root used for sores. *Pulmonary Aid* Compound decoction of root taken for "lung trouble." (54:130) **Meskwaki** *Burn Dressing* Poultice of pounded root applied to burns. *Dermatological Aid* Poultice of pounded root applied to sores. *Stimulant* Compound decoction of root "gives strength to one who is weak." (152:203) **Micmac** *Cough Medicine* Root used as a cough medicine. (40:54) **Mohegan** *Tonic* Complex compound infusion including sarsaparilla root taken as spring tonic. (174:266) Compound decoction of root taken as a spring tonic. (176:70, 128) **Montagnais** *Stimulant* Infusion of root taken for "weakness." *Tonic* Wine made from berries taken as a tonic. (as *A. medicalis* 156:315) **Montana Indian** *Cathartic* Root used as a cathartic. *Tonic* Root used as a tonic. (19:7) **Ojibwa** *Anticonvulsive* Infusion of leaves taken for fits. (135:231) *Blood Medicine* Infusion of plants taken as a blood medicine. (135:237) Infusion of leaves taken as a blood medicine. (135:231) *Dermatological Aid* Poultice of pounded root applied to boils and carbuncles. (153:356) *Hunting Medicine* Roots and sweet flag made into tea and used to soak gill nets before setting out to catch fish at night. (153:428) *Stimulant* Infusion of leaves taken for fainting. (135:231) **Okanagon** *Blood Medicine* Plant used as a blood purifier. *Dermatological Aid* Plant used for pimples. *Tonic* Plant used as a tonic. (125:42) **Penobscot** *Cough Medicine* Compound infusion of powdered root taken for coughs. (156:310) **Potawatomi** *Dermatological Aid* Poultice of pounded root applied to swellings and infections. *Disinfectant* Poultice of root applied to swellings and infections. (154:40, 41) **Thompson** *Blood Medicine* Plant used as a blood purifier. (125:42) Decoction of root taken "for the blood and pimples." (164:471) *Dermatological Aid* Plant used for pimples. (125:42) Decoction of root taken "for the blood and pimples." *Stimulant* Decoction of root taken for lassitude and general debility. (164:471) *Tonic* Plant used as a tonic. (125:42) Decoction of root taken as a tonic. (164:471)

- **Food**—**Algonquin, Quebec** *Beverage* Berries used to make wine. (18:115) **Bella Coola** *Beverage* Roots boiled and used as a beverage. (184:201) **Iroquois** *Beverage* Fruits used to make wine. (142:96) **Kwakiutl, Southern** *Unspecified* Roots roasted, broken into pieces, mixed with oulachen (candlefish) grease, and used for food. (183:277) **Micmac** *Beverage* Used to make a beverage. (159:258) **Montagnais** *Beverage* Dark berries fermented in cold water and used to make a wine. *Forage* Roots eaten by rabbits. (as *A. medicalis* 156:315)

Aralia racemosa L., American Spikenard

- **Drug**—**Algonquin, Quebec** *Misc. Disease Remedy* Infusion of roots and spurge taken for sugar diabetes. *Tuberculosis Remedy* Infusion of roots taken for tuberculosis. (18:204) **Cherokee** *Burn Dressing* Ooze of beaten roots used as wash for burns. *Cough Medicine* Taken for coughs. *Dermatological Aid* Astringent infusion taken for menstrual problems. Poultice of root ooze used on swellings, fresh wounds, and cuts. *Diaphoretic* Infusion of roots and berries taken as a diaphoretic. *Disinfectant* Infusion of roots and berries taken as an antiseptic. *Expectorant* Taken as an expectorant. *Gynecological Aid* Infusion taken for menstrual problems. *Orthopedic Aid* Taken for weak backs. *Pulmonary Aid* Taken for lung diseases. *Respiratory Aid* Taken for asthma. *Tonic* Infusion of roots and berries taken as a tonic. (80:57) **Chippewa** *Abortifacient* Compound decoction of root taken for "stoppage of periods." (53:358) *Cough Medicine* Decoction of root taken for cough. (53:340) *Dermatological Aid* Poultice of pounded root applied to "draw" and heal boils. (53:350) Poultice of roots applied to boils. (71:137) *Orthopedic Aid* Decoction of root or poultice of root applied to sprain or strained muscles. (53:362) Compound poultice of root or decoction of root applied to fractured bone. (53:366) **Choctaw** *Analgesic* Sweetened decoction of root given to children for "gripes, colic, etc." *Expectorant* Berries and root used as an expectorant. *Eye Medicine* Decoction of root used to steam sore eyes. *Gastrointestinal Aid* Sweetened decoction of root given to children for "gripes, colic, etc." *Pediatric Aid* Berries and root used for many children's complaints. Sweetened decoction of root given to children for "gripes, colic, etc." *Stimulant* Berries and root used as a stimulant. (28:287) **Iroquois** *Abortifacient* Plant used to promote menstruation when stopped by a cold. (87:393) *Anthelmintic* Chewed plant induced tapeworm to pass. (87:392) *Antidiarrheal* Infusion of smashed roots taken for diarrhea. *Antirheumatic (Internal)* Infusion of roots taken at night for rheumatism. (87:391) Compound used for rheumatism. *Blood Medicine* Compound decoction of roots and bark taken for watery blood and as purifier. *Cough Medicine* Decoction of roots taken for coughs. *Dermatological Aid* Decoction of roots used as wash and applied as poultice to deep cuts. *Gynecological Aid* Decoction of bark taken for prolapse of the uterus or fallen womb. (87:392) Compound decoction of roots taken for miscarriage. (87:393) *Kidney Aid* Compound decoction of roots and bark taken for dropsy or watery blood. *Liver Aid* Compound used for the liver. *Orthopedic Aid* Compound decoction of roots and bark taken for swellings on shins and calves. *Pulmonary Aid* Decoction of roots taken for whooping cough. *Strengthener* Cold infusion of roots taken for more strength. *Tonic* Compound decoction of roots taken by women as a tonic. *Tuberculosis Remedy* Decoction of roots taken for threatening consumption. *Urinary Aid* Compound decoction of bark taken

by old men with urinary problems. *Venereal Aid* Compound decoction of roots taken for venereal disease. (87:392) **Malecite** *Analgesic* Roots mixed with red osier dogwood and smoked for headaches. (116:248) Infusion of plant and snakeroot used by women with back and side pain. (116:257) *Cold Remedy* Infusion of roots used for head colds. (116:249) *Kidney Aid* Infusion of plant and snakeroot used for kidney trouble. (116:257) *Strengthener* Infusion of roots used for lassitude in spring. (116:248) *Tuberculosis Remedy* Infusion of roots used for tuberculosis. (116:251) *Venereal Aid* Infusion of plant used for gonorrhea. (116:257) **Menominee** *Analgesic* Root used to make a drink taken for stomachache. *Blood Medicine* Root used in cases of blood poisoning and as a poultice for sores. *Dermatological Aid* Poultice of root applied to sores and used for blood poisoning. *Gastrointestinal Aid* Root used to make a drink said to be good for stomachache. (151:24) **Meskwaki** *Adjuvant* Split root used as a seasoner for other medicines. *Gynecological Aid* Sprayed from the mouth upon women's heads, when they are giving birth. (152:203) **Micmac** *Analgesic* Root used for headaches and female pains. *Antihemorrhagic* Root used for spitting blood. *Cold Remedy* Root used for colds. (40:54, 55) Infusion of roots and angelica roots used for head colds. (116:259) *Cough Medicine* Root used for coughs. (40:54, 55) Infusion of roots and angelica roots used for coughs. (116:259) *Dermatological Aid* Root used for wounds. *Eye Medicine* Root used for sore eyes. *Gynecological Aid* Root used for female pains. *Kidney Aid* Root used for kidney troubles. *Stimulant* Root used for fatigue. *Throat Aid* Root used for sore throats. (40:54, 55) Infusion of roots and angelica roots used for sore throats. (116:259) *Tuberculosis Remedy* Root used for consumption. *Venereal Aid* Root used for gonorrhea. (40:54, 55) **Ojibwa** *Unspecified* Plant used for medicinal purposes. (135:237) **Penobscot** *Antihemorrhagic* Compound infusion of plant taken for "spitting up blood." *Kidney Aid* Compound infusion of plant taken for kidney trouble. *Tonic* Compound infusion of plant taken as a tonic. *Venereal Aid* Compound infusion of plant taken for gonorrhea. (156:311) **Potawatomi** *Dermatological Aid* Hot poultice of pounded root applied to inflammations. (154:41)
- *Food*—**Menominee** *Unspecified* An aboriginal Menomini dish was spikenard root, wild onion, wild gooseberry, and sugar. (151:62) **Potawatomi** *Soup* Young tips were relished in soups. Soup was a favorite aboriginal dish and still is among the Indians. Being expandable, it fits in well with the well-known Indian hospitality. After a meal is started, several more guests may arrive and they are always welcome. (154:96)
- *Other*—**Chippewa** *Protection* Decoction of root used to drive away "blue tailed swifts," a type of lizard. When they became troublesome, the Indians used this decoction to drive them away. (71:137)

Aralia sp.
- *Drug*—**Creek** *Antihemorrhagic* Root used for passing blood. (177:43)

Aralia spinosa L., Devil's Walking Stick
- *Drug*—**Cherokee** *Antirheumatic (Internal)* Used for rheumatism. *Carminative* Used as a carminative for "flatulent colic." *Dermatological Aid* Root used in salve for old sores. *Diaphoretic* Used as a diaphoretic. *Emetic* Infusion of roasted and pounded roots used as strong emetic. (80:31, 32) Decoction of roasted and pounded roots given as a very strong emetic. (203:74) *Orthopedic Aid* Ooze of root used as wash for paralysis. (80:31, 32) *Poison* Green roots considered poisonous. (203:74) *Tonic* Used as a tonic. *Toothache Remedy* Used for rheumatism and "ache of decaying teeth." *Venereal Aid* Used for venereal diseases. (80:31, 32) **Choctaw** *Dermatological Aid* Poultice of mashed, boiled roots applied to boils. *Poultice* Poultice of beaten roots applied to swollen leg veins. **Koasati** *Eye Medicine* Cold infusion of roots used as drops for sore eyes. (177:44) **Rappahannock** *Dermatological Aid* Decoction of root, sugar, and flour or bran used as a salve for boils and sores. *Febrifuge* Decoction of root, sugar, and flour or bran used as a salve for fever. (161:26)

Arbutus, Ericaceae
Arbutus menziesii Pursh, Pacific Madrone
- *Drug*—**Cahuilla** *Gastrointestinal Aid* Leaves used for stomach ailments. (15:40) **Concow** *Emetic* Plant eaten to cause vomiting. (41:374) **Cowichan** *Burn Dressing* Leaves used for burns. *Dermatological Aid* Infusion of bark used for cuts and wounds. *Misc. Disease Remedy* Infusion of bark used for diabetes. (182:82) **Karok** *Ceremonial Medicine* Leaves used in the puberty ceremony. (148:387) **Miwok** *Dietary Aid* Cider employed as an appetizer to create appetite. (12:161) *Gastrointestinal Aid* Cider used for stomach trouble. Manzanita cider was dipped with a plume stick from a hawk's tail feather; beverage was sucked from the feathers and was said to create appetite as well as cure stomach troubles. Leaves chewed for stomachache and cramps. (12:161, 162) **Pomo** *Dermatological Aid* Decoction of bark used as a wash for skin sores. (66:14) **Pomo, Kashaya** *Dermatological Aid* Decoction of bark used as a wash for sores and impetigo. Decoction of bark used by women as an astringent to close the pores and make the skin soft. *Love Medicine* Flowers used for love charm poisoning. *Throat Aid* Decoction of bark used as a gargle for sore throat and strep throat. (72:67) **Pomo, Little Lakes** *Cold Remedy* Infusion of leaves taken as a cold medicine. (41:374) **Saanich** *Cold Remedy* Fresh leaves chewed and juice swallowed for bad colds. (182:82) **Salish, Cowichan** *Throat Aid* Leaves chewed and juice swallowed for sore throat. (186:104) **Skokomish** *Cold Remedy* Infusion of leaves taken for colds. *Gastrointestinal Aid* Infusion of leaves taken for ulcerated stomach. *Throat Aid* Infusion of leaves taken for sore throats. (79:44) **Yuki** *Dermatological Aid* Infusion of leaves and bark taken for sores and cuts. (49:47) *Emetic* Plant eaten to cause vomiting. *Gastrointestinal Aid* Infusion of bark taken for stomachaches. (41:374) *Veterinary Aid* Infusion of leaves and bark given to horses with sore backs. (49:47)
- *Food*—**Costanoan** *Fruit* Fruit eaten in small quantities. (21:252) **Karok** *Dried Food* Berries steamed, dried, and stored for future use. *Frozen Food* Berries steamed, dried, stored, and soaked in warm water before eating. (148:387) *Fruit* Berries used for food. (6:17) **Mendocino Indian** *Forage* White, globular flowers eaten by doves, wild pigeons, and turkeys. Fruits eaten by deer. Leaves eaten by cows when green grass scarce. (41:374) **Miwok** *Beverage* Berries crushed for sweet, unfermented cider. *Winter Use Food* Dried berries stored for winter consumption, chewed but never swallowed. (12:161) **Pomo** *Fruit* Fruits eaten for food. (41:374) **Pomo, Kashaya** *Fruit* Berries eaten fresh or roasted. *Winter Use Food* Berries parched and stored for the winter. (72:67) **Wailaki** *Fruit* Fruits eaten for food. (41:374) **Yuki** *Fruit* Berries used for food. (49:87) **Yurok** *Fruit* Berries roasted over an open fire and eaten. (6:17)

- *Fiber*—**Karok** *Snow Gear* Bark used by children as sleds. *Unspecified Wood* used for carving, will not split when dry. (6:17) **Mendocino Indian** *Building Material* Wood used for lodge poles. (41:374) **Tolowa** *Clothing* Inner bark sewn together to make an "every day dress." *Snow Gear* Bark used by children as sleds. **Yurok** *Snow Gear* Bark used by children as sleds. (6:17)
- *Other*—**Cowichan** *Preservative* Bark boiled and used for tanning paddles and fishhooks. (182:82) **Hoh** *Smoke Plant* Leaves sometimes smoked. (137:66) **Karok** *Containers* Leaves used to cover stored madroño berries. Leaves placed over maple leaves in earth oven, forming last layer before oven covered with earth. (148:387) *Hunting & Fishing Item* Berries used as bait for steelhead. *Tools* Leaves used to test the temperature of pitch used in canoe construction. The pitch was ready to use when the leaf turned black. (6:17) **Mendocino Indian** *Decorations* Leaves and scarlet berries used for decorative purposes. *Stable Gear* Wood used to make stirrups. *Tools* Wood used to make saw handles and other tools. (41:374) **Pomo, Kashaya** *Fuel* Wood used for firewood. *Toys & Games* Leaves used for paper dolls for the younger children to play with. (72:67) **Quileute** *Smoke Plant* Leaves sometimes smoked. (137:66) **Saanich** *Preservative* Bark boiled and used for tanning paddles and fishhooks. **Salish, Coast** *Cooking Tools* Young branches used to make spoons. *Toys & Games* Young branches used to make gambling sticks. (182:82) **Tolowa** *Jewelry* Berries used to make necklaces. (6:17)

Arceuthobium, Viscaceae

Arceuthobium americanum Nutt. ex Engelm., American Dwarf Mistletoe

- *Drug*—**Bella Coola** *Antihemorrhagic* Decoction taken as a potent medicine for lung hemorrhages. *Pulmonary Aid* Decoction of plant taken as potent medicine for lung hemorrhages. **Carrier, Southern** *Antihemorrhagic* Decoction taken as a potent medicine for mouth hemorrhages. *Dietary Aid* Decoction of plant taken for emaciation and tuberculosis. *Tuberculosis Remedy* Decoction of plant taken for mouth hemorrhages and tuberculosis. (150:56)

Arceuthobium campylopodum Engelm., Western Dwarf Mistletoe

- *Drug*—**Navajo, Ramah** *Ceremonial Medicine* Cold infusion used internally and externally as ceremonial medicine. (191:23)

Arceuthobium occidentale Engelm., Digger Pine Dwarf Mistletoe

- *Drug*—**Mendocino Indian** *Gastrointestinal Aid* Decoction of plant taken for stomachaches. (as *Razoumofskya occidentalis* 41:345)

Arceuthobium vaginatum (Willd.) J. Presl, Pineland Dwarf Mistletoe

- *Drug*—**Navajo, Ramah** *Ceremonial Medicine* Decoction of plant used as a ceremonial medicine. (191:23)

Arctium, Asteraceae

Arctium lappa L., Greater Burrdock

- *Drug*—**Cherokee** *Antirheumatic (Internal)* Used for rheumatism. *Blood Medicine* Infusion of root or seed used to cleanse blood. *Dietary Aid* Used for scurvy. *Gynecological Aid* Used for "weakly females." *Urinary Aid* Used for "gravel." *Venereal Aid* Infusion of root or seed used for venereal diseases. (80:27) **Malecite** *Dermatological Aid* Roots smashed and used with gilead buds for sores. (116:247) *Venereal Aid* Infusion of buds used for chancre. (116:258) **Menominee** *Dermatological Aid* and *Tuberculosis Remedy* Poultice of boiled leaves applied to scrofulous sores on the neck. (54:132) **Micmac** *Dermatological Aid* Buds and roots used for sores. *Venereal Aid* Buds and roots used for chancre. (40:55) **Ojibwa** *Blood Medicine* Roots used as a blood medicine. (135:238)
- *Food*—**Iroquois** *Dried Food* Roots dried by the fire and stored away for winter use. *Soup* Dried roots soaked and boiled into a soup. (196:120) *Vegetable* Young leaves cooked and seasoned with salt, pepper, or butter. (196:118)

Arctium minus Bernh., Lesser Burrdock

- *Drug*—**Abnaki** *Analgesic* Used for headaches. (144:155) Poultice of leaves applied to the head for headaches. (144:173) *Antirheumatic (External)* Used as a medicine for rheumatism. (144:155) *Antirheumatic (Internal)* Decoction of roots taken for rheumatism. (144:173) *Febrifuge* Used for trembling fevers. (144:154) Used for trembling fevers. (144:173) *Misc. Disease Remedy* Used for grippe. (144:154) **Cherokee** *Dermatological Aid* Boiled to make "ooze" for leg ulcers and used to bathe swollen legs. (80:27) **Chippewa** *Cough Medicine* Infusion of leaves taken after a coughing spell for a hard, dry cough. (53:340) **Cowlitz** *Pulmonary Aid* Infusion of roots taken for whooping cough. (79:50) **Delaware** *Antirheumatic (Internal)* Infusion of roots used for rheumatism. *Blood Medicine* Infusion of roots used as a blood purifier. *Stimulant* Infusion of roots used as a stimulant. (176:36) **Delaware, Oklahoma** *Antirheumatic (Internal)* Infusion of root taken for rheumatism. *Blood Medicine* Infusion of root taken as a blood purifier. *Stimulant* Infusion of root taken as a stimulant. (175:31, 74) **Delaware, Ontario** *Analgesic* Poultice of leaves bound to body for pain. *Blood Medicine* Roots used as a blood purifier. (175:66, 82) **Hoh** *Unspecified* Leaves used to make a rubbing salve. Infusion of leaves and roots used for medicine. (137:70) **Iroquois** *Antirheumatic (External)* Poultice of wetted leaves and salt applied to swellings. (141:62) Roots and fruits used for rheumatism. (142:100) *Blood Medicine* Infusion of roots with other roots used to purify the blood. *Dermatological Aid* Poultice of crushed leaves and other leaves applied to blue swellings. (141:62) **Meskwaki** *Analgesic* Compound containing root used by women in labor. *Gynecological Aid* Compound containing root used by women in labor. (152:211) **Micmac** *Dermatological Aid* Roots used for boils and abscesses. (40:55) **Mohegan** *Analgesic* Poultice of leaves applied for rheumatic pains. (176:70) *Antirheumatic (External)* Poultice of leaves used for rheumatism. (174:269) Poultice of leaves applied for rheumatic pains. (176:70, 128) *Cold Remedy* Compound infusion of plants taken for wintertime colds. (174:266) **Nanticoke** *Dermatological Aid* Poultice of leaves applied to boils. (175:57, 84) **Ojibwa** *Analgesic* and *Gastrointestinal Aid* Compound containing root used for stomach pain. *Tonic* Root supposed to have tonic effect. (153:363) **Oto** *Pulmonary Aid* Decoction of root taken for pleurisy. (70:135) **Penobscot** *Dermatological Aid* Poultice of mashed, heated root applied to boils and abscesses. (156:309) **Potawatomi** *Blood Medicine* and *Tonic* Infusion of root taken as a general tonic and blood purifier. (154:49) **Quileute** *Unspecified* Leaves used to make a rubbing salve. Infusion of leaves and roots used for medicine. (137:70)

- *Fiber*—**Chippewa** *Clothing* Leaves used for head covering. (53:377)

Arctium sp., Burdock

- *Drug*—**Iroquois** *Analgesic* Compound poultice of smashed leaves applied to the head for headaches. Poultice of leaves applied to pain caused by sorcery. (87:474) *Antirheumatic (External)* Poultice of leaves applied to draw poison out and to sweat for rheumatism. (87:475) *Blood Medicine* Compound infusion of roots and leaves taken as a blood purifier. *Dermatological Aid* Compound decoction of bark or roots taken for sores or boils. Compound infusion of dried roots taken for face and neck pimples. Poultice of one smashed leaf applied to bee stings. (87:474) Poultice of powdered plant applied to cuts. *Diaphoretic* Poultice of leaves applied to draw poison out and to sweat for rheumatism. *Diuretic* Decoction of roots taken to increase urination. (87:475) *Ear Medicine* Decoction of one leaf used as drops for earaches. *Emetic* Decoction of roots taken to vomit for stomach troubles caused by sorcery. *Febrifuge* Plant used as a fever medicine. *Gastrointestinal Aid* Decoction of roots taken to vomit for stomach troubles caused by sorcery. *Gynecological Aid* Compound decoction of plants taken to induce pregnancy. (87:474) *Kidney Aid* Infusion of roots taken for the kidneys. (87:475) *Orthopedic Aid* Poultice of heated leaves applied to bruises or backaches. (87:474) *Veterinary Aid* Roots given to horses with sore legs and muscles. (87:475) *Witchcraft Medicine* Plant or roots used several ways for magical illnesses or sorcery. (87:474)

Arctostaphylos, Ericaceae

Arctostaphylos alpina (L.) Spreng., Alpine Bearberry

- *Drug*—**Ojibwa** *Antirheumatic (External)* Infusion of pounded plants used as wash for rheumatism. *Blood Medicine* Decoction of bark taken for internal blood diseases. (135:231) *Ceremonial Medicine* Leaves used for medicine ceremonies. *Narcotic* Leaves smoked to cause intoxication. (135:238) *Panacea* Infusion of pounded plants used as wash for general illnesses. (135:231) *Unspecified* Leaves used for medicinal purposes. (135:238)
- *Food*—**Alaska Native** *Fruit* Berries used for food. Berry was juicy but rather insipid in flavor. Not usually available in large quantities. Picked in poor berry years and mixed with blueberries. Flavor was much improved with cooking. (85:77) **Eskimo, Alaska** *Fruit* Fruit used for food. (4:715) **Eskimo, Arctic** *Forage* Berries eaten greedily by bears and ptarmigan. (128:23) **Eskimo, Inupiat** *Fruit* Berries, other berries, and sugar cooked and eaten. (98:108) **Koyukon** *Winter Use Food* Berries stored in grease or oil and eaten with fish or meat. (119:55)

Arctostaphylos canescens Eastw., Hoary Manzanita

- *Food*—**Karok** *Beverage* Berries used to make a drink. *Dried Food* Berries dried and stored in storage baskets for future use. *Fruit* Dried berries pounded, mixed with salmon eggs, cooked in a basket with a hot rock, and eaten. (148:388)
- *Other*—**Karok** *Containers* Wood used to make reels for string. *Cooking Tools* Wood used to make spoons and scraping sticks for acorn soup. *Walking Sticks* Wood used to make canes. (148:388)

Arctostaphylos ×*cinerea* T. J. Howell, Del Norte Manzanita

- *Food*—**Tolowa** *Bread & Cake* Berries mixed with salmon roe and sugar, formed into patties, and baked in rocks. **Yurok** *Fruit* Berries used for food. (6:18)

Arctostaphylos columbiana Piper, Hairy Manzanita

- *Drug*—**Pomo** *Antidiarrheal* Decoction of bark taken for diarrhea. (66:14) **Pomo, Kashaya** *Antidiarrheal* Decoction of bark used for diarrhea. (72:69)
- *Other*—**Pomo, Kashaya** *Lighting* Wood used on the fire at dances and ceremonials because it made a bright light to see by. *Tools* Wood used for making tools and awl handles. (72:69)

Arctostaphylos glandulosa Eastw., Eastwood's Manzanita

- *Drug*—**Cahuilla** *Antidiarrheal* Infusion of leaves used for diarrhea. *Dermatological Aid* Infusion of leaves used for poison oak rash. (15:40) **Pomo** *Unspecified* Plant used as medicine. (66:14) **Pomo, Kashaya** *Antidiarrheal* Decoction of bark taken for diarrhea and bleeding diarrhea. (72:68)
- *Food*—**Cahuilla** *Beverage* Mashed fruit mixed with water and strained into a drink. *Dried Food* Berries sun dried and stored for future use. *Fruit* Berries eaten fresh. *Porridge* Dried berries ground into flour and used to make mush. *Sauce & Relish* Berries used to make a gelatinous substance and eaten like aspic. *Staple* Seeds ground into a meal and used to make mush or cakes. (15:40) **Pomo, Kashaya** *Dried Food* Dried, pounded berries stored for later use and made into pinole, cakes, or mixed with water. (72:68)
- *Fiber*—**Cahuilla** *Building Material* Branches used in house construction. (15:40)
- *Other*—**Cahuilla** *Fuel* Wood, provided a hot fire and long-lasting coals, used for firewood. *Hunting & Fishing Item* Plant provided food for wild game and therefore a rich hunting opportunity. *Smoke Plant* Leaves mixed with tobacco. *Smoking Tools* Stems used to make pipes. *Tools* Wood used to make awl handles. Stems used to make small tools. (15:40) **Pomo, Kashaya** *Tools* Wood used for making tools and awl handles. (72:68)

Arctostaphylos glauca Lindl., Bigberry Manzanita

- *Drug*—**Cahuilla** *Antidiarrheal* Infusion of leaves used for diarrhea. *Dermatological Aid* Infusion of leaves used for poison oak rash. (15:40)
- *Food*—**Cahuilla** *Beverage* Mashed fruit mixed with water and strained into a drink. *Dried Food* Berries sun dried and stored for future use. *Porridge* Dried berries ground into flour and used to make mush. *Sauce & Relish* Berries used to make a gelatinous substance and eaten like aspic. *Staple* Seeds ground into a meal and used to make mush or cakes. (15:40) **Diegueño** *Fruit* Fruit used for food. (88:219) **Kawaiisu** *Beverage* Berries used to make a beverage. Berries were covered with a thin layer of dirt and sifted in a yaduci, a winnowing tray, so that the dirt fell through. Then they were sprinkled with water, kneaded with the hands, mashed, and soaked "in the sun" for about a half day. The yaduci was used as a sieve to remove the berry pulp from the infusion which could be drunk thus or mixed with chia. Water could be drained through the berry pulp a second time. The liquid was said to be sweet and fattening. *Fruit* Berries eaten fresh. (206:11)
- *Fiber*—**Cahuilla** *Building Material* Branches used in house construction. (15:40) **Diegueño** *Brushes & Brooms* Branches used to make a broom. (88:219)

- *Other*—**Cahuilla** *Fuel* Wood, provided a hot fire and long-lasting coals, used for firewood. *Hunting & Fishing Item* Plant provided food for wild game and therefore a rich hunting opportunity. *Smoke Plant* Leaves mixed with tobacco. *Smoking Tools* Stems used to make pipes. *Tools* Wood used to make awl handles. Stems used to make small tools. (15:40) **Diegueño** *Fuel* Used for firewood. (88:219)

Arctostaphylos manzanita Parry, Whiteleaf Manzanita
- *Drug*—**Concow** *Dermatological Aid* Poultice of chewed leaves applied to sores. *Veterinary Aid* Plant used for sore backs of horses. **Mendocino Indian** *Poison* Fruit considered poisonous. (41:375) **Miwok** *Dietary Aid* Cider employed as an appetizer to create appetite. *Gastrointestinal Aid* Cider used for stomach trouble. Leaves chewed for stomachache and cramps. (12:161, 162) **Pomo, Calpella** *Cold Remedy* Infusion of leaves taken for severe colds. **Pomo, Little Lakes** *Analgesic* Decoction of leaves used as a wash for headaches. *Antidiarrheal* Leaves used for diarrhea. (41:375)
- *Food*—**Karok** *Beverage* Berries used to make a drink. *Dried Food* Berries dried and stored in storage baskets for future use. *Fruit* Dried berries pounded, mixed with salmon eggs, cooked in a basket with a hot rock, and eaten. (148:388) **Mendocino Indian** *Beverage* Ripe berries used to make cider. *Forage* Fruits eaten by bears as forage. *Fruit* Green fruits eaten in small quantities to quench thirst. Ripe fruits eaten raw or cooked. *Unspecified* Globular, waxy flowers sucked or eaten by children. *Winter Use Food* Ripe berries stored as a winter use food. (41:375) **Miwok** *Beverage* Berries crushed for sweet, unfermented cider. *Winter Use Food* Dried berries stored for winter consumption, chewed but never swallowed. (12:161) **Numlaki** *Bread & Cake* Fruits made into bread and eaten. *Porridge* Fruits made into mush and eaten. *Staple* Fruits eaten like pinole. (41:375) **Yuki** *Beverage* Ripe fruits crushed, strained, and used to make cider. *Fruit* Ripe berries eaten raw. *Staple* Ripe berries parched and used in pinole. (49:85)
- *Other*—**Karok** *Containers* Wood used to make reels for string. *Cooking Tools* Wood used to make spoons and scraping sticks for acorn soup. *Walking Sticks* Wood used to make canes. (148:388) **Mendocino Indian** *Fuel* Wood used as an exceedingly fine fuel. **Pomo** *Soap* Leaves boiled and the yellowish red extract used as a cleansing body wash. **Yokia** *Containers* Two V-shaped branches used to carry wood on the back. (41:375)

Arctostaphylos nevadensis Gray, Pine Mat Manzanita
- *Drug*—**Karok** *Antidiarrheal* Leaves used for diarrhea. *Antidote* Plant used for poisoning from *Toxicodendron diversiloba*. (6:18)
- *Food*—**Karok** *Beverage* Berries pulverized and made into a drink. (6:18) Berries used to make a drink. *Dried Food* Berries dried and stored in storage baskets for future use. (148:388) *Fruit* Berries used for food. (6:18) Dried berries pounded, mixed with salmon eggs, cooked in a basket with a hot rock, and eaten. (148:388) **Paiute** *Fruit* Berries used for food. (111:101) **Tolowa** *Bread & Cake* Berries mixed with salmon roe and sugar, formed into patties, and baked in rocks. (6:18)
- *Other*—**Karok** *Containers* Wood used to make reels for string. *Cooking Tools* Wood used to make spoons and scraping sticks for acorn soup. *Walking Sticks* Wood used to make canes. (148:388) **Klamath** *Smoke Plant* Dried leaves mixed with tobacco and used for smoking. (45:102) **Paiute** *Smoke Plant* Roasted, dried leaves mixed with tobacco and smoked. (111:101)

Arctostaphylos parryana Lemmon, Parry Manzanita
- *Food*—**Luiseño** *Fruit* Ground berry pulp used for food. (as *A. parryi* 155:230)

Arctostaphylos patula Greene, Greenleaf Manzanita
- *Drug*—**Atsugewi** *Burn Dressing* Poultice of leaves applied to burns. Decoction of pounded leaves used for burns. *Dermatological Aid* Decoction of pounded leaves used for cuts. Poultice of leaves applied to cuts. (61:140) **Navajo, Kayenta** *Ceremonial Medicine* and *Emetic* Plant used as a ceremonial emetic. (205:35) **Shoshoni** *Venereal Aid* Decoction of leaves taken for venereal diseases. (180:38)
- *Food*—**Atsugewi** *Beverage* Berries made into cakes and eaten plain or put into water and drunk. Cider was made by adding water to pounded berries and was conveyed to the mouth with a deer tail sop. *Bread & Cake*, and *Winter Use Food* Berries made into flour, molded into cakes, and stored for later use. (61:138) **Karok** *Dried Food* Berries dried and eaten. (148:388) **Klamath** *Fruit* Berries used for food. (45:102) **Midoo** *Fruit* Berries used for food during an acorn crop failure. (117:308) **Navajo, Kayenta** *Fruit* Berries eaten raw. (205:35) **Paiute** *Forage* Berries eaten by bears and deer. (111:102) **Wintoon** *Fruit* Berries used for food. (117:263)
- *Other*—**Klamath** *Smoke Plant* Dried leaves mixed with tobacco and used for smoking. (45:102) **Paiute** *Smoke Plant* Fire dried, pulverized leaves smoked with other plants or alone. (111:102)

Arctostaphylos pringlei Parry, Pringle Manzanita
- *Food*—**Navajo** *Beverage* Crushed berries used to make a beverage. *Fruit* Berries eaten raw or cooked. *Porridge* Seeds ground into a mush. *Preserves* Berries used to make jelly. (110:23)

Arctostaphylos pumila Nutt., Sandmat Manzanita
- *Food*—**Costanoan (Olhonean)** *Beverage* Berries used to make cider. (117:373)

Arctostaphylos pungens Kunth, Pointleaf Manzanita
- *Drug*—**Cahuilla** *Antidiarrheal* Infusion of leaves used for diarrhea. *Dermatological Aid* Infusion of leaves used for poison oak rash. (15:40) **Navajo, Ramah** *Ceremonial Medicine* and *Emetic* Leaves used as a ceremonial emetic. (191:38)
- *Food*—**Cahuilla** *Beverage* Mashed fruit mixed with water and strained into a drink. *Dried Food* Berries sun dried and stored for future use. *Fruit* Berries eaten fresh. *Porridge* Dried berries ground into flour and used to make mush. *Sauce & Relish* Berries used to make a gelatinous substance and eaten like aspic. *Staple* Seeds ground into a meal and used to make mush or cakes. (15:40) **Yavapai** *Beverage* Fresh or stored pulverized berries put in mouth, solid matter spat out, and juice sucked. Sometimes the liquid was expressed by squeezing the moistened pulverized mass with the two hands. (63:213) Berries used to make a beverage. *Fruit* Berries chewed and used for food. (65:256)
- *Fiber*—**Cahuilla** *Building Material* Branches used in house construction. (15:40)
- *Other*—**Cahuilla** *Fuel* Wood, provided a hot fire and long-lasting coals, used for firewood. *Hunting & Fishing Item* Plant provided food for wild game and therefore a rich hunting opportunity. *Smoke Plant* Leaves mixed with tobacco. *Smoking Tools* Stems used to make pipes. *Tools* Wood used to make awl handles. Stems used to make

small tools. (15:40) **Navajo, Ramah** *Good Luck Charm* Dried leaves smoked with mountain tobacco to bring good luck. (191:38)

Arctostaphylos rubra (Rehd. & Wilson) Fern., Red Fruit Bearberry
- *Food*—**Eskimo, Arctic** *Forage* Berries eaten greedily by bears and ptarmigan. (128:23) **Eskimo, Inupiat** *Winter Use Food* Berries and salmonberries stored in barrels for future use. (98:109) **Tanana, Upper** *Fruit* Berries used for food. (102:10)
- *Other*—**Tanana, Upper** *Season Indicator* Red leaves indicated fattened moose and the time to hunt them. (102:10)

Arctostaphylos **sp.**, Manzanita
- *Drug*—**Costanoan** *Urinary Aid* Decoction of plant used for bladder ailments. (21:12) **Diegueño** *Kidney Aid* Infusion of leaves used for the kidneys. (84:15)
- *Food*—**Costanoan** *Beverage* Fruit steeped in cold water to produce a cider. *Dried Food* Fruit dried and stored for winter use. *Fruit* Raw fruit used for food. (21:252) **Diegueño** *Preserves* Berries used to make jelly. (84:15) **Hualapai** *Beverage* Berries used to make a drink. *Dried Food* Berries dried and stored for future use. *Fruit* Berries eaten fresh. (195:46) **Mahuna** *Fruit* Berries eaten mainly to quench the thirst. (140:70) **Paiute** *Dried Food* Fruit sun dried, stored in buckskin bags, and hung up for winter use. (167:245) **Tubatulabal** *Fruit* Berries used for food. (193:15)
- *Other*—**Miwok** *Cooking Tools* Wood used to make mush stirring paddles. (12:146)

Arctostaphylos tomentosa (Pursh) Lindl., Woollyleaf Manzanita
- *Drug*—**Costanoan (Olhonean)** *Antihemorrhagic* Infusion of bark powder taken for lung hemorrhages. (117:373) **Miwok** *Dietary Aid* Cider employed as an appetizer to create appetite. *Gastrointestinal Aid* Cider used for stomach trouble. Leaves chewed for stomachache and cramps. (12:161, 162)
- *Food*—**Mendocino Indian** *Beverage* Berries used to make cider. (41:377) **Miwok** *Beverage* Berries crushed for sweet, unfermented cider. *Winter Use Food* Dried berries stored for winter consumption, chewed but never swallowed. (12:161) **Pomo** *Dried Food* Seeds ground, molded into biscuits, and sun dried. *Porridge* Seeds ground into meal and rock boiled to make mush. (11:81)
- *Other*—**Hoh** *Smoke Plant* Leaves smoked. **Quileute** *Smoke Plant* Leaves smoked. (137:66)

Arctostaphylos uva-ursi (L.) Spreng., Kinnikinnick
- *Drug*—**Blackfoot** *Dermatological Aid* Infusion of plant, mixed with grease and boiled hoof, applied as a salve to itching and peeling scalp, as a salve to rashes and skin sores, and used as a wash for baby's head. (86:75) *Oral Aid* Infusion of plant used as a mouthwash for cankers and sore gums. (86:66) *Pediatric Aid* Infusion of plant, mixed with grease and boiled hoof, used as a wash for baby's head. (86:75) **Carrier** *Dermatological Aid* Poultice of ground leaves and stems applied to sores. (31:74) Leaves placed on a piece of wood, roasted to a powder, and placed on a cut for rapid healing. Leaves pounded into a paste and applied to boils and pimples. (89:12) **Cherokee** *Kidney Aid* Used for "dropsy." *Urinary Aid* Used for urinary diseases. (80:25) **Cheyenne** *Analgesic* Infusion of stems, leaves, and berries taken for back pain and sprained backs. Poultice of wetted leaves rubbed on the back for pain. (76:183) Infusion of leaves, stems, and berries taken for "persistent" back pain. Leaves wetted and used for pain relief. *Cold Remedy* Berries and other plants used for colds. *Cough Medicine* Berries and other plants used for coughs. (82:40) *Orthopedic Aid* Decoction of plant taken and leaves rubbed on back for painful or sprained back. (75:41) Infusion of stems, leaves, and berries taken for sprained backs. (75:183) Infusion of stems, leaves, and berries taken for sprained backs. (76:183) *Psychological Aid* Leaves burned to drive away bad spirits for people going crazy. *Unspecified* Berries used as an ingredient in medicinal mixtures. (83:25) **Chippewa** *Analgesic* Pulverized, dried leaves compounded and smoked for headache. (53:336) *Hunting Medicine* Roots smoked in pipes as charms to attract game. (53:376) **Cree, Woodlands** *Abortifacient* Infusion of whole plant and velvet leaf blueberry taken to bring menstruation. *Antidiarrheal* Fruit mixed with grease and used for children with diarrhea. *Gynecological Aid* Decoction of stems and blueberry stem taken to prevent miscarriage without causing damage to the baby, and to speed a woman's recovery after childbirth. Roots and several other herbs used to slow excessive menstrual bleeding. *Pediatric Aid* Fruit mixed with grease and used for children with diarrhea. (109:29) **Crow** *Oral Aid* Leaves pulverized and powder used for canker sores of the mouth. **Flathead** *Burn Dressing* Poultice of pulverized leaves used for burns. *Ear Medicine* Smoke from leaves used for earache. (82:40) **Hoh** *Unspecified* Leaves smoked as medicine. (137:66) **Kwakiutl** *Narcotic* Leaves smoked as a narcotic. (183:282) **Menominee** *Adjuvant* Dried leaves used as a seasoner to make certain female remedies taste good. (151:35) **Navajo, Ramah** *Ceremonial Medicine* and *Emetic* Leaves used as a ceremonial emetic. (191:38) **Ojibwa** *Antirheumatic (External)* Infusion of pounded plants used as wash for rheumatism. *Blood Medicine* Decoction of bark taken for internal blood diseases. (135:231) *Ceremonial Medicine* Leaves used for medicine ceremonies. *Narcotic* Leaves smoked to cause intoxication. (135:238) *Panacea* Infusion of pounded plants used as wash for general illnesses. (135:231) *Unspecified* Leaves used for medicinal purposes. (135:238) **Okanagan-Colville** *Antihemorrhagic* Decoction of leaves and stems taken for spitting of blood. *Blood Medicine* Decoction of leaves and stems taken as a blood tonic. *Eye Medicine* Decoction of leaves and stems used as a wash for sore eyes. *Kidney Aid* Decoction of leaves and stems taken as a tonic for the kidneys. *Urinary Aid* Decoction of leaves and stems taken as a tonic for the bladder. (188:101) **Okanagon** *Antihemorrhagic* Decoction of leaves and stems taken for blood-spitting. *Eye Medicine* Decoction of leaves and stems used as a wash for sore eyes. *Kidney Aid* Decoction of leaves and stems taken as a tonic for kidneys. *Tonic* Decoction of leaves and stems taken as a tonic for kidneys and bladder. *Urinary Aid* Decoction of leaves and stems taken as a tonic for bladder. (125:40) **Quileute** *Unspecified* Leaves smoked as medicine. (137:66) **Sanpoil** *Dermatological Aid* Green leaves dried, pulverized, and sprinkled on skin sores. Infusion of entire plant used as hair wash for dandruff and scalp diseases. Infusion of entire plant used as young girls' hair wash to insure growth. *Pediatric Aid* Infusion of entire plant used as young girls' hair wash to insure growth. (131:220) **Tanana, Upper** *Laxative* Raw berries eaten as a laxative. (102:10) **Thompson** *Antihemorrhagic* Decoction of leaves and stems taken for blood-spitting. (125:40) Decoction of root taken for "blood spitting." (164:458) *Dietary Aid* Raw leaves chewed to alleviate thirst. (187:211)

Diuretic Decoction of leaves and stems taken as a diuretic. *Eye Medicine* Decoction of leaves and stems used as a wash for sore eyes. (164:458) *Kidney Aid* Decoction of leaves and stems taken as a tonic for kidneys. (125:40) *Oral Aid* Infusion of leaves used as a mouthwash for canker sores and weak gums. *Orthopedic Aid* Infusion of plant taken and used as a wash for broken bones. (187:211) *Tonic* Decoction of leaves and stems taken as a tonic for kidneys and bladder. (125:40) Decoction of leaves and stems taken as a tonic for the kidneys and bladder. (164:458) *Urinary Aid* Decoction of leaves and stems taken as a tonic for bladder. (125:40) Decoction of leaves and stems taken as a tonic for the bladder. (164:458) Infusion of leaves used as a tonic, antiseptic, and astringent for bladder and urinary passage disorders. (187:211)

- *Food*—**Bella Coola** *Special Food* Berries formerly mixed with melted mountain goat fat and served to chiefs at feasts. (184:204) **Blackfoot** *Beverage* Crushed leaves used to make tea. *Dried Food* Berries dried and later soaked with sugar. *Fruit* Berries eaten fresh. (86:101) Berries eaten raw. (97:49) Berries eaten raw or mashed in fat and fried. (114:276) *Winter Use Food* Berries preserved for later use. (97:49) **Carrier** *Fruit* Berries mixed with salmon eggs as a palatable and nutritious food. *Soup* Berries used to make soup. (89:12) **Cherokee** *Fruit* Fruit used for food. (80:25) **Chinook, Lower** *Dried Food* Berries dried in bags, mixed with oil, and eaten. *Fruit* Berries eaten fresh. (79:44) **Chippewa** *Spice* Berries cooked with meat to season the broth. (53:318) **Coeur d'Alene** *Dried Food* Berries dried and used for food. *Fruit* Berries eaten fresh. *Soup* Berries dried, boiled with roots, and eaten as soup. (178:90) **Cree, Woodlands** *Fruit* Fruit cooked in grease, pounded, mixed with raw fish eggs, and eaten. Approximate proportions of ingredients were 1 tablespoon grease, 1½ cups fruit, and 2 tablespoons whitefish eggs separated from the adhering membranes. A little sugar was added for flavor. After the fruits were lightly cooked in grease, they were pounded until they were crumbly. They were then placed in a heavy cloth folded to make a sack and pounded with the back of an ax head. The fish eggs moistened the pounded fruit. (109:29) **Eskimo, Arctic** *Fruit* Berries cooked and eaten. (128:23) **Eskimo, Inupiat** *Frozen Food* Berries frozen for future use. *Fruit* Berries eaten with salmon eggs, to prevent the eggs from sticking to the teeth. Berries and oil eaten with dry meat. *Ice Cream* Berries stored in bear fat and cracklings or in seal oil and used to make ice cream. *Winter Use Food* Berries stored in seal oil, fish oil, or rendered bear fat. (98:99) **Flathead** *Sauce & Relish* Berries dried, powdered, and used as a condiment with deer liver. (82:40) **Hanaksiala** *Fruit* Berries mashed, mixed with grease, and eaten. *Special Food* Berries mixed with high-bush cranberries or Pacific crabapples and featured at winter feasts. (43:239) **Kimsquit** *Dried Food* Berries formerly dried, boiled, mixed with boiled dumplings, and used for food. (184:204) **Koyukon** *Winter Use Food* Berries stored in grease or oil and eaten with fish or meat. (119:55) **Kwakiutl, Southern** *Fruit* Dry, mealy berries formerly used for food. (183:282) **Makah** *Fruit* Berries used for food. (67:297) **Montana Indian** *Dried Food* Fruit eaten fresh and dried. *Fruit* Fresh fruit used for food. (19:7) *Soup* Berries boiled and used to make a broth. *Starvation Food* Berries eaten raw or fried during famines. (82:40) **Nitinaht** *Forage* Fruits eaten by grouse. *Fruit* Fruits formerly eaten fresh. (186:104) **Nuxalkmc** *Fruit* Berries used for food. (43:239) **Okanagan-Colville** *Fruit* Berries used for food. (188:101) **Okanagon** *Fruit* Insipid fruits eaten fresh. *Soup* Insipid fruits boiled in soups. (as *Arvtostaphylos uva-ursi* 125:38) *Staple* Berries used as a principal food. (178:239) **Oweekeno** *Fruit* Berries used for food. (43:239) **Salish, Coast** *Fruit* Berries eaten raw or cooked. (182:82) **Sanpoil & Nespelem** *Dried Food* Berries dried and stored for future use. (188:101) *Soup* Dried berries used in soups. (131:102) **Skokomish** *Fruit* Berries eaten with salmon eggs. (79:44) **Spokan** *Fruit* Berries used for food. (178:343) **Squaxin** *Fruit* Berries occasionally eaten. (79:44) **Tanana, Upper** *Fruit* Fruit used for food. (77:28) Raw berries mixed with grease, dried or fresh, raw whitefish eggs, and eaten. Berries warmed in grease and eaten. *Winter Use Food* Raw berries mixed with grease, dried or fresh, raw whitefish eggs, and stored for later use. (102:10) **Thompson** *Beverage* Leaves and young stems boiled and drunk as a tea. (164:493) *Forage* Fruits eaten by deer. (164:514) *Fruit* Insipid fruits eaten fresh. (as *Arvtostaphylos uva-ursi* 125:38) Drupes eaten fresh. (164:486) Dry, mealy fruits eaten with bear fat or fish oil because of the dryness. Washed berries fried in hot lard or salmon oil and used for food. The berries would crackle and pop "just like popcorn." They were the only berries prepared in this manner. (187:211) *Soup* Insipid fruits boiled in soups. (as *Arvtostaphylos uva-ursi* 125:38) Drupes boiled in soups. (164:486) **Tolowa** *Bread & Cake* Berries mixed with salmon roe and sugar, formed into patties, and baked in rocks. **Yurok** *Fruit* Berries used for food. (6:18)

- *Dye*—**Great Basin Indian** *Brown* Berries used to make a gray-brown dye. (121:49)

- *Other*—**Blackfoot** *Ceremonial Items* Leaves mixed with tobacco, dried cambium of red osier dogwood and used in all religious bundles. (86:14) Dried berries used in rattles. *Jewelry* Dried berries strung on necklaces. (86:107) *Season Indicator* Heavy fruit set taken by the medicine men as a sign of a severe winter to come. (97:49) *Smoke Plant* Crushed leaves smoked with tobacco. (86:101) Leaves dried and mixed with tobacco. (97:49) Dried leaves smoked as tobacco. (114:276) **Carrier** *Smoke Plant* Leaves and stems used to smoke. (31:74) Leaves mixed with tobacco and smoked. (89:12) **Cheyenne** *Smoke Plant* Leaves mixed with tobacco or red willow and used to smoke in a pipe. (76:183) Leaves mixed with skunkbush leaves in the absence of tobacco and smoked. (83:14) Leaves dried, mixed with red willow bark, and used for pipe smoking. (83:25) **Chippewa** *Smoke Plant* Used for smoking. (53:377) **Clallam** *Smoke Plant* Leaves pulverized and smoked before the introduction of tobacco and presently mixed with tobacco. (57:199) **Cree** *Smoke Plant* Leaves mixed with tobacco and smoked. (16:485) **Cree, Woodlands** *Smoke Plant* Dried leaves mixed with tobacco and smoked in a pipe. (109:29) **Eskimo, Arctic** *Smoke Plant* Leaves powdered, dried, used as a substitute for tobacco or mixed with the tobacco, and smoked. (128:23) **Eskimo, Inuktitut** *Smoke Plant* Leaves used as an additive to or substitute for tobacco. (202:191) **Great Basin Indian** *Smoke Plant* Leaves used as one of the elements in the tobacco mixture. (121:49) **Heiltzuk** *Smoke Plant* Leaves smoked like tobacco. (43:239) **Hesquiat** *Smoke Plant* Dried, toasted leaves mixed with tobacco for smoking. (185:64) **Hoh** *Ceremonial Items* Leaves smoked during religious ceremonies. *Smoke Plant* Leaves dried and smoked. (137:66) **Jemez** *Smoke Plant* Dried leaves smoked as tobacco. (44:20) **Keresan** *Smoke Plant* Leaves mixed with native-grown tobacco for smoking. (198:559) **Lakota**

Smoke Plant Leaves used as tobacco. (139:44) **Makah** *Cash Crop* Leaves dried and sold. (67:297) *Smoke Plant* Leaves used for smoking. (186:104) **Montana Indian** *Smoke Plant* Leaves mixed with tobacco and smoked. (19:7) **Navajo, Ramah** *Good Luck Charm* Dried leaves smoked with mountain tobacco to bring good luck. (191:38) **Nitinaht** *Smoke Plant* Leaves roasted, crushed, and smoked. (67:297) Leaves dried or roasted and smoked. **Nootka** *Smoke Plant* Leaves used for smoking. (186:104) **Nuxalkmc** *Cash Crop* Berries used for trade. (43:239) **Okanagan-Colville** *Smoke Plant* Leaves toasted and used as a tobacco. (188:101) **Okanagon** *Smoke Plant* Leaves mixed with other plant leaves and smoked. (125:39) **Oweekeno** *Cash Crop* Berries used for trade. (43:239) **Paiute** *Smoke Plant* Roasted, dried leaves mixed with tobacco and smoked. (111:100) **Pawnee** *Smoke Plant* Leaves smoked like tobacco. Leaves used for smoking, like tobacco. (as *Uva-ursi uva-ursi* 70:108) **Potawatomi** *Smoke Plant* Leaves mixed with tobacco. (154:118) **Quileute** *Ceremonial Items* Leaves smoked during religious ceremonies. *Smoke Plant* Leaves dried and smoked. (137:66) **Salish, Coast** *Smoke Plant* Leaves dried and smoked or mixed with tobacco and smoked. (182:82) **Shuswap** *Smoke Plant* Leaves roasted until dry, mashed, and mixed with tobacco. (123:62) **Skagit, Upper** *Smoke Plant* Leaves used as a tobacco substitute. (179:42) **Thompson** *Protection* Leaves placed in moccasins or shoes after the death of a husband or wife for protection. (187:211) *Smoke Plant* Leaves mixed with other plant leaves and smoked. (125:39) Dried, toasted leaves mixed with tobacco for smoking. (164:495) Dried or toasted leaves alone or mixed with tobacco and used for smoking. Too much smoking of these leaves was said to make one dizzy. *Waterproofing Agent* Mashed berries rubbed on the inside of coiled cedar root baskets to waterproof them. The berries were used to waterproof baskets such as those used for whipping soapberries. (187:211)

Arctostaphylos viscida Parry, Sticky Whiteleaf Manzanita
- *Drug*—**Miwok** *Dietary Aid* Cider employed as an appetizer to create appetite. *Gastrointestinal Aid* Cider used for stomach trouble. Leaves chewed for stomachache and cramps. (12:161, 162)
- *Food*—**Mewuk** *Beverage* Berries used to make cider. *Fruit* Berries used for food. (117:336) **Midoo** *Fruit* Berries pounded and eaten. (117:311) **Miwok** *Beverage* Berries crushed for sweet, unfermented cider. *Winter Use Food* Dried berries stored for winter consumption, chewed but never swallowed. (12:161) **Wintoon** *Fruit* Berries used for food. (117:263)

Arctostaphylos viscida ssp. *mariposa* (Dudley) P. V. Wells, Mariposa Manzanita
- *Food*—**Mewuk** *Beverage* Berries used to make cider. *Fruit* Berries used for food. (as *A. mariposa* 117:336) **Midoo** *Fruit* Berries pounded and eaten. (as *A. mariposa* 117:311)

Ardisia, Myrsinaceae
Ardisia escallonoides Schiede & Deppe ex Schlecht. & Cham., Island Marlberry
- *Food*—**Seminole** *Unspecified* Plant used for food. (169:470)
- *Other*—**Seminole** *Cooking Tools* Plant used to make meat roasting sticks. *Hunting & Fishing Item* Plant used to make arrows. (169:470)

Arecaceae, see *Cocos, Glaucothea, Sabal, Serenoa, Washingtonia*

Arenaria, Caryophyllaceae
Arenaria aculeata S. Wats., Prickly Sandwort
- *Drug*—**Shoshoni** *Eye Medicine* Decoction of root used as an eyewash. (180:38)

Arenaria congesta Nutt., Ballhead Sandwort
- *Drug*—**Gosiute** *Gastrointestinal Aid* Plant used as bowel medicine. (39:362) **Shoshoni** *Antirheumatic (External)* Poultice of steeped leaves applied for swellings. (118:42) *Blood Medicine* Infusion of flower heads and seeds taken as a blood purifier. (121:47) *Dermatological Aid* Poultice of steeped leaves and blossoms used for sun exposure. *Venereal Aid* Poultice of steeped leaves and blossoms used for gonorrheal ulcers. (118:47) **Washo** *Antirheumatic (External)* Poultice of steeped leaves applied for swellings. (118:42)

Arenaria eastwoodiae Rydb., Eastwood's Sandwort
- *Drug*—**Hopi** *Emetic* Plant used as an emetic for the stomach. (200:34, 75–76)

Arenaria fendleri Gray, Fendler's Sandwort
- *Drug*—**Navajo, Ramah** *Panacea* Root used, only in the summer, as a "life medicine." *Respiratory Aid* Powdered root used as snuff to cause sneezing for "congested nose."

Arenaria lanuginosa ssp. *saxosa* (Gray) Maguire, Spreading Sandwort
- *Drug*—**Navajo, Ramah** *Analgesic* Cold infusion used as lotion on forehead for headache. *Dermatological Aid* Infusion of plant used as lotion for pimples. *Eye Medicine* Infusion of plant used as eye drops. *Febrifuge* Cold infusion used as lotion on forehead for fever. *Respiratory Aid* Infusion of powdered root put in nose to cause sneezing for "congested nose." *Venereal Aid* Strong infusion of plant taken before sweat bath for venereal disease. (as *A. saxosa* 191:26)

Arenaria macradenia S. Wats., Mojave Sandwort
- *Drug*—**Kawaiisu** *Analgesic* Dried root smoke inhaled for headaches. Poultice of broken roots applied to the head for headaches. Root used as a salve for pain. *Antirheumatic (External)* Poultice of broken roots applied to area affected by rheumatism. *Dermatological Aid* Root used as a salve for pimples. *Respiratory Aid* Dried root smoke inhaled to clear the sinuses. (206:12)

Arenaria sp., Sandwort
- *Drug*—**Yavapai** *Cathartic* Decoction of pounded root, with cathartic qualities, taken for stomachaches. *Gastrointestinal Aid* Decoction of pounded root taken for stomachaches. (65:261)
- *Other*—**Shoshoni** *Smoke Plant* Leaves used to make tobacco mixture for smoking. (118:61)

Arenaria triflora var. *obtusa* S. Wats.
- *Drug*—**Gosiute** *Cathartic* Plant used as a purgative for babies and adults with intestinal disorders. (39:350) *Gastrointestinal Aid* Plant used as a purgative for babies and adults with intestinal disorders. (39:350) Used as a bowel medicine. (39:362) *Pediatric Aid* Plant used as a purgative for babies and adults with intestinal disorders. (39:350)

Argemone, Papaveraceae
Argemone hispida Gray, Rough Pricklypoppy

- *Drug*—**Paiute** *Dermatological Aid* Ground seeds used for sores. (118:44) **Shoshoni** *Cathartic* Infusion of seeds taken as a physic. (118:42)

Argemone munita Dur. & Hilg., Flatbud Pricklypoppy
- *Drug*—**Kawaiisu** *Burn Dressing* Roasted, ripe, and mashed seeds applied as a salve to burns. (206:12)

Argemone polyanthemos (Fedde) G. B. Ownbey, Crested Pricklypoppy
- *Drug*—**Comanche** *Eye Medicine* Sap used for sore eyes. (as *A. intermedia* 29:520) **Paiute** *Burn Dressing* and *Dermatological Aid* Salve of pulverized seeds used on burns, cuts, and sores. **Shoshoni** *Burn Dressing* Salve of pulverized seeds used on burns, sores, or cuts. *Cathartic* Roasted, mashed seeds taken as powder or pills to serve as a physic. *Dermatological Aid* Poultice of pulverized seed paste applied to bring boils to a head. Salve of moistened, pulverized seeds rubbed into hair to kill head lice. Salve of pulverized seeds used on burns, cuts, and sores. *Emetic* Roasted, mashed seeds taken as an emetic and physic. *Eye Medicine* Infusion of pulverized seeds used as a wash for sore eyes. *Toothache Remedy* Warmed root used various ways for toothache. (as *A. platyceras* 180:38, 39) **Tubatulabal** *Dermatological Aid* Poultice of pounded, ripe seeds applied to open sores. *Hemorrhoid Remedy* Poultice of pounded, ripe seeds applied to piles. (as *A. platyceras* 193:59) **Washo** *Burn Dressing* Salve of pulverized seeds used on burns, sores, or cuts. *Dermatological Aid* Salve of pulverized seeds used on burns, cuts, and sores. (as *A. platyceras* 180:38, 39)
- *Dye*—**Kiowa** *Unspecified* Leaf ash used under the skin for tattooing. (as *A. intermedia* 192:29) **Lakota** *Yellow* Used as a yellow dye for arrows. (139:53)
- *Other*—**Hopi** *Ceremonial Items* Plant used to whip children during initiation. (as *A. intermedia* 42:285)

Argentina, Rosaceae
Argentina anserina (L.) Rydb., Silverweed Cinquefoil
- *Drug*—**Blackfoot** *Antidiarrheal* Root used for diarrhea. (114:275) *Dermatological Aid* Poultice of chewed roots applied to sores and scrapes. (as *Potentilla anserina* 86:78) *Emetic* Plant soaked in water and the solution taken as an emetic for stomach disorders. (as *Potentilla anserina* 86:68) **Iroquois** *Antidiarrheal* Infusion of plant and another plant given to children for diarrhea. (as *Potentilla anserina* 141:49) *Diuretic* Infusion of leaves used as a diuretic. (as *Potentilla anserina* 142:92) *Pediatric Aid* Infusion of plant and another plant given to children for diarrhea. (as *Potentilla anserina* 141:49) **Kwakiutl** *Analgesic* and *Dermatological Aid* Decoction of root mixed with catfish oil and smeared on painful places. (20:382)
- *Food*—**Montana Indian** *Vegetable* Roots, tasted like sweet potatoes, used for food. (as *Potentilla anserina* 19:19) **Okanagon** *Staple* Roots used as a principal food. (as *Potentilla anserina* 178:238) *Unspecified* Roots eaten either raw or cooked. (as *Pontentilla anserina* 125:37) **Shuswap** *Unspecified* Roasted roots used for food. (as *Potentilla anserina* 123:66) **Thompson** *Unspecified* Roots eaten either raw or cooked. (as *Potentilla anserina* 164:480) Roots eaten raw, but more often cooked. (as *Potentilla anserina* ssp. *anserina* 187:262)
- *Fiber*—**Blackfoot** *Cordage* Runners used to fix leggings in place. The leggings were tied above the knee and then folded over to the ankle, like a boot. Runners used by girls to tie blankets. (as *Potentilla anserina* 86:119)

Argentina egedii (Wormsk.) Rydb. **ssp. *egedii***, Eged's Pacific Silverweed
- *Drug*—**Kwakiutl** *Dermatological Aid* Poultice of boiled roots and oil applied to sores and swellings. *Eye Medicine* Root juice used as a wash for inflamed eyes. (as *Potentilla pacifica* 183:289) **Tsimshian** *Unspecified* Roots used medicinally for unspecified purpose. (as *Potentilla anserina* ssp. *pacifica* 43:344)
- *Food*—**Alaska Native** *Vegetable* Roots eaten raw, boiled, or roasted like potatoes. (as *Potentilla pacifica* 85:127) **Bella Coola** *Unspecified* Roots used for food. (as *Potentilla pacifica* 184:209) **Haisla & Hanaksiala** *Unspecified* Roots used for food. (as *Potentilla anserina* ssp. *pacifica* var. 43:270) **Hesquiat** *Unspecified* Boiled or steamed roots eaten with oil or "stink salmon-eggs" (fermented eggs). (as *Potentilla pacifica* 185:73) **Kitasoo** *Unspecified* Roots mixed with sugar and oolichan (candlefish) grease and eaten. (as *Potentilla anserina* ssp. *pacifica* 43:344) **Kwakiutl, Southern** *Dried Food* Roots dried, steamed, and eaten with oil at large feasts. *Special Food* Roots dried, steamed, and eaten with oil at large feasts. (as *Potentilla pacifica* 183:289) **Makah** *Unspecified* Roots used for food. (as *Potentilla pacifica* 67:265) **Nitinaht** *Dessert* Long, fleshy roots steam cooked, dipped in oil, and eaten fresh like a dessert. *Dried Food* Roots steam cooked, dried for winter storage, soaked, briefly steamed, and eaten like dessert. (as *Potentilla pacifica* 186:118) *Unspecified* Peeled roots eaten raw or steamed. (as *Potentilla pacifica* 67:265) Roots eaten as accompaniments to cooked duck. (as *Potentilla pacifica* 186:131) Roots formerly used as an important food. (as *Potentilla pacifica* 186:63) **Oweekeno** *Unspecified* Roots used for food. (as *Potentilla anserina* ssp. *pacifica* 43:110) **Quileute** *Unspecified* Roots steamed, dipped in whale oil, and eaten. (as *Potentilla pacifica* 79:37) **Salish, Coast** *Unspecified* Fleshy taproots used for food. (as *Potentilla pacifica* 182:87)

Argyrochosma, Pteridaceae, fern
Argyrochosma fendleri (Kunze) Windham, Fendler's Falsecloak Fern
- *Drug*—**Tewa** *Dermatological Aid* Pulverized plant applied to lips for cold sores. (as *Notholaena fendleri* 138:67, 68)

Argythamnia, Euphorbiaceae
Argythamnia cyanophylla (Woot. & Standl.) Ingram, Charleston Mountain Silverbush
- *Drug*—**Navajo** *Panacea* Plant used as a "life medicine." (as *Ditaxis cyanophylla* 90:158) **Navajo, Ramah** *Panacea* Root or whole plant used in summer as "the head of the life medicine." (as *Ditaxis cyanophylla* 191:35)

Arisaema, Araceae
Arisaema dracontium (L.) Schott, Greendragon
- *Drug*—**Menominee** *Gynecological Aid* Plant used for "female disorders." (151:23)
- *Other*—**Menominee** *Sacred Items* Root used in sacred bundles and gave the power of supernatural dreams to the owner. (151:79)

Arisaema triphyllum (L.) Schott, Jack in the Pulpit

- **Drug**—**Cherokee** *Analgesic* Poultice of root used for headaches and various skin diseases. *Carminative* Used as a carminative. *Cold Remedy* Taken for colds. *Cough Medicine* Taken for dry coughs. *Dermatological Aid* Ointment used for "scald head," ringworm, tetterworm, and "scrofulous sores." Poultice of beaten, boiled roots mixed with meal and used on boils. *Diaphoretic* Used as a diaphoretic. *Expectorant* Used as an expectorant. *Orthopedic Aid* Used as a liniment. *Stimulant* Used as a stimulant. *Throat Aid* Infusion taken for throat irritations. *Tuberculosis Remedy* Given for "consumptions" and ointment used on "scrofulous sores." (80:41) **Chippewa** *Eye Medicine* Decoction of root used as a wash for sore eyes. (53:360) **Iroquois** *Analgesic* Compound snuff used for headaches. Decoction or infusion of roots taken for pains. *Antidiarrheal* Decoction of plant given to children for diarrhea. *Blood Medicine* Cold infusion of roots taken "for nonconception caused by cold blood." *Dermatological Aid* Plant used for face sores and hot poultice of plant applied to bruises. (87:276) *Eye Medicine* Steam from decoction of plant used for sore eyes. (87:277) *Febrifuge* Compound decoction steam used "when a person has cold sweats, not very sick." *Gynecological Aid* Cold infusion of roots taken "for nonconception caused by cold blood." (87:276) *Nose Medicine* Steam from decoction of plant used to "make you sneeze." (87:277) *Orthopedic Aid* Compound infusion of powdered plants taken for cramps. Hot poultice of plant applied to bruises and for lameness. (87:276) Compound of powdered plant and alcohol used as a liniment for sore joints. (87:277) *Pediatric Aid* Decoction of plant given to children for diarrhea. Infusion of roots used as a wash for listless babies. *Respiratory Aid* Compound of chopped root and whisky taken for bronchial colds. Compound snuff used for catarrh. *Stimulant* Infusion of roots used as a wash for listless babies. *Tuberculosis Remedy* Compound decoction used as poultice for infected and swollen tubercular glands. (87:276) *Veterinary Aid* Ground plant added to mare's feed to induce pregnancy and reduce listlessness. (87:275) **Malecite** *Dermatological Aid* Poultice of dried, pounded plants used for abscesses and boils. (116:247) **Menominee** *Eye Medicine* Poultice of pulverized root applied to sore eyes. (151:23) *Witchcraft Medicine* Compounded pulverized root used in lip incision to counteract "witchery" to face. (54:129) **Meskwaki** *Ceremonial Medicine* Seed used as a magical diagnostic medicine to predict recovery or death. *Poison* Finely chopped root put in meat for enemies, to cause pain and death. (152:202) Root cooked with meat used in abandoned vessels to poison enemy during war. (152:272) *Sedative* Compound used in very small doses for insomnia. *Snakebite Remedy* Root used for rattlesnake bite swellings. (152:202) **Micmac** *Dermatological Aid* Parts of plant used for boils and abscesses. *Orthopedic Aid* Parts of plant used as a liniment used for external use. (40:55) **Mohegan** *Analgesic* and *Antirheumatic (External)* Infusion of dried root used as a liniment for pain. (176:70, 128) *Orthopedic Aid* Infusion of plant used as a liniment. *Poison* Infusion of plant poisonous when taken internally. (174:269) Infusion of root, poisonous if swallowed, gargled for sore throat. (176:70) *Throat Aid* Infusion of root taken for sore throat. (174:269) Dilute infusion of root gargled for sore throat. (176:70, 128) **Ojibwa** *Eye Medicine* Root used for sore eyes. (153:356) *Unspecified* Plant used for medicinal purposes. (135:246) **Pawnee** *Analgesic* Crushed corm sprinkled on head and temples for headache and general pain. *Antirheumatic (External)* Poultice of pulverized corm applied as counterirritant for rheumatism. (70:69) **Penobscot** *Orthopedic Aid* Infusion of plant used as a liniment "for general external use." *Poison* Infusion of plant considered poisonous. (156:310) **Rappahannock** *Dermatological Aid* Compound dried root meal poultice applied for swelling and boils. (161:32)
- **Food**—**Potawatomi** *Unspecified* Thinly sliced roots cooked in a pit oven for 3 days to eliminate the poison. (154:95)
- **Other**—**Pawnee** *Toys & Games* Seeds used in gourd shells to make rattles. (70:69)

Arisaema triphyllum **ssp.** *quinatum* (Buckl.) Huttleston, Jack in the Pulpit
- **Drug**—**Choctaw** *Blood Medicine* Decoction of root taken "to make blood." (as *A. quinatum* 25:23) Decoction of plant taken to make blood. (as *A. quinatum* 177:6)

Arisaema triphyllum (L.) Schott **ssp.** *triphyllum*, Jack in the Pulpit
- **Drug**—**Iroquois** *Contraceptive* Infusion of rhizomes used by women for temporary sterility. (as *A. atrorubens* 141:69)

Aristida, Poaceae

Aristida divaricata Humb. & Bonpl. ex Willd., Poverty Threeawn
- **Drug**—**Keres, Western** *Burn Dressing* Grass ashes rubbed on burns. (171:27)

Aristida purpurea **var.** *longiseta* (Steud.) Vasey, Fendler Threeawn
- **Food**—**Navajo, Ramah** *Fodder* Used for horse feed, if better forage was not available. Recognized as poor sheep or horse feed. (as *A. longiseta* 191:15)
- **Fiber**—**Hopi** *Brushes & Brooms* Plant used for broom material. (as *A. longiseta* 42:286) **Navajo, Ramah** *Brushes & Brooms* Stems used to make hairbrushes. (as *A. longiseta* 191:15)
- **Other**—**Hopi** *Ceremonial Items* Plant used in ceremonials. *Decorations* Plant used to decorate the faces of the female kachina. (as *A. longiseta* 42:286)

Aristolochiaceae, see *Aristolochia*, *Asarum*, *Hexastylis*

Aristolochia, Aristolochiaceae

Aristolochia californica Torr., California Dutchman's Pipe
- **Drug**—**Miwok** *Cold Remedy* Decoction taken for colds. (12:167)

Aristolochia macrophylla Lam., Pipevine
- **Drug**—**Cherokee** *Dermatological Aid* Decoction of root applied externally for "swelling of feet and legs." *Urinary Aid* Compound infusion of "stalk chips" taken for "yellowish urine." (80:32)

Aristolochia serpentaria L. Virginia Snakeroot
- **Drug**—**Cherokee** *Abortifacient* Infusion taken for obstructions. *Analgesic* Infusion taken for sharp darting pains and used as an anodyne. Taken for sharp pains in the breast and used as a wash for headache. (80:55) Infusion of roots taken for breast pains. (177:20) *Antirheumatic (Internal)* Infusion taken for rheumatism. *Cold Remedy* Infusion taken or root chewed for colds and cold infusion of plant used for pain. *Cough Medicine* Taken for coughs. *Disinfectant* Used as an antiseptic. *Diuretic* Used as a diuretic. *Febrifuge* Taken for fever. *Gas-*

trointestinal Aid Used to "stop mortification and prevent putrefaction in the bowels," dyspepsia. *Misc. Disease Remedy* Used as a tonic for typhus fevers and taken for ague. Used for "black-yellow" diseases. *Nose Medicine* Used as a poultice for "nose made sore by constant blowing." *Pulmonary Aid* Infusion taken for pleurisy. *Snakebite Remedy* Root chewed and saliva spit on snakebite. *Stimulant* Used by "persons of weak, phlegmatic habits" and for dizziness or fainting. *Throat Aid* Used as a gargle for sore throat. *Tonic* Used as a tonic for typhus fevers. *Toothache Remedy* Poultice of bruised root applied to tooth for toothache. (80:55) **Choctaw** *Analgesic* Infusion of root taken for stomach pain. (25:24) Cold infusion of roots taken for stomach pains. (177:20) *Gastrointestinal Aid* Infusion of root taken for stomach pains. (25:24) Cold infusion of roots taken for stomach pains. (177:20) **Delaware** *Tonic* Used singly or combined with wintergreen to make a tonic. (176:39) **Delaware, Oklahoma** *Tonic* Root used alone or with wintergreen as a tonic. (175:32, 74) **Micmac** *Anticonvulsive* Root used for fits. (40:55) **Mohegan** *Snakebite Remedy* Poultice of plant applied to snakebites. (174:266) Poultice of pounded root applied to snakebite. (176:70, 128) **Nanticoke** *Anthelmintic* Infusion of plant taken for intestinal worms. (175:57, 84) **Natchez** *Febrifuge* Warm decoction of plant taken for fevers. (172:667) Decoction of plant taken for fevers. (177:20) **Penobscot** *Anticonvulsive* Infusion of root used for "fits." (156:310) **Rappahannock** *Dermatological Aid* Compound poultice with mashed roots used as salve for spider bites. *Febrifuge* Infusion of leaves taken for chills. *Snakebite Remedy* Compound poultice with mashed roots used as salve for snakebites. (161:27)

Aristolochia sp., Pipevine
- **Other**—**Wintoon** *Tools* Nuts used by children in play for blowing. (117:264)

Armillaria, Tricholomataceae, fungus
Armillaria ponderosa (PK.) Sacc.
- **Food**—**Karok** *Unspecified* Species used for food. **Yurok** *Unspecified* Species used for food. (6:18)

Armoracia, Brassicaceae
Armoracia rusticana P. G. Gaertn., B. Mey. & Scherb., Horseradish
- **Drug**—**Cherokee** *Abortifacient* Used for "obstructed menses." *Antirheumatic (Internal)* Used for rheumatism. *Cold Remedy* Used for colds. *Dietary Aid* Used to increase the appetite. *Diuretic* Used as a diuretic. *Gastrointestinal Aid* Used to aid digestion. *Oral Aid* Roots chewed for tongue and mouth diseases. *Respiratory Aid* Used for asthma. *Throat Aid* Infusion gargled for sore throat. *Tonic* Used as a tonic. *Urinary Aid* Used for "gravel." (80:39) **Delaware, Ontario** *Analgesic* Poultice of leaves applied for neuralgia. (as *Rorippa armoracia* 175:66, 82) **Iroquois** *Blood Medicine* Infusion of smashed roots taken for the blood. *Misc. Disease Remedy* Plant used for sugar diabetes. (as *A. lapathifolia* 87:342) **Mohegan** *Toothache Remedy* Poultice of leaves, with midrib removed, bound to cheeks for toothache. (as *Roripa amoracia* 174:266) Poultice of leaf bound to the face for toothaches. (as *Rorippa amoracia* 176:75, 132)
- **Food**—**Cherokee** *Sauce & Relish* Root used as a condiment. (as *A. lapathifolia* 126:36)

Arnica, Asteraceae
Arnica acaulis (Walt.) B.S.P., Common Leopardbane
- **Drug**—**Catawba** *Analgesic* Infusion of roots taken for back pain. *Orthopedic Aid* Infusion of roots taken for back pain. (157:189)

Arnica cordifolia Hook., Heartleaf Arnica
- **Drug**—**Okanagan-Colville** *Love Medicine* Roots used as a love medicine. Roots were mixed with a robin's heart and tongue and with ocher paint. The mixture was dried and powdered. The user went into the water and faced east, recited certain words, mentioned the name of the person he desired, and marked his face with the powdered arnica mixture. (188:75) **Shuswap** *Eye Medicine* Plant used for sore eyes. (123:58) **Thompson** *Antirheumatic (External)* Poultice of mashed plant used for swellings. *Dermatological Aid* Poultice of mashed plant used for bruises and cuts. *Tuberculosis Remedy* Infusion of plant taken for tuberculosis. (187:169)

Arnica latifolia Bong., Broadleaf Arnica
- **Drug**—**Okanagan-Colville** *Love Medicine* Roots used as a love medicine. (188:75) **Thompson** *Unspecified* Plant used medicinally for unspecified purpose. (164:473)

Arnoglossum, Asteraceae
Arnoglossum atriplicifolium (L.) H. E. Robins., Armoglossum
- **Drug**—**Cherokee** *Cancer Treatment* Poultice used for cancer and to draw out blood or poisonous matter. *Dermatological Aid* Poultice used for cuts and bruises. (as *Cacalia atriplicifolia* 80:58)
- **Food**—**Cherokee** *Spice* Powdered leaves used as seasoning. (as *Cacalia atriplicifolia* 80:58)

Aronia, Rosaceae
Aronia melanocarpa (Michx.) Ell., Black Chokeberry
- **Drug**—**Potawatomi** *Cold Remedy* Infusion of berry used for colds. (as *Pyrus melanocarpa* 154:76)
- **Food**—**Abnaki** *Fruit* Fruit used for food. (144:152) *Unspecified* Species used for food. (144:168) **Potawatomi** *Fruit* Berries used for food. (as *Pyrus melanocarpa* 154:107)

Artemisia, Asteraceae
Artemisia absinthium L., Absinth Sagewort
- **Drug**—**Chippewa** *Orthopedic Aid* Boiled plant top used as warm compress for sprain or strained muscles. (53:362) **Mohegan** *Anthelmintic* Infusion of leaves taken as a vermifuge. (176:70, 128) **Okanagan-Colville** *Cold Remedy* Decoction or infusion of twigs taken for head colds. Poultice of pounded leaves applied for chest colds. *Gastrointestinal Aid* Infusion of split roots taken for stomach ailments. *Gynecological Aid* Plant used as a sanitary napkin to "heal the mother's insides" after a baby's birth. *Misc. Disease Remedy* Poultice of pounded leaves applied for flu. *Orthopedic Aid* Poultice of mashed, boiled plant applied or decoction of plant used as a wash for broken limbs. *Tuberculosis Remedy* Decoction or infusion of twigs taken for tuberculosis. *Venereal Aid* Decoction or infusion of twigs taken for venereal disease. (188:75)
- **Other**—**Okanagan-Colville** *Insecticide* Branches used under mattresses as a repellent for bedbugs and other insect pests. (188:75)

Artemisia alaskana Rydb., Alaska Sagebrush
- *Drug*—**Tanana, Upper** *Cancer Treatment* Decoction taken for cancers. *Cold Remedy* Decoction taken for colds. *Cough Medicine* Decoction taken for coughs. *Eye Medicine* Cooled decoction used as a wash for eyes. *Misc. Disease Remedy* Decoction taken for diabetes. (102:17)

Artemisia arctica Less., Boreal Sagebrush
- *Drug*—**Tanana, Upper** *Cancer Treatment* Decoction taken for cancer. *Cold Remedy* Decoction taken for colds. *Cough Medicine* Decoction taken for coughs. *Eye Medicine* Cooled decoction used as a wash for eyes. *Misc. Disease Remedy* Decoction taken for diabetes. (102:17)

Artemisia australis Less., Oahu Wormwood
- *Drug*—**Hawaiian** *Febrifuge* Decoction of pounded leaves, trunk, and roots used as a steam bath and wash for high fevers. *Pulmonary Aid* Leaves used as an ingredient in a medicine for lung troubles. *Reproductive Aid* Leaves used as an ingredient for asthma medicine. (2:7)
- *Other*—**Hawaiian** *Decorations* Leaves woven into wreaths. (2:7)

Artemisia biennis Willd., Biennial Wormwood
- *Drug*—**Cherokee** *Analgesic* Used for cramps and painful menstruation. *Anthelmintic* Poultice applied to stomach for worms and seeds in molasses taken for worms. *Dermatological Aid* Used for sores and wounds. *Gastrointestinal Aid* Used for cramps and colic. *Gynecological Aid* Used for painful menstruation. (80:62)
- *Food*—**Gosiute** *Unspecified* Seeds formerly gathered and used for food. (39:362) **Iroquois** *Forage* Plants eaten by turkeys. (142:102)
- *Other*—**Navajo, Ramah** *Ceremonial Items* Plant ash used in blackening ceremonies. (191:48)

Artemisia californica Less., California Sagebrush
- *Drug*—**Cahuilla** *Cold Remedy* Leaves used for colds. *Gynecological Aid* Decoction of plant taken to start menstrual activity, for easy childbirth, and postnatal recovery, and to prevent dysmenorrhea and ease menopause trauma. *Pediatric Aid* Decoction of plant given to newborn babies 1 day after birth to flush out their system. *Unspecified* Plant used in the sweat houses for various cures. (15:42) **Costanoan** *Analgesic* Poultice of leaves applied to the tooth for pain. *Antirheumatic (External)* Decoction of plant used as a bath for rheumatism. *Cold Remedy* Decoction of plant used as a bath for colds. *Cough Medicine* Decoction of plant used as a bath for coughs. *Dermatological Aid* Poultice of leaves applied to wounds. *Respiratory Aid* Poultice of plant applied to the back or decoction of plant taken for asthma. *Toothache Remedy* Poultice of leaves applied to the tooth for pain. (21:25) **Mahuna** *Gynecological Aid* Infusion of plants taken for vaginal troubles. (140:14)
- *Other*—**Cahuilla** *Smoke Plant* Leaves chewed fresh or dried and smoked after mixing with tobacco and other leaves. (15:42) **Luiseño** *Ceremonial Items* Plant and white sage used to build a ceremonial hunting fire before hunting. The hunters stood around the fire and in its smoke before hunting because they believed that the fire and smoke would absolve them of any breach of social observances they might have committed which would otherwise have brought them bad luck. (155:199)

Artemisia campestris L., Field Sagewort
- *Drug*—**Blackfoot** *Abortifacient* Decoction of leaves taken to abort difficult pregnancies. (86:60) *Antirheumatic (External)* Chewed leaf spittle applied to rheumatic parts. (86:78) *Cough Medicine* Infusion of dried leaves taken for coughs. (86:71) *Dermatological Aid* Infusion of roots used, especially for children, as a hair tonic. Infusion of roots cleansed and used for scalp infections. (86:123) Infusion of leaves applied to eczema. (86:75) *Eye Medicine* Poultice of chewed leaves applied to sore eyes. (86:80) *Gastrointestinal Aid* Fresh leaves chewed for stomach troubles. (86:66) *Oral Aid* Leaves chewed by runners for the mentholating properties. (86:101) *Pediatric Aid* Infusion of roots used, especially for children, as a hair tonic. (86:123) *Veterinary Aid* Infusion of roots used for back sores on horses. (86:87) **Lakota** *Diuretic* Infusion of roots used to cause urination. *Gastrointestinal Aid* Infusion of roots used to cause bowel movements. *Gynecological Aid* Infusion of roots used by women for difficult births. *Sedative* Pulverized roots put on sleeping man's face so his horses could be stolen. (139:35) **Shuswap** *Cold Remedy* Decoction of plants taken for colds. *Cough Medicine* Decoction of plants taken for coughs. *Dermatological Aid* Poultice of steamed branches applied to bruises and sores. *Panacea* Decoction of plants taken as a medicine for everything. *Tuberculosis Remedy* Decoction of plants taken for tuberculosis. (123:58)
- *Other*—**Blackfoot** *Hide Preparation* Infusion of roots rubbed on a hide to be tanned. (86:109) *Soap* Infusion of roots cleansed and used for scalp infections. (86:123) **Lakota** *Incense & Fragrance* Pulverized roots used for perfume. (139:35)

Artemisia campestris ssp. *borealis* var. *borealis* (Pallas) M. E. Peck, Pacific Wormwood
- *Drug*—**Menominee** *Abortifacient* Compound infusion of leaf taken to restore menstrual flow. (as *A. canadensis* 151:29) **Meskwaki** *Burn Dressing* Poultice of leaves applied to bad burns. (as *A. canadensis* 152:211) **Okanagon** *Antidiarrheal* Decoction of whole plant used for diarrhea. *Gynecological Aid* Decoction of fresh or dried leaves taken by women after childbirth. **Thompson** *Antidiarrheal* Decoction of whole plant used for diarrhea. (as *A. canadensis* 125:41) Decoction of plant taken for diarrhea. (as *A. canadensis* 164:470) *Gynecological Aid* Decoction of fresh or dried leaves taken by women after childbirth. (as *A. canadensis* 125:41) Decoction of fresh or dried leaves taken postpartum to hasten recovery. *Unspecified* Decoction of root used medicinally. (as *A. canadensis* 164:470)

Artemisia campestris ssp. *caudata* (Michx.) Hall & Clements, Field Wormwood
- *Drug*—**Tewa** *Antiemetic* Leaves and stems chewed and juice swallowed when one was "sick at the stomach." *Febrifuge* Infusion of leaves and stems taken for "chills." (as *A. forwoodii* 138:53)
- *Other*—**Navajo, Ramah** *Ceremonial Items* Plant ash used in blackening ceremonies. (as *A. forwoodii* 191:48)

Artemisia campestris ssp. *pacifica* (Nutt.) Hall & Clements, Pacific Wormwood
- *Drug*—**Havasupai** *Unspecified* Sprays used in the sweat baths or infusion of leaves taken for sicknesses. (as *A. pacifica* 197:245) **Navajo, Kayenta** *Ceremonial Medicine* Plant used as a ceremonial fumigant ingredient. *Disinfectant* Plant used as a ceremonial fumigant ingredient. (as *A. pacifica* 205:45)

- *Food*—**Navajo, Kayenta** *Unspecified* Seeds made into mush and used for food. (as *A. pacifica* 205:45)

Artemisia cana Pursh, Silver Sagebrush

- *Drug*—**Lakota** *Unspecified* Used as a medicine. (139:35) **Montana Indian** *Dermatological Aid* Decoction of leaves used as a hair restorer. *Dietary Aid* Leaves chewed to allay thirst. *Tonic* Decoction of leaves used as a general tonic. *Unspecified* Decoction of leaves used for various complaints. (19:7)
- *Food*—**Blackfoot** *Forage* Plant used as fall and winter forage for horses. (97:56) **Lakota** *Forage* Best sage for winter browse by livestock and game. (139:35)
- *Other*—**Tubatulabal** *Cooking Tools* Used as brush beds for roasting pinyon cones. (193:12) *Fuel* Brush burned to roast cones. (193:17)

Artemisia carruthii Wood ex Carruth., Carruth's Sagewort

- *Drug*—**Navajo, Ramah** *Cough Medicine* Infusion of leaves taken for cough. *Dermatological Aid* Cold infusion used as lotion for sores. *Diaphoretic* Leaves used in sweat bath medicine. *Febrifuge* Infusion of leaves taken for fever. *Misc. Disease Remedy* Cold infusion of leaves taken for fever, influenza, and cough. *Panacea* Root used as a "life medicine." *Veterinary Aid* Cold infusion of leaves used as lotion for sheep's sore back. (191:48) **Zuni** *Analgesic* Seeds placed on coals and used as a sweat bath for body pains from a serious cold. (as *A. wrightii* 166:42, 43)
- *Food*—**Apache, White Mountain** *Unspecified* Species used for food. (as *A. wrightii* 136:155) **Navajo** *Bread & Cake* Seeds ground and made into bread and dumplings. *Porridge* Seeds ground and made into gruel. (as *A. wrightii* 165:223) *Unspecified* Seeds used for food. (as *A. wrightii* 55:82) **Zuni** *Bread & Cake* Ground seeds mixed with water, made into balls, steamed, and used for food. (as *A. wrightii* 166:65) *Unspecified* Seeds considered among the most important food plants when the Zuni reached this world. (*A. wrightii* 32:21)
- *Other*—**Hopi** *Ceremonial Items* Plant used to make prayer sticks. (42:287) **Navajo, Ramah** *Ceremonial Items* Plant ash used as Evilway and Hand Trembling blackenings. Branches used in Beautyway garment ceremony. (191:48)

Artemisia douglasiana Bess., Douglas's Sagewort

- *Drug*—**Costanoan** *Analgesic* and *Antirheumatic (External)* Decoction of plant used as a compress for rheumatism pain. *Dermatological Aid* Decoction of plant used as a compress for wounds. *Ear Medicine* Poultice of heated leaves applied to the ear for earaches. *Respiratory Aid* Decoction of plant used for asthma. *Urinary Aid* Decoction of plant used for urinary problems. (21:25) **Karok** *Antirheumatic (External)* Poultice of leaves applied for rheumatism and arthritis. (6:18) **Kawaiisu** *Abortifacient* Infusion of plant used when the menstrual flow had stopped. *Dermatological Aid* Infusion of plant used as a hair wash to prevent the hair from falling out. *Other* Infusion of plant used as a bath for mother and father after childbirth. *Pediatric Aid* Infusion of plant used to "prevent a girl from aging prematurely." (206:12) **Miwok** *Analgesic* Leaves worn in nostrils for headaches. *Antirheumatic (Internal)* Decoction of leaves taken for rheumatism. *Ceremonial Medicine* Leaves worn in nostrils by mourners when crying, the pungent odor clearing the head. *Witchcraft Medicine* Leaves rubbed on body to keep ghosts away. Plant worn on a necklace to prevent dreaming of the dead. Poisoned leaves carried to avoid personal injury. Leaves rubbed on corpse handlers to ward off ghosts of the deceased. (as *A. vulgaris* var. *heterophylla* 12:167) **Paiute** *Analgesic* Poultice of crushed, green leaves used as a compress for headaches. *Misc. Disease Remedy* Burning plant used as an inhalant for grippe. (180:39) **Paiute, Northern** *Cold Remedy* Branches put over a bed of ashes and slept on for colds. *Febrifuge* Branches put over a bed of ashes and slept on for fevers. (59:125) **Pomo, Kashaya** *Analgesic* Decoction or infusion of leaves taken for stomachache and cramps associated with diarrhea. *Dermatological Aid* Poultice of warmed leaves used on baby's severed umbilical cord. Decoction or infusion of leaves used for washing itching sores. *Gastrointestinal Aid* Decoction or infusion of leaves taken for stomachache and cramps associated with diarrhea. *Gynecological Aid* Decoction or infusion of leaves taken to stop excessive menstruation or to ease cramps. (72:119) **Tolowa** *Anthelmintic* Infusion of fresh leaves taken by children for "pin worms." *Antirheumatic (External)* Poultice of fresh leaves used for arthritis. Fresh leaves used as a liniment. *Orthopedic Aid* Fresh leaves used as a steamed herb for fractures. *Pediatric Aid* Infusion of fresh leaves taken by children for "pin worms." (6:18) **Washo** *Analgesic* Decoction of leaves used as a wash for headaches. *Antirheumatic (External)* Decoction of leaves used as a liniment for rheumatism. (180:39) **Yuki** *Analgesic* Decoction of leaves taken for pains or "troubles inside." Poultice of pounded leaves applied for rheumatism, arthritic or back pains. *Antidiarrheal* Decoction of leaves taken for dysentery. *Antirheumatic (External)* Poultice of pounded leaves applied for rheumatism or arthritic pains. *Dermatological Aid* Infusion used as wash and poultice of plant applied to cuts, bruises, and sores. Poultice of chewed leaves applied to spot affected by hunting accident. *Gynecological Aid* Infusion of plant used as a steam bath for difficulties attending childbirth. *Orthopedic Aid* Poultice of pounded plant applied for back pains. *Veterinary Aid* Infusion of plant given to injured animals. (49:45) **Yurok** *Anthelmintic* Infusion of fresh leaves taken by children for "pin worms." *Antirheumatic (External)* Poultice of fresh leaves used for arthritis. Fresh leaves used as a liniment. *Orthopedic Aid* Fresh leaves used as a steamed herb for fractures. *Pediatric Aid* Infusion of fresh leaves taken by children for "pin worms." (6:18)
- *Other*—**Costanoan** *Lighting* Burned branches used as torches during night fishing. *Tools* Burned branches used to smoke bees from nests. (21:254) **Karok** *Insecticide* Shoots used with drying salmon to keep "salmon beetles" away. (6:18) **Kawaiisu** *Ceremonial Items* Plant used for many different ceremonial purposes. Plant and meat chewed by a boy and his parents after his first kill. A boy did not eat the meat of his first kill but, together with his parents, chewed the meat mixed with California mugwort and then spit it into the fire. If this custom were neglected, the boy would never kill deer and would become a transvestite. *Incense & Fragrance* Plant used for the aromatic fragrance in baths and hair washes. (206:12) **Pomo, Kashaya** *Smoke Plant* Dried leaves used as tobacco. (72:119)

Artemisia dracunculus L., Wormwood

- *Drug*—**Costanoan** *Antidiarrheal* Decoction of roots used for dysentery. *Gastrointestinal Aid* Decoction of roots used for infants with colic. *Pediatric Aid* Decoction of roots used for infants with colic.

Urinary Aid Decoction of roots used for urinary problems. (21:26) **Crow** *Eye Medicine* Infusion of stems and leaves used as an eyewash for snow-blindness. Poultice of leaves used for the eyes. **Flathead** *Antirheumatic (External)* Infusion of foliage used lukewarm for swollen feet and legs. *Dermatological Aid* Foliage dried, powdered, and used for open sores. (82:45) **Kawaiisu** *Antirheumatic (External)* Infusion of leaves used as a wash for rheumatism. (206:13) **Luiseño** *Unspecified* Plant used for medicinal purposes. (155:228) **Okanagan-Colville** *Analgesic* Poultice of mashed, dampened leaves applied to the forehead for headaches. *Antirheumatic (External)* Leaves used in a steam bath for rheumatic or arthritic pain. *Dermatological Aid* Leaves used in diapers or used as a diaper for diaper rash and skin rawness. *Gynecological Aid* Leaves used as sanitary napkins. *Pediatric Aid* Leaves used in diapers or used as a diaper for diaper rash and skin rawness. *Tonic* Infusion of roots and yarrow roots taken as a general tonic. (188:76) **Sanpoil** *Cold Remedy* Cold infusion of root used for colds. (131:217) **Shuswap** *Gynecological Aid* Plant used by women at childbirth. *Stimulant* Decoction of leaves and roots used as a bath for tiredness. *Witchcraft Medicine* Plant used to keep away sickness and germs. (123:58) **Thompson** *Antirheumatic (External)* Decoction of plant used as a "liniment" for arthritis. *Cold Remedy* Plant used as a wash for colds.[15] *Dermatological Aid* Infusion or decoction of plant used as a bathing solution for swelling and discoloration of bruises. *Pediatric Aid* Plant used as a wash for colds, especially for babies and for chickenpox, to help the itching.[15] *Veterinary Aid* Decoction of plant used after injuries to wash horses' legs every day until healed. (187:169)
- *Food*—**Apache, Chiricahua & Mescalero** *Beverage* Leaves and young stems boiled to make a nonintoxicating beverage. (as *A. aromatica* 33:53) **Luiseño** *Unspecified* Seeds used for food. (155:228)
- *Fiber*—**Okanagan-Colville** *Mats, Rugs & Bedding* Leaves used in a baby's board or cradle as padding and kept the baby cool on hot days. (188:76)
- *Other*—**Costanoan** *Lighting* Burned branches used as torches during night fishing. (as *A. dranunculus* 21:254) **Keres, Western** *Insecticide* Plant used in the bed as a bedbug repellent. Crushed plant mixed with water and used on bed clothing as a bedbug repellent. (as *A. aromatica* 171:28) **Okanagan-Colville** *Cooking Tools* Branches with leaves used as spreaders for drying salmon and to separate stored layers of salmon. *Insecticide* Branches with leaves used as an insect repellent and stopped flies from laying eggs in stored salmon. Branches burned as a smudge to drive away mosquitoes. (188:76) **Shuswap** *Insecticide* Plant burned to keep away mosquitoes. (123:58)

Artemisia dracunculus L. ssp. *dracunculus*, Wormwood
- *Drug*—**Chippewa** *Abortifacient* Decoction of root taken for "stoppage of periods." Infusion or decoction of root or leaf and stalk used for "stoppage of periods." (as *A. dracunculoides* 53:356) *Antidiarrheal* Infusion of dried leaves and tops taken for chronic dysentery. (as *A. dracunculoides* 53:344) *Dermatological Aid* Poultice of chewed, fresh, or dried leaves and flowers applied to wounds. (as *A. dracunculoides* 53:356) Compound decoction of root used as wash to strengthen hair and make it grow. (as *A. dracunculoides* 53:350) *Gynecological Aid* Decoction of root taken for "excessive flowing." Decoction of whole plant taken to aid in difficult labor. (as *A. dracunculoides* 53: 356) *Heart Medicine* Infusion of leaf and flower taken or fresh leaf chewed for heart palpitations. (as *A. dracunculoides* 53:338) *Herbal Steam* Strong decoction of root used "for steaming old people to make them stronger." *Pediatric Aid* Strong decoction of root used as a strengthening bath for children. *Strengthener* Decoction of root used as strengthening bath for child and herbal steam for elderly. (as *A. dracunculoides* 53:362) **Navajo, Ramah** *Dermatological Aid* Cold infusion used as lotion for cuts. Used as hair rinse to make hair long and soft. (191:48) **Okanagon** *Analgesic* Infusion of whole plant used as a head wash for headaches. *Antirheumatic (External)* Plant used in the sweat house for rheumatism. *Diaphoretic* Plant used in the sweat house for rheumatism and stiff joints. *Gynecological Aid* Infusion of whole plant used as a bath for women after childbirth. *Orthopedic Aid* Plant used in the sweat house for stiff joints. (125:41) **Omaha** *Abortifacient* Decoction of plant taken for irregular menstruation. *Love Medicine* Chewed root put on clothes as a love charm. *Unspecified* Plant used in the smoke treatment of unspecified illnesses. (70:134) **Paiute** *Antirheumatic (External)* Hot decoction of branches used as a wash for rheumatism. Hot poultice of plant tops applied to sprains, swellings, or rheumatism. *Dermatological Aid* Poultice of plant tops applied for swellings. *Gynecological Aid* Decoction of whole plant taken as a tonic after childbirth. *Orthopedic Aid* Poultice of plant tops applied for sprains. *Tonic* Decoction of whole plant taken as a tonic after childbirth. (180:39, 40) **Pawnee** *Antirheumatic (External)* Decoction of tops used as a wash for rheumatism. *Unspecified* Plant used in the smoke treatment of unspecified illnesses. **Ponca** *Unspecified* Plant used in the smoke treatment of unspecified illnesses. (70:134) **Shoshoni** *Cathartic* Hot decoction of branches taken as a physic. *Cold Remedy* Hot decoction of branches taken for colds. *Dermatological Aid* Decoction of whole plant used as a wash for nettle stings. *Eye Medicine* Steam from boiling plant used for eye trouble. *Herbal Steam* Steam from boiling plant used for eye trouble. *Throat Aid* Poultice of pulped, green plant applied to sore throat or neck glands. *Venereal Aid* Decoction of whole plant taken or used as a wash for venereal diseases. (180:39, 40) **Thompson** *Analgesic* Infusion of whole plant used as a head wash for headaches. (125:41) Decoction of plant used to wash the head and temples for headache. Fresh plants used as a bed in the sweat bath for aching bones or muscles. (164:463) *Antirheumatic (External)* Plant used in the sweat house for rheumatism. (125:41) Fresh plants used as a bed in the sweat bath for rheumatism. (164:463) *Diaphoretic* Plant used in the sweat house for rheumatism and stiff joints. (125:41) Whole plant steamed to cause sweating for rheumatism and other aches. (164:463) *Gynecological Aid* Infusion of whole plant used as a bath for women after childbirth. (125:41) Decoction of plant used as a wash for women after childbirth. (164:463) *Orthopedic Aid* Plant used in the sweat house for stiff joints. (125:41) Plants used in sweat bath for sprains, stiff or aching joints, or muscles. (164:463) *Poison* Plant considered poisonous "if it enters the blood." (164:512) **Winnebago** *Febrifuge* Infusion of plant top sprinkled on the body for fevers. *Love Medicine* Chewed root put on clothes as a love charm. *Unspecified* Plant used in the smoke treatment of unspecified illnesses. (70:134)
- *Food*—**Gosiute** *Unspecified* Oily and nutritious seeds formerly used for food. (39:363) **Hopi** *Unspecified* Leaves boiled or roasted between hot, flat stones and eaten. (32:25) Leaves baked between hot stones,

dipped in salted water, and eaten. (56:19) **Shoshoni** *Spice* Steeped seeds added to dishes for flavoring. (118:29)
- *Fiber*—**Dakota** *Mats, Rugs & Bedding* Plant bunches used as towels in old times. **Omaha** *Mats, Rugs & Bedding* Plant bunches used as towels in old times. **Pawnee** *Brushes & Brooms* Plant tops bound into bundles and made into brooms used for sweeping the lodge floor. The plant was used for this purpose because of its agreeable, wholesome odor. *Mats, Rugs & Bedding* Plant bunches used as towels in old times. **Ponca** *Mats, Rugs & Bedding* Plant bunches used as towels in old times. **Winnebago** *Mats, Rugs & Bedding* Plant bunches used as towels in old times. (70:134)
- *Other*—**Paiute** *Ceremonial Items* Foliage used for medicine man's costume. (118:51)

Artemisia dracunculus ssp. *glauca* (Pallas ex Willd.) Hall & Clements, Dragon Wormwood
- *Drug*—**Ponca** *Burn Dressing* Decoction of plant taken and used as a wash for burns. (as *A. glauca* 94:152)

Artemisia filifolia Torr., Sand Sagebrush
- *Drug*—**Comanche** *Gynecological Aid* Padding of plants placed over hot coals as a bed after childbirth. (29:520) **Hopi** *Dermatological Aid* Plant used for boils. (42:288) Plant used for boils. (200:32, 94) *Gastrointestinal Aid* Infusion of plant and juniper branches taken for indigestion. (42:288) Simple or compound decoction of plant taken for indigestion. (200:33, 94) **Navajo, Ramah** *Snakebite Remedy* Strong infusion taken in large amounts and used as lotion for snakebites. (191:48) **Tewa** *Carminative* Plant chewed or decoction taken for indigestion and flatulence. (138:44) *Dermatological Aid* Plant used for boils. *Gastrointestinal Aid* Infusion of plant and juniper branches taken for indigestion. (42:288) Leaves chewed or decoction taken for indigestion or biliousness. Poultice of plant steeped in boiling water applied to stomach. (138:44)
- *Food*—**Navajo** *Fodder* Used as stock feed. (55:81)
- *Other*—**Hopi** *Ceremonial Items* Plant used for ritualistic purposes. (42:288) **Kiowa** *Paper* Used for drying hands and as a substitute for toilet paper. (192:56) **Navajo** *Paper* Very soft leaves used as a convenient substitute for toilet paper. (55:81) **Tewa** *Ceremonial Items* Plant used for ritualistic purposes. (42:288)

Artemisia frigida Willd., Fringed Sagewort
- *Drug*—**Arapaho** *Cough Medicine* Infusion of leaves taken as a cough medicine. (118:38) **Blackfoot** *Cold Remedy* Infusion of leaves taken for colds. *Cough Medicine* Infusion of leaves taken for coughs. (97:56) *Dermatological Aid* Poultice of chewed leaves applied to wounds to lessen the swelling. (86:83) *Febrifuge* Decoction of roots or tops taken for "mountain fever." (114:275) *Gastrointestinal Aid* Plant tops chewed and liquid swallowed for heartburn. (97:56) Plant chewed for heartburn. (114:275) *Gynecological Aid* Pad of the plant worn by women during menses to reduce skin irritation. (86:79) Infusion of plant taken by women during menses. (97:56) *Hemostat* Soft leaves used to stuff a bleeding nostril. (86:83) *Misc. Disease Remedy* Decoction of roots and tops taken for mountain fever. (97:56) *Stimulant* Crushed leaves used to revive gophers after children clubbed them while playing a game. (86:109) *Veterinary Aid* Horses rolled in patches of the plant to treat their wounds. Infusion of plant given to horses for coughing, sneezing, and to clean the sinuses. (86:87) **Cheyenne** *Ceremonial Medicine* Plant used in the Sun Dance ceremony. (83:18) *Gynecological Aid* Infusion of plant taken by women during menses. (97:56) *Hemostat* Plant braid tied around the head for nosebleed. (83:18) **Chippewa** *Anticonvulsive* Compound infusion or decoction of root taken for "fits." (51:63, 64) Compound decoction of root taken for convulsions. (53:336) *Disinfectant* Dried leaves burned to disinfect room of contagious patient. (53:366) *Gastrointestinal Aid* Infusion of leaves taken or leaf smoke inhaled for biliousness. (53:364) *Hemostat* Compound infusion or decoction of root used on bleeding wounds. (51:63, 64) Compound decoction of root used on bleeding wounds. (53:336) *Stimulant* Compound infusion or decoction of root taken or used externally as stimulant. (51:63, 64) Compound decoction of root taken as a stimulant. *Tonic* Compound infusion or decoction of root taken as a tonic. (53:364) **Cree, Woodlands** *Analgesic* Leaves used for headaches associated with fevers. *Febrifuge* Decoction of leaves taken for fevers. (109:30) **Dakota** *Abortifacient* Decoction of plant taken and used as wash for irregular menstruation. (70:134) **Delaware, Oklahoma** *Ceremonial Medicine* Leaves chewed as "ceremonial" medicine. (175:74) **Great Basin Indian** *Antihemorrhagic* Leaves used for stopping a hemorrhage. (121:50) **Isleta** *Gastrointestinal Aid* Infusion of plant used as a stomach medicine. (100:22) **Montana Indian** *Pulmonary Aid* Decoction used for lung troubles. *Tuberculosis Remedy* Decoction used for consumption. (19:7) **Navajo, Ramah** *Cough Medicine* Decoction of leaves taken for cough. *Panacea* Root used as a "life medicine." *Toothache Remedy* Hot poultice of leaves applied for toothache. (191:48) **Okanagan-Colville** *Cold Remedy* Infusion of leaves and branches taken for colds. *Misc. Disease Remedy* Infusion of leaves and branches taken for flu. (188:76) **Omaha** *Abortifacient* Decoction of plant taken and used as wash for irregular menstruation. **Pawnee** *Abortifacient* Decoction of plant taken and used as wash for irregular menstruation. **Ponca** *Abortifacient* Decoction of plant taken and used as wash for irregular menstruation. (70:134) **Potawatomi** *Stimulant* Leaves and flowers fumed on live coals to revive comatose patient. (154:49) **Sioux** *Abortifacient* Decoction used for menstrual irregularity. (82:45) **Tanana, Upper** *Cancer Treatment* Decoction taken for cancer. *Cold Remedy* Decoction taken for colds. *Cough Medicine* Decoction taken for coughs. *Eye Medicine* Cooled decoction used as a wash for eyes. *Misc. Disease Remedy* Decoction taken for diabetes. (102:17) **Tewa** *Carminative* Leaves chewed or decoction taken for indigestion or flatulence. *Gastrointestinal Aid* Leaves chewed or decoction taken for indigestion or biliousness. Poultice of plant steeped in boiling water applied to stomach for gastritis. (138:54) **Thompson** *Unspecified* Plant used medicinally for unspecified purpose. (164:465) *Venereal Aid* Decoction of plant taken as medicine, possibly for venereal disease. (187:170) **Zuni** *Cold Remedy* Infusion of whole plant taken as a cold remedy. (166:42)
- *Food*—**Blackfoot** *Spice* Crushed leaves mixed with stored meat to maintain a good odor. (86:101) **Hopi** *Spice* Used with sweet corn when roasting. (190:167) **Isleta** *Forage* Plant considered excellent grazing plant for sheep and cattle. (100:22)
- *Fiber*—**Dakota** *Mats, Rugs & Bedding* Plant bunches used as towels in old times. **Omaha** *Mats, Rugs & Bedding* Plant bunches used as towels in old times. **Pawnee** *Mats, Rugs & Bedding* Plant bunches

used as towels in old times. **Ponca** *Mats, Rugs & Bedding* Plant bunches used as towels in old times. **Winnebago** *Mats, Rugs & Bedding* Plant bunches used as towels in old times. (70:134)
- *Dye*—**Great Basin Indian** *Green* Leaves used to make a green dye. (121:50)
- *Other*—**Arapaho** *Ceremonial Items* Whole plant used for ceremonials. (118:51) **Blackfoot** *Ceremonial Items* Plant used in religious rituals. (97:56) Plant tied to articles that were sacrificed to the Sun. (114:275) *Hide Preparation* Infusion of plant used to treat hides. *Incense & Fragrance* Plant stuffed into saddles, women's pillows, hide bags, and quivers as a deodorant. *Insecticide* Plant put on campfire coals and the smoke repelled the mosquitoes. (86:109) Plant put on a fire to attract horses that run to the smoke because it kept flies and mosquitoes away. (86:87) *Paper* Leaves used as toilet paper. (86:124) *Soap* Plant used to clean paint applicators made from broken buffalo shoulder blades. *Tools* Plant made into balls and used as molds to make rattles. (86:109) **Chippewa** *Protection* Fresh leaves in nostrils and mouth as protection when "working over the dead." (53:366) **Cree, Woodlands** *Hunting & Fishing Item* Used as an ingredient in trap lures. (109:30) **Great Basin Indian** *Ceremonial Items* Whole plant used for all ceremonies. (121:50) **Hopi** *Ceremonial Items* Plant used to make pahos (prayer sticks). (42:289) Sprig attached to the prayer emblem and regarded as efficacious in petitions for water. (56:21) Used on prayer sticks. (190:167) Attached to prayer sticks. (200:94) **Kutenai** *Ceremonial Items* Plant burned and smoke used in religious ceremonies. *Insecticide* Plant burned and used as a smudge for mosquitoes. (82:45) **Navajo** *Paper* Very soft leaves used as a convenient substitute for toilet paper. (55:81) **Navajo, Ramah** *Ceremonial Items* Plant ash applied before painting Witcheryway prayer sticks. (191:48) **Shuswap** *Insecticide* Plant burned to keep away mosquitoes. (123:58) **Sioux** *Soap* Decoction used for bathing. (82:45) **Thompson, Upper (Nicola Band)** *Fuel* Used for fuel in smoking skins. (164:497) **Zuni** *Decorations* Sprigs and corn ears attached to decorated tablets and carried by female dancers in a drama. The sprigs and corn ears were carried by female dancers in the drama of "The Coming of the Corn Maidens." *Fertilizer* Sprigs dipped in water and planted with corn so that it would grow in abundance. (166:87)

Artemisia furcata **var.** *heterophylla* (Bess.) Hultén, Forked Sagewort
- *Drug*—**Luiseño** *Unspecified* Plant used for medicinal purposes. (as *A. heterophylla* 155:228) **Mendocino Indian** *Analgesic* Decoction of leaves taken for headaches. *Antidiarrheal* Decoction of leaves taken for diarrhea. *Antidote* Juice used as an antidote for effects of poison oak. *Antirheumatic (External)* Poultice of leaves applied or leaves used in sweat bath for rheumatism. *Cold Remedy* Decoction of leaves taken or bruised leaves placed in nostril for colds. *Eye Medicine* Decoction of leaves used as a wash for sore eyes. *Febrifuge* Decoction of leaves taken for fevers. *Gastrointestinal Aid* Decoction of leaves taken for colic and stomachaches. *Gynecological Aid* Poultice of leaves applied after childbirth to promote blood circulation. *Herbal Steam* Poultice of leaves applied or leaves used in sweat bath for rheumatism. *Respiratory Aid* Decoction of leaves taken for bronchitis. (as *A. heterophylla* Nutt. 41:392) **Paiute** *Febrifuge* and *Pediatric Aid* Steeped leaves put next to a baby's skin for fever. (as *A. heterophylla* 118:40)
- *Other*—**Luiseño** *Hunting & Fishing Item* Stems used to make small arrows. (as *A. heterophylla* 155:206) Plant used to make small boys' arrows. (as *A. heterophylla* 155:228)

Artemisia ludoviciana Nutt., Louisiana Sagewort
- *Drug*—**Blackfoot** *Dermatological Aid* Leaves put into moccasins as a foot deodorant. (86:124) Poultice of leaves applied to blisters and burst boils. (86:75) *Pediatric Aid* Leaves chewed, especially by children, during the sweat for respiratory disorders. *Pulmonary Aid* Infusion of leaves taken for chest constrictions. *Respiratory Aid* Leaves chewed, especially by children, during the sweat for respiratory disorders. *Throat Aid* Infusion of leaves taken for throat constrictions. (86:71) *Veterinary Aid* Infusion of plant given to horses for coughing, sneezing, and to clean the sinuses. (86:87) **Cheyenne** *Analgesic* Snuff of crushed leaves used for headache. (82:44) Crushed leaves used as snuff for headaches. *Ceremonial Medicine* Plant used in ceremonies. *Hemostat* Crushed leaves used as snuff for nosebleeds. (83:18) *Nose Medicine* Leaves crushed and used as a snuff for nosebleeds. *Respiratory Aid* Leaves crushed and used as a snuff for sinuses. (82:44) Crushed leaves used as snuff for sinus attacks. (83:18) *Unspecified* Plants rubbed on the body for immunity to sickness. (82:44) **Comanche** *Dermatological Aid* Leaves chewed and used for insect and spider bites. (99:7) **Crow** *Dermatological Aid* Salve of plants and neck fat used for sores. Infusion used as an astringent for eczema, and for underarm and foot perspiration and odor. **Flathead** *Cold Remedy* Infusion used for colds. *Dermatological Aid* Infusion used for bruises and itching. **Gros Ventre** *Febrifuge* Infusion used for high fevers. (82:44) **Havasupai** *Unspecified* Sprays used in the sweat baths or infusion of leaves taken for sicknesses. (197:245) **Kutenai** *Dermatological Aid* Decoction of plants used for sores. (82:44) **Lakota** *Antidiarrheal* Infusion of plant used for diarrhea. *Cold Remedy* Infusion used for colds. *Throat Aid* Infusion used for sore throats. (106:46) **Meskwaki** *Dermatological Aid* Poultice of leaves applied to old sores. Tincture of leaves used for old sores, especially scrofulous sores. *Throat Aid* Infusion of leaves taken for tonsillitis and sore throat. *Tuberculosis Remedy* Tincture of leaves used for old sores, especially scrofulous sores. *Veterinary Aid* Smudge of leaves used to "smoke ponies when they have the distemper." (152:211) **Mewuk** *Cathartic* Infusion of plant used as a mild cathartic. *Dermatological Aid* Poultice of bruised leaves applied to cuts and sores. (117:366) *Disinfectant* Infusion of plant used as a disinfectant to wash the bodies of the mourners after funerals. (117:353) *Gastrointestinal Aid* Infusion of plant used for indigestion. (117:366) *Other* Plant worn around the neck by orphans after parents death to keep the ghost away and prevent sickness. (117:353) *Panacea* Small bundles of plant made into necklaces to keep disease away. (117:336) *Unspecified* Plant used as a medicine. (as *A. lucoviciana* 117:338) Infusion of plant taken, used as a wash, or poultice applied as medicine. (117:353) *Veterinary Aid* Used for horses with sore backs. (117:366) **Navajo** *Unspecified* Used by the medicine men. (55:81) **Navajo, Ramah** *Panacea* Root used as a "life medicine." (191:48) **Ojibwa** *Veterinary Aid* Plant used as a horse medicine. (153:363) **Okanagan-Colville** *Carminative* Infusion of plant taken and splashed on the body during sweat bathing to "clear his wind." *Strengthener* Infusion of plant taken and splashed on body during sweat bathing by hunters, to walk long distances. (188:

Artemisia ludoviciana (Asteraceae)

78) **Poliklah** *Eye Medicine* Poultice of plant applied to sore eyes. (117:173) **Thompson** *Antirheumatic (External)* Decoction of plant used as a "liniment" for arthritis. *Cold Remedy* Decoction of plant used as a wash for colds. Hot decoction of plant taken, especially by the elderly, for colds. *Disinfectant* Plant used as incense to disinfect the house. *Gastrointestinal Aid* Plant used for overeating or indigestion. *Misc. Disease Remedy* Hot decoction of plant taken, especially by the elderly, for influenza. *Orthopedic Aid* Poultice of plant boiled with "any kind of weeds" and used on injured areas as a bone setter. (187:170) **Yokut** *Unspecified* Used as medicine. (as *A. lucoviciana* 117:437) **Yurok, South Coast (Nererner)** *Dermatological Aid* Infusion of plant used for itching skin and as a lotion for sores. *Eye Medicine* Infusion of plant used for sore eyes. (117:169)

- *Food*—**Apache, Chiricahua & Mescalero** *Spice* Sage used to flavor meats. (33:47) **Blackfoot** *Candy* Leaves chewed as a confection. (86:101)
- *Fiber*—**Blackfoot** *Mats, Rugs & Bedding* Used to cover the floor of the sweat lodge. (86:17) **Pomo** *Building Material* Used to thatch the sweat house. (117:282)
- *Other*—**Blackfoot** *Ceremonial Items* Used for cleansing in the sweat lodge. Chewed by assistants to the participants during the sweat lodge rituals to relieve thirst. This thirst resulted from taboos against the consumption of liquids. (86:17) Used to wipe the sweat from their bodies during the sweat lodge ceremonies. Used to cleanse the headdresses of the Horn Society before transferring to new owners. In the transferal, the headdress was laid on a bed of man sage. Used to cleanse the singers of the All Smoking ceremony. Packed into ceremonial moccasins for storage. Used as the Holy Offering of the All Smoking ceremony and for certain Horn Society rituals. Used to cleanse the body during participation in the All Smoking Ceremony. Worn around wrists and ankles by the Lodge dancers in the Okan (Holy Lodge) of the Sun dance. Used to wipe the black paint off the one man who fasted until the Okan center pole was raised. Used as a bed for the Buffalo Stones (Iniskim) during buffalo drive rituals. *Good Luck Charm* Used to cleanse a man entered in a horse race. (86:24) *Incense & Fragrance* Plant stuffed into saddles, women's pillows, hide bags, and quivers as a deodorant. (86:109) *Insecticide* Plant put on a fire to attract horses that run to the smoke because it kept flies and mosquitoes away. (86:87) *Malicious Charm* Stems used as a curse on horses to slacken in races. Short pieces were broken from the stem and a small, flat, white stone was collected. The man first sprayed the rock with juice from his mouth. Then the stems were placed on the ground; they represented the other horses. The stone was sprayed again and shoved toward the stems, without touching them, three times. At the fourth shove the stone was pressed into the stems. Horses so cursed were sure to slacken in the race. Leaves used in defense of the use of yellow angelica by an opponent in the hand game. The loser went to his lodge and asked for a small piece of meat, which he combined with man sage leaves. Carrying this charm in his pocket, he returned to the gambling place, where he walked abruptly among the participants and took the dice in his hands. Then he pretended to sing and perform with the dice–all the while secretly rubbing them with the meat and sage mixture. After he handed back the dice, the man who had been using the yellow angelica would get a headache and grow lazy, losing all he had won.

(86:40) *Paper* Leaves used as toilet paper. (86:124) *Protection* Used to cleanse a person afraid of a ghost. (86:24) **Cheyenne** *Ceremonial Items* Plant used extensively in ceremonies to drive away bad spirits, evil influences, and ominous dreams. Leaves burned as incense in ceremonies to purify implements, utensils, or persons. Plants wiped on persons who broke taboos for purification. Plants wrapped around Sundancers' eagle bone whistles for prevention of thirst. Sprigs used as ceremonial paintbrushes during the Sundance. *Tools* Sprigs used in sweat lodges to sprinkle water on the rocks. **Crow** *Ceremonial Items* Plants used extensively in ceremonies. (82:44) **Lakota** *Ceremonial Items* Used to make bracelets and head wreaths for the Sun Dance. (106:46) Plant used to make wreaths and bracelets for the sun dance and used in the sweat lodge. *Incense & Fragrance* Plant burned as incense. (139:36) *Protection* Smudged on the body and home to ward off evil spirits. (106:46) **Meskwaki** *Insecticide* Smudge of leaves used to drive away mosquitoes. (152:211) **Mewuk** *Decorations* Burned wood soot used for tattooing. (117:349) **Navajo, Ramah** *Ceremonial Items* Plant ash used as Evilway and Hand Tremblingway blackenings. Applied to unraveler strings (a woman's hair cord or buckskin string from her moccasins). (191:48) **Sioux** *Smoke Plant* Plant used in smoking tobacco. (153:417) **Thompson** *Fuel* Plant used on the fire in the first smoking of a hide during the curing process. The smoke from this plant was supposed to soften the hide. *Insecticide* Plant burned and the smoke used as a good smudge against mosquitoes. *Protection* Plant used as incense to protect one against evil influences. If one had to go into a large crowd, he could smoke himself with this plant to protect himself against strangers who might bring him harm. (187:170)

Artemisia ludoviciana ssp. *incompta* (Nutt.) Keck, Mountain Sagewort

- *Drug*—**Bella Coola** *Unspecified* Used as a medicine. **Carrier, Northern** *Analgesic* Hot infusion of plant taken for headache. **Carrier, Southern** *Dermatological Aid* Poultice of chewed plant applied to sprains and swellings. *Orthopedic Aid* Poultice of chewed leaves applied to sprains. (as *A. discolor* 150:65)
- *Food*—**Gosiute** *Unspecified* Seeds formerly used for food. (as *A. discolor* 39:362)

Artemisia ludoviciana Nutt. ssp. *ludoviciana*, Foothill Sagewort

- *Drug*—**Cheyenne** *Ceremonial Medicine* Plant used in ceremonies. *Psychological Aid* Plant used to drive away bad or ominous dreams. (as *A. gnaphalodes* 76:190) **Chippewa** *Antidote* Dried flowers placed on coals and the fumes used as an antidote to "bad medicine." (as *A. gnaphalodes* 53:376) Smoke of burned flowers inhaled as antidote for "bad medicine." (as *A. gnaphalodes* 53:366) **Keres, Western** *Antirheumatic (External)* Crushed plant rubbed on body as a liniment for soreness or stiffness. *Diaphoretic* Plant used as an ingredient in the sweat bath. *Other* Plant placed in shoes to keep feet from sweating. (as *A. gnaphaloides* 171:28) **Kiowa** *Gastrointestinal Aid* Infusion of plants taken for stomach troubles. *Pulmonary Aid* Infusion of plants taken for the lungs or to cut phlegm. (as *A. gnaphalodes* 192:56) **Omaha** *Ceremonial Medicine* Plant used in rites of lustration for humans or beasts. (as *A. gnaphalodes* 68:321) *Febrifuge* Decoction of leaves used as a bath for fevers. *Hemostat* Dried, powdered leaves applied to nostrils for nasal hemorrhage. (as *A. gnaphalodes* 68:334) **Paiute** *Analgesic* Decoction of plant used as a soaking bath for aching feet. Hot or cold decoction of whole plant or young growth taken for stomachaches. Poultice of steamed plants or bruised leaves used for rheumatism or other aches. Poultice of steeped leaves used as a compress for headaches. *Antidiarrheal* Decoction of plant taken for diarrhea. *Antirheumatic (External)* Branches used in a sweat bath for rheumatism. Poultice of steamed plants or bruised leaves used for rheumatism or other aches. *Dermatological Aid* Decoction of plant used as a wash for rashes, itching, or skin eruptions. Poultice of leaves or stems and leaves applied to swellings, boils, and sores. *Disinfectant* Branches used as a bed in a sweat bath to steam out infection of influenza. *Eye Medicine* Decoction of leaves used as an eyewash. *Febrifuge* Poultice of steeped leaves used, especially for babies, as a compress for fevers. *Gastrointestinal Aid* Decoction of whole plant or shoots taken for stomachaches. *Gynecological Aid* Decoction of root or entire plant taken as a tonic after childbirth. Infusion of leaves used as a regulator of menstrual disorders. *Herbal Steam* and *Misc. Disease Remedy* Branches used as a bed in a sweat bath to steam out infection of influenza. *Orthopedic Aid* Decoction of plant used as a soaking bath for aching feet. *Pediatric Aid* Poultice of steeped leaves used, especially for babies, as a compress for fevers. *Tonic* Decoction of root or entire plant taken as a tonic after childbirth. *Venereal Aid* Decoction of plant tops taken for venereal diseases. (as *A. gnaphalodes* 180:40–42) **Sanpoil** *Analgesic* Leaves placed in the nostrils for an hour for a headache. *Cold Remedy* Leaves placed in the nostrils for an hour for a cold. *Dermatological Aid* Pulverized leaves sprinkled on sores "to hasten their healing." (as *A. vulgaris* var. *ludoviciana* 131:217) **Shoshoni** *Analgesic* Decoction of leaves taken for headaches. Hot or cold decoction of whole plant or young growth taken for stomachaches. Poultice of steeped leaves used, especially for babies, as a compress for fevers. *Antidiarrheal* Decoction of plant tops taken for diarrhea. *Cathartic* Decoction of whole plant or shoots taken as a physic. *Cold Remedy* Compound decoction of plant tops taken for colds. Compound decoction of whole plant taken for heavy colds. Decoction of branches taken in small doses for colds. Decoction of leaves taken for colds. Decoction of tops alone, or sometimes with roots, taken for colds. *Cough Medicine* Compound decoction of plant tops taken for coughs. Compound decoction of whole plant or plant tops taken for coughs. Compound decoction of whole plant taken for coughs. Decoction of branches taken in small doses for coughs. Simple or compound decoction of leaves taken for coughs. *Dermatological Aid* Decoction of plant used as a wash for rashes, itching, or skin eruptions. *Disinfectant* Decoction of plant tops taken for severe infections. *Eye Medicine* Infusion of leaves used as an eyewash. *Febrifuge* Compound decoction of whole plant taken for fevers. Poultice of steeped leaves used, especially for babies, as a compress for fevers. *Gastrointestinal Aid* Decoction of whole plant or shoots taken for stomachaches. *Gynecological Aid* Infusion of leaves used as a regulator of menstrual disorders. *Misc. Disease Remedy* Decoction of branches taken for influenza. *Pediatric Aid* Poultice of steeped leaves used, especially for babies, as a compress for fevers. **Washo** *Analgesic* Decoction of leaves taken and used as a cooling, aromatic wash for headaches. *Cold Remedy* Decoction of leaves taken for "heavy colds" and head colds. *Cough Medicine* Decoction of leaves taken for colds, coughs, and headaches. (as *A. gnaphalodes* 180:40–42)

- *Fiber*—**Dakota** *Mats, Rugs & Bedding* Plant bunches used as towels in old times. **Omaha** *Mats, Rugs & Bedding* Plant bunches used as towels in old times. **Pawnee** *Mats, Rugs & Bedding* Plant bunches used as towels in old times. **Ponca** *Mats, Rugs & Bedding* Plant bunches used as towels in old times. **Winnebago** *Mats, Rugs & Bedding* Plant bunches used as towels in old times. (as *A. gnaphalodes* 70:134)
- *Other*—**Cheyenne** *Ceremonial Items* Plant used for many ceremonial purposes. On ceremonial occasions it is spread about the borders of a lodge in a special way. Other uses are to wipe off ceremonial paint; or to purify, by wiping off, with a bundle of the sage, the body of one who has committed some fault—violated some taboo. It is used by Contraries to wipe off the ground in a lodge where a Contrary had been seated. The dried leaves are burned on the coal to make a smoke used in purifying implements or utensils used in ceremony; or to smoke, and so purify, the body of an individual. This is to drive away bad spirits, and particularly to drive away a bad or ominous dream had by a sick person, which dream may remain in the mind of the person and trouble him. It may be mixed with "motsinists" (*Lomatium dissectum* var. *multifidum*)—a small pinch of each in about the same quantity—for the same purpose. The patient who is being smoked sits over the coal on which the sage is being sprinkled, with a blanket over the body and the coal in order to confine the smoke. (as *A. gnaphalodes* 76:190) **Lakota** *Ceremonial Items* Plant used to make wreaths and bracelets for the sun dance and used in the sweat lodge. *Incense & Fragrance* Plant burned as incense. (139:36) **Paiute** *Ceremonial Items* Plant used as wash by dancers after the Sun Dance. (as *A. gnaphalodes* 118:51)

Artemisia ludoviciana* ssp. *mexicana (Willd. ex Spreng.) Keck, Mexican White Sagebrush
- *Drug*—**Kiowa** *Dermatological Aid* Poultice of chewed leaves applied to sores. *Herbal Steam* Plant used as a purifying agent in the sweat house. *Throat Aid* Leaves chewed for sore throats. (as *A. mexicana* 192:56)
- *Other*—**Navajo, Ramah** *Ceremonial Items* Plant ash used in blackening ceremonies. (as *A. mexicana* 191:48)

Artemisia nova A. Nels., Black Sagebrush
- *Drug*—**Shoshoni** *Cold Remedy* Decoction of leaves taken for colds. *Cough Medicine* Decoction of leaves taken for coughs. (180:43)

***Artemisia* sp.**, Wormwood
- *Drug*—**Blackfoot** *Veterinary Aid* Bracts and flower heads used in horse medicine bundles. (97:56) **Dakota** *Ceremonial Medicine* Plant used in purificatory rites by women after menstruation. (69:369) Plant used to begin any ceremonial in order to drive away evil influences. *Disinfectant* Decoction of plant used as a wash for purification. (70:134, 135) *Gastrointestinal Aid* Decoction of plant taken for stomach troubles and many other ailments. (70:135) *Witchcraft Medicine* Plant used as incense to exorcise evil powers. (70:134, 135) **Eskimo, Alaska** *Antirheumatic (External)* Poultice of powdered plant used for swellings. *Cold Remedy* Infusion of dried plants used for colds. *Dermatological Aid* Poultice of powdered plant used for in-

juries. (4:716) **Eskimo, Kuskokwagmiut** *Cold Remedy* Plant taken for colds. *Unspecified* Poultice of plant used for unspecified ailments. (122:33) **Eskimo, Nunivak** *Cold Remedy* Infusion of plant used for colds. *Dermatological Aid* Poultice of powdered plant applied to injuries or swellings. *Unspecified* Dried plant used for medicine. (149:325) **Hualapai** *Analgesic* Infusion of leaves used as a bitter tonic for headaches. *Cold Remedy* Infusion of leaves used as a bitter tonic for colds. *Gastrointestinal Aid* Infusion of leaves used as a bitter tonic for indigestion. *Pulmonary Aid* Leaves used for cleansing the lungs in the sweat house. (195:41) **Isleta** *Laxative* Decoction of leaves thickened with sugar and used for constipation. (100:23) **Jemez** *Gastrointestinal Aid* Leaves used for all stomach troubles. (44:20) **Keres, Western** *Analgesic* Infusion of leaves used for headache. *Antirheumatic (External)* Plants beaten into a paste, rubbed on body parts and heated with hot stones for rheumatism. Infusion of plant used to bathe body parts for rheumatism. Compound decoction of leaves boiled into a thick paste used as a salve or liniment for athletes. *Diaphoretic* Plant used as an ingredient in the sweat bath. *Febrifuge* Infusion of leaves used for fevers. (171:27) *Kidney Aid* Six-inch piece of root eaten for kidney trouble. (171:29) *Strengthener* Infusion of plant taken or applied to a weak patient as a strengthener. *Veterinary Aid* Compound decoction of leaves boiled into a thick paste used as a salve or liniment for horses. (171:27) **Omaha** *Ceremonial Medicine* Plant used to begin any ceremonial in order to drive away evil influences. *Disinfectant* Decoction of plant used as a wash for purification. (70:134, 135) *Gastrointestinal Aid* Decoction of plant taken for stomach troubles and many other ailments. (70:135) *Witchcraft Medicine* Plant used as incense to exorcise evil powers. (70:134, 135) **Paiute** *Cathartic* Infusion taken as a physic. (118:42) **Pawnee** *Ceremonial Medicine* Plant used to begin any ceremonial in order to drive away evil influences. *Disinfectant* Decoction of plant used as a wash for purification. (70:134, 135) *Gastrointestinal Aid* Decoction of plant taken for stomach troubles and many other ailments. (70:135) *Witchcraft Medicine* Plant used as incense to exorcise evil powers. (70:134, 135) **Ponca** *Ceremonial Medicine* Plant used to begin any ceremonial in order to drive away evil influences. *Disinfectant* Decoction of plant used as a wash for purification. (70:134, 135) *Gastrointestinal Aid* Decoction of plant taken for stomach troubles and many other ailments. (70:135) *Witchcraft Medicine* Plant used as incense to exorcise evil powers. (70:134, 135) **Thompson** *Dermatological Aid* Bark used by young girls to wipe mouths after eating to prevent hair growth. (164:510) *Unspecified* Plant used medicinally for unspecified purpose. (164:468) **Winnebago** *Ceremonial Medicine* Plant used to begin any ceremonial in order to drive away evil influences. *Disinfectant* Decoction of plant used as a wash for purification. (70:134, 135) *Gastrointestinal Aid* Decoction of plant taken for stomach troubles and many other ailments. (70:135) *Witchcraft Medicine* Plant used as incense to exorcise evil powers. (70:134, 135)
- **Food**—**Havasupai** *Preserves* Seeds parched, ground, kneaded into seed butter, and eaten with fruit drinks or spread on bread. (197:67)
- **Dye**—**Navajo** *Yellow* Leaves used to make a soft yellow dye. (55:81)
- **Other**—**Apache, Chiricahua & Mescalero** *Ceremonial Items* Plant used in ceremonial contexts. (33:24) **Cheyenne** *Cooking Tools* Branches used to remove the spines of prickly pear cacti fruits. (83:16) **Dakota** *Toys & Games* Plant tops chewed and used for popgun wads. (70:116) **Navajo** *Ceremonial Items* Used as a wand when practicing for the Night Chant. Bunches of plant, with other plants, tied to corners of hoops used in unraveling ceremonial objects. (55:81) **Omaha** *Toys & Games* Plant tops chewed and used for popgun wads. **Pawnee** *Toys & Games* Plant tops chewed and used for popgun wads. **Ponca** *Toys & Games* Plant tops chewed and used for popgun wads. (70:116) **Thompson** *Stable Gear* Bark used to make saddle blankets. (164:500) **Winnebago** *Toys & Games* Plant tops chewed and used for popgun wads. (70:116)

Artemisia spinescens D. C. Eat., Bud Sagebrush
- **Drug**—**Paiute** *Analgesic* Decoction of branches taken for chronic stomach troubles and cramps. *Antirheumatic (External)* Decoction of stems and leaves used as a wash for rheumatism. *Cold Remedy* Decoction of root taken for colds. *Cough Medicine* Decoction of root taken for coughs. *Dermatological Aid* Green leaves rubbed on bed patients to prevent bedsores. Mashed, green leaves mixed with tobacco and used as a salve for sores or bruises. Poultice of mashed, green leaves applied to "draw out boils." Poultice of mashed green leaves or young branches applied to swellings. *Diuretic* Decoction of flowers and leaves taken for "stoppage of the bladder." *Gastrointestinal Aid* Decoction of branches taken for stomach trouble, cramps, or indigestion. *Pulmonary Aid* Decoction of root taken for chest congestion, coughs, or colds. Strained decoction of leaves and flowers taken for tubercular hemorrhage. (180:43, 44) *Urinary Aid* Plant juice heated and taken for bladder trouble. (118:41) **Shoshoni** *Dermatological Aid* Crushed, moistened leaves rubbed onto the skin for irritations and rashes. Poultice of whole plant, either fresh or boiled, applied to rash or itch. *Hemostat* Decoction of leaves sniffed for nosebleed. *Misc. Disease Remedy* Decoction of whole plant taken and used as a wash for influenza. *Pulmonary Aid* Decoction of branches taken for tubercular hemorrhage. (180:43, 44)

Artemisia suksdorfii Piper, Coastal Wormwood
- **Other**—**Nitinaht** *Incense & Fragrance* Plants dried and hung in houses for fresh scents. (186:97)

Artemisia tilesii Ledeb. Tilesius's Wormwood
- **Drug**—**Eskimo** *Cancer Treatment* Plant used as an antitumor agent. *Disinfectant* Plant used as an infection inhibitor. *Febrifuge* Plant used as a fever medicine. (149:326) **Eskimo, Alaska** *Antirheumatic (External)* Infusion of fresh or dried leaves used for arthritic-like ailments. Infusion of stems used for discomfort of swollen areas. *Hemostat* Poultice of leaves used to stop bleeding. *Laxative* Infusion of seed heads and plant tops used as a laxative. *Respiratory Aid* Plant boiled and the vapors inhaled for congestion. *Tonic* Infusion of plant taken daily as a tonic. (1:38) **Eskimo, Inuktitut** *Dermatological Aid* Poultice of plant applied to skin infections. (202:186) **Eskimo, Kuskokwagmiut** *Dermatological Aid* Poultice of dried, shredded plant applied to skin infections. *Orthopedic Aid* Plant used as switch during steam bath for a sprained or sore limb. (122:33) **Eskimo, Western** *Gastrointestinal Aid* Decoction of plant taken for stomachache. (108:13) *Orthopedic Aid* Poultice of heated leaves applied to painful joint and used internally. (108:5, 13) **Tanaina** *Antirheumatic (Internal)* Plant used for diseases from rheumatism to tuberculosis. *Misc. Disease Remedy* Plant used for diseases from rheumatism to tuberculo-

sis. *Tuberculosis Remedy* Plant used for diseases from rheumatism to tuberculosis. (149:329) **Tanana, Upper** *Antirheumatic (External)* Decoction of above-the-ground part of the plant used as a body wash for aches and pains. Poultice of leaves applied or decoction of leaves used as a wash for swellings and body aches. *Blood Medicine* Poultice of leaves applied or decoction of leaves used as a wash for blood poisoning. *Cough Medicine* Leaves chewed for coughs. *Dermatological Aid* Poultice of leaves applied or decoction of leaves used as a wash for skin rashes and cuts. *Disinfectant* Poultice of leaves applied or decoction of leaves used as a wash for infections. *Eye Medicine* Decoction of above-the-ground part of the plant used as an eyewash. *Oral Aid* Decoction of above-the-ground part of the plant taken for mouth sores. (102:17)

- *Food*—**Eskimo, Alaska** *Unspecified* Raw shoots peeled and eaten with seal oil. (1:38)
- *Fiber*—**Tanana, Upper** *Building Material* Wood used as a floor covering in the steam bath. (102:17)
- *Other*—**Eskimo, Alaska** *Cleaning Agent* Fresh, crushed leaves rubbed on hands to remove or mask odors after cleaning fish. (1:38) **Eskimo, Inuktitut** *Incense & Fragrance* Used to cover food odors and in the sweat lodges. *Smoke Plant* Used as a tobacco quid additive. (202:186) **Eskimo, Western** *Ceremonial Items* Plant used as a switch during the sweat bath. (108:39) **Tanana, Upper** *Containers* Used under fresh meat and fish to keep them clean. *Tools* Used as a steam bath switch. (102:17)

Artemisia tridentata Nutt., Big Sagebrush

- *Drug*—**Cahuilla** *Disinfectant* Dried leaves and stems burned, in the homes and sweat houses, as a disinfectant. *Respiratory Aid* Dried leaves and stems burned, in the homes and sweat houses, as an air purifier. (15:43) **Coahuilla** *Gastrointestinal Aid* Decoction of leaves taken for stomach complaints. (13:78) **Diegueño** *Cold Remedy, Cough Medicine,* and *Respiratory Aid* Infusion of fresh or dried leaves taken for a bad cold with coughing and bronchitis. (88:220) **Flathead** *Cold Remedy* Infusion taken for colds. *Pulmonary Aid* Infusion taken for pneumonia. (82:45) **Gosiute** *Antirheumatic (External)* and *Antirheumatic (Internal)* Plant used externally and internally for rheumatism. *Cold Remedy* Plant used for colds. *Cough Medicine* Plant used for coughs. *Febrifuge* Plant used for fevers. (39:351) Infusion of leaves used for febrile conditions. (39:363) *Panacea* Plant used as a panacea. (39:351) **Havasupai** *Cold Remedy* Stems and leaves used for colds.[16] *Cough Medicine* Stems and leaves used for coughs.[16] *Dermatological Aid* Decoction of leafy stems and leaves used as a wash for sores or pimples. *Gastrointestinal Aid* Stems and leaves used for intestinal upset.[16] *Nose Medicine* Stems and leaves used for runny noses.[16] *Throat Aid* Stems and leaves used for sore throats.[16] (197:246) **Hopi** *Gastrointestinal Aid* Plant used for digestive disorders. (200:34, 94) *Orthopedic Aid* Infusion of leaves taken for ailing ilium. (56:17) **Kawaiisu** *Analgesic* Decoction of plant fumes inhaled for headaches. *Cold Remedy* Decoction of plant fumes inhaled for head colds and chest colds. Decoction or infusion of leaves taken for bad colds. *Cough Medicine* Infusion of leaves taken for colds and coughs. *Herbal Steam* Decoction of plant fumes inhaled for head colds, chest colds, and headaches. *Misc. Disease Remedy* Decoction of leaves taken for influenza or bad colds. (206:13) **Klamath** *Antidiarrheal* Decoction of herbage used internally for diarrhea. (45:105) Decoction of herbs taken for diarrhea. (163:131) *Antirheumatic (External)* Poultice of herbage used as a substitute for liniment. *Eye Medicine* Decoction of herbage used as an eyewash. (45:105) Decoction of herbs used as an eyewash. *Orthopedic Aid* Smashed herbs used as substitute for liniment. (163:131) **Lakota** *Unspecified* Used as a medicine. (139:36) **Montana Indian** *Antidiarrheal* Decoction of herb taken for diarrhea. *Eye Medicine* Decoction of herb used externally as an eyewash. *Orthopedic Aid* Mashed herbs used as a substitute for liniment and as a poultice. (19:7) **Navajo** *Analgesic* Compound of plants used for headaches. (55:81) *Ceremonial Medicine* Plant used for religious and medicinal ceremonies. (90:158) *Cold Remedy* Plant used for colds. *Febrifuge* Plant used for fevers. *Gastrointestinal Aid* Decoction of plants taken for stomachaches. *Gynecological Aid* Infusion of plants taken by women as an aid for deliverance. *Sports Medicine* Plant taken before long hikes and athletic contests to rid the body of lingering, undesirable things. (55:81) **Navajo, Kayenta** *Laxative* Plant used for constipation. *Snakebite Remedy* Infusion of plant taken and used as a lotion for water snake bites. (205:45) **Navajo, Ramah** *Analgesic* Decoction of leaves taken for postpartum pain. *Cough Medicine* Decoction of leaves taken for "big cough." *Dermatological Aid* Poultice of wet leaves applied to swellings. *Diaphoretic* Plant used in a sweat bath medicine. *Gynecological Aid* Decoction of leaves taken for postpartum pain. *Veterinary Aid* Cold infusion of leaves used as lotion for cuts on sheep. (191:48) **Okanagan-Colville** *Cold Remedy* Decoction of leaves and branches taken for colds. Infusion of roots taken for colds. *Diaphoretic* Decoction of leaves and branches taken to cause sweating during a cold. *Oral Aid* Decoction of leaves and branches taken for tonsillitis. *Throat Aid* Decoction of leaves and branches taken for sore throats. Infusion of roots taken for sore throats. (188:78) **Paiute** *Analgesic* Burning plant used as an inhalant for headache. Decoction of branches taken for headache. Decoction of branches taken for stomachaches, especially children's. (180:44–47) *Ceremonial Medicine* Sagebrush used by dancers to pat themselves to be made spiritually clean, curing ceremonies. (111:119) *Cold Remedy* Plant chewed for colds. (167:317) Burning branches used as an inhalant for head colds. Compound decoction of plant tops taken for colds. Decoction of leaves taken or raw leaves eaten for colds. Poultice of mashed, green leaves applied for chest colds. (180:44–47) *Dermatological Aid* Poultice of mashed leaves applied to burns and sores. Infusion of leaves applied to the scalp as a hair tonic. (111:119) Decoction of leaves used as an antiseptic wash for cuts, wounds, or sores. Poultice of wet, steeped leaves applied to bullet wounds. Pulverized leaves used as a talcum powder for babies. (180:44–47) *Diaphoretic* Infusion of leaves taken to produce sweating during a fever. (167:317) *Disinfectant* Decoction of leaves used as an antiseptic wash for cuts, wounds, or sores. (180:44–47) *Emetic* Infusion of leaves taken as an emetic for respiratory diseases. (111:119) *Febrifuge* Infusion of leaves taken to produce sweating during a fever. (167:317) Decoction of leaves taken for malarial fever. (180:44–47) *Gastrointestinal Aid* Leaves chewed for indigestion. (118:45) Plant chewed for stomach disorders. (167:317) Decoction of branches used for stomachaches, especially children's. Raw leaves chewed for indigestion. *Gynecological Aid* Decoction of plant taken as a general tonic, especially after childbirth. *Misc. Disease Remedy* Decoction of

Artemisia tridentata (Asteraceae)

leaves taken for malarial fever. *Orthopedic Aid* Decoction of branches used as liniment for lumbago, muscular cramps and sore feet. *Pediatric Aid* Decoction of branches used for stomachaches, especially children's. Finely pulverized dried leaves used as a baby powder. *Pulmonary Aid* Compound decoction of leaves taken and poultice of decoction used for pneumonia. Poultice of mashed, green leaves applied for chest colds. (180:44–47) *Respiratory Aid* Infusion of leaves taken as an emetic for respiratory diseases. (111:119) *Tonic* Decoction of plant taken as a general tonic, especially after childbirth. (180:44–47) **Paiute, Northern** *Analgesic* Small plant pieces stuffed into the nostrils for headaches. Decoction of leaves taken for headaches. *Antidiarrheal* Decoction of leaves taken for diarrhea. *Antirheumatic (External)* Poultice of ground leaves and tobacco applied to swellings on adults or children. *Cold Remedy* Decoction of leaves taken for colds. Small plant pieces stuffed into the nostrils for colds. (59:128) Branches put over a bed of ashes and slept on for colds. (59:125) *Emetic* Decoction of leaves taken as an emetic. *Febrifuge* Poultice of ground leaves and cold water applied to the body for fevers. Poultice of ground leaves and tobacco applied to children for fevers. (59:128) Branches put over a bed of ashes and slept on for fevers. (59:125) *Pediatric Aid* Poultice of ground leaves and tobacco applied to children for fevers. Poultice of ground leaves and tobacco applied to swellings on adults or children. *Stimulant* Blossoms dipped in water and the blossomed branch used to comb the hair for fainting spells. (59:128) **Salish** *Cold Remedy* Plant used for colds. (178:294) **Sanpoil** *Cold Remedy* Infusion of pulverized leaves and stems taken for colds. *Diaphoretic* Infusion of stem tips and seedpods taken as a diaphoretic. *Gastrointestinal Aid* Infusion of stem tips and seedpods taken for indigestion and biliousness. *Laxative* Various infusions of leaves, stems, and seedpods taken as laxatives. *Misc. Disease Remedy* Infusion of pulverized leaves and stems taken for "la grippe." *Tuberculosis Remedy* Infusion of stem tips and seedpods taken for indigestion and tuberculosis. (131:217) **Shoshoni** *Analgesic* Decoction of branches taken for stomach cramps. Decoction of leaves used as a wash for headaches. Hot poultice of branches applied for various aches and pains. Poultice of crushed, moistened, green leaves applied to forehead for headache. *Antidote* Decoction of leaves taken or leaf chewed as an antidote for any poisoning. *Antirheumatic (External)* Poultice of boiled branches applied for aches and pains, especially rheumatism. *Cold Remedy* Compound decoction of plant tops taken for colds. Decoction of leaves taken or raw leaves eaten for colds. *Cough Medicine* Decoction of branches with salt taken for coughs. Decoction of leaves and salt taken for pneumonia coughs. *Dermatological Aid* Decoction of branches used for red ant bites. Decoction of leaves used as an antiseptic wash for cuts, wounds, or sores. Poultice of leaf decoction or powdered branches used for sores, cuts, or wounds. *Diaphoretic* Decoction of leaves taken to cause sweating and break a fever. *Disinfectant* Decoction of leaves used as an antiseptic wash for cuts, wounds, or sores. Warm decoction of leaves used as an antiseptic bath for newborns. *Emetic* Decoction of plant tops taken for colds and an overdose acts as an emetic. *Eye Medicine* Poultice of steeped leaves applied to inflamed eyes. *Febrifuge* Decoction of leaves taken to cause sweating and break a fever. (180:44–47) *Gastrointestinal Aid* Leaves chewed for indigestion. (118:45) Hot decoction of branches taken for stomach cramps. Raw leaves chewed for indigestion. *Gynecological Aid* Decoction of plant taken as a general tonic, especially after childbirth. *Orthopedic Aid* Decoction of branches used as a wash or liniment for lumbago or muscular cramps. *Pediatric Aid* Decoction of leaves used as a warm antiseptic bath for newborn babies. *Pulmonary Aid* Decoction of leaves with salt taken for cough of pneumonia. *Throat Aid* Strained decoction of leaves gargled for sore throat. *Tonic* Decoction of plant taken as a general tonic, especially after childbirth. *Toothache Remedy* Poultice of mashed leaves applied to cheek for toothache. (180:44–47) **Shuswap** *Cold Remedy* Decoction of plants inhaled for a bad cold. *Disinfectant* Plant used to fumigate the house and keep germs off. *Witchcraft Medicine* Plant used to fumigate the house and keep germs off. (123:58) **Tewa** *Carminative* Leaves eaten as a carminative. *Cough Medicine* Leaves eaten as a cough remedy and expectorant. *Expectorant* Leaves chewed and swallowed as an expectorant. *Gastrointestinal Aid* Leaves eaten for indigestion. (138:45) **Thompson** *Antirheumatic (External)* Decoction of plant used as a bath to "rest your bones" and relax you. Decoction of plant used as a bath for muscular ailments. (187:172) *Cold Remedy* Decoction of leaves taken and poultice or plugs of leaves used in nostrils for colds. (164:459) Decoction of leafless twigs taken for colds. Decoction of branches taken for colds. One informant's mother said that this sage was too strong and powerful to drink. She said, "you wouldn't have any more kids. No children." She said that it seems to close something up in one's system, that it is just too powerful. Weak decoction of plant used as a wash for colds. *Panacea* Dried branch smoke used to fumigate the house, to protect the inhabitants against sickness. (187:172) *Stimulant* Bruised leaves used as an inhalant to revive a patient. (164:459) *Throat Aid* Decoction of leafless twigs taken for laryngitis. (187:172) *Tuberculosis Remedy* Decoction of stems and leaves taken for consumption and colds. (164:459) **Ute** *Unspecified* Decoction of leaves used as a medicine. (38:32) **Washo** *Cold Remedy* Decoction of leaves taken for colds. *Disinfectant* Branches burned as fumigant for sickroom or for utensils used for childbirth. *Tonic* Decoction of plant taken as a general tonic. (180:44–47) **Zuni** *Antirheumatic (External)* Infusion of leaves used for body aches. *Cold Remedy* Infusion of leaves taken as a cold medicine. *Dermatological Aid* Leaves in shoes used for athlete's foot infection, fissures between toes, and foot deodorant. (27:374)

- *Food*—**Apache, White Mountain** *Beverage* Used to make tea. *Spice* Used as a seasoning. (136:155) **Paiute** *Staple* Seeds roasted, ground into flour, and eaten with water. *Starvation Food* Seeds used, generally mixed with other seeds, in times of food shortages. (167:243) **Paiute, Northern** *Candy* Gum chewed. (59:53)
- *Fiber*—**Cahuilla** *Building Material* Shoots laid across the rafters for roofing material or used in the construction of the walls. (15:43) **Havasupai** *Building Material* Plant used for thatch. (197:246) **Kawaiisu** *Clothing* Pounded bark used as a lining or wrapper inside winter shoes. (206:13) **Navajo** *Building Material* Used between the poles of the sweat house to prevent the sand from sifting through. (55:82) **Navajo, Ramah** *Brushes & Brooms* Leafy stems tied together with wire and used for brooms. (191:48) **Paiute** *Building Material* Wood used for drills, hearths, and tinder in the creation of fire by friction. *Clothing* Bark used to make cloth and sandals. Bark frayed and stuffed into moccasins for added warmth. *Cordage* Bark used to

make cordage. (111:119) **Thompson** *Clothing* Fibrous bark used in weaving bags and clothing. *Mats, Rugs & Bedding* Fibrous bark used in weaving mats. (187:172)
- *Other*—**Flathead** *Fuel* Used for firewood in absence of other wood. (82:45) **Gosiute** *Containers* Leaves used as a covering over berries and other foods preserved in caches. *Fuel* Wood used to produce fire by friction. (39:363) **Havasupai** *Containers* Bark used as a plug to keep water from spilling out of a water jug. (197:246) **Kawaiisu** *Ceremonial Items* Seeds thrown into a fire to explode "like firecrackers" during celebrations. *Containers* Used as preferred material for both hearth. Bunched bark used as a stopper for the basketry water bottle. *Fuel* Wood used to roast pinyon seeds. *Tools* Used as preferred material for foreshaft of composite drill for the fire making. Bitter wood used as a head scratcher by menstruating women. Any other kind of wood would cause the hair to fall out and the face to wrinkle. Tar-like lac gathered into a ball, softened in fire, and shaped into awl and knife handles. (206:13) **Klamath** *Fuel* Twig used as a twirling stick to produce fire by friction. (45:88) Wood used for fuel. *Tools* Dead stems used as twirling sticks. (45:105) **Montana Indian** *Tools* Dead twigs used for a twirling stick in fire making. (19:7) **Navajo** *Tools* Wood used in the end of the fire drill. (55:81) **Okanagan-Colville** *Fuel* Bark used as tinder and for making friction fires. Wood used as fuel. *Hide Preparation* Wood used for smoking hides during the tanning process. (188:78) **Okanagon** *Containers* Used to make quiver cases. *Hide Preparation* Used to smoke hides. *Stable Gear* Used to make saddle blankets. (125:40) **Paiute** *Decorations* Blossoms and leaves used as personal decorations in a spring dance. *Fuel* Wood used for tinder in the creation of fire by friction. *Lighting* Bark wound about a stick and used as a torch. *Tools* Wood used for drills, hearths, and tinder in the creation of fire by friction. (111:119) **Tewa** *Fuel* Dry bushes used for fuel in absence of other firewood. (138:45) **Thompson** *Containers* Used to make quiver cases. (125:40) Fibrous bark used in weaving mats, bags, and clothing. (187:172) *Hide Preparation* Used to smoke hides. *Stable Gear* Used to make saddle blankets. (125:40) **Tubatulabal** *Cooking Tools* Used as brush beds for roasting pinyon cones. (193:12) *Fuel* Brush burned to roast cones. (193:17) **Washo** *Ceremonial Items* Plant used for medicine man's costume. (118:51)

Artemisia tripartita Rydb., Threetip Sagebrush
- *Drug*—**Okanagan-Colville** *Cold Remedy* Infusion of roots taken for colds. Decoction of leaves and branches taken for colds. *Diaphoretic* Decoction of leaves and branches taken to cause sweating during a cold. *Oral Aid* Decoction of leaves and branches taken for tonsillitis. *Throat Aid* Infusion of roots taken for sore throats. Decoction of leaves and branches taken for sore throats. (188:79)
- *Other*—**Okanagan-Colville** *Fuel* Bark used as tinder and for making friction fires. Wood used as fuel. *Hide Preparation* Wood used for smoking hides during the tanning process. (188:79)

Artemisia tripartita Rybd. ssp. *tripartita*, Idaho Threetip Sagebrush
- *Drug*—**Navajo** *Analgesic* Plant used for headaches. (as *A. trifida* 55:97) *Ceremonial Medicine* Plant burned to charcoal and given to patient to blacken legs and forearms in Mountain Chant Ceremony. *Dermatological Aid* Infusion of plant used as a wash for wounds caused by removed corns. (as *A. trifida* 55:82)
- *Food*—**Gosiute** *Unspecified* Seeds formerly used for food. (as *A. trifida* 39:362)
- *Other*—**Navajo** *Containers* Bark used to stuff into the necks of bottles to keep the water from spilling out. (as *A. trifida* 55:82)

Artemisia vulgaris L., Common Wormwood
- *Drug*—**Karok** *Analgesic* Infusion of plant taken by women for the pains of afterbirth. *Cold Remedy* Poultice of branches applied for colds. *Gynecological Aid* Infusion of plant taken by women for the pains of afterbirth. *Panacea* Poultice of branches applied for any kind of sickness. (148:390) **Kiowa** *Anthelmintic* Plant used as a "worm" medicine. (192:57) **Miwok** *Analgesic* Leaves worn in nostrils for headaches. *Antirheumatic (Internal)* Decoction of leaves taken for rheumatism. *Ceremonial Medicine* Leaves worn in nostrils by mourners when crying, the pungent odor clearing the head. *Witchcraft Medicine* Leaves rubbed on body to keep ghosts away. Plant worn on a necklace to prevent dreaming of the dead. Poisoned leaves carried to avoid personal injury. Leaves rubbed on corpse handlers to ward off ghosts of the deceased. (12:167) **Paiute** *Cold Remedy* Poultice of crushed leaves applied to the chest for colds. (104:197) *Dermatological Aid* Decoction of tops applied to gonorrheal sores. *Orthopedic Aid* Plant used for female backache and knee-ache. *Venereal Aid* Decoction of tops applied to gonorrheal sores. (167:317) **Pomo** *Dermatological Aid* Poultice of heated leaves applied to newborn baby's navel. *Gynecological Aid* Decoction of leaves taken to stop excessive menstruation. Plant used in childbirth. *Pediatric Aid* Poultice of heated leaves applied to newborn baby's navel. (66:15) **Tlingit** *Herbal Steam* and *Pulmonary Aid* Plant taken or used in steam bath for pleurisy. (107:283)
- *Fiber*—**Kiowa** *Mats, Rugs & Bedding* Used to make cushions for the worshippers in the peyote ceremony. (192:57)
- *Other*—**Kiowa** *Cleaning Agent* Leaves rubbed on the face and hands as a purifying agent. (192:57)

Artemisia vulgaris var. *kamtschatica* Besser, Kamtschat's Wormwood
- *Drug*—**Aleut** *Antirheumatic (External)* Heated plant used externally as a "switch" for rheumatism. (as *A. unalaskensis* 8:426) *Dermatological Aid* Poultice of heated leaves applied to minor cuts. *Orthopedic Aid* Poultice of heated leaves applied to sore muscles. (as *A. unalaskensis* 8:425) *Tonic* Decoction of leaves taken as a tonic, especially good "for dying persons." (as *A. unalaskensis* 8:427)

Arthrocnemum, Chenopodiaceae
Arthrocnemum subterminale (Parish) Standl., Parish's Glasswort
- *Food*—**Cahuilla** *Staple* Seeds ground into a meal. (as *Salicornia subterminalis* 15:135)

Artocarpus, Moraceae
Artocarpus altilis (Parkinson) Fosberg, Breadfruit
- *Drug*—**Hawaiian** *Dermatological Aid* Milk and other plants used for skin diseases, boils, cuts, and cracked skin. *Oral Aid* Milk and other plants used for mouth sores. (as *A. incisa* 2:38)
- *Food*—**Hawaiian** *Fruit* Fruit cooked and eaten. (as *A. incisa* 2:38)

Aruncus, Rosaceae

Aruncus dioicus (Walt.) Fern., Bride's Feathers
- **Drug**—**Cherokee** *Dermatological Aid* Beaten root applied to bee stings on the face. *Eye Medicine* Beaten root applied to bee stings on the eye. *Gynecological Aid* Infusion of root used to prevent excessive bleeding at childbirth. *Orthopedic Aid* Infusion of root used to bathe swollen feet. *Urinary Aid* Infusion taken for excessive urination. (80:36) **Haihais** *Unspecified* Roots used medicinally for unspecified purpose. **Kitasoo** *Unspecified* Roots used medicinally for unspecified purpose. (43:342) **Thompson** *Gastrointestinal Aid* Decoction of washed roots taken for internal ailments. Infusion of plant taken for internal wounds and stomach problems. *Misc. Disease Remedy* Decoction of washed roots taken for influenza. It was said that too much of the decoction should not be taken because it would make you sick. (187:257)

Aruncus dioicus var. *acuminatus* (Rydb.) Rydb. ex Hara, Bride's Feathers
- **Drug**—**Thompson** *Cold Remedy* Decoction of root taken for colds and influenza. *Dermatological Aid* Decoction of root taken for "swellings" and stalk ashes and grease used as a salve. *Gastrointestinal Aid* Decoction of root taken for indigestion and general stomach disorders. *Misc. Disease Remedy* Decoction of roots taken for colds and influenza. *Orthopedic Aid* Salve of stalk ashes and grease used for paralysis. (as *A. acuminatus* 164:457)

Aruncus dioicus var. *vulgaris* (Maxim.) Hara, Bride's Feathers
- **Drug**—**Bella Coola** *Analgesic* Decoction of root taken for stomach pain. (as *A. sylvester* 150:59) *Antidiarrheal* Infusion of roots used for diarrhea. (as *A. sylvester* 184:208) *Diuretic* Decoction of root taken as a diuretic and for gonorrhea. *Gastrointestinal Aid* Decoction of root taken for stomach pain. (as *A. sylvester* 150:59) Infusion of roots used for stomach pain. (as *A. sylvester* 184:208) *Misc. Disease Remedy* Decoction of root in grease of mountain goat taken for smallpox. *Venereal Aid* Decoction of root taken for gonorrhea and stomach pain. (as *A. sylvester* 150:59) **Klallam** *Dermatological Aid* Salve of root ashes rubbed on sores. (as *A. sylvester* 79:33) **Kwakiutl** *Cough Medicine* Scraped roots held in the mouth for coughs. Dried root held in mouth for cough. (as *A. aruncus* 20:381) Dried, soaked root held in the mouth for coughing. *Love Medicine* Root used as a love charm. (as *A. sylvester* 183:288) **Lummi** *Dermatological Aid* Plant used for sores. *Misc. Disease Remedy* Raw leaves chewed for smallpox. (as *A. sylvester* 79:33) **Makah** *Antirheumatic (External)* Infusion of roots used for rheumatism. (as *A. sylvester* 67:261) *Dermatological Aid* Plant used for sores. (as *A. sylvester* 79:33) *Internal Medicine* Root juice taken for internal healing. (as *A. sylvester* 67:261) *Kidney Aid* Infusion of roots taken for kidney trouble. (as *A. sylvester* 79:33) Mixture of pounded roots taken for kidney pain. (as *A. sylvester* 186:116) *Unspecified* Used for medicine. Roots used to make a very good medicine. (as *A. sylvester* 67:261) *Venereal Aid* Infusion of roots taken for gonorrhea. (as *A. sylvester* 79:33) **Nitinaht** *Febrifuge* Infusion of pounded roots taken for bad fevers. *Misc. Disease Remedy* Infusion of pounded roots taken for measles-like illnesses. (as *A. sylvester* 186:116) **Quileute** *Dermatological Aid* Poultice of scraped roots applied to sores. *Tonic* Infusion of pounded roots taken as a general tonic. **Quinault** *Dermatological Aid* Plant used for sores. **Skagit** *Cold Remedy* Infusion of roots taken for colds. *Dermatological Aid* Plant used for sores. *Throat Aid* Infusion of roots taken for sore throats. Poultice of twig and root ashes with bear grease applied to throat swellings. (as *A. sylvester* 79:33) **Tlingit** *Blood Medicine* Infusion of root used for "diseases of the blood." (as *Spiraea aruncus* 107:283)
- **Dye**—**Makah** *Unspecified* Roots used to make dye. (as *A. sylvester* 67:261)

Arundinaria, Poaceae

Arundinaria gigantea (Walt.) Walt. ex Muhl., Giant Cane
- **Drug**—**Houma** *Kidney Aid* and *Stimulant* Decoction of root taken to stimulate the kidneys and "renew strength." (as *A. macrosperma* 158:61) **Seminole** *Cathartic* Decoction of root used as a cathartic. (169:275)
- **Fiber**—**Cherokee** *Basketry* Used to make burden baskets. (80:21) Used to make baskets. (80:28) *Building Material* Used to make cane webbing, plastered with mud, supported with wood, and used as a dwelling. (80:21) **Choctaw** *Basketry* Plant used in basketry. (as *A. macrosperma* 25:13)
- **Other**—**Cherokee** *Fuel* Used for fuel. *Hunting & Fishing Item* Wood used to make arrow shafts and blowguns with darts for small game. *Lighting* Used to make candles. *Musical Instrument* "Joint of reed" used to make flutes. *Weapon* Used as knives as last resort in committing suicide in 1738 smallpox epidemic. (80:28) **Choctaw** *Hunting & Fishing Item* Plant used to make blowguns and darts. (as *A. macrosperma* 25:18) **Houma** *Hunting & Fishing Item* Young shoots used to make arrow shafts. Stalks hollowed and used as blowguns. (as *A. macrosperma* 158:61) **Seminole** *Hunting & Fishing Item* Plant used to make blowguns, knives, arrows, and bows. *Musical Instrument* Plant used to make flutes. *Tools* Plant used to make blowing tubes used for working melted silver. *Weapon* Plant used to make war spears. (169:495)

Arundinaria gigantea ssp. *tecta* (Walt.) McClure, Switch Cane
- **Drug**—**Choctaw** *Analgesic* Decoction of roots taken for breast pain. (as *A. tecta* 177:6) **Houma** *Kidney Aid* and *Stimulant* Decoction of root taken to stimulate the kidneys and "renew strength." (as *A. tecta* 158:61)
- **Other**—**Houma** *Hunting & Fishing Item* Stalks hollowed and used as blowguns. Young shoots used to make arrow shafts. (as *A. tecta* 158:61)

Arundinaria sp., Bamboo
- **Other**—**Keres, Western** *Cash Crop* Cane made into cigarettes used to pay for medicine given by a cheani, a ceremonial leader. *Musical Instrument* Large stems used as flutes. *Toys & Games* Four pieces of hollow cane used as implements in the winter game of bish-i. (171:29)

Arundo, Poaceae

Arundo donax L., Giant Reed
- **Drug**—**Cahuilla** *Orthopedic Aid* Used as a splint for broken limbs. (15:102)
- **Dye**—**Papago** *Yellow* Pollen used as a yellow dye. (34:51)
- **Other**—**Cahuilla** *Musical Instrument* Used to make a flute, usually played by men. (15:102) **Navajo** *Ceremonial Items* Reed made into a

whistle and attached to the collar of an otter skin for the Night Chant. Reed, with tassels, used in the special pouch required for every chant. Reed used to make prayer sticks. (55:24)

Asarum, Aristolochiaceae

Asarum canadense L., Canadian Wildginger

- **Drug**—**Abnaki** *Cold Remedy* Decoction of plant and another plant used for colds. (144:166) *Cough Medicine* Used for coughs. (144:154) **Algonquin, Quebec** *Anticonvulsive* Infusion of roots given to infants for convulsions. *Febrifuge* Infusion of roots taken for fevers. *Pediatric Aid* Infusion of roots given to infants for convulsions. (18:159) **Cherokee** *Abortifacient* Used for "scant or painful menstruation" and infusion taken "to start periods." (80:35, 36) *Analgesic* Decoction of plant taken to cause vomiting for stomach pain. (177:21) *Anthelmintic* Root used as powerful stimulant and for worms. *Antidiarrheal* Infusion taken for "flux." *Blood Medicine* Compound infusion of root used "for blood." (80:35, 36) *Breast Treatment* Infusion of whole plant used as a wash for swollen breasts. (177:21) *Cold Remedy* Root used for colds. *Cough Medicine* Root used for coughs. *Dermatological Aid* Fresh leaves applied to wounds and liquid or salve used on sores. (80:35, 36) *Emetic* Infusion of plant taken as an emetic for swollen breasts and stomach pain. (177:21) *Eye Medicine* Snuff of dried leaves used for head and eyes. *Febrifuge* Taken for typhus fever, "ague and fever." *Gastrointestinal Aid* Compound infusion used for poor digestion. (80:35, 36) Decoction of plant taken to cause vomiting for stomach pain. (177:21) *Gynecological Aid* Used for "scant or painful menstruation" and infusion taken to start periods. *Heart Medicine* Compound infusion used for poor digestion and infusion of root used "for heart trouble." *Misc. Disease Remedy* Taken for typhus fever, "ague and fever." *Sedative* Leaves, roots, or blossoms used for hysterical or nervous debility. *Stimulant* Root used as powerful stimulant. (80:35, 36) **Chippewa** *Adjuvant* Root combined with other herbs to strengthen their action. (53:342) *Dermatological Aid* Compound poultice of chopped root applied to inflammations. (53:348) Roots used for bruises and contusions. (71:129) *Gastrointestinal Aid* Root cooked with foods to aid digestion. (53:342) *Orthopedic Aid* Compound poultice of root applied to fractured bones. (53:366) **Iroquois** *Adjuvant* "Plant may be added to all kinds of medicine to make them stronger." (87:309) *Analgesic* Cold infusion of roots given to children with headaches and fevers. (87:308) Infusion of plant taken for long-lasting headaches. (87:309) *Anticonvulsive* Compound decoction given to children with convulsions. (87:308) Plant and other plants given to children with convulsions and fevers. (141:41) *Antiemetic* Compound decoction taken for heaves. *Blood Medicine* Complex compound decoction taken as blood purifier. (87:310) *Cathartic* Infusion of root taken as a spring tonic by the old and works as a physic. (87:309) *Cold Remedy* Decoction of roots taken for scarlet fever, colds, and "peevies." Infusion of roots taken for colds and typhoid fever. (87:310) *Cough Medicine* Decoction of root taken for coughs and measles. (87:309) Compound decoction taken for coughs. (87:310) *Dermatological Aid* Compound used for boils. (87:311) *Diaphoretic* Cold infusion or decoction of roots taken for any kind of fever and sweating. *Dietary Aid* Decoction taken to become fit to visit the sick and for the lack of appetite. *Disinfectant* Infusion or decoction used as hand and face wash for ghost contamination. (87:309) *Febrifuge* Plant used several ways for adults and children with fevers. (87:308) Plant and other plants given to children with convulsions and fevers. (141:41) *Misc. Disease Remedy* Compound decoction of roots taken for typhoid, measles, and scarlet fever. *Pediatric Aid* Compound infusion given "when babies cry until they hold their breath." Infusion of roots given to children with headaches, fevers, or convulsions. (87:308) Plant and other plants given to children with convulsions and fevers. (141:41) *Psychological Aid* Infusion taken to prevent bad dreams caused by the dead. (87:308) *Pulmonary Aid* Infusion taken and used as a wash for fever and chest congestion, then vomit. *Respiratory Aid* Infusion of root taken for asthma. *Stimulant* Decoction of roots taken for fevers, colds, and as a stimulant. (87:310) Compound used for laziness. (87:311) *Throat Aid* Poultice used for sore throat. (87:310) *Tonic* Infusion of root taken as a spring tonic by the old and works as a physic. (87:309) Compound infusion taken for fevers and as a general tonic. (87:311) *Tuberculosis Remedy* Compound infusion of roots taken for tuberculosis. (87:310) *Urinary Aid* Decoction of roots taken for urinary disorders and urine stoppage. (87:309) *Venereal Aid* Complex compound decoction taken for venereal disease. *Veterinary Aid* Compound decoction given to horses for coughs or heaves. (87:310) Decoction with whisky given to horses that are sick from not being used. (87:311) *Witchcraft Medicine* Plant used several ways to detect or protect people from witchcraft. (87:308) Decoction or infusion used internally or externally before visiting the sick. (87:309) **Malecite** *Gastrointestinal Aid* and *Pediatric Aid* Infusion of small roots used by children with cramps. (116:255) **Menominee** *Gastrointestinal Aid* Decoction of root used for indigestion. (54:130) Fresh or dried root used as a mild stomachic. Root eaten to protect "weak stomach" so that desired food may be eaten. (151:24, 25) **Meskwaki** *Adjuvant* Used as a seasoner and for sore throats. *Analgesic* Compound used for stomach cramps. *Antidote* Root cooked with spoiled meat to prevent ptomaine poisoning. *Ear Medicine* Cooked root placed in ear for earache or sore ears. *Gastrointestinal Aid* Compound used for stomach cramps. *Hunting Medicine* Root chewed and spittle put on bait to enable fisherman to catch catfish. *Pulmonary Aid* Compound used for lung trouble. *Throat Aid* Used for sore throat and as a medicine used as a seasoning. (152:204) **Micmac** *Analgesic* and *Gastrointestinal Aid* Root used for cramps and as a stomachic. *Tonic* Root used for cramps and as a stomachic. (40:55) **Montagnais** *Panacea* Plant had "general medicinal properties." (156:314) **Ojibwa** *Dietary Aid* Root chewed by sick person as an appetite stimulant. (153:357) *Gastrointestinal Aid* Roots chewed or infusion of roots taken for stomach pain. (5:2250) **Potawatomi** *Antiemetic* Root used to help the appetite of persons who could not keep anything in their stomachs. (154:96)

- **Food**—**Chippewa** *Spice* Root used as an appetizer in all cooked foods. (53:318) **Meskwaki** *Spice* Root used as seasoning for mud catfish, to destroy the mud taste and to render them palatable. Root used to cook with an animal that had died, to remove the danger of ptomaine poisoning. (152:255) **Ojibwa** *Dietary Aid* Roots render any meat dish digestible by anyone, even if they are sick. *Spice* Roots processed in lye water and used to season food and take muddy taste away from fish. (153:397) **Potawatomi** *Spice* Root flavored meat or fish and rendered otherwise inedible food, palatable. (154:96)

- **Other**—**Cherokee** *Snuff* Dried leaves pounded and used for snuff.

(80:36) **Chippewa** *Incense & Fragrance* Slightly roasted roots made into a powder and sprinkled on clothing for perfume. (71:129)

Asarum caudatum Lindl., British Columbia Wildginger
- **Drug**—**Bella Coola** *Analgesic* Decoction of plant used externally for headache, intestinal pain, and knee pain. *Antirheumatic (External)* Decoction of plant used externally for knee pain. *Gastrointestinal Aid* Decoction of plant used externally for intestinal pain. Decoction of plant taken for stomach pain. (150:56) Decoction of roots taken for stomach pains. (184:201) **Okanagan-Colville** *Cold Remedy* Infusion of roots taken for colds. *Laxative* Infusion of roots taken as a laxative. (188:74) **Okanagon** *Gastrointestinal Aid* Decoction of rhizomes taken for stomach troubles, indigestion, and colic. (125:40) **Pomo** *Dermatological Aid* Poultice of heated leaves applied to boils. (66:13) **Pomo, Kashaya** *Dermatological Aid* Poultice of fresh, warmed leaves used to bring boils to a head. Decoction of leaves used to wash sores. *Toothache Remedy* Poultice of fresh, warmed leaves used for toothaches. (72:50) **Skagit** *Dietary Aid* Leaves eaten to increase appetite. *Tonic* Decoction of leaves taken as a tonic. *Tuberculosis Remedy* Dried leaves used for tuberculosis. (79:28) **Thompson** *Dermatological Aid* Dried, powdered leaves rubbed on the hands as a deodorant. (187:165) *Gastrointestinal Aid* Decoction of rhizomes taken for stomach troubles, indigestion, and colic. (125:40) Decoction of rhizome taken as a stomach tonic and for indigestion and colic. (164:460) *Pediatric Aid* and *Sedative* Whole plant or stems put in infant's bed to quiet baby and for illness. (164:508) *Tonic* Decoction of rhizomes taken as a tonic for the stomach. (164:460) *Unspecified* Fresh or dried leaves used as a medicine. (187:165) **Tolowa** *Disinfectant* Poultice of leaves applied for any infections. **Yurok** *Dermatological Aid* and *Pediatric Aid* Leaves used to keep a newborn baby's navel from becoming infected.[17] (6:19)
- **Fiber**—**Okanagon** *Mats, Rugs & Bedding* Plants mixed with sphagnum and used as bedding for infants. **Thompson** *Mats, Rugs & Bedding* Plants mixed with sphagnum and used as bedding for infants. (125:39) Plant used as a bedding for infants. (164:496)

Asarum sp., Wildginger
- **Drug**—**Cherokee** *Dermatological Aid* Poultice of fresh leaves applied to wounds. (203:74) **Karok** *Unspecified* Roots used for medicine. (117:209) **Lolahnkok** *Emetic* Pounded leaves soaked in cold water and taken to vomit for stomach pains. (117:190) **Modesse** *Dermatological Aid* Poultice of fresh, large, green leaves applied as an antiseptic to cuts and boils. (117:224)
- **Other**—**Cherokee** *Snuff* Dried and pounded leaves used for snuff. (203:74)

Asclepiadaceae, see *Asclepias, Gonolobus, Matelea, Sarcostemma*

Asclepias, Asclepiadaceae

Asclepias asperula ssp. *capricornu* (Woods.) Woods., Antelope Horns
- **Drug**—**Navajo, Kayenta** *Respiratory Aid* Plant used as a snuff for catarrh. (as *Asclepiodora decumbens* 205:37) **Navajo, Ramah** *Ceremonial Medicine* Plant used as ceremonial emetic. *Emetic* Plant used as ceremonial emetic. *Veterinary Aid* Infusion taken and used as lotion for mad dog or mad coyote bite on humans or animals. (as *Asclepiodora decumbens* 191:39)
- **Food**—**Gosiute** *Candy* Latex used as a chewing gum. (as *Asclepiodora decumbens* 39:363)

Asclepias auriculata (Engelm. ex Torr.) Holz., Eared Milkweed
- **Drug**—**Navajo, Kayenta** *Respiratory Aid* Plant used for nasal congestion from a cold. (as *Acerates auriculata* 205:36)

Asclepias californica Greene, California Milkweed
- **Drug**—**Kawaiisu** *Dermatological Aid* Dried, powdered plant applied to spider bites. (206:13)
- **Food**—**Kawaiisu** *Candy* Milky juice boiled until thick and chewed like chewing gum. Leaves roasted under hot ashes and chewed. (206:13)

Asclepias cordifolia (Benth.) Jepson, Heartleaf Milkweed
- **Drug**—**Miwok** *Unspecified* Root used as a medicine. (12:167)
- **Food**—**Karok** *Candy* Latex boiled, condensed, and chewed. **Yurok** *Candy* Dried latex chewed by the older people at their leisure. (6:19)

Asclepias cryptoceras S. Wats., Pallid Milkweed
- **Drug**—**Paiute** *Analgesic* Decoction of root used as a wash for headaches. *Dermatological Aid* Latex used for ringworm. (180:47) **Paiute, Northern** *Dermatological Aid* Poultice of dried, powdered roots applied to sores. (59:125) **Shoshoni** *Veterinary Aid* Juice of plant used for horse with sore back. (118:49)

Asclepias eriocarpa Benth., Woollypod Milkweed
- **Drug**—**Costanoan** *Cold Remedy* Decoction of plant and plant salve used for colds. *Dermatological Aid* Milky juice used to reduce corns. *Other* Powdered, dried roots inhaled to cause sneezing. *Respiratory Aid* Burning dried plant smoke inhaled for asthma. (21:12) **Mendocino Indian** *Dermatological Aid* Plant juice applied to cuts, sores, and warts. (41:379)
- **Food**—**Karok** *Candy* Milk stirred, heated, mixed with salmon fat or deer grease, and used for chewing gum. (148:388) **Luiseño** *Candy* Stem sap boiled in water until coagulation and used as chewing gum. (155:196) **Mendocino Indian** *Forage* Sweet-scented flowers used by bees as a source of nectar. (41:379)
- **Fiber**—**Concow** *Cordage* Used to make ropes and string. (41:379) **Luiseño** *Clothing* Plant used to make aprons. (155:201) Stem fiber made into twine and used to make front aprons worn by women. *Cordage* Stem fiber or decayed stem fiber used to make twine. The stems were soaked in boiling water or the decayed stems were basted with boiling water to separate the fiber. The fiber was then formed into a ball and made into twine. (155:202) **Pomo, Kashaya** *Clothing* Stem fibers shredded to make a woman's skirt. *Cordage* Stem fibers used to make two-ply string. (72:70)
- **Other**—**Luiseño** *Containers* Stem fiber made into twine and used to make large-meshed nets for carrying bulky or heavy articles. Stem fiber made into twine and used to make network sacks for carrying acorns and other small seeds. *Hunting & Fishing Item* Stem fiber made into twine and sometimes used to make bowstrings. Stem fiber made into twine, occasionally used to make long nets and draw nets for catching rabbits. Stem fiber made into twine and used to make slings. Stem fiber made into twine and used to make fishing nets.[14] (155:202) Plant fiber used to make bowstrings. (155:206)

Asclepias erosa Torr., Desert Milkweed

- *Food*—Coahuilla *Candy* Sap collected, set aside to solidify, heated over the fire, and used as a chewing gum. (13:75) **Tubatulabal** *Candy* Juice roasted until congealed and used as chewing gum. (193:19)

Asclepias exaltata L., Poke Milkweed
- *Drug*—Omaha *Gastrointestinal Aid* Raw root eaten for stomach trouble. **Ponca** *Gastrointestinal Aid* Root eaten raw for stomach trouble. (70:110)

Asclepias fascicularis Dcne., Mexican Whorled Milkweed
- *Drug*—California Indian *Snakebite Remedy* Poultice of fresh leaves used for snakebite. (as *A. mexicana* 118:47) **Mendocino Indian** *Poison* Flowers considered poisonous. (as *A. mexicana* 41:380)
- *Food*—Miwok *Cooking Agent* Boiled greens added to thicken manzanita cider. *Vegetable* Boiled greens used for food. (as *A. mexicana* 12:159) **Paiute** *Unspecified* Species used for food. (as *A. mexicana* 167:244) **Yokia** *Unspecified* Young blossoms occasionally eaten in small quantities. (as *A. mexicana* 41:380)
- *Fiber*—Costanoan *Cordage* Stem fiber used as cordage. (21:252) **Kawaiisu** *Cordage* Stems used as the principal source of cordage. (206:14)

Asclepias hallii Gray, Hall's Milkweed
- *Drug*—Navajo *Gynecological Aid* Infusion of plant used as tonic after deliverance. (55:69) **Navajo, Kayenta** *Veterinary Aid* Plant poisonous to livestock. (205:36)

Asclepias incarnata L., Swamp Milkweed
- *Drug*—Chippewa *Pediatric Aid* Infusion of root used as a strengthening bath for children. *Strengthener* Infusion of root used as a strengthening bath for children and adults. (53:364) **Iroquois** *Dermatological Aid* Cold infusion of roots applied to heal baby's navel. *Diuretic* Decoction of plants taken for too little urine. *Kidney Aid* Decoction of plants taken for the kidneys. *Orthopedic Aid* Decoction of plants taken for lame backs. (87:418) *Other* Compound decoction of roots taken and used as wash for stricture. (87:417) *Pediatric Aid* Cold infusion of roots applied to heal baby's navel. *Strengthener* Infusion of roots taken and used as wash to give strength. *Toothache Remedy* Dried stems made into cord and used for tooth extraction. *Urinary Aid* Decoction of plants taken for too much urine. *Witchcraft Medicine* Decoction of plant used to increase one's strength to be able to physically punish a witch. (87:418) **Meskwaki** *Anthelmintic* Infusion of root used to drive the tapeworms from a person in 1 hour. *Carminative* Root used as a carminative. *Cathartic* Root used as a cathartic. *Diuretic* Root used as a diuretic. *Emetic* Root used as an emetic. (152:205)
- *Food*—Menominee *Soup* Heads, deer broth, or fat used to make soup. *Unspecified* Heads added to cornmeal mush. *Winter Use Food* Cut, dried heads stored for winter use. (as *Ascepias incarnata* 151:62)
- *Fiber*—Chippewa *Cordage* Fiber used to make good twine. (71:140)

Asclepias involucrata Engelm. ex Torr., Dwarf Milkweed
- *Drug*—Keres, Western *Gastrointestinal Aid* Infusion of plant used as a stomach medicine. (171:30) **Navajo, Kayenta** *Toothache Remedy* Poultice of heated roots applied for toothaches. (205:36) **Zuni** *Unspecified* Dry powdered root and saliva used for unspecified illness. (27:373)
- *Food*—Zuni *Forage* Plant favored by jackrabbits. (166:65)

Asclepias latifolia (Torr.) Raf., Broadleaf Milkweed
- *Drug*—Isleta *Respiratory Aid* Ground leaf and stem powder inhaled for catarrh. (100:23)

Asclepias nyctaginifolia Gray, Mojave Milkweed
- *Drug*—Navajo, Kayenta *Antidiarrheal* and *Pediatric Aid* Infusion of plant given to infants with diarrhea. (205:37)

Asclepias perennis Walt., Aquatic Milkweed
- *Drug*—Cherokee *Analgesic* Infusion of root taken with root of "virgin's bower" for backache. *Dermatological Aid* Rubbed on warts to remove them. *Kidney Aid* Plant taken for "dropsy." *Laxative* Plant taken as a laxative. *Urinary Aid* Plant taken for "gravel." *Venereal Aid* Infusion of root taken for venereal diseases. *Veterinary Aid* Infusion given for "milksick (mastitis)." (80:44)
- *Other*—Cherokee *Hunting & Fishing Item* Plant fibers used to make bowstrings. (80:44)

Asclepias pumila (Gray) Vail, Plains Milkweed
- *Drug*—Lakota *Antidiarrheal* Infusion of leaves taken for diarrhea. *Pediatric Aid* Infusion of leaves taken by children with diarrhea. (139:34)

Asclepias quadrifolia Jacq., Fourleaf Milkweed
- *Drug*—Cherokee *Analgesic* Infusion of root taken with root of "virgin's bower" for backache. *Dermatological Aid* Rubbed on warts to remove them. *Kidney Aid* Plant taken for "dropsy." *Laxative* Plant taken for as a laxative. *Urinary Aid* Plant taken for "gravel." *Venereal Aid* Infusion of root taken for venereal diseases. *Veterinary Aid* Infusion given for "milksick (mastitis)." (80:44)
- *Other*—Cherokee *Hunting & Fishing Item* Plant fibers used to make bowstrings. (80:44)

Asclepias sp., Milkweed
- *Drug*—Cahuilla *Analgesic* Gum applied to insect sting pain. (15:43) **Delaware** *Misc. Disease Remedy* Infusion of pounded roots used for epilepsy in those born during certain phases of the moon. (176:39) **Delaware, Oklahoma** *Anticonvulsive* Infusion of pounded roots of five species taken for epileptic fits. (175:32, 74) **Natchez** *Kidney Aid* Infusion of root taken for "kidney trouble and Bright's disease." (172:667) Infusion of roots taken for kidney troubles and nephritis. (177:52) *Venereal Aid* Plant used for syphilis. (172:668) Plant used for syphilis. (177:52) **Navajo** *Gastrointestinal Aid* Infusion of crushed, dried leaves taken for stomach troubles. (55:69) **Tewa** *Gynecological Aid* Plant used for sore breasts. (138:54)
- *Food*—Cahuilla *Staple* Seeds ground into flour. *Vegetable* Parboiled leaves used as greens from May until June. (15:43) **Jemez** *Unspecified* Immature seeds used for food. (44:20) **Karok** *Candy* Juice used for chewing gum. (117:212) **Kiowa** *Fruit* Young fruits, after first removing the outer "hairy" surface, cooked and eaten. (192:47) **Navajo** *Unspecified* Plant eaten raw or boiled. (55:69) **Tewa** *Unspecified* Roots and immature pods eaten. (138:54)
- *Fiber*—Cahuilla *Cordage* Stem cordage used to make nets, slings, and snares to capture small game.[18] (15:43) **Modesse** *Cordage* Stems used for making string and cord. (117:224) **Neeshenam** *Cordage* Inner bark used to make strings and cords. (129:378) **Tewa** *Cordage* Mature plants used to make string and rope. (138:54) **Wintoon** *Cordage* Used to make string and ropes. (117:264)

- *Other*—Cahuilla *Hunting & Fishing Item* Stem cordage used to make nets, slings, and snares to capture small game.[18] (15:43) **Kiowa** *Cooking Tools* Dried pods used as spoons. *Decorations* Dried pods gathered for decorative purposes. (192:47) **Neeshenam** *Hunting & Fishing Item* Inner bark used to make nets. (129:378)

Asclepias speciosa Torr., Showy Milkweed

- *Drug*—Cheyenne *Eye Medicine* Decoction of plant tops strained and used as an eye medicine. (82:66) Decoction of plant tops used as an eyewash for blindness or snow-blindness. (83:14) **Flathead** *Gastrointestinal Aid* Fresh roots chewed or dried, pulverized, and boiled and used for stomachache. (82:66) **Lakota** *Unspecified* Used as a medicine. (139:34) **Miwok** *Dermatological Aid* Milk of plant applied to warts. *Venereal Aid* Decoction of root taken in small doses for venereal diseases. (12:167) **Navajo, Kayenta** *Ceremonial Medicine* and *Emetic* Plant used as an Eagleway, Female Shootingway, Beautyway, and Beadway emetic. (205:37) **Okanagan-Colville** *Antirheumatic (External)* Poultice of mashed roots applied for rheumatism. *Dermatological Aid* Latex rubbed on skin sores. (188:74) **Okanogon** *Analgesic* Decoction of roots taken for headaches and general debility. (125:42) **Paiute** *Antirheumatic (External)* Decoction of root used as a wash for rheumatism. *Cough Medicine* Decoction of root taken for cough, especially from tuberculosis. *Dermatological Aid* Latex or pulverized seeds used as an antiseptic and healing agent on sores. *Misc. Disease Remedy* Hot decoction of root taken to "bring out the rash of measles." *Snakebite Remedy* Decoction of seeds used to draw poison from snakebites. *Tuberculosis Remedy* Decoction of root taken for cough, especially from tuberculosis. **Shoshoni** *Antidiarrheal* Decoction of root taken for "bloody diarrhea." *Dermatological Aid* Latex applied to remove corns and calluses. Latex used as an antiseptic and healing agent on sores, cuts, and ringworm. Poultice of mashed root applied to swellings. *Venereal Aid* Latex used as an antiseptic and healing agent on syphilitic sores. (180:48) **Thompson** *Analgesic* Decoction of roots taken for headaches and general debility. (125:42) *Dermatological Aid* Milky juice from stem used as face cream. *Dietary Aid* Decoction of root taken for "general out-of-sorts feeling and emaciation." (164:470) *Poison* Root poisonous in large amounts. (164:513) *Tonic* Decoction of root taken for "general out-of-sorts feeling and emaciation." (164:470)
- *Food*—Acoma *Candy* Milky latex allowed to harden and used as chewing gum. (32:31) **Apache, Chiricahua & Mescalero** *Candy* "Milk" squeezed from leaves and stems and chewed as gum. (33:45) **Cheyenne** *Candy* Dried, hardened milk used for chewing gum. (76:184) Milky juice allowed to harden and used as chewing gum. (82:66) Milky juice hardened and chewed as gum. (83:14) *Fruit* Inner layer of fruit used for food. (76:184) Immature fruits peeled and inner layer eaten. (82:66) Green, immature fruits peeled and the inner layer eaten raw. (83:14) *Sauce & Relish* Flowers boiled with soup or meat, flour added, and eaten as a gravy. (82:66) *Soup* Whole buds boiled with meat or in water to make soup. (76:184) *Unspecified* Young, unopened buds boiled with meat, grease, gravy, or soup and used for food. Tender, spring shoots eaten raw. (83:46) Young, unopened buds boiled with meat, grease, gravy, or soup and eaten. Tender stalks cooked, sweetened, and used for food. (83:14) **Crow** *Sauce & Relish* Flowers boiled with soup or meat, flour added, and eaten as a gravy. (82:66) Flowers boiled for food. Seeds eaten raw. (19:7) **Hopi** *Unspecified* Boiled with meat. (190:164) **Keres, Western** *Candy* Milky juice used as chewing gum. (171:30) **Laguna** *Candy* Milky latex allowed to harden and used as chewing gum. (32:31) **Lakota** *Cooking Agent* Floral bud clusters used to thicken soup. *Preserves* Open flowers cut up for a sort of preserve. *Unspecified* Blossoms boiled, mixed with flour, and eaten. (139:34) **Paiute** *Candy* Dried sap chewed as gum. (111:105) *Unspecified* Species used for food. (167:242) **Shoshoni** *Candy* Milk rolled in hand and used for gum. (118:56)
- *Fiber*—Okanagan-Colville *Cordage* Stems used as a poor substitute for Indian hemp. (188:74) **Pomo, Kashaya** *Clothing* Stem fibers shredded to make a woman's skirt. *Cordage* Stem fibers used to make two-ply string. (72:70) **Thompson** *Clothing* Cottony seed pappus formerly used for infant diapers. *Cordage* Inner bark used as a substitute for Indian hemp in making thread used for tying and binding. (187:165)
- *Other*—Crow *Paint* Milky juice used for temporary branding of livestock. (82:66) **Montana Indian** *Tools* "Milk" from the broken stems used in cases of emergency for branding stock temporarily. (19:7) **Thompson** *Fasteners* Inner bark used for binding or tying. (164:498)

Asclepias stenophylla Gray, Slimleaf Milkweed

- *Drug*—Lakota *Dietary Aid* Root given to children to increase the appetite. (139:34)

Asclepias subulata Dcne., Rush Milkweed

- *Drug*—Pima *Cathartic* Plant used as a physic. *Emetic* Plant used as an emetic. *Eye Medicine* Plant used for sore eyes. *Gastrointestinal Aid* Plant used for stomach disorders. *Panacea* Plant used for many ailments. *Poison* Plant considered poisonous. (47:81)

Asclepias subverticillata (Gray) Vail, Whorled Milkweed

- *Drug*—Hopi *Gynecological Aid* Used by the mother to produce a flow of milk. (as *A. galioides* 190:164) Plant used to increase mother's milk flow. (as *A. galioides* 200:36, 87) **Keres, Western** *Gynecological Aid* Crushed leaves rubbed on mothers' breasts to produce more and richer milk. Infusion of plant used by mothers for more and richer milk. (as *A. galioides* 171:30)
- *Food*—Apache, White Mountain *Unspecified* First buds eaten by children. (as *A. galioides* 136:155) **Jemez** *Unspecified* Roots and unripe pods eaten raw. (as *A. galioides* 32:17) **Keres, Western** *Candy* Ripe seed silk mixed with grease and used as chewing gum. (as *A. galioides* 171:30) **Zuni** *Unspecified* Buds eaten by little boys. (as *A. galioides* 166:65)
- *Fiber*—Navajo, Ramah *Cordage* Spun seed hair made into string used in prayer sticks. (as *A. galioides* 191:39) **Zuni** *Clothing* Pods gathered when two-thirds ripe and the cotton used for weaving clothing. The cotton was used for weaving beautiful white dance kilts, women's belts, and other articles of clothing. (as *A. galioides* 166:77) *Cordage* Coma made into cords and used for fastening plumes to the prayer sticks.[19] (as *A. galioides* 166:88)
- *Other*—Zuni *Ceremonial Items* Coma made into cords and used for fastening plumes to the prayer sticks.[19] (as *A. galioides* 166:88)

Asclepias syriaca L., Common Milkweed

- *Drug*—Cherokee *Analgesic* Infusion of root taken with root of "virgin's bower" for backache. *Dermatological Aid* Rubbed on warts to

remove them. *Kidney Aid* Plant taken for "dropsy." *Laxative* Plant taken as a laxative. *Urinary Aid* Plant taken for "gravel." *Venereal Aid* Infusion of root taken for venereal diseases. *Veterinary Aid* Infusion given for "milksick (mastitis)." (80:44) **Chippewa** *Gynecological Aid* Cold decoction of root added to food to produce postpartum milk flow. (53:360) **Iroquois** *Antirheumatic (External)* Stalks cooked as greens and used for rheumatism. (124:93) *Contraceptive* Infusion of dried, pulverized roots and rhizomes taken by women for temporary sterility. (141:59) *Dermatological Aid* Milk used for warts, bee stings, and cuts. *Gastrointestinal Aid* Infusion of leaves taken as a stomach medicine. *Gynecological Aid* Compound decoction of plants taken to prevent hemorrhage after childbirth. *Kidney Aid* Compound used for dropsy. (87:417) *Other* Compound decoction of roots taken for stricture. (87:416) *Unspecified* Poultice of cotton applied to sick parts. (141:59) **Menominee** *Pulmonary Aid* Buds eaten or decoction of root used for chest discomfort. (54:130) **Ojibwa** *Gynecological Aid* Root used as a female remedy for unspecified ailment. (153:357) **Potawatomi** *Unspecified* Root used for unspecified ailments. (154:42) **Rappahannock** *Dermatological Aid* Milk of fresh plant applied to warts and ringworm. (161:32)
- *Food*—**Chippewa** *Appetizer* Plant eaten before a feast to increase the appetite. *Preserves* Flowers cut up, stewed, and eaten like preserves. (53:320) *Vegetable* Tender leaves, young green seedpods, sprouts, and tops cooked as greens. (71:140) **Dakota** *Unspecified* Sprouts used in early spring for food. (69:363) **Iroquois** *Vegetable* Stalks eaten as greens in spring. (124:93) Tender stems, leaves, and immature flower clusters cooked and seasoned with salt, pepper, or butter. (196:117) **Meskwaki** *Soup* Buds used in soups. *Vegetable* Buds cooked with meat or added to cornmeal mush, tastes like okra. *Winter Use Food* Dried buds stored away in paper bags for winter use. (152:256) **Mohegan** *Unspecified* Cooked and used for food. (176:83) **Ojibwa** *Unspecified* Young pods cooked with salt and vinegar. *Vegetable* Young shoots and flower buds cooked like spinach. (5:2205) Fresh flowers and shoot tips, mucilaginous like okra when cooked, used in meat soups. *Winter Use Food* Dried flowers, freshened in the wintertime, made into soup. (153:397) **Omaha** *Vegetable* Tender shoots boiled and eaten as a vegetable. (58:341) Young shoots used for food like asparagus. Inflorescence, before the flower buds opened, and young fruits used as greens. (68:325) Boiled young sprouts, floral bud clusters, and young, firm green fruits used for food. **Pawnee** *Vegetable* Boiled young sprouts, floral bud clusters, and young, firm green fruits used for food. **Ponca** *Vegetable* Boiled young sprouts, floral bud clusters, and young, firm green fruits used for food. (70:109) **Potawatomi** *Soup* Flowers and buds used to thicken meat soups and to impart a very pleasing flavor to the dish. (154:96) **Winnebago** *Vegetable* Boiled young sprouts, floral bud clusters, and young, firm green fruits used for food. (70:109)
- *Fiber*—**Menominee** *Cordage* Outer bark used for making cords. *Sewing Material* Outer bark used for sewing thread. (151:74) **Meskwaki** *Cordage* Outer rind or bark used for thread. (152:267) **Potawatomi** *Sewing Material* This and other species of milkweed used for thread materials. (154:111)
- *Other*—**Cherokee** *Hunting & Fishing Item* Plant fibers used to make bowstrings. (80:44) **Chippewa** *Hunting & Fishing Item* Roots applied to whistles used for calling deer. (53:376) **Mahuna** *Fasteners* Milky juice used to mount precious stones into necklaces, earrings, collars, and bracelets. (140:51) **Menominee** *Hunting & Fishing Item* Outer bark used for making cords for fish lines. (151:74) **Ojibwa** *Hunting & Fishing Item* Milk and Canada hawkweed milk used to put on a deer call to imitate the call of a hungry fawn. (153:428) **Omaha** *Toys & Games* Mature stalk fiber chewed and used for popgun wadding by little boys. **Pawnee** *Toys & Games* Mature stalk fiber chewed and used for popgun wadding by little boys. **Ponca** *Toys & Games* Mature stalk fiber chewed and used for popgun wadding by little boys. **Winnebago** *Toys & Games* Mature stalk fiber chewed and used for popgun wadding by little boys. (70:109)

Asclepias tuberosa L., Butterfly Milkweed

- *Drug*—**Cherokee** *Analgesic* Used for breast, stomach, and intestinal pains. *Antidiarrheal* Seeds boiled in "new milk" and used for diarrhea. *Expectorant* Used as an expectorant and taken for pleurisy. *Gynecological Aid* Infusion used for "bloody flux." *Heart Medicine* Infusion of root used for heart trouble. *Laxative* Seeds or root used as gentle laxative. *Pulmonary Aid* Used as an expectorant and taken for pleurisy and lung inflammations. (80:27) **Delaware** *Antirheumatic (External)* Roots used for rheumatism. *Gynecological Aid* Roots administered to women following childbirth. *Pulmonary Aid* Roots used for pleurisy. (176:37) **Delaware, Oklahoma** *Antirheumatic (Internal)* Root used for rheumatism. *Gynecological Aid* Root used to make a drink taken by women after childbirth. *Pulmonary Aid* Root used for pleurisy. (175:31, 74) **Iroquois** *Other* Infusion of roots used as a wash for arms, shoulders, and body for lifting. *Sports Medicine* Poultice of smashed roots applied to legs, and running shoes dampened or washed for running strength. (87:416) **Menominee** *Dermatological Aid* Poultice of root used or decoction taken for bruises and swellings. (54:132) Simple or compound poultice of pulverized root used on cuts, wounds, and bruises. (151:25) *Orthopedic Aid* Poultice of root used or decoction taken for lameness. *Tonic* Decoction of pounded root taken as a tonic. (54:132) **Mohegan** *Pulmonary Aid* Dried root used for pleurisy. (176:70, 128) **Navajo, Ramah** *Ceremonial Medicine* Plant used in ceremonial chant lotion. *Dermatological Aid* Decoction or infusion of various plant parts used for dog or coyote bites. *Misc. Disease Remedy* Plant used for influenza. (191:39) **Omaha** *Ceremonial Medicine* Ceremony connected with the obtaining and distribution of this prized root. *Dermatological Aid* Fresh and dried root used in several ways on wounds and sores. *Pulmonary Aid* and *Respiratory Aid* Root eaten raw for bronchial and pulmonary trouble. **Ponca** *Pulmonary Aid* Root eaten raw for pulmonary trouble. *Respiratory Aid* Root eaten raw for bronchial trouble. (70:109) **Rappahannock** *Snakebite Remedy* Poultice of bruised leaves bound to snakebites. (161:30)
- *Fiber*—**Cherokee** *Clothing* Stems used to make belts. (80:27)

Asclepias verticillata L., Whorled Milkweed

- *Drug*—**Choctaw** *Diaphoretic* Root used as a sudorific. *Snakebite Remedy* Root chewed, saliva swallowed, and strong decoction taken for snakebite. *Stimulant* Root used as a stimulant. (28:287) **Hopi** *Gynecological Aid* Infusion of entire plant taken by nursing mother with scanty flow of milk. (56:18) **Lakota** *Gynecological Aid* Used by mothers to increase their milk. (139:34) **Navajo** *Nose Medicine* Plant used

for nose troubles. *Throat Aid* Plant used for throat troubles. (55:96)
- **Food**—Hopi *Unspecified* Leaves and young shoots boiled with meat and eaten. (56:18)
- **Other**—Hopi *Tools* Used as a planting stick. (56:18)

Asclepias viridiflora Raf., Green Milkweed
- **Drug**—Blackfoot *Antirheumatic (External)* Poultice of chewed roots applied to swellings. *Dermatological Aid* Poultice of chewed roots applied to rashes. Poultice of chewed roots applied to diarrhea rash. (86:75) *Eye Medicine* Poultice of chewed roots applied to sore eyes. (as *A. viridis flora* 86:80) *Oral Aid* Poultice of chewed roots applied to nursing baby's sore gums. *Pediatric Aid* Poultice of chewed roots applied to diarrhea rash and nursing baby's sore gums. (86:75) *Throat Aid* Root chewed for sore throats. (86:71) **Lakota** *Antidiarrheal* Pulverized roots given to children with diarrhea. *Gynecological Aid* Infusion of whole plant taken by mothers to increase their milk. *Pediatric Aid* Pulverized roots given to children with diarrhea. (139:34)
- **Food**—Blackfoot *Soup* Root pieces stored for winter soups. *Spice* Plant used to spice soups. *Unspecified* Fresh roots used for food. (86:101)

Asimina, Annonaceae
Asimina triloba (L.) Dunal, Common Pawpaw
- **Food**—Cherokee *Fruit* Fruit used for food. (80:47) **Iroquois** *Bread & Cake* Fruit mashed, made into small cakes, and dried for future use. *Dried Food* Raw or cooked fruit sun or fire dried and stored for future use. *Fruit* Dried fruit taken as a hunting food. *Sauce & Relish* Dried fruit cakes soaked in warm water and cooked as a sauce or mixed with corn bread. (196:129)
- **Fiber**—Cherokee *Cordage* Inner bark used to make strong ropes and string. (80:47)

Asparagus, Liliaceae
Asparagus officinalis L., Garden Asparagus
- **Drug**—Cherokee *Dietary Aid* Infusion of plant taken for rickets. (80:24) **Iroquois** *Antirheumatic (External)* Compound decoction with roots used as a foot soak for rheumatism. (87:282) Stalks cooked as greens and used for rheumatism. (124:93) *Blood Medicine* Compound decoction with bark taken before meals for the blood. (87:282)
- **Food**—Cherokee *Vegetable* Species used for food. (80:24) **Iroquois** *Vegetable* Stalks eaten as greens in spring. (124:93) **Isleta** *Unspecified* Uncultivated but used as food when found in the wild. (32:17) *Vegetable* Boiled, seasoned spears used for food. (100:23)

Asplenium, Blechnaceae, fern
Asplenium horridum Kaulfuss, Lacy Spleenwort
- **Drug**—Hawaiian *Blood Medicine* Infusion of plant, other ingredients, and coconut milk taken for impure blood. *Oral Aid* Buds and burnt potato peel chewed for sore mouths. *Stimulant* Scraped wood, other ingredients, and water taken and used as wash for fainting spells and muscle stiffness. (2:14)
- **Dye**—Hawaiian *Red* Juice used as a red dye. (2:14)

Asplenium nidus L., Birdnest Fern
- **Drug**—Hawaiian *Dermatological Aid* Leaves and other plants pounded, squeezed, and resulting liquid used for ulcers or scrofulous sores. *Oral Aid* Shoots and other plants pounded, squeezed, and resulting liquid used for children with mouth sores. *Pediatric Aid* Shoots and other plants pounded, squeezed, and resulting liquid given to children for general weakness, and used for children with mouth sores. *Strengthener* Shoots and other plants pounded, squeezed, and resulting liquid given to children for general weakness, and taken for general body weakness. *Tuberculosis Remedy* Leaves and other plants pounded, squeezed, and resulting liquid used for scrofulous sores. (2:22)

Asplenium pseudofalcatum, Iwaiwa
- **Drug**—Hawaiian *Dermatological Aid* Leaf ashes, nut juice, and fruit milk mixed and used on sores. Infusion of leaves used as a bath to beautify children. *Oral Aid* Leaf ashes, nut juice, and fruit milk mixed and used on sores about the mouth. (2:29)

Asplenium rhizophyllum L., Walking Fern
- **Drug**—Cherokee *Breast Treatment* Compound used for swollen breasts. (80:61) Decoction of whole plant rubbed on swollen breast. *Emetic* Decoction of whole plant taken to induce vomiting for swollen breasts. (as *Camptosorus rhizophyllus* 177:3)

Asplenium sp., Spleenwort
- **Drug**—Tlingit *Pulmonary Aid* Infusion of plant used for chest inflammation due to catarrh. (107:283)

Asplenium trichomanes L., Maidenhair Spleenwort
- **Drug**—Cherokee *Abortifacient* Taken for irregular menses. *Breast Treatment* Infusion taken for "breast diseases" and "acrid humors." *Cough Medicine* Infusion taken for coughs. *Liver Aid* Taken for "liver complaints." (80:34)

Asteraceae, see *Acamptopappus, Achillea, Acourtia, Adenocaulon, Ageratina, Agoseris, Ambrosia, Amphiachyris, Anaphalis, Antennaria, Anthemis, Arctium, Arnica, Arnoglossum, Artemisia, Aster, Baccharis, Bahia, Baileya, Balsamita, Balsamorhiza, Bellis, Berlandiera, Bidens, Blennosperma, Brachyactis, Brickellia, Calycadenia, Carduus, Carthamus, Centaurea, Chaenactis, Chaetopappa, Chamaemelum, Chaptalia, Chloracantha, Chrysopsis, Chrysothamnus, Cichorium, Cirsium, Conyza, Coreopsis, Corethrogyne, Cosmos, Crepis, Dahlia, Dugaldia, Echinacea, Encelia, Enceliopsis, Ericameria, Erigeron, Eriophyllum, Eupatorium, Euthamia, Gaillardia, Gamochaeta, Glyptopleura, Gnaphalium, Grindelia, Gutierrezia, Haplopappus, Hazardia, Helenium, Helianthella, Helianthus, Heliomeris, Heliopsis, Hemizonia, Heterotheca, Hieracium, Holocarpha, Hymenoclea, Hymenopappus, Hymenoxys, Inula, Isocoma, Iva, Krigia, Lactuca, Lasthenia, Layia, Lessingia, Leucanthemum, Liatris, Lygodesmia, Machaeranthera, Madia, Malacothrix, Marshallia, Matricaria, Microseris, Mikania, Onopordum, Palafoxia, Parthenium, Pectis, Pentachaeta, Perezia, Pericome, Petasites, Petradoria, Picradeniopsis, Pityopsis, Pluchea, Polymnia, Porophyllum, Prenanthes, Psathyrotes, Psilostrophe, Pterocaulon, Pyrrhopappus, Ratibida, Rudbeckia, Sanvitalia, Schkuhria, Scorzonella, Senecio, Shinnersoseris, Silphium, Smallanthus, Solidago, Sonchus, Stenotus, Stephanomeria, Tagetes, Tanacetum, Taraxacum, Tetradymia, Tetraneuris, Thelesperma, Thymophylla, Townsendia, Tragopogon, Tussilago, Vanclevea, Verbesina, Vernonia, Wyethia, Xanthium, Xylorhiza, Zinnia*

Aster, Asteraceae

Aster carolinianus Walt., Climbing Aster
- *Drug*—Seminole *Dermatological Aid* Plant used for snake sickness: itchy skin. (169:166) Decoction of leaves taken and used as a body steam for snake sickness: itchy skin. (169:239)

Aster conspicuus Lindl., Showy Aster
- *Drug*—Okanagan-Colville *Dermatological Aid* Roots soaked in hot or cold water and used as a wash for sores, boils, wounds, and infections. Poultice of leaves applied to boils. *Hemorrhoid Remedy* Roots soaked in hot or cold water and used as a wash for hemorrhoids. Poultice of leaves applied to hemorrhoids. *Toothache Remedy* Roots applied to the tooth for toothaches. *Unspecified* Roots soaked in hot or cold water and taken for gonorrhea and other ailments. *Venereal Aid* Roots soaked in hot or cold water and taken for gonorrhea. *Veterinary Aid* Roots used for cuts with maggots on horses. (188:79)

Aster cordifolius L., Common Blue Wood Aster
- *Drug*—Ojibwa *Hunting Medicine* Root used to make a smoke or incense to attract deer near enough to shoot it with a bow and arrow. A number of the composites as well as plants from other families are used in the hunting charms. The deer carries its scent or spoor in between its toes, and wherever the foot is impressed into the ground, other animals can detect its presence. This allows dogs to track them. It is a peculiar scent and the Ojibwe tries successfully to counterfeit it with roots and herbs. The root of this aster is but one of [19] that can be used. They say that the white man drives the deer away when he smokes cigarettes or cigars, but the Indian brings them closer. (153:428)

Aster cusickii Gray, Cusick's Aster
- *Drug*—Cheyenne *Ear Medicine* Infusion of dried stems used as ear drops for earaches. (76:187)

Aster dumosus L., Rice Button Aster
- *Food*—Tewa *Fruit* Small fruits eaten. (as *Sericotheca dumosa* 138:49)

Aster ericoides L. **var. ericoides**, Heath Aster
- *Drug*—Meskwaki *Herbal Steam* Used in the sweat bath. *Stimulant* Used to revive an unconscious patient. (as *A. multiflorus* 152:212)

Aster falcatus var. commutatus (Torr. & Gray) A. G. Jones, Cluster Aster
- *Drug*—Zuni *Dermatological Aid* Ground blossoms mixed with yucca suds and used to wash newborn infants and make their hair grow. *Pediatric Aid* Ground blossoms mixed with yucca suds and used to wash newborn infants. This medicine was said to make the hair grow on the head and to give strength to the body. *Strengthener* Ground blossoms mixed with yucca suds and used as a strengthening wash for newborn infants. (as *A. incanopilosus* 166:84)

Aster falcatus var. crassulus (Rydb.) Cronq., Rough White Prairie Aster
- *Drug*—Navajo, Ramah *Snakebite Remedy* Compound decoction of plant taken and used as lotion for snakebite. (as *A. commutatus* var. *crassulus* 191:48)

Aster foliaceus Lindl. ex DC., Alpine Leafybract Aster
- *Drug*—Okanagan-Colville *Veterinary Aid* Decoction of whole plant used as wash for sores on a horse's back. (188:80) **Okanagon** *Dietary Aid* Decoction of roots taken for loss of appetite. *Gastrointestinal Aid* Decoction of roots taken for stomach swelling, dyspepsia, and indigestion. **Thompson** *Dietary Aid* Decoction of roots taken for loss of appetite. (125:41) Decoction of roots used to stimulate appetite. (164:461) *Gastrointestinal Aid* Decoction of roots taken for stomach swelling, dyspepsia, and indigestion. (125:41) Decoction of root taken for various stomach discomforts. *Venereal Aid* Strong decoction of root mixed with salmon oil and taken for syphilis. (164:461)

Aster furcatus Burgess, Forked Aster
- *Drug*—Potawatomi *Analgesic* Infusion of leaves rubbed on head for severe headache. (154:49, 50)

Aster laevis L., Smooth Aster
- *Drug*—Meskwaki *Herbal Steam* Entire plant used to furnish smoke in sweat bath. *Stimulant* Smoke forced into nostrils of unconscious patient to revive him. (152:211, 212)

Aster laevis var. geyeri Gray, Geyer's Smooth Aster
- *Food*—Keres, Western *Unspecified* Flowers mixed with parched corn and eaten. (as *A. geyeri* 171:30)
- *Dye*—Keres, Western *Unspecified* Flowers mixed with white clay and used to dye wool or eggs. (as *A. geyeri* 171:30)
- *Other*—Keres, Western *Paint* Petals mixed with whitewash. (as *A. geyeri* 171:30)

Aster lanceolatus ssp. hesperius (Gray) Semple & Chmielewski, Siskiyou Aster
- *Drug*—Zuni *Dermatological Aid* Decoction of plant used to dress arrow or bullet wounds. Dried, pulverized plant used for abrasions made by ceremonial mask. *Hemostat* Smoke from crushed blossoms inhaled for nosebleed. (as *A. hesperius* 166:43)

Aster lanceolatus Willd. **ssp. lanceolatus var. lanceolatus**, White Panicle Aster
- *Drug*—Iroquois *Febrifuge* Infusion of plant and another plant used for fevers. (as *A. paniculatus* 141:65)

Aster lateriflorus (L.) Britt., Calico Aster
- *Drug*—Meskwaki *Herbal Steam* Entire plant used as a smoke or steam in sweat bath. *Psychological Aid* Blossoms smudged "to cure a crazy person who has lost his mind." (152:212)

Aster linariifolius L., Ionactis
- *Drug*—Cherokee *Analgesic* Poultice of roots used for pain. *Antidiarrheal* Infusion of root taken for diarrhea. *Febrifuge* Infusion taken for fever. *Respiratory Aid* Ooze of roots sniffed for catarrh. (80:24)

Aster macrophyllus L., Bigleaf Aster
- *Drug*—Iroquois *Blood Medicine* Roots used as a blood medicine. *Laxative* and *Venereal Aid* Compound decoction of roots taken to loosen the bowels for venereal disease. (87:462) **Ojibwa** *Analgesic* Infusion of root used to bathe the head for headache. *Hunting Medicine* Plant used as a charm in hunting. (153:363) Plant smoked as one of the hunting charms to attract deer. (153:429) *Unspecified* Young and tender leaves eaten and act as a medicine at the same time that they are food. (153:398)
- *Food*—Algonquin, Quebec *Vegetable* Leaves used for greens. (18:

108) **Ojibwa** *Soup* Roots used as a soup material. *Unspecified* Young and tender leaves eaten and act as a medicine at the same time that they are food. (153:398)

Aster nemoralis Ait., Bog Aster
- *Drug*—**Chippewa** *Ear Medicine* Decoction of root used as drops or on a compress for sore ear. (53:360)

Aster novae-angliae L., New England Aster
- *Drug*—**Cherokee** *Analgesic* Poultice of roots used for pain. *Antidiarrheal* Infusion of root taken for diarrhea. *Febrifuge* Infusion taken for fever. *Respiratory Aid* Ooze of roots sniffed for catarrh. (80:24) **Chippewa** *Hunting Medicine* Roots smoked in pipes as a charm to attract game. (53:376) **Iroquois** *Dermatological Aid* Decoction of plants used for weak skin. *Febrifuge* Decoction of roots and leaves taken for all kinds of fevers. (87:463) Infusion of whole plant and rhizomes from another plant taken by mothers with intestinal fevers. (141:65) *Love Medicine* Plant used as a love medicine. (87:462) **Meskwaki** *Stimulant* Smudged and used to revive an unconscious patient. (152:212) **Potawatomi** *Stimulant* Plant used as a fumigating reviver by the Prairie Potawatomi. (154:50)

Aster oblongifolius Nutt., Aromatic Aster
- *Drug*—**Navajo, Ramah** *Witchcraft Medicine* Decoction used as lotion for protection from witches. (191:48)

Aster praealtus var. *coerulescens* (DC.) A. G. Jones, Willowleaf Aster
- *Drug*—**Navajo, Ramah** *Ceremonial Medicine* Decoction used ceremonially for snakebite. *Eye Medicine* Cold infusion of whole plant used as a ceremonial eyewash. *Gastrointestinal Aid* Cold infusion of whole plant used for stomachache. *Hunting Medicine* Dried leaves smoked for good luck in hunting. *Internal Medicine* Cold infusion of whole plant used for internal injury. *Snakebite Remedy* Decoction used ceremonially for snakebite. (as *A. coerulescens* 191:48)

Aster praealtus Poir. var. *praealtus*, Willowleaf Aster
- *Drug*—**Meskwaki** *Stimulant* Used to revive an unconscious patient. (as *A. salicifolius* 152:212)

Aster prenanthoides Muhl. ex Willd., Crookedstem Aster
- *Drug*—**Iroquois** *Cold Remedy* Compound decoction of roots taken for colds. *Febrifuge* Decoction of roots given to babies with fevers. *Kidney Aid* Compound decoction of roots taken for the kidneys. *Pediatric Aid* Decoction of roots given to babies with fevers. (87:463)

Aster puniceus L., Purplestem Aster
- *Drug*—**Chippewa** *Hunting Medicine* Root tendrils smoked with tobacco as a charm to attract game. (53:376) **Cree, Woodlands** *Abortifacient* Decoction of roots taken for failure to menstruate. *Diaphoretic* Decoction of roots taken to cause sweating and reduce a fever. *Febrifuge* Decoction of roots taken to cause sweating and reduce a fever. *Gynecological Aid* Decoction of roots taken to make a woman well after childbirth. *Orthopedic Aid* Roots used for facial paralysis. *Pediatric Aid* and *Toothache Remedy* Decoction of roots taken for teething sickness. Chewed root applied to tooth for toothache. (109:31) **Iroquois** *Cold Remedy* Infusion of roots taken for colds. *Febrifuge* Infusion of roots taken for fevers. *Misc. Disease Remedy* Infusion of roots taken for typhoid. *Pulmonary Aid* Infusion of roots taken for pneumonia. *Tuberculosis Remedy* Infusion of roots taken for consumption. (87:463)

Aster shortii Lindl., Short's Aster
- *Drug*—**Potawatomi** *Unspecified* Infusion of flowering tops used for unspecified ailments. (154:50)

Aster simmondsii Small, Simmonds's Aster
- *Drug*—**Seminole** *Other* Infusion of plant used for sunstroke. (169:303)

Aster sp., Prairie Aster
- *Drug*—**Blackfoot** *Laxative* and *Pediatric Aid* Infusion of plant used as an enema for babies with colic or intestinal troubles. (86:66) *Veterinary Aid* Infusion of plant put in the horse's or dog's nostril for nasal disorders. Infusion of plant used as an eyewash for a dog's infected eye. (86:88) **Iroquois** *Blood Medicine* Compound decoction of plants and roots taken as a blood remedy. *Gynecological Aid* Compound decoction of leaves and roots taken for leukorrhea (sick womb). *Hunting Medicine* Plant used as a "hunting medicine." *Love Medicine* Plant used as a love medicine. (87:462) **Pawnee** *Other* Moxa of burned stems used over affected part. (70:133) **Shuswap** *Analgesic* Decoction of leaves and roots used as a wash for pain. (123:59)
- *Food*—**Chippewa** *Unspecified* Leaves boiled with fish and eaten. (53:320)
- *Dye*—**Blackfoot** *Unspecified* Flowers rubbed by children on bouncing arrows for color. (86:109)
- *Other*—**Blackfoot** *Jewelry* Flowers used to make necklaces. (86:109) **Navajo** *Ceremonial Items* Used, with other plants, as a liniment for the Bead Chant. Used, with other plants, as the Bead Chant tobacco. (55:82)

Aster subulatus var. *ligulatus* Shinners, Annual Saltmarsh Aster
- *Drug*—**Kawaiisu** *Analgesic* Decoction of roots used as a wash for headaches. *Toothache Remedy* Mashed roots applied to the tooth for toothache. (as *A. exilis* 206:14)

Aster umbellatus P. Mill., Parasol Aster
- *Drug*—**Mohegan** *Gastrointestinal Aid* Infusion of leaves said to be good for the stomach. (176:70, 128) **Potawatomi** *Witchcraft Medicine* Flowers smudged to repel evil spirits from sickroom. (154:50)

Astragalus, Fabaceae

Astragalus adsurgens var. *robustior* Hook., Prairie Milkvetch
- *Drug*—**Cheyenne** *Dermatological Aid* Ground leaf and stem sprinkled on skin in cases of poison ivy. (as *A. nitidus* 75:40) Ground leaves and stems sprinkled on watery poison ivy rash. (as *A. nitidus* 76:179)

Astragalus allochrous Gray, Halfmoon Milkvetch
- *Drug*—**Navajo, Ramah** *Ceremonial Medicine* Leaves used as a ceremonial emetic. *Emetic* Leaves used as a ceremonial emetic. (191:31, 32)
- *Other*—**Navajo** *Ceremonial Items* Used in the Night Chant. (55:55)

Astragalus americanus (Hook.) M. E. Jones, American Milkvetch
- *Drug*—**Cree, Woodlands** *Gastrointestinal Aid* Roots chewed for

stomachaches, cramps, or stomach flu. *Misc. Disease Remedy* Roots chewed for stomach flu. (109:31)

Astragalus amphioxys Gray, Crescent Milkvetch
- *Drug*—Zuni *Snakebite Remedy* Fresh or dried root chewed by medicine man before sucking snakebite and poultice applied to wound. (27:376)

Astragalus australis (L.) Lam., Indian Milkvetch
- *Food*—Canadian Indian *Unspecified* Roots used for food. (as *A. aboriginum* 97:39)

Astragalus bisulcatus* var. *haydenianus (Gray) Barneby, Hayden's Milkvetch
- *Drug*—Navajo, Ramah *Ceremonial Medicine* and *Emetic* Fruit used as ceremonial emetic. *Eye Medicine* Infusion of plant used as an eyewash. *Toothache Remedy* Poultice of chewed leaves applied for toothache. (as *A. haydenianus* 191:32)

Astragalus calycosus* var. *scaposus (Gray) M. E. Jones, Torrey's Milkvetch
- *Drug*—Navajo, Kayenta *Dermatological Aid* Plant used as a lotion and poultice applied to injuries from hailstones. *Other* Plant used as a lotion for illness from exposure. *Poultice* Plant used as a lotion and poultice applied to injuries from water. (as *A. scaposus* 205:27) **Shoshoni** *Venereal Aid* Decoction of scraped roots taken for venereal disease. (as *A. scaposus* 180:49)

Astragalus canadensis L., Canadian Milkvetch
- *Drug*—Blackfoot *Antihemorrhagic* Roots chewed or infusion of root taken for spitting up blood. (86:71) *Dermatological Aid* Poultice of chewed roots applied to cuts. (86:83) *Pediatric Aid* and *Pulmonary Aid* Root boiled and the steam used to bathe a child's aching chest. (86:71) **Dakota** *Febrifuge* and *Pediatric Aid* Infusion of roots given to children with fevers. (69:365) **Lakota** *Analgesic* Roots pulverized and chewed for chest and back pains. *Antihemorrhagic* Roots and wild licorice roots used for spitting of blood. *Cough Medicine* Infusion of roots taken for coughs. *Pulmonary Aid* Roots pulverized and chewed for chest pains. (139:45)
- *Food*—Blackfoot *Staple* Root considered a staple. *Unspecified* Roots eaten fresh or boiled in blood or broth. (86:101) Roots eaten raw or boiled. (97:39) **Lakota** *Fodder* Seeds eaten by horses. (139:45)

Astragalus canadensis L. **var. *canadensis***, Canadian Milkvetch
- *Drug*—Dakota *Febrifuge* and *Pediatric Aid* Decoction of root used as a febrifuge for children. (as *A. caroliniana* 70:91)
- *Food*—Blackfoot *Unspecified* Roots eaten raw or boiled. (as *A. carolinianus* 114:278)
- *Other*—Omaha *Cooking Tools* Plant used as a mat to keep the meat free from dirt while butchering. *Toys & Games* Stalks with pods used by small boys as rattles in games in which they imitated tribal dances. **Ponca** *Cooking Tools* Plant used as a mat to keep the meat free from dirt while butchering. *Toys & Games* Stalks with pods used by small boys as rattles in games in which they imitated tribal dances. (as *A. caroliniana* 70:91)

Astragalus ceramicus Sheldon, Painted Milkvetch
- *Food*—Hopi *Candy* Sweet roots eaten by children. (42:291)

Astragalus ceramicus Sheldon **var. *ceramicus***, Painted Milkvetch
- *Food*—Hopi Roots eaten as a sweet. (as *A. pictus filifolius* 56:16)

Astragalus ceramicus* var. *filifolius (Gray) F. J. Herm., Painted Milkvetch
- *Food*—Hopi *Unspecified* Sweet roots dug up and eaten by children. (200:79)

Astragalus convallarius Greene **var. *convallarius***, Timber Milkvetch
- *Drug*—Gosiute *Veterinary Aid* Plant used as a horse medicine. (as *A. junceus* 39:363)

Astragalus crassicarpus Nutt., Groundplum Milkvetch
- *Drug*—Chippewa *Anticonvulsive* Compound infusion or decoction of root taken for "fits." (51:63, 64) Compound decoction of root taken for convulsions. (53:336) *Hemostat* Compound infusion or decoction of root used on bleeding wounds. (51:63, 64) Compound decoction of root used on bleeding wounds. (53:336) *Stimulant* Compound infusion or decoction of root taken or used externally as stimulant. (51: 63, 64) Compound decoction of root taken as a stimulant. *Tonic* Compound decoction of root taken as a tonic. (53:364) **Lakota** *Veterinary Aid* Used as medicine for horses. (139:46)
- *Food*—Dakota *Unspecified* Plant sometimes eaten raw and fresh. (69:365) **Lakota** *Fruit* Fruits eaten for food. (139:46)

Astragalus crassicarpus Nutt. **var. *crassicarpus***, Groundplum Milkvetch
- *Food*—Montana Indian *Unspecified* Fleshy, plum-like pods eaten raw, boiled, and used for pickles. (as *A. caryocarpus* 19:7)
- *Other*—Omaha *Ceremonial Items* Fruits gathered just before corn planting time and ceremonially soaked with seed corn.[20] **Ponca** *Ceremonial Items* Fruits gathered just before corn planting time and ceremonially soaked with seed corn.[20] (as *Geoprumnon crassicarpum* 70:91)

Astragalus cyaneus Gray, Cyanic Milkvetch
- *Food*—Keres, Western *Unspecified* Tubers eaten. (as *A. jemensis* 171:31)

Astragalus giganteus S. Wats., Giant Milkvetch
- *Food*—Thompson *Fodder* Used as a rich horse and deer feed. (164:514)

Astragalus gracilis Nutt., Slender Milkvetch
- *Drug*—Lakota *Gynecological Aid* Roots chewed by mothers with no milk. (139:46)

Astragalus humistratus* var. *sonorae (Gray) M. E. Jones, Groundcover Milkvetch
- *Drug*—Navajo, Ramah *Ceremonial Medicine* Plant used as a ceremonial chant lotion. *Dermatological Aid* Dried plant used as a dusting powder for sores. *Panacea* Leaves or whole plant used as "life medicine." (191:32)

Astragalus kentrophyta* var. *elatus S. Wats., Tall Spiny Milkvetch
- *Drug*—Navajo, Ramah *Ceremonial Medicine* Cold infusion of whole

plant used as a ceremonial chant lotion. *Panacea* Root used as a "life medicine." (as *A. impensus* 191:32)

Astragalus kentrophyta Gray **var. *kentrophyta***, Spiny Milkvetch

- **Drug**—Navajo *Misc. Disease Remedy* Plant used for rabies. (as *Kentrophyta montana* 55:56)

Astragalus lentiginosus **var. *diphysus*** (Gray) M. E. Jones, Speckledpod Milkvetch

- **Food**—Acoma *Unspecified* Fleshy roots eaten fresh. (as *A. diphysus* 32:17) **Apache, White Mountain** *Fruit* Pea fruit eaten raw and cooked. (as *A. diphysus* 136:155) **Jemez** *Unspecified* Pods eaten raw or cooked. **Laguna** *Unspecified* Fleshy roots eaten fresh. (as *A. diphysus* 32:17) **Zuni** *Dried Food* Pods dried for winter use. *Unspecified* Pods eaten fresh, boiled, and salted. (as *A. diphysus* 166:65)

Astragalus lentiginosus **var. *palans*** (M. E. Jones) M. E. Jones, Speckledpod Milkvetch

- **Other**—Navajo, Kayenta *Ceremonial Items* Plant used as a charm in some prayers. (as *A. palans* 205:27)

Astragalus lonchocarpus Torr., Rushy Milkvetch

- **Drug**—Navajo, Kayenta *Emetic* Plant used as an emetic. *Poultice* Poultice of plant applied to goiter. (205:27)

Astragalus miser Dougl., Weedy Milkvetch

- **Food**—Okanagan-Colville *Unspecified* Seeds used for food. (188:105) **Thompson** *Unspecified* Plant placed at the top of the cooking pit in the absence of black tree lichen and wild onion. (187:222)
- **Other**—Okanagan-Colville *Season Indicator* Blooming plant indicated that the lodgepole pine cambium was ready to harvest. (188:28) Blooms indicated that pine cambium was ready to eat. (188:105) *Tools* Plant used to wipe the juice from the lodgepole pine bark before the cambium was scraped off. (188:28) Used to wipe off the turpentine-like juice from the inside of stripped pine bark. (188:105)

Astragalus miser **var. *decumbens*** (Nutt. ex Torr. & Gray) Cronq., Prostrate Loco Milkvetch

- **Food**—Thompson *Fodder* Used as a rich horse and deer feed. (as *A. decumbens* 164:514)

Astragalus mollissimus Torr., Woolly Milkvetch

- **Drug**—Mahuna *Poison* Plant considered poisonous. (140:36)

Astragalus mollissimus **var. *matthewsii*** (S. Wats.) Barneby, Matthews's Woolly Milkvetch

- **Drug**—Navajo, Ramah *Ceremonial Medicine* and *Emetic* Leaves used as a ceremonial emetic. (as *A. matthewsii* 191:32)
- **Food**—Navajo, Ramah *Forage* Plant and roots eaten by sheep. (as *A. matthewsii* 191:32)
- **Other**—Navajo *Ceremonial Items* Used by the male and female shooters in the Lightning Chant. (as *A. matthewsii* 55:56)

Astragalus pachypus Greene, Thickpod Milkvetch

- **Drug**—Kawaiisu *Analgesic* Decoction of roots taken for menstrual pains. *Gynecological Aid* Decoction of roots taken for menstrual pains. (206:14)

Astragalus pattersonii Gray, Patterson's Milkvetch

- **Drug**—Navajo, Kayenta *Ear Medicine* Plant used for any disease of the ears. *Emetic* Plant used as an emetic. *Eye Medicine* Plant used for any disease of the eyes. *Misc. Disease Remedy* Plant used for mumps. *Other* Plant used for sore throats or swollen neck. *Throat Aid* Plant used for any disease of the throat. (205:27)

Astragalus polaris Benth., Polar Milkvetch

- **Food**—Eskimo, Alaska *Unspecified* Tiny peas eaten raw or cooked. (1:36)

Astragalus praelongus Sheldon, Stinking Milkvetch

- **Drug**—Navajo, Ramah *Ceremonial Medicine* and *Emetic* Leaves used as a ceremonial emetic. (191:32)

Astragalus purshii Dougl. ex Hook., Woollypod Milkvetch

- **Drug**—Thompson *Dermatological Aid* Decoction of whole plant used as a wash for the head, hair, and whole body. (164:473, 474) *Disinfectant* Decoction of roots taken and poured on head in sweat house for purification. (164:504) *Hunting Medicine* Decoction of plant poured onto hunting equipment which had "lost its luck." (164:507)
- **Food**—Thompson *Forage* Used as a common forage plant. (164:516)

Astragalus purshii **var. *tinctus*** M. E. Jones, Woollypod Milkvetch

- **Drug**—Kawaiisu *Analgesic* and *Gynecological Aid* Decoction of roots taken for menstrual pains. (206:15)

Astragalus racemosus Pursh, Alkali Poisonvetch

- **Drug**—Lakota *Poison* Plant poisonous to livestock. (139:46)

Astragalus sesquiflorus S. Wats., Sandstone Milkvetch

- **Drug**—Navajo, Kayenta *Ceremonial Medicine* Plant used as a ceremonial emetic. *Dermatological Aid* Plant used as a lotion and poultice of plant applied to ringworm. *Emetic* Plant used as a ceremonial emetic. (205:28)

Astragalus sp., Vetch

- **Drug**—Alaska Native *Poison* Plant considered poisonous. (85:159) **Cahuilla** *Poison* Plant poisonous to stock. (15:44) **Cheyenne** *Poison* Plant poisonous to horses. *Veterinary Aid* Plant applied as an ointment for animals with urination troubles. (83:28) **Hopi** *Ceremonial Medicine* and *Emetic* Plant used as a ceremonial emetic. (200:80) **Jemez** *Cathartic* Roots chewed as a cathartic. (44:20) **Keresan** *Veterinary Aid* Plant made horses crazy or killed them, if eaten. (198:562) **Navajo** *Poison* Plant considered poisonous. (55:55) **Navajo, Kayenta** *Gastrointestinal Aid* Plant used for stomach disorders. *Orthopedic Aid* Poultice of crushed leaves applied to lame back. *Throat Aid* Plant used as a gargle for sore throats. (205:27) **Shoshoni** *Dermatological Aid* Infusion of root used as a wash for sores. *Eye Medicine* Infusion or decoction of root used as a wash for granulated eyelids. *Toothache Remedy* Decoction of root used as a wash for toothaches. (180:50) **Thompson** *Dermatological Aid* Decoction of whole plant used as a wash for the head, hair, and whole body. (164:473, 474)
- **Food**—Cahuilla *Spice* Pounded seeds mixed with other foods and used as a spice. (15:44) **Havasupai** *Unspecified* Seeds used for food. (197:226) **Shoshoni** *Spice* Steeped seeds added to dishes for flavoring. (118:29)

Asyneuma, Campanulaceae

Asyneuma prenanthoides (Dur.) McVaugh, California Harebell
- *Food*—**Costanoan** *Winter Use Food* Bulbs eaten in winter and early spring. (as *Campanula prenanthoides* 21:254)

Athyrium, Dryopteridaceae, fern

Athyrium filix-femina (L.) Roth, Common Ladyfern
- *Drug*—**Chippewa** *Diuretic* Compound decoction of root taken for "stoppage of urine." (53:348) **Cowlitz** *Analgesic* Infusion of stems taken for body pains. (79:14) **Hesquiat** *Cancer Treatment* Young, unfurling fronds eaten for internal ailments, such as cancer of the womb. (185:29) **Makah** *Gynecological Aid* Decoction of pounded stems taken by women to ease labor. (79:14) **Meskwaki** *Analgesic* and *Gynecological Aid* Decoction of root taken by women for bosom pains caused by childbirth. (as *Asplenium filix-femina* 152:237) **Ojibwa** *Dermatological Aid* Grated dried root used as healing powder for sores. *Gynecological Aid* Infusion of root induced milk flow in patients with caked breast. (as *Asplenium filis-femina* 153:381) **Potawatomi** *Gynecological Aid* Infusion of root taken for caked breasts and other female disorders. (as *Asplenium filix-femina* 154:73) **Thompson** *Antihemorrhagic* Infusion of plant used for vomiting blood. (187:88)
- *Food*—**Quileute** *Unspecified* Rhizomes roasted, peeled, and the centers eaten. **Quinault** *Unspecified* Rhizomes roasted, peeled, and the centers eaten. (79:14) **Salish, Coast** *Unspecified* New shoots and rhizomes used for food. (182:68)
- *Other*—**Cowlitz** *Cooking Tools* Leaves used to cover camas while baking. (79:14) **Karok** *Cooking Tools* Leaves used to clean eel's blood from butchered eel. (6:20) **Kwakiutl, Southern** *Containers* Fronds used as covering for fungus placed on hot stones to make a red paint. (183:263) **Nitinaht** *Cooking Tools* Fronds placed in layers below and above food in steaming pits. (186:62) **Okanagan-Colville** *Water Indicator* Ferns considered to be a sign of water when traveling through the mountains. (188:18) **Quileute** *Cleaning Agent* Leaves used to wipe fish. (79:14) **Shuswap** *Protection* Used to cover berry baskets. (123:49)

Athyrium filix-femina ssp. *angustum* (Willd.) Clausen, Subarctic Ladyfern
- *Drug*—**Iroquois** *Febrifuge* Infusion of rhizomes and whole New England aster plant taken by mothers with intestinal fevers. *Reproductive Aid* Infusion of plant, vinegar bark, and flower stalks taken to prevent women's water from breaking. *Venereal Aid* Infusion of rhizomes and sensitive fern used by men with venereal diseases. (as *A. angustum* 141:34)

Athyrium filix-femina ssp. *cyclosorum* (Rupr.) C. Christens., Subarctic Ladyfern
- *Drug*—**Bella Coola** *Eye Medicine* Simple or compound decoction of root used as a wash for sore eyes. (as *Asplenium cyclosorum* 150:48)

Atriplex, Chenopodiaceae

Atriplex argentea Nutt., Silverscale Saltbush
- *Drug*—**Navajo, Ramah** *Analgesic* Leaves used as a fumigant for pain. *Dermatological Aid* Poultice of leaves applied to spider bites. *Other* Cold infusion used for sickness from drinking bad water and to purify water. (191:24) **Zuni** *Dermatological Aid* Poultice of chewed root applied to sores and rashes. *Gastrointestinal Aid* Infusion of root taken for stomachache. (27:374)
- *Food*—**Acoma** *Fruit* Fruits eaten for food. (32:18) **Hopi** *Unspecified* Leaves boiled and eaten with fat. (56:21) Boiled with meat. (190:160) *Vegetable* Young, tender leaves cooked and eaten as greens. (200:73) **Isleta** *Vegetable* Young leaves boiled and eaten as greens. (32:18) **Keres, Western** *Forage* Plant used as forage for cattle. *Unspecified* Seeds and expanded calyx eaten for food and the salty taste. (171:31) **Laguna** *Fruit* Fruits eaten for food. (32:18) **Paiute, Northern** *Porridge* Seeds parched, ground into a flour, and made into mush. (59:47) **Pueblo** *Spice* Boiled alone or with plant products and meats for flavoring. Boiled alone or with plant products and meats for flavoring. (as *A. cornuta* 32:18)

Atriplex argentea ssp. *expansa* (S. Wats.) Hall & Clements, Silverscale Saltbush
- *Food*—**Navajo** *Fodder* Plant used, for the salt, and stored for the winter as fodder. *Forage* Plant used, for the salt, to pasture sheep in the summer. (as *A. expansa* 55:43)

Atriplex canescens (Pursh) Nutt., Fourwing Saltbush
- *Drug*—**Havasupai** *Dermatological Aid* Leaves made into a soapy lather and used to wash the hair. *Misc. Disease Remedy* Leaves made into a soapy lather and used for itches or rashes, such as chickenpox or measles. (197:217) **Hopi** *Ceremonial Medicine* Plant used for kiva fires. (56:21) **Isleta** *Poison* Infectious wood used to make poison arrowheads for war purposes. (100:24) **Jemez** *Dermatological Aid* Poultice of crushed leaves applied to ant bites, probably to reduce the swelling and pain. *Stimulant* Leaves put unto a fire and smoke used to revive badly hurt, weak, and faint person. (44:20) **Navajo** *Dermatological Aid* Plant used for ant bites. (90:148) **Navajo, Kayenta** *Emetic* Plant used as an emetic. *Gastrointestinal Aid* Plant used for stomach disease. (205:20) **Navajo, Ramah** *Analgesic* Decoction of tops or roots taken as an emetic for gastric pain. *Ceremonial Medicine* Whole plant used as a ceremonial emetic. *Cough Medicine* Decoction of leaves or roots taken for bad cough. *Dermatological Aid* Poultice of leaves applied to ant bites. Leaf and stem ash rubbed on the scalp as a hair tonic. *Emetic* Whole plant used as a ceremonial emetic. *Gastrointestinal Aid* Decoction of tops or roots taken as an emetic for gastric pain. *Nose Medicine* Leaves used as snuff for nose trouble. *Toothache Remedy* Poultice of warm, pulverized root applied for toothache. *Veterinary Aid* Compound decoction given to sheep for bloating from overeating. (191:24) **Shoshoni** *Cathartic* Decoction of fresh roots with salt taken as a physic. (180:50) **Zuni** *Dermatological Aid* Poultice of fresh or dried flower used for ant bites. (27:374) Infusion of dried root and blossoms or poultice of blossoms used for ant bites. (166:44) *Hunting Medicine* Twigs attached to prayer plumes and sacrificed to the cottontail rabbit to ensure good hunting. (166:88)
- *Food*—**Gosiute** *Unspecified* Seeds used for food. (39:363) **Hopi** *Substitution Food* Ashes used instead of baking soda. (as *Calligonum canescens* 190:160) **Navajo** *Fodder* Plant used in the winter to provide salt for the sheep. *Forage* Plant used as forage for cattle, sheep, and goats, especially when other forage was scarce. (55:43) *Pie & Pudding* Flowers used to make puddings. (90:148) **Navajo, Ramah** *Fodder* Used for sheep feed. *Spice* Leaves placed on coals in pit for roasting

corn, to impart a salty taste. (191:24) **Tewa of Hano** *Cooking Agent* Ashes stirred into dough to give it a greenish blue color. (138:54)
- **Dye**—**Hopi** *Blue* Ashes used to maintain the blue coloring in blue cornmeal. (200:73) *Mordant* Ashes used as alkali to maintain blue coloring of piki. (42:292) **Navajo** *Yellow* Leaves and twigs used in coloring wool yellow. (55:43) **Navajo, Ramah** *Red* Leaf and twig ash used to intensify red color of buckskin dye. *Yellow* Young leaves and twigs used to dye wool yellow. (191:24) **Tewa** *Mordant* Ashes used as alkali to maintain blue coloring of piki. (42:292)
- **Other**—**Diegueño** *Soap* Leaves formerly used as soap. (88:217) **Hopi** *Ceremonial Items* Plant used to make pahos (prayer sticks). (42:292) **Isleta** *Weapon* Infectious wood used to carve arrowheads. The arrowheads were attached to light, swift bamboo shafts and always broke upon impact making them ideal for war purposes. (100:24) **Kawaiisu** *Hunting & Fishing Item* Hard wood used to make arrow points. (206:15) **Tewa** *Ceremonial Items* Plant used to make pahos (prayer sticks). (42:292) **Zuni** *Ceremonial Items* Twigs attached to prayer plumes and sacrificed to the cottontail rabbit to ensure good hunting. (166:88)

Atriplex confertifolia (Torr. & Frém.) S. Wats., Shadscale Saltbush
- **Drug**—**Hopi** *Anticonvulsive* Plant burned and smoke inhaled for epileptic medicine. (as *A. jonesii* 42:293) **Navajo** *Veterinary Aid* Plant rubbed on horses to repel gnats. (90:149) **Paiute, Northern** *Antirheumatic (External)* Leaves boiled and used as a liniment for sore muscles and aches. *Cold Remedy* Poultice of mashed leaves applied to the chest and decoction of leaves taken for colds. (59:125)
- **Food**—**Gosiute** *Unspecified* Seeds formerly used for food. (39:363) **Hopi** *Pie & Pudding* Scented leaves boiled and water mixed with cornmeal to make a pudding. (32:17) Leaves boiled in water, the water mixed with cornmeal, and baked into a pudding. (56:20) *Spice* Plant used as flavoring with meat or other vegetables. (as *A. jonesii* 42:293) *Unspecified* Boiled with meat. (as *Obione confertifolia* 190:160) *Vegetable* Plant used for greens. (as *A. jonesii* 42:293) Young, tender leaves cooked and eaten as greens. (200:73)
- **Other**—**Kawaiisu** *Hunting & Fishing Item* Hard wood used to make arrow points. (206:15)

Atriplex coronata S. Wats., Crownscale
- **Food**—**Pima** *Cooking Agent* Boiled with dried cane cactus to counteract its acidic flavor. (32:36) *Spice* Plants boiled with other foods for their salty flavor. *Unspecified* Plants roasted in pits with cactus fruits and eaten. (146:69)

Atriplex elegans (Moq.) D. Dietr., Wheelscale Saltbush
- **Food**—**Pima** *Cooking Agent* Boiled with dried cane cactus to counteract its acidic flavor. (32:36) *Spice* Plants boiled with other foods for their salty flavor. *Unspecified* Plants roasted in pits with cactus fruits and eaten. (146:69) **Pima, Gila River** *Unspecified* Leaves boiled and eaten. (133:7)

Atriplex garrettii Rydb., Garrett's Saltbush
- **Dye**—**Great Basin Indian** *Yellow* Whole plant used to make a yellow dye and set with bitter alum. (121:47)

Atriplex lentiformis (Torr.) S. Wats., Big Saltbush
- **Drug**—**Cahuilla** *Cold Remedy* Dried leaves smoked for head colds. Fresh leaves chewed for head colds. *Nose Medicine* Crushed flowers, stems, and leaves steamed and inhaled for nasal congestion. (15:45) **Pima** *Dermatological Aid* Poultice of powdered roots applied to sores. (47:66) Poultice of powdered root applied to sores. (146:80)
- **Food**—**Cahuilla** *Porridge* Seeds ground into a flour and used to make mush or small cakes. (15:45) **Papago** *Unspecified* Seeds used for food. (36:62) **Pima** *Dried Food* Seeds roasted, dried, parched, and stored. (32:23) *Porridge* Seeds pounded into meal, cooked, mixed with water, and eaten as mush. (95:263) Seeds pit roasted, dried, parched, added to water, and eaten as a thick gruel. (146:78) *Starvation Food* Tiny seeds formerly roasted and eaten during famines. (47:66) **Pima, Gila River** *Starvation Food* Seeds used as "starvation food." (133:6) *Unspecified* Seeds used for food. (133:7) **Yuma** *Porridge* Seeds boiled to make a mush. (37:187) *Unspecified* Seeds pounded, pit baked, ground, mixed with water to form stiff dough, and eaten raw. (37:200)
- **Other**—**Cahuilla** *Soap* Crushed leaves and roots used as a soap and rubbed into articles for cleaning. (15:45) **Pima** *Soap* Leaves rubbed in water and lather and used for washing clothing and baskets. (47:66)

Atriplex nuttallii S. Wats., Nuttall's Saltbush
- **Food**—**Pima** *Bread & Cake* Stems used as stuffing for roast rabbit. (146:77) *Spice* Young stems and flower heads used as flavoring. *Unspecified* Stems cut in short lengths and used as a stuffing in cooked rabbits. (32:18) Stems boiled with wheat and used for food. (146:77)

Atriplex obovata Moq., Mound Saltbush
- **Drug**—**Hopi** *Anticonvulsive* Plant burned and smoke inhaled for epileptic medicine. (42:293)
- **Food**—**Hopi** *Spice* Plant used as flavoring with meat or other vegetables. *Vegetable* Plant used for greens. (42:293) Young, tender leaves cooked and eaten as greens. (200:73)

Atriplex polycarpa (Torr.) S. Wats., Cattle Saltbush
- **Drug**—**Maricopa** *Antirheumatic (External)* Moxa of dried galls burned on the affected area for rheumatism. **Pima** *Antirheumatic (External)* Moxa of galls placed on area affected by rheumatism. (47:67)
- **Food**—**Pima** *Bread & Cake* Seeds made into bread and used for food. *Forage* Used as an important forage plant. *Starvation Food* Seeds formerly roasted, ground, and eaten during famines. (47:67) **Yuma** *Unspecified* Seeds separated from hulls and eaten. (37:187)

Atriplex powellii S. Wats., Powell's Saltweed
- **Food**—**Cochiti** *Vegetable* Young plants eaten as greens. **Hopi** *Unspecified* Salty leaves boiled and eaten with fat. (as *A. philonitra* 32:18) *Vegetable* Young, tender leaves cooked and eaten as greens. (200:73) **Keres, Western** *Vegetable* Young plants used for greens. (as *A. philonitra* 171:31) **Navajo, Kayenta** *Substitution Food* Used as a greens and salt substitute in foods. (205:20) **Pueblo** *Vegetable* Young plants eaten as greens. (as *A. philonitra* 32:18) **Zuni** *Porridge* Seeds mixed with ground corn to make a mush. (32:22) Seeds eaten raw before the presence of corn and afterwards, ground with cornmeal and made into a mush. (166:66)

Atriplex rosea L., Tumbling Saltweed
- **Food**—**Navajo, Ramah** *Fodder* Used for sheep and horse feed and harvested for winter use. *Porridge* Seeds of dried plants threshed on a blanket, winnowed, ground, and made into a mush or used like maize. (191:24)

- *Dye*—**Navajo, Ramah** *Black* Used as a black dye. (191:24)

Atriplex saccaria S. Wats., Sack Saltbush
- *Food*—**Hopi** *Vegetable* Young, tender leaves cooked and eaten as greens. (200:73)

Atriplex semibaccata R. Br., Australian Saltbush
- *Food*—**Cahuilla** *Fruit* Berries gathered and eaten fresh. (15:45)

Atriplex serenana A. Nels., Bractscale
- *Food*—**Kawaiisu** *Vegetable* Leaves boiled, fried in grease, and eaten. (206:15) **Pima** *Cooking Agent* Boiled with dried cane cactus to counteract its acidic flavor. (as *A. bracteosa* 32:36) *Spice* Plants boiled with other foods for their salty flavor. *Unspecified* Plants roasted in pits with cactus fruits and eaten. (as *A. bracteosa* 146:69)

***Atriplex* sp.**, Four Winged Salt Bush
- *Drug*—**Hualapai** *Antirheumatic (External)* Infusion of smaller leaves used as a wash for aching body, joints, and sore muscles. (195:11) **Keres, Western** *Blood Medicine* Infusion of plant used as a blood medicine. Infusion of any atriplex in which the stems or leaves have a red color was used for blood medicine. (171:32) **Navajo** *Dermatological Aid* Poultice of chewed plants applied to ant, bee, and wasp sting swellings. (55:43) **Yokut** *Cathartic* Infusion of leaves used as a cathartic. (117:437)
- *Food*—**Isleta** *Unspecified* Leaves characterized as having a salty taste. *Vegetable* Young, tender leaves boiled for greens. (100:24) **Paiute** *Unspecified* Species used for food. (167:244) **Pima** *Dried Food* Seeds dried, parched, ground, and eaten dry with sips of water. (146:73) *Forage* Herbaceous plants eaten by stock. (146:69) *Staple* Seeds dried, parched, ground, and eaten as pinole. (146:73)
- *Other*—**Pima** *Fuel* Woody plants used for fuel. (146:69)

Atriplex torreyi (S. Wats.) S. Wats., Torrey's Saltbush
- *Food*—**Kamia** *Staple* Pulverized seeds made into a meal. (62:24)

Atriplex truncata (Torr. ex S. Wats.) Gray, Wedgescale Saltbush
- *Food*—**Gosiute** *Unspecified* Seeds used for food. (39:363)

Atriplex wrightii S. Wats., Wright's Saltbush
- *Food*—**Papago** *Soup* Mixed with roasted cholla buds and eaten as a vegetable stew. (34:16) *Spice* Branches used as seasoning in cooking or in pit baking. (34:15) *Vegetable* Branches eaten as greens in summer. (34:14) Greens used for food. (36:61) **Pima** *Vegetable* Leaves boiled, strained, fried in grease, and eaten as greens. (47:69) **Pima, Gila River** *Unspecified* Leaves boiled and eaten. (133:7)

Aulacomniaceae, see *Aulacomnium*

Aulacomnium, Aulacomniaceae, moss

***Aulacomnium* sp.**, Moss
- *Fiber*—**Shuswap** *Building Material* Moss mixed with clay and used between the logs of a log house. (123:49)
- *Other*—**Shuswap** *Fertilizer* Moss mixed with house plant dirt as a fertilizer to make the plants healthier. (123:49)

Aureolaria, Scrophulariaceae

Aureolaria flava (L.) Farw., Smooth Yellow False Foxglove
- *Drug*—**Cherokee** *Antidiarrheal* Compound decoction taken for dysentery. *Other* Infusion taken while fasting for 4 days for apoplexy. (80:35)

Aureolaria laevigata (Raf.) Raf., Entireleaf Yellow False Foxglove
- *Drug*—**Cherokee** *Antidiarrheal* Compound decoction taken for dysentery. *Other* Infusion taken while fasting for 4 days for apoplexy. (80:35)

Aureolaria pedicularia (L.) Raf., Fernleaf Yellow False Foxglove
- *Drug*—**Cherokee** *Antidiarrheal* Compound decoction taken for dysentery. *Other* Infusion taken while fasting for 4 days for apoplexy. (80:35)

Aureolaria pedicularia (L.) Raf. **var. *pedicularia***, Fernleaf Yellow False Foxglove
- *Drug*—**Chickasaw** *Emetic* Plant used as an emetic. *Misc. Disease Remedy* Plant used as an antiscorbutic. (as *Dasystoma pedicularia* 28:289)

Aureolaria virginica (L.) Pennell, Downy Yellow False Foxglove
- *Drug*—**Cherokee** *Antidiarrheal* Decoction of plants taken for dysentery. (as *Gerardia virginica* 177:57)

Avena, Poaceae

Avena barbata Pott ex Link, Slender Oat
- *Food*—**Cahuilla** *Unspecified* Small seeds used for food. (15:46) **Miwok** *Porridge* Parched, stone boiled seeds pulverized and eaten as a mush. *Soup* Parched, stone boiled seeds pulverized and eaten as a soup. (12:152)

Avena fatua L., Wild Oat
- *Food*—**Cahuilla** *Porridge* Parched seeds ground into flour and used to make mush. (15:46) **Diegueño** *Porridge* Moistened, hulled kernels boiled and eaten as hot cereal. (84:15) **Kawaiisu** *Unspecified* Seeds pounded in a bedrock mortar hole, boiled, and eaten. (206:15) **Luiseño** *Staple* Seeds ground into a flour and used for food. (155:234) **Mendocino Indian** *Staple* Seeds parched, ground, and the flour eaten dry. (41:311) **Pomo** *Staple* Seeds used to make pinoles. (11:87) *Unspecified* Seeds used for food. (41:311) Parched, pounded seeds used for food. *Winter Use Food* Seeds stored for later use. (66:11)

Avena sativa L., Common Oat
- *Food*—**Haisla & Hanaksiala** *Unspecified* Grains used for food. (43:205) **Karok** *Unspecified* Species used for food. (148:380) **Navajo** *Fodder* Used for hay. (55:25) **Navajo, Ramah** *Fodder* Fed to horses without being thrashed and in a bad winter fed tfo sheep and goats. (191:15) **Pomo** *Unspecified* Seeds parched in a circular coiled basket and used for food. (66:11)

***Avena* sp.**, Oat
- *Food*—**Pomo, Kashaya** *Staple* Grain used in pinole, a very fine dry meal. (72:85) **Yuki** *Staple* Used to make pinole. (49:85)

Baccharis, Asteraceae

Baccharis douglasii DC., Saltmarsh Baccharis
- *Drug*—**Costanoan** *Dermatological Aid* Decoction of plant used as a wash for wounds. Dried, powdered stems applied as a disinfectant to wounds. Poultice of heated leaves and animal fat applied to boils. *Disinfectant* Dried, powdered stems applied as a disinfectant to wounds. *Kidney Aid* Infusion of plant taken for kidney ailments. (21:26) **Luiseño** *Dermatological Aid* Decoction of leaves used as a bath for sores and wounds. (155:228)
- *Other*—**Luiseño** *Tools* Wood used for drilling fires. A small hole was made in a flat, extremely dry stick. The drill, a short piece of wood, was inserted into the hole and twirled to form the dust which would ignite if conditions were dry and favorable. No tinder was used. (155:209)

Baccharis emoryi Gray, Emory's Baccharis
- *Fiber*—**Havasupai** *Basketry* Used in coil basketry. *Building Material* Used to make fence posts and in brush house construction. (197:246)
- *Other*—**Havasupai** *Fuel* Wood used for firewood. *Tools* Wood used to make planting sticks. *Toys & Games* "Down" put onto fires by children to produce a sudden burst of flame which spread rapidly. Pith used to make "peashooters" and stems and twigs used to make the shooter. (197:246)

Baccharis pilularis DC., Coyotebrush
- *Drug*—**Costanoan** *Panacea* Infusion of plant used as a general remedy. (21:26)
- *Other*—**Mendocino Indian** *Hunting & Fishing Item* Light and pithy wood formerly used for arrows. (as *B. consanguinea* 41:393)

Baccharis pteronioides DC., Yerba de Pasmo
- *Drug*—**Yavapai** *Antirheumatic (External)* Decoction of leaves and roots used as wash for rheumatism. *Venereal Aid* Decoction of leaves and roots used as wash for gonorrhea. (65:261)

Baccharis salicifolia (Ruiz & Pavón) Pers., Mule's Fat
- *Drug*—**Cahuilla** *Dermatological Aid* Leaves used in a hair wash solution to prevent baldness. *Gynecological Aid* Decoction of leaves and stems used as a female hygienic agent. (as *B. viminea* 15:46) **Coahuilla** *Eye Medicine* Infusion of leaves used as an eyewash. (as *B. glutinosa* 13:78) **Costanoan** *Dermatological Aid* Infusion of leaves and twigs used as wash for scalp and hair to encourage growth. (as *B. viminea* 21:26) **Diegueño** *Dermatological Aid* Infusion of leaves used as a wash or poultice of leaves applied to bruises, wounds, or insect stings. (as *B. glutinosa* 88:220) **Navajo, Kayenta** *Febrifuge* Compound infusion of plants used as a lotion for chills from immersion. (as *B. glutinosa* 205:45)
- *Food*—**Mohave** *Starvation Food* Young shoots roasted and eaten as a famine food. **Yuma** *Starvation Food* Young shoots roasted and eaten as a famine food. (as *B. glutinosa* 37:201)
- *Fiber*—**Cahuilla** *Building Material* Limbs and branches used in house construction. (as *B. viminea* 15:46)
- *Other*—**Kawaiisu** *Hunting & Fishing Item* Stems used to make one piece arrows for hunting small game. Plant burned into a black powder, mixed with another ingredient, and used for gun powder. (as *B. glutinosa* 206:15)

Baccharis sarothroides Gray, Desert Broom
- *Drug*—**Diegueño** *Cough Medicine* Infusion of plant taken for coughs. *Gastrointestinal Aid* Infusion of plant taken for stomachaches. (88:220)
- *Food*—**Papago** *Beverage* Seeds steeped and used as tea-like drinks for refreshment. (34:27)
- *Fiber*—**Pima** *Brushes & Brooms* Stalks used to make brooms. Green stalks cut, tied together with strings, and used as brooms. (47:65)
- *Other*—**Papago** *Hunting & Fishing Item* Wood used to make stone-tipped hunting arrows. *Weapon* Wood used to make stone-tipped war arrows. (34:71)

Baccharis sp., Seep Willow
- *Drug*—**Hualapai** *Antirheumatic (External)* Hot poultice of leaves applied to swellings and aches. (195:17) **Keres, Western** *Analgesic* Cold infusion used to bathe the head for headaches. *Antirheumatic (External)* Infusion of leaves used as a rub for swelling. (171:32)
- *Fiber*—**Hualapai** *Building Material* Long, straight stems used for ramada roofs. (195:17)
- *Other*—**Hualapai** *Fuel* Stems used for firewood. (195:17)

Baccharis wrightii Gray, Wright's Baccharis
- *Drug*—**Navajo, Ramah** *Ceremonial Medicine* and *Emetic* Plant used as a ceremonial emetic. *Venereal Aid* Strong decoction of plant taken in large amounts for sexual infection. (191:49)

Bacopa, Scrophulariaceae

Bacopa caroliniana (Walt.) B. L. Robins., Blue Waterhyssop
- *Drug*—**Seminole** *Cough Medicine* Plant used for turtle sickness: trembling, short breath, and cough. (as *Hydrotrida caroliniana* 169:237) *Other* Complex infusion of leaves taken for chronic conditions. (as *Hydrotrida caroliniana* 169:272) *Respiratory Aid* and *Sedative* Plant used for turtle sickness: trembling, short breath, and cough. (as *Hydrotrida caroliniana* 169:237)

Bahia, Asteraceae

Bahia dissecta (Gray) Britt., Ragleaf Bahia
- *Drug*—**Keres, Western** *Cathartic* Infusion of plant used as a cathartic. *Emetic* Infusion of plant used as an emetic. (171:32) **Navajo, Ramah** *Analgesic* Compound decoction taken for menstrual pain. *Antirheumatic (Internal)* Compound decoction of plant taken for arthritis. *Contraceptive* Compound decoction of plant taken as contraceptive. *Gynecological Aid* Compound decoction of plant taken for menstrual pain. (191:49) **Zuni** *Analgesic* Powdered plant rubbed on affected parts for headache. *Antirheumatic (External)* Powdered plant rubbed on affected parts for rheumatism. (as *Villanova dissecta* 166:62)

Baileya, Asteraceae

Baileya multiradiata Harvey & Gray ex Gray, Desert Marigold
- *Drug*—**Keres, Western** *Dermatological Aid* Plant rubbed under arms as deodorant. (171:32)
- *Fiber*—**Jemez** *Building Material* Plant mixed with clay, used in making adobes, and plant used in plaster. (44:20)

Balsaminaceae, see *Impatiens*

Balsamita, Asteraceae

Balsamita major Desf., Costmary
- **Drug**—**Iroquois** *Ear Medicine* Infusion of one smashed leaf used as drops for earaches. (as *Chrysanthemum balsamita* var. *tanacetoides* 87:472)

Balsamorhiza, Asteraceae

Balsamorhiza deltoidea Nutt., Deltoid Balsamroot
- **Drug**—**Kawaiisu** *Cold Remedy* Decoction of split roots taken for colds. *Cough Medicine* Decoction of split roots taken for coughing. (206:15)
- **Food**—**Atsugewi** *Bread & Cake* Parched, winnowed, ground seeds made into cakes and eaten without cooking. (61:139) **Karok** *Unspecified* Peduncles used for food. (6:20) **Klamath** *Unspecified* Roasted, ground seeds used for food. (45:106)

Balsamorhiza hookeri (Hook.) Nutt., Hooker's Balsamroot
- **Food**—**Atsugewi** *Bread & Cake* Parched, winnowed, ground seeds made into cakes and eaten without cooking. (61:139) **Gosiute** *Unspecified* Seeds used for food. (39:363) **Okanagan-Colville** *Unspecified* Roots pit cooked and eaten. (188:80) **Paiute, Northern** *Unspecified* Roots used for food. (59:43)

Balsamorhiza hookeri var. *hirsuta* (Nutt.) A. Nels., Hairy Balsamroot
- **Drug**—**Paiute** *Gastrointestinal Aid* Decoction of root considered good for severe stomach. *Urinary Aid* Decoction of root considered good for bladder troubles. **Washo** *Gynecological Aid* Decoction of root taken for female complaints. (as *B. hirsuta* 180:50)

Balsamorhiza incana Nutt., Hoary Balsamroot
- **Drug**—**Cheyenne** *Analgesic* Decoction of leaves, roots, and stems taken for stomach pains. Decoction of leaves, roots, and stems used as a steam bath for headaches. (76:189) *Cold Remedy* Plant used for colds. (76:189) Infusion of leaves, stems, and roots taken for colds. (83:20) *Gastrointestinal Aid* Decoction of leaves, roots and stems taken for stomach pains. (76:189) Infusion of leaves, stems, and roots taken for stomach pains. (83:20)
- **Food**—**Nez Perce** *Unspecified* Thick roots eaten raw. (19:7)

Balsamorhiza sagittata (Pursh) Nutt., Arrowleaf Balsamroot
- **Drug**—**Blackfoot** *Antirheumatic (Internal)* Root smudge smoke inhaled for body aches. (86:78) *Dermatological Aid* Poultice of chewed roots applied to blisters and sores. (86:75) **Cheyenne** *Analgesic* Infusion of leaves, roots, and stems taken for stomach pains and headaches. (76:189) Steam of decoction of plant inhaled for headache and used as wash on head. (75:38) Decoction of leaves, roots, and stems taken for stomach pains. (75:38, 39) *Cold Remedy* Plant used for colds. (76:189) Infusion of leaves, stems, and roots taken for colds. (83:20) *Febrifuge* Infusion of root taken for fever. (75:38) *Gastrointestinal Aid* Infusion of leaves, roots, and stems taken for stomach pains. (76:189) Decoction of leaves, roots, and stems taken for stomach pain. (75:38) Infusion of leaves, stems, and roots taken for stomach pains. (83:20) *Gynecological Aid* Decoction of root taken when labor begins, to insure easy delivery. (75:38) *Oral Aid* Root chewed and saliva allowed to run down throat for sore mouth and throat. *Panacea* Root chewed and rubbed over the body for any sickness. (75:38, 39) *Throat Aid* Root chewed and saliva allowed to run down throat for sore throat. *Toothache Remedy* Root chewed for toothaches. (75:38) **Flathead** *Burn Dressing* Poultice of coarse, large leaves used for burns. *Cathartic* Infusion of roots taken as a cathartic. *Pulmonary Aid* Infusion of roots taken for whooping cough. *Tuberculosis Remedy* Infusion of roots taken for tuberculosis. *Urinary Aid* Infusion of roots taken to increase urine. (82:20) **Gosiute** *Dermatological Aid* Poultice of pounded or chewed root paste applied to arrow or gunshot wounds. (39:348) Chewed roots or pounded root salve applied to fresh wounds. (39:363) *Hemostat* Poultice of plant applied to arrow or gunshot wound hemorrhages. (39:348) **Kutenai** *Dermatological Aid* Poultice of root infusion used for wounds, cuts, and bruises. (82:20) **Miwok** *Analgesic* Decoction of ground root cooled and taken for headaches. *Antirheumatic (Internal)* Decoction of ground root cooled and taken for rheumatism. *Diaphoretic* Decoction of root taken to produce profuse perspiration for rheumatism. (12:167) **Okanagan-Colville** *Burn Dressing* Poultice of dried, powdered leaves applied to severe skin burns. *Diaphoretic* Leaves placed on glowing coals and laid on to cause profuse sweating. (188:80) **Paiute** *Analgesic* Decoction of root taken for stomachaches. *Dermatological Aid* Poultice of mashed root applied to insect bites or swellings. Powdered, dried root applied to syphilitic sores. *Disinfectant* Root burned as a fumigant in the sickroom. *Gastrointestinal Aid* Decoction of root taken for stomachaches. *Tuberculosis Remedy* Root sap taken for consumption. *Venereal Aid* Decoction of root taken over a long period of time for venereal disease. Poultice of dry, powdered root applied to syphilitic sores. (180:50, 51) **Sanpoil** *Analgesic* Poultice of root prepared in various ways and applied to painful areas. *Dermatological Aid* Infusion of root rubbed into hair and scalp to help hair grow. Poultice of root prepared in various ways and applied to bruised areas. Pulverized root sprinkled on sores and boils. (131:219) **Shoshoni** *Dermatological Aid* Poultice of mashed root applied to insect bites or swellings, and to syphilitic sores. *Eye Medicine* Decoction of root used as an eyewash. *Venereal Aid* Poultice of mashed root applied to syphilitic sores. (180:50, 51) **Shuswap** *Dermatological Aid* Infusion of leaves used as a wash for poison ivy and running sores. (123:59) **Thompson** *Antidiarrheal* Seeds eaten for dysentery. (187:175) *Dietary Aid* Root sucked and chewed for hunger. (164:493) *Sedative* Young shoots, when eaten in great quantities, caused sleepiness like sleeping pills. (187:175) **Washo** *Disinfectant* Root burned as a fumigant in the sickroom. (180:50, 51)
- **Food**—**Atsugewi** *Bread & Cake* Parched, winnowed, ground seeds made into cakes and eaten without cooking. (61:139) **Flathead** *Unspecified* Young, immature flower stems peeled and eaten raw. Roots pit baked and used for food. (82:20) **Gosiute** *Cooking Agent* Seeds a highly prized source of oil. *Unspecified* Leaves and petioles boiled and eaten. Seeds a highly prized source of food. (39:363) **Klamath** *Unspecified* Roasted, ground seeds used for food. (45:106) **Kutenai** *Unspecified* Young, immature flower stems peeled and eaten raw. (82:20) **Miwok** *Unspecified* Cracked seeds pulverized, winnowed, and eaten. (12:152) **Montana Indian** *Staple* Roasted seeds ground into a flour. (19:8) *Unspecified* Plant heated, fermented, and eaten. (19:26) Roots eaten raw and cooked. *Vegetable* Young stems and

leaves eaten raw as a salad. (19:8) **Nez Perce** *Unspecified* Young, immature flower stems peeled and eaten raw. Seeds roasted, ground, grease added, and mixture eaten. (82:20) **Okanagan-Colville** *Dried Food* Seeds oven dried for future use. *Unspecified* Young shoots eaten raw or baked in the ground or oven. Flower bud stems peeled and succulent inner portion eaten raw or boiled. Powdered seeds eaten alone or mixed with deer grease, pine nuts, saskatoon berries, or fir sugar. (188:80) **Okanogon** *Staple* Seeds used as a principal food. (178:239) Roots used as a principal food. (178:238) *Unspecified* Young plants used for food. Old, large roots cooked and used for food. (125:36) Roots used as an important food. (178:237) Seeds roasted in baskets with hot stones and eaten. (178:240) **Paiute** *Beverage* Juice from the stems sucked when thirsty. *Candy* Root pitch chewed as gum. (111:117) *Pie & Pudding* Ground seed meal and juniper berries used to make a pudding. (118:26) *Porridge* Roasted, ground seeds made into flour and used to make mush. *Unspecified* Blooming stems peeled and eaten. *Winter Use Food* Roasted, ground seeds made into flour and stored for winter use. (111:117) **Sanpoil** *Special Food* Shoots mixed with chocolate tips and used in the "first roots" ceremony. (188:80) **Shuswap** *Unspecified* Roots steamed and eaten. (123:59) **Thompson** *Bread & Cake* Seeds mixed with deer fat or grease, boiled, cooled, and made into small cakes. (164:491) *Dessert* Dried roots cooked and eaten as a "sort of dessert" after meals.[21] *Dried Food* Cooked roots hung on strings, dried, and then stored on the strings or in baskets.[21] (187:175) *Staple* Seeds pounded and flour mixed with other foods. (164:491) *Starvation Food* Dried seed flour eaten as porridge, especially in times of famine. The seeds were laid on mats, sun dried, placed in buckskin bags, and pounded into a flour. The resulting flour was made into a porridge and eaten, especially in times of famine. One informant said that the seeds were "choky" and difficult to swallow if eaten alone. (187:175) *Unspecified* Young plants used for food. Old, large roots cooked and used for food. (125:36) Plant used for food. (164:480) Young stems eaten as a favorite food. Stalks soaked in water, peeled, and eaten raw. Crowns chewed or sucked. (164:484) Ripe seeds eaten raw. (164:491) Roots used as an important food. (178:237) Loose or skewered roots cooked overnight in a steaming pit and used for food. Young shoots chewed while eating fish. Young leafstalks, leaves, young bud stems, and fruits used for food. Root crown, with the young undeveloped leaves, used for food. (187:175) **Ute** *Unspecified* Young shoots, leaves, and roots used for food. (38:32)

- **Other**—**Blackfoot** *Cooking Tools* Leaves used in roasting camas roots. (114:277) *Incense & Fragrance* Roots used as incense during the preparatory rites for the ceremonial runner. The ceremonial runner, in pre-horse days, had the duty of herding the buffalo toward the piskun (buffalo jump). The runner bathed himself in the smoke from a smudge of the dried root; according to tradition, that would enable him to run long distances—more than 20 miles a day. The runner wore special moccasins, which were transferable annually. Roots used as incense during the Planting ceremonies of the Tobacco Society. A horse was encouraged to stand near a smudge of roots. Then a rider leapt on the horse and galloped across the planting grounds, stopping only to deposit small offerings to the Small People. Roots used as incense for the Crow feather headpiece during the transfer ceremony of Beaver bundle. (86:47) **Okanagan-Colville** *Tools* Leaves wrapped around young boy's feet to practice walking silently and carefully in the woods. **Sanpoil** *Containers* Leaves used under cleaned and washed salmon. (188:80) **Thompson** *Cash Crop* Roots strung on long strings and used in trading. (187:175)

Balsamorhiza sp.

- **Food**—**Coeur d'Alene** *Unspecified* Seeds used for food. (178:90) Growing stalks used for food. (178:91) **Spokan** *Unspecified* Roots used for food. (178:343) Seeds used for food. (178:344)

Balsamorhiza ×*terebinthacea* (Hook.) Nutt., Balsamroot

- **Food**—**Paiute** *Dried Food* Dried roots eaten raw. *Unspecified* Fresh roots roasted, ground, and pounded or eaten raw. (104:103)

Bambusa, Poaceae

Bambusa sp., Cultivated Bamboo

- **Other**—**Navajo** *Ceremonial Items* Stems made into whistles and used in certain ceremonies. (55:25)

Bangiaceae, see *Porphyra*

Baptisia, Fabaceae

Baptisia alba var. *macrophylla* (Larisey) Isely, Largeleaf Wild Indigo

- **Drug**—**Choctaw** *Dermatological Aid* Poultice of roots and leaves applied to swellings. **Koasati** *Antirheumatic (Internal)* Decoction of roots taken for rheumatism. (as *B. leucantha* 177:31) **Meskwaki** *Dermatological Aid* Root used for old sores and compound used on knife or ax wounds. *Hemorrhoid Remedy* Compound containing root used for piles. *Kidney Aid* Compound infusion taken for dropsy. *Respiratory Aid* Decoction of root used for catarrh. *Snakebite Remedy* Compound containing root used for rattlesnake bite. (as *B. leucantha* 152:228)

Baptisia australis (L.) R. Br. ex Ait. f., Blue Wild Indigo

- **Drug**—**Cherokee** *Antiemetic* Cold infusion used for vomiting. *Cathartic* Used as a purgative. *Emetic* Used as an emetic. *Gynecological Aid* Poultice used "to allay inflammation and stop mortification." *Toothache Remedy* Hot infusion of root or beaten root held against tooth for toothache. (80:40)
- **Dye**—**Cherokee** *Blue* Used to make a blue dye. (80:40)

Baptisia bracteata Muhl. ex Ell., Longbract Wild Indigo

- **Drug**—**Pawnee** *Gastrointestinal Aid* Pulverized seeds mixed with buffalo fat and applied to abdomen for colic. (70:90)

Baptisia sp., False Indigo

- **Drug**—**Cree** *Venereal Aid* Poultice of powdered rhizomes applied to syphilitic sores. Plant and catbriar used for syphilis. (16:495) **Creek** *Pediatric Aid* Decoction of root used as a wash and given to drowsy and lifeless children. (172:658) Decoction of roots used as a bath and given to drowsy and listless children. (177:30) *Stimulant* Decoction of root used as a wash and given to drowsy and lifeless children. (172:658) Decoction of roots used as a bath and given to drowsy and listless children. (177:30)

Baptisia tinctoria (L.) R. Br. ex Ait. f., Horseflyweed

- **Drug**—**Cherokee** *Antiemetic* Cold infusion used for vomiting. *Cathartic* Used as a purgative. *Emetic* Used as an emetic. *Gynecological Aid*

Poultice used "to allay inflammation and stop mortification." *Toothache Remedy* Hot infusion of root or beaten root held against tooth for toothache. (80:40) **Delaware** *Dermatological Aid* Infusion of roots used to clean cuts and ulcers. *Gynecological Aid* Infusion of roots used as a douche. (176:37) **Delaware, Oklahoma** *Dermatological Aid* Infusion of root used as a wash to clean cuts and ulcers. (175:31, 74) *Gynecological Aid* Infusion of root used as a douche. (175:31) **Iroquois** *Antirheumatic (Internal)* Compound used for rheumatism. *Gastrointestinal Aid* Decoction of roots rubbed on the stomach for cramps. *Liver Aid* Infusion of plant taken to concentrate bile. *Orthopedic Aid* Decoction of roots rubbed on the arms and legs for cramps. (87:363) **Micmac** *Antihemorrhagic* Root used for spitting blood. *Kidney Aid* Root used for kidney trouble. *Venereal Aid* Root used for gonorrhea. (40:55) **Mohegan** *Dermatological Aid* Infusion of plant used as wash for cuts and wounds. (174:266) Infusion of root used as a healing lotion for cuts or bruises. (176:70, 128) **Nanticoke** *Orthopedic Aid* Compound containing plant used as a lotion on sprains. (175:56, 84) **Ojibwa** *Unspecified* Plant used for medicinal purposes. (135:235) **Penobscot** *Antihemorrhagic* Compound infusion of plant taken for "spitting up blood." *Kidney Aid* Compound infusion of plant taken for kidney trouble. *Tonic* Compound infusion of plant taken as a tonic. *Venereal Aid* Compound infusion of plant taken for gonorrhea. (156:311)
- *Dye*—**Cherokee** *Blue* Used to make a blue dye. (80:40) Used as a blue dye. (203:74) **Ojibwa** *Unspecified* Plant used in the native coloring. (135:235)

Barbarea, Brassicaceae

Barbarea orthoceras Ledeb., American Yellowrocket
- *Food*—**Alaska Native** *Vegetable* Rosettes of dark green shiny leaves cooked as a green vegetable or eaten raw in a mixed salad. (85:17)

Barbarea verna (P. Mill.) Aschers., Early Yellowrocket
- *Food*—**Cherokee** *Unspecified* Plant boiled, fried, and eaten. (204:252) *Vegetable* Leaves parboiled, rinsed, seasoned with grease and salt, and cooked until tender as potherbs. Leaves used in salads. (126:36)

Barbarea vulgaris Ait. f., Garden Yellowrocket
- *Drug*—**Cherokee** *Blood Medicine* Cooked salad eaten to purify blood. (80:31) **Mohegan** *Cough Medicine* Infusion of leaves taken every half hour for coughs. **Shinnecock** *Cough Medicine* Infusion of leaves taken every half hour for coughs. (30:118)
- *Food*—**Cherokee** *Unspecified* Plant boiled, fried, and eaten. (204:252) *Vegetable* Leaves parboiled, rinsed, seasoned with grease and salt, and cooked until tender as potherbs. Leaves used in salads. (126:36)

Barbula, Pottiaceae, moss

Barbula unguiculata Hedw., Moss
- *Drug*—**Seminole** *Antirheumatic (External)* and *Febrifuge* Plant used for fire sickness: fever and body aches. (169:203)

Bazzania, Jungermanniaceae, liverwort

Bazzania trilobata (L.) S. F. Gray, Liverwort
- *Dye*—**Montana Indian** *Yellow* Used as a yellow dye. (19:8)

Beckmannia, Poaceae

Beckmannia syzigachne (Steud.) Fern., American Sloughgrass
- *Food*—**Klamath** *Unspecified* Seeds used for food. (as *B. erucaeformis* 45:91) **Montana Indian** *Unspecified* Seeds used for food. (as *B. erucaeformis* 19:8) **Navajo, Ramah** *Fodder* Used as sheep and horse feed. (191:15)
- *Fiber*—**Okanagan-Colville** *Mats, Rugs & Bedding* Used for bedding and pillows. (188:53)

Bellis, Asteraceae

Bellis perennis L., Lawndaisy
- *Drug*—**Iroquois** *Gastrointestinal Aid* Decoction of plant clusters taken for bad stomach. (87:461)

Berberidaceae, see *Achlys, Berberis, Caulophyllum, Diphylleia, Jeffersonia, Mahonia, Podophyllum, Vancouveria*

Berberis, Berberidaceae

Berberis canadensis P. Mill., American Barberry
- *Drug*—**Cherokee** *Antidiarrheal* Infusion of scraped bark taken for diarrhea. (80:48)

Berberis fendleri Gray, Colorado Barberry
- *Food*—**Jemez** *Fruit* Berries used for food. (32:19) Berries used for food. (44:21)

Berberis sp., Barberry
- *Food*—**Yana** *Porridge* Berries pounded into a flour and used to make mush. (147:251)

Berberis vulgaris L., Common Barberry
- *Drug*—**Micmac** *Oral Aid* Bark and root used for ulcerated gums. *Throat Aid* Bark and root used for sore throat. (40:55) **Mohegan** *Febrifuge* Cold, compound decoction of berries taken for fever. (174:269) Juice of berries mixed with water and used for fever. (176:70, 128) *Throat Aid* Berries used for sore throat and fever. (176:128) **Penobscot** *Oral Aid* Poultice of pounded root or bark applied to ulcerated gums. *Throat Aid* Pounded root or bark used for sore throat. (156:309) **Shinnecock** *Liver Aid* Decoction of leaves taken three times a day for jaundice. (30:119)

Berchemia, Rhamnaceae

Berchemia scandens (Hill) K. Koch, Alabama Supplejack
- *Drug*—**Choctaw** *Blood Medicine* Plant used for the blood. (177:40) **Houma** *Reproductive Aid* Decoction of leaf and bark taken for "impotency in male or female." Decoction of leaves and bark taken by males or females for impotency. (158:57) **Koasati** *Cough Medicine* Cold infusion of burned stems taken as a cough medicine. (177:40) **Seminole** *Other* Complex infusion of stems taken for chronic conditions. (169:272)
- *Other*—**Houma** *Fasteners* Stems used to fasten dugout canoes to the shore and for general fastening purposes. (158:57)

Berlandiera, Asteraceae

Berlandiera lyrata Benth., Lyreleaf Greeneyes
- *Drug*—**Keres, Western** *Psychological Aid* Dried roots burned, ground, and tossed on hot coals or smoke inhaled to give courage. *Sedative* Dried roots burned, ground, and tossed on hot coals or smoke inhaled for nervousness. (171:33)

- **Food**—**Acoma** *Spice* Flowers mixed with sausage as seasoning. (32:19) **Keres, Western** *Spice* Flowers mixed with sausage as seasoning. (171:33) **Laguna** *Spice* Flowers mixed with sausage as seasoning. (32:19)

Berula, Apiaceae

Berula erecta (Huds.) Coville, Cutleaf Waterparsnip

- **Drug**—**Apache, White Mountain** *Unspecified* Leaves and blossoms used for medicinal purposes. (136:155) **Zuni** *Antirheumatic (External)* Ingredient of "schumaakwe cakes" and used externally for rheumatism. (166:44) *Dermatological Aid* Infusion of whole plant used as wash for rashes and athlete's foot infection. (27:379) Ingredient of "schumaakwe cakes" and used externally for swelling. (166:44)
- **Food**—**Apache, White Mountain** *Unspecified* Leaves and blossoms used for food. (136:155)

Besseya, Scrophulariaceae

Besseya plantaginea (James) Rydb., White River Coraldrops

- **Drug**—**Navajo, Ramah** *Ceremonial Medicine* Plant used as ceremonial emetic. *Dermatological Aid* Dried root or leaf used as dusting powder on skin sores or infant's sore navel. *Diuretic* Cold infusion of plant taken by hunters for anuria. *Emetic* Plant used as ceremonial emetic. *Hunting Medicine* Lotion from plant applied to body for protection while hunting and in war. *Panacea* Root used as a "life medicine." *Pediatric Aid* Cold infusion of root used for birth injuries. Dried root or leaves used as dusting powder for skin sores or infant's sore navel. *Witchcraft Medicine* Lotion from plant applied to body for protection from witches. (191:43)

Beta, Chenopodiaceae

Beta vulgaris L., Common Beet

- **Food**—**Anticosti** *Beverage* Bulbs used to make wine. (143:65) **Cherokee** *Unspecified* Species used for food. (80:25)

Betulaceae, see *Alnus, Betula, Carpinus, Corylus, Ostrya*

Betula, Betulaceae

Betula alleghaniensis Britt., Yellow Birch

- **Drug**—**Ojibwa, South** *Diuretic* Compound decoction of inner bark taken as a diuretic. (as *B. excelsa* 91:199)

Betula alleghaniensis Britt. **var. *alleghaniensis***, Yellow Birch

- **Drug**—**Delaware, Oklahoma** *Cathartic* Decoction of bark taken as a cathartic. *Emetic* Decoction of bark taken as an emetic. (as *B. lutea* 175:25, 74) *Gastrointestinal Aid* and *Liver Aid* Decoction of bark taken "to remove bile from the intestines." (as *B. lutea* 175:25) **Iroquois** *Blood Medicine* Complex compound used as a blood purifier. *Dermatological Aid* Complex compound decoction used as wash for affected parts of "Italian itch." (as *B. lutea* 87:300) *Gynecological Aid* Decoction of plant used for lactation. (as *B. lutea* 87:301) **Micmac** *Other* Wood used as a hot-water bottle. (as *B. lutea* 40:55) **Ojibwa** *Blood Medicine* Decoction of bark taken for internal blood diseases. (as *B. lutea* 135:231) **Potawatomi** *Adjuvant* Infusion of twigs used as a seasoner for medicines. (as *B. lutea* 154:44)
- **Food**—**Algonquin, Quebec** *Substitution Food* Sap mixed with maple sap if the latter is not available in sufficient quantities. (as *B. lutea* 18:80) **Ojibwa** *Beverage* Sap and maple sap used for a pleasant beverage drink. (as *B. lutea* 153:397)
- **Fiber**—**Cherokee** *Building Material* Wood used for lumber. (as *B. lutea* 80:25) **Ojibwa** *Building Material* Bark used to build dwellings and lodges. *Canoe Material* Bark used to make birchbark canoes. (as *B. lutea* 135:241) **Potawatomi** *Building Material* Branches used as poles for the wigwam or medicine lodge. (as *B. lutea* 154:112)
- **Other**—**Ojibwa** *Ceremonial Items* Bark placed on the coffins when burying the dead. *Containers* Bark used to make storage containers, sap dishes, rice baskets, buckets, trays, and winnowing dishes. *Cooking Tools* Bark used to make dishes. (as *B. lutea* 135:241)

Betula lenta L., Sweet Birch

- **Drug**—**Algonquin, Quebec** *Unspecified* Infusion of plant used for many medicinal purposes. (18:151) **Cherokee** *Antidiarrheal* Leaves chewed or infusion taken for dysentery. *Cold Remedy* Infusion taken for colds. *Gastrointestinal Aid* Infusion of bark taken for the stomach. *Urinary Aid* Infusion of bark taken for "milky urine." (80:25) **Chippewa** *Antidiarrheal* Decoction of bark taken for diarrhea. *Pulmonary Aid* Bark used for pulmonary troubles and decoction of bark taken for pneumonia. (71:128) **Iroquois** *Blood Medicine* Compound decoction taken when the "blood gets bad and cold." *Cold Remedy* Compound infusion taken by women "when they catch cold with the menses." *Febrifuge* Compound decoction taken for fever. *Gynecological Aid* Compound decoction taken by women who have had gonorrhea and are pregnant. *Orthopedic Aid* Compound decoction taken for soreness. *Stimulant* Compound decoction taken "when a person tires." *Unspecified* "Highly valued medicine because it sustains the deer, the mainstay of life." (87:300) **Mohegan** *Tonic* Complex compound infusion including black birch bark taken as spring tonic. (174:266) Inner bark used to make a tonic. (176:70, 128)
- **Food**—**Iroquois** *Beverage* Twigs steeped into a beverage. (196:148)
- **Fiber**—**Ojibwa** *Building Material* Bark used to build dwellings and lodges. *Canoe Material* Bark used to make birchbark canoes. (135:241)
- **Other**—**Ojibwa** *Ceremonial Items* Bark placed on the coffins when burying the dead. *Containers* Bark used to make storage containers, sap dishes, rice baskets, buckets, trays, and winnowing dishes. *Cooking Tools* Bark used to make dishes. (135:241)

Betula nana L., Bog Birch

- **Drug**—**Eskimo, Western** *Analgesic* and *Gastrointestinal Aid* Compound decoction of leaves taken for stomachache and intestinal discomfort. (as *B. exilis* 108:5)
- **Fiber**—**Eskimo, Inuktitut** *Mats, Rugs & Bedding* Wood used for "springs" under skin bedding. (202:184)
- **Other**—**Eskimo, Alaska** *Fuel* Shrub burned to smoke fish. (1:35) **Eskimo, Inuktitut** *Fuel* Used as tinder, even when wet, and for cooking fires for lack of larger wood. (202:184)

Betula nigra L., River Birch

- **Drug**—**Cherokee** *Antidiarrheal* Leaves chewed or infusion taken for dysentery. *Cold Remedy* Infusion taken for colds. *Gastrointestinal Aid* Infusion of bark taken for the stomach. *Urinary Aid* Infusion of bark taken for "milky urine." (80:25) Decoction of inner bark taken for difficult urination with discharge. (177:15) **Chippewa** *Analgesic* and

Gastrointestinal Aid Decoction of bark taken for stomach pain. (53:342)

Betula occidentalis Hook., Water Birch

- **Drug**—Blackfoot *Abortifacient* Flowers and leaves included in two separate bundles and used for conception. Decoction of flowers and leaves taken when bundle to stop conception fails. (86:60)
- **Fiber**—Blackfoot *Building Material* Wands, sharpened at both ends, used to construct the dome shape of the sweat lodge. (86:17) **Okanagan-Colville** *Basketry* Bark used to make baskets. *Canoe Material* Bark used to make canoes. *Furniture* Bark used to make cradles. (188:89)
- **Dye**—Jemez *Red* Bark, mountain mahogany bark, and alder bark boiled together and used as red dye to paint moccasins. (as *B. fontinalis* 44:21) **Okanagan-Colville** *Brown* Inner bark used to make a brown dye. (188:89)
- **Other**—Blackfoot *Cooking Tools* Wood used to make bowls. (97:33) **Navajo** *Toys & Games* Branch used for the pole in the "hoop and pole" game. (as *B. fontinalis* 55:39) **Okanagan-Colville** *Fuel* Used for fuel. (188:89) **Paiute** *Stable Gear* Wood used to make a pack saddle with posts at both ends. (as *B. fontinalis* 111:64)

Betula papyrifera Marsh., Paper Birch

- **Drug**—Algonquin, Quebec *Dermatological Aid* and *Pediatric Aid* Bark powder used for diaper rash and other skin rashes. (18:152) **Chippewa** *Cathartic* Infusion of inner bark used as an enema. (53:364) **Cree, Woodlands** *Burn Dressing* Poultice of outer bark used to bandage a burn. *Dermatological Aid* Dried, finely powdered rotten wood used as baby powder to prevent rashes. Dried inner bark ground, added to pitch and grease, and used as ointment for persistent scabs and rashes. Decoction of inner bark used as a wash for skin rashes and other skin sores. *Diaphoretic* Decoction of wood taken to cause sweating. *Gynecological Aid* Decoction of wood taken to ensure an adequate supply of milk for breast feeding. Decoction of wood and inner bark used for "women's troubles." *Orthopedic Aid* Decoction of wood taken for back pain. *Pediatric Aid* Dried, finely powdered rotten wood used as baby powder to prevent rashes. Decoction of stems or branches taken for teething sickness. *Toothache Remedy* Decoction of stems or branches taken for teething sickness. *Venereal Aid* Wood mixed with other materials and used for gonorrhea. (109:32) **Iroquois** *Gynecological Aid* Burned bark ashes used to "shrivel the womb." (87:300) **Koyukon** *Unspecified* Plant spirit used by the shaman to heal sick people. (119:53) **Menominee** *Antidiarrheal* Decoction of inner bark used for dysentery. *Tonic* Decoction of branch tips used as a tonic. (54:131) **Ojibwa** *Adjuvant* Root used as a seasoner for medicines. *Analgesic* Root bark cooked with maple sugar as syrup for stomach cramps. (as *B. alba* var. *papyrifera* 153:358) *Blood Medicine* Decoction of bark taken for internal blood diseases. (135:231) *Gastrointestinal Aid* Compound decoction of root bark taken to alleviate stomach cramps. (as *B. alba* var. *papyrifera* 153:358) **Potawatomi** *Adjuvant* Infusion of twigs used as a seasoner for medicines. (as *B. alba* var. *papyrifera* 154:43, 44) **Shuswap** *Analgesic* Plant used for pain. (123:60) **Tanana, Upper** *Orthopedic Aid* Bark used as casts for broken limbs. A soft material such as a cloth was placed next to the skin on the broken limb over which birch bark was wrapped and tied. The birch bark was then heated until it shrank to fit the limb. (102:5) **Thompson** *Cold Remedy* Sap tapped from trees in early spring and taken for colds. *Contraceptive* Bark used for contraception. One informant recalled a case in which a woman in childbirth did not want any more children. An old woman told her to take the afterbirth, stick it with an old bone awl, wrap it in fish net and then in a piece of birch bark, and place it high up on a particular kind of tree. The patient was then given an infusion of bitter cherry or saskatoon wood and after that had no more children *Cough Medicine* Sap tapped from trees in early spring and taken for coughs. (187:189)
- **Food**—Algonquin, Quebec *Sauce & Relish* Sap used to make syrup. (18:80) **Cree, Woodlands** *Preservative* Soft, rotten wood burned to make a slow, smoky fire to smoke cure meat and fish. *Sauce & Relish* Sap collected, made into syrup, and eaten on bannock. *Substitution Food* Root bark used as a tea substitute. *Unspecified* Cambium eaten fresh from the tree trunk. (109:32) **Montagnais** *Dietary Aid* Inner bark grated and eaten to benefit the diet. (156:313) **Ojibwa** *Preservative* Birch bark keeps the food stored in it from spoiling. (as *B. alba* var. *papyrifera* 153:416) **Tanana, Upper** *Unspecified* Raw sap, sometimes mixed with fish grease, used for food. Sap used for food. (102:5)
- **Fiber**—Abnaki *Basketry* Bark used to make baskets. (144:156) *Canoe Material* Bark used to make canoes. (144:164) **Algonquin, Tête-de-Boule** *Basketry* Bark used to make baskets and containers. *Building Material* Plant used to make houses, tents, and shelters. *Canoe Material* Bark used to make canoes. (132:119) **Anticosti** *Snow Gear* Used to make snowshoes. (143:65) **Bella Coola** *Basketry* Bark occasionally used to make baskets. *Canoe Material* Bark occasionally used to make canoes. (184:202) **Carrier** *Basketry* Inner bark used to make baskets. *Canoe Material* Wood used to make canoes. *Sewing Material* Roots used with spruce roots to sew things together. *Snow Gear* Wood used to make toboggans. (31:67) **Chippewa** *Building Material* Used as coverings for dwellings. (53:377) **Cree, Woodlands** *Basketry* Bark used to make baskets for food storage and berry collection. *Building Material* Bark used to make bathtubs. Bark used to cover a tepee. Wood used as poles to frame a tepee. *Canoe Material* Bark used to make canoes. Wood used to make canoe paddles. *Snow Gear* Wood used to make toboggans and snowshoes. (109:32) **Gitksan** *Basketry* Bark used to make baskets. (73:154) **Koyukon** *Basketry* Bark used to make baskets and food storage containers. *Canoe Material* Wood used to make canoe ribs. *Snow Gear* Wood used to make sleds and snowshoe frames. (119:53) **Malecite** *Canoe Material* Bark used for canoes. (160:6) **Meskwaki** *Building Material* Bark strips used as the waterproof, top coverings of wigwams. *Canoe Material* Paper birch used to make canoes. (as *B. alba papyrifera* 152:267) **Micmac** *Basketry* Bark used to make baskets. *Building Material* Bark used to make house coverings. (159:258) *Canoe Material* Bark used to make canoes. (145:56) **Montana Indian** *Canoe Material* Bark used to make canoes. (19:8) **Ojibwa** *Basketry* Bark used for buckets and baskets. (as *B. alba* var. *papyrifera* 153:413) Bark stripped and used to make emergency trays or buckets in the woods. Baskets made for gathering and storing berries, maple sugar, dried fish, meat, or any food. (as *B. alba* var. *papyrifera* 153:416) *Building Material* Bark used to build dwellings and lodges. (135:241) Bark used for wigwam coverings. (as *B. alba* var. *papyrifera* 153:413) Sheets of bark sewn together, made into rolls and used as waterproof roofing for wigwams. (as *B. alba* var. *papyrifera* 153:416) *Canoe Material* Bark used to make

124 *Betula papyrifera* (Betulaceae)

birchbark canoes. (135:241) Bark used for canoes. (as *B. alba* var. *papyrifera* 153:413) Heavy pieces of bark used to make very durable canoes. (as *B. alba* var. *papyrifera* 153:414) **Okanagan-Colville** *Basketry* Bark used to make baskets. *Canoe Material* Bark used to make canoes. *Furniture* Bark used to make cradles. (188:89) **Potawatomi** *Building Material* Bark furnished a waterproof cover for the top of the wigwam. *Canoe Material* Bark furnished the outside cover of the birchbark canoe. (as *B. alba* var. *papyrifera* 154:112) **Shuswap** *Basketry* Gray-colored bark used to make baskets. (123:60) **Tanana, Upper** *Basketry* Bark used to make baskets. Bark used to make containers for cooking. To cook in a birch bark basket, clean rocks were made very hot and then placed in water in the basket. This process was repeated until the cooking was completed. *Building Material* Bark used as roofing material. Bark used in the construction of some buildings. *Canoe Material* Bark used to make canoes. *Clothing* Bark used to make hats. *Furniture* Bark used to make baby cradles. *Snow Gear* Wood used to make snowshoes, toboggans, drums, bowls, spoons, and wedges. (102:5) **Thompson** *Basketry* Tough, waterproof bark used as material for baskets. *Building Material* Tough, waterproof bark used as material for walls and roofing. *Canoe Material* Tough, waterproof bark used as material for canoes. *Furniture* Tough, waterproof bark used as material for cradles. (187:189)

- **Dye**—**Chippewa** *Red* Inner bark boiled, cedar ashes added and used to make a red dye. (53:370) **Ojibwa** *Red* Innermost bark boiled to extract a reddish dye. (as *B. alba* var. *papyrifera* 153:425) **Okanagan-Colville** *Brown* Inner bark used to make a brown dye. (188:89)
- **Other**—**Abnaki** *Containers* Bark used to wrap and store wood for a year. (144:164) **Algonquin, Tête-de-Boule** *Containers* Bark used to make canoes, baskets, and containers. *Designs* Bark folded, edges chewed, and resulting design transferred to baskets and moccasins. (132:119) **Blackfoot** *Cooking Tools* Wood used to make bowls. (97:33) **Carrier** *Cooking Tools* Inner bark made into dishes and used for processing fish, picking berries, and to eat with. (31:67) **Chippewa** *Cooking Tools* Used for utensils. *Designs* Used as patterns for work in decorative art. (53:377) **Cree, Woodlands** *Containers* Bark used to make baskets for food storage and berry collection, and to make dishes to collect birch sap and fresh cambium. *Cooking Tools* Bark used to make dippers for water, funnels, and cups. Wood used to make wooden spoons, stoppers for sturgeon skin jars, and hammers to pound fish eggs. Wood used for upright supports and cross bars of the smoke curing rack. *Fuel* Bark fragments ignited from coals or smoldering tinder and used to start a fire. *Hide Preparation* Dried rotten wood with other rotten woods used to smoke tan hides. *Hunting & Fishing Item* Bark used to make moose calls. (109:32) **Dakota** *Containers* Fine, shredded bark used as vessels to catch sap from trees in sugar-making time. *Cooking Tools* Fine, shredded bark used as household utensils. *Lighting* Fine, shredded bark bound into bundles and used for torches. (70:75) *Toys & Games* Papery bark chewed to a pulp and used for popgun wads. (70:116) **Flathead** *Containers* Bark used to line sacks and stiffen them into baskets. (19:8) **Gitksan** *Containers* Bark used to make containers and waterproof wrappings. *Lighting* Used to make torches. (73:154) **Koyukon** *Containers* Bark used to make baskets and food storage containers. *Fuel* Wood used for firewood. Bark used to start campfires or light the stove at home. *Hunting & Fishing Item* Wood used to make fish traps. (119:53) **Malecite** *Containers* Bark used for boxes and other containers. (160:6) **Micmac** *Containers* Bark used to make boxes, coffins, and other containers. *Cooking Tools* Bark used to make dishes and cooking utensils. (159:258) **Ojibwa** *Ceremonial Items* Bark placed on the coffins when burying the dead. *Containers* Bark used to make small vessels, pails, and trays, and to make storage containers, sap dishes, rice baskets, buckets, trays, and winnowing dishes. *Cooking Tools* Bark used to make dishes. (135:241) Nearly any kitchen utensil common to the white man, could be duplicated in birch bark by the Ojibwe. Bark used to make funnels for pouring hot lard, and to make all sorts of drying trays and shallow trays for winnowing wild rice. *Fuel* After stripping a felled tree of its bark, it was salvaged for firewood. Scraps of bark used by women to kindle or light fires. *Lighting* Bark rolled into a handy, burn-all-night torch. The Ojibwe often used a torch of rolled birch bark in lieu of candles. (as *B. alba* var. *papyrifera* 153:416) *Paper* Patterns for decorative art made upon the bark. (as *B. alba* var. *papyrifera* 153:413) Records of medicine lodge rituals kept on its virgin surface. There were many layers of bark ranging from the thinnest paper to quite heavy pieces. *Preservative* Wood had the property of protecting articles stored in it from decay. *Protection* Ojibwe claim that birch was never struck by lightning, hence offered a safe harbor in thunderstorms. *Sacred Items* Paper birch and cedar form the two most sacred trees of the Ojibwe, both of which were very useful. The Ojibwe regard the bark as a distinct "contribution from Winabojo." No birch was gathered by the Ojibwe without due offering of tobacco to Winabojo and Grandmother Earth. Families made a pilgrimage to birch groves during the latter part of June and in July to gather their supply of birch bark, because it peels most easily at that time. (as *B. alba* var. *papyrifera* 153:414) **Omaha** *Toys & Games* Papery bark chewed to a pulp and used for popgun wads. **Pawnee** *Toys & Games* Papery bark chewed to a pulp and used for popgun wads. **Ponca** *Toys & Games* Papery bark chewed to a pulp and used for popgun wads. (70:116) **Potawatomi** *Cooking Tools* Bark used to make many of the household utensils, storage vessels, and containers. (as *B. alba* var. *papyrifera* 154:112) **Tanana, Upper** *Containers* Bark used to make containers for storing food and picking berries, as lining in food storage pits, and as lining in storage pits. *Cooking Tools* Bark used to make containers for storing food, picking berries, and cooking. Wood used to make bowls and spoons. *Fuel* Wood used for firewood. Black stem growth used as tinder for kindling and starting fires with a fire drill. *Hunting & Fishing Item* Wood used to make spears for hunting bears and bows for hunting both large and small game. *Musical Instrument* Wood used to make drums. *Protection* Bark made into broad-rimmed hats used by young, menstruating girls to restrict their vision. The broad-rimmed hats prevented them from looking where they were not supposed to. *Tools* Wood used to make wedges and tool handles, the bow of the fire drill, and wedges. (102:5) **Thompson** *Containers* Tough, waterproof bark used as a material for lining storage caches.[22] *Cooking Tools* Tough, waterproof bark used as a material for wrapping food.[22] *Decorations* Bark used for decorations. *Paper* Bark used for paper and cards. (187:189) **Wet'suwet'en** *Containers* Bark used to make containers and waterproof wrappings. *Lighting* Used to make torches. (73:154) **Winnebago** *Toys & Games* Papery bark chewed to a pulp and used for popgun wads. (70:116)

Betula papyrifera Marsh. **var.** *papyrifera*, Paper Birch
- *Other*—**Haisla & Hanaksiala** *Decorations* Wood used for carving. (43:226)

Betula populifolia Marsh., Gray Birch
- *Drug*—**Iroquois** *Hemorrhoid Remedy* Decoction of bark taken for bleeding piles. (87:300) **Malecite** *Dermatological Aid* Inner bark scrapings used for swelling in infected cuts. (116:245) **Micmac** *Dermatological Aid* Inner bark used for infected cuts. *Emetic* Inner bark used as an emetic. (40:55)

Betula pubescens Ehrh. **ssp.** *pubescens*, Downy Birch
- *Drug*—**Cree, Hudson Bay** *Dermatological Aid* Boiled, powdered wood applied to chafed skin. (as *B. alba* 92:303)
- *Fiber*—**Chippewa** *Canoe Material* Bark used in boat building. The bark was stripped off at raspberry ripening time, laid away, and pressed flat until the next spring. When required for manufacture, especially in boat building, it was heated over a fire to make it pliable for shaping to the purpose. (as *B. alba* 71:128)

Betula pumila **var.** *glandulifera* Regel, Glandulose Birch
- *Drug*—**Ojibwa** *Gynecological Aid* Infusion of cones taken during menses and for strength after childbirth. *Respiratory Aid* Smoke of cones inhaled for catarrh. (153:358)
- *Fiber*—**Ojibwa** *Basketry* Twigs of this dwarf birch used for the ribs of baskets. (153:417)

Betula **sp.**, White Birch
- *Drug*—**Creek** *Tuberculosis Remedy* Bark used in medicine taken for pulmonary tuberculosis. (172:659) Decoction of bark taken for pulmonary tuberculosis. (177:15) **Malecite** *Unspecified* Wood heated and used like a hot-water bottle. (160:6)
- *Food*—**Malecite** *Beverage* Bark used to make tea. (160:6)
- *Fiber*—**Eskimo, Inuktitut** *Snow Gear* Wood used to make snowshoes. (202:192) **Malecite** *Brushes & Brooms* Used to make brooms. *Snow Gear* Used to make sled and toboggan runners. (160:6)
- *Other*—**Eskimo, Inuktitut** *Containers* Wood used to make containers. *Fuel* Bark used for tinder. *Hunting & Fishing Item* Bark used to make muskrat callers. (202:192) **Micmac** *Hunting & Fishing Item* Bark used to make trumpets for calling game. *Lighting* Bark used to make torches for night fishing. (159:258)

Bidens, Asteraceae

Bidens amplectens Sherff, Kokolau
- *Food*—**Hopi** *Beverage* Used to make coffee. (as *B. gracilis* 190:168)

Bidens bipinnata L., Spanish Needles
- *Drug*—**Cherokee** *Anthelmintic* Infusion taken for worms. *Throat Aid* Leaves chewed for sore throat. (80:57)

Bidens coronata (L.) Britt., Crowned Beggarticks
- *Drug*—**Seminole** *Analgesic* and *Antidiarrheal* Infusion of roots taken for sun sickness: eye disease, headache, high fever, and diarrhea. (169:206) *Antirheumatic (External)* Plant used for fire sickness: fever and body aches. (169:204) *Eye Medicine* Infusion of roots taken for sun sickness: eye disease, headache, high fever, and diarrhea. (169:206) Infusion of whole plant taken and used as a bath for mist sickness: eye disease, fever, and chills. (169:209) *Febrifuge* Plant used for fire sickness: fever and body aches. (169:204) Infusion of roots taken for sun sickness: eye disease, headache, high fever, and diarrhea. (169:206) Infusion of whole plant taken and used as a bath for mist sickness: eye disease, fever, and chills. (169:209)

Bidens laevis (L.) B.S.P., Smooth Beggartick
- *Food*—**Paiute** *Unspecified* Species used for food. (167:244)

Bidens **sp.**, Kookoolau
- *Drug*—**Hawaiian** *Dietary Aid* Infusion of pounded flowers, buds, leaves, and other plants taken to stimulate the appetite. *Gastrointestinal Aid* Infusion of pounded flowers, buds, leaves, and other plants taken for stomach troubles. *Pediatric Aid* Flowers and buds chewed by mothers and given to infants for general debility. Infusion of flowers, buds, and leaves given to infants for general debility. *Respiratory Aid* Infusion of pounded flowers, buds, leaves, and other plants strained and taken for asthma. *Strengthener* Flowers and buds chewed by mothers and given to infants for general debility. Infusion of flowers, buds, and leaves given to infants for general debility. Infusion of flowers, buds, and leaves taken for general debility. *Throat Aid* Infusion of pounded flowers, buds, leaves, and other plants taken for throat troubles. (2:53)

Bignoniaceae, see *Bignonia, Chilopsis*

Bignonia, Bignoniaceae

Bignonia capreolata L., Cross Vine
- *Drug*—**Cherokee** *Blood Medicine* Infusion of leaf taken to purify blood. (as *Anisostichus capreolata* 80:31) Infusion of leaves used to cleanse the blood. (203:74) **Choctaw** *Kidney Aid* Decoction of mashed bark used as a steam bath for dropsy. (177:57) **Creek** *Unspecified* Plant used medicinally for unspecified purpose. (as *B. crucigera* 172:670) **Houma** *Misc. Disease Remedy* Infusion of mashed root used as a gargle for diphtheria. (158:65) **Koasati** *Analgesic* Decoction of bark used as a bath and taken for headaches. *Antirheumatic (Internal)* Decoction of leaves taken for rheumatism. (177:57)

Bixaceae, see *Amoreuxia*

Blechnaceae, see *Asplenium, Blechnum, Sadleria, Woodwardia*

Blechnum, Blechnaceae, fern

Blechnum spicant (L.) Roth, Deer Fern
- *Drug*—**Hesquiat** *Cancer Treatment* Leaflets chewed for internal cancer. *Dermatological Aid* Fronds used as a good medicine for skin sores. This medicine was first learned about from watching the deer, who rub their antler stubs on this plant when their antlers break off. (185:29) **Kwakiutl** *Antidiarrheal* Compound decoction of root taken or root held in mouth for diarrhea. (as *Struthiopteris spicant* 20:381) Compound decoction of roots taken for diarrhea. (183:266) **Makah** *Gastrointestinal Aid* Green leaves chewed for stomach distress. *Pulmonary Aid* Green leaves chewed for lung trouble. **Quileute** *Orthopedic Aid* Poultice of fresh leaves applied to paralyzed parts of the body. *Panacea* Decoction of leaves taken for general ill health. **Quinault** *Gastrointestinal Aid* Raw leaves chewed for colic. (as *Struthiopteris spicant* 79:15)
- *Food*—**Haisla & Hanaksiala** *Forage* Plant eaten by mountain goats

126 *Blechnum spicant* (Blechnaceae)

and deer. (43:153) **Hesquiat** *Starvation Food* Young, tender stalks peeled and center portion eaten when hungry and there is nothing to eat. (185:29) **Makah** *Spice* Fronds used for flavor in cooking by placing them under the items to be cooked. (67:219) **Nitinaht** *Starvation Food* Fronds eaten to relieve hunger when lost in the bushes. (186:63)
- *Fiber*—**Yurok** *Mats, Rugs & Bedding* Leaves used for bedding. (6:20)
- *Other*—**Haisla & Hanaksiala** *Decorations* Plants transplanted to pots and used as house plants. (43:153) **Nitinaht** *Cooking Tools* Fronds placed below and above food in steaming pits. (186:63) **Quinault** *Cooking Tools* Leaves used with sword fern to cook baking camas. (as *Struthiopteris spicant* 79:15)

Blennosperma, Asteraceae

Blennosperma nanum (Hook.) Blake, Common Stickyseed
- *Food*—**Neeshenam** *Bread & Cake* Seeds parched, ground into flour, and used to make bread. *Porridge* Seeds parched, ground into flour, and used to make mush. *Staple* Seeds parched, ground into flour, and used for food. (as *B. californicum* 129:377)

Blephilia, Lamiaceae

Blephilia ciliata (L.) Benth., Downy Pagodaplant
- *Drug*—**Cherokee** *Analgesic* Poultice of leaves used for headache. (80:45)

Bloomeria, Liliaceae

Bloomeria crocea (Torr.) Coville, Common Goldenstars
- *Food*—**Cahuilla** *Unspecified* Corms eaten raw any time of the year. (15:47)
- *Other*—**Kawaiisu** *Fasteners* Corms rubbed on metate into an adhesive and spread on seed gathering baskets to close the interstices. (206:16)

Bloomeria crocea var. *aurea* (Kellogg) Ingram, Common Goldenstars
- *Food*—**Luiseño** *Unspecified* Bulb used for food. (as *B. aurea* 155:233)

Bobea, Rubiaceae

Bobea sp., Ahakea
- *Drug*—**Hawaiian** *Blood Medicine* Bark used as a blood purifier. *Dermatological Aid* Bark used for skin ulcers. (2:5)
- *Fiber*—**Hawaiian** *Canoe Material* Wood used to make canoes or canoe parts. (2:5)

Boerhavia, Nyctaginaceae

Boerhavia erecta L., Erect Spiderling
- *Other*—**Hopi** *Insecticide* Sticky leaves and stem hung in house to catch flies. (200:75)

Boisduvalia, Onagraceae

Boisduvalia densiflora (Lindl.) S. Wats., Denseflower Spike Primrose
- *Food*—**Mendocino Indian** *Bread & Cake* Seeds used to make bread. *Staple* Seeds eaten as a pinole. (41:370) **Miwok** *Unspecified* Parched, pulverized, dried seeds used for food. Stored, unparched seeds used for food. (12:152) **Pomo** *Staple* Seeds used to make pinoles. (11:86)

Boisduvalia stricta (Gray) Greene, Brook Spike Primrose
- *Food*—**Miwok** *Unspecified* Parched, pulverized seeds used for food. (12:152)

Boletaceae, see *Boletus*

Boletus, Boletaceae, fungus

Boletus edulis, Timber Mushroom
- *Food*—**Pomo, Kashaya** *Vegetable* Cooked on hot stones, baked in the oven, or fried. (72:132)

Boraginaceae, see *Amsinckia, Cryptantha, Cynoglossum, Echium, Hackelia, Heliotropium, Lappula, Lithospermum, Macromeria, Mertensia, Myosotis, Onosmodium, Plagiobothrys, Symphytum, Tiquilia*

Boschniakia, Orobanchaceae

Boschniakia glabra
- *Drug*—**Tlingit** *Dermatological Aid* Compound containing root used for sores. (107:284)

Boschniakia hookeri Walp., Vancouver Groundcone
- *Drug*—**Hesquiat** *Cough Medicine* Roots used for coughs. (185:70)
- *Food*—**Hesquiat** *Unspecified* Peeled roots eaten raw. (185:70) **Luiseño** *Unspecified* Roots used for food. (as *Orobanche tuberosa* 155:229)

Boschniakia rossica (Cham. & Schlecht.) Fedtsch., Northern Groundcone
- *Food*—**Tanana, Upper** *Fodder* Raw roots or above ground portion of plant diced, mixed with other food, and used for puppy and dog food. (102:15)

Boschniakia strobilacea Gray, California Groundcone
- *Food*—**Karok** *Unspecified* Eaten when young. **Yurok** *Unspecified* Species used for food. (6:20)

Bothriochloa, Poaceae

Bothriochloa saccharoides (Sw.) Rydb., Silver Bluestem
- *Drug*—**Kiowa** *Oral Aid* Stem used as a toothpick. Stem used as a toothpick. (192:13)

Botrychium, Ophioglossaceae, fern

Botrychium virginianum (L.) Sw., Rattlesnake Fern
- *Drug*—**Abnaki** *Other* Used as a demulcent. (144:155) *Pediatric Aid* and *Unspecified* Decoction of plant given to children with illness. (144:162) **Cherokee** *Emetic* Decoction of roots taken to cause vomiting. (177:4) *Snakebite Remedy* Decoction of root "boiled down to syrup" and rubbed on snakebite. (80:34) **Chickasaw** *Diaphoretic* Plant used as a diaphoretic. (28:288, 289) *Emetic* Decoction of root used as an emetic. (28:288, 289) *Expectorant* Decoction of root used as an emetic and plant used as an expectorant. (28:288, 289) **Chippewa** *Snakebite Remedy* Poultice of mashed, fresh root applied to snakebite and used as repellent. (53:352) **Ojibwa** *Pulmonary Aid* Plant said to be good for lung trouble. *Tuberculosis Remedy* Plant said to be good for consumption. (153:377) **Ojibwa, South** *Dermatological Aid* Poultice of bruised root applied to cuts. (91:201) **Potawatomi** *Unspecified* Compound containing root used medicinally. (154:67)

Botrychium virginianum (L.) Sw. **ssp.** *virginianum*, Rattlesnake Fern
- *Drug*—Iroquois *Cough Medicine* and *Tuberculosis Remedy* Cold infusion of root and liquor taken for the cough of consumption. (87:261)

Bouteloua, Poaceae

Bouteloua curtipendula (Michx.) Torr., Sideoats Grama
- *Fiber*—Tewa *Brushes & Brooms* Grass bundled, dried, made into brooms, and used to sweep floors, hearths and metates. Grass bundled, dried, and used as hairbrushes. (138:64)
- *Other*—Apache, Chiricahua & Mescalero *Containers* Moist grass laid onto hot stones to prevent steam from escaping. (33:36)

Bouteloua curtipendula (Michx.) Torr. **var.** *curtipendula*, Sideoats Grama
- *Food*—Kiowa *Fodder* Recognized as a good fodder. (as *Atheropogon curtipendula* and *Chloris curtipendula* 192:14)
- *Other*—Kiowa *Decorations* Worn by those, who in battle, had killed an enemy with a lance; grass resembled the feathered lance. (as *Atheropogon curtipendula* and *Chloris curtipendula* 192:14)

Bouteloua gracilis (Willd. ex Kunth) Lag. ex Griffiths, Blue Grama
- *Drug*—Navajo, Ramah *Antidote* Compound decoction of plant taken to counteract overdose of "life medicine." *Dermatological Aid* Roots chewed and blown on cuts. *Gynecological Aid* Decoction of whole plant taken as a postpartum medicine. *Panacea* Cold, compound infusion of root used internally and externally as "life medicine." *Veterinary Aid* Roots chewed and blown on incisions of castrated colts. (191:15, 16)
- *Food*—Apache, Western *Porridge* Seeds ground, mixed with cornmeal and water and made into a mush. (26:189) Apache, White Mountain *Bread & Cake* Seeds ground and used to make bread and pones. *Porridge* Seeds ground, mixed with meal, and water and eaten as mush. (136:149) Hopi *Forage* Used as an important forage grass. (200:64) Keres, Western *Forage* Grass used for grazing purposes. (as *B. cligostachya* 171:33) Montana Indian *Forage* Grass used for forage. (as *B. oligostachya* 19:8) Navajo, Ramah *Forage* Important forage grass. (191:15)
- *Fiber*—Apache, Western *Brushes & Brooms* Stem used as comb and broom material. (26:189) Apache, White Mountain *Brushes & Brooms* Blades bundled by a cord, the stiff end used as a hair comb and the other end used as a broom. (136:149) Hopi *Basketry* Used as the fill of coiled basketry. (200:64) Zuni *Brushes & Brooms* Grass bunches tied together and the severed end used as a hairbrush, the other as a broom. (166:83)
- *Other*—Apache, White Mountain *Cash Crop* Plant gathered and sold. (136:149) Lakota *Toys & Games* Most plants have two spikes: for sport, people would hunt for those with three. (139:29) Montana Indian *Season Indicator* Plant used to foretell winter, one fruit spike—mild winter, and more fruit spikes—severe winter. (as *B. oligostachya* 19:8) Navajo *Ceremonial Items* Tied to the end of the wand carried by the girl in the Squaw Dance. (55:25) Zuni *Cooking Tools* Grass bunches tied together and used to strain goat's milk. (166:83)

Bouteloua hirsuta Lag., Hairy Grama
- *Food*—Kiowa *Fodder* Very good fodder for horses and mules. (192:14)
- *Other*—Navajo *Sacred Items* Used to make sacred charcoal for certain ceremonies. (55:25)

Bouteloua simplex Lag., Matted Grama
- *Drug*—Navajo, Ramah *Dermatological Aid* Ashes applied to sores. *Throat Aid* Cold infusion used internally and externally for sore throat. (191:16)
- *Food*—Navajo, Ramah *Forage* Important forage grass for a short season. (191:16)

Bouteloua sp., Grama Grass
- *Food*—Navajo *Fodder* Used for sheep and horse feed. (55:25)
- *Other*—Apache, Chiricahua & Mescalero *Ceremonial Items* Plant used in ceremonial contexts. (33:24) Costanoan *Cooking Tools* Hollow stems used as straws. (21:255)

Bovista, Lycoperdaceae, fungus

Bovista pila Berk. & Curt., Puffball
- *Drug*—Haisla & Hanaksiala *Poison* Spores dangerous, especially harmful to the eyes. (43:134)
- *Other*—Chippewa *Paint* Used as paint for the dead. (53:377)

Bovista plumbea Pers., Puffball
- *Drug*—Haisla & Hanaksiala *Poison* Spores dangerous, especially harmful to the eyes. (43:134)

Bovistella, Lycoperdaceae, fungus

Bovistella sp., Puffball
- *Drug*—Haisla & Hanaksiala *Poison* Spores dangerous, especially harmful to the eyes. (43:134)

Boykinia, Saxifragaceae

Boykinia jamesii (Torr.) Engl., James's Telesonix
- *Drug*—Cheyenne *Antihemorrhagic* and *Pulmonary Aid* Infusion of finely powered plant taken for lung hemorrhage. (as *Saxifraga jamesi* 76:175)

Boykinia occidentalis Torr. & Gray, Coastal Brookfoam
- *Drug*—Quileute *Tuberculosis Remedy* Raw leaves eaten for tuberculosis. (as *B. elata* 79:31) Yuki *Unspecified* Roots used medicinally for unspecified purpose. (as *Therofon elatum* 41:353)
- *Other*—Karok *Incense & Fragrance* Dried leaves sometimes worn inside basket caps for the fragrance. (as *B. elata* 148:384) Makah *Decorations* Flowers used in bouquets. (as *B. elatea* 67:257)

Brachyactis, Asteraceae

Brachyactis frondosa (Nutt.) A. G. Jones, Leafy Rayless Aster
- *Drug*—Paiute *Antirheumatic (External)* Infusion of stems and flowers used as a wash for rheumatism. *Blood Medicine* and *Tonic* Infusion of dried stems taken as a general blood tonic. (as *A. frondosus* 180:48, 49)

Brassicaceae, see *Arabis, Armoracia, Barbarea, Brassica, Camelina, Capsella, Cardamine, Caulanthus, Cochlearia, Dentaria, Descurainia, Dimorphocarpa, Draba, Erysimum, Hirschfeldia, Iodanthus, Lepidium, Lesquerella, Lobularia, Moricandia, Parrya, Pennellia, Phoenicaulis, Physaria, Raphanus, Rorippa, Schoencrambe, Sinapis,*

Sisymbrium, Stanleya, Streptanthus, Thelypodiopsis, Thelypodium, Thlaspi, Thysanocarpus

Brassica, Brassicaceae

Brassica juncea (L.) Czern., India Mustard
- *Drug*—**Navajo, Ramah** *Gastrointestinal Aid* Plant used as a stomach medicine. (191:28)

Brassica napus L., Rape
- *Drug*—**Cherokee** *Dietary Aid* Taken to increase the appetite. *Febrifuge* Taken for fever and "nervous fever." *Kidney Aid* Taken for "dropsy." *Misc. Disease Remedy* Taken for "ague." *Orthopedic Aid* Taken for palsy. *Pulmonary Aid* Used as a poultice for croup. *Respiratory Aid* Given for "phthisic" or asthma. *Stimulant* Taken as a stimulant. *Tonic* Taken as a tonic. (80:46) **Iroquois** *Dermatological Aid* Poultice of hot, dried leaves applied to sores and boils. (87:341) **Micmac** *Cold Remedy* Bark used for colds. *Cough Medicine* Bark used for coughs. *Misc. Disease Remedy* Bark used for grippe and smallpox. (40:55)
- *Food*—**Cherokee** *Vegetable* Leaves boiled and served with drippings, or boiled, fried with other greens, and eaten. (as *B. napa* 204:253)

Brassica nigra (L.) W. D. J. Koch, Black Mustard
- *Drug*—**Cherokee** *Dietary Aid* Taken to increase appetite. *Febrifuge* Taken for fever and "nervous fever." *Kidney Aid* Taken for "dropsy." *Misc. Disease Remedy* Taken for "ague." *Orthopedic Aid* Taken for palsy. *Pulmonary Aid* Used as a poultice for croup. *Respiratory Aid* Given for "phthisic" or asthma. *Stimulant* Taken as a stimulant. *Tonic* Taken as a tonic. (80:46) **Hoh** *Unspecified* Plants used for medicine. (137:61) **Meskwaki** *Cold Remedy* Snuff of ground seeds used for head colds. (152:219) **Mohegan** *Analgesic* Poultice of mustard applied to body pains. Poultice of wilted leaves applied to the skin for headache. (30:120) Poultice of leaves bound to the skin for headache. (174:264) Poultice of fresh leaves applied to headaches. (176:71, 128) *Toothache Remedy* Poultice of wilted leaves applied to the skin for toothache. (30:120) Poultice of leaves bound to the skin for toothache. (174:264) Poultice of fresh leaves applied to toothaches. (176:71, 128) **Quileute** *Unspecified* Plants used for medicine. (137:61) **Shinnecock** *Analgesic* Poultice of mustard applied to body pains. Poultice of wilted leaves applied to the skin for headache. *Emetic* Mustard mixed with flour and water and taken to make "insides come up." *Toothache Remedy* Poultice of wilted leaves applied to the skin for toothache. (30:120)
- *Food*—**Diegueño** *Vegetable* Young, spring leaves boiled and eaten as greens. (84:15) **Hoh** *Spice* Used for flavoring. *Vegetable* Plants eaten as greens. (137:61) **Iroquois** *Vegetable* Cooked and seasoned with salt, pepper, or butter. (196:117) **Luiseño** *Vegetable* Plant used for greens. (155:232) **Mohegan** *Vegetable* Combined with pigweed, plantain, dock, and nettle and used as mixed greens. (176:83) **Quileute** *Spice* Used for flavoring. *Vegetable* Plants eaten as greens. (137:61)

Brassica oleracea L., Cabbage
- *Drug*—**Cherokee** *Dermatological Aid* Poultice of wilted leaf used for boils. (80:28) **Rappahannock** *Analgesic* Poultice of green leaves bound to head for headache. (161:25)
- *Food*—**Cherokee** *Vegetable* Leaves used for food. (80:28) **Haisla & Hanaksiala** *Vegetable* Species used for food. (43:227) **Kitasoo** *Vegetable* Leaves used for food. (43:328) **Lakota** *Vegetable* Leaves eaten as greens. (106:34) **Okanagan-Colville** *Vegetable* Heads used for food. (188:92) **Seminole** *Unspecified* Plant used for food. (169:485)
- *Other*—**Rappahannock** *Protection* Leaves worn under the hat to protect from a strong sun. (161:25)

Brassica rapa L., Rape Mustard
- *Food*—**Haisla & Hanaksiala** *Vegetable* Roots used for food. (43:227) **Kitasoo** *Vegetable* Roots used for food. (43:329) **Okanagan-Colville** *Vegetable* Roots used for food. (188:92) **Oweekeno** *Unspecified* Roots used for food. (43:89)

Brassica rapa L. var. *rapa*, Birdrape
- *Drug*—**Ojibwa** *Unspecified* Plant used for medicinal purposes. (as *B. campestris* 135:232)
- *Food*—**Cherokee** *Vegetable* Leaves cooked with turnip greens, creaseys (probably *Lepidium virginicum*), and sochan (*Rudbeckia laciniata*) and eaten. (as *B. campestris* 204:253) **Mendocino Indian** *Vegetable* Young leaves eaten as greens in imitation of the first white settlers who first ate them. (as *B. campestris* 41:352)

Brassica sp., Turnip
- *Drug*—**Rappahannock** *Poultice* Hot poultice of roasted roots used for chilblains. (161:32) **Shoshoni** *Burn Dressing* Poultice of pulverized ripe seeds applied to burns. (180:52)
- *Food*—**Iroquois** *Vegetable* Young plants boiled and eaten as greens. (124:93) **Malecite** *Unspecified* Species used for food. (160:6) **Pomo, Kashaya** *Unspecified* Flowers eaten raw or cooked and young leaves eaten boiled and fried. (72:76)

Brickellia, Asteraceae

Brickellia ambigens (Greene.) A. Nels.
- *Drug*—**Keres, Western** *Antirheumatic (External)* Dried, ground leaves mixed with water and used as a salve. *Carminative* and *Dietary Aid* Infusion of plant taken for flatulence and overeating. *Liver Aid* Infusion of plant used as liver medicine. (171:33)
- *Fiber*—**Keres, Western** *Brushes & Brooms* Tied bunches of plants used as rough brooms. (171:33)

Brickellia brachyphylla (Gray) Gray, Plumed Brickellbush
- *Drug*—**Navajo, Ramah** *Disinfectant* and *Pediatric Aid* Root used with "lizard figurine" for prenatal "lizard infection." (191:49)

Brickellia californica (Torr. & Gray) Gray, California Brickellbush
- *Drug*—**Diegueño** *Febrifuge* Infusion of leaves taken for fevers. (88:220) **Navajo, Kayenta** *Ceremonial Medicine* Plant used as a ceremonial emetic following clan incest. *Dermatological Aid* Plant used as a lotion on infant sores caused by prenatal infection. *Emetic* Plant used as a ceremonial emetic following clan incest. *Pediatric Aid* Plant used as a lotion on infant sores caused by prenatal infection. (205:45) **Navajo, Ramah** *Cough Medicine* Cold infusion of leaves taken and used as lotion for cough or fever. *Febrifuge* Infusion of leaves taken and used as lotion for cough or fever. (191:49)
- *Food*—**Sanel** *Substitution Food* Leaves used as a substitute for tea. (as *Coleosanthus californicus* 41:393)

Brickellia eupatorioides (L.) Shinners var. *eupatorioides*, False Boneset

- *Drug*—Navajo, Ramah *Cough Medicine* and *Other* Decoction of root taken for old injury or cough. (as *Kuhnia rosmarinifolia* 191:52)

Brickellia grandiflora (Hook.) Nutt., Tasselflower Brickellbush
- *Drug*—Gosiute *Poison* Seeds had poisonous effects. *Unspecified* Root used as medicine. (39:364) Keres, Western *Antirheumatic (External)* Dried, ground leaves mixed with water and used as a salve. *Carminative* Infusion of plant taken for flatulence. *Dietary Aid* Infusion of plant taken for overeating. *Liver Aid* Infusion of plant used as a liver medicine. (171:33) Navajo, Ramah *Analgesic* Cold infusion of dried leaves taken for headache. *Ceremonial Medicine* and *Emetic* Plant used as a ceremonial emetic. *Misc. Disease Remedy* Cold infusion of dried leaves taken for influenza. (191:49)
- *Food*—Gosiute *Cooking Agent* Seeds and other seeds made into a meal and used as "baking powder" to improve the cakes. (39:364)
- *Fiber*—Keres, Western *Brushes & Brooms* Tied bunches of plants used as rough brooms. (171:33)
- *Other*—Navajo *Ceremonial Items* Plant and other plants used as a ceremonial liniment for the Female Shooting Life Chant. (55:83)

Brickellia oblongifolia var. *linifolia* (D. C. Eat.) B. L. Robins., Narrowleaf Brickellbush
- *Drug*—Navajo, Kayenta *Dermatological Aid* and *Pediatric Aid* Plant lotion used on infant ear and finger sores caused by prenatal infection. (205:46) Shoshoni *Gastrointestinal Aid* Decoction of stems and leaves taken as a stomach medicine. (180:52)

Brodiaea, Liliaceae

Brodiaea coronaria (Salisb.) Engl., Harvest Clusterlily
- *Food*—Atsugewi *Unspecified* Roots boiled in water or cooked in earth oven and used for food. (61:138) Mendocino Indian *Forage* Corms eaten by sheep. *Unspecified* Corms roasted and used for food. (as *Hookera coronaria* 41:326) Miwok *Unspecified* Bulbs steamed in earth oven and eaten without salt. (12:156) Pomo *Unspecified* Corms roasted and eaten. (as *Hookera coronaria* 11:89) Pomo, Kashaya *Vegetable* Baked or boiled corms eaten like baked or boiled potatoes. (72:27)

Brodiaea elegans Hoover, Elegant Clusterlily
- *Food*—Yurok *Vegetable* Bulbs baked in sand with a fire built over them. (6:21)

Brodiaea minor (Benth.) S. Wats., Vernalpool Clusterlily
- *Food*—Yana *Unspecified* Roots steamed and eaten. (147:251)

Brodiaea sp., Grass Nut
- *Drug*—Mahuna *Dermatological Aid* Plant used as a shampoo for the hair. (140:40)
- *Food*—Costanoan *Unspecified* Parched or roasted bulbs used for food. (21:255) Miwok *Unspecified* Cooked bulbs used for food. Bulbs used for food. (12:156)

Bromeliaceae, see *Ananas, Catopsis, Guzmania, Tillandsia*

Bromus, Poaceae

Bromus anomalus Rupr. ex Fourn., Nodding Brome
- *Food*—Navajo, Ramah *Fodder* Used for horse feed. (191:16)

Bromus carinatus Hook. & Arn., California Brome
- *Drug*—Hesquiat *Poison* Long, sharp-awned fruit were said to be very dangerous if swallowed. (185:56)
- *Food*—Neeshenam *Bread & Cake* Seeds parched, ground into flour and used to make bread. *Porridge* Seeds parched, ground into flour, and used to make mush. *Staple* Seeds parched, ground into flour, and used for food. (as *Promus virens* 129:377)

Bromus catharticus Vahl, Rescue Grass
- *Food*—Kiowa *Fodder* Grass recognized as an important fodder. Grass recognized as an important fodder. (as *B. unioloides* 192:14)

Bromus ciliatus L., Fringed Brome
- *Other*—Iroquois *Fertilizer* Decoction of plant used as a soak for corn, a "corn planting medicine." (87:273)

Bromus diandrus Roth, Ripgut Brome
- *Food*—Karok *Porridge* Seeds parched, pounded into a meal, and mixed with water into a gruel. (as *B. rigidus* 148:380) Luiseño *Unspecified* Seeds used for food. (as *B. maximus* 155:234) Miwok *Unspecified* Pulverized seeds made into pinole. (as *B. rigidus* var. *gussonei* 12:152)

Bromus hordeaceus L., Soft Brome
- *Food*—Karok *Porridge* Seeds parched, pounded into a meal, and mixed with water into a gruel. (148:379)

Bromus marginatus Nees ex Steud., Mountain Brome
- *Food*—Mendocino Indian *Staple* Seeds formerly used for pinole. (41:312)
- *Fiber*—Keres, Western *Brushes & Brooms* Tied bunches of plants used as hairbrushes and light brooms. (171:34)

Bromus marginatus var. *breviaristatus* (Buckl.) Beetle, Mountain Brome
- *Food*—Gosiute *Unspecified* Seeds formerly eaten. (as *B. breviaristatus* 39:364)

Bromus sp., Brome Grass
- *Food*—Thompson *Forage* Plants used as a forage crop. (164:516)
- *Fiber*—Isleta *Brushes & Brooms* Tied bunches of stems used to make brooms and brushes. (100:25)

Bromus tectorum L., Cheat Grass
- *Drug*—Navajo, Kayenta *Ceremonial Medicine* Infusion of plant used as a face wash for God-Impersonators. (205:15)
- *Food*—Cahuilla *Starvation Food* Seeds, a famine food, cooked into a gruel during food shortages. (15:48) Navajo, Ramah *Fodder* Used for sheep and horse feed. (191:16)
- *Fiber*—Paiute *Mats, Rugs & Bedding* Leaves used under bedding when camping. (111:51)

Broussaisia, Hydrangeaceae

Broussaisia arguta Gaud., Kanawao
- *Drug*—Hawaiian *Pediatric Aid* Fruits eaten from conception until the child feeds itself to increase the child's survival rate. *Reproductive Aid* Fruits eaten with baked eggs to bring about conception by barren women. *Strengthener* Fruits eaten from conception until the child feeds itself to increase the child's survival rate. (as *Broussaisa pelluoida* 2:48)

Broussonetia, Moraceae
Broussonetia papyrifera (L.) L'Hér. ex Vent., Paper Mulberry
- *Drug*—**Apache, Chiricahua & Mescalero** *Narcotic* Plant used as a narcotic. (as *B. secundiflora* 33:54)
- *Other*—**Apache, Chiricahua & Mescalero** *Ceremonial Items* Seeds worn around the neck in a string during ceremonies. (as *B. secundiflora* 33:54)

Bryaceae, see *Bryum*

Bryoria, Alectoriaceae, lichen
Bryoria capillaris (Ach.) Brodo & D. Hawksw., Black Tree Lichen
- *Other*—**Haisla & Hanaksiala** *Paint* Plant burned into black powder and used to make wood paint. (43:141)

Bryoria glabra (Mot.) Brodo & D. Hawksw., Black Tree Lichen
- *Other*—**Haisla & Hanaksiala** *Paint* Plant burned into black powder and used to make wood paint. (43:141)

Bryoria trichodes (Ach.) Brodo & Hawksw., Black Tree Lichen
- *Other*—**Haisla & Hanaksiala** *Paint* Plant burned into black powder and used to make wood paint. (43:141)

Bryum, Bryaceae, moss
Bryum capillare (L.) Hedw., Moss
- *Drug*—**Seminole** *Antirheumatic (External)* and *Febrifuge* Plant used for fire sickness: fever and body aches. (169:203)

Buchloe, Poaceae
Buchloe dactyloides (Nutt.) Engelm., Buffalo Grass
- *Drug*—**Keres, Western** *Dermatological Aid* Stolons crushed with yucca root or soaked in water and used as a head bath to make the hair grow. (as *Bulbilis dactyloides* 171:34)
- *Food*—**Blackfoot** *Forage* Used as an excellent fall and winter pasture for horses. (97:20)

Burseraceae, see *Bursera*

Bursera, Burseraceae
Bursera laxiflora, Torote Prieto
- *Fiber*—**Seri** *Basketry* Coil for the warp of very light, shallow baskets made of a "whisp of splints" of several sorts of torote. (50:138)

Bursera microphylla Gray, Elephant Tree
- *Drug*—**Cahuilla** *Dermatological Aid* Sap used for skin diseases. *Panacea* Sap used for almost any disease. (15:48)
- *Fiber*—**Seri** *Basketry* Coil for the warp of very light, shallow baskets made of a "whisp of splints" of several sorts of torote. (50:138)
- *Other*—**Cahuilla** *Good Luck Charm* Sap used by players of peon, a popular gambling game, to acquire "power." (15:48)

Bursera simaruba (L.) Sarg., Gumbo Limbo
- *Other*—**Seminole** *Containers* Wood used to make medicine bowls. (169:95)

Butomaceae, see *Butomus*

Butomus, Butomaceae
Butomus umbellatus L., Flowering Rush
- *Drug*—**Iroquois** *Veterinary Aid* Decoction of whole plant and bark from another plant added to cow and horse feed for worms. (141:66)

Cactaceae, see *Carnegia, Cereus, Coryphantha, Echinocactus, Echinocereus, Escobaria, Ferocactus, Lophophora, Machaerocereus, Mammillaria, Myrtillocactus, Neomammillaria, Opuntia, Pachycereus, Peniocereus, Stenocereus*

Caesalpinia, Fabaceae
Caesalpinia bonduc (L.) Roxb., Yellow Nicker
- *Drug*—**Hawaiian** *Blood Medicine* Beans and other plants pounded, squeezed and the resulting liquid taken to purify the blood. *Laxative* and *Pediatric Aid* Beans ground and taken as a laxative by infants, children and adults. *Pulmonary Aid* Beans and other plants pounded, squeezed, and resulting liquid taken to clear the chest of tough phlegm. (2:47)

Caesalpinia gracilis, Baraprieta
- *Fiber*—**Seri** *Basketry* Splints used to make the woof for basketry. (50:138)

Caesalpinia jamesii (Torr. & Gray) Fisher, James's Holdback
- *Drug*—**Zuni** *Veterinary Aid* Infusion of plant given to sheep to make them "prolific." (as *Hoffmanseggia jamesii* 166:54)
- *Food*—**Comanche** *Unspecified* Raw or boiled tubers used for food. (as *Hoffmanseggia jamesii* 29:522)

Caesalpinia kavaiensis Mann, Uhiuhi
- *Drug*—**Hawaiian** *Blood Medicine* Bark, young leaves, and other plants pounded, squeezed, and resulting liquid taken to purify the blood. (as *Mezoneurum kauaiense* 2:38)

Calamagrostis, Poaceae
Calamagrostis canadensis (Michx.) Beauv., Blue Joint
- *Fiber*—**Cree, Woodlands** *Mats, Rugs & Bedding* Plant used to make mattresses when nothing else was available. (109:33)
- *Other*—**Cree, Woodlands** *Containers* Grass used to line and cover winter storage pits for potatoes. (109:33)

Calamagrostis rubescens Buckl., Pine Grass
- *Drug*—**Thompson** *Gynecological Aid* Dried grass rubbed until soft and used as sanitary napkins. (187:140)
- *Fiber*—**Okanagan-Colville** *Caulking Material* Leaves mixed with mud and used to chink log cabins and as mortar in making chimneys. *Clothing* Leaves dried, softened, and used as insoles in moccasins. (188:53) **Thompson** *Clothing* Grass, rubbed and softened, sometimes mixed with sagebrush bark, used to make socks. (187:140)
- *Other*—**Okanagan-Colville** *Containers* Leaves used over and under food in pit cooking. Leaves used at the bottom of berry baskets and in a layer over the berries to keep the berries clean. *Cooking Tools* Leaves tied to a stick and used as a beater for whipping soapberries.

(188:53) **Thompson** *Cooking Tools* Tied bunches of grass used as soapberry whips. Grass used in drying soapberries. The grass was washed, braided at the ends, and laid out on a rack upon which the soapberries were placed to dry. A small fire was lit under the racks and when the berries were dried, they were stored with the grass still attached. Then, for use, the berries and grass were soaked in water and hand mixed. The grass, which helped to whip the berries, eventually floated to the top after which it was removed. Any remaining grass was removed by the person eating the berry whip. (187:140)

Calamovilfa, Poaceae
Calamovilfa gigantea (Nutt.) Scribn. & Merr., Giant Sandreed
- *Other*—**Hopi** *Ceremonial Items* Plant used to make prayer sticks. Plant used to make a carrying case for a part of the wedding garments. *Decorations* Plumes used to decorate mask of kachina. *Hunting & Fishing Item* Plant used to make bows and arrows. (42:296) **Jemez** *Decorations* Plumes used to decorate masks. (200:65)

Calamovilfa longifolia (Hook.) Scribn., Prairie Sandreed
- *Other*—**Lakota** *Decorations* Plant top, instead of a feather, worn on the head as a war charm. *Smoking Tools* Used for pipe cleaning. (139:29)

Calandrinia, Portulacaceae
Calandrinia ciliata (Ruiz & Pavón) DC., Fringed Redmaids
- *Food*—**Costanoan** *Staple* Seeds, in great quantities, used for pinole. *Unspecified* Raw foliage used for food. (21:251) **Luiseño** *Unspecified* Seeds used for food. *Vegetable* Tender plant used for greens. (as *C. caulescens* 155:232) **Miwok** *Unspecified* Oily, pulverized seeds pressed into balls and cakes for eating. (as *C. caulescens* var. *mensiesii* 12:152) **Numlaki** *Staple* Tiny, jet-black seeds eaten as pinole. (as *C. elegans* 41:346)

Calla, Araceae
Calla palustris L., Water Arum
- *Drug*—**Cree, Woodlands** *Orthopedic Aid* Aerial stems used for sore legs. *Poison* Plant poisonous to touch and eat. (109:33) **Gitksan** *Antihemorrhagic* Decoction of root taken for hemorrhage. *Eye Medicine* Decoction of root taken for "cleaning the eyes of the blind." *Misc. Disease Remedy* Decoction of roots taken for influenza and hemorrhage. *Respiratory Aid* Decoction of root taken for shortness of breath. (150:53) *Tonic* Plant used in a spiritual spring tonic. (74:26) **Iroquois** *Snakebite Remedy* Compound decoction of roots and stems used as poultice on snakebites. (87:278) **Potawatomi** *Dermatological Aid* Poultice of pounded root applied to swellings. (154:40)

Calliandra, Fabaceae
Calliandra eriophylla Benth., Fairyduster
- *Drug*—**Yavapai** *Gynecological Aid* Decoction of leaves and stems taken after childbirth. (65:261)

Calliandra humilis Benth., Dwarf Stickpea
- *Drug*—**Navajo, Ramah** *Panacea* Plant used as "life medicine." (191:32) **Zuni** *Dermatological Aid* Powdered root used three times a day for rashes. (27:376)

Callicarpa, Verbenaceae
Callicarpa americana L., American Beautyberry
- *Drug*—**Alabama** *Antirheumatic (External)* Decoction of roots and branches used in sweat bath for rheumatism. *Diaphoretic* Decoction of root and branch used in sweat bath for malarial fever and rheumatism. *Febrifuge* Decoction of roots and branches used in sweat bath for malarial fevers. (172:663) Decoction of roots and leaves used as sweat bath for malarial fever. *Herbal Steam* and Decoction of roots and leaves used as sweat bath for malarial fever. **Choctaw** *Antidiarrheal* Decoction of roots taken for dysentery. (177:52) *Gastrointestinal Aid* Decoction of roots and berries taken for colic. (25:24) Decoction of roots and berries taken for colic. *Other* Decoction of roots taken during attacks of dizziness. **Koasati** *Gastrointestinal Aid* Decoction of roots taken for stomachaches. (177:52) **Seminole** *Dermatological Aid* Plant used for snake sickness: itchy skin. (169:166) Roots or stem bark used for snake sickness: itchy skin. (169:239) *Urinary Aid* Decoction of root bark taken for urine retention. (169:274)

Callirhoe, Malvaceae
Callirhoe involucrata (Torr. & Gray) Gray, Purple Poppymallow
- *Drug*—**Dakota** *Analgesic* Decoction of root taken for internal pains. Smoke of dried root used to bathe aching body parts. *Cold Remedy* Root smoke inhaled for head cold. (70:103)
- *Other*—**Kiowa** *Decorations* Flowers, because of their beauty, have an esthetic appeal. (as *Malva involucrata* 192:42)

Calocedrus, Cupressaceae, conifer
Calocedrus decurrens (Torr.) Florin, Incense Cedar
- *Drug*—**Klamath** *Herbal Steam* Branches and twigs used in administering a sweat bath. (as *Libocedrus decurrens* 45:88) **Mendocino Indian** *Gastrointestinal Aid* Decoction of leaves taken for stomach troubles. (as *Libocedrus decurrens* 41:306) **Paiute** *Cold Remedy* Infusion of leaves steam inhaled for colds. (as *Libocedrus decurrens* 111:46)
- *Food*—**Round Valley Indian** *Spice* Dense leaflets used as flavoring in leaching acorn meal. (as *Libocedrus decurrens* 41:306)
- *Fiber*—**Cahuilla** *Building Material* Bark used to make conical shaped houses for temporary use while camped to gather and process acorns. Wood used for permanent construction. (as *Libocedrus decurrens* 15:85) **Karok** *Brushes & Brooms* Boughs and twigs used to make brooms. *Building Material* Wood used to make boards. (as *Libocedrus decurrens* 148:379) **Klamath** *Basketry* Wood used for basket weaving. (as *Libocedrus decurrens* 45:88) **Maidu** *Basketry* Roots used as overlay twine warps and overlay twine weft bases in the manufacture of baskets. (as *Libocedrus decurrens* 173:71) **Paiute** *Basketry* Bark made into baskets used for picking huckleberries. *Building Material* Tree used for fence posts. (as *Libocedrus decurrens* 111:46)
- *Other*—**Klamath** *Fuel* Wood used for fire blocks. Twig used as a twirling stick to produce fire by friction. (as *Libocedrus decurrens* 45:88) **Round Valley Indian** *Cooking Tools* Leafy branches used to spread water gently over acorn meal. (as *Libocedrus decurrens* 41:337) Dense leaflets used to prevent sand from mixing with the meal in leaching acorn meal. *Hunting & Fishing Item* Small limbs used as bows. (as *Libocedrus decurrens* 41:306) **Washo** *Hunting & Fishing Item* Wood used to make bows. (as *Librocedrus decurrens* 118:52)

Calochortus, Liliaceae

Calochortus amabilis Purdy, Short Lily
- *Food*—**Pomo, Kashaya** *Vegetable* Baked or boiled bulbs eaten like baked or boiled potatoes. (72:32)

Calochortus aureus S. Wats., Golden Mariposa Lily
- *Drug*—**Hopi** *Ceremonial Medicine* Plant used in the Flute ceremony. (56:18) **Navajo, Ramah** *Panacea* Bulb used as "life medicine." (as *C. nuttallii* var. *aureus* 191:20)
- *Food*—**Hopi** *Unspecified* Roots eaten raw. (56:18) Bulbs and flowers eaten. (190:159) **Navajo** *Unspecified* Bulbs gathered in early spring, peeled, and eaten raw. (55:32) **Navajo, Ramah** *Unspecified* Bulbs used for food. (as *C. nuttallii* var. *aureus* 191:20)
- *Other*—**Hopi** *Ceremonial Items* Ceremonially used as the Yellow Flower associated with the northwest direction. *Toys & Games* Boys holding handfuls of this plant and larkspur above their heads chased by girls on occasions. (200:70)

Calochortus catalinae S. Wats., Santa Catalina Mariposa Lily
- *Food*—**Cahuilla** *Unspecified* Bulbs roasted in hot ash pits or steamed prior to eating. (15:50)

Calochortus concolor (Baker) Purdy, Goldenbowl Mariposa Lily
- *Food*—**Cahuilla** *Unspecified* Bulbs roasted in hot ash pits or steamed prior to eating. (15:50)

Calochortus flexuosus S. Wats., Winding Mariposa Lily
- *Food*—**Cahuilla** *Unspecified* Bulbs roasted in hot ash pits or steamed prior to eating. (15:50)

Calochortus gunnisonii S. Wats., Gunnison's Mariposa Lily
- *Drug*—**Cheyenne** *Unspecified* Dried, chopped bulbs used as an ingredient for a medicinal mixture. *Veterinary Aid* Root put into a horse's mouth before running the animal in a race. (83:12) **Keres, Western** *Antirheumatic (Internal)* Infusion of plant taken for swellings. (171:34) **Navajo, Ramah** *Ceremonial Medicine* Plant used as a ceremonial medicine. *Dermatological Aid* Juice of leaf applied to pimples. *Gynecological Aid* Decoction of whole plant taken to ease delivery of placenta. *Panacea* Bulb used as "life medicine." (191:20)
- *Food*—**Cheyenne** *Dried Food* Bulbs dried and used as a winter food. (83:12) *Porridge* Dried bulbs pounded fine and meal boiled into a sweet porridge or mush. (76:172) *Unspecified* Bulbs cooked fresh and used for food. Young buds used for food. (83:12) Species used for food. (83:45) *Winter Use Food* Dried bulbs stored for winter use. (76:172) **Navajo, Ramah** *Unspecified* Bulbs eaten raw or gathered in the fall and boiled. (191:20)

Calochortus leichtlinii Hook. f., Smoky Mariposa
- *Food*—**Paiute, Northern** *Dried Food* Roots dried and eaten. *Soup* Roots dried and eaten or ground and cooked in soup. *Vegetable* Roots and tubers peeled and eaten roasted or raw. (59:44)

Calochortus luteus Dougl. ex Lindl., Yellow Mariposa Lily
- *Food*—**Miwok** *Unspecified* Roasted bulbs used for food. (12:157) **Navajo** *Unspecified* Bulbs gathered in early spring, peeled, and eaten raw. (55:32) **Pomo, Kashaya** *Vegetable* Baked bulbs eaten like baked potatoes. (72:64)

Calochortus macrocarpus Dougl., Sagebrush Mariposa Lily
- *Drug*—**Okanagan-Colville** *Dermatological Aid* Poultice of mashed bulbs applied to the skin for poison ivy. (188:41) **Thompson** *Eye Medicine* Mashed bulbs placed in cheesecloth and used for the eyes. (187:119)
- *Food*—**Klamath** *Unspecified* Species used for food. (45:93) **Okanagan-Colville** *Unspecified* Bulbs eaten raw or pit cooked with other roots. (188:41) **Okanagon** *Staple* Roots used as a principal food. (178:238) *Unspecified* Corms formerly cooked and used for food. Sweet flower buds used for food. (125:37) **Paiute** *Unspecified* Bulbs skinned and eaten fresh in spring. (104:102) *Vegetable* Bulbs eaten raw, boiled, or roasted. (111:58) **Shuswap** *Forage* Plant eaten by cattle and sheep. *Unspecified* Roots eaten either raw or cooked. (123:54) **Thompson** *Unspecified* Corms formerly cooked and used for food. Sweet flower buds used for food. (125:37) Coated, starchy corms used for food. (164:481) Sweet buds eaten. Unopened flowers eaten raw. (164:483) Corms used for food, usually raw, but sometimes cooked. (187:119)

Calochortus nuttallii Torr. & Gray, Sego Lily
- *Food*—**Gosiute** *Dried Food* Bulbs formerly dried and preserved for winter use. *Unspecified* Bulbs formerly used for food. (39:364) **Great Basin Indian** *Winter Use Food* Bulbs used for food during the winter. (121:47) **Havasupai** *Unspecified* Bulbs eaten with bread and mescal. (197:212) **Hopi** *Candy* Raw roots filled with sugar and eaten by children in early spring. *Unspecified* Seeds and flowers ground to make yellow pollen. (42:295) **Navajo** *Baby Food* Bulbs, a children's food, eaten by children while playing. (55:32) *Starvation Food* Bulbs formerly used for food in times of scarcity. *Unspecified* Bulbs eaten raw. (as *Calohortus nuttali* 110:24) **Paiute, Nevada** *Unspecified* Bulbs used for food during the spring. (121:47) **Paiute, Northern** *Dried Food* Roots dried and eaten. *Soup* Roots dried and eaten or ground and cooked in soup. *Vegetable* Roots and tubers peeled and eaten roasted or raw. (59:44) **Ute** *Starvation Food* Bulbs used for food in starving times. (118:15) *Unspecified* Bulbs formerly used for food. (38:33)
- *Other*—**Hopi** *Ceremonial Items* Yellow flower used ceremonially. In the spring, the flower is collected in quantity together with larkspur. Boys holding handfuls of these two plants above their heads are chased by the girls upon certain occasions. (42:295)

Calochortus palmeri S. Wats., Palmer's Mariposa Lily
- *Food*—**Cahuilla** *Unspecified* Bulbs roasted in hot ash pits or steamed prior to eating. (15:50) **Tubatulabal** *Unspecified* Bulbs used for food. (193:15)

Calochortus pulchellus Dougl. ex Benth., Mount Diablo Globelily
- *Food*—**Karok** *Unspecified* Bulbs baked in the earth oven and eaten. (148:380) **Mendocino Indian** *Unspecified* Corms eaten raw or roasted. (41:323) **Pomo** *Unspecified* Corms eaten raw or roasted. (11:89)

Calochortus sp., Mariposa Lily
- *Food*—**Montana Indian** *Unspecified* Bulbs dug and eaten for the pleasant flavor. (19:8) **Pomo** *Unspecified* Corms eaten for food. (11:89)

Calochortus tolmiei Hook. & Arn., Tolmie Startulip
- *Food*—**Mendocino Indian** *Unspecified* Corms eaten mostly by children. (as *C. maweanus* 41:323) **Pomo, Kashaya** *Vegetable* Baked or boiled bulbs eaten like baked or boiled potatoes. (72:31) **Yuki** *Unspecified* Sweet corms eaten raw. (as *C. caeruleus* var. *maweanus* 49:85)

Calochortus venustus Dougl. ex Benth., Butterfly Mariposa Lily
- *Food*—**Miwok** *Unspecified* Roasted bulbs used for food. (12:157) **Tubatulabal** *Unspecified* Bulbs used for food. (193:15)

Calochortus vestae Purdy, Coast Range Mariposa Lily
- *Food*—**Pomo** *Unspecified* Bulbs eaten for food. (as *C. luteus* var. *oculatus* 11:90) **Pomo, Kashaya** *Vegetable* Baked bulbs eaten like baked potatoes. (72:63)

Caltha, Ranunculaceae

Caltha leptosepala DC. **ssp. *leptosepala* var. *leptosepala*,** White Marshmarigold
- *Drug*—**Okanagon** *Dermatological Aid* Poultice of chewed plant applied to inflamed wounds. **Thompson** *Dermatological Aid* Poultice of chewed plant applied to inflamed wounds. (as *C. rotundifolia* 125:42) Fresh plant chewed and spit on wounds and poultice of crushed plant used. (as *C. rotundifolia* 164:467)

Caltha palustris L., Yellow Marshmarigold
- *Drug*—**Abnaki** *Poison* Plant considered poisonous. (144:155) **Alaska Native** *Poison* Raw leaves considered poisonous. (85:19) **Chippewa** *Cold Remedy* Decoction of root taken as diaphoretic, expectorant, and emetic for colds. (53:340) *Dermatological Aid* Poultice of boiled and mashed roots applied to sores. (71:130) *Diaphoretic* Decoction of root taken as a diaphoretic and emetic for colds. (53:340) *Diuretic* Compound decoction of leaves and stalks taken as a diuretic. (53:348) *Emetic* Decoction of root taken as a diaphoretic and emetic for colds. (53:340) *Gynecological Aid* Compound decoction of root taken during "confinement." (53:360) *Tuberculosis Remedy* Poultice of mashed or powdered root applied to scrofula sores. (53:354) **Eskimo, Inupiat** *Poison* Young shoots poisonous, if not boiled. (98:143) **Eskimo, Western** *Laxative* Infusion of leaves taken for constipation. (108:14) **Iroquois** *Emetic* and *Love Medicine* Infusion of smashed roots taken to vomit against a love charm. (87:323)
- *Food*—**Abnaki** *Unspecified* Leaves boiled with lard and eaten. (144:166) *Vegetable* Seeds used for food. (144:152) **Alaska Native** *Unspecified* Leaves and thick fleshy smooth slippery stems cooked and eaten. Roots boiled and eaten. (85:19) **Chippewa** *Vegetable* Leaves cooked and used as greens. (71:130) **Iroquois** *Vegetable* Young plants boiled and eaten as greens. (124:93) Cooked and seasoned with salt, pepper, or butter. (196:117) **Menominee** *Vegetable* Leaves used as greens. (151:70) **Mohegan** *Unspecified* Cooked and used for food. (176:83) **Ojibwa** *Unspecified* Leaves cooked with pork in the springtime. (153:408)

Caltha palustris* var. *flabellifolia (Pursh) Torr. & Gray, Yellow Marshmarigold
- *Food*—**Eskimo, Alaska** *Unspecified* Leaves and stalks boiled and eaten with seal oil. The leaves and stalks were collected before the plants flowered because after flowering commenced, the plant was apparently inedible. But boiling the plant broke down the poisonous protoanemonin that it contained, rendering it edible. (1:35)

Caltha palustris L. **var. *palustris*,** Yellow Marshmarigold
- *Food*—**Eskimo, Alaska** Leaves eaten fresh. (as *C. asarifolia* 4:715)

***Caltha* sp.**
- *Drug*—**Aleut** *Unspecified* Used medicinally. (7:29)

Calvatia, Lycoperdaceae, fungus

***Calvatia* sp.**
- *Drug*—**Haisla & Hanaksiala** *Poison* Spores dangerous, especially harmful to the eyes. (43:134)

Calycadenia, Asteraceae

Calycadenia fremontii Gray, Frémont's Western Rosinweed
- *Drug*—**Yana** *Febrifuge* Cooked, dried, pounded seeds eaten for chills. (as *Hemizonia multiglandulosa* 147:252)

Calycanthaceae, see *Calycanthus*

Calycanthus, Calycanthaceae

Calycanthus floridus L., Eastern Sweetshrub
- *Drug*—**Cherokee** *Dermatological Aid* Bark ooze used on children's sores and infusion used for hives. *Emetic* Roots used as a strong emetic. (80:58) Roots used as very strong emetics. (203:74) *Eye Medicine* Cold infusion of bark used as eye drops for persons losing eyesight. *Pediatric Aid* Bark ooze used on children's sores and infusion used for hives. (80:58) *Poison* Seeds used to poison wolves. (203:74) *Urinary Aid* Roots used for urinary and bladder complaints. (80:58)
- *Other*—**Cherokee** *Incense & Fragrance* Used for perfume. (80:58)

Calycanthus floridus* var. *glaucus (Willd.) Torr. & Gray, Eastern Sweetshrub
- *Drug*—**Cherokee** *Urinary Aid* Infusion of bark taken for urinary troubles. (as *C. fertilis* 177:23)

Calycanthus occidentalis Hook. & Arn., Western Sweetshrub
- *Drug*—**Pomo** *Cold Remedy* Decoction of scraped bark taken for severe colds. (66:13) **Pomo, Kashaya** *Cold Remedy* Infusion of dried or fresh, peeled bark used for chest colds. *Expectorant* Infusion of dried or fresh, peeled bark used to cough up the phlegm in the chest. *Gastrointestinal Aid* Infusion of dried or fresh, peeled bark used for stomach problems. *Throat Aid* Infusion of dried or fresh, peeled bark used for sore throat. (72:109)
- *Fiber*—**Mendocino Indian** *Basketry* Wood and bark from fresh shoots used in basket work. (as *Butneria occidentalis* 41:348)
- *Other*—**Yokia** *Hunting & Fishing Item* Pithy shoots used to make arrows. (as *Butneria occidentalis* 41:348)

Calylophus, Onagraceae

Calylophus hartwegii* ssp. *fendleri (Gray) Towner & Raven, Hartweg's Sundrops
- *Drug*—**Navajo, Ramah** *Panacea* Plant used as "life medicine," especially for internal bleeding. (as *Oenothera hartwegii* var. *fendleri* 191:38)

Calylophus lavandulifolius (Torr. & Gray) Raven, Lavenderleaf Sundrops
- *Food*—**Apache, Chiricahua & Mescalero** *Unspecified* Pods cooked and eaten by children. (as *Galpinsia lavandulaefolia* 33:45)

Calypso, Orchidaceae

Calypso bulbosa (L.) Oakes, Fairyslipper Orchid
- *Drug*—**Thompson** *Anticonvulsive* Bulbs chewed or flowers sucked for mild epilepsy. (187:135)

Calypso bulbosa var. *americana* (R. Br. ex Ait. f.) Luer, Fairy-slipper Orchid
- **Drug**—**Thompson** *Unspecified* Plants used as charms for unspecified purpose. (as *Cytherea bulbosa* 164:506)

Calystegia, Convolvulaceae
Calystegia longipes (S. Wats.) Brummitt, Paiute False Bindweed
- **Drug**—**Kawaiisu** *Venereal Aid* Decoction of roots taken for gonorrhea. (206:17)

Calystegia occidentalis ssp. *fulcrata* (Gray) Brummitt, Chaparral False Bindweed
- **Drug**—**Karok** *Love Medicine* Plant used as a love medicine. (as *Convolvulus fulcratus* 148:388)

Camassia, Liliaceae
Camassia leichtlinii (Baker) S. Wats., Large Camas
- **Food**—**Cowichan** *Special Food* Bulbs formerly served to guests at potlatches or winter dances. (186:83) **Klamath** *Pie & Pudding* Bulbs used to make pies. (as *Quamasia leichtlinii* 45:93) **Kwakiutl, Southern** *Unspecified* Bulbs pit steamed and used for food. (183:272) **Nitinaht** *Dried Food* Bulbs steam cooked, flattened, and dried for future food use. *Vegetable* Bulbs formerly steam cooked, dipped in whale or seal oil, and eaten as vegetables. (186:83) **Round Valley Indian** *Unspecified* Bulbs roasted or cooked and used for food. (as *Quamasia leichtlinii* 41:326) **Salish** *Vegetable* Bulbs used for food. (185:55) **Salish, Coast** *Dried Food* Bulbs pit steamed, slightly dried, and used for food. *Vegetable* Bulbs pit steamed and eaten immediately as the most important vegetable food. (182:74) **Yuki** *Unspecified* Bulbs cooked and used for food. (as *Quamasia leichtlinii* 41:326)
- **Other**—**Salish, Halkomelem** *Cash Crop* Bulbs traded to the Nootka and Nitinaht tribes. **Salish, Straits** *Cash Crop* Bulbs traded to the Nootka and Nitinaht tribes. (186:83)

Camassia quamash (Pursh) Greene, Small Camas
- **Drug**—**Blackfoot** *Gynecological Aid* Decoction of roots taken to induce labor. Infusion of grass taken for vaginal bleeding after birth and to help expel the afterbirth. (86:60)
- **Food**—**Blackfoot** *Bread & Cake* Roots pit roasted and made into loaves. (97:24) *Unspecified* Bulbs roasted and eaten. (82:14) Roots pit roasted and boiled with meat. (97:24) *Special Food* Bulbs boiled and given in soup on special events. (86:101) *Winter Use Food* Roots kept dry and preserved for future use. (97:24) **Chehalis** *Soup* Bulbs smashed, pressed together like cheese, and boiled in a stew with salmon. (79:24) **Clallam** *Unspecified* Bulbs cooked in pits with meat. (57:196) **Cowichan** *Special Food* Bulbs formerly served to guests at potlatches or winter dances. (186:83) **Flathead** *Beverage* Boiled and used as a sweet, hot beverage. *Sauce & Relish* Boiled with flour and eaten as a thick gravy. *Soup* Simmered with moss in blood into a soup and used for food. *Unspecified* Bulbs roasted and eaten. (82:14) **Hesquiat** *Vegetable* Steamed or boiled bulbs dipped in dogfish oil or whale oil before being eaten. (185:54) **Hoh** *Unspecified* Bulbs pit baked and used for food. (as *Quamasia quamash* 137:59) **Karok** *Vegetable* Bulbs used for food. Bulbs were dug up with a stick and placed in a pit 2 feet in diameter. Leaves of *Vitis californica* were placed on the bottom, a layer of bulbs, and then another layer of *V. californica* leaves. Finally a layer of dirt was added and a fire built on top. The mush formed was pure white and eaten by itself. (6:21) **Klamath** *Dried Food* Bulbs stored for future use. *Unspecified* Steamed bulbs used for food. (as *Quamasia quamash* 45:93) **Kutenai** *Unspecified* Bulbs roasted and eaten. (82:14) **Kwakiutl, Southern** *Unspecified* Bulbs pit steamed and used for food. (183:272) **Makah** *Unspecified* Bulbs pit cooked and eaten. (67:338) **Montana Indian** *Bread & Cake* Oven baked bulbs squeezed into little cakes or pulverized, formed into round loaves, and stored. *Staple* Bulbs formerly fire baked and used as a sweet and nutritious staple. *Sweetener* Bulbs formerly used as a sweetening agent. (82:14) **Nisqually** *Dried Food* Bulbs cooked, sun dried, and stored for future use as food. (79:24) **Nitinaht** *Dried Food* Bulbs steam cooked, flattened, and dried for future food use. *Vegetable* Bulbs formerly steam cooked, dipped in whale or seal oil, and eaten as vegetables. (186:83) **Okanagan-Colville** *Dried Food* Bulbs pit cooked, dried, and stored for future use. *Sauce & Relish* Bulbs pit cooked, dried, ground, and mixed with water and butter to make a "gravy." *Unspecified* Bulbs pit cooked, boiled with dried bitterroots, and eaten. (188:41) **Okanogon** *Unspecified* Bulbs baked and used for food. (as *Quamasia quamash* 125:37) **Paiute** *Dried Food* Roots cooked overnight, dried, and used for food. (104:102) *Pie & Pudding* Dried, ground bulbs made into a pudding. (111:56) *Unspecified* Roots cooked overnight and eaten. (104:102) *Winter Use Food* Bulbs prepared, preserved in numerous ways, and stored for winter use. (111:56) **Quileute** *Unspecified* Bulbs pit baked and used for food. (as *Quamasia quamash* 137:59) **Salish** *Vegetable* Bulbs used for food. (185:54) **Salish, Coast** *Dried Food* Bulbs pit steamed, slightly dried, and used for food. *Vegetable* Bulbs pit steamed and eaten immediately as the most important vegetable food. (182:74) **Skagit, Upper** *Unspecified* Bulbs steamed in an earth oven and eaten. (179:40) **Thompson** *Unspecified* Bulbs baked and used for food. (as *Quamasia quamash* 125:37) Bulbs cooked and eaten. (as *Quamasia quamash* 164:481) **Yuki** *Unspecified* Bulbs pit cooked and eaten. (49:86)
- **Other**—**Kutenai** *Cash Crop* Traded to the Blackfeet. **Nez Perce** *Cash Crop* Traded to the Gros Ventre and Crow. **Pend d'Oreille, Upper** *Cash Crop* Traded to the Kutenai. (82:14) **Salish, Halkomelem** *Cash Crop* Bulbs traded to the Nootka and Nitinaht tribes. **Salish, Straits** *Cash Crop* Bulbs traded to the Nootka and Nitinaht tribes. (186:83) **Shoshoni** *Cash Crop* Traded to the Nez Perce. (82:14)

Camassia scilloides (Raf.) Cory, Atlantic Camas
- **Drug**—**Creek** *Unspecified* Plant used medicinally for unspecified purpose. (as *C. esculenta* 172:667)
- **Food**—**Blackfoot** *Unspecified* Roots baked and eaten. (as *C. esculenta* 114:278) **Coeur d'Alene** *Vegetable* Roots used as a principal vegetable food. (as *C. esculenta* 178:88) **Comanche** *Unspecified* Raw roots used for food. (as *C. esculenta* 29:520) **Gosiute** *Unspecified* Bulbs roasted in pits lined with hot stones and eaten. *Winter Use Food* Bulbs formerly preserved for winter use. (as *C. esculenta* 39:364) **Montana Indian** *Unspecified* Bulbs boiled for eating fresh or preserved. *Winter Use Food* Bulbs baked in the ground by hot stones and dried for winter use. (as *C. esculenta* 19:9) **Okanogon** *Staple* Roots used as a principal food. (as *C. esculenta* 178:238) *Unspecified* Roots used as an important food. (as *C. esculenta* 178:237) **Spokan** *Unspecified* Roots used for food. (as *C. esculenta* 178:343) **Thompson** *Unspecified* Roots

used as an important food. (as *C. esculenta* 178:237)

Camassia sp.
- *Food*—**Haisla** *Vegetable* Bulbs used for food. (camas 43:194) **Nitinaht** *Unspecified* Bulbs formerly used as an important food. (camas 186:63) **Sanpoil & Nespelem** *Bread & Cake* Raw or roasted bulbs pulverized, formed into small cakes or balls, and dried for storage. Raw or roasted root pulverized, formed into small cakes or balls, and dried for storage. *Unspecified* Skinless bulb roasted until tender and used for food. Steamed or raw roots used for food. A cooking basket was filled half full of water and hot rocks added to boil the water. Then small sticks were placed crisscross in the basket above the surface of the water to hold the roots. After the roots had been added the basket was covered with a flat rock or piece of cedar plank to keep the steam from escaping. Plant, raw or cooked in underground pits, used for food. (as *Quamasia*, meadow camas 131:99)

Camelina, Brassicaceae

Camelina microcarpa Andrz. ex DC., Littlepod Falseflax
- *Food*—**Apache, Chiricahua & Mescalero** *Bread & Cake* Seeds threshed, winnowed, ground, and the flour used to make bread. *Unspecified* Seeds boiled and eaten. (33:49)

Camellia, Theaceae

Camellia sinensis (L.) Kuntze, Tea
- *Drug*—**Makah** *Hemostat* Poultice of leaves applied to stop bleeding. (67:287)
- *Food*—**Haisla & Hanaksiala** *Beverage* Leaves used to make tea. (as *Thea sinensis* 43:294) **Oweekeno** *Beverage* Leaves used to make tea. (as *Thea sinensis* 43:119)

Camissonia, Onagraceae

Camissonia brevipes (Gray) Raven ssp. *brevipes*, Golden Suncup
- *Food*—**Mohave** *Unspecified* Seeds used for food. (as *Oenothera brevipes* 37:187)

Camissonia claviformis (Torr. & Frém.) Raven ssp. *claviformis*, Browneyes
- *Food*—**Cahuilla** *Vegetable* Leaves used for greens. (as *Oenothera claveaformis* 15:94)

Camissonia multijuga (S. Wats.) Raven, Froststem Suncup
- *Drug*—**Navajo, Kayenta** *Other* Infusion of plant used for injuries by water or hail or dreaming of it. (as *Oenothera multijuga* 205:33)

Camissonia ovata (Nutt. ex Torr. & Gray) Raven, Goldeneggs
- *Food*—**Costanoan** *Unspecified* Raw, boiled, or steamed foliage used for food. (as *Oenothera ovata* 21:250)

Camissonia tanacetifolia (Torr. & Gray) Raven ssp. *tanacetifolia*, Tansyleaf Eveningprimrose
- *Drug*—**Navajo** *Dermatological Aid* Plant rubbed on as a liniment for boils. (as *Oenothera tanacetifolia* 55:67)

Campanulaceae, see *Asyneuma, Campanula, Campanulastrum, Clermontia, Lobelia, Triodanis*

Campanula, Campanulaceae

Campanula aparinoides Pursh, Marsh Bellflower
- *Drug*—**Iroquois** *Gynecological Aid* Decoction of stems taken by young women to induce childbirth. (87:451)

Campanula divaricata Michx, Small Bonny Bellflower
- *Drug*—**Cherokee** *Antidiarrheal* Infusion of root taken for diarrhea. (80:37)

Campanula parryi Gray, Parry's Bellflower
- *Drug*—**Navajo, Kayenta** *Gynecological Aid* Plant taken by pregnant woman when female baby was desired. (205:44) **Navajo, Ramah** *Dermatological Aid* Dry plant used as a dusting powder for sores. (191:47) **Zuni** *Dermatological Aid* Blossoms chewed and saliva applied to skin as a depilatory. Poultice of chewed root applied to bruises. (166:44)

Campanula rotundifolia L., Bluebell Bellflower
- *Drug*—**Chippewa** *Ear Medicine* Infusion of root used as drops for sore ear. (53:362) **Cree, Woodlands** *Heart Medicine* Root chewed for heart ailments. (109:34) **Navajo, Ramah** *Analgesic* Plant used as ceremonial fumigant for head trouble. *Ceremonial Medicine* Plant used as ceremonial fumigant for various ailments. *Disinfectant* Plant used as ceremonial fumigant for deer infection. *Eye Medicine* Plant used as ceremonial fumigant for eye. *Hunting Medicine* Plant rubbed on body for protection while hunting. *Witchcraft Medicine* Plant rubbed on body for protection from witches. (191:47) **Ojibwa** *Pulmonary Aid* Compound containing root used for lung troubles. (153:360) **Thompson** *Eye Medicine* Decoction of plant taken or used as a wash for sore eyes. (187:196)

Campanula uniflora L., Arctic Bellflower
- *Other*—**Navajo** *Ceremonial Items* Pollen used for some of the sacred ceremonies. (55:79)

Campanulastrum, Campanulaceae

Campanulastrum americanum (L.) Small, American Bellflower
- *Drug*—**Iroquois** *Pulmonary Aid* Infusion of smashed roots taken for whooping cough. (87:451) **Meskwaki** *Cough Medicine* Leaves used for coughs. *Tuberculosis Remedy* Leaves used for consumption. (152:206)

Canavalia, Fabaceae

Canavalia galeata (Gaud.) Vogel, 'Awikiwiki
- *Drug*—**Hawaiian** *Dermatological Aid* Infusion of leaves, shoots, bark, and other plants used as a bath for itch, ringworm, and skin diseases. (2:21)

Candelaria, Parmeliaceae, lichen

Candelaria concolor (Dicks.) Stein
- *Other*—**Haisla & Hanaksiala** *Paint* Plant used to make paint for wooden items. (43:141)

Candelariella vitellina (Hoffm.) Muell. Arg.
- *Other*—**Haisla & Hanaksiala** *Paint* Plant used to make paint for wooden items. (43:141)

Cannabaceae, see *Cannabis, Humulus*

Cannaceae, see *Canna*

Canna, Cannaceae

Canna flaccida Salisb., Bandanna of the Everglades
- *Other*—Seminole *Musical Instrument* Plant used to make rattle pellets. (169:503)

Cannabis, Cannabaceae

Cannabis sativa L., Marijuana
- *Drug*—Iroquois *Psychological Aid* Used after patient gets well but does not think that he has recovered. *Stimulant* "This plant will get you going." (87:306)

Canotia, Celastraceae

Canotia holacantha Torr., Crucifixion Thorn
- *Food*—Apache, San Carlos *Fruit* Berries used for food. (95:258) Apache, Western *Fruit* Berries eaten raw. (26:191)

Cantharellaceae, see *Cantharellus*

Cantharellus, Cantharellaceae, fungus

Cantharellus cibarius, Chantarelle
- *Food*—Pomo, Kashaya *Vegetable* Baked on hot stones or fried with onions. (72:128)

Capparaceae, see *Cleome, Polanisia, Wislizenia*

Caprifoliaceae, see *Diervilla, Linnaea, Lonicera, Sambucus, Symphoricarpos, Triosteum, Viburnum*

Capsella, Brassicaceae

Capsella bursa-pastoris (L.) Medik., Shepherd's Purse
- *Drug*—Cheyenne *Analgesic* Infusion of powdered leaves and stems taken or small quantities of powder eaten for head pains. (as *Bursa bursa-pastoris* 76:174) Chippewa *Analgesic* Decoction of entire plant taken for dysentery cramps. *Antidiarrheal* Decoction of whole plant taken for dysentery and cramps. *Gastrointestinal Aid* Decoction of whole plant taken for stomach cramps and dysentery. (as *Bursa bursa-pastoris* 53:344) Costanoan *Antidiarrheal* Infusion of plant used for dysentery. (21:9) Mahuna *Antidiarrheal* Infusion of plants taken for dysentery and diarrhea. (140:7) Menominee *Dermatological Aid* Infusion of whole plant used as a wash for poison ivy. (as *Bursa bursa-pastoris* 54:134) Infusion of plant used as a wash for poison ivy. (151:33) Meskwaki *Unspecified* Used as a medicine. (152:219) Mohegan *Analgesic* Infusion of seedpods taken for stomach pains. (176:71) *Anthelmintic* Infusion of seedpods taken for stomach, the pungency killed internal worms. (as *Bursa bursa-pastoris* 174:265) Infusion of seedpods taken as a vermifuge. (176:71, 128) *Gastrointestinal Aid* Infusion of seedpods taken for stomach, the pungency killed internal worms. (as *Bursa bursa-pastoris* 174:265) Infusion of seedpods taken for stomach pains. (176:71)
- *Food*—Apache, Chiricahua & Mescalero *Bread & Cake* Seeds winnowed, dried, stored, ground into flour, and used to make bread. *Unspecified* Seeds roasted without grinding and combined with other foods. (33:48) *Vegetable* Tops cooked alone or with meat and used as greens. (33:47) Cahuilla *Unspecified* Seeds gathered for food. *Vegetable* Leaves used for greens. (15:51) Cherokee *Spice* Mixed into other greens for flavoring. (204:253) *Vegetable* Leaves cooked and eaten as greens. (80:54) Mendocino Indian *Staple* Seeds eaten as a pinole. (as *Bursa bursa-pastoris* 41:352) Thompson *Vegetable* Leaves soaked in water overnight and eaten raw or cooked as a green vegetable. (187:194)

Capsicum, Solanaceae

Capsicum annuum L., Cayenne Pepper
- *Food*—Hopi *Dried Food* Fruits strung and dried for winter use. *Spice* Dried peppers crushed and used as flavoring for food. (200:88) Keresan *Soup* Used in stews. (198:560) Navajo *Unspecified* Chile peppers used for food. (165:221) Papago *Unspecified* Species used for food. Pima *Unspecified* Species used for food. (36:121) Pima, Gila River *Fruit* Fruits eaten raw and boiled. *Staple* Fruits used as a staple food. (133:7) Sia *Unspecified* Cultivated and eaten almost daily or sometimes at more than one meal per day. (199:106)
- *Other*—Pima, Desert *Cash Crop* Used for trade. (133:6)

Capsicum annuum L. **var. *annuum***, Cayenne Pepper
- *Food*—Navajo, Ramah *Spice* Pepper used extensively as a condiment in soups and stews. (as *C. frutescens* var. *longum* 191:42) Pima *Spice* Used for seasoning. (as *C. frutescens* var. *baccatum* 36:121)

Capsicum annuum **var. *frutescens*** (L.) Kuntze, Cayenne Pepper
- *Drug*—Cherokee *Cold Remedy* Plant used for colds. *Febrifuge* Poultice applied to soles of feet "in nervous or low fevers." *Gastrointestinal Aid* Plant used for colics. *Poultice* Poultice used for gangrene and poultice applied to feet for fevers. *Stimulant* Plant used as a powerful stimulant. (as *C. frutescens* 80:48) Navajo, Ramah *Gynecological Aid* and *Pediatric Aid* Powdered chili pepper rubbed on breast to wean nursing child. (as *C. frutescens* var. *longum* 191:42)
- *Food*—Papago *Special Food* Added to meat and eaten as a delicacy. (as *C. frutescens* var. *baccatum* 34:47) *Spice* Berries used as a seasoning. (as *C. frutescens* 34:19)

Cardamine, Brassicaceae

Cardamine californica (Nutt.) Greene **var. *californica***, Milkmaids
- *Other*—Yurok *Season Indicator* First plant to flower in the spring. (as *Dentaria californica* 6:27)

Cardamine concatenata (Michx.) Sw., Cutleaf Toothwort
- *Drug*—Iroquois *Analgesic* Poultice of smashed roots applied to the head for headaches. *Cold Remedy* Used for colds. *Dietary Aid* Used to stimulate appetite and regulate stomach. *Gastrointestinal Aid* Plant used for colds, to stimulate appetite, and to regulate the stomach. (87:340) *Hallucinogen* Plant used to mesmerize. (87:339) *Heart Medicine* Roots used several ways for heart palpitations or other heart diseases. (87:340) *Hunting Medicine* Roots rubbed on guns, traps, fishing lines or hooks, a "hunting medicine." *Love Medicine* Roots or plant placed in pocket or mouth to attract women, a "love medicine." *Panacea* Compound infusion taken or placed on injured part, a "Little Water Medicine." (87:339)
- *Food*—Iroquois *Unspecified* Roots eaten raw with salt or boiled. (as *Dentaria laciniata* 196:120)
- *Other*—Iroquois *Ceremonial Items* Plant used for divination. (87:340)

Cardamine diphylla (Michx.) Wood, Crinkle Root
- *Drug*—**Algonquin, Quebec** *Febrifuge* Infusion of plant given to children for fevers. *Heart Medicine* Infusion of plant and sweet flag root taken for heart disease. *Pediatric Aid* Infusion of plant given to children for fevers. (as *Dentaria diphylla* 18:173) **Cherokee** *Analgesic* Poultice of root used for headache. *Cold Remedy* Root chewed for colds. *Throat Aid* Infusion gargled for sore throat and root chewed for colds. (80:59) **Delaware** *Gastrointestinal Aid* Roots used as a stomach medicine. (as *Dentaria diphylla* 176:37) *Venereal Aid* Infusion of roots combined with other plants and used for scrofula and venereal disease. (as *Dentaria diphylla* 176:34) **Delaware, Oklahoma** *Gastrointestinal Aid* Compound containing root used as a stomach remedy. *Tuberculosis Remedy* Compound containing root used for "scrofula." *Venereal Aid* Compound containing root used for venereal disease. (as *Dentaria diphylla* 175:31, 76) **Iroquois** *Breast Treatment* Infusion of whole plant taken to strengthen the breasts. (as *Dentaria diphylla* 141:45) *Carminative* Raw root chewed for stomach gas. *Dermatological Aid* Poultice of roots applied to swellings. *Febrifuge* Cold infusion of plant taken for fever. *Love Medicine* Infusion of roots taken when "love medicine is too strong." *Other* Cold infusion of plant taken for "summer complaint." *Psychological Aid* Infusion of roots taken when the "heart jumps and the head goes wrong." *Pulmonary Aid* Compound used for chest pains. (87:341) *Tuberculosis Remedy* Infusion of plant taken at the beginning of tuberculosis. (as *Dentaria diphylla* 141:45) **Malecite** *Pediatric Aid* Infusion of roots used as a tonic for children. (as *Dentaria diphylla* 116:252) *Throat Aid* Green or dried roots chewed and used for hoarseness. (as *Dentaria diphylla* 116:247) Green or dry roots chewed and used to clear the throat. *Tonic* Infusion of roots used as a tonic. (as *Dentaria diphylla* 116:252) **Micmac** *Sedative* Root used as a sedative. *Throat Aid* Root used to clear the throat and for hoarseness. *Tonic* Root used as a tonic. (as *Dentaria diphylla* 40:56)
- *Food*—**Abnaki** *Sauce & Relish* Used as a condiment. (*Dentaria diphylla* 144:152) *Unspecified* Species used for food. (as *Dentaria diphylla* 144:167) **Algonquin, Quebec** *Sauce & Relish* Ground root put into vinegar for use as a relish. (as *Dentaria diphylla* 18:86) **Cherokee** *Vegetable* Leaves and stems parboiled, rinsed, added to hot grease, salt, and water, and boiled until soft as potherbs. Leaves used in salads. (as *Dentaria diphylla* 126:37) **Iroquois** *Unspecified* Roots eaten raw with salt or boiled. (as *Dentaria diphylla* 196:120) **Ojibwa** *Sauce & Relish* Ground roots mixed with salt, sugar, or vinegar and used as a condiment or relish. (as *Dentaria diphylla* 5:2207)
- *Other*—**Cherokee** *Smoke Plant* Leaves used for smoking. (80:59)

Cardamine douglassii Britt., Limestone Bittercress
- *Drug*—**Iroquois** *Antidote* Infusion of smashed roots used to counteract any kind of poison. *Witchcraft Medicine* Infusion of smashed roots used to divine the perpetrator of witchcraft. (87:340)

Cardamine maxima (Nutt.) Wood, Large Toothwort
- *Drug*—**Menominee** *Gastrointestinal Aid* Good medicine for the stomach. (as *Dentaria maxima* 151:65) **Ojibwa** *Gastrointestinal Aid* Roots used as a good medicine for the stomach. (as *Dentaria maxima* 153:399)
- *Food*—**Menominee** *Vegetable* Roots fermented for 4 or 5 days to sweeten and cooked with corn. (as *Dentaria maxima* 151:65) **Ojibwa** *Vegetable* Favored wild potatoes cooked with corn and deer meat or beans and deer meat. (as *Dentaria maxima* 153:399)

Cardamine rhomboidea (Pers.) DC., Bulbous Bittercress
- *Drug*—**Iroquois** *Poison* Roots used as a poison to kill. (as *C. bulbosa* 87:340)

Carduus, Asteraceae
***Carduus* sp.**, Thistle
- *Food*—**Luiseño** *Unspecified* Raw buds eaten for food. *Vegetable* Plant used as greens. (155:228)

Carex, Cyperaceae
Carex aquatilis Wahlenb., Water Sedge
- *Food*—**Alaska Native** *Unspecified* Stem bases eaten raw. (85:129)

Carex aquatilis* var. *dives (Holm) Kukenth., Sitka Sedge
- *Fiber*—**Hesquiat** *Basketry* Leaves used to make strong handles for baskets and shopping baskets. (as *C. sitchensis* 185:53) **Makah** *Basketry* Used for the bottoms of trinket baskets. (as *C. sitchensis* 79:22)

Carex atherodes Spreng., Slough Sedge
- *Food*—**Thompson** *Fodder* Grass used for animal feed. *Forage* Roots sometimes eaten by muskrats. (187:114)
- *Fiber*—**Thompson** *Clothing* "Swamp hay" softened by rubbing and used as insoles for moccasins. (187:114)

Carex barbarae Dewey, Santa Barbara Sedge
- *Fiber*—**Maidu** *Basketry* Blades used as coil thread and overlay twine weft bases in the manufacture of baskets. (173:71) **Pomo** *Basketry* Woody root fibers used as the white background in baskets. (9:137) Rootstocks used to make the white or creamy groundwork for baskets. (41:315) **Pomo, Kashaya** *Cordage* White, woody center of the root used as a sewing element in coiled baskets and in twining. *Sewing Material* White, woody center of the root used as a sewing element in coiled baskets and in twining. (72:103)

Carex brevior (Dewey) Mackenzie ex Lunell, Fescue Sedge
- *Drug*—**Iroquois** *Gynecological Aid* Compound infusion of plant taken for evacuation of the placenta. (87:275)

Carex concinnoides Mackenzie, Northwestern Sedge
- *Other*—**Okanagan-Colville** *Containers* Used for pit cooking and various household purposes. (188:36)

Carex douglasii Boott, Douglas's Sedge
- *Food*—**Kawaiisu** *Unspecified* Raw stems used for food. (206:17)

Carex exsiccata Bailey, Western Inflated Sedge
- *Dye*—**Shoshoni** *Black* Roots used as black dye in basketry. (as *C. exsiccata* 118:8)

Carex inops* ssp. *heliophila (Mackenzie) Crins, Sun Sedge
- *Drug*—**Navajo, Ramah** *Disinfectant* Cold infusion of plant used as lotion for "eagle infections." *Gastrointestinal Aid* Cold infusion of plant taken to relieve discomfort from overeating. (as *C. pensylvanica* var. *digyna* 191:19)

Carex mendocinensis Olney, Mendocino Sedge
- *Fiber*—**Pomo** *Basketry* Roots used in basketry. (118:6)

Carex microptera Mackenzie, Smallwing Sedge
- *Drug*—**Navajo, Ramah** *Ceremonial Medicine* and *Emetic* Plant used as a ceremonial emetic. (as *C. festivella* 191:19)
- *Food*—**Navajo, Ramah** *Forage* Plant browsed by sheep. (as *C. festivella* 191:19)

Carex nebrascensis Dewey, Nebraska Sedge
- *Drug*—**Cheyenne** *Ceremonial Medicine* Plant used in the Sun Dance and Massaum ceremonies. (83:7)
- *Food*—**Blackfoot** *Forage* Leaves thought to be a favorite food of the buffalo. (97:22) Favorite grass of the buffalo. (114:277)
- *Other*—**Blackfoot** *Ceremonial Items* Leaves tied around the horn of the sacred buffalo skull used in ceremonials. (97:22) Grass tied by medicine men to horns of the sacred buffalo head used in the Sun Dance ceremony. (114:277) Plant tied around the horns of the buffalo head in the Sun Dance. (118:51)

Carex obnupta Bailey, Slough Sedge
- *Food*—**Thompson** *Fodder* Grass used for animal feed. *Forage* Roots sometimes eaten by muskrats. (187:114)
- *Fiber*—**Hesquiat** *Basketry* Leaves used to make fine, closely woven baskets. (185:18) Dried, split leaves used to make the finest baskets. (185:50) *Clothing* Leaves used to make fine, closely woven hats. (185:18) **Makah** *Basketry* Leaves used for the horizontal strands in basketry. (67:330) **Nitinaht** *Basketry* Leaves extensively used as wrapping and twining material for "grass" baskets. **Nootka** *Basketry* Leaves extensively used as wrapping and twining material for "grass" baskets. (186:79) **Thompson** *Clothing* "Swamp hay" softened by rubbing and used as insoles for moccasins. (187:114)
- *Other*—**Hesquiat** *Tools* Sharp edged leaves used by men for shaving. (185:50)

Carex oligosperma Michx., Fewseed Sedge
- *Drug*—**Iroquois** *Emetic* Compound decoction taken as an emetic before running or playing lacrosse. (87:275)

Carex plantaginea Lam., Plantainleaf Sedge
- *Drug*—**Menominee** *Snakebite Remedy* Root used as a charm to prevent snakebite and spittle from chewed root used on snakebite. (151:34)

Carex platyphylla Carey, Broadleaf Sedge
- *Drug*—**Iroquois** *Other* Used several ways to "wash the snowsnake," a snowsnake medicine. (87:274)

Carex prasina Wahlenb., Drooping Sedge
- *Drug*—**Iroquois** *Emetic* Decoction taken as an emetic. *Gastrointestinal Aid* Decoction taken "when stomach is bad from an unknown cause." *Veterinary Aid* Decoction given to hunting dogs "when stomach is bad from an unknown cause." (87:275)

Carex rostrata Stokes, Beaked Sedge
- *Food*—**Thompson** *Fodder* Grass used for animal feed. (187:114) *Forage* Used as a forage plant. (164:514) Roots sometimes eaten by muskrats. *Unspecified* Bulbs used for food. (187:114)

Carex sp., Sedge
- *Drug*—**Cherokee** *Antidiarrheal* Infusion of leaf taken to "check bowels." (80:54) **Gosiute** *Unspecified* Root used as medicine. (39:365)

Songish *Abortifacient* Leaves eaten to induce abortions. (182:73)
- *Food*—**Klamath** *Beverage* Pith juice used as beverage. *Unspecified* Fresh stems used for food. Tuberous base of stem used for food. (45:92) **Montana Indian** *Unspecified* Young stems used for food. (19:9) **Navajo, Kayenta** *Porridge* Seeds ground, cooked into a mush, and eaten. (205:16) **Thompson** *Forage* Used as a general forage plant. (164:515) Leaves eaten by goats, horses, and other animals. (187:114)
- *Fiber*—**Blackfoot** *Clothing* Grass used in moccasins to protect the feet during winter horse stealing expeditions. (97:22) **Costanoan** *Basketry* Roots of many species used in basketry. (21:255) **Klamath** *Mats, Rugs & Bedding* Leaves woven into mats. (45:92) **Mendocino Indian** *Basketry* Rootstocks used to make strong and durable baskets. *Cordage* Rootstocks formerly used to make rope. (41:314) **Montana Indian** *Mats, Rugs & Bedding* Leaves woven into mats. (19:9) **Pomo** *Basketry* Roots used for basket body material. Roots split finely and used for black design material in fine baskets. (117:296) *Cordage* Roots used as a sewing element in coiled baskets. (66:11) **Salish, Coast** *Basketry* Fibrous leaves used to make baskets. *Cordage* Fibrous leaves used to make twine. (182:73) **Thompson** *Brushes & Brooms* Leaves used as brushes for cleaning things. (187:114) **Wailaki** *Basketry* Roots and leaves used for baskets. *Clothing* Leaves used to weave hats. (41:315) **Yuki** *Basketry* Large roots used to make baskets. (49:93)
- *Other*—**Jemez** *Ceremonial Items* and *Sacred Items* Plant considered sacred and used in the kiva. (44:21) **Paiute** *Cooking Tools* Woven sedge used to make spoons. (111:52) **Pomo** *Hunting & Fishing Item* Used in tending "hako fish traps." *Lighting* Dried plant tied in tight bundles for torches and used for travel by night like a lantern. (66:12)

Carex utriculata Boott, Northwest Territory Sedge
- *Food*—**Gosiute** *Unspecified* Lower, tender stems and root parts eaten by children. (39:365)

Carex vicaria Bailey, Western Fox Sedge
- *Food*—**Mendocino Indian** *Forage* Foliage cut for hay and used for forage. (41:314)

Carex vulpinoidea Michx., Fox Sedge
- *Drug*—**Iroquois** *Other* Compound decoction of roots used as a "rooster fighting medicine." (87:275)

Caricaceae, see *Carica*

Carica, Caricaceae
Carica papaya L., Papaya
- *Drug*—**Hawaiian** *Breast Treatment* Infusion of fruit taken by mothers for dry breasts. *Dermatological Aid* Milk and other plants mixed and applied to deep cuts. (2:43)
- *Food*—**Hawaiian** *Fruit* Fruit used for food. (2:43) **Seminole** *Unspecified* Plant used for food. (169:486)

Carnegia, Cactaceae
Carnegia gigantea (Engelm.) Britt. & Rose, Saguaro
- *Drug*—**Pima** *Gynecological Aid* Plant used to make the milk flow after childbirth. *Orthopedic Aid* Dead ribs used as splints for broken bones. (47:53)
- *Food*—**Apache, Chiricahua & Mescalero** *Fruit* Fruit used for food. (33:40) *Substitution Food* Syrup used in the absence of sugar to

sweeten an intoxicating drink. (33:50) **Apache, San Carlos** *Bread & Cake* Fruits sun dried, made into large cakes, and used for food. *Fruit* Fruits eaten raw. (as *Cereus giganteus* 95:257) **Apache, Western** *Beverage* Juice used as a drink. *Bread & Cake* Squeezed pulp dried and made into cakes. *Dried Food* Seeds washed and dried. *Fruit* Fruit eaten raw. *Pie & Pudding* Seeds ground with corn into a pudding. *Porridge* Seeds roasted, ground, and mixed with water to make a mush. (26:178) **Apache, White Mountain** *Fruit* Fruit used for food. (as *Cereus gigantea* 136:156) *Preserves* Fruit used to make a kind of butter. (as *Cercus gigantea* 136:147) **Maricopa** *Beverage* Juice fermented to make an intoxicating drink. (as *Cereus gigantea* 37:204) **Papago** *Beverage* Pulp boiled with water, strained, boiled again, and used as a ceremonial drink. (34:20) Juice mixed with water, fermented, and used as an intoxicating drink in ceremonies to bring rain. (34:26) *Bread & Cake* Seeds parched, stored, and used to make meal cakes. *Cooking Agent* Seeds parched, ground, water added, and oil extracted. (34:20) *Dried Food* Fruits dried, stored in jars, and used as sweets. (34:46) *Fodder* Seeds parched and used as a chicken feed. (34:20) *Fruit* Fruits used as an important article of diet. *Preserves* Fruits made into a conserve. (32:19) Juice made into cactus jam and used as the most important sweet in the diet. Pulp boiled to a sweet, sticky mass and used like raspberry jam. (34:20) Fruits made into jam. (34:46) *Sauce & Relish* Fruits boiled to make a syrup. (32:19) Fruits made into a syrup. (34:46) *Staple* Seeds ground into flour. (32:19) Seeds made into flour and used for food. (34:20) Used as a staple food. (34:45) *Unspecified* Oil extracted from the seeds. (32:19) Pulp eaten fresh. (34:20) Fruits and seeds used for food. (36:59) **Papago & Pima** *Candy* Used to make candy. (35:17) *Fruit* Fruit used for food. (35:11) *Preserves* Fruit boiled, without sugar, to make preserves. (35:17) *Sauce & Relish* Fruit used to make syrup. (35:11) **Pima** *Beverage* Ripe, dried fruits shaped into balls, boiled, fermented, and used to make wine. (47:53) Fruits boiled, fermented, and used as an intoxicating liquor. *Bread & Cake* Seeds ground, put into water, meal combined with other meal and baked to make bread. (as *Cereus giganteus* 146:71) *Dessert* Pulp eaten as dessert. (47:53) *Dried Food* Ripe fruits made into balls and dried for future use. (32:20) Fruits dried in balls and used for food. (as *Cereus giganteus* 146:71) *Fodder* Seeds fed to chickens. (47:53) *Fruit* Ripe fruits eaten fresh. (32:20) Ripe fruits eaten raw. (as *Cereus giganteus* 146:71) *Porridge* Seeds dried, roasted, ground, and eaten as a moist and sticky mush. (47:53) Fresh or dried fruits boiled, residue ground into an oily paste and eaten. (as *Cereus giganteus* 146:71) *Sauce & Relish* Fresh or dried fruits boiled to make a syrup. (32:20) Pulp boiled, seeds strained, boiled again, and sealed in jars until thick as honey. Ripe, dried fruits shaped into balls, boiled, and used to make a syrup. (47:53) Fresh or dried fruits boiled and used as a syrup. *Staple* Seeds ground, put into water, and eaten as pinole. (as *Cereus giganteus* 146:71) *Substitution Food* Seeds ground, passed through a sieve or left mixed with husks, and used as a substitute for lard. (47:53) *Unspecified* Seeds eaten raw. (as *Cereus giganteus* 146:71) **Pima, Gila River** *Beverage* Pulp made into a syrup and fermented for the annual wine feast, an elaborate liturgical celebration intended to bring rain and to continue it through the growing season. Ripe fruit used to make a cold drink. (133:4) *Candy* Fruits used as sweets. (133:6) *Dried Food* Pulp dried whole for future use. *Porridge* Seeds ground, mixed with grains, and used to make a porridge. *Preserves* Pulp used to make jam. Seeds ground, mixed with grains, and used to make a paste resembling peanut butter. *Sauce & Relish* Pulp used to make syrup. (133:4) *Staple* Fruits used as a staple food. (133:7) **Seri** *Fruit* Fruits eaten for food. *Porridge* Seeds ground to a powder and made into a meal or paste. (50:134) **Southwest Indians** *Fruit* Fruit used for food. (17:15) **Yavapai** *Beverage* Fruit mixed with water and liquid scooped with hand. Dried fruit pressed into bricks and kept for later use, pieces broken off and stirred in water. *Bread & Cake* Dried, parched, seeds ground to consistency of peanut butter and squeezed into cakes. *Dried Food* Dried fruit smeared with fresh fruit juice, made into slabs, and dried for later use. Dried fruit pressed into bricks and kept for later use. *Fruit* Fruit used for food. (as *Cereus giganteus* 65:260)

- **Fiber—Papago** *Basketry* Ribs used as one of the chief warp materials. *Building Material* Ribs used for roofing. Slats joined with one or two rows of deer hide thong and used to make screen doors. (34:53) Used for the roofing of houses. (34:66) **Seri** *Caulking Material* Dried plant skeletons and sea lion oil used as a caulking compound. (50:136)

- **Other—Apache, Western** *Ceremonial Items* Whole fruit mashed, water added, and mixture drunk after a 2-day burial in a dry place. The drink was consumed immediately after the 2-day burial and said to be better than tiswin, a fermented beverage. *Containers* Burls used as containers. *Cooking Tools* Burls used as vessels or cups. (26:178) **Papago** *Cooking Tools* Ribs made into a drying rack for datil fruit. (34:23) *Hunting & Fishing Item* Ribs split, made into rough cages, and used to trap Gambel quail and morning doves. (34:43) *Musical Instrument* Ribs with shallow, close-spaced notches used as soft rattles for certain songs. (34:68) *Season Indicator* Saguaro harvest marked the beginning of the new year. (36:59) *Tools* Giant ribs split in two and used as wooden tongs for gathering cholla joints and buds. (34:15) Four needles tied in a row and used as piercing instruments for tattooing. (34:51) Ribs used to make several kinds of light tools. (34:53) Ribs used as the larger of the two pieces of twisters used to make rope. (34:62) Ribs used for the handles of skin scrapers. (34:69) Perforations bored in sticks and used to smooth rough cords. (34:70) **Papago & Pima** *Tools* Trunks used to make cactus fruit picking poles. (35:14) **Pima** *Season Indicator* Saguaro harvest marked the beginning of the new year. (36:59) **Seri** *Tools* Dried plant skeletons used as a straight, slender pole for knocking off ripe fruit. (50:136) **Southwest Indians** *Containers* Seeds spread on saguaro ribs to dry. (17:16)

Carpinus, Betulaceae

Carpinus caroliniana Walt., American Hornbeam

- **Drug—Cherokee** *Antidiarrheal* Compound infusion taken for flux. *Dermatological Aid* Compound infusion taken for "navel yellowness." Compound infusion of astringent inner bark taken for flux. *Urinary Aid* Compound infusion taken for cloudy urine. (80:39) Decoction of inner bark taken for difficult urination with discharge. (177:15) **Delaware, Ontario** *Gynecological Aid* Compound infusion of root or bark taken for "diseases peculiar to women." *Tonic* Compound infusion of root or bark taken for "general debility." (175:68) **Iroquois** *Antidiarrheal* Decoction used as a wash or infusion of vine given to babies with diarrhea. *Dermatological Aid* Complex compound decoction used as wash for affected parts of "Italian itch."

Gynecological Aid Compound decoction taken to facilitate childbirth and for parturition. *Other* Compound used for "big injuries." *Pediatric Aid* Decoction used as a wash or infusion of vine given to babies with diarrhea. *Tuberculosis Remedy* Compound decoction of bark chips taken for consumption. (87:299)
- **Fiber**—**Chippewa** *Building Material* Tree used for the main supporting posts for the ridge pole of the wigwam or tent. (71:127)

Carpobrotus, Aizoaceae

Carpobrotus aequilateralus (Haw.) N. E. Br., Baby Sunrose
- **Food**—**Luiseño** *Fruit* Fruit used for food. (as *Mesembryanthemum aequilaterale* 155:232) **Pomo** *Fruit* Raw fruit used for food. (as *Mesembryanthemum aequilaterale* 66:13) **Pomo, Kashaya** *Fruit* Fruit eaten raw. (as *Mesembryanthemum chilense* 72:48)

Carthamus, Asteraceae

Carthamus tinctorius L., Safflower
- **Food**—**Hopi** *Cooking Agent* Flowers used as a yellow coloring for paper bread. (56:20) Flowers used to color wafer bread yellow. (200:95)
- **Dye**—**Hopi** *Yellow* Flowers used to color bread yellow for certain dances. (190:167)

Carum, Apiaceae

Carum carvi L., Caraway
- **Drug**—**Abnaki** *Analgesic* Used as an analgesic. (144:155) **Cree, Woodlands** *Pediatric Aid* and *Sedative* Seed given to a crying child to quiet him or her. (109:34) **Iroquois** *Veterinary Aid* Rhizomes given to pigs to make them stronger. (141:55)
- **Food**—**Cree, Woodlands** *Spice* Seeds added as a flavoring to bannock. *Staple* Seeds ground into flour. (109:34)

Carya, Juglandaceae

Carya alba (L.) Nutt. ex Ell., Mockernut Hickory
- **Drug**—**Cherokee** *Abortifacient* Used for female obstructions. *Analgesic* Used for poliomyelitis pain. *Cold Remedy* Used for colds. *Dermatological Aid* Bark used as a dressing for cuts. Astringent and detergent inner bark used as dressing for cuts. *Diaphoretic* Used as a diaphoretic. *Emetic* Used as an emetic. *Gastrointestinal Aid* Used to invigorate the stomach. *Liver Aid* Used for bile. *Misc. Disease Remedy* Used for poliomyelitis pain. *Oral Aid* Bark chewed for sore mouth. (as *C. tomentosa* 80:38) Chewed inner bark used for sore mouth. (as *C. tomentosa* 177:14) *Orthopedic Aid* Infusion of bark taken by ballplayers to make limbs supple. (as *C. tomentosa* 80:38) **Delaware, Ontario** *Gynecological Aid* Compound infusion of bark taken for "female disorder." *Tonic* Compound infusion of bark taken as a tonic for general debility. (175:82)
- **Food**—**Cherokee** *Unspecified* Species used for food. (as *C. tomentosa* 80:38) **Choctaw** *Soup* Pounded nutmeat boiled, made into a paste, and eaten as a broth or soup. (as *Juglans squamosa* 25:8)
- **Fiber**—**Cherokee** *Basketry* Inner bark used to finish baskets. *Furniture* Inner bark used to make chair bottoms. (as *C. tomentosa* 80:38)
- **Other**—**Cherokee** *Containers* Wood used to make barrel hoops. *Cooking Tools* Wood used to make corn beaters. *Fuel* Wood used for firewood. *Hunting & Fishing Item* Wood used to make blowgun darts and arrow shafts. *Tools* Wood used to make tool handles. (as *C. tomentosa* 80:38) **Choctaw** *Insecticide* Leaves "scattered about to drive away fleas." (as *Juglans squamosa* 25:24)

Carya cordiformis (Wangenh.) K. Koch, Bitternut Hickory
- **Drug**—**Iroquois** *Dermatological Aid* Nutmeat oil formerly used for the hair, either alone or mixed with bear grease. (196:123) **Meskwaki** *Diuretic* Infusion of bark taken "to make the urine free." *Laxative* Infusion of bark taken "to make the bowels loose." *Panacea* Infusion of bark taken for "simple sicknesses." (152:224)
- **Food**—**Iroquois** *Beverage* Fresh nutmeats crushed, boiled, and liquid used as a drink. *Bread & Cake* Fresh nutmeats crushed and mixed with bread. (124:99) Nuts crushed, mixed with cornmeal and beans or berries, and made into bread. (196:123) *Pie & Pudding* Fresh nutmeats crushed and mixed with corn pudding. (124:99) *Sauce & Relish* Nuts pounded, boiled, resulting oil seasoned with salt and used as gravy. *Soup* Nutmeats crushed and added to corn soup. (196:123) *Special Food* Fresh nutmeats crushed, boiled, and oil used as a delicacy in corn bread and pudding. (124:99) Nutmeat oil added to the mush used by the False Face Societies. *Staple* Nutmeats crushed and added to hominy. *Unspecified* Nutmeats, after skimming off the oil, seasoned and mixed with mashed potatoes. (196:123)
- **Fiber**—**Iroquois** *Furniture* Bark used to make chairs. (141:39) **Omaha** *Snow Gear* Hickory rims used to make snowshoes tied with thongs of rawhide woven across. (as *Hicoria minima* 68:324)
- **Other**—**Iroquois** *Insecticide* Nutmeat oil mixed with bear grease and used as a preventive for mosquitoes. (196:123)

Carya glabra (P. Mill.) Sweet **var. glabra**, Pignut Hickory
- **Fiber**—**Omaha** *Snow Gear* Hickory rims used to make snowshoes tied with thongs of rawhide woven across. (as *Hicoria glabra* 68:324)

Carya illinoinensis (Wangenh.) K. Koch, Pecan
- **Drug**—**Comanche** *Dermatological Aid* Pulverized leaves rubbed on affected part for ringworm. (29:520) **Kiowa** *Tuberculosis Remedy* Decoction of bark taken for tuberculosis. (as *C. pecan* 192:20)
- **Food**—**Comanche** *Unspecified* Nuts used for food. (29:520) *Winter Use Food* Nuts stored for winter use. (29:531)

Carya laciniosa (Michx. f.) G. Don, Shellbark Hickory
- **Drug**—**Cherokee** *Abortifacient* Used for female obstructions. *Analgesic* Used for poliomyelitis pain. *Cold Remedy* Used for colds. *Dermatological Aid* Bark used as a dressing for cuts. Astringent and detergent inner bark used as dressing for cuts. *Diaphoretic* Used as a diaphoretic. *Emetic* Used as an emetic. *Gastrointestinal Aid* Used to invigorate the stomach. *Liver Aid* Used for bile. *Misc. Disease Remedy* Used for poliomyelitis pain. *Oral Aid* Bark chewed for sore mouth. *Orthopedic Aid* Infusion of bark taken by ballplayers to make limbs supple. (80:38)
- **Food**—**Cherokee** *Unspecified* Species used for food. (80:38)
- **Fiber**—**Cherokee** *Basketry* Inner bark used to finish baskets. *Furniture* Inner bark used to make chair bottoms. (80:38) **Omaha** *Snow Gear* Hickory rims used to make snowshoes tied with thongs of rawhide woven across. (as *Hicoria laciniosa* 68:324)
- **Other**—**Cherokee** *Containers* Wood used to make barrel hoops. *Cooking Tools* Wood used to make corn beaters. *Fuel* Wood used for firewood. *Hunting & Fishing Item* Wood used to make blowgun darts and arrow shafts. *Tools* Wood used to make tool handles. (80:38)

Carya ovata (P. Mill.) K. Koch, Shagbark Hickory
- *Drug*—**Chippewa** *Analgesic* Fresh, small shoots steamed as inhalant for headache. *Herbal Steam* Fresh small shoots placed on hot stones as herbal steam for headache. (as *Hicoria alba* 53:338) **Delaware, Ontario** *Gynecological Aid* Compound infusion of bark taken for "diseases peculiar to women." *Tonic* Compound infusion of bark taken for "general debility." (175:68) **Iroquois** *Anthelmintic* Compound decoction with white from inside bark taken by adults for worms. *Antirheumatic (External)* Decoction of bark applied as a poultice for arthritis. *Antirheumatic (Internal)* Decoction of bark taken for arthritis. (87:297) *Dermatological Aid* Nutmeat oil formerly used for the hair, either alone or mixed with bear grease. (196:123)
- *Food*—**Dakota** *Soup* Nuts used to make soup. *Sweetener* Sap used to make sugar. Hickory chips boiled to make sugar. *Unspecified* Nuts eaten plain or with honey. (as *Hicoria ovata* 70:74) **Iroquois** *Baby Food* Fresh nutmeats crushed, boiled, and oil used as a baby food. *Beverage* Fresh nutmeats crushed, boiled, and liquid used as a drink. *Bread & Cake* Fresh nutmeats crushed and mixed with bread. (124:99) Nuts crushed, mixed with cornmeal and beans or berries, and made into bread. (196:123) *Pie & Pudding* Fresh nutmeats crushed and mixed with corn pudding. (124:99) *Sauce & Relish* Nuts pounded, boiled, resulting oil seasoned with salt and used as gravy. *Soup* Nutmeats crushed and added to corn soup. (196:123) *Special Food* Fresh nutmeats crushed, boiled, and oil used as a delicacy in corn bread and pudding. (124:99) Nutmeat oil added to the mush used by the False Face Societies. *Staple* Nutmeats crushed and added to hominy. *Unspecified* Nutmeats, after skimming off the oil, seasoned and mixed with mashed potatoes. (196:123) **Lakota** *Unspecified* Nuts used for food. (139:49) **Meskwaki** *Winter Use Food* Nuts stored for winter use. (152:259) **Ojibwa** *Unspecified* Edible nuts were appreciated. (153:405) **Omaha** *Soup* Nuts used to make soup. *Sweetener* Sap used to make sugar. Hickory chips boiled to make sugar. *Unspecified* Nuts eaten plain or with honey. **Pawnee** *Soup* Nuts used to make soup. *Sweetener* Sap used to make sugar. Hickory chips boiled to make sugar. *Unspecified* Nuts eaten plain or with honey. **Ponca** *Soup* Nuts used to make soup. *Sweetener* Sap used to make sugar. Hickory chips boiled to make sugar. *Unspecified* Nuts eaten plain or with honey. (as *Hicoria ovata* 70:74) **Potawatomi** *Winter Use Food* Hickory nuts gathered for winter use. (154:103) **Winnebago** *Soup* Nuts used to make soup. *Sweetener* Sap used to make sugar. Hickory chips boiled to make sugar. *Unspecified* Nuts eaten plain or with honey. (as *Hicoria ovata* 70:74)
- *Fiber*—**Omaha** *Snow Gear* Hickory rims used to make snowshoes tied with thongs of rawhide woven across. (as *Hicoria alba* 68:324)
- *Other*—**Chippewa** *Hunting & Fishing Item* Used for bows. (as *Hicoria alba* 53:377) **Iroquois** *Insecticide* Nutmeat oil mixed with bear grease and used as a preventive for mosquitoes. (196:123) **Ojibwa** *Hunting & Fishing Item* Wood used for making bows. Some are quite particular about the piece of wood they select, choosing a billet from the tree that includes heartwood on one side and sapwood on the other. The heartwood is the front of the bow in use, while the sapwood is nearest the user. It is a wood of general utility. (153:419) **Potawatomi** *Hunting & Fishing Item* Strong, elastic wood used to make bows and arrows. (154:113)

Carya pallida (Ashe) Engl. & Graebn., Sand Hickory
- *Drug*—**Cherokee** *Abortifacient* Used for female obstructions. *Analgesic* Used for poliomyelitis pain. *Cold Remedy* Used for colds. *Dermatological Aid* Bark used as a dressing for cuts. Astringent and detergent inner bark used as dressing for cuts. *Diaphoretic* Used as a diaphoretic. *Emetic* Used as an emetic. *Gastrointestinal Aid* Used to invigorate the stomach. *Liver Aid* Used for bile. *Misc. Disease Remedy* Used for poliomyelitis pain. *Oral Aid* Bark chewed for sore mouth. *Orthopedic Aid* Infusion of bark taken by ballplayers to make limbs supple. (80:38)
- *Food*—**Cherokee** *Unspecified* Species used for food. (80:38)
- *Fiber*—**Cherokee** *Basketry* Inner bark used to finish baskets. *Furniture* Inner bark used to make chair bottoms. (80:38)
- *Other*—**Cherokee** *Containers* Wood used to make barrel hoops. *Cooking Tools* Wood used to make corn beaters. *Fuel* Wood used for firewood. *Hunting & Fishing Item* Wood used to make blowgun darts and arrow shafts. *Tools* Wood used to make tool handles. (80:38)

Carya sp., Hickory
- *Food*—**Cherokee** *Cooking Agent* Wood ash and water used as a lye to skin corn, corn ground into a fine meal and made into bread. *Dried Food* Nuts dried in the rafters for future use. *Soup* Nuts and shells ground into a fine meal and used to make soup. *Spice* Wood ash, salt, and black pepper used to cure pork. (126:40) **Seminole** *Unspecified* Plant used for food. (169:491)
- *Other*—**Cherokee** *Cooking Tools* Dried leaves used to wrap around ball of meal, boiled for 1 hour and used for bread. (126:40) **Seminole** *Hunting & Fishing Item* Plant used to make bows and arrows. (169:492)

Caryophyllaceae, see *Arenaria, Cerastium, Drymaria, Honckenya, Paronychia, Pseudostellaria, Saponaria, Silene, Stellaria*

Cassiope, Ericaceae

Cassiope mertensiana (Bong.) D. Don, Western Moss Heather
- *Drug*—**Thompson** *Tuberculosis Remedy* Decoction of plant taken over a period of time for tuberculosis and spitting up blood. (187:215)

Cassiope sp.
- *Drug*—**Thompson** *Tuberculosis Remedy* Decoction of several unidentified species of *Cassiope* taken over a period of time for tuberculosis and spitting up blood. (187:215)

Cassiope tetragona (L.) D. Don, White Arctic Mountain Heather
- *Fiber*—**Eskimo, Alaska** *Building Material* Plant, mosses, and lichens gathered for insulating houses against cold. (4:715)
- *Other*—**Eskimo, Inuktitut** *Fuel* Used for tinder, burned green and even frozen, and fuel. (202:185)

Cassytha, Lauraceae

Cassytha filiformis L., Devil's Gut
- *Drug*—**Hawaiian** *Gynecological Aid* Plant pounded, water added, and taken by women to remove blood from the womb while giving birth. *Respiratory Aid* Plant and other plants pounded, water added, and taken to remove phlegm causing congestion in the chest. (2:46)

Castanea, Fagaceae

Castanea dentata (Marsh.) Borkh., American Chestnut
- *Drug*—**Cherokee** *Cough Medicine* Compound decoction of leaves used as cough syrup. *Dermatological Aid* Leaves from young sprouts dipped in hot water and put on sores. *Gastrointestinal Aid* Infusion given for the stomach. *Gynecological Aid* Cold, compound infusion of bark used to stop bleeding after childbirth. *Heart Medicine* Infusion of year-old leaves taken for heart trouble. *Misc. Disease Remedy* Infusion given for typhoid. *Pediatric Aid* Warmed galls applied to make infant's navel recede. (80:29) **Iroquois** *Dermatological Aid* Compound decoction used as wash for parts affected by "Italian itch." Compound wood powder used for chafed babies. (87:302) Nutmeat oil formerly used for the hair, either alone or mixed with bear grease. (196:123) *Pediatric Aid* Compound wood powder used for chafed babies. *Veterinary Aid* Bark mixed into young dog's food for worms. (87:302) **Mohegan** *Antirheumatic (Internal)* Leaves used for rheumatism. *Cold Remedy* Leaves used for colds. (176:128) *Pulmonary Aid* Infusion of leaves used for whooping cough. (174:265) Infusion of leaves taken for whooping cough. (176:71, 128)
- *Food*—**Cherokee** *Bread & Cake* Nuts ground into a meal and used to make bread. (126:39) *Substitution Food* Used as a coffee substitute. *Unspecified* Nuts boiled, pounded with corn, kneaded, wrapped in a green corn blade, boiled, and eaten. (80:29) **Iroquois** *Beverage* Fresh nutmeats crushed, boiled, and liquid used as a drink. *Bread & Cake* Fresh nutmeats crushed and mixed with bread. (124:99) Nuts crushed, mixed with cornmeal and beans or berries, and made into bread. (196:123) *Pie & Pudding* Fresh nutmeats crushed and mixed with corn pudding. (124:99) *Sauce & Relish* Nuts pounded, boiled, resulting oil seasoned with salt and used as gravy. *Soup* Nutmeats crushed and added to corn soup. (196:123) *Special Food* Fresh nutmeats crushed, boiled, and oil used as a delicacy in corn bread and pudding. (124:99) Nutmeat oil added to the mush used by the False Face Societies. (196:123) *Spice* Meats dried, pounded into flour and mixed with bread for flavoring. (124:99) *Staple* Nutmeats crushed and added to hominy. *Unspecified* Nutmeats, after skimming off the oil, seasoned and mixed with mashed potatoes. (196:123)
- *Fiber*—**Cherokee** *Building Material* Wood used for lumber and fence rails. (80:29)
- *Dye*—**Cherokee** *Brown* Bark used to make a brown dye. (80:29)
- *Other*—**Cherokee** *Fuel* Wood used for firewood. (80:29) **Iroquois** *Insecticide* Nutmeat oil mixed with bear grease and used as a preventive for mosquitoes. (196:123)

Castanea pumila (L.) P. Mill. Allegheny Chinkapin
- *Drug*—**Cherokee** *Analgesic* Brittle leaves heated and blown on patient for headaches. *Dermatological Aid* Used for fever blisters. *Febrifuge* Used for "chills and cold sweats." (80:29) Infusion of dried leaves used as a wash for fevers, chills, and cold sweats. **Koasati** *Gastrointestinal Aid* Decoction of roots taken for stomach troubles. (177:16)

Castanopsis, Fagaceae

Castanopsis chrysophylla (Dougl. ex Hook.) A. DC., Golden Chinkapin
- *Food*—**Karok** *Unspecified* "Nuts" roasted in coals and eaten. (as *Chrysolepis chrysophylla* 6:24) Nuts used for food. *Winter Use Food* Nuts stored for winter use. (148:383) **Klamath** *Unspecified* Nuts used for food. (45:94) **Mendocino Indian** *Unspecified* Nuts sometimes used for food. (41:333) **Paiute** *Beverage* Leaves used to make tea. (111:65) **Pomo** *Dried Food* Nuts stored for later use. *Unspecified* Raw nuts used for food. (66:12–13) **Pomo, Kashaya** *Unspecified* Nuts eaten raw, roasted, or pounded into a meal. *Winter Use Food* Nuts stored in their shell for winter. (72:34) **Tolowa** *Unspecified* "Nuts" shaken out of the dried fruits, rolled over hot coals, and eaten. **Yurok** *Unspecified* "Nuts" used for food. (as *Chrysolepis chrysophila* 6:24)

Castanopsis sempervirens (Kellogg) Dudley ex Merriam, Sierran Chinkapin
- *Food*—**Kawaiisu** *Unspecified* Raw seeds eaten by hunters when in the field. (as *Chrysolepis sempervirens* 206:20)

Castela, Simaroubaceae

Castela emoryi (Gray) Moran & Felger, Thorn of Christ
- *Drug*—**Yavapai** *Dermatological Aid* Milky fluid of pulverized buds rubbed on face to stop pimples. (as *Holocantha emoryi* 65:261)

Castilleja, Scrophulariaceae

Castilleja affinis Hook. & Arn., Indian Paintbrush
- *Drug*—**Costanoan** *Dermatological Aid* and *Disinfectant* Decoction of plant used as a wash or powdered plant applied to infected sores. (21:15)
- *Other*—**Hopi** *Decorations* Used by maidens to deck their hair on holiday occasions. (56:19)

Castilleja angustifolia (Nutt.) G. Don, Northwestern Indian Paintbrush
- *Drug*—**Navajo** *Gastrointestinal Aid* Plant used for stomach troubles. (55:96) **Quileute** *Abortifacient* Infusion of whole plant taken to regulate menstruation. (79:46) **Shuswap** *Eye Medicine* Decoction of roots, stems, and leaves used for weak or sore eyes. (123:69)

Castilleja angustifolia var. dubia A. Nels., Northwestern Indian Paintbrush
- *Drug*—**Navajo, Kayenta** *Dermatological Aid* Plant used for spider bites. (as *C. chromosa* 205:41)

Castilleja applegatei var. pinetorum (Fern.) N. Holmgren, Wavyleaf Indian Paintbrush
- *Food*—**Miwok** *Beverage* Flowers sipped sporadically and as a pastime. (as *Castilleia pinetorum* 12:163)

Castilleja coccinea (L.) Spreng., Scarlet Indian Paintbrush
- *Drug*—**Cherokee** *Poison* Infusion used "to destroy your enemies." (80:40) **Chippewa** *Cold Remedy* Infusion of flower taken for colds. *Orthopedic Aid* Simple or compound decoction of flowers used for paralysis. (53:362) **Menominee** *Love Medicine* Herb secreted onto the person who is the object of the enamor, a love charm. (151:81)

Castilleja foliolosa Hook. & Arn., Texas Indian Paintbrush
- *Food*—**Cahuilla** *Sweetener* Flowers picked by children to suck the nectar. (15:51)

Castilleja hispida Benth., Harsh Indian Paintbrush
- *Drug*—**Okanagan-Colville** *Dermatological Aid* Plant pounded up and put into moccasins for "sweaty feet." (188:138)

- *Food*—Nitinaht *Candy* Sweet nectar sucked by children. (186:127)
- *Other*—Nitinaht *Hunting & Fishing Item* Bright flowers covered with snail slime and used to trap hummingbirds. (186:127)

Castilleja integra Gray, Wholeleaf Indian Paintbrush
- *Drug*—Navajo *Burn Dressing* Plant used for burns. (90:159) *Gastrointestinal Aid* Infusion of crushed leaves taken for stomach troubles. (55:76) **Navajo, Ramah** *Blood Medicine* Compound decoction of root used to "clean out the blood" after internal injury. *Burn Dressing* Poultice of leaves applied to burns. *Gynecological Aid* Decoction of leaf taken during pregnancy to keep baby small, for easy labor. (191:43, 44)
- *Dye*—Apache, White Mountain *Unspecified* Root bark used with other substances to color various kinds of skins, especially deerskin. (136:156) **Zuni** *Black* Root bark used with minerals to color deerskin black. (166:80)
- *Other*—Jemez *Preservative* Dried bracts mixed with chile seeds to prevent spoilage during storage. (44:21) **Keres, Western** *Ceremonial Items* Plant held by women for decoration during the harvest dance. (171:35)

Castilleja linariifolia Benth., Wyoming Indian Paintbrush
- *Drug*—Hopi *Contraceptive* Decoction of plant used to prevent conception. (42:297) Decoction of plant used as a contraceptive. (200:35, 91) *Gynecological Aid* Decoction of plant used for excessive menstrual discharge. (42:297) Decoction of plant used to ease menstrual difficulties. (200:35, 91) **Navajo, Ramah** *Analgesic* Plant used for stomachaches. *Gastrointestinal Aid* Plant used for stomachache. *Gynecological Aid* Decoction of leaf taken during pregnancy to keep baby small, for easy labor. (191:44) **Shoshoni** *Blood Medicine* Decoction of root taken as a blood purifier. *Cathartic* Decoction of root taken as a physic. *Emetic* Decoction of root taken as an emetic. *Venereal Aid* Decoction of root taken over a long period of time for venereal disease. (180:53) **Tewa** *Contraceptive* Decoction of plant used to prevent conception. *Gynecological Aid* Decoction of plant used for excessive menstrual discharge. (42:297)
- *Food*—Hopi *Unspecified* Flowers eaten as food. (190:166)
- *Other*—Hopi *Ceremonial Items* Used ceremonially as the "Red Flower" associated with the southeast direction. (42:297) Used ceremonially as the Red Flower associated with the southeast direction. (200:91) *Decorations* Used by maidens to deck their hair on holiday occasions. (56:19) *Paint* Root and juniper bark chewed, mixed with white clay, and used as ceremonial paint. (42:297) Root chewed, mixed with white clay, and the juice used to decorate artificial squash blossoms. (200:91) **Tewa** *Ceremonial Items* Used ceremonially as the "Red Flower" associated with the southeast direction. *Paint* Root and juniper bark chewed, mixed with white clay, and used as ceremonial paint. (42:297) **Tewa of Hano** *Decorations* Red flowers painted on pottery and carved in wood as decorations. (138:54)

Castilleja lineata Greene, Marshmeadow Indian Paintbrush
- *Drug*—Navajo *Gastrointestinal Aid* Infusion of crushed, dried leaves taken for stomach troubles. (55:76)
- *Food*—Navajo *Special Food* Flowers sucked for the honey, a delicacy. (55:76)

Castilleja miniata Dougl. ex Hook., Scarlet Indian Paintbrush
- *Drug*—Gitksan *Antihemorrhagic* Decoction of entire plant taken for bleeding, stiff lungs. *Cathartic* Decoction of entire plant taken as a purgative. *Cough Medicine* Decoction of seeds taken for coughs. *Diuretic* Decoction of entire plant taken as a diuretic. *Eye Medicine* Decoction of entire plant taken for sore eyes. *Kidney Aid* Decoction of entire plant taken for lame back, perhaps from kidney trouble. *Orthopedic Aid* Decoction of entire plant taken for lame back, stiff lungs, and sore eyes. *Pulmonary Aid* Decoction of entire plant taken for bleeding, stiff lungs. (150:63) **Navajo, Ramah** *Hunting Medicine* Plant used with any witchcraft plant to protect hunters. *Witchcraft Medicine* Plant used in a drink and lotion as protection from witches. (191:44) **Thompson** *Unspecified* Broken plant parts used in the house for decoration or for medicine. (187:284)
- *Food*—Kwakiutl, Southern *Unspecified* Flower nectar formerly sucked by children. (183:292) **Nitinaht** *Candy* Sweet nectar sucked by children. (186:127) **Thompson** *Fodder* Plant used as hummingbird feed. (187:284)
- *Dye*—Apache, White Mountain *Unspecified* Root bark used with other substances to color various kinds of skins, especially deerskin. (136:156)
- *Other*—Bella Coola *Toys & Games* Flowers used in young girls' games. (184:210) **Nitinaht** *Hunting & Fishing Item* Bright flowers covered with snail slime and used to trap hummingbirds. (186:127) **Paiute** *Malicious Magic* Touching the plant would cause an unwanted storm. (111:110) **Thompson** *Decorations* Broken plant parts used in the house for decoration or for medicine. (187:284)

Castilleja miniata Dougl. ex Hook. **var. miniata**, Green Indian Paintbrush
- *Other*—Keres, Western *Ceremonial Items* Plant held by women for decoration during the harvest dance. (as *C. confusa* 171:35)

Castilleja minor (Gray) Gray, Lesser Indian Paintbrush
- *Dye*—Apache, White Mountain *Unspecified* Root bark used with other substances to color various kinds of skins, especially deerskin. (136:156)

Castilleja parviflora Bong., Mountain Indian Paintbrush
- *Drug*—Ute *Gastrointestinal Aid* Roots used for bowel troubles. (38:33)
- *Dye*—Apache, White Mountain *Unspecified* Root bark used with other substances to color various kinds of skins, especially deerskin. (as *Castileia parriflora* 136:156)
- *Other*—Karok *Toys & Games* Flowers used by children as pretend woodpecker scalps, a form of money or wealth. (148:389)

Castilleja parviflora var. *douglasii* Jepson, Mountain Indian Paintbrush
- *Food*—Miwok *Beverage* Flowers sipped sporadically and as a pastime. (12:163)

Castilleja sessiliflora Pursh, Downy Paintedcup
- *Drug*—Menominee *Dermatological Aid* Flowers and leaves macerated in bear grease and used as invigorating hair oil. (151:53)
- *Food*—Cheyenne *Unspecified* Flower nectar sucked in spring. (83:39)

Castilleja sp.
- *Drug*—Blackfoot *Antihemorrhagic* Infusion of plant taken for vagi-

nal bleeding (not menses) and rubbed on the abdomens. (86:69) Infusion of plant taken or chest rubbed when spitting up blood. (86:71) *Diuretic* Plant mixed with warm water and taken as a diuretic. (86:69) **Nevada Indian** *Love Medicine* Infusion of flowers used for love medicine. (118:50) **Thompson** *Unspecified* Broken plant parts used in the house for decoration or for medicine. (187:284)
- *Food*—**Miwok** *Winter Use Food* Dried seeds stored for winter use then parched, pounded, and eaten dry. (12:153) **Paiute** *Sweetener* Plant bases sucked for the sweetness. (167:246)
- *Dye*—**Blackfoot** *Unspecified* Flowers rubbed by children on bouncing arrows for color and shine. *Yellow* Flowers used as a yellow dye for arrow feathers. (86:111) **Great Basin Indian** *Red-Brown* Blossoms used to make a red-tan dye. (121:50)
- *Other*—**Blackfoot** *Waterproofing Agent* Flowers used to shine and waterproof hides. (86:111) **Thompson** *Decorations* Broken plant parts used in the house for decoration or for medicine. (187:284)

Castilleja stenantha Gray, Largeflower Indian Paintbrush
- *Drug*—**Kawaiisu** *Dermatological Aid* Decoction of leaves used as a wash for sores. (206:17)

Castilleja thompsonii Pennell, Thompson's Indian Paintbrush
- *Drug*—**Okanagan-Colville** *Dermatological Aid* Plant tops dried, powdered, and placed on open cuts to draw out the germs. (188:138)

Castilleja unalaschcensis (Cham. & Schlecht.) Malte, Alaska Indian Paintbrush
- *Other*—**Bella Coola** *Toys & Games* Flowers used in young girls' games. (184:210)

Catabrosa, Poaceae
Catabrosa aquatica (L.) Beauv., Water Whorlgrass
- *Drug*—**Crow** *Ceremonial Medicine* Burned as incense during certain ceremonies. **Montana Indian** *Ceremonial Medicine* Burned as incense during certain ceremonies. (as *Glyceria aquatica* 19:12) **Shoshoni** *Stimulant* Decoction of plant taken as a stimulant. *Tonic* Decoction of plant taken as a tonic. (180:53)
- *Food*—**Crow** *Unspecified* Seeds used for food. (as *Glyceria aquatica* 19:12) **Gosiute** *Unspecified* Seeds used for food. (as *Glyceria aquatica* 39:370)

Catharanthus, Apocynaceae
Catharanthus roseus (L.) G. Don, Madagascar Periwinkle
- *Drug*—**Hawaiian** *Blood Medicine* Bark and other plants pounded, the resulting liquid heated and taken to purify the blood. (as *Lochnera rosea* 2:51)

Catopsis, Bromeliaceae
Catopsis sp., Airplant
- *Other*—**Seminole** *Cooking Tools* Plant used as a water supply for cooking during the dry season. (169:472)

Caulanthus, Brassicaceae
Caulanthus coulteri S. Wats., Coulter's Wild Cabbage
- *Food*—**Kawaiisu** *Vegetable* Leaves gathered in early spring before the flowers appear, boiled, salted, fried in grease, and eaten. (206:17)

Caulanthus crassicaulis (Torr.) S. Wats., Thickstem Wild Cabbage
- *Drug*—**Shoshoni** *Blood Medicine* Infusion of root taken as a blood tonic. (180:53)

Caulanthus inflatus S. Wats., Desert Candle
- *Food*—**Kawaiisu** *Unspecified* Soft upper section of the stem roasted in a pit oven covered with dirt and eaten. (206:17)

Caulophyllum, Berberidaceae
Caulophyllum thalictroides (L.) Michx., Blue Cohosh
- *Drug*—**Cherokee** *Anticonvulsive* Syrup or decoction of root given for "fits and hysterics." *Antirheumatic (Internal)* Root used for rheumatism. *Dermatological Aid* Leaves rubbed on "oak-poison." *Gastrointestinal Aid* Taken for "colics and nerves" or root ooze held in mouth for toothache. *Gynecological Aid* Plant promoted childbirth and used for womb inflammation. *Sedative* Syrup or decoction of root given for "fits and hysterics." *Toothache Remedy* Root ooze held in mouth for toothache. (80:30) **Chippewa** *Analgesic* Compound decoction of root taken for cramps. (53:344) *Antihemorrhagic* Infusion of scraped root taken for lung hemorrhages. *Emetic* Infusion of scraped root taken as an emetic. (53:346) *Gastrointestinal Aid* Compound infusion of root taken for indigestion. (53:342) Compound decoction of root taken for stomach cramps. (53:344) Infusion of scraped root taken for biliousness. (53:346) *Pulmonary Aid* Decoction of root taken for lung trouble. (53:340) **Iroquois** *Antirheumatic (External)* Infusion of roots used as a foot and leg bath for rheumatism. *Antirheumatic (Internal)* Compound decoction taken for rheumatism. *Emetic* Infusion of smashed roots taken to vomit for gallstones. *Febrifuge* Decoction of roots taken for any kind of fever. *Liver Aid* Infusion of smashed roots taken to vomit for gallstones. *Tonic* Roots used as a tonic. (87:333) **Menominee** *Gynecological Aid* Decoction of root taken to suppress profuse menstruation. (151:25) **Meskwaki** *Gynecological Aid* Decoction of root taken for profuse menstruation. *Urinary Aid* Decoction of root taken by men as a genitourinary remedy. (152:205) **Mohegan** *Kidney Aid* Root, "very rare," used for kidney disorders. (176:71, 128) **Ojibwa** *Analgesic* Root used for stomach cramps accompanying painful menstruation. *Emetic* Decoction of root taken as an emetic. *Gynecological Aid* Root used for stomach cramps accompanying painful menstruation. (153:358) **Omaha** *Febrifuge* Plant used as a fever medicine. (68:335) Decoction of root, considered highly effective, given for fevers. **Ponca** *Febrifuge* Decoction of root given for fevers. (70:83) **Potawatomi** *Gynecological Aid* Infusion of root taken to suppress profuse menstruation and aid in childbirth. (154:43)

Ceanothus, Rhamnaceae
Ceanothus americanus L., New Jersey Tea
- *Drug*—**Alabama** *Orthopedic Aid* Decoction of root used as a wash for injured legs or feet. (172:664) Decoction of roots used as a bath for injured feet and legs. (177:40) **Cherokee** *Gastrointestinal Aid* Infusion of root taken for "bowel complaint." *Toothache Remedy* Infusion of root held on aching tooth. (80:46) **Chippewa** *Gastrointestinal Aid* Infusion of roots taken for constipation with bloating and shortness of breath. *Laxative* Infusion of roots taken for constipation with bloating and shortness of breath. *Pulmonary Aid* Infusion of roots taken for pulmonary troubles. *Respiratory Aid* Infusion of roots taken for constipation with bloating and shortness of breath. (71:136)

Iroquois *Abortifacient* Decoction of roots taken for suppressed menses from catching cold. Decoction taken as an abortifacient when fetus is hurt within 2 or 3 months. (87:381) *Antidiarrheal* Compound decoction of plants taken for diarrhea. *Blood Medicine* Infusion of roots taken for the blood. *Cold Remedy* Infusion of roots taken for colds. (87:382) *Dermatological Aid* Powdered bark applied to open sores caused by venereal disease. (87:381) *Misc. Disease Remedy* Compound decoction of plants taken for sugar diabetes. *Oral Aid* Decoction of bark used as a wash for sore roof of the mouth. *Other and Pediatric Aid* Compound decoction of dried roots given to children with "summer complaint." (87:382) *Urinary Aid* Decoction of roots taken by women with urinating problems caused by colds. *Venereal Aid* Decoction of roots taken for venereal disease. (87:381) **Menominee** *Cough Medicine* Decoction of root taken for cough with a "tendency to consumption." (54:130) *Gastrointestinal Aid* Infusion of roots used as a cure-all for stomach troubles. (151:49) **Meskwaki** *Antidiarrheal* Boiled root chewed as main remedy for flux. *Dermatological Aid* Root and bark were strongly astringent. *Gastrointestinal Aid* Root, strongly astringent with great powers, used for bowel troubles. *Snakebite Remedy* Root used for snakebite. (152:240, 241)

- *Food*—**Dakota** *Beverage* Leaves used to make a tea-like beverage. (70:102) **Menominee** *Beverage* and *Substitution Food* Dried leaves used as a substitute for Ceylon black tea. (151:70) **Meskwaki** *Beverage* Leaves used as a beverage. (152:263) **Omaha** *Beverage* Leaves used to make tea. (58:342) Leaves used to make a hot, aqueous, tea-like beverage. (68:329) Leaves used to make a tea-like beverage. **Pawnee** *Beverage* Leaves used to make a tea-like beverage. **Ponca** *Beverage* Leaves used to make a tea-like beverage. **Winnebago** *Beverage* Leaves used to make a tea-like beverage. (70:102)
- *Other*—**Dakota** *Fuel* Woody roots used as fuel on the buffalo hunt during scarcities of timber. **Omaha** *Fuel* Woody roots used as fuel on the buffalo hunt during scarcities of timber. **Pawnee** *Fuel* Woody roots used as fuel on the buffalo hunt during scarcities of timber. **Ponca** *Fuel* Woody roots used as fuel on the buffalo hunt during scarcities of timber. **Winnebago** *Fuel* Woody roots used as fuel on the buffalo hunt during scarcities of timber. (70:102)

Ceanothus cuneatus (Hook.) Nutt., Buck Brush

- *Food*—**Mendocino Indian** *Forage* Leaves eaten by deer as forage. Seeds eaten by squirrels as forage. (41:367)
- *Fiber*—**Mewuk** *Basketry* Rods used to make burden baskets, broad shallow scoops, and deep spoon-shaped scoops with handles. (117:328) Used as rods for basketry. (117:329)
- *Other*—**Kawaiisu** *Fuel* Wood used for firewood. *Hunting & Fishing Item* Twigs used as foreshafts for the two-piece arrows. Straight stems were used by removing the leaves and bark and sharpening one end. The piece would then have been fitted into a section of hollow carizzo grass or "cane." (206:17) **Mendocino Indian** *Hunting & Fishing Item* Brushes used to build fish dams. (41:367) **Mewuk** *Fuel* Wood used for fuel. (117:339) *Tools* Rods used to make burden baskets, broad shallow scoops, and deep spoon-shaped scoops with handles. (117:328) **Modesse** *Tools* Wood made into the slender needle used in piercing the ear lobe of young girls. (117:223) **Paiute** *Tools* Wood used to make digging sticks. (167:244) **Tubatulabal** *Tools* Sharpened twigs used to pierce roasted pinyons. (193:17)

Ceanothus fendleri Gray, Fendler's Ceanothus

- *Drug*—**Keres, Western** *Oral Aid* Leaves chewed for sore mouth. (171:35) **Navajo** *Sedative* Compound infusion taken and poultice of plants applied for nervousness. (55:62) **Navajo, Kayenta** *Ceremonial Medicine* and *Emetic* Plant used as a Plumeway emetic. (205:31) **Navajo, Ramah** *Ceremonial Medicine* and *Emetic* Leaves and stems used as an emetic in various ceremonies. (191:36)
- *Food*—**Acoma** *Fruit* Berries sweetened with sugar and used for food. (32:21) **Keres, Western** *Fruit* Berries sweetened with sugar and used for food. (171:35) **Laguna** *Fruit* Berries sweetened with sugar and used for food. (32:21) **Navajo, Ramah** *Unspecified* Inner bark strips eaten in summer. (191:36)

Ceanothus greggii Gray, Desert Ceanothus

- *Other*—**Yavapai** *Fuel* Branches used for kindling. (65:259)

Ceanothus griseus (Trel. ex B. L. Robins.) McMinn, Carmel Ceanothus

- *Other*—**Pomo, Kashaya** *Ceremonial Items* Flowers used in dance wreaths at the Strawberry Festival. *Soap* Fresh or dried flowers mixed with water and used as a soap for washing hands, face, and body. (72:23)

Ceanothus herbaceus Raf., Jersey Tea

- *Drug*—**Chippewa** *Cough Medicine* Decoction of root taken as a cough remedy. (as *C. ovatus* 53:340)
- *Food*—**Lakota** *Beverage* Leaves used to make tea. (139:56)

Ceanothus integerrimus Hook. & Arn., Deerbrush

- *Drug*—**Karok** *Gynecological Aid* Plant used by women who have suffered an injury in childbirth. (148:386)
- *Food*—**Concow** *Staple* Seeds eaten as a pinole. (41:368) **Karok** *Forage* Plant eaten by deer. (148:386)
- *Fiber*—**Concow** *Basketry* Young, flexible shoots used for the circular withes of baskets. (41:368) **Karok** *Basketry* New shoots used to make baskets. (6:22) Young shoots used to make baskets. (148:386) **Maidu** *Basketry* Withes used as overlay twine weft bases in the manufacture of baskets. (173:71) **Mewuk** *Basketry* Rods used to make burden baskets, broad shallow scoops, and deep spoon-shaped scoops with handles. Rods used in the fine, coiled baskets. (117:328)
- *Other*—**Mewuk** *Tools* Rods used to make burden baskets, broad shallow scoops, and deep spoon-shaped scoops with handles. (117:328)

Ceanothus leucodermis Greene, Chaparral Whitethorn

- *Drug*—**Diegueño** *Dermatological Aid* Leaves picked when only the leaves were out, boiled, and used as a wash for itch. Leaves and cascara leaves boiled and used for poison oak. Decoction of berries, whole branch with berries or leaves used as bath for itch, sores, or impetigo. Blossom, leaf, or berry sap used by rubbing area affected by itch, sores, or impetigo. (84:15)
- *Other*—**Kawaiisu** *Soap* Viscid fruits dipped into water by children and used as soap by rubbing them between the hands. (206:18)

Ceanothus oliganthus Nutt. **ssp. *oliganthus***, Explorer's Bush

- *Other*—**Mahuna** *Protection* Trees used for protection from lightning. (as *C. divaricatus* 140:40) **Pomo** *Tools* Wood used to make the fire drill. (as *C. divaricatus* 117:288)

Ceanothus sanguineus Pursh, Redstem Ceanothus
- **Drug**—**Okanagan-Colville** *Burn Dressing* Poultice of dried, powdered bark applied to burns. (188:119) **Sanpoil** *Dermatological Aid* Poultice of "sap wood" sprinkled on grease or oil applied to sores or wounds. (131:217)
- **Food**—**Okanagan-Colville** *Forage* Buds and branches considered an important food for deer. (188:119)
- **Other**—**Okanagan-Colville** *Fuel* Wood used for fuel. *Preservative* Wood used to smoke deer meat. (188:119)

***Ceanothus* sp.**, California Lilac
- **Drug**—**Choctaw** *Antihemorrhagic* Decoction of roots taken in small doses for lung hemorrhage. (25:24) Decoction of roots taken for lung hemorrhages. (177:40) *Pulmonary Aid* Decoction of roots taken in small doses for "hemorrhage from the lungs." (25:24) Decoction of roots taken for lung hemorrhages. (177:40) **Costanoan** *Dermatological Aid* Decoction of plant used as a wash for facial blemishes and the hair. (21:22)
- **Other**—**Cahuilla** *Fuel* Used for firewood. (15:51) **California Indian** *Soap* Flowers used to make soapsuds. (118:57) **Costanoan** *Soap* Decoction used as a detergent. (21:250) **Mewuk** *Cooking Tools* Bark used as a filter to leach the bitter out of acorn meal. (117:362) **Paiute** *Smoke Plant* Dried, mashed leaves mixed with tobacco. (111:89)

Ceanothus thyrsiflorus Eschsch., Blueblossom Ceanothus
- **Drug**—**Poliklah** *Pediatric Aid* Decoction of leaves and twigs used to wash newborn babies. (117:173)
- **Other**—**Pomo, Kashaya** *Ceremonial Items* Flowers used in dance wreaths at the Strawberry Festival. *Soap* Fresh or dried flowers mixed with water and used as a soap for washing hands, face, and body. (72:23)

Ceanothus velutinus Dougl. ex Hook., Snowbrush Ceanothus
- **Drug**—**Great Basin Indian** *Other* Infusion of leaves taken for diagnosis and certain results mean certain things. (121:49) **Karok** *Dermatological Aid* Leaves used as a deodorant. (148:386) **Modesse** *Cough Medicine* Infusion of leaves taken for coughs. *Febrifuge* Infusion of leaves taken for fevers. (117:223) **Okanagan-Colville** *Ceremonial Medicine* Decoction of plant tops with leaves used as a cleansing solution in the sweat house. *Dermatological Aid* Decoction of plant tops with leaves used as a hair wash for dandruff, used to bathe babies to prevent diaper rash, and used to "condition" adult skin. Infusion of branches used to wash sores and eczema. Poultice of dried, powdered leaves applied or mixed with pitch and used as a salve for sores. Poultice of dried, powdered leaves used as a "baby powder." *Orthopedic Aid* Infusion of leaves taken for broken bones. *Pediatric Aid* Decoction of plant tops with leaves used to bathe babies to prevent diaper rash. Poultice of dried, powdered leaves used as a "baby powder." (188:120) **Okanagon** *Analgesic* Decoction of stems and leaves used internally and externally for dull pains. (125:40) **Oregon Indian, Warm Springs** *Unspecified* Infusion of leaves taken for puzzling illnesses. (118:40) **Shuswap** *Misc. Disease Remedy* Decoction of plants taken for the flu. (123:65) **Thompson** *Analgesic* Decoction of stems and leaves used internally and externally for dull pains. (125:40) Decoction of stem and leaf taken and used as a wash for dull, body pains. (164:457) *Antidiarrheal* Infusion of leaves and twigs used for diarrhea. *Antirheumatic (External)* Decoction of branches used as a wash for rheumatism. Decoction of leaves used as a bath or leaves used in a steam bath for rheumatism or arthritis. Infusion of leaves and twigs used for arthritis. *Antirheumatic (Internal)* Decoction of plant taken for arthritis. *Cancer Treatment* Decoction of plant used for cancer. *Dietary Aid* Decoction of branches taken for weight loss. *Orthopedic Aid* Infusion of leaves and twigs with Indian hellebore used for broken limbs. *Panacea* Decoction of branches taken for general illness. *Unspecified* Decoction of branches taken for an unspecified ailment. Plant used in sweat bath for an unspecified illness. (187:252) *Venereal Aid* Compound decoction of branches taken for mild forms of gonorrhea. (164:457)
- **Food**—**Okanagan-Colville** *Forage* Bush eaten by deer. (188:120) **Paiute** *Forage* Plant eaten by deer. (111:89) **Thompson** *Forage* Shrub extensively eaten by deer. (164:516) Plant considered a favorite food of deer. (187:252)
- **Other**—**Shuswap** *Insecticide* Smoke from plant used to kill bedbugs. (123:65) **Thompson** *Soap* Infusion of leaves and twigs with Indian hellebore used as a wash for bathing. (187:252)

Celastraceae, see *Canotia, Celastrus, Euonymus, Mortonia, Paxistima*

***Celastrus*,** Celastraceae

Celastrus scandens L., American Bittersweet
- **Drug**—**Cherokee** *Analgesic* Strong compound infusion used for pain of childbirth. *Antirheumatic (External)* Thorny branch used to scratch rheumatism. *Cough Medicine* Root chewed for cough. *Dermatological Aid* Decoction of highly astringent leaves taken for bowel complaint. Used as wash for "foul ulcers." *Gastrointestinal Aid* Infusion of bark used to settle stomach and decoction given for bowel complaint. *Gynecological Aid* Strong infusion combined with red raspberry leaves and used for childbirth pains. (80:25) **Chippewa** *Cancer Treatment* Boiled roots used as an ointment for cancer. (71:135) *Cathartic* Decoction of root used, especially for babies, as a physic. (53:344) *Dermatological Aid* Decoction of stalk applied to skin "eruptions." (53:350) Boiled roots used as an ointment for any obstinate sore. (71:135) *Diuretic* Decoction of root taken for "stoppage of urine." (53:348) *Pediatric Aid* Decoction of root used, especially for babies, as a physic. (53:344) **Creek** *Analgesic, Gynecological Aid,* and *Orthopedic Aid* Plant used by women with urinary trouble or pain in small of back. *Urinary Aid* Plant used by women with urinary trouble. (172:661) **Delaware** *Dermatological Aid* Poultice or salve of roots used for skin eruptions. *Liver Aid* Infusion of roots used to clear up liver spots. (176:37) **Delaware, Ontario** *Tuberculosis Remedy* Root taken for consumption. (175:66, 82) **Iroquois** *Abortifacient* Decoction of roots taken by young girls who catch cold and do not menstruate. Infusion of leaves and stems taken as a regulator by women. *Blood Medicine* Compound decoction of plants taken to make blood or for watery blood. (87:376) Infusion of root bark with another plant and wine taken for anemia. (141:54) *Cold Remedy* Decoction of roots taken by young girls who catch cold and do not menstruate. *Diuretic* Infusion of leaves and stems taken as a diuretic. *Febrifuge* Infusion of leaves and stems taken for fever and soreness from pregnancy. *Gynecological Aid* Infusion of leaves and stems taken for fever and soreness from pregnancy. *Kidney Aid* Compound

decoction of roots and bark taken for dropsy or watery blood. Infusion of roots taken for kidney trouble following childbirth. *Other* Decoction of roots used as a wash on lips of bad children. *Pediatric Aid* Decoction of roots used as wash on lips or gums of bad or teething children. *Poison* Berries considered poisonous. *Toothache Remedy* Decoction of roots used as a wash on lips and gums of teething child. *Urinary Aid* Infusion of leaves and stems taken for urine stoppage. (87:376) **Meskwaki** *Analgesic* and *Gynecological Aid* Compound containing root used for "the relief of women in labor." (152:208, 209) **Oglala** *Poison* Plant considered poisonous. (70:102) **Ojibwa** *Gastrointestinal Aid* Berries used for stomach trouble. (153:362) **Unspecified** Plant used for medicinal purposes. (135:233)
- *Food*—**Menominee** *Starvation Food* Palatable inner bark would sustain life when food was hard to get. (151:63) **Ojibwa** *Soup* Inner bark used to make a thick soup when other food unobtainable in the winter. The Ojibwe name of the bittersweet is "manidobima' kwit," which means "spirit twisted" and "refers to the twisted intestines of their culture hero, Winabojo." (153:398) **Potawatomi** *Starvation Food* Inner bark cooked in times of food scarcity, not highly commended as a food but valued. (154:97)
- *Other*—**Lakota** *Paint* Roots chewed and smeared on the body and be impervious to wounding. (139:43)

Celtis, Ulmaceae

Celtis laevigata Willd., Sugarberry
- *Drug*—**Houma** *Throat Aid* Decoction of bark taken for sore throat. *Venereal Aid* Compound decoction of bark with powdered shells taken for venereal disease. (158:57)
- *Food*—**Comanche** *Fruit* Fruits beaten to a pulp, mixed with fat, rolled into balls, and roasted over fire. (29:521) **Seminole** *Unspecified* Plant used for food. (169:489)
- *Other*—**Seminole** *Tools* Plant used to make squirting tubes. (169:489)

Celtis laevigata var. *brevipes* (S. Wats.) Sarg., Sugarberry
- *Food*—**Yavapai** *Unspecified* Ground, boiled, and used for food. (65:256)

Celtis laevigata var. *reticulata* (Torr.) L. Benson, Netleaf Hackberry
- *Drug*—**Navajo, Kayenta** *Gastrointestinal Aid* Plant used for indigestion. (as *C. reticulata* 205:18)
- *Food*—**Acoma** *Fruit* Berries extensively used as food. (as *C. reticulata* 32:21) **Apache, Chiricahua & Mescalero** *Bread & Cake* Fruit ground, caked, and dried for winter use. *Fruit* Fruit eaten fresh. *Preserves* Fruit used to make jelly. (as *C. reticulata* 33:46) **Hualapai** *Dried Food* Fruit dried for winter use. *Fruit* Fruit eaten fresh. (as *C. reticulata* 195:6) **Laguna** *Fruit* Berries extensively used as food. (as *C. reticulata* 32:21) **Navajo** *Fruit* Berries ground and eaten. (as *C. reticulata* 55:41) **Papago** *Fruit* Fruits eaten for food. **Pueblo** *Fruit* Berries used for food. (as *C. reticulata* 32:21) **Tewa** *Fruit* Berries eaten. (as *C. reticulata* 138:39)
- *Fiber*—**Papago** *Clothing* Bark used to make sandals. (as *C. reticulata* 34:49)
- *Dye*—**Navajo** *Red-Brown* Leaves and branches boiled into a dark brown or red dye for wool. (as *C. reticulata* 55:41)
- *Other*—**Havasupai** *Fuel* Wood used for firewood. (as *C. reticulata* 197:215) **Navajo** *Tools* Wood used to make tubes for bellows. (as *C. reticulata* 55:41) **Tewa** *Tools* Wood used to make handles for axes and hoes. (as *C. reticulata* 138:39)

Celtis occidentalis L., Common Hackberry
- *Drug*—**Houma** *Throat Aid* Decoction of bark taken for sore throat. *Venereal Aid* Compound decoction of bark with powdered shells taken for venereal disease. (158:57) **Iroquois** *Abortifacient* Decoction taken "for suppressed menses in girls, cause: working in the sun." *Cold Remedy* Compound decoction taken by "women when they catch cold with the menses." *Gynecological Aid* Decoction of bark used as "woman's medicine" and regulated menses. (87:306) **Meskwaki** *Veterinary Aid* Inner bark fed to ponies as a conditioner. (152:250)
- *Food*—**Dakota** *Spice* Dried fruit pounded to make a condiment used for seasoning meat in cooking. (69:362) Berries used to flavor meat. (70:76) **Keres, Western** *Fruit* Berries used extensively for food. (171:35) **Meskwaki** *Porridge* Ground, hard berries made into a mush. (152:265) **Omaha** *Fruit* Berries used occasionally for food. **Pawnee** *Fruit* Berries pounded fine, mixed with a little fat, and parched corn and used for food. (70:76)
- *Other*—**Kiowa** *Fuel* Wood used as fuel for the altar fire in the peyote ceremony. (192:22)

Celtis occidentalis L. var. *occidentalis*, Western Hackberry
- *Food*—**Kiowa** *Fruit* Berries pounded into a paste-like consistency, molded onto a stick, and baked over an open fire. (192:23)

Cenchrus, Poaceae

Cenchrus calyculatus, Ka-mano-mano
- *Drug*—**Hawaiian** *Dermatological Aid* Shoots, leaves, roots, and other plants pounded and resulting liquid used on fresh, deep cuts. *Tuberculosis Remedy* Shoots, leaves, roots, and other plants pounded and resulting liquid used on scrofulous sores. (2:48)

Centaurea, Asteraceae

Centaurea americana Nutt., American Star Thistle
- *Drug*—**Kiowa** *Dermatological Aid* Poultice of leaves applied to skin sores. (192:58)

Centaurea melitensis L., Maltese Star Thistle
- *Drug*—**Mahuna** *Kidney Aid* Plant used for the kidneys. (140:69)

Centaurea sp., Bachelor Buttons
- *Drug*—**Rappahannock** *Dermatological Aid* Bruised leaves used as a salve for sores. (161:33)

Centaurium, Gentianaceae

Centaurium exaltatum (Griseb.) W. Wight ex Piper, Desert Centaury
- *Drug*—**Miwok** *Analgesic* Decoction of stems and leaves taken for internal pains. *Gastrointestinal Aid* Decoction of stems and leaves taken for stomachache. *Toothache Remedy* Decoction of stems and leaves taken for toothaches. *Tuberculosis Remedy* Decoction of stems and leaves taken for consumption. (12:168)

Centaurium muehlenbergii (Griseb.) W. Wight ex Piper, Muhlenberg's Centaury
- ***Drug***—**Mahuna** *Febrifuge, Gastrointestinal Aid,* and *Laxative* Infusion of plants taken for constipation caused by stomach fevers. (as *Erythaea muehlenbergii* 140:8)

Centaurium venustum (Gray) B. L. Robins., Charming Centaury
- ***Drug***—**Luiseño** *Febrifuge* Infusion of plant taken for fevers. (as *Erythraea venusta* 155:230) **Miwok** *Febrifuge* Decoction of flowers and leaves taken for fever. *Misc. Disease Remedy* Decoction of flowers and leaves taken for ague. *Pulmonary Aid* Decoction of flowers, leaves, and brandy taken for pneumonia. (12:168)

Cephalanthus, Rubiaceae

Cephalanthus occidentalis L., Common Buttonbush
- ***Drug***—**Chickasaw** *Eye Medicine* Poultice of warmed roots applied to the head for eye troubles. (177:58) **Choctaw** *Antidiarrheal* Strong decoction of tree bark taken as a favorite medicine for dysentery. (28:287) *Eye Medicine* Decoction of bark used as wash for sore eyes. (25:24) Decoction of bark used as a bath for sore eyes. (177:58) *Febrifuge* Root bark and bark used as a febrifuge. *Tonic* Root bark and bark used as a tonic. (28:287) *Toothache Remedy* Bark chewed for tooth ache. (25:24) Bark chewed for toothaches. (177:58) **Kiowa** *Antihemorrhagic* Decoction of roots taken for hemorrhages. (192:51) **Koasati** *Antirheumatic (Internal)* Decoction of leaves taken for rheumatism. *Orthopedic Aid* Decoction of roots taken for enlarged muscles. (177:58) **Meskwaki** *Emetic* Inner bark, very important medicine, used as an emetic. (152:243) **Seminole** *Analgesic* Decoction of bark taken for headaches. (169:283) *Antidiarrheal* Decoction of plant taken for wolf ghost sickness: diarrhea and painful defecation. (169:228) *Antiemetic* Decoction of roots or berries used for horse sickness: nausea, constipation, and blocked urination. (169:189) *Blood Medicine* Decoction of roots taken for menstruation sickness: yellow eyes and skin, weakness, and shaking head.[13] (169:247) *Cathartic* Decoction of plant taken for wolf ghost sickness: diarrhea and painful defecation. (169:228) *Febrifuge* Decoction of bark taken for fevers. *Gastrointestinal Aid* Decoction of bark taken for stomachaches. (169:283) *Laxative* Decoction of roots or berries used for horse sickness: nausea, constipation, and blocked urination. (169:189) *Other* and *Strengthener* Decoction of roots taken for menstruation sickness: yellow eyes and skin, weakness, and shaking head.[13] (169:247) *Unspecified* Plant used for medicinal purposes. (169:162) *Urinary Aid* Decoction of roots or berries used for horse sickness: nausea, constipation, and blocked urination. (169:189) Plant taken for urine retention. (169:273)
- ***Other***—**Comanche** *Toys & Games* Wood used to make game sticks. (29:521)

Cerastium, Caryophyllaceae

Cerastium arvense L., Field Chickweed
- ***Drug***—**Iroquois** *Dermatological Aid* Decoction of plant used as an astringent. *Gynecological Aid* Decoction of plant taken for injuries and miscarriage. Decoction taken to "stop bleeding and stops child from passing through uterus." (87:317)

Cerastium beeringianum Cham. & Schlecht., Bering Chickweed
- ***Drug***—**Navajo, Ramah** *Veterinary Aid* Cold infusion of plant used for sheep or horses with eye troubles. (191:26)

Cerastium fontanum ssp. ***vulgare*** (Hartman) Greuter & Burdet, Big Chickweed
- ***Drug***—**Cherokee** *Anthelmintic* and *Pediatric Aid* Compound infusion of stem and root given to children for worms. (as *C. holosteoides* 80:29)

Cerasus, Rosaceae

Cerasus crenulata Greene, Wild Plum
- ***Dye***—**Navajo** *Purple* Roots used to color wool purple. (55:52)

Cercis, Fabaceae

Cercis canadensis L., Eastern Redbud
- ***Drug***—**Alabama** *Febrifuge* Cold infusion of roots and inner bark taken for fever. (177:31) *Pulmonary Aid* Infusion of root and inner bark taken for congestion. (172:665) *Respiratory Aid* Cold infusion of roots and inner bark taken for congestion. (177:31) **Cherokee** *Pulmonary Aid* Infusion of bark given for whooping cough. (80:52) **Delaware** *Antiemetic* Infusion of bark used as a cold drink for vomiting. *Febrifuge* Infusion of bark used as a cold drink for fever. (176:30) **Delaware, Oklahoma** *Antiemetic* Infusion of bark taken for vomiting. *Febrifuge* Infusion of bark taken for fever. (175:25, 74)
- ***Food***—**Cherokee** *Unspecified* Blossoms eaten by children. (203:74)
- ***Other***—**Kiowa** *Fuel* Slender stems used for fuel during the winter. *Season Indicator* Flower welcomed as a sign of spring. This plant was esteemed because it is one of the earliest shrubs to flower in the spring. The red or pink flowers that appear before the leaves gave rise to their saying "pink flowers form into leaves." Flowering branches were used in the homes to "drive winter out." (192:32)

Cercis canadensis var. ***texensis*** (S. Wats.) M. Hopkins, California Redbud
- ***Drug***—**Mendocino Indian** *Febrifuge* Bark used for chills and fever. (as *C. occidentalis* 41:356)
- ***Food***—**Navajo** *Unspecified* Pods roasted in ashes and seeds eaten. (as *C. occidentalis* 32:21) Seeds roasted and eaten. (as *C. occidentalis* 55:56) **Navajo, Kayenta** *Unspecified* Pods roasted in ashes and seeds eaten. (as *C. occidentalis* 205:28)
- ***Fiber***—**Havasupai** *Building Material* Wood used to make fence posts. (as *C. occidentalis* 197:226) **Maidu** *Basketry* Peeled withes used as coarse twine and coil thread in the manufacture of baskets. (as *Cercus occidentalis* 173:71) **Mendocino Indian** *Basketry* Wood used for withes in constructing basket skeletons. *Cordage* Bark and wood of young sprouts used like thread or woof to twine in and out of twined baskets. (as *Cercis occidentalis* 41:356) **Mewuk** *Basketry* Used as the outside strands in coiled basketry. (as *C. occidentalis* 117:328) **Modesse** *Basketry* Used for the red design in baskets. (as *C. occidentalis* 117:223) **Neeshenam** *Basketry* Wood used for the woof in basket making. (as *C. occidentalis* 129:375) **Pomo** *Basketry* Red bark used as design material for baskets. White, inner bark used in basketry. (as *C. occidentalis* 9:138) Split strands used for basket body material. (as *C. occidentalis* 117:296) **Pomo, Kashaya** *Basketry* Strips of switch bark used for brown design or bark peeled to show the white used to make baskets. (as *C. occidentalis* 72:96) **Yuki** *Basketry* Red bark used as design material on twined basketry. (as *C. occidentalis* 9:138) Used as

basket material. *Sewing Material* Wood, sapwood, and roots used for sewing material. (as *C. occidentalis* 103:423)
- *Other*—Havasupai *Hunting & Fishing Item* Wood used to make bows. *Tools* Wood used to make tool handles. (as *C. occidentalis* 197:226) **Maidu** *Decorations* Unpeeled withes used as decorative coil thread in the manufacture of baskets. (as *Cercus occidentalis* 173:71) **Navajo** *Incense & Fragrance* Leaves used as an incense in the Mountain Chant. (as *Cercis occidentalis* 55:56)

Cercis sp., Redbud
- *Fiber*—Wintoon *Basketry* Branches used in the making and decorating of baskets. (117:275)

Cercocarpus, Rosaceae

Cercocarpus intricatus S. Wats., Littleleaf Mountain Mahogany
- *Other*—Hopi *Ceremonial Items* and **Tewa** *Ceremonial Items* Plant used during midwinter ceremonial to make prayer sticks. (42:299)

Cercocarpus ledifolius Nutt., Curlleaf Mountain Mahogany
- *Drug*—Gosiute *Burn Dressing* Poultice of powdered green wood applied to burns. (39:350) Charred wood powder applied to burns. (39:365) **Kawaiisu** *Ear Medicine* Dried, powdered plant exudation applied for earaches. Dried exudation ground into a powder and applied to earaches. *Gynecological Aid* Decoction of bark and leaves taken for "women's disease." *Venereal Aid* Decoction of bark taken for gonorrhea. (206:18) **Paiute** *Analgesic* Decoction of bark taken for stomachaches. *Antidiarrheal* Compound infusion of scraped bark given to children for diarrhea. Decoction of bark taken for diarrhea. *Blood Medicine* Cold decoction of bark taken as a blood tonic. *Burn Dressing* Powder or paste of bark or wood applied to burns. (180:53–55) *Cold Remedy* Infusion of bark taken for colds. (118:38) Decoction or infusion of dried bark or leaves taken for colds and coughs. *Cough Medicine* Decoction of bark or infusion of bark or leaves taken for coughs and colds. *Dermatological Aid* Powder or paste of bark or wood applied to sores, cuts, or wounds. *Gastrointestinal Aid* Decoction of bark taken for stomachaches and stomach ulcers. *Heart Medicine* Decoction of leaves or bark taken for heart disorders. *Pediatric Aid* Compound infusion of scraped bark given to children for diarrhea. *Pulmonary Aid* Decoction of bark taken for pneumonia. *Tuberculosis Remedy* Simple or compound decoction of dried bark used for tuberculosis. *Venereal Aid* Decoction of bark taken for venereal diseases. Pulverized wood sprinkled on syphilitic sores. (180:53–55) **Paiute, Northern** *Antihemorrhagic* Decoction of dried bark taken for spitting up blood. *Tuberculosis Remedy* Decoction of dried bark taken for tuberculosis. (59:129) **Shoshoni** *Antidiarrheal* Compound infusion of scraped bark given to children for diarrhea. *Blood Medicine* Cold decoction of bark taken as a blood tonic. *Burn Dressing* Powder or paste of bark or wood applied to burns. *Cold Remedy* Decoction of bark taken for colds. *Cough Medicine* Decoction of bark taken for coughs. *Dermatological Aid* Poultice of pulverized leaves and bark applied to swellings. Powder or paste of bark or wood applied to sores, cuts, or wounds. *Eye Medicine* Strained decoction of inner bark used as a wash for eye diseases. *Heart Medicine* Decoction of leaves or bark taken for heart disorders. *Misc. Disease Remedy* Infusion of inner bark taken for diphtheria. *Pediatric Aid* Compound infusion of scraped bark given to children for diarrhea. (180:53–55) *Tuberculosis Remedy* Compound decoction of bark taken for tuberculosis. (180:122) *Unspecified* Decoction of soft inner bark taken for unspecified purpose. *Venereal Aid* Compound decoction of bark taken as an "unfailing cure for syphilis." (180:53–55)
- *Dye*—Havasupai *Red* Inner bark used as a red dye for buckskin. (197:222)
- *Other*—Gosiute *Hunting & Fishing Item* Wood used to make bows. (39:365) **Klamath** *Hunting & Fishing Item* Wood used for the heads of fish spears. *Tools* Used as a root digger or camas stick. Wood used to make root diggers or camas sticks. (45:98) **Montana Indian** *Hunting & Fishing Item* Hard, enduring wood used for making fish spear heads. *Tools* Hard, enduring wood used for making "camas sticks." (19:9) **Shoshoni** *Hunting & Fishing Item* Wood used to make arrow tips. (118:52)

Cercocarpus montanus Raf., True Mountain Mahogany
- *Drug*—Keres, Western *Strengthener* Infusion of leaves used as a strengthener. (171:35) **Navajo** *Gastrointestinal Aid* Roots and bark used for stomach troubles. (55:53) **Navajo, Ramah** *Gastrointestinal Aid* Compound decoction of leaves taken and used as lotion for sickness from overeating. *Gynecological Aid* Decoction of plant used to hasten postpartum recovery. *Hunting Medicine* Leaves from shrubs browsed by deer chewed by hunter for good luck in hunting. *Panacea* Root used as a "life medicine." (191:30) **Tewa** *Laxative* Cold infusion of plant or leaves taken as a laxative. (138:45)
- *Food*—Navajo *Forage* Whole plant used by sheep for forage. (55:53)
- *Fiber*—Keres, Western *Brushes & Brooms* Bunches of tied bushes used for rough brooms. (171:35) **Navajo** *Building Material* Wood used to make the handle of the weaving distaff, dice, and the sweat house for ceremonies. (55:53)
- *Dye*—Isleta *Red* Root bark, alder root bark, and wild plum root bark used to make a red dye for buckskin. (100:25) **Jemez** *Red* Bark, alder bark, and birch bark boiled together and used as red dye to paint moccasins. (44:20) **Keres, Western** *Red* Roots used as a red dye for buckskin. (171:35) **Navajo, Ramah** *Brown* Decoction of root bark used as a brown dye for buckskin and wool. *Red* Used as a red dye for baskets. (191:30)
- *Other*—Keres, Western *Containers* Dried root used as a hearth for fires. *Hide Preparation* Bark used to tan buckskin. *Hunting & Fishing Item* Wood made into arrow points. *Tools* Wood made into small tools. Dried sticks used as spindles for fire by friction. (171:35) **Navajo** *Ceremonial Items* Wood used to make the sweat house and male prayer sticks for ceremonies. *Tools* Wood used to make the handle of the weaving distaff. *Toys & Games* Wood used to make dice and the sweat house for ceremonies. (55:53) **Navajo, Ramah** *Ceremonial Items* Wood made into stirring sticks for Chiricahua Windway mixed decoction. *Tools* Wood used to make tool handles and weaving combs. (191:30) **Tewa** *Tools* Wood used to make rabbit sticks. (138:45)

Cercocarpus montanus var. *glaber* (S. Wats.) F. L. Martin, Birchleaf Mountain Mahogany
- *Drug*—Apache, White Mountain *Burn Dressing* Wood burned, the charcoal powdered and applied to burns. (as *C. parvifolius* 136:156) **Kawaiisu** *Cough Medicine* Decoction of roots used for coughing. *Internal Medicine* Decoction of roots used for internal ills. (as *C. betuloides* 206:18) **Mahuna** *Venereal Aid* Infusion of bark and roots taken

for venereal disease or gonorrhea gleet (urethral discharge). (as *C. betulaefolius* 140:70)
- **Other**—**Apache, White Mountain** *Hunting & Fishing Item* Wood used to make bows. (as *C. parvifolius* 136:156) **California Indian** *Smoking Tools* Root used to make pipe bowls. *Tools* Wood used for digging sticks. *Weapon* Wood used to make clubs. (as *C. betuloides* 118:62) **Hopi** *Ceremonial Items* Wood used to make pahos (prayer sticks). *Tools* Wood used to make implements. (as *C. betuloides* 42:298) **Karok** *Tools* Hard wood used to make digging sticks. (as *C. betuloides* 148:384) **Kawaiisu** *Smoking Tools* Wood carved into a pipe head and used with a hollowed section of a honeysuckle twig as a pipestem. (as *C. betuloides* 206:18) **Mendocino Indian** *Hunting & Fishing Item* Wood formerly used to make arrow tips. *Tools* Wood formerly used to make tools for digging Indian potatoes and worms out of the ground. *Weapon* Large sticks used for war spears and fighting clubs. (as *C. betuloides* 41:354) **Modesse** *Hunting & Fishing Item* Wood used for spear points. *Tools* Wood used for digging sticks. (as *C. parvifolius* 117:223) **Wintoon** *Tools* Wood used to make digging sticks. (as *C. betuloides* 117:264) **Yuki** *Hunting & Fishing Item* Wood used to make bows. *Walking Sticks* Wood used to make canes. (as *C. betuloides* 49:93)

Cercocarpus sp., Mountain Mahogany
- **Dye**—**Keresan** *Red* Used to make a red dye for staining moccasins. (198:562)

Cereus, Cactaceae
Cereus sp.
- **Food**—**Apache, White Mountain** *Fruit* Fruit used for food. *Preserves* Fruit used to make a kind of butter. (136:156)

Cetraria, Parmeliaceae, lichen
Cetraria crispa (Ach.) Nyl., Shield Lichen
- **Food**—**Eskimo, Inuktitut** *Spice* Used as a soup condiment. (202:183)

Cetraria cucullata (Bellard) Ach., Curled Shield Lichen
- **Food**—**Eskimo, Inuktitut** *Sauce & Relish* Used as a condiment for fish or duck soup. (202:188)

Chaenactis, Asteraceae
Chaenactis douglasii (Hook.) Hook. & Arn., Douglas's Dustymaiden
- **Drug**—**Gosiute** *Analgesic* and *Orthopedic Aid* Mashed plant rubbed on limbs for soreness or aching. (39:365) **Great Basin Indian** *Heart Medicine* and *Pediatric Aid* Infusion of whole plant given to children to slow their heartbeats. (121:50) **Okanagon** *Dermatological Aid* Infusion used as wash for chapped hands, pimples, boils, tumors, and swellings. *Snakebite Remedy* Infusion of plant used as wash for insect and snake bites. (125:42) **Paiute** *Analgesic* Infusion of young leaves taken or put on the hair for headaches. (111:118) *Cold Remedy* Decoction of plant or leaves taken for colds. *Cough Medicine* Decoction of plant or leaves taken for coughs. (180:55, 56) *Dermatological Aid* Poultice of crushed leaves applied to swellings. (104:196) Poultice of crushed, fresh plants or leaves applied to swellings. *Heart Medicine* Infusion of plant used as a heart depressant. (180:55, 56) *Orthopedic Aid* Poultice of crushed leaves applied to sprains. (104:196) *Snakebite Remedy* Poultice of pulped leaves and stems applied to rattlesnake bites. (180:55, 56) **Sanpoil** *Cathartic* Decoction of roots taken by family of dead one as a purge to avoid illness. *Tuberculosis Remedy* Decoction of roots taken by family of dead one to avoid taking consumption. (131:221) **Shoshoni** *Dermatological Aid* Poultice of crushed fresh plants or leaves applied to swellings. *Emetic* Decoction of plant taken as an emetic for indigestion. *Gastrointestinal Aid* Decoction of plant taken as an emetic for indigestion or sour stomach. *Kidney Aid* Decoction of plants used as a bath for swollen limbs or dropsical conditions. (180:55, 56) **Thompson** *Dermatological Aid* Infusion of plant used as wash for chapped hands, pimples, boils, and tumors. Infusion of whole plant taken for swellings. (125:42) Decoction of plant used on various skin conditions and insect bites. Decoction of whole plant taken for any kind of swellings. *Gastrointestinal Aid* Mild decoction taken as a tonic for the stomach. (164:473) *Snakebite Remedy* Infusion of plant used as wash for insect and snake bites. (125:42) Strong decoction of entire plant applied to insect and snake bites. *Stimulant* Mild decoction of entire plant taken as a tonic for the stomach and lassitude. *Tonic* Mild decoction taken as a tonic for the stomach and lassitude. (164:473) *Unspecified* Plant considered a good medicine. (187:178)

Chaenactis douglasii (Hook.) Hook. & Arn. **var. *douglasii*,** Douglas's Dustymaiden
- **Drug**—**Okanagan-Colville** *Eye Medicine* Infusion of roots used as an eyewash. (188:82)

Chaenactis glabriuscula DC., Yellow Chaenactis
- **Food**—**Cahuilla** *Porridge* Parched seeds ground into flour, mixed with other seeds, and used to form a mush. (15:52)

Chaenactis santolinoides Greene, Santolina Pincushion
- **Drug**—**Kawaiisu** *Analgesic* and *Orthopedic Aid* Decoction of roots taken for sore chest, sore shoulders, and internal soreness. (206:19)

Chaenactis stevioides Hook. & Arn., Steve's Dustymaiden
- **Drug**—**Nevada Indian** *Heart Medicine* and *Pediatric Aid* Infusion of plant used to slow down heartbeats of children with fevers. (118:40)
- **Other**—**Navajo, Kayenta** *Fasteners* Juice used as glue to mend broken ceremonial items. (205:46)

Chaerophyllum, Apiaceae
Chaerophyllum procumbens (L.) Crantz, Spreading Chervil
- **Drug**—**Chickasaw** *Emetic* and *Poison* Poisonous root used as an emetic. (28:289)

Chaetopappa, Asteraceae
Chaetopappa ericoides (Torr.) Nesom, Rose Heath
- **Drug**—**Havasupai** *Gastrointestinal Aid* Decoction of whole plant or roots taken or used as a wash for digestive troubles. *Pediatric Aid* Decoction of whole plant or roots given or used as a wash for children with digestive troubles. (as *Leucelene ericoides* 197:248) **Hopi** *Nose Medicine* Infusion of root used to "aid a sore nose." (as *Aster leucelene* 200:34, 95) *Panacea* Root used as a universal panacea. *Pediatric Aid* Infusion of herb used to "quiet the baby." (as *Aster leucelene* 200:95) *Reproductive Aid* Plant used to determine the sex of a child. This is quite an ambiguous reference. The text says this: "This plant is

used by the Hopi Indians as genetic factor among the Indian clans. Genetic factor refers to the choice of a small (female) or large (male) plant to assist in determining the sex of a child." It is, therefore, unclear if the plant is used to detect whether the fetus is male or female, or to cause the child to be one or the other. Elsewhere, this author tells us that the Hopi make a decoction of the leaves of juniper "which is said to be a laxative and is taken by women who desire a female child." This suggests that the second possibility may be the correct one, with administration of large plants if you want a son and small ones if you want a daughter. (as *Aster arenosus* 42:290) *Sedative* Infusion of root used to "quiet the baby." (as *Aster leucelene* 200:36, 95) *Stimulant* Plant used as a stimulant. (as *Aster leucelene* 200:31) **Keres, Western** *Antirheumatic (External)* Poultice or infusion of plant used for swellings. (as *Leucelene ericoides* 171:52) **Navajo, Kayenta** *Kidney Aid* Infusion of plant with sumac berries taken for kidney disease. *Urinary Aid* Infusion of plant with sumac berries taken for bladder disease. (as *Aster ericaefolius* 205:45) **Navajo, Ramah** *Nose Medicine* Dried pulverized plant used as snuff or cold infusion used as drops for "nose trouble." *Snakebite Remedy* Poultice of chewed leaves applied and infusion taken for snakebite. *Toothache Remedy* Leaves chewed for toothache. (as *Aster arenosus* 191:48) **Zuni** *Analgesic* Infusion of pulverized plant applied for pain from cold or rheumatism. *Antirheumatic (External)* Infusion of whole plant rubbed on body for swelling and rheumatic pain. *Cold Remedy* Infusion of whole plant rubbed on body for pain from a cold. *Dermatological Aid* Infusion of pulverized plant rubbed over body for swellings. *Gynecological Aid* Warm infusion of plant taken to "hasten parturition." (as *Leucelene ericoides* 166:55)

Chamaebatia, Rosaceae

Chamaebatia foliolosa Benth., Sierran Mountain Misery

- **Drug**—**Miwok** *Antirheumatic (Internal)* Infusion of leaves taken for rheumatism. *Cold Remedy* Decoction of leaves taken for colds. *Cough Medicine* Decoction of leaves taken for coughs. *Misc. Disease Remedy* Infusion of leaves taken for chickenpox, measles, and smallpox. *Venereal Aid* Leaves used as ingredient in medicines for venereal diseases. (12:168)

Chamaebatiaria, Rosaceae

Chamaebatiaria millefolium (Torr.) Maxim., Fernbush

- **Drug**—**Gosiute** *Venereal Aid* Plant used for gonorrhea. (39:365) Poultice of plant applied or plant used as wash for venereal diseases. (as *Spiraea millefolium* 39:351) **Navajo, Ramah** *Hunting Medicine* Leaves rolled in corn husk smoked for good luck in hunting. (191:30) **Paiute** *Orthopedic Aid* Compound decoction of young shoots taken for lumbago. **Shoshoni** *Analgesic* and *Gastrointestinal Aid* Decoction of fresh or dried leaves taken for stomachaches or cramps. (180:56, 57)
- **Food**—**Navajo, Ramah** *Fodder* Used as sheep, goat, and deer feed and not eaten by cattle. (191:30)

Chamaecrista, Fabaceae

Chamaecrista fasciculata (Michx.) Greene, Sleepingplant

- **Drug**—**Cherokee** *Sports Medicine* Root medicine used to keep ballplayers from tiring. *Stimulant* Compound infusion given for fainting spells. (as *Cassia fasciculata* 80:54) **Seminole** *Antiemetic* Cold decoction of plant used for nausea. (as *Chamaecrista brachista* 169:276)
- **Other**—**Seminole** *Tools* Plant used as a bed for ripening persimmons. (as *C. brachiata* 169:496)

Chamaecrista nictitans (L.) Moench **ssp. *nictitans* var. *nictitans*,** Partridge Pea

- **Drug**—**Cherokee** *Sports Medicine* Root medicine used to keep ballplayers from tiring. *Stimulant* Compound infusion given for fainting spells. (as *Cassia nictitans* 80:54)

Chamaecyparis, Cupressaceae, conifer

Chamaecyparis lawsoniana (A. Murr.) Parl., Port Orford Cedar

- **Fiber**—**Karok** *Brushes & Brooms* Branches used to make brooms. *Building Material* Wood made into planks and used to build sweat houses. Wood used as the main post in house construction. *Furniture* Wood used to make circular stools and headrests for the sweat house. *Mats, Rugs & Bedding* Wood used to make pillows for the sweat house. (148:379)
- **Other**—**Yurok** *Toys & Games* Fruits used by children to throw at each other. (6:23)

Chamaecyparis nootkatensis (D. Don) Spach, Alaska Cedar

- **Drug**—**Bella Coola** *Adjuvant* Soft bark used as cover for poultices of *Trautvetteria grandis* and *Ranunculus acris*. (150:49) **Kwakiutl** *Antirheumatic (External)* Plant used in sweat baths for arthritis and rheumatism. *Dermatological Aid* Infusion of branch tips used as a wash for sores and swellings. Poultice of chewed leaves applied to sores. Sharp boughs rubbed on sores and swellings until skin was broken. *Herbal Steam* Plant used in sweat baths for arthritis and rheumatism. *Kidney Aid* Compound decoction of leaves applied to swelling on woman's kidney. *Panacea* Infusion of branch tips taken for general illness. *Strengthener* Bark ash and oil used as a lotion to give strength to the very ill. (183:266) **Kwakiutl, Southern** *Antirheumatic (External)* Branches placed on top of burning sea wrack as part of a steam treatment for rheumatism. *Strengthener* Branches placed on top of burning sea wrack as part of a steam treatment for general sickness. (183:261)
- **Fiber**—**Bella Coola** *Clothing* Inner bark used for weaving capes. *Mats, Rugs & Bedding* Inner bark used for weaving mats and blankets. (184:197) **Haisla** *Clothing* Inner bark fiber used to make clothing for the nobility. (73:153) **Haisla & Hanaksiala** *Canoe Material* Wood used to make regular and racing paddles for canoes. Wood used to make boat ribs. *Clothing* Inner bark woven into capes and loincloths. *Mats, Rugs & Bedding* Inner bark woven into blankets. (43:159) **Hesquiat** *Canoe Material* Wood used for making wedge-shaped block for the back of a canoe, used to keep the feet dry. *Clothing* Wood used for making ornamental dishes and headdresses. Bark softened with special oil and used for weaving capes and other clothing of head chiefs. *Mats, Rugs & Bedding* Bark softened with special oil and used for weaving blankets. (185:33) **Hoh** *Canoe Material* Used to make canoes and paddles. (137:57) **Kitasoo** *Basketry* Inner bark used to make baskets. *Building Material* Wood used for construction. *Canoe Material* Wood used to make bows, adz handles, paddles, and storage containers. *Clothing* Inner bark used to make hats. Inner bark pounded and used to make fine quality clothing. *Cordage* Inner bark used to make cordage. *Mats, Rugs & Bedding* Inner bark used to make mats.

Inner bark pounded and used to make fine quality blankets. (43:313) **Kwakiutl, Southern** *Canoe Material* Wood used to make canoe paddles. (183:266) *Clothing* Inner bark used to make clothing. (183:296) *Furniture* Wood used to make chests. (183:266) *Mats, Rugs & Bedding* Inner bark used to make mats and blankets. (183:296) **Nitinaht** *Canoe Material* Wood used to make light paddles and canoes. *Clothing* Inner bark pounded and spun to make baby clothing, skirts, capes, and hats. Inner bark finely shredded and used to make face towels. (186:65) *Cordage* Bark fibers, nettle fibers, and dog hair used to make a stronger rope. (67:227) *Mats, Rugs & Bedding* Inner bark pounded and spun to make blankets. (186:65) **Oweekeno** *Canoe Material* Wood used to make canoes and canoe bailers. *Clothing* Inner bark used to make robes. *Cordage* Bark used to make cordage. *Mats, Rugs & Bedding* Inner bark used to make blankets. (43:61) **Quileute** *Canoe Material* Used to make canoes and paddles. (137:57)
- **Other**—**Bella Coola** *Decorations* Inner bark used for decorating masks. (184:197) **Haisla & Hanaksiala** *Decorations* Wood used to make carved items and masks. *Fasteners* Wooden pegs used to hold together bent boxes of red cedar. *Hunting & Fishing Item* Wood used to make bows. Rods struck with Pacific crabapple sticks or yew wood to make noise to herd animals while hunting. (43:159) **Hesquiat** *Cooking Tools* Wood used for making ornamental dishes and headdresses. (185:33) **Kitasoo** *Containers* Wood used to make storage containers. *Hunting & Fishing Item* Wood used to make bows and paddles. *Tools* Wood used to make adz handles. (43:313) **Kwakiutl, Southern** *Cooking Tools* Wood used to make dishes. *Hunting & Fishing Item* Wood used to make bows. (183:266) **Nitinaht** *Containers* Wood used to make small charcoal mixing boxes. *Fuel* Inner bark finely shredded and used as tinder. *Sacred Items* Wood used to carve totem pole models and talking sticks. (186:65) **Oweekeno** *Containers* Wood used to make containers. *Decorations* Wood carved into totem poles and masks. *Tools* Wood used to make implements. (43:61)

Chamaecyparis thyoides (L.) B.S.P., Atlantic White Cedar
- *Drug*—**Ojibwa, South** *Analgesic* Decoction of leaves used as herbal steam for headache and backache. Poultice of crushed leaves and bark applied for headache. (as *Cupressus thyoides* 91:198)

Chamaedaphne, Ericaceae
Chamaedaphne calyculata (L.) Moench, Leather Leaf
- *Drug*—**Potawatomi** *Dermatological Aid* Poultice of leaves applied to inflammations. *Febrifuge* Infusion of leaves used for fevers. (154:56)
- *Food*—**Ojibwa** *Beverage* Fresh or dried leaves used as a beverage tea. (153:400)

Chamaemelum, Asteraceae
Chamaemelum nobile (L.) All., Garden Dogfennel
- *Drug*—**Cherokee** *Abortifacient* Infusion of flower or herb used for "female obstructions." *Antiemetic* Used for vomiting. *Dermatological Aid* Used as poultice for ulcers and "hard swellings." *Gastrointestinal Aid* Used for colic and bowel complaints. *Sedative* Infusion of flower or herb used for "hysterical affections." (as *Anthemis nobilis* 80:28) **Mahuna** *Gastrointestinal Aid* Plant used to regulate unsettled stomachs or for babies suffering from colic. *Pediatric Aid* Plant used for babies suffering from colic. (as *Anthemis nobilis* 140:7)

Chamaesaracha, Solanaceae
Chamaesaracha coronopus (Dunal) Gray, Greenleaf Five Eyes
- *Drug*—**Navajo, Kayenta** *Dermatological Aid* Plant used for swellings. *Other* Compound containing plant used in cases of drowning. (205:41)

Chamaesyce, Euphorbiaceae
Chamaesyce albomarginata (Torr. & Gray) Small, Whitemargin Sandmat
- *Drug*—**Diegueño** *Dermatological Aid* Decoction of plant used to wash sores. (as *Euphorbia albinomarginata* 84:21) **Kawaiisu** *Snakebite Remedy* Ground leaves and flowers used as a salve for rattlesnake bites. *Veterinary Aid* Poultice applied or decoction of leaves given to animals with snakebites. (as *Euphorbia albomarginata* 206:31) **Keres, Western** *Eye Medicine* Crushed plant rubbed on sore eyes. *Gynecological Aid* Leaves rubbed on mothers' breasts to produce more and richer milk. (as *Euphorbia albomarginata* 171:44) **Navajo, Ramah** *Analgesic* Cold infusion used for stomachache. *Gastrointestinal Aid* Cold infusion of plant taken for stomachache. *Hemostat* Poultice of plant used as a hemostatic. (as *Euphorbia albomarginata* 191:35) **Shoshoni** *Snakebite Remedy* Poultice of crushed, whole plant applied to snakebites. *Tonic* Decoction of plant taken as a tonic for general debility. (as *Euphorbia albomarginata* 180:73, 74) **Zuni** *Gynecological Aid* Leaves and roots eaten to promote lactation. (as *Euphorbia albomarginata* 27:376)

Chamaesyce fendleri (Torr. & Gray) Small, Fendler's Sandmat
- *Drug*—**Hopi** *Dietary Aid* Young roots fed to sick baby whose mother's milk was failing. (200:84) *Oral Aid* Dried, ground plant used as soothing lip balm. (200:33, 83–84) *Pediatric Aid* Young roots fed to sick baby whose mother's milk was failing. (200:84)

Chamaesyce fendleri (Torr. & Gray) Small **var. *fendleri***, Fendler's Sandmat
- *Drug*—**Navajo** *Gastrointestinal Aid* Infusion of plant taken for stomachache. (as *Euphorbia fendleri* 90:151) **Navajo, Ramah** *Analgesic* Cold infusion or decoction used for stomachache and diarrhea. *Antidiarrheal* Cold infusion or decoction of plant taken for diarrhea. *Ceremonial Medicine* Plant used as a ceremonial medicine. *Dermatological Aid* Plant used topically for warts and poison ivy. *Gastrointestinal Aid* Cold infusion or decoction of plant taken for stomachache. *Gynecological Aid* Pulverized plant used topically as a galactagogue and for breast injuries. *Hemostat* Poultice of chewed plant applied to cuts as a hemostatic. *Toothache Remedy* Hot poultice of plant applied for toothache. *Veterinary Aid* Milky juice applied to snakebite in livestock. (as *Euphorbia fendleri* 191:35)

Chamaesyce geyeri (Engelm.) Small, Geyer's Sandmat
- *Drug*—**Lakota** *Preventive Medicine* Used as a medicine as protection for the head. (as *Euphorbia geyeri* 139:45)

Chamaesyce glyptosperma (Engelm.) Small, Ribseed Sandmat
- *Drug*—**Iroquois** *Gland Medicine* Compound decoction of stems taken and used as a wash for goiter. *Gynecological Aid* Plant used to stimulate lactation. (as *Euphorbia glyptosperma* 87:369) **Thompson** *Snakebite Remedy* Fresh plant rubbed on all snakebites, but especially rattlesnake bites. (as *Euphorbia glyptosperma* 164:462)

Chamaesyce hypericifolia (L.) Millsp., Graceful Sandmat
- ***Drug***—**Cherokee** *Urinary Aid* Infusion of bruised roots taken for yellow urine. (as *Euphorbia hypericifolia* 177:35)

Chamaesyce lata (Engelm.) Small, Hoary Sandmat
- ***Drug***—**Navajo** *Cathartic* Plant used as a purge. *Gastrointestinal Aid* Plant used for upset stomachs. (as *Euphorbia lata* 90:151)

Chamaesyce maculata (L.) Small, Spotted Sandmat
- ***Drug***—**Cherokee** *Cancer Treatment* Decoction prepared with herbs and taken for cancer. *Cathartic* Taken as a purgative. *Dermatological Aid* "Juice rubbed on skin eruptions, especially on children's heads." Juice used as ointment for "sores and sore nipples." *Gynecological Aid* Infusion taken for bleeding after childbirth. *Pediatric Aid* "Juice rubbed on skin eruptions, especially on children's heads." *Toothache Remedy* Root used for toothache. *Urinary Aid* Infusion of bruised root taken for urinary diseases. *Venereal Aid* Decoction taken for gonorrhea and "similar diseases." (as *Euphorbia maculata* 80:45) **Costanoan** *Blood Medicine* Infusion of plant taken to purify the blood. *Dermatological Aid* Decoction of plant used as a wash for cuts. Milky juice applied to pimples and infusion of foliage used as a hair wash. *Eye Medicine* Decoction of plant used as a wash for eyes. (as *Euphorbia maculata* 21:9)

Chamaesyce melanadenia (Torr.) Millsp., Squaw Sandmat
- ***Drug***—**Cahuilla** *Dermatological Aid* Sap used for bee stings and sores. *Ear Medicine* Sap used for earaches. (as *Euphorbia melanadenia* 15:73)

Chamaesyce multiformis (Hook. & Arn.) Croizat & Deg. **var. *multiformis***, Variable Sandmat
- ***Drug***—**Hawaiian** *Breast Treatment* Plant milk and other ingredients taken for dry breasts. *Dietary Aid* Buds or leaves chewed by the mother for the benefit of the nursing baby. Buds or leaves chewed by nursing mothers to stimulate the appetite, helpful in milk production. *Laxative* Buds or leaves chewed by nursing mothers as a laxative. *Pediatric Aid* Buds chewed by the mother and given to babies till the age of 6 months. Buds or leaves chewed by the mother for the benefit of the nursing baby. *Reproductive Aid* Buds, leaves, and other plants pounded and resulting liquid taken for female reproductive organ weakness. *Strengthener* Buds and leaves used for general debility of the body. *Tuberculosis Remedy* Poultice of plant milk and other ingredients applied to scrofulous sores. *Unspecified* Buds chewed by the mother and given to babies till the age of 6 months. (as *Euphorbia multiformis* 2:11)

Chamaesyce nutans (Lag.) Small, Eyebane
- ***Drug***—**Houma** *Dermatological Aid* Milk from stem rubbed on skin for itching and eczema. Poultice of crushed leaves applied to bad sores. (as *Euphorbia nutans* 158:65, 66) *Gastrointestinal Aid* Cool decoction of plant in milk given to babies for sickness from bad milk. (as *Euphorbia nutans* 158:65) *Pediatric Aid* Cool decoction of plant in milk given to babies for sickness from bad milk. (as *Euphorbia nutans* 158:65, 66)

Chamaesyce ocellata **ssp. *arenicola*** (Parish) Thorne, Contura Creek Sandmat
- ***Drug***—**Paiute** *Dermatological Aid* Poultice of mashed plant applied to swellings. (as *Euphorbia arenicola* 180:74) *Eye Medicine* Infusion of whole plant used as an eyewash. (as *Euphorbia arenicola* 118:39) Decoction of leaves used as an eyewash. (as *Euphorbia arenicola* 180:74)

Chamaesyce ocellata (Dur. & Hilg.) Millsp. **ssp. *ocellata***, Contura Creek Sandmat
- ***Drug***—**Miwok** *Blood Medicine* Decoction of leaves taken as a blood purifier. *Snakebite Remedy* Mashed leaves rubbed into snakebite to prevent swelling. (as *Euphorbia ocellata* 12:169)

Chamaesyce polycarpa (Benth.) Millsp. ex Parish **var. *polycarpa***, Smallseed Sandmat
- ***Drug***—**Luiseño** *Snakebite Remedy* Plant used for rattlesnake bites. (as *Euphorbia polycarpa* 155:231) **Pima** *Dermatological Aid* Poultice of plant applied to scorpion and snakebites. *Diaphoretic* Plant chewed to cause vomiting and sweating for snakebites. *Emetic* Roots chewed to vomit for stomach troubles, snakebites, and constipation. *Gastrointestinal Aid* Roots chewed to vomit and loosen bowels for stomach troubles. *Laxative* Roots chewed to loosen bowels for stomach troubles and constipation. *Poison* Plant considered poisonous. *Snakebite Remedy* Plant chewed to cause vomiting and sweating for snakebites. Plant juice used as wash and poultice of plant applied to snakebites. (as *Euphorbia polycarpa* 47:99) **Shoshoni** *Eye Medicine* Infusion of plant used as an eyewash. *Tonic* Infusion of plant taken as a tonic "for any general, indisposed feeling." (as *Euphorbia polycarpa* 180:74) **Zuni** *Gynecological Aid* Warm gruel made with plant and white cornmeal taken to promote milk flow. (as *Euphorbia polycarpa* 166:51)

Chamaesyce revoluta (Engelm.) Small, Threadstem Sandmat
- ***Drug***—**Navajo, Kayenta** *Dermatological Aid* Plant used as a lotion for chafing and sores. (as *Euphorbia revoluta* 205:30)

Chamaesyce serpyllifolia (Pers.) Small, Thymeleaf Sandmat
- ***Drug***—**Omaha** *Antidiarrheal* Dried leaves rubbed into abdominal scratches for children's dysentery. *Gastrointestinal Aid* Dried leaves rubbed into abdominal scratches for children's bloating. **Ponca** *Gynecological Aid* Decoction of plant taken to encourage milk flow in nursing mothers. (70:99) Decoction of plant taken by young mothers for scanty or lack of milk. (94:151)

Chamaesyce serpyllifolia (Pers.) Small **ssp. *serpyllifolia***, Thymeleaf Sandmat
- ***Drug***—**Apache, White Mountain** *Oral Aid* Plant chewed to sweeten the saliva. (as *Euphorbia serpyllifolia* 136:158) **Miwok** *Dermatological Aid* Decoction of leaves used as wash for running sores. *Snakebite Remedy* Poultice of plant applied, must be done immediately, to rattlesnake bites. (as *Euphorbia serpyllifolia* 12:170) **Navajo, Ramah** *Analgesic* Cold infusion or decoction of plant taken for stomachache. *Antidiarrheal* Cold infusion or decoction of plant taken for diarrhea. *Ceremonial Medicine* Plant used as a ceremonial medicine. *Dermatological Aid* Plant used topically for warts and poison ivy. *Gastrointestinal Aid* Cold infusion or decoction of plant taken for stomachache. *Gynecological Aid* Pulverized plant used topically as a galactagogue and for breast injuries. *Hemostat* Poultice of chewed plant applied to cuts as a hemostatic. *Toothache Remedy* Hot poultice of plant applied for toothache. *Veterinary Aid* Milky juice applied to

snakebite in livestock. (as *Euphorbia serpyllifolia* 191:35) **Zuni** *Cathartic* Plant used as a cathartic. *Emetic* Plant used as an emetic. *Gynecological Aid* Plant used to increase the flow of milk in nursing mother. (as *Euphorbia serpyllifolia* 166:51)
- *Food*—**Apache, White Mountain** *Beverage* Roots used to make a fermented, intoxicating drink. (as *Euphorbia serpyllifolia* 136:151) *Cooking Agent* Roots chewed and used as a yeast preparation for the wedding cake. *Dried Food* Roots dried for future use. (as *Euphorbia serpyllifolia* 136:148) **Zuni** *Candy* Leaves chewed for the pleasant taste. *Sweetener* Root pieces used to sweeten cornmeal. After the mouth had been thoroughly cleansed, the women who sweetened the corn placed a piece of it in their mouths. The root remained in the mouth for 2 days, except to take refreshment and to sleep. Each time the root was removed from the mouth, the mouth was cleansed with cold water before returning the root to it. Finally, when they began sweetening the corn, either yellow or black corn was used. The women, with their fingers, placed as much cornmeal as possible into their mouths and held it there, without chewing, until the accumulation of saliva forced ejection of the mass. (as *Euphorbia serpyllifolia* 166:67)

Chaptalia, Asteraceae
Chaptalia tomentosa Vent., Woolly Sunbonnets
- *Drug*—**Seminole** *Antirheumatic (External)* Decoction of leaves rubbed on body and body steamed for deer sickness: numb, painful limbs and joints. (169:192) *Urinary Aid* Decoction of roots taken for urine retention. (169:274)

Chasmanthium, Poaceae
Chasmanthium latifolium (Michx.) Yates, Indian Woodoats
- *Food*—**Cocopa** *Porridge* Seeds dried, ground, and made into mush. (as *Uniola palmeri* 37:187) *Unspecified* Seeds used for food. *Winter Use Food* Seeds stored for later use. (as *Uniola palmeri* 64:267)

Cheilanthes, Pteridaceae, fern
Cheilanthes covillei Maxon, Coville's Lipfern
- *Food*—**Kawaiisu** *Beverage* Stems and leaves used to make tea. (206:19)

Cheilanthes fendleri Hook., Fendler's Lipfern
- *Drug*—**Keres, Western** *Gynecological Aid* Infusion of plant used as a douche after childbirth. (171:36)
- *Food*—**Apache, Chiricahua & Mescalero** *Beverage* Leaves and young stems boiled to make a nonintoxicating beverage. (33:53)

Cheilanthes wootonii Maxon, Beaded Lipfern
- *Drug*—**Navajo, Ramah** *Dermatological Aid* Cold infusion of plant used as a lotion for gunshot wounds. *Panacea* Plant used as "life medicine." (191:11)

Cheirodendron, Araliaceae
Cheirodendron gaudicchaudii, Olapa
- *Drug*—**Hawaiian** *Respiratory Aid* Root bark and other plants pounded, squeezed, and the resulting liquid taken for asthma. (as *C. caudicchaudii* 2:33)
- *Other*—**Hawaiian** *Fuel* Wood used for fuel. (as *C. caudicchaudii* 2:33)

Chelidonium, Papaveraceae
Chelidonium majus L., Greater Celandine
- *Drug*—**Iroquois** *Veterinary Aid* Infusion of whole plant, another plant, and milk given to pigs that drool and have sudden movements. (141:45)

Chelone, Scrophulariaceae
Chelone glabra L., White Turtlehead
- *Drug*—**Algonquin, Quebec** *Unspecified* Infusion of roots and cedar bark used as a medicinal tea. (18:230) **Cherokee** *Anthelmintic* Infusion of blooms taken for worms. *Dermatological Aid* Used for sores or skin eruptions. *Dietary Aid* Taken to increase appetite. *Febrifuge* Infusion of blooms taken for fevers. *Laxative* Infusion of blooms taken as a laxative. (80:59) **Iroquois** *Liver Aid* Compound decoction of roots taken for too much gall. *Witchcraft Medicine* Infusion of smashed roots taken as an anti-witchcraft medicine. (87:434) **Malecite** *Contraceptive* Infusion of plants used to prevent pregnancy. (116:258) **Micmac** *Contraceptive* Herb used to prevent pregnancy. (40:55)
- *Food*—**Cherokee** *Unspecified* Young shoots and leaves boiled, fried, and eaten. (204:253)

Chelone sp., Deer Tongue
- *Food*—**Cherokee** *Vegetable* Leaves and stems parboiled, rinsed, and cooked in grease until tender. (126:55)

Chenopodiaceae, see *Allenrolfea, Arthrocnemum, Atriplex, Beta, Chenopodium, Cycloloma, Kochia, Krascheninnikovia, Monolepis, Salicornia, Salsola, Sarcobatus, Sarcocornia, Suaeda*

Chenopodium, Chenopodiaceae
Chenopodium album L., Lamb's Quarters
- *Drug*—**Carrier** *Blood Medicine* Decoction of plant taken to improve the blood. (31:86) **Cherokee** *Dietary Aid* Cooked salad greens eaten to "keep healthy." (80:42) **Cree, Woodlands** *Antirheumatic (External)* Decoction of plant used as wash for painful limbs. *Antirheumatic (Internal)* Decoction of plant taken for painful limbs. (109:35) **Eskimo, Inupiat** *Carminative* Leaves and stems cooked with beans to reduce the intestinal gas from eating the beans. (98:64) **Iroquois** *Antidiarrheal* Cold infusion of whole plant taken for diarrhea. (87:315) *Burn Dressing* Compound used as salve on burns. (87:316) *Gynecological Aid* Compound decoction used as wash and applied as poultice when bothered by milk flow. (87:315) **Mendocino Indian** *Gastrointestinal Aid* Leaves used for stomachaches. (41:346) **Meskwaki** *Dermatological Aid* Infusion of root used for urethral itching. (152:209) **Navajo** *Dietary Aid* Plant used as a nutrient. (90:149) **Navajo, Kayenta** *Burn Dressing* Poultice of plant applied to burns. (205:20) **Navajo, Ramah** *Antidote* Stem, 3 inches long, made into snake figurinefor snake infection. (191:24) **Paiute** *Emetic* Leaf chewed as an emetic. (as *C. alba* 167:317) **Potawatomi** *Misc. Disease Remedy* Plant considered a medicinal food used to prevent or cure scurvy. (154:47) Leaves included in the diet for scurvy or to prevent it. (154:98)
- *Food*—**Alaska Native** *Dietary Aid* Fresh leaves, properly cooked, furnished significant amounts of vitamins C and A. *Substitution Food* Young, tender leaves and stems used as a substitute for spinach or other greens. *Vegetable* Young, tender leaves and stems cooked in a small amount of boiling water and eaten. (85:21) **Apache** *Vegetable*

Young plants cooked as greens. (32:16) **Apache, Chiricahua & Mescalero** *Unspecified* Eaten without preparation or cooked with green chile and meat or animal bones. (as *C. alba* 33:46) **Cherokee** *Spice* Young growth mixed with mustard leaves, morning glory leaves, or potato leaves for flavoring. *Unspecified* Young growth parboiled, fried, and eaten. (204:253) *Vegetable* Leaves mixed with other leaves, parboiled, and cooked in grease until tender. Leaves mixed with other leaves and used for greens. (126:32) **Dakota** *Soup* Young, tender plant cooked as pottage. (70:78) *Unspecified* Young plants boiled for food. (as *C. albidum* 69:361) **Diegueño** *Vegetable* Leaves cooked and eaten as greens. (84:17) **Eskimo, Inupiat** *Dried Food* Leaves and stems dried for future use. *Frozen Food* Leaves and stems frozen for future use. *Vegetable* Leaves and stems eaten raw or cooked as hot greens with beans. (98:64) **Hopi** *Porridge* Ground seeds used to make mush. (190:160) *Unspecified* Leaves cooked with meat. (32:16) Leaves boiled and eaten with fat. (56:18) Boiled and eaten with other foods. (200:73) **Iroquois** *Vegetable* Cooked and seasoned with salt, pepper, or butter. (196:117) **Kawaiisu** *Vegetable* Upper leaves boiled, "rinsed" in cold water, and fried in grease and salt. (206:19) **Lakota** *Vegetable* Used as cooked greens. (139:43) **Luiseño** *Vegetable* Leaves used as greens. (155:233) **Mendocino Indian** *Vegetable* Young leaves boiled and eaten as greens. (41:346) **Miwok** *Dried Food* Boiled greens dried and stored for later use. *Vegetable* Boiled greens used for food. (12:159) **Mohegan** *Unspecified* Cooked and used for food. (176:83) **Montana Indian** *Staple* Seeds ground into flour and made into bread. *Vegetable* Young plant used as a potherb. (19:9) **Navajo** *Dried Food* Seeds dried and used like corn. (55:43) *Staple* Seeds ground and eaten as a nutrient. (90:149) *Vegetable* Young, tender plants eaten raw, boiled as herbs alone or with other foods. (55:43) **Navajo, Ramah** *Bread & Cake*, and *Special Food* Seeds winnowed, ground with maize, made into bread, and used as a ceremonial food in Nightway. *Winter Use Food* Seeds stored for winter use. (191:24) **Ojibwa** *Vegetable* Young plant cooked as greens. (5:2209) Leaves eaten as greens. (135:240) **Omaha** *Soup* Young, tender plant cooked as pottage. (70:78) **Paiute** *Staple* Seeds parched, ground, and eaten as meal. (104:98) *Unspecified* Species used for food. (167:244) **Papago** *Soup* Mixed with roasted cholla buds and eaten as a vegetable stew. (as *C. alba* 34:16) **Pawnee** *Soup* Young, tender plant cooked as pottage. (70:78) **Pima, Gila River** *Unspecified* Leaves used for food. (133:7) **Potawatomi** *Vegetable* Leaves used as a relish food for salads and spring greens. (154:98) **Pueblo** *Vegetable* Young plants cooked as greens. (32:16) **Shuswap** *Vegetable* Leaves boiled with butter, salt, and pepper and used for greens. (123:61) **Spanish American** *Vegetable* Young plants cooked as greens. (32:16) **Thompson** *Vegetable* Boiled leaves eaten as greens. (187:203) **Zuni** *Vegetable* Young plants cooked as greens. (32:16)

- *Other*—**Pawnee** *Paint* Plant formerly used for painting bows and arrows. (70:78)

Chenopodium ambrosioides L., Mexican Tea

- *Drug*—**Creek** *Febrifuge* Unspecified plant part used "in cases of fever." (172:657) Plant used as a fever medicine. (177:22) *Panacea* Plant used for "a great many ailments." *Tonic* Plant used as "a sort of spring tonic." (172:657) Plant used as a spring tonic. (177:22) **Houma** *Analgesic* Poultice of crushed leaves applied for headaches. *Anthelmintic* and *Pediatric Aid* Decoction of leaves in milk given to children for worms. (158:63) **Koasati** *Anthelmintic* Decoction of leaves taken for worms. (177:22) **Mahuna** *Abortifacient* Roots used for delayed menstrual period. (140:14) **Miwok** *Antirheumatic (External)* Plant used as wash for rheumatic parts. *Dermatological Aid* Poultice of boiled or raw plant applied to swellings. *Toothache Remedy* Plant used for toothache or an ulcerated tooth. *Venereal Aid* Plant used as wash for gonorrhea and injected into affected parts. (12:168) **Natchez** *Anthelmintic* Plant given to children for worms. *Febrifuge* Plant used as a fever medicine. *Pediatric Aid* Plant given to children for worms. (177:22) **Rappahannock** *Anthelmintic* Stewed seeds taken for worms. *Tonic* Stewed seeds taken as a tonic. (161:30) **Seminole** *Blood Medicine* Decoction of whole plant taken for worm sickness: pale skin and laziness. (169:241) *Gastrointestinal Aid* Infusion of root bark taken for stomach troubles. (169:276) *Pulmonary Aid* and *Sedative* Plant taken and rubbed on the body for lion disease: chest cramps, nervousness, and walking continually. (169:233) *Stimulant* Decoction of whole plant taken for worm sickness: pale skin and laziness. (169:241)

Chenopodium botrys L., Jerusalem Oak Goosefoot

- *Drug*—**Cherokee** *Analgesic* Cold infusion taken orally and used to moisten head for headache. *Anthelmintic* Decoction of any part of plant in sweet milk given for worms. *Cold Remedy* Cold infusion taken orally and used to moisten head for colds. *Misc. Disease Remedy* Warm infusion of root taken in winter for "fever diseases." (80:41)
- *Other*—**Thompson** *Incense & Fragrance* Plant wound in necklaces, stuffed in pillows, bags, baskets, or tied to clothes as a scent. (164:503)

Chenopodium californicum (S. Wats.) S. Wats., California Goosefoot

- *Drug*—**Cahuilla** *Gastrointestinal Aid* Decoction of entire plant used for stomach disorders. (15:52) **Costanoan** *Orthopedic Aid* Decoction of root applied as a poultice for numb or paralyzed limbs. (21:11) **Kawaiisu** *Dermatological Aid* Plant used as a hair wash. *Emetic* Decoction of leaves and stems taken as an emetic. *Poison* Plant considered poisonous. (206:19)
- *Food*—**Cahuilla** *Candy* Milky sap used to make gum. *Staple* Parched seeds ground into flour. *Vegetable* Boiled shoots and leaves eaten as greens. (15:52) **Luiseño** *Unspecified* Seeds used for food. (155:233)
- *Other*—**Costanoan** *Soap* Scraped roots and water used to produce detergent foam. (21:249) **Diegueño** *Soap* Mashed root used to clean and whiten clothes. (84:17) **Kawaiisu** *Soap* Grated root used as soap. (206:19) **Luiseño** *Soap* Grated root used as soap. (155:210)

Chenopodium capitatum (L.) Aschers., Blite Goosefoot

- *Drug*—**Navajo, Kayenta** *Dermatological Aid* Plant used as a lotion for head bruises. *Eye Medicine* Plant used as a lotion for black eyes. (205:21) **Potawatomi** *Pulmonary Aid* Juice of seeds and infusion of plant used for lung congestion. (154:47)
- *Food*—**Alaska Native** *Dietary Aid* Leaves properly cooked and used as a good source of vitamins C and A. *Vegetable* Young, tender leaves used in raw salad mixture or cooked like garden spinach. (85:23) **Gosiute** *Unspecified* Seeds used for food. (39:366)
- *Dye*—**Potawatomi** *Red* Fruit heads used as rouge to paint on clan marks or to heighten the color of cheeks and lips. (154:117)

- *Other*—Carrier *Paint* Berries used as red paint. (31:83) **Tanana, Upper** *Paint* Berries used by children as paint by rubbing it on what they wished to color. (102:13) **Thompson** *Paint* Plant tops mashed with a little water and used to make ink to write with. (187:203)

Chenopodium capitatum (L.) Aschers. **var. *capitatum***, Strawberry Spinach
- *Dye*—Thompson *Red* Calyx crushed and red stain used on the face, body, clothes, wood, and skins. (as *Blitum capitatum* 164:502)
- *Other*—Thompson *Paint* Calyx crushed and red paint used on the face, body, clothes, wood, and skins. (as *Blitum capitatum* 164:502)

Chenopodium carinatum R. Br., Tasmanian Goosefoot
- *Food*—Atsugewi *Bread & Cake* Parched, ground seeds made into cakes and eaten without cooking. (61:139)

Chenopodium fremontii S. Wats., Frémont's Goosefoot
- *Food*—Cahuilla *Vegetable* Boiled shoots and leaves eaten as greens. (15:52) **Cocopa** *Vegetable* Young shoots boiled as greens. (37:202) **Havasupai** *Bread & Cake* Seeds used to make bread. (197:66) **Hopi** *Porridge* Ground seeds used to make mush. (190:161) *Spice* Leaves used as flavoring with meat or other vegetables. *Vegetable* Leaves cooked alone as greens or boiled and eaten with a number of other foods. (42:300) **Klamath** *Unspecified* Roasted, ground seeds used for food. (45:95–96) **Mohave** *Vegetable* Young shoots boiled as greens. (37:202) **Navajo** *Bread & Cake* Seeds used to make tortillas and bread. (55:44) **Paiute** *Unspecified* Species used for food. (167:243) **Paiute, Northern** *Staple* Seeds ground into a meal and eaten. (59:48)
- *Other*—Hopi *Containers* Leaves packed around yucca fruit when baked in earth oven. (42:300)

Chenopodium graveolens Willd., Fetid Goosefoot
- *Drug*—Keres, Western *Emetic* Plant used as an emetic. (as *C. cornutum* 171:36) **Zuni** *Analgesic* and *Herbal Steam* Plant steeped in water and vapor inhaled for headache. (as *C. cornutum* 166:45)
- *Food*—Hopi *Bread & Cake* Seeds ground, mixed with cornmeal, and made into small dumplings wrapped in corn husks. (as *C. cornutum* 56:18)
- *Other*—Keres, Western *Protection* Plant used to protect people from lightning during thunder showers. (as *C. cornutum* 171:36) **Navajo** *Ceremonial Items* Used, with other herbs, in the liniment for the Mountain Chant. (as *C. cornutum* 55:44) **Navajo, Ramah** *Protection* Cold infusion taken to give protection in warfare. (191:25)

Chenopodium humile Hook., Marshland Goosefoot
- *Food*—Cahuilla *Vegetable* Boiled shoots and leaves eaten as greens. (15:52)

Chenopodium incanum (S. Wats.) Heller, Mealy Goosefoot
- *Drug*—Navajo, Ramah *Antidote* Stem, 3 inches long, made into snake figurine for snake infection. (191:25)
- *Food*—Apache, Western *Unspecified* Species used for food. (26:192) **Apache, White Mountain** *Unspecified* Seeds ground and used for food. Young sprouts boiled with meat and eaten. (136:156) **Hopi** *Vegetable* Young, tender leaves cooked and eaten as greens. (200:73) **Navajo, Ramah** *Bread & Cake*, and *Special Food* Seeds winnowed, ground with maize, made into bread, and used as a ceremonial food in Nightway. *Winter Use Food* Seeds stored for winter use. (191:25)

Chenopodium leptophyllum (Moq.) Nutt. ex S. Wats., Narrowleaf Goosefoot
- *Food*—Apache *Vegetable* Young plants cooked as greens. (32:16) **Apache, Western** *Unspecified* Species used for food. (26:192) **Apache, White Mountain** *Unspecified* Seeds ground and used for food. Young sprouts boiled with meat and eaten. (136:156) **Gosiute** *Unspecified* Seeds used for food. (39:366) **Hopi** *Porridge* Ground seeds used to make mush. (190:161) **Navajo, Ramah** *Unspecified* Seeds used for food. (191:25) **Pueblo** *Vegetable* Young plants cooked as greens. **Spanish American** *Vegetable* Young plants cooked as greens. (32:16) **Zuni** *Bread & Cake* Ground seeds mixed with cornmeal and salt, made into a stiff batter, formed into balls, and steamed. The Zuni say that upon reaching this world, the seeds were prepared without the meal because there was no corn. Now the young plants are boiled, either alone or with meat, and are greatly relished. (166:66) *Unspecified* Seeds considered among the most important food plants when the Zuni reached this world. (32:21) Young plants boiled alone or with meat and used for food. (166:66) *Vegetable* Young plants cooked as greens. (32:16)

Chenopodium murale L., Nettleleaf Goosefoot
- *Food*—Cahuilla *Vegetable* Boiled shoots and leaves eaten as greens. (15:52) **Mohave** *Vegetable* Young shoots boiled as greens. (37:202) **Papago** *Unspecified* Seeds used for food. (36:62) *Vegetable* Stalks eaten as greens in the summer. (34:14) **Pima** *Staple* Seeds parched, ground, and eaten as a pinole in combination with other meal. (32:23) Seeds parched, ground, and eaten as pinole. (146:73)

Chenopodium nevadense Standl., Nevada Goosefoot
- *Food*—Paiute *Staple* Seeds parched, ground, and eaten as meal. (104:98) **Paiute, Northern** *Unspecified* Seeds used for food. (59:48)

Chenopodium oahuense (Meyen) Aellen, Alaweo
- *Drug*—Hawaiian *Dermatological Aid* Plant used for beautifying the skin. Bark mixture eaten by nursing mother to beautify the skin of the child during growth and development. *Dietary Aid* Bark chewed by nursing mother to benefit the child. Juice mixed with other plants and given to children to fatten or add weight. *Pediatric Aid* Buds chewed by children with general weakness. Bark chewed by nursing mother to benefit the child. Bark mixture eaten by nursing mother to beautify the skin of the child during growth and development. Juice mixed with other plants and given to children to fatten or add weight. *Strengthener* Buds chewed by children with general weakness. (as *Chenepodium sandwicheum* 2:20)

Chenopodium pratericola Rydb., Desert Goosefoot
- *Food*—Pima, Gila River *Unspecified* Leaves boiled and eaten. (as *C. desiccatum* var. *leptophylloides* 133:7)

Chenopodium rubrum L., Red Goosefoot
- *Food*—Gosiute *Unspecified* Seeds used for food. (39:366)

Chenopodium sp., Goosefoot
- *Food*—Havasupai *Bread & Cake* Seeds parched, ground fine, boiled, thickened, made into balls, and eaten as dumplings. Seeds ground, kneaded into a thick paste, rolled into little balls, boiled, and eaten as marbles. (197:66) *Staple* Seeds ground and eaten as a ground or parched meal. (197:67) *Unspecified* Seeds used for food. (197:217)

Isleta *Vegetable* Leaves used as greens. (100:25) **Keresan** *Vegetable* Leaves used for greens. (198:560) **Malecite** *Unspecified* Species used for food. (160:6) **Navajo** *Bread & Cake* Seeds used to make bread. *Porridge* Seeds used to make a stiff porridge. *Staple* Seeds of several species ground and used like corn. (55:44) **Papago** *Vegetable* Leaves eaten as greens in midsummer. (34:14) Greens used for food. (36:61) **Pima** *Vegetable* Leaves boiled, salted, strained, fried in grease, and eaten as greens. **Yaqui** *Vegetable* Leaves eaten as greens. (47:70) **Yavapai** *Unspecified* Parched, ground, boiled seeds used for food. *Vegetable* Leaves and stems boiled for greens. (65:256)
- *Other*—**Navajo** *Ceremonial Items* Used, with other plants, as a liniment in the Mountain Chant. *Insecticide* Finely chopped plant used on the face and arms to keep the flies and mosquitoes from biting. (55:44)

Chenopodium watsonii A. Nels., Watson's Goosefoot
- *Food*—**Navajo, Ramah** *Unspecified* Seeds used for food. (191:25)

Chilopsis, Bignoniaceae
Chilopsis linearis (Cav.) Sweet, Desert Willow
- *Food*—**Cahuilla** *Unspecified* Blossoms and seedpods used for food. (15:53)
- *Fiber*—**Cahuilla** *Building Material* Wood used to make house frames and granaries. *Clothing* Bark used to make shirts and breechclouts. *Cordage* Bark used to make nets. (15:53) **Havasupai** *Basketry* Branches, with bark removed, used unsplit as rod foundations in coil basketry. (197:241) **Hualapai** *Furniture* Used to make the bed of the cradleboards. *Sewing Material* Used for making cloth woven "spudi." (195:8)
- *Other*—**Cahuilla** *Hunting & Fishing Item* Wood used to make bows. *Protection* Tree usually comfortable to camp under, providing some shade for the desert dweller. *Tools* Long limbs used as sticks to reach fruits and nuts too high to grasp by hand. (15:53)

Chimaphila, Ericaceae
Chimaphila maculata (L.) Pursh, Striped Prince's Pine
- *Drug*—**Cherokee** *Analgesic* Poultice of root used for pain. *Antirheumatic (Internal)* Tops and roots used for rheumatism. *Cancer Treatment* Used as a wash for cancer and ulcers. *Cold Remedy* Infusion of leaves used for colds. *Dermatological Aid* Decoction used for tetter and ringworm. *Emetic* Infusion given to make baby vomit and poultice of root used for pain. *Febrifuge* Infusion of leaves used for fevers. *Pediatric Aid* Infusion given to make baby vomit and poultice of root used for pain. *Poison* Infusion used to kill rats. *Tuberculosis Remedy* Used as a wash for scrofula. *Urinary Aid* Tops and roots used for urinary problems. *Veterinary Aid* Infusion given for "milksick." (80:62) **Nanticoke** *Misc. Disease Remedy* Plant used for ague. (175:84)
- *Food*—**Cherokee** *Snack Food* Leaves used for a nibble. (126:38)

Chimaphila menziesii (R. Br. ex D. Don) Spreng., Little Prince's Pine
- *Drug*—**Karok** *Gynecological Aid* Decoction of leaves taken for female troubles. *Kidney Aid* and *Urinary Aid* Decoction of leaves taken for kidney and bladder troubles. (6:23)

Chimaphila umbellata (L.) W. Bart., Pipsissewa
- *Drug*—**Abnaki** *Cold Remedy* Used for head colds. (144:154) *Nose Medicine* Powdered leaves mixed with bark from another plant and used as snuff for nasal inflammation. (144:170) **Catawba** *Analgesic* and *Orthopedic Aid* Plant used for backache. (177:47) **Chippewa** *Eye Medicine* Decoction of root used as drops for sore eyes. (53:360) *Venereal Aid* Plant used for gonorrhea. (71:138) **Cree, Woodlands** *Antihemorrhagic* Infusion or decoction of plant with other species used for coughing up blood. Decoction of plant taken for coughing up blood. *Heart Medicine* Used for pain and fever caused by chest ailments due to heart conditions such as angina pectoris. *Orthopedic Aid* Decoction of plant used for backaches. *Pulmonary Aid* Decoction of plant used for stabbing pain in the chest. (109:35) **Delaware** *Blood Medicine* Infusion of plant, mallow root, elder flowers, and dwarf elder bark used as a blood purifier. *Dermatological Aid* Infusion of plant used for blisters. *Pulmonary Aid* Infusion of plant, mallow root, elder flowers, and dwarf elder bark used to remove lung mucus. *Urinary Aid* Infusion of plant, mallow root, elder flowers, and dwarf elder bark used for bladder inflammation. *Venereal Aid* Infusion of plant, mallow root, elder flowers, and dwarf elder bark used for scrofula. (176:35) **Delaware, Oklahoma** *Blood Medicine* Compound containing plant taken as a blood purifier. *Dermatological Aid* Infusion of plant applied to blisters. *Expectorant* Compound containing plant taken to help remove mucus from the lungs. *Tuberculosis Remedy* Compound containing plant taken for "scrofula." *Urinary Aid* Compound containing plant used for bladder inflammation. (175:29, 74) **Flathead** *Eye Medicine* Solution of plant used as eye medicine. (82:34) **Karok** *Orthopedic Aid* Poultice of plant applied or infusion of leaves taken for backaches. (148:387) **Kutenai** *Eye Medicine* Solution of plant used as eye medicine. *Kidney Aid* Infusion of plant used for kidney trouble. (82:34) **Malecite** *Blood Medicine* Infusion of plants used as a blood purifier. (116:253) *Tuberculosis Remedy* Infusion of plants and juniper roots used for consumption. (116:251) **Menominee** *Adjuvant* Plant used as a seasoner to make female remedies taste good. (151:35) *Blood Medicine* Decoction of leaves taken to "clear the blood." (54:129) *Gynecological Aid* Compound decoction of root taken after childbirth to aid internal healing. (54:133) **Micmac** *Antirheumatic (Internal)* Herb used for rheumatism. *Blood Medicine* Herb used as a blood purifier. *Dermatological Aid* Herb used for blisters. *Gastrointestinal Aid* Herb used for stomach trouble. *Kidney Aid* Herb used for kidney trouble. (40:56) Used for kidney pains. (145:57) *Misc. Disease Remedy* Herb used for smallpox. *Tuberculosis Remedy* Herb used for consumption. *Urinary Aid* Herb used for "cold in bladder." (40:56) Infusion of roots, hemlock, parsley, and curled dock used for colds in the bladder. (116:259) **Mohegan** *Dermatological Aid* Infusion of plant applied to blisters. (174:265) Infusion of leaves applied to blisters. (176:71, 128) **Montagnais** *Diaphoretic* Decoction of plant taken to induce sweating. (156:316) **Montana Indian** *Febrifuge* Decoction of the herb or root used as a febrifuge for fevers. (19:9) Infusion taken for fever. (82:34) **Nanticoke** *Misc. Disease Remedy* Infusion of plant taken for ague. (175:56) **Ojibwa** *Gastrointestinal Aid* Infusion of plant used for stomach troubles. (153:368) **Okanagan-Colville** *Blood Medicine* Infusion of roots and leaves taken as a blood purifier. Decoction of whole plant taken as a blood purifier. *Cold Remedy* Infusion of roots and leaves taken for long-lasting colds. *Dietary Aid* Decoction of whole plant taken as an appetizer. *Kidney Aid* Infusion of roots and leaves taken to "clean

out" the kidneys. *Tuberculosis Remedy* Infusion of roots and leaves taken for tuberculosis. (188:101) **Okanagon** *Dermatological Aid* Poultice of crushed plant applied to leg and foot swellings. *Gynecological Aid* Plant chewed or infusion of leaves taken by women before and after childbirth. (125:42) **Penobscot** *Dermatological Aid* Infusion of plant applied to blisters. (156:309) **Rappahannock** *Blood Medicine* Infusion of roots taken to benefit blood. *Dietary Aid* Infusion of dried roots in brandy taken or chewed for the appetite. *Gastrointestinal Aid* Infusion of dried roots in brandy taken or chewed for the stomach. *Tonic* Infusion of dried roots in brandy used as a tonic for feeling low. *Veterinary Aid* Crushed and dried leaves mixed with the feed of mules to remove "bot worms." (161:26) **Saanich** *Antirheumatic (External)* Leaves put in bath water of sprinters and canoers as a liniment for sore muscles. (182:83) **Thompson** *Dermatological Aid* Poultice of crushed plant applied to leg and foot swellings. (125:42) *Gynecological Aid* Plant chewed or infusion of leaves taken by women before and after childbirth. (125:41) Plant chewed at childbirth to ease confinement. (164:462) Warm decoction of leaves taken before and after childbirth. *Orthopedic Aid* Poultice of crushed, fresh plant applied to leg and foot swellings. *Tonic* Decoction of leaves taken as a tonic for general indisposition. (164:477)
- *Other*—**Blackfoot** *Smoke Plant* Dried leaves used as the favorite smoking tobacco. (114:277) **Montana Indian** *Smoke Plant* Leaves dried and smoked. (82:34)

Chimaphila umbellata ssp. *cisatlantica* (Blake) Hultén, Pipsissewa
- *Drug*—**Iroquois** *Adjuvant* Plant mixed, as a medicine strengthener, with any medicine. (87:408) *Analgesic* Compound decoction of roots taken for urinating pain. *Anthelmintic* Infusion of plants given to babies with worms. (87:407) *Antirheumatic (Internal)* Compound decoction of plants taken for rheumatism. (87:408) *Blood Medicine* Decoction of roots or stems taken to purify bad blood or for blood chills. *Cancer Treatment* Decoction of stalks and roots taken for stomach cancer. *Dermatological Aid* Infusion of dried roots taken for pimples and sores on the face and neck. *Diuretic* Leaves and stems used as a diuretic. *Febrifuge* Infusion of plants taken by feverish and drowsy pregnant women. *Gynecological Aid* Compound decoction of bark and roots taken to induce pregnancy. Compound infusion of leaves and bark taken for miscarriage. Infusion of plants taken by feverish and drowsy pregnant women. (87:407) *Kidney Aid* Compound decoction of roots taken for the kidneys and dropsy. (87:408) *Laxative* Compound decoction of roots taken as a laxative. *Pediatric Aid* Infusion of plants given to babies with worms. *Stimulant* Infusion of plants taken by feverish and drowsy pregnant women. (87:407) *Tonic* Compound decoction of roots and bark taken as a tonic. (87:408) *Urinary Aid* Compound decoction of roots taken for urinating pain. *Venereal Aid* Compound decoction taken as blood purifier and for venereal disease. (87:407)

Chimaphila umbellata ssp. *occidentalis* (Rydb.) Hultén, Pipsissewa
- *Drug*—**Yurok** *Antirheumatic (External)* Leaves used for various aches and pains and to relax the muscles. *Kidney Aid* Leaves used for kidney ailments. (6:23)
- *Food*—**Thompson** *Beverage* Stem and roots boiled and drunk as a tea. Leaves made into a tea. (164:494)
- *Other*—**Blackfoot** *Smoke Plant* Leaves used in the tobacco mixture. (97:49)

Chionanthus, Oleaceae
Chionanthus virginicus L., White Fringetree
- *Drug*—**Choctaw** *Dermatological Aid* Decoction of bark used as wash or poultice as dressing for cuts or bruises. (25:23) Decoction of roots or bark used as a wash for infected sores and wounds. Poultice of beaten bark applied to cuts and bruises. *Disinfectant* Decoction of roots used as a wash for infected sores. **Koasati** *Dermatological Aid* Decoction of bark used as a wash for cuts. (177:50)

Chloracantha, Asteraceae
Chloracantha spinosa (Benth.) Nesom, Spiny Chloracantha
- *Food*—**Mohave** *Starvation Food* Young shoots roasted and eaten as a famine food. (as *Aster spinosus* 37:201) **Navajo** *Candy* Stems chewed for gum. (as *Aster spinosus* 55:83)

Chlorogalum, Liliaceae
Chlorogalum angustifolium Kellogg, Narrowleaf Soapplant
- *Other*—**Karok** *Soap* Bulbs pounded, mixed with water, and used as a detergent for washing clothes and buckskin blankets. (148:380)

Chlorogalum parviflorum S. Wats., Smallflower Soapplant
- *Food*—**Luiseño** *Unspecified* Bulb used for food. (155:233)

Chlorogalum pomeridianum (DC.) Kunth, Wavyleaf Soapplant
- *Drug*—**Cahuilla** *Dermatological Aid* Saponaceous material used as a dandruff shampoo. (15:54) **Costanoan** *Dermatological Aid* Pounded stalks used as a wash for dandruff. *Poison* Bulb used as a fish poison. (21:28) **Mendocino Indian** *Dermatological Aid* Plant used as a hair wash for dandruff. *Poison* Root considered poisonous. (41:319) **Mewuk** *Poison* Plant considered poisonous. (117:336) **Pomo** *Dermatological Aid* Plant juice rubbed on area affected by poison oak. (41:319) Bulb used as a soap for washing the hair and to prevent lice. (66:12) **Pomo, Kashaya** *Dermatological Aid* Bulb used as soap for washing body, hair, and utensils. (72:107) **Wailaki** *Analgesic* Bulb rubbed on body for cramps. *Antirheumatic (External)* Bulb rubbed on body for rheumatism. *Carminative* Decoction of bulbs taken for gas. *Dermatological Aid* Poultice of roasted bulbs used antiseptically for sores. *Disinfectant* Poultice of roasted bulbs used antiseptically for sores. *Diuretic* Decoction of bulbs taken as a diuretic. *Gastrointestinal Aid* Decoction of bulbs taken for stomachaches. *Laxative* Decoction of bulbs taken as a laxative. (41:319)
- *Food*—**Cahuilla** *Vegetable* Young, spring shoots used as a potherb. (15:54) **Costanoan** *Unspecified* Leaves of immature plant used for food. (21:255) **Karok** *Unspecified* Bulbs roasted and eaten. (148:380) **Mendocino Indian** *Unspecified* Young shoots used for food. (41:319) **Miwok** *Unspecified* Soaproot used for food. *Winter Use Food* Stored, dried bulbs used for food. (12:157) **Yuki** *Forage* Bulbs eaten by pigs. (49:93)
- *Fiber*—**Cahuilla** *Brushes & Brooms* Coarse husk fibers from the bulb tied together and used as a cleaning brush and as a hairbrush. (15:54) **Costanoan** *Brushes & Brooms* Fibrous bulb covers tied in bundles to make brushes. (21:255) **Karok** *Brushes & Brooms* Bulb fibers used

to make a small brush. (148:380) **Kawaiisu** *Brushes & Brooms* Upper fiber ends dipped in starch, tied together, and used as brushes. The brush was used in the following ways: (1) to clean out the bedrock mortar hole before tobacco leaves were pounded and to remove the pulverized tobacco afterwards—a usage that probably applied to other materials; (2) to sweep fine meal off the sifting tray; (3) to brush the hair—an old brush must be used for this purpose or "the hair ends will split"; and (4) when wet, to scrub things clean. (206:19) **Luiseño** *Brushes & Brooms* Bulb fiber made into small brushes used for sweeping up scattered meal after pounding acorns. (155:203) Bulb fibers used to make a brush. (155:233) **Maidu** *Brushes & Brooms* Plant made into brushes and used to sweep meal off the sifting baskets. (as *Chorogalum pomeridianum* 173:68) **Mendocino Indian** *Mats, Rugs & Bedding* Fiber used occasionally for bedding. (41:319) **Mewuk** *Caulking Material* Made into a white mucilaginous paste and used to coat baskets. (117:336) **Pomo, Kashaya** *Brushes & Brooms* Bulb fibers tied into bundles to make scrub brushes and hairbrushes. (72:107) **Round Valley Indian** *Brushes & Brooms* Outer fibers made into brushes and used to scrape meal off the ground. (41:336)

- *Other*—**Cahuilla** *Hunting & Fishing Item* Saponaceous material used as a stupefying agent and placed into streams to catch fish. *Soap* Crushed bulb used as soap. (15:54) **Costanoan** *Hunting & Fishing Item* Bulbs produced a detergent foam used as fish poison. *Soap* Bulbs produced a detergent foam used in washing. (21:255) **Karok** *Soap* Bulbs pounded, mixed with water, and used as a detergent for washing clothes and buckskin blankets. (148:380) **Kawaiisu** *Fasteners* Bulbs boiled into liquid starch and applied to twined seed gathering baskets to close the interstices. (206:19) **Luiseño** *Soap* Grated root used as soap. (155:210) **Mahuna** *Soap* Fibrous bulbs used as soap bars for washing clothes. (140:39) **Mendocino Indian** *Cooking Tools* Outer portion of the bulbs made into small brushes and used for grinding acorns. Fresh, green leaves used to cover acorn bread dough while cooking. *Decorations* Green leaves formerly pricked into the skin to form tattoo marks. *Fasteners* Bulbs roasted and the juice used as a substitute for glue in attaching feathers to arrows. *Hunting & Fishing Item* Bulbs formerly mashed, stirred in rivers, and used as a poison to stupefy fish and eels. *Paint* Juice diluted, smeared over the back of a bow, and soot applied to produce a permanent black color. (41:319) **Pomo, Kashaya** *Hunting & Fishing Item* Bulb used for fish poison. *Soap* Bulb used as soap for washing body, hair, and utensils. (72:107) **Yuki** *Hunting & Fishing Item* Bulbs crushed, placed in still waters, and used as a fish poison. *Soap* Bulbs pounded, rubbed between the hands, and used as hair shampoo. (49:93)

Chlorogalum pomeridianum var. *divaricatum* (Lindl.) Hoover, Wavyleaf Soapplant

- *Food*—**Neeshenam** *Unspecified* Eaten raw, roasted, or boiled. (as *C. divaricatum* 129:377)

Chlorogalum sp., Soap Root

- *Other*—**Mewuk** *Fasteners* Bulbs made into glue and used to fasten sinew on the backs of the sinew-backed bows. (117:352) **Pomo** *Cooking Tools* Long leaves used to line the ground ovens while baking acorn bread. (117:290)

Chorizanthe, Polygonaceae

Chorizanthe staticoides Benth., Turkish Rugging

- *Drug*—**Tubatulabal** *Dermatological Aid* Infusion of entire plant used as lotion for pimples. (193:59)

Chrysobalanaceae, see *Chrysobalanus*, *Licania*

Chrysobalanus, Chrysobalanaceae

Chrysobalanus icaco L., Icaco Coco Plum

- *Drug*—**Seminole** *Love Medicine* Decoction of wood ashes placed on the tongue to cleanse the body and strengthen the marriage. (169:250)
- *Food*—**Seminole** *Unspecified* Plant used for food. (169:484)
- *Other*—**Seminole** *Hunting & Fishing Item* Plant used to make arrows. (169:484)

Chrysophyllum, Sapotaceae

Chrysophyllum oliviforme L., Satinleaf

- *Drug*—**Seminole** *Love Medicine* Decoction of wood ashes placed on the tongue to cleanse the body and strengthen the marriage. (169:250)
- Food—Seminole Unspecified Plant used for food. (169:481)

Chrysopsis, Asteraceae

Chrysopsis mariana (L.) Ell., Maryland Goldenaster

- *Drug*—**Delaware** *Pediatric Aid* Infusion of roots used to quiet infants. Infusion of roots used as a tonic for sickly children. *Sedative* Infusion of roots used to quiet infants. *Tonic* Infusion of roots used as a tonic for sickly children. (176:33) **Delaware, Oklahoma** *Sedative* Infusion of root given to infants to quiet them. *Tonic* Infusion of root used as a tonic for "sickly children." (175:28, 74)

Chrysothamnus, Asteraceae

Chrysothamnus depressus Nutt., Longflower Rabbitbrush

- *Drug*—**Navajo, Ramah** *Gynecological Aid* Decoction of plant used to facilitate labor and delivery of placenta. (191:49)
- *Other*—**Hopi** *Decorations* Used as prayer stick decorations. (200:96)

Chrysothamnus greenei (Gray) Greene, Greene's Rabbitbrush

- *Drug*—**Navajo** *Dermatological Aid* and *Misc. Disease Remedy* Infusion of plant tops used as a wash for chickenpox and measles eruptions. (55:83)
- *Other*—**Hopi** *Decorations* Used as prayer stick decorations. (200:96)

Chrysothamnus nauseosus (Pallas ex Pursh) Britt., Rubber Rabbitbrush

- *Drug*—**Cahuilla** *Toothache Remedy* Decoction of twigs taken for toothaches. (15:54) **Cheyenne** *Cold Remedy* Infusion of flower parts taken or burning plant smoke inhaled for colds. *Cough Medicine* Infusion of flower parts taken for coughs. (83:20) *Dermatological Aid* Decoction of leaves and stem used as wash and taken for sores, especially smallpox. (as *C. nauseosa* 75:39, 40) Infusion of leaves and stems used as a wash or taken for eruptions or body sores. (76:187) Infusion of leaves and stems used for sores and skin eruptions. (83: 20) *Misc. Disease Remedy* Decoction of leaves and stem taken for smallpox. (as *C. nauseosa* 75:39, 40) Infusion of leaves and stems taken for smallpox. (76:187) *Psychological Aid* Burning leaf and branch smoke used to drive away the cause of nightmares. *Tuberculosis Remedy* Infusion of flower parts taken for tuberculosis. (83:20) **Coahuilla** *Analgesic* Infusion of twigs taken for chest pain. *Cough Medicine* Infusion of twigs taken for coughs. (as *Bigilovia graveolens* 13:79) **Klamath** *Dermatological Aid* Poultice of herbage used to raise blisters. (as *Chondrophora nauseosa* 45:106) Poultice of smashed herbs applied to blisters. (as *Chondrophora nauseosa* 163:131) **Okanagan-Colville** *Gynecological Aid* Leaves used as a sanitary napkin. Leaves used as a sanitary napkin particularly after childbirth, to "heal the insides." (188:83) **Thompson** *Antidiarrheal* Decoction of plant taken for diarrhea. *Cold Remedy* Decoction of plant taken for colds. *Gastrointestinal Aid* Decoction of plant taken for stomach cramps. *Other* Plant used as a medicine for "drinking and bathing." *Panacea* Decoction of plant taken for "all diseases." *Tuberculosis Remedy* Decoction of plant taken for tuberculosis. *Urinary Aid* Decoction of plant taken for colds, venereal diseases, bladder trouble, and tuberculosis. *Venereal Aid* Decoction of plant taken for venereal disease. (187:178)

- **Food**—**Blackfoot** *Forage* Plant used as a fall and winter forage for horses. (97:56) **Kawaiisu** *Spice* Sharpened twig, stripped of bark and leaves, threaded with pinyon nuts to improve their flavor. (206:20) **Navajo** *Forage* Plants browsed by animals. (90:159) **Paiute** *Candy* Roots used as chewing gum. (111:115) **Paiute, Northern** *Candy* Root bark chewed like gum. (59:53)
- **Fiber**—**Hopi** *Building Material* Brush used to make windbreaks. (42:303) **Keres, Western** *Brushes & Brooms* Plant used for rough brooms. (as *Chrysothamnus confinis* 171:37)
- **Dye**—**Hopi** *Green* Bark used to make green dye. *Yellow* Flowers used as yellow dye. (42:303)
- **Other**—**Hopi** *Ceremonial Items* Plant used to weave the wedding belt. Plant used on the Po-wa-mu altar and used in Caquol ceremony. *Decorations* Stems used to make wicker plaques. *Hunting & Fishing Item* Stems used to make arrows. (42:303) **Keres, Western** *Hunting & Fishing Item* Plant used for arrows. (as *C. confinis* 171:37) **Navajo** *Cash Crop* Source of a commercial volatile oil. *Fuel* Used as a kiva fuel. (90:159) **Paiute** *Tools* Peeled stems used for piercing ears in preparation for wearing earrings. (111:115)

Chrysothamnus nauseosus ssp. *albicaulis* (Nutt.) Hall & Clements, Rubber Rabbitbrush

- **Drug**—**Sanpoil** *Veterinary Aid* Container of brush lighted and held under horse's nostrils for distemper. Salve of branches and leaves used on horses to keep horseflies and gnats away. (131:217) **Shoshoni** *Antidiarrheal* Decoction of roots and tops taken for bloody diarrhea. *Cold Remedy* Compound decoction of stems taken for colds. Infusion of leaves taken as a cold medicine. *Cough Medicine* Compound decoction of stems taken for coughs. Decoction of stems and leaves taken for coughs. *Gastrointestinal Aid* Infusion of leaves taken for stomach disorders. *Tonic* Infusion of dried leaves and flowers taken as a general tonic. (180:57)
- **Other**—**Sanpoil** *Insecticide* Salve of branches and leaves used on horses to keep horseflies and gnats away. (131:217)

Chrysothamnus nauseosus ssp. *bigelovii* (Gray) Hall & Clements, Rubber Rabbitbrush

- **Drug**—**Navajo, Kayenta** *Ceremonial Medicine* and *Emetic* Plant used as a ceremonial emetic. (205:46) **Navajo, Ramah** *Analgesic* Leaves made into a lotion and used for headache and decoction of root used for menstrual pain. *Cathartic* Compound decoction of leaves used as a cathartic. (191:49, 50) *Ceremonial Medicine* Leaves used as a ceremonial emetic. (191:49) *Cold Remedy* Strong decoction of root taken for colds. *Cough Medicine* Strong decoction of root taken for cough. *Emetic* Leaves used as an emetic for several ceremonies. *Febrifuge* Strong decoction of root taken for fever. *Gynecological Aid* Strong decoction of root taken for menstrual pain. (191:49, 50) **Tewa** *Oral Aid* and *Pediatric Aid* White galls from plants hung around babies' necks to stop dribbling. (as *C. bigelovii* 138:45)
- **Food**—**Apache, White Mountain** *Unspecified* Seeds ground and used for food. (as *C. bigelovii* 136:156) **Navajo, Kayenta** *Unspecified* Cooked with cornmeal mush and used for food. (205:46)
- **Fiber**—**Navajo, Ramah** *Mats, Rugs & Bedding* Branches used to carpet the sweat house floor. (191:49) **Zuni** *Basketry* Stems used to make baskets. The outer bark was removed and the stems were covered with sand to render them more pliable. The stems were often dyed and some of the completed baskets were decorated with color. (as *C. bigelovii* 166:81)
- **Dye**—**Apache, White Mountain** *Yellow* Blossoms used as a yellow dye. (as *C. bigelovii* 136:156) **Navajo, Ramah** *Yellow* Flowers used as a yellow dye for wool. Flowers and buds boiled overnight and used as a yellow dye for basket material. (191:49) **Tewa** *Yellow* Flowers boiled to make a yellow dye for woolen yarn. (as *C. bigelovii* 138:45) **Zuni** *Yellow* Blossoms used to make a yellow dye. (as *C. bigelovii* 166:80)
- **Other**—**Hopi** *Protection* Used to make windbreaks and other shelters for melon plants and young peach trees. (as *C. bigelovii* 138:45) **Navajo, Ramah** *Ceremonial Items* Branch used to make Enemyway prayer stick. (191:49) **Tewa** *Protection* Used to make windbreaks and other shelters for melon plants and young peach trees. (as *C. bigelovii* 138:45)

Chrysothamnus nauseosus ssp. *consimilis* (Greene) Hall & Clements, Rubber Rabbitbrush

- **Food**—**Paiute** *Candy* Roots chewed until gummy as a "chewing gum." (104:104)
- **Dye**—**Hopi** *Green* Bark used to obtain a green dye. *Yellow* Yellow flowers used to make a yellow dye. (as *C. pinifolius* Greene 200:95)
- **Other**—**Hopi** *Decorations* Stems used to make wicker plaques. (as *C. pinifolius* 200:95)

Chrysothamnus nauseosus ssp. *graveolens* (Nutt.) Piper, Rubber Rabbitbrush

- **Dye**—**Hopi** *Green* Bark used to obtain a green dye. *Yellow* Yellow flowers used to make a yellow dye. (as *C. graveolens* 200:95) **Navajo** *Green* Immature flowers, leaves, or green bark boiled with heated alum and used as a green dye for wool. *Yellow* Mature flowering tops boiled with heated alum and used as a yellow dye for wool. (55:83)
- **Other**—**Hopi** *Decorations* Stems used to make wicker plaques. (as *C. graveolens* 200:95) **Navajo** *Ceremonial Items* Plant used to remove evil spells in the Witch, Wind, and other chants. (55:83)

Chrysothamnus nauseosus ssp. *hololeucus* (Gray) Hall & Clements, Rubber Rabbitbrush

- **Dye**—**Hopi** *Green* Bark used to obtain a green dye. *Yellow* Yellow flowers used to make a yellow dye. (as *C. speciosus gnaphalodes* 200:95)

- *Other*—Hopi *Decorations* Stems used to make wicker plaques. (as *C. speciosus gnaphalodes* 200:95)

Chrysothamnus nauseosus ssp. *latisquameus* (Gray) Hall & Clements, Rubber Rabbitbrush
- *Drug*—Isleta *Poison* Wood used to make poisonous war arrows. (as *C. latisquameus* 100:26)
- *Dye*—Navajo *Yellow* Twigs and flowers used as a yellow dye for wool. (as *C. latisquameus* 55:83)
- *Other*—Isleta *Lighting* Wood, burned slowly and brightly, used as candles. *Weapon* Wood used to make poisonous war arrows that broke upon impact and could not be shot back. (as *C. latisquameus* 100:26)

Chrysothamnus nauseosus var. *occidentalis*, Rubber Rabbitbrush
- *Other*—Karok *Decorations* Stems and flowers tied by girls onto the end of hair rolls as imitation mink skins. (148:389)

Chrysothamnus parryi (Gray) Greene, Parry's Rabbitbrush
- *Fiber*—Hopi *Building Material* Used in rows in the sand to act as a sand break. (as *Linosyris howardi* 190:167)

Chrysothamnus parryi ssp. *howardii* (Parry ex Gray) Hall & Clements, Howard's Rabbitbrush
- *Drug*—Hopi *Ceremonial Medicine* Plant used in initiatory ceremonials. (as *Bigelovia howardii* 56:20)
- *Fiber*—Hopi *Building Material* Used in rows in the sand to act as a sand break. (as *C. howardii* 190:167)
- *Dye*—Hopi *Green* Bark used to obtain a green dye. *Yellow* Yellow flowers used to make a yellow dye. (as *C. howardi* 200:95)
- *Other*—Hopi *Decorations* Stems used to make wicker plaques. (as *C. howardi* 200:95) *Fuel* Dried plant used as one of the four prescribed kiva fuels. *Paint* Flowers and chalky stones used as a bright yellow pigment for personal decoration in ceremonies. (as *Bigelovia howardii* 56:20)

Chrysothamnus sp., Rabbitbrush
- *Drug*—Isleta *Febrifuge* Decoction of entire plant, except the roots, used as a bath for fevers. *Toothache Remedy* Small bit of stem used in cavities for toothache. It was believed that if the cavity was not sealed in some manner after the stem was inserted, the tooth would break. *Venereal Aid* Decoction of entire plant, except the roots, taken for venereal diseases. (100:25) Jemez *Cold Remedy* Plant used for chest colds. (44:21) Plant used for chest colds. (100:25) *Oral Aid* Decoction of green parts tied with cedar boughs or twigs and used as a gargle. (44:21) *Throat Aid* Plant used as a gargle. (100:25) Keres, Western *Psychological Aid* Infusion of plant used to relieve fright. (171:36) Keresan *Emetic* Infusion taken as an emetic. *Gastrointestinal Aid* Infusion taken for stomach trouble. (198:563)
- *Dye*—Navajo *Yellow* Used as a yellow dye. (55:83)

Chrysothamnus viscidiflorus (Hook.) Nutt., Green Rabbitbrush
- *Drug*—Hopi *Dermatological Aid* Poultice of chewed plant tips applied to boils. (as *Bigelovia douglasii* 56:20) Paiute *Cold Remedy* Infusion of crushed leaves taken or used as a wash for colds. (104:197) Infusion of crushed leaves taken for colds. *Cough Medicine* Decoction of young growth taken for coughs. *Diaphoretic* Branches used as a bed in the sweat bath for rheumatism. Shoshoni *Antirheumatic (External)* Poultice of moistened, crushed stems and leaves applied for rheumatism. *Misc. Disease Remedy* Hot compound decoction of plant taken for influenza. *Toothache Remedy* Finely mashed leaves inserted in tooth cavities for toothaches. (180:57, 58)
- *Food*—Gosiute *Candy* Roots used as chewing gum. (as *Bigelovia douglasii* 39:364) Hopi *Spice* Plant used as an herb. (42:302) Paiute *Candy* Roots used as chewing gum. (111:115)
- *Fiber*—Hopi *Building Material* Used as a sand break to protect young corn and melons. (as *Bigelovia douglasii* var. *stenophylla* 190:167) Navajo *Building Material* Plant used as thatch to prevent the sand on top of the sweat house from sifting through. (55:84) Yavapai *Brushes & Brooms* Stems used to brush off spines on fruits. (65:257)
- *Dye*—Hopi *Yellow* Blossoms used as a yellow dye for wools and cotton yarn. (42:302) Navajo *Orange* Flowers boiled with roasted alum and used as a light orange dye for leather, wool, and basketry. *Yellow* Flowers boiled with roasted alum and used as a yellow dye for leather, wool, and basketry. (55:84)
- *Other*—Hopi *Ceremonial Items* Plant used for ceremonies. *Cooking Tools* Plant used for roasting corn. (42:302) Paiute *Tools* Peeled stems used for piercing ears in preparation for wearing earrings. (111:115)

Chrysothamnus viscidiflorus ssp. *viscidiflorus* var. *stenophyllus* (Gray) Hall, Green Rabbitbrush
- *Fiber*—Hopi *Building Material* Used as a sand break to protect young corn and melons. (as *C. stenophyllus* 190:167)
- *Other*—Hopi *Decorations* Used as prayer stick decorations. (as *C. stenophyllus* 200:96)

Chrysothamnus viscidiflorus (Hook.) Nutt. ssp. *viscidiflorus* var. *viscidiflorus*, Green Rabbitbrush
- *Drug*—Navajo *Emetic* Plant used to make a sick person vomit. (55:84)

Cibotium, Cyatheaceae, fern

Cibotium chamissoi Kaulfuss, Chamisso's Manfern
- *Drug*—Hawaiian *Analgesic* Infusion of powdered bark and other plants taken for chest pains. *Antirheumatic (External)* Very fine, downy hairs burned and the heat applied to hardened muscles and tired limbs. *Blood Medicine* Infusion of powdered bark and other plants taken to purify the blood. *Dietary Aid* Infusion of powdered bark and other plants taken to stimulate the appetite, and taken for weight loss. *Sedative* Very fine, downy hairs burned and the heat applied for nervousness. (as *Cibatium whamissoi* 2:43)
- *Fiber*—Hawaiian *Mats, Rugs & Bedding* Down used for pillows. (as *Cibatium whamissoi* 2:43)

Cicer, Fabaceae

Cicer arietinum L., Chick Pea
- *Food*—Papago *Dried Food* Threshed, dried on the ground or roofs, stored, and used for food. (34:33) *Unspecified* Species used for food. Pima *Unspecified* Species used for food. (36:120)

Cichorium, Asteraceae

Cichorium intybus L., Chicory
- *Drug*—Cherokee *Tonic* Infusion of root used as a tonic for nerves.

(80:29) **Iroquois** *Dermatological Aid* Decoction of roots used as a wash and poultice applied to chancres and fever sores. (87:476)

Cicuta, Apiaceae

Cicuta douglasii (DC.) Coult. & Rose, Western Water Hemlock

- *Drug*—**Alaska Native** *Poison* Roots considered poisonous. (85:153) **Bella Coola** *Cathartic* Roots used as a purgative. (150:61) *Emetic* Infusion of grated tubers taken as an emetic to "clean out the bile in the stomach." (184:200) **Haisla & Hanaksiala** *Poison* Roots considered poisonous. (43:212) **Kawaiisu** *Analgesic* and *Orthopedic Aid* "Mashed root put on a hot stone and sore limbs laid directly over it." *Poison* Plant considered poisonous. (206:20) **Kutenai** *Dermatological Aid* Roots pounded and used for sores. *Emetic* Infusion of roots taken with large amounts of warm water as an emetic. This remedy was used cautiously because of the poisonous effects of larger doses of water hemlocks. (82:71) **Kwakiutl** *Antidiarrheal* Cold, compound infusion of burned, pulverized bark taken for diarrhea. (183:270) *Cathartic* Poultice of soaked roots applied to the stomach as a purgative. *Dermatological Aid* Poultice of roots applied to draw out thorns and splinters. *Emetic* Root extract and grease taken to induce vomiting. *Misc. Disease Remedy* Roots smashed, steamed, peeled, powdered, mixed with oil, and taken for any serious disease. *Poison* Plant considered highly poisonous. (183:276) **Montana Indian** *Analgesic* Roots used for headaches. *Snakebite Remedy* Poultice of split roots used for rattlesnake bites. (82:71) **Okanagan-Colville** *Poison* Plant considered a very bad poison. (188:60) **Paiute** *Antirheumatic (External)* Decoction of mashed roots used as a soothing solution for "rheumatism" and tired and aching feet. (111:96) **Salish, Coast** *Cathartic* Used with caution as a purgative. *Emetic* Used with caution to induce vomiting. (182:89) **Shuswap** *Poison* Plant considered poisonous. (123:56) **Thompson** *Analgesic* Poultice of root used for severe pain in the legs and back. Decoction of plant used as a wash for aching bones. *Orthopedic Aid* Boiled, roots used by bedridden patients, or poultice of roots used for broken hips. The bedridden patient's back was splashed with water and the boiled, mashed roots placed on the back to help the patient recover. The informant warned that the poultice should not be left on for more than half an hour and that because of its toxicity it should never be taken internally. Decoction of plant used as a wash for broken bones. (187:150) *Panacea* Root eaten to protect against disease and give feeling of "perfect wellness." (as *Cicuta vagans* 164:476) *Poison* Roots known to be poisonous to both humans and animals. (as *Cicuta vagans* 164:513) Considered one of the most toxic plants in North America for people, horses, cattle, and sheep. (187:150) *Veterinary Aid* Roots known to be poisonous to animals. (as *Cicuta vagans* 164:513)
- *Other*—**Okanagan-Colville** *Weapon* Roots powdered and used as an arrow poison during warfare. (188:60)

Cicuta maculata L., Spotted Water Hemlock

- *Drug*—**Alaska Native** *Poison* Roots considered poisonous. (85:153) **Cherokee** *Ceremonial Medicine* Root chewed, if dizziness occurred person would die soon, if not, long life. *Contraceptive* Roots eaten for four consecutive days "to become sterile forever." (80:31) Roots chewed and swallowed by women as form of contraception and become sterile. (177:45) **Cree, Woodlands** *Antirheumatic (External)* Dried roots powdered, made into a liniment, and applied externally. (109:35) **Iroquois** *Dermatological Aid* Poultice of smashed roots applied for lameness, running sores, or cuts. *Disinfectant* Handling plants caused fits and decoction used as floor wash to prevent disease. *Orthopedic Aid* Decoction of plants used on bruises, sprains, sore joints, or broken bones. *Poison* Roots chewed to commit suicide. *Veterinary Aid* Poultice of smashed roots applied to horses for lameness or running sores. (87:398) **Klamath** *Poison* Poisonous roots mixed with rattlesnake poison or decomposed animal liver and used to poison arrows. (45:101) **Lakota** *Poison* Plant poisonous to humans. (139:33) **Ojibwa** *Hunting Medicine* Root used in hunting medicine smoked to attract the buck deer near enough to shoot with bow and arrow. (153:432) *Unspecified* Root used medicinally. (153:390)
- *Other*—**Cherokee** *Insecticide* Infusion of root used to soak corn before planting to repel insect pests. (80:31) **Chippewa** *Smoke Plant* Seeds mixed with tobacco and smoked. (53:377)

Cicuta maculata var. *angustifolia* Hook., Spotted Water Hemlock

- *Drug*—**Montana Indian** *Poison* Fleshy roots known as a virulent poison and sometimes used for suicide. (as *Cicuta occidentalis* 19:10) **Paiute** *Analgesic* Poultice of roasted roots applied to "deaden muscular pain." *Antirheumatic (External)* Poultice of roasted roots applied to rheumatic joints. *Dermatological Aid* Poultice of roasted roots applied for ordinary swellings. *Orthopedic Aid* Poultice of roasted roots applied to "deaden muscular pain." *Snakebite Remedy* Poultice of pulped root applied to rattlesnake bites for the swelling. **Shoshoni** *Eye Medicine* Cool decoction of root used as a wash for sore eyes or granulated lids. *Orthopedic Aid* Poultice of roots applied to "deaden muscular pain." *Poison* Root pulp considered poisonous for open wounds. (as *Cicuta occidentalis* 180:58, 59)

Cicuta maculata L. var. *maculata*, Spotted Water Hemlock

- *Drug*—**Seminole** *Febrifuge* Decoction of leaves, roots, and stems used as a bath for high fevers. (as *Cicuta curtissii* 169:282)

Cicuta sp., Water Hemlock

- *Drug*—**Eskimo, Arctic** *Poison* Roots considered poisonous. (128:17) **Mendocino Indian** *Poison* Plant considered poisonous. (41:372)

Cicuta virosa L., Mackenzie's Water Hemlock

- *Drug*—**Alaska Native** *Poison* Roots considered poisonous. (as *Cicuta mackenziana* 85:153) **Eskimo, Inupiat** *Poison* Whole plant considered poisonous. (as *Cicuta mackenziana* 98:137) **Eskimo, Kuskokwagmiut** *Poison* Root "considered to be poisonous to people." (as *Cicuta mackenziana* 122:21) **Eskimo, Western** *Poison* Roots considered poisonous. (as *Cicuta mackenziana* 108:17) **Haisla & Hanaksiala** *Poison* Roots considered poisonous. (as *Cicuta mackenzieana* 43:212)
- *Food*—**Eskimo, Inuktitut** *Unspecified* Leaves boiled with fresh fish. (as *Cicuta mackenziana* 202:192)

Cimicifuga, Ranunculaceae

Cimicifuga racemosa (L.) Nutt., Black Bugbane

- *Drug*—**Cherokee** *Abortifacient* Used to stimulate menstruation. *Analgesic* Used as an anodyne. (80:30) Infusion of root "in spirits" used for rheumatic pains. (203:73, 74) *Antirheumatic (Internal)* Infusion of roots in alcoholic spirits used for rheumatism. (80:30) Infusion of root "in spirits" used for rheumatic pains. (203:73, 74) *Cold Remedy* Infusion taken for colds. *Cough Medicine* Infusion taken for coughs. *Dermatological Aid* Given for hives. Infusion of slightly astringent plant taken for rheumatism. *Diuretic* Used as a diuretic. *Laxative* Infusion taken for constipation. *Pediatric Aid* and *Sedative* Given to make babies sleep. *Stimulant* Given for fatigue. *Tonic* Used as a tonic. *Tuberculosis Remedy* Infusion taken for colds, coughs, "consumption," and constipation. (80:30) **Delaware** *Tonic* Used with elecampane and stone root to make a tonic. (176:33) Combined with elecampane and stone roots to make a tonic. (176:39) **Delaware, Oklahoma** *Tonic* Compound containing root used as a tonic. (175:32, 74) **Iroquois** *Antirheumatic (External)* Decoction of roots or plants used as a soak and steam bath for rheumatism. (87:320) *Blood Medicine* Root used as a blood purifier. (87:321) *Gynecological Aid* Infusion of roots taken to "promote the flow of milk in women." *Orthopedic Aid* and *Pediatric Aid* Poultice of smashed leaves applied to babies with sore backs. (87:320) **Micmac** *Kidney Aid* Root used for kidney trouble. (40:56) **Penobscot** *Kidney Aid* Root used to make a medicine and taken for kidney trouble. (156:310, 311)

Cinchona, Rubiaceae

Cinchona calisaya, Quinine
- **Drug**—**Cherokee** *Reproductive Aid* Infusion of plant taken for impotence. *Tonic* Used to make a tonic. (as *C. ledgeriana* 80:49)

Cinna, Poaceae

Cinna arundinacea L., Sweet Woodreed
- **Drug**—**Iroquois** *Misc. Disease Remedy* Compound decoction of plants taken for sugar diabetes. (87:274)
- **Food**—**Gosiute** *Unspecified* Seeds used for food. (as *C. arundinaceae* var. *pendula* 39:366)

Circaea, Onagraceae

Circaea lutetiana ssp. *canadensis* (L.) Aschers. & Magnus, Broadleaf Enchanter's Nightshade
- **Drug**—**Iroquois** *Dermatological Aid* Plant used on wounds. *Other* Compound infusion taken and used as wash on injured part, a "Little Water Medicine." (87:391)

Cirsium, Asteraceae

Cirsium altissimum (L.) Hill, Tall Thistle
- **Drug**—**Cherokee** *Analgesic* Infusion of leaves taken for neuralgia. *Gastrointestinal Aid* Warm infusion of roots taken to help person who overeats. *Poultice* Roots used as poultice and decoction of bruised plant used to poultice sore jaw. (as *Carduus altissimus* 80:58)
- **Other**—**Cherokee** *Hunting & Fishing Item* Down used as the tail for blow darts. (as *Carduus altissimus* 80:58)

Cirsium arvense (L.) Scop., Canadian Thistle
- **Drug**—**Abnaki** *Anthelmintic* Used as a vermifuge. (144:155) Decoction of roots used by children for worms. *Pediatric Aid* Decoction of roots used by children for worms. (144:173) **Iroquois** *Oral Aid* Infusion of roots used for mouth sickness. (141:63) **Mohegan** *Oral Aid* and *Pediatric Aid* Infusion of leaves used as a mouthwash for infants. (176:71, 128) *Pulmonary Aid* Plant used for lung trouble. (176:128) *Tuberculosis Remedy* Decoction of plant taken for consumption. (174:269) **Montagnais** *Tuberculosis Remedy* Decoction of plant used for consumption. (156:314) Decoction of plant taken for consumption. (174:269) **Ojibwa** *Gastrointestinal Aid* Plant used as a "bowel tonic." (153:364)

Cirsium brevistylum Cronq., Clustered Thistle
- **Food**—**Cowichan** *Unspecified* Large taproots peeled and eaten raw or cooked. (182:81) **Hesquiat** *Unspecified* Flower heads chewed to get the nectar. (185:61) **Saanich** *Unspecified* Large taproots peeled and eaten raw or cooked. (182:81)
- **Fiber**—**Nitinaht** *Clothing* Down spun with yellow cedar inner bark and used for baby clothing. (186:97)
- **Other**—**Nitinaht** *Protection* Leaves or roots dried and placed around the house for protection from evil influences. (186:97) **Saanich** *Protection* Sharp leaves placed in bath water for protection from evil spirits, particularly before feasts. (182:81)

Cirsium calcareum (M. E. Jones) Woot. & Standl., Cainville Thistle
- **Drug**—**Hopi** *Anthelmintic* Plant used as a worm remedy. (as *C. pulchellum* 200:34, 95, 96) *Dermatological Aid* Plant used for itching. (as *C. pulchellum* 200:32, 95, 96) *Laxative* Plant used as a laxative. *Throat Aid* Decoction of plant used for tickling throat caused by a cold. (as *C. pulchellum* 200:34, 95–96) **Navajo, Ramah** *Eye Medicine* Cold infusion of root used as a wash for sore eyes. (191:50)

Cirsium californicum Gray, California Thistle
- **Food**—**Kawaiisu** *Unspecified* Spring stems skinned and eaten raw. (206:20)

Cirsium congdonii Moore & Frankton, Rosette Thistle
- **Food**—**Kawaiisu** *Unspecified* Spring stalks peeled and eaten raw. (206:20)

Cirsium discolor (Muhl. ex Willd.) Spreng., Field Thistle
- **Drug**—**Cree** *Dermatological Aid* Poultice of root paste applied to linen cloths and bound to the wound. (16:490) **Iroquois** *Dermatological Aid* Compound decoction of roots taken or poultice of roots applied to boils. *Hemorrhoid Remedy* Compound decoction of plants taken for piles. (87:475) **Meskwaki** *Analgesic* Infusion of root used for stomachaches. *Gastrointestinal Aid* Infusion of root taken for stomachache. (152:213)

Cirsium drummondii Torr. & Gray, Dwarf Thistle
- **Food**—**Atsugewi** *Unspecified* Young, raw stalks used for food. (61:139) **Cahuilla** *Unspecified* Bud, at the base of the thistle, used for food. (15:55) **Gosiute** *Unspecified* Stems formerly used for food. (as *Cnicus drummondi* 39:366)

Cirsium eatonii (Gray) B. L. Robins., Eaton's Thistle
- **Drug**—**Gosiute** *Dermatological Aid* Plant used for wounds, cuts, or sores. (as *Cnicus eatoni* 39:349, 366)
- **Food**—**Gosiute** *Unspecified* Stems used for food. (as *Cnicus eatoni* 39:366)

Cirsium edule Nutt., Edible Thistle
- **Food**—**Cheyenne** *Special Food* Young stems eaten raw as a "luxury food." (83:20) *Unspecified* Peeled stem used for food. (76:191) Tender, spring shoots eaten raw. (83:46) **Hoh** *Vegetable* Young shoots eaten

as greens. (as *Carduus edulis* 137:69) **Okanagon** *Unspecified* Roots boiled and used for food. (as *Carduus edulis* 125:36) **Quileute** *Vegetable* Young shoots eaten as greens. (as *Carduus edulis* 137:69) **Thompson** *Dried Food* Roots dried and stored for future use.[23] *Soup* Dried roots rehydrated, scraped, chopped, and cooked in stews.[23] (187:178) *Unspecified* Roots boiled and used for food. (as *Carduus edulis* 125:36) Fresh roots eaten cooked.[23] (187:178)

Cirsium hookerianum Nutt., White Thistle

- *Food*—**Okanagon** *Unspecified* Roots boiled and used for food. (as *Carduus hookerianus* 125:36) **Thompson** *Dried Food* Roots dried and stored for future use.[23] Dried roots rehydrated, scraped, chopped, and cooked in stews.[23] (187:178) *Unspecified* Roots boiled and used for food. (as *Carduus hookerianus* (125:36) Deep, thick roots cooked and eaten. (164:480) Fresh roots eaten cooked.[23] (187:178)

Cirsium horridulum Michx., Yellow Thistle

- *Drug*—**Houma** *Dermatological Aid* Infusion of leaves and root in whisky recognized as a strong astringent. *Expectorant* Infusion of leaf and root in whisky taken to clear phlegm from lungs and throat. *Throat Aid* Infusion of leaves and root in whisky taken to clear throat and lungs of phlegm. (158:57)
- *Food*—**Houma** *Unspecified* Tender, white hearts eaten raw. (158:57)
- *Other*—**Seminole** *Hunting & Fishing Item* Plant used to make blowgun darts. (169:507)

Cirsium neomexicanum Gray, New Mexico Thistle

- *Drug*—**Navajo** *Febrifuge* Plant used for chills and fevers. (55:96) **Navajo, Ramah** *Eye Medicine* Cold infusion of root used as a wash for eye diseases. *Panacea* Root used as a "life medicine." Cold infusion of plant taken when one "feels bad all over." *Veterinary Aid* Cold infusion of root used as a wash for livestock with eye diseases. (191:50)
- *Food*—**Yavapai** *Unspecified* Raw, peeled stems used for food. (65:256)

Cirsium occidentale (Nutt.) Jepson, Cobwebby Thistle

- *Food*—**Kawaiisu** *Unspecified* Spring stems skinned and eaten raw. (206:20) **Tubatulabal** *Unspecified* Stalks used extensively for food. (193:15)

Cirsium ochrocentrum Gray, Yellowspine Thistle

- *Drug*—**Kiowa** *Burn Dressing* Decoction of blossoms used as wash for burns. *Dermatological Aid* Decoction of blossoms used as wash for sores. (192:58) **Zuni** *Contraceptive* Infusion of root taken by both partners as a contraceptive. (27:374) *Diaphoretic* Infusion of whole plant taken as a diaphoretic for syphilis. *Diuretic* Infusion of whole plant taken as a diuretic for syphilis. *Emetic* Infusion of whole plant taken as an emetic for syphilis. (as *Carduus ochrocentrus* 166:44, 45) *Misc. Disease Remedy* Infusion of fresh or dried root taken three times a day for diabetes. *Venereal Aid* Infusion of whole plant taken for syphilis. (27:374) Infusion of whole plant taken for syphilis. (as *Carduus ochrocentrus* 166:44, 45)
- *Food*—**Kiowa** *Unspecified* Roots used for food. (192:58)
- *Other*—**Kiowa** *Protection* Blossoms used to cover graves of those recently buried to keep the wolves from digging up the body. (192:58)

Cirsium pallidum Woot. & Standl., Pale Thistle

- *Drug*—**Keres, Western** *Diuretic* Roots used as a diuretic. (171:37)
- *Food*—**Apache, Chiricahua & Mescalero** *Bread & Cake* Seeds threshed, winnowed, ground, and the flour used to make bread. *Unspecified* Seeds boiled and eaten. (33:49)

Cirsium parryi (Gray) Petrak, Parry's Thistle

- *Other*—**Keres, Western** *Unspecified* Plant known and named but no use was specified. (171:37)

Cirsium pastoris J. T. Howell, Snowy Thistle

- *Food*—**Paiute** *Unspecified* Stems peeled and eaten raw. (as *C. occidentale* var. *candidissimum* 104:103)

Cirsium remotifolium (Hook.) DC., Fewleaf Thistle

- *Drug*—**Kwakiutl** *Oral Aid* Root skins dried, soaked in water, and used as a wash for mouth rashes and cankers. Infusion of root used to wipe out child's mouth for rash and cankers. Root held in mouth for rash or cankers in mouth and infusion used for children. *Pediatric Aid* Infusion of root used to wipe out child's mouth for rash and cankers. (as *Carduus remaliflorus* 20:383)

Cirsium rothrockii (Gray) Petrak, Rothrock's Thistle

- *Drug*—**Navajo, Kayenta** *Febrifuge* Plant used for fevers caused by injuries. *Misc. Disease Remedy* Roots used as a lotion or eaten raw for smallpox. *Panacea* Plant used as a life medicine. (205:46)

Cirsium scariosum Nutt., Meadow Thistle

- *Food*—**Flathead** *Unspecified* Roots eaten raw or pit baked. **Kutenai** *Unspecified* Roots eaten raw or pit baked. **Nez Perce** *Unspecified* Roots eaten raw or pit baked. (82:13)

Cirsium sp., Thistle

- *Drug*—**Cherokee** *Poultice* Roots of various species used as poultices. (203:74) **Chippewa** *Analgesic* Compound decoction of root taken for back pain. *Gynecological Aid* Compound decoction of root taken for "female weakness." (53:356) Compound infusion of root taken to produce postpartum milk flow. (53:360) **Costanoan** *Analgesic* Raw stalks chewed for stomach pain. *Dermatological Aid* Pounded stalk pulp used for face sores. *Disinfectant* Pounded stalk pulp used to dry infections. *Gastrointestinal Aid* Raw stalks chewed for stomach pain. *Respiratory Aid* Decoction of roots taken for asthma. (21:26) **Kwakiutl** *Oral Aid* Dried outer root held in the mouth for cankers and gum sores. (183:278) **Lummi** *Gynecological Aid* Decoction of roots and tips taken by women at childbirth. (79:49)
- *Food*—**Blackfoot** *Unspecified* Flower head pedicels eaten fresh. (86:102) **Costanoan** *Unspecified* Raw or boiled stems used for food. (21:254) **Havasupai** *Starvation Food* Leaves held in flames to burn the spines off and eaten by hunting parties when food was scarce. (197:247)

Cirsium tioganum (Congd.) Petrak **var. *tioganum***, Tioga Thistle

- *Food*—**Paiute** *Unspecified* Roots eaten raw or roasted. (as *C. acaulescens* 104:103)

Cirsium undulatum (Nutt.) Spreng., Wavyleaf Thistle

- *Drug*—**Comanche** *Venereal Aid* Decoction of root used for gonorrhea. (29:521) **Navajo, Ramah** *Eye Medicine* Cold infusion of root used as a wash for eye diseases. *Panacea* Root used as a "life medicine." Cold infusion of plant taken when one "feels bad all over." *Veterinary Aid* Cold infusion of root used as a wash for livestock with eye diseases.

(191:50) **Shuswap** *Gastrointestinal Aid* Root used for the stomach and body. (123:59)
- *Food*—**Comanche** *Unspecified* Raw roots used for food. (29:521) **Gosiute** *Unspecified* Stems used for food. (as *Cnicus undulatus* 39:366) **Montana Indian** *Vegetable* Early spring roots eaten raw or cooked with meat. Young, summer stalks eaten like asparagus and greens. (as *Cnicus eriocephalus* 19:10) **Okanagon** *Staple* Roots used as a principal food. (as *Cnicus undulatus* 178:238) **Shuswap** *Unspecified* Young roots roasted and eaten. (123:59) **Spokan** *Unspecified* Roots used for food. (as *Cnicus undulatus* 178:343) **Thompson** *Dried Food* Dried roots rehydrated, scraped, chopped, and cooked in stews.[23] Roots dried and stored for future use.[23] (187:178) *Unspecified* Root cooked and eaten. (164:480) Fresh roots eaten cooked.[23] (187:178)

Cirsium vulgare (Savi) Ten., Bull Thistle
- *Drug*—**Cherokee** *Analgesic* Infusion of leaves taken for neuralgia. *Gastrointestinal Aid* Warm infusion of roots taken to help person who overeats. *Poultice* Roots used as poultice and decoction of bruised plant used to poultice sore jaw. (as *Carduus lanceolatus* 80:58) **Delaware** *Antirheumatic (External)* Hot infusion of roots or twigs used as a steam treatment for muscular swellings and stiff joints. (as *Cirsium lanceolatum* 176:36) **Delaware, Oklahoma** *Antirheumatic (External)* Infusion of whole plant used as herbal steam for rheumatism. (as *C. lanceolatum* 175:30, 74) *Herbal Steam* Infusion of roots or twigs used as herbal steam for rheumatism. (as *C. lanceolatum* 175:30) **Iroquois** *Cancer Treatment* Plant used for cancer. *Hemorrhoid Remedy* Plant used for bleeding piles. (as *C. lanceolatum* 87:475) Decoction of whole plant taken and poultice of plant and wool applied to hemorrhoids. (as *C. lanceolatum* 141:63) *Hemostat* Plant used for bleeding piles. (as *C. lanceolatum* 87:475) **Meskwaki** *Adjuvant* Root used as a seasoner for medicines. (as *C. lanceolatum* 152:213) **Navajo** *Emetic* Decoction of plant taken to induce vomiting. (as *C. lanceolatum* 55:84) **Ojibwa** *Analgesic* and *Gastrointestinal Aid* Root used by men and women for stomach cramps. (as *C. lanceolatum* 153:364) **Potawatomi** *Adjuvant* Fresh flower centers chewed to mask unpleasant flavors in medicines. (as *C. lanceolatum* 154:51)
- *Food*—**Hesquiat** *Unspecified* Flower heads chewed to get the nectar. (185:61) **Thompson** *Dried Food* Dried roots rehydrated, scraped, chopped, and cooked in stews.[23] Roots dried and stored for future use.[23] *Unspecified* Fresh roots eaten cooked.[23] (187:178)
- *Other*—**Cherokee** *Hunting & Fishing Item* Down used as the tail for blow darts. (as *Carduus lanceolatus* 80:58)

Cissus, Vitaceae
Cissus verticillata (L.) D. H. Nicols. & Jarvis, Seasonvine
- *Other*—**Seminole** *Ceremonial Items* Plant used at the busk ceremonies. *Hunting & Fishing Item* Plant used to make bow strings. (as *C. sicyoides* 169:474)

Cistaceae, see *Helianthemum, Hudsonia, Lechea*

Cistanthe, Portulacaceae
Cistanthe monandra (Nutt.) Hershkovitz, Common Pussypaws
- *Food*—**Kawaiisu** *Unspecified* Seeds used for food. (as *Calyptridium monandrum* 206:16)

Citrullus, Cucurbitaceae
Citrullus lanatus (Thunb.) Matsumura & Nakai, Watermelon
- *Food*—**Apalachee** *Fruit* Fruit used for food. (81:98)

Citrullus lanatus (Thunb.) Matsumura & Nakai **var. *lanatus***, Watermelon
- *Drug*—**Cherokee** *Kidney Aid* Infusion of seeds taken for kidney trouble. *Pediatric Aid* and *Urinary Aid* Seeds chewed for bed-wetting. (as *C. vulgaris* 80:61) **Cheyenne** *Diuretic* Decoction of seeds taken as a diuretic. (as *C. vulgaris* 83:24) **Chickasaw** *Urinary Aid* Decoction of mashed seeds taken for blood in the urine. (as *C. vulgaris* 177:59) **Iroquois** *Urinary Aid* Compound decoction of roots and seeds taken for urine stoppage. (as *C. vulgaris*. 87:451) **Kiowa** *Poison* Unripened plant considered poisonous. (as *C. vulgaris* 192:53) **Rappahannock** *Kidney Aid* Infusion of seeds taken for gravel (kidney stones). *Veterinary Aid* Infusion of seeds given to horses for gravel (kidney stones). (as *C. vulgaris* 161:30)
- *Food*—**Cahuilla** *Fruit* Eaten fresh. *Winter Use Food* Cut, peeled into strips, and dried for winter use. (as *C. vulgaris* 15:55) **Cherokee** *Unspecified* Species used for food. (as *C. vulgaris* 80:61) **Cocopa** *Dried Food* Dried, whole seeds used for food. *Fruit* Ripe melon scooped with fingers and used for food. *Winter Use Food* Ripe and green melons stored in pits and the green melons ripened in storage. (as *C. vulgaris* 64:266) **Havasupai** *Fruit* Fruit eaten fresh. (as *C. vulgaris* 197:66) *Porridge* Seeds parched and ground to make sumkwin (mush) and other dishes. (as *C. vulgaris* 197:243) **Hopi** *Cooking Agent* Seeds ground and used to oil the "piki" stones. *Staple* Eaten and considered to be almost a staple food. *Unspecified* Seeds parched and eaten with parched corn and "piki." (as *C. vulgaris* 200:92) **Iroquois** *Bread & Cake* Fresh or dried flesh boiled, mashed, and mixed into the paste when making corn bread. *Dried Food* Flesh cut into strips, dried, and stored away. *Special Food* Squash eaten at feasts of ceremonial importance and longhouse ceremonies. *Vegetable* Flesh boiled, baked in ashes or boiled, mashed with butter and sugar, and eaten. Flesh fried and sweetened or seasoned with salt, pepper, and butter. (as *Cucurbita citrullus* 196:113) **Kamia** *Unspecified* Species used for food. (as *Citrullus vulgaris* 62:21) **Meskwaki** *Unspecified* Melon used for food. (as *C. citrullus* 152:257) **Navajo** *Dried Food* Fruit cut into strips, wound upon sticks in the form of a rope, sun dried, and stored for months. (as *C. vulgaris* 165:222) **Navajo, Ramah** *Unspecified* Watermelon cultivated and used for food. *Winter Use Food* Watermelon sliced into strips, dried, and stored for winter use. (as *C. vulgaris* 191:46) **Okanagan-Colville** *Unspecified* Species used for food. (as *C. vulgaris* 188:98) **Pima** *Fruit* Fruits eaten as one of the most important foods. (as *C. vulgaris* 146:75) **Seminole** *Unspecified* Plant used for food. (as *C. vulgaris* 169:479) **Sia** *Unspecified* Cultivated watermelons used for food. (as *C. vulgaris* 199:106)
- *Other*—**Havasupai** *Planting Seeds* Seeds stored for future planting. (as *C. vulgaris* 197:243)

Citrus, Rutaceae
Citrus aurantifolia (Christm.) Swingle, Key Lime
- *Food*—**Seminole** Plant used for food. (169:513)

Citrus aurantium L., Sour Orange
- *Food*—**Seminole** Plant used for food. (169:511)

- *Other*—Seminole *Hunting & Fishing Item* Plant used to make bows. (169:511)

Citrus limon (L.) Burm. f., Lemon
- *Food*—Haisla & Hanaksiala *Fruit* Fruit used for food. (43:284) **Seminole** *Unspecified* Plant used for food. (169:512) **Thompson** *Fruit* Fruit much prized for food after it became available. (187:275)
- *Other*—Seminole *Hunting & Fishing Item* Plant used to make bows for boys. (169:512)

Citrus medica L., Citron
- *Food*—Thompson *Fruit* Fruit much prized for food after it became available. (187:275)

Citrus ×paradisi Macfad., Paradise Citrus
- *Food*—Seminole *Unspecified* Plant used for food. (169:512)

Citrus reticulata Blanco, Tangerine
- *Food*—Seminole *Unspecified* Plant used for food. (169:512)

Citrus sinensis (L.) Osbeck, Sweet Orange
- *Food*—Haisla & Hanaksiala *Fruit* Fruit used for food. (43:284) **Seminole** *Unspecified* Plant used for food. (169:513) **Thompson** *Fruit* Fruit much prized for food after it became available. (187:275)

Cladium, Cyperaceae

Cladium mariscus ssp. *jamaicense* (Crantz) Kukenth., Jamaica Swamp Sawgrass
- *Fiber*—Mewuk *Basketry* Roots used to make small baskets. (as *Mariscus cladium* 117:328)
- *Other*—Seminole *Tools* Stems used to make medicine tubes. (as *Mariscus jamaicensis* 169:172) Plant used to make medicine blowing tubes. (as *Mariscus jamaicensis* 169:498)

Cladium sp., Tulare
- *Fiber*—Pomo *Basketry* Roots used for basket body material. (117:296) Yokut *Basketry* Roots used to make baskets. (117:405)

Cladoniaceae, see *Cladonia*

Cladonia, Cladoniaceae, lichen

Cladonia rangiferina
- *Food*—Abnaki *Forage* Plant eaten by caribou. (144:152)

Cladrastis, Fabaceae

Cladrastis kentukea (Dum.-Cours.) Rudd, Kentucky Yellowwood
- *Fiber*—Cherokee *Building Material* Wood used for lumber. (as *C. lutea* 80:62)
- *Other*—Cherokee *Decorations* Wood used for carving. (as *C. lutea* 80:62)

Clarkia, Onagraceae

Clarkia amoena (Lehm.) A. Nels. & J. F. Macbr. ssp. *amoena*, Farewell to Spring
- *Food*—Miwok *Unspecified* Parched, pulverized dry seeds used for food. (as *Godetia amoena* 12:153)

Clarkia biloba (Dur.) A. Nels. & J. F. Macbr. ssp. *biloba*, Twolobe Fairyfan
- *Food*—Miwok Parched, pulverized seeds used for food. (as *Godetia biloba* 12:154)

Clarkia purpurea ssp. *quadrivulnera* (Dougl. ex Lindl.) H. F. & M. E. Lewis, Winecup Fairyfan
- *Drug*—Mendocino Indian *Eye Medicine* Decoction of leaves used as a wash for sore eyes. (as *Godetia albescens* 41:370)
- *Food*—Mendocino Indian *Staple* Seeds eaten as a pinole. (as *Godetia albescens* 41:370)

Clarkia purpurea ssp. *viminea* (Dougl. ex Hook.) H. F. & M. E. Lewis, Winecup Fairyfan
- *Food*—Miwok *Unspecified* Dried, pulverized, uncooked seeds used for food. (as *Godetia viminea* 12:154) *Vegetable* Seeds considered one of the most prized vegetable foods. (as *Godetia viminea* 12:137)

Clarkia rhomboidea Dougl. ex Hook., Diamond Fairyfan
- *Food*—Yana *Unspecified* Seeds eaten raw or parched and finely pounded. (147:251)

Clarkia unguiculata Lindl., Elegant Fairyfan
- *Food*—Miwok *Unspecified* Parched, pulverized seeds eaten dry with acorn mush. (as *C. elegans* 12:153)

Claytonia, Portulacaceae

Claytonia acutifolia Pallas ex J. A. Schultes, Bering Sea Springbeauty
- *Food*—Alaska Native *Unspecified* Fresh roots eaten raw or cooked with seal oil. (85:115) **Eskimo, Alaska** *Unspecified* Fleshy taproots used for food. (4:715)

Claytonia caroliniana Michx., Carolina Springbeauty
- *Food*—Gosiute *Unspecified* Bulbs used for food. (39:366) **Ute** *Unspecified* Bulbs formerly used for food. (38:33)

Claytonia cordifolia S. Wats., Heartleaf Springbeauty
- *Food*—Montana Indian *Sauce & Relish* Eaten raw as a relish. (as *Montia asarifolia* 19:16)

Claytonia lanceolata Pursh, Lanceleaf Springbeauty
- *Food*—Blackfoot *Vegetable* Tubers boiled and eaten. (114:278) **Montana Indian** *Fodder* Roots "better for fattening hogs than the best feed." (19:10) *Forage* Rootstocks eaten by marmots, ground squirrels, and grizzly bears. (82:29) *Unspecified* Roots eaten raw and roasted. (19:10) Crisp, tuber-like corms eaten fresh or boiled. (82:29) **Okanagan-Colville** *Unspecified* Corms used for food. *Winter Use Food* Corms stored for future use. (188:113) **Okanagon** *Unspecified* Corms boiled and used for food. (125:38) **Thompson** *Bread & Cake* Corms made into cakes and dried for future use. (187:239) *Unspecified* Corms boiled and used for food. (125:38) Small, oval corms eaten. (164:482) Corms rubbed clean, cooked in pits or steamed, and eaten. *Winter Use Food* Corms buried fresh in underground caches and stored for winter use. (187:239)

Claytonia lanceolata var. *sessilifolia* (Torr.) A. Nels., Lanceleaf Springbeauty
- *Food*—Okanagon *Staple* Roots used as a principal food. (as *C. sessilifolia* 178:238)

Claytonia multicaulis Nelson, Ground Nut
- *Food*—**Montana Indian** *Fodder* Roots "better for fattening hogs than the best feed." **Unspecified** Roots eaten raw and roasted. (19:10)

Claytonia parviflora Dougl. ex Hook., Streambank Springbeauty
- *Food*—**Montana Indian** *Sauce & Relish* Eaten raw as a relish. (as *Montia parviflora* 19:16)
- *Other*—**Karok** *Toys & Games* Shoot used to play a game.[24] (6:24)

Claytonia perfoliata Donn ex Willd., Miner's Lettuce
- *Drug*—**Shoshoni** *Analgesic* Poultice of mashed plants applied for rheumatic pains. *Antirheumatic (External)* Poultice of plants applied as a counterirritant for rheumatic pains. (180:59) **Thompson** *Eye Medicine* Plant used for sore eyes and for "helping someone to see the right." (187:241)
- *Food*—**Costanoan** *Unspecified* Raw foliage used for food in early spring and boiled or steamed when eaten later in the season. (21:251) **Kawaiisu** *Vegetable* Leaves eaten as greens. (206:21) **Mendocino Indian** *Unspecified* Plants eaten raw. *Vegetable* Plants cooked with salt and pepper and eaten as greens. (41:346) **Neeshenam** *Vegetable* Leaves eaten as greens. (129:377)
- *Other*—**Yurok** *Toys & Games* Shoot used to play a game.[24] (6:25)

Claytonia perfoliata Donn ex Willd. **ssp. *perfoliata***, Miner's Lettuce
- *Food*—**Diegueño** *Vegetable* Young leaves, picked in the spring before the flowers appear, boiled once and eaten as greens. (84:17)

Claytonia perfoliata Donn ex Willd. **ssp. *perfoliata* var. *perfoliata***, Miner's Lettuce
- *Drug*—**Mahuna** *Dietary Aid* Plant juice used as an appetite restorer. (as *Montia perfoliata* 140:62)
- *Food*—**Cahuilla** *Vegetable* Leaves eaten fresh or boiled as greens. (as *Montia perfoliata* 15:89) **Luiseño** *Vegetable* Plant used for greens or eaten raw. (as *Montia perfoliata* 155:232) **Miwok** *Unspecified* Raw stems, leaves, and blossoms used for food. (as *Montia perfoliata* 12:160) **Montana Indian** *Sauce & Relish* Eaten raw as a relish. (as *Montia perfoliata* 19:16) **Paiute, Northern** *Vegetable* Leaves eaten raw. (as *Montia perfoliata* 59:49)

Claytonia sibirica L., Siberian Springbeauty
- *Drug*—**Cowlitz** *Dermatological Aid* Cold infusion of stems used as a hair wash. **Quileute** *Dermatological Aid* Cold infusion of stems used as a hair wash for dandruff. *Eye Medicine* Juice from stems used as a wash for eyes. *Urinary Aid* Infusion of plants taken as an urinative. **Quinault** *Gynecological Aid* Whole plant chewed by women during pregnancy. **Skagit** *Throat Aid* Infusion of plants taken for sore throats. *Tonic* Infusion of plants taken as a general tonic. (79:29) **Skagit, Upper** *Throat Aid* Infusion of plant used as a general tonic for sore throats. (179:42) **Skokomish** *Dermatological Aid* Cold infusion of stems used as a hair wash. **Snohomish** *Dermatological Aid* Cold infusion of stems used as a hair wash. (79:29) **Tlingit** *Venereal Aid* Compound poultice of leaves applied for syphilis. (as *C. alsinoides* 107:284)
- *Food*—**Alaska Native** *Dietary Aid* Leaves properly prepared and used as a good source of vitamin C and provitamin A. *Vegetable* Leaves added raw to mixed salads or cooked as a green vegetable. (85:25) **Yurok** *Unspecified* Shoot tops eaten raw. (6:25)

Claytonia sibirica L. **var. *sibirica***, Siberian Springbeauty
- *Drug*—**Hesquiat** *Dermatological Aid* Poultice of chewed leaves used on cuts and sores. *Eye Medicine* Stem juice squeezed into the eye for sore, red eyes. (as *Montia sibirica* 185:71)
- *Other*—**Karok** *Toys & Games* Fresh flowers used by children to play a game. Each of the two players held a stem in his hand and tried to hook his flower around his opponent's flower. When the flowers were engaged, the players pulled and the one whose flower head came off lost. It was done over and over and a score was kept. (as *Montia sibirica* 148:383)

***Claytonia* sp.**
- *Food*—**Aleut** *Unspecified* Species used for food. (7:29) **Coeur d'Alene** *Vegetable* Roots used as a principal vegetable food. (178:89) **Eskimo, Alaska** *Soup* Corm added to duck or goose soup. *Unspecified* Corm eaten raw, alone or with seal oil. (1:35) **Spokan** *Unspecified* Roots used for food. (178:343)

Claytonia spathulata Dougl. ex Hook. **var. *spathulata***, Pale Springbeauty
- *Food*—**Cahuilla** *Vegetable* Leaves eaten fresh or boiled as greens. (as *Montia spathulata* 15:89)

Claytonia tuberosa Pallas ex J. A. Schultes, Tuberous Springbeauty
- *Food*—**Alaska Native** *Dietary Aid* Green, fresh leaves eaten raw as a source of vitamins C and A. *Soup* Corms cooked and added to stews. *Vegetable* Corms roasted and used for food. Basal leaves added to other greens and eaten raw or cooked. (85:117) **Eskimo, Arctic** *Vegetable* Tubers boiled and eaten. (128:31)

Claytonia umbellata S. Wats., Great Basin Springbeauty
- *Food*—**Paiute, Northern** *Unspecified* Roots roasted in the sand and eaten. (59:43)

Claytonia virginica L., Virginia Springbeauty
- *Drug*—**Iroquois** *Anticonvulsive* Cold infusion or decoction of powdered roots given to child with convulsions. (87:317) *Contraceptive* Eating raw plants permanently prevented conception. (87:318) *Pediatric Aid* Cold infusion or decoction of powdered roots given to child with convulsions. (87:317)
- *Food*—**Algonquin, Quebec** *Vegetable* Corm cooked and eaten like potatoes. (18:84) **Iroquois** *Unspecified* Roots used for food. (196:120)

***Clematis*, Ranunculaceae**

Clematis baldwinii Torr. & Gray **var. *baldwinii***, Baldwin's Clematis
- *Drug*—**Seminole** *Other* Infusion of plant used for sunstroke. (as *Viorna baldwinii* 169:303)

Clematis bigelovii Torr., Bigelow's Leather Flower
- *Other*—**Keres, Western** *Decorations* Flowers used for bouquets. (171:37)

Clematis columbiana (Nutt.) Torr. & Gray, Rock Clematis
- *Drug*—**Okanagan-Colville** *Dermatological Aid* Infusion of leaves alone or the stems and leaves used as a hair wash to prevent gray hair. Poultice of pounded, dampened leaves applied to the feet for sweaty feet. (188:117) **Thompson** *Dermatological Aid* Plant used as a head wash and for scabs and eczema. (164:459)

Clematis columbiana (Nutt.) Torr. & Gray **var. *columbiana***, Rock Clematis
- *Drug*—**Navajo, Ramah** *Orthopedic Aid* Cold infusion of plant used as a lotion for swollen knee or ankle. (as *C. pseudoalpina* 191:27)
- *Other*—**Isleta** *Decorations* Plant grown for ornamental and shade purposes. (as *Atragene pseudalpina* 100:24) **Keres, Western** *Unspecified* Plant known and named but no use was specified. (as *C. pseudoalpina* 171:37)

Clematis hirsutissima Pursh, Hairy Clematis
- *Drug*—**Montana Indian** *Analgesic* Decoction of leaves used for headaches. *Veterinary Aid* Scraped root held in nostril of fallen horse and acted as a stimulant to animal. (as *C. douglasii* 19:10) **Navajo, Ramah** *Analgesic* and *Respiratory Aid* Root used for congested nose pain. *Witchcraft Medicine* Cold infusion of plant or root taken and used as a lotion to protect from witches. (as *C. eriophora* 191:27)

Clematis lasiantha Nutt., Pipestem Clematis
- *Drug*—**Miwok** *Burn Dressing* Pulverized plant charcoal dusted onto burns. *Dermatological Aid* Pulverized plant charcoal dusted onto running sores. (12:168) **Shasta** *Cold Remedy* Decoction of pounded, whole stem or bark used as a steam bath for colds. Roots burned or chewed for colds. *Herbal Steam* Decoction of pounded, whole stem or bark used as a steam bath for colds. (93:340)

Clematis ligusticifolia Nutt., Western White Clematis
- *Drug*—**Costanoan** *Analgesic* Poultice of foliage applied for chest pains. (21:7) **Dakota** *Veterinary Aid* Plant used as a horse medicine. (97:35) **Gosiute** *Unspecified* Plant used as a medicine. (39:366) **Great Basin Indian** *Dermatological Aid* Roots used to make a shampoo. (121:47) **Lakota** *Analgesic* Infusion of roots taken for headaches. (139:55) **Mahuna** *Dermatological Aid* Infusion of plant used as a wash for skin eruptions. (140:17) **Mendocino Indian** *Cold Remedy* Stems and leaves chewed for colds. *Throat Aid* Stems and leaves chewed for sore throats. (41:347) **Montana Indian** *Cold Remedy* Chewed for colds. *Throat Aid* Chewed for sore throats. (19:10) **Navajo** *Analgesic* Plant used for pain. *Gynecological Aid* Plant used as tonic after deliverance. *Tonic* Plant used as tonic after deliverance. (55:47) **Navajo, Kayenta** *Dermatological Aid* Plant used for spider or sand cricket bites. (205:22) **Navajo, Ramah** *Analgesic* Cold infusion of plant used as lotion for backache. *Orthopedic Aid* Cold infusion of plant used as lotion for backache or swollen legs or arms. *Witchcraft Medicine* Cold infusion of plant or root taken and used as a lotion to protect against witches. (191:27) **Nevada Indian** *Dermatological Aid* Dried, powdered root used for shampoo. (118:57) **Nez Perce** *Veterinary Aid* Plant used as a horse medicine. (97:35) **Okanagan-Colville** *Contraceptive* Stalk and roots used to make a women's contraceptive. *Dermatological Aid* Decoction of mashed leaves and branches rubbed into the scalp as a shampoo to kill hair root "germs." (188:117) **Okanagon** *Tonic* Decoction of plants taken as a tonic for general or out-of-sorts feeling. (125:40) **Oregon Indian** *Dermatological Aid* Leaves and bark used as shampoo. (118:57) *Febrifuge* Infusion of white portion of bark used for fever. (118:40) **Paiute** *Kidney Aid* Decoction of leaves used as a wash or tub bath for dropsical conditions. (180:59, 60) **Sanpoil** *Dermatological Aid* Lather of leaves and water applied to sores or boils. *Veterinary Aid* Lather of leaves and water applied to animals for sores or boils. (as *C. lingusticifolia* 131:220) **Shoshoni** *Analgesic* Branches used to whip sore or painful areas as a counterirritant. Crushed dried leaves used as snuff or fresh leaves smelled for headaches. Decoction of roots taken for stomachaches or cramps. Poultice of mashed leaves applied for rheumatic pains. *Antirheumatic (External)* Poultice of mashed leaves applied for rheumatic pains. *Burn Dressing* Poultice of mashed, moistened seeds applied to severe burns. *Dermatological Aid* Simple or compound poultice of leaf for used swellings, bruises, wounds, or boils. *Gastrointestinal Aid* Decoction of leaves or roots taken for stomachaches or cramps. *Kidney Aid* Decoction of leaves used as a wash or tub bath for dropsical conditions. *Orthopedic Aid* Decoction of leaves used as a bath for tired feet. *Venereal Aid* Powdered leaves or decoction applied as a healing agent for syphilitic sores. (180:59, 60) **Thompson** *Dermatological Aid* Plant used as a head wash and for scabs and eczema. (164:459) Sap used for boils. *Other* Decoction of plant used to bathe babies if they seemed to take after animals or deceased people. *Pediatric Aid* Decoction of plant given to children who habitually wet their beds. Decoction of plant used to bathe babies if they seemed to take after animals or deceased people. (187:247) *Tonic* Decoction of plants taken as a tonic for general or out-of-sorts feeling. (125:40) Mild decoction of plant taken as a tonic or "remedy for general disorder." (164:459) *Toothache Remedy* Poultice of cut stem pieces applied to the tooth for toothache. *Urinary Aid* Decoction of plant given to children who habitually wet their beds. (187:247) **Yavapai** *Gastrointestinal Aid* Decoction of pulverized root taken for stomachaches. (65:261)
- *Food*—**Lakota** *Forage* Leaves eaten by horses. (139:55)
- *Fiber*—**Great Basin Indian** *Cordage* Stems used to make string. (121:47) **Thompson** *Clothing* Cottony seed fluff used in infant diapers. (187:247)
- *Other*—**Jemez** *Decorations* Plant used by the Koshares to decorate people and hats of performers in the summer and fall dances. (44:21) **Keres, Western** *Unspecified* Plant known and named but no use was specified. (171:37) **Montana Indian** *Decorations* Grown as an ornamental vine for houses and verandas. (19:10) **Okanagan-Colville** *Smoke Plant* Stem smoked by children. (188:117) **Shoshoni** *Containers* Used to make carrying nets for water bottles. *Hunting & Fishing Item* Used to make carrying nets for snares. Used to make bowstrings. (118:59) **Washo** *Hunting & Fishing Item* Wood used to make bows. (118:52)

Clematis occidentalis (Hornem.) DC. **var. *occidentalis***, Western Blue Virginsbower
- *Drug*—**Blackfoot** *Veterinary Aid* Infusion of plant given to horses as a diuretic. (as *C. verticellaris* 86:88)
- *Other*—**Blackfoot** *Ceremonial Items* Leaves used to remove "ghost bullets," supernatural objects shot into people by ghosts. The bullets were removed by a diviner either with a sucking tube or through a slit in the skin made with a flint. Then the leaves were boiled and applied to the place where the bullets were removed. Fainting was often the result of coming near a ghost and the victim was revived with a smudge of the stem of this plant. *Protection* Flowers worn by children in their hair at night to keep ghosts away. (as *C. verticellaris* 86:111)

Clematis viorna L., Vasevine
- *Drug*—**Meskwaki** *Panacea* Root used to make a drink taken for "any kind of common sickness." (152:239)

Clematis virginiana L., Devil's Darning Needles
- *Drug*—**Cherokee** *Analgesic* Infusion with milkweed used for backache. *Ceremonial Medicine* Used as an ingredient in green corn medicine. *Gastrointestinal Aid* Infusion of root taken for stomach trouble and infusion taken for nerves. *Kidney Aid* Infusion of root taken for kidneys. *Orthopedic Aid* Infusion with milkweed used for backache. (80:60) **Iroquois** *Dermatological Aid* Root powder and infusion of roots used on venereal disease sores. *Hallucinogen* Decoction of stems used as a wash to induce strange dreams. *Kidney Aid* Taken for burning kidney troubles. *Urinary Aid* Taken for burning kidney troubles. *Venereal Aid* Root powder and infusion of roots used on venereal disease sores. (87:330)

Cleome, Capparaceae

Cleome isomeris Greene, Bladderpod Spiderflower
- *Food*—**Diegueño** *Unspecified* Seeds and flowers used as food. (as *Isomeris arborea* var. *angustata* 88:217) **Kawaiisu** *Unspecified* Flowers eaten boiled or sun baked. (as *Isomeris arborea* 206:35)

Cleome lutea Hook., Yellow Spiderflower
- *Drug*—**Navajo, Kayenta** *Ceremonial Medicine* Plant used with ceremonial tobacco in some chants. *Dermatological Aid* Plant used for ant bites. (205:25)

Cleome multicaulis DC., Slender Spiderflower
- *Food*—**Navajo** *Sauce & Relish* Leaves used to make a gravy. *Soup* Leaves used to make a watery stew. *Special Food* Leaves made into tea and taken at a general feast after finishing the masks for the Night Chant. *Vegetable* Leaves used for greens. (as *C. sonorae* 55:51)
- *Other*—**Navajo** *Tools* Stalks used as a drill to start fires. The brittle stalks, about an inch in diameter, were used for the drills which were whirled between the palms of the hands and were made to revolve on the edge of a larger stalk into which a notch had been cut. A pinch of sand was sometimes placed under the point of the drill which caused the wood to become a fine powder. This powder then ran down the notch and formed a little pile on the ground. Smoke was produced in less than a minute, and in about 2 minutes tiny sparks dropped onto the pile of dry powder which took fire from them. By carefully feeding the fire with bits of dried bark and grass and with much blowing, a blaze was produced. (as *C. sonorae* 55:51)

Cleome serrulata Pursh, Rocky Mountain Beeplant
- *Drug*—**Gosiute** *Eye Medicine* Poultice of pounded, soaked leaves applied to sore eyes. (as *C. integrifolia* 39:366) **Navajo, Ramah** *Blood Medicine* Decoction of seeds used ceremonially to give "good blood." *Ceremonial Medicine* Decoction of seeds used ceremonially to improve voice and give "good blood." *Dermatological Aid* Cold infusion of leaves used as a body and shoe deodorant. *Throat Aid* Decoction of seeds used ceremonially to improve voice. (191:29) **Oregon Indian** *Febrifuge* Infusion of whole plant taken for fever. (118:40) **Tewa** *Gastrointestinal Aid* Infusion of plant taken for stomach disorders and poultice of plant used on abdomen. (as *Peritoma serrulata* 138:58, 59)
- *Food*—**Acoma** *Porridge* Seeds cooked well, dried, and made into mush before use. (32:22) **Apache, Western** *Vegetable* Leaves and whole, young plants used as greens. (26:192) **Havasupai** *Unspecified* Seeds used for food. (197:221) **Hopi** *Unspecified* Leaves and flowers boiled and used for food. (as *C. integrifolia* 56:16) Young plants boiled for food. (200:77) *Vegetable* Plants boiled and eaten like spinach.[25] (32:24) **Isleta** *Bread & Cake* Seeds made into a meal and used to make bread. (32:22) Large seeds formerly used to make a flour for bread. *Vegetable* Leaves used as greens. (100:26) **Jemez** *Bread & Cake* Green parts boiled, fibrous material removed, molded into cakes, and fried in grease, a delicacy. *Vegetable* Young and tender plants eaten as greens. (as *Peritoma serrulatum* 44:26) **Keres, Western** *Porridge* Dried seeds cooked into a mush and eaten. *Vegetable* Leaves and shoots used for food as greens. (171:37) **Keresan** *Unspecified* Seeds cooked and eaten. *Vegetable* Leaves cooked as greens. (as *Peritoma serrulatum* 198:559) **Laguna** *Porridge* Seeds cooked well, dried, and made into mush before use. (32:22) **Navajo** *Bread & Cake* Dried leaves and meat or tallow used to make dumplings. (55:50) *Dried Food* Young plants boiled, pressed, rolled into balls, dried, and stored for winter use. (32:24) Leaves dried and stored for winter use. (110:13) Young shoots boiled, rolled into small balls, and dried for winter use. (165:223) *Soup* Plant made into stew with wild onions, wild celery, tallow, or bits of meat. (32:24) Leaves, onions, wild celery, and tallow or meat used to make stew. (55:50) Dried leaves used to make stew. (110:13) *Spice* Used as a seasoning. (90:149) *Unspecified* Young plants boiled, pressed, rolled into balls, and eaten. (32:24) Pods used for food. (55:50) *Vegetable* Young plants boiled with a pinch of salt and eaten as greens. (32:24) Leaves boiled like spinach. Young plants boiled and rolled into balls and eaten. (55:50) Young shoots eaten as greens. Young shoots boiled, rolled into small balls, and eaten fresh with or without mutton. (165:223) *Winter Use Food* Young plants boiled, rolled into balls, dried, and stored for the winter. (55:50) **Navajo, Ramah** *Dried Food* Young plants boiled twice and dried in small balls for later use. *Fodder* Young plants used for sheep and horse feed. *Unspecified* Young plants boiled twice and meat added or plants removed and fried. (191:29) **Pueblo** *Staple* Used as one of the most important food plants. **San Felipe** *Unspecified* Flower buds salted and eaten as food. (32:24) **Sia** *Unspecified* Seeds used for food. *Vegetable* Leaves cooked as greens. (as *Peritoma serrulatum* 199:107) **Tewa** *Unspecified* Young plants boiled, dried, soaked in hot water, fried in grease, and used for food. (as *Peritoma serrulatum* 138:58) *Vegetable* Plants boiled and eaten like spinach.[25] (32:24) **Zuni** *Dried Food* Leaves gathered in large quantities and hung indoors to dry for winter use. (as *Peritoma serrulatum* 166:69) *Unspecified* Young plants cooked with corn strongly flavored with chile. (32:24) Tender leaves usually boiled with corn, on or off the cob, and highly seasoned with chile. (as *Peritoma serrulatum* 166:69)
- *Dye*—**Isleta** *Unspecified* Roots formerly used to make a dye. (100:26)
- *Other*—**Keresan** *Paint* Used to make the black paint for pottery decoration. (as *Peritoma serrulatum* 198:559) **Tewa** *Paint* Young plants boiled, dried, soaked in hot water, and used as black paint for pottery decorations. (as *Peritoma serrulatum* 138:58) **Zuni** *Ceremonial Items* Plant paste used with black mineral paint to color sticks of plume offerings to anthropic gods.[26] (as *Peritoma serrulatum* 166:96) *Decorations* Whole plant except for the root used in pottery decorations. The whole plant except for the root was boiled for a considerable amount of time and the water was allowed to evaporate. The firm paste secured from precipitation was used in conjunction with a black mineral paint for decorating pottery. (as *Peritoma serrulatum*

166:82) *Paint* Plant paste used with black mineral paint to color sticks of plume offerings to anthropic gods.[26] (as *Peritoma serrulatum* 166:96)

Clermontia, Campanulaceae

Clermontia arborescens (Mann) Hbd., 'Oha Wai Nui
- *Drug*—Hawaiian *Breast Treatment* Milk and other plants mixed, poured into a sweet potato, and eaten for restoring or producing milk. *Dermatological Aid* Milk, breadfruit milk, and a finely ground plant mixed and put into bad and deep cuts. *Respiratory Aid* Fruits with other plants pounded, squeezed, and the resulting liquid taken for asthma. (2:30)
- *Food*—Hawaiian *Fruit* Fruit used for food. (2:30)

Clethraceae, see *Clethra*

Clethra, Clethraceae

Clethra acuminata Michx., Mountain Sweetpepperbush
- *Drug*—Cherokee *Antiemetic* Decoction of bark scrapings taken for vomiting bile. (177:47) *Emetic* Inner bark used to make a drink taken to induce vomiting of "disordered bile." (80:22) Decoction of inner bark taken to cause vomiting when unable to retain food. (177:47) *Febrifuge* Decoction of bark and "wild cherry" bark taken to break high fever. *Gastrointestinal Aid* Hot infusion of bark taken for "bowel complaint." (80:22) Decoction of inner bark taken to cause vomiting when unable to retain food. *Liver Aid* Decoction of bark scrapings taken for vomiting bile. (177:47)

Clintonia, Liliaceae

Clintonia andrewsiana Torr., Western Bluebeadlily
- *Drug*—Pomo *Poison* Plant considered poisonous. (66:12) **Pomo, Kashaya** *Poison* Plant considered poisonous. (72:34)

Clintonia borealis (Ait.) Raf., Yellow Bluebeadlily
- *Drug*—Algonquin, Quebec *Dermatological Aid* Poultice of leaves applied to open wounds. *Disinfectant* Poultice of leaves applied to infections. (18:138) **Algonquin, Tête-de-Boule** *Dermatological Aid* Poultice of leaves applied to wounds and ulcers. (132:126) **Chippewa** *Burn Dressing* Poultice of fresh leaf applied to burns. *Dermatological Aid* Decoction of leaves applied externally to scrofulous sores. (53:354) **Iroquois** *Heart Medicine* Decoction of plant taken for the heart. *Misc. Disease Remedy* Compound decoction of smashed whole plants taken for sugar diabetes. (87:283) **Menominee** *Antidote* Plant put on bite of dog which has eaten plant, to draw out the poison. (151:40, 41) **Ojibwa** *Antidote* Root applied to draw poison from bite made by dog which has eaten the plant. (153:373) *Dermatological Aid* Poultice of roots used on wound caused by dog's northern clintonia-poisoned teeth. (153:430) *Gynecological Aid* Infusion of root used to aid parturition. (153:373) *Poison* Roots chewed by dogs to poison their teeth and kill animals they bite. (153:430) **Potawatomi** *Unspecified* Plant used as a medicine for unspecified ailments. (154:62)
- *Other*—Algonquin, Tête-de-Boule *Insecticide* Crushed leaves rubbed on the face and hands as protection from mosquitoes. (132:126) **Chippewa** *Toys & Games* Patterns bitten into leaves for entertainment. (53:377)

Clintonia umbellulata (Michx.) Morong, White Bluebeadlily
- *Drug*—Iroquois *Basket Medicine* Decoction of whole plant "makes people buy baskets," a basket medicine. *Febrifuge* Infusion of whole plant taken for chills. (87:283)

Clintonia uniflora (Menzies ex J. A. & J. H. Schultes) Kunth, Bride's Bonnet
- *Drug*—Bella Coola *Dermatological Aid* Decoction of plant used as a wash for the body. Poultice of toasted leaf applied to wounds. *Eye Medicine* Poultice of toasted leaf applied to eyes. (150:53) **Cowlitz** *Dermatological Aid* Juice from smashed plants used as a wash for cuts. *Eye Medicine* Juice from smashed plants used as a wash for sore eyes. (79:25) **Haisla & Hanaksiala** *Dermatological Aid* Poultice of plant applied to wounds and decoction of plant used to wash the body. *Eye Medicine* Poultice of plant applied to the eyes. (43:194) **Micmac** *Urinary Aid* Root juice taken with water for gravel. (as *Smilacrina borealis* 156:317)
- *Food*—Bella Coola *Forage* Berries eaten by wolves. (184:199)
- *Dye*—Thompson *Unspecified* Mashed, blue, berry-like fruits used as a dye or stain. Large quantities of the fruits had to be used in order for the dye or stain to be effective. (187:121)

Clitoria, Fabaceae

Clitoria mariana L., Atlantic Pigeonwings
- *Drug*—Cherokee *Oral Aid* Infusion held in mouth for 10 to 20 minutes for thrush. (80:47)

Clusiaceae, see *Hypericum*, *Triadenum*

Cneoridium, Rutaceae

Cneoridium dumosum (Nutt.) Hook. f. ex Baill., Bush Rue
- *Drug*—Luiseño *Unspecified* Plant used for medicinal purposes. (155:231)

Coccoloba, Polygonaceae

Coccoloba diversifolia Jacq., Tietongue
- *Food*—Seminole *Unspecified* Plant used for food. (as *Coccolobis laurifolia* 169:475)

Cocculus, Menispermaceae

Cocculus carolinus (L.) DC., Carolina Coralbead
- *Drug*—Houma *Blood Medicine* Root used to make a drink taken as a blood clarifier. (158:63)

Cochlearia, Brassicaceae

Cochlearia officinalis L., Common Scurvygrass
- *Food*—Alaska Native *Vegetable* Leaves eaten raw in mixed salads or cooked as greens. (85:27)

Cochlearia sp.
- *Food*—Aleut *Unspecified* Species used for food. (7:29)

Cocos, Arecaceae

Cocos nucifera L., Coconut Palm
- *Drug*—Hawaiian *Abortifacient* Fruit shells burned and the smoke used for swollen wombs. *Dermatological Aid* Young shoots used for deep cuts. *Other* Young meat applied as a rub for the brain. *Strength-*

ener Dried meat ash used for general debility of the body. *Unspecified* Oil used as a good rub. (2:73)
- **Food**—**Seminole** *Unspecified* Plant used for food. (169:503)
- **Other**—**Seminole** *Ceremonial Items* Plant used to make dance rattles. *Cooking Tools* Plant used to make spoons and cups. (169:503)

Coeloglossum, Orchidaceae

Coeloglossum viride var. virescens (Muhl. ex Willd.) Luer, Longbract Frog Orchid
- **Drug**—**Iroquois** *Gynecological Aid* Compound decoction with plant taken to bring away placenta after childbirth. (as *Habenaria viridis* ssp. *bracteata* 87:289) **Ojibwa** *Love Medicine* Root smuggled into another's food as an aphrodisiac. (as *Habenaria bracteata* 153:377) Plant used as a sort of love charm and often put to bad use. (as *Habenaria bracteata* 153:431)

Coix, Poaceae

Coix lacryma-jobi L., Job's Tears
- **Drug**—**Cherokee** *Pediatric Aid* Seeds strung around baby's neck for teething. *Toothache Remedy* Seeds strung around baby's neck for teething. *Unspecified* Long strands of seeds used for various unspecified medicinal purposes. (80:41)
- **Food**—**Cherokee** *Bread & Cake* Seeds used to make bread. (80:41)
- **Other**—**Cherokee** *Jewelry* Seeds used to make beads. (80:41)

Coleogyne, Rosaceae

Coleogyne ramosissima Torr., Blackbrush
- **Drug**—**Kawaiisu** *Venereal Aid* Decoction of bark taken for gonorrhea. (206:21)
- **Food**—**Havasupai** *Fodder* Plant used as a good feed for stock in the absence of grass. (197:223)

Collinsia, Scrophulariaceae

Collinsia parviflora Lindl., Smallflower Blue Eyed Mary
- **Drug**—**Navajo, Kayenta** *Veterinary Aid* Plant used to make a horse run fast. (205:42) **Ute** *Dermatological Aid* Plant used externally for sore flesh. (38:33)

Collinsia violacea Nutt., Violet Blue Eyed Mary
- **Drug**—**Creek** *Tuberculosis Remedy* Infusion of root taken for colds, coughs, consumption, and whooping cough. **Natchez** *Cold Remedy* Infusion of root taken for colds. *Cough Medicine* Infusion of root taken for coughs. (172:667) Infusion of roots taken for coughs. (177:56) *Pulmonary Aid* Infusion of root taken for consumption and whooping cough. (172:667) *Tuberculosis Remedy* Infusion of roots taken for consumption. (177:56)

Collinsonia, Lamiaceae

Collinsonia canadensis L., Rich Weed
- **Drug**—**Cherokee** *Breast Treatment* Compound used for swollen breasts. (80:52) Decoction of plant applied for swollen breasts. (177:53) *Dermatological Aid* Mashed flowers and leaves used as a deodorant. (80:52) *Emetic* Decoction of plant taken to cause vomiting. (177:53) *Veterinary Aid* Infusion used as a drench for horses with colic. (80:52) **Iroquois** *Analgesic* Poultice of powdered leaves applied to the forehead for headaches. *Antidiarrheal* Compound decoction of roots taken for diarrhea with blood. *Antirheumatic (External)* Decoction of roots taken and used as foot, back, and leg soak for rheumatism. *Antirheumatic (Internal)* Compound decoction of roots taken and used as foot soak for rheumatism. *Blood Medicine* Roots used as a blood medicine. *Dermatological Aid* Roots used for boils. *Heart Medicine* Decoction of roots taken for heart trouble. *Kidney Aid* Decoction of roots taken for kidney trouble. *Panacea* Decoction of roots taken for any ailment. *Pediatric Aid* Infusion of smashed roots given to strengthen listless children. *Stimulant* Infusion of smashed roots given to children for listlessness. (87:429) *Strengthener* Infusion of smashed roots used as a wash for babies to give them strength. (87:428)

Collomia, Polemoniaceae

Collomia grandiflora Dougl. ex Lindl., Largeflower Mountaintrumpet
- **Drug**—**Okanagan-Colville** *Febrifuge* Infusion of roots taken for high fevers. *Laxative* Infusion of leaves and stalks taken for constipation and to "clean out your system." Infusion of roots taken as a laxative. (188:111)
- **Other**—**Paiute** *Containers* Leaves used as a protective covering for filled berry containers. (111:106)

Collomia linearis Nutt., Narrowleaf Mountaintrumpet
- **Drug**—**Gosiute** *Dermatological Aid* Poultice of mashed plant applied to wounds and bruises. (as *Gilia linearis* 39:370)

Colocasia, Araceae

Colocasia esculenta (L.) Schott, Coco Yam
- **Drug**—**Hawaiian** *Laxative* Flesh and other plants pounded, squeezed, and resulting liquid taken as a laxative. (as *Coloccasia antiquorum* 2:47) *Unspecified* Plant used to make a draft and given to the sick. (as *Arum esculentum* 112:67)
- **Food**—**Hawaiian** *Unspecified* Roots beaten into poi and eaten. Plant mixed with coconut meat and eaten. *Vegetable* Leaves and stems cooked and eaten as greens. (as *Arum esculentum* 112:67) **Seminole** *Unspecified* Plant used for food. (169:465)

Comandra, Santalaceae

Comandra umbellata (L.) Nutt., Bastard Toadflax
- **Drug**—**Cherokee** *Dermatological Aid* Juice applied to cut or sore. *Kidney Aid* Compound infusion used for kidneys. (80:24) **Meskwaki** *Analgesic* Infusion of leaf taken for lung pains. *Cold Remedy* Medicine licked to ease labored breathing caused by a cold or other illness. *Pulmonary Aid* Infusion of leaf taken for lung pains. *Respiratory Aid* Medicine of immature florets licked to ease labored breathing from cold, etc. (152:246)
- **Food**—**Okanagan-Colville** *Sweetener* Flowers sucked by children for the sweet nectar. (188:138)

Comandra umbellata ssp. pallida (A. DC.) Piehl, Pale Bastard Toadflax
- **Drug**—**Navajo** *Dermatological Aid* Decoction of plant used as a foot bath for corns. (as *C. pallida* 90:150) **Navajo, Kayenta** *Eye Medicine* Plant used for sore eyes. *Narcotic* Plant used as a narcotic. *Oral Aid* Plant used as a mouthwash for canker sores. (as *Commandra pallida*

205:18) **Thompson** *Dermatological Aid* Decoction of plant used as a wash for sores. (187:281) *Eye Medicine* Fresh roots mixed with woman's milk and used as a wash for sore or inflamed eyes. (as *Comandra pallida* 164:459) **Ute** *Analgesic* Roots used for headaches. (as *Comandra pallida* 38:33)
- *Food*—**Navajo, Kayenta** *Unspecified* Seeds used for food. (as *Comandra pallida* 205:18) **Okanagon** *Staple* Seeds or nuts used as a principal food. (as *Comandra pallida* 178:239) **Paiute** *Unspecified* Small, round seeds eaten as nuts by children. (as *C. pallida* 111:66)
- *Dye*—**Arapaho** *Blue* Area next to the root bark used as a blue dye. (as *C. pallida* 121:50)

Comarum, Rosaceae
Comarum palustre L., Purple Marshlocks
- *Drug*—**Chippewa** *Antidiarrheal* Decoction of root taken for dysentery. (as *Potentilla palustris* 53:344) **Ojibwa** *Analgesic* Plant used alone for stomach cramps. *Gastrointestinal Aid* Plant used for stomach cramps. (as *Potentilla palustris* 153:384, 385)
- *Food*—**Eskimo, Alaska** *Beverage* Dried leaves used to make a hot, tea-like beverage before the availability of imported tea. (as *Potentilla palustris* 1:36)

Combretaceae, see *Conocarpus*

Commelinaceae, see *Commelina, Murdannia, Tradescantia*

Commelina, Commelinaceae
Commelina dianthifolia Delile, Birdbill Dayflower
- *Drug*—**Keres, Western** *Tuberculosis Remedy* Infusion of plant used as a strengthener for weakened tuberculosis patients. (171:38) **Navajo, Ramah** *Veterinary Aid* Cold simple or compound infusion given to livestock as an aphrodisiac. (191:19)

Commelina erecta var. *angustifolia* (Michx.) Fern., Whitemouth Dayflower
- *Drug*—**Seminole** *Other* Mucilaginous sap used to soothe irritations. (as *C. angustifolia* 169:303)

Comptonia, Myricaceae
Comptonia peregrina (L.) Coult., Sweet Fern
- *Drug*—**Algonquin, Quebec** *Analgesic* Infusion of leaves taken or crushed leaf perfume inhaled for headaches. (18:149) **Cherokee** *Anthelmintic* Infusion taken for roundworms. (80:58) **Chippewa** *Ceremonial Medicine* Burned, dried leaves used as incense in religious ceremonies. *Febrifuge* Infusion of leaves taken for fevers. *Unspecified* Leaves used for medicine. (as *Myrica asplenifolia* 71:127) **Delaware** *Blood Medicine* Infusion of plant, mallow root, elder flowers, and dwarf elder used as a blood purifier. *Dermatological Aid* Infusion of plant used for blisters. *Pulmonary Aid* Infusion of plant, mallow root, elder flowers, and dwarf elder used to remove mucus from the lungs. *Urinary Aid* Infusion of plant, mallow root, elder flowers, and dwarf elder used for bladder inflammation. *Venereal Aid* Infusion of plant, mallow root, elder flowers, and dwarf elder used for scrofula. (as *Myrica asplenifolia* 176:35) **Delaware, Oklahoma** *Blood Medicine* Infusion of plant taken as a blood purifier. *Dermatological Aid* Infusion of plant applied to blisters and leaves used for poison ivy rash. *Expectorant* Infusion of plant taken to help remove mucus from the lungs. *Tuberculosis Remedy* Plant used for scrofula. *Urinary Aid* Complex compound containing plant used for bladder inflammation. (as *Myrica asplenifolia* 175:29, 76) **Malecite** *Dermatological Aid* Infusion of plant and yarrow used as a liniment for swelling. (as *Myrica asplenifolia* 116:245) *Respiratory Aid* Plants smoked and used for catarrh. (as *Myrica asplenifolia* 116:248) **Menominee** *Adjuvant* Plant used as a seasoner and potent medicine in childbirth. *Gynecological Aid* Decoction of plants used as a potent medicine in childbirth. *Poison* Compound containing leaves sprinkled on medicine to kill a hated person. (as *Myrica asplenifolia* 151:42) *Tonic* Compound decoction of root taken as a mild tonic. (54:133) **Micmac** *Analgesic* Root used for headache and inflammation. *Dermatological Aid* Leaves used for sprains, swellings, poison ivy, and inflammation. (40:56) Leaves used for swellings and poison ivy. (as *Myrica asplenifolia* 40:58) *Orthopedic Aid* Leaves used for sprains. (40:56) Leaves used for sprains, swellings, and poison ivy. (as *Myrica asplenifolia* 40:58) *Respiratory Aid* Leaves used for catarrh. (40:56) Leaves used for catarrh and poison ivy. (as *Myrica asplenifolia* 40:58) *Stimulant* Berries, bark, and leaves used as an "exhilarant" and beverage. (40:56) **Mohegan** *Dermatological Aid* Infusion of leaves used as a wash for poison ivy. (as *Myrica asplenifolia* 174:264) Infusion of leaves used as poison ivy lotion. (as *Myrica asplenifolia* 176:74, 130) **Ojibwa** *Analgesic* Infusion of leaves taken for flux and stomach cramps. *Antidiarrheal* Infusion of leaves taken for stomach cramps and flux. *Gastrointestinal Aid* Decoction of leaves taken for stomach cramps and flux. (as *Myrica asplenifolia* 153:375) **Penobscot** *Dermatological Aid* Infusion of leaves rubbed on skin for poison ivy. (as *Myrica asplenifolia* 156:309) **Potawatomi** *Dermatological Aid* Infusion of leaves used for itch. (as *Myrica asplenifolia* 154:65) **Shinnecock** *Dermatological Aid* Infusion of leaves rubbed on the skin for itch. (30:119)
- *Food*—**Chippewa** *Beverage* Leaves used to make a hot, tea-like beverage. (as *Myrica asplenifolia* 71:127) **Ojibwa** *Preservative* Leaves used to line buckets when picking blueberries and cover them to prevent spoiling. (as *Myrica asplenifolia* 153:420)
- *Other*—**Chippewa** *Incense & Fragrance* Leaves used for perfume. *Tools* Sprigs used to sprinkle water on the hot stones of the vapor bath. (as *Myrica asplenifolia* 71:127) **Potawatomi** *Insecticide* Leaves thrown onto the fire to make a smudge and keep away mosquitoes. (as *Myrica asplenifolia* 154:121)

Condalia, Rhamnaceae
Condalia globosa I. M. Johnston, Bitter Snakewood
- *Food*—**Papago** *Fruit* Fruits eaten raw. (as *Condolis spathulata* 34:19)

Condalia hookeri M. C. Johnston var. *hookeri*, Hooker's Bluewood
- *Food*—**Maricopa** *Fruit* Fruits eaten raw. (as *C. obovata* 95:262) Black berries used for food. (as *C. obovata* 95:265) **Pima** *Fruit* Fruits eaten raw. Fruits roasted and used for food. *Sauce & Relish* Berries cooked, strained, and juice boiled to make syrup. (as *C. obovata* 95:262)

Conioselinum, Apiaceae
Conioselinum chinense (L.) B.S.P., Chinese Hemlockparsley
- *Drug*—**Micmac** *Urinary Aid* Infusion of roots, hemlock, prince's pine, and curled dock used for colds in the bladder. (116:259)

Conioselinum gmelinii (Cham. & Schlecht.) Steud., Pacific Hemlockparsley
- *Drug*—**Aleut** *Cold Remedy* Leaves used to make a tonic for colds. *Throat Aid* Leaves used to make a soothing drink for sore throats. (8: 427) **Kwakiutl** *Antirheumatic (External)* and *Herbal Steam* Plant used in sweat baths for arthritis and rheumatism. (as *C. pacificum* 183: 266) Plant used in steam bath for general weakness. *Stimulant* Plant used in steam bath for general weakness. (as *C. pacificum* 183:276)
- *Food*—**Haihais** *Unspecified* Roots used for food. **Haisla** *Unspecified* Roots used for food. (as *C. pacificum* 43:83) **Haisla & Hanaksiala** *Unspecified* Roots steamed and eaten with oolichan (candlefish) grease. (as *C. pacificum* 43:213) **Hanaksiala** *Unspecified* Roots used for food. **Heiltzuk** *Unspecified* Roots used for food. (as *C. pacificum* 43:83) **Kitasoo** *Vegetable* Roots used for food. (as *C. pacificum* 43:325) **Kwakwaka'wakw** *Unspecified* Roots used for food. **Nuxalkmc** *Unspecified* Roots used for food. **Oweekeno** *Unspecified* Roots used for food. (as *C. pacificum* 43:83)

Conioselinum scopulorum (Gray) Coult. & Rose, Rocky Mountain Hemlockparsley
- *Drug*—**Navajo, Kayenta** *Blood Medicine* Plant used as a postpartum blood purifier. *Gynecological Aid* Plant used as a postpartum blood purifier. *Respiratory Aid* Plant smoked for catarrh. *Snakebite Remedy* Infusion of plant used as a snake repellent. (205:34)
- *Food*—**Navajo, Kayenta** *Unspecified* Leaves cooked with meat and used for food. (205:34)

***Conioselinum* sp.**
- *Food*—**Aleut** *Unspecified* Species used for food. (7:29)

Conium, Apiaceae
Conium maculatum L., Poison Hemlock
- *Drug*—**Klallam** *Love Medicine* Roots rubbed on woman's body to attract the attention of a man. *Poison* Roots considered poisonous. (79: 42) **Lakota** *Poison* All plant parts very poisonous. (139:33) **Snohomish** *Poison* Roots considered poisonous. (79:42)

Conocarpus, Combretaceae
Conocarpus erectus L., Button Mangrove
- *Dye*—**Seminole** *Unspecified* Plant used as a buckskin dye. (169:468)
- *Other*—**Seminole** *Fuel* Wood used for barbecue firewood. (169:468)

Conocephalaceae, see *Conocephalum*

Conocephalum, Conocephalaceae, liverwort
Conocephalum conicum, Cone Headed Liverwort
- *Drug*—**Haisla & Hanaksiala** *Burn Dressing* Plant pulverized, mixed with mountain goat fat, and used for sunburns. (as *C. conicum* 43: 145) **Nitinaht** *Eye Medicine* Plant used as an eye medicine and for cataracts. *Kidney Aid* Plant formerly used for kidney troubles. *Psychological Aid* Plant eaten to stop recurring dreams of having sex with the deceased. (as *C. conicum* 186:58)
- *Other*—**Haisla & Hanaksiala** *Paint* Plant used to make a green paint for wood. (as *C. conicum* 43:145)

Conopholis, Orobanchaceae

Conopholis alpina* var. *mexicana (Gray ex S. Wats.) Haynes, Mexican Squawroot
- *Drug*—**Keres, Western** *Tuberculosis Remedy* Infusion of plant used as a strengthener for weakened tuberculosis patients. (as *C. mexicana* 171:38)
- *Other*—**Jemez** *Good Luck Charm* Dried plant used to rub the ground before the race to make runner more swift-footed. (as *C. mexicana* 44:21)

Consolida, Ranunculaceae
Consolida ajacis (L.) Schur, Doubtful Knight's Spur
- *Drug*—**Cherokee** *Heart Medicine* Infusion taken "for heart." *Poison* Root "makes cows drunk and kills them." (as *Delphinium ajacis* 80:42)

Convolvulaceae, see *Calystegia, Convolvulus, Evolvulus, Ipomoea, Jacquemontia, Merremia*

Convolvulus, Convolvulaceae
Convolvulus arvensis L., Field Bindweed
- *Drug*—**Navajo, Ramah** *Dermatological Aid* Cold infusion of plant taken and used as a lotion for spider bites. *Gastrointestinal Aid* Cold infusion taken with food after swallowing a spider. (191:39) **Pomo** *Gynecological Aid* Decoction of plant taken for excessive menstruation. (66:15) **Pomo, Kashaya** *Gynecological Aid* Decoction of stem with leaves taken for excessive menstruation. (72:73)
- *Fiber*—**Okanagan-Colville** *Cordage* Stems used as a "pack rope" for carrying birds and marmots home after hunting. (188:96)

Conyza, Asteraceae
Conyza canadensis (L.) Cronq., Canadian Horseweed
- *Drug*—**Cahuilla** *Antidiarrheal* Infusion of leaves used for diarrhea. (15:56)

Conyza canadensis (L.) Cronq. **var. *canadensis***, Canadian Horseweed
- *Drug*—**Blackfoot** *Antidiarrheal* Plant used for chronic diarrhea. *Antihemorrhagic* Plant used for childbirth hemorrhage. (as *Erigeron canadensis* 97:56) **Chippewa** *Analgesic* Decoction of root and leaves taken for stomach pain. *Gastrointestinal Aid* Decoction of root and leaves taken for stomach pain. (as *Erigeron canadensis* 53:342) *Gynecological Aid* Infusion of whole plant taken for "female weakness." (as *Erigeron canadensis* 53:356) **Cree, Hudson Bay** *Antidiarrheal* Plant used as a diarrhea medicine. (as *Erigeron canadensis* 92:303) **Hawaiian** *Antirheumatic (External)* Leaves and other plant parts pounded and resulting liquid applied to sore joints. *Orthopedic Aid* Leaves and other plant parts pounded and resulting liquid applied to sprains and backaches. Leaves, shoots, and other plants pounded and resulting liquid used for injuries caused by accidents. (as *Erigen canadense* 2:25) **Hopi** *Analgesic* Poultice of rubbed plant applied to temples for headache. (as *Erigeron canadensis* 200:33, 96) **Houma** *Gynecological Aid* Hot infusion of root taken for leukorrhea. (as *Erigeron canadense* 158:64) **Iroquois** *Anticonvulsive, Febrifuge*, and *Pediatric Aid* Infusion of whole plant and roots from another plant used for children with convulsions and fevers. (as *Erigeron canadensis* 141:65) **Keres, Western** *Burn Dressing* Crushed plant rubbed on

sunburns. *Dermatological Aid* Plant beaten into a paste and rubbed on the skin for blotches or liver spots. (as *Leptilon canadense* 171:51) **Meskwaki** *Diaphoretic* Used as a steaming agent in sweat bath. (as *Erigeron canadensis* 152:213) **Navajo, Kayenta** *Dermatological Aid* Plant used as a lotion for pimples. *Disinfectant* Hot poultice of plant applied to infants with prenatal infection. *Ear Medicine* Hot poultice of plant applied for earaches. *Gastrointestinal Aid* Plant used for stomachaches. *Pediatric Aid* Hot poultice of plant applied to infants with prenatal infection. (as *Erigeron canadensis* 205:47) **Navajo, Ramah** *Dermatological Aid* Poultice of crushed leaves or cold infusion of leaves used as a lotion for pimples. *Snakebite Remedy* Cold infusion taken or used as lotion for snakebite. (as *Erigeron canadensis* 191:50) **Ojibwa** *Hunting Medicine* Disk florets smoked as one of the hunting charms. (as *Erigeron canadensis* 153:429) **Potawatomi** *Veterinary Aid* Plant used as a medicine for horses. (as *Erigeron canadensis* 154:51, 52) **Seminole** *Cold Remedy* Infusion of leaves and bark taken and steam inhaled for runny nose, stuffy head, and sore throat. *Cough Medicine* Infusion of leaves and bark taken and steam inhaled for coughs. (as *Leptilon canadense* 169:279) *Love Medicine* Plant rubbed on the body by a doctor to rid himself of his wife. (as *Leptilon canadense* 169:401) *Respiratory Aid* Infusion of leaves and bark taken and steam inhaled for asthma. (as *Leptilon canadense* 169:279) **Zuni** *Respiratory Aid* Crushed flowers inserted in nostrils to cause sneezing, relieving "rhinitis." (as *Leptilon canadense* 166:55)
- **Food**—**Miwok** *Vegetable* Raw, pulverized leaves and tender tops, flavor similar to onions, used for food. (as *Erigeron canadensis* 12:159)

Coprosma, Rubiaceae

Coprosma sp., Maile-kaluhea
- **Drug**—**Hawaiian** *Dermatological Aid* Infusion of pounded vines, roots, and other plants strained and used as a wash for skin ulcers, and for skin diseases. *Tuberculosis Remedy* Infusion of pounded vines, roots, and other plants strained and used as a wash for scrofulous sores. (2:69) *Unspecified* Whole plant dried, stored, and used under the direction of a medicine man. (2:72)

Coptis, Ranunculaceae

Coptis macrosepala, Goldthread
- **Drug**—**Tlingit** *Pulmonary Aid* Compound infusion of plant used for lung inflammations. (107:283)

Coptis trifolia (L.) Salisb., Threeleaf Goldthread
- **Drug**—**Iroquois** *Anthelmintic* Compound decoction or infusion of roots taken for stomach cramps and worms. *Antiemetic* Infusion or decoction of roots taken, especially for babies, for vomiting. *Blood Medicine* Complex compound decoction taken as a blood purifier and blood remedy. *Emetic* Decoction of roots taken to vomit for jaundice, for the eyeballs, and dizziness. *Eye Medicine* Infusion or decoction of roots taken or used as drops for sore eyes. *Gastrointestinal Aid* Compound decoction of roots taken for stomach cramps and worms. Infusion of roots taken for vomiting, biliousness, and jaundice. *Oral Aid* Infusion of roots used for sore mouths of children and trench mouth. *Pediatric Aid* Decoction of roots given to "little babies when they throw up often." Infusion of roots used as a wash or poultice applied to sore mouths of children. (87:322) Decoction of plant given to babies with sickness caused by bad blood from mother. (87:323) *Throat Aid* Cold, compound infusion with plant taken for trench mouth and raw throat. (87:322) *Unspecified* Decoction of plant given to babies with sickness caused by bad blood from mother. (87:323) *Venereal Aid* Complex compound decoction taken as a blood purifier and for venereal disease. (87:322) **Malecite** *Oral Aid* and *Pediatric Aid* Infusion of plant used for children with sore mouths. (116:245) **Menominee** *Oral Aid* Astringent root used as a wash for oral cankers and babies' teething pains. *Pediatric Aid* Root yielded astringent mouthwash for sore throat and teething babies. *Throat Aid* Roots used in astringent mouthwash for babies with sore throats. *Toothache Remedy* Roots used in astringent mouthwash for teething babies. (151:48) **Micmac** *Oral Aid* Herb used for sore and diseased mouth. (40:56) **Mohegan** *Oral Aid* Infusion of plant used as a mouthwash for babies. (174:265) Infusion of leaves used as a mouthwash for infants. (176:72, 128) *Pediatric Aid* Infusion of plant used as a mouthwash for babies. (174:265) Infusion of leaves used as a mouthwash for infants. (176:72, 128) **Ojibwa** *Oral Aid* Decoction of root used as a wash for sore mouth and to soothe mouth of teething baby. *Pediatric Aid* Decoction of root used to soothe mouth of teething baby. (153:383) **Penobscot** *Oral Aid* Stems chewed for mouth sores and mouths irritated by tobacco smoking. (156:309) **Potawatomi** *Analgesic* Root used for babies with teething pains. *Oral Aid* Roots used for sore gums and especially for pain of teething babies. *Pediatric Aid* Roots used especially for pain of teething babies. (154:74)
- **Dye**—**Chippewa** *Yellow* Long, slender roots used to make a bright yellow dye. (53:374) Roots used to make a yellow dye. (71:130) **Ojibwa** *Yellow* Roots boiled to obtain a yellow dye. (96:114) Golden-colored roots added to other plant dyes to emphasize the yellow color. (153:426) **Potawatomi** *Yellow* Roots cooked with the cloth to dye an indelible yellow. (154:122)

Coptis trifolia ssp. *groenlandica* (Oeder) Hultén, Threeleaf Goldthread
- **Drug**—**Abnaki** *Cold Remedy* Decoction of plant and another plant used for colds. (as *C. groenlandica* 144:167) *Cough Medicine* Used for coughs. (as *C. groenlandica* 144:154) **Algonquin, Quebec** *Antidiarrheal* Infusion of rhizomes taken for diarrhea. *Eye Medicine* Infusion of rhizomes used as an eyewash. *Heart Medicine* Infusion of rhizomes taken for heart disease. *Toothache Remedy* Infusion of rhizomes taken for toothaches. (as *C. groenlandica* 18:167) **Algonquin, Tête-de-Boule** *Cold Remedy* Boiled roots used for serious colds. *Respiratory Aid* Boiled roots used for respiratory troubles. (as *C. groenlandica* 132:126) **Iroquois** *Ear Medicine* Infusion of plant, with another plant, used as ear drops for earaches. *Gastrointestinal Aid* Infusion of roots taken to ease digestion. (as *C. groenlandica* 141:42) *Oral Aid* Roots chewed for mouth pains. (as *C. groenlandica* 142:87) **Malecite** *Unspecified* Used for medicines. (as *C. groenlandica* 160:6) **Micmac** *Unspecified* Roots chewed for medicinal use. (as *C. groenlandica* 159:258)

Corallorrhiza, Orchidaceae

Corallorrhiza maculata (Raf.) Raf., Summer Coralroot
- **Drug**—**Navajo, Kayenta** *Dermatological Aid* Infusion of plant used as a lotion for ringworm or skin disease. (205:17) **Nevada Indian** *Cold Remedy* Infusion of dried, whole plant bits taken for colds. (118:

37) **Paiute** *Blood Medicine* Decoction of stalks used to "build up the blood" of pneumonia patients. *Pulmonary Aid* Infusion of dried stalks taken to "build up the blood" of pneumonia patients. **Shoshoni** *Blood Medicine* Decoction of stalks used to "build up the blood" of pneumonia patients. *Pulmonary Aid* Infusion of dried stalks taken to "build up the blood" of pneumonia patients. (180:60)

Corallorrhiza maculata (Raf.) Raf. **ssp. maculata**, Summer Coralroot
- **Drug**—**Iroquois** *Basket Medicine* Infusion of pounded root used as a basket medicine. (87:290) *Hunting Medicine* Root placed in a half cup of water and used to wash guns and clothes as a hunting medicine. *Love Medicine* Infusion of pounded roots used as a love medicine. *Tuberculosis Remedy* Compound infusion of roots taken for tuberculosis. (87:291) *Veterinary Aid* Infusion of whole plant added to horse's grain for heaves. *Witchcraft Medicine* Infusion of pounded roots used as an anti-witch medicine. (87:290)

Cordylanthus, Scrophulariaceae

Cordylanthus ramosus Nutt. ex Benth., Bushy Bird's Beak
- **Drug**—**Navajo** *Emetic* Infusion of plant taken to induce vomiting. *Gynecological Aid* Infusion of plant taken by menstruating women to stop menses. *Hemostat* Infusion of plant used by men for nosebleeds. (55:76) *Orthopedic Aid* Plant used to prevent broken ribs. (55:96) *Venereal Aid* Infusion of plant taken for syphilis. (55:76) **Shoshoni** *Venereal Aid* Decoction of plant taken for venereal disease, or "bad disease." (180:60)

Cordylanthus **sp.**, Sunflower
- **Drug**—**Luiseño** *Emetic* Plant used as an emetic. (155:229)
- **Food**—**Yavapai** *Unspecified* Parched, ground seeds eaten dry or dampened. (65:256)

Cordylanthus wrightii Gray, Wright's Bird's Beak
- **Drug**—**Navajo** *Venereal Aid* Decoction of plant used for syphilis. (55:76) **Navajo, Kayenta** *Ceremonial Medicine* Plant used for ceremonial purposes. *Gynecological Aid* Plant used for prolapse of the uterus. (205:42) **Navajo, Ramah** *Analgesic* Decoction of plant taken for menstrual pain and by men for leg or body aches. *Gynecological Aid* Decoction of plant taken for menstrual pain. (191:44)

Cordyline, Agavaceae

Cordyline fruticosa (L.) Chev., Tiplant
- **Drug**—**Hawaiian** *Febrifuge* Leaves applied to the head, chest, and abdomen for dry fevers. *Nose Medicine* Flowers and other plants pounded and resulting liquid fumes inhaled for nose growths. *Pulmonary Aid* Leaves, shoots, and other plants mixed with water and taken for chest congestion from tough phlegm. *Respiratory Aid* Flowers and other plants pounded, resulting liquid mixed with potato or poi and eaten for asthma. *Sedative* Leaves made into a wreath and worn to provide a restful condition of the nerves and body. (as *C. terminalis* 2:49)
- **Food**—**Hawaiian** *Beverage* Roots fermented into a very powerful alcohol. *Unspecified* Roots cooked and used for food. (as *C. terminalis* 2:49) Roots pit baked and eaten. (as *C. terminalis* 112:68)
- **Other**—**Hawaiian** *Containers* Leaves used for underground oven covers or as containers for cooking fish or pork. *Hunting & Fishing Item* Leaves used for fishing drags. (*C. terminalis* 2:49)

Coreopsis, Asteraceae

Coreopsis bigelovii (Gray) Hall, Bigelow's Tickseed
- **Food**—**Kawaiisu** *Sweetener* Stems chewed for the sweet juice. *Vegetable* Whole plant eaten fresh or cooked and fried in grease and salt. Raw, bruised leaves eaten boiled or with salt. (206:21) **Tubatulabal** *Unspecified* Leaves used extensively for food. (193:15)

Coreopsis leavenworthii Torr. & Gray, Leavenworth's Tickseed
- **Drug**—**Seminole** *Other* Infusion of plant used for heat prostration. (169:303)

Coreopsis palmata Nutt., Stiff Tickseed
- **Drug**—**Meskwaki** *Orthopedic Aid* Decoction of seeds used internally and as a poultice for one who is crippled. (152:213)

Coreopsis **sp.**, Tickseed
- **Dye**—**Cherokee** *Red* Whole plant used to give a red coloring. (203:74)

Coreopsis tinctoria Nutt., Golden Tickseed
- **Drug**—**Cherokee** *Antidiarrheal* Infusion of root taken for flux. (80:59)
- **Food**—**Lakota** *Beverage* Used to make tea. (139:37)
- **Dye**—**Cherokee** *Red* Used to make a red dye. (80:59)

Coreopsis tinctoria Nutt. **var. tinctoria**, Golden Tickseed
- **Drug**—**Navajo, Ramah** *Ceremonial Medicine* Plant used in ceremonial chant lotion. *Disinfectant* Cold infusion of dried plant taken with salt for "lightning infection." *Panacea* Root used as a "life medicine." *Venereal Aid* Plant used as fumigant for sexual infection. (as *C. cardaminefolia* 191:50) **Zuni** *Reproductive Aid* Infusion of whole plant, except for the root, taken by women desiring female babies. (as *C. cardaminefolia* 166:84)
- **Food**—**Zuni** *Beverage* Plant formerly used to make a hot beverage until the introduction of coffee by traders. The plant was folded while fresh, a number of folds being attached one below the other, and hung on the wall to dry. When the beverage was desired, a fold was detached from the wall and used to make a hot beverage. (as *C. cardaminefolia* 166:66)
- **Dye**—**Apache, White Mountain** *Red* Used as a dark, rich red dye. (as *C. cardaminefolia* 136:156) **Zuni** *Red* Blossoms used with other flowers as a mahogany red dye for yarn. (as *C. cardaminefolia* 166:80)

Coreopsis tripteris L., Tall Tickseed
- **Drug**—**Meskwaki** *Analgesic* Decoction of stems taken for internal pains. *Antihemorrhagic* Decoction of stems taken for internal bleeding. (152:213)

Corethrogyne, Asteraceae

Corethrogyne filaginifolia (Hook. & Arn.) Nutt., Common Sandaster
- **Drug**—**Kawaiisu** *Cold Remedy, Diaphoretic,* and *Herbal Steam* Infusion of twigs and leaves used as steam bath to induce sweating for colds. (206:22)

Coriandrum, Apiaceae

Coriandrum sativum L., Chinese Parsley

- **Food**—**Hopi** *Sauce & Relish* Plant dipped into a stew and eaten as a condiment. (56:20) *Spice* Used as flavoring in cooking. (200:86) *Unspecified* Dipped into water, eaten raw and green. (190:164) **Keresan** *Spice* Seeds used to flavor soups and stews. (198:560) **Zuni** *Sauce & Relish* Powdered seeds ground with chile and used a condiment with meat. *Vegetable* Leaves used as a salad. (166:66)

Cornaceae, see *Cornus, Nyssa*

Cornicularia, Alectoriaceae, lichen

Cornicularia divergens Ach., Caribou Moss
- **Food**—**Eskimo, Inuktitut** *Fodder* Plant given to fawns to try to get them to eat from their hands. (202:191)
- **Other**—**Eskimo, Inuktitut** *Fuel* Dried plant used for tinder. (202:191)

Cornus, Cornaceae

Cornus alternifolia L. f., Alternateleaf Dogwood
- **Drug**—**Cherokee** *Analgesic* Bark chewed for headache. *Anthelmintic* Compound infusion of bark and root used for childhood diseases like worms and measles. *Antidiarrheal* Compound infusion taken for diarrhea. *Antidote* Infusion of beaten bark used for bathing after "poisons of any kind." *Blood Medicine* Infusion taken "for blood." *Dermatological Aid* Root bark astringent and compound infusion taken for diarrhea. Root bark used for unspecified poultices and poultice of bark ooze applied to ulcers. *Diaphoretic* Infusion of flower taken "to sweat off flu." *Disinfectant* Root bark used as an antiseptic and astringent. *Febrifuge* Root bark used as a febrifuge. *Gastrointestinal Aid* Infusion of flower taken for colic. *Gynecological Aid* Infusion of bark used by women for backache. *Misc. Disease Remedy* Compound infusion of bark and root used for childhood diseases like worms and measles. Infusion of flower taken "to sweat off flu." *Pediatric Aid* Compound infusion of bark and root used for childhood diseases like worms and measles. *Stimulant* Root bark used as a stimulant. *Throat Aid* Infusion of inner bark taken for "lost voice." *Tonic* Root bark used as a tonic. (80:32) **Chippewa** *Cough Medicine* Inner bark used as a cough remedy. (71:138) *Eye Medicine* Compound decoction of root used as a wash or compress for sore eyes. Infusion of scraped root used as a wash or on a compress for sore eyes. (53:360) *Hunting Medicine* Roots used as a charm on muskrat traps. (53:376) **Iroquois** *Cold Remedy* Compound decoction of bark taken for colds. *Cough Medicine* Compound decoction of bark taken for coughing. (87:407) *Dermatological Aid* Poultice of powdered bark applied to heal the navel and blisters. (87:406) Infusion of bark applied as poultice to swollen areas. *Emetic* Compound decoction of bark taken as an emetic, especially for coughs. (87:407) *Eye Medicine* Plant used in a wash for eyes. *Gynecological Aid* Compound decoction of bark taken by pregnant women who have had gonorrhea. *Pediatric Aid* Poultice of powdered bark applied to heal navel. *Respiratory Aid* Decoction of bark taken to vomit for coughs or bronchial coughs. (87:406) *Tuberculosis Remedy* Compound decoction of bark taken for tuberculosis. (87:407) *Venereal Aid* Compound infusion used as wash on parts affected by venereal disease. (87:406) **Menominee** *Antidiarrheal* Bark liquid injected rectally and poultice of bark applied to anus for diarrhea. *Cancer Treatment* One reported case: poultice of bark plus something else cured facial cancer. *Hemorrhoid Remedy* Bark used to make a liquid and injected rectally for piles. (151:32, 33) **Ojibwa** *Emetic* Inner bark used as an emetic. (153:366) **Potawatomi** *Eye Medicine* Infusion of bark used as a wash for granulation of the eyelids. (154:54)
- **Fiber**—**Ojibwa** *Building Material* Twigs used for thatching and various other purposes. (153:417)
- **Other**—**Cherokee** *Decorations* Wood used to carve. *Tools* Wood used to make loom shuttles. (80:32) **Chippewa** *Weapon* Wood used to make awl handles, mauls, and war clubs because it would not split or check. (71:138) **Menominee** *Smoke Plant* Plant used for Indian tobacco, known as kinnikinnick. (151:32, 33) Toasted inner bark used for smoking tobacco. (151:80) **Ojibwa** *Hunting & Fishing Item* Root boiled to wash muskrat traps and make it lure the muskrat. (153:429) *Smoke Plant* Bark used for "kinnikinnick" smoking mixture. (153:417)

Cornus amomum P. Mill., Silky Dogwood
- **Drug**—**Iroquois** *Analgesic* Compound decoction of roots taken for urinating pain. *Dermatological Aid* Infusion of bark used as wash or powdered bark applied to gonorrhea sores. (87:402) Complex compound decoction used as wash for affected parts of "Italian itch." *Emetic* Decoction of bark taken as an emetic. (87:403) *Laxative* Compound decoction of roots taken as a laxative. *Pediatric Aid* Infusion of bark used as wash to make babies sleep. (87:402) *Poultice* Poultice of smashed bark applied for goiter. *Pulmonary Aid* Infusion of bark taken for chest congestion. (87:403) *Sedative* Infusion of bark used as wash to make babies sleep. *Urinary Aid* Compound decoction of roots taken for urinating pain. *Venereal Aid* Infusion of bark used as wash or powdered bark applied to gonorrhea sores. (87:402) **Menominee** *Antidiarrheal* Plant known as maimakwukwa and infusion of bark injected rectally for diarrhea. *Ceremonial Medicine* Plant known as kinnikinnick and bark smoked ceremonially. (151:32)
- **Other**—**Dakota** *Smoke Plant* Dried inner bark used for smoking. (69:367) Fragrant inner bark dried and used for smoking. (70:107) **Menominee** *Smoke Plant* Toasted inner bark used for smoking tobacco. (151:80) **Omaha** *Smoke Plant* Dried inner bark used either alone or with tobacco for smoking. (68:331) Fragrant inner bark dried and used for smoking. **Pawnee** *Smoke Plant* Fragrant inner bark dried and used for smoking. **Ponca** *Smoke Plant* Fragrant inner bark dried and used for smoking. (70:107) Inner bark used as an additive to tobacco. (94:47) **Winnebago** *Smoke Plant* Fragrant inner bark dried and used for smoking. (70:107)

Cornus amomum ssp. *obliqua* (Raf.) J. S. Wilson, Silky Dogwood
- **Drug**—**Iroquois** *Gastrointestinal Aid* Infusion of bark taken for dyspepsia. (as *C. obliqua* 141:54)
- **Other**—**Iroquois** *Tools* Branches used to stretch rat skins while drying. (as *C. obliqua* 141:54)

Cornus asperifolia Michx., Toughleaf Dogwood
- **Other**—**Comanche** *Hunting & Fishing Item* Stems used to make arrow shafts. (29:521) **Omaha** *Hunting & Fishing Item* Wood considered the favorite for arrow shafts. **Pawnee** *Hunting & Fishing Item* Wood considered the favorite for arrow shafts. **Ponca** *Hunting & Fishing Item* Wood considered the favorite for arrow shafts. **Winnebago** *Hunting & Fishing Item* Wood considered the favorite for arrow shafts. (70:107)

Cornus canadensis L., Bunchberry Dogwood
- **Drug**—**Abnaki** *Analgesic* Used for side pains. (144:155) Decoction of whole plant taken for side pains. (144:170) **Algonquin, Quebec** *Cathartic* Infusion of leaves used as a cathartic tea. (18:211) **Algonquin, Tête-de-Boule** *Cold Remedy* Decoction of plant and other plants used for colds. *Gynecological Aid* Plant mixed with other plants and used by women for stomachaches. (132:128) **Carrier, Northern** *Unspecified* Used as a medicine for unspecified malady. **Carrier, Southern** *Eye Medicine* Strong decoction of plant, without berries, used as an eyewash. (150:62) **Delaware** *Analgesic* Bark used for body pains. (176:31) **Delaware, Oklahoma** *Analgesic* Compound containing bark used for body pain. (175:26, 74) **Hoh** *Tonic* Infusion of bitter bark used as a tonic. (137:66) **Iroquois** *Cold Remedy* Decoction of whole plant taken for coughs. *Febrifuge* Decoction of whole plant taken for fevers. *Tuberculosis Remedy* Decoction of whole plant taken for tuberculosis. (87:402) **Malecite** *Anticonvulsive* Infusion of roots, leaves, and berries used for fits. (116:256) **Micmac** *Anticonvulsive* Berries, roots, and leaves used for fits. (40:56) **Montagnais** *Orthopedic Aid* Infusion of plant used as a medicine for paralysis. (156:315) **Ojibwa** *Gastrointestinal Aid* Infusion of root used for infant colic. *Pediatric Aid* Infusion of root used for infant colic. (153:366, 367) **Paiute** *Eye Medicine* Mashed roots strained through a clean cloth and used as an eyewash for the removal of foreign objects. Mashed roots strained through a clean cloth and used as an eyewash for eye soreness. (111:98) **Quileute** *Tonic* Infusion of bitter bark used as a tonic. (137:66) **Thompson** *Dermatological Aid* Leaf ash or powdered, toasted leaves sprinkled on sores. (164:458)
- **Food**—**Abnaki** *Fruit* Fruits eaten for food. (144:170) **Algonquin, Quebec** *Snack Food* Berries used as a nibble food. (18:102) **Chippewa** *Fruit* Berries eaten raw. (53:321) **Cree, Woodlands** *Snack Food* Fruit eaten as a fresh nibble. (109:36) **Eskimo, Alaska** *Fruit* Gathered and mixed with other berries. (4:715) **Haisla & Hanaksiala** *Dessert* Berries mashed, mixed with oolichan (candlefish) grease, and eaten as a dessert. *Dried Food* Berries dried for winter use. (43:234) **Hesquiat** *Special Food* Raw berries eaten with dogfish oil by the elders of the village at a big feast. (185:63) **Kwakiutl, Southern** *Fruit* Pulpy berries extensively used for food. (183:281) **Makah** *Fruit* Berries eaten fresh. (79:43) **Nitinaht** *Fruit* Berries eaten fresh and raw. (186:102) **Potawatomi** *Fruit* Berries used for food. (154:98) **Salish** *Fruit* Berries used for food. (182:81)
- **Other**—**Hoh** *Ceremonial Items* Berries used in ceremonies. *Smoke Plant* Leaves dried and smoked. **Quileute** *Ceremonial Items* Berries used in ceremonies. *Smoke Plant* Leaves dried and smoked. (137:66)

Cornus drummondii C. A. Mey., Roughleaf Dogwood
- **Drug**—**Iroquois** *Venereal Aid* Infusion of switches taken for gonorrhea. (87:403)

Cornus florida L., Flowering Dogwood
- **Drug**—**Cherokee** *Analgesic* Bark chewed for headache. *Anthelmintic* Compound infusion of bark and root used for childhood diseases like worms and measles. (80:32) Infusion of bark used as a bath and given to children with worms. (177:46) *Antidiarrheal* Compound infusion taken for diarrhea. *Antidote* Infusion of beaten bark used for bathing after "poisons of any kind." *Blood Medicine* Compound infusion taken for "for blood." *Dermatological Aid* Root bark astringent and compound infusion taken for diarrhea. Root bark used for unspecified poultices and poultice of bark ooze applied to ulcers. (80:32) Root bark used for wounds. (203:74) *Diaphoretic* Infusion of flower taken "to sweat off flu." *Disinfectant* Root bark used as an antiseptic and astringent. *Febrifuge* Root bark used as a febrifuge. *Gastrointestinal Aid* Infusion of flower taken for colic. *Gynecological Aid* Infusion of bark used by women for backache. *Misc. Disease Remedy* Compound infusion of bark and root used for childhood diseases like worms and measles. Infusion of flower taken "to sweat off flu." *Pediatric Aid* Compound infusion of bark and root used for childhood diseases like worms and measles. (80:32) Infusion of bark used as a bath and given to children with worms. (177:46) *Poultice* Root bark used in poultices. (203:74) *Stimulant* Root bark used as a stimulant. *Throat Aid* Infusion of inner bark taken for "lost voice." (80:32) Decoction of inner bark taken to loosen phlegm for hoarseness. (177:46) *Tonic* Root bark used as a tonic. (80:32) **Delaware** *Tonic* Roots used as a tonic. (176:31) **Delaware, Oklahoma** *Tonic* Compound containing root used as a tonic. (175:26, 74) **Houma** *Febrifuge* Decoction of root or bark scrapings taken for fever. *Misc. Disease Remedy* Decoction of root or bark scrapings taken for malaria. (158:55) **Iroquois** *Blood Medicine* Compound decoction of stems and roots taken for blood chills. (87:402) **Rappahannock** *Antidiarrheal* Infusion of root bark taken for diarrhea. *Blood Medicine* Decoction of dried bark from roots used to purify the blood. *Tonic* Decoction of dried bark from roots used as a tonic. (161:33)
- **Other**—**Cherokee** *Decorations* Wood used to carve. *Tools* Wood used to make loom shuttles. (80:32)

Cornus foemina P. Mill., Stiff Dogwood
- **Drug**—**Cherokee** *Throat Aid* Infusion used for "lost voice." (as *C. stricta* 80:30, 31) **Houma** *Febrifuge* Decoction of root or bark scrapings taken for fever. *Misc. Disease Remedy* Decoction of root or bark scrapings taken for malaria. (as *C. stricta* 158:55)
- **Other**—**Micmac** *Smoke Plant* Dried bark mixed with tobacco and used for smoking. (as *C. stricta* 156:317)

Cornus glabrata Benth., Brown Dogwood
- **Fiber**—**Modesse** *Basketry* Long shoots used for some of the baskets. (117:223) **Wintoon** *Basketry* Straight stems used for making baskets. (117:264) **Yuki** *Basketry* Branches used to make coarse baskets. (49:92)

Cornus nuttallii Audubon ex Torr. & Gray, Pacific Dogwood
- **Drug**—**Green River Group** *Cathartic* Plant used as a physic. *Emetic* Plant used as an emetic. (79:42) **Hoh** *Tonic* Infusion of bitter bark used as a tonic. (137:66) **Karok** *Herbal Steam* Boughs used in the fire of the sweat house. (148:387) **Lummi** *Laxative* Decoction of bark taken as a laxative. (79:42) **Quileute** *Tonic* Infusion of bitter bark used as a tonic. (137:66) **Thompson** *Blood Medicine* Decoction of two bark strips and two cascara bark strips taken as a "blood purifier." (187:204) *Gastrointestinal Aid* Decoction of wood or bark taken for stomach trouble. (164:461) Decoction of two bark strips and two cascara bark strips taken for ulcers. *Psychological Aid* Strained decoction of flower heads used to wash the skin for "seven year itch." *Pulmonary Aid* Decoction of two bark strips and two cascara bark strips taken to improve hunters lungs before hiking. (187:204)

- *Fiber*—Pomo, Kashaya *Furniture* Long, slender branches used in making baby baskets. (72:42)
- *Dye*—Thompson *Black* Bark and fir bark boiled into a black dye and used to dye bitter cherry bark for imbricating baskets. *Brown* Bark boiled to make an intense brown dye and used to color bitter cherry bark for imbricating baskets. (187:204)
- *Other*—Green River Group *Toys & Games* Wood used to make disks for gambling games. (79:42) **Hoh** *Ceremonial Items* Berries used in ceremonies. *Smoke Plant* Leaves dried and smoked. (137:66) **Karok** *Good Luck Charm* Plant used as a good luck charm. (148:387) **Klallam** *Toys & Games* Wood used to make disks for gambling games. (79:42) **Quileute** *Ceremonial Items* Berries used in ceremonies. *Smoke Plant* Leaves dried and smoked. (137:66) **Quinault** *Decorations* Charcoal used for tattooing. (79:42) **Saanich** *Preservative* Bark used to tan articles like cedar bark bailers. **Salish, Coast** *Hunting & Fishing Item* Wood used to make bows and arrows. *Smoke Plant* Bark occasionally mixed with tobacco and used for smoking. (182:81) **Skagit** *Hunting & Fishing Item* Dry wood used to make foreshafts of salmon harpoons. *Toys & Games* Wood used to make disks for gambling games. **Snohomish** *Cooking Tools* Sticks used to pound brake ferns after roasting. (79:42) **Thompson** *Hunting & Fishing Item* Wood used to make bows. *Tools* Wood used to make implement handles. (164:496)

Cornus racemosa Lam., Gray Dogwood

- *Drug*—Iroquois *Dermatological Aid* Decoction of bark applied as poultice to cuts. (as *C. paniculata* 87:405) *Gastrointestinal Aid* Compound decoction taken, used as wash, and poultice applied to swollen abdomen. *Orthopedic Aid* Compound poultice of bark applied to swollen legs after the birth of a baby. *Venereal Aid* Compound powder poultice "put in bag, place penis in bag and tie around waist." (as *C. paniculata* 87:406) *Veterinary Aid* Decoction of bark applied as poultice to cuts on horses. (as *C. paniculata* 87:405) **Meskwaki** *Analgesic* Infusion of bark held in mouth for neuralgia. *Antidiarrheal* Infusion of bark used, especially for children, as an enema for flux. *Oral Aid* Infusion of bark held in mouth for toothache. *Pediatric Aid* Infusion of bark given, often to children, as an enema for flux. *Stimulant* Smudged bark used to revive an unconscious patient. *Tuberculosis Remedy* Infusion of root used for consumption. (as *C. paniculata* 152:218, 219) **Ojibwa** *Antidiarrheal* Infusion of bark used for flux. *Hemorrhoid Remedy* Bark forced into the anus for piles. (as *C. paniculata* 153:367)
- *Other*—Meskwaki *Smoke Plant* Smoked at ceremonies. (as *C. paniculata* 152:272) **Ojibwa** *Smoke Plant* Bark used to make "kinnikinnick" for smoking. (as *C. paniculata* 153:418) Peeled, toasted, shredded twig bark used in the "kinnikinnick" or native smoking tobacco. (as *C. paniculata* 153:399)

Cornus rugosa Lam., Roundleaf Dogwood

- *Drug*—Iroquois *Cathartic* Bark taken as a general cathartic or emetic. *Emetic* Decoction of bark taken as an emetic. *Kidney Aid* Compound decoction of roots taken for the kidneys. *Tuberculosis Remedy* Compound infusion of smashed roots taken for tuberculosis. (87:405)
- *Other*—Chippewa *Smoke Plant* Used for smoking. (53:377)

Cornus sericea L. Red Osier Dogwood

- *Drug*—Cree, Hudson Bay *Cold Remedy* Decoction of bark taken as an emetic for colds. *Cough Medicine* Decoction of bark taken as an emetic for coughs. *Emetic* Decoction of bark taken as an emetic for coughs, colds, and fevers. *Febrifuge* Decoction of bark taken as an emetic for fevers. (92:303) **Ojibwa** *Ceremonial Medicine* Bark smoked for various ceremonies. *Unspecified* Bark used for medicinal purposes. (135:237) **Thompson** *Anthelmintic* Fruit considered a good "tonic," especially for intestinal worms. *Antidiarrheal* Decoction of branches, wild rose, and chokecherry branches taken for diarrhea. *Antiemetic* Decoction of branches, wild rose, and chokecherry branches taken for vomiting. *Cold Remedy* Decoction of branches taken for colds. *Panacea* Plant used as a medicine for anything by the elderly. *Pediatric Aid* Decoction of plant, squaw currant, branches, and fir or tamarack used as a baby bath.[27] *Poison* Sap used on arrowheads for the poisonous effect upon animals. *Strengthener* Decoction of plant, squaw currant branches, and fir or tamarack used as a baby bath.[27] (187:204)
- *Food*—Thompson *Dessert* Fruit eaten as dessert. *Dried Food* Berries and saskatoon berries smashed together, dried, rehydrated, and eaten in the winter. The berries were also pounded with chokecherries, seeds and all, and used for food. *Fruit* Bitter, seedy fruits eaten alone or mashed with dried, "white" saskatoon berries. (187:204)
- *Fiber*—Thompson *Furniture* Branches used to make the bow at the top of a baby's cradle. (187:204)
- *Other*—Thompson *Hunting & Fishing Item* Flexible branches used to make the frame of a cylindrical basketry trap. The frame was tied together with "gray willow" rope. Twisted branches used to make gill nets. Sap used on arrowheads for the poisonous effect upon animals. (187:204)

Cornus sericea ssp. *occidentalis* (Torr. & Gray) Fosberg, Western Dogwood

- *Drug*—Hoh *Tonic* Infusion of bitter bark used as a tonic. **Quileute** *Tonic* Infusion of bitter bark used as a tonic. (as *C. occidentalis* 137:66) **Snohomish** *Eye Medicine* Infusion of bark used for sore eyes. (as *C. pubescens* 79:42) **Thompson** *Gynecological Aid* Compound decoction of twigs taken by women after childbirth. (as *C. pubescens* 164:461) Simple or compound decoction of various plant parts taken after childbirth. (as *C. pubescens* 164:475)
- *Food*—Nitinaht *Fruit* Berries eaten fresh. (as *C. stolonifera* var. *occidentalis* 186:103) **Okanagon** *Staple* Berries used as a principal food. (as *C. pubescens* 178:238) **Sanpoil & Nespelem** *Fruit* Berries eaten fresh. (as *C. pubescens* 131:102) **Spokan** *Fruit* Berries used for food. (as *C. pubescens* 178:343) **Thompson** *Unspecified* Little, white drupes eaten occasionally. (as *C. pubescens* 164:490)
- *Other*—Hoh *Ceremonial Items* Berries used in ceremonies. *Smoke Plant* Leaves dried and smoked. **Quileute** *Ceremonial Items* Berries used in ceremonies. *Smoke Plant* Leaves dried and smoked. (as *C. occidentalis* 137:66) **Thompson** *Smoke Plant* Leaves occasionally smoked as tobacco. (as *C. pubescens* 164:495)

Cornus sericea L. ssp. *sericea*, Red Osier Dogwood

- *Drug*—Abnaki *Eye Medicine* Used for sore eyes. (as *C. stolonifera* 144:155) Decoction of bark and bark from two other plants used for eye pain. (as *C. stolonifera* 144:170) **Algonquin, Quebec** *Cold Remedy* Infusion of bark shavings taken for colds. *Hemorrhoid Remedy* Bark shavings used to stop bleeding. (as *C. stolonifera* 18:211) **Apache,**

White Mountain *Ceremonial Medicine* Plant used in medicine ceremonies. (as *Svida stolonifera* 136:161) **Bella Coola** *Eye Medicine* Infusion of inner bark used for sties and other eye infections. (as *C. stolonifera* 184:203) **Blackfoot** *Cold Remedy* Infusion of bark taken for chest colds. (as *C. stolonifera* 86:71) *Liver Aid* Infusion of cambium taken for liver troubles and related disorders. (as *C. stolonifera* 86:66) *Poison* Chewed berry spittle used on arrow points and musket balls to cause infections in the wound. (as *C. stolonifera* 86:84) **Carrier** *Analgesic* Poultice of water-soaked, inner bark applied with warmed ashes as a painkiller. *Pulmonary Aid* Bark scraped, mixed with tobacco, and smoked for lung sickness. (as *C. stolonifera* 31:71) **Carrier, Northern** *Dermatological Aid* Compound decoction of bark taken for body sores. *Orthopedic Aid* Compound decoction of bark taken for weakness or paralysis. *Stimulant* Compound decoction of bark taken for constitutional weakness. (as *C. stolonifora* 150:62) **Chippewa** *Antidiarrheal* Infusion of bark taken for diarrhea. *Dermatological Aid* Infusion of bark taken for eruptions caused by poison ivy. (as *C. stolonifera* 71:138) *Eye Medicine* Compound decoction of root used as a wash or compress for sore eyes. (as *C. stolonifera* 53:360) **Costanoan** *Febrifuge* Decoction of inner bark used for fevers. (as *C. californica* 21:24) **Cree, Woodlands** *Eye Medicine* Fruit or pith used to make a wash for snow-blindness. Pith used for cataracts. (as *C. stolonifera* 109:36) **Gosiute** *Narcotic* Plant used for the similar effect to opium. (as *C. stolonifera* 39:366) **Iroquois** *Analgesic* Infusion of bark taken for headaches. (as *C. alba* ssp. *stolonifera* 87:403) *Cold Remedy* Plant used for colds. (as *C. alba* ssp. *stolonifera* 87:405) *Cough Medicine* Infusion of bark taken for coughs. *Emetic* Decoction or infusion of bark taken as an emetic, especially for consumption. (as *C. alba* ssp. *stolonifera* 87:403) *Eye Medicine* Plant used in a wash for eyes. (as *C. alba* ssp. *stolonifera* 87:405) *Hemostat* Infusion of bark taken for nose or mouth hemorrhages. *Oral Aid* Infusion of bark taken for nose or mouth hemorrhages. (as *C. alba* ssp. *stolonifera* 87:403) *Panacea* Bark smoked for every ailment. *Psychological Aid* Compound decoction of bark taken for craziness. (as *C. alba* ssp. *stolonifera* 87:404) *Pulmonary Aid* Compound infusion of bark taken for pain or congestion of the chest. (as *C. alba* ssp. *stolonifera* 87:405) *Sports Medicine* Decoction of bark taken by lacrosse players and runners to vomit. (as *C. alba* ssp. *stolonifera* 87:404) *Tuberculosis Remedy* Compound decoction taken to vomit during the initial stages of consumption. (as *C. alba* ssp. *stolonifera* 87:403) *Witchcraft Medicine* Compound of plant and dried snake's blood used as a "witching medicine." (as *C. alba* ssp. *stolonifera* 87:405) **Malecite** *Analgesic* Plants mixed with spikenard roots and smoked for headaches. *Eye Medicine* Plants chewed or soaked in warm water and used for sore eyes. *Respiratory Aid* Plants smoked and used for catarrh. *Throat Aid* Infusion of plants used as a gargle for sore throats. (as *C. stolonifera* 116:248) **Micmac** *Analgesic* Herb used for headache. *Eye Medicine* Herb used for sore eyes. *Respiratory Aid* Herb used for catarrh. *Throat Aid* Herb used for sore throat. (as *C. stolonifera* 40:56) **Montana Indian** *Dermatological Aid* Decoction of bark used as a wash for ulcers. (as *C. stolonifera* 19:11) **Navajo, Kayenta** *Ceremonial Medicine* and *Emetic* Plant used as a Mountain-top-way emetic. (as *C. stolonifera* 205:35) **Navajo, Ramah** *Ceremonial Medicine* and *Emetic* Plant used as a ceremonial emetic. (as *C. stolonifera* 191:38) **Ojibwa** *Ceremonial Medicine* Bark smoked for various ceremonies. (as *C. stolonifera* 135:237) **Okanagan-Colville** *Blood Medicine* Decoction of inner bark and chokecherry bark or alder bark taken to clear the blood, taken to help circulation, and taken for the blood after childbirth. *Cold Remedy* Inner bark dried, mixed with kinnikinnick or tobacco, and smoked for colds. Poultice of inner bark alone or mixed with goose oil applied to babies for a chest cold. *Contraceptive* Decoction of inner bark and other bark taken after childbirth to prevent frequent pregnancies. *Dermatological Aid* Decoction of sticks taken for poison ivy rashes. Decoction of inner bark and chokecherry bark or alder bark taken for sores and rashes. Decoction of wood and bark used to wash skin, hair, and scalp for dandruff, falling hair, and itchy scalp. Decoction of wood, bark, cow parsnip roots, and chokecherry wood and bark used for sores and scabs. Berries rubbed into the scalp to prevent graying hair. *Gastrointestinal Aid* Decoction of sticks taken in the sweat house to cause vomiting for an upset stomach condition. *Gynecological Aid* Decoction of inner bark and chokecherry or alder bark taken to clean out the womb after childbirth. Poultice of inner bark applied to the back and belly to "heal a woman's insides" after childbirth. *Heart Medicine* Infusion of inner cambium taken for "heart conditions." *Other* Decoction of inner bark and chokecherry bark or alder bark taken to heal the body. *Panacea* Decoction of inner bark and chokecherry bark or alder bark taken for any kind of sickness. *Pediatric Aid* Poultice of inner bark alone or mixed with goose oil applied to babies for a chest cold. *Unspecified* Decoction of bark or entire branch used as a medicine. (as *C. stolonifera* 188:96) **Okanagon** *Gynecological Aid* Decoction of bark or leaves taken by women soon after childbirth. (as *C. stolonifera* 125:42) **Potawatomi** *Antidiarrheal* Root bark used for diarrhea and flux, "the most efficacious remedy." (as *C. stolonifera* 154:55) **Saanich** *Gastrointestinal Aid* Bark soaked in warm water and taken to induce vomiting to clean out the stomach. *Respiratory Aid* Bark soaked in warm water and taken to induce vomiting for improved breathing. (as *C. stolonifera* 182:81) **Shuswap** *Kidney Aid* Plant used for weak kidneys. *Pediatric Aid* and *Urinary Aid* Plant used for children for bed-wetting. (as *C. stolonifera* 123:61) **Thompson** *Gynecological Aid* Decoction of bark or leaves taken by women soon after childbirth. (as *C. stolonifera* 125:42) Simple or compound decoction of various plant parts taken after childbirth. (as *C. stolonifera* 164:475) **Wet'suwet'en** *Dermatological Aid* Bark used to make a skin wash. *Febrifuge* Bark used for fevers. *Hemorrhoid Remedy* Bark used for postpartum hemorrhaging. (as *C. stolonifera* 73:152)

- **Food**—**Blackfoot** *Fruit* Berries eaten ripe. (as *C. stolonifera* 86:102) *Snack Food* Berries and saskatoon berries used as a favorite snack reserved for men. (as *C. stolonifera* 86:100) **Flathead** *Fruit* Berries occasionally eaten raw. Berries mixed with serviceberries and sugar and eaten as a "sweet and sour" dish. (as *C. stolonifera* 82:21) **Haisla & Hanaksiala** *Forage* Berries eaten by bears. (43:233) **Hesquiat** *Dessert* Berries, sugar, and water whipped with salal branches until foamy and eaten as a confectionery dessert. (as *Shepherdia stolonifera* 185:64) **Kutenai** *Fruit* Berries occasionally eaten raw. Berries mixed with serviceberries and sugar and eaten as a "sweet and sour" dish. (as *C. stolonifera* 82:21) **Okanagan-Colville** *Forage* Berries eaten by black bears. *Fruit* Berries pounded, mixed with chokecherries or saskatoon berries, or boiled and eaten alone. (as *C. stolonifera* 188:96) **Shuswap** *Fruit* Berries used for food. *Preservative* Used with narrow leaf cot-

tonwood to smoke salmon. *Spice* Scraped wood, tasted like salt, used for barbecuing meat. (as *C. stolonifera* 123:61)
- *Fiber*—**Costanoan** *Basketry* Stems used in basketry. (as *C. californica* 21:252) **Cree, Woodlands** *Basketry* Stem used as a birch bark basket rim. (as *C. stolonifera* 109:36) **Gosiute** *Snow Gear* Wood used to make snowshoes. (as *C. stolonifera* 39:366) **Iroquois** *Basketry* Branches used to make baskets. (as *C. stolonifera* 142:95) **Okanagan-Colville** *Building Material* Large limbs used as frame poles. *Cordage* Bark twisted into rope and used to lash fish traps, raised caches, and other structures. (as *C. stolonifera* 188:96) **Paiute** *Furniture* Wood used for cradleboard frames. (as *C. stolonifera* 111:98) **Washo** *Basketry* Twigs used for the foundation of baskets. (as *C. stolonifera* 121:49)
- *Dye*—**Chippewa** *Black* Used with grindstone dust or black earth to make a black dye. (as *C. stolonifera* 53:372) *Red* Inner bark boiled, cedar ashes added, and used to make a red dye. Outer bark boiled, cedar ashes added, and used to make a red dye. (as *C. stolonifera* 53:370) *Yellow* Used with bloodroot and wild plum to make a yellow dye. (as *C. stolonifera* 53:374) **Costanoan** *Unspecified* Decoction of inner bark used as a dye. (as *C. californica* 21:252) **Cree, Woodlands** *Brown* Infusion of outer bark used to color leather from tan to brown. (as *C. stolonifera* 109:36)
- *Other*—**Abnaki** *Smoke Plant* Bark used for smoking. (as *C. stolonifera* 144:159) Shredded bark used for smoking. (as *C. stolonifera* 144:170) **Bella Coola** *Cooking Tools* Branches used for barbecue racks. (as *C. stolonifera* 184:203) **Blackfoot** *Ceremonial Items* Plant mixed with tobacco, kinnikinnick, or dried cambium, and used in all religious bundles. (as *C. stolonifera* 86:14) *Smoke Plant* Dried cambium greased, crushed, and mixed with smoking tobacco. (as *C. stolonifera* 86:102) Inner bark used in the tobacco mixture. (as *C. stolonifera* 97:49) *Smoking Tools* Stems used to make pipestems and tamps. (as *C. stolonifera* 86:111) *Toys & Games* Bark used to cover split beaver teeth for gambling wheels. (as *C. stolonifera* 82:21) Bark used to cover a circle of split beaver teeth and used as a gambling wheel. (as *C. stolonifera* 97:49) **Cheyenne** *Ceremonial Items* Plant smoked ceremonially. (as *C. stolonifera* 83:23) *Hunting & Fishing Item* Branches used to make arrows. (as *C. stolonifera* 82:21) *Smoke Plant* Dried, pulverized under bark mixed with tobacco, and used for smoking. (as *C. stolonifera* 76:183) Inner bark mixed with skunkbush leaves in the absence of tobacco and smoked. (as *C. stolonifera* 83:14) Bark mixed with dried "kinnikinnick" leaves and used for pipe smoking. (as *C. stolonifera* 83:25) **Cree** *Smoke Plant* Bark mixed with tobacco and used for smoking. (as *C. stolonifera* 16:485) **Cree, Woodlands** *Smoke Plant* Dried inner bark pulverized, mixed with tobacco, and used for smoking. (as *C. stolonifera* 109:36) **Crow** *Fasteners* Branches used to make tepee stakes and pins. *Musical Instrument* Branches used to make drumsticks. *Tools* Branches used to make forks for sweat lodge rocks. (as *C. stolonifera* 82:21) **Dakota** *Smoke Plant* Plant used for smoking. (as *C. stolonifera* 70:107, 108) **Flathead** *Hunting & Fishing Item* Branches used to make arrows. (as *C. stolonifera* 82:21) **Gosiute** *Smoke Plant* Inner bark smoked as a tobacco. (as *C. stolonifera* 39:366) **Great Basin Indian** *Smoke Plant* Winter bark used to make "kinnikinnick." (as *C. stolonifera* 121:49) **Haisla & Hanaksiala** *Cleaning Agent* Leaves used to wipe fish. **Hanaksiala** *Hunting & Fishing Item* Branches used to string fish. (43:233) **Jemez** *Hunting & Fishing Item* Tough branches used to make bows and arrows. (as *C. instolonea* 44:22) **Karok** *Hunting & Fishing Item* Branches used for arrows with tips of western serviceberry wood. (as *C. californica* 148:387) **Kutenai** *Tools* Wood used to make pelt stretchers. (as *C. stolonifera* 82:21) **Lakota** *Smoke Plant* Inner bark mixed with tobacco and smoked. (as *C. stolonifera* 106:43) Bark mixed with tobacco and used for smoking. (as *C. stolonifera* 139:44) **Montana Indian** *Ceremonial Items* Ribbons of inner bark roasted, mixed with tobacco, and used for ceremonial or religious pipe smoking. (as *C. stolonifera* 82:21) *Hunting & Fishing Item* Twisted branches used to make fish nets. (as *C. stolonifera* 19:11) Branches twisted and used to make fish nets. (as *C. stolonifera* 82:21) *Smoke Plant* Inner bark dried and mixed with tobacco as a substitute for kinnikinnick. (as *C. stolonifera* 19:11) **Okanagan-Colville** *Cooking Tools* Branches used to make spatulas. *Hunting & Fishing Item* Branches used to make fish traps. *Preservative* Old branches used in smoking hides. *Smoke Plant* Inner bark dried, mixed with kinnikinnick or tobacco, and smoked. (as *C. stolonifera* 188:96) **Okanagon** *Smoke Plant* Leaves mixed with other plant leaves and smoked. (as *C. stolonifera* 125:39) **Omaha** *Smoke Plant* Plant used for smoking. (as *C. stolonifera* 70:107, 108) **Oweekeno** *Hunting & Fishing Item* Branches used to make arrows. Wood used to make fishing hooks and salmon tethering poles. *Tools* Stems used to make drying frames for beaver skins. (43:92) **Ponca** *Smoke Plant* Plant used for smoking. (as *C. stolonifera* 70:107, 108) **Potawatomi** *Smoke Plant* Toasted, shredded bark used as kinnikinnick or smoking material. (as *C. stolonifera* 154:118) **Shuswap** *Cooking Tools* Sticks used to skewer a salmon flat for proper drying. *Smoke Plant* Scraped bark used with tobacco. (as *C. stolonifera* 123:61) **Thompson** *Smoke Plant* Leaves mixed with other plant leaves and smoked. (as *C. stolonifera* 125:39) Leaves occasionally smoked as tobacco. (as *C. stolonifera* 164:495) **Washo** *Designs* Bark used for the patterns in baskets. (as *C. stolonifera* 121:49)

Cornus sp., Dogwood
- *Drug*—**Alabama** *Antidiarrheal* Decoction of inner bark taken for dysentery. (177:46) **Algonquin, Quebec** *Unspecified* Valued as a medicine. Source states that a plant sample was identified by the common name "willow" but was actually *Cornus*. (18:213)
- *Fiber*—**Yuki** *Basketry* Used as basket material. (103:423)

Cornus suecica L., Lapland Cornel
- *Food*—**Eskimo, Alaska** *Fruit* Fresh, ripe berries used for food. (1:37) Gathered and mixed with other berries. (4:715)

Cornus unalaschkensis Ledeb., Western Cordilleran Bunchberry
- *Food*—**Bella Coola** *Fruit* Ripe berries eaten with sugar and grease. (184:204) **Haisla & Hanaksiala** *Dessert* Berries mashed, mixed with oolichan (candlefish) grease, and eaten as a dessert. *Dried Food* Berries dried for winter use. (43:234) **Kitasoo** *Fruit* Fruit used for food. (43:331) **Oweekeno** *Forage* Berries eaten by bears. *Fruit* Berries used for food. (43:93)

Coronilla, Fabaceae
Coronilla varia L., Purple Crownvetch
- *Drug*—**Cherokee** *Antirheumatic (External)* Plant crushed and rubbed on rheumatism. *Emetic* Used as an emetic. (80:60) Decoction of bark taken as an emetic. (177:31) *Orthopedic Aid* Plant crushed and rubbed on cramps. (80:60)

Corydalis, Papaveraceae

Corydalis aurea Willd., Scrambledeggs
- **Drug**—**Navajo, Kayenta** *Antidiarrheal* Plant used for diarrhea. *Dermatological Aid* Plant used for hand sores. *Disinfectant* and *Gynecological Aid* Plant used for puerperal infection. *Veterinary Aid* Plant sprinkled on livestock for snakebites. (205:23) **Ojibwa** *Stimulant* Root smoke inhaled to clear the head and revive the patient. (153:370)
- **Food**—**Navajo, Ramah** *Fodder* Used for sheep feed. (191:28)
- **Other**—**Navajo, Ramah** *Fertilizer* Cold infusion used to soak watermelon seeds to increase production. (191:28)

Corydalis aurea ssp. *occidentalis* (Engelm. ex Gray) G. B. Ownbey, Scrambledeggs
- **Drug**—**Navajo** *Antirheumatic (Internal)* Plant used as a rheumatic remedy. (as *C. montana* 55:96) **Navajo, Ramah** *Analgesic* Cold infusion taken for stomachache and used as lotion for backache. *Gastrointestinal Aid* Cold infusion of plant taken for stomachache. *Gynecological Aid* Compound decoction of plant used for menstrual difficulties. *Orthopedic Aid* Cold infusion of plant used as a lotion for backache. *Other* Plant used for injuries. *Throat Aid* Cold infusion of plant taken and used as a lotion for sore throat. (191:28)
- **Food**—**Navajo, Ramah** *Fodder* Used for sheep feed. (191:28)
- **Other**—**Keres, Western** *Unspecified* Plant known and named but no use was specified. (as *C. montanum* 171:38) **Navajo, Ramah** *Fertilizer* Cold infusion used to soak watermelon seeds to increase production. (191:28)

Corydalis sempervirens (L.) Pers., Rock Harlequin
- **Drug**—**Iroquois** *Hemorrhoid Remedy* Compound decoction of plants taken and used as a wash for piles. (87:339)

Corylus, Betulaceae

Corylus americana Walt., American Hazelnut
- **Drug**—**Cherokee** *Dermatological Aid* Infusion of scraped bark taken for hives. *Emetic* Compound of inner bark taken "to vomit bile." (80:37) Decoction of inner bark taken to induce vomiting when unable to retain food. (177:16) **Chippewa** *Analgesic* Compound containing charcoal pricked into temples with needles for headache. (53:338) **Iroquois** *Antidiarrheal* and *Antiemetic* Compound decoction taken for "summer disease—vomiting, diarrhea and cramps." *Antihemorrhagic* Raw nuts taken for hay fever, childbirth hemorrhage, and prenatal strength. (87:297) *Blood Medicine* Compound infusion taken as a blood purifier and for prenatal strength. (87:298) *Dermatological Aid* Nutmeat oil formerly used for the hair, either alone or mixed with bear grease. (196:123) *Gastrointestinal Aid* Compound decoction taken for summer disease: vomiting, diarrhea, and cramps. *Gynecological Aid* Decoction or raw nuts taken for childbirth hemorrhage and prenatal strength. *Pediatric Aid* Compound decoction of roots given when "baby's teeth are coming in." (87:297) Compound infusion taken or raw nuts eaten for prenatal strength. *Respiratory Aid* Compound decoction of buds taken for hay fever. (87:298) *Toothache Remedy* Compound decoction of roots given when "baby's teeth are coming in." (87:297) **Menominee** *Adjuvant* Inner bark used "with other herbs as a binder to cement the virtues of all." (151:26) **Ojibwa** *Dermatological Aid* Poultice of boiled bark applied to help close and heal cuts. (153:359)
- **Food**—**Cherokee** *Unspecified* Nuts used for food. (80:37) **Chippewa** *Unspecified* Nuts used for food in season. *Winter Use Food* Nuts stored for winter use. (71:127) **Dakota** *Soup* Nuts used as a body for soup. *Unspecified* Nuts eaten raw with honey. (70:74) **Iroquois** *Beverage* Fresh nutmeats crushed, boiled, and liquid used as a drink. *Bread & Cake* Fresh nutmeats crushed and mixed with bread. (124:99) Nuts crushed, mixed with cornmeal and beans or berries, and made into bread. (196:123) *Pie & Pudding* Fresh nutmeats crushed and mixed with corn pudding. (124:99) *Sauce & Relish* Nuts pounded, boiled, resulting oil seasoned with salt and used as gravy. *Soup* Nutmeats crushed and added to corn soup. (196:123) *Special Food* Fresh nutmeats crushed, boiled, and oil used as a delicacy in corn bread and pudding. (124:99) Nutmeat oil added to the mush used by the False Face Societies. *Staple* Nutmeats crushed and added to hominy. *Unspecified* Nutmeats, after skimming off the oil, seasoned and mixed with mashed potatoes. (196:123) **Menominee** *Unspecified* Nuts, in the milk stage, eaten. *Winter Use Food* Nuts, in the milk stage, dried for winter use. (151:63) **Meskwaki** *Unspecified* Nuts eaten in the milk stage or ripe. *Winter Use Food* Nuts stored for winter use. (152:256) **Ojibwa** *Unspecified* Nuts eaten as food and newly gathered nuts before the kernel had hardened were favored. (153:397) **Omaha** *Soup* Nuts used as a body for soup. (70:74) *Unspecified* Nuts eaten plain or mixed with honey. (68:326) Nuts eaten raw with honey. **Ponca** *Soup* Nuts used as a body for soup. *Unspecified* Nuts eaten raw with honey. **Winnebago** *Soup* Nuts used as a body for soup. *Unspecified* Nuts eaten raw with honey. (70:74)
- **Fiber**—**Meskwaki** *Basketry* Twigs used in making twig baskets. *Brushes & Brooms* Twigs made into brushes for cleaning the earthen floors of the wigwams. (152:267) **Ojibwa** *Basketry* Finer twigs used as ribs in making woven baskets for collecting/storing acorns or hard fruits. *Brushes & Brooms* Finer twigs bound into a bundle, with the tips sheared, to serve as a broom or brush. (153:417)
- **Dye**—**Chippewa** *Black* Boiled with butternut to make a black dye. Burs boiled with inner bark of bur oak, added to black earth and butternut, and used as a black dye. (53:372) **Ojibwa** *Mordant* Seed hulls and butternut boiled together and the hull tannic acid sat the black butternut color. (153:425)
- **Other**—**Chippewa** *Musical Instrument* Used for drumming sticks. (53:377) **Iroquois** *Insecticide* Nutmeat oil mixed with bear grease and used as a preventive for mosquitoes. (196:123) **Ojibwa** *Musical Instrument* Crooked stick with an enlarged base made the favorite drum stick. (153:417)

Corylus cornuta Marsh., Beaked Hazelnut
- **Drug**—**Abnaki** *Eye Medicine* Used for sore eyes. (144:155) Decoction of bark and bark from two other plants used for eye pain. (144:165) **Algonquin, Quebec** *Gastrointestinal Aid* Infusion of branches and leaves used for intestinal disorders. *Heart Medicine* Infusion of branches and leaves used for heart troubles. (18:151) **Algonquin, Tête-de-Boule** *Heart Medicine* Infusion of branch tips taken for heart problems. (132:128) **Iroquois** *Antirheumatic (External)* Poultice of branches applied for rheumatism. (142:85) *Emetic* Compound decoction of bark taken to vomit. *Pediatric Aid* Decoction of bark given to children for teething. (87:298) Infusion of stems and other plant stems made into a necklace used by children for teething pain. (141:

38) *Psychological Aid* Compound decoction of bark used as a wash for loneliness. *Toothache Remedy* Decoction of bark given to children for teething. (87:298) Infusion of stems and other plant stems made into a necklace used by children for teething pain. (141:38) **Thompson** *Throat Aid* Buds chewed to become a good singer. (187:190)
- *Food*—**Algonquin, Quebec** *Unspecified* Nuts used for food. (18:79) **Cree, Woodlands** *Unspecified* Nuts used for food. *Winter Use Food* Nuts collected in quantity to use at a later time. (109:37) **Iroquois** *Fruit* Fruit roasted and eaten. (142:85) **Salish, Coast** *Unspecified* Nuts used for food. (182:79) **Thompson** *Snack Food* Nuts eaten for refreshments. (187:190)
- *Other*—**Okanagan-Colville** *Cash Crop* Nuts used as a trading item. (188:90) **Salish, Coast** *Cash Crop* Nuts used as a common trade article. *Hunting & Fishing Item* Straight suckers used for arrows. (182:79) **Thompson** *Cash Crop* Nuts used for trade. *Cooking Tools* Young suckers used as salmon spreaders in the absence of saskatoon branches. *Hunting & Fishing Item* Young branch softened with urine, twisted, and used to make a dip net hoop. (187:190)

Corylus cornuta var. *californica* (A. DC.) Sharp, California Hazelnut
- *Food*—**Chehalis** *Unspecified* Nuts eaten fresh. *Winter Use Food* Nuts eaten during the winter. (as *C. californica* 79:27) **Costanoan** *Unspecified* Nuts used for food but only late in season. (21:248) **Cowlitz** *Winter Use Food* Nuts eaten during the winter. (as *C. californica* 79:27) **Karok** *Unspecified* Nuts used for food. (6:25) **Klamath** *Unspecified* Nuts used for food. (as *C. californica* 45:94) **Lummi** *Unspecified* Nuts eaten fresh. (as *C. californica* 79:27) **Okanagon** *Unspecified* Species used for food. (as *C. californica* 125:39) **Paiute** *Unspecified* Nutmeat eaten raw. *Winter Use Food* Nutmeat stored for future use. (as *C. californica* 111:64) **Sanpoil & Nespelem** *Unspecified* Nuts stored without removing the shells. Nutmeat eaten whole or pulverized before use. (as *C. californica* 131:104) **Shuswap** *Unspecified* Nuts used for food. (as *C. californica* 123:60) **Skagit** *Unspecified* Nuts cracked with stones and eaten fresh. (as *C. californica* 79:27) **Skagit, Upper** *Unspecified* Nuts eaten fresh. *Winter Use Food* Nuts stored for winter use. (as *C. californica* 179:42) **Snohomish** *Unspecified* Nuts eaten fresh. **Squaxin** *Unspecified* Nuts used for food. **Swinomish** *Unspecified* Nuts eaten fresh. (as *C. californica* 79:27) **Thompson** *Unspecified* Species used for food. (as *C. californica* 125:39) **Tolowa** *Dried Food* Nuts dried and stored for winter use. *Unspecified* Nuts eaten fresh. **Yurok** *Unspecified* Nuts eaten fresh. (6:25)
- *Fiber*—**Chehalis** *Cordage* Long twigs twisted and used to tie things. (as *C. californica* 79:27) **Costanoan** *Basketry* Wood used for basket rims. (21:248) **Karok** *Basketry* Stems used in basketry. (6:25) **Pomo** *Basketry* Stems used for warp for sedge baskets. (as *C. californica* 118:6) **Pomo, Kashaya** *Basketry* Switches used to weave large burden baskets, surf fish baskets, and other openwork baskets. Switches used as the foundation in coiled baskets. (72:55) **Round Valley Indian** *Basketry* Slender twigs used to make coarse sieve baskets and vertical withes of saw grass baskets. (as *C. californica* 41:333) **Skokomish** *Cordage* Long twigs twisted and used as rope. (as *C. californica* 79:27) **Tolowa** *Basketry* Dried shoots soaked in water and used to make baskets. (6:25) **Wintoon** *Basketry* Straight stems used for making baskets. (as *C. californica* 117:264) **Yurok** *Basketry* Stems used in basketry. (6:25)
- *Dye*—**Thompson** *Blue* Roots used to make a bluish dye. (as *C. californica* 164:501)
- *Other*—**Costanoan** *Hunting & Fishing Item* Wood used for arrow shafts. (21:248) **Okanagon** *Cash Crop* Traded with the Coast Indians. (as *C. californica* 125:39) **Pomo** *Cooking Tools* Stems used for sieves. *Hunting & Fishing Item* Stems used for fish traps. (as *C. californica* 118:6) **Pomo, Kashaya** *Hunting & Fishing Item* Straight branches used for arrows. (72:55) **Round Valley Indian** *Hunting & Fishing Item* Wands made into baskets and used as salmon traps. (as *C. californica* 41:333) **Thompson** *Cash Crop* Traded with the Coast Indians. (as *C. californica* 125:39)

Corylus cornuta Marsh. var. *cornuta*, Beaked Hazelnut
- *Drug*—**Ojibwa** *Anthelmintic* Hairs of husk used as a medicine to expel worms. *Dermatological Aid* Poultice of boiled bark applied to help close and heal cuts. (as *C. rostrata* 153:359) **Potawatomi** *Dermatological Aid* Inner bark used as an astringent. (as *C. rostrata* 154:44)
- *Food*—**Karok** *Unspecified* Nuts used for food. *Winter Use Food* Nuts stored for winter use. (as *C. rostrata* var. *californica* 148:382) **Miwok** *Unspecified* Nuts used for food. (as *C. rostrata* var. *californica* 12:153) **Ojibwa** *Unspecified* Species used for food. (as *C. rostrata* 135:242) Nuts eaten as a food. (as *C. rostrata* 153:398) **Potawatomi** *Winter Use Food* Mature or "in the milk" nut gathered and used as a favorite food during the winter. (as *C. rostrata* 154:97) **Yuki** *Unspecified* Nuts eaten raw. (as *C. rostrata* var. *californica* 49:87)
- *Fiber*—**Karok** *Basketry* Sticks used to make baskets. Young shoots used to make baby baskets and carrying baskets. *Cordage* Withes twisted to make rope. *Snow Gear* Wood used as the heavy part of the frame for snowshoes. (as *C. rostrata* var. *californica* 148:382) **Potawatomi** *Brushes & Brooms* Bunch of twigs bound together and used as a broom. (as *C. rostrata* 154:112) **Yuki** *Basketry* Used as basket material. (as *C. rostrata* var. *californica* 103:423)
- *Other*—**Karok** *Hunting & Fishing Item* Wood made into poles and used on the fish trigger or set net. (as *C. rostrata* var. *californica* 148:382)

Corylus sp., Hazel
- *Drug*—**Chippewa** *Antihemorrhagic* Compound infusion of root taken for lung hemorrhages. *Pulmonary Aid* Compound infusion of root taken for lung hemorrhage. (53:340)
- *Fiber*—**Hahwunkwut** *Basketry* Roots used to make carrying baskets, baby baskets, and other coarse baskets. (117:183) **Poliklah** *Basketry* Used to make baskets. (117:170)

Coryphantha, Cactaceae
Coryphantha sp., Cactus
- *Other*—**Comanche** *Weapon* Spines used to punish unfaithful wives. (29:521)

Cosmos, Asteraceae
Cosmos parviflorus (Jacq.) Pers., Southwestern Cosmos
- *Drug*—**Navajo, Ramah** *Ceremonial Medicine* Cold infusion of dried leaves used as ceremonial chant lotion. (191:50)

Cosmos sp., Jilla Flower
- *Drug*—**Navajo** *Burn Dressing* Plant used for burns. (90:150)

Costaria, Laminariaceae, alga

Costaria costata (Turner) Saunders, Short Kelp
- *Food*—**Hesquiat** *Dried Food* Stipes and fronds with attached herring eggs dried for later use.6 (185:24)
- *Fiber*—**Hesquiat** *Sporting Equipment* Dried stipes use as "pucks" and hitting sticks.5 (185:24)

Coursetia, Fabaceae

Coursetia glandulosa Gray, Rosary Babybonnets
- *Other*—**Papago** *Fasteners* Transparent, yellowish brown gum mixed with adobe and used to make jars of syrup air tight. (34:21)

Cowania, Rosaceae

Cowania sp.
- *Fiber*—**Southwest Indians** *Clothing* Bark used to make coiled sandals. (17:30)
- *Other*—**Walapai** *Toys & Games* Bark made into a ring used in a hoop and pole game. (17:54)

Crassulaceae, see *Dudleya, Sedum, Sempervivum*

Crataegus, Rosaceae

Crataegus calpodendron (Ehrh.) Medik., Pear Hawthorn
- *Drug*—**Meskwaki** *Analgesic* Infusion of twigs used for a pain in the side and bladder trouble. *Stimulant* Infusion of root bark used in cases of "general debility." *Urinary Aid* Fruit used for bladder ailments. (as *C. tomentosa* 152:241)
- *Food*—**Meskwaki** *Fruit* Fruit eaten raw and cooked. (as *C. tomentosa* 152:263)

Crataegus chrysocarpa Ashe, Fireberry Hawthorn
- *Drug*—**Blackfoot** *Laxative* Decoction of dried berries taken during the winter as a mild laxative. (86:66) **Ojibwa, South** *Antidiarrheal* Compound decoction of root taken for diarrhea. (as *C. coccinea* 91:200) **Potawatomi** *Gastrointestinal Aid* Fruit used for stomach complaints. (as *C. rotundifolia* var. *bicknellii* 154:76)
- *Food*—**Blackfoot** *Fruit* Berries used for food. Certain conditions had to be met before the berries were eaten. Otherwise, they would cause stomach cramps. The procedure was to offer the tree a gift, for boys a little bow and arrow made from the thorns, for girls a pair of miniature moccasins fashioned from the leaves. In return, the tree would not allow its berries to "bite" the stomach. The gifts were placed on the tree and the berries collected. (86:102) **Lakota** *Fruit* Fruits eaten for food. (139:56) **Ojibwa** *Fruit* Fruit used for food. (as *C. coccinea* 135:236) **Omaha** *Beverage* Twigs used to make a hot, aqueous, tea-like beverage. (as *C. coccinea* 68:329) *Fruit* Fruit eaten by children fresh from the hand. (as *C. coccinea* 68:326) Fruit sometimes used for food, but mostly as a famine food. (70:87) *Starvation Food* Fruit eaten by adults in times of famine. (as *C. coccinea* 68:326) Fruit sometimes used for food, but mostly as a famine food. **Ponca** *Fruit* and *Starvation Food* Fruit sometimes used for food, but mostly as a famine food. (70:87) **Potawatomi** *Fruit* Fruit eaten by deer, bears, and sometimes the Indians. (as *C. rotundifolia* var. *bicknellii* 154:107) **Winnebago** *Fruit* and *Starvation Food* Fruit sometimes used for food, but mostly as a famine food. (70:87)

Crataegus columbiana T. J. Howell, Columbian Hawthorn
- *Food*—**Montana Indian** *Fruit* Fruit eaten fresh. *Winter Use Food* Fruit mixed with chokecherries and serviceberries pressed into cakes and dried for winter use. (19:11) **Okanagan-Colville** *Bread & Cake* Berries mashed and formed into cakes, dried, and eaten like cookies. *Fruit* Berries eaten fresh. (188:123) **Okanagon** *Fruit* Fruits eaten for food. (125:38) **Oregon Indian** *Fruit* Fresh or dried fruit used for food. (118:22) **Sanpoil & Nespelem** *Fruit* Whole berries eaten fresh or mashed in a mortar. (131:103) **Thompson** *Fruit* Fruits eaten for food. (125:38) **Unspecified** Scarlet, pear-shaped pomes eaten. (164:487)
- *Other*—**Montana Indian** *Tools* Wood used for making "camas sticks" for digging these and other roots. (19:11) **Thompson** *Hunting & Fishing Item* Spines used as pins and fishhooks. (164:454) Spines used to make fishhooks. *Tools* Spines used as probes for ripe boils and ulcers. (164:497)

Crataegus douglasii Lindl., Black Hawthorn
- *Drug*—**Kwakiutl** *Dermatological Aid* Poultice of chewed leaves applied to swellings. (183:288) **Okanagan-Colville** *Antidiarrheal* Infusion of shoots given to children for diarrhea. *Antirheumatic (External)* Thorn used to pierce areas affected by arthritic pain. The upper end of the thorn was ignited and burned down to the point buried into the skin. This treatment was very painful, but after a scab had formed and disappeared, the arthritic pain also disappeared. *Oral Aid* Infusion of new shoots used to wash a baby's mouth for mouth sores. *Pediatric Aid* Infusion of new shoots used to wash a baby's mouth for mouth sores. Infusion of shoots given to children for diarrhea. (188:124) **Okanagon** *Dermatological Aid* Spines used as probes for boils and ulcers. *Gastrointestinal Aid* Decoction of sapwood, bark, and roots taken as a stomach medicine. (as *C. brevispina* 125:40) **Thompson** *Antidiarrheal* Fruit considered a good health food for diarrhea. Infusion of bark taken for diarrhea and dysentery. (187:258) *Dermatological Aid* Spines used as probes for boils and ulcers. (as *C. brevispina* 125:40) Spines used as probes "for ripe boils and ulcers." (164:457) *Gastrointestinal Aid* Decoction of sapwood, bark, and roots taken as a stomach medicine. (as *C. brevispina* 125:40) Decoction of sap, bark, wood, or root taken as stomach medicine. (164:457) *Panacea* Fruit considered a good health food for general sickness. (187:258)
- *Food*—**Bella Coola** *Fruit* Berries used for food. (184:208) **Cheyenne** *Dried Food* Fruits dried and used as a winter food. *Fruit* Fresh fruits cooked and used for food. (83:34) *Winter Use Food* Pulverized, dried berries saved for winter use. (76:176) **Kwakiutl, Southern** *Fruit* Berries used for food. (183:288) **Okanagan-Colville** *Bread & Cake* Berries mashed and dried into thin, hard cakes. Sometimes the cakes were decorated. The dried cakes were eaten as a snack on winter evenings and were used as crackers to dip into deer marrow soup to soak up the fat. *Forage* Berries eaten by bears and other animals. (188:124) **Okanagon** *Fruit* Fruits eaten for food. (125:38) **Paiute** *Dried Food* Berries formerly dried and eaten. *Fruit* Berries formerly eaten fresh. (104:100) Fruit eaten raw or boiled. (111:84) **Salish, Coast** *Fruit* Dry, sweetish fruits eaten in late fall. (182:86) **Sanpoil & Nespelem** *Dried Food* Berries boiled, dried, and stored. *Fruit* Berries eaten raw. (as *C. brevispina* 131:103) **Shuswap** *Preserves* Berries used to make jelly. (123:66) **Thompson** *Dried Food* Mashed fruit dried for winter use.

(187:258) *Fruit* Fruits eaten for food. (125:38) Fruit, without the seeds, eaten fresh or pureed. *Preserves* Fruit made into jam or jelly. (187:258) *Unspecified* Pomes eaten. (164:486)
- Fiber—Okanagan-Colville *Building Material* Wooden withes used to repair barbed wire fences. (188:124)
- Other—Okanagan-Colville *Season Indicator* Ripened berries indicated that the mountain huckleberries in Sanpoil areas were beginning to ripen. (188:124) **Thompson** *Hunting & Fishing Item* Spines used for fishhooks. (187:258) *Tools* Spines used to probe ripe boils and ulcers. (164:454) Strong wood used for digging sticks and ax handles. Spines used for piercing ears. (187:258)

Crataegus douglasii Lindl. var. *douglasii*, Douglas's Hawthorn
- Food—Haisla & Hanaksiala *Fruit* Fruit used for food. (43:263)

Crataegus erythropoda Ashe, Cerro Hawthorn
- Food—Apache, Chiricahua & Mescalero *Bread & Cake* Fruit pressed into pulpy cakes, dried, and stored. *Fruit* Fruits eaten fresh. *Winter Use Food* Fruit pressed into pulpy cakes, dried, and stored for winter use. (as *C. cerronis* 33:44)

Crataegus macrosperma Ashe, Bigfruit Hawthorn
- Food—Cherokee *Fruit* Fresh fruit used for food. (126:56)

Crataegus mollis Scheele, Arnold Hawthorn
- Food—Omaha *Beverage* Twigs used to make a hot, aqueous, tea-like beverage. (68:329) *Fruit* Fruit eaten by children fresh from the hand. *Starvation Food* Fruit eaten by adults in times of famine. (68:326)

Crataegus pruinosa (Wendl. f.) K. Koch, Waxyfruit Hawthorn
- Food—Iroquois *Bread & Cake* Fruit mashed, made into small cakes, and dried for future use. (196:128) Used to make bread. (196:82) *Dried Food* Raw or cooked fruit sun or fire dried and stored for future use. *Fruit* Dried fruit taken as a hunting food. *Sauce & Relish* Dried fruit cakes soaked in warm water and cooked as a sauce or mixed with corn bread. (196:128)

Crataegus punctata Jacq., Dotted Hawthorn
- Drug—Iroquois *Gastrointestinal Aid* Infusion of little branches without leaves and other plants taken for large stomachs. (141:46) *Gynecological Aid* Compound decoction of shoots and bark taken to stop menstrual flow. *Witchcraft Medicine* Decoction taken to prevent "breaking out like cancer" caused by witchcraft. (87:351)

Crataegus rivularis Nutt., River Hawthorn
- Drug—Mendocino Indian *Poison* Thorns considered poisonous. (41:355)
- Food—Okanagon *Staple* Berries used as a principal food. (178:238)
- Other—Mendocino Indian *Fuel* Wood used for fuel. (41:355)

Crataegus sp., Thorn Apple
- Drug—Chippewa *Analgesic* Compound decoction of root taken for back pain. *Gynecological Aid* Compound decoction of root taken for "female weakness." (53:356) *Tuberculosis Remedy* Roots used for consumption. (71:132) **Iroquois** *Gastrointestinal Aid* Infusion of little branches without leaves and other plants taken for large stomachs. (141:46) **Ojibwa** *Gynecological Aid* Fruit and bark used to make a medicine for women only. (153:384)
- Food—Abnaki *Fruit* Fruits eaten by children. (144:168) **Algonquin, Quebec** *Fruit* Fruit used for food. (18:91) **Chippewa** *Bread & Cake* Fruits squeezed, made into little cakes, dried, and stored for winter use. (53:321) **Coeur d'Alene** *Bread & Cake* Fresh berries boiled, spread on layers of grass, juice poured on them, dried, and made into cakes. Berries mashed, made into cakes, dried, and used for food. (178:93) *Fruit* Berries eaten fresh. (178:89) **Comanche** *Candy* Inner bark chewed as gum. *Fruit* Fruit used for food. (29:521) **Ojibwa** *Fruit* Haw apples used as a food in the fall. (153:409) **Spokan** *Fruit* Berries used for food. (178:343)
- Other—Chippewa *Tools* Thorns used as awls. (53:377) **Coeur d'Alene** *Tools* Wood used to make root diggers. (178:91) **Ojibwa** *Hunting & Fishing Item* Bark smoked to attract deer while hunting. (153:431) *Tools* Sharp thorns used for sewing awls on finer work such as buckskin sewing with sinew. (153:422)

Crataegus spathulata Michx., Littlehip Hawthorn
- Drug—Cherokee *Dietary Aid* Berries eaten "for appetite." *Heart Medicine* Infusion of bark taken for good circulation. *Preventive Medicine* Infusion of bark taken "to prevent current spasms." (80:37)
- Fiber—Cherokee *Sporting Equipment* Bark tea taken or bathed in by ballplayers "to ward off tacklers." (80:37)
- Other—Cherokee *Protection* Infusion of bark taken or bathed in by ballplayers "to ward off tacklers." (80:37)

Crataegus submollis Sarg., Quebec Hawthorn
- Drug—Iroquois *Witchcraft Medicine* Decoction and doll used to "make a person break out like cancer." (87:351)
- Food—Iroquois *Bread & Cake* Fruit mashed, made into small cakes, and dried for future use. (196:128) Used to make bread. (196:82) *Dried Food* Raw or cooked fruit sun or fire dried and stored for future use. *Fruit* Dried fruit taken as a hunting food. *Sauce & Relish* Dried fruit cakes soaked in warm water and cooked as a sauce or mixed with corn bread. (196:128)

Crepis, Asteraceae

Crepis acuminata Nutt., Longleaf Hawksbeard
- Drug—Shoshoni *Analgesic* Poultice of seeds or plant applied to sore breasts after childbirth. *Eye Medicine* Pulverized root sprinkled in eye to dislodge object and clear inflammation. *Gynecological Aid* Poultice of seeds or whole plant applied to sore breasts to induce milk flow. (180:62)
- Food—Karok *Vegetable* Stems peeled and eaten raw as greens. (148:389)

Crepis atribarba Heller, Slender Hawksbeard
- Drug—Okanagan-Colville *Orthopedic Aid* Infusion of pounded green tops used in a foot bath for a "sweaty feet" condition. (188:83)

Crepis modocensis Greene, Siskiyou Hawksbeard
- Drug—Paiute *Dermatological Aid* Latex applied to bee stings or insect bites. **Shoshoni** *Eye Medicine* Decoction of root used as a wash for sore eyes. *Gynecological Aid* Poultice of mashed plant applied to women's caked breasts. (as *C. scopulorum* 180:62)

Crepis occidentalis Nutt., Largeflower Hawksbeard
- Food—Paiute *Unspecified* Leaves eaten raw. (104:103)

Crepis runcinata (James) Torr. & Gray, Fiddleleaf Hawksbeard

Drug—**Meskwaki** *Cancer Treatment* and *Dermatological Aid* Poultice of whole plant applied "to open up a carbuncle or cancer." (152:213)

Crepis runcinata* ssp. *glauca (Nutt.) Babcock & Stebbins, Fiddleleaf Hawksbeard
- *Drug*—**Keres, Western** *Psychological Aid* Infusion of young plants drunk for homesickness and lonesomeness. (as *C. glauca* 171:40)
- *Food*—**Gosiute** *Unspecified* Leaves used for food. (as *C. glauca* 39:367)

Crossosomataceae, see *Glossopetalon*

Crotalaria, Fabaceae
Crotalaria rotundifolia Walt. ex J. F. Gmel., Rabbitbells
- *Drug*—**Seminole** *Throat Aid* Infusion of pods used for sore throats. (as *C. maritima* 169:281)

Crotalaria sagittalis L., Arrowhead Rattlebox
- *Drug*—**Delaware** *Narcotic* Roots used as a strong narcotic. *Venereal Aid* Roots used for venereal disease. (176:34) **Delaware, Oklahoma** *Narcotic* Root considered to be a very strong narcotic. *Venereal Aid* Root used for venereal disease. (175:29, 74) **Mohegan** *Blood Medicine* Root taken as a blood purifier. (176:72, 128)

Croton, Euphorbiaceae
Croton californicus Muell.-Arg., California Croton
- *Drug*—**Cahuilla** *Ear Medicine* and *Pediatric Aid* Warm decoction of mashed stems and leaves placed in the child's ear for earaches. *Poison* Toxic plant used only in small dosages for illnesses. *Respiratory Aid* Hot decoction of mashed stems and leaves taken for congestion caused by colds. (15:56) **Diegueño** *Cough Medicine* Decoction of whole plant taken for coughs. (88:218) **Luiseño** *Abortifacient* Plant used for abortions. (155:231)

Croton monanthogynus Michx., Prairie Tea
- *Other*—**Comanche** *Hide Preparation* Leaves mixed with animal brains and used for tanning. (29:521)

Croton pottsii (Klotzsch) Muell.-Arg. **var. *pottsii***, Potts's Leatherweed
- *Drug*—**Mahuna** *Kidney Aid* Infusion of plant taken for kidney infections. (as *C. corymbosus* 140:68)

Croton setigerus Hook., Croton
- *Drug*—**Concow** *Analgesic* Poultice of fresh, bruised leaves applied to chest for internal pains. *Febrifuge* Decoction of plant used as a bath or decoction taken for chills and fevers. *Misc. Disease Remedy* Decoction of plant used as a bath for typhoid and other fevers. (41:363) **Costanoan** *Antidiarrheal* Decoction of roots used for dysentery. (as *Eremocarpus setigerus*, turkey mullein 21:8) **Diegueño** *Veterinary Aid* Mashed stems and leaves placed in wormy, open wounds on horses to kill the worms and heal the sores. (as *Eremocarpus setigerus*, turkey mullein 84:20) **Kawaiisu** *Analgesic* Decoction of plant used as a wash or taken for headaches. *Heart Medicine* Infusion of plant taken for heart palpitations. (as *Eremocarpus setigerus*, turkey mullein 206:28) **Modesse** *Kidney Aid* Plant dried for a year to take on great power and used for dropsy. (as *Eremocarpus setigerus*, turkey mullein 117:224) **Neeshenam** *Misc. Disease Remedy* Decoction of plants taken for ague. (as *Eremocarpus setigerus*, turkey mullein 129:376) **Pomo** *Antidiarrheal* Decoction of smashed plant used for bleeding diarrhea. *Poison* Plant considered poisonous. (as *Eremocarpus setigerus*, turkey mullein 66:13) **Pomo, Kashaya** *Antidiarrheal* Decoction of mashed, boiled root taken for bleeding diarrhea. (as *Eremocarpus setigerus*, turkey mullein 72:75)
- *Food*—**Mendocino Indian** *Forage* Shiny, bean-like seeds eaten by wild mourning doves and turkeys. (41:363)
- *Other*—**Costanoan** *Hunting & Fishing Item* Pounded roots thrown into freshwater pools or dammed streams to stupefy fish. (as *Eremocarpus setigerus*, turkey mullein 21:249) **Mahuna** *Hunting & Fishing Item* Plants used as fish poison to make them easy to catch. (140:61) **Mendocino Indian** *Hunting & Fishing Item* Bruised leaves used as a substitute for soaproot to poison fish and make them easier to catch. (41:363) Used as a poison to stupefy fish. (41:321)

Croton texensis (Klotzsch) Muell.-Arg., Texas Croton
- *Drug*—**Apache, White Mountain** *Cathartic* Infusion of plant taken as a purgative. *Gastrointestinal Aid* Infusion of plant taken for stomach troubles. (136:156) **Hopi** *Emetic* Plant used as an emetic to "relieve the stomach." (200:34, 84) *Eye Medicine* Plant used in a very strong eyewash. (200:33, 84) **Isleta** *Ear Medicine* Seeds used in ears as a hearing aid in cases of partial deafness. *Laxative* Infusion of leaves taken or fresh leaves eaten as a laxative. (100:27) **Jemez** *Analgesic* Decoction of ground whole plant, roots, and salt taken for headaches. *Antirheumatic (Internal)* Decoction of ground whole plant, roots, and salt taken for body aches. *Misc. Disease Remedy* Decoction of ground whole plant, roots, and salt taken for grippe. (44:22) **Keres, Western** *Cathartic* Infusion of plant used as a cathartic. *Dermatological Aid* Ground seed powder used on open sores. *Hemorrhoid Remedy* Crushed roots used as a salve for piles. (171:40) **Lakota** *Gastrointestinal Aid* Infusion of leaves taken for stomach pains. (139:45) **Pawnee** *Pediatric Aid* Decoction of leaves used to bathe sick babies. (70:99) **Zuni** *Antiemetic* Decoction of plant taken for "sick stomach." *Cathartic* Decoction of plant taken as a purgative. *Diuretic* Decoction of plant taken as a diuretic. (166:45) *Gastrointestinal Aid* Infusion of leaves taken for stomachaches. (27:375) *Snakebite Remedy* Fresh or dried root chewed by medicine man before sucking snakebite and poultice applied to wound. (27:376) *Venereal Aid* Infusion of leaves taken for gonorrhea. Infusion of leaves taken for syphilis. (27:375)
- *Food*—**Hopi** *Fodder* Used as food for wild doves. (200:84)
- *Other*—**Navajo, Ramah** *Incense & Fragrance* Used on large fire to smoke clothes and remove skunk smell. (191:35) **Oglala** *Toys & Games* Plant used as a headdress by little boys while playing. (70:99)

Cryptantha, Boraginaceae
Cryptantha cinerea (Greene) Cronq. **var. *cinerea***, James's Catseye
- *Drug*—**Navajo, Kayenta** *Disinfectant* and *Pediatric Aid* Plant given to newborn infant for prenatal snake or toad infection. *Snakebite Remedy* Poultice of plant applied or plant used as lotion for snakebites. *Veterinary Aid* Poultice of plant applied or plant used as lotion for livestock with snakebites. (as *C. jamesii* var. *multicaulis* 205:40) **Navajo, Ramah** *Ceremonial Medicine* Root used as ceremonial medicine. *Dermatological Aid* Poultice of root or powdered root applied

to sores. *Panacea* Cold infusion of root used as "life medicine." *Pediatric Aid* Cold infusion of whole plant used for birth injury. (as *C. jamesii* var. *multicaulis* 191:40)

Cryptantha cinerea var. *jamesii* Cronq., James's Catseye
- **Drug**—**Hopi** *Analgesic* Poultice of pounded plant applied for body pains. (as *Cryptanthe jamesii* 200:32, 88) **Zuni** *Dermatological Aid* Powdered root used for a sore anus. (as *Cryptantha jamesii* 27:374)
- **Food**—**Navajo, Ramah** *Fodder* Used for sheep feed. (as *C. jamesii* 191:40)

Cryptantha crassisepala (Torr. & Gray) Greene, Thicksepal Catseye
- **Drug**—**Hopi** *Dermatological Aid* Plant used for boils or any swelling. (200:32, 33, 38) **Keres, Western** *Poison* Plant considered a bad, poisonous weed. (171:40) **Navajo, Kayenta** *Dermatological Aid* Plant used as a lotion for itching. (205:39) **Zuni** *Stimulant* Hot infusion of pulverized plant applied to limbs for fatigue. (166:45)

Cryptantha fendleri (Gray) Greene, Sanddune Catseye
- **Drug**—**Navajo, Ramah** *Cough Medicine* Decoction of plant taken for coughs. (191:40)
- **Food**—**Navajo, Ramah** *Fodder* Used for sheep feed, a nuisance because the ripe fruits stick in the wool. (191:40)

Cryptantha flava (A. Nels.) Payson, Brenda's Yellow Catseye
- **Drug**—**Hopi** *Cancer Treatment* Plant used for the cancer and growth in the throat. (42:301) **Navajo, Kayenta** *Eye Medicine* Plant used as a dusting powder for sore eyes. *Gastrointestinal Aid* Plant used for intestinal inflammation. *Gynecological Aid* Plant used for postpartum purification. (205:39)

Cryptantha fulvocanescens (S. Wats.) Payson, Tawny Catseye
- **Drug**—**Navajo** *Gynecological Aid* Decoction of plants taken at childbirth. (90:150) **Navajo, Ramah** *Snakebite Remedy* Cold infusion used as lotion for snakebite. *Toothache Remedy* Poultice of leaves applied for toothache. (191:40)

Cryptantha fulvocanescens (S. Wats.) Payson var. *fulvocanescens*, Tawny Catseye
- **Drug**—**Navajo** *Analgesic* Plant used for pain from a fall. *Cold Remedy* Plant chewed for colds. *Cough Medicine* Plant chewed for coughs. (as *Oreocarya fulvocanescens* 55:72)

Cryptantha sericea (Gray) Payson, Silky Catseye
- **Drug**—**Ute** *Gastrointestinal Aid* Roots used as a stomach medicine. (as *Krynitzkia sericea* 38:35)

Cryptantha sp., Hollowstomach
- **Drug**—**Navajo** *Dietary Aid* Infusion of plant taken to stay slender. (110:18) **Navajo, Kayenta** *Disinfectant* Plant used for coyote infection. (205:39)

Cryptogramma, Pteridaceae, fern
Cryptogramma sitchensis (Rupr.) T. Moore, Sitka Rockbrake
- **Drug**—**Thompson** *Eye Medicine* Infusion of washed, strained fronds used as an eyewash. *Liver Aid* Infusion of washed, strained fronds taken for gallstones. (as *C. crispa* 187:88)

Cucumis, Cucurbitaceae
Cucumis melo L., Cantaloupe
- **Food**—**Hopi** *Dried Food* Rind removed, meat pressed flat or stripped, wrapped into bundles, and dried. *Unspecified* Eaten fresh. (200:93) **Iroquois** *Bread & Cake* Fresh or dried flesh boiled, mashed, and mixed into the paste when making corn bread. *Dried Food* Flesh cut into strips, dried, and stored away. *Special Food* Squash eaten at feasts of ceremonial importance and longhouse ceremonies. *Vegetable* Flesh boiled, baked in ashes or boiled, mashed with butter and sugar, and eaten. Flesh fried and sweetened or seasoned with salt, pepper, and butter. (as *Cucurbita melo* 196:113) **Keresan** *Spice* Seeds ground on metate to remove the hulls and used to flavor various foods, especially rabbit stews. (198:560) **Navajo** *Dried Food* Fruit cut into strips, wound upon sticks in the form of a rope, sun dried, and stored for months. *Sauce & Relish* Dried fruit boiled with sugar and eaten like apple sauce. (165:222) **Navajo, Ramah** *Unspecified* Muskmelon cultivated and used for food. (191:46) **Okanagan-Colville** *Unspecified* Species used for food. (188:98) **Seminole** *Unspecified* Plant used for food. (169:480) **Sia** *Unspecified* Cultivated cantaloupes used for food. (199:106) **Thompson** *Fruit* Fruit used for food. (187:206)
- **Other**—**Hopi** *Ceremonial Items* Seeds mixed with juniper charcoal and water and made into a ceremonial body paint. (200:93)

Cucumis sativus L., Garden Cucumber
- **Food**—**Iroquois** *Bread & Cake* Fresh or dried flesh boiled, mashed, and mixed into the paste when making corn bread. *Dried Food* Flesh cut into strips, dried, and stored away. *Special Food* Squash eaten at feasts of ceremonial importance and longhouse ceremonies. *Vegetable* Flesh boiled, baked in ashes or boiled, mashed with butter and sugar, and eaten. Fruit preserved in brine made with salt and sheep sorrel. Flesh fried and sweetened or seasoned with salt, pepper, and butter. (as *Cucurbita sativus* 196:113) **Ojibwa** *Vegetable* Cucumbers eaten raw and sometimes flavored with maple sap vinegar and powdered maple sugar. (153:399) **Seminole** *Unspecified* Plant used for food. (169:478)

Cucumis sp., Muskmelon
- **Food**—**Cocopa** *Bread & Cake* Stored flesh washed in water, pounded, made into cakes, and sun dried. *Dried Food* Flesh sun dried and stored in pottery jars for later use. *Fruit* Fresh flesh used for food. *Unspecified* Seeds boiled with dried pumpkin, mashed, and eaten or whole seeds eaten without husking. (64:266) **Kamia** *Unspecified* Species used for food. (62:21) **Thompson** *Unspecified* Seeds eaten in quantities, especially by children. (164:492)

Cucurbitaceae, see *Citrullus*, *Cucumis*, *Cucurbita*, *Echinocystis*, *Lagenaria*, *Marah*, *Melothria*, *Sicyos*

Cucurbita, Cucurbitaceae
Cucurbita digitata Gray, Fingerleaf Gourd
- **Food**—**Pima, Gila River** *Snack Food* Seeds roasted and eaten as a snack food. (133:7)

Cucurbita foetidissima Kunth, Missouri Gourd
- **Drug**—**Apache, Western** *Veterinary Aid* Poultice of mashed stems, leaves, and roots soaked in hot water and applied to sores on horses' backs. (26:192) **Cahuilla** *Cathartic* Decoction of dried roots used as a

physic. *Dermatological Aid* Ground fruit shell used as a hair shampoo. Macerated roots applied to ulcers. Pulp used for open sores. *Emetic* Decoction of dried roots used as an emetic. *Other* Dried gourds used to make ladles, syringes for feminine hygiene, and rattles. *Veterinary Aid* Poultice of crushed pulp applied to saddle sores on horses. (15:57) **Coahuilla** *Veterinary Aid* Poultice of crushed root and sugar applied to saddle sores on horses. (as *Cucerbita perennis*, mock orange 13:80) **Dakota** *Panacea* Root used for any ailment, according to the doctrine of signatures. (as *Pepo foetidissima* 70:116, 117) **Isleta** *Pulmonary Aid* Decoction of roots used for chest pains. (100:27) **Keres, Western** *Dermatological Aid* Poultice of crushed roots applied to boils or other sores. (171:40) **Kiowa** *Emetic* Decoction of peeled roots taken as an emetic. (192:53) **Omaha** *Analgesic* Pulverized root mixed with water and taken for pains. (as *Cucurbita perennis* 58:584) *Dietary Aid* Plant used as an appetizer. (68:335) *Gynecological Aid* Root used in cases of protracted labor. (as *C. perennis* 58:585) *Panacea* Root used for any ailment, according to the doctrine of signatures. (as *Pepo foetidissima* 70:116, 117) *Tonic* Plant used as a tonic. (68:335) **Paiute** *Cathartic* Decoction of root taken as a physic for venereal disease. *Dermatological Aid* Decoction of root used to kill maggots in wounds. *Emetic* Decoction of root taken as an emetic for venereal disease. *Venereal Aid* Decoction of root taken as an emetic and physic for venereal diseases. Pulverized seeds sprinkled on venereal sores. (180:62, 63) **Pawnee** *Panacea* Root used for any ailment, according to the doctrine of signatures. **Ponca** *Panacea* Root used for any ailment, according to the doctrine of signatures. (as *Pepo foetidissima* 70:116, 117) **Shoshoni** *Cathartic* Decoction of root taken as a physic for venereal disease. *Emetic* Decoction of root taken as an emetic for venereal disease. (180:62, 63) *Venereal Aid* Infusion of plant taken for gonorrhea and syphilis. (118:47) Decoction of root taken as an emetic and physic for venereal diseases. (180:62, 63) *Veterinary Aid* Infusion of plant given to horses for bloat or worms. (118:48) **Tewa** *Laxative* Infusion of pulverized root taken as a laxative. (138:63) **Winnebago** *Panacea* Root used for any ailment, according to the doctrine of signatures. (as *Pepo foetidissima* 70:116, 117) **Zuni** *Antirheumatic (External)* Poultice of powdered seeds, flowers, and saliva applied to swellings. (27:375)
- *Food*—**Cahuilla** *Porridge* Seeds ground into a flour and used to make mush. (15:57) **Isleta** *Fruit* Fruit formerly used for food. (100:27) **Luiseño** *Unspecified* Seeds used for food. (155:229) **Pima** *Unspecified* Seeds roasted and eaten. (146:70)
- *Dye*—**Cahuilla** *Unspecified* Yellow blossoms used as a dye. (15:57)
- *Other*—**Apache, White Mountain** *Ceremonial Items* Leaves ground and used as "green paint" in making sand paintings. (as *C. perennis* 136:156) **Cahuilla** *Ceremonial Items* Dried gourds used to make rattles. *Cooking Tools* Dried gourds used to make ladles. *Soap* Root and pepo (fruit) useful as a soap and bleach. (15:57) **Diegueño** *Soap* Mashed pulp and seeds boiled in with the wash to whiten the clothes. (84:17) **Havasupai** *Toys & Games* Fruits used by girls for juggling. Roots made into a wooden ball and used in playing the "four hills" game. (197:243) **Kawaiisu** *Soap* Gourds used as soap when washing clothes. (206:22) **Keres, Western** *Ceremonial Items* Ripe gourds used as rattles in dances. *Toys & Games* Ripe gourds used as children's rattles. (171:40) **Kiowa** *Soap* Fruit used to cleanse hides and clothes by rubbing the dried fruit into the stains before washing. (as *Pepo foetidissimus* 192:53) **Luiseño** *Soap* Fruit used as soap to clean miscellaneous articles. (155:210) **Mahuna** *Soap* Gourds used for washing buckskin cloths and blankets. Roots used as soap bars for washing clothes. (140:39) **Papago** *Soap* Fruits dried, grated into soapy water, and used to bleach clothes. (34:52)

Cucurbita maxima Duchesne, Winter Squash

- *Drug*—**Hawaiian** *Dermatological Aid* Fruit meat and water taken for bad skin blotches. *Gynecological Aid* Fruits and shoots or buds with other plants pounded, mixed with water, and used for fallen wombs. *Laxative* Fruits mixed with water and used for internal cleaning of the bowels. *Psychological Aid* Leaves and young shoots eaten for partial insanity due to lack of sleep. (2:28) **Ojibwa** *Diuretic* Infusion of seeds taken as a diuretic. (153:367)
- *Food*—**Hopi** *Unspecified* Species used for food. (200:93) **Iroquois** *Bread & Cake* Fresh or dried flesh boiled, mashed and mixed into the paste when making corn bread. *Dried Food* Flesh cut into strips, dried, and stored away. *Special Food* Squash eaten at feasts of ceremonial importance and longhouse ceremonies. *Vegetable* Flesh boiled, baked in ashes or boiled, mashed with butter and sugar, and eaten. Flesh fried and sweetened or seasoned with salt, pepper, and butter. (196:113) **Navajo, Ramah** *Spice* Blossoms used as seasoning for soup. *Unspecified* Squash cultivated and used for food. *Winter Use Food* Squash peeled, cut into strips, sun dried, and stored in cellars or ground holes for winter use. (191:46) **Ojibwa** *Dried Food* Squash rings dried for winter use. (153:399) **Papago** *Fruit* Fruit grown for food. **Pima** *Fruit* Fruit grown for food. (36:101) **Sia** *Unspecified* Cultivated pumpkins used for food. (199:106)

Cucurbita moschata (Duchesne ex Lam.) Duchesne ex Poir., Crookneck Squash

- *Food*—**Cahuilla** *Dried Food* Cooked, cut into strips, and dried. *Unspecified* Cooked and eaten fresh. (15:58) **Havasupai** *Dried Food* Seeds dried, parched, shelled, and eaten. *Porridge* Seeds ground to form a paste or mixed with corn into a mush. (197:244) *Soup* Seeds parched, ground, and used to make soup or mush. (197:67) Flowers and amaranth leaves boiled, ground, and fresh or dried corn and water added to make soup. (197:74) *Vegetable* Fruit baked and the flesh eaten. (197:66) **Hopi** *Cooking Agent* Seeds used to oil the "piki" stones. *Dried Food* Meat cut spirally, wound into long bundles, tied in pairs, and dried for winter use. *Special Food* Flowers used to make special foods. *Unspecified* Meat boiled or baked. Seeds roasted and eaten. (200:93) **Iroquois** *Bread & Cake* Fresh or dried flesh boiled, mashed, and mixed into the paste when making corn bread. *Dried Food* Flesh cut into strips, dried, and stored away. *Special Food* Squash eaten at feasts of ceremonial importance and longhouse ceremonies. *Vegetable* Flesh boiled, baked in ashes or boiled, mashed with butter and sugar, and eaten. Flesh fried and sweetened or seasoned with salt, pepper, and butter. (196:113) **Maricopa** *Dried Food* Fruit peeled, cut spirally into strips, dried, and stored. Fruit cut spirally into strips, dried, and stored. *Fruit* Fruits cut into pieces and boiled with mesquite pods. *Vegetable* Pumpkin eaten as a cooked, mushy vegetable. (37:111) **Navajo** *Dried Food* Fruit cut into strips and dried for future use, could be kept for years. *Preserves* Dried fruit boiled with large amounts of sugar into a preserve. *Sauce & Relish* Dried fruit boiled and eaten with sugar as a sauce. (165:221) **Papago**

Dried Food Rind hung in long spirals from house roofs to dry, tied in bundles, stored, and used for food. (34:36) *Fruit* Fruit grown for food. (36:101) **Pima** *Dried Food* Seeds parched and eaten. (146:71) *Fruit* Fruit grown for food. (36:101) **Seminole** *Unspecified* Plant used for food. (169:490) **Sia** *Unspecified* Cultivated pumpkins used for food. (199:106) **Yuma** *Dried Food* Seed sun dried, parched, cracked, and the meat eaten. (37:111)

- **Other**—**Cocopa** *Hide Preparation* Seeds used to tan hides. (37:113) **Havasupai** *Containers* Rinds sun dried, filled with seeds and beans, and hidden from enemy raiders. (197:244) Rind baked, cleaned, dried, and used as a storage container for seeds. (197:66) *Planting Seeds* Seeds stored for future planting. (197:244) **Hopi** *Containers* Dried shell used by children to carry parched corn. *Musical Instrument* Shell dried and used as a sounding board for musical rasps. (200:93)

Cucurbita pepo L., Field Pumpkin

- **Drug**—**Cherokee** *Anthelmintic* Seeds eaten for worms. *Ceremonial Medicine* Used as an ingredient in green corn medicine. *Diuretic* Taken as a diuretic. *Kidney Aid* Taken for "dropsy." *Urinary Aid* Browned seeds eaten for bed-wetting. Taken for "gravel," "scalding of the urine," and spasms of urinary passage. (80:51) **Iroquois** *Diuretic* and *Pediatric Aid* Infusion of seeds given to children with reduced urination. (141:61) **Menominee** *Diuretic* Pulverized seeds taken in water "to facilitate the passage of urine." (151:33) **Meskwaki** *Gynecological Aid* Decoction of stem used for "female ills." (152:220) **Navajo** *Gastrointestinal Aid* Leaves used for upset stomachs. (90:150) **Pima** *Dermatological Aid* Ground seed paste used to cleanse and soften the skin. (47:72) **Zuni** *Antirheumatic (External)* and *Dermatological Aid* Ingredient of "schumaakwe cakes" and used externally for rheumatism and swelling. Poultice of seeds and blossoms applied to cactus scratches. (166:45, 46)

- **Food**—**Apache, White Mountain** *Bread & Cake* Blossoms baked as parts of certain kinds of cakes. *Unspecified* Flesh used for food. Blossoms used for food. (136:156) **Cherokee** *Unspecified* Species used for food. (80:51) *Vegetable* Flesh used for food. (80:21) **Cocopa** *Dried Food* Dried flesh strips stored and mixed with flesh of stored whole pumpkins to improve flavor. *Unspecified* Parched seeds used for food. *Vegetable* Fresh flesh boiled with rind on and sometimes mixed with maize meal. Roasted flesh eaten with fingers. (64:266) **Iroquois** *Bread & Cake* Fresh or dried flesh boiled, mashed, and mixed into the paste when making corn bread. Dried flesh pounded, sifted, soaked in cold water, sweetened, grease added, and baked into cakes. *Dried Food* Flesh cut into strips, dried and stored away. *Pie & Pudding* Flesh boiled, cornmeal, and sugar added and eaten as a pudding with sugar and milk. *Sauce & Relish* Dried flesh pounded into a fine meal or flour, boiled, sweetened, grease added, and used as a sauce. *Special Food* Squash eaten at feasts of ceremonial importance and longhouse ceremonies. *Vegetable* Flesh boiled, baked in ashes or boiled, mashed with butter and sugar, and eaten. Fresh or dried flesh boiled, mashed, and sweetened, or boiled with green beans, butter, and salt, and eaten. Flesh fried and sweetened or seasoned with salt, pepper, and butter. (196:113) **Kamia** *Unspecified* Species used for food. (62:21) **Meskwaki** *Winter Use Food* Squash sliced into rings, sun dried, pressed, and stored for winter use. (152:257) **Navajo** *Vegetable* Fruit pulp and seeds used for food. (90:150) **Navajo, Ramah** *Spice* Blossoms used as seasoning for soup. *Unspecified* Pumpkin cultivated and used for food. *Winter Use Food* Pumpkin peeled, cut into strips, sun dried, and stored in cellars or ground holes for winter use. (191:47) **Ojibwa** *Dried Food* Pumpkin rings dried for winter use. (153:400) **Okanagan-Colville** *Unspecified* Species used for food. (188:98) **Papago** *Dried Food* Rind hung in long spirals from house roofs to dry, tied in bundles, stored, and used for food. (34:36) *Fruit* Fruit grown for food. (36:101) *Staple* Seeds parched, sun dried, stored, ground into flour, and used as a staple food. (34:45) **Pima** *Fruit* Fruit grown for food. (36:101) *Unspecified* Seeds roasted, cracked, and the kernels eaten. (47:72) **Rappahannock** *Snack Food* Seeds eaten as "tid-bits." (161:30) **Zuni** *Dried Food* Fresh squash cut into spiral strips, folded into hanks, and hung up to dry for winter use. *Special Food* Blossoms cooked in grease and used as a delicacy in combination with other foods. *Unspecified* Fresh squash, either whole or in pieces, roasted in ashes and used for food. (166:67)

- **Other**—**Cocopa** *Protection* Oily kernels rubbed on hands as protection against cold. (64:266) **Zuni** *Ceremonial Items* Gourds worn in phallic dances symbolizing fructification or made into ceremonial rattles. The gourd rattles were used in ceremonies for both anthropic and zooic worship. *Containers* Gourds made into receptacles for storing precious articles. (166:88) *Cooking Tools* Gourds made into cups, ladles, and dippers and put to various uses. (166:67)

Cucurbita pepo var. *melopepo* (L.) Alef., Field Pumpkin

- **Food**—**Menominee** *Winter Use Food* Squash cut into strips or rings and dried for winter use. (151:65)

Cucurbita sp., Squash

- **Drug**—**Cheyenne** *Antirheumatic (Internal)* Infusion of rind taken for rheumatism and arthritis. *Diuretic* Infusion of rind taken as a diuretic. *Ear Medicine* Infusion of rind taken for earaches. *Febrifuge* Infusion of rind taken for fevers. *Heart Medicine* Infusion of rind taken for heart problems. *Kidney Aid* Infusion of rind taken for kidney troubles. *Laxative* Infusion of rind taken as a laxative. *Tuberculosis Remedy* Infusion of rind taken for tuberculosis. *Venereal Aid* Infusion of rind taken for venereal disease. (83:24)

- **Food**—**Seminole** *Unspecified* Plant used for food. (169:479)

Cuminum, Apiaceae

Cuminum cyminum L., Cumin

- **Drug**—**Apalachee** *Unspecified* Plant water used for medicinal purposes. (as *C. cymium* 81:98)

Cunila, Lamiaceae

Cunila marina L., Common Dittany

- **Drug**—**Cherokee** *Analgesic* Infusion taken for headache. *Cold Remedy* Infusion taken for colds. *Diaphoretic* Infusion taken to increase perspiration. *Febrifuge* Infusion taken for fever. *Gynecological Aid* Strong infusion taken to increase labor pains and aid in childbirth. *Snakebite Remedy* Used as a snakebite remedy. *Stimulant* Used as a stimulant. *Tonic* Used as a tonic. (as *C. origanoides* 80:32)

Cupressaceae, see *Calocedrus, Chamaecyparis, Cupressus, Juniperus, Libocedrus, Thuja*

Cupressus, Cupressaceae, conifer

Cupressus macrocarpa Hartw. ex Gord., Monterey Cypress
- *Drug*—**Costanoan** *Antirheumatic (Internal)* Decoction of foliage used for rheumatism. (21:6)

Cupressus nevadensis Abrams, Paiute Cypress
- *Drug*—**Kawaiisu** *Analgesic* Decoction of dried seeds taken for sore chest. *Cold Remedy* Decoction of dried seeds taken for colds. *Cough Medicine* Decoction of dried seeds taken for coughing. *Gynecological Aid* Hot or cold infusion of cones taken for menstruation problems. *Kidney Aid* Hot or cold infusion of cones taken for kidney problems. *Orthopedic Aid* Hot or cold infusion of cones taken for backaches. (206:22)

***Cupressus* sp.**, Cypress
- *Drug*—**Miwok** *Antirheumatic (Internal)* Decoction of stems taken for rheumatism. *Cold Remedy* Decoction of stems taken for colds. (12:168) **Navajo** *Strengthener* Infusion of branches taken by women to regain strength after childbearing. (as Cupressaceae sp. 110:22)
- *Food*—**Navajo** *Bread & Cake* Berries ground into a meal and mixed with bread dough. Leaf ash mixed with breads. *Fruit* Berries eaten raw or roasted. *Porridge* Leaf ash mixed with cornmeal mush. (as Cupressaceae sp. 110:22)
- *Other*—**Navajo** *Jewelry* Dried berries used to make necklaces. (as Cupressaceae sp. 110:22)

Curcuma, Zingiberaceae

Curcuma longa L., Common Turmeric
- *Drug*—**Hawaiian** *Blood Medicine* Bulbs, shoots, and other plants pounded, squeezed, and resulting liquid taken to cleanse the blood. *Nose Medicine* Bulbs and other plants pounded, squeezed, and the resulting liquid fumes inhaled for nose growths or odors. *Oral Aid* Bark and other plants pounded, squeezed, and the resulting liquid used to gargle. (as *C. louza* 2:33)

Cuscutaceae, see *Cuscuta*

Cuscuta, Cuscutaceae

Cuscuta californica Hook. & Arn., Chaparral Dodder
- *Drug*—**Diegueño** *Antidote* Infusion of plant, picked from buckwheat plants, taken for black widow spider bites. (84:17) **Kawaiisu** *Hemostat* Chewed stem juice or powdered plant snuffed up the nose for nosebleeds. (206:23)
- *Fiber*—**Cahuilla** *Scouring Material* Handfuls of plant used as scouring pads for cleaning. (15:59)

Cuscuta compacta Juss. ex Choisy, Compact Dodder
- *Drug*—**Pawnee** *Love Medicine* Vine used by girls to divine sincerity of suitors. (as *C. paradoxa* 70:110)
- *Dye*—**Pawnee** *Orange* Boiled vines used as an orange dye. (as *C. paradoxa* 70:110)

Cuscuta gronovii Willd. ex J. A. Schultes, Scaldweed
- *Drug*—**Cherokee** *Dermatological Aid* Used as poultice for bruises. (80:32)

Cuscuta megalocarpa Rydb., Bigfruit Dodder
- *Drug*—**Navajo, Ramah** *Ceremonial Medicine* Plant used as ceremonial emetic. *Emetic* Plant used as ceremonial emetic. (as *C. curta* 191:39)

***Cuscuta* sp.**, Dodder
- *Drug*—**Paiute** *Contraceptive* Plant, known as "woman without children," eaten by women as a contraceptive. (180:63) **Pima** *Poison* Plant considered poisonous. (47:66)
- *Food*—**Navajo** *Unspecified* Parched seeds used for food. (55:70)

Cyatheaceae, see *Cibotium*

Cycloloma, Chenopodiaceae

Cycloloma atriplicifolium (Spreng.) Coult., Winged Pigweed
- *Drug*—**Hopi** *Analgesic* Plant used for headache. (200:33, 74) *Antirheumatic (Internal)* Plant used for rheumatism. *Febrifuge* Plant used for fever. (200:32, 74)
- *Food*—**Apache, White Mountain** *Staple* Seeds used to make flour. (as *C. artriplicifolium* 136:156) **Hopi** *Porridge* Ground seeds used to make mush. (as *Salsola atriplicifolia* 190:161) **Zuni** *Porridge* Seeds mixed with ground corn to make a mush. (32:22) *Staple* Tiny seeds ground, mixed with cornmeal, and made into steamed cakes. (166:67)
- *Dye*—**Hopi** *Red* Seeds used to produce a pink dye. (200:74)
- *Other*—**Zuni** *Protection* Blossoms chewed and rubbed all over the hands for protection. This medicine belonged to the grandmother of the Gods of War. She gave it to the people with the instructions that, when near the enemy, they should chew the blossoms, eject the mass into their hand, rubbing their hands well together. As soon as the Gods of War had done this, a peculiar yellow light spread all over the world, preventing the enemy from seeing how to aim their arrows. (166:84)

Cycloloma cornutum
- *Food*—**Hopi** *Unspecified* Seeds and flowers used as food. (32:22)

Cymopterus, Apiaceae

Cymopterus acaulis (Pursh) Raf., Plains Springparsley
- *Food*—**Comanche** *Unspecified* Rootstocks used for food. (29:521) **Navajo** *Dried Food* Plant dried for future use. (as *C. glomeratus* 110:28) *Soup* Used with the Rocky Mountain bee plant to make stew. (as *C. glomeratus* 32:24) *Spice* Dried plant used as an herb for mutton stew. (as *C. glomeratus* 110:28)

Cymopterus acaulis* var. *fendleri (Gray) Goodrich, Fendler's Springparsley
- *Food*—**Apache, Chiricahua & Mescalero** *Spice* Leaves used with other green plant parts to flavor soups and meats. (as *C. fendleri* 33:48) *Unspecified* Roots eaten raw. (as *C. fendleri* 33:47) Raw roots eaten for food. (as *C. fendleri* 33:48) **Keres, Western** *Vegetable* Plant eaten much as celery. (as *C. fendleri* 171:40) **Navajo** *Dried Food* Leaves rubbed through hot ash to remove the strong taste and dried for winter use. *Spice* Leaves used as a seasoning for cornmeal mush, gruel, and boiled meat. *Vegetable* Leaves rubbed through hot ash to remove the strong taste and eaten fresh. (as *C. fendleri* 165:221)

Cymopterus bulbosus A. Nels., Bulbous Springparsley
- *Drug*—**Keres, Western** *Gastrointestinal Aid* Plant eaten as a stomach medicine. (as *Phellopterus bulbosus* 171:59) **Navajo, Ramah** *Panacea* Plant used as "life medicine." (191:38)
- *Food*—**Acoma** *Vegetable* Eaten like celery. **Cochiti** *Vegetable* Eaten

like celery. **Laguna** *Vegetable* Eaten like celery. (as *Phellopterus bulbosus* 32:39) **Navajo, Ramah** *Dried Food* Dried, ground root cooked with milk. *Unspecified* Root eaten raw or roasted in ashes. *Winter Use Food* Root dried for winter use. (191:38)

Cymopterus globosus (S. Wats.) S. Wats., Globe Springparsley
- **Other**—**Paiute** *Insecticide* Decoction of roots used as an insecticide. (180:63, 64)

Cymopterus longipes S. Wats., Longstalk Springparsley
- **Food**—**Gosiute** *Unspecified* Leaves boiled and used for food. (39:367) **Ute** *Unspecified* Leaves formerly boiled and eaten. (38:33)

Cymopterus montanus Nutt. ex Torr. & Gray, Mountain Springparsley
- **Food**—**Gosiute** *Unspecified* Seeds and underground parts used for food. (39:367) **Navajo** *Substitution Food* Roots peeled, baked, and ground as an occasional substitute for cornmeal. (as *Phellopterus montanus* 55:68) *Unspecified* Cooked with dried wild desert onions in the winter. (as *Phellopterus montanus* 32:15) Cooked in the winter with wild carrot roots. (as *Phellopterus montanus* 32:26) Raw roots used for food. Peeled stems used for food. (as *Phellopterus montanus* 55:68)

Cymopterus multinervatus (Coult. & Rose) Tidestrom, Purplenerve Springparsley
- **Food**—**Hopi** *Unspecified* Roots eaten in spring. (42:305)

Cymopterus newberryi (S. Wats.) M. E. Jones, Sweetroot Springparsley
- **Drug**—**Navajo, Kayenta** *Dermatological Aid* Infusion of plant taken and used as a lotion for wounds. (205:34)
- **Food**—**Hopi** *Unspecified* Sweet roots peeled and eaten by children. (200:86) **Navajo, Kayenta** *Vegetable* Eaten as greens with meat. (205:34)

Cymopterus purpurascens (Gray) M. E. Jones, Widewing Springparsley
- **Drug**—**Navajo, Kayenta** *Analgesic* Plant used for backache. *Antiemetic* and *Gastrointestinal Aid* Plant used to settle stomach after vomiting from swallowing a fly. *Orthopedic Aid* Plant used for backache. (205:34)
- **Other**—**Navajo, Kayenta** *Ceremonial Items* Used in paint for prayer sticks. (205:34)

Cymopterus purpureus S. Wats., Purple Springparsley
- **Food**—**Navajo** *Spice* Plant used as a potherb in seasoning mush and soup. (55:67)

Cymopterus sp., Wild Onion
- **Food**—**Hualapai** *Bread & Cake* Roots pit roasted, mashed, and fried into cakes. *Soup* Roots boiled for stew. *Unspecified* Roots eaten raw. (195:46)

Cynodon, Poaceae

Cynodon dactylon (L.) Pers. **var. *dactylon***, Bermuda Grass
- **Drug**—**Keres, Western** *Veterinary Aid* Infusion of diseased grass used as a wash in castrating domestic animals. (as *Capriola dactylon* 171:35)

Cynoglossum, Boraginaceae

Cynoglossum grande Dougl. ex Lehm., Pacific Hound's Tongue
- **Drug**—**Concow** *Burn Dressing* Grated roots used for inflamed burns and scalds. **Pomo, Potter Valley** *Gastrointestinal Aid* Grated roots used for stomachaches. *Venereal Aid* Root used for venereal diseases. (41:382)
- **Food**—**Yuki** *Unspecified* Roots cooked and used for food. (41:382)

Cynoglossum officinale L., Gypsyflower
- **Drug**—**Iroquois** *Antihemorrhagic* Compound infusion of plants taken for consumption with hemorrhage. (87:420) *Cancer Treatment* Decoction of plant used as a wash and applied as poultice to leg cancer. *Dermatological Aid* Compound infusion used as wash and applied as poultice to running sores. *Kidney Aid* Compound infusion used as wash and applied as poultice for dropsy. (87:421) *Tuberculosis Remedy* Compound infusion of plants taken for consumption with hemorrhage. (87:420) *Venereal Aid* Decoction of roots taken and used as a wash for internal venereal disease. (87:421)

Cynoglossum virginianum L., Wild Comfrey
- **Drug**—**Cherokee** *Cancer Treatment* Root used for cancer. *Ceremonial Medicine* Used as an ingredient in green corn medicine. *Dermatological Aid* Compound decoction of roots given for itch. (as *Gynoglossom virginianum* 80:30) Decoction of roots used as a bath and taken for itching genitals. (177:52) *Psychological Aid* Compound decoction taken every 4 days for "bad memory." *Urinary Aid* Root syrup taken for milky urine and compound decoction used for bad memory. (as *Gynoglossom virginianum* 80:30)

Cynoglossum virginianum var. *boreale* (Fern.) Cooperrider, Wild Comfrey
- **Drug**—**Ojibwa** *Analgesic* Plant burned on live coals and fumes inhaled for headaches. (as *C. boreale* 153:359, 360)

Cyperaceae, see *Carex, Cladium, Cyperus, Eleocharis, Eriophorum, Scirpus*

Cyperus, Cyperaceae

Cyperus erythrorhizos Muhl., Redroot Flatsedge
- **Food**—**Kamia** *Porridge* Pulverized seeds cooked as mush. (62:24)

Cyperus esculentus L., Chufa Flatsedge
- **Drug**—**Navajo, Ramah** *Ceremonial Medicine* and *Emetic* Plant used as a ceremonial emetic. (191:19) **Pima** *Cold Remedy* Roots chewed for colds. *Cough Medicine* Roots chewed for coughs. *Snakebite Remedy* Poultice of chewed roots applied to snakebites. *Veterinary Aid* Chewed roots placed in horse's nostrils as a stimulant. (47:98)
- **Food**—**Costanoan** *Unspecified* Tubers eaten. (21:255) **Paiute, Northern** *Dried Food* Roots dried, ground, and mixed with other foods. *Unspecified* Roots eaten raw. (59:44) **Pomo, Kashaya** *Vegetable* Tubers on the rootstock eaten raw, baked, or boiled like potatoes. (72:78)

Cyperus fendlerianus Boeckl., Fendler's Flatsedge
- **Food**—**Apache, Chiricahua & Mescalero** *Fodder* Flowers salted and fed to horses. Seeds salted and fed to horses. *Unspecified* Tubers eaten raw or peeled and cooked. (33:47)

Cyperus laevigatus L., Smooth Flatsedge
- **Drug**—**Hawaiian** *Cold Remedy* Stalks crushed into fine particles and

used as a snuff for hard head colds. *Dermatological Aid* Stalks crushed into fine particles and used for deep cuts, boils, skin ulcers, and other skin diseases. *Strengthener* Flower and stalk ashes and kukui nut juice mixed and rubbed on the tongue for general debility. *Venereal Aid* Stalks crushed into fine particles, mixed with water and clay, and taken for penis burning disease. (2:9)
- *Fiber*—Hawaiian *Cordage* Stem fibers woven into strings and ropes. (2:9)
- *Other*—Hawaiian *Tools* Stem fibers used to remove all impurities in liquids. (2:9)

Cyperus odoratus L., Fragrant Flatsedge
- *Food*—Cocopa *Unspecified* Seeds used for food. Mohave *Unspecified* Seeds used for food. (as *C. ferax* 37:192) Pima *Unspecified* Tubers eaten. (as *C. ferax* 47:99)

Cyperus rotundus L., Nut Grass
- *Food*—Paiute *Porridge* Tubers made into meal and cooked as cereal. *Unspecified* Raw tubers used for food. (118:16)

Cyperus schweinitzii Torr., Schweinitz's Flatsedge
- *Food*—Kiowa *Fodder* Considered an excellent fodder for fattening horses. (192:17)

Cyperus sp., Kaa
- *Drug*—Hawaiian *Antirheumatic (External)* Buds, leaves, roots, and other plants pounded and resulting liquid used as a bath for body aches. (2:45) Tanana, Upper *Cold Remedy* Infusion of underground stems taken for colds. (as Cyperaceae sp. 102:9) Yavapai *Cold Remedy* Decoction of dried, pulverized root taken for colds. *Dermatological Aid* Dried, pulverized root dusted on sores. *Gastrointestinal Aid* Decoction of dried, pulverized root taken for stomachaches. (65:261)
- *Food*—Pima, Gila River *Snack Food* Tubers eaten as snack food primarily by children. (133:5) *Unspecified* Roots boiled and eaten. (133:7)

Cyperus squarrosus L., Bearded Flatsedge
- *Food*—Acoma *Unspecified* Small, tuberous roots eaten as food. (as *C. inflexus* 32:25) Keres, Western *Vegetable* Tubers eaten. (as *C. inflexus* 171:41) Laguna *Unspecified* Small, tuberous roots eaten as food. (as *C. inflexus* 32:25)

Cypripedium, Orchidaceae

Cypripedium acaule Ait., Pink Lady's Slipper
- *Drug*—Algonquin, Quebec *Gynecological Aid* Roots used for menstrual disorders. *Venereal Aid* Roots used for venereal disease. (18:143) Algonquin, Tête-de-Boule *Gastrointestinal Aid* Infusion of roots used for stomachaches. *Kidney Aid* and *Pediatric Aid* Infusion of roots used by children for kidney troubles. *Urinary Aid* Infusion of roots used for urinary tract problems. (132:128) Cherokee *Analgesic* Root used for neuralgia and other pain and infusion taken for "rupture pains." *Anthelmintic* Infusion of root given for worms. *Anticonvulsive* Roots used for "spasms" and "fits." *Cold Remedy* Infusion of root taken for colds and hot infusion of root taken for flu. *Gastrointestinal Aid* Compound infusion taken for stomach cramps. *Gynecological Aid* Infusion taken for "female trouble." *Kidney Aid* Compound infusion taken for kidney trouble. *Misc. Disease Remedy* Infusion taken for diabetes and hot infusion of root taken for flu. *Sedative* Roots used for "fits" and "hysterical affections." (80:42) Iroquois *Analgesic* Decoction taken for pains all over the skin and body, caused by bad blood. *Dermatological Aid* Poultice of smashed leaves bound to bite of mad dog. (87:289) Menominee *Urinary Aid* Root used in "male disorders." (151:44) Meskwaki *Love Medicine* Compound containing root used as a love medicine. (152:233, 234) Micmac *Sedative* Root used for nervousness. (40:56) Penobscot *Sedative* Infusion of plant taken for "nervousness." (156:310) Rappahannock *Panacea* Compound of dried roots and whisky taken for general ailments. (161:32)

Cypripedium arietinum Ait. f., Ramhead Lady's Slipper
- *Drug*—Iroquois *Analgesic* and *Gastrointestinal Aid* Decoction of whole plant taken for intestinal trouble with inflation and pains. (87:288)

Cypripedium montanum Dougl. ex Lindl., Mountain Lady's Slipper
- *Drug*—Okanagan-Colville *Reproductive Aid* Infusion of leaves and stalks taken by a pregnant woman to have a small baby. (188:52)

Cypripedium parviflorum Salisb., Lesser Yellow Lady's Slipper
- *Drug*—Cherokee *Analgesic* Root used for neuralgia and other pain and infusion taken for "rupture pains." *Anthelmintic* Infusion of root given for worms. *Anticonvulsive* Roots used for "spasms" and "fits." *Cold Remedy* Infusion of root taken for colds and hot infusion of root taken for flu. *Gastrointestinal Aid* Compound infusion taken for stomach cramps. *Gynecological Aid* Infusion taken for "female trouble." *Kidney Aid* Compound infusion taken for kidney trouble. *Misc. Disease Remedy* Infusion taken for diabetes and hot infusion of root taken for flu. *Sedative* Roots used for "fits" and "hysterical affections." (as *C. calceolus* 80:42) Iroquois *Blood Medicine* Compound decoction taken as blood medicine when "blood is bad from scrofula." (as *C. calceolus* ssp. *parviflorum* 87:289) *Febrifuge* Compound decoction taken for fever. *Orthopedic Aid* Compound decoction for soreness. (as *C. calceolus* ssp. *parviflorum* 87:288) *Sedative* Used as a nerve medicine. Infusion of root taken for nervousness or lack of energy. *Stimulant* Root decoction or infusion taken for nervousness, tiredness, and lack of energy. *Tuberculosis Remedy* Compound decoction taken as blood medicine when "blood is bad from scrofula." (as *C. calceolus* ssp. *parviflorum* 87:289)

Cypripedium pubescens Willd., Greater Yellow Lady's Slipper
- *Drug*—Cherokee *Anthelmintic* Decoction of roots taken for worms. (as *C. calceolus* var. *pubescens* 177:11) Chippewa *Dermatological Aid* Poultice of chopped, moistened root applied to inflammations. (as *C. hirsutum* 53:348) *Gastrointestinal Aid* Infusion of root taken in small doses for indigestion. *Toothache Remedy* Dried, powdered root moistened and applied to decayed teeth for toothache. (as *C. hirsutum* 53:342) Iroquois *Analgesic* Roots used with another plant for lower back pain. *Pulmonary Aid* Infusion of roots used for "too much wind in the chest." *Tuberculosis Remedy* Roots used with another plant for tuberculosis. (as *C. calceolus* var. *pubescens* 141:69) Menominee *Gynecological Aid* Plant used in "female disorders." *Hallucinogen* Plant used in sacred bundles to induce dreams of the supernatural. (as *C. parviflorum* pubescens 151:44) Ojibwa *Gynecological Aid* Root used for all female troubles. (as *C. parviflorum* var. *pubescens* 153:377)

***Cypripedium* sp.**, Lady Slipper
- *Food*—**Lakota** *Unspecified* Species used for food. (139:28)

Cyrtandra, Gesneriaceae
***Cyrtandra* sp.**, Kanawao-keokeo
- *Drug*—**Hawaiian** *Pediatric Aid* Fruits eaten from conception until the child feeds itself to increase the child's survival rate. Fruits eaten by infants for a weak physical constitution. *Reproductive Aid* Fruits eaten with baked eggs to bring about conception by barren women. *Strengthener* Fruits eaten from conception until the child feeds itself to increase the child's survival rate. Fruits eaten by infants for a weak physical constitution. (2:49)

Cystopteris, Dryopteridaceae, fern
Cystopteris fragilis (L.) Bernh., Brittle Bladderfern
- *Drug*—**Navajo, Ramah** *Dermatological Aid* Cold, compound infusion of plant taken and used as lotion for injury. (191:11)

Cystopteris protrusa (Weatherby) Blasdell, Lowland Bladderfern
- *Drug*—**Cherokee** *Febrifuge* Compound infusion given for chills. (80:33)

Dahlia, Asteraceae
Dahlia pinnata Cav., Pinnate Dahlia
- *Dye*—**Navajo** *Orange-Yellow* Roots and flowers used as a yellow-orange dye. (55:85)

Dalea, Fabaceae
Dalea aurea Nutt. ex Pursh, Golden Prairieclover
- *Drug*—**Dakota** *Antidiarrheal* Infusion of leaves taken for dysentery. (as *Parosela aurea* 69:366) Decoction of leaves used for dysentery. (as *Parosela aurea* 70:94) *Gastrointestinal Aid* Infusion of leaves taken for stomachaches. (as *Parosela aurea* 69:366) Decoction of leaves used for colic. (as *Parosela aurea* 70:94)

Dalea candida Willd. **var. *candida*,** White Prairieclover
- *Drug*—**Navajo** *Analgesic* Compound of plants used for abdomen pain caused by colds and loose bowels. Roots chewed for pain. *Gastrointestinal Aid* Compound of plants used for abdomen pain caused by colds and loose bowels. *Toothache Remedy* Plant used as toothache medicine. (as *Petalostemon candidus* 55:57) **Navajo, Ramah** *Analgesic* Plant used for stomachache. *Disinfectant* Compound decoction used for "snake infection." Decoction of plant used for "snake infection." *Febrifuge* and *Gastrointestinal Aid* Plant used for stomachache. *Panacea* Plant used as "life medicine," especially for fever. *Veterinary Aid* Compound decoction used for "snake infection" in sheep. (as *Petalostemum candidum* 191:33) **Pawnee** *Panacea* Infusion of root taken as a prophylactic to keep away disease. (as *Petalostemum candidum* 70:94)
- *Food*—**Navajo** *Special Food* Roots eaten as a delicacy by little children and sheepherders. (as *Petalostemon candidus* 55:57) **San Ildefonso** *Unspecified* Roots eaten raw. **Santa Clara** *Special Food* Plant chewed by women and children as a delicacy. (as *Pentalostemum candidus* 138:58)
- *Other*—**Navajo** *Ceremonial Items* Ground plant mixed with other ingredients and used in the Wind Chant. (as *Petalostemon candidus* 55:57)

Dalea candida* var. *oligophylla (Torr.) Shinners, White Prairieclover
- *Drug*—**Hopi** *Emetic* Plant recognized as a strong emetic. (as *Petalostemon oligophyllum* 200:34, 80) **Keres, Western** *Dermatological Aid* Infusion of roots used as a hair wash to keep it from falling. (as *Petalostemon cliogophyllus* 171:58) **Navajo, Kayenta** *Dermatological Aid* Poultice of plant applied to arrow wounds. *Panacea* Plant used as a life medicine. *Veterinary Aid* Plant used for sheep with constipation. (as *Petalostemum oligophyllum* 205:29)
- *Food*—**Acoma** *Staple* Roots dried and ground into meal. (as *Petalostemon oligophyllum* 32:33) **Keres, Western** *Staple* Dried roots ground into flour. *Sweetener* Roots eaten for the sweetness. (as *Petalostemon cliogophyllus* 171:58) **Keresan** *Unspecified* Roots chewed in the spring before the leaves come out. (as *Petalostemon oligophyllus* 198:560) **Kiowa** *Unspecified* Peeled stems used for food. (as *Petalostemon oligophyllus* 192:33) **Laguna** *Staple* Roots dried and ground into meal. **San Felipe** *Staple* Roots dried and ground into meal. (as *Petalostemon oligophyllus* 32:33) **San Ildefonso** *Unspecified* Roots eaten raw. **Santa Clara** *Special Food* Plant chewed by women and children as a delicacy. (as *Pentalostemum oligophyllum* 138:58)
- *Fiber*—**Keres, Western** *Brushes & Brooms* Plant used for light brooms. (as *Petalostemon cliogophyllus* 171:58)

Dalea compacta Spreng., Compact Prairieclover
- *Drug*—**Zuni** *Dermatological Aid* Poultice of root applied to sores and rashes. *Gastrointestinal Aid* Infusion of root taken for stomachache. (27:376)

Dalea enneandra Nutt., Nineanther Prairieclover
- *Drug*—**Dakota** *Poison* Root considered poisonous. (as *Parosela enneandra* 69:366)
- *Other*—**Kiowa** *Hunting & Fishing Item*, and *Toys & Games* Erect, slender stems made into small arrows with a thorn in the end, used for games. (as *Parosela enneandra* 192:32, 33)

Dalea flavescens (S. Wats.) Welsh, Canyonlands Prairieclover
- *Other*—**Navajo, Kayenta** *Protection* Infusion of plant taken when lightning strikes near a hogan. (as *Petalostemum flavescens* 205:29)

Dalea formosa Torr., Featherplume
- *Drug*—**Jemez** *Cathartic* Decoction of leaves taken as a cathartic. (as *Parosela formosa* 44:25) **Keres, Western** *Emetic* Infusion of leaves used as an emetic before breakfast. *Strengthener* Infusion of leaves used by runners to increase endurance and long wind. (as *Parosela formosa* 171:57)
- *Other*—**Keres, Western** *Fuel* Plant used for firewood. (as *Parosela formosa* 171:57)

Dalea lanata Spreng., Woolly Prairieclover
- *Food*—**Hopi** *Candy* Scraped roots eaten as a sweet. (56:16) *Sweetener* Root eaten and regarded as sugar. (190:163)

Dalea lanata Spreng. **var. *lanata*,** Woolly Prairieclover
- ***Drug*—Navajo, Kayenta** *Dermatological Aid* Poultice of plant applied to centipede bites. (as *Parosela lanata* 205:28)

Dalea lasiathera Gray, Purple Prairieclover
- ***Food*—Zuni** *Candy* Root chewed, especially by children, and greatly enjoyed. *Spice* Flowers crushed by hand and sprinkled into meat stew as a flavoring after cooking. (as *Parosela lasianthera* 166:69)

Dalea nana Torr. ex Gray **var. *nana*,** Dwarf Prairieclover
- ***Drug*—Keres, Western** *Pediatric Aid* and *Strengthener* Infusion of plant used as a tonic for weak children. (as *Parosela nana* 171:57)

Dalea purpurea Vent., Purple Prairieclover
- ***Drug*—Montana Indian** *Dermatological Aid* Poultice of steeped, bruised leaves applied to fresh wounds. (as *Petalostemon violaceus* 19:17)
- ***Food*—Navajo** *Beverage* Leaves used to make tea. (as *Petalostemum purpureum* 90:154)

Dalea purpurea Vent. **var. *purpurea*,** Violet Prairieclover
- ***Drug*—Chippewa** *Heart Medicine* Decoction of leaves and blossoms taken for heart trouble. (as *Petalostemon purpureus* 53:338) **Meskwaki** *Antidiarrheal* Compound containing florets used for diarrhea. *Misc. Disease Remedy* Infusion of root taken for measles. (as *Petalostemum purpureum* 152:229) **Navajo** *Pulmonary Aid* Plant used for pneumonia. (as *Petalostemum purpureum* 90:154) **Pawnee** *Panacea* Infusion of root taken as a prophylactic to keep away disease. (as *Petalostemum purpureum* 70:94)
- ***Food*—Comanche** *Candy* Roots chewed for sweet flavor. (as *Petalostemum purpureum* 29:523) **Lakota** *Candy* Roots chewed as a gum. (as *Petalostemon purpureum* 139:47) **Oglala** *Beverage* Leaves sometimes used to make a tea-like beverage. **Ponca** *Candy* Root chewed for the pleasant taste. (as *Petalostemum purpureum* 70:94)
- ***Fiber*—Pawnee** *Brushes & Brooms* Tough, elastic stems made into brooms and used to sweep the lodge. (as *Petalostemum purpureum* 70:94)

Dalea sp.
- ***Food*—Paiute** *Unspecified* Species used for food. (as *Parosela* 167:243)

Dalea villosa (Nutt.) Spreng. **var. *villosa*,** Silky Prairieclover
- ***Drug*—Lakota** *Cathartic* Roots used as a purge. *Throat Aid* Leaves and blossoms eaten for swellings inside the throat. (as *Petalostemon villosum* 139:47)

***Dalibarda*,** Rosaceae

Dalibarda repens L., Robin Runaway
- ***Drug*—Iroquois** *Blood Medicine* Complex compound decoction of powdered plants taken as a blood purifier. *Venereal Aid* Complex compound decoction of powdered plants taken for venereal disease. (87:357)

***Daphne*,** Thymelaeaceae

Daphne mezereum L., Paradise Plant
- ***Drug*—Cherokee** *Analgesic* Root bark used for nocturnal venereal disease pains. Root bark used for venereal disease pains. *Diaphoretic* Used as a diaphoretic. *Stimulant* Used as a stimulant. *Venereal Aid* Root bark used for venereal disease pains. (80:32)

***Darlingtonia*,** Sarraceniaceae

Darlingtonia californica Torr., California Pitcherplant
- ***Other*—Yurok** *Insecticide* Plant ate insects. (6:27)

***Darmera*,** Saxifragaceae

Darmera peltata (Torr. ex Benth.) Voss, Indian Rhubarb
- ***Drug*—Karok** *Gynecological Aid* Infusion of roots taken by women to prevent fetus from getting too large. (as *Peltiphyllum peltatum* 148:384)
- ***Food*—Karok** *Vegetable* Young shoots eaten raw as green vegetables. (as *Peltiphyllum peltatum* 148:384) **Miwok** *Cooking Agent* Pulverized root mixed with acorn meal to whiten it. (as *Peltiphyllum peltatum* 12:144)

***Dasylirion*,** Agavaceae

Dasylirion durangensis
- ***Drug*—Tarahumara** *Ceremonial Medicine* Used in the annual ceremonial curing of animals and fields from sickness and lightning. (17:60)

Dasylirion texanum Scheele, Texas Sotol
- ***Food*—Southwest Indians** *Bread & Cake* Crowns pit baked, dried, pounded into flour and made into cakes. (17:57)

Dasylirion wheeleri S. Wats., Common Sotol
- ***Food*—Apache, Chiricahua & Mescalero** *Beverage* Crowns pit baked, removed, peeled, crushed, mixed with water, fermented, and used as a beverage. (33:52) *Bread & Cake* Crowns baked in pits, stripped, pounded to a pulp, spread out to dry, and eaten like cake. *Unspecified* Stalks roasted, boiled, or eaten raw. *Vegetable* Stalks boiled, dried, and stored to be used as vegetables. (33:38) **Apache, Mescalero** *Beverage* Pounded and used as a drink. *Bread & Cake* Plants pit cooked, formed into cakes, dried, and used for food. (14:41) *Dried Food* Crowns baked, pounded, and dried for winter use. (17:58) *Soup* Head hearts cooked with bones to make soup. *Unspecified* Fresh, young stalks used for food. (14:41) **Papago** *Vegetable* Crowns with leaves removed and central flowering stalks eaten as greens in May. (34:14) Flower stalks eaten as greens. (34:16) Flower stalks roasted in ashes and eaten as greens. (34:46) **Southwest Indians** *Bread & Cake* Crowns pit baked, dried, pounded into flour, and made into cakes. (17:57)
- ***Fiber*—Apache, Mescalero** *Furniture* Stalks used as cross pieces for cradleboard backs. (14:41) **Papago** *Basketry* Used as a source of basketry material. (34:16) Used to make two kinds of baskets. (34:55) Leaves used as foundation in coiled basketry. (34:57) *Clothing* Used to make headbands and head rings. (34:55) *Mats, Rugs & Bedding* Leaves twilled into mats. (17:60) Used to make large, tough sleeping mats, cradle mats, and back mats for the carrying frame. (34:55) **Pima** *Mats, Rugs & Bedding* Leaves twilled into mats. (17:60)
- ***Other*—Apache, Mescalero** *Ceremonial Items* Stalks used in the headdress of Mountain Spirit dancers. *Smoking Tools* Stalks and leaf base tissues used to make cigarette papers. *Tools* Stalks dried, split, drilled to make small holes, and used as fire drill hearths. (14:41)

Datiscaceae, see *Datisca*

Datisca, Datiscaceae

Datisca glomerata (K. Presl) Baill., Durango Root
- **Drug**—**Costanoan** *Throat Aid* Decoction of plant used for sore throat and swollen tonsils. (21:9) **Miwok** *Antirheumatic (External)* Decoction of pulverized root used as a wash for rheumatism. *Dermatological Aid* Decoction of pulverized root used as a wash for sores. (12:168)
- **Dye**—**Costanoan** *Red* Roots used as a red dye. *Yellow* Roots used as a yellow dye. (21:250) **Karok** *Yellow* Roots used as a yellow dye. (148:386) **Wintoon** *Yellow* Used as a yellow dye. (117:264)
- **Other**—**Mendocino Indian** *Hunting & Fishing Item* Leaves and roots used to kill trout. (41:370)

Datura, Solanaceae

Datura discolor Bernh., Desert Thornapple
- **Drug**—**Pima** *Analgesic* Infusion of leaves taken to mitigate the pains of childbirth. *Dermatological Aid* Poultice of pounded leaves applied to sores or to draw pus from a boil. *Ear Medicine* Poultice of heated flowers applied to ears for earaches. *Eye Medicine* Plant juice used as a wash for sore eyes. *Gastrointestinal Aid* Plant used for ulcers. *Gynecological Aid* Infusion of leaves taken to mitigate the pains of childbirth. *Hemorrhoid Remedy* Plant used for hemorrhoids. *Other* Roots chewed to become crazy for "bad disease." (47:85)
- **Food**—**Pima** *Beverage* Infusion of leaves and mescal used as a dangerously intoxicating brew. (47:85)

Datura ferox L., Chinese Thornapple
- **Drug**—**Keres, Western** *Psychological Aid* Roots eaten to see into the future. (as *D. quercifolia* 171:41)

Datura sp., Jimson Weed
- **Drug**—**Navajo** *Analgesic* Plant used as painkiller for headaches. *Ceremonial Medicine* Raw, dried roots chewed in a ceremony for chills and fevers. *Eye Medicine* Plant used for trachoma. *Febrifuge* Raw, dried roots chewed in a ceremony for chills and fevers. *Toothache Remedy* Plant used as painkiller for toothaches. *Veterinary Aid* Infusion of leaves used as a wash on castration wounds of sheep. (55:73)

Datura stramonium L., Jimson Weed
- **Drug**—**Cherokee** *Dermatological Aid* Poultice of wilted leaves used on boils. *Respiratory Aid* Smoked for asthma. (80:41) **Delaware** *Dermatological Aid* Poultice of crushed leaves applied to fresh wounds. *Hemorrhoid Remedy* Seeds pounded, mixed with tallow and salve, and used for piles. (176:37) **Delaware, Oklahoma** *Dermatological Aid* Poultice of seeds and leaves applied to wounds. *Hemorrhoid Remedy* Crushed seeds and tallow used as a salve for piles. (175:31, 74) **Iroquois** *Poison* Seeds considered poisonous. (141:56) **Mohegan** *Dermatological Aid* Poultice of crushed leaves, considered a "powerful plant," applied to cuts. (176:72, 128) **Rappahannock** *Dermatological Aid* Decoction of green or parched leaves used as salve on wounds. (161:27) Decoction of leaves applied to parts affected with inflammation. (161:28) *Febrifuge* Poultice of parched leaves bound to fevered part. (161:27) Decoction of leaves applied to parts affected with fever. *Poison* Seeds and leaves poisonous. *Pulmonary Aid* Poultice of decoction of leaves mash applied to the chest for pneumonia. (161:28) *Throat Aid* Compound poultice with crushed seeds rubbed on sore throat. (161:27)

Datura wrightii Regel, Sacred Thornapple
- **Drug**—**Apache, White Mountain** *Ceremonial Medicine* Powdered roots used in the religious-medicine ceremonies. *Disinfectant* Plant juice or ground flowers and roots used as a disinfectant. *Narcotic* Powdered roots used as a narcotic. (as *D. meteloides* 136:156) **Cahuilla** *Analgesic* Powdered leaves made into an ointment and applied as a painkiller in setting bones, and applied for pain in specific areas of the body. *Antidote* Plant paste used for poisonous tarantula, snake, spider, and insect bites. *Antirheumatic (External)* Powdered leaves made into an ointment and applied to swellings. *Hallucinogen* Most universally used hallucinogenic and medicinal plant known to humans. Used by the shaman to transcend reality and enter other worlds. *Datura* offered the shaman not only a means to transcend reality and come into contact with specific guardian spirits, it also enabled him to go on magical flights to other worlds or transform himself into other life forms such as the mountain lion or eagle. Such magical flights were a necessary and routine activity for Cahuilla shaman. A shaman might use the drug to visit the land of the dead, returning to the profane world with information useful to his people, or he might pursue a falling star to recapture a lost soul and return it to its owner. *Hunting Medicine* Used by hunters on long treks to increase strength, allay hunger, and gain power to capture game. *Other* Plant used to diagnose ailments and permitted the shamans to "see" the pain or disease. Plant used to divine cures for diseases. *Poison* An extremely poisonous plant. *Respiratory Aid* Leaves steamed and vapor inhaled for severe bronchial or nasal congestion. *Snakebite Remedy* Plant paste used for poisonous tarantula, snake, spider, and insect bites. *Sports Medicine* Used to enhance mental perception when playing "peon," the gambling game. *Toothache Remedy* Powdered leaves made into an ointment and applied for toothache pain. *Unspecified* Crushed leaves and roots, with other parts, mixed into a medicinal paste. Most universally used hallucinogenic and medicinal plant known to humans. *Veterinary Aid* Plant paste used for saddle sores on horses. **Chumash** *Hallucinogen* and *Unspecified* Most universally used hallucinogenic and medicinal plant known to humans. (as *D. meteloides* 15:60) **Coahuilla** *Hallucinogen* Plant used as a "delirient," but with extreme danger, as it can cause death. (as *D. meteloides* 13:80) *Poison* Pulverized plant given with water to cause death and now almost wholly avoided. (as *D. meteloides* 13:75) *Veterinary Aid* Crushed plant mixed with water and rubbed into saddle sores on horses. (as *D. meteloides* 13:80) **Costanoan** *Analgesic* Poultice of heated leaves applied for chest pains. *Cathartic* Dried leaves smoked as a purgative. *Dermatological Aid* Ground leaves used as a salve for boils. *Eye Medicine* Flower dew used as an eyewash. *Hallucinogen* Dried leaves smoked as a hallucinogen. *Love Medicine* Seeds and tobacco smoked as an aphrodisiac. *Respiratory Aid* Poultice of heated leaves applied for respiratory problems. (as *D. meteloides* 21:14) **Diegueño** *Hallucinogen* Most universally used hallucinogenic and medicinal plant known to humans. (as *D. meteloides* 15:60) Well known as a hallucinogenic plant used in rites marking boys' initiation into the toloache cult. *Poison* Plant considered poisonous. (as *D. meteloides* 84:17) *Unspecified* Most universally used hallucinogenic and medicinal plant known to humans. **Gabrielino** *Hallucinogen* and *Unspecified* Most universally used hallucinogenic and medicinal plant known to humans. (as *D. meteloides* 15:60) **Havasupai** *Derma-*

tological Aid Leaf folded several times and rubbed onto red ant bite. *Narcotic* Leaves or seeds, when eaten, made a person intoxicated for a day or more. (as *D. meteloides* 197:239) **Hopi** *Hallucinogen* Root chewed to induce visions by medicine man while making a diagnosis. (as *D. meteloides* 42:306) Roots chewed by doctor to induce visions while making diagnosis. (as *D. meteloides* 200:31, 89) *Narcotic* Plant used as a narcotic. (as *D. meteloides* 42:306) Plant well known for the narcotic properties. (as *D. meteloides* 200:89) *Other* Used to cure meanness. (as *D. meteloides* 200:37) *Poison* Plant sometimes fatal and given to a person "who is mean" to cure "meanness." (as *D. meteloides* 200:89) *Psychological Aid* Plant used as a cure for "meanness." (as *D. meteloides* 42:306) *Stimulant* Plant rarely used as a stimulant as it was sometimes fatal. (as *D. meteloides* 200:31, 89) **Kawaiisu** *Analgesic* Infusion of mashed roots taken for the pain of broken bones. Plant used for pain and swellings. *Antirheumatic (External)* Infusion of mashed roots used as a bath for rheumatic and arthritic limbs. *Ceremonial Medicine* Plant used for the puberty ceremony. *Dermatological Aid* Plant used for swellings and as a wash for cuts. *Hallucinogen* Plant used as a hallucinogen to induce dreams and visions. *Orthopedic Aid* Infusion of mashed roots taken for the mending of broken bones. *Pediatric Aid* Plant used for the puberty ceremony. *Poison* Plant considered poisonous. (206:23) **Keres, Western** *Dermatological Aid* Poultice of crushed leaves or roots used for boils. Poultice of burned, ground leaves used for boils. *Poison* Root killed any humans or other animals if eaten. (as *D. metefoides* 171:41) **Luiseño** *Ceremonial Medicine* Root juice used in boys' puberty ceremony to induce stupefaction. (as *D. meteloides* 155:229) *Hallucinogen* Most universally used hallucinogenic and medicinal plant known to humans. (as *D. meteloides* 15:60) *Narcotic* Root juice used in boys' puberty ceremony to induce stupefaction. (as *D. meteloides* 155:229) *Unspecified* Most universally used hallucinogenic and medicinal plant known to humans. (as *D. meteloides* 15:60) **Mahuna** *Dermatological Aid* Plant used as an antivenin for tarantula bites. *Narcotic* Smoked leaves or infusion of leaves taken as a narcotic. *Poison* Plant considered poisonous. *Snakebite Remedy* Plant used as an antivenin for rattlesnake bites. (as *D. meteloides* 140:43) **Miwok** *Hallucinogen* Root eaten to induce delirium which achieved supernatural power. Decoction of plant taken to induce delirium which achieved supernatural power. (as *D. meteloides* 12:169) **Navajo** *Veterinary Aid* Infusion of leaf used as wash for wounds of sheep after castration. (as *D. meteloides* 90:160) **Navajo, Kayenta** *Narcotic* Plant used as a narcotic. *Orthopedic Aid* Poultice of plant applied for sprains and fractures. (as *D. meteloides* 205:41) **Navajo, Ramah** *Analgesic* Cold infusion of root taken and used as a lotion for injury pain, a narcotic. *Ceremonial Medicine* Plant used as a ceremonial medicine. *Hallucinogen* Plant caused hallucinations and made "you drunk like from whisky." *Hunting Medicine* Plant mixed with pollen and smoked by hunters to make deer tame. *Narcotic* Cold infusion of root taken and used as a lotion for injury pain, a narcotic. *Veterinary Aid* Cold infusion of flower used as an eyewash for blindness in horses and lotion used for sores. *Witchcraft Medicine* Plant used by witches, but cannot harm one who knows how to use it. (as *D. meteloides* 191:42) **Paiute** *Blood Medicine* Decoction of ground, soaked roots taken for blood poisoning in the foot. *Hallucinogen* Decoction of ground, soaked roots taken to have visions, especially visitations from the dead. Seeds eaten to see dead relatives. (as *D. meteloides* 167:318) *Narcotic* Roots used to make a narcotic tea and not used medicinally. (as *D. meteloides* 180:66, 67) *Other* Plant enabled one to ascertain one's life span and those "whose days were numbered." Plant taken to find lost objects and remember where things were hidden. (as *D. meteloides* 167:318) **Paiute, Northern** *Hallucinogen* Roots eaten to discover things or see things that could not be seen with ordinary powers. *Poison* Plant poisonous if used incorrectly. (as *D. meteloides* 59:126) **Shoshoni** *Hallucinogen* Decoction of root taken to become unconscious and have visions. (as *D. meteloides* 118:50) *Narcotic* Roots used to make a narcotic tea and not used medicinally. (as *D. meteloides* 180:66, 67) **Tubatulabal** *Antirheumatic (Internal)* Infusion of plant taken for rheumatism. *Dermatological Aid* Plant used for wounds. Poultice of dried, pounded root applied to inflamed sores. *Gastrointestinal Aid* Plant used for bloat. *Laxative* Plant used for constipation. *Sedative* Cold infusion of plant taken to fall into a stupor. (as *D. meteloides* 193:59) **Ute** *Narcotic* Used as a narcotic. (as *D. meteloides* 34:26) **Yavapai** *Hunting Medicine* Decoction of leaves taken or leaves eaten for success in deer hunt. (as *D. meteloides* 65:261) **Yokut** *Ceremonial Medicine* Decoction of roots used as a ceremonial narcotic. *Gastrointestinal Aid* Decoction of roots taken for inflammation of the bowels (appendicitis). *Other* Decoction of roots taken for many different diseases. (as *D. meteloides* 117:423) **Yuma** *Narcotic* Used as a narcotic. (as *D. meteloides* 34:26) **Zuni** *Anesthetic* Powdered root given as an anesthetic for surgery. *Dermatological Aid* Poultice of root and flower meal applied to wounds to promote healing. (as *D. meteloides* 166:46, 48) *Narcotic* Used as a narcotic. (as *D. meteloides* 34:26) Powdered root given as a narcotic for surgery. (as *D. meteloides* 166:46, 48)

- *Food*—**Apache, White Mountain** *Beverage* Juice or powdered roots used to make a fermented, intoxicating drink. (as *D. meteloides* 136:151) **Navajo** *Dried Food* Fruits dried and used in the winter after soaking and boiling. *Fruit* Fruits ground and eaten without further preparation. (as *D. meteloides* 32:26) *Special Food* Seeds eaten in ceremonies. (as *D. meteloides* 55:74) **Papago** *Beverage* Roots ground, infused, and used as a beverage. (as *D. meteloides* 34:26)

- *Other*—**Cahuilla** *Ceremonial Items* Roots used to make a drink taken at rituals. Plant offered a means of coming into contact with the sacred world. Used ritually in male puberty ceremonies. *Good Luck Charm* Vial of the plant carried for "good luck." *Smoke Plant* Leaves used to smoke. (as *D. meteloides* 15:60) **Diegueño** *Ceremonial Items* Plant used to make a tea given to young men learning to become dancers for ceremonies. Well known as a hallucinogenic plant used in rites marking boys' initiation into the toloache cult. (as *D. meteloides* 84:17) **Keres, Western** *Smoke Plant* Dried leaves used as tobacco. (as *D. metefoides* 171:41) **Luiseño** *Ceremonial Items* Roots pounded in a mortar and used at boys' puberty ceremonies. (as *D. meteloides* 155:207) **Paiute** *Good Luck Charm* Seeds eaten for good luck while gambling and enabled the eater to guess correctly in the hand game. (as *D. meteloides* 167:318) **Zuni** *Ceremonial Items* Powdered root used by rain priests in a number of ways to ensure fruitful rains. Root pieces chewed by a robbery victim in order to find out the thief's identity. The root was given to the victim by a rain priest. The victim was told to chew the root and that the medicine would induce dreams of the thief. The rain priest would ask the victim to tell him everything he remembered in his dream so that he could identify the thief and recover the stolen property. (as *D. meteloides* 166:88)

Daucus, Apiaceae

Daucus carota L., Queen Anne's Lace
- **Drug**—**Cherokee** *Dermatological Aid* Infusion used as a wash for swelling. (80:51) **Delaware** *Misc. Disease Remedy* Infusion of full-blooming blossoms used for diabetes. (176:35) **Delaware, Oklahoma** *Misc. Disease Remedy* Infusion of fresh blossoms taken for diabetes. (175:29, 76) **Iroquois** *Blood Medicine* Decoction of roots taken by men for a blood disorder. *Dermatological Aid* Decoction of roots taken by men for pimples and paleness. *Dietary Aid* Decoction of roots taken by men with no appetite. *Diuretic* Decoction of roots taken for urine stoppage. (87:402) *Gynecological Aid* Plant used for fallen womb. (87:401) **Micmac** *Cathartic* Leaves used as a purgative. (40:56) **Mohegan** *Misc. Disease Remedy* Infusion of blossoms, must be full bloom, taken for diabetes. (174:269) Infusion of blossoms, must be in full bloom, taken for diabetes. (176:72, 130)
- **Food**—**Haisla & Hanaksiala** *Vegetable* Roots used for food. (43:214) **Kitasoo** *Vegetable* Roots used for food. (43:325) **Oweekeno** *Unspecified* Roots used for food. (43:83) **Sanpoil & Nespelem** *Dried Food* Dried roots stored for winter use. *Vegetable* Steamed or boiled root used for food. (131:100)

Daucus pusillus Michx., American Wild Carrot
- **Drug**—**Costanoan** *Blood Medicine* Decoction of plant taken to clean the blood. *Cold Remedy* Decoction of plant taken for colds. *Dermatological Aid* Decoction of plant taken for itching. *Febrifuge* Decoction of plant taken for fevers. *Snakebite Remedy* Decoction of plant taken for snakebites. (21:23) **Miwok** *Snakebite Remedy* Poultice of chewed plant applied to snakebite. (12:169)
- **Food**—**Clallam** *Unspecified* Carrots eaten raw or cooked in pits. (57:204) **Cowichan** *Unspecified* Roots eaten raw or steamed. (182:89) **Navajo** *Dried Food* Roots dried and cooked in the winter with wild celery. *Unspecified* Roots eaten fresh. (32:26) Roots eaten raw or cooked with or without wild celery. *Winter Use Food* Roots dried for winter use. (55:67) **Saanich** *Unspecified* Roots eaten raw or steamed. **Salish, Coast** *Unspecified* Roots eaten raw or steamed. (182:89)
- **Other**—**Mendocino Indian** *Good Luck Charm* Used as a talisman in gambling. (41:372)

Delphinium, Ranunculaceae

Delphinium bicolor Nutt., Little Larkspur
- **Drug**—**Blackfoot** *Antidiarrheal* Infusion of plant given to children with diarrhea. (86:82) *Dermatological Aid* Infusion of plant used by women to shine and straighten their hair. (86:124) *Oral Aid* Infusion of plant given to children with frothy mouth. *Pediatric Aid* Infusion of plant given to children with diarrhea, frothy mouth, and fainting spells. *Stimulant* Infusion of plant given to children with fainting spells. (86:82) **Gosiute** *Poison* Plant considered poisonous. (39:367)
- **Dye**—**Blackfoot** *Blue* Flowers used as a light blue dye for quills. (86:112)

Delphinium carolinianum ssp. *virescens* (Nutt.) Brooks, Carolina Larkspur
- **Drug**—**Lakota** *Poison* Plant poisonous to cattle. (as *D. virescens* 139:55)
- **Other**—**Kiowa** *Ceremonial Items* Small seeds used in peyote rattles. (as *D. virescens* 192:28)

Delphinium decorum Fisch. & C. A. Mey., Coastal Larkspur
- **Other**—**Karok** *Paint* Flowers pounded, mixed with salmon glue and fresh mountain grapes, and used to paint arrows and bows. (148:383)

Delphinium geraniifolium Rydb., Clark Valley Larkspur
- **Drug**—**Hopi** *Ceremonial Medicine* Plant taken as an emetic in the Po-wa-mu ceremony. *Gynecological Aid* Decoction of plant and juniper used to bathe mother during the lying-in period. (42:307)
- **Other**—**Hopi** *Ceremonial Items* Plant used ceremonially. (42:307)

Delphinium hesperium Gray, Foothill Larkspur
- **Drug**—**Mendocino Indian** *Poison* Plant poisonous to cattle. (41:347)
- **Food**—**Miwok** *Vegetable* Boiled leaves and flowers used for food. (12:159)

Delphinium menziesii DC., Menzies's Larkspur
- **Drug**—**Chehalis** *Dermatological Aid* Poultice of stalks and roots applied to sores. *Poison* Whole plant considered poisonous. (79:30) **Navajo** *Unspecified* Powdered petals sometimes used by the medicine man instead of larkspur petals. (55:47) **Thompson** *Love Medicine* Plant used as a charm by women "to help them obtain & hold affection of men." (164:506)
- **Dye**—**Thompson** *Blue* Flowers used as a blue dye for clothing. (164:502)
- **Other**—**Thompson** *Paint* Flowers used as a paint for clothing. (164:502)

Delphinium nudicaule Torr. & Gray, Red Larkspur
- **Drug**—**Mendocino Indian** *Narcotic* Plant had narcotic properties. (41:347)

Delphinium nuttallianum Pritz. ex Walp., Nuttall's Larkspur
- **Dye**—**Okanagan-Colville** *Blue* Flowers used to make a blue stain for coloring arrows and other items. (188:119)
- **Other**—**Shuswap** *Ceremonial Items* Used with roses and scattered by small girls in the Corpus Christi ceremony in the church. (123:65)

Delphinium nuttallianum Pritz. ex Walp. **var. *nuttallianum*,** Nuttall's Larkspur
- **Other**—**Navajo, Ramah** *Ceremonial Items* Petals and other blue flowers ground and used ceremonially. (as *D. nelsonii* 191:27)

Delphinium parryi Gray, San Bernardino Larkspur
- **Drug**—**Kawaiisu** *Orthopedic Aid* Dried, ground root used as a salve for swollen limbs. (206:25)

Delphinium scaposum Greene, Tall Mountain Larkspur
- **Drug**—**Hopi** *Ceremonial Medicine* Plant taken as an emetic in Po-wa-mu ceremony. (42:308) Plant used as a ceremonial emetic. (200:76) *Emetic* Plant used as a ceremonial emetic. (200:34, 76) *Gynecological Aid* Decoction of plant and juniper used to bathe mother during the lying-in period. (42:308) Compound decoction of plant used as a wash for mother after childbirth. (200:36, 76) **Navajo** *Unspecified* Powdered petals used by the medicine man. (55:47) **Navajo, Kayenta** *Gynecological Aid* Plant eaten by women to become prolific. *Veterinary Aid* Plant eaten by goats to become prolific. (205:22)

- *Dye*—Navajo *Blue* Petals used to make blue dye. (55:47)
- *Other*—Hopi *Ceremonial Items* Plant used ceremonially. (42:308) Petals and seeds ground into a very fine blue meal prescribed for the Flute altar. (56:16) *Toys & Games* Boys holding handfuls of this and mariposa lily above their heads chased by girls on occasions. (200:70) **Navajo** *Ceremonial Items* Pollen used extensively in many ceremonies. (55:47) **Navajo, Ramah** *Ceremonial Items* Petals and other blue flowers ground and used ceremonially. (191:27)

Delphinium scopulorum Gray, Rocky Mountain Larkspur
- *Other*—Navajo, Ramah *Hunting & Fishing Item* Petals and other blue flowers, ground and used for luck in hunting. (191:27)

Delphinium sp., Larkspur
- *Drug*—Blackfoot *Poison* Plant considered poisonous. *Unspecified* Plant used medicinally. (97:34)
- *Food*—Miwok *Vegetable* Young, boiled greens used for food. (12:159)
- *Dye*—Great Basin Indian *Blue* Flower blossoms used to make a blue dye. (121:47)

Delphinium tenuisectum Greene, Carrotleaf Larkspur
- *Other*—Navajo, Ramah *Ceremonial Items* Petals and other blue flowers ground and used ceremonially. (191:27)

Delphinium tricorne Michx., Dwarf Larkspur
- *Drug*—Cherokee *Heart Medicine* Infusion taken "for heart." *Poison* Root "makes cows drunk and kills them." (80:42)

Dendromecon, Papaveraceae

Dendromecon rigida Benth., Tree Poppy
- *Food*—Kawaiisu *Unspecified* Seeds used for food. (206:25)
- *Other*—Kawaiisu *Smoke Plant* One or two leaves put into the liquid and used in the preparation of the tobacco plug. (206:25)

Dennstaedtiaceae, see *Dennstaedtia, Microlepia, Pteridium*

Dennstaedtia, Dennstaedtiaceae, fern

Dennstaedtia punctilobula (Michx.) T. Moore, Eastern Hayscented Fern
- *Drug*—Cherokee *Febrifuge* Compound infusion taken for chills. (80:33, 34) **Mahuna** *Antihemorrhagic* Plant used for lung hemorrhages. (140:19)

Dentaria, Brassicaceae

Dentaria sp., Toothwort
- *Food*—Cherokee *Unspecified* Plants cooked with other greens and used for food. (204:252)

Dentinum, Hydnaceae, fungus

Dentinum repandum, Hedgehog Mushroom
- *Food*—Pomo, Kashaya *Vegetable* Baked on hot stones, in the oven, or fried. (72:130)

Deschampsia, Poaceae

Deschampsia cespitosa (L.) Beauv., Tufted Hairgrass
- *Food*—Gosiute *Unspecified* Seeds used for food. (39:367)

Deschampsia danthonioides (Trin.) Munro, Annual Hairgrass
- *Food*—Kawaiisu *Porridge* Seeds pounded, cooked into a mush and eaten. (206:26)

Descurainia, Brassicaceae

Descurainia incana (Bernh. ex Fisch. & C. A. Mey.) Dorn **ssp. incana**, Mountain Tansymustard
- *Food*—Paiute, Northern *Beverage* Seeds dried, cooked, ground, water added, kneaded, water added to make a fine batter, and drunk. *Unspecified* Seeds roasted, cooled, ground, mixed with cold water, and eaten. *Winter Use Food* Seeds stored for winter use. (as *D. richardsonii* 59:47)

Descurainia incana **ssp. incisa** (Engelm.) Kartesz & Gandhi, Mountain Tansymustard
- *Drug*—Gitksan *Dermatological Aid* Mashed and applied to bad cuts. (as *Sisymbrium incisum* 150:57) **Navajo, Kayenta** *Dermatological Aid* Plant used as a lotion for frozen body parts. *Throat Aid* Plant used as a lotion for sore throats. (as *D. incisa* 205:23)
- *Food*—Apache, Chiricahua & Mescalero *Bread & Cake* Seeds threshed, winnowed, ground, and the flour used to make bread. *Unspecified* Seeds boiled and eaten. (as *Sophia incisa* 33:49) **Klamath** *Unspecified* Parched, ground seeds used for food. (as *Sisymbrium incisum* 45:96–97) **Montana Indian** *Unspecified* Parched seeds ground for food. (as *Sisymbrium incisum* 19:24)

Descurainia obtusa (Greene) O. E. Schulz, Blunt Tansymustard
- *Food*—Cocopa *Vegetable* Young plants boiled as greens. (37:187) **Hopi** *Vegetable* Plant used as greens. (42:309)

Descurainia pinnata (Walt.) Britt., Western Tansymustard
- *Drug*—Cahuilla *Gastrointestinal Aid* Ground seeds used for stomach ailments. (15:66) **Navajo, Ramah** *Toothache Remedy* Poultice of plant applied for toothache. (191:28)
- *Food*—Cahuilla *Spice* Ground seeds used to flavor soups or used as a condiment with corn. *Vegetable* Leaves used as potherbs. (15:66) **Cocopa** *Staple* Seeds harvested, winnowed, parched, ground, and the meal eaten. (37:187) **Hopi** *Spice* Plant used as flavoring with meat or other vegetables. *Vegetable* Plant cooked alone as greens. (42:310) Greens pit baked, cooled and served in salted water with corn dumplings, boiled bread, or piki bread. (120:19) **Kawaiisu** *Beverage* Seeds parched, pounded, sifted, mixed with cold water, and taken as a nourishing beverage. *Winter Use Food* Pounded or raw seeds stored for future use. (206:26) **Navajo, Ramah** *Bread & Cake* Ground seeds used to make cakes. *Fodder* Used as sheep feed. (191:28) **Paiute, Northern** *Beverage* Seeds dried, cooked, ground, water added, kneaded, water added to make a fine batter, and drunk. *Unspecified* Seeds roasted, cooled, ground, mixed with cold water, and eaten. *Winter Use Food* Seeds stored for winter use. (59:47) **Pima, Gila River** *Beverage* Seeds mixed with water to make a drink. *Porridge* Seeds used to make a mucilaginous mass and eaten. *Staple* Seeds ground, parched and used to make pinole. (133:5) *Unspecified* Seeds mixed with water and eaten. (133:7)

Descurainia pinnata **ssp. halictorum** (Cockerell) Detling, Western Tansymustard
- *Food*—Navajo *Porridge* Parched seeds ground, made into a gruel, and used to dip bread in. (as *Sophia halictorum* 165:223) **Pueblo**

Dried Food Young plants boiled, pressed, rolled into balls, dried, and stored for winter use. *Soup* Plant made into a stew with wild onions, wild celery, tallow, or bits of meat. *Unspecified* Young plants boiled, pressed, rolled into balls, and eaten. *Vegetable* Young plants boiled with a pinch of salt and eaten as greens. (as *Sophia halictorum* 32:25)
- *Other*—Keres, Western *Fertilizer* Leaves buried with seed corn as a fertilizer or fungicide. Infusion of plant used to soak seed corn for faster maturity. *Preservative* Leaves stored with corn to prevent spoiling. (as *Sophia halictorum* 171:70)

Descurainia pinnata (Walt.) Britt. **ssp. *pinnata***, Western Tansymustard
- *Drug*—Pima *Dermatological Aid* Infusion of leaves used for sores. (as *Sophia pinnata* 146:77) Ute *Unspecified* Used as medicine. (as *Sisymbrium canescens* 38:36)
- *Food*—Atsugewi *Bread & Cake* Parched, winnowed, ground seeds made into cakes and eaten without cooking. (as *Sisymbrium pinnatum* 61:139) Gosiute *Porridge* Seeds used to make a mush. (as *Sisymbrium canescens* 39:382) Hopi *Unspecified* Leaves boiled or roasted between hot, flat stones and eaten. (as *Sisymbrium canescens* 32:25) Leaves boiled or roasted and eaten. (as *Sisymbrium canescens* 56:15) *Vegetable* Plant, salty in flavor, eaten as greens in the spring. (as *Sisymbrium canescens* 42:310) Eaten as greens in the spring. (as *Sophia pinnata* (200:77) Papago *Beverage* Seeds steeped and used as tea-like drinks for refreshment. (as *Sophia pinnata* 34:27) *Dried Food* Seeds basket winnowed, parched, sun dried, cooked, stored, and used for food. (as *Sophia pinnata* 34:24) *Unspecified* Seeds used for food. (as *Sophia pinnata* 36:62) Pima *Dried Food* Seeds parched, ground, and eaten mixed with hot or cold water. (as *Sophia pinnata* 95:263) *Staple* Seeds parched, ground, mixed with water, and eaten as pinole. (as *Sophia pinnata* 146:77)
- *Other*—Hopi *Paint* Plant used in the preparation of pottery paint. (as *Sisymbrium canescens* 42:310) Flowers mixed with dark iron pigment used as a black color for pottery decoration. (as *Sisymbrium canescens* 56:15)

Descurainia sophia (L.) Webb ex Prantl, Herb Sophia
- *Drug*—Navajo, Ramah *Toothache Remedy* Poultice of plant applied for toothache. (191:28) Paiute *Dermatological Aid* Poultice of ground seeds applied to burns and sores, including sores on horses. *Veterinary Aid* Poultice of ground seeds applied to sores on horses. (111:74)
- *Food*—Kawaiisu *Beverage* Seeds parched, pounded, sifted, mixed with cold water, and taken as a nourishing beverage. *Winter Use Food* Pounded or raw seeds stored for future use. (206:26) Navajo, Ramah *Bread & Cake* Ground seeds used to make cakes. *Fodder* Used as sheep feed. (191:28) Paiute *Beverage* Roasted, ground seeds mixed with water and used as a cooling beverage for hot weather. (111:74) *Ice Cream* Seeds mixed with snow and eaten as ice cream. *Staple* Seeds parched, ground, and eaten as meal. (as *Sisymbrium sophia* 104:98) Paiute, Northern *Beverage* Seeds dried, cooked, ground, water added, kneaded, water added to make a fine batter, and drunk. *Unspecified* Seeds roasted, cooled, ground, mixed with cold water, and eaten. *Winter Use Food* Seeds stored for winter use. (59:47) Pueblo *Dried Food* Young plants boiled, pressed, rolled into balls, dried, and stored for winter use. *Soup* Plant made into a stew with wild onions, wild celery, tallow, or bits of meat. *Unspecified* Young plants boiled, pressed, rolled into balls, and eaten. *Vegetable* Young plants boiled with a pinch of salt and eaten as greens. (as *Sophia sophia* 32:25)
- *Other*—Keres, Western *Fertilizer* Leaves buried with seed corn as a fertilizer or fungicide. Infusion of plant used to soak seed corn for faster maturity. *Preservative* Leaves stored with corn to prevent spoiling. (as *Sophia sophia* 171:70)

Descurainia sp., Tansymustard
- *Food*—Havasupai *Beverage* Seeds ground and added to water to make a refreshing, summer drink. (197:66) *Preserves* Seeds parched, ground, kneaded into seed butter, and eaten with fruit drinks or spread on bread. (197:67) *Staple* Seeds parched and ground into a flour. (197:220) Pima *Beverage* Seeds roasted, mixed with water, and eaten like atole. (47:84) Tewa of Hano *Unspecified* Cooked and eaten in the spring. (as *Sophia* 138:60)
- *Other*—Tewa of Hano *Paint* Plants moistened, steamed, liquid squeezed, and the remaining mass used as paint to decorate pottery. (as *Sophia* 138:60)

Desmanthus, Fabaceae

Desmanthus illinoensis (Michx.) MacM. ex B. L. Robins. & Fern., Prairie Bundleflower
- *Drug*—Paiute *Eye Medicine* Five seeds placed in eye at night for trachoma and washed out in morning. (180:67) Pawnee *Dermatological Aid* Decoction of leaves used as a wash for itch. (as *Acuan illinoensis* 70:89)
- *Other*—Keres, Western *Unspecified* Plant known and named but no use was specified. (as *Acuan illinoensis* 171:24) Omaha *Toys & Games* Mature plant, with the seed-filled pods, used as rattles by little boys. The little boys used the rattles as they mimicked some of the native dances in play. Pawnee *Toys & Games* Mature plant, with seed-filled pods, used as rattles by little boys, to mimic native dances. Ponca *Toys & Games* Mature plant, with seed-filled pods, used as rattles by little boys, to mimic native dances. (as *Acuan illinoensis* 70:89)

Desmodium, Fabaceae

Desmodium canadense (L.) DC., Showy Ticktrefoil
- *Drug*—Iroquois *Gastrointestinal Aid* Decoction of root taken for biliousness. (87:364)

Desmodium glutinosum (Muhl. ex Willd.) Wood, Pointedleaf Ticktrefoil
- *Drug*—Iroquois *Basket Medicine* Cold infusion of smashed roots used as a "basket medicine." (87:364)

Desmodium illinoense Gray, Illinois Ticktrefoil
- *Drug*—Meskwaki *Adjuvant* Plant in combination with others used as a powerful medicine. (152:228)

Desmodium incanum DC. **var. *incanum***, Zarzabacoa Comun
- *Drug*—Seminole *Analgesic* Plant used for adult's sickness caused by adultery: headache, body pains, and crossed fingers. (as *D. supina* 169:256) Decoction of leaves taken for headaches. *Febrifuge* Decoction of leaves taken for fevers. *Gastrointestinal Aid* Decoction of leaves taken for stomachaches. (as *D. supinum* 169:282)

Desmodium nudiflorum (L.) DC., Nakedflower Ticktrefoil
- *Drug*—**Cherokee** *Analgesic* Infusion of root used as a wash for cramps. *Oral Aid* Roots chewed for sore gums and mouth, including pyorrhea. (80:59)

Desmodium paniculatum (L.) DC., Panicledleaf Ticktrefoil
- *Drug*—**Houma** *Analgesic and Stimulant* Infusion of root in whisky taken for weakness or cramps. (158:63)

Desmodium perplexum Schub., Perplexed Ticktrefoil
- *Drug*—**Cherokee** *Analgesic* Infusion of root used as a wash for cramps. *Oral Aid* Roots chewed for sore gums and mouth, including pyorrhea. (80:59)

Desmodium sandwicense E. Mey., Hawaii Ticktrefoil
- *Drug*—**Hawaiian** *Cold Remedy* Leaves sun dried, crushed into a powder, mixed with tea, and taken for head colds. *Respiratory Aid* Leaves sun dried, crushed, and smoked in a pipe for asthma. *Strengthener* Leaves sun dried, crushed into a powder, mixed with tea, and taken for general debility. *Tuberculosis Remedy* Decoction of whole plant and other plants used as a wash for scrofulous sores. (as *D. uncinatum* 2:51)

***Desmodium* sp.**, Ticktrefoil
- *Drug*—**Alabama** *Cold Remedy* Infusion of plant taken to cause vomiting for bad lung cold. (177:31) Plant used as an emetic for bad colds. (as *Meibomia* 172:663) *Emetic* Infusion of plant taken to cause vomiting for bad lung cold. (177:31) Plant used as an emetic for lung trouble or bad colds. (as *Meibomia* 172:663) *Pulmonary Aid* Infusion of plant taken to cause vomiting for bad lung cold. (177:31) Plant used as an emetic for lung troubles. (as *Meibomia* 172:663)

Diapensiaceae, see *Galax*

***Dicentra*, Papaveraceae**

Dicentra chrysantha (Hook. & Arn.) Walp., Golden Eardrops
- *Drug*—**Kawaiisu** *Analgesic and Heart Medicine* Dried, mashed roots placed in heated bag and applied to the heart for heart pains. (206:26)

Dicentra cucullaria (L.) Bernh., Dutchman's Breeches
- *Drug*—**Iroquois** *Sports Medicine* Compound infusion of leaves used as liniment by runners to strengthen limbs. (87:339) **Menominee** *Love Medicine* Plant, most important love charm, thrown by young swain at his intended to hit her with it. Root chewed by young swain and breath attracts girl, even against her will. (151:81)

Dicentra formosa (Andr.) Walp., Pacific Bleedinghearts
- *Drug*—**Skagit** *Anthelmintic* Decoction of pounded roots taken as a worm medicine. *Dermatological Aid* Infusion of crushed plants used as a wash to make hair grow. *Toothache Remedy* Raw roots chewed for toothaches. (79:31) **Thompson** *Unspecified* Root used as some kind of medicine. (187:225)

***Dichanthelium*, Poaceae**

Dichanthelium laxiflorum (Lam.) Gould, Openflower Rosette Grass
- *Drug*—**Seminole** *Analgesic* Infusion of leaves rubbed on the abdomen for labor pains. (as *Panicum xalapense* 169:323) *Antirheumatic (External)* Whole plant used for rabbit sickness: muscular cramps. (as *Panicum xalapense* 169:194) *Cough Medicine, Pulmonary Aid, and Throat Aid* Infusion of plant taken and used as bath for gopher-tortoise sickness: cough, dry throat, noisy chest. (as *Panicum xalapense* 169:236)

Dichanthelium oligosanthes (J. A. Schultes) Gould **var. oligosanthes**, Heller's Rosette Grass
- *Drug*—**Lakota** *Poison* Plant poisonous to horses. (as *Panicum oligosanthes* 139:30)

Dichanthelium oligosanthes* var. *scribnerianum (Nash) Gould, Scribner's Rosette Grass
- *Drug*—**Navajo, Ramah** *Ceremonial Medicine and Disinfectant* Decoction of plant used ceremonially for "snake infection." (as *Panicum seribnerianum* 191:17)
- *Food*—**Kiowa** *Fodder* Used to fatten horses very quickly. (as *Panicum scribnerianum* 192:16)

Dichanthelium strigosum* var. *glabrescens (Griseb.) Freckmann, Roughhair Rosette Grass
- *Drug*—**Seminole** *Analgesic* Infusion of leaves rubbed on the abdomen for labor pains. (as *Panicum polycaulon* 169:323) *Antirheumatic (External)* Whole plant used for rabbit sickness: muscular cramps. (as *Panicum polycaulon* 169:194) *Cough Medicine, Pulmonary Aid, and Throat Aid* Infusion of plant taken and used as bath for gopher-tortoise sickness: cough, dry throat, noisy chest. (as *Panicum polycaulon* 169:236)

***Dichelostemma*, Liliaceae**

Dichelostemma ida-maia (Wood) Greene, Ida May's Snakelily
- *Other*—**Karok** *Decorations* Flowers used for bouquets. (6:27) Seedpods dried and hung up as ornaments. (as *Brodiaea ida-maia* 148:380)

Dichelostemma multiflorum (Benth.) Heller, Roundtooth Snakelily
- *Food*—**Atsugewi** *Unspecified* Cooked in earth oven and used for food. (as *Brodiaea multiflora* 61:138) **Karok** *Unspecified* Raw bulbs used for food. (6:27)

Dichelostemma pulchellum (Salisb.) Heller, Congested Snakelily
- *Food*—**Apache, San Carlos** *Unspecified* Bulbs eaten raw or cooked in spring. Blue flowers eaten raw. (as *Dichelostemma,* var. *brodiaca, capitata pauciflora* 95:258) **Cahuilla** *Unspecified* Corms eaten raw or cooked. (as *Brodiaea pulchella* 15:47) **Karok** *Unspecified* Bulbs used for food. (6:27) **Luiseño** *Unspecified* Bulb used for food. (as *Brodiaea capitata* 155:233) **Mendocino Indian** *Unspecified* Bulbs eaten raw or cooked. (as *D. capitatum* 41:323) **Miwok** *Unspecified* Bulbs steamed in earth oven and eaten. (as *Brodiaea pulchella* 12:156) **Neeshenam** *Unspecified* Eaten raw, roasted, or boiled. (as *Brodiaea congesta* 129:377) **Paiute** *Dried Food* Roots dried and stored. *Staple* Roots roasted and ground into flour. *Unspecified* Fresh roots eaten raw. (as *Brodiaea capitata* 167:245) **Papago** *Unspecified* Bulbs eaten raw in spring before other crops appeared. (as *Brodiaea capitata* var. *pauciflora* 34:17) Bulbs eaten raw in early spring before other foods available. (as *Brodiaea capitata* var. *pauciflora* 36:60) **Pima, Gila River** *Baby Food* Raw roots eaten primarily by children. (133:7) *Snack Food* Bulbs

eaten primarily by children as snack food. (133:5) **Pomo** *Unspecified* Bulbs eaten raw or cooked. (as *D. capitatum* 11:90) **Pomo, Kashaya** *Vegetable* Baked or boiled corms eaten like baked or boiled potatoes. (as *Brodiaea pulchella* 72:26) **Yuki** *Unspecified* Sweet bulbs used for food. (as *Brodiaea pulchella* 49:86)
- **Other**—**Kawaiisu** *Fasteners* Corms rubbed on metate into an adhesive and spread on seed gathering baskets to close the interstices. (206:16)

Dichelostemma volubile (Kellogg) Heller, Twining Snakelily
- **Food**—**Neeshenam** *Unspecified* Eaten raw, roasted, or boiled. (as *Brodiaea volubilis* 129:377) **Pomo** *Unspecified* Bulbs eaten for food. (as *Hookera volubilis* 11:90)

Dicranaceae, see *Dicranum*

Dicranum, Dicranaceae, moss
Dicranum bonjeanii De Not, Woodmoss
- **Fiber**—**Chippewa** *Other* Used as an absorbent. (53:377)

Dicranum sp., Moss
- **Fiber**—**Shuswap** *Building Material* Moss mixed with clay and used between the logs of a log house. (123:49)
- **Other**—**Shuswap** *Fertilizer* Moss mixed with house plant dirt as a fertilizer to make the plants healthier. (123:49)

Diervilla, Caprifoliaceae
Diervilla lonicera P. Mill., Northern Bush Honeysuckle
- **Drug**—**Algonquin, Tête-de-Boule** *Diuretic* Leaves used as a diuretic. (132:128) **Chippewa** *Analgesic* Compound decoction of leaves taken for stomach pain. (53:342) *Eye Medicine* Infusion of bark used as an eyewash. (71:141) *Gastrointestinal Aid* Compound decoction of leaves taken for stomach pain. (53:342) *Laxative* Infusion of bark taken for constipation. (71:141) **Cree, Woodlands** *Eye Medicine* Cooled infusion or decoction of roots or stems put into the eyes for soreness. *Gynecological Aid* Infusion of roots taken to ensure a good supply of breast milk. (109:37) **Iroquois** *Blood Medicine* Decoction of plant or roots taken as a blood medicine. *Gynecological Aid* Compound decoction of bark and plants taken for prolapse of the uterus. *Pediatric Aid* Decoction of plant or roots given to "spoiled babies with adulterous mother." *Urinary Aid* Compound decoction of bark and plants taken by old men who cannot retain urine. *Venereal Aid* Compound decoction of roots taken for gonorrhea. (87:442) **Menominee** *Blood Medicine* Compound decoction of stalk used to "clear the blood." (54:129) *Diuretic* Infusion of root taken as a mild diuretic. *Psychological Aid* Infusion of root used for senility and as a mild diuretic. (151:27) **Meskwaki** *Urinary Aid* Infusion of root taken by "one who is urinating blood." *Venereal Aid* Compound decoction of root taken for gonorrhea. (152:206, 207) **Ojibwa** *Urinary Aid* Compound containing root used as a valued urinary remedy. (153:360) **Potawatomi** *Diuretic* Simple or compound infusion of root taken as a diuretic. *Other* Compound infusion of twigs used for vertigo. *Venereal Aid* Infusion of root taken for gonorrhea. (154:45)

Digitalis, Scrophulariaceae
Digitalis purpurea L., Purple Foxglove
- **Other**—**Hoh** *Ceremonial Items* Flowers used for decorations in ceremonies. **Quileute** *Ceremonial Items* Flowers used for decorations in ceremonies. (137:68)

Digitaria, Poaceae
Digitaria cognata (J. A. Schultes) Pilger, Carolina Crabgrass
- **Food**—**Hopi** *Staple* Seeds ground into meal. (as *Panicum autumnale* 190:158)

Digitaria cognata (J. A. Schultes) Pilger **var. cognata**, Fall Witchgrass
- **Food**—**Hopi** *Staple* Seeds ground into meal. (as *Leptoloma cognatum* 190:158)

Digitaria setigera Roth ex Roemer & J. A. Schultes, East Indian Crabgrass
- **Drug**—**Hawaiian** *Antihemorrhagic* Grass pounded, mixed with water, strained, and taken for stomach and bowel hemorrhage. *Eye Medicine* Shoots chewed into a thick liquid and blown into the eye for cataracts. *Laxative* Leaves chewed by mothers and fed to children as a laxative. Shoot chewed and swallowed as a laxative. *Pediatric Aid* Shoots chewed by mothers and given to infants for run-down condition. Leaves chewed by mothers and fed to children as a laxative. *Strengthener* Shoots chewed by mothers and given to infants for run-down condition. Shoots chewed for run-down condition. (as *Panicum pruricus* 2:55)

Dimorphocarpa, Brassicaceae
Dimorphocarpa wislizeni (Engelm.) Rollins, Touristplant
- **Drug**—**Apache, White Mountain** *Ceremonial Medicine* Infusion of plant taken at medicine ceremonies. *Dermatological Aid* Infusion of plant used as wash for swellings. *Throat Aid* Infusion of plant used as wash for throat troubles. (as *Dithyraea wislizeni* 136:157) **Hopi** *Dermatological Aid* Pods ground and sprinkled on wounds. (as *Dithyrea wislizeni* 42:311) Dried, powdered leaves sprinkled on abrasions. (as *Biscutella wislizeni* 56:15) Ground stalk used as a salve for all kinds of sores. (as *Dithyraea wislizenii* 190:163) Powdered plant sprinkled on wounds. (as *Dithyrea wislizeni* 200:32, 77) **Keres, Western** *Nose Medicine* Crushed seeds and leaves inhaled for catarrh or sore nose. (as *Dithyraea wislizeni* 171:41) **Navajo, Kayenta** *Dermatological Aid* Infusion of plant taken and used as lotion for centipede or sand cricket bites. *Hemorrhoid Remedy* Poultice of plant applied to hemorrhoids. *Pediatric Aid* and *Toothache Remedy* Plant chewed by children to strengthen teeth. (as *Dithyraea wislizeni* 205:24) **Navajo, Ramah** *Dermatological Aid* Cold infusion of plant used as a lotion for itch. Poultice of leaves used to remove scabs. (as *Dithyrea wizlizenii* 191:28) **Zuni** *Dermatological Aid* Warm infusion of pulverized plant applied to swelling, especially the throat. (as *Dithyraea wislizeni* 166:48, 49) *Emetic* Flower and fruit eaten as an emetic for stomachaches. (as *Dithyraea wislizeni* 27:375) *Psychological Aid* Decoction of entire plant given for delirium. (as *Dithyraea wislizeni* 166:48, 49) Infusion of plant taken by men to "loosen their tongues so they may talk like fools & drunken men." It was said that this infusion should never be given to women because they "should not be made to talk too much." (as *Dithyraea wislizeni* 166:91)
- **Food**—**Navajo** *Forage* Plant used by sheep for forage. (as *Dithyrea wislizeni* 55:49)

- *Other*—**Hopi** *Toys & Games* Plant, a powerful irritant, placed in armpit as a practical joke. (as *Dithyrea wislizeni* 42:311) **Keres, Western** *Snuff* Crushed seeds and leaves used for snuff. (as *Dithyraea wislizeni* 171:41) **Navajo, Kayenta** *Ceremonial Items* Mixed with paint and used on prayer sticks or ceremonial figurines of water animals. (as *Dithyraea wislizeni* 205:24)

Dionaea, Droseraceae

Dionaea muscipula Ellis, Venus Flytrap
- *Other*—**Cherokee** *Hunting & Fishing Item* Small piece of plant chewed and spat on bait for fishing. (80:60)

Dioscoreaceae, see *Dioscorea*

Dioscorea, Dioscoreaceae

Dioscorea bulbifera L., Air Yam
- *Food*—**Hawaiian** *Fruit* Bitter fruit cooked, grated, washed several times, strained, and eaten. (112:68)

Dioscorea pentaphylla L., Fiveleaf Yam
- *Food*—**Hawaiian** *Unspecified* Tubers oven cooked and eaten. (112:68)

Dioscorea sp., Uhi-keokeo
- *Drug*—**Hawaiian** *Febrifuge* Bulb scrapings, other plant scrapings, and water taken for a heated body and excessive sweating. (2:37) *Unspecified* Tubers used to make a drink for the sick. (112:67)
- *Food*—**Hawaiian** *Unspecified* Bulbs used for food. (2:37) *Vegetable* Tubers baked or roasted and eaten. (112:67)

Dioscorea villosa L., Wild Yam
- *Drug*—**Meskwaki** *Analgesic* Root used by women for pain at childbirth. *Gynecological Aid* Root used by women for pain at childbirth. (152:220)

Diospyros, Ebenaceae

Diospyros texana Scheele, Texas Persimmon
- *Food*—**Comanche** *Fruit* Fruits eaten for food. (as *Brayodendron texanum* 29:520)

Diospyros virginiana L., Common Persimmon
- *Drug*—**Cherokee** *Antidiarrheal* Syrup taken for bloody discharge from bowels. *Dermatological Aid* Astringent plant used for sore throat and mouth. *Gastrointestinal Aid* Bark chewed for heartburn. Compound used in steam bath for indigestion or biliousness. *Hemorrhoid Remedy* Used as a wash for piles. *Liver Aid* Cold infusion of bark taken for bile and liver. *Oral Aid* Syrup used for thrush. *Throat Aid* Syrup used for sore throat. *Toothache Remedy* Compound infusion used for toothache. *Venereal Aid* Used for venereal diseases. (80:49) **Rappahannock** *Oral Aid* Infusion of inner bark used as a wash for thrash. *Throat Aid* Infusion of north side bark taken for sore throat. (161:25)
- *Food*—**Cherokee** *Fruit* Fruit used for food. *Pie & Pudding* Fruit used to make pudding. (126:38) **Comanche** *Fruit* Fruits eaten for food. (29:521) **Rappahannock** *Beverage* Fruits rolled in cornmeal, brewed in water, drained, baked, and mixed with hot water to make a beer. (161:25) **Seminole** *Unspecified* Plant used for food. (169:495)

Diphylleia, Berberidaceae

Diphylleia cymosa Michx., American Umbrellaleaf
- *Drug*—**Cherokee** *Diaphoretic* Infusion taken as a diaphoretic. *Disinfectant* Used as an antiseptic. *Diuretic* Taken as a diuretic. *Misc. Disease Remedy* Infusion taken for smallpox. (80:59, 60)

Diplacus, Scrophulariaceae

Diplacus aurantiacus (W. Curtis) Jepson, Orange Bush Monkeyflower
- *Drug*—**Tubatulabal** *Gastrointestinal Aid* Decoction of leaves and flowers taken for stomachaches. (193:59)

Diplacus aurantiacus (W. Curtis) Jepson **ssp. *aurantiacus***, Orange Bush Monkeyflower
- *Drug*—**Costanoan** *Kidney Aid* Decoction of plant used for kidney problems. *Urinary Aid* Decoction of plant used for bladder problems. (as *Mimulus aurantiacus* 21:15) **Mahuna** *Antidiarrheal* Infusion of leaves, flowers, and stems taken for diarrhea. (as *Dipsacus glutinosus* 140:6) **Pomo, Kashaya** *Eye Medicine* Strained decoction of flower, stem, and leaves used as an eyewash for sore eyes. (as *Mimulus aurantiacus* 72:72)

Diplacus longiflorus Nutt., Southern Bush Monkeyflower
- *Drug*—**Tubatulabal** *Gastrointestinal Aid* Decoction of leaves and flowers taken for stomachaches. (193:59)

Diplazium, Dryopteridaceae, fern

Diplazium meyenianum K. Presl, Meyen's Twinsorus Fern
- *Drug*—**Hawaiian** *Dermatological Aid* Young shoots powdered, mixed with milk and nuts, and applied to boils. (as *Dilazium arnottii* 2:44)
- *Food*—**Hawaiian** *Unspecified* Young shoots used for food. (as *Dilazium arnottii* 2:44)

Dipsacaceae, see *Dipsacus*

Dipsacus, Dipsacaceae

Dipsacus fullonum L., Fuller's Teasel
- *Drug*—**Iroquois** *Dermatological Aid* Infusion of leaves used as a wash for acne or "worms in the face." *Poison* Powdered roots considered poisonous. (as *D. sylvestris* 87:450)
- *Other*—**Navajo** *Tools* Used to card wool. (as *D. sylvestris* 55:78)

Dirca, Thymelaeaceae

Dirca palustris L., Eastern Leatherwood
- *Drug*—**Algonquin, Quebec** *Laxative* Infusion of inner bark taken as a laxative tea. (18:202) **Chippewa** *Cathartic* Infusion of stalk taken or green stalk chewed as a physic. (53:346) *Dermatological Aid* Compound decoction of root used as wash to strengthen hair and make it grow. (53:350) *Pulmonary Aid* Infusion of roots taken for pulmonary troubles. (71:137) **Iroquois** *Analgesic* Compound infusion of bark and roots taken for back pains. *Blood Medicine* Compound decoction or infusion of roots taken to purify the blood. (87:388) *Cathartic* Bark and wood used as a strong purgative. (141:50) *Dermatological Aid* Decoction of branches applied as poultice to swellings on the leg or limbs. (87:387) Compound infusion taken for dark circles and puffiness around the eyes. Compound used for neck sores. (87:388) *Emet-*

ic Infusion of bark and wood used as an emetic to remove yellow from the stomach. The yellow in the stomach was a sickness brought by the Europeans. As they introduced tea, butter, and tobacco, the yellow accumulated in the stomach and could not be evacuated. (141:50) *Eye Medicine* Compound infusion taken for dark circles and puffiness around the eyes. *Gynecological Aid* Compound decoction of bark and roots taken to induce pregnancy. *Internal Medicine* Root used for internal inflammation. *Kidney Aid* Compound infusion of bark and roots taken for kidney troubles. *Laxative* Decoction or infusion of smashed roots or bark taken as a laxative. (87:388) *Love Medicine* Decoction of stems used as an aphrodisiac. (87:389) *Misc. Disease Remedy* Infusion of smashed roots taken for typhoid fever. (87:388) *Orthopedic Aid* Infusion of roots taken for a strained back or back pains. (87:387) *Strengthener* Decoction of stems used to increase strength. (87:389) *Tonic* Compound decoction of roots taken as a blood medicine or tonic. *Tuberculosis Remedy* Whole plant used for consumption. *Urinary Aid* Compound infusion of roots taken for kidneys or for male urination problems. *Venereal Aid* Compound infusion of powdered roots taken for gonorrhea and syphilis. Decoction of roots taken to remove venereal germs from the blood. (87:388) **Menominee** *Diuretic* Infusion of roots taken as a diuretic. *Kidney Aid* Infusion of root taken as a diuretic for kidney troubles. (151:54) **Ojibwa** *Diuretic* Infusion of bark taken as a diuretic. (153:390) *Urinary Aid* Infusion of twigs taken for urinary infections. (5:2306) **Potawatomi** *Diuretic* Infusion of inner bark taken as a diuretic. (154:85)
- *Fiber*—**Iroquois** *Cordage* Bark twisted into cordage. (141:50) **Menominee** *Cordage* Bark or twigs used for cordage. (151:76) **Potawatomi** *Cordage* Tough, stringy bark made a good substitute for twine. (154:114)
- *Other*—**Chippewa** *Fasteners* Tough, pliant bark used for tying and binding. (71:137) **Ojibwa** *Fasteners* Used for withes. (135:240)

Disporum, Liliaceae

Disporum hookeri (Torr.) Nichols., Drops of Gold
- *Drug*—**Costanoan** *Kidney Aid* Fruit used for the kidneys. (21:28)
- *Food*—**Thompson** *Fruit* Fruit occasionally used for food, but not considered important. (187:121)

Disporum hookeri var. *oreganum* (S. Wats.) Q. Jones, Oregon Drops of Gold
- *Drug*—**Klallam** *Poison* Plant considered poisonous. **Makah** *Love Medicine* Plant used as a love medicine. (as *D. oreganum* 79:25)
- *Food*—**Nitinaht** *Forage* Berries eaten by wolves. (186:86)

Disporum smithii (Hook.) Piper, Largeflower Fairybells
- *Drug*—**Makah** *Love Medicine* Plant used as a love medicine. (79:25)
- *Food*—**Karok** *Forage* Berries eaten by squirrels. (148:381)

Disporum sp.
- *Drug*—**Thompson** *Analgesic* Compound decoction of roots taken for internal pains. (164:459)

Disporum trachycarpum (S. Wats.) Benth. & Hook. f., Roughfruit Fairybells
- *Drug*—**Blackfoot** *Eye Medicine* Fresh seed used to clear matter from the eye. A fresh seed was inserted and the closed eye rubbed until the seed was watered out with the matter clinging to it. Seeds placed in the eye overnight and infusion of bark used as an eyewash for snow-blindness. (86:80) **Okanagan-Colville** *Dermatological Aid* Infusion of leaves used as a wash for wounds. *Hemostat* Poultice of dampened, bruised leaves applied to bleeding wounds. *Unspecified* Berries used to make medicine. (188:44)
- *Food*—**Blackfoot** *Fruit* Berries used for food. (86:102) Berries eaten raw. (97:25) Berries eaten raw. (114:277) **Shuswap** *Fruit* Berries used for food. (123:54)

Distichlis, Poaceae

Distichlis spicata (L.) Greene, Inland Saltgrass
- *Drug*—**Kawaiisu** *Dermatological Aid* Decoction of plant taken for doodlebug bites that cause pimples. *Heart Medicine* Infusion of plant taken "when the heart beats fast." *Laxative* Infusion of plant taken as a laxative. *Venereal Aid* Cold infusion of plant taken for gonorrhea. (206:26) **Yokut** *Cold Remedy* Decoction of salt cooked into a gum, placed in the mouth, and allowed to melt for bad colds.[28] *Dietary Aid* Decoction of salt cooked into a gum, placed in the mouth, and allowed to melt for loss of appetite.[28] (117:423)
- *Food*—**Cahuilla** *Spice* Leaves burned into ashes to remove the salt and used as a condiment. (15:66) **Kawaiisu** *Beverage* Dried grass cakes used to make a beverage. Green grass immersed in cold water, strained, and used as a beverage. (206:26) **Tubatulabal** *Unspecified* Leaves and stems used extensively for food. (193:15)
- *Fiber*—**Cahuilla** *Scouring Material* Leaves used as a brushing material for cleaning implements or removing cactus thorns from objects. (15:66)

Dodecatheon, Primulaceae

Dodecatheon hendersonii Gray, Mosquito Bills
- *Drug*—**Pomo, Kashaya** *Sedative* Flowers hung on baby baskets to make the baby sleepy. (as *D. henersonii* 72:105)
- *Food*—**Yuki** *Unspecified* Leaves and roots formerly roasted in ashes and eaten. (41:378)
- *Other*—**Mendocino Indian** *Decorations* Flowers used by women as ornaments for dances. (41:378)

Dodecatheon jeffreyi Van Houtte, Tall Mountain Shootingstar
- *Drug*—**Thompson** *Love Medicine* Flowers used by women "to obtain the love of men and to help them control men." (164:506)
- *Other*—**Thompson** *Good Luck Charm* Flowers used as a charm "to obtain wealth & to make people give presents." (164:506)

Dodecatheon pulchellum (Raf.) Merr., Darkthroat Shootingstar
- *Other*—**Thompson** *Designs* Flower used as a pattern for beadwork on gloves, moccasins, vests, and jackets. (187:245)

Dodecatheon pulchellum ssp. *pauciflorum* (Greene) Hultén, Pride of Ohio
- *Drug*—**Okanagan-Colville** *Eye Medicine* Infusion of roots used as a wash for sore eyes. (as *D. pauciflorum* 188:117)
- *Dye*—**Okanagan-Colville** *Red* Flowers mashed and smeared on arrows to color them pink. (as *D. pauciflorum* 188:117)

Dodecatheon pulchellum (Raf.) Merr. ssp. *pulchellum*, Darkthroat Shootingstar
- *Drug*—**Blackfoot** *Eye Medicine* Cooled infusion of leaves used for

eye drops. (as *D. radicatum* 86:81) *Oral Aid* and *Pediatric Aid* Infusion of leaves gargled, especially by children, for cankers. (as *D. radicatum* 86:76)

Dodonaea, Sapindaceae

Dodonaea viscosa (L.) Jacq., Florida Hopbush
- **Drug**—**Hawaiian** *Ceremonial Medicine* Infusion of leaves and other plants used as a wash to keep evil influences away. *Dermatological Aid* Infusion of leaves and other plants used as a wash for rash and itch. *Misc. Disease Remedy* Infusion of leaves and other plants used as a wash for contagious diseases. (2:3)

Draba, Brassicaceae

Draba helleriana Greene, Heller's Whitlowgrass
- **Drug**—**Keres, Western** *Other* Plant made into a drink and taken when not feeling well. (171:42) **Navajo, Ramah** *Ceremonial Medicine* Plant used in various ways as a ceremonial medicine. *Cough Medicine* Decoction of leaves taken for bad cough, sore kidney, or gonorrhea. *Emetic* Whole plant used as a ceremonial emetic. *Eye Medicine* Cold infusion of leaves used as ceremonial eyewash. *Kidney Aid* Decoction of leaves taken for sore kidney, bad cough, or gonorrhea. *Panacea* Root used as a "life medicine." *Venereal Aid* Decoction of leaves taken for gonorrhea, sore kidney, or bad cough. *Witchcraft Medicine* Cold infusion of plant taken and used as a lotion to protect against witches. (191:28)

Draba incerta Payson, Yellowstone Whitlowgrass
- **Drug**—**Blackfoot** *Abortifacient* Plant used to cause abortions. (86:60) *Nose Medicine* Infusion of roots taken for nosebleeds. (86:71)

Draba rectifructa C. L. Hitchc., Mountain Whitlowgrass
- **Drug**—**Navajo** *Diuretic* Infusion of plants taken as a diuretic. (as *D. montana* 55:49)

Draba reptans (Lam.) Fern., Carolina Whitlowgrass
- **Drug**—**Navajo, Ramah** *Dermatological Aid* Poultice of crushed leaves applied to sores. (191:28)

Dracocephalum, Lamiaceae

Dracocephalum parviflorum Nutt., American Dragonhead
- **Drug**—**Navajo, Kayenta** *Antidiarrheal* Plant used for infants with diarrhea. *Panacea* Plant used as a life medicine. *Pediatric Aid* Plant used for infants with diarrhea. (as *Moldavica parviflora* 205:40) **Navajo, Ramah** *Analgesic* Cold, compound infusion of leaves taken for headache. *Eye Medicine* Cold infusion of leaves used as an eyewash. *Febrifuge* Cold, compound infusion of leaves taken for fever. (as *Moldavica parviflora* 191:41)
- **Food**—**Apache, Chiricahua & Mescalero** *Spice* Leaves used as flavoring. (33:47) **Gosiute** *Unspecified* Seeds used for food. (39:367)

Droseraceae, see *Dionaea, Drosera*

Drosera, Droseraceae

Drosera capillaris Poir., Pink Sundew
- **Drug**—**Seminole** *Dermatological Aid* Sticky, plant glands rubbed on ringworm sores. (169:211)

Drosera rotundifolia L., Roundleaf Sundew
- **Drug**—**Kwakiutl** *Dermatological Aid* Plant used for corns, warts, and bunions. *Love Medicine* Plant used as a "medicine to make women love-crazy," a love charm. (183:281)

Drymaria, Caryophyllaceae

Drymaria glandulosa K. Presl, Fendler's Drymary
- **Drug**—**Navajo, Ramah** *Dermatological Aid* Poultice of chewed plant applied to mouse bite. (as *D. fendleri* 191:26)

Dryopteridaceae, see *Athyrium, Cystopteris, Diplazium, Dryopteris, Gymnocarpium, Matteuccia, Onoclea, Polystichum, Woodsia*

Dryopteris, Dryopteridaceae, fern

Dryopteris arguta (Kaulfuss) Watt, Coastal Woodfern
- **Drug**—**Costanoan** *Dermatological Aid* Infusion of fronds used as a hair wash. (21:5) **Mewuk** *Antiemetic* Decoction of roots taken for vomiting. *Antihemorrhagic* Decoction of roots taken for spitting blood and other internal bleeding. (as *D. rigida arguta* 117:366)
- **Food**—**Costanoan** *Unspecified* Rhizomes gathered in spring and eaten. (21:247) **Thompson** *Unspecified* Rootstocks used for food. (187:88) **Yurok** *Cooking Agent* Leaves used to clean meats and to lay over meat to keep the flies off. (6:28)
- **Other**—**Karok** *Cooking Tools* Leaves used to clean eels. **Yurok** *Containers* Leaves used to clean meats and to lay over meat to keep the flies off. *Designs* Spores used to make designs on hands. (6:28)

Dryopteris campyloptera Clarkson, Mountain Woodfern
- **Drug**—**Eskimo, Western** *Analgesic* and *Gastrointestinal Aid* Compound decoction of leaves taken for stomachache and intestinal discomfort. (as *D. austriaca* 108:5) **Hesquiat** *Cancer Treatment* Young shoots used cancer of the womb. (as *D. austriaca* 185:29)
- **Food**—**Eskimo, Inuktitut** *Ice Cream* Boiled roots added to "Eskimo ice cream." (as *D. dilatata* ssp. *americana* 202:193) **Kwakiutl, Southern** *Unspecified* Rhizomes cooked in steaming pits and used for food. (as *D. austriaca* 183:264) **Salish, Coast** *Unspecified* Rhizomes used for food. (as *D. austriaca* 182:69)
- **Other**—**Kwakiutl, Southern** *Fuel* Thin, wiry roots used as the burning material in a "slow match." (as *D. austriaca* 183:264) **Nitinaht** *Cooking Tools* Fronds placed in layers below and above food in steaming pits. (as *D. austriaca* 186:62)

Dryopteris carthusiana (Vill.) H. P. Fuchs, Spinulose Woodfern
- **Drug**—**Bella Coola** *Antidote* Root eaten as an antidote for poison from eating shellfish in early summer. (as *Aspidium spinulosum* 150:48)
- **Food**—**Alaska Native** *Unspecified* Old leaf stalks on the underground stem roasted, peeled, and the inner portion eaten. *Vegetable* Young, curled fronds boiled or steamed and eaten like asparagus with butter, margarine, or cream sauce. (as *D. spinulosa* 85:29)

Dryopteris cristata (L.) Gray, Crested Woodfern
- **Drug**—**Ojibwa** *Gastrointestinal Aid* Infusion of root used for stomach trouble. (as *Aspidium cristatum* 153:381)

Dryopteris expansa (K. Presl) Fraser-Jenkins & Jermy, Spreading Woodfern

- *Drug*—**Klallam** *Dermatological Aid* Poultice of pounded roots applied to cuts. **Snohomish** *Dermatological Aid* Infusion of leaves used as a hair wash. (as *D. dilatata* 79:14)
- *Food*—**Clallam** *Unspecified* Rhizomes used for food. (as *D. dilatata* 57:194) **Cowlitz** *Unspecified* Rhizomes pit baked overnight and the insides used for food. (as *D. dilatata* 79:14) **Eskimo, Alaska** *Soup* Fiddleheads, with the chaffy coverings removed, added to soups. *Unspecified* Fiddleheads, with the chaffy coverings removed, boiled and eaten with seal oil and dried fish. (as *D. dilatata* 1:34) **Thompson** *Unspecified* Rootstocks used for food. (as *D. assimilis* 187:88)

Dryopteris filix-mas (L.) Schott, Male Fern
- *Drug*—**Bella Coola** *Antidote* Rhizomes eaten raw to neutralize plant and shellfish poisoning. (184:197) **Cherokee** *Anthelmintic* Infusion of root taken for worms. (80:34)
- *Food*—**Bella Coola** *Dietary Aid* Rhizomes eaten raw for losing weight. *Unspecified* Rhizomes eaten raw or steamed. (184:197)

Dryopteris fragrans (L.) Schott, Fragrant Woodfern
- *Fiber*—**Eskimo, Inuktitut** *Mats, Rugs & Bedding* Used for bedding. (202:184)

Dryopteris marginalis (L.) Gray, Marginal Woodfern
- *Drug*—**Cherokee** *Antirheumatic (Internal)* Infusion of root used alone or in a compound for rheumatism. *Emetic* Infusion of root taken as an emetic. *Toothache Remedy* Warm infusion held in mouth for toothache. (80:34)

Dryopteris sp., Woodfern
- *Drug*—**Kitasoo** *Antidote* Rhizomes used as an antidote for food poisoning. (43:312)
- *Food*—**Haisla & Hanaksiala** *Forage* Rootstocks eaten by mountain goats. **Hanaksiala** *Unspecified* Rhizomes, rootstocks, and stipe bases steamed and eaten. (43:149) **Kitasoo** *Vegetable* Rhizomes and stipe bases used for food. (43:312) **Oweekeno** *Unspecified* Roots used for food. (43:53) **Thompson** *Unspecified* Rootstocks used for food. (187:88)
- *Fiber*—**Iroquois** *Mats, Rugs & Bedding* Made into pillows and used by children under their lower backs to prevent bed-wetting. (142:82)
- *Other*—**Haisla & Hanaksiala** *Containers* Fronds used to pack freshly caught salmon to prevent them from drying out. (43:149) **Oweekeno** *Ceremonial Items* Roots used as a shamanistic device in the tsaika ritual. (43:53)

Dudleya, Crassulaceae
Dudleya lanceolata (Nutt.) Britt. & Rose, Lanceleaf Liveforever
- *Food*—**Neeshenam** *Vegetable* Leaves eaten raw as greens. (as *Echeveris lanceolata* 129:377)

Dudleya pulverulenta (Nutt.) Britt. & Rose, Chalk Liveforever
- *Drug*—**Diegueño** *Dermatological Aid* Fleshy leaves used to remove corns and calluses. Three descriptions of its use, essentially the same but differing in detail, were given by the three consultants: (1) Prick the leaf all over with a pin or needle, put it on the stove, and bake it on one side, then the other. Place the leaf over the corn or callus and leave it there to remove the growth. (2) Heat the leaf over the fire, peel the skin off one side, place the leaf over the corn or callus—peeled side down—and leave it there to remove the corn or callus. (3) Cook the leaf over a flame, peel one side, and prick the peeled side with a needle and bind the leaf over the corn or callus to remove it. *Respiratory Aid* Decoction of roots taken for asthma. (84:19)
- *Food*—**Diegueño** *Unspecified* Leaves chewed, with or without salt, by children. (84:19)

Dudleya sp., Pygmy Weed
- *Food*—**Cahuilla** *Unspecified* Leaves and flowering stems eaten raw. (15:67)

Dugaldia, Asteraceae
Dugaldia hoopesii (Gray) Rydb., Owlsclaws
- *Drug*—**Great Basin Indian** *Analgesic* Snuff of crushed blossoms and string plant leaves inhaled for headaches. *Respiratory Aid* Snuff of crushed blossoms inhaled for hay fever. (as *Helenium hoopesii* 121:50) **Navajo** *Antiemetic* Plant used to inhibit vomiting. (as *Helenium hoopesii* 55:87)
- *Food*—**Navajo** *Candy* Roots used as a chewing gum. (as *Helenium hoopesii* 55:87)
- *Dye*—**Navajo** *Yellow* Crushed flowers boiled with juniper ash and used as a yellow dye. (as *Helenium hoopesii* 55:87)

Dyssodia, Asteraceae
Dyssodia papposa (Vent.) A. S. Hitchc., Fetid Marigold
- *Drug*—**Dakota** *Veterinary Aid* Plant given to horses for coughs. (as *Boebera papposa* 69:369) Compound decoction of plant used for horses with coughs. (as *Boebera papposa* 70:132, 133) **Keres, Western** *Febrifuge* Infusion of fresh or dried plants taken or used as a rub for fever. *Other* Plant smoked for epileptic fits. (as *Boebera papposa* 171:33) **Lakota** *Analgesic* Plant breathed in for headaches. *Antihemorrhagic* Decoction of plant and gumweed blossoms taken for the spitting of blood. *Reproductive Aid* Pulverized leaves used for breathing difficulties. (139:37) **Navajo, Ramah** *Dermatological Aid* Poultice of chewed leaves applied to ant bites. *Gastrointestinal Aid* Cold infusion of plant taken after swallowing a red ant. (191:50) **Omaha** *Analgesic* Leaves stuffed up nostrils to cause nosebleed for headache. (as *Boebera papposa* 70:132, 133)
- *Food*—**Apache, Chiricahua & Mescalero** *Bread & Cake* Seeds winnowed, dried, stored, ground into flour, and used to make bread. *Unspecified* Seeds roasted without grinding and combined with other foods. (33:48) *Vegetable* Tops cooked alone or with meat and used as greens. (33:47) **Dakota** *Forage* Plant considered a choice prairie dog food. (as *Boebera papposa* 69:369). Plant eaten by prairie dogs. (as *Boebera papposa* 70:132)

Ebenaceae, see *Diospyros*

Echeandia, Liliaceae
Echeandia flavescens (J. A. & J. H. Schultes) Cruden, Torrey's Craglily
- *Drug*—**Navajo, Ramah** *Gynecological Aid* Cold infusion of root taken to ease delivery of placenta. *Veterinary Aid* Cold infusion of root used as a lotion on sheep's swollen leg. Cold simple or compound

infusion given to livestock as an aphrodisiac. (as *Anthericum torreyi* 191:20)

Echinacea, Asteraceae

Echinacea angustifolia DC., Blacksamson Echinacea

- **Drug—Blackfoot** *Toothache Remedy* Roots chewed to cause mouth numbness for toothaches. (97:56) **Cheyenne** *Analgesic* Infusion of powdered leaves and roots used as a wash for sore and painful necks. *Oral Aid* Infusion of powdered leaves and roots taken or root chewed for sore mouth or gums. Root used to stimulate the flow of saliva. *Orthopedic Aid* Infusion of powdered leaves and roots used as a wash for sore and painful necks. (76:188) *Throat Aid* Infusion of leaves and roots taken for sore throat. (75:188) Infusion of powdered leaves and roots taken or root chewed for sore throat. (76:188) *Toothache Remedy* Root juice or infusion of powdered leaves and roots used for toothaches. (76:188) **Dakota** *Analgesic* Juice used as wash for pain from burns. Plant used in smoke treatment for headache. *Antidote* Plant used as an antidote for many poisonous conditions. *Burn Dressing* Juice used as wash for pain from burns. Juice used by jugglers as wash for arms, to protect against boiling water. *Misc. Disease Remedy* Poultice of plant applied to enlarged glands, as in mumps. *Other* Plant used in the steam bath to "render the great heat endurable." *Snakebite Remedy* Plant used for snake and other venomous bites and stings in unspecified ways. *Toothache Remedy* Plant applied to tooth for toothache. *Veterinary Aid* Plant used in smoke treatment for horses with distemper. (70:131) **Lakota** *Antirheumatic (External)* Poultice of chewed roots applied to swellings. (139:37) *Dermatological Aid* Poultice used for wounds and sores. (106:47) *Gastrointestinal Aid* Plant chewed for stomachaches. *Oral Aid* Plant chewed when thirsty. *Other* Plant chewed for overperspiring. (139:37) *Throat Aid* Roots chewed for tonsillitis. *Toothache Remedy* Roots chewed or powdered and used for toothache. (106:47) Plant chewed for toothaches. (139:37) **Montana Indian** *Oral Aid* Dried root with "smarting, acrid taste" caused a profuse flow of saliva. *Snakebite Remedy* Root used as an antidote for rattlesnake bites. (19:11) **Omaha** *Analgesic* Juice used as wash for pain from burns. Plant used in smoke treatment for headache. (70:131) *Anesthetic* Poultice of smashed roots applied as an anesthetic to arms and hands. *Antidote* Poultice of smashed roots applied to septic diseases. (68:333) Plant used as an antidote for many poisonous conditions. (70:131) *Blood Medicine* Poultice of smashed roots applied to septic diseases. (68:333) *Burn Dressing* Juice used as wash for pain from burns. Juice used by jugglers as wash for arms, to protect against boiling water. (70:131) *Dermatological Aid* Poultice of smashed roots applied to stings and septic diseases. (68:333) *Eye Medicine* Plant used for sore eyes and roots used for eye troubles. (68:335) *Misc. Disease Remedy* Poultice of plant applied to enlarged glands, as in mumps. *Other* Plant used in the steam bath to "render the great heat endurable." (70:131) *Snakebite Remedy* Poultice of smashed roots applied to snakebites. (68:333) Plant used for snake and other venomous bites and stings in unspecified ways. *Toothache Remedy* Plant applied to tooth for toothache. *Veterinary Aid* Plant used in smoke treatment for horses with distemper. **Pawnee** *Analgesic* Juice used as wash for pain from burns. Plant used in smoke treatment for headache. *Antidote* Plant used as an antidote for many poisonous conditions. *Burn Dressing* Juice used as wash for pain from burns. Juice used by jugglers as wash for arms, to protect against boiling water. *Misc. Disease Remedy* Poultice of plant applied to enlarged glands, as in mumps. *Other* Plant used in the steam bath to "render the great heat endurable." *Snakebite Remedy* Plant used for snake and other venomous bites and stings in unspecified ways. *Toothache Remedy* Plant applied to tooth for toothache. *Veterinary Aid* Plant used in smoke treatment for horses with distemper. **Ponca** *Analgesic* Juice used as wash for pain from burns. Plant used in smoke treatment for headache. *Antidote* Plant used as an antidote for many poisonous conditions. *Burn Dressing* Juice used as wash for pain from burns. Juice used by jugglers as wash for arms, to protect against boiling water. *Misc. Disease Remedy* Poultice of plant applied to enlarged glands, as in mumps. *Other* Plant used in the steam bath to "render the great heat endurable." *Snakebite Remedy* Plant used for snake and other venomous bites and stings in unspecified ways. *Toothache Remedy* Plant applied to tooth for toothache. *Veterinary Aid* Plant used in smoke treatment for horses with distemper. (70:131) **Sioux, Teton** *Analgesic* Root used for bowel pain. *Gastrointestinal Aid* Root used for bowel pain. *Throat Aid* Root used for tonsillitis. *Toothache Remedy* Root used as toothache remedy. (52:270) **Winnebago** *Analgesic* Juice used as wash for pain from burns. Plant used in smoke treatment for headache. *Antidote* Plant used as an antidote for many poisonous conditions. *Burn Dressing* Juice used as wash for pain from burns. Juice used by jugglers as wash for arms, to protect against boiling water. Plant used to make mouth insensitive to hot coals put in mouth for show. *Misc. Disease Remedy* Poultice of plant applied to enlarged glands, as in mumps. *Other* Plant used in the steam bath to "render the great heat endurable." *Snakebite Remedy* Plant used for snake and other venomous bites and stings in unspecified ways. *Toothache Remedy* Plant applied to tooth for toothache. *Veterinary Aid* Plant used in smoke treatment for horses with distemper. (70:131)

Echinacea angustifolia DC. **var.** *angustifolia*, Blacksamson Echinacea

- **Drug—Kiowa** *Cough Medicine* Ground root chewed for coughs. *Throat Aid* Ground root chewed for sore throats. (as *Brauneria angustifolia* 192:57) **Meskwaki** *Analgesic* Root used for stomach cramps and fits. *Anticonvulsive* Compound containing root used for stomach cramps and fits. *Gastrointestinal Aid* Root used for stomach cramps and fits. (as *Brauneria angustifolia* 152:212)

Echinacea pallida (Nutt.) Nutt., Pale Purple Coneflower

- **Drug—Cheyenne** *Antirheumatic (Internal)* Decoction of roots and leaves taken for rheumatism and arthritis. (82:38) Decoction of roots taken for rheumatism and arthritis. (83:20) *Burn Dressing* Decoction of roots used for burns. (82:38) Decoction of roots used as a wash for burns. *Cold Remedy* Root chewed for colds. (83:20) *Dermatological Aid* Roots mixed with puffball mushroom spores and skunk oil and used for boils. (82:38) Roots used for boils. *Dietary Aid* Root chewed to increase the flow of saliva and prevent thirst. *Febrifuge* Decoction of roots used as a wash for fevers. (83:20) *Misc. Disease Remedy* Decoction of roots and leaves taken for mumps and measles. (82:38) Decoction of roots taken for smallpox, mumps, and measles. (83:20) *Oral Aid* Infusion of powdered roots and leaves taken for sore mouth and gums. *Throat Aid* Infusion of powdered roots and leaves taken

for sore throat. *Toothache Remedy* Roots chewed for toothache. (82: 38) Root chewed for toothaches, especially cavities. (83:20) **Crow** *Cold Remedy* Roots chewed and used for colds. *Gastrointestinal Aid* Infusion of roots taken for colic. *Toothache Remedy* Roots chewed for toothache. (82:38) **Dakota** *Anthelmintic* Decoction of roots taken as a vermifuge. (as *Brauneria pallida* 69:361) *Dermatological Aid* Poultice of roots applied to inflammation to relieve the burning sensation. (as *Brauneria pallida* 69:368) *Eye Medicine* Decoction of roots used for sore eyes. (as *Brauneria pallida* 69:367) *Snakebite Remedy* Plant used as an antidote for snakebites. (as *Brauneria pallida* 69: 368) **Sioux** *Analgesic* Plant used in the smoke treatment for headache. *Antidote* Used as an antidote for rattlesnake and other venomous bites, stings, and poisonous conditions. *Burn Dressing* Decoction of roots used for burns. *Toothache Remedy* Roots chewed for toothache. *Veterinary Aid* Plant used in the smoke treatment for horses with distemper. (82:38)
- *Other*—**Sioux** *Protection* Juice used by "jugglers" for protection in handling hot meat. (82:38)

Echinacea purpurea (L.) Moench, Eastern Purple Coneflower
- *Drug*—**Choctaw** *Cough Medicine* Root chewed, saliva swallowed, and tincture of root used for cough. *Gastrointestinal Aid* Root chewed, saliva swallowed, and tincture of root used for dyspepsia. (28:288) **Delaware** *Venereal Aid* Roots combined with staghorn sumac roots and used for venereal disease. (as *Brauneria purpurea* 176:33) Infusion of roots used for advanced cases of venereal disease. (as *Brauneria purpurea* 176:35) **Delaware, Oklahoma** *Venereal Aid* Simple or compound infusion of root, highly effective, taken for gonorrhea. (as *Brauneria purpurea* 175:29, 74)

Echinacea sp., Purple Coneflower
- *Drug*—**Comanche** *Throat Aid* Decoction of root taken for sore throat. *Toothache Remedy* Root held against tooth for toothaches. (29:521)

Echinocactus, Cactaceae
Echinocactus polycephalus Engelm. & Bigelow, Cottontop Cactus
- *Food*—**Cahuilla** *Staple* Berries and stems were an important and dependable food source. (15:49)
- *Other*—**Kawaiisu** *Tools* Spines used as awls in the making of coiled basketry. (206:27) **Panamint** *Tools* Thorns formerly used as awls in basket making. (105:83)

Echinocactus sp., Water Barrel Cactus
- *Drug*—**Mahuna** *Oral Aid* Plant used for the prevention of salivary gland swelling. (140:47)
- *Food*—**Havasupai** *Dried Food* Seeds dried for future use. *Porridge* Fresh or dried seeds parched, ground, and made into mush. (197:232)
- *Other*—**Havasupai** *Containers* Pieces of plant used as tray for baked mescal. *Cooking Tools* Used as improvised cooking vessels particularly on hunting expeditions. *Jewelry* Red spines fire warmed and bent into finger rings. (197:232) **Mahuna** *Hunting & Fishing Item* Thorns used as fishhooks for deep-water fishing. (140:47)

Echinocactus williamsii Lem., Mescal
- *Drug*—**Omaha** *Other* Plant used for alcohol addiction. (68:318)

Echinocereus, Cactaceae
Echinocereus coccineus Engelm., Scarlet Hedgehog Cactus
- *Drug*—**Navajo** *Heart Medicine* Plant used as a heart stimulant. *Poison* Plant considered poisonous. (55:64)
- *Food*—**Apache, Chiricahua & Mescalero** *Fruit* Raw fruit used for food. (33:41) **Navajo, Ramah** *Fruit* Fruit used for food. (191:37)
- *Other*—**Keres, Western** *Unspecified* Plant known and named but no use was specified. (171:42)

Echinocereus engelmannii (Parry ex Engelm.) Lem., Saints Cactus
- *Food*—**Pima** *Fruit* Ripe fruits freed from spines and eaten raw. (47:57) **Pima, Gila River** *Snack Food* Fruit eaten primarily by children as a snack food. (133:5) **Yavapai** *Fruit* Raw fruit used for food. *Unspecified* Boiled buds used for food. (as *Cereus engelmanni* 65:256)

Echinocereus engelmannii var. *chrysocentrus* (Engelm. & Bigelow) Rumpl., Saints Cactus
- *Food*—**Apache, Chiricahua & Mescalero** *Fruit* Raw fruit used for food. (33:41)

Echinocereus fendleri (Engelm.) F. Seitz, Pinkflower Hedgehog Cactus
- *Food*—**Apache, Chiricahua & Mescalero** *Fruit* Raw fruit used for food. (33:41) **Cochiti** *Unspecified* Stems pit roasted and eaten. (32:26) **Hopi** *Sweetener* Fruits dried and used as a source of sweetening. (200:85)

Echinocereus polyacanthus Engelm., Mojave Mound Cactus
- *Food*—**Apache, Chiricahua & Mescalero** *Fruit* Raw fruit used for food. (33:41)

Echinocereus rigidissimus (Engelm.) Haage f., Rainbow Hedgehog Cactus
- *Food*—**Apache, Chiricahua & Mescalero** *Fruit* Raw fruit used for food. (33:41)

Echinocereus sp., Coccinia
- *Drug*—**Isleta** *Antirheumatic (External)* Poultice of roasted leaves used for swellings. (100:28)
- *Food*—**Apache, Mescalero** *Fruit* Fruits eaten for food. (14:45) **Navajo** *Fruit* Fruits eaten for food. (55:64)

Echinocereus triglochidiatus Engelm., Kingcup Cactus
- *Food*—**Isleta** *Beverage* Water extracted from pulp in emergencies. (100:27) *Bread & Cake* Pulp macerated and cooked with sugar to make cakes. (32:26) Pulp baked with sugar and used to make cakes. (100:27) *Candy* Pulp baked with sugar to make candy. (32:26) Pulp baked with sugar and used to make candy. (100:27) *Fruit* Fruits, with spines removed by burning, eaten fresh. (32:26) Fruit eaten fresh. (100:27) *Preserves* Fruits, with spines removed by burning, made into a conserve. (32:26) Fruit eaten as conserves. (100:27) *Sauce & Relish* Pulp baked with sugar to make sweet pickles. *Unspecified* Pulp sliced and baked like squash. (32:26) *Vegetable* Sliced pulp baked like squash and used for food. Pulp baked with sugar and used to make a sweet pickle. (100:27)
- *Other*—**Isleta** *Lighting* Dried pulp used as candles. (100:27)

Echinocereus triglochidiatus Engelm. **var. *triglochidiatus*,** Kingcup Cactus
- ***Food*—Cochiti** *Unspecified* Stems pit roasted and eaten. **Isleta** *Bread & Cake* Pulp macerated and cooked with sugar to make cakes. *Candy* Pulp baked with sugar to make candy. *Fruit* Fruits, with spines removed by burning, eaten fresh. *Preserves* Fruits, with spines removed by burning, made into a conserve. *Sauce & Relish* Pulp baked with sugar to make sweet pickles. *Unspecified* Pulp sliced and baked like squash. (as *E. gonocanthus* 32:26) **Keres, Western** *Fruit* Tunas used for food. (as *E. gonacanthus* 171:42)

***Echinochloa*,** Poaceae

Echinochloa colona (L.) Link, Jungle Rice
- ***Food*—Cocopa** *Porridge* Seeds parched, ground, and the flour cooked into a mush. *Staple* Seeds parched, ground, and the flour eaten dry. (as *E. colonum* 37:187)

Echinochloa crus-galli (L.) Beauv., Barnyard Grass
- ***Food*—Cocopa** *Unspecified* Seeds used for food. *Winter Use Food* Seeds stored for later use. (as *E. crusgalli* 64:267) **Paiute** *Unspecified* Species used for food. (as *E. crusgalli* 167:243) **Tubatulabal** *Unspecified* Used extensively for food. (193:15) **Yuma** *Porridge* Seeds pounded, winnowed, ground, made into mush, and used to cook with fish. *Staple* Seeds pounded, winnowed, parched, and ground into a meal. (as *E. crusgalli* 37:187) *Unspecified* Wild seeds eaten for food. (as *E. crusgalli* 37:173)

Echinochloa crus-pavonis* var. *macera (Wieg.) Gould, Gulf Cockspur Grass
- ***Drug*—Navajo, Ramah** *Ceremonial Medicine* and *Emetic* Plant used as a ceremonial emetic. (as *E. crusgalli* var. *zelayensis* 191:16)
- ***Food*—Navajo, Ramah** *Fodder* Used for horse feed. (as *E. crusgalli* var. *zelayensis* 191:16)

***Echinocystis*,** Cucurbitaceae

Echinocystis brandegei Cogn.
- ***Drug*—Tubatulabal** *Dermatological Aid* Burned, ripe seeds rubbed on pimples and newborn baby's navel. *Pediatric Aid* Burned, ripe seeds rubbed on newborn baby's navel. (193:59)

Echinocystis lobata (Michx.) Torr. & Gray, Wild Cucumber
- ***Drug*—Cherokee** *Abortifacient* Taken for "obstructed menses." *Antirheumatic (Internal)* Infusion taken for rheumatism. *Febrifuge* Taken for chills and fevers. *Kidney Aid* Infusion taken for kidneys. (80:40) **Menominee** *Analgesic* Poultice of pulverized root used for headache. *Love Medicine* Root used in love potions. *Panacea* Plant considered to be "the greatest of all medicines" and always useful. *Tonic* Decoction of root taken as a bitter tonic. (151:33, 34) **Meskwaki** *Analgesic* Poultice of pounded root applied for headache. *Panacea* Compound containing root used as a universal remedy for all sicknesses. (152:220) **Ojibwa** *Gastrointestinal Aid* Infusion of root used as a bitter medicine for stomach troubles. *Tonic* Infusion of root used as a tonic. (153:367, 368)
- ***Other*—Oglala** *Jewelry* Seeds used for beads. (as *Micrampelis lobata* 70:129)

***Echinocystis* sp.,** Wild Cucumber
- ***Drug*—California Indian** *Kidney Aid* Roasted seeds eaten for kidney troubles. (118:41)

***Echium*,** Boraginaceae

Echium vulgare L., Common Vipersbugloss
- ***Drug*—Cherokee** *Urinary Aid* Compound taken for milky urine. (80:60) **Iroquois** *Gynecological Aid* Compound infusion of roots taken for the evacuation of the placenta. (87:421) **Mohegan** *Kidney Aid* Infusion of root taken for the kidneys. (174:266) Leaves or root, "a rare plant," used for kidney disorders. (176:72, 130)
- ***Other*—Cherokee** *Jewelry* Seeds used to make beads. (80:60)

***Egregia*,** Alariaceae, alga

Egregia menziesii (Turner) Areschoug, Boa Kelp
- ***Food*—Bella Coola** *Dried Food* Whole plants with attached herring spawn dried and used as a winter food. *Unspecified* Whole plants with attached herring spawn eaten fresh. (184:195) **Haisla & Hanaksiala** *Dried Food* Plant used to collect herring roe, dried, and eaten with the roe. (43:125) **Kitasoo** *Unspecified* Plant eaten with herring roe. (43:302) **Oweekeno** *Dried Food* Plant and herring eggs dried for future use. *Unspecified* Plant eaten with herring roe. *Winter Use Food* Plant and herring eggs salted and stored for future use. (43:44)
- ***Other*—Bella Coola** *Hunting & Fishing Item* Used to collect herring spawn. (184:195) **Hesquiat** *Fertilizer* Considered a good fertilizer for potatoes. (185:23)

Elaeagnaceae, see *Elaeagnus, Shepherdia*

***Elaeagnus*,** Elaeagnaceae

Elaeagnus commutata Bernh. ex Rydb., Silverberry
- ***Drug*—Blackfoot** *Dermatological Aid* and *Pediatric Aid* Strong decoction of bark and grease used as a salve for children with frostbite. (86:85) **Thompson** *Venereal Aid* Decoction of roots and sumac roots taken for syphilis. This medicine was considered very poisonous and had to be taken with fish head soup to counteract the poison. One informant who was treated with this medicine recovered from syphilis but, afterwards, could never have children. (187:207)
- ***Food*—Alaska Native** *Fruit* Berries fried in moose fat and eaten. (85:144) **Blackfoot** *Candy* Peeled berries mixed with grease, stored in a cool place, and eaten as a confection. *Fruit* Peeled berries used for food. *Soup* Peeled berries used to make soups and broths. (86:102) **Cree** *Beverage* Berries used to make wine. *Fruit* Berries used for food. (as *Eleagnus argentea* 16:485) **Montana Indian** *Fruit* Fruit used for food. (as *Elaeagnus argentea* 19:11) **Okanagan-Colville** *Fruit* Berries used for food. (188:99) **Okanagon** *Staple* Seeds used as a principal food. (as *E. argentea* 178:239) **Paiute** *Unspecified* Seeds used for food. (as *Eleagnus asgentae* 167:244) **Tanana, Upper** *Fruit* Berries mixed with grease, cooked, and eaten. Berries eaten raw. *Soup* Berries used to make soup. (102:13)
- ***Fiber*—Blackfoot** *Cordage* Bark used to make strong rope. (as *Elaeagnus argentea* 114:277) **Cree** *Cordage* Bark used to cordage. (as *Elaeagnus argentea* 16:485) **Okanagan-Colville** *Clothing* Inner bark twisted to make clothing. *Cordage* Inner bark twisted to make ropes. *Mats, Rugs & Bedding* Inner bark twisted to make blankets. (188:99) **Thompson** *Clothing* Inner bark stripped off, shredded, and

the fiber used to make clothing. (as *Elaeagnus argentea* 164:496) Bark made into two-ply twine and used for twining bags, capes, skirts, and other clothing.[29] *Cordage* Bark made into two-ply twine and used for twining mats, bags, capes, skirts, and other clothing.[29] (187:207) *Mats, Rugs & Bedding* Inner bark stripped off, shredded, and the fiber used to make mats. (as *E. argentea* 164:496) Bark made into two-ply twine and used for twining mats.[29] (187:207)

- **Other**—**Arapaho** *Jewelry* Berries used to make necklaces. Boiled seeds strung on sinew and used as necklaces. (121:49) **Blackfoot** *Jewelry* Seeds cleaned, boiled, threaded, dried, greased, and made into necklaces. (86:112) Seeds used to make necklaces. (as *Eleagnus argentea* 118:56) *Soap* Berries used to make soap. (as *Elaeagnus argentea* 114:277) *Toys & Games* Braided bark used to whip stone tops spun by children on the ice during the winter. (86:112) **Cree** *Jewelry* Seeds used to make beads. (as *Eleagnus argentea* 16:485) **Okanagan-Colville** *Containers* Inner bark twisted to make sacks. Bark used to imbricate clematis bags. *Cooking Tools* Inner bark twisted to make soapberry beaters. (188:99) **Tanana, Upper** *Jewelry* Seeds sometimes used as beads. (102:13) **Thompson** *Ceremonial Items* Bark made into a headband and worn by the man chosen to sing when newborn twins first cried. (as *Elaeagnus argentea* 164:508) *Containers* Bark made into two-ply twine and used for twining mats, bags, capes, skirts, and other clothing.[29] *Cooking Tools* Bark twine used in drying roots and bunches of tied bark used as soapberry whippers. The bark was peeled off in as long strips as possible in the spring or fall when it was "kind of dry" and split with a knife (originally of stone). The grayish outer bark was removed and the inner bark scraped, cleaned, and cut into desired widths. At this stage, the bark could be dried for future use. The long, even strands of the fresh or dried bark, after it was soaked, could be spun on the bare leg into a strong, two-ply twine used for many different purposes. The bark wine was used in twining mats, bags, capes, skirts, and other clothing and also to thread bitterroots, avalanche lily corms, and other roots for drying. *Decorations* Seeds, sometimes painted gold, used to decorate Christmas cards. (187:207) *Jewelry* Seeds strung and worn as necklaces. (as *E. argentea* 164:496) Cleaned, polished seeds, with holes drilled in them, strung and used as necklaces. (187:207)

Eleocharis, Cyperaceae

Eleocharis equisetoides (Ell.) Torr., Jointed Spikesedge
- **Other**—**Seminole** *Jewelry* Plant used to make beads. (169:498)

Eleocharis geniculata (L.) Roemer & J. A. Schultes, Canada Spikesedge
- **Drug**—**Seminole** *Analgesic* Infusion of leaves taken as an emetic for rainbow sickness: fever, stiff neck, and backache. (as *E. caribaea* 169:210) Infusion of leaves taken as an emetic for thunder sickness: fever, dizziness, headache, and diarrhea. *Antidiarrheal* Infusion of leaves taken as an emetic for thunder sickness: fever, dizziness, headache, and diarrhea. (as *E. caribaea* 169:213) *Antirheumatic (External)* Plant used for fire sickness: fever and body aches. (as *E. caribaea* 169:203) *Emetic* Infusion of leaves taken as an emetic for rainbow sickness: fever, stiff neck, and backache. (as *E. caribaea* 169:210) Infusion of leaves taken as an emetic for thunder sickness: fever, dizziness, headache, and diarrhea. (as *E. caribaea* 169:213) *Febrifuge* Plant used for fire sickness: fever and body aches. (as *E. caribaea* 169:203) Infusion of leaves taken as an emetic for rainbow sickness: fever, stiff neck, and backache. (as *E. caribaea* 169:210) Infusion of leaves taken as an emetic for thunder sickness: fever, dizziness, headache, and diarrhea. (as *E. caribaea* 169:213) *Unspecified* Plant used for medicinal purposes. (as *E. caribaea* 169:162) *Urinary Aid* Decoction of plant taken for urine retention. (as *E. caribaea* 169:274) *Vertigo Medicine* Infusion of leaves taken as an emetic for thunder sickness: fever, dizziness, headache, and diarrhea. (as *E. caribaea* 169:213)

Eleocharis montevidensis Kunth, Sand Spikerush
- **Drug**—**Navajo, Ramah** *Ceremonial Medicine* and *Emetic* Plant used as a ceremonial emetic. (191:19)

Eleocharis palustris (L.) Roemer & J. A. Schultes, Common Spikerush
- **Food**—**Paiute, Northern** *Unspecified* Sap eaten. (59:49)
- **Fiber**—**Okanagan-Colville** *Mats, Rugs & Bedding* Used as bedding, for pillows, and for sitting on in the sweat house. (188:36)

Eleocharis rostellata (Torr.) Torr., Beaked Spikerush
- **Drug**—**Navajo, Ramah** *Ceremonial Medicine* and *Emetic* Plant used as a ceremonial emetic. (191:19)

Eleocharis sp., Spikerush
- **Food**—**Paiute** *Unspecified* Bulbs used for food. (167:245)
- **Fiber**—**Cheyenne** *Basketry* Rushes made into large baskets and used to hold small utensils or carry loads on the back. (76:170) *Other* Used in weaving. (83:46)
- **Other**—**Cheyenne** *Cooking Tools* Rushes made into small baskets and used as dishes to serve food. *Protection* Rushes fastened to the head of a baby's board to form a shade for the face. *Toys & Games* Rushes made into small baskets and used as children's playthings. (76:170)

Elliottia, Ericaceae

Elliottia pyroliflorus (Bong.) S. W. Brim & P. F. Stevens, Copperbush
- **Drug**—**Kitasoo** *Dietary Aid* Decoction of plant used as an appetite stimulant. (as *Cladothamnus pyroliflorus* 43:332)

Elodea, Hydrocharitaceae

Elodea canadensis Michx., Canadian Waterweed
- **Drug**—**Iroquois** *Emetic* Infusion of plant taken as a strong emetic. (as *Anacharis canadensis* 141:66)

Elymus, Poaceae

Elymus canadensis L., Canada Wildrye
- **Drug**—**Iroquois** *Ceremonial Medicine* Decoction of plant with other plants used as medicine to soak corn seeds before planting. (196:19) *Kidney Aid* Compound decoction of roots taken for the kidneys. *Other* Compound decoction of plants taken for stricture. (87:274)
- **Food**—**Gosiute** *Unspecified* Seeds formerly used for food. (39:368) **Ute** *Unspecified* Seeds formerly gathered for food. (38:34)
- **Other**—**Iroquois** *Fertilizer* Decoction of roots used as a soak for "corn medicine." (87:274)

Elymus canadensis L. **var. *canadensis***, Canada Wildrye
- **Food**—**Kiowa** *Fodder* Foliage and lemmas used as a palatable fodder for livestock. (as *E. robustus* 192:15)

Elymus elymoides (Raf.) Swezey, Bottlebrush Squirreltail
- **Food**—**Navajo, Ramah** *Fodder* Young plants used for sheep and horse feed, mature plants made animals' mouths sore. (as *Sitanion hystrix* 191:17)

Elymus glaucus Buckl., Blue Wildrye
- **Drug**—**Karok** *Other* Plant used as a medicine to settle quarrels between families or individuals. (148:380)
- **Food**—**Karok** *Porridge* Seeds parched, pounded into a flour, and mixed with water into a paste. (148:380)
- **Other**—**Keres, Western** *Unspecified* Plant known and named but no use was specified. (171:42)

Elymus glaucus Buckl. **ssp. *glaucus***, Blue Wildrye
- **Food**—**Costanoan** *Staple* Seeds used in pinole. (21:255)

Elymus hystrix L. **var. *hystrix***, Eastern Bottlebrush Grass
- **Drug**—**Iroquois** *Ceremonial Medicine* Decoction of leaves and reed grass rootstocks used as medicine to soak corn seeds before planting. (as *Hystrix patula* 196:18)

Elymus multisetus M. E. Jones, Big Squirreltail
- **Food**—**Kawaiisu** *Porridge* Seeds parched, pounded, and cooked into a thin mush. (as *Sitanion jubatum* 206:64)

Elymus sibiricus L., Siberian Wildrye
- **Food**—**Gosiute** *Unspecified* Seeds formerly used for food. (39:368)

Elymus sp., Rye Grass
- **Drug**—**Comanche** *Eye Medicine* Single looped blades used as eye scrapers for removing cataracts. (99:5)
- **Food**—**Pomo, Kashaya** *Staple* Grain grounded into a fine powder and used in pinole. (72:101)
- **Fiber**—**Aleut** *Other* Used for weaving. (7:29)

Elymus trachycaulus (Link) Gould ex Shinners **ssp. *trachycaulus***, Slender Wheatgrass
- **Drug**—**Navajo, Ramah** *Veterinary Aid* Leaves eaten by dogs, causing emesis. (as *Agropyron trachycaulum* 191:15)
- **Food**—**Navajo, Ramah** *Fodder* Used for horse feed and sometimes harvested for winter use. (as *Agropyron trachycaulum* 191:15) **Thompson** *Fodder* Cut and fed to horses as hay. (as *Agropyron tenerum* 164:515)

Elytrigia, Poaceae

Elytrigia repens (L.) Desv. ex B. D. Jackson **var. *repens***, Quack Grass
- **Drug**—**Cherokee** *Orthopedic Aid* Decoction used to wash swollen legs and infusion taken for "gravel." *Urinary Aid* Infusion taken for "gravel" and "[in]continence and bedwetting." (as *Agropyron repens* 80:31) **Iroquois** *Anthelmintic* Used as a worm remedy. (as *Agropyron repens* 87:274) *Urinary Aid* Infusion of rhizomes and stems from another plant used for thick urine. (as *Agropyron repens* 141:67)
- **Food**—**Apache, White Mountain** *Fodder* Plant used for hay. *Unspecified* Seeds used for food. (as *Agropyron repens* 136:155) **Gosiute** *Unspecified* Seeds formerly eaten. (as *Agropyron repens* 39:360)
- **Other**—**Okanagan-Colville** *Containers* Used under and over the food in pit cooking. (as *Agropyron repens* 188:52)

Empetraceae, see *Empetrum*

Empetrum, Empetraceae

Empetrum nigrum L., Black Crowberry
- **Drug**—**Bella Coola** *Cathartic* Decoction of green leaves taken as a purgative. (150:60) **Cree, Woodlands** *Diuretic* and *Pediatric Aid* Leafy branches used, especially for children with a fever, as a diuretic.[30] (109:38) **Tanana, Upper** *Antidiarrheal* Decoction or infusion of stems taken for diarrhea. Berries cooked and eaten for diarrhea. *Cold Remedy* Decoction of leaves, stems, Hudson bay tea, and young spruce tree tip used for colds. *Eye Medicine* Cooled decoction of roots used as an eyewash to remove a growth. *Kidney Aid* Decoction of leaves and stems taken for kidney troubles. (102:12)
- **Food**—**Alaska Native** *Pie & Pudding* Berries mixed with other berries and used to make pie. *Preserves* Berries mixed with other berries and used to make jelly. (as *Empertrum nigrum* 85:79) **Cree, Woodlands** *Fruit* Fruit eaten in the fall. (109:38) **Eskimo, Alaska** *Fruit* Fruit used for food. (4:715) **Eskimo, Arctic** *Frozen Food* Berries stored frozen and eaten with seal blubber or oil. *Fruit* Berries eaten fresh. (128:21) **Eskimo, Inupiat** *Dessert* Berries cooked with sourdock and eaten as a dessert. *Fruit* Berries eaten with oil and sugar or mixed with other berries, sourdock, ice cream, or fish livers. This was a favorite food made just the same way and still just as good as it had been for centuries. It was one food one could eat all one wanted, for any meal, day after day, and still like it. It was good fresh or leftover and as a main meal, side dish, or dessert. The sweet acidic berries and fat fish livers balanced each other and also were exceptionally nutritious. The only limiting factor in how much one ate was picking enough berries and catching enough fish to have the ingredients. The recipe was as follows: pick clean, ripe blackberries, at least 1 gallon. Save the livers from four large, freshly caught fall trout. Pinch out the bile sack, without breaking it, and discard. Soak the livers in a bowl of cold water while you finish caring for the fish. Rinse the livers, throw out the soak water. Simmer the livers in clean water until just done, 5 to 10 minutes. Lift the livers out to drain and cool. They could be stored a few days this way in the refrigerator. Skim the oil off the broth and save to add. Mash the livers thoroughly in a bowl, every tiny lump, using your hand or a fork. Mix in a little water as you mash to make a smooth paste, like thick hot cake batter. Stir in the whole blackberries until all the paste was taken up coating the berries. No salt or sugar was ever used or needed. Newcomers would prefer trout livers, which were mild, but after a while began to crave the stronger taste and more satisfying oiliness of tom cod livers. *Pie & Pudding* Berries, cornstarch, water, and butter used to make pie. Berries cooked with blueberries and used to make pie and ice cream. *Winter Use Food* Berries mixed with salmonberries and stored for winter use. Berries stored in seal oil, a sealskin poke, or plastic bag for future use. (98:92) **Koyukon** *Beverage* Berries eaten by hunters to quench their thirst in the waterless high country. (119:55) **Ojibwa** *Fruit* Fruit used for food. (135:243) **Tanana, Upper** *Frozen Food* Berries frozen for future use. (102:12) *Fruit* Fruit used for food. (77:28) Berries eaten raw,

plain, or mixed raw with sugar, grease, or the combination of the two. Berries fried in grease with sugar or dried fish eggs. Berries boiled with sugar and flour to thicken. *Pie & Pudding* Berries used to make pies. *Preserves* Berries used to make jam and jelly. *Winter Use Food* Berries preserved alone or in grease and stored in a birch bark basket in an underground cache. (102:12) **Tsimshian** *Fruit* Berries used for food. (43:332)

Empetrum nigrum ssp. *hermaphroditum* (Lange ex Hagerup) Böcher, Black Crowberry

- *Food*—**Eskimo, Alaska** *Fruit* Berries eaten alone. *Ice Cream* Berries added to ice cream. *Winter Use Food* Berries preserved in seal oil for use in fall and winter. (1:37)

Encelia, Asteraceae

Encelia farinosa Gray ex Torr., Goldenhills

- *Drug*—**Cahuilla** *Toothache Remedy* Decoction of blossoms, leaves, and stems held in the mouth for toothaches. (15:69) **Pima** *Analgesic* Poultice of plant applied for pain. (47:102)
- *Food*—**Papago** *Candy* Gum secretions chewed by children. (34:28) **Pima** *Candy* Resin used as a primitive chewing gum. (47:102) Amber-colored gum used for chewing gum. (95:265)
- *Other*—**Papago** *Fasteners* "Gum" used to fasten arrow points to the slit ends of arrow twigs. (34:71) *Waterproofing Agent* Gum-like secretions formerly smeared on tall, slender water bottles. (34:59) **Pima** *Fuel* Resinous branches used to make quick fires. *Paint* Resin melted and used as a varnish. (47:102)

Encelia frutescens var. *resinosa* M. E. Jones, Button Brittlebush

- *Drug*—**Navajo, Kayenta** *Dermatological Aid* Plant used for shingles. (205:47)
- *Food*—**Navajo, Kayenta** *Spice* Used as a seasoning for broth. (205:47)

Encelia virginensis var. *actonii* (Elmer) B. L. Turner, Acton's Brittlebush

- *Drug*—**Kawaiisu** *Analgesic* and *Antirheumatic (External)* Decoction of leaves and flowers used as a wash for rheumatic pains. *Veterinary Aid* Decoction of leaves used as a wash for horses with cuts and bruises. (206:27) **Tubatulabal** *Antirheumatic (External)* Compound decoction of roots used as a wash for rheumatism. (as *E. actoni* 193:59)

Enceliopsis, Asteraceae

Enceliopsis nudicaulis (Gray) A. Nels., Nakedstem Sunray

- *Drug*—**Shoshoni** *Antidiarrheal* Decoction of root taken for bloody diarrhea. *Cough Medicine* Decoction of leaves taken for coughs. *Venereal Aid* Decoction of root taken for venereal disease. (180:67, 68)

Enteromorpha, Ulvaceae, alga

Enteromorpha intestinalis (L.) Link, Tubular Green Alga

- *Food*—**Hesquiat** *Forage* Plants float upright during high tide and the brant geese like to pick at them. (185:23)

Epacridaceae, see *Styphelia*

Ephedraceae, see *Ephedra*

Ephedra, Ephedraceae, conifer

Ephedra antisyphilitica Berl. ex C. A. Mey., Clapweed

- *Drug*—**Pima** *Venereal Aid* Plant used for syphilis. (146:80) **Tewa** *Antidiarrheal* Leaves and stems chewed or decoction taken for diarrhea. (138:46)

Ephedra californica S. Wats., California Jointfir

- *Drug*—**Diegueño** *Blood Medicine* Infusion of branches taken to purify the blood. *Dietary Aid* Infusion of branches taken to improve the appetite. (84:19) *Gastrointestinal Aid* Infusion of plant taken for stomachaches caused by eating too much food or eating bad food. (88:216) *Kidney Aid* Infusion of branches taken for the kidneys. (84:19)
- *Food*—**Kawaiisu** *Beverage* Stems used to make tea. *Unspecified* Seeds formerly used for food. (206:27)
- *Other*—**Kawaiisu** *Tools* Wood provided the best charcoal for tattooing. The charcoal was mashed, a little water added, and a design made with it on the skin. (206:27)

Ephedra fasciculata A. Nels., Arizona Jointfir

- *Drug*—**Pima** *Dermatological Aid* Poultice of dried, powdered roots applied to sores. *Venereal Aid* Poultice of dried, powdered roots applied for syphilis. (47:76)
- *Food*—**Havasupai** *Beverage* Upper portions of plant boiled into tea. (197:207) **Pima** *Beverage* Ends of branches boiled and made into a beverage. **Pima, Lehi** *Beverage* Roots used as a tea. (47:76)

Ephedra nevadensis S. Wats., Nevada Jointfir

- *Drug*—**Apache, White Mountain** *Venereal Aid* Infusion of stems and leaves taken for gonorrhea or first stages of syphilis. (136:157) **Cahuilla** *Blood Medicine* Decoction of fresh or dried twigs used to purify the blood. *Other* Decoction of fresh or dried twigs used to "clear the system." (15:70) **Coahuilla** *Unspecified* Plant used for unspecified medicinal purposes. (13:73, 74) **Navajo** *Kidney Aid* Infusion of stems and leaves taken for kidney troubles. *Venereal Aid* Infusion of stems and leaves taken for venereal troubles. (136:157) **Paiute** *Adjuvant* Twigs used in medicines "to lessen disagreeable flavors." *Burn Dressing* Compound decoction of plant used as a salve for burns. *Venereal Aid* Decoction of twigs and branches taken for venereal diseases. **Shoshoni** *Dermatological Aid* Poultice of powdered twigs and branches applied to sores. *Diuretic* Decoction of twigs and branches taken to stimulate urination. *Venereal Aid* Decoction of twigs and branches taken for venereal diseases. (180:68) **Zuni** *Venereal Aid* Infusion of stems and leaves taken for venereal troubles. (136:157) Infusion of whole plant, except root, taken for syphilis. (166:49)
- *Food*—**Apache, White Mountain** *Beverage* Stems used to make tea. (136:157) **Cahuilla** *Beverage* Fresh or dried twigs boiled to make tea. (15:70) **Coahuilla** *Beverage* Used to make a pleasant and refreshing beverage. (13:73) **Havasupai** *Beverage* Upper portions of plant boiled into tea. (197:207) **Kawaiisu** *Beverage* Stems used to make tea. *Unspecified* Seeds formerly used for food. (206:27) **Papago** *Beverage* Seeds steeped and used as tea-like drinks for refreshment. (34:27) **Zuni** *Beverage* Plant without the root occasionally used to make a hot, tea-like beverage. (166:67)
- *Other*—**Kawaiisu** *Tools* Wood provided the best charcoal for tattooing. (206:27)

Ephedra sp., Indian Tea

- *Drug*—**Hualapai** *Unspecified* Infusion of green branches taken for medicinal purposes. (195:34) **Mahuna** *Analgesic* Infusion of roots

taken for painful bloating of the stomach. (140:68) *Blood Medicine* Infusion of plants taken to purify the blood. (140:21) *Gastrointestinal Aid* Infusion of roots taken for painful bloating of the stomach. (140:68) *Kidney Aid* Infusion of plants taken to flush the kidneys. (140:21) Infusion of leaves and blossoms taken for pleurisy of the kidneys. *Venereal Aid* Infusion of roots taken for gonorrhea. (140:68)
- **Food**—**Paiute** *Beverage* Dried twigs made into an aromatic tea. **Shoshoni** *Beverage* Parched, ground seeds used for coffee. (118:17)
- **Fiber**—**Havasupai** *Mats, Rugs & Bedding* Twigs used to make drying mat for pulp. (17:17)

Ephedra torreyana S. Wats., Torrey's Jointfir
- **Drug**—**Hopi** *Venereal Aid* Plant used for syphilis. (200:35, 63) **Isleta** *Dermatological Aid* Decoction of leaves and stems used to make a lotion for itching skin. (100:28) **Keres, Western** *Cough Medicine* Infusion of stems used as a cough medicine. *Diaphoretic* Stems used as an ingredient in the sweat bath. *Kidney Aid* Infusion of stems taken or stems chewed for kidney trouble. *Urinary Aid* Infusion of stems taken or stems chewed for bladder trouble. (171:42) **Navajo, Ramah** *Cough Medicine* Decoction of whole plant taken for bad cough. *Gastrointestinal Aid* Decoction of whole plant taken for stomachache. (191:14)
- **Food**—**Havasupai** *Beverage* Upper portions of plant boiled into tea. (197:207) **Navajo** *Beverage* Branches used to make tea. (55:24)

Ephedra trifurca Torr. ex S. Wats., Longleaf Jointfir
- **Drug**—**Cocopa** *Dermatological Aid* Poultice of pulverized or boiled stems and leaves applied to sores. (64:268) **Navajo** *Gastrointestinal Aid* Infusion of dried plants taken for stomach troubles. *Kidney Aid* Infusion of dried plants taken for kidney affections. *Venereal Aid* Infusion of dried plants taken for venereal disease. Wood burned for venereal disease. Wood burned with charcoal, buffalo hair, wood rat hair, and bat hair in a hole in the middle of the hogan. The person with venereal disease sits over the hole and the smudge covers his exposed parts and cures him. (55:24) **Pima** *Dermatological Aid* Moxa used to burn the boils caused by "bad disease." Poultice of plant applied to bleeding sores caused by venereal disease. *Other* Plant used as an "antileuretic." *Venereal Aid* Poultice of plant applied to bleeding sores caused by venereal disease. (47:76)

Ephedra viridis Coville, Mormon Tea
- **Drug**—**Havasupai** *Emetic* Used to make a draft and taken to vomit for bowel complaints. *Laxative* Used to make a draft and taken to clear the bowels. (162:285) **Hopi** *Tonic* Dried flowers and stems taken as a tonic. *Venereal Aid* Plant used for syphilis. (42:312) Plant used for syphilis. (200:64) **Kawaiisu** *Blood Medicine* Infusion of stems taken for anemia. *Orthopedic Aid* Infusion of stems taken for backaches. (206:27) **Navajo** *Cough Medicine* Decoction of plant tops taken as a cough medicine. (55:24) *Venereal Aid* Strong infusion of plant used for syphilis. (110:19) **Paiute** *Antidiarrheal* Compound infusion of plant given to children for diarrhea. *Antirheumatic (Internal)* Infusion or decoction of twigs or branches taken for rheumatism. *Blood Medicine* Infusion or decoction of twigs or branches taken as a blood purifier. *Cold Remedy* Infusion or decoction of twigs or branches taken for colds. *Dermatological Aid* Dried, powdered stems applied to sores. *Gastrointestinal Aid* Infusion or decoction of twigs or branches used for stomach ulcers and disorders. *Kidney Aid* Infusion or decoction of twigs or branches taken as a kidney regulator. *Pediatric Aid* Compound infusion of plant given to children for diarrhea. *Tonic* Infusion or decoction of twigs or branches taken as a tonic. *Urinary Aid* Infusion or decoction of twigs or branches used as a kidney regulator and for bladder. *Venereal Aid* Decoction of twigs taken for syphilis or gonorrhea. (180:68–70) **Paiute, Northern** *Dermatological Aid* Poultice of dried, powdered stems applied to sores. (59:128) **Shoshoni** *Antidiarrheal* Compound infusion of plant given to children for diarrhea. *Blood Medicine* Infusion or decoction of twigs or branches taken as a blood purifier. *Burn Dressing* Poultice of moistened, powdered stems applied to burns. *Cathartic* Decoction of root or salted decoction of stems taken as a physic. *Cold Remedy* Infusion or decoction of twigs or branches taken for colds. *Dermatological Aid* Dried, powdered stems alone or mixed with pitch and used as a salve for sores. *Gastrointestinal Aid* Infusion or decoction of twigs or branches taken for stomach disorders. *Kidney Aid* Infusion or decoction of twigs or branches taken as a kidney regulator. *Pediatric Aid* Compound infusion of plant given to children for diarrhea. *Tonic* Infusion or decoction of twigs or branches taken as a tonic. *Urinary Aid* Infusion or decoction of twigs or branches used as a kidney regulator and for bladder. *Venereal Aid* Simple or compound decoctions of plant parts taken for syphilis or gonorrhea. (180:68–70) **Tewa** *Tonic* Dried flowers and stems taken as a tonic. *Venereal Aid* Plant used for syphilis. (42:312) **Tubatulabal** *Blood Medicine* Decoction of stalks and leaves used for the blood. *Venereal Aid* Decoction of stalks and leaves used for syphilis. (193:59) **Washo** *Gynecological Aid* Infusion or decoction of twigs or branch used for delayed or difficult menstruation. (180:68–70)
- **Food**—**Havasupai** *Beverage* Upper portions of plant boiled into tea. (197:207) Twigs boiled into a tea. (197:66) **Kawaiisu** *Beverage* Stems used to make tea. *Unspecified* Seeds formerly used for food. (206:27) **Navajo** *Beverage* Roasted stems used to make tea. Stems chewed to relieve thirst when on the move and away from water supplies. (110:19) **Paiute** *Beverage* Leafless needles boiled into a drink. (167:245) **Paiute, Northern** *Beverage* Stems used to make tea. (59:128) Stalks boiled in water to make tea. (59:53) **Tubatulabal** *Unspecified* Leaves and stalks used for food. (as *E. viridis* 193:15)
- **Dye**—**Navajo** *Brown* Twigs and leaves boiled with alum and used as a light tan dye. (110:19)
- **Other**—**Kawaiisu** *Tools* Wood provided the best charcoal for tattooing. (206:27)

Epicampes, Poaceae
Epicampes sp.
- **Fiber**—**Shoshoni** *Basketry* Grass used to make grass coil baskets. (117:446)

Epifagus, Orobanchaceae
Epifagus virginiana (L.) W. Bart., Beech Drops
- **Drug**—**Iroquois** *Antidiarrheal* Infusion of plants taken for diarrhea caused by menstruating women. (87:437)

Epigaea, Ericaceae
Epigaea repens L., Trailing Arbutus
- **Drug**—**Algonquin, Quebec** *Kidney Aid* Infusion of leaves used for

kidney disorders. (18:216) **Cherokee** *Analgesic* Decoction of plant taken to cause vomiting for abdominal pains. *Antidiarrheal* Infusion of plant given to children with diarrhea. *Emetic* Decoction of plant taken to cause vomiting for abdominal pains. (177:48) *Gastrointestinal Aid* Compound infusion taken for indigestion. (80:23) Decoction of plant taken to cause vomiting for abdominal pains. (177:48) *Kidney Aid* Infusion taken for kidneys and "chest ailment." (80:23) *Pediatric Aid* Infusion of plant given to children with diarrhea. (177:48) *Pulmonary Aid* Infusion used for "chest ailment." (80:23) **Iroquois** *Analgesic* Compound used for labor pains in parturition. *Antirheumatic (Internal)* Compound decoction of plant taken for rheumatism. *Gastrointestinal Aid* Decoction of leaves taken to aid digestion. *Gynecological Aid* Compound used for labor pains in parturition. *Kidney Aid* Decoction of whole plant or roots, stalks, and leaves taken for the kidneys. (87:410)
- *Other*—**Potawatomi** *Sacred Items* Tribal flower of Forest Potawatomi and considered these flowers came directly from their divinity. (154:118)

Epilobium, Onagraceae
Epilobium angustifolium L., Fireweed
- *Drug*—**Abnaki** *Cough Medicine* Roots used for coughs. (144:154) **Algonquin, Tête-de-Boule** *Dermatological Aid* Poultice of boiled roots applied to "sick" skin. (132:128) **Bella Coola** *Dermatological Aid* Poultice of roasted and mashed root applied to boils. (150:60) Poultice of roasted and mashed roots applied to boils. (184:207) **Blackfoot** *Dermatological Aid* Powdered inner cortex rubbed on the hands and face to protect them from the cold during the winter. (86:112) *Laxative* and *Pediatric Aid* Infusion of roots and inner cortex given to babies as an enema for constipation. (86:66) **Chippewa** *Dermatological Aid* Poultice of moistened fresh or dried leaf used for bruises or to remove a sliver. (53:352) **Cree, Woodlands** *Dermatological Aid* Poultice of barked, macerated roots applied to boils, abscesses, or wounds to draw out the infection. Poultice of leaves applied to bruises. Poultice of barked, chewed roots applied to cuts and wounds. (109:38) **Eskimo, Alaska** *Laxative* Infusion of old, dry leaves used as a laxative. (1:36) **Eskimo, Western** *Analgesic* and *Gastrointestinal Aid* Compound decoction of leaves taken for stomachache and intestinal discomfort. (108:5) **Iroquois** *Analgesic* Infusion of bark applied as poultice for pain anywhere in the body. *Basket Medicine* Infusion of plant used as a "basket medicine." *Internal Medicine* Decoction of roots taken for internal injuries from lifting. (87:389) *Kidney Aid* Compound infusion of roots taken for kidneys or for male urination problems. *Orthopedic Aid* Poultice of smashed roots applied to swollen knees. (87:390) *Panacea* Compound infusion of twigs and roots taken as a panacea for pain. *Tuberculosis Remedy* Compound decoction of roots taken for consumption. *Urinary Aid* Infusion of roots taken for burning urination and other urination problems. *Witchcraft Medicine* Compound decoction of plants and a doll used for black magic. (87:389) **Kwakiutl** *Cancer Treatment* Poultice of seeds, down, and oil applied to wound after cutting open the tumor. *Dermatological Aid* Poultice of seeds, down, and oil applied to wound after cutting open the tumor. (183:287) **Menominee** *Dermatological Aid* Root used to make a wash for swellings. (151:43) **Navajo, Kayenta** *Gastrointestinal Aid* Plant used for gastritis. (205:32) **Ojibwa** *Dermatological Aid* Poultice of pounded root applied to boils and carbuncles. (153:376) **Potawatomi** *Unspecified* Plant used as a medicine for unspecified ailments. (154:66) **Skokomish** *Tuberculosis Remedy* Infusion of roots taken for tuberculosis. **Snohomish** *Throat Aid* Infusion of roots taken for sore throats. **Swinomish** *Other* Decoction of whole plants used as a bath for invalids. *Poison* Infusion of plant considered poisonous. (79:41) **Thompson** *Dermatological Aid* Decoction of plant used as a wash for sores. (187:235)
- *Food*—**Alaska Native** *Dietary Aid* Young, tender greens, properly prepared, used as a good source of vitamin C and provitamin A. *Unspecified* Young stems peeled and eaten raw. *Vegetable* Young shoots mixed with other greens and eaten. (85:31) **Bella Coola** *Vegetable* Young shoots eaten as greens in spring. (184:207) **Blackfoot** *Unspecified* Fresh roots used for food. (86:102) **Clallam** *Beverage* Roots boiled and used as a drink. (57:201) **Eskimo, Alaska** *Unspecified* Young shoots eaten raw or blanched, with seal oil. (1:36) Young shoots gathered, boiled, and mixed with other plants and sometimes bacon. (4:715) **Eskimo, Inuktitut** *Unspecified* Young shoots used for food. (202:192) **Eskimo, Inupiat** *Sweetener* Pith used as a berry sweetener and eaten by children. *Vegetable* Violet stems, with dark purple leaves, used in salads. Pink stems with leaves boiled and eaten or steamed and served with cream sauce or cheese sauce. *Winter Use Food* Violet stems, with dark purple leaves, preserved in seal oil. (98:23) **Gitksan** *Unspecified* Shoots and stems used for food in summer. **Haisla** *Unspecified* Shoots and stems used for food in summer. (73:154) **Okanagan-Colville** *Forage* Plant eaten by horses and deer. (188:110) **Okanagon** *Unspecified* Young shoots eaten raw. (125:38) **Saanich** *Beverage* Young leaves boiled to make a refreshing tea. (182:85) **Tanana, Upper** *Fodder* Shoots cooked with fish for dog feed. *Preservative* Used for smoking fish and as a mosquito repellent. *Unspecified* Shoots eaten raw or boiled. (102:16) **Thompson** *Fodder* Stalks used for pig feed. (187:235) *Unspecified* Young shoots eaten raw. (125:38) Young shoots peeled and eaten. Stalks eaten raw like celery, boiled, or steamed. (187:235) **Wet'suwet'en** *Unspecified* Shoots and stems used for food in summer. (73:154)
- *Fiber*—**Clallam** *Clothing* Puffs mixed with dog hair and used for weaving cloth. (57:201) **Cree, Woodlands** *Sewing Material* Stem fibers used for thread. (109:38) **Quinault** *Mats, Rugs & Bedding* Cotton combined with duck feathers and used as blankets. **Skokomish** *Mats, Rugs & Bedding* Cotton combined with duck feathers and used as blankets. (79:41)
- *Other*—**Blackfoot** *Waterproofing Agent* Flowers rubbed on rawhide thongs and mittens for waterproofing. (86:112) **Cree, Woodlands** *Season Indicator* Flowering coincided with the season of moose fattening. Comas coincided with the moose mating season which followed fattening. (109:38) **Kitasoo** *Hunting & Fishing Item* Fibers used to make fishing nets. (43:339) **Shuswap** *Decorations* Used in bouquets with roses. (123:64) **Tanana, Upper** *Insecticide* Used for smoking fish and as a mosquito repellent. *Tools* Stem used to draw the pus out of a boil or cut. (102:16) **Thompson** *Containers* Leaves put in cooking pits by old-timers, to protect the food while cooking. *Hunting & Fishing Item* Plant blooms indicated that the deer would be fat. (187:235)

Epilobium angustifolium L. ssp. *angustifolium*, Fireweed
- *Drug*—**Cheyenne** *Antihemorrhagic* Infusion of dried, pulverized

leaves taken for bowel hemorrhage. *Gastrointestinal Aid* Infusion of dried leaves or dried roots taken for bowel hemorrhages. (as *Chamoenerion angustifolium* 76:181)

Epilobium angustifolium* ssp. *circumvagum Mosquin, Fireweed
- *Food*—**Haisla & Hanaksiala** *Unspecified* Young shoots used for food. (43:257) **Oweekeno** *Unspecified* Young shoots eaten as a spring food. (43:106)
- *Fiber*—**Haisla** *Cordage* Stem fibers possibly used to make cordage. (43:106) **Haisla & Hanaksiala** *Cordage* Plant used to make twine, cordage, and binding. (43:257) **Hanaksiala** *Cordage* Stem fibers possibly used to make cordage. **Heiltzuk** *Cordage* Stem fibers possibly used to make cordage. **Oweekeno** *Cordage* Stem fibers possibly used to make cordage. (43:106)
- *Other*—**Haisla** *Cleaning Agent* Leaves used to wipe hands, especially after handling slimy fish. **Haisla & Hanaksiala** *Hunting & Fishing Item* Plant used to make oolichan (candlefish) nets. (43:257)

Epilobium brachycarpum K. Presl, Autumn Willowweed
- *Drug*—**Okanagan-Colville** *Dermatological Aid* Infusion of plant tops applied to the hair as a conditioner for dandruff and hair manageability. (as *E. paniculatum* 188:111)

Epilobium canum* ssp. *angustifolium (Keck) Raven, Hummingbird Trumpet
- *Drug*—**Costanoan** *Dermatological Aid* Decoction of plant used for infected sores. *Disinfectant* Decoction of plant used for infected sores. *Febrifuge* Decoction of plant used for infant's fever. *Panacea* Decoction of plant used as a general remedy. *Pediatric Aid* Decoction of plant used for infant's fever. *Urinary Aid* Decoction of plant used for urinary problems. (as *Zauschneria californica* 21:22) **Miwok** *Antihemorrhagic* Used by women after parturition for hemorrhages. *Cathartic* Decoction of leaves taken as a cathartic. *Gynecological Aid* Used by women after parturition for hemorrhages. *Kidney Aid* Decoction of leaves taken for kidney trouble. *Tuberculosis Remedy* Decoction of leaves taken for tuberculosis. *Urinary Aid* Decoction of leaves taken for bladder trouble. *Venereal Aid* Used for syphilis. (as *Zauschneria californica* 12:174)

Epilobium canum* ssp. *latifolium (Hook.) Raven, Hummingbird Trumpet
- *Food*—**Karok** *Unspecified* Blossoms sucked for the nectar. (as *Zauschneria latifolia* 148:386)

Epilobium ciliatum Raf. **ssp. *ciliatum*,** Coast Willowweed
- *Drug*—**Hopi** *Analgesic* Plant used for leg pains. (as *E. adenocaulon* 200:33, 86) **Navajo, Kayenta** *Orthopedic Aid* Infusion used as lotion and poultice of roots applied to muscular cramps. (as *E. adenocaulon* 205:32) **Potawatomi** *Antidiarrheal* Infusion of root used "to check diarrhea." (as *E. adenocaulon* 154:66)

Epilobium latifolium L., Dwarf Fireweed
- *Drug*—**Eskimo, Inupiat** *Dermatological Aid* Leaves, rich in vitamins A and C, eaten for healthy, beautiful skin. *Eye Medicine* Leaves, rich in vitamins A and C, eaten for healthy, beautiful eyes. (98:26)
- *Food*—**Alaska Native** *Vegetable* Young, tender greens, properly prepared, used as a good source of vitamin C and provitamin A. (85:33)

Eskimo, Arctic *Vegetable* Flowers eaten raw as a salad. Leaves cooked and eaten. **Eskimo, Greenland** *Vegetable* Flowers and leaves eaten raw with seal blubber. (128:25) **Eskimo, Inuktitut** *Unspecified* Young shoots used for food. (202:192) **Eskimo, Inupiat** *Unspecified* Leaves preserved in seal oil and eaten within 48 hours with walrus blubber. (98:26)

Epilobium minutum Lindl. ex Lehm., Small Willowweed
- *Drug*—**Okanagan-Colville** *Antidiarrheal* and *Pediatric Aid* Infusion of roots and stems given to children for diarrhea. (188:111)

***Epilobium sp.*,** Willowweed
- *Drug*—**Abnaki** *Cough Medicine* Infusion of roots and bark from other plants taken for persistent coughs. (144:164) **Thompson** *Unspecified* Plant used medicinally for unspecified purpose. (164:468)
- *Fiber*—**Mewuk** *Building Material* Used to line acorn caches. (117:362)
- *Other*—**Thompson** *Good Luck Charm* Plant used as an "especially efficacious" charm for good luck in gambling. (164:506)

Epipactis, Orchidaceae

Epipactis gigantea Dougl. ex Hook., Giant Helleborine
- *Drug*—**Navajo, Kayenta** *Ceremonial Medicine* Plant used in girl's puberty rite. *Other* Plant used for general body disease. *Pediatric Aid* Plant used to purify a newborn infant and plant used in girl's puberty rite. (205:17)
- *Other*—**Karok** *Decorations* Flowers used for their prettiness. (148:381)

Epixiphium, Scrophulariaceae

Epixiphium wislizeni (Engelm. ex Gray) Munz, Balloonbush
- *Drug*—**Keres, Western** *Antirheumatic (External)* Leaves heated with stones and rubbed on swellings. *Snakebite Remedy* Leaves heated with stones and rubbed onto snakebites. (as *Maurandia wislizeni* 171:53)
- *Food*—**Apache, Chiricahua & Mescalero** *Unspecified* Pods eaten fresh or boiled. (as *Maurandia wislizeni* 33:45)

Equisetaceae, see *Equisetum*

Equisetum, Equisetaceae, horsetail

Equisetum arvense L., Field Horsetail
- *Drug*—**Blackfoot** *Dermatological Aid* Poultice of stem pieces applied to rash under the arm and in the groin. (86:76) *Diuretic* Infusion of fertile stem roots used as a powerful diuretic. (86:69) *Orthopedic Aid* Powdered stems put in moccasins to avoid foot cramps when traveling long distances. (86:112) *Veterinary Aid* Infusion of fertile stem roots given to horses as a diuretic. Infusion of fertile stem roots rubbed on the groins of horses. Powdered stems and water given to perk a horse up. (86:88) **Cherokee** *Kidney Aid* Infusion taken for kidneys. *Laxative* Strong infusion taken for constipation. (80:39) **Cheyenne** *Veterinary Aid* Infusion of leaves and stems given to horses with a hard cough. (76:169) **Chippewa** *Urinary Aid* Decoction of stems taken for dysuria. (71:122) **Iroquois** *Analgesic* Used for headaches and pains. *Antirheumatic (Internal)* Used for rheumatism. *Orthopedic Aid* Used for joint aches. *Pediatric Aid* Raw stems chewed by

teething babies. (87:261) Infusion of rhizomes and hazel stems given to children for teething. (141:33) *Toothache Remedy* Raw stems chewed by teething babies. (87:261) Infusion of rhizomes and hazel stems given to children for teething. (141:33) **Kwakiutl** *Dermatological Aid* Poultice of rough leaves and stems applied to cuts and sores. (183:263) **Ojibwa** *Kidney Aid* Infusion of whole plant used for dropsy. (153:368) **Okanagan-Colville** *Antirheumatic (Internal)* Infusion of stems taken for lumbago. *Dermatological Aid* Plant pounded, mixed with water, and used to wash areas of the body affected by poison ivy. *Diuretic* Infusion of stems taken as a diuretic to stimulate the kidneys. *Orthopedic Aid* Infusion of stems taken for backaches. *Stimulant* Infusion of stems taken for sluggishness due to a cold. *Venereal Aid* Decoction of plant and false box taken or used as a bath for syphilis and gonorrhea. *Veterinary Aid* Given to thin, old horses with diarrhea after eating fresh grass in spring. (188:17) **Pomo, Kashaya** *Dermatological Aid* Decoction of plant used as a wash for itching or open sores. (72:58) **Potawatomi** *Analgesic* Infusion of whole plant used for lumbago. *Kidney Aid* Infusion of plant used for kidney trouble. *Orthopedic Aid* Infusion of plant used for lumbago. *Urinary Aid* Infusion of plant used for bladder trouble. (154:55, 56) **Saanich** *Blood Medicine* Tender, young shoots eaten raw or boiled and thought to be "good for the blood." (182:68) **Thompson** *Dermatological Aid* Decoction or infusion of stems used after childbirth to expel the afterbirth more quickly.[31] *Urinary Aid* Decoction of new plant tops taken for "stoppage of urine." (187:86)
- *Food*—**Chinook, Lower** *Unspecified* Young shoots used as food. (79:15) **Eskimo, Alaska** *Unspecified* Black, edible nodules attached to roots used for food. The effort of collecting the nodules was considerable and therefore rarely done. However, these nodules were often obtained from underground caches of roots and tubers collected by lemmings and other tundra rodents. The caches were raided by the people and the "mouse nuts" were used for food. (1:33) **Haisla & Hanaksiala** *Forage* Plant eaten by geese. (43:156) **Hesquiat** *Vegetable* Tender, young, vegetative shoots peeled and eaten raw. These shoots were green but had not yet branched out, and the segments were still very close together. The leaf sheaths were peeled off two at a time and the succulent stems eaten raw. They were "nothing but juice." The Hesquiat people traveled up towards Esteven Point especially to get these shoots, and sometimes they would collect 20 or more kilograms of them at a time. When they returned home, the harvesters would call together all their relatives and friends and have a feast of horsetail shoots. The white, fertile shoots were apparently not eaten, although they are in other areas of the Northwest Coast. (185:28) **Meskwaki** *Fodder* Plant fed to captive wild geese to make them fat in a week. (152:272) **Ojibwa** *Fodder* Plant gathered to feed domesticated ducks and fed to ponies to make their coats glossy. (153:400) **Okanagan-Colville** *Fodder* Used in winter for fodder during hay shortage. (188:17) **Saanich** *Unspecified* Tender, young shoots eaten raw or boiled. (182:68) **Tanana, Upper** *Unspecified* Tubers eaten. (102:9) **Tewa** *Forage* Plant eaten by horses. (138:68)
- *Fiber*—**Costanoan** *Basketry* Roots used in basketry. (21:247) **Kwakiutl, Southern** *Scouring Material* Rough leaves and stems used for polishing canoes and other wooden articles. (183:264) **Okanagan-Colville** *Scouring Material* Stems used as sandpaper to polish bone tools and soapstone pipes. Used to polish fingernails. (188:17)
- *Dye*—**Blackfoot** *Red* Crushed stems used as a light pink dye for porcupine quills. (86:112)
- *Other*—**Blackfoot** *Soap* Plant used by children to shine their bouncing arrows. (86:112) **Chippewa** *Malicious Charm* Plant pieces carried in men's pockets to prevent their rivals from having good luck. (71:122) **Okanagan-Colville** *Containers* Hollow stems used to administer medicines to babies. (188:17) **Shoshoni** *Musical Instrument* Plant used for whistles. (118:57) **Shuswap** *Tools* Used as a file. (123:49)

Equisetum hyemale L., Scouringrush Horsetail

- *Drug*—**Blackfoot** *Veterinary Aid* Infusion used as a drench for horse medicine. (82:58) Decoction of foliage used in horse medicine as a drench. (97:16) Decoction of plant used as a horse medicine. (114:276) **Carrier** *Kidney Aid* Decoction of plant taken for kidney problems. *Urinary Aid* Decoction of plant taken for the inability to pass water. (31:84) **Cherokee** *Kidney Aid* Infusion taken for kidneys. *Laxative* Strong infusion taken for constipation. (80:39) **Cheyenne** *Veterinary Aid* Plant used as a medicine for horses. (83:4) **Chippewa** *Disinfectant* Leaves burned as a disinfectant. (53:366) **Cree** *Abortifacient* Used for irregular menstruation. (82:58) Decoction of plant and two unknown roots used to correct menstrual irregularities. (97:16) **Crow** *Analgesic* Poultice used for bladder and prostate pains. *Diuretic* Infusion of stems used as a diuretic. **Flathead** *Diuretic* Infusion of stems used as a diuretic. (82:58) **Hoh** *Ceremonial Medicine* Rootstocks eaten during medicinal ceremonies. (137:57) **Iroquois** *Urinary Aid* Infusion of rhizomes taken by old people "when the urine is too red." (141:33) **Karok** *Ceremonial Medicine* Plant used in ceremonial cleansing for the priests in First Salmon ceremony. *Eye Medicine* Decoction of plant used as a wash or poultice of stalks applied for sore eyes. (148:378) **Mahuna** *Urinary Aid* Infusion of dried plants taken for prostate gland troubles. (140:21) **Makah** *Antidiarrheal* Raw shoots chewed for diarrhea. (79:15) **Menominee** *Gynecological Aid* Decoction of rushes taken after childbirth "to clear up the system." *Kidney Aid* Decoction of rushes taken for kidney troubles. (151:34) **Meskwaki** *Venereal Aid* Infusion of whole plant taken by both men and women for gonorrhea. (152:220) **Ojibwa** *Unspecified* Plant used as a medicine. (153:418) **Okanagan-Colville** *Antirheumatic (Internal)* Infusion of stems taken for lumbago. *Dermatological Aid* Decoction of stems used as a wash on children for skin sores. Plant pounded, mixed with water, and used to wash areas of the body affected by poison ivy. *Diuretic* Infusion of stems taken as a diuretic to stimulate the kidneys. *Eye Medicine* Stem fluid used as an eyewash. *Orthopedic Aid* Infusion of stems taken for backaches. *Pediatric Aid* Decoction of stems used as a wash on children for skin sores. *Stimulant* Infusion of stems taken for sluggishness due to a cold. *Venereal Aid* Decoction of plant and false box taken or used as a bath for syphilis and gonorrhea. *Veterinary Aid* Given to thin, old horses with diarrhea after eating fresh grass in spring. (188:17) **Quileute** *Ceremonial Medicine* Rootstocks eaten during medicinal ceremonies. (137:57) *Sports Medicine* Plant rubbed on swimmers to make them feel strong. **Quinault** *Abortifacient* Decoction taken to regulate menses, informant insisted not an abortive. *Eye Medicine* Infusion of roots or root juice used as a wash for sore eyes. (79:15) **Sanpoil** *Adjuvant* Used as a drinking tube for medicine and used for giving medicine to infants. *Pediatric Aid* Used as a drinking tube for medicine and used for giv-

ing medicine to infants. (131:218) **Thompson** *Eye Medicine* Stem liquid used for sore eyes or decoction of stems used for sore, itchy eyes or cataracts.32 *Gynecological Aid* Decoction of roots taken during difficult childbirth, to accelerate it. Decoction or infusion of stems taken after childbirth to expel the afterbirth more quickly.31 *Urinary Aid* Decoction of new growths taken for bladder trouble. (187:86)
- *Food*—**Blackfoot** *Beverage* Blades boiled to make a drink. (114:276) **Cowlitz** *Dried Food* Stalk tops dried, mashed, mixed with salmon eggs, and eaten. (79:15) **Hoh** *Dried Food* Rootstocks dried and used for food. *Special Food* Rootstocks eaten during puberty ceremonies. *Unspecified* Rootstocks used for food. (137:57) **Lakota** *Fodder* Plant given to horses to fatten them. (139:25) **Meskwaki** *Fodder* Plant fed to ponies to make them fat in a week. (152:273) **Okanagan-Colville** *Fodder* Used in winter for fodder during hay shortage. (188:17) **Quileute** *Dried Food* Rootstocks dried and used for food. *Special Food* Rootstocks eaten during puberty ceremonies. *Unspecified* Rootstocks used for food. (137:57)
- *Fiber*—**Bella Coola** *Scouring Material* Stems formerly used for sandpaper to smooth wooden objects. (184:196) **Chippewa** *Scouring Material* Used for scouring. (53:377) **Costanoan** *Basketry* Roots used in basketry. (21:247) **Cowlitz** *Scouring Material* Used to polish arrow shafts. (79:15) **Klamath** *Scouring Material* Used to smooth arrow shafts. (45:88) **Menominee** *Scouring Material* Used as a scouring rush for pots and pans. (151:75) **Meskwaki** *Scouring Material* Used to scour pots and pans. (152:268) **Missouri River Indian** *Mats, Rugs & Bedding* Used to make mats. **Montana Indian** *Scouring Material* Abrasive stems used to polish pipes, bows, and arrows and formerly used to scrub tins and floors. (82:58) **Ojibwa** *Scouring Material* Handful of stems used to scour the kettles and pans. (153:418) **Okanagan-Colville** *Scouring Material* Stems used as sandpaper to polish bone tools and soapstone pipes. Used to polish fingernails. (188:17) **Quinault** *Scouring Material* Used to polish arrow shafts. (79:15) **Sanpoil & Nespelem** *Basketry* Roots used to imbricate woven bags and baskets. (188:17) **Thompson** *Scouring Material* Rough, silicon-impregnated stems used to smooth and polish implements of wood, bone, and steatite.33 (187:86)
- *Other*—**Cowlitz** *Insecticide* Decoction of stalks used as a wash for hair infested with vermin. (79:15) **Gosiute** *Toys & Games* Used by children as whistles. (39:368) **Havasupai** *Tools* Joints pulled apart and used by children to produce a whistling sound. (197:204) **Okanagan-Colville** *Containers* Hollow stems used to administer medicines to babies. (188:17) **Sioux** *Toys & Games* Stems formerly used by children to make whistles. (82:58) **Thompson** *Fertilizer* Stem liquid used to kill any type of weed.34 (187:86)

Equisetum hyemale var. *affine* (Engelm.) A. A. Eat., Scouring-rush Horsetail

- *Drug*—**Iroquois** *Diuretic* Decoction of plant taken for urinating too infrequently. *Eye Medicine* Infusion of whole plant used as an eyewash for white spot on the eye. *Kidney Aid* Decoction of plant taken for kidney trouble. *Other* Decoction of plant taken for backache or "summer complaint." *Urinary Aid* Decoction of plant taken for urinating too much. Decoction used by women with excessive urination who are ruptured. *Venereal Aid* Compound decoction of roots taken for gonorrhea. (87:262)
- *Fiber*—**Karok** *Scouring Material* Dried stalks used to sharpen mussel shell scrapers and for polishing arrows. (148:378)

Equisetum laevigatum A. Braun, Smooth Horsetail

- *Drug*—**Costanoan** *Abortifacient* Decoction of plant used for retarded menstruation. *Contraceptive* Decoction of plant used as a contraceptive. *Dermatological Aid* Decoction of stalks used as a hair wash. *Urinary Aid* Decoction of plant used for bladder ailments. (21:4) **Diegueño** *Hypotensive* Infusion of stems taken for high blood pressure. (84:19) **Hoh** *Ceremonial Medicine* Rootstocks eaten during medicinal ceremonies. (137:57) **Hopi** *Ceremonial Medicine* Dried, ground plant used for ceremonial bread. (56:17) **Keres, Western** *Hemorrhoid Remedy* Plant chewed before meals for piles. (171:42) **Navajo, Kayenta** *Analgesic* and *Orthopedic Aid* Infusion of plant taken or cold infusion used as a lotion for backaches. (as *E. kansanum* 205:15) **Navajo, Ramah** *Disinfectant* Compound decoction of plant used for "lightning infection." (191:11) **Okanagan-Colville** *Antirheumatic (Internal)* Infusion of stems taken for lumbago. *Cold Remedy* Decoction of plant and chokecherry twigs given to children for colds. *Dermatological Aid* Plant pounded, mixed with water, and used to wash areas of the body affected by poison ivy. *Diuretic* Infusion of stems taken as a diuretic to stimulate the kidneys. *Orthopedic Aid* Infusion of stems taken for backaches. *Pediatric Aid* Decoction of plant and chokecherry twigs given to children for colds. *Stimulant* Infusion of stems taken for sluggishness due to a cold. *Venereal Aid* Decoction of plant and false box taken or used as a bath for syphilis and gonorrhea. *Veterinary Aid* Given to thin, old horses with diarrhea after eating fresh grass in spring. (188:17) **Okanagon** *Burn Dressing* Poultice of plant ash and grease applied to burns. (125:41) **Pomo, Kashaya** *Kidney Aid* Decoction of whole plant taken for kidney trouble and associated back trouble. (as *E. funstoni* 72:59) **Quileute** *Ceremonial Medicine* Rootstocks eaten during medicinal ceremonies. (137:57) **Thompson** *Burn Dressing* Poultice of plant ash and grease applied to burns. (125:41) *Eye Medicine* Stem liquid used for sore eyes or decoction of stems used for sore, itchy eyes or blindness.32 *Gynecological Aid* Decoction of roots taken to accelerate a difficult childbirth. Decoction or infusion of stems taken after childbirth to expel the afterbirth more quickly.31 *Urinary Aid* Decoction of new growths taken for bladder trouble. (187:86)
- *Food*—**Hoh** *Dried Food* Rootstocks dried and used for food. *Special Food* Rootstocks eaten as a delicacy. Rootstocks eaten during puberty ceremonies. (137:57) **Isleta** *Fodder* Plant used for horse feed. (100:28) **Okanagan-Colville** *Fodder* Used in winter for fodder during hay shortage. *Unspecified* Heads used for food. (188:17) **Quileute** *Dried Food* Rootstocks dried and used for food. *Special Food* Rootstocks eaten as a delicacy. Rootstocks eaten during puberty ceremonies. (137:57) **San Felipe** *Porridge* Plant dried and ground to make mush. (32:27)
- *Fiber*—**Karok** *Scouring Material* Stems used to sandpaper madrone spoons. (6:29) **Okanagan-Colville** *Scouring Material* Stems used as sandpaper to polish bone tools and soapstone pipes. Used to polish fingernails. (188:17) **Okanagon** *Scouring Material* Used for sharpening and polishing bone tools. **Thompson** *Scouring Material* Used for sharpening and polishing bone tools. (125:39) Rough, silicon-impregnated stems used to smooth and polish implements of wood, bone, and steatite.33 (187:86)

- **Other**—Hopi *Ceremonial Items* Dried, ground with cornmeal, and used to make a ceremonial bread. (56:17) **Okanagan-Colville** *Containers* Hollow stems used to administer medicines to babies. (188:17) **Thompson** *Fertilizer* Stem liquid used to kill any type of weed.[34] (187:86) **Ute** *Toys & Games* Used by children as whistles. (38:34)

Equisetum palustre L., Marsh Horsetail
- **Drug**—Ojibwa *Gastrointestinal Aid* Infusion or decoction of plants taken for stomach or bowel troubles. *Laxative* Decoction of plants taken for sick stomach, bowels, or for constipation. (135:231)

Equisetum pratense Ehrh., Meadow Horsetail
- **Food**—Eskimo, Inupiat *Unspecified* Raw roots eaten with seal oil. *Winter Use Food* Roots stored in oil for future use. (98:121)

Equisetum scirpoides Michx., Dwarf Scouringrush
- **Food**—Haisla & Hanaksiala *Forage* Plant eaten by grizzly bears. (43:156)

Equisetum sp., Mare's Tail
- **Drug**—Aleut *Poison* Decoction of plant fed to hated guest as a magical poison. (8:428) **Costanoan** *Abortifacient* Decoction of plant used for retarded menstruation. *Contraceptive* Decoction of plant used as a contraceptive. *Urinary Aid* Decoction of plant used for bladder ailments. (21:4) **Miwok** *Unspecified* Stems used for medicine. (12:169) **Modesse** *Cough Medicine* Infusion of plant taken for coughs. *Urinary Aid* Infusion of plant taken for bladder troubles. (117:224) **Shoshoni** *Kidney Aid* Decoction of plant taken for kidney trouble. (180:70) **Thompson** *Burn Dressing* Poultice of stem ash of several species used alone or with grease on burns. (164:462) *Eye Medicine* Stem liquid used for sore eyes or decoction of stem used for sore, itchy eyes or blindness.[32] *Gynecological Aid* Decoction or infusion of stems taken after childbirth to expel the afterbirth more quickly.[31] *Urinary Aid* Decoction of new growths taken for bladder trouble. (187:86)
- **Food**—Skagit, Upper *Unspecified* Tender shoots eaten. (179:42)
- **Fiber**—Cahuilla *Scouring Material* Used for a cleaning pad as a cleansing agent. (15:70) **Koyukon** *Mats, Rugs & Bedding* Blades used as dog bedding. (119:56) **Modesse** *Scouring Material* Used to polish arrows. (117:224) **Omaha** *Scouring Material* Plant used like sandpaper for polishing. **Pawnee** *Scouring Material* Plant used like sandpaper for polishing. **Ponca** *Scouring Material* Plant used like sandpaper for polishing. (70:63) **Thompson** *Scouring Material* Used to smooth and finish soapstone pipes. (164:497) Rough, silicon-impregnated stems used to smooth and polish implements of wood, bone, and steatite.[33] (187:86)
- **Other**—Koyukon *Fuel* Blades used to produce smoke in smudge fires. (119:56) **Paiute** *Toys & Games* Stalk sections used by children to make whistles. (111:37) **Thompson** *Ceremonial Items* Stem used to hold lice found in girls' hair and thrown in a stream during puberty ceremonies. (164:510) *Fertilizer* Stem liquid used to kill any type of weed.[34] (187:86) *Tools* Stems used to sharpen and polish bone. (164:497) **Winnebago** *Toys & Games* Stems used by children to make whistles. The elders warned children not to use the stems as whistles as they might cause the appearance of snakes. (70:63)

Equisetum sylvaticum L., Woodland Horsetail
- **Drug**—Eskimo, Alaska *Antihemorrhagic* Infusion of branches and stems used for internal bleeding. Green plants could be used, but a stronger medicine could be made from plants collected in autumn. The plant was also dried for future use, but only the stems and branches were used. The tea from this plant was strong and bitter. (1:33) **Menominee** *Hemostat* Poultice of pulverized stem applied to stop bleeding. *Kidney Aid* Infusion of stems used for dropsy. (151:35) **Ojibwa** *Kidney Aid* Infusion of plant used for kidney trouble and dropsy. (153:368)

Equisetum telmateia Ehrh., Giant Horsetail
- **Drug**—Kwakiutl *Dermatological Aid* Poultice of rough leaves and stems applied to cuts and sores. (183:263) **Saanich** *Blood Medicine* Tender, young shoots eaten raw or boiled and thought to be "good for the blood." (182:68) **Thompson** *Urinary Aid* Decoction of new plant tops used for "stoppage of urine." (187:86) **Yuki** *Diuretic* Decoction of plant taken as a diuretic. (49:47)
- **Food**—Clallam *Unspecified* Sprouts peeled and eaten raw or pit baked and eaten. (57:193) **Cowlitz** *Unspecified* Root stock bulbs cooked and eaten. Bulbs eaten raw. (79:15) **Klallam** *Unspecified* Reproductive and vegetative sprouts used for food. (78:197) **Makah** *Unspecified* Young stems peeled and eaten raw. (79:15) **Nitinaht** *Substitution Food* Hollow, water-filled stem segments used when water scarce. *Unspecified* Young shoots eaten in spring. (186:60) **Quileute** *Fodder* Used as fodder for horses. *Unspecified* Young stems peeled and eaten raw. **Quinault** *Fodder* Used as fodder for horses. *Unspecified* Young stems peeled and eaten raw. Roots eaten with whale or seal oil. (79:15) **Saanich** *Unspecified* Tender, young shoots eaten raw or boiled. (182:68) **Swinomish** *Unspecified* Bulbs eaten raw. (79:15)
- **Fiber**—Cowlitz *Basketry* Black roots used for imbrication on coiled baskets. (79:15) **Kwakiutl, Southern** *Scouring Material* Rough leaves and stems used for polishing canoes and other wooden articles. (183:264) **Quileute** *Basketry* Black roots used for imbrication on coiled baskets. (79:15) **Salish, Coast** *Basketry* Stems used for black imbrication in basket making. (182:68) **Skokomish** *Scouring Material* Used with dogfish as sandpaper. **Swinomish** *Basketry* Black roots used for imbrication on coiled baskets. *Scouring Material* Used to polish arrow shafts. (79:15) **Yuki** *Scouring Material* Stalks used to smooth Indian hemp stems and to polish arrows. (49:92)

Equisetum telmateia var. *braunii* (Milde) Milde, Giant Horsetail
- **Drug**—Pomo, Kashaya *Gynecological Aid* Decoction of stem taken for menstrual cramps. (72:58) **Tolowa** *Oral Aid* and *Pediatric Aid* Stem rubbed on child's teeth to keep them from gritting their teeth. (6:29)
- **Food**—Makah *Unspecified* Young, sterile or fertile shoots peeled, washed, or soaked in cold water and eaten raw. Strobili boiled in water for 10 minutes and eaten. **Nitinaht** *Beverage* Vegetative shoots used as a source of drinking water when traveling. *Unspecified* Fertile and sterile shoots used for food. (67:215) **Yurok** *Unspecified* Very small, fresh sprouts used for food. (6:29)
- **Other**—Pomo, Kashaya *Fasteners* Plant used as binding to fasten feathers onto the coat of a wale-pu. *Tools* Leafless, fertile stems used as sandpaper in smoothing arrow shafts and drill shafts. (72:58)

Equisetum variegatum Schleich. ex F. Weber & D. M. H. Mohr, Variegated Scouringrush
- **Drug**—Yuki *Eye Medicine* Plant used for sore eyes. (41:304)

- *Food*—Mendocino Indian *Forage* Used as an occasional forage food for horses. (41:304)
- *Fiber*—Mendocino Indian *Scouring Material* Silicious stems used as a substitute for sandpaper in finishing off arrows and other woodwork. (41:304)

Eragrostis, Poaceae
Eragrostis mexicana (Hornem.) Link, Mexican Lovegrass
- *Food*—Cocopa *Porridge* Seeds parched, ground, and the flour cooked into a mush. *Staple* Seeds parched, ground, and the flour eaten dry. (37:187)

Eragrostis secundiflora J. Presl, Red Lovegrass
- *Food*—Paiute *Unspecified* Species used for food. (167:243)

Eremalche, Malvaceae
Eremalche exilis (Gray) Greene, White Mallow
- *Food*—Pima, Gila River *Unspecified* Leaves boiled and eaten. (133:7) *Vegetable* Leaves boiled, or boiled, strained, refried, and eaten as greens.[11] (133:5)

Eremocrinum, Liliaceae
Eremocrinum albomarginatum (M. E. Jones) M. E. Jones, Lonely Lily
- *Drug*—Navajo, Kayenta *Snakebite Remedy* Plant used for snakebites. (205:17)

Eriastrum, Polemoniaceae
Eriastrum densifolium (Benth.) Mason, Giant Woolstar
- *Drug*—Kawaiisu *Dermatological Aid* and *Venereal Aid* Dried, pounded flowers and roots used as a salve for venereal sores. (206:28)

Eriastrum eremicum (Jepson) Mason, Desert Woolstar
- *Drug*—Paiute *Antidiarrheal* Decoction of plant taken for diarrhea. *Gastrointestinal Aid* Decoction of plant taken as a stomach medicine. *Pediatric Aid* Infusion of plant used for children with tuberculosis. *Tuberculosis Remedy* Infusion of plant used for children with tuberculosis. (as *Gilia eremica* var. *arizonica* 180:80)

Eriastrum filifolium (Nutt.) Woot. & Standl., Lavender Woolstar
- *Drug*—Paiute *Cathartic* Decoction of plant taken as a physic. *Emetic* Decoction of plant taken as an emetic. Shoshoni *Analgesic* and *Antirheumatic (External)* Decoction of plant used as a bath for rheumatic pains. *Cathartic* Decoction of plant taken as a physic. *Emetic* Decoction of plant taken as an emetic. *Venereal Aid* Decoction of plant taken for venereal disease. (as *Gilia filifolia* var. *sparsiflora* 180:80, 81)

Eriastrum sparsiflorum (Eastw.) Mason, Great Basin Woolstar
- *Drug*—Paiute, Northern *Gastrointestinal Aid* Decoction of stalks taken as an emetic for stomach troubles. (59:128)

Eriastrum virgatum (Benth.) Mason, Wand Woolstar
- *Drug*—Paiute *Throat Aid* Decoction of entire plant used as a gargle for sore throats. (as *Hugelia virgata* 167:317)

Ericaceae, see *Andromeda, Arbutus, Arctostaphylos, Cassiope, Chamaedaphne, Chimaphila, Elliottia, Epigaea, Gaultheria, Gaylussacia, Kalmia, Ledum, Leiophyllum, Leucothoe, Lyonia, Menziesia, Moneses, Monotropa, Orthilia, Oxydendrum, Phyllodoce, Pterospora, Pyrola, Rhododendron, Vaccinium*

Ericameria, Asteraceae
Ericameria arborescens (Gray) Greene, Goldenfleece
- *Drug*—Miwok *Antirheumatic (External)* Decoction of leaves used for rheumatism. Poultice of twigs and leaves applied to rheumatic parts. *Dermatological Aid* Poultice of leaves applied to bring boils to a head. *Gastrointestinal Aid* Decoction of leaves taken for stomach trouble. *Gynecological Aid* Decoction of leaves taken during menstruation and after parturition for pain. *Orthopedic Aid* Poultice of leaves applied to foot sores. (as *Haplopappus aborescens* 12:170)

Ericameria bloomeri (Gray) J. F. Macbr., Rabbitbush Heathgoldenrod
- *Drug*—Klamath *Dermatological Aid* Poultice of leaves used to draw blisters. (as *Chrysothamnus bloomeri* 45:106) Poultice of smashed leaves applied to blisters. (as *Chrysothamnus bloomeri* 163:131)

Ericameria brachylepis (Gray) Hall, Chaparral Heathgoldenrod
- *Drug*—Diegueño *Dermatological Aid* Decoction of fresh or dried, entire plant used as a wash for wounds. *Misc. Disease Remedy* Decoction of fresh or dried, entire plant taken for "pasmo," a malady with chills. (as *Aplopappus propinquus* 88:219)

Ericameria cooperi (Gray) Hall, Cooper's Heathgoldenrod
- *Drug*—Tubatulabal *Antirheumatic (External)* Compound decoction of stalks and flowers used as a wash for rheumatism. (as *E. monactis* 193:59)

Ericameria cuneata (Gray) McClatchie **var. cuneata**, Cliff Heathgoldenrod
- *Drug*—Miwok *Cold Remedy* Decoction of stems taken for colds. (as *Haplopappus cuneatus* 12:170)

Ericameria linearifolia (DC.) Urbatsch & Wussow, Narrowleaf Heathgoldenrod
- *Drug*—Kawaiisu *Antirheumatic (External)* Decoction of leaves and flowers applied to limbs for rheumatism. *Dermatological Aid* Decoction of leaves and flowers applied to soreness, bruises, and cuts. Decoction of roots used as a hair wash to make the hair grow. *Orthopedic Aid* Decoction of roots used as a wash for tired feet. *Veterinary Aid* Decoction of leaves and flowers applied to sore backs of horses. (as *Haplopappus linearifolius* var. *interior* 206:33) Tubatulabal *Antirheumatic (External)* Compound decoction of leaves and flowers used as a wash for rheumatism. (as *Stenotopsis linearifolius* 193:59)

Ericameria nana Nutt., Dwarf Heathgoldenrod
- *Drug*—Paiute *Antidiarrheal* Decoction of plant taken for diarrhea. *Cold Remedy* Decoction of flowering heads and stems or stems alone used for colds. *Cough Medicine* Decoction of flowering heads and stems used for coughs and colds. *Eye Medicine* Decoction of roots used as a wash for sore eyes. *Febrifuge* Decoction of whole plant taken for high fevers. *Gastrointestinal Aid* Decoction of whole plant taken for stomach troubles. *Misc. Disease Remedy* Decoction of whole plant taken for grippe and high fever. Shoshoni *Analgesic* Decoction of flowering tops taken for stomachaches or stomach cramps. *Cold Remedy* Decoction of flowering heads and stems or whole plant used

for colds. *Cough Medicine* Decoction of flowering heads and stems used for coughs and colds. *Eye Medicine* Decoction of roots used as a wash for sore eyes. *Gastrointestinal Aid* Decoction of flowering tops taken for stomachaches or cramps. (as *Aplopappus nanus* 180:36)

Ericameria palmeri var. *pachylepis* (Hall) Nesom, Palmer's Heathgoldenrod
- **Other**—**Kawaiisu** *Hunting & Fishing Item* Straight stems used as arrow foreshafts. (as *Haplopappus palmeri* ssp. *pachylepus* 206:34)

Ericameria palmeri (Gray) Hall var. *palmeri*, Palmer's Heathgoldenrod
- **Drug**—**Cahuilla** *Dermatological Aid* Poultice of boiled leaves applied to sores. *Throat Aid* Leaves soaked in a pan of boiling water and steam inhaled for sore throats. (as *Haplopappus palmeri* 15:75) **Coahuilla** *Analgesic* Poultice of leaves and twigs applied to feet for swelling and pain. *Orthopedic Aid* Hot poultice of leaves and twigs bound to feet for swelling and pain. (as *Aplopappus palmeri* 13:78)
- **Fiber**—**Cahuilla** *Building Material* Plant used to build fences as a protection from cold winds. (as *Haplopappus palmeri* 15:75)

Ericameria parishii (Greene) Hall, Parish's Heathgoldenrod
- **Drug**—**Luiseño** *Unspecified* Plant used for medicinal purposes. (as *Bigelovia parishii* 155:228)
- **Food**—**Luiseño** *Unspecified* Seeds used for food. (as *Haplopappus parishii* 155:228)

Erigenia, Apiaceae
Erigenia bulbosa (Michx.) Nutt., Harbinger of Spring
- **Drug**—**Cherokee** *Toothache Remedy* Chewed for toothache. (80:48)

Erigeron, Asteraceae
Erigeron aphanactis (Gray) Greene var. *aphanactis*, Rayless Shaggy Fleabane
- **Drug**—**Paiute** *Analgesic* Decoction of whole plant taken for stomachaches and cramps. *Cathartic* Decoction of whole plant, a violent remedy, taken as a physic. *Emetic* Decoction of whole plant, a violent remedy, taken as an emetic. *Gastrointestinal Aid* Decoction of plant taken for stomachaches and cramps. **Shoshoni** *Analgesic* Decoction of whole plant taken for stomachaches and cramps. *Eye Medicine* Decoction of plant used as an eyewash. *Gastrointestinal Aid* Decoction of plant taken for stomachaches and cramps. (as *E. concinnus* var. *aphanactis* 180:70, 71)

Erigeron asper Nutt., Rough Fleabane
- **Fiber**—**Keres, Western** *Brushes & Brooms* Tied bunches of plants used for brooms. (171:42)

Erigeron bellidiastrum Nutt., Western Daisy Fleabane
- **Drug**—**Navajo, Ramah** *Ceremonial Medicine* Cold infusion of dried leaves used as ceremonial chant lotion. (191:50)

Erigeron caespitosus Nutt., Tufted Fleabane
- **Drug**—**Paiute** *Antidiarrheal* Strong decoction of root taken for diarrhea. *Eye Medicine* Cool decoction of root used as an eyewash. (180:70)

Erigeron canus Gray, Hoary Fleabane
- **Drug**—**Navajo, Ramah** *Ceremonial Medicine* Plant used in ceremonial chant lotion. *Disinfectant* Plant used for "deer infection." (191:50)

Erigeron compositus Pursh, Cutleaf Daisy
- **Drug**—**Thompson** *Dermatological Aid* Plant chewed, possibly taken internally, and spit on sores. (164:465) *Orthopedic Aid* Decoction of plant and any kind of "weeds" used for broken bones. (187:180)

Erigeron concinnus (Hook. & Arn.) Torr. & Gray, Navajo Fleabane
- **Drug**—**Navajo, Ramah** *Analgesic* Cold infusion taken and used as lotion for general body pain. *Disinfectant* Plant used for "antelope infection." (191:50) **Shoshoni** *Venereal Aid* Infusion of whole plant taken for gonorrhea. (118:47)

Erigeron concinnus var. *condensatus* D. C. Eat., Navajo Fleabane
- **Drug**—**Navajo, Kayenta** *Analgesic* Plant used as a lotion for headaches. *Gynecological Aid* Plant used for difficult labor. (205:47)

Erigeron divergens Torr. & Gray, Spreading Fleabane
- **Drug**—**Navajo** *Gynecological Aid* Infusion of plant taken by women as an aid for deliverance. (55:85) **Navajo, Kayenta** *Analgesic* Plant used as a snuff for headaches. (205:47) **Navajo, Ramah** *Ceremonial Medicine* Plant used ceremonially in several ways. *Disinfectant* Cold infusion of plant taken and used as a lotion for "lightning infection." *Eye Medicine* Cold, compound infusion of plant used as an eyewash. *Panacea* Root used as a "life medicine." *Snakebite Remedy* Compound used for snakebites. (191:50)
- **Other**—**Kiowa** *Good Luck Charm* Plant considered an omen of good fortune and brought into the home. (192:60)

Erigeron eximius Greene, Sprucefir Fleabane
- **Drug**—**Navajo, Ramah** *Ceremonial Medicine* Cold infusion of plant taken and used ceremonially as a lotion for various ills. *Cough Medicine* Cold infusion of plant taken and used as a lotion for cough. *Febrifuge* Cold infusion of plant taken and used as a lotion for fever. *Hunting Medicine* Infusion of plant used internally and externally for protection in warfare or hunting. *Misc. Disease Remedy* Cold infusion of plant taken and used as a lotion for influenza. *Witchcraft Medicine* Cold infusion of plant taken and used as a lotion for protection from witches. (as *E. superbus* 191:51)

Erigeron filifolius (Hook.) Nutt., Threadleaf Fleabane
- **Drug**—**Thompson** *Dermatological Aid* Toasted, powdered stems and leaves sprinkled on sores, cuts, and wounds. (164:473) Plant chewed, possibly taken internally, and spit on sores. (164:465) *Unspecified* Decoction of plant and any kind of "weeds" used for broken bones. (187:180)

Erigeron flagellaris Gray, Trailing Fleabane
- **Drug**—**Navajo, Ramah** *Ceremonial Medicine* Cold infusion of leaves used ceremonially as a medicine and as a fumigant. (as *E. nudiflorus* 191:50, 51) *Dermatological Aid* Poultice of chewed leaves applied to spider bites and used as a hemostat. (191:50) *Disinfectant* Cold infusion of leaves used ceremonially for "lightning infection." (as *E. nudiflorus* 191:50, 51) *Hemostat* Poultice of chewed leaves applied as a hemostatic. *Snakebite Remedy* Compound poultice of plant applied to snakebite. *Veterinary Aid* Cold infusion of leaves used as eyewash for livestock. (as *E. nudiflorus* 191:50, 51)

- *Fiber*—**Keres, Western** *Brushes & Brooms* Tied bunches of plants used for brooms. (171:42)

Erigeron foliosus var. *stenophyllus* (Nutt.) Gray, Leafy Fleabane

- *Drug*—**Kawaiisu** *Analgesic* Infusion of roots used as a wash for headaches. *Toothache Remedy* Root held between the teeth for toothache. (206:28) **Miwok** *Dermatological Aid* Decoction of root and vinegar weed used as a bath for pustules and skin eruptions from smallpox. (12:173, 174) *Febrifuge* Decoction of washed and pounded root taken for fever. *Misc. Disease Remedy* Decoction of washed and pounded root taken for ague. *Toothache Remedy* Root chewed and placed in cavity. (12:169)

Erigeron formosissimus Greene, Beautiful Fleabane

- *Drug*—**Navajo, Ramah** *Hunting Medicine* Cold, compound infusion taken and used as lotion for good luck in hunting. (191:50)

Erigeron grandiflorus Hook., Largeflower Fleabane

- *Drug*—**Gosiute** *Poison* Roots used for arrow poison. (39:368)

Erigeron linearis (Hook.) Piper, Desert Yellow Fleabane

- *Drug*—**Okanagan-Colville** *Tuberculosis Remedy* Decoction of whole plant taken for tuberculosis. (188:83)

Erigeron neomexicanus Gray, New Mexico Fleabane

- *Drug*—**Navajo, Kayenta** *Dermatological Aid* Powdered plant applied to dog or bear bite sores. *Gastrointestinal Aid* Plant used for stomachaches caused by eating unripe fruit. (205:47)

Erigeron peregrinus ssp. *callianthemus* (Greene) Cronq., Subalpine Fleabane

- *Drug*—**Cheyenne** *Analgesic* and *Orthopedic Aid* Infusion of dried, pulverized roots, stems, and flowers used as a steam bath or taken for backaches. *Other* Infusion of dried, pulverized roots, stems, and flowers used as a steam bath or taken for dizziness. *Stimulant* Infusion of roots, stems and flowers used as steam bath when dizzy and drowsy. (as *E. salsuginosus* 76:187) Infusion of dried, pulverized roots, stems, and flowers used as a steam bath or taken for drowsiness. (as *E. salsuginosus* 76:187)
- *Fiber*—**Thompson** *Basketry* Plant used as a pattern in basketry. (as *E. salsuginosus* 164:497)

Erigeron philadelphicus L., Philadelphia Fleabane

- *Drug*—**Blackfoot** *Antidiarrheal* Plant used for chronic diarrhea. *Antihemorrhagic* Plant used for childbirth hemorrhage. (97:56) **Cherokee** *Abortifacient* Taken for "suppressed menstruation." *Analgesic* Used as a poultice for headache. *Anticonvulsive* and *Antihemorrhagic* Used for hemorrhages, "spitting of blood," and epilepsy. *Cold Remedy* Cold infusion of root taken and root chewed for colds. *Cough Medicine* Taken for coughs. *Dermatological Aid* Astringent plant boiled, mixed with tallow, and used on sores. Boiled plant mixed with tallow and used on sores. *Diaphoretic* Used as a sudorific. *Diuretic* Used as a diuretic. *Eye Medicine* Taken for "dimness of sight." *Kidney Aid* Taken for gout and infusion taken "for kidneys." (80:35) **Houma** *Gynecological Aid* Decoction of root taken for "menstruation troubles." (158:62) **Iroquois** *Dermatological Aid* Compound infusion of plants used as wash for poison ivy and itch. Poultice of plants applied to running sores. *Pulmonary Aid* Decoction of whole plant taken to open the lungs. (87:464) **Meskwaki** *Analgesic* Snuff of powdered florets used for sick headaches. (152:213, 214) *Cold Remedy* Powdered disk florets used as snuff to make one sneeze for cold or catarrh. (152:213) *Respiratory Aid* Snuff of powdered florets used to make patient sneeze for catarrh. (152:213, 214) **Ojibwa** *Cold Remedy* Smoke of dried flowers inhaled for head cold. Snuff of pulverized flowers used to cause sneezing to loosen head colds. *Febrifuge* Infusion of flowers used to break fevers. (153:364) **Okanagan-Colville** *Analgesic* Infusion of leaves and blossoms taken for headaches. (188:83)
- *Food*—**Ojibwa** *Forage* Plant eaten by deer and cows. (153:398)
- *Other*—**Ojibwa** *Hunting & Fishing Item* Disk florets smoked to attract the buck deer. They say that cows and deer eat the blossoms. (153:429) *Smoke Plant* Plant used in the smoking tobacco or "kinnikinnick" mixture. (153:398)

Erigeron pulchellus Michx., Robin's Plantain

- *Drug*—**Cherokee** *Abortifacient* Taken for "suppressed menstruation." *Analgesic* Used as a poultice for headache. *Anticonvulsive* and *Antihemorrhagic* Used for hemorrhages, "spitting of blood," and epilepsy. *Cold Remedy* Cold infusion of root taken and root chewed for colds. *Cough Medicine* Taken for coughs. *Dermatological Aid* Astringent plant boiled, mixed with tallow, and used on sores. Boiled plant mixed with tallow and used on sores. *Diaphoretic* Used as a sudorific. *Diuretic* Used as a diuretic. *Eye Medicine* Taken for "dimness of sight." *Kidney Aid* Taken for gout and infusion taken "for kidneys." (80:35) **Iroquois** *Cold Remedy* Decoction of roots taken for colds. *Cough Medicine* Decoction of roots taken for coughs. *Tuberculosis Remedy* Decoction of plants and flowers taken for consumption. (87:464)

Erigeron pumilus Nutt., Shaggy Fleabane

- *Drug*—**Okanagan-Colville** *Eye Medicine* Infusion of roots used to wash the eyes as an "eye tonic." (188:84)

Erigeron sp., Fleabane

- *Drug*—**Thompson** *Analgesic* Salve of toasted, crushed plant and grease rubbed on painful area. (164:468) Decoction of plant taken for backache, stomachache, or menstrual cramps. (187:180) *Dermatological Aid* Salve of toasted, crushed plant and grease rubbed on painful, swollen areas. (164:468) Toasted, powdered stems and leaves sprinkled on sores, ulcers, cuts, and wounds. (164:473) *Gastrointestinal Aid* Decoction of plant taken for stomachache. (187:180) *Gland Medicine* Salve of toasted, crushed plant and grease rubbed on swollen glands. (164:468) *Gynecological Aid* Decoction of plant taken for menstrual cramps. (187:180) *Throat Aid* Salve of crushed plant used on throat or plant chewed for sore throat. (164:468)

Erigeron speciosus var. *macranthus* (Nutt.) Cronq., Aspen Fleabane

- *Drug*—**Navajo, Ramah** *Analgesic* Compound decoction taken for menstrual pain and as a contraceptive. *Contraceptive* Compound decoction of plant used as a contraceptive. *Gynecological Aid* Compound decoction of plant used for menstrual pain. (as *E. macranthus* 191:50)

Erigeron strigosus Muhl. ex Willd. var. *strigosus*, Prairie Fleabane

- *Drug*—**Catawba** *Heart Medicine* Infusion of roots taken for heart troubles. (as *E. ramosus* 157:191) Infusion of roots taken for heart troubles. (as *E. ramosus* 177:61) **Ojibwa** *Analgesic* Plant used for sick headache. (as *E. ramosus* 153:364)

Eriochloa, Poaceae

Eriochloa aristata Vasey, Bearded Cupgrass
- **Food**—**Cocopa** *Porridge* Seeds parched, ground, and the flour cooked into a mush. *Staple* Seeds parched, ground, and the flour eaten dry. (37:187)

Eriodictyon, Hydrophyllaceae

Eriodictyon angustifolium Nutt., Narrowleaf Yerbasanta
- **Drug**—**Hualapai** *Dermatological Aid* Decoction of leaves used as a wash for cuts. *Gastrointestinal Aid* Decoction of leaves taken for indigestion. *Laxative* Decoction of leaves taken as a laxative. *Orthopedic Aid* Decoction of leaves used as a wash for tired feet. (195:48) **Paiute** *Antidiarrheal* Decoction of leaves taken for diarrhea. *Antiemetic* Decoction of leaves taken for vomiting. *Cold Remedy* Decoction of leaves or shoots taken for colds. *Cough Medicine* Decoction of leaves or shoots taken for coughs. *Expectorant* Decoction of leaves or plant tops used as an expectorant for lungs or tuberculosis. *Pulmonary Aid* Infusion of leaves or tops taken as an expectorant for pulmonary troubles. *Tuberculosis Remedy* Decoction of leaves or plant tops used as an expectorant for lungs or tuberculosis. **Shoshoni** *Analgesic* Infusion of leaves taken for stomachaches. Poultice of decoction of stems, leaves, and flowers applied for rheumatic pains. *Antirheumatic (External)* Decoction of plant used in hot compresses for rheumatic pains. *Cold Remedy* Decoction of leaves or shoots taken for colds. *Cough Medicine* Decoction of leaves or shoots taken for coughs. *Expectorant* Decoction of leaves or plant tops used as an expectorant for lungs or tuberculosis. *Gastrointestinal Aid* Decoction of leaves taken for stomachaches. *Pulmonary Aid* Infusion of leaves or tops taken as an expectorant for pulmonary troubles. *Tuberculosis Remedy* Decoction of leaves or plant tops used as an expectorant for lungs or tuberculosis. *Venereal Aid* Decoction of leaves taken for venereal disease. (180:71, 72)

Eriodictyon californicum (Hook. & Arn.) Torr., California Yerbasanta
- **Drug**—**Atsugewi** *Antirheumatic (External)* Branches and leaves used in a steam bath for rheumatism. Steam of burned branches and leaves used for rheumatism. *Cold Remedy* Juice of chewed plant used for colds. Plant chewed and juice swallowed for colds. *Herbal Steam* Branches and leaves used in a steam bath for rheumatism. *Pulmonary Aid* Plant chewed and juice swallowed for whooping cough. Juice of chewed plant used for whooping cough. (61:140) **Coahuilla** *Analgesic* Decoction of leaves used as a wash for "sore parts" or painful, fatigued limbs. *Dermatological Aid* Poultice of leaves applied to men and animals with sores. *Orthopedic Aid* Decoction of leaves used as a wash for painful or fatigued limbs. *Veterinary Aid* Poultice of leaves applied to men and animals with sores. (13:78) **Costanoan** *Analgesic* Poultice of heated leaves applied to the forehead for headaches. *Antirheumatic (External)* Decoction of plant used for rheumatism. *Blood Medicine* Decoction of plant used to purify the blood. *Cold Remedy* Infusion of plant used for colds. *Dermatological Aid* and *Disinfectant* Plant combined with other herbs and used for infected sores. *Eye Medicine* Infusion of plant used as an eyewash. *Respiratory Aid* Decoction of plant used or leaves chewed or smoked for asthma. *Tuberculosis Remedy* Decoction of plant used for tuberculosis. (21:13) **Karok** *Cold Remedy* Decoction of leaves taken for colds. *Pulmonary Aid* Decoction of leaves taken for pleurisy. *Tuberculosis Remedy* Decoction of leaves taken for tuberculosis. (148:388) **Kawaiisu** *Cold Remedy* Infusion of leaves taken as a cold medicine. *Gastrointestinal Aid* Infusion of leaves used for stomach problems. *Venereal Aid* Infusion of leaves taken for gonorrhea. (206:29) **Mahuna** *Antirheumatic (Internal)* Plant used for rheumatism. *Cough Medicine* Plant used for coughs. *Pulmonary Aid* Plant used for pneumonia. *Respiratory Aid* Plant used for asthma. (140:19) **Mendocino Indian** *Cold Remedy* Leaves used for colds and asthma. *Respiratory Aid* Leaves used for inflammation of the bronchial tubes or asthma. (41:381) **Miwok** *Antirheumatic (External)* Poultice of leaves used as plasters on aching or sore spots. *Antirheumatic (Internal)* Infusion of leaves and flowers taken for rheumatism. Leaves chewed for rheumatism. *Cold Remedy* Infusion of leaves and flowers taken for colds. Leaves chewed for colds. Leaves smoked in form of cigarette for colds. *Cough Medicine* Infusion of leaves and flowers taken for coughs. Leaves chewed for coughs. Leaves smoked in form of cigarette for coughs. *Dermatological Aid* Poultice of mashed leaves applied to cuts, wounds, and abrasions. *Gastrointestinal Aid* Infusion of leaves and flowers taken for stomachaches. Leaves chewed for stomachache. *Orthopedic Aid* Poultice of mashed leaves applied to fractured bones in order to keep down the swelling, aid the knitting, and for pain. (12:169) **Pomo** *Expectorant* Infusion of gummy leaf taken as an expectorant. (118:38) **Pomo, Kashaya** *Blood Medicine* Decoction of leaves used as a blood purifier. *Cough Medicine* Decoction of leaves used as a cough medicine. *Dermatological Aid* Decoction of leaves used to make a wash for sores. *Febrifuge* Decoction of leaves used to bring down the fever of a cold. (72:74) **Round Valley Indian** *Antirheumatic (Internal)* Infusion of leaves taken for rheumatism. *Blood Medicine* Infusion of leaves taken as a blood purifier. *Febrifuge* Infusion of leaves used as a wash for fevers. *Misc. Disease Remedy* Leaves used for grippe. *Respiratory Aid* Infusion of leaves taken or used as wash for catarrh. *Tuberculosis Remedy* Infusion of leaves taken or used as wash for consumption. (41:381) **Yokut** *Diaphoretic* Infusion of plant taken and used as a steam for sweating. (yerba santa 117:437) **Yuki** *Cough Medicine* Plant used in cough syrup. *Dermatological Aid* Poultice of leaves applied to scabby sores. (49:47) **Yurok** *Cold Remedy* Infusion of leaves taken for colds. *Cough Medicine* Infusion of leaves taken for coughs. (6:30)
- **Food**—**Karok** *Beverage* Decoction of leaves and *Pinus lambertiana* pitch or leaves chewed and water taken as soothing drink. (6:30)
- **Fiber**—**Costanoan** *Clothing* Leaves woven into skirts and aprons. (21:253)

Eriodictyon crassifolium Benth., Thickleaf Yerbasanta
- **Drug**—**Luiseño** *Unspecified* Plant used for medicinal purposes. (155:230)

Eriodictyon lanatum (Brand) Abrams, San Diego Yerbasanta
- **Drug**—**Diegueño** *Cold Remedy* Decoction of leaves, with or without honey, taken for colds. *Cough Medicine* Decoction of leaves, with or without honey, taken for coughs. (as *E. trichocalyx* ssp. *lanatum* 84:21) Decoction of leaves taken for coughs. (88:219)

- *Food*—Diegueño *Candy* Decoction of leaves and honey boiled down into a syrup or candy and used by children. *Dried Food* Leaves dried, stored indefinitely, and used for colds, coughs, candy, and with soap to wash hair. (as *E. trichocalyx* ssp. *lanatum* 84:21)
- *Other*—Diegueño *Incense & Fragrance* Leaves used with soap to wash the hair. (as *E. trichocalyx* ssp. *lanatum* 84:21)

Eriodictyon tomentosum Benth., Woolly Yerbasanta
- *Drug*—Luiseño *Unspecified* Plant used for medicinal purposes. (155:230)

Eriodictyon trichocalyx Heller, Hairy Yerbasanta
- *Drug*—Cahuilla *Antirheumatic (External)* Poultice of fresh, pounded leaves applied to sore or fatigued limbs for rheumatism. *Antirheumatic (Internal)* Decoction of leaves taken for rheumatism. *Blood Medicine* Decoction of leaves used as a blood purifier. *Cold Remedy* Decoction of leaves used for colds. *Cough Medicine* Decoction of leaves used for coughs. Decoction of three leaves and ½ teaspoon of sugar taken for coughs, 1 teaspoon every 4 hours. *Febrifuge* Decoction of leaves applied as a liniment for fevers. *Oral Aid* Fresh leaves chewed as a thirst quencher. *Respiratory Aid* Decoction of leaves used for asthma and catarrh. *Throat Aid* Decoction of leaves used for sore throats. *Tuberculosis Remedy* Decoction of leaves used for tuberculosis. (15:71)
- *Food*—Cahuilla *Beverage* Fresh or dried leaves boiled into tea. (15:71)

Eriogonum, Polygonaceae

Eriogonum abertianum Torr. **var. *abertianum***, Abert's Buckwheat
- *Drug*—Navajo, Ramah *Dermatological Aid* Decoction of plant used as lotion for skin cuts. *Veterinary Aid* Decoction of plant used as lotion for skin cuts on horses. (191:23)

Eriogonum alatum Torr., Winged Buckwheat
- *Drug*—Navajo *Analgesic* Plant used for pain. (55:42) Navajo, Kayenta *Dermatological Aid* Plant used as a lotion for rashes. *Panacea* Plant used as a life medicine. (205:19) Navajo, Ramah *Antidiarrheal* Cold infusion of root used for diarrhea. *Ceremonial Medicine* Cold infusion of root used as a ceremonial medicine. *Cough Medicine* Cold infusion of root used for bad cough. *Dermatological Aid* Powdered root mixed with tallow and used as ointment for infant's sore navel. *Oral Aid* Cold infusion of root used as a mouthwash for sore gums. *Panacea* Cold infusion of root used as an important "life medicine." *Pediatric Aid* Powdered root mixed with tallow and used as ointment for infant's sore navel. (191:23) Zuni *Emetic* Root eaten as an emetic for stomachaches. (27:378) *Other* Infusion of powdered root taken after a fall and relieve general misery. (166:49)
- *Food*—Navajo *Unspecified* Roots used for food. (55:42) Navajo, Ramah *Porridge* Ground seeds made into a mush with milk. *Unspecified* Root chewed by children. (191:23)
- *Other*—Navajo *Ceremonial Items* Plant used in the Life or Knife Chant. (55:42)

Eriogonum androsaceum Benth., Rockjasmine Buckwheat
- *Drug*—Thompson *Analgesic* Decoction of plant taken for internal pains, especially stomach pain. Plants used in steam bath for rheumatism, stiff and aching joints and muscles. *Antirheumatic (External)* Plants used in steam bath for aching and rheumatic joints. Plants steamed in sweat bath for rheumatism and various aches and stiffness. *Dermatological Aid* Salve of dry leaves or leaf ash mixed with grease used for swellings. *Gastrointestinal Aid* Decoction of plant taken for stomach pain. *Herbal Steam* Plants steamed in sweat bath for rheumatism and various aches and stiffness. *Orthopedic Aid* Plants used in steam bath for sprains, stiff and aching joints and muscles. *Other* Mild or medium decoction taken for general indisposition. *Venereal Aid* Strong decoction of plant used for syphilis. (164:470)

Eriogonum angulosum Benth., Anglestem Buckwheat
- *Food*—Kawaiisu *Unspecified* Seeds pounded in a bedrock mortar hole and eaten "without boiling." (206:29)

Eriogonum annuum Nutt., Annual Buckwheat
- *Drug*—Lakota *Diuretic* Infusion of plant used for urination problems. *Oral Aid* and *Pediatric Aid* Infusion of plant used for children with sore mouths. (139:54) Navajo, Ramah *Ceremonial Medicine* Plant used as a ceremonial medicine. *Dermatological Aid* Cold infusion taken or used as lotion for red ant bite. *Disinfectant* Cold infusion taken for "lightning infection." *Other* Cold infusion taken or used as lotion for sickness from swallowing an ant. *Panacea* Plant used as a "life medicine," "the boss of all medicines." *Witchcraft Medicine* Cold infusion used for protection against witches. (191:23)
- *Dye*—Kiowa *Unspecified* Leaves rubbed on buffalo or deer hides in the process of staining and tanning. (192:24) Lakota *White* Blossoms, brains, liver or gall and spleen rubbed into hides to bleach them. (139:54)
- *Other*—Kiowa *Hide Preparation* Leaves rubbed on buffalo or deer hides in the process of staining and tanning. The leaves were applied fresh, or, if the hide was dry, the leaves were moistened before application. (192:24)

Eriogonum baileyi S. Wats., Bailey's Buckwheat
- *Drug*—Tubatulabal *Dermatological Aid* Infusion of entire plant used as lotion for pimples. (193:59)
- *Food*—Kawaiisu *Beverage* Seeds pounded into a meal, mixed with water, and used as a beverage. *Staple* Seeds pounded into a meal and eaten dry. (206:29)

Eriogonum cernuum Nutt., Nodding Buckwheat
- *Drug*—Navajo, Kayenta *Dermatological Aid* Plant used for rashes. *Kidney Aid* Infusion of plant used for kidney disease. (205:19) Navajo, Ramah *Dermatological Aid* Poultice of chewed leaves applied to red ant bite. (191:23)
- *Food*—Navajo, Kayenta *Porridge* Seeds made into a mush and used for food. (205:19)

Eriogonum compositum Dougl. ex Benth., Arrowleaf Buckwheat
- *Drug*—Okanagan-Colville *Cold Remedy* Decoction of roots and stems taken for colds. *Dermatological Aid* Decoction of roots and stems used to wash infected cuts. Poultice of mashed leaves applied to cuts or infusion of leaves used as a wash for cuts and sores. (188:112) Sanpoil *Antidiarrheal* Decoction of root taken for diarrhea. (131:218)
- *Other*—Okanagan-Colville *Toys & Games* Stems used by children to play a game.[35] (188:112)

Eriogonum corymbosum Benth., Crispleaf Buckwheat
- *Drug*—**Havasupai** *Analgesic* Decoction of leaves taken three times a day for headaches. (197:216)
- *Food*—**Hopi** *Bread & Cake* Leaves boiled, mixed with water and cornmeal, and baked into a bread. (56:21) *Dried Food* Boiled stalks pressed into cakes, dried, and eaten with salt. (190:159)

Eriogonum davidsonii Greene, Davidson's Buckwheat
- *Food*—**Kawaiisu** *Staple* Seeds pounded into a meal and eaten dry. (206:29)

Eriogonum divaricatum Hook., Divergent Buckwheat
- *Drug*—**Navajo, Kayenta** *Ceremonial Medicine* Plant used for "Big Snake chant." *Orthopedic Aid* Poultice of plant applied to back for leg paralysis. *Snakebite Remedy* Plant smoked for snakebites. (205:19)

Eriogonum elatum Dougl. ex Benth., Tall Woolly Buckwheat
- *Drug*—**Mahuna** *Cathartic* Branch chewed or infusion of plant taken as a physic. (140:20)
- *Other*—**Paiute** *Toys & Games* Stems used in game similar to wishbone pulling. When the plant has matured and the stems are dry and brittle, the stout, long-branched flowering stems of this plant are used to play a game. Two children fashion a hook for themselves from the branching portion of the stem. These hooks are interlocked, and the participants pull on them until one of the hooks is broken. The person with the unbroken hook is considered to be the winner. (111:68)

Eriogonum elongatum Benth., Longstem Buckwheat
- *Drug*—**Mahuna** *Blood Medicine* Plant used as a blood tonic. *Hypotensive* Plant used for high blood pressure and hardening of arteries. (140:22)
- *Other*—**Tubatulabal** *Cooking Tools* Tubular sections of jointed stalks used to collect and roast juice for chewing gum. (193:19)

Eriogonum fasciculatum Benth., Eastern Mojave Buckwheat
- *Drug*—**Coahuilla** *Analgesic* Decoction of leaves taken for headache and stomach pain. *Eye Medicine* Infusion of flower used as an eyewash. *Gastrointestinal Aid* Decoction of leaves taken for stomach pain and headache. (13:78) **Costanoan** *Urinary Aid* Decoction of plant used for urinary problems. (21:11) **Diegueño** *Antidiarrheal* Decoction of flowers given to babies for diarrhea. *Emetic* Decoction of flowers taken to "throw up badness in the stomach." (84:21) *Heart Medicine* Decoction of dried flowers or dried roots taken for a healthy heart. (88:216) *Pediatric Aid* Decoction of flowers given to babies for diarrhea. (84:21) **Navajo** *Witchcraft Medicine* Decoction of plants used as an anti-witchcraft medicine. (55:42) **Omaha** *Dermatological Aid* Poultice of powdered root applied to wounds. (58:49) **Zuni** *Dermatological Aid* Poultice of powdered root applied to cuts and arrow or bullet wounds. *Gynecological Aid* Decoction of root taken after parturition to heal lacerations. *Throat Aid* Decoction of root taken for hoarseness and colds involving the throat. (166:49)

Eriogonum fasciculatum* var. *polifolium (Benth.) Torr. & Gray, Eastern Mojave Buckwheat
- *Drug*—**Tubatulabal** *Antidiarrheal* Decoction of dried flowers given to children for bloody flux. Infusion of dried heads taken for diarrhea. *Gastrointestinal Aid* Infusion of dried heads taken for stomachaches. *Pediatric Aid* Decoction of dried flowers given to children for bloody flux. (193:59)
- *Other*—**Kawaiisu** *Containers* Leaves used to line the acorn granary to prevent the acorns from getting wet. *Tools* Wood used to pierce ears. (206:29)

Eriogonum flavum Nutt., Yellow Eriogonum
- *Food*—**Blackfoot** *Unspecified* Roots used for food. (97:33)

Eriogonum gracillimum S. Wats., Rose and White Buckwheat
- *Drug*—**Tubatulabal** *Dermatological Aid* Infusion of entire plant used as lotion for pimples. (193:59)

Eriogonum heracleoides Nutt., Parsnipflower Buckwheat
- *Drug*—**Okanagan-Colville** *Cold Remedy* Decoction of roots and stems taken for colds. *Dermatological Aid* Decoction of roots and stems used to wash infected cuts. Poultice of mashed leaves applied to cuts or infusion of leaves used as a wash for cuts and sores. (188:112) **Sanpoil** *Antidiarrheal* Decoction of root taken for diarrhea. (131:218) **Thompson** *Analgesic* Decoction of plant taken for internal pains, especially stomach pain. Plants used in steam bath for rheumatism, stiff and aching joints and muscles. *Antirheumatic (External)* Plants used in steam bath for aching and rheumatic joints. Plants steamed in sweat bath for rheumatism and various aches and stiffness. (164:470) *Ceremonial Medicine* Decoction of whole plant used as a purifying ceremonial wash in the sweat house. (164:505) *Dermatological Aid* Salve of dry leaves or leaf ash mixed with grease used for swellings. (164:470) Infusion of plant used as a wash for sores. (187:237) *Disinfectant* Decoction of whole plant used as a purifying ceremonial wash in the sweat house. (164:505) *Eye Medicine* Decoction of leaves used as a wash for sore eyes. (187:237) *Gastrointestinal Aid* Decoction of plant taken for stomach pain. *Herbal Steam* Plants steamed in sweat bath for rheumatism and various aches and stiffness. *Orthopedic Aid* Plants used in steam bath for sprains, stiff and aching joints and muscles. *Other* Mild or medium decoction taken for general indisposition. (164:470) *Pulmonary Aid* Decoction of washed, clean plant taken for sickness on the lung. *Tuberculosis Remedy* Decoction of washed, clean plant taken for tuberculosis. Infusion of plant taken in large quantities for tuberculosis. *Unspecified* Decoction of plant taken or used as a wash for an unspecified illness. (187:237) *Venereal Aid* Strong decoction of plant used for syphilis. (164:470)
- *Other*—**Okanagan-Colville** *Toys & Games* Stems used by children to play a game.[35] (188:112)

Eriogonum hookeri S. Wats., Hooker's Buckwheat
- *Food*—**Hopi** *Spice* Boiled with mush for flavor. (190:160)

Eriogonum inflatum Torr. & Frém., Native American Pipeweed
- *Drug*—**Navajo, Kayenta** *Dermatological Aid* Plant used as a lotion for bear or dog bite. (205:19)
- *Food*—**Havasupai** *Vegetable* Leaves boiled for 5 to 10 minutes and eaten. (197:216) Young, fresh, tender leaves boiled, drained, balled into individual portions, and served. (197:66) **Kawaiisu** *Porridge* Seeds pounded into a meal and eaten mixed with water. *Staple* Seeds pounded into a meal and eaten dry. (206:29)
- *Other*—**Havasupai** *Cooking Tools* Stems cut at both ends and used as drinking tubes. (197:216) **Yavapai** *Smoking Tools* Dried stem used as tobacco pipe if pottery pipe lacking, burned with tobacco. (65:263)

Eriogonum insigne S. Wats., Ladder Buckwheat
- **Other**—**Kawaiisu** *Smoking Tools* Stems used as a substitute for "cane" as a pipe. (206:30)

Eriogonum jamesii Benth., James's Buckwheat
- **Drug**—**Apache, White Mountain** *Ceremonial Medicine* Plant used in medicine ceremonies. *Oral Aid* Plant chewed to sweeten the saliva. *Unspecified* Plant used for medicinal purposes. (136:157) **Keres, Western** *Heart Medicine* Roots chewed for a heart medicine. *Psychological Aid* Infusion of roots used for despondency. (171:43) **Navajo, Kayenta** *Psychological Aid* Plant smoked when disturbed by dreaming of tobacco worms. (205:19) **Navajo, Ramah** *Analgesic* Decoction of whole plant taken to ease labor pains. *Contraceptive* Root used as a contraceptive. *Gastrointestinal Aid* Cold infusion of whole plant taken to kill a swallowed red ant. *Gynecological Aid* Decoction of whole plant taken to ease labor pains. *Panacea* Root used as a "life medicine." (191:23) **Zuni** *Eye Medicine* Root soaked in water and used as a wash for sore eyes. *Gastrointestinal Aid* Fresh or dried root eaten for stomachaches. (27:378) *Oral Aid* Root carried in mouth for sore tongue, then buried in river bottom. (166:50)
- **Other**—**Zuni** *Ceremonial Items* Ground blossom powder given to ceremonial dancers impersonating anthropic gods to bring rain. The blossom powder was given to the dancers after they were dressed for the ceremony. The dance director placed it in the mouth of each dancer so that the dance would bring rain. Each dancer ejected the medicine from his mouth over his body and apparel. (166:91)

Eriogonum latifolium Sm., Seaside Buckwheat
- **Drug**—**Costanoan** *Cold Remedy* Decoction of root, stalk, and leaves taken for colds. *Cough Medicine* Decoction of root, stalk, and leaves taken for coughs. (21:11) **Round Valley Indian** *Analgesic* Decoction of leaves, stems, and roots taken for stomach pains and headaches. *Eye Medicine* Decoction of roots used for sore eyes. *Gastrointestinal Aid* Decoction of leaves, stems, and roots taken for stomach pains. *Gynecological Aid* Decoction of leaves, stems, and roots taken for female complaints. (41:345)
- **Food**—**Mendocino Indian** *Unspecified* Young stems eaten by children in early summer. (41:345)

Eriogonum leptophyllum (Torr. & Gray) Woot. & Standl., Slenderleaf Buckwheat
- **Drug**—**Navajo, Ramah** *Analgesic* Decoction of whole plant used for postpartum pain. *Gynecological Aid* Decoction of whole plant used for postpartum pain and to aid in placenta delivery. *Panacea* Decoction of whole plant used as a "life medicine." *Snakebite Remedy* Decoction of whole plant used internally and externally for snakebite. (191:23)

Eriogonum longifolium Nutt., Longleaf Buckwheat
- **Drug**—**Comanche** *Gastrointestinal Aid* Infusion of root taken for stomach trouble. (29:521)
- **Food**—**Kiowa** *Unspecified* Root used for food. (192:25)

Eriogonum microthecum Nutt., Slender Buckwheat
- **Drug**—**Paiute** *Tuberculosis Remedy* Decoction of roots or tops used for tuberculosis. *Urinary Aid* Decoction of stems and leaves used for bladder trouble. **Shoshoni** *Antirheumatic (External)* Decoction of plant used in hot compresses or as a wash for lameness or rheumatism. *Cough Medicine* Decoction of roots and sometimes tops taken for tubercular cough. *Orthopedic Aid* Decoction of whole plant used as a wash or as a compress for lameness. *Tuberculosis Remedy* Decoction of roots or tops used for tuberculosis. (180:72)
- **Food**—**Havasupai** *Beverage* Used to make tea. (197:217)

Eriogonum niveum Dougl. ex Benth., Snow Buckwheat
- **Drug**—**Okanagan-Colville** *Cold Remedy* Decoction of roots and stems taken for colds. *Dermatological Aid* Decoction of roots and stems used to wash infected cuts. Poultice of mashed leaves applied to cuts or infusion of leaves used as a wash for cuts and sores. (188:112)
- **Other**—**Okanagan-Colville** *Toys & Games* Stems used by children to play a game.[35] (188:112)

Eriogonum nudum Dougl. ex Benth., Naked Buckwheat
- **Food**—**Karok** *Vegetable* Sour tasting, young stems eaten raw as greens. (148:383) **Miwok** *Vegetable* Raw greens, sour flavor, used for food. (12:159)
- **Fiber**—**Miwok** *Brushes & Brooms* Twigs and leaves used as a brush to clear ground under manzanita bushes before knocking off berries. (12:161)
- **Other**—**Karok** *Toys & Games* Stems used by children to play a game by hooking each other's plant. (148:383) **Kawaiisu** *Cooking Tools* Hollow stems used as drinking tubes. *Smoking Tools* Hollow stems used as pipes for smoking. (206:30)

Eriogonum nudum* var. *oblongifolium S. Wats., Naked Buckwheat
- **Drug**—**Karok** *Gastrointestinal Aid* Roots used for abdominal ailments. (6:30)
- **Food**—**Karok** *Unspecified* Young shoots used for food. (6:30) *Vegetable* Sour tasting, young stems eaten raw as greens. (148:383)
- **Other**—**Karok** *Toys & Games* Stems used by children to play a game by hooking each other's plant. (148:383)

Eriogonum nudum* var. *pauciflorum S. Wats., Naked Buckwheat
- **Drug**—**Kawaiisu** *Cold Remedy* Infusion of roots taken for colds. *Cough Medicine* Infusion of roots taken for coughs. (206:30)
- **Other**—**Kawaiisu** *Cooking Tools* Hollow stems used as drinking tubes. *Smoking Tools* Hollow stems used as pipes for smoking. (206:30)

Eriogonum nudum* var. *pubiflorum Benth., Naked Buckwheat
- **Food**—**Kawaiisu** *Unspecified* Dried flowers mixed with valley stickweed seeds and eaten. (206:30)

Eriogonum ovalifolium Nutt., Cushion Buckwheat
- **Drug**—**Gosiute** *Eye Medicine* Plant used as an eye medicine. *Gastrointestinal Aid* Plant used for stomachaches. (39:369) *Venereal Aid* Poultice of plant applied or plant used as wash for venereal diseases. (39:351) **Paiute** *Cold Remedy* Decoction of root taken as a cold remedy. **Shoshoni** *Cold Remedy* Decoction of root taken as a cold remedy. (180:72) **Ute** *Unspecified* Used as medicine. (38:34)

Eriogonum plumatella Dur. & Hilg., Yucca Buckwheat
- **Food**—**Kawaiisu** *Porridge* Seeds pounded, cooked into a mush, and eaten. (206:30)

Eriogonum pusillum Torr. & Gray, Yellowturbans
- *Food*—Kawaiisu *Staple* Seeds pounded and eaten dry. *Unspecified* Flowers mixed with valley stickweed seeds and eaten. (206:30)

Eriogonum racemosum Nutt., Redroot Buckwheat
- *Drug*—**Navajo, Kayenta** *Analgesic* Plant used for backaches and side aches. *Orthopedic Aid* Plant used for backaches and side aches. (205:19) **Navajo, Ramah** *Blood Medicine* Cold infusion of whole plant taken for blood poisoning or internal injuries. (191:23)
- *Food*—**Navajo, Kayenta** *Unspecified* Leaves and stems eaten raw. (205:19)

Eriogonum roseum Dur. & Hilg., Wand Buckwheat
- *Drug*—**Tubatulabal** *Dermatological Aid* Infusion of entire plant used as lotion for pimples. (as *E. virgatum* 193:59)
- *Food*—**Kawaiisu** *Beverage* Seeds pounded into a meal, mixed with water, and used as a beverage. *Staple* Seeds pounded into a meal and eaten dry. (206:30)

Eriogonum rotundifolium Benth., Roundleaf Buckwheat
- *Drug*—**Keres, Western** *Emetic* Infusion of roots used as an emetic to eliminate the ozone in cases of lightning shock. *Psychological Aid* Plant eaten by children to become good looking. *Sedative* Infusion of roots used for lightning shock. (171:43) **Navajo** *Emetic* Plant taken to vomit after swallowing ants. (55:42) *Throat Aid* Leaves used for sore throats. *Unspecified* Roots used as medicine. (90:150)
- *Food*—**Navajo** *Unspecified* Stems used for food. (90:150)

***Eriogonum* sp.**, Buckwheat
- *Drug*—**Cahuilla** *Cathartic* Leaves, growing near the root, used as a physic. *Eye Medicine* Infusion of flowers used as an eyewash. *Gastrointestinal Aid* Infusion of flowers used to clean out the intestines. *Gynecological Aid* Infusion of plant taken to shrink the uterus and reduce dysmenorrhea. (15:72) **Hopi** *Analgesic* Plant used for severe pain in hips and back. Plant used for severe pain in hips and back, especially in pregnant state. (42:314) Plant used for pain in hips and back, especially during pregnancy. (200:35) *Antihemorrhagic* Plant used for hemorrhage. *Gynecological Aid* Plant used as menstruation medicine. Plant used to expedite childbirth. (42:314) Plant used for pain in hips and back, especially during pregnancy. (200:35) Plant used to ease menstrual difficulties and ease childbirth. (200:35, 73) *Orthopedic Aid* Plant used for pain in hips and back. (200:35) **Kawaiisu** *Antidiarrheal* Decoction of roots taken for diarrhea. *Heart Medicine* Root used as heart medicine. (206:30) **Mahuna** *Oral Aid* Infusion of flowers and leaves used as a mouthwash for pyorrhea. (140:25) **Navajo** *Antidiarrheal* Cold infusion of roots taken for diarrhea. (55:42) *Gynecological Aid* Plant used during confinement after childbirth. (90:150) **Navajo, Kayenta** *Gastrointestinal Aid* Plant used for stomach disease. (205:18) **Thompson** *Analgesic* Decoction of plant taken for internal pains, especially stomach pain. Plants used in steam bath for rheumatism, stiff and aching joints and muscles. Plants steamed in sweat bath for rheumatism and various aches and stiffness. *Antirheumatic (External)* Plants used in steam bath for aching and rheumatic joints. *Dermatological Aid* Salve of dry leaves or leaf ash mixed with grease used for swellings. *Gastrointestinal Aid* Decoction of plant taken for stomach pain. *Herbal Steam* Plants steamed in sweat bath for rheumatism and various aches and stiff-

ness. *Orthopedic Aid* Plants used in steam bath for sprains, stiff and aching joints and muscles. *Other* Mild or medium decoction taken for general indisposition. *Venereal Aid* Strong decoction of plant used for syphilis. (164:470)
- *Food*—**Cahuilla** *Unspecified* Shoots and seeds used for food. (15:72) **Tubatulabal** *Unspecified* Seeds used for food. (193:15)
- *Fiber*—**Yavapai** *Brushes & Brooms* Stems used to brush off spines on fruits. (65:257)

Eriogonum sphaerocephalum Dougl. ex Benth., Rock Buckwheat
- *Drug*—**Paiute** *Antidiarrheal* Decoction of root taken for diarrhea. *Cold Remedy* Decoction of root taken for colds. **Shoshoni** *Antidiarrheal* Decoction of root taken for diarrhea. (180:73)

Eriogonum tenellum Torr., Tall Buckwheat
- *Drug*—**Keres, Western** *Febrifuge* Hot infusion of plant given to mothers after childbirth for fever. *Gynecological Aid* Infusion of roots or raw roots given to women during difficult labor. Infusion of plant used as a douche after childbirth. (171:43)

Eriogonum umbellatum Torr., Sulphur Wildbuckwheat
- *Drug*—**Kawaiisu** *Dermatological Aid* and *Venereal Aid* Mashed flowers used as a salve for gonorrheal sores. (206:30) **Mahuna** *Gastrointestinal Aid* Infusion of blossoms taken for ptomaine poisoning. (140:49) **Navajo, Kayenta** *Disinfectant* Plant used as a fumigant for biliousness. *Emetic* Plant used as an emetic for biliousness. *Gastrointestinal Aid* Plant used as a fumigant or emetic for biliousness. (205:20) **Nevada Indian** *Cold Remedy* Infusion of roots taken for colds. (118:37) **Paiute** *Analgesic* Decoction of roots taken for stomachaches. *Antirheumatic (External)* Poultice of mashed leaves, often with roots, used for lameness or rheumatism. (180:73) *Cold Remedy* Infusion of roots taken for colds. (167:317) Hot decoction of roots taken for colds. *Gastrointestinal Aid* Decoction of root taken for stomachaches. *Orthopedic Aid* Poultice of leaves, and sometimes roots, applied for lameness or rheumatism. **Shoshoni** *Antirheumatic (External)* Poultice of mashed leaves, often with roots, used for lameness or rheumatism. *Cold Remedy* Hot decoction of roots taken for colds. *Orthopedic Aid* Poultice of leaves, and sometimes roots, applied for lameness or rheumatism. (180:73)

Eriogonum umbellatum* var. *majus Hook., Sulphurflower Buckwheat
- *Drug*—**Cheyenne** *Gynecological Aid* Infusion of powdered stems and flowers taken for lengthy menses. (as *E. subalpinum* 76:172) Stems and flowers powdered, made into a tea, and used for menses that ran too long. (83:32)
- *Food*—**Blackfoot** *Beverage* Leaves boiled to make tea. (as *E. subalpinum* 97:33)

Eriogonum umbellatum* var. *stellatum (Benth.) M. E. Jones, Sulphurflower Buckwheat
- *Drug*—**Klamath** *Burn Dressing* Leaves placed on burns to soothe the pain. (as *E. stellatum* 45:95) Poultice of leaves applied to burns. (as *E. stellatum* 163:131)

Eriogonum wrightii Torr. ex Benth., Bastardsage
- *Drug*—**Navajo, Kayenta** *Emetic* Plant used as an emetic. (205:20)

- *Food*—Kawaiisu *Beverage* Seeds pounded into a meal, mixed with water, and used as a beverage. *Staple* Seeds pounded into a meal and eaten dry. (206:30)

Erioneuron, Poaceae
Erioneuron pulchellum (Kunth) Tateoka, Low Woollygrass
- *Drug*—Havasupai *Laxative* Decoction of blades taken as a laxative. (as *Tridens pulchellus* 197:210)

Eriophorum, Cyperaceae
Eriophorum angustifolium Honckeny, Tall Cottongrass
- *Drug*—Eskimo, Kuskokwagmiut *Panacea* Raw stems eaten to restore good health to persons in generally poor health. (122:27)
- *Food*—Alaska Native *Unspecified* Stem bases eaten raw with seal oil. "Mouse nuts" found in mice caches, cooked and eaten with seal oil. (85:131) Eskimo, Inupiat *Unspecified* Roots eaten raw or boiled. *Winter Use Food* Roots stored in seal oil for future use. (98:119)
- *Fiber*—Eskimo, Alaska *Mats, Rugs & Bedding* Dried leaves and stems woven into soft mats or covers for coarse grass mattresses. (1:34)

Eriophorum angustifolium ssp. *subarcticum* (Vassiljev) Hultén, Tall Cottongrass
- *Drug*—Eskimo, Inuktitut *Unspecified* "Female" stems used medicinally. (202:184)
- *Food*—Eskimo, Inuktitut *Unspecified* "Female" stems used for food. (202:184)
- *Fiber*—Eskimo, Inuktitut *Clothing* "Female" stems dried, split, and inserted into boot welts to seal them. *Mats, Rugs & Bedding* "Female" stems dried, split, and used for weaving. (202:184)

Eriophorum callitrix Cham. ex C. A. Mey., Arctic Cottongrass
- *Drug*—Ojibwa *Hemostat* Matted fuzz used as a "hemostatic." (153:368)

Eriophorum russeolum Fries ex Hartman, Red Cottongrass
- *Drug*—Eskimo, Western *Dermatological Aid* Poultice of "cotton" from plant applied to boils to absorb the pus. (108:17) *Eye Medicine* "Cotton" from plant put in corner of eye to absorb fluid from "watery eyes." (108:22)

Eriophorum scheuchzeri Hoppe, White Cottongrass
- *Drug*—Eskimo, Western *Dermatological Aid* Poultice of "cotton" from plant applied to boils to absorb the pus. (108:17)
- *Fiber*—Eskimo, Inuktitut *Clothing* Dried stems used in summer for boot insoles. (202:186)

Eriophorum sp., Cottongrass
- *Other*—Eskimo, Inuktitut *Lighting* Seed head bristles used to make wicks for oil lamps. (202:184) Tanana, Upper *Hunting & Fishing Item* Flower used as a lure for catching grayling. (102:9)

Eriophyllum, Asteraceae
Eriophyllum ambiguum (Gray) Gray, Beautiful Woollysunflower
- *Food*—Kawaiisu *Dried Food* Seeds parched, pounded, and eaten dry. (206:30)

Eriophyllum confertiflorum (DC.) Gray, Yellow Yarrow
- *Food*—Cahuilla *Staple* Parched seeds ground into flour. (15:72)

Eriophyllum lanatum (Pursh) Forbes, Woolly Eriophyllum
- *Drug*—Chehalis *Love Medicine* Dried flowers used as a love charm. Skagit *Dermatological Aid* Leaves rubbed on the face to prevent chapping. (79:49)

Eriophyllum lanatum var. *leucophyllum* (DC.) W. R. Carter, Common Woollysunflower
- *Drug*—Miwok *Antirheumatic (External)* Poultice of leaves bound on body over aching parts. (as *E. caespitosum* 12:169)

Erodium, Geraniaceae
Erodium cicutarium (L.) L'Hér. ex Ait., Redstem Stork's Bill
- *Drug*—Costanoan *Misc. Disease Remedy* Infusion of leaves used for typhoid fever. (21:8) Jemez *Gynecological Aid* Plant and roots eaten by women to produce more milk for the nursing children. (44:22) Navajo, Kayenta *Dermatological Aid* Plant used for wildcat, bobcat, or mountain lion bites. *Disinfectant* Plant used for infections. (205:29) Zuni *Dermatological Aid* Poultice of chewed root applied to sores and rashes. *Gastrointestinal Aid* Infusion of root taken for stomachache. (27:376)
- *Food*—Costanoan *Unspecified* Raw stems used for food. (21:252) Diegueño *Vegetable* Leaves picked early in the spring before the flowers appeared and cooked as greens. (84:21) Hopi *Candy* Roots chewed by children, sometimes as gum. (42:313) Isleta *Forage* High moisture content of leaves and stems made it a good grazing plant for livestock. (100:28) Kawaiisu *Forage* Plant eaten by horses, cows, and rabbits. (206:31) Navajo, Ramah *Fodder* Used for sheep feed. (191:34)
- *Other*—Jemez *Protection* Dried plant powder mixed with watermelon seeds during storage and planting stops watermelon disease. (44:22) Navajo, Kayenta *Ceremonial Items* Used on prayer sticks. (205:29)

Erodium sp., Filaree
- *Drug*—Hualapai *Gastrointestinal Aid* Roots used for stomach disorders. (195:43)
- *Food*—Cahuilla *Vegetable* Used as a potherb. Leaves eaten fresh or cooked. (15:72)

Eryngium, Apiaceae
Eryngium alismifolium Greene, Modoc Eryngo
- *Drug*—Paiute *Antidiarrheal* Infusion of whole plant used for diarrhea. (118:42) Infusion of plant taken for diarrhea. (180:73)

Eryngium aquaticum L., Rattlesnakemaster
- *Drug*—Alabama *Emetic* Infusion of plant taken as an emetic. Cherokee *Emetic* and *Gastrointestinal Aid* Infusion of plant taken to cause vomiting for nausea. (177:45) Choctaw *Antidote* Root used as an "anti-poison"; especially good for snakebite. *Diuretic* Root used as a powerful diuretic. *Expectorant* Root used as a powerful expectorant. *Snakebite Remedy* Root used as an "anti-poison"; especially good for snakebite. *Stimulant* Root used as a powerful stimulant. *Venereal Aid* Plant used for gonorrhea. (28:287) Delaware *Anthelmintic* Used for intestinal tape and "pin" worms. (176:35) Delaware, Oklahoma *Anthelmintic* Root used for tapeworms and pinworms. *Venereal Aid* Root used alone and in compound for venereal disease. (175:29, 76) Koasati *Emetic* Decoction of roots taken as an emetic. (177:45)

Eryngium yuccifolium Michx., Button Eryngo
- **Drug**—**Cherokee** *Pulmonary Aid* Decoction used to prevent whooping cough. *Snakebite Remedy* Used as snakebite remedy. *Toothache Remedy* Infusion held in mouth for toothaches. (80:27) **Creek** *Analgesic* Cold infusion of root taken for neuralgia and kidney troubles. (172:655, 656) Cold infusion of pounded roots taken for neuralgia. (177:45) *Antirheumatic (Internal)* Plant used with "deer potato" for rheumatism. *Blood Medicine* Infusion of root taken "to cleanse the system and purify the blood." (172:655, 656) Plant used to cleanse the system and purify the blood. (177:45) *Cathartic* Plant used as a physic called "the war physic." *Gastrointestinal Aid* Plant used for diseases of the spleen. *Kidney Aid* Cold infusion of root taken for kidney troubles and neuralgia. (172:655, 656) Cold infusion of pounded roots taken for kidney troubles. (177:45) *Panacea* Infusion of root used to produce "an access of health." *Sedative* Plant used to produce a feeling of peace and tranquillity. *Snakebite Remedy* Infusion of root used for snakebite. (172:655, 656) Plant used for snakebites. (177:45) *Venereal Aid* Compound infusion of root taken for "the clap." (172:655, 656) **Meskwaki** *Antidote* Root used as an antidote for poisons and for bladder trouble. *Ceremonial Medicine* Leaves and fruit formerly introduced into rattlesnake medicine song and dance. *Snakebite Remedy* Root used for rattlesnake bites and bladder trouble. *Urinary Aid* Root used for bladder trouble and as an antidote for poisons. (152:248) **Natchez** *Antidiarrheal* Infusion of parched leaves taken for dysentery. *Hemostat* Stem and leaves chewed for nosebleeds. (177:45)

Eryngium yuccifolium var. *synchaetum* Gray ex Coult. & Rose, Button Eryngo
- **Drug**—**Seminole** *Analgesic* Plant used for pains. (as *E. synchaetum* 169:161) Decoction of roots used for cow sickness: lower chest pain, digestive disturbances, and diarrhea. (as *E. synchaetum* 169:191) Infusion of plant taken by men for menstruation sickness: stomachache, headache, and body soreness. (as *E. synchaetum* 169:248) Decoction of plant taken for dead people's sickness.[36] (as *E. synchaetum* 169:257) *Antidiarrheal* Decoction of roots used for cow sickness: lower chest pain, digestive disturbances, and diarrhea. (as *E. synchaetum* 169:191) Infusion of plant taken for otter sickness: severe diarrhea, bloody stools, and severe stomachache. *Antihemorrhagic* Infusion of plant taken for otter sickness: severe diarrhea, bloody stools, and severe stomachache. (as *E. synchaetum* 169:223) *Antirheumatic (External)* Cold infusion of plant used as a bath for body aches. (as *E. synchaetum* 169:215) *Antirheumatic (Internal)* Infusion of plant taken by men for menstruation sickness: stomachache, headache, and body soreness. (as *E. synchaetum* 169:248) *Ceremonial Medicine* Plant used as a ceremonial emetic. (as *E. synchaetum* 169:161) Roots used as an emetic in purification after funerals, at doctor's school, and after death of patient. (as *E. synchaetum* 169:167) Plant used to make a medicine taken by students in medical training. (as *E. synchaetum* 169:95) *Dermatological Aid* Plant used for sores. (as *E. synchaetum* 169:161) Plant used for snake sickness: itchy skin. (as *E. synchaetum* 169:166) Decoction of plant taken and used as a body steam for snake sickness: itchy skin. (as *E. synchaetum* 169:238) *Dietary Aid* Decoction of plant taken for dead people's sickness.[36] (as *E. synchaetum* 169:257) *Emetic* Decoction of plant and other plants taken as an emetic by doctors to strengthen his internal medicine. (as *E. synchaetum* 169:145) Roots used as an emetic to "clean the insides." (as *E. synchaetum* 169:167) Plant used as an emetic by the doctor to prevent the next patient from getting worse.[37] (as *E. synchaetum* 169:184) Infusion of roots taken as an emetic during religious ceremonies. (as *E. synchaetum* 169:408) *Febrifuge* Decoction of plant taken for dead people's sickness.[36] (as *E. synchaetum* 169:257) *Gastrointestinal Aid* Decoction of roots used for cow sickness: lower chest pain, digestive disturbances, and diarrhea. (as *E. synchaetum* 169:191) Infusion of plant taken for otter sickness: severe diarrhea, bloody stools, and severe stomachache. (as *E. synchaetum* 169:223) Infusion of plant taken by men for menstruation sickness: stomachache, headache, and body soreness. (as *E. synchaetum* 169:248) Decoction of plant taken and root chewed for stomachaches. (as *E. synchaetum* 169:276) *Heart Medicine* Roots eaten as a heart medicine. (as *E. synchaetum* 169:304) *Orthopedic Aid* Decoction of plant taken for dead people's sickness.[36] (as *E. synchaetum* 169:257) Decoction of roots applied to foot swellings. (as *E. synchaetum* 169:288) *Panacea* Plant used medicinally for everything. (as *E. synchaetum* 169:161) *Respiratory Aid* Decoction of plant taken for dead people's sickness.[36] (as *E. synchaetum* 169:257) *Snakebite Remedy* Plant used for snakebites. (as *E. synchaetum* 169:295) *Stimulant* Decoction of plant taken for dead people's sickness.[36] (as *E. synchaetum* 169:257) *Unspecified* Plant used for medicinal purposes. (as *E. synchaetum* 169:161) Plant used medicinally. (as *E. synchaetum* 169:164)

Erysimum, Brassicaceae

Erysimum asperum (Nutt.) DC., Plains Wallflower
- **Drug**—**Okanagan-Colville** *Dermatological Aid* Poultice of pounded, whole plant applied to open, fresh wounds. (188:92) **Sioux, Teton** *Analgesic* Infusion of crushed seed taken and used externally for stomach or bowel cramps. *Gastrointestinal Aid* Infusion of crushed seeds used for stomach or bowel cramps. (as *Cheirinia aspera* 52:269)

Erysimum capitatum (Dougl. ex Hook.) Greene, Sanddune Wallflower
- **Drug**—**Hopi** *Tuberculosis Remedy* Plant used for advanced cases of tuberculosis. (42:315) **Navajo, Ramah** *Analgesic* Crushed leaves "smelled" for headache. *Ceremonial Medicine* and *Emetic* Whole plant used as a ceremonial emetic. *Gynecological Aid* Whole plant chewed and blown over patient to aid in difficult labor. *Respiratory Aid* Pulverized pods snuffed to cause sneezing for "congested nose." *Toothache Remedy* Poultice of warmed root applied for toothache. (191:28, 29) **Zuni** *Antirheumatic (External)* Infusion of whole plant used for muscle aches. *Emetic* Flower and fruit eaten as an emetic for stomachaches. (27:375)

Erysimum capitatum (Dougl. ex Hook.) Greene var. *capitatum*, Sanddune Wallflower
- **Drug**—**Keres, Western** *Antirheumatic (External)* Poultice of chewed leaves applied to swellings. (as *E. wheeleri* 171:43)
- **Other**—**Keres, Western** *Paint* Ground flowers used as yellow paint. (as *E. wheeleri* 171:43)

Erysimum cheiranthoides L., Wormseed Wallflower
- **Drug**—Chippewa *Dermatological Aid* Decoction of root applied to skin eruptions. (53:350)

Erysimum inconspicuum (S. Wats.) MacM., Shy Wallflower
- **Drug**—Hopi *Tuberculosis Remedy* Plant used for tuberculosis. (42:316)

Erysimum sp.
- **Drug**—Zuni *Analgesic* Infusion of whole plant applied to forehead and temples for headache from heat. *Dermatological Aid* Infusion of whole plant rubbed over body to prevent sunburn. (166:50)
- **Other**—Zuni *Ceremonial Items* Plant used ceremonially to insure the coming of rain so that the corn and all vegetation would grow. (166:92)

Erythrina, Fabaceae

Erythrina herbacea L., Redcardinal
- **Drug**—Alabama *Gastrointestinal Aid* Cold infusion of root taken by women for bowel pain. (172:666) **Choctaw** *Tonic* Decoction of leaves taken as a general tonic. (25:23) **Creek** *Analgesic* Cold infusion of root taken by women for bowel pain. (172:666) **Seminole** *Antiemetic* Decoction of roots or berries used for horse sickness: nausea, constipation, and blocked urination. (169:188) *Antirheumatic (External)* Decoction of "beans" or inner bark used as a body rub and steam for deer sickness: numb, painful limbs and joints. (169:192) *Laxative* and *Urinary Aid* Decoction of roots or berries used for horse sickness: nausea, constipation, and blocked urination. (169:188)

Erythrina sandwicensis O. Deg., Wili Wili
- **Drug**—Hawaiian *Venereal Aid* Flowers used for venereal diseases. Infusion of pounded bark taken for sexual organ diseases. (as *E. monosperma* 2:74)

Erythronium, Liliaceae

Erythronium americanum Ker-Gawl., American Troutlily
- **Drug**—Cherokee *Dermatological Aid* Warmed leaves crushed and juice poured over "wound that won't heal." *Febrifuge* Infusion of root given for fever. *Hunting Medicine* Root chewed and spit into river to make fish bite. *Stimulant* Compound infusion given for fainting. (80:43) **Iroquois** *Contraceptive* Raw plants, except the roots, taken by young girls to prevent conception. *Dermatological Aid* Poultice of smashed roots used for swellings and removing slivers. (87:282)

Erythronium grandiflorum Pursh, Dogtooth Lily
- **Drug**—Montana Indian *Dermatological Aid* Poultice of crushed bulb-like roots applied to boils. (19:11) **Okanagan-Colville** *Cold Remedy* Corms used for bad colds. (188:45)
- **Food**—Blackfoot *Soup* Bulbs eaten with soup. *Unspecified* Bulbs eaten fresh. (86:102) **Montana Indian** *Forage* Plants eaten by bears and ground squirrels. (82:24) *Unspecified* Bulb-like roots used for food. (19:11) Bulbs occasionally eaten raw or boiled. (82:24) **Okanagan-Colville** *Dried Food* Corms dried for future use. *Unspecified* Corms eaten fresh. (188:45) **Okanagon** *Staple* Roots used as a principal food. (178:238) *Unspecified* Steamed and eaten as a sweet, mealy, and starchy food. (125:37) Roots used as an important food. (178:237) **Shuswap** *Winter Use Food* Roots dried for winter use. (123:54) **Thompson** *Candy* Small root ends of corms eaten as candy by children. *Dried Food* Raw corms dried for future use in soups or stews. The corms were laid out loosely on a scaffold and allowed to partially dry until they had wilted so that they would not split when strung. Then they were strung with needles onto long strings or thin sticks and allowed to dry completely. The strings were tied at the ends to make a large necklace-like loop which could be hung up for storage. *Pie & Pudding* Corms used to make a traditional kind of pudding. The pudding was made by boiling together such traditional ingredients as dried black tree lichen, dried saskatoon berries, cured salmon eggs, tiger lily bulbs, or bitterroot and deer fat. Some of these ingredients, including avalanche lily corms, were optional. Nowadays flour is often used as a substitute for black tree lichen and sugar is added. *Soup* Raw, dried corms used in soups and stews. (187:121) *Unspecified* Steamed and eaten as a sweet, mealy, and starchy food. (125:37) Corms cooked and eaten. (164:481) Roots used as an important food. (178:237) Corms considered an important traditional food source. Because raw corms were considered poisonous, most of the corms were pit cooked, either immediately after harvesting or at a later date after they had been strung and dried. In the latter case, they were soaked for a few minutes in lukewarm water until they had regained about two-thirds of their moisture before being placed in the cooking pit. They could be eaten immediately or redried for later use, when they could be could again very quickly. One informant confirmed that the corms cooked and eaten immediately after harvesting were not as sweet and good as those that had been stored first. Corms eaten with meat and fish as the vegetable portion of a meal, like potatoes. Deep-fried corms used for food. (187:121)
- **Other**—Thompson *Cash Crop* Strings of dried corms used as trading items. *Toys & Games* Corms used as wagers in gambling. Some of the women used to climb up the valley sides to dig sacks of corms which they used as wagers in gambling. The winners would stagger down the hillside with several sacks of corms, while others, who had worked just as hard, would return home empty-handed, having lost in the gambling. (187:121)

Erythronium grandiflorum Pursh ssp. *grandiflorum*, Yellow Avalanche Lily
- **Food**—Thompson *Unspecified* Corms cooked and eaten. (164:481)

Erythronium mesochoreum Knerr, Midland Fawnlily
- **Food**—Winnebago *Unspecified* Raw plant, freshly dug in springtime, eaten avidly by children. (70:71)

Erythronium oregonum Applegate, Giant White Fawnlily
- **Food**—Kwakiutl *Dried Food* Bulbs dried and used for food. *Unspecified* Bulbs eaten raw, baked, or boiled. (182:75)

Erythronium oregonum Applegate ssp. *oregonum*, Oregon Fawnlily
- **Drug**—Wailaki *Dermatological Aid* Poultice of crushed plant applied to boils. (as *E. giganteum* 41:320)

Erythronium revolutum Sm., Mahogany Fawnlily
- **Food**—Kwakiutl, Southern *Dried Food* Bulbs sun dried, boiled, mixed with grease, and eaten at large feasts. *Unspecified* Bulbs eaten raw, baked, or steamed. (183:272)

Eschscholzia, Papaveraceae

Eschscholzia californica Cham., California Poppy
- *Drug*—**California Indian** *Toothache Remedy* Leaves used for toothache. (as *Escholtzia californica* 118:45) **Costanoan** *Pediatric Aid* Flowers laid underneath bed to put child to sleep. *Poison* "Plant avoided by pregnant or lactating women as smell may be poisonous." *Sedative* Flowers laid underneath bed to put child to sleep. (21:9) **Mahuna** *Poison* Plant considered poisonous. (140:34) **Mendocino Indian** *Analgesic* Root juice used as a wash for headaches. *Dermatological Aid* Root juice used as a wash for suppurating sores. *Emetic* Root juice taken as an emetic. *Gastrointestinal Aid* Root juice taken for stomachaches. *Gynecological Aid* Root juice used as a wash by women to stop the secretion of milk. *Narcotic* Root used for the stupefying effect. *Toothache Remedy* Root placed in cavity of tooth for toothaches. *Tuberculosis Remedy* Root juice taken for consumption. (as *Eschscholtzia douglasii* 41:351) **Pomo, Kashaya** *Gynecological Aid* Mashed seedpod rubbed on a nursing mother's breast to dry up her milk. Decoction of mashed seedpod rubbed on a nursing mother's breast to dry up her milk. (72:94)
- *Food*—**Luiseño** *Candy* Flowers chewed with chewing gum. *Vegetable* Leaves used as greens. (155:232) **Mendocino Indian** *Vegetable* Leaves eaten as greens. (as *Eschscholtzia douglasii* 41:351) **Neeshenam** *Vegetable* Leaves boiled or roasted, laid in water, and eaten as greens. (as *Escholtzia californica* 129:377)
- *Other*—**Costanoan** *Insecticide* Decoction of flowers rubbed into the hair to kill lice. (21:9)

Eschscholzia parishii Greene, Parish's Goldenpoppy
- *Drug*—**Kawaiisu** *Dermatological Aid* Poultice of dried, ground roots applied to venereal sores. *Venereal Aid* Root used for gonorrhea and syphilis. (206:31)

Eschscholzia sp., California Poppy
- *Drug*—**Cahuilla** *Pediatric Aid* and *Sedative* Plant used as a sedative for babies. (15:73)
- *Other*—**Cahuilla** *Decorations* Pollen used by women as a facial cosmetic. (15:73)

Escobaria, Cactaceae

Escobaria missouriensis (Sweet) D. R. Hunt **var. missouriensis**, Missouri Foxtail Cactus
- *Food*—**Crow** *Fruit* Red, ripe fruit eaten. (as *Mamillaria missouriensis* 19:15)

Escobaria vivipara (Nutt.) Buxbaum **var. vivipara**, Spinystar
- *Drug*—**Blackfoot** *Antidiarrheal* Fruit eaten in small amounts for diarrhea. (as *Mamillaria vivipara* 86:67) *Eye Medicine* Seed inserted into the eye to remove matter. (as *Mammilaria vivpara* 86:81)
- *Food*—**Blackfoot** *Candy* Fruit eaten as a confection. (as *Mammilaria vivipara* 86:103) *Fruit* Fruits eaten for food. (as *Mamillaria viviparia* 97:45) **Cheyenne** *Dried Food* Fruits dried, boiled, and eaten. *Fruit* Fruits boiled fresh and eaten. (as *Coryphantha vivipara* 83:16)
- *Other*—**Blackfoot** *Toys & Games* Plant used to play a joke on people by placing it under the covers. (as *Mamillaria vivipara* 86:115)

Eucalyptus, Myrtaceae

Eucalyptus sp., Eucalyptus
- *Drug*—**Cahuilla** *Cold Remedy* Leaves used in steam treatments for colds. The leaves were boiled in water and the patient held his head over the bowl. A blanket was then placed over the patient, who inhaled the steam to relieve sinus congestion. (15:73) **Hawaiian** *Analgesic* Oil used for rubbing over backaches and rheumatic pain. *Antirheumatic (External)* Oil used for rubbing over rheumatic pain. *Dermatological Aid* Oil used for rubbing over sores and cuts. *Febrifuge* Leaves used in sweat baths for fevers. *Orthopedic Aid* Oil used for rubbing over sprains. (2:73)

Eugenia, Myrtaceae

Eugenia axillaris (Sw.) Willd., White Stopper
- *Other*—**Seminole** *Hunting & Fishing Item* Plant used to make bows. (169:467)

Euonymus, Celastraceae

Euonymus americana L., American Strawberrybush
- *Drug*—**Cherokee** *Analgesic* Infusion taken for stomachache. *Antihemorrhagic* Taken for "breast complaints" or "spitting blood." *Dermatological Aid* Astringent infusion of bark sniffed for sinus. Used for "white swelling." *Disinfectant* Used as an antiseptic. *Expectorant* Used as an expectorant. *Gynecological Aid* Infusion of root used for "falling of the womb." *Orthopedic Aid* Infusion of bark rubbed on cramps in veins. *Other* Compound infusion taken for "bad disease." *Respiratory Aid* Infusion sniffed for sinus. *Tonic* Used as a tonic. *Urinary Aid* Compound decoction used for "irregular urination." Infusion of bark taken for urinary troubles. (177:38) *Venereal Aid* Infusion of root used for "claps." (80:38) **Iroquois** *Abortifacient* Decoction of plants taken for suppressed menses. Decoction taken to stimulate suppressed menses, not taken when pregnant. *Gynecological Aid* Decoction of vine taken for excessive menstrual flow. *Urinary Aid* Compound decoction of plant taken for difficult urination due to excess gall. (87:375)

Euonymus atropurpurea Jacq., Eastern Wahoo
- *Drug*—**Meskwaki** *Dermatological Aid* Poultice of pounded, fresh trunk bark applied to old facial sores. *Eye Medicine* Infusion or decoction of bark used as a wash for weak or sore eyes. (as *Evonymus atropurpurea* 152:209) **Mohegan** *Cathartic* Infusion of plant used as a physic. (as *Evonymus atropurpurea* 174:265) Infusion of leaves taken as a physic. (176:72, 130) **Winnebago** *Gynecological Aid* Decoction of inner bark taken for uterine trouble. (70:102)

Euonymus europaea L., European Spindletree
- *Drug*—**Iroquois** *Anthelmintic* Compound decoction of plant taken to remove worms caused by solid food taboo. Decoction of bark given to children with worms. *Cathartic* Infusion of roots taken as a physic. *Dietary Aid* Infusion of roots taken to stimulate the appetite. *Pediatric Aid* Decoction of bark given to children with worms. *Urinary Aid* Decoction of roots taken for bloody urine. (87:374)

Euonymus obovata Nutt., Running Strawberrybush
- *Drug*—**Iroquois** *Other* Compound decoction of plants taken for stricture caused by bad blood. (87:375) *Urinary Aid* Decoction of vines taken for difficult urination. (87:376) *Witchcraft Medicine*

Compound infusion of plants taken by people who are bewitched. (87:375)

Euonymus sp.
- **Drug**—Iroquois *Kidney Aid* Compound taken for dropsy. *Other* Compound decoction taken for stricture caused by a menstruating woman. *Venereal Aid* Compound decoction of roots taken for gonorrhea and syphilis. (87:374)

Eupatorium, Asteraceae
Eupatorium maculatum L., Spotted Joepyeweed
- **Drug**—Algonquin, Quebec *Gynecological Aid* Used for menstrual disorders and to facilitate the recovery of women after childbirth. *Venereal Aid* White flowered plant used for males and pink flowered plant used for females for venereal disease. (18:238) **Cherokee** *Adjuvant* Section of stem used to blow or spray medicine. *Antirheumatic (Internal)* Root used for rheumatism. *Diuretic* Root used as a diuretic. *Gynecological Aid* Root used for "female problems." *Kidney Aid* Root used for "dropsy" and infusion taken for kidney trouble. *Misc. Disease Remedy* Root used for gout and "dropsy." *Other* Infusion of root used as a wash "after becoming sick from odor of corpse." *Tonic* Infusion of root used as a tonic during pregnancy. *Urinary Aid* Compound decoction of root taken for "difficult urination." (80:41, 42) **Chippewa** *Antirheumatic (External)* Decoction of root used as a wash for joint inflammations. (53:348) *Pediatric Aid* and *Sedative* Decoction of root used as a quieting bath for fretful child. (53:364) **Iroquois** *Antidiarrheal* Compound decoction of smashed plants taken for diarrhea. *Antirheumatic (External)* Compound plants used for the liver and rheumatism. *Carminative* Compound decoction of plants taken for stomach gas. *Cold Remedy* Infusion of roots taken for colds. *Febrifuge* Infusion of roots taken for chills and fever. *Gastrointestinal Aid* Decoction of dried roots taken for dried stomach. (87:456) *Gynecological Aid* Infusion of roots taken for soreness of womb and abdomen after childbirth. *Kidney Aid* Infusion of roots taken for kidney trouble. (87:455) *Liver Aid* Compound infusion of roots taken for liver sickness. *Love Medicine* Compound decoction of roots used as a wash for anti-love medicine. *Tuberculosis Remedy* Decoction of dried roots taken for consumption. *Venereal Aid* Decoction of roots taken for gonorrhea. (87:456)
- **Other**—Cherokee *Cooking Tools* Stems used as a straw in sucking water from low springs. (80:41)

Eupatorium perfoliatum L., Common Boneset
- **Drug**—Abnaki *Orthopedic Aid* Used to mend bones. (144:154) **Cherokee** *Cathartic* Used as a purgative. *Cold Remedy* Infusion taken for colds. *Diaphoretic* Used as a sudorific. *Disinfectant* and *Diuretic* Used as a tonic, sudorific, stimulant, emetic, and diuretic. *Emetic* Used as an emetic. *Febrifuge* Taken for fever. *Gastrointestinal Aid* Taken for the "biliary system." *Misc. Disease Remedy* Infusion taken for "ague," colds and flu. *Stimulant* Used as a stimulant. *Throat Aid* Infusion taken for sore throat. *Tonic* Used as a tonic. (80:26) **Chippewa** *Abortifacient* Root used to correct irregular menses. *Antirheumatic (External)* Poultice of boiled plant tops applied for rheumatism. (71:142) *Hunting Medicine* Root fibers applied to whistles and used as a charm to attract deer. (53:376) *Snakebite Remedy* Poultice of chewed plants applied to rattlesnake bites. (71:142) **Delaware** *Febrifuge* Infusion of roots and occasionally the leaves used for chills and fever. (176:33) **Delaware, Oklahoma** *Febrifuge* Infusion of root, sometimes with leaves, used for chills and fever. (175:28, 76) **Delaware, Ontario** *Gastrointestinal Aid* Infusion of leaves, considered a powerful herb, taken as a stomach medicine. (175:67, 82) **Iroquois** *Analgesic* Infusion of roots taken for pains in the stomach and on the left side. Poultice of smashed plants applied for headaches. (87:457) *Cold Remedy* Compound decoction of roots taken for colds. Infusion of stems with leaves taken during the onset of a cold. (87:458) *Dermatological Aid* Infusion of roots used as a wash and applied as poultice to syphilitic chancres. (87:456) *Febrifuge* Infusion of whole plant, plant tops, or roots taken for fevers. *Gastrointestinal Aid* Infusion of roots taken for pains in the stomach. (87:457) *Hemorrhoid Remedy* Plant used for piles. *Kidney Aid* Compound decoction of roots taken for the kidneys. (87:458) *Laxative* Compound decoction of flowers and leaves taken as a laxative. (87:456) *Misc. Disease Remedy* Decoction of smashed plants and roots taken for typhoid. (87:458) *Orthopedic Aid* Cold, compound infusion of leaves applied as poultice to broken bones. *Other* Decoction of roots taken for stricture caused by menstruating girls. *Poison* Plant put in enemy's liquor flask to kill him. *Psychological Aid* Decoction of smashed roots taken to stop the liquor habit. (87:457) *Pulmonary Aid* Decoction of roots taken for pneumonia and pleurisy. (87:458) *Venereal Aid* Infusion of roots used as a wash and applied as poultice to syphilitic chancres. (87:456) *Veterinary Aid* Infusion of whole plant given to horses with fevers. *Witchcraft Medicine* Plant used for sorcery. (87:457) **Koasati** *Emetic* Decoction of leaves taken as an emetic. *Urinary Aid* Decoction of roots taken for urinary troubles. (177:61) **Menominee** *Febrifuge* Infusion of whole plant used for fever. (151:30) **Meskwaki** *Anthelmintic* Infusion of leaves and blossoms used to expel worms. *Snakebite Remedy* Root used for snakebite. (152:214) **Micmac** *Kidney Aid* Parts of plant used for kidney trouble. *Venereal Aid* Parts of plant used for persons spitting blood and gonorrhea. (40:56) **Mohegan** *Cold Remedy* Bitter infusion taken for colds. (30:118) Infusion taken for colds. (174:265) Simple or compound infusion of leaves taken in small doses for colds. (176:72, 130) *Febrifuge* Bitter infusion taken for fever. (30:118) Infusion taken for fevers. (174:265) Infusion of leaves taken in small doses for colds and fever. (176:72, 130) *Gastrointestinal Aid* Leaves used for stomach trouble and colds. (176:130) *Panacea* Infusion taken for many ailments and general illness. (174:265) Infusion of leaves taken in small doses for "general debility." (176:72, 130) *Tonic* Complex compound infusion including boneset taken as spring tonic. (174:266) **Nanticoke** *Febrifuge* Compound infusion of whole plant taken for chills and fever. (175:56, 84) **Penobscot** *Antihemorrhagic* Compound infusion of plant taken for "spitting up blood." *Kidney Aid* Compound infusion of plant taken for kidney trouble. *Tonic* Compound infusion of plant taken as a tonic. *Venereal Aid* Compound infusion of plant taken for gonorrhea. (156:311) **Rappahannock** *Tonic* Infusion of dried leaves, picked before flowers matured, taken as a tonic. (161:34) **Seminole** *Emetic* Decoction of plant used as a gentle emetic. *Febrifuge* Plant used as a fever medicine. (169:283) **Shinnecock** *Cold Remedy* Bitter infusion taken for colds. *Diaphoretic* Infusion taken cold, then a hot cup before bed to cause perspiring. *Febrifuge* Bitter infusion taken for fever. (30:118)
- **Other**—Iroquois *Ceremonial Items* Plant used for divination. (87:457)

Eupatorium pilosum Walt., Rough Boneset
- **Drug**—**Cherokee** *Breast Treatment* Used for "breast complaints." *Cold Remedy* Used for colds. *Laxative* Used as a laxative. *Respiratory Aid* Used for "phthisic." *Tonic* Used as a tonic. *Urinary Aid* Taken to increase urination. (80:38)

Eupatorium purpureum L., Sweetscented Joepyeweed
- **Drug**—**Cherokee** *Adjuvant* Section of stem used to blow or spray medicine. *Antirheumatic (Internal)* Root used for rheumatism. *Diuretic* Root used as a diuretic. *Gynecological Aid* Root used for "female problems" and infusion of root used as a tonic during pregnancy. *Kidney Aid* Root used for "dropsy" and infusion taken for kidney trouble. *Misc. Disease Remedy* Root used for gout and "dropsy." *Other* Infusion of root used as a wash "after becoming sick from odor of corpse." *Tonic* Root used for "female problems" and infusion of root used as a tonic during pregnancy. *Urinary Aid* Compound decoction of root taken for "difficult urination." (80:41, 42) **Chippewa** *Cold Remedy* Vapors from infusion of plant tops inhaled for colds. *Gynecological Aid* Plant used to counteract the bad effects of a miscarriage. (71:142) **Mahuna** *Laxative* Infusion of roots taken as a laxative. (140:18) **Menominee** *Gynecological Aid* Compound decoction of root taken after childbirth "for internal healing." (54:133) *Urinary Aid* Plant used for diseases of the genitourinary canal. (151:30) **Meskwaki** *Love Medicine* Root kept in mouth and nibbled when wooing women. (152:214) **Navajo** *Antidote* Plant used as an antidote for poison. *Dermatological Aid* Decoction of plant taken for arrow wounds. (55:85) **Ojibwa** *Pediatric Aid* Strong solution of root used as strengthening wash for infants. (153:364) **Potawatomi** *Burn Dressing* Poultice of fresh leaves applied to burns. *Gynecological Aid* Root used "to clear up afterbirth." (154:52) **Rappahannock** *Blood Medicine* An ingredient of a blood medicine. (161:31)
- **Food**—**Cherokee** *Spice* Root ash used as salt. (126:33)
- **Other**—**Cherokee** *Cooking Tools* Stems used as a straw in sucking water from low springs. (80:41) **Potawatomi** *Good Luck Charm* Flowering tops used as a good luck talisman for gambling. (154:117)

Eupatorium purpureum L. **var. *purpureum***, Sweetscented Joepyeweed
- **Drug**—**Iroquois** *Other* Compound infusion used as wash on injured parts, a "Little Water Medicine." (as *E. falcatum* 87:455)

Eupatorium serotinum Michx., Lateflowering Thoroughwort
- **Drug**—**Houma** *Febrifuge* Decoction of flowers taken for typhoid fever. *Misc. Disease Remedy* Decoction of flowers taken for typhoid fever. (158:64)

Euphorbiaceae, see *Acalypha, Aleurites, Antidesma, Argythamnia, Chamaesyce, Croton, Euphorbia, Jatropha, Manihot, Reverchonia, Ricinus, Sapium, Sebastiania, Stillingia, Tragia*

Euphorbia, Euphorbiaceae
Euphorbia corollata L., Flowering Spurge
- **Drug**—**Cherokee** *Cancer Treatment* Decoction prepared with herbs and taken for cancer. *Cathartic* Taken as a purgative. *Dermatological Aid* "Juice rubbed on skin eruptions, especially on children's heads." Juice used as an ointment for "sores and sore nipples." *Gynecological Aid* Infusion taken for bleeding after childbirth. *Pediatric Aid* "Juice rubbed on skin eruptions, especially on children's heads." *Toothache Remedy* Root used for toothache. *Urinary Aid* Infusion of bruised root taken for urinary diseases. (80:45) Infusion of bruised roots taken for yellow urine. (177:35) *Venereal Aid* Decoction taken for gonorrhea and "similar diseases." (80:45) **Meskwaki** *Anthelmintic* Compound infusion of root used to expel pinworms. *Antirheumatic (Internal)* Decoction of root taken for rheumatism. *Cathartic* Decoction of root or compound taken before breakfast as a physic. (152:220, 221) **Micmac** *Emetic* Root used as an emetic. (40:56) **Ojibwa** *Cathartic* Infusion of pounded root taken before eating as a physic. (153:369)

Euphorbia dentata Michx., Toothed Spurge
- **Drug**—**Keres, Western** *Gynecological Aid* Plant eaten by mothers to produce more milk. (as *Poinsettia dentata* 171:62)

Euphorbia helioscopia L., Madwoman's Milk
- **Drug**—**Iroquois** *Gastrointestinal Aid* and *Pediatric Aid* Infusion of plant given to babies with stomachaches. (141:41)

Euphorbia incisa Engelm., Mojave Spurge
- **Drug**—**Navajo, Kayenta** *Gynecological Aid* Plant used to increase fertility. *Veterinary Aid* Plant used to increase fertility in livestock. (205:30)

Euphorbia ipecacuanhae L., American Ipecac
- **Drug**—**Cherokee** *Diaphoretic* Used as a diaphoretic. *Emetic* Used as an emetic. *Expectorant* Used as an expectorant. *Gynecological Aid* and *Pulmonary Aid* "Stops violent hemorrhaging from lungs and womb when given in small doses." (as *Cephaelis ipecacuanha* 80:40)

Euphorbia lurida Engelm., San Francisco Mountain Spurge
- **Drug**—**Navajo, Ramah** *Gynecological Aid* Poultice of root applied to hard areas of "caked breast." (191:35)
- **Other**—**Navajo, Ramah** *Good Luck Charm* Root tasted, rubbed on the clothing so that opponents smell it and used for good luck in gambling. (191:35)

Euphorbia marginata Pursh, Snow on the Mountain
- **Drug**—**Lakota** *Antirheumatic (External)* Infusion of crushed leaves used as a liniment for swellings. *Gynecological Aid* Infusion of plant used by mothers without milk. (139:45) **Pawnee** *Poison* Plant considered poisonous. (as *Dichrophyllum marginatum* 70:99)
- **Food**—**Kiowa** *Candy* Used for chewing gum. (as *Lepadena marginata* 192:36)

Euphorbia robusta (Engelm.) Small, Rocky Mountain Spurge
- **Drug**—**Navajo** *Cathartic* Compound infusion of plants taken for purging. *Dermatological Aid* Plant rubbed as a liniment or poultice of plant applied to boils and pimples. *Gynecological Aid* Compound infusion of plants taken for confinement. (as *E. montana* 55:60) **Navajo, Kayenta** *Analgesic* Plant used for injuries and pain. *Witchcraft Medicine* Plant used for bewitchment. (205:30)

Euphorbia **sp.**, Spurge
- **Drug**—**Algonquin, Quebec** *Misc. Disease Remedy* Infusion of leaves taken for sugar diabetes. (18:192) **Cahuilla** *Febrifuge* Decoction of plant used as a bath for fevers. *Misc. Disease Remedy* Decoction of plant used as a bath for chickenpox and smallpox. *Oral Aid* Infusion

of plant taken for mouth sores. (15:73) **Creek** *Cathartic* Infusion of roots, "a very violent remedy," taken to make the bowels act. (172:\661) **Mahuna** *Dermatological Aid* Infusion of plant used as a wash for minor skin eruptions and scalp diseases. (140:13) **Yavapai** *Dermatological Aid* Decoction used for sores. *Venereal Aid* Decoction used for gonorrhea. (65:261)

Euthamia, Asteraceae

Euthamia graminifolia (L.) Nutt., Flattop Goldentop
- *Drug*—**Chippewa** *Analgesic* Decoction of root taken for chest pain. *Pulmonary Aid* Decoction of root taken for lung trouble, especially chest pain. (53:340)

Euthamia graminifolia (L.) Nutt. **var. *graminifolia***, Flattop Goldentop
- *Drug*—**Ojibwa** *Analgesic* Infusion of flowers taken for chest pain. *Hunting Medicine* Plant used in a hunting medicine. (as *Solidago graminifolia* 153:366) Flowers used in the hunting medicine and smoked to simulate the odor of a deer's hoof. (as *Solidago graminifolia* 153:429) **Potawatomi** *Febrifuge* Infusion of blossoms used for some kinds of fevers. (as *Solidago graminifolia* 154:53)

Evernia, Usneaceae, lichen

Evernia sp., Yellow Lichen
- *Drug*—**Modesse** *Poison* Plant used for poison on stone arrow tips. (117:225)

Evernia vulpina (L.) Acharius
- *Drug*—**Blackfoot** *Dermatological Aid* Plant blackened in a fire and rubbed on rashes, eczema, and wart sores. *Gastrointestinal Aid* Infusion of plant and marrow taken for stomach disorders like ulcers. (86:76) **Wailaki** *Dermatological Aid* Used for drying running sores. **Yuki** *Dermatological Aid* Used for drying running sores. (118:44)
- *Fiber*—**Montana Indian** *Clothing* Used for making clothing. *Mats, Rugs & Bedding* Used for making bedding. (tree moss 19:12)
- *Dye*—**Blackfoot** *Yellow* Plant pieces used as a yellow dye for porcupine quills. (86:113) **Montana Indian** *Yellow* Used for making clothing, bedding, yellow dye, and yellow paint. (tree moss 19:12) **Thompson** *Yellow* Used to make a bright yellow dye. (wolf's moss 164:501)
- *Other*—**Montana Indian** *Paint* Plant used to make a yellow paint. (tree moss 19:12) **Thompson** *Paint* Used as a paint on wood and skin. (wolf's moss 164:501)

Evolvulus, Convolvulaceae

Evolvulus nuttallianus J. A. Schultes, Shaggy Dwarf Morningglory
- *Drug*—**Navajo, Kayenta** *Nose Medicine* Plant used as a snuff for itching in the nose and sneezing. (as *E. pilosus* 205:37)

Exobasidiaceae, see *Exobasidium*

Exobasidium, Exobasidiaceae, fungus

Exobasidium sp., Ghost Ear Fungus
- *Food*—**Haisla & Hanaksiala** *Unspecified* Galls used for food. (43:135)

Fabaceae, see *Acacia, Amorpha, Amphicarpaea, Apios, Arachis, Astragalus, Baptisia, Caesalpinia, Calliandra, Canavalia, Cercis, Chamaecrista, Cicer, Cladrastis, Clitoria, Coronilla, Coursetia, Crotalaria, Dalea, Desmanthus, Desmodium, Erythrina, Galactia, Gleditsia, Glycyrrhiza, Gymnocladus, Hedysarum, Hoffmannseggia, Hoita, Lathyrus, Lens, Lespedeza, Lotus, Lupinus, Medicago, Melilotus, Mucuna, Olneya, Orbexilum, Oxytropis, Parkinsonia, Parryella, Pediomelum, Peteria, Phaseolus, Pisum, Prosopis, Psoralidium, Psorothamnus, Robinia, Senna, Sophora, Strophostyles, Stylosanthes, Tephrosia, Thermopsis, Trifolium, Vicia, Vigna*

Fagaceae, see *Castanea, Castanopsis, Fagus, Lithocarpus, Quercus*

Fagopyrum, Polygonaceae

Fagopyrum esculentum Moench, Fagopyrum
- *Drug*—**Iroquois** *Pediatric Aid* Decoction of plant given when "baby is sick because of mother's adultery." *Witchcraft Medicine* Decoction taken by the mother "who is running around, making baby sick." (87:313)

Fagus, Fagaceae

Fagus grandifolia Ehrh., American Beech
- *Drug*—**Cherokee** *Anthelmintic* Nuts chewed for worms. (80:25) **Chippewa** *Pulmonary Aid* Bark used for pulmonary troubles. (71:128) **Iroquois** *Abortifacient* Bark used for abortions, only when mother was suffering. *Blood Medicine* Complex compound used as a blood purifier. *Burn Dressing* Compound decoction of leaves applied as poultice to burns or scalds. *Dermatological Aid* Compound decoction taken when "skin becomes thin." (87:302) Nutmeat oil formerly used for the hair, either alone or mixed with bear grease. (196:123) *Liver Aid* Compound decoction taken for yellow skin and gall. *Tuberculosis Remedy* Compound decoction of bark taken for consumption. (87:302) **Malecite** *Dermatological Aid* Leaves used for sores. (116:246) **Menominee** *Unspecified* Inner bark of the trunk and root used in medicinal compounds. (151:36) **Micmac** *Venereal Aid* Leaves used for chancre. (40:56) **Potawatomi** *Burn Dressing* Decoction of leaves used for burned or scalded wounds. *Dermatological Aid* Decoction of leaves used to restore frostbitten extremities. (154:58) **Rappahannock** *Dermatological Aid* Compound infusion of north side bark used as a wash for poison ivy. (161:34)
- *Food*—**Algonquin, Quebec** *Unspecified* Nuts used, mainly by men working in the bush, for food. (18:80) **Chippewa** *Unspecified* Nuts used for food. People sought stores of beechnuts that had been put away by chipmunks. These hoards saved the labor not only of gathering, but also of shucking, and were certain to contain only sound nuts. The people had observed that chipmunks never stored any that were not good. (71:128) **Iroquois** *Beverage* Fresh nutmeats crushed, boiled, and liquid used as a drink. *Bread & Cake* Fresh nutmeats crushed and mixed with bread. (124:99) Nuts crushed, mixed with cornmeal and beans or berries, and made into bread. (196:123) *Pie & Pudding* Fresh nutmeats crushed and mixed with corn pudding. (124:99) *Sauce & Relish* Nuts pounded, boiled, resulting oil seasoned with salt, and used as gravy. *Soup* Nutmeats crushed and added to

corn soup. (196:123) *Special Food* Fresh nutmeats crushed, boiled, and oil used as a delicacy in corn bread and pudding. (124:99) Nutmeat oil added to the mush used by the False Face Societies. *Staple* Nutmeats crushed and added to hominy. *Unspecified* Nutmeats, after skimming off the oil, seasoned and mixed with mashed potatoes. (196:123) **Menominee** *Winter Use Food* Beechnuts stored for winter use. (151:66) **Ojibwa** *Unspecified* Sweet nuts much appreciated and never enough to store for winter. (153:401) **Potawatomi** *Unspecified* Beechnuts used for food. The hidden stores of the small deer mouse was what the Indians relied upon. The deer mouse is outdone by no other animal in laying up winter stores. Its favorite food is the beechnut. It will lay up, in some safe log or hollow tree, from 4 to 8 quarts, shelled in the most careful manner. The Indians easily found the stores, when the snow was on the ground, by the refuse on the snow. (154:100)
- **Fiber**—**Cherokee** *Building Material* Wood used for lumber. (80:25) **Micmac** *Snow Gear* Used to make snowshoe frames. (159:258)
- **Other**—**Cherokee** *Fasteners* Wood used to make buttons. (80:25) **Iroquois** *Insecticide* Nutmeat oil mixed with bear grease and used as a preventive for mosquitoes. (196:123) **Potawatomi** *Cooking Tools* Wood used to make food or chopping bowls. (154:113)

Fallugia, Rosaceae

Fallugia paradoxa (D. Don) Endl. ex Torr., Apache Plume
- **Drug**—**Navajo, Kayenta** *Witchcraft Medicine* Plant used as witchcraft to cause insanity. (205:26) **Navajo, Ramah** *Ceremonial Medicine* Cold infusion of leaves used as a ceremonial lotion and leaves used as a ceremonial emetic. *Emetic* Leaves used as an emetic in various ceremonies. (191:30, 31) **Tewa** *Dermatological Aid* Infusion of leaves used as shampoo, to promote growth of hair. (138:46, 47)
- **Fiber**—**Havasupai** *Basketry* Used for the top ring of baskets. *Furniture* Used for the ladder-back rungs of the cradleboards. (197:223) **Hualapai** *Furniture* Branches used to make cradleboard hoods and beds. (195:37) **Jemez** *Brushes & Brooms* Branches bound together and used as a broom for outdoor sweeping and to separate chaff from the wheat. (44:22) **Keres, Western** *Brushes & Brooms* Brush used for rough brooms, especially in Acoma. (171:44) **Keresan** *Brushes & Brooms* Used to make arrows and brooms. (198:558) **Tewa** *Brushes & Brooms* Slender branches bound together and used as brooms for rough outdoor sweeping. (138:46)
- **Other**—**Hopi** *Hunting & Fishing Item* Stems used for arrows. (200:78) **Isleta** *Hunting & Fishing Item* Slender, smooth, straight branches used to make arrow shafts. (100:29) **Keres, Western** *Hunting & Fishing Item* Straight sticks used for arrow shafts. (171:44) **Keresan** *Hunting & Fishing Item* Used to make arrows and brooms. (198:558) **Tewa** *Hunting & Fishing Item* Straight, slender branches used to make arrows. (138:46)

Fendlera, Hydrangeaceae

Fendlera rupicola Gray, Cliff Fendlerbush
- **Drug**—**Navajo** *Gastrointestinal Aid* Infusion of inner bark taken for swallowed ants. (55:51) **Navajo, Kayenta** *Cathartic* Plant used as a cathartic. *Ceremonial Medicine* Plant used for Plumeway, Nightway, Male Shootingway, and Windway ceremonies. (205:25)
- **Other**—**Havasupai** *Hunting & Fishing Item* Wood used to make arrow foreshafts. (197:221) **Hopi** *Ceremonial Items* Used in religious ceremonies. (42:318) **Navajo** *Ceremonial Items* Notched stick rubbed with a smooth stick instead of beating a drum in the Mountain Chant Ceremony. Used by the Home God in the Mountain Chant Ceremony. *Hunting & Fishing Item* Wood used to make arrow shafts. *Insecticide* Plant used to kill hair lice. *Tools* Wood used to make weaving forks, planting sticks, and knitting needles. (55:51) **Navajo, Kayenta** *Ceremonial Items* Boiled with juniper berries, pinyon buds and cornmeal, and used in mush-eating ceremonies. (205:25)

Ferocactus, Cactaceae

Ferocactus coulteri, Barrelcactus
- **Food**—**Seri** *Beverage* Plant provided drinking water. (50:136)

Ferocactus cylindraceus (Engelm.) Orcutt **var. *cylindraceus***, California Barrelcactus
- **Food**—**Cahuilla** *Beverage* Plant used to obtain water. The barrel cactus provided a desert reservoir, one which had long been familiar to many desert travelers at times of emergency. To obtain water, the top of the cactus was sliced off, a portion of the pulp was removed to create a depression, and then the pulp was squeezed by hand in the depression until water was released from the spongy mass. *Dried Food* Buds sun dried for storage. Flowers sun dried for storage. (as *Echinocactus acanthodes* 15:67) *Staple* Berries and stems were an important and dependable food source. (as *Echinocactus acanthodes* 15:49) *Unspecified* Buds eaten fresh, parboiled, or baked in a pit. Flowers eaten fresh, parboiled, or baked in a pit. (as *Echinocactus acanthodes* 15:67)
- **Other**—**Cahuilla** *Cooking Tools* Body of the plant used as a cooking vessel. The top was cut off of the cactus and the interior was dug out. Water was then put into the depression and heated with hot stones. (as *Echinocactus acanthodes* 15:67)

Ferocactus cylindraceus **var. *lecontei*** (Engelm.) H. Bravo, Leconte's Barrelcactus
- **Food**—**Pima** *Beverage* Juice extracted from pulp and used to quench thirst. *Candy* Used to make cactus candy. (as *Echinocactus lecontei* 47:55) *Unspecified* Plants sliced, cut into small pieces, boiled with mesquite beans, and eaten as a sweet dish. (as *Echinocactus lecontei* 47:56)

Ferocactus sp., Barrelcactus
- **Food**—**Hualapai** *Dried Food* Fruits dried, pounded, and eaten. *Fruit* Fruits eaten for food. (195:42) **Pima, Gila River** *Special Food* Flesh prepared as a special dish with mesquite pods. (133:5) *Unspecified* Pulp boiled and eaten. (133:7)

Ferocactus wislizeni (Engelm.) Britt. & Rose, Candy Barrelcactus
- **Food**—**Apache, San Carlos** *Beverage* Juice used for extreme thirst. *Porridge* Small, black seeds parched, ground, boiled, and eaten as mush. (as *Echinocereus wislizeni* 95:257) **Papago** *Beverage* Plant tops pounded and the juice used as a drink. (as *Echinocactus wislizeni* 34:17) *Vegetable* Pulp eaten as greens in May. (as *Echinocactus wislizeni* 34:14) **Pima** *Beverage* Juice extracted from pulp and used to quench thirst. *Candy* Used to make cactus candy. (as *Echinocactus wislizeni* 47:55) *Substitution Food* Pulp used in lieu of water for thirst. (as *Echinocactus wislizeni* 146:77) *Unspecified* Plants sliced, cut into

small pieces, boiled with mesquite beans, and eaten as a sweet dish. (as *Echinocactus wislizeni* 47:56) Pulp cut in strips, boiled, and used for food. (as *Echinocactus wislizeni* 146:77) **Seri** *Beverage* Plant provided drinking water. (50:136)
- *Other*—**Pima** *Hunting & Fishing Item* Thorns used to make fishhooks. (as *Echinocactus wislizeni* 47:56) **Yuma** *Hunting & Fishing Item* Spines heated and bent to make fishing hooks. (as *Echinocactus wislizeni* 37:222)

Ferula, Apiaceae
Ferula dissoluta
- *Food*—**Okanagon** *Staple* Roots used as a principal food. (178:238)

Festuca, Poaceae
Festuca brachyphylla J. A. Schultes ex J. A. & J. H. Schultes ssp. *brachyphylla*, Alpine Fescue
- *Food*—**Gosiute** *Unspecified* Seeds used for food. (as *F. ovina* var. *brevifolia* 39:369)

Festuca idahoensis Elmer, Idaho Fescue
- *Fiber*—**Navajo, Ramah** *Scouring Material* Bunch about a foot long, tied with string or yucca fiber, used as a brush for cleaning metates. (191:16)

Festuca sp.
- *Food*—**Costanoan** *Staple* Seeds used in pinole. (21:255)

Festuca subverticillata (Pers.) Alexeev, Nodding Fescue
- *Drug*—**Iroquois** *Heart Medicine* Decoction of smashed roots taken for heart disease. *Other* Compound used as a "corn medicine." (as *F. obtusa* 87:273)

Ficus, Moraceae
Ficus aurea Nutt., Florida Strangler Fig
- *Drug*—**Seminole** *Dermatological Aid* Poultice of mashed bark applied to cuts and sores. (169:300)
- *Food*—**Seminole** *Candy* Plant used for chewing gum. *Unspecified* Plant used for food. (169:481)
- *Other*—**Seminole** *Cooking Tools* Plant used for meat stringing. *Fasteners* Plant used to make house lashings and cane mill lashings. (169:481) Root bark twisted and used to bind together the pole frame of the medical training school house. (169:95) *Hunting & Fishing Item* Plant used to make arrows, bowstrings, and fish line. (169:481)

Ficus carica L., Common Fig
- *Food*—**Havasupai** *Beverage* Plant used to make a drink. (197:66) *Dried Food* Fruit sun dried and stored in sacks for winter use. *Fruit* Fruit eaten fresh. *Winter Use Food* Fallen fruit ground, mixed with water into a thick paste, dried in sheets, and eaten during the winter. (197:216)

Filipendula, Rosaceae
Filipendula rubra (Hill) B. L. Robins., Queen of the Prairie
- *Drug*—**Meskwaki** *Heart Medicine* Root used as an important medicine for various heart troubles. *Love Medicine* Compound containing root used as a love medicine. (152:241, 242)

Foeniculum, Apiaceae
Foeniculum vulgare P. Mill., Sweet Fennel
- *Drug*—**Cherokee** *Carminative* Used for colic and given to children for flatulent colic. *Cold Remedy* Used as a tonic and given for colds and to children for flatulence. *Gastrointestinal Aid* Used for colic and given to children for flatulent colic. *Gynecological Aid* Used as a tonic and given to women in labor. *Pediatric Aid* Used for colic and given to children for flatulent colic. *Tonic* Used as a tonic. (80:33) **Pomo, Kashaya** *Eye Medicine* Strained decoction of seeds used as an eyewash. *Gastrointestinal Aid* Seeds chewed for upset stomach, indigestion, and heartburn. (72:44)
- *Other*—**Hopi** *Smoke Plant* Plant used as a substitute for tobacco. (as *F. officinale* 56:20) Plant used as a substitute for tobacco. (as *F. officinale* 200:86, 87)

Fomes, Polyporaceae, fungus
Fomes igniarius (L. ex Fries) Kickx, Shelf Fungus
- *Drug*—**Eskimo, Inuktitut** *Laxative* Infusion of plant taken as a laxative. (202:187)
- *Other*—**Eskimo, Inuktitut** *Cash Crop* Plant collected and widely traded. *Insecticide* Plant added to tobacco as a mosquito repellent. (202:187)

Fomes sp.
- *Fiber*—**Cowlitz** *Unspecified* Used to draw pictures on. (bracket fungus 79:50)
- *Other*—**Chehalis** *Toys & Games* Used as a target for archery. (bracket fungus 79:50) **Keres, Western** *Unspecified* Fungus known and named but no use was specified. (bracket fungus 171:44) **Haisla & Hanaksiala** *Ceremonial Items* Plant used for some aspects of the secret society rituals. *Protection* Plant used as hand protection for handling live coals during a secret society ritual. Plant placed at entrance of special shamanistic dance house as protection from bad spirits and ghosts. *Toys & Games* Plant thrown into the stream and used as a target by young boys for spear practice. Plant used as a ball in a women's and children's game. (shelf fungus 43:135) **Oweekeno** *Paint* Whole plant made into a white powder, sometimes mixed with coloring, and used to make a paint. (shelf fungus 43:49) **Snohomish** *Toys & Games* Used as a target for archery. (bracket fungus 79:50)

Fomitopsis, Polyporaceae, fungus
Fomitopsis officinalis (Vill ex Fries) Bond. & Sing.
- *Drug*—**Haisla & Hanaksiala** *Tuberculosis Remedy* Decoction of ground plant taken for tuberculosis. (43:138)

Fomitopsis pinicola (Sw. ex Fr.) Karst., Shelf Fungus
- *Other*—**Nitinaht** *Protection* Used for protection against people with ill feelings and wishes toward others. (186:56)

Fomitopsis sp., Shelf Fungus
- *Other*—**Haisla & Hanaksiala** *Ceremonial Items* Plant used for some aspects of the secret society rituals. *Protection* Plant placed at entrance of special shamanistic dance house as protection from bad spirits and ghosts. Plant used as hand protection for handling live coals during a secret society ritual. *Toys & Games* Plant used as a ball in a women's and children's game. Plant thrown into the stream and used as a target by young boys for spear practice. (43:135)

Forestiera, Oleaceae

Forestiera acuminata (Michx.) Poir., Eastern Swampprivet
- **Drug**—Houma *Panacea* Decoction of roots and bark taken as a "health beverage." (158:63)

Forestiera pubescens Nutt. **var. pubescens**, Stretchberry
- **Drug**—Navajo, Ramah *Ceremonial Medicine* Leaves used as a ceremonial emetic. *Disinfectant* Plant used for "bear infection." *Emetic* Leaves used as a ceremonial emetic. (as *F. neomexicana* 191:39)
- **Food**—Apache, Chiricahua & Mescalero *Fruit* Raw fruits occasionally eaten as food. (as *F. neomexicana* 33:44)
- **Other**—Hopi *Ceremonial Items* Used to make pahos (prayer sticks). *Tools* Used for digging stick. (as *F. neomexicana* 42:319) Wood used for digging sticks. (as *F. neomexicana* 200:87) **Isleta** *Water Indicator* Large shrubs considered water indicators because wells dug where plants grew always produced water. (as *F. neomexicana* 100:29) **Jemez** *Ceremonial Items* Berry juice mixed with white clay and used as purple body paint for summer dances. (as *F. neomexicana* 44:22) **Navajo** *Ceremonial Items* Used to make prayer sticks. (as *F. neomexicana* 55:68) **Navajo, Ramah** *Ceremonial Items* Stem used to make Evilway big hoop. (as *F. neomexicana* 191:39)

Forestiera segregata (Jacq.) Krug & Urban **var. segregata**, Florida Swampprivet
- **Other**—Seminole *Hunting & Fishing Item* Plant used to make arrows. (as *F. porulosa* 169:492)

Fortunella, Rutaceae

Fortunella sp., Kumquat
- **Food**—Seminole *Unspecified* Plant used for food. (169:513)

Fouquieriaceae, see *Fouquieria*

Fouquieria, Fouquieriaceae

Fouquieria sp., Ocotillo
- **Drug**—Hualapai *Orthopedic Aid* Roots used in a soothing bath for swollen feet. (195:22)
- **Fiber**—Hualapai *Building Material* Branches used to construct huts. (195:22)

Fouquieria splendens Engelm., Ocotillo
- **Drug**—Mahuna *Blood Medicine* Plant used as a blood specific, purifier, and tonic. (140:28)
- **Food**—Cahuilla *Beverage* Fresh blossoms soaked in water and used to make a summer drink. *Porridge* Parched seeds ground into a flour and used to make mush or cakes. *Unspecified* Fresh blossoms used for food. (as *Fourquieria splendens* 15:74) **Papago** *Special Food* Nectar pressed out of blossoms, hardened like rock candy, and chewed as a delicacy. (34:28) **Yavapai** *Snack Food* Flowers sucked by children for nectar. (65:256)
- **Fiber**—Cahuilla *Building Material* Wood used to make fences to prevent rodents from attacking cultivated crops. (as *Fourquieria splendens* 15:74) **Papago** *Building Material* Used for the warp of wrapped weaving in house frames. (34:53) Withes used to bind together the house dome ribs. (34:66) Used for house construction. **Pima** *Building Material* Stalks freed from thorns, bound together with rawhide or wire, and used as shelves. (47:89) **Seri** *Building Material* Branches used to make sun and wind shelters. (50:136)
- **Other**—Cahuilla *Fuel* Wood used for firewood. (as *Fourquieria splendens* 15:74) **Papago** *Ceremonial Items* Flexible rods used as the basis of ceremonial structures representing clouds or mountains. (34:54) *Tools* Thorns used to pierce the ears of both sexes. (34:51) **Pima** *Decorations* Plants grown around gardens for decorations. (47:89)

Fragaria, Rosaceae

Fragaria ×ananassa var. cuneifolia (Nutt. ex T. J. Howell) Staudt, Hybrid Strawberry
- **Food**—Chehalis *Fruit* Berries eaten fresh. (as *F. cuneifolia* 79:36) **Klallam** *Fruit* Berries used for food. Fruits eaten fresh in early summer. (as *F. cuneifolia* 78:197) Berries eaten fresh. **Squaxin** *Fruit* Berries eaten fresh. (as *F. cuneifolia* 79:36)

Fragaria chiloensis (L.) P. Mill., Beach Strawberry
- **Drug**—Quileute *Burn Dressing* Poultice of chewed leaves applied to burns. (79:36)
- **Food**—Alaska Native *Dietary Aid* Berries used as a rich source of vitamin C. *Fruit* Berries eaten raw. *Preserves* Berries made into a jam. (85:81) **Clallam** *Fruit* Berries eaten fresh. (57:202) **Hesquiat** *Fruit* Berries used for food. (185:72) **Hoh** *Fruit* Fruits eaten raw. Fruits stewed and used for food. (137:63) **Kitasoo** *Fruit* Fruit used for food. (43:342) **Makah** *Fruit* Fruit eaten fresh. (67:262) Berries eaten fresh immediately after picking. (79:36) *Preserves* Fruit used to make jams and jellies. (67:262) **Nitinaht** *Fruit* Berries eaten fresh. (186:117) **Oweekeno** *Fruit* Berries eaten fresh. *Preserves* Berries used to make jam. (43:108) **Pomo, Kashaya** *Fruit* Berries eaten fresh. (72:109) **Quileute** *Fruit* Berries eaten after fish. (79:36) Fruits eaten raw. Fruits stewed and used for food. (137:63) **Quinault** *Special Food* Berries served by young women to their guests at parties. (79:36) **Salish, Coast** *Beverage* Leaves dried and used to make tea. *Fruit* Fruits eaten fresh. (182:86) **Tolowa** *Fruit* Fresh fruit used for food. **Yurok** *Fruit* Fresh fruit used for food. (6:31)
- **Other**—Pomo, Kashaya *Ceremonial Items* Berry used in the flower dance at the Strawberry Festival, danced by young girls.[38] (72:109)

Fragaria chiloensis ssp. lucida (Vilm.) Staudt, Beach Strawberry
- **Food**—Haisla & Hanaksiala *Fruit* Berries eaten fresh. *Preserves* Berries used to make jam. (43:264)

Fragaria chiloensis ssp. pacifica Staudt, Pacific Beach Strawberry
- **Food**—Haisla & Hanaksiala *Fruit* Berries eaten fresh. *Preserves* Berries used to make jam. (43:264)

Fragaria sp., Wild Strawberry
- **Drug**—Blackfoot *Antidiarrheal* Infusion of plant used for diarrhea. (86:66) **Carrier** *Antihemorrhagic* Decoction of stems taken for stomach bleeding. (31:78) **Iroquois** *Abortifacient* Infusion of whole plant taken by women to regulate menses. *Antidiarrheal* Compound decoction of roots taken for diarrhea with blood. *Blood Medicine* Compound decoction of roots taken by all ages as blood remedy. *Eye Medicine* Decoction of roots used as a wash for sties. *Gastrointestinal Aid* Plant used for babies with colic. *Gynecological Aid* Plant used for gonorrhea. *Heart Medicine* Compound used for stroke. *Oral Aid* Decoction of roots used as a wash for chancre sores. *Pediatric Aid* and

Toothache Remedy Compound decoction of plant given when "baby's teeth are coming in." (87:352) **Skokomish** *Antidiarrheal* Decoction of whole plant taken for diarrhea. (79:36)
- *Food*—**Carrier** *Fruit* Berries used for food. (31:78) **Chinook, Lower** *Fruit* Berries eaten fresh. (79:36) **Costanoan** *Fruit* Raw fruits used for food. (21:249) **Montana Indian** *Fruit* Berries eaten fresh. *Winter Use Food* Berries dried and stored for winter use. (19:12) **Nisqually** *Fruit* Berries mashed and eaten. (79:36) **Paiute** *Fruit* Berries eaten fresh. (111:80) **Puyallup** *Fruit* Berries mashed and eaten. (79:36) **Skagit, Upper** *Dried Food* Fruit pulped and dried for winter use. *Fruit* Fruit eaten fresh. (179:38) **Skokomish** *Fruit* Berries mashed and eaten. (79:36)

Fragaria vesca L., Woodland Strawberry
- *Drug*—**Okanagan-Colville** *Dermatological Aid* Poultice of leaf powder and deer fat applied to sores. *Disinfectant* Leaf powder applied to any open sore as a disinfectant. *Oral Aid* and *Pediatric Aid* Leaf powder dusted into baby's sore mouth. (188:125) **Potawatomi** *Gastrointestinal Aid* Root used for stomach complaints. (154:76, 77) **Thompson** *Antidiarrheal* Infusion of roots or whole plant taken for diarrhea or dysentery. Decoction of leaves taken for diarrhea. *Pediatric Aid* Infusion of roots or whole plant bottle fed to babies for diarrhea or dysentery. Decoction of leaves given to children for diarrhea. (187:259)
- *Food*—**Bella Coola** *Fruit* Berries used for food. (184:208) **Clallam** *Fruit* Berries eaten fresh. (57:202) **Gosiute** *Fruit* Berries used for food in season. (39:370) **Hesquiat** *Fruit* Berries used for food. (185:72) **Kitasoo** *Fruit* Fruit used for food. (43:342) **Lakota** *Fruit* Fruits eaten fresh. Fruits eaten with other foods. (106:37) Fruits eaten for food. (139:56) **Nitinaht** *Fruit* Berries eaten fresh. (186:117) **Okanagan-Colville** *Fruit* Berries eaten fresh. *Winter Use Food* Berries canned for future use. (188:125) **Oweekeno** *Fruit* Berries eaten fresh. *Preserves* Berries used to make jam. (43:108) **Potawatomi** *Dried Food* Berries sometimes dried for winter use. *Winter Use Food* Berries sometimes dried and at other times preserved for winter use. (154:107) **Salish, Coast** *Beverage* Leaves dried and used to make tea. *Fruit* Fruits eaten fresh. (182:86) **Thompson** *Dried Food* Berries, if plentiful, dried for future use. *Fruit* Berries eaten fresh. (187:259)

Fragaria vesca ssp. *americana* (Porter) Staudt, Woodland Strawberry
- *Food*—**Chippewa** *Fruit* Strawberries considered an important part of the diet. (as *F. americana* 71:132) **Dakota** *Fruit* Fruit used for food. (as *F. americana* 70:84) **Iroquois** *Bread & Cake* Fruit mashed, made into small cakes and dried for future use. *Dried Food* Raw or cooked fruit sun or fire dried and stored for future use. *Fruit* Dried fruit taken as a hunting food. *Sauce & Relish* Dried fruit cakes soaked in warm water and cooked as a sauce or mixed with corn bread. (196:127) **Omaha** *Fruit* Fruit used for food. **Pawnee** *Fruit* Fruit used for food. **Ponca** *Fruit* Fruit used for food. (as *F. americana* 70:84) **Thompson** *Fruit* Large, wild berries eaten as a favorite food. (as *F. vesca americana* 164:487) **Winnebago** *Beverage* Young leaves used to make a tea-like beverage. *Fruit* Fruit used for food. (as *F. americana* 70:84)

Fragaria vesca ssp. *bracteata* (Heller) Staudt, Woodland Strawberry
- *Food*—**Apache** *Special Food* Fruits eaten as a delicacy. (as *F. bracteata* 32:29) **Apache, Chiricahua & Mescalero** *Fruit* Raw fruits occasionally eaten as food. (as *F. bracteata* 33:44) **Cochiti** *Special Food* Fruits eaten as a delicacy. (as *F. bracteata* 32:29) **Cowlitz** *Beverage* Leaves used for a beverage. *Dried Food* Berries dried and used for food. *Fruit* Berries eaten fresh. (as *F. bracteata* 79:36) **Haisla & Hanaksiala** *Fruit* Berries eaten fresh. *Preserves* Berries used to make jam. (43:264) **Isleta** *Fruit* Flavorful fruit considered a delicacy. (as *F. bracteata* 100:29) *Special Food* Fruits eaten as a delicacy. **Navajo** *Special Food* Fruits eaten as a delicacy. (as *F. bracteata* 32:29) Fruits used for food and considered a delicacy. (as *F. bracteata* 55:53) **Swinomish** *Fruit* Berries eaten fresh. (as *F. bracteata* 79:36)

Fragaria vesca ssp. *californica* (Cham. & Schlecht.) Staudt, California Strawberry
- *Drug*—**Diegueño** *Antidiarrheal* Decoction of leaves taken for diarrhea. (84:21) **Navajo, Ramah** *Panacea* Whole plant used as "life medicine." (as *F. californica* 191:31)
- *Food*—**Cahuilla** *Fruit* Fruit always eaten fresh. (as *F. californica* 15:74) **Coeur d'Alene** *Fruit* Berries eaten fresh. Berries mashed and eaten. (as *F. californica* 178:90) **Diegueño** *Fruit* Fruit eaten fresh with cream. (84:21) **Karok** *Fruit* Fresh fruit used for food. (6:31) Berries used for food. (as *F. californica* 148:384) **Mendocino Indian** *Fruit* Berries eaten fresh by children. (as *F. californica* 41:354) **Navajo, Ramah** *Unspecified* Very small fruit, hard to find, used for food. (as *F. californica* 191:31) **Okanagon** *Staple* Berries used as a principal food. (as *F. californica* 178:239) **Pomo** *Fruit* Raw berries used for food. (as *F. californica* 66:13) **Pomo, Kashaya** *Fruit* Berries eaten fresh. (as *F. californica* 72:110) **Spokan** *Fruit* Berries used for food. (as *F. californica* 178:343) **Thompson** *Dried Food* Berries washed, dried, and stored for winter use. *Fruit* Berries eaten fresh. (as *F. californica* 164:488) *Spice* Flowers and stems used to flavor roots. (as *F. californica* 164:478) **Yurok** *Fruit* Fresh fruit used for food. (6:31)
- *Other*—**Pomo, Kashaya** *Ceremonial Items* Berry used in the flower dance at the Strawberry Festival, danced by young girls.[38] (as *F. californica* 72:110) **Thompson** *Incense & Fragrance* Leaves made into pads and worn under the armpits to make them smell sweet. (as *F. californica* 164:509)

Fragaria virginiana Duchesne, Virginia Strawberry
- *Drug*—**Blackfoot** *Antidiarrheal* Decoction of roots used for diarrhea. (97:38) **Cherokee** *Antidiarrheal* Infusion taken for dysentery. *Gastrointestinal Aid* Taken for visceral obstructions. *Kidney Aid* Taken for disease of the kidneys. *Liver Aid* Taken for jaundice. *Misc. Disease Remedy* Taken for scurvy. *Psychological Aid* Kept in home to insure happiness. *Sedative* Infusion taken to calm nerves. *Toothache Remedy* Fruit held in mouth to remove tartar from teeth. *Urinary Aid* Taken for disease of the bladder. (80:57) **Chippewa** *Misc. Disease Remedy* and *Pediatric Aid* Infusion of root given for "cholera infantum." (53:346) **Iroquois** *Unspecified* Fruits eaten as a spring medicine. (124:96) **Malecite** *Abortifacient* Infusion of plant and dwarf raspberry used for irregular menstruation. (116:258) **Micmac** *Abortifacient* Parts of plant used for irregular menstruation. (40:56) **Ojibwa** *Analgesic* Infusion taken for stomachaches. *Gastrointestinal Aid* and *Pediatric Aid* Infusion of root used, especially for babies, for stomachache. (153:384) **Okanagan-Colville** *Dermatological Aid* Poultice of leaf powder and deer fat applied to sores. *Disinfectant* Leaf powder applied to any

open sore as a disinfectant. *Oral Aid* and *Pediatric Aid* Leaf powder dusted into baby's sore mouth. (188:125) **Thompson** *Antidiarrheal* Infusion of roots or whole plant taken for diarrhea or dysentery. Decoction of leaves taken for diarrhea. *Dermatological Aid* Berries used as deodorant. *Pediatric Aid* Infusion of roots or whole plant bottle fed to babies for diarrhea or dysentery. Decoction of leaves given to children for diarrhea. (187:259)

- *Food*—**Abnaki** *Fruit* Fruits eaten for food. (144:169) **Algonquin, Quebec** *Fruit* Fruit gathered, cultivated, and eaten fresh. *Preserves* Fruit gathered, cultivated, and preserved. (18:91) **Algonquin, Tête-de-Boule** *Fruit* Berries used for food. (132:128) **Bella Coola** *Fruit* Berries used for food. (184:208) **Blackfoot** *Beverage* Leaves used to make tea. *Fruit* Fruits eaten raw. (97:38) **Cherokee** *Fruit* Fruit eaten raw. (80:57) Fresh berries used for food. *Preserves* Berries used to make jam. *Sauce & Relish* Berries used on shortcake. (126:56) **Cheyenne** *Fruit* Fruits formerly used for food. (83:34) **Chippewa** *Fruit* Berries eaten raw. (53:321) Strawberries considered an important part of the diet. (71:132) **Clallam** *Fruit* Berries eaten fresh. (57:202) **Cree, Woodlands** *Snack Food* Fresh fruit eaten on sight as a nibble. (109:38) **Dakota** *Fruit* Fruit used for food. (70:84) **Hesquiat** *Fruit* Berries used for food. (185:72) **Iroquois** *Bread & Cake* Fruit mashed, made into small cakes, and dried for future use. *Dried Food* Raw or cooked fruit sun or fire dried and stored for future use. (196:127) *Fruit* Fruits eaten raw. (124:96) Dried fruit taken as a hunting food. *Sauce & Relish* Dried fruit cakes soaked in warm water and cooked as a sauce or mixed with corn bread. (196:127) **Kitasoo** *Fruit* Fruit used for food. (43:342) **Klamath** *Fruit* Fresh fruit used for food. (45:98) **Menominee** *Fruit* Berries eaten fresh. (151:71) **Meskwaki** *Preserves* Berries cooked into a jam for winter use. (152:263) **Nitinaht** *Fruit* Berries eaten fresh. (186:117) **Ojibwa** *Fruit* Berries used fresh or preserved. (5:2220) Berries used in season. *Preserves* Berries used to make preserves for winter use. (153:409) *Winter Use Food* Berries used fresh or preserved. (5:2220) **Okanagan-Colville** *Fruit* Berries eaten fresh. *Winter Use Food* Berries canned for future use. (188:125) **Omaha** *Dried Food* Fruit dried for winter use. *Fruit* Fruit eaten fresh. (68:326) Fruit used for food. (70:84) **Oweekeno** *Fruit* Berries eaten fresh. *Preserves* Berries used to make jam. (43:108) **Pawnee** *Fruit* Fruit used for food. **Ponca** *Fruit* Fruit used for food. (70:84) **Salish, Coast** *Beverage* Leaves dried and used to make tea. *Fruit* Fruits eaten fresh. (182:86) **Shuswap** *Dried Food* Dried berries used for food. (123:66) **Thompson** *Dried Food* Berries, if plentiful, dried for future use. *Fruit* Berries eaten fresh. (187:259) **Winnebago** *Beverage* Young leaves used to make a tea-like beverage. *Fruit* Fruit used for food. (70:84)
- *Other*—**Iroquois** *Ceremonial Items* Fruits used as symbols of the Creator's beneficence in the Strawberry Thanksgiving ceremony. (124:96)

Fragaria virginiana ssp. *platypetala* (Rydb.) Staudt, Virginia Strawberry

- *Food*—**Haisla & Hanaksiala** *Fruit* Berries eaten fresh. *Preserves* Berries used to make jam. (43:264) **Ojibwa** *Fruit* Fruit used for food. (135:235) **Sanpoil & Nespelem** *Fruit* Berries eaten fresh. (131:102)

Frangula, Rhamnaceae

Frangula betulifolia (Greene) V. Grub. ssp. *betulifolia*, Beech-leaf Frangula

- *Drug*—**Navajo, Kayenta** *Ceremonial Medicine* Plant used in a hoop for the emetic ceremony of Mountain-top-way. *Emetic* Plant used in a hoop for the emetic ceremony of Mountain-top-way. (as *Rhamnus betulaefolia* 205:31)

Frangula californica (Eschsch.) Gray, California Buckthorn

- *Drug*—**Neeshenam** *Toothache Remedy* Heated root held in the mouth for toothaches. (129:376)

Frangula californica (Eschsch.) Gray ssp. *californica*, California Buckthorn

- *Drug*—**Costanoan** *Cathartic* Decoction of inner bark used as a purgative. *Dermatological Aid* Decoction of leaves used for poison oak dermatitis. *Laxative* Dried, ground inner bark used as a laxative. (as *Rhamnus californica* 21:22) **Kawaiisu** *Antidote* Crushed berries used to counteract poisoning. *Burn Dressing* Crushed leaves and berries rubbed into burns. *Dermatological Aid* Crushed berries applied to infected sores. Crushed leaves and berries used for wounds. *Disinfectant* Crushed berries applied to infected sores. *Hemostat* Crushed berries used to stop the flow of blood. *Laxative* Ripe berries eaten as a laxative. *Unspecified* Roots or dried seeds used as medicine. (as *Rhamnus californica* ssp. *tomentella* 206:58) **Mahuna** *Cathartic* Powdered bark used as a cathartic for constipation. (as *Rhamnus californica* 140:21) **Mendocino Indian** *Cathartic* Bark used as a cathartic. *Kidney Aid* Bark used for kidney troubles. *Misc. Disease Remedy* Bark used for grippe. *Psychological Aid* Decoction of bark taken for mania. (as *Rhamnus californica* 41:368) **Mewuk** *Cathartic* Infusion of bark and leaves used as a cathartic. (as *Rhamnus californica* 117:366) **Modesse** *Antirheumatic (External)* Used as a medicine for rheumatism. *Cathartic* Used as a cathartic. (as *Rhamnus californica* 117:224) **Pomo** *Laxative* Decoction of bark taken for constipation. *Poison* Berries considered poisonous. (as *Rhamnus californica* 66:14) **Pomo, Kashaya** *Laxative* Decoction of bark stored for a whole year and taken for constipation. Fresh berries eaten as a laxative. (as *Rhamnus californica* 72:39) **Yokia** *Misc. Disease Remedy* Decoction of bark taken for grippe. (as *Rhamnus californica* 41:368)
- *Food*—**Costanoan** *Fruit* Raw berries used for food. (as *Rhamnus californica* 21:250) **Paiute** *Dried Food* Fruit sun dried, stored in buckskin bags, and hung up for winter use. *Fruit* Fruit eaten fresh. (as *Rhamnus californica* 167:245)

Frangula californica ssp. *occidentalis* (T. J. Howell) Kartesz & Gandhi, California Buckthorn

- *Drug*—**Cahuilla** *Laxative* Infusion of berries taken as a laxative. Dried, ground bark used for constipation. *Tonic* Infusion of berries taken as a tonic. (as *Rhamnus californica* ssp. *occidentalis* 15:131)

Frangula californica ssp. *tomentella* (Benth.) Kartesz & Gandhi, California Buckthorn

- *Drug*—**Diegueño** *Cathartic* Decoction of bark used as a physic. *Dermatological Aid* Decoction of bark and salt used as a bath for poison oak. (as *Rhamnus californica* ssp. *tomentella* 84:37)
- *Food*—**Kawaiisu** *Fruit* Fruit eaten fresh. (as *Rhamnus californica* ssp. *tomentella* 206:58)

Frangula caroliniana (Walt.) Gray, Carolina Buckthorn

- *Drug*—**Creek** *Liver Aid* Infusion of wood taken for jaundice. (as *Rhamnus caroliniana* 172:667) **Delaware, Oklahoma** *Cathartic* Decoction of bark taken as a cathartic. *Emetic* Decoction of bark taken as an emetic. (as *Rhamnus caroliniana* 175:25, 78) *Gastrointestinal Aid* and *Liver Aid* Decoction of bark taken "to remove bile from the intestines." (as *Rhamnus caroliniana* 175:25)

Frangula purshiana (DC.) Cooper, Pursh's Buckthorn
- *Drug*—**Bella Coola** *Laxative* Infusion of bark taken as a strong laxative. (as *Rhamnus purshiana* 184:208) **Clallam** *Dermatological Aid* Poultice of bark used for wounds. (as *Rhamnus purshiana* 57:201) **Cowlitz** *Laxative* Bark used as a laxative. (as *Rhamnus purshiana* 79:40) **Flathead** *Cathartic* Infusion of bark used as a purgative. (as *Rhamnus purshiana* 82:56) *Poison* Fruit considered poisonous. (as *Rhamnus purshiana* 19:21) **Green River Group** *Laxative* Bark used as a laxative. (as *Rhamnus purshiana* 79:40) **Haisla & Hanaksiala** *Laxative* Infusion of bark used as a laxative. (as *Rhamnus purshianus* 43:262) **Hesquiat** *Anthelmintic* Decoction of bark, infusion of bark or chewed bark used by children for worms. *Gastrointestinal Aid* Decoction of bark, infusion of bark or chewed bark used for general stomach upset. *Laxative* Decoction of bark, infusion of bark or chewed bark used as a laxative. It was believed that the bigger the tree, the stronger the medicine. Thick bark from the larger trees was used if a very strong dose was required; thin bark from young trees was used for a mild dose. (as *Rhamnus purshiana* 185:71) **Karok** *Cathartic* Infusion of bark taken as a physic. (as *Rhamnus purshiana* 148:385) **Klallam** *Laxative* Bark used as a laxative. (as *Rhamnus purshiana* 79:40) **Klamath** *Emetic* Infusion of foliage, twigs, and bark taken as an emetic. Berries used as an emetic. (as *Rhamnus purshiana* 45:100) Infusion of leaves, twigs, bark, and berries taken as an emetic. (as *Rhamnus purshiana* 163:131) **Kutenai** *Cathartic* Infusion of bark used as a purgative. (as *Rhamnus purshiana* 82:56) **Kwakiutl** *Gastrointestinal Aid* Decoction of dried bark taken for biliousness. *Laxative* Decoction of dried bark taken as a laxative. (as *Rhamnus purshiana* 183:288) **Lummi** *Laxative* Bark used as a laxative. (as *Rhamnus purshiana* 79:40) **Makah** *Adjuvant* Bark mixed with crab apple bark to prevent the crab apple from constipating the user. *Laxative* Used as a laxative. (as *Rhamnus purshiana* 67:286) Bark used as a laxative. (as *Rhamnus purshiana* 79:40) **Montana Indian** *Emetic* Decoction of leaves, bark, and fruit used as an emetic. *Unspecified* Bark used as a source of medicine. (as *Rhamnus purshiana* 19:21) **Nitinaht** *Disinfectant* Infusion of spring or early summer bark used as a disinfectant for cuts, wounds, and sores. *Gastrointestinal Aid* Infusion of spring or early summer bark taken as a tonic for bowel regularity. *Laxative* Infusion of spring or early summer bark taken as a mild but effective laxative. (as *Rhamnus purshiana* 186:115) **Okanagan-Colville** *Antirheumatic (Internal)* Infusion of bark taken for rheumatism and arthritis. *Blood Medicine* Infusion of bark taken as a blood purifier. *Laxative* Infusion of bark taken as a mild laxative. (as *Rhamnus purshiana* 188:120) **Paiute** *Gastrointestinal Aid* Decoction of bark taken for "any trouble in the stomach." (as *Rhamnus purshiana* 111:89) **Quileute** *Laxative* Bark used as a laxative. (as *Rhamnus purshiana* 79:40) *Panacea* Infusion of bark used for "any sort of disease." *Venereal Aid* Infusion of bark taken for gonorrhea. (as *Rhamnus purshiana* 137:65) **Quinault** *Laxative* Bark used as a laxative. (as *Rhamnus purshiana* 79:40) **Salish, Coast** *Tonic* Bark soaked in cold water and used as an excellent tonic. (as *Rhamnus purshiana* 182:86) **Sanpoil** *Cathartic* Decoction of bark used as a cathartic. (as *Rhamnus purshiana* 131:221) **Shuswap** *Laxative* Decoction of bark taken as a laxative. (as *Rhamnus purshiana* 123:65) **Skagit** *Antidiarrheal* Decoction of inner bark taken for dysentery. *Dermatological Aid* Salve of bark ashes and grease rubbed on swellings. *Laxative* Bark used as a laxative. (as *Rhamnus purshiana* 79:40) **Skagit, Upper** *Laxative* Decoction of bark used as a laxative. (as *Rhamnus purshiana* 179:42) **Squaxin** *Dermatological Aid* Poultice of chewed bark applied or infusion of bark used as a wash for sores. *Laxative* Bark used as a laxative. **Swinomish** *Laxative* Bark used as a laxative. (as *Rhamnus purshiana* 79:40) **Thompson** *Analgesic* Decoction of four bark strips used as a skin wash for sciatica. (as *Rhamnus purshiana* 187:253) *Cathartic* Strong decoction of bark or wood used as a physic. (as *Rhamnus purshiana* 164:473) Strong or mild decoction of bark and sometimes wood used as a physic. *Gastrointestinal Aid* Decoction of two bark strips and flowering dogwood bark taken for ulcers. (as *Rhamnus purshiana* 187:253) *Laxative* Mild decoction of bark or wood used as a laxative. (as *Rhamnus purshiana* 164:473) *Liver Aid* Infusion of bark and red elderberry roots taken for liver diseases. (as *Rhamnus purshiana* 187:253) **Tolowa** *Laxative* Bark used as a laxative. (as *Rhamnus purshiana* 6:50) **West Coast Indian** *Cathartic* Infusion of root bark or bark taken as a cathartic. *Panacea* Infusion of root bark or bark taken for most any sort of disease. *Poison* Infusion of root bark or bark taken in large doses caused death. *Venereal Aid* Infusion of root bark or bark taken for gonorrhea. (as *Rhamus purshiana* 134:133) **Yurok** *Laxative* Decoction of bark or bark chewed as a laxative. (as *Rhamnus purshiana* 6:50) **Yurok, South Coast (Nererner)** *Cathartic* Decoction of bark used as a cathartic medicine. (as *Rhamnus purshiana* 117:169)
- *Food*—**Makah** *Fruit* Berries eaten fresh in the summer. (as *Rhamnus purshiana* 79:40)
- *Dye*—**Skagit** *Green* Bark boiled and used as a green dye for mountain goat wool. (as *Rhamnus purshiana* 79:40)
- *Other*—**Hesquiat** *Tools* Wood used to make implement handles, especially D-adz handles. (as *Rhamnus purshiana* 185:71) **Nitinaht** *Tools* Wood used to make D-adz handles. (as *Rhamnus purshiana* 186:115)

Frangula rubra (Greene) V. Grub. **ssp. *rubra***, Red Buckthorn
- *Drug*—**Miwok** *Cathartic* Decoction of bark taken as a cathartic. (as *Rhamnus rubra* 12:172)
- *Food*—**Atsugewi** *Fruit* Fresh berries used for food. (as *Rhamnus rubra* 61:139)

Frasera, Gentianaceae

Frasera albomarginata S. Wats., Desert Elkweed
- *Drug*—**Navajo, Kayenta** *Dermatological Aid* Poultice of plant applied to gunshot wounds. (205:36)

Frasera albomarginata var. *induta* (Tidestrom) Card, Desert Elkweed
- *Drug*—**Shoshoni** *Eye Medicine* Decoction of root used as an eyewash. (180:75)

Frasera caroliniensis Walt., American Columbo
- *Drug*—**Cherokee** *Antidiarrheal* Root used as tonic and taken for

dysentery. *Antiemetic* Used to "check vomiting." *Dietary Aid* Taken for dysentery and given for "want of appetite." *Disinfectant* Used as an antiseptic. *Gastrointestinal Aid* Root used for indigestion, colics, and cramps. *Tonic* Root used as tonic. (as *Swertia caroliniensis* 80:30)

Frasera montana Mulford, White Elkweed
- **Drug**—**Okanagan-Colville** *Tuberculosis Remedy* Infusion of roots taken for tuberculosis. (188:106)

Frasera sp., American Deer Ears
- **Drug**—**Mahuna** *Dermatological Aid* Infusion of plant used for infected sores. (140:15)
- **Food**—**Yavapai** *Unspecified* Leaves and roots boiled and eaten. (65:256)

Frasera speciosa Dougl. ex Griseb., Showy Frasera
- **Drug**—**Apache** *Unspecified* Root used to make a medicine. (121:49) **Cheyenne** *Antidiarrheal* Infusion of dried, pulverized leaves or roots taken for diarrhea. (76:184) **Havasupai** *Cold Remedy* Cooled decoction of roots taken for colds and similar troubles. *Gastrointestinal Aid* Cooled decoction of roots taken for digestive upsets and similar troubles. *Venereal Aid* Cooled decoction of roots used in conjunction with the sweat bath for gonorrhea. (as *Swertia radiata* 197:236) **Isleta** *Analgesic* Poultice of large, salted leaves applied to the head for headaches. *Pulmonary Aid* Decoction of large, fleshy root used as a lung medicine for asthma. *Throat Aid* Decoction of large, fleshy root used as a throat medicine for asthma. (100:29) **Navajo** *Sedative* Plant used for alarm and nervousness. (as *Swertia radiata* 55:97) **Navajo, Kayenta** *Panacea* Plant used as a life medicine. *Veterinary Aid* Ground plant sprinkled on incision when castrating livestock. (205:36) **Navajo, Ramah** *Psychological Aid* Dried leaves mixed with mountain tobacco and smoked to "clear the mind if lost." *Strengthener* Cold, compound infusion rubbed on hunters to strengthen them. *Veterinary Aid* Cold, compound infusion rubbed on hunters' horses to strengthen them. (191:39) **Shoshoni** *Panacea* Decoction of roots taken as a tonic for any general weakness or illness. *Tonic* Decoction of root taken as a tonic for any general weakness or illness. (180:76)
- **Food**—**Apache** *Unspecified* Roots used for food. (32:29) Root used for food. (121:49) **Arapaho** *Sweetener* Nectar used for honey. (118:17)
- **Other**—**Apache** *Hunting & Fishing Item* Large stems used to make an elk call. (121:49)

Fraxinus, Oleaceae
Fraxinus americana L., White Ash
- **Drug**—**Abnaki** *Abortifacient* Used as an emmenagogue. (144:154) Infusion of bark taken by women to provoke menses. (144:172) **Cherokee** *Gastrointestinal Aid* Tonic of inner bark taken for liver and stomach. *Gynecological Aid* Infusion of bark used to "check discharge." (80:23) **Delaware, Oklahoma** *Cathartic* Decoction of bark taken as a cathartic. *Emetic* Decoction of bark taken as an emetic. (175:25, 76) *Gastrointestinal Aid* and *Liver Aid* Decoction of bark taken "to remove bile from the intestines." (175:25) **Iroquois** *Blood Medicine* Compound used for bad blood. (87:412) *Cathartic* Decoction of bark taken as a physic. (87:411) *Dermatological Aid* Compound infusion of bark taken and applied as poultice to syphilitic lumps. Compound used for neck sores. (87:412) *Ear Medicine* Infusion of plant, with another plant, used as ear drops for earaches. Branch sap used for earaches. (141:60) *Emetic* Bark chewed to cause vomiting and clean out the insides, as a hunting medicine for deer. *Gastrointestinal Aid* Compound decoction of bark taken for stomach cramps. *Hunting Medicine* Bark chewed to cause vomiting and clean out the insides, as a hunting medicine for deer. *Laxative* Compound decoction of bark taken as a laxative. (87:412) *Reproductive Aid* Compound decoction of roots and bark taken to induce pregnancy. *Snakebite Remedy* Decoction of roots taken and applied as poultice to snakebites. (87:411) *Venereal Aid* Compound infusion of bark taken and applied as poultice to syphilitic lumps. *Veterinary Aid* Compound decoction of plants mixed with feed as a laxative for horses. (87:412) **Meskwaki** *Dermatological Aid* Infusion of bark used for sores, itch, and vermin on the scalp. *Snakebite Remedy* Decoction of flowers taken as an antidote for a bite, probably a snakebite. (152:233) **Micmac** *Gynecological Aid* Leaves used for cleansing after childbirth. (40:56) **Ojibwa** *Unspecified* Root bark used for medicinal purposes. (135:245) **Penobscot** *Gynecological Aid* Strong decoction of leaves taken after childbirth for cleansing. (156:310)
- **Fiber**—**Abnaki** *Snow Gear* Wood used to make the frames of snowshoes. (144:160) Wood used to make snowshoes. (144:172) **Iroquois** *Basketry* Plant used to make baskets. *Furniture* Plant used to make chair backs. (141:60) **Malecite** *Canoe Material* Used to make boat frames. *Snow Gear* Used to make snowshoes. (160:6) **Meskwaki** *Basketry* Wood splints used for weaving baskets. (152:268) **Ojibwa** *Canoe Material* Used to make canoes. *Snow Gear* Used to make snowshoes. (135:245)
- **Other**—**Chippewa** *Hunting & Fishing Item* Wood used to make handles for fishing spears. (71:139) **Micmac** *Tools* Used to make ax and knife handles. (159:258)

Fraxinus anomala Torr. ex S. Wats., Singleleaf Ash
- **Other**—**Hopi** *Ceremonial Items* Used for prayer sticks. **Navajo, Kayenta** *Ceremonial Items* Seeds used in prayer for rain. (205:35)

Fraxinus cuspidata Torr., Fragrant Ash
- **Other**—**Navajo** *Hunting & Fishing Item* Stems used to make arrows. *Stable Gear* Wood used to make a fair imitation of the Mexican saddle. (55:68) **Navajo, Ramah** *Hunting & Fishing Item* Wood used to make arrow shafts and bows. *Tools* Wood used to make weaving tools. (191:39)

Fraxinus latifolia Benth., Oregon Ash
- **Drug**—**Costanoan** *Febrifuge* Cold infusion of twigs used for fevers. *Snakebite Remedy* Leaves placed in sandals as a snake repellent. (21:12) **Cowlitz** *Anthelmintic* Infusion of bark taken for worms. (as *F. oregana* 79:45) **Karok** *Preventive Medicine* Bark used to prevent bad effect on medicine by ceremonially impure person. (as *F. oregana* 148:388) **Yokia** *Dermatological Aid* Mashed roots used for wounds. (as *F. oregana* 41:378)
- **Fiber**—**Cowlitz** *Canoe Material* Wood used to make canoe paddles. (as *F. oregana* 79:45) **Karok** *Basketry* Roots used to make baskets. (as *F. oregana* 148:388)
- **Other**—**Costanoan** *Protection* Leaves placed in sandals as a snake repellent. (21:250) **Cowlitz** *Tools* Wood used to make digging sticks. (as *F. oregana* 79:45) **Kawaiisu** *Tools* Peeled pole, 10 or more feet in length, used to knock down pinyon cones. (206:32) **Mendocino Indian** *Tools* Wood used to make handles and small tools. *Walking Sticks*

Wood used to make canes. **Yuki** *Fuel* Used for fuel. *Smoking Tools* Used to make tobacco pipes. (as *F. oregana* 41:378) Wood used to make straight pipes. (as *F. oregona* 49:93)

Fraxinus nigra Marsh., Black Ash

- **Drug**—**Cherokee** *Gastrointestinal Aid* Tonic of inner bark taken for liver and stomach. *Gynecological Aid* Infusion of bark used to "check discharge." (80:23) **Iroquois** *Analgesic* Infusion of bark taken for painful urination. *Antirheumatic (External)* Compound infusion of roots and bark used as foot soak for rheumatism. (87:412) *Ear Medicine* Compound infusion of roots used as drops for earaches. *Laxative* Compound decoction of bark taken as a laxative. *Other* Compound decoction of bark taken for stricture. *Reproductive Aid* Compound decoction of roots and bark taken to induce pregnancy. (87:413) *Urinary Aid* Infusion of bark taken for painful urination. (87:412) *Veterinary Aid* Compound decoction of plants mixed with feed given to horses as a laxative. (87:413) **Menominee** *Adjuvant* Inner bark used as a seasoner for medicines. (151:43) **Meskwaki** *Laxative* Compound infusion of wood used to loosen the bowels. *Panacea* Inner bark of trunk considered a remedy for any internal ailments. (152:233) **Ojibwa, South** *Eye Medicine* Infusion of inner bark applied to sore eyes. (as *F. sambucifolia* 91:200)
- **Fiber**—**Abnaki** *Basketry* Wood used to make baskets. (144:157) Used to make baskets. (144:172) **Chippewa** *Basketry* Wood logs beaten with mauls to separate the growth layers, cut into strips, and woven into baskets. The wood logs were beaten with mauls until the growth layers were loosened so that they could be separated. The thin sheets of wood were then cut into strips of the desired size and woven into baskets. (71:139) *Building Material* Bark used to cover wigwams. (53:377) **Malecite** *Basketry* Used to make basket splints. (160:6) **Meskwaki** *Basketry* Inner bark and wood used to make baskets. (152:269) **Micmac** *Basketry* Used to make basket ware. (159:258) **Ojibwa** *Basketry* Wood used for basketry splints. (153:420)
- **Dye**—**Chippewa** *Blue* Bark used to make a blue dye in a manner similar to that of blue ash. (71:139)
- **Other**—**Chippewa** *Fuel* Wood used for fuel for quiet fires because it did not crackle and shoot sparks like other woods. (71:139) **Menominee** *Hunting & Fishing Item* Wood used for bows and arrows. (151:75) **Meskwaki** *Hunting & Fishing Item* Wood used to make bows and arrows. (152:269)

Fraxinus pennsylvanica Marsh., Green Ash

- **Drug**—**Algonquin, Tête-de-Boule** *Psychological Aid* Infusion of inner bark taken for depression. *Stimulant* Infusion of inner bark taken for fatigue. (132:128) **Ojibwa** *Tonic* Compound containing inner bark used as a tonic. (153:376) **Omaha** *Ceremonial Medicine* Plant used in various rituals. (as *F. viridis* 68:322)
- **Food**—**Ojibwa** *Unspecified* Cambium layer scraped down in long, fluffy layers and cooked. They say it tastes like eggs. (153:407)
- **Fiber**—**Cherokee** *Building Material* Wood used for firewood and lumber. *Sporting Equipment* Wood used to make handles, ball bats, and butter paddles. (80:23) **Cheyenne** *Building Material* Wood used to make Sundance lodges. (82:20) Trunks used to construct the medicine lodge for the Sun Dance ceremony. (83:46) Wood used to make tent poles, pegs, and tepee pins. Wood used to make posts for the Sun Dance lodge. (83:31) **Havasupai** *Building Material* Wood used for house and fence construction. *Furniture* Wood used to make the oval frame for the cradleboard. (197:235) **Ojibwa** *Basketry* All ash wood quite valuable and used for basketry splints. *Furniture* All ash wood quite valuable and used for cradleboards. *Snow Gear* All ash wood quite valuable and used for snowshoe frames and sleds. (153:420) **Omaha** *Building Material* Wood and cottonwood used to make the sacred pole. **Ponca** *Building Material* Wood and cottonwood used to make the sacred pole. (70:108) **Potawatomi** *Basketry* Wood rings used for making woven wooden baskets. (154:113)
- **Other**—**Cherokee** *Cooking Tools* Wood used to make butter paddles. *Fuel* Wood used for firewood and lumber. *Tools* Wood used to make handles and ball bats. (80:23) **Cheyenne** *Cooking Tools* Wood used to make racks for drying meat. (83:31) *Hunting & Fishing Item* Used to make bows. (83:46) Wood used to make bows and arrows. *Smoking Tools* Wood used to make pipestems. (83:31) **Dakota** *Hunting & Fishing Item* Wood used to make bows. Young stems used to make arrow shafts. *Smoking Tools* Wood used to make pipestems. (70:108) **Havasupai** *Fuel* Wood used for fuel. *Hunting & Fishing Item* Wood used to make bows. *Tools* Wood used for handles of various tools, such as hoes or axes. *Toys & Games* Wood used to make the hoop for hoop and pole game. (197:235) **Lakota** *Hunting & Fishing Item* Wood used to make bows. *Smoking Tools* Wood used to make pipestems. (139:52) **Ojibwa** *Hunting & Fishing Item* All ash wood quite valuable and used for bows and arrows. (153:420) **Omaha** *Ceremonial Items* Wood and cottonwood used to make the sacred pole. *Hunting & Fishing Item* Wood used to make bows. Young stems used to make arrow shafts. *Smoking Tools* Wood used to make pipestems. **Pawnee** *Hunting & Fishing Item* Wood used to make bows. Young stems used to make arrow shafts. *Smoking Tools* Wood used to make pipestems. **Ponca** *Ceremonial Items* Wood and cottonwood used to make the sacred pole. *Hunting & Fishing Item* Wood used to make bows. Young stems used to make arrow shafts. *Smoking Tools* Wood used to make pipestems. (70:108) **Potawatomi** *Cooking Tools* Wood used for making wooden spoons. (154:113) **Sioux** *Ceremonial Items* Wood used as a part of sacred poles. (82:20) *Hunting & Fishing Item* Wood used for bows. (as *F. viridis* 19:12) **Winnebago** *Hunting & Fishing Item* Wood used to make bows. Young stems used to make arrow shafts. *Smoking Tools* Wood used to make pipestems. (70:108)

Fraxinus sp., Ash

- **Drug**—**Algonquin, Quebec** *Ear Medicine* Sap used for earaches. Medicine was made by placing the end of a fresh log or branch in a fire. The sap was collected as it appeared from the opposite end. (18:218) **Chippewa** *Cathartic* Decoction of root used as an enema. *Stimulant* Decoction of inner bark taken as a stimulant. *Tonic* Decoction of inner bark taken as a tonic. (53:364)
- **Fiber**—**Chippewa** *Snow Gear* Used to make snowshoe frames and sleds. (53:377) **Iroquois** *Furniture* Bark used to make chair backs. (142:99)

Fraxinus velutina Torr., Velvet Ash

- **Other**—**Hualapai** *Hunting & Fishing Item* Wood used to make bows. *Tools* Wood used to make a sharp tool for gathering mescal agave. Wood made into long prongs used to pick from saguaro cacti and pinyon pine trees. *Walking Sticks* Wood used to make canes and staffs. (as *F. velutina* 195:25)

Fremontodendron, Sterculiaceae

Fremontodendron californicum (Torr.) Coville, California Flannelbush
- *Drug*—**Kawaiisu** *Cathartic* Infusion of inner bark taken as a physic. (206:32)
- *Fiber*—**Kawaiisu** *Building Material* Bark made into cordage and tied in a loop to upper ends of poles to make a winter house smoke hole. *Cordage* Bark make into a twine and used to sting pinyon seeds for winter storage. *Furniture* Wood used to make cradles. (206:32) **Yokut** *Cordage* Bark made into ropes and used to bound acorn caches. (117:420)
- *Other*—**Kawaiisu** *Containers* Bark made into cordage and used to make heavy load carrying nets. *Hunting & Fishing Item* Bark made into cordage and used to make rabbit nets. (206:32) **Shoshoni** *Fasteners* Bark used to bind bundles of fine brush for acorn caches. (117:447)

Fremontodendron sp.
- *Fiber*—**Shoshoni** *Cordage* Tough bark used to make cord. (as *Fremontia* 117:440)

Freycinetia, Pandanaceae

Freycinetia arborea Gaud., 'Ie'ie
- *Drug*—**Hawaiian** *Analgesic* Shoots and leaves laid over the sheets in bed for severe body pain. *Gynecological Aid* Stems and other plants pounded, squeezed, and resulting juice taken for excessive menses. *Pediatric Aid* Shoots and other plants pounded, squeezed, and the resulting juice given to children with general debility. *Strengthener* Shoots and other plants pounded, squeezed, and the resulting juice given to children with general debility, or taken for general debility. (as *F. arnotti* 2:22)

Fritillaria, Liliaceae

Fritillaria atropurpurea Nutt., Spotted Missionbells
- *Drug*—**Lakota** *Cancer Treatment* Plant pulverized into a salve and applied to scrofulous swellings. (139:27) **Ute** *Poison* Decoction of bulbs and roots in large quantities regarded dangerously poisonous. *Unspecified* Decoction of bulbs and roots used as medicine. (38:34)

Fritillaria camschatcensis (L.) Ker-Gawl., Kamchatka Missionbells
- *Food*—**Alaska Native** *Dried Food* Bulbs dried and used in fish and meat stews. *Soup* Bulbs dried and used in fish and meat stews. *Staple* Bulbs pounded into a flour. (85:119) **Bella Coola** *Unspecified* Bulbs formerly boiled and eaten with sugar and grease. (184:199) **Haisla & Hanaksiala** *Unspecified* Bulbs eaten with western dock. (43:196) **Hesquiat** *Dried Food* Bulbs dried for winter use. *Forage* The first horse seen in the Hesquiat area was said to have eaten mission bells. *Vegetable* Boiled bulbs eaten with oil. (185:55) **Kitasoo** *Unspecified* Bulbs used for food. (43:320) **Kwakiutl, Southern** *Dried Food* Bulbs sun dried, steamed, covered with oil, and eaten at feasts. (183:273) **Oweekeno** *Unspecified* Bulbs boiled, mixed with oolichan (candlefish) grease and sugar, and eaten. (43:77) **Salish, Straits** *Unspecified* Roots formerly used for food. (186:85)
- *Other*—**Hanaksiala** *Ceremonial Items* Flowers used on costumes for the New Year "flower dance." *Season Indicator* Flower appearance signals the "Indian New Year." (43:196)

Fritillaria lanceolata Pursh, Rice Root
- *Food*—**Okanagon** *Staple* Roots used as a principal food. (178:238) *Unspecified* Cooked and used for food. (125:37) Roots used as an important food. (178:237) **Saanich** *Unspecified* Bulbs used for food. (182:75) **Salish, Coast** *Unspecified* Bulbs used for food. (183:300) **Shuswap** *Unspecified* Roasted roots and stems used for food. (123:54) **Thompson** *Dried Food* Washed bulbs dried for future use. *Spice* Bulbs used in flavoring soups. (187:125) *Unspecified* Cooked and used for food. (125:37) Thick, scaly bulbs cooked and eaten. (164:481) Roots used as an important food. (178:237) Roots steam cooked with a little water and put in puddings or pit cooked and used for food. (187:125)

Fritillaria pudica (Pursh) Spreng., Yellow Missionbells
- *Food*—**Blackfoot** *Soup* Bulbs eaten with soup. *Unspecified* Bulbs eaten fresh. (86:102) **Flathead** *Unspecified* Bulbous, underground corms boiled and used for food. (82:25) **Gosiute** *Unspecified* Bulbs formerly used for food. (39:370) **Montana Indian** *Forage* Bulbous, underground corms eaten by bears, gophers, and ground squirrels. Leafy tops eaten by deer. (82:25) *Unspecified* Bulb used for food. (19:12) **Okanagan-Colville** *Dried Food* Bulbs pit cooked, dried, and stored for future use. *Unspecified* Small bulbs eaten raw. (188:46) **Okanagon** *Staple* Roots used as a principal food. (178:238) *Unspecified* Small bulbs steamed and used for food. (125:37) **Paiute** *Unspecified* Bulb gathered, boiled, and eaten. (111:57) **Shuswap** *Unspecified* Root used for food. (123:54) **Spokan** *Unspecified* Roots used for food. (178:343) **Thompson** *Unspecified* Small bulbs steamed and used for food. (125:37) Bulbs used for food. (164:482) Bulbs eaten when available. (187:125) **Ute** *Unspecified* Bulbs formerly used for food. (38:34)
- *Other*—**Okanagan-Colville** *Season Indicator* Flowers used as a sign that spring had arrived. (188:46) **Shuswap** *Decorations* Flowers used to make a bouquet. (123:54)

Fritillaria recurva Benth., Scarlet Missionbells
- *Food*—**Shasta** *Unspecified* Bulbs boiled or roasted in ashes and eaten. (93:308)

Fritillaria sp., Fritillary
- *Food*—**Similkameen** *Unspecified* Bulbs used for food. (164:481) **Yana** *Unspecified* Roots roasted and eaten. (147:251)

Fucaceae, see *Fucus*

Fucus, Fucaceae, alga

Fucus gardneri Silva, Sea Wrack
- *Drug*—**Kwakiutl, Southern** *Analgesic* Poultice applied for aches and pains. *Antirheumatic (External)* and *Strengthener* Plants used to make a steam bath for general sickness. *Venereal Aid* Fresh plants rubbed on legs and feet for locomotor ataxia. (183:260)
- *Food*—**Haisla & Hanaksiala** *Dried Food* Plant used to collect herring roe, dried and eaten with the roe. (43:125)
- *Other*—**Haisla & Hanaksiala** *Tools* Plant used with an open fire to steam heat kerfed boards to bend into red cedar bentwood boxes. (43:125) **Hesquiat** *Fertilizer* Seaweed used to fertilize potatoes. (185:24) **Nitinaht** *Ceremonial Items* Plants rubbed on body by pregnant women expecting their unborn baby boys to become whalers. *Good*

Luck Charm Plants rubbed on body until receptacles broke by whalers for good luck in hunting. (186:51) **Oweekeno** *Toys & Games* Bladders squeezed and popped by children for entertainment. (43:44)

Fucus sp.
- *Food*—**Eskimo, Alaska** *Unspecified* Seaweed, densely covered with herring eggs, eaten raw or cooked. (1:33) **Yurok, South Coast (Ner-erner)** *Dried Food* Plant dried and eaten without cooking. (117:169)
- *Other*—**Bella Coola** *Fuel* Used in steaming pits to generate steam for cooking. (184:195)

Gaillardia, Asteraceae

Gaillardia aristata Pursh, Common Gaillardia
- *Drug*—**Blackfoot** *Breast Treatment* Infusion of plant rubbed on nursing mother's sore nipples. *Dermatological Aid* Poultice of chewed, powdered roots applied to skin disorders. (86:76) *Eye Medicine* Infusion of plant used as an eyewash. (86:81) *Gastrointestinal Aid* Infusion of roots taken for gastroenteritis. (86:66) *Nose Medicine* Infusion of plant used as nose drops. (86:71) *Orthopedic Aid* Infusion of flower heads used as a foot wash. (86:124) *Veterinary Aid* Infusion of roots rubbed on saddle sores and places where the hair was falling out. Infusion of roots used for horses as an eyewash for minor lacerations. (86:88) **Okanagan-Colville** *Analgesic* Flowers used to "paint" the body for pain. *Kidney Aid* Decoction of plant taken for kidney problems. *Orthopedic Aid* Poultice of mashed plant applied for backaches. *Venereal Aid* Infusion or decoction of whole plant used as a bath for venereal disease. (188:84) **Thompson** *Analgesic* Decoction of plant taken for headache and general indisposition. (164:469) *Cancer Treatment* Infusion of whole plant used for cancer. *Misc. Disease Remedy* Poultice of lightly toasted, pounded plant mixed with bear grease and used for "mumps." *Tuberculosis Remedy* Decoction of plant taken for tuberculosis. (187:181)
- *Food*—**Blackfoot** *Soup* Flower heads used to absorb soups and broth. (86:113)
- *Other*—**Blackfoot** *Cooking Tools* Flower heads served as spoons for the sick and invalid. *Waterproofing Agent* Flower heads rubbed on rawhide bags for waterproofing. (86:113)

Gaillardia pinnatifida Torr., Red Dome Blanketflower
- *Drug*—**Hopi** *Analgesic* Plant used as a diuretic for painful urination. (200:96) *Diuretic* Taken as a diuretic. (42:320) Plant used as a diuretic for painful urination. (200:35, 96) **Keres, Western** *Gynecological Aid* Plant rubbed on mothers' breasts to wean infant. *Psychological Aid* Infusion of plant used to become good drummers. (171:44) **Navajo** *Misc. Disease Remedy* Infusion of leaves taken and poultice of leaves applied for gout. (55:86) **Navajo, Kayenta** *Other* Plant used for the effects of immersion. *Witchcraft Medicine* Plant used for bewitchment. (205:48) **Navajo, Ramah** *Antiemetic* Two cupfuls of decoction taken for heartburn and nausea. *Gastrointestinal Aid* Decoction of plant taken for heartburn and nausea. *Respiratory Aid* Plant used as snuff for "congested nose." (191:51)
- *Food*—**Havasupai** *Preserves* Seeds parched, ground, kneaded into seed butter, and eaten with fruit drinks or spread on bread. (197:67)

Gaillardia pulchella Foug., Firewheel
- *Other*—**Kiowa** *Decorations* Flowers used for ornaments in the homes. *Good Luck Charm* Flowers believed to bring luck. (192:60)

Gaillardia pulchella Foug. **var.** *pulchella*, Firewheel
- *Drug*—**Keres, Western** *Gynecological Aid* Plant rubbed on mothers' breasts to wean infant. *Psychological Aid* Infusion of plant used to become good drummers. (as *G. neo-mexicana* 171:44)

Galactia, Fabaceae

Galactia volubilis (L.) Britt., Downy Milkpea
- *Drug*—**Seminole** *Analgesic* Roots used for baby sickness caused by adultery: appetite loss, fever, headache, and diarrhea. (169:253) Roots used for adult's sickness caused by adultery: headache, body pains and crossed fingers. (169:256) *Antidiarrheal* Roots used for baby sickness caused by adultery: appetite loss, fever, headache, and diarrhea. (169:253) Cold infusion of roots taken for baby's sickness: vomiting, diarrhea, and grogginess. *Antiemetic* Cold infusion of roots taken for baby's sickness: vomiting, diarrhea, and grogginess. (169:306) *Dietary Aid* Roots used for baby sickness caused by adultery: appetite loss, fever, headache, and diarrhea. (169:253) Roots and mother's milk or canned milk used for baby's sickness: refusal to suckle. (169:255) *Febrifuge* and *Pediatric Aid* Roots used for baby sickness caused by adultery: appetite loss, fever, headache, and diarrhea. (169:253) Roots and mother's milk or canned milk used for baby's sickness: refusal to suckle. (169:255) Cold infusion of roots taken for baby's sickness: vomiting, diarrhea, and grogginess. (169:306) *Reproductive Aid* Infusion of roots taken and rubbed on the body for protracted labor. (169:323) *Stimulant* Cold infusion of roots taken for baby's sickness: vomiting, diarrhea, and grogginess. (169:306) *Unspecified* Plant used for medicinal purposes. (169:162)

Galax, Diapensiaceae

Galax urceolata (Poir.) Brummitt, Beetleweed
- *Drug*—**Cherokee** *Kidney Aid* Infusion of root taken for kidneys. *Sedative* Infusion taken for "nerves." (as *G. aphylla* 80:35)

Galeopsis, Lamiaceae

Galeopsis tetrahit L., Brittlestem Hempnettle
- *Drug*—**Iroquois** *Emetic* and *Witchcraft Medicine* Infusion of roots taken to vomit as a cure for bewitching. (87:425) **Potawatomi** *Pulmonary Aid* Infusion of plant used for pulmonary troubles. (154:61)

Galium, Rubiaceae

Galium aparine L., Stickywilly
- *Drug*—**Cherokee** *Laxative* Infusion taken to "move bowels." (80:36) **Chippewa** *Dermatological Aid* Cold infusion of stems rubbed on skin troubles. (71:141) **Cowlitz** *Love Medicine* Infusion of plant used as a bath by women to be successful in love. *Poison* Plant considered poisonous. (79:46) **Gosiute** *Veterinary Aid* Plant used as a horse medicine. (39:370) **Iroquois** *Dermatological Aid* Compound infusion of plants used as wash for poison ivy and itch. (87:439) **Meskwaki** *Emetic* Decoction of whole plant taken as an emetic. (152:243) **Micmac**

Antihemorrhagic Parts of plant used for persons spitting blood and gonorrhea. *Kidney Aid* Parts of plant used for kidney trouble. *Venereal Aid* Parts of plant used for gonorrhea. (40:56) **Nitinaht** *Dermatological Aid* Plant good for the hair, making it grow long. (67:316) **Ojibwa** *Diuretic* Infusion of whole plant used as a diuretic. *Kidney Aid* Infusion of whole plant used for kidney trouble. *Urinary Aid* Infusion of whole plant used for gravel, urine stoppage, and allied ailments. (153:386) **Penobscot** *Antihemorrhagic* Compound infusion of plant taken for "spitting up blood." *Kidney Aid* Compound infusion of plant taken for kidney trouble. *Tonic* Compound infusion of plant taken as a tonic. *Venereal Aid* Compound infusion of plant taken for gonorrhea. (156:311)
- *Other*—**Nitinaht** *Soap* Used as a hair wash. (186:125)

Galium asprellum Michx., Rough Bedstraw
- *Drug*—**Choctaw** *Diaphoretic* Whole plant used as a diaphoretic. *Diuretic* Whole plant used as a diuretic. *Misc. Disease Remedy* Whole plant used for measles. (28:287)

Galium boreale L., Northern Bedstraw
- *Drug*—**Choctaw** *Abortifacient* Decoction of whole plant used as a "deobstruent." *Contraceptive* Decoction of whole plant used to prevent pregnancy. *Diaphoretic* Whole plant used as a diaphoretic. *Diuretic* Whole plant used as a diuretic. (28:287) **Cree, Hudson Bay** *Diuretic* Leaves used as a diuretic. (92:303) **Shuswap** *Poison* Plant considered poisonous. (123:68)
- *Dye*—**Cree** *Red* Decoction of roots used as a red dye for porcupine quills. (97:53) **Great Basin Indian** *Red* Root used as a red dye and set with alum. (121:50)

Galium circaezans Michx., Licorice Bedstraw
- *Drug*—**Cherokee** *Cough Medicine* Taken for coughs. *Expectorant* Used as an expectorant. *Respiratory Aid* Taken for asthma. *Throat Aid* Taken for hoarseness. (80:43)

Galium concinnum Torr. & Gray, Shining Bedstraw
- *Drug*—**Meskwaki** *Kidney Aid* Infusion of whole plant used for kidney trouble. *Misc. Disease Remedy* Infusion of whole plant used for ague. *Urinary Aid* Infusion of whole plant used for bladder trouble. (152:244)

Galium fendleri Gray, Fendler's Bedstraw
- *Drug*—**Navajo, Ramah** *Analgesic* Infusion of plant taken and used as lotion for headache. *Ceremonial Medicine* and *Emetic* Plant used as a ceremonial emetic. *Misc. Disease Remedy* Infusion of plant taken and used as lotion for influenza. (191:45)

Galium sp., Bedstraw
- *Drug*—**Costanoan** *Antidiarrheal* Decoction of plant taken for dysentery. *Antirheumatic (External)* Decoction of plant used externally for rheumatism. *Dermatological Aid* Decoction of plant used externally for wounds. (21:24) **Iroquois** *Basket Medicine* Cold infusion of smashed roots used as a "basket or peddler's medicine." *Eye Medicine* Compound of plants used for blindness. *Urinary Aid* Compound decoction of roots and seeds taken for urine stoppage. *Venereal Aid* Infusion of plants used as wash for parts affected by venereal disease. (87:439) **Neeshenam** *Antirheumatic (External)* Poultice of heated leaves and stems applied for rheumatism. (129:376)
- *Other*—**Cowichan** *Cleaning Agent* Plants rubbed on the hands to take pitch off. *Fuel* Dried plants used for lighting fires. (182:88)

Galium tinctorium (L.) Scop., Stiff Marsh Bedstraw
- *Drug*—**Ojibwa** *Pulmonary Aid* Infusion of whole plant used for "beneficial effect upon the respiratory organs." (153:386, 387)
- *Dye*—**Micmac** *Red* Roots used to make a red dye for porcupine quills. (159:254)

Galium trifidum L., Threepetal Bedstraw
- *Drug*—**Ojibwa** *Dermatological Aid* Infusion of plant used for skin diseases like eczema and ringworm. *Tuberculosis Remedy* Infusion of plant used for skin diseases like scrofula. (153:387)

Galium triflorum Michx., Fragrant Bedstraw
- *Drug*—**Cherokee** *Gastrointestinal Aid* Infusion taken for gallstones. (80:25) **Iroquois** *Love Medicine* Compound used as love medicine. *Orthopedic Aid* and *Pediatric Aid* Poultice of whole plant applied to babies for backaches. *Urinary Aid* Compound decoction taken and poultice applied to swollen testicles or ruptures. (87:440) **Karok** *Love Medicine* Plant placed in women's bed as a love medicine. (148:389) **Klallam** *Dermatological Aid* Poultice of smashed plants applied to the hair to make it grow. (79:46) **Kwakiutl** *Analgesic* Nettles or vines and then hellebore used to rub the chest for chest pains. (20:379) Plant rubbed on the skin for chest pains. (183:291) **Makah** *Dermatological Aid* Poultice of smashed plants applied to the hair to make it grow. (79:46) **Menominee** *Kidney Aid* Infusion of herb used "to clear up kidney troubles." (151:51) **Miwok** *Kidney Aid* Decoction of plant taken as a tea for dropsy. (12:170) **Quileute** *Love Medicine* Plant used by women to attract men. **Quinault** *Dermatological Aid* Poultice of smashed plants applied to the hair to make it grow. (79:46)
- *Other*—**Makah** *Incense & Fragrance* Plant crushed and used as a perfume. (67:316) **Nitinaht** *Soap* Used as a hair wash. (186:125) **Omaha** *Incense & Fragrance* Plant gathered in green state and used only by women as perfume by tucking into the girdle. (68:323) Plant tucked under women's girdles for the delicate fragrance given off during withering. **Ponca** *Incense & Fragrance* Plant tucked under women's girdles for the delicate fragrance given off during withering. (70:115)

Galium uniflorum Michx., Oneflower Bedstraw
- *Drug*—**Choctaw** *Dermatological Aid* Whole plant used as an astringent. *Diaphoretic* Whole plant used as an diaphoretic. *Diuretic* Whole plant used as an diuretic. (28:287)

Gamochaeta, Asteraceae

Gamochaeta purpurea (L.) Cabrera, Spoonleaf Purple Everlasting
- *Drug*—**Houma** *Cold Remedy* Decoction of dried plant taken for colds. *Misc. Disease Remedy* Decoction of dried plant taken for grippe. (as *Gnaphalium purpureum* 158:64)

Ganoderma, Polyporaceae, fungus

Ganoderma applanatum (Pers. ex Wallr.) Pat., Bracket Fungus
- *Other*—**Nitinaht** *Protection* Used for protection against people with ill feelings and wishes toward others. (186:56)

Ganoderma sp., Shelf Fungus
- *Other*—**Haisla & Hanaksiala** *Ceremonial Items* Plant used for some aspects of the secret society rituals. *Protection* Plant used as hand

protection for handling live coals during a secret society ritual. Plant placed at entrance of special shamanistic dance house as protection from bad spirits and ghosts. *Toys & Games* Plant thrown into the stream and used as a target by young boys for spear practice. Plant used as a ball in a women's and children's game. (43:135)

Garryaceae, see *Garrya*

Garrya, Garryaceae

Garrya elliptica Dougl. ex Lindl., Wavyleaf Silktassel
- **Drug**—**Pomo, Kashaya** *Abortifacient* Infusion of leaves taken to bring on a woman's period. (72:106)
- **Other**—**Yurok, South Coast (Nererner)** *Tools* Wood hardened by fire and used for mussel bars to pry the mussels off the rocks. (117:169)

Garrya flavescens ssp. *pallida* (Eastw.) Dahling, Pallid Silktassel
- **Drug**—**Kawaiisu** *Cold Remedy* Decoction of plant taken for colds. *Gastrointestinal Aid* Infusion of leaves taken for stomachaches. *Laxative* Infusion of leaves taken as a laxative. *Venereal Aid* Infusion of leaves taken for gonorrhea. (206:32)

Garrya sp., Fever Bush
- **Drug**—**Poliklah** *Cold Remedy* Infusion of plant taken for colds. (117:172) **Wintoon** *Unspecified* Used for medicine. (117:264)
- **Other**—**Havasupai** *Toys & Games* Straight, thick stocks used to make whistles. (197:235)

Gaultheria, Ericaceae

Gaultheria hispidula (L.) Muhl. ex Bigelow, Creeping Snowberry
- **Drug**—**Algonquin, Quebec** *Gastrointestinal Aid* Infusion of leaves used as a tonic for overeating. (18:216) **Anticosti** *Sedative* Used to facilitate sleeping. (as *Chiogenes hispidula* 143:68) **Micmac** *Unspecified* Decoction of leaves or whole plant taken for unspecified purpose. (as *Vaccinium hispidotum* 156:317)
- **Food**—**Algonquin, Quebec** *Fruit* Fruit used for food. (18:102) **Chippewa** *Beverage* Leaves used to make a beverage. (as *Chiogenes hispidula* 53:317)

Gaultheria humifusa (Graham) Rydb., Alpine Spicywintergreen
- **Dye**—**Navajo** *Black* Used to make a black dye. (55:68)

Gaultheria ovatifolia Gray, Western Teaberry
- **Food**—**Hoh** *Fruit* Fruits eaten fresh. *Preserves* Fruits stewed and made into jelly. *Sauce & Relish* Fruits stewed and made into a sauce. **Quileute** *Fruit* Fruits eaten fresh. *Preserves* Fruits stewed and made into jelly. *Sauce & Relish* Fruits stewed and made into a sauce. (137:67)

Gaultheria procumbens L., Eastern Teaberry
- **Drug**—**Algonquin, Quebec** *Analgesic* Infusion of plant used for headaches and general discomforts. *Cold Remedy* Infusion of plant used for colds. (18:216) *Unspecified* Used to make tea and medicinal tea. (18:116) Infusion of plant used as a medicinal tea. (18:216) **Algonquin, Tête-de-Boule** *Cold Remedy* Poultice of whole plant applied to the chest for colds. Infusion of leaves used for colds. *Gastrointestinal Aid* Infusion of leaves used for stomachaches. *Misc. Disease Remedy* Infusion of leaves used for grippe. (132:129) **Cherokee** *Antidiarrheal* Leaves chewed for dysentery. *Cold Remedy* Infusion taken for colds. *Gastrointestinal Aid* Infusion of root taken with trailing arbutus for chronic indigestion. *Oral Aid* Leaves chewed for tender gums. (80:61) **Chippewa** *Blood Medicine* Decoction of plants taken as spring and fall tonic to keep blood in good order. *Cold Remedy* Plant used for colds. *Tonic* Decoction of plants taken as spring and fall tonic to keep blood in good order. (71:138) **Delaware** *Antirheumatic (External)* Plants used with poke root, mullein leaves, wild cherry, and black cohosh barks for rheumatism. (176:33) *Kidney Aid* Infusion of plant used for kidney disorders. (176:36) *Tonic* Plants used with poke root, mullein leaves, wild cherry, and black cohosh barks as a tonic. (176:33) **Delaware, Oklahoma** *Antirheumatic (Internal)* Complex compound containing entire plant taken for rheumatism. *Tonic* Complex compound containing entire plant taken as a tonic. (175:28, 76) **Iroquois** *Anthelmintic* Compound infusion of roots taken as blood remedy and for tapeworms. *Antirheumatic (Internal)* Plant used for rheumatism and arthritis. (87:410) *Blood Medicine* Compound decoction or infusion taken as blood purifier or blood remedy. *Cold Remedy* Decoction of leaves taken for colds. (87:409) *Kidney Aid* Compound decoction of plants taken for the kidneys. (87:410) *Venereal Aid* Compound decoction taken as blood purifier and for venereal disease. (87:409) **Menominee** *Antirheumatic (Internal)* Infusion of leaf and berry taken for rheumatism. (151:35) **Mohegan** *Kidney Aid* Infusion taken for kidney trouble. (30:121) Infusion of leaves taken as a kidney medicine. (176:72, 130) **Ojibwa** *Antirheumatic (Internal)* Infusion of leaves taken for rheumatism and "to make one feel good." (153:369) Young, tender leaves used as a beverage tea and rheumatic medicine. (153:400) **Potawatomi** *Analgesic* Infusion of leaves used for lumbago and rheumatism. *Antirheumatic (Internal)* Infusion of leaves taken for rheumatism and lumbago. *Febrifuge* Infusion of leaves used for fevers. (154:56, 57) **Shinnecock** *Kidney Aid* Infusion taken for kidney trouble. (30:121)
- **Food**—**Abnaki** *Beverage* Used to make tea. (144:152) Leaves used to make tea. (144:171) **Algonquin, Quebec** *Beverage* Used to make tea and medicinal tea. (18:116) *Snack Food* Berries used as a nibble food. (18:102) **Cherokee** *Beverage* Leaves used to make tea. *Fruit* Berries used for food. (126:38) *Unspecified* Species used for food. (80:61) **Chippewa** *Beverage* Leaves used to make a beverage. (53:317) Leaves used to make a pleasant, tea-like beverage. *Spice* Leaves used as a cooking flavor. (71:138) **Iroquois** *Bread & Cake* Fruit mashed, made into small cakes, and dried for future use. *Dried Food* Raw or cooked fruit sun or fire dried and stored for future use. *Fruit* Dried fruit taken as a hunting food. *Sauce & Relish* Dried fruit cakes soaked in warm water and cooked as a sauce or mixed with corn bread. (196:128) **Ojibwa** *Beverage* Leaves used to make tea. (96:17) Young, tender leaves used as a beverage tea and rheumatic medicine. (153:400) *Fruit* Fruit used for food. (135:239) Berries used for food. (153:400)
- **Other**—**Cherokee** *Smoke Plant* Dried leaves used as a substitute for chewing tobacco. (80:61)

Gaultheria shallon Pursh, Salal
- **Drug**—**Bella Coola** *Dermatological Aid* Poultice of toasted, pulverized leaves applied to cuts. (150:63) **Klallam** *Burn Dressing* Poultice of chewed leaves applied to burns. (79:43) **Makah** *Oral Aid* Leaves used to dry the mouth. **Nitinaht** *Gastrointestinal Aid* Infusion of leaves used as a stomach tonic. (67:299) *Reproductive Aid* Large

leaves eaten by both newly wed husband and wife for a firstborn baby boy. (186:104) **Quileute** *Dermatological Aid* Poultice of chewed leaves applied to sores. **Quinault** *Antidiarrheal* Decoction of leaves taken for diarrhea. *Gastrointestinal Aid* Leaves chewed for heartburn and colic. **Samish** *Cough Medicine* Infusion of leaves taken for coughs. *Tuberculosis Remedy* Infusion of leaves taken for tuberculosis. **Skagit** *Tonic* Infusion of leaves taken as a convalescent tonic. (79:43) **Skagit, Upper** *Other* Infusion of leaves taken as a convalescent tea. (179:38) **Swinomish** *Cough Medicine* Infusion of leaves taken for coughs. *Tuberculosis Remedy* Infusion of leaves taken for tuberculosis. (79:43)

- *Food*—**Alaska Native** *Fruit* Berry-like fruits used for food. (85:83) **Bella Coola** *Bread & Cake* Berries dried in cakes and used as a winter food. (184:204) **Clallam** *Bread & Cake* Berries mashed, dried in cakes, soaked, dipped in oil, and eaten. (57:200) **Haisla & Hanaksiala** *Fruit* Berries used for food. (43:240) **Hesquiat** *Dried Food* Dried, caked berries rehydrated and eaten with oil. *Spice* Branches, with leaves attached, layered between fish heads and fish for flavoring. (185:65) **Karok** *Fruit* Berries used for food. (148:387) **Kitasoo** *Fruit* Fruit used for food. (43:333) **Klallam** *Bread & Cake* Berries mashed, dried, made into cakes, dipped in whale or seal oil, and eaten. (79:43) **Kwakiutl, Southern** *Dried Food* Berries mashed, dried over fire, and resulting cakes used as a winter food. (183:282) *Special Food* Berries mashed with stink currant berries and eaten by chiefs and their wives. (183:286) Ripe berries dipped into oil and eaten fresh at feasts. (183:282) **Makah** *Beverage* Leaves used as a remedy for thirst. *Bread & Cake* Berries mashed, formed into cakes, and sun or air dried for winter use. (67:299) Berries mashed, dried, made into cakes, dipped in whale or seal oil, and eaten. (79:43) *Dried Food* Berries dried for future use. *Fruit* Fruit used for food. Berries eaten fresh. *Pie & Pudding* Berries used to make pies. *Preserves* Berries used to make jellies. *Spice* Leaves used to flavor smoked fish. Leaves steamed with halibut heads for flavoring. **Nitinaht** *Dried Food* Berries dried for future use. (67:299) Berries mashed, dried into rectangular cakes, soaked, boiled, and eaten in winter. *Frozen Food* Berries frozen and used for food. (186:104) *Fruit* Fruit used for food. (67:299) Berries eaten fresh. *Preserves* Berries made into jam and used for food. *Spice* Branches and leaves used in steam cooking pits to flavor the cooking food. *Starvation Food* Leaves chewed by those lost in the bushes to alleviate hunger. (186:104) **Okanagon** *Bread & Cake* Fruits pressed into cakes and used as a winter food. *Fruit* Fruits eaten fresh. (125:39) **Oweekeno** *Fruit* Berries eaten as fresh fruit. Berries mixed with stink currants, sugar, and oolichan (candlefish) grease and eaten. *Preserves* Berries used to make jam or jelly. (43:96) **Pomo** *Fruit* Raw or cooked berries used for food. (66:14) **Pomo, Kashaya** *Fruit* Berries eaten fresh from the vine. *Pie & Pudding* Berries used in pies. (72:101) **Quileute** *Bread & Cake* Berries mashed, dried, made into cakes, dipped in whale or seal oil, and eaten. *Fruit* Berries dipped in whale oil, and eaten fresh. **Quinault** *Bread & Cake* Berries mashed, dried, made into cakes, dipped in whale or seal oil and eaten. (79:43) **Salish, Coast** *Bread & Cake* Berries boiled, poured into frames, sun or fire dried into cakes, and used as a winter food. *Fruit* Berries eaten fresh. (182:83) **Samish** *Bread & Cake* Berries mashed, dried, made into cakes, dipped in whale or seal oil, and eaten. **Skagit** *Bread & Cake* Berries mashed, dried, made into cakes, dipped in whale or seal oil, and eaten. (79:43) **Skagit, Upper** *Dried Food* Fruit pulped and dried for winter use. *Fruit* Fruit eaten fresh. (179:38) **Skokomish** *Bread & Cake* Berries mashed, dried, made into cakes, dipped in whale or seal oil, and eaten. **Snohomish** *Bread & Cake* Berries mashed, dried, made into cakes, dipped in whale or seal oil, and eaten. **Swinomish** *Bread & Cake* Berries mashed, dried, made into cakes, dipped in whale or seal oil, and eaten. (79:43) **Thompson** *Bread & Cake* Fruits pressed into cakes and used as a winter food. (125:39) Berries picked with the stems attached, washed, destemmed, dried, and made into cakes for later use. (187:213) *Fruit* Fruits eaten fresh. (125:39) Berries picked with the stems attached, washed, destemmed, and eaten fresh with other berries. *Pie & Pudding* Berries made into pies. *Preserves* Berries made into jams. (187:213) **Tolowa** *Fruit* Fresh fruit used for food. **Yurok** *Fruit* Fresh fruit used for food. (6:31)
- *Dye*—**Karok** *Black* Berries rubbed over basket caps as a black stain. (148:387) **Nitinaht** *Yellow* Infusion of leaves used as a greenish yellow dye. (186:104)
- *Other*—**Hesquiat** *Cooking Tools* Branches, with leaves attached, used as beaters for whipping soapberries. Branches, with leaves attached, layered between fish heads and fish to prevent sticking. Leaves, folded around like a cone, made a good drinking cup. (185:65) **Makah** *Cooking Tools* Branches used to whip soapberries into a froth. (67:299) *Smoke Plant* Leaves dried, pulverized, and smoked with kinnikinnick. (79:43) **Nitinaht** *Cooking Tools* Branches and leaves used in steam cooking pits to circulate steam and keep food from burning. *Paint* Leaves crushed, mixed with salmon roe, and used as paint for masks and wooden item designs. (186:104) **Oweekeno** *Cooking Tools* Branches tied into a bunch and used for whipping soapberries. (43:96)

Gaura, Onagraceae

Gaura coccinea Nutt. ex Pursh, Scarlet Beeblossom
- *Drug*—**Navajo, Ramah** *Antiemetic* Cold infusion given to settle child's stomach after vomiting. *Panacea* Plant used as "life medicine," especially for serious internal injury. *Pediatric Aid* Cold infusion given to settle child's stomach after vomiting. (191:37)
- *Other*—**Lakota** *Stable Gear* Plant chewed and rubbed on hands to catch horses. (139:52)

Gaura hexandra ssp. *gracilis* (Woot. & Standl.) Raven & Gregory, Harlequinbush
- *Drug*—**Navajo, Ramah** *Gastrointestinal Aid* Infusion of plant taken for stomachache. (as *G. gracilis* 191:37)

Gaura parviflora Dougl. ex Lehm., Velvetweed
- *Drug*—**Hopi** *Snakebite Remedy* Decoction of root taken for snakebite. (200:86) **Isleta** *Dermatological Aid* Fresh, soft leaves worn as a headband for their cooling effect in hot weather. (100:30) **Keres, Western** *Febrifuge* Leaves used for fever and the cooling effect. *Sedative* Fresh leaves used in pillows to overcome insomnia. (171:44) **Navajo** *Burn Dressing* Infusion of plant used for burns. *Dermatological Aid* Infusion of plant used for inflammation. (55:66) **Navajo, Kayenta** *Disinfectant* Plant used as a fumigant. *Gynecological Aid* Poultice of plant applied for postpartum sore breast. (205:33) **Zuni** *Snakebite Remedy* Fresh or dried root chewed by medicine man before sucking snakebite and poultice applied to wound. (27:377)

- *Food*—**Navajo, Kayenta** *Unspecified* Roots stewed with meat or roasted and used for food. (205:33)
- *Other*—**Navajo** *Protection* Plant used to keep the dancers from burning themselves during the Fire Dance at the Mountain Chant. (55:66)

Gaylussacia, Ericaceae

Gaylussacia baccata (Wangenh.) K. Koch, Black Huckleberry
- *Drug*—**Cherokee** *Antidiarrheal* Infusion of leaves and infusion of bark taken for dysentery. *Kidney Aid* Infusion of leaves taken for Bright's disease and dysentery. (80:39) **Iroquois** *Blood Medicine* Berries considered "good" for the blood. (124:96) *Ceremonial Medicine* Berries used ceremonially by those desiring health and prosperity for the coming season. (196:142) *Liver Aid* Berries considered "good" for the liver. (124:96)
- *Food*—**Cherokee** *Bread & Cake* Berries mixed with flour or cornmeal, soda, and water and made into bread. *Frozen Food* Berries frozen for future use. *Fruit* Berries used for food. *Pie & Pudding* Berries used to make cobblers and pies. *Preserves* Berries used to make jam and canned for future use. (126:39) *Unspecified* Species used for food. (80:39) **Iroquois** *Bread & Cake* Fruits dried, soaked in water and used in bread. (124:96) Fruit mashed, made into small cakes, and dried for future use. *Dried Food* Raw or cooked fruit sun or fire dried and stored for future use. (196:128) *Fruit* Fruits eaten raw. (124:96) Dried fruit taken as hunting food. (196:128) *Pie & Pudding* Fruits dried, soaked in water and used in pudding. *Porridge* Berries dried, soaked in cold water, heated slowly, and mixed with bread meal or hominy in winter. *Sauce & Relish* Fruits dried, soaked in water, and used as a sauce. Berries dried, soaked in cold water, heated slowly, and used as a winter sauce. (124:96) Dried fruit cakes soaked in warm water and cooked as a sauce or mixed with corn bread. (196:128) *Soup* Fruits dried, soaked in water, and used in soups. (124:96) **Ojibwa** *Unspecified* Species used for food. (as *G. resinosa* 135:238)

Gaylussacia sp., Huckleberry
- *Drug*—**Chickasaw** *Psychological Aid* and *Sedative* Roots used for delirium. (177:48) **Rappahannock** *Gastrointestinal Aid* Infusion of dried or fresh roots taken for the stomach. (161:34)

Gaylussacia ursina (M. A. Curtis) Torr. & Gray ex Gray, Bear Huckleberry
- *Food*—**Cherokee** *Preserves* Berries made into jelly or canned for future use. (126:39)

Gayophytum, Onagraceae

Gayophytum ramosissimum Torr. & Gray, Pinyon Groundsmoke
- *Drug*—**Navajo, Kayenta** *Dermatological Aid* Plant used as a lotion for cuts. *Psychological Aid* Plant used for the effects of a dream of a spider bite. (205:33) **Navajo, Ramah** *Hemostat* Poultice applied to cuts as a hemostatic. (191:37)

Geastrum, Lycoperdaceae, fungus

Geastrum sp., Earth Star
- *Drug*—**Isleta** *Dermatological Aid* Spores used as baby powder similar to talcum. (100:30) **Keres, Western** *Ear Medicine* Spores used in the ear for running ear. (171:45)

Gelsemium, Loganiaceae

Gelsemium sempervirens St.-Hil., Evening Trumpetflower
- *Drug*—**Delaware** *Blood Medicine* Roots used as a blood purifier. (176:33) **Delaware, Oklahoma** *Blood Medicine* Root used as a blood purifier. *Dermatological Aid* Compound containing root used as a salve. (175:28, 76)

Gentianaceae, see *Centaurium, Frasera, Gentiana, Gentianella, Gentianopsis, Obolaria, Sabatia*

Gentiana, Gentianaceae

Gentiana affinis Griseb., Pleated Gentian
- *Drug*—**Navajo** *Analgesic* Plant used as a snuff for headaches. *Stimulant* Plant used for fainting. *Witchcraft Medicine* Plant used as an antidote for witchcraft. (55:69)
- *Other*—**Blackfoot** *Decorations* Flowers used for their attractiveness. (97:49)

Gentiana alba Muhl. ex Nutt., Plain Gentian
- *Drug*—**Potawatomi** *Alterative* Infusion of root taken as an alterative. (as *G. flavida* 154:58, 59)

Gentiana andrewsii Griseb., Closed Bottle Gentian
- *Drug*—**Iroquois** *Analgesic* Infusion of roots used as a wash and taken for pain and headaches. *Eye Medicine* Infusion of roots used as drops for sore eyes. (87:413) *Febrifuge* Infusion of roots taken for chills. *Liver Aid* Compound used as a liver medicine. *Orthopedic Aid* Poultice of roots applied for muscular soreness. *Psychological Aid* Compound infusion of roots taken and used as wash for lonesomeness and craziness. *Witchcraft Medicine* Dried root hung in house as an anti-witch charm. Infusion of dried roots taken for headaches and to cure jealous witchcraft. (87:414) **Meskwaki** *Gynecological Aid* Root used for "caked breast." *Snakebite Remedy* Root used for snakebite. (152:222)

Gentiana douglasiana Bong., Swamp Gentian
- *Food*—**Hanaksiala** *Candy* Flowers sucked by children for the sweet nectar. (43:252)

Gentiana saponaria L., Harvestbells
- *Drug*—**Dakota** *Tonic* Simple or compound decoction of root taken as a tonic. **Winnebago** *Tonic* Simple or compound decoction of root taken as a tonic. (as *Dasystephana puberula* 70:109)

Gentianella, Gentianaceae

Gentianella propinqua (Richards.) J. Gillett, Fourpart Dwarfgentian
- *Drug*—**Tanana, Upper** *Cold Remedy* Decoction of leaves, stems, and flowers taken for colds. *Cough Medicine* Decoction of leaves, stems, and flowers taken for coughs. (102:17)

Gentianella quinquefolia (L.) Small, Agueweed
- *Drug*—**Cherokee** *Cathartic* Root used as a cathartic. *Gastrointestinal Aid* Used for "dyspepsy," "weak stomach and hysterical affections." *Laxative* Root used as a laxative. *Sedative* Used for "weak stomach and hysterical affections." *Stimulant* Root used as a stimulant. *Tonic* Root used as a tonic. (80:35) **Iroquois** *Anthelmintic* Plant used for worms. *Antidiarrheal* Infusion of plants taken for diarrhea. *Gastrointestinal Aid* Plant used for stomachaches. *Pulmonary Aid* Infusion of

plants taken for sore chest. (87:414) **Meskwaki** *Hemostat* Liquid from root used for hemorrhages. (152:222)

Gentianopsis, Gentianaceae

Gentianopsis crinita (Froel.) Ma, Greater Fringedgentian
- **Drug**—**Delaware** *Blood Medicine* Infusion of roots used as a blood purifier. *Gastrointestinal Aid* Infusion of roots used as a stomach strengthener. (176:39) **Delaware, Oklahoma** *Blood Medicine* Infusion of root taken as a blood purifier. *Gastrointestinal Aid* Infusion of root taken as a "stomach strengthener." (175:32, 76) **Rappahannock** *Blood Medicine* An ingredient of a blood medicine. (161:31)

Geocaulon, Santalaceae

Geocaulon lividum (Richards.) Fern., False Toadflax
- **Drug**—**Cree, Hudson Bay** *Cathartic* Decoction of chewed leaves and bark taken as a purgative. *Dermatological Aid* Poultice of chewed leaves and bark applied to wounds. *Emetic* Decoction of chewed leaves and bark taken as an emetic. *Unspecified* Plant used as medicine. (as *Comandra livida* 92:303)
- **Food**—**Alaska Native** *Fruit* Fruit used for food. (85:144)

Geraniaceae, see *Erodium*, *Geranium*

Geranium, Geraniaceae

Geranium atropurpureum Heller, Western Purple Cranesbill
- **Drug**—**Navajo, Kayenta** *Other* Plant used for overexertion. (205:29) **Navajo, Ramah** *Panacea* Plant used as "life medicine." (191:34)
- **Fiber**—**Jemez** *Cordage* Split epidermis used to sew moccasins. (44:22)
- **Other**—**Navajo, Kayenta** *Ceremonial Items* Blossoms dipped into white shell containing sea water, dipped into white shell containing salt, then dipped into white shell containing plants that grow near water, to bring rain. (205:29)

Geranium caespitosum James, Pineywoods Geranium
- **Drug**—**Keres, Western** *Dermatological Aid* Roots bruised into a paste for sores. (171:45)
- **Food**—**Keres, Western** *Fodder* Considered good turkey food. (171:45)

Geranium caespitosum var. fremontii (Torr. ex Gray) Dorn, Frémont's Geranium
- **Drug**—**Gosiute** *Antidiarrheal* Decoction of roots used for diarrhea. *Dermatological Aid* Plant used as an astringent. (as *G. fremontii* 39:370)

Geranium erianthum DC., Woolly Geranium
- **Drug**—**Aleut** *Throat Aid* Leaves used in a gargle for sore throat. (8:428)

Geranium lentum Woot. & Standl., Mogollon Geranium
- **Drug**—**Navajo, Ramah** *Dermatological Aid* Poultice of moist leaves and root applied to injuries, a "life medicine." *Panacea* Decoction of plant taken for internal injury, a "life medicine." (191:34)

Geranium maculatum L., Spotted Geranium
- **Drug**—**Cherokee** *Dermatological Aid* Astringent, compound decoction used as a wash for thrush in child's mouth. Used for open wounds and to remove canker sores. *Hemostat* Used as a styptic.

Oral Aid and *Pediatric Aid* Decoction and fox grapes used to wash children's mouths for "thrush." (80:35) **Chippewa** *Antidiarrheal* Infusion of roots taken for diarrhea. (71:134) *Oral Aid* Dried, pulverized root put in mouth, especially by children, for sores. *Pediatric Aid* Dried, powdered root placed in mouth, especially by children, for soreness. (53:342) **Choctaw** *Dermatological Aid* Root used as powerful astringent. *Venereal Aid* Root used for "the venereal." (28:287) **Iroquois** *Antidiarrheal* Infusion of plant taken for diarrhea. (87:367) *Dermatological Aid* Decoction of roots used as a wash for face sores or parts infected with itch. Poultice of powdered or chewed roots applied to unhealed navel of babies. (87:366) *Emetic* Decoction of roots taken as an emetic. *Heart Medicine* Decoction of roots taken for heart trouble. *Laxative* Infusion of plant used to "clean out the innards." *Love Medicine* Root placed in victim's tea to counteract a love medicine. (87:367) *Oral Aid* Roots used several ways for sore mouth, trench mouth, and chancre sores. *Pediatric Aid* Decoction of roots given to child with sore mouth, trench mouth, or chancre sores. (87:366) Poultice of chewed or powdered roots applied to severed umbilical cord. *Throat Aid* Decoction of smashed roots used as a wash for sore throats. (87:367) *Venereal Aid* Compound infusion of root used as wash on parts affected by venereal disease. Decoction of roots taken for venereal disease. (87:366) **Menominee** *Antidiarrheal* Root used for "flux and like troubles." (151:36, 37) **Meskwaki** *Analgesic* Infusion of root used for neuralgia and toothache. *Antidiarrheal* Compound containing root used for diarrhea. *Burn Dressing* Poultice of decoction of root applied to burns and infusion used for toothache. *Hemorrhoid Remedy* Poultice of pounded root bound on the anus to cause piles to recede. *Oral Aid* Infusion of root used for pyorrhea, sore gums, and toothache. *Toothache Remedy* Infusion of root used for aching teeth and sore gums. (152:222, 223) **Ojibwa** *Antidiarrheal* Root used for flux and sore mouth. *Oral Aid* Root used for sore mouths and flux. (153:370, 371)

Geranium oreganum T. J. Howell, Oregon Geranium
- **Drug**—**Miwok** *Antirheumatic (External)* Decoction of root rubbed on aching joints. (as *G. incisum* 12:170) **Montana Indian** *Antidiarrheal* Root used for diarrhea. *Dermatological Aid* Root used as an astringent. (as *G. incisum* 19:12) **Salish** *Oral Aid* Leaf held between lips for sore lips. (as *G. incisum* 178:293)

Geranium richardsonii Fisch. & Trautv., Richardson's Geranium
- **Drug**—**Cheyenne** *Hemostat* Infusion of dried roots taken or powdered leaves used as snuff for nosebleed. *Nose Medicine* Pulverized leaf rubbed on the nose and powder snuffed up the nostrils for nosebleeds. Infusion of powdered roots taken for nosebleeds. (76:179) **Navajo, Ramah** *Panacea* Plant used as "life medicine." (191:34) **Thompson** *Unspecified* Plant used medicinally for unspecified purpose. (164:461)

Geranium viscosissimum Fisch. & C. A. Mey. ex C. A. Mey., Sticky Geranium
- **Drug**—**Blackfoot** *Cold Remedy* Infusion of leaves and simple sweat bath taken for colds. (86:72) *Eye Medicine* Infusion of leaves used for sore eyes. (86:81) *Other* Infusion of leaves applied to the head and eaten for large head, from dropsy or severe malnutrition. Two cases were described of a young girl and a woman whose heads became

abnormally large, as with dropsy or severe malnutrition. An infusion of the leaves was applied to the head and consumed, effecting temporary recovery, but both lost their hair and died later. (86:82) **Okanagan-Colville** *Unspecified* Poultice of pounded, heated roots applied medicinally. (188:106) **Sanpoil** *Dermatological Aid* Poultice of crushed leaves applied to sores. *Eye Medicine* Decoction of roots used as a wash for sore eyes. (131:219) **Thompson** *Gynecological Aid* Plant used as a medicine for women. *Love Medicine* Plant used as a love charm or love potion. (187:225) *Unspecified* Plant used medicinally for unspecified purpose. (164:461) Roots used for medicine. *Witchcraft Medicine* Flowers possibly used for witchcraft. (187:225)

- *Food*—**Blackfoot** *Spice* Leaves kept in food storage bags to mask the spoiling of the contents. (86:103)

Gesneriaceae, see *Cyrtandra*

Geum, Rosaceae

Geum aleppicum Jacq., Yellow Avens

- *Drug*—**Cree, Woodlands** *Diaphoretic* Decoction of root and other herbs used to make a person sweat. *Panacea* Powdered roots used in a many herb remedy for various ailments. *Pediatric Aid* Decoction of root alone or with other herbs used for teething sickness. *Throat Aid* Decoction of root used for sore throats. *Toothache Remedy* Decoction of root alone or with other herbs used for teething sickness. Decoction of root used for sore teeth. (109:39) **Iroquois** *Anticonvulsive* Decoction of roots taken for convulsions. *Antidiarrheal* Compound infusion or decoction of roots taken for diarrhea. *Emetic* Compound decoction of roots taken to vomit as cure for love medicine. *Febrifuge* Decoction of roots taken for high fevers. *Love Medicine* Compound decoction of roots taken to vomit as cure for love medicine. *Veterinary Aid* Compound decoction mixed with horse feed and used as nose drops for cramps. (87:353) **Malecite** *Cough Medicine* Infusion of one root used by children with coughs. *Pediatric Aid* Infusion of one root used by children with croup. Infusion of one root used by children with coughs. *Pulmonary Aid* Infusion of one root used by children with croup. (as *G. strictum* 116:249) **Micmac** *Cough Medicine* Roots used for coughs and croup. *Pulmonary Aid* Root used for coughs and croup. (as *G. strictum* 40:57) **Ojibwa, South** *Analgesic* Weak decoction of root taken for chest soreness. *Cough Medicine* Weak decoction of root taken for cough. *Pulmonary Aid* Weak decoction of root taken for chest soreness. (as *G. strictum* 91:200)

Geum calthifolium Menzies ex Sm., Calthaleaf Avens

- *Drug*—**Aleut** *Cold Remedy* Decoction of root taken as a tonic for colds and sore throat. *Dermatological Aid* Poultice of plant applied to sores "that refused to heal." *Throat Aid* Decoction of root taken as a tonic for colds and sore throats. *Tonic* Decoction of root taken as a tonic. (8:427)

Geum canadense Jacq., White Avens

- *Drug*—**Chippewa** *Gynecological Aid* Root used for "female weakness." (53:356) **Iroquois** *Love Medicine* Decoction of whole plant used as a love medicine. *Panacea* Compound infusion taken or placed on injured part, a "Little Water Medicine." (87:353)

Geum macrophyllum Willd., Largeleaf Avens

- *Drug*—**Bella Coola** *Analgesic* Decoction of root taken for stomach pain. *Dermatological Aid* Poultice of chewed or bruised leaves applied to boils. *Gastrointestinal Aid* Decoction of root taken for stomach pain. **Carrier, Southern** *Dermatological Aid* Poultice of boiled leaves applied to bruises. *Panacea* Decoction of leaves taken for any sickness. (150:59) **Chehalis** *Contraceptive* Infusion of leaves taken to avoid conception. (79:37) **Clallam** *Dermatological Aid* Leaves used for boils. (57:202) **Gosiute** *Unspecified* Decoction of roots used as a medicine. (39:370) **Hesquiat** *Gastrointestinal Aid* Entire plant, including the roots, eaten as a medicine for stomach pains or excess acid. *Gynecological Aid* Young, small leaves chewed after childbirth to heal the womb. (185:72) **Klallam** *Gynecological Aid* Raw leaves chewed during labor. (79:37) **Ojibwa** *Gynecological Aid* Plant used as a female remedy. (153:384) **Okanagan-Colville** *Gynecological Aid* Infusion of roots taken by women after childbirth. (188:126) **Quileute** *Dermatological Aid* Poultice of leaves applied to boils. **Quinault** *Dermatological Aid* Poultice of smashed leaves applied to cuts. *Gynecological Aid* Raw leaves chewed during labor. (79:37) *Panacea* Leaves chewed as a universal remedy, "good for everything." (201:276) **Snohomish** *Dermatological Aid* Poultice of leaves applied to boils. (79:37)

- *Other*—**Salish, Coast** *Protection* Leaves eaten before seeing a dying person for protection from germs. (182:86)

Geum macrophyllum var. *perincisum* (Rydb.) Raup, Largeleaf Avens

- *Drug*—**Cree, Woodlands** *Pediatric Aid* and *Toothache Remedy* Decoction of root with other herbs used for teething sickness. (109:39)

Geum rivale L., Purple Avens

- *Drug*—**Algonquin, Tête-de-Boule** *Antihemorrhagic* Decoction of roots boiled four times and used for the spitting of blood. (132:129) **Iroquois** *Antidiarrheal* Infusion of roots taken for diarrhea. *Febrifuge* Infusion of roots taken for fevers. (87:354) **Malecite** *Antidiarrheal* and *Pediatric Aid* Infusion of one root used by children with diarrhea. (116:255) **Micmac** *Antidiarrheal* Root used for diarrhea or dysentery. (40:57) Decoction of root taken, especially by children, for dysentery. *Cold Remedy* Decoction of root taken, especially by children, for colds. *Cough Medicine* Decoction of root taken, especially by children, for coughs. *Pediatric Aid* Decoction of root taken, especially by children, for dysentery, coughs, and colds. (as *G. nivale* 156:316)

Geum sp., Avens

- *Drug*—**Thompson** *Analgesic* Plants used in sweat bath for sprains, stiff or aching joints or muscles. *Antirheumatic (External)* Fresh plants used as a bed in the sweat bath for rheumatism. *Herbal Steam* Plants used in sweat bath for sprains, stiff or aching joints or muscles. (164:464) *Love Medicine* Plant used by men as a charm "for gaining a woman's affection." (164:507) *Misc. Disease Remedy* Decoction of root taken for diseases with rash: measles, chickenpox, and smallpox. (164:476) *Orthopedic Aid* Plants used in sweat bath for sprains, stiff or aching joints or muscles. (164:464) *Unspecified* Plants used as charms for unspecified purpose. (164:506)

Geum triflorum Pursh, Prairie Smoke

- *Drug*—**Blackfoot** *Cough Medicine* Infusion of plant taken as a general tonic for severe coughs. (86:72) *Dermatological Aid* Infusion of roots and grease applied as a salve to sores, rashes, blisters, and flesh wounds. (86:76) Infusion of roots applied to wounds. (86:84) *Eye Med-*

icine Decoction of roots used for sore or swollen eyes. (97:38) *Oral Aid* Infusion of roots used as a mouthwash for cankers. (86:66) *Respiratory Aid* Scraped roots mixed with tobacco and smoked to "clear the head." (86:79) *Throat Aid* Infusion of roots used as a mouthwash for sore throats. (86:66) *Tonic* Leaves dried, crushed, mixed with other medicines, and used as a tonic. (97:38) *Veterinary Aid* Infusion of roots used for bleeding and promoted rapid healing on horse boils and castration wounds. (86:88) **Okanagan-Colville** *Cold Remedy* Infusion of roots taken for colds. *Dietary Aid* Infusion of roots taken for the lack of appetite due to "poor blood." *Febrifuge* Infusion of roots taken for fevers. *Gynecological Aid* Infusion of roots taken by women to "heal her insides" from a vaginal yeast infection. *Love Medicine* Infusion of roots taken as a love potion by a woman who wanted to win back the affections of a man. *Misc. Disease Remedy* Infusion of roots taken for flu. (188:126) **Thompson** *Analgesic* Decoction of root used as a wash for pain. Plants used in the sweat bath for aching joints and sore, stiff muscles. *Antirheumatic (External)* Plants used in the sweat bath for rheumatism and stiff joints and muscles. (164:466) *Disinfectant* Decoction of whole plant used as a wash after the purifying sweat bath. (164:504) *Herbal Steam* Plants used in the sweat bath for rheumatism and stiff joints and muscles. Plants steamed in the sweat bath for rheumatism and joint and muscle stiffness. *Orthopedic Aid* Decoction of root used as a wash for body stiffness. Plants used in the sweat bath for sprains, aching and stiff joints and muscles. *Tonic* Decoction of root taken as a tonic. (164:466)
- *Food*—**Thompson** *Beverage* Roots boiled and drunk as tea. (164:493)
- *Other*—**Blackfoot** *Incense & Fragrance* Ripe seeds crushed and used as perfume. (97:38)

Geum triflorum **var.** *ciliatum* (Pursh) Fassett, Old Man's Whiskers
- *Drug*—**Blackfoot** *Blood Medicine* Infusion of roots taken to build the blood. (as *Sieversia ciliata* 121:48) *Eye Medicine* Decoction of plant used as a wash for sore and inflamed eyes. (as *Sieversia ciliata* 114:275) Decoction of root applied to eyes. (as *Sieversia ciliata* 118:39) Infusion of roots taken for sore eyes. (as *Sieversia ciliata* 121:48) **Chippewa** *Gastrointestinal Aid* Compound decoction of root taken for indigestion. (as *Sieversia ciliata* 53:342) *Stimulant* Dried root chewed as strong stimulant before feats of endurance. (as *Sieversia ciliata* 53:364) *Veterinary Aid* Dried, powdered root added to horse's feed as a stimulant before a race. (as *Sieversia ciliata* 53:366) **Paiute** *Veterinary Aid* Decoction of roots given to stimulate tired horses. (as *G. ciliatum* 111:81)
- *Other*—**Blackfoot** *Incense & Fragrance* Crushed seedpods used for perfume. (as *Sieversia ciliata* 118:57)

Gilia, Polemoniaceae

Gilia capitata **ssp.** *staminea* (Greene) V. Grant, Bluehead Gilia
- *Food*—**Luiseño** *Unspecified* Seeds used for food. (as *G. staminea* 155:230)

Gilia inconspicua (Sm.) Sweet, Shy Gilia
- *Drug*—**Navajo, Ramah** *Febrifuge* Cold, compound infusion of plant taken and used as lotion for fever. (191:40)

Gilia leptomeria Gray, Sand Gilia
- *Drug*—**Navajo, Kayenta** *Dermatological Aid* Poultice of plant applied to scorpion stings or worm bites. *Sedative* Plant used as a soporific. *Tonic* Infusion of plant taken or plant smoked as a tonic. (205:38)

Gilia rigidula **ssp.** *acerosa* (Gray) Wherry, Bluebowls
- *Drug*—**Keres, Western** *Antirheumatic (External)* Crushed plant used to massage the muscles for cramps. (as *G. acerosa* 171:45)

Gilia sinuata Dougl. ex Benth., Rosy Gilia
- *Food*—**Havasupai** *Preserves* Seeds parched, ground, kneaded into seed butter, and eaten with fruit drinks or spread on bread. (197:67)

Gilia **sp.**, Blue Gilia
- *Drug*—**Gosiute** *Dermatological Aid* Plant used for wounds, cuts, or sores. (39:349) **Shoshoni** *Pediatric Aid* Infusion of whole plant taken by children for colds. (118:37) **Zuni** *Analgesic* Infusion of fresh or dried plant taken and applied to head for headache. *Diuretic* Warm infusion of plant taken as a diuretic. *Emetic* Warm infusion of plant taken as an emetic. *Febrifuge* Infusion of fresh or dried plant taken and rubbed on body for fever. *Laxative* Warm infusion of plant taken as a laxative. *Throat Aid* Infusion of plant taken and applied to neck for swollen throat. (166:52, 53)

Gilia subnuda Torr. ex Gray, Coral Gilia
- *Drug*—**Navajo, Kayenta** *Gynecological Aid* Ground flowers eaten to insure healthy pregnancy and ease labor. (205:38)

Glandularia, Verbenaceae

Glandularia bipinnatifida (Nutt.) Nutt., Dakota Mock Vervain
- *Drug*—**Keres, Western** *Snakebite Remedy* Leaves crushed with rocks and rubbed on snakebites. *Throat Aid* Infusion of leaves used as a gargle for sore throat. (as *Verbena lupinnatifida* 171:73)

Glandularia wrightii (Gray) Umber, Davis Mountain Mock Vervain
- *Drug*—**Navajo, Ramah** *Panacea* Plant used as "life medicine." (as *Verbena wrightii* 191:41)

Glaucothea, Arecaceae

Glaucothea armata (S. Wats.) O. F. Cook, Blue Palm
- *Other*—**Cocopa** *Musical Instrument* Seeds used in gourd rattles. (as *Glaucotheca armata* 37:187)

Glaux, Primulaceae

Glaux maritima L., Sea Milkwort
- *Drug*—**Kwakiutl** *Sedative* Boiled roots eaten to make one very sleepy. (183:288)
- *Food*—**Kwakiutl, Southern** *Unspecified* Fleshy roots boiled, dipped in oil, and used for food. (183:288) **Salish, Coast** *Unspecified* Roots eaten in spring. (182:86)

Glechoma, Lamiaceae

Glechoma hederacea L., Ground Ivy
- *Drug*—**Cherokee** *Cold Remedy* Infusion used for colds. *Dermatological Aid* Infusion used for babies' hives. *Misc. Disease Remedy* Infusion used for measles. *Pediatric Aid* Infusion used for babies' hives. (80:37)

Gleditsia, Fabaceae

Gleditsia triacanthos L., Honey Locust

- *Drug*—**Cherokee** *Adjuvant* and *Anthelmintic* Pods used to sweeten worm medicine. *Gastrointestinal Aid* Compound taken for "dyspepsia from overeating." (80:43) Infusion of bark taken and used as a bath for dyspepsia. (177:32) *Misc. Disease Remedy* Infusion of pod taken for measles. *Pulmonary Aid* Compound infusion of bark taken for whooping cough. (80:43) **Creek** *Misc. Disease Remedy* Decoction of sprigs, thorns, and branches used as a bath to prevent smallpox. (177:32) *Panacea* Pod considered a good antidote for the complaints of children. *Pediatric Aid* Pod considered a good antidote for the complaints of children. (172:669) **Delaware** *Blood Medicine* Bark mixed with bark of prickly ash, wild cherry, and sassafras and used as a tonic to purify blood. *Cough Medicine* Bark combined with bark of prickly ash, wild cherry, and sassafras and used as a tonic for coughs. (176:30) **Delaware, Oklahoma** *Blood Medicine* Compound containing bark used as a blood purifier. *Cough Medicine* Compound containing bark used for a severe cough. (175:25, 76) *Tonic* Compound containing bark used as a general tonic. (175:25) **Meskwaki** *Cold Remedy* Infusion of twig bark used for bad colds. *Febrifuge* Infusion of bark used for fevers. *Misc. Disease Remedy* Infusion of bark used for measles and especially smallpox. *Tonic* Decoction of bark taken by patient to help regain flesh and strength. (152:228, 229) **Rappahannock** *Cold Remedy* Infusion of roots and bark used as a cold medicine. *Cough Medicine* Infusion of roots and bark used as a cough medicine. (161:31)
- *Food*—**Cherokee** *Beverage* Seed pulp used to make a drink. (80:43) Pod juice, water, and sugar or pods soaked in water used as a beverage. *Unspecified* Ripe, raw pods used for food. (126:45)
- *Fiber*—**Cherokee** *Building Material* Wood used to make fence posts. (80:43)
- *Other*—**Cherokee** *Protection* Compound infusion of bark used by ballplayers "to ward off tacklers." (80:43) **Keres, Western** *Unspecified* Plant known and named but no use was specified. (171:45)

Glossopetalon, Crossosomataceae

Glossopetalon spinescens var. *aridum* M. E. Jones, Spiny Greasebush
- *Drug*—**Shoshoni** *Tuberculosis Remedy* Decoction of shrub taken regularly for tuberculosis. (as *Forsellesia nevadensis* 180:75)

Glyceria, Poaceae

Glyceria canadensis (Michx.) Trin., Rattlesnake Mannagrass
- *Drug*—**Ojibwa** *Gynecological Aid* Root used as a female remedy. (153:371)

Glyceria fluitans (L.) R. Br., Water Mannagrass
- *Drug*—**Crow** *Ceremonial Medicine* Burned as incense during certain ceremonies. **Montana Indian** *Ceremonial Medicine* Burned as incense during certain ceremonies. (19:12)
- *Food*—**Crow** *Unspecified* Seeds used for food. (19:12) **Klamath** *Unspecified* Seeds used for food. (as *Panicularia fluitans* 45:91)

Glyceria obtusa (Muhl.) Trin., Atlantic Mannagrass
- *Drug*—**Catawba** *Analgesic* Infusion of beaten roots taken for backaches. *Orthopedic Aid* Infusion of beaten roots taken for backaches. (177:6)

Glycyrrhiza, Fabaceae

Glycyrrhiza glabra L., Cultivated Licorice
- *Drug*—**Cherokee** *Cough Medicine* Used for coughs. *Expectorant* Used as an expectorant. *Respiratory Aid* Used for asthma. *Throat Aid* Used for hoarseness. (80:43) **Meskwaki** *Gynecological Aid* Compound containing root, not a native plant, used for female trouble. (152:229)

Glycyrrhiza lepidota Pursh, American Licorice
- *Drug*—**Bannock** *Throat Aid* Root chewed for strong throat for singing. Root boiled into a tonic and taken for sore throat. (118:38) **Blackfoot** *Analgesic* Infusion of roots taken for chest pains. (86:72) *Antirheumatic (External)* Infusion of roots applied to swellings. (86:76) *Cough Medicine* Infusion of roots taken for coughs. (86:72) *Oral Aid* Burs kept in the mouth by buffalo runners to protect against thirst. (86:113) *Throat Aid* Infusion of roots taken for sore throats. (86:72) *Veterinary Aid* Roots used for horse windgalls. (86:88) **Cheyenne** *Antidiarrheal* Infusion of roots or leaves used for diarrhea. (82:35) Infusion of roots or leaves taken for diarrhea. (83:28) *Ceremonial Medicine* Roots chewed to cool the body in the Sweatlodge and Sundance Ceremonies. (82:35) Roots chewed in the Sun Dance ceremony for the cooling effect. (83:28) *Gastrointestinal Aid* Infusion of roots or leaves used for stomachache. (82:35) Infusion of roots or leaves taken for upset stomach. (83:28) **Dakota** *Ear Medicine* Infusion of leaves applied to ears for earaches. (69:365) Poultice of steeped leaves applied to ears for earache. *Febrifuge* Decoction of root used for children with fevers. *Pediatric Aid* Decoction of root used as a febrifuge for children. (70:92) *Toothache Remedy* Root held in the mouth for toothaches. (69:365) Root chewed and held in mouth for toothache. (70:92) *Veterinary Aid* Poultice of chewed leaves applied to sores on horses. (69:365) Poultice of chewed leaves applied to sore backs of horses. (70:92) **Great Basin Indian** *Throat Aid* Roots chewed or decoction of roots taken for sore throats. (121:48) **Isleta** *Dermatological Aid* Leaves used in shoes to absorb moisture. (100:30) **Keres, Western** *Cough Medicine* Roots used as cough drops by singers or talkers. (171:45) **Keresan** *Febrifuge* Infusion of plant used as a wash for chills. (198:561) **Lakota** *Antihemorrhagic* Roots and Canadian milk vetch roots used for spitting of blood. *Misc. Disease Remedy* Roots chewed for the flu. (139:46) *Toothache Remedy* Roots chewed for toothache. *Unspecified* Roots used for "doctoring the sick." (106:40) **Montana Indian** *Throat Aid* Roots chewed and juice swallowed to strengthen the throat for singing. *Tonic* Infusion of roots taken as a tonic. (82:35) **Navajo, Ramah** *Cathartic* Decoction of root used as a cathartic. (191:32) **Paiute** *Other* Infusion of plant used for some sicknesses. (167:317) **Pawnee** *Ear Medicine* Poultice of steeped leaves applied to ears for earache. *Febrifuge* Decoction of root used for children with fevers. *Pediatric Aid* Decoction of root used as a febrifuge for children. *Toothache Remedy* Root chewed and held in mouth for toothache. (70:92) **Sioux** *Ear Medicine* Infusion of leaves used for earache. *Febrifuge* and *Pediatric Aid* Infusion of roots used for children with fever. *Toothache Remedy* Roots chewed and used for toothache. *Veterinary Aid* Poultice of chewed root leaves applied to sore horse backs. (82:35) **Zuni** *Oral Aid* Root chewed to keep the mouth sweet and moist. (27:376)
- *Food*—**Cheyenne** *Unspecified* Shoots eaten raw. (76:178) Young shoots eaten raw in spring. (82:35) Tender, spring shoots eaten raw.

(83:46) **Montana Indian** *Unspecified* Roots used for food. **Northwest Indian** *Unspecified* Roots used for food. (19:12)
- *Other*—Blackfoot *Weapon* Burs believed to be shot by ghosts inflicting disease in their victims. (86:113)

Glyptopleura, Asteraceae
Glyptopleura marginata D. C. Eat., Carveseed
- *Food*—Paiute *Vegetable* Raw leaves eaten as greens. (118:23) **Paiute, Northern** *Vegetable* Leaves and stems eaten raw. (59:49)

Gnaphalium, Asteraceae
Gnaphalium californicum DC., Ladies' Tobacco
- *Drug*—Costanoan *Analgesic* Infusion of plant taken for stomach pain. *Cold Remedy* Infusion of plant taken for colds. *Gastrointestinal Aid* Infusion of plant taken for stomach pain. (21:26)

Gnaphalium canescens DC., Wright's Cudweed
- *Drug*—Keres, Western *Cold Remedy* Ground, white flowers inhaled for head colds. *Dermatological Aid* Bruised leaves made into a paste and used as a liniment. (as *G. wrightii* 171:46)

Gnaphalium microcephalum Nutt., Smallhead Cudweed
- *Drug*—Karok *Eye Medicine* Cold infusion of plant used as a wash for sore eyes. (148:390)

Gnaphalium obtusifolium L., Rabbit Tobacco
- *Drug*—Alabama *Sedative* Compound decoction of plant used many ways for nervousness or sleeplessness. (172:663, 664) Decoction of plant used as a face wash for nerves and insomnia. (177:61) **Cherokee** *Analgesic* Compound used for local pains, muscular cramps, and twitching. (80:51, 52) Infusion of plant rubbed into scratches made over muscle cramp pain. (177:61) *Antirheumatic (Internal)* Used with Carolina vetch for rheumatism. *Cold Remedy* Decoction taken for colds. *Cough Medicine* Used as a cough syrup. *Misc. Disease Remedy* Used in a sweat bath for various diseases. Warm liquid blown down throat for clogged throat (diphtheria). *Oral Aid* Chewed for sore mouth. *Orthopedic Aid* Compound used for muscular cramps and twitching. (80:51, 52) Infusion of plant rubbed into scratches made over muscle cramp pain. (177:61) *Respiratory Aid* Smoked for asthma. *Throat Aid* Chewed for sore throat. (80:51, 52) **Choctaw** *Analgesic* Decoction of leaves and blossoms taken for lung pain. (as *G. polycephalum* 25:24) Decoction of leaves and blossoms taken for lung pain. (177:61) *Cold Remedy* Decoction of leaves and blossoms taken for colds. (as *G. polycephalum* 25:24) Decoction of leaves and blossoms taken for colds. (177:61) *Pulmonary Aid* Decoction of leaves and blossoms taken for lung pain. (as *G. polycephalum* 25:24) Decoction of leaves and blossoms taken for lung pain. (177:61) **Creek** *Adjuvant* Leaves added to medicines as a perfume. *Antiemetic* Decoction of leaves taken for vomiting. *Cold Remedy* Compound decoction of plant tops taken and used as inhalant for colds. *Misc. Disease Remedy* Poultice of decoction of leaves applied to throat for mumps. (172:661) Decoction of leaves used as a throat wash for mumps. (177:61) *Psychological Aid* Decoction of plant used as a wash for persons who "wanted to run away." (172:663, 664) *Sedative* Decoction of plant tops taken and used as a wash for old people unable to sleep. (172:661) *Witchcraft Medicine* Decoction of plant used as a wash for persons afflicted by ghosts. (172:663, 664) **Koasati** *Febrifuge* Decoction of leaves taken for fevers. *Pediatric Aid* Decoction of leaves used as a bath and given to children with fevers. (177:61) **Menominee** *Analgesic* Dried leaves steamed as an inhalant for headache. (54:129) *Disinfectant* Smudge of leaves used to fumigate premises to dispel ghost of a dead person. (as *G. polycephalum* 151:30) *Psychological Aid* Dried leaves steamed as an inhalant for "foolishness." (54:129) *Stimulant* Leaf smoke blown into nostrils to revive one who had fainted. *Witchcraft Medicine* Smudge of leaves used to fumigate premises to dispel ghost of a dead person. (as *G. polycephalum* 151:30) **Meskwaki** *Psychological Aid* Smudge of herb used to "bring back a loss of mind." *Stimulant* Smudged and used to revive an unconscious patient. (as *G. polycephalum* 152:214, 215) **Montagnais** *Cough Medicine* Decoction of plant taken for coughing. *Tuberculosis Remedy* Decoction of plant taken for consumption. (as *G. popycephalum* 156:314) **Rappahannock** *Febrifuge* Infusion of roots taken for chills. *Respiratory Aid* Infusion of dried stems or dried leaves smoked in a pipe for asthma. (161:29)
- *Food*—Rappahannock *Candy* Leaves chewed for "fun." (161:29)

Gnaphalium sp., Everlasting
- *Drug*—Karok *Unspecified* Used for medicine. (117:209) **Pomo** *Dermatological Aid* and *Pediatric Aid* Poultice of crushed leaves applied to the baby's navel string. (117:284) **Weyot** *Unspecified* Plant used to make a strong medicine. (117:180)

Gnaphalium stramineum Kunth, Cotton Batting Plant
- *Drug*—Kawaiisu *Analgesic* Hot poultice of leaves or stems applied to parts of body affected by pain. (as *G. chilense* 206:33) **Navajo, Ramah** *Ceremonial Medicine* and *Emetic* Plant used as a ceremonial emetic. *Panacea* Plant used as "life medicine." (as *G. chilense* 191:51) **Pomo** *Dermatological Aid* Poultice of boiled plant applied to a swollen face. (as *G. chilense* 66:15)
- *Other*—Pomo, Kashaya *Hunting & Fishing Item* Cottony flower tops used like stuffing to line deer antler head disguises. (as *G. chilense* 72:43)

Gnaphalium uliginosum L., Marsh Cudweed
- *Drug*—Iroquois *Orthopedic Aid* Plants used for bruises. *Respiratory Aid* Compound infusion of plants used for asthma. (87:465)

Gnaphalium viscosum Kunth, Winged Cudweed
- *Drug*—Miwok *Antirheumatic (External)* Poultice of leaves used for swelling. *Cold Remedy* Decoction of leaves taken for colds. *Gastrointestinal Aid* Decoction of leaves taken for stomach trouble. (as *G. decurrens* var. *californicum* 12:170)

Gonolobus, Asclepiadaceae
Gonolobus sp., Milk Vine
- *Drug*—Houma *Antiemetic* Infusion of root taken for "sick stomach." (158:63)

Goodeniaceae, see *Scaevola*

Goodyera, Orchidaceae
Goodyera oblongifolia Raf., Western Rattlesnake Plantain
- *Drug*—Cowlitz *Tonic* Infusion of plants taken as a tonic. (as *Peramium decipiens* 79:26) **Okanagan-Colville** *Dermatological Aid* Poultice of softened leaves applied to cuts and sores. *Reproductive Aid* Leaves

split open and blown on several times by women wishing to become pregnant. (188:52) **Okanagon** *Gynecological Aid* Plant chewed by women before and at the time of childbirth. (as *Peramium decipiens* 125:41) **Saanich** *Antirheumatic (External)* Infusion of leaves used in the bath water of sprinters and canoers as a liniment for stiff muscles. (182:77) **Thompson** *Gynecological Aid* Plant chewed by women before and at the time of childbirth. (as *Peramium decipiens* 125:41) Plant chewed at childbirth to ease confinement. (as *Peramium decipiens* 164:462) Leaves chewed prenatally to determine the sex of a baby and to insure an easy delivery. If the mother could swallow the chewed leaf, the baby was going to be a girl, but if she could not, then it was going to be a boy. (187:136)

Goodyera pubescens (Willd.) R. Br. ex Ait. f., Downy Rattlesnake Plantain
- **Drug**—**Cherokee** *Blood Medicine* Compound decoction taken as a blood tonic. *Burn Dressing* Poultice of wilted leaves applied "to draw out burn." *Cold Remedy* Cold infusion of leaf taken for colds. *Dietary Aid* Cold infusion of leaf taken with whisky to improve the appetite. Compound decoction taken to build the appetite. *Emetic* Taken with whisky to improve the appetite and as an emetic. *Eye Medicine* Ooze dripped into sore eyes. *Kidney Aid* Cold infusion of leaf taken for kidneys. *Toothache Remedy* Infusion held in mouth for toothache. (80:50) **Delaware** *Antirheumatic (External)* Used as a medicine for rheumatism. *Gynecological Aid* Administered to women following childbirth. *Pulmonary Aid* Used for pleurisy. (as *Epipactis pubescens* 176:37) **Delaware, Oklahoma** *Antirheumatic (Internal)* Root used for rheumatism. *Gynecological Aid* Root given to women after childbirth. *Pulmonary Aid* Root used for pleurisy. (as *Epipactis pubescens* 175:31, 76) *Unspecified* Poultice of leaves used for unspecified purpose. (as *Epipactis pubescens* 175:76) **Mohegan** *Oral Aid* Poultice of mashed leaves used for babies with sore mouths. (as *Epipactis pubescens* 174:265) Mashed leaves used to wipe out infants' mouths to prevent soreness. (as *Epipactis pubescens* 176:72) *Pediatric Aid* Poultice of mashed leaves used for babies with sore mouths. (as *Epipactis pubescens* 174:265) Mashed leaves used to wipe out infants' mouths to prevent soreness. (as *Epipactis pubescens* 176:72)

Goodyera repens (L.) R. Br. ex Ait. f., Lesser Rattlesnake Plantain
- **Drug**—**Cherokee** *Blood Medicine* Compound decoction taken as a blood tonic. *Burn Dressing* Poultice of wilted leaves applied "to draw out burn." *Cold Remedy* Cold infusion of leaf taken for colds. *Dietary Aid* Cold infusion of leaf taken with whisky to improve the appetite. Compound decoction taken to build the appetite. *Emetic* Taken with whisky to improve the appetite and as an emetic. *Eye Medicine* Ooze dripped into sore eyes. *Kidney Aid* Cold infusion of leaf taken for kidneys. *Toothache Remedy* Infusion held in mouth for toothache. (80:50) **Potawatomi** *Gastrointestinal Aid* Root and leaves used for stomach diseases. *Gynecological Aid* Root and leaves used for female disorders. *Snakebite Remedy* Poultice of chewed leaves and swallowed juice used for snakebite, reference from 1796. *Urinary Aid* Root and leaves used for bladder diseases. (as *Epipactis repens* var. *ophioides* 154:67)

Gossypium, Malvaceae

Gossypium herbaceum L., Levant Cotton
- **Drug**—**Koasati** *Gynecological Aid* Decoction of roots taken to ease childbirth. (177:42)

Gossypium hirsutum L., Upland Cotton
- **Food**—**Pima, Gila River** *Unspecified* Seeds used for food. (133:5)
- **Fiber**—**Zuni** *Clothing* Cotton used to make ceremonial garments. (166:77) *Cordage* Fuzz made into cords and used ceremonially. (166:92)
- **Other**—**Zuni** *Ceremonial Items* Cotton used to make ceremonial garments. (166:77) Fuzz used alone or made into cords and used ceremonially in a number of ways. The cotton cords were tied loosely around the wrists and ankles of the newborn child while supplications were offered that the rainmakers would provide enough rain to insure proliferative crops so that the child would have full nourishment its whole life. Cotton down was used to cover the heads of rain priests after their deaths, symbolizing their duties in this world and also their obligations in the undermost world. Crowns and certain masks were also covered with raw cotton to indicate that the gods represented were rainmakers or were specially associated with the rainmakers. (166:92)

Gossypium sp., Cotton
- **Drug**—**Tewa** *Dermatological Aid* and *Pediatric Aid* Poultice of chewed kernels applied to child's head for baldness. (138:102, 103)
- **Food**—**Papago** *Bread & Cake* Seeds made into flour and baked on hot sands as browned cakes. (34:37) Seed flour mixed with saguaro seed flour, baked on sand, and eaten as browned cakes. (34:46) **Pima** *Dried Food* Seeds formerly parched and eaten without grinding. (146:77)
- **Fiber**—**Isleta** *Clothing* Cotton used to make belts, sashes, and red bands for the hair. (100:30) **Navajo** *Cordage* Used to make string for many different ceremonies. *Sewing Material* Used to make fabrics. (55:62)
- **Other**—**Havasupai** *Fuel* Used as a strike-a-light. (162:105) *Tools* Cotton twisted into thread, braided into a thick cord, and used in the strike-a-light. (197:231) **Navajo** *Tools* Twisted, soaked in mutton tallow, and used as a lampwick for soldering. (55:62) **Santa Clara** *Ceremonial Items* Formerly used to weave large ceremonial blankets. **Tewa of Hano** *Ceremonial Items* Used to make the strings for prayer feathers. (138:102)

Gossypium thurberi Todaro, Thurber's Cotton
- **Fiber**—**Papago** *Unspecified* Used as a source of fiber. (as *Thurberia thespesioides* 36:106)

Grindelia, Asteraceae

Grindelia camporum Greene, Great Valley Gumweed
- **Drug**—**Costanoan** *Dermatological Aid* Decoction of plant used for dermatitis, poison oak, boils, and wounds. (21:26) **Kawaiisu** *Analgesic* and *Orthopedic Aid* Decoction of leaves and flowers applied to sore parts of the body. (206:33) **Mewuk** *Blood Medicine* Fresh buds used extensively as a medicine for blood disorders. (117:338)

Grindelia decumbens Greene, Reclined Gumweed
- **Drug**—**Keres, Western** *Gastrointestinal Aid* Infusion of plant used for severe stomachache. (171:46)

Grindelia fastigiata Greene, Pointed Gumweed

- *Drug*—Keres, Western *Gastrointestinal Aid* Infusion of plant used for severe stomachache. (171:46)

Grindelia hallii Steyermark ex Rothrock, Hall's Gumweed
- *Drug*—Diegueño *Blood Medicine* Decoction of leaves and stems taken as a blood tonic. (84:23)

Grindelia humilis Hook. & Arn., Hairy Gumweed
- *Drug*—Mahuna *Dermatological Aid* Plant used for itching skin eruption caused by poison oak. (140:11)

Grindelia integrifolia var. *macrophylla* (Greene) Cronq., Puget Sound Gumweed
- *Other*—Pomo, Kashaya *Fasteners* Sticky sap used like glue. (as *G. stricta venulosa* 72:55)

Grindelia nana Nutt., Idaho Gumweed
- *Drug*—Paiute *Cough Medicine* Decoction of plant said to be a good cough medicine. *Expectorant* Decoction of plant said to be a good expectorant. *Pulmonary Aid* Hot decoction of young shoots taken for pneumonia. *Urinary Aid* Infusion of plant taken for bladder trouble. (180:81, 82) Sanpoil *Tuberculosis Remedy* Decoction of roots used for tuberculosis. (131:218) Shoshoni *Analgesic* Decoction of plant taken for stomachaches. *Cough Medicine* Decoction of plant said to be a good cough medicine. *Dermatological Aid* Poultice of boiled plant applied to swellings. *Disinfectant* Decoction of plant used as an antiseptic wash to help heal broken bones. *Emetic* Infusion of plant taken as an emetic. *Expectorant* Decoction of plant said to be a good expectorant. *Gastrointestinal Aid* Decoction of plant taken for stomachaches. *Misc. Disease Remedy* Decoction of plant taken for smallpox and measles. *Orthopedic Aid* Poultice of boiled plant applied to broken leg bones. *Urinary Aid* Infusion of plant taken for bladder trouble. *Venereal Aid* Decoction of plant taken for venereal disease. (180:81, 82)

Grindelia nuda var. *aphanactis* (Rydb.) Nesom, Curlytop Gumweed
- *Drug*—Navajo, Ramah *Dermatological Aid* Plant used to hold cuts together until they heal. *Emetic* Plant used as an emetic. *Gastrointestinal Aid* and *Pediatric Aid* Cold infusion of plant given to children to kill a swallowed ant. *Veterinary Aid* Cold infusion of plant given to lambs to kill a swallowed ant. (as *G. aphanactis* 191:51) Zuni *Poultice* Poultice of flower applied to ant bites. (as *G. aphanactis* 27:375) *Snakebite Remedy* Fresh or dried root chewed by medicine man before sucking snakebite and poultice applied to wound. (as *G. aphanactis* 27:374)
- *Other*—Navajo, Ramah *Insecticide* Strong infusion of plant poured on anthill to kill ants. (as *G. aphanactis* 191:51)

Grindelia robusta Nutt., Great Valley Gumweed
- *Drug*—Miwok *Dermatological Aid* Decoction of leaves used to wash running sores. Infusion of pulverized leaves applied to sores. (12:170)
- *Food*—Karok *Vegetable* Leaves eaten raw as greens. (148:389)
- *Other*—Karok *Insecticide* Decoction of roots used as a shampoo to kill hair lice. (148:389)

Grindelia sp., Gum Plant
- *Drug*—Jemez *Dermatological Aid* Decoction of dried, ground plant used as a wash for cuts. *Veterinary Aid* Decoction of dried, ground plant used as a wash for cuts on horses. (44:23) Mendocino Indian *Blood Medicine* Decoction of plant taken as a blood purifier. *Cold Remedy* Decoction of plant taken for colds. *Gastrointestinal Aid* Decoction of plant taken for colic. *Laxative* Decoction of plant taken to open bowels. (41:394)
- *Food*—Mendocino Indian *Substitution Food* Leaves used as a substitute for tea. (41:394)

Grindelia squarrosa (Pursh) Dunal, Curlycup Gumweed
- *Drug*—Blackfoot *Liver Aid* Infusion taken for the "liver." (82:32) Decoction of roots taken for liver troubles. (97:56) Decoction of root taken for liver trouble. (114:276) Cheyenne *Dermatological Aid* Decoction of flowering tops applied to skin diseases, scabs, and sores. *Eye Medicine* Gum rubbed on the outside of eyes for snow-blindness. (83:21) Cheyenne, Northern *Disinfectant* Decoction of flowering tops used to wash sores and other skin lesions. *Eye Medicine* Sticky, flower heads used for snow-blindness. (82:32) Cree *Abortifacient* Used to prevent childbearing. (16:485) *Gynecological Aid* Infusion of buds and flowers taken to ease and lessen menses. *Kidney Aid* Plant and camomile used for kidney pains. (16:494) *Venereal Aid* Used for gonorrhea. (16:485) Crow *Cold Remedy* Taken for colds. *Cough Medicine* Taken for coughs. *Pulmonary Aid* Taken for whooping cough and pneumonia. *Respiratory Aid* Infusion sniffed up the nose for catarrh. Taken for bronchitis and asthma. (82:32) Dakota *Gastrointestinal Aid* and *Pediatric Aid* Infusion of plant tops given to children for stomachaches. (69:368) Decoction of plant given to children for colic. (70:133) Flathead *Cold Remedy* Taken for colds. *Cough Medicine* Taken for coughs. *Pulmonary Aid* Taken for whooping cough and pneumonia. *Respiratory Aid* Taken for bronchitis and asthma. *Tuberculosis Remedy* Infusion taken for tuberculosis. *Veterinary Aid* Flower heads rubbed on horses' hooves for protection against injury. (82:32) Gosiute *Cough Medicine* Roots used as a cough medicine. (39:371) Lakota *Antihemorrhagic* Decoction of blossoms and fetid marigold taken for the spitting of blood. (139:37) Mahuna *Dermatological Aid* Poultice of plants applied to cuts. *Disinfectant* Infusion used as a disinfectant wash. (140:15) Montana Indian *Venereal Aid* Decoction used as an antisyphilitic. (19:12) Paiute *Cough Medicine* Decoction of plant said to be a good cough medicine. *Expectorant* Decoction of plant said to be a good expectorant. *Pulmonary Aid* Hot decoction of young shoots taken for pneumonia. *Urinary Aid* Infusion of plant taken for bladder trouble. (180:81, 82) Pawnee *Veterinary Aid* Decoction of tops and leaves used as a wash for saddle galls and sores on horses. Ponca *Tuberculosis Remedy* Decoction of plant taken for consumption. (70:133) Shoshoni *Analgesic* Decoction of plant taken for stomachaches. (180:81, 82) *Cough Medicine* Dried buds used for coughs. (118:37) Decoction of plant said to be a good cough medicine. *Dermatological Aid* Poultice of boiled plant applied to swellings. *Disinfectant* Decoction of plant used as an antiseptic wash to help heal broken bones. *Emetic* Infusion of plant taken as an emetic. *Expectorant* Decoction of plant said to be a good expectorant. *Gastrointestinal Aid* Decoction of plant taken for stomachaches. *Misc. Disease Remedy* Decoction of plant taken for smallpox and measles. *Orthopedic Aid* Poultice of boiled plant applied to broken legs. *Urinary Aid* Infusion of plant taken for bladder trouble. *Venereal Aid* Decoction of plant taken for venereal disease. (180:81, 82) Sioux *Gastroin-

testinal Aid Infusion taken for colic. (82:32) *Kidney Aid* Infusion taken for kidney trouble. (19:12) **Ute** *Cold Remedy* Used as a cough medicine. (38:34)

Grindelia squarrosa* var. *serrulata (Rydb.) Steyermark, Curlycup Gumweed
- **Drug**—**Blackfoot** *Liver Aid* Infusion of root taken as a liver aid. (as *G. squarrosa serrulata* 118:45) **Shoshoni** *Kidney Aid* Dried upper third of plant and buds taken for dropsy. *Misc. Disease Remedy* Dried upper third of plant and buds taken for smallpox. (as *G. squarrosa serrulata* 118:43)

Grossulariaceae, see *Ribes*

Guajacum, Zygophyllaceae

Guajacum coulteri
- **Drug**—**Seri** *Unspecified* Berries used for medicine. (as *Guaiacum coulteri* 50:136)

Gutierrezia, Asteraceae

Gutierrezia californica (DC.) Torr. & Gray, San Joaquin Snakeweed
- **Drug**—**Kawaiisu** *Orthopedic Aid* Poultice of heated plant applied to aching back or limbs. (206:33)
- **Fiber**—**Kawaiisu** *Building Material* Plant used as wall filler in the construction of the winter house. Built on a circular ground plan of vertical and horizontal poles, the house had matchweed packed in so tightly between the poles that one could not see through. However, there was apparently an outer covering usually of tule mats. (206:33)

Gutierrezia microcephala (DC.) Gray, Threadleaf Snakeweed
- **Drug**—**Cahuilla** *Toothache Remedy* Infusion of plant used as a gargle or plant placed inside the mouth for toothaches. (15:75) **Hopi** *Carminative* Used for "gastric disturbances." (42:323) **Navajo** *Veterinary Aid* Poultice of plant applied to the back and legs of horses. (as *G. lucida* 90:151) **Tewa** *Carminative* Used for "gastric disturbances." (42:323)
- **Other**—**Hopi** *Cooking Tools* Used in roasting sweet corn. *Decorations* Used as paho (prayer stick) decorations. (42:323) Used as prayer stick decorations. (as *G. lucida* 200:96) **Tewa** *Cooking Tools* Used in roasting sweet corn. *Decorations* Used as paho (prayer stick) decorations. (42:323)

Gutierrezia sarothrae (Pursh) Britt. & Rusby, Broom Snakeweed
- **Drug**—**Blackfoot** *Herbal Steam* Roots used in herbal steam for unspecified ailments. (as *G. diversifolia* 114:276) *Respiratory Aid* Roots placed in boiling water and steam inhaled for respiratory ailments. (97:56) **Comanche** *Pulmonary Aid* Compound containing leaves used for whooping cough. (29:522) **Dakota** *Veterinary Aid* Decoction of flowers given to horses as a laxative. (69:368) **Diegueño** *Antidiarrheal* Decoction of fresh flowers or fresh roots taken for diarrhea. (88:220) **Isleta** *Dermatological Aid* Poultice of moistened leaves used for bruises. *Febrifuge* Infusion of leaves used as a bath for fevers. *Venereal Aid* Infusion of leaves used for venereal diseases. (as *G. furfuracea* 100:31) **Jemez** *Dermatological Aid* Decoction of plant used for sores. *Gynecological Aid* Decoction of plant taken by women after childbirth following the cedar decoction. (as *G. furfuracea* 44:23) **Keres, Western** *Antirheumatic (External)* Strong, black infusion of plant used as a rub for rheumatism. *Cathartic* Infusion of plant used as a cathartic. *Diaphoretic* Plant used as an ingredient in the sweat bath. *Emetic* Infusion of plant used as an emetic. *Eye Medicine* Infusion of plant used as an eyewash. *Snakebite Remedy* Chewed leaf juice taken for and rubbed on rattlesnake bites. *Veterinary Aid* Infusion of leaves used as a wash for horses after castration. (as *G. longifolia* 171:46) **Lakota** *Cold Remedy* Decoction of plant taken for colds. *Cough Medicine* Decoction of plant taken for coughs. *Vertigo Medicine* Decoction of plant taken for dizziness. (139:37) **Navajo** *Analgesic* Plant ashes rubbed on the body for headaches. (55:86) Plant used for headaches. (90:151) *Ceremonial Medicine* Wood made into charcoal used in the medicines applied to the ailing gods. Two kinds of charcoal were used in the medicines which were applied to the ailing gods. The first was made from the bark of the pine and willow. The second was made from this plant and three-lobed sagebrush, to which were added the feathers dropped from a live crow and a live buzzard. (55:86) *Dermatological Aid* Plant used for wounds. (55:97) Poultice of chewed plant applied to ant, bee, and wasp sting swellings. (55:86) *Sedative* Plant used for "nervousness." (90:151) *Snakebite Remedy* Plant used for snakebites. *Veterinary Aid* Decoction of ground plant applied as poultice to sheep bitten by a snake. (55:86) **Navajo, Kayenta** *Antidiarrheal* Plant used for bloody diarrhea. *Ceremonial Medicine* and *Disinfectant* Plant used as a ceremonial fumigant ingredient. *Gastrointestinal Aid* Plant used for gastrointestinal disease. (205:48) **Navajo, Ramah** *Analgesic* Decoction of root taken for painful urination and stomachache. *Antidote* Compound decoction of plant used as an antidote for taking too much medicine. *Ceremonial Medicine* Decoction used ceremonially for snake infection or snakebite. *Dermatological Aid* Poultice or infusion of flowers and leaves applied to red ant bite and bee sting. *Disinfectant* Decoction used ceremonially for snake infection. *Febrifuge* Cold infusion of leaves applied to forehead for fever. *Gastrointestinal Aid* Decoction of root taken for stomachache. *Gynecological Aid* Decoction of root taken to hasten delivery of placenta. *Panacea* Root used as a "life medicine." *Snakebite Remedy* Decoction used ceremonially for snakebite. *Urinary Aid* Decoction of root taken for painful urination. *Veterinary Aid* Cold infusion of leaves used as a lotion on incisions and bites on lambs or colts. (191:51) **Paiute** *Antirheumatic (External)* Poultice of boiled leaves in cloth applied as a heat pack for rheumatism. *Hemostat* Poultice of boiled leaves applied to top of head for nosebleed. *Orthopedic Aid* Poultice of boiled leaves applied for sprains. **Shoshoni** *Cold Remedy* Decoction of plant taken for colds. *Dermatological Aid* and *Disinfectant* Compound decoction of plant used as an antiseptic wash for measles and other rashes. *Misc. Disease Remedy* Compound decoction of plant used as an antiseptic wash for measles. (180:82, 83) **Tewa** *Analgesic* Plant used on hot coals to fumigate patient with painful menstruation. *Disinfectant* Plant used on hot coals to fumigate mother and newborn child. *Ear Medicine* Chopped, fresh plant rubbed around ear for earache. *Gastrointestinal Aid* Decoction of plant used for gastric disturbances, especially "gastric influenza." *Gynecological Aid* Compound containing plant used as snuff and as a fumigant for painful periods, and for women in labor. *Misc. Disease Remedy* Decoction of plant taken for gastric influenza. *Pediatric Aid*

Plant used on hot coals to fumigate mother and newborn child. (as *G. longifolia* 138:56) **Zuni** *Antirheumatic (External)* Infusion of whole plant used for muscle aches. (as *G.* cf. *sarothrae* 27:375) *Diaphoretic* Infusion of blossoms taken as a diaphoretic. *Diuretic* Infusion of blossoms taken as a diuretic for "obstinate cases." *Strengthener* Infusion of blossoms taken to "make one strong in the limbs and muscles." (as *G. filifolia* 166:53) *Urinary Aid* Infusion of whole plant taken to increase strength for urinary retention. (as *G.* cf. *sarothrae* 27:375)
- *Food*—**Tewa** *Forage* Plant eaten by livestock. (as *G. longifolia* 138:56)
- *Fiber*—**Comanche** *Brushes & Brooms* Stems used to make brooms. (29:522)
- *Dye*—**Navajo** *Yellow* Tops used to make a yellow dye. (55:86)
- *Other*—**Hopi** *Ceremonial Items* Sprig attached to the paho (prayer emblem). (as *G. euthamiae* 56:15) Sprigs tied on prayer sticks during the December ceremonies. (as *G. longifolia* 138:56) Tied onto the prayer stick. (as *Solidago sarothrae* 190:168) *Decorations* Used as prayer stick decorations. (200:96) **Isleta** *Soap* Infusion of leaves used as pleasant and refreshing bath. (as *G. furfuracea* 100:31) **Jemez** *Insecticide* Plant chewed and juice spit upon bees to kill the insects. Plant placed upon a slow fire and smoke destroyed bees. (as *G. furfuracea* 44:23) **Keres, Western** *Insecticide* Chewed leaf juice had an intoxicating effect upon bees. (as *G. longifolia* 171:46) **Navajo** *Ceremonial Items* Leaves, grama grass, sagebrush, and unidentified leaves burned to charcoal for blackening ceremony. Wood ash and pitch used to cover the oak bull-roarer for the Female Shooting Life Chant. *Tools* Stems used for whirls when making fire by friction. (55:86) **Navajo, Kayenta** *Ceremonial Items* Plant placed on top of most ceremonial prayer sticks and figurines. (205:48) **Navajo, Ramah** *Ceremonial Items* Plant ash used as Evilway, Holyway, and Hand Tremblingway blackenings. Fresh branches used to make Evilway unravelers. Fresh branches used to make cactus prayer sticks for Chiricahua Windway and Enemyway prayer sticks. (191:51) **Tewa of Hano** *Ceremonial Items* Sprigs tied on prayer sticks during the December ceremonies. (as *G. longifolia* 138:56)

Gutierrezia sp., Snakeweed
- *Drug*—**Hopi** *Gastrointestinal Aid* Plant used for disorders of the digestive system. (200:34) **Keresan** *Ceremonial Medicine* and *Emetic* Infusion of plant taken as a ceremonial emetic. (198:563)
- *Fiber*—**Hualapai** *Brushes & Brooms* Used as a utilitarian brush to remove stickers off prickly pear fruits and for sweeping the floor. (195:16)
- *Other*—**Apache, Chiricahua & Mescalero** *Ceremonial Items* Plant used in ceremonial contexts. (33:24) **Hualapai** *Ceremonial Items* Used as an important plant in rain ceremonies. (195:16)

Guzmania, Bromeliaceae
Guzmania sp., Airplant
- *Other*—**Seminole** *Cooking Tools* Plant used as a water supply for cooking during the dry season. (169:472)

Gymnocarpium, Dryopteridaceae, fern
Gymnocarpium disjunctum (Rupr.) Ching, Pacific Oakfern
- *Drug*—**Abnaki** *Other* Used as a demulcent. (as *Dryopteris disjuncta* 144:155)

Gymnocarpium dryopteris (L.) Newman, Western Oakfern
- *Other*—**Okanagan-Colville** *Water Indicator* Ferns considered to be a sign of water when traveling through the mountains. (188:18)

Gymnocladus, Fabaceae
Gymnocladus dioicus (L.) K. Koch, Kentucky Coffee Tree
- *Drug*—**Dakota** *Laxative* Infusion of root used as an enema and infallible remedy for constipation. *Stimulant* Pulverized root bark used as snuff to cause sneezing in comatose patient. (as *G. dioica* 70:89, 90) **Meskwaki** *Psychological Aid* Wax of pods "fed to a patient to cure him of lunacy." (as *G. dioica* 152:229) **Omaha** *Dietary Aid* Bark used as an appetizer. (as *G. dioica* 68:335) *Gynecological Aid* Powdered root mixed with water and given to women during protracted labor. *Hemostat* Root bark used for hemorrhages, especially from nose and during childbirth. *Kidney Aid* Root used "when kidneys failed to act." (as *G. canadensis* 58:584) *Laxative* Infusion of root used as an enema and infallible remedy for constipation. *Stimulant* Pulverized root bark used as snuff to cause sneezing in comatose patient. (as *G. dioica* 70:89, 90) *Tonic* Bark used as a tonic. (as *G. dioica* 68:335) **Oto** *Laxative* Infusion of root used as an enema and infallible remedy for constipation. **Pawnee** *Analgesic* Pulverized pod sniffed to cause sneezing for headaches. *Stimulant* Pulverized root bark used as snuff to cause sneezing in comatose patient. **Ponca** *Laxative* Infusion of root used as an enema and infallible remedy for constipation. *Stimulant* Pulverized root bark used as snuff to cause sneezing in comatose patient. **Winnebago** *Laxative* Infusion of root used as an enema and infallible remedy for constipation. *Stimulant* Pulverized root bark used as snuff to cause sneezing in comatose patient. (as *G. dioica* 70:89, 90)
- *Food*—**Meskwaki** *Beverage* Roasted, ground seeds boiled to make coffee. *Unspecified* Roasted seeds eaten. (as *G. dioica* 152:260) **Pawnee** *Unspecified* Roasted seeds eaten like chestnuts. **Winnebago** *Unspecified* Seeds pounded in a mortar and used for food. (as *G. dioica* 70:89)
- *Dye*—**Dakota** *Black* Root sometimes used with another component to make a black dye. The root was not very highly esteemed for making a dye and alone was considered useless, but was occasionally used with another component to make a black dye. (as *G. dioica* 70:89)
- *Other*—**Winnebago** *Toys & Games* Seeds used as counters or tally checks in gambling. (as *G. dioica* 70:89)

Habenaria, Orchidaceae
Habenaria odontopetala Reichenb. f., Toothpetal False Reinorchid
- *Drug*—**Seminole** *Strengthener* Plant used to make a medicine and given to students in medical training to make the body strong. (as *H. strictissima* var. *odontopetala* 169:102)

Hackelia, Boraginaceae
Hackelia diffusa (Lehm.) I. M. Johnston, Spreading Stickseed
- *Food*—**Thompson** *Forage* Plant eaten by sheep. The plant was not

used by people as it was considered a noxious weed because the burred fruits stuck to fur and clothing. (187:192)

Hackelia floribunda (Lehm.) I. M. Johnston, Manyflower Stickseed
- *Drug*—**Isleta** *Poison* Prickles from fruit caused skin irritation and swelling. (as *Lappula floribunda* 100:33) **Navajo, Ramah** *Orthopedic Aid* Root of this or any poisonous plant used for serious injury such as fracture. *Poison* Plant considered poisonous. (191:40, 41)
- *Other*—**Navajo, Ramah** *Good Luck Charm* Leaves and pollen used various ways for good luck in gambling and trading. (191:40, 41)

Hackelia hispida (Gray) I. M. Johnston **var. *hispida***, Showy Stickseed
- *Drug*—**Thompson** *Unspecified* Plant used medicinally for unspecified purpose. (as *Lappula hispida* 164:474)

Hackelia virginiana (L.) I. M. Johnston, Beggarslice
- *Drug*—**Cherokee** *Cancer Treatment* Bruised root with bear oil used as ointment for cancer. *Dermatological Aid* Compound decoction of root given for itch. (80:25) Decoction of roots used as a bath and taken for itching genitals. (as *Lappula virginiana* 177:52) *Kidney Aid* Decoction used for kidney trouble. *Love Medicine* Used for love charms. *Psychological Aid* Used for "good memory." (80:25)
- *Other*—**Iroquois** *Insecticide* Plants used around potatoes to keep bugs off. (as *Lappula virginiana* 87:420)

Haemodoraceae, see *Lachnanthes*

Halesia, Styracaceae

Halesia carolina L., Carolina Silverbell
- *Fiber*—**Cherokee** *Building Material* Wood used for lumber. (80:55)

Haloragaceae, see *Myriophyllum*

Halosaccion, Palmariaceae, alga

Halosaccion glandiforme (Gmelin) Ruprecht, Bladder Seaweed
- *Drug*—**Hesquiat** *Unspecified* Seaweed used as a medicine. (185:24) **Nitinaht** *Reproductive Aid* Sacs chewed by newly wed women wanting their first baby to be a boy. (186:51)

Hamamelidaceae, see *Hamamelis, Liquidambar*

Hamamelis, Hamamelidaceae

Hamamelis virginiana L., American Witchhazel
- *Drug*—**Cherokee** *Analgesic* Infusion taken for periodic pains. *Cold Remedy* Infusion taken for colds. *Dermatological Aid* Infusion used as wash for sores and skinned places and leaves rubbed on scratches. *Febrifuge* Compound infusion taken for fevers. *Gynecological Aid* Infusion taken for periodic pains. *Throat Aid* Infusion taken for sore throat. *Tuberculosis Remedy* Infusion of bark taken for tuberculosis. (80:62) **Chippewa** *Dermatological Aid* Infusion of inner bark used as lotion for skin troubles. *Emetic* Inner bark used, especially in cases of poisoning, as an emetic. *Eye Medicine* Infusion of inner bark used as a wash for sore eyes. (71:131) **Iroquois** *Antidiarrheal* Infusion of twig bark taken for bloody dysentery. *Antiemetic* Poultice of branches applied to body part affected by colds and heaves. (87:346) *Antirheumatic (Internal)* Compound used for arthritis. (87:348) *Blood Medicine* Compound decoction of tips and sprouts taken as a blood purifier. (87:347) *Cold Remedy* Decoction of young branches taken or poultice applied for colds. *Cough Medicine* Decoction of young branches taken as medicine for coughs and colds. (87:348) *Dermatological Aid* Bark used as an astringent. (87:347) *Dietary Aid* Decoction of bark taken "when one can't eat," to stimulate the appetite. *Emetic* Decoction of bark taken as an emetic. (87:346) *Gynecological Aid* Compound decoction taken to prevent hemorrhage after childbirth. Decoction of shoots taken by a pregnant woman who has fallen or been hurt. (87:347) *Heart Medicine* Decoction of leaves and twigs taken for "cold around the heart." (87:346) *Kidney Aid* Decoction of twigs taken and poultice of bark used to regulate the kidneys. (87:348) *Misc. Disease Remedy* Infusion of twig bark taken for cholera. (87:346) *Orthopedic Aid* Decoction of shoots taken and poultice of bark used for bruises. *Panacea* Compound decoction of roots taken as a panacea. (87:347) *Pulmonary Aid* Decoction of bark taken for lung troubles or for spots and scars on lungs. *Respiratory Aid* Decoction of new growth twigs taken for chest colds and asthma. (87:348) *Toothache Remedy* Plant used as toothache medicine. *Tuberculosis Remedy* Compound decoction of roots or bark taken for consumption. (87:346) *Venereal Aid* Compound decoction of bark taken for venereal disease. (87:347) **Menominee** *Ceremonial Medicine* Seeds used as the sacred bead in the medicine ceremony. *Orthopedic Aid* Decoction rubbed on legs during sports, to keep legs limber. Infusion of twigs used to "cure a lame back." (151:37) *Other* Dried seeds used in a test to tell whether sick person would recover. (54:120) **Mohegan** *Dermatological Aid* Infusion of twigs and leaves used as a lotion for cuts, bruises, and insect bites. (176:72, 130) **Potawatomi** *Orthopedic Aid* Twigs used to create steam in the sweat bath for sore muscles. (154:59, 60)
- *Food*—**Cherokee** *Beverage* Leaves and twigs used to make tea. (126:44)
- *Other*—**Mohegan** *Water Indicator* Crotched sticks used to locate underground water or buried treasure. (176:87)

Haplopappus, Asteraceae

***Haplopappus* sp.**
- *Drug*—**Paiute** *Antidiarrheal* Infusion of plant taken for diarrhea. *Gastrointestinal Aid* Infusion of plant taken for stomach troubles. (167:317)
- *Food*—**Paiute** *Unspecified* Species used for food. (as *Haploppapus* 167:243)

Hastingsia, Liliaceae

Hastingsia alba (Dur.) S. Wats., White Rushlily
- *Other*—**Karok** *Toys & Games* Leaves put over the teeth to make a snapping sound for amusement. (as *Schoenolirion album* 148:380)

Hazardia, Asteraceae

Hazardia squarrosa (Hook. & Arn.) Greene **var. *squarrosa***, Sawtooth Goldenbush
- *Drug*—**Diegueño** *Antirheumatic (External)* Decoction of plant used for bathing the aches and pains of the body. (as *Aplopappus squarrosus* ssp. *grindelioides* 88:220)

Hedeoma, Lamiaceae

Hedeoma drummondii Benth., Drummond's Falsepennyroyal

- *Drug*—Navajo *Analgesic* Plant used for pain. (55:72) **Navajo, Ramah** *Misc. Disease Remedy* Infusion of plant taken in large quantities for influenza. (191:41)
- *Food*—Lakota *Soup* Leaves used to make soup. (139:49)

Hedeoma hispida Pursh, Rough Falsepennyroyal
- *Drug*—Dakota *Cold Remedy* Infusion of leaves used for colds. *Dietary Aid* Infusion of leaves used as a flavor and tonic appetizer in diet for the sick. (70:112)

Hedeoma nana (Torr.) Briq., Falsepennyroyal
- *Drug*—Navajo *Ceremonial Medicine* Used by assistant during the War Dance. At noon of the third day of the War Dance, the body of the patient was painted black. Medicine was then made of yarrow, red juniper, pine needles, and meadow rue, which were previously pulverized, then thrown into a bowl of water and stirred. This was then dabbed all over the patient who sipped the mixture before bathing his whole body in it. Foxtail grass and mock pennyroyal were then chewed by the assistant and sputtered on the patient. (55:72) **Shoshoni** *Cathartic* Decoction of plant taken as a physic. *Gastrointestinal Aid* Decoction of plant taken for indigestion. (180:83)
- *Food*—Apache, Chiricahua & Mescalero *Beverage* Leaves and young stems boiled to make a nonintoxicating beverage. (33:53) *Spice* Leaves used as flavoring. (33:47) **Isleta** *Spice* Leaves chewed for the mint flavor. (100:31) *Unspecified* Leaves chewed for their pleasing flavor. (32:30)
- *Other*—Keres, Western *Soap* Infusion of plant used as a hair wash and body bath. (171:47)

Hedeoma pulegioides (L.) Pers., American Falsepennyroyal
- *Drug*—Catawba *Cold Remedy* Decoction of roots used for colds. (157:188) Decoction of roots taken for colds. (177:53) **Cherokee** *Abortifacient* Infusion taken for "obstructed menses." *Analgesic* Poultice of leaves used for headaches. *Antidiarrheal* Infusion taken for "flux" and leaves rubbed on body as insect repellent. *Cold Remedy* Taken for colds. *Cough Medicine* Taken for coughs. *Diaphoretic* Decoction taken as a diaphoretic. *Expectorant* Decoction taken as an expectorant. *Febrifuge* Infusion taken for fever. *Pulmonary Aid* Taken for whooping cough. *Stimulant* Decoction taken as a stimulant. *Toothache Remedy* Beaten leaves held in mouth for toothache. (80:48) **Chickasaw** *Eye Medicine* Cold infusion of roots applied to forehead for itching eyes. (177:53) **Delaware** *Gastrointestinal Aid* Infusion of leaves used for stomach pains. (176:35) **Delaware, Oklahoma** *Analgesic* and *Gastrointestinal Aid* Infusion of leaves taken for stomach pains. (175:29, 76) **Iroquois** *Analgesic* Infusion of plants taken for headaches. (87:426) **Mahuna** *Antidiarrheal* Plant used for dysentery. (140:7) **Mohegan** *Gastrointestinal Aid* Infusion of plant taken to warm the stomach. (174:265) Infusion of leaves said to be "warming and good for stomach." (176:72) **Nanticoke** *Diaphoretic* Whole plant used as a sudorific. (175:58, 84) *Kidney Aid* Plant considered an excellent remedy for kidney troubles. *Liver Aid* Plant considered an excellent remedy for liver troubles. (175:58) **Ojibwa** *Febrifuge* Infusion of plant taken for cold fevers. *Gastrointestinal Aid* Infusion of plant taken for upset stomachs. (5:2274) **Rappahannock** *Gynecological Aid* Infusion of fresh or dried plants taken for menstruation pains. (161:33) **Shinnecock** *Analgesic* Infusion of leaves taken for pains. (30:121)
- *Other*—Cherokee *Insecticide* Infusion taken for "flux" and leaves rubbed on body as insect repellent. (80:48) **Rappahannock** *Insecticide* Leaves dried and used indoors for fleas. (161:33)

Hedeoma sp., Pennyroyal
- *Drug*—Dakota *Cold Remedy* Infusion of plants taken for colds. *Dietary Aid* and *Tonic* Infusion of plants taken as a tonic appetizer in diet of sick. (69:363)

Hedophyllum, Laminariaceae, alga
Hedophyllum sessile (C. Agardh) Setchell
- *Food*—Nitinaht *Unspecified* Plants eaten with herring spawn. (186:51)
- *Other*—Nitinaht *Hunting & Fishing Item* Fronds used to catch herring spawn. (186:51)

Hedysarum, Fabaceae
Hedysarum alpinum L., Alpine Sweetvetch
- *Food*—Alaska Native *Unspecified* Roots eaten raw, boiled, or roasted. (85:121) **Eskimo, Arctic** *Forage* Root tubers eaten by brown and black bears and meadow mice. *Vegetable* Tubers located in mice "caches" by specially trained dogs and eaten. (128:30) **Eskimo, Inupiat** *Frozen Food* Roots frozen for future use. *Vegetable* Roots, always with some kind of oil, eaten raw or cooked. *Winter Use Food* Roots stored in buried sacks for winter use. Roots stored in seal oil, fish oil, or bear fat for winter use. (98:115) **Tanana, Upper** *Beverage* Fried roots, with or without grease, used to make tea. *Vegetable* Roots eaten raw, roasted over a fire, fried, or boiled. Roots dipped in or mixed with grease and eaten. *Winter Use Food* Used in the winter during times of food shortage. A large fire was set over an area where the Indians knew the roots to be abundant. By thawing the ground this way, they were able to dig them out. Roots stored, with or without grease, in a birch bark basket in an underground cache. (102:14)

Hedysarum boreale Nutt., Northern Sweetvetch
- *Food*—Eskimo, Arctic *Forage* Roots eaten by the brown bears, meadow mice, and lemmings. *Vegetable* Roots located in mice "caches" by dogs and eaten. (127:1)

Hedysarum boreale ssp. *mackenziei* (Richards.) Welsh, Mackenzie's Sweetvetch
- *Drug*—Alaska Native *Poison* Plant considered poisonous. (as *H. mackenzii* 85:155) **Eskimo, Inupiat** *Poison* Roots considered poisonous. (as *H. mackenzii* 98:142) **Tanana, Upper** *Poison* Plant considered poisonous. (as *H. mackenzii* 102:14) **Ute** *Unspecified* Roots used as medicine. (as *H. mackenzii* 38:35)
- *Food*—Tanana, Upper *Unspecified* Roots used for food. (as *H. mackenzii* 77:28) Roots eaten fresh and boiled. *Winter Use Food* Fresh roots stored underground in brush-lined caches for future use. (as *H. mackenzii* 115:36)

Hedysarum sp.
- *Food*—Eskimo, Alaska *Vegetable* Fleshy roots used the same as potatoes. (4:715)

Helenium, Asteraceae
Helenium amarum (Raf.) H. Rock var. *amarum*, Yellowdicks

- *Drug*—**Koasati** *Dermatological Aid* Decoction of entire plant used as a sweat bath for swellings. *Herbal Steam* Decoction of entire plant used as a sweat bath for dropsy and swellings. *Kidney Aid* Decoction of entire plant used as a sweat bath for dropsy. (as *H. tenuifolium* 177:62)

Helenium autumnale L., Common Sneezeweed
- *Drug*—**Cherokee** *Gynecological Aid* Compound infusion of roots given to prevent menstruation after childbirth. *Nose Medicine* Powdered, dry leaves used to induce sneezing. (80:56) **Comanche** *Febrifuge* Infusion of stems used as a wash for fever. (29:522) **Mahuna** *Respiratory Aid* Plant used for catarrh. (140:24) **Menominee** *Alterative* Compound infusion of flower heads taken "for its alterative effects." (151:30, 31) *Analgesic* Compound of dried flowers applied to small cuts made on temples for headache. Snuff of compounded flowers used to cause sneezing for headaches. (54:129) *Cold Remedy* Simple or compound snuff of flowers caused sneezing to clear a stuffy head cold. (151:30, 31) **Meskwaki** *Cold Remedy* Disk florets used as snuff for colds or catarrh. *Gastrointestinal Aid* Infusion of florets taken for stomach catarrh. *Poison* Plant known to be poisonous to cattle. *Respiratory Aid* Snuff of dried disk florets inhaled for catarrh or colds. (152:215)

Helenium microcephalum DC., Littlehead Tarweed
- *Drug*—**Comanche** *Gynecological Aid* Pulverized flowers inhaled to cause sneezing and expulsion of afterbirth. (29:522) *Heart Medicine* Flowers dried, crushed, and inhaled for "heart flutter." *Hypotensive* Flowers dried, crushed, and inhaled for low blood pressure. (99:4) *Respiratory Aid* Pulverized flowers inhaled to cause sneezing and clear nasal passages. (29:522) Flowers dried, crushed, and inhaled for sinus congestion. (99:4)

Helenium puberulum DC., Rosilla
- *Drug*—**Costanoan** *Cold Remedy* Dried, powdered plant rubbed on the forehead and nose for colds. *Dermatological Aid* Dried, powdered plant applied to wounds. (21:26) **Mendocino Indian** *Venereal Aid* Plant used for venereal diseases. (41:394)
- *Food*—**Mendocino Indian** *Unspecified* Leaves and heads eaten raw. (41:394)
- *Other*—**Costanoan** *Snuff* Dried, powdered plant used as a snuff to induce sneezing. (21:26)

Helianthella, Asteraceae
Helianthella californica Gray, California Helianthella
- *Food*—**Yana** *Unspecified* Flowers cooked and eaten. (147:251)

Helianthella parryi Gray, Parry's Dwarfsunflower
- *Drug*—**Navajo, Ramah** *Dermatological Aid* Decoction of root taken and used as a lotion on arrow or bullet wound, a "life medicine." *Panacea* Decoction of root used as "life medicine," especially for arrow or bullet wounds. (191:51)

Helianthella uniflora (Nutt.) Torr. & Gray, Oneflower Helianthella
- *Drug*—**Paiute** *Dermatological Aid* Hot poultice of mashed root applied for swellings and sprains. *Orthopedic Aid* Poultice of mashed root applied for swellings and sprains. **Shoshoni** *Analgesic* Infusion of root used as a wash or compress for headaches. *Antirheumatic (External)* Poultice of mashed root applied for rheumatism of the shoulder or knee. (180:83, 84)

Helianthemum, Cistaceae
Helianthemum canadense (L.) Michx., Longbranch Frostweed
- *Drug*—**Cherokee** *Kidney Aid* Infusion of leaf taken for kidneys. (80:35) **Delaware** *Throat Aid* Poultice of roots applied to sore throats. (176:32) **Delaware, Oklahoma** *Analgesic* Infusion of plant taken and poultice of root applied for sore throat. (175:27, 76) *Tonic* Plant used as a "strengthening" medicine. (175:76)

Helianthus, Asteraceae
Helianthus annuus L., Common Sunflower
- *Drug*—**Apache, White Mountain** *Snakebite Remedy* Poultice of crushed plants applied to snakebites. (136:158) **Dakota** *Analgesic* Infusion of flowers used for chest pains. (69:369) *Pulmonary Aid* Decoction of flower heads taken for pulmonary troubles. (70:130) **Gros Ventre** *Ceremonial Medicine* Oil from seeds used "to lubricate or paint the face or body." *Stimulant* Dried, powdered seeds mixed into cakes and taken on war party to combat fatigue. (19:12, 13) **Hopi** *Dermatological Aid* Plant used as a "spider bite medicine." (200:32, 96) **Jemez** *Dermatological Aid* Juice applied to cuts. (44:23) **Kiowa** *Oral Aid* Coagulated sap chewed, by the elders, to diminish thirst. (192:60) **Mandan** *Ceremonial Medicine* Oil from seeds used "to lubricate or paint the face or body." *Stimulant* Dried, powdered seeds mixed into cakes and taken on war party to combat fatigue. (19:12, 13) **Navajo** *Ceremonial Medicine* Plant, double bladderpod, sumac, and mistletoe used in the liniment for the War Dance. (55:87) *Dietary Aid* Seeds eaten to give appetite. (90:152) **Navajo, Kayenta** *Ceremonial Medicine* Plant used for sun sand-painting ceremony. *Disinfectant* Plant used for prenatal infection caused by solar eclipse. *Pediatric Aid* Plant used for prenatal infection caused by solar eclipse. (205:48) **Navajo, Ramah** *Dermatological Aid* Moxa of pith used on scratched wart for removal. *Other* Salve of pulverized seed and root used on injury from horse falling on person. (191:51) **Paiute** *Antirheumatic (External)* Decoction of root used as a warm wash for rheumatism. (180:84) **Pawnee** *Gynecological Aid* Dry seed compound eaten by pregnant nursing women to protect suckling child. (70:130) **Pima** *Anthelmintic* Poultice of warm ashes applied to stomach for worms. *Febrifuge* Decoction of leaves taken for high fevers. *Veterinary Aid* Decoction of leaves used as a wash for horses with sores caused by screwworms. (47:103) **Ree** *Ceremonial Medicine* Oil from seeds used "to lubricate or paint the face or body." *Stimulant* Dried, powdered seeds mixed into cakes and taken on war party to combat fatigue. (19:12, 13) **Thompson** *Dermatological Aid* Powdered leaves alone or in ointment used on sores and swellings. (as *H. lenticularis* 164:469) **Zuni** *Snakebite Remedy* Fresh or dried root chewed by medicine man before sucking snakebite and poultice applied to wound. (27:375) Compound poultice of root applied with much ceremony to rattlesnake bite. (166:53, 54)
- *Food*—**Apache, Chiricahua & Mescalero** *Bread & Cake* Seeds ground, sifted, made into dough, and baked on hot stones. *Sauce & Relish* Seeds ground into flour and used to make a thick gravy. (33:48) **Apache, White Mountain** *Staple* Seeds used to make flour. (136:158) **Cahuilla** *Staple* Dried seeds ground and mixed with flour from other

seeds. (15:76) **Costanoan** *Unspecified* Seeds used for food, usually not in pinole. (21:254) **Gosiute** *Cooking Agent* Seeds a highly prized source of oil. *Unspecified* Seeds a highly prized source of food. (39:371) **Gros Ventre** *Staple* Powdered seed meal boiled or made into cakes with grease. *Unspecified* Seeds eaten raw. (19:12) **Havasupai** *Dried Food* Seeds sun dried and stored for winter use. (197:248) *Preserves* Seeds parched, ground, kneaded into seed butter, and eaten with fruit drinks or spread on bread. *Staple* Seeds ground and eaten as a ground or parched meal. (197:67) **Hopi** *Fodder* Used as an important food for summer birds. (200:96) **Kawaiisu** *Staple* Roasted seeds pounded, ground into a meal, and eaten dry. (206:34) **Kiowa** *Unspecified* Seeds ground into a paste-like consistency and eaten. (192:60) **Luiseño** *Unspecified* Seeds used for food. (155:228) **Mandan** *Staple* Powdered seed meal boiled or made into cakes with grease. *Unspecified* Seeds eaten raw. (19:12) **Mohave** *Staple* Seeds winnowed, parched, ground, and eaten as pinole. *Winter Use Food* Seeds stored in gourds or ollas. (37:187) **Montana Indian** *Bread & Cake* Seeds dried, powdered, and grease added to make cakes. *Porridge* Seeds dried, powdered, and boiled to make gruel. (82:30) **Navajo** *Bread & Cake* Seeds mixed with corn, ground into a meal, and made into cakes. (55:87) Seeds ground and made into bread and dumplings. *Porridge* Seeds ground and made into gruel. (165:223) **Navajo, Ramah** *Fodder* Used for livestock feed. *Unspecified* Roasted, ground seeds made into cakes. (191:51) **Paiute** *Porridge* Roasted, ground seeds made into flour and used to make mush. (111:117) *Staple* Seeds parched, ground, and eaten as meal. (104:98) *Winter Use Food* Roasted, ground seeds made into flour and stored for winter use. (111:117) **Paiute, Northern** *Staple* Seeds ground into a meal and eaten. (59:47) **Pima** *Candy* Inner pulp of stalks used as chewing gum. Petals used by children as chewing gum. *Staple* Seeds ground into meal and used as food. *Unspecified* Seeds eaten raw or roasted. (47:103) **Pueblo** *Unspecified* Seeds used for food. (32:30) **Ree** *Staple* Powdered seed meal boiled or made into cakes with grease. *Unspecified* Seeds eaten raw. (19:12) **Sanpoil & Nespelem** *Dried Food* Dried roots stored for winter use. (131:100) *Unspecified* Stems eaten raw. (131:103) Seeds parched until brown, pulverized, and eaten. *Winter Use Food* Seeds parched until brown, pulverized, and stored in salmon skins. (131:104)
- **Fiber**—**Jemez** *Building Material* Sunflower mixed with clay, to hold the particles together, and used for plaster. (44:23)
- **Dye**—**Navajo** *Red* Outer seed coatings boiled and used as a dull, dark red dye. (55:87)
- **Other**—**Hopi** *Decorations* Petals dried, ground, mixed with yellow cornmeal, and used as a face powder in women's basket dance. (200:96) **Isleta** *Ceremonial Items* Pith used to light the ceremonial cigarettes. (100:31) **Jemez** *Decorations* Flowers used by the Koshares as a decoration for dances. *Soap* Seeds boiled and water used to wash in. (44:23) **Navajo** *Ceremonial Items* Hollow stalk used in the illusion of swallowing the arrow during the Mountain Chant. Stalk made into flute used in an ancient custom of timing the grinding of the corn at the War Dance. *Hunting & Fishing Item* Stalks used to make bird snares. Bird snares were made of stalks in which were drilled two small holes. In one of these holes was inserted a twig of greasewood and at the end of this was fastened a sliding loop of horsehair. The greasewood twig was then bent in a bow and the loop passed through the upper hole, across which was laid a small piece of reed. The small stick below the loop was placed so that one end rested on the rim of the stalk and the other end on the reed. When a bird alighted on this, the small piece of reed was disturbed and the greasewood twig straightened, drawing the horsehair loop with the bird's foot in it into the stalk. (55:87) **Navajo, Ramah** *Ceremonial Items* Stem used to make Holyway Prayer stick. (191:51) **Pima** *Lighting* Inner pulp of dried stalks strung and used to make quick-burning candles. (47:103) **Tewa** *Smoking Tools* Dried stalks made into fire sticks and used to light cigarettes. (138:56) **Zuni** *Ceremonial Items* Blossoms used ceremonially for anthropic worship. (166:93)

Helianthus anomalus Blake, Western Sunflower
- **Drug**—**Hopi** *Dermatological Aid* Plant used as a "spider medicine." (200:32, 96)
- **Food**—**Hopi** *Fodder* Used as an important food for summer birds. (200:96)
- **Other**—**Hopi** *Decorations* Petals dried, ground, mixed with yellow cornmeal, and used as a face powder in women's basket dance. (200:96)

Helianthus bolanderi Gray, Serpentine Sunflower
- **Food**—**Paiute** *Unspecified* Species used for food. (167:243)

Helianthus cusickii Gray, Cusick's Sunflower
- **Drug**—**Paiute** *Heart Medicine* Infusion of roots taken for heart troubles. *Tuberculosis Remedy* Infusion of roots taken for tuberculosis. (111:116) **Shasta** *Analgesic* Pounded roots used in a steam bath for internal pain. *Carminative* Decoction of smashed roots taken for gas. *Dermatological Aid* Poultice of roots applied to swellings. *Disinfectant* Root burned in the house after a death. *Febrifuge* Poultice of roots applied for chills and fever. Root burned in the house for long, slow sickness with chills and fever. *Herbal Steam* Pounded roots used in a steam bath for internal pain. *Preventive Medicine* Roots burned to keep away disease. (93:340)
- **Food**—**Paiute, Northern** *Unspecified* Roots peeled and eaten raw. (59:43)

Helianthus decapetalus L., Thinleaf Sunflower
- **Drug**—**Meskwaki** *Dermatological Aid* Poultice of macerated root applied to sores of long standing. (152:215)

Helianthus giganteus L., Giant Sunflower
- **Drug**—**Cherokee** *Nose Medicine* Dry powder sprinkled to induce sneezing. (80:58)

Helianthus grosseserratus Martens, Sawtooth Sunflower
- **Drug**—**Meskwaki** *Burn Dressing* Poultice of blossoms used for burns. (152:215)

Helianthus maximiliani Schrad., Maximilian Sunflower
- **Food**—**Sioux** *Unspecified* Tubers were dug and eaten. (19:13)

Helianthus niveus ssp. *canescens* (Gray) Heiser, Showy Sunflower
- **Drug**—**Keres, Western** *Hemostat* Stem juice applied to open bleeding wounds. (as *H. canus* 171:47)

Helianthus nuttallii Torr. & Gray, Nuttall's Sunflower
- **Drug**—**Navajo** *Gastrointestinal Aid* Infusion of dried, crushed leaves taken for stomach troubles. (55:87)

Helianthus occidentalis Riddell, Fewleaf Sunflower
- **Drug**—**Ojibwa, South** *Dermatological Aid* Poultice of crushed root applied to "bruises and contusions." (91:199)

Helianthus petiolaris Nutt., Prairie Sunflower
- **Drug**—**Hopi** *Dermatological Aid* Plant used as a "spider medicine." (200:32, 96) *Other* Used as a spider medicine. (42:324) **Navajo, Ramah** *Hunting Medicine* Cold infusion of flowers sprinkled on clothing for good luck in hunting. *Panacea* Cold infusion of whole plant used as "life medicine." (191:52) **Thompson** *Dermatological Aid* Powdered leaves alone or in ointment used on sores and swellings. (164:469)
- **Food**—**Havasupai** *Dried Food* Seeds sun dried and stored for winter use. (197:248) *Preserves* Seeds parched, ground, kneaded into seed butter, and eaten with fruit drinks or spread on bread. *Staple* Seeds ground and eaten as a ground or parched meal. (197:67) **Hopi** *Fodder* Used as an important food for summer birds. (200:96)
- **Other**—**Hopi** *Ceremonial Items* Dried petals ground and mixed with cornmeal to make yellow face powder for women's basket dance. *Decorations* Whole plant used in the decoration of flute priests in the Flute ceremony. (42:324) Petals dried, ground, mixed with yellow cornmeal, and used as a face powder in women's basket dance. (200:96) *Season Indicator* Amount of flowers present used as a sign that there will be copious rains and abundant harvest. (42:324)

***Helianthus* sp.,** Sunflower
- **Drug**—**Cheyenne** *Ceremonial Medicine* Flower heads used in the Massaum ceremony. (83:21)
- **Food**—**Apache, Western** *Bread & Cake* Seeds ground, mixed with cornmeal, put into hot water and eaten as a pasty bread. *Candy* Seeds parched and ground with mescal to taste like candy. *Porridge* Seeds made into meal, mixed with cornmeal, and boiled with salt into a cereal. *Special Food* Seeds ground and used by army scouts as rations. (26:184) **Havasupai** *Bread & Cake* Seeds ground, made into small cakes, and baked for a short time. (197:65) **Hopi** *Dried Food* Seeds dried, cracked, and eaten like nuts after dyes were obtained from them. (200:97) **Hualapai** *Unspecified* Seeds used for food. *Winter Use Food* Seeds stored for winter use. (195:2) **Thompson** *Unspecified* Seeds eaten in quantities, especially by children. (164:492)
- **Dye**—**Hopi** *Black* Seeds used to make a black textile and basketry dye. *Purple* Seeds used to make a purple dye for basketry and textiles. (200:97) **Hualapai** *Black* Seeds used to make a black dye. *Purple* Seeds used to make a purple dye. (195:2)
- **Other**—**Hopi** *Ceremonial Items* Seeds used to make a ceremonial body paint. (200:97)

Helianthus strumosus L., Paleleaf Woodland Sunflower
- **Drug**—**Iroquois** *Anthelmintic* Decoction of roots given to children and adults with worms. *Pediatric Aid* Decoction of roots given to children with worms. (87:469) **Meskwaki** *Pulmonary Aid* Infusion of root taken for lung troubles. (152:215)

Helianthus tuberosus L., Jerusalem Artichoke
- **Food**—**Cherokee** *Vegetable* Root used as a vegetable food. (126:34) **Cheyenne** *Unspecified* Species used for food. (83:45) **Chippewa** *Unspecified* Roots eaten raw like a radish. (53:319) **Dakota** *Unspecified* Tubers boiled and sometimes fried after boiling for food. Overuse of these tubers was said to cause flatulence. (69:369) **Hopi** *Unspecified* Tubers eaten in the spring. (200:97) **Huron** *Starvation Food* Roots used with acorns during famine. (3:63) **Iroquois** *Unspecified* Roots used raw, boiled, or fried. (196:120) **Lakota** *Starvation Food* Dried and eaten during famines. *Unspecified* Eaten fresh. (106:47) Stalks and tubers used for food. (139:38) **Malecite** *Unspecified* Species used for food. (160:6) **Micmac** *Unspecified* Tubers eaten. (159:258) **Omaha** *Fruit* Fruits eaten raw. (58:341) *Unspecified* Tubers used as a common food article. (68:325) Noncultivated tubers eaten raw, boiled, or roasted. **Pawnee** *Unspecified* Noncultivated, raw tubers used for food. **Ponca** *Unspecified* Noncultivated tubers eaten raw, boiled, or roasted. (70:131) **Potawatomi** *Unspecified* Roots gathered for foodstuffs. (154:98) **Winnebago** *Unspecified* Noncultivated tubers eaten raw, boiled, or roasted. (70:131)

***Heliomeris*,** Asteraceae

Heliomeris longifolia* var. *annua (M. E. Jones) Yates, Longleaf Falsegoldeneye
- **Drug**—**Navajo, Ramah** *Panacea* Plant used as "life medicine." (as *Viguiera annua* 191:54)
- **Food**—**Navajo, Ramah** *Fodder* Used for sheep and goat feed. (as *Viguiera annua* 191:54)

Heliomeris longifolia (Robins. & Greenm.) Cockerell **var. *longifolia*,** Longleaf Falsegoldeneye
- **Drug**—**Navajo, Ramah** *Panacea* Plant used as "life medicine." (as *Viguiera longifolia* 191:54)
- **Food**—**Navajo, Ramah** *Fodder* Used for sheep feed. (as *Viguiera longifolia* 191:54)

Heliomeris multiflora Nutt., Showy Goldeneye
- **Food**—**Gosiute** *Unspecified* Seeds formerly used for food. (as *Gymnolomia multiflora* 39:371)

Heliomeris multiflora Nutt. **var. *multiflora*,** Showy Goldeneye
- **Drug**—**Navajo, Ramah** *Witchcraft Medicine* Plant designated as a witchcraft plant. (as *Viguiera multiflora* 191:54)
- **Food**—**Navajo, Ramah** *Fodder* Used for sheep and deer feed. (as *Viguiera multiflora* 191:54)

***Heliopsis*,** Asteraceae

Heliopsis helianthoides* var. *scabra (Dunal) Fern., Smooth Oxeye
- **Drug**—**Chippewa** *Stimulant* Decoction of dried root or chewed fresh root spit on limbs as stimulant. (as *H. scabra* 53:364) **Meskwaki** *Pulmonary Aid* Root used for lung troubles. (as *H. scabra* 152:215)

***Heliotropium*,** Boraginaceae

Heliotropium convolvulaceum (Nutt.) Gray, Phlox Heliotrope
- **Food**—**Navajo, Kayenta** *Porridge* Seeds made into mush and used for food. (205:40)

Heliotropium curassavicum L., Salt Heliotrope
- **Drug**—**Paiute** *Antidiarrheal* Plant used as a diarrhea medicine. (167:317) *Diuretic* Decoction of plant or roots taken in cases of "retention of urine." *Emetic* Decoction of root taken as an emetic. *Throat Aid* Decoction of root gargled for sore throat. (180:84, 85) **Pima** *Dermatological Aid* Poultice of dried, pulverized root applied to sores and

wounds. (146:79) **Shoshoni** *Diuretic* Decoction of plant or roots taken in cases of "retention of urine." *Emetic* Decoction of root taken as an emetic. *Misc. Disease Remedy* Decoction of plant tops taken to aid in "bringing out" measles. *Throat Aid* Decoction of root gargled for sore throat. *Venereal Aid* Decoction of plant taken for venereal disease. (180:84, 85) **Tubatulabal** *Antidiarrheal* Decoction of entire plant taken for bloody flux. (193:59)
- *Food*—Tubatulabal *Unspecified* Seeds used extensively for food. (193:15)

Helvellaceae, see *Morchella*

Hemizonia, Asteraceae

Hemizonia clevelandii Greene, Cleveland's Tarweed
- *Food*—Pomo *Staple* Seeds used to make pinoles. (11:86)

Hemizonia corymbosa (DC.) Torr. & Gray, Coastal Tarweed
- *Food*—Costanoan *Staple* Seeds eaten as a pinole. (21:254)
- *Other*—Costanoan *Hunting & Fishing Item* Foliage burned to drive ground squirrels from burrows. (21:254)

Hemizonia fasciculata (DC.) Torr. & Gray, Clustered Tarweed
- *Food*—Cahuilla *Starvation Food* Whole plant, including the seeds, used as a famine plant. (15:77)

Hemizonia fitchii Gray, Fitch's Tarweed
- *Food*—Miwok *Unspecified* Seeds used to make mush. (as *Centromadia fitchii* 12:153)

Hemizonia luzulifolia DC., Hayfield Tarweed
- *Food*—Mendocino Indian *Staple* Seeds used as an important source of pinole. (41:394) **Pomo** *Staple* Seeds used to make pinoles. (11:86)

***Hemizonia* sp.**, Tarweed
- *Food*—Wintoon *Staple* Seeds used to make pinole. (117:274)

Hepatica, Ranunculaceae

Hepatica nobilis* var. *acuta (Pursh) Steyermark, Sharplobe Hepatica
- *Drug*—Cherokee *Analgesic* Infusion of plant taken as an emetic for abdominal pains. (as *H. acutiloba* 177:22) *Breast Treatment* Compound used for swollen breasts. (as *H. acutiloba* 80:38) *Emetic* Infusion of plant taken as an emetic for abdominal pains. (as *H. acutiloba* 177:22) *Gastrointestinal Aid* Compound decoction used for poor digestion. (as *H. acutiloba* 80:38) Infusion of plant taken as an emetic for abdominal pains. (as *H. acutiloba* 177:22) *Laxative* Infusion used as a laxative. *Liver Aid* Infusion used for the liver. (as *H. acutiloba* 80:38) **Iroquois** *Analgesic* Decoction of plants taken by pregnant women with side or labor pains. *Blood Medicine* Plant used as a blood purifier. *Contraceptive* Infusion of plants taken to prevent conception. *Gynecological Aid* Compound used for labor pains in parturition. Decoction of plants taken by middle-aged women to induce childbirth, and by pregnant women for sore abdomen or side pains. *Orthopedic Aid* Compound decoction of plants taken for stiff muscles. *Other* Compound decoction of roots given to children with "summer complaint." Roots used to tell fortune. *Pediatric Aid* Compound decoction of roots given to children with "summer complaint." (87:328) *Pulmonary Aid* Infusion of whole plant and another plant taken by forest runners with shortness of breath. (as *H. acutiloba* 141:42) *Witchcraft Medicine* "Chewed by women to bewitch men and make them crazy by affecting their hearts." (87:328) **Menominee** *Gynecological Aid* Compound containing root used for female maladies, especially leukorrhea. (as *H. acutiloba* 151:48, 49) **Meskwaki** *Eye Medicine* and *Other* Infusion of root taken and used as a wash for twisted mouth or crossed eyes. (as *H. acutiloba* 152:239)

Hepatica nobilis* var. *obtusa (Pursh) Steyermark, Roundlobed Hepatica
- *Drug*—Chippewa *Abortifacient* Decoction of roots taken for amenorrhea. (as *H. triloba* 71:129) *Anticonvulsive* Decoction of root taken, especially by children, for convulsions. (as *H. americana* 53:336) *Dermatological Aid* Poultice of plants applied to inflammations and bruises. (as *H. triloba* 71:129) *Hunting Medicine* Roots used as charms on traps for fur-bearing animals. (as *H. triloba* 53:376) *Liver Aid* Plant used for liver ailments. (as *H. triloba* 71:129) *Pediatric Aid* Decoction of root taken for convulsions, "used chiefly for children." (as *H. americana* 53:336) **Menominee** *Antidiarrheal* Compound decoction of root used for dysentery. (as *H. triloba* 54:131) **Nanticoke** *Febrifuge* Petals chewed "to prevent fever in summer." (as *H. americana* 175:56, 84) **Potawatomi** *Other* Infusion of root and leaves taken for vertigo. (as *H. triloba* 154:74)
- *Dye*—Potawatomi *Unspecified* Roots used to make a dye for mats and baskets. (as *H. triloba* 154:123)

Heracleum, Apiaceae

Heracleum maximum Bartr., Common Cow Parsnip
- *Drug*—Aleut *Cold Remedy* Leaves used to make a tonic for colds. (as *H. lanatum* 8:427) *Dermatological Aid* Poultice of heated leaves applied to minor cuts. *Orthopedic Aid* Poultice of heated leaves applied to sore muscles. (as *H. lanatum* 8:425) *Throat Aid* Leaves used to make a soothing drink for sore throats. (as *H. lanatum* 8:427) **Bella Coola** *Analgesic* Poultice of compound containing roots used for lung or hip pains. *Antirheumatic (External)* Compound infusion of root used as poultice for pains like rheumatism. *Dermatological Aid* Poultice of crushed, boiled root, baked root, or raw root applied to boils. (as *H. lanatum* 150:61) Poultice of crushed and cooked roots applied to boils. (as *H. lanatum* 184:201) *Orthopedic Aid* Poultice of compound containing roots used for hip pains. *Pulmonary Aid* Poultice of compound containing roots used for lung pains. (as *H. lanatum* 150:61) **Blackfoot** *Antidiarrheal* Infusion of fresh, young stems taken for diarrhea. (as *H. lanatum* 86:67) *Dermatological Aid* Infusion of young stems applied in the removal of warts. (as *H. lanatum* 86:76) Poultice of roots applied to bruises and chronic swellings. (as *H. lanatum* 97:48) **California Indian** *Antirheumatic (Internal)* Strong decoction of root used for rheumatism. (as *H. lanatum* 19:13) **Carrier** *Antirheumatic (External)* Poultice of ground roots applied for rheumatism. (as *H. lanatum* 31:82) **Carrier, Northern** *Dermatological Aid* Poultice of root applied to swellings and bruises. (as *H. lanatum* 150:61) **Chippewa** *Dermatological Aid* Poultice of boiled or dried root and flowers applied to boils. (as *H. lanatum* 53:350) *Throat Aid* Decoction of root gargled or dried root chewed for ulcerated sore throat. (as *H. lanatum* 53:342) **Cree** *Dermatological Aid* Powdered roots and lard used as an ointment or poultice of root paste applied

to boils and swellings. (as *H. lanatum* 16:492) *Poison* Plant considered poisonous. *Toothache Remedy* Root held on the sore tooth for toothaches. (as *H. lanatum* 16:491) *Venereal Aid* Powdered roots and lard used as ointment or root paste poultice applied to venereal disease chancres. (as *H. lanatum* 16:494) **Cree, Woodlands** *Analgesic* Poultice of ground root, calamus, and yellow pond lily applied to the head for severe headaches. Decoction of root, calamus, and yellow pond lily used as a wash for severe headaches. *Antirheumatic (External)* Poultice of ground root, calamus, and yellow pond lily applied to painful limbs. Decoction of root, calamus, and yellow pond lily used as a wash for painful limbs. *Dermatological Aid* Poultice of root, calamus, root and yellow pond lily root applied to mancos, worms in the flesh. (as *H. lanatum* 109:40) **Gitksan** *Antirheumatic (External)* Poultice of fresh roots used for rheumatism. (as *H. lanatum* 74:25) Poultice of mashed root applied to rheumatic or other swellings. *Dermatological Aid* Poultice of mashed root applied to boils and other swellings. (as *H. lanatum* 150:61) *Witchcraft Medicine* Roots, red elder bark, and juniper boughs used as a smudge for evil witchcraft victims. (as *H. lanatum* 74:25) **Haisla** *Dermatological Aid* Poultice of roots, Indian hellebore, and Sitka spruce pitch applied to wounds. (as *H. lanatum* 43:214) **Iroquois** *Analgesic* Compound infusion of plant used as steam bath for headaches. *Antirheumatic (External)* Compound infusion of plant used as steam bath for rheumatism. *Dermatological Aid* Compound decoction used as wash or poultice applied to chancres or lumps on penis. *Diaphoretic* Infusion of plant used as steam bath to sweat out rheumatism and headaches. *Gastrointestinal Aid* Compound decoction of roots taken for bruises on the back of the stomach. *Hunting Medicine* Decoction of roots used as wash for rifles, a "hunting medicine." (as *H. lanatum* 87:400) *Misc. Disease Remedy* Plant used for influenza. (as *H. lanatum* 141:56) **Karok** *Poison* Roots poisonous to cattle. (as *H. lanatum* 148:387) **Klamath** *Unspecified* Roots used medicinally for unspecified purpose. (as *H. lanatum* 45:102) **Kwakiutl** *Dermatological Aid* Dried, pounded roots and oil used as a hair ointment. *Gynecological Aid* and *Pediatric Aid* Dried, pounded roots and oil rubbed on face and waist of girl at puberty. (as *H. lanatum* 183:276) **Makah** *Eye Medicine* Heated poultice of leaves applied for eye problems. *Tonic* Used as a spring tonic. *Unspecified* Central stalk considered strong medicine. (as *H. lanatum* 67:293) **Malecite** *Misc. Disease Remedy* Infusion of root shoots used for smallpox. Infusion used for cholera. (as *H. lanatum* 116:256) **Menominee** *Hunting Medicine* Herb used in the hunting bundle and smudged for 4 days to remove the charm. (as *H. lanatum* 151:81) *Witchcraft Medicine* An evil medicine used by sorcerers. (as *H. lanatum* 151:55) **Meskwaki** *Analgesic* Seeds used for severe headache and root used for stomach cramps. *Dermatological Aid* Poultice of stems applied to wounds. *Gastrointestinal Aid* Root used for colic or any kind of stomach cramps. *Misc. Disease Remedy* Infusion of root used for erysipelas. (as *H. lanatum* 152:249) **Mewuk** *Antirheumatic (External)* Poultice of mashed roots applied to swellings. *Misc. Disease Remedy* Used for mumps. (as *H. lanatum* 117:366) **Micmac** *Misc. Disease Remedy* Root used for smallpox and cholera. (as *H. lanatum* 40:57) **Ojibwa** *Dermatological Aid* Poultice of pounded, fresh root applied to sores. (as *H. lanatum* 153:390) *Hunting Medicine* Root or seeds used to smudge a fire and drive away a bad spirit from the camp of the hunter. There is a bad spirit who is always present trying to steal away one's luck in hunting game. He must be driven away from the camp of the hunter by smudging a fire with the roots or seeds. This gets into the spirit's eyes and he cannot see the hunter leave the camp, so naturally does not follow and bother him. (as *H. lanatum* 153:432) **Okanagan-Colville** *Dermatological Aid* Decoction of roots, red willow, and chokecherry branches used as a cleansing medicine for the scalp. Decoction of branches used as a hair tonic to prevent gray hair and dandruff. *Orthopedic Aid* Heated poultice of sliced, pounded roots applied to sore backs. (as *H. lanatum* 188:62) **Okanagon** *Cathartic* Decoction of roots taken as a purgative. *Tonic* Decoction of roots taken as a tonic. (as *H. lanatum* 125:40) **Omaha** *Analgesic* Decoction of root taken for intestinal pains. *Cathartic* Decoction of root taken as a physic. *Gastrointestinal Aid* Decoction of root taken for intestinal pains. (as *H. lanatum* 70:107) **Paiute** *Antirheumatic (External)* Poultice of mashed root applied for rheumatism. (as *H. lanatum* 180:85, 86) *Cold Remedy* Decoction of roots taken for colds. (as *H. lanatum* 104:197) *Dermatological Aid* Roots used as a salve for sores. (as *H. lanatum* 104:196) Salve made from root applied to wounds. (as *H. lanatum* 180:85, 86) **Paiute, Northern** *Antirheumatic (External)* Poultice of roasted, split plants applied to aching joints for rheumatism. (as *H. lanatum* 59:130) **Pawnee** *Dermatological Aid* Poultice of scraped, boiled root applied to boils. (as *H. lanatum* 70:107) **Pomo** *Antirheumatic (External)* Decoction of plant used as a wash for rheumatism. Poultice of pounded, raw, or heated roots applied to rheumatism. *Dermatological Aid* Decoction of plant used as a wash for swellings. Poultice of pounded, raw, or heated roots applied to swellings. (as *H. lanatum* 66:14) **Pomo, Kashaya** *Antirheumatic (External)* Poultice of baked, pounded root used for rheumatism, arthritis, and other muscular pains. (as *H. lanatum* 72:87) **Quinault** *Analgesic* Poultice of warmed leaves applied to sore limbs. *Orthopedic Aid* Poultice of warmed leaves applied to sore limbs. (as *H. lanatum* 79:42) **Salish, Coast** *Dermatological Aid* Roots pounded, roasted, mixed with dogfish oil and used as a hair lotion to make hair grow long. (as *H. lanatum* 182:89) **Sanpoil** *Analgesic* Poultice of roots applied overnight to "painful parts, sore eyes, etc." *Dermatological Aid* Pounded root mixed with water and used as a hair wash for dandruff. *Eye Medicine* Poultice of roots applied overnight to "painful parts, sore eyes, etc." (as *H. lanatum* 131:220) **Shoshoni** *Cold Remedy* Decoction of root in whisky taken and smoke of root compound inhaled for colds. *Cough Medicine* Decoction of root in whisky taken for colds and coughs. *Throat Aid* Infusion of mashed root gargled and poultice applied for sore throat. *Toothache Remedy* Raw root placed in cavities for toothaches. *Tuberculosis Remedy* Decoction of root taken for tuberculosis. (as *H. lanatum* 180:85, 86) **Shuswap** *Dermatological Aid* Infusion of roots taken for sores. *Internal Medicine* Infusion of roots taken to kill all the internal germs. *Urinary Aid* Infusion of roots taken for the bladder. (as *Meracleum lanatum* 123:56) **Sikani** *Analgesic* Poultice of mashed roots applied to swellings of neuralgia or rheumatism. *Antirheumatic (External)* Poultice of mashed roots applied to swellings of rheumatism. *Dermatological Aid* Poultice of mashed roots applied to swellings of neuralgia. (as *H. lanatum* 150:61) **Tanaina** *Unspecified* Root used as a medicine. (as *H. lanatum* 149:329) **Thompson** *Cathartic* Decoction of roots taken as a purgative. (as *H. lanatum* 125:40) Decoction of root used as a purgative and tonic. (as *H. lanatum*

164:457) Decoction of roots taken by warriors and hunters as a purgative. (as *H. lanatum* 164:504) *Ceremonial Medicine* and *Disinfectant* Decoction of root used ceremonially as a wash for purification. (as *H. lanatum* 164:457) Decoction of roots taken by warriors and hunters as a purifier. (as *H. lanatum* 164:504) *Tonic* Decoction of roots taken as a tonic. (as *H. lanatum* 125:40) Decoction of root used as a tonic and purgative. (as *H. lanatum* 164:457) *Unspecified* Plant used medicinally. (as *H. lanatum* 187:152) *Venereal Aid* Strong decoction of root used for syphilis. (as *H. lanatum* 164:457) **Washo** *Antidiarrheal* Decoction of root taken for diarrhea. *Toothache Remedy* Raw root placed in cavities for toothaches. (as *H. lanatum* 180:85, 86) **Winnebago** *Anticonvulsive* Plant tops used in smoke treatment for convulsions. *Stimulant* Plant tops used in smoke treatment for fainting. (as *H. lanatum* 70:107)

- **Food**—**Alaska Native** *Unspecified* Inner stem pulp eaten raw and often dipped in seal oil. (as *H. lanatum* 85:133) **Anticosti** *Forage* Whole plant eaten by cows. (as *H. lanatum* 143:67) **Bella Coola** *Unspecified* Young stems peeled and eaten with grease. (as *H. lanatum* 184:201) **Blackfoot** *Soup* Stem pieces dipped in blood, stored, and used to make soup and broths. (as *H. lanatum* 86:103) *Unspecified* Stalks roasted over hot coals and eaten. (as *H. lanatum* 114:277) *Vegetable* Young plant stems peeled and eaten like celery. (as *H. lanatum* 86:103) **California Indian** *Vegetable* Young, raw shoots eaten like celery. (as *H. lanatum* 19:13) **Carrier** *Unspecified* Young growth used for food. (as *H. lanatum* 31:82) **Coeur d'Alene** *Unspecified* Growing stalks used for food. (as *H. lanatum* 178:91) **Costanoan** *Unspecified* Boiled roots and foliage used for food. (as *H. spondylium* ssp. *montanum* 21:251) **Cree, Woodlands** *Unspecified* Leaf petiole peeled and eaten fresh. Pith scraped out of the roasted, main stem and eaten. (as *H. lanatum* 109:40) **Gitksan** *Unspecified* Stems used for food in spring. (as *H. lanatum* 73:154) Stalks eaten in spring. (as *H. lanatum* 74:25) **Haisla** *Unspecified* Stems used for food in spring. (as *H. lanatum* 73:154) **Haisla & Hanaksiala** *Unspecified* Petioles considered "the main food in spring." (as *H. lanatum* 43:214) **Hesquiat** *Forage* Young shoots eaten by cattle. *Unspecified* Raw stalks of young leaves and flower buds eaten with sugar or honey. (as *H. lanatum* 185:60) **Hoh** *Vegetable* Young shoots eaten raw as greens. (as *H. lanatum* 137:66) **Karok** *Unspecified* Fresh shoot used for food. (as *H. lanatum* 148:387) **Kitasoo** *Vegetable* Young stems and petioles eaten as a spring vegetable. (as *H. lanatum* 43:326) **Klamath** *Unspecified* Young shoots used for food. (as *H. lanatum* 45:102) **Kwakiutl, Southern** *Unspecified* Young stems and petioles peeled and eaten raw like celery. (as *H. lanatum* 183:276) **Makah** *Unspecified* Fresh petioles peeled, mixed with oil, and used for food. Stems considered a favored food. Plant eaten after peeling. (as *H. lanatum* 67:293) Young tops eaten raw in the spring. Stems used for food. (as *H. lanatum* 79:42) **Mendocino Indian** *Vegetable* Tender leaf and flower stalks eaten as green food in spring and early summer. (as *H. lanatum* 41:373) **Meskwaki** *Vegetable* Potatoes cooked like the rutabaga. (as *H. lanatum* 152:265) **Mewuk** *Unspecified* Young stems peeled and eaten raw. (as *H. lanatum* 117:366) **Montana Indian** *Vegetable* Young, raw shoots eaten like celery. (as *H. lanatum* 19:13) **Nitinaht** *Unspecified* Hollow and solid leafstalks peeled and used for food. (as *H. lanatum* 186:91) **Ojibwa** *Vegetable* Leaves used as greens. (as *H. lanatum* 135:237) **Okanagan-Colville** *Vegetable* Flower stalks and leaf stems peeled and eaten fresh. (as *H. lanatum* 188:62) **Okanagon** *Staple* Growing stalks used as a principal food. (as *H. lanatum* 178:239) *Unspecified* Young flower stalks peeled and eaten raw. (as *H. lanatum* 125:38) **Oweekeno** *Unspecified* Stems and petioles peeled and used for food. (as *H. lanatum* 43:84) **Pomo, Kashaya** *Unspecified* New shoots peeled and eaten raw. (as *H. lanatum* 72:87) **Quileute** *Unspecified* Stems dipped in seal oil and eaten. (as *H. lanatum* 79:42) *Vegetable* Young shoots eaten raw as greens. (as *H. lanatum* 137:66) **Quinault** *Unspecified* Stems dipped in seal oil and eaten. (as *H. lanatum* 79:42) **Salish, Coast** *Unspecified* Young stems and leaf stalks eaten raw or boiled. (as *H. lanatum* 182:89) **Shuswap** *Unspecified* Young stems eaten raw. (as *Meracleum lanatum* 123:56) **Spokan** *Unspecified* Stalks used for food. (as *H. lanatum* 178:344) **Thompson** *Dried Food* Plant formerly dried for storage. (as *H. lanatum* 187:152) *Forage* Stalks used as a common food for cattle. (as *H. lanatum* 164:482) *Frozen Food* Plant frozen for future use. (as *H. lanatum* 187:152) *Unspecified* Young flower stalks peeled and eaten raw. (as *H. lanatum* 125:38) Young stalks peeled and eaten raw. (as *H. lanatum* 164:482) *Vegetable* Peeled shoots eaten as vegetables with meat or fish. Peeled, raw, or cooked leaf stalks and flower stalks used for food. The stalks were ready to use around May and June, but after a while, they became tough, dry, or sticky and were no longer good to eat. The raw stalks would cause a burning like pepper if eaten in too great a quantity; it was better to eat cooked stalks. *Winter Use Food* Plant canned for future use. (as *H. lanatum* 187:152) **Tolowa** *Unspecified* Stem inner layers eaten raw. (as *H. lanatum* 6:32) **Wet'suwet'en** *Unspecified* Stems used for food in spring. (as *H. lanatum* 73:154) **Yuki** *Unspecified* Tender, young stems peeled and eaten raw. (as *H. lanatum* 49:87) **Yurok** *Unspecified* Stem inner layers eaten raw. (as *H. lanatum* 6:32)

- **Fiber**—**Makah** *Basketry* Large blossom stems twined with seaweed, made into baskets, and used by girls for playing. **Quileute** *Basketry* Large blossom stems twined with seaweed, made into baskets, and used by girls for playing. (as *H. lanatum* 79:42)

- **Dye**—**Karok** *Yellow* Roots used as a yellow dye for porcupine quills. (as *H. lanatum* 148:387)

- **Other**—**Blackfoot** *Ceremonial Items* Stalks placed on the altar of the Sun Dance ceremonial. (as *H. lanatum* 114:277) Stalk placed on altar of Sun Dance ceremonial. (as *H. lanatum* 118:50) *Cooking Tools* Hollow stems used by infirm people to suck soup and stew without raising up. *Musical Instrument* Hollow stems used to make children's flutes. *Toys & Games* Hollow stems used to make children's toy blowguns. (as *H. lantum* 86:113) **Carrier, Southern** *Insecticide* Infusion of blossoms rubbed on body to keep off flies and mosquitoes. (as *H. lanatum* 150:61) **Cheyenne** *Musical Instrument* Hollow stems made into whistles and used for romantic purposes at night. (as *H. lanatum* 83:40) **Haisla & Hanaksiala** *Toys & Games* Stems used to make whistles. Plant used to play a game by throwing the plant into a pot. (as *H. lanatum* 43:214) **Menominee** *Protection* Plant smudged to drive away the evil spirit, whose special mission was to steal one's hunting luck. (as *H. lanatum* 151:81) **Nitinaht** *Toys & Games* Swollen leaf sheaths and small, unexpanded leaves used in children's games. (as *H. lanatum* 186:91) **Ojibwa** *Hunting & Fishing Item* Roots boiled and sprinkled on the fishing nets to lure fish. (as *H. lanatum* 153:432) **Omaha** *Ceremonial Items* Pounded, dried roots mixed with beaver dung and planted in the same hole as the sacred pole. (as *H. lana-*

tum 70:107) **Pomo, Kashaya** *Containers* Hollow stems used to carry water. *Toys & Games* Dried, hollow stems used as toy blowguns to shoot berries or small pebbles. (as *H. lanatum* 72:87) **Shuswap** *Containers* Leaves used to cover a basket of berries. (as *Meracleum lanatum* 123:56) **Tsimshian** *Hunting & Fishing Item* Petioles rubbed on fishing gear to insure success in fishing. (as *H. lanatum* 43:326)

Heracleum sp.
- **Food**—**Aleut** *Unspecified* Species used for food. (7:29)

Heracleum sphondylium L., Eltrot
- **Drug**—**Micmac** *Gynecological Aid* Green and light-color plant used as medicine for women. (194:30) *Unspecified* Part of plant considered "good medicine." (40:57) *Urinary Aid* Dark and ripe plant used as medicine for men. (194:30)

Hericium, Hydnaceae, fungus

Hericium coralloides, Coral Mushroom
- **Food**—**Pomo, Kashaya** *Vegetable* Baked on hot stones, in the oven, or fried. (72:129)

Hesperocallis, Liliaceae

Hesperocallis undulata Gray, Desert Lily
- **Food**—**Cahuilla** *Unspecified* Bulbs eaten raw or oven pit baked. (15:77) **Yuma** *Unspecified* Bulbs eaten raw, baked, or boiled. (37:207)

Heteranthera, Pontederiaceae

Heteranthera reniformis Ruiz & Pavón, Kidneyleaf Mudplantain
- **Drug**—**Cherokee** *Dermatological Aid* Hot poultice of root applied to inflamed wounds and sores. (80:45)

Heteromeles, Rosaceae

Heteromeles arbutifolia (Lindl.) M. Roemer, Toyon
- **Drug**—**Costanoan (Olhonean)** *Abortifacient* Infusion of leaves taken "for suppression of menses or irregular menses of girls." (117:373) **Diegueño** *Dermatological Aid* Infusion of bark and leaves used as wash for infected wounds. (88:217) **Mendocino Indian** *Analgesic* Decoction of leaves taken for various aches and pains. *Gastrointestinal Aid* Decoction of leaves taken for stomachaches. (41:355)
- **Food**—**Cahuilla** *Fruit* Berries eaten cooked and raw. (15:77) **Costanoan** *Fruit* Fruits eaten toasted or dried. (21:249) **Diegueño** *Fruit* Fruit used for food. (88:217) **Karok** *Fruit* Berries roasted over an open fire and eaten. (6:32) **Luiseño** *Dried Food* Parched berries used for food. (155:194) **Mahuna** *Fruit* Berries eaten mainly to quench the thirst. (140:70) **Mendocino Indian** *Fruit* Fruits eaten fresh. Fruits boiled or roasted and used for food. (41:355) **Pomo, Kashaya** *Fruit* Berries wilted in hot ashes and winnowed in a basket plate. (72:115) **Yurok** *Fruit* Berries roasted over an open fire and eaten by children. (6:32)

Heteromeles arbutifolia M. Roem. var. *arbutifolia*, Toyon
- **Food**—**Karok** *Fruit* Berries put on a basket plate in front of the fire, turned until wilted and eaten. (as *Photinia arbutifolia* 148:385) **Neeshenam** *Fruit* Bright, red berries used for food. (as *Photinea arbutifolia* 129:375) **Pomo** *Fruit* Wilted, winnowed berries used for food. (as *Photinia arbutifolia* 66:13)
- **Other**—**Karok** *Toys & Games* Leaves thrown into the fire by children to hear them crack like firecrackers. (as *Photinia arbutifolia* 148:385)

Heterotheca, Asteraceae

Heterotheca grandiflora Nutt., Telegraphweed
- **Other**—**Luiseño** *Hunting & Fishing Item* Tall stems sometimes used to make arrow main shafts. (155:228) Stems used to make small arrows. (155:206)

Heterotheca villosa var. *hispida* (Hook.) Harms, Bristly Hairy Goldaster
- **Drug**—**Isleta** *Poison* Plant, when touched, caused a skin irritation similar to ant bites. (as *Chrysopsis hirsutissima* 100:25) **Navajo, Ramah** *Dermatological Aid* Poultice of leaves applied to ant bites. *Nose Medicine* Poultice of leaves applied to sore nose. *Toothache Remedy* Poultice of root applied for toothache. (as *Chrysopsis hispida* 191:49)

Heterotheca villosa (Pursh) Shinners var. *villosa*, Hairy Goldenaster
- **Drug**—**Cheyenne** *Disinfectant* Plant burned as incense to remove evil spirits from the house. (as *Chrysopsis villosa* 83:20) *Sedative* Infusion of tops and stems taken for feeling poorly and made one sleepy. (as *Chrysopsis foliosa* 76:187) **Hopi** *Analgesic* Infusion of leaves and flowers used for chest pain. (as *Chrysopsis villosa* 200:95) **Navajo, Kayenta** *Ceremonial Medicine* Plant used in the corral dance. (as *Chrysopsis villosa* 205:46) **Navajo, Ramah** *Ceremonial Medicine* Plant used as a ceremonial emetic and chant lotion. *Emetic* Plant used alone as a sweat house emetic. Plant used as a ceremonial and sweat house emetic for various ailments. *Gastrointestinal Aid* Cold infusion of leaves used to kill a swallowed red ant. Plant used as a sweat house emetic for indigestion. *Heart Medicine* Plant used as an "aorta medicine." *Panacea* Root used as a "life medicine." *Toothache Remedy* Poultice of heated root applied for toothache. *Venereal Aid* Plant used as a sweat house emetic for sexual infection. (as *Chrysopsis villosa* 191:49)
- **Food**—**Navajo, Ramah** *Fodder* Used for sheep feed. (as *Chrysopsis villosa* 191:49)
- **Other**—**Navajo, Ramah** *Insecticide* Infusion thrown on anthill to kill red ants. (as *Chrysopsis foliosa* 191:49)

Heuchera, Saxifragaceae

Heuchera americana L., American Alumroot
- **Drug**—**Cherokee** *Antidiarrheal* Taken for dysentery. *Dermatological Aid* Powdered root used on malignant ulcers and infusion sprinkled on bad sores. Infusion of astringent root taken for bowel complaints. *Gastrointestinal Aid* Infusion of root taken for bowel complaints. *Gynecological Aid* Used for "immoderate flow of menses." *Hemorrhoid Remedy* Infusion of root taken for piles. *Oral Aid* Infusion used for "thrash" and sore mouth and root chewed to take coat off tongue. (80:23) **Chickasaw** *Dermatological Aid* Root used as a powerful astringent. *Tonic* Root used as a tonic. **Choctaw** *Dermatological Aid* Root used as a powerful astringent. *Tonic* Root used as a tonic. **Creek** *Dermatological Aid* Root used as a powerful astringent. *Tonic* Root used as a tonic. (28:286, 287) **Menominee** *Analgesic* Compound decoction of root used for stomach pain. *Gastrointestinal Aid* Raw root eaten for "disordered stomach." (54:130) **Meskwaki** *Dermatological Aid* Foliage used as an astringent for sores. *Panacea* Compound containing root used as a "healer." (152:246)

Heuchera americana var. hispida (Pursh) E. Wells, Hairy Alumroot
- **Drug**—**Chippewa** *Analgesic* Dried root chewed and juice swallowed for stomach pain. (as *H. hispida* 53:344) *Eye Medicine* Decoction of root used as a wash for sore eyes. (as *H. hispida* 53:360) *Gastrointestinal Aid* Dried root chewed and juice swallowed for stomach pain. (as *H. hispida* 53:344) **Menominee** *Antidiarrheal* Infusion of root used for diarrhea. (as *H. hispida* 151:53) **Sioux, Teton** *Antidiarrheal* Root, very powerful and small dose for children, used for chronic diarrhea. (as *H. hispida* 52:269)

Heuchera bracteata (Torr.) Ser., Bracted Alumroot
- **Drug**—**Navajo** *Gastrointestinal Aid* Plant chewed for indigestion. *Oral Aid* Plant chewed for sore gums. *Toothache Remedy* Compound poultice of crushed leaves applied to toothaches. (55:52)
- **Dye**—**Navajo** *Red-Brown* Stems used to make a pinkish tan dye. (55:52)

Heuchera cylindrica Dougl. ex Hook., Roundleaf Alumroot
- **Drug**—**Blackfoot** *Antidiarrheal* Decoction of roots used for diarrhea. *Dermatological Aid* Decoction of roots used as an astringent. (97:36) **Flathead** *Antidiarrheal* Roots infused or chewed for diarrhea. *Gastrointestinal Aid* Roots infused or chewed for stomach cramps. **Kutenai** *Antirheumatic (External)* Decoction of roots used for "aching bones." *Tuberculosis Remedy* Decoction of roots taken for tuberculosis. (82:31) **Okanagan-Colville** *Blood Medicine* Decoction of roots and Oregon grape roots used as a tonic for the "changing of the blood." *Dermatological Aid* Poultice of mashed, peeled roots applied to sores and cuts. Infusion of roots used to wash sores and cuts. Roots mixed with puffball spores and used as a salve for diaper rash. *Pediatric Aid* Decoction of roots used, especially for children and babies, to rinse out the mouth for sore throats. Roots mixed with puffball spores and used as a salve for diaper rash. *Throat Aid* Fresh root held in the mouth and sucked for sore throats. Decoction of roots used, especially for children and babies, to rinse out the mouth for sore throats. (188:138) **Shuswap** *Antidiarrheal* Decoction of leaves and roots taken for diarrhea. *Dermatological Aid* Decoction of leaves and roots used as a wash for sores. (123:68) **Thompson** *Dermatological Aid* Chewed leaves and roots spat on sores or wounds. Poultice of root with Douglas fir pitch used for wounds. *Liver Aid* Infusion of root taken for liver trouble. *Oral Aid* Small, peeled, cleaned root piece chewed for mouth sores and gum boils. *Throat Aid* Infusion of root taken for sore throats. *Unspecified* Root used for medicine. (187:282)

Heuchera cylindrica var. alpina Sw., Alpine Alumroot
- **Drug**—**Cheyenne** *Antirheumatic (External)* Powdered roots rubbed on the skin for rheumatism or sore muscles. *Antirheumatic (Internal)* Infusion of powdered plant tops taken for rheumatism or sore muscles. (as *H. ovalifolia* 76:176) *Dermatological Aid* Poultice of powdered roots applied for poison ivy and other skin rashes. (as *H. ovalifolia* 83:38) *Orthopedic Aid* Infusion of roots taken or powdered roots rubbed on skin for sore muscles. (as *H. ovalifolia* 76:176)

Heuchera cylindrica var. glabella (Torr. & Gray) Wheelock, Beautiful Alumroot
- **Drug**—**Arapaho** *Unspecified* Roots used medicinally for unspecified purpose. **Blackfoot** *Dermatological Aid* Pounded roots used for sores. (as *H. glabella* 121:47) *Eye Medicine* Infusion of root used as an eyewash. (as *H. glabella* 118:39) *Snakebite Remedy* Poultice of mashed, raw root applied to snakebites. *Veterinary Aid* Poultice of mashed, raw root applied to horses for snakebites. (*H. glabella* 118:49)
- **Dye**—**Blackfoot** *Mordant* Root added to dye baths to set the color in native dress. (as *H. glabella* 121:47)

Heuchera flabellifolia Rydb., Bridger Mountain Alumroot
- **Drug**—**Blackfoot** *Gastrointestinal Aid* Decoction of roots taken for stomach troubles and cramps. (97:36)

Heuchera glabra Willd. ex Roemer & J. A. Schultes, Alpine Heuchera
- **Drug**—**Tlingit** *Venereal Aid* Plant used for inflammation of testicles from syphilis. (as *H. devaricata* 107:284)

Heuchera micrantha Dougl. ex Lindl., Crevice Alumroot
- **Drug**—**Skagit** *Dermatological Aid* Pounded plants rubbed on hair to make it grow or applied to cuts. (79:31) **Thompson** *Dermatological Aid* Poultice of mashed root with Douglas fir pitch used for wounds. The poultice was covered with a cloth and when it was taken off, all the poison was extracted from the open wound. Chewed leaves and roots spat on sores or wounds. *Liver Aid* Infusion of roots taken for liver trouble. *Oral Aid* Small, peeled, cleaned root piece chewed for mouth sores and gum boils. *Throat Aid* Infusion of root taken for sore throat. *Unspecified* Root used as medicine. (187:282)
- **Food**—**Miwok** *Dried Food* Steamed leaves dried and stored. *Vegetable* Boiled or steamed leaves eaten in spring. (12:159)

Heuchera novomexicana Wheelock, New Mexico Alumroot
- **Drug**—**Navajo, Ramah** *Analgesic* Decoction of root taken as needed for internal pain. *Dermatological Aid* Poultice of split root applied to infected sores and swellings. *Disinfectant* Poultice of split root applied to infected sores. *Orthopedic Aid* Poultice of split root applied to infected sores, swellings, and fractures. *Panacea* Plant used as "life medicine." (191:29)

Heuchera parviflora Bartl., Littleflower Alumroot
- **Drug**—**Blackfoot** *Dermatological Aid* Poultice of pounded root applied to sores and swellings. (114:274)

Heuchera parvifolia Nutt. ex Torr. & Gray, Littleleaf Alumroot
- **Drug**—**Blackfoot** *Antirheumatic (External)* Pounded, wetted root used for rheumatism. *Dermatological Aid* Pounded, wetted roots used for sores. (118:43) *Eye Medicine* Infusion of root used as an eyewash. (118:39) *Hemostat* Poultice of chewed roots applied to wounds and sores as a styptic. (86:76) Poultice of chewed roots applied to wounds and sores as a styptic. (86:84) *Oral Aid* and *Pediatric Aid* Poultice of chewed roots applied to cold sores and children's mouth cankers. (86:76) *Veterinary Aid* Infusion of roots given to horses for respiratory troubles. (86:88) **Navajo, Kayenta** *Dermatological Aid* Plant used for rat bites. (205:25) **Navajo, Ramah** *Analgesic* Decoction of root taken for stomachache. *Gastrointestinal Aid* Decoction of root taken for stomachache. *Gynecological Aid* Decoction of split root taken to ease delivery of placenta. *Panacea* Root used as a "life medicine." *Venereal Aid* Infusion of root used as a lotion for venereal disease. (191:29, 30)

Heuchera richardsonii R. Br., Richardson's Alumroot
- **Drug**—**Blackfoot** *Antidiarrheal* Rootstocks chewed for diarrhea. (97: 37) **Cree, Woodlands** *Antidiarrheal* Decoction of root or root chewed for diarrhea. *Eye Medicine* Infusion of root used to wash sore eyes. (109:40) **Lakota** *Antidiarrheal* Infusion of roots taken for diarrhea. *Dermatological Aid* Poultice of powdered roots applied to sores. (139:58)

Heuchera rubescens Torr., Pink Alumroot
- **Drug**—**Gosiute** *Dermatological Aid* Decoction of roots used as an astringent. *Gastrointestinal Aid* and *Pediatric Aid* Decoction of roots used for babies and children with colic. (39:371) **Paiute** *Eye Medicine* Infusion of root used as an eyewash. *Venereal Aid* Decoction of root taken for venereal disease. **Shoshoni** *Antidiarrheal* Infusion of root taken for diarrhea. *Febrifuge* Decoction of root taken for high fever. *Heart Medicine* Decoction of root taken for heart trouble. *Liver Aid* Infusion of roots taken for liver trouble or biliousness. *Tonic* Decoction of root taken as a tonic for general debility. *Venereal Aid* Decoction of root taken for venereal disease. *Veterinary Aid* Mashed, boiled leaves used as a wash for horses' saddle sores. Soaked roots given to horses and cows for cramps. (180:87, 88)

***Heuchera* sp.**, Alumroot
- **Drug**—**Blackfoot** *Veterinary Aid* Plant used for saddle sores on horses. (97:36) **Chippewa** *Antidiarrheal* Compound decoction of root taken for dysentery. (53:344) *Gastrointestinal Aid* Compound decoction of root taken for indigestion. *Oral Aid* and *Pediatric Aid* Compound decoction of root used as mouthwash for teething children. (53:342) **Gosiute** *Cathartic*, *Gastrointestinal Aid*, and *Pediatric Aid* Plant used as a purgative for babies and adults with intestinal disorders. (39:350)

Hexastylis, Aristolochiaceae

Hexastylis arifolia (Michx.) Small, Littlebrownjug
- **Drug**—**Catawba** *Analgesic* and *Gastrointestinal Aid* Infusion of leaves taken for stomach pains. *Heart Medicine* Infusion of leaves taken for heart troubles. (as *Asarum apiifolia* 157:190)

Hexastylis arifolia (Michx.) Small **var. *arifolia***, Littlebrownjug
- **Drug**—**Catawba** *Analgesic* Leaves used for severe pain in the heart from heart disease. (as *Asarum arifolia* 157:188) Infusion of leaves taken for stomach pains or backaches. (as *Asarum arifolium* 177:20) *Gastrointestinal Aid* Leaves used for stomach trouble. (as *Asarum arifolia* 157:188) Infusion of leaves taken for stomach pains. (as *Asarum arifolium* 177:20) *Heart Medicine* Leaves used for severe pain in the heart from heart disease. (as *Asarum arifolia* 157:188) Infusion of leaves taken for heart troubles. *Orthopedic Aid* Plant used for backache. (as *Asarum arifolium* 177:20) **Rappahannock** *Febrifuge* Infusion of leaves taken for fever. *Pulmonary Aid* Infusion of leaves taken for whooping cough. *Respiratory Aid* Decoction of leaves with alcohol taken for asthma. (as *Asarum arifolium* 161:25)

Hexastylis virginica (L.) Small, Virginia Heartleaf
- **Drug**—**Cherokee** *Gastrointestinal Aid* Infusion taken "to stop blood from passing." (80:37)

Hibiscus, Malvaceae

Hibiscus moscheutos L. **ssp. *moscheutos***, Crimsoneyed Rosemallow
- **Drug**—**Shinnecock** *Urinary Aid* Infusion of dried stalks applied for inflammation of the bladder. (as *H. palustris* 30:120)

***Hibiscus* sp.**, Kokio
- **Drug**—**Hawaiian** *Blood Medicine* Infusion of pounded roots and other plants strained and taken to purify the blood. (2:54) *Laxative* Flower bases chewed by the mother and given to infants as a laxative. (2:40) Buds chewed by mothers and given to children as a laxative. Leaves chewed and swallowed as a laxative. (2:54) *Pediatric Aid* Flower bases chewed by the mother and given to infants as a laxative. Seeds chewed and swallowed by children with general weakness of the body. (2:40) Buds chewed by mothers and given to children as a laxative, and for general debility and run-down conditions. (2:54) *Strengthener* Seeds chewed and swallowed by children with general weakness of the body. (2:40) Buds chewed by mothers and given to children for general debility and run-down conditions. Leaves chewed and swallowed for general debility and run-down conditions. (2:54)

Hibiscus tiliaceus L., Sea Hibiscus
- **Drug**—**Hawaiian** *Gynecological Aid* Slimy substance from inner bark and water taken before or between the pain accompanying childbirth. *Laxative* Slimy substance from bark or the flower bases used as a laxative for adults and children. *Pediatric Aid* Slimy substance from bark or the flower bases used as a laxative for adults and children. *Pulmonary Aid* Bark and other plants crushed, water added, strained, and resulting liquid taken for congested chest. *Throat Aid* Shoots and buds chewed and swallowed for dry throat. (2:39)

Hieracium, Asteraceae

Hieracium canadense Michx., Canadian Hawkweed
- **Drug**—**Ojibwa** *Hunting Medicine* Flowers used to make a hunting lure and mixed with other hunting charms. Roots nibbled when hunting to attract a doe. (153:429)

Hieracium cynoglossoides Arv.-Touv., Houndstongue Hawkweed
- **Drug**—**Okanagan-Colville** *Tonic* Infusion of leaves and roots taken as a general tonic. (188:84)

Hieracium fendleri Schultz-Bip., Yellow Hawkweed
- **Drug**—**Navajo, Ramah** *Diuretic* Cold infusion of plant taken by hunters for anuria. *Hunting Medicine* Leaves chewed for good luck in hunting. (191:52)

Hieracium pilosella L., Mouseear Hawkweed
- **Drug**—**Iroquois** *Antidiarrheal* Infusion of plants taken for diarrhea. (87:480)

Hieracium scabrum Michx., Rough Hawkweed
- **Drug**—**Rappahannock** *Antidiarrheal* Infusion of leaves taken or chewed for diarrhea. (161:27)

Hieracium scouleri Hook., Woollyweed
- **Drug**—**Okanagan-Colville** *Tonic* Infusion of leaves and roots taken as a general tonic. (188:84)

***Hieracium* sp.**, Hawkweed

- **Drug**—Iroquois *Dermatological Aid* Poultice of roots applied to sores close to the bone. *Tuberculosis Remedy* Decoction of plants taken for consumption. (87:480) **Thompson** *Oral Aid* Gummy juice chewed to cleanse the mouth. (164:492) *Unspecified* Root used as a charm for unspecified purpose. (164:506)
- **Food**—Thompson *Unspecified* Chewed for pleasure. (164:492)

Hieracium venosum L., Rattlesnakeweed
- **Drug**—Cherokee *Gastrointestinal Aid* Compound infusion of root given for bowel complaints. (80:37)

Hierochloe, Poaceae

Hierochloe alpina (Sw. ex Willd.) Roemer & J. A. Schultes ssp. *alpina*, Alpine Sweet Grass
- **Fiber**—Haisla & Hanaksiala *Basketry* Blades used to make baskets. (43:207)

Hierochloe hirta (Schrank) Borbás ssp. *hirta*, Northern Sweet Grass
- **Fiber**—Haisla & Hanaksiala *Basketry* Blades used to make baskets. (as *H. odorata* ssp. *hirta* 43:207)

Hierochloe occidentalis Buckl., California Sweet Grass
- **Drug**—Karok *Gynecological Aid* Infusion of plant taken by women after miscarriage or to arrest fetus growth. *Veterinary Aid* Plant given to sick dogs. (as *Torresia macrophylla* 148:380)

Hierochloe odorata (L.) Beauv., Sweet Grass
- **Drug**—Blackfoot *Cold Remedy* Smoke from burning leaves used for colds. (82:28) Burning leaf smoke inhaled for colds. (97:20) *Cough Medicine* Infusion of plant taken for coughs. (86:72) *Dermatological Aid* Leaves and boiled hoof sticky substance used as a hair tonic. (86:124) Stems soaked in water and used for chapping and windburn. (86:77) *Eye Medicine* Stems soaked in water and used as an eyewash. (86:81) *Strengthener* Grass chewed as a means of extended endurance in ceremonies involving prolonged fasting. (86:9) *Throat Aid* Infusion of plant taken for sore throats. (86:72) *Venereal Aid* Infusion of blades taken by men for venereal infections. (86:69) *Veterinary Aid* Leaves used for saddle sores on horses. (97:20) **Cheyenne** *Ceremonial Medicine* Plant used as a ceremonial incense for purification. *Witchcraft Medicine* Plant burned in homes to prevent evil. (83:9) **Flathead** *Analgesic* Infusion used for "sharp pains inside." *Cold Remedy* Infusion used for colds. *Febrifuge* Infusion used for fevers. *Respiratory Aid* Infusion mixed with meadow rue seeds and used for congested nasal passages. (82:28) **Kiowa** *Dermatological Aid* Dried foliage employed as a perfume. (192:15) **Menominee** *Dermatological Aid* Grass used in basketry and as a perfume. (151:75) **Plains Indian** *Veterinary Aid* Leaves given to horses to make them long-winded on the chase. (97:20) **Thompson** *Dermatological Aid* Infusion or decoction of plant used as a wash for the hair and body. (164:476)
- **Fiber**—Iroquois *Basketry* Plant used to make baskets. (141:67) **Kiowa** *Mats, Rugs & Bedding* Fragrant leaves used as stuffing for pillows and mattresses. (192:15) **Malecite** *Basketry* Used to make baskets. (160:6) **Menominee** *Basketry* Grass used in basketry and as a perfume. *Sewing Material* Wet grass used for sewing, dried tight, and resin used over the stitches. (151:75) **Micmac** *Basketry* Used to make baskets. *Mats, Rugs & Bedding* Used to make mats. (159:258)
- **Other**—Blackfoot *Ceremonial Items* Smoke from burning grass used to purify Sundance dancers. Leaves ceremonially smoked with tobacco. (82:28) Stems burned and prayers said during every ceremony. Grass essential to the raising of a fallen dancer or fallen paraphernalia. Grass water used to bathe the mother 34 days after giving birth and before returning home. Grass braids tied vertically around the base of the Horn Society staffs. Grass braids strung on the inside of headbands of the Motokiks headdresses. (86:9) Used in the Sun Dance ceremony and burned on a small altar found in many lodges. (97:20) *Decorations* Used to decorate women's hair. (82:28) *Incense & Fragrance* Grass packed into saddles to keep them smelling good. (86:114) Leaves mixed with red ocher to make it smell good. (86:124) Stems bound at the lower end with other stems, braided, and used as incense during ceremonies. Grass used by everyone as incense during daily prayers. Grass water used as incense smudge by the mother 34 days after giving birth and before returning home. (86:9) Used as an incense, natural sachet, or perfume. (97:20) *Soap* Leaves soaked in water and used as a hair wash. (82:28) Leaves soaked in water and used as a hair wash. (97:20) **Cheyenne** *Ceremonial Items* Smoke from burning grass used for purification of rattles, sacred shields, and Sundance dancers. (82:28) *Paint* Used to paint pipes in the Sun Dance and the Sacred Arrow ceremonies. (83:9) *Protection* Burned for protection from lightning and thunder. **Flathead** *Decorations* Used to decorate women's hair. *Insecticide* Used as an incense to "keep the bugs away." **Gros Ventre** *Soap* Leaves soaked in water and used as a hair wash. (82:28) **Kiowa** *Incense & Fragrance* Dried leaves sprinkled over the fire to yield incense and used during the peyote ceremony. (192:15) **Lakota** *Ceremonial Items* Strands of grass burned to bring guardian spirits. (106:49) Grass used in religious ceremonies. (139:30) *Incense & Fragrance* Used as a perfume. (106:49) **Menominee** *Sacred Items* Grass used to burn as an oblation to the deities. (151:75) **Montana Indian** *Ceremonial Items* Burned as incense for spiritual protection and purification. *Incense & Fragrance* Used as a clothes and body perfume. (82:28) **Okanagan-Colville** *Incense & Fragrance* Blades braided together and packed among clothes to give them a nice smell. (188:55) **Sioux** *Ceremonial Items* Smoke from burning grass used to purify Sundance dancers. Leaves ceremonially smoked with tobacco. (82:28) **Thompson** *Incense & Fragrance* Grass tied in the hair and on neck and arm ornaments as a scent. (164:503) *Toys & Games* Easily braided grass used in play by children. (187:141)

Hierochloe odorata (L.) Beauv. ssp. *odorata*, Vanilla Grass
- **Drug**—Blackfoot *Ceremonial Medicine* Plant burned as ceremonial incense. (as *Savastana odorata* 114:273, 274) *Dermatological Aid* Used to make a hair tonic. (as *Savastana odorata* 114:273) Leaves used as a hair wash and incense. (as *Sevastana odorata* 114:278) **Cheyenne** *Ceremonial Medicine* Dried plant burned in many ceremonies. (as *Torresia odorata* 76:170) **Dakota** *Ceremonial Medicine* Plant used in propitiatory rites. (as *Savastana odorata* 69:359) Plant used as incense in ceremony to invoke good powers and in peace ceremony. (as *Savastana odorata* 70:66) **Kiowa** *Dermatological Aid* Dried foliage employed as a perfume. (as *Torresia odorata* 192:15) **Omaha** *Ceremonial Medicine* Plant used in various rituals. (as *Savastana odorata* 68:322) Plant used as incense in ceremony to invoke good powers and in peace ceremony. **Pawnee** *Ceremonial Medicine*

Plant used as incense in ceremony to invoke good powers and in peace ceremony. **Ponca** *Ceremonial Medicine* Plant used as incense in ceremony to invoke good powers and in peace ceremony. **Winnebago** *Ceremonial Medicine* Plant used as incense in ceremony to invoke good powers and in peace ceremony. (as *Savastana odorata* 70:66)

- *Fiber*—**Kiowa** *Mats, Rugs & Bedding* Fragrant leaves used as stuffing for pillows and mattresses. (as *Torresia odorata* 192:15)
- *Other*—**Blackfoot** *Ceremonial Items* Braided plant put up on Sun Dance alters and used in religious services. (as *Savastana odorata* 118:51) *Incense & Fragrance* Leaves braided and placed with the clothes or carried in small bags as perfume. Leaves used as a hair wash and incense. (as *Sevastana odorata* 114:278) **Cheyenne** *Ceremonial Items* Dried leaves burned over coals in many ceremonies. *Incense & Fragrance* Dried leaves used as a perfume by wrapping the article in the leaves. (as *Torresia odorata* 76:170) **Chippewa** *Ceremonial Items* Used for ceremonial, economic, and pleasurable purposes. (as *Torresia odorata* 53:378) **Dakota** *Ceremonial Items* Plant used in religious ceremonies. (as *Savastana odorata* 70:91) **Kiowa** *Incense & Fragrance* Dried leaves sprinkled over the fire to yield incense and used during the peyote ceremony. (as *Torresia odorata* 192:15) **Omaha** *Ceremonial Items* Plant used as incense. (as *Savastana odorata* 68:320) *Incense & Fragrance* Plant used as perfume. (as *Savastana odorata* 68:323)

Hilaria, Poaceae

Hilaria jamesii (Torr.) Benth., Galleta
- *Drug*—**Navajo, Ramah** *Dietary Aid* and *Pediatric Aid* Cold infusion given to babies to make them "want to eat a lot." (191:16)
- *Food*—**Navajo, Ramah** *Forage* Used as horse and sheep feed and able to withstand trampling and close grazing. (191:16)
- *Fiber*—**Hopi** *Basketry* Grass used by the women to make coil trays. (56:17) Used as the fill of coiled basketry. (200:65) *Brushes & Brooms* Culms used as a floor and hairbrush. **Tewa** *Brushes & Brooms* Culms used as a floor and hairbrush. (42:325)
- *Other*—**Hopi** *Ceremonial Items* Used as the artificial arm worn by the manipulator of the serpent effigy. (200:65) *Decorations* Stems used to form the base of the coils for manufactured plaques. (17:33) Used to make plaques. **Tewa** *Decorations* Used to make plaques. (42:325)

Hippocastanaceae, see *Aesculus*

Hippuridaceae, see *Hippuris*

Hippuris, Hippuridaceae

Hippuris tetraphylla L. f., Fourleaf Marestail
- *Food*—**Eskimo, Alaska** *Vegetable* Small, young leaves eaten as greens. (4:715)

Hippuris vulgaris L., Common Marestail
- *Food*—**Alaska Native** *Soup* Whole plant used to make soup. *Winter Use Food* Leaves piled on high ground and stored for winter use. (85:135) **Eskimo, Alaska** *Soup* Plant added to seal blood soup and tom cod-liver soup. (1:37) **Eskimo, Inuktitut** *Ice Cream* Used to make "Eskimo ice cream." *Soup* Used as a condiment for soups. *Unspecified* Eaten raw or with seal oil and salmon eggs. (202:191)

Hirschfeldia, Brassicaceae

Hirschfeldia incana (L.) Lagreze-Fossat, Shortpod Mustard
- *Food*—**Cahuilla** *Porridge* Seeds ground into a mush. *Vegetable* Leaves eaten fresh or boiled. *Winter Use Food* Leaves and seeds used as an important winter food. (as *Brassica geniculata* 15:47)

Hoffmannseggia, Fabaceae

Hoffmannseggia glauca (Ortega) Eifert, Indian Rushpea
- *Food*—**Apache** *Unspecified* Potatoes roasted and eaten much more commonly in the past than currently. (as *H. densiflora* 32:52) **Apache, Chiricahua & Mescalero** *Unspecified* Roots eaten either raw or cooked. (as *H. densiflora* 33:42) **Cocopa** *Unspecified* Tuberous roots utilized as food. (as *H. densiflora* 37:207) **Pima** *Unspecified* Bulbs eaten raw or boiled. (as *H. stricta* 95:262) *Vegetable* Tubers boiled and eaten like potatoes. (as *H. densiflora* 47:92) **Pima, Gila River** *Unspecified* Roots boiled or roasted and eaten. (as *H. densiflora* 133:7) Tubers eaten. (133:5) **Pueblo** *Unspecified* Potatoes roasted and eaten much more commonly in the past than currently. (as *H. densiflora* 32:52)

Hoita, Fabaceae

Hoita macrostachya (DC.) Rydb., Large Leatherroot
- *Drug*—**Luiseño** *Dermatological Aid* Plant used for ulcers and sores. (as *Psoralea macrostachya* 155:231)
- *Fiber*—**California Indian** *Cordage* Root fiber used to make rope. *Sewing Material* Inner bark used for thread. (as *Psoralea macrostachya* 118:59) **Concow** *Sewing Material* Fine, strong inner bark formerly used for thread. **Mendocino Indian** *Cordage* Root fibers used to make rope. **Yokia** *Sewing Material* Fine, strong inner bark formerly used for thread. (as *Psoralea macrostachya* 41:358)
- *Dye*—**Cahuilla** *Yellow* Roots boiled with basket weeds as a yellow dye. (as *Psoralea macrostachya* 15:121) **Luiseño** *Yellow* Roots boiled to make a yellow dye. (as *Psoralea macrostachya* 155:209)
- *Other*—**California Indian** *Containers* Root fibers used to make bags. (as *Psoralea macrostachya* 118:59) **Mendocino Indian** *Hunting & Fishing Item* Root fibers used to make hunting bags. (as *Psoralea macrostachya* 41:358)

Hoita orbicularis (Lindl.) Rydb., Roundleaf Leatherroot
- *Drug*—**Costanoan** *Blood Medicine* Decoction of plant used for the blood. *Febrifuge* Decoction of plant used for fevers. (as *Psoralea orbicularis* 21:19)
- *Food*—**Luiseño** *Vegetable* Plant used for greens. (as *Psoralea orbicularis* 155:231)

Holocarpha, Asteraceae

Holocarpha virgata (Gray) Keck, Yellowflower Tarweed
- *Drug*—**Miwok** *Febrifuge* Decoction of plant used as a bath for fevers. *Misc. Disease Remedy* Decoction of plant used as a bath for measles. (as *Hemizonia virgata* 12:170, 171)

Holodiscus, Rosaceae

Holodiscus discolor (Pursh) Maxim., Ocean Spray
- *Drug*—**Chehalis** *Misc. Disease Remedy* Infusion of seeds taken for smallpox, black measles, and chickenpox. **Lummi** *Antidiarrheal*

Holodiscus discolor (Rosaceae)

Blossoms used for diarrhea. *Eye Medicine* Infusion of inner bark used as an eyewash. *Oral Aid* Poultice of leaves applied to sore lips. *Orthopedic Aid* Poultice of leaves applied to sore feet. **Makah** *Tonic* Decoction of bark taken as a tonic by convalescents and athletes. (79:33) *Unspecified* Used to make medicine. (67:263) **Navajo, Ramah** *Misc. Disease Remedy* Decoction of leaves taken for influenza. (191:31) **Okanagan-Colville** *Burn Dressing* Bark dried, powdered, mixed with Vaseline, and used on burns. (188:126) **Sanpoil** *Dermatological Aid* Powder of dried leaves used for sores. (131:221) **Squaxin** *Blood Medicine* Seeds used as a blood purifier. (79:33)

- **Fiber**—**Nitinaht** *Sewing Material* Wood used to make knitting needles and long needles for mat making. (186:117) **Okanagan-Colville** *Furniture* Wood used to make baby cradle covers. (188:126) **Pomo, Kashaya** *Furniture* Long branches used to make baby baskets and arrows. (72:40) **Salish, Cowichan** *Sewing Material* Wood used to make knitting needles. (186:117) **Squaxin** *Canoe Material* Wood used to make canoe paddles. (79:33)
- **Other**—**Chehalis** *Cooking Tools* Wood used to make roasting tongs. *Hunting & Fishing Item* Wood used to make shafts and spear prongs. (79:33) **Hesquiat** *Tools* Plant used to make needles for sewing tule and basket sedge. (185:72) **Karok** *Toys & Games* Shoots used to make "Indian cards." A set of little sticks was prepared by scraping 10-inch lengths of shoot clean of bark. All the sticks were straight and one was marked with a black mark. The dealer took the set of sticks in his hands, shuffled the sticks, sang, and held his hands behind his back. He tried to prevent his opponents from guessing where the black-marked stick was. If the dealer kept the other side from guessing where the ace was for 10 times, he won. This was a gambling game. (148:384) **Klallam** *Cooking Tools* Wood used to make roasting tongs. *Hunting & Fishing Item* Wood used to make the prongs of duck spears. (79:33) **Kwakiutl, Southern** *Hunting & Fishing Item* Wood used to make arrows. *Tools* Wood used to make digging sticks. (183:288) **Lummi** *Cooking Tools* Wood used to make roasting tongs. *Hunting & Fishing Item* Wood used to make the prongs of duck spears. **Makah** *Cooking Tools* Wood used to make roasting tongs. *Hunting & Fishing Item* Wood used to make the prongs of duck spears. (79:33) **Nitinaht** *Cooking Tools* Branches used for holding fish while barbecuing because they do not burn. (67:263) Wood used to make barbecue sticks. *Hunting & Fishing Item* Wood used with a yew wood barb tied on the end as an octopus spear. (186:117) *Tools* Wood used to make knitting needles. *Toys & Games* Wood used to make practice bows and arrows for children. (67:263) Wood used to make practice bows for children. (186:117) **Nootka** *Tools* Plant used to make needles and harpoons. (185:72) **Okanagan-Colville** *Fasteners* Wood used to make tepee pins. *Hunting & Fishing Item* Wood used to make arrows, fishing spear heads, and bows. *Musical Instrument* Wood used to make drum hoops. *Tools* Wood used to make digging sticks. *Toys & Games* Wood used to make gambling game sticks. (188:126) **Poliklah** *Hunting & Fishing Item* Used to make arrows. (as *H. ariaefolius* 117:173) **Pomo** *Hunting & Fishing Item* Used for arrows. (66:13) **Pomo, Kashaya** *Hunting & Fishing Item* Long branches used to make baby baskets and arrows. (72:40) **Saanich** *Cooking Tools* Wood used to make salmon barbecuing sticks. *Hunting & Fishing Item* Wood used to make bows, arrows, harpoon shafts, and halibut hooks. *Tools* Wood used to make camas bulb digging sticks and cambium scrapers, and knitting needles and cattail mat needles. (182:86) **Salish** *Hunting & Fishing Item* Wood used to make arrows. *Tools* Wood used to make digging sticks. (183:288) **Salish, Coast** *Cooking Tools* Wood used to make salmon barbecuing sticks. *Hunting & Fishing Item* Wood used to make bows, arrows, harpoon shafts, and halibut hooks. *Tools* Wood used to make camas bulb digging sticks, and cambium scrapers. Wood used to make knitting needles and cattail mat needles. (182:86) **Salish, Cowichan** *Cooking Tools* Wood used to make skewers for roasting and drying clams. (186:117) **Skagit** *Cooking Tools* Wood used to make roasting tongs. *Hunting & Fishing Item* Wood used to make the prongs of duck spears. **Snohomish** *Cooking Tools* Wood used to make roasting tongs. *Hunting & Fishing Item* Wood used to make the prongs of duck spears. **Squaxin** *Cooking Tools* Wood used to make roasting tongs. *Hunting & Fishing Item* Wood used to make the prongs of duck spears. **Swinomish** *Cooking Tools* Wood used to make roasting tongs. *Hunting & Fishing Item* Wood used to make flounder spears and the prongs of duck spears. (79:33) **Thompson** *Hunting & Fishing Item* Wood used to make arrows. (as *Sericotheca discolor* 164:497) *Protection* Extremely hard wood used to make cuirasses and other types of armor. (187:261) *Weapon* Wood used for cuirasses and armor in general. (as *Sericotheca discolor* 164:497)

Holodiscus dumosus (Nutt. ex Hook.) Heller, Rockspirea

- **Drug**—**Paiute** *Antidiarrheal* Decoction of root taken for diarrhea. *Cold Remedy* Decoction of stems taken for colds. *Gastrointestinal Aid* Decoction of root taken for stomach disorders. **Shoshoni** *Disinfectant* Decoction of leaves, flowers, and stems used as an antiseptic wash. *Emetic* Decoction of leaves taken as an emetic. *Gastrointestinal Aid* Decoction of leaves and stems taken for stomachaches. *Unspecified* Decoction of leaf, flower, and stem taken for illnesses of "undefined cause." *Venereal Aid* Decoction of leaves or stems taken for venereal disease. (as *H. discolor* var. *dumosus* 180:88, 89)
- **Food**—**Isleta** *Beverage* Leaves steeped to make a beverage. (100:32)

Honckenya, Caryophyllaceae

Honckenya peploides (L.) Ehrh., Seaside Sandplant

- **Food**—**Eskimo, Inupiat** *Vegetable* Sour leaves and shoots eaten with seal oil and sugar. *Winter Use Food* Leaves and shoots boiled many times and stored in a large wooden barrel for winter use. (98:42)

Honckenya peploides ssp. *major* (Hook.) Hultén, Seaside Sandplant

- **Food**—**Eskimo, Alaska** *Unspecified* Leaves and stems boiled and eaten with seal oil. (1:35) Leaves used for food. (as *Ammodenia peploides major* 4:715)

Honckenya peploides (L.) Ehrh. ssp. *peploides*, Seaside Sandplant

- **Food**—**Alaska Native** *Dietary Aid* Fresh and raw leaves eaten as a good source of vitamins A and C. *Ice Cream* Leaves chopped, cooked in water, soured, and mixed with reindeer fat and berries into Eskimo ice cream. *Unspecified* Leaves eaten with dried fish. *Vegetable* Leaves eaten raw or mixed with other greens. Leaves mixed with other greens and made into a kraut. (as *Arenaria peploides* 85:15) **Eskimo, Arctic** *Vegetable* Young stems and leaves pickled as "sauerkraut" or eaten as a potherb. (as *Arenaria peploides* 128:29)

Hordeum, Poaceae

Hordeum jubatum L., Foxtail Barley
- *Drug*—**Chippewa** *Eye Medicine* Dry root wrapped, moistened, and used as a compress for sties or inflammation of lid. (53:360) **Navajo, Ramah** *Poison* Plant considered poisonous and children taught to avoid it. (191:16) **Potawatomi** *Unspecified* Root used for unspecified ailments. (154:59)
- *Food*—**Kawaiisu** *Unspecified* Seeds pounded and eaten dry. (206:34)
- *Other*—**Iroquois** *Toys & Games* Used by children to place in the sleeves of playmates as a joke. (142:106) **Kawaiisu** *Tools* Used to rub the skin off yucca stalks. (206:34)

Hordeum marinum ssp. *gussonianum* (Parl.) Thellung, Mediterranean Barley
- *Food*—**Mendocino Indian** *Fodder* Green grass used for fodder. (41:313)

Hordeum murinum L., Mouse Barley
- *Food*—**Mendocino Indian** *Staple* Seeds used for pinole. (41:313)

Hordeum murinum ssp. *glaucum* (Steud.) Tzvelev, Smooth Barley
- *Drug*—**Costanoan** *Urinary Aid* Decoction of plant used for bladder ailments. (as *H. glaucum* 21:30)
- *Food*—**Cahuilla** *Unspecified* Seeds eaten, when other foods were scarce. (as *H. stebbinsi* 15:78) **Costanoan** *Staple* Seeds used for pinole. (as *H. glaucum* 21:255)

Hordeum sp., Foxtail Grass
- *Food*—**Pomo, Kashaya** *Unspecified* Seeds used in pinole. (72:53)

Hordeum vulgare L., Common Barley
- *Food*—**Cahuilla** *Unspecified* Cultivated and used for food. (15:78) **Papago** *Unspecified* Species used for food. **Pima** *Unspecified* Species used for food. (36:117) **Yuki** *Bread & Cake* Seeds ground into flour and used to make bread. *Substitution Food* Seeds parched and used as a substitute for coffee. (41:313)

Horkelia, Rosaceae

Horkelia californica Cham. & Schlecht., California Honeydew
- *Drug*—**Pomo, Kashaya** *Blood Medicine* Decoction of root used as a blood purifier. (72:57)

Hosta, Liliaceae

Hosta lancifolia Engl., Narrowleaf Plantainlily
- *Drug*—**Cherokee** *Antihemorrhagic* Warm infusion of root taken for spitting blood. *Cough Medicine* Warm infusion of root taken for coughing. *Dermatological Aid* Leaf rubbed on swollen legs and feet after scratching insect bites. (as *H. japonica* 80:50)

Houstonia, Rubiaceae

Houstonia caerulea L., Azure Bluet
- *Drug*—**Cherokee** *Urinary Aid* Infusion given for bed-wetting. (80:26)

Houstonia rubra Cav., Red Bluet
- *Drug*—**Keres, Western** *Eye Medicine* Infusion of plant used for sore eyes. *Gastrointestinal Aid* Infusion of plant used for the stomach.

(171:48) **Navajo, Kayenta** *Gynecological Aid* Decoction of plant used for menstrual troubles. (205:43)

Houstonia wrightii Gray, Pygmy Bluet
- *Drug*—**Navajo, Ramah** *Ceremonial Medicine* Plant used as a ceremonial fumigant for "deer infection." *Dermatological Aid* Cold, compound infusion of plant taken and used as lotion for poison ivy rash. Dried, pulverized root used as dusting powder for sores on humans or livestock. *Disinfectant* Plant used as a ceremonial fumigant for "deer infection." *Panacea* Root used as a "life medicine." (191:45)

Hudsonia, Cistaceae

Hudsonia tomentosa Nutt., Woolly Beachheather
- *Drug*—**Montagnais** *Blood Medicine* Decoction of plant taken by women to "purge the blood." (156:313)

Humulus, Cannabaceae

Humulus lupulus L., Common Hop
- *Drug*—**Cherokee** *Analgesic* "Alleviates pain and produces sleep." *Antirheumatic (Internal)* Taken for rheumatism. *Breast Treatment* Used for breast and womb problems. *Gynecological Aid* Used for "breast & female complaints where womb is debilitated." *Kidney Aid* Taken for inflamed kidneys. *Sedative* "Alleviates pain and produces sleep." *Urinary Aid* Taken for "gravel" and the bladder. (80:39) **Dakota** *Analgesic* Decoction of fruits taken for intestinal pains. *Febrifuge* Decoction of fruits taken for fevers. *Gastrointestinal Aid* Decoction of fruits taken for intestinal pains. (69:362) **Delaware** *Ear Medicine* Poultice of heated plants in small bags applied for earache. *Sedative* Blossoms used for nervousness. *Stimulant* Infusion of plants used as a tonic stimulant. *Toothache Remedy* Poultice of heated plants in small bags applied for toothache. (176:31) **Delaware, Oklahoma** *Ear Medicine* Poultice of heated herb in bag applied for earache. *Sedative* Blossoms used in medicine for "nervousness." *Stimulant* Infusion of plant taken as a tonic and stimulant. (175:26, 76) *Tonic* Infusion of plant taken as a "tonic-stimulant." (175:26) *Toothache Remedy* Poultice of heated herb in bag applied for toothache. (175:26, 76) **Meskwaki** *Sedative* Root used for insomnia. (152:250) **Mohegan** *Analgesic* Blossoms used for pain. (176:130) *Ear Medicine* Dried blossoms applied to earache. (174:266) *Sedative* Used in making "nerve medicine." (30:120) Infusion of blossoms used for nerves. (174:266) Infusion of blossoms taken for nervous tension. (176:72, 130) *Toothache Remedy* Dried blossoms applied to toothache. (174:266) **Ojibwa** *Diuretic* Infusion of herb taken as a diuretic and to reduce acidity of urine. (153:391) **Omaha** *Dermatological Aid* Root used for wounds. (58:584) **Round Valley Indian** *Dermatological Aid* Poultice of soaked hops applied to swellings and bruises. (41:344) **Shinnecock** *Pulmonary Aid* Poultice of dried hops heated in a cloth bag applied for pneumonia. *Sedative* Used in making "nerve medicine" and used as a poultice for pneumonia. (30:120)
- *Food*—**Algonquin, Quebec** *Bread & Cake* Hops used to make bread. (18:83) **Lakota** *Cooking Agent* Used to make bread swell. (139:51) **Ojibwa** *Cooking Agent* Hop fruit often used as a substitute for baking soda. (153:411)

Humulus lupulus var. *lupuloides* E. Small, Common Hop
- *Drug*—**Dakota** *Analgesic* Infusion of fruit taken for intestinal pains.

Dermatological Aid Simple or compound poultice of chewed root applied to wounds. *Febrifuge* Infusion of fruit taken for fevers. *Gastrointestinal Aid* Infusion of fruits taken "to allay fevers and intestinal pains." (as *H. americana* 70:77) **Navajo, Ramah** *Cough Medicine* Plant used for bad cough. *Hunting Medicine* Plant used as a "big medicine" for "good luck" in hunting. *Misc. Disease Remedy* Plant used for influenza. *Witchcraft Medicine* Plant used for protection against witches. (as *H. americanus* 191:22)

Humulus lupulus* var. *neomexicanus A. Nels. & Cockerell, Common Hop
- *Food*—**Apache, Chiricahua & Mescalero** *Spice* Hops boiled and used to flavor wheat flour and potatoes. (33:47) Flower used to flavor drinks and make them stronger. (33:51) **Navajo** *Unspecified* Hops used for cooking. (55:41)

Huperzia, Lycopodiaceae, club moss

Huperzia lucidula (Michx.) Trevisan, Shining Clubmoss
- *Drug*—**Iroquois** *Blood Medicine* Compound used when "blood is bad." *Cold Remedy* Decoction used when woman catches cold due to suppressed menses. *Dermatological Aid* Compound used for neck sores. (as *Lycopodium lucidulum* 87:263)

Huperzia selago (L.) Bernh. ex Mart. & Schrank **var. *selago***, Fir Clubmoss
- *Drug*—**Nitinaht** *Cathartic* Plant used as a purgative. *Emetic* Plant used as a fast acting emetic. *Gastrointestinal Aid* Branches used to "clean . . . out" the insides. (as *Lycopodium selago* 186:60) **Tanana, Upper** *Analgesic* Poultice of the whole plant applied to the head for headaches. (as *Lycopodium selago* 102:18)

Hybanthus, Violaceae

Hybanthus concolor (T. F. Forst.) Spreng., Eastern Greenviolet
- *Drug*—**Iroquois** *Veterinary Aid* Infusion of roots and stems mixed with feed for mare with injured fetus. (87:386)

Hydnaceae, see *Dentinum*, *Hericium*

Hydrangeaceae, see *Broussaisia*, *Fendlera*, *Hydrangea*, *Jamesia*, *Philadelphus*

Hydrangea, Hydrangeaceae

Hydrangea arborescens L., Wild Hydrangea
- *Drug*—**Cherokee** *Abortifacient* Compound infusion taken for menstrual period. *Antiemetic* Used as an antiemetic and cold infusion of bark used as antiemetic for children. *Burn Dressing* Poultice of scraped bark used for burns. *Cancer Treatment* Used for tumors. *Cathartic* Used as a purgative. *Dermatological Aid* Used for ulcers and poultice of scraped bark used for "risings." *Disinfectant* Used as an antiseptic. *Emetic* Infusion of bark given to induce vomiting to "throw off disordered bile." *Gastrointestinal Aid* Bark chewed for stomach trouble. *Hypotensive* Bark chewed for high blood pressure. *Liver Aid* Infusion of bark given to induce vomiting to "throw off disordered bile." *Orthopedic Aid* Used for sprains and as poultice for sore or swollen muscles. *Pediatric Aid* Used as an antiemetic and cold infusion of bark used as antiemetic for children. *Stimulant* Inner bark and leaves used as a stimulant. (80:54) **Delaware** *Kidney Aid* Roots and blue flag roots used for gallstones. (176:36) **Delaware, Oklahoma** *Liver Aid* Root combined with root of *Iris versicolor* and used for gallstones. (as *Hydranga aborescens* 175:30, 76)
- *Food*—**Cherokee** *Beverage* Peeled branches and twigs boiled to make tea. (126:54) *Unspecified* New growth of young twigs peeled, boiled thoroughly, fried, and eaten. (204:253) *Vegetable* Peeled branches and twigs cooked in grease like green beans. (126:54)

Hydrangea cinerea Small, Ashy Hydrangea
- *Drug*—**Cherokee** *Antiemetic* Infusion of bark scrapings taken for vomiting bile. *Cathartic* Infusion of roots taken as a cathartic by women during menses. *Emetic* Infusion of roots taken as an emetic by women during menses. *Gynecological Aid* Infusion of roots taken as an emetic and cathartic by women during menses. *Liver Aid* Infusion of bark scrapings taken for vomiting bile. (177:25)

Hydrastis, Ranunculaceae

Hydrastis canadensis L., Golden Seal
- *Drug*—**Cherokee** *Cancer Treatment* Used for cancer. *Dermatological Aid* Used as a tonic and wash for local inflammations. *Dietary Aid* Used to improve the appetite. *Gastrointestinal Aid* Used for "general debility" and "dyspepsy." *Stimulant* Used for cancer, general debility, dyspepsia, and to improve appetite. *Tonic* Used as a tonic and wash for local inflammations. (80:36) **Iroquois** *Antidiarrheal* Decoction of roots taken for whooping cough and diarrhea. *Carminative* Infusion of powdered root taken for gas. *Ear Medicine* Compound infusion with roots used as drops for earaches. *Emetic* Infusion of roots taken as an emetic for biliousness. *Eye Medicine* Compound decoction of plants taken for scrofula and used as drops for sore eyes. *Febrifuge* Infusion or decoction of roots taken for fevers. *Gastrointestinal Aid* Infusion of powdered root taken for sour stomach. *Heart Medicine* Infusion of roots with whisky taken for heart trouble. *Liver Aid* Infusion of powdered root taken for liver trouble, gall, sour stomach, and gas. *Pulmonary Aid* Infusion or decoction of roots taken for pneumonia. *Stimulant* Infusion of roots with whisky taken for run-down system. *Tuberculosis Remedy* Infusion or decoction of plants taken for tuberculosis, especially scrofula. (87:324) **Micmac** *Dermatological Aid* Root used for chapped or cut lips. (40:57)
- *Dye*—**Cherokee** *Unspecified* Used to make a dye. (80:36)

Hydrocharitaceae, see *Elodea*

Hydrocotyle, Apiaceae

Hydrocotyle poltata, Po-he-po-he
- *Drug*—**Hawaiian** *Pulmonary Aid* Plant used for lung troubles. *Strengthener* Plant used for general body weakness. *Venereal Aid* Plant used for diseases of the sexual organs. (2:74)

***Hydrocotyle* sp.**, Marsh Pennywort
- *Food*—**Cahuilla** *Vegetable* Plant used for greens. (15:79)

Hydrocotyle umbellata L., Manyflower Marshpennywort
- *Drug*—**Seminole** *Cough Medicine*, *Respiratory Aid*, and *Sedative* Roots, or whole plant, used for turtle sickness: trembling, short breath and cough. (169:237)

Hydrophyllaceae, see *Eriodictyon*, *Hydrophyllum*, *Nama*, *Nemophila*, *Phacelia*, *Pholistoma*, *Turricula*

Hydrophyllum, Hydrophyllaceae

Hydrophyllum canadense L., Bluntleaf Waterleaf
- **Drug**—Iroquois *Antidote* Compound infusion of roots taken as antidote for poisons. (87:420)

Hydrophyllum fendleri var. *albifrons* (Heller) J. F. Macbr., White Waterleaf
- **Food**—Okanagon *Forage* Thick roots eaten by cattle. *Unspecified* Thick roots cooked and eaten. **Thompson** *Forage* Thick roots eaten by cattle. *Unspecified* Thick roots cooked and eaten. (as *H. albifrons* 125:37)

Hydrophyllum occidentale (S. Wats.) Gray, Western Waterleaf
- **Food**—Okanagon *Staple* Roots used as a principal food. (178:238) **Thompson** *Forage* Roots eaten by cattle. *Unspecified* Root cooked and eaten. (164:480)

Hydrophyllum tenuipes Heller, Pacific Waterleaf
- **Food**—Cowlitz *Unspecified* Roots broken and eaten. (79:45)

Hydrophyllum virginianum L., Shawnee Salad
- **Drug**—Iroquois *Oral Aid* Decoction or chewed roots used as wash for cracked lips and mouth sores. (87:420) **Menominee** *Analgesic* Compound decoction of root used for chest pain. (54:130) *Antidiarrheal* Astringent root used for flux. (151:37) **Ojibwa** *Antidiarrheal* and *Pediatric Aid* Root used by men, women, or children to "keep flux in check." (153:371)
- **Food**—Iroquois *Vegetable* Young plants or leaves cooked and seasoned with salt, pepper, or butter. (196:117) **Menominee** *Vegetable* Leaves wilted in maple sap vinegar, simmered, and boiled in fresh water with pork and fine meal. (151:68) **Ojibwa** *Fodder* Roots fed to ponies to make them fatten rapidly. (153:405) Root chopped and put into pony feed to make them grow fat and have glossy hair. (153:419)

Hylocomiaceae, see *Hylocomium*

Hylocomium, Hylocomiaceae, moss

Hylocomium splendens (Hedw.) B.S.G.
- **Fiber**—Bella Coola *Mats, Rugs & Bedding* Used for padding and bedding. (184:196)

Hymenoclea, Asteraceae

Hymenoclea monogyra Torr. & Gray, Singlewhorl Burrobush
- **Food**—Seri *Unspecified* Seeds used for food. (50:136)

Hymenoclea sp., Burrobush
- **Fiber**—Hualapai *Brushes & Brooms* Used to make brushes and brooms. (195:47)
- **Other**—Hualapai *Fuel* Used as kindling to ignite sparks from the friction of fire sticks. *Hunting & Fishing Item* Used to make arrow shafts. (195:47)

Hymenopappus, Asteraceae

Hymenopappus filifolius Hook., Fineleaf Hymenopappus
- **Drug**—Zuni *Dermatological Aid* Poultice of chewed root with lard applied to swellings. *Emetic* Warm decoction of root taken as an emetic. (166:54, 55)
- **Food**—Hopi *Bread & Cake* Leaves boiled, rubbed with cornmeal, and baked into bread. (32:29) **Zuni** *Candy* Root used as chewing gum. (166:68)

Hymenopappus filifolius var. *cinereus* (Rydb.) I. M. Johnston, Fineleaf Hymenopappus
- **Drug**—Navajo, Ramah *Cough Medicine* Decoction of plant taken for cough. *Panacea* Cold infusion of root used as "life medicine." (191:52)
- **Other**—Keres, Western *Unspecified* Plant known and named but no use was specified. (as *H. arenosus* 171:48)

Hymenopappus filifolius var. *lugens* (Greene) Jepson, Idaho Hymenopappus
- **Drug**—Hopi *Ceremonial Medicine* and *Emetic* Compound containing plant used as a ceremonial emetic. (as *H. lugens* 200:97) *Toothache Remedy* Root chewed for decaying teeth. (as *H. lugens* 200:33, 97) **Navajo** *Blood Medicine* Decoction of whole plant taken for blood poisoning. (as *H. nudatus* 55:88) **Navajo, Kayenta** *Dermatological Aid* Poultice of plant applied to sores caused by bird infections. *Other* Plant used for illness caused by lunar eclipse. (as *H. lugens* 205:48) **Navajo, Ramah** *Dermatological Aid* Infusion or decoction of plant taken and used as a lotion for arrow or bullet wound. (as *H. lugens* 191:52)

Hymenopappus filifolius var. *pauciflorus* (I. M. Johnston) B. L. Turner, Fineleaf Hymenopappus
- **Food**—Hopi *Beverage* Used to make tea and coffee. (as *H. pauciflorus* 42:326)
- **Dye**—Hopi *Unspecified* Used for dye. (as *H. pauciflorus* 42:326)

Hymenopappus newberryi (Gray) I. M. Johnston, Newberry's Hymenopappus
- **Drug**—Isleta *Gastrointestinal Aid* Infusion of plant taken for stomachache. Dried, ground plants made into a powder and used on the stomach for stomachaches. *Pediatric Aid* Dried, ground plants made into a powder and used on children's stomachs for stomachaches. (as *Leucampyx newberri* 100:34)

Hymenopappus sp.
- **Food**—Isleta *Beverage* Leaves and stems used to make a beverage. Plant kept well in storage and used to make a beverage in all seasons. (100:32) **Jemez** *Beverage* Little bundles of plant steeped into tea. (44:24)

Hymenopappus tenuifolius Pursh, Chalk Hill Hymenopappus
- **Drug**—Lakota *Veterinary Aid* Plant made into a tea and salve used for horses' hooves. (139:38)

Hymenoxys, Asteraceae

Hymenoxys bigelovii (Gray) Parker, Bigelow's Rubberweed
- **Drug**—Hopi *Antirheumatic (External)* Used for severe pains in hips and back. *Cathartic* Used as a purge. *Gynecological Aid* Used for severe pains in hips and back, especially in pregnant state. *Stimulant* Used as a stimulant. *Unspecified* Infusion of plant used for medicinal tea. (42:328)

Hymenoxys cooperi (Gray) Cockerell, Cooper's Hymenoxys
- **Food**—Hopi *Beverage* Used to make tea. (42:329)
- **Dye**—Hopi *Unspecified* Used for a dye. (42:329)

- *Other*—**Hopi** *Ceremonial Items* Used for peach tree pahos (prayer sticks). (42:329)

Hymenoxys richardsonii (Hook.) Cockerell, Pingue Hymenoxys
- *Drug*—**Zuni** *Dermatological Aid* Poultice of chewed root applied to sores and rashes. *Gastrointestinal Aid* Infusion of root taken for stomachache. (as *H. richarsonii* 27:375)
- *Food*—**Navajo** *Candy* Plant used as a chewing gum. (as *Actinella richardsoni* 55:80)

Hymenoxys richardsonii var. *floribunda* (Gray) Parker, Colorado Rubberweed
- *Drug*—**Isleta** *Psychological Aid* Leaves characterized as making cattle crazy. (as *H. floribunda* 100:32) **Keres, Western** *Poison* Plant considered poisonous to sheep. (as *H. floribunda* 171:48)
- *Food*—**Isleta** *Candy* Roots used as chewing gum. (as *H. floribunda* 100:32) **Keres, Western** *Candy* Root used as chewing gum. (as *H. floribunda* 171:48) **Spanish American** *Candy* Roots chewed as chewing gum. (as *H. floribunda* 32:30) **Tewa** *Candy* Root skins pounded and the gummy material chewed as gum. (as *H. floribunda* 138:56)
- *Dye*—**Navajo** *Yellow* Flowers used as a yellow dye for wool. (as *H. metcalfei* 55:88)

Hymenoxys richardsonii (Hook.) Cockerell **var. *richardsonii***, Pingue Hymenoxys
- *Drug*—**Navajo, Ramah** *Ceremonial Medicine* Plant used as a ceremonial emetic. *Dermatological Aid* Decoction of plant taken and used as lotion for red ant bites. *Emetic* Plant used as a ceremonial emetic. *Poison* Toxic to livestock, especially sheep. (as *Actinea richardsoni* 191:47)

Hymenoxys sp.
- *Drug*—**Isleta** *Venereal Aid* Decoction of leaves used for gonorrhea. (100:32)

Hypericum, Clusiaceae
Hypericum ascyron L. Great St. John's Wort
- *Drug*—**Menominee** *Kidney Aid* Compounded with blackcap raspberry root and used for kidney troubles. *Pulmonary Aid* Compound containing root used for weak lungs and as a specific for consumption. *Tuberculosis Remedy* Root, thought to be a "specific," used in the first stages of consumption. (151:37, 38) **Meskwaki** *Snakebite Remedy* Powder of boiled root applied to draw poison from water moccasin bite. *Tuberculosis Remedy* Compound containing root used for consumption in the first stages. (152:223)

Hypericum brachyphyllum (Spach) Steud., Coastalplain St. John's Wort
- *Drug*—**Seminole** *Cathartic* Plant used as a cathartic. (as *H. aspalathoides* 169:275)

Hypericum concinnum Benth., Gold Wire
- *Drug*—**Miwok** *Dermatological Aid* Decoction of plant used as a wash for running sores. (12:171)

Hypericum crux-andreae (L.) Crantz, St. Peter's Wort
- *Drug*—**Choctaw** *Analgesic* Decoction of root used for colic. *Eye Medicine* Decoction of leaves used as wash for sore eyes. (as *Aseyrum crux andreae* 25:23)

Hypericum ellipticum Hook., Pale St. John's Wort
- *Drug*—**Iroquois** *Abortifacient* Decoction of stems taken after the remedy for the suppression of menses. (87:386)

Hypericum fasciculatum Lam., Peelbark St. John's Wort
- *Drug*—**Seminole** *Cathartic* Infusion of roots taken for rat sickness: blocked urination and bowels. (169:231) Plant used as a cathartic. (169:275) *Urinary Aid* Infusion of roots taken for rat sickness: blocked urination and bowels. (169:231)

Hypericum gentianoides (L.) B.S.P., Orangegrass
- *Drug*—**Cherokee** *Abortifacient* Compound decoction taken "to promote menstruation." *Antidiarrheal* Infusion taken for bloody flux and bowel complaint. *Dermatological Aid* Milky substance rubbed on sores. *Febrifuge* Infusion taken for fever. *Gastrointestinal Aid* Infusion taken for bloody flux and bowel complaint. *Hemostat* Crushed plant sniffed for nosebleed. *Snakebite Remedy* Root chewed, a portion swallowed, and rest used as poultice for snakebite. *Strengthener* Infusion of root used as wash to give infants strength. *Venereal Aid* Milky substance used for venereal disease. (80:53)

Hypericum hypericoides (L.) Crantz, St. Andrew's Cross
- *Drug*—**Cherokee** *Abortifacient* Compound decoction taken "to promote menstruation." *Antidiarrheal* Infusion taken for bloody flux and bowel complaint. *Dermatological Aid* Milky substance rubbed on sores. *Febrifuge* Infusion taken for fever. *Gastrointestinal Aid* Infusion taken for bloody flux and bowel complaint. *Hemostat* Crushed plant sniffed for nosebleed. *Snakebite Remedy* Root chewed, a portion swallowed, and rest used as poultice for snakebite. *Strengthener* Infusion of root used as wash to give infants strength. *Venereal Aid* Milky substance used for venereal disease. (80:53)

Hypericum hypericoides (L.) Crantz **ssp. *hypericoides***, St. Andrew's Cross
- *Drug*—**Alabama** *Antidiarrheal* Infusion of entire plant taken for dysentery. *Eye Medicine* Infusion of plant used as an eyewash. *Orthopedic Aid* Decoction of mashed plants used as a bath for children too weak to walk. *Pediatric Aid* Decoction of mashed plants used as a bath for children too weak to walk. **Choctaw** *Eye Medicine* Infusion of leaves used as a wash for sore eyes. *Gastrointestinal Aid* Decoction of roots taken for colic. (as *Ascyrum hypericoides* 177:42) **Houma** *Analgesic* Decoction of root taken, especially in childbirth, for severe pain. *Febrifuge* Decoction of scraped roots and bark taken for fever. *Gynecological Aid* Decoction of root taken, especially in childbirth, for severe pain. *Toothache Remedy* Bark used to pack aching tooth. (as *Ascyrum hypericoides* 158:55) **Koasati** *Antirheumatic (Internal)* Decoction or infusion of leaves taken for rheumatism. (as *Ascyrum linifolium* 177:43) **Natchez** *Pediatric Aid* and *Urinary Aid* Infusion of plant given to children unable to urinate. (as *Ascyrum hypericoides* 177:42)

Hypericum multicaule, St. Peter's Wort
- *Drug*—**Alabama** *Antidiarrheal* Decoction of whole plant taken for dysentery. *Eye Medicine* Infusion of plant used as an eyewash. (as *Ascyrum multicaule* 172:664)

Hypericum perforatum L., Common St. John's Wort
- *Drug*—**Cherokee** *Abortifacient* Compound decoction taken "to pro-

mote menstruation." *Antidiarrheal* Infusion taken for bloody flux and bowel complaint. *Dermatological Aid* Milky substance rubbed on sores. *Febrifuge* Infusion taken for fever. *Gastrointestinal Aid* Infusion taken for bloody flux and bowel complaint. *Hemostat* Crushed plant sniffed for nosebleed. *Snakebite Remedy* Root chewed, a portion swallowed, and rest used as poultice for snakebite. *Strengthener* Infusion of root used as wash to give infants strength. *Venereal Aid* Milky substance used for venereal disease. (80:53) **Iroquois** *Febrifuge* Plant used as a fever medicine. *Reproductive Aid* Roots used to prevent sterility. (87:385) **Montagnais** *Cough Medicine* Decoction of plant used as a cough medicine. (156:314)

Hypericum punctatum Lam., Spotted St. John's Wort
- **Drug**—**Meskwaki** *Unspecified* Compound containing root used as a medicine. (152:223)

Hypericum scouleri Hook., Scouler's St. John's Wort
- **Drug**—**Paiute** *Analgesic* Decoction of plant used as a bath for aching feet. (180:89) *Dermatological Aid* Flowers used for perfume. (111:90) *Orthopedic Aid* Decoction of plant used as a bath for aching feet. **Shoshoni** *Analgesic* Decoction of plant used as a bath for aching feet. *Dermatological Aid* Plant used several ways as a poultice for sores, swellings, wounds, and cuts. *Orthopedic Aid* Decoction of plant used as a bath for aching feet. *Toothache Remedy* Dried root used for toothache. *Venereal Aid* Infusion of tops taken over a long period of time for venereal disease. (180:89)

Hypericum scouleri Hook. **ssp. *scouleri***, Scouler's St. John's Wort
- **Food**—**Miwok** *Dried Food* and *Staple* Eaten fresh, dried, or ground into flour and used like acorn meal. (as *H. formosum* var. *scouleri* 12:158)

Hypericum sp., St. Andrew's Cross
- **Drug**—**Natchez** *Diuretic* and *Pediatric Aid* Infusion of plant given to children unable to pass urine. (as *Ascyrum* 172:666)

Hypoxis, Liliaceae

Hypoxis hirsuta (L.) Coville, Common Goldstar
- **Drug**—**Cherokee** *Heart Medicine* Infusion taken for the heart. (80:57)

Hyptis, Lamiaceae

Hyptis emoryi Torr., Desert Lavender
- **Drug**—**Cahuilla** *Antihemorrhagic* Infusion of blossoms and leaves taken for hemorrhages. (15:79)

Hyptis pectinata (L.) Poit., Comb Bushmint
- **Drug**—**Seminole** *Dermatological Aid* Infusion of roots applied to sores and ulcers on the legs and feet. (169:307) *Psychological Aid* Leaves and fruit used for insanity. (169:293)

Hyssopus, Lamiaceae

Hyssopus officinalis L., Hyssop
- **Drug**—**Cherokee** *Abortifacient* Infusion taken to "bring on menses." *Cold Remedy* Syrup taken for colds. *Cough Medicine* Syrup taken for coughs. *Febrifuge* Infusion taken for fevers. *Pulmonary Aid* Syrup taken for "asthma and other lung and breast diseases." *Respiratory Aid* Syrup taken for "asthma and other lung and breast diseases." (80:40)

Ilex, Aquifoliaceae

Ilex aquifolium L., English Holly
- **Drug**—**Micmac** *Cough Medicine* Root used for cough. *Febrifuge* Part of plant used for fevers and root used for consumption. *Tuberculosis Remedy* Root used for consumption. *Urinary Aid* Root used for gravel. (40:57)

Ilex cassine L., Dahoon
- **Drug**—**Cherokee** *Diaphoretic* Infusion, "black drink," caused sweating to purify physically and morally. *Emetic* Strong decoction called "black drink" induced vomiting for purification. *Kidney Aid* Used for "dropsy and gravel." *Urinary Aid* Plant prepared in unspecified manner and taken for "dropsy and gravel." (80:12, 62)
- **Other**—**Seminole** *Soap* Plant used as soap. (169:466)

Ilex opaca Ait., American Holly
- **Drug**—**Alabama** *Eye Medicine* Decoction of bark used as a wash for sore eyes. (177:37) **Catawba** *Dermatological Aid* Infusion of leaves taken for sores. *Misc. Disease Remedy* Infusion of leaves taken for measles. (157:188) Decoction of leaves taken for measles. (177:37) **Cherokee** *Gastrointestinal Aid* Berries chewed for "colics" and "dyspepsia." (80:38) Berries used for colics. (203:74) *Orthopedic Aid* Leaves used to scratch cramped muscles. (80:38) **Choctaw** *Eye Medicine* Decoction of leaves used as drops for sore eyes. **Koasati** *Dermatological Aid* Infusion of bark rubbed on areas affected by itching. (177:37)
- **Dye**—**Cherokee** *Unspecified* Berries used to make a dye. (80:38)
- **Other**—**Cherokee** *Cooking Tools* Wood used to make spoons. (203:74) *Decorations* Wood used to carve. Whole plant used for Christmas trees. (80:38)

Ilex sp., Holly
- **Drug**—**Alabama** *Eye Medicine* Decoction of inner bark used as an eyewash. (172:665)
- **Food**—**Comanche** *Beverage* Leaves used to make a beverage. (29:522)

Ilex verticillata (L.) Gray, Common Winterberry
- **Drug**—**Iroquois** *Cathartic* Decoction of bark taken as a physic. *Emetic* Decoction of bark taken as an emetic. *Gastrointestinal Aid* Plant taken for biliousness. *Other* Taken to retain vigor. *Psychological Aid* Decoction of bark taken as an emetic for craziness. (87:373) *Respiratory Aid* Compound decoction of roots taken for hay fever. (87:374) **Ojibwa** *Antidiarrheal* Bark used for diarrhea. (153:355)

Ilex vomitoria Ait., Yaupon
- **Drug**—**Alabama** *Ceremonial Medicine* Plant taken to "clear out the system and produce ceremonial purity." (172:666) *Emetic* Decoction of toasted leaves taken as an emetic. **Cherokee** *Emetic* Infusion of leaves taken as an emetic. (177:38) *Hallucinogen* Used to "evoke ecstasies." (80:12, 62) **Creek** *Cathartic* "Black drink" used to "clear out the system." (172:666) *Emetic* Decoction of leaves and shoots taken as an emetic. **Natchez** *Emetic* Plant used as an emetic. (177:38) **Seminole** *Psychological Aid* Bark used as medicine for old people's dance sickness: nightmarish dreams and waking up talking. (169:261)
- **Other**—**Seminole** *Hunting & Fishing Item* Plant used to make arrows and ramrods. (169:476)

Impatiens, Balsaminaceae

Impatiens capensis Meerb., Jewelweed
- ***Drug***—**Cherokee** *Ceremonial Medicine* Used as an ingredient in green corn medicine. *Dermatological Aid* Juice rubbed on "ivy poisoning" and infusion of root used for babies with hives. *Gastrointestinal Aid* Crushed leaves rubbed on "child's sour stomach." *Gynecological Aid* Decoction taken and used to "bathe private parts" to aid in delivery. (80:41) Decoction of stems taken to ease childbirth. (as *I. biflora* 177:40) *Misc. Disease Remedy* Infusion of leaf taken for measles. *Pediatric Aid* Infusion of root used for babies with "bold hives" and leaves used for "child's sour stomach." (80:41) **Chippewa** *Dermatological Aid* Poultice of bruised stems applied to rashes or other skin troubles. (as *I. biflora* 71:136) **Iroquois** *Dermatological Aid* Compound decoction of plants taken and used as a wash for liver spots. *Diuretic* Infusion of roots taken to increase urination. *Eye Medicine* Poultice of smashed stems applied to sore or raw eyelids. *Febrifuge* Cold infusion of plants taken for fevers. *Kidney Aid* Decoction of plants taken for kidney problems and dropsy. *Liver Aid* Compound decoction of plants taken and used as a wash for liver spots. *Urinary Aid* Decoction of plants taken for stricture or for difficult urination. (as *I. biflora* 87:380) **Malecite** *Liver Aid* Infusion of leaves used for jaundice. (as *I. biflora* 116:256) **Meskwaki** *Dermatological Aid* Poultice of fresh plant applied to sores and juice used for nettle stings. (as *I. biflora* 152:205) **Micmac** *Liver Aid* Herbs used for jaundice. (as *I. biflora* 40:57) **Mohegan** *Burn Dressing* Compound of balsam buds and rum used as ointment for burns. (as *I. biflora* 174:269) Poultice of crushed buds applied to burns. (as *I. biflora* 176:72) Poultice of crushed flower buds applied to burns. (as *I. biflora* 176:72, 130) *Dermatological Aid* Compound of balsam buds and rum used as ointment for cuts. (as *I. biflora* 174:269) Poultice of crushed flower buds applied to cuts and bruises. (as *I. biflora* 176:72, 130) *Orthopedic Aid* Compound of balsam buds and rum used as ointment for bruises. **Nanticoke** *Burn Dressing* Compound of balsam buds and rum used as ointment for burns. (as *I. biflora* 174:269) Infusion of plant taken and poultice of leaves applied to burns. (as *I. biflora* 175:57, 84) *Dermatological Aid* Compound of balsam buds and rum used as ointment for cuts. *Orthopedic Aid* Compound of balsam buds and rum used as ointment for bruises. (as *I. biflora* 174:269) **Ojibwa** *Analgesic* Juice of fresh plant rubbed on head for headache. *Unspecified* Infusion of leaves used medicinally for unspecified purpose. (as *I. biflora* 153:357, 358) **Omaha** *Dermatological Aid* Poultice of crushed stems and leaves applied to skin for rash and eczema. (as *I. biflora* 70:101) **Penobscot** *Burn Dressing* Compound of balsam buds and rum used as ointment for burns. *Dermatological Aid* Compound of balsam buds and rum used as ointment for cuts. *Orthopedic Aid* Compound of balsam buds and rum used as ointment for bruises. (as *I. biflora* 174:269) **Potawatomi** *Analgesic* Infusion of whole plant taken for stomach cramps and used as a liniment for soreness. *Dermatological Aid* Fresh juice of plant used as a wash on nettle stings or poison ivy rash. Infusion of whole plant used as a liniment for sprains and bruises. *Gastrointestinal Aid* Infusion of whole plant taken for stomach cramps. *Orthopedic Aid* Decoction of plant used as a liniment for sprains, bruises, and soreness. *Pulmonary Aid* Infusion of whole plant taken for chest cold. (as *I. biflora* 154:42) **Shinnecock** *Dermatological Aid* Salve made of balsam buds and Vaseline. (as *I. biflora* 30:122)
- ***Dye***—**Menominee** *Orange-Yellow* Whole plant used to make an orange-yellow dye. (as *I. biflora* 151:78) **Ojibwa** *Yellow* Whole plant used to make a yellow dye, the material boiled in the mixture with rusty nails. (as *I. biflora* 153:425) **Potawatomi** *Orange* Material placed in pot of boiling plant juice to dye it orange. *Yellow* Material placed in pot of boiling plant juice to dye it yellow. (as *I. biflora* 154:116)

Impatiens pallida Nutt., Pale Touchmenot
- ***Drug***—**Cherokee** *Ceremonial Medicine* Used as an ingredient in green corn medicine. *Dermatological Aid* Juice rubbed on "ivy poisoning" and infusion of root used for babies with hives. *Gastrointestinal Aid* Crushed leaves rubbed on "child's sour stomach." *Gynecological Aid* Decoction taken and used to "bathe private parts" to aid in delivery. *Misc. Disease Remedy* Infusion of leaf taken for measles. *Pediatric Aid* Infusion of root used for babies with "bold hives" and leaves used for "child's sour stomach." (80:41) **Iroquois** *Dermatological Aid* Smashed stalks and juice rubbed on poison ivy blisters and mosquito bites. *Febrifuge* Cold infusion of plants taken for fevers. *Gynecological Aid* Infusion of plant taken to induce childbirth. Infusion of stalks taken to stop suffering while having a baby. Poultice of mashed plants applied to women's breast injury. (87:379) **Ojibwa** *Dermatological Aid* Juice rubbed on sores. (5:2311) **Omaha** *Dermatological Aid* Poultice of crushed stems and leaves applied to skin for rash and eczema. (70:101)

***Impatiens* sp.**
- ***Drug***—**Creek** *Kidney Aid* Compound decoction of plant taken and used as a wash for dropsy. (172:663)

Inula, Asteraceae

Inula helenium L., Elecampane Inula
- ***Drug***—**Cherokee** *Cough Medicine* Root used for coughs. *Gynecological Aid* "For female obstructions and pregnant women with weak bowels and wombs." *Pulmonary Aid* Root used for lung disorders. *Respiratory Aid* Root used for asthma. *Tuberculosis Remedy* Root used for "consumption." (80:33) **Delaware** *Gastrointestinal Aid* Roots made into a tonic and used to strengthen digestive organs. Roots made into a tonic and used to remove intestinal mucus. (176:37) *Tonic* Roots used with black snakeroot and stone root as a tonic. (176:33) **Delaware, Oklahoma** *Gastrointestinal Aid* Root used in a tonic to strengthen the digestive organs. *Laxative* Root used in a tonic to remove mucus from the intestines. (175:31) *Tonic* Root used in tonic taken for "strengthening digestive organs." (175:31, 76) **Delaware, Ontario** *Cold Remedy* Compound decoction of root or bark taken for colds. (175:67, 82) **Iroquois** *Analgesic* Compound roots used for chest pains. (87:465) *Antirheumatic (External)* Poultice of leaves or roots applied for rheumatism and arthritic sores. (87:467) *Carminative* Decoction or cold infusion of powdered roots taken for stomach gas. (87:465) *Cathartic* Compound roots used to clean out the intestines. (87:467) *Cold Remedy* Plant used for colds. *Cough Medicine* Compound decoction of roots taken for coughs or heaves. (87:466) *Dermatological Aid* Poultice of smashed plants applied to sores, cuts, or arthritic sores. (87:465) *Diuretic* Compound decoction of leaves and roots taken as a diuretic. *Febrifuge* Decoction of powdered plants or dried roots taken for fevers. *Gastrointestinal Aid* Compound decoction of roots and flowers taken for bruise on back of stomach. (87:

466) Infusion of roots and other plant branches taken for large stomachs. (141:64) *Gynecological Aid* Compound decoction of dried roots taken by girls who "leak rotten." *Heart Medicine* Decoction of dried roots taken for fevers, tuberculosis, and heart troubles. (87:466) Decoction of roots taken for stroke. (87:467) *Misc. Disease Remedy* Compound decoction of powdered plants taken for fevers or typhoid. *Panacea* Plant used as medicine for anything. (87:466) *Pediatric Aid* Dried leaves given to children for asthma. (87:467) *Pulmonary Aid* Compound roots used for chest pains. (87:465) *Respiratory Aid* Infusion of roots taken for asthma. (87:466) Dried leaves given to children for asthma. (87:467) *Tuberculosis Remedy* Decoction of leaf or root or infusion of one root taken for consumption. (87:466) *Veterinary Aid* Powdered roots mixed with horse's feed or decoction of root given for heaves. (87:465) **Malecite** *Analgesic* Dried roots finely powdered and snuffed for headaches. *Cold Remedy* Infusion of roots used for colds. *Heart Medicine* Infusion of roots used for heart trouble. (116:248) **Micmac** *Analgesic* Root used for headaches. *Cold Remedy* Root used for colds. *Heart Medicine* Root used for heart trouble. (40:57) **Mohegan** *Pulmonary Aid* Infusion of plant taken for lungs. (174:266) Leaves used for lung trouble. (176:130) *Tuberculosis Remedy* Infusion of leaves taken for tuberculosis. (176:72, 130) *Veterinary Aid* Infusion of plant given to horses for colic. (174:266) Infusion of leaves used for horses with colic. (176:72, 130)

Iodanthus, Brassicaceae

Iodanthus pinnatifidus (Michx.) Steud., Purple Rocket
- *Drug*—**Meskwaki** *Love Medicine* Decoction of root used as a paint by women for a love medicine. (152:219, 220) *Poultice* Poultice used on head of old man who is cold, to bring warmth to whole body. (152:219)

Ipomoea, Convolvulaceae

Ipomoea batatas (L.) Lam., Sweet Potato
- *Food*—**Cherokee** *Vegetable* Potatoes used for food. (as *Impomoea batatas* 80:51) **Seminole** *Vegetable* Tubers eaten. (169:465)

Ipomoea cairica (L.) Sweet, Mile a Minute Vine
- *Food*—**Hawaiian** *Unspecified* Tubers grated, roasted, and eaten. (as *I. tuberculata* 112:69)

Ipomoea indica (Burm. f.) Merr., Oceanblue Morningglory
- *Drug*—**Hawaiian** *Analgesic* Poultice of pounded flowers, leaves, and salt applied to the back for pain. *Dermatological Aid* Poultice of pounded roots, other plants, and the resulting liquid applied to flesh wounds. *Laxative* Whole plant and other plants baked and eaten as a laxative. *Orthopedic Aid* Poultice of pounded roots, other plants, and the resulting liquid applied to broken bones. *Pediatric Aid* and *Strengthener* Flowers chewed by mothers and given to infants for general weakness. (as *Impomea insularis* 2:52)

Ipomoea leptophylla Torr., Bush Morningglory
- *Drug*—**Keres, Western** *Gastrointestinal Aid* Infusion of staminate cones used as a stomach tonic. (171:48) **Keresan** *Veterinary Aid* Dried, ground root added to water and given to colts to cause them to become large horses. Infusion of dried, pulverized root used for fertility of mares and growth of colts. (198:559) **Lakota** *Gastrointestinal Aid* Root scraped and eaten raw for stomach troubles. (139:43) **Pawnee** *Analgesic* Pulverized root dusted on body for pain. *Sedative* Root used in smoke treatment for nervousness and bad dreams. *Stimulant* Pulverized root used to revive "one who had fainted." (70:110) **Sia** *Veterinary Aid* Infusion of ground roots used to promote the fertility of horses and the growth of the colts. (199:284)
- *Food*—**Arapaho** *Starvation Food* Root roasted for food when pressed by hunger. **Cheyenne** *Starvation Food* Root roasted for food when pressed by hunger. **Kiowa** *Starvation Food* Root roasted for food when pressed by hunger. (19:13)
- *Other*—**Keres, Western** *Unspecified* Plant known and named but no use was specified. (171:48) **Keresan** *Ceremonial Items* Root used to make hindquarters of little hobbyhorse "ridden" by saints in ceremonial impersonations. (198:559) **Lakota** *Fuel* Roots used in place of matches. It is said that in olden days when there were no matches, they used to start a fire in the root, wrap it up, and hang it outside. The fire would keep for 7 months. (139:43)

Ipomoea pandurata (L.) G. F. W. Mey., Man of the Earth
- *Drug*—**Cherokee** *Antirheumatic (External)* Poultice of root applied to rheumatism. *Cough Medicine* Taken for coughs. *Diuretic* Taken as a diuretic. *Expectorant* Taken as an expectorant. *Kidney Aid* Taken for "dropsy." *Laxative* Taken as a laxative. *Misc. Disease Remedy* Infusion of root taken for cholera morbis. *Respiratory Aid* Taken for asthma. *Tuberculosis Remedy* Taken for consumption. *Urinary Aid* Taken for "gravel" and "suppression of urine." (80:51) **Creek** *Diuretic* Plant used as a diuretic. *Kidney Aid* Plant used "nephritic complaints." (172:670) **Iroquois** *Analgesic* Decoction of roots taken for abdominal pains. Infusion of powdered plants taken for headaches. *Blood Medicine* Compound infusion of bark, roots, and leaves taken as blood purifier. *Cough Medicine* Decoction of roots taken as a cough medicine. *Gastrointestinal Aid* Decoction of roots taken for abdominal pains. Infusion of powdered plants taken for upset stomachs. *Liver Aid* Plant used for the liver. *Misc. Disease Remedy* Infusion of dried roots taken for all kinds of diseases. *Other* Compound infusion taken or injured parts washed, a "Little Water Medicine." *Tuberculosis Remedy* Decoction of roots taken for initial stages of tuberculosis. *Witchcraft Medicine* Plant had magical potency. (87:419)
- *Food*—**Cherokee** *Unspecified* Roots used for food. (80:51) *Vegetable* Potatoes used for food. (as *Impomoea pandurata* 80:21)
- *Other*—**Cherokee** *Insecticide* Infusion of vine used for soaking sweet potatoes to keep away bugs and moles. (80:51)

Ipomoea pes-caprae (L.) R. Br., Bayhops
- *Drug*—**Hawaiian** *Reproductive Aid* Plant used by expectant mothers. (as *Impomea pes-caprae* 2:73)

Ipomoea sagittata Poir., Saltmarsh Morningglory
- *Drug*—**Houma** *Blood Medicine* Decoction of root taken to remove poison from the blood or heart. *Dermatological Aid* Poultice of boiled leaves applied to swellings. *Heart Medicine* Hot decoction of roots taken to "take poison out of the blood or heart." *Snakebite Remedy* Leaf chewed and juice swallowed or poultice of chewed leaves used on snakebite. (158:62, 63)

Ipomoea sp., Uwala
- *Drug*—**Hawaiian** *Emetic* Infusion of tubers and other plants strained and taken to vomit stomach contents that caused vomiting. *Gynecol-*

ogical Aid Tubers and other plants pounded, resulting liquid strained, heated, and taken for womb troubles. *Laxative* Tubers and other plants pounded, resulting liquid strained and taken for constipation. Tubers cooked with taro leaves and nuts and used as a laxative for children and adults. *Pediatric Aid* Tubers used to strengthen children. Tubers cooked with taro leaves and nuts and used as a laxative for children and adults. *Respiratory Aid* Tubers and other plants pounded, resulting liquid heated and taken for asthma. *Sedative* Tubers and other plants pounded, resulting liquid strained and taken for lack of sleep. *Strengthener* Tubers used to strengthen children. (2:35)
- *Food*—Hawaiian *Vegetable* Tubers eaten. (2:35)

Ipomoea tiliacea (Willd.) Choisy, Darkeye Morningglory
- *Drug*—Hawaiian *Laxative* Plant used to make a laxative. (as *Argyreia tiliaefolia* 2:73)

Ipomopsis, Polemoniaceae

Ipomopsis aggregata (Pursh) V. Grant, Skyrocket Gilia
- *Drug*—Great Basin Indian *Blood Medicine* Infusion of whole plant used for blood disease. (as *Gilia appregata* 121:49)

Ipomopsis aggregata (Pursh) V. Grant **ssp. *aggregata***, Skyrocket Gilia
- *Drug*—Hopi *Gynecological Aid* Plant used after birth when the mother lied in bed for 15 or 20 days. (as *Gilia aggregata* 42:321) **Navajo, Kayenta** *Cathartic* Plant used as a cathartic. *Dermatological Aid* Plant used for spider bites. *Emetic* Plant used as an emetic. *Gastrointestinal Aid* Plant used for stomach disease. (as *Gilia aggregata* 205:37) **Navajo, Ramah** *Hunting Medicine* Cold infusion taken and applied to body of hunter and weapons for good luck. (as *Gilia aggregata* 191:39) **Okanagan-Colville** *Febrifuge* Infusion of roots taken for high fevers. *Laxative* Infusion of leaves and stalks taken for constipation and to "clean out your system." Infusion of roots taken as a laxative. (as *Gilia aggregata* 188:111) **Paiute** *Cathartic* Simple or compound decoction of plant or root taken as a physic. *Cold Remedy* Decoction of root taken as a cold remedy. *Emetic* Simple or compound decoction of plant or root taken as an emetic. (as *Gilia aggregata* 180:76, 77) **Salish** *Dermatological Aid* Decoction of plants used as a face and hair wash by adolescent girls. *Eye Medicine* Decoction of plants used as an eyewash. *Pediatric Aid* Decoction of plants used as a face and hair wash by adolescent girls. (as *Gilia aggregata* 178:294) **Shoshoni** *Analgesic* Poultice of crushed, whole plant applied for rheumatic aches. *Antirheumatic (External)* Poultice of crushed plant applied for rheumatic aches. *Blood Medicine* Decoction of plant taken as a blood tonic. *Cathartic* Decoction of plant or root taken as a physic. *Dermatological Aid* Decoction of whole plant used as a disinfecting wash for the itch. *Disinfectant* Decoction of whole plant used as a disinfectant wash for the itch. *Emetic* Compound decoction of roots used to induce vomiting. Decoction of plant or root taken as an emetic. *Tonic* Simple or compound decoction of whole plant taken as a blood tonic. *Venereal Aid* Simple or compound decoction of plant taken and used as a wash for gonorrhea and syphilis. (as *Gilia aggregata* 180:76, 77)
- *Food*—Hopi *Beverage* Boiled for a drink. (as *Gilia aggregata* 42:321) **Klamath** *Snack Food* Nectar sucked from flowers by children. (as *Gilia aggregata* 45:103)
- *Dye*—Hopi *Unspecified* Plant used for dye. (as *Gilia aggregata* 42:321)
- *Other*—Hopi *Decorations* Plant used for decoration. (as *Gilia aggregata* 42:321) **Ute** *Fasteners* Whole plant boiled for glue. (as *Gilia aggregata* 118:56)

Ipomopsis aggregata ssp. *attenuata* (Gray) V. & A. Grant, Scarlet Skyrocket
- *Drug*—Navajo *Gastrointestinal Aid* Infusion of crushed, dried leaves taken for stomach troubles. (as *Gilia attenuata* 55:70)
- *Food*—Navajo *Forage* Used as a browse plant. (as *Gilia attenuata* 90:160)
- *Other*—Navajo *Decorations* Cultivated as an ornamental flower. (as *Gilia attenuata* 90:160)

Ipomopsis aggregata ssp. *candida* (Rydb.) V. & A. Grant, Scarlet Skyrocket
- *Other*—Keres, Western *Unspecified* Plant known and named but no use was specified. (as *Gilia greeneana* 171:45)

Ipomopsis congesta (Hook.) V. Grant **ssp. *congesta***, Ballhead Gilia
- *Drug*—Great Basin Indian *Analgesic* Poultice of dried, powdered blossoms applied for pain. (as *Gilia congesta* 121:49) **Paiute** *Antidiarrheal* Decoction of plant taken for diarrhea. *Cathartic* Decoction of plant taken as a physic. *Cold Remedy* Decoction of plant taken for colds. *Emetic* Decoction of plant taken as an emetic. *Gastrointestinal Aid* Decoction of plant taken for indigestion and stomach trouble. *Venereal Aid* Plant used in many ways for venereal diseases. **Shoshoni** *Antidiarrheal* Decoction of plant taken for diarrhea. *Cathartic* Decoction of plant taken as a physic. *Cold Remedy* Decoction of plant taken for colds. *Dermatological Aid* Decoction of plant used as an antiseptic wash for wounds, cuts, sores, and bruises. *Disinfectant* Decoction of plant used as an antiseptic wash for skin problems. *Emetic* Decoction of plant taken as an emetic. *Eye Medicine* Decoction or infusion of plant used as an eyewash. *Gastrointestinal Aid* Decoction of plant taken for indigestion and stomach trouble. *Kidney Aid* Decoction of plant taken for "kidney complaint." *Liver Aid* Decoction of plant taken for liver trouble. *Misc. Disease Remedy* Decoction of plant taken for influenza. Poultice of boiled, drained, and mashed plant applied for erysipelas. *Venereal Aid* Plant used in many ways for venereal diseases. *Veterinary Aid* Poultice of crushed, raw plants applied to back sores on horses. **Washo** *Antidiarrheal* Decoction of plant taken for diarrhea. *Cathartic* Decoction of plant taken as a physic. *Cold Remedy* Decoction of plant taken for colds. *Emetic* Decoction of plant taken as an emetic. *Gastrointestinal Aid* Decoction of plant taken for indigestion and stomach trouble. *Kidney Aid* Infusion of crushed plant taken and poultice applied for dropsy. (as *Gilia congesta* 180:77–80)

Ipomopsis gunnisonii (Torr. & Gray) V. Grant, Sanddune Skyrocket
- *Drug*—Navajo, Kayenta *Blood Medicine* Plant used as a blood purifier. *Dermatological Aid* Poultice of plant applied to sores. (as *Gilia gunnisoni* 205:37)

Ipomopsis laxiflora (Coult.) V. Grant, Iron Skyrocket
- *Drug*—Keres, Western *Emetic* Infusion of roots used as an emetic to

eliminate the ozone in cases of lightning shock. (as *Gilia laxiflora* 171:45)

Ipomopsis longiflora (Torr.) V. Grant **ssp. *longiflora***, Flax-flowered Gilia
- **Drug**—**Hopi** *Analgesic* Decoction of leaves used for stomachache. (as *Gilia longiflora* 200:87) *Gastrointestinal Aid* Decoction of leaves taken for stomachache. (as *Gilia longiflora* 200:33, 87) **Keres, Western** *Emetic* Infusion of roots used as an emetic to eliminate the ozone in cases of lightning shock. (as *Gilia longiflora* 171:45) **Navajo** *Ceremonial Medicine* Plant used as medicine in the Wind and Female Shooting Chants. *Emetic* Decoction of pounded plant taken to vomit. *Gastrointestinal Aid* Decoction of pounded plant taken for the bowels. *Veterinary Aid* Infusion of flowers mixed with feed and given to sheep for stomach troubles. (as *Gilia longiflora* 55:70) **Navajo, Kayenta** *Blood Medicine* and *Gynecological Aid* Plant used for postpartum septicemia. (as *Gilia longiflora* 205:38) **Navajo, Ramah** *Analgesic* Plant used for stomachache and arthritis. *Antirheumatic (Internal)* Plant used for arthritis. *Ceremonial Medicine* Plant used as ceremonial eyewash and chant lotion. *Dermatological Aid* Infusion of plant used as hair tonic to lengthen hair and prevent baldness. *Disinfectant* Plant used for "deer infection" and "snake infection." *Eye Medicine* Plant used as ceremonial eyewash and chant lotion. *Gastrointestinal Aid* Plant chewed with salt for heartburn. Plant used for stomachache. *Gynecological Aid* Plant used to facilitate delivery of placenta. *Panacea* Plant used as "life medicine." *Veterinary Aid* Cold infusion of plant applied daily to heal incision in castrated colt. (as *Gilia longiflora* 191:40) **Tewa** *Analgesic* Infusion of pulverized flowers and leaves used for headache. *Dermatological Aid* Infusion of pulverized flowers and leaves used on sores. (as *Gilia longiflora* 138:55) **Zuni** *Dermatological Aid* and *Pediatric Aid* Poultice of dried, powdered flowers and water applied to remove hair on newborns and children. (as *Gilia longiflora* 27:378)
- **Other**—**Navajo** *Ceremonial Items* Used to make prebreakfast drink and taken to make the person "bark" or sing loudly for Squaw Dance. (as *Gilia longiflora* 55:70)

Ipomopsis multiflora (Nutt.) V. Grant, Manyflowered Gilia
- **Drug**—**Navajo, Ramah** *Ceremonial Medicine* Decoction of plant used as a ceremonial medicine. (as *Gilia multiflora* 191:40) **Zuni** *Analgesic* Powdered, whole plant applied to face for headache. *Dermatological Aid* Powdered plant applied to wounds. *Pulmonary Aid* Crushed blossoms smoked in corn husks to "relieve strangulation." (as *Gilia multiflora* 166:52)

Ipomopsis polycladon (Torr.) V. Grant, Manybranched Gilia
- **Drug**—**Navajo, Kayenta** *Sedative* Plant used as a soporific. *Tonic* Plant used as a tonic. (as *Gilia polycladon* 205:38)

Ipomopsis pumila (Nutt.) V. Grant, Dwarf Gilia
- **Other**—**Keres, Western** *Unspecified* Plant known and named but no use was specified. (as *Gilia pulmila* 171:45)

Iresine, Amaranthaceae
Iresine diffusa Humb. & Bonpl. ex Willd., Juba's Bush
- **Drug**—**Houma** *Pulmonary Aid* Syrup of leaves and stems taken for whooping cough. (as *I. paniculata* 158:66)

Iridaceae, see *Iris*, *Sisyrinchium*

Iris, Iridaceae
Iris cristata Ait., Dwarf Crested Iris
- **Drug**—**Cherokee** *Dermatological Aid* Compound decoction of pulverized root used as salve for ulcers. *Liver Aid* Infusion taken for liver. *Urinary Aid* Compound decoction of root used for "yellowish urine." (80:41)

Iris douglasiana Herbert, Douglas Iris
- **Drug**—**Yokia** *Oral Aid* and *Pediatric Aid* Leaves used to wrap babies during berry gathering trips to retard perspiration and prevent thirst. (41:330)
- **Other**—**Mendocino Indian** *Hunting & Fishing Item* Leaf edges made into nets and ropes used to make snares for catching deer. (41:330) **Pomo, Kashaya** *Ceremonial Items* Flowers used in dance wreaths at the Strawberry Festival. (72:62)

Iris innominata Henderson, Del Norte County Iris
- **Fiber**—**Tolowa** *Cordage* Roots and leaves used to make cordage. (6:33)

Iris macrosiphon Torr., Bowltube Iris
- **Drug**—**Pomo** *Gynecological Aid* Roots used to hasten the birth of a baby. (117:284)
- **Fiber**—**Karok** *Cordage* Leaves dried, scraped, and used to make string or cord. Used to make rope. (148:381)
- **Other**—**Karok** *Hunting & Fishing Item* Leaves dried, scraped, and used to make fish nets, camping bags, deer snares, traps, and woodpecker nets. (148:381) **Pomo** *Hunting & Fishing Item* Used to make the strongest deer snares. (117:284)

Iris missouriensis Nutt., Rocky Mountain Iris
- **Drug**—**Great Basin Indian** *Toothache Remedy* Root put in a hollow tooth for toothaches. (121:47) **Klamath** *Emetic* Dried rootstocks used by medicine men as smoking material to cause nausea. Dried rootstocks sometimes used by medicine men as a smoking material, mixed with poison camas and a little tobacco, to give a person a severe nausea, in order to secure a heavy fee for making him well again. (45:93) **Montana Indian** *Emetic* Decoction of rootstocks used by medicine men to induce vomiting. (19:13) **Navajo, Ramah** *Ceremonial Medicine* and *Emetic* Decoction of plant used as a ceremonial emetic. (191:21) **Nevada Indian** *Kidney Aid* Infusion of roots taken for kidney troubles. *Urinary Aid* Infusion of roots taken for bladder troubles. (121:47) **Paiute** *Analgesic* Decoction of root taken for stomachaches. *Dermatological Aid* Paste of ripe seeds applied to sores. *Ear Medicine* Warm decoction of root dropped into ear for earache. *Gastrointestinal Aid* Decoction of root taken for stomachaches. *Toothache Remedy* Raw root placed in cavity or against gum for toothache. *Urinary Aid* Decoction of root taken for bladder troubles. *Venereal Aid* Decoction of root used for gonorrhea. **Shoshoni** *Analgesic* Decoction of root taken for stomachaches. Poultice of mashed roots applied for rheumatic pains. *Antirheumatic (External)* Poultice of mashed roots applied to rheumatic pains. *Burn Dressing* Paste of ripe seeds applied to burns. *Dermatological Aid* Pulped root applied as a salve for venereal sores. *Ear Medicine* Warm decoction of root dropped into ear for earache. *Gastrointestinal Aid* Decoction of root

taken for stomachaches. *Toothache Remedy* Raw root placed in cavity or against gum for toothache. *Venereal Aid* Decoction of root taken for gonorrhea and root salve used for venereal sores. (180:89, 90) **Yavapai** *Cathartic* Decoction of root taken as a purgative. (65:261) **Zuni** *Pediatric Aid* Poultice of chewed root applied to increase strength of newborns and infants. *Strengthener* Poultice of chewed root used for newborns and infants to increase strength. (27:373)
- *Other*—**Jemez** *Decorations* Flower used as a decoration for dances. (44:24)

Iris setosa Pallas ex Link, Beachhead Iris
- *Drug*—**Aleut** *Laxative* Decoction of root taken as a laxative. (8:428) **Eskimo, Inupiat** *Poison* Whole plant considered poisonous. (98:140)
- *Food*—**Eskimo, Alaska** *Beverage* Roasted, ground seeds used for coffee. (4:715)
- *Dye*—**Eskimo, Alaska** *Unspecified* Petals made into a dye and used for staining strands of grass for weaving colored patterns on baskets. (1:34)

Iris sp., Iris
- *Drug*—**Seminole** *Dermatological Aid* Plant used for alligator bites. (169:298) **Yana** *Cough Medicine* Raw roots chewed for coughs. (147:253)
- *Fiber*—**Wintoon** *Cordage* Used to make cord for fish nets. (117:264)
- *Dye*—**Navajo** *Green* Used to make a green dye. (55:37)

Iris tenax ssp. *klamathensis* Lenz, Klamath Iris
- *Fiber*—**Tolowa** *Cordage* Used to make cordage. (6:33)

Iris tenuissima Dykes, Longtube Iris
- *Drug*—**Pomo** *Gynecological Aid* Roots used to hasten the birth of a baby. (117:284)
- *Other*—**Pomo** *Hunting & Fishing Item* Used to make the strongest deer snares. (117:284)

Iris verna L., Dwarf Violet Iris
- *Drug*—**Cherokee** *Dermatological Aid* Compound decoction of pulverized root used as salve for ulcers. *Liver Aid* Infusion taken for liver. *Urinary Aid* Compound decoction of root used for "yellowish urine." (80:41) **Creek** *Cathartic* Plant used as a powerful cathartic. (172:669, 670)

Iris versicolor L., Harlequin Blueflag
- *Drug*—**Abnaki** *Poison* Plant considered poisonous. (144:155) Plant considered poisonous. (144:175) **Algonquin, Tête-de-Boule** *Burn Dressing* Poultice of smashed roots applied to burns. *Dermatological Aid* Poultice of smashed roots applied to wounds. (132:129) **Chippewa** *Dermatological Aid* Poultice of root, very strong, applied to swellings. (53:366) Poultice of roots applied to scrofulous sores. *Tuberculosis Remedy* Poultice of roots applied to scrofulous sores. (71:126) **Cree, Hudson Bay** *Cathartic* Plant used as a purgative. *Liver Aid* Plant used to increase the flow of bile. (92:303) **Creek** *Cathartic* Plant used as a powerful cathartic. (172:669, 670) Plant used as a cathartic. (177:10) **Delaware** *Antirheumatic (External)* Roots used for rheumatism. *Kidney Aid* Roots used for kidney disorders. *Liver Aid* Roots used for liver disorders. *Venereal Aid* Roots used for scrofula. (176:36) **Delaware, Oklahoma** *Antirheumatic (Internal)* Root taken for rheumatism. (175:30, 76) *Kidney Aid* Root used for disorders of the kidneys. *Liver Aid* Root used for disorders of the liver. (175:30) Root combined with root of *Hydrangea arborescens* and used for gallstones. *Tuberculosis Remedy* Root taken for "scrofula." (175:30, 76) **Iroquois** *Blood Medicine* Poultice of crushed rhizomes applied for blood poisoning caused by contusions. (141:67) *Cathartic* Infusion of plant taken as a physic. *Gynecological Aid* Infusion of smashed roots taken at menses to induce pregnancy. *Orthopedic Aid* Infusion of smashed roots taken to induce paralysis. *Respiratory Aid* Compound decoction with roots taken for hay fever. (87:287) **Malecite** *Throat Aid* Infusion of plants and bulrush used as a gargle for sore throats. (116:248) **Meskwaki** *Burn Dressing* Poultice of freshly macerated root applied to burns. *Cold Remedy* Root used for colds. *Dermatological Aid* Poultice of freshly macerated root applied to sores. *Pulmonary Aid* Root used for lung trouble. (152:224) **Micmac** *Dermatological Aid* Root used for wounds and herb used for sore throat. *Misc. Disease Remedy* Root used for cholera and the prevention of disease. *Panacea* Root used as a "basic medical cure" and for cholera. *Throat Aid* Herbs used for sore throat and root used for wounds. (40:57) **Mohegan** *Analgesic* Poultice of pulverized root mixed with flour applied to pain. (176:72, 130) **Montagnais** *Analgesic* Poultice of crushed plant mixed with flour applied to any pain. (156:315) Compound poultice of plant and flour applied to pain. (174:268) **Ojibwa** *Cathartic* Decoction of root taken as a "quick physic." *Emetic* Decoction of root taken as an emetic. (153:371) **Omaha** *Dermatological Aid* Paste of pulverized rootstock applied to sores and bruises. *Ear Medicine* Pulverized rootstock mixed with water or saliva and dropped in ear for earache. *Eye Medicine* Rootstock used to medicate "eye-water." (70:72) **Penobscot** *Herbal Steam* Plant steamed throughout the house to keep away "disease in general." (156:311) *Misc. Disease Remedy* Infusion of root taken for cholera. (156:308, 309) *Preventive Medicine* Root chewed to keep disease away; the plant is thought to "kill" sickness. (156:311) **Ponca** *Dermatological Aid* Paste of pulverized rootstock applied to sores and bruises. *Ear Medicine* Pulverized rootstock mixed with water or saliva and dropped in ear for earache. *Eye Medicine* Rootstock used to medicate "eye-water." (70:72) **Potawatomi** *Dermatological Aid* Poultice of root used to allay inflammation. (154:60) **Rappahannock** *Panacea* Infusion of dried roots taken for "every complaint." (161:28)
- *Fiber*—**Potawatomi** *Basketry* Leaves used to weave baskets. *Mats, Rugs & Bedding* Leaves used to weave mats. (154:120)
- *Other*—**Ojibwa** *Protection* Used as a charm against snakes. When blueberry picking, everyone carries a piece of this plant in his clothes and will handle it every little while to perpetuate the scent. They believe that snakes will shun them while so protected. They say that the Arizona Indians use it when they hold their snake dances and are never struck as long as their clothes are fumigated with it. They also chew it to get the odor into their mouths, preparatory to taking rattlesnakes into their teeth. The rattlesnake never offers to bite them so long as the scent of the blue flag persists. (153:430)

Iris virginica L., Virginia Iris
- *Drug*—**Cherokee** Dermatological Aid Compound decoction of pulverized root used as salve for ulcers. Liver Aid Infusion taken for liver. Urinary Aid Compound decoction of root used for "yellowish urine." (80:41)

Isocoma, Asteraceae

Isocoma acradenia (Greene) Greene **var. *acradenia***, Alkali Goldenbush
- *Drug*—Cahuilla *Dermatological Aid* Poultice of boiled leaves applied to sores. *Throat Aid* Leaves soaked in a pan of boiling water and steam inhaled for sore throats. (as *Haplopappus acradenius* 15:75)
- *Fiber*—Cahuilla *Building Material* Plant used to build fences as a protection from cold winds. (as *Haplopappus acradenius* 15:75)

Isocoma pluriflora (Torr. & Gray) Greene, Southern Jimmyweed
- *Drug*—Navajo, Kayenta *Dermatological Aid* and *Pediatric Aid* Plant used as a lotion to heal infant's navel. (as *Aplopappus heterophyllus* 205:44) Pima *Analgesic* Poultice of plant applied for muscular pain. *Cough Medicine* Leaves chewed for coughs. *Orthopedic Aid* Poultice of plant applied for muscular pain. (as *Aplopappus heterophyllus* 47:101)
- *Other*—Pima *Fuel* Dried plants used for kindling. (as *Aplopappus heterophyllus* 47:101)

Iva, Asteraceae

Iva axillaris Pursh, Poverty Weed
- *Drug*—Mahuna *Abortifacient* Plant used to cause abortions. *Contraceptive* Plant used to prevent conception. (140:67) Paiute *Dermatological Aid* Leaves used as a plaster or infusion used as a wash for sores or skin irritations. Shoshoni *Analgesic* Infusion or decoction of plant taken, especially by children, for stomachaches or cramps. *Antidiarrheal* Decoction of plant taken for diarrhea. *Cold Remedy* Infusion or decoction of plant used by children for colds. (180:90, 91) *Gastrointestinal Aid* Root soaked in cold water for tea and taken for bowel disorders. (118:42) Infusion or decoction of plant taken, especially by children, for stomachaches or cramps. Raw, roasted, or boiled root eaten for indigestion. *Pediatric Aid* Infusion or decoction of plant taken, especially by children, for stomachaches or cramps. Infusion or decoction of plant used by children for colds. (180:90, 91) Ute *Unspecified* Used as medicine. (38:35)

Iva xanthifolia Nutt., Giant Sumpweed
- *Drug*—Navajo, Kayenta *Dermatological Aid* Poultice of plant applied to boils. *Veterinary Aid* Plant used to heal castration incision in sheep. (205:48) Navajo, Ramah *Cough Medicine* Infusion or decoction taken and used as lotion for cough. *Misc. Disease Remedy* Infusion or decoction taken and used as lotion for influenza. *Witchcraft Medicine* Infusion or decoction taken and used as lotion for protection from witches. (191:52)

Ivesia, Rosaceae

Ivesia gordonii (Hook.) Torr. & Gray, Gordon's Ivesia
- *Drug*—Arapaho *Tonic* Infusion of root used as a tonic. (as *Horkelia gordonii* 118:40) Infusion of resinous roots used as a general tonic. (121:48)

Jacquemontia, Convolvulaceae

Jacquemontia ovalifolia* ssp. *sandwicensis (Gray) Robertson, Ovalleaf Clustervine
- *Drug*—Hawaiian *Dermatological Aid* Plant mixed with taro leaves and salt and used for cuts. *Pediatric Aid* and *Strengthener* Plant used for babies with general body weakness. (as *J. sandwicensis* 2:73)
- *Food*—Hawaiian *Beverage* Dried leaves and stems used to make tea. *Unspecified* Dried leaves and stems eaten with coconut. (as *J. sandwicensis* 2:73)

Jacquinia, Theophrastaceae

Jacquinia pungens
- *Food*—Seri *Unspecified* Nuts used for food. (50:136)
- *Other*—Seri *Ceremonial Items* Dried nuts used as favorite rattle beads. (50:136)

Jamesia, Hydrangeaceae

Jamesia americana Torr. & Gray, Cliffbush
- *Food*—Apache, Chiricahua & Mescalero *Unspecified* Seeds occasionally eaten fresh. (33:45)

Jatropha, Euphorbiaceae

Jatropha cardiophylla (Torr.) Muell.-Arg., Sangre de Cristo
- *Fiber*—Papago *Basketry* Thick, rubbery stems used to make bulky baskets. (as *Tatropolia cordiophylla*, bloodroot bush 34:57). Seri *Basketry* Coil for the warp of very light, shallow baskets made of a "whisp of splints" of several sorts of torote. (50:138)

Jatropha spatulata, Torote Amarillo
- *Fiber*—Seri *Basketry* Coil for the warp of very light, shallow baskets made of a "whisp of splints" of several sorts of torote. (50:138)

Jeffersonia, Berberidaceae

Jeffersonia diphylla (L.) Pers., Twinleaf
- *Drug*—Cherokee *Dermatological Aid* Poultice used for sores, ulcers, and inflamed parts. *Kidney Aid* Infusion taken for dropsy. *Urinary Aid* Infusion taken for gravel and urinary problems. (80:59) Iroquois *Antidiarrheal* Decoction of whole plant taken by adults and children with diarrhea. *Liver Aid* Decoction of whole plant taken for gall. *Pediatric Aid* Decoction of whole plant taken by adults and children with diarrhea. (87:332)

Juglandaceae, see *Carya*, *Juglans*

Juglans, Juglandaceae

Juglans californica S. Wats., California Walnut
- *Drug*—Costanoan *Blood Medicine* Infusion of leaves taken for thin blood. (21:20)
- *Food*—Costanoan *Unspecified* Nuts used for food. (21:248)

Juglans cinerea L., Butternut
- *Drug*—Cherokee *Antidiarrheal* Infusion of bark taken to check bowels. *Cathartic* Pills from inner bark used as a cathartic and compound infusion used for toothache. (80:61) Pills prepared from inner bark and used as a cathartic. (203:75) *Toothache Remedy* Pills from

inner bark taken as a cathartic and compound infusion used for toothache. (80:61) **Chippewa** *Cathartic* Decoction of plant sap taken as a cathartic. (71:127) **Iroquois** *Analgesic* Compound decoction of plants taken for urinating pain. *Anthelmintic* Compound decoction with bark taken to kill worms in adults. *Blood Medicine* Compound decoction taken as a blood purifier and for venereal disease. (87:295) *Cathartic* Decoction of bark taken as a physic and cathartic. (87:296) *Dermatological Aid* Compound decoction taken when skin becomes thin. Infusion or chewed bark applied to bleeding wounds. (87:295) Nutmeat oil formerly used for the hair, either alone or mixed with bear grease. (196:123) *Emetic* Infusion of plant and other plant wood and bark used as an emetic to remove yellow from the stomach. (141:39) *Gynecological Aid* Compound decoction with bark taken to induce pregnancy. (87:294) *Hemostat* Infusion or chewed bark applied to bleeding wounds. *Laxative* Compound decoction of bark or shoots taken as a laxative. *Liver Aid* Compound decoction taken for yellow skin and too much gall. *Oral Aid* Compound infusion of buds used as mouthwash for mouth ulcers. *Psychological Aid* Compound decoction with plant taken for "loss of senses during menses." *Toothache Remedy* Juice used for toothache. *Tuberculosis Remedy* Compound decoction used as poultice for infected and swollen tubercular glands. *Urinary Aid* Compound decoction of plants taken for urinating pain. *Venereal Aid* Decoction of shoots taken as a laxative and for venereal disease. (87:295) **Malecite** *Cathartic* Infusion of bark used as a purgative. (116:254) **Menominee** *Cathartic* Syrup from sap used as a standard "physic." (151:38, 39) **Meskwaki** *Cathartic* Decoction of twig bark or decoction of wood and bark taken as a cathartic. (152:224) **Micmac** *Cathartic* Bark used as a purgative. (40:57) **Potawatomi** *Cathartic* Bark used as a physic and infusion of inner bark taken as a tonic. *Tonic* Infusion of inner bark taken as a tonic and bark used as a physic. (154:60, 61)
- **Food**—**Algonquin, Quebec** *Unspecified* Nuts used for food. (18:78) **Cherokee** *Unspecified* Nuts used for food. (80:61) Raw nut used for food. (126:42) **Iroquois** *Baby Food* Fresh nutmeats crushed, boiled, and oil used as a baby food. *Beverage* Fresh nutmeats crushed, boiled, and liquid used as a drink. *Bread & Cake* Fresh nutmeats crushed and mixed with bread. (124:99) Nuts crushed, mixed with cornmeal and beans or berries, and made into bread. (196:123) *Pie & Pudding* Fresh nutmeats crushed and mixed with corn pudding. (124:99) *Sauce & Relish* Nuts pounded, boiled, resulting oil seasoned with salt and used as gravy. *Soup* Nutmeats crushed and added to corn soup. (196:123) *Special Food* Fresh nutmeats crushed, boiled, and oil used as a delicacy in corn bread and pudding. (124:99) Nutmeat oil added to the mush used by the False Face Societies. *Staple* Nutmeats crushed and added to hominy. *Unspecified* Nutmeats, after skimming off the oil, seasoned and mixed with mashed potatoes. (196:123) **Menominee** *Unspecified* Used in the same way that the white man did. (151:68) **Meskwaki** *Winter Use Food* Nuts stored for winter use. (152:259) **Ojibwa** *Unspecified* Nuts used for food. (153:405) **Potawatomi** *Winter Use Food* Butternuts gathered for their edible quality and furnished a winter supply of food. (154:103)
- **Fiber**—**Cherokee** *Building Material* Wood used for lumber. (80:61)
- **Dye**—**Cherokee** *Black* Young roots used to make a black dye. *Brown* Bark used to make a brown dye. (80:61) **Chippewa** *Black* Boiled with hazel to make a black dye. Inner bark and a little of the root boiled with black earth and ocher to make a black dye. Used with black earth to make a black dye. (53:372) *Brown* Root bark used to make a brown dye which did not need a mordant. (71:127) **Menominee** *Black* Bark boiled with blue clay to obtain a deep black color. *Brown* Juice of nut husk used as a brown dye for deerskin shirts. (151:78) **Ojibwa** *Brown* Nut hulls used as best brown dye, because it was attained from the tree at any time of the year. Butternut was usually used in other combinations for brown and black colors. (153:425)
- **Other**—**Iroquois** *Insecticide* Nutmeat oil mixed with bear grease and used as a preventive for mosquitoes. (196:123)

Juglans hindsii (Jepson) Jepson ex R. E. Sm., Hinds's Black Walnut
- **Food**—**Pomo, Kashaya** *Dried Food* Sweet nutmeat dried and stored for later use. *Unspecified* Sweet nutmeat eaten fresh. (72:117)
- **Dye**—**Pomo, Kashaya** *Black* Nut husk used in dying bulrush root a black color for making basket design. (72:117)

Juglans major (Torr.) Heller, Arizona Walnut
- **Food**—**Apache, Chiricahua & Mescalero** *Unspecified* Nutmeats eaten raw. *Winter Use Food* Nutmeats mixed with mesquite gravy or ground with roasted mescal and stored. (33:42) **Apache, Mescalero** *Unspecified* Nutmeats mixed with mescal, datil, sotol, or mesquite and used for food. (14:46) **Hualapai** *Unspecified* Nuts used for food. (195:13) **Navajo** *Unspecified* Nuts gathered and eaten on a fairly large scale. (55:39) **Yavapai** *Beverage* Decoction of pulverized nut juice dipped up and sucked. (63:209) Meat pulverized in mescal syrup and used as a beverage. *Unspecified* Nutmeat used for food. *Winter Use Food* Nuts stored for later use. (65:256)
- **Fiber**—**Apache, Mescalero** *Building Material* Trees used to construct dome-shaped lodges when away from home. (14:46)
- **Dye**—**Hualapai** *Unspecified* Nutshells boiled and used as a dye. (195:13) **Navajo** *Brown* Nut hulls used as a golden brown dye. Young twigs used as a light brown dye. (55:39)
- **Other**—**Apache, Mescalero** *Paint* Outer shell coverings soaked in water to make a black paint. (14:46)

Juglans nigra L., Black Walnut
- **Drug**—**Cherokee** *Dermatological Aid* Infusion used as a wash for sores. *Misc. Disease Remedy* Infusion of inner bark taken for smallpox and infusion of leaves used for goiter. *Poison* "Bark used cautiously in medicine because it is poisonous." *Toothache Remedy* Bark chewed for toothache. (80:61) **Comanche** *Dermatological Aid* Pulverized leaves rubbed on affected part for ringworm. (29:522) **Delaware** *Anthelmintic* Juice from green hulls of fruits rubbed over areas infected by ringworm. *Dermatological Aid* Sap used in applications for inflammations. *Gastrointestinal Aid* Three bundles of bark boiled to make a strong tea and used for 2 days to remove intestinal bile. (176:29) **Delaware, Oklahoma** *Cathartic* Strong decoction of bark taken as a cathartic. *Dermatological Aid* Juice from green hull of fruit rubbed on skin for ringworm. Sap applied to any inflammation. *Emetic* Strong decoction of bark taken as an emetic. (175:24, 76) *Gastrointestinal Aid* and *Liver Aid* Decoction of bark taken "to remove bile from the intestines." (175:24) **Houma** *Dermatological Aid* Infusion of nutshells used as a wash for "the itch." *Hypotensive* Decoction of mashed leaves taken for relief from "blood pressure." (158:66) **Iroquois** *Analgesic* Poultice of bark applied for headache. *Blood Medicine* Compound decoction with brandy taken as a blood purifier.

(87:296) *Dermatological Aid* Nutmeat oil formerly used for the hair, either alone or mixed with bear grease. (196:123) *Laxative* Compound decoction of bark taken as a laxative. *Psychological Aid* Poultice of bark applied for "craziness." *Witchcraft Medicine* Infusion of bark used as a medicine for rain. (87:296) **Kiowa** *Anthelmintic* Decoction of root bark taken to kill "worms." (192:21) **Meskwaki** *Cathartic* Inner bark used as a very strong physic. *Snakebite Remedy* Coiled and charred twig bark and old bark applied in water for snakebite. (152:224, 225) **Rappahannock** *Antidiarrheal* Infusion of root bark taken to prevent dysentery. (161:32) *Febrifuge* Compound with north side bark used as a poultice for chills. (161:31) *Gastrointestinal Aid* Infusion of root bark taken to "roughen the intestines." (161:32)
- *Food*—**Cherokee** *Dried Food* Nuts dried in the rafters for future use. *Porridge* Nuts mixed with skinned hominy corn, water, and pinto beans. (126:43) *Unspecified* Nuts used for food. (80:61) **Comanche** *Unspecified* Nuts used for food. (29:522) *Winter Use Food* Nuts stored for winter use. (29:531) **Dakota** *Soup* Nuts used to make soup. *Unspecified* Nuts eaten plain or with honey. (70:74) **Iroquois** *Beverage* Fresh nutmeats crushed, boiled, and liquid used as a drink. *Bread & Cake* Fresh nutmeats crushed and mixed with bread. (124:99) Nuts crushed, mixed with cornmeal and beans or berries, and made into bread. (196:123) *Pie & Pudding* Fresh nutmeats crushed and mixed with corn pudding. (124:99) *Sauce & Relish* Nuts pounded, boiled, resulting oil seasoned with salt and used as gravy. *Soup* Nutmeats crushed and added to corn soup. (196:123) *Special Food* Fresh nutmeats crushed, boiled, and oil used as a delicacy in corn bread and pudding. (124:99) Nutmeat oil added to the mush used by the False Face Societies. *Staple* Nutmeats crushed and added to hominy. *Unspecified* Nutmeats, after skimming off the oil, seasoned and mixed with mashed potatoes. (196:123) **Kiowa** *Unspecified* Nuts used for food. (192:20) **Lakota** *Unspecified* Nuts used for food. (139:49) **Meskwaki** *Unspecified* Nuts were relished. (152:259) **Omaha** *Soup* Nuts used to make soup. (70:74) *Unspecified* Nuts eaten plain or mixed with honey. (68:326) Nuts eaten plain or with honey. **Pawnee** *Soup* Nuts used to make soup. *Unspecified* Nuts eaten plain or with honey. **Ponca** *Soup* Nuts used to make soup. *Unspecified* Nuts eaten plain or with honey. **Winnebago** *Soup* Nuts used to make soup. *Unspecified* Nuts eaten plain or with honey. (70:74)
- *Fiber*—**Cherokee** *Furniture* Wood used to make furniture. (80:61)
- *Dye*—**Cherokee** *Brown* Bark, roots, and husks used to make a brown dye. *Green* Leaves used to make a green dye. (80:61) **Chippewa** *Black* Bark used to make a black dye. *Brown* Bark used to make a dark brown dye. (71:127) **Dakota** *Black* Roots used to make a black dye. (69:367) Nuts used to make a black dye. (70:74) **Kiowa** *Blue-Black* Roots boiled to make a bluish, black dye for buffalo hides. (192:20) **Meskwaki** *Black* Wood and bark charred to make the best black dye. (152:271) **Omaha** *Black* Nuts used to make a black dye. **Pawnee** *Black* Nuts used to make a black dye. **Ponca** *Black* Nuts used to make a black dye. **Winnebago** *Black* Nuts used to make a black dye. (70:74)
- *Other*—**Cherokee** *Decorations* Wood used to carve. *Hunting & Fishing Item* Wood used to make gunstocks. (80:61) **Delaware** *Insecticide* Leaves scattered about the house to dispel fleas. (176:29) **Delaware, Oklahoma** *Insecticide* Leaves scattered about house to "dispel fleas." (175:24) **Iroquois** *Insecticide* Nutmeat oil mixed with bear grease and used as a preventive for mosquitoes. (196:123)

Juglans regia L., English Walnut
- *Dye*—**Navajo** *Brown* Nut hulls used as a golden brown dye. (55:39)

Juglans sp., Walnut
- *Drug*—**Apache, Western** *Dermatological Aid* Juice used to clear maggots from wounds. *Veterinary Aid* Juice given to dogs for worms. (26:187)
- *Food*—**Apache, Western** *Sauce & Relish* Walnuts pulverized, mixed with mescal juice, and used as dip for corn bread. *Unspecified* Nuts parched with corn, ground, and eaten by the pinch. (26:187)

Juncaceae, see *Juncus, Luzula*

Juncaginaceae, see *Triglochin*

Juncus, Juncaceae

Juncus acutus var. *sphaerocarpus* Engelm., Leopold's Rush
- *Fiber*—**Cahuilla** *Basketry* Rushes made into baskets used for collecting foods, leaching acorn meal, and finely woven baskets. (15:80)

Juncus balticus Willd., Baltic Rush
- *Food*—**Paiute** *Candy* Sugar, formed along tops of plants, gathered and eaten as candy. *Unspecified* Seeds used for food. (167:246) Species used for food. (167:243) **Paiute, Northern** *Beverage* Stems used to make a fermented drink. (59:53)
- *Fiber*—**Cheyenne** *Basketry* Stems used to weave baskets. (76:171) Stems formerly used in basket weaving. (83:12) *Other* Used in weaving. (83:46) **Kawaiisu** *Basketry* Split stems used in weaving coiled baskets. (206:35) **Klamath** *Basketry* Stems used in the weaving of baskets. *Mats, Rugs & Bedding* Stems used in the weaving of mats. (45:92) **Montana Indian** *Basketry* Used for weaving light baskets. *Mats, Rugs & Bedding* Used for weaving mats. (19:13) **Pomo** *Basketry* Used by girls to simulate basket making. (66:12)
- *Other*—**Cheyenne** *Decorations* Rootlets used to sew patterns for ornamentations on robes or other leather. (76:171) **Hopi** *Ceremonial Items* Ceremonially associated with water. (200:70) **Panamint** *Decorations* Basal portions of stems used as light yellow-brown decorations for baskets. (105:78) **Pomo** *Tools* Used to hold drilled clamshell beads in place when rolled on a stone slab to smooth them. (66:12) **Pomo, Kashaya** *Jewelry* Blades used to string clamshell beads to hold them together when being smoothed. *Toys & Games* Blades used by children to make play baskets. (72:100)

Juncus bufonius L., Toad Rush
- *Drug*—**Iroquois** *Emetic* Infusion of plant taken as an emetic by runners. The runner drank about 2 quarts the first time, vomited, drank the same quantity, and vomited again. The face and body were also washed with the liquid. This was done about three times during the week before the race. (196:89)

Juncus bufonius L. var. *bufonius*, Toad Rush
- *Drug*—**Iroquois** *Dermatological Aid* Compound decoction used as wash for the entire body. *Emetic* Compound decoction taken as an emetic. *Strengthener* Compound decoction taken to "give strength to runners and other athletes." (87:279)

Juncus dudleyi Wieg., Dudley's Rush
- *Fiber*—**Ojibwa** *Mats, Rugs & Bedding* Tiny rush used in the finest mat work and for small pieces. (153:419)

Juncus effusus L., Common Rush
- **Drug**—**Cherokee** *Emetic* Decoction of plant taken as an emetic. (177: 7) *Oral Aid* Decoction used "to dislodge spoiled saliva." *Orthopedic Aid* Infusion given to babies to prevent lameness. *Pediatric Aid* Infusion used as a wash to strengthen babies and given to babies to prevent lameness. *Strengthener* Infusion used as a wash to strengthen babies. (80:53) **Karok** *Unspecified* Stems and leaves placed in the fire and the medicine man prayed over it. (6:33)
- **Food**—**Mendocino Indian** *Forage* Plants eaten by cows and horses in early spring. (41:318) **Okanagan-Colville** *Fodder* Plant used to feed horses. (188:38) **Snuqualmie** *Unspecified* Early sprouts eaten raw. (79:23)
- **Fiber**—**Cherokee** *Cordage* Used to make string to bind up dough in oak leaves for cooking bread. (80:53) **Chippewa** *Mats, Rugs & Bedding* Rushes used for weaving small table mats and other larger mats. (71:125) **Hesquiat** *Cordage* Tough, round stems dried, twisted or braided, and used for tying and binding. (185:54) **Karok** *Basketry* Stems used by young females to practice making baskets. (6:33) **Mendocino Indian** *Basketry* Wiry stalks used to make temporary baskets. *Cordage* Wiry stalks used for tying. **Pomo** *Sporting Equipment* Formerly used to make a device for trapping and catching salmon and trout as a sport. (41:318) **Snuqualmie** *Cordage* Stalks used for tying things. (79:23) **Tolowa** *Basketry* Stems used by young females to practice making baskets. **Yurok** *Basketry* Stems used by young females to practice making baskets. (6:33)
- **Other**—**Chippewa** *Containers* Rushes used for weaving little bags and pouches. (71:125)

Juncus effusus* var. *pacificus Fern. & Wieg., Pacific Rush
- **Fiber**—**Cahuilla** *Basketry* Rushes made into baskets used for collecting foods, leaching acorn meal, and finely woven baskets. (15:80)

Juncus ensifolius Wikstr., Swordleaf Rush
- **Drug**—**Hoh** *Unspecified* Used as a medicine. **Quileute** *Unspecified* Used as a medicine. (137:59)
- **Food**—**Paiute** *Fodder* Rushes used as food for livestock. (111:53) **Swinomish** *Unspecified* Bulbs used for food. (as *J. xiphioides* var. *triandrus* 79:23)
- **Fiber**—**Karok** *Basketry* Used in teaching little girls to make baskets. (148:380)

Juncus lesueurii Boland., Salt Rush
- **Fiber**—**Cahuilla** *Basketry* Rushes made into baskets used for collecting foods, leaching acorn meal, and finely woven baskets. (15:80)

Juncus mertensianus Bong., Mertens's Rush
- **Drug**—**Okanagan-Colville** *Witchcraft Medicine* Plant used for "witchcraft" or "plhax." (188:38)
- **Fiber**—**Luiseño** *Basketry* Rushes used to make woven and twined baskets. (155:204)
- **Other**—**Luiseño** *Containers* Rushes made into woven or twined baskets and used as gathering containers for acorns and cacti. *Cooking Tools* Rushes made into woven and twined baskets and used as sifters or to leach acorn meal. (155:204)

Juncus mexicanus Willd. ex J. A. & J. H. Schultes, Mexican Rush
- **Dye**—**Shoshoni** *Green* Roots used as green dye in basketry. (118:8)

***Juncus* sp.**, Rush
- **Fiber**—**Costanoan** *Basketry* Leaves used in basketry. *Clothing* Stems and leaves used as raw textile material. *Cordage* Stems and leaves used as cordage. *Mats, Rugs & Bedding* Stems and leaves used as stuffing. (21:255) **Isleta** *Building Material* Plant used for thatch in building houses. (100:32) **Neeshenam** *Clothing* Used to make breech cloths. (129:378)
- **Other**—**Navajo** *Tools* Used as a sandpaper for smoothing bows. (55:31)

Juncus stygius L., Moor Rush
- **Fiber**—**Ojibwa** *Mats, Rugs & Bedding* Used to weave mats. (135:245)

Juncus tenuis Willd., Poverty Rush
- **Drug**—**Cherokee** *Oral Aid* Decoction used "to dislodge spoiled saliva." *Orthopedic Aid* Infusion given to babies to prevent lameness. *Pediatric Aid* Infusion used as a wash to strengthen babies and given to babies to prevent lameness. *Strengthener* Infusion used as a wash to strengthen babies. (80:53) **Iroquois** *Emetic* Decoction or infusion of plant taken by lacrosse players and runners to vomit. *Sports Medicine* Infusion of plant taken to vomit and used as a wash by lacrosse players. *Veterinary Aid* Infusion of plant given to "colt that has had too much feed." (87:279)
- **Fiber**—**Cherokee** *Cordage* Used to make string to bind up dough in oak leaves for cooking bread. (80:53)

Juncus textilis Buch., Basket Rush
- **Fiber**—**Cahuilla** *Basketry* Rushes made into baskets used for collecting foods, leaching acorn meal, and finely woven baskets. (15:80) **Diegueño** *Basketry* Split stems used in basket making. Allowed to dry, the stems were split three or four ways into splints and used as wrapping material for coiled baskets, or sometimes as a foundation material in openwork, coiled leaching baskets. Only the lower 2 feet of the plant, which grows up to 8 feet tall, was gathered and used. The plant was collected at any time during the year, but if the centers of the stems were brown, it was not as good for basket making as when the centers were white. Basket designs were formed with the various natural shades of green, tan, and brown found in the plant or it was sometimes dyed black. (84:23)

Juncus torreyi Coville, Torrey's Rush
- **Other**—**Hopi** *Ceremonial Items* Ceremonially associated with water. (200:70)

Jungermanniaceae, see *Bazzania*

Juniperus, Cupressaceae, conifer
Juniperus californica Carr., California Juniper
- **Drug**—**Apache, White Mountain** *Anticonvulsive* Scorched twigs rubbed on body for fits. *Cold Remedy* Infusion of leaves taken for colds. *Cough Medicine* Infusion of leaves taken for coughs. *Gynecological Aid* Infusion of leaves taken by women previous to childbirth to relax muscles. (136:158) **Costanoan** *Analgesic* Decoction of leaves taken for pain. *Diaphoretic* Decoction of leaves taken to cause sweating. (21:6) **Diegueño** *Analgesic* Infusion of leaves and bark taken for hangovers. *Hypotensive* Infusion of leaves and bark taken for high blood pressures. (88:216) **Gosiute** *Cold Remedy* Infusion of leaves

used for colds. *Cough Medicine* Infusion of leaves used for coughs. (39:372) **Mahuna** *Febrifuge* and *Misc. Disease Remedy* Infusion of berries taken or berries chewed for grippe fevers. (140:9)
• **Food**—**Cahuilla** *Dried Food* Berries sun dried and preserved for future use. *Fruit* Berries eaten fresh. *Porridge* Dried berries ground into a flour and used to make mush or bread. (15:81) **Costanoan** *Fruit* Berries used for food. (21:248) **Diegueño** *Fruit* Fruit eaten, informally only. *Starvation Food* Fruit eaten in times of starvation. (88:216) **Kawaiisu** *Bread & Cake* Berries seeded, pounded into a meal, moistened, molded into cakes, and dried. *Dried Food* Unseeded berries dried and stored. *Fruit* Berries boiled fresh and eaten cold. *Staple* Berries seeded, pounded into a meal, and eaten. (206:35) **Mendocino Indian** *Dried Food* Dried fruits boiled and eaten. (41:306)
• **Fiber**—**Kawaiisu** *Building Material* Bark used as a house covering. (206:35)
• **Other**—**Kawaiisu** *Cooking Tools* Wood used to make acorn mush stirrers and ladles. *Hunting & Fishing Item* Wood used as the primary material for making bows, either self-bows or sinew-backed. (206:35)

Juniperus communis L., Common Juniper
• **Drug**—**Algonquin** *Other* Used for "cold" conditions, since plant was regarded as "hot." (22:142) **Bella Coola** *Analgesic* Decoction of roots, leaves, branches, and bark taken for stomach pain. *Cough Medicine* Decoction of root, leaves, branches, and bark taken for "cough from the lungs." *Gastrointestinal Aid* Decoction of roots, leaves, branches, and bark taken for stomach pain. (150:49) Infusion of roots, leaves, branches, and bark taken for stomach pains, for ulcers, and for heartburn. *Pulmonary Aid* Infusion of roots, leaves, branches, and bark taken for lung cough. (184:197) **Blackfoot** *Pulmonary Aid* Used for lung diseases. (82:37) Decoction of berries used for lung diseases. (97:17) *Venereal Aid* Used for venereal diseases. (82:37) Decoction of berries used for venereal diseases. (97:17) **Carrier** *Tuberculosis Remedy* Decoction of berries and kinnikinnick leaves or balsam strained and taken for tuberculosis. (31:71) **Carrier, Northern** *Cathartic* Decoction of tips taken as a purgative and for coughs. *Cough Medicine* Decoction of tips taken for coughs and as a purgative. **Carrier, Southern** *Analgesic* Steam from boiling branches inhaled for headache and chest pain. (150:49) **Cheyenne** *Ceremonial Medicine* Leaves burned as incense in ceremonies, especially to remove fear of thunder. *Cold Remedy* Cones chewed, infusion of boughs or cones taken or used as steam bath for colds. *Cough Medicine* Infusion of boughs or fleshy cones taken for coughing. *Febrifuge* Infusion of boughs or fleshy cones taken for high fevers. *Gynecological Aid* Leaves burned at childbirth to promote delivery. *Herbal Steam* Cones chewed, infusion of boughs or cones taken or used as steam bath for colds. *Love Medicine* Wood flutes used to "charm a girl whom a man loved to make her love him." *Sedative* Infusion of boughs or fleshy cones taken as a sedative. *Throat Aid* Infusion of boughs or cones taken for tickles in the throat or tonsillitis. (83:4) **Chippewa** *Respiratory Aid* Decoction of twigs and leaves taken for asthma. (71:124) **Cree, Hudson Bay** *Dermatological Aid* Poultice of bark applied to wounds. *Disinfectant* Plant used for the antiseptic qualities on wounds. (92:302) **Cree, Woodlands** *Antidiarrheal* Decoction of barked procumbent stem or branch used for diarrhea. *Cough Medicine* Decoction of branch or wood and other herbs used for coughs. *Febrifuge* Decoction of branch or wood and other herbs used for fevers. *Gynecological Aid* Decoction of branch or wood and other herbs used for "woman's troubles," and for sickness after giving birth. *Kidney Aid* Decoction of green berries taken for sore backs from kidney troubles. *Pediatric Aid* Decoction of branch or wood and other herbs used for teething sickness. *Pulmonary Aid* Decoction of barked procumbent stem or branch used for sore chest from lung infections. *Respiratory Aid* Blue berries smoked in a pipe for asthma. *Toothache Remedy* Decoction of branch or wood and other herbs used for teething sickness. (109:41) **Delaware, Ontario** *Gynecological Aid* Compound infusion of bark taken for women's diseases. (176:110) *Tonic* Compound infusion of bark taken as a tonic. (175:68, 82) **Eskimo, Inupiat** *Cold Remedy* Infusion of berries taken or one berry a day eaten to prevent colds, and for colds. *Cough Medicine* Decoction of berries, needles, and twigs taken one cup a day for coughing. *Misc. Disease Remedy* Infusion of berries taken or one berry a day eaten to prevent flu, and for the flu. *Respiratory Aid* Decoction of berries, needles, and twigs taken one cup a day for respiratory ailments. (98:110) **Gitksan** *Unspecified* Plant used for many medicinal applications. (43:314) **Hanaksiala** *Dermatological Aid* Heated poultice of branch and "berry" paste applied to wounds and cuts. (43:160) **Iroquois** *Cold Remedy* Decoction taken for colds caused by overheating and chills. *Cough Medicine* Decoction taken for coughs caused by overheating and chills. (87:271) *Kidney Aid* Infusion of boughs used for kidney pain. *Tonic* Infusion of boughs used as a tonic. (142:83) **Kwakiutl** *Antidiarrheal* Compound decoction of berries taken for diarrhea. *Blood Medicine* Decoction of wood and bark taken to purify the blood. *Respiratory Aid* Decoction of wood and bark taken for short breath. (183:266) **Malecite** *Dermatological Aid* Infusion of boughs used as a hair wash. (116:250) *Tonic* Infusion of boughs used as a tonic. *Tuberculosis Remedy* Infusion of roots and prince's pine used for consumption. (116:252) **Micmac** *Antirheumatic (Internal)* Part of plant used for rheumatism and bark used for tuberculosis. *Dermatological Aid* Stems used in hair wash, gum used for wounds, and cones used for ulcers. *Orthopedic Aid* Gum used for sprains and bark used for tuberculosis. *Tonic* Stems used in a tonic and bark used for tuberculosis. *Tuberculosis Remedy* Root or bark used for consumption and stems used as a tonic. (40:57) **Navajo, Ramah** *Emetic* Used as an emetic for all ceremonials. (191:11) **Okanagan-Colville** *Cold Remedy* Infusion of bark and needles taken for colds. *Tonic* Infusion of bark and needles taken as a tonic before entering the sweat house. *Tuberculosis Remedy* Infusion of bark and needles taken for consumption. (188:18) **Okanagon** *Eye Medicine* Infusion of twigs used as a wash for sore eyes. *Kidney Aid* Berries eaten for kidney disorders. *Tonic* Decoction of small branches used as a tonic. (125:42) **Potawatomi** *Urinary Aid* Compound containing berries used for urinary tract diseases. (154:69) **Shuswap** *Diaphoretic* Used in the sweat house. *Panacea* Decoction of stems and needles taken for any sickness. (123:50) **Tanana, Upper** *Antirheumatic (External)* Decoction of branches and fruit used as a wash for body aches and pains. *Cold Remedy* Decoction of branches and fruit taken for colds. Raw fruit eaten for colds. Decoction of branches taken for colds. Decoction of berries taken for colds. *Cough Medicine* Decoction of branches and fruit taken for coughs. Raw fruit eaten for coughs. *Kidney Aid* Decoction of branches and fruit taken for kidney problems. Raw fruit eaten

for kidney problems. *Panacea* Branches burned on top of the wood stove to keep sickness away. *Throat Aid* Decoction of branches taken for sore throats. *Tuberculosis Remedy* Decoction of branches taken for tuberculosis. (102:4) **Thompson** *Antirheumatic (Internal)* Infusion of branches taken for aching muscles. *Cathartic* Decoction of branches taken as a physic. *Cold Remedy* Decoction of branches used for colds. (187:92) *Eye Medicine* Infusion of twigs used as a wash for sore eyes. (125:42) Infusion or decoction of twigs used as a wash for sore eyes. *Gastrointestinal Aid* Decoction of twigs taken as a tonic for the stomach. (164:474) Infusion of three 10-cm-long branches taken to "make your insides nice." *Heart Medicine* Infusion of boughs taken for "leakage of the heart." *Hypotensive* Infusion of branches taken for high blood pressure. The branches were steeped in boiling water until the water cooled. The cool infusion was taken for 2 weeks after which the blood pressure returned to normal. (187:92) *Kidney Aid* Berries eaten for kidney disorders. (125:42) Decoction of branches used for kidney ailments. (187:92) *Tonic* Decoction of small branches used as a tonic. (125:42) Decoction of twigs taken as a tonic for the stomach. (164:474) Decoction of branches taken as a tonic. *Tuberculosis Remedy* Branches used for tuberculosis. It was said that for the medicine to be really effective, the boughs should be taken from a plant growing all by itself. (187:92)
- *Food*—**Anticosti** *Beverage* Fruits, branches, potatoes, yeast, and water boiled into a drink. (143:64) **Thompson** *Beverage* Small pieces of branches used to make a tea-like beverage. (187:92)
- *Other*—**Gitksan** *Ceremonial Items* Plant used for rituals. (43:314) **Haisla** *Ceremonial Items* Wood used to make rattles worn on belts by shamans. (43:160) **Heiltzuk** *Ceremonial Items* Plant used as a part of a process of preparation undergone by shamanistic initiates. (43:62) **Kitasoo** *Good Luck Charm* Plant rubbed on the back for good luck. (43:314) **Navajo, Ramah** *Good Luck Charm* Used as a "good luck" smoke for hunters. (191:11) **Okanagan-Colville** *Ceremonial Items* Used in the sweat house during the winter. *Protection* Decoction of branches used as a wash for the body to protect a person from evil influences. (188:18) **Oweekeno** *Ceremonial Items* Plant used for ritualistic purposes. (43:62) **Tanana, Upper** *Incense & Fragrance* Branches used on rocks in the steam bath for the aromatic properties. (102:4) **Thompson** *Soap* Stems and leaf whorls boiled and used as a body wash by hunters, warriors, and widowers. (164:505)

Juniperus communis var. *montana* Ait., Common Juniper
- *Drug*—**Arapaho** *Disinfectant* Needles burned as a disinfectant. *Gastrointestinal Aid* Infusion of needles taken for bowel troubles. (as *J. sibirica* 121:46) *Misc. Disease Remedy* Ground needles scent used to drive smallpox away. (as *J. sibirica* 118:50) **Cheyenne** *Cough Medicine* Infusion of leaves used for coughs. One or two berries chewed and the juice swallowed for bad coughs. *Throat Aid* Infusion of leaves used for a tickling in the throat. (as *J. sibirica* 76:169) **Gitksan** *Witchcraft Medicine* Boughs, red elder bark, and cow parsnip roots used for evil witchcraft victims. (74:25) **Navajo, Ramah** *Ceremonial Medicine* Decoction of plant used as a ceremonial emetic. *Cough Medicine* Decoction taken and used as lotion for fever or "big cough." *Emetic* Decoction of plant used as a ceremonial emetic. *Febrifuge* Decoction of plant used internally and externally for fever. (as *J. sibirica* 191:12) **Paiute** *Blood Medicine* Seeds from dried fruit eaten as a blood tonic. *Orthopedic Aid* Seeds from dried fruit eaten for lumbago. *Tonic* Seeds from dried fruit eaten as a blood tonic. *Venereal Aid* Cold decoction of twigs taken for venereal disease. **Shoshoni** *Blood Medicine* and *Tonic* Decoction of branches taken as a blood tonic. (180:91, 92)
- *Food*—**Jemez** *Beverage* Leaves boiled into a beverage similar to coffee. (as *J. sibirica* 44:24)
- *Fiber*—**Ojibwa** *Building Material* Bark used to build houses, wigwams, and wickiups. Split strips or stakes used to make a pen to enclose graves. *Furniture* Wood used to make cradleboards. *Mats, Rugs & Bedding* Inner bark crushed and used to pad cradleboards. Bark used to make mats. (135:245)
- *Other*—**Arapaho** *Incense & Fragrance* Needles ground and used for their scent. (as *J. sibirica* 121:46) **Gitksan** *Ceremonial Items* Boughs burned as a fumigant to purify dwellings. (74:25) **Navajo, Ramah** *Smoke Plant* Dried fruits added to flavor tobacco. (as *J. sibirica* 191:12) **Ojibwa** *Cash Crop* Pulp wood and wood posts sold to make paper and fencing. *Ceremonial Items* Split strips thatched and placed on graves. (135:245) **Tolowa** *Decorations* Dried berries used to decorate dresses. *Jewelry* Dried berries used for beads to make necklaces. **Yurok** *Decorations* Dried berries used to decorate dresses. (6:34)

Juniperus deppeana Steud., Alligator Juniper
- *Food*—**Apache** *Fruit* Berries boiled for food. (as *J. pachyphloea* 32:32) **Apache, Chiricahua & Mescalero** *Fruit* Raw fruit eaten fresh. *Preserves* Berries boiled and made into jelly or preserves. (as *J. pachyphloea* 33:45) **Isleta** *Fruit* Berries boiled for food. (as *J. pachyphloea* 32:32) Large fruit boiled and eaten as food. (as *J. pachyphloea* 100:33) **Navajo, Ramah** *Fruit* Fruit eaten raw or boiled and ground. *Winter Use Food* Fruit stored for winter use. (as *J. pachyphloea* 191:12) **San Felipe** *Fruit* Berries boiled for food. (as *J. pachyphloea* 32:32) **Yavapai** *Beverage* Pulverized berries soaked in water, put in mouth, and juice sucked, the solid matter spat out. (as *J. pachyphloea* 63:212) Ground berries made into a meal, water added and used as a beverage. *Bread & Cake* Ground berries made into a meal, stored in baskets, and later made into a cake by dampening. *Staple* Ground berries made into a meal, water added and used as a beverage. (as *J. pachyphloea* 65:257)
- *Other*—**Yavapai** *Fuel* Dead wood used for fuel. (as *J. pachyphloea* 65:259)

Juniperus horizontalis Moench, Creeping Juniper
- *Drug*—**Blackfoot** *Kidney Aid* Used for kidney problems. (82:37) *Veterinary Aid* Roots soaked in water and used as a bath on horses for shiny hair. (86:89) **Cheyenne** *Ceremonial Medicine* Leaves burned as incense in ceremonies, especially to remove fear of thunder. *Cold Remedy* Cones chewed, infusion of boughs or cones taken or used as steam bath for colds. *Cough Medicine* Infusion of boughs or fleshy cones taken for coughing. *Febrifuge* Infusion of boughs or fleshy cones taken for high fevers. *Gynecological Aid* Leaves burned at childbirth to promote delivery. *Herbal Steam* Cones chewed, infusion of boughs or cones taken or used as steam bath for colds. *Love Medicine* Wood flutes used to "charm a girl whom a man loved to make her love him." *Sedative* Infusion of boughs or fleshy cones taken as a sedative. *Throat Aid* Infusion of boughs or cones taken for tickles in the throat or tonsillitis. (83:4) **Crow** *Ceremonial Medicine*

Young twigs and leaves burned as incense during incantations. **Montana Indian** *Kidney Aid* Infusion of seeds taken for kidney trouble. (as *J. sabina procumbens* 19:13)
- **Food**—**Ojibwa** *Beverage* Leaves used to make tea. (as *J. prostrata* 96:17)
- **Fiber**—**Blackfoot** *Mats, Rugs & Bedding* Branches used to form a carpet for the Holy Lodge dancer of the Sun Dance. (86:33) **Ojibwa** *Building Material* Bark used to build houses, wigwams, and wickiups. Split strips or stakes used to make a pen to enclose graves. *Furniture* Wood used to make cradleboards. *Mats, Rugs & Bedding* Inner bark crushed and used to pad cradleboards. Bark used to make mats. (as *J. sabina* var. *procumbens* 135:245)
- **Other**—**Blackfoot** *Ceremonial Items* Branch held in the right hand and the wing of an owl in the other by the Okan dancer. *Decorations* Seven berries, representing the Bunched Stars, used to make headpieces worn by some dancers. Sprigs used symbolically to decorate the altar of the Marten designed tepee. (86:33) **Ojibwa** *Cash Crop* Pulp wood and wood posts sold to make paper and fencing. *Ceremonial Items* Split strips thatched and placed on graves. (as *J. sabina* var. *procumbens* 135:245)

Juniperus monosperma (Engelm.) Sarg., Oneseed Juniper
- **Drug**—**Apache, White Mountain** *Anticonvulsive* Scorched twigs rubbed on body for fits. *Cold Remedy* Infusion of leaves taken for colds. *Cough Medicine* Infusion of leaves taken for coughs. *Gynecological Aid* Infusion of leaves taken by women previous to childbirth to relax muscles. (136:158) **Hopi** *Antirheumatic (External)* Poultice of heated twigs bound over a bruise or sprain for swelling. *Gastrointestinal Aid* Decoction of plant and sagebrush taken for indigestion. *Gynecological Aid* Infusion of leaves taken and used for many purposes.[41] *Laxative* Decoction of leaves taken as a laxative. *Pediatric Aid* Plant ashes rubbed on newborn baby.[42] *Reproductive Aid* Decoction of leaves taken by women who desire a female child. (42:330) **Isleta** *Antirheumatic (External)* Infusion of cedar bark used for bathing and washing sore feet. *Emetic* Strong infusion of leaves given in large quantities as an emetic. *Gynecological Aid* Infusion of leaves given to mothers after childbirth. (100:32) **Jemez** *Gastrointestinal Aid* Decoction of leaves taken for stomach or bowel disorders. *Gynecological Aid* Decoction of leaves taken by women after the birth of an infant. (44:24) **Keres, Western** *Antidiarrheal* Infusion of staminate cones used for diarrhea. *Dermatological Aid* Chewed bark taken for or applied to spider bites. *Diaphoretic* Plant used as an ingredient in the sweat bath. *Ear Medicine* Ground leaves mixed with salt and used in ears to eliminate bugs. *Emetic* Infusion of twigs or chewed twigs used as an emetic before breakfast. *Gastrointestinal Aid* Infusion of staminate cones used as a stomach tonic. *Laxative* Infusion of staminate cones used as a laxative. Bark chewed as a laxative. (171:48) **Navajo, Ramah** *Analgesic* Decoction used for postpartum or menstrual pain and cold infusion used for stomachache. *Ceremonial Medicine* Decoction used in "bath for purification of burial party." *Cough Medicine* Compound decoction, sometimes salted, taken for cough. (191:11, 12) *Dermatological Aid* Bark highly prized as a medicine for burns. (191:11) *Diaphoretic* Compound used as sweat bath medicine. *Emetic* Infusion of inner bark given to newborns "to clean out impurities." *Febrifuge* Cold infusion of plant used for fever. *Gastrointestinal Aid* Cold infusion of plant used for stomachache. *Gynecological Aid* Decoction or smoke of various plant parts used for childbirth difficulties. *Pediatric Aid* Infusion of inner bark used as an emetic for newborn "to clean out all impurities." Plant used as bed and coverlet for baby, "to make him strong and healthy." *Stimulant* Wet twigs or pulverized needles used as stimulant in postpartum fainting. *Veterinary Aid* Decoction given to sheep for bloating from eating "chamiso." (191:11, 12) **Paiute** *Cold Remedy* Decoction of twigs taken and fumes from burning branches inhaled for colds. *Dermatological Aid* and *Misc. Disease Remedy* Heated twigs rubbed on measles eruptions to relieve the discomfort. **Shoshoni** *Cold Remedy* Decoction of twigs taken and fumes from burning branches inhaled for colds. *Dermatological Aid* and *Misc. Disease Remedy* Heated twigs rubbed on measles eruptions to relieve the discomfort. (180:92) **Tewa** *Analgesic* Poultice of toasted leafy twigs applied to bruise or sprain pains. (138:39, 40) *Antirheumatic (External)* Poultice of heated twigs bound over a bruise or sprain for swelling. (42:330) *Dermatological Aid* Poultice of leafy twigs used for the pain and swellings of bruises or sprains. *Disinfectant* Leaves placed on hot coals as an herbal steam to "fumigate" new mother. *Diuretic* Berries used as an "active diuretic." (138:39, 40) *Gastrointestinal Aid* Decoction of plant and sagebrush taken for indigestion. *Gynecological Aid* Infusion of leaves taken and used for many purposes.[41] (42:330) Decoction of leaves taken and used as a postpartum wash. *Herbal Steam* Leaves placed on hot coals as an herbal steam to "fumigate" new mother. *Internal Medicine* Berries eaten or decoction of berries used "for every kind of internal chill." (138:39, 40) *Laxative* Decoction of leaves taken as a laxative. (42:330) *Orthopedic Aid* Poultice of toasted leafy twigs applied to bruise or sprain pains. (138:39, 40) *Pediatric Aid* Plant ashes rubbed on newborn baby.[42] *Reproductive Aid* Decoction of leaves taken by women who desire a female child. (42:330) *Toothache Remedy* Gum used as a filling for decayed teeth. (138:39, 40) **Zuni** *Antirheumatic (External)* Infusion of leaves used for muscle aches. *Contraceptive* Infusion of leaves taken to prevent conception. *Gynecological Aid* Infusion of leaves taken postpartum to prevent uterine cramps and stop vaginal bleeding. (27:373) Simple or compound infusion of twigs used to promote muscular relaxation at birth, and after childbirth to stop blood flow. *Hemostat* Simple or compound infusion of twigs taken after childbirth to stop blood flow. (166:55)
- **Food**—**Acoma** *Fruit* Fruits mixed with chopped meat, put into a clean deer stomach, and roasted. *Spice* Fruits used to season meats. *Starvation Food* Fruits eaten when other foods became scarce. (32:31) **Apache, Chiricahua & Mescalero** *Sauce & Relish* Fruit roasted, water added, and the mixture made into a gravy. (33:45) **Apache, White Mountain** *Fruit* Berries boiled and eaten. (136:158) **Cochiti** *Fruit* Fresh or cooked berries used for food. (32:31) **Hopi** *Fruit* Berries eaten with piki or cooked with stew. (42:330) **Jemez** *Fruit* Fresh or cooked berries used for food. (32:31) **Keres, Western** *Spice* Berries used to season meat.[40] *Starvation Food* Berries eaten in the fall or when food was scarce. (171:48) **Keresan** *Fruit* Berries used for food. (198:561) **Laguna** *Fruit* Fruits mixed with chopped meat, put into a clean deer stomach, and roasted. *Spice* Fruits used to season meats. *Starvation Food* Fruits eaten when other foods became scarce. (32:31) **Navajo** *Fodder* Branches cut off and given to the sheep to eat when

the snow was deep. *Fruit* Berries eaten ripe. (55:19) *Starvation Food* Inner bark chewed in times of food shortage to obtain the juice. (32:31) Inner bark chewed in times of food shortage. (55:19) **Navajo, Ramah** *Unspecified* Berry-like cones eaten roasted or boiled. Pinyon nuts used for food. (191:11) **San Ildefonso** *Fruit* Berries eaten. (138:40) **Tewa** *Fruit* Fruits eaten fresh or heated. (32:31) Berries eaten with piki. (42:330) Berries eaten by children and young people. **Tewa of Hano** *Special Food* Gum chewed as a delicacy. (138:40)

- **Fiber**—**Hopi** *Building Material* Used for construction. (42:330) **Jemez** *Building Material* Trunks used as uprights, beams, and fence posts. Limbs and boughs placed across corrals or enclosures as shelters for livestock. (44:24) **Keres, Western** *Basketry* Larger twigs used for basket frames.[39] *Building Material* Tree used to make posts and lumber. *Clothing* Bark rubbed fine and used to make baby clothes. (171:48) **Navajo** *Building Material* Wood used to make fence posts and hogan roofs. Wood used to make a canopy to protect a newborn child from the sparks of the fire. (55:19) **Navajo, Ramah** *Building Material* Wood used for fence posts and hogan poles. Boughs used for the sides and roofs of shade houses or special hogans for the Enemyway ceremonial. Bark used as lining in sweat houses. *Clothing* Bark used in the winter as a lining for moccasins to absorb moisture. *Furniture* Sticks used as frame for baby cradles. (191:11) **Tewa** *Building Material* Used for construction. (42:330) **Tewa of Hano** *Building Material* Bark used to chink the walls and roofs of log houses built after the Navajo fashion. (138:39)

- **Dye**—**Great Basin Indian** *Mordant* Whole plant ash added to various dye baths as a mordant. *Yellow* Whole plant used to make a yellow dye. (121:46) **Keres, Western** *Green* Green twigs rubbed on moccasins as a green dye. (171:48) **Navajo** *Green* Bark and berries used as a green dye for wool. (55:19) **Navajo, Ramah** *Unspecified* Needle ashes burned on rocks or in a pan and used as an ingredient for buckskin dye. (191:11)

- **Other**—**Hopi** *Ceremonial Items* Charcoal of plant, chewed melon seeds, and water used to make a ceremonial body paint. Branches used in the kachina dances. *Cleaning Agent* Boiled branch used as wash by men returning from burying a corpse. *Cooking Tools* Twigs used to separate corn dumplings while boiling. *Decorations* Seeds strung for beads. *Fuel* Wood used for firewood and tinder. *Malicious Charm* Plant used to do away with evil spirits after a death. *Tools* Used as a rake for clearing brush from the fields. *Toys & Games* Berries used in rattles. (42:330) **Isleta** *Fuel* Wood used in open ovens to produce very hot fires. (100:32) **Jemez** *Ceremonial Items* Ornamental branches and twigs used as decorations in nearly all of the dances. (44:24) **Keres, Western** *Ceremonial Items* Branches used in ceremonial dances. Infusion of plant taken by all household members for 4 days after a death. Infusion of cedar twigs taken with wafer bread by mothers of infants who died during birth. Cedar wood fire smoke used to fumigate property of the deceased. Cedar purge kept one from getting tired, but did not preclude sleepiness. *Fuel* Considered an important source of firewood for steady, even fires. *Hunting & Fishing Item* Larger trees used to make bows backed with sinew. *Preservative* Twigs mixed with commercial dyes to prevent them from fading. *Unspecified* Plant known and named but no use was specified. (171:48) **Navajo** *Ceremonial Items* Wood used to make prayer sticks. *Decorations* Used to make bows for the canopy of the baby's cradle. *Fuel* Wood used for firewood. Wood made into charcoal and used for smelting silver. *Good Luck Charm* Leaves chewed and spat out for better luck. *Weapon* Wood used to make bows, formerly carried in war. (55:19) **Navajo, Ramah** *Containers* Bark used as platform for sun drying roasted corn. Bark used as lining in corn storage pits. *Fuel* Wood used as one of the main sources of fuel. Bark used as tinder for making ceremonial fire with fire drill. *Hunting & Fishing Item* Wood used to make hunting bows. *Lighting* Bark used as a torch in the "Fire Dance." (191:11) **Tewa** *Ceremonial Items* Charcoal of plant, chewed melon seeds, and water used to make a ceremonial body paint. Branches used in the kachina dances. (42:330) Branches used in a few ceremonies and dances. (138:40) *Cleaning Agent* Boiled branch used as wash by men returning from burying a corpse. *Cooking Tools* Twigs used to separate corn dumplings while boiling. *Decorations* Seeds strung for beads. *Fuel* Wood used for firewood and tinder. (42:330) Used largely for firewood. *Hunting & Fishing Item* Wood used to make bows. *Lighting* Bark formerly shredded, bound into bundles, and used as torches to give light in houses. (138:39) *Malicious Charm* Plant used to do away with evil spirits after a death. *Tools* Used as a rake for clearing brush from the fields. *Toys & Games* Berries used in rattles. (42:330) **Tewa of Hano** *Fuel* Used largely for firewood. (138:39) **Zuni** *Fuel* Wood used as a favorite firewood, but more importantly in ceremonies. *Tools* Shredded, fibrous bark used as tinder to ignite the fire sticks used for the New Year fire. The bark was also used to make firebrands carried by personators of certain gods. (166:93)

Juniperus occidentalis Hook., Western Juniper

- **Drug**—**Apache, White Mountain** *Anticonvulsive* Scorched twigs rubbed on body for fits. *Cold Remedy* Infusion of leaves taken for colds. *Cough Medicine* Infusion of leaves taken for coughs. *Gynecological Aid* Infusion of leaves taken by women previous to childbirth to relax muscles. (136:158) **Paiute** *Analgesic* Decoction of berries taken for menstrual cramps. Decoction of young twigs taken for stomachaches. Fumes from burning twigs or leaves inhaled for headaches. *Antihemorrhagic* Decoction of young twigs taken for hemorrhages. *Antirheumatic (External)* Branches used in the sweat bath for rheumatism. Decoction of berries taken or poultice of decoction applied for rheumatism. Poultice of boiled twigs applied and cooled decoction used as a wash for rheumatism. *Blood Medicine* Decoction of berries or young twigs taken as a blood tonic. (180:92–96) *Cold Remedy* Infusion of leaves taken as a cold medicine. (111:47) Branches used in the sweat bath for "heavy colds." Fumes from burning twigs or leaves inhaled for colds. Simple or compound decoction of twigs or berries taken for colds. *Cough Medicine* Decoction of twigs or berries taken for coughs. *Dermatological Aid* Compound poultice of twigs used as a drawing agent for boils or slivers. Poultice of mashed twigs applied for swellings or rheumatism. Strong decoction used as an antiseptic wash for sores. *Disinfectant* Branches burned as a fumigant after illness. Decoction of twigs used as an antiseptic wash for sores. *Diuretic* Decoction of berries taken to induce urination. *Febrifuge* Simple or compound decoction of young twigs taken for fevers. *Gastrointestinal Aid* Decoction of young twigs taken for stomachaches. *Gynecological Aid* Decoction of berries taken for menstrual cramps. *Kidney Aid* Decoction of berries taken for kidney ailments. Simple or compound decoction of young twigs taken for kidney trouble. (180:92–96) *Misc. Disease Remedy* Bed of hot coals and branches used for malaria and other diseases. (111:47) Compound decoction of young twigs taken for smallpox. Decoction of young twigs taken for influenza. (180:92–96) *Pulmonary Aid* Bed of hot coals and branches used for pneumonia. (111:47) Compound decoction of twigs taken for fever, pneumonia, and influenza. *Tonic* Decoction of berries taken as a blood tonic. Decoction of young twigs taken as a blood tonic. *Venereal Aid* Decoction of shaved root taken for venereal disease. Decoction of twig or compound decoction of berry taken for venereal disease. (180:92–96) *Veterinary Aid* Boughs placed in a pan of coals and fumes inhaled by horses that have eaten poison camas. (111:47) **Shoshoni** *Anthelmintic* Strained cold water infusion of pulverized terminal twigs taken for worms. *Burn Dressing* Poultice of mashed twigs applied to burns. *Cold Remedy* Simple or compound decoction of twigs or berries taken for colds. *Cough Medicine* Decoction of twigs taken for coughs. *Dermatological Aid* Poultice of mashed twigs applied to swellings. Strong decoction used as an antiseptic wash for measles and smallpox. *Disinfectant* Branches burned as a fumigant after illness. Decoction of twigs used as an antiseptic wash for measles and smallpox. *Diuretic* Decoction of berries taken to induce urination. *Heart Medicine* Decoction of berries taken for heart trouble. *Kidney Aid* Simple or compound decoction of twigs or decoction of berry used for kidney trouble. *Misc. Disease Remedy* Compound decoction of young twigs taken for influenza or smallpox. Decoction of twigs used as an antiseptic wash for measles and smallpox. *Oral Aid* Poultice of pounded, moistened leaves applied to jaw for swollen and sore gums. *Throat Aid* Poultice of twigs applied to neck for sore throat. *Tonic* Decoction of young twigs taken as a general tonic. *Toothache Remedy* Poultice of leaves applied to jaw for toothache. *Venereal Aid* Decoction of twigs taken for venereal disease. **Washo** *Analgesic* Fumes from burning twigs inhaled for headaches. *Cold Remedy* Fumes from burning twigs inhaled for colds. *Disinfectant* Branches burned as a fumigant after illness. (180:92–96)

- **Food**—**Apache, White Mountain** *Fruit* Berries boiled and eaten. (136:158) **Atsugewi** *Dried Food* Berries dried, pounded into flour, and stored for later use. *Fruit* Fresh berries used for food. (61:139) **Miwok** *Unspecified* Ripe nuts used for food. (12:151) **Paiute** *Fruit* Berries and roasted, mashed deer liver combination used for food. *Winter Use Food* Berries stored without drying in a grass-lined hole in the ground for winter use. (111:47) **Paiute, Northern** *Fruit* Berries roasted, mixed with warm water, crushed, and eaten. (59:50)
- **Fiber**—**Paiute** *Building Material* Used as a material for housing. *Clothing* Bark rubbed between hands until soft and fibers woven into clothing. Bark rolled into rope, coiled, and sewn with sinew to form sandal soles. (111:47) **Pomo** *Basketry* Root fiber used to make twined baskets. (9:139)
- **Dye**—**Navajo** *Red* Wood ash, mountain mahogany, and black alder used as a red dye for buckskin. (55:19)
- **Other**—**Klamath** *Hunting & Fishing Item* Used for bows. (45:88) **Navajo** *Ceremonial Items* Wood used to make the wand for the War Dance. Branchlets, with needles, used to make prayer sticks of the west. (55:19) **Paiute** *Fuel* Wood was one of the principal sources of fuel and material. Bark mixed with dirt to use as tinder. *Hunting & Fishing Item* Wood and sinew strips used to make laminated bows. *Lighting* Bark wound around a stick and used as a torch to provide light and carry a fire to a new campsite. *Musical Instrument* Wood strips used for drum frames. *Tools* Wood hearth board used as a base for a fire drill. *Toys & Games* Leaves stuffed into buckskin and used as a ball in a game like lacrosse or hockey. (111:47)

Juniperus osteosperma (Torr.) Little, Utah Juniper

- **Drug**—**Havasupai** *Cold Remedy* Green branches used singly or together with other plants for colds. (197:206) **Hopi** *Gynecological Aid* Decoction of branches used especially by women during confinement. (as *J. utahensis* 190:157) *Other* Misbehaving youngsters held in a blanket over a smoldering fire of plant. (as *J. utahensis* 200:37) **Navajo** *Analgesic* Seeds eaten for headaches. *Dermatological Aid* Used to wash the hair. (90:152) **Paiute** *Analgesic* Decoction of berries taken for menstrual cramps. Decoction of young twigs taken for stomachaches. Fumes from burning twigs inhaled for headaches and colds. *Antihemorrhagic* Decoction of young twigs taken for hemorrhages. *Antirheumatic (External)* Branches used in the sweat bath for rheumatism. Decoction of berries taken or poultice of decoction applied for rheumatism. Poultice of boiled twigs applied and cooled decoction used as a wash for rheumatism. *Blood Medicine* Decoction of berries or young twigs taken as a blood tonic. *Cold Remedy* Branches used in the sweat bath for heavy colds. Fumes from burning twigs or leaves inhaled for headaches and colds. Simple or compound decoction of twigs or berries taken for colds. *Cough Medicine* Decoction of twigs or berries taken for coughs. *Dermatological Aid* Compound poultice of twigs used as a drawing agent for boils or slivers. Poultice of mashed twigs applied for swellings or rheumatism. Strong decoction used as an antiseptic wash for sores. *Disinfectant* Branches burned as a fumigant after illness. Strong decoction of twigs used as an antiseptic wash for sores. *Diuretic* Decoction of berries taken to induce urination. *Febrifuge* Simple or compound decoction of young twigs taken for fevers. *Gastrointestinal Aid* Decoction of young twigs taken for stomachaches. *Gynecological Aid* Decoction of berries taken for menstrual cramps. *Kidney Aid* Decoction of berries taken for kidney ailments. Simple or compound decoction of young twigs taken for kidney trouble. *Misc. Disease Remedy* Compound decoction of twigs taken for smallpox. Decoction of young twigs taken for influenza. *Pulmonary Aid* Compound decoction of twigs taken for fevers, pneumonia, and influenza. *Tonic* Decoction of young twigs or berries taken as a blood tonic. *Venereal Aid* Decoction of shaved root taken for venereal disease. Decoction of twig or compound decoction of berry taken for venereal disease. (as *J. utahensis* 180:93–96) **Paiute, Northern** *Analgesic* Roasted berry steam used for pains. *Antirheumatic (External)* Roasted berry steam used for rheumatism. *Cold Remedy* Infusion of leaves taken and burning leaf scent inhaled for colds. (59:130) **Shoshoni** *Anthelmintic* Strained cold water infusion of pulverized terminal twigs taken for worms. *Burn Dressing* Poultice of mashed twigs applied to burns. *Cold Remedy* Simple or compound decoction of twigs or berries taken for colds. *Cough Medicine* Decoction of twigs taken for coughs. *Dermatological Aid* Poultice of mashed twigs applied to swellings. Strong decoction used as an antiseptic wash for measles and smallpox. *Disinfectant* Branches burned as a fumigant after illness. Strong

decoction of twigs used as an antiseptic wash for measles and smallpox. *Diuretic* Decoction of berries taken to induce urination. *Heart Medicine* Decoction of berries taken for heart trouble. *Kidney Aid* Compound decoction of twigs or decoction of berry taken for kidney ailments. Simple or compound decoction of young twigs taken for kidney trouble. *Misc. Disease Remedy* Compound decoction of twigs taken for influenza and smallpox. Strong decoction of twigs used as an antiseptic wash for measles and smallpox. *Oral Aid* Poultice of pounded leaves held to the jaw for swollen and sore gums. *Throat Aid* Poultice of twigs applied to neck for sore throat. *Tonic* Decoction of young twigs taken as a general tonic. *Toothache Remedy* Poultice of leaves applied to jaw for toothache. *Venereal Aid* Decoction of twigs taken for venereal disease. **Washo** *Analgesic* Fumes from burning twigs inhaled for headaches. *Cold Remedy* Fumes from burning twigs inhaled for colds. *Disinfectant* Branches burned as a fumigant after illness. (as *J. utahensis* 180:93–96)
- *Food*—**Acoma** *Soup* Berries cooked in a stew. (as *J. utahensis* 200:63) **Apache, White Mountain** *Fruit* Berries boiled and eaten. (as *J. californica* var. *utahensis* (*J. utahensis*) 136:158) **Gosiute** *Fruit* Berries eaten in the fall and winter after proper boiling. (as *J. californica* var. *utahensis* 39:372) **Havasupai** *Beverage* Dried berries used to make a drink. *Dried Food* Berries sun dried and stored for winter use. (197:206) **Hopi** *Fruit* Berries eaten with piki bread. (as *J. utahensis* 120:18) Berries used for food. (as *J. utahensis* 200:63) **Tubatulabal** *Fruit* Berries used extensively for food. Berries used extensively for food. (as *J. utaliensis* 193:15) **Yavapai** *Beverage* Ground berries made into a meal, water added, and used as a beverage. *Bread & Cake* Ground berries made into a meal, stored in baskets, and later made into a cake by dampening. *Staple* Ground berries made into a meal, water added, and used as a beverage. (as *J. utahensis* 65:257)
- *Fiber*—**Gosiute** *Building Material* Wood used in the construction of winter lodges. Bark used for thatching and as a floor covering. Small branches used as a floor covering. (as *J. californica* var. *utahensis* 39:372) **Havasupai** *Building Material* Logs and brush, covered with dirt, used to make winter houses. Bark used on top of the brush covering of the winter houses to keep the dirt from falling through. (197:206) **Navajo** *Building Material* Green timber used to make corrals. (as *J. utahensis* 200:62)
- *Other*—**Gosiute** *Containers* Bark used to line and cover the fruit storing pits. (as *J. californica* var. *utahensis* 39:372) **Havasupai** *Fuel* Wood used for firewood. Crushed bark used for tinder. Crushed bark used as a "slow match." The crushed bark was twisted into a rope, tied at intervals with yucca, and wrapped into a coil. The free end was set on fire and kept smoldering by blowing on it at intervals. Fire could be carried in this fashion from early dawn until noon. *Toys & Games* Wood used to make the pole of the hoop and pole game. (197:206) **Hopi** *Fuel* Used for firewood. (as *J. utahensis* 200:62) *Jewelry* Seeds pierced and strung for beads in ancient times. (as *J. utahensis* 200:63) **Yavapai** *Fuel* Dead wood used for fuel. *Lighting* Bark used as a torch. (as *J. utahensis* 65:259)

Juniperus pinchotii Sudworth, Pinchot's Juniper
- *Drug*—**Comanche** *Analgesic* Dried leaves sprinkled on live coals and smoke inhaled for headache. *Ceremonial Medicine* Dried leaves sprinkled on live coals and smoke inhaled for ghost sickness. *Gynecological Aid* Decoction of dried and pulverized roots taken for menstrual complaints. *Other* Dried leaves sprinkled on live coals and smoke inhaled for vertigo. (99:3)

Juniperus scopulorum Sarg., Rocky Mountain Juniper
- *Drug*—**Blackfoot** *Antiemetic* Infusion of berries taken for vomiting. (97:17) Infusion of berries taken for vomiting. (114:276) *Antirheumatic (External)* Leaves boiled, turpentine added, mixture cooled and used for arthritis and rheumatism. (82:36) Decoction of leaves and turpentine rubbed on parts affected by arthritis and rheumatism. (97:17) **Cheyenne** *Ceremonial Medicine* Leaves burned as incense in ceremonies, especially to remove fear of thunder. (83:4) *Cold Remedy* Infusion of boughs, branches, and cones used for colds. Fleshy cones chewed for colds. (82:36) Cones chewed, infusion of boughs or cones taken or used as steam bath for colds. (83:4) *Cough Medicine* Infusion of leaves taken for constant coughing. (76:170) Infusion taken for coughs. (82:36) Infusion of boughs or fleshy cones taken for coughing. (83:4) *Febrifuge* Infusion of boughs, branches, and cones used for fevers. (82:36) Infusion of boughs or fleshy cones taken for high fevers. *Gynecological Aid* Leaves burned at childbirth to promote delivery. *Herbal Steam* Cones chewed, infusion of boughs or cones taken or used as steam bath for colds. *Love Medicine* Wood flutes used to "charm a girl whom a man loved to make her love him." (83:4) *Pulmonary Aid* Infusion of boughs, branches, and cones used for pneumonia. *Sedative* Infusion used for sedating hyperactive persons. (82:36) Infusion of boughs or fleshy cones taken as a sedative. (83:4) *Throat Aid* Infusion of leaves taken for a tickling in the throat. (76:170) Infusion taken for "tickling of the throat." (82:36) Infusion of boughs or cones taken for tickles in the throat or tonsillitis. (83:4) **Crow** *Antidiarrheal* Infusion taken for diarrhea. *Antihemorrhagic* Infusion taken for lung or nose hemorrhages. *Dietary Aid* Fleshy cones chewed to increase the appetite. *Gastrointestinal Aid* Fleshy cones chewed for upset stomach. *Gynecological Aid* Infusion taken after birth for cleansing and healing. **Flathead** *Ceremonial Medicine* Plant burned and smoke used to purify the air and ward off illness. *Cold Remedy* Infusion of boughs, branches, and cones used for colds. *Febrifuge* Infusion of boughs, branches, and cones used for fevers. *Pulmonary Aid* Infusion of boughs, branches, and cones used for pneumonia. *Veterinary Aid* Leaves placed on hot coals and smoke used for sick horses. **Kutenai** *Cold Remedy* Infusion of boughs, branches, and cones used for colds. Plant burned and smoke used for colds. *Febrifuge* Infusion of boughs, branches, and cones used for fevers. *Misc. Disease Remedy* Infusion taken for sugar diabetes. *Pulmonary Aid* Infusion of boughs, branches, and cones used for pneumonia. **Montana Indian** *Antirheumatic (External)* Concoction used for arthritis and rheumatic pain. (82:36) *Ceremonial Medicine* Aromatic twigs burned as incense. (19:14) **Navajo** *Ceremonial Medicine* Plant taken as a "War Dance medicine." *Dermatological Aid* Plant rubbed on the hair for dandruff. *Unspecified* Pounded mixture of herbs given to patient during the blackening ceremony of the War Dance. (55:20) **Navajo, Kayenta** *Analgesic* Plant used for pain. (205:15) **Navajo, Ramah** *Analgesic* Decoction of needles taken and used as lotion for headache and stomachache. *Ceremonial Medicine* Cold infusion used as a ceremonial medicine to protect from enemies and witches. *Cold Remedy* Decoction of needles taken and used as lotion for colds.

Febrifuge Decoction of needles taken and used as lotion for fever. *Gastrointestinal Aid* Decoction of needles taken and used as lotion for stomachache. *Kidney Aid* Decoction of needles taken and used as lotion for kidney trouble. *Witchcraft Medicine* Cold infusion taken and used as lotion in ceremony for protection from witches. (191:12) **Nez Perce** *Cold Remedy* Infusion of boughs, branches, and cones used for colds. *Febrifuge* Infusion of boughs, branches, and cones used for fevers. *Pulmonary Aid* Infusion of boughs, branches, and cones used for pneumonia. (82:36) **Okanagan-Colville** *Antihemorrhagic* Decoction of branch tips and needles taken for internal hemorrhaging. *Antirheumatic (External)* Poultice of mashed and dampened branches applied to arthritic joints. *Dermatological Aid* Poultice of mashed and dampened branches applied to skin sores. *Misc. Disease Remedy* Decoction of sap used for the flu and colds. Five strips of bark each about 5 cm by 10 cm were boiled in about 2 liters of water in order to obtain the sap. Only bark from the bottom part of the tree could be used. *Other* Decoction of branch tips and needles considered a good emergency medicine. *Poison* Berries believed to be poisonous. (188:19) **Okanagon** *Urinary Aid* Fruit eaten for bladder troubles. (125:41) **Saanich** *Misc. Disease Remedy* Scented branches hung around the house during disease epidemics to "drive the germs away." (182:70) **Sanpoil** *Dermatological Aid* Decoction of leaves, stems, and berries used as a wash for sores. *Tuberculosis Remedy* Berries eaten or decoction taken for tuberculosis. *Unspecified* Berries eaten or decoction taken for general illnesses. (131:221) **Shoshoni** *Venereal Aid* Decoction of twigs taken over a long period of time for venereal disease. (180:92) **Shuswap** *Diaphoretic* Used in the sweat house. *Misc. Disease Remedy* Decoction of plant used as a steam bath for the flu. *Panacea* Decoction of stems and needles taken for any sickness. (123:50) **Sioux** *Cold Remedy* Infusion of boughs, branches, and cones used for colds. Plant burned and smoke used for colds. *Febrifuge* Infusion of boughs, branches, and cones used for fevers. *Misc. Disease Remedy* Infusion of leaves formerly used for cholera. *Pulmonary Aid* Infusion of boughs, branches, and cones used for pneumonia. **Stony Indian** *Antihemorrhagic* Infusion taken for hemorrhages. (82:36) **Swinomish** *Antirheumatic (External)* Infusion of roots used as a foot soak for rheumatism. *Disinfectant* Decoction of leaves used to disinfect the house. *Panacea* Infusion of leaves used as a wash for all ailments. *Tonic* Infusion of leaves taken as a general tonic. (79:21) **Thompson** *Antirheumatic (External)* Decoction of berries used externally for rheumatism. *Cold Remedy* Decoction of branches and berries taken for colds. *Dermatological Aid* Decoction of berries used as a wash for all types of bites and stings. Decoction of boughs taken or used as a wash for hives or sores. The informant said that she used a decoction of mashed boughs and Douglas fir to bathe her children when they had the "seven year itch" and that it worked, but not as well as modern medicine. *Disinfectant* Decoction or infusion of plant used to disinfect the house after an illness or death. The decoction was used to scrub the floors, walls, and furniture after an illness or death in the house. It was also used to wash the deceased person's bedding and clothing as well as serving as a protective wash for other members of the household. The steam from the infusion was also said to have a disinfecting effect. If they knew that an illness was going to arrive, they broke the branches and burned them in the house for the strong smoke which they said would keep the air fresh so that the sickness would not affect them. They also burned the branches after a death in the house to freshen the air. (187:92) *Diuretic* Fresh berries eaten as a diuretic. (164:465) Fresh or dried berries eaten as a diuretic. *Gastrointestinal Aid* Decoction of berries used externally for stomach ailments. *Gynecological Aid* Decoction of branches and berries taken every morning just before childbirth. The decoction was taken every morning just before childbirth to promote muscular relaxation. *Heart Medicine* Decoction of branches and berries taken for heart trouble. *Kidney Aid* Infusion of plant taken for kidney trouble. *Misc. Disease Remedy* Decoction of boughs used for "black measles" or chickenpox. *Other* Plant considered effective in combating evil "spirits" associated with illness and death. *Tuberculosis Remedy* Decoction of branches and berries taken for tuberculosis. (187:92) *Urinary Aid* Fruit eaten for bladder troubles. (125:41) Fresh berries eaten as a medicine for the bladder. (164:465) Fresh or dried berries eaten for bladder trouble. (187:92) *Veterinary Aid* Strong decoction of berries used to kill ticks on horses. (164:512)

- *Food*—**Apache, Chiricahua & Mescalero** *Fruit* Berries mixed with mescal and eaten. (33:37) **Jemez** *Fruit* Berries eaten raw or stewed. (44:24) **Keresan** *Fruit* Berries eaten raw by hunters while out in the mountains, but better when cooked. (198:561) **Okanagan-Colville** *Beverage* Berries made into a drink and taken in the sweat house. This drink could only be taken with great caution, because the berries were believed to be poisonous. (188:19) **Tewa** *Fruit* Fruits eaten fresh or heated. (32:32)
- *Fiber*—**Montana Indian** *Building Material* Wood used for fence posts. (19:14)
- *Other*—**Blackfoot** *Ceremonial Items* Plant used in the Sun Dance ceremony, the summer festival of the Blackfoot. (97:17) Used on the altar of the sacred woman at the Sun Dance. (114:276) **Cheyenne** *Cooking Tools* Knots used to make bowls. (83:13) *Hunting & Fishing Item* Used to make bows. (83:46) Wood used as the best material for bows. (83:5) *Incense & Fragrance* Burned as an incense when making medicine. (83:13) *Musical Instrument* Wood used to make courting flutes. (83:5) *Protection* Plant burned and smoke used for protection from thunder and lightning. (82:36) **Hoh** *Ceremonial Items* Twigs and berries used in ceremonies. (137:57) **Montana Indian** *Hunting & Fishing Item* Wood used to make lance shafts and bows. Wood used to make bows. (82:36) **Okanagan-Colville** *Hunting & Fishing Item* Tough wood used to make bows. Pounded branches, berries, and water used to soak arrowheads and render them poisonous. Arrowheads, soaked overnight in a solution of pounded juniper branches (with berries) and water, were said to cause a deer's blood to coagulate when it was wounded so that it could not run far. This "poison" worked effectively, even if the deer were only nicked with the arrowhead. It was said not to affect the edibility of the meat. *Protection* Boughs considered an extremely powerful medicine for combating evil spirits associated with death. When a person died, his family used the boughs to fumigate the house. All the doors and windows were closed and the boughs were burned and the smoke allowed to fill all the rooms. This treatment was made even more effective by adding rose branches to the juniper. After the smoke treatment, rose and juniper branches were boiled together and the water used to wash the entire house—lights, windows, floors, walls, and ceilings.

This wash water was then taken outside and splashed all around the house and along the trails leading to the outbuildings to prevent the spirit of the dead person from coming back to the house. *Stable Gear* Tough wood used to make double yokes for horses. *Toys & Games* Tough wood made into a spoked wheel and used in a throwing game. The wheel was rolled along a trough and contestants threw spear-like sticks at it, trying to stop it by having their stick enter the center of the wheel, thus making it fall over. Lesser points were made by getting the stick part way through the spokes, each of which gave a different value according to its color. The winner had to get 20 points. This game was played by men and was often accompanied by betting. *Weapon* Pounded branches, berries, and water used as a poison to kill people quickly in warfare. Pounded branches, berries, and water used as a poison on bullets. (188:19) **Quileute** *Ceremonial Items* Twigs and berries used in ceremonies. (137:57) **Shuswap** *Hunting & Fishing Item* Used to make bows. *Insecticide* Plant used to keep earwigs and bedbugs out of the house. (123:50) **Thompson** *Good Luck Charm* Tree used to bring good luck. (187:92) *Hunting & Fishing Item* Wood used to make bows and clubs. (164:498) Wood used to make the two outer prongs of a leister, the center from "ironwood" or saskatoon wood. Hunters rubbed the boughs on themselves as protection against grizzlies. (187:92) *Insecticide* Strong decoction of berries used to kill ticks on horses. (164:512) *Musical Instrument* Wood used to make drums. (164:498) *Protection* Boughs used as protection against illnesses and death. One informant said that, formerly, when a person died the branches were broken and laid in the coffin to keep the germs away and to keep the spirit or "ghost" of the deceased person from harming or scaring the living. The informant also said that the branches could be placed on the stove in a little dish and the scent allowed to permeate the room. The branches could also be placed around the edges of the family's bedrooms as a disinfectant. They were left there until they lost their strong, pungent odor. Hunters rubbed the boughs on themselves as protection against grizzlies. (187:92)

Juniperus sp., Juniper

- **Drug—Apache, Western** *Pulmonary Aid* Poultice of heated, wrapped branches applied to pneumonia patients' backs. (26:187) **Blackfoot** *Antirheumatic (External)* Infusion of roots and poplar leaves applied like a liniment to stiff backs or backaches. (86:78) *Dermatological Aid* Infusion of plant used to soothe the face after whiskers were plucked. *Orthopedic Aid* Infusion of plant used as a foot wash. (86:124) *Tonic* Infusion of roots used as a general tonic. (86:83) **Creek** *Analgesic* and *Antirheumatic (External)* Poultice of warm sprigs and leaves applied to rheumatic aches and pains. *Blood Medicine* and *Tonic* Plant used as a spring tonic, to thin the blood. (172:657) **Gitksan** *Antihemorrhagic* Plant used for mouth hemorrhages. (43:314) Strong decoction of whole plant taken for hemorrhage and kidney trouble. (150:49) *Cathartic* Plant used as a purgative. (43:314) Strong decoction of whole plant taken as a purgative and diuretic. (150:49) *Dermatological Aid* Plant used for cuts. *Diuretic* Plant used as a diuretic. (43:314) Strong decoction of whole plant taken as a diuretic and purgative. (150:49) *Kidney Aid* Plant used for kidney troubles. (43:314) Strong decoction of whole plant used for kidney trouble and hemorrhage. (150:49) *Strengthener* Plant used as a strengthener. (43:314) **Hopi** *Dermatological Aid* Poultice of heated twigs applied to bruise or sprain for swelling. *Disinfectant* Decoction of branch used as wash to disinfect persons after corpse burial. (200:62, 63) *Gastrointestinal Aid* Compound decoction of plant taken for indigestion. (200:33, 62) *Gynecological Aid* Plant used several ways to ease pregnancy and childbirth. (200:35, 36, 62) *Orthopedic Aid* Poultice of heated twigs bound on bruise or sprain for swelling. (200:32, 62) *Pediatric Aid* Plant ashes rubbed on newborn baby. Plant smoke used to make child behave by holding the child over the fire. (200:62, 63) **Hualapai** *Dermatological Aid* Poultice of leaf ash applied to sores. *Panacea* Decoction of leaves taken for various disorders. (195:32) **Navajo** *Misc. Disease Remedy* Decoction of berries taken for influenza. (55:17) **Shoshoni** *Contraceptive* Infusion of berries taken on three successive days for birth control. (118:46)

- **Food—Apache, Mescalero** *Fruit* Berries boiled, ground, or mashed and used with other foods. (14:43) **Apache, Western** *Beverage* Berries soaked, pounded with yucca fruit, mixed with water, and drained to make a drink. *Dried Food* Unseasoned berries dried and boiled. *Sauce & Relish* Berries pounded with yucca fruit to make a gravy. *Spice* Ashes mixed with corn mush for color and flavor. (26:187) **Hualapai** *Starvation Food* Berries considered a starvation food because of their abundance. (195:32) **Navajo** *Forage* Plant eaten by sheep during droughts. (55:17)

- **Fiber—Apache, Mescalero** *Building Material* Used for tepee poles. (14:43) **Navajo** *Building Material* Boughs used to build the corral for public exhibitions at the close of a ceremony. Boughs used to make the summer shelters where the women weave. Bark used in the construction of hogans. *Clothing* Bark woven into garments and used to make sandals. Dry bark mixed with mud and worn as clothing during hard times. *Mats, Rugs & Bedding* Bark used to make blankets and passageway curtains. (55:17)

- **Dye—Navajo** *Unspecified* Bark, berries, and twigs used for dye purposes. (55:17)

- **Other—Apache, Chiricahua & Mescalero** *Fuel* Wood used to heat cooking stones. (33:36) **Apache, Mescalero** *Fuel* Bark used as tinder for fire drills. *Hunting & Fishing Item* Used to make bows. *Tools* Used to make handles for scrapers. (14:43) **Apache, Western** *Lighting* Dried bark made into a torch. (26:187) **Blackfoot** *Hide Preparation* Greased leaves used to smoke hides yellow. *Jewelry* Dried, smoked berries used to make necklaces, wristlets, or clothing decorations. (86:114) **Micmac** *Fuel* Wood used for kindling and fuel. (159:258) **Navajo** *Ceremonial Items* Wood, struck by lightning, used as the two parts of the fire drill for the Night Chant. Branches made into a fagot and used by the personator of the Black God, owner of all fire. Shredded bark carried by the dancers in the Fire Dance during the last night of the Mountain Chant. Wood burned into charcoal, ground, and used for black in sand paintings. Branches made into wands and used in certain ceremonies. Wood used to make prayer sticks. *Containers* Concave bark used to make improvised trays for the sand painting powders. *Fuel* Light bark used as tinder to catch the spark from the fire drill. Wood burned into charcoal and used as a fuel. *Jewelry* Seeds used to make necklaces, bracelets, anklets, and wristlets. *Toys & Games* Wood used to make dice. (55:17)

Juniperus virginiana L., Eastern Red Cedar

- **Drug**—**Cherokee** *Abortifacient* Used for "female obstructions." *Anthelmintic* Decoction of berries given for worms. *Antirheumatic (Internal)* Used for rheumatism and "female obstructions." *Cold Remedy* Infusion taken for colds. *Dermatological Aid* Used as an ointment for itch, skin diseases, and "white swelling." *Diaphoretic* Used as a diaphoretic. *Misc. Disease Remedy* Used as a diaphoretic and for measles. (80:28) **Chippewa** *Antirheumatic (External)* Compound decoction of twigs used as herbal steam for rheumatism. *Antirheumatic (Internal)* Compound decoction of twigs taken for rheumatism. *Herbal Steam* Compound decoction of twigs taken or used as herbal steam for rheumatism. (53:362) **Comanche** *Disinfectant* Smoke from leaves inhaled for purifying effect. (29:522) **Cree, Hudson Bay** *Diuretic* Leaves used as a diuretic. (92:303) **Dakota** *Cold Remedy* Smoke from burned twigs inhaled as a cold remedy. *Cough Medicine* Decoction of fruits and leaves taken for coughs. *Misc. Disease Remedy* Decoction of leaves taken and used as a wash for cholera. *Veterinary Aid* Decoction of fruits and leaves given to horses for coughs. (70:63, 64) **Delaware, Oklahoma** *Antirheumatic (External)* Infusion of twigs used as herbal steam for rheumatism. (175:30, 76) *Herbal Steam* Infusion of roots or twigs used as herbal steam for rheumatism. (175:30) **Iroquois** *Antirheumatic (Internal)* Compound decoction taken for rheumatism. *Cold Remedy* Compound decoction taken for colds. *Cough Medicine* Compound decoction taken for coughs. *Diuretic* Compound decoction taken as a diuretic. (87:271) **Kiowa** *Oral Aid* Berries chewed for canker sores in the mouth. (192:13) **Lakota** *Cold Remedy* Leaves burned and smoke inhaled for head colds. (106:30) **Meskwaki** *Adjuvant* Wood prepared in warm water and used as a seasoner for other medicines. *Other* Decoction of leaves taken by convalescent patients. *Stimulant* Decoction of leaves taken for weakness and as a convalescent medicine. (152:234) **Ojibwa, South** *Analgesic* Bruised leaves and berries used internally for headache. (91:198) **Omaha** *Ceremonial Medicine* Plant used in the sun dance ceremony and various rituals. Plant used in the vapor bath of the purificatory rites. (68:320) *Cold Remedy* Smoke from burned twigs inhaled as a cold remedy. *Cough Medicine* Decoction of fruits and leaves taken for coughs. (70:63, 64) *Diaphoretic* Plant used in the vapor bath of the purificatory rites. (68:320) *Herbal Steam* Twigs used on hot stones in vapor bath, especially in purification rites. *Veterinary Aid* Decoction of fruits and leaves given to horses for coughs. **Pawnee** *Cold Remedy* Smoke from burned twigs inhaled as a cold remedy. *Cough Medicine* Decoction of fruits and leaves taken for coughs. *Sedative* Smoke from burning twigs used for nervousness and bad dreams. *Veterinary Aid* Decoction of fruits and leaves given to horses for coughs. **Ponca** *Cold Remedy* Smoke from burned twigs inhaled as a cold remedy. *Cough Medicine* Decoction of fruits and leaves taken for coughs. *Herbal Steam* Twigs used on hot stones in vapor bath, especially in purification rites. *Veterinary Aid* Decoction of fruits and leaves given to horses for coughs. (70:63, 64) **Rappahannock** *Other* Infusion of bark taken for summer complaint (summer cholera). *Pulmonary Aid* Compound infusion with berries taken for shortness of breath. (161:30) *Respiratory Aid* Compound infusion with berries taken for asthma. (161:33) **Salish** *Disinfectant* Plant used for fumigation. (178:294)
- **Food**—**Comanche** *Fruit* Fruits eaten for food. (29:522) **Lakota** *Beverage* Berries eaten to relieve thirst. *Spice* Berries crushed and used to flavor soups, meats, and stews. (106:30)
- **Fiber**—**Cherokee** *Building Material* Wood used to make fence posts. *Furniture* Wood used to make furniture. (80:28) **Chippewa** *Mats, Rugs & Bedding* Used for mats. (53:377) **Ojibwa** *Building Material* Bark used to build houses, wigwams, and wickiups. Split strips or stakes used to make a pen to enclose graves. *Furniture* Wood used to make cradleboards. *Mats, Rugs & Bedding* Inner bark crushed and used to pad cradleboards. Bark used to make mats. (135:245)
- **Dye**—**Chippewa** *Red-Brown* Bark used to make a mahogany-colored dye for coloring cedar strips in mats. (53:371)
- **Other**—**Cherokee** *Decorations* Wood used to carve. *Insecticide* Used for mothproofing. (80:28) **Dakota** *Protection* Boughs put on tepee poles to ward off lightning. (70:63) **Kiowa** *Incense & Fragrance* Needles thrown into the fire and used as incense during prayers in the peyote meeting. *Musical Instrument* Red, aromatic heartwood used to make "love flutes." (192:13) **Lakota** *Incense & Fragrance* Leaves and twigs burned as incense in funerals. (106:30) **Navajo** *Ceremonial Items* Wood used to make the wand carried in the War Dance Ceremony. (55:20) **Ojibwa** *Cash Crop* Pulp wood and wood posts sold to make paper and fencing. *Ceremonial Items* Split strips thatched and placed on graves. (135:245) **Omaha** *Incense & Fragrance* Twigs used as incense. (68:320) *Protection* Boughs put on tepee poles to ward off lightning. **Pawnee** *Protection* Boughs put on tepee poles to ward off lightning. **Ponca** *Protection* Boughs put on tepee poles to ward off lightning. (70:63) **Thompson** *Fuel* Used as a fuel to make a heavy smoke for smoking skins. **Thompson, Upper (Nicola Band)** *Fuel* Used in combination with sagebrush as a fuel to make a heavy smoke when desiring very dark skins. (164:500)

Juniperus virginiana **var.** *silicicola* (Small) J. Silba, Southern Red Cedar
- **Drug**—**Seminole** *Analgesic* Infusion of leaves taken as an emetic for rainbow sickness: fever, stiff neck, and backache. (as *J. silicicola* 169:210) Infusion of leaves taken as an emetic for thunder sickness: fever, dizziness, headache, and diarrhea. (as *J. silicicola* 169:213) Leaves used for scalping sickness: severe headache, backache, and low fever. (as *J. silicicola* 169:262) *Antidiarrheal* Infusion of leaves taken as an emetic for thunder sickness: fever, dizziness, headache, and diarrhea. (as *J. silicicola* 169:213) *Antirheumatic (External)* Decoction of leaves used as a body rub and steam for joint swellings. (as *J. silicicola* 169:193) *Cold Remedy* Infusion of leaves and bark taken and steam inhaled for runny nose, stuffy head, and sore throat. *Cough Medicine* Infusion of leaves and bark taken and steam inhaled for coughs. (as *J. silicicola* 169:279) *Emetic* Infusion of leaves taken as an emetic for rainbow sickness: fever, stiff neck, and backache. (as *J. silicicola* 169:210) Infusion of leaves taken as an emetic for thunder sickness: fever, dizziness, headache, and diarrhea. (as *J. silicicola* 169:213) Plant used as an emetic during religious ceremonies. (as *J. silicicola* 169:409) *Eye Medicine* Infusion of leaves taken and used as a bath for mist sickness: eye disease, fever, and chills. *Febrifuge* Infusion of leaves taken and used as a bath for mist sickness: eye disease, fever, and chills. (as *J. silicicola* 169:209) Infusion of leaves taken as an emetic for rainbow sickness: fever, stiff neck, and backache. (as *J. silicicola* 169:210) Infusion of leaves taken as an emetic for thunder sickness: fever, dizziness, headache, and diarrhea. (as *J. silicicola*

169:213) Leaves used for scalping sickness: severe headache, backache, and low fever. *Orthopedic Aid* Leaves used for scalping sickness: severe headache, backache, and low fever. (as *J. silicicola* 169:262) Leaves used to smoke the body for eagle sickness: stiff neck or back. Leaves used to smoke the body for fawn sickness: swollen legs and face. (as *J. silicicola* 169:305) *Other* Leaves burned for ghost sickness: dizziness and staggering. (as *J. silicicola* 169:260) *Pediatric Aid* Plant and other plants used as a baby's charm for fear from dreams about raccoons or opossums. (as *J. silicicola* 169:221) *Psychological Aid* Plant burned to smoke the body for insanity. (as *J. silicicola* 169:293) *Sedative* Plant and other plants used as a baby's charm for fear from dreams about raccoons or opossums. (as *J. silicicola* 169:221) *Stimulant* Decoction of leaves used as a bath for hog sickness: unconsciousness. (as *J. silicicola* 169:229) *Unspecified* Plant used for medicinal purposes. (as *J. cilicicola* 169:161) Leaves used as medicine. (as *J. silicicola* 169:156) Plant used as medicine. (as *J. silicicola* 169:158) Plant used medicinally. (as *J. silicicola* 169:164) *Vertigo Medicine* Infusion of leaves taken as an emetic for thunder sickness: fever, dizziness, headache, and diarrhea. (as *J. silicicola* 169:213) *Witchcraft Medicine* Leaves used to make a witchcraft medicine. (as *J. bilicicola* 169:394)
- *Other*—**Seminole** *Protection* Leaves kept with eagle tail feathers to prevent the feathers from causing sickness. (as *J. silicola* 169:404)

Juniperus virginiana L. **var.** *virginiana*, Eastern Red Cedar
- *Other*—**Kiowa** *Incense & Fragrance* Needles thrown into the fire and used as incense during prayers in the peyote meeting. *Musical Instrument* Red, aromatic heartwood used to make "love flutes." (as *Sabina virginiana* 192:13)

Justicia, Acanthaceae
Justicia californica (Benth.) D. Gibson, Beloperone
- *Food*—**Diegueño** *Sweetener* Flower sucked for the nectar. (as *Beloperone californica* 15:47)

Justicia crassifolia (Chapman) Chapman ex Small, Thickleaf Waterwillow
- *Drug*—**Seminole** *Reproductive Aid* Plant used to restore virility. (169:319)

Kallstroemia, Zygophyllaceae
Kallstroemia californica (S. Wats.) Vail, California Caltrop
- *Drug*—**Tewa** *Antidiarrheal* Root used for diarrhea. *Dermatological Aid* Poultice of chewed leaves applied to sores or swellings. (as *K. brachystylis* 138:56, 57)

Kalmia, Ericaceae
Kalmia angustifolia L., Sheep Laurel
- *Drug*—**Abnaki** *Cold Remedy* Used for head colds. (144:154) *Nose Medicine* Powdered leaves mixed with bark from another plant and used as snuff for nasal inflammation. (144:170) **Algonquin, Quebec** *Cold Remedy* Singed, crushed leaves used like snuff for colds. *Poison* Plant considered poisonous. (18:215) **Algonquin, Tête-de-Boule** *Analgesic* Leaves boiled and used for headaches. *Poison* Infusion of leaves taken in great quantities caused death. (132:129) **Cree, Hudson Bay** *Gastrointestinal Aid* Decoction of twigs with leaves and flowers taken for bowel complaints. *Tonic* Decoction of twigs with leaves and flowers taken as a tonic. (92:303) **Malecite** *Dermatological Aid* Salve of pounded, fresh plant used for swellings. *Orthopedic Aid* Salve of pounded, fresh plant used for sprains. (116:244) **Micmac** *Analgesic* Herb used for pain, swellings, and sprains. (40:57) Poultice of crushed leaves bound to head for headache. (156:316) *Dermatological Aid* and *Orthopedic Aid* Herb used for swellings, pain, and sprains. (40:57) *Panacea* Infusion of leaves considered valuable as a "non-specific remedy." *Poison* Plant considered very poisonous. (156:316) **Montagnais** *Analgesic* Infusion of leaves taken sparingly for backache. *Cold Remedy* Infusion of leaves taken sparingly for colds. (156:314) *Gastrointestinal Aid* Infusion of leaves taken, poisonous if too strong, for stomach complaints. (156:317) *Poison* Leaves considered poisonous. (156:314) **Penobscot** *Analgesic* Compound poultice of plant used on cuts made in painful area to treat pain. *Panacea* Compound poultice of plant applied "for all kinds of trouble." (156:311)

Kalmia latifolia L., Mountain Laurel
- *Drug*—**Cherokee** *Analgesic* Infusion of leaves put on scratches made over location of the pain. (177:48) *Antirheumatic (External)* "Bristly edges of ten to twelve leaves" rubbed over skin for rheumatism. *Dermatological Aid* Crushed leaves used to "rub brier scratches." *Disinfectant* Infusion used as a wash "to get rid of pests" and as a liniment. *Orthopedic Aid* "Leaf ooze rubbed into scratched skin of ball players to prevent cramps." Compound used as liniment. *Panacea* Leaf salve used "for healing." (80:42) **Cree, Hudson Bay** *Antidiarrheal* Decoction of leaves taken for diarrhea. *Poison* Plant considered poisonous. (92:303) **Mahuna** *Dermatological Aid* Plant used as a body deodorizer. *Poison* Plant considered poisonous. (140:52)
- *Food*—**Mahuna** *Forage* Plants eaten by deer. (140:52)
- *Other*—**Cherokee** *Decorations* Wood used to carve. (80:42)

Kalmia microphylla (Hook.) Heller, Alpine Laurel
- *Drug*—**Kwakwaka'wakw** *Antiemetic* and *Antihemorrhagic* Decoction of leaves used for vomiting and spitting blood. (43:241)
- *Food*—**Hanaksiala** *Beverage* Leaves used to make tea. Leaves used to make tea. (43:241)

Kalmia polifolia Wangenh., Bog Laurel
- *Drug*—**Gosiute** *Unspecified* Leaves used as medicine. (as *K. glauca* 39:373) **Hesquiat** *Poison* Leaves could be poisonous and should never be used to make tea. (185:65) **Kwakiutl** *Antihemorrhagic* Decoction of leaves taken for "spitting of blood." (as *K. glauca* 20:380) Decoction of leaves taken for blood-spitting. (183:283) *Dermatological Aid* Decoction of leaves used as a wash for open sores and wounds that do not heal. (as *K. glauca* 20:380, 382) Decoction of leaves used as a wash for open sores and wounds. (183:283) **Thompson** *Unspecified* Decoction of plant used medicinally. (164:465) **Tlingit** *Dermatological Aid* Infusion of whole plant used for skin ailments. (as *K. glauca* 107:284)

Keckiella, Scrophulariaceae

Keckiella breviflora (Lindl.) Straw **ssp. *breviflora***, Bush Beardtongue
- *Drug*—**Miwok** *Cold Remedy* Infusion taken for colds. (as *Pentstemon breviflorus* 12:171) **Paiute** *Dermatological Aid* Ground, dry leaves used for running sores. (as *Pentstemon berviflorus* 118:44)

Keckiella cordifolia (Benth.) Straw, Heartleaf Penstemon
- *Drug*—**Mahuna** *Dermatological Aid* Infusion used as a wash or poultice of plant applied for fistulas and ulcers. (as *Pentstemon cordifolius* 140:12)

Kochia, Chenopodiaceae

Kochia americana S. Wats., Greenmolly
- *Drug*—**Navajo, Kayenta** *Venereal Aid* Plant used for venereal disease. (205:21)

Kochia scoparia (L.) Schrad., Common Kochia
- *Drug*—**Navajo** *Ceremonial Medicine* Used by the medicine man for painting a patient during a healing ceremony. *Dermatological Aid* Plant used for sores. (as *K. trichophylla* 90:152)
- *Food*—**Navajo** *Forage* Plant used as sheep forage, especially in the winter. (as *K. trichophylla* 90:152)

Koeleria, Poaceae

Koeleria macrantha (Ledeb.) J. A. Schultes, Prairie Junegrass
- *Drug*—**Cheyenne** *Ceremonial Medicine* Plant used in the Sun Dance ceremony. *Dermatological Aid* Plant used for cuts. *Stimulant* Plant tied to Sun Dancers head to prevent him from getting tired. (as *K. cristata* 83:10)
- *Food*—**Havasupai** *Bread & Cake* Seeds used to make bread. (as *K. cristata* 197:66) *Forage* Plant grazed by livestock. *Unspecified* Seeds used for food. *Winter Use Food* Seeds stored in blankets or bags of skin in caves. (as *K. cristata* 197:209) **Isleta** *Bread & Cake* Seeds made into a meal and used to make bread. *Porridge* Seeds made into a meal and used to make mush. (as *K. cristata* 32:22) *Staple* Considered a very important source of food before the introduction of wheat. Seeds used to make flour for bread and mush. (as *K. cristata* 100:33) **Okanagan-Colville** *Fodder* Used as a good feed for cattle and horses. (as *K. cristata* 188:55)
- *Fiber*—**Cheyenne** *Brushes & Brooms* Plants used as paintbrushes to paint ceremonial participants. (as *K. cristata* 83:10) **Jemez** *Brushes & Brooms* Blades tied together and used as a broom. (as *K. cristata* 44:25) **Navajo, Ramah** *Scouring Material* Bunch about a foot long, tied with string or yucca fiber, used as a brush for cleaning metates. (as *K. cristata* 191:16)
- *Other*—**Isleta** *Fasteners* Straw mixed with adobe to give strength and adhesion. (as *K. cristata* 100:33)

Krameriaceae, see *Krameria*

Krameria, Krameriaceae

Krameria erecta Willd. ex J. A. Schultes, Littleleaf Ratany
- *Drug*—**Pima** *Dermatological Aid* Poultice of powdered root applied to sores. (as *K. parvifolia* 146:80)
- *Dye*—**Papago** *Red* Roots used as a red dye for garments. (as *K. glandulosa* 34:48) Used to dye cotton red. (as *K. glandulosa* 34:60) Roots peeled, cut, split, boiled, and used as a red dye for buckskins. (as *K. glandulosa* 34:69)

Krameria grayi Rose & Painter, White Ratany
- *Drug*—**Paiute** *Dermatological Aid* Infusion of root used as a wash for gonorrheal sores. Root powder or decoction of root applied to sores. *Disinfectant* and *Eye Medicine* Infusion of root used as a wash for gonorrheal eye infections. *Venereal Aid* Decoction of root taken and used as a wash for gonorrheal eye infections and sores. (180:96) **Pima** *Analgesic* Infusion of roots taken for pain. *Cough Medicine* Infusion of roots taken for coughs. *Dermatological Aid* Poultice of powdered roots applied to prevent infection on newborn's navel. Decoction of roots applied as poultice to sores caused by "bad disease." *Disinfectant* Poultice of powdered roots applied to prevent infection on newborn's navel. *Eye Medicine* Infusion of twigs used for sore eyes. *Febrifuge* Infusion of roots taken for fevers. *Pediatric Aid* Poultice of powdered roots applied to prevent infection on newborn's navel. *Throat Aid* Root chewed for sore throats. (47:91) **Shoshoni** *Dermatological Aid* Infusion of pulverized root used as a wash for swellings. (180:96)
- *Dye*—**Pima** *Brown* Dry roots ground, boiled in water, and used as a brown dye for basket making. (47:91)

Krascheninnikovia, Chenopodiaceae

Krascheninnikovia lanata (Pursh) Guldenstaedt, Winter Fat
- *Drug*—**Gosiute** *Febrifuge* Plant used for intermittent fevers. (as *Eurotia lanata* 39:369) **Hopi** *Burn Dressing* Powdered root used for burns. *Febrifuge* Decoction of leaves used for fever. (as *Eurotia lanata* 42:317) Compound containing plant used for fever. (as *Eurotia lanata* 200:32, 74) *Orthopedic Aid* Plant used for sore muscles. (as *Eurotia lanata* 200:32) **Navajo** *Antihemorrhagic* Decoction of leaves taken for blood-spitting. (as *Eurotia lanata* 55:44) *Dermatological Aid* Plant used for sores and boils. *Misc. Disease Remedy* Plant used for smallpox. (as *Eurotia lanata* 90:151) **Navajo, Ramah** *Antidote* Cold infusion of plant taken as needed for *Datura* poisoning. *Dermatological Aid* Poultice of chewed leaves applied to poison ivy rash. *Panacea* Cold infusion of plant used as "life medicine." (as *Eurotia lanata* 191:25) **Paiute** *Dermatological Aid* Decoction of plant used as a head and scalp tonic and prevents graying. (as *Eurotia lanata* 180:74, 75) *Eye Medicine* Decoction of leaves alone or with stems used as a wash or compress for sore eyes. (as *Eurotia lanata* 180:74, 75) **Shoshoni** *Dermatological Aid* Decoction of plant used as a head and scalp tonic. *Eye Medicine* Decoction of leaves alone or with stems used as a wash or compress for sore eyes. (as *Eurotia lanata* 180:74, 75) **Tewa** *Burn Dressing* Powdered root used for burns. *Febrifuge* Decoction of leaves used for fever. (as *Eurotia lanata* 42:317) **Zuni** *Burn Dressing* Poultice of ground root applied to burns and bound with cotton cloth. (as *Eurotia lanata* 166:51)
- *Food*—**Havasupai** *Fodder* Plant used for horse feed. (as *Ceratoides lanata* 197:218) **Keres, Western** *Forage* Considered a good forage plant. (as *Eurotia lanata* 171:44) **Navajo** *Forage* Plant used as winter forage for the sheep. (as *Eurotia lanata* 55:44)
- *Other*—**Hopi** *Ceremonial Items* Used in ceremonials to produce steam. (as *Eurotia lanata* 42:317) **Navajo** *Ceremonial Items* Armful of stems with leaves used on heated stones in the sweat house for the

Mountain Chant. (as *Eurotia lanata* 55:44) **Tewa** *Ceremonial Items* Used in ceremonials to produce steam. (as *Eurotia lanata* 42:317)

Krigia, Asteraceae
Krigia biflora (Walt.) Blake, Twoflower Dwarfdandelion
- **Other**—**Menominee** *Hunting & Fishing Item* Stem used by hunters to make a wail that simulated a fawn in distress and lured the doe to the hunter. (as *Krigia amplexicaulis* 151:80)

Lachnanthes, Haemodoraceae
Lachnanthes caroliana (Lam.) Dandy, Carolina Redroot
- **Drug**—**Catawba** *Veterinary Aid* Infusion of dried, powdered root given to horses as a tonic. (as *Gyrotheca capitata* 157:188) **Cherokee** *Antihemorrhagic* Taken for "spitting blood." *Cancer Treatment* Strong decoction used as a wash for cancer. *Dermatological Aid* Strong decoction of astringent root used as a wash for cancer. *Gastrointestinal Aid* Taken for bowel complaints. *Gynecological Aid* Taken for "flooding." *Hemorrhoid Remedy* Used for bloody piles. *Oral Aid* Used for sore mouth. *Throat Aid* Used for sore throat. *Venereal Aid* Compound decoction taken for venereal disease. (80:52)

Lactuca, Asteraceae
Lactuca biennis (Moench) Fern., Tall Blue Lettuce
- **Drug**—**Bella Coola** *Analgesic* Decoction of root taken for body pain, but not pain in the limbs. *Antidiarrheal* Decoction of root taken for diarrhea. *Antiemetic* Decoction of root taken for vomiting. *Antihemorrhagic* Decoction of root taken for hemorrhage, body pain, and heart trouble. *Heart Medicine* Decoction of root taken for heart trouble, hemorrhage, and pain. (as *L. spicata* 150:65) **Ojibwa** *Gynecological Aid* Infusion of plant used for caked breast and to ease lactation. (as *L. spicata* 153:364, 365) **Potawatomi** *Unspecified* Plant used as a medicine for unspecified illnesses. (as *L. spicata* 154:52)
- **Other**—**Ojibwa** *Hunting & Fishing Item* Plant used in the same manner as the Canada hawkweed to attract a doe to them for a close shot. (as *L. spicata* 153:429)

Lactuca canadensis L., Canada Lettuce
- **Drug**—**Cherokee** *Analgesic* Used for pain and infusion given "for calming nerves." *Ceremonial Medicine* Used as an ingredient in a green corn medicine. *Sedative* Infusion given "for calming nerves" and "produces sleep." *Stimulant* Infusion used as a stimulant. *Veterinary Aid* Infusion given for "milksick." (80:42) **Chippewa** *Dermatological Aid* Milky sap from fresh plant rubbed on warts. (53:350) **Iroquois** *Analgesic* Compound infusion of roots and bark taken for back pain. *Eye Medicine* Compound infusion of roots and bark taken for dark circles and puffy eyes. *Hemostat* Poultice of smashed roots applied to severe bleeding from a cut. *Kidney Aid* Compound infusion of roots and bark taken for kidney trouble. *Orthopedic Aid* Compound infusion of roots and bark taken for back pain. (87:478) **Menominee** *Dermatological Aid* Milky juice of plant rubbed on poison ivy eruptions. (151:31)
- **Food**—**Cherokee** *Vegetable* Leaves cooked and eaten as greens. (80:42)

Lactuca ludoviciana (Nutt.) Riddell, Biannual Lettuce
- **Food**—**Gosiute** *Unspecified* Leaves used for food. (39:373)

Lactuca sativa L., Garden Lettuce
- **Drug**—**Isleta** *Gastrointestinal Aid* Fresh leaves eaten for stomachaches. (as *L. integrata* 100:33) **Meskwaki** *Gynecological Aid* Infusion of leaves taken after childbirth to hasten the flow of milk. (as *L. scariola* var. *integrata* 152:215)
- **Food**—**Acoma** *Vegetable* Young, tender plants eaten as greens. (as *L. integrata* 32:32) **Keres, Western** *Vegetable* Young, tender plants used as lettuce. (as *L. integrata* 171:51) **Laguna** *Vegetable* Young, tender plants eaten as greens. (as *L. integrata* 32:32)

Lactuca serriola L., Prickly Lettuce
- **Drug**—**Navajo, Ramah** *Ceremonial Medicine* and *Emetic* Compound decoction of plant used as a ceremonial emetic. (191:52)

Lactuca tatarica var. *pulchella* (Pursh) Breitung, Blue Lettuce
- **Drug**—**Iroquois** *Hemorrhoid Remedy* Poultice of plants applied to piles. (as *L. pulchella* 87:478) **Okanagan-Colville** *Antidiarrheal* and *Pediatric Aid* Infusion of roots and stems given to children for diarrhea. (as *L. pulchella* 188:84)
- **Food**—**Apache, White Mountain** *Candy* Gummy substance from the root used for chewing gum. **Navajo** *Candy* Gummy substance from the root used for chewing gum. **Zuni** *Candy* Gummy substance from the root used for chewing gum. (as *L. pulchella* 136:158) Dried root gum used as chewing gum. (as *L. pulchella* 166:68)

Lactuca virosa L., Bitter Lettuce
- **Drug**—**Navajo** *Antidiarrheal*, *Antiemetic*, and *Gastrointestinal Aid* Plant used for gastroenteritis (nausea, vomiting, and diarrhea). (90:152)

Lagenaria, Cucurbitaceae
Lagenaria siceraria (Molina) Standl., Bottle Gourd
- **Drug**—**Cherokee** *Dermatological Aid* Poultice of soaked seeds used for boils. (as *L. vulgaris* 80:37) **Houma** *Analgesic* Poultice of crushed leaves applied to the forehead for headaches. (158:62) **Seminole** *Analgesic* Seeds used for adult's sickness caused by adultery: headache, body pains, and crossed fingers. (169:256) *Psychological Aid* Seeds burned to smoke the body for insanity. (169:293)
- **Food**—**Cherokee** *Unspecified* Species used for food. (as *L. vulgaris* 80:37) **Ojibwa** *Vegetable* Gourds eaten young, before the rind had hardened. (as *L. vulgaris* 153:400)
- **Other**—**Cherokee** *Ceremonial Items* Fruit used to make ceremonial rattles. *Cooking Tools* Fruit used to make dippers. (as *L. vulgaris* 80:37) **Cocopa** *Musical Instrument* Fruit made into a rattle and used to provide rhythm for singing and dancing. (37:115) **Dakota** *Ceremonial Items* Gourds made into rattles and used for ritualistic music.[43] (as *Cucurbita lagenaria* 70:117) **Havasupai** *Containers* Rinds made into containers used for carrying water on foot or on horseback trips away from home. *Musical Instrument* Rinds used to make rattles. (as *L. vulgaris* 197:244) **Hopi** *Ceremonial Items* Used as prayer sticks. Covered with a cord net to be used as water containers in ceremonies

and buried with the dead. *Containers* Used as containers for sacred honey, cups, seed bottles, and medicine holders. *Cooking Tools* Used as dippers, canteens, and spoons. *Decorations* Used to make noses, horns, and flowers for masks. *Hunting & Fishing Item* Used in hunting to imitate the sound of a deer. *Musical Instrument* Used as trumpets or megaphones to represent the bellowing of the plumed serpent in ceremonies. Used as rattles. *Tools* Used as pottery scrapers. (as *L. vulgaris* 200:93) **Houma** *Cooking Tools* Used for water dippers, cups, and bowls. *Musical Instrument* Used for drums and rattles. (158:62) **Iroquois** *Ceremonial Items* Fruit made into rattles used by the Medicine Societies. (as *L. vulgaris* 196:113) **Keres, Western** *Cooking Tools* Gourds made into dippers. *Toys & Games* Gourds made into rattles. (as *L. vulgaris* 171:51) **Keresan** *Cooking Tools* Used to make dippers. *Toys & Games* Used to make rattles. (198:561) **Mohave** *Musical Instrument* Fruit made into a rattle and used to provide rhythm for singing and dancing. (37:115) **Navajo** *Ceremonial Items* Used to make rattles for various ceremonies. *Containers* Used to make cups for preparing medicines. *Cooking Tools* Used to make dippers. (as *L. vulgaris* 55:79) **Navajo, Ramah** *Ceremonial Items* Used to make chant rattles. *Containers* Used to make water dippers. *Tools* Used to make pottery scrapers. (191:47) **Ojibwa** *Ceremonial Items* Gourds used to make rattles for the medicine lodge. *Cooking Tools* Gourds used to make drinking and dipping cups. (as *L. vulgaris* 153:400) **Omaha** *Ceremonial Items* Gourds made into rattles and used for ritualistic music.[43] (as *Cucurbita lagenaria* 70:117) **Papago** *Cooking Tools* Used for a drinking and eating vessel. (as *L. vulgaris* 34:17) *Musical Instrument* Fruits dried, freed of seeds and pulp, and used as rattles. (as *L. vulgaris* 34:68) **Pima** *Ceremonial Items* Gourds dried, filled with gravel, and used in ceremonial songs. (as *L. vulgaris* 47:72) **Ponca** *Ceremonial Items* Gourds made into rattles and used for ritualistic music.[43] (as *Cucurbita lagenaria* 70:117) **Seminole** *Cooking Tools* Plant used to make dippers, dishes, and water bottles. (169:484) **Yuma** *Containers* Fruit contents removed, shells cleaned and dried, and used as water and food containers. *Musical Instrument* Fruit made into a rattle and used to provide rhythm for singing and dancing. (37:115)

Lamarckia, Poaceae
Lamarckia aurea (L.) Moench, Goldentop
- **Drug**—**Diegueño** *Analgesic* Decoction of grass with "small tassels" taken for headaches. (84:23)

Lamiaceae, see *Agastache, Blephilia, Collinsonia, Cunila, Dracocephalum, Galeopsis, Glechoma, Hedeoma, Hyptis, Hyssopus, Leonurus, Lepechinia, Lycopus, Marrubium, Melissa, Mentha, Monarda, Monardella, Nepeta, Perilla, Physostegia, Piloblephis, Pogogyne, Poliomintha, Prunella, Pycnanthemum, Salvia, Satureja, Scutellaria, Stachys, Thymus, Trichostema*

Laminariaceae, see *Costaria, Hedophyllum, Laminaria*

Laminaria, Laminariaceae, alga
Laminaria groenlandica Rosenvinge
- **Food**—**Nitinaht** *Unspecified* Plants eaten with herring spawn. (186:51)
- **Other**—**Nitinaht** *Hunting & Fishing Item* Fronds used to catch herring spawn. (186:51)

Laportea, Urticaceae
Laportea canadensis (L.) Weddell, Canadian Woodnettle
- **Drug**—**Houma** *Febrifuge* Decoction of plant taken for fever. (158:60) **Iroquois** *Antidote* Decoction taken to counteract poison made from menstrual blood and fruit. *Emetic* Decoction of roots taken to vomit to neutralize a love medicine. (87:307) *Gynecological Aid* Infusion of smashed roots taken to facilitate childbirth. (87:306) *Psychological Aid* Decoction taken to counteract loneliness because your woman has left. *Tuberculosis Remedy* Compound infusion of smashed roots taken for tuberculosis. *Witchcraft Medicine* Decoction taken "when your woman goes off and won't come back." (87:307) **Meskwaki** *Diuretic* and *Urinary Aid* Root used as a "diurient" and for urine incontinence. (152:250, 251) **Ojibwa** *Diuretic* Infusion of root taken as a diuretic. *Urinary Aid* Infusion of root used for various urinary ailments. (153:391, 392)
- **Fiber**—**Abnaki** *Basketry* Used to make baskets. (144:156) **Chippewa** *Cordage* Used for twine. (as *Urticastrum divaricatum* 53:378) **Menominee** *Basketry* Plant made into hemp twine and used to make fiber bags. (151:77) **Meskwaki** *Cordage* Inner bark braided to make cords. (152:270) **Ojibwa** *Sewing Material* Rind of this nettle used by the old people as a sewing fiber. (153:423)

Lappula, Boraginaceae
Lappula occidentalis (S. Wats.) Greene, Flatspine Stickseed
- **Other**—**Keres, Western** *Unspecified* Plant known and named but no use was specified. (171:51)

Lappula occidentalis var. *cupulata* (Gray) Higgins, Flatspine Stickseed
- **Drug**—**Navajo** *Gynecological Aid* Parts of the plant used at confinement. *Hemostat* Parts of the plant used for nosebleeds. (as *L. texana* 90:153) **Navajo, Kayenta** *Dermatological Aid* Plant used as a lotion for itching. (as *L. texana* 205:40)

Lappula occidentalis (S. Wats.) Greene var. *occidentalis*, Desert Stickseed
- **Drug**—**Navajo, Kayenta** *Dermatological Aid* Poultice of plant applied to sores caused by insects. (as *L. redowskii* 205:40) **Navajo, Ramah** *Dermatological Aid* Cold infusion used as lotion for sores or swellings. (as *L. redowskii* 191:41)
- **Food**—**Navajo, Ramah** *Fodder* Used for sheep feed. (as *L. redowskii* 191:41)

Lappula squarrosa (Retz.) Dumort., European Stickseed
- **Drug**—**Ojibwa, South** *Analgesic* Roots on hot stones use as an inhalant or snuff of raw root used for headache. (as *Echinospermum lappula* 91:201)

Larix, Pinaceae, conifer
Larix americana
- **Drug**—**Micmac** *Dermatological Aid* Poultice of boiled inner bark applied to sores and swellings. *Diuretic* Decoction of boughs taken as a diuretic. (as *Pinus microcarpa* 156:317)

Larix laricina (Du Roi) K. Koch, Tamarack
- **Drug**—**Abnaki** *Cough Medicine* Used for coughs. (144:154) Decoction of plant and bark from another plant used for coughs. Infusion

of bark and roots from other plants taken for persistent coughs. (144: 163) **Algonquin, Quebec** *Cough Medicine* Needles and inner bark used for cough medicine. *Disinfectant* Poultice of needles and inner bark applied to infections. *Unspecified* Used with ground pine as a medicinal tea. (18:127) **Algonquin, Tête-de-Boule** *Laxative* Infusion of young branches used as a laxative. (132:129) **Anticosti** *Kidney Aid* Decoction of bark and bark from another plant taken for kidney troubles. (143:63) **Chippewa** *Blood Medicine* Infusion of bark taken for anemic conditions. (71:123) *Burn Dressing* Poultice of chopped inner bark applied to burns. (53:352) **Cree, Woodlands** *Antiemetic* Used with eight different trees for vomiting. *Dermatological Aid* Decoction used as a wash and poultice of boiled inner bark and wood applied to frostbite or deep cuts. Poultice of warm, boiled inner bark applied to wounds to draw out infection. (109:41) **Iroquois** *Analgesic* Fermented compound decoction taken for soreness. *Antirheumatic (Internal)* Compound decoction taken for rheumatism. *Blood Medicine* Fermented compound decoction taken when "blood gets bad and cold." *Cold Remedy* Compound decoction taken for colds. *Cough Medicine* Compound decoction taken for coughs. *Febrifuge* Fermented compound decoction taken for fever. *Stimulant* Fermented compound decoction taken when one is tired from complaint. *Venereal Aid* Compound decoction taken for gonorrhea. Compound powder poultice "put in bag, place penis in bag and tie around waist." (87: 268) **Malecite** *Cold Remedy* Infusion of bark used for colds. *Strengthener* Infusion of bark used for general debility. *Tuberculosis Remedy* Infusion of bark used for consumption. (116:249) *Unspecified* Bark used as medicine. (160:6) *Venereal Aid* Infusion of bark, spruce bark, and balsam bark used for gonorrhea. (116:257) **Menominee** *Dermatological Aid* Poultice of bark used for unspecified ailments. *Other* Infusion of bark "drives out inflammation and generates heat." *Veterinary Aid* Infusion of bark given to horses "to better their condition from distemper." (151:45) **Micmac** *Cold Remedy* Bark used for colds. *Dermatological Aid* Bark used for "suppurating wounds" and colds. *Stimulant* Bark used for physical weakness. *Tuberculosis Remedy* Bark used for consumption. *Venereal Aid* Bark used for gonorrhea. (40:58) **Montagnais** *Expectorant* Infusion of buds and bark taken as an expectorant. (24:14) **Ojibwa** *Disinfectant* Dried leaves used as an inhalant and fumigator. (153:378, 379) *Unspecified* Infusion of roots and bark used as a general medicine. (as *L. americana* 135:244) **Ojibwa, South** *Analgesic* Boiled, crushed leaves, and bark used as herbal steam for headache and backache. Poultice of crushed leaves and bark applied for headache. *Herbal Steam* Boiled, crushed leaves and bark used as herbal steam for headache and backache. (as *L. americana* 91:198) **Potawatomi** *Dermatological Aid* Poultice of fresh inner bark applied to wounds and inflammations. *Other* Infusion of bark taken to drive out inflammation and to warm body. *Veterinary Aid* Shredded inner bark mixed with feed to make horse's hide loose. (154:69, 70)

- **Food**—**Anticosti** *Beverage* Branches and needles used to make tea. (143:63) **Potawatomi** *Fodder* Shredded inner bark mixed with oats and fed to horses to make the hide of the animal loose. (154:122)
- **Fiber**—**Cree, Woodlands** *Snow Gear* Wood used to make toboggans. (109:41) **Ojibwa** *Basketry* Root fibers used to make durable bags. (153:421) *Canoe Material* Roots used to sew canoes and used as the strong upper wrappings over the canoe edges. (as *L. americana* 135: 244) Roots used to sew canoes. *Sewing Material* Roots used as a sewing material. (153:421)
- **Other**—**Chippewa** *Containers* Roots used to weave bags. (53:377) **Cree, Woodlands** *Hide Preparation* Rotten wood used to smoke tan and yellow tint hides. (109:41) **Malecite** *Hunting & Fishing Item* Used to make arrow shafts. (160:6) **Micmac** *Fuel* Wood used for kindling and fuel. (159:258)

Larix occidentalis Nutt., Western Larch

- **Drug**—**Kutenai** *Dermatological Aid* Gum used for cuts and bruises. *Tuberculosis Remedy* Infusion of bark used for tuberculosis. **Nez Perce** *Cold Remedy* Infusion of bark used for colds. *Cough Medicine* Infusion of bark used for coughs. *Throat Aid* Sap chewed for sore throat. (82:22) **Okanagan-Colville** *Antirheumatic (External)* Decoction of plant tops used to soak arthritic limbs. *Antirheumatic (Internal)* Decoction of plant taken for severe arthritis. *Blood Medicine* Decoction of plant tops taken to help "changing of the blood." Decoction of plant tops and Oregon grapes used as a blood purifier. *Cancer Treatment* Decoction of plant taken for cancer. *Dermatological Aid* Decoction of plant tops used as an antiseptic wash for cuts and sores, and to soak severe skin sores. (188:25) **Thompson** *Burn Dressing* Poultice of pitch mixed with fat or Vaseline and used for sores, cuts, and burns. *Cancer Treatment* Decoction of small pieces of branches and tops used for cancer. A decoction of plant tops was used to wash the areas affected by cancer. A second decoction of branch pieces was taken internally. It made the emaciated patient get better and gain weight. This treatment was used after a "western" doctor diagnosed the breast cancer patient as being terminal. *Cough Medicine* Branches used for dry coughs. *Dermatological Aid* Decoction of bark used as a wash for wounds, such as bullet wounds. Poultice of pitch used for sores, cuts, and burns. The pitch was mixed with tallow and used as a poultice for sores or it was mixed with fat or Vaseline and used for cuts and burns. *Dietary Aid* Decoction of small pieces of branches and bark used to stimulate the appetite. *Gastrointestinal Aid* Decoction of small pieces of branches and bark used for ulcers. *Gynecological Aid* Decoction of small pieces of branches and bark taken as a form of birth control after childbirth. *Orthopedic Aid* Pitch considered a valuable bone setter for broken bones that would not heal. Branches used for broken bones. *Panacea* Branches used as a medicine for any type of illness. (187:99) *Pediatric Aid* Decoction of leaves used as a healthful, strengthening wash for infants. Decoction of plant used as a wash to make babies strong and healthy. (164:475) Decoction of bark used as a wash or bath for babies, to make them strong and healthy. *Respiratory Aid* Poultice of pitch used or infusion of pitch taken for respiratory diseases. *Strengthener* Decoction of bark used as a wash or bath for babies, to make them strong and healthy. *Tuberculosis Remedy* Poultice of pitch used or infusion of pitch taken for tuberculosis. (187:99)
- **Food**—**Flathead** *Candy* Solidified pitch chewed as gum. *Sauce & Relish* Sap used to make a sweet syrup. *Unspecified* Cambium layer eaten in spring. **Kutenai** *Sauce & Relish* Sap used to make a sweet syrup. (82:22) **Okanagan-Colville** *Candy* Sap hardened and eaten like candy. *Forage* Buds eaten by blue grouse. (188:25) **Paiute** *Candy* Syrup or "dark sugar" gathered as a confection. (111:43) **Sanpoil & Nespelem** *Unspecified* Gum collected on stump of a burned or fallen larch and

used for food. (131:105) **Thompson** *Candy* Gum from trunk and branches chewed for pleasure. (187:99) *Unspecified* Gum chewed for pleasure. (164:493)
- *Other*—**Kutenai** *Ceremonial Items* Used for the center pole of the religious Sundance. *Hide Preparation* Rotten wood used for smoking buckskins. **Nez Perce** *Hunting & Fishing Item* Wood used to make bows. (82:22) **Okanagan-Colville** *Paint* Pitch heated, rubbed into a fine powder, mixed with grease, and used as a red paint for girls' faces. *Season Indicator* Leaf color changes used to indicate fall and pregnant bears going into their dens for the winter. (188:25) **Thompson** *Paint* Pitch burned until dry to make a reddish pigment and used as a face paint for women and men. (187:99)

Larrea, Zygophyllaceae

Larrea tridentata (Sessé & Moc. ex DC.) Coville, Creosote Bush
- *Drug*—**Coahuilla** *Gastrointestinal Aid* Infusion of leaves taken for bowel complaints and consumption. *Tuberculosis Remedy* Infusion of leaves taken for consumption and bowel complaints. (as *L. mexicana* 13:78) *Veterinary Aid* Plant given to horses for colds, distemper, or runny nose. (as *L. mexicana* 13:79) **Diegueño** *Antirheumatic (External)* Decoction of leaves used as a bath for rheumatism and painful arthritis. *Orthopedic Aid* Decoction of leaves used as a bath for aching bones and sprains. (84:23) **Hualapai** *Cold Remedy* Infusion of leaves taken or leaves steamed for colds. *Disinfectant* Infusion of leaves used as a disinfecting skin cleanser. *Respiratory Aid* Infusion of leaves taken or leaves steamed for congestion and asthma. (195:28) **Kawaiisu** *Analgesic* Decoction of leaves used as a wash for sore and aching parts of the body. *Disinfectant* Plant used for the antiseptic properties. *Orthopedic Aid* Decoction of leaves used as a wash for sore and aching parts of the body. Poultice of heated leaves applied to aching limbs. *Unspecified* Plant used for medicinal purposes. *Veterinary Aid* Decoction of leaves used for collar sores on draft animals. (206:36) **Mahuna** *Dermatological Aid* Infusion of plant used for dandruff. *Disinfectant* Infusion of plant used as a disinfectant and deodorizer. (as *L. mexicana* 140:37) *Gastrointestinal Aid* Plant used for stomach cramps from delayed menstruation. (as *L. mexicana* 140:14) **Paiute** *Cold Remedy* Infusion of leaves taken as a cold medicine. (118:37) **Pima** *Analgesic* Decoction of twigs taken for gas pains or headaches caused by upset stomachs. Infusion of leaves taken for pain or used as bath and rub for rheumatic pains. Poultice of heated branches and leaves applied for pain. *Antidiarrheal* Plant gum chewed and swallowed as an antidysenteric and intestinal antispasmodic. *Antirheumatic (External)* Infusion of leaves used as bath and rub or poultice applied to rheumatic pains. *Antirheumatic (Internal)* Infusion of plant taken for rheumatism. *Carminative* Decoction of branches taken for gas caused by upset stomach or gas pains. *Cold Remedy* Decoction of gum taken for colds. *Dermatological Aid* Infusion of plant used as wash for impetigo sores or dandruff. Poultice of leaves applied to prevent feet from perspiring or as a deodorant. Poultice of leaves applied to scratches, wounds, sores, and bruises. *Emetic* Decoction of leaves taken as an emetic for high fevers. (47:61) Decoction of leaves taken as an emetic. (as *L. mexicana* 146:79) *Febrifuge* Decoction of leaves taken as an emetic for high fevers. *Gastrointestinal Aid* Decoction of plant taken for stomachaches and cramps. Plant gum chewed and swallowed as an intestinal antispasmodic. *Oral Aid* Decoction of gum used as a gargle. *Panacea* Plant used to cure everything. *Strengthener* Smoke from plant used for weakness and laziness. *Toothache Remedy* Infusion of plant held in the mouth for toothaches. *Tuberculosis Remedy* Decoction of gum taken for tuberculosis. (47:61) *Unspecified* Poultice of boiled leaves used for unspecified purpose. (as *L. mexicana* 146:79) *Urinary Aid* Infusion of leaves taken for dysuria (difficulty in passing urine). (47:61) **Yavapai** *Antirheumatic (External)* Decoction of leaves and stems used as wash for rheumatism. *Dermatological Aid* Decoction of leaves and stems used as a wash for cuts and sores. Dried, pulverized leaves used for sores. *Throat Aid* Decoction of leaves and stems taken for sore throat. *Venereal Aid* Decoction of leaves and stems used as a wash for gonorrhea. Whole leaves used on penis for gonorrhea. (as *L. mexicana* 65:261)
- *Other*—**Kawaiisu** *Tools* Wood used to make a pointed digging stick. Gum-like substance gathered into a ball, softened in fire, and shaped into awl and knife handles. (206:36)

Larrea tridentata (Sessé & Moc. ex DC.) Coville **var. *tridentata*,** Creosote Bush
- *Drug*—**Cahuilla** *Antirheumatic (External)* Plant made into liniment used by elderly people for swollen limbs caused by poor blood circulation. *Cancer Treatment* Infusion of stems and leaves used for cancer. *Cold Remedy* Infusion of stems and leaves used for colds. *Dermatological Aid* Decoction or poultice of leaves used on open wounds. Crushed leaf powder applied to sores and wounds. *Disinfectant* Decoction or poultice of leaves used to draw out poisons and for infections. *Emetic* Infusion of stems and leaves used, in heavy doses, to induce vomiting. *Gastrointestinal Aid* Infusion of stems and leaves used for bowel complaints. *Gynecological Aid* Infusion of stems and leaves used for stomach cramps from delayed menstruation. *Pulmonary Aid* Infusion of stems and leaves used for chest infections, and as a decongestant for clearing lungs. *Respiratory Aid* Leaves boiled or heated and the steam inhaled for congestion. *Tonic* Infusion of stems and leaves mixed with honey and used as a general health tonic before breakfast. (as *L. divaricata* 15:83) **Isleta** *Antirheumatic (External)* Decoction of leaves used as a body bath for rheumatism. *Dermatological Aid* Leaves used in shoes to absorb moisture. *Disinfectant* Decoction of leaves used as a disinfectant. (as *Covillea glutinosa* 100:26) **Paiute** *Analgesic* Decoction of leaves taken for bowel cramps. *Antirheumatic (External)* Infusion of leaves used as a wash for rheumatism. *Burn Dressing* Compound decoction of leaves with badger oil used as a salve for burns. *Cold Remedy* Decoction of leaves taken for colds. *Dermatological Aid* Dried, powdered leaves sprinkled on sores. *Gastrointestinal Aid* Decoction of leaves taken for bowel cramps. *Misc. Disease Remedy* Infusion of leaves used as a wash for chickenpox. *Panacea* Plant used for many different illnesses and considered a "cure-all." *Venereal Aid* Compound decoction of leaves taken for gonorrhea. (as *L. divaricata* 180:96, 97) **Papago** *Analgesic* Branches used as bed for women with menstrual cramps or after childbirth. *Antirheumatic (External)* Poultice of heated branches applied for rheumatism. Poultice of heated branches applied to joints. *Dermatological Aid* Plant used for poisonous bites and sores. Dried, powdered leaf rubbed on infant's navel to promote healing. Poultice of chewed leaves placed on insect bites, snakebites, and

sores. Poultice of chewed plant applied to spider or scorpion bites. Poultice of dried, powdered leaves applied to infant's navel. *Emetic* Decoction of leaves taken as an emetic. *Gynecological Aid* Branches used as bed for women with menstrual cramps or after childbirth. Infusion of leaves rubbed on breasts to start milk flow. Infusion of leaves used as wash on breasts to start milk flow. Poultice of heated branches applied to facilitate childbirth. *Orthopedic Aid* Plant used for stiff limbs. Green branches laid on ashes, aching feet, and stiff limbs held in smoke. Smoke from smoldering green branches used for sore feet. *Pediatric Aid* Dried, powdered leaf rubbed on infant's navel to promote healing. Poultice of dried, powdered leaves applied to infant's navel. *Snakebite Remedy* Poultice of chewed leaves placed on snakebites, insect bites, and sores. Poultice of chewed plant applied to snakebites. (as *Covillea glutinosa* 34:64, 65) **Shoshoni** *Cold Remedy* Decoction of leaves taken for colds. *Diuretic* Decoction of leaves taken to "stimulate urination." *Venereal Aid* Decoction of leaves taken for venereal disease. (as *L. divaricata* 180:96, 97)
- *Fiber*—**Papago** *Brushes & Brooms* Branches used to brush off the spines of prickly pears. (as *Covillea glutinosa* 34:23) *Building Material* Piled on top of saguaro ribs to strengthen house roofs. Tops tied together and used to thatch menstruation huts. (as *Covillea glutinosa* 34:67)
- *Other*—**Panamint** *Tools* Lac used to make awl handles. (as *L. divaricata* 105:84) **Papago** *Decorations* Charcoal used in tattooing as a permanent greenish blue color. (as *Covillea glutinosa* 34:51) *Hunting & Fishing Item* Wood used to make arrows for hunting small animals. (as *Covillea glutinosa* 34:42) Twigs cut, peeled, straightened, dried, split and used as the foreshafts of hunting arrows. (as *Covillea glutinosa* 34:71) *Musical Instrument* Wood thrust through gourd rattles and used as the handles. (as *Covillea glutinosa* 34:68) *Protection* Branches stuck in the ground to shade tobacco plants. (as *Covillea glutinosa* 34:37) *Tools* Short transverse sticks affixed to poles and used to dislodge saguaro fruits from the shafts. (as *Covillea glutinosa* 34:20) Wood used to make handles for basketry awls. (as *Covillea glutinosa* 34:59) Smooth sticks used as the shuttles in weaving cotton. (as *Covillea glutinosa* 34:60) Wood used as the smaller of the two pieces of twisters used to make rope. (as *Covillea glutinosa* 34:62) Stakes used to stake out sheep hides so they can be made more pliable. (as *Covillea glutinosa* 34:69) *Weapon* Twigs cut green, peeled, straightened, dried, split, and used as the foreshafts of war arrows. (as *Covillea glutinosa* 34:71)

Lasthenia, Asteraceae

Lasthenia californica DC. ex Lindl., California Goldfields
- *Food*—**Cahuilla** *Porridge* Parched seeds ground into flour and used to make mush. (as *Baeria chrysostoma* 15:46)

Lasthenia glabrata Lindl., Yellowray Goldfields
- *Food*—**Cahuilla** *Dried Food* Parched seeds eaten dry. *Porridge* Parched seeds ground into flour and used to make mush. (15:84)

Lathyrus, Fabaceae

Lathyrus brachycalyx Rydb. ssp. *brachycalyx*, Bonneville Peavine
- *Food*—**Omaha** *Unspecified* Roasted pods eaten by children in sport, but not considered of any importance. **Ponca** *Unspecified* Roasted seedpods eaten by children in sport, but not considered of any importance. (as *L. ornatus* 70:98)

Lathyrus eucosmus Butters & St. John, Bush Vetchling
- *Drug*—**Navajo, Kayenta** *Gynecological Aid* Plant used to remove placenta. (205:28) **Navajo, Ramah** *Disinfectant* Cold infusion taken and used as a lotion for "deer infection." *Veterinary Aid* Cold infusion used as lotion on horses for swellings or injuries. (191:32)

Lathyrus graminifolius (S. Wats.) White, Grassleaf Peavine
- *Food*—**Karok** *Vegetable* Tender plant eaten as greens in the spring. (148:385)

Lathyrus japonicus var. *maritimus* (L.) Kartesz & Gandhi, Sea Peavine
- *Drug*—**Eskimo, Inupiat** *Poison* Peas considered poisonous. (as *L. maritimus* 98:141) **Iroquois** *Antirheumatic (External)* Stalks cooked as greens and used for rheumatism. (as *L. maritimus* 124:93)
- *Food*—**Eskimo, Alaska** *Beverage* Roasted seeds used to make coffee. (as *L. maritimus* 4:715) **Iroquois** *Vegetable* Stalks eaten as greens in spring. (as *L. maritimus* 124:93) **Makah** *Vegetable* Immature seeds eaten as peas. (67:281)

Lathyrus jepsonii ssp. *californicus* (S. Wats.) C. L. Hitchc., California Peavine
- *Drug*—**Mendocino Indian** *Orthopedic Aid* Poultice of boiled plants applied to swollen joints. (as *L. watsoni* 41:357)
- *Food*—**Mendocino Indian** *Fodder* Cut for hay and used as fodder for horses and cattle. **Yokia** *Vegetable* Cooked and eaten as greens when 3 inches high. (as *L. watsoni* 41:357)

Lathyrus lanszwertii var. *leucanthus* (Rydb.) Dorn, Aspen Peavine
- *Food*—**Apache, Chiricahua & Mescalero** *Dried Food* Ripe pods dried, stored, and soaked and boiled when needed. *Unspecified* Ripe pods cooked and eaten. (as *L. leucanthus* 33:49)

Lathyrus nevadensis ssp. *lanceolatus* var. *nuttallii* (S. Wats.) C. L. Hitchc., Nuttall's Peavine
- *Food*—**Thompson** *Forage* Used as a general forage for animals. (as *L. nuttallii* 164:516)

Lathyrus ochroleucus Hook., Cream Peavine
- *Drug*—**Ojibwa** *Gastrointestinal Aid* Plant used for stomach trouble. *Veterinary Aid* Foliage fed to a pony to make him lively for a race. (153:372, 373)
- *Food*—**Ojibwa** *Fodder* Leaves and roots used to put spirit into a pony just before they expected to race him. (153:419) *Vegetable* Peas used for food. (135:235) Roots used as a sort of Indian potato and stored in deep garden pits, like regular potatoes. (153:406)

Lathyrus palustris L., Slenderstem Peavine
- *Drug*—**Ojibwa** *Veterinary Aid* Plant fed to a sick pony to make him fat. (153:373)
- *Food*—**Chippewa** *Unspecified* Full grown peas shelled and cooked for food. (71:133) **Ojibwa** *Fodder* Foliage was specially fed to a pony to make it grow fat. (153:419) *Vegetable* Peas used for food. (135:235)
- *Other*—**Meskwaki** *Hunting & Fishing Item* Root used as a lure to trap beaver and other game. (152:273)

Lathyrus polymorphus Nutt. **ssp. *polymorphus* var. *polymorphus***, Manystem Peavine
- *Food*—**Acoma** *Unspecified* Whole pods used for food. **Cochiti** *Unspecified* Whole pods used for food. (as *L. decaphyllus* 32:32) **Keres, Western** *Vegetable* Peas used for food. (as *L. decaphyllus* 171:51) **Laguna** *Unspecified* Whole pods used for food. (as *L. decaphyllus* 32:32)
- *Other*—**Keres, Western** *Decorations* Flowers used for bouquets. (as *L. decaphyllus* 171:51)

***Lathyrus* sp.**, Wild Pea
- *Drug*—**Weyot** *Antidiarrheal* Plant used as a diarrhea medicine. (117:180)
- *Food*—**Carrier** *Forage* Plant eaten by cows and horses. (31:81)
- *Fiber*—**Aleut** *Other* Used for weaving. (7:29)

Lathyrus venosus Muhl. ex Willd., Veiny Peavine
- *Drug*—**Chippewa** *Anticonvulsive* Simple or compound decoction of root taken or applied to chest for convulsions. (53:336) *Emetic* Decoction of root taken as an emetic for internal blood accumulation. *Hemostat* Poultice of boiled root applied to bleeding wounds. (53:356) *Stimulant* Decoction of root taken as a stimulant. *Tonic* Decoction of root taken as a tonic. (53:364)
- *Other*—**Chippewa** *Protection* Roots carried as a charm to insure successful outcomes of difficulties. (53:376)

Lathyrus vestitus Nutt., Pacific Peavine
- *Drug*—**Costanoan** *Emetic* Decoction of roots used as an emetic for internal injuries. *Panacea* Decoction of roots used as a general remedy. (21:19)
- *Food*—**Miwok** *Unspecified* Raw seeds used for food. *Vegetable* Greens used for food. (12:159)

Lauraceae, see *Cassytha, Lindera, Ocotea, Persea, Sassafras, Umbellularia*

Layia, Asteraceae
Layia glandulosa (Hook.) Hook. & Arn., Whitedaisy Tidytips
- *Food*—**Cahuilla** *Porridge* Seeds ground into flour and used with other ground seeds in a mush. (15:84) **Luiseño** *Unspecified* Seeds used for food. (155:228)

Layia platyglossa (Fisch. & C. A. Mey.) Gray, Coastal Tidytips
- *Food*—**Cahuilla** *Porridge* Seeds ground into flour and used with other ground seeds in a mush. (15:85) **Costanoan** *Staple* Seeds eaten in pinole. (21:254) **Mendocino Indian** *Staple* Seeds used to make a pinole. (as *Blepharipappus platyglossus* 41:393)

Leathesiaceae, see *Leathesia*

Leathesia, Leathesiaceae, alga
Leathesia difformis (L.) Areschoug, Bubble Seaweed
- *Drug*—**Hesquiat** *Unspecified* Used for some kind of medicine. (185:24)

Lecanoraceae, see *Candelariella*

Lechea, Cistaceae
Lechea minor L., Thymeleaf Pinweed
- *Drug*—**Seminole** *Analgesic* Decoction of leaves taken for headaches. (169:282) *Antidiarrheal* Plant used as an astringent for diarrhea. (169:168) Decoction of plant taken by babies and adults for bird sickness: diarrhea, vomiting, and appetite loss. (169:234) Plant used as a diarrhea medicine. (169:275) *Antiemetic* and *Dietary Aid* Decoction of plant taken by babies and adults for bird sickness: diarrhea, vomiting, and appetite loss. (169:234) *Febrifuge* Decoction of leaves taken for fevers. *Gastrointestinal Aid* Decoction of leaves taken for stomachaches. (169:282) *Pediatric Aid* Decoction of plant taken by babies and adults for bird sickness: diarrhea, vomiting, and appetite loss. (169:234)

***Lechea* sp.**, Pinweed
- *Drug*—**Catawba** *Dermatological Aid* Infusion of beaten roots applied to sores. (177:43)

Ledum, Ericaceae
Ledum ×columbianum Piper, Coast Labradortea
- *Food*—**Pomo, Kashaya** *Beverage* Leaves used to make a beverage tea. (as *L. glandulosum* ssp. *columbianum* 72:113)

Ledum glandulosum Nutt., Western Labradortea
- *Food*—**Tolowa** *Beverage* Leaves simmered to make tea. **Yurok** *Beverage* Leaves simmered to make a most prized tea. (6:34)

Ledum groenlandicum Oeder, Bog Labradortea
- *Drug*—**Abnaki** *Cold Remedy* Used for head colds. (144:154) *Nose Medicine* Powdered leaves mixed with bark from another plant and used as snuff for nasal inflammation. (144:170) **Algonquin, Quebec** *Analgesic* Infusion of plant used for headaches. *Ceremonial Medicine* Infusion of plant taken for colds. *Tonic* Infusion of plant used as a tonic. (18:214) *Unspecified* Leaves used to make tea and medicinal tea. (18:116) **Anticosti** *Unspecified* Infusion of plant used medicinally. (143:68) **Bella Coola** *Analgesic* and *Gastrointestinal Aid* Decoction of leaves taken for stomach pain. (150:63) **Chippewa** *Burn Dressing* Powder containing powdered root applied to burns. *Dermatological Aid* Powder containing powdered root applied to ulcers. (53:354) **Cree** *Burn Dressing* Poultice of powdered leaf ointment applied to burns and scalds. (16:492) *Diuretic* Infusion of leaves used as a diuretic. (16:493) *Emetic* Used as an emetic. (16:484) **Cree, Hudson Bay** *Analgesic* Infusion of flowering tops used for insect sting pain. *Antirheumatic (External)* Infusion of flowering tops used for rheumatism. *Dermatological Aid* Boiled, powdered wood applied to chafed skin. Poultice of fresh, chewed leaves applied to wounds. *Orthopedic Aid* Infusion of flowering tops used for tender feet. (as *L. latifolium* 92:303) **Cree, Woodlands** *Breast Treatment* Poultice of leaves applied to cracked nipples. *Burn Dressing* Poultice of leaves and grease applied to burns. Decoction of plant used to wash burns before application of burn ointment. *Dermatological Aid* Decoction of plant used as a wash for itchy skin, hand sores, and chapped skin. Poultice of leaves and fish oil applied to the umbilical scab. Powdered leaves applied directly to a baby's skin for rashes in the skin folds. *Diuretic* Plant used as a diuretic. *Pediatric Aid* Poultice of leaves and fish oil applied to the umbilical scab. Powdered leaves applied directly to a baby's skin for rashes in the skin folds. *Pulmonary Aid* Decoction of plant used for pneumonia. Decoction of plant and calamus used for whooping cough. (109:42) **Gitksan** *Diuretic* Decoction of leaves used

as a diuretic and beverage. (150:63) **Haisla & Hanaksiala** *Cold Remedy* Infusion of leaves taken for colds. *Dietary Aid* Decoction of leaves and small branches taken to increase the appetite. *Tuberculosis Remedy* Infusion of leaves taken for tuberculosis. (43:241) **Kitasoo** *Cold Remedy* and *Respiratory Aid* Decoction of dried leaves used for colds and other respiratory ailments. (43:333) **Kwakiutl** *Narcotic* Leaves considered narcotic. (183:283) **Makah** *Blood Medicine* Infusion of leaves taken as a blood purifier. (79:43) *Gynecological Aid* Infusion of leaves and sugar given to mothers after childbirth to gain their strength. *Kidney Aid* Infusion of leaves used as a kidney medicine. (67:301) *Unspecified* Infusion of fresh or dried plant used as a medicine. (186:106) **Malecite** *Kidney Aid* Infusion of plant used for kidney trouble. (116:257) **Micmac** *Cold Remedy* Leaves used for the common cold. (as *L. latifolium* 40:58) *Diuretic* Decoction of leaves taken as a diuretic. (as *L. latifolium* 156:317) *Kidney Aid* Leaves used for kidney trouble and to make a beverage. *Misc. Disease Remedy* Leaves used for scurvy and as a beverage. *Respiratory Aid* Leaves used for asthma. (as *L. latifolium* 40:58) *Tonic* Infusion of leaves taken for a "beneficial effect on the system." (156:316) **Montagnais** *Blood Medicine* Infusion of leaves and twigs taken to purify the blood. (156:313) *Febrifuge* Poultice of plant applied or infusion taken for fever. (24:14) Infusion of leaves and twigs "taken in case of chill." (156:313) *Liver Aid* and *Pediatric Aid* Poultice of plant applied or infusion given to children for jaundice. (24:14) **Nitinaht** *Dietary Aid* Infusion of leaves used as an appetite stimulant. (67:301) Infusion of fresh or dried plant taken as a tonic for increased appetite. *Strengthener* Infusion of fresh or dried plant taken as a tonic when "run down." **Nootka** *Unspecified* Infusion of fresh or dried plant used as a medicine. (186:106) **Okanagan-Colville** *Kidney Aid* Infusion of leaves and twigs taken for the kidneys. (188:102) **Oweekeno** *Cold Remedy* Infusion of leaves taken as a cold medicine. *Throat Aid* Infusion of leaves taken for sore throat. (43:96) **Potawatomi** *Unspecified* Compound containing leaves used to correct unspecified ailment. (154:57) **Quinault** *Antirheumatic (Internal)* Infusion of leaves taken for rheumatism. (79:43) **Salish** *Tonic* Decoction of plants taken as a tonic. (178:294) **Shuswap** *Dermatological Aid* Infusion or decoction of leaves taken for poison ivy. (123:56) *Eye Medicine* Decoction of plants used for blindness, sore eyes, and poison ivy. (123:62)

- *Food*—**Alaska Native** *Beverage* Strongly, aromatic leaves used to make tea. (as *L. palustre* ssp. *groenlandicum* 85:35) **Algonquin, Quebec** *Beverage* Leaves used to make tea and medicinal tea. (18:116) **Anticosti** *Beverage* Used to make tea. (143:68) **Bella Coola** *Beverage* Leaves boiled and used as a beverage. (184:205) **Chippewa** *Beverage* Leaves used to make a beverage. (53:317) **Cree** *Beverage* Used to make tea. (16:484) **Cree, Woodlands** *Beverage* Plant, with flower tops removed, used to make a tea. (109:42) **Eskimo, Arctic** *Beverage* Leaves dried and used as a substitute for tea. (128:31) **Haisla & Hanaksiala** *Beverage* Leaves used to make tea. (43:241) **Hesquiat** *Beverage* Toasted, dried leaves brewed or steeped to make tea. (185:65) **Kitasoo** *Beverage* Leaves used to make a beverage. (43:333) **Kwakiutl, Southern** *Beverage* Leaves used to make a hot, refreshing drink. (183:293) Leaves used to make tea. (183:283) **Makah** *Beverage* Leaves used to make a beverage tea. (67:301) Leaves steeped and drunk as a beverage tea. (79:43) **Malecite** *Beverage* Used to make tea. (160:6) **Micmac** *Beverage* Used to make a beverage. (159:258) **Nitinaht** *Beverage* Fresh or dried plant used to make a hot tea beverage. (186:106) **Ojibwa** *Beverage* Leaves used to make tea. (96:17) Tender leaves used for beverage tea, a well-known tea, and sometimes eaten with the tea. (153:401) **Okanagan-Colville** *Beverage* Leaves and twigs used to make tea. (188:102) **Oweekeno** *Beverage* Leaves used to make tea. (43:96) **Potawatomi** *Beverage* Leaves used to make a beverage and also used as a brown dye material. (154:120) Leaves used to make a beverage. (154:99) **Saanich** *Beverage* Fresh or dried leaves made into tea. **Salish, Coast** *Beverage* Fresh or dried leaves made into tea. (182:83) **Shuswap** *Beverage* Dried leaves mixed with tea or mint. (123:62) **Thompson** *Beverage* Leaves made into a tea-like beverage. Leaves and twigs made into a tea-like beverage and used in place of coffee. (187:214)

- *Dye*—**Iroquois** *Brown* Plant used as a dark brown dye for wool. (142:96) **Potawatomi** *Brown* Leaves used to make a beverage and also used as a brown dye material. (154:120)

Ledum palustre L., Marsh Labradortea

- *Drug*—**Eskimo, Inupiat** *Poison* Plant contains ledol, a poisonous substance known to cause cramps, diarrhea, and paralysis. *Unspecified* Infusion of young, dried, stored leaves used as a medicinal tea. (98:60) **Tanana, Upper** *Antirheumatic (External)* Decoction of leaves and stems used for arthritis. *Antirheumatic (Internal)* Fresh or dried leaves chewed for body aches. *Blood Medicine* Decoction of leaves and stems used for weak blood. *Cold Remedy* Decoction of stems and leaves taken for colds. *Cough Medicine* Decoction of stems and leaves taken for coughs. *Dermatological Aid* Decoction of stems and leaves used as a wash for rashes and dandruff. Dried leaves ground into a powder or leaf ash used on sores. *Disinfectant* Decoction of stems and leaves used as a wash for infections. Fresh or dried leaves chewed for infections. *Gastrointestinal Aid* Decoction of leaves and stems used for heartburn. *Misc. Disease Remedy* Decoction of leaves and stems used for flu. *Panacea* Decoction of stems and leaves, blackberry leaves, and spruce inner bark taken for sickness in general. *Respiratory Aid* Fresh or dried leaves chewed for congestion. *Throat Aid* Decoction of stems and leaves taken for sore throats. *Vertigo Medicine* Decoction of leaves and stems used for dizziness. (102:16) **Tlingit** *Venereal Aid* Compound infusion of sprouts and bark taken for syphilis. (107:283)

- *Food*—**Eskimo, Inupiat** *Beverage* Young, dried, stored leaves used to make tea. (98:60) **Tanana, Upper** *Beverage* Leaves and stems used to make tea. *Spice* Leaves used as a spice for strong tasting meat. (102:16)

- *Other*—**Tanana, Upper** *Tools* Used as a switch in the sweat house. (102:16)

Ledum palustre ssp. *decumbens* (Ait.) Hultén, Marsh Labradortea

- *Drug*—**Eskimo, Alaska** *Antihemorrhagic* Infusion of plant used for spitting up blood. *Gastrointestinal Aid* Infusion of plant used for upset stomach. (1:37) *Unspecified* Infusion of leaves used for medicinal purposes. (as *L. decumbens* 4:715) **Eskimo, Inupiat** *Poison* Plant contains ledol, a poisonous substance known to cause cramps, diarrhea, and paralysis. *Unspecified* Infusion of young, dried, stored leaves used as a medicinal tea. (as *L. decumbens* 98:60) **Eskimo, Kuskokwagmiut** *Pediatric Aid* Burning dried stalk shaken around head and shoulders of sick child. (as *L. decumbens* 122:32) **Eskimo, Nunivak**

Unspecified Infusion of stems and leaves used for the medicinal value. (as *L. decumbens* 149:325) **Eskimo, Western** *Analgesic* and *Gastrointestinal Aid* Compound decoction of stem and leaf used for stomachache and intestinal discomfort. (as *L. decumbens* 108:5)
- *Food*—**Eskimo, Alaska** *Beverage* Leaves used for tea. (as *L. decumbens* 4:715) *Spice* Sprigs added to tea to give it flavor. (1:37) **Eskimo, Arctic** *Beverage* Leaves dried and used as a substitute for tea. (as *L. decumbens* 128:31) **Eskimo, Inupiat** *Beverage* Young, dried, stored leaves used to make tea. (as *L. decumbens* 98:60)
- *Other*—**Eskimo, Inuktitut** *Fuel* Wood used for firewood. (as *L. decumbens* 202:190)

Leiophyllum, Ericaceae
Leiophyllum buxifolium (Berg.) Ell., Sand Myrtle
- *Drug*—**Nanticoke** *Tonic* Berries used to make spring tonic. (175:58, 84)

Lemnaceae, see *Lemna*

Lemna, Lemnaceae
Lemna trisulca L., Star Duckweed
- *Drug*—**Iroquois** *Antirheumatic (External)* Poultice of wetted plant and another plant applied to swellings. (141:71)

Lennoaceae, see *Pholisma*

Lens, Fabaceae
Lens culinaris Medik., Lentil
- *Food*—**Papago** *Dried Food* Threshed, dried on the ground or roofs, stored, and used for food. (as *L. esculenta* 34:33) *Unspecified* Species used for food. **Pima** *Unspecified* Species used for food. (as *L. esculenta* 36:120)

Lentibulariaceae, see *Pinguicula*

Leonurus, Lamiaceae
Leonurus cardiaca L., Common Motherwort
- *Drug*—**Cherokee** *Gastrointestinal Aid* Taken for "disease of the stomach." *Sedative* Given for "nervous and hysterical affections" and taken as a stimulant. *Stimulant* Taken as a stimulant for fainting. (80:45) **Delaware** *Gynecological Aid* Infusion of leaves used for female diseases. (176:38) **Delaware, Oklahoma** *Gynecological Aid* Infusion of leaves taken for "female diseases." (175:32, 76) **Iroquois** *Gastrointestinal Aid* Infusion of dried plant taken to facilitate digestion. *Sedative* Infusion of dried plant taken as a tonic for nerves. *Tonic* Infusion of dried plant taken as a tonic. (142:98) **Micmac** *Gynecological Aid* Part of plant used for obstetric cases. (40:58) **Mohegan** *Gynecological Aid* Infusion taken for "female ills." (30:121) Infusion of plant taken for peculiar ills of women. (174:265) Infusion of leaf, highly regarded herb, taken for diseases peculiar to women. (176:72, 130) *Tonic* Complex compound infusion including motherwort taken as spring tonic. (174:266) **Shinnecock** *Gynecological Aid* Infusion taken for "female ills." (30:121)

Lepechinia, Lamiaceae
Lepechinia calycina (Benth.) Epling ex Munz, Woodbalm
- *Drug*—**Miwok** *Analgesic* Decoction of leaves taken for headaches. *Febrifuge* Decoction of leaves taken for fever. *Misc. Disease Remedy* Decoction of leaves taken for ague. (as *Sphacele calycina* 12:173)

Lepidium, Brassicaceae
Lepidium campestre (L.) Ait. f., Field Pepperweed
- *Food*—**Cherokee** *Unspecified* Young plants boiled, fried, and eaten. (204:252)

Lepidium densiflorum Schrad., Common Pepperweed
- *Drug*—**Isleta** *Analgesic* Leaves chewed for headaches. (as *L. apetalum* 100:34) **Lakota** *Kidney Aid* Infusion of plant used for the kidneys. (139:41) **Mahuna** *Dietary Aid* Infusion of plant taken as a reducing aid. (as *L. epetalum* 140:66) **Navajo, Kayenta** *Gastrointestinal Aid* Plant used for effects of swallowing an ant. *Pediatric Aid* and *Sedative* Plant rubbed on baby's face to put infant to sleep. (205:24)
- *Other*—**Keres, Western** *Unspecified* Plant known and named but no use was specified. Crushed plant applied to sunburns that dancers got during the harvest, mask, and other dances. (as *L. epitalum* 171:51)

Lepidium fremontii S. Wats., Desert Pepperweed
- *Food*—**Kawaiisu** *Beverage* Seeds pounded, mixed with water, and used as a beverage. (206:36)

Lepidium lasiocarpum Nutt., Shaggyfruit Pepperweed
- *Drug*—**Navajo** *Disinfectant* Plant used as a "disinfectant." (90:153)
- *Food*—**Havasupai** *Bread & Cake* Seeds used to make bread. (197:66) *Preserves* Seeds parched, ground, kneaded into seed butter, and eaten with fruit drinks or spread on bread. *Staple* Seeds ground and eaten as a ground or parched meal. (197:67) *Unspecified* Seeds used in a variety of ways. (197:220)

Lepidium montanum Nutt., Mountain Pepperweed
- *Drug*—**Navajo, Kayenta** *Gastrointestinal Aid* Plant used for biliousness and gastrointestinal disorders. *Other* Plant used for palpitations and dizziness. (205:24)
- *Food*—**Havasupai** *Unspecified* Seeds used in a variety of ways. (197:220) **Navajo, Ramah** *Fodder* Used for sheep and horse feed. (191:29)

Lepidium nitidum Nutt., Shining Pepperweed
- *Drug*—**Cahuilla** *Dermatological Aid* Decoction of leaves used to wash hair, kept the scalp clean and prevented baldness. (15:85) **Diegueño** *Gastrointestinal Aid* Tablespoon of seeds in water used, followed the next day by a physic, for indigestion. (84:23)
- *Food*—**Diegueño** *Vegetable* Plant tops and flowers boiled and eaten as greens. (84:23) **Luiseño** *Unspecified* Seeds used for food. *Vegetable* Leaves used for greens. (155:232)

Lepidium sp., Pepper Grass
- *Food*—**Cherokee** *Vegetable* Tender plant and roots eaten as potherbs. (126:37)

Lepidium thurberi Woot., Thurber's Pepperweed
- *Food*—**Papago** *Dried Food* Seeds basket winnowed, parched, sun dried, cooked, stored, and used for food. (34:24)

Lepidium virginicum L., Virginia Pepperweed
- *Drug*—**Cherokee** *Dermatological Aid* Poultice of bruised root applied to "draw blister quickly." *Pulmonary Aid* Used as a poultice for

croup. *Veterinary Aid* Infusion given to sick chickens and mixed with feed to make chickens lay. (80:48) **Houma** *Tuberculosis Remedy* Compound decoction of plant with whisky taken for tuberculosis. (158:64) **Menominee** *Dermatological Aid* Infusion of plant used as a wash or bruised plant used for poison ivy. (151:33)
- *Food*—**Cherokee** *Unspecified* Species used for food. (80:48) Young plants boiled, fried, and eaten. (204:252)

Lepidium virginicum var. *menziesii* (DC.) C. L. Hitchc., Menzies's Pepperweed
- *Food*—**Hoh** *Unspecified* Eaten raw. *Vegetable* Leaves eaten as greens. **Quileute** *Unspecified* Eaten raw. *Vegetable* Leaves eaten as greens. (as *L. menziesii* 137:62)

Leptarrhena, Saxifragaceae

Leptarrhena pyrolifolia (D. Don) R. Br. ex Ser., Fireleaf Leptarrhena
- *Drug*—**Aleut** *Misc. Disease Remedy* Infusion of leaves taken for "sicknesses such as influenza." (8:427) **Thompson** *Dermatological Aid* Poultice of chewed, fresh leaves applied to wounds and sores. (as *Leptarrhenia amplexifolia* 164:465)

Leptodactylon, Polemoniaceae

Leptodactylon pungens (Torr.) Torr. ex Nutt., Granite Pricklygilia
- *Drug*—**Navajo, Kayenta** *Dermatological Aid* Plant used for scorpion stings. *Kidney Aid* Plant used for kidney disease. (as *Gilia pungens* 205:38) **Navajo, Ramah** *Disinfectant* Decoction of plant used for "snake infection." *Gynecological Aid* Decoction of plant taken during pregnancy keeps baby small, for easy labor. (as *Gilia pungens* 191:40) **Shoshoni** *Eye Medicine* Decoction or infusion of plant used as a wash for sore or swollen eyes. (as *Gilia pungens* 180:81)

Lespedeza, Fabaceae

Lespedeza capitata Michx., Roundhead Lespedeza
- *Drug*—**Meskwaki** *Antidote* Root used as antidote for poison. (152:229) **Omaha** *Analgesic* Moxa of stems used in cases of neuralgia. *Antirheumatic (External)* Moxa of stems used in cases of rheumatism. **Ponca** *Analgesic* Moxa of stems used in cases of neuralgia. *Antirheumatic (External)* Moxa of stems used in cases of rheumatism. (70:97, 98)
- *Food*—**Comanche** *Beverage* Leaves boiled for tea. (29:522)

Lespedeza sp.
- *Drug*—**Iroquois** *Other* Compound decoction of plant taken for stricture caused by bad blood. (87:364)

Lesquerella, Brassicaceae

Lesquerella douglasii S. Wats., Douglas's Bladderpod
- *Drug*—**Okanagan-Colville** *Antidiarrheal* Roots chewed, juice swallowed, and pulp spat out for diarrhea. *Gastrointestinal Aid* Roots chewed, juice swallowed, and pulp spat out for "heartburn." (188:92) **Shuswap** *Dermatological Aid* Poultice of mashed plants applied to sores. *Diaphoretic* Plant used to produce sweating. (123:61)

Lesquerella fendleri (Gray) S. Wats., Fendler's Bladderpod
- *Drug*—**Keres, Western** *Antirheumatic (External)* Bruised plant mixed with salt and used as a rub for swellings. *Emetic* Infusion of plant used as an emetic. (171:52) **Navajo** *Dermatological Aid* Infusion of plants taken to counteract the effects of spider bites. (55:49)
- *Other*—**Navajo, Ramah** *Ceremonial Items* Tied to ceremonial rattle string and wetted with infusion of the plant. (191:29)

Lesquerella intermedia (S. Wats.) Heller, Mid Bladderpod
- *Drug*—**Hopi** *Ceremonial Medicine* and *Emetic* Infusion of root taken as a ceremonial emetic. (200:77) *Gynecological Aid* Root rubbed on abdomen when uterus failed to contract after childbirth. (200:36, 77) *Snakebite Remedy* Root eaten and poultice of chewed root used for snakebite. (200:32, 77) **Navajo, Kayenta** *Ceremonial Medicine* Plant used as a Nightway medicine. *Eye Medicine* Poultice of roots applied to sore eyes. (205:24)

Lesquerella rectipes Woot. & Standl., Straight Bladderpod
- *Drug*—**Navajo, Ramah** *Ceremonial Medicine* Cold infusion used as a ceremonial eyewash and pulverized plant used as a ceremonial snuff. *Eye Medicine* Cold infusion of plant used as a ceremonial eyewash. *Respiratory Aid* Finely ground leaves used ceremonially as snuff to clear nasal passages. *Toothache Remedy* Poultice of crushed or chewed leaves applied for toothache. (191:29)

Lessingia, Asteraceae

Lessingia glandulifera Gray var. *glandulifera*, Valley Vinegarweed
- *Drug*—**Kawaiisu** *Analgesic* Moxa of dried stem bark used for pain. (as *L. germanorum* var. *vallicola* 206:36)

Lessoniaceae, see *Lessoniopsis, Macrocystis, Nereocystis, Postelsia*

Lessoniopsis, Lessoniaceae, alga

Lessoniopsis littoralis (Farlow & Setchell) Reinke, Short Kelp
- *Drug*—**Nitinaht** *Strengthener* Burned stipes made into a salve and used to strengthen young boys. (186:51)
- *Food*—**Hesquiat** *Dried Food* Stipes and fronds with attached herring eggs dried for later use.[6] (185:24)
- *Fiber*—**Hesquiat** *Sporting Equipment* Dried stipes use as "pucks" and hitting sticks.[5] (185:24)
- *Other*—**Nitinaht** *Toys & Games* Flattish, hardened stipes used for "beach hockey." (186:51)

Letharia, Usneaceae, lichen

Letharia vulpina
- *Dye*—**Cheyenne** *Yellow* Boiled in water and used as a yellow dye for porcupine quills. (as *L. vulpina* 83:3) **Karok** *Unspecified* Used as a dye for porcupine quills. **Oweekeno** *Yellow* Thalli used to make a yellow dye. (as *L. vulpina*, wolf lichen 43:49) **Yurok** *Unspecified* Used as a dye for porcupine quills. (as *L. vulpina* 6:34)

Leucanthemum, Asteraceae

Leucanthemum vulgare Lam., Ox Eye Daisy
- *Drug*—**Iroquois** *Eye Medicine* Infusion of flowers and roots with other plants used as an eyewash. (as *Chrysanthemum leucanthemum* var. *pinnatifidum* 141:64) **Menominee** *Febrifuge* Plant used as a fever medicine. (as *Chrysanthemum leucanthemum* 151:29) **Mohegan** *Tonic* Dandelion and white daisy used to make wines and taken as tonics. (as *Chrysanthemum leucanthemum* 30:121) Compound de-

coction or infusion of plants taken as a spring tonic. (as *Chrysanthemum leucanthemum* 174:266) Flowers used to make a tonic. (as *Chrysanthemum leucanthemum* 176:72, 128) **Quileute** *Dermatological Aid* Decoction of dried flowers and stems used as a wash for chapped hands. (as *Chrysanthemum leucanthemum* 79:49) **Shinnecock** *Tonic* Dandelion and white daisy used to make wines and taken as tonics. (as *Chrysanthemum leucanthemum* 30:121)

Leucobryaceae, see *Octoblephorum*

Leucocrinum, Liliaceae

Leucocrinum montanum Nutt. ex Gray, Common Starlily
- **Drug**—**Paiute** *Dermatological Aid* Poultice of pulverized roots applied to sores or swellings. **Shoshoni** *Dermatological Aid* Poultice of pulverized roots applied to sores or swellings. (180:100)
- **Food**—**Crow** *Unspecified* Roots used for food. (19:14)

Leucodontaceae, see *Antitrichia*

Leucothoe, Ericaceae

Leucothoe axillaris (Lam.) D. Don, Coastal Doghobble
- **Drug**—**Cherokee** *Analgesic* Infusion used for "shifting pains." *Antirheumatic (External)* Infusion rubbed on for rheumatism. *Antirheumatic (Internal)* Compound decoction of leaf used for rheumatism. *Dermatological Aid* Infusion of leaf and stem used to bathe itch. (80:32) *Other* Decoction of leaves rubbed into scratches made on legs as preliminary treatment. (as *L. catesbaei* 177:49) *Stimulant* Infusion rubbed on for "languor." *Veterinary Aid* Root ooze applied to mangy dog. (80:32)

Lewisia, Portulacaceae

Lewisia columbiana (T. J. Howell ex Gray) B. L. Robins., Columbian Bitterroot
- **Food**—**Okanagon** *Winter Use Food* Steamed or boiled and used as a winter food. (125:36) **Thompson** *Unspecified* Fleshy roots eaten. (164:480) *Winter Use Food* Steamed or boiled and used as a winter food. (125:36) **Thompson, Upper (Nicola Band)** *Unspecified* Fleshy roots eaten. (164:480)
- **Other**—**Okanagon** *Cash Crop* Traded with other tribes for dried salmon and other items. **Thompson** *Cash Crop* Traded with other tribes for dried salmon and other items. (125:36)

Lewisia pygmaea (Gray) B. L. Robins., Pigmy Bitterroot
- **Drug**—**Thompson** *Psychological Aid* Some believed that eating the roots caused insanity. (164:479)
- **Food**—**Blackfoot** *Dried Food* Roots dried for future use. (97:34) **Thompson** *Unspecified* Roots used for food. (164:479)
- **Other**—**Thompson** *Good Luck Charm* Plant used as a charm for good luck in gambling. (164:507)

Lewisia rediviva Pursh, Oregon Bitterroot
- **Drug**—**Blackfoot** *Throat Aid* Pounded, dry root chewed for sore throat. (118:38) **Flathead** *Breast Treatment* Roots eaten for increased milk flow after childbirth. *Gynecological Aid* Infusion of roots taken for increased milk flow after childbirth. *Heart Medicine* Infusion of roots taken for heart pain. *Pulmonary Aid* Infusion of roots taken for pleurisy pain. **Nez Perce** *Blood Medicine* Plant used for impure blood. *Gynecological Aid* Roots eaten for increased milk flow after childbirth. Infusion of roots taken for increased milk flow after childbirth. (82:46) **Okanagan-Colville** *Dermatological Aid* Poultice of raw roots applied to sores. Raw roots eaten for poison ivy rashes. *Misc. Disease Remedy* Dried or fresh roots eaten for diabetes. *Witchcraft Medicine* "Hearts" used in some type of witchcraft. (188:114)
- **Food**—**Blackfoot** *Unspecified* Plant boiled and eaten. (114:278) **Coeur d'Alene** *Vegetable* Roots used as a principal vegetable food. (178:88) **Kutenai** *Cooking Agent* Roots steamed and used to thicken gravy. *Dessert* Roots steamed, added to camas bulbs, and eaten as a "sweet treat." *Dried Food* Roots dried, stored, and used for food. *Unspecified* Roots used for food as the most important root crop. (82:46) **Montana Indian** *Unspecified* Small pieces of bitterroot steeped, boiled in water, and eaten. (19:14) Roots boiled or steamed and eaten plain, mixed with berries, or added to meat or bone marrow. (82:46) **Okanagan-Colville** *Dried Food* Roots peeled and dried for future use. *Unspecified* Fresh or dried roots steamed or boiled and eaten. (188:114) **Okanagon** *Staple* Roots used as a principal food. (178:238) *Unspecified* Roots used as an important food. (178:237) *Winter Use Food* Steamed or boiled and used as a winter food. (125:36) **Oregon Indian, Warm Springs** *Unspecified* Roots used for food. **Paiute** *Dried Food* Roots dried and used for food. *Unspecified* Roots boiled "like macaroni." (104:102) *Winter Use Food* Roots peeled and dried for winter use and boiled and eaten with salmon. (111:70) **Paiute, Northern** *Unspecified* Roots peeled, boiled, or roasted and eaten without grinding. *Vegetable* Leaves boiled like spinach and eaten. (59:43) **Sanpoil & Nespelem** *Porridge* Roots mixed with serviceberries, grease or fat added, and boiled into a congealed mass. (131:100) **Shuswap** *Unspecified* Roots cooked with serviceberries. (123:65) **Spokan** *Unspecified* Roots used for food. (178:343) **Thompson** *Bread & Cake* Roots used as an ingredient in fruit cake. *Dried Food* Peeled roots dried loose or large roots stored on strings for future use. The roots were dried on strings in order to determine the market value or trade worth. The dried roots were eaten with saskatoon berries and salmon eggs. *Pie & Pudding* Roots cooked with black tree lichen, dough, and fresh salmon and made into a pudding. Sometimes the roots were cooked with black tree lichen, fermented salmon eggs, yellow avalanche lily corms, saskatoon berries, and deer fat to make a similar kind of pudding. *Special Food* Dried roots cooked in soups such as fish head soup, but only served on special occasions. Because the roots were so valuable, they were only served on special occasions. (187:243) *Unspecified* Used as an important food. (164:478) Fleshy taproot eaten. (164:479) Roots used as an important food. (178:237) Fresh roots pit cooked or boiled in watertight baskets using red-hot stones. (187:243) *Winter Use Food* Steamed or boiled and used as a winter food. (125:36)
- **Other**—**Okanagan-Colville** *Cash Crop* Roots formerly an important article of trade. (188:114) **Okanagon** *Cash Crop* Traded with other tribes for dried salmon and other items. (125:36) Roots traded to the Lower Thompson for dried salmon. (164:479) **Thompson** *Cash Crop* Traded with other tribes for dried salmon and other items. (125:36) Strung, dried roots used as a trade item. *Plant Indicator* Presence of plant indicated the growth of another plant type. (187:243) **Thompson, Upper (Lytton Band)** *Cash Crop* Fleshy taproot traded to the Lower Thompson band. (164:479)

Leymus, Poaceae

Leymus cinereus (Scribn. & Merr.) Á. Löve, Basin Wildrye
- *Drug*—**Okanagan-Colville** *Antihemorrhagic* Decoction of roots taken for internal hemorrhaging. *Dermatological Aid* Decoction of roots used as a wash to stimulate hair growth. *Venereal Aid* Infusion of mashed roots taken for gonorrhea. (as *Elymus cinereus* 188:55) **Thompson** *Veterinary Aid* Hollow straw used to clear the blocked nipple of a cow. The udder was splashed with warm water, massaged, and the straw poked into it to clear the blockage. (as *Elymus cinereus* 187:140)
- *Food*—**Blackfoot** *Forage* Used for grazing during the winter. (as *Elymus cinereus* 97:20) **Okanagan-Colville** *Fodder* Leaves used as bedding and horse feed. (as *Elymus cinereus* 188:55)
- *Fiber*—**Blackfoot** *Mats, Rugs & Bedding* Grass used for beds in lodges made from sticks when on war parties. (as *Elymus cinereus* 97:20) **Okanagan-Colville** *Mats, Rugs & Bedding* Leaves used as bedding and horse feed. Leaves used to cover the floor of sweat house. (as *Elymus cinereus* 188:55) **Thompson** *Basketry* Culms used for basket imbrication as a substitute for another plant or other swamp grasses. (as *Elymus cinereus* 187:140)
- *Dye*—**Cheyenne** *Black* Plants tied in bunches, burned, ash mixed in blood, and used as a permanent black dye. (as *Elymus cinereus* 83:8) *Unspecified* Used to make a dye. (as *Elymus cinerus* 83:46)
- *Other*—**Cheyenne** *Ceremonial Items* Plants used to make bedding for various ceremonies. (as *Elymus cinereus* 83:8) **Okanagan-Colville** *Containers* Leaves used over and under the food in the cooking pits. *Hunting & Fishing Item* Stems straightened, notched, fixed with wooden tips into arrows, and used for hunting. *Toys & Games* Stems straightened, notched, fixed with wooden tips into arrows, and used in games. (as *Elymus cinereus* 188:55) **Thompson** *Ceremonial Items* Grass used to line old-style graves. *Cooking Tools* Stout culms broken into lengths and poked into edges of cut fish to hold it flat while drying. (as *Elymus cinereus* 187:140)

Leymus condensatus (J. Presl) Á. Löve, Giant Wildrye
- *Drug*—**Paiute** *Eye Medicine* Dried leaves used to scrape pimples from the under side of the eyelid. (as *Elymus condensatus* 111:51) Decoction or infusion of leaves used as a wash for sore eyes. Sharp edges of leaf blades used to scrape granulated eyelids. **Shoshoni** *Eye Medicine* Decoction or infusion of leaves used as a wash for sore eyes. Sharp edges of leaf blades used to scrape granulated eyelids. (as *Elymus condensatus* 180:67)
- *Food*—**Klamath** *Unspecified* Grains used for food. (as *Elymus condensatus* 45:91) **Montana Indian** *Unspecified* Seeds used for food. (as *Elymus condensatus* 19:11) **Paiute** *Unspecified* Species used for food. (as *Elymus condensatus* 167:244) **Paiute, Southern** *Unspecified* Species used for food. (as *Elymus condensatus* 15:69) **Shoshoni** *Starvation Food* Seeds stored for times of famine. (as *Elymus condensatus* 118:17)
- *Fiber*—**Cahuilla** *Building Material* Stalks used for roof thatching. (as *Elymus condensatus* 15:69) **Paiute** *Brushes & Brooms* Roots tied together and used as hair combs. (as *Elymus condensatus* 111:51)
- *Other*—**Cahuilla** *Ceremonial Items* Stems made into painted arrows and used in ceremonial dances. *Hunting & Fishing Item* Fire hardened stems used as the main shaft in arrow making. **Luiseño** *Hunting & Fishing Item* Fire hardened stems used as the main shaft in arrow making. (as *Elymus condensatus* 15:69) Plant used to make arrow main shafts. (as *Elymus condensatus* 155:205)

Leymus mollis (Trin.) Hara **ssp. *mollis***, American Dunegrass
- *Drug*—**Makah** *Unspecified* Bundles of roots used to rub the body after bathing. (as *Elymus mollis* 79:21) **Nitinaht** *Strengthener* Rootstocks twisted together and rubbed on bodies of young men while bathing for strength. (as *Elymus mollis* var. *mollis* 186:88) **Quileute** *Unspecified* Roots braided, tied into bundles, and used to rub the body after bathing. (as *Elymus mollis* 79:21)
- *Fiber*—**Eskimo, Alaska** *Basketry* Dried, brown leaves woven into mats, baskets, and tote sacks. *Cordage* Dried, brown leaves woven into ropes for hanging herring and other fish. *Mats, Rugs & Bedding* Dried, brown leaves woven into mats, baskets, and tote sacks. (as *Elymus arenarius* ssp. *mollis* 1:34) **Hesquiat** *Basketry* Tough, coarse leaves used to make handles for bags, but not the bags themselves. (as *Elymus mollis* 185:58) **Kwakiutl, Southern** *Basketry* Fibrous leaves used to make baskets. *Clothing* Fibrous leaves used to make hats. (as *Elymus mollis* 183:275) **Nitinaht** *Sewing Material* Tough, sharply pointed leaves used as "needle-and-thread" for sewing and tying material. (as *Elymus mollis* var. *mollis* 186:88)
- *Other*—**Eskimo, Alaska** *Cash Crop* Dried, brown leaves woven into mats and other marketable products and sold for cash. The sale of baskets, mats, tote sacks, and ropes provided a significant supplementary cash income. *Cooking Tools* Dried, brown leaves woven into mats, baskets, tote sacks, and ropes for hanging herring and other fish. (as *Elymus arenarius* ssp. *mollis* 1:34) **Eskimo, Inuktitut** *Hunting & Fishing Item* Plant used as an indicator of marmot burrows. (as *Elymus arenarius* ssp. *mollis* 202:185) **Haisla & Hanaksiala** *Containers* Blades used to line oolichan (candlefish) ripening pits. *Fasteners* Blades used to tie Pacific silverweed roots together before steaming. (as *Elymus mollis* var. *mollis* 43:205) **Kwakiutl, Southern** *Cooking Tools* Leaves used with skunk cabbage leaves to line steaming boxes for cooking lupine roots. (as *Elymus mollis* 183:285) **Nitinaht** *Fasteners* Tough, sharply pointed leaves used as "needle-and-thread" for sewing and tying material. (as *Elymus mollis* var. *mollis* 186:88) **Quinault** *Containers* Leaves placed under drying salal berries. (as *Elymus mollis* 79:21)

Leymus triticoides (Buckl.) Pilger, Beardless Wildrye
- *Food*—**Kawaiisu** *Forage* Plant eaten by cows. *Porridge* Seeds pounded in a bedrock mortar hole, cooked into a thick mush and eaten. (as *Elymus triticoides* 206:27) **Mendocino Indian** *Fodder* Foliage used as fodder in late summer. *Staple* Seeds used for pinole. (as *Elymus triticoides* 41:312)
- *Fiber*—**Thompson** *Basketry* Culms used as a substitute in making basketry. (as *Elymus triticoides* 164:499)

Liatris, Asteraceae

Liatris acidota Engelm. & Gray, Sharp Gayfeather
- *Drug*—**Koasati** *Antirheumatic (Internal)* Decoction of roots taken for rheumatism. (177:62)

Liatris laxa Small, Rattlesnake Master
- *Drug*—**Seminole** *Analgesic* Decoction of roots used for cow sickness: lower chest pain, digestive disturbances, and diarrhea. *Antidiarrheal*

Decoction of roots used for cow sickness: lower chest pain, digestive disturbances, and diarrhea. (169:191) Decoction of roots taken by babies and adults for bird sickness: diarrhea, vomiting, and appetite loss. *Antiemetic* Decoction of roots taken by babies and adults for bird sickness: diarrhea, vomiting, and appetite loss. (169:234) *Antirheumatic (External)* Decoction of plant used as a body rub and steam for joint swellings. (169:193) *Dietary Aid* Decoction of roots taken by babies and adults for bird sickness: diarrhea, vomiting, and appetite loss. (169:234) *Gastrointestinal Aid* Decoction of roots used for cow sickness: lower chest pain, digestive disturbances, and diarrhea. (169:191) *Pediatric Aid* Decoction of roots taken by babies and adults for bird sickness: diarrhea, vomiting, and appetite loss. (169:234)

Liatris punctata Hook., Dotted Gayfeather
- **Drug**—**Blackfoot** *Dermatological Aid* Poultice of boiled roots applied to swellings. *Gastrointestinal Aid* Infusion of roots taken for stomachaches. (97:59) **Comanche** *Urinary Aid* Root chewed and juice swallowed for swollen testes. (29:522) **Meskwaki** *Dermatological Aid* Infusion of root applied locally for itch. *Urinary Aid* Infusion of root used for bloody urine and by women for bladder trouble. *Venereal Aid* Infusion of root used for gonorrhea. *Veterinary Aid* Infusion of root used for ponies to make them spirited for hunting in hot weather. (152:216)
- **Food**—**Blackfoot** *Unspecified* Roots used for food. (97:59) **Kiowa** *Unspecified* Springtime, sweet roots baked over a fire and eaten. (192:61) **Lakota** *Dietary Aid* Roots pulverized and eaten to improve the appetite. (139:38)

Liatris punctata Hook. var. *punctata*, Dotted Gayfeather
- **Drug**—**Blackfoot** *Analgesic* Infusion of root taken for stomachache. *Dermatological Aid* Boiled root applied to swellings. *Gastrointestinal Aid* Infusion of root taken for stomachache. (as *Lacinaria punctata* 114:274)
- **Food**—**Blackfoot** *Unspecified* Plant eaten raw. (as *Lacinaria punctata* 114:274) **Kiowa** *Unspecified* Springtime, sweet roots baked over a fire and eaten. (as *Lacinaria punctata* 192:61) **Tewa** *Unspecified* Roots eaten as food. (as *Laciniaria punctata* 138:57)

Liatris scariosa (L.) Willd., Devil's Bite
- **Drug**—**Meskwaki** *Kidney Aid* Used for kidney troubles. *Urinary Aid* Used for bladder troubles. (152:216)

Liatris scariosa (L.) Willd. var. *scariosa*, Devil's Bite
- **Drug**—**Chippewa** *Veterinary Aid* Decoction of root used as a horse stimulant before a race. (as *Laciniaria scariosa* 53:366) **Omaha** *Dermatological Aid* Poultice of powdered plants applied to external inflammation. *Dietary Aid* Roots used as an appetizer. *Gastrointestinal Aid* Plant taken for abdominal troubles. *Tonic* Roots used as a tonic. (as *Laciniaria scariosa* 68:335) *Veterinary Aid* Chewed corm blown into nostrils of horses to strengthen them and help them. **Pawnee** *Antidiarrheal* and *Pediatric Aid* Decoction of leaves and corms given to children for diarrhea. (as *Laciniaria scariosa* 70:133, 134)

Liatris sp., Deer's Potato
- **Drug**—**Creek** *Antirheumatic (External)* Simple or compound infusion of root rubbed on affected rheumatic part. *Antirheumatic (Internal)* Simple or compound infusion of root taken for affected rheumatic part. (as *Lacinaria* 172:660)

Liatris spicata (L.) Willd., Dense Gayfeather
- **Drug**—**Cherokee** *Analgesic* Used as an anodyne and decoction or tincture used for backache and limb pains. *Carminative* Used as a carminative. *Diaphoretic* Root used as a sudorific. *Diuretic* Root used as a diuretic. *Expectorant* Root used as an expectorant. *Gastrointestinal Aid* Decoction or tincture taken for colic. *Kidney Aid* Root used for dropsy. *Stimulant* Root used as a stimulant. (80:27) **Menominee** *Heart Medicine* Compound decoction of root used for a "weak heart." (as *Lacinaria spicata* 54:129)

Libocedrus, Cupressaceae, conifer
Libocedrus sp.
- **Fiber**—**Mewuk** *Building Material* Boughs placed on top of the acorn caches. (117:362)
- **Other**—**Mewuk** *Cooking Tools* Boughs used to line the leach. (117:362)

Licania, Chrysobalanaceae
Licania michauxii Prance, Gopher Apple
- **Drug**—**Seminole** *Analgesic* Plant used for wolf sickness: vomiting, stomach pain, diarrhea, and frequent urination. Infusion of plant taken for wolf sickness: vomiting, stomach pain, diarrhea, and frequent urination. Roots and leaves used for labor pains and hasten the birth. (as *Chrysobalanus oblongifolius* 169:165, 227, 323) *Antidiarrheal* Plant used for wolf sickness: vomiting, stomach pain, diarrhea, and frequent urination. Infusion of plant taken for wolf sickness: vomiting, stomach pain, diarrhea, and frequent urination. (as *Chrysobalanus oblongifolius* 169:165, 227) *Antiemetic* Plant used for wolf sickness: vomiting, stomach pain, diarrhea, and frequent urination. Infusion of plant taken for wolf sickness: vomiting, stomach pain, diarrhea, and frequent urination. (as *Chrysobalanus oblongifolius* 169:166, 227) *Gastrointestinal Aid* Plant used for wolf sickness: vomiting, stomach pain, diarrhea, and frequent urination. Infusion of plant taken for wolf sickness: vomiting, stomach pain, diarrhea, and frequent urination. (as *Chrysobalanus oblongifolius* 169:165, 227) *Other* Complex infusion of leaves and roots taken for chronic conditions. (as *Chrysobalanus oblongifolia* 169:272) *Psychological Aid* Infusion of plant used to steam and bathe the body for insanity. (as *Chrysobalanus oblongifolia* 169:292) *Reproductive Aid* Roots and leaves used for labor pains and hasten the birth. (as *Chrysobalanus oblongifolius* 169:323) *Unspecified* Plant used for medicinal purposes. (as *Chrysobalanus oblongifolius* 169:162) *Urinary Aid* Plant used for wolf sickness: vomiting, stomach pain, diarrhea, and frequent urination. Infusion of plant taken for wolf sickness: vomiting, stomach pain, diarrhea, and frequent urination. (as *Chrysobalanus oblongifolius* 169:165, 227)
- **Food**—**Seminole** *Forage* Berries eaten by gophers. (as *Chrysobalanus oblongifolius* 169:434)

Ligusticum, Apiaceae
Ligusticum apiifolium (Nutt. ex Torr. & Gray) Gray, Celeryleaf Licoriceroot
- **Drug**—**Karok** *Dietary Aid* Infusion of roots taken by person who lacks an appetite. (as *L. apiodorum* 148:387) **Pomo** *Antihemorrhagic* and *Pulmonary Aid* Decoction of roots taken for lung hemorrhages.

(as *L. apiodorum* 66:14) **Pomo, Kashaya** *Blood Medicine* Decoction of root taken for anemia. *Pulmonary Aid* Decoction of root taken for lung hemorrhage. *Tuberculosis Remedy* Decoction of root taken for the beginning of tuberculosis. (72:64)

Ligusticum californicum Coult. & Rose, California Licoriceroot
- **Food**—**Tolowa** *Unspecified* Roots used for food. (6:34)

Ligusticum canadense (L.) Britt., Canadian Licoriceroot
- **Drug**—**Cherokee** *Gastrointestinal Aid* Roots chewed or smoked for all stomach disorders. **Creek** *Gastrointestinal Aid* Roots chewed or smoked for all stomach disorders. (as *Lingusticum canadense* 169:276)
- **Food**—**Cherokee** *Dried Food* Fresh greens gathered into a bundle, dried, and hung until needed. (126:58) *Unspecified* Young growth boiled, fried with ramps (*Allium tricoccum*?) and eaten. (204:252) *Vegetable* Leaves cooked and eaten as greens. (80:61) Leaves and stalks boiled, rinsed, and fried with grease and salt until soft as a potherb. *Winter Use Food* Leaves and stalks blanched, boiled in a can, and stored for future use. (126:58)

Ligusticum canbyi Coult. & Rose, Canby's Licoriceroot
- **Drug**—**Cree** *Heart Medicine* Used for heart troubles. **Crow** *Cold Remedy* Roots chewed for colds. *Cough Medicine* Roots chewed for coughs. *Ear Medicine* Infusion of roots used for earache. *Respiratory Aid* Root shavings added to boiling water and steam inhaled for sinus infection and congestion. **Flathead** *Anticonvulsive* Roots chewed, rubbed on the body, or smoked for seizures. *Unspecified* Plants used as herbal medicine. **Kutenai** *Unspecified* Plants used as herbal medicine. (82:24) **Okanagan-Colville** *Ceremonial Medicine* Roots burned and smoke used to revive singers from a trance, considered ceremonially dead. Roots burned and smoke used to revive a subdued person possessed by the "bluejay spirit." *Cold Remedy* Root tied in a cheesecloth and kept near a baby's face to prevent a cold. *Internal Medicine* Roots used as a good general internal medicine. *Other* Roots burned and smoke used for unconsciousness, trances, or "possession" by spirits. *Pediatric Aid* Root tied in a cheesecloth and kept near a baby's face to prevent a cold. (188:64) **Thompson** *Unspecified* Roots used medicinally whenever obtainable. (187:153)
- **Other**—**Crow** *Incense & Fragrance* Root shavings sprinkled on live coals for incense. *Smoke Plant* Root shavings added to tobacco and kinnikinnick and smoked. **Flathead** *Soap* Roots and buckbrush leaves used to make a special hair rinse. (82:24) **Okanagan-Colville** *Smoke Plant* Roots mixed with tobacco or rolled in cigarettes to give the smoke a pleasant menthol taste. (188:64)

Ligusticum filicinum S. Wats., Fernleaf Licoriceroot
- **Drug**—**Menominee** *Panacea* Root used for many ailments. (151:55) **Paiute** *Cough Medicine* Root used in a cough remedy. (180:100, 101)

Ligusticum grayi Coult. & Rose, Gray's Licoriceroot
- **Drug**—**Atsugewi** *Analgesic* Roots used to avoid pains. *Cold Remedy* Infusion of root taken or roots chewed for colds. *Cough Medicine* Infusion of root taken or roots chewed for coughs. *Gastrointestinal Aid* Infusion of root taken or roots chewed for children's stomachaches. *Panacea* Infusion of root taken or roots chewed for ailments. *Pediatric Aid* Infusion of root taken or roots chewed for children's stomachaches. (61:140)
- **Food**—**Atsugewi** *Substitution Food* Tender leaves soaked in water, cooked, and used as a meat substitute when acorns were eaten. *Vegetable* Tender leaves soaked in water, cooked, and used for food. *Winter Use Food* Tender leaves soaked in water, cooked, and stored for later use. (61:139)
- **Other**—**Atsugewi** *Hunting & Fishing Item* Pulverized root used for poisoning fish. (61:137)

Ligusticum porteri Coult. & Rose, Porter's Licoriceroot
- **Drug**—**Zuni** *Antirheumatic (External)* Infusion of root used for body aches. *Ceremonial Medicine* Root chewed by medicine man and patient during curing ceremonies for various illnesses. *Throat Aid* Crushed root and water used as wash and taken for sore throat. (27:379)
- **Food**—**Apache, Chiricahua & Mescalero** *Unspecified* Eaten without preparation or cooked with green chile and meat or animal bones. (33:46)
- **Other**—**Yuki** *Protection* Roots used to ward off rattlesnakes. (49:44)

Ligusticum scothicum L., Scottish Licoriceroot
- **Drug**—**Bella Coola** *Unspecified* Leaves spread over hot stones and used as medicinal bed for the sick. (150:61)
- **Food**—**Anticosti** *Spice* Used to season fish or salads. (143:67) **Eskimo, Inupiat** *Spice* Leaves used as a spice for soups. *Vegetable* Leaves stored in oil or cooked and eaten with dried meat or boiled fish. Leaves used as greens in salads. (98:13)

Ligusticum scothicum ssp. *hultenii* (Fern.) Calder & Taylor, Hultén's Licoriceroot
- **Drug**—**Eskimo, Western** *Poison* Mature plant in late summer considered mildly poisonous. (as *L. hultenii* 108:60)
- **Food**—**Alaska Native** *Dietary Aid* Fresh leaves used as a good source for vitamins C and A. *Substitution Food* Leaves and stalks used as a substitute for celery. *Unspecified* Leaves and stalks eaten raw with seal oil. Leaves and stalks used in cooking fish. *Vegetable* Leaves and stalks used as a cooked vegetable. *Winter Use Food* Leaves and stalks stored in seal oil for winter use. (as *L. hultenii* 85:37) **Eskimo, Alaska** *Vegetable* Young leaves and stems eaten raw or cooked and often mixed with other wild greens. (1:37) *Winter Use Food* Cut, mixed with fish, and boiled for winter use. (as *L. hultenii* 4:715)

Liliaceae, see *Aletris, Allium, Amianthium, Asparagus, Bloomeria, Brodiaea, Calochortus, Camassia, Chlorogalum, Clintonia, Dichelostemma, Disporum, Echeandia, Eremocrinum, Erythronium, Fritillaria, Hastingsia, Hesperocallis, Hosta, Hypoxis, Leucocrinum, Lilium, Maianthemum, Medeola, Ornithogalum, Polygonatum, Smilacina, Stenanthium, Streptopus, Trillium, Triteleia, Uvularia, Veratrum, Xerophyllum, Zephyranthes, Zigadenus*

Lilium, Liliaceae

Lilium canadense L., Canadian Lily
- **Drug**—**Algonquin, Quebec** *Gastrointestinal Aid* Root used for stomach disorders. (18:138) **Cherokee** *Antidiarrheal* Infusion of root given for "flux." *Antirheumatic (Internal)* Infusion of root used in various ways for rheumatism. *Dietary Aid* and *Pediatric Aid* Decoction of boiled tubers given "to make child fleshy and fat." (80:43) **Chippewa** *Snakebite Remedy* Decoction of root applied to snakebites. (53:352)

Malecite *Abortifacient* Infusion of plant and sweet viburnum roots used for irregular menstruation. (116:258) **Micmac** *Abortifacient* Parts of plant used for irregular menstruation. (40:58)
- **Food**—**Cherokee** *Starvation Food* Roots made into flour and used to make bread for famine times. (80:43) **Huron** *Starvation Food* Roots used with acorns during famine. (3:63)

Lilium columbianum hort. ex Baker, Columbian Lily
- **Drug**—**Okanagan-Colville** *Witchcraft Medicine* Bulbs dried, mashed with "stink bugs," powdered, and used against "plhax," that is, witchcraft. (188:46)
- **Food**—**Clallam** *Unspecified* Bulbs steamed in pits and used for food. (57:196) **Klallam** *Unspecified* Corms steamed and eaten. **Lummi** *Unspecified* Corms steamed and eaten. (79:25) **Nitinaht** *Unspecified* Bulbs formerly steamed and eaten cold with oil. (186:85) **Okanagan-Colville** *Bread & Cake* Bulbs dried into cakes and stored for winter use. *Spice* Bulbs dried into cakes and used as seasoning in meat soups. *Unspecified* Bulbs eaten raw or boiled alone or with saskatoon berries. (188:46) **Okanagon** *Staple* Roots used as a principal food. (178:238) *Unspecified* Roots used as an important food. (178:237) Roots used extensively for food. (178:89) **Quileute** *Unspecified* Corms steamed and eaten. **Quinault** *Unspecified* Corms steamed and eaten. **Samish** *Unspecified* Corms steamed and eaten. (79:25) **Shuswap** *Unspecified* Roasted roots used for food. (123:54) Roots used extensively for food. (178:89) **Skagit** *Unspecified* Corms steamed and eaten. (79:25) **Skagit, Upper** *Unspecified* Bulbs baked or steamed in an earth oven and eaten. (179:40) **Skokomish** *Unspecified* Corms steamed and eaten. **Swinomish** *Unspecified* Corms steamed and eaten. (79:25) **Thompson** *Dried Food* Pit cooked bulbs dried for future use and usually cooked with meat. *Soup* Bulbs used to make a soup like clam chowder. A vegetable soup was made with salmon heads, bitterroot, tiger lily bulbs, water horehound roots, chocolate lily bulbs, the "dry" variety of saskatoon berries, dried powdered bracken fern rhizome, and chopped wild onions. *Spice* Thick, scaly bulbs eaten mainly as a condiment or cooked with food to add a pepper-like flavoring. (187:126) *Unspecified* Bulbs mixed with salmon roe and panther lily, boiled, and eaten as a favorite dish. Thick, scaly bulbs mixed with salmon roe, boiled, and eaten as a favorite dish. (as *L. parviflorum* 164:482) Roots used as an important food. (178:237) Roots used extensively for food. (178:89)

Lilium occidentale Purdy, Eureka Lily
- **Food**—**Karok** *Unspecified* Bulbs baked in the earth oven and eaten. (148:381)

Lilium pardalinum Kellogg, Leopard Lily
- **Food**—**Atsugewi** *Unspecified* Bulbs cooked in earth oven and used for food. (61:138) **Karok** *Unspecified* Bulbs baked in the earth oven and eaten. (148:381) **Yana** *Unspecified* Roots steamed and eaten. (147:251)

Lilium parvum Kellogg, Sierran Tiger Lily
- **Food**—**Paiute** *Unspecified* Roots used for food. (167:244)

Lilium philadelphicum L., Wood Lily
- **Drug**—**Algonquin, Quebec** *Gastrointestinal Aid* Root used for stomach disorders. (18:138) **Chippewa** *Dermatological Aid* Poultice of boiled bulbs applied to wounds, contusions, and dog bites. *Witchcraft Medicine* Poultice of bulbs applied to dog bites and caused dog's fangs to drop out. (71:125) **Iroquois** *Gynecological Aid* Decoction of whole plant taken "to bring away placenta after childbirth." *Love Medicine* "Dry plants in sun, if twists together, wife is unfaithful; determines love." Decoction of roots taken by wife as emetic and used as a wash "if husband is unfaithful." (87:282) **Malecite** *Adjuvant* Roots used to strengthen other medicines. (116:245) *Cough Medicine* Roots used with roots of blackberry and mountain raspberry, staghorn sumac for coughs. (116:251) *Dermatological Aid* Poultice of ground roots used for swellings and bruises. (116:245) *Febrifuge* Roots used with roots of blackberry and mountain raspberry, staghorn sumac for fevers. *Tuberculosis Remedy* Roots used with roots of blackberry and mountain raspberry, staghorn sumac for consumption. (116:251) **Menominee** *Dermatological Aid* Poultice of boiled, mashed root applied to sores. (54:132) **Micmac** *Cough Medicine* Roots used for coughs. *Dermatological Aid* Roots used for swellings and bruises. *Febrifuge* Roots used for fever. *Tuberculosis Remedy* Roots used for consumption and fever. (40:58)
- **Food**—**Blackfoot** *Soup* Bulbs eaten with soup. *Unspecified* Bulbs eaten fresh. (86:103) **Meskwaki** *Vegetable* Straight roots gathered for potatoes. (152:262)

Lilium philadelphicum var. *andinum* (Nutt.) Ker-Gawl., Wood Lily
- **Drug**—**Dakota** *Dermatological Aid* Pulverized or chewed flowers applied as antidote for spider bites. (as *L. umbellatum* 70:71)
- **Food**—**Cree, Woodlands** *Snack Food* Bulb segments eaten dried as a nibble. *Unspecified* Bulb segments eaten fresh. Seeds and underground bulbs used for food. (109:43)

Lilium rubescens S. Wats., Redwood Lily
- **Other**—**Karok** *Decorations* Used for bouquets. (6:34)

Limonium, Plumbaginaceae

Limonium californicum (Boiss.) Heller, California Sealavender
- **Drug**—**Costanoan** *Blood Medicine* Decoction of plant used to clean the blood. *Respiratory Aid* Powdered plant placed in nostrils to cause sneezing for congestion. *Urinary Aid* Decoction of plant used for internal injuries or urinary problems. *Venereal Aid* Decoction of plant used for venereal disease. (21:11)

Limonium carolinianum (Walt.) Britt., Carolina Sealavender
- **Drug**—**Micmac** *Tuberculosis Remedy* Roots pounded, ground, added to boiling water, and used for consumption with hemorrhage. (116:259)

Limonium vulgare P. Mill., Mediterranean Sealavender
- **Drug**—**Micmac** Roots used for consumption with hemorrhage. (as *Statice limonium* 40:62)

Limosella, Scrophulariaceae

Limosella aquatica L., Water Mudwort
- **Drug**—**Navajo, Ramah** *Ceremonial Medicine* Leaves ceremonially rubbed on body for protection in hunting and from witches. *Dermatological Aid* Roll of washed leaves used to plug a bullet or arrow wound. *Hunting Medicine* Leaves ceremonially rubbed on body for

protection while hunting. *Witchcraft Medicine* Leaves rubbed on body ceremonially for protection against witches. (191:44)

Linaceae, see *Linum*

Linanthus, Polemoniaceae

Linanthus ciliatus (Benth.) Greene, Whiskerbrush
- *Drug*—Pomo, Calpella *Blood Medicine* Cold decoction of plant taken to purify the blood. *Cold Remedy* Infusion of plant given to children for colds. *Cough Medicine* Infusion of plant given to children for coughs. *Pediatric Aid* Infusion of plant given to children for coughs and colds. (41:381)
- *Food*—Yuki *Substitution Food* Flowering heads used in the summer as a substitute for coffee. (41:381)

Linanthus nuttallii (Gray) Greene ex Milliken **ssp. *nuttallii***, Nuttall's Desert Trumpets
- *Drug*—Navajo, Ramah *Panacea* Plant used as "life medicine." (191:40)

Linaria, Scrophulariaceae

Linaria vulgaris P. Mill., Butter and Eggs
- *Drug*—Iroquois *Antidiarrheal* Cold infusion of leaves taken for diarrhea. *Emetic* Infusion of plants taken to vomit as an anti-love medicine and remove bewitching. *Love Medicine* Compound infusion of smashed plants taken to vomit as an anti-love medicine. *Pediatric Aid* and *Sedative* Compound infusion of plants and flowers given to babies that cry too much. *Witchcraft Medicine* Compound infusion of smashed plants taken to vomit and remove bewitching. (87:433) **Ojibwa** *Herbal Steam* Compound containing plant used as a bronchial inhalant in the sweat lodge. *Respiratory Aid* Compound containing dried plant used as a bronchial inhalant in the sweat lodge. (153:389)

Lindera, Lauraceae

Lindera benzoin (L.) Blume, Northern Spicebush
- *Drug*—Cherokee *Abortifacient* Taken for female obstructions. *Blood Medicine* Taken for the blood and female obstructions. *Cold Remedy* Taken for colds. *Cough Medicine* Taken for coughs. *Dermatological Aid* Infusion of bark taken for "bold hives." *Diaphoretic* Compound decoction of bark taken as a diaphoretic, any part is a diaphoretic. *Misc. Disease Remedy* Infusion of bark taken to break out measles. *Pulmonary Aid* Taken for croup. *Respiratory Aid* Taken for phthisic. *Tonic* Used for "white swellings" and infusion taken as a spring tonic. (80:56) **Creek** *Analgesic* and *Diaphoretic* Infusion of branches taken and steam bath used to cause perspiring for aches. *Emetic* Plant used as an emetic. (177:24) **Iroquois** *Cold Remedy* Decoction or infusion of stripped leaves and twigs taken for colds. *Febrifuge* Compound decoction of plants used as a steam bath for cold sweats. *Misc. Disease Remedy* Decoction of stripped leaves and twigs taken for colds and measles. *Panacea* Compound decoction of roots taken as a panacea. (87:335) *Venereal Aid* Compound decoction of roots taken for gonorrhea and syphilis. Compound poultice of bark applied to lumps that remain after having syphilis. (87:334)
- *Food*—Cherokee *Beverage* Used to make a beverage. (80:56) Stems used to make tea. (126:44) *Spice* Used to flavor opossum or ground hog. (80:56)

Lindera benzoin (L.) Blume **var. *benzoin***, Northern Spicebush
- *Drug*—Chippewa *Unspecified* Leaves used medicinally. (as *Benzoin aestivale* 71:131) **Creek** *Antirheumatic (Internal)* Infusion of branches taken or used as herbal steam to cause sweating for pains. *Blood Medicine* Compound containing plant taken to cause vomiting which purifies the blood. *Diaphoretic* Infusion of branches taken or used as herbal steam to cause sweating for pains. *Emetic* Compound containing plant taken to cause vomiting which purifies the blood. *Herbal Steam* Infusion of branches used as herbal steam for aches and pains. (as *Benzoin aestivale* 172:657) **Rappahannock** *Abortifacient* Infusion taken to correct delayed menses. *Gynecological Aid* Infusion taken for menstruation pains. (as *Benzoin aestivale* 161:33)
- *Food*—Chippewa *Beverage* Leaves used to make a pleasant, tea-like beverage. *Spice* Leaves used as a flavor for masking or modifying the taste of naturally strong flavored meats. (as *Benzoin aestivale* 71:131)

Lindera sp., Spice Wood
- *Drug*—Mohegan *Anthelmintic* Chewed or infusion of leaves used by children for worms. (as *Benzoin* 174:265) Fresh leaves chewed or infusion taken by children and adults as a vermifuge. (as *Benzoin* 176:70) *Pediatric Aid* Chewed or infusion of leaves used by children for worms. (as *Benzoin* 174:265) Fresh leaves chewed or infusion taken by children and adults as a vermifuge. (as *Benzoin* 176:70)

Linnaea, Caprifoliaceae

Linnaea borealis L., Twinflower
- *Drug*—Montagnais *Orthopedic Aid* Mashed plant used for "inflammation of the limbs." (156:314) **Tanana, Upper** *Analgesic* Poultice of the whole plant applied to the head for headaches. *Pediatric Aid* and *Psychological Aid* Poultice of the whole plant applied to the child's head to insure him a long life. (102:18) **Thompson** *Unspecified* Decoction of plant used as a medicine for unspecified purpose. (164:458)
- *Food*—Carrier *Unspecified* Species used for food. (31:74)

Linnaea borealis ssp. *longiflora* (Torr.) Hultén, Longtube Twinflower
- *Drug*—Algonquin, Quebec *Gynecological Aid* Infusion of entire plant used for menstrual difficulties. Infusion of entire plant used by pregnant women to insure the good health of the child. (18:235) **Iroquois** *Febrifuge* Decoction of twigs given to children with fever. *Gastrointestinal Aid* Decoction of twigs given to children with cramps. *Pediatric Aid* Decoction of twigs given to children with cramps, fever, or for crying. *Sedative* Decoction of twigs given to children for crying. (87:444) **Potawatomi** *Gynecological Aid* Entire plant used for unspecified female troubles. (154:45, 46) **Snohomish** *Cold Remedy* Decoction of leaves used for colds. (79:47)

Linum, Linaceae

Linum australe Heller, Southern Flax
- *Drug*—Hopi *Gastrointestinal Aid* Infusion of plant taken for stomach disorders. (200:83) *Gynecological Aid* Decoction of plant taken and used as a wash to ease protracted labor. (200:36, 83)

Linum australe Heller **var. *australe***, Southern Flax
- *Drug*—Navajo, Kayenta *Kidney Aid* Infusion of plant used for kidney disease. (as *L. aristatum* var. *australe* 205:30)

Linum lewisii Pursh, Prairie Flax
- **Drug**—**Gosiute** *Dermatological Aid* Infusion of flax used for bruise swellings. (39:349) Poultice of plant applied to bruises for the swelling. (39:373) **Great Basin Indian** *Eye Medicine* Seeds used to make an eye medicine. (121:48) **Navajo, Kayenta** *Analgesic* Plant used for headaches. *Disinfectant* Plant used as a fumigant. (205:30) **Navajo, Ramah** *Gastrointestinal Aid* Decoction of leaves taken for heartburn. (191:33, 34) **Okanagon** *Dermatological Aid* and *Pediatric Aid* Infusion of flowers, leaves, and stem used as skin and hair wash by young females. (125:42) **Paiute** *Carminative* Infusion of whole stem used for gas. (118:44) *Dermatological Aid* Poultice of leaves alone or stems and leaves applied to swellings. *Eye Medicine* Infusion or decoction of plant parts used as an eyewash. (180:101, 102) *Gastrointestinal Aid* Infusion of whole stem used for a disordered stomach. (118:44) *Poultice* Poultice of leaves applied for goiter. (180:101, 102) *Unspecified* Infusion of roots used as an medicinal tea. (121:48) **Shoshoni** *Dermatological Aid* Poultice of leaves applied to swellings. (180:101, 102) *Eye Medicine* Infusion of root used as an eye medicine. (118:39) Infusion or decoction of plant parts used as an eyewash. *Liver Aid* Poultice of crushed leaves applied for "gall trouble." (180:101, 102) **Thompson** *Dermatological Aid* Infusion of flowers, leaves, and stem used as skin and hair wash by young females. (125:42) Infusion of flowers, leaves, and stems used as wash for adolescents' skin and hair. (164:467) Decoction of stems and flowers used as wash by girls for beautiful hair and face. (164:507) Decoction of whole plant with the roots used to wash the hair and scalp if hair loss occurred. (187:234) *Pediatric Aid* Infusion of flowers, leaves, and stem used as skin and hair wash by young females. (125:42)
- **Food**—**Dakota** *Fodder* Seeds used to flavor feed. (121:48) *Unspecified* Seeds used in cooking for the nutritive value and agreeable flavor. **Omaha** *Unspecified* Seeds used in cooking for the nutritive value and agreeable flavor. **Pawnee** *Unspecified* Seeds used in cooking for the nutritive value and agreeable flavor. **Ponca** *Unspecified* Seeds used in cooking for the nutritive value and agreeable flavor. **Winnebago** *Unspecified* Seeds used in cooking for the nutritive value and agreeable flavor. (70:98)
- **Fiber**—**Great Basin Indian** *Cordage* Roots and stems used to make string. (121:48) **Klamath** *Basketry* Stems made into strings and cords used to make baskets. *Cordage* Stems fiber used to make strings and cords. *Mats, Rugs & Bedding* Stems made into strings and cords used to make mats. *Snow Gear* Stem made into strings and cords used to make mesh on snowshoes. (45:99) **Montana Indian** *Basketry* Bark fibers used in baskets. *Cordage* Bark fibers used for cordage. *Mats, Rugs & Bedding* Bark fibers used as the warp for mats. *Snow Gear* Bark fibers used as the mesh for snowshoes. (19:14)
- **Other**—**Klamath** *Hunting & Fishing Item* Stems made into strings and cords used for weaving fish nets. (45:99) **Montana Indian** *Hunting & Fishing Item* Bark fibers used in fish nets. (19:14) **Thompson** *Soap* Flowers, leaves, and stems soaked in cold water and used by girls at puberty as a head and face wash. (164:504)

Linum puberulum (Engelm.) Heller, Plains Flax
- **Drug**—**Apache, White Mountain** *Eye Medicine* Berry juice used as an eye medicine. (136:158) **Navajo, Ramah** *Gastrointestinal Aid* Decoction of leaves taken for heartburn. Infusion of plant taken to kill a swallowed red ant. *Panacea* Plant used as "life medicine." (191:35) **Zuni** *Eye Medicine* Berry juice squeezed into eye for inflammation. (166:56)
- **Other**—**Keres, Western** *Paint* Flowers made into yellow paint. (as *Cathartolinum puberulum* 171:35)

Linum rigidum Pursh, Stiffstem Flax
- **Drug**—**Keres, Western** *Psychological Aid* Infusion of plant used by racers to make them speedy. (171:52)

Linum usitatissimum L., Common Flax
- **Drug**—**Cherokee** *Cold Remedy* and *Cough Medicine* Taken for "violent colds, coughs and diseases of lungs." *Febrifuge* Decoction poured over body for "fever attacks." (80:34) Decoction of plant poured over patient for fevers. (177:34) *Pulmonary Aid* Taken for "violent colds, coughs and diseases of lungs." *Urinary Aid* Seeds used for "gravel" or burning during urination. (80:34)

Liparis, Orchidaceae
Liparis loeselii (L.) L. C. Rich., Yellow Widelip Orchid
- **Drug**—**Cherokee** Compound infusion of root taken for urinary problems. (80:59)

Liquidambar, Hamamelidaceae
Liquidambar styraciflua L., Sweet Gum
- **Drug**—**Cherokee** *Antidiarrheal* Rosin or inner bark used for diarrhea, flux, and dysentery. *Dermatological Aid* Salve used for wounds, sores, and ulcers and mixed with sheep or cow tallow for itch. *Gynecological Aid* Infusion of bark taken to stop flooding. *Other* Gum used as drawing plaster and compound infusion of bark taken for bad disease. *Sedative* Infusion of bark given to nervous patients. (80:58) Gum used as a "drawing plaster" and infusion of inner bark used for nervous patients. (203:74) **Choctaw** *Dermatological Aid* Compound decoction of root used as a dressing for cuts and wounds. (25:23) Decoction of plant used as a poultice for cuts and bruises. (177:26) **Houma** *Dermatological Aid* Decoction of root applied to skin sores thought to be caused by worms. *Diaphoretic* and *Febrifuge* Decoction of Spanish moss from this tree taken as a diaphoretic for fever. (158:61, 62) **Koasati** *Other* Decoction of bark taken for "night sickness." (177:26) **Rappahannock** *Antidiarrheal* Compound infusion with dried bark taken for dysentery. (161:34) *Veterinary Aid* Rolled, hardened sap placed in dog's nose for distemper. (161:27)
- **Food**—**Cherokee** *Beverage* Bark, hearts-a-bustin-with-love (*Euonymus americana*), and summer grapes used to make tea. *Candy* Hardened gum used for chewing gum. (80:58)

Liriodendron, Magnoliaceae
Liriodendron tulipifera L., Tuliptree
- **Drug**—**Cherokee** *Anthelmintic* Infusion of bark taken for pinworms. Used for cholera infantum and infusion of bark given for pinworms. *Antidiarrheal* Bark used for cholera infantum, "dyspepsy, dysentery and rheumatism." *Antirheumatic (Internal)* Bark used for "dyspepsy, dysentery and rheumatism." *Cough Medicine* Bark used in cough syrup. *Dermatological Aid* Decoction blown onto wounds and boils. *Febrifuge* Infusion of root bark taken for fever. (80:50) Infusion of root bark taken for fevers. (203:74) *Gastrointestinal Aid* Bark used for

"dyspepsy, dysentery and rheumatism." Compound used as steam bath for indigestion or biliousness. *Misc. Disease Remedy* Given for cholera infantum and infusion of bark given for pinworms. *Orthopedic Aid* Decoction blown onto fractured limbs. *Pediatric Aid* Given for cholera infantum and infusion of bark given for pinworms. *Poultice* Used as a poultice and infusion of bark given for pinworms. (80:50) Infusion of root bark used in poultices. (203:74) *Sedative* Used for "women with hysterics and weakness." *Snakebite Remedy* Decoction used as a wash for snakebite. *Stimulant* Used for "women with hysterics and weakness." (80:50) **Rappahannock** *Analgesic* Poultice of bruised leaves bound to head for neuralgic pain. *Love Medicine* Raw, green bark chewed as a sex invigorant. *Stimulant* Raw, green bark chewed as a stimulant. (161:25)
- *Food*—**Cherokee** *Sauce & Relish* Used to make honey. (80:50)
- *Fiber*—**Cherokee** *Building Material* Wood used for lumber. *Canoe Material* Wood used to make 30- to 40-foot-long canoes. *Furniture* Wood used to make cradles. (80:50)
- *Other*—**Cherokee** *Paper* Wood used for pulpwood. (80:50)

Lithocarpus, Fagaceae
Lithocarpus densiflorus (Hook. & Arn.) Rehd., Tan Oak
- *Drug*—**Costanoan** *Dermatological Aid* Infusion of bark used as a wash for face sores. *Toothache Remedy* Infusion of bark held in the mouth to tighten loose teeth. (21:20) **Pomo, Kashaya** *Cough Medicine* Acorns, the tannin soothed the cough, used as cough drops. (72:83) **Yurok** *Strengthener* Acorn mush taken by old people on their death bed to survive the day. (6:35)
- *Food*—**Costanoan** *Unspecified* Acorns used for food. (21:248) **Hahwunkwut** *Bread & Cake* Acorns used to make bread. *Porridge* Acorns used to make mush. *Staple* Acorns used to make a meal. (117:187) **Hupa** *Bread & Cake* Acorns used to make bread, biscuits, pancakes, and cake. *Porridge* Acorns used to make mush. *Staple* Acorns used to make meal. *Unspecified* Acorns roasted and eaten. (117:200) **Karok** *Bread & Cake* Acorn paste made into patties and baked in hot coals.[44] *Porridge* Acorn flour used to make paste and gruel and flavored with venison and herbs.[44] (6:35) Acorns shelled, dried, pounded into a meal, leached, and used to make gruel. (148:382) *Staple* Acorns considered the main staple.[44] Acorns used to make flour.[44] *Winter Use Food* Acorn flour stored in large storage baskets.[44] (6:35) Acorns stored for winter use. (148:382) **Mendocino Indian** *Unspecified* Acorns leached and used for food. (as *Quercus densiflora* 41:342) **Poliklah** *Bread & Cake* Acorns used to make bread. (117:172) *Porridge* Acorns used to make mush. (117:170, 172) *Staple* Acorns form one of the principal foods. (117:168) **Pomo** *Bread & Cake* Acorns used to make black bread. (as *Quercus densiflora* 11:67) Acorns used to make bread. (117:290) *Porridge* Acorns used to make mush and gruel. (as *Quercus densiflora* 11:67) Moldy acorns mixed with whitened dried acorns and made into a mush. Leached acorns used for mush. (66:12) Acorns used to make mush. (117:290) *Soup* Acorns used to make soup. (as *Quercus densiflora* 11:67) Leached acorns used for soup. *Unspecified* Pulverized, leached acorns used for food. (66:12) **Pomo, Kashaya** *Dried Food* Acorns sun dried before storing. *Forage* Acorns collected by woodpeckers. *Porridge* Acorns used as flour for pancakes, bread, mush, or soup.[45] (72:83) **Shasta** *Bread & Cake* Acorns pounded, winnowed, leached, and made into bread. *Porridge* Acorns pounded, winnowed, leached and made into mush. *Soup* Acorns pounded, winnowed, leached, and made into thin soup. *Staple* Acorns used as the basic staple. (as *Quercus densiflora* 93:308) **Tolowa** *Staple* Acorns considered the main staple. (6:35) **Yuki** *Bread & Cake* Acorns used to make pancakes. *Porridge* Acorns used to make mush. *Soup* Acorns used to make soup. (49:88) **Yurok** *Bread & Cake* Acorns used to make dough. *Soup* Acorns used to make soup. *Staple* Acorns considered the main staple. Acorns leached and ground into flour. (6:35) **Yurok, South Coast (Nererner)** *Staple* Acorns form one of the principal foods. (117:168)
- *Dye*—**Costanoan** *Unspecified* Bark used to prepare dye. (21:248) **Tolowa** *Unspecified* Bark used to dye baskets and fishing nets so the fish could not see them. (6:35)
- *Other*—**Pomo, Kashaya** *Ceremonial Items* Acorns used in a first fruits ceremony in October after the first rainfall. Bark used by a wale-pu (a ceremonial figure) as tinder to create flashes of light. *Musical Instrument* Strung acorns twirled in a special way to make music. (72:83)

Lithophragma, Saxifragaceae
Lithophragma affine Gray, Common Woodlandstar
- *Drug*—**Mendocino Indian** *Cold Remedy* Root chewed for colds. *Gastrointestinal Aid* Root chewed for stomachaches. (as *Tellima affinis* 41:353)

Lithospermum, Boraginaceae
Lithospermum canescens (Michx.) Lehm., Hoary Puccoon
- *Drug*—**Menominee** *Sedative* Compound infusion taken and rubbed on body to quiet person near convulsions. (54:128)
- *Food*—**Omaha** *Cooking Agent* Root chewed with gum by children, to color it red. Flowers chewed with gum by children, to color it yellow. **Ponca** *Cooking Agent* Root chewed with gum by children, to color it red. Flowers chewed with gum by children, to color it yellow. (70:111)
- *Other*—**Menominee** *Sacred Items* White, ripened seed used as a sacred bead in the Midewewin ceremony. (151:80)

Lithospermum caroliniense (Walt. ex J. F. Gmel.) MacM., Hairy Puccoon
- *Drug*—**Lakota** *Pulmonary Aid* Root powder taken for chest wounds. (139:40)
- *Dye*—**Chippewa** *Red* Dried or pulverized roots boiled and used to make a red dye. (53:371)
- *Other*—**Chippewa** *Paint* Used for face paint. (53:377)

Lithospermum incisum Lehm., Narrowleaf Gromwell
- *Drug*—**Blackfoot** *Ceremonial Medicine* Dried tops burned as incense in ceremonials. (as *L. linearifolium* 114:277) **Cheyenne** *Orthopedic Aid* Finely ground leaves rubbed on paralyzed part. (as *L. linearifolium* 75:40, 41) Leaf, root, and stem powder rubbed on body for paralysis. *Psychological Aid* Infusion of roots, leaves, and stems rubbed on head and face for irrational behavior from any illness. (as *L. linearifolium* 76:185) Infusion of stems, leaves, and roots used as a wash for "irrationalness." (as *L. linearifolium* 83:15) *Sedative* Infusion of plant parts rubbed on face when irrational from illness. (as *L. linearifolium* 76:185) Decoction of root rubbed on one who was "irrational by reason of illness." *Stimulant* Plant chewed, then spit and blown in

face or rubbed on chest as a stimulant. (as *L. linearifolium* 75:40, 41) Chewed plant spit and blown into face and rubbed over the heart by the doctor for sleepiness. (as *L. linearifolium* 76:185) Chewed plant spit and blown onto face to keep a very sleepy person awake. (as *L. linearifolium* 83:15) **Great Basin Indian** *Unspecified* Roots used medicinally for unspecified purpose. (as *L. angustifolium* 121:50) **Hopi** *Antihemorrhagic* Plant used for hemorrhages. *Blood Medicine* Plant used for building up the blood. (42:331) *Unspecified* Used as a medicinal plant. Used as a medicinal plant. (as *L. linearifolium* 190:165) **Navajo** *Cold Remedy* Plant chewed for colds. (as *L. angustifolium* 55:71) Plant used for colds. *Contraceptive* Plant used as an oral contraceptive. (90:161) *Cough Medicine* Plant chewed for coughs. (as *L. angustifolium* 55:71) Plant used for coughs. (as *L. angustifolium* 90:161) *Dermatological Aid* Roots used for soreness at the attachment of the umbilical cord. *Pediatric Aid* Roots used for soreness at the attachment of the umbilical cord. (as *L. angustifolium* 55:71) **Navajo, Ramah** *Eye Medicine* Cold infusion of pulverized root and seeds used as an eyewash. *Panacea* Root used as a "life medicine" and considered a "big medicine." (191:41) **Sioux, Teton** *Pulmonary Aid* Fragrant herb used for lung hemorrhages. (as *L. linearifolium* 52:269, 270) **Zuni** *Ceremonial Medicine* Salve of powdered root applied ceremonially to swelling of any body part. (as *L. linearifolium* 166:56) *Dermatological Aid* Powdered root mixed with bum branch resin and used for abrasions and skin infections. (27:374) Poultice of root used and decoction of plant taken for swelling. (as *L. linearifolium* 166:56) *Gastrointestinal Aid* Infusion of root taken for stomachache. *Kidney Aid* Infusion of root taken for kidney problems. (27:374) *Throat Aid* Poultice of root used and decoction of plant taken for sore throat. (as *L. linearifolium* 166:56)
- *Food*—**Blackfoot** *Beverage* Roots used to make tea. (as *L. angustifolium* 121:50) *Unspecified* Roots eaten boiled or roasted. (as *L. linearifolium* 114:278) **Okanagon** *Unspecified* Plants boiled and used for food. (as *Lithosperum angustofolium* 125:37) **Shoshoni** *Beverage* Roots used to make tea. (as *Lithospermum angustifolium* 121:50) **Thompson** *Unspecified* Plants boiled and used for food. (as *Lithosperum angustofolium* 125:37) Root cooked and eaten. (as *Lithospermum angustifolium* 164:480)
- *Dye*—**Great Basin Indian** *Blue* Roots used to make a blue dye. (as *L. angustifolium* 121:50) **Montana Indian** *Purple* Root used to produce a violet-colored dye. (as *L. angustifolium* 19:14)
- *Other*—**Blackfoot** *Ceremonial Items* Plant top dried and burned ceremonially. (as *L. linearifolium* 118:50) **Navajo** *Ceremonial Items* Used in the Life or Knife Chant. (as *L. angustifolium* 55:71) **Okanagon** *Toys & Games* Seeds used as beads by children. (as *Lithosperum angustofolium* 125:37) **Thompson** *Paint* Roots dipped in hot grease and used as a red paint on the face and on dressed skins. (as *Lithospermum angustifolium* 164:502) *Toys & Games* Seeds used as beads by children. (as *Lithosperum angustofolium* 125:37) **Zuni** *Weapon* Leaves bound to arrow shafts, close to the point, obscured by sinew wrapping and used in wartime. The leaves were said to be so deadly poisonous that they would cause the immediate death of anyone pierced by them. (as *Lithospermum linearifolium* 166:93)

Lithospermum multiflorum Torr. ex Gray, Manyflowered Gromwell

- *Drug*—**Navajo, Ramah** *Panacea* Root used as a "life medicine" and considered a "big medicine." (191:41)
- *Food*—**Gosiute** *Unspecified* Seeds formerly used for food. (39:373)

Lithospermum officinale L., European Gromwell

- *Drug*—**Iroquois** *Diuretic* and *Pediatric Aid* Infusion of dried, powdered seeds given to children as a diuretic. (141:56)

Lithospermum ruderale Dougl. ex Lehm., Western Gromwell

- *Drug*—**Cheyenne** *Analgesic* Poultice of dried, powdered leaves and stems applied for rheumatic pains. *Antirheumatic (External)* Poultice of dried, powdered leaves and stems applied for rheumatic pains. (76:185) Poultice of dried, pulverized leaves and stems applied for rheumatic pains. (76:185) **Gosiute** *Diuretic* Infusion or decoction of roots used as a diuretic. (as *L. pilosum* 39:351) Roots used as a diuretic for kidney troubles. (as *L. pilosum* 39:373) *Kidney Aid* Infusion or decoction of roots used for kidney troubles. (as *L. pilosum* 39:351) Roots used as a diuretic for kidney troubles. (as *L. pilosum* 39:373) **Navajo** *Contraceptive* Plant used as an oral birth control. (97:51) **Okanagan-Colville** *Antihemorrhagic* Infusion of roots taken for internal hemorrhaging. (188:91) **Shoshoni** *Antidiarrheal* Infusion or decoction of root taken for diarrhea. (180:102) *Contraceptive* Plant used as an oral birth control. (97:51) Infusion of root taken for 6 months as a permanent birth control. (118:46) Cold water infusion of root taken daily for 6 months as a contraceptive. (180:102) **Shuswap** *Dermatological Aid* Plant used for sores. *Disinfectant* Decoction of plant used as a bath to "clean the germs off the body." (123:60) **Thompson** *Witchcraft Medicine* Root used to "inflict sickness or bad luck on persons." (as *L. pilosum* 164:508) **Ute** *Diuretic* Decoction of roots used as a diuretic. (as *L. pilosum* 38:35)
- *Food*—**Gosiute** *Unspecified* Seeds formerly used for food. (as *L. pilosum* 39:373)
- *Other*—**Blackfoot** *Toys & Games* Stems used by children to make headpieces for playing and mimicking the affairs of the Holy Woman. (86:114) **Okanagan-Colville** *Good Luck Charm* Plant rubbed on fishing line for good luck. *Water Indicator* Plant used as a charm to make it rain. (188:91) **Thompson** *Decorations* Hard, white, shiny seeds formerly used to make beads. *Protection* Plant used as a charm to stop a thunderstorm. (187:192)

Lithospermum sp.

- *Food*—**Pima** *Unspecified* Leaves eaten raw. (146:77)
- *Other*—**Keres, Western** *Paint* Ground flowers used to make yellow paint. (171:52)

Loasaceae, see *Mentzelia*

Lobelia, Campanulaceae

Lobelia cardinalis L., Cardinal Flower

- *Drug*—**Cherokee** *Analgesic* Compound given for pain and poultice of crushed leaves used for headache. *Anthelmintic* Infusion of root taken for worms. *Antirheumatic (Internal)* Infusion given for rheumatism. *Cold Remedy* Infusion of leaf taken for colds. *Dermatological Aid* Poultice of root used for "risings" and infusion used for "sores hard to heal." *Febrifuge* Infusion of leaf taken for fever. *Gastrointestinal Aid* Infusion of root taken for stomach trouble. *Hemostat* Cold infusion "snuffed" for nosebleed. *Pulmonary Aid* Used for croup.

Venereal Aid Used for syphilis. (80:28) **Delaware** *Misc. Disease Remedy* Infusion of roots used for typhoid. (176:34) **Delaware, Oklahoma** *Misc. Disease Remedy* Infusion of root, considered to be very strong medicine, taken for typhoid. (175:28, 76) **Iroquois** *Adjuvant* Plant strengthened all medicines. (87:453) *Analgesic* Infusion of roots taken or poultice applied for pain. *Anticonvulsive* Compound decoction of plants taken by women for epilepsy. (87:452) *Basket Medicine* Infusion of stalks and flowers used as wash for baskets, a "basket medicine." (87:453) *Dermatological Aid* Decoction of roots used as a wash and poultice applied to chancres and fever sores. (87:452) *Febrifuge* Plant used as a fever medicine. *Gastrointestinal Aid* Compound decoction of plants taken for bad stomach caused by consumption. (87:453) *Gynecological Aid* Decoction or infusion taken and poultice or wash used for breast troubles. (87:452) *Love Medicine* Decoction of plants or infusion of roots used as a wash for love medicine. (87:453) *Other* Compound infusion of plants taken for stricture caused by menstruating women. Decoction of roots, plants, and blossoms taken for cramps. (87:452) Infusion of roots taken or washed on injured parts, "Little Water Medicine." *Panacea* Plant used for every ailment. (87:453) *Psychological Aid* Compound decoction of whole plant taken for sickness caused by grieving. (87:452) *Tuberculosis Remedy* Compound decoction of plants taken for bad stomach caused by consumption. (87:453) *Witchcraft Medicine* Infusion of roots taken or poultice applied for trouble caused by witchcraft. (87:452) **Meskwaki** *Love Medicine* Roots used as a love medicine. (152:231) Ground roots used in food to end quarrels, a love medicine and anti-divorce remedy. (152:273) **Pawnee** *Love Medicine* Compound containing roots and flowers used as a love charm. (70:129)
- *Other*—**Meskwaki** *Ceremonial Items* Ceremonial "tobacco" not smoked, but used to ward off storms and strewn onto graves. (152:273)

Lobelia cardinalis ssp. *graminea* var. *propinqua* (Paxton) Bowden, Cardinal Flower
- *Drug*—**Zuni** *Antirheumatic (External)* Ingredient of "schumaakwe cakes" and used externally for rheumatism. *Dermatological Aid* Ingredient of "schumaakwe cakes" and used externally for swelling. (as *L. splendens* 166:56)
- *Other*—**Jemez** *Ceremonial Items* Flowers used in the rain dance. (as *L. splendens* 44:25)

Lobelia inflata L., Indian Tobacco
- *Drug*—**Cherokee** *Analgesic* Poultice of root used for body aches and leaves rubbed on aches and stiff neck. *Dermatological Aid* Used for bites and stings and roots and leaves used on boils and sores. *Emetic* Plant used as a strong emetic. *Gastrointestinal Aid* Tincture in small doses prevented colics. *Orthopedic Aid* Poultice of root used for body aches and leaves rubbed on aches and stiff neck. *Other* Smoked "to break tobacco habit." *Pulmonary Aid* Given for croup and tincture in small doses prevented croup. *Respiratory Aid* Plant taken for "asthma and phthisic." *Throat Aid* Chewed for sore throat. (80:40) **Crow** *Ceremonial Medicine* Used in religious ceremonies. (19:14, 15) **Iroquois** *Cathartic* Cold infusion of whole plant taken as a physic. (87:455) *Dermatological Aid* Infusion of roots or leaves used as a wash or poultice on abscesses or sores. *Emetic* Infusion of plant taken to vomit and cure tobacco or whisky habit. (87:454) Cold infusion of whole plant taken as an emetic. (87:455) *Love Medicine* Infusion of plant taken as a love or anti-love medicine. (87:454) *Other* Cold infusion of whole plant used as a divining agent. (87:455) *Psychological Aid* Infusion of plant taken to vomit and cure tobacco or whisky habit. *Venereal Aid* Infusion of smashed roots used as wash and poultice for venereal disease sores. *Witchcraft Medicine* Decoction of plant taken to counteract sickness produced by witchcraft. (87:454)
- *Other*—**Cherokee** *Insecticide* Used to smoke out gnats. (80:40)

Lobelia kalmii L., Ontario Lobelia
- *Drug*—**Cree, Hudson Bay** *Emetic* Plant used as an emetic. (92:303) **Iroquois** *Dermatological Aid* Infusion of smashed plants used as drops for abscesses. *Ear Medicine* Infusion of smashed plants used as drops for earaches. *Emetic* Infusion of plants taken to vomit to remove the effect of a love medicine. *Love Medicine* Infusion of plants taken to vomit to remove the effect of a love medicine. (87:455)

Lobelia siphilitica L., Great Blue Lobelia
- *Drug*—**Cherokee** *Analgesic* Compound given for pain and poultice of crushed leaves used for headache. *Anthelmintic* Infusion of root taken for worms. *Antirheumatic (Internal)* Infusion given for rheumatism. *Cold Remedy* Infusion of leaf taken for colds. *Dermatological Aid* Poultice of root used for "risings" and infusion used for "sores hard to heal." *Febrifuge* Infusion of leaf taken for fever. *Gastrointestinal Aid* Infusion of root taken for stomach trouble. *Hemostat* Cold infusion "snuffed" for nosebleed. *Pulmonary Aid* Used for croup. *Venereal Aid* Used for syphilis. (80:28) **Iroquois** *Cough Medicine* Plant used as a gargle for coughs. (87:454) *Witchcraft Medicine* Infusion of smashed plants taken for anti-bewitchment. (87:453) **Meskwaki** *Love Medicine* Finely chopped roots eaten by couple to avert divorce and renew love. (152:231, 232) Ground roots used in food to end quarrels, a love medicine and anti-divorce remedy. (152:273)

Lobelia sp.
- *Drug*—**Iroquois** *Other* Plant used for "divining sickness." (87:451) **Yokut** *Cold Remedy* Dried, pulverized leaves used as snuff for head colds. *Emetic* Infusion of leaves used as a potent emetic. (117:437)

Lobelia spicata Lam., Palespike Lobelia
- *Drug*—**Cherokee** *Orthopedic Aid* Compound taken for "arm shakes and trembles." (80:43) Cold infusion of roots put into scratches made for shaking arms. (177:60) **Iroquois** *Blood Medicine* Decoction of stalks used as wash for bad blood. *Dermatological Aid* Decoction of stalks used as wash for neck and jaw sores. *Emetic* and *Love Medicine* Infusion of plants taken as an emetic for the lovelorn. (87:454)

Lobularia, Brassicaceae
Lobularia maritima (L.) Desv., Seaside Lobularia
- *Food*—**Costanoan** *Unspecified* Raw stems used for food. (21:252)

Loeseliastrum, Polemoniaceae
Loeseliastrum matthewsii (Gray) Timbrook, Desert Calico
- *Drug*—**Tubatulabal** *Cold Remedy* Decoction of plant taken for colds. (as *Langloisia matthewsii* 193:59)

Loganiaceae, see *Gelsemium*, *Spigelia*

Lolium, Poaceae

Lolium temulentum L., Darnel Ryegrass
- **Food**—**Pomo** *Staple* Seeds formerly used for pinole. **Yuki** *Staple* Seeds formerly used for pinole. (41:314)

Lomatium, Apiaceae

Lomatium ambiguum (Nutt.) Coult. & Rose, Wyeth Biscuitroot
- **Drug**—**Okanagan-Colville** *Cold Remedy* Infusion of flowers and upper leaves taken for colds and sore throats. *Throat Aid* Infusion of flowers and upper leaves taken for sore throats. (188:70)
- **Food**—**Montana Indian** *Staple* Spring roots reduced to flour. *Unspecified* Spring roots eaten. (19:15) **Okanagan-Colville** *Dried Food* Flowers and upper leaves dried for future use. *Spice* Dried flowers and upper leaves used to flavor meats, stews, and salads. *Substitution Food* Flowers and upper leaves sometimes used as a substitute food. (188:70)

Lomatium bicolor var. *leptocarpum* (Torr. & Gray) Schlessman, Wasatch Desertparsley
- **Food**—**Paiute** *Dried Food* Roots dried and used for food. *Unspecified* Roots eaten fresh. (as *L. leptocarpum* 104:101)

Lomatium californicum (Nutt.) Mathias & Constance, California Lomatium
- **Drug**—**Karok** *Dietary Aid* Decoction of roots taken by person who does not feel like eating. (as *Leptotaenia californica* 148:387) **Kawaiisu** *Cold Remedy* and *Emetic* Decoction of dried roots taken for colds, but caused vomiting. *Gastrointestinal Aid* Pounded root rubbed on the stomach for stomachaches. *Throat Aid* Root chewed for sore throat. (206:37) **Yuki** *Analgesic* and *Antirheumatic (External)* Root moxa used for arthritic pains. *Cold Remedy* Dried root smoked or decoction of roots taken for colds. *Other* Root dried, ground, and smoked in a pipe for severe colds; this occasionally caused dizziness. (49:44)
- **Food**—**Karok** *Unspecified* Roots eaten raw. (as *Leptotaenia californica* 148:387) **Kawaiisu** *Vegetable* Spring leaves eaten raw as greens. (206:37) **Yuki** *Unspecified* Young stems eaten raw. Shoots cooked and used for food. (49:87)
- **Other**—**Karok** *Smoke Plant* Root chewed and smoked in the pipe. (6:37) **Poliklah** *Sacred Items* Plant considered the most sacred plant of the tribe. (as *Leptotaenia californica* 117:173) **Yuki** *Good Luck Charm* Root bits placed in pockets for good luck in gambling. *Hunting & Fishing Item* Chewed while hunting to prevent deer from detecting human scent. (48:44) *Protection* Poultice of roots applied to neck to ward off sickness and rattlesnakes. (49:44) **Yurok** *Ceremonial Items* Thrown into the fire at ceremonies. (6:37)

Lomatium canbyi (Coult. & Rose) Coult. & Rose, Canby's Biscuitroot
- **Food**—**Klamath** *Dried Food* Dried roots used for food. *Porridge* Mashed and boiled roots made into mush. **Modoc** *Unspecified* Roots used for food. (as *Peucedanum canbyi* 45:102) **Okanagan-Colville** *Dried Food* Roots dried for future use. *Unspecified* Roots eaten raw or pit cooked and boiled. (188:64) **Paiute** *Bread & Cake* Peeled, mashed roots formed into cakes and allowed to dry, "Indian bread." (111:94) *Dried Food* Dried roots cooked and used for food. *Unspecified* Fresh roots cooked and used for food. (104:101)

Lomatium cous (S. Wats.) Coult. & Rose, Cous Biscuitroot
- **Food**—**Montana Indian** *Bread & Cake* Roots pulverized, moistened, partially baked, and made into different sized cakes. *Dried Food* Whole roots sun dried and stored for future food use. *Porridge* Roots pulverized and made into a gruel. *Soup* Roots pulverized, moistened, partially baked, mixed in water, and eaten as soup. (82:26) *Staple* Spring roots eaten or reduced to flour. *Unspecified* Spring roots eaten or reduced to flour. (as *L. montanum* 19:15) Peeled roots eaten raw or boiled. (82:26) **Okanagan-Colville** *Dried Food* Roots dried for future use. *Unspecified* Roots used for food. (188:65) **Oregon Indian** *Soup* Roots and fish used to make stew. These roots were eaten at the first feast of the new year. This was called the Root Feast. (as *Cogswellia cous* 118:12)

Lomatium dissectum (Nutt.) Mathias & Constance, Fernleaf Biscuitroot
- **Drug**—**Nez Perce** *Dermatological Aid* Root oil used for sores. *Dietary Aid* Infusion of cut roots taken to increase the appetite. *Eye Medicine* Root oil used for sore eyes. *Respiratory Aid* Roots mixed with tobacco and smoked for sinus trouble. *Tuberculosis Remedy* Infusion of cut roots taken for tuberculosis. (82:26) **Okanagan-Colville** *Antirheumatic (Internal)* Infusion or decoction of roots taken for arthritis. *Blood Medicine* Shoots used to "change the blood" to adapt to the summer's heat. *Dermatological Aid* Poultice of peeled, pounded roots applied to open cuts, sores, boils, or bruises. Roots soaked in cold water and used for dandruff. *Dietary Aid* Infusion of roots taken to increase the appetite. *Orthopedic Aid* Poultice of peeled, pounded roots applied to sore backs. *Other* Infusion or decoction of roots taken for some general illness. *Poison* Purple shoots considered poisonous. Mature tops and roots considered poisonous. Strong infusion or decoction of roots considered poisonous. *Tuberculosis Remedy* Infusion or decoction of roots taken for tuberculosis. *Veterinary Aid* Plant tops rubbed on cattle to kill lice. (188:66) **Paiute, Northern** *Analgesic* Roots smoked and decoction of roots taken or used as a head wash for head pains. Roots mixed with tobacco and smoked for headaches. *Antirheumatic (External)* Decoction of plant rubbed on the joints for rheumatism and aches. *Cold Remedy* Decoction of roots taken and roots smoked for colds. Root chips thrown on the fire and fumes inhaled or roots sliced, boiled, and eaten for colds. Roots mixed with tobacco and smoked for colds. *Dermatological Aid* Poultice of roasted roots applied to sores. Decoction of plant rubbed on pimples and sores. *Diuretic* Decoction of plant taken to pass water stopped by venereal disease. *Panacea* Plant used for all the common ailments and injuries. *Throat Aid* Roots chewed and juice swallowed for sore throats. *Vertigo Medicine* Roots smoked and decoction of roots taken or used as a head wash for dizziness. (59:129) **Shuswap** *Dermatological Aid* Poultice of mashed roots or decoction of mashed roots applied to sores. (123:56) **Thompson** *Cold Remedy* Infusion of dried root used for colds. *Orthopedic Aid* Poultice of washed, pounded root used for sprains and as a bone setter for broken bones. (187:154)
- **Food**—**Nez Perce** *Unspecified* Roots pit baked and eaten. (82:26) **Okanagan-Colville** *Sauce & Relish* Young shoots eaten raw as a relish, alone or with meat. **Sanpoil** *Special Food* Shoots mixed with balsamroot and featured in the "first roots" ceremony. (188:66) **Shuswap**

Unspecified Root of the young plant roasted and eaten. (123:56) **Thompson** *Unspecified* Roots dug in the early spring, pit cooked until soft, like balsamroots, and used for food. (187:154)
- *Other*—Okanagan-Colville *Hunting & Fishing Item* Roots pounded, steeped in water overnight, and used to poison fish. (188:66) **Paiute, Northern** *Hunting & Fishing Item* Plant used to stupefy fish. *Smoke Plant* Roots smoked for pleasure. (59:129)

Lomatium dissectum (Nutt.) Mathias & Constance **var. *dissectum***, Fernleaf Biscuitroot

- *Drug*—**Thompson** *Burn Dressing* Root powder mixed with grease and used as a salve for burns. *Dermatological Aid* Dried root powder sprinkled on wounds and sores to aid healing. (as *Leptotaenia dissecta* 164:472) *Veterinary Aid* Dried, crushed root sprinkled on horses' sores or wounds. (as *Leptotaenia dissecta* 164:513)
- *Food*—**Okanagon** *Dried Food* Thick, fleshy roots split, dried, and cooked for food. **Thompson** *Dried Food* Thick, fleshy roots split, dried, and cooked for food. (as *Leptotaenia dissecta* 125:37) Roots split, strung, dried, and cooked as needed. (as *Leptotaenia dissecta* 164:480)

Lomatium dissectum var. *multifidum* (Nutt.) Mathias & Constance, Carrotleaf Biscuitroot

- *Drug*—**Blackfoot** *Ceremonial Medicine* Pulverized root burned as incense. *Dietary Aid* Root used to make a drink taken as a tonic to help weakened people gain weight. *Stimulant* and *Tonic* Root used to make a drink taken as a tonic for "people in a weakened condition." *Veterinary Aid* Root smoke inhaled by horses for distemper. (as *Leptotaenia multifida* 114:274) **Cheyenne** *Analgesic* Infusion of dried, powdered roots taken for stomach pains or any internal disorder. Infusion of pulverized stems and leaves taken for stomach pains or any internal disorder. *Gastrointestinal Aid* Infusion of dried, powdered roots taken for stomach pains or any internal disorder. Infusion of pulverized stems and leaves taken for stomach pains or any internal disorder. *Tonic* Infusion of dried, powdered roots taken as a tonic. Infusion of pulverized stems and leaves used as a tonic. (as *Leptotaenia multifida* 76:182) **Gosiute** *Dermatological Aid* Poultice of roots applied to wounds, cuts, or bruises. (as *Ferula multifida* 39:348) Poultice of roots applied to wounds and bruises. (as *Ferula multifida* 39:369) *Disinfectant* and *Orthopedic Aid* Poultice of roots applied to infected, compound fractures. (as *Ferula multifida* 39:348) Poultice of roots applied to foot crushed under the wheel of a wagon. *Veterinary Aid* Burning root smoke inhaled by horse for distemper. (as *Ferula multifida* 39:369) **Great Basin Indian** *Cold Remedy* Decoction of roots taken or steam inhaled for colds. *Cough Medicine* Infusion of roots taken for coughs. *Dermatological Aid* Poultice of powdered roots used on sores. *Misc. Disease Remedy* Decoction of roots taken or steam inhaled for flu. Decoction of roots taken for colds and flu. *Other* Plant used to make a scent for a sick person. *Respiratory Aid* Dried roots burned on coals and smoke inhaled for asthma or bronchial troubles. *Tonic* Infusion of roots taken as a tonic. *Unspecified* Decoction of root water used to sponge a sick person. Dried, pounded roots and grease massaged on affected parts. (as *Leptotaenia multifida* 121:49) **Kawaiisu** *Analgesic* Pounded roots applied as a salve to sore limbs. *Dermatological Aid* Infusion of roots used as a wash or poultice applied to cuts and open wounds. Pounded roots applied as a salve to cuts and wounds. *Orthopedic Aid* Pounded roots applied as a salve to sore limbs. (206:37) **Montana Indian** *Poison* Young sprouts eaten, but poisonous to stock in early spring. (as *Leptotaenia multifida* 19:14) **Navajo** *Ceremonial Medicine* Infusion of dried, ground plant mixed with other plants and taken by patients for Mountain Top Chant. (as *Leptotaenia dissecta* var. *multifida* 55:67) **Nevada Indian** *Veterinary Aid* Poultice of ground chips given to horses for head trouble and sores. **Oregon Indian** *Veterinary Aid* Roots used in a wash for horse ticks and dandruff. (as *Leptotaenia multifida* 118:49) **Paiute** *Antirheumatic (External)* Poultice of root applied and decoction of root used as a wash for rheumatism. *Cold Remedy* Compound decoction of root taken for colds. Compound of pulverized roots smoked for colds. *Cough Medicine* Decoction of root taken as a cough remedy. (as *Leptotaenia multifida* 180:97–100) *Dermatological Aid* Smashed roots used for sores and swellings. (as *Leptotaenia multifida* 104:196) Poultice of root applied and decoction used as a wash for swellings. Poultice of root applied or decoction used as a wash for rashes, cuts, or sores. Sap from cut roots or oil from decoction used as a salve on cuts and sores. *Disinfectant* Root used as "the basis of a number of antiseptics." *Herbal Steam* Compound of roots used as herbal steam for lung or nasal congestion and asthma. *Misc. Disease Remedy* Decoction of dried root taken for influenza. Decoction of root and sometimes leaves used as an antiseptic wash for smallpox. *Orthopedic Aid* Poultice of root applied and decoction used as a wash for sprains. *Panacea* Root used for a wide variety of ailments, usually as a decoction. *Pulmonary Aid* Compound of roots used as herbal steam for lung congestion. Pulverized roots smoked to clear lungs and nasal passages. Root used in various ways for pneumonia. *Respiratory Aid* Compound of roots used as herbal steam for nasal congestion and asthma. Decoction of dried root taken for hay fever, bronchitis, and pneumonia. Pulverized roots smoked alone or in compound for asthma. *Throat Aid* Raw root chewed and decoction of root taken for sore throat. (as *Leptotaenia multifida* 180:97–100) *Tuberculosis Remedy* Decoction of roots taken for tuberculosis. (as *Leptotaenia multifida* 104:198) Compound of pulverized roots smoked for tuberculosis. Decoction of dried root taken for tuberculosis. *Venereal Aid* Simple or compound decoction of root taken for venereal diseases. *Veterinary Aid* Root smoke inhaled by horses for distemper. **Shoshoni** *Antirheumatic (External)* Poultice of root applied and decoction of root used as a wash for rheumatism. *Cold Remedy* Compound decoction of root taken for colds. *Cough Medicine* Decoction of root taken as a cough remedy. *Dermatological Aid* Poultice of root applied and decoction used as a wash for swellings. Sap from cut roots or oil from decoction used as a salve on cuts and sores. *Disinfectant* Root used as "the basis of a number of antiseptics." (as *Leptotaenia multifida* 180:97–100) *Eye Medicine* Root oil used as eye drops for "trachoma." (97:48) Oil of root used for trachoma. (as *Leptotaenia multifida* 118:39) Sap from cut roots used as eye drops for trachoma or gonorrheal eye infections. *Herbal Steam* Compound of roots used as herbal steam for lung or nasal congestion and asthma. *Misc. Disease Remedy* Decoction of dried root taken for influenza. Decoction of root and sometimes leaves used as an antiseptic wash for smallpox. *Orthopedic Aid* Poultice of root applied and decoction used as a wash for sprains. *Panacea* Root used for a wide variety of ailments, usually as a decoc-

tion. *Pulmonary Aid* Compound of roots used as herbal steam for lung congestion. Decoction of dried root taken for pneumonia. Pulverized roots smoked to clear lungs and nasal passages. *Respiratory Aid* Compound of roots used as herbal steam for nasal congestion and asthma. Decoction of dried root taken for hay fever and bronchitis. Pulverized roots smoked for asthma. *Throat Aid* Raw root chewed for sore throat. *Tuberculosis Remedy* Compound of pulverized roots smoked for tuberculosis. Decoction of dried root taken for tuberculosis. *Venereal Aid* Simple or compound decoction of root taken for venereal diseases. *Veterinary Aid* Smoke from root alone or in compound inhaled by horses for distemper. (as *Leptotaenia multifida* 180:97–100) **Ute** *Dermatological Aid* Poultice of root pulp applied to wounds and bruises. *Veterinary Aid* Roots burned in a pan and held beneath the horse's nose for distemper. (as *Ferula multifida* 38:34) **Washo** *Cold Remedy* Decoction of root taken as a cold remedy. *Cough Medicine* Decoction of root taken as a cough remedy. *Dermatological Aid* Pulverized root applied as powder to infant's severed umbilical cord. Sap from cut roots or oil from decoction used as a salve on cuts and sores. *Herbal Steam* Compound of roots used as herbal steam for lung or nasal congestion and asthma. *Misc. Disease Remedy* Decoction of dried root taken for influenza. *Panacea* Root used for a wide variety of ailments, usually as a decoction. *Pediatric Aid* Poultice of fresh root pulp applied to severed umbilical cord. *Pulmonary Aid* Compound of roots used as herbal steam for lung congestion. Decoction of dried root taken for pneumonia. Pulverized roots smoked to clear lungs and nasal passages. *Respiratory Aid* Compound of roots used as herbal steam for nasal congestion and asthma. Decoction of dried root taken for hay fever and bronchitis. Pulverized roots smoked for asthma. *Throat Aid* Raw root chewed for sore throat. *Tuberculosis Remedy* Decoction of dried root taken for tuberculosis. (as *Leptotaenia multifida* 180:97–100)
- *Food*—**Gosiute** *Unspecified* Young shoots, seeds used for food. (as *Ferula multifida* 39:369) **Great Basin Indian** *Beverage* Roots boiled to make a drink. *Vegetable* Long, young shoots cooked in the spring for greens. (as *Leptotaenia multifida* 121:49) **Montana Indian** *Unspecified* Young sprouts eaten, but poisonous to stock in early spring. (as *Leptotaenia multifida* 19:14)
- *Other*—**Blackfoot** *Hide Preparation* Plant mixed with brains and used in soft tanning. (as *Leptotaenia multifida* 114:274) **Great Basin Indian** *Smoke Plant* Dried root chips and Bull Durham used as a friendly smoke. (as *Leptotaenia multifida* 121:49) **Oregon Indian** *Hide Preparation* Root used in tanning hides. (as *Leptotaenia multifida* 118:55)

Lomatium farinosum (Hook.) Coult. & Rose, Northern Biscuitroot
- *Food*—**Okanagan-Colville** *Unspecified* Roots boiled and eaten fresh. (188:68)

Lomatium foeniculaceum ssp. daucifolium (Torr. & Gray) Theobald, Desert Biscuitroot
- *Drug*—**Dakota** *Love Medicine* Compound containing seeds used by men as a love charm. **Omaha** *Love Medicine* Compound containing seeds used by men as a love charm. **Pawnee** *Love Medicine* Compound containing seeds used by men as a love charm. **Ponca** *Love Medicine* Compound containing seeds used by men as a love charm. **Winnebago** *Love Medicine* Compound containing seeds used by men as a love charm. (as *Cogswellia daucifolia* 70:107)

Lomatium geyeri (S. Wats.) Coult. & Rose, Geyer's Biscuitroot
- *Food*—**Okanagan-Colville** *Unspecified* Roots peeled, cooked, and eaten with bitterroot. (188:68)

Lomatium graveolens (S. Wats.) Dorn & Hartman **var. graveolens**, King Desertparsley
- *Drug*—**Gosiute** *Cold Remedy* Compound decoction of roots taken for biliousness with severe colds. (as *Peucedanum graveolens* 39:351) Decoction of plant used for severe colds. (as *Peucedanum graveolens* 39:376) *Gastrointestinal Aid* Compound decoction of roots taken for biliousness with severe colds. (as *Peucedanum graveolens* 39:351) Decoction of plant used for biliousness. (as *Peucedanum graveolens* 39:376) *Throat Aid* Decoction of roots used or poultice of root pulp applied for sore throats. (as *Peucedanum graveolens* 39:351) Poultice of mashed plant applied to throat for a sore throat. *Unspecified* Root used as medicine. (as *Peucedanum graveolens* 39:376)

Lomatium grayi (Coult. & Rose) Coult. & Rose, Gray's Biscuitroot
- *Food*—**Paiute** *Starvation Food* Roots eaten when hungry in the winter. *Unspecified* Tender, young stems eaten raw. (111:95)

Lomatium macrocarpum (Nutt. ex Torr. & Gray) Coult. & Rose, Bigseed Biscuitroot
- *Drug*—**Blackfoot** *Strengthener* Infusion of roots taken for weakness. *Veterinary Aid* Smoke from burning roots or decoction of roots inhaled by horses for distemper. **Crow** *Antirheumatic (External)* Poultice of boiled root shavings used for swellings. *Cold Remedy* Infusion of root shavings mixed with animal fat and used for colds. *Throat Aid* Roots chewed and juice used for sore throats. (82:26) **Okanagan-Colville** *Cold Remedy* Dried roots soaked overnight and chewed for colds. *Misc. Disease Remedy* Dried roots soaked overnight and chewed for flu. *Oral Aid* and *Pediatric Aid* Poultice of pounded roots applied to the inside of babies mouths for mouth sores or "thrush." *Respiratory Aid* Dried roots soaked overnight and chewed for bronchitis. (188:69) **Thompson** *Pediatric Aid* Leaves used in babies' bath water to make them sleep a lot. (187:155) *Reproductive Aid* Root eaten by childless women for infertility. (as *Peucedanum macrocarpum* 164:508) Roots eaten by elderly couples to help them conceive. *Sedative* Leaves used in babies' bath water to make them sleep a lot. Leaves used as padding, especially in children's cradles, to cause them to sleep a lot. (187:155)
- *Food*—**Flathead** *Dried Food* Roots dried and stored for future use. *Unspecified* Roots eaten raw. (82:26) **Okanagan-Colville** *Unspecified* Roots peeled and eaten raw or boiled. (188:69) **Okanagon** *Staple* Roots used as a principal food. (as *Peucedanum macrocarpum* 178:238) *Unspecified* Thick roots, tiger lily bulbs, and salmon roe boiled and eaten. (125:36) **Paiute** *Dried Food* Roots dried and used for food. *Unspecified* Roots eaten fresh. (104:101) Peeled roots eaten raw or baked. (111:95) **Paiute, Northern** *Unspecified* Roots roasted in the sand and eaten. (59:43) **Pomo, Kashaya** *Spice* Sweet seed used to flavor tea and pinole. *Unspecified* Young leaves used for food. (72:31) **Sanpoil** *Unspecified* Roots pit cooked and eaten. (188:69) **Shuswap** *Spice* Roots used to flavor dried salmon heated with dried bread over an open fire. *Unspecified* Roots roasted and eaten. (123:57) **Thompson** *Dried Food* Roots dug in the springtime, peeled, and dried for

later use. *Pie & Pudding* Roots used in puddings. *Spice* Roots cooked with meat stews, saskatoon berries, or tiger lily bulbs as a flavoring. (187:155) *Unspecified* Thick roots, tiger lily bulbs, and salmon roe boiled and eaten. (125:36) Thick roots combined with salmon roe, boiled, and eaten. (as *Peucedanum macrocarpum* 164:479) Boiled roots used for food. (187:155)
- *Fiber*—**Thompson** *Mats, Rugs & Bedding* Leaves finely divided and used as a padding in child carriers. (as *Peucedanum macrocarpum* 164:496) Leaves used as padding, especially in children's cradles, to cause them to sleep a lot. (187:155)
- *Other*—**Crow** *Ceremonial Items* Root shavings sprinkled on live coals to produce a ceremonial incense. *Incense & Fragrance* Root shavings sprinkled on live coals to deodorize and purify the air. (82:26) **Thompson, Upper (Lytton Band)** *Cash Crop* Plant traded to the Lower Thompson band. (as *Peucedanum macrocarpum* 164:479)

Lomatium nevadense (S. Wats.) Coult. & Rose, Nevada Biscuitroot
- *Food*—**Paiute, Northern** *Unspecified* Roots eaten raw or cooked in the sand. (59:44)

Lomatium nevadense* var. *parishii (Coult. & Rose) Jepson, Parish's Biscuitroot
- *Food*—**Paiute** *Vegetable* Peeled roots eaten fresh like radishes. (111:95)

Lomatium nudicaule (Pursh) Coult. & Rose, Barestem Biscuitroot
- *Drug*—**Cowichan** *Cold Remedy* Seeds chewed for colds. *Throat Aid* Seeds chewed for sore throats. (182:89) **Kwakiutl** *Analgesic* Poultice of chewed seeds applied or chewed seeds blown on head for headaches. (183:276) *Antirheumatic (External)* Poultice of chewed seeds applied to back for sore places, pains, or itching. (as *Peucedanum lecocarpum* 20:382) *Cold Remedy* Poultice of chewed seeds applied for colds. (183:276) *Cough Medicine* Seeds kept in the mouth and the saliva swallowed to loosen the phlegm for hoarseness and coughs. (as *Penoedanum leiocarpum* 20:381) Seeds sucked for coughs. *Dermatological Aid* Poultice of chewed seeds applied to carbuncles. *Gastrointestinal Aid* Poultice of chewed seeds applied for stomachaches. *Gynecological Aid* Infusion of seeds taken by pregnant women to insure an easy delivery. Poultice of chewed seeds applied for swelling of a woman's breasts. *Herbal Steam* Compound with seeds used in a steam bath for general sickness. *Hunting Medicine* Seeds used by hunters for protection. *Laxative* Seeds eaten for constipation. *Orthopedic Aid* Poultice of chewed seeds applied for backaches and swollen knees and feet. *Panacea* Compound with seeds used in a steam bath for general sickness. (183:276) *Throat Aid* Seeds kept in the mouth and the saliva swallowed to loosen the phlegm for hoarseness and coughs. (as *Penoedanum leiocarpum* 20:381) Seeds sucked for sore throats. (183:276) **Nitinaht** *Ceremonial Medicine* Seeds burned as a protective fumigant against bad spirits and illness. *Cold Remedy* Poultice of warm, soaked seeds applied to the chest for colds. (186:92) **Saanich** *Cold Remedy* Seeds chewed for colds. *Throat Aid* Seeds chewed for sore throats. **Salish, Coast** *Internal Medicine* Seeds swallowed for internal complaints. **Songish** *Cold Remedy* Seeds chewed for colds. *Throat Aid* Seeds chewed for sore throats. (182:89) **Thompson** *Cold Remedy* Strong decoction of whole plant or stems and leaves taken for colds. (as *Peucedanum nudicaulis* 164:473) Decoction of leaves, strawberry leaves, and ginger root used as a vitamin supplement for colds. *Diaphoretic* Infusion of 2 teaspoons of dried seeds used to "sweat the cold out." (187:156) *Febrifuge* Strong decoction of whole plant or stems and leaves taken for fevers. (as *Peucedanum nudicaulis* 164:473)
- *Food*—**Atsugewi** *Unspecified* Raw leaves and tender stems used for food. (61:139) **Okanagon** *Vegetable* Stalks used like celery. (125:38) **Paiute** *Vegetable* Stem eaten raw like celery. (111:96) **Thompson** *Beverage* Flowers, leaves, and stems dried, brought to a boil, and used as a drink. (as *Peucedanum leiocarpum* 164:494) Dried leaves used to make a tea-like beverage. Mature fruits, leaves and other plant parts preserved and used all year to make a tea-like beverage. Young, green fruits used to make tea. *Dried Food* Leaves frozen or canned for future use or dried and used to flavor stews or other dishes. *Frozen Food* Leaves frozen, canned, or dried for future use, and used to flavor stews or other dishes. *Fruit* Green, undeveloped fruits chewed raw. *Spice* Leaves used as a flavoring in soups and stews. Green, undeveloped fruits used as a flavoring. (187:156) *Unspecified* Stalks used for food. (as *Peucedanum leiocarpum* 164:484) Roots formerly used as food. (as *Peucedanum leiocarpum* 164:479) Stalks peeled and eaten as celery. (as *Peucedanum leiocarpum* 164:483) *Vegetable* Stalks used like celery. (125:38) Leaves eaten raw or cooked as a potherb. *Winter Use Food* Leaves frozen or canned for future use or dried and used to flavor stews and other dishes. (187:156)
- *Other*—**Cowichan** *Ceremonial Items* Seeds burned to fumigate homes and to "drive away ghosts." (182:89) **Nitinaht** *Hunting & Fishing Item* Leaves or seeds used for devil's club codfish lures. *Incense & Fragrance* Leaves or seeds used as scents or charms. (186:92) **Saanich** *Ceremonial Items* Seeds burned to fumigate homes and to "drive away ghosts." **Songish** *Ceremonial Items* Seeds burned to fumigate homes and to "drive away ghosts." (182:89) **Thompson** *Incense & Fragrance* Plant sometimes used as a scent. Stems used as a scent. (as *Peucedanum leiocarpum* 164:503)

Lomatium nuttallii (Gray) J. F. Macbr., Nuttall's Biscuitroot
- *Drug*—**Creek** *Poison* Plant considered poisonous if eaten in winter. *Unspecified* Used as a medicine in summer. (172:667)

Lomatium orientale Coult. & Rose, Northern Idaho Biscuitroot
- *Drug*—**Cheyenne** *Analgesic* Infusion of roots and leaves used by children and adults for bowel pain and diarrhea. (as *Cogswellia orientalis* 76:182) Infusion of roots and leaves used or dried roots and leaves eaten for bowel pain. (82:26) *Antidiarrheal* Infusion of pounded roots and leaves taken for diarrhea. (as *Cogswellia orientalis* 76:182) Roots and leaves infused or eaten dry for diarrhea. (82:26) *Gastrointestinal Aid* Infusion of pounded roots and leaves taken for bowel pain. *Pediatric Aid* Infusion of pounded roots and leaves given to children for bowel pain or diarrhea. (as *Cogswellia orientalis* 76:182)
- *Food*—**Lakota** *Unspecified* Roots used for food. (139:33) **Navajo** *Unspecified* Roots used for food. (55:68) Roots rubbed in hot ash to remove the strong taste and eaten raw or baked. (as *Cogswellia orientalis* 165:221)

Lomatium piperi Coult. & Rose, Indian Biscuitroot
- *Food*—**Paiute** *Unspecified* Roots used for food. (111:94)

Lomatium simplex (Nutt.) J. F. Macbr., Narrowleaf Lomatium

- *Food*—Montana Indian *Vegetable* Fusiform root eaten baked, roasted, or raw. (19:15)

Lomatium simplex var. leptophyllum (Hook.) Mathias, Narrowleaf Lomatium
- *Food*—Blackfoot *Unspecified* Roots eaten raw or roasted. (97:48)

Lomatium simplex (Nutt.) J. F. Macbr. **var. simplex**, Great Basin Desertparsley
- *Food*—Montana Indian *Staple* Spring roots reduced to flour. *Unspecified* Spring roots eaten. (as *L. platycarpum* 19:15)

Lomatium sp., Biscuitroot
- *Drug*—Cheyenne *Antirheumatic (External)* Infusion of dried pulverized roots applied to swellings. *Dermatological Aid* Infusion of dried roots applied to swellings. (as *Cogswellia* 76:182) **Oregon Indian** *Love Medicine* Aromatic seeds carried by men as a love charm. (as *Cogswellia* 118:57)

Lomatium triternatum (Pursh) Coult. & Rose, Nineleaf Biscuitroot
- *Drug*—Blackfoot *Panacea* Chewed roots blown onto affected part by the diviner. The healing qualities of the spray were believed to penetrate the body at that place. (86:83) *Pulmonary Aid* Infusion of roots and leaves taken for chest troubles. (86:72) *Strengthener* Fruit chewed by long distance runners to avoid side aches. (86:67) **Okanagan-Colville** *Cold Remedy* Infusion of flowers and upper leaves taken for colds. *Throat Aid* Infusion of flowers and upper leaves taken for sore throats. (188:70)
- *Food*—Atsugewi *Unspecified* Roots cooked in earth oven and used for food. (61:138) **Blackfoot** *Unspecified* Flowers used to make pemmican. (86:103) Roots eaten raw or roasted. (97:49) **Montana Indian** *Staple* Spring roots reduced to flour. *Unspecified* Spring roots eaten. (19:15) Roots eaten raw, roasted, or baked. (82:26) *Vegetable* Fusiform root eaten baked, roasted, or raw. (19:15) **Okanagan-Colville** *Dried Food* Flowers and upper leaves dried for future use. *Spice* Dried flowers and upper leaves used to flavor meats, stews, and salads. *Substitution Food* Flowers and upper leaves sometimes used as a substitute food. (188:70)
- *Other*—Blackfoot *Good Luck Charm* Fruits stuffed into a porcupine foot and tied on a young girl's hair as a good luck charm. *Hide Preparation* Fruits used during the tanning process of animal pelts to keep them from smelling. (86:115) **Paiute** *Malicious Magic* When broken, it brought the cold wind. (111:96)

Lomatium utriculatum (Nutt. ex Torr. & Gray) Coult. & Rose, Common Lomatium
- *Drug*—Kawaiisu *Dermatological Aid* Decoction of plant used as a wash for swollen limbs. *Orthopedic Aid* Decoction of plant used as a wash for broken limbs. (206:38) **Salish, Coast** *Analgesic* Roots chewed or soaked in water and taken for headaches. *Gastrointestinal Aid* Roots chewed or soaked in water and taken for stomach disorders. (182:89)
- *Food*—Atsugewi *Unspecified* Raw leaves used for food. (61:139) **Kawaiisu** *Vegetable* Leaves, sometimes with flowers, cooked, fried in grease and salt, and eaten. (206:38) **Mendocino Indian** *Unspecified* Young leaves eaten raw in early summer. (41:373)

Lomatium watsonii (Coult. & Rose) Coult. & Rose, Watson's Desertparsley
- *Food*—Paiute *Winter Use Food* Peeled roots dried for winter use, ground, and boiled into a mush or used to flavor dried crickets. (111:94)

Lonicera, Caprifoliaceae

Lonicera arizonica Rehd., Arizona Honeysuckle
- *Drug*—Navajo, Ramah *Ceremonial Medicine* and *Emetic* Leaves used as a ceremonial emetic. (191:45)

Lonicera canadensis Bartr. ex Marsh., American Fly Honeysuckle
- *Drug*—Iroquois *Blood Medicine* Complex compound taken as blood purifier. *Dermatological Aid* Decoction of shoots taken for chancres caused by syphilis. *Pediatric Aid* Infusion of bark given to children who cry all night. *Psychological Aid* Infusion of bark taken for homesickness. *Sedative* Infusion of bark given to children who cry all night. *Venereal Aid* Decoction of shoots taken for chancres caused by syphilis. (87:443) **Menominee** *Urinary Aid* Bark used for urinary diseases. *Venereal Aid* Compound containing bark used for gonorrhea. (151:27) **Montagnais** *Diuretic* Infusion of vines taken as a diuretic. (156:315) **Potawatomi** *Diuretic* Compound infusion of bark used as a diuretic. (154:46)

Lonicera ciliosa (Pursh) Poir. ex DC., Orange Honeysuckle
- *Drug*—Chehalis *Contraceptive* Infusion of leaves taken as a contraceptive. *Dermatological Aid* Infusion of crushed leaves used as a hair wash to make it grow. (79:48) **Cowichan** *Unspecified* Leaves used for medicine. (182:79) **Klallam** *Dermatological Aid* Poultice of chewed leaves applied to bruises. **Lummi** *Tuberculosis Remedy* Decoction of leaves taken for tuberculosis. **Skagit** *Tonic* Decoction of leaves applied to the body as a strengthening tonic. **Squaxin** *Gynecological Aid* Infusion of leaves taken for womb trouble. **Swinomish** *Cold Remedy* Infusion of bark or chewed leaf juice taken for colds. *Gynecological Aid* Infusion of leaves used as a steam bath to stimulate lacteal flow. *Throat Aid* Infusion of bark taken for colds and sore throats. (79:48) **Thompson** *Anticonvulsive* Infusion of woody part of vine taken in small amounts or used as a bath for epilepsy. Infusion of woody part of vine taken in small amounts or used as a bath for children with epilepsy. Flowers sucked for epilepsy. *Reproductive Aid* Decoction of chopped, cooked vine stems taken by women who could not become pregnant. *Sedative* Vine pieces used under the pillow to induce sound sleep. (187:196) *Tonic* Decoction of peeled stems taken as a tonic. (164:471)
- *Food*—Nitinaht *Candy* Tubes formerly sucked by children for sweet nectar. (186:99) **Okanagan-Colville** *Forage* Flower nectar sucked by hummingbirds. (188:93) **Saanich** *Candy* Flower nectar sucked by children. (182:79) **Thompson** *Candy* Nectar sucked from flowers by children. (187:196) *Forage* Flower nectar eaten by bees and hummingbirds. (164:516)
- *Fiber*—Thompson *Building Material* Vines used with other plants as building materials. The vines were used with willow withes to reinforce suspension bridges across canyons and rivers. The vines were also twisted with coyote willow to lash together the framing poles of underground pit houses and to make a pliable ladder on the outside of the pit house, running from the opening down to the ground. (187:

196) *Cordage* Fiber obtained from stems used as twine. *Sewing Material* Fiber obtained from stems used as thread. (164:499)
- **Dye**—Thompson *Black* Stems used as a black dye for bitter cherry. (187:196)

Lonicera conjugialis Kellogg, Purpleflower Honeysuckle
- **Food**—Klamath *Fruit* Fresh berries used for food. (45:104)

Lonicera dioica L., Limber Honeysuckle
- **Drug**—Algonquin, Quebec *Cathartic* Infusion of bark used as a cathartic. *Gynecological Aid* Infusion of bark used for menstrual difficulties. *Kidney Aid* Infusion of bark used for kidney stones. (18:234) **Chippewa** *Diuretic* Infusion of stems taken as a diuretic. *Urinary Aid* Infusion of stems taken for dysuria. (as *L. divica* 71:141) **Iroquois** *Emetic* Decoction of vines taken as an emetic to throw off effects of love medicine. *Love Medicine* Decoction of vines taken as an emetic, love or anti-love medicine. *Venereal Aid* Compound decoction of roots taken for gonorrhea. (87:443) **Meskwaki** *Anthelmintic* Infusion of berry and root bark taken by pregnant women for worms. (152:207)

Lonicera dioica var. *glaucescens* (Rydb.) Butters, Limber Honeysuckle
- **Drug**—Cree, Woodlands *Diuretic* Infusion of inner bark, either scraped from or attached to the stem, used as a diuretic. Infusion of peeled, internodal stem lengths used for urine retention. *Gynecological Aid* Decoction of stems used for blood clotting after childbirth. *Misc. Disease Remedy* Infusion of peeled, internodal stem lengths used for flu. *Venereal Aid* Decoction of stems used for venereal disease. (109:43) **Iroquois** *Febrifuge* Decoction of plants given to children for fevers and sickness. *Gynecological Aid* Decoction taken by pregnant women for internal and leg soreness. *Pediatric Aid* Decoction of plants given to children for fevers and sickness. *Tuberculosis Remedy* Compound decoction of roots taken for consumption. (87:444)
- **Other**—Cree, Woodlands *Cooking Tools* Hollow stems used as straws by children. *Smoking Tools* Hollow stem sections used as pipestems for corn cob pipes and toy rose hip pipes. (109:43)

Lonicera hispidula var. *californica* Jepson, Pink Honeysuckle
- **Other**—Pomo *Smoking Tools* Stems used for clay elbow pipes. (66:15)

Lonicera hispidula var. *vacillans* (Benth.) Gray, Pink Honeysuckle
- **Other**—Pomo, Kashaya *Designs* Burned wood ashes made into a paste for tattooing. *Smoking Tools* Hollow stems used for pipestems. (72:56)

Lonicera interrupta Benth., Chaparral Honeysuckle
- **Drug**—Mendocino Indian *Eye Medicine* Infusion of leaves taken and used as a wash for sore eyes. (41:388) **Shoshoni** *Antirheumatic (External)* Poultice of raw root applied to swellings. **Yuki** *Dermatological Aid* Leaves used in a wash for sores. (118:44)
- **Food**—Mendocino Indian *Unspecified* Nectar sucked out of long, yellow flowers by children. (41:388)
- **Fiber**—Mendocino Indian *Basketry* Long, flexible stems used for the circular withes of baskets. (41:388)
- **Other**—Kawaiisu *Smoking Tools* Hollowed stem section used as a cigarette-type pipe. (206:38)

Lonicera involucrata Banks ex Spreng., Twinberry Honeysuckle
- **Drug**—Bella Coola *Cough Medicine* Decoction of bark taken for cough. *Dermatological Aid* Poultice of chewed leaves used on itch, boils, and gonorrheal sores and bark used for sores. (150:63) Leaves chewed and used for itchy skin and boils. (184:203) *Venereal Aid* Poultice of chewed leaves or toasted bark applied to gonorrheal sores. (150:63) **Blackfoot** *Cathartic* and *Emetic* Infusion of berries used as a cathartic and emetic to cleanse the body. *Gastrointestinal Aid* Infusion of berries used for stomach troubles. *Pulmonary Aid* Infusion of berries used for chest troubles. (86:67) **Carrier** *Dermatological Aid* Poultice of crushed leaves applied to open sores. *Eye Medicine* Decoction of leaves used to bathe sore eyes. (31:77) **Carrier, Northern** *Dermatological Aid* Compound decoction of stems taken for body sores. *Orthopedic Aid* and *Stimulant* Compound decoction of stems taken for constitutional weakness or paralysis. **Carrier, Southern** *Eye Medicine* Decoction of bark used daily as an eyewash. (150:63) **Gitksan** *Eye Medicine* Fruit juice used for sore eyes. Infusion of inner bark used for sore eyes. (43:229) Fresh juice of berries or infusion of inner bark used in sore eyes. (150:63) **Kwakiutl** *Analgesic* Compound infusion of bark used as foot bath for painful legs and feet. (20:380) Chewed leaves with yellow cedar rubbed on painful places. (20:382) *Antirheumatic (External)* Plant used in sweat baths for arthritis and rheumatism. (183:266) *Dermatological Aid* Leaves chewed with yellow cedar and rubbed on painful places. (20:380) Chewed leaves rubbed on sores. Poultice of bark, berries, or leaves and grease applied to swellings or sores. *Gynecological Aid* Decoction of bark applied to women's breasts to make milk flow. (183:279) *Herbal Steam* Plant used in sweat baths for arthritis and rheumatism. (183:266) *Orthopedic Aid* Compound decoction of bark used as foot bath for leg and foot pains. (20:380) Compound infusion of bark used as a soak for sore feet and legs. Infusion of leaves or roots applied as poultice to swollen shoulders and feet. (183:279) **Kwakiutl, Southern** *Analgesic* Poultice of bark and berries or leaves, fresh sea wrack, and alder bark applied for aches and pains. (183:260) **Makah** *Dermatological Aid* Mashed fruit applied to the scalp for dandruff. *Emetic* Fruit used as an emetic. (67:317) *Gynecological Aid* Leaves chewed by women during confinement. (79:48) *Poison* Fruit considered poisonous. (67:317) **Navajo, Ramah** *Ceremonial Medicine* and *Emetic* Leaves used as a ceremonial emetic. (191:45) **Nitinaht** *Psychological Aid* Buds eaten in spring or bark rubbed on body as a tonic for nervous breakdowns. **Nootka, Manhousat** *Love Medicine* Decoction of bark or fresh bark eaten by whalers to relieve effects of sexual abstinence. (186:99) **Nuxalkmc** *Cough Medicine* Leaves or bark used for coughs. *Dermatological Aid* Leaves or bark used for boils and itchy areas. *Venereal Aid* Leaves or bark used for gonorrhea. (43:229) **Okanagan-Colville** *Gynecological Aid* Branches used to make a medicine for mothers after childbirth. *Poison* Berries considered poisonous. (188:94) **Poliklah** *Poison* Berries considered poisonous. (117:173) **Quileute** *Antidote* and *Emetic* Leaves chewed as an emetic when poisoned. (79:48) **Quinault** *Dermatological Aid* Leaves chewed or rubbed on sores. (201:276) *Gynecological Aid* Leaves chewed by women during confinement. (79:48) *Oral Aid* Leaves chewed for sore

mouth. (201:276) **Thompson** *Dermatological Aid* Poultice of boiled leaves applied to swellings. (164:457) Decoction of sticks, leaves and all, used for scabs and sores. *Dietary Aid* Decoction of stems and leaves taken as a tonic "for vitamins." (187:197) *Orthopedic Aid* Decoction of leaves and twigs used as a liniment. (164:457) Decoction of sticks, leaves and all, used for broken bones. (187:197) *Poison* Berries considered poisonous if more than two or three eaten. (164:489) *Throat Aid* Decoction of sticks, leaves and all, taken for sore throat. *Urinary Aid* Decoction of sticks, leaves and all, taken for bladder trouble. (187:197) **Tolowa** *Poison* "Not good to eat, poison." (6:37) **Wet'suwet'en** *Burn Dressing* Bark used for burns. *Dermatological Aid* Bark used for wounds. *Disinfectant* Bark used for infections. (73:152)
- *Food*—**Bella Coola** *Forage* Berries eaten by birds. (184:203) **Hesquiat** *Forage* Berries eaten by crows and other birds. (185:63) **Montana Indian** *Winter Use Food* Fruit dried and stored for winter use. (19:15) **Okanagan-Colville** *Forage* Berries eaten by bears. (188:94) **Okanagon** *Fruit* Fruits occasionally used for food. (125:39) **Oweekeno** *Fruit* Berries used for food. (43:89) **Thompson** *Forage* Berries eaten by grizzly bears. (187:197) *Fruit* Fruits occasionally used for food. (125:39) Berries eaten, but not commonly exploited as a food source. One informant ate the berries, but was told by her mother not to eat them. (187:197)
- *Dye*—**Hesquiat** *Purple* Mashed berries boiled to make a purple paint. (185:63) **Makah** *Unspecified* Fruit used as a dye for basketry materials. (67:317)
- *Other*—**Quileute** *Paint* Juice used to paint the faces of dolls. (79:48)

Lonicera japonica Thunb., Japanese Honeysuckle
- *Fiber*—**Cherokee** *Basketry* Vines used to make baskets. (80:38)

Lonicera oblongifolia (Goldie) Hook., Swamp Fly Honeysuckle
- *Drug*—**Iroquois** *Analgesic* Poultice of hot bark applied to abdomen for urinating pain. *Gynecological Aid* Compound decoction of branches taken for falling of the womb. *Psychological Aid* Infusion of bark taken for loneliness. *Sedative* Infusion of bark taken for restlessness. *Urinary Aid* Poultice of hot bark applied to abdomen for urinating pain. (87:443)

Lonicera sp., Honeysuckle
- *Drug*—**Chippewa** *Pulmonary Aid* Compound decoction of root taken for lung trouble. (53:340) **Costanoan** *Cough Medicine* Decoction of dried fruit used as a cough syrup. *Dermatological Aid* and *Disinfectant* Decoction of plant used for infected sores. *Orthopedic Aid* Decoction of plant used as a bath for swollen feet. (21:24) **Iroquois** *Antidiarrheal* and *Pediatric Aid* Compound decoction of twigs given to babies with diarrhea. (87:442)

Lonicera subspicata var. *johnstonii* Keck, Johnston's Honeysuckle
- *Drug*—**Diegueño** *Veterinary Aid* Decoction of plant used to wash sores on horses. (84:24)

Lonicera utahensis S. Wats., Utah Honeysuckle
- *Drug*—**Navajo, Ramah** *Hunting Medicine* Chewed leaves blown on weapons for good luck in hunting. (191:45) **Okanagan-Colville** *Blood Medicine* Infusion of branches taken as a tonic to "change the blood" in the spring and fall. *Dermatological Aid* Infusion of branches and leaves used to wash sores and infections. *Laxative* Infusion of branches taken as a mild laxative. (188:94)
- *Food*—**Okanagan-Colville** *Fruit* Berries used for food. (188:94)

Lophophora, Cactaceae
Lophophora williamsii (Lem. ex Salm-Dyck) Coult., Peyote
- *Drug*—**Comanche** *Ceremonial Medicine* and *Narcotic* Plant used in ceremonies as a narcotic. (29:522) **Delaware** *Panacea* Carried in small beaded bags and worn around the neck to protect against illness. *Tuberculosis Remedy* Used for tuberculosis. (176:39) **Kiowa** *Analgesic* and *Antirheumatic (External)* Poultice of plants applied for rheumatic pains. *Cold Remedy* Decoction of plants taken for colds. *Dermatological Aid* Poultice of plants applied for cuts. *Febrifuge* Decoction of plants taken for fevers. *Gastrointestinal Aid* Decoction of plants taken for intestinal ills. *Misc. Disease Remedy* Decoction of plants taken for grippe and scarlet fever. *Narcotic* Plant used as a narcotic. *Orthopedic Aid* Poultice of plants applied for bruises. *Panacea* Decoction of plants taken as a panacea. *Pulmonary Aid* Decoction of plants taken for pneumonia and scarlet fever. *Tuberculosis Remedy* Decoction of plants taken for tuberculosis. *Venereal Aid* Decoction of plants taken for venereal disease. (192:43) **Omaha** *Ceremonial Medicine* Plant revered and used in important ritual and ceremonial sacraments. (70:104, 110) **Paiute** *Unspecified* Plant used by one shaman for curing. (111:91) **Ponca** *Hallucinogen* Dried flesh "buttons" eaten to cause auditory and visual hallucinations. *Unspecified* Decoction of dried flesh "buttons" taken for illness. (94:48) **Winnebago** *Ceremonial Medicine* Plant revered and used in important ritual and ceremonial sacraments. (70:104, 110)
- *Other*—**Blackfoot** *Ceremonial Items* Plant used in ceremonial rites of the Native American Church. (97:45) **Ponca** *Ceremonial Items* Flesh dried into "buttons" and eaten during religious ceremonies. (94:48)

Lotus, Fabaceae
Lotus humistratus Greene, Foothill Deervetch
- *Drug*—**Karok** *Gynecological Aid* Infusion of plant taken and used as a wash by women in labor. (148:385)

Lotus mearnsii (Britt.) Greene, Mearns's Birdsfoot Trefoil
- *Food*—**Havasupai** *Unspecified* Species used for food. (197:226)

Lotus procumbens (Greene) Greene, Silky Deerweed
- *Food*—**Kawaiisu** *Spice* Plant added to the dry pine needles spread as a layer in the pit roasting of the yucca. (206:38)
- *Fiber*—**Kawaiisu** *Building Material* Plant used as wall filler in the construction of the winter house. (206:38)

Lotus scoparius (Nutt.) Ottley, Common Deerweed
- *Drug*—**Costanoan** *Cough Medicine* Decoction of foliage used for coughs. (21:19)
- *Food*—**Diegueño** *Fodder* Leaves fed to domesticated animals. (88:218) **Tubatulabal** *Unspecified* Leaves used for food. (193:15)
- *Fiber*—**Cahuilla** *Building Material* Plant used as a material in house construction. (15:87) **Costanoan** *Building Material* Foliage used for house thatching. (21:250)
- *Other*—**Diegueño** *Soap* Roots used for soap. (88:218)

Lotus scoparius (Nutt.) Ottley var. *scoparius*, Common Deerweed

- **Drug**—Mahuna *Blood Medicine* Infusion of plant taken to build the blood. (as *Hosackia glabra* 140:34)

Lotus strigosus (Nutt.) Greene, Bishop Lotus
- **Food**—Luiseño *Vegetable* Plant used for greens. (155:231)

Lotus unifoliolatus (Hook.) Benth. **var. *unifoliolatus***, Prairie Trefoil
- **Food**—Kawaiisu *Spice* Plant used as a mat for the juniper cake which improves the taste of the cake. (as *L. purshianus* 206:39) **Miwok** *Cooking Agent* Green leaves pounded with oily acorns, to absorb some of the oil. (as *L. americanus* 12:144)

Lotus wrightii (Gray) Greene, Wright's Deervetch
- **Drug**—Navajo, Ramah *Analgesic* Decoction of leaves used for stomachache. *Cathartic* Decoction of leaves used as a cathartic. *Disinfectant* Decoction of leaves used for "deer infection." *Gastrointestinal Aid* Decoction of leaves used for stomachache. *Panacea* Plant used as "life medicine." (191:32) **Zuni** *Witchcraft Medicine* Poultice of chewed root applied to swellings caused by being witched by a bullsnake. (27:376)
- **Food**—Isleta *Forage* Considered an excellent grazing plant for sheep. (100:34)
- **Other**—Keres, Western *Unspecified* Plant known and named but no use was specified. (171:52)

Ludwigia, Onagraceae

Ludwigia bonariensis (M. Micheli) Hara, Carolina Primrosewillow
- **Drug**—Hawaiian *Blood Medicine* Infusion of pounded whole plants and other plants strained and taken to purify the blood. *Dermatological Aid* and *Pediatric Aid* Seeds or root pulp used by children on small cuts or scratches. (as *Jussiaea villosa* 2:48)

Ludwigia virgata Michx., Savannah Primrosewillow
- **Drug**—Seminole *Dermatological Aid* Decoction of root taken and used as a body steam for snake sickness: itchy skin. (169:239)

Luetkea, Rosaceae

Luetkea pectinata (Pursh) Kuntze, Partridgefoot
- **Drug**—Okanagon *Analgesic* Decoction of plant taken for abdominal pains. *Dermatological Aid* Poultice of crushed plant applied to sores. *Gastrointestinal Aid* Decoction of plant taken for abdominal pains. *Gynecological Aid* Decoction of plant taken by women for profuse or prolonged menstruation. (125:42) **Thompson** *Analgesic* Decoction of plant taken for abdominal pains. (164:472) *Dermatological Aid* Poultice of crushed plant applied to sores. (125:42) Poultice of crushed, fresh plant applied to sores. (164:472) *Gastrointestinal Aid* Decoction of plant taken for abdominal pains. (125:42) Decoction of plant taken for abdominal pain. (164:472) *Gynecological Aid* Decoction of plant taken by women for profuse or prolonged menstruation. (125:42) Decoction of plant taken for profuse or prolonged menstruation. (164:472)

Lupinus, Fabaceae

Lupinus affinis J. G. Agardh, Fleshy Lupine
- **Food**—Mendocino Indian *Vegetable* Young leaves formerly roasted and eaten as greens. (as *L. carnosulus* 41:357)

- **Other**—Pomo, Kashaya *Ceremonial Items* Flowers used in wreaths for the Flower Dance performed at the Strawberry Festival in May. (72:65)

Lupinus albifrons Benth. ex Lindl., Silver Lupine
- **Drug**—Karok *Gastrointestinal Aid* Decoction of plant taken and used as a steam bath for stomach troubles. (148:385)
- **Other**—Pomo, Kashaya *Ceremonial Items* Flowers used in wreaths for the Flower Dance performed at the Strawberry Festival in May. (72:65)

Lupinus arboreus Sims, Bush Lupine
- **Fiber**—Pomo *Cordage* Root fibers used for string. (66:13) **Pomo, Kashaya** *Cordage* Root fibers used to make string for fish nets, deer and rabbit nets, gill nets, and carrying nets. (72:65)
- **Other**—Pomo, Kashaya *Ceremonial Items* Flowers used in wreaths for the Flower Dance performed at the Strawberry Festival in May. (72:65)

Lupinus arcticus S. Wats., Arctic Lupine
- **Drug**—Eskimo, Inupiat *Poison* Seeds considered poisonous. (98:143)

Lupinus argenteus* ssp. *ingratus (Greene) Harmon, Silvery Lupine
- **Drug**—Navajo, Ramah *Dermatological Aid* Poultice of crushed leaves applied to poison ivy blisters. (as *L. ingratus* 191:32)

Lupinus brevicaulis S. Wats., Shortstem Lupine
- **Drug**—Navajo *Dermatological Aid* Plant rubbed on as a liniment for boils. *Reproductive Aid* Plant used for sterility. (55:56)
- **Other**—Navajo *Ceremonial Items* Used in the female shooters branch of the Lightning Chant. (55:56)

Lupinus caudatus* ssp. *argophyllus (Gray) L. Phillips, Kellogg's Spurred Lupine
- **Drug**—Navajo, Ramah *Ceremonial Medicine* Leaves used as a ceremonial emetic. *Dermatological Aid* Cold infusion of leaves used as a lotion on poison ivy blisters. *Emetic* Leaves used as a ceremonial emetic. (as *L. aduncus* 191:32)

Lupinus densiflorus Benth., Whitewhorl Lupine
- **Food**—Miwok *Unspecified* Steamed leaves and flowers eaten with acorn soup. (12:159)
- **Other**—Pomo, Kashaya *Ceremonial Items* Flowers used in wreaths for the Flower Dance performed at the Strawberry Festival in May. (72:65)

Lupinus kingii S. Wats., King's Lupine
- **Drug**—Hopi *Eye Medicine* Plant used as an eye medicine. (200:33, 80) **Navajo, Ramah** *Dermatological Aid* Poultice of crushed leaves used for poison ivy blisters and other skin irritations. *Panacea* Leaves used as "life medicine." (191:32)

Lupinus latifolius Lindl. ex J. G. Agardh, Broadleaf Lupine
- **Food**—Miwok *Sauce & Relish* Steamed, dried leaves and flowers boiled and used as a relish with manzanita cider. *Winter Use Food* Steamed leaves and flowers dried and stored for winter use. (12:159)
- **Other**—Miwok *Containers* Leaves used to line acorn leaching basket, to prevent meal from running through the interstices. (12:146)

Lupinus littoralis Dougl., Seashore Lupine
- **Drug**—Kwakiutl *Pediatric Aid* and *Sedative* Root ash rubbed into a newborn baby's cradle to make infant sleep well. (183:284)
- **Food**—Haisla & Hanaksiala *Unspecified* Roots peeled and eaten raw. (43:249) **Kwakiutl, Southern** *Unspecified* Fleshy taproots eaten raw, boiled, or steamed in spring. If eaten raw, these roots caused dizziness. Therefore, they were usually eaten raw only before bedtime in the evening. (183:284)

Lupinus luteolus Kellogg, Pale Yellow Lupine
- **Food**—Mendocino Indian *Forage* Succulent tops eaten sparingly by horses in early summer. *Vegetable* Plant tops eaten as greens. (41:358)
- **Other**—Pomo, Kashaya *Ceremonial Items* Flowers used in wreaths for the Flower Dance performed at the Strawberry Festival in May. (72:65)

Lupinus lyallii Gray, Dwarf Mountain Lupine
- **Drug**—Navajo *Dermatological Aid* Plant used for boils. (55:97)
- **Dye**—Navajo *Blue* Flowers used to make a blue dye. *Green* Used to make a green dye. (55:57)

Lupinus nanus* ssp. *latifolius (Benth. ex Torr.) D. Dunn, Sky Lupine
- **Other**—Pomo, Kashaya *Ceremonial Items* Flowers used in wreaths for the Flower Dance performed at the Strawberry Festival in May. (72:65)

Lupinus nootkatensis Donn ex Sims, Nootka Lupine
- **Drug**—Alaska Native *Poison* Roots considered poisonous. (85:157)
- **Food**—Alaska Native *Unspecified* Roots peeled and inner portion eaten raw or boiled. (85:157)

Lupinus nootkatensis* var. *fruticosus Sims, Nootka Lupine
- **Food**—Haisla & Hanaksiala *Unspecified* Roots peeled and eaten raw. (43:249) **Kimsquit** *Unspecified* Roots formerly roasted and used for food. (184:205)

Lupinus nootkatensis Donn ex Sims **var. *nootkatensis***, Nootka Lupine
- **Food**—Haisla & Hanaksiala *Unspecified* Roots peeled and eaten raw. (43:249)

Lupinus perennis L., Sundial Lupine
- **Drug**—Cherokee *Antiemetic* Cold infusion taken and used as wash "to check hemorrhage and vomiting." *Antihemorrhagic* Cold infusion taken and used as wash "to check hemorrhage and vomiting." (80:43, 44) **Menominee** *Veterinary Aid* Plant used to fatten a horse and make him spirited and full of fire. *Witchcraft Medicine* Plant rubbed on hands or body to give person power to control horses. (151:40)

Lupinus polyphyllus Lindl., Bigleaf Lupine
- **Drug**—Salish *Tonic* Decoction of plants used as a tonic. (178:293) **Thompson** *Poison* Plant considered poisonous. (187:224) *Unspecified* Plant used medicinally for unspecified purpose. (164:461) *Veterinary Aid* Plant eaten by horses as medicine. (187:224)
- **Food**—Kwakiutl *Unspecified* Roots eaten fresh or steamed. (182:84)
- **Other**—Pomo, Kashaya *Ceremonial Items* Flowers used in wreaths for the Flower Dance performed at the Strawberry Festival in May. (72:65)

Lupinus pusillus Pursh, Rusty Lupine
- **Drug**—Hopi *Ear Medicine* Plant used as an ear medicine. *Eye Medicine* Plant used as an eye medicine. (42:333)
- **Other**—Hopi *Ceremonial Items* Juice used as holy water in the Po-wa-mu ceremony. (42:333)

Lupinus pusillus* ssp. *intermontanus (Heller) D. Dunn, Intermountain Lupine
- **Drug**—Navajo, Kayenta *Disinfectant* Plant used as a fumigant ingredient. *Ear Medicine* Plant used for earaches. *Hemostat* Plant used for nosebleeds. (205:28)

Lupinus rivularis Dougl. ex Lindl., Riverbank Lupine
- **Drug**—Thompson *Unspecified* Plant used medicinally for unspecified purpose. (164:461)

Lupinus sericeus Pursh, Silky Lupine
- **Drug**—Okanagan-Colville *Eye Medicine* Seeds pounded, mixed with water, strained, and resulting liquid used as an eye medicine. (188:105)
- **Food**—Okanagan-Colville *Forage* Plant considered the marmot's favorite food. (188:105)
- **Fiber**—Okanagan-Colville *Mats, Rugs & Bedding* Plants used for bedding and as flooring in the sweat house. (188:105)
- **Other**—Okanagan-Colville *Season Indicator* Blooms indicated that groundhogs were fat enough to eat. (188:105)

Lupinus sericeus Pursh **var. *sericeus***, Silky Lupine
- **Drug**—Thompson *Poison* Plant considered poisonous. *Veterinary Aid* Plant eaten by horses as medicine. (187:224)

***Lupinus* sp.**, Lupine
- **Drug**—Blackfoot *Gastrointestinal Aid* Infusion of plant taken for indigestion and gas. (86:67) *Misc. Disease Remedy* Infusion of roots taken and rubbed on mumps. (86:77) *Respiratory Aid* Infusion of plant taken for hiccups. (86:72) *Veterinary Aid* Infusion of leaves applied to wounds caused by small biting flies, especially on the chest and udder. (86:89) **Paiute** *Diuretic* Plant used for failure to urinate. (167:317) Plant used for "failure in urination." (180:102) *Urinary Aid* Plant used for bladder trouble. (167:317) **Shoshoni** *Diuretic* Plant used for "failure in urination." (180:102) **Thompson** *Poison* Plant considered poisonous. *Veterinary Aid* Plant eaten by horses as medicine. (187:224)
- **Food**—Costanoan *Staple* Seeds used for pinole. (21:250) **Kitasoo** *Unspecified* Roots used for food. (43:337) **Luiseño** *Unspecified* Leaves used for food. (15:87) *Vegetable* Plant used for greens. (155:231) **Paiute** *Forage* Plants used for horse and cattle food. (111:86) **Thompson** *Fodder* Used as a fodder for horses and cattle. (164:514) **Yavapai** *Vegetable* Boiled leaves used for greens. (65:257) **Yuki** *Vegetable* Young plants roasted and eaten as greens. (49:88)
- **Other**—Blackfoot *Ceremonial Items* Leaves chewed by ceremonialist, to reinforce his powers, before he undertook any face painting. *Incense & Fragrance* Used as incense in the Ghost Dance. (86:38) **Navajo** *Ceremonial Items* Used in the Male Shooting Chant. (55:56)

Lupinus succulentus Dougl. ex K. Koch, Hollowleaf Annual Lupine
- **Other**—Pomo, Kashaya *Ceremonial Items* Flowers used in wreaths

for the Flower Dance performed at the Strawberry Festival in May. (72:65)

Lupinus sulphureus Dougl. ex Hook., Sulphur Lupine
- *Drug*—Okanagan-Colville *Eye Medicine* Seeds pounded, mixed with water, strained, and resulting liquid used as an eye medicine. (188:105)
- *Food*—Okanagan-Colville *Forage* Plant considered the marmot's favorite food. (188:105)
- *Fiber*—Okanagan-Colville *Mats, Rugs & Bedding* Plants used for bedding and as flooring in the sweat house. (188:105)
- *Other*—Okanagan-Colville *Season Indicator* Blooms indicated that groundhogs were fat enough to eat. (188:105)

Lupinus versicolor Lindl., Manycolored Lupine
- *Other*—Pomo, Kashaya *Ceremonial Items* Flowers used in wreaths for the Flower Dance performed at the Strawberry Festival in May. (as *L. variicolor* 72:65)

Lupinus wyethii S. Wats., Wyeth's Lupine
- *Drug*—Okanagan-Colville *Eye Medicine* Seeds pounded, mixed with water, strained, and resulting liquid used as an eye medicine. (188:105)
- *Food*—Okanagan-Colville *Forage* Plant considered the marmot's favorite food. (188:105)
- *Fiber*—Okanagan-Colville *Mats, Rugs & Bedding* Plants used for bedding and as flooring in the sweat house. (188:105)
- *Other*—Okanagan-Colville *Season Indicator* Blooms indicated that groundhogs were fat enough to eat. (188:105)

Luzula, Juncaceae

Luzula multiflora (Ehrh.) Lej., Common Woodrush
- *Drug*—Navajo, Ramah *Ceremonial Medicine* Plant used as a ceremonial emetic and for other ceremonial purposes. *Emetic* Plant used as a ceremonial emetic. (191:20)

Luzula sp., Woodrush
- *Drug*—Iroquois *Other* Decoction used to increase strength to physically punish one of bewitchment. (87:279)

Lycium, Solanaceae

Lycium andersonii (Pursh) Nutt., Anderson's Wolfberry
- *Food*—Paiute, Northern *Dried Food* Berries dried in the sand for winter use. (59:50)

Lycium andersonii Gray, Anderson's Wolfberry
- *Food*—Cahuilla *Dried Food* Dried berries boiled into mush or ground into flour and mixed with water. *Fruit* Berries eaten fresh. (15:87) **Kawaiisu** *Beverage* Fruit juice used as a beverage. *Dried Food* Fruit mashed, dried, soaked in warm water for an hour, and eaten. *Fruit* Fruit eaten fresh. (206:39) **Mohave** *Beverage* Berries crushed, strained, and used as a drink. *Dried Food* Berries dried like raisins. (37:205) **Paiute, Northern** *Fruit* Berries eaten fresh and crushed or mixed with water. *Porridge* Berries dried, mashed, and eaten like a mush. (59:50)

Lycium exsertum Gray, Arizona Desertthorn
- *Food*—Yuma *Beverage* Berries gathered, washed, boiled, ground, mixed with water, and used as a beverage. *Dried Food* Berries sun dried, stored, and eaten without preparation. Berries washed, boiled, dried, and stored. *Porridge* Berries washed, boiled, strained, mashed, and wheat added to make mush. (37:204)

Lycium fremontii Gray, Frémont's Desertthorn
- *Food*—Cahuilla *Dried Food* Dried berries boiled into mush or ground into flour and mixed with water. *Fruit* Berries eaten fresh. (15:87) **Maricopa** *Fruit* Black berries used for food. (95:265) **Papago** *Dried Food* Berries dried and eaten like raisins. (34:19) *Fruit* Berries used for food. (36:62) **Pima** *Beverage* Red berries boiled, mashed, and the liquid used as a beverage. (47:87) *Fruit* Red berries cooked and eaten warm or cold with sugar. (95:262) Red berries boiled and eaten. (146:75) **Yuma** *Beverage* Berries gathered, washed, boiled, ground, mixed with water, and used as a beverage. *Dried Food* Berries sun dried, stored, and eaten without preparation. Berries washed, boiled, dried, and stored. *Porridge* Berries washed, boiled, strained, mashed, and wheat added to make mush. (37:204)
- *Other*—Papago *Hunting & Fishing Item* Used to make bows. (34:70)

Lycium pallidum Miers, Pale Wolfberry
- *Drug*—Hopi *Ceremonial Medicine* Plant used at the annual "Nimankatcina" ceremony. (56:19) **Navajo, Kayenta** *Toothache Remedy* Ground root placed in cavity for toothaches. (205:41) **Navajo, Ramah** *Ceremonial Medicine* and *Emetic* Leaves or root used as ceremonial emetic. *Misc. Disease Remedy* Plant used for chickenpox and poultice of plant used for toothaches. *Panacea* Bark and dried berries used as "life medicine." *Toothache Remedy* Poultice of heated root applied for toothache. (191:42)
- *Food*—Acoma *Sauce & Relish* Berries cooked into a syrup. (32:33) **Havasupai** *Beverage* Dried berries ground and mixed with water to make a drink. *Dried Food* Berries sun dried for future use. (197:239) **Hopi** *Fruit* Berries eaten fresh from the shrub. (56:19) Berries eaten. (138:47) *Porridge* Ground berries mixed with "potato clay" and eaten. (42:332) *Preserves* Berries cooked to make a jam-like food and served with fresh piki bread. (120:19) *Starvation Food* Berries boiled, ground, mixed with "potato clay," and eaten during past famines. (200:89) *Unspecified* Seeds eaten. (190:166) **Isleta** *Fruit* Fresh, summer berries eaten for food. (100:34) **Jemez** *Fruit* Ripe or cooked berries used for food. *Special Food* Unripe berries stewed, sweetened, and eaten as a delicacy. (32:33) **Keres, Western** *Sauce & Relish* Cooked berries made into a syrup. (171:52) **Laguna** *Sauce & Relish* Berries cooked into a syrup. (32:33) **Navajo** *Beverage* Berries mashed in water and used as a beverage. (110:32) *Dried Food* Fruits boiled, dried, stored for winter use, and eaten dry. (55:74) Sun dried berries used for food. (110:32) *Fruit* Fruits eaten fresh. (55:74) Berries used for food. (90:153) Berries eaten fresh off the bush. (110:32) Fresh, mashed berries mixed with powdered clay to counteract astringency and used for food. (165:222) *Soup* Fruits boiled, dried, stored for winter use, and made into a soup. (55:74) Berries used to make soup and stew. (110:32) *Special Food* Fruit sacrificed to the gods. (55:74) *Winter Use Food* Fresh berries soaked, boiled until tender, ground with clay, and stored for winter use. (165:222) **Navajo, Ramah** *Dried Food* Berries dried and boiled with clay, sugar, or wild potatoes. *Fruit* Berries eaten raw or boiled with clay. (191:42) **Zuni** *Fruit* Berries eaten raw when perfectly ripe or boiled and sometimes sweetened. (166:68)

- *Other*—Hopi *Ceremonial Items* Whole shrub used in Niman kachina dance. (42:332) **Navajo** *Sacred Items* Plant considered to be a sacred plant. (56:19) **Navajo, Ramah** *Ceremonial Items* Thorn ash used for Evilway blackening. Stem used to make Evilway big hoop. (191:42) **Zuni** *Protection* Ground leaves, twigs, and flowers given to warriors for protection during war. A pinch of the mixture was given to each warrior. The warriors placed it in their mouths, ejected the mass into their hands, and rubbed in on their faces, arms, and bodies so that the enemy's arrows could not harm them. (166:94)

Lycium sp., Wolfberry
- *Food*—**Pima, Gila River** *Fruit* Fruits eaten raw and boiled. (133:7)

Lycium torreyi Gray, Squawthorn
- *Drug*—**Navajo, Ramah** *Ceremonial Medicine* and *Emetic* Leaves or root used as ceremonial emetic. *Misc. Disease Remedy* Plant used for chickenpox and poultice of plant used for toothaches. *Panacea* Bark and dried berries used as "life medicine." *Toothache Remedy* Poultice of heated root applied for toothache. (191:42)
- *Food*—**Navajo, Ramah** *Dried Food* Berries dried and boiled with clay, sugar, or wild potatoes. *Fruit* Berries eaten raw or boiled with clay. (191:42) **Tubatulabal** *Fruit* Berries used extensively for food. (193:15)
- *Other*—**Navajo, Ramah** *Ceremonial Items* Stem used to make Evilway big hoop. Thorn ash used for Evilway blackening. (191:42)

Lycoperdaceae, see *Bovista, Bovistella, Calvatia, Geastrum, Lycoperdon*

Lycoperdon, Lycoperdaceae, fungus
Lycoperdon sp., Puffball
- *Drug*—**Blackfoot** *Antihemorrhagic* Spores mixed with water and taken for internal hemorrhage. *Hemostat* Poultice of spores applied to wounds as a styptic. Plant pieces held to the nose for nosebleeds. (86:84) *Veterinary Aid* Plant pieces applied as a styptic on castration wounds and other cuts. (86:89) **Haisla & Hanaksiala** *Poison* Spores dangerous, especially harmful to the eyes. (43:134) **Pomo, Kashaya** *Poison* Plant considered poisonous. (72:132)
- *Other*—**Blackfoot** *Ceremonial Items* Puffballs figured into religious life. Puffballs were thought to be stars that had fallen to earth during supernatural events. There was a story about the woman who married Morning Star and had a child by this supernatural being. When she returned to earth with the Natoas bundles and her child, she was directed by the star personage to keep her baby from touching the ground for 14 days. She managed all right until the day she went for wood and left the child in the care of a grandmother. The grandmother was careless and the baby touched the ground. It turned into a large puffball and returned to the heavens as the Fixed Star (North Star), plugging the hole left by the woman when she pulled out the Holy Turnip. Used in the Firelighters bundle of the Horn Society for use as punk to light a fire easily. *Designs* Small, painted circles at the base of the tepee represented puffballs to insure fire to those within. *Incense & Fragrance* Puffballs used as incense to keep ghosts away. (86:38)

Lycopersicon, Solanaceae
Lycopersicon esculentum P. Mill., Garden Tomato
- *Food*—**Haisla & Hanaksiala** *Fruit* Fruit used for food. (43:291) **Seminole** *Unspecified* Plant used for food. (169:508)

Lycopodiaceae, see Huperzia, Lycopodium

Lycopodium, Lycopodiaceae, club moss
Lycopodium annotinum L., Stiff Clubmoss
- *Fiber*—**Shuswap** *Building Material* Moss mixed with clay and used between the logs of a log house. (123:49)
- *Other*—**Cree, Woodlands** *Cooking Tools* Plant used to separate raw fish eggs from the membranes. (109:44) **Shuswap** *Fertilizer* Moss mixed with house plant dirt as a fertilizer to make the plants healthier. (123:49)

Lycopodium clavatum L., Running Clubmoss
- *Drug*—**Aleut** *Analgesic* and *Gynecological Aid* Infusion of plant taken for postpartum pain. (8:427) **Carrier, Southern** *Analgesic* Moss inserted into the nose to cause bleeding for headaches. (150:48) **Montagnais** *Febrifuge* "Brew" from plant used for weakness and fever. *Stimulant* Compound containing plant used for weakness and fever. (156:315) **Potawatomi** *Hemostat* Spores of fruiting spikes used as a styptic and coagulant. (154:64)
- *Fiber*—**Hanaksiala** *Clothing* Plant used as a belt for the blankets that were worn. (43:157)
- *Other*—**Bella Coola** *Decorations* Used to make wreaths. (184:196) **Hesquiat** *Decorations* Used by children to make Christmas decorations. (185:29) **Oweekeno** *Jewelry* Plant used to make a decorative necklace worn during festive occasions. (43:59) **Thompson** *Decorations* Plant used as a Christmas decoration. One informant used it as a Christmas decoration but was told not to use it by her chief as it was considered to bring bad luck. He said, "that's for the devil." (187:87)

Lycopodium complanatum L., Groundcedar
- *Drug*—**Blackfoot** *Dermatological Aid* Spores applied as an antiseptic dust on wounds. *Hemostat* Spores snuffed for nosebleed. *Pulmonary Aid* Decoction of plant used for lung diseases. *Venereal Aid* Decoction of plant used for venereal diseases. (97:16) **Iroquois** *Reproductive Aid* Compound decoction taken to induce pregnancy. (87:263) **Ojibwa** *Stimulant* Dried leaves used as a reviver. (153:375)
- *Dye*—**Blackfoot** *Mordant* Whole plant used as a mordant to set certain dyes. (97:16)

Lycopodium dendroideum Michx., Tree Groundpine
- *Drug*—**Montagnais** *Cathartic* and *Gastrointestinal Aid* Decoction of plant taken as a purgative "in case of biliousness." (156:316) **Penobscot** *Unspecified* Plant thought to have "some medicinal value." (156:309)

Lycopodium obscurum L., Rare Clubmoss
- *Drug*—**Chippewa** *Antirheumatic (External)* Compound decoction of moss used as herbal steam for rheumatism. (53:362) **Iroquois** *Blood Medicine* Cold, compound decoction taken for weak blood. *Gynecological Aid* Decoction of root taken for change of life and resulting blindness and deafness. (87:263) **Ojibwa** *Diuretic* Plant combined with *Diervilla lonicera* and taken as a diuretic. (153:375) **Potawatomi** *Hemostat* Spores of fruiting spikes used as a styptic and coagulant. (154:64)

- *Other*—Cree, Woodlands *Cooking Tools* Plant used to separate raw fish eggs from the membranes. (109:44)

Lycopodium sabinifolium Willd., Savinleaf Groundpine
- *Drug*—Iroquois *Venereal Aid* Compound decoction with plant taken for gonorrhea. (87:263)

Lycopodium sp., Clubmoss
- *Drug*—Algonquin, Quebec *Gynecological Aid* Used to make a medicinal tea for inducing labor and making childbirth easier. *Pediatric Aid* and *Urinary Aid* Used to make a medicinal tea for children with bladder trouble. (18:120) Iroquois *Anticonvulsive* Compound decoction taken by men for epilepsy. *Diuretic* Compound decoction used as a diuretic. *Hemostat* Smoke (spores) from plant sprinkled on nosebleed. (87:262) Micmac *Febrifuge* Herb used for fever. (as *Lycopodia* 40:58)

Lycopus, Lamiaceae

Lycopus americanus Muhl. ex W. Bart., American Waterhorehound
- *Drug*—Meskwaki *Analgesic* and *Gastrointestinal Aid* Compound containing entire plant used for stomach cramps. (152:225)

Lycopus asper Greene, Rough Bugleweed
- *Drug*—Iroquois *Laxative* and *Pediatric Aid* Decoction of plants given to children as a laxative. *Poison* Plant considered poisonous. (as *L. lucidus* ssp. *americanus* 87:427)
- *Food*—Chippewa *Dried Food* Dried, boiled, and used for food. (53:320)

Lycopus sp., Waterhorehound
- *Drug*—Blackfoot *Cold Remedy* and *Pediatric Aid* Plant compounded with other plants and used for children's colds. (97:51)

Lycopus uniflorus Michx., Northern Bugleweed
- *Food*—Okanagon *Staple* Roots used as a principal food. (as *Cycopus uniflorus* 178:238) Thompson *Dessert* Cooked tuberous root eaten for dessert. (187:232) *Unspecified* Roots eaten. (164:480) Tuberous root steamed or baked and used for food. (187:232)

Lycopus virginicus L., Virginia Waterhorehound
- *Drug*—Cherokee *Ceremonial Medicine* Infusion taken at green corn ceremony. *Other* and *Pediatric Aid* Chewed root given to infants to give them "eloquence of speech." *Snakebite Remedy* Root chewed, a portion swallowed, the rest applied to snakebite wound. *Veterinary Aid* Decoction fed to snakebitten dog. (80:39) Iroquois *Poison* Roots and leaves considered poisonous. (87:426)

Lygodesmia, Asteraceae

Lygodesmia grandiflora (Nutt.) Torr. & Gray, Largeflower Skeletonplant
- *Drug*—Gosiute *Veterinary Aid* Plant used as a horse medicine. (39:374) Hopi *Gynecological Aid* Leaves chewed to increase mother's milk supply. (200:36, 97) Navajo, Kayenta *Dermatological Aid* Plant milk applied to sores caused by sunburn. (205:48)
- *Food*—Hopi *Spice* Boiled with a certain kind of mush for flavor. (190:168) *Unspecified* Leaves boiled with meats and eaten. (56:19) Leaves boiled with meat. (200:97) Navajo, Kayenta *Vegetable* Used for greens in foods. (205:48)

Lygodesmia juncea (Pursh) D. Don ex Hook., Rush Skeletonplant
- *Drug*—Blackfoot *Cough Medicine* Infusion of stems taken for burning coughs. (86:72) *Dermatological Aid* Infusion of stems mixed with grease and applied as a hair tonic. (86:124) *Diuretic* Infusion of powdered galls taken as a diuretic. (86:70) *Gastrointestinal Aid* Decoction of plant taken for heartburn. (97:61) *Gynecological Aid* Decoction of plant taken for symptoms resembling heartburn caused by pregnancy. (86:61) *Kidney Aid* Infusion of plant taken for kidney trouble. (86:70) *Orthopedic Aid* Crushed stems used as foot pads in moccasins. (86:115) *Pediatric Aid* and *Tonic* Infusion of plant used as a general tonic for children. (86:67) *Veterinary Aid* Infusion of plant rubbed on saddle sores and leg wounds. Infusion of plant given to horses with coughs. (86:89) Cheyenne *Breast Treatment* Infusion of dried stems taken to increase milk flow. (82:27) *Gynecological Aid* Simple or compound decoction of plant taken to increase maternal milk flow. (75:41) Infusion of plant taken by women after childbirth to increase milk flow. (76:191) Infusion of stems taken by pregnant and nursing mothers to increase milk flow. (83:22) *Misc. Disease Remedy* Infusion of leaves taken for smallpox and measles. (76:191) *Pediatric Aid* Infusion of stems taken by pregnant and nursing mothers for a healthy baby. (83:22) *Psychological Aid* Infusion of dried stems taken to bring feelings of contentment to mothers. (82:27) Keres, Western *Antirheumatic (External)* Poultice of crushed plant, warmed with rocks, applied to swellings. (171:53) Lakota *Antidiarrheal* and *Pediatric Aid* Infusion of whole plant used for children with diarrhea. (139:38) Omaha *Eye Medicine* Infusion of stems used as a wash for sore eyes. *Gynecological Aid* Infusion of stems taken by nursing mothers to increase milk flow. (70:136) Ponca *Antidiarrheal* Infusion of plant taken for diarrhea. (94:152) *Eye Medicine* Infusion of stems used as a wash for sore eyes. (70:136) Infusion of plant applied to sore eyes. (94:152) *Gynecological Aid* Infusion of stems taken by nursing mothers to increase milk flow. (70:136) Infusion of plant used to increase milk flow of mothers. (94:152) Sioux *Eye Medicine* Infusion of dried stems used for sore eyes. *Gynecological Aid* Infusion of dried stems taken to increase milk flow. (82:27)
- *Food*—Lakota *Unspecified* Plant chewed. (139:38) Navajo, Ramah *Candy* Roots left in the sun until gum came out and hardened and used for chewing gum. (191:52) Sioux *Unspecified* Hardened juice chewed for the flavor. (82:27)
- *Other*—Blackfoot *Waterproofing Agent* Stems used to waterproof newly tanned buffalo hides. (86:115) Keres, Western *Decorations* Plant used for decorations in ceremonial dances. (171:53)

Lygodesmia sp., Desert Gum
- *Food*—Washo *Candy* Plant gum used for chewing gum. (118:56)

Lyonia, Ericaceae

Lyonia fruticosa (Michx.) G. S. Torr., Coastalplain Staggerbush
- *Other*—Seminole *Smoking Tools* Plant used to make pipe bowls. (as *Xolisma fruticosa* 169:466)

Lyonia mariana (L.) D. Don, Piedmont Staggerbush
- *Drug*—Cherokee *Dermatological Aid* Infusion used for toe itch, "ground-itch," and ulcers. (80:57)

Lysichiton, Araceae

Lysichiton americanus Hultén & St. John, American Skunk-cabbage

- ***Drug***—**Bella Coola** *Gastrointestinal Aid* Decoction of root taken for stomach trouble. (as *L. kamtschatcense* 150:52, 53) **Clallam** *Dermatological Aid* Poultice of roots used for sores. (57:196) **Cowlitz** *Antirheumatic (External)* Poultice of heated blossoms applied to the body for rheumatism. (79:22) **Gitksan** *Antihemorrhagic* Compound containing root used as plaster on the chest for lung hemorrhages. *Antirheumatic (External)* Compound containing root used for rheumatism. Poultice applied or leaves sat on or lain on in sweat bath for rheumatism. *Antirheumatic (Internal)* Smoke of root inhaled for influenza, rheumatism, and bad dreams. *Dermatological Aid* Simple or compound poultice of mashed root applied for blood poisoning and boils. *Misc. Disease Remedy* Smoke from burning roots inhaled for influenza and bad dreams. *Poison* Roots considered poisonous. *Pulmonary Aid* Compound containing root used as plaster on the chest for lung hemorrhages. *Sedative* Smoke of root inhaled for bad dreams, influenza, and rheumatism. (as *L. kamtschatcense* 150:52, 53) **Haisla & Hanaksiala** *Burn Dressing* Poultice of pounded root paste applied to burns. *Urinary Aid* Roots used experimentally for bloody urine. (43:189) **Hesquiat** *Burn Dressing* Poultice of cold and fresh leaves applied for burns. (186:78) *Dermatological Aid* Poultice of cool leaves used for bad burns. *Unspecified* Roots used as a medicine. (185:48) **Klallam** *Dermatological Aid* Poultice of baked roots applied to carbuncles. *Tuberculosis Remedy* Poultice of leaves applied to parts of the body sore with scrofula. (79:22) **Kwakiutl** *Dermatological Aid* Poultice of leaf, oil, down, and Douglas fir bark applied to carbuncles. (183:270) Heated leaves used to draw out thorns and splinters. Poultice of steamed, mashed roots applied to swellings. Poultice of washed, heated leaves applied to boils, carbuncles, and sores. Pulverized root rubbed into a child's head to make his hair grow. *Herbal Steam* Leaves used in a sweat bath for general weakness or undefined sickness. *Other* Leaves used in a sweat bath for undefined sickness. *Pediatric Aid* Pulverized root rubbed into a child's head to make his hair grow. *Stimulant* Leaves used in a sweat bath for general weakness. (183:271) **Makah** *Abortifacient* Raw root chewed by women to effect an abortion. (79:22) Roots chewed to induce an abortion. (186:78) *Analgesic* Poultice of warmed leaves applied for chest pain. (79:22) Warmed leaves applied to chest for pain. (186:78) *Antirheumatic (External)* Roots used for arthritis. (67:336) *Blood Medicine* Decoction of roots taken as a blood purifier. *Gastrointestinal Aid* Root chewed to soothe the stomach after taking an emetic. *Pulmonary Aid* Poultice of warmed leaves applied for chest pain. (79:22) *Unspecified* Roots used medicinally for unspecified purpose. (67:336) **Nitinaht** *Burn Dressing* Poultice of one leaf applied for severe burns. (186:78) **Quileute** *Analgesic* Leaves used for headaches. *Dermatological Aid* Poultice of leaves applied to cuts and swellings. *Febrifuge* Leaves used for fevers. *Gynecological Aid* Decoction of pounded root taken to bring about easy delivery. **Quinault** *Panacea* Poultice of leaves applied for many ailments. *Urinary Aid* Decoction of roots taken to clean out the bladder. **Samish** *Other* Infusion of roots used as a wash for invalids. (79:22) **Shuswap** *Analgesic* Poultice of leaves applied for pain, particularly pains in the knees. *Dermatological Aid* Poultice of leaves applied to sores. *Orthopedic Aid* Poultice of leaves applied for pain, particularly pains in the knees. *Panacea* Cold infusion of roots taken for any sickness. (123:53) **Skokomish** *Analgesic* Leaves used for headaches. *Cathartic* Infusion of roots taken as a physic. *Dermatological Aid* Poultice of leaves applied to cuts and swellings. *Febrifuge* Leaves used for fevers. **Swinomish** *Other* Infusion of roots used as a wash for invalids. (79:22) **Thompson** *Dermatological Aid* Powdered, charred rhizome mixed with bear grease, used as an ointment for animal bites and infections. Charcoal used for wounds. The charcoal was applied four times, the fourth time being mixed with bear grease. *Psychological Aid* Leaves placed under pillows during sleep or the head washed with charcoal to induce "power dreams." (187:113) **Tolowa** *Antirheumatic (External)* Roots used in a steam for arthritis and lumbago. *Other* Roots used in a steam for stroke. **Yurok** *Antirheumatic (External)* Roots used in a steam for arthritis and lumbago. *Misc. Disease Remedy* Roots used in a steam for stroke. (6:38)
- ***Food***—**Cowlitz** *Unspecified* Blossoms cooked overnight and eaten no more than two or three at a time, otherwise one became sick. (79:22) **Haisla & Hanaksiala** *Forage* Roots eaten by black and grizzly bears after hibernation, to cleanse and strengthen their stomachs. (43:189) **Hesquiat** *Forage* Roots eaten by deer and bear. (185:48) **Hoh** *Forage* Plants eaten by bears in spring. *Spice* Leaves placed over roasting camas, wild onion, or garlic for flavoring. (as *L. camtschatcense* 137:59) **Okanagan-Colville** *Forage* Flower stalks sucked by grizzly and black bears. (188:35) **Oweekeno** *Forage* Roots eaten by bears after emerging from hibernation. (43:76) **Quileute** *Forage* Plants eaten by bears in spring. *Spice* Leaves placed over roasting camas, wild onion, or garlic for flavoring. (as *L. camtschatcense* 137:59) *Unspecified* Root cooked and eaten. **Skokomish** *Unspecified* Young leaves steamed and eaten. (79:22) **Tolowa** *Unspecified* Root centers eaten after boiling eight times. **Yurok** *Unspecified* Root centers eaten after boiling eight times. (6:38)
- ***Other***—**Bella Coola** *Containers* Large leaves folded and used as berry containers. *Cooking Tools* Large leaves folded and used as drinking cups, as covering for drying cakes, and to line pits. (184:198) **Haisla & Hanaksiala** *Cooking Tools* Leaves used to wrap western hemlock cambium, bear meat, and porcupine meat while cooking. Leaves used as a mat when drying berries. (43:189) **Hesquiat** *Containers* Leaves used as sheets to dry berries. (185:48) **Hoh** *Containers* Leaves wrapped around cooked fruits and buried in swampy regions for preservation. *Cooking Tools* Leaves used to wrap red elderberries during baking. (as *L. camtschatcense* 137:59) **Kitasoo** *Toys & Games* Spadices on sticks thrown by children in distance contests. (43:320) **Kwakiutl, Southern** *Containers* Leaves used to cover baskets of stink currants. (183:286) Leaves used to cover baskets of freshly picked berries. (183:271) *Cooking Tools* Leaves used for drying salal berry cakes. (183:282) Leaves used with green grass leaves to line steaming boxes for cooking lupine roots. Leaves used to wrap wild clover roots for baking, boiling, and steaming. (183:285) Leaves used for steam cooking salmon. (183:271) **Makah** *Containers* Leaves used for drying salal berries and to line berry baskets. (79:22) *Cooking Tools* Leaves used to cover sprouts while cooking. (67:336) **Nitinaht** *Containers* Large, waxy leaves used as berry containers and for wrapping leftover food. (186:78) *Cooking Tools* Leaves used to make rectangular drying frames for drying mashed salal berries. (186:105) Large, waxy leaves used as plates, drinking cups, berry drying racks, and steam

pit covers. (186:78) **Okanagan-Colville** *Containers* Leaves placed over and under the food in steaming pit cooking. (188:35) **Oweekeno** *Containers* Leaves used to line berry baskets to prevent the berries from falling through holes in the baskets. *Cooking Tools* Leaves folded into makeshift cups. (43:76) **Quileute** *Containers* Leaves used to wrap salal berries and elderberries while drying. (79:22) Leaves wrapped around cooked fruits and buried in swampy regions for preservation. *Cooking Tools* Leaves used to wrap red elderberries during baking. (as *L. camtschatcense* 137:59) *Good Luck Charm* Leaves placed under canoe bow pieces to make seals easier to catch. (79:22) **Salish, Coast** *Cooking Tools* Leaves used to make water dippers on camping trips. (182:73) **Samish** *Containers* Large leaves doubled or rolled and used as cups for drinking or picking berries. **Swinomish** *Containers* Large leaves doubled or rolled and used as cups for drinking or picking berries. (79:22) **Thompson** *Protection* Charcoal used as protection against "witchcraft." (187:113) **Tolowa** *Containers* Leaves used as a vessel to drive water from streams. *Cooking Tools* Leaves used to wrap sturgeon eggs baked in ashes. (6:38) **Tsimshian** *Containers* Leaves used to line cooking pits. Leaves used to cover and wrap foods during collection, transit, storage, or cooking. *Cooking Tools* Leaves used as an underlay for drying berries. (43:320) **Yurok** *Containers* Leaves used as a vessel to drive water from streams. *Cooking Tools* Leaves used to wrap sturgeon eggs baked in ashes. (6:38)

Lysichiton sp., Skunkcabbage
- **Other**—**Poliklah** *Containers* Leaves used as a temporary lining for openwork baskets when used to hold berries. (117:173)

Lysimachia, Primulaceae
Lysimachia quadrifolia L., Whorled Yellow Loosestrife
- **Drug**—**Cherokee** *Gastrointestinal Aid* Decoction of root taken for "bowel trouble" and "kidney problems." *Gynecological Aid* Infusion taken for "female trouble." *Kidney Aid* Taken "for kidney problems." (80:43) *Urinary Aid* Infusion of roots taken for urinary troubles. (177:50) **Iroquois** *Emetic* Infusion of roots taken as an emetic. (87:411)

Lysimachia thyrsiflora L., Tufted Loosestrife
- **Drug**—**Iroquois** *Gynecological Aid* Compound decoction used as wash and applied as poultice to stop milk flow. (87:411)

Lythraceae, see *Ammannia, Lythrum*

Lythrum, Lythraceae
Lythrum alatum var. *lanceolatum* (Ell.) Torr. & Gray ex Rothrock, Winged Lythrum
- **Drug**—**Cherokee** *Kidney Aid* Infusion taken "for kidneys." (as *L. lanceolatum* 80:43)

Lythrum californicum Torr. & Gray, California Loosestrife
- **Drug**—**Kawaiisu** *Dermatological Aid* Plant used for washing hair. *Unspecified* Plant used medicinally. (206:40)

Lythrum salicaria L., Purple Loosestrife
- **Drug**—**Iroquois** *Febrifuge* and *Witchcraft Medicine* Compound decoction of plants taken for fever and sickness caused by the dead. (87:389)

Machaeranthera, Asteraceae
Machaeranthera alta A. Nels., Purple Aster
- **Drug**—**Navajo** *Dermatological Aid* Plant used as a rub on pimples. (55:88)

Machaeranthera canescens (Pursh) Gray ssp. *canescens* var. *canescens*, Cutleaf Goldenweed
- **Drug**—**Navajo** *Nose Medicine* Dried and pulverized plant used as a snuff for nose troubles. *Throat Aid* Dried and pulverized plant used as a snuff for throat troubles. (as *Aster canescens* 55:82) **Okanagan-Colville** *Witchcraft Medicine* Used for witchcraft. (as *Aster canescens* 188:79) **Zuni** *Emetic* Infusion of whole plant taken and rubbed on abdomen as an emetic. (as *M. glabella* 166:56)

Machaeranthera canescens ssp. *canescens* var. *leucanthemifolia* (Greene) Welsh, Whiteflower Machaeranthera
- **Drug**—**Paiute** *Throat Aid* Poultice of mashed leaves applied to swollen jaw or neck glands. **Shoshoni** *Analgesic* Decoction of fresh or dried leaves taken for headaches. *Blood Medicine* Decoction of whole plant taken as a blood tonic. *Cathartic* Warm infusion of plant tops taken as a physic. *Eye Medicine* Infusion of scraped roots used as an eyewash. *Tonic* Decoction of whole plant taken as a blood tonic. (as *Aster leucanthemifolius* 180:49)

Machaeranthera canescens ssp. *glabra* var. *aristata* (Eastw.) B. L. Turner, Hoary Tansyaster
- **Drug**—**Hopi** *Gynecological Aid* Decoction of plant taken by parturient women for any disorder. (as *Aster cichoriaceus* 200:36, 94) *Stimulant* Decoction of plant taken as a strong stimulant. (as *Aster cichoriaceus* 200:31, 94)

Machaeranthera gracilis (Nutt.) Shinners, Slender Goldenweed
- **Drug**—**Navajo, Ramah** *Dermatological Aid* Cold infusion used as lotion for pimples, boils, and sores. *Eye Medicine* Cold, compound infusion of plant used as an eyewash. *Internal Medicine* Decoction of plant taken for internal injury. *Respiratory Aid* Plant used as snuff to cause sneezing, clearing congested nose. (as *Aplopappus gracilis* 191:47)
- **Food**—**Navajo, Ramah** *Dried Food* Dried seeds used for food. (as *Aplopappus gracilis* 191:47)

Machaeranthera grindelioides (Nutt.) Shinners var. *grindelioides*, Rayless Aster
- **Drug**—**Hopi** *Cough Medicine* Decoction of root taken for cough. (as *Aplopappus nuttallii* 200:35, 94)

Machaeranthera parviflora Gray, Smallflower Tansyaster
- **Drug**—**Navajo** *Cathartic* Plant used as a purgative. (as *Aster parviflorus* 90:148)

Machaeranthera pinnatifida (Hook.) Shinners ssp. *pinnatifida*, Lacy Tansyaster
- **Drug**—**Navajo** *Analgesic* Plant or some part of it used for headaches. (as *Haplopappus spinulosus* ssp. *typicus* 90:151)

Machaeranthera tanacetifolia (Kunth) Nees, Tanseyleaf Aster
- **Drug**—**Hopi** *Gynecological Aid* Decoction of plant taken by parturi-

ent women for any disorder. (as *Aster tanacetifolius* 200:36, 94) *Stimulant* Decoction of plant taken as a strong stimulant. (as *Aster tanacetifolius* 200:31, 94) **Navajo, Ramah** *Gastrointestinal Aid* Decoction of whole plant taken for stomachache. *Respiratory Aid* Dried root used as snuff to cause sneezing to relieve congested nose. (as *Aster tanacetifolius* 191:49) **Zuni** *Unspecified* Infusion of flowers taken with other flowers for unspecified illnesses. (27:375)

Machaerocereus, Cactaceae

Machaerocereus eruca (T. Brandeg.) Britt. & Rose
- *Food*—**Papago & Pima** *Fruit* Fruit used for food. (35:42)

Machaerocereus gummosus (Engelm.) Britt. & Rose, Pitahaya Agria
- *Food*—**Papago & Pima** *Fruit* Fruit used for food. (35:40)

Maclura, Moraceae

Maclura pomifera (Raf.) Schneid., Osage Orange
- *Drug*—**Comanche** *Eye Medicine* Decoction of root used as a wash for sore eyes. (29:522)
- *Dye*—**Kiowa** *Yellow* Outer portion of the roots yielded a yellow dye. (192:23) **Pima** *Unspecified* Inner wood and large roots formerly used as dyes. (47:83)
- *Other*—**Comanche** *Hunting & Fishing Item* Branches used to make bows. (29:522) **Kiowa** *Ceremonial Items* Wood used as the favorite material for the staff held by singer in the peyote ceremony. *Hunting & Fishing Item* Wood used for making bows. (192:23) **Omaha** *Hunting & Fishing Item* Wood used to make bows. **Pawnee** *Hunting & Fishing Item* Wood used to make bows. (as *Toxylon pomiferum* 70:76) **Pima** *Hunting & Fishing Item* Used to make bows. (47:83) **Ponca** *Hunting & Fishing Item* Wood used to make bows whenever obtainable. (as *Toxylon pomiferum* 70:76) **Seminole** *Hunting & Fishing Item* Plant used to make bows. (169:467) **Tewa** *Hunting & Fishing Item* Wood used to make bows. This plant grew outside Tewa country. It was considered to be better for making bows than any native wood. It was brought from the east by the Tewa, or obtained from the Comanche or other eastern tribes. (as *M. aurantiaca* 138:68)

Macranthera, Scrophulariaceae

Macranthera sp.
- *Drug*—**Jemez** *Veterinary Aid* Decoction of plant given to horses with blood poisoning. (44:25)

Macrocystis, Lessoniaceae, alga

Macrocystis integrifolia Bory, Giant Kelp
- *Food*—**Haisla & Hanaksiala** *Unspecified* Plant used to collect herring roe, dried and eaten with the roe. (43:127) **Kitasoo** *Unspecified* Plant eaten with herring roe. (43:303) **Oweekeno** *Dried Food* Plant and herring eggs dried for future use. *Unspecified* Plant eaten with herring roe. *Winter Use Food* Plant and herring eggs preserved in brine for future use. (43:45)
- *Other*—**Heiltzuk** *Cash Crop* Plant traded for oolichan (candlefish) grease and smoked oolichans. (43:127) **Hesquiat** *Toys & Games* Children threw dried, little floats from blade base onto fire to make them explode; firecrackers. (185:24) **Kitkatla** *Cash Crop* Plant traded for oolichan (candlefish) grease and smoked oolichans. (43:127) **Kwakiutl, Southern** *Hunting & Fishing Item* Blades weighted, placed underwater at river mouths, and used to catch herring spawn. (183:261)

Macrocystis luetkeana, Giant Kelp
- *Food*—**Pomo** *Unspecified* Plant chewed raw. (as *M. lütkeana* 11:94)

Macromeria, Boraginaceae

Macromeria viridiflora DC., Gianttrumpets
- *Drug*—**Hopi** *Anticonvulsive* Dried plant and mullein smoked for "fits," craziness, and witchcraft. (as *Onosmodium thurberi* 200:33, 88) *Psychological Aid* Compound of plant smoked by persons not in their "right mind." *Witchcraft Medicine* Compound of plant smoked as a cure for persons with "power to charm." (as *Onosmodium thurberi* 200:88)

Madia, Asteraceae

Madia capitata Nutt., Coast Tarweed
- *Food*—**Pomo** *Staple* Seeds used to make pinoles. (11:87)

Madia elegans D. Don ex Lindl., Common Madia
- *Food*—**Hupa** *Staple* Seeds parched and pounded into a flour. (148:390) **Mewuk** *Staple* Seeds roasted with hot coals, pounded or rolled into flour, and eaten dry. (117:338) **Miwok** *Staple* Pulverized seeds eaten as a dry meal. (12:154) **Neeshenam** *Bread & Cake* Seeds parched, ground into flour, and used to make bread. *Porridge* Seeds parched, ground into flour, and used to make mush. *Staple* Seeds parched, ground into flour, and used for food. (as *Madaria*, tarry smelling weed 129:377) **Pomo** *Staple* Seeds used to make pinoles. (11:87) **Pomo, Kashaya** *Staple* Seeds used to make pinole. (72:112) **Shoshoni** *Unspecified* Seeds roasted and eaten alone or mixed with manzanita berries, acorns, and pine nuts. (117:440)

Madia elegans ssp. *densifolia* (Greene) Keck, Showy Tarweed
- *Food*—**Pomo** *Staple* Seeds used to make pinoles. (as *M. densifolia* 11:87)

Madia glomerata Hook., Mountain Tarweed
- *Drug*—**Cheyenne** *Ceremonial Medicine* Dried plant used in special ceremony for perverted, oversexed people. *Herbal Steam* Infusion of stems and leaves taken and used as a steam bath for venereal disease. *Love Medicine* Dried plant aroma used as a love medicine to attract a woman. *Psychological Aid* Dried plant used for perverted, oversexed people. *Venereal Aid* Infusion of stems and leaves taken and used as a steam bath for venereal disease. (83:22) **Crow** *Ceremonial Medicine* Dried herbs burned as incense in some ceremonies. (19:15)
- *Food*—**Crow** *Unspecified* Seeds used for food. (19:15) **Klamath** *Unspecified* Seeds used for food. (45:106)

Madia gracilis (Sm.) Keck & J. Clausen ex Applegate ssp. *gracilis*, Grassy Tarweed
- *Food*—**Mendocino Indian** *Staple* Seeds used to make pinole. (as *M. dissitiflora* 41:395) **Miwok** *Staple* Parched, pulverized seeds made into oily meal and readily picked up in lumps. (as *M. dissitiflora* 12:154) **Pomo** *Staple* Seeds used to make pinoles. (as *M. dissitiflora* 11:87)

Madia sativa Molina, Coast Tarweed
- *Food*—**Mendocino Indian** *Cooking Agent* Oil from seeds used for

cooking. (41:395) **Miwok** *Unspecified* Seeds used for food. (12:154) **Pomo** *Porridge* Parched, pulverized seeds eaten as pinole and meal moistened to keep people from choking on dry meal. (66:15) *Staple* Seeds used to make pinoles. (11:87) *Winter Use Food* Raw seeds stored for later use, parched, and pounded when used for food. (66:15) **Pomo, Kashaya** *Staple* Seeds used to make pinole. (72:111)

Madia sp., Tarweed
- *Food*—**Wintoon** *Staple* Seeds used to make pinole. (117:274)

Magnoliaceae, see *Liriodendron, Magnolia*

Magnolia, Magnoliaceae

Magnolia acuminata (L.) L., Cucumber Tree
- *Drug*—**Cherokee** *Analgesic* Infusion of bark taken for stomachache or cramps. *Antidiarrheal* Compound medicine containing bark taken for "bloody flux." *Gastrointestinal Aid* Infusion of bark used for toothache. Used in steam bath for "indigestion or biliousness with swelling abdomen." *Respiratory Aid* Hot infusion of bark snuffed for sinus and used for toothache. *Toothache Remedy* Warm compound decoction of bark held in mouth for toothache. (80:44) **Iroquois** *Anthelmintic* Compound decoction taken by men for worms caused by venereal disease. *Toothache Remedy* Infusion of inner bark "chewed" for toothaches. *Venereal Aid* Compound decoction taken by men for worms caused by venereal disease. (87:330)
- *Fiber*—**Cherokee** *Building Material* Wood used for lumber. *Furniture* Wood used to make furniture. (80:44)
- *Other*—**Cherokee** *Paper* Wood used for pulpwood. (80:44)

Magnolia grandiflora L., Southern Magnolia
- *Drug*—**Choctaw** *Dermatological Aid* Decoction of bark used as wash for prickly heat itching. (25:23) Decoction of plant used as a bath for prickly heat. *Kidney Aid* Infusion of mashed bark used as a steam bath for dropsy. **Koasati** *Dermatological Aid* Decoction of bark used as a wash for sores. (177:23)

Magnolia macrophylla Michx., Bigleaf Magnolia
- *Drug*—**Cherokee** *Analgesic* Infusion of bark taken for stomachache or cramps. *Antidiarrheal* Compound medicine containing bark taken for "bloody flux." *Gastrointestinal Aid* Infusion of bark taken for stomachache or cramps. Used in steam bath for "indigestion or biliousness with swelling abdomen." *Respiratory Aid* Hot infusion of bark snuffed for sinus and used for toothache. *Toothache Remedy* Warm compound decoction of bark held in mouth for toothache. (80:44)
- *Fiber*—**Cherokee** *Building Material* Wood used for lumber. *Furniture* Wood used to make furniture. (80:44)
- *Other*—**Cherokee** *Paper* Wood used for pulpwood. (80:44)

Magnolia virginiana L., Sweet Bay
- *Drug*—**Houma** *Blood Medicine* Decoction of leaves and twigs taken to warm the blood. *Cold Remedy* Decoction of leaves and twigs taken for colds. *Febrifuge* Decoction of leaves and twigs taken for chills. (158:56) **Rappahannock** *Hallucinogen* Leaves or bark placed in cupped hands over nose and inhaled as "mild dope." (as *M. glauca*, laurel 161:28)

Mahonia, Berberidaceae

Mahonia aquifolium (Pursh) Nutt., Hollyleaved Barberry
- *Drug*—**Blackfoot** *Antihemorrhagic* Decoction of root used for hemorrhages. *Gastrointestinal Aid* Decoction of root used for stomach trouble. (as *Berberis aquifolium* 114:275) **Karok** *Misc. Disease Remedy* Leaves and roots used as a steam bath for "yellow fever." (as *Berberis aquifolium* 148:383) *Other* Fruits, if eaten, caused diarrhea. (6:38) *Panacea* Decoction of roots taken as a good medicine for all kinds of sickness. *Poison* Plant considered poisonous. (as *Berberis aquifolium* 148:383) **Keres, Western** *Other* Plant chewed for sickness that occurred during hunting when approached by a dying deer. *Preventive Medicine* Infusion of leaves used to prevent sickness that occurred while hunting and approached by dying deer (as *Berberis aquifolium* 171:32) **Nitinaht** *Laxative* Used as a laxative. (186:98) *Tuberculosis Remedy* Used with hemlock and alder as drink for tuberculosis. *Unspecified* Bark used medicinally. (as *Berbaris aquifolium* 67:254) **Okanagan-Colville** *Blood Medicine* Decoction of branches and chokecherry branches taken for the "changing of the blood." Infusion of branches taken as a blood tonic. *Eye Medicine* Infusion of plant used to wash out blurry or bloodshot eyes. *Kidney Aid* Roots used to make a tonic for the kidneys. Decoction of roots and chokecherry or kinnikinnick branches taken for bad kidneys. (as *Berberis aquifolium* 188:85) **Samish** *Tonic* Infusion of roots taken as a general tonic. (as *Berberis aquifolium* 79:30) **Sanpoil** *Antiemetic* Decoction of stem tips taken for vomiting. *Eye Medicine* Infusion of root parts used as a wash for the eyes. *Gastrointestinal Aid* Decoction of stem tips taken "to relieve a disturbed stomach." *Tuberculosis Remedy* Decoction of roots used for tuberculosis. (as *Berberis aquifolium* 131:219) **Squaxin** *Blood Medicine* Infusion of roots taken to purify the blood. *Throat Aid* Infusion of roots taken as a gargle for sore throats. **Swinomish** *Tonic* Infusion of roots taken as a general tonic. (as *Berberis aquifolium* 79:30) **Thompson** *Antirheumatic (External)* Decoction of peeled, chopped root bark used as a wash for arthritis. *Antirheumatic (Internal)* Decoction of peeled, chopped root bark taken for arthritis. *Blood Medicine* Decoction of peeled, chopped root bark taken as a blood tonic. *Eye Medicine* Infusion of stems and bark used to make an eyewash for red, itchy eyes.[46] *Laxative* Fruit considered an "excellent laxative." *Tonic* Fruit eaten as a "tonic." *Venereal Aid* Decoction of peeled, chopped roots taken for syphilis. (187:187)
- *Food*—**Klallam** *Fruit* Berries used for food. (as *Berberis aquifolium* 78:197) **Kwakiutl, Southern** *Fruit* Sour berries occasionally used for food. (as *Berberis aquifolium* 183:279) **Makah** *Preserves* Fruit used to make preserves. (as *Berbaris aquifolium* 67:254) **Okanagan-Colville** *Fruit* Berries eaten raw. (as *Berberis aquifolium* 188:85) **Salish, Coast** *Preserves* Berries used to make jelly. (as *Berberis aquifolium* 182:78) **Samish** *Fruit* Berries eaten fresh. (as *Berberis aquifolium* 79:30) **Sanpoil** *Fruit* Berries eaten fresh. *Preserves* Berries boiled into a jam. (as *Berberis aquifolium* 188:85) **Sanpoil & Nespelem** *Fruit* Berries eaten fresh. (as *Berberis aquifolium* 131:102) **Skagit, Upper** *Dried Food* Berries pulped, dried, and stored in cakes for winter use. (as *Berberis aquifolium* 179:37) *Fruit* Fruit eaten raw or mashed. (as *Berberis aquifolium* 179:38) **Snohomish** *Fruit* Berries eaten fresh. **Squaxin** *Fruit* Berries eaten. **Swinomish** *Fruit* Berries eaten fresh. (as *Berberis aquifolium* 79:30) **Thompson** *Dried Food* Fruit dried in the absence of any other fruit. *Fruit* Fruit eaten fresh, a few at a time. *Preserves* Fruit used to make jelly. (187:187)

- **Dye**—**Chehalis** *Yellow* Roots used to make a yellow dye. (as *Berberis aquifolium* 79:30) **Makah** *Yellow* Roots or possibly the leaves used for yellow dye. (as *Berbaris aquifolium* 67:254) **Nitinaht** *Yellow* Bark scrapings steeped and used as a yellow dye. (186:98) **Okanagan-Colville** *Yellow* Stem and root inner bark used as bright yellow dye for basket materials, wool, and porcupine quills. (as *Berberis aquifolium* 188:85) **Salish, Coast** *Yellow* Root bark shredded, boiled, and used as a yellow dye for basketry. (as *Berberis aquifolium* 182:78) **Skagit** *Yellow* Roots used to make a yellow dye. **Snohomish** *Yellow* Roots used to make a yellow dye. (as *Berberis aquifolium* 79:30) **Thompson** *Yellow* Outer bark boiled to make a bright yellow dye used for basket materials. (187:187) **Yurok** *Yellow* Root used to dye porcupine quills yellow. (6:38)
- **Other**—**Karok** *Paint* Fruit mixed with salmon glue and pounded larkspur flowers and used to paint arrows and bows. (as *Berberis aquifolium* 148:383)

Mahonia dictyota (Jepson) Fedde, Shining Netvein Barberry
- **Drug**—**Kawaiisu** *Venereal Aid* Decoction of roots taken for gonorrhea. (as *Berberis dictyota* 206:15)

Mahonia fremontii (Torr.) Fedde, Frémont's Mahonia
- **Drug**—**Apache, White Mountain** *Ceremonial Medicine* Plant used for ceremonial purposes. (as *Berberis fremontii* 136:155) **Hopi** *Oral Aid* Plant used for gums. (as *Odostemon fremontii* 200:33, 76) **Hualapai** *Gastrointestinal Aid* Roots used as a bitter tonic to promote digestion. *Laxative* Roots made into a bitter tonic and used as a laxative. *Liver Aid* Roots used as a bitter tonic for the liver. (as *Berberis fermontii* 195:5)
- **Food**—**Hualapai** *Beverage* Berries used to make a beverage. *Fruit* Berries used for food. (as *Berberis fermontii* 195:5) **Yavapai** *Fruit* Raw berries used for food. (65:257)
- **Dye**—**Havasupai** *Yellow* Roots used as a yellow buckskin dye. (as *Berberis fremontii* 197:219) **Hualapai** *Yellow* Roots used to make a brilliant yellow dye. (as *Berberis fermontii* 195:5) **Navajo** *Yellow* Roots and bark used as a yellow dye for buckskin. (as *Berberis fremontii* 55:48) **Walapai** *Yellow* Roots used as a yellow basket dye. (as *Berberis fremontii* 197:219)
- **Other**—**Hopi** *Tools* Wood used to make various tools. (as *Odostemon fremontii* 200:76) **Zuni** *Ceremonial Items* and *Paint* Crushed berries used as purple coloring for the skin and for objects employed in ceremonies. (as *Berberis fremontii* 166:88)

Mahonia haematocarpa (Woot.) Fedde, Red Barberry
- **Drug**—**Apache, Mescalero** *Eye Medicine* Inner wood shavings soaked in water and used as an eyewash. (as *Berberis haematocarpa* 14:49)
- **Food**—**Apache** *Fruit* Berries eaten fresh. (as *Berberis haematocarpa* 32:19) **Apache, Chiricahua & Mescalero** *Preserves* Fruit cooked with a sweet substance, strained, and eaten as jelly. (as *Berberis haematocarpa* 33:46) **Apache, Mescalero** *Fruit* Berries eaten fresh. (as *Berberis haematocarpa* 14:49) **Pueblo** *Preserves* Berries used to make jelly. **Spanish American** *Preserves* Berries used to make jelly. (as *Berberis haematocarpa* 32:19)
- **Dye**—**Apache, Mescalero** *Yellow* Root shavings used to make a yellow dye for hides. (as *Berberis haematocarpa* 14:49)

Mahonia nervosa (Pursh) Nutt., Cascade Oregongrape
- **Drug**—**Nitinaht** *Laxative* Used as a laxative. (186:98) **Thompson** *Antirheumatic (External)* Decoction of peeled, chopped root bark used as wash for arthritis. *Antirheumatic (Internal)* Decoction of peeled, chopped root bark taken for arthritis. *Blood Medicine* Decoction of peeled, chopped root bark taken as a blood tonic. *Eye Medicine* Infusion of woody stems and bark used as an eyewash for red, itchy eyes.[46] *Laxative* Fruit considered an "excellent laxative." *Psychological Aid* Plant induced dreams of someone sleeping when brought into the house. *Venereal Aid* Decoction of peeled, chopped roots taken for syphilis. (187:187)
- **Food**—**Thompson** *Preserves* Berries used to make jelly. (187:187)
- **Dye**—**Nitinaht** *Yellow* Bark scrapings steeped and used as a yellow dye. (186:98) **Thompson** *Yellow* Root bark boiled to make a bright yellow dye used for basket materials. (187:187)

Mahonia nervosa (Pursh) Nutt. **var. *nervosa*,** Cascade Oregongrape
- **Drug**—**Hoh** *Blood Medicine* Infusion of roots used as a blood remedy. (as *Berberis nervosa* 137:61) **Nitinaht** *Tuberculosis Remedy* Used with hemlock and alder as drink for tuberculosis. *Unspecified* Bark used medicinally. (as *Berbaris nervosa* 67:254) **Paiute** *Blood Medicine* Infusion of roots and leaves taken as a general tonic "to make the blood good." *Hemostat* Infusion of roots and leaves taken as a general tonic for nosebleeds. (as *Berberis nervosa* 111:72) **Quileute** *Blood Medicine* Infusion of roots used as a blood remedy. (as *Berberis nervosa* 137:61) **Skagit** *Venereal Aid* Decoction of roots taken for venereal disease. (as *Berberis nervosa* 79:30)
- **Food**—**Clallam** *Fruit* Sour berries used for food. (as *Berberis nervosa* 57:197) **Hoh** *Preserves* Berries used to make jelly. (as *Berberis nervosa* 137:61) **Klallam** *Fruit* Berries used for food. (as *Berberis nervosa* 78:197) **Kwakiutl, Southern** *Fruit* Sour berries occasionally used for food. (as *Berberis nervosa* 183:279) **Makah** *Preserves* Fruit used to make preserves. (as *Berbaris nervosa* 67:254) **Quileute** *Preserves* Berries used to make jelly. (as *Berberis nervosa* 137:61) **Salish, Coast** *Preserves* Berries used to make jelly. (as *Berberis nervosa* 182:78) **Skagit** *Fruit* Ripe berries formerly used for food. *Preserves* Ripe berries used to make jam. (as *Berberis nervosa* 79:30) **Skagit, Upper** *Dried Food* Berries pulped, dried, and stored in cakes for winter use. (as *Berberis nervosa* 179:37) *Fruit* Fruit eaten raw or mashed. (as *Berberis nervosa* 179:38)
- **Dye**—**Klallam** *Unspecified* Roots used to dye basketry material. (as *Berberis nervosa* 79:30) **Makah** *Yellow* Roots or possibly the leaves used for yellow dye. (as *Berbaris nervosa* 67:254) **Skagit** *Unspecified* Roots used to dye basketry material. **Snohomish** *Unspecified* Roots used to dye basketry material. (as *Berberis nervosa* 79:30)

Mahonia pinnata (Lag.) Fedde **ssp. *pinnata*,** California Barberry
- **Drug**—**Miwok** *Antirheumatic (Internal)* Decoction of root taken for rheumatism. *Dermatological Aid* Chewed root liquid placed on cuts, wounds, and abrasions to prevent swelling. Decoction of root used as a wash for cuts and bruises. (as *Berberis pinnata* 12:168) *Gastrointestinal Aid* Decoction of root taken for heartburn. *Misc. Disease Remedy* Decoction of root taken for ague. (as *Berberis pinnata* 12:167) Leaves chewed as a preventative of ague. *Tuberculosis Remedy* Decoction of roots taken for consumption. (as *Berberis pinnata* 12:168)

Mahonia pumila (Greene) Fedde, Dwarf Barberry
- **Drug**—Karok *Tonic* Root used in a tonic. **Tolowa** *Blood Medicine* Roots used in a concoction for blood purification. *Cough Medicine* Roots used in a concoction for coughs. (6:38)
- **Dye**—Karok *Yellow* Root used to dye porcupine quills yellow. (6:38)

Mahonia repens (Lindl.) G. Don, Oregongrape
- **Drug**—**Blackfoot** *Antihemorrhagic* Decoction of roots used for hemorrhages. (as *Berberis repens* 118:45) *Dermatological Aid* Poultice of fresh berries applied to boils. Infusion of roots applied to boils. (as *Berberis repens* 86:75) *Disinfectant* Infusion of roots applied as an antiseptic to wounds. (as *Berberis repens* 86:83) *Gastrointestinal Aid* Decoction of roots used for stomach trouble. (as *Berberis repens* 118:45) *Kidney Aid* Infusion of berries used for kidney troubles. (as *Berberis repens* 86:66) *Veterinary Aid* Berries mixed with water and given to horses with coughs. Infusion or roots used for body sores on horses. (as *Berberis repens* 86:88) **Cheyenne** *Unspecified* Fruit used in medicinal preparations. (as *Berberis repens* 83:15) **Flathead** *Antirheumatic (External)* Infusion of roots used for rheumatism. *Contraceptive* Infusion of roots used as a contraceptive. *Cough Medicine* Infusion of roots used for coughs. *Dermatological Aid* Roots crushed and used for wounds and cuts. *Gynecological Aid* Infusion of roots used for delivery of the placenta. *Venereal Aid* Infusion of roots taken for gonorrhea and syphilis. (as *Berberis repens* 82:18) **Great Basin Indian** *Antidiarrheal* Infusion of roots taken for dysentery. *Blood Medicine* Infusion of roots taken to thicken the blood for bleeders. (as *Berberis repens* 121:47) **Havasupai** *Analgesic* Cooled decoction of roots taken three times a day for headaches. *Antirheumatic (External)* Cooled decoction of roots used as a wash for aches. *Cold Remedy* Cooled decoction of roots used as a wash for colds. *Gastrointestinal Aid* Cooled decoction of roots taken three times a day for stomach upsets. *Laxative* Cooled decoction of roots taken as a laxative for colds and stomach ailments. *Pediatric Aid* and *Unspecified* Cooled decoction of roots given to sick babies. (as *Berberis repens* 197:219) **Kutenai** *Blood Medicine* Taken to "enrich" the blood. *Kidney Aid* Infusion of roots taken for kidney trouble. (as *Berberis repens* 82:18) **Mendocino Indian** *Blood Medicine* Decoction of root bark taken as a blood purifier. *Gastrointestinal Aid* Decoction of root bark taken for stomach troubles. (as *Berberis repens* 41:348) **Montana Indian** *Febrifuge* Decoction of root bark used for mountain fever. *Gastrointestinal Aid* Decoction of root bark used for stomach trouble. *Kidney Aid* Decoction of root bark used for kidney trouble. *Tonic* Decoction of root bark used as a tonic. (as *Berberis repens* 19:8) **Navajo** *Antirheumatic (Internal)* Decoction of leaves and twigs taken for rheumatic stiffness. (as *Berberis repens* 55:48) **Navajo, Kayenta** *Panacea* Infusion of plant taken and poultice of plant applied as a cure-all. (as *Berberis repens* 205:23) **Navajo, Ramah** *Ceremonial Medicine* Whole plant used as a ceremonial emetic. *Dermatological Aid* Cold infusion of plant used as a lotion on scorpion bites. *Emetic* Whole plant used as a ceremonial emetic. *Laxative* Decoction of root used for constipation. (as *Berberis repens* 191:28) **Paiute** *Antidiarrheal* Decoction of root taken to prevent or stop bloody dysentery. *Blood Medicine* Decoction of root taken as a blood tonic or purifier. Decoction of root taken to "thicken the blood of haemophilic persons." *Cough Medicine* Decoction of root, sometimes with whisky, taken for coughs. *Gastrointestinal Aid* Decoction of stems taken as a tonic for stomach troubles. *Urinary Aid* Decoction of root taken for bladder difficulties. (as *Berberis repens* 180:51, 52) *Venereal Aid* Decoction of roots taken for venereal disease. (as *Berberis repens* 104:198) Decoction of roots taken for venereal diseases. **Shoshoni** *Analgesic* Decoction of leaves taken or root used for general aches or rheumatic pains. *Antidiarrheal* Decoction of root taken to prevent or stop bloody dysentery. *Antirheumatic (Internal)* Decoction of roots or leaves taken for general aches or rheumatic pains. *Blood Medicine* Infusion or decoction of root taken as a blood tonic or purifier. *Cough Medicine* Decoction of root, sometimes with whisky, taken for coughs. *Kidney Aid* Decoction of root taken as a kidney medicine. *Venereal Aid* Decoction of roots taken for venereal diseases. (as *Berberis repens* 180:51, 52) **Shuswap** *Blood Medicine* Decoction of leaves and stems taken as a blood tonic. (as *Berberis repens* 123:59)
- **Food**—**Blackfoot** *Fruit* Berries eaten when nothing else was available. (as *Berberis repens* 86:101) Fruit used for food. (as *Berberis repens* 97:35) **Cheyenne** *Fruit* Fruits eaten for food. (as *Berberis repens* 83:15) **Flathead** *Dessert* Berries mashed, sugar and milk added, and eaten as a dessert. *Fruit* Berries roasted and used for food. **Kutenai** *Appetizer* Root tea taken as an appetizer. *Dessert* Berries mashed, sugar and milk added, and eaten as a dessert. (as *Berberis repens* 82:18) **Montana Indian** *Beverage* Fruit used to make wine and "lemonade." (as *Berberis repens* 19:8) Berries crushed, mixed with sugar and water, and made into a refreshing beverage. (as *Berberis repens* 82:18) *Fruit* Fruit eaten raw. *Preserves* Fruit used to make jelly. (as *Berberis repens* 19:8) Berries used to make jams and jellies. (as *Berberis repens* 82:18) **Shuswap** *Fruit* Ripe berries used for food. (as *Berberis repens* 123:59)
- **Dye**—**Blackfoot** *Yellow* Roots used to make a yellow dye. (as *Berberis repens* 97:35) **Great Basin Indian** *Orange* Plant used to make an orange dye. (as *Berberis repens* 121:47) **Montana Indian** *Yellow* Bark shredded, boiled, and used as a brilliant yellow dye. (as *Berberis repens* 82:18)
- **Other**—**Hopi** *Ceremonial Items* Yellow root and leaves used for ceremonial purposes in the Home Dance. The yellow root is the most important part. The leaf is sometimes used to represent a mountain lion's paw. (as *Berberis repens* 42:294) **Navajo, Kayenta** *Ceremonial Items* Sprinkled on grass where lightning struck near livestock. (as *Berberis repens* 205:23)

Mahonia sp., Oregongrape
- **Drug**—**Cowlitz** *Dermatological Aid* Infusion of bark used as a wash for skin sores. *Oral Aid* Infusion of bark used as a wash for mouth sores. (as *Berberis* 79:30) **Modesse** *Blood Medicine* Infusion of plant taken as a blood medicine. (as *Berberis* 117:224) **Quinault** *Cough Medicine* Decoction of roots taken for coughs. *Gastrointestinal Aid* Decoction of roots taken for stomach disorders. (as *Berberis* 79:30)
- **Food**—**Coeur d'Alene** *Fruit* Berries eaten fresh. (as *Berberis* 178:90) **Cowlitz** *Fruit* Berries eaten raw. Berries boiled and eaten. **Lummi** *Fruit* Berries eaten. (as *Berberis* 79:30) **Modesse** *Preserves* Berries used to make jelly. (as *Berberis* 117:224) **Okanagon** *Staple* Berries used as a principal food. (as *Berberis* 178:239) **Spokan** *Fruit* Berries used for food. (as *Berberis* 178:343)
- **Dye**—**Cowlitz** *Unspecified* Roots used to make a dye. **Makah** *Unspec-*

ified Roots used to make a dye. (as *Berberis* 79:30) **Nitinaht** *Yellow* Formerly used to make yellow basket dyes. (186:79) **Wintoon** *Yellow* Used to make a yellow dye. (as *Berberis* 117:264)

Maianthemum, Liliaceae

Maianthemum canadense Desf., Canada Beadruby

- *Drug*—**Iroquois** *Kidney Aid* Compound decoction of roots taken for the kidneys. (87:284) **Montagnais** *Analgesic* Infusion of plant taken for headache. (156:314) **Ojibwa** *Analgesic* Plant used for headache and sore throat. *Gynecological Aid* and *Kidney Aid* Plant used "to keep the kidneys open during pregnancy." *Throat Aid* Plant used for sore throat and headache. *Unspecified* Smoke inhaled for unspecified purpose. (153:373, 374) **Potawatomi** *Throat Aid* Root used for sore throat. (154:62, 63)
- *Food*—**Potawatomi** *Fruit* Berries eaten, but the preparation as a food was not discovered. (154:105)
- *Other*—**Potawatomi** *Good Luck Charm* Root used as a good luck charm to win a game. (154:121)

Maianthemum dilatatum (Wood) A. Nels. & J. F. Macbr., Twoleaf False Solomon's Seal

- *Drug*—**Hesquiat** *Dermatological Aid* Poultice of whole or mashed leaves used for boils and cuts. *Tuberculosis Remedy* Fruit used as a good medicine for tuberculosis. (185:55) **Makah** *Reproductive Aid* Chewed roots taken to correct sterility. (79:25) **Nitinaht** *Burn Dressing* Poultice of leaves used for minor burns. *Dermatological Aid* Poultice of leaves used for sores, boils, cuts, and wounds. (186:86) **Oweekeno** *Dermatological Aid* Poultice of water-soaked, bruised leaves applied to wounds. (43:78) **Quinault** *Eye Medicine* Infusion of pounded roots used as a wash for sore eyes. (79:25) Poultice of chewed roots applied to sore eyes. (as *M. bifolium* 201:276)
- *Food*—**Bella Coola** *Fruit* Ripe berries occasionally eaten by hunters and berry pickers. (184:199) **Haisla & Hanaksiala** *Fruit* Fruits eaten for food. (43:198) **Hesquiat** *Fruit* Raw fruit eaten with oil. (185:55) **Kitasoo** *Fruit* Berries eaten fresh. (43:321) **Kwakiutl, Southern** *Fruit* Berries occasionally eaten raw. (183:273) **Oweekeno** *Forage* Berries eaten by frogs. (43:78) **Salish, Coast** *Fruit* Berries occasionally eaten raw. (182:76)

Maianthemum racemosum ssp. *amplexicaule* (Nutt.) LaFrankie, Western Solomon's Seal

- *Drug*—**Karok** *Dermatological Aid* and *Pediatric Aid* Poultice of root applied to the severed umbilical cord of child. (as *Smilacina amplexicaulis* 148:381)
- *Food*—**Tewa** *Fruit* Ripe berries eaten. (as *Vagnera amplexicaulis* 138:70)

Maianthemum racemosum (L.) Link ssp. *racemosum*, Feather Solomon's Seal

- *Drug*—**Abnaki** *Antihemorrhagic* Used by men for spitting up blood. Used for spitting up blood. (as *Smilacina racemosa* 144:154) Decoction of plant taken for spitting up blood. (as *Smilacina racemosa* 144:174) **Algonquin, Quebec** *Antirheumatic (Internal)* Infusion of plant used as a tea for sore backs. (as *Smilacina racemosa* 18:139) **Cherokee** *Eye Medicine* Cold infusion of root used as a wash for sore eyes. (as *Smilacina racemosa* 80:56) **Chippewa** *Analgesic* Compound decoction of root taken for back pain. (as *Vagnera racemosa* 53:356) Burning root fumes inhaled for headaches and pain. (as *Smilacina racemosa* 71:125) *Gynecological Aid* Compound decoction of root taken for "female weakness." (as *Vagnera racemosa* 53:356) **Costanoan** *Contraceptive* Decoction of leaves used as a contraceptive. (as *Smilacina racemosa* 21:28) **Delaware, Oklahoma** *Tonic* Compound containing root used as a tonic. (as *Smilacina racemosa* 175:80) **Gitksan** *Antirheumatic (Internal)* Decoction of root, a very strong medicine, taken for rheumatism. *Cathartic* Decoction of root, a very strong medicine, taken as a purgative. *Dermatological Aid* Poultice of mashed roots bound on cuts. *Kidney Aid* Decoction of root taken for sore back and kidney trouble. *Orthopedic Aid* Decoction of roots taken as a very strong medicine for sore back. (as *Smilacina racemosa* 150:53) **Iroquois** *Anthelmintic* Compound infusion with whisky taken for tapeworms. (as *Smilacina racemosa* 87:284) *Antidote* Compound infusion taken as remedy for poison. *Antirheumatic (External)* Compound infusion used as foot soak. *Antirheumatic (Internal)* Compound decoction taken for rheumatism. (as *Smilacina racemosa* 87:283) *Blood Medicine* Compound infusion with whisky taken as a blood remedy. (as *Smilacina racemosa* 87:284) *Dermatological Aid* Poultice used for swellings. (as *Smilacina racemosa* 87:283) *Gynecological Aid* Compound decoction of roots used "when woman has miscarriage." (as *Smilacina racemosa* 87:284) *Hunting Medicine* Cold, compound infusion used to "get a fish on each hook, every cast." *Other* Compound decoction of roots used as a "rooster fighting medicine." *Snakebite Remedy* Roots bound to spoiled snakebites. (as *Smilacina racemosa* 87:283) *Witchcraft Medicine* Compound used for witching. (as *Smilacina racemosa* 87:284) **Kitasoo** *Unspecified* Plant used for medicinal purposes. (as *Smilacina racemosa* 43:321) **Malecite** *Dermatological Aid* Leaves and stalks boiled and used for rashes or itching. (as *Smilacina racemosa* 116:250) **Menominee** *Herbal Steam* and *Respiratory Aid* Root used in herbal steam inhaled for catarrh. (as *Smilacina racemosa* 151:41) **Meskwaki** *Anticonvulsive* Smudge of root used in cases of a fit, to bring back to normal. *Ceremonial Medicine* Root used in meeting when medicine man wants to perform trick or cast spells. *Laxative* Compound containing root used to loosen the bowels. *Misc. Disease Remedy* Root cooked in kettle to prevent sickness during time of plague. *Pediatric Aid* Smudge used "to hush a crying child." *Psychological Aid* Smudge of root used in cases of insanity, to bring back to normal. *Sedative* Smudge used "to hush a crying child." *Stimulant* Smudge used to "smoke patient for five minutes" and revive him. *Veterinary Aid* Root mixed with food fed to hogs to prevent hog cholera. (as *Smilacina racemosa* 152:230, 231) **Micmac** *Dermatological Aid* Leaves and stems used for rashes and itch. (as *Smilacina racemosa* 40:62) **Mohegan** *Cough Medicine* Infusion of leaves used for cough. (as *Smilacina racemosa* 174:265) Infusion of leaves used as a cough remedy. (as *Smilacina racemosa* 176:75, 132) *Gastrointestinal Aid* Infusion of root used for "a stronger stomach." (as *Smilacina racemosa* 174:265) Infusion of root taken for stomach disorders. (as *Smilacina racemosa* 176:75, 132) *Tonic* Complex compound infusion including spikenard root taken as spring tonic. (as *Smilacina racemosa* 174:266) **Ojibwa** *Analgesic* Compound containing root used for headache. *Gynecological Aid* and *Kidney Aid* Compound containing root taken "to keep kidneys open during pregnancy." *Stimulant* Root used as a reviver. *Throat*

Aid Compound containing root used for sore throat. (as *Smilacina racemosa* 153:374) **Ojibwa, South** *Analgesic* Roots used as an inhalant for headache. *Gynecological Aid* Decoction of leaves used by "lying-in women." *Hemostat* Poultice of crushed, fresh leaves applied to bleeding cuts. (as *Smilacina racemosa* 91:199) **Okanagan-Colville** *Cold Remedy* Decoction of rhizomes taken for colds. *Dietary Aid* Decoction of rhizomes taken to increase the appetite. (as *Smilacina racemosa* 188:48) **Potawatomi** *Stimulant* Root smudged on coals and used to revive comatose patient. (as *Smilacina racemosa* 154:63) **Shuswap** *Blood Medicine* Decoction of roots taken as a blood purifier. (as *Smilacina racemosa* 123:55) **Thompson** *Analgesic* Compound decoction of root taken for internal pains. (as *Vagnera racemosa* 164:459) *Antirheumatic (Internal)* Decoction of leaves taken two or three times a day for rheumatism. *Cancer Treatment* Decoction of rhizomes taken in several doses over a period of several days for cancer. (as *Smilacina racemosa* 187:127) *Gastrointestinal Aid* Decoction of rhizomes taken as a stomach medicine. *Gynecological Aid* Decoction of rhizomes taken during the menstrual period. (as *Vagnera racemosa* 164:458) Decoction of leaves and roots taken during pregnancy for internal soreness. *Heart Medicine* Decoction of rhizomes taken in several doses over a period of several days for heart trouble. *Throat Aid* Decoction of rhizomes taken for a sore or ulcerated throat. The decoction was taken in several doses over a period of several days. (as *Smilacina racemosa* 187:127)

- *Food*—**Costanoan** *Fruit* Fruits eaten for food. (as *Smilacina racemosa* 21:255) **Hanaksiala** *Beverage* Juice mixed with Pacific crabapples and high-bush cranberries and drunk. (as *Smilacina racemosa* 43:200) **Ojibwa** *Fodder* Roots added to oats to make a pony grow fat. *Vegetable* Roots soaked in lye water, parboiled to get rid of the lye, and cooked like potatoes. (as *Smilacina racemosa* 153:407) **Okanagan-Colville** *Unspecified* Rhizomes dried, soaked, pit cooked with camas, and eaten. (as *Smilacina racemosa* 188:48) **Okanagon** *Fruit* Bright-colored berries used for food. (as *Vagnera racemosa* 125:38) **Skagit, Upper** *Fruit* Berries used for food. (as *Smilacina racemosa* 179:38) **Thompson** *Forage* Rhizomes eaten by bears. (as *Smilacina racemosa* 187:127) *Fruit* Bright-colored berries used for food. (as *Vagnera racemosa* 125:38) Berries eaten in large quantities. (as *Vagnera racemosa* 164:486) *Spice* Leafy shoots cooked as a flavoring for meat. *Unspecified* Roots used for food. *Vegetable* Young shoots cooked and eaten like asparagus. (as *Smilacina racemosa* 187:127)

Maianthemum stellatum (L.) Link, Starry False Solomon's Seal
- *Drug*—**Delaware** *Cathartic* Roots combined with others and used to cleanse the system. *Gastrointestinal Aid* Roots combined with others and used to stimulate the stomach. *Gynecological Aid* Roots combined with others and used for leukorrhea. (as *Smilacina stellata* 176:38) **Delaware, Oklahoma** *Cathartic* Compound containing root taken to "cleanse the system." (as *Smilacina stellata* 175:32, 80) *Gastrointestinal Aid* Compound containing root taken to stimulate the stomach and cleanse the system. (as *Smilacina stellata* 175:32) *Gynecological Aid* Compound containing root taken for leukorrhea. *Stimulant* Compound containing root taken to "stimulate the stomach." (as *Smilacina stellata* 175:32, 80) *Tuberculosis Remedy* Root used alone or in compound for scrofula. *Venereal Aid* Root used alone or in compound for venereal disease. (as *Smilacina stellata* 175:80) **Gosiute** *Antirheumatic (External)* Pounded roots rubbed on limbs affected by rheumatism. (as *Smilacina stellata* 39:382) **Iroquois** *Gynecological Aid* Compound infusion taken for stricture caused when a woman has her changes. (as *Smilacina stellata* 87:283) **Navajo, Kayenta** *Ceremonial Medicine* Plant used in the Fire Dance. (as *Smilacina stellata* 205:17) **Navajo, Ramah** *Ceremonial Medicine* and *Emetic* Decoction of plant used as a ceremonial emetic. (as *Smilacina stellata* 191:20) **Paiute** *Cough Medicine* Exudate from plant used as a cough syrup. *Dermatological Aid* Poultice of fresh or dried roots applied to boils or swellings. *Ear Medicine* Pulped root squeezed into ear for earache. *Eye Medicine* Infusion of root used as a wash for eye inflammations. *Gastrointestinal Aid* Decoction of root taken for stomach trouble. *Gynecological Aid* Decoction of root taken to regulate menstrual disorders. *Orthopedic Aid* Poultice of fresh or dried roots applied to sprains. **Shoshoni** *Eye Medicine* Infusion of root used as a wash for eye inflammations. *Gastrointestinal Aid* Decoction of root taken for stomach trouble. *Gynecological Aid* Decoction of leaf taken daily for a week by women as a contraceptive. Decoction of root taken to regulate menstrual disorders. *Venereal Aid* Decoction of root taken for venereal disease. (as *Smilacina stellata* 180:139, 140) **Thompson** *Analgesic* Compound decoction of roots taken for internal pains. (as *Vagnera stellata* 164:459) *Antirheumatic (Internal)* Decoction of leaves taken two or three times a day for rheumatism. *Cold Remedy* Decoction of crushed, dried leaves and fruits taken for colds. (as *Smilacina stellata* 187:129) **Washo** *Blood Medicine* and *Dermatological Aid* Infusion of root used as an antiseptic wash in cases of blood poisoning. *Hemostat* Powdered root applied to bleeding wounds. *Tonic* Decoction of root taken as a tonic. (as *Smilacina stellata* 180:139, 140)
- *Food*—**Bella Coola** *Fruit* Berries chewed and juice swallowed. (as *Smilacina stellata* 184:199) **Okanagon** *Fruit* Bright-colored berries used for food. **Thompson** *Fruit* Bright-colored berries used for food. (as *Vagnera stellata* 125:38) Berries eaten in large quantities. (as *Vagnera stellata* 164:486)
- *Other*—**Kawaiisu** *Hunting & Fishing Item* Mashed roots used as a fish stupefier. When put in a stream it stunned the fish, which floated on the water. They were caught in the basket winnower and thrown to the bank. (as *Smilacina stellata* 206:64)

Malacothrix, Asteraceae
Malacothrix californica DC., California Desertdandelion
- *Food*—**Luiseño** *Unspecified* Seeds used for food. (155:228)

Malacothrix fendleri Gray, Fendler's Desertdandelion
- *Drug*—**Navajo, Ramah** *Dermatological Aid* Poultice of leaves applied to sores. *Eye Medicine* Cold infusion used as wash for sore eyes. (191:52)

Malacothrix glabrata (D. C. Eat. ex Gray) Gray, Smooth Desertdandelion
- *Drug*—**Apache, White Mountain** *Blood Medicine* Roots used as a blood medicine. (136:158)

Malacothrix sonchoides (Nutt.) Torr. & Gray, Sowthistle Desertdandelion
- *Drug*—**Navajo, Kayenta** *Antiemetic* Plant used for vomiting. (205:49)

Malaxis, Orchidaceae

Malaxis unifolia Michx., Green Addersmouth Orchid
- **Drug**—**Ojibwa** *Diuretic* Compound containing root used as a diuretic. (as *Microstylis unifolia* 153:377)

Malus, Rosaceae

Malus angustifolia (Ait.) Michx., Southern Crabapple
- **Food**—**Cherokee** *Dried Food* Sun dried, sliced fruit used for food. *Preserves* Fruit used to make clear jelly. (126:56)

Malus coronaria (L.) P. Mill., Sweet Crabapple
- **Drug**—**Cherokee** *Gastrointestinal Aid* Infusion of bark taken for gallstones and piles and infusion used for sore mouth. *Hemorrhoid Remedy* Infusion of bark taken for gallstones and "piles" and infusion used for sore mouth. *Oral Aid* Infusion of bark taken for gallstones and infusion used as a wash for sore mouth. (80:31)
- **Food**—**Cherokee** *Fruit* Fruit used for food. (80:31)

Malus coronaria (L.) P. Mill. **var. *coronaria***, Sweet Crabapple
- **Drug**—**Iroquois** *Abortifacient* Decoction of roots taken for suppressed menses. (as *Pyrus coronaria* 87:351) *Dermatological Aid* Cold infusion of bark used as wash for black eyes. *Eye Medicine* Cold infusion of bark used as wash for snow-blindness and black or sore eyes. *Gynecological Aid* Cold, compound infusion of twig bark taken for difficult birth. *Tuberculosis Remedy* Compound decoction of roots taken for consumption. (as *Pyrus coronaria* 87:350)
- **Food**—**Iroquois** *Bread & Cake* Fruit mashed, made into small cakes, and dried for future use. *Dried Food* Raw or cooked fruit sun or fire dried and stored for future use. *Fruit* Dried fruit taken as a hunting food. *Sauce & Relish* Dried fruit cakes soaked in warm water and cooked as a sauce or mixed with corn bread. (as *Pyrus coronaria* 196:129) **Ojibwa** *Fruit* Fruit used for food. (as *Pyrus coronaria* 135:236)

Malus fusca (Raf.) Schneid., Oregon Crabapple
- **Drug**—**Bella Coola** *Eye Medicine* Compound decoction of bark or root used as an eyewash for soreness. (as *Pyrus diversifolia* 150:60) **Cowichan** *Panacea* Infusion of bark and wild cherry bark taken as a cure-all tonic. (as *Pyrus fusca* 182:87) **Gitksan** *Antirheumatic (Internal)* Decoction of trunk and branch or inner bark taken for rheumatism. *Dietary Aid* Decoction of trunk, branches, or inner bark taken as a "fattening medicine." *Diuretic* Decoction of trunk and branches or inner bark taken as a laxative and diuretic. *Eye Medicine* Juice, scraped from peeled trunk, used as an eye medicine. *Laxative* Decoction of trunk and branch or inner bark taken as a laxative and diuretic. *Tuberculosis Remedy* Decoction of trunk and branch or inner bark taken for consumption. (as *Pyrus diversifolia* 150:60) **Haisla & Hanaksiala** *Antirheumatic (Internal)* Fruit eaten after a long day of hunting, "kills poison in muscles." *Ceremonial Medicine* Afterbirth of a child tied to a young tree to ensure the child would grow up strong. (43:265) **Hoh** *Venereal Aid* Infusion used for gonorrhea. (as *Pyrus diversifolia* 137:64) **Klallam** *Eye Medicine* Infusion of bark used as an eyewash. (as *Pyrus diversifolia* 79:38) **Kwakiutl** *Antihemorrhagic* Shredded bark used for blood-spitting. *Dermatological Aid* Tree used for eczema or skin troubles. (as *Pyrus fusca* 183:290) **Makah** *Antidiarrheal* Infusion of bark taken for dysentery and diarrhea. (as *Pyrus diversifolia* 79:38) *Blood Medicine* Bark used as a "blood purifier" and it "puts something in your blood that cuts down the clots." (as *Pyrus fusca* 67:268) *Dermatological Aid* Poultice of chewed bark applied to wounds. (as *Pyrus diversifolia* 79:38) *Gastrointestinal Aid* Bark used for ulcers. (as *Pyrus fusca* 67:268) Infusion of bark taken for intestinal disorders. (as *Pyrus diversifolia* 79:38) *Heart Medicine* Bark used for the heart. *Internal Medicine* Bark used for any internal ailment. Bark used for internal organs. *Laxative* Bark of larger trees used as a laxative. *Orthopedic Aid* Used for fractures. *Panacea* Bark used for any illness and considered a complete medicine, all in itself. (as *Pyrus fusca* 67:268) *Pulmonary Aid* Soaked leaves chewed for lung trouble. (as *Pyrus diversifolia* 79:38) Leaves chewed for lung trouble. (as *Pyrus fusca* 186:121) *Tonic* Used as a tonic. (as *Pyrus fusca* 67:268) Infusion of bark used as a tonic. (as *Pyrus fusca* 186:121) *Tuberculosis Remedy* Used for tuberculosis. *Unspecified* Bark used to make the most popular and healing medicine. (as *Pyrus fusca* 67:268) **Nitinaht** *Cough Medicine* Infusion of bark taken for coughs. *Dietary Aid* Infusion of bark taken for losing weight. *Panacea* Infusion of bark taken for "any kind of sickness." (as *Pyrus fusca* 186:121) *Tonic* Bark and roots used as a tonic for young men in training. Used as a tonic to "repair the damage done by the elder" during puberty rites. *Unspecified* Used as a medicine. (as *Pyrus fusca* 67:268) **Oweekeno** *Oral Aid* Bark chewed by hunters to suppress thirst. (43:109) **Quileute** *Pulmonary Aid* Infusion of bark taken for lung trouble. (as *Pyrus diversifolia* 79:38) *Venereal Aid* Infusion used for gonorrhea. (as *Pyrus diversifolia* 137:64) **Quinault** *Analgesic* and *Blood Medicine* Infusion of bark taken for "soreness inside, for it is throughout the blood." *Eye Medicine* Infusion of bark used as an eyewash. (as *Pyrus diversifolia* 79:38) **Saanich** *Panacea* Infusion of bark and wild cherry bark taken as a cure-all tonic. (as *Pyrus fusca* 182:87) **Samish** *Dermatological Aid* Decoction of bark used as a wash for cuts. *Gastrointestinal Aid* Decoction of bark taken for stomach disorders. **Swinomish** *Dermatological Aid* Decoction of bark used as a wash for cuts. *Gastrointestinal Aid* Decoction of bark taken for stomach disorders. (as *Pyrus diversifolia* 79:38) **Thompson** *Analgesic* Decoction of bark and cascara bark taken for sciatica. (187:262)
- **Food**—**Alaska Native** *Cooking Agent* Used as a source of pectin for jelly making. (85:85) **Bella Coola** *Fruit* Berries used for food. (as *Pyrus fusca* 184:209) **Chinook, Lower** *Fruit* Fruits stored in baskets until soft and used for food. (as *Pyrus diversifolia* 79:38) **Clallam** *Fruit* Fruit softened in baskets and eaten. (as *Pyrus fusca* 57:202) **Cowlitz** *Fruit* Fruits cooked, stored in baskets until soft and used for food. (as *Pyrus diversifolia* 79:38) **Haisla & Hanaksiala** *Fruit* Fruit used for food. *Winter Use Food* Fruit boiled and stored in the cooking water or oil for winter use. (43:265) **Hesquiat** *Dried Food* Sour fruit dried for future use. (as *Pyrus fusca* 185:73) **Hoh** *Fruit* Fruits eaten for food. (as *Pyrus diversifolia* 137:64) **Kitasoo** *Special Food* Fruit used as a food item associated with ceremonial situations. *Winter Use Food* Fruit stored in water and topped with mammal or fish grease or oil. (43:342) **Kwakiutl, Southern** *Special Food* Fruits boiled until soft and eaten with oil at large feasts. (as *Pyrus fusca* 183:290) **Makah** *Fruit* Ripe fruit used for food. (as *Pyrus fusca* 67:268) Fruits stored in baskets until soft and used for food. (as *Pyrus diversifolia* 79:38) *Preserves* Ripe fruit used to make jelly. (as *Pyrus fusca* 67:268) **Nitinaht** *Forage* Fruits eaten by grouse. *Fruit* Fruits eaten for food. (as *Pyrus fusca*

186:121) **Oweekeno** *Fruit* Overripe fruit cooked with sugar and eaten. Fruit boiled and stored under grease in special boxes for future use. (43:109) **Quileute** *Fruit* Fruits eaten raw. (as *Pyrus diversifolia* 79:38) Fruits eaten for food. (as *Pyrus diversifolia* 137:64) **Quinault** *Fruit* Fruits stored in baskets until soft and used for food. (as *Pyrus diversifolia* 79:38) **Salish, Coast** *Fruit* Berries used for food. (as *Pyrus fusca* 182:87) **Samish** *Fruit* Fruits eaten raw. (as *Pyrus diversifolia* 79:38) **Skagit, Upper** *Fruit* Fruit ripened in storage and then eaten. (as *Pyrus diversifolia* 179:38) **Swinomish** *Fruit* Fruits eaten raw. (as *Pyrus diversifolia* 79:38) **Thompson** *Fruit* Fruit picked in fall when still green, allowed to ripen in a basket, and eaten with oulachen (candlefish) oil. (187:262)
- *Other*—**Haisla & Hanaksiala** *Hunting & Fishing Item* Wood used to make bows. Sticks used to strike rods of cedar to make noise while driving animals to be killed. *Tools* Wood used to make sledgehammer handles and mallet heads. Wood used to make root digging sticks. **Hanaksiala** *Cooking Tools* Sticks used to retrieve special cooking stones used for cooking edible seaweed. (43:265) **Hesquiat** *Tools* Wood used to make ax handles. (as *Pyrus fusca* 185:73) **Kwakiutl, Southern** *Cash Crop* Fruits used as a common article of trade. (as *Pyrus fusca* 183:290) **Nitinaht** *Hunting & Fishing Item* V-shaped branches used as gaffs for salmon. *Tools* Wood used to make digging sticks. (as *Pyrus fusca* 186:121) **Oweekeno** *Cooking Tools* Wood used to make spoons. (43:109) **Quileute** *Hunting & Fishing Item* Wood used to make seal spear prongs and as bait lure on sea bass hooks. *Tools* Wood used to make mauls for driving stakes. (as *Pyrus diversifolia* 79:38) **Salish, Coast** *Hunting & Fishing Item* Wood used to make halibut hooks, bows, and fishing floats. *Tools* Wood used to make digging sticks and adz handles. (as *Pyrus fusca* 182:87)

Malus ioensis (Wood) Britt., Prairie Crabapple
- *Food*—**Omaha** *Fruit* Fruit used for food. **Ponca** *Fruit* Fruit used for food. (70:86)

Malus ioensis (Wood) Britt. **var. ioensis**, Prairie Crabapple
- *Drug*—**Meskwaki** *Misc. Disease Remedy* Used 50 years ago for smallpox. (as *Pyrus ioensis* 152:242)
- *Food*—**Meskwaki** *Preserves* Fruit reduced to jelly. *Winter Use Food* Fruit dried for winter use. (as *Pyrus ioensis* 152:263)

Malus pumila P. Mill., Cultivated Apple
- *Drug*—**Cherokee** *Gastrointestinal Aid* and *Hemorrhoid Remedy* Infusion of bark taken for gallstones and "piles." *Pulmonary Aid* Used with other ingredients to give ballplayers wind during the game. *Throat Aid* Infusion of inner bark used for lost voice and used as an ingredient in drink for dry throat. (80:23)
- *Dye*—**Cherokee** *Yellow* Bark used to make a yellow dye. (80:23)

Malus sp., Wild Crabapple
- *Drug*—**Creek** *Herbal Steam* and *Misc. Disease Remedy* Strong decoction of plant taken, used as wash and herbal steam for rabies. (172:659)

Malus sylvestris P. Mill., Apple
- *Drug*—**Cherokee** *Throat Aid* Decoction of inner bark taken to loosen phlegm for hoarseness. (as *Pyrus malus* 177:29) **Iroquois** *Ear Medicine* Decoction of bark used as drops for earaches. *Eye Medicine* Compound poultice of bark and fruit peelings used for black eyes. Compound infusion of bark and leaves used as drops for blindness. *Orthopedic Aid* Compound poultice of bark and fruit peelings used for bruises. (as *Pyrus malus* 87:350)
- *Food*—**Haisla & Hanaksiala** *Fruit* Fruit used for food. (43:270) **Hopi** *Unspecified* Species used for food. (200:79) **Iroquois** *Bread & Cake* Fruit mashed, made into small cakes, and dried for future use. *Dried Food* Raw or cooked fruit sun or fire dried and stored for future use. *Fruit* Dried fruit taken as a hunting food. *Sauce & Relish* Dried fruit cakes soaked in warm water and cooked as a sauce or mixed with corn bread. (as *Pyrus malus* 196:129) **Oweekeno** *Fruit* Fruit used for food. (43:110)
- *Dye*—**Navajo** *Red-Yellow* Bark used to make a red yellow dye. (as *Pyrus malus* 55:55)
- *Other*—**Mohegan** *Water Indicator* Crotched sticks used to locate underground water. (as *Pyrus malus* 176:87)

Malvaceae, see *Abutilon, Alcea, Callirhoe, Eremalche, Gossypium, Hibiscus, Malva, Malvella, Modiola, Napaea, Sida, Sidalcea, Sphaeralcea*

Malva, Malvaceae

Malva moschata L., Musk Mallow
- *Drug*—**Iroquois** *Febrifuge* Infusion of plant taken for chills. *Stimulant* Infusion of plant taken for lassitude. (87:385)

Malva neglecta Wallr., Common Mallow
- *Drug*—**Cherokee** *Dermatological Aid* Flowers put in oil and mixed with tallow for use on sores. (80:44) **Iroquois** *Dermatological Aid* Compound infusion of plants applied as poultice to swellings of all kinds. (87:384) *Emetic* Infusion of smashed plant taken to vomit for a love medicine. *Gastrointestinal Aid* Compound decoction of plants applied as poultice to baby's swollen stomach. *Love Medicine* Infusion of smashed plant taken to vomit for a love medicine. *Orthopedic Aid* Cold, compound infusion of leaves applied as poultice to broken bones. Compound decoction of plants applied as poultice to baby's sore back. *Pediatric Aid* Compound decoction of plants applied to baby's swollen stomach or sore back. (87:385) **Mahuna** *Analgesic* Plant used for painful congestion of the stomach. *Gastrointestinal Aid* Plant used for painful congestion of the stomach. (as *M. rotundifolia* 140:8) **Navajo, Ramah** *Other* Cold infusion of plant taken and used as a lotion for injury or swelling. (191:36)

Malva nicaeensis All., Bull Mallow
- *Drug*—**Costanoan** *Analgesic* Decoction of plant taken for migraine headaches. Poultice of heated leaves applied to the stomach or head for pain. *Dermatological Aid* Decoction of roots used as a hair rinse. *Emetic* Decoction of plant taken as an emetic. *Febrifuge* Decoction of roots used, especially for children, for fevers. *Gastrointestinal Aid* Poultice of heated leaves applied to stomach or head for pain. *Other* Decoction of plant taken for migraine headaches. *Pediatric Aid* Decoction of roots used, especially for children, for fevers. (21:8)
- *Food*—**Pima** *Unspecified* Leaves cooked, mixed with white flour, cooked again, and used for food. (as *M. borealis* 95:264)

Malva parviflora L., Cheeseweed Mallow
- *Drug*—**Diegueño** *Dermatological Aid* Decoction of leaves or roots used as a rinse for dandruff and to soften the hair after the hair

wash. *Febrifuge* Decoction of leaves or roots used as an enema and bath for babies with fevers. (84:24) **Miwok** *Antirheumatic (External)* Infusion of leaves, soft stems, and flowers used as poultice on swellings. *Dermatological Aid* Infusion of leaves, soft stems, and flowers used as poultice on running sores and boils. (12:171) **Pima** *Dermatological Aid* Decoction of plant used as a shampoo. (47:79)
• *Food*—**Pima** *Forage* Seeds eaten by hogs. (47:79)

Malva sp., Mallow
• *Drug*—**Mahuna** *Kidney Aid* Plant used for the kidneys. (as *Malvacea rubra*, creeping rock mallow 140:70) **Tewa** *Analgesic* Poultice of pulverized plant paste applied to head for headache. (mallow 138:70, 71)
• *Food*—**Cahuilla** *Unspecified* Seeds eaten fresh. (15:88) **Papago** *Starvation Food* Plants boiled and liquid used to make pinole during famine. (146:76)

Malvella, Malvaceae
Malvella leprosa (Ortega) Krapov., Alkali Mallow
• *Drug*—**Choctaw** *Antidiarrheal, Burn Dressing,* and *Gastrointestinal Aid* Root used for "dysentery, diarrhea, inflammation of the bowels, burns, etc." (as *Sida hederacea* 28:287)

Mammillaria, Cactaceae
Mammillaria dioica K. Brandeg., Strawberry Cactus
• *Food*—**Diegueño** *Fruit* Small fruits eaten raw. (84:25)

Mammillaria grahamii Engelm., Graham's Nipple Cactus
• *Food*—**Apache, Chiricahua & Mescalero** *Dried Food* Dried fruit cooked and eaten. *Fruit* Raw fruit used for food. Raw fruit used for food. (as *Neomammillaria olivia* 33:41) **Apache, San Carlos** *Fruit* Fruits eaten for food. (95:257)

Mammillaria grahamii Engelm. **var. *grahamii*,** Graham's Nipple Cactus
• *Drug*—**Pima** *Ear Medicine* Plant boiled and placed warm in the ear for earaches and suppurating ears. (as *M. microcarpa* 47:57)
• *Food*—**Pima, Gila River** *Baby Food* Raw pulp eaten primarily by children. (as *M. microcarpa* 133:7) *Snack Food* Pulp eaten, primarily by children, as a snack food. (as *M. microcarpa* 133:5)

Mammillaria mainiae K. Brandeg., Counterclockwise Nipple Cactus
• *Food*—**Apache, Chiricahua & Mescalero** *Fruit* Raw fruit used for food. (as *Neomammillaria mainae* 33:41)

Mammillaria sp., Cactus
• *Food*—**Apache, White Mountain** *Unspecified* Flesh used for food. (136:158) **Gosiute** *Unspecified* Skinned inner portion of plant used for food. (39:374) **Navajo** *Unspecified* Flesh used for food. (55:64) **Tewa** *Unspecified* Spines burned off and the entire plant eaten raw. (138:62)

Mammillaria wrightii Engelm., Wright's Nipple Cactus
• *Food*—**Navajo, Ramah** *Unspecified* Stems and ripe fruits used for food. (as *Neomammillaria wrightii* 191:37)

Manfreda, Agavaceae
Manfreda virginica (L.) Salisb. ex Rose, False Aloe
• *Drug*—**Catawba** *Kidney Aid* Infusion of pounded roots taken and used externally for dropsy. (as *Agave virginica* 157:191) Infusion of pounded roots taken and used as a wash for dropsy. (as *Agave virginica* 177:10) *Snakebite Remedy* Infusion of roots taken and used externally for snakebite. (as *Agave virginica* 157:191) Infusion of roots taken and used as a wash for snakebites. (as *Agave virginica* 177:10) **Cherokee** *Anthelmintic* Root chewed for worms. *Antidiarrheal* Root chewed for diarrhea. (as *Agave virginica* 80:23) Root, a very strong medicine, chewed for persistent diarrhea. (as *Agave virginica* 203:74) *Liver Aid* Root chewed for the liver. (as *Agave virginica* 80:23) **Creek** *Snakebite Remedy* Decoction of root in sweet milk taken or used as wash for rattlesnake bite. Root chewed and swallowed or used externally for rattlesnake bite. (28:289) **Seminole** *Snakebite Remedy* Plant used for snakebites. (as *Agave virginica* 169:297)

Mangifera, Anacardiaceae
Mangifera indica L., Mango
• *Food*—**Seminole** *Unspecified* Plant used for food. (169:491)

Manihot, Euphorbiaceae
Manihot esculenta Crantz, Tapioca
• *Food*—**Seminole** *Unspecified* Plant used for food. (169:490)

Marah, Cucurbitaceae
Marah fabaceus (Naud.) Naud. ex Greene, California Manroot
• *Drug*—**Pomo** *Dermatological Aid* Pounded nuts and grease rubbed on the head for falling hair. (as *Echinocystis fabacea* 66:14) **Pomo, Kashaya** *Dermatological Aid* Raw, pounded root mixed with pounded pepper nuts and skunk grease applied to head to prevent baldness. (72:41)
• *Other*—**Pomo** *Hunting & Fishing Item* Root used as a fish drug. (as *Echinocystis fabacea* 66:14) **Pomo, Kashaya** *Hunting & Fishing Item* Mashed root used in pools in the river and in tide pools at the beach for poisoning fish. (72:41)

Marah horridus (Congd.) S. T. Dunn, Sierran Manroot
• *Drug*—**Kawaiisu** *Dermatological Aid* Mashed, roasted seeds used as a salve for skin irritations and sores, and as a salve to eliminate baldness. *Ear Medicine* Roasted seeds placed in the ear for earaches. (206:40) **Tubatulabal** *Dermatological Aid* Burned, ripe seeds rubbed on pimples and newborn baby's navel. *Pediatric Aid* Burned, ripe seeds rubbed on newborn baby's navel. (as *Echinocystis horrida* 193:59)

Marah macrocarpus (Greene) Greene, Cucamonga Manroot
• *Drug*—**Costanoan** *Dermatological Aid* Seeds used as a paste for pimples and skin sores. (21:24) **Luiseño** *Cathartic* Roots used as a purgative. (as *Echinocystis macrocarpa* 155:229)
• *Other*—**Costanoan** *Soap* Root used to make detergent lather. (21:251) **Luiseño** *Paint* Seeds mixed with iron oxide and turpentine to make a red paint. (as *Echinocystis macrocarpa* 155:210)

Marah macrocarpus **var. *major*** (S. T. Dunn) Stocking, Cucamonga Manroot
• *Drug*—**Mahuna** *Dermatological Aid* Plant juices rubbed on parts afflicted by ringworm. Seed oil rubbed on the head for diseased scalps and hair roots. (as *Micrapelis micracarpa* 140:38)

Marah oreganus (Torr. ex S. Wats.) T. J. Howell, Coastal Manroot
- *Drug*—**Chehalis** *Dermatological Aid* and *Tuberculosis Remedy* Salve of root ash and grease applied to scrofula sores. (as *Echinocystis oregana* 79:48) **Karok** *Dermatological Aid* Poultice of roots applied to bruises and boils. (6:39) *Poison* Plant considered poisonous. (as *Echinocystis oregana* 148:386) **Mendocino Indian** *Antirheumatic (External)* Seeds and roots used for rheumatism or root rubbed on rheumatic joints. *Dermatological Aid* Root rubbed on rheumatic joints, boils, and swellings. *Poison* Roots and seeds considered poisonous. *Urinary Aid* Seeds eaten for urinary troubles. *Venereal Aid* Seeds and roots used for rheumatism and venereal disease. (as *Micrampelis marah* 41:390) **Paiute** *Eye Medicine* Decoction of peeled, sliced, and dried root used for "sore eyes." (as *Echinocystis oreganas* 111:113) **Squaxin** *Analgesic* and *Orthopedic Aid* Infusion of smashed stalks used as a soak for aching hands. *Poison* Plant considered poisonous. (as *Echinocystis oregana* 79:48)
- *Food*—**Yurok** *Beverage* Young shoots and *Polypodium* rhizomes used to make tea. (6:39)
- *Other*—**Mendocino Indian** *Hunting & Fishing Item* Roots formerly used as fish poison. (as *Micrampelis marah* 41:390) **Yurok** *Toys & Games* Fruit used by children to construct representations of animals by inserting twigs. Fruit tossed by children at one another in play. (6:39)

Marantaceae, see *Thalia*

Marattiaceae, see *Marattia*

Marattia, Marattiaceae, fern
Marattia sp., Pala Fern
- *Food*—**Hawaiian** *Unspecified* Leaf stem bases overcooked and eaten. (112:68)

Marrubium, Lamiaceae
Marrubium vulgare L., Horehound
- *Drug*—**Cahuilla** *Kidney Aid* Infusion of whole plant used for flushing the kidneys. (15:88) **Cherokee** *Breast Treatment* Infusion used for "breast complaints." *Cold Remedy* Taken for colds. *Cough Medicine* Mixed with sugar to make cough syrup. *Pediatric Aid* Infusion given to babies. *Throat Aid* Taken for hoarseness. (80:39) **Costanoan** *Cough Medicine* Decoction of leaves used for coughs. *Dermatological Aid* Heated leaf salve used on boils. *Pulmonary Aid* Decoction of leaves used for whooping cough. (21:16) **Diegueño** *Cold Remedy* Infusion of leaves taken for colds. *Pediatric Aid* Infusion of leaves mixed with honey and given to children for colds and whooping cough. *Pulmonary Aid* Infusion of leaves taken for whooping cough. (84:25) **Hopi** *Unspecified* Used as a medicinal plant. (190:165) **Isleta** *Antirheumatic (External)* Poultice of crushed leaves used for swellings. (100:34) **Kawaiisu** *Cold Remedy* Hot or cold infusion of leaves and flowering tops taken for colds. *Cough Medicine* Hot or cold infusion of leaves and flowering tops taken for coughs. *Respiratory Aid* Plant used as a syrup for respiratory ailments. (206:40) **Mahuna** *Cough Medicine* Infusion of leaves and flowers taken for coughs. *Throat Aid* Infusion of leaves and flowers taken for sore throats. (140:18) **Navajo** *Throat Aid* Infusion of plant taken for sore throats. (55:73) **Navajo, Ramah** *Analgesic* Decoction of plant used for stomachache and influenza. *Disinfectant* Strong infusion used for "lightning infection." *Gastrointestinal Aid* Decoction of plant taken for stomachache. *Gynecological Aid* Root used before and after childbirth. *Misc. Disease Remedy* Decoction of plant taken for influenza. (191:41) **Paiute** *Analgesic* Branches used to whip aching body parts to stimulate circulation. (180:103) **Rappahannock** *Cold Remedy* Infusion of roots taken for colds. *Cough Medicine* Compound decoction taken for coughs. (161:27) **Round Valley Indian** *Antidiarrheal* Decoction of leaves taken for diarrhea. *Cold Remedy* Decoction of leaves taken for colds. (41:383) **Yuki** *Cough Medicine* Infusion of plant taken for coughs. (49:47)
- *Food*—**Diegueño** *Candy* Infusion of leaves mixed with honey and made into candy. (84:25) **Navajo, Ramah** *Fodder* Used for sheep feed, made the meat bitter. (191:41)

Marshallia, Asteraceae
Marshallia obovata (Walt.) Beadle & F. E. Boynt., Spoonshape Barbara's Buttons
- *Drug*—**Catawba** *Misc. Disease Remedy* Plant used in certain diseases. (157:191)

Martynia, Pedaliaceae
Martynia sp., Devil's Claw
- *Food*—**Apache, Western** *Beverage* Seeds cracked and chewed for the juice. *Winter Use Food* Seeds stored in pottery, gourd, or water-basket receptacles. (26:189)
- *Fiber*—**Shoshoni** *Basketry* Pods used to make the black design in basketry. (117:445)
- *Other*—**Santa Clara** *Ceremonial Items* Open seed vessels used to make artificial flowers for dancers' headdresses. **Tewa of Hano** *Ceremonial Items* Open seed vessels used to make artificial flowers for dancers' headdresses. (138:57)

Matelea, Asclepiadaceae
Matelea biflora (Raf.) Woods., Star Milkvine
- *Drug*—**Comanche** *Ceremonial Medicine* Decoction of thick, white roots used for ghost sickness. *Dermatological Aid* Decoction of thick, white roots used for bruises. *Gastrointestinal Aid* Root paste used for severe stomach pains. *Gynecological Aid* Decoction of thick, white roots used for menstrual cramps. *Misc. Disease Remedy* Root paste used for diphtheria and other throat closing ailments in children. *Orthopedic Aid* Decoction of thick, white roots used for broken bones. *Pediatric Aid* Root paste used for diphtheria and other throat closing ailments in children. (99:9)

Matelea cynanchoides (Engelm.) Woods., Prairie Milkvine
- *Drug*—**Comanche** *Ceremonial Medicine* Decoction of thick, white roots used for ghost sickness. *Dermatological Aid* Decoction of thick, white roots used for bruises. *Gastrointestinal Aid* Root paste used for severe stomach pains. *Gynecological Aid* Decoction of thick, white roots used for menstrual cramps. *Misc. Disease Remedy* Root paste used for diphtheria and other throat closing ailments in children. *Orthopedic Aid* Decoction of thick, white roots used for broken bones. *Pediatric Aid* Root paste used for diphtheria and other throat closing ailments in children. (99:9)

Matelea producta (Torr.) Woods., Texas Milkvine

- *Food*—**Apache, Chiricahua & Mescalero** *Unspecified* Seeds eaten fresh or boiled. (as *Vincetoxicum productum* 33:45)

Matricaria, Asteraceae
Matricaria discoidea DC., Disc Mayweed
- *Drug*—**Aleut** *Analgesic* and *Carminative* Infusion of leaves taken for stomach pains, especially from gas. *Gastrointestinal Aid* Infusion of leaves taken for stomach pain, especially for gas on the stomach. *Laxative* Infusion of leaves taken as a laxative. *Panacea* Plant used as a cure-all. *Tonic* Plant used to make a tonic. (as *M. matricarioides* 8:426) **Blackfoot** *Antidiarrheal* Decoction of plant and flowers used for diarrhea. (as *M. matricarioides* 97:61) **Cahuilla** *Antidiarrheal* Infusion of plant used for diarrhea. *Gastrointestinal Aid* Infusion of plant used for colic and to settle upset stomachs. (as *M. matricarioides* 15:88) **Cherokee** *Gastrointestinal Aid* Infusion taken "to keep regular." (as *M. matricarioides* 80:49) **Cheyenne** *Ceremonial Medicine* Plant used in the Sun Dance ceremony. (as *M. matricarioides* 83:22) *Dermatological Aid* Dried, pulverized flowers, leaves, sweet grass, horse mint, and sweet pine used as a perfume. (as *M. matricarioides* 76:189) *Unspecified* Plant tops used as an ingredient in many medicines. (as *M. matricarioides* 83:22) **Costanoan** *Analgesic* Decoction of plant taken for stomach pain. *Anticonvulsive* Decoction of plant used for infant convulsions. *Dermatological Aid* and *Disinfectant* Seeds used as salve for infected sores. *Febrifuge* Decoction of plant taken for fever. *Gastrointestinal Aid* Decoction of plant taken for stomach pain and indigestion. *Pediatric Aid* Decoction of plant used for infant convulsions. (as *M. matricarioides* 21:27) **Diegueño** *Febrifuge* Decoction of plant mixed with mallow and elderberry blossoms and used as an enema for babies with fever. *Gynecological Aid* Decoction of whole plant taken by women following childbirth. *Pediatric Aid* Decoction of plant mixed with mallow and elderberry blossoms and used as an enema for babies with fever. (as *M. matricarioides* 84:25) **Eskimo, Alaska** *Antihemorrhagic* Plant tops chewed for spitting up blood. (as *M. matricarioides* 1:38) **Eskimo, Inuktitut** *Unspecified* Used for medicinal purposes. (as *M. matricarioides* 202:183) **Eskimo, Kuskokwagmiut** *Adjuvant* Plant used in the steam bath for the pleasant odor. *Cold Remedy* Decoction of dried seed heads taken for colds. *Gastrointestinal Aid* Decoction of dried seed heads taken for indigestion. (as *M. suaveolens* 122:22, 23) **Eskimo, Western** *Adjuvant* Plants added to sweat bath water container to impart fragrance. (as *M. suaveolens* 108:39) *Cold Remedy* and *Gastrointestinal Aid* Decoction of dried seed heads used "for either indigestion or a cold." (as *M. suaveolens* 108:13) **Flathead** *Antidiarrheal* Infusion of herb used for diarrhea. *Cold Remedy* Infusion of herb used for children with colds. *Gastrointestinal Aid* Infusion of herb used for upset stomach. *Pediatric Aid* Infusion of herb used for children with colds. (as *M. matricarioides* 82:23) **Montana Indian** *Antidiarrheal* Decoction of herbs and flowers used for diarrhea. (19:15) *Gynecological Aid* Infusion of herb used for building up blood at childbirth and delivering the placenta. Infusion of herb used by young girls for menstrual cramps. (as *M. matricarioides* 82:23) **Okanagan-Colville** *Love Medicine* Tops and human hair buried on range to prevent loved ones or relations from going away. *Veterinary Aid* Tops, horse and human hair, and musk gland material buried on range to prevent horses from running away. *Witchcraft Medicine* Plant used for witchcraft. (as *M. matricarioides* 188:84) **Shuswap** *Cold Remedy* Dried plants used for colds. *Heart Medicine* Dried plants used for the heart. (as *M. matricarioides* 123:59) **Ute** *Unspecified* Used as a medicine. (38:35) **Yokia** *Antidiarrheal* Decoction of leaves and flowers taken for diarrhea. (41:395)
- *Food*—**Eskimo, Alaska** *Candy* Plant tops chewed by children for the pleasant flavor. (as *M. matricarioides* 1:38) **Flathead** *Preservative* Plants dried, pulverized, and used to preserve meat and berries. **Kutenai** *Unspecified* Small, yellowish green flower heads eaten occasionally. **Montana Indian** *Unspecified* Occasionally used for food. (as *M. matricarioides* 82:23) **Okanagan-Colville** *Unspecified* Flower heads eaten by children. (as *M. matricarioides* 188:84)
- *Fiber*—**Crow** *Mats, Rugs & Bedding* Plants dried, crushed, and used to line baby cradles. (as *M. matricarioides* 82:23)
- *Other*—**Blackfoot** *Incense & Fragrance* Blossoms dried and used for perfume. (as *M. matricarioides* 114:278) *Insecticide* Dried blossoms used as an insect repellent. (as *M. matricarioides* 97:61) **Cheyenne** *Incense & Fragrance* Leaves dried, powdered, mixed with fir or sweet grass, and used as perfume. (as *M. matricarioides* 82:23) **Eskimo, Inuktitut** *Incense & Fragrance* Used as an aromatic in sweat lodges. *Season Indicator* Plant used as an indicator of salmonberry picking time. (as *M. matricarioides* 202:183) **Kutenai** *Incense & Fragrance* Leaves dried, powdered, and used as perfume. *Jewelry* Dried heads used to make necklaces. **Montana Indian** *Incense & Fragrance* Used as a perfume. *Insecticide* Used as a bug repellent. (as *M. matricarioides* 82:23)

Matricaria sp.
- *Drug*—**Aleut** *Unspecified* Used medicinally. (7:29)

Matteuccia, Dryopteridaceae, fern
Matteuccia struthiopteris (L.) Todaro, Ostrich Fern
- *Drug*—**Cree, Woodlands** *Gynecological Aid* Decoction of leaf stalk base from the sterile frond taken to speed expulsion of the afterbirth. *Orthopedic Aid* Decoction of leaf stalk base from the sterile frond taken for back pain. (109:44)

Medeola, Liliaceae
Medeola virginiana L., Indian Cucumberroot
- *Drug*—**Iroquois** *Anticonvulsive* Infusion of crushed dried berries and leaves given to babies with convulsions. *Panacea* Compound infusion taken or placed on injured part, a "Little Water Medicine." *Pediatric Aid* Infusion of crushed dried berries and leaves given to babies with convulsions. *Witchcraft Medicine* Raw root chewed and spit on hook to "make fish bite." (87:285)

Medicago, Fabaceae
Medicago polymorpha L., Bur Clover
- *Food*—**Cahuilla** *Porridge* Parched, ground seeds used to make mush. (as *M. hispida* 15:88) **Mendocino Indian** *Forage* Seeds and leaves used as a forage plant. Dried seedpods eaten by sheep in summer. (as *M. denticulata* 41:358)

Medicago sativa L., Alfalfa
- *Drug*—**Costanoan** *Ear Medicine* Poultice of heated leaves applied to the ear for earaches. (21:19)
- *Food*—**Navajo, Ramah** *Fodder* Plant cultivated, harvested, dried,

stacked, or stored in hogans, and fed to livestock in winter. (191:32) **Okanagan-Colville** *Spice* Plants placed above and below black tree lichen and camas in cooking pits for the sweet flavor. (188:105) **Shuswap** *Fodder* Used for horse feed. (123:64)
- *Other*—Keres, Western *Unspecified* Plant known and named but no use was specified. (171:53)

Melampyrum, Scrophulariaceae

Melampyrum lineare Desr., Narrowleaf Cowheat
- *Drug*—Ojibwa *Eye Medicine* Infusion of plant used as a "little medicine for the eyes." (153:389)

Melastomataceae, see *Rhexia*

Meliaceae, see *Melia*

Melia, Meliaceae

Melia azedarach L., Chinaberrytree
- *Drug*—Cherokee *Anthelmintic* Infusion of root and bark given for worms. *Dermatological Aid* Used for "scald head," ringworm, and "tetterworm." (80:29)
- *Other*—Cherokee *Insecticide* Crushed leaves used to drive out "house insects." (80:29) **Omaha** *Good Luck Charm* Fruits used as good luck charm beads. *Jewelry* Fruits used as beads. **Ponca** *Good Luck Charm* Fruits used as good luck charm beads. *Jewelry* Fruits used as beads. (70:98)

Melica, Poaceae

Melica bulbosa Geyer ex Porter & Coult., Oniongrass
- *Food*—Pomo *Porridge* Raw roots pounded like pinole. *Unspecified* Raw roots used for food. (66:11)

Melica imperfecta Trin., Smallflower Melicgrass
- *Drug*—Kawaiisu *Toothache Remedy* Plant clump used to discard children's milk teeth, "then another one will grow in." (206:40)
- *Food*—Kawaiisu *Porridge* Seeds winnowed, pounded in a bedrock mortar, and cooked into a mush. (206:40)

Melicope, Rutaceae

Melicope cinerea Gray, Manena
- *Drug*—Hawaiian *Venereal Aid* Plant used for venereal diseases. (as *Pelea cinerea* 2:72)

Melilotus, Fabaceae

Melilotus indicus (L.) All., Annual Yellow Sweetclover
- *Drug*—Pomo, Kashaya *Laxative* Decoction of whole plant taken as a purgative, a very strong laxative. (72:37)
- *Other*—Isleta *Insecticide* Plant used in beds as a bedbug repellant. (100:34) **Pima** *Toys & Games* Used in target shooting games. (47:131)

Melilotus officinalis (L.) Lam., Yellow Sweetclover
- *Drug*—Iroquois *Dermatological Aid* Infusion of flowers and rhizomes from another plant applied to the face for pimples and sunburn. (as *M. alba* 141:49) *Febrifuge* Infusion taken for typhoid-like fever caused by odor from killed snake. (as *M. alba* 87:364) **Navajo, Ramah** *Cold Remedy* Cold infusion taken and used as lotion for colds caused by becoming chilled. (as *M. alba* 191:33)
- *Food*—Jemez *Forage* Plant very nutritious food for horses. (as *M. alba* 44:25)
- *Other*—Dakota *Incense & Fragrance* Grass hung in houses for the pleasant fragrance. (as *M. alba* 69:365) Bunches of plants hung in the home for the fragrance. (as *M. alba* 70:91) **Iroquois** *Incense & Fragrance* Flowers used in a bouquet to perfume the house. (as *M. alba* 142:93) **Keres, Western** *Insecticide* Plant used in beds as a bedbug repellent. (as *M. alba* 171:53)

Melilotus sp., Sweetclover
- *Other*—Havasupai *Incense & Fragrance* Leaves dried, ground, placed in a small bundle, and tied onto women's clothes as a perfume. (197:227)

Melissa, Lamiaceae

Melissa officinalis L., Common Balm
- *Drug*—Cherokee *Cold Remedy* Used for old colds. *Febrifuge* Used for typhus fevers, chills, and fevers. *Misc. Disease Remedy* Plant used for typhus fevers, chills, and fevers. *Stimulant* Used as a stimulant. *Tonic* Used as a tonic. (80:24) **Costanoan** *Gastrointestinal Aid* Decoction of plant used for infants' colic and stomachaches. *Pediatric Aid* Decoction of plant used for infants with colic. (21:16)

Melothria, Cucurbitaceae

Melothria pendula L., Guadeloupe Cucumber
- *Drug*—Houma *Snakebite Remedy* Poultice of pulverized leaves and gunpowder applied to moccasin bite. (158:64)

Menispermaceae, see *Cocculus*, *Menispermum*

Menispermum, Menispermaceae

Menispermum canadense L., Common Moonseed
- *Drug*—Cherokee *Antidiarrheal* Taken for weak stomachs and bowels. *Dermatological Aid* Root used for skin diseases. *Gastrointestinal Aid* Taken for weak stomachs and bowels. *Gynecological Aid* Taken by "weakly females." *Laxative* Root used as a laxative. *Stimulant* Taken by "weakly females." *Venereal Aid* Taken for venereal diseases and as a laxative. (80:54) **Delaware, Oklahoma** *Dermatological Aid* Salve containing plant used on chronic sores. (175:27)

Menodora, Oleaceae

Menodora scabra Engelm. ex Gray, Rough Menodora
- *Drug*—Navajo, Ramah *Analgesic* Decoction of root used for backbone pain. *Gastrointestinal Aid* Cold infusion taken for heartburn. *Gynecological Aid* Decoction of plant taken to facilitate labor. *Orthopedic Aid* Decoction of root used for "pain in backbone." *Panacea* Plant used as "life medicine." (191:39)

Mentha, Lamiaceae

Mentha arvensis L., Wild Mint
- *Drug*—California Indian *Kidney Aid* Infusion of leaves taken for kidney complaint. (118:41) **Cherokee** *Febrifuge* Infusion given for fever. (80:45) **Cheyenne** *Antiemetic* Infusion of ground leaves and stems taken for vomiting. (82:64) *Ceremonial Medicine* Plant used in the Sun Dance ceremony. *Dermatological Aid* Decoction of plant used as a hair oil. *Heart Medicine* Infusion of ground leaves and stems taken

to strengthen heart muscles. (83:27) *Love Medicine* Leaves chewed and placed on body for improved love life. (82:64) Infusion of ground leaves and stems used to improve one's love life. *Stimulant* Infusion of ground leaves and stems taken to stimulate vital organs. (83:27) **Flathead** *Cold Remedy* Infusion taken for colds. *Cough Medicine* Infusion taken for coughs. *Febrifuge* Infusion taken for fevers. *Tonic* Infusion taken as a tonic. *Toothache Remedy* Leaves used for carious teeth. **Gros Ventre** *Analgesic* Infusion taken for headaches. (82:64) **Iroquois** *Antidote* Compound decoction of plants taken to vomit as cure for poison. (87:428) **Kawaiisu** *Analgesic* Poultice of leaves and stems applied to areas of pain. *Dermatological Aid* Poultice of leaves and stems applied to areas of swelling. (206:40) **Kutenai** *Antirheumatic (External)* Poultice of leaves used for rheumatism and arthritis. *Cold Remedy* Infusion taken for colds. *Cough Medicine* Infusion taken for coughs. *Febrifuge* Infusion taken for fevers. *Kidney Aid* Infusion taken for kidney problems. *Tonic* Infusion taken as a tonic. (82:64) **Menominee** *Pulmonary Aid* Compound infusion taken and poultice applied to chest for pneumonia. (151:39) **Navajo, Kayenta** *Dermatological Aid* Plant used as a lotion for swellings. *Disinfectant* and *Pediatric Aid* Roots used for prenatal snake infection. (205:40) **Navajo, Ramah** *Febrifuge* Cold infusion taken and used as lotion for fever. *Misc. Disease Remedy* Cold infusion taken and used as lotion for influenza. *Stimulant* Cold infusion given to counteract effects of being struck by a whirlwind. (191:41) **Okanagan-Colville** *Analgesic* Infusion of stems taken for pains. *Antirheumatic (Internal)* Infusion of stems taken for swellings. *Cold Remedy* Infusion of stems taken for colds. *Febrifuge* Infusion of stems taken for fevers. *Gastrointestinal Aid* and *Pediatric Aid* Infusion of stems taken for colic in children. (188:109) **Paiute** *Other* Plant chewed or infusion of entire plant, except root, taken to keep cool. (167:317) **Thompson** *Cold Remedy* Infusion of plant taken for colds. *Misc. Disease Remedy* Infusion of plant taken to prevent influenza. One informant said that during the flu epidemic after the First World War, her grandmother made a big potful of mint tea. She and her family drank this and did not get sick. (187:233)

- **Food**—**Blackfoot** *Beverage* Dried plant used to make tea. *Spice* Dried plant used to spice pemmican and soups. (86:103) **Cherokee** *Unspecified* Species used for food. (80:45) **Cheyenne** *Beverage* Leaves and stems made into a tea and used as a beverage. (83:27) **Kawaiisu** *Beverage* Green leaves brewed into an nonmedicinal beverage tea. (206:40) **Lakota** *Beverage* Used to make tea. (139:49) **Navajo, Kayenta** *Spice* Used as flavoring with meats or cornmeal mush. (205:40) **Okanagan-Colville** *Beverage* Stems used to make tea. (188:109) **Paiute** *Beverage* Dried leaves used to make a tea. (104:103) Leaves boiled into a refreshing drink. (167:245) **Saanich** *Spice* Leaves used for flavoring food. (182:84) **Sanpoil** *Beverage* Stems used to make tea. (188:109) **Shuswap** *Beverage* Leaves used in tea. (123:64) **Thompson** *Unspecified* Greens warmed over an open fire and eaten with dried fish. (187:233)

- **Other**—**Cheyenne** *Incense & Fragrance* Leaves and stems used as perfume and deodorizers in houses. (83:27) **Flathead** *Insecticide* Leaves powdered and sprinkled on meat and berries as a bug repellent. **Kutenai** *Insecticide* Leaves powdered and sprinkled on meat and berries as a bug repellent. **Montana Indian** *Incense & Fragrance* Used as a home fragrance. **Sioux** *Hunting & Fishing Item* Plant solution used for concealing human scent from traps. (82:64) **Thompson** *Containers* Plant tops used as a liner for dried fish platters, to counteract the strong odor. *Incense & Fragrance* Whole plant soaked in warm water to make a solution used to scent feather pillows. *Insecticide* Plant used all over the house for bedbugs and other insect pests. *Soap* Whole plant soaked in warm water to make a solution used as a hair dressing. (187:233)

Mentha canadensis L., Canadian Mint

- **Drug**—**Abnaki** *Panacea* Used by children for maladies. *Pediatric Aid* Used by children for maladies. (144:155) Used for crying babies. *Sedative* Used for crying babies. (144:171) **Algonquin, Tête-de-Boule** *Febrifuge* Plant used as a fever medicine. (132:129) **Bella Coola** *Analgesic* Decoction of entire plant taken for stomach pain. *Gastrointestinal Aid* Decoction of entire plant taken for stomach pain. (150:63) **Blackfoot** *Analgesic* Dried leaves chewed and swallowed for chest pains. *Heart Medicine* Dried leaves chewed and swallowed for heart ailments. (as *M. arvensis* var. *villosa* 97:51) **Carrier, Southern** *Cold Remedy* Decoction of entire plant taken for colds and the stomach. *Gastrointestinal Aid* Decoction of entire plant taken for the stomach and various ailments. *Pulmonary Aid* Decoction of entire plant taken for "lung affections" and colds. (150:63) **Cheyenne** *Antiemetic* Decoction of finely ground leaves and stems taken to prevent vomiting. (75:39) Decoction of ground stems and leaves taken to prevent vomiting. (76:186) **Chippewa** *Carminative* Plant used as a carminative. (71:140) **Cree, Hudson Bay** *Gastrointestinal Aid* Infusion of plant used as a stomachic. (92:303) **Cree, Woodlands** *Analgesic* Infusion of leaves taken for headaches. *Antihemorrhagic* Infusion of plant taken for coughing up blood. *Cold Remedy* Infusion of plant taken to prevent the onset of a cold and for prolonged colds. Infusion of leaves taken for colds. *Febrifuge* Infusion of leaves taken for fevers. *Hemostat* Leafy stems and flowers inserted into the nostril for serious nosebleeds. *Oral Aid* Ground flowers and yarrow placed in a cloth, moistened, and rubbed on infected gums to remove pus. *Toothache Remedy* Poultice of ground leaves or leafy stems applied to the gums for toothaches. (as *M. arvensis* var. *villosa* 109:45) **Dakota** *Carminative* Sweetened infusion taken as a carminative or beverage. (70:112, 113) **Gosiute** *Analgesic, Cold Remedy,* and *Cough Medicine* Decoction of plant taken for coughs and colds with headaches. (39:351) **Great Basin Indian** *Gastrointestinal Aid* Infusion of whole plant taken for indigestion. (121:50) **Hoh** *Unspecified* Used as smelling and rubbing medicine. (137:68) **Iroquois** *Emetic* Compound decoction of plants taken to vomit as cure for poison. (as *M. arvensis* var. *canadensis* 87:428) *Febrifuge* Infusion of plant given to children for fevers. (141:58) *Hemorrhoid Remedy* Compound decoction of roots taken and used as a wash for piles. (as *M. arvensis* var. *canadensis* 87:428) *Pediatric Aid* Infusion of plant given to children for fevers. (141:58) **Isleta** *Eye Medicine* Poultice of moistened, crushed leaves used for eye trouble. (as *M. penardi* 100:34) **Keres, Western** *Analgesic* Infusion of dried plants used for headaches. *Febrifuge* Infusion of dried plants used for fevers. (171:53) **Keresan** *Febrifuge* Infusion of plant used for fever. (198:562) **Mahuna** *Sedative* Plant used as a sedative. (140:23) **Malecite** *Gastrointestinal Aid* Infusion of plants used by children with stomach trouble. *Pediatric Aid* Infusion of plants used to quiet children suffering from croup, and used by children with stomach trouble. *Pulmonary Aid* Infusion of plants used by children with croup. *Sedative* Infusion of plants used to quiet children suffering from croup. (116:250) **Menominee** *Febrifuge* Infusion of whole plant used for fever. (54:132) **Micmac** *Antiemetic* and *Pediatric Aid* Herb used for children with an upset stomach. *Pulmonary Aid* Herb used for croup. (40:58) **Mohegan** *Gastrointestinal Aid* Infusion of leaves considered beneficial to the stomach. (176:73, 130) **Montana Indian** *Unspecified* Infusion of leaves used for various complaints. (19:15) **Ojibwa** *Blood Medicine* Infusion of entire plant taken as a blood remedy. *Diaphoretic* Plant used in the sweat bath. *Febrifuge* Infusion of leaves taken for fevers. (as *M. arvensis* var. *canadensis* 153:371, 372) *Gastrointestinal Aid* Infusion of plants taken for stomach troubles. (135:231) **Okanagon** *Analgesic* Infusion of leaves and plant tips given to children with colicky pains. Infusion of leaves and plant tips taken for pains. *Cold Remedy* Infusion of leaves and plant tips taken for colds. *Dermatological Aid* Infusion of leaves and plant tips taken for swellings. *Gastrointestinal Aid* and *Pediatric Aid* Infusion of leaves and plant tips given to children with colicky pains. (125:42) **Omaha** *Carminative* Plant used as a carminative. (68:334) Sweetened infusion taken as a carminative or beverage. (70:112, 113) **Paiute** *Analgesic* Decoction of various plant parts taken for headaches. Leaves used in several ways for headaches. (180:104, 105) *Carminative* Infusion of leaves and stems taken for gas pains. (as *M. penardi* 118:45) *Cold Remedy* Infusion of fresh or dried leaves taken for colds. (as *M. arvensis* var. *glabrata* 111:107) Decoction of various plant parts taken for colds. *Dermatological Aid* Poultice of crushed leaves applied to swellings. *Febrifuge* Decoction of various plant parts taken and used as a wash for fevers. *Gastrointestinal Aid* and *Pediatric Aid* Infusion and/or decoction of plant parts used for stomachache, indigestion, and babies' colic. *Throat Aid* Leaves chewed for sore throats. (180:104, 105) **Paiute, Northern** *Cold Remedy* Fresh leaves put in the nostrils for colds. Plant spread out on the ground and lied on for a cold. *Febrifuge* Plant spread out on the ground and lied on for a fever. (59:129) **Pawnee** *Carminative* Sweetened infusion taken as a carminative or beverage. **Ponca** *Carminative* Sweetened infusion taken as a carminative or beverage. (70:112, 113) **Potawatomi** *Febrifuge* Leaves or plant top used for fevers. *Pulmonary Aid* Decoction of leaves used for pleurisy. (as *M. arvensis* var. *canadensis* 154:61) **Quileute** *Unspecified* Used as smelling and rubbing medicine. (137:68) **Salish** *Unspecified* Decoction of plants used as a medicine. (as *M. borealis* 178:294) **Sanpoil** *Cold Remedy* Decoction of leaves taken by adults for colds and infusion given to children. Decoction of plant given to infants for colds. *Panacea* Decoction of leaves taken by adults and given to children for "illnesses of a general nature." *Pediatric Aid* Decoction of leaves taken by adults for colds and infusion given to children. Decoction of plant given to infants for colds. Infusion of leaves given to children for "illnesses of a general nature." (131:218) **Shoshoni** *Carminative* Infusion of leaves and stems taken for gas pains. (as *M. penardi* 118:45) *Cold Remedy* Decoction of various plant parts taken for colds. *Febrifuge* Decoction of various plant parts taken and used as a wash for fevers. *Gastrointestinal Aid* Decoction of plant parts used for stomachache, indigestion, or babies' colic. (180:104, 105) *Pediatric Aid* Decoction of plant parts used for stomachache, indigestion, or babies' colic. (180:104, 105) **Sia** *Febrifuge* Infusion of leaves taken for fevers. (199:284) **Thompson** *Analgesic* Infusion of leaves and plant tips given to children with colicky pains. Infusion of leaves and plant tips taken for pains. (125:42) Decoction of leaves and tops taken for pains. *Antirheumatic (External)* Leaves used in the sweat bath for rheumatism. Plant steamed in the sweat bath for rheumatism and severe colds. (164:475) *Cold Remedy* Infusion of leaves and plant tips taken for colds. (125:42) Decoction of leaves and tops taken and used as herbal steam for colds. Plant steamed in the sweat bath for severe colds. (164:475) *Dermatological Aid* Infusion of leaves and plant tips taken for swellings. (125:42) Decoction of leaves and tops taken for swellings. (164:475) *Gastrointestinal Aid* Infusion of leaves and plant tips given to children with colicky pains. (125:42) *Herbal Steam* Plant steamed in the sweat bath for rheumatism and severe colds. (164:475) *Pediatric Aid* Infusion of leaves and plant tips given to children with colicky pains. (125:42) *Unspecified* Plant used as a charm for unspecified purpose. (164:507) **Washo** *Antidiarrheal* Decoction of various plant parts taken for diarrhea. *Cold Remedy* Decoction of various plant parts taken for colds. *Febrifuge* Decoction of various plant parts taken for fevers. *Gastrointestinal Aid* and *Pediatric Aid* Decoction of plant parts used for stomachache, indigestion, and babies' colic. (180:104, 105) **Winnebago** *Carminative* Sweetened infusion taken as a carminative or beverage. (70:112, 113)

- **Food**—**Apache, Chiricahua & Mescalero** *Spice* Leaves used as flavoring. (as *M. penardi* 33:47) **Blackfoot** *Beverage* Leaves used to make tea. *Spice* Leaves placed in parfleches to flavor dried meat. (114:278) **Chippewa** *Beverage* Leaves used to make a pleasant, tea-like beverage. *Spice* Leaves used to add flavor to certain meats in cooking. (71:140) **Cree, Woodlands** *Beverage* Leaves added to store bought tea to improve the flavor. *Spice* Leaves added to sturgeon oil to sweeten the odor. (as *M. arvensis* var. *villosa* 109:45) **Dakota** *Beverage* Plant used to make a hot, tea-like beverage. (69:363) Plant used to make a tea-like beverage enjoyed for its pleasing, aromatic flavor. (70:112) *Spice* Plant used as a flavor for meat. (69:363) Plant used as a flavor in cooking meat. (70:112) *Unspecified* Plant laid in alternate layers with dried meat in the packing case. (69:363) Plant parts packed in alternate layers with dried meat for storage. (70:112) **Gosiute** *Beverage* Leaves formerly used to make tea. (39:374) **Hopi** *Sauce & Relish* Plant eaten as a relish. (56:19) *Spice* Boiled with mush for flavor. (190:165) **Klamath** *Beverage* Herbage used for tea. (45:104) **Malecite** *Spice* Plant used as a flavoring in soup. (116:250) **Ojibwa** *Beverage* Foliage used to make a beverage tea. (as *M. arvensis* var. *canadensis* 153:405) **Omaha** *Beverage* Leaves used to make a hot, aqueous, tea-like beverage. (68:329) Plant used to make a tea-like beverage enjoyed for its pleasing, aromatic flavor. (70:112) **Paiute** *Beverage* Fresh or dried leaves made into tea. (as *M. arvensis* var. *glabrata* 111:107) **Pawnee** *Beverage* Plant used to make a tea-like beverage enjoyed for its pleasing, aromatic flavor. **Ponca** *Beverage* Plant used to make a tea-like beverage enjoyed for its pleasing, aromatic flavor. (70:112) **Sanpoil & Nespelem** *Beverage* Leaves and stems boiled, liquid strained and drunk. (as *M. arvensis* var. *lanata* 131:104) **Winnebago** *Beverage* Plant used to make a tea-like beverage enjoyed for its pleasing, aromatic flavor. (70:112)

- **Other**—**Blackfoot** *Hunting & Fishing Item* Plant boiled with traps to destroy the human scent. (as *M. arvensis* var. *villosa* 97:51) **Thompson** *Incense & Fragrance* Plant used extensively as a scent. (164:503) **Winnebago** *Hunting & Fishing Item* Plant boiled with traps to deodorize them so that the smell of blood would not deter the animals. *Incense & Fragrance* Plant boiled with traps to deodorize them so that the smell of blood would not deter the animals. (70:112)

Mentha ×*piperita* L.

- **Drug**—**Cherokee** *Adjuvant* Plant used to flavor medicine and foods. *Analgesic* Taken for colic pains, cramps, and used for nervous headache. *Antiemetic* Taken for vomiting. *Carminative* Taken to "dispel flatulence and remove colic pains." *Cold Remedy* Infusion taken for colds. *Febrifuge* Infusion taken for fevers. *Gastrointestinal Aid* Taken for "affections of stomach and bowels" and infusion used for upset stomach. Taken for bowel problems. *Hemorrhoid Remedy* Tincture applied externally to piles. *Misc. Disease Remedy* Given for cholera infantum. *Pediatric Aid* Given for cholera infantum. *Sedative* Taken for hysterics. *Stimulant* Used as a stimulant. *Urinary Aid* Taken for "suppression of urine and gravelly affection." (water mint 80:48, 49) **Delaware, Oklahoma** *Tonic* Compound containing leaves used as a tonic. (water mint 175:76) **Hoh** *Unspecified* Used as smelling and rubbing medicine. **Quileute** *Unspecified* Used as smelling and rubbing medicine. (peppermint 137:68) **Iroquois** *Cold Remedy* Infusion of whole plant taken for colds. *Febrifuge* Infusion of whole plant taken for fevers. *Other* Compound infusion used as wash on injured parts, a "Little Water Medicine." *Witchcraft Medicine* Infusion of plant will throw off witchcraft. (water mint 87:428) **Menominee** *Pulmonary Aid* Compound infusion taken and poultice applied to chest for pneumonia. (water mint 151:39) **Mohegan** *Anthelmintic* Infusion of plant given to babies for worms. (water mint 174:265) Infusion of leaves taken by children and adults as a vermifuge. (water mint 176:73, 130) *Pediatric Aid* Infusion given to babies for worms. (water mint 174:265) Infusion of leaves taken as a vermifuge for children and adults. (water mint 176:73)
- **Food**—**Cherokee** *Spice* Used to flavor foods. *Unspecified* Species used for food. (water mint 80:48)

Mentha sp., Mint

- **Drug**—**Chehalis** *Cold Remedy* Infusion of leaves taken as a cold medicine. **Cowlitz** *Cold Remedy* Infusion of leaves taken as a cold medicine. (79:45) **Kiowa** *Gastrointestinal Aid* Decoction of leaves taken for stomach troubles. (192:48) **Navajo** *Ceremonial Medicine* Used with sage, red penstemon, red willow, scrub oak, and chokecherry as medicine for Shooting Chant. (55:73)
- **Food**—**Kiowa** *Candy* Fresh leaves frequently chewed. (192:48)

Mentha spicata L., Spearmint

- **Drug**—**Cherokee** *Adjuvant* Plant used to flavor medicine and foods. *Analgesic* Taken for colic pains, cramps, and used for nervous headache. *Antiemetic* Taken for vomiting. *Carminative* Taken to "dispel flatulence and remove colic pains." *Cold Remedy* Infusion taken for colds. *Febrifuge* Infusion taken for fevers. *Gastrointestinal Aid* Taken for "affections of stomach and bowels" and infusion used for upset stomach. Taken for bowel problems. *Hemorrhoid Remedy* Tincture applied externally to piles. *Misc. Disease Remedy* and *Pediatric Aid* Given for cholera infantum. *Sedative* Taken for hysterics. *Stimulant* Used as a stimulant. *Urinary Aid* Taken for "suppression of urine and gravelly affection." (80:48, 49) **Iroquois** *Analgesic* Cold infusion applied to forehead or powdered plant snuffed for headaches. (87:427) *Cold Remedy* Infusion of whole plant taken as a cold remedy. (87:428) *Emetic* Infusion of plants given to children as an emetic. *Febrifuge* Compound infusion of powdered plants taken for fevers. *Gastrointestinal Aid* Infusion of plants given to children for a bad stomach. *Misc. Disease Remedy* Compound infusion of powdered plants taken for typhoid. *Other* Compound infusion used as wash on injured parts, a "Little Water Medicine." *Pediatric Aid* Infusion of plants given to children as an emetic or for a bad stomach. *Respiratory Aid* Compound decoction of roots and berries taken for hay fever. (87:427) **Mahuna** *Sedative* Plant used as a sedative. (140:23) **Miwok** *Antidiarrheal* Infusion of leaves taken for diarrhea. *Gastrointestinal Aid* Infusion of leaves taken for stomach trouble. (12:171) **Mohegan** *Anthelmintic* Infusion of plant taken as a worm medicine. (174:265) Infusion of leaves taken as a vermifuge. (176:73, 130)
- **Food**—**Cherokee** *Spice* Used to flavor foods. *Unspecified* Species used for food. (80:48) **Kawaiisu** *Beverage* Leaves brewed into an nonmedicinal beverage tea. (206:41) **Miwok** *Beverage* Leaves used for tea. (12:171) **Yuki** *Beverage* Used to make a beverage. (49:88)
- **Other**—**Yuki** *Incense & Fragrance* Used as body and garment perfume. (49:93)

Mentzelia, Loasaceae

Mentzelia affinis Greene, Yellowcomet

- **Food**—**Kawaiisu** *Preserves* Seeds parched and ground into a "peanut butter"-like substance. *Winter Use Food* Seeds stored for future use. (206:41)

Mentzelia albicaulis (Dougl. ex Hook.) Dougl. ex Torr. & Gray, Whitestem Blazingstar

- **Drug**—**Gosiute** *Burn Dressing* Seeds used for burns. (39:375) **Hopi** *Toothache Remedy* Plant used as toothache medicine. (42:335) **Navajo, Ramah** *Snakebite Remedy* Compound containing leaves used for snakebite. *Toothache Remedy* Poultice of crushed, soaked seeds applied for toothache. (191:36)
- **Food**—**Cahuilla** *Porridge* Parched seeds ground into flour and used to make mush. (15:88) **Havasupai** *Preserves* Seeds parched, ground, kneaded into seed butter, and eaten with fruit drinks or spread on bread. (197:67) *Soup* Seeds and Indian millet seeds ground and used to make soup or mush. (197:73) *Unspecified* Seeds formerly used for food. (197:232) **Hopi** *Staple* Seeds parched, ground into a fine, sweet meal, and eaten in pinches. (56:20) *Unspecified* Mashed seeds rolled into sticks and eaten. (190:164) **Kawaiisu** *Preserves* Seeds parched and ground into a "peanut butter"-like substance. *Winter Use Food* Seeds stored for future use. (206:41) **Klamath** *Unspecified* Seeds used for food. (45:100) **Montana Indian** *Unspecified* Seeds used for food. (19:15) **Paiute** *Sauce & Relish* Fried seeds and water used for gravy. (118:27) *Staple* Seeds parched, ground, and eaten as meal. (104:98) **Paiute, Northern** *Dried Food* Seeds dried and stored for winter use. *Porridge* Seeds dried, roasted, ground into a flour, and used to make mush. (59:46) **Tubatulabal** *Unspecified* Used extensively for food. (193:15)
- **Other**—**Hopi** *Smoke Plant* Plant used as substitute for tobacco. (42:335)

Mentzelia albicaulis var. *veatchiana* (Kellogg) Urban & Gilg, Whitestem Blazingstar

- *Drug*—Kawaiisu *Burn Dressing* Pounded seeds made into a salve and rubbed on burned skin. (as *M. veatchiana* 206:41)
- *Food*—Cahuilla *Porridge* Parched seeds ground into flour and used to make mush. (as *M. veatchiana* 15:88) Kawaiisu *Preserves* Seeds parched and ground into a "peanut butter"-like substance. *Winter Use Food* Seeds stored for future use. (206:41)

Mentzelia congesta Nutt. ex Torr. & Gray, United Blazingstar
- *Food*—Kawaiisu *Preserves* Seeds parched and ground into a "peanut butter"-like substance. *Winter Use Food* Seeds stored for future use. (206:41)

Mentzelia dispersa S. Wats., Bushy Blazingstar
- *Food*—Kawaiisu *Preserves* Seeds parched and ground into a "peanut butter"-like substance. *Winter Use Food* Seeds stored for future use. (206:41)

Mentzelia gracilenta (Nutt.) Torr. & Gray, Grass Blazingstar
- *Food*—Tubatulabal *Unspecified* Used extensively for food. (193:15)

Mentzelia involucrata S. Wats., Whitebract Blazingstar
- *Food*—Cahuilla *Porridge* Parched seeds ground into flour and used to make mush. (15:88)
- *Other*—Kawaiisu *Toys & Games* Leaves thrown by children at one another because they stick and were hard to remove. (206:41)

Mentzelia laciniata (Rydb.) J. Darl., Cutleaf Blazingstar
- *Drug*—Navajo, Ramah *Eye Medicine* Infusion of flowers used as an eyewash. (191:36)

Mentzelia laevicaulis (Dougl. ex Hook.) Torr. & Gray, Smoothstem Blazingstar
- *Drug*—Cheyenne *Antirheumatic (Internal)* Roots used for rheumatism and arthritis. *Dietary Aid* Roots chewed for thirst prevention. *Ear Medicine* Roots used for earaches. *Febrifuge* Roots used for fevers. *Misc. Disease Remedy* Infusion of roots taken for mumps, measles, and smallpox. *Unspecified* Plant used as an ingredient in medicinal preparations. Roots used for complicated illnesses. (83:30) **Gosiute** *Dermatological Aid* Infusion of roots used for bruise swellings. (39:349) **Mendocino Indian** *Dermatological Aid* Decoction of leaves used as a wash for skin diseases. *Gastrointestinal Aid* Decoction of leaves taken for stomachaches. (41:369) **Montana Indian** *Dermatological Aid* Decoction of leaves applied as a lotion for certain skin diseases. *Gastrointestinal Aid* Decoction of leaves taken for stomach trouble. (19:16) **Thompson** *Unspecified* Plant used medicinally for unspecified purpose. (164:474)
- *Food*—Paiute *Sauce & Relish* Fried seeds and water used for gravy. (118:27)

Mentzelia multiflora (Nutt.) Gray, Manyflowered Mentzelia
- *Drug*—Keres, Western *Diuretic* Infusion of plant used as a diuretic. *Psychological Aid* Plants used to make infants good horseback riders. Plants used to whip 3- or 4-month-old infants, or ground leaves rubbed on their thighs so that they will become good horseback riders when they grow up. *Tuberculosis Remedy* Leaves and roots used as the strongest tuberculosis medicine by the strongest patients. (171:54) **Navajo** *Emetic* Plant used as an emetic. (90:161)
- *Food*—Navajo *Unspecified* Seeds used for food. (55:63)
- *Other*—Navajo *Ceremonial Items* Leaves chewed and sprayed with the mouth on offerings before and after making prayer sticks. (55:63)

Mentzelia multiflora (Nutt.) Gray **var. *multiflora***, Adonis Blazingstar
- *Drug*—Navajo, Kayenta *Ceremonial Medicine* Plant used as fumigant for collared lizard ceremony. *Dermatological Aid* Plant used to keep smallpox sores from pitting. *Disinfectant* Plant used as fumigant for collared lizard ceremony. *Gastrointestinal Aid* Plant used for abdominal swellings. (as *M. pumila* var. *multiflora* 205:32) **Navajo, Ramah** *Eye Medicine* Infusion of flowers used as an eyewash. (as *M. pumila* var. *multiflora* 191:37)
- *Food*—Navajo, Ramah *Staple* Seeds parched with hot coals in an old basket, ground lightly with a special rock. (as *M. pumila* var. *multiflora* 191:37)
- *Other*—Navajo, Ramah *Ceremonial Items* Used to make Big Snake's prayer stick in Beautyway. (as *M. pumila* var. *multiflora* 191:37)

Mentzelia nuda (Pursh) Torr. & Gray **var. *nuda***, Bractless Blazingstar
- *Drug*—Dakota *Febrifuge* Boiled, strained sap applied externally for fever. (as *Nuttallia nuda* 70:103)

Mentzelia puberula J. Darl., Roughstem Blazingstar
- *Food*—Cahuilla *Porridge* Parched seeds ground into flour and used to make mush. (15:88)

Mentzelia pumila Nutt. ex Torr. & Gray, Dwarf Mentzelia
- *Drug*—Apache, White Mountain *Laxative* Powdered roots used for constipation. (136:158) **Hopi** *Toothache Remedy* Plant used as "a toothache medicine." (200:85) **Zuni** *Laxative* Powdered root inserted into rectum as a suppository for constipation. (166:57) *Pediatric Aid and Strengthener* Plant used to whip children to make them strong so they could hold on to a horse without falling. (166:84)
- *Other*—Hopi *Smoke Plant* Plant used as a substitute for tobacco. (200:85)

Mentzelia sp., Buena Mujer
- *Drug*—Miwok *Poultice* Pulverized seeds mixed with water, fox grease, or wildcat grease and applied as a poultice. (12:171)
- *Food*—Hualapai *Dried Food* Mature seeds parched and stored for winter use. *Porridge* Green seeds pounded into a gruel and cooked. *Staple* Seeds considered an important staple. (195:52) **Miwok** *Unspecified* Pulverized seeds made into pinole. (12:155)

Menyanthaceae, see *Menyanthes, Nymphoides*

Menyanthes, Menyanthaceae
Menyanthes sp.
- *Drug*—Aleut *Unspecified* Used medicinally. (7:29)

Menyanthes trifoliata L., Common Buckbean
- *Drug*—Aleut *Analgesic* Infusion of roots taken for gas pains, constipation, and rheumatism. *Antirheumatic (Internal)* Compound containing roots taken as a tonic for gas pains and rheumatism. *Carminative* Compound containing roots taken as a tonic for gas pains. *Laxative* Compound containing roots taken as a tonic for constipation. *Tonic* Roots used as a powerful ingredient in a tonic. (8:427) **Kwakiutl** *Antiemetic* Decoction of roots or leaves taken when sick to

the stomach. (183:287) *Antihemorrhagic* Decoction of root and stem used for "spitting of blood and other internal diseases." (20:380) Decoction of ground stem and roots taken for blood-spitting. *Dietary Aid* Decoction of roots or leaves taken to put on weight. *Gastrointestinal Aid* Decoction of roots or leaves taken when sick to the stomach. *Misc. Disease Remedy* Decoction of roots or leaves taken to put on weight during the flu. (183:287) **Menominee** *Unspecified* Plant used in medicines. (151:36) **Micmac** *Unspecified* Strong decoction of root taken for unspecified purpose. (as *Merganthes trifolia* 156:317) **Tlingit** *Unspecified* Plant used for the medicinal value. (149:330)
- *Food*—**Alaska Native** *Bread & Cake* Rootstocks dried, ground, leached, dried, ground into flour, and used to make bread. *Dried Food* Rootstocks dried, ground, leached, dried, and used for food. *Starvation Food* Rootstocks used in the past as an emergency food. (85:145) **Hesquiat** *Forage* Deer put their heads under the surface of the water to get at the long, green rhizomes. (185:69)

Menziesia, Ericaceae
Menziesia ferruginea Sm., Rusty Menziesia
- *Drug*—**Hesquiat** *Oral Aid* Nectar sucked from flowers to sweeten the mouth. (185:65) **Kwakiutl** *Analgesic* Leaves chewed for heart pain. *Dermatological Aid* Poultice of heated leaves applied to sores and swellings. *Gastrointestinal Aid* Leaves chewed for stomach troubles. *Heart Medicine* Leaves chewed for heart pain. (183:283) **Nitinaht** *Witchcraft Medicine* Bark used to counteract evil spells and doctor remedies. (186:107) **Quinault** *Love Medicine* Forked twig waved in the air by a woman to make a man fall in love with her. (79:43)
- *Fiber*—**Quileute** *Canoe Material* Twigs woven together with cedar bark and used for grills on the bottom of canoes. (79:43)
- *Other*—**Nitinaht** *Toys & Games* Forked branches used by children to make sling shots. (186:107)

Merremia, Convolvulaceae
Merremia dissecta (Jacq.) Hallier f., Noyau Vine
- *Drug*—**Hawaiian** *Analgesic* Poultice of pounded flowers, leaves, and salt applied to the back for pain. *Dermatological Aid* Poultice of pounded roots, other plants, and the resulting liquid applied to flesh wounds. *Laxative* Roots and other plants pounded, mixed with water and an egg, and taken as a laxative. *Orthopedic Aid* Poultice of pounded roots, other plants, and the resulting liquid applied to broken bones. *Pediatric Aid* and *Strengthener* Flowers chewed by mothers and given to infants for general weakness. (as *Impomea dissecta* 2:52)

Mertensia, Boraginaceae
Mertensia ciliata (James ex Torr.) G. Don, Mountain Bluebells
- *Drug*—**Cheyenne** *Breast Treatment* Infusion of plant used to increase milk flow of mothers. (as *Mestensia ciliata* 83:16) *Dermatological Aid* Infusion of powdered roots taken for itching from smallpox. (76:184) *Gynecological Aid* Infusion of plant taken by women after childbirth to increase milk flow. *Misc. Disease Remedy* Infusion of leaves taken for smallpox and measles. (76:184) Infusion of leaves used for measles and smallpox. (as *Mestensia ciliata* 83:16)

Mertensia maritima (L.) S. F. Gray, Oysterleaf
- *Food*—**Eskimo, Alaska** *Unspecified* Long, leafy stems boiled, cooked briefly, and eaten with seal oil. (1:38) Rootstock used for food. (4:715)

Mertensia virginica (L.) Pers. ex Link, Virginia Bluebells
- *Drug*—**Cherokee** *Pulmonary Aid* Taken for whooping cough. *Tuberculosis Remedy* Taken for consumption. (80:26) **Iroquois** *Antidote* Compound infusion of roots taken as an antidote for poisons. *Venereal Aid* Decoction of roots taken for venereal disease. (87:421)

Metrosideros, Myrtaceae
Metrosideros polymorpha Gaud. **var.** *polymorpha*, 'Ohi'a
- *Drug*—**Hawaiian** *Analgesic* Flowers and other plants rubbed, squeezed, and the resulting liquid taken for severe childbirth pain. (as *Metrosideros collins polym* 2:31)

Microlepia, Dennstaedtiaceae, fern
Microlepia setosa (Sm.) Alston, Pa-la-pa-la-i
- *Drug*—**Hawaiian** *Psychological Aid* Plant used for insanity. (as *M. strigosa* 2:73)

Microseris, Asteraceae
Microseris laciniata (Hook.) Schultz-Bip., Cutleaf Silverpuffs
- *Food*—**Mendocino Indian** *Substitution Food* Milky juice exposed to the sun and used by school children as a substitute for gum. *Unspecified* Roots formerly used for food. (as *Scorzonella maxima* 41:391)

Microseris nutans (Hook.) Schultz-Bip., Nodding Microseris
- *Food*—**Montana Indian** *Unspecified* Bitter, milky root juice eaten raw. (19:16)

Mikania, Asteraceae
Mikania batatifolia DC., Southern Hempvine
- *Drug*—**Seminole** *Dermatological Aid* Plant used for snake sickness: itchy skin. (169:166) Decoction of plant taken and used as a body steam for snake sickness: itchy skin. (169:239)

Mimulus, Scrophulariaceae
Mimulus cardinalis Dougl. ex Benth., Crimson Monkeyflower
- *Drug*—**Karok** *Pediatric Aid* Infusion of plant used as a wash for newborn baby. (148:389)
- *Food*—**Kawaiisu** *Unspecified* Tender stalks eaten raw. (206:41)

Mimulus eastwoodiae Rydb., Eastwood's Monkeyflower
- *Drug*—**Navajo, Kayenta** *Anticonvulsive* Plant used for hiccups. (205:42)
- *Food*—**Navajo, Kayenta** *Fruit* Berries eaten raw. Berries stewed and used for food. (205:42)

Mimulus glabratus **var.** *jamesii* (Torr. & Gray ex Benth.) Gray, James's Monkeyflower
- *Drug*—**Potawatomi** *Unspecified* Leaves used as treatment for unspecified ailments. (154:83)
- *Food*—**Isleta** *Vegetable* Tender shoots slit and eaten as a salad. (as *M. geyeri* 32:34) Salted, tender, young leaves used for salad. (as *M. geyeri* 100:35)
- *Other*—**Keres, Western** *Water Indicator* Plant used as an indication of surface water. (as *M. geyeri* 171:54)

Mimulus guttatus DC., Seep Monkeyflower
- **Drug**—**Kawaiisu** *Analgesic, Herbal Steam,* and *Orthopedic Aid* Decoction of stems and leaves used as steam bath for chest and back soreness. (206:41) **Shoshoni** *Dermatological Aid* Poultice of crushed leaves applied to wounds or rope burns. (180:105) **Yavapai** *Gastrointestinal Aid* Decoction taken as tea for stomachache. (as *M. nasutus* 65:261)
- **Food**—**Mendocino Indian** *Substitution Food* Plants used as a substitute for lettuce. (41:387) **Miwok** *Vegetable* Boiled leaves used for food. (12:160)

Mimulus moschatus Dougl. ex Lindl., Musk Monkeyflower
- **Food**—**Miwok** *Vegetable* Boiled, young plant used for food. (12:160)

Mimulus ringens L., Ringen Monkeyflower
- **Drug**—**Iroquois** *Anticonvulsive* Compound decoction of roots taken by women for epilepsy. *Antidote* Compound decoction of plants used as wash to counteract poison. (87:435)

***Mimulus* sp.**, Monkeyflower
- **Drug**—**Miwok** *Antidiarrheal* Infusion of root used for diarrhea. (12:171)

Mimulus tilingii* var. *caespitosus (Greene) A. L. Grant, Subalpine Monkeyflower
- **Food**—**Neeshenam** *Vegetable* Leaves eaten as greens. (as *M. luteus* 129:377)

Mirabilis, Nyctaginaceae

Mirabilis alipes (S. Wats.) Pilz, Winged Four o'Clock
- **Drug**—**Paiute** *Analgesic* Decoction of root taken and used as a wash for headaches. Decoction of root used as a wash for neuralgia. Poultice of crushed, fresh leaves applied for headaches. *Antiemetic* Decoction of root taken or used as a wash for fainting spells and nausea. *Burn Dressing* Dried root powder moistened and used as a salve for burns. *Cathartic* Decoction of root taken as a physic. *Dermatological Aid* Dried root powder sprinkled on sores or made into a wash for impetigo. Poultice of mashed leaves applied to swellings. *Psychological Aid* Decoction of root used as a wash for "delirium," neuralgia, and dizziness. *Stimulant* Decoction of root taken or used as a wash for fainting spells and dizziness. (as *Hermidium alipes* 180:86, 87) **Paiute, Northern** *Dermatological Aid* Poultice of dried, powdered roots applied and decoction of powdered roots used as a wash for sores. (as *Hermidium alipes* 59:125)

Mirabilis bigelovii* var. *retrorsa (Heller) Munz, Bigelow's Four o'Clock
- **Drug**—**Paiute** *Dermatological Aid* Powdered root used as shampoo and infusion of root taken simultaneously. (as *Hesperonia retrorsa* 118:46) Powdered root used as shampoo and infusion taken simultaneously. (as *Hesperonia retrorsa* 118:57)

Mirabilis californica Gray, California Four o'Clock
- **Drug**—**Luiseño** *Cathartic* Decoction of leaves taken as a purgative. (155:232) **Mahuna** *Febrifuge* Plant used for eruptive fevers. (140:10)

Mirabilis coccineus (Torr.) Benth. & Hook. f., Scarlet Four o'Clock
- **Drug**—**Hopi** *Dermatological Aid* Decoction of plant used as a wash for wounds. (as *Oxybaphus coccinea* 200:32, 75) **Yavapai** *Venereal Aid* Pounded, boiled root taken for gonorrhea. (as *Allionia coccinea* 65:260)

Mirabilis greenei S. Wats., Greene's Four o'Clock
- **Drug**—**Karok** *Pediatric Aid* Plant used to make a newborn baby healthy. (148:383)

Mirabilis linearis (Pursh) Heimerl, Narrowleaf Four o'Clock
- **Drug**—**Lakota** *Diuretic* Infusion of roots taken for urinating difficulties. (139:52) **Navajo, Kayenta** *Gastrointestinal Aid* Root used for stomach disorders. *Gynecological Aid* Infusion of roots used for postpartum treatment. *Panacea* Plant used as a life medicine. (as *Oxybaphus linearis* 205:21) **Navajo, Ramah** *Burn Dressing* Poultice of soaked, split root applied to burns. *Cough Medicine* Decoction of plant used for coughs. *Hunting Medicine* Cold infusion of plant used as a lotion for good luck in trading or hunting. *Veterinary Aid* Decoction of plant used for sheep and horses with coughs. (as *Oxybaphus linearis* 191:26) **Zuni** *Diuretic* Root eaten to induce urination. *Emetic* Root eaten to induce vomiting. *Gastrointestinal Aid* Infusion of root taken for stomachache. (as *Oxybaphus linearis* 27:377)
- **Food**—**Navajo, Kayenta** *Fruit* Berries stewed and used for food. *Unspecified* Seeds roasted and used for food. (as *Oxybaphus linearis* 205:21)

Mirabilis multiflora (Torr.) Gray, Colorado Four o'Clock
- **Drug**—**Hopi** *Gynecological Aid* Used to push up the blood in the woman during the pregnant stage. *Hallucinogen* Root chewed by medicine man to induce visions while making a diagnosis. *Veterinary Aid* Used as antiseptic to wash out wounds in horses. (as *Nirabilis multiflora* 42:334) **Navajo** *Antirheumatic (Internal)* Plant used for rheumatism. *Dermatological Aid* Plant used for "swellings." *Oral Aid* Plant used for various mouth disorders. (90:161) **Navajo, Ramah** *Dermatological Aid* Poultice of root applied to swellings. (191:26) **Zuni** *Dietary Aid* Powdered root mixed with flour, made into a bread and used to decrease appetite. (27:377)
- **Food**—**Navajo, Ramah** *Beverage* Used to make tea. (191:26)
- **Other**—**Hopi** *Hunting & Fishing Item* Heavy root used to anchor the bird trap string. (as *Nirabilis multiflora* 42:334) **Keres, Western** *Smoke Plant* Dried leaves used as tobacco. (171:54)

Mirabilis multiflora (Torr.) Gray **var. *multiflora***, Colorado Four o'Clock
- **Drug**—**Hopi** *Hallucinogen* Roots chewed by doctor to induce visions while making diagnosis. (as *Quamoclidion multiflorum* 200:31, 75) **Tewa** *Kidney Aid* Infusion of pulverized root taken for swellings "of dropsical origin." (as *Quamoclidion multiflorum* 138:60) **Zuni** *Dietary Aid* Infusion of root taken and rubbed on abdomen of hungry adults and children. (as *Quamoclidion multiflorum* 166:58, 59) *Gastrointestinal Aid* and *Pediatric Aid* Infusion of powdered root taken by adults or children after overeating. (as *Quamoclidion multiflorum* 166:58)

Mirabilis nyctaginea (Michx.) MacM., Heartleaf Four o'Clock
- **Drug**—**Cherokee** *Dermatological Aid* Poultice of beaten root used for boils and milk poured over leaves used as fly poison. (80:34, 35) **Chippewa** *Orthopedic Aid* Decoction of root or poultice of root applied to sprain or strained muscles. (as *Allionia nyctaginea* 53:362) **Dakota** *Anthelmintic* Decoction of roots taken as a vermifuge. (as

Allionia nyctaginea 69:361) Compound decoction of root taken as a vermifuge. *Dermatological Aid* Compound decoction of root used as wash for swollen arms or legs. (as *Allionia nyctaginea* 70:78) *Febrifuge* Decoction of roots taken as a febrifuge. (as *Allionia nyctaginea* 69:361) Decoction of root taken for fever. (as *Allionia nyctaginea* 70:78) **Meskwaki** *Burn Dressing* Poultice of macerated root applied to burns. *Urinary Aid* Root or whole herb used for bladder troubles. (as *Oxybaphus nyctaginea* 152:232) **Ojibwa** *Orthopedic Aid* Root used for sprains and swellings. (as *Oxybaphus nyctagineus* 153:375) **Pawnee** *Gynecological Aid* Decoction of root taken after childbirth for abdominal swelling. *Oral Aid* Dried, ground root applied to baby's sore mouth. *Pediatric Aid* Pulverized dried root applied to babies for sore mouth. **Ponca** *Dermatological Aid* Chewed root blown into wounds. (as *Allionia nyctaginea* 70:78) **Sioux, Teton** *Dermatological Aid* Moistened, grated root rubbed on skin for swelling. *Orthopedic Aid* Herb used externally for broken bones. (as *Allionia nyctaginea* 52:270)
- *Other*—**Cherokee** *Insecticide* Poultice of beaten root used for boils and milk poured over leaves used as fly poison. (80:34, 35) **Keres, Western** *Smoke Plant* Leaves used as tobacco. (as *Allionia nyctaginea* 171:25)

Mirabilis oblongifolia (Gray) Heimerl, Mountain Four o'Clock
- *Drug*—**Navajo, Ramah** *Burn Dressing* Poultice of soaked, split root applied to burns. (as *Oxybaphus comatus* 191:26)

Mirabilis oxybaphoides (Gray) Gray, Smooth Spreading Four o'Clock
- *Drug*—**Navajo, Kayenta** *Dermatological Aid* Plant used for spider bites or as a hair lotion for dandruff. (205:21) **Navajo, Ramah** *Orthopedic Aid* Poultice of whole plant applied to fractures. (191:26)
- *Food*—**Navajo, Kayenta** *Vegetable* Used for greens in foods. (205:21) **Navajo, Ramah** *Beverage* Used to make tea. (191:26)

Mirabilis pumila (Standl.) Standl., Dwarf Four o'Clock
- *Drug*—**Navajo, Kayenta** *Dermatological Aid* Plant used as a lotion for sores and skin eruptions. (as *Oxybaphus pumilus* 205:21)

Mirabilis sp., Four o'Clock
- *Dye*—**Navajo** *Brown* Petals boiled and used as a light brown dye for wool. *Purple* Petals boiled and used as a purple dye for wool. *Red* Petals boiled for about 15 minutes and used as a light red dye. *Yellow* Petals boiled for about 15 minutes and used as a muddy yellow dye. (55:46)

Mitchella, Rubiaceae

Mitchella repens L., Partridge Berry
- *Drug*—**Abnaki** *Antirheumatic (External)* Used for swellings. (144:155) Poultice of plant applied to swellings. (144:173) **Cherokee** *Analgesic* Taken for "monthly period pains." *Antidiarrheal* Decoction made with milk taken for dysentery. *Dermatological Aid* Infusion taken for hives and used for sore nipples. *Diaphoretic* Taken as a diaphoretic. *Dietary Aid* Given to baby before it "takes the breast." *Diuretic* Taken as a diuretic. *Gastrointestinal Aid* Infusion of root taken with rattlesnake weed for bowel complaint. *Gynecological Aid* Used to facilitate childbirth, for sore nipples, and taken for menstrual cramps. *Hemorrhoid Remedy* Decoction made with milk taken for piles. *Pediatric Aid* Given to baby before it "takes the breast." *Veterinary Aid* Given to pregnant cat and her kittens. (80:47) **Chippewa** *Unspecified* Infusion of whole plant used for medicinal purposes. (71:141) **Delaware** *Abortifacient* Infusion used for suppressed menstruations to strengthen the female generative organs. *Antirheumatic (External)* Hot infusion of roots or twigs used as a steam treatment for muscular swellings and stiff joints. (176:36) **Delaware, Oklahoma** *Abortifacient* Infusion of plant taken for "suppressed menses." *Antirheumatic (External)* Strong infusion of roots or twigs used as herbal steam for rheumatism. (175:30, 76) *Gynecological Aid* Infusion of plant taken to strengthen the "female generative organs." *Herbal Steam* Infusion of roots or twigs used as herbal steam for rheumatism. (175:30) **Iroquois** *Analgesic* Compound decoction of roots taken for urinating pain. Decoction of plants taken by pregnant women with side or labor pains. (87:441) Compound infusion of roots and bark taken for back pain. (87:442) Berries used to prevent severe labor pains. (124:96) *Anticonvulsive* Compound infusion given to children with convulsions. (87:440) *Antiemetic* Compound infusion of roots and bark taken for vomiting. (87:442) *Blood Medicine* Compound decoction of plants taken as blood purifier. Decoction of plants taken to "remove chill from the blood." (87:441) *Carminative* Decoction of plant taken for stomach gas. (87:442) *Cathartic* Decoction of roots given to newborn babies as a physic for stomachaches. *Dermatological Aid* Decoction of plants given to babies with rashes. Poultice of smashed plant applied to bleeding cuts. (87:441) *Febrifuge* Compound infusion given to children with inward fever. Compound infusion of plants taken for typhoid-like fever or inward fever. (87:440) Poultice of hot plant applied to chest for fevers. (87:442) *Gastrointestinal Aid* Compound infusion taken for inward fever from stomach trouble. (87:440) Decoction of roots given to newborn babies as a physic for stomachaches. Poultice applied or decoction of vines given to babies with swollen abdomens. *Gynecological Aid* Compound decoction of leaves and roots taken for leukorrhea (sick womb). Compound decoction of plants taken by pregnant women with side pains. (87:441) Compound of plants taken for labor pain in parturition. (87:442) Berries used to prevent severe labor pains and to facilitate delivery. (124:96) *Hemostat* Poultice of smashed plant applied to bleeding cuts. (87:441) *Kidney Aid* Infusion or decoction of roots and bark taken for kidney troubles. (87:442) *Love Medicine* Compound of plant used as love medicine. (87:441) *Misc. Disease Remedy* Decoction of plant taken by pregnant mother to prevent rickets in baby. *Orthopedic Aid* Compound infusion of roots and bark taken for back pain. (87:442) *Pediatric Aid* Compound infusion given to children with inward fever and convulsions. Decoction of vines given to babies when they will not suckle. (87:440) Decoction of plants given to babies with rashes. Decoction of roots given to newborn babies as a physic for stomachaches. Poultice applied or decoction of vines given to babies with swollen abdomens. (87:441) Decoction of plant taken by pregnant mother to prevent rickets in baby. (87:442) *Psychological Aid* Compound infusion of plants taken for typhoid-like fever or craziness. (87:440) *Urinary Aid* Compound decoction of plants taken for swollen testicles or ruptures. Compound decoction of roots taken for urinating pain. Decoction of plants taken for bladder stricture. *Venereal Aid* Compound decoction of plants taken for venereal disease. (87:441) **Menominee** *Gynecological Aid* Decoction of leaves used for "diseases of women."

(54:133) *Sedative* Infusion of leaves taken "to cure insomnia." (151:51) **Montagnais** *Febrifuge* Berries cooked into a jelly and used for fevers. (156:313) **Ojibwa** *Ceremonial Medicine* Leaves smoked during ceremonies. (135:239) **Penobscot** *Unspecified* Infusion of leaves used as a medicine for unspecified purpose. (156:309) **Seminole** *Kidney Aid* Plant used for kidney disorders. (169:274)

- **Food**—**Cherokee** *Fruit* Fruit used for food. (80:47) **Iroquois** *Bread & Cake* Fruit mashed, made into small cakes, and dried for future use. *Dried Food* Raw or cooked fruit sun or fire dried and stored for future use. (196:128) *Fruit* Berries eaten by women. (124:96) Dried fruit taken as a hunting food. *Sauce & Relish* Dried fruit cakes soaked in warm water and cooked as a sauce or mixed with corn bread. (196:128) **Micmac** *Beverage* Used to make a beverage. (159:258)

Mitella, Saxifragaceae

Mitella diphylla L., Twoleaf Miterwort

- **Drug**—**Iroquois** *Emetic* Decoction of whole plants taken to vomit and counteract bad luck. *Eye Medicine* Infusion of plant used as drops for sore eyes. (87:345)
- **Other**—**Iroquois** *Good Luck Charm* Decoction of whole plants used as body and rifle wash to counteract bad luck. (87:345) **Menominee** *Sacred Items* Seed used as the sacred bead and swallowed in the medicine dance, during the reinstatement ceremony. (151:81)

Mitella nuda L., Naked Miterwort

- **Drug**—**Cree, Woodlands** *Ear Medicine* Crushed leaf wrapped in a cloth and inserted in the ear for earaches. (109:45)

Mitella sp.

- **Drug**—**Gosiute** *Cathartic* Infusion of root juice taken to hasten elimination and purging. (39:348) Plant used as a purgative for babies and adults with intestinal disorders. *Gastrointestinal Aid* Plant used as a purgative for babies and adults with intestinal disorders. (39:350) *Laxative* Infusion of root juice taken to hasten elimination and purging. (39:348) *Pediatric Aid* Plant used as a purgative for babies and adults with intestinal disorders. (39:350)

Mitella trifida Graham, Threeparted Miterwort

- **Drug**—**Gosiute** *Gastrointestinal Aid* and *Pediatric Aid* Infusion of roots used for babies with colic. (39:375)

Mniaceae, see *Mnium, Plagiomnium, Rhizomnium*

Mnium, Mniaceae, moss

Mnium affine

- **Drug**—**Carrier, Southern** *Antirheumatic (External)* Decoction of plant used to bathe a swollen face. (43:53)

Mnium punctatum L.

- **Drug**—**Makah** *Antirheumatic (External)* Leaves used for swellings. (43:53)

Modiola, Malvaceae

Modiola caroliniana (L.) G. Don, Carolina Bristlemallow

- **Drug**—**Houma** *Misc. Disease Remedy* Compound infusion gargled and decoction taken for sore throat or diphtheria. *Throat Aid* Compound infusion gargled and decoction taken for tonsillitis or sore throat. (158:64)

Monarda, Lamiaceae

Monarda citriodora Cerv. ex Lag., Lemon Beebalm

- **Food**—**Hopi** *Unspecified* Plant boiled and eaten only with hares. (56:19)

Monarda didyma L., Scarlet Beebalm

- **Drug**—**Cherokee** *Abortifacient* Used for female obstructions. *Analgesic* Poultice of leaves used for headache. *Carminative* Used carminative for colic and flatulence. *Cold Remedy* Poultice of leaves used for colds. *Diaphoretic* Used as a diaphoretic. *Diuretic* Used as a diuretic. *Febrifuge* Hot infusion of leaf used to "bring out measles" and infusion used as febrifuge. *Gastrointestinal Aid* Infusion of leaf and plant top taken for weak bowels and stomach. *Heart Medicine* Infusion used for heart trouble. *Hemostat* Infusion of leaf or root taken orally and wiped on head for nosebleed. *Misc. Disease Remedy* Infusion of leaf used to "bring out measles" and infusion used to "sweat off flu." *Sedative* Used for hysterics and restful sleep. (80:39)
- **Food**—**Cherokee** *Unspecified* Species used for food. (80:39)

Monarda fistulosa L., Wildbergamot Beebalm

- **Drug**—**Blackfoot** *Cough Medicine* Infusion of plant taken for coughs. (86:72) *Dermatological Aid* Poultice of a flower head applied to a burst boil and removed after the wound healed. (86:77) Poultice of plant pieces applied to cuts. (86:84) *Emetic* Infusion of plant and another plant taken and used as a steam to serve as an emetic. (86:65) *Eye Medicine* Used to make a solution for sore eyes. (82:70) *Kidney Aid* Infusion of plant taken for aching kidneys. (86:67) *Throat Aid* Root chewed for swollen neck glands. (86:77) **Cherokee** *Abortifacient* Used for female obstructions. *Analgesic* Poultice of leaves used for headache. *Carminative* Used as a carminative for colic and flatulence. *Cold Remedy* Poultice of leaves used for colds. *Diaphoretic* Used as a diuretic, diaphoretic, and especially for "sweating off flu." *Diuretic* Used as a carminative for colic, diuretic, and diaphoretic. *Febrifuge* Hot infusion of leaf used to "bring out measles" and infusion used as febrifuge. *Gastrointestinal Aid* Infusion of leaf and plant top taken for weak bowels and stomach. *Heart Medicine* Infusion used for heart trouble. *Hemostat* Infusion of leaf or root taken orally and wiped on head for nosebleed. *Misc. Disease Remedy* Infusion of leaf used to "bring out measles" and infusion used to "sweat off flu." *Sedative* Used for hysterics and restful sleep. (80:39) **Chippewa** *Analgesic* Chewed leaves placed in nostrils for headaches. *Cold Remedy* Plant tops used for colds. (71:140) **Choctaw** *Analgesic* Plant rubbed on child's chest for pain. *Cathartic* Infusion of leaves taken as a cathartic. *Pediatric Aid* Plant rubbed on child's chest for pain. (177:54) **Crow** *Respiratory Aid* Infusion taken for respiratory problems. (82:70) **Dakota** *Analgesic* Infusion of flowers and leaves taken for abdominal pains. (69:363) Decoction of flowers and leaves taken for abdominal pains. (70:111) *Gastrointestinal Aid* Infusion of flowers and leaves taken for abdominal pains. (69:363) Decoction of leaves and flowers taken for abdominal pains. (70:111) **Flathead** *Cold Remedy* Infusion taken for colds. Plants hung on walls for colds. *Cough Medicine* Used for coughs. *Eye Medicine* Used to make a solution for sore eyes. *Febrifuge* Infusion taken for fevers. *Misc. Disease Remedy* Infusion taken for flu and chills. *Pulmonary Aid* Infusion taken for pneumonia. *Toothache Remedy* Used for toothache. (82:70) **Koasati** *Febrifuge* Decoction of leaves used as a bath for chills. (177:54) **Kutenai** *Kidney Aid* Infu-

sion used for kidney problems. (82:70) **Lakota** *Cough Medicine* Infusion of leaves used as a cough remedy. *Eye Medicine* Infusion of leaves used on a cloth placed on sore eyes overnight. *Hemostat* Poultice of chewed leaves applied to stop the flow of blood. *Pulmonary Aid* Infusion of leaves used for whooping cough. *Stimulant* Infusion of leaves used for fainting. (139:50) **Menominee** *Pediatric Aid* Decoction of stem and leaves used as strengthening bath for infants. (54:133) *Respiratory Aid* Simple or compound infusion of leaves and flowers used as a universal remedy for catarrh. (151:39) **Meskwaki** *Cold Remedy* Compound used for colds. (152:225) **Montana Indian** *Gynecological Aid* Infusion taken for expulsion of the afterbirth. (82:70) **Navajo** *Analgesic* Cold infusion of plant used as a wash for headaches. (55:73) **Ojibwa** *Anticonvulsive* Infusion of plant taken or used as a bath for infant convulsions. *Febrifuge* Infusion of flowers taken for fevers. *Pediatric Aid* Infusion of plant taken or used as a bath for infant convulsions. (5:2274) *Respiratory Aid* Plant boiled and steam inhaled "to cure catarrh and bronchial affections." (153:372) **Ojibwa, South** *Analgesic* and *Gastrointestinal Aid* Decoction of root taken for "pain in the stomach and intestines." (91:201) **Sioux** *Gastrointestinal Aid* Infusion taken for stomach pains. (82:70) **Sioux, Teton** *Cold Remedy* Infusion of blossoms used for "hard cold." *Febrifuge* Infusion of blossoms used for fever. (52:270) **Winnebago** *Dermatological Aid* Decoction of leaves used on pimples and other skin eruptions on the face. (70:111)
- *Food*—**Cherokee** *Unspecified* Species used for food. (80:39) **Flathead** *Preservative* Leaves pulverized and sprinkled on meats as a preservative. (82:70) **Iroquois** *Beverage* Used to make a beverage. (196:149) **Lakota** *Special Food* Leaves chewed while people were singing and dancing. (139:50)
- *Other*—**Blackfoot** *Cooking Tools* Dried flowerheads used by invalids for sucking broth and soup. *Tools* Dried flowerheads used to apply water to a green hide to make it easier to scrape the hide. (86:115) **Cheyenne** *Incense & Fragrance* Leaves chewed and used as horse perfume. **Crow** *Incense & Fragrance* Plants mixed with other plants and beaver castor oil and used as a hair, body, or clothing perfume. (82:70) **Dakota** *Incense & Fragrance* Bunches of plants carried in bachelors' coats for the pleasant fragrance. Plant used for perfume. (69:363) **Flathead** *Insecticide* Leaves pulverized and sprinkled on meats as a bug repellent. **Kutenai** *Incense & Fragrance* Leaves placed on hot rocks in the sweat house as incense. (82:70) **Omaha** *Incense & Fragrance* Leaves used as a perfume for hair oil. (68:323)

Monarda fistulosa **ssp.** *fistulosa* **var.** *menthifolia* (Graham) Fern., Mintleaf Beebalm
- *Drug*—**Cheyenne** *Ceremonial Medicine* Plant used in ceremonies. (as *M. menthoefolia* 76:186) *Dermatological Aid* and *Veterinary Aid* Chewed or dried leaves used as a perfume for horses, bodies, and clothing. (as *M. menthoefolia* 76:186) **Navajo, Ramah** *Dermatological Aid* Cold infusion taken and used as lotion for gunshot or arrow wounds. (as *M. menthaefolia* 191:41) **Tewa** *Analgesic* Pulverized plant rubbed on the head for headache. *Eye Medicine* Plant used for sore eyes. *Febrifuge* Dried plant or leaves rubbed over body and infusion taken for fever. *Throat Aid* Dried leaves worn around neck and decoction taken for sore throat. (as *M. menthaefolia* 138:57, 58)
- *Food*—**Acoma** *Spice* Leaves ground and mixed with sausage for seasoning. (as *M. menthaefolia* 32:34) **Apache, Chiricahua & Mescalero** *Beverage* Leaves and young stems boiled to make a nonintoxicating beverage. (as *M. menthaefolia* 33:53) *Spice* Leaves used as flavoring. (as *M. menthaefolia* 33:47) **Hopi** *Dried Food* Dried in bundles for winter use. (as *M. menthaefolia* 200:91) **Isleta** *Spice* Leaves used for seasoning soups and stews. (as *M. menthaefolia* 100:35) **Laguna** *Spice* Leaves ground and mixed with sausage for seasoning. **Pueblo** *Dried Food* Dried and stored for winter use. *Spice* Cooked with meats and soups as a flavoring. (as *M. menthaefolia* 32:34) **San Ildefonso** *Spice* Used to flavor meat during cooking. (as *M. menthaefolia* 138:57) **Spanish American** *Dried Food* Dried and stored for winter use. *Spice* Cooked with meats and soups as a flavoring. (as *M. menthaefolia* 32:34) **Tewa of Hano** *Unspecified* Plant cooked and eaten. (as *M. menthaefolia* 138:57)
- *Fiber*—**Cheyenne** *Mats, Rugs & Bedding* Stems and flowers used as fragrant pillow stuffing by young girls from puberty to marriage. (as *M. menthoefolia* 76:186)
- *Other*—**Blackfoot** *Incense & Fragrance* Flowers dried and used as a perfume. (as *Monardo fisiulosa* var. *menthaefolia* 97:51) **Cheyenne** *Incense & Fragrance* Dried leaves and pine needles burned over coal for a fragrance. (as *Monarda menthoefolia* 76:186) **Keres, Western** *Hunting & Fishing Item* Plant chewed by hunters while hunting. (as *M. menthaefolia* Graham 171:54)

Monarda fistulosa **ssp.** *fistulosa* **var.** *mollis* (L.) Benth., Oswego Tea
- *Drug*—**Blackfoot** *Eye Medicine* Infusion of blossoms used as an eyewash to allay inflammation. (as *M. scabra* 114:275) **Chippewa** *Anthelmintic* Decoction of root and blossoms taken for worms. (as *M. mollis* L. 53:346) *Burn Dressing* Poultice of moistened, dry flowers and leaves applied to scalds and burns. (as *M. mollis* 53:354) *Dermatological Aid* Infusion of flower and leaf used as a wash, especially for children, for "eruptions." (as *M. mollis* 53:350) **Flathead** *Gynecological Aid* Infusion of leaves used by women after confinement. **Sioux** *Gynecological Aid* Infusion of leaves used by women after confinement. (as *M. scabra* 19:16)

Monarda pectinata Nutt., Pony Beebalm
- *Drug*—**Kiowa** *Dermatological Aid* Flowers gathered, placed in water, and the liquid sprinkled on the hair as a perfume. Infusion of flowers used as a wash for insect bites and stings. Flowers gathered, placed in water, and the liquid sprinkled on the hair as a perfume. (192:49) **Navajo** *Analgesic* Plant used for headaches. (90:153) **Navajo, Kayenta** *Gastrointestinal Aid* Plant used for stomach disease. (205:41) **Navajo, Ramah** *Analgesic* Cold infusion taken and used as poultice for headache. *Ceremonial Medicine* Plant used in a ceremonial lotion. *Cough Medicine* Cold infusion taken and used as poultice for cough. *Febrifuge* Cold infusion taken and used as poultice for fever. *Misc. Disease Remedy* Cold infusion taken and used as poultice for influenza. (191:41)
- *Food*—**Acoma** *Spice* Leaves ground and mixed with sausage for seasoning. (32:34) **Keres, Western** *Spice* Ground leaves mixed with sausage for seasoning. (171:54) **Laguna** *Spice* Leaves ground and mixed with sausage for seasoning. (32:34)

Monarda punctata L., Spotted Beebalm
- *Drug*—**Delaware** *Dermatological Aid* Infusion of plant used to bathe patients' faces. *Febrifuge* Infusion of plant used for fever. (176:35)

Delaware, Oklahoma *Febrifuge* Infusion of whole plant taken and used as a face wash for fever. (175:29, 76) **Meskwaki** *Analgesic* Compound containing leaves snuffed up nostrils for sick headache. Compound containing leaves used for stomach cramps. *Cold Remedy* Compound used as a snuff for head colds and catarrh. *Gastrointestinal Aid* Compound containing leaves used for stomach cramps. *Respiratory Aid* Compound used as a snuff for catarrh and used for head cold. *Stimulant* Compound applied at nostrils of patient to rally him when at point of death. (152:225, 226) **Mohegan** *Febrifuge* Infusion of leaves used for fevers. (176:73, 130) **Nanticoke** *Cold Remedy* Infusion of whole plant taken as a cold remedy. (175:55, 84) **Navajo, Ramah** *Analgesic* Cold infusion taken and used as poultice for headache. *Cough Medicine* Cold infusion taken and used as poultice for cough. *Febrifuge* Cold infusion taken and used as poultice for fever. (191:42) **Ojibwa** *Gastrointestinal Aid* Decoction of plants taken for stomach or bowel troubles. *Laxative* Decoction of plants taken for sick stomach, bowels, or for constipation. (135:231) *Unspecified* Plant used as a rubbing medicine. (135:240)
- *Other*—**Navajo** *Incense & Fragrance* Plant hung in the hogan for the pleasing odor. (55:73)

Monarda sp., Mint

- *Drug*—**Bannock** *Cold Remedy* Infusion of seed heads taken for colds. (118:38) **Blackfoot** *Veterinary Aid* Infusion of plant pieces and *Allium* applied to saddle sores. (86:87) **Creek** *Antirheumatic (External)* Compound decoction of plant used as a wash for rheumatism. *Antirheumatic (Internal)* Compound decoction of plant taken for rheumatism. *Dermatological Aid* Compound decoction of plant taken and used as a wash for swollen legs. *Diaphoretic* Infusion of entire plant taken to induce perspiration. (172:657) Decoction of entire plant used to promote perspiration. (177:54) *Ear Medicine* Compound decoction of plant taken and used as wash to prevent deafness from ghosts. *Kidney Aid* Compound decoction of plant taken and used as a wash for dropsy. (172:657) Infusion of plant used as a bath and taken for dropsy. (177:54) *Psychological Aid* Compound decoction of plant administered to delirious person. (172:657) Decoction of plant used for delirium. *Sedative* Decoction of plant used for delirium. (177:54) *Witchcraft Medicine* Compound decoction of plant taken and used as a wash to protect from ghosts. (172:657) **Iroquois** *Analgesic* Infusion of roots given to children for headaches. (87:426) *Emetic* Infusion of plants taken to "soften the brain" or as an emetic. (87:425) *Febrifuge* Compound decoction of leaves and roots taken for recurring chills from fevers. Decoction of roots taken for fevers and sickness caused by dead people. *Laxative* Infusion of roots given to children for constipation. *Other* Compound decoction of roots given to children with "summer complaint." Plant used in "Little Water Medicine." *Pediatric Aid* Compound decoction of roots given to children with "summer complaint." Infusion of roots given to children for headaches and constipation. *Stimulant* Infusion of plants taken for general lassitude. (87:426)
- *Food*—**Bannock** *Appetizer* Infusion of seed heads used as an appetizer. (118:38)

Monardella, Lamiaceae

Monardella candicans Benth., Sierran Mountainbalm

- *Food*—**Tubatulabal** *Beverage* Fresh or dried plants boiled, sugar added, and tea used as a beverage. (193:19) *Unspecified* Leaves and stalks used for food. (193:15)

Monardella lanceolata Gray, Mustang Mountainbalm

- *Drug*—**Diegueño** *Unspecified* Infusion of plant used as a medicinal tea and beverage. (84:25) **Luiseño** *Unspecified* Infusion of plant used for medicinal purposes. (155:229) **Miwok** *Analgesic* Decoction of leaves, upper stems, and flowers taken for headaches. *Cold Remedy* Decoction of leaves, upper stems, and flowers taken for colds. (12:171)
- *Food*—**Diegueño** *Beverage* Infusion of plant used as a medicinal tea and beverage. (84:25) **Luiseño** *Beverage* Plant used to make a tea. (155:211)

Monardella linoides Gray, Narrowleaf Monardella

- *Food*—**Kawaiisu** *Beverage* Leaves and flowers used to make a nonmedicinal tea. (206:42)

Monardella odoratissima Benth., Pacific Monardella

- *Drug*—**Karok** *Diaphoretic* Plant used as a sweat medicine. *Love Medicine* Plant used as a love medicine by women. (148:389) **Miwok** *Cold Remedy* Decoction of stems and flower heads taken for colds. *Febrifuge* Decoction of stems and flower heads taken for fevers. (12:171) **Okanagan-Colville** *Cold Remedy* and *Pediatric Aid* Infusion of leaves and stems given to children and adults for the common cold. (188:109) **Paiute** *Analgesic* Decoction of plant taken for gas pains. *Cold Remedy* Decoction of plant taken for colds. *Eye Medicine* Decoction of branches used as an eyewash for soreness or inflammation. *Gastrointestinal Aid* Decoction of plant taken for indigestion, gas pain, or minor digestive upset. (180:105, 106) **Sanpoil** *Cold Remedy* and *Pediatric Aid* Decoction of stems and leaves taken by adults and children for severe colds. (as *M. adoratissima* 131:218) **Shoshoni** *Analgesic* Decoction of plant taken for gas pains. *Blood Medicine* Decoction of branches taken as a blood tonic. *Cathartic* Decoction of branches taken as a physic. *Cold Remedy* Decoction of plant taken for colds. *Gastrointestinal Aid* Decoction of plant taken for indigestion, gas pain, or minor digestive upset. *Tonic* Decoction of branches taken as a general tonic. **Washo** *Analgesic* Decoction of plant taken for gas pains. *Cold Remedy* Decoction of plant taken for colds. *Gastrointestinal Aid* Decoction of plant taken for indigestion, gas pain, or minor digestive upset. (180:105, 106)
- *Food*—**Kawaiisu** *Beverage* Leaves and flowers used to make a nonmedicinal tea. (206:42) **Miwok** *Beverage* Decoction of stems and flower heads used as a beverage. (12:171) **Okanagan-Colville** *Beverage* Leaves and stems used to make a hot or cold tea. (188:109) **Sanpoil & Nespelem** *Beverage* Leaves and stems boiled, liquid strained and used as a hot or cold beverage. (131:105)
- *Other*—**Karok** *Incense & Fragrance* Plant put inside the hat for the nice smell when going on a journey. (148:389) **Okanagan-Colville** *Hunting & Fishing Item* Plant used to wipe fishing hooks and lines, spears, harpoons, animal snares, and arrows to remove scent. (188:109)

Monardella sp., Aromatic Mint

- *Drug*—**Karok** *Unspecified* Used for medicine. (117:209)

Monardella villosa Benth., Coyote Mint

- ***Drug*—Cahuilla** *Gastrointestinal Aid* Infusion of leaves taken for stomachaches. (15:89) **Costanoan** *Blood Medicine* and *Pulmonary Aid* Poultice of plant applied to back cuts to draw out "bad blood" for pneumonia. *Respiratory Aid* Decoction of plant and plant salve used for respiratory conditions. (21:16) **Mahuna** *Gastrointestinal Aid* Plant used for stomachaches. (140:8)

Monardella villosa ssp. *sheltonii* (Torr.) Epling, Shelton's Mountainbalm
- ***Drug*—Mendocino Indian** *Blood Medicine* Infusion of dried leaves taken to purify the blood. *Gastrointestinal Aid* Infusion of dried leaves taken for colic. (as *M. sheltonii* 41:384)
- ***Food*—Mendocino Indian** *Substitution Food* Aromatic, sweet-scented leaves used dried or fresh as a substitute for tea. (as *M. sheltonii* 41:384)

Monardella viridis Jepson, Green Mountainbalm
- ***Food*—Kawaiisu** *Beverage* Leaves and flowers used to make a nonmedicinal tea. (206:42)

Moneses, Ericaceae
Moneses uniflora (L.) Gray, Single Delight
- ***Drug*—Cowichan** *Dermatological Aid* Poultice of leaves applied to draw out the pus from boils or abscesses. (182:83) **Eskimo, Alaska** *Cold Remedy* Infusion of dried plants used for colds. *Cough Medicine* Infusion of dried plants used for coughs. (149:331) **Haisla & Hanaksiala** *Throat Aid* Plant chewed for sore throats. (43:261) **Kwakiutl** *Analgesic* Poultice of chewed or pounded plant applied to pains. (183:283) *Dermatological Aid* Poultice of chewed plant applied to swellings to draw blisters. (20:382) Plant used to draw blisters. Poultice of chewed or pounded plant applied to swellings. (183:283) **Kwakwaka'wakw** *Dermatological Aid* Poultice of chewed plant applied to swellings and blisters. (43:261) **Montagnais** *Orthopedic Aid* Infusion of plant used as a medicine for paralysis. (as *Monensis uniflora* 156:314) **Salish, Coast** *Dermatological Aid* Poultice of leaves applied to draw out the pus from boils or abscesses. (182:83)
- ***Food*—Montana Indian** *Fruit* Fruit used for food. (19:16)

Moneses uniflora ssp. *reticulata* (Nutt.) Calder & Taylor, Single Delight
- ***Drug*—Haisla & Hanaksiala** *Throat Aid* Plant chewed for sore throats. **Kwakwaka'wakw** *Dermatological Aid* Poultice of chewed plant applied to swellings and blisters. (43:261)

Monolepis, Chenopodiaceae
Monolepis nuttalliana (J. A. Schultes) Greene, Nuttall's Povertyweed
- ***Drug*—Navajo, Ramah** *Ceremonial Medicine* Plant used as ceremonial emetic. *Dermatological Aid* Poultice of moist leaves applied to skin abrasions. *Emetic* Plant used as ceremonial emetic. *Hunting Medicine* Pinch of dried plant eaten by hunters to prevent "buck fever." (191:25)
- ***Food*—Hopi** *Porridge* Ground seeds used to make mush. Ground seeds used to make mush. Ground seeds used to make mush. (190:161) **Navajo, Ramah** *Fodder* Used for sheep feed. (191:25) **Papago** *Dried Food* Seeds basket winnowed, parched, sun dried, cooked, stored, and used for food. (34:24) *Unspecified* Roots used for food. (36:60) **Pima** *Staple* Seeds boiled, partially dried, parched, ground, and eaten as pinole. *Unspecified* Roots boiled, cooled, mixed with fat or lard and salt, cooked, and eaten with tortillas. (as *M. chenopoides* 146:70) *Vegetable* Leaves boiled until tender, salted, fried in lard or fat, and eaten as greens. (47:70) **Pima, Gila River** *Vegetable* Leaves boiled or boiled, strained, refried, and eaten as greens. (133:5)

Monotropa, Ericaceae
Monotropa hypopithys L., Pinesap
- ***Drug*—Kwakiutl** *Love Medicine* Plant used in a love potion. (as *Hypopites monotropa* 183:283)

Monotropa uniflora L., Indian Pipe
- ***Drug*—Cherokee** *Anticonvulsive* Pulverized root given to children for fits, epilepsy, and convulsions. *Dermatological Aid* Crushed plant rubbed on bunions or warts. *Eye Medicine* Juice and water used to wash sore eyes. *Pediatric Aid* Pulverized root given to children for fits, epilepsy, and convulsions. (80:40) **Cree, Woodlands** *Toothache Remedy* Flower chewed for toothaches. (109:46) **Mohegan** *Analgesic* and *Cold Remedy* Infusion of root or leaves taken for pain due to colds. (176:73, 130) *Febrifuge* Leaves used for colds and fever. (176:130) **Potawatomi** *Gynecological Aid* Infusion of root taken for female troubles. (154:57) **Thompson** *Dermatological Aid* Poultice of plant used for sores that would not heal. Dried powdered stems applied to sores or burned stalk rubbed on sores. (187:215)
- ***Other*—Thompson** *Plant Indicator* Abundance of plant in woods indicated many mushrooms in the coming season. (187:215)

Monroa, Poaceae
Monroa squarrosa (Nutt.) Torr., False Buffalograss
- ***Other*—Navajo, Kayenta** *Decorations* Pollen mixed with corn pollen and mineral pigments and painted on masks of the God Impersonators. (205:16)

Moraceae, see *Artocarpus, Broussonetia, Ficus, Maclura, Morus*

Morchella, Helvellaceae, fungus
Morchella sp., Morelle
- ***Other*—Crow** *Soap* Used as a substitute for soap. (19:16)

Moricandia, Brassicaceae
Moricandia arvensis (L.) DC., Purple Mistress
- ***Drug*—Hoh** *Unspecified* Plants used for medicine. **Quileute** *Unspecified* Plants used for medicine. (as *Brassica arvensis* 137:61)
- ***Food*—Hoh** *Spice* Used for flavoring. *Vegetable* Plants eaten as greens. **Quileute** *Spice* Used for flavoring. *Vegetable* Plants eaten as greens. (as *Brassica arvensis* 137:61)
- ***Other*—Keres, Western** *Unspecified* Plant known and named but no use was specified. (as *Brassica arvensis* 171:33)

Morinda, Rubiaceae
Morinda citrifolia L., Indian Mulberry
- ***Drug*—Hawaiian** *Dermatological Aid* Young fruit thoroughly pounded with salt or fruit juice and used for broken bones and deep cuts. *Orthopedic Aid* Young fruit thoroughly pounded with salt or fruit juice used for broken bones and deep cuts. *Unspecified* Leaves used to make medicine. (2:73)

Mortonia, Celastraceae

Mortonia sempervirens ssp. scabrella (Gray) Prigge, Rio Grande Saddlebush
- *Other*—Havasupai *Hunting & Fishing Item* Wood used to make the foreshaft on the reed arrow. (as *M. scabrella* 197:230)

Morus, Moraceae

Morus alba L., White Mulberry
- *Drug*—Cherokee *Anthelmintic* Infusion of bark taken for worms. *Antidiarrheal* Infusion of bark taken to "check dysentery." *Cathartic* Infusion of bark used as a purgative. *Laxative* Infusion of bark taken as a laxative. (80:45)
- *Food*—Cherokee *Fruit* Fruit used for food. (80:45)

Morus microphylla Buckl., Texas Mulberry
- *Food*—Apache, Chiricahua & Mescalero *Bread & Cake* Fruit pressed into pulpy cakes, dried, and stored. *Fruit* Fruits eaten fresh. *Winter Use Food* Fruit pressed into pulpy cakes, dried, and stored for winter use. (33:44) **Apache, Mescalero** *Fruit* Berries eaten fresh. *Sauce & Relish* Berries dried and used as a spread on mescal. (14:47) **Yavapai** *Fruit* Raw berries used for food. (65:257)
- *Fiber*—Papago *Basketry* Twigs split in half lengthwise and used to make serviceable baskets. (34:57)
- *Other*—Apache, Mescalero *Weapon* Wood used to make the best war bows. (14:47) **Papago** *Hunting & Fishing Item* Long shoots peeled of bark, laid in hot ashes, dried, and used as bows. (34:70)

Morus nigra L., Black Mulberry
- *Drug*—Delaware, Oklahoma *Cathartic* Decoction of bark taken as a cathartic. *Emetic* Decoction of bark taken as an emetic. *Gastrointestinal Aid* and *Liver Aid* Decoction of bark taken "to remove bile from the intestines." (175:25, 76)

Morus rubra L., Red Mulberry
- *Drug*—Alabama *Urinary Aid* Decoction of roots taken for passing yellow urine. (177:19) **Cherokee** *Anthelmintic* Infusion of bark taken for worms. *Antidiarrheal* Infusion of bark taken to "check dysentery." *Cathartic* Infusion of bark used as a purgative. *Laxative* Infusion of bark taken as a laxative. (80:45) **Creek** *Emetic* Roots used as an emetic. *Stimulant* Infusion of root taken for weakness. *Urinary Aid* Infusion of root taken for urinary problems. (172:659) **Meskwaki** *Panacea* Root bark used as a medicine for any sickness. (152:251) **Rappahannock** *Dermatological Aid* Tree sap rubbed on skin for ringworm. (161:30)
- *Food*—Cherokee *Beverage* Berries used to make juice. *Bread & Cake* Berries and pokeberries crushed, strained, mixed with sugar and cornmeal, and made into dumplings. (126:48) *Fruit* Fruit used for food. (as *M. ruba* 80:21) Fresh berries used for food. *Preserves* Berries used to make jam. *Winter Use Food* Berries canned for future use. (126:48) **Comanche** *Fruit* Fruits eaten for food. (29:523) **Iroquois** *Bread & Cake* Fruit mashed, made into small cakes, and dried for future use. *Dried Food* Raw or cooked fruit sun or fire dried and stored for future use. *Fruit* Dried fruit taken as a hunting food. *Sauce & Relish* Dried fruit cakes soaked in warm water and cooked as a sauce or mixed with corn bread. (196:128) **Omaha** *Dried Food* Fruit dried for winter use. *Fruit* Fruit eaten fresh. (68:326) **Seminole** *Unspecified* Plant used for food. (169:475)
- *Other*—Seminole *Hunting & Fishing Item* Plant used to make bows. (169:475)

Morus sp., Wild Mulberry
- *Food*—Hualapai *Fruit* Berries used for food. (195:50) **Lakota** *Fruit* Fruits eaten for food. (139:51)
- *Fiber*—Hualapai *Basketry* Used to make the frame for coil weaving. (195:50)
- *Other*—Hualapai *Hunting & Fishing Item* Used to make bows. *Walking Sticks* Used to make canes. (195:50)

Mucuna, Fabaceae

Mucuna gigantea (Willd.) DC., Seabean
- *Drug*—Hawaiian *Laxative* Fruit meat and other plants chewed, mixed with salt water, and injected with an enema as a laxative. (2:45)

Muhlenbergia, Poaceae

Muhlenbergia andina (Nutt.) A. S. Hitchc., Foxtail Muhly
- *Fiber*—Navajo, Ramah *Brushes & Brooms* Stems used to make hairbrushes. (191:16)

Muhlenbergia cuspidata (Torr. ex Hook.) Rydb., Plains Muhly
- *Fiber*—Navajo *Brushes & Brooms* Stems used to make hairbrushes and brooms for sweeping out the hogan. (55:25)

Muhlenbergia dubia Fourn. ex Hemsl., Pine Muhly
- *Drug*—Navajo *Veterinary Aid* Compound poultice with roots applied to make sheep's blood cake. (90:153)

Muhlenbergia filiformis (Thurb. ex S. Wats.) Rydb., Pullup Muhly
- *Food*—Navajo, Ramah *Fodder* Used for sheep, horse, and cow feed. (191:16)

Muhlenbergia mexicana (L.) Trin. **var. mexicana**, Mexican Muhly
- *Food*—Navajo, Ramah *Fodder* Used for sheep and horse feed. (as *M. foliosa* 191:16)
- *Fiber*—Navajo, Ramah *Brushes & Brooms* Stems used to make hairbrushes. (as *M. foliosa* 191:16)

Muhlenbergia pauciflora Buckl., New Mexico Muhly
- *Other*—Apache, Chiricahua & Mescalero *Containers* Moist grass laid onto hot stones to prevent steam from escaping. (as *M. neomexicana* 33:36)

Muhlenbergia pungens Thurb., Sandhill Muhly
- *Fiber*—Hopi *Brushes & Brooms* Used to make a hairbrush and broom. (56:16) Used to make brooms. (190:158) Culms stripped of leaves and bound together to form a broom and hairbrush. (200:65) **Navajo** *Brushes & Brooms* Stems, pulled out of their sheaths when dry, tied with string and used as brooms and brushes. (55:26)

Muhlenbergia richardsonis (Trin.) Rydb., Mat Muhly
- *Drug*—Blackfoot *Veterinary Aid* Roots used as a horse medicine. (97:22)
- *Food*—Blackfoot *Forage* Plant eaten by horses. (97:22)

Muhlenbergia rigens (Benth.) A. S. Hitchc., Deer Grass
- *Food*—Apache, Western *Porridge* Seeds ground, mixed with cornmeal and water, and made into a mush. (as *Epicompes rigens* 26:189)

Apache, White Mountain *Bread & Cake* Seeds ground and used to make bread and pones. (as *Epicompes rigens* 136:149) *Fodder* Plant used for hay. (as *Epicampes rigens* 136:157) *Porridge* Seeds ground, mixed with meal and water, and eaten as mush. (as *Epicompes rigens* 136:149) *Unspecified* Seeds used for food. (as *Epicampes rigens* 136:157) **Hopi** *Bread & Cake* Ground seed meal used to make bread. Ground seed meal used to make bread. (190:158)
- *Fiber*—**Cahuilla** *Basketry* Stalks used as the horizontal or foundation around which the coils were wrapped in basket making. (15:89) **Diegueño** *Basketry* Seed stems used as the foundation material for coiled baskets. (84:25) **Kawaiisu** *Basketry* Stems used as the multiple rod foundation material in coiled basketry. (206:42) **Luiseño** *Basketry* Long grass used to make coiled baskets. *Clothing* Long grass made into coiled, conical baskets and used as hats. (as *Epicampes rigens californica* 155:204) **Shoshoni** *Sewing Material* Grass used as thread, very white, long, and fine. (as *Epicampes rigens* 118:8)
- *Other*—**Apache, White Mountain** *Cash Crop* Plant gathered and sold. (as *Epicompes rigens* 136:149) **Kawaiisu** *Tools* Stems used in pierced ears to keep the hole from growing together. (206:42) **Luiseño** *Containers* Long grass made into large, coiled baskets and used for storing food. *Cooking Tools* Long grass made into coiled, conical baskets and used as eating and drinking vessels. Long grass made into nearly flat, coiled baskets and used for winnowing and cleaning seeds. (as *Epicampes rigens californica* 155:204) **Zuni** *Ceremonial Items* Grass attached to sticks of plume offerings to anthropic gods. This grass was used only by Galaxy and Shu'maakwe fraternities. The sticks designated the god to whom the offerings were made and the plumes of the eagle and of other birds conveyed the breath prayers to the gods. (as *Epicampes rigens* 166:91)

Muhlenbergia sp.
- *Food*—**Apache, Chiricahua & Mescalero** *Bread & Cake* Seeds threshed, winnowed, ground, and the flour used to make bread. (33:48)
- *Fiber*—**Navajo** *Brushes & Brooms* Used in the making of brushes and brooms. (55:25)

Muhlenbergia wrightii Vasey ex Coult., Spike Muhly
- *Other*—**Apache, Chiricahua & Mescalero** *Containers* Moist grass laid onto hot stones to prevent steam from escaping. (33:36)

Murdannia, Commelinaceae
Murdannia nudiflora (L.) Brenan, Nakedstem Dewflower
- *Drug*—**Hawaiian** *Blood Medicine* Infusion of pounded leaves and other plants strained and taken to purify the blood. (as *Commelina nudiflora* 2:70)
- *Other*—**Hawaiian** *Containers* Leaves used as a covering for underground ovens. (as *Commelina nudiflora* 2:70)

Musaceae, see *Musa*

Musa, Musaceae
Musa ×*paradisiaca* L., Paradise Banana
- *Drug*—**Hawaiian** *Strengthener* Bud or young flower juice rubbed on tongue and mouth interior for body weakness from stomach disorder. (as *M. sapientum* 2:65)

Musa sp., Banana
- *Food*—**Seminole** *Unspecified* Plant used for food. (169:509)

Musineon, Apiaceae
Musineon divaricatum (Pursh) Raf. **var. *divaricatum*,** Leafy Wildparsley
- *Food*—**Blackfoot** *Unspecified* Roots eaten raw. (114:278)

Musineon divaricatum var. *hookeri* (Torr. & Gray) Mathias, Hooker's Wildparsley
- *Food*—**Crow** *Unspecified* Fleshy root used for food. (as *Museneon hookeri* 19:16)

Myosotis, Boraginaceae
Myosotis laxa Lehm., Bay Forget Me Not
- *Drug*—**Makah** *Dermatological Aid* Plant rubbed on the hair to act like a hair spray. (67:312)

Myosotis sp., Forget Me Not
- *Drug*—**Iroquois** *Veterinary Aid* Compound decoction of plants mixed with feed to aid cows with birthing. (87:421)

Myosurus, Ranunculaceae
Myosurus aristatus Benth., Bristle Mousetail
- *Drug*—**Navajo, Ramah** *Panacea* Plant used as "life medicine." *Witchcraft Medicine* Cold infusion of plant taken to protect from witches. (191:27)

Myosurus cupulatus S. Wats., Arizona Mousetail
- *Drug*—**Navajo, Ramah** *Dermatological Aid* Cold infusion used internally or externally for ant bites. *Gastrointestinal Aid* Cold infusion of plant taken and used as a lotion for effects of swallowing an ant. (191:27)

Myosurus minimus L., Tiny Mousetail
- *Drug*—**Navajo, Ramah** *Dermatological Aid* Cold infusion taken or poultice of chewed plant applied to ant bite. (191:27)

Myrcianthes, Myrtaceae
Myrcianthes fragrans (Sw.) McVaugh, Twinberry
- *Other*—**Seminole** *Cooking Tools* Plant used to make food paddles. *Hunting & Fishing Item* Plant used to make blowgun darts. *Tools* Plant used to make pestles and ax handles. (as *Eugenia simpsoni* 169:468)

Myricaceae, see *Comptonia*, *Myrica*

Myrica, Myricaceae
Myrica cerifera L., Southern Bayberry
- *Drug*—**Choctaw** *Febrifuge* Decoction of leaves and stems taken "during attacks of fever." (25:23) Decoction of leaves and stems taken for fevers. *Throat Aid* Decoction of roots used as a gargle for inflamed tonsils. (177:13) **Houma** *Anthelmintic* Decoction of leaves taken as a vermifuge. (158:56) **Koasati** *Gastrointestinal Aid* and *Pediatric Aid* Decoction of roots given to children with stomachaches. (177:13) **Micmac** *Analgesic* Roots used for headaches. (40:58) *Antirheumatic (External)* Hot poultice of pounded, water-soaked roots applied to inflammations. (194:30) *Dermatological Aid* Roots used for inflam-

mations. *Stimulant* Berries, bark, and leaves used as an exhilarant and beverage. (40:58) **Seminole** *Analgesic* Decoction of leaves taken for headaches. *Febrifuge* Decoction of leaves taken for fevers. *Gastrointestinal Aid* Decoction of leaves taken for stomachaches. (169:282) *Love Medicine* Decoction of wood ashes placed on the tongue to cleanse the body and strengthen the marriage. (169:250)
- *Other*—**Houma** *Lighting* Berries boiled and wax used to make candles. (158:56) **Seminole** *Cleaning Agent* Plant used to make lye. *Smoke Plant* Plant used as a tobacco substitute. (169:480)

Myrica gale L., Sweet Gale
- *Drug*—**Bella Coola** *Diuretic* Decoction of pounded branches taken as a diuretic and for gonorrhea. *Venereal Aid* Decoction of pounded branches taken for gonorrhea and as a diuretic. (150:55) Infusion of pounded branches and fruits taken as a diuretic for gonorrhea. (184:206)
- *Food*—**Potawatomi** *Preservative* Plant used to line the blueberry pail to keep the berries from spoiling. (154:121)
- *Dye*—**Ojibwa** *Brown* In the fall, the branch tips grow into an abortive scale and boiled to yield a brown dye stuff. (153:425) *Yellow* Seeds boiled to obtain a yellow dye. (96:114)
- *Other*—**Cree, Woodlands** *Hunting & Fishing Item* Pistillate catkins used as an ingredient in lures. (109:46) **Potawatomi** *Insecticide* Plant thrown onto the fire to make a smudge and keep away mosquitoes. (154:121)

Myrica heterophylla Raf., Evergreen Bayberry
- *Other*—**Houma** *Lighting* Berries boiled and wax used to make candles. (as *M. caroliniensis* 158:56)

Myrica sp., Sweet Bay
- *Drug*—**Creek** *Emetic* Compound decoction of leaves taken after a burial as an emetic before eating. (172:664) **Delaware** *Blood Medicine* Bark used as a blood purifier. *Kidney Aid* Bark used for kidney troubles. (176:35) **Delaware, Oklahoma** *Blood Medicine* Bark used as a blood purifier. *Gynecological Aid* Compound containing root taken for "female generative organs." *Kidney Aid* Bark used for "kidney trouble." (175:29, 76) **Mohegan** *Kidney Aid* Infusion of bark taken for kidney disorders. (176:74, 130)

Myriophyllum, Haloragaceae

Myriophyllum sibiricum Komarov, Shortspike Watermilfoil
- *Drug*—**Iroquois** *Blood Medicine* Infusion of whole plant and two other plants taken by adolescents for poor blood circulation. (as *M. exalbescens* 141:51) *Emetic* Infusion of plant taken as a strong emetic. (as *M. exalbescens* 141:51). *Pediatric Aid* Infusion of whole plant and two other plants taken by adolescents for poor blood circulation. (as *M. exalbescens* 141:51)

Myriophyllum spicatum L., Spike Watermilfoil
- *Drug*—**Menominee** *Unspecified* Some plants of this class used in medicines. (151:37)
- *Food*—**Tanana, Upper** *Frozen Food* Rhizomes frozen for future use. *Unspecified* Rhizomes eaten raw, fried in grease, or roasted. Rhizomes were sweet and crunchy and a much relished food. They were said to have been an important food during periods of low food supply as they were usually able to obtain them when needed. People give accounts of how they saved people's lives during such times. (102:14)

Myriophyllum verticillatum L., Whorlleaf Watermilfoil
- *Drug*—**Iroquois** *Other* Compound infusion of plants used as a "snowsnake medicine." *Pediatric Aid* and *Stimulant* Decoction of plant given to children when they lie very quiet and never move. (87:391)

Myrsinaceae, see *Ardisia*

Myrtaceae, see *Eucalyptus, Eugenia, Metrosideros, Myrcianthes, Psidium, Syzygium*

Myrtillocactus, Cactaceae

Myrtillocactus cochal (Orcutt) Britt. & Rose, Cochal
- *Food*—**Papago & Pima** *Fruit* Fruit used for food. (35:42)

Nama, Hydrophyllaceae

Nama demissum Gray **var. *demissum***, Purplemat
- *Food*—**Kawaiisu** *Porridge* Seeds pounded in a bedrock mortar and boiled into a mush. (206:43)

Nama hispidum Gray, Bristly Nama
- *Drug*—**Navajo, Kayenta** *Dermatological Aid* Plant used as a lotion for spider or tarantula bites. (205:39)

Napaea, Malvaceae

Napaea dioica L., Glade Mallow
- *Drug*—**Meskwaki** *Dermatological Aid* Poultice of root applied to keep old sores soft and boiled root used for swellings. *Gynecological Aid* Root used to ease childbirth and for female troubles. *Hemorrhoid Remedy* Roots used as a special remedy for piles. *Hunting Medicine* Root used as a hunting charm and for female troubles. (152:232)

Navarretia, Polemoniaceae

Navarretia atractyloides (Benth.) Hook. & Arn., Hollyleaf Pincushionplant
- *Drug*—**Costanoan** *Burn Dressing* Toasted, powdered plant or plant ash applied to burns. (21:13)

Navarretia cotulifolia (Benth.) Hook. & Arn., Cotulaleaf Pincushionplant
- *Drug*—**Miwok** *Antirheumatic (External)* Decoction applied to swelling. (12:171)

Navarretia sp., Skunk Weed
- *Food*—**Miwok** *Unspecified* Dried, stored, parched, pulverized seeds used for food. (12:155)

Nelumbonaceae, see *Nelumbo*

Nelumbo, Nelumbonaceae

Nelumbo lutea Willd., American Lotus

- *Food*—**Comanche** *Unspecified* Boiled roots used for food. (29:523) **Dakota** *Soup* Hard, nut-like seeds cracked, freed from the shells, and used with meat to make soup. *Unspecified* Peeled tubers cooked with meat or hominy and used for food. (70:79) **Huron** *Starvation Food* Roots used with acorns during famine. Roots used with acorns during famine. (as *Nelumbium luteim* 3:63) **Meskwaki** *Unspecified* Seeds cooked with corn. *Winter Use Food* Terminal shoots cut crosswise, strung on string, and dried for winter use. (152:262) **Ojibwa** *Unspecified* Hard chestnut-like seeds roasted and made into a sweet meal. Shoots cooked with venison, corn, or beans. The terminal shoots are cut off at either end of the underground creeping rootstock and the remainder is their potato. These shoots are similar in shape and size to a banana, and form the starchy storage reservoirs for future growth. They have pores inside, but have more substance to them than the stems. They are cut crosswise and strung upon basswood strings, to hang from the rafters for winter use. (153:407) **Omaha** *Soup* Hard, nut-like seeds cracked, freed from the shells, and used with meat to make soup. *Unspecified* Peeled tubers cooked with meat or hominy and used for food. (70:79) *Vegetable* Roots boiled and eaten as vegetables. (as *Nelumbium luteum* 58:341) **Pawnee** *Soup* Hard, nut-like seeds cracked, freed from the shells, and used with meat to make soup. *Unspecified* Peeled tubers cooked with meat or hominy and used for food. **Ponca** *Soup* Hard, nut-like seeds cracked, freed from the shells, and used with meat to make soup. *Unspecified* Peeled tubers cooked with meat or hominy and used for food. (70:79) **Potawatomi** *Unspecified* Seeds gathered and roasted like chestnuts. *Winter Use Food* Roots gathered, cut, and strung for winter use. (154:105) **Winnebago** *Soup* Hard, nut-like seeds cracked, freed from the shells, and used with meat to make soup. *Unspecified* Peeled tubers cooked with meat or hominy and used for food. (70:79)
- *Other*—**Dakota** *Ceremonial Items* Plant characterized as having mystic powers. **Omaha** *Ceremonial Items* Plant characterized as having mystic powers. **Pawnee** *Ceremonial Items* Plant characterized as having mystic powers. **Ponca** *Ceremonial Items* Plant characterized as having mystic powers. **Winnebago** *Ceremonial Items* Plant characterized as having mystic powers. (70:79)

Nemopanthus, Aquifoliaceae
Nemopanthus mucronatus (L.) Loes., Catberry
- *Drug*—**Malecite** *Cough Medicine* Used with blackberry roots, staghorn sumac, lily roots, and mountain raspberry roots for coughs. *Febrifuge* Used with blackberry roots, staghorn sumac, lily roots, and mountain raspberry roots for fevers. (116:251) *Kidney Aid* Infusion of root scrapings used for gravel. (116:257) *Tuberculosis Remedy* Used with blackberry roots, staghorn sumac, lily roots, and mountain raspberry roots for consumption. (116:251) **Ojibwa** *Unspecified* Berries used medicinally for unspecified purpose. (153:355) **Potawatomi** *Panacea* Compound decoction boiled down to syrup and used for many kinds of diseases. *Tonic* Decoction of small branches reduced to syrup and taken as a tonic. (154:39)
- *Food*—**Potawatomi** *Sour* Berries edible, but quite bitter and kept for a food. (154:95)

Nemophila, Hydrophyllaceae
Nemophila menziesii Hook. & Arn., Menzies's Baby Blue Eyes
- *Food*—**Kawaiisu** *Forage* Plant eaten by the cows. (206:43)

Neomammillaria, Cactaceae
Neomammillaria sp., Fishhook Cactus
- *Food*—**Navajo** *Unspecified* Spines removed and used for food. (55:64)

Nepeta, Lamiaceae
Nepeta cataria L., Catnip
- *Drug*—**Cherokee** *Abortifacient* Infusion used for female obstructions. *Anthelmintic* Infusion used for worms. *Anticonvulsive* Infusion used for spasms. *Cold Remedy* "Syrup" and honey used for colds and infusion used for babies' colds. *Cough Medicine* "Syrup" and honey used for coughs. *Dermatological Aid* Infusion taken for hives and poultice of leaf used for boils and swellings. *Febrifuge* Infusion taken for fevers and poultice of leaf used for boils. *Gastrointestinal Aid* Infusion used for colic and infusion of leaves used for stomach. *Pediatric Aid* Infusion used for babies' colds. *Sedative* Infusion used for hysterics. *Stimulant* Infusion of leaf used as a stimulant. *Tonic* Infusion of leaf used as a tonic. (80:28) **Chippewa** *Febrifuge* Simple or compound decoction of leaves taken for fever. (53:354) **Delaware** *Pediatric Aid* Leaves used with peach seeds to make a beneficial syrup for children. (176:37) **Delaware, Oklahoma** *Pediatric Aid* Leaves used with peach pits to make a tonic for children. (175:31, 82) *Tonic* Leaves prepared with peach pit and given as a pediatric tonic. (175:31, 76) **Delaware, Ontario** *Pediatric Aid* and *Sedative* Infusion of leaves given to soothe infants. (175:67) **Hoh** *Pediatric Aid* Infusion used as medicine for infants. (137:68) **Iroquois** *Analgesic* Cold infusion of plants taken for headaches. Decoction of plant tips given to children with headaches. (87:423) *Antidiarrheal* Compound decoction of stems given to children with diarrhea. (87:422) Plant used as a laxative and astringent for diarrhea. *Antiemetic* Infusion of plants taken for vomiting from fever or unknown cause. *Cold Remedy* Infusion of stems taken for colds. *Cough Medicine* Infusion of stems taken for coughs. *Febrifuge* Cold infusion of plants taken or applied to forehead for fevers and chills. *Gastrointestinal Aid* Infusion of plant given to children for stomachaches due to colds. Infusion of plants taken for trouble caused by eating rich foods. *Laxative* Plant used as a laxative and astringent for diarrhea. (87:423) *Oral Aid* Flowers and roots used for excess saliva. (141:58) *Other* Cold infusion of plants taken for fever and "summer complaint." (87:423) *Pediatric Aid* Compound decoction of stems given to children with diarrhea. (87:422) Decoction of plant tips given to children when peevish or with headaches. Infusion of plant given to children for stomachaches due to colds. Infusion of plant tips given to babies who are restless and cannot sleep. Plant used for babies with fevers. *Sedative* Decoction of plant tips given to children when peevish. *Throat Aid* Infusion of stems taken for sore throats and chills. (87:423) **Keres, Western** *Stimulant* Infusion of plant used as a bath for tiredness. (171:55) **Menominee** *Diaphoretic* Decoction of whole plant, except root, taken to produce perspiration. (54:132) *Pulmonary Aid* Compound infusion taken and poultice applied to chest for pneumonia. (151:39) *Sedative* Decoction of whole plant, except root, taken to produce restful sleep. (54:132) **Mohegan** *Gastrointestinal Aid* Infusion of plant given to babies for colic. (174:266) Infusion of leaves taken for the stomach and given to infants for colic. (176:74, 130) *Pediatric Aid* Infusion given to babies

for colic. (174:266) Infusion of leaves given to infants for colic. (176: 74, 13) **Ojibwa** *Blood Medicine* Infusion of leaves taken as a blood purifier. *Other* Infusion of leaves used to bathe a patient to raise the body temperature. (153:372) **Okanagan-Colville** *Cold Remedy* Infusion of plant tops taken for colds. (188:110) **Quileute** *Pediatric Aid* Infusion used as medicine for infants. (137:68) **Rappahannock** *Analgesic* Infusion of leaves given to babies for pains. *Antirheumatic (Internal)* Infusion of leaves given to babies for rheumatism. *Pediatric Aid* Infusion of leaves given to babies for pains, rheumatism, and measles. Weak infusion of plant given to children as a tonic. *Tonic* Weak infusion of plant given to children as a tonic. (161:25) **Shinnecock** *Antirheumatic (Internal)* Dried leaves smoked in a pipe for rheumatism. (30:119)
- *Food*—**Ojibwa** *Beverage* Leaves used to make a beverage tea. (153:405) **Okanagan-Colville** *Forage* Plant eaten by skunks. (188:110)

Nephroma, Peltigeraceae, lichen

Nephroma arcticum (L.) Tross., Arctic Kidney Lichen
- *Drug*—**Eskimo, Inuktitut** *Strengthener* Infusion of plant used for "weakness." (202:187)
- *Food*—**Eskimo, Inuktitut** *Unspecified* Plant boiled and eaten with fish eggs. (202:187)

Nereocystis, Lessoniaceae, alga

Nereocystis luetkeana (Mertens) Postels & Ruprecht
- *Drug*—**Kwakiutl, Southern** *Burn Dressing* Leaves used for burns. *Dermatological Aid* Leaves used for scabs and nonpigmented spots. Leaves dried, pulverized, and rubbed into children's heads to make hair grow long. *Orthopedic Aid* Leaves used for swollen feet. *Pediatric Aid* Leaves dried, pulverized, and rubbed into children's heads to make hair grow long. (common kelp 183:261) **Nitinaht** *Dermatological Aid* Bulbs dried, melted, hardened, and used as skin cream for protection from sun, wind, and cold. (common kelp 186:52) **Pomo, Kashaya** *Expectorant* Dried, salty stalk strips sucked for colds with sore throats and to clear mucus. *Throat Aid* Dried, salty stalk strips sucked for colds with sore throats. (as *N. luetkeama*, bull kelp 72:124)
- *Food*—**Oweekeno** *Unspecified* Plant eaten with herring roe. *Winter Use Food* Plant and herring eggs salted for storage. (bull kelp 43:46) **Pomo, Kashaya** *Unspecified* Thick part of the stalk cooked in an oven or in hot ashes. *Winter Use Food* Stalks cut into lengthwise strips and dried for winter use. (bull kelp 72:124)
- *Fiber*—**Hesquiat** *Cordage* Long stipes used to make fishing lines and anchor ropes. Long stipes were dried, then soaked in dogfish or whale oil so they would not lose their flexibility. Kelp ropes were very strong and could be plaited or spliced together to make them longer. (common kelp 185:25) **Makah** *Cordage* Solid stipes used for tying. (bull kelp 67:206) **Nitinaht** *Cordage* Lower stipes used for ropes and fishing lines. (bull kelp 67:206) **Pomo, Kashaya** *Cordage* Dried, shredded stems used as cordage or fish line. (bull kelp 72:124)
- *Other*—**Alaska Native** *Hunting & Fishing Item* Long, hollow stalks used to make fishing lines for deep-sea fishing. (giant kelp 85:139) **Bella Coola** *Containers* Used to store eulachon (candlefish) grease. (bull kelp 184:195) **Haisla & Hanaksiala** *Designs* Plant heads used as the design for the horizontal timber ends of the Killer Whale house. (bull kelp 43:128) **Kitasoo** *Hunting & Fishing Item* Stipes used to make fishing lines. (bull kelp 43:303) **Hesquiat** *Containers* Hollow floats and upper stipes used to store oil and other liquids. Hollow floats and upper stipes used as molds for cottonwood resin and deer fat skin "cream." This ointment was poured in hot and melted. After it had solidified, the kelp mold was cut open and the ball of ointment removed. (common kelp 185:25) **Kwakiutl, Southern** *Ceremonial Items* Tubes used in special pre-war rituals. *Containers* Long, hollow stipes used to store oulachen (candlefish) grease and other oils. *Cooking Tools* Tubes put into steaming pits so that water could be poured directly on hot rocks at the bottom. *Hunting & Fishing Item* Blades weighted, placed underwater at river mouths, and used to catch herring spawn. (common kelp 183:261) **Makah** *Containers* Bottle ends used to carry fish oil and molasses. *Hunting & Fishing Item* Kelp used as fishing culture for fish line. (kelp 79:50) Solid stipes used for fish lines. (bull kelp 67:206) *Toys & Games* Plants and stems used by children as beach games. (kelp 79:50) Used by children to make "kelp cars" to tow around on the beach. (bull kelp 67:206) **Nitinaht** *Containers* Enlarged upper portion of the stipes used as molds for cosmetics. Enlarged upper portion of stipes dried and rinsed with fresh water and used for oil storage bottles. Blades used to cover fish in the boat, while at sea, to prevent the fish from drying out. (bull kelp 67:206) Hollow stipes used as containers for storing oil. Leafy fronds used to cover fish to prevent spoiling or drying out. (common kelp 186:52) *Cooking Tools* Enlarged upper portion of stipes used as funnels for pouring water onto hot rocks in pit cooking. *Hunting & Fishing Item* Lower stipes used for ropes and fishing lines. (bull kelp 67:206) Stipes partially dried, rubbed with oil until saturated, and used as fishing line. (common kelp 186:52) *Tools* Enlarged upper portion of the stipes used as steam boxes for making halibut hooks. (bull kelp 67:206) Bulbs used for curving and molding halibut hooks. (common kelp 186:52) **Oweekeno** *Hunting & Fishing Item* Stipes used to make fishing lines. (bull kelp 43:46) **Pomo, Kashaya** *Hunting & Fishing Item* Dried, shredded stems used as cordage or fish line. (bull kelp 72:124) **Quileute** *Containers* Bottle ends used to carry fish oil and molasses. *Hunting & Fishing Item* Kelp used as fishing culture for fish line. **Quinault** *Containers* Bottle ends used to carry fish oil and molasses. *Hunting & Fishing Item* Kelp used as fishing culture for fish line. (kelp 79:50) **Tsimshian** *Hunting & Fishing Item* Stipes used to make fishing lines. (bull kelp 43:303)

Nicotiana, Solanaceae

Nicotiana attenuata Torr. ex S. Wats., Coyote Tobacco
- *Drug*—**Apache, White Mountain** *Ceremonial Medicine* Plant smoked in the medicine ceremonies. (136:158) **Hopi** *Ceremonial Medicine* Plant smoked for all ceremonial occasions. (56:19) **Navajo, Kayenta** *Hemostat* Plant used for nosebleed. *Narcotic* Plant used as a narcotic. (205:41) **Navajo, Ramah** *Analgesic* Leaves smoked in corn husks for headache. *Ceremonial Medicine* Plant smoked in corn husks for ceremonial purposes. *Cough Medicine* Leaves smoked in corn husks for cough. *Veterinary Aid* Plant used to heal castration cuts on a young racehorse. (191:43) **Paiute** *Anthelmintic* Decoction of leaves taken sparingly to expel worms. (180:106, 107) *Antirheumatic (External)* Infusion of stems and leaves used as a wash for aches and pains. (111:108) Crushed seeds used as a liniment for rheumatic swellings. Poultice of crushed leaves applied to swellings, especially from rheu-

matism. *Cathartic* Weak decoction of leaves taken as a physic. (180: 106, 107) *Cold Remedy* Leaves smoked during sweat house bathing and prayer, connected with spiritual power. (111:108) Dried leaves smoked alone or in a compound for colds and asthma. *Dermatological Aid* Decoction of leaves used as a healing wash for hives or other skin irritations. Poultice of chewed leaves applied to cuts. Poultice of crushed leaf applied or crushed seed used as a liniment for swellings. Poultice of crushed leaves applied to eczema or other skin infections. Pulverized dust of plant sprinkled on sores. (180:106, 107) *Emetic* Infusion of stems and leaves taken as an emetic. (111:108) Weak decoction of leaves taken as an emetic. *Kidney Aid* Decoction of leaves used as a wash for "dropsical conditions." (180:106, 107) *Misc. Disease Remedy* Infusion of plant taken for measles. *Respiratory Aid* Infusion of stems and leaves taken for respiratory diseases. (111:108) Compound containing dried leaves smoked for asthma. *Snakebite Remedy* Poultice of chewed leaves bound on snakebite after removing poison. (180:106, 107) *Tuberculosis Remedy* Infusion of stems and leaves taken for tuberculosis. (111:108) Compound containing dried leaves smoked for tuberculosis. (180:106, 107) **Salish** *Dermatological Aid* Plant used as a head wash for dandruff. (178:294) **Shoshoni** *Anthelmintic* Decoction of leaves taken sparingly to expel worms. *Antirheumatic (External)* Poultice of crushed leaves applied to swellings, especially from rheumatism. *Cathartic* Weak decoction of leaves taken as a physic. *Dermatological Aid* Decoction of leaves used as a healing wash for hives or other skin irritations. Poultice of chewed leaves applied to cuts. Poultice of crushed leaf applied to reduce swellings. *Emetic* Weak decoction of leaves taken as an emetic. *Toothache Remedy* Poultice of crushed leaves applied to gum for toothache. *Tuberculosis Remedy* Compound of dried leaves smoked for tuberculosis. (180:106, 107) **Shuswap** *Urinary Aid* Plant used for the bladder. (123:69) **Tewa** *Ceremonial Medicine* Dried leaves and other plant parts smoked ceremonially. (138:103, 104) *Cough Medicine* Poultice of leaves mixed with oil and soot applied to neck and chest for cough. *Gynecological Aid* Snuff containing leaves used by women in labor. (138:106) *Nose Medicine* Snuff of leaves used for "a discharge from the nose." (138:103, 104) *Toothache Remedy* Leaves placed on or in a tooth for toothache. (138:106) **Thompson** *Dermatological Aid* Decoction of plant used as a wash to remove dandruff and prevent falling hair. (164:467, 468) **Zuni** *Snakebite Remedy* Smoke blown over body for throbbing from rattlesnake bite. (166:54)

- **Other**—**Blackfoot** *Smoke Plant* Leaves used for smoking. (97:52) **Coahuilla** *Smoke Plant* Leaves pounded, mixed with water, chewed, and used as a smoking material. (13:74) **Gosiute** *Smoke Plant* Leaves dried and smoked. (39:375) **Havasupai** *Smoke Plant* Leaves smoked for pleasure. (197:240) **Hopi** *Ceremonial Items* Plant smoked in pipes for ceremonial purposes only. (200:90) *Smoke Plant* Used for smoking. (36:109) Used very much. (190:166) **Kawaiisu** *Protection* Leaves and lime placed in the camp fire to prevent supernatural beings from bothering you. *Smoke Plant* Leaves dried, pulverized, powdered, mixed with liquid, formed into cakes, and smoked in hollow stems. (206:43) **Klamath** *Smoke Plant* Leaves used for smoking. (45:104) **Mewuk** *Smoke Plant* Leaves dried, pounded, and used for smoking. (117:356) **Midoo** *Smoke Plant* Leaves dried, pounded, and used for smoking. (117:319) **Navajo** *Ceremonial Items* Smoked after the feast following the completion of the masks for the Night Chant. Used for filling ceremonial prayer sticks in the Night Chant. (55:75) **Navajo, Ramah** *Smoke Plant* Plant used as substitute for commercial tobacco. (191:43) **Okanagan-Colville** *Smoke Plant* Leaves used for smoking. (188:140) **Okanagon** *Smoke Plant* Leaves dried, greased, mixed with leaves of other plants, and smoked. (125:39) **Paiute** *Smoke Plant* Roasted, dried leaves and small twigs used for smoking. (111:108) Leaves dried, ground, moistened, and made into balls for preservation. (167:319) **Papago** *Smoke Plant* Used for smoking. **Pima** *Smoke Plant* Used for smoking. (36:109) **Shoshoni** *Cash Crop* Leaves pulverized and made into large cakes and sold. (117:446) **Shuswap** *Smoke Plant* Mixed with kinnikinnick and red willow and smoked at ceremonies. (123:69) **Tewa** *Smoke Plant* Dried leaves and other plant parts smoked in pipes and cigarettes. (138:103, 104) **Thompson** *Smoke Plant* Leaves dried, greased, mixed with leaves of other plants, and smoked. (125:39) Dried, toasted leaves considered the most important source of tobacco. (164:495) **Yavapai** *Smoke Plant* Dried stems and leaves used for smoking. (65:263) **Zuni** *Ceremonial Items* and *Smoke Plant* Leaves smoked ceremonially. (166:95)

Nicotiana clevelandii Gray, Cleveland's Tobacco

- **Drug**—**Cahuilla** *Dermatological Aid* Poultice of leaves applied to cuts, bruises, swellings, and other wounds. *Ear Medicine* Leaf smoke blown into the ear and covered with a warm pad for earaches. *Emetic* Infusion of leaves used as an emetic. *Hunting Medicine* Leaves smoked as part of a hunting ritual. (15:90)
- **Food**—**Cahuilla** *Beverage* Leaves chewed, smoked, or used in a drinkable decoction. (15:90)
- **Other**—**Cahuilla** *Ceremonial Items* Used as an integral part of every ritual.[47] Used by shamans to control rain, increase crop production, divining, and improve health of community. Used by shamans, at community gatherings, to drive away malevolent powers. *Protection* Leaves smoked by travelers to clear away all danger and ensure blessing from spiritual guides. *Smoke Plant* Leaves chewed, smoked, or used in a drinkable decoction. (15:90)

Nicotiana glauca Graham, Tree Tobacco

- **Drug**—**Cahuilla** *Dermatological Aid* Poultice of leaves applied to cuts, bruises, swellings, and other wounds. *Ear Medicine* Leaf smoke blown into the ear and covered with a warm pad for earaches. *Emetic* Infusion of leaves used as an emetic. *Hunting Medicine* Leaves smoked as part of a hunting ritual. (15:90) **Hawaiian** *Dermatological Aid* Plant used for removing the pus from scrofulous sores or boils. Plant used for sores and the smoke used for cuts. *Tuberculosis Remedy* Plant used for removing the pus from scrofulous sores. (2:73) **Mahuna** *Antirheumatic (External)* Infusion of leaves used as a steam bath for rheumatism. *Throat Aid* Poultice of leaves applied to inflamed throat glands. *Tuberculosis Remedy* Poultice of leaves applied for scrofula. (140:60)
- **Food**—**Cahuilla** *Beverage* Leaves chewed, smoked, or used in a drinkable decoction. (15:90)
- **Other**—**Cahuilla** *Ceremonial Items* Used as an integral part of every ritual.[47] Used by shamans to control rain, increase crop production, divining, and improve health of community. Used by shamans, at community gatherings, to drive away malevolent powers. *Protection* Leaves smoked by travelers to clear away all danger and ensure blessing from spiritual guides. *Smoke Plant* Leaves chewed, smoked,

or used in a drinkable decoction. (15:90) **Diegueño** *Smoke Plant* Leaves used for smoking. (84:27)

Nicotiana plumbaginifolia Viviani, Tex-mex Tobacco

- *Other*—**Neeshenam** *Smoke Plant* Leaves sun dried, finely cut, and smoked. (129:378)

Nicotiana quadrivalvis Pursh, Indian Tobacco

- *Other*—**Blackfoot** *Ceremonial Items* Plants planted, harvested ceremonially, and smoked as an important part of every ritual. Leaves mixed with kinnikinnick, dried cambium, or red osier dogwood and used in all religious bundles. *Smoke Plant* Leaves used for ritual smoking. Ritual smoking was begun by an orderly, who filled the pipe and passed it, unlit, to the man sitting next to the officiating ceremonialist. This man had the favored position because of his wealth in bundle ownership. He drew on the unlit pipe four times and then passed it back to the orderly, who lit the pipe and gave it to the man next to the distinguished bundle owner. This man drew on the pipe four times (not inhaling) and blew the smoke upward. Then the pipe was passed sunwise (clockwise) to each participant until it reached the door of the tepee, whence it was returned to the orderly. The pipe was not passed across the door to the other side of the lodge, where women and children were seated. If the pipe went out during the smoke, it was given to the orderly, who cleaned and refilled it. The manner in which the participant received the pipe varied according to bundle ownership. Thus, a Medicine Pipe bundle owner would grasp the pipe roughly with both hands half clenched, imitating the actions of a bear. A ceremony in which smoking had special significance was the Big Smoke, or All Smoking, ceremony. This ceremony was confined to ceremonialists, diviners, and bundle owners. They gathered for the single purpose of recounting their prestigious and wealthy positions in the tribe. The Big Smoke commenced at sundown and continued until daybreak, and there was continuous use of many pipes. Four songs were allowed to be sung for each bundle owned; participants would often qualify for 16 songs or more. (as *N. multivalvis* 86:14) **Dakota** *Smoke Plant* Plant cultivated and used for smoking. (70:113, 114) **Montana Indian** *Smoke Plant* Used in compound for smoking and the delicate tobacco from flower was never chewed. (19:16) **Omaha** *Smoke Plant* Plant preferred for smoking because of the mild quality. (68:331) Plant cultivated and used for smoking. **Pawnee** *Smoke Plant* Plant cultivated and used for smoking. **Ponca** *Smoke Plant* Plant cultivated and used for smoking. (70:113, 114) **Tsimshian** *Smoke Plant* Leaves used for smoking. (43:350) **Winnebago** *Smoke Plant* Plant cultivated and used for smoking. (70:113, 114)

Nicotiana quadrivalvis var. *bigelovii* (Torr.) DeWolf, Bigelow's Tobacco

- *Drug*—**Costanoan** *Cathartic* Leaves smoked as a general purgative in social and ritual contexts. *Ceremonial Medicine* Leaves smoked as a general purgative in social and ritual contexts. *Ear Medicine* Plant smoke blown into the ear for earaches. *Emetic* Fresh leaves chewed as an emetic. (as *N. bigelovii* 21:14) **Karok** *Unspecified* Used for medicine. (as *N. bigelovi* 117:209) **Kawaiisu** *Analgesic* Chewed plant put in the nostril for headaches. Poultice of plant applied to the chest for internal pains. *Dermatological Aid* Poultice of plant applied to itchy bites. Poultice of plant applied to bleeding cuts. *Ear Medicine* Chewed plant put in the ear for earaches. *Emetic* Plant eaten to cause vomiting. Plant induced vomiting. *Gastrointestinal Aid* Plant eaten to clean out the stomach. *Gynecological Aid* Poultice of plant applied to woman's stomach during parturition. *Hallucinogen* Plant eaten to cause dreams. *Hemostat* Poultice of plant applied to bleeding cuts. *Other* Plant used for infanticide and suicide. *Poison* Plant considered poisonous. *Psychological Aid* Plant blown in the air to prevent bad dreams. *Respiratory Aid* Plant used as snuff for stuffy noses. *Sedative* Plant used as a soporific. *Stimulant* Plant used for fatigue. *Toothache Remedy* Plant held between the teeth for toothaches. (as *N. bigelovii* 206:43)

- *Other*—**Karok** *Smoke Plant* Leaves dried, powdered, and smoked. (as *N. bigelovii* var. *exaltata* 148:389) **Kawaiisu** *Protection* Leaves and lime placed in the camp fire to prevent supernatural beings from bothering you. *Smoke Plant* Leaves dried, pulverized, powdered, mixed with liquid, formed into cakes, and smoked in hollow stems. (as *N. bigelovii* 206:43) **Mendocino Indian** *Smoke Plant* Light green leaves used for smoking. (as *N. bigelovii* 41:386) **Mewuk** *Smoke Plant* Leaves dried, pounded, and used for smoking. (as *N. bigelovi* 117:356) **Midoo** *Smoke Plant* Leaves dried, pounded, and used for smoking. (as *N. bigelovi* 117:319) **Pomo, Kashaya** *Smoke Plant* Dried leaves used as tobacco. (as *N. bigelovii* 72:115) **Tolowa** *Smoke Plant* Leaves smoked for leisure. (as *N. bigelovii* 6:41)

Nicotiana rustica L., Aztec Tobacco

- *Drug*—**Cherokee** *Analgesic* Used for cramps and sharp pains. (80:59) Chewed plant used for headaches. (177:56) *Anthelmintic* Used as an anthelmintic. *Anticonvulsive* Used as an antispasmodic. *Cathartic* Used as a cathartic. *Ceremonial Medicine* Used extensively in rituals. *Dermatological Aid* Poultice of beaten plant used for boils and applied to insect bites. *Diaphoretic* Used as a sudorific. *Diuretic* Used as a diuretic. *Emetic* Used as an emetic. *Expectorant* Used as an expectorant. *Gastrointestinal Aid* Taken for colic. *Kidney Aid* Taken for dropsy. *Misc. Disease Remedy* Decoction of leaf used for ague, "locked-jaw," and "black-yellow disease." *Other* Compound used for apoplexy, dizziness, and fainting. (80:59) Decoction of leaves rubbed in scratches made on the patient for apoplexy. (177:56) *Snakebite Remedy* Juice applied to snakebite. *Toothache Remedy* Smoke blown on toothache. *Vertigo Medicine* Used for dizziness and fainting. (80:59) **Iroquois** *Antidote* Compound decoction of plants used as wash to counteract poison. *Dermatological Aid* Poultice of chewed plant applied to all insect bites. *Psychological Aid* Decoction of plants taken for insanity caused by masturbation. *Tuberculosis Remedy* Plant used for consumption. (87:430)

- *Other*—**Apalachee** *Cash Crop* Plant cultivated and sold. *Smoke Plant* Plant smoked in the pre-ballgame rituals. (81:97) **Cherokee** *Smoke Plant* Leaves used for smoking. (80:59) **Iroquois** *Ceremonial Items* Leaves used for ceremonial purposes. (196:4) *Smoke Plant* Smashed roots used for smoking. (141:57) Leaves used to smoke and if the smoke blew in a streak to one side, it would rain in 24 hours. (196:30) Leaves used for smoking purposes. (196:4) **Pima** *Smoke Plant* Used for smoking. (36:111)

Nicotiana sp., Tobacco

- *Drug*—**Cheyenne** *Dermatological Aid* Poultice of wet leaves applied to sores. (as *Nicotina* 83:39) **Eskimo, Inuktitut** *Veterinary Aid* Used medicinally for dogs. (202:184) **Eskimo, Kuskokwagmiut** *Veterinary Aid* Leaves considered to be "a good cure for a dog in a weakened condition." (122:32) **Kwakiutl** *Dermatological Aid* Poultice of chewed leaves applied or compound rubbed on sores and swellings. (183:266) Poultice of leaves applied to sores, lumps, and cuts. *Herbal Steam* Leaves used in steam baths. (183:292) **Kwakiutl, Southern** *Analgesic* Poultice of dried plant and fresh sea wrack, alder, and black twinberry applied for aches and pains. (183:260) **Navajo** *Ceremonial Medicine* Infusion of leaves given to the patient in a painted turtle shell during the Raven Chant. *Dermatological Aid* Plant used for sores caused by the handling or burning a raven's nest. (55:74)

- *Other*—**Eskimo, Inuktitut** *Smoke Plant* Plant chewed rather than smoked. (202:184) **Luiseño** *Smoke Plant* Plant formerly used as tobacco. (155:229) **Yuki** *Smoke Plant* Leaves crushed, stuffed into pipes, and smoked. (49:91)

Nicotiana tabacum L., Cultivated Tobacco

- *Drug*—**Cherokee** *Analgesic* Used for cramps and sharp pains. *Anthelmintic* Used as an anthelmintic. *Anticonvulsive* Used as an antispasmodic. *Cathartic* Used as a cathartic. *Ceremonial Medicine* Used extensively in rituals. *Dermatological Aid* Poultice of beaten plant used for boils and applied to insect bites. *Diaphoretic* Used as a sudorific. *Diuretic* Used as a diuretic. *Emetic* Used as an emetic. *Expectorant* Used as an expectorant. *Gastrointestinal Aid* Taken for colic. *Kidney Aid* Taken for dropsy. *Misc. Disease Remedy* Decoction of leaf used for ague, "locked-jaw," and "black-yellow disease." *Other* Compound used for apoplexy, dizziness, and fainting. *Snakebite Remedy* Juice applied to snakebite. *Toothache Remedy* Smoke blown on toothache. *Vertigo Medicine* Used for dizziness and fainting. (80:59) **Haisla & Hanaksiala** *Antirheumatic (External)* Juice rubbed on the hands and feet for rheumatism. (43:291) **Hawaiian** *Dermatological Aid* Plant used for removing the pus from scrofulous sores or boils. Plant used for sores and the smoke used for cuts. *Tuberculosis Remedy* Plant used for removing the pus from scrofulous sores. (2:73) **Hesquiat** *Dermatological Aid* Chewed leaves used as a poultice or rubbed on bruises and cuts. (185:76) **Micmac** *Ear Medicine* Leaves used for earache. *Hemostat* Leaves used for bleeding. (40:58) **Mohegan** *Ear Medicine* Smoke blown into the ear for an earache. **Montauk** *Toothache Remedy* Tobacco placed in tooth for toothache. **Rappahannock** *Ear Medicine* Smoke blown into the ear for an earache. *Toothache Remedy* Tobacco placed in tooth for toothache. **Shinnecock** *Ear Medicine* Smoke blown into the ear for an earache. *Toothache Remedy* Tobacco placed in tooth for toothache. (30:120) **Thompson** *Dermatological Aid* Poultice of plant used on cuts and sores. (187:288)
- *Other*—**Cherokee** *Smoke Plant* Leaves used for smoking. (80:59) **Haisla & Hanaksiala** *Smoke Plant* Leaves used for chewing and smoking. (43:291) **Hesquiat** *Smoke Plant* Leaves mixed with kinnikinnick and smoked. *Snuff* Leaves mixed with kinnikinnick and chewed. (185:76) **Iroquois** *Smoke Plant* Smashed roots used for smoking. (141:57) **Navajo** *Sacred Items* Sacred plant depicted with beans, corn, and squash in the first sacred painting of the Mountain Chant. (55:75) **Oweekeno** *Smoke Plant* Leaves chewed and smoked. (43:118) **Papago** *Smoke Plant* Leaves half or fully dried and smoked. (34:36) Used for smoking. **Pima** *Smoke Plant* Used for smoking. (36:108) Used for smoking. (36:110) **Thompson** *Ceremonial Items* Plant used as an offering in the sweat house. (187:288) **Tsimshian** *Smoke Plant* Leaves used for chewing. (43:350)

Nicotiana trigonophylla Dunal, Desert Tobacco

- *Drug*—**Cahuilla** *Dermatological Aid* Poultice of leaves applied to cuts, bruises, swellings, and other wounds. *Ear Medicine* Leaf smoke blown into the ear and covered with a warm pad for earaches. *Emetic* Infusion of leaves used as an emetic. *Hunting Medicine* Leaves smoked as part of a hunting ritual. (15:90)
- *Food*—**Cahuilla** *Beverage* Leaves chewed, smoked, or used in a drinkable decoction. (15:90)
- *Other*—**Cahuilla** *Ceremonial Items* Used as an integral part of every ritual.[47] Used by shamans to control rain, increase crop production, divining, and improve health of community. Used by shamans, at community gatherings, to drive away malevolent powers. *Protection* Leaves smoked by travelers to clear away all danger and ensure blessing from spiritual guides. *Smoke Plant* Leaves chewed, smoked, or used in a drinkable decoction. (15:90) **Havasupai** *Smoke Plant* Leaves used for smoking. (162:105) Leaves smoked for pleasure. (197:240) **Hopi** *Ceremonial Items* Plant smoked in pipes for ceremonial purposes only. (200:90) *Smoke Plant* Used for smoking. (36:109) **Hualapai** *Smoke Plant* Used to smoke in ceremonials. (195:54) **Mohave** *Smoke Plant* Wild tobacco smoked. (37:120) **Papago** *Smoke Plant* Leaves dried and smoked. (34:27) Leaves half or fully dried and smoked. (34:36) Used for smoking. **Pima** *Smoke Plant* Used for smoking. (36:108) **Yuma** *Smoke Plant* Wild tobacco smoked. (37:120)

Nicotiana trigonophylla var. *palmeri* (Gray) M. E. Jones, Palmer's Tobacco

- *Other*—**Navajo** *Ceremonial Items* Used for filling ceremonial prayer sticks. (as *N. palmeri* 55:75)

Nolina, Agavaceae

Nolina bigelovii (Torr.) S. Wats., Bigelow's Nolina

- *Food*—**Cahuilla** *Unspecified* Stalk baked in a rock-lined roasting pit and eaten. (15:94)
- *Fiber*—**Papago** *Basketry* Bleached or green grass used for basketry. (118:9)

Nolina bigelovii var. *parryi* (S. Wats.) L. Benson, Parry's Nolina

- *Food*—**Hualapai** *Fruit* Fruit used for food. (as *N. parryi* 195:26) **Tubatulabal** *Unspecified* Stalks used for food. (as *N. parryi* 193:15)
- *Fiber*—**Hualapai** *Basketry* Used to make coil baskets. (as *N. parryi* 195:26)

Nolina erumpens (Torr.) S. Wats., Foothill Beargrass

- *Fiber*—**Papago** *Basketry* Leaves sun dried, split into strands, and used as foundation in coiled basketry. (34:57) Used as the warp element of baskets. (34:59)

Nolina microcarpa S. Wats., Sacahuista

- *Drug*—**Isleta** *Antirheumatic (Internal)* Decoction of root taken for rheumatism. *Pulmonary Aid* Decoction of root taken for pneumonia and lung hemorrhages. (100:35)
- *Food*—**Apache, Chiricahua & Mescalero** *Unspecified* Stalks roasted, boiled, or eaten raw. *Vegetable* Stalks boiled, dried, and stored to be used as vegetables. (33:38) **Apache, Western** *Unspecified* Young stalks

placed in fire, peeled, and eaten. (26:183) **Isleta** *Bread & Cake* Seeds made into a meal and used to make bread. (32:22) *Fruit* Fruit eaten fresh. (100:35) *Porridge* Seeds made into a meal and used to make mush. (32:22) *Preserves* Fruit eaten preserved. *Staple* Seeds used to make flour. (100:35)

- **Fiber**—**Apache, Mescalero** *Mats, Rugs & Bedding* Grass used as tepee ground covering. (14:51) **Apache, Western** *Building Material* Grass used as a thatching material for wickiup or ramada. (26:183) **Havasupai** *Building Material* Leaves used for thatch. *Mats, Rugs & Bedding* Leaves woven into a coarse mat and used for drying mescal. (197:212) **Isleta** *Basketry* Leaf fibers formerly used in basketry. *Brushes & Brooms* Leaf fibers used to make brushes. *Cordage* Leaf fibers used to make cords, ropes, and whips. (100:35) **Jemez** *Basketry* Leaves used to make baskets for storage and washing of grains. (17:34) Leaves woven together into baskets. (44:25) **Keres, Western** *Basketry* Plant used to make baskets. *Mats, Rugs & Bedding* Plant used to make mats. (171:55) **Keresan** *Basketry* Used to make baskets. (198:559) **Papago** *Basketry* Grass used as the foundation in coiled basketry. (17:34) Leaves dried, split, and made into baskets. **Pima** *Basketry* Leaves used to fashion coils for storage baskets. (17:62) **Southwest Indians** *Basketry* Beargrass used to make basketry. Made into baskets and used for storage containers. Leaves used as foundation element in coiled basketry. *Cordage* Leaves used as tying material. *Mats, Rugs & Bedding* Used to make matting to cover the dead. (17:61) **Yavapai** *Building Material* Leaves used to thatch dwellings. (17:62)
- **Dye**—**Navajo** *Unspecified* Plant used to make a dye for blankets. (100:35)
- **Other**—**Apache, Chiricahua & Mescalero** *Containers* Moist grass laid onto hot stones to prevent steam from escaping. (33:36) **Apache, Mescalero** *Cooking Tools* Grass woven into trays and used for processing datil and mescal. (14:51) **Apache, Western** *Cooking Tools* Dried leaves fashioned into spoons. *Protection* Grass used as wrapping material for foods to be transported or stored. (26:183) *Soap* Roots used as soap. (26:182) **Havasupai** *Cooking Tools* Lower stalks split open to form an alternative base for drying mescal. (197:212) **Isleta** *Ceremonial Items* Seeds used in dried gourd shells to make ceremonial rattles. *Stable Gear* Leaf fibers used to make brushes, cords, ropes, and whips. (100:35)

Nolina sp., Beargrass

- **Fiber**—**Hualapai** *Building Material* Used for tying and thatching houses. (195:51) **Southwest Indians** *Basketry* Leaves used for making coarse forms of basketry. (17:60) *Mats, Rugs & Bedding* Grass made into mats and used in cradles. (17:48)
- **Other**—**Apache, Western** *Soap* Crowns and bases of leaves pounded together and mixed with water to make soap. (26:183) **Hualapai** *Hunting & Fishing Item* Used to make snares. (195:51)

Nothochelone, Scrophulariaceae

Nothochelone nemorosa (Dougl. ex Lindl.) Straw, Woodland Beardtongue

- **Drug**—**Paiute** *Dermatological Aid* Green, mashed plant juice applied to sores "like iodine." (as *Penstemon nemorosus* 111:109)

Nuphar, Nymphaeaceae

Nuphar lutea (L.) Sm., Yellow Pondlily

- **Food**—**Lakota** *Unspecified* Roots boiled and eaten. (as *N. luteum* 139:52)

Nuphar lutea ssp. *advena* (Ait.) Kartesz & Gandhi, Yellow Pondlily

- **Drug**—**Iroquois** *Analgesic* Compound decoction taken for pain between shoulder blades. *Anticonvulsive* Compound decoction with plant taken by men with epilepsy. *Blood Medicine* Compound infusion of dried roots taken for blood disease. *Febrifuge* Compound decoction with roots taken for recurring chills followed by fever. (as *N. luteum* ssp. *macrophyllum* 87:319) *Gastrointestinal Aid* Poultice of roots or decoction taken and used as wash for swollen abdomen. (as *N. luteum* ssp. *macrophyllum* 87:318) *Heart Medicine* Infusion of dried, grated plant taken for heart trouble. (as *N. luteum* ssp. *macrophyllum* 87:319) *Misc. Disease Remedy* Infusion of roots taken to dry up smallpox. (as *N. luteum* ssp. *macrophyllum* 87:318) *Other* Cold infusion of roots used as a "ghost medicine." *Pulmonary Aid* Compound decoction taken for swollen lungs. *Witchcraft Medicine* "Hung up inside to keep witches away" as an anti-witch remedy. Compound used to "detect bewitchment." Poultice of roots applied to sore areas caused by witchcraft diseases. (as *N. luteum* ssp. *macrophyllum* 87:319) **Menominee** *Dermatological Aid* Poultice of dried, powdered root applied to cuts and swellings. (as *Nymphaea advena* 151:42, 43) **Micmac** *Dermatological Aid* Poultice of bruised root with flour or meal applied to swellings and bruises. (as *Nuphar advena* 156:317) *Orthopedic Aid* Leaves used for limb swellings. (as *Nymphaea advena* 40:58) **Ojibwa** *Dermatological Aid* Poultice of grated root applied to sores and powdered root used for cuts and swellings. (as *Nymphaea advena* 153:376) **Penobscot** *Dermatological Aid* Poultice of mashed leaves applied to swollen limbs. (as *Nymphaea advena* 156:310) **Potawatomi** *Dermatological Aid* Poultice of pounded root applied for "many inflammatory diseases." (as *Nymphaea advena* 154:65) **Rappahannock** *Dermatological Aid* Poultice of parched and bruised leaf used to remove fever and inflammation from sores. Warmed leaves applied to boils. *Febrifuge* Poultice of parched and bruised leaf used to remove fever and inflammation from sores. (as *Nymphaea advena* 161:32) **Sioux** *Hemostat* Dry, porous rhizomes ground fine and applied to wounds as a styptic. (as *Nuphar advena* 19:17) **Thompson** *Analgesic* Cold decoction of stems or roots taken for internal pains. *Dermatological Aid* Poultice of fresh or dried leaves applied to wounds, cuts, or sores. (as *Nymphaea advena* 164:460)
- **Food**—**Comanche** *Unspecified* Boiled roots used for food. (as *Nymphaea advena* 29:523) **Menominee** *Vegetable* Rhizomes cooked in the same manner as rutabagas. (as *Nymphaea advena* 151:69) **Montana Indian** *Snack Food* Parched seeds eaten like popcorn. *Soup* Seeds ground into meal used for thickening soups. *Unspecified* Thick, fleshy rhizomes boiled with fowl or other meat. Mucilaginous seedpods were well flavored and nutritious. (as *Nuphar advena* 19:17) **Pawnee** *Unspecified* Cooked seeds used for food. (as *Nymphaea advena* 70:79)

Nuphar lutea ssp. *polysepala* (Engelm.) E. O. Beal, Rocky Mountain Pondlily

- **Drug**—**Bella Coola** *Analgesic* Decoction of root taken for pain in any part of the body. *Antirheumatic (Internal)* Decoction of root taken for rheumatic pain. *Blood Medicine* Decoction of root considered "good for the blood." *Heart Medicine* Decoction of root taken for heart disease pain. *Tuberculosis Remedy* Decoction of root taken for consumption pain. (as *Nymphaea polysepala* 150:56) Rhizomes used for tuberculosis. (as *Nuphar polysepalum* 184:206) *Venereal Aid* Decoction of root taken for gonorrheal pain. **Gitksan** *Antihemorrhagic* Infusion of toasted root scrapings taken for lung hemorrhages. *Gynecological Aid* Infusion of toasted root or decoction of root heart used as a contraceptive. *Pulmonary Aid* Decoction of root taken for lung hemorrhage. (as *Nymphaea polysepala* 150:56) **Haisla & Hanaksiala** *Tuberculosis Remedy* Decoction of rhizomes taken for tuberculosis. *Unspecified* Decoction of rhizomes taken as medicine. (43:256) **Hesquiat** *Unspecified* Pond lily was a good medicine. (as *Nuphar polysepalum* 185:70) **Kitasoo** *Gynecological Aid* Decoction of plant and devil's club used for unspecified woman's illness. (43:339) **Kwakiutl** *Analgesic* Poultice of heated leaves applied or rhizome extract taken for chest pains. *Orthopedic Aid* Rhizomes used as a medicine for internal swellings or sickness in the bones. *Respiratory Aid* Rhizome extract taken for asthma and chest pains. (as *N. polysepalum* 183:287) **Nitinaht** *Other* Large rhizomes placed in hot water and liquid taken to prevent sickness during epidemics. (as *N. polysepalum* 186:114) *Unspecified* Decoction of rhizomes used as a medicinal drink. (as *N. polysepalum* 67:251) Rhizomes used as a medicine. (as *N. polysepalum* 185:70) **Okanagan-Colville** *Poison* Roots considered poisonous. *Toothache Remedy* Stems placed directly on the tooth for toothaches. (as *N. polysepalum* 188:110) **Quinault** *Analgesic* Poultice of heated roots applied for pain. *Antirheumatic (External)* Poultice of heated roots applied for rheumatism. (as *Nymphozanthus polysepalus* 79:29) **Shuswap** *Analgesic* Infusion of mashed roots applied for sore back pain. *Antirheumatic (External)* Infusion of mashed roots applied for rheumatism. *Dermatological Aid* Infusion of mashed roots taken for sores. *Orthopedic Aid* Infusion of mashed roots applied for sore back pain. (as *Nuphar polysepalum* 123:64) **Tanana, Upper** *Analgesic* Poultice of sliced, warmed rhizomes applied for pain. (as *N. polysepalum* 102:17) **Thompson** *Antirheumatic (External)* Powdered, dried leaves mixed with bear grease and used as an ointment for swellings. *Dermatological Aid* Powdered, dried leaves mixed with bear grease and used as an ointment for bites and infections. *Gastrointestinal Aid* Large rhizomes chewed for ulcers. (187:235)

- **Food**—**Alaska Native** *Vegetable* Rootstocks boiled or roasted and eaten as a vegetable. (as *Nuphar polysepalum* 85:145) **Cheyenne** *Unspecified* Roots eaten raw or boiled. (as *Nymphea polysepala* 76:173) Species used for food. (as *Nuphar polysepala* 83:45) **Klamath** *Bread & Cake* Ground seeds used for bread. *Dried Food* Seeds stored for later use. *Porridge* Ground seeds used for porridge. (as *Nymphaea polysepala* 45:96) *Special Food* Used as a delicacy. *Staple* Used as a staple food in primitive times. (as *Nymphaea polysepala* 46:728) Dried, roasted seeds used as cereal. (as *Nuphar polysepalum* 118:29) *Unspecified* Roasted seeds, tasted like popcorn, used for food. (as *Nymphaea polysepala* 45:96) **Mendocino Indian** *Forage* Fleshy roots eaten as a favorite food by deer. *Unspecified* Seeds used for food. (as *Nymphaea polysepala* 41:347) **Thompson** *Dried Food* Rhizomes sliced and dried like apples. (187:235) **Tolowa** *Unspecified* Seeds used for food. (as *Nuphar polysepalum* 6:41)

Nuphar lutea* ssp. *variegata (Dur.) E. O. Beal, Variegated Yellow Pondlily

- **Drug**—**Algonquin, Quebec** *Dermatological Aid* Poultice of mashed rhizomes applied to swellings. *Disinfectant* Poultice of mashed rhizomes applied to infections. (as *N. variegatum* 18:163) **Cree, Woodlands** *Analgesic* Poultice of grated rhizome, calamus, water or grease, and sometimes cow parsnip applied for headaches. *Antirheumatic (External)* Poultice of grated rhizomes and other ingredients applied to sore joints, swellings, and painful limbs. Poultice consisted of grated rhizomes, grated calamus rootstocks, and occasionally cow parsnip with the addition of water or grease. *Dermatological Aid* Poultice of roots with calamus and cow parsnip roots applied to mancos, worms in the flesh. Poultice of fresh or rehydrated, dried rhizome slice applied to infected skin lesions. Poultice of sliced, dried roots soaked in water and applied to infected wounds. *Panacea* Powdered rhizomes added to a many herb remedy for various ailments. (as *N. variegatum* 109:46) **Flathead** *Antirheumatic (External)* Decoction of rootstocks added to bath water for rheumatism. *Dermatological Aid* Poultice of baked rootstocks used for sores. *Venereal Aid* Infusion of rootstocks taken for venereal disease. *Veterinary Aid* Poultice of rootstocks used for horses with cuts and bruises. (as *N. variegatum* 82:33) **Iroquois** *Blood Medicine* and *Pediatric Aid* Infusion of rhizomes and two other plants taken by adolescents for poor blood circulation. *Veterinary Aid* Infusion of plant, other plant fragments, and milk given to pigs that drool and have sudden movements. (as *N. variegatum* 141:43) **Kutenai** *Dermatological Aid* Poultice of baked rootstocks used for sores. **Sioux** *Hemostat* Rootstocks powdered and used as a styptic for wounds. (as *N. variegatum* 82:33)
- **Food**—**Algonquin, Tête-de-Boule** *Beverage* Petiole sucked to relieve thirst. *Unspecified* Grains used for food. (as *N. variegatum* 132:129) **Cree, Woodlands** *Dried Food* Sliced roots dried for food and eaten dried or cooked. (as *N. variegatum* 109:46) **Montana Indian** *Cooking Agent* Thin slices of rootstocks dried, ground, or pulverized into meal or gruel, and used to thicken soups. *Porridge* Seeds parched, ground into meal, and used for mush or gruel. *Unspecified* Root stocks eaten raw or boiled with meat. (as *N. variegatum* 82:33)

***Nuphar* sp.**, Yellow Pondlily

- **Drug**—**Abnaki** *Love Medicine* Used as an anaphrodisiac. (144:154) *Psychological Aid* Infusion of roots taken by men to inhibit sexual drives for 2 months. (144:167) **Iroquois** *Analgesic* Compound roots used for "pains in the chest which make you hold your breath." (87:320)

Nyctaginaceae, see *Abronia, Allionia, Boerhavia, Mirabilis, Tripterocalyx*

Nymphaeaceae, see *Nuphar, Nymphaea*

Nymphaea, Nymphaeaceae

Nymphaea odorata Ait., American White Waterlily

- **Drug**—**Chippewa** *Oral Aid* Dried, pulverized root put in the mouth for sores. (as *Castalia odorata* 53:342) **Micmac** *Cold Remedy* Leaves used for colds. (40:58) *Cough Medicine* Juice of root taken for coughs. *Dermatological Aid* Poultice of boiled root applied to swellings. (156:317) *Gland Medicine* Roots used for suppurating glands and leaves used for colds. *Misc. Disease Remedy* Leaves used for grippe. *Orthopedic Aid* Leaves used for limb swellings and colds. (40:58) **Ojibwa** *Cough Medicine* Root used as a cough medicine for tuberculosis.

Tuberculosis Remedy Root used as a cough medicine for tuberculosis. (as *Castalia odorata* 153:376) **Okanagan-Colville** *Poison* Roots considered poisonous. *Toothache Remedy* Stems placed directly on the tooth for toothaches. (188:110) **Penobscot** *Dermatological Aid* Compound poultice of mashed leaves applied to limb swellings. (as *Castalia odorata* 156:310) **Potawatomi** *Unspecified* Poultice of pounded root used for unspecified ailments. (as *Castalia odorata* 154:65)
- *Food*—**Ojibwa** *Unspecified* Buds eaten before opening. (as *Castalia odorata* 153:407)

Nymphaea sp., Waterlily
- *Drug*—**Abnaki** *Psychological Aid* Infusion of roots taken by men to inhibit sexual drives for 2 months. (144:167) **Seminole** *Cough Medicine, Respiratory Aid,* and *Sedative* Plant used for turtle sickness: trembling, short breath, and cough. (as *Castalia,* Pond Lily 169:237)
- *Food*—**Kiowa** *Vegetable* Tubers stewed or prepared like potatoes. (192:27) **Klamath** *Unspecified* Seeds used for food. (117:224) **Lakota** *Unspecified* Roots boiled and eaten. (139:52)

Nymphoides, Menyanthaceae
Nymphoides cordata (Ell.) Fern., Little Floatingheart
- *Drug*—**Seminole** *Cough Medicine* Plant used for turtle sickness: trembling, short breath, and cough. (as *N. lacunosum* 169:237) *Other* Complex infusion of roots taken for chronic conditions. (as *N. lacunosum* 169:272) *Respiratory Aid* and *Sedative* Plant used for turtle sickness: trembling, short breath, and cough. (as *N. lacunosum* 169:237)

Nyssa, Cornaceae
Nyssa aquatica L., Water Tupelo
- *Dye*—**Choctaw** *Red* Burned bark and red oak ash added to water and used as a red dye. (25:14)

Nyssa sp., Black Gum
- *Drug*—**Creek** *Tuberculosis Remedy* Decoction of bark and chips taken and used as a wash for tuberculosis. (172:659)

Nyssa sylvatica Marsh., Black Gum
- *Drug*—**Cherokee** *Anthelmintic* Compound given for worms. (80:26) Infusion of bark used as a bath and given to children with worms. (177:47) *Antidiarrheal* Compound decoction given for diarrhea. *Emetic* Inner bark used as part of "drink to vomit bile." (80:26) Decoction of inner bark taken to cause vomiting when unable to retain food. (177:47) *Eye Medicine* Strong ooze from root dripped into eyes. (80:26) *Gastrointestinal Aid* Decoction of inner bark taken to cause vomiting when unable to retain food. (177:47) *Gynecological Aid* Infusion given for childbirth and infusion of bark given for "flooding." *Other* Compound infusion of bark used for "bad disease." (80:26) *Pediatric Aid* Infusion of bark used as a bath and given to children with worms. (177:47) *Urinary Aid* Used as ingredient in drink for "milky urine." (80:26) **Creek** *Tuberculosis Remedy* Decoction of bark used as a bath and taken for pulmonary tuberculosis. (177:47) **Houma** *Anthelmintic* Decoction of root or bark taken for worms. (158:55) **Koasati** *Dermatological Aid* Decoction of bark taken and applied to gun wounds. (177:47)
- *Other*—**Chippewa** *Tools* Wood used to make awl handles, mauls, and war clubs because it would not split or check. (71:138)

Obolaria, Gentianaceae
Obolaria virginica L., Virginia Pennywort
- *Drug*—**Cherokee** *Cold Remedy* Taken for colds. *Cough Medicine* Taken for coughs. *Diaphoretic* Taken as a diaphoretic. *Gastrointestinal Aid* Taken for colic. (80:48) **Choctaw** *Dermatological Aid* Simple or compound decoction of root used as a wash or dressing for cuts and bruises. (25:23) Decoction of root used as bath or poultice of root applied to cuts and bruises. (177:51)

Ochrosia, Apocynaceae
Ochrosia compta K. Schum., Ho-le-i
- *Drug*—**Hawaiian** *Herbal Steam* Infusion of bark and leaves used for steam in the sweat bath. *Pediatric Aid* and *Strengthener* Nuts and other plants chewed and fed to infants for general debility. (as *Ochorosia sandwicensis* 2:44)
- *Food*—**Hawaiian** *Unspecified* Nuts used for food. (as *Ochorosia sandwicensis* 2:44)

Ocotea, Lauraceae
Ocotea coriacea (Sw.) Britt., Lance Wood
- *Other*—**Seminole** *Hunting & Fishing Item* Plant used to make bows. (169:469)

Octoblephorum, Leucobryaceae, moss
Octoblephorum albidum Hedw., Moss
- *Drug*—**Seminole** *Antirheumatic (External)* and *Febrifuge* Plant used for fire sickness: fever and body aches. (169:203)

Odontoglossum, Orchidaceae
Odontoglossum chinensis, Pa-la-a
- *Drug*—**Hawaiian** *Laxative* Infusion of plant used for softening the bowels and constipation. (as *Odonto-glossum chineusis* 2:73)

Oemleria, Rosaceae
Oemleria cerasiformis (Torr. & Gray ex Hook. & Arn.) Landon, Indian Plum
- *Drug*—**Kwakiutl** *Analgesic* and *Dermatological Aid* Poultice of chewed, burned plant and oil applied to sore places. (as *Osmaronia cerasiformis* 183:289) **Makah** *Laxative* Bark used as a mild laxative. *Tuberculosis Remedy* Decoction of bark taken for tuberculosis. *Unspecified* Bark used as a healing agent. (67:264)
- *Food*—**Cowlitz** *Dried Food* Berries dried and eaten in the winter. *Fruit* Berries eaten fresh. (as *Osmaronia cerasiformis* 79:37) **Karok** *Forage* Berries eaten by ground squirrels. (as *Osmaronia cerasiformis* 148:384) **Kitasoo** *Fruit* Fruit used for food. (43:343) **Kwakiutl, Southern** *Fruit* Fruits eaten fresh with oil at family meals or large feasts. *Special Food* Fruits eaten fresh with oil at large feasts. (as *Osmaronia cerasiformis* 183:289) **Lummi** *Fruit* Berries eaten fresh. (as *Osmaronia cerasiformis* 79:37) **Nitinaht** *Fruit* Fruits formerly cooked and used for food. (as *Osmaronia cerasiformis* 186:118) **Quinault** *Fruit* Berries eaten fresh. (as *Osmaronia cerasiformis* 79:37) **Saanich** *Fruit* Berries eaten ripe. (as *Osmaronia cerasiformis* 182:86) **Samish** *Fruit* Berries eaten fresh. (as *Osmaronia cerasiformis* 79:37) **Shasta** *Fruit*

Catalog of Plants 361

Berries eaten raw with wild currants. (as *Osmaronia cerasiformis* 93:308) **Skagit** *Fruit* Berries eaten fresh. (as *Osmaronia cerasiformis* 79:37) **Skagit, Upper** *Fruit* Fruit eaten fresh. (as *Osmaronia cerasiformis* 179:38) **Snohomish** *Fruit* Berries eaten fresh. **Squaxin** *Fruit* Berries eaten. **Swinomish** *Fruit* Berries eaten fresh. (as *Osmaronia cerasiformis* 79:37) **Thompson** *Bread & Cake* Smashed fruit made into bread. *Fruit* Fruit eaten fresh. It was cautioned that if too much fruit was eaten, one would get "bleeding lungs." (187:262) **Tolowa** *Fruit* Fruit used for food. This was called the "wood that lies" because it was the first to bloom in the spring and the last to set fruit. (6:41)
- *Other*—**Makah** *Fasteners* Inner bark strips used to bind harpoons. (67:264)

Oenanthe, Apiaceae
Oenanthe sarmentosa K. Presl ex DC., Water Parsley
- *Drug*—**Haisla & Hanaksiala** *Poison* Plant considered highly toxic. (43:216) **Kitasoo** *Cathartic* Roots used as a purgative. *Emetic* Roots used as an emetic. (43:326) **Kwakiutl** *Emetic* Seeds and roots used as an emetic. (183:277) **Kwakwaka'wakw** *Emetic* Plant used as an emetic. (43:216) Roots used as an emetic. (43:326) **Makah** *Laxative* Pounded roots used as a laxative. (79:42) **Nitinaht** *Gynecological Aid* Roots squashed and swallowed to facilitate and speed up delivery. (186:93) **Nuxalkmc** *Emetic* Plant used as an emetic. (43:216) **Tsimshian** *Ceremonial Medicine* Roots eaten as an emetic to seek supernatural powers and purify. (43:326)
- *Food*—**Costanoan** *Unspecified* Raw or cooked stems used for food. (21:251) **Cowlitz** *Unspecified* Young, tender stems used for food. (79:42) **Hesquiat** *Unspecified* Stems formerly eaten. (185:61) **Skokomish** *Unspecified* Young, tender stems used for food. **Snuqualmie** *Unspecified* Young, tender stems used for food. (79:42)
- *Other*—**Makah** *Toys & Games* Stalks cut and used as whistles by children. **Quileute** *Toys & Games* Stalks cut and used as whistles by children. (79:42)

Oenanthe sp., Twiggy Water Dropwort
- *Food*—**Miwok** *Unspecified* Raw stems used for food. (12:160)

Oenothera, Onagraceae
Oenothera albicaulis Pursh, Whitest Eveningprimrose
- *Drug*—**Hopi** *Ceremonial Medicine* Used to ward out the cold through prayer. (42:336) **Keres, Western** *Antirheumatic (External)* Poultice of plant used for swellings. (as *Anogra albicaulis* 171:27) **Navajo, Ramah** *Ceremonial Medicine* Dried flowers used as ceremonial medicine. (as *O. ctenophylla* 191:37, 38) *Orthopedic Aid* Decoction of root taken and used as a lotion for strain from carrying heavy load. *Panacea* Decoction of root taken and used as a lotion for muscle strain, a "life medicine." (191:37) Root used as a "life medicine." (as *O. ctenophylla* 191:37, 38) *Throat Aid* Compound poultice of plant applied for "throat trouble." (191:37)
- *Food*—**Apache** *Fruit* Fruits eaten for food. (as *Anogra albicaulis* 32:17) **Apache, Chiricahua & Mescalero** *Sauce & Relish* Seeds ground and made into a gravy. *Soup* Seeds boiled in soups. *Special Food* Fruit chewed as a delicacy without preparation. (as *Anogra albicaulis* 33:45)
- *Other*—**Hopi** *Ceremonial Items* Flower used ceremonially as the "white flower." (42:336) *Decorations* Flowers used by marriageable maids in their hair on holidays. (as *O. pinnatifida* 56:16) *Smoke Plant* Plant used for tobacco. (42:336) **Zuni** *Ceremonial Items* Chewed blossoms rubbed on the bodies of young girls so that they could dance well and ensure rain. The blossoms were given by the High Priest and the Sun Priest of the Corn Maidens. The girls chewed the blossoms, ejected the mass into their hands, and rubbed it on the neck, breast, arms, and hands, ensuring that they would dance well so that it would rain and the corn would grow. (as *Anogra albicaulis* 166:87)

Oenothera biennis L., Common Eveningprimrose
- *Drug*—**Cherokee** *Dietary Aid* Infusion taken for "overfatness." *Hemorrhoid Remedy* Hot root poultice used for piles. (80:33) **Iroquois** *Dermatological Aid* Compound used for boils. *Hemorrhoid Remedy* Compound decoction of roots taken and used as a wash for piles. *Stimulant* Compound used for laziness. *Strengthener* Chewed roots rubbed on arms and muscles to provide athletes great strength. (87:390) **Ojibwa** *Dermatological Aid* Poultice of soaked, whole plant applied to bruises. (153:376) **Potawatomi** *Unspecified* Tiny seeds used as a valuable medicine for unspecified ailment. (154:66, 67)
- *Food*—**Cherokee** *Vegetable* Leaves cooked and eaten as greens. (80:33) Roots boiled like potatoes. (126:49) **Gosiute** *Unspecified* Seeds used for food. (39:375)
- *Other*—**Lakota** *Incense & Fragrance* Seeds aromatic. (139:52)

Oenothera brachycarpa Gray, Shortfruit Eveningprimrose
- *Drug*—**Navajo, Kayenta** *Dermatological Aid* Plant used as a lotion for sores. (205:33)

Oenothera cespitosa Nutt., Tufted Eveningprimrose
- *Drug*—**Blackfoot** *Dermatological Aid* Wet poultice of crushed roots applied to sores and swellings. (97:48) **Gosiute** *Unspecified* Root used as medicine. (39:375) **Isleta** *Dermatological Aid* Poultice of dried, ground leaves used on sores for rapid healing. (as *Pachylophus hirsutus* 100:36) **Navajo, Kayenta** *Ceremonial Medicine* Plant used in various ceremonies. *Dermatological Aid* Plant used as dusting powder for chafing. *Gynecological Aid* Poultice of ground plant applied for prolapse of the uterus. (205:33)
- *Other*—**Navajo, Kayenta** *Decorations* Mixed with cornmeal and placed on Nightway sand painting figures. (205:33)

Oenothera cespitosa Nutt. ssp. *cespitosa*, Tufted Eveningprimrose
- *Drug*—**Blackfoot** *Dermatological Aid* Poultice of pounded root applied to inflamed sores and swellings. (as *Pachylobus caespitosus* 114:274) Poultice of pounded, wetted root applied to inflamed sores. (as *Pachylophus caespitosus* 118:44)

Oenothera cespitosa ssp. *marginata* (Nutt. ex Hook. & Arn.) Munz, Tufted Eveningprimrose
- *Drug*—**Hopi** *Eye Medicine* Plant used with Kachina ears for sore eyes. *Toothache Remedy* Plant used as toothache medicine. (42:337) **Navajo, Ramah** *Panacea* Poultice of plant or root used only for large swellings, a "life medicine." (191:37)
- *Other*—**Hopi** *Ceremonial Items* Flowers used ceremonially as "white flower." *Smoke Plant* Plant used as substitute for tobacco. (42:337)

Oenothera coronopifolia Torr. & Gray, Crownleaf Eveningprimrose

- **Drug**—Navajo, Ramah *Adjuvant* Dried leaves added to improve the flavor of wild tobacco. *Analgesic* and *Gastrointestinal Aid* Cold infusion of leaves taken for stomachache. *Panacea* Poultice of plant or root used only for large swellings, a "life medicine." (191:37) **Zuni** *Antirheumatic (External)* Poultice of powdered flower and saliva applied at night to swellings. (27:377)

Oenothera elata ssp. *hookeri* (Torr. & Gray) W. Dietr. & W. L. Wagner, Hooker's Eveningprimrose
- **Drug**—Navajo, Kayenta *Ceremonial Medicine* Plant used as a Plumeway emetic. *Cold Remedy* Plant used for colds. *Dermatological Aid* Poultice of plant applied to sores. *Emetic* Plant used as a Plumeway emetic. *Misc. Disease Remedy* Hot poultice of plant applied for mumps. (as *O. hookeri* 205:33) **Navajo, Ramah** *Panacea* Poultice of root used only for large swellings, a "life medicine." (as *O. hookeri* var. *hirsutissima* 191:37) **Zuni** *Antirheumatic (External)* Poultice of powdered flower and saliva applied at night to swellings. (as *O. hookeri* 27:377)
- **Food**—Paiute *Unspecified* Species used for food. (as *O. hookeri* 167:243)
- **Dye**—Pomo, Kashaya *Yellow* Flowers chewed with gum to make gum yellow. (as *O. hookeri* 72:95)
- **Other**—Jemez *Good Luck Charm* Root carried by deer hunters as a charm. (as *O. hookeri* 44:25) **Paiute** *Hunting & Fishing Item* Root rubbed on hunter's moccasins and body to attract deer. *Protection* Root rubbed on hunter's moccasins and body to repel snakes. (as *O. hookeri* 118:50)

Oenothera flava (A. Nels.) Garrett, Yellow Eveningprimrose
- **Drug**—Navajo, Ramah *Burn Dressing* Seedpod ashes applied to burns. *Panacea* Poultice of plant or root used only for large swellings, a "life medicine." Poultice of root used for swellings and internal injuries, a "life medicine." *Throat Aid* Poultice of compound containing whole plant used for "throat trouble." (191:38)

Oenothera fruticosa L., Narrowleaf Eveningprimrose
- **Food**—Cherokee *Vegetable* Leaves parboiled, rinsed, and cooked in hot grease as a potherb. (126:49)

Oenothera pallida Lindl., Pale Eveningprimrose
- **Drug**—Navajo, Kayenta *Ceremonial Medicine* Plant used as a Beadway emetic. *Dermatological Aid* Plant used as dusting powder for venereal disease sores. Poultice of plant applied to spider bites. *Emetic* Plant used as a Beadway emetic. *Kidney Aid* Infusion of plant used for kidney disease. *Venereal Aid* Plant used as dusting powder for venereal disease sores. *Veterinary Aid* Plant used for livestock with colic. (205:34)

Oenothera pallida Lindl. ssp. *pallida*, Pale Eveningprimrose
- **Other**—Hopi *Ceremonial Items* Used ceremonially as the White Flower associated with the northeast direction. (as *Anogra pallida* 200:86)

Oenothera pallida ssp. *runcinata* (Engelm.) Munz & W. Klein, Pale Eveningprimrose
- **Drug**—Navajo, Ramah *Ceremonial Medicine* and *Emetic* Plant used as a ceremonial emetic. *Snakebite Remedy* Decoction of root and leaves used as a lotion for snakebites. *Throat Aid* Soaked plant rubbed on throat and infusion taken for sore throat. (as *O. runcinata* 191:38)
- **Other**—Hopi *Ceremonial Items* Used ceremonially as the White Flower associated with the northeast direction. (as *Anogra runcinata* 200:86)

Oenothera perennis L., Small Eveningprimrose
- **Drug**—Iroquois *Orthopedic Aid* Decoction of plants taken for paralysis. (87:390)

Oenothera primiveris Gray, Desert Eveningprimrose
- **Drug**—Navajo, Ramah *Ceremonial Medicine* Fresh or dried flowers used as ceremonial lotion and medicine. *Dermatological Aid* Poultice of plant applied to swellings. (191:38)

Oenothera rhombipetala Nutt. ex Torr. & Gray, Fourpoint Eveningprimrose
- **Other**—Kiowa *Decorations* Yellow flowers picked and brought into the house. (as *Raimannia rhombipetala* 192:45)

Oenothera sp., Eveningprimrose
- **Drug**—Navajo *Dermatological Aid* Compound infusion of plants used as a wash for sore skin. (55:66)
- **Food**—Cherokee *Unspecified* Leaves boiled, fried, and often eaten with greens. (204:253)

Oenothera triloba Nutt., Stemless Eveningprimrose
- **Drug**—Zuni *Antirheumatic (External)* Ingredient of "schumaakwe cakes" and used externally for rheumatism. *Dermatological Aid* Ingredient of "schumaakwe cakes" and used externally for swelling. (as *Lavauxia triloba* 166:55)
- **Food**—Zuni *Unspecified* Roots ground and used for food. (as *Lavauxia triloba* 136:158)

Oenothera villosa ssp. *strigosa* (Rydb.) W. Dietr. & Raven, Hairy Eveningprimrose
- **Drug**—Navajo, Ramah *Disinfectant* Cold infusion of dried root taken for "deer infection." *Hunting Medicine* Dried leaves and tobacco smoked for good luck in hunting. (as *O. procera* 191:38)

Olacaceae, see *Ximenia*

Olneya, Fabaceae

Olneya tesota Gray, Desert Ironwood
- **Food**—Cahuilla *Staple* Roasted pods and seeds ground into flour. (15:94) **Cocopa** *Porridge* Seeds roasted, ground, and made into mush. **Mohave** *Bread & Cake* Seeds parched, ground lightly, roasted, and the meal made into thin loaves and baked. *Dried Food* Seeds parched, ground lightly, roasted, and eaten. (37:187) **Papago** *Dried Food* Seeds basket winnowed, parched, sun dried, cooked, stored and used for food. (34:24) Beans flailed, winnowed, parched, and used for food. (34:25) *Staple* Beans parched, sun dried, stored, ground into flour, and used as a staple food. (34:45) *Unspecified* Ground, leached seeds used for food. (36:60) **Pima** *Dried Food* Beans formerly pit roasted, parched, and eaten whole. (47:93) Seeds formerly dried, roasted, ground coarsely, and used for food. (95:263) Nuts parched and eaten. (146:70) *Staple* Beans formerly pit roasted, ground, mixed with water, and eaten as pinole. (47:93) **Pima, Gila River** *Unspecified* Seeds leached, roasted, and eaten. (133:5) Seeds parched and eaten. (133:7)

Seri *Porridge* Beans ground into a meal, mixed with water or sea lion oil, and eaten. (50:136) **Yavapai** *Bread & Cake*, and *Staple* Dried, mashed, parched seeds ground into a meal and used to make greasy cakes. (63:211) **Yuma** *Bread & Cake* Seeds parched, ground lightly, roasted, and the meal made into thin loaves and baked. *Dried Food* Seeds parched, ground lightly, roasted, and eaten. (37:187)
- *Fiber*—**Papago** *Building Material* Posts of wood, forked at the top, used for the core of the house frame. (34:66)
- *Other*—**Cahuilla** *Fuel* Wood used for firewood. *Tools* Wood used to make implements requiring extreme hardness: throwing sticks and clubs. (15:94) **Papago** *Musical Instrument* Concave sticks with far-spaced, deep notches used as loud rattles for scraping stick songs. (34:68) *Tools* Four-foot sticks with sharp points used as digging sticks. (34:31) Wooden stakes driven into the ground and used for weaving cotton. (34:60) **Pima** *Fuel* Wood used for firewood. *Tools* Wood used to make tool handles. Formerly used to make shovels. (47:93)

Onagraceae, see *Boisduvalia, Calylophus, Camissonia, Circaea, Clarkia, Epilobium, Gaura, Gayophytum, Ludwigia, Oenothera*

Onoclea, Dryopteridaceae, fern
Onoclea sensibilis L., Sensitive Fern
- *Drug*—**Iroquois** *Antirheumatic (Internal)* Used for arthritis and infection. (87:256) *Blood Medicine* Fermented compound decoction taken before meals and bed to "make blood." (87:254) Compound decoction of roots taken for "cold in blood." Decoction used as a hair wash and taken for the blood which caused the hair to fall out. (87:255) Infusion of rhizomes given to children when "the blood doesn't have a determined path." (141:34) *Dermatological Aid* Poultice of plant top used for deep cuts. (87:256) *Gastrointestinal Aid* Used "for trouble with the intestines, when you catch cold and get inflated and sore." (87:255) *Gynecological Aid* Infusion of root taken for pain after childbirth. Infusion of whole plant or roots applied to full, nonflowing breasts. (87:254) Decoction of roots taken for fertility in women and the blood, to give strength after childbirth, and to start menses and for swellings, cramps, and sore abdomen. (87:255) *Pediatric Aid* Infusion of rhizomes given to children when "the blood doesn't have a determined path." (141:34) *Tuberculosis Remedy* Compound decoction of roots taken during the early stages of consumption. (87:256) *Venereal Aid* Compound decoction used for venereal disease. (87:254) Cold, compound infusion of plant washed on sores and taken for gonorrhea. (87:255) Infusion of plant and female fern rhizomes used by men for venereal diseases. (141:34) **Ojibwa** *Gynecological Aid* Decoction of powdered, dried root used by patients with caked breast for milk flow. (153:382)
- *Food*—**Iroquois** *Vegetable* Cooked and seasoned with salt, pepper, or butter. (196:118)

Onoclea sp.
- *Fiber*—**Iroquois** *Mats, Rugs & Bedding* Made into pillows and used by children under their lower backs to prevent bed-wetting. (142:82)

Onopordum, Asteraceae
Onopordum acanthium L., Scotch Cottonthistle
- *Drug*—**Iroquois** *Emetic* Decoction used as an emetic to counter witchcraft. *Poison* Decoction used for witchcraft poison. *Witchcraft Medicine* Decoction used for witchcraft poison and as an emetic to counter witchcraft. (87:476)

Onosmodium, Boraginaceae
Onosmodium molle Michx., Smooth Onosmodium
- *Drug*—**Lakota** *Antirheumatic (External)* Infusion of roots and seeds used by men for swellings. *Veterinary Aid* Infusion of roots and seeds given to horses or used as a rubbing solution. (139:41)

Onosmodium molle ssp. *hispidissimum* (Mackenzie) Boivin, Softhair Marbleseed
- *Other*—**Chippewa** *Good Luck Charm* Seeds used as a love charm and to attract money and worldly goods. (as *O. hispidissimum* 53:376)

Onosmodium molle ssp. *occidentale* (Mackenzie) Cochrane, Western Onosmodium
- *Drug*—**Cheyenne** *Antirheumatic (External)* Pulverized leaves and stems mixed with grease and used as a rub for lumbago. (as *O. occidentale* 76:185) *Dermatological Aid* Pulverized leaves and stems mixed with grease and rubbed on numb skin. *Orthopedic Aid* Smashed leaves and stems rubbed on back for lumbago. (as *O. occidentale* 76:185)

Ophioglossaceae, see *Botrychium*

Oplopanax, Araliaceae
Oplopanax horridus (Sm.) Torr. & Gray ex Miq., Devil's Club
- *Drug*—**Bella Coola** *Antirheumatic (Internal)* Decoction of root bark and stems taken for rheumatism. *Cathartic* Root bark chewed as purgative and decoction of root bark and stems taken. (as *Fatsia horrida* 150:62) *Emetic* Inner bark chewed as an emetic. (as *O. horridum* 184:201) **Carrier** *Analgesic* Poultice of bark scrapings applied or bark pills taken for pain. (as *O. horridum* 31:82) **Carrier, Northern** *Analgesic* Inner bark taken for stomach and bowel cramps. *Cathartic* Inner bark acts as a purgative, especially if taken with hot water. *Gastrointestinal Aid* Inside layer of inner bark taken for stomach and bowel cramps. **Carrier, Southern** *Cathartic* and *Gynecological Aid* Decoction of bark taken as a purgative before and after childbirth. (as *Fatsia horrida* 150:62) **Cheyenne** *Analgesic* Root mixed with tobacco and smoked for headache. (as *Fatsia horrida* 19:12) **Cowlitz** *Antirheumatic (External)* Infusion of bark used as wash for rheumatism. *Cold Remedy* Infusion of bark taken for colds. *Poison* Plant considered poisonous. (as *O. horridum* 79:41) **Crow** *Analgesic* Root mixed with tobacco and smoked for headache. (as *Fatsia horrida* 19:12) **Gitksan** *Analgesic* Fresh or dried inner bark used for stomach ulcers and pain. Infusion of dried bark used for stomach pain and ulcers. (as *O. horridum* 74:16) *Antihemorrhagic* Poultice of compound containing bark used as a chest plaster for lung hemorrhage. (as *Fatsia horrida* 150:62) *Antirheumatic (External)* Bark and other plants used for arthritis. (73:152) Poultice of compound containing bark used for rheumatism. (as *Fatsia horrida* 150:62) *Cancer Treatment* Bark and other plants used for cancer. (73:152) *Cathartic* Compound decoction taken as a diuretic and purgative for "strangury." (as *Fatsia horrida* 150:62) *Cold Remedy* Decoction of inner bark used for colds. *Cough Medicine* Decoction of inner bark used for coughs. *Dermatological Aid* Poultice of inner bark applied to wounds. Bark

and other plants used as a skin wash. (73:152) Poultice of compound containing bark applied to boils and ulcers. *Diuretic* Compound decoction taken as a diuretic and purgative for "strangury." (as *Fatsia horrida* 150:62) *Gastrointestinal Aid* Bark used as a "cleanser." *Misc. Disease Remedy* Bark and other plants used for diabetes. Decoction of inner bark used for flu. (73:152) *Orthopedic Aid* Decoction taken to aid the knitting of broken bones. *Other* Compound decoction taken continuously for rupture. *Pulmonary Aid* Poultice of compound containing bark used as a chest plaster for lung hemorrhage. (as *Fatsia horrida* 150:62) *Respiratory Aid* Bark and other plants used for bronchitis. *Tonic* Decoction of inner bark used as a tonic. *Tuberculosis Remedy* Bark and other plants used for tuberculosis. (73:152) *Venereal Aid* Decoction taken as a purgative for gonorrhea. (as *Fatsia horrida* 150:62) **Green River Group** *Cold Remedy* Infusion of roots taken for colds. *Dermatological Aid* Dried bark used as a deodorant. (as *O. horridum* 79:41) **Haisla** *Antirheumatic (External)* Bark and other plants used for arthritis. *Cancer Treatment* Bark and other plants used for cancer. *Cold Remedy* Decoction of inner bark used for colds. *Cough Medicine* Decoction of inner bark used for coughs. *Dermatological Aid* Poultice of inner bark applied to wounds. Bark and other plants used as a skin wash. *Gastrointestinal Aid* Bark used as a "cleanser." *Misc. Disease Remedy* Bark and other plants used for diabetes. Decoction of inner bark used for flu. *Respiratory Aid* Bark and other plants used for bronchitis. *Tonic* Decoction of inner bark used as a tonic. *Tuberculosis Remedy* Bark and other plants used for tuberculosis. (73:152) **Haisla & Hanaksiala** *Antirheumatic (External)* Infusion of pounded leaves applied to arthritic joints. *Dermatological Aid* Inner bark placed in wounds followed by an application of Sitka spruce pitch. Juice used for sores. *Emetic* Decoction or infusion of plant and sea water taken as an emetic. *Eye Medicine* Decoction of stem bark, stems, or winter roots used as an eyewash for cataracts. Decoction of plant used as an eyewash for cataracts. *Gastrointestinal Aid* Juice taken for stomach sickness. *Laxative* Decoction or infusion taken as a laxative. **Hanaksiala** *Cold Remedy* Infusion or decoction of plant taken for winter colds. (43:217) **Hoh** *Unspecified* Used as medicine. (as *Echinopanax horridum* 137:65) **Kwakiutl** *Analgesic* Bark used in a steam bath for body pains. Root held in the mouth and juice swallowed for stomach pains. *Dermatological Aid* Rotten stem ash and oil rubbed on swellings. *Gastrointestinal Aid* Root held in the mouth and juice swallowed for stomach pains. *Herbal Steam* Bark used in a steam bath for body pains. *Laxative* Root held in the mouth and juice swallowed for constipation. *Poison* Spines considered poisonous. *Tuberculosis Remedy* Bark extract taken for tuberculosis. *Witchcraft Medicine* Plant used for the magical powers. (as *O. horridum* 183:278) **Lummi** *Gynecological Aid* Poultice of bark applied to woman's breast to stop an excessive flow of milk. (as *O. horridum* 79:41) **Makah** *Antirheumatic (External)* Poultice of cooked or boiled plant applied to sore spots. Plant used for arthritis. *Unspecified* Bark and roots used medicinally. (as *O. horridum* 67:289) **Montana Indian** *Ceremonial Medicine* Used by medicine men in their incantations. (as *Fatsia horrida* 19:12) **Nitinaht** *Antirheumatic (Internal)* Infusion of bark taken for arthritis and rheumatism. (as *O. horridum* 67:289) Infusion of stem pieces taken for arthritis. (186:95) *Orthopedic Aid* Infusion of bark taken for bone ailments. *Unspecified* Used for medicine. (as *O. horridum* 67:289) **Okanagan-Colville** *Cough Medicine* Infusion of roots and stems taken for dry coughs. *Tuberculosis Remedy* Infusion of roots and stems taken for consumption. (as *O. horridum* 188:73) **Okanagon** *Blood Medicine* Infusion of crushed stems taken as a blood purifier. *Dermatological Aid* Burned stems and grease salve rubbed on swollen parts. *Gastrointestinal Aid* Infusion of crushed stems taken for stomach troubles and indigestion. *Tonic* Infusion of crushed stems taken as a tonic. (as *Fatsia horrida* 125:40) **Oweekeno** *Analgesic* Decoction of inner bark from young spring growth taken for general aches and pains. *Antirheumatic (External)* Roots used in a bath for rheumatism. *Cold Remedy* Decoction of inner bark from young spring growth taken for colds. Plant boiled and vapor inhaled for colds. *Dermatological Aid* Berries mashed into a foam and rubbed into the scalp for head lice. *Panacea* Decoction of inner bark from young spring growth taken for any kind of sickness. *Poison* Berries considered poisonous. *Tonic* Decoction of inner bark from young spring growth used as a tonic. (43:85) **Quileute** *Unspecified* Used as medicine. (as *Echinopanax horridum* 137:65) **Salish, Coast** *Analgesic* Poultice of pounded, boiled roots used for rheumatism and other aches and pains. *Antirheumatic (External)* Poultice of pounded, boiled roots used for rheumatism. Prickly stems beaten against the skin as a counterirritant for sore limbs. (as *O. horridum* 182:78) **Sanpoil** *Cold Remedy* Infusion of inner pith of stalk taken for colds. (as *Echinopanax horridum* 131:220) **Skagit** *Gynecological Aid* Decoction of bark taken by women to start menstrual flow after childbirth. *Tuberculosis Remedy* Decoction of bark taken for tuberculosis. (as *O. horridum* 79:41) **Thompson** *Blood Medicine* Infusion of crushed stems taken as a blood purifier. (as *Fatsia horrida* 125:40) Decoction of stems taken as a blood purifier. (as *Echinopanax horridum* 164:459) *Dermatological Aid* Burned stems and grease salve rubbed on swollen parts. (as *Fatsia horrida* 125:40) Stem ash with grease used as an ointment for swellings. (as *Echinopanax horridum* 164:459) *Dietary Aid* Infusion of sticks, with the spines and outer bark removed, taken to cease weight loss. The infusion was taken in doses of about ½ cup before meals, to replace milk and other beverages. It was noted that if the infusion was taken for too great a period of time, one could gain too much weight. Infusion of whole plant taken to give one a good appetite. (187:164) *Gastrointestinal Aid* Infusion of crushed stems taken for stomach troubles and indigestion. (as *Fatsia horrida* 125:40) Infusion of crushed stems taken for indigestion and stomach troubles. (as *Echinopanax horridum* 164:459) Infusion of whole plant taken for ulcers. (187:164) *Laxative* Decoction of stems taken as a laxative. (as *Echinopanax horridum* 164:459) *Misc. Disease Remedy* Infusion of sticks, with the spines and outer bark removed, taken for influenza and other illnesses.[48] Infusion of roots taken for diabetes. *Panacea* Infusion of sticks, with the spines and outer bark removed, taken for everything.[48] (187:164) *Tonic* Infusion of crushed stems taken as a tonic. (as *Fatsia horrida* 125:40) Decoction of stems taken as a tonic. (as *Echinopanax horridum* 164:459) **Tlingit** *Dermatological Aid* Compound containing plant ash used for sores. (as *Panax horridum* 107:284) **Wet'suwet'en** *Antirheumatic (External)* Bark and other plants used for arthritis. *Cancer Treatment* Bark and other plants used for cancer. *Cold Remedy* Decoction of inner bark used for colds. *Cough Medicine* Decoction of inner bark used for coughs. *Dermatological Aid* Poultice of inner bark applied to wounds. Bark and other plants used as a skin wash.

Gastrointestinal Aid Bark used as a "cleanser." *Misc. Disease Remedy* Bark and other plants used for diabetes. Decoction of inner bark used for flu. *Respiratory Aid* Bark and other plants used for bronchitis. *Tonic* Decoction of inner bark used as a tonic. *Tuberculosis Remedy* Bark and other plants used for tuberculosis. (73:152)
- *Food*—**Oweekeno** *Unspecified* Young, spring buds boiled and eaten. (43:85)
- *Dye*—**Hesquiat** *Unspecified* Bark shavings and berries made into paint and used to color basket materials and other objects. (as *O. horridum* 185:61)
- *Other*—**Bella Coola** *Protection* Spiny stems used as protective charms against supernatural powers. (as *O. horridum* 184:201) **Clallam** *Hunting & Fishing Item* Sticks peeled, cut into pieces, fastened to bass fishing line, and used to attract fish. Wood made into fishing lures. (as *O. horridum* 57:197) **Gitksan** *Ceremonial Items* Inner bark chewed during pre-hunting purification rituals. **Haisla** *Ceremonial Items* Inner bark chewed during pre-hunting purification rituals. (73:152) *Paint* Plant made into black face paint and used by warriors. **Haisla & Hanaksiala** *Ceremonial Items* Bark used for ritual purification. *Good Luck Charm* Bark used for the acquisition of luck by hunters, fishers, and shamanistic initiates. Plant used by bathing black bear and brought observer good luck. Plant used to bring good luck. *Paint* Bark charred, mixed with pounded salmon eggs, and used as black face paint for dancing. *Protection* Bark made into face paint and used by shamans to repel enemy spirits from the shaman's patient. Plant used to cleanse areas where people had died. *Soap* Decoction or infusion of bark used to wipe one's body after bathing. (43:217) **Hesquiat** *Hunting & Fishing Item* Spiny stems used as spears for catching octopus and carved into fishing lures. Because the wood is light, it spins around when pulled through the water and helps to attract fish. (as *O. horridum* 185:61) **Kitasoo** *Hunting & Fishing Item* Plant used by hunters in a bath to remove the human smell. (43:327) **Klallam** *Hunting & Fishing Item* Sticks peeled, cut, fastened to bass lines, and used as fish lures. (as *O. horridum* 79:41) **Kwakiutl** *Protection* Stem used as a protective charm. (as *O. horridum* 183:278) **Lummi** *Paint* Sticks burned, mixed with grease or Vaseline, and used as a reddish brown face paint. (as *O. horridum* 79:41) **Lummi, Washington** *Paint* Stems charred, mixed with grease or Vaseline, and used as a black face paint. (as *O. horridum* 182:78) **Makah** *Hunting & Fishing Item* Wood used to make lures and hooks for bass fishing. **Nitinaht** *Ceremonial Items* Plant burned to make charcoal used as a protective face paint for ceremonial dancers. A person wearing this kind of paint would have so much power you could not look them in the eye. *Hunting & Fishing Item* Wood used to make lures for cod fishing. (as *O. horridum* 67:289) Wood used to make codfish and sea or black bass lures. *Paint* Wood charcoal used as a special ceremonial paint for dancers. (186:95) **Tsimshian** *Ceremonial Items* Plant used by shamans, novices, and warriors for power seeking. (43:327) **Wet'suwet'en** *Ceremonial Items* Inner bark chewed during pre-hunting purification rituals. (73:152)

Opuntia, Cactaceae

Opuntia acanthocarpa Engelm. & Bigelow, Buckhorn Cholla
- *Drug*—**Cahuilla** *Burn Dressing* Stem ash applied to burns. *Dermatological Aid* Stem ash applied to cuts. (15:95) **Pima** *Gastrointestinal Aid* Plant used for stomach troubles. (47:58)
- *Food*—**Cahuilla** *Dried Food* Fruit gathered in the spring and dried for storage. *Fruit* Fruit gathered in the spring and eaten fresh. (15:95) *Staple* Berries and stems were an important and dependable food source. (15:49) **Pima, Gila River** *Dried Food* Calyxes pit roasted with inkweed and dried for future use. (133:4) *Staple* Flowers pit roasted and eaten as a staple. (133:7) *Unspecified* Calyxes pit roasted with inkweed and eaten fresh. (133:4)

Opuntia acanthocarpa var. *ramosa* Peebles, Cholla
- *Food*—**Maricopa** *Unspecified* Flower buds pit baked and used for food. (37:201)
- *Other*—**Pima** *Cash Crop* Dry, woody joints made into canes, napkin rings, and other tourist souvenirs. (47:58)

Opuntia arbuscula Engelm., Arizona Pencil Cholla
- *Food*—**Pima** *Fruit* Green fruits boiled with saltbush and used for food. (47:59) **Pima, Gila River** *Dried Food* Calyxes pit roasted with inkweed and dried for future use. *Unspecified* Calyxes pit roasted with inkweed and eaten fresh. (133:4) Flowers pit roasted and eaten. (133:7)

Opuntia basilaris Engelm. & Bigelow, Beavertail Pricklypear
- *Food*—**Cahuilla** *Dried Food* Buds cooked and dried for indefinite storage. *Porridge* Seeds ground into mush. *Unspecified* Buds cooked and eaten. *Vegetable* Joints boiled and mixed with other foods or eaten as greens. (15:95) **Diegueño** *Dried Food* Fruit cleaned of thorns, dried, and eaten. (84:27) **Kawaiisu** *Unspecified* Buds cooked and eaten. (206:46) **Tubatulabal** *Unspecified* Species used for food. (as *O. basilans* 193:16)
- *Other*—**Diegueño** *Cash Crop* Fruit dried and sold by children in small sacks for 10 cents. (84:27)

Opuntia basilaris var. *aurea* (E. M. Baxter) W. T. Marsh., Golden Pricklypear
- *Drug*—**Shoshoni** *Analgesic* Poultice of inner pulp applied to cuts and wounds and for the pain. *Dermatological Aid* Fuzz-like spines rubbed into warts or moles to remove them. Poultice of inner pulp applied to cuts and wounds and for the pain. (180:107, 108)

Opuntia bigelovii Engelm., Teddybear Cholla
- *Food*—**Cahuilla** *Dried Food* Buds cooked and dried for indefinite storage. (15:96) *Staple* Berries and stems were an important and dependable food source. (15:49) *Unspecified* Buds cooked and eaten. (15:96)

Opuntia caseyi var. *magenta* Pursh
- *Food*—**Tubatulabal** *Unspecified* Species used for food. (193:16)

Opuntia chlorotica Engelm. & Bigelow, Dollarjoint Pricklypear
- *Food*—**Yavapai** *Fruit* Raw fruit used for food. (65:257)

Opuntia clavata Engelm., Club Cholla
- *Drug*—**Keres, Western** *Dermatological Aid* Dried joints ground or burned into a powder and used on open sores or bad wounds. *Veterinary Aid* Dried joints ground or burned into a powder and used on open sores or bad wounds on horses. (171:56)
- *Food*—**Acoma** *Starvation Food* Stems and fruits roasted and eaten in times of food shortage. Joints roasted and eaten during famines. (32:

35) **Keres, Western** *Starvation Food* Roasted joints used for food in times of famine. (171:56) **Laguna** *Starvation Food* Stems and fruits roasted and eaten in times of food shortage. Joints roasted and eaten during famines. (32:35)

Opuntia echinocarpa Engelm. & Bigelow, Staghorn Cholla
- *Food*—**Cocopa** *Fruit* Fruits rolled on ground to remove spines and eaten raw. **Maricopa** *Fruit* Fruits rolled on ground to remove spines and eaten raw. **Mohave** *Fruit* Fruits rolled on ground to remove spines and eaten raw. (37:204) **Papago** *Staple* Buds and joints used as a staple crop. (34:15) Pit baked buds, fruits, and joints considered a staple food. (36:59) *Vegetable* Buds eaten as greens in May. (34:14) **Yavapai** *Fruit* Fruit boiled and eaten without mashing. (65:257)

Opuntia engelmannii Salm-Dyck, Cactus Apple
- *Drug*—**Pima** *Gynecological Aid* Poultice of heated plant applied to breasts to encourage the flow of milk. (47:60)
- *Food*—**Acoma** *Fruit* Ripe tunas eaten fresh. *Porridge* Tunas split, dried, ground, and the meal mixed with cornmeal to make a mush for winter use. (32:35) **Cocopa** *Fruit* Fruits rolled on ground to remove spines and eaten raw. (37:204) **Keres, Western** *Cooking Agent* Tunas used as a red dye for corn mush. *Fruit* Fresh tunas used for food. *Winter Use Food* Ground, dried tunas mixed in equal proportions with cornmeal and made into a mush for winter food. (171:56) **Laguna** *Fruit* Ripe tunas eaten fresh. *Porridge* Tunas split, dried, ground, and the meal mixed with cornmeal to make a mush for winter use. (32:35) **Maricopa** *Fruit* Fruits rolled on ground to remove spines and eaten raw. **Mohave** *Fruit* Fruits rolled on ground to remove spines and eaten raw. (37:204) **Papago** *Beverage* Fruits formerly fermented and used for a beverage. (34:26) *Sauce & Relish* Fruits used to make syrup. (146:75) *Staple* Fruits and joints used as a staple food. (36:60) *Vegetable* Leaves with thorns scraped off sliced in strips and eaten as greens in summer. (34:14) **Pima** *Fruit* Fruits freed from thorns, peeled, and eaten. (146:75) *Unspecified* Tender leaves sliced, cooked, seasoned like string beans and used for food. (47:60) **Pima, Gila River** *Fruit* Fruits eaten raw. (133:7) **San Felipe** *Fruit* Ripe tunas eaten fresh. *Porridge* Tunas split, dried, ground, and the meal mixed with cornmeal to make a mush for winter use. Seeds ground with white corn and meal eaten as mush. (32:35)
- *Other*—**Keres, Western** *Paint* Tunas used for red paint. *Tools* Thorns used for needles. (171:56)

Opuntia engelmannii Salm-Dyck **var. *engelmannii***, Cactus Apple
- *Food*—**Cahuilla** *Fruit* Fruit used for food. (as *O. megacarpa* 15:97) *Staple* Berries and stems were an important and dependable food source. (as *O. occidentalis* var. *megacarpa* 15:49) *Unspecified* Diced joints used for food. (as *O. megacarpa* 15:97) **Diegueño** *Fruit* Fruit eaten raw. *Vegetable* Pads boiled like cabbage or string beans with tomatoes, onions, and peppers, like a stew. (as *O. phaeacantha* var. *discata* 84:27)
- *Other*—**Diegueño** *Lubricant* Pad juice used to lubricate oxcart wheels. (as *O. phaeacantha* var. *discata* 84:27)

Opuntia engelmannii var. *lindheimeri* (Engelm.) Parfitt & Pinkava, Texas Pricklypear
- *Food*—**Keresan** *Unspecified* Plant, with thorns burned off, roasted in damp sand and eaten with chili. (as *O. lindheimeri* 198:560) **Sia** *Unspecified* Roasted in damp sand and eaten with chili. (as *O. lindheimeri* 199:107)

Opuntia erinacea Engelm. & Bigelow ex Engelm., Grizzlybear Pricklypear
- *Food*—**Yavapai** *Fruit* Raw fruit used for food. (65:257)

Opuntia erinacea var. *hystricina* (Engelm. & Bigelow) L. Benson, Grizzlybear Pricklypear
- *Food*—**Hopi** *Fruit* Fruits cooked, freed from thorns, and served with cornmeal boiled bread. (as *O. hystricina* 120:18) *Unspecified* Joints boiled, dipped into syrup, and eaten after thorn removal. (as *O. hystricina* 200:85)

Opuntia ficus-indica (L.) P. Mill., Tuna Cactus
- *Drug*—**Cahuilla** *Cathartic* Boiled fruit used as a purgative. *Dermatological Aid* Plugs made from the plant and inserted into wounds as healing agents. *Laxative* Boiled fruit used for constipation. (as *O. megacantha* 15:96)
- *Food*—**Cahuilla** *Dried Food* Diced pads dried and stored for later use. Buds dried for future use. *Fruit* Peeled, cool fruit eaten as a refreshing early morning meal. (as *O. megacantha* 15:96) *Staple* Berries and stems were an important and dependable food source. (as *O. megacantha* 15:49) *Unspecified* Diced pads boiled and eaten. Buds eaten fresh. (as *O. megacantha* 15:96)

Opuntia fragilis (Nutt.) Haw., Brittle Pricklypear
- *Drug*—**Okanagan-Colville** *Dermatological Aid* Poultice of flesh applied to skin sores and infections. *Diuretic* Flesh eaten to cause urination. (188:92) **Shuswap** *Dermatological Aid* Poultice of heated quills applied to cuts, sores, and boils. *Throat Aid* Poultice of heated quills applied to swollen throats. (123:60)
- *Food*—**Okanagan-Colville** *Soup* Flesh and fat boiled into a soup. *Unspecified* Flesh pit cooked or roasted and eaten. (188:92) **Shuswap** *Unspecified* Stems used for food. (123:60) **Thompson** *Dessert* Stems roasted over a fire, peeled, and eaten as dessert by children. *Starvation Food* Stems used for food during times of famine.[49] *Unspecified* Stems steam cooked in pits, the outer, spiny skin peeled off, and the insides used for food.[50] *Winter Use Food* Stems mixed with berry juice and canned for future use. (187:194)
- *Other*—**Okanagan-Colville** *Hunting & Fishing Item* Spines used to make fishhooks. *Season Indicator* Blooms indicated saskatoon berries ready to be picked. (188:92) **Thompson** *Fasteners* Mucilaginous material from cut stems used for glue by some people, but not considered very good. (187:194)

Opuntia fulgida Engelm., Jumping Cholla
- *Food*—**Papago** *Staple* Pit baked buds, fruits, and joints considered a staple food. (36:59) *Vegetable* Young shoots and buds eaten as greens in summer. (34:14)

Opuntia humifusa (Raf.) Raf., Pricklypear
- *Drug*—**Dakota** *Dermatological Aid* Poultice of peeled stems bound on wounds. (70:104) **Lakota** *Snakebite Remedy* Cut stems used for rattlesnake bites. (106:32) **Nanticoke** *Dermatological Aid* Juice of fruit rubbed on warts. (175:56, 84) **Pawnee** *Dermatological Aid* Poultice of peeled stems bound on wounds. (70:104)

- **Food**—**Dakota** *Dried Food* Fruit dried for winter use. (69:366) Fruits, with bristles removed, dried for winter use. (70:104) *Fruit* Fruit eaten raw or stewed. (69:366) Fruits, with bristles removed, eaten fresh and raw or stewed. (70:104) *Starvation Food* Stems, cleared of spines, roasted, and used for food in times of scarcity. (69:366) Stems, with spines removed, roasted during food scarcities. (70:104) **Lakota** *Beverage* Fruit "insides" eaten for thirst. *Dried Food* Fruits dried and stored for winter use food. *Fruit* Fruits eaten fresh. Fruits stewed and used for food. (106:32) **Pawnee** *Dried Food* Fruits, with bristles removed, dried for winter use. *Fruit* Fruits, with bristles removed, eaten fresh and raw or stewed. *Starvation Food* Stems, with spines removed, roasted during food scarcities. (70:104)
- **Dye**—**Dakota** *Mordant* Mucilaginous stem juice used to fix the colors painted on hides or receptacles made from hides.[51] **Pawnee** *Mordant* Mucilaginous stem juice used to fix the colors painted on hides or receptacles made from hides.[51] (70:104)
- **Other**—**Dakota** *Toys & Games* Plant used by small boys in playing games. The "cactus game" was played on the prairie where the cactus abounded. One boy was chosen to be "it" and he would take a stick, place a cactus plant upon it, and hold it up it the air. The other boys would attempt to shoot at it with their bows and arrows and the target holder would run after the boy who hit the target and strike him with the spiny cactus making him "it." (70:104)

Opuntia imbricata (Haw.) DC. **var.** *imbricata*, Tree Cholla
- **Drug**—**Keres, Western** *Dermatological Aid* Ground needle coverings made into a paste and used for boils. *Ear Medicine* Dried stem pith used for earache and running ear. *Strengthener* Thorn coverings eaten by men in times of war to make them tough. (as *O. arborescens* 171:55)
- **Food**—**Acoma** *Dried Food* Young joints split lengthwise, dried, and stored for winter use. *Unspecified* Joints roasted and eaten. (as *O. arborescens* 32:35) **Apache, White Mountain** *Dried Food* Fruit dried for winter use. *Fruit* Fruit eaten raw or stewed. (as *O. arborescens* 136:159) **Keres, Western** *Starvation Food* Roasted joints used for food during times of famine. *Winter Use Food* Young, dried joints stored for winter food. (as *O. arborescens* 171:55) **Laguna** *Dried Food* Young joints split lengthwise, dried, and stored for winter use. *Unspecified* Joints roasted and eaten. (as *O. arborescens* 32:35) **Papago** *Vegetable* Eaten as greens in summer. (as *O. arborescens* 34:14) **Pima** *Dried Food* Fruits pit baked overnight, dried, and stored. (as *O. arborescens* 32:36) Fruits pit cooked, dried, boiled, salted, and eaten with pinole. (as *O. arborescens* 146:71) *Fruit* Fruits roasted in pits and eaten. (as *O. arborescens* 146:69) **Tewa of Hano** *Fruit* Fruits boiled and eaten with sweetened cornmeal porridge. (as *O. arborescens* 138:62)
- **Fiber**—**Keres, Western** *Sewing Material* Thorns used as sewing material and for tattooing. (as *O. arborescens* 171:55)
- **Other**—**Keres, Western** *Lighting* Dried woody stems used for candles and torches before the presence of other forms of lighting. *Season Indicator* Red flowers used as an indicator of when to plant beans. *Tools* Thorns used as sewing needles and for tattooing. (as *O. arborescens* 171:55) **Zuni** *Ceremonial Items* Plant used ceremonially. (as *O. arborescens* 166:95)

Opuntia ×*kelvinensis* V. & K. Grant, Kelvin's Pricklypear
- **Food**—**Pima, Gila River** *Unspecified* Flowers pit roasted and eaten. (133:7)

Opuntia leptocaulis DC., Christmas Cactus
- **Drug**—**Apache, Chiricahua & Mescalero** *Narcotic* Fruits crushed and mixed with a beverage to produce narcotic effects. (33:55)
- **Food**—**Pima** *Fruit* Fruits freed from thorns and eaten raw. (47:60) Small fruits eaten raw. (95:261) **Pima, Gila River** *Fruit* Fruits eaten raw. (133:7)

Opuntia macrorhiza Engelm. **var.** *macrorhiza*, Twistspine Pricklypear
- **Drug**—**Navajo, Ramah** *Dermatological Aid* Cactus spines formerly used to pierce ears and lance small skin abscesses. *Gynecological Aid* Stem roasted and material used to lubricate midwife's hand for placenta removal. (as *O. plumbea* 191:37)
- **Food**—**Navajo, Ramah** *Dried Food* Fruit dried and boiled. *Fruit* Fruit eaten raw. *Winter Use Food* Fruit harvested for winter use. (as *O. plumbea* 191:37)

Opuntia ×*occidentalis* Engelm. & Bigelow, Pricklypear
- **Food**—**Cahuilla** *Fruit* Fruit used for food. *Unspecified* Diced joints used for food. (15:97)

Opuntia parryi Engelm., Brownspined Pricklypear
- **Food**—**Cahuilla** *Staple* Berries and stems were an important and dependable food source. (15:49)

Opuntia phaeacantha Engelm., Tulip Pricklypear
- **Drug**—**Pima** *Gynecological Aid* Poultice of heated plant applied to breasts to encourage the flow of milk. (47:60)
- **Food**—**Havasupai** *Beverage* Plant used to make a drink. (197:66) *Bread & Cake* Dried fruit pounded into cakes for storage or pieces of cake eaten without further preparation. *Dried Food* Fruits sun dried for future use. *Fruit* Fruits eaten fresh. (197:233) **Navajo** *Beverage* Plant used to make fruit juice. *Bread & Cake* Pad pulp formed into cakes, dried, stored for later use, and fried or roasted. *Candy* Pads peeled, sliced, roasted, boiled in sugar water, dried, and eaten like candy. Pad strips peeled, parboiled, boiled, and used as chewing gum. *Cooking Agent* Seed flour used to thicken soups, puddings, or fruit dishes. *Dried Food* Plant eaten dried. *Fruit* Fruit eaten raw. *Preserves* Plant used to make jelly. Pads peeled, sliced, roasted, boiled in sugar water until dissolved into a syrup, and eaten like jelly. *Staple* Dried seeds ground into flour. *Unspecified* Plant eaten fresh. Pads parboiled, peeled, sliced, boiled in salted water, and eaten. (110:14) **Pima** *Unspecified* Tender leaves sliced, cooked, seasoned like string beans, and used for food. (47:60)
- **Other**—**Havasupai** *Containers* Used in preparing pottery clay. *Tools* Spines used to prick the design into the skin for tattooing. (197:233)

Opuntia phaeacantha **var.** *camanchica* (Engelm. & Bigelow) L. Benson, Tulip Pricklypear
- **Food**—**Keres, Western** *Fruit* Mountain tunas used for food. (as *O. camanchica* 171:56) **Tewa** *Fruit* Fruits eaten for food. (as *O. camanchica* 138:62)

Opuntia polyacantha Haw., Plains Pricklypear
- **Drug**—**Flathead** *Analgesic* Stems smashed and used for backache. *Antidiarrheal* Infusion of stems taken for diarrhea. (82:39) **Navajo** *Poison* Plant used as a poison for hunting. (55:65) **Okanagan-Colville** *Dermatological Aid* Poultice of flesh applied to skin sores and

infections. *Diuretic* Flesh eaten to cause urination. (188:92) **Sioux** *Dermatological Aid* Stems peeled and used for wounds. (82:39)

- **Food**—**Cheyenne** *Cooking Agent* Pulp dried and used to thicken soups and stews. (82:39) *Dried Food* Fruits dried and used as a winter food. (83:16) *Fruit* Fruit eaten raw or dried for winter use. (76:180) Fruits eaten raw. (83:16) *Soup* Fruit stewed with meat and game into a soup. *Winter Use Food* Fruit dried for winter use. (76:180) **Gosiute** *Unspecified* Joints roasted in hot coals and eaten. (39:375) **Hopi** *Fruit* Fruits cooked, freed from thorns, and served with cornmeal boiled bread. (120:18) *Unspecified* Joints boiled, dipped into syrup, and eaten after thorn removal. (200:85) **Keres, Western** *Winter Use Food* Joints singed in hot coals, boiled with dried sweet corn, and used as a winter food. (171:57) **Montana Indian** *Dried Food* Fruits dried and stored for winter use. (82:39) *Fodder* In times of scarcity, spines were singed off and fed to stock. *Fruit* Ripe fruit eaten raw. (19:17) Fruits eaten raw. (82:39) *Preserves* Fruit made into preserves. *Unspecified* Young joint pulp boiled and fried. (19:17) Stems occasionally used for food. (82:39) **Okanagan-Colville** *Soup* Flesh and fat boiled into a soup. *Unspecified* Flesh pit cooked or roasted and eaten. (188:92) **Okanagon** *Unspecified* Insides of plants oven roasted and used for food. (125:36) **Paiute, Northern** *Unspecified* Flesh peeled and eaten roasted or uncooked and fresh. (59:49) **San Felipe** *Unspecified* Joints formerly roasted and eaten. *Winter Use Food* Joints singed in hot coals, boiled and dried with sweet corn to make a winter use food. (32:36) **Sanpoil & Nespelem** *Dried Food* Berry pits roasted, after spines burned off and removed, and used for food. (131:103) **Thompson** *Dessert* Stems roasted over a fire, peeled, and eaten as dessert by children. *Starvation Food* Stems used for food during times of famine.[49] (187:194) *Unspecified* Insides of plants oven roasted and used for food. (125:36) Roots and little bulbs cooked, peeled, and the inside eaten. Stalks cooked, peeled, and the inside eaten. (164:480) Stems steam cooked in pits, the outer, spiny skin peeled off, and the insides used for food.[50] *Winter Use Food* Stems mixed with berry juice and canned for future use. (187:194)
- **Dye**—**Crow** *Mordant* Stems peeled and used to fix color on hides. (82:39) **Navajo** *Red* Fruit used to dye wool pink. Dead, ripe fruits used to make a cardinal dye. (55:65) **Sioux** *Mordant* Stems peeled and used to fix color on hides. (82:39)
- **Other**—**Navajo** *Fasteners* Juice used to adhere buckskin cuttings and trimmings to the buckskin war shirt. (55:65) **Okanagan-Colville** *Hunting & Fishing Item* Spines used to make fishhooks. *Season Indicator* Blooms indicated saskatoon berries ready to be picked. (188:92) **Thompson** *Fasteners* Mucilaginous material from cut stems used for glue by some people, but not considered very good. (187:194) *Jewelry* Seeds strung and worn as necklaces. (164:498)

Opuntia polyacantha var. *rufispina* (Engelm. & Bigelow ex Engelm.) L. Benson, Hairspine Pricklypear

- **Food**—**Gosiute** *Unspecified* Joints roasted in hot coals and eaten. (as *O. rutila* 39:375)

Opuntia ramosissima Engelm., Branched Pencil Cholla

- **Food**—**Cahuilla** *Dried Food* Fruit dried for later use. Stalks, with thorns removed, dried for future use. *Fruit* Fruit eaten fresh. *Soup* Stalks, with thorns removed, boiled into a soup. (15:97) *Staple* Berries and stems were an important and dependable food source. (15:49)

Opuntia sp., Pricklypear

- **Drug**—**Apache, Mescalero** *Dermatological Aid* and *Disinfectant* Stems scorched, split, and used for infections and cuts. *Eye Medicine* Needles used for scraping infected eyelids and tattoos. (14:38) **Apache, Western** *Antidiarrheal* Liquid extract of boiled roots used for thin and frequent bowel movements. *Burn Dressing* Poultice of peeled stalks applied to burns. *Laxative* and *Pediatric Aid* Boiled roots used as laxative for babies and small children. (26:180) **Costanoan** *Antirheumatic (External)* Poultice of warm fruit applied and warm fruit juice rubbed on for rheumatism. (21:10) **Hualapai** *Burn Dressing* Inner pad juice applied to burns. *Dermatological Aid* Inner pad juice applied to cuts. (195:4) **Jemez** *Dermatological Aid* Hot poultice of baked pear skin applied to boils, probably to remove the swelling and pain. (44:25) **Kiowa** *Dermatological Aid* Thorns used to puncture the skin for boils. *Hemostat* Poultice of peeled stems applied as a hemostat. (192:45) **Lakota** *Abortifacient* Roots and soapweed roots used as "medicine for not give birth." *Diuretic* Decoction of roots taken for urinary problems. *Gynecological Aid* Roots and soapweed roots used by mothers when they cannot give birth. (139:42) **Lummi** *Gynecological Aid* Infusion of smashed plants taken to facilitate childbirth. (79:41) **Mahuna** *Dermatological Aid* Leaves inserted into wound. (140:16) **Navajo** *Dermatological Aid* Plant used for boils. (90:161)
- **Food**—**Apache, Mescalero** *Dried Food* Unpeeled fruits split, covered with juice, sun dried, and stored for future food use. *Fruit* Tunas eaten fresh. (14:38) **Apache, San Carlos** *Porridge* Seeds parched, ground, boiled, and eaten as mush. *Staple* Seeds parched, ground, and flour eaten with drafts of water. (95:257) **Apache, Western** *Porridge* Seeds roasted, mixed with corn and meal, moistened with water and salt before eating. *Soup* Fruit pit baked, dried, and boiled with fat or in soups. (26:180) **Comanche** *Fruit* Fruits eaten for food. (29:523) **Costanoan** *Fruit* Fruits eaten for food. (21:251) **Hopi** *Unspecified* Stems, with spines removed, boiled and eaten. (56:17) **Hualapai** *Beverage* Fruit made into a drink. (195:4) *Dried Food* Fruits sun dried and used for food. (195:10) Fruit dried for future use. (195:4) *Fruit* Fruits pit baked and eaten. (195:10) Fruit eaten fresh. (195:4) **Isleta** *Fruit* Fruit eaten fresh. *Preserves* Fruit eaten as conserves. (100:35) **Jemez** *Fruit* Pears used for food. (44:25) **Kiowa** *Candy* Ripe fruits gathered in large quantities and employed in making candy. *Fruit* Ripe fruits gathered in large quantities and used fresh in jams. *Preserves* Ripe fruits gathered in large quantities and used fresh in jams. (192:45) **Luiseño** *Dried Food* Fruit eaten dried. *Fruit* Fruit eaten fresh. *Staple* Seeds ground into a meal. (155:230) **Maricopa** *Fruit* Fruits eaten raw. (95:265) **Navajo** *Dried Food* Fruits split, sun dried, and used for food. (55:64) Fruit with thorns rubbed off, dried and used for food. *Fruit* Fruit boiled and eaten plain or boiled with dried peaches. *Sauce & Relish* Juice mixed with sugar and used to make syrup. (165:222) *Unspecified* Tunas stewed with dried peaches and eaten. (32:37) **Okanagon** *Staple* Used as a principal food. (178:239) **Papago** *Dried Food* Pulp spread on grass, sun dried for 2 days, stored, and used for food. (34:22) Fruits dried, stored in jars, and used as sweets. (34:46) Fruits pit cooked, dried, boiled, salted, and eaten with pinole. (95:262) *Fruit* Fruits eaten fresh. *Sauce & Relish* Pulp mashed with sticks, juice squeezed, strained, boiled, strained again, and used as a syrup. (34:22) Fruits made into a syrup. (34:46) *Unspecified* Large, waxy

flowers fried in grease or lard and used for food. (34:16) *Vegetable Joints* pit baked and used as greens. Joints roasted in ashes and eaten as greens. (34:46) **Pima** *Unspecified* Pulp sliced, cooked with mesquite bean pods, and used for food. (95:257) Cooked and used for food. (95:262) **Pima, Gila River** *Unspecified* Buds used for food. (133:6) **Southwest Indians** *Unspecified* Buds used for food. (17:15) **Spokan** *Unspecified* Species used for food. (178:344) **Thompson, Upper (Spences Bridge)** *Unspecified* Peeled stems baked or steamed. (164:484) **Yavapai** *Bread & Cake* Ground fruit made into cakes. *Dried Food* Fruit dried in cakes or opened and dried without expressing juice. *Fruit* Salty fruit eaten only out of necessity and the seeds spat out. Juice used as a beverage. (65:257)
- *Other*—**Havasupai** *Containers* Juice used to mix with pottery clay. (197:234) **Kiowa** *Hunting & Fishing Item* Sharp thorns used as points for small arrows to kill birds. *Protection* Cut stem secretion applied to buckskin moccasins as a varnish. (192:45) **Navajo** *Designs* Plant shape used as form for figures in the sand painting of the Cactus People for the Wind Chant. (55:64) **Papago** *Protection* Used between fence posts to protect tobacco plants from marauding animals. (34:37)

Opuntia spinosior (Engelm.) Toumey, Walkingstick Cactus
- *Food*—**Papago** *Staple* Pit baked buds, fruits, and joints considered a staple food. (36:60)

Opuntia tunicata (Lehm.) Link & Otto, Thistle Cholla
- *Drug*—**Hawaiian** *Laxative* Leaf juice and roots used for constipation. *Reproductive Aid* Leaf juice and roots used by expectant mothers. (as *O. tuna* 2:73)

Opuntia versicolor Engelm. ex Coult., Staghorn Cholla
- *Food*—**Papago** *Staple* Pit baked buds, fruits, and joints considered a staple food. (36:60) *Vegetable* Young shoots and buds eaten as greens in summer. (34:14) **Pima** *Fruit* Green fruits boiled with saltbush and used for food. (47:59) Fruits eaten raw. (146:78)

Opuntia whipplei Engelm. & Bigelow, Whipple Cholla
- *Drug*—**Hopi** *Antidiarrheal* Root chewed or compound decoction taken for diarrhea. (200:34, 86)
- *Food*—**Apache, White Mountain** *Dried Food* Fruit dried for winter use. *Fruit* Fruit eaten raw or stewed. (136:159) **Hopi** *Unspecified* Buds boiled and eaten with cornmeal boiled bread. (120:19) **Zuni** *Dried Food* Fruit, with the spines rubbed off, dried for winter use. (166:69) *Fruit* Spineless fruits eaten raw or stewed. (32:36) Fruit, with the spines rubbed off, eaten raw or stewed. *Porridge* Dried fruit ground into a flour, mixed with parched cornmeal and made into a mush. (166:69)
- *Other*—**Navajo, Ramah** *Ceremonial Items* Used to make cactus prayer stick, Chiricahua Windway. Branches made into a wand and used in Red Antway. The Antway wand consisted of five cactus branches with branches of rabbitbrush and other plants wrapped around their combined bases. The base was wrapped with yucca fiber. A small colored wooden disk was attached to each branch by a yucca fiber, each disk a different color. (191:37)

Orbexilum, Fabaceae
Orbexilum pedunculatum (P. Mill.) Rydb. **var. *pedunculatum***, Sampson's Snakeroot

- *Drug*—**Catawba** *Dermatological Aid* Root used as salve for boils, sores, and wounds. (as *Psorglea pedunculata* 157:188) Poultice of boiled roots applied to sores. (as *Psoralea pedunculata* 177:32) *Orthopedic Aid* Root used as salve for broken bones. (as *Psorglea pedunculata* 157:188)

Orbexilum pedunculatum var. ***psoralioides*** (Walt.) Isely, Sampson's Snakeroot
- *Drug*—**Cherokee** *Abortifacient* Infusion taken for obstructed menstruation. *Diaphoretic* Taken as a diaphoretic. *Gastrointestinal Aid* Taken for colic and indigestion. *Gynecological Aid* Infusion taken to "check discharge." *Tonic* Taken as a tonic. (as *Psoralea psoralioides* 80:55)

Orchidaceae, see *Aplectrum, Calypso, Coeloglossum, Corallorrhiza, Cypripedium, Epipactis, Goodyera, Habenaria, Liparis, Malaxis, Odontoglossum, Piperia, Platanthera, Spiranthes, Zeuxine*

Oreoxis, Apiaceae
Oreoxis alpina (Gray) Coult. & Rose **ssp. *alpina***, Alpine Oreoxis
- *Drug*—**Navajo** *Ceremonial Medicine* Plant, greasewood, and wild privet used as a medicine for the Coyote Chant. (as *Cymopterus alpinus* 55:67)

Ornithogalum, Liliaceae
Ornithogalum umbellatum L., Sleepydick
- *Other*—**Thompson** *Decorations* Plant used only ornamentally. (187:127)

Orobanchaceae, see *Boschniakia, Conopholis, Epifagus, Orobanche*

Orobanche, Orobanchaceae
Orobanche californica Cham. & Schlecht., California Broomrape
- *Drug*—**Paiute** *Cold Remedy* Decoction of plant taken for colds. *Pulmonary Aid* Decoction of plant taken for pneumonia or pulmonary trouble. (180:108)

Orobanche cooperi (Gray) Heller, Desert Broomrape
- *Food*—**Pima, Gila River** *Unspecified* Stalk, below the ground, eaten cooked or raw. (133:5) Roots eaten raw or roasted. (133:7)

Orobanche cooperi (Gray) Heller **ssp. *cooperi***, Cooper's Broomrape
- *Food*—**Cahuilla** *Unspecified* Root peeled and eaten. (as *O. ludoviciana* var. *cooperi* 15:97)

Orobanche fasciculata Nutt., Clustered Broomrape
- *Drug*—**Blackfoot** *Dermatological Aid* Chewed root blown onto wounds by medicine men. (as *Thalesia fasciculata* 114:276) **Keres, Western** *Pulmonary Aid* Roots eaten as a lung medicine. (as *Thalesia fasciculata* 171:72) **Montana Indian** *Cancer Treatment* Parasite (cancer root) on sweet sage roots used for cancer. *Poison* Plant poisonous to stock. (as *Aphyllon fasciculatum* 19:6) **Navajo** *Dermatological Aid* Infusion of leaves used as wash for sores. (55:77) Poultice of plant applied to wounds and open sores. (90:153) **Navajo, Ramah** *Panacea* Plant used as "life medicine." *Pediatric Aid* Plant used for birth injuries. (191:45) **Zuni** *Hemorrhoid Remedy* Powdered plant inserted

into rectum as a specific for hemorrhoids. (as *Thalesia fasciculata* 166:61)
- **Food**—**Gosiute** *Unspecified* Entire plant sometimes eaten. (as *Aphyllon fasciculatum* 39:361) **Navajo, Ramah** *Unspecified* Roasted in ashes, the skin peeled off and eaten like a baked potato. (191:45) **Paiute, Northern** *Unspecified* Stems eaten raw or boiled. (59:49)

Orobanche ludoviciana Nutt., Louisiana Broomrape
- **Drug**—**Blackfoot** *Dermatological Aid* Plant chewed by medicine men and blown upon wounds. (97:53) **Pima** *Dermatological Aid* Poultice of stems applied to ulcerated sores. (47:49)
- **Food**—**Pima** *Unspecified* Young sprouts covered with hot ashes, baked and the lower parts eaten. (47:49)

Orobanche ludoviciana Nutt. **ssp.** *ludoviciana*, Sand Broomrape
- **Drug**—**Navajo, Kayenta** *Dermatological Aid* Powdered plant applied to gunshot wounds. (as *O. multiflora* var. *arenosa* 205:43)

Orobanche ludoviciana **ssp.** *multiflora* (Nutt.) Collins, Manyflowered Broomrape
- **Food**—**Pima** *Unspecified* Plants cooked and used for food. (as *Orobranche multiflora* 95:264)

Orobanche **sp.**, Broomrape
- **Drug**—**Navajo** *Other* Plant used for infections. (90:153)

Orthilia, Ericaceae

Orthilia secunda (L.) House, Sidebells Wintergreen
- **Drug**—**Carrier, Southern** *Eye Medicine* Strong decoction of root used as an eyewash. (as *Pyrola secunda* 150:62)

Orthocarpus, Scrophulariaceae

Orthocarpus attenuatus Gray, Attenuate Indian Paintbrush
- **Food**—**Miwok** *Unspecified* Dried, parched, pulverized seeds used for food. (12:155)

Orthocarpus densiflorus Benth., Denseflower Indian Paintbrush
- **Other**—**Pomo, Kashaya** *Ceremonial Items* Flowers used in dance wreaths at the Strawberry Festival in May. (72:35)

Orthocarpus faucibarbatus Gray, Triphysaria
- **Other**—**Pomo, Kashaya** *Ceremonial Items* Flowers used in dance wreaths at the Strawberry Festival in May. (72:35)

Orthocarpus lithospermoides Benth., Indian Paintbrush
- **Food**—**Mendocino Indian** *Forage* Plants eaten sparingly by horses. (41:387)

Orthocarpus luteus Nutt., Yellow Owlclover
- **Dye**—**Blackfoot** *Red* Leaves crushed and pressed firmly into skins, horsehair, and feathers as a red dye. (97:53) Plant pounded and pressed firmly into the gopher skin as a red dye. (114:276) *Red-Brown* Whole, blooming plant pressed firmly into skins, horsehair, and feathers as a reddish tan dye. (97:53) **Great Basin Indian** *Yellow* Whole plant used to make a yellow dye. (121:50)

Orthocarpus purpurascens Benth., Exserted Indian Paintbrush
- **Other**—**Pomo, Kashaya** *Ceremonial Items* Flowers used in dance wreaths at the Strawberry Festival in May. (72:35)

Orthocarpus purpureoalbus Gray ex S. Wats., Purplewhite Owlclover
- **Drug**—**Navajo, Ramah** *Cathartic* Decoction of whole plant taken as a cathartic. *Ceremonial Medicine* Compound decoction used as ceremonial medicine. *Gastrointestinal Aid* Cold infusion of plant taken for heartburn. (191:44)

Orthocarpus **sp.**
- **Drug**—**Costanoan** *Cough Medicine* Decoction of foliage used for coughs. (21:15)

Oryza, Poaceae

Oryza sativa L., Rice
- **Food**—**Haisla & Hanaksiala** *Staple* Grains used for food. (43:207) **Kitasoo** *Unspecified* Grains used for food. (43:324) **Oweekeno** *Unspecified* Grains used for food. (43:81) **Seminole** *Unspecified* Seeds used for food. (169:470)

Oryzopsis, Poaceae

Oryzopsis hymenoides (Roemer & J. A. Schultes) Ricker ex Piper, Indian Ricegrass
- **Food**—**Apache, Western** *Porridge* Seeds ground, mixed with cornmeal and water, and made into a mush. (as *Eriocoma cuspidata* 26:189) **Apache, White Mountain** *Bread & Cake* Seeds ground and used to make bread and pones. (as *Eriocoma cuspidata* 136:149) *Fodder* Plant used for hay. (as *Eriocoma cuspidata* 136:157) *Porridge* Seeds ground, mixed with meal and water, and eaten as mush. (as *Eriocoma cuspidata* 136:149) *Unspecified* Seeds used for food. (as *Eriocoma cuspidata* 136:157) **Gosiute** *Unspecified* Seeds formerly used for food. (as *O. cuspidata* 39:375) **Havasupai** *Bread & Cake* Seeds parched, ground fine, boiled, thickened, made into balls, and eaten as dumplings. (197:66) *Soup* Seeds and Indian millet seeds ground and used to make soup or mush. (197:73) *Staple* Seeds ground and eaten as a ground or parched meal. (197:67) **Hopi** *Bread & Cake* Seeds ground with corn into fine meal and used to make tortilla bread. (120:20) *Staple* Ground seeds used to make meal. (as *Stipa hymenoides* 190:158) *Starvation Food* Seeds eaten, especially in time of famine. (42:338) Plants formerly used for food during famines. (as *O. membranacea* 101:43) Seeds used during famines. (200:65) **Kawaiisu** *Staple* Seeds pounded into a meal and eaten dry. (206:46) **Montana Indian** *Unspecified* Seeds used for food. (as *Ericooma cuspidata* 19:11) **Navajo** *Bread & Cake* Ground seeds made into cakes. (55:26) Seeds ground and made into bread and dumplings. (165:223) *Fodder* Plant used as a fodder for both wild and domesticated animals. *Forage* Plant used as a forage for both wild and domesticated animals. (90:154) *Porridge* Seeds ground and made into gruel. (165:223) *Staple* Ground seeds used for food. (90:154) *Unspecified* Seeds used for food. (as *Eriocoma cuspidata* 32:27) **Navajo, Ramah** *Fodder* Young plants used as horse feed. *Porridge* Seeds finely ground and cooked into a mush with milk or water. (191:16) **Paiute** *Porridge* Seeds ground into a meal for mush. (118:26–27) *Sauce & Relish* Ground seeds used for sauce. *Staple* Ground seeds used for flour. (118:32) Roasted and ground into flour. (167:244) **Paiute, Northern** *Porridge* Seeds dried, winnowed, ground into a flour, and used to make mush. *Soup* Seeds dried, winnowed, ground into a flour, and used to make soup. *Special Food* Seeds considered a good food to eat when suffering from stomachaches, colic,

or aching bones. When a person was suffering from any of these sicknesses, Indian ricegrass seeds should have been the only food eaten. *Staple* Seeds used as a staple food. *Winter Use Food* Seeds stored for winter use. (59:46) **Zuni** *Staple* Ground seeds used as a staple before the availability of corn. After the introduction of corn, the ground seeds were mixed with cornmeal and made into steamed balls or pats. (as *Eriocoma cuspidata* 166:67) *Unspecified* Used especially in earlier times as an important source of food. (as *Eriocoma cuspidata* 32:27)

- **Other**—**Apache, White Mountain** *Cash Crop* Plant gathered and sold. (as *Eriocoma cuspidata* 136:149)

Osmorhiza, Apiaceae

Osmorhiza berteroi DC., Sweetcicely

- **Drug**—**Bella Coola** *Cathartic* Infusion of ground root pieces taken as a purgative. *Emetic* Infusion of ground root pieces taken as an emetic. (as *O. chilensis* 184:201) **Blackfoot** *Cold Remedy* Hot drink containing root taken for colds. *Throat Aid* Hot drink containing root taken for tickling in throat. (as *Washingtonia divaricata* 114:276) *Veterinary Aid* Whole plant fed to mares in the winter to put them into condition for foaling. (as *O. chilensis* 97:49) Plant given to mares to put them in good foaling condition. (as *Washingtonia divaricata* 114:276) Roots placed in mares' mouths and chewed to put them in good condition for foaling. (as *Washingtonia divaricata* 118:49) **Cheyenne** *Adjuvant* Plant used as an ingredient in all medicines. *Cold Remedy* Root chewed or infusion of leaves taken for colds. *Stimulant* Root chewed to "bring one around." (as *O. chilensis* 83:40) **Karok** *Analgesic* Root chewed for headaches. *Panacea* Roots used for any illness. *Preventive Medicine* Roots placed under pillow to prevent sickness. *Psychological Aid* Infusion of roots used as a bath for grieving person. (as *Osmorrhiza nuda* var. *brevipes* 148:386) **Kwakiutl** *Emetic* Seeds and roots used as an emetic. *Poison* Plant "sure to kill" if eaten. (as *Osmorhiza chilensis* 183:277) **Swinomish** *Love Medicine* Roots chewed and used as powerful love charms. (as *O. chilensis* 79:41)
- **Food**—**Karok** *Vegetable* Young tops eaten raw as greens. (as *Osmorrhiza nuda* var. *brevipes* 148:386) **Miwok** *Vegetable* Boiled leaves used for food. (as *Osmorrhiza nuda* 12:160) **Okanagon** *Unspecified* Thick, aromatic roots used for food. (as *Washingtonia nuda* 125:36) **Thompson** *Unspecified* Thick, aromatic roots eaten. (as *Osmorrhiza nuda* 164:480) Thick, aromatic roots used for food. (as *Washingtonia nuda* 125:36)
- **Other**—**Karok** *Good Luck Charm* Plant growing in a place where it had never been seen before was very good luck. (as *Osmorrhiza nuda* var. *brevipes* 148:386)

Osmorhiza brachypoda Torr., California Sweetcicely

- **Drug**—**Kawaiisu** *Cold Remedy* Decoction of roots taken for colds. *Cough Medicine* Decoction of roots taken for coughs. *Dermatological Aid* Infusion of mashed plant used as a hair wash to kill fleas. (206:47)
- **Other**—**Kawaiisu** *Insecticide* Infusion of mashed plant used as a hair wash to kill fleas. (206:47)

Osmorhiza claytonii (Michx.) C. B. Clarke, Clayton's Sweetroot

- **Drug**—**Chippewa** *Dermatological Aid* Poultice of moistened, pulverized root used for ulcers, especially running sores. (53:354) *Throat Aid* Decoction of root gargled or root chewed for sore throat. (53:342) **Menominee** *Dietary Aid* Root eaten "to enable one to put on flesh." (151:55) Branch or piece of root eaten cautiously for losing flesh, a fattener. (151:72) *Eye Medicine* Decoction of root used as an eyewash for sore eyes. (54:133) **Ojibwa** *Gynecological Aid* Infusion of root used to ease parturition. *Throat Aid* Infusion of root taken for sore throat. (153:391) **Tlingit** *Cough Medicine* Warm infusion of whole plant taken for coughs. (as *Osmorhyza brevistyla* 107:283)

Osmorhiza depauperata Phil., Bluntseed Sweetroot

- **Food**—**Isleta** *Beverage* Roots and stems boiled to make a beverage. (as *Washingtonia obtusa* 100:45)

Osmorhiza longistylis (Torr.) DC., Longstyle Sweetroot

- **Drug**—**Cheyenne** *Gastrointestinal Aid* Infusion of pulverized leaves, stems, and roots taken for bloated stomachs or disordered stomachs. *Kidney Aid* Infusion of pulverized leaves, stems, and roots taken for kidney troubles. (76:181) **Chippewa** *Gynecological Aid* Infusion of roots taken for amenorrhea. *Veterinary Aid* Decoction of roots used as nostril wash to increase dog's sense of scent. (71:137) **Meskwaki** *Dietary Aid* Compound infusion of leaves taken "to regain flesh and strength." *Eye Medicine* Used chiefly as an eye remedy. *Panacea* Used as a "good medicine for everything." *Veterinary Aid* Grated root mixed with salt for distemper in horses. (152:249) **Ojibwa** *Gynecological Aid* Infusion of root used to ease parturition. *Throat Aid* Infusion of root taken for sore throat. (153:391) **Omaha** *Dermatological Aid* Poultice of pounded root applied to boils. **Pawnee** *Stimulant* Decoction of root taken for weakness and general debility. (as *Washingtonia longistylis* 70:107) **Potawatomi** *Eye Medicine* Root used to make an eye lotion. *Gastrointestinal Aid* Infusion of root used as a stomachic. (154:86) **Winnebago** *Dermatological Aid* Poultice of pounded root applied to wounds. (as *Washingtonia longistylis* 70:107)
- **Food**—**Omaha** *Fodder* Root used to attract horses and catch them.[52] **Ponca** *Fodder* Root used to attract horses and catch them.[52] (as *Washingtonia longistylis* 70:107) **Potawatomi** *Fodder* Chopped roots added to oats or other seeds to fatten the ponies. (154:124)

Osmorhiza occidentalis (Nutt. ex Torr. & Gray) Torr., Western Sweetroot

- **Drug**—**Blackfoot** *Breast Treatment* Infusion of roots applied to swollen breasts. (86:77) *Cough Medicine* Infusion of plant taken for coughs. (86:72) *Dermatological Aid* Infusion of roots applied to sores. (86:77) *Eye Medicine* Infusion of roots used for eye troubles. (86:81) *Gynecological Aid* Roots used by women as a feminine deodorant. (86:124) Infusion of roots taken by women to induce labor. (86:61) *Nose Medicine* Infusion of roots used for nose troubles. (86:81) **Okanagan-Colville** *Febrifuge* Infusion of roots taken for fevers. *Toothache Remedy* Roots applied to the tooth for toothaches. (188:70) **Paiute** *Analgesic* Decoction of root taken for stomachaches and gas pains. *Cathartic* Decoction of root taken as a physic. *Cold Remedy* Decoction of root taken for colds. *Dermatological Aid* Decoction of root used as a wash for venereal sores and skin rashes. Hot decoction of root applied to kill head lice. Poultice of pulped roots applied to cuts, sores, swellings, and bruises. *Disinfectant* Decoction of root used as an antiseptic wash for venereal sores. *Eye Medicine* Decoction of root used as an eyewash. *Febrifuge* Decoction of root taken for fever, and

for chills. *Gastrointestinal Aid* Decoction of root taken for stomachaches, indigestion, or gas pains. *Misc. Disease Remedy* Decoction of root taken for colds and influenza. (180:109, 110) *Pulmonary Aid* Infusion of roots taken "when sick in the chest, not feeling well." (111:93) Decoction of root taken for pulmonary disorders and pneumonia. *Snakebite Remedy* Poultice of pulped roots applied to snakebites. *Throat Aid* Root chewed for sore throat. *Venereal Aid* Decoction of root used as an antiseptic wash for venereal sores. Simple or compound decoction of root taken for venereal disease. (180:109, 110) **Paiute, Northern** *Analgesic* Roots chewed for chest pains. *Antirheumatic (External)* Roots used to make a liniment. *Cold Remedy* Roots chewed for colds. *Dermatological Aid* Decoction of roots taken for sores. *Eye Medicine* Decoction of roots taken for sore eyes. (59:128) **Shoshoni** *Analgesic* Decoction of root taken for stomachaches and gas pains. Fresh root inserted into nostrils for headaches. *Antidiarrheal* Decoction of root taken for diarrhea. (180:109, 110) *Cathartic* Infusion of roots taken as a general physic. (118:41) Decoction of root taken as a physic. (180:109, 110) *Cold Remedy* Infusion of aromatic roots with Indian balsam taken for heavy colds. (118:38) Simple or compound decoction of root taken or pulverized root smoked for colds. *Cough Medicine* Compound decoction of roots taken for coughs, "heavy colds," and fevers. *Dermatological Aid* Hot decoction of root applied to kill head lice. Poultice of pulped roots applied to cuts, sores, swellings, and bruises. *Disinfectant* Decoction of root used as an antiseptic wash for measles. *Eye Medicine* Decoction of root used as an eyewash. *Febrifuge* Simple or compound decoction of root taken for fevers. *Gastrointestinal Aid* Decoction of root taken for stomachaches, indigestion, or gas pains. *Gynecological Aid* Decoction of root taken to regulate menstrual disorders. *Misc. Disease Remedy* Decoction of root taken for colds and influenza. Decoction of root used as an antiseptic wash for measles. (180:109, 110) *Pulmonary Aid* Infusion of aromatic roots with Indian balsam taken for pneumonia. (118:38) Decoction of root used for pulmonary disorders, pneumonia, and whooping cough. *Snakebite Remedy* Poultice of pulped roots applied to snakebites. *Throat Aid* Root chewed for sore throat. *Tonic* Decoction of root taken as a tonic to protect against illness. *Toothache Remedy* Poultice of raw root applied for toothaches. *Venereal Aid* Infusion or decoction of root taken for venereal disease. (180:109, 110) **Thompson** *Cold Remedy* Root chewed for colds. (187:158) **Washo** *Analgesic* Decoction of root taken for stomachaches and gas pains. *Cathartic* Decoction of root taken as a physic. *Cold Remedy* Decoction of root taken for colds. *Gastrointestinal Aid* Decoction of root taken for stomachaches, indigestion, or gas pains. *Misc. Disease Remedy* Decoction of root taken for colds and influenza. *Pulmonary Aid* Decoction of root taken for pulmonary disorders and pneumonia. (180:109, 110)
- *Food*—**Blackfoot** *Candy* Root chewed, especially during the winter, as a confection. (86:103) **Shoshoni** *Spice* Steeped seeds added to dishes for flavoring. (118:29)
- *Dye*—**Blackfoot** *Unspecified* Stems mixed with ocher and applied to robes. (86:115)
- *Other*—**Blackfoot** *Incense & Fragrance* Root pieces kept in quivers and clothing as a deodorant. (86:115) Infusion of roots used to sweeten diapers. (86:124) **Okanagan-Colville** *Hunting & Fishing Item* Roots carried by hunters for the "good smell" to cover the human scent. (188:70) **Paiute** *Insecticide* Decoction of root used as a dip to kill chicken lice. (180:109, 11)

Osmorhiza purpurea (Coult. & Rose) Suksdorf, Purple Sweetroot
- *Drug*—**Songish** *Love Medicine* Roots used by girls as love charms. (182:89)

Osmorhiza sp.
- *Drug*—**Bella Coola** *Emetic* Infusion of ground root taken as an emetic and sometimes acted as a purgative. *Pulmonary Aid* Infusion of ground root used for pneumonia. (as *Osmorhiza*, sweet cicely 150:61) **Iroquois** *Analgesic* Infusion of roots taken for fevers and headaches. *Antidiarrheal* Infusion of plants taken for diarrhea. *Blood Medicine* Complex compound decoction of powdered roots taken as blood purifier. (as *Osmorrhiza*, sweet cicely 87:397) *Cathartic* Infusion of roots taken as a physic. (as *Osmorrhiza*, sweet cicely 87:398) *Febrifuge* Infusion of roots given to children or adults with intermittent fevers. *Hunting Medicine* Infusion of roots used as a soak for muskrat traps, a "hunting medicine." Roots chewed and spit on bait, a "fishing medicine." *Laxative* Compound decoction of roots taken for venereal disease or to loosen bowels. *Love Medicine* Roots chewed for an anti-love medicine. *Pediatric Aid* Infusion of roots given to children with intermittent fevers. (as *Osmorrhiza*, sweet cicely 87:397) *Tonic* Infusion of roots taken as a tonic. (as *Osmorrhiza*, sweet cicely 87:398) *Venereal Aid* Compound decoction of roots or powdered roots taken for venereal disease. (as *Osmorrhiza*, sweet cicely 87:397) **Miwok** *Snakebite Remedy* Chewed and put on snakebite. (snake root 12:171) **Sanpoil** *Dermatological Aid* Poultice of inner portion of root applied to sores, wounds, boils, or bruises. *Unspecified* Infusion taken instead of water "when a person was generally ill." *Veterinary Aid* Infusion of crushed root used as a wash for horse's sore back or scabby body. (snake root 131:221)
- *Other*—**Iroquois** *Fertilizer* Decoction of roots and whole plant used as wash for seeds, a "seed medicine." (as *Osmorrhiza*, sweet cicely 87:397)

Osmundaceae, see *Osmunda*

Osmunda, Osmundaceae, fern
Osmunda cinnamomea L., Cinnamon Fern
- *Drug*—**Cherokee** *Antirheumatic (External)* Compound decoction of root applied with warm hands for rheumatism. (80:8) Decoction of roots rubbed on area affected by rheumatism. (177:4) *Febrifuge* Compound decoction used for chills. *Snakebite Remedy* Root chewed, a portion swallowed, and remainder applied to the wound. *Tonic* Cooked fronds eaten as "spring tonic." (80:33) **Iroquois** *Analgesic* Decoction taken for headache and joint pain. *Antirheumatic (External)* Decoction taken for rheumatism. *Cold Remedy* Decoction taken for colds. *Gynecological Aid* Roots used for "woman's troubles." *Orthopedic Aid* Decoction taken for joint pain. *Panacea* Decoction taken for malaise. *Venereal Aid* Cold, compound infusion used as a wash and decoction taken for affected parts. *Veterinary Aid* Compound chopped and added to cows food for difficult birth of a calf. (87:261) **Menominee** *Gynecological Aid* Used to promote the flow of milk and for caked breasts. (151:70)
- *Food*—**Abnaki** *Snack Food* Used as a nibble. (144:152) *Unspecified*

White base of plant eaten raw. (144:162) **Menominee** *Soup* Frond tips simmered to remove the ants, added to soup stock, and thickened with flour. (151:70)
- **Other**—Menominee *Hunting & Fishing Item* Shoots eaten by hunters to have same scent as shoot-eating deer; deer will not be frightened away. (151:70)

Osmunda claytoniana L., Interrupted Fern
- **Drug**—Iroquois *Blood Medicine* Cold, compound decoction taken for weak blood. *Venereal Aid* Compound decoction taken for gonorrhea. (87:260)

Osmunda regalis L., Royal Fern
- **Drug**—Iroquois *Anticonvulsive* Infusion of fronds and wild ginger rhizomes used by children with convulsions from intestinal worms. (141:33) *Blood Medicine* Decoction taken by women for watery blood. *Gynecological Aid* Decoction taken by women for strong menses. Decoction used when "girls leak rotten; affected women can't raise children." *Kidney Aid* Decoction taken by women for cold in kidneys. (87:260) *Pediatric Aid* Infusion of fronds and wild ginger rhizomes used by children with convulsions from intestinal worms. (141:33) **Menominee** *Unspecified* Roots used medicinally for unspecified purpose. (151:44) **Seminole** *Other* Complex infusion of roots taken for chronic conditions. (169:272) *Pediatric Aid* Plant used for chronically ill babies. (169:329) *Psychological Aid* Infusion of plant used to steam and bathe the body for insanity. (169:292) *Unspecified* Plant used for medicinal purposes. (169:162)

Osmunda sp.
- **Fiber**—Iroquois *Mats, Rugs & Bedding* Made into pillows and used by children under their lower backs to prevent bed-wetting. (142:82)

Osteomeles, Rosaceae
Osteomeles anthyllidifolia (Sm.) Lindl., Hawaii Hawthorn
- **Drug**—Hawaiian *Dermatological Aid* Leaves, root bark and salt pounded and the resulting liquid used on deep cuts. *Laxative* Seeds and buds chewed and given to children as a laxative for general debility of the body. (2:38)

Ostrya, Betulaceae
Ostrya virginiana (P. Mill.) K. Koch, Eastern Hop Hornbeam
- **Drug**—Cherokee *Blood Medicine* Infusion of bark taken to build up blood. *Orthopedic Aid* Decoction of bark used to bathe sore muscles. *Toothache Remedy* Infusion of bark held in mouth for toothache. (80:39) **Chippewa** *Antihemorrhagic* Compound infusion of heartwood taken for lung hemorrhages. (53:340) *Antirheumatic (External)* Compound decoction of heartwood used as herbal steam for rheumatism. (53:362) *Cough Medicine* Compound liquid made from wood taken as a cough syrup. (53:340) *Kidney Aid* Decoction of wood taken for kidney trouble. (53:346) *Pulmonary Aid* Compound infusion of inner wood taken for lung hemorrhage. (53:340) **Delaware, Ontario** *Gynecological Aid* Compound containing root used for "female weakness." *Tonic* Compound containing root used as a tonic. (175:82) **Iroquois** *Cancer Treatment* Decoction of bark used for rectum cancer. (87:299) *Cough Medicine* Decoction of heart chips taken for catarrh coughs. (87:298) *Dermatological Aid* Infusion used for swellings. (87:299) *Tuberculosis Remedy* Compound decoction of bark taken for consumption. (87:298) **Potawatomi** *Antidiarrheal* Infusion of bark used for flux. *Antihemorrhagic* Compound decoction of heartwood chips taken for hemorrhages. (154:44)
- **Fiber**—Chippewa *Building Material* Used as frames for dwellings. (53:377)
- **Other**—Lakota *Hunting & Fishing Item* Wood used to make bows. *Paint* Blossoms used for painting the face. (139:40) **Malecite** *Hunting & Fishing Item* Used to make bows. *Tools* Used to make utensil handles. (160:6)

Oxalidaceae, see *Oxalis*

Oxalis, Oxalidaceae
Oxalis corniculata L., Creeping Woodsorrel
- **Drug**—Cherokee *Anthelmintic* Infusion taken and used as a wash for children with hookworms. *Antiemetic* Cold infusion of leaf taken to stop vomiting. *Blood Medicine* Infusion taken for blood. *Cancer Treatment* Used for cancer "when it is first started." *Dermatological Aid* Salve of infusion of leaf mixed with sheep grease used for sores. *Oral Aid* Leaves chewed for "disordered saliva" and sore mouth. *Pediatric Aid* Infusion taken and used as a wash for children with hookworms. *Throat Aid* Chewed for sore throat. (80:56)
- **Food**—Cherokee *Unspecified* Species used for food. (80:56) **Iroquois** *Vegetable* Eaten raw, sometimes with salt. (196:118)
- **Dye**—Menominee *Yellow* Boiled whole plant used as a yellow dye. (151:78)

Oxalis drummondii Gray, Drummond's Woodsorrel
- **Drug**—Navajo, Ramah *Analgesic* Decoction of bulb used for pain. *Dermatological Aid* Poultice of bulbs, alone or in compound, applied to sores. (as *O. amplifolia* 191:34)

Oxalis montana Raf., Mountain Woodsorrel
- **Food**—Potawatomi *Dessert* Plant gathered, cooked, and sugar added to make a dessert. (as *O. acetosella* 154:106)
- **Dye**—Menominee *Yellow* Boiled whole plant used as a yellow dye. (as *O. acetosella* 151:78)

Oxalis oregana Nutt., Oregon Oxalis
- **Drug**—Cowlitz *Eye Medicine* Fresh juice from plant applied to sore eyes. (79:39) **Karok** *Dietary Aid* Plant used by anyone who does not feel like eating. (148:385) **Makah** *Other* Decoction of plants taken for "summer complaint." (79:39) **Pomo** *Antirheumatic (External)* Decoction of whole plant used as a wash for rheumatism. (66:13) **Pomo, Kashaya** *Antirheumatic (External)* Decoction of whole plant used to wash parts of the body afflicted with rheumatism. (72:108) **Quileute** *Dermatological Aid* Poultice of wilted leaves applied to boils. **Quinault** *Eye Medicine* Chewed root juice applied to sore eyes. (79:39) **Tolowa** *Antirheumatic (External)* Poultice of plant applied to swollen areas on the skin. *Dermatological Aid* Poultice of plant applied to sores. *Disinfectant* Poultice of plant applied to draw out infections. (6:42)
- **Food**—Cowlitz *Unspecified* Leaves eaten fresh or cooked. (79:39) **Makah** *Unspecified* Leaves eaten fresh. (67:284) **Pomo, Kashaya** *Sour* Flowering plant leaves and stem chewed for the sour taste. (72:108) **Quileute** *Unspecified* Leaves eaten by hunters or by those traveling

in the woods. **Quinault** *Unspecified* Leaves cooked with grease and used for food. (79:39) **Tolowa** *Unspecified* Plant eaten with dried fish. **Yurok** *Unspecified* Plant eaten with dried fish. (6:42)
- *Other*—**Pomo, Kashaya** *Toys & Games* Children ate as many leaves as they could without making an awful face; a children's game. (72:108)

Oxalis sp., Sorrel
- *Drug*—**Iroquois** *Alterative* Sprouts used as an alterative. (124:93)
- *Food*—**Iroquois** *Unspecified* Sprouts eaten raw. (124:93)

Oxalis stricta L., Common Yellow Oxalis
- *Drug*—**Iroquois** *Blood Medicine* Compound decoction of roots taken as a blood medicine. (as *O. europaea* 87:366) *Febrifuge* Infusion of plant taken for fever. *Gastrointestinal Aid* Infusion of plant taken for cramps and nausea. (as *O. europaea* 87:365) *Oral Aid* Infusion of plant used as wash to refresh the mouth. (as *O. europaea* 87:366) *Other* Infusion of plant taken for "summer complaint." *Witchcraft Medicine* Compound used as an anti-witch medicine. (as *O. europaea* 87:365) **Kiowa** *Oral Aid* Leaves chewed on long walks to relieve thirst. The Kiowa name means "salt weed." This name may indicate that there was an early realization that the loss of salt through perspiration may be counteracted by chewing the leaves of this plant. (192:35) **Omaha** *Dermatological Aid* Poultice of plants applied for swellings. (68:335)
- *Food*—**Cherokee** *Vegetable* Leaves used for food. (126:49) **Meskwaki** *Sour* Eaten for its acidity. (152:271) **Omaha** *Fodder* Pounded bulbs fed to horses to make them fleet. *Unspecified* Leaves, flowers, scapes, and bulbs used for food by children. **Pawnee** *Fodder* Pounded bulbs fed to horses to make them fleet. *Forage* Plant much esteemed by buffalo. *Unspecified* Leaves, flowers, scapes, and bulbs used for food by children. Plant considered to have a salty and sour taste. **Ponca** *Fodder* Pounded bulbs fed to horses to make them fleet. *Unspecified* Leaves, flowers, scapes, and bulbs used for food by children. (as *Xanthoxalis stricta* 70:98)
- *Dye*—**Menominee** *Yellow* Boiled whole plant used as a yellow dye. (151:78) **Meskwaki** *Orange* Whole plant boiled to obtain an orange dye. (152:271)

Oxalis violacea L., Violet Woodsorrel
- *Drug*—**Cherokee** *Anthelmintic* Infusion taken and used as a wash for children with hookworms. *Antiemetic* Cold infusion of leaf taken to stop vomiting. *Blood Medicine* Infusion taken for blood. *Cancer Treatment* Used for cancer "when it is first started." *Dermatological Aid* Salve of infusion of leaf mixed with sheep grease used for sores. *Oral Aid* Leaves chewed for "disordered saliva" and sore mouth. *Pediatric Aid* Infusion taken and used as a wash for children with hookworms. *Throat Aid* Chewed for sore throat. (80:56) **Pawnee** *Veterinary Aid* Pounded bulbs fed to horses to make them run faster. (as *Ionoxalis violacea* 70:98)
- *Food*—**Apache, Chiricahua & Mescalero** *Unspecified* Mixed with other leaves and cooked or eaten raw. Bulbs eaten raw or boiled. (33:47) **Cherokee** *Unspecified* Species used for food. (80:56) **Omaha** *Fodder* Pounded bulbs fed to horses to make them fleet. *Unspecified* Leaves, flowers, scapes, and bulbs used for food by children. **Pawnee** *Fodder* Pounded bulbs fed to horses to make them fleet. *Unspecified* Plant considered to have a salty and sour taste. Leaves, flowers, scapes, and bulbs used for food by children. **Ponca** *Fodder* Pounded bulbs fed to horses to make them fleet. *Unspecified* Leaves, flowers, scapes, and bulbs used for food by children. (as *Ionoxalis violacea* 70:98)

Oxydendrum, Ericaceae
Oxydendrum arboreum (L.) DC., Sourwood
- *Drug*—**Catawba** *Gynecological Aid* Cold infusion of plant taken by women to check excessive flow of blood. Cold infusion of plant taken by women when very sick from the change of life. (157:188) Cold infusion of plant taken to regulate flow of blood during menopause. (177:49) **Cherokee** *Antidiarrheal* Compound infusion taken for diarrhea. *Dermatological Aid* Bark ooze used for itch. *Gastrointestinal Aid* Taken as a tonic for "dyspepsy." *Oral Aid* Bark chewed for mouth ulcers. *Pulmonary Aid* Taken as a tonic for lung diseases. *Respiratory Aid* Taken as a tonic for phthisic and asthma. *Sedative* Infusion taken for nerves. (80:56)
- *Food*—**Cherokee** *Sauce & Relish* Used to make honey. (80:56)
- *Fiber*—**Cherokee** *Snow Gear* Wood used to make sled runners. (80:56)
- *Other*—**Cherokee** *Cooking Tools* Wood used to make butter paddles. *Decorations* Wood used to carve. *Fuel* Wood used for firewood. *Hunting & Fishing Item* Wood used to make arrow shafts. *Smoking Tools* Wood used to make pipestems. (80:56)

Oxypolis, Apiaceae
Oxypolis rigidior (L.) Raf., Stiff Cowbane
- *Food*—**Cherokee** *Unspecified* Roots baked and eaten. (80:51)

Oxyria, Polygonaceae
Oxyria digyna (L.) Hill, Alpine Mountainsorrel
- *Food*—**Alaska Native** *Dietary Aid* Leaves used as a good source of vitamin C. *Unspecified* Leaves eaten fresh and raw. (85:39) **Eskimo, Alaska** *Unspecified* Leaves and stems eaten raw or cooked with seal oil. (1:35) Leaves eaten fresh, soured, boiled or in oil, and root also utilized. (4:715) Fresh leaves mixed with seal blubber and eaten. **Eskimo, Arctic** *Unspecified* Leaves and young stems eaten raw and cooked. **Eskimo, Greenland** *Unspecified* Juice sweetened, thickened with a small amount of rice or potato flour and eaten. Fresh leaves mixed with seal blubber and eaten. (128:24) **Eskimo, Inuktitut** *Unspecified* Leaves eaten with seal oil. (202:190) **Eskimo, Inupiat** *Vegetable* Leaves eaten raw, with seal oil, cooked, or fermented. (98:65) **Montana Indian** *Vegetable* Acid-tasting leaves used as a salad. (19:17)

Oxytropis, Fabaceae
Oxytropis campestris (L.) DC., Cold Mountain Crazyweed
- *Drug*—**Thompson** *Disinfectant* Decoction of roots taken and poured on head in sweat house for purification. (164:504)
- *Food*—**Thompson** *Forage* Used as a common forage plant. (164:516)

Oxytropis lagopus Nutt., Haresfoot Pointloco
- *Drug*—**Blackfoot** *Dermatological Aid* Plant chewed to allay swelling. *Throat Aid* Plant chewed for sore throat. (as *Aragallus lagopus* 114:274) Plant chewed for sore throat and swelling. (as *Aragallus lagopus* 118:38)

Oxytropis lambertii Pursh, Lambert's Crazyweed
- *Drug*—**Hopi** *Poison* Plant poisonous to cattle. (200:80) **Lakota** *Poison*

Plant, in quantities, poisonous to livestock and horses. (139:47) **Navajo, Kayenta** *Laxative* Plant used for constipation. (205:28)
- *Food*—**Lakota** *Forage* Whole plant and roots eaten by horses. (139:47) **Navajo, Kayenta** *Porridge* Used to make a mush or parched and used for food. (205:28)
- *Other*—**Navajo** *Ceremonial Items* Plant offered to the bighorn at the Night Chant. (55:57)

Oxytropis maydelliana Trautv., Maydell's Oxytrope
- *Food*—**Eskimo, Inupiat** *Frozen Food* Roots frozen for future use. *Vegetable* Roots, always with some kind of oil, eaten raw or cooked. *Winter Use Food* Roots stored in buried sacks for winter use. Roots stored in seal oil, fish oil, or bear fat for winter use. (98:122)

Oxytropis monticola Gray, Yellowflower Locoweed
- *Drug*—**Thompson** *Dermatological Aid* Decoction of whole plant used as a wash for the head, hair, and whole body. (164:473, 474)

Oxytropis nigrescens (Pallas) Fisch. ex DC., Blackish Oxytrope
- *Food*—**Alaska Native** *Unspecified* Roots used for food. (85:159)

Oxytropis sericea Nutt., Silvery Oxytrope
- *Drug*—**Blackfoot** *Dermatological Aid* Infusion of leaves applied to sores. (86:77) *Ear Medicine* Infusion of leaves used for ear troubles. (86:81)
- *Other*—**Blackfoot** *Jewelry* Stems used by children to make headdresses. The headdresses were worn during a single-file dance to different spots while singing a song. The seeds would rattle in their pods. Then the leader would suddenly stop and look behind him while the others dropped to the ground. The leader struck whoever remained standing with a smoking stick which he carried. Then he took some of the seeds of the plant, chewed them, and applied the mixture to the burn. Shortly, the irritation stopped and the game continued with changing leaders. (86:115)

Oxytropis sp., Locoweed
- *Drug*—**Alaska Native** *Poison* Plant considered poisonous. (85:159) **Blackfoot** *Poison* Plant poisonous to horses. *Throat Aid* Leaves chewed and liquid swallowed for sore throats. (97:40) **Cheyenne** *Gynecological Aid* Powdered root used to increase flow of milk. *Pediatric Aid* Powdered roots taken by women when milk does not agree with the child. (76:179) **Navajo** *Misc. Disease Remedy* Plant used for rabies. *Respiratory Aid* Infusion of crushed leaves taken for bronchial and esophagus troubles. (55:57) **Thompson** *Analgesic* and *Antirheumatic (External)* Plant used in the sweat bath for rheumatism, stiff and aching muscles and joints. *Dermatological Aid* Plant used in steam bath for swellings and rheumatism. (164:468) Decoction of whole plant used as a wash for the head, hair, and whole body. (164:473, 474) *Diaphoretic* Plant used in the sweat bath for rheumatism and various aches. *Herbal Steam* Plants steamed in the sweat bath for rheumatism and other joint and muscle aches. *Orthopedic Aid* Plant used in the sweat bath for sprains, stiff and aching muscles and joints. (164:468) **Tlingit** *Gastrointestinal Aid* Root extract used for colic. (107:283)
- *Food*—**Navajo** *Forage* Plant used by sheep, in the spring, for forage. (55:57)

Pachycereus, Cactaceae
Pachycereus pringlei (S. Wats.) Britt. & Rose
- *Food*—**Papago & Pima** *Cooking Agent* Pulp made into a flavoring substance and used for jellies. (cardon 35:37) **Seri** *Fruit* Fruits eaten for food. *Porridge* Seeds ground to a powder and made into a meal or paste. (giant cactus 50:134)
- *Fiber*—**Papago & Pima** *Building Material* Ribs used for building and fence materials. *Furniture* Ribs used as frames and springs for beds and couches. (cardon 35:37) **Seri** *Caulking Material* Dried plant skeletons and sea lion oil used as a caulking compound. (giant cactus 50:136)
- *Other*—**Papago & Pima** *Hunting & Fishing Item* Ribs used as shafts for fish spears. (cardon 35:37) **Seri** *Tools* Dried plant skeletons used as a straight, slender pole for knocking off ripe fruit. (giant cactus 50:136)

Paeoniaceae, see *Paeonia*

Paeonia, Paeoniaceae
Paeonia brownii Dougl. ex Hook., Brown's Peony
- *Drug*—**Costanoan** *Gastrointestinal Aid* Decoction of roots or plants used for stomachaches and indigestion. *Laxative* Decoction of plants taken for constipation. *Pulmonary Aid* Decoction of roots used for pneumonia. (21:7) **Mahuna** *Febrifuge* and *Pulmonary Aid* Infusion of roots taken for complicated lung fevers. (140:9) **Paiute** *Cough Medicine* Seeds used to make cough medicine. (15:98) Cold infusion of seeds used as a cough medicine. (104:197) Decoction of root taken as a cough remedy. *Dermatological Aid* Decoction of root used as a liniment for swellings. (180:110, 111) *Dietary Aid* Decoction of sun dried roots taken to make people grow fat. (111:71) *Eye Medicine* Infusion of root used as a wash for sore eyes. (180:110, 111) *Heart Medicine* Fresh roots chewed and the juice swallowed for heart trouble. (111:71) *Kidney Aid* Decoction of root taken for kidney trouble. *Tuberculosis Remedy* Decoction of root taken for tuberculosis. (180:110, 111) *Veterinary Aid* Decoction of sun dried roots given to make horses grow fat. (111:71) **Shoshoni** *Antidiarrheal* Decoction of root taken for diarrhea. *Burn Dressing* Powder of pulverized, dried roots applied to burns. *Cough Medicine* Decoction of root taken as a cough remedy. *Dermatological Aid* Powdered, dried roots used on cuts, wounds, sores, and burns. Poultice of mashed root applied to boils and deep cuts or wounds. *Eye Medicine* Infusion or decoction of root used as a wash for sore eyes. *Kidney Aid* Decoction of root taken for kidney trouble. *Throat Aid* Decoction of root gargled for sore throat. *Tuberculosis Remedy* Decoction of root taken for tuberculosis. *Venereal Aid* Decoction of root taken for venereal disease. **Washo** *Analgesic* Decoction of root used as a lotion for headaches. *Tuberculosis Remedy* Decoction of root taken for tuberculosis. (180:110, 111)
- *Other*—**Paiute** *Jewelry* Seeds strung by children and worn as beads. (111:71)

Paeonia californica Nutt., California Peony
- *Drug*—**Diegueño** *Gastrointestinal Aid* Infusion of sliced, oven baked roots taken for indigestion. (84:28)
- *Food*—**Diegueño** *Vegetable* Leaves cooked as greens. Young leaves were picked before the blossoms appeared in the spring. They were

prepared by boiling, placing the boiled leaves in a cloth sack and weighting the sack down in the river with a stone, allowing the water to flow through the greens overnight to remove the bitterness in them. Alternatively, the boiled leaves could be soaked in a pan and the water changed until the bitterness was removed. The leaves were then cooked as greens, with onions, and eaten as a vegetable with acorn mush. The greens could also be prepared by boiling them twice, rather than letting them wash in the river. Buds cooked as vegetables. (84:28)

Palafoxia, Asteraceae

Palafoxia arida B. L. Turner & Morris, Desert Palafox
- **Dye**—Cahuilla *Yellow* Used as a yellow dye. (as *P. linearis* 15:98)

Palmariaceae, see *Halosaccion*

Panax, Araliaceae

Panax quinquefolius L., American Ginseng
- **Drug**—Cherokee *Analgesic* Used for headache. *Anticonvulsive* Root used for convulsions and palsy. *Expectorant* Root used as an expectorant. *Gastrointestinal Aid* Root chewed for colic. *Gynecological Aid* Used for "weakness of the womb and nervous affections." *Oral Aid* Infusion used for "thrush." *Other* Root used as a tonic and expectorant and for palsy and vertigo. *Tonic* Root used as a tonic. (80:36) **Creek** *Dermatological Aid* Poultice of plant applied to bleeding cuts. *Diaphoretic* and *Febrifuge* Decoction of plant taken to produce sweating for fevers. *Hemostat* Poultice of plant applied to bleeding cuts. *Pulmonary Aid* Decoction of roots taken for short-windedness. (177:44) **Delaware** *Tonic* Roots and other plant parts used as a general tonic. *Unspecified* Plants used to effect cures where all others have failed. (176:32) **Delaware, Oklahoma** *Panacea* Infusion of root used in any severe illness as a cure when others have failed. *Tonic* Infusion of root and other plant parts taken as a general tonic. (175:27, 76) **Houma** *Antiemetic* Decoction of root taken for vomiting. *Antirheumatic (Internal)* Decoction of root with whisky taken for rheumatism. (158:61) **Iroquois** *Anthelmintic* Compound infusion of roots taken for tapeworms. *Antiemetic* Decoction of roots taken for vomiting from cholera morbus. Infusion of roots taken for vomiting gall. *Blood Medicine* Compound infusion of roots taken as a blood remedy. *Dermatological Aid* Infusion of roots taken for sores on body and boils. *Dietary Aid* Infusion of roots taken for a bad appetite. *Ear Medicine* Compound infusion of roots used as drops for earaches. *Eye Medicine* Infusion of roots used as a wash for 2-year-old children with sore eyes. (87:395) *Febrifuge* Infusion of roots used for night fevers. (141:55) *Gastrointestinal Aid* Decoction of roots taken for upset stomach and vomiting gall. *Gynecological Aid* Compound infusion of seedpods taken by women having difficulty with labor. *Liver Aid* Decoction or infusion of roots taken to stop vomiting gall. *Misc. Disease Remedy* Decoction of roots taken for vomiting from cholera morbus. *Panacea* Compound decoction of roots taken or dried roots smoked as a panacea. (87:395) Dried roots smoked for every ailment or fainting spells. (87:396) *Pediatric Aid* Infusion of roots used as a wash for 2-year-old children with sore eyes. *Respiratory Aid* Smashed root smoked for asthma. (87:395) *Stimulant* Plant used for laziness and as a stimulant. *Tonic* Plant used as a tonic. (87:396) *Tuberculosis Remedy* Compound infusion of roots taken for tuberculosis. (87:395) *Venereal Aid* Plant used for gonorrhea. (87:396) **Menominee** *Hunting Medicine* Root used in some war bundles and hunting bundles. (151:80) *Psychological Aid* and *Tonic* Plant acted as a tonic and "strengthener of mental powers." (151:24) **Meskwaki** *Adjuvant* Used chiefly as a seasoner to render other remedies powerful. *Love Medicine* Compound called a "bagger" and used by a woman to get a husband. *Panacea* and *Pediatric Aid* Used as a "universal remedy for children and adults." (152:204) **Micmac** *Blood Medicine* Roots used as a "detergent for the blood." (40:58) **Mohegan** *Panacea* Herb highly valued as a cure-all. (176:74, 130) *Tonic* Complex compound infusion including ginseng root taken as spring tonic. (174:266) Root used alone or in combination to make a tonic. (176:74, 130) **Pawnee** *Love Medicine* Compound containing root used by men as a love charm. (70:106) **Penobscot** *Reproductive Aid* Infusion of root taken by women to increase fertility. (156:310) **Potawatomi** *Adjuvant* Root used as a seasoner in many powdered medicines. *Ear Medicine* Poultice of pounded root applied to earache. *Eye Medicine* Infusion of pounded root used as wash for sore eyes. (154:41) **Seminole** *Antirheumatic (External)* Decoction of roots used as a body rub and steam for joint swellings. (169:193) *Dermatological Aid* Poultice of roots applied to burst and drain boils and carbuncles. (169:243) Plant used for gunshot wounds. (169:302) Plant used in medicine bundles for bullet wounds. (169:422) *Love Medicine* Plant rubbed on the body and clothes to get back a divorced wife. (169:401) *Pediatric Aid* Plant and other plants used as a baby's charm for fear from dreams about raccoons or opossums. (169:221) *Respiratory Aid* Infusion of roots taken for shortness of breath. Four bits of root were floated on water in a container. If none sank, the water was drunk and the patient would recover immediately. If one or two sank, the recovery would be slower, and if all four sank the patient was certain to die. (169:278) *Sedative* Plant and other plants used as a baby's charm for fear from dreams about raccoons or opossums. (169:221) *Tonic* Plant used as a general tonic. (169:318) *Unspecified* Plant used for medicinal purposes. (169:161) Plant used as medicine. (169:158) Plant used medicinally. (169:164) *Witchcraft Medicine* Plant used to make a witchcraft medicine. (169:394)
- **Other**—Cherokee *Cash Crop* Roots sold in large quantities to traders in the late nineteenth century for 50 cents a pound. (80:36) **Chippewa** *Cash Crop* Root became a money commodity because of the white traders' demand for it. *Good Luck Charm* Root considered a good luck charm if carried in the pocket. (71:137) **Menominee** *Hunting & Fishing Item* Root chewed by hunters to impart a lure to the breath and to attract deer. *Protection* Root used in some war bundles and hunting bundles. (151:80)

Panax sp., Ginseng
- **Drug**—Creek *Adjuvant* Healthful root added to many compounds. *Ceremonial Medicine* Plant used to keep ghosts away and for religious occasions. *Diaphoretic* and *Febrifuge* Compound decoction of root taken to induce sweating, reducing fever. *Hemostat* Infusion of root applied to bleeding cuts or wounds. *Pediatric Aid* and *Pulmonary Aid* Decoction of root taken for shortness of breath and by children for croup. (172:656)

Panax trifolius L., Dwarf Ginseng
- **Drug**—Cherokee *Analgesic* An "ingredient to relieve sharp pains in

the breast." (80:36) Chewed plant used for headaches. Infusion of plant taken for breast pains and chewed plant used for headaches. (177:44) *Antirheumatic (Internal)* Used for rheumatism. (80:36) *Breast Treatment* Infusion of plant taken for breast pains. (177:44) *Dermatological Aid* Compound infusion given to children for "bold hives." *Gastrointestinal Aid* Root chewed for short breath and colic and infusion taken for colic. Used for "nervous debility," "dyspepsia," and apoplexy. *Kidney Aid* Root used for "dropsy" and gout. *Liver Aid* Used for the liver. *Misc. Disease Remedy* Used for "diseases induced by mercury" and pox. *Other* Taken for "nervous debility, dyspepsia and apoplexy." (80:36) Decoction of roots rubbed into scratches made for apoplexy. (177:44) *Pediatric Aid* Compound infusion given to children for "bold hives." *Pulmonary Aid* Root chewed for short breath and colic and infusion used for colic. *Stimulant* Cold, compound infusion of beaten roots given for fainting. *Tuberculosis Remedy* Infusion used for tuberculosis and "scrofulous sores." *Venereal Aid* Root used for stubborn venereal disease. (80:36) **Iroquois** *Analgesic* Compound used for chest pains. *Hunting Medicine* Infusion of roots used as a wash for fishing equipment, a "fishing medicine." *Pulmonary Aid* Compound used for chest pains. *Sports Medicine* Decoction of plant rubbed on the arms and legs of lacrosse players. (87:396) **Ojibwa, South** *Hemostat* Poultice of chewed root applied to cuts as a coagulant. (as *Aralia trifolia* 91:201)

Pandanaceae, see *Freycinetia, Pandanus*

Pandanus, Pandanaceae

Pandanus tectorius Parkinson ex Zucc., Tahitian Screwpine
- *Drug*—**Hawaiian** *Analgesic* Roots and other plants pounded, squeezed, resulting liquid heated and taken for chest pains. *Laxative* and *Pediatric Aid* Flowers chewed by the mothers and given to infants with constipation. *Strengthener* Roots and other plants pounded, resulting liquid heated and taken for weakness from too many births. (as *P. odoratissimus* 2:41)
- *Fiber*—**Hawaiian** *Clothing* Leaves used to make hats. *Mats, Rugs & Bedding* Leaves used to make mats. (as *P. odoratissimus* 2:41)

Panicum, Poaceae

Panicum bulbosum Kunth, Bulb Panicgrass
- *Food*—**Apache, Chiricahua & Mescalero** *Bread & Cake* Seeds threshed, winnowed, ground, and the flour used to make bread. *Sauce & Relish* Seeds ground, made into gravy, and mixed with meat. (33:48)

Panicum capillare L., Witch Grass
- *Drug*—**Keres, Western** *Emetic* Infusion of leaves used as an emetic before breakfast. (171:57) **Mahuna** *Dietary Aid* Infusion of plant taken as a reducing aid. (140:66)
- *Food*—**Hopi** *Bread & Cake* Ground seed meal used to make bread. (190:159) *Staple* Seeds ground and mixed with cornmeal. (56:17) **Navajo** *Unspecified* Seeds used for food. (55:26) **Navajo, Ramah** *Fodder* Used for sheep and horse feed. (191:17)
- *Fiber*—**Tewa** *Brushes & Brooms* Grass made into brooms and used to clean metates and metate boxes. (as *P. barbipulvinatum* 138:64)
- *Other*—**Keres, Western** *Unspecified* Plant known and named but no use was specified. (171:57)

Panicum hirticaule J. Presl, Mexican Panicgrass
- *Food*—**Cocopa** *Bread & Cake* Seeds ground into a meal and used to make bread. *Sauce & Relish* Seeds ground into a meal and used to make gravy. *Winter Use Food* Seeds stored in ollas for future use. (37:175) **Yuma** *Staple* Seeds parched, winnowed, and ground into flour. (37:190)

Panicum obtusum Kunth, Obtuse Panicgrass
- *Drug*—**Isleta** *Dermatological Aid* Grass characterized as making the hair grow rapidly. (100:36)
- *Food*—**Apache, Chiricahua & Mescalero** *Sauce & Relish* Seeds ground, made into gravy, and mixed with meat. (33:48) **Navajo, Ramah** *Fodder* Cut for hay. *Forage* Good forage. (191:17)
- *Other*—**Isleta** *Soap* Ground stolons mixed with soapweed and used in washing hair. (100:36)

Panicum sonorum Beal, Sauwi
- *Food*—**Cocopa** *Bread & Cake* Seeds ground, mixed with water, and dried to make cakes. *Winter Use Food* Seeds harvested, winnowed, and stored for winter use. **Warihio** *Beverage* Seeds ground into flour and mixed with milk to make a nourishing drink. *Staple* Seeds ground into flour and seasoned with salt and sugar. (37:170)

Panicum sp., Panicgrass
- *Drug*—**Creek** *Misc. Disease Remedy* Warm infusion of leaves taken for fevers, especially malaria. **Natchez** *Febrifuge* Warm infusion of leaves taken for fevers, especially malaria. (172:667) *Misc. Disease Remedy* Infusion of plant taken for malaria. (177:6) **Seminole** *Antirheumatic (External)* Plant used for rabbit sickness: muscular cramps. (169:166) *Cough Medicine, Pulmonary Aid,* and *Throat Aid* Infusion of plant taken and used as bath for gopher-tortoise sickness: cough, dry throat, noisy chest. (169:236)
- *Fiber*—**Cherokee** *Clothing* Stems used for padding inside of moccasins. (80:47)

Panicum urvilleanum Kunth, Desert Panicgrass
- *Food*—**Cahuilla** *Porridge* Singed seeds boiled and made into a gruel. (15:98)

Papaveraceae, see *Argemone, Chelidonium, Corydalis, Dendromecon, Dicentra, Eschscholzia, Papaver, Platystemon, Romneya, Sanguinaria*

Papaver, Papaveraceae

Papaver somniferum L., Opium Poppy
- *Drug*—**Cherokee** *Analgesic* Large dose used to produce sleep, for pain and cramps. *Anticonvulsive* Large dose given as antispasmodic, to produce sleep and for pain. *Sedative* Large dose used for pain, to produce sleep and soothe and tranquilize the system. *Stimulant* Small doses given as a stimulant and larger dose produced sleep and used for pain. (80:50, 51)

Parkinsonia, Fabaceae

Parkinsonia aculeata L., Jerusalem Thorn
- *Food*—**Papago** *Dried Food* Seeds basket winnowed, parched, sun dried, cooked, stored, and used for food. (34:24) Beans flailed, winnowed, parched, and used for food. (34:25) *Unspecified* Seeds used for food. (36:60)

Parkinsonia florida (Benth. ex Gray) S. Wats., Blue Paloverde
- *Food*—Cahuilla *Porridge* Dried beans ground into flour and used to make mush or cakes. (as *Cercidium floridum* 15:52) **Cocopa** *Porridge* Seeds roasted, ground, and made into mush. **Mohave** *Starvation Food* Seeds parched until almost burned and eaten as a famine food. (as *Cercidium floridum* 37:187) **Pima** *Unspecified* Green pods eaten raw in summer. (as *Cercidium floridum* 47:90) Beans formerly eaten fresh. (as *P. torreyana* 146:75) **Yuma** *Starvation Food* Seeds parched until almost burned and eaten as a famine food. (as *Cercidium floridum* 7:187)
- *Other*—Cahuilla *Protection* Trees large enough to shelter campers. (as *Cercidium floridum* 15:52) **Pima** *Cooking Tools* Trunk and larger branches used to make ladles. (as *Cercidium floridum* 47:90)

Parkinsonia microphylla Torr., Yellow Paloverde
- *Food*—Cocopa *Porridge* Seeds roasted, ground, and made into mush. **Mohave** *Starvation Food* Seeds parched until almost burned and eaten as a famine food. (as *Cercidium microphyllum* 37:187) **Papago** *Dried Food* Seeds basket winnowed, parched, sun dried, cooked, stored, and used for food. (34:24) Beans flailed, winnowed, parched, and used for food. (34:25) *Staple* Beans parched, sun dried, stored, ground into flour, and used as a staple food. (34:45) *Unspecified* Seeds used for food. (as *Cercidium microphyllum* 36:60) **Pima** *Unspecified* Beans formerly eaten fresh. (146:75) **Pima, Gila River** *Unspecified* Peas eaten raw or cooked. (as *Cercidium microphyllum* 133:5) Seeds eaten raw and boiled. (as *Cercidium microphyllum* 133:7) **Yuma** *Starvation Food* Seeds parched until almost burned and eaten as a famine food. (as *Cercidium microphyllum* 37:187)

Parmeliaceae, see *Candelaria, Cetraria, Parmelia*

Parmelia, Parmeliaceae, lichen
Parmelia physodes (L.) Ach., Lichen
- *Food*—Potawatomi *Soup* and *Vegetable* Cooked into a soup, material swelled and afforded a pleasant flavor. (154:107)

Parmelia saxatilis (L.) Ack., Flat Lichen
- *Fiber*—Eskimo, Inuktitut *Canoe Material* Used to stuff caribou skins for rafts. (202:190)

Parnassia, Saxifragaceae
Parnassia fimbriata Koenig, Rocky Mountain Parnassia
- *Drug*—Cheyenne *Gastrointestinal Aid* Infusion of powdered leaves given to small babies for dullness or sick to the stomach. *Pediatric Aid* Infusion of powdered leaves given to small babies for dullness or sick to the stomach. *Stimulant* Infusion of powdered leaves given to babies when dull. (76:176) **Gosiute** *Venereal Aid* Poultice of plant applied or plant used as wash for venereal diseases. (39:351)

Paronychia, Caryophyllaceae
Paronychia jamesii Torr. & Gray, James's Nailwort
- *Food*—Kiowa *Beverage* Used as a "tea" plant. (192:27)

Parrya, Brassicaceae
Parrya nudicaulis (L.) Boiss., Nakedstem Wallflower
- *Food*—Alaska Native *Soup* Roots cooked and added to fish and meat stews. *Unspecified* Roots cooked and used for food. New, young leaves used for food. *Winter Use Food* Leaves stored raw in seal oil for winter use. (85:123)

Parryella, Fabaceae
Parryella filifolia Torr. & Gray ex Gray, Common Dunebroom
- *Drug*—Hopi *Toothache Remedy* Beans used for toothaches. (42:339) Beans used for toothaches. (200:33, 80) **Navajo, Ramah** *Ceremonial Medicine* and *Emetic* Leaves used as a ceremonial emetic. (191:33) **Tewa** *Toothache Remedy* Beans used for toothaches. (42:339)
- *Fiber*—Hopi *Basketry* Plant used as basketry material. (42:339) Used as an important basketry material. (200:80) **Navajo, Ramah** *Basketry* Stems used as material for small baskets. (191:33) **Tewa** *Basketry* Plant used as basketry material. (42:339) **Zuni** *Basketry* Pleasantly fragrant plant used for weaving baskets. (166:81)
- *Other*—Hopi *Ceremonial Items* Plant used to weave kachina masks. *Stable Gear* Roots made into hooks and used to secure packs on burros during salt expeditions. **Tewa** *Ceremonial Items* Plant used to weave kachina masks. *Stable Gear* Roots made into hooks and used to secure packs on burros during salt expeditions. (42:339)

Parthenium, Asteraceae
Parthenium hysterophorus L., Santa Maria Feverfew
- *Drug*—Koasati *Antidiarrheal* Decoction of roots taken for dysentery. (177:63)

Parthenium incanum Kunth, Mariola
- *Food*—Apache, Chiricahua & Mescalero *Beverage* Fresh leaves boiled and used similarly to coffee. (33:53)

Parthenium integrifolium L., Wild Quinine
- *Drug*—Catawba *Burn Dressing* Poultice of fresh leaves applied to burns. *Veterinary Aid* Leaf ash rubbed on horse's sore back. (157:191)

Parthenocissus, Vitaceae
Parthenocissus quinquefolia (L.) Planch., Virginia Creeper
- *Drug*—Cherokee *Liver Aid* Infusion taken for yellow jaundice. (80:60) **Iroquois** *Antidote* Compound decoction of twigs taken and used as wash to counteract poison sumac. *Orthopedic Aid* Poultice of vines applied to bunches (swellings) on wrists. *Other* Compound decoction of bark taken for stricture caused by menstruating woman. *Poison* Plant considered poisonous. *Urinary Aid* Compound decoction of plants taken for difficult urination. (87:382)
- *Food*—Chippewa *Unspecified* Stalks cut, boiled, peeled, and the sweetish substance between the bark and the wood used for food. (53:320)
- *Dye*—Kiowa *Pink* Fruits used as pink paint for skin and feathers worn in war dance. (192:41)

Parthenocissus quinquefolia (L.) Planch. **var. *quinquefolia***, Virginia Creeper
- *Drug*—Houma *Dermatological Aid* Hot decoction of stems and leaves applied to reduce swellings. Poultice of crushed leaves and vinegar applied to wounds. *Misc. Disease Remedy* Poultice of crushed leaves and vinegar applied for lockjaw. (as *Psedera quinquefolia* 158:63) **Meskwaki** *Antidiarrheal* Decoction of root taken for diarrhea. (as *Psedera quinquefolia* 152:252)

- *Food*—**Montana Indian** *Fruit* Ripe fruit collected and eaten like grapes. (as *Ampelopsis quinquefolia* 19:6) **Ojibwa** *Special Food* Root cooked and given as a special food by Winabojo. *Unspecified* Root cooked and eaten. (as *Psedera quinquefokia* 153:411)
- *Dye*—**Kiowa** *Pink* Fruits used as pink paint for skin and feathers worn in war dance. (as *Psedera quinquefolia* 192:41)

Parthenocissus sp., Woodbine
- *Drug*—**Creek** *Venereal Aid* Root used for gonorrhea. (172:662) Roots used for gonorrhea. (177:41)

Parthenocissus vitacea (Knerr) A. S. Hitchc., Woodbine
- *Drug*—**Iroquois** *Urinary Aid* Compound decoction of plants taken for difficult urination caused by gall. (87:382) **Navajo** *Ceremonial Medicine* Used as part of the medicine the patient takes in the Mountain Chant Ceremony. (55:62) **Navajo, Ramah** *Dermatological Aid* Infusion of leaves and berries used as a lotion for swollen arm or leg. (191:36)
- *Food*—**Navajo, Ramah** *Fruit* Berries used for food. (191:36)
- *Fiber*—**Navajo** *Building Material* Used on ramadas for shade. (55:62)
- *Other*—**Jemez** *Ceremonial Items* Berry juice mixed with white clay and used as a purple body paint for the summer dance. (44:26) **Keres, Western** *Unspecified* Plant known and named but no use was specified. (171:58)

Pascopyrum, Poaceae
Pascopyrum smithii (Rydb.) Á. Löve, Western Wheatgrass
- *Food*—**Lakota** *Forage* Heads eaten by horses. (as *Agropyron smithii* 139:28) **Montana Indian** *Fodder* and *Forage* Most valuable forage grass and cultivated for hay, good keeping qualities, and high nutritional value. (as *Agropyron occidentale* 19:5)
- *Other*—**Keres, Western** *Insecticide* Grass and water used to soak melon seeds before planting to keep worms away and make melons sweeter. (as *Agropyron smithii* 171:25)

Paspalidium, Poaceae
Paspalidium geminatum var. *paludivagum* (A. S. Hitchc. & Chase) Gould, Egyptian Panicum
- *Drug*—**Seminole** *Dermatological Aid* Plant used for snake sickness: itchy skin. (as *Panicum paludivagum* 169:166) Decoction of whole plant taken and used as a body steam for snake sickness: itchy skin. (as *Panicum paludivagum* 169:239)

Paspalum, Poaceae
Paspalum setaceum Michx., Thin Paspalum
- *Food*—**Kiowa** *Fodder* Used as a valuable fodder plant. *Forage* Used as a valuable pasture plant. (as *P. stramineum* 192:16)

Passifloraceae, see *Passiflora*

Passiflora, Passifloraceae
Passiflora incarnata L., Purple Passionflower
- *Drug*—**Cherokee** *Dermatological Aid* Compound infusion of root used for boils. Pounded root applied to "draw out inflammation" of brier or locust wounds. *Dietary Aid* Infusion of root given to babies to aid in weaning. *Ear Medicine* Warm infusion of beaten root dropped into ear for earache. *Liver Aid* Infusion taken for liver and compound infusion of root used for boils. *Pediatric Aid* Infusion of root given to babies to aid in weaning. (80:47) **Houma** *Blood Medicine* Infusion of roots taken as a blood tonic. (158:63)
- *Food*—**Cherokee** *Beverage* Used to make a social drink. (80:47) Crushed fruit strained into a juice, mixed with flour or cornmeal to thicken, and used as a beverage. (126:50) *Fruit* Fruit used for food. (80:47) Fruit eaten raw. (126:50) *Unspecified* Young shoots and leaves boiled, fried, and often eaten with other greens. (204:253) *Vegetable* Leaves parboiled, rinsed, and cooked in hot grease with salt as a potherb. (126:50)

Pastinaca, Apiaceae
Pastinaca sativa L., Wild Parsnip
- *Drug*—**Cherokee** *Analgesic* Taken for "sharp pains." (80:47) **Iroquois** *Dermatological Aid* Compound decoction used as wash or poultice applied to chancres or lumps on penis. (87:399) **Ojibwa** *Gynecological Aid* Compound infusion of minute quantity of root taken for female troubles. *Poison* Root powerful in small amounts and poisonous in large amounts. (153:391) **Paiute** *Tuberculosis Remedy* Root medicine taken "for their all over inside—when they have this awful, maybe TB." (111:93) **Potawatomi** *Dermatological Aid* Poultice of root applied to inflammation and sores. *Poison* Root considered poisonous when taken internally. (154:86)

Paxistima, Celastraceae
Paxistima myrsinites (Pursh) Raf., Boxleaf Myrtle
- *Drug*—**Bella Coola** *Unspecified* Formerly used for medicine. (184:203) **Navajo, Ramah** *Ceremonial Medicine* and *Emetic* Plant used as an emetic in various ceremonies. (191:36) **Okanagan-Colville** *Cold Remedy* Decoction of branches taken for colds. *Kidney Aid* Decoction of branches taken for kidney troubles. *Tuberculosis Remedy* Decoction of branches taken for tuberculosis. (188:95) **Thompson** *Analgesic* Poultice of leaves used for pain in any part of the body. *Dermatological Aid* Poultice of boiled leaves applied to swellings and body pains. (164:468) *Internal Medicine* Decoction or infusion of plant used for internal ailments. *Orthopedic Aid* Decoction or infusion of plant used for broken bones. *Tuberculosis Remedy* Decoction or infusion of branches taken for tuberculosis. (187:202)
- *Food*—**Karok** *Fruit* Berries used for food. (148:385) **Okanagan-Colville** *Forage* Plant used by deer as a good winter food. (188:95) **Thompson** *Forage* Long, narrow leaves eaten by cattle when other foods scarce. (164:515)
- *Other*—**Paiute** *Decorations* Plant used for grave decorations. (111:88) **Saanich** *Cash Crop* Branches sold to local florist shops. (182:80)

Pectis, Asteraceae
Pectis angustifolia Torr., Narrowleaf Pectis
- *Drug*—**Keres, Western** *Gastrointestinal Aid* Blossoms and salt eaten for stomach trouble. *Psychological Aid* Infusion of plant used as an emetic before breakfast to relieve sadness and worry. (171:58) **Navajo** *Carminative* Plant used as a carminative. *Ceremonial Medicine* Plant used in the liniment for the Chiricahua Apache Wind Chant. *Gastrointestinal Aid* Crushed leaves used for stomachaches. (55:88)
- *Food*—**Acoma** *Spice* Used as seasoning to counteract the taste of tainted meat. (32:38) **Havasupai** *Sauce & Relish* Plant used as a con-

diment. (197:74) **Hopi** *Dried Food* Plants dried and eaten with fresh roasted corn, dried parched corn, or corn dumplings. (120:20) Dried, stored, and used for food. *Spice* Used as a flavoring. *Unspecified* Eagerly eaten raw. (200:97) **Keres, Western** *Spice* Plant used as a seasoning for meat, to kill the tainted taste. (171:58) **Laguna** *Spice* Used as seasoning to counteract the taste of tainted meat. **Pueblo** *Spice* Used as seasoning. (32:38)
- *Dye*—**Hopi** *Unspecified* Used to make an inferior dye. (200:97)

Pectis papposa Harvey & Gray, Cinchweed Fetidmarigold
- *Drug*—**Pima** *Laxative* Decoction of plant or dried plant taken as a laxative. (47:104) **Zuni** *Carminative* Infusion of whole plant taken as a carminative. *Eye Medicine* Infusion of blossoms used as eye drops for snow-blindness. (166:57)
- *Food*—**Havasupai** *Porridge* Seeds parched, ground, and used to make mush. (197:67) *Sauce & Relish* Fresh plant dipped in salted water and eaten with mush or cornmeal as a condiment. (197:249) *Soup* Seeds parched, ground, and used to make soup. (197:67) **Pueblo** *Spice* Used as seasoning. (32:38)
- *Other*—**Zuni** *Ceremonial Items*, and *Incense & Fragrance* Chewed blossoms used as perfume before a dance in ceremonies of the secret fraternities.[53] (166:83)

Pedaliaceae, see *Martynia, Proboscidea, Sesamum*

Pedicularis, Scrophulariaceae
Pedicularis attollens Gray, Attol Lousewort
- *Drug*—**Washo** *Dermatological Aid* Poultice of plant applied to cuts, sores, and swellings. *Tonic* Decoction of leaves taken as a tonic. (180:112)

Pedicularis bracteosa Benth., Bracted Lousewort
- *Drug*—**Thompson** *Unspecified* Plant used medicinally for unspecified purpose. (164:467)
- *Other*—**Thompson** *Designs* Leaves used as designs on baskets. (164:500)

Pedicularis canadensis L., Canadian Lousewort
- *Drug*—**Catawba** *Analgesic* Infusion of roots used for stomach pains. *Gastrointestinal Aid* Infusion of roots used for stomach pains and disorders. (157:190) **Cherokee** *Antidiarrheal* Taken "for bloody discharge from bowels" and used to rid sheep of lice. *Cough Medicine* Used as an ingredient in cough medicine. *Dermatological Aid* Infusion of root rubbed on sores. *Gastrointestinal Aid* Decoction of root taken for stomachache and infusion taken for "flux." *Veterinary Aid* Put in dog bed to delouse pups and used to rid sheep of lice. (80:43) **Chippewa** *Blood Medicine* Infusion of dried roots used for anemic conditions. (71:140) **Iroquois** *Emetic* and *Gastrointestinal Aid* Decoction taken to vomit for stomachaches caused by menstruating women. *Heart Medicine* Infusion of smashed roots taken for heart troubles. *Orthopedic Aid* Compound decoction of plants used as steam bath for sore legs or knees. *Tuberculosis Remedy* Compound infusion of whole plants taken for consumption with bad hemorrhage. (87:436) **Menominee** *Love Medicine* Root carried on the person who is contemplating making love advances. (151:81) *Veterinary Aid* Chopped root added to feed to make pony fat and vicious to all but its owner. (151:53) **Meskwaki** *Cancer Treatment* Poultice of root applied to tumors. *Dermatological Aid* Poultice of root applied to external swellings. *Internal Medicine* Decoction of plant taken for internal swelling and poultice applied for external swelling. (152:247) *Love Medicine* Root used in food to make estranged married people congenial and a love medicine used to return love (152:273) **Mohegan** *Abortifacient* Infusion of leaves taken to induce abortion. (176:74, 130) **Ojibwa** *Gastrointestinal Aid* Infusion of roots taken for stomach ulcers. (5:2304) *Love Medicine* Finely cut root secretly added to another's food as an aphrodisiac. (153:389, 390) Chopped root added to food as a love charm. The root was added to some dish of food that was cooking, without the knowledge of the people who were going to eat it, and if they had been quarrelsome, then they became lovers again. However, the informant said that it was too often put to bad uses. (153:432) *Throat Aid* Infusion of fresh or dried leaves taken for sore throats. (5:2304) **Potawatomi** *Cathartic* Root used as a physic. *Internal Medicine* Root used by Prairie Potawatomi for both internal and external swellings. (154:83)
- *Food*—**Cherokee** *Vegetable* Cooked leaves and stems used for food. (126:54) **Iroquois** *Vegetable* Cooked and seasoned with salt, pepper, or butter. (196:118) **Potawatomi** *Fodder* Roots mixed with oats to fatten the ponies. (154:123)

Pedicularis centranthera Gray, Dwarf Lousewort
- *Drug*—**Shoshoni** *Gastrointestinal Aid* and *Pediatric Aid* Decoction of root given to children for stomachaches. (180:112)

Pedicularis densiflora Benth. ex Hook., Indian Warrior
- *Food*—**Mendocino Indian** *Forage* Flower nectar used by yellowhammer birds. *Unspecified* Honey sucked out of the flowers by children. (41:388)

Pedicularis groenlandica Retz., Elephanthead Lousewort
- *Drug*—**Cheyenne** *Cough Medicine* Infusion of powdered leaves and stems taken to stop or loosen a long-lasting cough. (76:187) Infusion of smashed leaves and stems taken for coughs. (83:39)

Pedicularis lanata Cham. & Schlecht., Woolly Lousewort
- *Food*—**Alaska Native** *Unspecified* Flowers with water added allowed to ferment. Roots boiled or roasted. (85:125) **Eskimo, Arctic** *Unspecified* Roots eaten either raw or cooked. Flowers sucked by children for the sweet nectar. *Vegetable* Flowering stems boiled and eaten as a potherb. (128:23) **Eskimo, Inupiat** *Dessert* Fermented, frozen greens mashed, creamed, and mixed with sugar and oil for a dessert. *Unspecified* Raw shoots and roots used for food. *Vegetable* Fermented young flower tops eaten with oil and sugar, like sauerkraut. (98:56)

Pedicularis lanata Cham. & Schlecht. **ssp. *lanata*,** Woolly Lousewort
- *Food*—**Eskimo, Alaska** *Unspecified* Raw roots eaten with seal oil. (as *P. kanei* 1:38)

Pedicularis lanceolata Michx., Swamp Lousewort
- *Food*—**Iroquois** *Vegetable* Cooked and seasoned with salt, pepper, or butter. (196:118)

Pedicularis procera Gray, Giant Lousewort
- *Other*—**Navajo, Ramah** *Ceremonial Items* Used in Mountaintopway. (as *P. grayi* 191:44)

Pedicularis racemosa Dougl. ex Benth., Sickletop Lousewort
- **Drug**—Thompson *Unspecified* Plant used medicinally for unspecified purpose. (164:467)

Pedicularis **sp.**, Bumblebee Plant
- **Food**—Eskimo, Alaska *Sour* Soured leaves used for food. (4:716) *Unspecified* Nectar rich flowers eaten by children. (1:38) Root used for food. (4:716)

Pediomelum, Fabaceae

Pediomelum argophyllum (Pursh) J. Grimes, Silverleaf Scurfpea
- **Drug**—Cheyenne *Febrifuge* Decoction of plant taken for fever and salve of plant used for high fever. (as *Psoralea argophylla* 75:40) Infusion of ground leaves and stems taken for fevers. (as *Psoralea argophylla* 76:178) Ground leaf and stem powder mixed with grease and rubbed over the body for high fevers. (as *Psoralea argophylla* 76:178) Chippewa *Veterinary Aid* Compound infusion of root applied to chest and legs of horse as a stimulant. (as *Psoralea argophylla* 53:366) Lakota *Unspecified* Plant used as a medicine. *Veterinary Aid* Roots fed to tired horses. (as *Psoralea argophylla* 139:47) Meskwaki *Laxative* Infusion of root used for chronic constipation. (as *Psoralea argophylla* 152:230) Montana Indian *Dermatological Aid* Decoction of plant used as a wash for wounds. (as *Psoralea argophylla* 19:20)
- **Fiber**—Lakota *Basketry* Tough, green stems made into a basket to carry meat home. (as *Psoralea argophylla* 139:47)

Pediomelum canescens (Michx.) Rydb., Buckroot
- **Drug**—Seminole *Analgesic* Poultice of warmed roots applied externally as an analgesic. (as *Psoralea canescens* 169:167) Poultice of wet, warm tuberous roots applied for pains. *Antirheumatic (External)* Poultice of wet, warm tuberous roots applied for rheumatism. (as *Psoralea canescens* 169:285) *Cold Remedy* Infusion of tuberous roots taken and steam inhaled for runny nose, stuffy head, and sore throat. *Cough Medicine* Infusion of tuberous roots taken and steam inhaled for coughs. (as *Psoralea canescens* 169:279) *Unspecified* Plant used as medicine. (as *Psoralea canescens* 169:158)

Pediomelum cuspidatum (Pursh) Rydb., Largebract Indian Breadroot
- **Drug**—Lakota *Unspecified* Used as a medicine. (as *Psoralea cuspidata* 139:47)

Pediomelum esculentum (Pursh) Rydb., Breadroot Scurfpea
- **Drug**—Blackfoot *Antirheumatic (External)* Poultice of chewed roots applied to sprains. (as *Psoralea esculenta* 86:80) *Ear Medicine* Chewed root spittle used for earaches. *Eye Medicine* Chewed root spittle applied to the eye to remove matter. (as *Psoralea esculenta* 86:82) *Gastrointestinal Aid* Infusion of dried roots taken for gastroenteritis. Chewed roots blown into a baby's rectum for colic. (as *Psoralea esculenta* 86:68) Roots chewed by children for bowel complaints. (as *Psoralea esculenta* 97:41) *Orthopedic Aid* Poultice of chewed roots applied to fractures. (as *Psoralea esculenta* 86:80) *Pediatric Aid* Chewed roots blown into a baby's rectum for colic. (as *Psoralea esculenta* 86:68) *Pulmonary Aid* Infusion of roots taken for chest troubles. (as *Psoralea esculenta* 86:73) *Throat Aid* Infusion of dried roots taken for sore throats. (as *Psoralea esculenta* 86:68) Roots chewed for sore throat. (as *Psoralea esculenta* 86:73) *Toothache Remedy* Roots chewed by teething children. (as *Psoralea esculenta* 97:41) *Unspecified* Root pieces dried and attached to clothing and robes as ornamentation and medicine. (as *Psoralea esculenta* 86:119) Cheyenne *Antidiarrheal* Plant used as a diarrhea medicine. *Burn Dressing* Plant used as a burn medicine. *Unspecified* Plant used as an ingredient for medicinal mixtures. (as *Psoralea esculenta* 83:29)
- **Food**—Blackfoot *Dried Food* Peeled roots dried and added to winter supplies. *Unspecified* Peeled roots eaten fresh. (as *Psoralea esculenta* 86:104) Cheyenne *Cooking Agent* Dried plant pieces powdered and used as a thickening for soups, gravy, and dry meat. (as *Psoralea esculenta* 83:29) *Dried Food* Roots dried and eaten as a winter food. (as *Psoralea esculenta* 83:30) Roots formerly cut into thin, lengthwise slices and dried for winter use. *Pie & Pudding* Dried plant slices boiled, a sweetener added, and eaten as a sweet pudding. (as *Psoralea esculenta* 83:29) *Unspecified* Roots eaten fresh. (as *Psoralea esculenta* 83:30) Species used for food. (as *Psoralea esculenta* 83:45) Roots formerly eaten raw or cooked as one of the most important foods. (as *Psoralea esculenta* 83:29) Dakota *Dried Food* Roots dried for winter use. The roots were peeled and braided into festoons by their tapering roots or were split into halves or quarters and after drying were stored in any convenient container. (as *Psoralea esculenta* 69:365) Peeled roots braided and dried for winter use. (as *Psoralea esculenta* 70:92) *Unspecified* Roots eaten fresh. (as *Psoralea esculenta* 69:365) Peeled roots eaten fresh and uncooked or cooked. (as *Psoralea esculenta* 70:92) Lakota *Dried Food* Roots peeled, dried, and used as a winter food. *Soup* Roots cooked in soups and stews. *Unspecified* Roots peeled and eaten raw. (as *Psoralea esculenta* 106:41) Montana Indian *Bread & Cake* Roots dried, mashed, and used to make cakes or breads. *Cooking Agent* Roots dried, mashed, and used to thicken soups. *Dried Food* Roots shredded, dried, and stored for future use. *Porridge* Roots dried, mashed, and used to make mush and gruel. *Unspecified* Inner root core eaten raw, roasted, or boiled. (as *Psoralea esculenta* 82:61) *Vegetable* Roots, similar to yams, roasted in ashes. *Winter Use Food* Peeled, sliced roots dried for winter use. (as *Psoralea esculenta* 19:20) Omaha *Dried Food* Thickened root eaten dried. (as *Psoralea esculenta* 68:325) Peeled roots braided and dried for winter use. (as *Psoralea esculenta* 70:92) *Soup* Thickened root cooked with soup. *Unspecified* Thickened root eaten fresh and raw. (as *Psoralea esculenta* 68:325) Peeled roots eaten fresh and uncooked or cooked. Pawnee *Dried Food* Peeled roots braided and dried for winter use. *Unspecified* Peeled roots eaten fresh and uncooked or cooked. Ponca *Dried Food* Peeled roots braided and dried for winter use. *Unspecified* Peeled roots eaten fresh and uncooked or cooked. (as *Psoralea esculenta* 70:92) Sioux *Soup* Boiled or roasted roots eaten or dried and ground into meal and used in soups. *Winter Use Food* Plant gathered and hung up for winter use. (as *Psoralea esculenta* 118:13) Winnebago *Dried Food* Peeled roots braided and dried for winter use. *Unspecified* Peeled roots eaten fresh and uncooked or cooked. (as *Psoralea esculenta* 70:92)
- **Other**—Blackfoot *Decorations* Root pieces dried and attached to clothing and robes as ornamentation and medicine. (as *Psoralea esculenta* 86:119)

Pediomelum hypogaeum (Nutt. ex Torr. & Gray) Rydb. var. *hypogaeum*, Scurfpea

- **Food**—**Cheyenne** *Dried Food* Roots dried and eaten as a winter food. (as *Psoralea hypogaea* 83:30) *Unspecified* Root eaten fresh. (as *Psoralea hypogeae* 76:178) Roots eaten fresh. (as *Psoralea hypogaea* 83:30) Species used for food. (as *Psoralea hypogea* 83:46) *Winter Use Food* Root dried for winter use. (as *Psoralea hypogeae* 76:178) **Comanche** *Unspecified* Raw roots used for food. (as *Psoralea hypogeae* 29:523)

Pelea, Rutaceae

Pelea sp., Alani-kuahiwi
- **Drug**—**Hawaiian** *Blood Medicine* Infusion of pounded bark and other ingredients taken to purify the blood. *Ceremonial Medicine* Leaves placed on the bed as a beauty remedy for king, queens, and their sons and daughters. The alani was the Hawaiian beauty remedy and was dedicated to the exclusive use of the kings and queens and their sons and daughters. The leaves, in sufficient quantity, were taken and laid on the bed, covering the space, from the neck to the feet. A sheeting of tapa, tightly drawn, was laid over the leaves. In the meantime, 20 leaves were allowed to remain in the water overnight and placed in the sun during the day. This was for bathing. Towards evening, the royal child, or the one chosen for beauty, was given a bath of this water. In it were put the alani flowers. After the bath the child was fed a fattening ration. After feeding, and when the child became sleepy, it was placed in the bed covered with the alani leaves. This was repeated for five consecutive days. The bedding was then changed, the old alani leaves were removed and new ones took their place, and the process continued from that point on for 5 days more. Not only did this treatment improve the appearance, but it made the skin immune to certain diseases, especially skin diseases. *Dermatological Aid* Leaves placed in beds of kings, queens, their sons, and daughters to make the skin immune to diseases. *Pediatric Aid* and *Strengthener* Young shoots or buds used for children with general debility. (2:15)

Pellaea, Pteridaceae, fern

Pellaea atropurpurea (L.) Link, Purple Cliffbrake
- **Drug**—**Mahuna** *Blood Medicine* Infusion of plants taken to tone and thin the blood. *Kidney Aid* Infusion of plants taken to flush the kidneys. *Preventive Medicine* Infusion of plants taken as a preventative against sunstroke. (140:22)

Pellaea mucronata (D. C. Eat.) D. C. Eat., Birdfoot Cliffbrake
- **Drug**—**Costanoan** *Antihemorrhagic* and *Blood Medicine* Decoction of plant used for internal injuries to cough up "bad blood." *Dermatological Aid* Infusion of leaves used as a wash for facial sores. *Emetic* Decoction of plant used for internal injuries to cough up "bad blood." *Febrifuge* Infusion of sprouts taken for fevers. (21:5) **Yavapai** *Dermatological Aid* Dried, pulverized leaves dusted on sores. *Gynecological Aid* Decoction taken as tea by women after childbirth. (65:261)
- **Food**—**Diegueño** *Beverage* Used to make tea. (88:215) **Kawaiisu** *Beverage* Ferns brewed to make a nonmedicinal tea. (206:47) **Luiseño** *Beverage* Fronds used to make a beverage. (155:234)

Pellaea mucronata ssp. *californica* (Lemmon) Windham, California Cliffbrake
- **Food**—**Tubatulabal** *Unspecified* Leaves and stalks used for food. (as *P. compacta* 193:15)

Pellaea mucronata (D. C. Eat.) D. C. Eat. **ssp. *mucronata***, Birdfoot Cliffbrake
- **Drug**—**Diegueño** *Antihemorrhagic* Decoction of rhizomes taken for hemorrhage. (as *P. mucronata* var. *mucronata* 84:28) **Luiseño** *Unspecified* Decoction of fronds used for medicinal purposes. (as *P. ornithopus* 155:234) **Miwok** *Antihemorrhagic* Infusion taken for nosebleed. *Blood Medicine* Infusion taken as a blood purifier. *Other* Infusion taken as a spring medicine. (as *P. ornithopus* 12:171)
- **Food**—**Luiseño** *Beverage* Plant used to make a tea. (as *P. ornithopus* 155:211)
- **Other**—**Diegueño** *Good Luck Charm* Rhizome pieces scattered about the home to keep animals and enemies away and encourage friends to visit. (as *P. mucronata* var. *mucronata* 84:28)

Peltandra, Araceae

Peltandra virginica (L.) Schott, Green Arrow Arum
- **Drug**—**Nanticoke** *Pediatric Aid* and *Unspecified* Grated root in milk given to babies for unspecified purpose. (175:58)
- **Food**—**Seminole** *Unspecified* Plant used for food. (169:494)

Peltigeraceae, see *Nephroma*, *Peltigera*

Peltigera, Peltigeraceae, lichen

Peltigera aphthosa (L.) Willd.
- **Drug**—**Nitinaht** *Tuberculosis Remedy* Plants chewed and eaten for tuberculosis. *Urinary Aid* Formerly used to facilitate urination. (186:55)
- **Fiber**—**Eskimo, Inuktitut** *Canoe Material* Used to stuff caribou skins for rafts. (as *P. apthosa*, flat lichen 202:190)

Peltigera canina Willd., Dogtooth Lichen
- **Drug**—**Nitinaht** *Urinary Aid* Formerly used to facilitate urination. (186:55)

Peltigera sp., Veined Lichen
- **Drug**—**Oweekeno** *Dermatological Aid* Poultice of whole, pounded plants and spruce pitch applied to wounds. *Veterinary Aid* Poultice of chewed, whole plants and pitch applied by black and grizzly bears to wounds. (43:50)

Peniocereus, Cactaceae

Peniocereus greggii (Engelm.) Britt. & Rose **var. *greggii***, Night-blooming Cereus
- **Drug**—**Nevada Indian** *Heart Medicine* Infusion of root taken as a cardiac stimulant. (as *Cereus greggii* 118:40) **Papago** *Dermatological Aid* Seedpod and deer grease salve rubbed on sores. Seedpod mixed with deer grease as salve for sores. (as *Cereus greggii* 34:65) **Pima** *Misc. Disease Remedy* Decoction of roots taken for diabetes. (as *Cereus greggii* 47:55)
- **Food**—**Apache, San Carlos** *Unspecified* Red, doughnut-like fruits and flowers used for food. (as *Cereus greggii* 95:257) **Papago** *Beverage* Roots chewed for thirst. *Unspecified* Roots baked whole in ashes, peeled, and eaten. (as *Cereus greggii* 34:18) *Vegetable* Stalks eaten as greens. (as *Cereus greggii* 34:14) Shoots eaten as greens. (as *Cereus greggii* 34:16)

Peniocereus striatus (Brandeg.) Buxbaum, Gearstem Cactus

- *Food*—**Papago & Pima** *Fruit* Fruit used for food. (as *Wilcoxia striata* 35:42)

Pennellia, Brassicaceae
Pennellia micrantha (Gray) Nieuwl., Mountain Mock Thelypody
- *Drug*—**Navajo, Ramah** *Gynecological Aid* Decoction of root taken to expedite delivery. *Toothache Remedy* Poultice of crushed, heated roots applied for toothache. (191:29)

Pennisetum, Poaceae
Pennisetum glaucum (L.) R. Br., Pearl Millet
- *Drug*—**Navajo, Ramah** *Dermatological Aid* Fruits rubbed on open facial pimples. (as *Setaria lutescens* 191:17)

Penstemon, Scrophulariaceae
Penstemon acuminatus Dougl. ex Lindl., Sharpleaf Penstemon
- *Drug*—**Blackfoot** *Analgesic* Decoction of plant taken for stomach pain. (as *Pentstemon acuminatus* 114:276) *Antiemetic* Infusion of leaves taken for vomiting. (97:53) Decoction of plant taken for vomiting. *Gastrointestinal Aid* Decoction of plant taken for cramps and stomach pain. (as *Pentstemon acuminatus* 114:276)

Penstemon ambiguus Torr., Gilia Beardtongue
- *Drug*—**Keres, Western** *Emetic* Infusion of plant used as an emetic. (as *Pentstemon ambiguus* 171:58) **Navajo, Kayenta** *Dermatological Aid* Plant used for solpugid (wind scorpion) bites or poultice of plant applied to eagle bites. *Disinfectant* and *Veterinary Aid* Plant used as a fumigant for livestock with snakebites. (205:42)
- *Other*—**Hopi** *Ceremonial Items* Plant, associated with east direction, used in the Po-wa-mu ceremony. *Season Indicator* Flowers used to indicate when watermelon planting was over. **Tewa** *Ceremonial Items* Plant, associated with east direction, used in the Po-wa-mu ceremony. *Season Indicator* Flowers used to indicate when watermelon planting was over. (42:340)

Penstemon angustifolius Nutt. ex Pursh, Broadbeard Beardtongue
- *Other*—**Lakota** *Paint* Blossoms used to make blue paint for moccasins. (139:59)

Penstemon barbatus (Cav.) Roth, Beardlip Penstemon
- *Drug*—**Navajo, Ramah** *Analgesic* Decoction of root taken for menstrual pain and stomachache. *Burn Dressing* Cold infusion or powdered plant applied to burns. *Cough Medicine* Decoction of plant taken for cough. *Dermatological Aid* Poultice of root applied to swellings, gun wounds, and arrow wounds, a "life medicine." *Gastrointestinal Aid* Simple or compound decoction of root taken for stomachache. *Gynecological Aid* Honey sucked from flower by pregnant woman to keep baby small for easy labor. Simple or compound decoction of root taken for menstrual pain. *Panacea* Decoction of plant taken for internal injuries, a "life medicine." Poultice of root applied to gun wounds, arrow wounds, and swellings, a "life medicine." *Veterinary Aid* Poultice of plant applied to sheep for fractured legs. (191:44)

Penstemon barbatus ssp. torreyi (Benth.) Keck, Torrey's Penstemon
- *Drug*—**Apache, White Mountain** *Witchcraft Medicine* Plant used as a magic medicine. (as *Pentstemon torreyi* 136:159) **Navajo** *Diuretic* Infusion of plants taken as a diuretic. (as *Pentstemon torreyi* 55:77) **Tewa** *Dermatological Aid* Plant used as a dressing for sores. (as *Pentstemon torreyi* 138:58) **Zuni** *Hunting Medicine* Chewed root rubbed over the rabbit stick to insure success in the hunt.[54] (as *Pentstemon torreyi* 166:95)
- *Other*—**Keres, Western** *Decorations* Flowers used for bouquets and decorations in dances. (as *Pentstemon torreyi* 171:58) **Zuni** *Ceremonial Items* Chewed root rubbed over the rabbit stick to insure success in the hunt.[54] (as *Pentstemon torreyi* 166:95)

Penstemon centranthifolius Benth., Scarlet Bugler
- *Drug*—**Costanoan** *Dermatological Aid* and *Disinfectant* Poultice of plant applied to deep, infected sores. (as *Pentstemon centranthifolius* 21:15)
- *Food*—**Diegueño** *Unspecified* Flowers sucked for the good taste. (88:219)
- *Other*—**Cahuilla** *Decorations* Used as decorations at funerals or church affairs. (15:99)

Penstemon confertus Dougl. ex Lindl., Yellow Penstemon
- *Drug*—**Thompson** *Cathartic* Decoction of root taken as a purgative. (as *Pentstemon confertus* 164:467) Decoction used as a beverage, but if too strong acted as a purgative. (as *Pentstemon confertus* 164:493) *Dermatological Aid* Toasted, powdered stems and leaves sprinkled on sores, cuts, and wounds. (as *Pentstemon confertus* 164:473) *Gastrointestinal Aid* Decoction of outer bark taken for stomach troubles. (as *Pentstemon confertus* 164:467)
- *Food*—**Thompson** *Beverage* Dried stems and leaves boiled for a short time and drunk as a tea. (as *Pentstemon confertus* 164:493)
- *Dye*—**Okanagan-Colville** *Blue* Flowers boiled and rubbed on arrows and other items to give them a blue, indelible coloring. (188:139)

Penstemon deustus Dougl. ex Lindl., Scabland Penstemon
- *Drug*—**Paiute** *Dermatological Aid* Poultice of smashed leaves applied to sores. (104:196) Poultice of mashed, fresh leaves applied to boils, mosquito bites, tick bites, and open sores. (111:109) Poultice of green leaves or leaf powder applied to various skin problems. Poultice of green or dried plant applied for swellings. *Eye Medicine* Decoction of plant used as an eyewash. *Gastrointestinal Aid* and *Pediatric Aid* Decoction of plant taken for stomachaches, especially children's. (180:112, 113) **Paiute, Northern** *Dermatological Aid* Poultice of dried, ground leaves and stalks applied to chapped and cracked skin. (59:129) **Shoshoni** *Analgesic* Decoction of plants used as a hot bath for sore feet and swollen legs and veins. Decoction of plants taken for colds and rheumatic aches. *Antirheumatic (Internal)* Decoction of whole plant taken for rheumatic aches. *Cold Remedy* Decoction of whole plant taken for colds. (180:112, 113) *Dermatological Aid* Powdered root used for sores. (as *Pentstemon deustus* 118:44) Poultice of green leaves or leaf powder applied to various skin problems. *Disinfectant* Decoction of stems and leaves dropped into the ear for ear infections. *Ear Medicine* Strong decoction of stems and leaves dropped into ear for ear infection. *Eye Medicine* Decoction of plant used as an eyewash. *Gastrointestinal Aid* Decoction of plant taken for stomachaches, especially children's. *Orthopedic Aid* Decoction of plants used as a hot bath for sore feet and swollen legs and

veins. *Pediatric Aid* Decoction of plant taken for stomachaches, especially children's. (180:112, 113) *Venereal Aid* Juice of mashed, raw leaves used as wash for venereal disease. (as *Pentstemon deustus* 118:47) Compound infusion of plant used as a wash for gonorrheal sores. Plant used in various ways both internally and externally for venereal disease. (180:112, 113)

Penstemon eatonii Gray, Eaton's Penstemon
- **Drug**—**Navajo, Kayenta** *Dermatological Aid* Plant used for spider bites. *Disinfectant* Plant used as a fumigant and Lightning infection emetic. *Emetic* Plant used as a Lightning infection emetic. *Gastrointestinal Aid* Plant used for stomach troubles. *Hemostat* Plant used as a hemostatic. *Orthopedic Aid* Plant used for backache. *Snakebite Remedy* Poultice of plant applied to snakebites. *Veterinary Aid* Plant used for livestock with colic. (205:42) **Shoshoni** *Analgesic* Decoction of whole plant used as a wash for pain and healing of burns. *Burn Dressing* Decoction of whole plant used as a wash for pain and healing of burns. (180:114)
- **Other**—**Hopi** *Ceremonial Items* Plant, associated with east direction, used in the Po-wa-mu ceremony. *Season Indicator* Flowers used to indicate when watermelon planting was over. (42:341)

Penstemon fendleri Torr. & Gray, Fendler's Penstemon
- **Drug**—**Navajo, Ramah** *Dermatological Aid* Plant used for arrow or gunshot wounds. (191:44)

Penstemon fruticosus (Pursh) Greene, Bush Penstemon
- **Drug**—**Iroquois** *Emetic* Compound decoction of plants taken as an emetic to cure a love medicine. *Gynecological Aid* Compound decoction used as wash by women who are bothered by milk flow. *Love Medicine* Compound decoction of plants taken as an emetic to cure a love medicine. (as *P. pubescens* 87:434) **Okanagan-Colville** *Analgesic* Infusion of plant tops taken for headaches. *Cold Remedy* Infusion of plant tops taken for colds. *Dermatological Aid* Infusion of plant tops used for sore and itchy scalp and to bathe the skin for acne and pimples. *Gastrointestinal Aid* Infusion of plant tops taken for internal disorders. *Misc. Disease Remedy* Infusion of plant tops taken for flu. *Toothache Remedy* Raw roots placed on the tooth for severe toothaches. *Veterinary Aid* Infusion of plant tops used on animals for skin problems. (188:139) **Salish** *Unspecified* Decoction of plants used as a medicine. (as *Pentstemon douglassii* 178:294) **Shuswap** *Urinary Aid* Plant used for the bladder. (123:69) **Thompson** *Antirheumatic (External)* Whole plant used to make bathing water for rheumatism. Decoction of plant used as a wash for arthritis or as a bath for any kind of aches and sores. (187:286) *Eye Medicine* Infusion of fresh plant used as a wash for sore eyes. (as *Pentstemon douglasii* 164:468) Infusion of plant used as an eyewash. The informant used the infusion as an eyewash after she had gotten glass splinters in her eye. Decoction of leaves used as an eyewash for sore, red eyes. *Gastrointestinal Aid* Decoction of plant taken for ulcers and "to clean you out." (187:286) *Kidney Aid* Decoction of stems, flowers, and leaves taken for kidney trouble. *Orthopedic Aid* Decoction of stems, flowers, and leaves taken for sore back. (as *Pentstemon douglasii* 164: 468) Decoction of plant with other "weeds" used as a poultice for broken bones. *Veterinary Aid* Decoction of plant used on horses' legs. The decoction was used to wash a horse's leg, and after just a couple of days the horse was able to walk again. (187:286)
- **Food**—**Thompson** *Forage* Plant frequented by bees and hummingbirds for the nectar. *Spice* Plant used in pit cooking nodding onions. (187:286)
- **Fiber**—**Okanagan-Colville** *Clothing* Mashed leaves placed inside moccasins for insoles. (188:139)
- **Dye**—**Thompson** *Unspecified* Plant used in making a dye for basket designs. (187:286)
- **Other**—**Thompson** *Decorations* Plant used as a yellow-flowered garden shrub frequented by hummingbirds. (187:286)

Penstemon fruticosus var. *scouleri* (Lindl.) Cronq., Littleleaf Bush Penstemon
- **Drug**—**Okanagon** *Eye Medicine* Decoction of stems, flowers, and leaves used as a wash for inflamed eyes. *Kidney Aid* Decoction of stems, flowers, and leaves taken for kidney troubles. *Orthopedic Aid* Decoction of stems, flowers, and leaves taken for sore back. **Thompson** *Eye Medicine* Decoction of stems, flowers, and leaves used as a wash for inflamed eyes. *Kidney Aid* Decoction of stems, flowers, and leaves taken for kidney troubles. *Orthopedic Aid* Decoction of stems, flowers, and leaves taken for sore back. (as *Pentstemon scouleri* 125:41)

Penstemon gracilis Nutt., Lilac Penstemon
- **Other**—**Lakota** *Protection* Roots used against snakebite. (139:59)

Penstemon grandiflorus Nutt., Large Beardtongue
- **Drug**—**Dakota** *Analgesic* Decoction of roots used for chest pains. (as *Pentstemon grandiflorus* 69:363) **Kiowa** *Gastrointestinal Aid* Decoction of roots taken for stomachaches. (192:51) **Pawnee** *Febrifuge* Decoction of leaves taken for chills and fever. (as *Pentstemon grandiflorus* 70:114)

Penstemon jamesii Benth., James's Beardtongue
- **Drug**—**Navajo, Kayenta** *Emetic* and *Pediatric Aid* Plant used as an emetic and lotion to purify a newborn infant before nursing. (205: 43) **Navajo, Ramah** *Analgesic* Cold, compound infusion of plant taken for headache caused by hunting. *Ceremonial Medicine* Plant used ceremonially for headache and sore throat. *Throat Aid* Cold infusion taken and used as lotion for sore throat. (191:44)

Penstemon laetus Gray, Mountain Blue Penstemon
- **Drug**—**Karok** *Psychological Aid* Infusion of plant taken and used as a steam bath by grieving person. (as *Pentstemon laetus* 148:389)

Penstemon laevigatus Ait., Eastern Smooth Beardtongue
- **Drug**—**Cherokee** *Gastrointestinal Aid* Infusion taken for cramps. (80:25)

Penstemon linarioides ssp. *coloradoensis* (A. Nels.) Keck, Colorado Penstemon
- **Drug**—**Navajo, Ramah** *Gynecological Aid* Decoction of plant taken early in pregnancy to insure birth of female child. Decoction of plant taken to facilitate labor and delivery of placenta. (191:44)

Penstemon pachyphyllus Gray ex Rydb., Thickleaf Beardtongue
- **Other**—**Havasupai** *Hunting & Fishing Item* Leaves folded lengthwise and place in the mouth to produce a baby deer sound while hunting. (197:241)

Penstemon palmeri Gray, Palmer's Penstemon
- *Drug*—**Navajo, Kayenta** *Snakebite Remedy* Poultice of plant applied to snakebite sores. (205:43)

Penstemon pruinosus Dougl. ex Lindl., Chilean Beardtongue
- *Dye*—**Okanagan-Colville** *Blue* Flowers boiled and rubbed on arrows and other items to give them a blue, indelible coloring. (188:139)

Penstemon richardsonii Dougl. ex Lindl., Cutleaf Beardtongue
- *Drug*—**Okanagan-Colville** *Misc. Disease Remedy* Infusion of stalks with leaves and flowers taken for typhoid fever. (188:139) **Paiute** *Dermatological Aid* Poultice of crushed leaves applied to sores. (111:109)

Penstemon rostriflorus Kellogg, Bridge Penstemon
- *Drug*—**Kawaiisu** *Orthopedic Aid* Poultice of mashed roots applied to swollen limbs. (as *P. bridgesii* 206:47)

Penstemon sp., Penstemon
- *Drug*—**Creek** *Tuberculosis Remedy* Infusion of root taken for colds, coughs, consumption, and whooping cough. **Natchez** *Cold Remedy* Infusion of root taken for colds. *Cough Medicine* Infusion of root taken for coughs. *Pulmonary Aid* Infusion of root taken for consumption and whooping cough. (172:667) **Navajo** *Snakebite Remedy* Infusion taken and poultice of pounded leaves applied to rattlesnake bites. (55:77) **Paiute** *Toothache Remedy* Chewed root inserted into the tooth cavity for pain. (180:114, 115) **Thompson** *Eye Medicine* Infusion of plants used as a wash for sore eyes. *Kidney Aid* Decoction of plants taken for kidney trouble. *Orthopedic Aid* Decoction of plants taken for sore back. (164:465)
- *Food*—**Navajo** *Beverage* Used to make tea. (55:77) Used to make beverages. *Forage* Plant browsed by animals. (90:162)

Penstemon utahensis Eastw., Utah Penstemon
- *Other*—**Hopi** *Ceremonial Items* Plant, associated with east direction, used in the Po-wa-mu ceremony. *Decorations* Flowers used for personal decoration. *Season Indicator* Flowers used to indicate when watermelon planting was over. (42:342)

Penstemon virgatus Gray, Upright Blue Beardtongue
- *Drug*—**Navajo, Ramah** *Panacea* Whole plant or root used as "life medicine." (191:45)

Pentachaeta, Asteraceae

Pentachaeta aurea Nutt., Golden Chaetopappa
- *Other*—**Cahuilla** *Decorations* Pollen used as a cosmetic for women. (as *Chaetopappa aurea* 15:52)

Pentagrama, Pteridaceae, fern

Pentagrama triangularis (Kaulfuss) Yatskievych, Windham & Wollenweber **ssp. *triangularis***, Western Goldfern
- *Drug*—**Karok** *Analgesic* and *Gynecological Aid* Plant used to mitigate the afterpains of childbirth. (as *Gymnogramme triangularis* 148:377) **Miwok** *Toothache Remedy* Chewed for toothache. (as *Gymnogramma triangularis* 12:170)
- *Other*—**Yurok** *Designs* Spores used by children to make a design on their hands. (as *Pityrogramma triangularis* 6:45)

Pentaphylloides, Rosaceae

Pentaphylloides floribunda (Pursh) Á. Löve, Shrubby Cinquefoil
- *Drug*—**Cheyenne** *Ceremonial Medicine* Dried, powdered leaves rubbed over hands, arms, and body for Contrary dance. (as *Dasiphora fruticosa* 76:176) *Other* Plant used as a medicine against an enemy. *Poison* Plant considered poisonous. (as *Potentilla fruticosa* 83:35) **Tanana, Upper** *Gynecological Aid* Branches placed under the mattress to lessen first menstruation and number of years of menstruation. (as *Potentilla fruticosa* 102:8)
- *Food*—**Blackfoot** *Spice* Leaves mixed with dried meat as a deodorant and spice. (as *Potentilla fruticosa* 86:104) **Eskimo, Alaska** *Beverage* Dried leaves used to make tea. (as *Potentilla fruticosa* 4:715) **Eskimo, Arctic** *Beverage* Leaves dried and used as a substitute for tea. (as *Potentilla fruticosa* 128:31)
- *Fiber*—**Blackfoot** *Mats, Rugs & Bedding* Leaves used to fill pillows. (as *Potentilla fruticosa* 86:119)
- *Other*—**Blackfoot** *Fuel* Dry, flaky bark used as tinder when starting a fire with twirling sticks. (as *Potentilla fruticosa* 97:39) **Cheyenne** *Protection* Powdered leaves or infusion rubbed over body to protect hands from hot soup during Contrary dance. (as *Dasiphora fruticosa* 76:176) **Jemez** *Ceremonial Items* Yellow flowers used for the summer dances. (as *Dasiophora fruticosa* 44:22)

Penthorum, Saxifragaceae

Penthorum sedoides L., Ditch Stonecrop
- *Drug*—**Meskwaki** *Cough Medicine* Seeds used to make a cough syrup. (152:219)
- *Food*—**Cherokee** *Vegetable* Leaves used as a potherb. (126:50)

Peperomia, Piperaceae

Peperomia sp., Ala-ala-waionui-pehu
- *Drug*—**Hawaiian** *Gynecological Aid* Whole plant with other ingredients and coconut milk taken by women with sexual organ afflictions. *Laxative* Buds chewed by the mother and given to the newborn infant as a laxative. *Other* Stems and other ingredients pounded and the resulting liquid taken for wasting away of the body. *Pediatric Aid* Buds chewed by the mother and given to the newborn infant as a laxative. *Pulmonary Aid* Leaves with other ingredients and water taken for pulmonary diseases. (2:13) *Respiratory Aid* Buds and other plants pounded, the resulting liquid heated and taken for asthma. (2:57) *Strengthener* Stems and other ingredients taken for general debility. Stems and other ingredients pounded and the resulting liquid taken for general weakness. (2:13)

Perezia, Asteraceae

Perezia sp.
- *Drug*—**Yavapai** *Dermatological Aid* and *Pediatric Aid* Cotton-like material at root base placed on baby's umbilicus. (65:261)

Pericome, Asteraceae

Pericome caudata Gray, Mountain Leaftail
- *Drug*—**Navajo, Ramah** *Analgesic* Decoction of root taken for general body pain. Fresh leaves "smelled" for headache. *Ceremonial Medicine* Cold infusion of leaves used as a ceremonial chant lotion and emetic. *Cough Medicine* Decoction of root taken for cough. *Dermatological Aid* Compound containing stems used as shampoo to pre-

vent falling hair. *Diaphoretic* Decoction of root used as a sweat bath medicine. *Emetic* Cold infusion of leaves used as a ceremonial chant lotion and emetic. *Febrifuge* Cold infusion of leaves taken for fever. *Gynecological Aid* Decoction of root taken to facilitate delivery of placenta. *Misc. Disease Remedy* Cold infusion of leaves taken for influenza. *Toothache Remedy* Poultice of heated root applied for toothache. *Witchcraft Medicine* Decoction of root used for protection from witches. (191:52)

Perideridia, Apiaceae

Perideridia bolanderi (Gray) A. Nels. & J. F. Macbr., Bolander's Yampah

- *Food*—Atsugewi *Bread & Cake* Stored, dried roots pounded and made into bread. *Dried Food* Dried roots stored for winter use. Stored, dried roots pounded and made into bread or dried and stored for later use. *Soup* Stored, dried roots pounded and made into soup. *Unspecified* Fresh, ground roots used for food. (as *Pteridendia bolanden* 61:138) Miwok *Substitution Food* Served as substitution food when acorn supply was reduced. *Unspecified* Eaten raw or cooked in baskets by stone boiling, becoming mealy like potatoes. (as *Eulophus bolanderi* 12:157)

Perideridia gairdneri (Hook. & Arn.) Mathias, Gairdner's Yampah

- *Drug*—Blackfoot *Antidiarrheal* and *Antiemetic* Infusion of roots taken to counteract cathartic and emetic effects of another infusion. (86:67) *Breast Treatment* Infusion of roots used to massage sore breasts with warm stones. (86:77) *Cough Medicine* Infusion of roots or roots chewed for coughs. Root smudge smoke inhaled for nagging coughs. (86:72) *Dermatological Aid* Infusion of roots applied to sores and wounds. (86:77) *Diuretic* Roots eaten in quantity as a diuretic. (86:67) Infusion of roots taken as a diuretic. (86:70) *Laxative* Roots eaten in quantity as a mild laxative. (86:67) *Panacea* Chewed roots sprayed onto affected part by the diviner. A diviner, like Dog Child, would find the root mysteriously during the rituals. While he sang, often with a drum, he would dig the ground with a special bear claw, coming up with the root every time and anywhere. (86:83) *Respiratory Aid* Infusion of roots used as a nostril wash for catarrh. (86:72) *Strengthener* Roots chewed by buffalo runners to extend their endurance. (86:116) *Throat Aid* Infusion of roots or roots chewed for sore throats. (86:72) *Veterinary Aid* Infusion of roots given to horses as a diuretic. Roots chewed by lazy horses to enliven them. Infusion of roots used for horses with nasal gleet. (86:89) Cheyenne *Unspecified* Roots used as an ingredient in medicines. (83:41)
- *Food*—Blackfoot *Snack Food* Roots eaten as snacks by children while playing on the prairie. *Soup* Roots stored for use in soups. *Staple* Root considered a staple. *Unspecified* Roots eaten fresh. (86:103) Cheyenne *Dried Food* Roots scraped, dried, and stored for winter use. *Porridge* Roots cooked, dried, pulverized, and eaten as mush. (82:65) *Unspecified* Species used for food. (83:45) Flathead *Bread & Cake* Roots smashed, formed into small, round cakes, sun dried, and stored for winter use. Montana Indian *Unspecified* Roots eaten raw or boiled. (82:65) Okanagan-Colville *Dried Food* Roots eaten raw, boiled, or cooked, sliced, dried, and mixed with dried, powdered deer meat. *Pie & Pudding* Roots mixed with flour or black tree lichen into a pudding. *Unspecified* Roots boiled alone or with saskatoon berries. *Winter Use Food* Roots stored in pits for future use. (188:71) Paiute *Dried Food* Dried, mashed corms used in a mush or gravy. *Unspecified* Corms eaten raw or boiled. *Winter Use Food* Roots mixed with dirt and buried for winter use. (111:97) Paiute, Northern *Porridge* Roots dried, pounded, ground, and used to make mush. *Soup* Roots dried, pounded, ground, and used to make soup. *Staple* Roots ground into flour. *Unspecified* Roots peeled and eaten fresh, boiled, or roasted. (59:43) Skagit, Upper *Unspecified* Roots steamed in an earth oven and eaten. (179:40)
- *Other*—Blackfoot *Waterproofing Agent* Plant rubbed on arrows for shine and waterproofing. (86:116)

Perideridia gairdneri (Hook. & Arn.) Mathias **ssp.** *gairdneri*, Gairdner's Yampah

- *Drug*—Blackfoot *Dermatological Aid* Root used to draw inflammation from swellings. *Throat Aid* Root used for sore throat. (as *Carum gairdneri* 114:274) Cheyenne *Unspecified* Decoction of roots, stems, and leaves used as a medicine. (as *Carum gairdneri* 76:182)
- *Food*—Blackfoot *Spice* Used to flavor stews. *Vegetable* Eaten raw or boiled as a vegetable. (as *Carum gairdneri* 114:274) Cheyenne *Porridge* Dried roots cooked and used as a mush by pouring soup over them. *Unspecified* Roots eaten fresh. *Winter Use Food* Roots dried for winter use. (as *Carum gairdneri* 76:182) Gosiute *Unspecified* Roots roasted in a pit lined with hot stones and eaten. *Winter Use Food* Roots preserved in quantity for winter use. (as *Carum gairdneri* 39:365) Great Basin Indian *Unspecified* Root eaten raw or cooked. (as *Carum gairdneri* 121:49) Karok *Unspecified* Roots dried, peeled, cooked in an earth oven, and eaten. (as *Carum gairdneri* 148:387) Klamath *Unspecified* Roots used for food. (as *Carum gairdneri* 45:101) Miwok *Unspecified* Boiled and eaten like a potato. (as *Carum gairdneri* 12:157) Montana Indian *Vegetable* Roots boiled like potatoes. (as *Carum gairdneri* 19:9) Nevada Indian *Pie & Pudding* Roots ground into flour for puddings. *Unspecified* Boiled roots used for food. *Winter Use Food* Roots stored for winter use. (as *Carum gairdneri* 118:16) Pomo *Staple* Seeds used for pinole. Tubers eaten raw, cooked, or used for pinole. *Vegetable* Fresh tops eaten as greens. (as *Carum gairdneri* 11:89) Umatilla *Unspecified* Roots used for food. Ute *Unspecified* Roots used for food. (as *Carum gairdneri* 45:101) Yana *Unspecified* Roots roasted and eaten. Leaves eaten raw. (as *Carum gairdneri* 147:251)

Perideridia kelloggii (Gray) Mathias, Kellogg's Yampah

- *Drug*—Pomo *Antiemetic* Decoction of flowers taken for vomiting. (as *Carum kelloggii* 66:14) Pomo, Kashaya *Antiemetic* Decoction of flowers used for vomiting. (72:89)
- *Food*—Mendocino Indian *Spice* Seeds used to flavor pinole. *Staple* Tubers and semifleshy roots eaten as pinole. *Unspecified* Leaves eaten raw. Tubers and semifleshy roots eaten raw or cooked like acorn bread. (as *Carum kelloggii* 41:372) Miwok *Unspecified* Species used for food. (as *Carum kelloggii* 12:157) Pomo *Staple* Seeds used to make pinoles. (as *Carum kelloggii* 11:86) Pomo, Kashaya *Unspecified* Young greens eaten raw. (72:89) Yuki *Unspecified* Young plants eaten raw. (as *Carum kelloggii* 49:88)
- *Fiber*—Mendocino Indian *Brushes & Brooms* Rigid, root fibers made into compact cylindrical brushes and used "for combs." (as *Carum kelloggii* 41:372) Pomo, Kashaya *Brushes & Brooms* Roots bundled

together to make a hairbrush or scrub brush. (72:89) **Yuki** *Brushes & Brooms* Strong, fibrous roots used to make hairbrushes. (as *Carun kelloggii* 49:91)

Perideridia oregana (S. Wats.) Mathias, Squaw Potato
- **Food**—**Klamath** *Dried Food* Dried roots eaten raw. (as *Carum oreganum* 45:101) **Paiute** *Dried Food* Roots sun dried and used for food. *Unspecified* Roots eaten raw or boiled. (as *Carum oreganum* 104:101)

Perideridia pringlei (Coult. & Rose) A. Nels. & J. F. Macbr., Adobe Yampah
- **Food**—**Kawaiisu** *Vegetable* Peeled roots boiled like potatoes and eaten. (206:47) **Yana** *Unspecified* Roots roasted and eaten. (as *Eulophus pringlei* 147:251)

Perilla, Lamiaceae

Perilla frutescens (L.) Britt., Beefsteakplant
- **Drug**—**Rappahannock** *Blood Medicine* An ingredient of a blood medicine. (as *P. fructescens* 161:31)

Persea, Lauraceae

Persea borbonia (L.) Spreng., Red Bay
- **Drug**—**Seminole** *Abortifacient* Infusion of leaves taken to abort a fetus up to about 4 months old. (169:320) *Analgesic* Infusion of leaf taken for bear sickness: fever, headache, thirst, constipation, and blocked urination. (169:198) Infusion of leaves taken for sun sickness: eye disease, headache, high fever, and diarrhea. (169:206) Infusion of leaves taken as an emetic for rainbow sickness: fever, stiff neck, and backache. (169:210) Infusion of leaves taken as an emetic for thunder sickness: fever, dizziness, headache, and diarrhea. (169:213) Infusion of plant taken for wolf sickness: vomiting, stomach pain, diarrhea, and frequent urination. (169:227) Leaves used for baby sickness caused by adultery: appetite loss, fever, headache, and diarrhea. (169:253) Leaves used for adult's sickness caused by adultery: headache, body pains, and crossed fingers. (169:256) Decoction of leaves taken for dead people's sickness.[36] (169:257) Leaves used for scalping sickness: severe headache, backache, and low fever. (169:262) Decoction of leaves taken for headaches. (169:282) *Antidiarrheal* Leaves used for bird sickness: diarrhea, vomiting, and appetite loss. (169:156) Infusion of leaves taken for sun sickness: eye disease, headache, high fever, and diarrhea. (169:206) Infusion of leaves taken as an emetic for thunder sickness: fever, dizziness, headache, and diarrhea. (169:213) Leaves burned and smoke "smelled" by the baby for raccoon sickness: diarrhea. (169:218) Infusion of leaves taken by babies and adults for otter sickness: diarrhea and vomiting. (169:222) Infusion of plant taken for wolf sickness: vomiting, stomach pain, diarrhea, and frequent urination. (169:227) Decoction of leaves taken by babies and adults for bird sickness: diarrhea, vomiting, and appetite loss. (169:234) Leaves used for baby sickness caused by adultery: appetite loss, fever, headache, and diarrhea. (169:253) *Antiemetic* Leaves used for bird sickness: diarrhea, vomiting, and appetite loss. (169:156) Infusion of leaves taken by babies and adults for otter sickness: diarrhea and vomiting. (169:222) Infusion of leaves taken as an emetic and rubbed on the body for cat sickness: nausea. (169:224) Infusion of plant taken for wolf sickness: vomiting, stomach pain, diarrhea, and frequent urination. (169:227) Decoction of leaves taken by babies and adults for bird sickness: diarrhea, vomiting, and appetite loss. (169:234) Infusion of leaves taken by children for buzzard sickness: vomiting. (169:305) *Antirheumatic (External)* Decoction of leaves rubbed on body and body steamed for deer sickness: numb, painful limbs and joints. (169:192) Plant used for fire sickness: fever and body aches. (169:203) *Ceremonial Medicine* Leaf used as an emetic in purification after funerals, at doctor's school, and after death of patient. (169:167) Infusion of leaves added to food after a recent death. (169:342) *Dietary Aid* Leaves used for bird sickness: diarrhea, vomiting, and appetite loss. (169:156) Infusion of leaves taken by babies for opossum sickness: appetite loss, and drooling. (169:220) Decoction of leaves taken by babies and adults for bird sickness: diarrhea, vomiting, and appetite loss. (169:234) Leaves used for baby sickness caused by adultery: appetite loss, fever, headache, and diarrhea. (169:253) Decoction of leaves taken for dead people's sickness.[36] (169:257) Infusion of leaves taken as emetic for ghost sickness: grief, lung cough, appetite loss, and vomiting. (169:260) *Emetic* Decoction of plant and other plants taken as an emetic by doctors to strengthen his internal medicine. (169:145) Leaves used as an emetic to "clean the insides." (169:167) Leaves used as an emetic by the doctor to prevent the next patient from getting worse.[37] (169:184) Infusion of leaves taken as an emetic for rainbow sickness: fever, stiff neck, and backache. (169:210) Infusion of leaves taken as an emetic for thunder sickness: fever, dizziness, headache, and diarrhea. (169:213) Infusion of leaves taken as an emetic and rubbed on the body for cat sickness: nausea. (169:224) Plant used as an emetic during religious ceremonies. (169:409) *Eye Medicine* Infusion of leaves taken for sun sickness: eye disease, headache, high fever, and diarrhea. (169:206) Infusion of leaves taken and used as a bath for mist sickness: eye disease, fever, and chills. (169:208) *Febrifuge* Infusion of leaf taken for bear sickness: fever, headache, thirst, constipation, and blocked urination. (169:198) Plant used for fire sickness: fever and body aches. (169:203) Infusion of leaves taken for sun sickness: eye disease, headache, high fever, and diarrhea. (169:206) Infusion of leaves taken and used as a bath for mist sickness: eye disease, fever, and chills. (169:208) Infusion of leaves taken as an emetic for rainbow sickness: fever, stiff neck, and backache. (169:210) Infusion of leaves taken as an emetic for thunder sickness: fever, dizziness, headache, and diarrhea. (169:213) Leaves used for baby sickness caused by adultery: appetite loss, fever, headache, and diarrhea. (169:253) Decoction of leaves taken for dead people's sickness.[36] (169:257) Leaves used for scalping sickness: severe headache, backache, and low fever. (169:262) Decoction of leaves taken for fevers. (169:282) *Gastrointestinal Aid* Infusion of plant taken for wolf sickness: vomiting, stomach pain, diarrhea, and frequent urination. (169:227) Infusion of leaves taken as emetic for ghost sickness: grief, lung cough, appetite loss, and vomiting. (169:260) Decoction of leaves taken for stomachaches. (169:282) *Laxative* Infusion of leaf taken for bear sickness: fever, headache, thirst, constipation, and blocked urination. (169:198) *Love Medicine* Leaves sung over to get the love of a particular girl. (169:400) *Oral Aid* Infusion of leaf taken for bear sickness: fever, headache, thirst, constipation, and blocked urination. (169:198) Infusion of leaves taken by babies for opossum sickness: appetite loss and drooling. (169:220) *Orthopedic Aid* Decoction of leaves taken for dead people's sickness.[36] (169:257) Leaves

used for scalping sickness: severe headache, backache and low fever. (169:262) *Other* Leaves burned for ghost sickness: dizziness and staggering. (169:260) Infusion of leaves taken and rubbed on the body for "mythical wolf" sickness. (169:306) *Panacea* Leaves used medicinally for everything and could be added to any medicine. (169:161) *Pediatric Aid* Leaves burned and smoke "smelled" by the baby for raccoon sickness: diarrhea. (169:218) Infusion of leaves taken by babies for opossum sickness: appetite loss and drooling. (169:220) Leaves with other plants used as a baby's charm for fear from dreams about raccoons or opossums. (169:221) Infusion of leaves taken by babies and adults for otter sickness: diarrhea and vomiting. (169:222) Decoction of leaves taken by babies and adults for bird sickness: diarrhea, vomiting, and appetite loss. (169:234) Leaves used for baby sickness caused by adultery: appetite loss, fever, headache, and diarrhea. (169:253) Infusion of leaves taken by children for buzzard sickness: vomiting. (169:305) Whole plant used for chronically ill babies. (169:329) *Psychological Aid* Infusion of leaves taken as emetic for ghost sickness: grief, lung cough, appetite loss, and vomiting. (169:260) Infusion of leaves used to steam and bathe the body for insanity. (169:292) Plant burned to smoke the body for insanity. (169:293) *Pulmonary Aid* Infusion of leaves taken as emetic for ghost sickness: grief, lung cough, appetite loss, and vomiting. (169:260) *Reproductive Aid* Infusion of leaves taken and rubbed on the body for protracted labor. (169:323) *Respiratory Aid* Decoction of leaves taken for dead people's sickness.[36] (169:257) *Sedative* Leaves with other plants used as a baby's charm for fear from dreams about raccoons or opossums. (169:221) Infusion of leaves taken and used as a steam for turkey sickness: dizziness or "craziness." (169:236) *Stimulant* Decoction of leaves used as a bath for hog sickness: unconsciousness. (169:229) Decoction of leaves taken for dead people's sickness.[36] (169:257) *Unspecified* Leaves used for medicinal purposes. (169:161) Plant used medicinally. (169:164) *Urinary Aid* Infusion of leaf taken for bear sickness: fever, headache, thirst, constipation, and blocked urination. (169:198) Infusion of plant taken for wolf sickness: vomiting, stomach pain, diarrhea, and frequent urination. (169:227) *Vertigo Medicine* Infusion of leaves taken as an emetic for thunder sickness: fever, dizziness, headache, and diarrhea. (169:213)
- *Other*—**Seminole** *Ceremonial Items* Leaves used in funeral ceremonies. (169:161) Leaves carried by every member of the burial party and placed on top of the casket. (169:338) Leaves burned to keep the soul of the recently deceased from returning home. (169:342) *Cooking Tools* Plant used to make spoons. (169:507)

Persea palustris (Raf.) Sarg., Swamp Bay
- *Drug*—**Creek** *Alterative* Root used as a "hydragogue" and alterant. *Diaphoretic* and *Febrifuge* Decoction of root used as a diaphoretic in "fevers of all descriptions." *Kidney Aid* Decoction of root used for dropsy. (as *P. pubescens* 28:289)

Persea planifolia, American Avocado
- *Drug*—**Mahuna** *Oral Aid* Powdered seeds used for pyorrhea. *Toothache Remedy* Infusion used for toothaches. (140:25)

Petasites, Asteraceae

Petasites frigidus (L.) Fries, Arctic Sweet Coltsfoot
- *Drug*—**Eskimo, Inupiat** *Cold Remedy* Infusion of dried, stored leaves used for colds and head congestion. *Respiratory Aid* Infusion of dried, stored leaves used for chest congestion. (98:62)
- *Food*—**Alaska Native** *Vegetable* Leaves mixed with other greens. (85:41) **Eskimo, Alaska** *Vegetable* Leaves used for greens. (4:716) **Eskimo, Arctic** *Vegetable* Young leaves and flowering stems eaten raw as salad, cooked as a potherb, or made into a "sauerkraut." (128:26)
- *Fiber*—**Eskimo, Alaska** *Mats, Rugs & Bedding* Cotton-like seed heads formerly used for mattress stuffing with duck and goose feathers. (1:38)
- *Other*—**Alaska Native** *Containers* Large, mature leaves used to cover berries and other greens stored in kegs for winter use. (85:41) **Eskimo, Alaska** *Containers* Leaves used by children to make cone-shaped buckets to hold the picked berries. *Cooking Tools* Leaves occasionally used to form makeshift funnels. *Smoke Plant* Dried, burned leaves added to chewing tobacco for flavoring. *Snuff* Dried, burned leaves added to snuff for flavoring. (1:38) **Eskimo, Inuktitut** *Smoke Plant* Dried, burned plant ashes added to chewing tobacco. (202:189) **Eskimo, Inupiat** *Containers* Large, mature leaves used to cover barrels of rhubarb and blueberries, to prevent mold from growing. (98:62)

Petasites frigidus var. *nivalis* (Greene) Cronq., Arctic Sweet Coltsfoot
- *Drug*—**Eskimo, Inupiat** *Cold Remedy* Infusion of dried, stored leaves used for colds and head congestion. *Respiratory Aid* Infusion of dried, stored leaves used for chest congestion. (as *P. hyperboreus* 98:62)
- *Fiber*—**Eskimo, Alaska** *Mats, Rugs & Bedding* Cotton-like seed heads formerly used for mattress stuffing with duck and goose feathers. (as *P. hyperboreus* 1:38)
- *Other*—**Eskimo, Alaska** *Containers* Leaves used by children to make cone-shaped buckets to hold the picked berries. Leaves occasionally used to form makeshift funnels. *Smoke Plant* Dried, burned leaves added to chewing tobacco for flavoring. *Snuff* Dried, burned leaves added to snuff for flavoring. (as *P. hyperboreus* 1:38) **Eskimo, Inupiat** *Containers* Large, mature leaves used to cover barrels of rhubarb and blueberries, to prevent mold from growing. (as *P. hyperboreus* 98:62)

Petasites frigidus var. *palmatus* (Ait.) Cronq., Arctic Sweet Coltsfoot
- *Drug*—**Concow** *Dermatological Aid* Dried, grated roots applied to boils and running sores. *Misc. Disease Remedy* Root used for the first stages of grippe. *Tuberculosis Remedy* Root used for the first stages of consumption. (as *P. palmata* Gray. 41:395) **Delaware** *Cough Medicine* Combined with great mullein, plum root, and glycerin and used as a syrup for coughs. *Pulmonary Aid* Combined with great mullein, plum root, and glycerin and used as a syrup for lung trouble. *Respiratory Aid* Combined with great mullein, plum root, and glycerin and used as a syrup for catarrh. (as *P. palmatus* 176:36) **Delaware, Oklahoma** *Cough Medicine* Compound decoction of leaves taken for coughs. *Pulmonary Aid* Compound decoction of leaves taken for catarrh and lung trouble. *Respiratory Aid* Compound containing plant taken for catarrh, coughs, and lung trouble. (as *P. palmata* 175:30, 31) **Karok** *Panacea* Plant used for sickly babies. *Pediatric Aid* Plant used for sickly babies. (as *P. palmata* 148:390) **Lummi** *Emetic* Decoction of roots taken as an emetic. (as *P. speciosus* 79:49) **Menominee** *Dermatological Aid* Decoction of root used for itch. (as *P. palmatus* 151:31) **Quileute** *Cough Medicine* Decoction of roots or raw roots

eaten as a cough medicine. **Quinault** *Dermatological Aid* Infusion of smashed roots used as a wash for swellings. *Eye Medicine* Infusion of smashed roots used as a wash for sore eyes. **Skagit** *Antirheumatic (External)* Poultice of warmed leaves applied to parts afflicted with rheumatism. *Tuberculosis Remedy* Decoction of roots taken for tuberculosis. (as *P. speciosus* 79:49) **Tanaina** *Antirheumatic (Internal)* and *Misc. Disease Remedy* Plant used for diseases from rheumatism to tuberculosis. *Tuberculosis Remedy* Plant used for diseases from rheumatism to tuberculosis. (as *P. palmatus* 149:329) **Tlingit** *Dermatological Aid* Compound containing plant used for sores. (as *Nardosmia palmata* 107:284) **Tolowa** *Antirheumatic (External)* Leaves placed in hot water and used for arthritic joints. (as *P. palmatum* 6:42)
- *Food*—**Concow** *Unspecified* Leaves and young stems used for food. (as *P. palmata* 41:395) **Nitinaht** *Forage* Plants eaten by elk. (186:98) **Sanpoil & Nespelem** *Unspecified* Petioles eaten raw after removal of integumental fibers. (as *Pedasites speciosa* 131:103)
- *Other*—**Quinault** *Cooking Tools* Leaves used to cover berries cooking in pits. (as *Petasites speciosus* 79:49)

Petasites sagittatus (Banks ex Pursh) Gray, Arrowleaf Sweetcoltsfoot
- *Drug*—**Cree, Woodlands** *Dermatological Aid* Poultice of leaves applied to worms eating the flesh and itchy skin. (109:48)

Peteria, Fabaceae

Peteria scoparia Gray, Rush Peteria
- *Drug*—**Navajo, Ramah** *Ceremonial Medicine* Cold infusion of root used by family to protect hogan and livestock. *Dermatological Aid* Plant used as a lotion for injury inflicted by porcupine. *Misc. Disease Remedy* Compound infusion of tops taken for influenza. *Veterinary Aid* Smoke from dried tops inhaled by sheep for cough. *Witchcraft Medicine* Compound infusion of tops taken for protection from witches. (191:33)

Petradoria, Asteraceae

Petradoria pumila (Nutt.) Greene, Grassy Rockgoldenrod
- *Dye*—**Navajo** *Yellow* Flowering tops mixed with wild rhubarb and used as a yellow dye. (as *Solidago pumila* 55:89)

Petradoria pumila (Nutt.) Greene **ssp. *pumila***, Grassy Rockgoldenrod
- *Drug*—**Hopi** *Analgesic* Plant considered a good remedy for breast pain. (as *Solidago petradoria* 200:98) *Breast Treatment* Plant used for breast pain and to dry up flow of milk. (as *Solidago petradoria* 42:361) *Gynecological Aid* Plant used to decrease milk flow and ease breast pain. (as *Solidago petradoria* 200:36, 98) **Navajo, Kayenta** *Dermatological Aid* Plant used for ant bites. (as *Solidago petradoria* 205:50) **Navajo, Ramah** *Cathartic* Strong decoction taken as a cathartic. *Ceremonial Medicine* Plant used as a ceremonial emetic. *Dermatological Aid* Cold infusion used as a lotion for injuries. *Emetic* Plant used as a ceremonial emetic. (as *Solidago petradoria* 191:53)
- *Other*—**Hopi** *Decorations* Used as prayer stick decorations. (as *Solidago petradoria* 200:96)

Petrophyton, Rosaceae

Petrophyton caespitosum (Nutt.) Rydb. **var. *caespitosum***, Rocky Mountain Rockspirea
- *Drug*—**Gosiute** *Burn Dressing* Poultice of boiled roots applied to burns. (as *Spiraea caespitosa* 39:349) Boiled roots used as a salve for burns. *Gastrointestinal Aid* Leaves used as a bowel medicine. (as *Spiraea caespitosa* 39:382) **Navajo, Kayenta** *Ceremonial Medicine* Plant used as a charm or prayer in the "Pleiades rite." *Narcotic* Plant used as a narcotic. (as *Spiraea caespitosa* 205:27)

Petroselinum, Apiaceae

Petroselinum crispum (P. Mill.) Nyman ex A. W. Hill, Parsley
- *Drug*—**Cherokee** *Abortifacient* Infusion of top and root taken as an abortive for "female obstructions." *Gynecological Aid* Infusion taken by "lying-in women whose discharges are too scant." *Kidney Aid* Infusion of top and root taken for kidneys and "dropsy." *Urinary Aid* Infusion of top and root taken for the bladder. (80:47) **Micmac** *Urinary Aid* Herb used for "cold in the bladder." (as *P. sativum* 40:58)

Peucedanum, Apiaceae

Peucedanum sandwicense Hbd., Makou
- *Drug*—**Hawaiian** *Laxative* and *Pediatric Aid* Bark eaten by children and adults as a mild laxative. *Reproductive Aid* Bark taken by expectant mother for the healthy effect on the growing life. (2:71)

Peucedanum sp., Wild Celery
- *Food*—**Coeur d'Alene** *Unspecified* Growing stalks used for food. (178:91) **Okanagon** *Staple* Growing stalks used as a principal food. (178:239) **Thompson** *Unspecified* Roots and stalks eaten. (164:482) **Thompson, Upper (Nicola Band)** *Unspecified* Roots and stems eaten. (164:479)

Pezizaceae, see *Peziza*

Peziza, Pezizaceae, fungus

Peziza aurantia, Orange Peel Mushroom
- *Food*—**Pomo, Kashaya** *Vegetable* Cooked on hot stones, coals, or eaten fresh. (72:131)

Phacelia, Hydrophyllaceae

Phacelia californica Cham., California Scorpionweed
- *Drug*—**Costanoan** *Febrifuge* Decoction of root used for fevers. (21:13) **Kawaiisu** *Cold Remedy* Infusion of roots taken for colds. *Cough Medicine* Infusion of roots taken for coughs. *Gastrointestinal Aid* Infusion of roots taken for stomach problems. *Stimulant* Infusion of roots taken when weak and not feeling good. (206:48) **Pomo, Kashaya** *Dermatological Aid* Fresh, crushed leaf juice rubbed on cold sores and impetigo. (72:48)

Phacelia crenulata Torr. ex S. Wats., Cleftleaf Wildheliotrope
- *Drug*—**Hopi** *Veterinary Aid* Plant used for injury in animals, especially horses. (42:344)

Phacelia crenulata var. ***corrugata*** (A. Nels.) Brand, Cleftleaf Wildheliotrope
- *Drug*—**Hopi** Plant used for injury in animals, especially horses. (as *P. corrugata* 42:343) **Keres, Western** *Antirheumatic (External)* Infusion of root used as a rub for swellings. *Throat Aid* Infusion of plant used for sore throat. (as *P. corrugata* 171:59)

Phacelia distans Benth., Distant Phacelia
- *Food*—Kawaiisu *Vegetable* Leaves steam cooked and eaten as greens. (206:48)

Phacelia dubia (L.) Trel., Smallflower Scorpionweed
- *Food*—Cherokee *Unspecified* Young growth boiled, fried, and eaten. (204:252) *Vegetable* Leaves cooked and eaten as greens. (80:49)

Phacelia hastata Dougl. ex Lehm. **var. *hastata*,** Silverleaf Phacelia
- *Drug*—Thompson *Gynecological Aid* Decoction of plant taken for difficult menstruation. (as *P. leucophylla* 164:470)

Phacelia heterophylla Pursh, Varileaf Phacelia
- *Drug*—Miwok *Dermatological Aid* Poultice of pulverized, dried plant put in fresh wounds. (12:171, 172)
- *Food*—Navajo, Kayenta *Vegetable* Used for greens in foods. (205:39)

Phacelia linearis (Pursh) Holz., Threadleaf Phacelia
- *Drug*—Shuswap *Cold Remedy* Infusion of plant taken for a bad cold. (123:64) Thompson *Unspecified* Decoction of plant used medicinally. (as *P. menziesii* 164:468)

Phacelia neomexicana Thurb. ex Torr., New Mexico Scorpionweed
- *Drug*—Zuni *Dermatological Aid* Powdered root mixed with water and used for rashes. (27:376)

Phacelia purshii Buckl., Miami Mist
- *Drug*—Cherokee *Antirheumatic (External)* Poultice of plant used for swollen joints. (80:49)

Phacelia ramosissima Dougl. ex Lehm., Branching Phacelia
- *Drug*—Kawaiisu *Emetic* Decoction of roots taken to cause vomiting. *Gastrointestinal Aid* Decoction of roots taken to clear the "bad stomach." *Venereal Aid* Decoction of roots taken for gonorrhea. (206:48)
- *Food*—Kawaiisu *Vegetable* Leaves steam cooked and eaten as greens. (206:48) Luiseño *Vegetable* Plant used for greens. (155:230)

***Phacelia* sp.,** Wildheliotrope
- *Food*—Pima, Gila River *Unspecified* Leaves boiled and eaten. (133:7) *Vegetable* Leaves boiled or boiled, strained, refried, and eaten as greens.[11] (133:5)

***Phalaris*,** Poaceae

Phalaris arundinacea L., Reed Canarygrass
- *Other*—Okanagan-Colville *Ceremonial Items* Used to make peaked hats suitable for wearing by Indian doctors. *Cooking Tools* Used to make eating mats and mats for drying roots and berries. *Hunting & Fishing Item* Used to make fishing weirs. (188:57)

Phalaris caroliniana Walt., Carolina Canarygrass
- *Food*—Pima, Gila River *Unspecified* Seeds parched and eaten. (133:7)

Phalaris minor Retz., Littleseed Canarygrass
- *Food*—Pima, Gila River *Unspecified* Seeds parched and eaten. (133:7)

***Phaseolus*,** Fabaceae

Phaseolus acutifolius Gray, Tepary Bean
- *Drug*—Papago *Toothache Remedy* Plant bitten and held between teeth for toothache. (34:65)
- *Food*—Havasupai *Soup* Beans parched, ground, and added to hot water to make a soup. *Vegetable* Beans cooked with fresh corn, cooked in hot ashes under a fire, or boiled. *Winter Use Food* Beans stored in granaries or in frame houses for later use. (197:227) **Sia** *Vegetable* Cultivated beans used for food. (199:106)
- *Other*—Keresan *Ceremonial Items* Beans made into a flour by the Koshairi and used for ritual purposes. Prayer meal ground from beans was exceedingly unusual; it was almost always made from corn. (198:558)

Phaseolus acutifolius* var. *latifolius Freeman, Tepary Bean
- *Food*—Cocopa *Staple* Parched, ground, boiled beans and unparched maize made into a meal. *Winter Use Food* Beans stored in pots for later use. (64:264) Kamia *Unspecified* Species used for food. (62:21) Papago *Dried Food* Beans threshed, dried on the ground or roofs, stored, and used for food. *Staple* Used as a staple crop. (34:32)

Phaseolus angustissimus Gray, Slimleaf Bean
- *Drug*—Zuni *Pediatric Aid* and *Strengthener* Crushed leaves, blossoms, and powdered root rubbed on a child's body as a strengthener.[55] (166:85)

Phaseolus coccineus L., Scarlet Runner
- *Food*—Iroquois *Bread & Cake* Seeds cooked, mixed with corn bread paste, and again cooked in the making of the bread. *Dried Food* Seedpods boiled, dried in evaporating baskets or on flat boards, and stored away in bags or barrels. *Soup* Seedpods cooked and used to make soup. Beans boiled with green sweet corn, meat, and seasoned with salt, pepper, and butter or fat. Dried seedpods soaked, boiled, seasoning and butter added, and eaten as a soup. Seeds washed with hot water, cooked until soft, and sugar added to make a sweet soup. Ripe seeds boiled with beef or venison, mashed until thoroughly mixed, and eaten as soup. *Vegetable* Seedpods cooked and eaten whole or cooked with butter, squash, or meat. Seeds boiled or fried in bear or sunflower oil, seasoned, and eaten. Seeds cooked "like potatoes" and mashed or pounded. (as *P. multiflorus* 196:103)
- *Other*—Iroquois *Ceremonial Items* Seeds intimately associated with the annual ceremonies of planting time and the harvest thanksgiving. (as *P. multiflorus* 196:103)

Phaseolus lunatus L., Sieva Bean
- *Food*—Cherokee *Bread & Cake* Beans used to make bean bread. *Soup* Beans used to make hickory nut soup. *Vegetable* Beans used for food. (80:24) Havasupai *Soup* Beans parched, ground, and added to hot water to make a soup. *Vegetable* Beans cooked with fresh corn, cooked in hot ashes under a fire, or boiled. *Winter Use Food* Beans stored in granaries or in frame houses for later use. (197:227) **Iroquois** *Bread & Cake* Seeds cooked, mixed with corn bread paste, and again cooked in the making of the bread. *Dried Food* Seedpods boiled, dried in evaporating baskets or on flat boards, and stored away in bags or barrels. *Soup* Seedpods cooked and used to make soup. Beans boiled with green sweet corn, meat, and seasoned with salt, pepper, and butter or fat. Dried seedpods soaked, boiled, seasoning and butter added, and eaten as a soup. Seeds washed with hot water, cooked until soft, and sugar added to make a sweet soup. Ripe seeds boiled with beef or venison, mashed until thoroughly mixed, and eaten as soup. *Vegetable* Seedpods cooked and eaten

whole or cooked with butter, squash, or meat. Seeds boiled or fried in bear or sunflower oil, seasoned, and eaten. Seeds cooked "like potatoes" and mashed or pounded. (196:103) **Navajo, Ramah** *Unspecified* Large, white bean cultivated for local use. Small, white lima bean cultivated for local use. (191:33) **Ojibwa** *Vegetable* The Ojibwe claim to have originally had the lima bean, but that is doubtful. (153:406)
- **Other**—**Iroquois** *Ceremonial Items* Seeds intimately associated with the annual ceremonies of planting time and the harvest thanksgiving. (196:103)

Phaseolus sp., Bean
- **Food**—**Havasupai** *Porridge* Seeds parched, ground, and used to make mush. *Soup* Seeds parched, ground, and used to make soup. (197:67) **Okanagan-Colville** *Unspecified* Seeds used for food. (188:105)

Phaseolus vulgaris L., Kidney Bean
- **Food**—**Abnaki** *Vegetable* Beans used for food. (144:169) **Apache, White Mountain** *Vegetable* Beans used for food. (136:159) **Cherokee** *Bread & Cake* Beans used to make bean bread. *Soup* Beans used to make hickory nut soup. (80:24) *Vegetable* Beans used for food. (80:21) **Havasupai** *Soup* Beans parched, ground, and added to hot water to make a soup. *Vegetable* Beans cooked with fresh corn, cooked in hot ashes under a fire, or boiled. *Winter Use Food* Beans stored in granaries or in frame houses for later use. (197:227) **Iroquois** *Bread & Cake* Seeds cooked, mixed with corn bread paste, and again cooked in the making of the bread. *Dried Food* Seedpods boiled, dried in evaporating baskets or on flat boards, and stored away in bags or barrels. *Soup* Seedpods cooked and used to make soup. Beans boiled with green sweet corn, meat, and seasoned with salt, pepper, and butter or fat. Dried seedpods soaked, boiled, seasoning and butter added, and eaten as a soup. Seeds washed with hot water, cooked until soft, and sugar added to make a sweet soup. Ripe seeds boiled with beef or venison, mashed until thoroughly mixed, and eaten as soup. *Vegetable* Seedpods cooked and eaten whole or cooked with butter, squash, or meat. Seeds boiled or fried in bear or sunflower oil, seasoned, and eaten. Seeds cooked "like potatoes" and mashed or pounded. (196:103) **Menominee** *Staple* Berry used as a staple article of food. (151:69) **Navajo** *Soup* Beans boiled and used in stews. *Vegetable* Beans formed a large part of the vegetable diet. (165:221) **Navajo, Ramah** *Fodder* Plants, after harvesting the beans, used as stock feed. *Winter Use Food* Beans cultivated and stored for use during the winter. (191:33) **Ojibwa** *Vegetable* Similar to the white man's navy bean. Original source of all best commercial pole beans, used alone or in many peculiar combinations. (153:406) **Papago** *Dried Food* Beans threshed, dried on the ground or roofs, stored, and used for food. (34:32) *Unspecified* Beans grown for food. (36:99) **Potawatomi** *Vegetable* A great number of varieties of beans were used. (154:104) **Sia** *Vegetable* Cultivated beans used for food. (199:106) **Tewa** *Staple* Used as a staple food. (138:100) **Zuni** *Vegetable* Beans boiled and fried or crushed, boiled beans mixed with mush, baked in corn husks, and used for food. Boiled and fried beans used for food. (166:69)
- **Other**—**Diegueño** *Cash Crop* Plant grown and traded to the Cocopa Indians. (37:108) **Iroquois** *Ceremonial Items* Seeds intimately associated with the annual ceremonies of planting time and the harvest thanksgiving. (196:103) **Kiliwa** *Cash Crop* Plant grown and traded to the Cocopa Indians. (37:108) **Navajo, Ramah** *Cash Crop* Beans cultivated as a commercial crop. (191:33) **Paipai** *Cash Crop* Plant grown and traded to the Cocopa Indians. (37:108)

Phegopteris, Thelypteridaceae, fern
Phegopteris sp., Ako-lea
- **Drug**—**Hawaiian** *Dietary Aid* Young shoots or buds and bark mixed with cooked leaves and eaten to restore the lost of appetite. *Gynecological Aid* Inner bark scraped or buds mixed with cooked taro leaves and water and eaten during childbirth. (2:12)
- **Food**—**Hawaiian** *Unspecified* Roots cooked for food. (2:12)

Philadelphus, Hydrangeaceae
Philadelphus lewisii Pursh, Lewis's Mockorange
- **Drug**—**Okanagan-Colville** *Cathartic* Decoction of plant taken as a physic in the morning and evening. (188:108) **Thompson** *Antirheumatic (External)* Powdered, burned wood mixed with pitch or bear grease and rubbed on the skin for swellings. Dried, powdered leaves mixed with pitch or bear grease and rubbed on the skin for swellings. *Breast Treatment* Poultice of bruised leaves used by women for infected breasts. *Dermatological Aid* Powdered, burned wood mixed with pitch or bear grease and rubbed on the skin for sores. Dried, powdered leaves mixed with pitch or bear grease and rubbed on the skin for sores. Strained decoction of branches, sometimes with the blossoms, used as a soaking solution for eczema.[56] *Hemorrhoid Remedy* Strained decoction of branches, sometimes with the blossoms, used to soak bleeding hemorrhoids.[56] *Pulmonary Aid* Strained decoction of branches taken for sore chest. (187:230)
- **Fiber**—**Mewuk** *Basketry* Rods used in the fine, coiled baskets. (117:328) **Okanagan-Colville** *Furniture* Wood used to make cradle hoops. *Snow Gear* Wood used to make snowshoes. (188:108) **Paiute** *Brushes & Brooms* Sticks split and used as hair combs. (111:77) **Shuswap** *Snow Gear* Wood used for fish spears and snowshoes. (123:63) **Thompson** *Basketry* Sticks used as edging for birch bark baskets. (187:230) *Brushes & Brooms* Wood used to make combs. (164:499) *Furniture* Sticks used as edging for birch bark cradle hoods. (187:230)
- **Other**—**Coeur d'Alene** *Tools* Wood used to make root diggers. (178:91) **Klamath** *Hunting & Fishing Item* Stems used in the manufacture of arrows for war purposes or large game. Stems used to make arrows for war purposes or large game. (45:97) **Montana Indian** *Hunting & Fishing Item* Stems used for making arrows. (19:17) **Okanagan-Colville** *Hunting & Fishing Item* Wood used to make harpoon shafts, bows and arrows, arrow tips, and clubs. *Protection* Wood used to make breast plate armor. *Season Indicator* Blooming bushes indicated the groundhogs were fat. *Smoking Tools* Wood used to make pipestems. *Soap* Leaves rubbed with water, made into a frothy lather, and used to wash the hands and hair. *Tools* Wood used to make digging sticks. (188:108) **Paiute** *Tools* Sticks used as digging sticks, one of the principal tools. (111:77) **Poliklah** *Hunting & Fishing Item* Used to make arrows. (117:173) **Saanich** *Hunting & Fishing Item* Wood occasionally used to make bows and arrows. (182:84) **Shuswap** *Hunting & Fishing Item* Wood used for fish spears and snowshoes. *Soap* Bark soaked in warm water and used for washing the face. (123:63) **Thompson** *Tools* Hard wood used for making knitting needles. (187:230)

Philadelphus lewisii* var. *gordonianus (Lindl.) Jepson, Gordon's Mockorange
- *Drug*—**Snohomish** *Dermatological Aid* Soapy lather from bruised leaves rubbed on sores. (as *P. gordonianus* 79:31)
- *Fiber*—**Cowlitz** *Brushes & Brooms* Wood used to make combs. **Lummi** *Brushes & Brooms* Wood used to make combs. (as *P. gordonianus* 79:31) **Mendocino Indian** *Furniture* Pithy stems used to make light baskets for carrying babies. (as *P. gordonianus* 41:352)
- *Other*—**California Indian** *Hunting & Fishing Item* Wood used to make bows. Shoots used to make arrows. (as *P. gordonianus* 118:52) **Karok** *Hunting & Fishing Item* Young shoots used to make arrow shafts. *Smoking Tools* Twigs, with the pithy center removed, used to make tobacco pipes. (148:384) **Lummi** *Tools* Wood used for netting shuttles and knitting needles. **Skagit** *Hunting & Fishing Item* Wood used to make arrow shafts. (as *P. gordonianus* 79:31) **Wailaki** *Hunting & Fishing Item* Older, less pithy wood formerly used to make bows. Young, very pithy shoots used to make arrows. **Yuki** *Hunting & Fishing Item* Older, less pithy wood formerly used to make bows. Young, very pithy shoots used to make arrows. (as *P. gordonianus* 41:352) Straight branches used to make arrows. (49:91)

Philadelphus microphyllus Gray, Littleleaf Mockorange
- *Food*—**Isleta** *Fruit* Fruits formerly used for food. (32:30) Fruit formerly eaten as food. (100:36)

***Philadelphus* sp.,** Wild Syringa
- *Other*—**Midoo** *Hunting & Fishing Item* Wood used to make the best arrow shafts. (117:320) **Modesse** *Hunting & Fishing Item* Used to make spear tips. (117:224) **Washo** *Hunting & Fishing Item* Wood used to make shafts. (118:51)

Phlebodium, Polypodiaceae, fern

Phlebodium aureum (L.) J. Sm., Golden Polypody
- *Drug*—**Seminole** *Other* Complex infusion of tuberous roots taken for chronic conditions. (169:272) *Pediatric Aid* Plant used for chronically ill babies. (169:329) *Psychological Aid* Infusion of plant used to steam and bathe the body for insanity. (169:292) *Unspecified* Plant used for medicinal purposes. (169:162)

Phleum, Poaceae

Phleum pratense L., Timothy
- *Food*—**Shuswap** *Fodder* Plant used as feed for cows. (123:55)
- *Fiber*—**Navajo, Ramah** *Brushes & Brooms* Stems used to make hairbrushes. (191:17)
- *Other*—**Okanagan-Colville** *Containers* Used in pit cooking. (188:57)

Phlox, Polemoniaceae

Phlox austromontana Coville, Desert Phlox
- *Drug*—**Havasupai** *Antirheumatic (External)* Decoction of pounded roots rubbed all over the body for aches. *Cold Remedy* Decoction of pounded roots rubbed all over the body for colds. *Gastrointestinal Aid* and *Pediatric Aid* Decoction of pounded roots given to babies with stomachaches. (197:238) **Navajo, Kayenta** *Toothache Remedy* Crushed plant placed in cavity for toothaches. (205:38)

Phlox caespitosa Nutt., Tufted Phlox
- *Drug*—**Navajo** *Burn Dressing* Plant used for burns. *Cathartic* Plant used as a cathartic. (as *P. douglasii* var. *diffusa* 90:162) *Ceremonial Medicine* Crushed plant and other plants used to make the Night Chant liniment. (55:70) Plant used in medicine ceremonies. *Diuretic* Plant used as a diuretic. *Gynecological Aid* Plant used for childbirth. *Toothache Remedy* Plant used as toothache medicine. (as *P. douglasii* var. *diffusa* 90:162)

Phlox gracilis (Hook.) Greene **ssp. *gracilis*,** Slender Phlox
- *Drug*—**Gosiute** *Dermatological Aid* Poultice of mashed plant applied to wounds and bruises. (as *Gilia gracilis* 39:370) **Navajo, Ramah** *Dermatological Aid* Poultice of plant applied to sores on body. *Oral Aid* Cold infusion used as mouthwash for mouth sores. (as *Gilia gracilis* 191:40) **Ute** *Dermatological Aid* Poultice of plant applied to bruised or sore legs. (as *Gilia gracilis* 38:34)

Phlox hoodii Richards., Spiny Phlox
- *Drug*—**Blackfoot** *Laxative* and *Pediatric Aid* Infusion of plant given to children as a mild laxative. (86:67) *Pulmonary Aid* Infusion of plant taken for chest pains. (86:73)
- *Dye*—**Blackfoot** *Yellow* Plant used to make a yellow dye. (86:116)

Phlox longifolia Nutt., Longleaf Phlox
- *Drug*—**Havasupai** *Antirheumatic (External)* Decoction of pounded roots rubbed all over the body for aches. *Cold Remedy* Decoction of pounded roots rubbed all over the body for colds. *Gastrointestinal Aid* and *Pediatric Aid* Decoction of pounded roots given to babies with stomachaches. (197:238) **Okanagan-Colville** *Blood Medicine* and *Pediatric Aid* Infusion of whole plant given to "anemic" children. (188:112) **Paiute** *Cathartic* Decoction of root taken as a physic. *Eye Medicine* Infusion or decoction of root used as an eyewash. *Gastrointestinal Aid* and *Pediatric Aid* Decoction of root given to children for stomachaches. *Venereal Aid* Decoction of root taken for venereal disease. **Shoshoni** *Antidiarrheal* Infusion of mashed root taken for diarrhea. (180:115) *Dermatological Aid* Decoction of leaves used for boils. (118:44) *Eye Medicine* Infusion or decoction of root used as an eyewash. *Gastrointestinal Aid* Decoction of entire plant taken for stomach disorders. Infusion of root given to children for stomachaches. *Pediatric Aid* Infusion of root given to children for stomachaches. **Washo** *Eye Medicine* Infusion or decoction of root used as an eyewash. (180:115)

Phlox maculata L., Wild Sweetwilliam
- *Drug*—**Cherokee** *Dietary Aid* and *Pediatric Aid* Infusion of root used as a wash to make children grow and fatten. (80:58)

Phlox multiflora A. Nels., Flowery Phlox
- *Drug*—**Cheyenne** *Stimulant* Infusion of pulverized leaves and flowers used as a wash and taken as a stimulant for body numbness. (76:184)

Phlox pilosa L., Downy Phlox
- *Drug*—**Meskwaki** *Blood Medicine* Infusion of leaves taken to cure and purify blood. *Dermatological Aid* Infusion of leaves used as wash for eczema. *Love Medicine* Compound containing root used as a love medicine. (152:235)

Phlox stansburyi (Torr.) Heller, Colddesert Phlox
- *Drug*—**Navajo, Ramah** *Contraceptive* Decoction of leaves taken dur-

ing menstruation as a contraceptive. *Dermatological Aid* Decoction of leaves used as a lotion for sores. *Disinfectant* Plant used as a fumigant for "deer infection." *Gynecological Aid* Decoction of leaves taken during pregnancy to insure birth of female baby. Decoction of leaves taken to facilitate delivery of placenta. (191:40)
- *Food*—**Navajo, Kayenta** *Vegetable* Eaten as greens with meat or as an emergency food. (205:38)
- *Other*—**Hopi** *Protection* Infusion used to keep grasshoppers, rabbits, and pack rats from eating corn. **Navajo, Kayenta** *Protection* Infusion used to keep grasshoppers, rabbits, and pack rats from eating corn. (205:38)

Phlox subulata L., Moss Phlox
- *Drug*—**Mahuna** *Antirheumatic (Internal)* Plant used for rheumatism. (140:59)

Phoenicaulis, Brassicaceae
Phoenicaulis cheiranthoides Nutt., Wallflower Phoenicaulis
- *Drug*—**Paiute** *Gynecological Aid* and *Tonic* Decoction of root taken as a tonic after childbirth. (as *Parrya menziesii* 180:112)

Pholisma, Lennoaceae
Pholisma arenarium Nutt. ex Hook., Desert Christmas Tree
- *Food*—**Kawaiisu** *Unspecified* Stems eaten raw, "roasted," or baked below the fire "like mushrooms." (206:48)

Pholisma sonorae (Torr. ex Gray) Yatskievych, Sandfood
- *Food*—**Cocopa** *Dried Food* Roots baked, dried, boiled, and eaten. *Unspecified* Roots baked and eaten after stripping off the thin bark. (as *Ammobroma sonorae* 37:207) **Papago** *Dried Food* Surplus of roots sun dried on roofs and used for food. *Staple* Used as a staple root crop. (as *Ammobroma sonorae* 34:17) *Unspecified* Species used for food. (*Ammobroma sonorae* 32:7) Roots used for food. (as *Ammobroma sonorae* 36:60)

Pholistoma, Hydrophyllaceae
Pholistoma membranaceum (Benth.) Constance, White Fiestaflower
- *Food*—**Tubatulabal** *Unspecified* Rolled in palm of hand with salt grass leaves and stems and eaten. (as *Ellisea membranacea* 193:19) Leaves used for food. (as *Ellisia membranacea* 193:15)

Phoradendron, Viscaceae
Phoradendron californicum Nutt., Mesquite Mistletoe
- *Drug*—**Pima** *Cathartic* Decoction of berries taken as a purge. *Dermatological Aid* Infusion of plant used as a wash for sores. *Gastrointestinal Aid* Decoction of berries taken for stomachaches. (47:82)
- *Food*—**Maricopa** *Porridge* Berries boiled to produce liquid and combined with wheat mush. (37:204) **Papago** *Dried Food* Berries sun dried, stored, and used for food. (34:19) **Pima** *Fruit* Berries boiled and eaten. (32:39) Berries boiled and eaten. (146:71) **Pima, Gila River** *Fruit* Berries eaten cooked or raw. (133:5) *Snack Food* Fruits eaten raw or boiled as a snack food. (133:7)

Phoradendron juniperinum Engelm., Juniper Mistletoe
- *Drug*—**Hopi** *Gastrointestinal Aid* Plant used as "medicine for the stomach." (200:34, 72) *Unspecified* Plant used medicinally. (42:345)

Witchcraft Medicine Plant used as "medicine for the stomach and bad medicine of wizards." (200:72) **Keres, Western** *Antidiarrheal* Crushed plant given to children for diarrhea. *Antirheumatic (External)* Crushed plant used as a rub for rheumatism. *Pediatric Aid* Crushed plant given to children for diarrhea. (171:59) **Navajo** *Dermatological Aid* Plant used for warts. (90:162) **Navajo, Ramah** *Gastrointestinal Aid* Cold infusion taken to relieve distress caused by eating too much meat. (191:23) **Tewa** *Gastrointestinal Aid* Infusion of pulverized plant taken for "chill in the stomach." (138:47) **Zuni** *Emetic* Infusion of whole plant taken as an emetic for stomachaches. (27:377) *Gynecological Aid* Compound infusion of plant taken to promote muscular relaxation at birth. Simple or compound infusion of twigs taken after childbirth to stop blood flow. *Hemostat* Simple or compound infusion of twigs taken after childbirth to stop blood flow. (166:55)
- *Food*—**Acoma** *Starvation Food* Berries eaten when other foods became scarce. (32:39) **Havasupai** *Unspecified* Plant pounded and boiled for food. (197:216) **Keres, Western** *Fodder* Plant used as sheep and goat feed, to produce good milk. *Starvation Food* Berries eaten when other food was scarce. (171:59) **Laguna** *Starvation Food* Berries eaten when other foods became scarce. (32:39) **Navajo** *Beverage* Stems used to make tea. *Fruit* Berries used for food. (55:42)

Phoradendron juniperinum Engelm. **ssp.** **juniperinum**, Juniper Mistletoe
- *Drug*—**Navajo** *Dermatological Aid* Plant used for warts. (as *P. ligatum* 90:162)

Phoradendron leucarpum (Raf.) Reveal & M. C. Johnston, Oak Mistletoe
- *Drug*—**Cherokee** *Analgesic* "Tea ooze" used to bathe head for headache. *Anticonvulsive* Dried and pulverized plant "good for epilepsy or fits, best if from oak." *Gynecological Aid* Hot infusion used as "medicine for pregnant women." *Hypotensive* Infusion used for high blood pressure. *Love Medicine* Infusion taken after vomiting for 4 days, to cure "love sickness." (as *P. serotinum* 80:45) **Creek** *Pulmonary Aid* Compounds containing leaves and branches used for lung trouble. (as *P. flavescens* 172:659) Leaves and branches used for lung troubles. (as *P. flavescens* 177:20) *Tuberculosis Remedy* Compounds containing leaves and branches used for consumption. (as *P. flavescens* 172:659) **Houma** *Orthopedic Aid* Decoction of plant taken for debility and paralytic weakness. *Panacea* Decoction of plant said to be good for sickness in general, a panacea. (as *P. flavescens* 158:58) **Mendocino Indian** *Abortifacient* Infusion of roots taken for abortions. *Poison* Plant considered poisonous. *Toothache Remedy* Root chewed for toothaches. (as *P. flavescens* 41:344) **Seminole** *Antirheumatic (External)* Decoction of leaves rubbed on body and body steamed for deer sickness: numb, painful limbs and joints. (as *P. flavescens* 169:192) *Emetic* Plant used as an emetic during religious ceremonies. (as *P. flavescens* 169:409) *Pediatric Aid* Plant used for chronically ill babies. (as *P. flavescens* 169:329)

Phoradendron macrophyllum (Engelm.) Cockerell **ssp.** **macrophyllum**, Colorado Desert Mistletoe
- *Drug*—**Diegueño** *Dermatological Aid* Decoction of entire, fresh plant used for dandruff. (as *P. tomentosum* ssp. *macrophyllum* 84:28)

Phoradendron sp., Mistletoe

- **Drug**—**Cahuilla** *Disinfectant* and *Eye Medicine* Powdered berries mixed with water and used to bathe sore or infected eyes. *Unspecified* Leaves used to make tea, which may have had a medicinal use. (15:101) **Hopi** *Unspecified* Plant growing on cottonwood used medicinally for unspecified purpose. (200:72) **Papago** *Analgesic* Bed of heated branches used by women for menstrual cramps. Decoction of leaves taken for stomach cramps and menstrual cramps. *Gastrointestinal Aid* Decoction of leaves taken for stomach cramps. *Gynecological Aid* Decoction of leaves taken for menstrual cramps. Decoction of leaves taken and bed of heated branches used for menstrual cramps. (34:65)
- **Food**—**Cahuilla** *Beverage* Leaves used to make tea, which may have had a medicinal use. *Fruit* Ground berries mixed with a small amount of ashes, boiled in a pot, and eaten. (15:101)
- **Dye**—**Cahuilla** *Black* Leaves used to dye basket weeds permanently black. (15:101)
- **Other**—**Navajo** *Ceremonial Items* Used in the War Dance liniment. *Protection* Twigs hung over the doorway of a hogan for protection from lightning. (55:41)

Phoradendron villosum (Nutt.) Nutt., Pacific Mistletoe

- **Drug**—**Kawaiisu** *Abortifacient* Infusion of plant taken first 2 months of pregnancy to cause an abortion. *Antirheumatic (External)* Infusion of plant used as a wash on limbs affected by rheumatism. (as *P. flavescens* var. *villosum* 206:49) **Pomo** *Abortifacient* Decoction of leaves taken to bring on delayed menstruation. (66:13)

Phoradendron villosum (Nutt.) Nutt. **ssp. *villosum*,** Pacific Mistletoe

- **Drug**—**Pomo, Kashaya** *Abortifacient* Decoction of leaves used for delayed menstruation. (as *P. flavescens* var. *villosum* 72:72)

Phragmites, Poaceae

Phragmites australis (Cav.) Trin. ex Steud., Common Reed

- **Drug**—**Apache, White Mountain** *Antidiarrheal* Root used for diarrhea and kindred diseases. *Gastrointestinal Aid* Root used for stomach troubles and kindred diseases. (as *P. communis* 136:159) **Blackfoot** *Emetic* Decoction of whole plant taken as an emetic. (as *P. communis* 97:22) **Cahuilla** *Orthopedic Aid* Used as a splint for broken limbs. (as *P. communis* var. *berlandieri* 15:101) **Iroquois** *Ceremonial Medicine* Decoction of rootstocks and bottle brush grass used as medicine to soak corn seeds before planting. (as *P. communis* 196:18) *Other* Compound used as a "corn medicine." (87:273) **Keres, Western** *Pediatric Aid* Crushed plant given to children for diarrhea. (as *P. communis* 171:59) **Paiute** *Analgesic* Sugary sap taken by pneumonia patients to loosen phlegm and soothe lung pain. *Expectorant* Sugary sap taken by pneumonia patients to loosen phlegm. *Pulmonary Aid* Sugary sap taken by pneumonia patients to loosen phlegm and soothe lung pain. (as *P. communis* 180:116) **Seminole** *Dermatological Aid* Poultice of plant applied to dry up boils and carbuncles. (as *P. communis* 169:243)
- **Food**—**Kawaiisu** *Sweetener* Stems dried and beaten with sticks to remove the sugar crystals. (206:49) **Klamath** *Unspecified* Seeds used for food. (as *P. phragmites* 45:91) **Montana Indian** *Unspecified* Seeds used for food. (as *P. communis* 19:17) **Paiute** *Sweetener* Dried sap made into balls, softened by fire, and eaten like sugar. (as *P. communis* 167:245) **Paiute, Northern** *Candy* Sap crystallized, gathered, and eaten like candy. (as *P. communis* 59:53) **Thompson** *Forage* Used as a forage plant only in absence of other foods. (as *P. communis* 164:516) **Yuma** *Unspecified* Honeydew obtained from grass. (as *P. communis* 37:218)
- **Fiber**—**Havasupai** *Mats, Rugs & Bedding* Stems used to make mats for drying yucca fruit pulp, baked mescal, peaches, or figs. (as *P. communis* 197:209) **Hopi** *Building Material* Used for roofing, tubular pipes, pipestems, and weaving rods. (as *P. communis* 200:66) **Klamath** *Basketry* Stems used for surface finish of baskets. (as *P. phragmites* 45:91) **Montana Indian** *Basketry* Hard, hollow culms used for pipestems, arrow shafts, and in making baskets. (as *P. communis* 19:17) **Okanagon** *Basketry* Extensively used for basketry. *Mats, Rugs & Bedding* Extensively used to make mats. (as *P. communis* 125:39) **Pima** *Mats, Rugs & Bedding* Used to make mats. (as *P. communis* 47:75) **Seri** *Canoe Material* Canes used to make rafts. (as *P. communis* 50:134) **Thompson** *Basketry* Extensively used for basketry. (as *P. communis* 125:39) Stems used commonly as basketry material. Reed used for weaving baskets. (as *P. communis* 164:497) Stems used in basket imbrication. (187:142) *Mats, Rugs & Bedding* Extensively used to make mats. (as *P. communis* 125:39) Stems twined together to make food drying mats similar to those of tule stems. (187:142)
- **Other**—**Apache, White Mountain** *Hunting & Fishing Item* Reeds used as an arrow shaft for hunting small birds with arrows. *Smoke Plant* Reeds filled with tobacco and used as a cigarette. *Smoking Tools* Reeds used to make pipestems. (as *P. communis* 136:159) **Cahuilla** *Musical Instrument* Used to make a flute, usually played by men. (as *P. communis* var. *berlandieri* 15:101) **Chippewa** *Cooking Tools* Used to weave frames for drying berries. (as *P. communis* 53:378) **Cocopa** *Smoking Tools* Tubular internodes used to smoke tobacco. (as *P. communis* 37:122) **Havasupai** *Hunting & Fishing Item* Stems used for arrow shafts. *Smoking Tools* Stems used to make pipestems. (as *P. communis* 197:209) **Hopi** *Ceremonial Items* Associated ceremonially with the bow and arrow. (as *P. communis* 200:66) **Hualapai** *Hunting & Fishing Item* Shoots used to make arrow shafts. *Musical Instrument* Shoots used to make flutes. (as *P. comminus* 195:7) **Kawaiisu** *Hunting & Fishing Item* Straight, rigid, hollow, bamboo-like stems used in the making of arrows. *Musical Instrument* Reed arrows used to play the musical bow. Stems used as a clapper to maintain the rhythm in singing and dancing. *Smoking Tools* Straight, rigid, hollow, bamboo-like stems used in the making of pipes. *Tools* Straight, rigid, hollow, bamboo-like stems used in the making of fire drills. Stem split and the sharp edge used at birth to cut the navel cord. *Toys & Games* Small stem sections used in the dice game and ring and pin game. (206:49) **Keres, Western** *Unspecified* Plant known and named but no use was specified. (as *P. communis* 171:59) **Klamath** *Hunting & Fishing Item* Stems used for arrow shafts. (as *P. phragmites* 45:92) **Maricopa** *Smoking Tools* Tubular internodes used to smoke tobacco. (as *P. communis* 37:122) **Montana Indian** *Hunting & Fishing Item* Hard, hollow culms used for arrow shafts and in making baskets. *Smoking Tools* Hard, hollow culms used for pipestems. (as *P. communis* 19:17) **Navajo** *Ceremonial Items* Reeds used to make prayer sticks for the Mountain Chant Ceremony. The reeds were first rubbed with a polishing stone to remove the siliceous surface in order that the paint might adhere well. The reeds were then rubbed

with finely powdered tobacco or sometimes with snakeweed. Afterwards the reed was cut into four pieces (or 10 pieces for the second ceremony). When this was finished, the sticks were colored and yucca inserted to serve as handles. The sections were then filled with some kind of tobacco. These had to be kept in order. The section growing nearest the ground was segment number 1, the next number 2, and so on. It was also important that the side of the reed growing toward the east be indicated, so the painting would be done on the side having that exposure. This made it more potent. Fifty-two prayer sticks were made for the evening of the third day of the Night Chant. Of these, four were made of sections of reed, 12 of mountain mahogany, 12 of Russian olive, 12 of sierra juniper, and 12 of cherry. The first people, according to the Navajo, were supposed to have come up to this earth on a reed. Reeds made into frames, like kite frames, and carried by dancers on last night of Mountain Chant. *Hunting & Fishing Item* Stems used to make arrow shafts. (as *P. communis* 55:26) **Okanagon** *Decorations* Extensively used to make fringe for dresses. (as *P. communis* 125:39) **Paiute, Northern** *Hunting & Fishing Item* Stalks used to make arrow shafts. (60:75) **Papago** *Smoking Tools* Six-inch stems used as smoking tubes. (as *P. communis* 34:27) **Pima** *Musical Instrument* Used to make flutes. (as *P. communis* 47:75) **Seminole** *Tools* Stems used to make medicine tubes. (as *P. communis* 169:172) Plant used to make medicine blowing tubes. (as *P. communis* 169:495) **Tewa** *Hunting & Fishing Item* Plant used to make arrows. *Toys & Games* Plant used to make game sticks for the canute game. (as *P. phragmites* 138:66) **Thompson** *Cooking Tools* Stems twined together to make food drying mats similar to those of tule stems. (187:142) *Decorations* Extensively used to make fringe for dresses. (as *P. communis* 125:39) Whitish culms valued for the use in decoration of coiled split cedar root baskets. The culms were harvested while still green and soft, warmed over the coals of a fire, and broken at the nodes. They were then split open, flattened, and used together with dyed and undyed bitter cherry bark to create patterns on coiled cedar root baskets. (187:142) *Designs* Used to design ornamenting baskets. (as *P. communis* 164:499) *Jewelry* Reed cut in different lengths, dyed, and used interspersed with seed beads for necklaces. (as *P. communis* 164:497) **Yuma** *Smoking Tools* Tubular stalk internodes used to smoke tobacco. (as *P. communis* 37:122)

Phryma, Verbenaceae

Phryma leptostachya L., American Lopseed
- *Drug*—**Chippewa** *Throat Aid* Decoction of root gargled or root chewed for sore throat. (53:342) **Ojibwa, South** *Antirheumatic (Internal)* Decoction of root taken for rheumatic leg pains. (91:201)

Phyla, Verbenaceae

Phyla cuneifolia (Torr.) Greene, Fogfruit
- *Drug*—**Navajo, Kayenta** *Dermatological Aid* Poultice of plants applied to spider bites. (as *Lippia cuneifolia* 205:40)

Phyla lanceolata (Michx.) Greene, Lanceleaf Fogfruit
- *Drug*—**Mahuna** *Antirheumatic (Internal)* Plant used for rheumatism. (as *Lippia lanceolata* 140:60)

Phyla nodiflora (L.) Greene, Turkey Tangle Fogfruit
- *Drug*—**Houma** *Orthopedic Aid* and *Pediatric Aid* Decoction of plant used as a wash to make weak, lazy babies walk. (as *Lippia nodiflora* 158:65)

Phyllodoce, Ericaceae

Phyllodoce empetriformis (Sm.) D. Don, Pink Mountainheath
- *Drug*—**Thompson** *Tuberculosis Remedy* Decoction of plant taken over a period of time for tuberculosis and spitting up blood. (187:215)

Phyllodoce sp.
- *Drug*—**Thompson** *Tuberculosis Remedy* Decoction of plant taken over a period of time for tuberculosis and spitting up blood. (187:215)

Phyllospadix, Zosteraceae

Phyllospadix scouleri Hook., Scouler's Surfgrass
- *Food*—**Hesquiat** *Unspecified* Leaves occasionally cooked and eaten when it had herring eggs on it. (185:58) **Makah** *Unspecified* Rhizomes chewed or eaten raw. (67:328) Roots eaten raw in the spring. (79:21)
- *Other*—**Quileute** *Toys & Games* Grass used by boys as arrow target practice. (79:21)

Phyllospadix serrulatus Rupr. ex Aschers., Toothed Surfgrass
- *Food*—**Makah** *Unspecified* Rhizomes chewed or eaten raw. (67:328)

Phyllospadix sp., Seaweed
- *Food*—**Makah** Species used for food. (67:205)

Phyllospadix torreyi S. Wats., Torrey's Surfgrass
- *Drug*—**Kwakiutl** *Pediatric Aid* and *Strengthener* Leaves placed in the bottom of child's cradle to make him grow strong. (183:274)
- *Food*—**Hesquiat** *Dried Food* Leaves, with herring eggs on it, dried for later use. (185:58) **Makah** *Unspecified* Rhizomes chewed or eaten raw. (67:328)
- *Fiber*—**Hesquiat** *Basketry* Bleached leaves used to make baskets. (185:58) **Makah** *Basketry* White, sun bleached leaves used in basketry. (67:328) **Nitinaht** *Basketry* Leaves dried, split, and used in basketry. (186:89)
- *Other*—**Hesquiat** *Toys & Games* Dried, curly leaves used by children to make wigs. (185:58) **Nitinaht** *Hunting & Fishing Item* Leaves used to trap herring spawn. (186:89)

Physalis, Solanaceae

Physalis acutifolia (Miers) Sandw., Sharpleaf Groundcherry
- *Food*—**Pima, Gila River** *Baby Food* Fruits eaten raw, primarily by children. (133:7) *Snack Food* Fruit eaten primarily by children as a snack food. (133:5)

Physalis hederifolia var. *fendleri* (Gray) Cronq., Fendler's Groundcherry
- *Food*—**Apache, White Mountain** *Fruit* Fruit eaten raw and cooked. (as *P. fendleri* 136:159) **Mohave** *Fruit* Fruits eaten fresh by children. **Yuma** *Fruit* Fruits eaten fresh by children. (as *P. fendleri* 37:207) **Zuni** *Sauce & Relish* Fruit boiled in small quantities of water, crushed, and used as a condiment. (as *P. fendleri* 166:70)

Physalis heterophylla Nees, Clammy Groundcherry
- *Drug*—**Iroquois** *Burn Dressing* Compound infusion of dried leaves

and roots used as wash for scalds and burns. *Emetic* and *Gastrointestinal Aid* Compound infusion of leaves and roots taken to vomit for bad stomachaches. *Venereal Aid* Compound infusion of dried leaves and roots used as wash for venereal disease. (87:430) **Lakota** *Dietary Aid* Three or five berries used for lack of appetite. (139:60) **Meskwaki** *Unspecified* Root used as a medicine. (152:247)
- *Food*—**Cherokee** *Fruit* Fruit used for food. (80:37) **Cheyenne** *Fruit* Ripe fruits eaten in fall. (83:39) **Dakota** *Dried Food* Fruits, when in sufficient quantity, dried for winter use. (70:113) *Sauce & Relish* Fruit made into a sauce. (69:362) Fruits made into a sauce for food.˙(70:113) *Winter Use Food* Fruit dried and stored for winter use. (69:362) **Meskwaki** *Fruit* Berries eaten raw. (152:264) **Omaha** *Dried Food* Fruits, when in sufficient quantity, dried for winter use. *Sauce & Relish* Fruits made into a sauce for food. **Pawnee** *Dried Food* Fruits, when in sufficient quantity, dried for winter use. *Sauce & Relish* Fruits made into a sauce for food. **Ponca** *Dried Food* Fruits, when in sufficient quantity, dried for winter use. *Sauce & Relish* Fruits made into a sauce for food. (70:113)

Physalis lanceolata Michx., Lanceleaf Groundcherry
- *Drug*—**Omaha** *Analgesic* Decoction of root used for headache and stomach trouble. *Dermatological Aid* Root used as a dressing for wounds. *Gastrointestinal Aid* Decoction of root used for stomach trouble and headache. *Unspecified* Root used in smoke treatment for unspecified ailments. **Ponca** *Analgesic* Decoction of root used for headache and stomach trouble. *Dermatological Aid* Root used as a dressing for wounds. *Gastrointestinal Aid* Decoction of root used for stomach trouble and headache. *Unspecified* Root used in smoke treatment for unspecified ailments. **Winnebago** *Analgesic* Decoction of root used for headache and stomach trouble. *Dermatological Aid* Root used as a dressing for wounds. *Gastrointestinal Aid* Decoction of root used for stomach trouble and headache. *Unspecified* Root used in smoke treatment for unspecified ailments. (70:113)
- *Food*—**Dakota** *Unspecified* Bud clusters used in the spring for food. Firm, young, green seedpods boiled with meat in the spring. (69:363) **Navajo** *Fruit* Berries used for food. (90:154)
- *Other*—**Dakota** *Toys & Games* Large calyx of plant inflated by children in play and popped by striking it on the forehead or hand. (69:362)

Physalis longifolia Nutt., Longleaf Groundcherry
- *Food*—**Keres, Western** *Fruit* Berries used for food. (171:59) **Pueblo** *Fruit* Berries eaten fresh or boiled. **San Felipe** *Fruit* Berries eaten fresh or boiled. (32:40) **Zuni** *Fruit* Berries boiled, ground in a mortar with raw onions, chile, and coriander seeds and used for food. (166:70)

Physalis philadelphica Lam., Mexican Groundcherry
- *Drug*—**Diegueño** *Eye Medicine* Berries squeezed and the juice used as an eyewash. (84:28)

Physalis pubescens L., Husk Tomato
- *Drug*—**Navajo, Ramah** *Panacea* Dried leaves and root used as "life medicine." (191:43)
- *Food*—**Mohave** *Fruit* Fruits eaten fresh by children. (37:207) **Navajo** *Fruit* Sour berries mixed with honey and eaten. *Preserves* Sour berries used to make jam. *Staple* Berries dried, ground into a flour, and stored for winter use. (110:17) **Navajo, Ramah** *Fruit* Fruit eaten raw or boiled. (191:43) **Yuma** *Fruit* Fruits eaten fresh by children. (37:207)

Physalis sp., Groundcherry
- *Food*—**Cherokee** *Fruit* Fresh fruit used for food. (126:55) **Hualapai** *Fruit* Berries eaten fresh from the vine. *Preserves* Berries used to make preserves. *Sauce & Relish* Berries used to make relish. (195:9) **Iroquois** *Bread & Cake* Fruit mashed, made into small cakes, and dried for future use. *Dried Food* Raw or cooked fruit sun or fire dried and stored for future use. *Fruit* Dried fruit taken as a hunting food. *Sauce & Relish* Dried fruit cakes soaked in warm water and cooked as a sauce or mixed with corn bread. (196:129) **Isleta** *Fruit* Fresh berries eaten for food. (100:36)

Physalis subulata var. *neomexicana* (Rydb.) Waterfall ex Kartesz & Gandhi, New Mexican Groundcherry
- *Food*—**Apache, Chiricahua & Mescalero** *Special Food* Fresh fruit eaten by children as a delicacy. (as *P. neomexicana* 33:45) **Keres, Western** *Fruit* Berries used for food. (as *P. neo-mexicana* 171:59) **Navajo** *Fruit* Raw fruit used for food. (as *P. neomexicana* 165:222) **Pueblo** *Fruit* Berries eaten fresh or boiled. (as *P. neo-mexicana* 32:39) **Tewa** *Fruit* Berries used for food. (as *P. neomexicana* 138:59)

Physalis virginiana P. Mill., Virginia Groundcherry
- *Drug*—**Meskwaki** *Stimulant* Infusion of whole plant taken for dizziness. (152:247, 248)
- *Food*—**Meskwaki** *Fruit* Berries, touched by frost, eaten raw. (152:264)

Physalis viscosa L., Starhair Groundcherry
- *Drug*—**Omaha** *Dermatological Aid* Root used to dress wounds. (as *P. viscora* 58:584)

Physaria, Brassicaceae

Physaria chambersii Rollins, Chambers's Twinpod
- *Drug*—**Paiute** *Eye Medicine* Infusion or decoction of various plant parts used as a wash for sore eyes or sties. **Shoshoni** *Eye Medicine* Infusion or decoction of various plant parts used as a wash for sore eyes and sties. (180:116)

Physaria didymocarpa (Hook.) Gray, Common Twinpod
- *Drug*—**Blackfoot** *Abortifacient* Infusion of plant taken in small amounts to abort. (86:61) *Analgesic* Plant chewed for cramps and stomach trouble. (114:274) *Antirheumatic (External)* Infusion of roots applied to aching parts of the body. (86:78) Strong infusion of plant used as a liniment on sprains. (86:79) Decoction of plant used for swellings. (97:35) *Dermatological Aid* Weak decoction of plant used for diaper rash. Weak decoction of leaf used on newborn's umbilical. (86:77) Infusion of plant applied to wounds to heal with less irritation. (86:84) Infusion of plant used to allay swelling. (114:274) *Dietary Aid* Decoction of plant taken slowly to gradually expand the stomach until food was eaten without pain. This decoction was used by a person who had not eaten for a long time. (86:104) *Ear Medicine* Infusion of leaves used as drops for ear infections. *Eye Medicine* Infusion of leaves used as drops for bloodshot eyes. (86:81) *Gastrointestinal Aid* Plant chewed for cramps and stomach troubles. (97:35) Plant chewed for cramps, stomach trouble, and sore throat. (114:274) Infusion of leaves taken for stomach trouble. (118:38) *Orthopedic Aid*

Strong infusion of plant used as a liniment on dislocations. (86:79) *Pediatric Aid* Weak decoction of leaf used on newborn's umbilical. (86:77) *Throat Aid* Plant chewed for sore throats. (97:35) Plant chewed for sore throat. (114:274) Infusion of leaves taken for sore throat. (118:38) *Toothache Remedy* Leaf clenched between the teeth for toothache. (86:77) Plant used as toothache medicine. (86:78) *Veterinary Aid* Infusion of plant applied as a liniment to the shoulders of work and wagon horses. (86:89)

Physaria didymocarpa var. *lanata* A. Nels., Common Twinpod
- **Drug**—**Blackfoot** *Gastrointestinal Aid* Plant chewed for stomach troubles. *Throat Aid* Plant chewed for sore throats. (121:47)

Physaria newberryi Gray, Newberry's Twinpod
- **Drug**—**Hopi** *Antidote* and *Ceremonial Medicine* Plant taken as an antidote after the snake dance. (56:16) **Navajo** *Respiratory Aid* Plant used as a snuff for catarrh. (55:49)

Physocarpus, Rosaceae

Physocarpus capitatus (Pursh) Kuntze, Pacific Ninebark
- **Drug**—**Bella Coola** *Emetic* Decoction of 3-foot stick taken alternatively with large amounts of water as an emetic. (184:208) **Green River Group** *Emetic* Young shoots, peeled of bark, used as an emetic. (79:33) **Hesquiat** *Antidote* Decoction of bark taken as an antidote for poisoning, caused vomiting. *Antirheumatic (External)* Decoction of bark used as a wash or soaking solution for rheumatic pain. *Antirheumatic (Internal)* Decoction of bark taken for rheumatic fever. *Emetic* Decoction of bark taken as an antidote for poisoning, caused vomiting. Bark chewed and juice swallowed to induce vomiting. *Laxative* Decoction of bark taken in small doses as a laxative. (185:73) **Kwakiutl** *Cathartic* Root extract used as a purgative. *Emetic* Decoction of bark taken to induce vomiting. *Laxative* Decoction of bark taken for constipation. *Venereal Aid* Root extract used for locomotor ataxia. (183:289) **Saanich** *Laxative* Infusion of macerated roots taken as a quick laxative. (182:86)
- **Food**—**Miwok** *Fruit* Raw berries used for food. (12:162)
- **Dye**—**Hesquiat** *Brown* Bark soaked with cedar bark to darken the cedar. (185:73)
- **Other**—**Hesquiat** *Toys & Games* Wood used for making children's bows. (185:73) **Karok** *Hunting & Fishing Item* Shoots used to make arrows. (148:384) **Tolowa** *Toys & Games* Seeds squeezed and popped. (6:43) **Wintoon** *Hunting & Fishing Item* Straight stems used to make arrows. (117:264)

Physocarpus malvaceus (Greene) Kuntze, Mallow Ninebark
- **Drug**—**Okanagan-Colville** *Hunting Medicine* Infusion of bark used to wash arrows and other hunting equipment to protect them from spells. (188:126)
- **Food**—**Okanagan-Colville** *Unspecified* Roots steam cooked and eaten. (188:126)
- **Other**—**Okanagan-Colville** *Malicious Charm* Plant used to cause other people bad luck. (188:126)

Physocarpus opulifolius (L.) Maxim., Common Ninebark
- **Drug**—**Bella Coola** *Analgesic* and *Emetic* Decoction of inner bark taken as an emetic by persons "dizzy with pain." *Laxative* Decoction of inner bark taken as a laxative for gonorrhea. *Tuberculosis Remedy* Decoction of inner bark taken and used as a wash for scrofulous glands in neck. *Venereal Aid* Decoction of inner bark taken and used as wash for gonorrhea. **Carrier, Southern** *Cathartic* and *Emetic* Decoction of bark taken as an emetic, a large dose fatal. (150:59) **Chippewa** *Emetic* Infusion of roots taken as an emetic. (71:132) **Iroquois** *Gynecological Aid* Poultice used when women swell after copulation; caused by bad medicine. (87:349) **Menominee** *Gynecological Aid* Bark used to make a drink for female maladies, to cleanse system and enhance fertility. (151:49)

Physostegia, Lamiaceae

Physostegia parviflora Nutt. ex Gray, Western False Dragonhead
- **Drug**—**Meskwaki** *Cold Remedy* Infusion of leaves taken for bad cold. (152:226)

Phytolaccaceae, see *Phytolacca*

Phytolacca, Phytolaccaceae

Phytolacca americana L., American Pokeweed
- **Drug**—**Cherokee** *Antirheumatic (Internal)* Infusion of berry taken for arthritis. Roots and berries or berry wine used for rheumatism. *Blood Medicine* Cooked greens eaten or infusion of root taken to build the blood. *Dermatological Aid* Poultice used for ulcers and swellings and infusion of root used for eczema. Salve used on "ulcerous sores" and dried, crushed roots sprinkled on old sores. *Febrifuge* Poultice used for "nervous fevers, ulcers and swellings." *Kidney Aid* Cold infusion of powdered root taken for kidneys. (80:50) *Laxative* Plant used in a side dish with laxative properties. (204:251) *Other* Compound used for "white swelling." (80:50) *Poison* Roots and berries considered poisonous. *Unspecified* Berries used for medicine. (126:51) **Delaware** *Antirheumatic (External)* Roasted, crushed roots used with sarsaparilla and mountain grape barks for rheumatism. *Blood Medicine* Roasted, crushed roots used with sarsaparilla and mountain grape barks as a blood purifier. *Dermatological Aid* Roots roasted and the salve used for chronic sores. *Gland Medicine* Roots roasted and the salve used for glandular swellings. *Stimulant* Roasted, crushed roots used with sarsaparilla and mountain grape barks as a stimulant. (as *P. decandra* 176:32) **Delaware, Oklahoma** *Antirheumatic (External)* Strong infusion of roots or twigs used as herbal steam for rheumatism. (as *P. decandra* 175:30, 78) *Antirheumatic (Internal)* Compound containing root used for rheumatism. *Blood Medicine* Compound containing root used as a blood purifier. (as *P. decandra* 175:27, 78) *Herbal Steam* Infusion of roots or twigs used as herbal steam for rheumatism. (as *P. decandra* 175:30) *Stimulant* Compound containing root used as a stimulant. (as *P. decandra* 175:27, 78) **Iroquois** *Antirheumatic (External)* Stalks cooked as greens and used for rheumatism. (as *P. decandra* 124:93) *Cathartic* Plant used as a cathartic. (87:316) *Cold Remedy* Decoction of stems taken for chest colds. (87:317) *Dermatological Aid* Compound with undried roots applied as a salve on bunions. Poultice of crushed roots applied to bruises. Raw berries rubbed on skin lumps. *Emetic* Plant used as an emetic. *Expectorant* Plant used as an expectorant. (87:316) *Liver Aid* Compound infusion of whole roots used for liver sickness. (87:317) *Love Medicine* "Tie in a poplar tree, then place among roots," as a love medicine. *Orthopedic Aid* Decoction of roots applied as poultice to sprains, bruises, and swollen joints. *Witchcraft Medi-*

cine Plant used for bewitchment. (87:316) **Mahuna** *Analgesic* Roots used for severe, neuralgic pains. *Dermatological Aid* Leaves used for skin diseases or to remove pimples and blackheads. *Poison* Plant considered poisonous. (as *P. decandra* 140:65) **Micmac** *Hemostat* Leaves used for bleeding wounds. (as *P. decandra* 40:59) **Mohegan** *Gynecological Aid* Poultice of mashed berries applied to sore breasts. (as *P. decandra* 176:74, 130) *Poison* Root considered poisonous. (as *P. decandra* 176:74) **Rappahannock** *Antidiarrheal* Infusion of berries taken for dysentery. *Antirheumatic (Internal)* Fermented infusion of leaves taken for rheumatism. *Dermatological Aid* Compound infusion with roots applied to ivy poison. Poultice of mashed root applied to wart until it bleeds. *Hemorrhoid Remedy* Steam from decoction of roots used for piles. (161:29) **Seminole** *Analgesic* Berries eaten as an analgesic. (169:167) Berries eaten for pains. *Antirheumatic (Internal)* Berries eaten for rheumatism. (169:285)
- **Food**—**Cherokee** *Beverage* Crushed berries and sour grapes strained, mixed with sugar and cornmeal, and used as a beverage. (126:51) *Cooking Agent* Crushed berries used to add color to canned fruit. (80:50) Berries used to color canned fruit. *Dried Food* Leaves gathered into bundle and dried for future use. (126:51) *Unspecified* Young shoots cut, cooked, and eaten. (204:251) *Vegetable* Shoots, leaves, and stems parboiled, rinsed, and cooked alone or mixed with other greens and eggs. Peeled stalks cut lengthwise, parboiled, dipped in egg, rolled in cornmeal, and fried like fish. (126:51) **Iroquois** *Vegetable* Stalks eaten as greens in spring. (as *P. decandra* 124:93) **Malecite** *Unspecified* Shoots used for food. (160:6) **Mohegan** *Unspecified* Cooked and used for food. (as *P. decandra* 176:83)
- **Dye**—**Mahuna** *Unspecified* Berries used to make dyes and inks. (as *P. decandra* 140:65)
- **Other**—**Kiowa** *Jewelry* Dark, dry fruits used by girls to make necklaces. (192:26) **Pawnee** *Paint* Fruit made into a red stain used in painting horses and various articles of adornment. (70:78)

Picea, Pinaceae, conifer

Picea abies (L.) Karst., Norway Spruce
- **Drug**—**Mohegan** *Analgesic* Poultice of sap or gum applied for boil and abscess pains. (176:74) *Dermatological Aid* Sap or gum applied to boil or abscess pains. (176:74, 130)

Picea engelmannii Parry ex Engelm., Engelmann's Spruce
- **Drug**—**Navajo, Ramah** *Ceremonial Medicine* and *Emetic* Plant used as a ceremonial emetic. (191:12) **Okanagan-Colville** *Respiratory Aid* Infusion of bark used for respiratory ailments. *Tuberculosis Remedy* Infusion of bark used for tuberculosis. (188:27) **Thompson** *Cancer Treatment* Decoction of needles and gum taken for cancer. It was said that if this treatment did not work, nothing would work. The decoction was taken with a spoon directly from the bark blisters and in concentrated form. *Cough Medicine* Decoction of needles and gum taken for coughs. (187:100) *Dermatological Aid* Twig ashes mixed with grease and used as an ointment or salve. (164:475) Pitch used for eczema. *Psychological Aid* Tree and red cedar tree caused vivid dreams for anyone who slept under it. (187:100)
- **Food**—**Okanagan-Colville** *Beverage* Branches used by mountain travelers to make a tea. (188:27) **Thompson** *Unspecified* Sap considered edible. (187:100)
- **Fiber**—**Hoh** *Building Material* Timber used to make shakes, clapboards, and framing timbers. *Cordage* Limbs and roots shredded, pounded, and used to make cord and rope. (137:59) **Paiute** *Mats, Rugs & Bedding* Boughs used on the floor of sweat houses and for camping beds. (111:44) **Quileute** *Building Material* Timber used to make shakes, clapboards, and framing timbers. *Cordage* Limbs and roots shredded, pounded, and used to make cord and rope. (137:59) **Thompson** *Basketry* Bark used to make baskets. *Building Material* Bark used to thatch the roofs of lodges. *Canoe Material* Bark used to cover canoes. (164:499)
- **Other**—**Hoh** *Toys & Games* Timber used to make toys. **Quileute** *Toys & Games* Timber used to make toys. (137:59) **Thompson** *Cooking Tools* Bark used to make utensils of all kinds. (164:499) *Good Luck Charm* Tree and red cedar tree provided good luck and wishes for those who asked for it. (187:100) *Soap* Branch tips and needles boiled and used as a wash by hunters, warriors, and boys at puberty. (164:505) See also *P. glauca × engelmannii* under *P. glauca*.

Picea glauca (Moench) Voss, White Spruce
- **Drug**—**Abnaki** *Urinary Aid* Infusion of cones taken for urinary troubles. (144:164) **Algonquin, Quebec** *Cough Medicine* Inner bark chewed and infusion of inner bark taken for coughs. *Dermatological Aid* Gum used as a salve. *Gynecological Aid* Used in the sudatory, this is taken by women after childbirth and for other complaints. *Internal Medicine* Infusion of branch tips taken to "heal the insides." *Laxative* Gum chewed as a laxative. (18:126) Resin chewed as a laxative. (18:73) *Unspecified* Used in the sudatory, this is taken by women after childbirth and for other complaints. (18:126) **Chippewa** *Antirheumatic (External)* Compound decoction of twigs used as herbal steam for rheumatism. (as *P. canadensis* 53:362) **Cree, Woodlands** *Antirheumatic (Internal)* Decoction of inner bark used for arthritis. *Blood Medicine* Poultice of gum and lard applied for blood poisoning. *Dermatological Aid* Pitch and grease used as an ointment for skin rashes, scabies, persistent scabs, and growing boils. Rotten, dried, finely powdered wood used as baby powder and for skin rashes. Poultice of gum and lard applied to infections. Rotten wood used in baby dusting powder. *Pediatric Aid* Rotten, dried, finely powdered wood used as baby powder and for skin rashes. Rotten wood used in baby dusting powder. (109:48) **Eskimo, Alaska** *Dermatological Aid* Poultice of resin applied to wounds. *Unspecified* Infusion of needles used as medicine for all purposes. (as *P. canadensis* 4:716) **Eskimo, Inuktitut** *Dermatological Aid* Poultice of gum and grease applied to pustulant wounds. *Respiratory Aid* Decoction of gum or needles taken for respiratory infections. (202:188) **Eskimo, Kuskokwagmiut** *Cough Medicine* Decoction of green needles taken or raw needles chewed as cough medicine. (122:28, 29) **Eskimo, Nunivak** *Dermatological Aid* Resin applied to wounds. *Panacea* Infusion of needles used as a medicine for all purposes. (as *P. canadensis* 149:325) **Eskimo, Western** *Cough Medicine* Decoction of needles taken or raw needles chewed as a cough medicine. (108:24) **Gitksan** *Cold Remedy* Decoction of bark or inner bark used for colds. *Cough Medicine* Decoction of bark or inner bark used for coughs. *Misc. Disease Remedy* Decoction of bark or inner bark used for flu. *Tonic* Decoction of bark or inner bark used as a tonic. (as *P. glauca × engelmannii* 73:152) **Iroquois** *Gastrointestinal Aid* Gum chewed to facilitate digestion. (142:83) **Koyukon**

Ceremonial Medicine Pitch, swan feathers, and slender grass tops burned by shamans when making medicine for a sick person. (119:50) *Dermatological Aid* Infusion of needles used as a rub or bath for dry skin or sores. (119:49) Pitch applied to sores and cuts. (119:50) *Hunting Medicine* Tops put in animal track by girls before stepping over it, to avoid alienating animals from hunters. *Kidney Aid* Infusion of needles taken for kidney problems. *Panacea* Infusion of needles taken to promote general good health. *Unspecified* Tree tops used by the shamans to brush people and remove their sickness. (119:49) **Malecite** *Unspecified* Pitch used in medicines. (160:6) **Menominee** *Dermatological Aid* Poultice of cooked, beaten inner bark applied to wounds, cuts, or swellings. *Internal Medicine* Infusion of inner bark taken for "inward troubles for either man or woman." (as *P. canadensis* 151:45) **Micmac** *Cough Medicine* Bark used as a cough remedy. *Dermatological Aid* Bark used to prepare a salve for cuts and wounds. Gum used for scabs and sores. *Gastrointestinal Aid* Parts of plant used for stomach trouble. *Misc. Disease Remedy* Bark, leaves, and stems used for scurvy. (as *P. canadensis* 40:59) **Montagnais** *Tonic* Infusion of twigs taken "for generally beneficial effects." (as *P. canadensis* 156:314) **Ojibwa** *Disinfectant* Dried leaves used as an inhalant and fumigator. (as *P. canadensis* 153:379) **Ojibwa, South** *Antidiarrheal* Compound containing outer bark taken for diarrhea. (as *Abies canadensis* 91:198) **Okanagan-Colville** *Respiratory Aid* Infusion of bark used for respiratory ailments. *Tuberculosis Remedy* Infusion of bark used for tuberculosis. (188:27) **Shuswap** *Dermatological Aid* Poultice of soft pitch applied to sores. *Panacea* Decoction of bark taken for tuberculosis and other sickness. *Toothache Remedy* Hard pitch chewed to clean the teeth. *Tuberculosis Remedy* Decoction of bark taken for tuberculosis. (123:51) **Tanana, Upper** *Antirheumatic (Internal)* Decoction of tree top, young birch tip, and Hudson Bay tea taken for body aches. *Cold Remedy* Decoction of young tips, Hudson Bay tea, and blackberry stems taken for colds. Raw cambium chewed for colds. Decoction of tree top, young birch tip, and Hudson Bay tea taken for colds. Decoction of tree tip, Hudson Bay tea, and blackberry stems used for colds. *Cough Medicine* Raw cambium chewed for coughs. *Dermatological Aid* Poultice of raw or boiled cambium applied to sores and infected areas or used to bandage cuts. Decoction of tree tip used as a wash for rashes and sores. Pitch and moose fat warmed into an ointment and used for sores. Pitch boiled in water and applied to sores. Soft pitch, sometimes mixed with grease, used as an ointment for sores. *Disinfectant* Decoction of tree top and cottonwood taken for infections. Soft pitch, sometimes mixed with grease, used as an ointment for external infections. *Hemorrhoid Remedy* Chewed pitch applied to bleeding cuts. *Oral Aid* Decoction of young tips, Hudson Bay tea, and blackberry stems taken for mouth sores. Decoction of tree tip, Hudson Bay tea, and blackberry stems used for mouth sores. *Pulmonary Aid* Decoction of wood ash taken for chest problems. *Respiratory Aid* Decoction of tree top, young birch tip, and Hudson Bay tea taken for congestion. *Throat Aid* Pitch chewed for sore throats. *Tuberculosis Remedy* Raw cambium chewed for tuberculosis. Decoction of wood ash taken for tuberculosis. (102:2) **Tlingit** *Antidiarrheal* Sap mixed with mountain goat tallow and used for diarrhea. (as *Pinus canadensis* 107:283) **Wet'suwet'en** *Cold Remedy* Decoction of bark or inner bark used for colds. *Cough Medicine* Decoction of bark or inner bark used for coughs. *Misc. Disease Remedy* Decoction of bark or inner bark used for flu. *Tonic* Decoction of bark or inner bark used as a tonic. (as *Picea glauca × engelmannii* 73:152)

- *Food*—**Algonquin, Quebec** *Candy* Resin chewed like chewing gum. (18:73) **Cree, Woodlands** *Candy* Gum chewed for pleasure. Gum chewed as a confection. (109:48) **Eskimo, Alaska** *Candy* Resin chewed for pleasure. (as *P. canadensis* 4:716) **Eskimo, Inuktitut** *Unspecified* Cambium eaten in the spring. (202:188) **Gitksan** *Unspecified* Cambium eaten fresh. (as *P. glauca × engelmannii* 73:151) **Koyukon** *Candy* Pitch chewed like gum. (119:50) **Micmac** *Beverage* Bark used to make a beverage. (159:258) **Okanagan-Colville** *Beverage* Branches used by mountain travelers to make a tea. (188:27) **Tanana, Upper** *Candy* Hard pitch used for chewing gum. *Fodder* Rotten wood mixed with poque and fed to puppies. *Starvation Food* Cambium used as a food during periods of food shortage. *Unspecified* Fresh sap eaten as food during the summer. (102:2) **Wet'suwet'en** *Unspecified* Cambium eaten fresh. (as *P. glauca × engelmannii* 73:151)

- *Fiber*—**Algonquin, Tête-de-Boule** *Basketry* Roots used to sew baskets. *Canoe Material* Roots used to sew canoes. *Snow Gear* Roots used to sew snowshoes. (132:129) **Cree, Woodlands** *Basketry* Wood used for the edging of a birch bark sewing basket base and lid. *Building Material* Bark sheets used for roofing on buildings. *Canoe Material* Wood used to make canoe paddles. Wood used to make ribs and gunwales for birchbark canoes. *Caulking Material* Pitch used as a sealant for birchbark canoes. *Mats, Rugs & Bedding* Bark sheets used for tent flooring. *Sewing Material* Roots used to sew birch bark baskets and canoes. (109:48) **Eskimo, Inuktitut** *Building Material* Wood used to make cabins and caches. *Canoe Material* Split wood used to make fish traps and canoe or kayak stringers. *Cordage* Split, inner root bark or small rootlets used as fishing lines and cord for making and repairing tools. *Mats, Rugs & Bedding* Needles used as flooring in tents. (202:188) **Iroquois** *Brushes & Brooms* Wood used to make scrub brushes. (142:83) **Koyukon** *Building Material* Wood used for the building log houses. Wood split or ripsawed and used as lumber for house construction, caches, and tent frames. *Canoe Material* Wood split or ripsawed and used to make boats and canoes. *Caulking Material* Pitch used to caulk boat seams. *Snow Gear* Wood split or ripsawed and used to make sleds. (119:50) **Malecite** *Building Material* Bark used for hut roofing. *Canoe Material* Bark used for canoes. *Mats, Rugs & Bedding* Needles and branches used for pillows and bedding. *Sewing Material* Roots used for thread. (160:6) **Micmac** *Mats, Rugs & Bedding* Boughs used to make beds. (159:258) **Tanana, Upper** *Basketry* Roots used to sew birch bark baskets.[57] *Brushes & Brooms* Twigs used by young menstruating girls to clean their teeth and to scratch their heads. *Building Material* Bark used as siding and roofing material for steam bath houses and other structures. Wood used for fuel and building logs. *Canoe Material* Roots used for the bow of a canoe.[57] Wood used to make boats, boat paddles, shovels, skin stretchers, and wedges for chopping wood. *Cordage* Split or whole roots used to make line.[57] *Mats, Rugs & Bedding* Boughs used for camp mattresses and dog bedding. Boughs used on the floor of camp buildings to sit on. *Snow Gear* Boughs used as temporary snowshoes by securing with line. (102:2)

- *Dye*—**Cree, Woodlands** *Yellow-Brown* Rotten wood used as a yellow brown dye for white goods. (109:48)

- *Other*—Cree, Woodlands *Cooking Tools* Split log hollowed out to make a dish to feed fish broth to puppies. *Fasteners* Wood pegs used to fasten the tabs on the bottom and top pieces to the basket body during construction. *Hide Preparation* Rotten, dried wood burned in a slow fire to smoke tan hides. *Hunting & Fishing Item* Wood used to make floats for fishing nets. *Tools* Dead, standing trees used to make a moose hide stretcher. (109:48) **Eskimo, Inuktitut** *Ceremonial Items* Roots used to make headgear and masks. *Containers* Roots used to make trays and buckets. *Cooking Tools* Roots used to make spoons, dippers, and bowls. *Fuel* Wood used for fires. *Hunting & Fishing Item* Split wood used to make fish traps and canoe or kayak stringers. Roots used to make floats. (202:188) **Iroquois** *Fuel* Bark used to start fires. (142:83) **Koyukon** *Ceremonial Items* Knot ring worn by children on a string around the neck to make them skilled with their hands. *Cooking Tools* Roots used for carving spoons and bowls. *Fasteners* Roots skinned and used to lash birch bark baskets. *Fuel* Wood used for household heating fuel. *Good Luck Charm* Golden needles conferred good luck to those who found them. *Hide Preparation* Rotten wood pulverized, mixed with rotten willow and used to smoke hides. (119:50) *Protection* Trees protected those who slept beneath them, especially from malevolent spirits. (119:49) Trees nullified dangerous spiritual forces. Boughs taken home as talismans for protection. (119:50) **Malecite** *Waterproofing Agent* Pitch used for waterproofing seams in canoes. (160:6) **Micmac** *Fuel* Wood used for kindling and fuel. (159:258) **Tanana, Upper** *Containers* Roots woven into waterproof containers.[57] Small, dead tree used to dry fish on. *Cooking Tools* Rough bark used to cut fish on, prevented the fish from slipping. Bark made into a container and used to roast waterfowl eggs. The spruce bark was cut large enough to surround the eggs, tied around the eggs, and the ends plugged with moss. *Fasteners* Warmed pitch used as glue to patch birchbark canoes and to attach feathers to arrows. *Fuel* Wood used for fuel and building logs. *Hide Preparation* Rotten, reddish colored wood smoke used to tan moose skins. *Hunting & Fishing Item* Roots woven into dip nets.[57] White, inner side of bark used in the bottom of a weir to act as a reflector. Reflector used in order to more easily spot fish as they swam through the weir. Wood used to build weirs, fish traps, fish racks, fish rafts, and boat poles. *Insecticide* Needles burned to keep mosquitoes away. *Tools* Wood used to make boats, boat paddles, shovels, skin stretchers, and wedges for chopping wood. (102:2)

Picea mariana (P. Mill.) B.S.P., Black Spruce

- *Drug*—**Algonquin, Quebec** *Dermatological Aid* Gum used as a salve. *Internal Medicine* Infusion of branch tips used for "healing the insides." *Unspecified* Used in the medicinal sudatory. (18:127) **Cree, Woodlands** *Antidiarrheal* Decoction of cones taken for diarrhea. *Burn Dressing* Pitch mixed with grease and used as ointment for bad burns. *Dermatological Aid* Pitch mixed with grease and used as ointment for skin rashes, scabies, and persistent scabs. *Oral Aid* Cone chewed for a sore mouth. *Throat Aid* Decoction of cones used as a gargle for sore throats. *Toothache Remedy* Cone chewed for toothaches. *Venereal Aid* Decoction of cones and other herbs taken for venereal disease. (109:49) **Eskimo, Inuktitut** *Dermatological Aid* Poultice of gum and grease applied to pustulant wounds. *Respiratory Aid* Decoction of gum or needles taken for respiratory infections. (202: 188) **Iroquois** *Gastrointestinal Aid* Gum chewed to facilitate digestion. (142:83) **Koyukon** *Dermatological Aid* Infusion of needles used as a rub or bath for dry skin or sores. *Hunting Medicine* Tops put in animal track by girls before stepping over it, to avoid alienating animals from hunters. *Kidney Aid* Infusion of needles taken for kidney problems. *Panacea* Infusion of needles taken to promote general good health. *Unspecified* Tree tops used by the shamans to brush people and remove their sickness. (119:49) **Malecite** *Unspecified* Pitch used in medicines. (160:6) **Montagnais** *Cough Medicine* Decoction of twigs taken for coughs. (156:314) **Ojibwa** *Analgesic* Infusion of roots and bark used for stomach pain. *Anticonvulsive* Infusion of roots and bark used for trembling and fits. *Gastrointestinal Aid* Infusion of roots and bark used for stomach pain. (as *P. nigra* 135:244) *Stimulant* Leaves used as a reviver and bark used as a medicinal salt. *Unspecified* Bark used as a medicinal salt. (153:379) **Ojibwa, South** *Unspecified* Decoction of leaves and crushed bark taken for unspecified ailments. (as *Abies nigra* 91:198) **Potawatomi** *Dermatological Aid* Poultice of inner bark applied to infected inflammations. *Disinfectant* Poultice of inner bark applied to infected inflammations. (154:70)

- *Food*—**Anticosti** *Beverage* Branches used to make beer. (143:63) **Carrier** *Candy* Pitch used to chew. (31:69) **Cree, Woodlands** *Candy* Gum chewed for pleasure. (109:49) **Eskimo, Inuktitut** *Unspecified* Cambium eaten in the spring. (202:188) **Micmac** *Beverage* Bark used to make a beverage. (159:258)

- *Fiber*—**Algonquin, Tête-de-Boule** *Basketry* Roots used to sew baskets. *Canoe Material* Roots used to sew canoes. *Snow Gear* Roots used to sew snowshoes. (132:129) **Carrier** *Snow Gear* Young wood used to make snowshoes. (31:69) **Cree, Woodlands** *Building Material* Peeled, wood poles preferred for tepee framing because of their straightness and lack of taper. Small trees and boughs used in the construction of shelters made of brush and shelters for storing moss. *Canoe Material* Wood used to make canoe paddles. *Caulking Material* Pitch used to seal seams on a birchbark canoe. *Cordage* Roots used to tie and secure the stick and bundle game made from black spruce boughs. Roots used to tie and secure the arched roof trees of the shelter for storing moss. Roots used to tie and secure the ends of a birch bark dish. *Mats, Rugs & Bedding* Boughs used on the ground as flooring in tepees and in front of the tent door as a door mat. *Sewing Material* Roots used to sew sheets of birch bark together for a tepee cover. Roots used to stitch birch bark basket and canoe seams. (109: 49) **Eskimo, Inuktitut** *Building Material* Wood used to make cabins and caches. *Canoe Material* Split wood used to make fish traps and canoe or kayak stringers. *Cordage* Split, inner root bark or small rootlets used as fishing lines and cord for making and repairing tools. *Mats, Rugs & Bedding* Needles used as flooring in tents. (202:188) **Iroquois** *Brushes & Brooms* Wood used to make scrub brushes. (142: 83) **Koyukon** *Building Material* Wood used to make poles for tent frames. (119:49) Wood used for the building log houses. Wood split or ripsawed and used as lumber for house construction, caches, and tent frames. *Canoe Material* Wood split or ripsawed and used to make boats and canoes. *Snow Gear* Wood split or ripsawed and used to make sleds. (119:50) **Malecite** *Mats, Rugs & Bedding* Needles and branches used for pillows and bedding. *Sewing Material* Roots used for thread. (160:6) **Micmac** *Mats, Rugs & Bedding* Boughs used to make beds. *Sewing Material* Roots used as sewing material for canoe

birch bark products. (159:258) **Ojibwa** *Canoe Material* Roots used to sew canoes. *Caulking Material* Boiled resin and tallow used to make pitch for caulking canoes. (153:421)
- *Other*—**Carrier** *Cooking Tools* Wood made into drying poles and used for smoking and drying meat. (31:69) **Cree, Woodlands** *Designs* Roots used to make designs on baskets. *Hide Preparation* Dry cones mixed with rotten white spruce wood and used to smoke tan hides a golden brown color. *Toys & Games* Roots used to tie and secure the stick and bundle game made from black spruce boughs. (109:49) **Eskimo, Inuktitut** *Ceremonial Items* Roots used to make headgear and masks. *Containers* Roots used to make trays and buckets. *Cooking Tools* Roots used to make spoons, dippers, and bowls. *Fuel* Wood used for fires. *Hunting & Fishing Item* Split wood used to make fish traps and canoe or kayak stringers. Roots used to make floats. (202:188) **Iroquois** *Fuel* Bark used to start fires. (142:83) **Koyukon** *Cooking Tools* Roots used for carving spoons and bowls. *Fasteners* Roots skinned and used to lash birch bark baskets. (119:50) *Fuel* Wood used for firewood. (119:49) *Hide Preparation* Rotten wood pulverized, mixed with rotten willow, and used to smoke hides. (119:50) *Tools* Wood used to make drying racks. (119:49) **Malecite** *Waterproofing Agent* Pitch used for waterproofing seams in canoes. (160:6) **Micmac** *Fuel* Wood used for kindling and fuel. (159:258)

Picea parryana Parry, Spruce
- *Drug*—**Keres, Western** *Antirheumatic (External)* Infusion of leaves used as a bath for rheumatism. *Cold Remedy* Infusion of leaves used for colds. *Gastrointestinal Aid* Infusion of plant used to clean the stomach. (171:60)
- *Other*—**Keres, Western** *Ceremonial Items* Plant used in ceremonies. *Good Luck Charm* Twig bits given as presents to bring good luck. (171:60)

Picea pungens Engelm., Blue Spruce
- *Other*—**Navajo** *Ceremonial Items* Branches used for the Chant of the Sun's House. (55:20)

Picea rubens Sarg., Red Spruce
- *Drug*—**Cherokee** *Cold Remedy* Infusion of bough taken for colds. *Misc. Disease Remedy* Infusion of bough taken to break out measles. (80:57) **Montagnais** *Pulmonary Aid* Compound decoction of bark taken for lung trouble. *Throat Aid* Compound decoction of bark taken for throat trouble. (156:315)
- *Food*—**Chippewa** *Beverage* Leaves used to make a beverage. (as *P. rubra* 53:317)
- *Fiber*—**Cherokee** *Basketry* Bark used to make baskets. *Building Material* Bark used to make lumber. (80:57) **Chippewa** *Canoe Material* Roots used in sewing canoes. *Caulking Material* Gum used to make pitch. (as *P. rubra* 53:377)

Picea sitchensis (Bong.) Carr., Sitka Spruce
- *Drug*—**Bella Coola** *Analgesic* Decoction of cones taken for pain and bark used as steam bed for backache. *Antirheumatic (External)* Poultice of compound containing gum applied to the arms for rheumatism. Steam bed of ripe cones or bark on hot stones used by rheumatics. (150:51, 52) Sapling bark and ripe cones used to make steam baths for rheumatism. (184:198) *Burn Dressing* "Branches used to whip a burned arm or leg until the blood came." (150:51, 52) *Ceremonial Medicine* Boughs used ritually for protection from death and illness. (184:198) *Dermatological Aid* Gum applied to "small cuts, broken skin and suppurating sores." (150:51, 52) Poultice of gum applied to cuts. *Disinfectant* Poultice of gum applied to infections. (184:198) *Diuretic* Decoction of gum taken as a diuretic for gonorrhea. (150:51, 52) *Gastrointestinal Aid* Sapling bark and ripe cones used to make steam baths for stomach troubles. (184:198) *Heart Medicine* Poultice of compound containing gum applied to the chest for heart trouble. *Laxative* Sap from peeled trunk taken in large doses as a laxative. (150:51, 52) Cambium eaten as a laxative. (184:198) *Venereal Aid* Compound decoction of tips of small spruces taken for gonorrhea. Decoction of gum taken as a diuretic for gonorrhea. (150:51, 52) Decoction of branch tips and other herbs taken for gonorrhea. Gum taken for gonorrhea. (184:198) **Carrier, Southern** *Analgesic* Decoction of new shoots and bark taken for stomach pain. *Eye Medicine* Gum from new shoots and small branches placed in the eyes for snow-blindness. *Gastrointestinal Aid* Decoction of new shoots and bark taken for stomach pain. **Gitksan** *Antirheumatic (Internal)* Compound decoction of twigs with leaves and bark taken for rheumatism. *Tuberculosis Remedy* Compound containing gum taken before meals for consumption. (150:51, 52) **Haisla & Hanaksiala** *Antirheumatic (External)* Poultice of pitch, Indian hellebore roots and rhizomes applied to sore areas. *Cold Remedy* Inner bark chewed for colds. *Cough Medicine* Inner bark chewed for coughs. *Dermatological Aid* Poultice of boiled pitch applied to cuts, sores, boils, and wounds. *Oral Aid* Pitch chewed as a breath freshener. *Tuberculosis Remedy* Pitch chewed everyday for tuberculosis. **Hanaksiala** *Laxative* Infusion of dried bough tips taken for constipation. (43:175) **Hesquiat** *Analgesic* Boughs used to scrub skin, until it bled, for aches and pains. *Dermatological Aid* Rendered pitch and deer oil used as salve for sores and sunburn. (185:41) **Kwakiutl** *Analgesic* Head struck with branches until it bled for headaches. (183:269) *Antidiarrheal* Compound decoction of roots taken for diarrhea. (183:264) Decoction of roots used for diarrhea. *Cold Remedy* Bud extract taken for coughs and colds. *Cough Medicine* Bud extract or pitch and grease taken for coughs. *Dermatological Aid* Poultice of pitch applied to boils, swellings, cuts, and abrasions. *Disinfectant* Branches in house of sick person to prevent anything unclean from entering. *Kidney Aid* Compound poultice of boiled root bark applied to woman's kidney swellings. *Other* Branch tips rubbed to cleanse person contaminated with menstrual blood. (183:269) **Makah** *Blood Medicine* Decoction of plants used to "take out bad blood." *Dermatological Aid* Compound poultice of ashes applied to infant's navel. (79:17) *Gastrointestinal Aid* Pitch used as a stomach medicine. (67:234) *Pediatric Aid* Compound poultice of ashes applied to infant's navel. *Strengthener* Decoction of plants used as a strengthening bath. (79:17) **Oweekeno** *Antirheumatic (External)* Decoction of bark used as a soak for soreness. Pitch mixed with badge moss and used for arthritic joints. *Dermatological Aid* Pitch boiled and used for dermatitis. Pitch mixed with pounded dog-tooth lichens and used for wounds. *Gastrointestinal Aid* Decoction of bark used for gastrointestinal difficulties. *Unspecified* Pitch eaten as medicine. (43:68) **Quinault** *Dermatological Aid* Poultice of gum applied to cuts and wounds. *Throat Aid* Infusion of inner bark taken for throat problems. (79:17) **Sikani** *Cough Medicine* Inner bark chewed for a cough. (150:51, 52) **Thompson** *Antidiarrheal* Decoction

of burned cone ashes taken for dysentery. *Eye Medicine* Needles used to restore eyesight. A blind person, or one with poor eyesight, rubbed his hands with the needles and then rubbed his eyes with his hands to restore his eyesight. *Panacea* Decoction of boughs used for any kind of illness. *Unspecified* Infusion of bark taken as a medicine. Decoction of inner bark taken as a medicine. Evergreen tops considered good medicine. (187:100) **Tlingit** *Toothache Remedy* Compound containing warmed seeds used for toothache. *Venereal Aid* Compound poultice of sap applied for syphilis. (107:284) **Tsimshian** *Hunting Medicine* Boughs used by shamans, hunters, and fishers during preparatory and purification rituals. (43:317)

- *Food*—**Haisla & Hanaksiala** *Candy* Pitch chewed like chewing gum. (43:175) **Hesquiat** *Candy* Cooled, rendered pitch chewed like gum. (185:41) **Kitasoo** *Dried Food* Inner bark cooked and dried for later use. (43:317) **Kwakiutl, Southern** *Candy* Pitch used as chewing gum. (183:293) **Makah** *Candy* Pitch used as chewing gum. (67:234) Pitch chewed as gum for pleasure. (79:17) *Unspecified* "Little cones" and buds used for food. (67:234) Young shoots eaten raw. (79:17) **Oweekeno** *Candy* Pitch boiled and used for chewing. (43:68) **Quinault** *Candy* Pitch chewed as gum for pleasure. (79:17)

- *Fiber*—**Bella Coola** *Basketry* Long roots split and used to make finely woven baskets. (184:198) **Hahwunkwut** *Basketry* Roots used to make cooking bowls, mush baskets, and other small baskets. (117:183) **Haisla & Hanaksiala** *Basketry* Roots mixed with red cedar bark and used to make baskets. (43:175) **Hesquiat** *Building Material* Wood sometimes used as lumber. (185:41) **Hoh** *Building Material* Timber used to make shakes, clapboards, and framing timbers. *Cordage* Limbs and roots shredded, pounded, and used to make cord and rope. (137:59) **Kitasoo** *Building Material* Branches used by hunters as shelter. (43:317) **Kwakiutl, Southern** *Basketry* Roots used to make baskets. (183:296) Roots burned over a fire, freed from root bark, dried, split, and used to make baskets. (183:269) *Clothing* Roots used to make hats. (183:296) Roots burned over a fire, freed from root bark, dried, split, and used to make hats. *Cordage* Roots burned over a fire, freed from root bark, dried, split, and used to make ropes. *Mats, Rugs & Bedding* Roots burned over a fire, freed from root bark, dried, split, and used to make mats. *Sewing Material* Roots burned over a fire, freed from root bark, dried, split, and used for "sewing wood." (183:269) **Nitinaht** *Basketry* Split roots used for basketry. (67:234) Roots soaked, split in quarters, and used to make sturdy pack baskets. (186:71) *Caulking Material* Pitch used to fill cracks and knot holes in canoes. (67:234) **Oweekeno** *Basketry* Roots used for structural elements in basketry. (43:68) **Poliklah** *Basketry* Roots used to make baskets. The body material of baskets was spruce roots, which were dug out and cut off in lengths of 2½ to 3 feet and from ½ to 1 inch in diameter. These were at once (while full of sap and soft) split into broad flat bands, and these in turn were subdivided by knife and teeth until the desired size was obtained—a little larger than coarse thread, about like small twine. The vertical rods were hazel. The overlay was bear grass. The design was commonly of black maidenhair fern stem or salmon red strands made by dying the stem bundles of *Woodwardia* fern with chewed alder bark. (117:170) **Quileute** *Basketry* Roots used for basketry. (79:17) *Building Material* Timber used to make shakes, clapboards, and framing timbers. (137:59) *Caulking Material* Pitch used for caulking canoes. *Clothing* Roots used for rain hats. (79:17) *Cordage* Limbs and roots shredded, pounded, and used to make cord and rope. (137:59) **Quinault** *Caulking Material* Pitch used for caulking canoes. (79:17) **Yurok** *Basketry* Roots used to make the horizontal weave in coarse baskets used for drying foods in the smokehouse. (6:43)

- *Other*—**Hahwunkwut** *Cooking Tools* Roots used to make cooking bowls, mush baskets, and other small baskets. (117:183) **Haisla** *Lighting* Pitch applied to the ends of red cedar torches used in night fishing. **Haisla & Hanaksiala** *Containers* Roots and red cedar bark used to make bag-like implement for the oolichan (candlefish) grease rendering process. *Fasteners* Branches and roots made into pegs, dipped in pitch, and used as nails to hold together bent boxes. *Hunting & Fishing Item* Wood made into rake-like implements and used to catch herring and oolichans (candlefish). Wood used to make oolichan spears. Wood used to make arrows. *Tools* Wood used to make digging sticks. Wood used to make a bark peeling tool. **Hanaksiala** *Ceremonial Items* Boughs used to hit and rub boys as part of a ritual treatment to increase their strength and tolerance. (43:175) **Hesquiat** *Ceremonial Items* Boughs used at girl's puberty potlatch to brush with sweeping motions and scare away bad influences. *Fasteners* Rendered pitch used as a glue for arrows and harpoons before they were tied. *Fuel* Knots used as fuel to keep the fire burning all night. *Weapon* Sharpened knots used to make a weapon. (185:41) **Hoh** *Toys & Games* Timber used to make toys. (137:59) **Kitasoo** *Protection* Prickly leaves used to discourage and repel animals. (43:317) **Kwakiutl, Southern** *Fasteners* Roots burned, dried, split, and used as strings to tie nets, hooks, and harpoons together. *Hunting & Fishing Item* Roots burned over a fire, freed from root bark, dried, split, and used to make fish nets. *Sacred Items* Branch tips used as sacred items. (183:269) **Makah** *Fasteners* Pitch used as glue to repair items such as harpoons. (67:234) *Hunting & Fishing Item* Warmed pitch used to protect the harpoon point. (79:17) **Nitinaht** *Ceremonial Items* Branches used ceremonially to initiate the children. (67:234) Branches used in winter dances and to make traditional costumes for initiation ceremonies. *Fasteners* Roots used for binding gaff implement joints. *Hunting & Fishing Item* Wood used to make the longer prong in the two-pronged salmon harpoon. Pitch used as a protective coating for fishing spears and whaling harpoon heads. (186:71) *Paint* Pitch used like shellac on harpoons. The pitch was ignited and caught with a mussel shell as it melted. The whale hunter's entire family would join in and chew the pitch until it was the right consistency. Then the hunter would put the pitch on his harpoon, smooth it over, and then burn off the excess. Finally, he would shine it until it was smooth like shellac. *Waterproofing Agent* Pitch used to waterproof boxes. These boxes were used only for cold materials, as hot water would melt the pitch. (67:234) **Quileute** *Hunting & Fishing Item* Roots used to tie the tines of salmon spears. Saplings used for the spring poles of snares for deer, elk, and other game animals. (79:17) *Toys & Games* Timber used to make toys. (137:59) **Quinault** *Hunting & Fishing Item* Roots used to tie the tines of salmon spears. (79:17) **Thompson** *Protection* Branches rubbed on skin to protect one against evil or "witchcraft." The protective powers were attributed to the prickly needles. (187:100) **Tsimshian** *Ceremonial Items* Boughs used by shamans, hunters, and fishers during preparatory and purification rituals. (43:317)

Picea sp., Spruce
- **Drug**—**Carrier** *Cold Remedy* Inner bark chewed for colds. *Tuberculosis Remedy* Inner bark chewed for tuberculosis. (31:69) **Cree** *Throat Aid* Cones chewed for sore throats. (97:18) *Venereal Aid* Gum used as a salve for syphilitic sores. (16:495) **Iroquois** *Antirheumatic (External)* Compound decoction applied to parts affected by rheumatism. *Antirheumatic (Internal)* Compound decoction taken for rheumatism. *Cold Remedy* Compound decoction or infusion of plants taken for colds. *Cough Medicine* Compound decoction taken for coughs. *Dermatological Aid* Gum applied to ingrown nails and cuts and used as a chewing gum. *Emetic* Infusion taken for colds and to vomit in the spring. (87:268) *Tuberculosis Remedy* Gum used for tuberculosis. (141:36) **Navajo** *Ceremonial Medicine* Plant used for "Shooting, Witch, Lightning and Night Chant" ceremonies. *Stimulant* Used to make an arrow and shot over the person to revive them from fainting. (55:21) **Penobscot** *Dermatological Aid* Poultice of soft gum or pine pitch applied to boils and abscesses. (156:309)
- **Food**—**Penobscot** *Candy* Gum extensively chewed as a "pastime." (156:309)
- **Fiber**—**Carrier** *Mats, Rugs & Bedding* Needles used to make tent floor coverings. (31:69) **Eskimo, Alaska** *Building Material* Wood used to support buildings and to build the framework for sod-covered buildings. *Canoe Material* Logs and poles used for making kayak parts, weapon and tool handles, and other utilitarian objects. (1:34) **Hoh** *Basketry* Limbs and roots split, pared, scraped, and used to make baskets. **Quileute** *Basketry* Limbs and roots split, pared, scraped, and used to make baskets. (137:59)
- **Other**—**Eskimo, Alaska** *Cooking Tools* Wood used to build fish drying racks and legs for elevated caches. *Fuel* Logs considered an important source of fuel for heating the homes and steam baths of the village. *Tools* Logs and poles used for making kayak parts, tool handles, and other utilitarian objects. *Weapon* Logs and poles used for making and weapon handles. (1:34) **Iroquois** *Smoke Plant* Dried roots used to make cigars and smoked. (141:36) **Navajo** *Ceremonial Items* Used to make hoops, dresses, collars, bows, and arrows for many different ceremonies. *Tools* Twigs used as beaters to make a high, stiff, lasting lather of yucca roots and water. (55:21)

Picradeniopsis, Asteraceae
Picradeniopsis oppositifolia (Nutt.) Rydb. ex Britt., Oppositeleaf Bahia
- **Drug**—**Navajo, Ramah** *Dermatological Aid* Poultice of chewed leaves applied to red ant bites. *Gastrointestinal Aid* Cold infusion of leaves taken after swallowing a red ant. *Herbal Steam* Steam from compound containing plant used medicinally. *Panacea* Plant used as "life medicine." (as *Bahia oppositifolia* 191:49)

Picradeniopsis woodhousei (Gray) Rydb., Woodhouse's Bahia
- **Drug**—**Zuni** *Dermatological Aid* Poultice of chewed root applied to sores and rashes. (as *Bahia woodhousei* 27:374) *Emetic* Infusion of whole plant taken, vomiting ensued, for "sick stomach." (as *Bahia woodhousei* 166:44) *Gastrointestinal Aid* Infusion of root taken for stomachache. (as *Bahia woodhousei* 27:374)

Pilea, Urticaceae
Pilea pumila (L.) Gray, Canadian Clearweed
- **Drug**—**Cherokee** *Dermatological Aid* Stems rubbed between the toes for itching. *Dietary Aid* and *Pediatric Aid* Infusion given to children to reduce excessive hunger. (80:52, 53) **Iroquois** *Respiratory Aid* "Squeeze water out of stem and inhale for sinus problems." (87:308)

Piloblephis, Lamiaceae
Piloblephis rigida (Bartr. ex Benth.) Raf., Wild Pennyroyal
- **Drug**—**Seminole** *Ceremonial Medicine* Infusion of leaves added to food after a recent death. (as *Pycnothymus rigidus* 169:342) *Cold Remedy* Infusion of plant taken for colds. (as *Pycnothymus rigidus* 169:283) *Dermatological Aid* Infusion of roots applied to sores and ulcers on the legs and feet. (as *Pycnothymus rigidus* 169:307) *Emetic* Plant used as an emetic during religious ceremonies. (as *Pycnothymus rigidus* 169:409) *Febrifuge* Infusion of plant taken for fevers. (as *Pycnothymus rigidus* 169:283) *Pediatric Aid* Plant used for chronically ill babies. (as *Pycnothymus rigidus* 169:329) *Stimulant* Decoction of whole plant minus the roots used as a bath for hog sickness: unconsciousness. (as *Pycnothymus rigidus* 169:229)
- **Food**—**Seminole** *Spice* Plant used for soup flavoring. (as *Pycnothymus rigidus* 169:482)

Pimpinella, Apiaceae
Pimpinella anisum L., Anise Burnet Saxifrage
- **Drug**—**Cherokee** *Respiratory Aid* Infusion of half a teaspoonful in a cup of hot water taken for catarrh. (80:23) **Delaware** *Cathartic* Roots used as a cathartic. *Gastrointestinal Aid* Roots used as a stomach tonic. (176:33) **Delaware, Oklahoma** *Cathartic* Root used as a cathartic. (175:28, 78) *Gastrointestinal Aid* and *Tonic* Root used as a stomach tonic. (175:28)

Pinaceae, see *Abies, Larix, Picea, Pinus, Pseudotsuga, Tsuga*

Pinguicula, Lentibulariaceae
Pinguicula lutea Walt., Yellow Butterwort
- **Drug**—**Seminole** *Analgesic* and *Gastrointestinal Aid* Infusion of whole plant taken for raw meat sickness: severe abdominal pains. (169:276, 277)

Pinguicula pumila Michx., Small Butterwort
- **Drug**—**Seminole** *Analgesic* and *Gastrointestinal Aid* Infusion of whole plant taken for raw meat sickness: severe abdominal pains. (169:276, 277)

Pinguicula vulgaris L., Common Butterwort
- **Other**—**Oweekeno** *Good Luck Charm* Dried roots kept for good luck. (43:106)

Pinus, Pinaceae, conifer
Pinus albicaulis Engelm., Whitebark Pine
- **Food**—**Coeur d'Alene** *Unspecified* Nutlets used for food. (178:90) Nutlets cooked in hot ashes and used for food. (178:93) **Montana Indian** *Unspecified* Inner bark used for food. Nuts were an important article of food. (19:18) **Okanagan-Colville** *Unspecified* Seeds used for food. *Winter Use Food* Seeds gathered and stored for winter use. (188:27) **Spokan** *Unspecified* Nutlets used for food. (178:344) **Thompson** *Dried Food* Dried nuts kept alone in sacks or mixed with dried serviceberries and stored for future use. *Porridge* Parched seeds pound-

ed in a mortar to make a flour and mixed with water to form a mush. (187:101) *Unspecified* Seeds oven cooked or fire roasted. (164:492) Seeds eaten roasted or raw, but often considered bitter. If too many raw seeds were eaten, it would cause constipation. Roasted seeds were therefore preferred to raw seeds. (187:101) *Winter Use Food* Seeds cooked, crushed, mixed with dried serviceberries, and preserved for winter use. (164:492) Cooked, crushed seeds mixed with dried berries and preserved for winter use. (187:101)

Pinus aristata Engelm., Bristlecone Pine
- *Drug*—Shoshoni *Dermatological Aid* Poultice of heated pitch applied to sores and boils. (180:117)

Pinus attenuata Lemmon, Knobcone Pine
- *Other*—Karok *Decorations* Nuts used as beads and ornaments for dresses. (as *P. tuberculata* 148:379)

Pinus banksiana Lamb., Jack Pine
- *Drug*—Cree, Woodlands *Dermatological Aid* Poultice of inner bark applied to deep cuts. (109:50) **Menominee** *Unspecified* Every part of tree used as a medicine. (151:45) **Ojibwa** *Anticonvulsive* Plant used for fits. *Stimulant* Plant used for fainting. (as *P. baksiana* 135:244) Leaves used as a reviver. (153:379) **Potawatomi** *Dermatological Aid* Pitch from boiled cones used as an ointment for unspecified ailment. *Pulmonary Aid* Leaves used as a fumigant to clear congested lungs. *Stimulant* Leaves used as a fumigant to revive a comatose patient. (154:70)
- *Food*—Cree, Woodlands *Unspecified* Inner bark used for food. (109:50)
- *Fiber*—Algonquin, Tête-de-Boule *Caulking Material* Gum used to caulk canoes. (132:130) **Menominee** *Sewing Material* Small, boiled roots used as cords to sew birchbark canoe and stitching sealed with pitch or resin. (151:75) **Ojibwa** *Mats, Rugs & Bedding* Boughs used on the ground or floor, covered with blankets and other bedding, and used as a bed. (as *P. baksiana* 135:244) *Sewing Material* Roots used as fine sewing material for canoes and other coarse and durable sewing. (153:421) **Potawatomi** *Sewing Material* Roots used as a heavy sewing material. The roots extend near the surface of the ground through the sandy soil for 30 to 35 feet and were easy to pull out of the ground in their entire length. When they were gathered they were made into coils and sunk beneath the surface of the lake until the outer bark had loosened from the root. Then, they were peeled and split in half, each half being a serviceable cord for sewing together canoes and bark strips intended for the roofs of wigwams and for other purposes. (154:113)
- *Other*—Cree, Woodlands *Hide Preparation* Dry, open cones mixed with rotten white spruce wood used to smoke tan hides. (109:50) **Potawatomi** *Lighting* Pine pitch and cedar used to make torches and attached to the canoe bow for night hunting. (154:122) *Waterproofing Agent* Pitch from the cones used to waterproof sewn seams. (154:113)

Pinus contorta Dougl. ex Loud., Lodgepole Pine
- *Drug*—Bella Coola *Antirheumatic (External)* Compound containing gum used as poultice on arms for rheumatism. *Dermatological Aid* Gum applied to cuts and chewed gum applied to broken skin. *Heart Medicine* Compound containing gum used as poultice on chest for heart trouble. *Tuberculosis Remedy* Decoction of gum taken for consumption. (150:49, 50) **Blackfoot** *Tuberculosis Remedy* Infusion of pitch taken for tubercular coughs. Here is a fine example of the origin and use of a "personal medicine" which was later expanded to include general therapeutic practice. There was once a woman named Last Calf who was riddled with tuberculosis. While she and her husband were camped near a beaver lodge, she noticed the animal's tracks in the mud and left some food for it. The beaver took the gift and returned the favor by appearing to her in a vision. He gave her a cure for tuberculosis. She was to collect the pitch of the lodgepole pine, boil it in water, and drink the infusion while uttering a special song. (The song had no words.) Last Calf's husband was alarmed at this treatment and cautioned her against poisoning but she went ahead and drank the brew. She said she felt as though she were going to die and began vomiting profusely. She drank again with the same result, but the next morning her chest was cleared as never before. (86:73) **Carrier, Northern** *Dermatological Aid* Compound decoction of needle tips taken for paralysis and body sores. *Eye Medicine* Gum painted on eye "to remove white scum" and for snow-blindness. *Orthopedic Aid* Compound decoction of needle tips taken for paralysis, weakness, or sores. *Stimulant* Compound decoction of needle tips taken for constitutional weakness. **Carrier, Southern** *Analgesic* Decoction of new shoots taken for stomach pain. *Gastrointestinal Aid* Decoction of new shoots taken for stomach pain. (150:49, 50) **Eskimo, Alaska** *Cold Remedy* Juice taken for colds. *Cough Medicine* Juice taken for coughs. *Unspecified* Sap used as a medicine. (149:331) **Flathead** *Burn Dressing* Poultice of heated sap and bone marrow used for burns. *Dermatological Aid* Poultice of sap, red axle grease, and Climax chewing tobacco used for boils. (82:52) **Gitksan** *Blood Medicine* Inner bark eaten as a blood purifier and used as a cathartic. *Cathartic* Needles or inner bark eaten or decoction of inner bark taken as a purgative. *Diuretic* Needles eaten or decoction of inner bark taken as a purgative and diuretic. (150:49, 50) *Tonic* Decoction of bark used as a tonic. (73:152) *Tuberculosis Remedy* Decoction of inner bark taken for consumption and gonorrhea. (150:49, 50) *Unspecified* Bark used for medicines. (73:152) *Venereal Aid* Decoction of inner bark taken for gonorrhea and other "serious ailments." (150:49, 50) **Kutenai** *Tuberculosis Remedy* Inner bark eaten for tuberculosis. (82:52) **Kwakiutl** *Cough Medicine* Decoction of buds and pitch taken for coughs. *Gastrointestinal Aid* Decoction of buds and pitch taken for stomachaches. (183:269) **Okanagan-Colville** *Gastrointestinal Aid* Cambium layer eaten for stomach troubles such as ulcers. Decoction of sap taken for ulcers. *Throat Aid* Pitch sucked and juice swallowed for sore throats. (188:28) **Okanogon** *Cold Remedy* Gum used for colds. *Cough Medicine* Gum used for coughs. *Orthopedic Aid* Decoction of bark gum and fat rubbed on the body for muscle and joint aches. *Throat Aid* Gum used for sore throats. (125:40) **Quinault** *Dermatological Aid* Poultice of pitch applied to open sores. *Throat Aid* Buds chewed for sore throats. (79:17) **Salish, Coast** *Dermatological Aid* Sap mixed with deer tallow and used for psoriasis and other diseases. *Misc. Disease Remedy* Sap mixed with deer tallow and used for psoriasis and other diseases. (182:69, 70) **Shuswap** *Cough Medicine* Infusion of inner bark taken for coughs. *Tuberculosis Remedy* Infusion of inner bark taken for tuberculosis. (123:51) **Sikani** *Cough Medicine* Pitch chewed and saliva swallowed for a cough. (150:49, 50) **Thompson** *Analgesic* Salve of boiled sap and grease applied for pains. *Anti-*

rheumatic (External) Salve of boiled sap and grease applied for rheumatism. (164:461) *Cold Remedy* Gum used for colds. (125:40) Salve of boiled sap and grease used for colds. (164:461) *Cough Medicine* Gum used for coughs. (125:40) Salve of boiled sap and grease used for coughs. (164:461) *Dermatological Aid* Pitch mixed with bear tallow, rose petals, and red ocher and used as face cream or for blemishes. (187:102) *Disinfectant* Salve of resin with animal fat applied to body as a purifier after sweat bath. (164:504) Pitch used as a sort of "cold cream" with disinfectant properties. *Misc. Disease Remedy* Infusion of twigs with needles attached used for influenza. (187:102) *Orthopedic Aid* Decoction of bark gum and fat rubbed on the body for muscle and joint aches. (125:40) Salve of boiled sap and grease applied for muscle and joint soreness. (164:461) *Pediatric Aid* Pitch mixed with bear tallow, rose petals, and red ocher and rubbed on the skin of newborn babies. (187:102) *Pulmonary Aid* Salve of boiled sap and grease applied to back and chest for congestion. (164:461) *Throat Aid* Gum used for sore throats. (125:40) Salve of boiled sap and grease used for sore throats. (164:461) **Tlingit** *Venereal Aid* Compound infusion of sprouts and bark taken for syphilis. (as *P. inops* 107:283)

- **Food**—**Blackfoot** *Candy* Pitch chewed like gum. (86:104) **Coeur d'Alene** *Unspecified* Cambium layer used for food. (178:91) **Flathead** *Candy* Pitchy secretions chewed as gum. *Unspecified* Inner bark used for food. Seeds used for food. (82:52) **Gitksan** *Unspecified* Sap eaten fresh. (73:151) **Hesquiat** *Candy* Pitch chewed like gum. (185:44) **Kutenai** *Unspecified* Inner bark used for food. (82:52) **Okanagan-Colville** *Forage* Cambium layer eaten by grizzly bears. *Unspecified* Cambium layer used for food. (188:28) **Okanagon** *Staple* Cambium layer used as a principal food. (178:239) *Unspecified* Cambium layer and sap used for food. (125:38) **Salish, Coast** *Bread & Cake* Juicy inner bark dried in cakes and used for food. *Unspecified* Juicy inner bark eaten fresh. (182:70) **Shuswap** *Unspecified* Inner bark used for food. (123:51) **Spokan** *Unspecified* Cambium used for food. (178:344) **Thompson** *Beverage* Needles used to make a tea-like beverage. Twigs with needles attached used to make a tea-like beverage. *Candy* Young shoots of branches chewed for the honey. *Dried Food* Cambium and adjacent phloem tissue dried for winter use. (187:102) *Unspecified* Cambium layer and sap used for food. (125:38) Cambium and adjacent phloem tissue eaten fresh. (187:102) **Wet'suwet'en** *Dried Food* Inner bark strips dried and stored for future food use. *Unspecified* Sap eaten fresh. (73:151)
- **Fiber**—**Blackfoot** *Building Material* Wood used to make travois and tepee poles. *Furniture* Wood used to make backrest poles and bed supports. The backrest poles were cut about five forearms in length and dried over a fire of rotten logs. One end of the pole was perforated and the other end sharpened. Then a stick was inserted through the hole and the pole etched. Later it was painted red and blue with buffalo shoulder blade applicators. Backrest poles were often notched to record the number of camp moves. (86:116) **Cheyenne** *Building Material* Used for tepee poles. (83:46) Trunks used for tepee poles. (83:6) **Montana Indian** *Building Material* Poles used to make the foundations for tepees. (82:52) **Thompson** *Building Material* Delimbed trunks used as framework poles for traditional sleeping platforms. (187:102)
- **Other**—**Blackfoot** *Cooking Tools* Wood burls scraped with a rough stone, grease applied to prevent cracking and made into a bowl. *Fasteners* Sticks notched to act as fasteners on designated food storage bags. Resin boiled with buffalo phallus and used as a glue for headdresses and bows. *Musical Instrument* Used to make wind chimes and presented to newly married couples. *Toys & Games* Wood used to make story sticks. Story sticks were prepared by older men and presented to children in return for favors. The sticks were notched to count the number of stories that the man would tell the child. They were often varnished with a solution of boiled hoof and steer phallus and sometimes red ocher was added. Then the stick was polished with a piece of rawhide. Story sticks were sometimes used to hang tepee doors. *Waterproofing Agent* Resin boiled with buffalo phallus and applied to moccasins for waterproofing. (86:116) **Hesquiat** *Fasteners* Pitch used on joints of implements, arrows, and harpoons, before bound with twine. *Preservative* Chewed pitch sprayed onto mats to preserve them. (185:44) **Kwakiutl, Southern** *Cooking Tools* Wood used to make fire tongs. *Tools* Wood used to make cedar bark peelers, digging sticks, and board bending tools. (183:296) **Nitinaht** *Ceremonial Items* Wood used to make small totem poles and model canoes. (186:73) **Okanagan-Colville** *Ceremonial Items* Wood placed in a basket of water to bring rain and pine cones burned to stop rain. *Containers* Bark used to make temporary, berry picking containers. *Season Indicator* Pollen cone ripening used as an indication that the cambium was ready to harvest. (188:28) **Salish, Coast** *Fasteners* Pitch used to fasten arrowheads onto shafts. (182:70) **Tolowa** *Hunting & Fishing Item* Branches rubbed on the hunters' bodies to hide the human scent. (6:44)

Pinus contorta Dougl. ex Loud. **var. *contorta*,** Lodgepole Pine

- **Drug**—**Haisla & Hanaksiala** *Antirheumatic (External)* Smoldering twigs applied to arthritic or injured joints for the pain and swelling. (43:178) **Kwakwaka'wakw** *Unspecified* Buds and pitch used medicinally. (43:70) **Nitinaht** *Dermatological Aid* Pitch mixed with melted deer tallow and used as a skin cosmetic. (186:73)
- **Fiber**—**Tsimshian** *Building Material* Branches used by hunters as shelter to discourage and repel animals. (43:318)
- **Other**—**Haisla** *Tools* Wood used to make maul heads. **Haisla & Hanaksiala** *Paint* Twigs burned and used as a pigment material for tattoos. *Tools* Smoldering twigs used to singe and trim hair. (43:178) **Kwakwaka'wakw** *Fuel* Wood used for firewood. *Hunting & Fishing Item* Pitch used to trap hummingbirds and the bird's hearts used in love medicines. *Tools* Wood used to make implements. (43:70) **Nitinaht** *Hunting & Fishing Item* Pitch used as protective coating for whaling and fishing equipment. (186:73) **Tsimshian** *Protection* Branches used by hunters as shelter to discourage and repel animals. (43:318)

Pinus contorta **var. *latifolia*** Engelm. ex S. Wats., Tall Lodgepole Pine

- **Fiber**—**Blackfoot** *Building Material* Wood used to make tepee frames. (97:18)
- **Other**—**Thompson** *Hunting & Fishing Item* Wood used to make a leister pole. (187:102)

Pinus contorta **var. *murrayana*** (Grev. & Balf.) Engelm., Murray Lodgepole Pine

- **Drug**—**Klamath** *Eye Medicine* Pitch placed inside the lid for sore eyes. (as *P. murrayana* 45:89)

- **Food**—**Montana Indian** *Starvation Food* Inner cambium layer of the bark eaten in times of scarcity. (as *P. murrayana* 19:18) **Okanagon** *Unspecified* Cambium layer and sap used for food. **Thompson** *Unspecified* Cambium layer and sap used for food. (125:38) Sap eaten especially in the spring. (as *P. murrayana* 164:483)
- **Fiber**—**Dakota** *Building Material* Tree used for tepee poles. (as *P. murrayana* 70:63) **Klamath** *Canoe Material* Trunk used to make poles to push boats through shallow water. (as *P. murrayana* 45:89) Peeled sapling used to make poles to propel canoes. (as *P. murrayana* 46:728) **Montana Indian** *Basketry* Young bark used to make baskets. *Building Material* Smaller trees stripped of the bark and used for lodge poles and extensively for lumber. (as *P. murrayana* 19:18) **Paiute** *Building Material* Barkless trunks used as tepee poles. (111:41) **Thompson** *Building Material* Trunk used as a favorite for building. *Scouring Material* Pitch mixed with grease and used for smoothing and polishing steatite pipes. (as *P. murrayana* 164:496)
- **Other**—**Klamath** *Containers* Bark used to make buckets for gathering berries. (as *P. murrayana* 45:89)

Pinus coulteri D. Don, Coulter's Pine
- **Fiber**—**Diegueño** *Basketry* Needles used in making baskets. (84:29)

Pinus echinata P. Mill., Shortleaf Pine
- **Drug**—**Choctaw** *Anthelmintic* Cold infusion of buds taken for worms. (177:5) **Nanticoke** *Analgesic* "Pellets of tar" considered "beneficial for soreness of the back." (175:55) *Cathartic* "Pellets of tar" used as a cathartic. (175:55, 84) *Orthopedic Aid* "Pellets of tar" considered "beneficial for soreness of the back." (175:55) **Rappahannock** *Dermatological Aid* Compound infusion or decoction of top branches used as wash for swellings. *Emetic* Compound with grated dried bark taken to induce vomiting. *Veterinary Aid* Compound with dried bark fed to dogs with distemper to induce vomiting. (161:27)
- **Fiber**—**Cherokee** *Building Material* Wood used for lumber. *Canoe Material* Wood used to make 30- to 40-foot-long canoes. (80:49)
- **Other**—**Cherokee** *Decorations* Wood used to carve. (80:49)

Pinus edulis Engelm., Twoneedle Pinyon
- **Drug**—**Apache, Mescalero** *Cold Remedy* Needles burned and smoke inhaled for colds. (14:35) **Apache, Western** *Dermatological Aid* Heated pitch applied to the face to remove facial hair. (26:185) **Apache, White Mountain** *Venereal Aid* Leaves chewed for venereal diseases. (136:159) **Havasupai** *Dermatological Aid* Poultice of melted gum applied to cuts. *Veterinary Aid* Poultice of melted gum applied to horses for cuts. (197:205) **Hopi** *Dermatological Aid* Poultice of gum used to exclude air from cuts and sores. (200:32) *Disinfectant* Gum smoke used as disinfectant for family of dead person. (200:63) *Tuberculosis Remedy* Plant used for "consumption." (200:35, 63) *Witchcraft Medicine* Gum applied to forehead as a protection from sorcery. (200:63) **Hualapai** *Expectorant* Decoction of inner bark taken as an expectorant tea. *Other* Fresh, white pitch burned to purify the air. (195:35) **Isleta** *Dermatological Aid* Gum mixed with tallow and used as a salve for cuts and open sores. (100:37) **Keres, Western** *Dermatological Aid* Pitch used on open sores. *Gastrointestinal Aid* Infusion of foliage used as an emetic to clean the stomach. (171:60) **Navajo** *Ceremonial Medicine* Needles used in the medicine for the "War Dance." Pitch painted all over the patient in the War Dance. (55:21) *Dermatological Aid* Plant used for cuts and sores. (55:97) Gum with tallow and red clay and used as a salve on open cuts and sores. (55:21) *Emetic* Resin used as an emetic. (90:162) **Navajo, Ramah** *Analgesic* Compound decoction used for headache. *Burn Dressing* Poultice of chewed buds applied to burns. *Ceremonial Medicine* Decoction of wood or needles used as ceremonial emetic. *Cold Remedy* Compound decoction used for colds. Fumes from burning resin inhaled for head colds. *Cough Medicine* Compound decoction used for cough. *Ear Medicine* Pulverized, dried buds used as fumigant for earache. *Emetic* Decoction of wood or needles used as ceremonial emetic. *Febrifuge* Compound decoction used for fever. *Misc. Disease Remedy* Compound decoction used for influenza. *Other* Compound containing inner bark used for injuries. (191:12, 13) **Tewa** *Dermatological Aid* Poultice of gum used to exclude air from cuts and sores. (42:347) Resin applied to cuts and sores to keep out the air. (138:41) **Zuni** *Dermatological Aid* Powdered resin sprinkled in opened abscess or mixed with lard or Vaseline and placed in abscess. Powdered resin used for skin infections. (27:373) *Diaphoretic* Needles chewed and swallowed as a diaphoretic. *Disinfectant* Powdered gum sprinkled on lanced groin swellings as an antiseptic. *Diuretic* Needles eaten and infusion of twigs used as a diuretic and diaphoretic for syphilis. *Venereal Aid* Needles eaten and infusion of twigs used as a diuretic and diaphoretic for syphilis. Powdered gum sprinkled on scraped syphilitic ulcers. (166:57, 58)
- **Food**—**Apache, Chiricahua & Mescalero** *Pie & Pudding* Seeds mixed with yucca fruit pulp to make a pudding. *Special Food* Seeds ground, rolled into balls and eaten as a delicacy. (33:43) *Unspecified* Secretion from the trunk chewed. (33:45) **Apache, Mescalero** *Dried Food* Nuts parched, ground, mixed with datil fruit, mescal, mesquite beans, or sotol and used for food. *Special Food* Nuts used as an essential food during girls' puberty ceremonies. (14:35) **Apache, Western** *Candy* Pitch used as chewing gum. *Porridge* Pinyon and corn flour mixed and cooked into a mush. *Staple* Used as a staple food. Nuts eaten raw, roasted, or ground into flour. *Winter Use Food* Nuts stored in baskets or pottery jars. (26:185) **Apache, White Mountain** *Unspecified* Nuts eaten raw. (136:159) **Gosiute** *Unspecified* Nuts used for food. (39:377) **Havasupai** *Preserves* Seeds parched, ground, kneaded into seed butter, and eaten with fruit drinks or spread on bread. (197:67) *Soup* Nuts ground with the shells and used to make soup. (197:73) *Spice* Sprigs placed in the cooking pit with porcupine, bobcat, or badger to improve the taste of the meat. *Unspecified* Nuts formerly used as an important food source. (197:205) **Hopi** *Special Food* Nuts roasted and eaten as an after supper luxury. (120:18) *Unspecified* Nuts used for food. (42:347) Nuts eaten for food. (200:63) **Hualapai** *Beverage* Needles used to make a tea. *Bread & Cake* Nuts formed into cakes. *Candy* Pitch chewed as a gum. *Porridge* Nuts used to make a paste. *Soup* Nuts used to make a soup. *Unspecified* Nuts eaten raw or roasted. (195:35) **Isleta** *Staple* Nuts formerly used as a staple food. (100:37) *Unspecified* Seeds formerly considered an important food. (32:40) *Winter Use Food* Nuts gathered and stored for winter use. (100:37) **Jemez** *Unspecified* Nuts gathered in large quantities to save and sell. (44:26) **Keres, Western** *Unspecified* Raw or roasted nuts used for food. (171:60) **Keresan** *Winter Use Food* Nuts gathered in large quantities, roasted, and eaten during the winter. (198:562) **Navajo** *Bread & Cake* Ground nuts formed into cakes. (110:21) *Candy* Sap

used as a chewing gum. (55:21) *Porridge* Nuts boiled into a gruel. (110:21) *Preserves* Roasted nuts mashed into a butter. (55:21) Nuts roasted, cracked and shelled on a metate, ground fine, made into butter, and used with bread. (165:222) *Special Food* Ground nuts rolled into balls and eaten as a delicacy. (110:21) *Staple* Nuts hulled, parched, and ground with cornmeal to make a flour. (32:40) *Unspecified* Hardened resinous secretions chewed. (32:32) Nuts hulled, roasted, and eaten without further preparation. (32:40) Seeds used for food. (90:162) Nuts eaten raw or roasted directly from the shell. (110:21) **Navajo, Ramah** *Candy* Resin used for chewing gum. *Preserves* Roasted, ground nuts made into butter and spread on corn cakes or mixed with roasted, ground corn. *Starvation Food* Inside bark used as an emergency ration, when food was scarce. *Winter Use Food* Nuts gathered and stored for winter use. Roasted, ground nuts made into sun dried cakes and stored for winter. (191:12) **Pueblo** *Unspecified* Hardened resinous secretions chewed. (32:32) Seeds formerly considered an important food. (32:40) **Sia** *Unspecified* Nuts gathered in considerable quantities, roasted, and used for food. (199:107) **Tewa** *Unspecified* Fresh or roasted seeds formerly considered an important food. (32:40) Nuts used for food. (42:347) Nuts formerly roasted and used for food. (138:41) **Zuni** *Winter Use Food* Nuts gathered in great quantities, toasted, and stored for winter use. (166:70)

- **Fiber**—**Apache, Mescalero** *Furniture* Young trees used for the main hoop of infant cradleboards. (14:35) **Havasupai** *Building Material* Wood used for house construction. *Caulking Material* Melted gum used to plug a leaky canteen or other containers. (197:205) **Navajo** *Building Material* Logs used to make hogans for ordinary and ceremonial purposes. Boughs used to build the corral for public exhibitions at the close of a ceremony. *Furniture* Wood used to make various parts of the cradle. (55:21) **Navajo, Ramah** *Basketry* Resin used in pottery and basketry making. *Building Material* Rot- and wood-eating-beetle-resistant logs used as the chief building material for hogans. Wood used to make summer shade houses. Branches used to cover a sweat house. Wood used for fence posts and corral construction. (191:12)

- **Dye**—**Hopi** *Unspecified* Gum used in the preparation of certain dyes. (42:347) Gum used to prepare certain dyes. (200:63) **Jemez** *Red* Gum from old and new trees used as a red paint for jars and bowls. (44:26) **Navajo** *Black* Gum used to make black dye. A black dye was made from pinyon gum, the leaves and twigs of sumac, and a native yellow ocher. The sumac leaves were put in water and allowed to boil until the mixture became strong. While this was boiling, the ocher was powdered and roasted. Pinyon gum was then added to the ocher and the whole roasted again. As roasting proceeded, the gum melted, and finally the mixture was reduced to a black powder. This was cooled and thrown into the sumac mixture, forming a rich blue-black fluid which was essentially an ink. When this process was finished the wool was put in and allowed to boil until it was dyed the right shade. This same dye was also used to color leather and buckskin. (55:21) **Navajo, Ramah** *Black* Resin used as an ingredient of black dye for wool or basketry. (191:12) **Tewa** *Unspecified* Gum used in the preparation of certain dyes. (42:347)

- **Other**—**Apache, Mescalero** *Ceremonial Items* Pollen used instead of cattail pollen in ceremonies. *Waterproofing Agent* Resin used for waterproofing woven water jugs. (14:35) **Apache, Western** *Waterproofing Agent* Pitch used to waterproof baskets. (26:185) **Apache, White Mountain** *Waterproofing Agent* Pitch warmed and applied inside and out to waterproof water jugs. (136:150) **Havasupai** *Cash Crop* Nuts sold in considerable quantities to stores. *Fuel* Wood used for firewood. *Paint* Gum used in the paint used on the base of arrows. *Tools* Wood used to make the knife for trimming mescal heads. *Waterproofing Agent* Gum used to waterproof basketry water jugs and basketry drinking cups. (197:205) **Hopi** *Ceremonial Items* Gum put on hot coals and fumes used to smoke people and their clothes after a funeral. Pollen used for the Snake Ceremonial. *Fasteners* Gum used in making turquoise mosaics. (42:347) Gum used in making turquoise mosaics. (200:63) *Protection* Gum put on forehead when going outside of house as protection against sorcery. *Waterproofing Agent* Gum used to waterproof and repair pottery vessels. Gum used to prevent absorption of moisture and warping. (42:347) Gum used in waterproofing and repairing pottery vessels. (200:63) **Hualapai** *Fasteners* Pitch used to glue arrows and cradleboards. Pitch spread on the palms of the hand to make gripping rope easier. *Waterproofing Agent* Melted pitch used for waterproofing baskets. (195:35) **Jemez** *Cash Crop* Nuts gathered in large quantities to save and sell. (44:26) **Keres, Western** *Cash Crop* Nuts used for trade. *Cooking Tools* Pitch rubbed on stone to blacken and the stone used to make paper bread from black corn. *Paint* Pitch mixed with ground lichens or mineral colors to make a paint medium. (171:60) **Navajo** *Cash Crop* Seeds gathered in large quantities and sold or traded. (32:40) Nuts gathered and sold or traded. (55:21) Seeds used as a commercial crop. (90:162) Nuts sold to the nearest trading posts. (110:21) Nuts sold to the Hano, Jemez, and the Keresan Pueblos. (138:41) *Ceremonial Items* Pitch smeared on burier's body before burying person and on forehead and under the eyes during mourning. Wood used to make ceremonial pokers and wands. Wood charcoal used to make the best black for sand paintings. Sapling, stripped of its branches, carried by the Talking God on the fourth day of the Night Chant. Tree used for ceremonial purposes. On the ninth day of the Night Chant, The Slayer of Alien Gods and The Child of the Water deposit their cigarettes in the shade of a tree, preferably a pinyon, while The Shooting Divinity lays hers on the ground in a cluster of snakeweed. Branches used to make the circle of branches for the Mountain Chant. Bunches of needles carried in each hand by dancers on the last night of the Mountain Chant. (55:21) *Fasteners* Resin used to cement turquoise in jewelry. (90:162) *Fuel* Wood used for firewood. *Incense & Fragrance* Dried gum, together with parts of different birds, used as an incense for ceremonial fumigation. *Jewelry* Dried seeds used to make necklaces, bracelets, anklets, and wristlets. *Tools* Wood used to make loom poles, beams, and uprights used in the construction of looms. *Waterproofing Agent* Gum used to make water bottles watertight. The gum was heated and poured into the jar, and by turning the jar, the melted gum was brought in contact with the entire inner surface, after which the surplus was poured off. The outside was also covered with the gum to which a red clay had been added so that the bottle, when finished, had a reddish hue. (55:21) Resin used to waterproof containers. (90:162) **Navajo, Ramah** *Cash Crop* Nuts gathered and sold to make up a considerable portion of the cash income of many families. *Ceremonial Items* Branches, preferably one broken from a lightning struck tree, used in Evilway ceremonials as pokers. Needles used in Evilway ceremonials as pokers. *Containers* Resin used in pottery and basketry making. *Fuel* Wood used for fires because it throws fewer sparks. *Stable Gear* Wood used to make saddle horns. *Tools* Wood used to make sharp sticks for perforating buckskin and various other tools. *Toys & Games* Wood used to make tops for spinning and sticks used in the moccasin game. (191:12) **Pueblo** *Cash Crop* Seeds gathered in large quantities and sold or traded. **Spanish American** *Cash Crop* Seeds gathered in large quantities and sold or traded. (32:40) **Tewa** *Ceremonial Items* Gum put on hot coals and fumes used to smoke people and their clothes after a funeral. Pollen used for the Snake Ceremonial. *Fasteners* Gum used in making turquoise mosaics. (42:347) *Fuel* Used extensively for firewood. (138:41) *Protection* Gum put on forehead when going outside of house as protection against sorcery. *Waterproofing Agent* Gum used to waterproof and repair pottery vessels. Gum used to prevent absorption of moisture and warping. (42:347) **Tewa of Hano** *Waterproofing Agent* Resin used to mend cracked water jars. (138:41)

Pinus elliottii Engelm., Slash Pine
- **Drug—Seminole** *Analgesic* Decoction of wood bits or bark applied externally as an analgesic. (as *P. caribaea* 169:167) *Antirheumatic (External)* Decoction of wood or bark used as a bath for aches and pains. (as *P. caribaea* 169:286) *Dermatological Aid* Decoction of roots and buds used for ballgame sickness: sores, back or limb pains, and hemorrhoids. (as *P. caribaea* 169:269) Decoction of wood or bark used as a bath for sores and cuts. (as *P. caribaea* 169:286) *Hemorrhoid Remedy* and *Orthopedic Aid* Decoction of roots and buds used for ballgame sickness: sores, back or limb pains, and hemorrhoids. (as *P. caribaea* 169:269)
- **Fiber—Seminole** *Basketry* Plant used to make baskets. *Building Material* Plant used to make houses. *Furniture* Plant used to make Bighouse seats. (as *P. caribaea* 169:480)
- **Other—Seminole** *Ceremonial Items* Plant used for religious scarification. *Fasteners* Plant used to make arrow point glue. *Hide Preparation* Plant used for tanning. *Lighting* Plant used to make torches. *Toys & Games* Plant used to make ball poles. (as *P. caribaea* 169:480)

Pinus flexilis James, Limber Pine
- **Drug—Navajo, Ramah** *Ceremonial Medicine* Plant used as a ceremonial emetic. *Cough Medicine* Plant used as a cough medicine. *Emetic* Plant used as a ceremonial emetic. *Febrifuge* Plant used as medicine for fever. *Hunting Medicine* Plant smoked by hunters for "good luck." (191:13)
- **Food—Apache, Chiricahua & Mescalero** *Unspecified* Seeds roasted and hulled or sometimes the seeds ground, shell and all, and eaten. (33:43) **Montana Indian** *Unspecified* Nuts were an important article of food. (19:18)
- **Other—Navajo** *Ceremonial Items* Wood used to make the small bow and arrow used in the Witch and Shooting Chants. (55:23)

Pinus glabra Walt., Spruce Pine
- **Drug—Cherokee** *Anthelmintic* Given for worms. *Antidiarrheal* Bark chewed "to check bowels." *Antirheumatic (External)* Oil used to bathe painful joints. *Antirheumatic (Internal)* Syrup taken for chronic rheumatism. *Cold Remedy* Infusion, steam, and oil used in various ways as cold remedy. *Cough Medicine* Syrup taken by pregnant women with cough. *Dermatological Aid* Poultice of tar used on scald head, tetterworm, stone bruises, and ulcers. *Febrifuge* Compound infusion of needles to "break out fever." *Gastrointestinal Aid* Taken for colics and gout. *Gynecological Aid* Compound infusion of needles for "child-bed-fevers." Syrup taken by pregnant women with cough and poultice used for swollen breasts. *Hemorrhoid Remedy* Compound infusion of root taken for piles. *Kidney Aid* Taken for "weak back or kidneys." *Laxative* Taken as a gentle laxative. *Misc. Disease Remedy* Taken for gout, to break out measles, and for complications from mumps. *Orthopedic Aid* Taken for "weak back or kidneys." *Other* Skim turpentine off root decoction and use on deer's skin for drawing plaster. *Respiratory Aid* Compound infusion of needles in apple juice taken by ballplayers "for wind." Syrup taken for "catarrh (ulcer of the lungs)." *Sedative* Given for hysterics. *Stimulant* Compound infusion of root taken as a stimulant. *Tuberculosis Remedy* Tar used for consumption. *Urinary Aid* Syrup used as poultice for swollen testicles caused by mumps. *Venereal Aid* Syrup taken for chronic rheumatism and venereal disease. (80:49)
- **Other—Cherokee** *Incense & Fragrance* Needles or gum used to scent soap. (80:49)

Pinus jeffreyi Grev. & Balf., Jeffrey Pine
- **Food—Paiute, Northern** *Candy* Sap crystallized, gathered, and eaten like candy. *Winter Use Food* Sap crystallized, gathered, and stored for winter use. (59:53)
- **Fiber—Diegueño** *Basketry* Needles used in making baskets. Butt ends of the needle clusters were left protruding on the outside of the basket as a decorative touch. *Building Material* Bark used to make shelters for those gathering acorns in the mountains. (84:29)
- **Other—Paiute, Northern** *Cash Crop* Sap crystallized, gathered, and sold. (59:53)

Pinus lambertiana Dougl., Sugar Pine
- **Drug—Kawaiisu** *Carminative* Dried sap powder eaten for stomach gas. *Eye Medicine* Powdered sap and milk used as drops for sore eyes and gives infants good eyes. *Laxative* Dried sap powder eaten to loosen the bowels. *Pediatric Aid* Powdered sap and milk used as drops for sore eyes and gives infants good eyes. (206:50) **Mendocino Indian** *Cathartic* Plant used as a cathartic. (41:306) **Miwok** *Eye Medicine* Sugar pine sugar used as a wash for sore or blind eyes. (12:151) **Pomo** *Unspecified* Sugar found in bark wounds ground up, molded into cakes, and used as a medicine. (11:79)
- **Food—Karok** *Unspecified* Roasted seeds used for food.[58] (6:44) Coagulated sap gathered from hollow trees and eaten without preparation or mixing with other foods. Nuts roasted and used for food. (148:378) *Winter Use Food* Roasted seeds stored for winter use.[58] (6:44) Nuts roasted and stored for winter use. (148:378) **Kawaiisu** *Sweetener* Sap, drained through a hole cut into the tree, dried into a "powdered sugar" and eaten. *Unspecified* Seeds eaten raw, roasted, parched, boiled, or pounded and mixed with cold water. (206:50) **Klamath** *Unspecified* Seeds used for food. (45:88) **Mendocino Indian** *Unspecified* Nuts used for food. (41:306) **Miwok** *Sweetener* Sugar pine sugar eaten as a delicacy. (12:151) *Unspecified* Shelled nutmeats used for food. (12:150) Pulverized nutshells and meat made into peanut butter and used for feasts. (12:151) **Pomo** *Unspecified* Nuts rarely used for food. (11:79) Nuts used for food. Pitch used for food. (66:11) **Pomo,**

Kashaya *Candy* Pitch chewed for gum. *Sweetener* Pitch tasted sweet like candy. *Unspecified* Nuts, inside the cone, eaten fresh. *Winter Use Food* Nuts, inside the cone, dried for winter use. (72:93) **Shasta** *Bread & Cake* Nuts dried, powdered, made into small cakes, and eaten with a very thin mush made of grass seeds. *Dried Food* Nuts dried and eaten. *Unspecified* Whole nuts mixed with powdered salmon and eaten. (93:308) **Yuki** *Candy* Sweet exudation chewed as gum. (49:88)
- **Fiber**—**Karok** *Building Material* Wood used for building sweat houses. (148:378)
- **Other**—**Karok** *Fasteners* Pitch used as an adhesive. (148:378) *Jewelry* Seeds used as beads in jewelry. (6:44) **Pomo, Kashaya** *Toys & Games* Pitch used in whistles. (72:93)

Pinus monophylla Torr. & Frém., Singleleaf Pinyon
- **Drug**—**Apache, Western** *Dermatological Aid* Heated pitch applied to the face to remove facial hair. (26:185) **Cahuilla** *Dermatological Aid* Pitch used as a face cream by girls to prevent sunburn. (15:102) **Gosiute** *Anthelmintic* Decoction of gum taken for worms or other intestinal parasites. (39:350) **Havasupai** *Dermatological Aid* Poultice of melted gum applied to cuts. *Veterinary Aid* Poultice of melted gum applied to horses for cuts. (197:205) **Hopi** *Dermatological Aid* Poultice of gum used to exclude air from cuts and sores. *Disinfectant* Gum smoke used as disinfectant for family of dead person. *Witchcraft Medicine* Gum applied to forehead as a protection from sorcery. (200:63) **Kawaiisu** *Contraceptive* Cooked pitch taken by women who wanted no more children. *Dermatological Aid* Pitch used as salve on cuts. *Gynecological Aid* Cooked pitch taken by women to stop menstruation. *Pediatric Aid* Cooked pitch given to adolescent girls to keep youthful and increase life span. (206:50) **Paiute** *Analgesic* Poultice of heated resin applied for general muscular soreness. *Antidiarrheal* Decoction of resin or simple or compound pills of resin taken for diarrhea. *Antiemetic* Decoction of resin taken for nausea. *Antirheumatic (Internal)* Decoction of resin taken for rheumatism. *Cold Remedy* Compound poultice of heated resin applied for chest congestion from colds. Simple or compound decoction of several plant parts taken for colds. *Dermatological Aid* Compound poultice of pitch applied to sores, cuts, swellings, and insect bites. Simple or compound poultice of heated resin applied to draw boils or slivers. *Febrifuge* Decoction of resin taken for fevers. *Gastrointestinal Aid* Decoction of resin taken for indigestion, nausea, or bowel troubles. *Gynecological Aid* Decoction of resin taken as a tonic after childbirth and for general debility. *Misc. Disease Remedy* Decoction of resin taken for influenza. *Orthopedic Aid* Poultice of heated resin applied to treat any general muscular soreness. *Pulmonary Aid* Compound poultice of heated resin applied for chest congestion from colds. Poultice of heated resin applied for pneumonia. *Throat Aid* Resin chewed or pulverized resin applied with swab for sore throat. *Tonic* Decoction of resin taken as a tonic after childbirth and for general debility. *Tuberculosis Remedy* Decoction of resin taken for tuberculosis and influenza. *Venereal Aid* Compound decoction of resin taken for venereal disease. Decoction of resin taken, resin chewed or used as pills for venereal disease. Pulverized resin dusted on syphilitic sores. (180:117, 118) **Shoshoni** *Analgesic* Poultice of heated resin applied for sciatic pains or muscular soreness. Poultice of heated resin used for sciatic pains and general muscular soreness. *Antiemetic* Decoction of resin taken for nausea. *Cold Remedy* Compound decoction of several plant parts taken for colds. Smoke of pitch compound inhaled for colds. *Cough Medicine* Compound decoction of pitch taken for coughs. *Dermatological Aid* Compound decoction of needles used as antiseptic wash for rashes. Compound poultice of pitch applied to sores, cuts, swellings, and insect bites. Poultice of heated resin applied to draw boils or imbedded slivers. *Disinfectant* Compound decoction of needles used as antiseptic wash for measles and other rashes. *Febrifuge* Decoction of resin taken for fevers. *Gastrointestinal Aid* Decoction of resin taken for indigestion, nausea, or bowel troubles. *Kidney Aid* Compound decoction of resin taken for the kidneys. *Misc. Disease Remedy* Compound decoction of needles used as antiseptic wash for measles. Compound decoction of pitch taken for smallpox. *Orthopedic Aid* Poultice of heated resin applied for sciatic pains or muscular soreness. *Poultice* Poultice of heated resin applied for ruptures. *Pulmonary Aid* Poultice of heated resin applied for pneumonia. *Venereal Aid* Decoction of resin taken for venereal disease. *Veterinary Aid* Smoke from root compound inhaled by horses for distemper. (180:117, 118) **Washo** *Cold Remedy* Compound decoction of resin taken for colds. *Venereal Aid* Decoction of needles or wood taken or fresh resin used as pills for gonorrhea. (180:117, 118)
- **Food**—**Apache, Western** *Candy* Pitch used as chewing gum. *Porridge* Pinyon and corn flour mixed and cooked into a mush. *Staple* Used as a staple food. Nuts eaten raw, roasted, or ground into flour. *Winter Use Food* Nuts stored in baskets or pottery jars. (26:185) **Cahuilla** *Baby Food* Nuts used as one of the few foods fed to babies instead of a natural milk diet. *Beverage* Ground nuts mixed with water and used as a drink. *Dried Food* Cooked, unshelled nuts stored for future use. *Porridge* Roasted, shelled nuts eaten whole or ground and made into mush. (15:102) **Cocopa** *Unspecified* Pinyons eaten in the mountains away from home. (37:188) **Diegueño** *Unspecified* Nuts used for food. (84:30) Seeds used for food. (88:215) **Gosiute** *Unspecified* Nuts used for food. (39:377) **Havasupai** *Spice* Sprigs placed in the cooking pit with porcupine, bobcat, or badger to improve the taste of the meat. *Unspecified* Nuts formerly used as an important food source. (197:205) **Hopi** *Unspecified* Nuts eaten for food. (200:63) **Kawaiisu** *Porridge* Roasted, steamed seeds pounded into a meal, mixed with cold water, and eaten. *Unspecified* Roasted, steamed seeds eaten hulled or unhulled. Roasted, steamed seeds hulled, the kernels boiled and eaten. *Winter Use Food* Unhulled seeds strung on cord, dried, and stored in sacks for winter use. (206:50) **Paiute** *Dried Food* Nuts sun dried or roasted and stored for future use. *Porridge* Roasted nuts ground into a flour and mixed with water into a paste or mush. *Soup* Roasted nuts ground into a flour and mixed with water into a soup. *Staple* Roasted nuts ground into flour. (167:241) *Unspecified* Nuts eaten and obtained from Nevada. Seeds used for food. (111:42) Nuts roasted and eaten whole. *Winter Use Food* Nuts gathered in great quantity and stored for future use. (167:241) **Paiute, Northern** *Bread & Cake* Nuts roasted, winnowed, dried, ground into a meal, made into a stiff flour dough, and eaten. (59:51) *Candy* Gum chewed as gum. (59:53) *Dried Food* Nuts dried and stored for future use. *Ice Cream* Nuts roasted, dried, ground into a meal, made into a stiff dough, frozen, and eaten like ice cream. *Soup* Nuts roasted, ground into a fine flour, and cooked into a thick soup. (59:51) **Shoshoni** *Unspecified* Nuts formed an important part of the diet. (117:443) **Tubat-

ulabal Unspecified Nuts used extensively for food. (193:15) **Washo** *Dried Food* Roasted nuts eaten fresh or stored for later use. (10:13) *Porridge* Nuts used to make mush. (10:14)
- **Fiber**—**Cahuilla** *Basketry* Needles and roots used to make baskets. *Building Material* Bark used as roofing material in house construction. (15:102) **Havasupai** *Building Material* Wood used for house construction. *Caulking Material* Melted gum used to plug a leaky canteen or other containers. (197:205)
- **Dye**—**Hopi** *Unspecified* Gum used to prepare certain dyes. (200:63)
- **Other**—**Apache, Western** *Waterproofing Agent* Pitch used to waterproof baskets. (26:185) **Cahuilla** *Cash Crop* Nuts used as an important trade item. *Fasteners* Pitch used as an adhesive for mending pottery and baskets and attaching arrow points to shafts. *Fuel* Wood, high combustibility, used for firewood and kindling. *Incense & Fragrance* Wood, gave off a pleasant odor, used for firewood. (15:102) **Havasupai** *Cash Crop* Nuts sold in considerable quantities to stores. *Fuel* Wood used for firewood. *Paint* Gum used in the paint used on the base of arrows. *Tools* Wood used to make the knife for trimming mescal heads. *Waterproofing Agent* Gum used to waterproof basketry water jugs and basketry drinking cups. (197:205) **Hopi** *Fasteners* Gum used in making turquoise mosaics. *Waterproofing Agent* Gum used in waterproofing and repairing pottery vessels. (200:63) **Kawaiisu** *Jewelry* Cracked shells pinched by children onto the ear lobes and worn as ornaments. *Waterproofing Agent* Hot pitch applied to waterproof the inside and outside of a basketry water bottle. (206:50)

Pinus monticola Dougl. ex D. Don, Western White Pine

- **Drug**—**Hoh** *Cough Medicine* Gum used for coughs. (137:58) **Kwakiutl** *Cough Medicine* Pitch used for coughs. *Dermatological Aid* Pitch used for sores. *Gastrointestinal Aid* Pitch used for stomachaches. *Reproductive Aid* Gum chewed by women for fertility and by girls to become pregnant without sex. (183:270) **Lummi** *Tuberculosis Remedy* Infusion of bark taken for tuberculosis. (79:16) **Mahuna** *Antirheumatic (Internal)* Plant used for rheumatism. (140:60) **Nitinaht** *Dermatological Aid* Pitch mixed with melted deer tallow and used as a skin cosmetic. (186:73) **Quileute** *Cough Medicine* Gum used for coughs. (137:58) **Quinault** *Blood Medicine* Infusion of bark taken to purify the blood. *Gastrointestinal Aid* Infusion of bark taken for stomach disorders. (79:16) **Shuswap** *Tuberculosis Remedy* Bark used for tuberculosis. (123:51) **Skagit** *Antirheumatic (External)* Decoction of young shoots used as a soak for rheumatism. *Dermatological Aid* Decoction of bark used for cuts and sores. *Tuberculosis Remedy* Infusion of bark taken for tuberculosis. (79:16) **Thompson** *Panacea* Infusion of boughs used for any kind of illness by old people. *Unspecified* Pitch used medicinally. (187:103)
- **Food**—**Paiute** *Unspecified* Nuts served as a minor source of subsistence. (111:40) **Salish, Coast** *Dried Food* Inner bark dried in cakes and used for food. *Unspecified* Inner bark eaten fresh. (182:71) **Shuswap** *Unspecified* Cones used for food. (123:51) **Thompson** *Unspecified* Gummy substance collected from trunk and branches and chewed. (164:493)
- **Fiber**—**Okanagan-Colville** *Canoe Material* Bark used to make sturgeon nosed canoes. (188:29) **Skagit** *Canoe Material* Used rarely to make light dugouts. (79:16)
- **Other**—**Nitinaht** *Ceremonial Items* Wood used to make small totem poles and model canoes. *Hunting & Fishing Item* Pitch used as protective coating for whaling and fishing equipment. (186:73) **Paiute** *Ceremonial Items* Green branch thrown into the fire and rain will come. (111:40) **Salish, Coast** *Fasteners* Pitch used to fasten arrowheads onto shafts. (182:71)

Pinus muricata D. Don, Bishop Pine

- **Food**—**Pomo, Kashaya** *Unspecified* Nuts eaten fresh. *Winter Use Food* Nuts dried for winter use. (72:92)
- **Fiber**—**Pomo** *Basketry* Root used in basketry. (66:11)
- **Other**—**Pomo** *Fuel* Wood used for firewood. (66:11) **Pomo, Kashaya** *Fasteners* Pitch used like glue. *Hunting & Fishing Item* Roots used in making fish traps. (72:92)

Pinus ponderosa P. & C. Lawson, Ponderosa Pine

- **Drug**—**Cheyenne** *Dermatological Aid* Pitch used to hold the hair in place. (82:50) Gum used as a salve or ointment for sores and scabby skin. (83:6) **Flathead** *Analgesic* Poultice of pitch and melted animal tallow or lard used for backache. *Antirheumatic (External)* Poultice of pitch and melted animal tallow or lard used for rheumatism. Boughs used in sweat lodges for muscular pain. *Dermatological Aid* Pitch warmed and used for boils and carbuncles. Needles jabbed into the scalp for dandruff. *Gynecological Aid* Needles heated and used for faster delivery of the placenta. (82:50) **Navajo** *Ceremonial Medicine* Pollen used in the "Night Chant" medicine. (55:23) **Navajo, Ramah** *Ceremonial Medicine* Cones with seeds removed used as a ceremonial medicine. Needles used as a ceremonial emetic. *Cough Medicine* Compound decoction of needles taken for bad coughs and fever. *Emetic* Needles used as a ceremonial emetic. *Febrifuge* Compound decoction of needles taken for fever and bad cough. (191:13, 14) **Okanagan-Colville** *Abortifacient* Green buds never chewed by pregnant women because it would cause a miscarriage. *Antihemorrhagic* Decoction of plant tops taken for internal hemorrhaging. *Dermatological Aid* Poultice of pitch applied to boils. *Eye Medicine* Infusion of dried buds used as an eyewash. *Febrifuge* Decoction of plant tops taken for high fevers. *Gastrointestinal Aid* Good medicine for the stomach. *Witchcraft Medicine* Needles spread on the floor of the sweat house to fight off "plhax," witchcraft. (188:29) **Okanogon** *Eye Medicine* Decoction of gum used as an ointment for sore eyes. (125:41) **Paiute** *Dermatological Aid* Poultice of dry, chewed pitch used on boils. (111:40) **Shuswap** *Dermatological Aid* Plant used to remove underarm odors. *Panacea* and *Pediatric Aid* Infusion of plant used as a wash for sick babies. *Stimulant* Used in the sweat house to hit oneself at the hottest point. (123:52) **Thompson** *Antirheumatic (External)* Pitch used for aching backs, joints, and limbs. (187:104) *Dermatological Aid* Boiled gum mixed with grease and used as an ointment for sores. (164:466) Decoction of tops used in washing the face and head by girls who want fair and smooth skin. (164:508) Poultice of gum applied to boils, sores, and chapped skin. White gum was used as a poultice with buckskin on boils and chronic sores while reddish gum was used on hard, red sores. The reddish gum was mixed with any kind of lard, such as deer fat, strained, and used on sores. Pitch made into a salve and used for boils or cuts. The pitch ointment was left on the skin for 3 or 4 days. It was said to get quite itchy, but after a while, the pitch was removed with the bandage and then took effect. If the pitch stuck to the skin, it was not ready to remove. *Ear*

Medicine Poultice of warmed gum applied to the ear for earache. (187:104) *Eye Medicine* Decoction of gum used as an ointment for sore eyes. (125:41) Boiled gum mixed with grease and used as an ointment for inflamed eyes. (164:466) *Pediatric Aid* Gum used on babies' skin like baby oil. The ointment caused the baby to sleep all the time, just like aspirin. *Sedative* Gum used on babies' skin like baby oil causing them to sleep all the time. (187:104) *Veterinary Aid* Hot gum and animal fat poured on horses' sores or wounds. (164:514)

- *Food*—**Blackfoot** *Unspecified* Inner bark used for food. (97:18) **Cheyenne** *Candy* Pitch chewed as a gum. (82:50) *Unspecified* Seeds used for food. Young male cones chewed for the juice. (83:6) **Coeur d'Alene** *Unspecified* Nutlets used for food. (178:90) Cambium layer used for food. (178:91) **Havasupai** *Unspecified* Nuts roasted and eaten. (197:206) **Kawaiisu** *Unspecified* Kernels eaten raw. (206:51) **Klamath** *Sauce & Relish* Cambium layer scraped off and eaten as a relish. *Starvation Food* Cambium layer scraped off and eaten in time of famine. *Unspecified* Sweet layer between bark and sapwood scraped and used for food. (45:89) **Miwok** *Dried Food* Cones' extracted nuts gathered, dried in the sun, and eaten. (12:150) **Montana Indian** *Unspecified* Inner bark eaten in the spring. (19:18) Inner bark formerly used for food. (82:50) **Navajo, Ramah** *Unspecified* Bark eaten raw. (191:13) **Okanagan-Colville** *Candy* Pitch used as chewing gum. Green buds chewed and the juice sucked by children. *Frozen Food* Cambium frozen for future use. *Unspecified* Cambium used for food. Seeds eaten like nuts. *Winter Use Food* Seeds stored for winter use. (188:29) **Okanagon** *Staple* Nutlets or seeds used as a principal food. Cambium layer used as a principal food. (178:239) *Unspecified* Seeds used for food. (125:39) **Paiute** *Candy* Dried pitch used as chewing gum. *Dried Food* Inner bark sun dried and stored. *Unspecified* Inner bark eaten fresh and raw. Seeds used for food. (111:40) **Sanpoil & Nespelem** *Unspecified* Cambium layer eaten raw. This was an important food. The bark was removed in sections with the aid of wooden wedges. Sap scrapers were made from the rib of the deer by cutting it to an appropriate length, sharpening the edges, and rounding the working end. (as *P. poderosa* 131:103) Pine nuts eaten without special preparation. (131:104) **Shasta** *Bread & Cake* Nuts dried, powdered, made into small cakes, and eaten with a very thin mush made of grass seeds. *Dried Food* Nuts dried and eaten. *Unspecified* Whole nuts mixed with powdered salmon and eaten. (93:308) **Spokan** *Unspecified* Nutlets used for food. Cambium used for food. (178:344) **Thompson** *Porridge* Seeds and whitebark pine seeds placed in a bag, pounded into a powder, mixed with water, and eaten. (187:104) *Unspecified* Seeds used for food. (125:39) Cambium of young twigs eaten. (164:484) Seeds eaten in small quantities. (164:491)
- *Fiber*—**Diegueño** *Basketry* Needles used in making baskets. *Building Material* Bark used to make shelters for those gathering acorns in the mountains. (84:29) **Hopi** *Building Material* Used for large roof timbers. (200:63) **Isleta** *Building Material* Wood used to furnish the beams or "vega (or viga) poles" of the houses. (100:37, 51) **Karok** *Basketry* Bigger roots used for basketry. (6:45) Root fibers used to make baskets. (148:378) **Kawaiisu** *Building Material* Needles used as an outer covering for the winter house. (206:51) **Klamath** *Canoe Material* Logs used to make boats. (45:89) Single logs used to make dugout canoes. (46:728) **Maidu** *Basketry* Roots used as the overlay twine warps and overlay twine weft bases in the manufacture of baskets. (173:71) **Mendocino Indian** *Building Material* Wood used for lodge poles. (41:307) **Mewuk** *Building Material* Branches with tips down used to hang from the top of acorn caches to keep out the rain in winter. (117:346) **Montana Indian** *Building Material* Most important lumber tree in the state. *Canoe Material* Trunks hollowed by fire to make dugouts. (19:18) **Navajo, Ramah** *Building Material* Wood used for hogans, fence posts, and corral construction. Branches often used to cover a sweat house. *Furniture* Wood used to make boards and cradle bow of the two-board type of baby cradle. A young tree, in an area where few people go and therefore not likely to be cut down, is selected, corn pollen is sprinkled on it from the bottom upward, and a solid piece is taken from the east side. As the cradle is made, prayers are said but no songs sung. If the first baby is a boy, the top tips of the boards are truncated, if it is a girl, they are pointed; thereafter either kind can be used for either sex and the cradle is saved for later children unless the baby dies. The cradle is rubbed with red ocher and tallow to protect if from evil spirits who never use red paint. Formerly, a buckskin covering was used over the top but now a blanket is considered better. The footboard is moved down as the baby grows and the cradle is discarded when the baby begins to walk. Small branches of a tree from which squirrels have gnawed the bark are tied together in a row about 5 inches long and tied to the cradle to keep the baby from hurting himself (until he is 3 years old). Dirt from a spot where a squirrel has landed on the ground is placed in a buckskin bag and attached to the sticks as an additional precaution (effective even when the baby is grown). *Snow Gear* Wood slabs tied together with yucca fiber used as snowshoes. (191:13) **Okanagan-Colville** *Building Material* Needles used as insulation for underground storage pits. *Canoe Material* Wood used to make dugout canoes. (188:29) **Paiute** *Building Material* Bark used to make houses. (111:40) **Thompson** *Building Material* Needles used as insulation on the roofs of pit houses. (187:104) **Thompson, Upper (Fraser Band)** *Canoe Material* Used to make dugout canoes. **Thompson, Upper (Lytton Band)** *Canoe Material* Used to make dugout canoes. (164:499) **Wintoon** *Basketry* Straight stems used for making baskets. (117:264)
- *Dye*—**Cheyenne** *Blue* Roots used to make a blue dye. (83:6) *Unspecified* Used to make a dye. (83:46)
- *Other*—**Blackfoot** *Tools* Twigs used as twirling sticks in fire making. (97:18) **Cheyenne** *Musical Instrument* Pitch used to make bone and wooden whistles and flutes. (82:50) Gum placed inside whistles and flutes to improve their sounds. (83:6) **Crow** *Fasteners* Pitch used as glue. (82:50) **Hopi** *Ceremonial Items* Plant parts smoked ceremonially. *Tools* Used to make ladders. (200:63) **Hualapai** *Cash Crop* Trees considered a main economic resource for the tribe. (195:21) **Isleta** *Fuel* Wood used as principal source of firewood. (100:37) **Kawaiisu** *Ceremonial Items* Branch used to hang the outgrown cradle of a male child so the boy will grow strong like the tree. *Containers* Needles used to form a layer in the roasting of the yucca "heart." (206:51) **Klamath** *Fuel* Dried needles stuffed loosely between cross sticks and lighted to ignite them. (46:735) **Mendocino Indian** *Fasteners* Pitch used for the adhesive qualities. *Fuel* Wood used for fuel. (41:307) **Montana Indian** *Tools* Twigs used for twirling sticks in fire production. (19:18) **Navajo, Ramah** *Containers* Bark used to make containers for sand painting pigments. *Fuel* Wood used for firewood. *Stable*

Gear Wood used to make saddle horns, pommel, and back. (191:13) **Nez Perce** *Fasteners* Pitch used as glue. *Lighting* Pitch used to make torches. (82:50) **Okanagan-Colville** *Ceremonial Items* Smoldering cones thrown into the air in the direction of rain clouds to make the rain stop. *Fasteners* Pitch used to cement feathers onto arrow shafts. *Hide Preparation* Rotten wood used for smoking deer hides. (188:29) **Paiute** *Fasteners* Pitch used as glue in arrow making and other manufactures. *Preservative* Pitch used to protect pictures painted on rocks. *Waterproofing Agent* Melted pitch used to waterproof the outside of water jugs woven of willow. (111:40) **Shuswap** *Fuel* Bark used as fuel because it cooled quickly and enemies cannot tell how long ago camp was broken. *Hide Preparation* Wood used for smoking buckskin. (123:52) **Thompson** *Containers* Needles used to line food caches and cellars. *Cooking Tools* Needles supported on a framework of poles used for drying cooked berries. The needles were interspersed between layers of dried salmon or any other food being stored. They kept the food dry but allowed air to circulate around it to prevent spoiling. (187:104) *Incense & Fragrance* Needles inserted into the flesh under the arms by girls who wish their armpits to smell sweet. (164:508) *Smoking Tools* Plant tops hollowed out with mock orange sticks and used to make the stems of smoking pipes. *Waterproofing Agent* Pitch used to waterproof moccasins and other items. (187:104) **Thompson, Upper (Fraser Band)** *Fuel* Dry cones mixed with fir bark to make the best smoke for smoking skins. **Thompson, Upper (Lytton Band)** *Fuel* Dry cones mixed with fir bark to make the best smoke for smoking skins. (164:499)

Pinus ponderosa P. & C. Lawson **var. *ponderosa***, Ponderosa Pine

- *Other*—**Tewa of Hano** *Ceremonial Items* Leaves attached to prayer feathers prepared during December ceremonies. (as *P. brachyptera* 138:41)

Pinus ponderosa **var. *scopulorum*** Engelm., Ponderosa Pine

- *Food*—**Apache, Chiricahua & Mescalero** *Bread & Cake* Inner bark scraped off and baked in the form of cakes. *Starvation Food* Seeds ground, rolled into balls, and eaten raw only in times of food scarcity. *Unspecified* Bark boiled or eaten raw. (as *P. scopulorum* 33:43)
- *Fiber*—**Jemez** *Building Material* Used as timbers for roofs. (as *P. scopulorum* 44:26) **Keres, Western** *Building Material* Wood used for logs. (as *P. scopulorum* 171:61)
- *Other*—**Hopi** *Ceremonial Items* Plant smoked ceremonially. Plant used in the Su-ya-lung ceremony. *Season Indicator* Needles attached to prayer sticks to bring cold. *Tools* Wood used to make kiva ladders. (42:348)

Pinus pungens Lamb., Table Mountain Pine

- *Fiber*—**Cherokee** *Building Material* Wood used for lumber. *Canoe Material* Wood used to make 30- to 40-foot-long canoes. (80:49)
- *Other*—**Cherokee** *Decorations* Wood used to carve. (80:49)

Pinus quadrifolia Parl. ex Sudworth, Parry Pinyon

- *Drug*—**Cahuilla** *Dermatological Aid* Pitch used as a face cream by girls to prevent sunburn. (15:102)
- *Food*—**Cahuilla** *Baby Food* Nuts used as one of the few foods fed to babies instead of a natural milk diet. *Beverage* Ground nuts mixed with water and used as a drink. *Dried Food* Cooked, unshelled nuts stored for future use. *Porridge* Roasted, shelled nuts eaten whole or ground and made into mush. (15:102) **Diegueño** *Unspecified* Nuts used for food. (84:30) Seeds used for food. (88:215)
- *Fiber*—**Cahuilla** *Basketry* Needles and roots used to make baskets. *Building Material* Bark used as roofing material in house construction. (15:102)
- *Other*—**Cahuilla** *Cash Crop* Nuts used as an important trade item. *Fasteners* Pitch used as an adhesive for mending pottery and baskets and attaching arrow points to shafts. *Fuel* Wood, high combustibility, used for firewood and kindling. *Incense & Fragrance* Wood, gave off a pleasant odor, used for firewood. (15:102)

Pinus resinosa Soland., Red Pine

- *Drug*—**Algonquin, Tête-de-Boule** *Cold Remedy* Poultice of wetted, inner bark applied to the chest for strong colds. (132:129) **Ojibwa** *Stimulant* Powdered, dried leaves used as a reviver or inhalant. *Unspecified* Bark and cones used medicinally. (153:379) **Ojibwa, South** *Analgesic* Decoction of leaves and bark used as herbal steam for headache and backache. Poultice of crushed leaves and bark applied for headache. (91:198) **Potawatomi** *Stimulant* Leaves used as a fumigant to revive a comatose patient. (154:70)
- *Fiber*—**Ojibwa** *Building Material* Resin boiled twice, added to tallow, and used for mending roof rolls of birch bark. *Caulking Material* Resin boiled twice, added to tallow, and used for caulking canoes. (153:421)
- *Other*—**Chippewa** *Toys & Games* Used for toys. (53:378) **Ojibwa** *Waterproofing Agent* Resin boiled twice and added to tallow to make a serviceable waterproof pitch. (153:421)

Pinus rigida P. Mill., Pitch Pine

- *Drug*—**Iroquois** *Antirheumatic (Internal)* Pitch taken for rheumatism. *Burn Dressing* Pitch used for burns. *Dermatological Aid* Compound infusion applied as a poultice to break open boils. Pitch applied to cuts in joints and boils. *Laxative* Pitch taken as a laxative. (87:267) **Shinnecock** *Dermatological Aid* Poultice applied to boils and abscesses. (30:121)
- *Fiber*—**Cherokee** *Building Material* Wood used for lumber. *Canoe Material* Wood used to make 30- to 40-foot-long canoes. (80:49)
- *Other*—**Cherokee** *Decorations* Wood used to carve. (80:49) **Iroquois** *Insecticide* Smoke from burning leaves used to get rid of fleas. (87:267)

Pinus sabiniana Dougl. ex Dougl., California Foothill Pine

- *Drug*—**Costanoan** *Antirheumatic (Internal)* Pitch chewed for rheumatism. (21:6) **Mendocino Indian** *Burn Dressing* Pitch applied to burns. *Dermatological Aid* Pitch applied to sores. (41:307) **Miwok** *Burn Dressing* Crushed nuts' charcoal applied to burns. *Dermatological Aid* Crushed nuts' charcoal applied to sores and abrasions. (12:149) **Wailaki** *Antirheumatic (Internal)* Gum chewed for rheumatism. **Yokia** *Analgesic* Burning twigs and leaves used as sweat bath for rheumatism pain. *Antirheumatic (External)* Burning twigs and leaves used as sweat bath for rheumatism pain. *Dermatological Aid* Burning twigs and leaves used as sweat bath for bruises. *Diaphoretic* Burning twigs and leaves used as sweat bath for rheumatism pain and bruises. *Tuberculosis Remedy* Infusion of bark taken for consumption. (41:307)
- *Food*—**Costanoan** *Unspecified* Pine nuts used for food. (21:248) **Kawaiisu** *Porridge* and *Unspecified* Seeds eaten fresh, roasted,

boiled, or pounded, and mixed with cold water. (206:52) **Mendocino Indian** *Starvation Food* Fresh, inner bark formerly used for food during prolonged winters when other foods were scarce. (41:307) **Mewuk** *Unspecified* Nuts used for food. (117:333) **Miwok** *Unspecified* Nuts and cone pith eaten for food. (12:149) **Pomo** *Unspecified* Nuts rarely used for food. (11:79) **Pomo, Kashaya** *Staple* Dried nut eaten whole or pounded into a flour and mixed with pinole. *Unspecified* Nuts eaten fresh. *Winter Use Food* Nuts dried for winter use. (72:92) **Shasta** *Bread & Cake* Nuts dried, powdered, made into small cakes, and eaten with a very thin mush made of grass seeds. *Dried Food* Nuts dried and eaten. *Unspecified* Whole nuts mixed with powdered salmon and eaten. (93:308) **Tubatulabal** *Unspecified* Nuts used extensively for food. (193:15) **Wailaki** *Candy* Gum chewed by children for pleasure. (41:307)

- **Fiber**—**Kawaiisu** *Building Material* Needles used as an outer covering for the winter house. (206:52) **Mewuk** *Basketry* Sprouts used to make coiled bowls. (117:335) **Miwok** *Basketry* Twigs and rootlets used as sewing material for coiled basket. *Building Material* Needles used for thatch. Bark used for house coverings. *Mats, Rugs & Bedding* Needles used for bedding and floor covering. (12:149) **Pomo** *Basketry* Root fiber used to make twined baskets. (9:138) Root wood used to make V-shaped baskets for carrying acorns. (41:307) Young growth split into ribbon-like strands and used for basket body material. (117:296) **Wintoon** *Building Material* Wood used to make planks for houses. (117:273)

- **Other**—**Karok** *Decorations* Nuts used as beads to decorate dance dresses. (148:378) **Kawaiisu** *Containers* Needles used to form a layer in the roasting of the yucca "heart." *Smoke Plant* Seeds put into the liquid used to moisten dry tobacco meal and shaped into plugs. (206:52) **Mahuna** *Protection* Trees used for protection from lightning. (140:40) **Mendocino Indian** *Decorations* Pitch burned and the resulting soot used for tattooing. *Musical Instrument* Logs formerly hollowed out by fire and used as drums for dances. **Pomo** *Fasteners* Pitch exudations used to fasten feathers on arrows. (41:307) *Hunting & Fishing Item* Pitch used to make the eyes for deer hunting masks. (117:284) **Yuki** *Fasteners* Formerly used like a glue to hold feathers on the body in times of war. (41:307)

Pinus **sp.**, Pine
- **Drug**—**Alabama** *Antidiarrheal* Decoction of inner bark taken for dysentery. (177:5) **Cherokee** *Other* Decoction of root used on deerskin for a "drawing plaster." (203:74) **Navajo** *Ceremonial Medicine* Needles, in water, used ceremonially. In the first ceremony of the fourth day of the Mountain Chant, the medicine man carried a bowl of pine needles in water in which the patient washed both hands. He then drank some of it and finally bathed his feet and legs to the thighs, his arms and shoulders, his body, and then his face and head, before he emptied the remainder over his back. (55:23) **Sanpoil** *Dermatological Aid* Poultice of pitch applied to bring boils to a head. (131:221)

- **Food**—**Apache, Western** *Unspecified* Inner bark used for food. (26:192) **Paiute** *Dried Food* Roasted, dried seeds stored for later use without further cooking. (111:43)

- **Fiber**—**Navajo** *Building Material* Bark used as a covering for summer shelters. (55:23)

- **Other**—**Navajo** *Ceremonial Items* Gum mixed with gypsum and used as a white paste on the "spirits of the fire" in the Fire Dance. Wood used to make the bull-roarer for some ceremonies. *Containers* Wood used to make a tinderbox for fire by friction. Bark used to make the trays for the colored powders used in the sand paintings. *Fuel* Wood used extensively for firewood. *Toys & Games* Wood used to make the ball for the game, shinny. *Waterproofing Agent* Gum used to make water bottles watertight. (55:23) **Paiute** *Ceremonial Items* Pine nuts used around the wrist when dancing, the number of nuts increased magically. (111:43)

Pinus strobus L., Eastern White Pine
- **Drug**—**Abnaki** *Cough Medicine* Decoction of bark and another plant used for coughs. (144:163) **Algonquin, Tête-de-Boule** *Cold Remedy* Poultice of wetted, inner bark applied to the chest for strong colds. (132:129) **Chippewa** *Dermatological Aid* Compound poultice of trunk of young tree applied to cuts and wounds. (53:352) Poultice of pitch applied to felons and similar inflammations. (71:123) **Delaware, Ontario** *Analgesic* Poultice of pitch applied to draw out the poison and pain from boils. (175:68) *Dermatological Aid* Pitch applied to boils to "draw out the poison and reduce the pain." *Kidney Aid* Infusion of twigs taken for kidney disorders. (175:68, 82) *Pediatric Aid* Powder from decayed plant used on babies "because of its healing properties." (175:68) *Pulmonary Aid* Infusion of twigs taken for pulmonary diseases. (175:68, 82) *Unspecified* Powder from decayed plant used on babies "because of its healing properties." (175:68) **Iroquois** *Antirheumatic (Internal)* Decoction of raw bark taken for rheumatism. (87:264) *Blood Medicine* Infusion of young trees taken as a blood tonic, "don't vomit." (87:266) *Cold Remedy* Compound decoction taken for colds, coughs, and rheumatism. (87:264) Steam from decoction of bark inhaled for head cold. (87:267) *Cough Medicine* Compound decoction or infusion taken for colds, coughs, or rheumatism. (87:264) *Dermatological Aid* Powdered wood used on chafed babies, sores, and improperly healed navels. (87:265) Compound decoction used as wash and compound poultice applied to deep cuts. Decoction of bark used for skin eruptions and scabs. Poultice used for drawing thorns and slivers. Strained, compound poultice used for cuts, bruises, sores, and scabs on face. (87:266) Compound decoction used as salve for cuts and wounds. Decoction of shaved knots, "better than penicillin," used for poison ivy. (87:267) *Dietary Aid* Decoction of knots taken to increase the appetite. *Emetic* Decoction of one knot taken to vomit for regular consumption. (87:265) Decoction of branches taken as a spring emetic. Decoction used as an emetic "when someone dies and you can't forget it." (87:266) *Gastrointestinal Aid* Raw bark taken for the stomach and cramps. (87:265) Compound decoction taken to clean the stomach. *Liver Aid* Compound decoction taken for gall. (87:266) *Misc. Disease Remedy* Raw bark taken to prevent typhoid. (87:265) *Orthopedic Aid* Decoction of leaves used as a wash for nonwalking 2- or 3-year-old infants. (87:264) Compound poultice of leaves bound to broken bones. Poultice of gum applied to broken coccyx. Raw bark taken for rheumatism and stiff limbs. (87:265) *Other* Compound used as a sugar medicine. (87:267) *Panacea* Leaves burned in spring and fall, smoke used to fill the house and prevent all sickness. (87:265) *Pediatric Aid* Decoction of leaves used as a wash for nonwalking 2- or 3-year-old infants. (87:

264) Powdered wood used on chafed babies, sores, and improperly healed navels. (87:265) *Psychological Aid* Decoction used as an emetic "when someone dies and your can't forget it." (87:266) *Pulmonary Aid* Compound decoction of bark taken by fat people for breathing difficulties. (87:264) Compound decoction taken for shortness of wind. (87:266) Salve applied to chest for chest cold. *Throat Aid* Taken for infections and sore throats. (87:267) *Tuberculosis Remedy* Decoction of knots taken for consumption. Decoction of one knot taken to vomit for regular consumption. (87:265) Decoction of twigs taken for scrofula. (87:267) *Venereal Aid* Compound decoction used as a salve on dry, cracked, venereal diseased penis. (87:265) *Veterinary Aid* Decoction of twigs used for boils on horses' necks. (87:267) *Witchcraft Medicine* Burning branch smoke drove away ghosts from the house of returning people. Smoke from plant used as a wash for a person who has seen a dead person. (87:266) **Menominee** *Analgesic* Infusion of bark, an important medicine, taken for chest pain. (151:46) *Dermatological Aid* Poultice of pounded inner bark applied to sores. (54:132) Poultice of pounded bark applied to wounds, sores, or ulcers. (151:46) **Micmac** *Cold Remedy* Bark, leaves, and stems used for colds. *Cough Medicine* Bark, leaves, and stems used for coughs. *Dermatological Aid* Bark used for wounds and sap used for hemorrhaging. (40:59) Boiled inner bark used for sores and swellings. (156:317) *Hemostat* Sap used for hemorrhaging. *Kidney Aid* Plant parts used for kidney trouble. *Misc. Disease Remedy* Bark, leaves, and stems used for grippe. Inner bark, bark, and leaves used for scurvy. (40:59) **Mohegan** *Analgesic* Poultice of sap or gum applied for boil and abscess pains. (176:74) *Cold Remedy* Cold infusion of bark taken for colds. (174:264) Bark, sap, or gum used for coughs, colds, and boils. (176:130) *Cough Medicine* Infusion of bark used for stubborn cough and pitch chewed for cough. (30:121) Infusion of bark taken for coughs and colds. (174:269) Infusion of dried inner bark used as a cough remedy. *Dermatological Aid* Sap or gum applied to boil or abscess pains. (176:74, 130) **Montagnais** *Cold Remedy* Boiled gum taken for colds. *Throat Aid* Boiled gum taken for sore throats. *Tuberculosis Remedy* Boiled gum taken for consumption. (156:315) **Ojibwa** *Stimulant* Dried leaves used as a reviver or inhalant. (153:379) *Unspecified* Plant used for medicinal purposes. (135:244) Bark and cones used medicinally. (153:379) **Ojibwa, South** *Analgesic* Boiled, crushed leaves used as herbal steam for headache and backache. Poultice of crushed leaves applied for headaches. *Herbal Steam* Boiled, crushed leaves used as herbal steam for headache and backache. (91:198) **Potawatomi** *Dermatological Aid* Pitch or resin of wood and bark used as the base for a salve. (154:70) **Shinnecock** *Cough Medicine* Infusion of bark used for stubborn cough and pitch chewed for cough. (30:121)
- *Food*—**Iroquois** *Unspecified* Species used for food. (196:119) **Micmac** *Beverage* Bark used to make a beverage. (159:258) **Ojibwa** *Unspecified* Young staminate catkins of this pine cooked for food and stewed with meat. One might think this would taste rather like pitch, but they assured the writer that is was sweet and had no pitchy flavor. (153:407)
- *Fiber*—**Cherokee** *Building Material* Wood used for lumber. *Canoe Material* Wood used to make 30- to 40-foot-long canoes. (80:49) **Ojibwa** *Caulking Material* Pitch from boiled cones and resin used for caulking and waterproofing. (153:421) *Mats, Rugs & Bedding* Boughs used on the ground or floor, covered with blankets and other bedding, and used as a bed. (135:244) **Potawatomi** *Caulking Material* Pitch rendered from the bark or cone and used to caulk boats and canoes. (154:122)
- *Other*—**Cherokee** *Decorations* Wood used to carve. (80:49) **Micmac** *Fuel* Wood used for kindling and fuel. (159:258) **Ojibwa** *Waterproofing Agent* Pitch from boiled cones and resin used for caulking and waterproofing. (153:421)

Pinus taeda L., Loblolly Pine
- *Fiber*—**Cherokee** *Building Material* Wood used for lumber. *Canoe Material* Wood used to make 30- to 40-foot-long canoes. (80:49)
- *Other*—**Cherokee** *Decorations* Wood used to carve. (80:49)

Pinus virginiana P. Mill., Virginia Pine
- *Drug*—**Cherokee** *Anthelmintic* Given for worms. *Antidiarrheal* Bark chewed "to check bowels." *Antirheumatic (External)* Oil used to bathe painful joints. *Antirheumatic (Internal)* Syrup taken for chronic rheumatism. *Ceremonial Medicine* Branches burned and ashes thrown on hearth fire after a death in the home. *Cold Remedy* Infusion, steam, and oil used in various ways as cold remedy. *Cough Medicine* Syrup taken by pregnant women with cough. *Dermatological Aid* Poultice of tar used on scald head, tetterworm, stone bruises, and ulcers. *Febrifuge* Compound infusion of needles used to "break out fever." *Gastrointestinal Aid* Taken for colics and gout. *Gynecological Aid* Compound infusion of needles used for "child-bed-fevers." Syrup taken by pregnant women with cough and poultice used for swollen breasts. *Hemorrhoid Remedy* Compound infusion of root taken for piles. *Kidney Aid* Taken for "weak back or kidneys." *Laxative* Taken as a gentle laxative. *Misc. Disease Remedy* Taken for gout, to break out measles, and for complications from mumps. *Orthopedic Aid* Taken for "weak back or kidneys." *Other* Skim turpentine off root decoction and use on deer's skin for drawing plaster. *Respiratory Aid* Compound infusion of needles in apple juice taken by ballplayers "for wind." Syrup taken for "catarrh (ulcer of the lungs)." *Sedative* Given for hysterics. *Stimulant* Compound infusion of root taken as a stimulant. *Tuberculosis Remedy* Tar used for consumption. *Urinary Aid* Syrup used as poultice for swollen testicles caused by mumps. *Venereal Aid* Syrup taken for chronic rheumatism and venereal disease. (80:49) **Choctaw** *Anthelmintic* Infusion of buds taken for worms. (as *P. mitis* 25:24) **Rappahannock** *Kidney Aid* Rolled, hardened sap used as pills and taken for kidney trouble. (161:27)
- *Other*—**Cherokee** *Incense & Fragrance* Needles or gum used to scent soap. (80:49)

Piperaceae, see *Peperomia*, *Piper*

Piper, Piperaceae

Piper methysticum G. Forst., Kava
- *Drug*—**Hawaiian** *Analgesic* Roots chewed for sharp, blinding headaches. *Cold Remedy* Plant and other plants pounded, squeezed, the resulting juice heated and taken for chills and hard colds. *Dermatological Aid* and *Eye Medicine* Roots chewed to prevent contagious diseases of all sorts, especially skin diseases and eye troubles. *Gastrointestinal Aid* Plant ashes and other ashes rubbed on children for a disorderly stomach. *Gynecological Aid* Plant pieces and other plants mixed with water and taken for weaknesses arising from vir-

ginity. Plant pieces and other plants mixed with coconut milk and taken for displacement of the womb. *Misc. Disease Remedy* Roots chewed to prevent contagious diseases of all sorts, especially skin diseases and eye troubles. *Oral Aid* Plant ashes and other ashes rubbed on children for thick white coatings on the tongue. *Pediatric Aid* Buds chewed by children for general debility. Plant ashes and other ashes rubbed on children with general weakness of the body, rubbed on children for a disorderly stomach, and for thick white coatings on the tongue. *Pulmonary Aid* Decoction of whole plant and other plants taken for lung and kindred troubles. *Sedative* Decoction of plant and other plants taken for sleeplessness. *Stimulant* Buds chewed by children for general debility. *Strengthener* Plant ashes and other ashes rubbed on children with general weakness of the body. *Urinary Aid* Infusion of plant, other plants, and coconut milk taken for difficulty in passing urine. (2:17)

Piper nigrum L., Black Pepper
- *Drug*—**Cherokee** *Dermatological Aid* Used as an astringent. *Stimulant* Used as a stimulus. (80:48)
- *Food*—**Cherokee** *Spice* Used to season food. (80:48) **Haisla & Hanaksiala** *Spice* Used for seasoning. (43:259)

Piperia, Orchidaceae

Piperia elegans (Lindl.) Rydb., Hillside Bogorchid
- *Food*—**Pomo, Kashaya** *Vegetable* Baked bulbs eaten like baked potatoes. (as *Habenaria elegans* 72:62)

Piperia sp.
- *Drug*—**Mahuna** *Dermatological Aid* Plant used for black widow and scorpion bites. *Eye Medicine* Plant used for trachoma. *Snakebite Remedy* Plant used for rattlesnake bites. (140:34)

Piperia unalascensis (Spreng.) Rydb., Alaska Rein Orchid
- *Food*—**Pomo, Kashaya** *Vegetable* Baked bulbs eaten like baked potatoes. (as *Habenaria unalascensis* 72:62)

Piptatherum, Poaceae

Piptatherum miliaceum (L.) Coss., Smilograss
- *Food*—**Paiute** *Staple* Roasted and ground into flour. (as *Oryzopsis miliacea* 167:244)

Pipturus, Urticaceae

Pipturus sp., Mamaki
- *Drug*—**Hawaiian** *Pediatric Aid* Seeds eaten by infants for general debility of the body. *Strengthener* Seeds eaten by expectant mothers for general debility of the body. Seeds eaten by infants for general debility of the body. (2:71)
- *Other*—**Hawaiian** *Tools* Wood made into clubs and used to beat the tapa. (2:71)

Pisum, Fabaceae

Pisum sativum L., Garden Pea
- *Food*—**Cherokee** *Vegetable* Peas used for food. (80:48) **Navajo, Ramah** *Vegetable* Used as a garden vegetable. (191:34) **Okanagan-Colville** *Unspecified* Seeds used for food. (188:106) **Papago** *Unspecified* Species used for food. **Pima** *Unspecified* Species used for food. (36:120)

Pittosporaceae, see *Pittosporum*

Pittosporum, Pittosporaceae

Pittosporum sp., Ho-a-wa
- *Drug*—**Hawaiian** *Tuberculosis Remedy* Fruit and other plants pounded and resulting liquid used as a wash for scrofulous neck swellings. (2:44)

Pityopsis, Asteraceae

Pityopsis graminifolia (Michx.) Nutt. var. *graminifolia*, Narrowleaf Silkgrass
- *Drug*—**Choctaw** *Oral Aid* Burned plant ashes used for mouth sores. (as *Chrysopsis graminea* 25:24) Plant ashes used as powder for mouth sores. (as *Chrysopsis graminifolia* 177:60) **Seminole** *Cold Remedy* Infusion of plant taken for colds. *Febrifuge* Infusion of plant taken for fevers. (as *Chrysopsis graminifolia* 169:283)

Plagiobothrys, Boraginaceae

Plagiobothrys arizonicus (Gray) Greene ex Gray, Arizona Popcornflower
- *Dye*—**Diegueño** *Red* Red coating on outside leaves and lower stems used as a red pigment to paint the body and face. (84:30)

Plagiobothrys fulvus var. *campestris* (Greene) I. M. Johnston, Fulvous Popcornflower
- *Food*—**Mendocino Indian** *Staple* Seeds used to make pinole. Seeds winnowed, parched, and flour eaten dry. *Unspecified* Crisp, tender shoots and flowers used as a sweet and aromatic food. (as *P. campestris* 41:382)
- *Dye*—**Mendocino Indian** *Red* Matter at the base of young leaves used by women and children to stain their cheeks crimson. (as *P. campestris* 41:382)
- *Other*—**California Indian** *Paint* Red matter around root used as rouge. (as *P. campestris* 118:55)

Plagiobothrys jonesii Gray, Mojave Popcornflower
- *Other*—**Kawaiisu** *Paint* Juice from the stem and root juncture used to paint the face "just for fun." (206:52)

Plagiobothrys nothofulvus (Gray) Gray, Rusty Popcornflower
- *Food*—**Yuki** *Vegetable* Young leaves eaten as greens. (49:85)
- *Other*—**Kawaiisu** *Paint* Juice from the stem and root juncture used to paint the face "just for fun." (206:52)

Plagiomnium, Mniaceae, moss

Plagiomnium insigne (Mitt.) Koponen, Badge Mnium
- *Drug*—**Bella Coola** *Antirheumatic (External)* Poultice of crushed "leaves" applied to infections and swellings. *Blood Medicine* Poultice of "leaves" used for blood blisters. *Breast Treatment* Poultice of "leaves" used for breast abscesses in women. *Dermatological Aid* Poultice of "leaves" used for boils. (184:196) **Oweekeno** *Antirheumatic (External)* Heated or cooled poultice of boiled plant and Sitka spruce pitch applied to sore and swollen joints. *Dermatological Aid* Poultice of boiled plant and Sitka spruce pitch applied to cuts for the swelling. *Internal Medicine* Plant dried, chewed for the juice, and swallowed for internal ailments. (43:52)

Plagiomnium juniperinum, Hair Cap Moss
- *Drug*—**Heiltzuk** *Antirheumatic (External)* Plant used as an anti-swelling medicine. (43:53)

Plantaginaceae, see *Plantago*

Plantago, Plantaginaceae

Plantago aristata Michx., Largebracted Plantain
- *Drug*—**Cherokee** *Analgesic* Poultice of leaf applied for headache. *Antidiarrheal* Infusion of root taken for dysentery and infusion given to check babies' bowels. *Antidote* Infusion used for poisonous bites and stings. *Burn Dressing* Poultice of wilted or scalded leaf applied to burns. *Dermatological Aid* Poultice used for blisters, ulcers, insect stings, and infusion used orally and as wash. *Eye Medicine* Juice used for sore eyes. *Gastrointestinal Aid* Taken for bowel complaints. *Gynecological Aid* Infusion taken to "check discharge" and used as a douche. *Orthopedic Aid* Compound infusion of leaf given to strengthen a child learning to walk. *Pediatric Aid* Compound infusion of leaf given to strengthen a child learning to walk. Infusion given to check babies' bowels. *Snakebite Remedy* Infusion used for snakebites. *Urinary Aid* Taken for "bloody urine." (80:50)

Plantago australis* ssp. *hirtella (Kunth) Rahn, Mexican Plantain
- *Drug*—**Tolowa** *Dermatological Aid* Poultice of leaves applied to cuts and boils. **Yurok** *Dermatological Aid* Poultice of steamed leaves applied to boils. (as *P. hirtella* 6:45)

Plantago cordata Lam., Heartleaf Plantain
- *Drug*—**Houma** *Burn Dressing* and *Dermatological Aid* Poultice of raw leaves and oil or grease used on cuts, sores, burns, and boils. (158:62)

Plantago lanceolata L., Narrowleaf Plantain
- *Drug*—**Cherokee** *Analgesic* Poultice of leaf applied for headache. *Antidiarrheal* Infusion of root taken for dysentery and infusion given to check babies' bowels. *Antidote* Infusion used for poisonous bites, stings, and snakebites. *Burn Dressing* Poultice of wilted or scalded leaf applied to burns. *Dermatological Aid* Poultice used for blisters, ulcers, insect stings, and infusion used orally and as wash. *Eye Medicine* Juice used for sore eyes. *Gastrointestinal Aid* Taken for bowel complaints. *Gynecological Aid* Infusion taken to "check discharge" and used as a douche. *Orthopedic Aid* Compound infusion of leaf given to strengthen a child learning to walk. *Pediatric Aid* Compound infusion of leaf given to strengthen a child learning to walk. Infusion given to check babies' bowels. *Snakebite Remedy* Infusion used for snakebites. *Urinary Aid* Taken for "bloody urine." (80:50) **Kawaiisu** *Ear Medicine* Infusion of leaves put in the ear for earaches. (206:52)
- *Food*—**Mendocino Indian** *Fodder* Plant used as fodder for cattle. (41:388)

Plantago macrocarpa Cham. & Schlecht., Seashore Plantain
- *Drug*—**Aleut** *Tonic* Decoction of root taken as a tonic. (8:428)
- *Food*—**Alaska Native** *Vegetable* Young, tender leaves used raw in salads or cooked as spinach. (85:43)

Plantago major L., Common Plantain
- *Drug*—**Abnaki** *Analgesic* Poultice of leaves applied for pain. *Antirheumatic (External)* Poultice of leaves applied to the foot for rheumatism or swellings. (144:172) **Algonquin, Quebec** *Poultice* Leaves used as poultices. (18:231) **Algonquin, Tête-de-Boule** *Burn Dressing* Poultice of leaves applied to burns. *Dermatological Aid* Poultice of leaves applied to wounds and contusions. (132:130) **Carrier** *Cough Medicine* Decoction of plant taken for coughs. *Gastrointestinal Aid* Decoction of plant taken for stomach problems. *Laxative* Decoction of plant taken as a laxative. (31:86) **Cherokee** *Analgesic* Poultice of leaf applied for headache. *Antidiarrheal* Infusion of root taken for dysentery and infusion given to check babies' bowels. *Antidote* Infusion used for poisonous bites, stings, and snakebites. *Burn Dressing* Poultice of wilted or scalded leaf applied to burns. *Dermatological Aid* Poultice used for blisters, ulcers, insect stings, and infusion used orally and as wash. *Eye Medicine* Juice used for sore eyes. *Gastrointestinal Aid* Taken for bowel complaints. *Gynecological Aid* Infusion taken to "check discharge" and used as a douche. *Orthopedic Aid* Compound infusion of leaf given to strengthen a child learning to walk. *Pediatric Aid* Compound infusion of leaf given to strengthen a child learning to walk. Infusion given to check babies' bowels. *Snakebite Remedy* Infusion used for snakebites. *Urinary Aid* Taken for "bloody urine." (80:50) **Chippewa** *Antirheumatic (External)* Poultice of chopped, fresh leaves applied for rheumatism. (53:362) *Dermatological Aid* Simple or compound poultice of chopped root or fresh leaf used for inflammations. (53:348) *Snakebite Remedy* Poultice of chopped, fresh leaves, and root applied to snakebites. (53:353) **Costanoan** *Febrifuge* Decoction of roots taken for fever. *Laxative* Decoction of roots taken for constipation. (21:11) **Delaware** *Gynecological Aid* Plant combined with other plant parts and used for female diseases. *Unspecified* Poultice of crushed leaves used medicinally. (176:37) **Delaware, Oklahoma** *Gynecological Aid* Compound containing plant used for "female diseases." *Unspecified* Poultice of crushed leaves used for unspecified ailments. (175:31, 82) **Delaware, Ontario** *Dermatological Aid* Poultice of crushed leaves applied to bruises. (175:66, 82) **Hesquiat** *Dermatological Aid* Poultice of leaves used for drawing out the pus from sores, cuts, and infections. (185:70) **Iroquois** *Blood Medicine* Infusion of roots with other roots used to purify the blood. (141:59) *Dermatological Aid* Poultice of leaves applied to sores. Poultice of boiled flower stems applied to abscesses. *Gastrointestinal Aid* Infusion of seeds taken to facilitate digestion and reduce intestinal inflammation. *Gynecological Aid* Infusion of seeds taken to regulate and shorten menses. (142:98) *Respiratory Aid* Infusion of roots with other roots used for difficult breathing caused by lower chest pains. (141:59) **Isleta** *Gastrointestinal Aid* Infusion of leaves used as a stomach tonic. (100:38) **Kawaiisu** *Ear Medicine* Infusion of leaves put in the ear for earaches. (206:52) **Keres, Western** *Blood Medicine* Roots used for blood medicine. (171:61) **Kwakiutl** *Dermatological Aid* Poultice of leaves applied to draw blisters on sores and swellings. (183:287) **Mahuna** *Dermatological Aid* Plant used to dislodge and draw out poisonous thorns and splinters. (140:16) **Meskwaki** *Burn Dressing* Infusion of leaves used for burns. *Dermatological Aid* Fresh leaf used for swellings. *Diuretic* Infusion of leaves taken for bowel troubles and as a urinary. *Gastrointestinal Aid* Infusion of leaves used for bowel troubles and as a urinary. (152:234, 235) **Mohegan** *Burn Dressing* Leaves bound over burns. *Dermatological Aid* Leaves bound over bruises. (174:266) Poultice of fresh leaves applied to insect bites, cuts, or swellings. (176:74, 130) *Snakebite*

Remedy Leaves bound over snakebites to draw out poison. (174:266) Poultice of fresh leaves applied to snake and insect bites to remove poison. (176:74) **Navajo, Ramah** *Disinfectant* Cold infusion of plant taken for "lightning infection." *Panacea* Root used as a "life medicine." (191:45) **Nitinaht** *Dermatological Aid* Poultice of moist leaves placed on cuts, boils, and open sores. *Disinfectant* Poultice of moist leaves placed on infections and boils. *Gastrointestinal Aid* Leaves chewed and swallowed for stomach ulcers. (186:115) **Ojibwa** *Burn Dressing* Poultice of soaked leaves bound on burns, scalds, and snakebites. *Dermatological Aid* Poultice of soaked leaves bound on bruises, sprains, sores, and bee stings. *Orthopedic Aid* Poultice of soaked leaves bound on bruises, sprains, or sores. *Snakebite Remedy* Poultice of soaked leaves bound on snakebites. (153:380, 381) *Unspecified* Poultice of pounded leaves applied for medicinal purposes. (5:2317) **Okanagan-Colville** *Dermatological Aid* Poultice of mashed leaves applied to kill the germs of sores. (188:111) **Paiute** *Cold Remedy* Decoction of root taken for colds. *Dermatological Aid* Poultice of leaves bound onto cuts and wounds to promote healing without scars. *Pulmonary Aid* Decoction of root taken for pneumonia. (180:119, 120) **Ponca** *Dermatological Aid* Hot leaves applied to foot to draw out thorns or splinters. (70:115) **Potawatomi** *Dermatological Aid* Poultice of heated leaf bound on swellings and inflammations. *Throat Aid* Decoction of root taken to lubricate throat for removal of lodged bone. (154:71, 72) **Rappahannock** *Febrifuge* Poultice of ground leaves used for fever. (161:25) **Shinnecock** *Dermatological Aid* Poultice of pounded leaves applied to draw out inflammation from sore spots. *Eye Medicine* Infusion of leaves used as a wash for sore eyes. (30:119) **Shoshoni** *Antirheumatic (External)* Compound poultice of leaves applied to wounds, swellings, and rheumatism. (180:119, 120) *Dermatological Aid* Poultice and infusion of whole plant used for battle bruises. Poultice of raw leaves with wild clematis applied to wounds. (118:43) Compound poultice of leaves applied for wounds, bruises, swellings, and boils. Poultice of mashed leaves applied to dropsical swellings and infections. *Disinfectant* Poultice of mashed leaves applied to dropsical swellings and infections. *Gastrointestinal Aid* Decoction of root taken for stomach trouble. *Kidney Aid* Poultice of mashed leaves applied to dropsical swellings and infections. (180:119, 120) **Shuswap** *Dermatological Aid* Poultice of mashed plants applied to sores. *Diaphoretic* Leaf rubbed on the body to produce sweating. (123:64) **Thompson** *Dermatological Aid* Poultice of chewed leaves used for sores and carbuncles. *Hemorrhoid Remedy* Poultice of chewed leaves used for hemorrhoids. (187:236) **Yurok** *Dermatological Aid* Poultice of steamed leaves applied to boils. (6:46)

- *Food*—**Acoma** *Unspecified* Young leaves used for food. (32:42) **Cherokee** *Vegetable* Leaves cooked and eaten as greens. (80:50) Cut leaves and stems cooked with fatback. (126:52) **Keres, Western** *Unspecified* Tender shoots used for food. (171:61) **Laguna** *Unspecified* Young leaves used for food. (32:42) **Mohegan** *Vegetable* Combined with pigweed, mustard, dock, and nettle and used as mixed greens. (176:83)
- *Other*—**Chippewa** *Protection* Powdered roots carried as protection against snakebites. (53:376) **Ojibwa** *Protection* Ground root always carried in the pockets to ward off snakes. (153:431)

Plantago maritima L., Goose Tongue

- *Food*—**Alaska Native** *Unspecified* Plant eaten fresh or cooked. *Winter Use Food* Plant canned for winter use. (85:45)

Plantago ovata Forsk., Desert Indianwheat

- *Drug*—**Pima** *Antidiarrheal* Cold infusion of seeds taken for diarrhea. (as *P. fastigiata* 47:96)
- *Food*—**Pima** *Fodder* Herbs used for fodder. (as *P. fastigiata* 47:96) **Pima, Gila River** *Unspecified* Seeds used for food. (as *P. insularis* 133:7)

Plantago patagonica Jacq., Woolly Plantain

- *Drug*—**Hopi** *Psychological Aid* Plant given to a person to make him more agreeable. (as *P. purshii* 42:349) Used to make a person more agreeable. (as *P. purshii* 200:37) Plant given to a person "to make him more agreeable." (as *P. purshii* 200:92) **Keres, Western** *Analgesic* Infusion of plant used for headaches. *Antidiarrheal* Infusion of plant used for diarrhea. (as *P. purshii* 171:61) **Navajo** *Gastrointestinal Aid, Laxative,* and *Pediatric Aid* Infusion of seeds given to babies when they "spoil" (colic or constipation). (as *P. purshii* 90:154) **Navajo, Ramah** *Dietary Aid* Cold infusion of plant parts taken to reduce appetite and prevent obesity. (as *P. purshii* 191:45) **Okanagan-Colville** *Dermatological Aid* Poultice of mashed leaves applied to sores. (188:111) **Zuni** *Antidiarrheal* Infusion of whole plant taken three times a day for bloody diarrhea. (27:377)
- *Food*—**Havasupai** *Porridge* Seeds ground and made into mush. (as *P. purshii* 197:242) **Navajo, Kayenta** *Porridge* Seeds made into mush and used for food. (as *P. purshii* 205:43) **Pima, Gila River** *Unspecified* Seeds used for food. (as *P. purshii* 133:7)
- *Other*—**Hopi** *Ceremonial Items* Plant used for participants of the clowning crew. (as *P. purshii* 42:349) **Navajo, Ramah** *Ceremonial Items* Infusion of root, leaves, and seeds blown over rattle at beginning of hoof rattle song. (as *P. purshii* 191:45)

Plantago rugelii var. *asperula* Farw., Blackseed Plantain

- *Drug*—**Menominee** *Burn Dressing* Poultice of fresh leaves applied to burn or any inflammation. (54:132) *Dermatological Aid* Poultice of fresh leaves applied to any inflammation. (54:131) Poultice of specific sides of leaf applied to swellings and other ailments. (151:46, 47)

Plantago sp., Plantain

- *Drug*—**Cree** *Burn Dressing* Chewed leaves used for burns and scalds. (16:492) **Iroquois** *Analgesic* Decoction of whole plant taken for chest and stomach pains. Poultice of hot leaves applied to bruises, swellings, or for pain. (87:438) *Antidiarrheal* Compound decoction of bark and roots taken for diarrhea. *Antiemetic* Compound decoction of bark and roots taken for vomiting. (87:437) *Antirheumatic (External)* Poultice of leaves applied for arthritis. (87:439) *Blood Medicine* Decoction of roots taken for the blood and pain. (87:438) *Burn Dressing* Poultice of leaves applied to burns or sores. *Dermatological Aid* Poultice of leaves applied to burns, sores, cuts, bruises, and wounds. Poultice of smashed leaves applied to swollen face caused by erysipelas. (87:437) Poultice of leaves applied to spider bites and for drawing sores. (87:438) *Febrifuge* Infusion of roots taken for fevers. *Gastrointestinal Aid* Compound decoction of bark and roots taken for cramps. (87:437) Decoction of whole plant taken for stomach pains. (87:438) *Gynecological Aid* Decoction of roots taken by women who cannot have babies. (87:437) Compound decoction of roots

taken for falling of the womb. Decoction of roots taken as an annual female medicine. *Heart Medicine* Compound of plants taken for heart trouble. (87:438) *Hemostat* Poultice of heated leaves applied for bleeding or cuts. *Orthopedic Aid* Decoction of roots used as wash for sore joints and sprains. (87:437) *Psychological Aid* Decoction of roots taken for nervous breakdown. *Pulmonary Aid* Decoction of whole plant taken for chest pains. (87:438) *Strengthener* Compound decoction of plants taken by runners and athletes for strength. (87:437) *Urinary Aid* Decoction of roots taken for the bladder and caused urine to flow. (87:438)
- *Food*—Pima, Gila River *Beverage* Seeds used to make a mucilaginous drink. *Porridge* Seeds used to make a mucilaginous mass and eaten. (133:5)

Plantago virginica L., Virginia Plantain
- *Other*—Kiowa *Ceremonial Items* Used to make garlands or wreaths worn by old men around their heads for dances as symbol of health. (192:51)

Platanaceae, see *Platanus*

Platanthera, Orchidaceae

Platanthera ciliaris (L.) Lindl., Yellow Fringed Orchid
- *Drug*—Cherokee *Analgesic* Cold infusion used for headache. *Antidiarrheal* Infusion of root taken every hour for "flux." (as *Habenaria ciliaris* 80:47) **Seminole** *Snakebite Remedy* Roots used for snakebites. (as *Habenaria ciliaris* 169:297)
- *Other*—Cherokee *Hunting & Fishing Item* Root used on hooks to make the fish bite. (as *Habenaria ciliaris* 80:47)

Platanthera dilatata (Pursh) Lindl. ex Beck **var. dilatata**, Scent-bottle
- *Drug*—Micmac *Urinary Aid* Root juice taken with water for gravel. (as *Havernaria dilatata* 156:317) **Shuswap** *Poison* Leaves considered poisonous. (as *Habenaria dilatata* 123:55)

Platanthera grandiflora (Bigelow) Lindl., Greater Purple Fringed Orchid
- *Other*—Iroquois *Protection* Decoction of smashed, dried roots taken to frighten away ghosts. (as *Habenaria fimbriata* 87:290)

Platanthera leucostachys Lindl., Bog Orchid
- *Drug*—Thompson *Analgesic* Plant used in the sweat bath for rheumatism and various joint and muscle aches. *Antirheumatic (External)* Heated decoction of plant used as a wash for rheumatism. Plant used in the sweat bath for rheumatism and various joint and muscle aches. (as *Habenaria leucostachys* 164:467) *Disinfectant* Decoction of whole plant used as a wash after the purifying sweat bath. (as *Habenaria leucostachys* 164:504) *Herbal Steam* Plant used in the sweat bath for rheumatism and various joint and muscle aches. Plant steamed in the sweat bath for rheumatism and other joint and muscle aches. (as *Habenaria leucostachys* 164:467) *Hunting Medicine* Plant used to wash guns "to insure good luck when hunting." *Love Medicine* Plant used in wash by women "hoping to gain a mate and have success in love." (as *Habenaria leucostachys* 164:506) *Orthopedic Aid* Plant used in the sweat bath for sprains, stiff and aching muscles and joints. (as *Habenaria leucostachys* 164:467)
- *Other*—Thompson *Good Luck Charm* Plant used in a wash by both sexes "to obtain riches and property." Plant used in wash to make young men "lucky, good looking and sweet smelling." (as *Habenaria leucostachys* 164:506)

Platanthera × *media* (Rydb.) Luer, Bog Orchid
- *Drug*—Potawatomi *Love Medicine* Women rubbed this plant on their cheek as a love charm to enable them to secure a good husband. (as *Habenaria dilatata* var. *media* 154:121)

Platanthera orbiculata (Pursh) Lindl. **var. orbiculata**, Large Roundleaved Orchid
- *Drug*—Iroquois *Dermatological Aid* Poultice of leaves applied to sores caused by scrofula and cuts. *Tuberculosis Remedy* Poultice of leaves bound to sore caused by scrofula. (as *Habenaria orbiculata* 87:289) **Montagnais** *Dermatological Aid* Poultice of leaves applied to blisters on hands or feet. (as *Habenaria orbiculata* 156:313)

Platanthera psycodes (L.) Lindl., Lesser Purple Fringed Orchid
- *Drug*—Iroquois *Analgesic* Infusion of single root and flowers given to young children for cramps. (as *Habenaria psycodes* 87:289) *Gynecological Aid* Infusion of root taken as a parturition medicine. *Panacea* Compound infusion taken or placed on injured part, a "Little Water Medicine." (as *Habenaria psycodes* 87:290) *Pediatric Aid* Infusion of single root and flowers given to young children for cramps. (as *Habenaria psycodes* 87:289)

Platanthera sparsiflora (S. Wats.) Schlechter **var. sparsiflora**, Canyon Bog Orchid
- *Food*—San Felipe *Starvation Food* Plant used as food in times of food shortage. (as *Habenaria sparsiflora* 32:30)

Platanthera stricta Lindl., Modoc Bog Orchid
- *Drug*—Kwakiutl *Love Medicine* Compound with plant used as a love charm. (as *Habenaria saccata* 183:275)

Platanus, Platanaceae

Platanus occidentalis L., American Sycamore
- *Drug*—Cherokee *Abortifacient* Compound decoction taken "for menstrual period." *Antidiarrheal* Taken for dysentery. (80:58) Infusion of inner bark taken for dysentery. (177:26) *Cathartic* Taken as a purgative. (80:58) Decoction of roots taken by women during menses as a cathartic. (177:26) *Cough Medicine* Infusion of inner bark taken for cough. *Dermatological Aid* Bark ooze used as wash for infected sores and infusion given for infant rash. *Emetic* Taken as an emetic. (80:58) Decoction of roots taken by women during menses as an emetic. (177:26) *Gastrointestinal Aid* Compound used in steam bath for indigestion or biliousness. *Gynecological Aid* Compound decoction used to aid in expelling afterbirth. (80:58) Decoction of roots taken by women during menses as an emetic and cathartic. (177:26) *Misc. Disease Remedy* Infusion of inner bark taken for measles. *Other* Compound infusion of bark taken for "bad disease." *Pediatric Aid* Infusion given for infant rash. *Urinary Aid* Taken for milky urine. (80:58) Infusion of inner barks taken for difficult urination with yellow discharge. (177:26) **Creek** *Tuberculosis Remedy* Bark used in medicine taken for pulmonary tuberculosis. (172:659) Decoction of bark taken for pulmonary tuberculosis. (177:26) **Delaware** *Cold Remedy* Infusion of three chips from east side of honey locust and sycamore trees

used as drinks for colds. *Throat Aid* Infusion of bark mixed with honey locust bark and used as a gargle for hoarseness and sore throat. (176:30) **Delaware, Oklahoma** *Cold Remedy* Compound infusion of bark taken for colds. (as *P. occidentalis* 175:25) *Throat Aid* Compound infusion of bark gargled for hoarseness and sore throat. (as *P. occidentalis* 175:25, 78) **Iroquois** *Antirheumatic (External)* Compound infusion of bark and roots used as foot soak for rheumatism. *Dermatological Aid* Compound decoction used for skin eruptions, scabs, and eczema. (87:348) **Mahuna** *Gastrointestinal Aid* Plant used for internal ulcers. *Respiratory Aid* Plant used for catarrh. (140:24) **Meskwaki** *Analgesic* Bark eaten for internal pains. *Blood Medicine* Infusion of bark taken as a blood purifier and used for colds. *Cold Remedy* Infusion of bark used for colds and to purify the blood. *Dermatological Aid* Compound containing bark used on knife or ax wounds. *Dietary Aid* Bark eaten to become fat. *Hemostat* Bark used for hemorrhages and lung troubles. *Misc. Disease Remedy* Infusion of bark used as wash to dry smallpox pustules and prevent scars. *Pulmonary Aid* Bark used for lung troubles and hemorrhages. (152:235)
- *Fiber*—**Cherokee** *Building Material* Wood used for lumber. (80:58)
- *Other*—**Cherokee** *Fasteners* Wood used to make buttons. (80:58)

Platanus racemosa Nutt., California Sycamore
- *Drug*—**Costanoan** *Panacea* Infusion of plant used as a general remedy. (21:20) **Diegueño** *Blood Medicine* Decoction of bark used as a tonic for the blood. (88:217) *Respiratory Aid* Decoction of bark taken for a week for asthma. (84:30) **Kawaiisu** *Unspecified* Infusion of bark taken for the relief of some indisposition. Decoction of small bark pieces taken for an unrecorded indisposition. (206:53)
- *Food*—**Costanoan** *Unspecified* Inner bark used for food. (21:249) **Kawaiisu** *Beverage* Small bark pieces boiled in water and drunk warm with sugar. (206:53)
- *Fiber*—**Cahuilla** *Building Material* Limbs and branches used in house construction. (15:105)
- *Other*—**Costanoan** *Cooking Tools* Leaves used to wrap bread during baking. (21:249)

Platystemon, Papaveraceae

Platystemon californicus Benth., California Creamcups
- *Food*—**Mendocino Indian** *Vegetable* Green leaves eaten as greens. (41:351)

Pleomele, Agavaceae

Pleomele aurea (Mann) N. E. Br., Golden Hala Pepe
- *Drug*—**Hawaiian** *Febrifuge* Buds, bark, root bark, and other plants pounded and resulting liquid taken for chills and high fever. *Pulmonary Aid* Leaves, bark, root bark, and other plants pounded and resulting liquid taken for lung troubles. *Respiratory Aid* Leaves, bark, root bark, and other plants pounded and resulting liquid taken for asthma. (as *Dracaena aurea* 2:42)

Pleopeltis, Polypodiaceae, fern

Pleopeltis polypodioides (L.) Andrews & Windham **ssp. *polypodioides***, Resurrection Fern
- *Drug*—**Houma** *Analgesic* Decoction of fronds taken for headaches. *Oral Aid* Cold decoction of fronds used as a wash for babies' sore mouth or thrush. Decoction of fronds used for bleeding gums. *Pediatric Aid* Cold decoction of fronds used as a wash for babies' sore mouth or thrush. *Vertigo Medicine* Decoction of fronds taken for dizziness. (as *Marginaria polypodioides* 158:55)

Pleurotus, Tricholomataceae, fungus

Pleurotus ostreatus, Oyster Mushroom
- *Food*—**Pomo, Kashaya** *Vegetable* Cooked on hot stones, baked in the oven, or fried. (72:131)

Pluchea, Asteraceae

Pluchea foetida (L.) DC., Stinking Camphorweed
- *Drug*—**Choctaw** *Febrifuge* Decoction of leaves taken "during attacks of fever." (25:23) Decoction of leaves taken for fevers. (177:63)

Pluchea sericea (Nutt.) Cav., Arrow Weed
- *Drug*—**Havasupai** *Throat Aid* Leaves chewed or boiled for throat irritations. (162:285) **Paiute** *Antidiarrheal* Decoction of root taken for diarrhea, especially bloody diarrhea. *Gastrointestinal Aid* Raw root chewed or decoction of root taken for indigestion. (180:120) **Pima** *Antidiarrheal* Decoction of roots taken for diarrhea. (47:105) *Dermatological Aid* Infusion of root bark used as a wash for the face and sore eyes. (as *P. borealis* 146:79) *Eye Medicine* Decoction of roots used for sore eyes. (47:105) Infusion of bark used as a wash for the face and sore eyes. (as *P. borealis* 146:79) *Gastrointestinal Aid* Decoction of roots taken for stomachaches. *Pediatric Aid* and *Sedative* Poultice of roots applied to soothe nervous child that cried while sleeping. *Veterinary Aid* Poultice of chewed roots applied to snakebites on horses. (47:105)
- *Food*—**Cahuilla** *Unspecified* Roots roasted and eaten. (15:105) **Pima** *Forage* Plants browsed by deer, horses, and cattle. (47:105)
- *Fiber*—**Cahuilla** *Building Material* Long, slender, pliable stems with leaves used as a roofing material. Long, slender, pliable stems with leaves interwoven with stronger materials into walls of houses. Used in the construction of ramadas, windbreaks, fences, and granaries. (15:105) **Havasupai** *Building Material* Plant used for thatch in building houses. *Furniture* Straight stems used to form the back of the cradleboard. (as *Tessaria sericea* 197:249) **Hualapai** *Building Material* Shafts used to make thatch for houses. *Furniture* Shafts used to make cradleboard beds. (195:27) **Luiseño** *Building Material* Plant formerly used to roof houses. (as *P. borealis* 155:228)
- *Other*—**Cahuilla** *Hunting & Fishing Item* Used to make arrow shafts. (15:105) **Havasupai** *Hunting & Fishing Item* Stems used to make arrows, especially for children. *Tools* Sticks used to peg a hide to the ground while stretching it during tanning. (as *Tessaria sericea* 197:249) **Hualapai** *Hunting & Fishing Item* Shafts used to make cradleboard beds, arrow shafts, and thatch for houses. (195:27) **Luiseño** *Hunting & Fishing Item* Plant used to make arrows. (as *P. borealis* 155:206) Plant sometimes used to make arrows. (as *P. borealis* 155:228) **Papago** *Tools* Used as a spindle for spinning cotton. (34:59)

***Pluchea* sp.**, Indian Balm
- *Drug*—**Houma** *Febrifuge* Decoction of dried plant taken for fevers. *Hemorrhoid Remedy* Decoction of dried plant taken for piles. (158:63)

Plumbaginaceae, see *Limonium, Plumbago*

Plumbago, Plumbaginaceae

Plumbago zeylanica L., Wild Leadwort
- **Drug**—**Hawaiian** *Antirheumatic (External)* Poultice of pounded bark, leaves, and roots applied to swollen parts of the body. *Dermatological Aid* Leaves, stems, other plant parts, and fruit juice made into a paste and applied to sores. (as *Chunbago zelanica* 2:25)

Poaceae, see *Agropyron, Agrostis, Alopecurus, Andropogon, Anthoxanthum, Aristida, Arundinaria, Arundo, Avena, Bambusa, Beckmannia, Bothriochloa, Bouteloua, Bromus, Buchloe, Calamagrostis, Calamovilfa, Catabrosa, Cenchrus, Chasmanthium, Cinna, Coix, Cynodon, Deschampsia, Dichanthelium, Digitaria, Distichlis, Echinochloa, Elymus, Elytrigia, Epicampes, Eragrostis, Eriochloa, Erioneuron, Festuca, Glyceria, Hierochloe, Hilaria, Hordeum, Koeleria, Lamarckia, Leymus, Lolium, Melica, Monroa, Muhlenbergia, Oryza, Oryzopsis, Panicum, Pascopyrum, Paspalidium, Paspalum, Pennisetum, Phalaris, Phleum, Phragmites, Piptatherum, Poa, Polypogon, Pseudoroegneria, Puccinellia, Saccharum, Schedonnardus, Schizachyrium, Sitanion, Sorghum, Spartina, Sporobolus, Stenotaphrum, Stipa, Tridens, Trisetum, Triticum, Vulpia, Zea, Zizania*

Poaceae sp., Grass
- **Fiber**—**Tanana, Upper** *Clothing* Blades rubbed until soft, peat moss and squirrel's nest material placed in a cradle for a diaper. Used as baby diapers. *Mats, Rugs & Bedding* Dried plant placed on top of spruce boughs and used as a mattress and dog bedding. Blades placed on the floor of sweat houses and camp shelters to sit on. Used for bedding and insulation in footgear. Used to weave mats. (as Gramineae 102:8)
- **Other**—**Tanana, Upper** *Hunting & Fishing Item* Blades placed under a trap to keep it from freezing to the ground and over it to hide the trap. Blades used to make duck-hunting blinds. Used to conceal traps. Used in the construction of waterfowl blinds. *Protection* Used for bedding and insulation in footgear. *Tools* Bunches placed on trees as trail markers. Used as trail markers. (as Gramineae 102:8)

Poa, Poaceae

Poa arida Vasey, Plains Bluegrass
- **Food**—**Gosiute** *Unspecified* Seeds used for food. (as *P. californica* 39:377)

Poa fendleriana (Steud.) Vasey, Mutton Grass
- **Drug**—**Hopi** *Ceremonial Medicine* Pollen used in prayer medicine. (42:350)
- **Food**—**Havasupai** *Bread & Cake* Seeds parched, ground fine, boiled, thickened, made into balls, and eaten as dumplings. Seeds ground, kneaded into a thick paste, rolled into little balls, boiled, and eaten as marbles. (197:66) *Staple* Seeds ground and eaten as a ground or parched meal. (197:67) *Unspecified* Seeds used for food. (197:210)

Poa fendleriana (Steud.) Vasey **ssp. *fendleriana***, Skyline Bluegrass
- **Food**—**Navajo, Ramah** *Fodder* Used for sheep and horse feed. (as *P. longiligula* 191:17)

Poa secunda J. Presl, Sandberg Bluegrass
- **Food**—**Gosiute** *Unspecified* Seeds used for food. (as *P. tenuifolia* 39:377)

Poa sp., Bluegrass
- **Fiber**—**Eskimo, Alaska** *Clothing* Fine leaves and stems used in the past to line skin boots. (1:34) **Eskimo, Inuktitut** *Clothing* Dried leaves used for boot insoles. *Mats, Rugs & Bedding* Dried, split leaves used for weaving. Dried leaves used for winter bedding for dogs. (202:189)
- **Other**—**Eskimo, Inuktitut** *Paper* Leaves used to dry hands. (202:189)

Podophyllum, Berberidaceae

Podophyllum peltatum L., May Apple
- **Drug**—**Cherokee** *Anthelmintic* Root used as anthelmintic. *Antirheumatic (Internal)* Root soaked in whisky and taken for rheumatism and as a purgative. *Cathartic* Boiled root eaten as a purgative. (80:44) Decoction of root boiled into a syrup, made into pills, and given as a purgative. (203:74) *Dermatological Aid* Powdered root used on ulcers and sores. *Ear Medicine* "Drop of juice of fresh root" put in ear for deafness. (80:44) Juice of fresh root dropped into the ear for deafness. (203:74) *Laxative* Powdered root eaten "to correct constipation." *Poison* Root joints considered poisonous. (80:44) **Delaware** *Laxative* Roots used to make a laxative. *Tonic* Roots used to make a spring tonic. (176:38) **Delaware, Oklahoma** *Laxative* Root used to make a laxative. *Love Medicine* Plant used as a love charm. *Tonic* Root used to make a spring tonic. (175:32, 78) **Iroquois** *Cathartic* Cold infusion of smashed root taken or raw root chewed for a strong physic. (87:331) *Ceremonial Medicine* Decoction of leaves with other plants used as medicine to soak corn seeds before planting. (196:19) *Dermatological Aid* Compound decoction of roots taken for boils. *Laxative* Decoction or infusion of roots taken or raw root chewed as a laxative. *Poison* Root considered poisonous. *Strengthener* Compound decoction of plants taken to increase strength. (87:331) *Veterinary Aid* Seeds and pulp of fruit placed in cut of atrophied shoulder muscle of horse. (87:330) Decoction of plant used as a laxative for horses bound up on green grass. (87:331) **Meskwaki** *Antirheumatic (Internal)* Root used for rheumatism and as a physic. *Cathartic* Compound containing root used as a physic and for rheumatism. *Emetic* Decoction of root taken as an emetic. (152:206)
- **Food**—**Cherokee** *Fruit* Fruit used for food. (80:44) Ripe fruit used for food. (126:32) **Chippewa** *Fruit* Fruit considered very palatable. (71:130) **Iroquois** *Bread & Cake* Fruit mashed, made into small cakes, and dried for future use. *Dried Food* Raw or cooked fruit sun or fire dried and stored for future use. *Fruit* Dried fruit taken as a hunting food. *Sauce & Relish* Dried fruit cakes soaked in warm water and cooked as a sauce or mixed with corn bread. (196:129) **Menominee** *Fruit* Fresh, ripe fruits eaten. *Preserves* Fresh, ripe fruits preserved. (151:62) **Meskwaki** *Fruit* Fresh fruits eaten raw. *Preserves* Fruits cooked into a conserve. (152:256)
- **Other**—**Cherokee** *Insecticide* Root ooze used to soak corn before planting to keep off crows and insects. (80:44) **Iroquois** *Fertilizer* Root mixed with water for sprouting corn, a "corn medicine." (87:331) **Menominee** *Insecticide* Decoction of whole plant sprinkled on potato plants to kill potato bugs. (151:25, 26)

Pogogyne, Lamiaceae

Pogogyne douglasii* ssp. *parviflora (Benth.) J. T. Howell, Douglas's Mesamint

- **Drug**—Concow *Analgesic* and *Gastrointestinal Aid* Leaves used as a counterirritant for stomach and bowel pains. (as *P. parviflora* 41:384)
- **Food**—Concow *Substitution Food* Leaves used as a substitute for tea. **Numlaki** *Staple* Seeds used as a sweet, aromatic ingredient of wheat and barley pinole. **Yuki** *Staple* Seeds used as a sweet, aromatic ingredient of wheat and barley pinole. (as *P. parviflora* 41:384)
- **Other**—Mendocino Indian *Insecticide* Plant used for fleas. (as *P. parviflora* 41:384)

Polanisia, Capparaceae

Polanisia dodecandra ssp. *trachysperma* (Torr. & Gray) Iltis, Sandyseed Clammyweed

- **Food**—Pueblo *Dried Food* Young plants boiled, pressed, rolled into balls, dried, and stored for winter use. *Soup* Plant made into a stew with wild onions, wild celery, tallow, or bits of meat. *Unspecified* Young plants boiled, pressed, rolled into balls, and eaten. *Vegetable* Young plants boiled with a pinch of salt and eaten as greens. (as *P. trachysperma* 32:25)
- **Other**—Isleta *Ceremonial Items* Dried, rubbed leaves rolled in corn husks to make ceremonial cigarettes. *Smoke Plant* Dried, rubbed leaves rolled in corn husks to make ceremonial cigarettes. (as *P. trachycarpa* 100:38) **Zuni** *Ceremonial Items* Switches, roots, and blossoms used ceremonially. When the Cactus fraternity returned to their chamber from the last dance at sunset, they were whipped with switches and then roots and blossoms were chewed and ejected over the bodies of the whipped people. (as *P. trachysperma* 166:96)

Polemoniaceae, see *Collomia, Eriastrum, Gilia, Ipomopsis, Leptodactylon, Linanthus, Loeseliastrum, Navarretia, Phlox, Polemonium*

Polemonium, Polemoniaceae

Polemonium elegans Greene, Elegant Jacobsladder

- **Drug**—Thompson *Dermatological Aid* Compound decoction containing several species used as a wash for the head and hair. (164:467)

Polemonium pulcherrimum ssp. *lindleyi* (Wherry) V. Grant, Lindley's Polemonium

- **Drug**—Thompson *Dermatological Aid* Decoction of plant used as a wash for the head and hair. (as *P. humile* 164:467)

Polemonium reptans L., Greek Valerian

- **Drug**—Meskwaki *Cathartic* Compound containing root used as powerful physic. *Diuretic* Compound containing root used as powerful urinary. (152:235, 236)

Poliomintha, Lamiaceae

Poliomintha incana (Torr.) Gray, Hoary Rosemarymint

- **Drug**—Comanche *Adjuvant* Leaves chewed by medicine woman retaining other drugs in her mouth to sweeten the taste. Used to increase the efficacy of other medicine plants. (as *Poliomentha ineana* 99:7) **Hopi** *Antirheumatic (External)* Plant used for rheumatism. *Ear Medicine* Plant used for ear trouble. (42:351) **Navajo, Kayenta** *Dermatological Aid* Plant used for sores. (205:41) **Tewa** *Antirheumatic (External)* Plant used for rheumatism. *Ear Medicine* Plant used for ear trouble. (42:351)
- **Food**—Hopi *Dried Food* Dried plant stored for winter use. (42:351) Dried for winter use. (200:91) *Spice* Flowers boiled with a certain mush to give it a flavor. (190:165) Flowers used as flavoring. (200:91) *Unspecified* Plant eaten raw or boiled. (42:351) Plant dipped in salted water and eaten. (56:19) Flowers eaten. Flowers eaten and also boiled with a certain mush to give it a flavor. (190:165) Eaten raw or boiled. (200:91) **Tewa** *Dried Food* Dried plant stored for winter use. *Spice* Flowers used as flavoring. *Unspecified* Plant eaten raw or boiled. (42:351)

Polygalaceae, see *Polygala*

Polygala, Polygalaceae

Polygala alba Nutt., White Milkwort

- **Drug**—Sioux *Ear Medicine* Decoction of root used for earache. (19:18)

Polygala cornuta Kellogg, Sierran Milkwort

- **Drug**—Miwok *Analgesic* Decoction used for pains. *Cold Remedy* Decoction used for colds. *Cough Medicine* Decoction used for coughs. *Emetic* Decoction used as an emetic. (12:172)

Polygala lutea L., Orange Milkwort

- **Drug**—Choctaw *Dermatological Aid* Poultice of dried blossoms mixed with water applied to swellings. (25:24) Poultice of dried blossoms applied to swellings. (177:35) **Seminole** *Antirheumatic (External), Blood Medicine,* and *Heart Medicine* Plant used for sapiyi sickness: heart palpitations, yellow skin, body swelling, and short breath. (169:264) *Other* Complex infusion of whole plant taken for chronic conditions. (169:272) *Respiratory Aid* Plant used for sapiyi sickness: heart palpitations, yellow skin, body swelling, and short breath. (169:264)

Polygala paucifolia Willd., Gaywings

- **Drug**—Iroquois *Dermatological Aid* Infusion of plant taken and poultice of leaves applied to abscesses on limbs. (87:368) Decoction of plant used as a wash for boils and syphilitic sores. (87:369) *Orthopedic Aid* Compound poultice of plants applied to sore legs. (87:368) *Pediatric Aid* and *Venereal Aid* Decoction of plant used as a wash for babies with syphilitic sores. (87:369)

Polygala polygama Walt., Racemed Milkwort

- **Drug**—Montagnais *Cough Medicine* Decoction of plant used as a cough medicine. (156:314)

Polygala rugelii Shuttlw. ex Chapman, Yellow Milkwort

- **Drug**—Seminole *Antirheumatic (External), Blood Medicine, Heart Medicine,* and *Respiratory Aid* Plant used for sapiyi sickness: heart palpitations, yellow skin, body swelling, and short breath. (169:264) *Snakebite Remedy* Infusion of plant taken for snakebites. (169:297)

Polygala senega L., Seneca Snakeroot

- **Drug**—Blackfoot *Respiratory Aid* Decoction of roots used for respiratory diseases. (97:42) **Cherokee** *Abortifacient* Used as an emmenagogue. *Antirheumatic (Internal)* Used for rheumatism. *Cathartic* Infusion of root or root powder taken as a cathartic. *Cold Remedy* Used for colds. *Diaphoretic* Used as a sudorific. *Diuretic* Used as a diuretic and cathartic. *Expectorant* Infusion of root or root powder taken as an expectorant. *Kidney Aid* Used for "dropsy" and infusion of root or root powder used as an expectorant. *Other* Used for "inflammatory complaints" and swellings. *Pulmonary Aid* Used for pleurisy and croup. *Snakebite Remedy* Root chewed, "sufficient quantity" swallowed, and wound poulticed with the rest. (80:55) **Chippewa** *Anticonvulsive* Compound infusion or decoction of root taken for "fits." (51:63, 64) Compound decoction of root taken for convulsions. (53:336) *Heart Medicine* Compound decoction of root prepared ceremonially and taken for heart trouble. (53:338) *Hemostat* Compound infusion or decoction of root used on bleeding wounds. (51:63, 64) Compound decoction of root used on bleeding wounds. (53:336) *Stimulant* Compound infusion or decoction of root taken or used externally as stimulant. (51:63, 64) Compound decoction of root taken as a stimulant. (53:364) *Tonic* Compound decoction of root or dried root alone taken as a tonic. (53:336) *Unspecified* Roots carried for general health and safe journeys. (53:376) **Cree, Woodlands** *Blood Medicine* Infusion of blooms taken as a blood medicine. *Oral Aid* Root used for sore mouths. *Panacea* Powdered roots added to a many herb remedy and used for various ailments. *Throat Aid* Chewed root juice swallowed for sore throats. *Toothache Remedy* Root applied directly to a tooth for toothache. (109:51) **Malecite** *Cold Remedy* Dried roots chewed for colds. (116:244) **Meskwaki** *Heart Medicine* Decoction of root taken for heart trouble. (152:236) **Micmac** *Cold Remedy* Root used for colds. (40:59) **Ojibwa** *Unspecified* Plant used for medicinal purposes. (135:235) **Ojibwa, South** *Cold Remedy* Decoction of root used for colds. *Cough Medicine* Decoction of root used for cough. *Gastrointestinal Aid* Infusion of leaves taken to "destroy water bugs that have been swallowed." *Throat Aid* Infusion of leaves taken for sore throat. (91:199)

Polygala verticillata L., Whorled Milkwort
- **Drug**—**Cherokee** *Other* Infusion taken for "summer complaint." (80:45) **Iroquois** *Other* and *Pediatric Aid* Infusion of plant given to babies with "summer complaint." (87:369)

Polygonaceae, see *Chorizanthe, Coccoloba, Eriogonum, Fagopyrum, Oxyria, Polygonum, Rheum, Rumex*

Polygonatum, Liliaceae

Polygonatum biflorum (Walt.) Ell., King Solomon's Seal
- **Drug**—**Cherokee** *Antidiarrheal* Taken for dysentery. *Breast Treatment* Used for breast diseases. *Dermatological Aid* Hot poultice of bruised root used to draw risings or carbuncles. *Gastrointestinal Aid* Infusion of roasted roots taken for stomach trouble. *Gynecological Aid* Taken by females afflicted with "whites or profuse menstruation." *Pulmonary Aid* Used for lung diseases. *Tonic* Root used as a mild tonic for "general debility." (80:56) **Chippewa** *Sedative* Plant used to insure sound sleep. (71:126) **Menominee** *Analgesic* Compound poultice of boiled, mashed root applied for sharp pains. (54:134) *Stimulant* Smudge of compound containing dried root used to revive unconscious patient. (151:41) **Meskwaki** *Stimulant* Root heated on coals and fumes inhaled by unconscious patient to revive him. (152:230) **Ojibwa** *Cathartic* Root used as a physic and decoction used as cough remedy. *Cough Medicine* Decoction of root used as a cough remedy and root used as a physic. (153:374) **Rappahannock** *Dermatological Aid* Decoction of roots applied as a salve to cuts, bruises, and sores. *Orthopedic Aid* Decoction of roots applied as a salve to cuts and bruises. (161:30)
- **Food**—**Cherokee** *Bread & Cake* Roots dried, beaten into flour, and used to make bread. (80:56) Roots used to make bread. (126:47) *Spice* Roots ground and used as salt. (80:56) *Unspecified* Young growth boiled, fried, and eaten. (204:252) *Vegetable* Leaves cooked and eaten as greens. (80:56) Stems and leaves parboiled, rinsed, fried with grease and salt until soft, and eaten as a potherb. *Winter Use Food* Stems and leaves mixed with bean salad and wanegedum (angelico, *Ligusticum canadense*), blanched, and boiled for 3 hours in a can. (126:47) Rhizomes boiled and eaten especially during winter. (204:252)
- **Other**—**Chippewa** *Incense & Fragrance* Root burned for the fragrance. When one burned it in a room before going to bed, it insured sound sleep and caused one to awaken refreshed, rested, and feeling young, so it is said. (71:126)

Polygonatum biflorum var. *commutatum* (J. A. & J. H. Schultes) Morong, King Solomon's Seal
- **Drug**—**Chippewa** *Analgesic* Decoction of root steamed and inhaled for headache. *Herbal Steam* Decoction of root sprinkled on hot stones and used as an herbal steam for headache. (as *P. commutatum* 53:336) *Misc. Disease Remedy* Roots used to prevent measles and other diseases. (as *P. commutatum* 71:125)
- **Other**—**Chippewa** *Incense & Fragrance* Root burned, especially in the house, for the pleasant fragrance. (as *P. commutatum* 71:125)

Polygonatum pubescens (Willd.) Pursh, Hairy Solomon's Seal
- **Drug**—**Abnaki** *Antihemorrhagic* Used by women for spitting up blood. Used for spitting up blood. (144:154) Decoction of plant taken for spitting up blood. (144:174) **Iroquois** *Carminative* Compound decoction of plant taken for "gas on the stomach." (87:284) *Eye Medicine* Infusion of roots used to wash the eyes for snow-blindness. *Hunting Medicine* Compound infusion of roots used to "get a fish on each hook, every cast." (87:285)

Polygonum, Polygonaceae

Polygonum alpinum All., Alaska Wild Rhubarb
- **Drug**—**Tanana, Upper** *Cold Remedy* Raw roots and stem bases chewed for colds. *Cough Medicine* Raw roots and stem bases chewed for coughs. (as *P. alaskanum* 102:15)
- **Food**—**Alaska Native** *Pie & Pudding* Chopped leaves and stems added to a thick pudding of flour and sugar and eaten. *Vegetable* Young stems cut into small pieces and used in the same manner as domesticated rhubarb. Young, tender leaves mixed with other greens and cooked in boiling water. (as *P. alaskanum* 85:47) **Eskimo, Arctic** *Beverage* Juice sweetened and used to make a beverage. *Pie & Pudding* Stems stewed and used as pie filling. *Unspecified* Stems stewed and eaten. (as *P. alaskanum* 128:26) **Eskimo, Inupiat** *Beverage* Stalks boiled, strained, and juice used as a beverage. *Dessert* Raw stalks eaten in a garden rhubarb dessert. Stored stalks boiled, mixed with cranberries, raisins, dried apples, or peaches and eaten as a dessert. *Sauce & Relish* Stalks boiled into a sauce and used on cooked fish. Stalks boiled, mixed with oil and sugar, and used as a sauce for dumplings, cake, or sweet breads. *Unspecified* Fresh, chopped stalks mixed with whitefish or pike eggs and livers, oil, and sugar and eaten. *Vegetable* Fresh stalks eaten raw with seal oil and meat or fish. Raw stalks eaten with peanut butter or cut up in a salad. Leaves boiled and eaten as

hot greens. *Winter Use Food* Stalks boiled and stored in a barrel for winter use. (as *P. alaskanum* 98:45) **Koyukon** *Unspecified* Plant cooked and eaten. (as *P. alaskanum* 119:56) **Tanana, Upper** *Frozen Food* Leaves and stems frozen for future use. *Vegetable* Stems and leaves boiled with sugar and flour or in whitefish broth, grease, and sugar and eaten. Stems and leaves eaten raw. (as *P. alaskanum* 102:15)

Polygonum amphibium L., Water Knotweed
- *Drug*—**Cree, Woodlands** *Oral Aid* Poultice of fresh roots applied directly to blisters in the mouth. *Panacea* Powdered roots added to a many herb remedy and used for various ailments. (109:51) **Okanagan-Colville** *Cold Remedy* Infusion of dried, pounded roots taken or raw root eaten for chest colds. (188:113)
- *Other*—**Thompson** *Hunting & Fishing Item* Flowers used as bait for trout. (164:515)

Polygonum amphibium var. *emersum* Michx., Longroot Smartweed
- *Drug*—**Meskwaki** *Antidiarrheal* Infusion of leaves and stems used for children with flux. *Gynecological Aid* Compound decoction of root taken for injured womb. *Oral Aid* Root used for mouth sores. *Pediatric Aid* Infusion of leaves and stems used for children with flux. (as *P. muhlenbergii* 152:236) **Ojibwa** *Analgesic* Infusion of plant taken for stomach pain. *Gastrointestinal Aid* Infusion of plant used for stomach pain. *Hunting Medicine* Plant used as hunting medicine. (as *P. muhlenbergii* 153:381) Dried flowers included in the hunting medicine and smoked to attract deer to the hunter. (as *P. muhlenbergii* 153:431)
- *Food*—**Lakota** *Unspecified* Species used for food. (as *P. coccineum* 139:54) **Sioux** *Sauce & Relish* Young shoots eaten in the spring as a relish. (as *P. emersum* 19:18)

Polygonum amphibium var. *stipulaceum* Coleman, Water Smartweed
- *Drug*—**Potawatomi** *Unspecified* Root used for unspecified ailments. (154:72)

Polygonum arenastrum Jord. ex Boreau, Ovalleaf Knotweed
- *Drug*—**Iroquois** *Gynecological Aid* Decoction of whole plant used for miscarriage injuries. *Love Medicine* Powdered, dry root placed in other person's tea as a love medicine. *Orthopedic Aid* Decoction of whole plant used for lame back. *Veterinary Aid* Decoction of plant mixed with feed and given to heifers to restore their milk. (87:314)

Polygonum argyrocoleon Steud. ex Kunze, Silversheath Knotweed
- *Food*—**Cocopa** *Dried Food* Seeds parched, ground, and eaten. (37:187)

Polygonum aviculare L., Prostrate Knotweed
- *Drug*—**Cherokee** *Analgesic* Taken for painful urination. Infusion mixed with meal used as poultice for pain. *Antidiarrheal* Infusion of root given to children for diarrhea. *Dermatological Aid* Used for "scaldhead." *Pediatric Aid* Infusion of root given to children for diarrhea. Leaves rubbed on children's thumb to prevent thumb sucking. *Poison* Used to poison fish and infusion mixed with meal used as poultice for pain. *Poultice* Used as a poultice for "swelled and inflamed parts." *Urinary Aid* Taken for "gravel," painful urination, and bloody urine. (80:55) **Choctaw** *Gynecological Aid* Strong infusion of whole plant taken freely to prevent abortion. (28:286) **Iroquois** *Antidiarrheal* Infusion of plant and another plant given to children for diarrhea. (141:40) *Dermatological Aid* Compound poultice of raw plants applied to cuts and wounds. *Orthopedic Aid* and *Pediatric Aid* Compound decoction taken and poultice used for baby's broken coccyx. (87:313) Infusion of plant and another plant given to children for diarrhea. (141:40) **Mendocino Indian** *Dermatological Aid* Decoction of whole plant used as an astringent. (41:345) **Navajo, Ramah** *Analgesic* and *Gastrointestinal Aid* Warm infusion of plant taken for stomachache. (191:23) **Thompson** *Antidiarrheal* and *Pediatric Aid* Decoction of whole plant taken, especially by children, for diarrhea. (187:238)

Polygonum bistorta L., Meadow Bistort
- *Drug*—**Aleut** *Tonic* Root used as a tonic. (189:263)
- *Food*—**Eskimo, Inupiat** *Unspecified* Roots eaten raw and cooked. *Vegetable* Leaves preserved in seal oil and eaten with any meat or eaten raw in salads. (98:19)

Polygonum bistorta var. *plumosum* (Small) Boivin, Meadow Bistort
- *Food*—**Alaska Native** *Dietary Aid* Leaves rich in vitamin C and provitamin A. *Soup* Roots boiled and added to stews. *Unspecified* Roots boiled, mixed with seal oil, and eaten. *Vegetable* Leaves mixed with other greens, cooked, and eaten. (85:49)

Polygonum bistortoides Pursh, American Bistort
- *Drug*—**Miwok** *Dermatological Aid* Poultice of root used on sores and boils. (12:172)
- *Food*—**Blackfoot** *Soup* Roots used in soups and stews. (97:33) Roots used in soups and stews. (114:278) **Cheyenne** *Unspecified* Fresh roots boiled with meat. (as *Bistorta bistortoides* 76:173) Species used for food. (83:46) Roots formerly used for food. (83:32)

Polygonum careyi Olney, Carey's Smartweed
- *Drug*—**Potawatomi** *Cold Remedy* and *Febrifuge* Infusion of entire plant taken for cold accompanied by fever. (154:72)

Polygonum cuspidatum Sieb. & Zucc., Japanese Knotweed
- *Food*—**Cherokee** *Vegetable* Cooked leaves used for food. (126:53)

Polygonum densiflorum Meisn., Denseflower Knotweed
- *Drug*—**Hawaiian** *Blood Medicine* Infusion of pounded whole plants and other plants strained and taken to purify the blood. (as *P. glabrum* 2:48)

Polygonum douglasii Greene, Douglas's Knotweed
- *Food*—**Klamath** *Porridge* Ground, parched seeds used to make meal and eaten dry or mixed with water and boiled. (45:95) **Montana Indian** *Staple* Parched seeds made into meal. (19:18)

Polygonum douglasii ssp. *johnstonii* (Munz) Hickman, Johnston's Knotweed
- *Drug*—**Navajo, Ramah** *Nose Medicine* Plant used as snuff for nose troubles. (as *P. sawatchense* 191:24)

Polygonum hydropiper L., Marshpepper Knotweed
- *Drug*—**Cherokee** *Analgesic* Taken for painful urination. Infusion

mixed with meal used as poultice for pain. *Antidiarrheal* Infusion of root given to children for diarrhea. *Dermatological Aid* Used for "scaldhead." *Pediatric Aid* Infusion of root given to children for diarrhea. Leaves rubbed on children's thumb to prevent thumb sucking. *Poison* Used to poison fish and infusion mixed with meal used as poultice for pain. *Poultice* Used as a poultice for "swelled and inflamed parts." *Urinary Aid* Taken for "gravel," painful urination, and bloody urine. (80:55) **Iroquois** *Analgesic* Poultice of wetted plant applied to the forehead for headaches. (141:40) *Febrifuge* Decoction of plant taken for fever, chills, and "when cold." *Gastrointestinal Aid* Decoction of small piece of plant taken for indigestion. Whole plant used for children with swollen stomachs. *Pediatric Aid* Whole plant used for children with swollen stomachs. (87:314) **Malecite** *Kidney Aid* Infusion of dried leaves used for dropsy. (116:244)
- *Food*—**Cherokee** *Unspecified* Young growth boiled, fried, and eaten. (as *P. hydopiper* 204:253) **Iroquois** *Spice* Whole plant, except the roots, used by older people as pepper. (141:40)

Polygonum lapathifolium L., Curlytop Knotweed
- *Drug*—**Apache, White Mountain** *Unspecified* Plant used for medicinal purposes. (136:159) **Keres, Western** *Gastrointestinal Aid* Infusion of plant taken for stomach trouble. (171:62) **Navajo, Ramah** *Ceremonial Medicine* Cold infusion of plant used as ceremonial chant lotion. (191:23, 24) **Potawatomi** *Febrifuge* Infusion of whole plant used for fever. (154:72) **Zuni** *Cathartic* Decoction of plant taken as an emetic and purgative. *Emetic* Decoction of root taken as an emetic and purgative. (166:58)

Polygonum pensylvanicum L., Pennsylvania Smartweed
- *Drug*—**Chippewa** *Anticonvulsive* Infusion of plant tops taken for epilepsy. (71:129) **Iroquois** *Veterinary Aid* Decoction of plant given to horses for colic and "when urine is bound up." (87:314) **Menominee** *Antihemorrhagic* Infusion of leaf taken for "hemorrhage of blood from the mouth." *Gynecological Aid* Compound infusion of leaf taken to aid postpartum healing. (151:47) **Meskwaki** *Antidiarrheal* Used to wipe anus for bloody flux. *Hemorrhoid Remedy* Used for piles. (152:236, 237)

Polygonum persicaria L., Spotted Ladysthumb
- *Drug*—**Cherokee** *Analgesic* Decoction mixed with meal and used as poultice for pain. *Dermatological Aid* Crushed leaves rubbed on poison ivy. *Urinary Aid* Infusion taken for "gravel." (80:26) **Chippewa** *Analgesic* Decoction of leaves and flowers taken for stomach pain. *Gastrointestinal Aid* Simple or compound decoction of flowers and leaves taken for stomach pain. (53:344) **Iroquois** *Antirheumatic (External)* Decoction of plant used as a foot and leg soak for rheumatism. *Heart Medicine* Plant used for heart trouble. *Veterinary Aid* Plant rubbed over horses to keep flies away. (87:315)

Polygonum punctatum Ell., Dotted Smartweed
- *Drug*—**Chippewa** *Analgesic* and *Gastrointestinal Aid* Compound decoction of leaves and flowers taken for stomach pain. (53:344) **Houma** *Analgesic* and *Orthopedic Aid* Decoction of root taken for pains and swellings in the legs and joints. (158:58) **Iroquois** *Psychological Aid* Compound decoction taken for "loss of senses during menses." (87:315)

Polygonum ramosissimum Michx., Bushy Knotweed
- *Drug*—**Navajo, Ramah** *Analgesic* and *Gastrointestinal Aid* Warm infusion of plant taken for stomachache. *Panacea* Plant used as a "life medicine." (191:24)

Polygonum sp., Smartweed
- *Drug*—**Algonquin, Quebec** *Hemostat* Leaves used for bleeding. (18:161)
- *Food*—**Paiute** *Unspecified* Species used for food. (167:244)

Polygonum virginianum L., Jumpseed
- *Drug*—**Cherokee** *Pulmonary Aid* Hot infusion of leaves with bark of honey locust given for whooping cough. (as *Tovara virginiana* 80:42)

Polymnia, Asteraceae
Polymnia canadensis L., Whiteflower Leafcup
- *Drug*—**Houma** *Dermatological Aid* Poultice of crushed leaves applied to swellings. (158:63) **Iroquois** *Toothache Remedy* Plant used as toothache medicine. (87:468)

Polypodiaceae, see *Phlebodium, Pleopeltis, Polypodium*

Polypodium, Polypodiaceae, fern
Polypodium californicum Kaulfuss, California Polypody
- *Drug*—**Mendocino Indian** *Eye Medicine* Infusion of roots used as a wash for sore eyes. **Wailaki** *Antirheumatic (External)* Root juice rubbed on areas affected by rheumatism. *Dermatological Aid* Root juice rubbed on sores. (41:303) **Yurok** *Disinfectant* Rhizomes used as an "antibiotic" for infections. (6:46)

Polypodium glycyrrhiza D. C. Eat., Licorice Fern
- *Drug*—**Bella Coola** *Oral Aid* Rhizomes chewed to flavor the mouth. *Throat Aid* Rhizomes chewed and juice swallowed for sore throat. (184:196) **Haisla** *Analgesic* Rhizomes used for chest pains. *Respiratory Aid* Rhizomes used for shortness of breath. **Haisla & Hanaksiala** *Cold Remedy* Rhizomes used for colds. *Cough Medicine* Rhizomes used for coughs. *Throat Aid* Rhizomes chewed or sucked for sore throats. (43:158) **Hesquiat** *Carminative* Rhizomes growing on the wild crabapple used for gas. *Cough Medicine* Long, slender rhizomes eaten as a medicine for coughs. *Oral Aid* Long, slender rhizomes eaten raw to sweeten the mouth. *Throat Aid* Long, slender rhizomes eaten as a medicine for sore throats. (185:30) **Kitasoo** *Cough Medicine* Rhizomes used for coughs. *Throat Aid* Rhizomes used for sore throats. (43:312) **Kwakiutl** *Antidiarrheal* Compound decoction of plants or roots taken for diarrhea. (183:266) *Antiemetic* and *Antihemorrhagic* Roots sucked and juice swallowed for vomiting blood. (183:264) **Makah** *Unspecified* Rhizomes used for internal ailments. (67:220) **Nitinaht** *Cough Medicine* Licorice flavored rhizomes chewed and juice swallowed for coughs. *Respiratory Aid* Licorice flavored rhizomes chewed and juice swallowed for sore chest. (186:64) **Nootka** *Alterative* and *Venereal Aid* Plant used as an excellent alterative for venereal complaints. (as *P. falcatum* 170:80, 81) **Oweekeno** *Cough Medicine* Rhizomes chewed for coughs. *Throat Aid* Rhizomes chewed for sore throats. (43:59) **Thompson** *Cold Remedy* Rhizomes chewed or infusion of rhizomes taken for colds. *Oral Aid* Rhizomes used as medicine for sore gums. *Throat Aid* Rhizomes chewed or infusion of rhizomes taken for sore throats. (187:91)

Food—**Hesquiat** *Vegetable* Long, slender rhizomes eaten raw as a food and to sweeten the mouth. (185:30) **Kwakiutl, Southern** *Dietary Aid* Roots kept in the mouth to prevent hunger and thirst. *Starvation Food* Roots dried, steamed, and eaten during famines. *Unspecified* Roots scorched, pounded, cut in bite-size pieces, dipped in oil, and chewed and sucked by old people. (183:264) **Makah** *Dietary Aid* Rhizomes chewed, on hunting trips, to curb the appetite. *Unspecified* Rhizomes eaten raw, especially by children, because of the licorice flavor. (67:220) **Thompson** *Candy* Rhizomes chewed for the pleasant, sweet, licorice flavor. (187:91)

Polypodium hesperium Maxon, Western Polypody
- *Drug*—**Thompson** *Cold Remedy* Rhizomes chewed or infusion of rhizomes taken for colds. *Oral Aid* Rhizomes used as medicine for sore gums. *Throat Aid* Rhizomes chewed or infusion of rhizomes taken for sore throats. (187:91)
- *Food*—**Thompson** *Candy* Rhizomes chewed for the pleasant, sweet, licorice taste. (187:91)

Polypodium incanum Sw., Resurrection Fern
- *Drug*—**Seminole** *Other* Complex infusion of leaves taken for chronic conditions. (169:272) *Psychological Aid* Infusion of plant used to steam and bathe the body for insanity. (169:291)

Polypodium scouleri Hook. & Grev., Leathery Polypody
- *Food*—**Hesquiat** *Candy* Children chewed the thick rhizomes. (185:30) **Makah** *Unspecified* Species used for food. (67:221)

Polypodium virginianum L., Rock Polypody
- *Drug*—**Abnaki** *Gastrointestinal Aid* Used for stomachaches. (144:154) Decoction of whole plant used for stomachaches. (144:162) **Algonquin, Quebec** *Heart Medicine* Used to make a medicinal tea for heart disease. (18:123) **Bella Coola** *Analgesic* Simple or compound decoction taken for stomach pains. *Gastrointestinal Aid* Compound decoction of root used for stomach pain, not vomiting or diarrhea. *Throat Aid* Roots chewed for swollen, sore throat and compound decoction used for stomach pain. (as *P. vulgare* 150:48) **Cherokee** *Dermatological Aid* Poultice used for inflamed swellings and wounds and infusion taken for hives. (80:60, 61) **Cowichan** *Cold Remedy* Rhizomes used for colds. *Gastrointestinal Aid* Rhizomes used for stomach ailments. *Throat Aid* Rhizomes used for sore throat. (as *P. vulgare* 182:69) **Cowlitz** *Misc. Disease Remedy* Infusion of crushed stems taken for the measles. (as *P. vulgare* 79:13) **Cree, Woodlands** *Tuberculosis Remedy* Decoction of leaf taken for tuberculosis. (109:51) **Green River Group** *Cough Medicine* Baked or raw roots used as a cough medicine. (as *P. vulgare* 79:13) **Iroquois** *Misc. Disease Remedy* Compound decoction taken for cholera. (87:260) **Klallam** *Cough Medicine* Baked or raw roots used as a cough medicine. **Makah** *Cough Medicine* Peeled stems chewed for coughs. (as *P. vulgare* 79:13) **Malecite** *Pulmonary Aid* Infusion of pounded roots used for pleurisy. (as *P. vulgare* 116:252) **Micmac** *Diuretic* Infusion of plant used for urine retention. (145:55) *Pulmonary Aid* Roots used for pleurisy. (40:59) **Quinault** *Cough Medicine* Baked or raw roots used as a cough medicine. (as *P. vulgare* 79:13) **Saanich** *Cold Remedy* Rhizomes used for colds. *Gastrointestinal Aid* Rhizomes used for stomach ailments. *Throat Aid* Rhizomes used for sore throat. (as *P. vulgare* 182:69) **Skagit, Upper** *Dermatological Aid* Plant used to make a demulcent. *Expectorant* Plant used to make an expectorant. *Laxative* Plant used to make a laxative. (as *P. vulgare* 179:42)
- *Food*—**Salish, Coast** *Dried Food* Rhizomes sun dried and used as a winter food. *Substitution Food* Rhizomes formerly used as a substitute for sugar. *Unspecified* Rhizomes eaten fresh. (as *P. vulgare* 182:69)

Polypogon, Poaceae

Polypogon monspeliensis (L.) Desf., Annual Rabbitsfoot Grass
- *Drug*—**Navajo, Kayenta** *Heart Medicine* Infusion of ashes taken for palpitations. (205:16)
- *Food*—**Tubatulabal** *Unspecified* Used extensively for food. (193:15)
- *Other*—**Navajo, Kayenta** *Soap* Used as lotion to wash a snake figurine before painting it. (205:16)

Polyporaceae, see *Fomes, Fomitopsis, Ganoderma, Polyporus*

Polyporus, Polyporaceae, fungus

Polyporus harlowii, Bracket Fungus
- *Food*—**Isleta** *Unspecified* Fungi boiled or baked and eaten. *Winter Use Food* Fungi stored for winter use. **Pueblo** *Special Food* Fungi boiled and eaten as a delicacy. (32:33)

Polyporus sp.
- *Drug*—**Blackfoot** *Antidiarrheal* Infusion of plant taken for diarrhea and dysentery. (86:67)
- *Other*—**Blackfoot** *Ceremonial Items* Used in the Firelighters bundle of the Horn Society for use as punk to light a fire easily. *Incense & Fragrance* Used as incense to keep ghosts away. (86:38) **Haisla & Hanaksiala** *Ceremonial Items* Plant used for some aspects of the secret society rituals. *Protection* Plant placed at entrance of special shamanistic dance house as protection from bad spirits and ghosts. Plant used as hand protection for handling live coals during a secret society ritual. *Toys & Games* Plant used as a ball in a women's and children's game. Plant thrown into the stream and used as a target by young boys for spear practice. (43:135)

Polystichum, Dryopteridaceae, fern

Polystichum acrostichoides (Michx.) Schott, Christmas Fern
- *Drug*—**Cherokee** *Antirheumatic (External)* Compound decoction of root applied with warm hands for rheumatism. (as *Aspidium acrostichoides* 80:8) Roots used in "medicine rubbed on skin for rheumatism after scratching." (80:33) Decoction of roots rubbed on area affected by rheumatism. (177:3) *Antirheumatic (Internal)* Infusion taken for rheumatism. *Emetic* Roots used as an ingredient in an emetic and infusion taken for rheumatism. *Febrifuge* Compound decoction taken for chills and infusion taken for fever. *Gastrointestinal Aid* Cold infusion of root used for "stomachache or bowel complaint." *Pulmonary Aid* Infusion taken for pneumonia. *Toothache Remedy* Compound decoction used for toothache and chills. (80:33) **Iroquois** *Analgesic* Decoction of plant used by children for cramps. (87:256) *Anticonvulsive* Poultice of wet, smashed roots used on children's back and head for convulsions. *Antidiarrheal* Compound decoction taken for diarrhea. (87:257) *Antirheumatic (External)* Infusion of smashed roots used as a foot soak for "rheumatism" in back and legs. (87:256) *Blood Medicine* Cold, compound decoction taken for weak

blood and as a blood purifier. *Dermatological Aid* Poultice of wet, smashed roots used on children's back and head for red spots. (87: 257) *Emetic* Infusion of roots taken as an emetic for dyspepsia and consumption. *Febrifuge* Decoction of vine with small leaves used for children with fevers. (87:256) *Gynecological Aid* Plant taken before and after baby to clean womb. Roots used as a "Lady's medicine" for the insides. (87:257) *Orthopedic Aid* Poultice applied to back and feet for spinal trouble and sore back of babies. *Pediatric Aid* Decoction of plant used by children for cramps. Decoction of vine with small leaves used for children with fevers. Poultice applied to back and feet for spinal trouble and sore back of babies. (87:256) Poultice of wet, smashed roots used on children's back and head for convulsions, and for red spots. *Stimulant* Decoction of plant given to children (sometimes mother too) for listlessness. *Throat Aid* Powder inhaled and coughed up by a man who cannot talk. *Tuberculosis Remedy* Infusion of roots taken as an emetic for consumption. *Venereal Aid* Compound decoction used as a blood purifier and for venereal disease. (87:257) **Malecite** *Throat Aid* Roots chewed and used for hoarseness. (116:247) **Micmac** *Throat Aid* Roots used for hoarseness. (40:59)
- **Food**—**Cherokee** *Unspecified* Fiddle heads used for food. (80:33)

Polystichum munitum (Kaulfuss) K. Presl, Western Swordfern
- **Drug**—**Cowlitz** *Dermatological Aid* Infusion of stems used as a wash for sores. (79:13) **Hesquiat** *Cancer Treatment* Young shoots or fiddleheads chewed for cancer of the womb. (185:32) **Kwakiutl** *Gynecological Aid* Boughs placed under bed of young girl to have as many children as plants. (183:265) **Lummi** *Gynecological Aid* Leaves chewed by women to facilitate childbirth. **Quileute** *Dermatological Aid* Poultice of chewed leaves applied to sores and boils. **Quinault** *Burn Dressing* Poultice of spore sacs from the leaves applied to burns. *Dermatological Aid* Decoction of roots used as a wash for dandruff. **Swinomish** *Throat Aid* Raw plant chewed and eaten for sore throats or tonsillitis. (79:13) **Thompson** *Hunting Medicine* Plant rubbed on the hands to bring luck in whaling and sturgeon fishing. (187:89)
- **Food**—**Klallam** *Unspecified* Rhizomes boiled and eaten. (79:13) **Kwakiutl, Southern** *Unspecified* Basal leaves and rhizomes steamed, peeled, and used for food. (183:265) **Makah** *Spice* Leaves used to steam salmonberry sprouts on hot rocks, to give the sprouts flavor. *Unspecified* Roots steamed or cooked in a pit. (67:221) Rhizomes boiled and eaten. (79:13) **Nitinaht** *Unspecified* Large rootstocks steam cooked and eaten in summer. (186:62) **Quileute** *Unspecified* Rhizomes peeled, pit baked, and eaten with fresh or dried salmon eggs. **Quinault** *Unspecified* Rhizomes pit baked on hot rocks and used for food. (79:13) **Thompson** *Unspecified* Rootstocks used for food. (187:89)
- **Fiber**—**Cowlitz** *Mats, Rugs & Bedding* Leaves tied with maple bark and used for mattresses. (79:13) **Hesquiat** *Clothing* Long, straight fronds worn as head decoration when visiting another place and bringing gifts. *Mats, Rugs & Bedding* Long, straight fronds used as bedding before mats or mattresses were used. (185:32) **Nitinaht** *Mats, Rugs & Bedding* Fronds laid side by side several layers thick and used as a "place mat" for food at feasts. (186:62) **Oweekeno** *Clothing* Leaves used to make a "hula hula skirt" as part of the costume of some male, tsaika dancers. *Mats, Rugs & Bedding* Leaves used as a mat under fish when cleaning and cutting. (43:56) **Quileute** *Mats, Rugs & Bedding* Leaves used for mattresses. (79:13) **Salish, Coast** *Mats, Rugs & Bedding* Large, fleshy leaves used to cover floors. (182:69) **Yurok** *Mats, Rugs & Bedding* Leaves used for bedding. (6:46)
- **Other**—**Chehalis** *Cooking Tools* Leaves used to line pits when baking camas. (79:13) **Clallam** *Toys & Games* Leaves used in a children's game. (57:194) **Cowlitz** *Cooking Tools* Leaves used to line pits when baking camas or wapatoo. (79:13) **Hesquiat** *Containers* Long, straight fronds used on the ground under fish and other foods to keep them clean. (185:32) **Karok** *Toys & Games* Fronds used in a game played by adults of both sexes to see who had the longest wind. Beginning at the bottom of the frond, the player touched each leaflet, first on one side of the stem and then the other and said "tiip" each time he touched a leaflet. Whoever went the farthest up the frond won. There was no gambling on this game. (148:378) **Klallam** *Toys & Games* Leaves pulled off the plants by children playing an endurance game. (79:13) **Kwakiutl** *Ceremonial Items* Plant used as a charm to call the northwest wind. **Kwakiutl, Southern** *Containers* Leaves used to line food storage boxes, berry picking baskets, and berry drying racks. *Cooking Tools* Leaves used to line steaming pits. (183:265) **Makah** *Cooking Tools* Leaves used for lining cooking pits, both above and below the foods. Leaves used to wipe salmon. (67:221) Leaves used to line pits when steaming sprouts. (79:13) *Toys & Games* Fronds used in the game, pile pile.[59] (67:221) Leaves pulled off the plants by children playing an endurance game. (79:13) **Nitinaht** *Cooking Tools* Leaves used for lining cooking pits, both above and below the foods. (67:221) Fronds placed below and above food in steaming pits. (186:62) *Toys & Games* Fronds used in the game, pile pile.[59] (67:221) Fronds used in games. (186:62) **Oweekeno** *Containers* Leaves used as a liner for oolichan (candlefish) bins and pits. (43:56) **Paiute** *Decorations* Plant used to decorate graves. *Toys & Games* Kids played with it as a feather. (111:36) **Pomo** *Cooking Tools* Fronds used for lining the top and bottom of an earth oven in baking acorn bread. Fronds used as a lining for an acorn-leaching basin. (66:11) **Pomo, Kashaya** *Cooking Tools* Fronds used to line an earth baking oven or sand leaching basin. (72:47) **Quileute** *Cooking Tools* Leaves used to line pits when baking camas. **Quinault** *Cooking Tools* Leaves used to line pits when baking rhizomes or camas. (79:13) **Saanich** *Ceremonial Items* Fronds used during initiation dances. **Salish, Coast** *Cooking Tools* Large, fleshy leaves used for laying food on and for spreading on berry drying racks. (182:69) **Squaxin** *Containers* Leaves spread on racks for berries to dry. (79:13) **Thompson** *Containers* Rootstocks used to line the steaming pits for cooking "Indian potatoes" and other root-type foods. (187:89)

Polystichum munitum (Kaulf.) K. Presl. **ssp. *munitum***, Western Swordfern
- **Food**—**Costanoan** *Unspecified* Rhizomes eaten, boiled or baked in coals. (21:247)
- **Other**—**Costanoan** *Cooking Tools* Fronds used to line earth ovens. (21:247)

Polytaenia, Apiaceae
Polytaenia nuttallii DC., Nuttall's Prairie Parsley
- **Drug**—**Meskwaki** *Antidiarrheal* and *Gynecological Aid* Decoction of seeds taken by women for severe diarrhea. (152:249)

Polytrichaceae, see *Polytrichum*

Polytrichum, Polytrichaceae, moss
Polytrichum commune Hedw., Hair Moss
- **Drug**—**Nitinaht** *Gynecological Aid* Plant chewed by women in labor to speed up the process. (186:59)

Pontederiaceae, see *Heteranthera, Pontederia*

Pontederia, Pontederiaceae
Pontederia cordata L., Pickerel Weed
- **Drug**—**Malecite** *Contraceptive* Infusion of plant used to prevent pregnancy. (116:259) **Micmac** *Contraceptive* Herbs used to prevent pregnancy. (40:59) **Montagnais** *Panacea* "Brew" from plant used for "illness in general." (156:315)

Populus, Salicaceae
Populus × *acuminata* Rydb., Lanceleaf Cottonwood
- **Food**—**Lakota** *Fodder* Boughs and bark fed to horses during winter. (106:33)
- **Fiber**—**Lakota** *Building Material* Used for posts and other building materials. (106:33) **Navajo, Ramah** *Building Material* Used for hogan construction. (191:22)
- **Other**—**Lakota** *Fuel* Used for fuel. (106:33) **Navajo, Ramah** *Ceremonial Items* Root used to make snake figurines. *Tools* Used to make fire drills. (191:22)

Populus alba L., White Poplar
- **Drug**—**Iroquois** *Cold Remedy* Infusion of inner bark taken to "cleans you out after a cold." *Love Medicine* Compound decoction of branches used as a body wash for anti-love medicine. *Tonic* Infusion of inner bark taken as a tonic. (87:293) **Ojibwa** *Antirheumatic (External)* Infusion of pounded plants used as wash for rheumatism. *Blood Medicine* Infusion of bark and root or decoction of bark taken for internal blood diseases. *Panacea* Infusion of pounded plants used as wash for general illnesses. (135:231) *Unspecified* Roots and bark used for medicinal purposes. (135:243)
- **Other**—**Ojibwa** *Paper* Wood used for pulpwood. (135:243)

Populus angustifolia James, Narrowleaf Cottonwood
- **Food**—**Apache, White Mountain** *Candy* Buds used as chewing gum. *Unspecified* Buds used for food. (136:159) **Montana Indian** *Fodder* Young twigs fed to horses when other food was not obtainable. *Unspecified* Inner bark considered a valuable mucilaginous food. (19:19) **Navajo** *Candy* Buds used as chewing gum. *Unspecified* Buds used for food. **Zuni** *Candy* Buds used as chewing gum. *Unspecified* Buds used for food. (136:159)
- **Fiber**—**Gosiute** *Basketry* Shoots used to make baskets. (39:378) **Montana Indian** *Building Material* Wood used for fire and shelter during the winter. (19:19) **Navajo** *Furniture* Soft wood used for parts of the cradle. (55:37)
- **Other**—**Montana Indian** *Fuel* Wood used for fire and shelter during the winter. *Smoke Plant* Inner bark used for "kinnikinnick." (19:19)

Populus balsamifera L., Balsam Poplar
- **Drug**—**Algonquin, Quebec** *Dermatological Aid* Spring buds used to make a salve. Poultice of steeped, root scrapings applied to open sores. Buds used to make a salve and applied to open sores. *Disinfectant* Poultice of steeped, root scrapings applied to infected wounds. Buds used to make a salve and applied to infected wounds. (18:148) **Bella Coola** *Analgesic* Branches with leaves used in a sweat bath for pains similar to rheumatism. Decoction of rotten leaves used as a bath for body pain. Poultice of compound containing buds applied for lung or hip pains. *Antirheumatic (External)* Leaves used in sweat bath for pains similar to rheumatism. *Diaphoretic* Branches with leaves used in a sweat bath for pains similar to rheumatism. *Orthopedic Aid* Poultice of compound containing buds applied to lung or hip pain. *Pulmonary Aid* Poultice of compound containing buds applied for lung pains. **Carrier, Northern** *Dermatological Aid* Poultice of chewed, green roots applied to bleeding wounds. *Eye Medicine* Decoction of inner bark used as an eyewash. **Carrier, Southern** *Cough Medicine* and *Pulmonary Aid* Decoction of buds taken for "coughs and lung affections." (150:54) **Chippewa** *Analgesic* Compound decoction of root taken for back pain. (53:356) *Dermatological Aid* Decoction of buds used as salve for frostbite, sores, and inflamed wounds. (71:126) *Gynecological Aid* Compound decoction of root taken for "female weakness." (53:356) Compound infusion of root taken for excessive flowing during confinement. (53:358) *Heart Medicine* Compound decoction of root, bud, and blossom prepared ceremonially and used for the heart. (53:338) *Orthopedic Aid* Poultice of infusion or decoction of used buds for sprains or strained muscles. (53:362) **Cree, Woodlands** *Disinfectant* Poultice of fresh leaves applied to a sore to draw out the infection. *Hemostat* Poultice of sticky buds applied directly to the nostril for a nosebleed. (109:52) **Malecite** *Dermatological Aid* Bud balm and burdock roots used for sores. (116:247) **Micmac** *Dermatological Aid* Buds and other parts of plant used as salve for sores. *Venereal Aid* Buds and other parts of plant used as salve for chancre. (40:59) **Ojibwa** *Antirheumatic (External)* Infusion of pounded plants used as wash for rheumatism. *Blood Medicine* Decoction of bark taken for internal blood diseases. (135:231) *Cold Remedy* Buds cooked in grease and rubbed in nostrils for cold. *Dermatological Aid* Buds cooked in grease and used as salve for cuts, wounds, or bruises. (153:387) *Panacea* Infusion of pounded plants used as wash for general illnesses. (135:231) *Respiratory Aid* Buds cooked in grease and rubbed in nostrils for catarrh or bronchitis. (153:387) **Paiute** *Gastrointestinal Aid* Decoction of sap taken for stomach disorders. (180:121, 122) **Potawatomi** *Dermatological Aid* Buds melted with tallow and used as ointment for sores or eczema. (154:80, 81) **Shoshoni** *Analgesic* Decoction of root used as a lotion for headaches. *Blood Medicine* Compound decoction of bark taken as a blood tonic. *Tonic* Compound decoction of bark taken as a blood tonic and for general debility. *Tuberculosis Remedy* Simple or compound decoction of bark taken for tuberculosis. *Venereal Aid* Simple or compound decoction of bark taken for venereal disease. (180:121, 122) **Tanana, Upper** *Cold Remedy* Decoction of buds taken for colds. Buds heated on top of a wood stove and the aroma inhaled for colds. Decoction of mashed buds used for colds. *Cough Medicine* Decoction of buds taken for coughs. *Dermatological Aid* Mashed buds cooked in grease or dried, powdered buds used as a salve for rashes and sores. *Panacea* Decoction of buds taken for colds, coughs, and other illnesses. (102:4)
- **Food**—**Montana Indian** *Fodder* Young twigs fed to horses when oth-

Populus balsamifera (Salicaceae)

er food was not obtainable. *Unspecified* Inner bark considered a valuable mucilaginous food. (19:19) **Tanana, Upper** *Unspecified* Sap used for food. (102:4)

- *Fiber*—**Klamath** *Clothing* Bark used in the manufacture of cloth. (45:94) **Montana Indian** *Building Material* Wood used for fire and shelter during the winter. (19:19)
- *Other*—**Carrier, Southern** *Insecticide* Resin from buds used to repel mosquitoes, blackflies, and gadflies. (150:54) **Cree, Woodlands** *Hunting & Fishing Item* Buds mixed with other ingredients to make a trap lure. (109:52) **Eskimo, Inuktitut** *Hunting & Fishing Item* Bark used for fishing floats. *Insecticide* Bark burned for a mosquito repelling smoke. *Smoke Plant* Leaf galls used with or as tobacco. (202:188) **Montana Indian** *Fuel* Wood used for fire and shelter during the winter. *Smoke Plant* Inner bark used for "kinnikinnick." (19:19) **Ojibwa** *Paper* Wood used for pulpwood. (135:243) **Tanana, Upper** *Cooking Tools* Soft wood used to make bowls and other items. *Fuel* Wood used for fuel. *Preservative* Wood used to smoke fish. *Smoke Plant* Wood ashes mixed with tobacco and smoked. *Toys & Games* Bark used to make toys. (102:4)

Populus balsamifera L. **ssp. *balsamifera*,** Balsam Poplar

- *Drug*—**Anticosti** *Dermatological Aid* Poultice of buds and alcohol applied to wounds. (as *P. tacamahacca* 143:65) **Cherokee** *Antirheumatic (Internal)* Tincture of buds used for chronic rheumatism. *Dermatological Aid* Juice of buds used on sores. *Gastrointestinal Aid* Tincture of buds used for colic and bowels. *Stimulant* Given to "persons of phlegmatic habits." *Toothache Remedy* Used for aching teeth. *Venereal Aid* Tincture of buds used for old venereal complaints. (as *P. candicans* 80:24) **Iroquois** *Anthelmintic* Compound decoction with bark taken to kill worms in adults. (87:291) *Antirheumatic (Internal)* Used for arthritis. *Blood Medicine* Compound with whisky taken as blood remedy and for tapeworms. (87:292) *Dermatological Aid* Compound decoction used for skin eruptions and scabs. *Laxative* Compound decoction of bark taken as a laxative. *Veterinary Aid* Decoction of bark taken by people and given to horses for worms. (87:291) **Menominee** *Cold Remedy* Decoction of resinous buds in fat used in the nostrils for a head cold. *Dermatological Aid* Decoction of resinous buds in fat used as a salve for wounds. (as *P. candicans* 151:52)
- *Other*—**Ojibwa** *Paper* Wood used for pulpwood. (135:243) **Thompson** *Stable Gear* Wood used to make the sides of riding and pack saddles. (as *P. tacamahacca* 164:497)

Populus balsamifera ssp. *trichocarpa* (Torr. & Gray ex Hook.) Brayshaw, Black Cottonwood

- *Drug*—**Bella Coola** *Dermatological Aid* Buds mixed with chewed and warmed mountain goat kidney fat and used as a face cream. Infusion of buds mixed with eulachon (candlefish) grease or sockeye salmon oil and rubbed on scalp for baldness. *Orthopedic Aid* Poultice of buds applied for hip pains. *Pulmonary Aid* Infusion of buds and animal fat taken for whooping cough. Poultice of buds applied for lung and hip pains. *Throat Aid* Infusion of buds in dogfish oil taken for sore throat. *Tuberculosis Remedy* Buds mixed with balsam sap and used for tuberculosis. (as *P. trichocarpa* 184:210) **Flathead** *Cold Remedy* Bark eaten for colds. *Dermatological Aid* Poultice of leaves used for bruises, sores, and boils. *Venereal Aid* Infusion of young branches, buds, and other plants taken for syphilis. (as *P. trichocarpa* 82:68) **Haisla & Hanaksiala** *Burn Dressing* Buds cooked with mountain goat fat and rubbed on the face for sunburn. *Dermatological Aid* Buds used to make a hair dressing. Buds cooked with mountain goat fat and rubbed on the body to soften the skin. (43:284) **Hesquiat** *Dermatological Aid* Decoction of buds mixed with deer fat and used to make a fragrant salve. (185:75) **Hoh** *Unspecified* Infusion of bark used for medicine. (as *P. trichocarpa* 137:60) **Karok** *Love Medicine* Leaves used as a love medicine. (as *P. trichocarpa* 148:381) **Klallam** *Eye Medicine* Infusion of buds used as an eyewash. (as *P. trichocarpa* 79:26) **Kutenai** *Dermatological Aid* Poultice of leaves used for bruises, sores, and boils. *Respiratory Aid* Infusion of bark taken for tuberculosis. Infusion of bark taken for whooping cough. (as *P. trichocarpa* 82:68) **Kwakiutl** *Dermatological Aid* Buds and oil used as a hair tonic. Buds rubbed on the skin to prevent sunburn. (as *P. trichocarpa* 183:292) **Nez Perce** *Antirheumatic (External)* Leaves used for aching muscles. *Veterinary Aid* Poultice of leaves used for horses' sores. (as *P. trichocarpa* 82:68) **Nitinaht** *Dermatological Aid* Resin used as a salve for wounds and cuts. (186:126) **Okanagan-Colville** *Venereal Aid* Infusion of buds taken for gonorrhea. (188:134) **Oweekeno** *Dermatological Aid* Buds boiled with grease, strained, combined with an unknown ingredient, and used as hair dressing. (43:116) **Quileute** *Unspecified* Infusion of bark used for medicine. (as *P. trichocarpa* 137:60) **Quinault** *Dermatological Aid* and *Disinfectant* Gum of burls applied as an antiseptic to cuts and wounds. *Tuberculosis Remedy* Infusion of bark taken for tuberculosis. (as *P. trichocarpa* 79:26) **Shuswap** *Dermatological Aid* Poultice of pitch applied to sores. (as *P. trichocarpa* 123:68) **Squaxin** *Dermatological Aid* and *Disinfectant* Infusion of bruised leaves applied as an antiseptic for cuts. *Throat Aid* Infusion of bark used as a gargle for sore throats. (as *P. trichocarpa* 79:26) **Thompson** *Ceremonial Medicine* Decoction of bark taken "for your health" after childbirth if someone close had passed away. *Dermatological Aid* Poultice of mashed buds mixed with pitch used for ringworm. Infusion of white inner bark used for washing sores and especially itchy skin. *Gynecological Aid* Infusion of white inner bark taken by women after childbirth. *Orthopedic Aid* Concoction of wood, willow, soapberry branches, and "anything weeds" used for broken bones. *Unspecified* Decoction of buds taken for "some kind of disease." It was cautioned that one should not drink too much of this decoction because it would kill you. (187:276) **Yurok** *Unspecified* Decoction of shoot tips used for medicine. (as *P. trichocarpa* 6:47)
- *Food*—**Bella Coola** *Dried Food* Inner cambium "slime" sun dried and eaten with grease. *Unspecified* Inner cambium "slime" eaten fresh. (as *P. trichocarpa* 184:210) **Blackfoot** *Unspecified* Inner bark and sap used for food. **Cheyenne** *Fodder* Twigs and bark fed to horses and other livestock. (as *P. trichocarpa* 82:68) **Clallam** *Dried Food* Sap eaten dried. *Unspecified* Sap eaten fresh. (as *P. trichocarpa* 57:203) **Flathead** *Unspecified* Inner bark and sap used for food. (as *P. trichocarpa* 82:68) **Haisla** *Unspecified* Cambium eaten fresh in spring. (73:151) **Haisla & Hanaksiala** *Unspecified* Inner bark used for food. (43:284) **Kutenai** *Unspecified* Inner bark and sap used for food. (as *P. trichocarpa* 82:68) **Kwakiutl, Southern** *Unspecified* Cambium eaten in early spring. (as *P. trichocarpa* 183:292) **Oweekeno** *Unspecified* Cambium used for food. (43:116) **Thompson** *Forage* Leaves and twigs eaten by moose. (187:276)

- *Fiber*—**Carrier** *Canoe Material* Wood used to make canoes. (as *P. trichocarpa* 31:69) **Cheyenne** *Building Material* Trunks used as framework for Sundance lodges. (as *P. trichocarpa* 82:68) **Hanaksiala** *Mats, Rugs & Bedding* Seed "wool" spun and used to make blankets and toques. (43:284) **Karok** *Basketry* Roots used to make baskets. (as *P. trichocarpa* 148:381) **Nisga** *Canoe Material* Wood used to make canoes. (43:349) **Nitinaht** *Cordage* Fibers, dog hair, and nettles used to make stronger ropes. (as *P. trichocarpa* 67:241) Inner bark shredded, spun together with red or yellow cedar inner bark and used as twine. (186:126) **Okanagan-Colville** *Canoe Material* Wood used to make dugout canoes. (188:134) **Quinault** *Building Material* Bark used for house coverings. (as *P. trichocarpa* 79:26) **Salish, Coast** *Canoe Material* Wood used to make canoes. (as *P. trichocarpa* 182:89) **Shuswap** *Canoe Material* Wood used to make dugout canoes. (as *P. trichocarpa* 123:68) **Squaxin** *Building Material* Young shoots used to make sweat lodges. *Cordage* Young shoots used as lashings or tying thongs. (as *P. trichocarpa* 79:26) **Thompson** *Canoe Material* Wood used for dugout canoes. *Mats, Rugs & Bedding* Cottony seed fluff used for stuffing mattresses and pillows. *Scouring Material* Inner bark used as a scouring pad. (187:276)
- *Dye*—**Missouri River Indian** *Yellow* Buds used to make a yellow dye. (as *P. trichocarpa* 82:68)
- *Other*—**Blackfoot** *Protection* Sap used to conceal human scent when stealing enemy horses. **Cheyenne** *Paint* Fruits used to make red, green, yellow, purple, and white paint for suitcases and tepees. (as *P. trichocarpa* 82:68) **Haisla & Hanaksiala** *Preservative* Wood dried and used for smoking salmon. (43:284) **Karok** *Fasteners* Leaf buds, in spring, used as glue to stick feathers to arrows. (as *P. trichocarpa* 117:209) **Montana Indian** *Fuel* Branches used for firewood. (as *P. trichocarpa* 82:68) **Nitinaht** *Hunting & Fishing Item* Inner bark shredded, spun with stinging nettle fiber, and used for fishing lines and duck nets. Knots used to make molded halibut hooks. *Incense & Fragrance* Sweet smelling, yellow resin used as a scent in deer fat skin cosmetics. *Paint* Sweet smelling, yellow resin used as a base for paints. (186:126) **Okanagan-Colville** *Containers* Bark used to make food storage containers and to line food storage pits. *Fasteners* Bud scale resin used to glue arrowhead onto shafts and in making spears, fishhooks, and canoes. *Season Indicator* Leaves trembling or shimmering with no perceptible wind was a signal that bad weather was coming. *Tools* Thin board of wood placed at the top of a cradle to flatten a child's head. (188:134) **Oweekeno** *Fasteners* Buds used as a binding agent to glue duck feathers to red cedar hoops used for festive applications. (43:116) **Paiute** *Fuel* Wood used for fuel. (as *P. trichocarpa* 111:61) **Shuswap** *Hide Preparation* Wood used for smoking buckskin. (as *P. trichocarpa* 123:68) **Thompson** *Fuel* Rotten wood used as a fuel in smoking hides. *Soap* Dried inner bark packaged into small, fist-sized bundles and used as a soap substitute. Each person carried his own package with him. Men's and women's were packaged differently. The soap was also used as a laundry soap. *Tools* Dried root used as a drill in making friction fires. (187:276) **Yurok** *Tools* Pitch used to apply soot in the tattooing process. (as *P. trichocarpa* 6:47)

Populus deltoides Bartr. ex Marsh., Eastern Cottonwood
- *Drug*—**Delaware** *Strengthener* Bark combined with black haw and wild plum barks and used by women for weakness and debility. (176:31) **Delaware, Oklahoma** *Gynecological Aid* Simple or compound infusion taken for "weakness and debility in women." (175:26, 78) **Flathead** *Cold Remedy* Bark eaten for colds. *Dermatological Aid* Poultice of leaves used for bruises, sores, and boils. *Venereal Aid* Infusion of young branches, buds, and other plants taken for syphilis. (82:68) **Iroquois** *Anthelmintic* Decoction of bark used for intestinal worms. *Veterinary Aid* Poultice of dried bark flour and water applied to horses with bumps containing worms. (141:39) **Kutenai** *Dermatological Aid* Poultice of leaves used for bruises, sores, and boils. *Respiratory Aid* Infusion of bark taken for whooping cough. *Tuberculosis Remedy* Infusion of bark taken for tuberculosis. (82:68) **Nanticoke** *Orthopedic Aid* Compound containing bark used as a lotion for sprains. (175:56, 84) **Nez Perce** *Antirheumatic (External)* Leaves used for aching muscles. *Veterinary Aid* Poultice of leaves used for horses' sores. (82:68)
- *Food*—**Blackfoot** *Unspecified* Inner bark and sap used for food. **Cheyenne** *Fodder* Twigs and bark fed to horses and other livestock. (82:68) Bark and twigs formerly used to feed horses in winter. *Unspecified* Inner bark scraped and eaten in spring. (83:36) **Flathead** *Unspecified* Inner bark and sap used for food. **Kutenai** *Unspecified* Inner bark and sap used for food. (82:68) **Lakota** *Forage* Bark eaten by horses. (139:57) **Montana Indian** *Fodder* Young twigs fed to horses when other food was not obtainable. *Unspecified* Inner bark considered a valuable mucilaginous food. (19:19) **Pima** *Unspecified* Catkins eaten raw. (146:69)
- *Fiber*—**Cheyenne** *Building Material* Trunks used as framework for Sundance lodges. (82:68) Trunks used to construct the medicine lodge for the Sun Dance ceremony. (83:46) **Kiowa** *Building Material* Used to make poles for the ceremonial tepee. (192:19) **Montana Indian** *Building Material* Wood used for fire and shelter during the winter. (19:19)
- *Dye*—**Cheyenne** *Green* Brown, gummy leaf buds scratched and used to make a green dye. *Purple* Brown, gummy leaf buds scratched and used to make a purple dye. *Red* Brown, gummy leaf buds scratched and used to make a red dye. *Unspecified* Used to make a dye. (83:46) *White* Brown, gummy leaf buds scratched and used to make a white dye. (83:36) **Missouri River Indian** *Yellow* Buds used to make a yellow dye. (82:68)
- *Other*—**Blackfoot** *Protection* Sap used to conceal human scent when stealing enemy horses. **Cheyenne** *Paint* Fruits used to make red, green, yellow, purple, and white paint for suitcases and tepees. (82:68) **Kiowa** *Ceremonial Items* Used to make smoke sticks for the peyote ceremony. *Fuel* Used for fuel. (192:19) **Montana Indian** *Fuel* Wood used for fire and shelter during the winter. (19:19) Branches used for firewood. (82:68) *Smoke Plant* Inner bark used for "kinnikinnick." (19:19) **Pima** *Containers* Used moistened to line pits for roasting saltbush overnight. (32:23)

Populus deltoides Bartr. ex Marsh. **ssp. *deltoides***, Eastern Cottonwood
- *Drug*—**Choctaw** *Dermatological Aid* Decoction of leaves and bark steam used for wounds. (as *P. angulata* 25:24) *Herbal Steam* Steam from decoction of stems, bark, and leaves used for snakebites. *Snakebite Remedy* Steam from decoction of stems, bark, and leaves used for snakebites. (as *P. angulata* 25:23) Decoction of stems, bark, and leaves used as a steam bath for snakebites. (as *P. angulata* 177:12)

Populus deltoides ssp. monilifera (Ait.) Eckenwalder, Plains Cottonwood

- *Drug*—**Ojibwa, South** *Dermatological Aid* Cotton down used as an absorbent on open sores. (as *P. monilifera* 91:199) **Omaha** *Ceremonial Medicine* Plant used in various rituals. (as *P. sargentii* 68:322)
- *Food*—**Dakota** *Candy* Fruit used as chewing gum by children. *Forage* Branches used as forage for horses. *Unspecified* Inner bark eaten for its pleasant, sweet taste and nutritive value. (as *P. sargentii* 70:72) **Ojibwa** *Unspecified* Buds and seed capsules used for food. (as *P. monilifera* 135:243) **Omaha** *Candy* Fruit used as chewing gum by children. **Pawnee** *Candy* Fruit used as chewing gum by children. **Ponca** *Candy* Cottony fruits used as chewing gum by children. (as *P. sargentii* 70:72)
- *Fiber*—**Omaha** *Building Material* Plant used to make the sacred pole. (as *P. sargentii* 70:72)
- *Dye*—**Dakota** *Yellow* Waxy leaf buds boiled to make a yellow dye. Seed vessels boiled to make a yellow dye for pluming arrow feathers. (as *P. sargentii* 70:72) **Omaha** *Yellow* Leaf buds used to make a yellow dye. (as *P. sargentii* 68:324) Seed vessels boiled to make a yellow dye for pluming arrow feathers. **Pawnee** *Yellow* Waxy leaf buds boiled to make a yellow dye. Seed vessels boiled to make a yellow dye for pluming arrow feathers. **Ponca** *Yellow* Seed vessels boiled to make a yellow dye for pluming arrow feathers. (as *P. sargentii* 70:72)
- *Other*—**Dakota** *Musical Instrument* Leaves used by girls and young women to make a flute-like instrument. *Toys & Games* Leaves used by children to make toy tepees. Leaves used by little girls to make toy moccasins. Green, unopened fruits used by children as beads and ear pendants in play. (as *P. sargentii* 70:72) **Ojibwa** *Paper* Wood used for pulpwood. (as *P. monilifera* 135:243) **Omaha** *Ceremonial Items* Wood used for the sacred pole and for the poles of the "buffalo tent." (as *P. sargentii* 68:321) *Fuel* Bark used as fuel for roasting the clays in making skin paints. *Musical Instrument* Leaves used by girls and young women to make a flute-like instrument. *Sacred Items* Plant used to make the sacred pole. *Toys & Games* Leaves used by children to make toy tepees. Leaves used by little girls to make toy moccasins. Green, unopened fruits used by children as beads and ear pendants in play. **Pawnee** *Musical Instrument* Leaves used by girls and young women to make a flute-like instrument. *Toys & Games* Leaves used by children to make toy tepees. Leaves used by little girls to make toy moccasins. Green, unopened fruits used by children as beads and ear pendants in play. **Ponca** *Musical Instrument* Leaves used by girls and young women to make a flute-like instrument. *Toys & Games* Leaves used by children to make toy tepees. Leaves used by little girls to make toy moccasins. Green, unopened fruits used by children as beads and ear pendants in play. (as *P. sargentii* 70:72)

Populus deltoides ssp. wislizeni (S. Wats.) Eckenwalder, Rio Grande Cottonwood

- *Food*—**Acoma** *Candy* Cotton from the pistillate catkins used as chewing gum. (as *P. wislizeni* 32:31) **Apache, Chiricahua & Mescalero** *Candy* Buds used as chewing gum. (as *P. wislizeni* 33:45) **Apache, White Mountain** *Candy* Buds used as chewing gum. *Unspecified* Buds used for food. (as *P. wislizeni* 136:159) **Isleta** *Candy* Fruit used by children for chewing gum. (as *P. wislizeni* 100:39) *Unspecified* Catkins eaten raw. **Jemez** *Unspecified* Catkins eaten raw. (as *P. wislizeni* 32:43) **Keres, Western** *Candy* Cotton used by children for chewing gum. (as *P. wislizenii* 171:62) **Laguna** *Candy* Cotton from the pistillate catkins used as chewing gum. (as *P. wislizeni* 32:31) **Navajo** *Candy* Sap or catkins, alone or mixed with animal fat, used for chewing gum. (as *P. wislizeni* 55:38) Buds used as chewing gum. *Unspecified* Buds used for food. (as *P. wislizeni* 136:159) **Pima** *Candy* Buds used for chewing gum in early spring. (as *P. fremontii wislizeni* 95:265) **Zuni** *Candy* Buds used as chewing gum. *Unspecified* Buds used for food. (as *P. wislizeni* 136:159)
- *Fiber*—**Isleta** *Building Material* Smaller limbs and leaves used for thatching houses. *Canoe Material* Wood formerly used in making small boats and rafts. (as *P. wislizeni* 100:39) **Navajo** *Building Material* Wood used for firewood, fence posts, vigas (heavy rafters), and tinder boxes. *Furniture* Wood used to make cradles. (as *P. wislizeni* 55:38)
- *Other*—**Isleta** *Cash Crop* Limbs used to make small bows and arrows for sale to tourists. (as *P. wislizeni* 100:39) **Keres, Western** *Ceremonial Items* Twigs mixed with spruce branches the day after the mask dance. *Fuel* Wood used for fuel. (as *P. wislizenii* 171:62) **Navajo** *Ceremonial Items* Wood used to carve dolls and images of some animals for ceremonial purposes. *Containers* Wood used to make tinder boxes. *Fuel* Wood used for firewood. *Tools* Used to make wooden tubes for the bellows used in silversmithing. (as *P. wislizeni* 55:38) **Tewa** *Decorations* Wood used to make many artifacts. (as *P. wislizeni* 138:42)

Populus fremontii S. Wats., Frémont's Cottonwood

- *Drug*—**Cahuilla** *Analgesic* Infusion of bark and leaves used to wet a handkerchief and tie it around the head for headaches. *Antirheumatic (External)* Poultice of boiled bark and leaves applied to swellings caused by muscle strain. *Dermatological Aid* Infusion of bark and leaves used as a wash for cuts. *Veterinary Aid* Infusion of bark and leaves used on horses for saddle sores and swollen legs. (15:106) **Diegueño** *Dermatological Aid* Infusion of leaves used as a wash or poultice of leaves applied to bruises, wounds, or insect stings. (88:216) **Kawaiisu** *Orthopedic Aid* Decoction of inner bark used to wash broken limbs. *Poultice* Poultice of inner bark applied to injured areas. (206:53) **Mendocino Indian** *Dermatological Aid* Decoction of bark used as a wash for bruises and cuts. *Veterinary Aid* Decoction of bark used as a wash for horse sores caused by chafing. (41:330) **Pima** *Dermatological Aid* Decoction of plant used as a wash for sores. (47:109) **Yuki** *Cold Remedy* Infusion of bark or leaves taken for colds. *Dermatological Aid* Decoction of bark used as a wash for sores. Infusion of bark or leaves taken for cuts and sores. *Throat Aid* Infusion of bark or leaves taken for sore throats. (49:46)
- *Food*—**Havasupai** *Candy* "Berries" eaten or chewed like gum. (197:213) **Pima** *Candy* Young, green pods chewed as gum. (47:109) **Pima, Gila River** *Snack Food* Catkins eaten as a snack food by all age groups. (133:5) Flowers eaten as a snack food. (133:7)
- *Fiber*—**Havasupai** *Basketry* Peeled stems split and used to make baskets. *Building Material* Wood used for fence posts and in the construction of shades and houses. (197:213) **Pima** *Basketry* Twigs used for basket making. *Building Material* Used to make fence posts. (47:109)
- *Other*—**Cahuilla** *Cooking Tools* Trunks used to make wooden mor-

tars. (15:106) **Diegueño** *Fuel* Used for firewood. (88:216) **Havasupai** *Cooking Tools* Wood used to make bowls and plates. *Fuel* Wood used for firewood. *Musical Instrument* Hollowed logs used to make drums. *Season Indicator* Falling seeds indicate the time to plant. (197:213) **Mendocino Indian** *Fuel* Wood used occasionally for fuel. (41:330) **Pima** *Fuel* Used as a poor source of fuel. (47:109)

Populus fremontii S. Wats. **var. *fremontii*,** Frémont's Cottonwood
- *Drug*—**Diegueño** *Orthopedic Aid* Decoction of green leaves used as a bath or poultice of hot leaves applied to breaks or sprains. (84:30)

Populus grandidentata Michx., Bigtooth Aspen
- *Drug*—**Cree** *Abortifacient* Used to prevent childbearing. (16:485) *Gynecological Aid* Infusion of bark taken to ease and lessen menses. (16:494) **Iroquois** *Dermatological Aid* Dust from the bark applied to parts affected by itch. (87:293) **Malecite** *Dietary Aid* Infusion of bark used for stimulating the appetite. (116:253) **Ojibwa** *Hemostat* Infusion of young root used as a "hemostatic." (153:387, 388)
- *Food*—**Ojibwa** *Unspecified* Cambium layer scraped, boiled, and eaten, something like eggs. (153:410)
- *Other*—**Ojibwa** *Paper* Wood used for pulpwood. (135:243)

Populus ×jackii Sarg., Jack's Poplar
- *Drug*—**Iroquois** *Oral Aid* Compound infusion used as a wash for mouth ulcers. (as *P. gileadensis* 87:292) **Malecite** *Unspecified* Used to make medicines. (as *P. gileadensis* 160:6)
- *Other*—**Malecite** *Hunting & Fishing Item* Used for trap scents. (as *P. gileadensis* 160:6)

Populus nigra L., Lombardy Poplar
- *Drug*—**Cherokee** *Antirheumatic (Internal)* Tincture of buds used for chronic rheumatism. *Dermatological Aid* Juice of buds used on sores. *Gastrointestinal Aid* Tincture of buds used for colic and bowels. *Stimulant* Given to "persons of phlegmatic habits." *Toothache Remedy* Used for aching teeth. *Venereal Aid* Tincture of buds used for old venereal complaints. (80:24)

Populus sp., Poplar
- *Drug*—**Blackfoot** *Antirheumatic (External)* Infusion of leaves and juniper roots applied like a liniment to stiff backs or backaches. (86:78) *Liver Aid* Infusion of bark taken for liver troubles. (86:67) **Carrier** *Anthelmintic* Fresh bark growth scraped and given to children with worms. *Dermatological Aid* Poultice of boiled inner bark applied to wounds. *Pediatric Aid* Fresh bark growth scraped and given to children with worms. (31:68) **Chickasaw** *Antidiarrheal* Decoction of roots taken for dysentery. (177:11) **Costanoan** *Orthopedic Aid* Decoction of bark made into a syrup and used to set broken bones. (21:21) **Cree, Hudson Bay** *Cathartic* Inner bark eaten as a purgative. *Cough Medicine* Decoction of bark taken for coughs. *Dermatological Aid* Bark used for the astringent qualities. (92:303) **Creek** *Kidney Aid* Decoction of root used as a wash in cases of dropsy. *Orthopedic Aid* Decoction of bark poured over fractured limbs and inner bark used to make splints. Decoction of bark poured over sprained ankles or other joints. (172:660) Decoction of bark used as a wash for a broken arm. (177:11)
- *Food*—**Blackfoot** *Fodder* Bark fed to horses during war parties. (97:28) **Coeur d'Alene** *Unspecified* Cambium layer occasionally used for food. (178:91) **Costanoan** *Unspecified* Inner bark used for food. (21:248) **Dakota** *Candy* Fruit seeds used by children as chewing gum. (69:361) *Fodder* Bark, similar to oats, used for horse feed. *Sweetener* Inner bark eaten in the spring and winter for the sweet taste and agreeable flavor. In the winter, the inner bark was chewed to extract the sweetness, but the fiber was rejected. (69:360) **Hopi** *Candy* "Berries" chewed as gum, particularly with chili. (42:346) Berries chewed as gum with chili. (200:71)
- *Fiber*—**Blackfoot** *Building Material* Used to make the center poles of the ceremonial lodges. Branches used to complete the building of the lodge. (86:29) **Carrier** *Clothing* Rotten wood used to wrap babies in at night as a diaper. (31:68) **Hopi** *Building Material* Trunks used as beams in construction of houses. (200:71) **Hualapai** *Basketry* New shoots used in basketry. (195:3) **Navajo** *Building Material* Boughs used to make the circular or oval summer shelter. (55:37)
- *Other*—**Blackfoot** *Ceremonial Items* Pole used as Sun Dance ceremony centerpiece to symbolize the axis between people and world beyond. Inner bark used as punk in the ceremonial lighting of pipes. (97:28) *Decorations* Used to make head wreaths by the Motokiks. (86:29) *Fuel* Branches used for firewood. *Incense & Fragrance* Sap rubbed on the bodies of horse thieves to disguise the human scent. *Preservative* Branches used for drying meat. (97:28) **Dakota** *Ceremonial Items* Plant used in the tree burial of old times. The body was either placed in hollow tree trunks or laid on a support placed across branches. *Fuel* Wood used for fuel. (69:360) **Eskimo, Inuktitut** *Cooking Tools* Wood used to make carved utensils. *Fuel* Wood used for firewood. (202:186) **Hopi** *Ceremonial Items* Peeled shoots used to make pahos (prayer sticks). Leafy branches used during Snake Dance and related ceremonials. (42:346) Peeled shoots used to make prayer sticks. Leafy branches used in the Snake Dance and related ceremonies. (200:71) *Containers* Roots carved into boxes for sacred feathers and various ceremonial objects. (42:346) Roots carved into boxes for sacred feathers and other ceremonial objects. (200:71) *Musical Instrument* Hollowed, rotten logs used to make drums. (42:346) Hollowed sections of rotten logs made into drums. (200:71) *Tools* Wood used to make fire spindle and sometimes the hearth. *Toys & Games* Roots carved into kachina dolls for children. (42:346) Roots carved into kachina dolls for children and tourists. (200:71) **Hualapai** *Musical Instrument* Trunk hollowed out and used as a drum. (195:3) **Navajo** *Ceremonial Items* Wood used to make prayer sticks. Wood used to carve the image of a duck for the Water Chant. *Containers* Wood used to make tinderboxes. *Fuel* Sticks used in making fire by friction and fiber used for tinder. *Tools* Wood used to make the frame of the loom. *Toys & Games* Wood used to make dice. Wood used to make clubs for the moccasin game. (55:37)

Populus tremuloides Michx., Quaking Aspen
- *Drug*—**Abnaki** *Anthelmintic* Used as a vermifuge. (144:155) Infusion of bark taken as a vermifuge. (144:165) **Algonquin, Tête-de-Boule** *Antirheumatic (External)* Poultice of shredded roots applied to joints for rheumatism. (132:130) **Bella Coola** *Venereal Aid* Decoction of root bark taken for gonorrhea with urethral hemorrhage. (150:54) **Blackfoot** *Gastrointestinal Aid* Infusion of bark used for heartburn. (86:83) *Gynecological Aid* Infusion of bark scrapings taken by women about to give birth. (86:61) *Panacea* Infusion of bark used for general dis-

comfort. (86:83) **Carrier, Southern** *Analgesic* and *Gastrointestinal Aid* Decoction of bark taken for stomach pain. (150:54) **Chippewa** *Dermatological Aid* Poultice of chewed bark or root applied to cuts. (53:350) *Gynecological Aid* Compound infusion of root taken for "excessive flowing" during confinement. (53:358) *Heart Medicine* Compound decoction of inner bark prepared ceremonially for heart trouble. (53:338) **Cree, Woodlands** *Dermatological Aid* Poultice of crushed leaf applied to bee stings to reduce the irritation. Poultice of inner bark used as a wound dressing. *Hemostat* White, powdery substance on the outer bark surface scraped off and used as a styptic. *Venereal Aid* Bark outer surface scraped and used for venereal disease. Decoction of bark taken for venereal disease. (109:52) **Delaware, Ontario** *Cold Remedy* Compound containing bark taken for colds. (175:82) **Flathead** *Other* Infusion of bark used for ruptures. (82:37) **Gitksan** *Cathartic* Bark used as a purgative. (73:152) Decoction of bark taken as a purgative. *Dermatological Aid* Poultice of chewed or mashed root bark applied to cuts. (150:54) **Haisla** *Laxative* Decoction of bark taken as a laxative. **Haisla & Hanaksiala** *Oral Aid* Leaves used for mouth abscesses. (43:286) **Iroquois** *Anthelmintic* Decoction of bark taken for worms. *Gastrointestinal Aid* Infusion of bark taken for cramps caused by worms. *Misc. Disease Remedy* Cold, compound infusion of bark taken for measles. Poultice of wetted bark applied for pleurisy. *Pediatric Aid* Infusion of bark from young trees given when "baby cries, but is not sick." (87:292) Infusion of bark or decoction of young shoots given to children for worms. *Urinary Aid* Compound used for bed-wetting. (87:293) *Venereal Aid* Infusion of roots or bark used internally and externally for venereal diseases. *Veterinary Aid* Decoction of bark mixed with feed for horses for worms. (87:292) Decoction of bark given to dogs and cats with fits caused by worms. (87:293) **Isleta** *Orthopedic Aid* Bark used as casts in setting broken limbs. (as *P. aurea* 100:38) **Meskwaki** *Cold Remedy, Cough Medicine,* and *Pediatric Aid* Decoction of buds used as a nasal salve by children and adults for coughs and colds. (152:245) **Micmac** *Cold Remedy* Bark used for colds. *Dietary Aid* Bark used to stimulate the appetite. (40:59) **Montagnais** *Anthelmintic* Infusion of dried bark given to children suffering from worms. (156:315) **Ojibwa** *Dermatological Aid* Poultice of bark applied to cuts and wounds. *Orthopedic Aid* Poultice of inner bark applied to sore arm or leg and used as a splint for broken limb. (153:388) **Okanagan-Colville** *Dermatological Aid* Bark powder used on the feet and underarms as a deodorant and antiperspirant. *Eye Medicine* Infusion of young growth used as a bath for bruised eyes. (188:134) **Okanagon** *Antirheumatic (External)* Decoction of stems and branches used as a wash for rheumatism. *Antirheumatic (Internal)* Decoction of stems and branches taken for rheumatism. *Gastrointestinal Aid* Decoction of stems and branches taken for dyspepsia. (125:41) **Paiute** *Cough Medicine* Infusion of inner bark taken for coughs from pneumonia. *Febrifuge* Infusion of inner bark taken for fever with excessive perspiration. (111:61) **Penobscot** *Cold Remedy* and *Diaphoretic* Infusion of bark taken as a diaphoretic for colds. (156:310) **Potawatomi** *Veterinary Aid* Burned bark ashes mixed with lard and used as a salve for sores on horses. (154:81) **Salish** *Venereal Aid* Decoction of rootlets and stems taken for syphilis. (178:294) **Shoshoni** *Venereal Aid* Decoction of bark taken over a long period of time for venereal disease. (180:120, 121) **Sikani** *Anthelmintic* Infusion of scraped bark taken for worms and caused a stool immediately. *Dermatological Aid* Poultice of pulverized bark and water applied as paste to wounds. (150:54) **Tanana, Upper** *Cold Remedy* Decoction of inner bark, outer bark, and Hudson Bay tea used for colds. *Cough Medicine* Decoction of inner bark, outer bark, and Hudson Bay tea used for coughs. (102:5) **Tewa** *Urinary Aid* Decoction of leaves taken for urinary trouble. (138:42) **Thompson** *Antirheumatic (External)* Decoction of stems and branches used as a wash for rheumatism. *Antirheumatic (Internal)* Decoction of stems and branches taken for rheumatism. (125:41) *Dermatological Aid* Wood ash mixed with water or grease and used as a salve on swellings. (164:464) Powdery substance from bark rubbed on girls' armpits so that they would not grow underarm hair. The powder was rubbed on girls' armpits after their first menstrual period. Young men, too, rubbed the powdery substance on their arms and faces to prevent the growth of hair. Wood ashes rubbed on men's faces and arms to prevent the growth of hair. (187:277) *Disinfectant* Decoction of bark rubbed on adolescents' bodies for purification. (164:504) *Gastrointestinal Aid* Decoction of stems and branches taken for dyspepsia. (125:41) *Pediatric Aid* Decoction of bark rubbed on adolescents' bodies for purification. (164:504) *Psychological Aid* Decoction of branches taken by people suffering from insanity through excessive drinking. (187:277) *Venereal Aid* Decoction of branches or roots taken and used as a wash for syphilis. (164:464)

- **Food**—**Apache, Chiricahua & Mescalero** *Bread & Cake* Inner bark scraped off and baked in the form of cakes. *Unspecified* Bark boiled or eaten raw. (33:43) **Apache, Mescalero** *Spice* Sap used as flavoring for wild strawberries. (14:50) **Blackfoot** *Fodder* Bark made an excellent winter food for horses. (86:89) Bark fed to horses during the winter. (97:28) *Snack Food* Cambium used as a snack food by children. *Special Food* Bark sucked by anyone observing a liquid taboo. *Unspecified* Cambium used for food. (86:104) Inner bark eaten in the spring. (97:28) **Cree, Woodlands** *Preservative* Dry, rotted wood used to make a fire to smoke cure whitefish and moose meat. *Unspecified* Cambium eaten fresh in early summer. Inner bark used for food in the spring. (109:52) **Montana Indian** *Fodder* Young twigs fed to horses when other food was not obtainable. *Unspecified* Inner bark considered a valuable mucilaginous food. (19:19) **Navajo, Ramah** *Starvation Food* Inner bark eaten raw as an emergency ration. (191:22) **Tanana, Upper** *Preservative* Wood used to smoke fish. *Unspecified* Sap and cambium used for food. (102:5) **Thompson** *Forage* Bark eaten by beavers. (187:277)

- **Fiber**—**Cheyenne** *Building Material* Logs used to make Sundance lodges. (82:37) Trunks used to construct the medicine lodge for the Sun Dance ceremony. (83:46) Trunks used to build the Sun Dance Lodge. (83:37) **Cree, Woodlands** *Building Material* Poles placed upon the birch bark cover of a tepee to secure it. Poles used to frame a tepee. (109:52) **Crow** *Building Material* Logs used to make Sundance lodges. (82:37) **Klamath** *Clothing* Bark used to make hats. (45:94) **Montana Indian** *Building Material* Wood used for fire and shelter during the winter. *Cordage* Bark sometimes employed as cordage. (19:19) **Paiute** *Building Material* Young trees used for tent and tepee poles. (111:61) **Thompson, Upper (Fraser Band)** *Canoe Material* Wood used to make dugout canoes. (164:497)

- **Other**—**Blackfoot** *Toys & Games* Bark or moistened leaves used to make whistles. (86:119) **Cree, Woodlands** *Hunting & Fishing Item*

Trees used to make a deadfall trap for bear. *Toys & Games* Short section of fresh branch used to make a toy whistle. (109:52) **Hopi** *Ceremonial Items* Plant smoked ceremonially. (as *P. aurea* 200:71) **Montana Indian** *Fuel* Wood used for fire and shelter during the winter. (19:19) Used occasionally for firewood. (82:37) *Smoke Plant* Inner bark used for "kinnikinnick." (19:19) **Navajo** *Ceremonial Items* Tree important to the Sun's House Chant. This tree, according to legend, has the distinction of being the first tree against which the bear rubs his back in the Sun's House Chant. The others are red willow, fir, and chokecherry. (55:38) **Navajo, Ramah** *Ceremonial Items* Stem used to make Evilway hoop. *Cooking Tools* Knots used to make wooden cups. (191:22) **Ojibwa** *Paper* Wood used for pulpwood. (135:243) **Okanagan-Colville** *Tools* Logs used to scrape deer hides. (188:134) **Paiute** *Fuel* Dry limbs used as a source of fuel. (111:61) **Shuswap** *Toys & Games* Branches used by boys to make whistles. (123:68) **Tanana, Upper** *Fuel* Wood used for fuel. *Smoke Plant* Wood ashes mixed with tobacco and used for chewing tobacco. (102:5) **Thompson** *Hunting & Fishing Item* Decoction of branches used to wash traps, guns, buckskins, and hunters. The decoction was used to wash humans such as hunters who desired to be exceptionally "clean." *Protection* Decoction of branches used as a protective bath against witches. (187:277)

Porophyllum, Asteraceae

Porophyllum gracile Benth., Slender Poreleaf

- **Drug**—**Havasupai** *Analgesic* Decoction of pounded plant taken for pain. *Antirheumatic (External)* Decoction of pounded plant rubbed in as a liniment. *Antirheumatic (Internal)* Decoction of pounded plant taken for aches. *Dermatological Aid* Decoction of pounded plant used as a wash on sores. *Gastrointestinal Aid* Decoction of pounded plant taken for abdominal pain. (197:249) **Paiute** *Abortifacient* Decoction of root taken as "a regulator for delayed menstruation." (as *P. leucospermum* 180:122) **Shoshoni** *Abortifacient* Plant used to regulate delayed menstruation. (as *P. leucospermum* 118:46)

Porphyra, Bangiaceae, alga

Porphyra abbottae Krishnamurthy, Edible Seaweed

- **Drug**—**Hanaksiala** *Gastrointestinal Aid* Decoction of plant taken or poultice applied for any kind of sickness in the stomach or body. *Orthopedic Aid* Poultice of plant applied to broken collarbones. *Panacea* Decoction of plant taken or poultice applied for any kind of sickness in the stomach or body. (43:131)
- **Food**—**Haisla & Hanaksiala** *Dried Food* Plant gathered and dried for winter use. *Sauce & Relish* Plant dried, crushed, and sprinkled on various foods as a condiment. (43:131) **Kitasoo** *Bread & Cake* Plant pressed into boxes to form compressed cakes, dried, and stored for future use. *Unspecified* Plant eaten with salmon roe or butter clams. (43:304) **Oweekeno** *Bread & Cake* Whole plant formed into flat sheets, pressed in boxes, dried, and made into cakes. *Dried Food* Whole plant dried for future use. *Unspecified* Whole plant cooked and eaten with salmon eggs, cooked salmon, clams, herring eggs, and other foods. (43:47)
- **Other**—**Tsimshian** *Cash Crop* Plant used for trade. (43:304)

Porphyra laciniata (Lightfoot) Agardh, Seaweed

- **Food**—**Alaska Native** *Dried Food* Leaves sun dried, chopped, dried, and stored in closed containers. *Snack Food* Leaves sun dried, chopped, dried, and eaten raw like popcorn. *Soup* Leaves used in fish stews and soups. (85:141)

Porphyra lanceolata (Setchell) G. M. Smith

- **Food**—**Pomo, Kashaya** *Dried Food* Fresh seaweed dried for later use. *Unspecified* Fresh seaweed baked and eaten. (72:126) **Tolowa** *Unspecified* Species used for food. **Yurok** *Unspecified* Species used for food. (6:47)

Porphyra perforata J. Agardh

- **Food**—**Hesquiat** *Unspecified* Boiled with herring spawn and eaten with dogfish oil or eulachon (candlefish) oil. (edible seaweed 185:25) **Pomo** *Bread & Cake* Weeds stacked like cakes and dried until needed. *Dried Food* Seaweed sun dried and used for food. (11:94) *Winter Use Food* Plant made into a cake, cooked in earth oven, and stored for winter consumption. (66:10) **Pomo, Kashaya** *Dried Food* Fresh seaweed dried for later use. *Unspecified* Fresh seaweed baked and eaten. (72:125)
- **Other**—**Nitinaht** *Cash Crop* Plants formerly sold to the Chinese. (edible seaweed 186:54)

Porphyra sp., Red Laver

- **Food**—**Bella Coola** *Dried Food* Sun dried and eaten alone or cooked with other food. (184:195) **Kwakiutl, Southern** *Bread & Cake* Plants allowed to rot, shaped into cakes, sun dried, and eaten with dried salmon at feasts. *Dessert* Mixed with water, beaten until frothy and white, and eaten as a dessert. *Dried Food* Plants sun dried, boiled with clams or creamed corn, and eaten with large amounts of oulachen (candlefish) grease. *Snack Food* Plants sun dried and eaten as a snack. *Unspecified* Strips lightly browned, pulverized, cooked, and used for food. (183:262) **Nitinaht** *Unspecified* Plant used for food. (67:210)

Porteranthus, Rosaceae

Porteranthus sp.

- **Drug**—**Creek** *Gynecological Aid* Plant thought to be used by women during their menstrual periods. (172:667)

Porteranthus stipulatus (Muhl. ex Willd.) Britt., Indian Physic

- **Drug**—**Cherokee** *Antirheumatic (External)* Roots used for rheumatism. *Cold Remedy* Mild infusion used in slight doses for colds. *Dermatological Aid* Cold infusion of root given or root chewed "for bee and other stings." *Emetic* Mild infusion taken as an emetic. (as *Gillenia stipulata* 80:40) Decoction or strong infusion of whole plant taken a pint at a time as an emetic. (as *Gillenia stipulata* 203:74) *Kidney Aid* Infusion taken for kidneys. *Liver Aid* Compound taken for liver. *Orthopedic Aid* Poultice used for "leg swelling." *Respiratory Aid* Mild infusion used in slight doses for asthma. *Toothache Remedy* Infusion used for toothache. *Veterinary Aid* Tincture of root given for "milksick."

Porteranthus trifoliatus (L.) Britt., Bowman's Root

- **Drug**—**Cherokee** *Antirheumatic (External)* Poultice used for rheumatism. *Cold Remedy* Mild infusion used in slight doses for colds. *Dermatological Aid* Cold infusion of root given or root chewed "for bee and other stings." (as *Gillenia trifoliata* 80:40) Infusion of roots used as a wash for leg scratches. (as *Gillenia trifoliata* 177:27) *Emetic*

Mild infusion taken as an emetic. (as *Gillenia trifoliata* 80:40) Decoction or strong infusion of whole plant taken a pint at a time as an emetic. (as *Gillenia trifoliata* 203:74) *Kidney Aid* Infusion taken for kidneys. *Liver Aid* Compound taken for liver. *Orthopedic Aid* Poultice used for "leg swelling." *Respiratory Aid* Mild infusion used in slight doses for asthma. *Toothache Remedy* Infusion used for toothache. *Veterinary Aid* Tincture of root given for "milksick." (as *Gillenia trifoliata* 80:40) **Iroquois** *Antidiarrheal* Compound decoction of leaves and switches taken for diarrhea. (as *Gillenia trifoliata* 87:349) *Cathartic* Plant used as a physic. *Cold Remedy* Infusion of roots taken for colds, fevers, and chills caused by fever and sore throats. *Diaphoretic* Decoction of roots taken to cause sweating. *Febrifuge* Infusion of roots taken for colds, fevers, and chills caused by fever and sore throats. *Misc. Disease Remedy* Decoction of roots taken for grippe. *Throat Aid* Infusion of roots taken for colds, fevers, and chills caused by fever and sore throats. (as *Gillenia trifoliata* 87:350)

Portulacaceae, see *Calandrinia, Cistanthe, Claytonia, Lewisia, Portulaca, Talinum*

Portulaca, Portulacaceae

Portulaca oleracea L., Little Hogweed

- **Drug**—**Cherokee** *Anthelmintic* Compound decoction taken for worms. *Ear Medicine* Juice used for earache. (80:51) **Hawaiian** *Strengthener* Plant and other plants pounded, squeezed, and resulting liquid taken to check run-down conditions. (2:24) **Iroquois** *Antidote* Good medicine to cure you if someone has given you some bad medicine. *Burn Dressing* Poultice of mashed plant used on burns. *Dermatological Aid* Poultice of entire plant used on bruises. (87:318) **Keres, Western** *Antidiarrheal* Infusion of leaf stems used for diarrhea. *Blood Medicine* Infusion of leaf stems used as an antiseptic wash for blood clots. *Oral Aid* Raw leaves rubbed in mouth for difficulty in opening the mouth. (171:62) **Navajo** *Analgesic* Plant used for pain. (55:97) *Gastrointestinal Aid* Plant taken for stomachaches. *Panacea* Plant used to "cure sick people." (55:47) **Rappahannock** *Dermatological Aid* Compound decoction of bruised leaves applied as salve for "footage" trouble. (161:28)
- **Food**—**Acoma** *Vegetable* Plants cooked with meat and eaten like spinach. (32:43) **Apache, Chiricahua & Mescalero** *Unspecified* Eaten without preparation or cooked with green chile and meat or animal bones. (33:46) **Hopi** *Unspecified* Cooked in a gravy. (200:75) **Iroquois** *Vegetable* Cooked and seasoned with salt, pepper, or butter. (196:118) **Isleta** *Vegetable* Plants oven dried, stored, and used as greens during the winter. (32:43) *Winter Use Food* Plants dried in ovens, stored, and used as greens in the winter. (100:39) **Keres, Western** *Vegetable* Plant cooked with meat as greens. (171:62) **Laguna** *Vegetable* Plants cooked with meat and eaten like spinach. (32:43) **Luiseño** *Vegetable* Plant used for greens. (155:232) **Navajo** *Unspecified* Seeds used for food. (55:47) **Navajo, Ramah** *Spice* Leaves used as a potherb. *Vegetable* Leaves boiled as greens with meat. (191:26) **Pima, Gila River** *Unspecified* Leaves boiled and eaten. (133:7) **Tewa** *Unspecified* Fleshy plant tops boiled and eaten. (138:59)

Portulaca oleracea L. ssp. *oleracea*, Little Hogweed

- **Drug**—**Navajo, Kayenta** *Misc. Disease Remedy* Plant used as a lotion for scarlet fever. (as *P. retusa* 205:22)

- **Food**—**Hopi** *Unspecified* Plant boiled with meats and eaten. (as *P. retusa* 56:15) Plant formerly cut up fine and eaten in gravy. (as *P. retusa* 138:60) **Navajo** *Forage* Plant used as a good sheep forage. *Unspecified* Seeds used for food. (as *P. retusa* Engelm. 55:47) Plants used for food. (as *P. retusa* 90:154) **Pima, Gila River** *Unspecified* Leaves boiled and eaten. (as *P. retusa* 133:7) **San Felipe** *Unspecified* Young plants fried or boiled and mixed with young peas. *Vegetable* Young plants used as greens. (as *P. retusa* 32:43)

Portulaca sp., Common Purslane

- **Food**—**Pima, Gila River** *Vegetable* Leaves used as greens. (133:5)

Postelsia, Lessoniaceae

Postelsia palmaeformis Ruprecht, Sea Palm

- **Drug**—**Hesquiat** *Strengthener* Whalers rubbed four or eight pieces of plant on their arms to make them as strong as the plant. (185:26) **Nitinaht** *Anticonvulsive* Plants burned and ashes used for convulsions. *Psychological Aid* Plants burned and ashes used for craziness. *Strengthener* Stipes dried, burned, powdered, mixed with raccoon marrow, and salve used to strengthen young boys. **Nootka** *Strengthener* Stipes dried, burned, powdered, mixed with raccoon marrow, and salve used to strengthen young boys. (186:54)
- **Food**—**Hesquiat** *Dried Food* Stipes and fronds with attached herring eggs dried for later use.[6] (185:24) **Pomo** *Unspecified* Cooked stalks used for food. Raw stalks chewed like sugar cane. (as *P. palmiformis* 66:10) Plant chewed raw. (11:95) **Pomo, Kashaya** *Dried Food* Stems cut into long strips and sun dried for winter use. *Unspecified* Fresh stem chewed raw or baked in an oven or hot ashes. (72:126)
- **Fiber**—**Hesquiat** *Sporting Equipment* Dried stipes use as "pucks" and hitting sticks.[5] (185:24)
- **Other**—**Nitinaht** *Toys & Games* Tough, rubbery holdfasts carved into "beach hockey" balls. (186:54)

Postelsia sp., Short Beach Kelp

- **Fiber**—**Hesquiat** *Sporting Equipment* Dried stems used as "pucks" and sticks for "beach hockey." (185:18)

Potamogetonaceae, see *Potamogeton, Ruppia*

Potamogeton, Potamogetonaceae

Potamogeton diversifolius Raf., Waterthread Pondweed

- **Fiber**—**Kawaiisu** *Cordage* Dried stem fibers used to make a strong cord. (206:53)
- **Other**—**Kawaiisu** *Containers* Dried stem fibers made into strong cords and used to make carrying nets. *Hunting & Fishing Item* Dried stem fibers made into strong cords and used to make rabbit nets. (206:53)

Potamogeton natans L., Floating Pondweed

- **Drug**—**Navajo, Ramah** *Ceremonial Medicine* and *Emetic* Decoction of plant taken as ceremonial emetic. (191:15)

Potamogeton sp., Pondweed

- **Drug**—**Iroquois** *Witchcraft Medicine* Compound poultice bound to "soreness all over in men from being witched." (87:272)
- **Food**—**Hesquiat** *Forage* Plant browsed by deer. (185:17) Deer wade into the water and put their heads under the surface to eat this plant. (185:56)

Potentilla, Rosaceae

Potentilla arguta Pursh, Tall Cinquefoil
- **Drug**—**Okanagan-Colville** *Gynecological Aid* Infusion of roots taken by women after childbirth. (188:126)

Potentilla arguta Pursh **ssp. *arguta***, Tall Cinquefoil
- **Drug**—**Chippewa** *Analgesic* Dry, pulverized root pricked into temples or placed in nostrils for headache. (as *Drymocallis arguta* 53:338) *Antidiarrheal* Simple or compound decoction of root taken for dysentery. (as *Drymocallis arguta* 53:344) *Dermatological Aid* Poultice of moistened, dried, powdered root applied to cuts. (as *Drymocallis arguta* 53:350)

Potentilla canadensis L., Dwarf Cinquefoil
- **Drug**—**Iroquois** *Antidiarrheal* Infusion of pounded roots taken for diarrhea. (87:353) **Natchez** *Witchcraft Medicine* Plant given to one who was bewitched. (172:667)

Potentilla crinita Gray, Bearded Cinquefoil
- **Drug**—**Navajo, Ramah** *Panacea* Cold infusion of whole plant taken as "life medicine." (191:31)

Potentilla glandulosa Lindl., Gland Cinquefoil
- **Drug**—**Gosiute** *Dermatological Aid* Plant taken and poultice of plant applied to swollen parts. *Unspecified* Root used as medicine. (39:378) **Okanagon** *Stimulant* Infusion of whole plant taken as a stimulant. *Tonic* Infusion of whole plant taken as a tonic. **Thompson** *Stimulant* Infusion of whole plant taken as a stimulant. (125:42) Weak decoction of leaves taken as a stimulant. (164:469) Decoction of leaves or whole plant said to be slightly stimulant. (164:494) *Tonic* Infusion of whole plant taken as a tonic. (125:42) Decoction of plant taken as a tonic for "general out-of-sorts feeling." (164:469)

Potentilla gracilis Dougl. ex Hook., Northwest Cinquefoil
- **Drug**—**Okanagan-Colville** *Analgesic* Infusion of pounded roots taken as a general tonic for pains. *Antidiarrheal* Infusion of pounded roots taken for diarrhea. *Antirheumatic (Internal)* Infusion of pounded roots taken as a general tonic for aches. *Blood Medicine* Infusion of pounded roots taken as a blood tonic. *Dermatological Aid* Infusion of pounded roots used to wash sores. *Venereal Aid* Infusion of pounded roots taken for gonorrhea. (188:127) **Thompson** *Dermatological Aid* Poultice of mashed leaves, roots, and subalpine fir pitch used on wounds, to draw out the pain. (187:263)

Potentilla hippiana Lehm. **var. *hippiana***, Woolly Cinquefoil
- **Drug**—**Navajo, Kayenta** *Burn Dressing* Plant used as a lotion for burns. *Dermatological Aid* Powdered plant applied to sores caused by a bear. *Gynecological Aid* Plant used to expedite childbirth. (205:26) **Navajo, Ramah** *Dermatological Aid* Poultice of fresh leaves applied to injury. *Panacea* Cold infusion of root taken as "life medicine." (191:31)

Potentilla nana Willd. ex Schlecht., Arctic Cinquefoil
- **Drug**—**Eskimo** *Unspecified* Root eaten for the medicinal value. (as *P. hyparctica* 149:325)

Potentilla norvegica **ssp. *monspeliensis*** (L.) Aschers. & Graebn., Norwegian Cinquefoil
- **Drug**—**Chippewa** *Throat Aid* Decoction of root gargled or root chewed for sore throat. (as *P. monspeliensis* 53:342) **Navajo, Ramah** *Analgesic* Cold infusion of whole plant used for pain. *Venereal Aid* Fumes from plant used for sexual infection. (as *P. monspeliensis* 191:31) **Ojibwa** *Cathartic* Plant known to be a physic, even by the very young. (as *P. monspeliensis* 153:384) **Potawatomi** *Unspecified* Root used for unspecified malady. (as *P. monspeliensis* 154:77)

Potentilla pensylvanica L., Pennsylvania Cinquefoil
- **Drug**—**Navajo, Ramah** *Panacea* Root used as a "life medicine." (191:31)

Potentilla recta L., Sulphur Cinquefoil
- **Drug**—**Okanagan-Colville** *Dermatological Aid* Poultice of pounded leaves and stems applied to open sores and wounds. *Internal Medicine* Infusion of leaves taken for all types of internal troubles. (188:127)

Potentilla simplex Michx., Common Cinquefoil
- **Drug**—**Cherokee** *Antidiarrheal* Infusion of root taken for dysentery. *Dermatological Aid* Infusion of astringent root used as a mouthwash for "thrash." *Febrifuge* Used for "fevers and acute diseases with great debility." *Oral Aid* Infusion of root used as a mouthwash for "thrash." *Pulmonary Aid* Root eaten and infusion of root used by ballplayers "for wind" and safety. (as *P. simplex* 80:29)

Potentilla sp., Five Finger
- **Drug**—**Cherokee** *Febrifuge* Infusion given for fever. (203:74) **Paiute** *Laxative* Whole plant used as a laxative. (118:42)

Pottiaceae, see *Barbula*

Prenanthes, Asteraceae

Prenanthes alata (Hook.) D. Dietr., Western Rattlesnakeroot
- **Drug**—**Bella Coola** *Analgesic* Poultice of chewed root applied to any painful part of body. *Burn Dressing* Poultice of chewed root applied to burns. *Cold Remedy* Decoction of root taken and small dose given to babies for colds. *Pediatric Aid* Decoction of roots taken daily and small dose given to babies for colds. (150:65)

Prenanthes alba L., White Rattlesnakeroot
- **Drug**—**Chippewa** *Gynecological Aid* Dried, powdered root added to food to produce postpartum milk flow. (53:360) **Iroquois** *Dermatological Aid* Poultice of roots applied to dog bites. (87:479) *Snakebite Remedy* Poultice of roots applied to rattlesnake bites. *Stimulant* Infusion of smashed roots used as wash for weakness. (87:478) **Ojibwa** *Diuretic* Milk of lettuce used, especially in female diseases, as a diuretic. *Gynecological Aid* Milk of plant used as a diuretic for female diseases and root used as a female remedy. (153:365)

Prenanthes altissima L., Tall Rattlesnakeroot
- **Drug**—**Iroquois** *Snakebite Remedy* Poultice of smashed roots applied to rattlesnake bites. (87:480)

Prenanthes aspera Michx., Rough Rattlesnakeroot
- **Drug**—**Choctaw** *Analgesic* Decoction of roots and plant tops used as an anodyne. *Diuretic* Decoction of roots and plant tops used as a stimulating diuretic. *Other* Plant used as a "secernant" (which induces secretion). *Stimulant* Plant used as a stimulant. (as *Nabalus asper* 28:288)

Prenanthes serpentaria Pursh, Cankerweed
- *Drug*—Cherokee *Analgesic* Roots used in stomachache medicine. (80:35)
- *Food*—Cherokee *Vegetable* Leaves eaten as cooked salad. (80:35)

Prenanthes trifoliolata (Cass.) Fern., Gall of the Earth
- *Drug*—Cherokee *Analgesic* Roots used in stomachache medicine. (80:35) **Iroquois** *Dermatological Aid* Poultice of roots applied to skin swellings or bristles on a toe or foot. (87:480) *Eye Medicine* Decoction of roots used as drops for sore eyes. *Hunting Medicine* Compound infusion of root used as wash for rifles, a "deer hunting medicine." *Love Medicine* Root chewed and rubbed on hands and face as a love medicine. *Snakebite Remedy* Roots used for rattlesnake bites. (87:479)
- *Food*—Cherokee *Vegetable* Leaves eaten as cooked salad. (80:35)

Primulaceae, see *Anagallis, Androsace, Dodecatheon, Glaux, Lysimachia, Trientalis*

Proboscidea, Pedaliaceae

Proboscidea althaeifolia (Benth.) Dcne., Devilshorn
- *Drug*—Pima *Analgesic* and *Antirheumatic (External)* Plant moxa used for rheumatic pains. (as *Martynia arenaria* 47:107)
- *Food*—Cahuilla *Unspecified* Seeds used for food. (15:107) **Papago** *Unspecified* Young pods used for food. (as *Martynia arenaria* 47:107)
- *Fiber*—Pima *Basketry* Used for basket making. (as *Martynia arenaria* 47:107)
- *Other*—Cahuilla *Tools* Hooked thorns used as a tool in mending baskets and broken pottery. (15:107)

Proboscidea louisianica (P. Mill.) Thellung, Common Devilsclaw
- *Fiber*—Kawaiisu *Basketry* Long, pointed seed capsule horns used as the black pattern material in coiled basketry. (206:53)
- *Dye*—Shoshoni *Black* Dried pods used as black dye, pieces buried in wood ashes to deepen the shade. (as *Martynia proboscides* 118:7)
- *Other*—Panamint *Decorations* Used as black decorations for baskets. (as *Martynia proboscidea* 105:78)

Proboscidea louisianica ssp. *fragrans* (Lindl.) Bretting, Ram's Horn
- *Food*—Papago *Dried Food* Seeds basket winnowed, parched, sun dried, cooked, stored, and used for food. (as *Martynia fragrans* 34:24) *Unspecified* Seeds boiled and eaten. (as *Martynia fragrans* 34:25)
- *Fiber*—Papago *Basketry* Pods softened with lye and water, split, bent, and used as sewing withes in coiled basketry. (as *Martynia fragrans* 34:57)

Proboscidea louisianica (P. Mill.) Thellung **ssp. *louisianica***, Louisiana Ram's Horn
- *Food*—Apache, Chiricahua & Mescalero *Unspecified* Seeds eaten by prisoners of war in Oklahoma. (as *Martynia louisiana* 33:45)
- *Other*—Hopi *Ceremonial Items* Used in the preparation of ceremonial paraphernalia. (as *Martynia louisiana* Mill. 200:92)

Proboscidea parviflora (Woot.) Woot. & Standl., Doubleclaw
- *Food*—Havasupai *Dried Food* Fruit sun dried for future use. *Unspecified* Seeds used for food. (197:241) **Hualapai** *Unspecified* Young pods used for food. (195:38)
- *Fiber*—Havasupai *Basketry* Spines stored and used as decorative elements in basketry. (197:241) **Hualapai** *Basketry* Mature pods used for the black designs in baskets. (195:38)

Proboscidea parviflora (Woot.) Woot. & Standl. **ssp. *parviflora***, Doubleclaw
- *Drug*—Pima *Analgesic* and *Antirheumatic (External)* Plant moxa used for rheumatic pains. (as *Martynia parviflora* 47:107)
- *Food*—Papago *Unspecified* Young pods used for food. **Pima** *Unspecified* Seeds dried, cracked, and eaten like pine nuts. (as *Martynia parviflora* 47:107)
- *Fiber*—Pima *Basketry* Used extensively for basket making. (as *Martynia parviflora* 47:107)
- *Other*—Pima *Designs* Seedpods used as black designs for coiled baskets. (as *Martynia parviflora* 47:116)

Proboscidea sp., Devilsclaw
- *Food*—Pima, Gila River *Snack Food* Seeds eaten raw as a snack food. (133:7) *Unspecified* Oily seeds used for food. (133:5)

Prosopis, Fabaceae

Prosopis chilensis, Algarrobo
- *Food*—Kiowa *Fodder* Leaves used for fodder. *Vegetable* Pounded beans and pods used for food. (as *Ceratonia chilensis* 192:33)

Prosopis glandulosa Torr., Honey Mesquite
- *Drug*—Apache, Mescalero *Eye Medicine* Juice from leaves used for irritated eye lids. *Pediatric Aid* and *Urinary Aid* Infusion of bark used for children with enuresis. (14:37) **Comanche** *Gastrointestinal Aid* Leaves chewed and juice swallowed to neutralize acid stomach. (29:523) **Isleta** *Eye Medicine* Decoction of leaves and pods without beans used as an eye medicine. (100:39) **Keres, Western** *Eye Medicine* Leaves made into an eyewash. (171:63)
- *Food*—Acoma *Porridge* Beans formerly ground into flour and prepared as mush. *Unspecified* Beans eaten raw or cooked as string beans. (32:43) **Apache** *Bread & Cake* Seeds ground into flour and used in pancakes. *Preserves* Beans boiled, pounded or ground, hand kneaded, and made into a jam. (as *P. prosopis* 32:45) **Apache, Chiricahua & Mescalero** *Beverage* Cooked pods and seeds ground, water added, mixture allowed to ferment, and used as a beverage. (33:53) *Bread & Cake* Bean flour made into pancakes and bread. Beans were gathered, boiled, pounded on a hide or ground on a metate, placed in a pan, and worked with the hands until a thick consistency was attained. *Pie & Pudding* Pods boiled in water, taken out, mashed, boiled again, and eaten as pudding. (33:41) *Spice* Root used to flavor drinks and make them stronger. (33:51) *Substitution Food* Flour used in the absence of sugar to sweeten an intoxicating drink. (33:50) *Unspecified* Beans cooked with meat and seed coats spit out when eaten. (33:41) **Apache, Mescalero** *Beverage* Beans boiled, strained, and used as a drink. *Staple* Beans ground into flour, mixed with other plant foods, and eaten. (14:37) **Comanche** *Staple* Pods made into a meal and used for food. (29:523) **Isleta** *Bread & Cake* Beans ground into a flour and used to make bread. (100:39) *Candy* Beans toasted and eaten as a confection by sucking out the juice. (32:43) Roasted beans eaten as a confection. (100:39) **Keres, Western** *Porridge* Beans ground into a flour, made into a mush, and used for food. *Vegetable*

Beans eaten raw for the sweet taste or cooked like string beans. (171:63) **Kiowa** *Fodder* Leaves used for fodder. *Vegetable* Pounded beans and pods used for food. (192:33) **Laguna** *Porridge* Beans formerly ground into flour and prepared as mush. *Unspecified* Beans eaten raw or cooked as string beans. (32:43) **Pima** *Candy* White resinous secretions used to make candy. (as *P. prosopis* 32:45) **Yavapai** *Staple* Pods pulverized and made into a meal for transporting. (65:257)
- *Other*—**Apache, Mescalero** *Hunting & Fishing Item* Resin used for fletching arrows. (14:37) **Isleta** *Hunting & Fishing Item* Limbs used to make shafts for hunting arrows. (100:39)

Prosopis glandulosa Torr. **var. *glandulosa***, Honey Mesquite
- *Food*—**Apache, Western** *Beverage* Pounded bean pulp squeezed for the juice and drunk just like milk. *Bread & Cake* Dried seeds pounded into flour, moistened, allowed to harden into cakes, and stored. *Candy* Dried beans pounded into flour and eaten as candy. *Dried Food* Pods dried and stored. *Porridge* Dried beans pounded into flour and mixed into a mush. *Staple* Fresh pods pounded into a flour. *Substitution Food* Pitch chewed as a substitute for gum. (as *P. chilensis* 26:176) **Havasupai** *Beverage* Plant used to make a drink. (as *P. juliflora* 197:66) *Candy* Pods eaten raw like a stick of candy. (as *P. juliflora* 197:228) **Kamia** *Unspecified* Pod used for food. (as *P. juliflora* 62:23) **Kiowa** *Fodder* Leaves used for fodder. Leaves used for fodder. Leaves used for fodder. *Vegetable* Pounded beans and pods used for food. Pounded beans and pods used for food. Pounded beans and pods used for food. (as *P. juliflora glandulosa* 192:33) **Luiseño** *Staple* Ground beans made into a flour and used for food in some places. (as *P. juliflora* 155:231) **Mahuna** *Bread & Cake* Bean pods ground into flour and used to make cakes and tarts. *Dried Food* Dried bean pods eaten raw. *Porridge* Bean pods ground into flour, mixed with hot or cold water, and eaten as porridge. (as *P. juliflora* 140:57) **Maricopa** *Beverage* Unripe beans pounded and mixed with water to make a drink. (as *P. juliflora* 37:181) **Mohave** *Bread & Cake* Dried bean pods ground into a meal and used to make cakes. *Dried Food* Beans dried and stored in giant basket granaries for winter use. *Vegetable* Beans eaten raw or roasted. (as *P. juliflora* 168:46) **Paiute** *Unspecified* Pounded beans used for food. (as *P. juliflora* 118:27) **Papago** *Staple* Fruits and seeds used for food. (as *P. chilensis* 36:60) **Seri** *Porridge* Beans ground into a meal, mixed with water or sea lion oil, and eaten. (as *P. juliflora* 50:136) **Southwest Indians** *Unspecified* Seeds used for food. (as *P. chilensis* 17:15) **Yuma** *Beverage* Dried pods boiled to make a beverage. Pods crushed and steeped in water to make a beverage. *Bread & Cake* Meal molded into cakes for storage. *Dried Food* Pods dried on roof tops and stored. *Staple* Pods crushed or ground into a meal. (as *P. juliflora* 37:181)
- *Fiber*—**Havasupai** *Furniture* Wood used to make the base frame of the cradleboard. (as *P. juliflora* 197:228) **Seri** *Cordage* Outer root tissues pounded, split, worked between the hand and the mouth, and twisted into cords. *Mats, Rugs & Bedding* Root strips made into doughnut-shaped head pads used to balance earthen water jars on the heads. *Sewing Material* Outer root tissues woven into rough fabric. (as *P. juliflora* 50:134)
- *Other*—**Apache, Western** *Fasteners* Pitch used to attach arrow points to shafts. *Fuel* Used for firewood. (as *P. chilensis* 26:176) **Havasupai** *Fuel* Wood used for firewood. (as *P. juliflora* 197:228) **Navajo** *Hunting & Fishing Item* Wood used to make bows. (as *P. chilensis* 55:58) **Seri** *Hunting & Fishing Item* Fiber made into cord used for bows. (as *P. juliflora* 50:138)

Prosopis glandulosa var. *torreyana* (L. Benson) M. C. Johnston, Western Honey Mesquite
- *Drug*—**Cahuilla** *Dermatological Aid* Gum diluted with water and used as a wash for open wounds and sores. *Eye Medicine* Gum diluted with water and used as a wash for sore eyes. (as *P. juliflora* var. *torreyana* 15:107) **Diegueño** *Eye Medicine* Infusion of leaves used as an eyewash. *Febrifuge* Infusion of leaves taken for fevers. (as *P. juliflora* var. *torreyana* 88:218)
- *Food*—**Cahuilla** *Beverage* Blossoms used to make tea. Pods crushed into a pulpy juice and used to make a beverage. Pod meal and water used to make a beverage. *Bread & Cake* Pod meal and water used to make cakes. *Porridge* Pod meal and water used to make mush. *Staple* Pods dried and ground into a meal. *Unspecified* Roasted blossoms stored in pottery vessels and cooked in boiling water when needed. Pods eaten fresh. (as *P. juliflora* var. *torreyana* 15:107) **Cocopa** *Unspecified* Pods used for food. *Winter Use Food* Pods stored for later use. (as *P. odorata* 64:267) **Diegueño** *Bread & Cake* Beans ground into a meal and used to make cakes. (84:32) **Kawaiisu** *Bread & Cake* Seeds pounded and molded into a cake without cooking. Pods crushed into a meal, molded into a dry cake, and stored and eaten at a later time. *Porridge* Pods crushed into a meal and eaten with water. (206:54) **Pima** *Beverage* Beans sun dried, pounded into meal, mixed with cold water, and used as a drink. (as *P. odorata* 95:261) **Yavapai** *Unspecified* Seeds used for food. (as *Cercidium torreyana* 63:211) Parched, ground seeds dampened, sometimes mixed with ground saguaro seed, and used for food. (as *Cercidium torreyanum* 65:256) **Yuma** *Beverage* Pods crushed and steeped in water to make a beverage. *Dried Food* Pit cooked pods dried and stored in baskets. Pods dried on roof tops and stored. *Staple* Beans dried thoroughly and pounded into meal. *Unspecified* Pit cooked pods pounded in a mortar and prepared as food. (as *P. odorata* 37:181)
- *Fiber*—**Cahuilla** *Building Material* Large limbs used as corner posts for houses, as rafters, and granary posts. Leaves used for roofing houses. *Clothing* Pounded, rubbed, and pulled bark used as a soft fiber for weaving skirts and making diapers for babies. *Cordage* Pounded, rubbed, and pulled bark used as a soft fiber to make a carrying net for pottery. (as *P. juliflora* var. *torreyana* 15:107)
- *Other*—**Cahuilla** *Cooking Tools* Trunk used to make wooden mortars. *Fasteners* Gum used as an adhesive for arrows. Gum used to secure foreshafts to arrows and baskets to mortars. *Fuel* Wood used as firewood for cooking, baking pottery, and warmth. Bark used as kindling for cooking and firewood in sweat houses. *Hunting & Fishing Item* Smaller limbs used for bow making. Fire hardened branches used as the foreshaft inserted into the main shaft of an arrow. *Paper* Bark used as a wrapping. (as *P. juliflora* var. *torreyana* 15:107) *Protection* Trees used by women as shaded working areas, out of the direct rays of the sun, for grinding food. (as *P. juliflora* var. *torreyana* 15:114) *Tools* Thorns used to puncture the skin for tattooing. (as *P. juliflora* var. *torreyana* 15:107) **Diegueño** *Cooking Tools* Wood used to make wooden spoons and other utensils. (84:32) *Fuel* Wood used for firewood. (as *P. juliflora* var. *torreyana* 88:218) *Tools* Wood used to

make knives for cutting yucca stalks, pottery paddles, and tools for digging pottery clay. (84:32)

Prosopis pubescens Benth., Screwbean Mesquite
- ***Drug***—**Apache, Mescalero** *Ear Medicine* Pods soaked in water and used for earache. (as *Strombocarpa pubescens* 14:44) **Apache, Western** *Ear Medicine* Bean placed in ear for earache. (26:178) **Cahuilla** *Unspecified* Roots and bark had medicinal value. (15:118) **Paiute** *Eye Medicine* Infusion of gummy exudate on bark used as an eyewash. (180:123) **Pima** *Dermatological Aid* Decoction of roots used as a wash or powdered roots applied to sores. (as *Strombocarpa pubescens* 47:96) Powdered root bark or decoction used to dress wounds. (146:79) *Gynecological Aid* Infusion of roots taken for troubles with menses. (as *Strombocarpa pubescens* 47:96) **Tewa** *Ear Medicine* Pods twisted into the ear for an earache. (138:69)
- ***Food***—**Apache, Chiricahua & Mescalero** *Beverage* Fruit ground and sugar added to make a thick drink. (as *Strombocarpa pubescens* 33:53) *Bread & Cake* Pods dried, washed, ground into flour, and made into bread. *Dried Food* Fruits gathered, dried, and stored in sacks. *Special Food* Raw pods chewed and eaten as a delicacy. (as *Strombocarpa pubescens* 33:41) **Cahuilla** *Beverage* Pods crushed into a pulpy juice and used to make a beverage. Pod meal and water used to make a beverage. *Bread & Cake* Pod meal and water used to make cakes. *Dried Food* Ripe pods allowed to dry or picked after fully dried and ground into meal. *Staple* Pods used as one of the important food staples. Ripe pods allowed to dry or picked after fully dried and ground into meal. Pod meal and water used to make mush. (15:118) **Hualapai** *Dried Food* Pods dried and stored for later use. *Unspecified* Pods used for food. (195:45) **Isleta** *Unspecified* Pods chewed for the starch content and agreeable taste. (as *Strombocarpa pubescens* 100:43) **Kamia** *Unspecified* Coiled pod used for food. (62:23) **Mohave** *Beverage* Bean pods rotted in a pit for a month, dried, ground into a flour, and used to make a drink. *Vegetable* Bean pods used for food. (168:46) **Paiute** *Unspecified* Pounded beans used for food. (118:27) **Pima** *Beverage* Beans ground, mixed with water, and made into a nourishing and sweet beverage. (as *Strombocarpa pubescens* 47:96) Beans sun dried, pounded into meal, mixed with cold water, and used as a drink. (95:261) *Candy* Fresh, sugary pods chewed by children. *Forage* Pods and foliage eaten by grazing animals. (as *Strombocarpa pubescens* 47:96) *Staple* Beans pit roasted for several days, dried, and ground into a pinole. (as *Strombocarpa pubescens* 32:45) Beans pit cooked, dried, pounded, and eaten as pinole. (146:75) **Pima, Gila River** *Snack Food* Sap eaten as a snack food by all age groups. Catkins eaten as a snack food by all age groups. *Staple* Beans used to make flour. (133:5) Fruit used as a staple food. (133:7)
- ***Fiber***—**Cahuilla** *Building Material* Large limbs used in construction. (15:118) **Hualapai** *Furniture* Roots used to make cradleboard frames. (195:45) **Pima** *Building Material* Wood used for fence posts. (as *Strombocarpa pubescens* 47:96)
- ***Other***—**Cahuilla** *Hunting & Fishing Item* Small limbs used to make bows. *Tools* Long branch made into a mescal cutter to sever agave leaves. (15:118) **Pima** *Fuel* Wood used for fuel. (as *Strombocarpa pubescens* 47:96) **Pima, Gila River** *Season Indicator* Leaves used as a sign that planted crops would be safe from freezing weather. (133:6)

***Prosopis* sp.**, Mesquite
- ***Food***—**Hualapai** *Dried Food* Beans dried and stored for winter use. (195:44)
- ***Fiber***—**Hualapai** *Furniture* Roots used to make cradleboards. (195:44)
- ***Dye***—**Hualapai** *Unspecified* Black sap used to make hair dye. (195:44)

Prosopis velutina Woot., Velvet Mesquite
- ***Drug***—**Papago** *Dermatological Aid* Poultice of chewed leaves applied for red ant stings. Poultice of chewed leaves applied to red ant stings. Poultice of pulverized gum applied to sores and impetigo. Poultice of pulverized plant applied to sores and impetigo pustules. (34:65) **Pima** *Analgesic* Cold infusion of leaves taken for headaches. Decoction of gum held in mouth for painful gums or applied to painful burns. *Antidiarrheal* Infusion of roots taken for diarrhea. *Burn Dressing* Decoction of gum applied to burns to prevent soreness. *Cathartic* Decoction of gum taken to cleanse the system. (47:93) Decoction of inner bark taken as a cathartic. (146:79) *Dermatological Aid* Decoction of beans used as bleach for severe sunburn. Decoction of gum applied to chapped and cracked fingers or sore lips. Poultice of dried gum applied to prevent infection in newborn's navel. Resin used for sores. (47:93) Decoction of black gum used as a wash for open wounds. (146:79) *Disinfectant* Poultice of dried gum applied to prevent infection in newborn's navel. (47:93) *Emetic* Decoction of inner bark taken as an emetic. (146:79) *Eye Medicine* Decoction of leaves applied as poultice to pink eye. (47:93) Decoction of black gum used as a wash for sore eyes. (146:79) *Gastrointestinal Aid* Cold infusion of leaves taken for stomach troubles. *Oral Aid* Decoction of gum held in the mouth for painful gums or sore lips. *Other* Decoction of gum applied as a lotion for "bad disease." *Pediatric Aid* Poultice of dried gum applied to prevent infection in newborn's navel. (47:93)
- ***Food***—**Cocopa** *Unspecified* Pods used for food. *Winter Use Food* Pods stored for later use. (64:267) **Maricopa** *Unspecified* Beans formerly eaten as an important food. (32:44) **Papago** *Candy* Gum-like secretions found on branches and chewed. Gum-like secretions found on branches, dried, ground, boiled in gruel, cooled, and eaten like candy. (34:28) *Dried Food* Seeds basket winnowed, parched, sun dried, cooked, stored, and used for food. (34:24) *Preserves* Gum-like secretions found on branches, dried, ground, mixed with saguaro syrup, and eaten like jam. (34:28) *Staple* Beans ground into flour and used for food. (34:25) Beans pounded in mortars and used as a staple food. (34:45) *Unspecified* Pods eaten fresh. Beans and pods pounded into a pulpy mass, boiled, and used for food. (34:25) **Pima** *Beverage* Beans pounded, added to cold water, strained, and used as a sweet drink. (47:93) *Bread & Cake* Seeds ground into flour and used to make bread. (32:44) Beans boiled, cooled, pressed out into dumplings, and eaten. *Candy* Gum formerly eaten raw as a sweet. (47:93) White gum used to make candy. (146:74) *Porridge* Beans used to make mush. (47:93) *Staple* Seeds ground into flour and eaten as a pinole. (32:44) Beans parched, ground, and eaten as pinole. *Substitution Food* Inner bark used as a substitute for rennet. (146:74) *Sweetener* Seeds ground into flour and used to sweeten pinole. (32:44) *Unspecified* Sugary bean pods relished as food. Catkins sucked for their sweet taste. (47:93) Beans and pods pounded, ground, and used for food. Catkins eaten raw. (146:74) **Pima, Gila River** *Beverage* Pods crushed in a wooden mortar, soaked in water, and used to make vau

(a drink). *Bread & Cake* Pods made into flour and used to make an uncooked cake or loaf. (133:4) *Snack Food* Sap eaten as a snack food by all age groups. Catkins eaten as a snack food by all age groups. *Staple* Beans used to make flour. (133:5) Fruit used as a staple food. (133:7) *Winter Use Food* Pods stored in great quantities in large arrowweed baskets or bins. (133:4)

- *Fiber*—**Papago** *Basketry* Used as the warp element of baskets. (34:59) *Building Material* Posts of wood, forked at the top, used for the core of the house frame. (34:66) *Other* Roots used for curved structures in wrapped weaving. (34:53) **Pima** *Building Material* Wood used for fence posts. *Furniture* Roots used to make cradle frames. (47:93)
- *Dye*—**Pima** *Black* Decoction of gum applied to gray hair and used with black clay or mud as a black hair dye. (47:93)
- *Other*—**Papago** *Cooking Tools* Green sticks used to turn roasting ears of corn. (34:35) *Fasteners* "Gum" used to fasten handles to gourds. (34:68) *Fuel* Used to heat stones for baking cholla buds and joints. (34:15) *Protection* Posts used to make a fence to protect tobacco plants from marauding animals. (as *P. veluntina* 34:37) *Tools* Two-foot-long, sword-shaped slabs sharpened and used as weed hoes. (34:32) **Pima** *Cooking Tools* Green wood used to make ladles. (47:93) *Fuel* Stumps used as fuel for pit baking. (47:58) Wood used for fuel. *Paint* Resin boiled and used as a pottery paint. *Toys & Games* Sticks used in gambling games. Wood made into balls and used in racing games. (47:93) **Pima, Gila River** *Cash Crop* Wood cut and sold. (133:9) *Season Indicator* Leaves used as a sign that planted crops would be safe from freezing weather. (133:6)

Prunella, Lamiaceae

Prunella vulgaris L., Common Selfheal

- *Drug*—**Algonquin, Quebec** *Febrifuge* Infusion of leaves used for fevers. (18:224) **Bella Coola** *Heart Medicine* Weak decoction of roots, leaves, and blossoms taken for the heart. (150:63) **Blackfoot** *Dermatological Aid* Infusion of plant used to wash a burst boil. Infusion of plant applied to neck sores. (86:78) *Eye Medicine* Infusion of plant used as an eyewash to keep the eyes moist on cold or windy days. (86:82) *Veterinary Aid* Infusion of plant used for saddle and back sores on horses. Infusion of plant used as an eyewash for horses. (86:90) **Catawba** *Misc. Disease Remedy* Plant used in certain diseases. (157:191) **Cherokee** *Adjuvant* Used to flavor other medicines. *Burn Dressing* Cold infusion used as a wash for burns. *Dermatological Aid* Infusion of root used as wash for bruises, diabetic sores, cuts, and acne. (80:54) **Chippewa** *Cathartic* Compound decoction of root taken as a physic. (53:346) **Cree, Hudson Bay** *Throat Aid* Plant used or herb chewed for sore throats. (92:303) **Delaware** *Febrifuge* Plant tops used to make a cooling drink and body wash for fevers. (176:37) **Delaware, Oklahoma** *Febrifuge* Liquid made from plant tops taken and used as a wash for fever. (175:31, 78) **Iroquois** *Analgesic* Infusion of plant taken for backaches. (87:425) *Antidiarrheal* Decoction of plants taken for diarrhea. (87:423) *Antiemetic* Decoction of plants taken for vomiting and diarrhea. *Blood Medicine* Compound decoction of roots and shoots taken as a blood purifier. (87:424) *Cold Remedy* Decoction of plants taken for colds. *Cough Medicine* Decoction of plants taken for coughs. (87:423) *Emetic* Decoction of whole plant taken as an emetic. (87:425) *Febrifuge* Decoction of plants taken for fevers and shortness of breath. (87:424) *Gastrointestinal Aid* Infusion of plants taken for stomach cramps and biliousness. (87:423) Compound infusion of roots and plants taken for upset stomachs. (87:424) Infusion of plant taken for biliousness. *Gynecological Aid* Decoction of whole plant taken to strengthen the womb. (87:425) *Hemorrhoid Remedy* Compound decoction of roots taken and used as wash for piles. (87:424) *Misc. Disease Remedy* Plant used for sugar diabetes. (87:425) *Orthopedic Aid* Compound decoction of plants used as steam bath for sore legs or stiff knees. (87:424) Infusion of plant taken for backaches. *Panacea* Infusion of plant taken for any ailment. (87:425) *Pediatric Aid* Compound infusion of plants given to babies that cry too much. *Psychological Aid* Compound infusion of plants taken for sickness caused by grieving. *Pulmonary Aid* Decoction or infusion of roots taken for shortness of breath. *Respiratory Aid* Infusion of roots taken for heaves or shortness of breath. *Sedative* Compound infusion of plants given to babies that cry too much. *Tuberculosis Remedy* Compound decoction of roots taken for consumption. *Venereal Aid* Compound decoction of roots and shoots taken for venereal disease. (87:424) **Menominee** *Antidiarrheal* and *Pediatric Aid* Infusion of stalk used, especially good for babies, for dysentery. (54:131) **Mohegan** *Febrifuge* "Drink" made from leaves taken and used as a wash for fevers. (176:74, 130) **Ojibwa** *Gynecological Aid* Compound containing root used as a female remedy. (153:372) *Hunting Medicine* Root, sharpened the powers of observation, used to make a tea to drink before going hunting. (153:430) **Quileute** *Dermatological Aid* Plant used for boils. **Quinault** *Dermatological Aid* Plant juice rubbed on boils. (79:45) **Salish, Coast** *Dermatological Aid* Leaves used for boils, cuts, bruises, and skin inflammations. (182:84) **Thompson** *Tonic* Hot or cold infusion of plant taken as a tonic for general indisposition. (164:471)
- *Food*—**Cherokee** *Vegetable* Leaves cooked and eaten as greens. (80:54) Small leaves used as a potherb. (as *Frunella vulgaris* 126:44) Leaves cooked with sochan (*Rudbeckia laciniata*), creaseys (probably *Lepidium virginicum*), and other potherbs and eaten. (204:253) **Thompson** *Beverage* Plant soaked in cold water and used as one of the most common drinks. (164:494)

Prunus, Rosaceae

Prunus americana Marsh., American Plum

- *Drug*—**Cherokee** *Cough Medicine* Bark used to make cough syrup. *Kidney Aid* Infusion of bark taken for the kidneys. *Urinary Aid* Infusion of bark taken for the bladder. (80:50) **Cheyenne** *Ceremonial Medicine* Branches used for the Sun Dance ceremony. *Oral Aid* Smashed fruits used for mouth disease. (83:35) **Chippewa** *Anthelmintic* Compound decoction of root taken for worms. (53:346) *Dermatological Aid* Compound poultice of inner bark applied to cuts and wounds. (53:352) *Disinfectant* Compound decoction of inner bark used as a disinfectant wash. (53:366) Decoction of bark used as a disinfecting wash. (53:376) **Meskwaki** *Oral Aid* Root bark used as an astringent medicine for mouth cankers. (152:242) **Mohegan** *Respiratory Aid* Infusion of twigs taken for asthma. (174:270) Infusion of twigs taken for asthma. (176:74, 130) **Ojibwa, South** *Antidiarrheal* Compound decoction of small rootlets taken for diarrhea. (91:200) **Omaha** *Dermatological Aid* Poultice of boiled root bark applied to skin abrasions. (70:87) **Rappahannock** *Unspecified* "An ingredient of a medicine made after diagnosis." (161:31)

- *Food*—**Apache, Mescalero** *Dried Food* Fruits dried and stored for future food use. (14:50) **Cherokee** *Beverage* Fruit used to make juice. *Fruit* Fruit used for food. *Preserves* Fruit used to make jelly. (80:50) **Cheyenne** *Pie & Pudding* Fruits, sugar, and flour used to make a pudding. *Special Food* Fruits pulverized, sun dried, boiled, and eaten as a delicacy. (83:35) *Winter Use Food* Sun dried plums stored for winter use. (76:177) **Chippewa** *Bread & Cake* Berries cooked, spread on birch bark into little cakes, dried, and stored for winter use. *Fruit* Berries eaten raw. (53:321) **Crow** *Fruit* Ripe plums used fresh. *Winter Use Food* Ripe plums dried for winter use. (19:19) **Dakota** *Dried Food* Fruit boiled, pitted, and dried for winter use. (69:364) Highly valued fruit pitted and dried for winter use. (70:87) *Fruit* Fruit eaten fresh. (69:364) Highly valued fruit eaten fresh and raw. (70:87) *Sauce & Relish* Fruit made into a sauce. (69:364) Highly valued fruit cooked as a sauce. (70:87) **Iroquois** *Beverage* Fruit sun dried and boiled in water to make coffee. (196:145) *Bread & Cake* Fruit mashed, made into small cakes, and dried for future use. *Dried Food* Raw or cooked fruit sun or fire dried and stored for future use. *Fruit* Dried fruit taken as a hunting food. *Sauce & Relish* Dried fruit cakes soaked in warm water and cooked as a sauce or mixed with corn bread. (196:128) **Isleta** *Fruit* Fruits eaten for food. Fruits eaten fresh. (32:46) Fruit eaten for food. (100:40) **Kiowa** *Fruit* Fruit gathered in great quantities and used immediately. *Winter Use Food* Fruit gathered in great quantities, dried, and stored for winter use. (192:29) **Lakota** *Fruit* Fruits eaten fresh. (106:37) Fruits eaten for food. (139:56) *Starvation Food* Fruits dried and eaten during famines. (106:37) **Meskwaki** *Fruit* Plums eaten fresh. *Preserves* Plums made into plum butter for winter use. (152:263) **Ojibwa** *Dried Food* Fruit dried for winter use. *Fruit* Fruit eaten fresh. *Soup* Dried fruit ground into a flour and used to make soup. (135:235) **Omaha** *Dried Food* Fruit pitted and dried for winter use. (68:326) Highly valued fruit pitted and dried for winter use. (70:87) *Fruit* Fruit eaten fresh in season. (68:326) Highly valued fruit eaten fresh and raw. *Sauce & Relish* Highly valued fruit cooked as a sauce. **Pawnee** *Dried Food* Highly valued fruit eaten fresh and raw, cooked as a sauce, or dried with the pits for winter use. *Fruit* Highly valued fruit eaten fresh and raw. *Sauce & Relish* Highly valued fruit cooked as a sauce. **Ponca** *Dried Food* Highly valued fruit pitted and dried for winter use. *Fruit* Highly valued fruit eaten fresh and raw. *Sauce & Relish* Highly valued fruit cooked as a sauce. **Winnebago** *Dried Food* Highly valued fruit pitted and dried for winter use. *Fruit* Highly valued fruit eaten fresh and raw. *Sauce & Relish* Highly valued fruit cooked as a sauce. (70:87)
- *Fiber*—**Dakota** *Brushes & Brooms* Tough, elastic twigs bound into bundles and used as brooms for sweeping the floor. **Omaha** *Brushes & Brooms* Tough, elastic twigs bound into bundles and used as brooms for sweeping the floor. **Pawnee** *Brushes & Brooms* Bound bundles of tough, elastic twigs used a brooms for sweeping the floor. **Ponca** *Brushes & Brooms* Tough, elastic twigs bound into bundles and used as brooms for sweeping the floor. **Winnebago** *Brushes & Brooms* Tough, elastic twigs bound into bundles and used as brooms for sweeping the floor. (70:87)
- *Dye*—**Chippewa** *Mordant* Inner bark scraped and used to set the color of a yellow dye. (53:374) *Red* Inner bark boiled with other inter barks and bloodroot and used to make a red dye. (53:371) *Yellow* Single handful of shredded roots boiled with bloodroot to make a dark yellow dye. (53:374) **Isleta** *Red* Root bark, alder root bark, and mountain mahogany root bark used to make a red dye for buckskin. (100:40) **Navajo** *Red* Roots used as a red dye for wool. (55:54)
- *Other*—**Dakota** *Ceremonial Items* Sprout or young growth made into a wand and used ceremonially. The sprout or young growth was made into a wand by peeling it and painting it with emblematic colors and designs. An offering which consisted of tobacco or anything acceptable to higher powers was attached to the top of the wand and usually made for the benefit of the sick. The offering could be made anywhere and by anyone as long as it was executed with appropriate ceremony, but was most efficiently performed if an altar were prepared at which the wand was placed upright with the offering fastened near the top. (70:87) *Hide Preparation* Fruit dried with buffalo hides in preparing them for tanning. *Toys & Games* Seeds used to make playing pieces of a game similar to dice. (69:364) **Mohegan** *Water Indicator* Crotched sticks used to locate underground water. (176:87) **Omaha** *Season Indicator* Blossoms used as an indicator of when to plant corn, beans, and squashes. (70:87)

Prunus andersonii Gray, Desert Peach

- *Drug*—**Paiute** *Antidiarrheal* Decoction of stems, leaves, or roots taken for diarrhea. *Antirheumatic (Internal)* Weak decoction of bark taken for rheumatism. (180:123) *Cold Remedy* Infusion of branches taken for colds. (118:38) Hot infusion of leaves or decoction of branches taken for colds. *Misc. Disease Remedy* Decoction of dried bark strips taken as a winter tonic to ward off influenza. *Tonic* Decoction of dried bark strips taken as a winter tonic to ward off influenza. *Tuberculosis Remedy* Decoction of bark taken or twigs chewed for tuberculosis. (180:123)
- *Food*—**Cahuilla** *Fruit* Fruit considered a great delicacy, important food and a highly prized food source. *Preserves* Fruit boiled, sweetened with sugar, and used to make jelly. (15:119)

Prunus angustifolia Marsh., Chickasaw Plum

- *Food*—**Comanche** *Fruit* Fresh fruits used for food. *Winter Use Food* Stored fruits used for food. (29:523)

Prunus armeniaca L., Apricot

- *Food*—**Hopi** *Unspecified* Species used for food. (200:79) **Keresan** *Dried Food* Fruit dried for winter use. *Fruit* Fruit eaten fresh. (198:558)

Prunus cerasus L., Sour Cherry

- *Drug*—**Cherokee** *Blood Medicine* Compound used as a blood tonic. *Cold Remedy* Infusion of bark taken for colds. *Cough Medicine* Infusion of bark taken for coughs. *Dermatological Aid* Astringent root bark used in a wash for old sores and ulcers. Root bark used as a wash for old sores and ulcers. *Febrifuge* Infusion or decoction of bark used for fevers, including the "great chill." *Gastrointestinal Aid* Boiled fruit used for "blood discharged from bowels." Used in steam bath for indigestion, biliousness, and jaundice. *Gynecological Aid* Warm infusion given when labor pains begin. *Misc. Disease Remedy* Compound of barks added to corn whisky and used to break out measles. *Oral Aid* Infusion of bark used for "thrash." *Throat Aid* Decoction of inner bark used for laryngitis. (80:28, 29)
- *Food*—**Cherokee** *Fruit* Fruit used for food. (80:28)
- *Fiber*—**Cherokee** *Building Material* Wood used for lumber. *Furni-*

ture Wood used to make furniture. (80:28)
- *Other*—**Cherokee** *Decorations* Wood used to carve. (80:28)

Prunus dulcis (P. Mill.) D. A. Webber, Sweet Almond
- *Food*—**Hopi** *Unspecified* Species used for food. (as *P. amygdalus* 200:79)

Prunus emarginata (Dougl. ex Hook.) Walp., Bitter Cherry
- *Drug*—**Bella Coola** *Heart Medicine* Decoction of root and inner bark taken daily for heart trouble. (150:58) Infusion of bark used for heart trouble. *Tuberculosis Remedy* Infusion of bark used for tuberculosis. (184:209) **Cowichan** *Cold Remedy* Infusion of bark and crabapple bark used as a cure-all tonic for colds. *Panacea* Infusion of bark and crabapple bark used as a cure-all tonic for numerous ailments. (182:87) **Hoh** *Blood Medicine* Decoction of bark used as a blood remedy. (137:64) **Kwakiutl** *Cancer Treatment* Bark used to wrap lint after treating tumors. (20:383) *Dermatological Aid* Bark used to cover poultice on swellings. (20:382) Bark ash rubbed on newborn's chest to protect against rash and sore mouth. (20:383) Bark ash rubbed on chest of baby as protection from rashes. Infusion of bark taken for eczema. (183:290) *Dietary Aid* Roots applied to nipples of mother to induce the infant to nurse. (20:386) Roots applied to the nipples of a mother to induce her infant to nurse. (183:290) *Gynecological Aid* Roots applied to nipples of mother to induce the infant to nurse. (20:386) Decoction of split roots taken for blood discharge. *Heart Medicine* Infusion of bark taken for heart trouble. (183:290) *Hemostat* Poultice of bark strips used for holding down all kinds of plasters applied to bleeding wounds. (as *P. emarginata* var. *villosa* 20:384) *Oral Aid* Bark ash rubbed on newborn's chest to protect against rash and sore mouth. Poultice of rubbed root applied to sores in child's mouth. (20:383) Bark ash rubbed on chest of baby as protection from mouth sores. Roots held in the mouth by children with canker sores. (183:290) *Pediatric Aid* Plant used as part of charm worn by children to ward off disease. (20:379) Bark ash rubbed on newborn's chest to protect against rash and sore mouth. Poultice of rubbed root applied to sores in child's mouth. (20:383) Roots applied to nipples of mother to induce the infant to nurse. (20:386) Bark ash rubbed on chest of baby as protection from rashes and mouth sores. Roots applied to the nipples of a mother to induce her infant to nurse. Roots held in the mouth by children with canker sores. (183:290) *Preventive Medicine* Plant used as part of charm worn by children to ward off disease. (20:379) *Tuberculosis Remedy* Infusion of bark taken for tuberculosis. (183:290) **Lummi** *Gynecological Aid* Bark chewed to facilitate childbirth. Infusion of rotten wood taken as a contraceptive. (79:37) **Makah** *Blood Medicine* Bark used as a blood purifier. *Laxative* Bark used as a laxative. *Tonic* Bark used as a tonic. (67:266) **Nitinaht** *Panacea* Infusion of bark taken as a general tonic for healing any sickness. (186:120) **Okanagan-Colville** *Gastrointestinal Aid* Berries eaten as a laxative for "sour stomach." (188:127) **Paiute** *Eye Medicine* Infusion of bark used as an eye medicine. (111:85) **Quileute** *Blood Medicine* Decoction of bark used as a blood remedy. (137:64) **Quinault** *Gynecological Aid* Infusion of rotten wood taken as a contraceptive. *Laxative* Decoction of bark taken as a laxative. (79:37) **Saanich** *Cold Remedy* Infusion of bark and crabapple bark used as a cure-all tonic for colds. *Panacea* Infusion of bark and crabapple bark used as a cure-all tonic for numerous ailments. *Psychological Aid* Concoction of roots and gooseberry roots used to make children intelligent and obedient. (182:87) **Skagit** *Cold Remedy* Decoction of bark taken for colds. *Gynecological Aid* Infusion of rotten wood taken as a contraceptive. **Skokomish** *Cold Remedy* Decoction of bark taken for colds. *Gynecological Aid* Infusion of rotten wood taken as a contraceptive. (79:37) **Thompson** *Orthopedic Aid* Bark used to wrap splints for broken limbs. *Unspecified* Infusion of branches taken for an unspecified illness. (187:263)
- *Food*—**Cahuilla** *Fruit* Fruit considered a great delicacy, important food and a highly prized food source. (15:119) **Coeur d'Alene** *Fruit* Berries occasionally eaten fresh. (178:90) **Hoh** *Preserves* Fruits used to make jelly. (137:64) **Klamath** *Fruit* Fruit used for food. (45:98–99) **Quileute** *Preserves* Fruits used to make jelly. (137:64) **Shuswap** *Fruit* Cherries used for food. (123:67) **Thompson** *Dessert* Fruits sometimes eaten as a dessert. *Fruit* Fruits eaten occasionally because of the bitter taste. (187:263)
- *Fiber*—**Bella Coola** *Basketry* Bark formerly used for imbricating baskets. (184:209) **Hanaksiala** *Basketry* Bark used to make baskets. (43:272) **Hesquiat** *Basketry* Bark used in basket decoration and in weaving the large part of the berry picking baskets. (185:73) **Nitinaht** *Basketry* Bark strips used to make baskets. (67:266) **Okanagon** *Basketry* Bark split and used to make baskets. *Mats, Rugs & Bedding* Bark split and used to make mats. (125:40) **Salish, Coast** *Basketry* Bark used for imbrication in cedar bark baskets. (182:87) **Shuswap** *Basketry* Bark used for basket trim. (123:67) **Snohomish** *Basketry* Bark used in the imbricated designs of baskets. (79:37) **Thompson** *Basketry* Bark split and used to make baskets. (125:40) Bark used extensively in basketry. (164:498) Tough, waterproof bark used with grass stems for imbrication of coiled split cedar root baskets. The basket was either left a natural light reddish brown color or was dyed by burying it in damp earth or letting it sit in a rusty tin can. After being buried a short time, it became a dark brown color and when kept for a longer time, it became black. *Building Material* Bark made into twine and used for reinforcement of old suspension bridges. *Cordage* Bark used to make twine. (187:263) *Mats, Rugs & Bedding* Bark split and used to make mats. (125:40) Bark softened and used to make mats. (164:497)
- *Other*—**Bella Coola** *Containers* Bark formerly used for wrapping implements. (184:209) **Clallam** *Hunting & Fishing Item* Bark fashioned into twine and used as fishing line. (57:202) **Haisla & Hanaksiala** *Decorations* Wood used for carving. (43:272) **Hesquiat** *Fasteners* Bark used to wrap the joints of implements such as harpoons, where the head is fixed to the shaft. First some pitch was smeared over the joint, then the cherry bark was wrapped around and bound tightly with twine or sinew. Finally more pitch was plastered over to make the joint completely watertight. Cherry bark is both strong and flexible and is decorative as well. (185:73) **Klamath** *Weapon* Branches used for whips. (45:98–99) **Nitinaht** *Ceremonial Items* Bark used to make the reed for a ceremonial wolf whistle. *Fasteners* Smooth, tough bark used for binding and wrapping joints of fishing and hunting implements. (186:120) **Okanagan-Colville** *Decorations* Bark used to imbricate split cedar root baskets. Bark used to decorate bows, "tomahawk" handles, and pipestems. Root bark used to imbricate cedar root baskets. (188:127) **Okanagon** *Containers* Bark split and used to make bags. (125:40) **Quinault** *Fasteners* Bark used to tie the prongs of

fish spears. (79:37) **Salish, Coast** *Containers* Wood used for the hearth. *Fuel* Wood used for fuel and the hearth and drill in making friction fires. *Hunting & Fishing Item* Bark used to make harpoons, spears, fishing lines, nets, and other hunting gear. *Tools* Wood used for the drill to make friction fires. (182:87) **Thompson** *Containers* Bark split and used to make bags. (125:40) Bark softened and used to make bags. (164:497) *Decorations* Bark used to bind bows considered a decorative contrast to the wood of the bows. *Fasteners* Bark used to bind bows in the middle and ends for strength. (164:498)

Prunus fasciculata (Torr.) Gray, Desert Almond
- **Food**—**Cahuilla** *Fruit* Fruit considered a great delicacy, important food and a highly prized food source. (15:119)
- **Other**—**Kawaiisu** *Hunting & Fishing Item* Straight twigs, fitted into the hollow stems of carizzo grass main shafts, used as arrow foreshafts. *Tools* Used as a drill in fire making. (206:55)

Prunus fremontii S. Wats., Desert Apricot
- **Food**—**Cahuilla** *Fruit* Fruit considered a great delicacy, important food and a highly prized food source. (15:119)

Prunus gracilis Engelm. & Gray, Oklahoma Plum
- **Food**—**Kiowa** *Candy* Dried fruit used as an ingredient in making candy. *Winter Use Food* Dried fruit stored for winter use, eaten uncooked, or pounded and made into cakes. (192:30)

Prunus ilicifolia (Nutt. ex Hook. & Arn.) D. Dietr., Holly Leaf Cherry
- **Drug**—**Diegueño** *Cough Medicine* Infusion of leaves taken as a cough medicine. (88:217) **Mahuna** *Cough Medicine* Infusion of bark or roots taken for coughs. (140:18)
- **Food**—**Cahuilla** *Fruit* Fruit considered a great delicacy, important food and a highly prized food source. (15:119) **Costanoan** *Fruit* Fruits eaten for food. *Unspecified* Soaked, roasted inner kernels used for food. (21:249) **Diegueño** *Bread & Cake* Large seed cracked, the kernel extracted, pounded into a meal, and made into patties and roasted. *Fruit* Fruit eaten fresh. (84:32) Fruit used for food. *Porridge* Seeds ground, leached, and used to make atole. (88:217) **Luiseño** *Fruit* Fruit used for food. (as *Cerasus ilicifolia* 155:232) Fruit, similar to plums or cherries, formerly used to some extent as food. *Porridge* Sun dried fruit kernels made into a flour and cooked in an earthen vessel. The sun dried fruit kernels were extracted from the shells, made into a flour, and then leached to remove the bitterness. The flour was either leached with hot water, placed in a rush basket, and warm water poured over it, or placed in a sand hole and warm water poured over it to remove the bitterness. (155:194) *Staple* Kernels ground into a flour and used for food. (as *Cerasus ilicifolia* 155:232) *Unspecified* Pulp eaten for food. (155:194) **Mahuna** *Fruit* Berries eaten mainly to quench the thirst. (140:70)
- **Other**—**Costanoan** *Hunting & Fishing Item* Wood used for bows. (21:249)

Prunus nigra Ait., Canadian Plum
- **Drug**—**Algonquin, Quebec** *Cough Medicine* Infusion of inner bark taken for coughs. *Unspecified* Infusion of roots used as a medicinal tea. (18:184) **Meskwaki** *Antiemetic* Infusion of bark used to settle stomach when it will not retain food. (152:242)
- **Food**—**Algonquin, Quebec** *Fruit* Fruit eaten. *Preserves* Fruit made into preserves. (18:95) **Iroquois** *Bread & Cake* Fruit mashed, made into small cakes and dried for future use. *Dried Food* Raw or cooked fruit sun or fire dried and stored for future use. *Fruit* Dried fruit taken as a hunting food. *Sauce & Relish* Dried fruit cakes soaked in warm water and cooked as a sauce or mixed with corn bread. (196:128) **Meskwaki** *Fruit* Plums eaten fresh. *Preserves* Plums made into plum butter for winter use. (152:263) **Ojibwa** *Fruit* Large quantities of plums found in thickets and gathered for food. *Preserves* Large quantities of plums found in thickets and gathered for preserves. (153:409)
- **Dye**—**Ojibwa** *Mordant* Inner bark used as an astringent color fixative in dyeing with other plant dyes. (153:426)

Prunus pensylvanica L. f., Pin Cherry
- **Drug**—**Algonquin, Quebec** *Blood Medicine* Infusion of bark taken for blood poisoning. *Cough Medicine* Infusion of bark taken for coughs. *Disinfectant* Infusion of bark taken for infections. *Pulmonary Aid* Infusion of bark taken for bronchitis. (18:184) **Algonquin, Tête-de-Boule** *Hemostat* and *Pediatric Aid* Poultice of boiled, shredded inner bark applied to bleeding umbilical cord. (132:130) **Cherokee** *Blood Medicine* Compound used as a blood tonic. *Cold Remedy* Infusion of bark taken for colds. *Cough Medicine* Infusion of bark taken for coughs. *Dermatological Aid* Astringent root bark used in a wash for old sores and ulcers. Root bark used as a wash for old sores and ulcers. *Febrifuge* Infusion or decoction of bark used for fevers, including the "great chill." *Gastrointestinal Aid* Boiled fruit used for "blood discharged from bowels." Used in steam bath for indigestion, biliousness, and jaundice. *Gynecological Aid* Warm infusion given when labor pains begin. *Misc. Disease Remedy* Compound of barks added to corn whisky and used to break out measles. *Oral Aid* Infusion of bark used for "thrash." *Throat Aid* Decoction of inner bark used for laryngitis. (80:28, 29) **Cree, Woodlands** *Eye Medicine* Infusion of inner bark used for sore eyes. (109:53) **Gitksan** *Unspecified* Bark used for medicine. (73:152) **Iroquois** *Burn Dressing* Compound of roots applied as a salve to burns. (87:359) *Cough Medicine* Bark and another bark used to make cough syrup. (142:91) **Malecite** *Dermatological Aid* Outer layer of dried trees used as a powder for prickly heat. (116:250) Infusion of bark used for erysipelas. (116:256) *Pediatric Aid* Outer layer of dried trees used for chafed babies. (116:250) **Micmac** *Dermatological Aid* Wood used for chafed skin and prickly heat. *Misc. Disease Remedy* Bark used for erysipelas. (40:59) **Ojibwa** *Cough Medicine* Inner bark used as a cough remedy. (153:385) **Ojibwa, South** *Analgesic* Decoction of crushed root taken for stomach pains. *Gastrointestinal Aid* Decoction of crushed root taken for stomach disorders. (91:199) **Potawatomi** *Analgesic* and *Cough Medicine* Infusion of inner bark taken internal pain and cough. (154:77) **Wet'suwet'en** *Cough Medicine* Bark used for coughs. (73:152)
- **Food**—**Algonquin, Quebec** *Fruit* Fruit eaten fresh. *Preserves* Fruit made into jelly. (18:95) **Cherokee** *Fruit* Fruit used for food. (80:28) *Pie & Pudding* Fruit used to make pies. *Preserves* Fruit used to make jam. (126:58) **Cree, Woodlands** *Preserves* Juice used to make jelly. (109:53) **Iroquois** *Bread & Cake* Fruit mashed, made into small cakes, and dried for future use. *Dried Food* Raw or cooked fruit sun or fire dried and stored for future use. (196:128) *Fruit* Fruit used for food. (141:46) Dried fruit taken as a hunting food. *Sauce & Relish* Dried fruit cakes

soaked in warm water and cooked as a sauce or mixed with corn bread. (196:128) **Ojibwa** *Dried Food* Fruit dried for winter use. *Fruit* Fruit eaten fresh. (135:235) Berries used for food. The pin cherry was abundant around the Flambeau Reservation and the Ojibwe were fond of it. It was an education in itself to see a group of Ojibwe women working on mats with a supply of fruit laden branches beside them. With one hand they would start a stream of berries into the mouth and the stream of cherry stones ejected from the other corner of the mouth seemed ceaseless. The Pillager Ojibwe also had the tree and used it is the same manner. (153:409) *Soup* Dried fruit ground into a flour and used to make soup. (135:235) **Potawatomi** *Fruit* Cherries eaten as the women worked making baskets. (154:108)

- *Fiber*—**Cherokee** *Building Material* Wood used for lumber. *Furniture* Wood used to make furniture. (80:28)
- *Other*—**Cherokee** *Decorations* Wood used to carve. (80:28)

Prunus persica (L.) Batsch, Peach

- *Drug*—**Cherokee** *Anthelmintic* Decoction or teaspoon of parched seed kernels taken for worms. *Antiemetic* Infusion of scraped bark taken for vomiting. *Cathartic* Infusion of any part taken as a purgative. *Dermatological Aid* Used for skin diseases and leaves wrung in cold water used to bathe swelling. *Febrifuge* Strong infusion taken for fever. *Gastrointestinal Aid* Infusion of leaves taken for sick stomach. (80:47, 48) **Delaware** *Anthelmintic* Infusion of leaves used to expel pinworms. *Antiemetic* and *Pediatric Aid* Infusion of leaves used by children for vomiting. (176:31) **Koasati** *Orthopedic Aid* Leaves rubbed on the scratches of tired legs. (177:27) **Navajo** *Cathartic* Plant used as a purgative. (55:96) Dried fruit used as a purgative. (55:54) **Rappahannock** *Kidney Aid* Infusion of fresh or dried leaves taken for kidney trouble. (161:33)
- *Food*—**Cherokee** *Fruit* Fruit used for food. (80:47) **Havasupai** *Beverage* Dried fruits pounded, stewed, and the water drunk. *Dried Food* Fruit split open, pitted, and sun dried for later consumption. (197:224) **Hopi** *Dried Food* Fruits split open and dried for winter use. *Fruit* Fruits eaten fresh. (200:79) **Iroquois** *Bread & Cake* Fruit mashed, made into small cakes, and dried for future use. *Dried Food* Raw or cooked fruit sun or fire dried and stored for future use. *Fruit* Dried fruit taken as a hunting food. *Sauce & Relish* Dried fruit cakes soaked in warm water and cooked as a sauce or mixed with corn bread. (196:129) **Keres, Western** *Fruit* Fresh peaches eaten for food. *Winter Use Food* Peaches dried for winter use. (171:63) **Keresan** *Dried Food* Fruit dried for winter use. *Fruit* Fruit eaten fresh. (198:562) **Navajo, Ramah** *Fruit* Favorite fruit used for food. (191:31) **Seminole** *Unspecified* Plant used for food. (169:507)
- *Dye*—**Navajo** *Yellow* Leaves used as a yellow dye. (55:54)
- *Other*—**Hopi** *Tools* Wood used to make weaving batons. (200:79)

Prunus pumila L., Sandcherry

- *Food*—**Menominee** *Fruit* Berries eaten fresh. *Preserves* Berries sometimes gathered and preserved. (151:71) **Ojibwa** *Dried Food* Fruit used dried. *Fruit* Fruit used fresh. (5:2221) This species was plentiful on sandy openings in the forest and the fruit gathered for food. (153:409)

Prunus pumila var. *besseyi* (Bailey) Gleason, Western Sandcherry

- *Food*—**Dakota** *Dried Food* Pitted fruit dried for winter use. (as *P. besseyi* 69:364) Fruit dried for winter use. (as *P. besseyi* 70:88) *Fruit* Pitted fruit eaten fresh. (as *P. besseyi* 69:364) *Sauce & Relish* Fruit used to make a sauce during the fruiting season. (as *P. besseyi* 70:88) **Lakota** *Fruit* Fruits eaten for food. (as *P. besseyi* 139:56) **Omaha** *Dried Food* Fruit dried for winter use. (as *P. besseyi* 70:88) *Fruit* Fruit eaten fresh. (as *P. besseyi* 68:326) *Sauce & Relish* Fruit used to make a sauce during the fruiting season. **Pawnee** *Dried Food* Fruit dried for winter use. *Sauce & Relish* Fruit used to make a sauce during the fruiting season. **Ponca** *Dried Food* Fruit dried for winter use. *Sauce & Relish* Fruit used to make a sauce during the fruiting season. (as *P. besseyi* 70:88)
- *Other*—**Lakota** *Paint* Fruit used to paint the face. (as *P. besseyi* 139:56)

Prunus pumila var. *susquehanae* (hort. ex Willd.) Jaeger, Sesquehana Sandcherry

- *Food*—**Potawatomi** *Beverage* Cherries used to improve the flavor of whisky. (as *P. cuneata* 154:107)

Prunus serotina Ehrh., Black Cherry

- *Drug*—**Cherokee** *Blood Medicine* Compound used as a blood tonic. *Cold Remedy* Infusion of bark taken for colds. *Cough Medicine* Infusion of bark taken for coughs. *Dermatological Aid* Astringent root bark used in a wash for old sores and ulcers. Root bark used as a wash for old sores and ulcers. *Febrifuge* Infusion or decoction of bark used for fevers, including the "great chill." (80:28, 29) Decoction of bark used as a wash for chills and fevers. (177:28) Infusion of bark taken for fevers. (203:74) *Gastrointestinal Aid* Boiled fruit used for "blood discharged from bowels." Used in steam bath for indigestion, biliousness, and jaundice. *Gynecological Aid* Warm infusion given when labor pains begin. (80:28, 29) Infusion of bark taken for childbirth. (177:28) *Misc. Disease Remedy* Compound of barks added to corn whisky and used to break out measles. (80:28, 29) Decoction of bark used as a wash for ague. (177:28) *Oral Aid* Infusion of bark used for "thrash." *Throat Aid* Decoction of inner bark used for laryngitis. (80:28, 29) **Chippewa** *Anthelmintic* Compound decoction of root taken for worms. (53:346) *Burn Dressing* Powder containing powdered root applied to burns. (53:354) *Dermatological Aid* Compound poultice of inner bark applied to cuts and wounds. (53:352) Poultice of fresh roots or decoction of bark used as a wash for "scrofulous neck." Powder containing powdered root applied to ulcers. (53:354) *Disinfectant* Compound decoction of inner bark used as a disinfectant wash. (53:366) *Misc. Disease Remedy* and *Pediatric Aid* Decoction of root given for "cholera infantum." (53:346) *Tuberculosis Remedy* Poultice of root applied or decoction of bark used as a wash for scrofulous, neck sores. (53:354) **Delaware** *Antidiarrheal* Bark used for diarrhea. *Cough Medicine* Fruits used to make cough syrup. *Tonic* Bark combined with other roots and used as a tonic. (176:32) **Delaware, Oklahoma** *Antidiarrheal* Bark used as a diarrhea remedy. *Cough Medicine* Fruit used to make cough syrup. *Tonic* Compound containing bark taken as a tonic. (175:27, 78) **Delaware, Ontario** *Gynecological Aid* Compound infusion of bark taken for "diseases peculiar to women." *Tonic* Compound infusion of bark taken as a tonic for general debility. (175:68, 82) **Iroquois** *Analgesic* Decoction of bark taken or poultice applied to forehead and neck for headaches. (87:362) *Blood Medicine* Compound infusion of bark and roots taken as a blood purifier. (87:361) Infusion of roots and other roots taken by young mothers for thick blood. (141:46) *Burn Dressing* Compound of roots applied as a salve to burns. (87:362) *Cold Remedy* Infusion or decoction of bark taken or inhaled for colds or sore throats. (87:361) *Cough Medicine* Decoction of bark taken for consumption or an "old cough." (87:360) *Dermatological Aid* Compound decoction taken for "sores all over the body caused by bad blood." Compound poultice of bark applied to chancres caused by syphilis or cuts. (87:361) Compound decoction used as wash for parts affected by "Italian itch." (87:362) *Emetic* Compound decoction of plants taken to vomit for sleepiness and weakness. (87:361) *Febrifuge* Decoction of bark taken for colds and fever. (87:360) *Gynecological Aid* Compound decoction taken when a woman has a miscarriage. (87:361) *Liver Aid* Decoction of bark taken for too much gall. *Pediatric Aid* Decoction of bark used as a steam bath for babies with bronchitis. *Pulmonary Aid* Decoction of bark taken for soreness and lung inflammation. *Respiratory Aid* Decoction of bark used as a steam bath for babies with bronchitis. (87:360) *Stimulant* Compound decoction of plants taken to vomit for sleepiness and weakness. *Throat Aid* Infusion of bark taken for colds and sore throats. (87:361) *Tuberculosis Remedy* Decoction of bark or roots taken for consumption or for an "old cough." (87:360) *Venereal Aid* Compound poultice of bark applied to chancres caused by syphilis. (87:361) **Mahuna** *Cough Medicine* Infusion of bark or roots taken for coughs. (140:18) **Malecite** *Cold Remedy* Infusion of bark, "beaver castor," and gin used for colds. Castor or castorecum is a strong-smelling, brown, concrete substance from the preputial follicles of the beaver, *Castor fiber*. It has long been used in medicine as a stimulant and antispasmodic, and also In the manufacture of perfume. *Cough Medicine* Infusion of bark, beaver castor, and gin used for coughs. *Tuberculosis Remedy* Infusion of bark, beaver castor, and gin used by men for consumption. (116:249) **Micmac** *Cold Remedy* Bark used for colds. *Cough Medicine* Bark used for coughs. *Misc. Disease Remedy* Bark used for smallpox. *Tonic* Fruit used as a tonic. *Tuberculosis Remedy* Bark used for consumption. (40:60) **Mohegan** *Antidiarrheal* Ripe fruit fermented 1 year and used for dysentery. (174:264) Liquid from fermented fruit taken for dysentery. (176:74, 130) *Cold Remedy* Compound infusion of leaves and boneset taken with molasses for colds. Infusion of buds, leaves, or bark taken with sugar for colds. (30:118) Compound infusion taken, hot at night and cold in the morning, for colds. (174:264) Compound infusion of bark taken for colds. (176:74, 13) *Gastrointestinal Aid* Fruit put in a bottle and allowed to stand, then taken for stomach trouble. (30:118) *Tonic* Complex compound infusion including wild cherry bark taken as spring tonic. (174:266) **Narraganset** *Cold Remedy* Infusion of buds, leaves, or bark taken with sugar for colds. (30:118) **Ojibwa** *Cold Remedy* Infusion of bark used for colds. *Cough Medicine* Infusion of bark used for coughs. (153:385) **Ojibwa, South** *Analgesic* Infusion of inner bark taken for chest pain and soreness. *Dermatological Aid* Poultice of boiled, bruised, or chewed inner bark applied to sores. *Pulmonary Aid* Infusion of inner bark taken for chest pain and soreness. (91:199) **Penobscot** *Cough Medicine* Infusion of bark taken for coughs. *Tonic* Infusion of berries taken as a "fine bitter tonic." (156:310) **Potawatomi** *Adjuvant* Inner bark used as a seasoner for medicines. (154:77) **Rappahannock** *Cold Remedy* Infusion of buds, leaves, or bark taken with sugar for colds. (30:118) *Cough Medicine* Infusion of bark or berries with honey used for coughs, if stale it is poisonous. *Dietary Aid* Infusion of fresh or dried bark taken as an appetizer. *Poison* Infusion of bark or berries with honey used for coughs, if stale it is poisonous. *Tonic* Infusion of fresh or dried bark taken as a tonic. (161: 26) **Shinnecock** *Cold Remedy* Compound infusion of leaves and boneset taken with molasses for colds. Infusion of buds, leaves, or bark taken with sugar for colds. *Gastrointestinal Aid* Fruit put in a bottle and allowed to stand, then taken for stomach trouble. (30:118)

- **Food**—**Cherokee** *Fruit* Fruit used for food. (80:28) **Chippewa** *Beverage* Twigs used to make a beverage. (53:317) *Bread & Cake* Berries cooked, spread on birch bark into little cakes, dried, and stored for winter use. *Fruit* Berries eaten raw. (53:321) **Iroquois** *Bread & Cake* Fruit mashed, made into small cakes, and dried for future use. *Dried Food* Raw or cooked fruit sun or fire dried and stored for future use. *Fruit* Dried fruit taken as a hunting food. *Sauce & Relish* Dried fruit cakes soaked in warm water and cooked as a sauce or mixed with corn bread. (196:128) **Mahuna** *Fruit* Berries eaten mainly to quench the thirst. (140:70) **Menominee** *Fruit* Cherries, if eaten when picked and allowed to stand some time, said to make the Indian drunk. Cherries eaten fresh. (151:71) **Ojibwa** *Beverage* Ripe cherries used to make whisky. (153:409) *Dried Food* Fruit dried for winter use. (135: 235) This cherry was preferred to all other wild cherries and dried for winter use. (153:409) *Fruit* Fruit eaten fresh. *Soup* Dried fruit ground into a flour and used to make soup. (135:235) **Potawatomi** *Beverage* Cherries mostly used in wine or whisky. *Fruit* Cherries used for food. (154:108)
- **Fiber**—**Cherokee** *Building Material* Wood used for lumber. *Furniture* Wood used to make furniture. (80:28)
- **Other**—**Cherokee** *Decorations* Wood used to carve. (80:28)

Prunus sp., Chokecherry

- **Drug**—**Apache, Mescalero** *Antidiarrheal* Berries used for diarrhea. *Burn Dressing* Ripe berries mashed and used for burns. (14:48) **Chippewa** *Dermatological Aid* Poultice of fresh root or decoction of dried root applied to ulcers. (53:354) *Gynecological Aid* Poultice of fresh root or decoction of dried root applied to "broken breast." (53:360) **Creek** *Antidiarrheal* Decoction of root taken for dysentery. (172:659) Infusion of roots taken for dysentery. **Koasati** *Gastrointestinal Aid* Infusion of inner bark taken for dyspepsia. (177:27) **Malecite** *Unspecified* Used to make medicines. (160:6)
- **Food**—**Apache, Mescalero** *Bread & Cake* Berries ground, formed into cakes, and dried. *Fruit* Berries eaten fresh. (14:48) **Coeur d'Alene** *Fruit* Berries eaten fresh. (178:89) **Comanche** *Dried Food* Fruits eaten dried and stored for later use. (29:523) **Malecite** *Fruit* Fruits eaten for food. (160:6) **Micmac** *Beverage* Bark used to make a beverage. (159: 258) **Navajo** *Fruit* Fruits eaten as soon as they were picked. (55:54) **Oweekeno** *Fruit* Fruit used for food. (43:111) **Thompson** *Preserves* Berries collected in large quantities and cured. (178:237)
- **Dye**—**Navajo** *Green* Fruits used to make a green dye. *Purple* Roots used to make a purple dye. (55:54)
- **Other**—**Apache, Mescalero** *Hunting & Fishing Item* Small shoots used to make arrow shafts. (14:48) **Malecite** *Smoking Tools* Wood used to make pipes. (160:6) **Navajo** *Ceremonial Items* Wood used to make prayer sticks. Wood used to make a staff carried by the Humpback in the Night Chant. (55:54)

Prunus subcordata Benth., Klamath Plum

- *Food*—**Atsugewi** *Bread & Cake* Seeds removed, pulp pounded, and stored for winter in small cakes. (61:139) **Klamath** *Dried Food* Fresh or dried fruit used for food. *Fruit* Dried or fresh fruit used for food. (45:99) **Mendocino Indian** *Dried Food* Fruits dried and eaten. *Fruit* Fruits eaten for food. (41:356) **Modesse** *Fruit* Fruit used for food. (117:223) **Wintoon** *Fruit* Fruit used for food. (117:264)

Prunus subcordata var. *oregana* (Greene) W. Wight ex M. E. Peck, Oregon Klamath Plum
- *Food*—**Wintoon** *Fruit* Fruit used for food. (as *P. oregana* 117:264)

Prunus virginiana L., Common Chokecherry
- *Drug*—**Algonquin, Quebec** *Cough Medicine* Infusion of bark and sweet flag taken for coughs. (18:185) **Blackfoot** *Antidiarrheal* Berry juice used for diarrhea. *Cathartic* Infusion of cambium and saskatoon taken as a purge. *Pediatric Aid* Infusion of cambium and saskatoon taken by nursing mothers to pass medicinal qualities to baby. *Throat Aid* Berry juice used for sore throats. (86:68) **Cherokee** *Blood Medicine* Compound used as a blood tonic. *Cold Remedy* Infusion of bark taken for colds. *Cough Medicine* Infusion of bark taken for coughs. *Dermatological Aid* Astringent root bark used in a wash for old sores and ulcers. Root bark used as a wash for old sores and ulcers. *Febrifuge* Infusion or decoction of bark used for fevers, including the "great chill." (80:28, 29) Decoction of bark used as a wash for chills and fevers. (177:28) *Gastrointestinal Aid* Boiled fruit used for "blood discharged from bowels." Used in steam bath for indigestion, biliousness, and jaundice. *Gynecological Aid* Warm infusion given when labor pains begin. *Misc. Disease Remedy* Compound of barks added to corn whisky and used to break out measles. (80:28, 29) Decoction of bark used as a wash for ague. (177:28) *Oral Aid* Infusion of bark used for "thrash." *Throat Aid* Decoction of inner bark used for laryngitis. (80:28, 29) Decoction of inner bark taken to loosen phlegm for hoarseness. (177:28) **Cheyenne** *Antidiarrheal* Unripened berries pulverized and used for diarrhea. (82:42) Unripened fruits eaten by children for diarrhea. *Dietary Aid* and *Pediatric Aid* Dried, smashed, ripe berries given to children with loss of appetite. Unripened fruits eaten by children for diarrhea. *Unspecified* Dried, smashed, ripe berries used as an ingredient for medicines. (83:35) **Chippewa** *Analgesic* Decoction of inner bark taken for cramps. (53:344) *Antihemorrhagic* Compound infusion of inner bark taken for hemorrhages from the lungs. (53:340) *Blood Medicine* Compound decoction of inner bark used as cathartic blood cleanser for scrofula. *Cathartic* Compound decoction of bark used as a blood-cleansing cathartic for scrofula sores. (53:354) *Dermatological Aid* Decoction of bark used as a wash to strengthen the hair and make it grow. (53:350) *Disinfectant* Compound decoction of inner bark used as a disinfectant wash. (53:366) *Gastrointestinal Aid* Decoction of inner bark taken for stomach cramps. (53:344) *Pulmonary Aid* Compound infusion of inner bark taken for lung hemorrhage. (53:340) *Throat Aid* Decoction of inner bark gargled for sore throat. (53:342) *Tuberculosis Remedy* Compound decoction of bark used as a blood-cleansing cathartic for scrofula sores. Compound decoction of inner bark used as cathartic blood cleanser for scrofula. (53:354) **Cree, Hudson Bay** *Antidiarrheal* Decoction of fresh bark taken for diarrhea. (92:303) **Cree, Woodlands** *Antidiarrheal* Decoction of roots taken for diarrhea. (109:53) **Crow** *Antidiarrheal* Infusion of bark used for diarrhea and dysentery. *Burn Dressing* Infusion of bark used for cleansing burns. *Dermatological Aid* Infusion of bark used for cleansing sores. **Flathead** *Anthelmintic* Infusion used for intestinal worms. *Antidiarrheal* Infusion of bark used for diarrhea and dysentery. *Eye Medicine* Bark resin warmed, strained, cooled, and used for sore eyes. **Gros Ventre** *Antidiarrheal* Infusion of bark used for diarrhea and dysentery. (82:42) **Iroquois** *Antidiarrheal* Bark used for diarrhea. *Antihemorrhagic* Stalk used for hemorrhages. *Blood Medicine* Stalk used as a blood purifier. (87:359) *Cough Medicine* Decoction of plant taken as a cough syrup. (87:360) *Dermatological Aid* Inner bark used for wounds. (87:359) *Gynecological Aid* Compound decoction of stalks taken to prevent hemorrhage after childbirth. *Misc. Disease Remedy* Compound decoction of plants and bark taken for cholera. (87:360) *Pediatric Aid* Stalk used for prenatal care. *Tuberculosis Remedy* Compound decoction of roots taken for consumption. *Veterinary Aid* Decoction of branches, leaves, and berries given to horses for diarrhea. (87:359) **Kutenai** *Antidiarrheal* Infusion of bark used for diarrhea and dysentery. (82:42) **Menominee** *Antidiarrheal* Infusion of inner bark or decoction of berries taken for diarrhea. *Dermatological Aid* Poultice of pounded inner bark applied to man or beast for wounds or galls. *Pediatric Aid* Sweetened infusion of inner bark given to children for diarrhea. *Veterinary Aid* Poultice of inner bark applied to heal a wound or gall on humans or beasts. (151:49, 50) **Meskwaki** *Dermatological Aid* Decoction of bark used as an astringent and spoken of as "a puckering." *Gastrointestinal Aid* Infusion of root bark used for stomach troubles and as a sedative. *Hemorrhoid Remedy* Decoction of root bark used as an astringent, rectal douche for piles. *Sedative* Infusion of root bark used as a sedative and for stomach trouble. (152:242) **Micmac** *Antidiarrheal* Bark used for diarrhea. (40:60) **Navajo, Ramah** *Analgesic* Cold infusion of dried fruit taken for stomachache. *Ceremonial Medicine* and *Emetic* Leaves used as an emetic in various ceremonies. *Gastrointestinal Aid* Cold infusion of dried fruit taken for stomachache. *Panacea* Dried fruit used as "life medicine." (191:31) **Ojibwa** *Pulmonary Aid* Infusion of inner bark taken for lung trouble. (153:385) **Ojibwa, South** *Gynecological Aid* "Branchlets" used in unspecified manner during gestation. (91:199) **Okanagan-Colville** *Antidiarrheal* Decoction of wood, branches and bark taken for diarrhea. *Cold Remedy* Decoction of wood, branches, and bark taken for colds. *Cough Medicine* Decoction of wood, branches, and bark taken for coughs. *Dermatological Aid* Poultice of wood scraped until pasty and applied to woman's stomach to eliminate the "stretch marks." *Gastrointestinal Aid* Mashed seeds taken as a stomach medicine. *Tonic* Decoction of branches and red willow roots used as a general tonic for any type of sickness. (188:127) **Penobscot** *Antidiarrheal* Infusion of bark taken for diarrhea. (156:310) **Potawatomi** *Eye Medicine* Bark used in an eyewash and berries used to make tonic drink. *Tonic* Berries used to make tonic drink and bark used in an eyewash. (154:77, 78) **Sanpoil** *Antidiarrheal* Decoction of bark taken for diarrhea. (131:221) **Thompson** *Antidiarrheal* Decoction of twigs taken for diarrhea. *Cold Remedy* Decoction of broken sticks taken for colds. Decoction of branches, sometimes with red willow branches and wild rose roots, taken for colds. *Cough Medicine* Decoction of branches, sometimes with red willow branches and wild rose roots, taken for coughs. *Laxative* Decoction of branches, sometimes with red willow branches and wild rose roots, taken as a laxative. *Misc. Disease Remedy* De-

coction of branches, sometimes with red willow branches and wild rose roots, taken for influenza. *Unspecified* Decoction of broken sticks taken for a sick feeling. (187:264)
- **Food**—**Abnaki** *Fruit* Fruits eaten for food. (144:168) **Algonquin, Quebec** *Beverage* Fruits used to make a wine. (18:113) Cherries used to make wine. *Fruit* Cherries eaten fresh. *Preserves* Cherries made into preserves. (18:96) **Apache, Western** *Fruit* Berries eaten raw. (26:190) **Blackfoot** *Beverage* Juice given as a special drink to husbands or the favorite child. *Dried Food* Berries greased, sun dried, and stored for future use. *Fruit* Crushed berries mixed with back fat, and used to make pemmican. *Soup* Crushed berries mixed with back fat, and used to make soup. (86:104) *Special Food* Berry soup used for most ceremonial events.10 (86:26) *Spice* Peeled sticks inserted into roasting meat as a spice. *Staple* Berries considered a staple. (86:104) **Cherokee** *Fruit* Fruit used for food. (80:28) **Cheyenne** *Bread & Cake* Fruits pounded, formed into flat cakes, sun dried, and used as a winter food. *Pie & Pudding* Berries boiled, sugar and flour added, and eaten as a pudding. (83:35) **Chippewa** *Beverage* Twigs used to make a beverage. (53:317) *Dried Food* Fruits pounded, dried, and used for food. (53:321) **Cree, Woodlands** *Fruit* Fruit and pits, sometimes with fish eggs, crushed, mixed with grease, and eaten. *Sauce & Relish* Fruit used to make pancake syrup. (109:53) **Iroquois** *Bread & Cake* Fruit mashed, made into small cakes, and dried for future use. (196:128) *Dried Food* Fruits dried and used as a winter food. (124:95) Raw or cooked fruit sun or fire dried and stored for future use. *Fruit* Dried fruit taken as a hunting food. *Sauce & Relish* Dried fruit cakes soaked in warm water and cooked as a sauce or mixed with corn bread. (196:128) *Soup* Fruits pulverized, mixed with dried meat flour, and eaten as soup. (124:95) **Lakota** *Beverage* Leaves used to make a tea during the Sun Dance. *Fruit* Berries eaten fresh. (106:38) Fruits eaten for food. (139:57) *Pie & Pudding* Berries mixed with cornstarch and sugar to make a pudding. *Special Food* Small branches sucked or chewed for thirst during the Sun Dance. (106:38) **Menominee** *Beverage* Bark boiled into regular tea and drunk with meals. *Fruit* Cherries eaten fresh. (151:71) **Meskwaki** *Beverage* Bark made into a beverage. *Fruit* Cherries eaten raw. (152:263) **Montana Indian** *Bread & Cake* Berries pulverized, shaped into round cakes, sun dried, and stored for winter use. *Fruit* Berries eaten raw. Berries pulverized, shaped into round cakes, sun dried, and used to make pemmican. *Pie & Pudding* Berries mixed with sugar and flour and used to make a pudding. *Soup* Berries pulverized, shaped into round cakes, sun dried, and used in soups and stews. (82:42) **Ojibwa** *Dried Food* Berries used dried. (5:2222) *Fruit* dried for winter use. (135:235) *Fruit* Berries used fresh. (5:2222) *Fruit* eaten fresh. (135:235) Fruit of this cherry was liked, especially after the fruit had been frosted. (153:409) *Soup* Dried berry powder mixed with dried meat flour for soup. (5:2222) Dried fruit ground into a flour and used to make soup. (135:235) **Okanagan-Colville** *Bread & Cake* Berries mashed, seeds and all, and sun dried into thin cakes. *Fruit* Berries eaten fresh. *Winter Use Food* Berries stored for winter use. (188:127) **Omaha** *Dried Food* Fruit pounded with the pits, made into thin cakes, and dried for winter use.60 *Fruit* Fruit eaten fresh.60 (68:326) **Potawatomi** *Fruit* Cherry used for food and for seasoning or flavoring wine. (154:108) **Thompson** *Beverage* Fruit used to make wine and juice. *Dried Food* Fruit, with the pit, dried for future use. *Fruit* Fruit used for food. *Sauce & Relish* Fruit used to make syrup. *Winter Use Food* Fruit, with the pit, canned for future use. (187:264)
- **Fiber**—**Blackfoot** *Furniture* Straight branches used to make backrests. (86:119) **Cherokee** *Building Material* Wood used for lumber. *Furniture* Wood used to make furniture. (80:28)
- **Other**—**Blackfoot** *Containers* Hard wood used to make incense tongs. *Cooking Tools* Hard wood used to make roasting skewers. (86:119) *Tools* Sticks used to dig roots. (86:104) **Cherokee** *Decorations* Wood used to carve. (80:28) **Cheyenne** *Hunting & Fishing Item* Limbs used to make arrow shafts and bows. (82:42) Used to make arrow shafts. Used to make bows. (83:46) Wood used to make arrow shafts and bows. (83:35) **Crow** *Fasteners* Wood used to make tepee stakes and pins. Sap mixed with the neck portion of certain animals and used to make a glue. (82:42) **Lakota** *Ceremonial Items* Branch bundles tied to sacred Sun Dance poles. (106:38) *Hunting & Fishing Item* Stems used to make arrows. (139:57) **Montana Indian** *Paint* Sap mixed with different colored clays and used as paint for Indian designs. (82:42) **Okanagan-Colville** *Season Indicator* Ripened fruit indicated that the spring salmon were coming up the river to spawn. (188:127)

Prunus virginiana var. *demissa* (Nutt.) Torr., Western Chokecherry

- **Drug**—**Atsugewi** *Dermatological Aid* Poultice of leaves applied to cuts, sores, bruises, and black eyes. Decoction of bark used for bathing wounds. (as *P. demissa* 61:140) **Blackfoot** *Unspecified* Decoction of bark and roots of western sweet cicely, northern valerian, and horehound taken internally. (as *P. demissa* 114:277) **Gosiute** *Blood Medicine* Decoction of bark used as a blood medicine for nose hemorrhages. (as *P. demissa* 39:378) *Gastrointestinal Aid* Decoction of wood scrapings used by children and adults for bowel troubles. (as *P. demissa* 39:350) *Hemostat* Decoction of bark used as a blood medicine for nose hemorrhages. (as *P. demissa* 39:378) *Pediatric Aid* Decoction of wood scrapings used by children and adults for bowel troubles. (as *P. demissa* 39:350) **Haisla & Hanaksiala** *Oral Aid* Poultice of mashed leaves applied to oral abscesses. (43:273) **Karok** *Cold Remedy* and *Pediatric Aid* Bark scrapings placed beside the nose of babies for colds. (as *P. demissa* 148:384) **Kawaiisu** *Laxative* Ripe berries had a laxative effect. (as *P. demissa* 206:54) **Mendocino Indian** *Antidiarrheal* Inner bark used for diarrhea. *Sedative* Inner bark used for nervous excitability. *Tonic* Inner bark used in a tonic. (as *Cerasus demissa* 41:356) **Menominee** *Pulmonary Aid* Decoction of inner bark used for lung trouble. (as *P. demissa* 54:130) **Oregon Indian** *Antidiarrheal* Pounded, dried cherries mixed with dry salmon and sugar and used for dysentery. (as *P. demissa* 118:42) **Paiute** *Analgesic* Dried, pulverized bark smoked for headache or head cold. *Cold Remedy* Decoction of peeled bark or root taken for colds and bark smoked for head colds. *Cough Medicine* Decoction of peeled bark or root taken for coughs and colds. *Dermatological Aid* Pulverized, dried bark used as a drying powder on sores. *Eye Medicine* and *Herbal Steam* Steam from boiling bark allowed to rise into the eyes for snow-blindness. *Tuberculosis Remedy* Decoction of leaves, bark, or roots taken for tuberculosis. **Shoshoni** *Antiemetic* Decoction of bark taken for indigestion or upset stomach. *Eye Medicine* Steam from boiling bark allowed to rise into the eyes for snow-blindness. *Gastrointestinal Aid*

Decoction of bark taken for indigestion or upset stomach. *Herbal Steam* Steam from boiling bark allowed to rise into the eyes for snow-blindness. (180:123, 124) **Sioux** *Adjuvant* Wood used to make "medicine-spoons" for use in ceremonial dog feasts. *Antidiarrheal* Infusion of bark used for dysentery. *Ceremonial Medicine* Wood used to make "medicine-spoons" for use in ceremonial dog feasts. *Hemostat* Dried roots chewed and placed in bleeding wounds. (as *P. demissa* 19:19) **Thompson** *Gynecological Aid* Decoction of bark taken after childbirth as a strengthening tonic. *Tonic* Decoction of bark taken as a tonic. (as *P. demissa* 164:477)

- *Food*—**Atsugewi** *Porridge* Ripe, mashed fruit added to water to form a paste and eaten without cooking. (as *P. demissa* 61:139) **Blackfoot** *Fruit* Berries eaten raw. Berries pounded, mixed with meat, and eaten. *Soup* Berries used for soups. (as *P. demissa* 114:277) **Cahuilla** *Dried Food* Fruit sun dried for future use. *Fruit* Fruit considered a great delicacy, important food and a highly prized food source. Fruit eaten fresh. *Staple* Ground pit used as a meal. (15:119) **Coeur d'Alene** *Dried Food* Berries dried and used for food. *Fruit* Berries eaten fresh. Berries mashed and eaten. *Soup* Berries dried, boiled with roots, and eaten as soup. (as *P. demissa* 178:89) **Costanoan** *Fruit* Fruits used for food, late in season only. (21:249) **Gosiute** *Dried Food* Fruit mashed, sun dried, and stored for winter use. *Fruit* Fruit used for food. *Porridge* Fruit mashed, sun dried, stored for winter, and used to make a mush. (as *P. demissa* 39:378) **Haisla & Hanaksiala** *Forage* Fruit eaten by bears. *Fruit* Fruit used for food. (43:273) **Karok** *Fruit* Berries used for food. (as *P. demissa* 148:384) **Kawaiisu** *Fruit* Berries eaten fresh. *Preserves* Berries used to make jelly. (as *P. demissa* 206:54) **Kiowa** *Fruit* Fruit eaten fresh. *Winter Use Food* Fruit dried in large quantities for winter use. (as *Cerasus demissa* 192:30) **Klamath** *Dried Food* Fruit dried for later use. (as *P. demissa* 45:98) **Luiseño** *Fruit* Fruit used for food. (as *P. demissa* 155:232) **Mendocino Indian** *Preserves* Fruits made into a jelly and used for food. (as *Cerasus demissa* 41:356) **Modesse** *Fruit* Fruit used for food. (as *Cerasus demissa* 117:223) **Montana Indian** *Beverage* Ripe fruit collected each fall and made into wine. *Fruit* Fruit eaten fresh.[61] *Preserves* Ripe fruit collected each fall and made into marmalade. *Staple* Fruit used as an important ingredient in the preparation of "pemmican."[61] *Winter Use Food* Crushed, dried fruit strips stored for winter use. (as *P. demissa* 19:19) **Ojibwa** *Dried Food* Fruit dried for winter use. *Fruit* Fruit eaten fresh. *Soup* Dried fruit ground into a flour and used to make soup. (as *P. demissa* 135:235) **Okanagon** *Staple* Berries used as a principal food. (as *P. demissa* 178:238) **Paiute** *Beverage* Fruits added to hot water and used as a beverage. Stems used to make tea. (as *P. demissa* 104:99) Bark and twigs made into a tea and taken with meals. (as *P. demissa* 111:84) *Dried Food* Fruits broken, molded into cakes, hardened, ground, boiled, dried, and used for food. *Fruit* Fruits eaten fresh. (as *P. demissa* 104:99) Chokecherry cakes ground and boiled with flour, sugar, and occasionally roasted deer liver. *Winter Use Food* Chokecherries made into cakes for winter use. (as *P. demissa* 111:84) **Paiute, Northern** *Bread & Cake* Berries mashed, made into round cakes, and eaten dry. *Dried Food* Berries dried, cooked, and eaten. *Fruit* Berries eaten ripe. *Porridge* Berries dried, ground, and boiled into a mush. *Winter Use Food* Berries dried and stored for winter use. (as *P. demissa* 59:49) **Round Valley Indian** *Dried Food* Fruits dried and used for food. *Fruit* Fruits eaten. (as *Cerasus demissa* 41:356) **Shuswap** *Beverage* Boiled roots used to make beer. Dried berries used to make wine. *Winter Use Food* Berries dried for winter use. (123:67) **Spokan** *Fruit* Berries used for food. (as *P. demissa* 178:343) **Thompson** *Unspecified* Dark purple drupe used as part of the diet. (as *P. demissa* 164:490) **Wintoon** *Fruit* Fruit used for food. (as *P. demissa* 117:264) **Yuki** *Fruit* Berries eaten raw. Ripe berries cooked and eaten. (as *P. demissa* 49:87) **Yurok** *Fruit* Fruit used for food. (6:48)
- *Fiber*—**Maidu** *Basketry* Withes used as overlay twine weft bases in the manufacture of baskets. (as *P. demissa* 173:71)
- *Other*—**Karok** *Fasteners* Gum used to fasten foreshafts to the end of arrows. *Paint* Gum applied to the surface of bows and arrows before painting the design. (as *P. demissa* 148:384) **Kawaiisu** *Hunting & Fishing Item* Straight stems used to make arrows. *Tools* Straight stems used to make gun cleaners. (as *P. demissa* 206:54) **Montana Indian** *Hunting & Fishing Item* Straight shoots used to make arrow shafts. (as *P. demissa* 19:19) **Navajo** *Ceremonial Items* Wood used to make dance implements, prayer sticks, and square hoops for ceremonies. *Sacred Items* Tree sacred to the Navajo. (as *P. demissa* 55:54) **Paiute** *Hunting & Fishing Item* Limbs used for arrow shafts. (as *P. demissa* 111:84) **Paiute, Northern** *Smoke Plant* Berries mashed, made into little cakes, dried, and used like chewing tobacco. (as *P. demissa* 59:49) **Shuswap** *Paint* Berries mixed with bear grease and used to make paint for painting pictographs. (123:67) **Thompson** *Designs* Shredded bark used to ornament the rims of baskets. *Tools* Wood used to make handles for root diggers. (as *P. demissa* 164:500)

Prunus virginiana var. *melanocarpa* (A. Nels.) Sarg., Black Chokecherry

- *Drug*—**Keres, Western** *Cough Medicine* Bark made into a cough medicine. (as *P. melanocarpa* 171:63) **Navajo** *Unspecified* Fruit and seeds ground raw, patted into a cake, sun dried, and used for medicinal purposes. (as *P. melanocarpa* 165:222) **Sanpoil & Nespelem** *Unspecified* Decoction of branches taken as medicine. (131:104)
- *Food*—**Acoma** *Dried Food* Fruits dried for winter use. *Fruit* Fruits eaten fresh. (as *P. melanocarpa* 32:46) **Apache** *Bread & Cake* Berries ground and meal made into sweet, blackish cakes. (as *Padus melanocarpa* 138:47) **Apache, Chiricahua & Mescalero** *Fruit* Fruit eaten fresh. *Preserves* Fruit cooked to make a preserve. *Winter Use Food* Fruits ground, pressed, and saved for winter. (as *Prunus melanocarpa* 33:46) **Cheyenne** *Fruit* Fresh or pounded, dried berries and pits used to make berry pemmican. *Winter Use Food* Pounded berries and pits made into flat cakes and sun dried for winter use. (as *P. melanocarpa* 76:177) **Cochiti** *Dried Food* Fruits dried for winter use. *Fruit* Fruits eaten fresh. (as *P. melanocarpa* 32:46) **Dakota** *Fruit* Fresh fruit used for food. *Winter Use Food* Fruit pounded to a pulp, made into small cakes, dried in the sun, and stored for winter use. (as *P. melanocarpa* 69:364) **Great Basin Indian** *Dried Food* Mashed berries dried for winter use. (as *P. melanocarpa* 121:48) **Keres, Western** *Fruit* Fruit used for food. *Winter Use Food* Fruit dried for winter use. (as *P. melanocarpa* 171:63) **Kiowa** *Fruit* and *Winter Use Food* Fruit eaten fresh and dried in large quantities for winter use. (192:30) **Laguna** *Dried Food* Fruits dried for winter use. *Fruit* Fruits eaten fresh. **Navajo** *Porridge* Fruits cooked into a gruel with cornmeal. (as *P. melanocarpa* 32:46) **Navajo, Ramah** *Bread & Cake* Fruit ground and made into small cakes. *Fruit* Fruit eaten fresh. *Winter Use Food*

Fruit dried for winter use. (191:31) **Pueblo** *Fruit* Fruits eaten fresh or cooked. **San Felipe** *Dried Food* Fruits dried for winter use. *Fruit* Fruits eaten fresh. (as *P. melanocarpa* 32:46) **Sanpoil & Nespelem** *Beverage* Branches used to make a beverage. (131:104) *Bread & Cake* Berries mashed, mixed with dried salmon into a pemmican, formed into cakes, dried, and stored. *Fruit* Fruit eaten fresh or dried. (131:101) **Spanish American** *Preserves* Fruits made into jelly and jam. (as *P. melanocarpa* 32:46) **Tewa** *Fruit* Berries boiled and eaten. Berries eaten raw. (as *Padus melanocarpa* 138:47)
- **Dye**—**Great Basin Indian** *Red* Fruit used to make a dark red dye. *Red-Brown* Inner bark used to make a red-brown dye. (as *Prunus melanocarpa* 121:48)
- **Other**—**Isleta** *Hunting & Fishing Item* Strong, supple, straight-grained limbs used to make bows. (as *P. melanocarpa* 100:40) **Keres, Western** *Hunting & Fishing Item* Wood, backed with sinew, made into bows. (as *P. melanocarpa* 171:63) **Navajo, Ramah** *Ceremonial Items* Stems used to make Evilway and Mountaintopway big hoops and Bear's prayer stick in Mountaintopway. (191:31) **Paiute** *Designs* Flower used as the favorite basket pattern. (as *P. melanocarpa* 121:48) **Tewa** *Hunting & Fishing Item* Wood used to make bows. (as *Padus melanocarpa* 138:47)

Prunus virginiana L. var. *virginiana*, Chokecherry
- **Drug**—**Dakota** *Ceremonial Medicine* Fruit prepared in unspecified way and used in old-time ceremonies. **Omaha** *Ceremonial Medicine* Fruit prepared in unspecified way and used in old-time ceremonies. **Pawnee** *Ceremonial Medicine* Fruit prepared in unspecified way and used in old-time ceremonies. **Ponca** *Antidiarrheal* Infusion of fruit or decoction of bark taken for diarrhea. *Ceremonial Medicine* Fruit prepared in unspecified way and used in old-time ceremonies. **Winnebago** *Ceremonial Medicine* Fruit prepared in unspecified way and used in old-time ceremonies. (as *Padus nana* 70:88, 89)
- **Food**—**Dakota** *Bread & Cake* Fruit and pits pounded to a pulp, formed into small cakes, sun dried, and stored for winter use. *Dried Food* Fruit dried for winter use. *Fruit* Fruit eaten fresh. **Omaha** *Bread & Cake* Fruit and pits pounded to a pulp, formed into small cakes, sun dried, and stored for winter use. *Dried Food* Fruit dried for winter use. *Fruit* Fruit eaten fresh. **Pawnee** *Bread & Cake* Fruit and pits pounded to a pulp, formed into small cakes, sun dried, and stored for winter use. *Dried Food* Fruit dried for winter use. *Fruit* Fruit eaten fresh. **Ponca** *Bread & Cake* Fruit and pits pounded into small cakes, sun dried, and stored for winter use. *Dried Food* Fruit dried for winter use. *Fruit* Fruit eaten fresh. (as *Padus nana* 70:88)
- **Other**—**Dakota** *Season Indicator* Ripe fruit used as an indicator for the time of the Sun dance. **Omaha** *Season Indicator* Ripe fruit used as an indicator for the time of the Sun dance. **Pawnee** *Season Indicator* Ripe fruit used as an indicator for the time of the Sun dance. **Ponca** *Hunting & Fishing Item* Boiled bark water used as a wash for traps to remove the scent of the former captures. *Season Indicator* Ripe fruit used as an indicator for the time of the Sun dance. (as *Padus nana* 70:88)

Psathyrotes, Asteraceae
Psathyrotes annua (Nutt.) Gray, Annual Psathyrotes
- **Drug**—**Paiute** *Eye Medicine* Decoction of dried plant used as an eyewash. *Toothache Remedy* Dried leaves chewed for toothache. **Shoshoni** *Gastrointestinal Aid* and *Pediatric Aid* Decoction of entire plant used for stomachaches, especially children's. *Urinary Aid* Decoction of whole plant taken for urinary troubles. (180:124)

Psathyrotes pilifera Gray, Hairybeast Turtleback
- **Drug**—**Navajo, Kayenta** *Ceremonial Medicine* Plant used as a ceremonial emetic. *Dermatological Aid* Plant used as a lotion for chilblains. *Emetic* Plant used as a ceremonial emetic. (205:49)

Psathyrotes ramosissima (Torr.) Gray, Velvet Turtleback
- **Drug**—**Paiute** *Analgesic* Decoction of plant used as a head wash for headaches. *Antidiarrheal* Decoction of plant taken for diarrhea. *Cathartic* Decoction of plant taken as a physic. *Dermatological Aid* Compound poultice of crushed plants applied to draw boils and imbedded slivers. Compound poultice of plant applied to sores, cuts, swellings, and insect bites. *Emetic* Decoction of plant taken as an emetic. *Gastrointestinal Aid* Decoction of plant taken for stomachaches, bowel disorders, and biliousness. *Laxative* Decoction of plant taken for constipation. *Liver Aid* Decoction of plant taken for liver trouble. *Snakebite Remedy* Poultice of crushed, green plant or moistened dried plant used on snakebites. (180:125, 126) *Toothache Remedy* Dry bits chewed for toothache. (118:45) *Venereal Aid* Decoction of plant taken for venereal diseases. **Shoshoni** *Analgesic* Decoction of plant used as a head wash for headaches. *Antidiarrheal* Decoction of plant taken for diarrhea. *Cathartic* Decoction of plant taken as a physic. *Cough Medicine* Decoction of plant taken for tubercular cough. *Dermatological Aid* Compound poultice of crushed plants applied to draw boils and imbedded slivers. Compound poultice of plant applied to sores, cuts, swellings, and insect bites. *Emetic* Decoction of plant taken as an emetic. *Gastrointestinal Aid* Decoction of plant taken for stomachaches, bowel disorders, and biliousness. *Laxative* Decoction of plant taken for constipation. *Liver Aid* Decoction of plant taken for liver trouble. *Snakebite Remedy* Poultice of crushed, green plant applied to snakebites. *Tuberculosis Remedy* Decoction of plant taken for tubercular cough. *Venereal Aid* Decoction of plant taken for venereal diseases. (180:125, 126)

Pseudephebe, Alectoriaceae, lichen
Pseudephebe pubescens (L.) Choisy
- **Other**—**Haisla & Hanaksiala** *Paint* Plant used to make black paint for wood. (43:143)

Pseudocymopterus, Apiaceae
Pseudocymopterus montanus (Gray) Coult. & Rose, Alpine False Springparsley
- **Drug**—**Navajo, Kayenta** *Ceremonial Medicine* and *Emetic* Plant used as a ceremonial emetic. (205:35) **Navajo, Ramah** *Ceremonial Medicine* and *Emetic* Whole plant used as a ceremonial emetic. *Gastrointestinal Aid* Infusion or decoction taken after swallowing an ant. (191:38)
- **Food**—**Hopi** *Vegetable* Plant used for greens. (42:352) **Navajo, Ramah** *Unspecified* Ground root cooked with meat. Leaves boiled with cornmeal. (191:38)

Pseudocymopterus sp.
- **Food**—**Isleta** *Beverage* Leaves and stems boiled to make a beverage. (100:40)

Pseudoroegneria, Poaceae

Pseudoroegneria spicata (Pursh)Á. Löve **ssp. *spicata***, Bluebunch Wheatgrass
- **Drug**—Okanagan-Colville *Antirheumatic (External)* Decoction of leaves used for bathing sore, swollen, crippled, or paralyzed limbs caused by arthritis. (as *Agropyron spicatum* 188:53)
- **Food**—Okanagan-Colville *Forage* Plant used as grazing grass for livestock and deer. (as *Agropyron spicatum* 188:53)

Pseudostellaria, Caryophyllaceae

Pseudostellaria jamesiana (Torr.) W. A. Weber & R. L. Hartman, Tuber Starwort
- **Drug**—Navajo, Kayenta *Ceremonial Medicine* Plant chewed for corral dance. *Dermatological Aid* Poultice of plant applied to hailstone injuries. (as *Stellaria jamesiana* 205:22)

Pseudotsuga, Pinaceae, conifer

Pseudotsuga menziesii (Mirbel) Franco, Douglas Fir
- **Drug**—Apache, White Mountain *Cough Medicine* Pitch used for coughs. (as *P. mucronata* 136:159) Haisla *Unspecified* Poultice of pitch and roasted, pounded frogs applied for unspecified serious illness. Hanaksiala *Gastrointestinal Aid* Infusion of green bark taken for bleeding bowels and stomach troubles. *Gynecological Aid* Infusion of green bark taken for excessive menstruation. *Throat Aid* Pitch chewed for sore throats. (43:179) Havasupai *Unspecified* Leaves boiled and used as medicine. (197:206) Isleta *Antirheumatic (External)* Infusion of leaves used for rheumatism. *Orthopedic Aid* Infusion of leaves used for paralysis. (as *P. mucronata* 100:41) Karok *Cold Remedy* Infusion of young sprouts used for colds. (6:48) Kwakiutl *Antidiarrheal* Cold, compound infusion of burned, pulverized bark taken for diarrhea. *Dermatological Aid* Poultice of bark, oil, down, and skunk cabbage leaf applied to carbuncles. *Unspecified* Pitch used as a "good medicine." (183:270) Montana Indian *Antirheumatic (External)* Leaves used in the sweat bath for rheumatism. *Venereal Aid* Decoction of spring buds used for certain venereal diseases. (as *P. mucronata* 19:20) Okanagon *Antirheumatic (External)* Moxa of ashes or ash and fat salve used for rheumatism. *Kidney Aid* Decoction of twigs or shoots taken as a kidney remedy. *Urinary Aid* Decoction of twigs or shoots taken as a bladder remedy. (as *P. douglasii* 125:42) Pomo, Little Lakes *Antirheumatic (External)* and *Herbal Steam* Leaves used as a sweat bath for rheumatism. *Venereal Aid* Decoction of buds used for venereal disease. (as *P. mucronata* 41:309) Thompson *Antirheumatic (External)* Moxa of ashes or ash and fat salve used for rheumatism. (as *P. douglasii* 125:42) Heated branches or moxa of bough tips and needles used for rheumatism. (as *P. mucronata* 164:475) *Cold Remedy* Infusion of plant top used for colds. (187:107) *Dermatological Aid* Twig ashes mixed with grease and used as a general ointment or salve. (as *P. mucronata* 164:475) Shoots used in the tips of moccasins to keep the feet from perspiring and to prevent athlete's foot. Poultice of pitch used for cuts, boils, and other skin ailments. (187:107) *Disinfectant* Decoction of branches and twigs used as a purifying body wash in sweat house. (as *P. mucronata* 164:505) *Diuretic* Decoction of twigs taken as a diuretic. (as *P. mucronata* 164:475) *Kidney Aid* Decoction of twigs or shoots taken as a kidney remedy. (as *P. douglasii* 125:42) *Oral Aid* Peeled plant tops chewed, especially by young people at puberty, as a mouth freshener. *Orthopedic Aid* Poultice of pitch used for injured or dislocated bones. (187:107) *Tonic* Decoction of young twigs and leaves used for the tonic properties. (as *P. mucronata* 164:494) *Urinary Aid* Decoction of twigs or shoots taken as a bladder remedy. (as *P. douglasii* 125:42)
- **Food**—Apache, White Mountain *Candy* Pitch used as gum. (as *P. mucronata* 136:159) Gosiute *Candy* Gum used for chewing gum. (as *P. douglasii* 39:378) Karok *Beverage* Young sprouts used to make tea. (6:48) Montana Indian *Substitution Food* Leaves sometimes used as a substitute for coffee. (as *P. mucronata* 19:20) Paiute *Spice* Boughs or branches used for flavoring barbecued bear meat. (as *P. mucronata* 111:44) Round Valley Indian *Substitution Food* Fresh leaves used as a substitute for coffee. (as *P. mucronata* 41:309) Shuswap *Sweetener* Sap used as a sugar-like food. (123:52) Thompson *Sweetener* Wild sugar gathered and eaten whenever possible. (187:107) Yuki *Substitution Food* Fresh leaves used as a substitute for coffee. (as *P. mucronata* 41:309) Yurok *Beverage* Young sprouts used to make tea. *Candy* Young sprouts used to chew. (6:48)
- **Fiber**—Klamath *Canoe Material* Single logs used to make dugout canoes. (as *P. mucronata* 46:728) Montana Indian *Building Material* Wood used extensively for lumber. (as *P. mucronata* 19:20) Paiute *Mats, Rugs & Bedding* Boughs used on the floor of sweat houses and for camping beds. (as *P. mucronata* 111:44) Round Valley Indian *Basketry* Small roots used to make baskets. (as *P. mucronata* 41:309) Salish, Coast *Caulking Material* Pitch used to patch canoes or water vessels. (182:71) Thompson *Building Material* Logs considered important for construction. Young, second growth boughs used to make early summer lodges. *Caulking Material* Pitch used for caulking canoes. *Mats, Rugs & Bedding* Boughs used as floor coverings for lodges and sweat houses. The boughs were generally mixed with juniper and sagebrush branches for the sweat house floor coverings. Boughs used in the sweat lodge as a mat for scrubbing the skin. The scrubbing mats prevented them from having body odor and made them feel fresh and clean. *Snow Gear* Wood from young trees used to make snowshoe frames. (187:107)
- **Other**—Bella Coola *Fuel* Bark used as a valuable fuel. (184:198) Blackfoot *Hunting & Fishing Item* Wood used to make bows. (97:18) Clallam *Fuel* Bark and wood used for firewood. *Hunting & Fishing Item* Wood used to make spear and harpoon shafts. (57:195) Haisla *Hunting & Fishing Item* Wood used to make herring and oolichan (candlefish) rakes. Haisla & Hanaksiala *Fasteners* Pitch used for bindings. (43:179) Havasupai *Ceremonial Items* Branches used ceremonially. (197:206) Hesquiat *Fuel* Pitch laden bark and limbs used as an excellent fuel. In Hesquiat mythology, Black Bear used to break off Douglas fir bark with one swipe of his paw and pile it on end in the fire. Raven wanted to have a meal with Bear and he tried to imitate Bear in collecting fuel, but he could not break off the bark; he only hurt himself. When the fire was going, Bear put his paws up to the fire and oil dripped out of them into a dish. Raven watched him doing this, and when Bear went over to eat at Raven's house, Raven tried to produce oil in a similar manner. But no oil came out of his feet, and his claws burned and shriveled up into their present state. (185:44) Hopi *Ceremonial Items* Branches considered important in many of the ceremonies. (as *P. mucronata* 200:63) Isleta *Ceremonial Items* Trees cut, brought to the pueblo, and used in the Easter cere-

monies. Boughs used in the Easter and Evergreen dances. (as *P. mucronata* 100:41) **Jemez** *Ceremonial Items* Branches used by the Koshares for dances. (as *P. mucronata* 44:26) **Keres, Western** *Ceremonial Items* Plant sometimes used in ceremonies. (as *P. mucronata* 171:64) **Keresan** *Ceremonial Items* Used to make costumes for dancers, prayer sticks, and other ceremonial items. (as *P. mucronata* 198:563) **Kitasoo** *Fuel* Wood used as an excellent fuel source. (43:318) **Navajo** *Cash Crop* Bartered with the Hano for corn and meal. (as *P. mucronata* 138:42) **Oweekeno** *Fuel* Wood used for firewood, especially for cooking fish. (43:70) **Pomo, Kashaya** *Fasteners* Pitch used like glue. *Fuel* Wood used for firewood. (72:49) **Round Valley Indian** *Lighting* Small branches used as torches in fishing. (as *P. mucronata* 41:309) **Salish, Coast** *Fuel* Bark used as a top-quality fuel. *Hunting & Fishing Item* Knots steamed, placed in hollow kelp stems overnight and bent to make halibut and cod hooks. (182:71) **Tewa** *Ceremonial Items* Branches used in almost all dances. Twig used as part of the headdress worn in dances. **Tewa of Hano** *Ceremonial Items* Branches and twigs used in almost all the winter dances. (as *P. mucronata* 138:42) **Thompson** *Ceremonial Items* Boughs used for scrubbing and purification by girls at puberty. *Cooking Tools* Dry wood broken and placed in a thick layer above and below the food in the cooking pit and used as fuel. Peeled twig bundles used as whippers for soapberries. *Fuel* Rotten wood used as fuel for smoking hides. *Good Luck Charm* Boughs used as scrubbers by boys and girls at puberty because it would bring good luck. The branches were boiled for good luck and good health, either to drink or used as a wash. *Hunting & Fishing Item* Young saplings used to make dip net hoops and handles. Boughs used by hunters to scrub themselves before hunting so that the deer could not smell them. (187:107)

Pseudotsuga menziesii var. *glauca* (Beissn.) Franco, Rocky Mountain Douglas Fir

- **Drug**—**Okanagan-Colville** *Blood Medicine* Decoction of first-year-growth shoots taken as an emetic for anemia. *Dermatological Aid* Infusion of boughs with other plants taken and used to wash the skin and hair during sweat baths. Decoction of bark used for allergies caused by touching water hemlock. *Emetic* Decoction of first-year-growth shoots taken as an emetic for high fevers and anemia. *Febrifuge* Decoction of first-year-growth shoots taken as an emetic for high fevers. (188:34)
- **Food**—**Okanagan-Colville** *Unspecified* White, crystalline sugar exuded from branches and eaten alone or with balsamroot seeds. *Winter Use Food* White, crystalline sugar exuded from branches and stored for future use. (188:34)
- **Fiber**—**Okanagan-Colville** *Building Material* Wood used to make tepee poles and spear shafts. Boughs used to cover the floor of the sweat house. (188:34)
- **Other**—**Okanagan-Colville** *Ceremonial Items* Boughs used by bereaved persons to scrub themselves as a purification ritual. *Containers* Boughs used under a freshly killed deer while butchering it. *Hunting & Fishing Item* Wood used to make tepee poles and spear shafts. *Season Indicator* Pollen shedding cones used as an indication that ponderosa pine cambium was ripe. *Soap* Boughs used to scrub the body during a sweat bath to impart a nice smell and make the skin clean. (188:34)

Pseudotsuga menziesii (Mirbel) Franco var. *menziesii*, Douglas Fir

- **Drug**—**Bella Coola** *Analgesic* Gum mixed with dogfish oil and taken as emetic and purgative for intestinal pains. *Antidiarrheal* Gum mixed with dogfish oil and taken as emetic and purgative for diarrhea. *Antirheumatic (Internal)* Gum mixed with dogfish oil and taken as emetic and purgative for rheumatism. *Cathartic* Gum and dogfish oil taken as a purgative for many ailments. *Cold Remedy* Gum mixed with dogfish oil and taken as emetic and purgative for colds. *Diuretic* Decoction of gum taken warm as a diuretic for gonorrhea. *Emetic* Gum and dogfish oil taken as an emetic for many ailments. *Gastrointestinal Aid* Gum mixed with dogfish oil and taken as emetic and purgative for intestinal pains. *Laxative* Gum mixed with dogfish oil and taken as emetic and purgative for constipation. *Venereal Aid* Decoction of gum taken warm as a diuretic for gonorrhea. Gum mixed with dogfish oil and taken as emetic and purgative for gonorrhea. (as *P. taxifolia* 150:51) **Cowlitz** *Cold Remedy* Decoction of plant taken as a cold medicine. *Dermatological Aid* Poultice of pitch applied to sores. (as *P. taxifolia* 79:19) **Karok** *Disinfectant* Boughs used as an antiseptic. *Herbal Steam* Plant used as sweat house fuel. (as *P. taxifolia* 148:379) **Navajo, Kayenta** *Analgesic* Plant used for headaches. *Disinfectant* Plant used for fumigation. *Gastrointestinal Aid* Plant used for stomach disease. (as *P. taxifolia* 205:15) **Navajo, Ramah** *Ceremonial Medicine* Needles used as a ceremonial emetic. *Emetic* Needles used as a ceremonial emetic. (as *P. taxifolia* 191:14) **Quinault** *Dermatological Aid* Poultice of pitch applied to sores. **Skagit** *Dermatological Aid* Poultice of pitch applied to sores. *Disinfectant* Decoction of bark used as an antiseptic for infections. **Squaxin** *Cold Remedy* Decoction of plant taken as a cold medicine. **Swinomish** *Analgesic* Decoction of needles applied as poultice to the chest to draw out the pain. *Cold Remedy* Decoction of root bark taken for colds. *Oral Aid* Bud tips chewed for mouth sores. *Throat Aid* Bud tips chewed for sore throats. *Tonic* Decoction of needles taken as a general tonic. (as *P. taxifolia* 79:19)
- **Food**—**Apache, White Mountain** *Candy* Pitch used as gum. (as *P. taxifolia* 136:159) **Cowlitz** *Candy* Pitch chewed as a gum. (as *P. taxifolia* 79:19) **Karok** *Spice* Boughs used as "seasoning" for barbecued elk or deer meat. (as *P. taxifolia* 148:379) **Klallam** *Candy* Pitch chewed as a gum. **Quinault** *Candy* Pitch chewed as a gum. (as *P. taxifolia* 79:19)
- **Fiber**—**Hoh** *Building Material* Formerly used to make shakes and clapboards for houses. (as *P. taxifolia* 137:58) **Karok** *Building Material* Wood used to make planks. (as *P. taxifolia* 148:379) **Maidu** *Basketry* Twigs used as coarse twine warp in the manufacture of baskets. (as *P. taxifolia* 173:71) **Quileute** *Building Material* Formerly used to make shakes and clapboards for houses. (as *P. taxifolia* 137:58) **Yuki** *Building Material* Branches used to make camp shelters. (as *P. taxifolia* 49:93)
- **Dye**—**Swinomish** *Brown* Bark boiled and used on fish nets as a light brown dye to make them invisible to the fish. (as *P. taxifolia* 79:19)
- **Other**—**Chehalis** *Fuel* Bark used for firewood. *Good Luck Charm* Warmed cones used as charms to stop the rain. *Hunting & Fishing Item* Used to make salmon spears and harpoons. **Cowlitz** *Fuel* Bark used for firewood. *Good Luck Charm* Cones placed near the fire and used as charms to bring sunshine. *Hunting & Fishing Item* Used to make salmon spears and harpoons. **Green River Group** *Fuel* Bark

used for firewood. *Hunting & Fishing Item* Used to make salmon spears and harpoons. (as *P. taxifolia* 79:19) **Hopi** *Ceremonial Items* Plant used for the Ni-man and Po-wa-me ceremony. (as *P. taxifolia* 42:353) **Karok** *Decorations* Soot from burned pitch rubbed into the punctures made in tattooing a girl's chin. *Good Luck Charm* Boughs considered "good luck." *Hunting & Fishing Item* Wood used to make the salmon harpoon shaft and dip net poles. Boughs used in fire to pass the bow and arrow through, to prevent the deer from smelling the hunter. *Tools* Wood used to make the hook for climbing sugar pine trees. (as *P. taxifolia* 148:379) **Klallam** *Fuel* Bark used for firewood. *Hunting & Fishing Item* Used to make salmon spears and harpoons. **Lummi** *Fuel* Bark used for firewood. *Hunting & Fishing Item* Used to make salmon spears and harpoons. (as *P. taxifolia* 79:19) **Navajo** *Ceremonial Items* Branches used in the Shooting Chant. (as *P. taxifolia* 55:23) **Navajo, Ramah** *Ceremonial Items* Used to make garment for garment ceremony of Evilway. Used to make bow and chant arrow for overshooting ceremony of Evilway. Branches used to make Holyway big hoop. Used to make unravelers for several ceremonials. Branches attached to masks and carried in hands by god impersonators in Nightway. (as *P. taxifolia* 191:14) **Nitinaht** *Fuel* Bark and wood used for fuel. *Hunting & Fishing Item* Hard knots used to make molded halibut hooks. Wood used to make sea urchin spear shafts, seal spears, and poles for placing codfish lures. (186:73) Used as a pole to place fishing lures in the water. (186:95) **Pomo** *Fuel* Wood used for firewood. (as *P. taxifolia* 66:11) **Quinault** *Fuel* Bark used for firewood. *Hunting & Fishing Item* Used to make shafts for harpoons, salmon spears, and handles of dip nets. **Skagit** *Fuel* Bark used for firewood. *Hunting & Fishing Item* Used to make salmon spears and harpoons. **Swinomish** *Fuel* Bark used for firewood. *Hunting & Fishing Item* Used to make salmon spears and harpoons. (as *P. taxifolia* 79:19)

Psidium, Myrtaceae

Psidium guajava L., Guava

- *Drug*—**Hawaiian** *Antidiarrheal* Buds chewed by mothers and given to infants for diarrhea. Buds chewed for diarrhea. *Antihemorrhagic* Infusion of pounded roots and other plants strained and taken for bowel or intestinal hemorrhage. *Dermatological Aid* Buds and other plants pounded and resulting liquid applied to deep cuts. *Orthopedic Aid* Buds and other plants pounded and resulting liquid applied to sprains. *Pediatric Aid* Buds chewed by mothers and given to infants for diarrhea. (as *P. guayava* 2:55)
- *Food*—**Seminole** *Unspecified* Plant used for food. (169:464)
- *Other*—**Seminole** *Hunting & Fishing Item* Plant used to make bows. (169:464)

Psilostrophe, Asteraceae

Psilostrophe sparsiflora (Gray) A. Nels., Greenstem Paperflower

- *Drug*—**Hopi** *Adjuvant* Plant used with other plants to make medicine stronger. (42:354) **Navajo, Kayenta** *Antidiarrheal* Plant used as a diarrhea medicine. *Blood Medicine* Plant used as a postpartum blood purifier. *Dermatological Aid* Poultice of plant applied to wounds. *Gynecological Aid* Plant used as a postpartum blood purifier. *Panacea* Plant used as a life medicine. (205:49)
- *Other*—**Hopi** *Ceremonial Items* Plant used in the Snake Dance ceremonials. (42:354)

Psilostrophe tagetina (Nutt.) Greene, Woolly Paperflower

- *Drug*—**Navajo, Ramah** *Analgesic* Strong infusion taken for stomachache or as a cathartic. *Cathartic* Strong infusion of plant taken as a cathartic. *Ceremonial Medicine* Cold infusion of leaves used as ceremonial eyewash. *Dermatological Aid* Infusion of plant used as lotion for itching. *Eye Medicine* Cold infusion of leaves used as ceremonial eyewash. *Gastrointestinal Aid* Strong infusion of plant taken for stomachache. *Throat Aid* Cold infusion gargled or poultice of leaves applied for sore throat. (191:52) **Zuni** *Snakebite Remedy* Compound poultice of root applied with much ceremony to rattlesnake bite. (166:53)
- *Dye*—**Apache, White Mountain** *Yellow* Blossoms used to make a yellow dye. (136:160) **Keres, Western** *Yellow* Boiled, crushed flowers used for yellow paint or dye. (171:64) **Zuni** *Yellow* Blossoms used to make a yellow dye. (166:80)
- *Other*—**Keres, Western** *Paint* Boiled, crushed flowers used for yellow paint or dye. (171:64) **Zuni** *Ceremonial Items* and *Paint* Blossoms used by personators of anthropic gods for painting masks and for coloring bodies yellow.[62] (166:97)

Psoralidium, Fabaceae

Psoralidium lanceolatum (Pursh) Rydb., Lemon Scurfpea

- *Drug*—**Arapaho** *Analgesic* Snuff of leaves and sneezeweed blossoms inhaled for headaches. Infusion of leaves used on the head for headaches. *Dermatological Aid* Oily leaves rubbed on the skin for dryness. (121:48) *Throat Aid* Fresh leaves chewed for sore throat and voice. (118:38) Root chewed for hoarseness. (121:48) **Cheyenne** *Ceremonial Medicine* Plant used for certain ceremonies. (76:178) **Navajo, Kayenta** *Dermatological Aid* Plant used as a lotion and poultice of plant applied to itch and sores. (205:29) **Navajo, Ramah** *Analgesic* Cold infusion of plant taken for stomachache and menstrual pain. *Ceremonial Medicine* Cold infusion of plant used as ceremonial chant lotion. *Gastrointestinal Aid* Cold infusion of plant taken for stomachache. *Gynecological Aid* Cold infusion of plant taken for menstrual cramps. *Venereal Aid* Compound decoction of root used for venereal disease. *Witchcraft Medicine* Cold infusion of plant used as a lotion for protection from witches. (191:34) **Zuni** *Gastrointestinal Aid* Fresh flower eaten for stomachaches. (27:376)
- *Fiber*—**Great Basin Indian** *Cordage* Roots used to make string and nets. (121:48)
- *Other*—**Cheyenne** *Ceremonial Items* Only used in certain ceremonies. (76:178)

Psoralidium tenuiflorum (Pursh) Rydb., Slimflower Scurfpea

- *Drug*—**Dakota** *Tuberculosis Remedy* Decoction of plants taken for consumption. (as *Psoralea floribunda* 69:366) Compound decoction of root taken for consumption. (70:93) **Lakota** *Analgesic* Infusion of roots used for headaches. (139:48) **Navajo, Ramah** *Misc. Disease Remedy* Infusion of plant taken or leaves smoked for influenza. *Veterinary Aid* Plant used several ways for sheep with coughs. (191:34) **Zuni** *Disinfectant* Poultice of moistened leaves applied to any body part for purification. (166:58)
- *Food*—**Yavapai** *Beverage* Used as ingredient of modern intoxicant made from mescal. (65:257)
- *Fiber*—**Dakota** *Clothing* Plant tops used to make garlands worn like

hats in hot weather. (as *Psoralea floribunda* 69:366) Plant tops used to make garlands worn on the head as protection from the sun on very hot days. (70:93)
- **Other**—**Dakota** *Protection* Plant tops used to make garlands worn on the head as protection from the sun on very hot days. (70:93) **Keres, Western** *Unspecified* Plant known and named but no use was specified. (171:64) **Kiowa** *Cooking Tools* Stout stem used as a fork to eat buffalo steak. (192:34) **Lakota** *Insecticide* Plant smudge used against mosquitoes. (139:48) **Navajo** *Smoke Plant* Leaves smoked by the masker after the feast celebrating the completion of masks for the Night Chant. (55:58)

Psorothamnus, Fabaceae

Psorothamnus fremontii (Torr. ex Gray) Barneby var. *fremontii*, Frémont's Dalea
- **Drug**—**Paiute** *Antihemorrhagic* Decoction of root or plant tops taken for internal hemorrhages. **Shoshoni** *Gastrointestinal Aid* Decoction of roots taken for stomach trouble. (as *Dalea fremontii* 180:64)

Psorothamnus polydenius (Torr. ex S. Wats.) Rydb., Nevada Smokebush
- **Drug**—**Paiute, Northern** *Cathartic* Infusion of bark taken as a physic. *Cold Remedy* Infusion of bark taken for light colds. *Misc. Disease Remedy* Decoction of brush taken for the flu. *Pulmonary Aid* Decoction of brush taken for pneumonia. *Throat Aid* Decoction of brush taken for sore throats. (59:126)

Psorothamnus polydenius (Torr. ex S. Wats.) Rydb. var. *polydenius*, Nevada Smokebush
- **Drug**—**Paiute** *Analgesic* Decoction of stem taken for muscular pain and stem chewed for facial neuralgia. Decoction of stems taken for stomachaches. *Antidiarrheal* Strong infusion of plant taken for diarrhea. *Antirheumatic (External)* Hot decoction of plant used as a wash for rheumatism. *Cold Remedy* Plant used in a variety of ways for colds. *Cough Medicine* Plant used in a variety of ways for coughs. *Dermatological Aid* Powdered, dried stems or crushed fresh stems used for sores. *Disinfectant* Simple or compound decoction of stems and tops used as an antiseptic bath for measles. Simple or compound decoction of stems and tops used as an antiseptic bath for smallpox. *Diuretic* Decoction of plant tops taken to induce urination. *Gastrointestinal Aid* Decoction of plant taken for stomachaches. *Kidney Aid* Compound decoction of stems and tops taken for kidney ailments. *Misc. Disease Remedy* Decoction of plant taken and used as an antiseptic wash for measles. Decoction of plant used as an antiseptic wash for smallpox. Decoction of stems taken for influenza. *Orthopedic Aid* Infusion of plant taken for muscular pains. *Pulmonary Aid* Plant used in several ways for pneumonia. Sweetened decoction of stems taken for whooping cough. *Toothache Remedy* Stems chewed for toothache or face neuralgia. *Tuberculosis Remedy* Plant used in several ways for tuberculosis. *Venereal Aid* Decoction of plant taken for venereal diseases. **Shoshoni** *Analgesic* Decoction of plant tops taken for pains in the back over the kidneys. Decoction of stems taken for stomachaches. *Antidiarrheal* Strong infusion of plant taken for diarrhea. *Cold Remedy* Plant used in a variety of ways for colds. *Cough Medicine* Plant used in a variety of ways for coughs. *Disinfectant* Simple or compound decoction of stems and tops used as an antiseptic bath for smallpox. *Gastrointestinal Aid* Decoction of plant taken for stomachaches. *Kidney Aid* Simple or compound decoction of plant tops used for kidney pain and urine incontinence. *Misc. Disease Remedy* Decoction of stems taken for influenza. Simple or compound decoction of plant taken and used as a wash for smallpox. *Pulmonary Aid* Plant used in several ways for pneumonia. *Tuberculosis Remedy* Plant used in several ways for tuberculosis. *Urinary Aid* Decoction of plant tops taken for kidney pain and urine incontinence. *Venereal Aid* Decoction of plant taken for venereal diseases. (as *Dalea polyadenia* 180:64–66)

Psorothamnus scoparius (Gray) Rydb., Broom Dalea
- **Drug**—**Keres, Western** *Dermatological Aid* Infusion of plant rubbed on spider bites. *Emetic* Infusion of plant used as an emetic. *Gastrointestinal Aid* Infusion of plant used for stomach trouble. (as *Parosela scoparia* 171:57)

Ptelea, Rutaceae

Ptelea trifoliata L., Common Hoptree
- **Drug**—**Menominee** *Adjuvant* Root bark used as a seasoner and to render other medicines potent. *Panacea* Root considered a sacred medicine and credited with all sorts of cures. (151:51) **Meskwaki** *Adjuvant* Root often added to other medicines to make them potent. *Pulmonary Aid* Compound infusion of pounded root used for lung troubles, a good medicine. (152:244)

Ptelea trifoliata ssp. *pallida* var. *pallida* (Greene) V. Bailey, Pallid Hoptree
- **Drug**—**Havasupai** *Gastrointestinal Aid* and *Pediatric Aid* Decoction of leaves rubbed on a child's abdomen for stomachaches. *Poison* Leaves made into poison and used on arrow tips for hunting large game and in warfare.[63] (as *P. pallida* 197:229)
- **Other**—**Havasupai** *Fuel* Wood used for firewood. *Hunting & Fishing Item* Leaves made into poison and used on arrow tips for hunting large game.[63] *Weapon* Leaves made into poison and used on arrow tips in warfare.[63] (as *P. pallida* 197:229)

Ptelea trifoliata ssp. *trifoliata* var. *mollis* Torr. & Gray, Common Hoptree
- **Food**—**San Felipe** *Fruit* Fruits commonly eaten by children. (as *P. tomentosa* 32:47)

Pteridaceae, see *Acrostichum, Adiantum, Argyrochosma, Cheilanthes, Cryptogramma, Pellaea, Pentagrama*

Pteridium, Dennstaedtiaceae, fern

Pteridium aquilinum (L.) Kuhn, Western Brackenfern
- **Drug**—**Alaska Native** *Poison* Full grown fronds poisonous to cattle. (85:51) **Cherokee** *Antiemetic* Root used as a tonic and antiemetic and given for "cholera-morbus." *Disinfectant* Root used as an antiseptic. *Misc. Disease Remedy* Root used as a tonic, antiseptic, antiemetic, and for "cholera-morbus." *Tonic* Root used as a tonic. (80:33) **Costanoan** *Dermatological Aid* Decoction of root used as hair rinse or roots rubbed on scalp for hair growth. (21:5) **Hesquiat** *Cancer Treatment* Young shoots eaten as medicine for "troubles with one's insides," such as cancer of the womb. (185:32) **Iroquois** *Antidiarrheal* Decoction taken for diarrhea. *Antirheumatic (Internal)* Compound

used for rheumatism. *Blood Medicine* Cold, compound decoction of roots taken for weak blood. *Gynecological Aid* Compound decoction taken for prolapse of uterus. Decoction taken when suffering after birth. Decoction used to make "good blood" after menses, taken after baby's birth. *Liver Aid* Used as a liver and rheumatism medicine. *Tuberculosis Remedy* Compound decoction taken during the early stages of consumption. Decoction of plant taken by women for tuberculosis. *Urinary Aid* Compound decoction taken by men to retain urine. *Venereal Aid* Compound used for infection, probably venereal disease. *Witchcraft Medicine* Ingredients placed in coffin with root shaped into a person and the person dies in 10 days. (87:259) **Koasati** *Analgesic* Decoction of ground roots taken for chest pain. (177:4) **Mendocino Indian** *Veterinary Aid* Plant used as a diuretic for horses. (41:304) **Menominee** *Gynecological Aid* Decoction of root taken for "caked breast" and a dog whisker used to pierce teat. (as *Pteris aquilina* 151:48) **Micmac** *Pediatric Aid* and *Stimulant* Fronds of plant used for weak babies and old people. (as *Pteris aquilina* 40:60) **Montagnais** *Orthopedic Aid* Fronds used as a bed to strengthen babies' backs and old people. *Pediatric Aid* Fronds used as a bed to strengthen babies' backs. (as *Pteris aquilina* 156:315) **Ojibwa** *Analgesic* Infusion of root taken by women to allay stomach cramps. Smoke from dried leaves on coals used for headaches. *Gynecological Aid* Infusion of root taken by women to allay stomach cramps. (as *Pteris aquilina* 153:382) **Okanagan-Colville** *Poison* Fronds considered poisonous when mature and known to contain carcinogenic substances. (188:18) **Thompson** *Antihemorrhagic* Infusion of rhizomes taken for vomiting blood, possibly from internal injuries. *Antirheumatic (External)* Leaves used in a steam bath for arthritis. The leaves were placed over red-hot rocks in a steaming pit, a little water was added, and the person laid on top of the fronds. *Cold Remedy* Decoction of rhizomes taken for colds. *Dermatological Aid* Poultice of pounded fronds and leaves applied to sores of any type. Fronds, pounded with a rock, mixed with leaves and melted pine pitch, strained, and applied to sores from one to several days. *Dietary Aid* Decoction of rhizomes taken for lack of appetite. *Orthopedic Aid* Poultice of boiled, pounded fronds mixed with leaves, and used to set broken bones in place. Decoction of leaves used as a bath for broken bones or poultice of leaves used to bind broken bones. (187:90) **Yana** *Burn Dressing* Poultice of pounded, heated roots applied to burns. (as *Pteris aquilina* 147:253)

- *Food*—**Alaska Native** *Substitution Food* Young fiddlenecks peeled, boiled, or steamed and eaten as a substitute for asparagus. *Winter Use Food* Young fiddlenecks canned for winter use. (85:51) **Atsugewi** *Unspecified* Raw leaves and tender stems used for food. (61:139) **Bella Coola** *Unspecified* Rhizomes toasted and eaten in summer. (184:197) **Clallam** *Staple* Rhizomes roasted, pounded into a flour, and eaten. (57:194) **Costanoan** *Unspecified* Young fronds eaten raw or cooked. (21:247) **Hahwunkwut** *Unspecified* Roots cooked in ground ovens. (as *Pteris aquilina* 117:185) **Hesquiat** *Vegetable* Long, mashed rhizomes eaten boiled or steamed. (185:32) **Kwakiutl, Southern** *Unspecified* Rhizomes roasted, beaten until soft, broken into pieces, and used for food. (183:265) **Mahuna** *Unspecified* Young shoots cut, cooked, and eaten like asparagus. (as *Pteris aquilina* 140:58) **Montana Indian** *Unspecified* Peeled root roasted for food. (as *Pteris aquilina* 19:20) **Nitinaht** *Unspecified* Rhizomes roasted, pounded, and inner portions used for food. *Vegetable* Long, thick rhizomes formerly steamed, dried, and used as a vegetable food in winter. (186:63) **Ojibwa** *Soup* Young fern sprouts used as a soup material. The tips were thrown into hot water for an hour to rid them of ants, then put into soup stock and thickened with flour. The flavor resembles wild rice. Hunters were very careful to live wholly upon this when stalking does in the spring. The doe feeds upon the fronds and the hunter does also, so that his breath does not betray his presence. He claims to be able to approach within 20 feet without disturbing the deer, from which distance he can easily make a fatal shot with his bow and arrow. After killing the deer, the hunter will eat whatever strikes his fancy. *Unspecified* Young fern tips, with coiled fronds, were like asparagus tips, only not stringy like asparagus. (as *Pteris aquilina* 153:408) **Okanagon** *Unspecified* Rootstocks boiled or roasted and used for food. (125:38) **Salish, Coast** *Bread & Cake* Rhizomes pounded into flour and baked to make bread. *Unspecified* Rhizomes eaten fresh in late fall or winter. Young shoots used for food. (182:69) **Sierra** *Staple* Roots used as a staple food. (41:304) **Skagit, Upper** *Unspecified* Roots roasted in ashes, peeled, and eaten. (179:40) **Thompson** *Staple* Cooked, inner rhizome pounded into a flour and used for food. (187:90) *Unspecified* Rootstocks boiled or roasted and used for food. (125:38) Rootstocks cooked and eaten. Rootstocks used as a nutritious food. (164:482) Dried, toasted rhizomes beaten with a stick to remove the bark and the white insides used for food. The rhizomes were usually eaten with fish and were said to be very sweet, but one informant's father said it would give her worms. Fiddleheads broken off and the stem portion of the shoot used for food, often with fish. (187:90)

- *Fiber*—**Costanoan** *Basketry* Roots used in basketry. (21:247) **Mendocino Indian** *Basketry* Root wood split into flat bands and used for the black strands of cheap baskets. (41:304) **Mewuk** *Basketry* Split roots used for the black design in coiled basketry. (as *Pteris aquilina* 117:328) **Nitinaht** *Mats, Rugs & Bedding* Fronds used for bedding while camping. (186:63) **Pomo** *Basketry* Root fiber made into coils and used in basketry. (9:139) **Round Valley Indian** *Basketry* Root wood, not frequently used, split and used for the black strands of cheap baskets. (41:304) **Shuswap** *Mats, Rugs & Bedding* Used for bedding in camp. (123:49) **Ukiah** *Basketry* Root wood, not frequently used, split and used for the black strands of cheap baskets. (41:304)

- *Other*—**Clallam** *Cooking Tools* Fronds used to cover berry baskets and to wipe fish before hanging up to smoke. (57:194) **Costanoan** *Containers* Fronds used to line acorn-leaching pits and earth ovens. *Protection* Fronds used as sunshades. (21:247) **Okanagan-Colville** *Containers* Fronds dipped in water and used in pit cooking to place over and under the food. *Water Indicator* Ferns considered to be a sign of water when traveling through the mountains. (188:18) **Oweekeno** *Fuel* Rhizomes chewed, used as punk in a clamshell, and placed in a fire. (43:58) **Shuswap** *Protection* Used to cover berry baskets. (123:49)

Pteridium aquilinum* var. *latiusculum (Desv.) Underwood ex Heller, Western Brackenfern

- *Drug*—**Iroquois** *Veterinary Aid* Rhizomes, raspberry leaves, and wheat flour given to cows at birthing. (as *P. latiusculum* 141:34)

Pteridium aquilinum* var. *pubescens Underwood, Hairy Brackenfern

- **Drug**—**Makah** *Toothache Remedy* Fiddleheads placed on each side of the gums adjacent to the affected tooth for toothaches. (67:224) **Pomo, Kashaya** *Dermatological Aid* Young, curled frond juice used as a body deodorant. (72:44) **Tolowa** *Anticonvulsive* Poultice of pulverized leaves applied for *Toxicodendron* poisoning. (6:48)
- **Food**—**Chehalis** *Unspecified* Rhizomes roasted, peeled, and the starchy centers eaten. **Cowlitz** *Unspecified* Rhizomes roasted, peeled, and the starchy centers eaten. Young plant tops eaten raw. **Green River Group** *Unspecified* Rhizomes roasted, peeled, and the starchy centers eaten. **Klallam** *Unspecified* Rhizomes roasted, peeled and the starchy centers eaten. **Lummi** *Unspecified* Rhizomes roasted, peeled, and the starchy centers eaten. (79:14) **Makah** *Unspecified* Steamed rhizomes used for food. (67:224) Rhizomes roasted, peeled, and the starchy centers eaten. (79:14) **Nitinaht** *Unspecified* Steamed rhizomes used for food. (67:224) **Quileute** *Unspecified* Rhizomes roasted, peeled, and the starchy centers eaten. **Quinault** *Unspecified* Rhizomes roasted, peeled, and the starchy centers eaten. **Skagit** *Unspecified* Rhizomes roasted, peeled, and the starchy centers eaten. **Skokomish** *Unspecified* Rhizomes roasted, peeled, and the starchy centers eaten. **Snohomish** *Unspecified* Rhizomes roasted, peeled, and the starchy centers eaten. **Squaxin** *Unspecified* Rhizomes roasted, peeled, and the starchy centers eaten. **Swinomish** *Unspecified* Rhizomes roasted, peeled, and the starchy centers eaten. (79:14)
- **Fiber**—**Kawaiisu** *Basketry* Leaf midrib or root used for the black pattern material in coiled basketry. (206:55) **Pomo** *Basketry* Root used in basketry. (as *Pteris aquilina* var. *lanuginosa* 66:11) **Pomo, Kashaya** *Basketry* Root pounded to remove bark, bark core split into layers and used as material for basket design. (72:44) **Squaxin** *Mats, Rugs & Bedding* Leaves used for camp bedding. (79:14)
- **Other**—**Karok** *Containers* Leaves used under draining fish. (as *Pteris aquilina* var. *lanuginosa* 148:377) *Cooking Tools* Leaves used to clean eels and salmon. (6:48) **Maidu** *Decorations* Roots used as decorative coil thread and decorative overlay twine in the manufacture of baskets. (as *Pteridum quilina* var. *lanuginosa* 173:71) **Makah** *Cooking Tools* Leaves placed beneath fish being cleaned and used to wipe the fish. (79:14) **Paiute** *Preservative* Plant used to cover berry baskets to keep the berries fresh. (111:36) **Quileute** *Cooking Tools* Leaves placed beneath fish being cleaned and used to wipe the fish. **Squaxin** *Cooking Tools* Leaves placed beneath fish being cleaned and used to wipe the fish. **Swinomish** *Cooking Tools* Leaves placed beneath fish being cleaned and used to wipe the fish. (79:14) **Yurok** *Containers* Leaves used in layers to dry food. *Cooking Tools* Fronds used as plates to serve fish, to put over fish to keep the flies off, and to clean fish. (6:48)

Pteridium caudatum (L.) Maxon, Southern Brackenfern
- **Drug**—**Seminole** *Orthopedic Aid* Plant used for turkey sickness: permanently bent toes and fingers. Roots used for turkey sickness: permanently bent toes and fingers. (as *Pteris caudata* 169:166, 236)

Pteridium sp., Fern
- **Fiber**—**Yokut** *Basketry* Roots used to make the black designs on baskets. (117:405)

Pterocaulon, Asteraceae
Pterocaulon virgatum (L.) DC., Wand Blackroot
- **Drug**—**Seminole** *Abortifacient* Infusion of plant used "to correct irregularities and to relieve menstrual pain." (as *P. undulatum* 169:284) *Antidiarrheal* and *Antihemorrhagic* Infusion of plant taken for otter sickness: severe diarrhea, bloody stools, and severe stomachache. (as *P. undulatum* 169:223) *Cold Remedy* Infusion of plant taken for colds. *Febrifuge* Infusion of plant taken for fevers. (as *P. undulatum* 169:283) *Gastrointestinal Aid* Infusion of plant taken for otter sickness: severe diarrhea, bloody stools, and severe stomachache. (as *P. undulatum* 169:223) Plant used for water bison sickness: digestive difficulties. (as *P. undulatum* 169:306) *Gynecological Aid* Decoction of roots used for backache and excessive bleeding following childbirth. (as *P. undulatum* 169:271) Infusion of plant used "to correct irregularities and to relieve menstrual pain." (as *P. undulatum* 169:284) Decoction of roots taken for backache and excessive bloody discharge after giving birth. (as *P. undulatum* 169:326) *Orthopedic Aid* Decoction of roots used for persistent backache. *Other* Decoction of roots used for chronic sickness. (as *P. undulatum* 169:271) Infusion of plant taken for beaver sickness. (as *P. undulatum* 169:304) *Pulmonary Aid* Plant used for pulmonary disorders. (as *P. undulatum* 169:282) *Unspecified* Plant used medicinally. (as *P. undulatum* 169:164)

Pterospora, Ericaceae
Pterospora andromedea Nutt., Woodland Pinedrops
- **Drug**—**Cheyenne** *Antihemorrhagic* Cold infusion of ground stems and berries taken for lung hemorrhages. (76:183) Cold infusion of stem and berries taken for "bleeding at the lungs." (75:39) *Dermatological Aid* Cold infusion of ground stems and berries used as an astringent. (76:183) Plant used as an astringent. (75:39) *Disinfectant* Infusion of ground berries and stems used as an astringent. (76:183) *Hemostat* Cold infusion of ground stems and berries used as snuff for nosebleeds. (76:183) Decoction of stem and berries snuffed and used as wash to prevent nosebleed. (75:39) *Nose Medicine* Infusion of ground berries and stems snuffed up the nose and put on the head for nosebleed. (76:183) *Pulmonary Aid* Cold infusion of ground stems and berries taken for lung hemorrhages. (76:183) Decoction of plant taken for bleeding "at the lungs." (75:39) Infusion of ground berries and stems taken for lung hemorrhage. (76:183) **Keres, Western** *Emetic* Boiled plant used as an emetic. (171:64) **Okanagan-Colville** *Venereal Aid* Infusion of roots taken for gonorrhea. (188:102)
- **Food**—**Kawaiisu** *Unspecified* Stems eaten raw, "roasted," or baked below the fire "like mushrooms." (206:55)
- **Other**—**Jemez** *Ceremonial Items* Leaves smoked in the kiva. (44:26)

Pteryxia, Apiaceae
Pteryxia terebinthina (Hook.) Coult. & Rose **var. terebinthina**, Turpentine Wavewing
- **Drug**—**Okanagan-Colville** *Dermatological Aid* Poultice of dried, pounded, peeled roots, alone or with Vaseline, applied to sores. *Tonic* Infusion of dried, pounded roots taken as a general tonic. (as *Cymopteris terebinthinus* var. *terebinthinus* 188:60)

Puccinellia, Poaceae
Puccinellia distans (Jacq.) Parl., Weeping Alkaligrass
- **Food**—**Gosiute** *Unspecified* Seeds formerly used for food. (as *Glyceria distans* 39:370)

Pulsatilla, Ranunculaceae

Pulsatilla occidentalis (S. Wats.) Freyn, White Pasqueflower
- **Drug**—**Okanagon** *Gastrointestinal Aid* Decoction of roots or plants taken for stomach and bowel troubles. (as *Anemone occidentalis* 125:40) **Thompson** *Antirheumatic (External)* Infusion of plant used as a wash for rheumatism. *Eye Medicine* Infusion of plant used as an eyewash. (187:249) *Gastrointestinal Aid* Decoction of roots or plants taken for stomach and bowel troubles. (as *Anemone occidentalis* 125:40) Decoction of root or whole plant taken for stomach and bowel troubles. (as *Anemone occidentalis* 164:459) *Unspecified* Plant used medicinally for unspecified purpose. (as *Anemone occidentalis* 164:466)

Pulsatilla patens (L.) P. Mill., American Pasqueflower
- **Drug**—**Omaha** *Analgesic* Poultice of crushed, fresh leaves applied for rheumatism and neuralgia. (70:80, 81) *Antirheumatic (External)* Poultice of crushed, fresh leaves applied to affected part for rheumatism. (70:80–82)

Pulsatilla patens* ssp. *multifida (Pritz.) Zamels, Cutleaf Anemone
- **Drug**—**Blackfoot** *Dermatological Aid* Poultice of crushed leaves applied to affected parts as a counterirritant. (as *Anemone patens* var. *wolfgangiana* 97:35) *Gynecological Aid* Decoction of plant taken to speed delivery. (as *Anemone patens* 86:60) **Cheyenne** *Poison* Plant considered poisonous. *Stimulant* Smashed root used symbolically by passing over the body to revive a person. *Unspecified* Plant used for the medicinal properties. (as *Anemone nuttalliana* 83:34) **Chippewa** *Analgesic* Dried, pulverized leaves "smelled" for headache. (as *P. hirsutissima* 53:336) *Pulmonary Aid* Compound decoction of root taken for lung trouble. (as *P. hirsutissima* 53:340) **Great Basin Indian** *Antirheumatic (External)* Poultice of fresh leaves used as a counterirritant for rheumatism. (as *P. hirsutissima* 121:47)
- **Other**—**Blackfoot** *Toys & Games* Leaves acted as a vesicant and given to unsuspecting people as toilet paper as a prank. (as *Anemone patens* 86:107)

***Pulsatilla* sp.**, Pasqueflower
- **Drug**—**Blackfoot** *Antirheumatic (External)* Poultice of leaves applied for rheumatism. (118:43)

Punicaceae, see *Punica*

Punica, Punicaceae

Punica granatum L., Pomegranate
- **Other**—**Navajo** *Jewelry* Blossoms used to make necklaces. (55:66)

Purshia, Rosaceae

Purshia glandulosa Curran, Desert Bitterbrush
- **Drug**—**Kawaiisu** *Analgesic* Plant used for menstrual cramps. *Emetic* Decoction of inner bark and leaves used as an emetic. *Gynecological Aid* Plant used for menstrual cramps. *Laxative* Decoction of inner bark and leaves used as a laxative. *Venereal Aid* Decoction of inner bark and leaves used for gonorrhea. (206:55)

Purshia mexicana (D. Don) Henrickson, Mexican Cliffrose
- **Drug**—**Apache, White Mountain** *Unspecified* Leaves used for medicinal purposes. (as *Cowania mexicana* 136:156) **Gosiute** *Unspecified* Leaves used as a medicine. (as *Cowania mexicana* 39:367) **Havasupai** *Cold Remedy* Decoction of green branches, sagebrush, and juniper used for colds to loosen the mucus. *Laxative* Decoction of green branches, sagebrush, and juniper used as a laxative for colds. (as *Cowania mexicana* 197:223) **Hualapai** *Antirheumatic (Internal)* Leaves chewed for arthritis. *Dermatological Aid* Leaves made into a tea for bathing and cleansing the skin. (as *Cowania mexicana* 195:31) **Paiute** *Cathartic* Decoction of leaves and stems or flowers taken as a physic. *Cold Remedy* Decoction of leaves and stems or flowers taken for colds. *Venereal Aid* Decoction of leaves and stem or flowers taken for venereal diseases. **Shoshoni** *Cathartic* Decoction of leaves and stems or flowers taken as a physic. *Dermatological Aid* and *Disinfectant* Decoction of plant parts used as an antiseptic wash for smallpox or measles. *Kidney Aid* Decoction of leaves and stems or flowers taken for pains over the kidneys. *Misc. Disease Remedy* Plant used in several compounds taken and used as washes for smallpox and measles. *Venereal Aid* Decoction of leaves and stem or flowers taken for venereal diseases. (as *Cowania mexicana* 180:61)
- **Fiber**—**Havasupai** *Clothing* Bark crushed, rubbed into softness, and stuffed into overshoes for warmth. Soft bark used as an absorbent diaper for children. *Mats, Rugs & Bedding* Soft bark used in a thick layer in infants' cradleboards. Bark made into loosely twisted ropes and used to make sleeping mats. (as *Cowania mexicana* 197:223) **Hualapai** *Building Material* Inner bark used for huts and the lining of sweat houses. *Clothing* Inner bark used for diapers, clothing, and sandals. (as *Cowania mexicana* 195:31)
- **Other**—**Havasupai** *Fuel* Fine, soft bark used as tinder for the fire drill. (as *Cowania mexicana* 197:223)

Purshia stansburiana (Torr.) Henrickson, Stansbury Cliffrose
- **Drug**—**Hopi** *Dermatological Aid* Plant used in a wash for wounds. (as *Cowania mexicana* var. *stansburiana* 42:304) Plant used in a wash for wounds. (as *Cowania stansburiana* 200:32, 78) *Emetic* Plant used as an emetic. (as *Cowania mexicana* var. *stansburiana* 42:304) Bark used as an emetic. (as *Cowania stansburiana* 200:34, 78) **Navajo, Ramah** *Ceremonial Medicine* Leaves used as an emetic in various ceremonies. *Cough Medicine* Decoction of leaves taken for bad cough. *Emetic* Leaves used as an emetic in various ceremonies. (as *Cowania stansburiana* 191:30)
- **Food**—**Navajo** *Forage* Plant used for deer and livestock forage. (as *Cowania stansburiana* 90:159)
- **Fiber**—**Hopi** *Clothing* Bark spun and woven into kilts worn by the snake priests. (as *Cowania mexicana* var. *stansburiana* 42:304) Bark spun and woven into kilts worn by the snake priests. (as *Cowania stansburiana* 200:78) *Mats, Rugs & Bedding* Bark used as padding for the cradleboard. (as *Cowania mexicana* var. *stansburiana* 42:304) Bark from large stems used as the padding for cradleboards. (as *Cowania stansburiana* 200:78) **Navajo** *Mats, Rugs & Bedding* Softened bark used as backing for cradleboards and as stuffing for pillows. (as *Cowania stansburiana* 55:53) **Navajo, Ramah** *Clothing* Shredded bark used for bedding or diaper for cradleboard. *Mats, Rugs & Bedding* Shredded bark used for bedding or stuffed into a sack for pillows. (as *Cowania stansburiana* 191:30)
- **Dye**—**Navajo** *Brown* Pounded leaves and stems mixed with pounded juniper and used to make a tan dye. *Yellow-Brown* Pounded leaves

and stems mixed with pounded juniper and used to make a yellow brown dye. (as *Cowania stansburiana* 55:53)
- **Other**—**Hopi** *Ceremonial Items* Plant used ceremonially on the Po-wa-mu altar. (as *Cowania mexicana* var. *stansburiana* 42:304) *Hunting & Fishing Item* Used for arrows. (as *Cowania stansburiana* 200:78) **Keres, Western** *Hunting & Fishing Item* Plant sometimes used for arrow shafts. (as *Cowania stansburiana* 171:38) **Navajo** *Ceremonial Items* Wood used to make female prayer sticks for the Night Chant. Wood used to make arrows for the Mountain Chant Ceremony. *Toys & Games* Softened bark used to stuff baseballs. (as *Cowania stansburiana* 55:53)

Purshia tridentata (Pursh) DC., Antelope Bitterbrush
- **Drug**—**Klamath** *Cough Medicine* Infusion of root taken for coughs. (as *Kunzia tridentata* 45:98) Infusion of roots taken for coughs. (as *Kunzia tridentata* 163:131) *Emetic* Ripe fruit mashed in cold water and taken as an emetic. (as *Kunzia tridentata* 45:98) Infusion of smashed, dried, ripe, bitter fruits taken as an emetic. (as *Kunzia tridentata* 163:131) *Pulmonary Aid* Infusion of root taken for lung and bronchial troubles. (as *Kunzia tridentata* 45:98) Infusion of roots taken for lung troubles. (as *Kunzia tridentata* 163:131) *Respiratory Aid* Infusion of root taken for lung and bronchial troubles. (as *Kunzia tridentata* 45:98) Infusion of roots taken for bronchial troubles. (as *Kunzia tridentata* 163:131) **Montana Indian** *Emetic* Dry, ripe fruits mashed in cold water and used as an emetic. *Pulmonary Aid* Infusion of roots taken for lung troubles. (19:20) **Navajo** *Gynecological Aid* Plant taken during confinement. (90:154) **Navajo, Ramah** *Ceremonial Medicine* Leaves used as an emetic in various ceremonies. *Disinfectant* Leaves chewed by hunters for "deer infection." *Emetic* Leaves used as an emetic in various ceremonies. *Febrifuge* Decoction of root used for fever. *Gynecological Aid* Decoction of root used to facilitate delivery of placenta. *Hunting Medicine* Leaves chewed by hunters for good luck in hunting. (191:31) **Paiute** *Analgesic* Compound infusion of twigs and leaves taken for tubercular lung pain. *Cathartic* Decoction of leaves or twigs taken as a physic. *Cold Remedy* Decoction of leaves taken for colds. *Dermatological Aid* Poultice of leaf, decoction of plant or leaf powder used for skin problems. *Emetic* Decoction of leaves or twigs taken as an emetic. *Liver Aid* Decoction of leaf taken for liver trouble. *Misc. Disease Remedy* Decoction of plant taken and used as a wash for smallpox, chickenpox, and measles. *Pulmonary Aid* Decoction of leaf taken for pneumonia. *Tonic* Decoction of leaf taken as a "blood or general tonic." *Tuberculosis Remedy* Compound decoction of bark taken for tuberculosis and tubercular lung pain. *Venereal Aid* Decoction of leaves or inner bark taken for venereal diseases. (180:126–128) **Paiute, Northern** *Anthelmintic* Decoction of sun dried leaves taken for intestinal worms. *Gastrointestinal Aid* Decoction of sun dried leaves taken to vomit and move the bowels for stomachaches and constipation. *Laxative* Decoction of sun dried leaves taken to vomit and move the bowels for stomachaches and constipation. (59:126) **Sanpoil** *Antihemorrhagic* Infusion of crushed berries used for hemorrhage. *Laxative* Infusion of crushed berries used for constipation. (131:217) **Shoshoni** *Antihemorrhagic* Decoction of inner bark taken to aid the healing of an internal rupture. *Cathartic* Decoction of leaves or twigs taken as a physic. *Dermatological Aid* Decoction of leaf used as a wash for the swelling of milk leg. Poultice of leaf, decoction of plant or leaf powder used for skin problems. *Emetic* Decoction of leaves or twigs taken as an emetic. *Misc. Disease Remedy* Simple or compound decoction of plant taken and used as a wash for diseases with rashes. *Other* Bundle of inner bark strips sucked and decoction of leaf used as a wash for milk leg. *Tonic* Decoction of leaf taken as a "blood or general tonic." Decoction of leaf taken as a blood tonic or general tonic. *Venereal Aid* Compound decoction of inner bark taken specifically for gonorrhea. Decoction of leaves or roots taken for venereal diseases. **Washo** *Cathartic* Decoction of ripe, whole seeds taken as a physic. (180:126–128)
- **Food**—**Navajo** *Forage* Considered an important browse plant. (90:154) **Okanagan-Colville** *Forage* Plant eaten by deer. (188:128)
- **Fiber**—**Navajo** *Clothing* Bark used for diapers. (90:154) **Navajo, Ramah** *Mats, Rugs & Bedding* Shredded bark used as bedding for cradleboard. (191:31) **Paiute** *Clothing* Bark used to make moccasins. (111:82)
- **Dye**—**Great Basin Indian** *Purple* Seed coats used to make a violet dye. (121:48) **Klamath** *Purple* Outer seed coat used as a purple stain to produce temporary color on arrows, bows, and other objects. (as *Kunzia tridentata* 45:98) **Montana Indian** *Purple* Outer seed coats used to make a purple stain for wood. (19:20)
- **Other**—**Navajo** *Hunting & Fishing Item* Used to make arrows. (55:54) **Navajo, Ramah** *Ceremonial Items* Twig and leaf ash used for Evilway blackening. (191:31) **Okanagan-Colville** *Fuel* Branches used for fuel. Branches used to make the initial fire for pit cooking. (188:128) **Paiute** *Fuel* Large branches that grow close to the roots used as firewood. (111:82)

Pycnanthemum, Lamiaceae
Pycnanthemum albescens Torr. & Gray, Whiteleaf Mountainmint
- **Drug**—**Choctaw** *Cold Remedy* Hot decoction of leaves taken as diaphoretic for colds. (25:24) Decoction of leaves taken to cause sweating for colds. (177:54) *Diaphoretic* Hot decoction of leaves taken as diaphoretic for colds. (25:24) Decoction of leaves taken to cause sweating for colds. (177:54)

Pycnanthemum californicum Torr., Sierra Mint
- **Drug**—**Miwok** *Cold Remedy* Decoction taken for colds. (12:172)

Pycnanthemum flexuosum (Walt.) B.S.P., Appalachian Mountainmint
- **Drug**—**Cherokee** *Analgesic* Poultice of leaves used for headache. *Antidiarrheal* Infusion taken with "green corn" to prevent diarrhea. *Cold Remedy* Infusion taken for colds. *Dermatological Aid* Warm infusion used to bathe inflamed penis and infusion taken for upset stomach. *Febrifuge* Infusion taken for fevers. *Gastrointestinal Aid* Infusion used for upset stomach. *Heart Medicine* Infusion of leaves taken for "heart trouble." (80:45)
- **Food**—**Cherokee** *Unspecified* Species used for food. (80:45)

Pycnanthemum incanum (L.) Michx., Hoary Mountainmint
- **Drug**—**Cherokee** *Analgesic* Poultice of leaves used for headache. *Antidiarrheal* Infusion taken with "green corn" to prevent diarrhea. *Cold Remedy* Infusion taken for colds. *Dermatological Aid* Warm infusion used to bathe inflamed penis and infusion taken for upset stomach. *Febrifuge* Infusion taken for fevers. *Gastrointestinal Aid* In-

fusion used for upset stomach. *Heart Medicine* Infusion of leaves taken for "heart trouble." (80:45) **Choctaw** *Analgesic* Infusion of mashed leaves taken and used as a wash for headaches. *Panacea* Infusion of mashed leaves blown on sickly patient. **Koasati** *Analgesic* Decoction of roots taken for headaches. *Hemostat* Soaked plants put up the nose for nosebleeds. *Stimulant* Cold infusion of leaves taken and used as a bath for laziness. (177:55)
- **Food**—**Cherokee** *Unspecified* Species used for food. (80:45)

Pycnanthemum virginianum (L.) T. Dur. & B. D. Jackson ex B. L. Robins. & Fern., Virginia Mountainmint
- **Drug**—**Chippewa** *Abortifacient* Decoction of powdered root taken for "stoppage of periods." (as *Koellia virginiana* 53:358) *Febrifuge* Compound decoction of leaves taken for chills and fever. (as *Koellia virginiana* 53:354) **Lakota** *Cough Medicine* Infusion of plant taken for coughs. (139:50) **Meskwaki** *Alterative* Infusion of leaf used as an alterative "when a person is all run down." *Febrifuge* Infusion of plant tops taken for chills. *Misc. Disease Remedy* Infusion of plant tops taken for ague. *Stimulant* Compound containing florets applied at nostrils to rally a dying patient. (152:226, 227)
- **Food**—**Chippewa** *Spice* Buds and flowers used to season meat or broth. (as *Koellia virginiana* 53:318)
- **Other**—**Meskwaki** *Hunting & Fishing Item* Leaves used to scent mink traps. (152:273)

Pyrola, Ericaceae
Pyrola americana Sweet, American Wintergreen
- **Drug**—**Cherokee** *Dermatological Aid* "Stick on cuts and sores to heal them." (as *P. rotundifolia* 80:55) **Ojibwa** *Hunting Medicine* Dried leaves used to make tea and drunk as good luck potion in the morning before the hunt started. (153:430)

Pyrola asarifolia Michx., Liverleaf Wintergreen
- **Drug**—**Carrier, Southern** *Eye Medicine* Decoction of leaves or leaves and roots used as an eyewash. (150:62) **Cree, Woodlands** *Antihemorrhagic* Decoction of plant taken for coughing up blood. *Eye Medicine* Infusion of leaves used for sore eyes. (109:54) **Karok** *Ceremonial Medicine*, *Pediatric Aid*, and *Sedative* Decoction of plant used as a steam bath for "goofy" child in the Brush Dance. (148:387) **Shoshoni** *Liver Aid* Decoction of root taken for liver trouble. (180:128)

Pyrola asarifolia Michx. ssp. *asarifolia*, Liverleaf Wintergreen
- **Drug**—**Micmac** *Antihemorrhagic* Parts of plant used for spitting blood. *Kidney Aid* Parts of plant used for kidney trouble. *Venereal Aid* Parts of plant used for gonorrhea. (as *P. uliginosa* 40:60) **Montagnais** *Panacea* Decoction of leaves taken for any ailment. (as *P. uliginosa* 156:314) **Penobscot** *Antihemorrhagic* Compound infusion of plant taken for "spitting up blood." *Kidney Aid* Compound infusion of plant taken for kidney trouble. *Tonic* Compound infusion of plant taken as a tonic. *Venereal Aid* Compound infusion of plant taken for gonorrhea. (as *P. uliginosa* 156:311)

Pyrola chlorantha Sw., Greenflowered Wintergreen
- **Drug**—**Navajo, Kayenta** *Antidiarrheal* Plant used for infants with bloody diarrhea. *Gynecological Aid* Plant used for menorrhagia. *Hemostat* Plant used as a hemostatic. *Pediatric Aid* Plant used for infants with bloody diarrhea. (205:35)
- **Other**—**Navajo, Kayenta** *Ceremonial Items* Used in paint for the God Impersonators. (205:35)

Pyrola elliptica Nutt., Waxflower Shinleaf
- **Drug**—**Cherokee** *Dermatological Aid* "Stick on cuts and sores to heal them." (80:55) **Chippewa** *Unspecified* Poultice of plant and two other plants used medicinally. (71:138) **Iroquois** *Anticonvulsive* Decoction of roots and leaves given to babies with fits or epileptic seizures. (87:408) *Antirheumatic (Internal)* Compound infusion of plants taken for rheumatism. *Blood Medicine* Compound used for bad blood. *Dermatological Aid* Compound used for neck sores. *Eye Medicine* Decoction of plant used as drops for sore eyes, sties, and inflamed lids. *Gastrointestinal Aid* Compound decoction of whole plants taken for indigestion. *Orthopedic Aid* Compound infusion of smashed plants applied as poultice to sore legs. (87:409) *Pediatric Aid* Decoction of roots and leaves given to babies with fits or epileptic seizures. (87:408) **Mohegan** *Oral Aid* Infusion of leaves used as a gargle for sores or cankers in the mouth. (174:264) Infusion of leaves used as a mouthwash for canker sores. *Throat Aid* Infusion of leaves used as a wash for mouth sores and sore throat. (176:74, 130) **Montagnais** *Stimulant* Decoction of root taken for "weakness." (156:315) **Nootka** *Cancer Treatment* Poultice of bruised plant applied to "tumors." (170:79)

Pyrola picta Sm., Whiteveined Wintergreen
- **Drug**—**Karok** *Panacea* and *Pediatric Aid* Infusion of plant used as a wash for sick child. (148:387)

Pyrola sp., Wintergreen
- **Drug**—**Blackfoot** *Cough Medicine* Infusion of flowers given to children with coughs. (86:73) *Dermatological Aid* Poultice of chewed roots applied to wounds. (86:84) *Diuretic* Infusion of leaves and/or roots taken as a diuretic. (86:70) *Ear Medicine* Poultice of chewed roots applied to ear disorders. *Eye Medicine* Poultice of chewed roots applied to eye disorders. (86:82) *Gynecological Aid* Infusion of leaves and roots taken to expel the afterbirth. (86:61) *Laxative* Infusion of leaves taken as a laxative. (86:68) *Pediatric Aid* Infusion of flowers given to children with coughs. *Throat Aid* Root pieces used for throat lozenges. (86:73) Infusion of roots applied to swollen neck glands. (86:78) **Nootka** *Gastrointestinal Aid* Plant used for "derangement" of stomach and bowels. (170:81) **Thompson** *Gynecological Aid* Plant chewed at childbirth to ease confinement. (164:462)
- **Food**—**Malecite** *Beverage* Used to make tea. (160:6)

Pyrrhopappus, Asteraceae
Pyrrhopappus carolinianus (Walt.) DC., Carolina Desertchicory
- **Drug**—**Cherokee** *Blood Medicine* Infusion taken to purify blood. (80:31)
- **Food**—**Kiowa** *Unspecified* Autumn, sweet roots used for food. (as *Sitilias caroliniana* 192:61)

Pyrrhopappus pauciflorus (D. Don) DC., Desertchicory
- **Drug**—**Navajo, Ramah** *Ceremonial Medicine* and *Emetic* Flower stalks used as a ceremonial emetic. (as *P. multicaulis* 191:52)
- **Other**—**Keres, Western** *Unspecified* Plant known and named but no use was specified. (as *Sitilias multicaulis* 171:69)

Pyrularia, Santalaceae

Pyrularia pubera Michx., Buffalo Nut
- **Drug**—**Cherokee** *Dermatological Aid* Used as a salve for "old sores." *Emetic* Nut chewed to "make vomit for colic." (80:27)

Pyrus, Rosaceae

Pyrus communis L., Common Pear
- **Food**—**Cherokee** *Fruit* Fruit used for food. (80:48) **Hopi** *Unspecified* Species used for food. (200:79) **Iroquois** *Bread & Cake* Fruit mashed, made into small cakes and dried for future use. *Dried Food* Raw or cooked fruit sun or fire dried and stored for future use. *Fruit* Dried fruit taken as a hunting food. *Sauce & Relish* Dried fruit cakes soaked in warm water and cooked as a sauce or mixed with corn bread. (196:129) **Seminole** *Unspecified* Plant used for food. (169:496)

Pyrus sp., Service Tree
- **Drug**—**Cree, Hudson Bay** *Dermatological Aid* Branch bark used for its astringent qualities. *Misc. Disease Remedy* Branch bark used for inflammatory diseases. *Pulmonary Aid* Branch bark used for pleurisy. (92:303)
- **Food**—**Iroquois** *Dried Food* Fruits cut up in thin slices, strung on twine, dried, and used for food. *Fruit* Fruits eaten raw. Fruits boiled or baked and eaten. *Sauce & Relish* Fruits cut up and used as sauce. (124:94)

Quercus, Fagaceae

Quercus agrifolia Née, California Live Oak
- **Drug**—**Mahuna** *Hemostat* and *Pediatric Aid* Plant used for newborns with bleeding navels. (140:56)
- **Food**—**Cahuilla** *Bread & Cake* Acorns ground into a fine meal and used to make bread. *Dried Food* Dried acorns stored for a year or more in granaries. *Porridge* Cooked acorns used to make mush. *Special Food* Acorn meat considered a delicacy and favored at social and ceremonial occasions. (15:121) **Costanoan** *Unspecified* Acorns used for food. (21:248) **Luiseño** *Porridge* Acorns leached, ground into a meal, cooked in an earthen vessel, and eaten. (155:194) *Staple* Acorns eaten as a staple food. (155:193) Acorns from storage granaries pounded in a mortar and pestle to make a flour.[64] *Winter Use Food* Acorns formerly gathered for storage in acorn granaries. (155:194) **Pomo** *Unspecified* Acorns used for food. (66:12) **Pomo, Kashaya** *Dried Food* Acorns sun dried before storing. *Porridge* Acorns used as flour for pancakes, bread, mush, or soup.[45] (72:80)
- **Other**—**Cahuilla** *Cash Crop* Acorn meal exchanged for pinyon nuts, mesquite beans, and palm tree fruit. Acorn meal used as payment to a shaman for special services. *Fuel* Dried wood considered an ideal firewood for heating and cooking. *Hunting & Fishing Item* Acorns used as bait in trigger traps to capture small animals. *Jewelry* Unhusked acorns dried and strung as necklaces. *Musical Instrument* Acorns gathered on a cord and swung against the teeth to produce music. *Toys & Games* Acorns used by children in a game like jacks and for juggling. (15:121)

Quercus agrifolia Née **var. *agrifolia***, California Live Oak
- **Drug**—**Diegueño** *Dermatological Aid* Decoction of chipped bark used as a wash for sores. (84:33)
- **Food**—**Diegueño** *Porridge* Acorns shelled, pounded, leached, and cooked into a mush or gruel. (84:33)
- **Other**—**Diegueño** *Fuel* Bark used as fuel for firing pottery. (84:33)

Quercus alba L., White Oak
- **Drug**—**Cherokee** *Antidiarrheal* Bark used for chronic dysentery. *Dermatological Aid* Astringent bark chewed for mouth sores. Infusion of bark applied to sore, chapped skin. *Disinfectant* Bark used as an antiseptic. *Emetic* Bark used as an emetic. (80:46) Bark used as an emetic. (203:74) *Febrifuge* Bark used after long, intermittent fevers and as a wash for chills and fevers. *Gastrointestinal Aid* Bark used for indigestion and "any debility of the system." *Oral Aid* Bark chewed for mouth sores. *Respiratory Aid* Infusion of bark taken for asthma. *Throat Aid* Decoction of inner bark used for "lost voice." *Tonic* Bark used as a tonic. *Urinary Aid* Unspecified liquid preparation taken for "milky urine." (80:46) **Delaware** *Cough Medicine* Infusion of bark used for severe coughs. *Disinfectant* Infusion of bark used as a disinfectant. *Gynecological Aid* Infusion of bark used as a douche. *Throat Aid* Infusion of bark used for sore throats. (176:30) **Delaware, Oklahoma** *Cough Medicine* Infusion of bark taken for severe cough. (175:25, 78) *Dermatological Aid* Strong infusion of bark used to cleanse bruises and ulcers. (175:25) *Disinfectant* Compound containing bark used as an antiseptic. (175:78) *Gynecological Aid* Infusion of bark used as an excellent douche. *Panacea* Bark used in many medicinal compounds. (175:25, 78) *Throat Aid* Strong infusion of bark gargled for a sore throat. (175:25) **Delaware, Ontario** *Gynecological Aid* Compound infusion of bark taken for "diseases peculiar to women." *Tonic* Compound infusion of bark taken as a tonic. (175:68, 82) **Houma** *Antirheumatic (External)* Crushed root mixed with whisky and used as liniment on rheumatic parts. (158:56) **Iroquois** *Psychological Aid* Compound decoction used to counteract loneliness. *Tuberculosis Remedy* Compound decoction of bark taken for consumption. *Veterinary Aid* Bark used for horses with distemper. *Witchcraft Medicine* Compound decoction used "when your woman goes off and won't come back." (87:303) **Menominee** *Unspecified* Inner bark used in compounds. (151:36) **Meskwaki** *Antidiarrheal* Compound containing bark used for diarrhea. *Pulmonary Aid* Decoction of inner bark taken to "throw up phlegm from the lungs." (152:221) **Micmac** *Dietary Aid* Nuts used to induce thirst. *Hemorrhoid Remedy* Plant parts used for bleeding piles. (40:60) **Mohegan** *Analgesic* Infusion of bark used as liniment for muscular pains. (30:121) Infusion of inner bark used as liniment for humans and horses with pain. (176:75) *Antirheumatic (External)* Infusion of inner bark used as a liniment for pain. (176:75, 132) *Cold Remedy* Bark used for colds. (176:132) *Orthopedic Aid* Infusion of bark used as liniment for muscular pains. (30:121) Infusion of bark used as a liniment for people. *Veterinary Aid* Infusion of bark used as a liniment for horses. (174:266) Infusion of inner bark used as a liniment for horses with pain. (176:75, 132) **Ojibwa, South** *Antidiarrheal* Decoction of root bark and inner bark taken for diarrhea. (91:198) **Penobscot** *Dietary Aid* Acorns eaten to induce thirst and plenty of water thought to be beneficial. (156:309) *Hemorrhoid Remedy* Infusion of bark taken for bleeding piles. (156:

310) **Shinnecock** *Analgesic* Infusion of bark used as liniment for muscular pains. *Orthopedic Aid* Infusion of bark used as liniment for muscular pains. (30:121)
- *Food*—**Iroquois** *Unspecified* Acorns used for food. (196:123) **Menominee** *Pie & Pudding* Acorns boiled, simmered to remove lye, ground, sifted, and made into pie. *Porridge* Acorns boiled, simmered to remove lye, ground, sifted, and made into mush with bear oil seasoning. *Staple* Acorns boiled, simmered to remove lye, ground, sifted, cooked in soup stock to flavor, and eaten. (151:66) **Meskwaki** *Beverage* Ground, scorched acorns made into a drink similar to coffee. *Porridge* Dried acorns made into mush. (152:257) **Ojibwa** *Soup* Acorns soaked in lye water to remove bitter tannin taste, dried for storage, and used to make soup. Lye for leaching acorns was obtained by soaking wood ashes in water. Acorns were put in a net bag and then soaked in the lye, then rinsed several times in warm water. The acorns were then dried for storage, and when wanted, pounded into a coarse flour which was used to thicken soups or form a sort of mush. (153:401)
- *Fiber*—**Cherokee** *Basketry* Used to make baskets. *Building Material* Wood used for lumber, railroad ties, wagon spokes and rims. *Furniture* Wood used to make furniture. Used to make woven chair bottoms. (80:46) **Ojibwa** *Building Material* Wood used in making wigwams and for several other things. (153:418)
- *Other*—**Cherokee** *Cooking Tools* Wood used to make corn beaters and mortars. Leaves used to wrap dough for bread making. *Fuel* Wood used for firewood. (80:46) **Ojibwa** *Tools* Wood was of much value, especially for making awls to punch holes in birch bark. (153:418)

Quercus bicolor Willd., Swamp White Oak
- *Drug*—**Iroquois** *Misc. Disease Remedy* Compound decoction of bark taken for cholera. *Orthopedic Aid* Compound decoction of bark taken for broken bones. *Respiratory Aid* Compound of leaves smoked and exhaled through the nostrils for catarrh. *Tuberculosis Remedy* Compound decoction of bark chips taken for consumption. *Witchcraft Medicine* Used "when wife runs around, takes away lonesomeness." (87:303)
- *Food*—**Iroquois** *Unspecified* Acorns used for food. (196:123)
- *Other*—**Chippewa** *Cleaning Agent* Bark boiled with hemlock and soft maple bark and the liquid used to clean the rust from traps. The solution was believed to prevent the trap from becoming rusty again. (71:128)

Quercus chrysolepis Liebm., Canyon Live Oak
- *Drug*—**Mendocino Indian** *Poison* Nuts considered poisonous. (41:342)
- *Food*—**Cahuilla** *Bread & Cake* Acorns ground into a fine meal and used to make bread. *Dried Food* Dried acorns stored for a year or more in granaries. *Porridge* Cooked acorns used to make mush. *Special Food* Acorn meat considered a delicacy and favored at social and ceremonial occasions. (15:121) **Diegueño** *Porridge* Acorns shelled, pounded, leached, and cooked into a mush or gruel. (84:33) **Karok** *Fruit* Fruit buried from 1 to 4 years to kill the bugs and worms and used for food. (6:49) *Unspecified* Acorns used for food. (148:382) **Kawaiisu** *Bread & Cake* Acorns made into a fine meal, cooked into a mush, and allowed to stand and harden into a "cake." *Staple* Acorns dried, pounded, sifted into a fine meal, and leached. *Winter Use Food* Acorns stored for future use.[65] (206:56) **Luiseño** *Porridge* Acorns leached, ground into a meal, cooked in an earthen vessel, and eaten. *Staple* Acorns from storage granaries pounded in a mortar and pestle to make a flour.[64] (155:194) *Substitution Food* Acorns used as a substitution during a scarcity of common live oak or black oak. (155:193) *Winter Use Food* Acorns formerly stored in acorn granaries. (155:194) **Pomo** *Bread & Cake* Acorns used to make bread. *Porridge* Acorns used to make mush. (117:290) **Shasta** *Bread & Cake* Acorns pounded, winnowed, leached, and made into bread. *Porridge* Acorns pounded, winnowed, leached, and made into mush. *Soup* Acorns pounded, winnowed, leached, and made into thin soup. *Staple* Acorns used as the basic staple. (93:308) **Tubatulabal** *Unspecified* Acorns used extensively for food. (193:15) **Wintoon** *Dried Food* Acorns dried and preserved for future use. *Unspecified* Acorns leached all winter in cold, wet, swampy ground, boiled or roasted, and eaten in the spring. (117:265)
- *Fiber*—**Kawaiisu** *Building Material* Logs used in house construction. (206:56)
- *Dye*—**Diegueño** *Black* Acorn cups soaked in water containing iron and used as a black dye to color basket materials. (84:33)
- *Other*—**Cahuilla** *Cash Crop* Acorn meal exchanged for pinyon nuts, mesquite beans, and palm tree fruit. Acorn meal used as payment to a shaman for special services. *Fuel* Dried wood considered an ideal firewood for heating and cooking. *Hunting & Fishing Item* Acorns used as bait in trigger traps to capture small animals. *Jewelry* Unhusked acorns dried and strung as necklaces. *Musical Instrument* Acorns gathered on a cord and swung against the teeth to produce music. *Toys & Games* Acorns used by children in a game like jacks and for juggling. (15:121) **Kawaiisu** *Fasteners* Acorn meal used to mend cracks in clay pots. *Toys & Games* Acorn cupule used to make a top for children. (206:56)

Quercus douglasii Hook. & Arn., Blue Oak
- *Drug*—**Kawaiisu** *Burn Dressing* Poultice of ground galls and salt applied to burns. *Dermatological Aid* Poultice of ground galls and salt applied to sores and cuts. *Eye Medicine* Ground gall powder and salt wrapped in a small piece of cloth and dipped in water applied to sore eyes. (206:56) **Midoo** *Throat Aid* Leaves chewed for sore throats. (117:310)
- *Food*—**Kawaiisu** *Bread & Cake* Acorns made into a fine meal, cooked into a mush, and allowed to stand and harden into a "cake." *Staple* Acorns dried, pounded, sifted into a fine meal, and leached. *Winter Use Food* Acorns stored for future use.[65] (206:56) **Mendocino Indian** *Bread & Cake* Thick acorns used to make bread. *Soup* Thick acorns used to make soup. (41:342) **Miwok** *Bread & Cake* Acorns ground into a meal and used to make bread and biscuits. *Porridge* Acorns considered a staple food and used to make mush. *Soup* Acorns ground into a meal and used to make soup. *Winter Use Food* Whole acorns stored for winter use. (12:142) **Tubatulabal** *Unspecified* Acorns used extensively for food. (193:14) **Yana** *Bread & Cake* Acorn flour used to make bread. *Dried Food* Acorns dried for winter use. *Porridge* Acorn flour used to make mush. *Staple* Dried acorns ground into flour. (147:249) **Yokut** *Unspecified* Acorns used for food. (117:420)
- *Fiber*—**Kawaiisu** *Basketry* Branches used to make rims for twined

work baskets. *Building Material* Logs used in house construction. (206:56)
- *Other*—Kawaiisu *Cooking Tools* Wood used to carve a ladle about a foot long. A branch with a bulge was sought and the bulge was hollowed out by burning to form the bowl. Such a utensil was used for stirring and dipping out foods. *Fasteners* Acorn meal used to mend cracks in clay pots. *Fuel* Wood preferred as firewood for roasting yucca bulbs. Spongy pith material used for starting fires. *Toys & Games* Acorn cupule used to make a top for children. (206:56) **Miwok** *Cash Crop* Acorns gathered in large quantities and traded for other foods. (12:142)

Quercus dumosa Nutt., California Scrub Oak
- *Drug*—Diegueño *Eye Medicine* Decoction of broken galls used as an eyewash. (84:33) **Luiseño** *Dermatological Aid* Gallnuts used for sores and wounds and as an astringent. (155:233)
- *Food*—Cahuilla *Bread & Cake* Acorns ground into a fine meal and used to make bread. *Dried Food* Dried acorns stored for a year or more in granaries. *Porridge* Cooked acorns used to make mush. *Special Food* Acorn meat considered a delicacy and favored at social and ceremonial occasions. (15:121) **Diegueño** *Porridge* Acorns shelled, pounded, leached, and cooked into a mush or gruel. (84:33) **Kawaiisu** *Bread & Cake* Acorns made into a fine meal, cooked into a mush, and allowed to stand and harden into a "cake." *Staple* Acorns dried, pounded, sifted into a fine meal, and leached. *Winter Use Food* Acorns stored for future use.[65] (206:56) **Luiseño** *Porridge* Acorns leached, ground into a meal, cooked in an earthen vessel, and eaten. *Staple* Stored acorns pounded in a mortar and pestle to make a flour.[64] (155:194) *Substitution Food* Acorns used only when more preferred species could not be obtained. (155:193) *Winter Use Food* Acorns formerly stored in acorn granaries. (155:194) **Pomo, Kashaya** *Forage* Acorns not used by people but eaten as a favorite food by deer, squirrels, chipmunks, quail, and jays. (72:82) **Tubatulabal** *Unspecified* Acorns used extensively for food. (193:15)
- *Fiber*—Diegueño *Basketry* Branches, with willow branches, used to make acorn storage baskets. *Furniture* Branches used as framework material for cradles. (84:33) **Kawaiisu** *Building Material* Logs used in house construction. (206:56)
- *Other*—Cahuilla *Cash Crop* Acorn meal exchanged for pinyon nuts, mesquite beans, and palm tree fruit. Acorn meal used as payment to a shaman for special services. *Fuel* Dried wood considered an ideal firewood for heating and cooking. *Hunting & Fishing Item* Acorns used as bait in trigger traps to capture small animals. *Jewelry* Unhusked acorns dried and strung as necklaces. *Musical Instrument* Acorns gathered on a cord and swung against the teeth to produce music. *Toys & Games* Acorns used by children in a game like jacks and for juggling. (15:121) **Kawaiisu** *Fasteners* Acorn meal used to mend cracks in clay pots. *Toys & Games* Acorn cupule used to make a top for children. (206:56)

Quercus dunnii Kellogg, Palmer Oak
- *Food*—Diegueño *Fruit* Fruit formerly used for food. (as *Q. palmeri* 88:216) **Paiute** *Porridge* Acorns boiled into mush. *Winter Use Food* Acorns stored for future use in pits lined and covered with sage bark. (as *Q. palmeri* 167:246)

Quercus ellipsoidalis E. J. Hill, Northern Pin Oak
- *Drug*—Menominee *Abortifacient* Compound decoction of inner bark taken for suppressed menses caused by cold. (54:133)
- *Food*—Menominee *Beverage* Roasted acorn ground for coffee. (151:66)

Quercus emoryi Torr., Emory's Oak
- *Food*—Apache, Western *Unspecified* Acorns eaten whole and raw, ground on a metate, or boiled. (26:174) **Papago** *Candy* Acorns chewed as a confection. (34:47) *Unspecified* Acorns eaten fresh from the shell. (34:19) Acorns used for food. (36:61) **Yavapai** *Cooking Agent* Ground meat used as thickening for venison stew. *Winter Use Food* Nuts stored for later use. (65:257)

Quercus engelmannii Greene, Engelmann's Oak
- *Food*—Diegueño *Candy* Bark gum pounded, washed, and chewed like chewing gum. *Porridge* Acorns shelled, pounded, leached, and cooked into a mush or gruel. (84:33) **Luiseño** *Porridge* Acorns leached, ground into a meal, cooked in an earthen vessel, and eaten. *Staple* Stored acorns pounded in a mortar and pestle to make a flour.[64] (155:194) *Substitution Food* Acorns used only when more preferred species could not be obtained. (155:193) *Winter Use Food* Acorns formerly stored in acorn granaries. (155:194)

Quercus falcata Michx., Southern Red Oak
- *Drug*—Cherokee *Antidiarrheal* Bark used for chronic dysentery. *Dermatological Aid* Astringent bark chewed for mouth sores. Infusion of bark applied to sore, chapped skin. *Disinfectant* Bark used as an antiseptic. *Emetic* Bark used as an emetic. *Febrifuge* Bark used after long, intermittent fevers and as a wash for chills and fevers. *Gastrointestinal Aid* Bark used for indigestion and "any debility of the system." *Oral Aid* Bark chewed for mouth sores. *Respiratory Aid* Infusion of bark taken for asthma. *Throat Aid* Decoction of inner bark used for "lost voice." *Tonic* Bark used as a tonic. *Urinary Aid* Unspecified liquid preparation taken for "milky urine." (80:46)
- *Fiber*—Cherokee *Basketry* Used to make baskets. *Building Material* Wood used for lumber, railroad ties, wagon spokes and rims. *Furniture* Wood used to make furniture. Used to make woven chair bottoms. (80:46)
- *Other*—Cherokee *Cooking Tools* Wood used to make corn beaters and mortars. Leaves used to wrap dough for bread making. *Fuel* Wood used for firewood. (80:46)

Quercus gambelii Nutt., Gambel's Oak
- *Drug*—Isleta *Reproductive Aid* Acorns eaten to give greater sexual potency. (32:47) Consumption of acorns believed to give greater sexual potency. (100:41) **Navajo, Ramah** *Analgesic* Decoction of root bark used for postpartum pain. *Cathartic* Decoction of root bark used as a cathartic. *Ceremonial Medicine* and *Emetic* Leaves used as a ceremonial emetic. *Gynecological Aid* Decoction of root bark used for postpartum pain and to help in delivery of placenta. *Panacea* Root bark used as a "life medicine." (191:22)
- *Food*—Acoma *Staple* Acorns ground into meal. *Unspecified* Acorns boiled and eaten. (32:47) **Apache, Chiricahua & Mescalero** *Fruit* Raw fruit used for food. *Winter Use Food* Acorns roasted slightly, pounded, mixed with dried meat, and stored away in hide containers. (33:42) **Apache, Western** *Unspecified* Acorns eaten whole and raw, ground on a metate, or boiled. (26:174) **Apache, White Mountain** *Unspecified*

Acorns used for food. (136:160) **Cochiti** *Staple* Acorns ground into meal. *Unspecified* Acorns boiled and eaten. (32:47) **Havasupai** *Porridge* Acorns parched, ground, and used to make mush. *Soup* Acorns parched, ground, and used to make soup. (197:67) *Spice* Acorns ground and added to flavor beef or deer soups. (197:74, 215) *Unspecified* Acorns parched on a tray or eaten raw. (197:215) **Hualapai** *Soup* Acorns used to make soup. *Unspecified* Acorns roasted and used for food. (195:12) **Isleta** *Staple* Acorns formerly used as a staple food. (100:41) **Laguna** *Staple* Acorns ground into meal. *Unspecified* Acorns boiled and eaten. (32:47) **Navajo, Ramah** *Staple* Acorns eaten raw, boiled, roasted in ashes, or dried, ground, and cooked like cornmeal. (191:22) **Neeshenam** *Bread & Cake* Acorns ground into flour, soaked in water, and baked to make a bread. *Porridge* Acorns ground into flour, soaked in water, and cooked to make mush. (129:374) **Pueblo** *Unspecified* Acorns formerly used extensively for food. **San Felipe** *Staple* Acorns ground into meal. *Unspecified* Acorns boiled and eaten. (32:47) **Yavapai** *Cooking Agent* Acorns sometimes added as thickening to venison stews. *Unspecified* Uncooked acorns used for food. (65:257)
- *Fiber*—**Navajo, Ramah** *Building Material* Whole trees used for shade house construction. *Furniture* Wood used to make frames for baby cradles. (191:22)
- *Dye*—**Navajo, Ramah** *Unspecified* Red leaf galls and red clay or gum used to make stripes on arrow shafts between and below the feathers. (191:22)
- *Other*—**Apache, White Mountain** *Hide Preparation* Bark used to tan skins. (136:160) **Havasupai** *Tools* Wood used to make handles for implements, such as hoes and axes. (197:215) **Hopi** *Ceremonial Items* Plant used in Oaqol ceremony. (42:355) **Isleta** *Tools* Wood used to make handles and other wooden portions of various implements. (100:41) **Jemez** *Hunting & Fishing Item* Hard, tough wood made into clubs and used in rabbit hunts. (44:27) **Navajo, Ramah** *Ceremonial Items* Wood used to make ceremonial bull-roarers. *Tools* Wood used to make ax handles, hoe handles, digging sticks, and weaving tools. Wood sticks notched by sheepherders to keep track of the days they have worked. (191:22)

Quercus gambelii Nutt. var. *gambelii*, Gambel's Oak

- *Drug*—**Isleta** *Reproductive Aid* Acorns eaten to give greater sexual potency. (as *Q. utahensis* 32:47) **Keres, Western** *Emetic* Infusion of ground leaves and oak galls used as an emetic. *Oral Aid* and *Pediatric Aid* Velvet pubescence rubbed on babies' tongues to remove milk coating. (as *Q. utahensis* 171:64)
- *Food*—**Acoma** *Staple* Acorns ground into meal. *Unspecified* Acorns boiled and eaten. **Cochiti** *Staple* Acorns ground into meal. *Unspecified* Acorns boiled and eaten. (as *Q. utahensis* 32:47) **Keres, Western** *Staple* Acorns ground into flour. *Unspecified* Acorns boiled and eaten. (as *Q. utahensis* 171:64) **Laguna** *Staple* Acorns ground into meal. *Unspecified* Acorns boiled and eaten. (as *Q. utahensis* 32:47) **Navajo** *Unspecified* Acorns seldom used for food. (as *Q. utahensis* 165:222) **Pueblo** *Unspecified* Acorns formerly used extensively for food. **San Felipe** *Staple* Acorns ground into meal. *Unspecified* Acorns boiled and eaten. (as *Q. utahensis* 32:47) **Tewa** *Unspecified* Acorns used for food. (as *Q. utahensis* 138:44)
- *Fiber*—**Tewa of Hano** *Sewing Material* Wood used to make embroidery stretchers. (as *Q. utahensis* 138:44)
- *Other*—**Tewa** *Tools* Woods used to make digging sticks. *Weapon* Woods used to make bows and war clubs. **Tewa of Hano** *Tools* Wood used to make rabbit sticks and other utensils. (as *Q. utahensis* 138:44)

Quercus garryana Dougl. ex Hook., Oregon White Oak

- *Drug*—**Cowlitz** *Tuberculosis Remedy* Decoction of bark taken for tuberculosis. (79:27) **Karok** *Gynecological Aid* Infusion of plant taken by mother before her first baby comes. Pounded bark rubbed on abdomen and sides of mother before her first delivery. (148:382)
- *Food*—**Chehalis** *Unspecified* Acorns roasted and eaten. **Cowlitz** *Unspecified* Acorns buried in the mud for leaching and used for food. (79:27) **Karok** *Unspecified* Acorns used for food. (148:382) **Mendocino Indian** *Bread & Cake* Acorns used to make bread. *Soup* Acorns used to make soup. (41:343) **Nisqually** *Unspecified* Acorns used for food. (79:27) **Paiute** *Unspecified* Autumn acorns buried in mud to ripen and eaten. (111:65) **Pomo** *Bread & Cake* Acorns used to make bread. *Porridge* Acorns used to make mush. (117:290) *Unspecified* Acorns used for food. (66:12) **Pomo, Kashaya** *Dried Food* Acorns sun dried before storing. *Porridge* Acorns used as flour for pancakes, bread, mush, or soup.[45] (72:81) **Salish, Coast** *Unspecified* Acorns steamed, roasted, or boiled and used for food. (182:84) **Shasta** *Bread & Cake* Acorns pounded, winnowed, leached, and made into bread. *Porridge* Acorns pounded, winnowed, leached, and made into mush. *Soup* Acorns pounded, winnowed, leached, and made into thin soup. *Staple* Acorns used as the basic staple. (93:308) **Squaxin** *Unspecified* Acorns roasted on hot rocks and eaten. (79:27)
- *Fiber*—**Cowlitz** *Brushes & Brooms* Wood used to make combs. (79:27)
- *Other*—**Cowlitz** *Fuel* Wood used as a fuel. *Tools* Wood used to make digging sticks. (79:27) **Paiute** *Cooking Tools* Wood made into dishes used to pound roots. (111:65) **Pomo** *Hunting & Fishing Item* Branches used to make arrows. (66:12)

Quercus garryana var. *semota* Jepson, Oregon White Oak

- *Food*—**Kawaiisu** *Bread & Cake* Acorns made into a fine meal, cooked into a mush, and allowed to stand and harden into a "cake." *Staple* Acorns dried, pounded, sifted into a fine meal, and leached. *Winter Use Food* Acorns stored for future use.[65] (206:56)
- *Fiber*—**Kawaiisu** *Building Material* Logs used in house construction. (206:56)
- *Other*—**Kawaiisu** *Fasteners* Acorn meal used to mend cracks in clay pots. *Toys & Games* Acorn cupule used to make a top for children. (206:56)

Quercus grisea Liebm., Gray Oak

- *Food*—**Apache, Chiricahua & Mescalero** *Fruit* Raw fruit used for food. (33:42) *Spice* Shaved root chips used to flavor drinks. (33:51) *Winter Use Food* Ripe acorns roasted slightly, pounded, and mixed with dried meat and stored. (33:42) **Navajo, Ramah** *Unspecified* Acorns used for food. (191:22)
- *Other*—**Navajo, Ramah** *Protection* Used to protect new or ceremonial hogans from lightning, ghosts, and witches. (191:22)

Quercus ilicifolia Wangenh., Bear Oak

- *Drug*—**Iroquois** *Gynecological Aid* Used as a wash and taken for "female troubles." *Other* Used as a "sugar medicine." (87:302)

Quercus imbricaria Michx., Shingle Oak

- **Drug**—Cherokee *Antidiarrheal* Bark used for chronic dysentery. *Dermatological Aid* Astringent bark chewed for mouth sores. Infusion of bark applied to sore, chapped skin. *Disinfectant* Bark used as an antiseptic. *Emetic* Bark used as an emetic. *Febrifuge* Bark used after long, intermittent fevers and as a wash for chills and fevers. *Gastrointestinal Aid* Bark used for indigestion and "any debility of the system." *Oral Aid* Bark chewed for mouth sores. *Respiratory Aid* Infusion of bark taken for asthma. *Throat Aid* Decoction of inner bark used for "lost voice." *Tonic* Bark used as a tonic. *Urinary Aid* Unspecified liquid preparation taken for "milky urine." (80:46)
- **Fiber**—Cherokee *Basketry* Used to make baskets. *Building Material* Wood used for lumber, railroad ties, wagon spokes and rims. *Furniture* Wood used to make furniture. Used to make woven chair bottoms. (80:46)
- **Other**—Cherokee *Cooking Tools* Wood used to make corn beaters and mortars. Leaves used to wrap dough for bread making. *Fuel* Wood used for firewood. (80:46)

Quercus kelloggii Newberry, California Black Oak

- **Food**—Cahuilla *Bread & Cake* Acorns ground into a fine meal and used to make bread. *Dried Food* Dried acorns stored for a year or more in granaries. *Porridge* Cooked acorns used to make mush. *Special Food* Acorn meat considered a delicacy and favored at social and ceremonial occasions. (15:121) **Diegueño** *Porridge* Acorns shelled, pounded, leached, and cooked into a mush or gruel. (84:33) **Karok** *Fruit* Fruit soaked in mud for a year and used for food. (6:49) *Unspecified* Acorns made into "houm" and eaten. (148:382) **Kawaiisu** *Bread & Cake* Acorns made into a fine meal, cooked into a mush, and allowed to stand and harden into a "cake." *Staple* Acorns dried, pounded, sifted into a fine meal, and leached. *Winter Use Food* Acorns stored for future use.[65] (206:56) **Luiseño** *Porridge* Acorns leached, ground into a meal, cooked in an earthen vessel, and eaten. (as *Q. californica* 155:194) *Staple* Acorns eaten as a staple food. (as *Q. californica* 155:193) Acorns from storage granaries pounded in a mortar and pestle to make a flour.[64] *Winter Use Food* Acorns formerly gathered for storage in acorn granaries. (as *Q. californica* 155:194) **Mendocino Indian** *Bread & Cake* Acorns used to make bread. *Soup* Acorns used to make soup. (as *Q. californicum* 41:342) **Mewuk** *Bread & Cake* Acorns used to make bread. *Porridge* Acorns used to make mush. (as *Q. californicus* 117:327) *Unspecified* Acorns used for food. (as *Q. californicus* 117:333) **Miwok** *Bread & Cake* Acorns ground into a meal and used to make bread and biscuits. *Porridge* Acorns considered a staple food and used to make mush. *Soup* Acorns ground into a meal and used to make soup. *Winter Use Food* Whole acorns stored for winter use. (12:142) **Modesse** *Staple* Acorns used as the principal vegetable food. (as *Q. californica* 117:223) **Neeshenam** *Unspecified* Acorns occasionally used for food. (as *Q. sonomensis* 129:374) **Paiute** *Porridge* Acorns boiled into mush. *Winter Use Food* Acorns stored for future use in pits lined and covered with sage bark. (167:246) **Paiute, Northern** *Staple* Acorns ground into flour, leached, and eaten. (59:52) **Pomo** *Bread & Cake* Acorns used to make white bread. (as *Q. californica* 11:67) Acorns used to make bread. (as *Q. californica* 117:290) *Porridge* Acorns used to make gruel and mush. (as *Q. californica* 11:67) Acorns used to make mush. (as *Q. californica* 117:290) *Soup* Acorns used to make soups. (as *Q. californica* 11:67) *Unspecified* Acorns used for food. (66:12) **Pomo, Kashaya** *Dried Food* Acorns sun dried before storing. *Porridge* Acorns used as flour for pancakes, bread, mush, or soup.[45] (72:79) **Shasta** *Bread & Cake* Acorns pounded, winnowed, leached, and made into bread. *Porridge* Acorns pounded, winnowed, leached, and made into mush. *Soup* Acorns pounded, winnowed, leached, and made into thin soup. *Staple* Acorns used as the basic staple. (as *Q. californica* 93:308) **Tolowa** *Fruit* Fruit used for food. (6:49) **Tubatulabal** *Unspecified* Acorns used extensively for food. (as *Q. californica* 193:15) **Yokut** *Unspecified* Acorns used for food. (as *Q. californica* 117:420) **Yuki** *Bread & Cake* Nutmeats pounded into fine meal, winnowed, and made into bread. *Porridge* Nutmeats pounded into fine meal, winnowed, boiled, and eaten as mush. (49:89) **Yurok** *Fruit* Fruit used for food. (6:49)
- **Fiber**—Kawaiisu *Building Material* Logs used in house construction. (206:56) **Maidu** *Basketry* Withes used to make basket rims. (173:71) **Mewuk** *Basketry* Shoots split into strands and used for twining the rods of baskets and scoops. (as *Q. californicus* 117:328)
- **Dye**—Pomo, Kashaya *Black* Round, fleshy insect galls made into a dark hair dye. (72:79)
- **Other**—Cahuilla *Cash Crop* Acorn meal exchanged for pinyon nuts, mesquite beans, and palm tree fruit. Acorn meal used as payment to a shaman for special services. *Fuel* Dried wood considered an ideal firewood for heating and cooking. *Hunting & Fishing Item* Acorns used as bait in trigger traps to capture small animals. *Jewelry* Unhusked acorns dried and strung as necklaces. *Musical Instrument* Acorns gathered on a cord and swung against the teeth to produce music. *Toys & Games* Acorns used by children in a game like jacks and for juggling. (15:121) **Kawaiisu** *Fasteners* Acorn meal used to mend cracks in clay pots. *Toys & Games* Acorn cupule used to make a top for children. (206:56) **Mewuk** *Tools* Shoots split into strands and used for twining the rods of baskets and scoops. (as *Q. californicus* 117:328) **Miwok** *Cooking Tools* Branch used to make mush stirring paddles. (12:146) **Pomo** *Fuel* Small square of thick dry bark used to carry fire from one place to another. (as *Q. californica* 117:283)

Quercus laurifolia Michx., Laurel Oak

- **Other**—Choctaw *Paint* Bark, red oak, and live oak boiled and used for paint. (as *Q. obtusiloba* 25:14)

Quercus lobata Née, California White Oak

- **Drug**—Kawaiisu *Burn Dressing* Poultice of ground galls and salt applied to burns. *Dermatological Aid* Poultice of ground galls and salt applied to sores and cuts. *Eye Medicine* Ground gall powder and salt wrapped in a small piece of cloth and dipped in water applied to sore eyes. (206:56) **Miwok** *Cough Medicine* Decoction of bark taken as a cough medicine. *Dermatological Aid* and *Pediatric Aid* Pulverized, outer bark dusted on running sores and particularly used for babies with sore umbilicus. (12:172) **Yuki** *Antidiarrheal* Bark used for diarrhea. (41:343)
- **Food**—Kawaiisu *Bread & Cake* Acorns made into a fine meal, cooked into a mush, and allowed to stand and harden into a "cake." *Staple* Acorns dried, pounded, sifted into a fine meal, and leached. *Winter Use Food* Acorns stored for future use.[65] (206:56) **Mendocino Indian** *Bread & Cake* Large acorns used to make bread. (41:343) **Miwok** *Bread & Cake* Acorns ground into a meal and used to make bread and bis-

cuits. *Soup* Acorns ground into a meal and used to make soup. *Staple* Acorns considered a staple food and used to make mush. *Winter Use Food* Whole acorns stored for winter use. (12:142) **Pomo** *Bread & Cake* Acorns used to make white and black bread. (11:67) Acorns used to make bread. (117:290) *Porridge* Acorns used to make gruel and mush. (11:67) Acorns used to make mush. (117:290) *Soup* Acorns used to make soup. (11:67) **Pomo, Kashaya** *Porridge* Acorns used to make mush or soup rather than bread. (72:84) **Tubatulabal** *Unspecified* Acorns used extensively for food. (193:15) **Wintoon** *Unspecified* Roasted seeds used for food. (117:274) **Yokut** *Unspecified* Acorns used for food. (117:420) **Yuki** *Bread & Cake* Nutmeats pounded into fine meal, winnowed, and made into bread. *Porridge* Nutmeats pounded into fine meal, winnowed, boiled, and eaten as mush. (49:89)
- *Fiber*—**Kawaiisu** *Building Material* Logs used in house construction. (206:56)
- *Dye*—**Concow** *Black* Bark used to blacken strands of red buds for basket making. (41:343)
- *Other*—**Kawaiisu** *Fasteners* Acorn meal used to mend cracks in clay pots. *Toys & Games* Acorn cupule used to make a top for children. (206:56)

Quercus macrocarpa Michx., Bur Oak
- *Drug*—**Chippewa** *Analgesic* Decoction of root or inner bark taken for cramps. *Gastrointestinal Aid* Decoction of inner bark used for cramps. (53:340) *Heart Medicine* Compound decoction of inner bark prepared ceremonially for heart trouble. (53:338) *Pulmonary Aid* Compound decoction of inner bark taken for lung trouble. (53:340) **Iroquois** *Antidiarrheal* Infusion of bark chips taken for diarrhea. *Antidote* "Plant will stop the effects of the laxative made from V[iburnum] opulus." *Dermatological Aid* Complex compound decoction used as wash for affected parts of "Italian itch." (87:303) **Menominee** *Abortifacient* Compound decoction of inner bark taken for suppressed menses caused by cold. (54:133) **Meskwaki** *Anthelmintic* Compound containing wood and inner bark used to expel pinworms. (152:221, 222) **Ojibwa** *Dermatological Aid* Bark used as an astringent medicine. *Orthopedic Aid* Bark used to bandage a broken foot or leg. (153:369)
- *Food*—**Cheyenne** *Unspecified* Acorns formerly used for food. (83:26) **Chippewa** *Unspecified* Acorns roasted in ashes or boiled, mashed, and eaten with grease or duck broth. *Vegetable* Acorns boiled, split open, and eaten like a vegetable. (53:320) **Dakota** *Unspecified* Acorns leached with basswood ashes to remove the bitter taste and used for food. (70:75) **Lakota** *Soup* Acorns chopped and cooked in soups and meats. *Unspecified* Acorns chopped, cooked over fire, and eaten. (106:31) **Ojibwa** *Unspecified* Acorns treated with lye to remove bitterness and eaten. (153:402) **Omaha** *Unspecified* Acorns leached with basswood ashes to remove the bitter taste and used for food. **Pawnee** *Unspecified* Acorns leached with basswood ashes to remove the bitter taste and used for food. **Ponca** *Unspecified* Acorns leached with basswood ashes to remove the bitter taste and used for food. **Winnebago** *Unspecified* Acorns leached with basswood ashes to remove the bitter taste and used for food. (70:75)
- *Dye*—**Chippewa** *Black* Boiled with black earth and ocher to make a black dye. Inner bark boiled with green hazel burs, added to black earth and butternut, and used as a black dye. (53:372) **Ojibwa** *Mordant* Bark used in combination with other materials to set color. (153:425)
- *Other*—**Dakota** *Toys & Games* Young growths used to make popgun pistons. **Omaha** *Toys & Games* Young growths used to make popgun pistons. **Pawnee** *Toys & Games* Young growths used to make popgun pistons. **Ponca** *Toys & Games* Young growths used to make popgun pistons. **Winnebago** *Toys & Games* Young growths from this or another plant used to make popgun pistons. (70:116)

Quercus marilandica Muenchh., Blackjack Oak
- *Drug*—**Choctaw** *Analgesic* Infusion of tree bark coal taken to remove the afterbirth and ease cramps. *Gynecological Aid* Infusion of tree bark coal taken to aid in childbirth, and to remove the afterbirth and ease cramps. (177:17)
- *Food*—**Comanche** *Starvation Food* Boiled acorns used for food in times of scarcity. (29:524)
- *Other*—**Comanche** *Smoking Tools* Leaves used as cigarette wrappers. (29:524)

Quercus muehlenbergii Engelm., Chinkapin Oak
- *Drug*—**Delaware, Ontario** *Antiemetic* Infusion of bark taken for vomiting. (175:68, 82)

Quercus nigra L., Water Oak
- *Food*—**Choctaw** *Staple* Pounded acorns boiled and made into a meal. Pounded acorns used as cornmeal. (as *Q. aquatica* 25:8) **Kiowa** *Beverage* Acorns used to make a beverage. *Unspecified* Acorns used for food. (192:21)
- *Other*—**Kiowa** *Fuel* Wood burned in the home and in the peyote ceremony. *Smoke Plant* Leaves used as a substitute for paper in rolling cigarettes. (192:21)

Quercus oblongifolia Torr., Mexican Blue Oak
- *Food*—**Papago** *Unspecified* Acorns used for food. (36:61) **Pima** *Staple* Hulls removed, acorns parched, ground into meal, and used for food. (146:78)

Quercus pagoda Raf., Cherrybark Oak
- *Drug*—**Houma** *Antidiarrheal* Compound decoction of bark taken for dysentery. *Orthopedic Aid* Strong decoction of root or bark applied to swollen joints. *Throat Aid* Decoction of bark and roots taken for sore throat or hoarseness. *Tonic* Decoction of bark and root taken as a tonic for "run-down health." (as *Q. pagodaefolia* 158:56)

Quercus palustris Muenchh., Pin Oak
- *Drug*—**Delaware** *Gastrointestinal Aid* Infusion of inner bark used for intestinal pains. (176:30) **Delaware, Oklahoma** *Analgesic* and *Gastrointestinal Aid* Infusion of inner bark taken for intestinal pains. (175: 25, 78)
- *Other*—**Cherokee** *Fasteners* Wood used to make pins or small pegs. (80:47)

Quercus ×pauciloba Rydb., Wavyleaf Oak
- *Drug*—**Navajo, Kayenta** *Sedative* Plant used for nervousness. (as *Q. undulata* 205:18) **Navajo, Ramah** *Analgesic* Decoction of root bark used for internal pains. Decoction of root bark taken for internal pains. (as *Q. undulata* 191:22)
- *Food*—**Apache, Western** *Unspecified* Acorns eaten whole and raw, ground on a metate, or boiled. (as *Q. undulata* 26:174) **Apache,**

White Mountain *Beverage* Acorns used to make "coffee." *Bread & Cake* Acorns ground into flour and used to make bread. *Unspecified* Acorns eaten raw. (as *Q. undulata* 136:148) **Gosiute** *Unspecified* Acorns used only in season for food. (as *Q. undulata* 39:378) **Navajo, Ramah** *Unspecified* Acorns used for food. (as *Q. undulata* 191:22)
- *Fiber*—**Navajo** *Furniture* Wood used to make batten sticks and bows for the baby's cradle. (as *Q. undulata* 55:41)
- *Dye*—**Navajo, Ramah** *Black* Charcoal used as a black pigment for sand paintings. (as *Q. undulata* 191:22)
- *Other*—**Apache, White Mountain** *Hide Preparation* Bark used to tan skins. (as *Q. undulata* 136:160) **Navajo** *Tools* Wood used to make batten sticks and bows for the baby's cradle. (as *Q. undulata* 55:41) **Navajo, Ramah** *Protection* Used to protect new or ceremonial hogans from lightning, ghosts, and witches. (as *Q. undulata* 191:22)

Quercus peninsularis Trel., Peninsula Oak
- *Food*—**Diegueño** *Staple* Acorns pounded, sun dried, ground, and leached. (88:216)

Quercus phellos L., Willow Oak
- *Drug*—**Seminole** *Analgesic* Decoction of wood bits or bark applied externally as an analgesic. (169:167) *Antirheumatic (External)* Decoction of wood or bark used as a bath for aches and pains. (169:286) *Dermatological Aid* Decoction of bark used for ballgame sickness: sores, back or limb pains, and hemorrhoids. Decoction of wood or bark used as a bath for sores and cuts. (169:269, 286) *Hemorrhoid Remedy* Decoction of bark used for ballgame sickness: sores, back or limb pains, and hemorrhoids. (169:269) *Love Medicine* Decoction of wood ashes placed on the tongue to cleanse the body and strengthen the marriage. (169:250) *Orthopedic Aid* Decoction of bark used for ballgame sickness: sores, back or limb pains, and hemorrhoids. (169:269)
- *Food*—**Seminole** *Unspecified* Plant used for food. (169:471)
- *Other*—**Seminole** *Cleaning Agent* Plant used to make lye. *Toys & Games* Plant used to make ball sticks. (169:471)

Quercus prinus L., Chestnut Oak
- *Food*—**Iroquois** *Unspecified* Acorns used for food. (196:123)
- *Fiber*—**Cherokee** *Building Material* Wood used to make railroad ties. (80:46)
- *Dye*—**Cherokee** *Brown* Bark used to make a tan dye. (80:46)

Quercus pumila Walt., Running Oak
- *Food*—**Seminole** *Unspecified* Plant used for food. (169:493)

Quercus pungens Liebm., Pungent Oak
- *Food*—**Navajo** *Candy* Gum used for chewing gum. (55:41)
- *Dye*—**Navajo** *Brown* Bark exudation used as a tan dye. (55:41)
- *Other*—**Keres, Western** *Tools* Wood used for tool handles. Twigs used to tie warp of rugs to beam while weaving. (171:64) **Navajo** *Ceremonial Items* Wood charcoal used as the black for sand paintings. *Paint* Gum used for painting arrows between the feathers. *Tools* Wood used to make batten for weaving. (55:41)

Quercus rubra L., Northern Red Oak
- *Drug*—**Cherokee** *Antidiarrheal* Bark used for chronic dysentery. *Dermatological Aid* Astringent bark chewed for mouth sores. Infusion of bark applied to sore, chapped skin. *Disinfectant* Bark used as an antiseptic. *Emetic* Bark used as an emetic. *Febrifuge* Bark used after long, intermittent fevers and as a wash for chills and fevers. *Gastrointestinal Aid* Bark used for indigestion and "any debility of the system." *Oral Aid* Bark chewed for mouth sores. *Respiratory Aid* Infusion of bark taken for asthma. *Throat Aid* Decoction of inner bark used for "lost voice." (80:46) Decoction of inner bark taken for hoarseness. (177:17) *Tonic* Bark used as a tonic. *Urinary Aid* Unspecified liquid preparation taken for "milky urine." (80:46) **Chippewa** *Heart Medicine* Compound decoction of inner bark prepared ceremonially for heart trouble. (53:338) **Delaware** *Cough Medicine* Infusion of bark used for severe coughs. *Throat Aid* Infusion of bark used for hoarseness. (176:30) **Delaware, Oklahoma** *Cough Medicine* Infusion of bark taken for severe cough. *Throat Aid* Infusion of bark taken for hoarseness. (175:25, 78) **Mahuna** *Toothache Remedy* Plant juice used for straightening and setting loose teeth. (140:25) **Malecite** *Antidiarrheal* Infusion of plant and fir buds or cones used for diarrhea. (116:244) Infusion of bark or roots used for diarrhea. (116:255) **Micmac** *Antidiarrheal* Bark and roots used for diarrhea. (40:60) **Ojibwa** *Blood Medicine* Decoction of bark taken for internal blood diseases. (as *Q. ruba* 135:231) *Heart Medicine* and *Respiratory Aid* Bark used for "heart troubles and bronchial affections." (153:369, 370) *Unspecified* Plant used for medicinal purposes. (135:242) *Venereal Aid* Infusion of root bark taken for gonorrhea. (as *Q. ruba* 135:231) **Ojibwa, South** *Antidiarrheal* Decoction of root bark and inner bark taken for diarrhea. (91:198) **Potawatomi** *Antidiarrheal* Inner bark used for flux. (154:58) **Rappahannock** *Dietary Aid* Infusion of north side bark taken as an appetizer. *Tonic* Decoction of bark and leaves taken as a beneficial beverage (bitters). (161:26)
- *Food*—**Dakota** *Unspecified* Acorns leached with basswood ashes to remove the bitter taste and used for food. (70:75) **Iroquois** *Unspecified* Acorns used for food. (196:123) **Ojibwa** *Staple* Acorns leached with lye and used as of the most important starchy foods. (153:402) **Omaha** *Unspecified* Acorns freed from tannic acid by boiling with wood ashes and used for food. (68:327) Acorns leached with basswood ashes to remove the bitter taste and used for food. **Pawnee** *Unspecified* Acorns leached with basswood ashes to remove the bitter taste and used for food. **Ponca** *Unspecified* Acorns leached with basswood ashes to remove the bitter taste and used for food. (70:75) **Potawatomi** *Porridge* Dried, ground acorns used as a flour to make gruel. Hardwood ashes and water furnished the lye for soaking the acorns, to swell them and remove the tannic acid. A bark bag or reticule served to hold the acorns while they were washed through a series of hot and cold water to remove the lye. Then they were dried in the sun and became perfectly sweet and palatable. They were ground on depressions of rocks which served as a mortar with a stone pestle, to a flour, which was cooked as a gruel, sometimes called samp. (154:100)
- *Fiber*—**Cherokee** *Basketry* Used to make baskets. *Building Material* Wood used for lumber, railroad ties, wagon spokes and rims. *Furniture* Wood used to make furniture. Used to make woven chair bottoms. (80:46)
- *Dye*—**Ojibwa** *Unspecified* Bark used in tanning and coloring. (135:242) **Omaha** *Black* Bark used to make a black dye for porcupine quills. (68:325) **Potawatomi** *Red-Brown* Rushes gathered for mat weaving and boiled with bark to impart a brownish red dye. (154:120)

- **Other**—**Cherokee** *Cooking Tools* Wood used to make corn beaters and mortars. Leaves used to wrap dough for bread making. *Fuel* Wood used for firewood. (80:46) **Ojibwa** *Hide Preparation* Bark used in tanning and coloring. (135:242) **Potawatomi** *Designs* Leaves used to furnish a design for beadwork. (154:120)

Quercus rubra L. var. *rubra*, Northern Red Oak
- **Drug**—**Alabama** *Dermatological Aid* Decoction of bark used as a wash for bad smelling sores on the head or feet. *Orthopedic Aid* and *Pediatric Aid* Infusion of bark given to child old enough to walk but too weak to do so. *Pulmonary Aid* Infusion of bark taken for lung trouble. *Throat Aid* Decoction of bark taken for sore throats. **Cherokee** *Antidiarrheal* Infusion of twig juice taken for dysentery. (as *Q. borealis* var. *maxima* 177:16) **Iroquois** *Dermatological Aid* Complex compound decoction used as wash for affected parts of "Italian itch." Poultice of powdered bark bound to ruptured or improperly healed navels. (as *Q. borealis* var. *maxima* 87:304)
- **Other**—**Kiowa** *Fuel* Wood used as a favorite fuel for the altar fire in the peyote ceremony. (as *Q. borealis* var. *maxima* 192:21)

Quercus sadleriana R. Br. Campst., Deer Oak
- **Food**—**Karok** *Unspecified* Acorns shelled, parched, and eaten. (148:382)

Quercus sp., Oak
- **Drug**—**Alabama** *Dermatological Aid* Decoction of bark used as a wash for bad smelling sores on the head or feet. *Emetic* and *Pulmonary Aid* Decoction of bark taken as emetic for lung troubles. *Throat Aid* Boiled bark used for sore throat. (172:665) **Atsugewi** *Blood Medicine* Decoction taken by women to prevent blood poisoning. *Cold Remedy* Decoction taken by women to prevent catching cold during the birth ordeal. (as *Cambrium* 61:140) **Chippewa** *Hemostat* Poultice of chewed, fresh or dry root applied to wounds as a styptic. (53:356) **Costanoan** *Antidiarrheal* Infusion of acorns used for diarrhea. *Toothache Remedy* Decoction of bark used for toothaches and to tighten loose teeth. (21:20) **Creek** *Orthopedic Aid* and *Pediatric Aid* Compound decoction of bark used as a wash to strengthen children unable to walk. (172:665) **Dakota** *Gastrointestinal Aid* and *Pediatric Aid* Decoction of root bark given for bowel trouble, especially in children. (70:75) **Iroquois** *Other* Infusion of bark and elm bark taken for ruptures caused by violence. (141:38) *Throat Aid* Poultice of inner bark used for "sore throats that will not heal." *Tuberculosis Remedy* Compound used for tuberculosis. (87:302) **Malecite** *Unspecified* Used to make medicines. (160:6) **Mendocino Indian** *Dietary Aid* Plant used for fattening. (41:26) **Neeshenam** *Antirheumatic (Internal)* Burning pitch smoke inhaled for rheumatism. *Burn Dressing* Poultice of powdered acorns applied to burns or scalds. *Cold Remedy* Burning pitch smoke inhaled for colds. *Cough Medicine* Burning pitch smoke inhaled for coughs. (129:374) *Dermatological Aid* Pitch rubbed on wounds, sores, or arrow wounds. *Psychological Aid* Poultice of hot pitch and powdered, burned acorns applied to mourning widows. (129:375) **Omaha** *Gastrointestinal Aid* and *Pediatric Aid* Decoction of root bark given for bowel trouble, especially in children. **Pawnee** *Gastrointestinal Aid* and *Pediatric Aid* Decoction of root bark given for bowel trouble, especially in children. **Ponca** *Gastrointestinal Aid* and *Pediatric Aid* Decoction of root bark given for bowel trouble, especially in children. **Winnebago** *Gastrointestinal Aid* and *Pediatric Aid* Decoction of root bark given for bowel trouble, especially in children. (70:75)
- **Food**—**Apache, Mescalero** *Unspecified* Acorns boiled, pounded, and mixed with mescal. Acorns eaten raw. (14:41) **Chippewa** *Unspecified* Acorns, with the tannin removed by using wood ash lye and leached out with water, used for food. (71:129) **Comanche** *Unspecified* Acorns used for food. (29:524) **Concow** *Bread & Cake* Acorns made into bread and eaten. *Porridge* Acorns made into mush and eaten. (41:333) **Costanoan** *Unspecified* Acorns used for food. (21:248) **Iroquois** *Beverage* Fresh nutmeats crushed, boiled, and liquid used as a drink. *Bread & Cake* Fresh nutmeats crushed and mixed with bread. *Pie & Pudding* Fresh nutmeats crushed and mixed with corn pudding. *Soup* Acorns boiled, roasted, pounded, mixed with meal or meat, and eaten as soup. *Special Food* Fresh nutmeats crushed, boiled, and oil used as a delicacy in corn bread and pudding. *Unspecified* Acorns eaten raw by children. (124:99) **Malecite** *Unspecified* Acorns baked and used for food. (160:6) **Miwok** *Bread & Cake* Acorns ground into a meal and used to make bread and biscuits. *Soup* Acorns ground into a meal and used to make soup. (12:142) **Navajo** *Staple* Dried acorns ground into flour. *Unspecified* Acorns boiled like beans and roasted over coals. (55:40) **Round Valley Indian** *Bread & Cake* Nuts dried, cracked, pulverized, water added, and the dough made into bread. *Porridge* Nuts dried, cracked, pulverized, water added, and the dough made into brownish red mush. (41:333)
- **Fiber**—**Apache, Mescalero** *Building Material* Used for poles in dome-shaped lodges and as tepee stakes. *Furniture* Used as footrests for cradleboards. (14:41) **Comanche** *Building Material* Trunks used for fence posts. (29:524) **Navajo** *Basketry* Twigs used as the framework of a temporary carrying basket. A temporary carrying basket was made of two staves or bows of oak twigs crossed in the center and brought upwards to the hoop. This framework was then covered with sheep or goatskin. These carrying baskets were usually made in the field for carrying yucca fruits. (55:40)
- **Dye**—**Chippewa** *Black* Used with grindstone dust or black earth to make a black dye. (53:372) *Red* Inner bark boiled, cedar ashes added, and used to make a red dye. (53:370)
- **Other**—**Apache, Chiricahua & Mescalero** *Fuel* Wood used on fire to heat cooking stones. (33:36) *Tools* Branches used to dig out crowns of the mescal plants. (33:35) **Apache, Mescalero** *Cooking Tools* Used to make platters and shelves for mescal cakes. *Stable Gear* Used to make stirrups. *Tools* Used to make digging sticks and wooden tweezers. *Toys & Games* Used to make toy bows. (14:41) **Cherokee** *Hunting & Fishing Item* Wood used to make bows. (80:21) **Chippewa** *Tools* Used for awls. (53:378) **Costanoan** *Cooking Tools* Wood used for bowls and mortars. *Fuel* Bark used as tinder. (21:248) **Miwok** *Cooking Tools* Wood used to make mush stirring paddles. (12:146) **Navajo** *Ceremonial Items* Used to make digging sticks for the Female Shooting Life Chant for digging medicinal roots. Sticks inserted in crevice above door during the dedication and purification of the hogan. Curled twig used as a drum stick in the War Dance Ceremony. Wood used, because of its hardness and great resisting power, in nearly all of the ceremonies. *Containers* Acorn shells used to hold medicine and a hummingbird was made to sip from each shell. *Hunting & Fishing Item* Used to make throwing sticks. The Navajo throwing

stick, which was of oak, was made by whittling the piece down to the shape of a batten and then heating it and bending it over the knee to give it a slight curve. *Tools* Used to make batten stick for weaving. Concave hole in wood used as a die to make metallic hemispheres for beads and sunflower blossoms. Used to make hoes and digging sticks. *Toys & Games* Sticks kicked out of the ground while playing "football." Stick curved in hot ashes to make a J-shaped stick or bat for shinny and other games. *Weapon* Wood used to make the bow carried into war. Branches used to make clubs. In warfare, clubs were used by some of the warriors. The older type consisted of a grooved stone, which was hafted by twisting a small branch from an oak twice around the grooved section of the stone and tying the free ends together. (55:40) **Pima, Desert** *Cash Crop* Acorns used for trade. (133:6)

Quercus stellata Wangenh., Post Oak

- *Drug*—**Cherokee** *Antidiarrheal* Bark used for chronic dysentery. (80:46) Infusion of twig juice taken for dysentery. (177:18) *Dermatological Aid* Astringent bark chewed for mouth sores. Infusion of bark applied to sore, chapped skin. *Disinfectant* Bark used as an antiseptic. *Emetic* Bark used as an emetic. *Febrifuge* Bark used after long, intermittent fevers and as a wash for chills and fevers. *Gastrointestinal Aid* Bark used for indigestion and "any debility of the system." *Oral Aid* Bark chewed for mouth sores. *Respiratory Aid* Infusion of bark taken for asthma. *Throat Aid* Decoction of inner bark used for "lost voice." *Tonic* Bark used as a tonic. *Urinary Aid* Unspecified liquid preparation taken for "milky urine." (80:46) Infusion of inner bark taken for difficult urination with discharge. **Choctaw** *Gastrointestinal Aid* Decoction of bark taken for stomachaches. (177:18) **Creek** *Antidiarrheal* Bark used to make a drink taken for dysentery. (172:659) Infusion of bark taken for dysentery. (177:18)
- *Food*—**Kiowa** *Beverage* Acorns used to make a drink similar to coffee. *Dried Food* Dried, pounded acorns used for food. (192:22)
- *Fiber*—**Cherokee** *Basketry* Used to make baskets. *Building Material* Wood used for lumber, railroad ties, wagon spokes and rims. *Furniture* Wood used to make furniture. Used to make woven chair bottoms. (80:46)
- *Other*—**Cherokee** *Cooking Tools* Wood used to make corn beaters and mortars. Leaves used to wrap dough for bread making. *Fuel* Wood used for firewood. (80:46) **Kiowa** *Fuel* Wood used for firewood. *Smoke Plant* Leaves used as cigarette wrappers for the peyote ceremony. (192:22)

Quercus texana Buckl., Nuttall Oak

- *Dye*—**Choctaw** *Red* Burned bark and black gum ash added to water and used as a red dye. (25:14)
- *Other*—**Choctaw** *Paint* Bark, post oak, and live oak boiled and used for paint. (25:14)

Quercus turbinella Greene, Shrub Live Oak

- *Food*—**Hualapai** *Bread & Cake* Acorns used to make bread. *Soup* Acorns used to make stew. *Unspecified* Acorns roasted like pinyons. (195:11) **Mohave** *Porridge* Acorns used to make mush. (37:187) **Pima, Gila River** *Snack Food* Fruits eaten raw as a snack food. (133:7)
- *Other*—**Cocopa** *Cash Crop* Acorns gathered and traded with the Paipai for wild sheep skins. (37:187) **Havasupai** *Tools* Wood used to make the hoe and ax handles. (197:215)

Quercus velutina Lam., Black Oak

- *Drug*—**Cherokee** *Antidiarrheal* Bark used for chronic dysentery. *Dermatological Aid* Astringent bark chewed for mouth sores. Infusion of bark applied to sore, chapped skin. *Disinfectant* Bark used as an antiseptic. *Emetic* Bark used as an emetic. *Febrifuge* Bark used after long, intermittent fevers and as a wash for chills and fevers. *Gastrointestinal Aid* Bark used for indigestion and "any debility of the system." *Gynecological Aid* Compound infusion of bark of black oak taken for "female trouble." *Oral Aid* Bark chewed for mouth sores. *Respiratory Aid* Infusion of bark taken for asthma. *Throat Aid* Decoction of inner bark used for "lost voice." *Tonic* Bark used as a tonic. *Urinary Aid* Unspecified liquid preparation taken for "milky urine." (80:46) **Delaware** *Cold Remedy* Infusion of inner bark used as a gargle for colds. *Throat Aid* Infusion of inner bark used as a gargle for hoarseness. (176:30) **Delaware, Oklahoma** *Cold Remedy* Infusion of inner bark taken and used as a gargle for colds. *Throat Aid* Decoction of inner bark taken and used as a gargle for hoarseness. (175:25, 78) **Menominee** *Eye Medicine* Decoction of crushed bark used as a wash for sore eyes. (151:36) **Meskwaki** *Pulmonary Aid* Compound containing inner bark used for lung troubles. (152:222)
- *Food*—**Lakota** *Staple* Acorns used to make flour. (139:49) **Ojibwa** *Unspecified* Acorns, with tannic acid extracted, equally as good as other acorns. (153:402)
- *Fiber*—**Cherokee** *Basketry* Used to make baskets. *Building Material* Wood used for lumber, railroad ties, wagon spokes and rims. *Furniture* Wood used to make furniture. Used to make woven chair bottoms. (80:46)
- *Dye*—**Ojibwa** *Mordant* Bark used for a reddish yellow dye and to set its own color. *Red-Yellow* Bark used for a reddish yellow dye and to set its own color. (153:425)
- *Other*—**Cherokee** *Cooking Tools* Wood used to make corn beaters and mortars. Leaves used to wrap dough for bread making. *Fuel* Wood used for firewood. (80:46)

Quercus virginiana P. Mill., Live Oak

- *Drug*—**Houma** *Antidiarrheal* Decoction of bark taken for dysentery. (158:56) **Mahuna** *Unspecified* Bark used for medicine. (140:55) **Seminole** *Analgesic* Decoction of wood bits or bark applied externally as an analgesic. (169:167) *Antirheumatic (External)* Decoction of wood or bark used as a bath for aches and pains. (169:286) *Dermatological Aid* Decoction of bark used for ballgame sickness: sores, back or limb pains, and hemorrhoids. Decoction of wood or bark used as a bath for sores and cuts. (169:269, 286) *Hemorrhoid Remedy* Decoction of bark used for ballgame sickness: sores, back or limb pains, and hemorrhoids. (169:269) *Love Medicine* Decoction of wood ashes placed on the tongue to cleanse the body and strengthen the marriage. (169:250) *Orthopedic Aid* Decoction of bark used for ballgame sickness: sores, back or limb pains, and hemorrhoids. (169:269)
- *Food*—**Mahuna** *Dessert* Acorns ground into a fine meal, sun dried, made into porridge, cooked, and eaten as a dessert. *Unspecified* Acorns ground into a fine meal, sun dried, made into porridge, and eaten with deer meat. (140:55) **Seminole** *Fodder* Acorns used as hog food. (169:493)
- *Fiber*—**Mahuna** *Mats, Rugs & Bedding* Leaves used to make mattress bedding. (140:55) **Seminole** *Caulking Material* Plant used to make mortar. (169:493)

- *Dye*—**Houma** *Red* Roots and bark boiled to make a red basket dye. (158:56) **Mahuna** *Black* Bark blended with other oak barks and roots and used to make a black dye for buckskins. *Brown* Bark blended with other oak barks and roots and used to make a light or dark brown dye for buckskin. *Gray* Bark blended with other oak barks and roots and used to make a gray dye for buckskins. *Red* Bark blended with other oak barks and roots and used to make a red dye for buckskins. *White* Bark blended with other oak barks and roots and used to make a white dye for buckskins. *Yellow* Bark blended with other oak barks and roots and used to make a yellow dye for buckskins. (140:55)
- *Other*—**Choctaw** *Paint* Bark, red oak, and post oak boiled and used for paint. (as *Q. virens* 25:14) **Seminole** *Cleaning Agent* Plant used to make lye. *Fuel* Wood used to burn out mortar hollow. *Hide Preparation* Plant used for tanning. *Tools* Plant used to make pestles and cane mills. (169:493)

Quercus wislizeni A. DC., Interior Live Oak
- *Drug*—**Miwok** *Cough Medicine* Decoction of bark taken as a cough medicine. *Dermatological Aid* and *Pediatric Aid* Pulverized, outer bark dusted on running sores and particularly used for babies with sore umbilicus. (12:172)
- *Food*—**Luiseño** *Porridge* Acorns leached, ground into a meal, cooked in an earthen vessel, and eaten. (155:194) *Staple* Stored acorns pounded in a mortar and pestle to make a flour.[64] *Substitution Food* Acorns used only when more preferred species could not be obtained. (155:193) *Winter Use Food* Acorns formerly stored in acorn granaries. (155:194) **Miwok** *Bread & Cake* Acorns ground into a meal and used to make bread and biscuits. *Porridge* Acorns considered a staple food and used to make mush. *Soup* Acorns ground into a meal and used to make soup. *Winter Use Food* Whole acorns stored for winter use. (12:142) **Neeshenam** *Unspecified* Acorns occasionally used for food. (129:374) **Tubatulabal** *Unspecified* Acorns used extensively for food. (193:15)

Quercus wislizeni var. *frutescens* Engelm., Interior Live Oak
- *Drug*—**Kawaiisu** *Antirheumatic (Internal)* Decoction of inner bark taken for arthritis. *Burn Dressing* Ground plant applied to burns. (206:56)
- *Food*—**Diegueño** *Porridge* Acorns shelled, pounded, leached, and cooked into a mush or gruel. (84:33) **Kawaiisu** *Bread & Cake* Acorns made into a fine meal, cooked into a mush, and allowed to stand and harden into a "cake." *Staple* Acorns dried, pounded, sifted into a fine meal, and leached. *Winter Use Food* Acorns stored for future use.[65] (206:56)
- *Fiber*—**Kawaiisu** *Basketry* Branches used to make rims for twined work baskets. *Building Material* Logs used in house construction. (206:56)
- *Other*—**Kawaiisu** *Fasteners* Acorn meal used to mend cracks in clay pots. *Toys & Games* Acorn cupule used to make a top for children. (206:56)

Quincula, Solanaceae
Quincula lobata (Torr.) Raf., Chinese Lantern
- *Drug*—**Kiowa** *Misc. Disease Remedy* Decoction of roots taken or poultice of pounded roots applied for grippe. (as *Physalis lobata* 192:50)
- *Food*—**Kiowa** *Preserves* Berries gathered to make jelly. Berries gathered to make jelly. (192:50)
- *Other*—**Kiowa** *Toys & Games* Bladdery envelope blown up by children and busted on the forehead. (192:50)

Ramalina, Usneaceae, lichen
Ramalina menziesii, Spanish Moss
- *Fiber*—**Pomo, Kashaya** *Clothing* Used for baby diapers and other sanitary purposes. (72:123)

Ranunculaceae, see *Aconitum, Actaea, Anemone, Aquilegia, Caltha, Cimicifuga, Clematis, Consolida, Coptis, Delphinium, Hepatica, Hydrastis, Myosurus, Pulsatilla, Ranunculus, Thalictrum, Trautvetteria, Trollius, Xanthorhiza*

Ranunculus, Ranunculaceae
Ranunculus abortivus L., Littleleaf Buttercup
- *Drug*—**Cherokee** *Dermatological Aid* Used as poultice for abscesses. *Oral Aid* Infusion used for "thrash." *Sedative* Juice used as sedative. *Throat Aid* Infusion gargled for sore throat. (80:31) **Iroquois** *Anticonvulsive* Compound decoction of roots taken by men and women for epilepsy. (87:325) *Antidote* Decoction of smashed root taken to counteract poison. (87:320) *Blood Medicine* Compound infusion of roots taken for blood disease which caused fainting. (87:325) *Emetic* Decoction of root taken to vomit for stomach trouble and to counteract poison. (87:320) Compound decoction of plants taken to vomit after taking epilepsy medicine. *Eye Medicine* Decoction of roots and leaves used as wash for sore eyes from catching cold. (87:325) *Gastrointestinal Aid* Decoction of smashed root taken for stomach trouble. *Misc. Disease Remedy* Decoction of roots taken to dry up smallpox. (87:320) *Orthopedic Aid* Compound decoction of roots taken for stiff muscles. *Psychological Aid* Compound decoction of plants taken for "loss of senses during menses." (87:325) *Snakebite Remedy* Decoction of smashed root used as a wash for snakebites. (87:320) *Toothache Remedy* Plant used for sore teeth. (87:325) **Meskwaki** *Hemostat* Root used as a styptic for nosebleeds. (152:239)
- *Food*—**Cherokee** *Vegetable* Leaves cooked and eaten as greens. (80:31)

Ranunculus acris L., Tall Buttercup
- *Drug*—**Abnaki** *Analgesic* Used for headaches. (144:155) Flowers and leaves smashed and sniffed for headaches. (144:166) **Bella Coola** *Dermatological Aid* Poultice of pounded roots applied to boils. (as *R. arcis* 150:57) **Cherokee** *Dermatological Aid* Used as poultice for abscesses. *Oral Aid* Infusion used for "thrash." *Sedative* Juice used as sedative. *Throat Aid* Infusion gargled for sore throat. (80:31) **Iroquois** *Analgesic* Poultice of smashed plant applied to chest for pains. *Antidiarrheal* Infusion of roots taken for diarrhea. *Antihemorrhagic* Poultice of smashed plant applied to chest for colds. (87:320) *Blood Medicine* Poultice of plant fragments with another plant applied to the skin for excess water in the blood. (141:42) **Micmac** *Analgesic* Leaves used for headaches. (40:60) **Montagnais** *Analgesic* Crushed leaves inhaled for headaches. (156:315)

- *Food*—Cherokee *Vegetable* Leaves cooked and eaten as greens. (80:31)

Ranunculus aquatilis L., Whitewater Crowfoot
- *Food*—Gosiute *Unspecified* Entire plant boiled and eaten. (39:379)

Ranunculus bulbosus L., St. Anthony's Turnip
- *Drug*—Iroquois *Toothache Remedy* Root placed in cavity to break up the tooth for a toothache. *Venereal Aid* Decoction of plants taken for venereal disease. (87:320)

Ranunculus californicus Benth., California Buttercup
- *Food*—Miwok *Unspecified* Dried, stored, parched, pulverized seeds used for food. (12:155) **Neeshenam** *Bread & Cake* Seeds parched, ground into flour and used to make bread. *Porridge* Seeds parched, ground into flour, and used to make mush. *Staple* Seeds parched, ground into flour, and used for food. (129:377)
- *Other*—Pomo, Kashaya *Toys & Games* Flower put under the chin by a child, if yellow was reflected, the child would like butter. (72:30)

Ranunculus californicus var. *cuneatus* Greene, California Buttercup
- *Other*—Karok *Musical Instrument* Stems split and sucked on to make sounds. (6:49)

Ranunculus cymbalaria Pursh, Alkali Buttercup
- *Drug*—Kawaiisu *Dermatological Aid* Mashed leaves and flowers applied as salve to sores and cuts. (206:58) **Navajo** *Venereal Aid* Plant used for syphilis. (55:96) **Navajo, Kayenta** *Panacea* Plant used as a life medicine. (205:22)
- *Food*—Iroquois *Unspecified* Mature leaves used for food. (142:87)

Ranunculus cymbalaria var. *saximontanus* Fern., Rocky Mountain Buttercup
- *Drug*—Navajo, Ramah *Ceremonial Medicine* and *Emetic* Plant used as an emetic in various ceremonies. (191:27)

Ranunculus flabellaris Raf., Yellow Water Buttercup
- *Drug*—Meskwaki *Cold Remedy* Flower stigmas used as snuff to induce sneezing for catarrh and head colds. *Respiratory Aid* Compound containing leaves used as a snuff for catarrh and head cold. (as *R. delphinifolius* 152:239, 240)

Ranunculus glaberrimus Hook., Sagebrush Buttercup
- *Drug*—Okanagan-Colville *Analgesic* Poultice of mashed and dampened whole plants applied to pains of any kind. *Antirheumatic (External)* Poultice of mashed and dampened whole plants applied to sore joints. *Poison* Dried or mashed, fresh whole plant placed on a piece of meat as poisoned bait for coyotes. (188:119) **Thompson** *Dermatological Aid* Poultice of mashed flowers used for warts. (187:249) *Poison* Flowers or whole plant rubbed on arrow points as a poison. (164:512) Plant considered a skin irritant. (187:249)

Ranunculus hispidus var. *nitidus* (Chapman) T. Duncan, Bristly Buttercup
- *Drug*—Iroquois *Toothache Remedy* Root placed in cavity to break up the tooth for a toothache. (as *R. septentrionalis* 87:320)

Ranunculus inamoenus Greene, Graceful Buttercup
- *Drug*—Navajo, Ramah *Hunting Medicine* Cold infusion of plant taken and used as a lotion to protect hunters from animals. (191:27)
- *Food*—Acoma *Unspecified* Roots used for food. (32:48) **Keres, Western** *Unspecified* Roots considered good to eat. (171:65) **Laguna** *Unspecified* Roots used for food. (32:48)

Ranunculus lapponicus L., Lapland Buttercup
- *Drug*—Eskimo, Kuskokwagmiut *Dietary Aid* Plants soaked and eaten by starving persons before eating other food. (122:23)
- *Food*—Eskimo, Inuktitut *Soup* Leaves and stems stewed with duck and fresh fish. (202:183)

Ranunculus occidentalis Nutt., Western Buttercup
- *Drug*—Aleut *Poison* Flower juice slipped into food to cause a person "to waste away to nothing." (8:428)
- *Other*—Shasta *Season Indicator* Plant blooms indicated the coming of the summer salmon. (93:310)

Ranunculus occidentalis var. *eisenii* (Kellogg) Gray, Western Buttercup
- *Food*—Mendocino Indian *Staple* Smooth, flat, and orbicular seeds used alone or mixed with other seeds to make pinole. (as *R. eisenii* 41:347) **Pomo** *Staple* Seeds used to make pinoles. (11:87)

Ranunculus occidentalis var. *rattanii* Gray, Western Buttercup
- *Food*—Pomo *Staple* Seeds used to make pinoles. (11:87)

Ranunculus pallasii Schlecht., Pallas's Buttercup
- *Drug*—Eskimo, Inupiat *Poison* Young shoots poisonous, if not boiled. (98:143)
- *Food*—Alaska Native *Unspecified* Young, tender shoots cooked and eaten. (85:53) **Eskimo, Alaska** *Unspecified* Shoots and stems boiled until tender and eaten with seal oil. (1:35) Rootstocks used as food, but became bitter after leaves developed. (4:715)

Ranunculus pensylvanicus L. f., Pennsylvania Buttercup
- *Drug*—Ojibwa *Hunting Medicine* Seeds used as a hunting medicine. (153:383) Seeds smoked in hunting medicine to lure buck deer near enough for a shot with bow and arrow. (153:431) **Potawatomi** *Dermatological Aid* Plant used as an astringent medicine for unspecified diseases. (154:75)
- *Dye*—Ojibwa *Red* Entire plant boiled to yield a red coloring dye and bur oak added to set the color. (153:426) **Potawatomi** *Yellow* Entire plant boiled with rushes or flags to dye them yellow; used to make mats or baskets. (154:123)

Ranunculus recurvatus Poir., Blisterwort
- *Drug*—Cherokee *Dermatological Aid* Used as poultice for abscesses. *Oral Aid* Infusion used for "thrash." *Sedative* Juice used as sedative. *Throat Aid* Infusion gargled for sore throat. (80:31) **Iroquois** *Laxative* Compound decoction of roots taken to loosen bowels and for venereal disease. *Toothache Remedy* Decoction of roots taken to "kill the worms" in sore and hollow teeth. *Venereal Aid* Decoction of roots taken for venereal disease. (87:320)
- *Food*—Cherokee *Vegetable* Leaves cooked and eaten as greens. (80:31)
- *Dye*—Menominee *Red* Boiled root used for red coloring. (151:79)

Ranunculus repens L., Creeping Buttercup
- *Drug*—Hesquiat *Analgesic* and *Antirheumatic (External)* Poultice of

chewed leaves used for muscular aches and rheumatic pains. *Dermatological Aid* Poultice of chewed leaves used for sores. *Gynecological Aid* Three or four leaves eaten to help heal the insides after childbirth. *Other* Chewed leaves swallowed for general sickness. (185:71) **Thompson** *Poison* Plant considered a skin irritant. (187:249)
- *Food*—**Hesquiat** *Forage* Eaten by cows and deer. (185:71)

Ranunculus reptans L., Buttercup
- *Food*—**Makah** *Unspecified* Roots cooked on hot rocks, dipped in whale or seal oil, and eaten with dried salmon eggs. **Quileute** *Unspecified* Roots cooked on hot rocks, dipped in whale or seal oil, and eaten with dried salmon eggs. (79:29)

Ranunculus sceleratus L., Celeryleaf Buttercup
- *Drug*—**Thompson** *Poison* Flowers or whole plant rubbed on arrow points as a poison. (164:512)

Ranunculus sceleratus* var. *multifidus Nutt., Blister Buttercup
- *Drug*—**Keres, Western** *Poison* Plant considered poisonous. (as *R. eremogenes* 171:65)

***Ranunculus* sp.**, Buttercup
- *Drug*—**Costanoan** *Dermatological Aid* Decoction of plant used as a wash for wounds. (21:8) **Iroquois** *Veterinary Aid* Compound decoction given to cows when bearing a calf and womb comes out. (87:324) **Thompson** *Poison* Flowers of several species used as a poison on arrowheads. (164:512) *Tuberculosis Remedy* Poultice of mashed root mixed with pitch applied to the chest for tuberculosis. (187:249)
- *Food*—**Cherokee** *Unspecified* Leaves boiled and used for food. (204:253) **Skokomish** *Winter Use Food* Roots eaten as winter food. (79:30)

Ranunculus uncinatus D. Don ex G. Don, Hooked Buttercup
- *Drug*—**Thompson** *Antirheumatic (External)* Decoction of plant used as a wash or steam for stiff, sore muscles, and rheumatism. (as *R. douglasii* 164:473) *Disinfectant* Decoction of whole plant used as a body wash for purification in sweat house. (as *R. douglasii* 164:505) *Herbal Steam* Decoction of plant used as a sweat bath wash for stiff, sore muscles and bones. Decoction of plant used as a wash or steam for stiff, sore muscles and rheumatism. *Orthopedic Aid* Decoction of plant used as a sweat bath wash for stiff, sore muscles and bones. (as *R. douglasii* 164:473) *Poison* Plant considered a skin irritant. (187:249)

***Raphanus*,** Brassicaceae

Raphanus sativus L., Wild Radish
- *Food*—**Costanoan** *Unspecified* Raw stems used for food. (as *R. sativum* 21:252)

***Ratibida*,** Asteraceae

Ratibida columnifera (Nutt.) Woot. & Standl., Upright Prairie Coneflower
- *Drug*—**Cheyenne** *Analgesic* Decoction of leaves and stems used as wash for pain. *Dermatological Aid* Decoction of leaves and stems used for poison ivy rash. *Snakebite Remedy* Decoction of leaves and stems used as wash to draw out poison of a rattlesnake's bite. (as *R. columnaris* 76:188) Leaves and stems boiled and solution used for rattlesnake bites. (83:23) **Dakota** *Analgesic* Flowers used for chest pains and other ailments. *Dermatological Aid* Flowers used for wounds. *Panacea* Flowers used for chest pains and other ailments. (as *R. columnaris* 69:368) **Keres, Western** *Gynecological Aid* Crushed leaves rubbed on mothers' breast to wean child. (171:65) **Lakota** *Analgesic* Infusion of plant tops taken for headaches. *Gastrointestinal Aid* Infusion of plant tops taken for stomachaches. *Veterinary Aid* Plant given to horses for urinary problems. (139:39) **Navajo, Ramah** *Febrifuge* Cold infusion used for fever. *Veterinary Aid* Cold infusion given to sheep that are "out of their minds." (as *R. columnaris* var. *pulcherrima* 191:52) **Zuni** *Emetic* Infusion of whole plant taken as an emetic. (as *R. columnaris* 166:59)
- *Food*—**Dakota** *Beverage* Leaves used to make a hot, tea-like beverage. (as *R. columnaris* 69:368) **Oglala** *Beverage* Leaves and cylindrical heads used to make a tea-like beverage. (as *R. columnaris* 70:131)
- *Other*—**Lakota** *Cooking Tools* Plant top used as a nipple. (139:39)

Ratibida tagetes (James) Barnh., Green Prairie Coneflower
- *Drug*—**Keres, Western** *Hunting Medicine* Roots carried while deer hunting to prevent craziness. Chewing the root prevented craziness which occurred if a wounded deer blew his breath into one's face. *Sedative* Infusion of plant used for epileptic fits. *Unspecified* Plant considered strong medicine. (171:65) **Navajo, Ramah** *Analgesic* Strong infusion of leaves used for stomachache or as a cathartic. *Cathartic* Strong infusion of leaves used as a cathartic. *Ceremonial Medicine* Plant used in ceremonial chant lotion. *Cough Medicine* Cold infusion of leaves taken for coughs. *Febrifuge* Cold infusion of leaves taken for fever. *Gastrointestinal Aid* Strong infusion of leaves taken for stomachache. *Panacea* Plant used as "life medicine." *Pediatric Aid* Decoction of root used for "birth injuries." *Venereal Aid* Plant used as a fumigant for sexual infection. (191:52)

***Reverchonia*,** Euphorbiaceae

Reverchonia arenaria Gray, Sand Reverchonia
- *Drug*—**Hopi** *Gynecological Aid* Plant used for postpartum hemorrhage. (200:36, 84) **Navajo, Kayenta** *Veterinary Aid* Plant used for livestock bloat. (205:31)

Rhamnaceae, see *Berchemia, Ceanothus, Condalia, Frangula, Rhamnus, Ziziphus*

***Rhamnus*,** Rhamnaceae

Rhamnus alnifolia L'Hér., Alderleaf Buckthorn
- *Drug*—**Iroquois** *Antidote* Infusion of plant taken and poultice applied to swelling caused by poison. *Blood Medicine* Infusion of bark given to children as a blood purifier. *Cathartic* Infusion of bark given to children as a physic. *Dermatological Aid* Infusion of plant taken and poultice applied to swelling caused by poison. *Orthopedic Aid* Infusion given and used as a wash for bad backs. *Pediatric Aid* Infusion given and used as a wash for peevish children. Infusion of bark given to children as a tonic, physic, and as blood purifier. *Sedative* Infusion given and used as a wash for peevish children. *Tonic* Infusion of bark given to children as a tonic. *Venereal Aid* Compound decoction of roots taken for gonorrhea. (87:381) **Meskwaki** *Laxative* Decoction of bark taken for constipation. (152:241) **Potawatomi** *Cathartic* Inner bark used as a physic. (154:75)

Rhamnus cathartica L., Common Buckthorn
- *Drug*—**Cherokee** *Cathartic* Bark and fruit used as a cathartic. *Der-

matological Aid Decoction of bark used for itch. *Eye Medicine* Used as wash for sore and inflamed eyes. (80:27)

Rhamnus crocea Nutt., Redberry Buckthorn
- *Food*—Cahuilla *Fruit* Berries used for food. (15:131)

Rhamnus crocea ssp. *ilicifolia* (Kellogg) C. B. Wolf, Hollyleaf Redberry
- *Drug*—Kawaiisu *Analgesic* Decoction of roots and bark taken for internal soreness. Plant smoke inhaled for headaches. *Antirheumatic (Internal)* Plant smoke inhaled for rheumatism. *Blood Medicine* Infusion of roots and bark taken as a blood medicine. Plant used by women for blood shortages. *Cold Remedy* Decoction of roots and bark taken for colds. *Cough Medicine* Decoction of roots and bark taken for coughs. *Dermatological Aid* Plant used for boils and carbuncles. *Diuretic* Decoction of roots used to increase urination. *Gastrointestinal Aid* Plant used for stomach disorders and the spleen. *Gynecological Aid* Plant used by women for blood shortages. *Kidney Aid* Plant used for the kidneys. *Laxative* Decoction of roots used to loosen the bowels and as a laxative. *Liver Aid* Plant used for the liver and spleen. *Stimulant* Infusion of roots and bark taken when "one feels tired and run down." *Venereal Aid* Decoction of roots used for gonorrhea. (as *R. ilicifolia* 206:58) **Yuki** *Unspecified* Inner bark used as a "good medicine." (as *R. ilicifolia* 41:369)
- *Other*—Wintoon *Lighting* Wood used for torches because it burned a long time. (as *R. crocea ilicifolia* 117:276)

Rhamnus sp., Buckthorn
- *Drug*—Costanoan (Olhonean) *Cathartic* Infusion of bark taken as a cathartic. (as *Rhamus* 117:374) **Iroquois** *Dermatological Aid* Poultice of ripe berry jelly applied to sores. (as *Rhamus* 117:374) *Emetic* Bark used as an emetic. (87:380)
- *Food*—Cahuilla *Preserves* Plant boiled into a jelly-like substance and eaten. (15:131)

Rheum, Polygonaceae
Rheum rhabarbarum L., Garden Rhubarb
- *Food*—Haisla & Hanaksiala *Unspecified* Petioles used for food. (43:259) **Kitasoo** *Pie & Pudding* Stalks used to make pie. *Preserves* Stalks used to make jam. (43:340)

Rheum rhaponticum L., False Rhubarb
- *Drug*—Cherokee *Antidiarrheal* Used for dysentery. *Cathartic* Used as a mild purgative. *Dermatological Aid* Used as an astringent. *Laxative* Infusion of plant taken for constipation. *Poison* Leaves considered poisonous. *Strengthener* Used for strengthening. (80:52)
- *Food*—Cherokee *Unspecified* Species used for food. (80:52)

Rhexia, Melastomataceae
Rhexia virginica L., Handsome Harry
- *Drug*—Micmac *Throat Aid* Leaves and stems used as a throat cleanser. (40:60) **Montagnais** *Throat Aid* "Brew" from leaves and stems used for cleaning the throat. (156:314)
- *Food*—Montagnais *Beverage* Leaves and stems used to make a sour drink. (156:314)

Rhizomnium, Mniaceae, moss

Rhizomnium glabrescens (Kindb.) Koponen
- *Drug*—Bella Coola *Antirheumatic (External)* Poultice of crushed "leaves" applied to infections and swellings. *Blood Medicine* Poultice of "leaves" used for blood blisters. *Breast Treatment* Poultice of "leaves" used for breast abscesses in women. *Dermatological Aid* Poultice of "leaves" used for boils. (184:196)

Rhizophoraceae, see *Rhizophora*

Rhizophora, Rhizophoraceae
Rhizophora mangle L., American Mangrove
- *Dye*—Seminole *Unspecified* Plant used as a buckskin dye. (169:468)

Rhododendron, Ericaceae
Rhododendron albiflorum Hook., Cascade Azalea
- *Drug*—Okanagon *Dermatological Aid* Poultice of powdered, burned wood and grease applied to swellings. *Gastrointestinal Aid* Decoction of bark taken as a stomach remedy. (125:40) **Skokomish** *Cold Remedy* Decoction of buds taken for colds. *Dermatological Aid* Poultice of chewed buds applied to cuts. *Gastrointestinal Aid* Chewed buds eaten for an ulcerated stomach. *Throat Aid* Decoction of buds taken for sore throats. (79:43) **Thompson** *Dermatological Aid* Poultice of powdered, burned wood and grease applied to swellings. (125:40) Wood ash mixed with grease and used as a salve for swellings. (164:460) *Gastrointestinal Aid* Decoction of bark taken as a stomach remedy. (125:40)
- *Food*—Okanagan-Colville *Beverage* Leaves used to make tea. (188:102) **Thompson** *Preservative* Branches used in bottoms of berry baskets and on top of the berries to keep them fresh. (187:216)
- *Other*—Thompson *Incense & Fragrance* Used as a scent. (164:502)

Rhododendron calendulaceum (Michx.) Torr., Flame Azalea
- *Drug*—Cherokee *Antirheumatic (External)* Peeled and boiled twig rubbed on rheumatism. *Gynecological Aid* Infusion taken by women. (80:24)
- *Food*—Cherokee *Unspecified* Fungus "apple" formed on the stem eaten to appease thirst. (80:24)
- *Other*—Cherokee *Decorations* Flowers used to decorate the home. (80:24)

Rhododendron macrophyllum D. Don ex G. Don, Pacific Rhododendron
- *Drug*—Karok *Ceremonial Medicine* Plant used in the luck-getting ceremony of the sweat house. (as *R. californicum* 148:387)
- *Other*—Pomo *Decorations* Flowers used for dance wreaths. (as *R. californicum* 66:14) **Pomo, Kashaya** *Ceremonial Items* Flowers used in dance wreathes at the Strawberry Festival. (72:98)

Rhododendron maximum L., Great Laurel
- *Drug*—Cherokee *Analgesic* Compound used as liniment for pains and poultice of leaves used for headache. (80:52) Infusion of leaves put on scratches made over the location of the pain. (177:49) *Antirheumatic (Internal)* Compound decoction of leaf taken for rheumatism. *Ceremonial Medicine* "Throw clumps of leaves into a fire and dance around it to bring cold weather." *Heart Medicine* Infusion of leaf taken for heart trouble. (80:52) *Other* Decoction of leaves rubbed into scratches made on legs as preliminary treatment. (177:49)

- *Other*—**Cherokee** *Cooking Tools* Wood used to make spoons. *Decorations* Wood used to carve. *Smoking Tools* Wood used to make pipes. (80:52)

Rhododendron occidentale (Torr. & Gray ex Torr.) Gray, Western Azalea
- *Other*—**Pomo** *Decorations* Flowers used for dance wreaths. (66:14) **Pomo, Kashaya** *Ceremonial Items* Flowers used in dance wreaths at the Strawberry Festival. (72:21)

Rhododendron occidentale (Torr. & Gray ex Torr.) Gray **var. *occidentale***, Western Azalea
- *Drug*—**Modesse** *Antidote* Used for poisoning. (as *Azalea occidentalis* 117:224)

Rhodymeniaceae, see *Rhodymenia*

Rhodymenia, Rhodymeniaceae, alga

Rhodymenia palmata (L.) Greville, Dulse
- *Food*—**Alaska Native** *Dried Food* Leaves air dried and stored for winter use. *Soup* Leaves air dried and added to soups and fish head stews. *Unspecified* Leaves eaten fresh or singed on a hot stove or griddle. (85:143)

Rhus, Anacardiaceae

Rhus aromatica Ait., Fragrant Sumac
- *Drug*—**Natchez** *Dermatological Aid* Poultice of root applied to boils. (172:667) **Ojibwa** *Ceremonial Medicine* Bark and berries used in medicine ceremonies. *Unspecified* Bark and berries used in medicinal purposes. (135:234) **Ojibwa, South** *Antidiarrheal* Compound decoction of root taken for diarrhea. (91:201)
- *Food*—**Midoo** *Fruit* Berries pounded and eaten. (117:312)
- *Other*—**Lakota** *Smoke Plant* Leaves mixed with tobacco and smoked. (139:32)

Rhus copallinum L., Flameleaf Sumac
- *Drug*—**Cherokee** *Antiemetic* Red berries eaten for vomiting. *Burn Dressing* Infusion poured over sunburn blisters. (80:57) *Dermatological Aid* Decoction of bark used as a wash for blisters. (177:36) *Gynecological Aid* Infusion of bark taken "to make human milk flow abundantly." *Urinary Aid* Red berries chewed for bed-wetting. (80:57) **Creek** *Antidiarrheal* Decoction of root taken for dysentery. (172:659) Decoction of roots taken for dysentery. (177:36) **Delaware** *Dermatological Aid* Poultice of roots applied to sores and skin eruptions. Infusion of leaves used to cleanse and purify skin eruptions. *Oral Aid* Berries used to make mouthwash. *Venereal Aid* Infusion of roots used for venereal disease. (176:32) **Delaware, Oklahoma** *Ceremonial Medicine* Leaves and root used in "ceremonial tobacco mixture." *Dermatological Aid* Poultice of roots or infusion of leaves used for sores and skin eruptions. *Oral Aid* Berries used to make mouthwash. *Venereal Aid* Infusion of root taken for venereal disease. (175:26, 78) **Koasati** *Orthopedic Aid* and *Pediatric Aid* Decoction of leaves used as a bath and given to babies to make them walk. (177:36) **Ojibwa** *Ceremonial Medicine* Bark and berries used in medicine ceremonies. *Unspecified* Bark and berries used for medicinal purposes. (135:234)
- *Food*—**Cherokee** *Fruit* Berries used for food. (80:57)
- *Dye*—**Cherokee** *Black* Berries used to make black dye. *Red* Berries used to make red dye. (80:57)
- *Other*—**Delaware, Oklahoma** *Ceremonial Items* Leaves and root used in ceremonial tobacco mixture. (175:78)

Rhus copallinum* var. *leucantha (Jacq.) DC., Winged Sumac
- *Drug*—**Seminole** *Dermatological Aid* Poultice of plant applied for ant sickness: boils and infections. (as *R. leucantha* Jacq. 169:304) *Urinary Aid* Decoction of root bark taken for urine retention. *Venereal Aid* Decoction of bark taken for gonorrhea. (as *R. leucantha* 169:274)

Rhus glabra L., Smooth Sumac
- *Drug*—**Cherokee** *Antiemetic* Red berries eaten for vomiting. *Burn Dressing* Infusion poured over sunburn blisters. (80:57) *Dermatological Aid* Decoction of bark used as a wash for blisters. (177:36) *Gynecological Aid* Infusion of bark taken "to make human milk flow abundantly." *Urinary Aid* Red berries chewed for bed-wetting. (80:57) **Chippewa** *Antidiarrheal* Decoction of "growth which sometimes appears on the tree" used for dysentery. (53:344) *Cold Remedy* Infusion of roots taken for colds. *Emetic* Infusion of roots taken as an emetic. (71:135) *Oral Aid* Compound decoction of blossoms used as mouthwash for teething children. (53:342) Blossoms chewed for sore mouth. (71:135) *Pediatric Aid* Compound decoction of flower used as a mouthwash for teething child. (53:342) *Respiratory Aid* Infusion of plants taken for asthma. (71:135) **Creek** *Antidiarrheal* Decoction of root taken for dysentery. (172:659) Decoction of roots taken for dysentery. (177:36) *Other* Leaves mixed with tobacco and smoked for "all cephalic and pectoral complaints." (172:659) **Flathead** *Cathartic* Fruits used as a purgative. *Tuberculosis Remedy* Infusion of green or dried branches taken for tuberculosis. (82:55) **Iroquois** *Alterative* Sprouts used as an alterative. (124:93) **Kiowa** *Other* Plant used to "purify" the body and mind. *Tuberculosis Remedy* Plant used for tuberculosis. (192:37) **Kutenai** *Throat Aid* Roots squeezed and juice swallowed for sore throat. (82:55) **Meskwaki** *Dermatological Aid* Root bark used as a rubefacient, to raise a blister on the patient. *Dietary Aid* Decoction of root taken as an appetizer by invalids. (152:200) **Micmac** *Ear Medicine* Parts of plant used for earaches. (40:60) **Nez Perce** *Dermatological Aid* Leaves moistened and used for skin rashes. (82:55) **Ojibwa** *Ceremonial Medicine* Bark and berries used in medicine ceremonies. (135:234) *Dermatological Aid* Inner bark of trunk or twig used in compounds as astringents. *Eye Medicine* Infusion of blossoms used as a wash for sore eyes. *Hemostat* Infusion of root bark used as a "hemostatic." (153:354) *Unspecified* Bark and berries used for medicinal purposes. (135:234) Poultice of leaves used for unspecified conditions. (153:354) **Okanagan-Colville** *Dermatological Aid* Decoction of branches with seed heads used for an itchy scalp condition. Milky latex used as a salve on sores. *Gynecological Aid* Decoction of seed heads taken by women during childbirth. *Heart Medicine* Infusion of bark and/or roots taken and applied externally to the chest for a "tight chest." *Other* Decoction of branches with seed heads used as bathing water for frost-bitten limbs. *Venereal Aid* Decoction of seed heads used as bathing water for gonorrhea. (188:59) **Okanagon** *Oral Aid* Root chewed for sore mouth or tongue. (125:41) **Omaha** *Analgesic* Decoction of root taken for painful urination and retention of urine. (70:99, 100) *Antidote* Poultice of plants applied for poisoning. (68:335) *Dermatological Aid* Infusion used as wash for sores and powdered plants applied to

wounds and open sores. (68:334) Poultice of leaves or fruits applied "in case of poisoning of the skin." *Diuretic* Decoction of root taken "in case of retention of urine." *Gynecological Aid* Decoction of root used as a postpartum styptic wash. *Hemostat* Decoction of fruits used as a postpartum styptic wash. *Urinary Aid* Decoction of root taken for painful urination and retention of urine. **Pawnee** *Antidiarrheal* Decoction of fruit used for "bloody flux." *Gynecological Aid* Decoction of fruit used for dysmenorrhea. (70:99, 100) **Sanpoil** *Dermatological Aid* Mashed leaves rubbed on sore lips. *Oral Aid* Leaves chewed and held in the mouth for sore gums. (131:219) **Sioux** *Antihemorrhagic* Decoction of fruits used by women for hemorrhaging after parturition. *Dermatological Aid* Poultice of bruised and wetted leaves or fruits used for poisoned skin. *Urinary Aid* Infusion of roots used for urine retention and painful urination. (82:55) **Thompson** *Gastrointestinal Aid* Decoction of shredded bark with another plant taken for ulcers. *Internal Medicine* Infusion of plant used after internal surgery, to make the wounds heal faster. (187:149) *Oral Aid* Root chewed for sore mouth or tongue. (125:41) Fresh root chewed for sore mouth or tongue. (164:466) *Poison* Decoction of plant considered poisonous if too strong or taken in large dose. (164:512) *Venereal Aid* Decoction of stems and roots taken for syphilis. (164:466)

- **Food**—**Apache, Chiricahua & Mescalero** *Special Food* Bark eaten by children as a delicacy. (as *R. cismontana* 33:44) **Cherokee** *Fruit* Berries used for food. (80:57) **Comanche** *Fruit* Fruits eaten by children. (29:524) **Gosiute** *Fruit* Berries used for food. (39:379) **Iroquois** *Beverage* Bobs boiled and used as a drink in winter. (124:96) *Unspecified* Sprouts eaten raw. (124:93) Fresh shoots peeled and eaten raw. (196:119) **Meskwaki** *Beverage* Berries and sugar used to make a cooling drink in the summertime and stored for winter use. (152:255) **Ojibwa** *Beverage* Fresh or dried berries sweetened with maple sugar and made into a hot or cool beverage like lemonade. (153:397) **Okanagan-Colville** *Beverage* Seed heads used to make tea. (188:59)
- **Fiber**—**Thompson** *Basketry* Leaves used as basket covers. (187:149)
- **Dye**—**Cherokee** *Black* Berries used to make black dye. *Red* Berries used to make red dye. (80:57) **Chippewa** *Red* Fruit used to make a dull, red dye. (71:135) *Yellow* Stalk pulp used to make a light yellow dye. (53:373) Inner bark, bloodroot, and wild plum inner bark used to make a yellow dye. (53:374) **Kiowa** *Orange-Yellow* Spring roots used as a yellow, orange dye. (192:37) **Meskwaki** *Yellow* Root used to dye rush mats and woven bark mats yellow. (152:271) **Ojibwa** *Orange* Inner bark and central pith of the stem mixed with bloodroot and used for the orange color. (153:424) **Omaha** *Yellow* Inner bark used to make a yellow dye. (68:325) Roots used to make a yellow dye. (70:99) **Plains Indian** *Black* Leaves, bark, and roots used to make a black dye. *Gray* Leaves, bark, and roots used to make a gray dye. *Yellow* Leaves, bark, and roots used to make a yellow-tan dye. (82:55) **Thompson** *Unspecified* Juice used as a stain. (as *R. glabra* 164:502) **Winnebago** *Yellow* Roots used to make a yellow dye. (70:99)
- **Other**—**Cheyenne** *Smoke Plant* Leaves mixed with tobacco and used for smoking. (76:180) **Chippewa** *Smoke Plant* Leaves dried and smoked. (71:135) **Comanche** *Smoke Plant* Leaves added to tobacco for smoking. (29:524) **Dakota** *Smoke Plant* Scarlet leaves gathered in the fall and dried for smoking. (69:367) Red leaves dried and used for smoking. (70:99, 100) **Gosiute** *Smoke Plant* Leaves formerly used to smoke. (39:379) **Kiowa** *Smoke Plant* Dried leaves smoked in a mixture of tobacco. Dried leaves smoked in a mixture of tobacco. (192:37) **Lakota** *Smoke Plant* Red, autumn leaves used to smoke. (139:33) **Okanagan-Colville** *Season Indicator* Leaves changing color used as an indication that the sockeye salmon were spawning. (188:59) **Omaha** *Smoke Plant* Dried, deveined, red leaves broken fine and used for smoking in the absence of kinnikinnick. (68:331) Red leaves dried and used for smoking. (70:99, 100) **Pawnee** *Smoke Plant* Red leaves dried and used for smoking. (70:99, 100) **Plains Indian** *Smoke Plant* Leaves used to make a tobacco mixture. (82:55) **Ponca** *Smoke Plant* Red leaves dried and used for smoking. (70:99, 100) **Tewa** *Smoke Plant* Leaves smoked as tobacco. (as *R. cismontana* 138:47) **Winnebago** *Smoke Plant* Red leaves dried and used for smoking. (70:99, 100)

Rhus hirta (L.) Sudworth, Staghorn Sumac

- **Drug**—**Algonquin, Quebec** *Antirheumatic (Internal)* Infusion of plant with chokecherry, oak, yellow birch, and dogwood used for rheumatism. *Dietary Aid* Infusion of fruits used as tonic to improve the appetite. *Unspecified* Root used as a medicine. (as *R. typhina* 18:192) **Cherokee** *Antiemetic* Red berries eaten for vomiting. *Burn Dressing* Infusion poured over sunburn blisters. *Gynecological Aid* Infusion of bark taken "to make human milk flow abundantly." *Urinary Aid* Red berries chewed for bed-wetting. (as *R. typhina* 80:57) **Chippewa** *Analgesic* Decoction of flowers taken for stomach pain. *Gastrointestinal Aid* Decoction of flowers taken for stomach pain. (53:344) **Delaware** *Venereal Aid* Roots combined with purple coneflower roots and used for venereal disease. (as *R. typhina* 176:33) **Delaware, Oklahoma** *Venereal Aid* Compound containing root used for venereal disease. (as *R. typhina* 175:28, 78) **Delaware, Ontario** *Antidiarrheal* Infusion of berries taken for diarrhea. (as *R. typhina* 175:69, 82) **Iroquois** *Dietary Aid* Wood pieces eaten by mothers to improve the milk. *Gynecological Aid* Infusion of bark, buds, and branches from another plant taken before giving birth. *Reproductive Aid* Infusion of bark and flowers taken to prevent the water from breaking too early during the pregnancy. (as *R. typhina* 141:51) **Malecite** *Blood Medicine* Infusion of roots or berries used as a blood purifier. (as *R. typhina* 116:253) *Cough Medicine* Used with blackberry roots, mountain holly, lily roots, and mountain raspberry roots for coughs. *Febrifuge* Used with blackberry roots, mountain holly, lily roots, and mountain raspberry roots for fevers. *Tuberculosis Remedy* Used with blackberry roots, mountain holly, lily roots, and mountain raspberry roots for consumption. (as *Thus typhina* 116:251) **Menominee** *Cough Medicine* Decoction of "red top" sweetened, strained, "boiled down," and used for coughs. (54:130) *Dermatological Aid* Inner bark considered astringent and used as a valuable pile remedy. *Gastrointestinal Aid* Infusion of root bark taken for "inward troubles." *Gynecological Aid* Hairy twigs of smaller shrubs used for various "female diseases." *Hemorrhoid Remedy* Astringent, inner bark of trunk considered a valuable pile remedy. *Pulmonary Aid* and *Tuberculosis Remedy* Compound containing berries used for consumption and pulmonary troubles. (as *R. typhina* 151:22) **Meskwaki** *Anthelmintic* Compound containing berries used for pinworms. (as *R. typhina* 152:201) **Micmac** *Dietary Aid* Berries and roots used for loss of appetite. *Throat Aid* Parts of plant used for sore throats. (as *R. typhina* 40:60) Used for sore throats. Used for sore throats. (as *R. typbira* 194:25) **Mohegan** *Throat Aid* Berries used to make a gargle for sore throat. (as *R. typhina* 176:75, 132) **Natchez** *Dermatological Aid* Poultice of roots applied to boils. (as *R. typhina* 177:37) **Ojibwa** *Hemostat* Root used for hemorrhages. (as *R. typhina* 153:354) *Oral Aid* Infusion of gall-infected leaves taken for mouth sores. *Throat Aid* Infusion of gall-infected leaves taken for sore throat. (as *R. typhina* 5:2244) **Potawatomi** *Anthelmintic* Compound containing berries used to expel worms. *Hemostat* Root bark used as a "hemostatic." *Misc. Disease Remedy* Infusion of leaves used as gargle for sore throat, tonsillitis, and erysipelas. *Throat Aid* Infusion of leaves gargled for sore throat, tonsillitis, and erysipelas. (as *R. typhina* 154:38) **Rappahannock** *Panacea* Decoction of stems, leaves, or berries used for complaint. (as *R. typhina* 161:30)

- *Food*—**Algonquin, Quebec** *Beverage* Berries steeped in water, sweetened with sugar, and drunk like lemonade. (as *R. typhina* 18:114) **Cherokee** *Fruit* Berries used for food. (as *R. typhina* 80:57) **Menominee** *Beverage* Infusion of dried berries used as a beverage very similar to lemonade. *Winter Use Food* Berries dried for winter use. (as *R. typhina* 151:62) **Ojibwa** *Beverage* Fresh or dried berries sweetened with maple sugar and made into a hot or cool beverage like lemonade. *Winter Use Food* Seed heads dried for winter use. (as *R. typhina* 153:397) **Potawatomi** *Sour* Berries eaten to satisfy a natural craving for something acid or tart. (as *R. typhina* 154:95)
- *Dye*—**Cherokee** *Black* Berries used to make black dye. *Red* Berries used to make red dye. (as *R. typhina* 80:57) **Menominee** *Yellow* Roots boiled for yellow dye. (as *R. typhina* 151:77) **Ojibwa** *Orange* Inner bark and central pith of the stem mixed with bloodroot and used for the orange color. (as *R. typhina* 153:424)
- *Other*—**Potawatomi** *Smoke Plant* Leaves mixed with tobacco to cause it to smoke pleasantly. (as *R. typhina* 154:116)

Rhus integrifolia (Nutt.) Benth. & Hook. f. ex Brewer & S. Wats., Lemonade Sumac
- *Food*—**Cahuilla** *Beverage* Berries soaked in water and used as a beverage. (15:131) **Diegueño** *Beverage* Leaf wad kept in the mouth to assuage thirst on long journeys by foot. (84:37) **Mahuna** *Fruit* Berries eaten mainly to quench the thirst. (140:70)

Rhus microphylla Engelm. ex Gray, Littleleaf Sumac
- *Food*—**Apache** *Fruit* Fruits eaten for food. (32:49) **Apache, Chiricahua & Mescalero** *Preserves* Dried fruits ground, pulp mixed with water and sugar and cooked to make jam. (33:46)

Rhus ovata S. Wats., Sugar Sumac
- *Drug*—**Cahuilla** *Cold Remedy* Infusion of leaves taken for colds. *Cough Medicine* Infusion of leaves taken for coughs. (15:131) **Coahuilla** *Analgesic* Infusion of leaves taken for chest pain. *Cough Medicine* Infusion of leaves taken for coughs. (13:78) **Diegueño** *Gynecological Aid* Infusion of leaves taken just before the birth for an easy delivery. (88:218)
- *Food*—**Cahuilla** *Dried Food* Berries dried. *Fruit* Berries eaten fresh. *Porridge* Berries ground into a flour for mush. *Sweetener* Fruit sap used as a sweetener. (15:131) **Yavapai** *Fruit* Mashed, raw berries used for food. (63:212)

***Rhus* sp.,** Sumac
- *Drug*—**Iroquois** *Abortifacient* Decoction of roots taken for irregular menses. *Analgesic* Warm poultice of bark applied to abdomen for pain when urinating. *Anthelmintic* and *Anticonvulsive* Infusion of root given to children with worms that cause convulsions. *Carminative* Compound decoction of roots taken for stomach gas. (87:370) *Cathartic* Infusion of bark given to newborn babies as a physic. (87:372) *Cold Remedy* Berries used for colds. (87:373) *Cough Medicine* Compound decoction of roots taken for coughs. (87:370) *Dermatological Aid* Decoction of inner bark used for itching, bruises, bumps, or black eyes. (87:371) Decoction of roots applied as poultice to red swellings and wounds. Plant used for cuts and blisters. *Dietary Aid* Taken as an appetizer by a sick person. (87:372) *Emetic* Infusion of seed cluster taken to vomit for consumption. (87:370) Decoction of berries taken as an emetic for sugar diabetes. (87:373) *Expectorant* Compound decoction of roots taken for phlegm. (87:370) *Eye Medicine* Compound poultice of bark applied to bruises or black eyes. (87:372) *Febrifuge* Compound poultice applied for fever. Decoction of bark taken for mild fevers. *Gastrointestinal Aid* Poultice of smashed bark applied for intestinal pains. (87:371) *Gynecological Aid* Compound decoction of plants taken for a fallen womb. Compound infusion of roots taken for evacuation of the placenta. (87:370) Compound poultice applied for lumps and swellings of the breast. *Hemorrhoid Remedy* Compound decoction of bark taken for piles. (87:371) *Laxative* Infusion of bark taken as a laxative. (87:370) *Liver Aid* Decoction of inner bark taken for gall caused by diphtheria. (87:371) *Misc. Disease Remedy* Infusion of bark taken to prevent smallpox. (87:370) Decoction of inner bark taken for gall caused by diphtheria. (87:371) Compound infusion of berries and poultice of bark used for mumps. Decoction of bark or infusion of cones taken for scarlet fever. (87:372) Decoction of berries taken as an emetic for sugar diabetes and measles. (87:373) *Oral Aid* Infusion of cones used as a wash for chancres and sore throats. (87:371) *Orthopedic Aid* Poultice of bark applied to dislocated elbows or shoulders. (87:372) *Pediatric Aid* Infusion of bark given to children with throat rashes. Infusion of root given to children with worms that cause convulsions. (87:370) Compound infusion of berries given to children with measles. (87:371) Infusion of bark given to newborn babies as a physic. (87:372) *Pulmonary Aid* Berries used for whooping cough. (87:373) *Throat Aid* Infusion of bark given to children with throat rashes. (87:370) Infusion of cones or berries used as a wash for chancres and sore throats. *Tuberculosis Remedy* Compound decoction or infusion of roots or berries taken for consumption. (87:371) *Urinary Aid* Warm poultice of bark applied to abdomen for pain when urinating. (87:370) Compound decoction of bark taken for urine retention. (87:371) *Venereal Aid* Compound poultice of bark applied to lumps that remain after having syphilis. (87:369) *Veterinary Aid* Infusion of berries given to horses with bellyaches. (87:370) **Winnebago** *Antidiarrheal* Compound decoction of leaves taken as an antidiarrheal. (130:265)
- *Dye*—**Rappahannock** *Unspecified* Stems, leaves, or berries used to make a dark dye. (161:30)

Rhus trilobata Nutt., Skunkbush Sumac
- *Drug*—**Blackfoot** *Misc. Disease Remedy* Dried berries ground and dusted onto smallpox pustules. (97:42) **Cheyenne** *Burn Dressing* Plant used to protect the hands when removing dog meat from a boiling pot. *Cold Remedy* Leaves used for head colds. *Diuretic* Decoction of leaves taken as a diuretic. *Hemostat* Plant used for bleeding.

474 Rhus trilobata (Anacardiaceae)

Reproductive Aid "Old man took this medicine and bore a child (an aphrodisiac?)." *Toothache Remedy* Fruit chewed for toothaches. *Veterinary Aid* Fruit used for horses with urinary troubles and to prevent tiredness. (83:14) **Comanche** *Cold Remedy* Bark chewed and juice swallowed for colds. (29:524) **Hopi** *Ceremonial Medicine* Twigs used for ceremonial purposes. (56:16) *Dermatological Aid* Roots used as deodorant. (42:356) Buds used on the body as a medicinal deodorant or perfume. (200:84) *Tuberculosis Remedy* Compound containing root used for "consumption." (200:35, 84) *Unspecified* Roots used medicinally for unspecified purpose. (42:356) **Jemez** *Oral Aid* Bark chewed for sore gums. (44:27) **Keres, Western** *Emetic* Infusion of leaves used as an emetic. *Gastrointestinal Aid* Infusion of leaves used as a stomach wash. *Gynecological Aid* Infusion of bark used as a douche after childbirth. *Oral Aid* Berries used for a mouthwash. (171:66) **Kiowa** *Gastrointestinal Aid* Berries eaten for stomach troubles. *Misc. Disease Remedy* Berries eaten for grippe. (192:39) **Mahuna** *Dietary Aid* and *Gastrointestinal Aid* Plant used as appetite restorative for inactive stomach that refused food. (140:63) **Montana Indian** *Misc. Disease Remedy* Powdered fruit applied as a lotion or dusted on the affected surface in cases of smallpox. (19:21) **Navajo, Kayenta** *Dermatological Aid* Plant used as a lotion for poison ivy dermatitis. *Gastrointestinal Aid* Plant used for bowel troubles. (205:31) **Navajo, Ramah** *Analgesic* Leaves chewed for stomachache. *Contraceptive* Decoction of leaves taken to induce impotency, as a means of contraception. *Dermatological Aid* Poultice of leaves used for itch and decoction of fruits used to prevent falling hair. *Gastrointestinal Aid* Leaves chewed for stomachache. *Gynecological Aid* Decoction of root bark used to facilitate delivery of placenta. (191:35, 36) **Paiute** *Dermatological Aid* Dried, powdered fruits used as an astringent for smallpox sores. (180:129) **Round Valley Indian** *Dermatological Aid* and *Misc. Disease Remedy* Dried, powdered berries used for smallpox sores. **Yokia** *Dermatological Aid* and *Misc. Disease Remedy* Dried, powdered berries used for smallpox sores. (41:365)

- **Food**—**Acoma** *Appetizer* Fruits eaten fresh as appetizers. *Spice* Fruits mixed with various foods as seasoning. **Apache** *Fruit* Fruits eaten fresh. *Staple* Fruits ground into meal. (32:48) **Apache, Chiricahua & Mescalero** *Dried Food* Fruits ground with mescal, dried, and stored. (33:37) *Preserves* Dried fruits ground, pulp mixed with water, and sugar and cooked to make jam. (33:46) Fruits formerly used to make jam. (33:49) **Apache, Western** *Beverage* Berries stirred in warm water to make a nonintoxicating drink. *Fruit* Berries ground or chewed raw for the juice. (26:190) **Apache, White Mountain** *Fruit* Berries used for food. (136:160) **Atsugewi** *Beverage* Berries pounded into flour, mixed with manzanita flour and water, and used as a beverage. *Dried Food* Berries washed, dried, and stored for later use. *Preserves* Berries mixed with sugar and made into jam. (61:139) **Cahuilla** *Beverage* Berries soaked in water and used as a beverage. *Fruit* Berries eaten fresh. *Soup* Berries ground into a flour and used to make soup. (15:131) **Gosiute** *Fruit* Berries used for food. (39:379) **Great Basin Indian** *Beverage* Berries used to make an acid drink. (121:48) **Havasupai** *Beverage* Berries crushed, soaked in water, ground, more water added, and used as a drink. *Dried Food* Berries sun dried and kept in sacks for future use. (197:229) **Hopi** *Beverage* Berries used to make "lemonade." (42:356) Berries pounded, soaked in water, and used to make a refreshing drink. (120:20) Berries made into lemonade. (200:84) *Fruit* Berries eaten by young people. (56:16) **Hualapai** *Beverage* Berries used to make a drink. *Fruit* Berries used for food. (195:15) **Isleta** *Sauce & Relish* Sour, acid flavored fruits eaten as an appetizer or relish. (100:41) **Jemez** *Fruit* Berries used for food. (44:27) **Keres, Western** *Beverage* Berries used to make a beverage. *Fruit* Raw berries eaten as an appetizer. *Spice* Berries used as a lemon flavored seasoning for food. (171:66) **Keresan** *Fruit* Berries used for food. (198:563) **Kiowa** *Beverage* Berries boiled into a "tea." *Fruit* Berries mixed with cornmeal and eaten. Berries mixed with cornmeal and eaten. (as *Schmaltzia bakeri* 192:39) **Laguna** *Appetizer* Fruits eaten fresh as appetizers. *Spice* Fruits mixed with various foods as seasoning. (32:48) **Luiseño** *Porridge* Berries ground into a meal and used for food. (155:195) **Mahuna** *Fruit* Berries eaten mainly to quench the thirst. (140:70) **Montana Indian** *Fruit* Berries used for food. (19:21) **Navajo** *Beverage* Berries used to make juice. (110:26) Berries ground, washed, mixed with water, and used as a beverage. *Bread & Cake* Berries used to make cakes. (165:222) *Dried Food* Berries dried for future use. (110:26) *Fruit* Fruits eaten fresh. (32:48) Fruits eaten as they come off the bush. Fruits ground with sugar in a little water and eaten. (55:60) Berries boiled with meat. (165:222) *Porridge* Fruits cooked into a gruel with cornmeal. (32:48) Fruits ground into a meal, cooked with cornmeal, and eaten as a gruel. (55:60) Berries ground, mixed with flour and sugar, and made into a mush. (165:222) *Staple* Fruits ground into a meal and eaten. (55:60) Berries ground into a flour. (110:26) **Navajo, Ramah** *Dried Food* Fruit dried for later use, ground, and soaked before using. *Fruit* Fruit eaten raw, with sugar, sometimes ground, and used with other foods, especially roasted corn. *Unspecified* Inner bark used for food. (191:35) **Shoshoni** *Beverage* Berries used to make a cooling drink. *Winter Use Food* Berries kept in large quantities for future use. (117:440) **Tewa** *Fruit* Fruits eaten whole or ground. (as *Schmaltzia bakeri* 138:49) **Wintoon** *Fruit* Berries used for food. (117:264) **Yokut** *Fruit* Sour berries gathered and used for food. (117:420)

- **Fiber**—**Apache, Western** *Basketry* Stalks split or peeled off the bark and used to make pitched water baskets. (26:190) **Apache, White Mountain** *Basketry* Used in basket weaving. (136:160) **Cahuilla** *Basketry* Thin, pliable stems used for the woof in baskets. (15:131) **Great Basin Indian** *Basketry* Branches used to make baskets. (121:48) **Havasupai** *Basketry* Stems used as an important basketry material. (197:229) **Hopi** *Basketry* Twigs used in basketry. (42:356) Twigs used for coarse basketry. (56:16) *Furniture* Twigs used to make cradles. (42:356) **Hualapai** *Basketry* Limbs used to make baskets. (195:15) **Jemez** *Basketry* Twigs and small branches used to make baskets. (44:27) **Keres, Western** *Basketry* Branches woven into rough baskets. (171:66) **Luiseño** *Basketry* Grass used as splints for wrapping the basket coils. Rushes used as splints for wrapping the basket coils, to give a brown color. (155:204) **Mahuna** *Basketry* Vine fibers used to make baskets. (140:63) **Mewuk** *Basketry* Rods used in the fine, coiled baskets. (117:328) **Navajo** *Basketry* Split stems used to make baskets. Used to make carrying baskets. *Sewing Material* Used to sew water bottles. (55:60) **Navajo, Kayenta** *Basketry* Used as basket material. (205:31) **Navajo, Ramah** *Basketry* Split stems used to make baskets, water bottles, and basket sacks. *Clothing* Small stems used to make sunshades or hats. (191:35) **Panamint** *Basketry* Formerly used as withes and splints for baskets. (105:78) **Pomo** *Basketry* Rods with

bark used for coarse baskets, fish baskets, and the coarsest kind of burden baskets. (117:296) **Tewa** *Basketry* Stems used to make baskets. (as *Schmaltzia bakeri* 138:49) **Wintoon** *Basketry* Long shoots used in making large storehouse baskets. (117:264) **Zuni** *Basketry* Stems, with the bark removed, used in making fine "Apache" and other baskets. The bark covered stems were used to form patterns in the weave. (166:81)
- *Dye*—**Dakota** *Mordant* Fruits used for the mordant effect. *Red* Ripe, red fruits boiled with another plant to make a red dye. (69:367) **Great Basin Indian** *Black* Twigs and pine gum used to make a black dye. *Red-Brown* Bark and leaves used to make a red-brown dye. Berries used to make a pink-tan dye. (121:48) **Hopi** *Mordant* Berries used as a mordant in dying wool and in the preparation of body paint. (42:356) **Hualapai** *Unspecified* Roots boiled and used to make a dye. (195:15) **Navajo** *Black* Leaves used to make black dye for baskets and leather. *Blue* Used to make a blue dye. *Mordant* Ashes used in setting dyes. (55:60) **Navajo, Ramah** *Black* Leaves boiled to dye basketry and wool black. (191:35)
- *Other*—**Cheyenne** *Smoke Plant* Dried leaves mixed with tobacco and used for smoking. (76:180) Leaves dried, mixed with tobacco, and smoked. (83:14) **Hopi** *Ceremonial Items* Plant used for ceremonial equipment and prayer sticks. (42:356) Twigs used for many ceremonial purposes. (56:16) Used for ceremonial equipment. Used to make prayer sticks. (200:84) *Fuel* Dry shrub used as one of the four prescribed fuels for the kivas. (56:16) *Incense & Fragrance* Roots used as perfume. (42:356) **Hualapai** *Insecticide* Leaves used on a person's body as an insect repellent. *Protection* Leaves used on a person's body as a snake repellent. (195:15) **Jemez** *Tools* Branches used to make hoe handles. (44:27) **Keres, Western** *Hunting & Fishing Item* Larger bushes sometimes used for bows. *Paint* Crushed berry juice used as a vehicle for paint. (171:66) **Keresan** *Smoke Plant* Dried leaves mixed with native tobacco and used for smoking. (198:563) **Kiowa** *Smoke Plant* Leaves mixed with tobacco and used for smoking. Leaves mixed with tobacco and used for smoking. (as *Schmaltzia bakeri* 192:39) **Luiseño** *Tools* Twigs made into a seed fan and used to beat the seeds off plants. (155:231) **Navajo** *Ceremonial Items* Twigs used to make a light frame for the bag carried by the Hunchback in the Night Chant. Branch, with eagle down attached, carried by the dancers on the last night of the Mountain Chant. Pollen used in some ceremonies. Wood tied with yucca and used to make circle prayer sticks. *Containers* Used to make water bottles. Used to make "bugaboos" to subdue insubordinate children. *Decorations* Twigs painted white and used to decorate masks for the Fringe Mouths in the Night Chant. *Hunting & Fishing Item* Small, sharpened stick driven into the reed shaft of an arrow and fastened with sinew. Wood used to make bows. *Sacred Items* Used to make sacred baskets to hold sacred meal for rites. (55:60) **Navajo, Ramah** *Ceremonial Items* Used to make small hoops on cactus prayer stick of Chiricahua Windway. *Containers* Split stems used to make baskets, water bottles, and basket sacks. *Hunting & Fishing Item* Large stems used to make bows. *Weapon* Six-foot stems made into spear shafts used for thrusting in warfare, not thrown or used in hunting. (191:35) **Santa Clara** *Hunting & Fishing Item* Wood used to make bows. (as *Schmaltzia bakeri* 138:49)

Rhus trilobata* var. *pilosissima Engelm., Pubescent Squawbush
- *Drug*—**Diegueño** *Eye Medicine* and *Pediatric Aid* Infusion of leaves used as a wash for babies' eyes. (84:37)
- *Food*—**Yavapai** *Beverage* Mashed berries mixed with water or mescal syrup and used as a beverage. *Unspecified* Seeds used for food. (as *R. emoryi* 65:257)
- *Fiber*—**Diegueño** *Basketry* Stems split into thin splints, dried, and used as wrapping material for baskets. (84:37) **Kawaiisu** *Basketry* Split stem strands used to make twined and coiled baskets. (206:59) **Southwest Indians** *Basketry* Stems used to make the warp and weft of baskets. (as *R. emoryi* 17:35)

Rhus trilobata Nutt. **var. *trilobata*,** Skunkbush Sumac
- *Food*—**Kiowa** *Beverage* Berries boiled into a "tea." *Fruit* Berries mixed with cornmeal and eaten. (as *Schmaltzia trilobata* 192:39) **Ute** *Fruit* Berries used for food. (as *R. aromatica* var. *trilobata* 38:36)
- *Other*—**Kiowa** *Smoke Plant* Leaves mixed with tobacco and used for smoking. (as *Schmaltzia trilobata* 192:39) **Modesse** *Tools* Wood made into the large plug used to keep the pierced ear lobe open on young girls. (as *R. aromatica trilobata* 117:223)

Rhytidiaceae, see *Rhytidiadelphus*

Rhytidiadelphus, Rhytidiaceae, moss
Rhytidiadelphus loreus (Hedw.) Warnst.
- *Fiber*—**Bella Coola** *Mats, Rugs & Bedding* Used for padding and bedding. (184:196)

Ribes, Grossulariaceae
Ribes amarum McClatchie, Bitter Gooseberry
- *Food*—**Mahuna** *Fruit* Berries eaten mainly to quench the thirst. (140:70)

Ribes americanum P. Mill., American Black Currant
- *Drug*—**Blackfoot** *Gynecological Aid* Decoction of roots taken by women for uterine troubles. *Kidney Aid* Decoction of roots taken for kidney ailments. (97:37) **Iroquois** *Antidote* Compound infusion of branches taken as remedy for poison. (87:345) *Antiemetic* Compound infusion of roots and bark taken for vomiting. (87:346) *Dermatological Aid* Poultice of bark used for swellings. (87:345) *Kidney Aid* Compound infusion of roots and bark taken for kidney trouble. *Orthopedic Aid* Compound infusion of roots and bark taken for back pains. *Other* Compound decoction of bark taken for fortune-telling or divination. Decoction of leaves taken by women hurting inside from lifting. (87:346) **Meskwaki** *Anthelmintic* Root bark used to expel intestinal worms. (as *R. floridum* 152:246) **Ojibwa** *Unspecified* Root and bark used for medicinal purposes. (as *R. floridum* 135:236) **Omaha** *Kidney Aid* Strong decoction of root taken for kidney trouble. **Winnebago** *Gynecological Aid* Root used by women for "uterine trouble." (70:84)
- *Food*—**Chippewa** *Dried Food* Dried fruit used for food. *Fruit* Fresh fruit used for food. (as *R. floridum* 71:131) **Iroquois** *Bread & Cake* Fruit mashed, made into small cakes, and dried for future use. *Dried Food* Raw or cooked fruit sun or fire dried and stored for future use. *Fruit* Dried fruit taken as a hunting food. *Sauce & Relish* Dried fruit cakes soaked in warm water and cooked as a sauce or mixed with corn bread. (as *R. floridum* 196:128) **Lakota** *Fruit* Fruits eaten fresh.

(106:35) Fruits eaten for food. (139:58) *Starvation Food* Berries dried and eaten during famines. (106:35) **Meskwaki** *Fruit* Currants used for food. (as *R. floridum* 152:264) **Montana Indian** *Fruit* Fruit highly esteemed as an article of diet. (as *R. floridum* 19:21) **Ojibwa** *Dried Food* Fruit dried for future use. (as *R. floridum* 135:236) Berries dried for winter use. (153:410) *Fruit* Fruit eaten fresh. (as *R. floridum* 135:236) Berries eaten fresh. In the winter, a favorite dish was wild currants cooked with sweet corn. *Preserves* Berries used to make jams and preserves. (153:410)

Ribes aureum Pursh, Golden Currant

- **Drug**—**Paiute** *Dermatological Aid* Dried, pulverized inner bark sprinkled on sores. *Orthopedic Aid* Decoction of inner bark taken for leg swellings. **Shoshoni** *Orthopedic Aid* Decoction of inner bark taken for leg swellings. (180:129) *Unspecified* Poultice of second bark applied medicinally. (118:43)
- **Food**—**Blackfoot** *Fruit* Berries used for food. (86:119) **Cheyenne** *Winter Use Food* Pounded, dried berries formed into cakes for winter use. (76:175) **Gosiute** *Dried Food* Berries dried and stored for winter use. *Fruit* Berries used for food. (39:379) **Klamath** *Fruit* Berries used for food. (45:97) **Montana Indian** *Fruit* Fruit highly esteemed as an article of diet. (19:21) **Okanagan-Colville** *Bread & Cake* Dried berries mixed with other berries and made into cakes. *Dried Food* Berries dried and stored for future use. *Fruit* Berries eaten fresh. (188:106) **Paiute** *Dried Food* Berries dried for later use. (111:78) *Fruit* sun dried, stored in buckskin bags, and hung up for winter use. (167:245) *Fruit* Ripe, crushed berries eaten with sugar. (111:78) *Fruit* eaten fresh. (167:245) **Paiute, Northern** *Dried Food* Berries eaten dried. *Fruit* Berries eaten fresh. *Porridge* Berries dried, ground, mixed with seed flour, and used to make mush. (59:50) **Ute** *Fruit* Berries used for food. (38:36) **Yavapai** *Fruit* Raw berries used for food. (65:258)

Ribes aureum var. villosum DC., Golden Currant

- **Drug**—**Kiowa** *Snakebite Remedy* Poultice of plant parts applied to snakebites. (as *R. odoratum* 192:29)
- **Food**—**Comanche** *Fruit* Fruits eaten for food. (as *R. odoratum* 29:524) **Kiowa** *Fruit* Fruit eaten raw. *Preserves* Fruit made into jelly. (as *Chrysobotrya odorata* 192:29) **Lakota** *Fruit* Fruits eaten fresh. (as *R. odoratum* 106:36) Fruits eaten for food. (as *R. odoratum* 139:58) *Starvation Food* Fruits dried and eaten during famines. (as *R. odoratum* 106:36)
- **Other**—**Lakota** *Hunting & Fishing Item* Stems used to make arrows. (as *R. odoratum* 139:58)

Ribes bracteosum Dougl. ex Hook., Stink Currant

- **Drug**—**Bella Coola** *Unspecified* Berries and "cane" used for medicine. (184:206) *Venereal Aid* Compound decoction of berry taken for gonorrhea. Compound decoction taken for gonorrhea. (150:57) **Haisla** *Dermatological Aid* Plant used for impetigo. *Reproductive Aid* Roots used as a birthing aid. **Hanaksiala** *Unspecified* Roots used for unspecified type of medicine. (43:253) **Nitinaht** *Laxative* Berries eaten in quantity as a laxative. (186:113) **Sanpoil** *Cold Remedy* and *Pediatric Aid* Infusion of stems given to children for colds. (131:218)
- **Food**—**Bella Coola** *Fruit* Berries used for food. (184:206) **Hanaksiala** *Winter Use Food* Fruit cooked and stored underground in barrels with elderberries and cooked western dock for winter use. (43:253) **Hesquiat** *Fruit* Berries eaten with oil and could cause stomachache,

if too many were eaten. *Preserves* Berries made excellent jam. (185:68) **Kitasoo** *Fruit* Fruit used for food. (43:338) **Kwakiutl, Southern** *Fruit* Fruits eaten raw with large quantities of oil. *Winter Use Food* Fruits boiled, mixed with powdered skunk cabbage leaves, dried, and eaten with oil in winter. (183:286) **Makah** *Fruit* Fruit used for food. (67:257) Berries eaten fresh. (79:32) **Nitinaht** *Dessert* Berries boiled, mixed in molasses, and eaten as dessert. (186:113) **Oweekeno** *Fruit* Berries mixed with salal berries, oolichan (candlefish) grease, and sugar, and eaten. (43:103) **Paiute** *Forage* Berries eaten only by bears. (111:78) **Salish, Coast** *Bread & Cake* Berries boiled, dried into rectangular cakes, and used as a winter food. (182:84)
- **Other**—**Quileute** *Containers* Large leaves used to line and cover hemlock bark containers. *Tools* Pithless stems used as tubes to inflate seal paunches made into oil containers. (79:32)

Ribes californicum Hook. & Arn., Hillside Gooseberry

- **Food**—**Kawaiisu** *Dried Food* Berries dried in the shade for about a week and stored. *Fruit* Berries eaten fresh. *Preserves* Fresh berries boiled into a jelly. (206:59) **Mendocino Indian** *Fruit* Fruits eaten directly from the bushes by children. (41:353) **Pomo, Kashaya** *Fruit* Singed berries eaten whole. (72:51)

Ribes cereum Dougl., Wax Currant

- **Drug**—**Hopi** *Gastrointestinal Aid* Used for stomach pains. (190:163) **Okanagan-Colville** *Eye Medicine* Infusion of inner bark used to wash sore eyes. (188:107) **Shoshoni** *Emetic* Fruit used as an emetic. (121:48) **Thompson** *Antidiarrheal* Berries eaten for diarrhea. *Pediatric Aid* and *Strengthener* Decoction of branches with many other branches used to wash babies to make them strong.[66] (187:226)
- **Food**—**Hopi** *Fruit* Berries used for food. (56:16) **Keres, Western** *Fruit* Berries used for food. (171:66) **Klamath** *Fruit* Berries used for food. (45:97) **Montana Indian** *Fruit* Fruit highly esteemed as an article of diet. (19:21) **Okanagan-Colville** *Forage* Berries eaten by grouse and pheasant. *Fruit* Berries eaten fresh. (188:107) **Okanagon** *Fruit* Insipid, bright orange-red fruits used for food. (125:38) *Staple* Berries used as a principal food. (178:239) **Paiute** *Fruit* Fruits eaten fresh. (104:100) **Sanpoil & Nespelem** *Fruit* Berries eaten raw. Only currants from the bushes growing along the Columbia River were eaten. Berries from bushes growing in the hills were not eaten because it was thought that they caused headaches, nosebleeds, and sore eyes. (131:102) **Thompson** *Fruit* Insipid, bright orange-red fruits used for food. (125:38) Insipid, rubbery berries used for food. (187:226)
- **Other**—**Havasupai** *Hunting & Fishing Item* Stems made into arrow shafts and used in hunting large game. *Weapon* Stems made into arrow shafts and used in war. (197:221) **Okanagan-Colville** *Season Indicator* First plant to sprout green leaves in spring. (188:107)

Ribes cereum var. pedicellare Brewer & S. Wats., Whisky Currant

- **Drug**—**Navajo, Kayenta** *Ceremonial Medicine* Plant used as an Evilway, Nightway, and Mountain-top-way emetic. *Dermatological Aid* Poultice of plant applied to sores. *Emetic* Plant used as an Evilway, Nightway, and Mountain-top-way emetic. *Pediatric Aid* Plant used to purify a child who has seen a forbidden sand painting. (as *R. inebrians* 205:26)
- **Food**—**Acoma** *Fruit* Fruits eaten fresh. *Preserves* Fruits preserved and eaten. (as *R. inebrians* 32:49) **Apache, White Mountain** *Fruit* Fruit eaten raw and cooked. (as *R. inebrians* 136:160) **Cheyenne**

Dried Food Pounded berries formed into cakes, dried, and stewed with buffalo hide chips. (as *R. inebrians* 76:175) **Hopi** *Fruit* Berries eaten with fresh piki bread. (as *R. inebrians* 120:18) **Isleta** *Fruit* Fruit eaten fresh. *Preserves* Fruit eaten preserved. (as *R. inebrians* 100:42) **Laguna** *Fruit* Fruits eaten fresh. *Preserves* Fruits preserved and eaten. (as *R. inebrians* 32:49) **Navajo** *Fruit* Fruits eaten for food. (as *R. inebrians* 55:52) **Navajo, Ramah** *Fruit* Berries eaten raw. (as *R. inebrians* 191:30) **Tewa** *Fruit* Fruits eaten for food. (as *R. inebrians* 138:48) **Zuni** *Fruit* Highly relished berries used for food. (as *R. inebrians* 166:70) *Unspecified* Leaves eaten with uncooked mutton fat or deer fat. (as *R. inebrians* 32:49) Fresh leaves eaten with uncooked mutton fat or with deer fat. (as *R. inebrians* 166:70)
- *Other*—**Havasupai** *Hunting & Fishing Item* Stems made into arrow shafts and used in hunting large game. *Weapon* Stems made into arrow shafts and used in war. (as *R. inebrians* 197:221) **Hopi** *Hunting & Fishing Item* Wood used for arrows. (as *R. inebrians* 200:78) **Navajo** *Hunting & Fishing Item* Wood used to make arrow shafts. *Tools* Wood used to make the distaff used in spinning. (as *R. inebrians* 55:52) **Navajo, Ramah** *Hunting & Fishing Item* Stems used to make arrow shafts. *Season Indicator* Green plant indicated time for plowing and leafy plant indicated time to plant maize. (as *R. inebrians* 191:30) **Tewa** *Hunting & Fishing Item* Wood used to make bows. (as *R. inebrians* 138:48)

Ribes cruentum Greene, Shinyleaf Currant
- *Food*—**Karok** *Fruit* Fresh fruits used for food. (6:50)

Ribes cruentum Greene **var. *cruentum***, Shinyleaf Currant
- *Food*—**Karok** Berries eaten raw. (as *R. roezlii* var. *cruentum* 148:384)

Ribes cynosbati L., Eastern Prickly Gooseberry
- *Drug*—**Meskwaki** *Gynecological Aid* Root used for uterine trouble caused by having too many children. (152:246) **Potawatomi** *Eye Medicine* Infusion of root used for sore eyes. *Gynecological Aid* Root bark used by the Prairie Potawatomi as a uterine remedy. (154:82)
- *Food*—**Algonquin, Quebec** *Fruit* Fruit used fresh. *Preserves* Fruit preserved. (18:87) **Cherokee** *Winter Use Food* Berries canned for future use. (126:54) **Chippewa** *Fruit* Berries used for food. (71:131) **Menominee** *Fruit* Berries used in favorite aboriginal Menomini dish. *Preserves* Berries preserved. *Winter Use Food* Berries stored for winter use. (151:71) **Meskwaki** *Dessert* Berries cooked with sugar as a dessert. (152:264) **Ojibwa** *Fruit* Berries relished when ripe. *Preserves* Berries made into preserves for winter use. (153:410) **Potawatomi** *Fruit* Berries used for food. *Preserves* Berries made into jams and jellies. (154:109)

Ribes divaricatum Dougl., Spreading Gooseberry
- *Drug*—**Bella Coola** *Cold Remedy* Inner bark chewed and juice swallowed for colds. (184:206) *Eye Medicine* Simple or compound decoction of bark or root used as an eyewash for soreness. (150:58) *Throat Aid* Inner bark chewed and juice swallowed for sore throats. (184:206) **Cowlitz** *Dermatological Aid* Burned stems rubbed on neck sores. **Klallam** *Eye Medicine* Infusion of bark used as an eyewash. **Makah** *Eye Medicine* Infusion of bark used as an eyewash. (79:32) **Saanich** *Psychological Aid* Roots used with wild cherry roots to wash newborn children for intelligence and obedience. **Salish, Coast** *Other* Infusion of roots rubbed on the skin for a charley horse. (182:84) **Skagit, Upper** *Throat Aid* Roots boiled for sore throats. (179:38) **Swinomish** *Throat Aid* Infusion of roots taken for sore throats. *Tuberculosis Remedy* Infusion of roots taken for tuberculosis. *Venereal Aid* Infusion of roots taken for venereal disease. (79:32)
- *Food*—**Bella Coola** *Fruit* Ripe, black berries used for food. *Sauce & Relish* Green berries boiled into a thick sauce and used for food. (184:206) **Clallam** *Fruit* Berries eaten fresh. (57:200) **Cowlitz** *Dried Food* Green berries dried and stored for winter use. *Fruit* Green berries eaten fresh. (79:32) **Gosiute** *Dried Food* Berries dried and stored for winter use. *Fruit* Berries used for food. (39:379) **Haisla & Hanaksiala** *Fruit* Fruit used for food. (43:254) **Hesquiat** *Fruit* Raw, fresh berries eaten with oil. (185:69) **Karok** *Fruit* Berries eaten raw. (148:384) **Makah** *Fruit* Fruit eaten fresh. (67:258) **Mendocino Indian** *Fruit* Black, juicy berries used for food. (41:353) **Nitinaht** *Fruit* Fruit used for food. (67:258) Berries formerly eaten fresh. (186:114) **Oweekeno** *Fruit* Berries used for food. (43:104) **Quinault** *Preservative* Berries mixed with elderberries and buried with them for preservation. (79:32) **Salish, Coast** *Bread & Cake* Berries boiled, dried into rectangular cakes, and used as a winter food. (182:84) **Skagit, Upper** *Fruit* Berries eaten fresh and never stored. (179:38) **Swinomish** *Fruit* Berries eaten fresh. (79:32) **Thompson** *Beverage* Berries made into juice. *Fruit* Berries eaten fresh or cooked. *Pie & Pudding* Berries made into pies. (187:227)
- *Fiber*—**Cowichan** *Cordage* Roots boiled with cedar and wild rose roots, pounded, and woven into rope. **Saanich** *Cordage* Roots boiled with cedar and wild rose roots, pounded, and woven into rope. (182:84)
- *Other*—**Bella Coola** *Smoking Tools* "Canes" hollowed out and used for pipestems. (184:206) **Cowichan** *Hunting & Fishing Item* Roots used to make reef nets. **Saanich** *Hunting & Fishing Item* Roots used to make reef nets. **Salish, Coast** *Tools* Stiff, sharp thorns used as probes for boils, for removing splinters, and for tattooing. (182:84)

Ribes glandulosum Grauer, Skunk Currant
- *Drug*—**Chippewa** *Analgesic* Compound decoction of root taken for back pain. *Gynecological Aid* Compound decoction of root taken for "female weakness." (53:356) **Cree, Woodlands** *Gynecological Aid* Decoction of stem used alone or with wild red raspberry to prevent blood clotting after birth. (109:54)
- *Food*—**Algonquin, Quebec** *Fruit* Fruit used for food. (18:88) **Cree, Woodlands** *Beverage* Stem used to make a bitter tea. *Fruit* Fresh berries eaten in considerable quantity. (109:54)

Ribes hirtellum Michx., Hairystem Gooseberry
- *Food*—**Klamath** *Dried Food* Dried berries used for food. *Fruit* Fresh berries used for food. (as *R. oxyacanthoides saxosum* 45:97)

Ribes hudsonianum Richards., Northern Black Currant
- *Drug*—**Cree, Woodlands** *Gynecological Aid* Decoction of stem sections alone or with wild gooseberry stems used for sickness after childbirth. (109:55) **Ojibwa** *Unspecified* Root and bark used for medicinal purposes. (135:236) **Tanana, Upper** *Cold Remedy* Raw currants eaten for colds. *Panacea* Decoction of leaves and berries taken for sickness in general. (102:11) **Thompson** *Cold Remedy* Decoction of stems and leaves taken for colds. (164:471) Infusion of branches possibly taken for colds. (187:227) *Gastrointestinal Aid* Decoction of stems and leaves taken for stomach troubles. (164:471) *Panacea*

Roots used for any kind of sickness. (187:227) *Pediatric Aid* and *Sedative* Sprigs placed in baby's carrier to quiet child. (164:509) *Throat Aid* Decoction of stems and leaves taken for sore throats. (164:471) *Tuberculosis Remedy* Roots used for tuberculosis. (187:227)

- **Food**—**Cree, Woodlands** *Preserves* Berries used to make jam and eaten with fish, meat, or bannock. (109:55) **Ojibwa** *Dried Food* Fruit dried for future use. *Fruit* Fruit eaten fresh. (135:236) **Salish, Coast** *Bread & Cake* Berries boiled, dried into rectangular cakes, and used as a winter food. (182:84) **Shuswap** *Fruit* Berries used for food. (123:63) **Tanana, Upper** *Fruit* Currants mixed with moose grease and dried whitefish eggs and eaten. *Preserves* Currants used to make jam. (102:11) **Thompson** *Forage* Berries eaten by bears. (164:514) *Fruit* Berries eaten, sparingly by some. (164:489) Fresh berries used for food. (187:227)
- **Other**—**Thompson** *Hunting & Fishing Item* Plant growth around a lake indicated the presence of fish. If there were no black currant plants around the lake, then there would be no fish. (187:228)

Ribes hudsonianum var. *petiolare* (Dougl.) Jancz., Western Black Currant

- **Food**—**Montana Indian** *Fruit* Fruit highly esteemed as an article of diet. (as *R. petiolare* 19:21)

Ribes indecorum Eastw., Whiteflower Currant

- **Drug**—**Luiseño** *Toothache Remedy* Roots used for toothaches. (155:232)

Ribes inerme Rydb., Whitestem Gooseberry

- **Food**—**Apache** *Fruit* Berries used for food. (32:49) **Keres, Western** *Fruit* Berries used for food. (as *R. inverme* 171:66) **Navajo** *Fruit* Berries eaten during the winter. (90:155) **Pueblo** *Fruit* Berries used for food. (32:49) **Thompson** *Beverage* Berries used to make juice. *Fruit* Berries eaten fresh or cooked. *Pie & Pudding* Berries used to make pies. (187:227)

Ribes lacustre (Pers.) Poir., Prickly Currant

- **Drug**—**Bella Coola** *Dermatological Aid* Leaves or bark chewed and cud tied on sores caused by the prickers of plant. *Laxative* Decoction of root taken many times a day for constipation. **Gitksan** *Unspecified* Decoction of bark used for "some unspecified malady." (150:58) **Lummi** *Analgesic* Decoction of twigs taken for general body aches. (79:32) **Okanagan-Colville** *Antidiarrheal* Decoction of dried branches taken for diarrhea. *Cold Remedy* Decoction of dried branches taken for colds. (188:107) **Oweekeno** *Poison* Plant considered poisonous. (43:104) **Saanich** *Psychological Aid* Roots used with wild cherry roots to wash newborn children for intelligence and obedience. **Salish, Coast** *Other* Infusion of roots rubbed on the skin for a charley horse. (182:84) **Shuswap** *Panacea* Berries used for health and strength. (123:63) **Skagit** *Eye Medicine* Decoction of bark used as a wash for sore eyes. *Gynecological Aid* Decoction of bark taken during childbirth. (79:32) **Skagit, Upper** *Gynecological Aid* Decoction of bark taken by women during childbirth. (179:38) **Swinomish** *Poison* Thorns considered poisonous. (79:32) **Thompson** *Eye Medicine* Infusion of cambium layer used as a wash for sore eyes. *Gastrointestinal Aid* Decoction of wood taken as a tonic for the stomach. (164:469) *Gynecological Aid* Berries considered good medicine for women. (187:229) *Other* Decoction of roots and scraped stems taken for "general indisposition."

Tonic Decoction of wood taken as a tonic for the stomach. (164:469)

- **Food**—**Bella Coola** *Fruit* Berries used for food. (184:206) **Cheyenne** *Dried Food* Berries dried for future use. *Fruit* Berries eaten fresh. (76:175) **Gosiute** *Dried Food* Berries dried and stored for winter use. *Fruit* Berries used for food. (39:379) **Haisla & Hanaksiala** *Fruit* Berries used for food. (43:254) **Montana Indian** *Fruit* Fruit highly esteemed as an article of diet. (19:21) **Okanagon** *Staple* Berries used as a principal food. (178:239) **Paiute** *Sauce & Relish* Fruit eaten raw or boiled into a sauce. (111:77) **Salish, Coast** *Bread & Cake* Berries boiled, dried into rectangular cakes, and used as a winter food. (182:84) **Shuswap** *Fruit* Berries used for food. (123:63) **Skagit, Upper** *Fruit* Berries eaten fresh. (179:38) **Thompson** *Dried Food* Berries dried or sometimes buried fresh in the ground for future use. *Frozen Food* Berries stored in the freezer for future use. *Preserves* Berries used to make jam. (187:229)
- **Fiber**—**Cowichan** *Cordage* Roots boiled with cedar and wild rose roots, pounded, and woven into rope. **Saanich** *Cordage* Roots boiled with cedar and wild rose roots, pounded, and woven into rope. (182:84)
- **Other**—**Bella Coola** *Protection* Used as a deterrent against snakes. (184:206) **Cowichan** *Hunting & Fishing Item* Roots used to make reef nets. **Saanich** *Hunting & Fishing Item* Roots used to make reef nets. **Salish, Coast** *Tools* Stiff, sharp thorns used as probes for boils, for removing splinters, and for tattooing. (182:84)

Ribes laxiflorum Pursh, Trailing Black Currant

- **Drug**—**Bella Coola** *Eye Medicine* Simple or compound decoction of root used each day as an eyewash to remove matter. (150:57) Infusion of roots and branches used as an eyewash. (184:206) **Lummi** *Tonic* Decoction of leaves and twigs taken as a general tonic. **Skagit** *Cold Remedy* Decoction of bark taken as a cold medicine. (79:32) **Skagit, Upper** *Cold Remedy* Bark boiled and used as a cold medicine. (179:38) **Skokomish** *Tuberculosis Remedy* Decoction of bark and roots taken for tuberculosis. (79:32)
- **Food**—**Bella Coola** *Fruit* Berries used for food. (184:206) **Haisla & Hanaksiala** *Fruit* Berries used for food. (43:255) **Hesquiat** *Fruit* Raw or cooked berries eaten with oil or sugar. (185:69) **Kwakiutl, Southern** *Fruit* Fruits eaten for food. (183:286) **Lummi** *Fruit* Berries eaten fresh. (79:32) **Makah** *Fruit* Fruit eaten fresh. (67:260) Berries eaten fresh. (79:32) *Preserves* Fruit used to make jelly. (67:260) **Oweekeno** *Fruit* Berries used for food. (43:105) **Skagit** *Fruit* Berries eaten fresh. (79:32) **Skagit, Upper** *Fruit* Berries eaten fresh. (179:38) **Tanana, Upper** *Fruit* Fruit used for food. (77:28)
- **Other**—**Hesquiat** *Smoking Tools* Stems used to make pipestems. (185:69)

Ribes leptanthum Gray, Trumpet Gooseberry

- **Food**—**Apache, Chiricahua & Mescalero** *Bread & Cake* Fruit made into cakes for use during winter. *Fruit* Raw fruit eaten fresh. (33:44) **Isleta** *Fruit* Berries eaten fresh. **Jemez** *Fruit* Berries eaten fresh. (32:49) Berries eaten fresh. (as *Grossularia leptantha* 44:23) **Spanish American** *Beverage* Fruits used to make wine. *Preserves* Fruits used to make jelly. (32:49)

Ribes lobbii Gray, Gummy Gooseberry

- **Drug**—**Kwakiutl** *Antidiarrheal* Roots used for diarrhea. *Dermatological Aid* Poultice of roots and salt water applied to sores, blisters,

and carbuncles. Root ash and oil used as salve on boils. *Oral Aid* Poultice of roots and salt water applied to mouth sores. (183:286) **Saanich** *Psychological Aid* Roots used with wild cherry roots to wash newborn children for intelligence and obedience. **Salish, Coast** *Other* Infusion of roots rubbed on the skin for a charley horse. (182:84)
- *Food*—**Klallam** *Fruit* Berries used for food. (78:197) **Kwakiutl, Southern** *Fruit* Green berries occasionally eaten raw. *Special Food* Berries boiled, cooled, and eaten with oulachen (candlefish) oil at feasts. (183:286) **Salish, Coast** *Bread & Cake* Berries boiled, dried into rectangular cakes, and used as a winter food. (182:84) **Tolowa** *Fruit* Fresh fruits used for food. (6:50)
- *Fiber*—**Cowichan** *Cordage* Roots boiled with cedar and wild rose roots, pounded, and woven into rope. **Saanich** *Cordage* Roots boiled with cedar and wild rose roots, pounded, and woven into rope. (182:84)
- *Other*—**Cowichan** *Hunting & Fishing Item* Roots used to make reef nets. **Saanich** *Hunting & Fishing Item* Roots used to make reef nets. **Salish, Coast** *Tools* Stiff, sharp thorns used as probes for boils, for removing splinters, and for tattooing. (182:84)

Ribes malvaceum Sm., Chaparral Currant
- *Drug*—**Luiseño** *Toothache Remedy* Roots used for toothaches. (155:232)

Ribes malvaceum var. *viridifolium* Abrams, Chaparral Currant
- *Food*—**Cahuilla** *Fruit* Berries eaten fresh. (15:133)

Ribes mescalerium Coville, Mescalero Currant
- *Food*—**Apache, Chiricahua & Mescalero** *Fruit* Raw fruit eaten without preparation. (33:44)

Ribes missouriense Nutt., Missouri Gooseberry
- *Food*—**Chippewa** *Fruit* Berries used for food. (as *R. gracile* 71:131) **Dakota** *Fruit* Berries used for food in season. (as *Grossularia missouriensis* 70:84) **Lakota** *Fruit* Fruits eaten fresh. *Starvation Food* Fruits dried and eaten during famines. (106:35) **Ojibwa** *Dried Food* Fruit dried for future use. *Fruit* Fruit eaten fresh. (as *R. gracile* 135:236) **Omaha** *Dried Food* Fruit dried for winter use. *Fruit* Fruit eaten fresh. (68:326) Berries used for food in season. **Ponca** *Fruit* Berries used for food in season. **Winnebago** *Fruit* Berries used for food in season. (as *Grossularia missouriensis* 70:84)
- *Other*—**Omaha** *Toys & Games* Unripe, acidulous berries used in children's games. The children were divided into two teams and each child was given a portion of the unripe berries which were to be eaten without making a grimace. The team less successful in completing the task had to forfeit to the winners and execute some performances for their amusement such as hopping backwards on one foot. (as *Grossularia missouriensis* 70:84)

Ribes montigenum McClatchie, Gooseberry Currant
- *Food*—**Cahuilla** *Fruit* Berries eaten fresh. (15:133)
- *Other*—**Keres, Western** *Unspecified* Plant known and named but no use was specified. (171:66)

Ribes nevadense Kellogg, Sierran Currant
- *Food*—**Miwok** *Fruit* Raw berries used for food. (12:162) **Tolowa** *Spice* Leaves placed between seaweed patties to keep them from sticking and flavors the patties. (6:50)

Ribes oxyacanthoides L., Canadian Gooseberry
- *Drug*—**Cree, Woodlands** *Gynecological Aid* Decoction of stems and wild black currant stems used for sickness after childbirth. (109:55) **Ojibwa** *Unspecified* Root and bark used for medicinal purposes. (135:236)
- *Food*—**Blackfoot** *Fruit* Berries eaten fresh. *Soup* Berries added to soups. (86:104) **Cree, Woodlands** *Fruit* Berries eaten fresh. (109:55) **Gosiute** *Dried Food* Berries dried and stored for winter use. *Fruit* Berries used for food. (39:379) **Ojibwa** *Dried Food* Fruit dried for future use. *Fruit* Fruit eaten fresh. (135:236) Berries gathered for fresh food. Berries often cooked with sweet corn. *Preserves* Berries used to make preserves for winter use. (153:410)
- *Other*—**Blackfoot** *Toys & Games* Berries used by children to play a game. The children sat in a circle and began counting to 10, each child counting one number around the circle. The tenth child would take five berries and eat them at once, trying his best not to show a bitter face. If he was successful, the child next to him would do the same and this would continue until one grimaced at the sour taste. He was then struck on the thigh with a knuckle punch, thus giving the name "punctured berry" to the plant. The child who never grimaced won all the others' berry supplies. (86:122)

Ribes oxyacanthoides ssp. *irriguum* (Dougl.) Sinnott, Idaho Gooseberry
- *Drug*—**Thompson** *Gastrointestinal Aid* and *Tonic* Decoction of root taken as a tonic for the stomach. (as *Grossularia irrigua* 164:472)
- *Food*—**Okanagan-Colville** *Bread & Cake* Berries, alone or mixed with other berries, used to make cakes. *Forage* Berries eaten by bears. *Fruit* Berries eaten green. *Winter Use Food* Berries canned for future use. (as *R. irriguum* 188:107) **Sanpoil & Nespelem** *Fruit* Berries eaten fresh. (as *R. irriguum* 131:102) **Thompson** *Beverage* Berries used to make juice. (as *R. irriguum* 187:227) *Fruit* Berries mainly eaten fresh. (as *Grossularia irrigua* 164:489) Berries eaten fresh or cooked. *Pie & Pudding* Berries used to make pies. (as *R. irriguum* 187:227)
- *Other*—**Okanagan-Colville** *Cash Crop* Berries dried, mixed with bitterroot, and sold to the Thompson. (as *R. irriguum* 188:107)

Ribes oxyacanthoides L. ssp. *oxyacanthoides*, Canadian Gooseberry
- *Drug*—**Chippewa** *Analgesic* and *Gynecological Aid* Compound decoction of berry taken for back pain and "female weakness." (as *Grossularia oxyacanthoides* 53:356)

Ribes oxyacanthoides ssp. *setosum* (Lindl.) Sinnott, Inland Gooseberry
- *Food*—**Cheyenne** *Fruit* Fruit eaten raw or cooked. *Winter Use Food* Dried fruit formed into little cakes and used for winter food. (as *Grossularia setosa* 76:175) **Montana Indian** *Fruit* Fruit highly esteemed as an article of diet. (as *R. setosum* 19:21)

Ribes pinetorum Greene, Orange Gooseberry
- *Drug*—**Navajo, Ramah** *Ceremonial Medicine* Leaves used as emetics in various ceremonies. *Emetic* Leaves used as a ceremonial emetic. (191:30)
- *Food*—**Apache, Chiricahua & Mescalero** *Bread & Cake* Fruit ground and compressed into cakes for winter use. (33:44) **Navajo, Ramah** *Fruit* Berries used for food. (191:30)

- *Other*—**Navajo, Ramah** *Hunting & Fishing Item* Stems used to make arrow shafts. Thorns used to make arrow points. (191:30)

Ribes quercetorum Greene, Rock Gooseberry
- *Food*—**Kawaiisu** *Dried Food* Berries dried in the shade for about a week and stored. *Fruit* Berries eaten fresh. *Preserves* Fresh berries boiled into a jelly. (206:59) **Tubatulabal** *Fruit* Berries used extensively for food. (193:15)

Ribes roezlii Regel, Sierran Gooseberry
- *Food*—**Atsugewi** *Fruit* Fresh berries used for food. (61:139) **Karok** *Fruit* Fresh berries used for food. (6:50) **Kawaiisu** *Dried Food* Berries dried in the shade for about a week and stored. *Fruit* Berries eaten fresh. *Preserves* Fresh berries boiled into a jelly. (206:59) **Miwok** *Fruit* Pulverized, raw berries used for food. (12:162) **Yurok** *Fruit* Fresh berries used for food. (6:50)

Ribes rotundifolium Michx., Appalachian Gooseberry
- *Drug*—**Cherokee** *Antidiarrheal* Infusion of bark taken "to check bowels." *Misc. Disease Remedy* Infusion taken for measles. *Sedative* Infusion of leaf taken for nerves. (80:36) **Iroquois** *Other* Compound decoction of bark taken for fortune-telling or divination. (87:345)
- *Food*—**Cherokee** *Winter Use Food* Berries canned for future use. (126:54)

Ribes rubrum L., Cultivated Currant
- *Drug*—**Ojibwa** *Unspecified* Root and bark used for medicinal purposes. (135:236)
- *Food*—**Chippewa** *Fruit* Fresh or dried fruit used for food. (71:131)

Ribes rubrum var. *subglandulosum* Maxim., Cultivated Currant
- *Food*—**Ojibwa** *Dried Food* Fruit dried for future use. *Fruit* Fruit eaten fresh. (135:236)

Ribes sanguineum Pursh, Redflower Currant
- *Food*—**Chehalis** *Fruit* Berries eaten by children. (79:32) **Hoh** *Fruit* Fruits eaten raw. Fruits stewed and used for food. *Winter Use Food* Fruits canned and saved for future food use. (137:62) **Klallam** *Fruit* Berries used for food. Fruits eaten fresh. (78:197) **Paiute** *Fruit* Berries used for food. (111:78) **Quileute** *Fruit* Fruits eaten raw. Fruits stewed and used for food. *Winter Use Food* Fruits canned and saved for future food use. (137:62) **Salish, Coast** *Bread & Cake* Berries boiled, dried into rectangular cakes, and used as a winter food. (182:84) **Skagit, Upper** *Fruit* Fruit eaten fresh. (179:38) **Squaxin** *Fruit* Berries eaten by children. (79:32) **Thompson** *Dried Food* Berries sometimes dried and used in soups as flavoring. *Fruit* Berries eaten fresh. *Spice* Berries sometimes dried and used in soups as flavoring. (187:229) **Thompson, Upper (Lytton Band)** *Fruit* Grayish black berries eaten. (164:487)

Ribes sanguineum var. *glutinosum* (Benth.) Loud., Blood Currant
- *Food*—**Mahuna** *Fruit* Berries eaten mainly to quench the thirst. (as *R. glutinosum* 140:70)

Ribes sp., Gooseberry
- *Drug*—**Blackfoot** *Laxative* Berries eaten as a mild laxative. (86:68) **Carrier, Northern** *Dermatological Aid* Compound decoction of inner bark taken for body sores. *Orthopedic Aid* Compound decoction of inner bark taken for paralysis. *Stimulant* Compound decoction of inner bark taken for constitutional weakness. (150:58) **Chippewa** *Diuretic* Compound decoction of leaves and stalk taken as a diuretic. (53:348) **Kwakiutl** *Antidiarrheal* Compound decoction of berries taken for diarrhea. (183:264) *Dermatological Aid* Juice used as wash for open wounds. Poultice of rubbed root applied to open, running sores. Root "rubbed on sores or around the mouths of children." Ground roots and salt water used as a wash for opened blisters. Roots rubbed on sandstone and placed on open, running sores. *Oral Aid* and *Pediatric Aid* Roots scraped on sandstone and "rubbed on sores or around the mouths of children." (20:382)
- *Food*—**Anticosti** *Preserves* Berries used to make jam. (143:67) **Carrier** *Preserves* Berries used to make jelly. (31:73) **Coeur d'Alene** *Fruit* Berries eaten fresh. (178:90) **Costanoan** *Fruit* Fruits eaten for food. (21:250) **Okanagon** *Staple* Berries used as a principal food. (178:239) **Paiute** *Fruit* Berries eaten raw. (111:79) **Sanpoil & Nespelem** *Fruit* Fresh or dried berries sweetened with serviceberries in water and whipped to a froth. Berries eaten raw. Only currants found south of the Columbia River were eaten raw without ill results. Those found on the north side were eaten only if mixed with other foods. Otherwise illness resulted. (131:103) **Spokan** *Fruit* Berries used for food. (178:343) **Wintoon** *Fruit* Berries used for food. (117:264)
- *Other*—**Cheyenne** *Hunting & Fishing Item* Used to make arrow shafts. (83:46)

Ribes triste Pallas, Red Currant
- *Drug*—**Chippewa** *Abortifacient* Compound decoction of stalk taken for "stoppage of periods." (53:358) *Urinary Aid* Decoction of root and stalk taken for "gravel." (53:348) **Ojibwa** *Gynecological Aid* Leaves used as some sort of female remedy. (153:389) **Tanana, Upper** *Eye Medicine* Decoction of stems, without the bark, used as a wash for sore eyes. *Unspecified* Decoction of stems, without the bark, taken as a medicine. (102:11)
- *Food*—**Alaska Native** *Fruit* Berries used raw. *Preserves* Berries made into jams and jellies. (85:87) **Chippewa** *Bread & Cake* Berries cooked, spread on birch bark into little cakes, dried, and stored for winter use. *Fruit* Berries eaten raw. (53:321) **Eskimo, Alaska** *Unspecified* Species used for food. (4:715) **Eskimo, Inupiat** *Dessert* Berries mixed with other berries and used to make traditional dessert. *Fruit* Berries eaten raw or cooked. *Sauce & Relish* Berries mixed with rose hips and high-bush cranberries and boiled into a catsup or syrup. (98:105) **Iroquois** *Bread & Cake* Fruit mashed, made into small cakes, and dried for future use. *Dried Food* Raw or cooked fruit sun or fire dried and stored for future use. *Fruit* Dried fruit taken as a hunting food. *Sauce & Relish* Dried fruit cakes soaked in warm water and cooked as a sauce or mixed with corn bread. (196:128) **Ojibwa** *Dried Food* Berries dried for winter use. *Fruit* In the winter, a favorite dish was wild currants cooked with sweet corn. Berries eaten fresh. *Preserves* Berries used to make jams and preserves. (153:410) **Tanana, Upper** *Fruit* Berries used for food. (102:11)

Ribes uva-crispa var. *sativum* DC., European Gooseberry
- *Drug*—**Micmac** *Cathartic* Bark and roots used as a physic. (as *R. grossularia* 40:61)

Ribes velutinum Greene, Desert Gooseberry
- *Food*—**Gosiute** *Dried Food* Berries dried and stored for winter use.

Fruit Berries used for food. (as *R. leptanthum* var. *brachyanthum* 39:379)

Ribes velutinum Greene **var. *velutinum*,** Desert Gooseberry
- *Food*—**Kawaiisu** *Dried Food* Berries dried in the shade for about a week and stored. *Fruit* Berries eaten fresh. *Preserves* Fresh berries boiled into a jelly. (206:59)

Ribes viscosissimum Pursh, Sticky Currant
- *Food*—**Montana Indian** *Fruit* Fruit highly esteemed as an article of diet. (19:21)

Ribes wolfii Rothrock, Wolf's Currant
- *Food*—**Apache, Chiricahua & Mescalero** *Bread & Cake* Fruit ground, dried, and pressed into cakes for storage. *Fruit* Raw fruit eaten without preparation. *Preserves* Fruit used to make jelly. (33:44)

Ricinus, Euphorbiaceae

Ricinus communis L., Castor Bean
- *Drug*—**Cahuilla** *Dermatological Aid* Seeds crushed into a greasy substance and used for sores. *Poison* Seeds and leaves considered poisonous. (15:133) **Cherokee** *Cathartic* Infusion of beans used as a purgative. *Dermatological Aid* Poultice of beans used for boils. (80:24) **Diegueño** *Dermatological Aid* Bean nuts mashed to make an ointment, similar to cold cream, and used for acne and pimples. (84:37) **Hawaiian** *Analgesic* Leaves used for severe headaches. *Febrifuge* Poultice of leaves applied to the head and body for strong fevers. *Pediatric Aid* Poultice of leaves applied to children's heads and bodies for strong fevers. (2:55) **Navajo** *Contraceptive* Plant used by women to become sterile. (55:60) **Pima** *Analgesic* Beans eaten for headaches. *Cathartic* Beans eaten as a purge. *Dermatological Aid* Beans dried, ground, and sprinkled on sores. *Laxative* Beans eaten for constipation. *Poison* Plant considered poisonous. (47:100) **Seminole** *Cathartic* Beans used as a cathartic. (169:167) Infusion of seeds taken as a cathartic for constipation. (169:275)
- *Other*—**Navajo** *Protection* Plant used in an unknown manner as a protection from the spirit of the bear. (55:60)

Robinia, Fabaceae

Robinia hispida L., Bristly Locust
- *Drug*—**Cherokee** *Emetic* Root bark chewed as an emetic. *Toothache Remedy* Beaten root held on tooth for toothache. *Veterinary Aid* Infusion given to cows as a "tonic." (80:43)
- *Fiber*—**Cherokee** *Building Material* Wood used to make fence posts and sills for houses. (80:43)
- *Other*—**Cherokee** *Fasteners* Wood used to make pegs for log cabins. *Hunting & Fishing Item* Wood used to make bows and blowgun darts. (80:43)

Robinia neomexicana Gray, New Mexico Locust
- *Drug*—**Hopi** *Emetic* Used as an emetic to purify the stomach. (200:83)
- *Food*—**Apache, Chiricahua & Mescalero** *Vegetable* Raw pods eaten as food. *Winter Use Food* Pods cooked and stored. (33:42) **Apache, Mescalero** *Dried Food* Flowers boiled, dried, and stored for winter food use. *Unspecified* Fresh flowers cooked with meat or bones and used for food. (14:47) **Apache, White Mountain** *Vegetable* Beans and pods used for food. (136:160) **Jemez** *Unspecified* Large clusters of flowers eaten without preparation. (32:49) Flowers eaten as food. (44:27)
- *Fiber*—**Hualapai** *Furniture* Branches used to make cradleboards. (195:34)
- *Other*—**Apache, Mescalero** *Hunting & Fishing Item* Wood used to make high-quality bows. (14:47) **Hualapai** *Hunting & Fishing Item* Wood, cured for a year, used to make hunting bows. (195:34) **Jemez** *Hunting & Fishing Item* Tough, elastic branches used to make bows. (44:27) **Keres, Western** *Hunting & Fishing Item* Branches used in making arrow shafts. (171:67) **Tewa** *Hunting & Fishing Item* Wood used to make bows. (138:48)

Robinia pseudoacacia L., Black Locust
- *Drug*—**Cherokee** *Emetic* Root bark chewed as an emetic. *Toothache Remedy* Beaten root held on tooth for toothache. *Veterinary Aid* Infusion given to cows as a "tonic." (80:43) **Menominee** *Adjuvant* Trunk bark used as a seasoner to give flavor to medicines. (151:40)
- *Food*—**Cherokee** *Beverage* Bark steeped into tea. (126:46) **Mendocino Indian** *Forage* Leaves eaten by horses as forage. **Wailaki** *Forage* Seeds eaten by chickens as forage. (41:359)
- *Fiber*—**Cherokee** *Building Material* Wood used to make fence posts and sills for houses. (80:43)
- *Other*—**Cherokee** *Fasteners* Wood used to make pegs for log cabins. *Hunting & Fishing Item* Wood used to make bows and blowgun darts. (80:43)

Robinia sp., Black Locust
- *Drug*—**Chickasaw** *Analgesic* Roots used for headaches. (177:33)

Romneya, Papaveraceae

Romneya coulteri Harvey, Coulter's Matilija Poppy
- *Food*—**Cahuilla** *Beverage* Watery substance in the stalk used as a beverage. (15:133)

Rorippa, Brassicaceae

Rorippa alpina (S. Wats.) Rydb., Alpine Yellowcress
- *Drug*—**Navajo** *Gynecological Aid* Infusion of plants taken as a tonic after deliverance. (as *Radicula alpina* 55:49)

Rorippa curvisiliqua (Hook.) Bess. ex Britt. **var. *curvisiliqua*,** Curvepod Yellowcress
- *Food*—**Paiute** *Unspecified* Species used for food. (as *Radicula curvisiliqua* 167:242)

Rorippa islandica (Oeder) Borbas, Northern Marsh Yellowcress
- *Food*—**Eskimo, Inuktitut** *Spice* Used as a condiment in fish soup. (202:185)

Rorippa nasturtium-aquaticum (L.) Hayek, Water Cress
- *Drug*—**Costanoan** *Febrifuge* Cold infusion of plants taken for fevers. *Kidney Aid* Decoction of plant used as a kidney remedy. *Liver Aid* Decoction of plant used as a liver remedy. (as *Nasturtium officinale* 21:10) **Mahuna** *Liver Aid* Plant used for torpid liver, cirrhosis of the liver, and gallstones. (140:65) **Okanagan-Colville** *Analgesic* Poultice of fresh, whole plants applied to the forehead for headaches. *Vertigo Medicine* Poultice of fresh, whole plants applied to the forehead for dizziness. (188:92)

- **Food**—**Algonquin, Quebec** *Vegetable* Used as a salad plant. (as *Nasturtium officinale* 18:86) **Cahuilla** *Vegetable* Eaten fresh in the spring, cooked like spinach, or mixed with less flavorful greens into a salad. (as *Nasturtium officinale* 15:90) **Cherokee** *Vegetable* Leaves eaten cooked or raw as greens. (as *Nasturtium officinale* 80:61) Leaves boiled and eaten with bacon grease as potherbs. Leaves used in salads. (as *␣␣␣␣␣␣␣␣␣␣* *Nasturtium officinale* 126:37) **Diegueño** *Vegetable* Leaves boiled and eaten as greens. (84:37) **Gosiute** *Unspecified* Species used for food. (as *Nasturtium palustre* 39:375) **Havasupai** *Unspecified* Species used for food. (197:220) **Iroquois** *Vegetable* Eaten raw, sometimes with salt. (as *Radicula nasturtium-aquaticum* 196:118) **Karok** *Unspecified* Young plants boiled and eaten. (6:51) **Kawaiisu** *Unspecified* Leaves eaten raw, usually with salt, or boiled and fried in grease and salt. (206:60) **Luiseño** *Vegetable* Plant used for greens. (as *Nasturtium officinale* 155:232) **Mendocino Indian** *Sauce & Relish* Leaves eaten as a relish. (as *Roripa nasturtium* 41:352) **Okanagan-Colville** *Starvation Food* Leaves used as a good emergency food. *Vegetable* Leaves eaten raw as salad greens. (188:92) **Saanich** *Unspecified* Young leaves eaten raw. (182:82) **Tubatulabal** *Vegetable* Leaves and stems boiled as greens. (as *Radicula nasturtium-aquaticum* 193:16)

Rorippa palustris ssp. *hispida* (Desv.) Jonsell, Hispid Yellowcress
- **Drug**—**Navajo, Ramah** *Ceremonial Medicine* and *Eye Medicine* Plant used in ceremonial eyewash. (as *R. hispida* 191:29)

Rorippa sinuata (Nutt.) A. S. Hitchc., Spreading Yellowcress
- **Drug**—**Zuni** *Eye Medicine* Infusion of plant used as a wash and smoke from blossoms used for inflamed eyes. (as *Radicula sinuata* 166:59)
- **Other**—**Navajo, Ramah** *Fertilizer* Cold infusion used to soak watermelon seeds to increase productivity. (191:29)

Rorippa sylvestris (L.) Bess., Creeping Yellowcress
- **Drug**—**Iroquois** *Febrifuge* and *Pediatric Aid* Decoction of plant taken by mother for fever in baby. (87:342)

Rorippa teres (Michx.) R. Stuckey, Southern Marsh Yellowcress
- **Food**—**Navajo, Ramah** *Fodder* Used for sheep feed. (as *R. obtusa* 191:29)

Rosaceae, see *Adenostoma, Agrimonia, Amelanchier, Argentina, Aronia, Aruncus, Cerasus, Cercocarpus, Chamaebatia, Chamaebatiaria, Coleogyne, Comarum, Cowania, Crataegus, Dalibarda, Fallugia, Filipendula, Fragaria, Geum, Heteromeles, Holodiscus, Horkelia, Ivesia, Luetkea, Malus, Oemleria, Osteomeles, Pentaphylloides, Petrophyton, Physocarpus, Porteranthus, Potentilla, Prunus, Purshia, Pyrus, Rosa, Rubus, Sorbus, Spiraea, Waldsteinia*

Rosa, Rosaceae
Rosa acicularis Lindl., Prickly Rose
- **Drug**—**Cree, Woodlands** *Cough Medicine* Decoction of roots taken as a cough remedy. *Eye Medicine* Infusion of roots used for sore eyes. (109:55) **Iroquois** *Eye Medicine* Compound infusion of rose leaves and bark used as drops for blindness. (87:358) *Gynecological Aid* Cold, compound infusion of twig bark taken for difficult birth. (87:359) *Witchcraft Medicine* Compound decoction of plants and doll used for black magic. (87:358) **Okanagan-Colville** *Ceremonial Medicine* Decoction of leaves, branches, and other boughs taken and used as body and hair wash by sweat bathers. *Dermatological Aid* Poultice of chewed leaves applied to bee stings. (188:131) **Tanana, Upper** *Blood Medicine* Decoction of stems and branches taken for weak blood. *Cold Remedy* Decoction of stems and branches taken for colds. *Emetic* Infusion of bark strained and taken to induce vomiting. *Febrifuge* Decoction of stems and branches taken for fevers. *Gastrointestinal Aid* Decoction of stems and branches taken for stomach troubles. (102:12) **Thompson** *Antidiarrheal* Decoction of branches, chokecherry, and red willow taken for diarrhea. *Antiemetic* Decoction of branches, chokecherry, and red willow taken for vomiting. *Dermatological Aid* Leaves placed in moccasins for athlete's foot and possibly for protection. *Gynecological Aid* Hips chewed by women in labor, to hasten the delivery. Decoction of roots taken by women after childbirth. Decoction of branches, chokecherry, and red willow taken for women's illnesses. *Venereal Aid* Decoction of roots taken for syphilis. (187:267)
- **Food**—**Alaska Native** *Beverage* Rose hips cooked, juice extracted, pasteurized, and mixed with other fruit juices. Leaves used to make tea. *Dietary Aid* Rose hips used as one of the richest known food sources of vitamin C. *Preserves* Rose hip juice used to make jellies. Rose hip pulp, with seeds and skins removed, used to make jams and marmalades. *Sauce & Relish* Rose hip juice used to make syrups. Rose hip pulp, with seeds and skins removed, used to make ketchups. (85:89) **Cree, Woodlands** *Snack Food* Ripe hips eaten as a nibble. (109:55) **Eskimo, Inupiat** *Beverage* Used to make juice. *Frozen Food* Frozen and stored for future use. *Ice Cream* Used with oil and water to make ice cream. *Preserves* Used to make jam or jelly.[67] *Sauce & Relish* Used to make syrup. *Unspecified* Eaten fresh or cooked. (98:101) **Koyukon** *Fruit* Berries used for food. (119:55) **Montana Indian** *Fruit* Fruit and hips used for food. (19:21) **Okanagan-Colville** *Forage* Hips eaten by coyotes. *Spice* Leaves placed under and over food while pit cooking to add flavor and prevent burning. *Unspecified* Orange, outer rind of the hips used for food. (188:131) **Tanana, Upper** *Beverage* Leaves boiled into tea. (102:12) *Fruit* Rose hips used for food. (as *R. acidularis* 77:28) Rose hips eaten raw or cooked with grease and sugar. *Preserves* Rose hips used to make jelly. *Unspecified* Raw petals used for food. *Winter Use Food* Leaves dried and saved for later use. (102:12) **Thompson** *Beverage* Shoots or hips[67] or leaves and young twigs used to make a tea-like beverage. *Forage* Hips eaten by bears before hibernation. *Preserves* Hips used to make jelly. *Sauce & Relish* Hips used to make syrup.[67] *Unspecified* Hips used only sparingly for food because of the seeds and the insipid taste.[68] Young, tender shoots peeled and eaten in the spring. (187:267)
- **Fiber**—**Thompson** *Furniture* Heavy, split wood used to make cradle hoops. (187:267)
- **Other**—**Cree, Woodlands** *Jewelry* Firm hips strung on a string to make a necklace. *Toys & Games* Halved, fresh hip hollowed out to make a bowl for a toy pipe. (109:55) **Okanagan-Colville** *Ceremonial Items* Branches used by an Indian doctor to sweep out the grave before the corpse was lowered into it.[69] *Containers* Leaves placed under and over food while pit cooking to add flavor and prevent burning. *Good Luck Charm* Branches boiled in water and used to soak fishing lines and nets to obtain good luck. *Hunting & Fishing Item* Branches used in a wash by hunters to get rid of the human scent. *Protection* Branches made into tea and taken as protection from bad spirits and

ghosts. Branches made into tea and used as washing water for one who was being jinxed by some bad person. Branches placed around the house and yard of the deceased to keep his or her spirit from returning. (188:131) **Thompson** *Good Luck Charm* Plant asked "for good luck." *Hunting & Fishing Item* Plant used to wipe dip net hoops, to improve the chances of a good catch. *Paint* Petals mixed with pine pitch, grease, and red ocher paint to make a cosmetic. *Protection* Leaves placed in moccasins for athlete's foot and possibly for protection. Branches placed around the body and house of a dead person to protect other people from its spirit.[70] (187:267)

Rosa acicularis ssp. *sayi* (Schwein.) W. H. Lewis, Prickly Rose

- **Drug**—**Blackfoot** *Antidiarrheal* and *Pediatric Aid* Root used to make a drink given to children for diarrhea. (as *R. sayi* 114:275)
- **Food**—**Blackfoot** *Fruit* Berries eaten raw. (as *R. sayi* 114:275) **Montana Indian** *Fruit* Fruit used for food. (as *R. sayi* 19:21) **Ojibwa** *Unspecified* Buds used for food. (as *R. sayi* 135:236)

Rosa arkansana Porter, Prairie Rose

- **Drug**—**Chippewa** *Anticonvulsive* Compound infusion or decoction of root taken for "fits." (51:63, 64) Compound decoction of root taken for convulsions. (53:336) *Hemostat* Compound infusion or decoction of root used on bleeding wounds. (51:63, 64) Compound decoction of root used on bleeding wounds. (53:336) *Stimulant* Compound infusion or decoction of root taken or used externally as stimulant. (51:63, 64) Compound decoction of root taken as a stimulant. *Tonic* Compound decoction of root taken as a tonic. (53:364) **Omaha** *Eye Medicine* Roots used for eye troubles. (68:336)
- **Food**—**Lakota** *Beverage* Petals used to make tea. Roots used to make a strong tea. *Dried Food* Hips dried and used for food. *Preserves* Petals used to make jam. *Soup* Hips dried, added to soups or stews, and used for food. *Starvation Food* Hips eaten during famines. (106:39)
- **Other**—**Omaha** *Incense & Fragrance* Petals used as a perfume for hair oil. (68:323)

Rosa arkansana var. *suffulta* (Greene) Cockerell, Prairie Rose

- **Drug**—**Omaha** *Eye Medicine* Infusion of fruit used as a wash for inflamed eyes. **Pawnee** *Burn Dressing* Poultice of charred, crushed hypertrophied stem growths applied to burns. (as *R. pratincola* 70:85)
- **Food**—**Dakota** *Starvation Food* Fruit sometimes eaten in times of food scarcity. **Omaha** *Starvation Food* Fruit sometimes eaten in times of food scarcity. **Pawnee** *Starvation Food* Fruit sometimes eaten in times of food scarcity. **Ponca** *Starvation Food* Fruit sometimes eaten in times of food scarcity. (as *R. pratincola* 70:85)
- **Other**—**Dakota** *Smoke Plant* Inner bark, alone or mixed with tobacco, used for smoking. **Omaha** *Smoke Plant* Inner bark, alone or mixed with tobacco, used for smoking. **Pawnee** *Smoke Plant* Inner bark, alone or mixed with tobacco, used for smoking. **Ponca** *Smoke Plant* Inner bark, alone or mixed with tobacco, used for smoking. (as *R. pratincola* 70:85)

Rosa blanda Ait., Smooth Rose

- **Drug**—**Meskwaki** *Dermatological Aid* Decoction of fruit used for itching piles or any other itch. *Gastrointestinal Aid* Rose hip skin used for stomach troubles and decoction of fruit used for piles. *Hemorrhoid Remedy* Decoction of fruit used for itching piles. (152:242, 243) **Ojibwa** *Gastrointestinal Aid* Dried, powdered flowers used for heartburn. Rose hip skin used for stomach trouble and indigestion. (153:385) **Ojibwa, South** *Eye Medicine* Infusion of root used as a wash for inflamed eyes. (91:200) **Potawatomi** *Analgesic* Infusion of root taken for headache or lumbago. *Orthopedic Aid* Infusion of root taken for lumbago. *Unspecified* Rose hip skin used as medicine by the Prairie Potawatomi. (154:78)

Rosa californica Cham. & Schlecht., California Wild Rose

- **Drug**—**Cahuilla** *Analgesic* and *Pediatric Aid* Infusion of blossoms used for infant pain. (15:133) **Costanoan** *Antirheumatic (Internal)* Decoction of fruit "hips" used for rheumatism. *Cold Remedy* Decoction of fruit "hips" used for colds. *Dermatological Aid* Decoction of fruit "hips" used as a wash for scabs and sores. *Febrifuge* Decoction of fruit "hips" used for fevers. *Gastrointestinal Aid* Decoction of fruit "hips" used for indigestion. *Kidney Aid* Decoction of fruit "hips" used for kidney ailments. *Throat Aid* Decoction of fruit "hips" used for sore throats. (21:18) **Diegueño** *Febrifuge* and *Pediatric Aid* Infusion of petals given to babies with fever. (84:39) **Mahuna** *Analgesic* Infusion of seeds taken for painful congestion. *Febrifuge* Infusion of seeds taken for stomach fevers. *Gastrointestinal Aid* Infusion of seeds taken for stomach fevers and painful congestion. (140:8) **Miwok** *Analgesic* Infusion of leaves and berries taken for pains. *Gastrointestinal Aid* Infusion of leaves and berries taken for colic. (12:172)
- **Food**—**Cahuilla** *Beverage* Blossoms soaked in water to make a beverage. *Unspecified* Buds eaten fresh. (15:133) **Gosiute** *Fruit* Berries used for food. (39:379) **Kawaiisu** *Fruit* Fruit, a "fleshy hip," eaten ripe. (206:60) **Pomo, Kashaya** *Fruit* Fresh fruit used for food. (72:99)
- **Fiber**—**Kawaiisu** *Basketry* Unsplit stems used as rims in twined basketry. (206:60)

Rosa carolina L., Carolina Rose

- **Drug**—**Menominee** *Gastrointestinal Aid* Fruit skin eaten for stomach troubles. (as *R. humilis* 151:50)

Rosa eglanteria L., Sweetbriar Rose

- **Drug**—**Iroquois** *Gastrointestinal Aid* Plant used as an intestinal astringent by forest runners. (141:47) *Urinary Aid* Compound decoction of plants taken for difficult urination. (as *R. rubiginosa* 87:358)

Rosa gallica L., French Rose

- **Drug**—**Mahuna** *Febrifuge* Plant used for high fevers and chills. (140:26)

Rosa gymnocarpa Nutt., Dwarf Rose

- **Drug**—**Okanagan-Colville** *Ceremonial Medicine* Decoction of leaves, branches, and other boughs taken and used as body and hair wash by sweat bathers. *Dermatological Aid* Poultice of chewed leaves applied to bee stings. (188:131) **Thompson** *Eye Medicine* Decoction of bark used as a wash for sore eyes. (164:466) *Hunting Medicine* Decoction of plant poured onto hunting equipment that had "lost its luck." (164:507) *Poison* Spines considered poisonous as they caused swelling and irritation if touched. Hips considered poisonous and would give one an itchy bottom if eaten. (187:266) *Tonic* Decoction of stems taken for "general indisposition" and as a tonic. (164:466)
- **Food**—**Okanagan-Colville** *Forage* Hips eaten by coyotes. *Spice* Leaves placed under and over food while pit cooking to add flavor and prevent burning. *Unspecified* Orange, outer rind of the hips used for food. (188:131) **Pomo, Kashaya** *Fruit* Fresh fruit used for food.

(72:99) **Thompson** *Beverage* Young leaves and stalks boiled and drunk as a tea. (164:493) Shoots used to make a tea-like beverage. (187:267) *Fruit* Small fruits occasionally eaten. (164:488) Fruits eaten, but not in large quantities. (164:489)
- ***Fiber***—**Thompson** *Furniture* Wood used for the hoops of baby carriers. (164:498) *Mats, Rugs & Bedding* Twigs put in the beds of widows and widowers during the period of their widowhood. (164:504)
- ***Other***—**Okanagan-Colville** *Ceremonial Items* Branches used by an Indian doctor to sweep out the grave before the corpse was lowered into it.[69] *Containers* Leaves placed under and over food while pit cooking to add flavor and prevent burning. *Good Luck Charm* Branches boiled in water and used to soak fishing lines and nets to obtain good luck. *Hunting & Fishing Item* Branches used in a wash by hunters to get rid of the human scent. *Protection* Branches made into tea and taken as protection from bad spirits and ghosts. Branches made into tea and used as washing water for one who was being jinxed by some bad person. Branches placed around the house and yard of the deceased to keep his or her spirit from returning. (188:131) **Okanogan** *Smoke Plant* Leaves mixed with other plant leaves and smoked. (125:39) *Toys & Games* Hips used as beads by children. (125:38) **Thompson** *Ceremonial Items* Large branches used for sweeping evil influences out of graves before burial. (164:504) *Hunting & Fishing Item* Wood used to make arrows. (164:498) *Jewelry* Small fruits strung and used as beads by children. (164:488) *Smoke Plant* Leaves mixed with other plant leaves and smoked. (125:39) Dried, toasted, powdered leaves and bark occasionally used for smoking. (164:495) *Tools* Wood used to make handles. (164:498) *Toys & Games* Hips used as beads by children. (125:38)

Rosa nutkana K. Presl, Nootka Rose
- ***Drug***—**Bella Coola** *Diaphoretic* Roots and sprouts used in steam baths. *Eye Medicine* Infusion of roots and sprouts used as an eyewash. (184:209) **Carrier** *Eye Medicine* Decoction of roots applied to sore eyes. (31:86) **Chehalis** *Analgesic* Decoction of bark taken by women to ease labor pains in childbirth. *Gynecological Aid* Decoction of bark taken by women to ease labor pains in childbirth. **Cowlitz** *Pediatric Aid* and *Strengthener* Decoction of leaves used as a wash to strengthen babies. (79:34) **Keres, Western** *Misc. Disease Remedy* and *Pediatric Aid* Crushed petals rubbed on children's bodies to prevent smallpox. (as *R. nelina* 171:67) **Nitinaht** *Unspecified* Infusion of leaves used for medicine. (186:123) **Okanagan-Colville** *Ceremonial Medicine* Decoction of leaves, branches, and other boughs taken and used as body and hair wash by sweat bathers. *Dermatological Aid* Poultice of chewed leaves applied to bee stings. (188:131) **Quileute** *Dermatological Aid* Poultice of haw ashes applied to "swellings." **Quinault** *Dermatological Aid* and *Venereal Aid* Poultice of twig ashes and skunk oil applied to syphilitic sores. **Skagit** *Eye Medicine* Infusion of roots used as an eyewash. *Throat Aid* Decoction of roots taken for sore throats. (79:34) **Skagit, Upper** *Throat Aid* Decoction of roots and sugar used for sore throats. (179:42) **Thompson** *Antidiarrheal* Decoction of branches, chokecherry, and red willow taken for diarrhea. *Antiemetic* Decoction of branches, chokecherry, and red willow taken for vomiting. *Dermatological Aid* Leaves placed in moccasins for athlete's foot and possibly for protection. *Gynecological Aid* Decoction of roots taken by women after childbirth. Decoction of branches, chokecherry, and red willow taken for women's illnesses. *Venereal Aid* Decoction of roots taken for syphilis. (187:267)
- ***Food***—**Bella Coola** *Fruit* Fruits used for food in late fall. (184:209) **Blackfoot** *Fruit* Raw berries used for food. *Preserves* Berries used to make jelly. (118:22) **Carrier** *Preserves* Berries used to make jam. (31:86) **Cowichan** *Fruit* Hips eaten raw in fall. (182:87) **Hesquiat** *Forage* Eaten by deer. *Fruit* Outside of the fruit, or hip, eaten with oil. (185:74) **Lummi** *Beverage* Twigs peeled, boiled, and used as a beverage. *Dried Food* Hips dried and used for food. (79:34) **Montana Indian** *Fruit* Fruit and hips used for food. (19:21) **Nitinaht** *Fruit* Hips eaten raw in fall. (186:123) **Okanagan-Colville** *Forage* Hips eaten by coyotes. *Spice* Leaves placed under and over food while pit cooking to add flavor and prevent burning. *Unspecified* Orange, outer rind of the hips used for food. (188:131) **Quinault** *Unspecified* Hips used for food. (79:34) **Rocky Boy** *Fruit* Raw berries used for food. *Preserves* Berries used to make jelly. (118:22) **Saanich** *Fruit* Hips eaten raw in fall. (182:87) **Salish** *Unspecified* Hips used for food. (183:290) **Salish, Coast** *Unspecified* Young shoots used for food. (182:87) **Shuswap** *Beverage* Stems and flowers used to make tea. (123:67) **Skagit** *Beverage* Leaves used to make tea. *Unspecified* Hips mixed with dried salmon eggs and used for food. (79:34) **Skagit, Upper** *Beverage* Leaves used to make tea. *Spice* Hips mixed with dried salmon eggs to enhance the flavor. (179:42) **Skokomish** *Unspecified* Hips eaten in the fall. **Snohomish** *Unspecified* Hips used for food. **Swinomish** *Unspecified* Hips used for food. (79:34) **Thompson** *Beverage* Shoots used to make a tea-like beverage. Leaves and young twigs used to make a tea-like beverage. *Unspecified* Young, tender shoots peeled and eaten in the spring. (187:267) **Washo** *Fruit* Raw berries used for food. (118:22)
- ***Fiber***—**Thompson** *Furniture* Heavy, split wood used to make cradle hoops. (187:267)
- ***Other***—**Cowichan** *Hunting & Fishing Item* Roots used with gooseberry and cedar roots to make reef nets. (182:87) **Okanagan-Colville** *Ceremonial Items* Branches used by an Indian doctor to sweep out the grave before the corpse was lowered into it.[69] *Containers* Leaves placed under and over food while pit cooking to add flavor and prevent burning. *Good Luck Charm* Branches boiled in water and used to soak fishing lines and nets to obtain good luck. *Hunting & Fishing Item* Branches used in a wash by hunters to get rid of the human scent. *Protection* Branches made into tea and taken as protection from bad spirits and ghosts. Branches made into tea and used as washing water for one who was being jinxed by some bad person. Branches placed around the house and yard of the deceased to keep his or her spirit from returning. (188:131) **Shuswap** *Protection* Branches broken and left in the house after removal of corpse to keep the disease in the body. (123:67) **Thompson** *Good Luck Charm* Plant asked "for good luck." *Hunting & Fishing Item* Plant used to wipe dip net hoops, to improve the chances of a good catch. *Paint* Petals mixed with pine pitch, grease, and red ocher paint to make a cosmetic. *Protection* Branches placed around the body and house of a dead person to protect other people from its spirit.[70] Leaves placed in moccasins for athlete's foot and possibly for protection. (187:267)

Rosa nutkana var. *hispida* Fern., Bristly Nootka Rose
- ***Food***—**Blackfoot** *Fruit* Raw berries used for food. *Preserves* Berries used to make jelly. **Rocky Boy** *Fruit* Raw berries used for food. *Pre-*

serves Berries used to make jelly. **Washo** *Fruit* Raw berries used for food. (as *R. spaldingii* 118:22)

Rosa nutkana K. Presl var. *nutkana*, Nootka Rose
• **Food**—**Haisla & Hanaksiala** *Beverage* Leaves dried and mixed with American red raspberry leaves to make tea. *Unspecified* Hips mixed with oolichan (candlefish) grease and sugar and eaten. (43:273) **Makah** *Beverage* Leaves used to make tea. *Unspecified* Hips used for food. **Nitinaht** *Unspecified* Hips and petals used for food. (67:270)
• **Other**—**Haisla & Hanaksiala** *Ceremonial Items* Flowers used in "flower dance" costume. (43:273) **Makah** *Jewelry* Hips used to make necklaces. (67:270) **Oweekeno** *Ceremonial Items* Plant, wild parsnip, salmonberry, gooseberry, and mask represented a child in a ceremonial dance. (43:111)

Rosa palustris Marsh., Swamp Rose
• **Drug**—**Cherokee** *Anthelmintic* Infusion of bark and root used for worms. *Antidiarrheal* Decoction of roots taken for dysentery. (80:53)

Rosa pisocarpa Gray, Cluster Rose
• **Drug**—**Snohomish** *Throat Aid* Decoction of roots taken for sore throats. **Squaxin** *Gynecological Aid* Infusion of bark taken after childbirth. (79:34) **Thompson** *Antidiarrheal* Decoction of branches, chokecherry, and red willow taken for diarrhea. *Antiemetic* Decoction of branches, chokecherry, and red willow taken for vomiting. *Dermatological Aid* Leaves placed in moccasins for athlete's foot and possibly for protection. *Gynecological Aid* Decoction of roots taken by women after childbirth. Decoction of branches, chokecherry, and red willow taken for women's illnesses. *Venereal Aid* Decoction of roots taken for syphilis. (187:267) **Yurok** *Unspecified* Fruit used to make a medicinal tea. (6:51)
• **Food**—**Hoh** *Fruit* Fruits eaten fresh. *Winter Use Food* Fruits eaten in winter. (137:63) **Paiute** *Unspecified* Haws pounded with deer tallow and eaten. (104:103) Species used for food. (167:244) **Quileute** *Fruit* Fruits eaten fresh. *Winter Use Food* Fruits eaten in winter. (137:63) **Squaxin** *Unspecified* Hips eaten fresh. (79:34) **Thompson** *Beverage* Shoots used to make a tea-like beverage. Leaves and young twigs used to make a tea-like beverage. *Unspecified* Young, tender shoots peeled and eaten in the spring. (187:267)
• **Fiber**—**Thompson** *Furniture* Heavy, split wood used to make cradle hoops. (187:267)
• **Other**—**Thompson** *Good Luck Charm* Plant asked "for good luck." *Hunting & Fishing Item* Plant used to wipe dip nets, to improve the chances of a good catch. *Paint* Petals mixed with pine pitch, grease, and red ocher paint to make a cosmetic. *Protection* Branches placed around the body and house of a dead person to protect other people from its spirit.[70] Leaves placed in moccasins for athlete's foot and possibly for protection. (187:267)

Rosa sp., Wild Rose
• **Drug**—**Bella Coola** *Analgesic* Decoction of roots and branches taken as a purgative for stomach pain. *Cathartic* and *Gastrointestinal Aid* Decoction of roots and branches taken as a purgative for stomach pain. (150:59) **Blackfoot** *Antidiarrheal* Infusion of stems or root bark taken for diarrhea. (82:62) Infusion of roots used for diarrhea. (97:39) *Gastrointestinal Aid* Infusion of stems or root bark taken for stomach maladies. (82:62) **Carrier, Northern** *Dermatological Aid* Compound decoction of inner bark taken for body sores. *Orthopedic Aid* Compound decoction of inner bark taken for weakness or paralysis. *Stimulant* Compound decoction of inner pulp taken for constitutional weakness. (150:59) **Cherokee** *Antidiarrheal* Decoction of roots taken for dysentery. (203:74) **Cheyenne** *Antidiarrheal* Infusion of stems or root bark taken for diarrhea. (82:62) Infusion of inner bark used for diarrhea. (83:36) *Eye Medicine* Infusion of petals, stem bark, or root bark used for sore eyes. *Gastrointestinal Aid* Infusion of stems or root bark taken for stomach maladies. (82:62) Infusion of inner bark used for stomach troubles. *Unspecified* Flowers and roots used as an ingredient for medicines. (83:36) **Chippewa** *Eye Medicine* Infusion of root bark used as a wash to reduce inflammation from cataracts. (53:360) **Crow** *Antidiarrheal* Infusion of stems or root bark taken for diarrhea. *Antirheumatic (External)* Infusion of crushed roots used for swelling. *Gastrointestinal Aid* Infusion of stems or root bark taken for stomach maladies. *Nose Medicine* Vapor from infusion of crushed roots used for nosebleeding. *Oral Aid* Infusion of crushed roots taken for mouth bleeding. *Throat Aid* Infusion of crushed roots taken for tonsillitis and sore throat. **Flathead** *Eye Medicine* Infusion of petals, stem bark, or root bark used for sore eyes. **Gros Ventre** *Antidiarrheal* Infusion of stems or root bark taken for diarrhea. *Gastrointestinal Aid* Infusion of stems or root bark taken for stomach maladies. **Kutenai** *Antidiarrheal* Infusion of stems or root bark taken for diarrhea. *Gastrointestinal Aid* Infusion of stems or root bark taken for stomach maladies. (82:62) **Paiute** *Dermatological Aid* Poultice of chewed rose galls applied to boils. *Gastrointestinal Aid* Decoction of bark and small stems taken for stomach troubles or "when they don't feel very good." *Stimulant* Decoction of bark and small stems applied to the scalp for fainting or dizzy sensations. (111:81) **Sikani** *Eye Medicine* Infusion of crushed roots used as an eyewash. (150:59)
• **Food**—**Blackfoot** *Fruit* Crushed rose hips used to make pemmican. (86:105) Fruits eaten fresh or roasted. *Starvation Food* Dried fruits, hanging on the bushes in the winter, used as a famine food. (97:39) **Cheyenne** *Pie & Pudding* Hips boiled, sugar and flour added, and eaten as a pudding. *Unspecified* Petals used for food. (83:36) **Coeur d'Alene** *Fruit* Berries eaten fresh. (178:90) **Montana Indian** *Starvation Food* Hips used for food during famines. *Unspecified* Hips used for food. **Nez Perce** *Starvation Food* Hips used for food during famines. (82:62) **Okanagon** *Staple* Berries used as a principal food. (178:239) **Paiute** *Fruit* Rose hips eaten fresh. (111:81) **Spokan** *Fruit* Berries used for food. (178:343) **Washo** *Beverage* Roots used to make a rose-colored tea. (118:17)
• **Other**—**Blackfoot** *Jewelry* Fruits used to make necklaces. (97:39) **Cheyenne** *Hunting & Fishing Item* Used to make arrow shafts. (83:46) **Flathead** *Protection* Stems used for the howling dead. **Nez Perce** *Protection* Sprigs hung on cradleboards to keep ghosts from babies. (82:62) **Paiute** *Jewelry* Rose hips strung for beads and worn by children. *Protection* Sprig of rose carried to keep the ghost away at a funeral. (111:81) **Tewa** *Incense & Fragrance* Dried petals used as a house perfume. (138:48)

Rosa virginiana P. Mill., Virginia Rose
• **Drug**—**Cherokee** *Anthelmintic* and *Pediatric Aid* Decoction of roots used as a bath and given to children with worms. (177:29) **Ojibwa** *Dermatological Aid* Infusion of roots taken and used as a wash for

bleeding foot cuts. *Hemostat* Infusion of roots taken and used as a wash for bleeding foot cuts. (as *R. lucida* 135:231) *Unspecified* Root and bark used for medicinal purposes. (as *R. lucida* 135:236) **Ojibwa, South** *Eye Medicine* Infusion of root used as a wash for sore eyes. (as *R. lucida* 91:201)
- *Food*—**Ojibwa** *Unspecified* Buds used for food. (as *R. lucida* 135:236)

Rosa woodsii Lindl., Woods's Rose
- *Drug*—**Arapaho** *Antirheumatic (External)* Seeds used to produce a drawing effect for muscular pains. (121:48) **Okanagan-Colville** *Ceremonial Medicine* Decoction of leaves, branches, and other boughs taken and used as body and hair wash by sweat bathers. *Dermatological Aid* Poultice of chewed leaves applied to bee stings. (188:131) **Paiute** *Antidiarrheal* Decoction of root taken by adults and children for diarrhea. *Burn Dressing* Poultice of various plant parts applied to burns. *Cold Remedy* Decoction of root or inner bark taken for colds. *Dermatological Aid* Poultice of mashed fungus galls applied to opened boils. Poultice of various plant parts applied to sores, cuts, swellings, and wounds. *Misc. Disease Remedy* and *Pediatric Aid* Decoction of roots given to children for intestinal influenza. *Tonic* Infusion of leaves taken as a spring tonic. **Shoshoni** *Antidiarrheal* Decoction of root taken for diarrhea. *Blood Medicine* Compound decoction of roots taken as a blood tonic and for general debility. *Burn Dressing* Poultice of various plant parts applied to burns. *Cold Remedy* Decoction of root or inner bark taken for colds. *Dermatological Aid* Poultice of various plant parts applied to sores, cuts, swellings, and wounds. *Diuretic* Decoction of root taken for "failure of urination." *Tonic* Compound decoction of roots taken as a blood tonic and for general debility. (180:129–131) **Thompson** *Antidiarrheal* Decoction of branches, chokecherry, and red willow taken for diarrhea. *Antiemetic* Decoction of branches, chokecherry, and red willow taken for vomiting. *Cough Medicine* Infusion of one handful of washed hips taken for coughs. Infusion of sticks taken for coughs. *Dermatological Aid* Leaves placed in moccasins for athlete's foot and possibly for protection. *Gynecological Aid* Hips chewed by women in labor, to hasten the delivery. Decoction of roots taken by women after childbirth. Decoction of branches, chokecherry, and red willow taken for women's illnesses. *Pediatric Aid* Infusion of one handful of washed hips taken, especially by babies, for coughs. Infusion of sticks taken for coughs, especially babies' coughs. *Throat Aid* Infusion of one handful of washed hips taken for sore, itchy throats. Infusion of sticks taken for sore, itchy throats. *Venereal Aid* Decoction of roots taken for syphilis. (187:267) **Washo** *Cold Remedy* Decoction of root or inner bark taken for colds. (180:129–131)
- *Food*—**Arapaho** *Beverage* Bark used to make tea. (121:48) **Lakota** *Fruit* Rose hips used for food. (139:57) **Montana Indian** *Fruit* Fruit used for food. (19:21) **Okanagan-Colville** *Forage* Hips eaten by coyotes. *Spice* Leaves placed under and over food while pit cooking to add flavor and prevent burning. *Unspecified* Orange, outer rind of the hips used for food. (188:131) **Thompson** *Beverage* Shoots or hips[67] or leaves and young twigs used to make a tea-like beverage. *Forage* Hips eaten by bears before hibernation. *Preserves* Hips used to make jelly.[67] *Sauce & Relish* Hips used to make syrup.[67] *Unspecified* Hips eaten only sparingly because of the seeds and the insipid taste.[68] Young, tender shoots peeled and eaten in the spring. (187:267)
- *Fiber*—**Thompson** *Furniture* Heavy, split wood used to make cradle hoops. (187:267)
- *Dye*—**Arapaho** *Orange* Root used to make an orange dye. (121:48)
- *Other*—**Okanagan-Colville** *Ceremonial Items* Branches used by an Indian doctor to sweep out the grave before the corpse was lowered into it.[69] *Containers* Leaves placed under and over food while pit cooking to add flavor and prevent burning. *Good Luck Charm* Branches boiled in water and used to soak fishing lines and nets to obtain good luck. *Hunting & Fishing Item* Branches used in a wash by hunters to get rid of the human scent. *Protection* Branches made into tea and taken as protection from bad spirits and ghosts. Branches made into tea and used as washing water for one who was being jinxed by some bad person. Branches placed around the house and yard of the deceased to keep his or her spirit from returning. (188:131) **Thompson** *Good Luck Charm* Plant asked "for good luck." *Hunting & Fishing Item* Plant used to wipe dip net hoops, to improve the chances of a good catch. *Paint* Petals mixed with pine pitch, grease, and red ocher paint to make a cosmetic. *Protection* Branches placed around the body and house of a dead person to protect other people from its spirit.[70] Leaves placed in moccasins for athlete's foot and possibly for protection. (187:267)

Rosa woodsii var. *ultramontana* (S. Wats.) Jepson, Woods's Rose
- *Food*—**Hopi** *Fruit* Fruits occasionally eaten by children. (as *R. arizonica* 200:78) **Kawaiisu** *Fruit* Fruit, a "fleshy hip," eaten ripe. (206:60) **Sanpoil & Nespelem** *Starvation Food* Pips eaten in times of famine. (as *R. californica* var. *ultramontana* 131:108)
- *Fiber*—**Kawaiisu** *Basketry* Unsplit stems used as rims in twined basketry. (206:60)

Rosa woodsii Lindl. var. *woodsii*, Woods's Rose
- *Drug*—**Isleta** *Pediatric Aid* Rose petals soaked in water and the liquid given to newborn babies before the mother's milk. (as *R. fendleri* 100:42) **Navajo, Ramah** *Ceremonial Medicine* and *Emetic* Leaves used as a ceremonial emetic. (as *R. fendleri* 191:31)
- *Food*—**Apache, Chiricahua & Mescalero** *Fruit* Rose hips eaten fresh. *Preserves* Rose pulps squeezed into water and boiled to make jelly. (as *R. fendleri* 33:46) **Cheyenne** *Fruit* Berries not to be eaten too freely. (as *R. fendleri* 76:177) **Gosiute** *Fruit* Berries used for food. (as *R. fendleri* 39:379) **Klamath** *Fruit* Fruit used for food. (as *R. fendleri* 45:99) **Navajo** *Fruit* Fruits eaten for food. (as *R. fendleri* 55:55) **Navajo, Ramah** *Fruit* Fruit eaten raw. Fruit eaten raw. (as *R. neomexicana* 191:31) **Ute** *Fruit* Berries used for food. (as *R. fendleri* 38:36)
- *Other*—**Klamath** *Hunting & Fishing Item* Stems used for light arrow shafts. *Smoking Tools* Stems used for pipestems. (as *R. fendleri* 45:99) **Navajo** *Ceremonial Items* Used as a medicine in the Sun's House Chant. *Tools* Wood used to make needles for leather work. (as *R. fendleri* 55:55) **Navajo, Ramah** *Ceremonial Items* Stem used to make Holyway big hoop. (as *R. fendleri* 191:31)

Rubiaceae, see *Bobea, Cephalanthus, Cinchona, Coprosma, Galium, Houstonia, Mitchella, Morinda, Sherardia*

Rubus, Rosaceae
Rubus aculeatissimus, Red Raspberry
- *Food*—**Iroquois** *Unspecified* Fresh shoots peeled and eaten. (196:119)

Rubus allegheniensis Porter, Allegheny Blackberry
- **Drug**—**Cherokee** *Antidiarrheal* Infusion of root or leaf used for diarrhea. *Antirheumatic (Internal)* Infusion given for rheumatism. (80:26) *Dermatological Aid* Compound, astringent, and tonic infusion of root used as a wash for piles. (80:25, 26) *Hemorrhoid Remedy* Compound infusion of root used for piles. *Oral Aid* Washed root chewed for coated tongue. *Stimulant* Used as a stimulant. *Throat Aid* Used with honey as a wash for sore throat. *Tonic* Used as a tonic. *Urinary Aid* Compound decoction taken to regulate urination. (80:26) Infusion of bark taken for urinary troubles. (177:29) *Venereal Aid* Used for venereal disease. (80:26) **Chippewa** *Antidiarrheal* Infusion of roots taken for diarrhea. *Gynecological Aid* Infusion of roots taken by pregnant women threatened with miscarriage. (71:133) **Iroquois** *Analgesic* Compound of plant used as a snuff for headaches. (87:357) *Antidiarrheal* Plant used as a diarrhea medicine. *Blood Medicine* Compound decoction of roots taken by all ages as blood remedy. (87:356) *Cold Remedy* Compound decoction of roots taken for colds. *Cough Medicine* Compound decoction of roots taken for coughs. *Dermatological Aid* Poultice of smashed roots applied to baby's sore navel after birth. *Other* Root used for "summer complaint." *Pediatric Aid* Poultice of smashed roots applied to baby's sore navel after birth. *Respiratory Aid* Compound of plant used as a snuff for catarrh. (87:357) *Tuberculosis Remedy* Compound decoction of roots taken for tuberculosis. *Witchcraft Medicine* Infusion of roots used to make dogs good hunters and ensure them from theft. (87:356) **Menominee** *Eye Medicine* Infusion of root used as a wash for sore eyes. *Unspecified* Poultice of infusion of root used for unspecified ailments. (151:50) **Meskwaki** *Antidote* Decoction of root used as an antidote for poison. *Eye Medicine* Root extract used for sore eyes and stomach trouble. *Gastrointestinal Aid* Root extract used for stomach trouble and sore eyes. (152:243) **Ojibwa** *Antidiarrheal* Infusion of root used to "arrest flux." *Diuretic* Decoction of canes taken as a diuretic. (153:385, 386) **Potawatomi** *Eye Medicine* Root bark used by the Prairie Potawatomi for sore eyes. (154:79)
- **Food**—**Cherokee** *Beverage* Fruit used to make juice. *Fruit* Fruit used for food. (80:26) **Chippewa** *Dried Food* Fruit dried for winter use. *Fruit* Fruit eaten fresh. (71:133) **Menominee** *Fruit* Berries eaten fresh. *Pie & Pudding* Berries made into pies. *Winter Use Food* Berries dried for winter use. (151:71) **Meskwaki** *Fruit* Berries eaten fresh. *Pie & Pudding* Berries made into pies. *Preserves* Berries made into jams. *Winter Use Food* Berries sun dried for winter use. (152:264) **Ojibwa** *Preserves* Berries used to make jam for winter use. (153:409) **Potawatomi** *Fruit* Blackberries only used for food. (154:108)

Rubus allegheniensis Porter **var.** ***alleghaniensis***, Allegheny Blackberry
- **Drug**—**Delaware** *Antidiarrheal* Vine combined with wild cherry bark and used for diarrhea. (as *R. nigrobaccus* 176:33) **Delaware, Oklahoma** *Antidiarrheal* Compound containing vine and wild cherry bark used for dysentery. (as *R. nigrobaccus* 175:28, 78)

Rubus arcticus L., Arctic Blackberry
- **Food**—**Alaska Native** *Preserves* Fruit used to make a superior jelly. (85:91) **Eskimo, Alaska** *Fruit* Berries sometimes used for food, but not considered a significant food source. (1:36) **Eskimo, Inuktitut** *Winter Use Food* Berries added to stored salmonberries. (202:189) **Eskimo, Inupiat** *Dessert* Berries used to make traditional dessert. *Fruit* Berries eaten fresh. *Winter Use Food* Berries mixed with salmonberries and stored in a barrel for future use. (98:103) **Koyukon** *Fruit* Berries used for food. (119:55) **Tanana, Upper** *Frozen Food* Berries frozen for future use. *Fruit* Berries eaten raw, plain, or mixed raw with sugar, grease, or the combination of the two. Berries fried in grease with sugar or dried fish eggs. Berries boiled with sugar and flour to thicken. *Pie & Pudding* Berries used to make pies. *Preserves* Berries used to make jam and jelly. *Winter Use Food* Berries preserved alone or in grease and stored in a birch bark basket in an underground cache. (102:12)

Rubus arcticus ssp. ***acaulis*** (Michx.) Focke, Dwarf Raspberry
- **Drug**—**Shuswap** *Antidiarrheal* Leaves used for diarrhea. (as *R. acaulis* 123:67)
- **Food**—**Cree, Woodlands** *Fruit* Fruit eaten fresh. (109:56)

Rubus argutus Link, Sawtooth Blackberry
- **Drug**—**Cherokee** *Antidiarrheal* Infusion of root or leaf used for diarrhea. *Antirheumatic (Internal)* Infusion given for rheumatism. (80:26) *Dermatological Aid* Compound, astringent, and tonic infusion of root used as a wash for piles. (80:25, 26) *Hemorrhoid Remedy* Compound infusion of root used for piles. *Oral Aid* Washed root chewed for coated tongue. *Stimulant* Used as a stimulant. *Throat Aid* Used with honey as a wash for sore throat. *Tonic* Used as a tonic. *Urinary Aid* Compound decoction taken to regulate urination. *Venereal Aid* Used for venereal disease. (80:26)
- **Food**—**Cherokee** *Beverage* Fruit used to make juice. *Fruit* Fruit used for food. (80:26)

Rubus arizonensis Focke, Arizona Dewberry
- **Food**—**Apache, Chiricahua & Mescalero** *Bread & Cake* Fruit pressed into pulpy cakes, dried, and stored. *Fruit* Fruits eaten fresh. *Winter Use Food* Fruit pressed into pulpy cakes, dried, and stored for winter use. (33:44) **Navajo** *Fruit* Fruits eaten for food. (as *R. arizonicus* 55:55)

Rubus canadensis L., Smooth Blackberry
- **Drug**—**Delaware, Oklahoma** *Antidiarrheal* Vine and berries used for dysentery. (175:78) **Iroquois** *Unspecified* Berries, maple sap, and water used to make a medicine. (196:142) **Menominee** *Antidiarrheal* Simple or compound decoction of root used for dysentery. (54:131)
- **Food**—**Chippewa** *Fruit* Fruit used for food. (71:133) **Iroquois** *Beverage* Berries, water, and maple sugar used to make a drink for home consumption and longhouse ceremonies. *Bread & Cake* Fruit mashed, made into small cakes, and dried for future use. *Dried Food* Raw or cooked fruit sun or fire dried and stored for future use. *Fruit* Dried fruit taken as a hunting food. *Sauce & Relish* Dried fruit cakes soaked in warm water and cooked as a sauce or mixed with corn bread. (196:127) **Ojibwa** *Fruit* Berries used fresh. *Winter Use Food* Berries used preserved. (5:2223)

Rubus chamaemorus L., Cloudberry
- **Drug**—**Cree, Woodlands** *Gynecological Aid* Decoction of roots used as a "woman's medicine." Decoction of plant used for hard labor. *Reproductive Aid* Decoction of root and lower stem used by barren women. (109:56) **Micmac** *Cough Medicine* Roots used for cough. *Febrifuge* Roots used for fever. *Tuberculosis Remedy* Roots used for consumption. (40:61)

- **Food**—Alaska Native *Dietary Aid* Berries used as a very rich source of vitamin C. *Frozen Food* Fruit stored in seal pokes, kegs, or barrels and buried in the frozen tundra for future use. *Fruit* Berries eaten raw with sugar, seal oil, or both. *Pie & Pudding* Berries used to make berry shortcakes and pies. *Winter Use Food* Fruit stored in large quantities for winter use. (85:93) **Anticosti** *Preserves* Fruits used to make jelly. *Winter Use Food* Fruits stored for winter use. (143:67) **Cree, Woodlands** *Fruit* Fruit eaten fresh. (109:56) **Eskimo** *Fruit* Fruits eaten for food. (181:233) **Eskimo, Alaska** *Winter Use Food* Berries stored with seal oil in barrels or sealskin pokes for winter use. (1:36) **Eskimo, Arctic** *Ice Cream* Berries mixed with seal oil and chewed caribou tallow, beaten, and eaten as "Eskimo ice cream." (128:21) **Eskimo, Inuktitut** *Frozen Food* Berries frozen for future use. *Fruit* Berries eaten fresh or mixed with oil or fat. (202:183) **Eskimo, Inupiat** *Dessert* Berries mixed with sugar and seal oil and eaten as a dessert. *Frozen Food* Berries mixed with blueberries and frozen for future use. *Fruit* Berries eaten fresh. Berries mixed with blueberries and eaten fresh. *Ice Cream* Berries added to fluffy fat and eaten as ice cream. *Winter Use Food* Berries mixed with blackberries, preserved in a poke or barrel, and stored for winter use. (98:73) **Koyukon** *Fruit* Berries used for food. (119:55) **Tanana, Upper** *Frozen Food* Berries frozen for future use. *Fruit* Berries eaten raw, plain, or mixed raw with sugar, grease, or the combination of the two. Berries fried in grease with sugar or dried fish eggs. Berries boiled with sugar and flour to thicken. *Pie & Pudding* Berries used to make pies. *Preserves* Berries used to make jam and jelly. *Winter Use Food* Berries preserved alone or in grease and stored in a birch bark basket in an underground cache. (102:12)

Rubus cuneifolius Pursh, Sand Blackberry
- **Drug**—Seminole *Other* Complex infusion of roots taken for chronic conditions. (169:272)

Rubus discolor Weihe & Nees, Himalayan Blackberry
- **Food**—Haisla & Hanaksiala *Fruit* Fruit used for food. (43:278) **Kitasoo** *Fruit* Fruit used for food. (43:348) **Makah** *Fruit* Fruit used for food. (67:272)

Rubus flagellaris Willd., Northern Dewberry
- **Drug**—Cherokee *Antidiarrheal* Infusion of root or leaf used for diarrhea. *Antirheumatic (Internal)* Infusion given for rheumatism. (80:26) *Dermatological Aid* Compound, astringent, and tonic infusion of root used as a wash for piles. (80:25, 26) *Hemorrhoid Remedy* Compound infusion of root used for piles. *Oral Aid* Washed root chewed for coated tongue. *Stimulant* Used as a stimulant. *Throat Aid* Used with honey as a wash for sore throat. *Tonic* Used as a tonic. *Urinary Aid* Compound decoction taken to regulate urination. *Venereal Aid* Used for venereal disease. (80:26)
- **Food**—Cherokee *Beverage* Fruit used to make juice. *Fruit* Fruit used for food. (80:26)

Rubus frondosus Bigelow, Yankee Blackberry
- **Drug**—Chippewa *Abortifacient* Decoction of root taken for "stoppage of periods." (53:358) *Pulmonary Aid* Compound decoction of root taken for lung trouble. (53:340)
- **Food**—Chippewa *Bread & Cake* Berries cooked, spread on birch bark into little cakes, dried, and stored for winter use. *Fruit* Berries eaten raw. (53:321)

Rubus fruticosus L., Shrubby Blackberry
- **Drug**—Micmac *Antidiarrheal* and *Pediatric Aid* Bark and roots used for children's diarrhea. (40:61)

Rubus hawaiensis Gray, Hawaii Blackberry
- **Drug**—Hawaiian *Antiemetic* Plant ashes mixed with poi or potatoes and eaten for vomiting. *Dermatological Aid* Infusion of plant ashes and tobacco leaf ashes used as a wash for scaly scalps. *Gastrointestinal Aid* Plant ashes mixed with "papaia meat" and eaten for burning in the chest. (2:8)

Rubus hispidus L., Bristly Dewberry
- **Drug**—Micmac *Cough Medicine* Roots used for cough. *Febrifuge* Roots used for fever. *Tuberculosis Remedy* Roots used for consumption. (40:61) **Mohegan** *Anthelmintic* Infusion of berries taken as a vermifuge. (174:265) *Antidiarrheal* Juice of plant taken for dysentery. (174:269) Juice of berries taken for dysentery. (176:75, 132) **Rappahannock** *Antidiarrheal* Infusion taken for diarrhea. (30:119) Fermented decoction of berries or roots taken for dysentery. *Dermatological Aid* Infusion of leaves taken for boils. *Tonic* Fermented decoction of berries taken for dysentery and as a tonic. (161:32) **Shinnecock** *Antidiarrheal* Infusion taken for diarrhea and fruit used to check dysentery. (30:119)

Rubus idaeus L., American Red Raspberry
- **Drug**—Algonquin, Quebec *Antidiarrheal* Root used for diarrhea. *Unspecified* Root had medicinal value. (18:180) **Algonquin, Tête-de-Boule** *Urinary Aid* Decoction of roots used for bloody urine. (132:130) **Cherokee** *Analgesic* Strong infusion of red raspberry leaves used for childbirth pains. *Antirheumatic (External)* Thorny branch used to scratch rheumatism. *Cathartic* Taken as a purgative. *Cough Medicine* Root chewed for cough. *Dermatological Aid* Infusion taken as a tonic for boils. Leaves highly astringent and decoction taken for bowel complaint. Used as wash for old and foul sores and infusion taken as tonic for boils. *Emetic* Taken as an emetic. *Gastrointestinal Aid* Decoction taken for bowel complaint. *Gynecological Aid* Strong infusion used for childbirth pains and decoction used for menstrual period. *Tonic* Infusion taken as a tonic for boils. *Toothache Remedy* Roots used for toothache. (80:52) **Cree, Woodlands** *Gynecological Aid* Decoction of stem and upper part of the roots used to help a woman recover after childbirth, and to slow menstrual bleeding. *Heart Medicine* Fruit used as a heart medicine. *Pediatric Aid* and *Toothache Remedy* Decoction of stem and upper part of the roots used for teething sickness. (109:57) **Iroquois** *Analgesic* Decoction of leaves taken for "burning and pain when passing water." *Blood Medicine* Compound used when the "blood is bad and sores break out on the neck." Decoction of roots taken as a blood purifier. *Cathartic* Decoction of leaves taken as a physic. *Dermatological Aid* Compound used for boils. Compound used when the "blood is bad and sores break out on the neck." *Emetic* Decoction of leaves taken as an emetic. *Gynecological Aid* Compound decoction taken by "ladies who are run down from period sickness." *Hypotensive* Decoction of roots taken for low or high blood pressure. *Kidney Aid* Decoction of leaves taken for the kidneys. *Liver Aid* Decoction of leaves taken for bile. *Stimulant* Compound decoction taken by "ladies who are run down from period sickness." Compound used for laziness. *Tonic* Plant used as a tonic. *Urinary Aid* Decoction of leaves taken for "burning and pain

when passing water." *Venereal Aid* Compound decoction of roots taken for gonorrhea. (87:355) *Veterinary Aid* Leaves, rhizomes from another plant, and wheat flour given to cows at birthing. (141:48) **Menominee** *Adjuvant* Root used as a seasoner for medicines. (151:50) **Okanagan-Colville** *Antidiarrheal* Decoction of branches taken for diarrhea. *Cathartic* Decoction of branches taken as a physic. *Gastrointestinal Aid* Decoction of branches taken for heartburn. *Laxative* Decoction of roots taken for constipation. (188:131)
- *Food*—**Abnaki** *Fruit* Fruits eaten for food. (144:169) **Alaska Native** *Fruit* Berries eaten raw. *Preserves* Berries made into jams and jellies. (85:93) **Algonquin, Quebec** *Fruit* Fruit eaten fresh. *Preserves* Fruit preserved. (18:92) **Algonquin, Tête-de-Boule** *Fruit* Fruits eaten for food. (132:130) **Bella Coola** *Fruit* Berries eaten fresh. *Preserves* Berries cooked into jam. (184:209) **Cherokee** *Fruit* Fruit used for food. (80:52) **Cree, Woodlands** *Fruit* Fruit eaten with dried fish flesh and fish oil. *Unspecified* Young, leafy shoots peeled and the tender inner part eaten. (109:57) **Eskimo, Inupiat** *Dessert* Berries used to make traditional dessert. (98:107) **Koyukon** *Fruit* Berries used for food. (119:55) **Okanagan-Colville** *Dried Food* Berries dried for future use. *Frozen Food* Berries frozen for future use. *Fruit* Berries eaten fresh. *Winter Use Food* Berries canned for future use. (188:131) **Tanana, Upper** *Frozen Food* Berries frozen for future use. (102:12) *Fruit* Berries used for food. (77:28) Berries eaten raw, plain, or mixed raw with sugar, grease, or the combination of the two. Berries fried in grease with sugar or dried fish eggs. Berries boiled with sugar and flour to thicken. *Pie & Pudding* Berries used to make pies. *Preserves* Berries used to make jam and jelly. *Winter Use Food* Berries preserved alone or in grease and stored in a birch bark basket in an underground cache. (102:12) **Thompson** *Bread & Cake* Fruit steamed, dried, and made into a cake. *Dried Food* Fruit sun dried loose on mats. *Frozen Food* Fruit frozen or made into a jam. *Fruit* Fruit eaten fresh. *Preserves* Fruit frozen or made into jam. (187:269)

Rubus idaeus* ssp. *strigosus (Michx.) Focke, Grayleaf Red Raspberry
- *Drug*—**Cherokee** *Cathartic* Infusion of roots taken as a cathartic by women during menses. *Emetic* Infusion of roots taken as an emetic by women during menses. *Gynecological Aid* Infusion of roots taken as an emetic and cathartic by women during menses. (177:30) **Chippewa** *Antidiarrheal* Decoction of root taken for dysentery. (as *R. strigosus* 53:344) *Eye Medicine* Infusion of root bark used as a wash for cataracts. (as *R. strigosus* 53:360) *Misc. Disease Remedy* Decoction of roots or stems taken for measles. (as *R. strigosus* 71:132) **Meskwaki** *Adjuvant* Root used as a seasoner in medicines. (as *R. idaeus-aculeatissi* 152:243) **Ojibwa** *Adjuvant* Berries used as a seasoner for medicines. *Eye Medicine* Infusion of root bark used for sore eyes. (153:386) **Ojibwa, South** *Analgesic* Decoction of crushed root taken for stomach pain. *Gastrointestinal Aid* Decoction of crushed root taken for stomach pain. (as *R. strigosus* 91:199) **Omaha** *Gastrointestinal Aid* and *Pediatric Aid* Decoction of scraped root given to children for bowel trouble. (as *R. strigosus* 70:84, 85) **Potawatomi** *Eye Medicine* Infusion of root used as an eyewash. (154:79) **Thompson** *Antihemorrhagic* Decoction of leaves taken for spitting or vomiting blood. *Gastrointestinal Aid* and *Tonic* Decoction of root taken as a tonic for the stomach. (as *R. strigosus* 164:466)
- *Food*—**Cheyenne** *Fruit* Berries always eaten fresh. (as *R. melanolasius* 76:177) **Chippewa** *Beverage* Twigs used to make a beverage. (as *R. strigosus* 53:317) *Bread & Cake* Berries cooked, spread on birch bark into little cakes, dried, and stored for winter use. (as *R. strigosus* 53:321) *Dried Food* Fruit dried for winter use. (as *R. strigosus* 71:132) *Fruit* Berries eaten raw. (as *R. strigosus* 53:321) Fruit eaten fresh. (as *R. strigosus* 71:132) **Dakota** *Beverage* Young leaves steeped to make a tea-like beverage. *Dried Food* Fruit dried for winter use. *Fruit* Fruit eaten fresh. (as *R. strigosus* 70:84) **Haisla & Hanaksiala** *Beverage* Leaves mixed with young Nootka rose leaves to make tea. *Fruit* Fruit used for food. (43:274) **Hoh** *Fruit* Fruits eaten raw. Fruits stewed and used for food. *Winter Use Food* Fruits canned and saved for future food use. (as *R. strigosus* 137:63) **Iroquois** *Bread & Cake* Fruit mashed, made into small cakes, and dried for future use. *Dried Food* Raw or cooked fruit sun or fire dried and stored for future use. *Fruit* Dried fruit taken as a hunting food. *Sauce & Relish* Dried fruit cakes soaked in warm water and cooked as a sauce or mixed with corn bread. (196:127) **Kitasoo** *Fruit* Fruit used for food. (43:345) **Menominee** *Fruit* Berries eaten fresh. (as *R. idaeus aculeatissimus* 151:71) **Montana Indian** *Fruit* Fruit was highly esteemed. (as *R. strigosus* 19:22) **Ojibwa** *Dried Food* Berries used dried. (5:2223) Fruit dried for winter use. (as *R. strigosus* 135:235) *Fruit* Berries used fresh. (5:2223) Fruit eaten fresh. (as *R. strigosus* 135:235) This was a favorite fresh fruit. *Preserves* Berries used to make jam for winter use. (153:410) **Omaha** *Beverage* Young leaves steeped to make a tea-like beverage. *Dried Food* Fruit dried for winter use. *Fruit* Fruit eaten fresh. (as *R. strigosus* 70:84) **Oweekeno** *Fruit* Fruit used for food. (43:111) **Pawnee** *Beverage* Young leaves steeped to make a tea-like beverage. *Dried Food* Fruit dried for winter use. *Fruit* Fruit eaten fresh. **Ponca** *Beverage* Young leaves steeped to make a tea-like beverage. *Dried Food* Fruit dried for winter use. *Fruit* Fruit eaten fresh. (as *R. strigosus* 70:84) **Potawatomi** *Fruit* Berries, a favorite article of food, eaten fresh. *Preserves* Berries, a favorite article of food, made into jams and jellies. (154:109) **Quileute** *Fruit* Fruits eaten raw. Fruits stewed and used for food. *Winter Use Food* Fruits canned and saved for future food use. (as *R. strigosus* 137:63) **Shuswap** *Fruit* Berries used for food. (123:67)

Rubus laciniatus Willd., Cutleaf Blackberry
- *Food*—**Hoh** *Fruit* Fruits stewed and used for food. Fruits eaten raw. *Winter Use Food* Fruits canned and saved for future food use. (137:63) **Makah** *Fruit* Fresh fruit used for food. *Pie & Pudding* Fruit used to make pies. *Preserves* Fruit used to make jam. (67:272) **Quileute** *Fruit* Fruits stewed and used for food. Fruits eaten raw. *Winter Use Food* Fruits canned and saved for future food use. (137:63)

Rubus lasiococcus Gray, Roughfruit Berry
- *Food*—**Hoh** *Fruit* Fruits eaten raw. Fruits stewed and used for food. *Winter Use Food* Fruits canned and saved for future food use. **Quileute** *Fruit* Fruits eaten raw. Fruits stewed and used for food. *Winter Use Food* Fruits canned and saved for future food use. (137:62)

Rubus leucodermis Dougl. ex Torr. & Gray, Whitebark Raspberry
- *Drug*—**Pomo, Kashaya** *Antidiarrheal* Infusion of leaves or root taken for diarrhea. *Gastrointestinal Aid* Infusion of leaves or root taken for upset stomach. *Other* Infusion of leaves or root taken for weak bowels. (72:96) **Shoshoni** *Dermatological Aid* Poultice of powdered stems

applied to wounds and cuts. (180:131) **Thompson** *Misc. Disease Remedy* Mild infusion of washed roots taken for influenza. (187:269)
- **Food**—**Bella Coola** *Bread & Cake* Berries formerly dried in cakes and used for food. *Fruit* Berries eaten fresh. *Preserves* Berries cooked into jam. (184:209) **Cahuilla** *Beverage* Berries soaked in water to make a beverage. *Dried Food* Berries dried for later use. *Fruit* Berries eaten fresh. (15:134) **Coeur d'Alene** *Fruit* Berries eaten fresh. Berries mashed and eaten. (178:90) **Costanoan** *Fruit* Raw fruits used for food. (21:250) **Cowlitz** *Dried Food* Berries sun or fire dried and stored for winter use. *Fruit* Berries eaten fresh. (79:35) **Gosiute** *Fruit* Berries used for food. (39:380) **Green River Group** *Fruit* Berries eaten fresh. (79:35) **Hesquiat** *Fruit* Berries eaten with oil. (185:74) **Hoh** *Fruit* Fruits eaten raw. Fruits stewed and used for food. *Winter Use Food* Fruits canned and saved for future food use. (137:63) **Karok** *Fruit* Fruit eaten fresh. (6:51) Berries used for food. (148:384) **Klallam** *Dried Food* Berries dried and used for food. *Fruit* Fruits eaten fresh. (78:197) Berries eaten fresh. (79:35) *Unspecified* Leaves and sprouts used for food. (78:197) Sprouts and young leaves eaten. (79:35) **Klamath** *Dried Food* Berries dried for later use. *Fruit* Berries used for food. (45:99) **Makah** *Fruit* Fruit eaten fresh. *Pie & Pudding* Fruit used to make pies. *Preserves* Fruit used to make jam. (67:273) **Mendocino Indian** *Dried Food* Fruits dried or canned for winter use. *Fruit* Fruits eaten fresh. *Winter Use Food* Fruits dried or canned for winter use. (41:355) **Montana Indian** *Fruit* Fruit was highly esteemed. (19:22) **Nisqually** *Fruit* Berries eaten fresh. (79:35) **Nitinaht** *Fruit* Fruits eaten fresh and raw. (186:123) **Okanagan-Colville** *Dessert* Dried berries soaked in water or boiled and eaten as a dessert. *Dried Food* Berries dried and stored for winter use. *Fruit* Dried berries eaten with dried meat or fish. (188:132) **Okanagon** *Staple* Berries used as a principal food. (178:238) *Unspecified* Used commonly for food. (125:38) **Pomo** *Fruit* Raw berries used for food. *Winter Use Food* Berries cooked, bottled, and stored for later use. (66:13) **Pomo, Kashaya** *Fruit* Berries eaten fresh. *Winter Use Food* Berries canned. (72:96) **Puyallup** *Fruit* Berries eaten fresh. (79:35) **Quileute** *Fruit* Fruits eaten raw. Fruits stewed and used for food. *Winter Use Food* Fruits canned and saved for future food use. (137:63) **Salish, Coast** *Bread & Cake* Berries mashed, dried in rectangular frames, and cakes used as a winter food. *Fruit* Berries eaten fresh. (182:87) **Shuswap** *Fruit* Berries used for food. (123:67) **Skagit, Upper** *Dried Food* Berries pulped and dried for winter use. *Fruit* Berries eaten fresh. (179:38) **Spokan** *Fruit* Berries used for food. (178:343) **Thompson** *Dried Food* Fruit dried for winter use. (187:269) *Fruit* Reddish purple berries eaten. (164:487) Fruit used for food. (187:269) *Unspecified* Used commonly for food. (125:38) Young shoots peeled, cooked over a fire, and eaten alone or with fish. (187:269) **Yuki** *Dried Food* Berries dried and cooked as a winter food. *Fruit* Berries eaten fresh. (49:87) **Yurok** *Fruit* Fruit eaten fresh. (6:51)
- **Dye**—**Thompson** *Unspecified* Juice squeezed from dark reddish purple fruits and used as a stain. (164:502)

Rubus leucodermis Dougl. ex Torr. & Gray **var. *leucodermis*,** Whitebark Raspberry

- **Food**—**Haisla & Hanaksiala** *Fruit* Fruit used for food. (43:275) **Kitasoo** *Fruit* Fruit used for food. (43:348) **Oweekeno** *Fruit* Fruit used for food. (43:112)

Rubus macraei Gray, 'Akala

- **Drug**—**Hawaiian** *Antiemetic* Plant ashes mixed with poi or potatoes and eaten for vomiting. *Dermatological Aid* Infusion of plant ashes and tobacco leaf ashes used as a wash for scaly scalps. *Gastrointestinal Aid* Plant ashes mixed with "papaia meat" and eaten for burning in the chest. (2:8)

Rubus nivalis Dougl. ex Hook., Snow Raspberry

- **Food**—**Hoh** *Fruit* Fruits eaten raw. Fruits stewed and used for food. *Winter Use Food* Fruits canned and saved for future food use. **Quileute** *Fruit* Fruits eaten raw. Fruits stewed and used for food. *Winter Use Food* Fruits canned and saved for future food use. (137:63)

Rubus occidentalis L., Black Raspberry

- **Drug**—**Cherokee** *Analgesic* Strong infusion of red raspberry leaves used for childbirth pains. *Antirheumatic (External)* Thorny branch used to scratch rheumatism. *Cathartic* Taken as a purgative. (80:52) Infusion of roots taken as a cathartic by women during menses. (177:30) *Cough Medicine* Root chewed for cough. *Dermatological Aid* Infusion taken as a tonic for boils. Leaves highly astringent and decoction taken for bowel complaint. Used as wash for old and foul sores and infusion taken as tonic for boils. *Emetic* Taken as an emetic. (80:52) Infusion of roots taken as an emetic by women during menses. (177:30) *Gastrointestinal Aid* Decoction taken for bowel complaint. *Gynecological Aid* Strong infusion used for childbirth pains and decoction used for menstrual period. (80:52) Infusion of roots taken as an emetic and cathartic by women during menses. (177:30) *Tonic* Infusion taken as a tonic for boils. *Toothache Remedy* Roots used for toothache. (80:52) **Chippewa** *Analgesic* Compound decoction of root taken for back pain. (53:356) *Eye Medicine* Decoction of roots used as a wash for sore eyes. (71:133) *Gynecological Aid* Compound decoction of root taken for "female weakness." (53:356) **Iroquois** *Antidiarrheal* Compound decoction of roots taken for diarrhea with blood. *Cathartic* Leaves used as a physic. *Emetic* Leaves used as an emetic. *Liver Aid* Leaves used for removing bile. *Pediatric Aid* and *Pulmonary Aid* Decoction of roots, stalks, and leaves given to children with whooping cough. *Venereal Aid* Compound decoction of roots taken for gonorrhea. *Witchcraft Medicine* Decoction taken by a hunter and his wife to prevent her from fooling around. (87:356) **Menominee** *Tuberculosis Remedy* Root used with *Hypericum* sp. for consumption in the first stages. (151:50) **Ojibwa, South** *Analgesic* and *Gastrointestinal Aid* Decoction of crushed root taken for stomach pain. (91:199) **Omaha** *Gastrointestinal Aid* and *Pediatric Aid* Decoction of scraped root given to children for bowel trouble. (70:84, 85)
- **Food**—**Cherokee** *Fruit* Fruit used for food. (80:52) Fresh fruit used for food. *Pie & Pudding* Fruit used to make pies. *Preserves* Fruit used to make jelly. *Winter Use Food* Fruit canned for future use. (126:57) **Chippewa** *Dried Food* Fruit dried for winter use. *Fruit* Fruit eaten fresh. (71:133) **Dakota** *Beverage* Young leaves steeped to make a tea-like beverage. *Dried Food* Fruit dried for winter use. *Fruit* Fruit eaten fresh. (70:84) **Iroquois** *Bread & Cake* Fruits dried, soaked in water, and used in bread. (124:95) Fruit mashed, made into small cakes, and dried for future use. *Dried Food* Raw or cooked fruit sun or fire dried and stored for future use. *Fruit* Dried fruit taken as a hunting food. (196:127) *Pie & Pudding* Fruits dried, soaked in water, and used in pudding. *Porridge* Berries dried, soaked in cold water, heated

slowly, and mixed with bread meal or hominy in winter. *Sauce & Relish* Fruits dried, soaked in water, and used as a sauce. Berries dried, soaked in cold water, heated slowly, and used as a winter sauce. (124:95) Dried fruit cakes soaked in warm water and cooked as a sauce or mixed with corn bread. (196:127) *Soup* Fruits dried, soaked in water, and used in soups. (124:95) **Lakota** *Fruit* Fruits eaten for food. (139:57) **Menominee** *Fruit* Berries eaten fresh, not important as a fresh fruit. (151:71) **Meskwaki** *Beverage* Root bark used to make tea. *Fruit* Berries eaten fresh. *Winter Use Food* Berries sun dried for winter use. (152:264) **Ojibwa** *Fruit* Berries used fresh. *Winter Use Food* Berries used preserved. (5:2224) **Omaha** *Beverage* Leaves used to make a hot, aqueous, tea-like beverage. (68:329) Young leaves steeped to make a tea-like beverage. (70:84) *Dried Food* Fruit dried for winter use. *Fruit* Fruit eaten fresh. (68:326) **Pawnee** *Beverage* Young leaves steeped to make a tea-like beverage. *Dried Food* Fruit dried for winter use. *Fruit* Fruit eaten fresh. **Ponca** *Beverage* Young leaves steeped to make a tea-like beverage. *Dried Food* Fruit dried for winter use. *Fruit* Fruit eaten fresh. (70:84) **Thompson** *Unspecified* Sprouts or young shoots eaten like rhubarb. (164:484)

Rubus odoratus L., Purpleflowering Raspberry

- **Drug**—**Cherokee** *Analgesic* Strong infusion of red raspberry leaves used for childbirth pains. *Antirheumatic (External)* Thorny branch used to scratch rheumatism. *Cathartic* Taken as a purgative. *Cough Medicine* Root chewed for cough. *Dermatological Aid* Infusion taken as a tonic for boils. Leaves highly astringent and decoction taken for bowel complaint. Used as wash for old and foul sores and infusion taken as tonic for boils. *Emetic* Taken as an emetic. *Gastrointestinal Aid* Decoction taken for bowel complaint. *Gynecological Aid* Strong infusion used for childbirth pains and decoction used for menstrual period. *Tonic* Infusion taken as a tonic for boils. *Toothache Remedy* Roots used for toothache. (80:52) **Iroquois** *Antidiarrheal* Decoction of scraped bark or roots taken for diarrhea. *Blood Medicine* Decoction taken as blood medicine and blood purifier. (87:354) *Cold Remedy* Roots used for colds. (87:355) *Dermatological Aid* Compound decoction taken and used as wash for venereal disease chancres and sores. (87:354) *Diuretic* Berries eaten in late summer or dried in winter and used as a diuretic. (124:96) *Gastrointestinal Aid* Decoction or infusion of branches taken to settle the stomach. (87:355) *Gynecological Aid* Compound infusion of plants taken by women who have a miscarriage. *Kidney Aid* Compound decoction of stalks and leaves taken as kidney medicine. *Laxative* and *Pediatric Aid* Decoction given as blood medicine and for the bowels of newborn babies. *Venereal Aid* Compound decoction taken and used as wash for venereal disease chancres and sores. (87:354)
- **Food**—**Algonquin, Quebec** *Fruit* Fruit used for food. (18:92) **Cherokee** *Fruit* Fruit used for food. (80:52) Fresh fruit used for food. *Pie & Pudding* Fruit used to make pies. *Preserves* Fruit used to make jelly. *Winter Use Food* Fruit canned for future use. (126:57) **Chippewa** *Dried Food* Fruit dried for winter use. *Fruit* Fruit eaten fresh. (71:133) **Iroquois** *Bread & Cake* Fruit mashed, made into small cakes, and dried for future use. *Dried Food* Raw or cooked fruit sun or fire dried and stored for future use. *Fruit* Dried fruit taken as a hunting food. *Sauce & Relish* Dried fruit cakes soaked in warm water and cooked as a sauce or mixed with corn bread. (196:127)
- **Other**—**Iroquois** *Protection* Leaves placed inside the shoes of forest runners to protect the feet. (141:48)

Rubus parviflorus Nutt., Thimbleberry

- **Drug**—**Blackfoot** *Pulmonary Aid* Berries given by diviners to patients to eat for chest disorders. (86:74) **Cowlitz** *Burn Dressing* Poultice of dried leaves applied to burns. (79:34) **Karok** *Dietary Aid* Infusion of roots taken by thin people as an appetizer or tonic. *Tonic* Infusion of roots taken by thin people as a tonic. (148:384) **Kwakiutl** *Antiemetic* Decoction of leaves taken for vomiting. *Antihemorrhagic* Decoction of leaves taken for blood-spitting. *Dermatological Aid* Dried, powdered leaves applied to wounds. *Gynecological Aid* Leaves used when a woman's period was unduly long. *Internal Medicine* Dried, powdered leaves eaten for internal disorders. (183:291) **Makah** *Blood Medicine* Decoction of leaves taken for anemia and to strengthen the blood. (79:34) **Montana Indian** *Alterative* Young sprouts considered a valuable alterative. *Misc. Disease Remedy* Young sprouts considered a valuable antiscorbutic. (as *R. nutkanus* 19:21) **Okanagan-Colville** *Dermatological Aid* Decoction of roots taken by young people with pimples and blackheads. Leaves rubbed on the face of young people with pimples and blackheads. *Gastrointestinal Aid* Infusion of roots taken for stomach ailments. (188:132) **Saanich** *Antidiarrheal* Leaves dried and chewed for diarrhea. *Gastrointestinal Aid* Leaves dried and chewed for stomachaches. (182:87) **Skagit** *Dermatological Aid* Poultice of leaf ashes and grease applied to swellings. (79:34) **Thompson** *Dermatological Aid* and *Pediatric Aid* Green insect galls found on stems burned and the ashes rubbed on babies' navels if they did not heal. (187:270)
- **Food**—**Alaska Native** *Fruit* Berries used for food. (85:97) **Bella Coola** *Preserves* Berries cooked with wild raspberries and other fruits into a thick jam, dried, and used for food. *Unspecified* Young sprouts peeled and eaten in spring. (184:209) **Blackfoot** *Fruit* Ripe fruit used for food. (86:105) **Cahuilla** *Beverage* Berries soaked in water to make a beverage. *Dried Food* Berries dried for later use. *Fruit* Berries eaten fresh. (15:134) **Chehalis** *Fruit* Berries eaten fresh. (79:34) **Clallam** *Fruit* Berries eaten fresh. (57:203) **Cowlitz** *Fruit* Berries eaten fresh. (79:34) **Gosiute** *Fruit* Berries used for food. (as *R. nutkanus* 39:380) **Hesquiat** *Dried Food* Berries dried for future use. *Fruit* Berries eaten fresh. *Preserves* Berries made into jam. *Spice* Fish boiled with leaves as flavoring and kept the fish from sticking to the pot. (185:74) **Hoh** *Fruit* Fruits eaten raw. Fruits stewed and used for food. *Winter Use Food* Fruits canned and saved for future food use. (137:63) **Isleta** *Fruit* Berries grown in the mountains, considered a strawberry and used for food. (as *Bossekia parvifolia* 100:25) *Special Food* Fruits eaten as a delicacy. (as *Bossekia parviflora* 32:19) **Karok** *Fruit* Berries used for food. (148:384) **Klallam** *Dried Food* Berries dried and used for food. *Unspecified* Sprouts used for food. (78:197) Sprouts eaten in early spring. (79:34) **Kwakiutl, Southern** *Dried Food* Berries dried in cakes and used as a winter food. *Fruit* Berries eaten fresh. (183:291) **Luiseño** *Fruit* Fruit used for food. (155:232) **Makah** *Fruit* Fruit eaten fresh. *Preservative* Fruit used to make jam and jelly. *Unspecified* Raw sprouts used for food. (67:273) Sprouts eaten in early spring. (79:34) **Montana Indian** *Unspecified* Young sprouts eaten raw or tied into bundles and steamed. (as *R. nutkanus* 19:21) **Nitinaht** *Fruit* Berries eaten raw. *Unspecified* Young, tender sprouts peeled and eaten raw

in spring. (186:124) **Okanagan-Colville** *Fruit* Berries eaten fresh. (188:132) **Paiute** *Fruit* Berries eaten ripe and fresh. (111:83) **Pomo** *Fruit* Raw berries used for food. (66:13) **Pomo, Kashaya** *Fruit* Berries eaten fresh. (72:113) **Quileute** *Fruit* Berries eaten fresh. (79:34) Fruits eaten raw. Fruits stewed and used for food. *Winter Use Food* Fruits canned and saved for future food use. (137:63) **Quinault** *Fruit* Berries eaten fresh. (79:34) **Salish, Coast** *Bread & Cake* Berries dried into cakes and used for food. *Fruit* Berries eaten fresh or boiled. *Unspecified* Young, tender shoots eaten in spring. (182:87) **Samish** *Fruit* Berries eaten fresh. *Unspecified* Sprouts eaten in early spring with half-dried salmon eggs. (79:34) **Sanpoil & Nespelem** *Fruit* Berries eaten fresh. (131:102) **Shuswap** *Fruit* Berries used for food. (123:67) **Skagit** *Fruit* Berries eaten fresh. *Unspecified* Sprouts eaten in early spring. (79:34) **Skagit, Upper** *Fruit* Berries eaten fresh. *Unspecified* Tender shoots peeled and eaten in spring and early summer. (179:38) **Snohomish** *Fruit* Berries eaten fresh. **Squaxin** *Dried Food* Berries dried, stored in soft or hard baskets, and used for food. *Fruit* Berries mixed with blackberries and eaten fresh. **Swinomish** *Fruit* Berries eaten fresh. *Unspecified* Sprouts eaten in early spring with half-dried salmon eggs. (79:34) **Thompson** *Fruit* Berries eaten fresh, often with fish. *Sweetener* Roots used for sugar. *Unspecified* Toasted shoots eaten alone or with meat and fish. (187:270) **Tsimshian** *Fruit* Berries used for food. (43:346) **Wintoon** *Fruit* Berries used for food. (117:264) **Yurok** *Fruit* Fruit eaten fresh. (6:51)
- **Dye**—**Blackfoot** *Unspecified* Berries used to dye tanned robes. (86:122)
- **Other**—**Blackfoot** *Protection* Berries applied to quivers to strengthen them. (86:122) **Carrier** *Cooking Tools* Leaves used to dry berries on. (31:74) **Cowlitz** *Soap* Bark boiled and used for soap. (79:34) **Kwakiutl, Southern** *Cooking Tools* Leaves placed above and below seaweed in steaming pits. (183:264) **Okanagan-Colville** *Containers* Leaves used to line steam cooking pits, berry baskets, and placed between layers of fresh berries. (188:132) **Pomo** *Cooking Tools* Leaves used to wrap meat for baking. (66:13) **Pomo, Kashaya** *Cooking Tools* Leaves used to wrap food for baking. (72:113) **Quileute** *Containers* Leaves used to wrap cooked elderberries for storage. **Quinault** *Containers* Leaves used with skunk cabbage leaves to line baskets in preserving elderberries. (79:34) **Shuswap** *Containers* Leaves used to cover huckleberries, to prevent them from spilling over when they fall. (123:67) **Tsimshian** *Cleaning Agent* Leaves used to wipe slime from salmon. (43:346)

Rubus parviflorus Nutt. ssp. *parviflorus*, Thimbleberry
- **Food**—**Haisla** *Bread & Cake* Berries used to make dried berry cakes for winter use. (43:276) **Oweekeno** *Beverage* Fall, brown leaves used to make tea. *Fruit* Fruit eaten fresh. *Preserves* Fruit used to make jelly. *Unspecified* Sprouts used for food. (43:112)
- **Other**—**Hanaksiala** *Cooking Tools* Leaves used to whip soapberries. (43:276)

Rubus parviflorus var. *velutinus* (Hook. & Arn.) Greene, Western Thimbleberry
- **Food**—**Mendocino Indian** *Fruit* Berries eaten fresh. (41:354)

Rubus pedatus Sm., Strawberryleaf Raspberry
- **Food**—**Alaska Native** *Preserves* Fruit used to make an excellent jelly. (85:99) **Haisla & Hanaksiala** *Forage* Berries eaten by porcupines and groundhogs. *Fruit* Berries eaten in small quantities. (43:278) **Kitasoo** *Fruit* Berries eaten fresh. (43:346) **Thompson** *Fruit* Small fruits rarely eaten. (187:272)

Rubus procumbens, Wild Blackberry
- **Drug**—**Mahuna** *Antidiarrheal* Infusion of roots taken for diarrhea. (as *R. villosus* 140:7)
- **Food**—**Mahuna** *Fruit* Berries eaten mainly to quench the thirst. (as *R. villosus* 140:70)

Rubus pubescens Raf., Dwarf Red Blackberry
- **Drug**—**Okanagon** *Antiemetic* and *Antihemorrhagic* Decoction of leaves taken for vomiting of blood and blood-spitting. *Gastrointestinal Aid* Decoction of leaves taken as a stomach tonic. *Tonic* Decoction of leaves taken as a stomach tonic. **Thompson** *Antiemetic* Decoction of leaves taken for vomiting of blood and blood-spitting. *Antihemorrhagic* Decoction of leaves taken for vomiting of blood and blood-spitting. (125:41) Decoction of leaves taken for spitting or vomiting blood. (164:466) *Gastrointestinal Aid* Decoction of leaves taken as a stomach tonic. (125:41) Decoction of root taken as a tonic for the stomach. (164:466) *Tonic* Decoction of leaves taken as a stomach tonic. (125:41) Decoction of root taken as a tonic for the stomach. (164:466)
- **Food**—**Cree, Woodlands** *Fruit* Fruit eaten fresh. (109:57) **Iroquois** *Fruit* Fruit used for food. (142:92)

Rubus pubescens Raf. var. *pubescens*, Dwarf Red Blackberry
- **Drug**—**Malecite** *Abortifacient* Infusion of plant and wild strawberry used for irregular menstruation. (as *R. triflorus* 116:258) **Micmac** *Abortifacient* Parts of plant used for irregular menstruation. (as *R. triflorus* 40:61)
- **Food**—**Chippewa** *Fruit* Delicate, delicious fruit used for food. (as *R. triflorus* 71:133) **Iroquois** *Bread & Cake* Fruit mashed, made into small cakes, and dried for future use. *Dried Food* Raw or cooked fruit sun or fire dried and stored for future use. *Fruit* Dried fruit taken as a hunting food. *Sauce & Relish* Dried fruit cakes soaked in warm water and cooked as a sauce or mixed with corn bread. (as *R. triflorus* 196:127)

Rubus sp., Raspberry
- **Drug**—**Algonquin, Tête-de-Boule** *Respiratory Aid* Infusion of shredded branches taken for bronchial troubles. (132:130) **Carrier** *Abortifacient* Decoction of stems taken by women with sickness in their womb. (31:79) **Carrier, Northern** *Dermatological Aid* Compound decoction of inner bark taken for body sores. *Orthopedic Aid* Compound decoction of inner bark taken for paralysis. *Stimulant* Compound decoction of inner bark taken for constitutional weakness. (150:58) **Cherokee** *Cough Medicine* Root chewed for coughs. (203:74) **Choctaw** *Antidiarrheal* Infusion of roots taken for dysentery. *Tonic* Decoction of roots taken as a tonic to improve circulation. (177:29) **Eskimo, Western** *Antidiarrheal* Berries eaten for diarrhea. (108:15) **Iroquois** *Blood Medicine* Tender, new shoots used as a blood remedy. (124:95) Infusion of roots and other roots taken by young mothers for thick blood. (141:48) *Dermatological Aid* Roots used as an effectual astringent. (124:95) **Klallam** *Cold Remedy* Roots used for colds. (79:36) **Malecite** *Antidiarrheal* and *Pediatric Aid* Infusion of 1-foot section of tree used by children with diarrhea. (116:255) **Rappahan-

nock *Anesthetic* and *Antidiarrheal* Root or berry infusion taken for diarrhea, an overdose would cause numbness. *Gastrointestinal Aid* Infusion of dried, brown runners taken for dyspepsia. (161:29)

- *Food*—**Abnaki** *Preserves* Fruits used to make jelly. (144:169) **Algonquin, Quebec** *Preserves* Fruit used to make preserves. (18:94) **Carrier** *Preserves* Berries used to make jelly. (31:79) **Cherokee** *Cooking Agent* Berries mixed with apples to color the jelly red. (126:58) *Pie & Pudding* Berries used to make pies and cobblers. (126:57) Fruit used to make pies. (126:58) *Preserves* Berries used to make jelly and jam. (126:57) Fruit used to make jelly. (126:58) *Unspecified* Tips of new, young shoots boiled thoroughly, fried, and eaten. (204:253) *Vegetable* Shoots used in salads and as potherbs. *Winter Use Food* Berries canned for future use. (126:57) **Coeur d'Alene** *Dried Food* Berries dried and used for food. *Fruit* Berries eaten fresh. Berries mashed and eaten. *Soup* Berries dried, boiled with roots, and eaten as soup. (178:89) **Iroquois** *Fruit* Fruits eaten raw. *Sauce & Relish* Fruits dried, soaked in sugared water, cooked, and eaten as a sauce. *Special Food* Dried berries soaked in honey and water and used as a ceremonial food by the Bear Society. (124:95) **Okanagon** *Staple* Berries used as a principal food. (178:238) **Paiute** *Fruit* Berries used for food. (111:83) **Sanpoil & Nespelem** *Dried Food* Berries eaten dried. *Fruit* Berries eaten raw or dried. (131:102) **Spokan** *Fruit* Berries used for food. (178:343)

Rubus spectabilis Pursh, Salmon Berry

- *Drug*—**Bella Coola** *Gastrointestinal Aid* Decoction of root bark taken for stomach troubles. (150:58) **Kwakiutl** *Burn Dressing* Powdered bark applied to burns. *Dermatological Aid* Powdered bark applied to sores. *Pediatric Aid* Chewed sprouts applied to the head of a child to make him grow. (183:291) **Makah** *Analgesic* and *Dermatological Aid* Poultice of bark applied to wounds for the pain. *Toothache Remedy* Poultice of bark applied to aching tooth. **Quileute** *Burn Dressing* Poultice of chewed leaves or bark applied to burns. **Quinault** *Analgesic* Decoction of bark taken to lessen labor pains. *Burn Dressing*, *Dermatological Aid*, and *Disinfectant* Decoction of bark used to clean infected wounds, especially burns. *Gynecological Aid* Decoction of bark taken to lessen labor pains. (79:35)

- *Food*—**Alaska Native** *Fruit* Fruit eaten raw. *Preserves* Fruit made into jams and jellies. (85:101) **Bella Coola** *Bread & Cake* Berries cooked, dried in cakes, and used for food. *Fruit* Berries eaten raw. *Unspecified* Sprouts peeled and eaten in spring. (184:209) **Carrier** *Fruit* Berries used for food. (31:77) **Chehalis** *Fruit* Berries eaten fresh. *Unspecified* Sprouts cooked in a pit and eaten with dried salmon. **Chinook, Lower** *Fruit* Berries eaten fresh. *Unspecified* Sprouts cooked in a pit and eaten with dried salmon. (79:35) **Clallam** *Fruit* Berries eaten fresh. (57:203) **Cowlitz** *Fruit* Berries eaten fresh. *Unspecified* Sprouts cooked in a pit and eaten with dried salmon. **Green River Group** *Fruit* Berries eaten fresh. *Unspecified* Sprouts cooked in a pit and eaten with dried salmon. (79:35) **Haisla & Hanaksiala** *Beverage* Berries used to make homemade wine. *Dried Food* Berries dried for winter use. *Fruit* Berries eaten fresh. *Special Food* Young sprouts peeled and served as a featured item at salmonberry sprout feasts. (43:279) **Hesquiat** *Unspecified* Young, fresh shoots eaten with oil. (185:74) **Hoh** *Fruit* Fruits stewed and used for food. Fruits eaten raw. *Winter Use Food* Fruits canned and saved for future food use. (137:63) **Kitasoo** *Fruit* Berries eaten fresh. *Unspecified* Sprouts peeled and eaten fresh or steamed with oolichan (candlefish) grease, salmon, or salmon roe. (43:347) **Kwakiutl, Southern** *Dried Food* Fruits boiled, mashed, dried, and used as a winter food. *Fruit* Fruits eaten fresh. *Unspecified* Young shoots eaten in spring. (183:291) **Lummi** *Fruit* Berries eaten fresh. *Unspecified* Sprouts cooked in a pit and eaten with dried salmon. (79:35) **Makah** *Fruit* Fruit eaten fresh. (67:275) Berries eaten fresh. (79:35) *Special Food* Sprouts available in large amounts often the occasion for sprout parties. Makah women would collect canoe loads of sprouts and pit steam them on the beach. People would sing and dance while waiting for the steaming sprouts to finish cooking. *Unspecified* Sprouts peeled and eaten raw, boiled, or steamed on hot rocks. (67:275) Sprouts cooked in a pit and eaten with dried salmon. (79:35) *Winter Use Food* Fruit canned for winter use. Sprouts eaten with fermented salmon eggs collected during the previous autumn. (67:275) **Nitinaht** *Dessert* Sprouts eaten raw or steam cooked like a dessert. *Fruit* Berries eaten fresh. (186:124) **Okanagon** *Fruit* Yellow fruits used for food. *Unspecified* Young, sweet shoots used for food. (125:38) **Oweekeno** *Fruit* Berries eaten fresh. *Preserves* Berries used to make jam. *Unspecified* Sprouts used for food. *Winter Use Food* Berries preserved for winter use. (43:113) **Paiute** *Fruit* Berries eaten ripe and fresh. (111:82) **Pomo** *Fruit* Raw berries used for food. (66:13) **Pomo, Kashaya** *Fruit* Berries eaten fresh. (72:102) **Quileute** *Fruit* Berries eaten fresh. (79:35) Fruits stewed and used for food. Fruits eaten raw. (137:63) *Unspecified* Sprouts cooked in a pit and eaten with dried salmon. (79:35) *Winter Use Food* Fruits canned and saved for future food use. (137:63) **Quinault** *Fruit* Berries eaten fresh. *Unspecified* Sprouts cooked in a pit and eaten with dried salmon. (79:35) **Salish, Coast** *Fruit* Berries eaten fresh in summer. *Unspecified* Sprouts peeled and eaten raw in early spring. (182:88) **Skagit, Upper** *Fruit* Berries eaten fresh. *Unspecified* Green sprouts peeled and eaten or cooked in an earth oven. (179:38) **Squaxin** *Fruit* Berries eaten fresh. *Unspecified* Sprouts cooked in a pit and eaten with dried salmon. **Swinomish** *Fruit* Berries eaten fresh. *Unspecified* Sprouts cooked in a pit and eaten with dried salmon. (79:35) **Thompson** *Dried Food* Fruit eaten dried. (187:272) *Fruit* Yellow fruits used for food. (125:38) Fruits eaten for food. (164:486) Fruit eaten fresh. (187:272) *Unspecified* Young, sweet shoots used for food. (125:38) Young shoots eaten. (164:482) **Tolowa** *Fruit* Berries eaten fresh. *Unspecified* Young sprouts eaten with seaweed and dried eels. **Yurok** *Fruit* Berries eaten fresh. (6:51)

- *Other*—**Haisla & Hanaksiala** *Ceremonial Items* Flower used in "flower dance" costume and in shamanistic performances. *Cooking Tools* Leaves used to whip soapberries. *Season Indicator* Plant used as an indicator for picking edible seaweed. (43:279) **Hesquiat** *Cooking Tools* Sticks used to make salmon spreaders and for stringing clams for cooking and smoking. Leaves spread at bottom of wooden cooking containers to prevent the hot rocks from burning the wood. *Smoking Tools* Roots used to make pipe bowls. (185:74) **Hoh** *Ceremonial Items* Sprouts formerly used in courting ceremonies. (137:63) **Kwakiutl, Southern** *Cooking Tools* Leaves placed above and below seaweed in steaming pits. (183:264) **Nitinaht** *Toys & Games* Stems used to make children's practice bows. (186:124) **Nuxalkmc** *Ceremonial Items* Plant, wild parsnip, gooseberry, and rose used in the dance of Winwina. **Oweekeno** *Ceremonial Items* Plant, wild parsnip, gooseberry, rose, and mask represented a child in a ceremonial dance. *Containers* Leaves used as a mat under any kind of berries and hemlock cambium when drying. (43:113) **Quileute** *Ceremonial Items* Sprouts formerly used in courting ceremonies. (137:63) *Hunting & Fishing Item* Wood made into a plug stopper for seal hair floats used for whaling. (79:35)

Rubus trivialis Michx., Southern Dewberry
- *Drug*—**Cherokee** *Antidiarrheal* Infusion of root or leaf used for diarrhea. *Antirheumatic (Internal)* Infusion given for rheumatism. (80:26) *Dermatological Aid* Compound, astringent, and tonic infusion of root used as a wash for piles. (80:25, 26) *Hemorrhoid Remedy* Compound infusion of root used for piles. *Oral Aid* Washed root chewed for coated tongue. *Stimulant* Used as a stimulant. *Throat Aid* Used with honey as a wash for sore throat. *Tonic* Used as a tonic. *Urinary Aid* Compound decoction taken to regulate urination. *Venereal Aid* Used for venereal disease. (80:26) **Seminole** *Gastrointestinal Aid* Infusion of herbage used for stomach troubles. (169:276)
- *Food*—**Cherokee** *Beverage* Fruit used to make juice. *Fruit* Fruit used for food. (80:26)

Rubus ursinus Cham. & Schlecht., California Blackberry
- *Drug*—**Diegueño** *Antidiarrheal* Decoction of roots taken or fresh fruit eaten for diarrhea. (84:39) **Hesquiat** *Gastrointestinal Aid* Decoction of the entire vine taken for stomach troubles. *Other* Decoction of the entire vine taken for a general sick feeling. (185:75) **Kwakiutl** *Antidiarrheal* Compound decoction of vines taken for diarrhea. (183:264) *Antiemetic* Decoction of vines and roots taken for vomiting. *Antihemorrhagic* Decoction of vines and roots taken for blood-spitting. (183:291) **Pomo, Kashaya** *Antidiarrheal* Decoction of root taken for diarrhea. *Gynecological Aid* Berries not to be eaten by pregnant women or fathers to be; if berries eaten, the baby would be dark. (72:22)
- *Food*—**Clallam** *Fruit* Berries used for food. (57:203) **Diegueño** *Dried Food* Fruit dried and cooked. *Fruit* Fruit eaten fresh. (84:39) **Hesquiat** *Fruit* Berries eaten and well liked. (185:75) **Makah** *Fruit* Fruit eaten fresh. *Pie & Pudding* Fruits used to make pies. *Preserves* Fruits used to make jam. (67:278) **Pomo, Kashaya** *Fruit* Fresh berries eaten whole or mashed with bread. *Pie & Pudding* Berries cooked as pie filling. *Sauce & Relish* Fresh berries mashed as topping for ice cream. Berries cooked as sauce for dumplings. (72:22) **Saanich** *Beverage* Old, dry leaves used to make tea. **Salish, Coast** *Bread & Cake* Berries mashed, dried in cakes, placed in hot water, and used for food. *Fruit* Berries eaten fresh. (182:88) **Thompson** *Dried Food* Berries sun dried on mats. (187:272)
- *Other*—**Saanich** *Ceremonial Items* Stems used in purification rituals before dancing. (182:88)

Rubus ursinus ssp. *macropetalus* (Dougl. ex Hook.) Taylor & MacBryde, California Blackberry
- *Drug*—**Skagit** *Gastrointestinal Aid* Infusion of leaves taken for stomach trouble. (as *R. macropetalus* 79:35) **Skagit, Upper** *Gastrointestinal Aid* Infusion of leaves taken for stomach trouble. (as *R. macropetalus* 179:38)
- *Food*—**Cowlitz** *Dried Food* Berries dried and eaten. *Fruit* Berries eaten fresh. (as *R. macropetalus* 79:35) **Haisla & Hanaksiala** *Fruit* Berries used for food. (43:282) **Hoh** *Fruit* Fruits stewed and used for food. Fruits eaten raw. *Winter Use Food* Fruits canned and saved for future food use. (as *R. macropetalus* 137:63) **Nitinaht** *Fruit* Berries used for food. (186:125) **Okanagon** *Unspecified* Used commonly for food. Used commonly for food. (as *R. macropetalus* 125:38) **Quileute** *Beverage* Fresh or dried vines and leaves used to make a beverage tea. *Dried Food* Berries dried and eaten. *Fruit* Berries eaten fresh. (as *R. macropetalus* 79:35) Fruits stewed and used for food. Fruits eaten raw. *Winter Use Food* Fruits canned and saved for future food use. (as *R. macropetalus* 137:63) **Skagit** *Dried Food* Berries dried and eaten. *Fruit* Berries eaten fresh. (as *R. macropetalus* 79:35) **Skagit, Upper** *Dried Food* Berries pulped and dried for winter use. *Fruit* Berries eaten fresh. (as *R. macropetalus* 179:38) **Thompson** *Fruit* Berries eaten. (as *R. macropetalus* 164:487) *Unspecified* Used commonly for food. (as *R. macropetalus* 125:38)

Rubus vitifolius Cham. & Schlecht., Pacific Dewberry
- *Drug*—**Costanoan** *Antidiarrheal* Decoction of roots used for dysentery and diarrhea. *Dermatological Aid* and *Disinfectant* Roots used for infected sores. (21:19) **Mendocino Indian** *Antidiarrheal* Infusion of roots taken for diarrhea. (41:355)
- *Food*—**Cahuilla** *Beverage* Berries soaked in water to make a beverage. *Dried Food* Berries dried for later use. *Fruit* Berries eaten fresh. (15:134) **Costanoan** *Fruit* Fruits eaten for food. (21:250) **Karok** *Fruit* Fruit eaten fresh. (6:52) Berries used for food. (148:384) **Klamath** *Unspecified* Species used for food. (45:99) **Luiseño** *Fruit* Fruit used for food. (155:232) **Mendocino Indian** *Dried Food* Black, juicy berries dried for winter use. *Fruit* Black, juicy berries eaten fresh. (41:355) **Wintoon** *Fruit* Berries used for food. (117:264) **Yurok** *Beverage* Young shoots boiled with other vine shoots into a tea. *Fruit* Berries eaten fresh. (6:52)
- *Dye*—**Luiseño** *Unspecified* Berry juice used to stain wood. (155:232)

Rudbeckia, Asteraceae

Rudbeckia fulgida Ait., Orange Coneflower
- *Drug*—**Cherokee** *Anthelmintic* Used as wash for "swelling caused by worms." *Dermatological Aid* Warm infusion of root used to bathe sores. *Ear Medicine* Root ooze used for earache. *Gynecological Aid* Taken for "flux and some private diseases." *Kidney Aid* Infusion taken for dropsy. *Snakebite Remedy* Used as wash for snakebites. *Venereal Aid* Taken for "flux and some private diseases."

Rudbeckia hirta L., Black Eyed Susan
- *Drug*—**Cherokee** *Anthelmintic* Used as wash for "swelling caused by worms." *Dermatological Aid* Warm infusion of root used to bathe sores. *Ear Medicine* Root ooze used for earache. *Gynecological Aid* Taken for "flux and some private diseases." *Kidney Aid* Infusion taken for dropsy. *Snakebite Remedy* Used as wash for snakebites. *Venereal Aid* Taken for "flux and some private diseases." (80:30) **Chippewa** *Pediatric Aid* Poultice of blossoms and another plant used for babies. (71:143) **Iroquois** *Anthelmintic* Infusion of roots given to children with worms. *Heart Medicine* Decoction of plants taken for the heart. *Pediatric Aid* Infusion of roots given to children with worms. (87:469) **Potawatomi** *Cold Remedy* Infusion of root taken for colds. (154:52, 53) **Shuswap** *Eye Medicine* Plant used for sore eyes. (123:59)
- *Dye*—**Potawatomi** *Yellow* Disk florets boiled with rushes to dye them yellow. Rushes used to make woven mats. (154:117)

Rudbeckia hirta* var. *angustifolia (T. V. Moore) Perdue, Black Eyed Susan
- **Drug**—**Seminole** *Analgesic* Cold infusion of cone flowers used for headaches. *Febrifuge* Cold infusion of cone flowers used for fevers. (as *R. divergens* 169:283)

Rudbeckia laciniata L., Cutleaf Coneflower
- **Drug**—**Cherokee** *Dietary Aid* Cooked spring salad eaten to "keep well." (80:30) **Chippewa** *Burn Dressing* Compound poultice of blossoms applied to burns. (53:352) *Gastrointestinal Aid* Compound infusion of root taken for indigestion. (53:342) *Veterinary Aid* Compound infusion of root applied to chest and legs of horse as a stimulant. (53:366)
- **Food**—**Cherokee** *Dried Food* Leaves and stems tied together and hung up to dry or sun dried and stored for future use. *Frozen Food* Tender leaves and stems frozen in early spring. (126:34) *Unspecified* Young shoots and leaves boiled, fried with fat, and eaten. (204:251) *Vegetable* Leaves used as cooked spring salad to keep well. (80:30) Leaves and stems parboiled, rinsed, and boiled in hot grease until soft. Leaves and stems cooked alone or with poke, eggs, dock, cornfield creasy (probably *Barbarea verna*), or any other greens. *Winter Use Food* Leaves and stems preserved by blanching, then boiling in a can, with or without salt. (126:34) **San Felipe** *Vegetable* Young stems eaten like celery. (32:50)

Rumex, Polygonaceae

Rumex acetosa L., Garden Sorrel
- **Drug**—**Apalachee** *Unspecified* Plant water used for medicinal purposes. (81:98) **Eskimo, Kuskokwagmiut** *Antidiarrheal* Decoction of leaves and stems taken for diarrhea. (122:24)
- **Food**—**Cherokee** *Vegetable* Leaves used for food. (126:53)

Rumex acetosella L., Common Sheep Sorrel
- **Drug**—**Aleut** *Dermatological Aid* Poultice of steamed leaves applied to warts and bruises. (8:427) **Cherokee** *Dermatological Aid* Poultice of bruised leaves and blossoms applied to old sores. (80:56) **Mohegan** *Gastrointestinal Aid* Fresh leaves chewed as a stomach aid. (176:75, 132) **Squaxin** *Tuberculosis Remedy* Raw leaves eaten for tuberculosis. (79:29)
- **Food**—**Anticosti** *Unspecified* Leaves eaten fresh by children. (143:65) **Bella Coola** *Unspecified* Leaves eaten raw. (184:207) **Chehalis** *Unspecified* Leaves eaten raw or boiled. (79:29) **Cherokee** *Unspecified* Species used for food. (80:56) *Vegetable* Leaves used for food. (126:53) **Delaware** *Unspecified* Plant used as filling for pies. (176:59) **Hanaksiala** *Unspecified* Leaves eaten by children. (43:260) **Hesquiat** *Sour* Tart, tangy leaves chewed by children. (185:71) **Iroquois** *Spice* Used with salt in a brine for cucumbers. (196:113) *Vegetable* Eaten raw, sometimes with salt. (196:118) **Miwok** *Vegetable* Moistened, pulverized leaves eaten with salt, tasted sour like vinegar. (12:160) **Okanagan-Colville** *Vegetable* Leaves eaten raw. (188:113) **Saanich** *Vegetable* Acid-tasting leaves eaten like lettuce. (182:85) **Thompson** *Unspecified* Leaves chewed by children for the tangy, sour taste. (187:239)

Rumex altissimus Wood, Pale Dock
- **Drug**—**Dakota** *Dermatological Aid* Poultice of green leaves applied to boils. (69:361) **Lakota** *Antidiarrheal* Used for diarrhea. *Antihemorrhagic* Used for hemorrhages. *Gastrointestinal Aid* Used for stomach cramps. (139:55) **Ojibwa** *Unspecified* Plant used for medicinal purposes. (135:240)

Rumex aquaticus* var. *fenestratus (Greene) Dorn, Western Dock
- **Drug**—**Bella Coola** *Analgesic* Leaves used in a sweat bath for pains similar to rheumatism all over body. *Antirheumatic (External)* Leaves used in sweat bath for pains similar to rheumatism. *Dermatological Aid* Poultice of leaves and mashed, roasted roots applied to boils and wounds. (as *R. occidentalis* 150:56) Poultice of long, yellow roots, and "klondike soap" applied to boils, cuts, and scrapes. (as *R. occidentalis* 184:207) *Herbal Steam* Leaves used in a sweat bath for pains similar to rheumatism all over body. (as *R. occidentalis* 150:56) **Haisla** *Laxative* Plant used as a laxative. **Hanaksiala** *Antirheumatic (External)* Poultice of pounded leaves and stems applied to sores and other painful areas. *Dermatological Aid* Roots cooked and inserted into wounds. Poultice of root paste applied to cuts and boils. Poultice of pounded leaves and stems applied to sores and other painful areas. (as *R. occidentalis* var. *procerus* 43:260) **Kwakiutl** *Dermatological Aid* Root extract used to wash sores and swellings. *Gastrointestinal Aid* Boiled roots eaten and applied as poultice for stomachaches. (as *R. occidentalis* 183:287)
- **Food**—**Apache, Chiricahua & Mescalero** *Unspecified* Leaves eaten without preparation or cooked with green chile and meat or animal bones. (as *R. occidentalis* 33:46) **Bella Coola** *Vegetable* Young leaves mashed, cooked, mixed with grease, and eaten like spinach. (as *R. occidentalis* 184:207) **Haisla & Hanaksiala** *Vegetable* Leaves and stems eaten with oolichan (candlefish) grease and sugar. **Hanaksiala** *Beverage* Plant formerly used to make a type of home-brew or wine. *Winter Use Food* Plant cooked and stored underground in barrels with stink currants and red elderberries for winter use. (as *R. occidentalis* var. *procerus* 43:260) **Heiltzuk** *Unspecified* Stems, leaves, sprouts, and shoots eaten with sugar and grease. (as *R. occidentalis* var. *procerus* 43:107) **Kitasoo** *Vegetable* Leaves cooked and eaten. (as *R. occidentalis* 43:340) **Klallam** *Unspecified* Plants used for food. (as *R. occidentalis* 78:197) **Montana Indian** *Unspecified* Seeds used for food. *Vegetable* Spring leaves used for "greens." (as *R. occidentalis* 19:22) **Oweekeno** *Unspecified* Stems, leaves, sprouts, and shoots used for food. (as *R. occidentalis* var. *procerus* 43:107) **Tanana, Upper** *Frozen Food* Leaves and stems frozen for future use. *Vegetable* Leaves and stems eaten raw or boiled with sugar. (as *R. fenestratus* 102:15)

Rumex arcticus Trautv., Arctic Dock
- **Drug**—**Eskimo, Inuktitut** *Antidiarrheal* Leaves and stems used for diarrhea. (202:186)
- **Food**—**Alaska Native** *Dietary Aid* Fresh, green leaves used as a source for vitamins A and C. *Vegetable* Leaves used as salad greens and cooked as vegetables. *Winter Use Food* Leaves cooked, chopped, mixed with other greens, and stored in kegs or barrels for winter use. (85:55) **Eskimo, Alaska** *Unspecified* Young, tender leaves boiled and eaten either hot or cold with seal oil and sometimes with sugar. The cooked leaves were sometimes served with a sauce-like coating of imported milk. (1:35) Leaves eaten fresh, soured, boiled or in oil. Root also utilized. (4:715) *Winter Use Food* Boiled leaves mixed with seal oil and preserved for months. (1:35) **Eskimo, Arctic** *Vegetable* Leaves from young stems eaten raw as a salad or cooked like spinach.

(128:26) **Eskimo, Inuktitut** *Ice Cream* Leaves and stems boiled, cooled, and added to "Eskimo ice cream." *Unspecified* Leaves used for food. (202:186) **Eskimo, Inupiat** *Dessert* Leaves eaten cold with seal oil and sugar, like a rhubarb dessert. *Vegetable* Leaves eaten raw in a salad or boiled and eaten hot with seal oil, blubber, or butter. *Winter Use Food* Leaves chopped, cooked with blubber, and stored in a 10- to 30-gallon wooden barrel for winter use. (98:35) **Koyukon** *Unspecified* Plant cooked and eaten. (119:56) **Tanana, Upper** *Frozen Food* Leaves and stems frozen for future use. *Vegetable* Leaves and stems eaten raw or boiled with sugar. (102:15)

Rumex conglomeratus Murr., Clustered Dock
- **Drug**—**Karok** *Herbal Steam* Decoction of plant used as a steam bath medicine. (148:383) **Miwok** *Dermatological Aid* Decoction of root taken for boils. Poultice of boiled root applied to boils. (12:172)
- **Food**—**Miwok** *Vegetable* Cooked leaves eaten as greens. (12:160)

Rumex crispus L., Curly Dock
- **Drug**—**Blackfoot** *Antirheumatic (External)* Mashed root pulp used for swellings. *Dermatological Aid* Mashed root pulp used for sores. (118:43) **Cherokee** *Antidiarrheal* Infusion taken for dysentery. *Blood Medicine* Infusion of root taken "for blood." *Dermatological Aid* Juice and infusion of root used as poultice and salve for various skin problems. *Kidney Aid* Infusion of root taken "to correct fluids." *Laxative* Infusion of root used for constipation. *Throat Aid* Leaves rubbed in mouth for sore throat. *Veterinary Aid* Beaten roots fed to horses for "sick stomach." (80:32) **Cheyenne** *Antihemorrhagic* Infusion of dried roots taken for lung hemorrhages. *Dermatological Aid* Wet poultice of pounded, dried root applied to wounds or sores. *Pulmonary Aid* Infusion of dried, pulverized root taken for lung hemorrhage. (76:172) Infusion of pulverized roots used for lung hemorrhages. (83:32) **Chippewa** *Dermatological Aid* Poultice of moistened, dried, powdered root applied to cuts or itches. (53:350) Poultice of dried, pounded root applied to ulcers and swellings. (53:354, 366) **Costanoan** *Urinary Aid* Decoction of plant used for urinary problems. (21:11) **Dakota** *Dermatological Aid* Poultice of crushed, green leaves applied to boils to draw out pus. (70:77) **Delaware** *Blood Medicine* Root used as a blood purifier. *Liver Aid* Root used for jaundice. (176:33) **Delaware, Oklahoma** *Blood Medicine* Root used as a blood purifier. *Liver Aid* Root used for jaundice. (175:28, 78) **Iroquois** *Antidiarrheal* Compound decoction of roots taken for diarrhea with blood. (87:311) *Antirheumatic (External)* Stalks cooked as greens and used for rheumatism. (124:93) *Blood Medicine* Compound decoction taken as blood medicine when "blood is bad from scrofula." *Cathartic* Decoction taken as a physic for general bowel trouble and intestinal colds. *Dermatological Aid* Infusion taken and poultice used for swellings caused by blood catching cold. (87:312) *Dietary Aid* Decoction of roots taken "when one can't eat." *Emetic* Decoction of plants taken as an emetic before running or playing lacrosse. *Gastrointestinal Aid* Decoction of roots taken for intestinal colds, cramps, or abdominal pains. Decoction of roots taken for upset stomach. Poultice used for swollen body from yellow fever, abdominal cramps, and pains. (87:311) *Hemorrhoid Remedy* Compound decoction of plant taken and used as a wash for piles. *Hemostat* Used for bleeding. (87:312) *Kidney Aid* Decoction of plants taken for kidney trouble. (87:311) *Love Medicine* Decoction used as a wash for face, hands, and clothes as love medicine. (87:312) *Misc. Disease Remedy* Decoction of roots taken and poultice applied for yellow fever. (87:311) *Orthopedic Aid* Used for strained muscles. *Panacea* "Good for all illnesses." (87:312) *Reproductive Aid* Compound decoction of roots taken to induce pregnancy. *Strengthener* Decoction taken to makes muscles strong for running or playing lacrosse. (87:311) *Tonic* Compound decoction of roots taken by women as a tonic. *Venereal Aid* Compound used for gonorrhea. (87:312) **Isleta** *Gastrointestinal Aid* Leaves eaten as greens for the beneficial effect upon the stomach. (100:42) **Micmac** *Cathartic* Roots used as a purgative. (40:61) Infusion of roots used as a purgative. (116:259) *Urinary Aid* Roots used "cold in bladder." (40:61) Infusion of roots, hemlock, parsley, and prince's pine used for colds in the bladder. (116:259) **Mohegan** *Blood Medicine* Cooked leaves said to "purify the blood." *Tonic* Root used to make a tonic. (176:75, 132) **Navajo** *Stimulant* Plant used for fainting. (90:155) **Navajo, Ramah** *Ceremonial Medicine* Whole plant used as a ceremonial emetic. *Dermatological Aid* Dried, powdered leaves dusted on sores. *Emetic* Whole plant used as a ceremonial emetic. *Oral Aid* Cold infusion of leaf used on mouth sores. *Panacea* Root used as a "life medicine." (191:24) **Nevada Indian** *Veterinary Aid* Poultice of root applied to saddle sores. (118:49) **Ojibwa** *Antidiarrheal* Boiled seeds used for diarrhea. (5:2289) Seeds boiled and used for diarrhea. (5:2318) *Dermatological Aid* Root used to close and heal cuts. (153:381) *Hunting Medicine* Dried seeds smoked as a favorable lure to game when mixed with kinnikinnick. (153:431) **Ojibwa, South** *Dermatological Aid* Poultice of bruised or crushed root applied to sores and abrasions. (91:200) **Paiute** *Analgesic* Poultice of pulped root applied to rheumatic pains. *Antidiarrheal* Boiled seeds eaten alone or in a compound for diarrhea. (180:131, 132) *Antirheumatic (External)* Poultice of chewed roots used for pain and swelling of sprained or swollen areas. (111:67) Poultice of pulped root applied to rheumatic swellings. *Blood Medicine* Decoction of root taken as a blood purifier. *Burn Dressing* Poultice of pulped root applied to burns. (180:131, 132) *Dermatological Aid* Roots used for the astringent properties. (167:317) Poultice of pulped root applied to bruises and swellings. (180:131, 132) *Gastrointestinal Aid* Decoction of roots taken, or raw, peeled roots eaten for stomach disorders. *Tonic* Roots used for the tonic properties. (167:317) Decoction of root taken as a general tonic. *Venereal Aid* Decoction of root taken for venereal disease. (180:131, 132) **Paiute, Northern** *Dermatological Aid* Roots ground into a powder and used on sores and cuts. (59:128) **Rappahannock** *Blood Medicine* An ingredient of a blood medicine. (161:31) **Shoshoni** *Analgesic* Poultice of pulped root applied to rheumatic pains. *Antirheumatic (External)* Poultice of pulped root applied to rheumatic swellings. *Blood Medicine* Decoction of root taken as a blood purifier. *Burn Dressing* Poultice of pulped root applied to burns. *Cathartic* Decoction of root taken as a physic. *Dermatological Aid* Poultice of pulped root applied to bruises and swellings. *Liver Aid* Decoction of root taken for liver complaints. *Tonic* Decoction of root taken as a general tonic. *Venereal Aid* Decoction of root taken for venereal disease. (180: 131, 132) **Thompson** *Cough Medicine* Plant used as a cough medicine. (187:239) **Yavapai** *Cough Medicine* Decoction of tubers taken for coughs. *Dermatological Aid* Dried, pulverized tubers used for sores. *Gastrointestinal Aid* Decoction of tubers taken for stomachache. *Pediatric Aid* Dried, pulverized tubers used for babies with

chafed skin. *Throat Aid* Decoction of tubers gargled for sore throat. *Toothache Remedy* Fresh or boiled tuber placed against gum or tooth or decoction held in mouth for toothaches. *Venereal Aid* Decoction of tuber used as wash and powder applied for gonorrhea. (65:261) **Yuki** *Antidiarrheal* Infusion of seeds taken by adults and babies for dysentery. *Dermatological Aid* Decoction of leaves and seeds applied to sores. *Pediatric Aid* Infusion of seeds taken by adults and babies for dysentery. (49:46) **Zuni** *Dermatological Aid* Poultice of powdered root applied to sores, rashes, and skin infections. Infusion of root used for athlete's foot infection. (27:378)

- *Food*—**Cherokee** *Vegetable* Leaves and stems mixed with other greens, parboiled, rinsed, and cooked in hot grease as a potherb. (126:53) Young leaves cooked with sochan (*Rudbeckia laciniata*), creaseys (probably *Lepidium virginicum*), and other greens and eaten. (204:253) **Cheyenne** *Unspecified* Stems peeled and inner portions eaten raw. (83:32) **Cocopa** *Unspecified* Seeds gathered and eaten. (37:188) **Costanoan** *Staple* Seeds used for pinole. *Vegetable* Leaves used for greens. (21:249) **Havasupai** *Vegetable* Leaves boiled and eaten. (197:217) Young, fresh, tender leaves boiled, drained, balled into individual portions, and served. (197:66) **Iroquois** *Vegetable* Stalks eaten as greens in spring. (124:93) Young leaves, before the stem appeared, cooked and seasoned with salt, pepper, or butter. (196:117) **Isleta** *Vegetable* Leaves eaten as greens. (100:42) **Kawaiisu** *Porridge* Seeds parched with hot coals, pounded, and cooked to the consistency of "thick gravy." *Unspecified* Stems boiled with sugar or roasted, inner pulp pushed out of the burned skin, and eaten hot or cold. (206:60) **Mendocino Indian** *Porridge* Seeds used to make mush. *Vegetable* Leaves used as greens in food. (41:345) **Mohave** *Vegetable* Leaves boiled and eaten as greens. (37:201) **Mohegan** *Vegetable* Combined with pigweed, mustard, plantain, and nettle and used as mixed greens. (176:83) **Montana Indian** *Unspecified* Seeds used for food. *Vegetable* Spring leaves used for "greens." (19:22) **Omaha** *Unspecified* Boiled leaves used for food. (70:77) **Paiute, Northern** *Bread & Cake* Seeds soaked in water, ground into a doughy flour, and baked in the sand. *Starvation Food* Roots pit baked in the winter when food was scarce. (59:48) **Pima** *Vegetable* Leaves eaten as greens. (47:51) **Pima, Gila River** *Unspecified* Leaves boiled and eaten. (133:7) **Yavapai** *Starvation Food* Upper stalk roasted during food shortage. (65:258)
- *Dye*—**Cheyenne** *Unspecified* Used to make a dye. (83:46) *Yellow* Leaves and stems boiled and used as a yellow dye. (76:172) **Choctaw** *Yellow* Pounded, dry roots boiled and used as a yellow dye. (25:14) **Pima** *Yellow* Roots pounded, boiled and used to make a yellow dye. (47:51)

Rumex giganteus Ait., Pa-wale

- *Drug*—**Hawaiian** *Blood Medicine* Plant used for purifying the blood. *Heart Medicine* Plant used for heart disease. *Misc. Disease Remedy* Plant used for leprosy. *Reproductive Aid* Plant used to condition the mother for pregnancy. *Strengthener* Plant used for general body weakness. *Tuberculosis Remedy* Plant used for consumption. (as *Punex gigauteus* 2:73)

Rumex hymenosepalus Torr., Canaigre Dock

- *Drug*—**Arapaho** *Dermatological Aid* Stems and leaves used in a wash for sores. (118:44) **Hopi** *Cold Remedy* Plant used for colds. (200:34, 73) *Dermatological Aid* Plant used for ant bites and infected cuts. (42:

357) **Mahuna** *Cough Medicine* Infusion of roots used as a gargle for coughs. *Throat Aid* Infusion of roots used as a gargle for sore throats. (140:17) **Navajo** *Unspecified* Plant used for medicine. (110:30) **Navajo, Kayenta** *Panacea* Plant used as a life medicine. (205:20) **Navajo, Ramah** *Ceremonial Medicine* Cold infusion of root used as a ceremonial medicine. *Gynecological Aid* Cold infusion of root used as a lactagogue on breasts. *Veterinary Aid* Cold infusion of root used as a galactagogue on breasts of goats. (191:24) **Paiute** *Burn Dressing* Dried, powdered root used on burns. *Dermatological Aid* Dried, powdered root used on sores. (118:44) **Papago** *Dermatological Aid* Poultice of pulverized, dried root applied to sores. (34:64, 65) Poultice of dried and ground roots applied to sores. (34:65) *Throat Aid* Powdered root eaten or piece of root held in mouth for sore throat. (34:64, 65) Dried and pounded root taken for sore throats. (34:64) **Pawnee** *Antidiarrheal* Root used for diarrhea. (70:77) **Pima** *Cold Remedy* Root chewed for colds. *Cough Medicine* Root chewed for coughs. *Dermatological Aid* Poultice of roots applied or decoction of roots used as wash for skin sores. (47:51) Poultice of dried, powdered root applied to sores. (146:80) *Oral Aid* Root held in the mouth for sore gums. *Throat Aid* Root chewed or decoction of roots used as a gargle for sore throats. (47:51) *Unspecified* Roots used for medicine. (95:264)

- *Food*—**Cahuilla** *Vegetable* Crisp, juicy stalks eaten as greens. (15:134) **Hualapai** *Beverage* Stems, before the buds bloom, boiled into a drink. (195:53) **Kawaiisu** *Porridge* Seeds parched with hot coals, pounded, and cooked to the consistency of "thick gravy." *Unspecified* Stems boiled with sugar, or roasted, inner pulp pushed out of the burned skin and eaten hot or cold. (206:60) **Navajo** *Porridge* Seeds used to make mush. *Unspecified* Leaves roasted in ashes or boiled and served with butter or chopped and fried with mutton grease. (110:30) *Vegetable* Stems baked and eaten. (55:43) **Papago** *Vegetable* Leaves eaten as greens in spring. (34:14) Roasted in ashes and eaten as greens. (34:46) Greens used for food. (36:61) **Pima** *Bread & Cake* Seeds formerly roasted, ground, added to water to form flat cakes, baked, and eaten. (47:51) *Candy* Roots used for chewing gum by school girls. (95:265) *Pie & Pudding* Stems boiled, strained, flour added, combined with sugar, filled into pie crusts, baked, and eaten. (47:51) *Unspecified* Stalks formerly cooked or roasted, peeled, and insides eaten. (95:264) Stems roasted or stewed and used for food. Roots eaten raw by children in early spring. (146:77) *Vegetable* Young, succulent leaves boiled or roasted and eaten as greens in spring. (47:51)
- *Dye*—**Hopi** *Unspecified* Root used for dye. (42:357) Root used as an important source of dye. (200:73) **Hualapai** *Unspecified* Roots used as a dye. (195:53) **Navajo** *Brown* Roots boiled and used to make a medium brown dye for yarn. (47:51) Dried, ground roots used as a brown dye.[71] *Green* Dried, ground roots used as a green dye.[71] *Orange* Dried, ground roots used as an orange dye.[71] *Red* Dried, ground roots used as a red dye.[71] *Yellow* Fresh, crushed roots mixed with alum, made into soft paste, and rubbed into wool as a gold dye. Dried, ground roots used as a yellow dye.[71] (55:43) **Navajo, Ramah** *Yellow-Brown* Root used as a yellow-brown dye for wool. (191:24) **Pima** *Brown* Dry roots crushed, placed in water, and used as a brown dye for basket making. *Red-Brown* Dry roots crushed, placed in water, and used as a brownish red dye for tanning hides. *Yellow* Dry roots crushed, placed in water, and used as a yellow dye for basket making. (47:51)

Rumex maritimus L., Golden Dock
- *Drug*—**Navajo, Kayenta** *Gastrointestinal Aid* Infusion of plant taken for bloat. (as *R. fueginus* 205:20)
- *Food*—**Navajo, Kayenta** *Porridge* Seeds made into a mush and used for food. (as *R. fueginus* 205:20)

Rumex obtusifolius L., Bitter Dock
- *Drug*—**Chippewa** *Dermatological Aid* and *Pediatric Aid* Infusion of root applied, especially to children, for skin eruptions. (53:350) **Delaware** *Blood Medicine* Root used as a blood purifier. *Liver Aid* Root used for jaundice. (176:33) **Delaware, Oklahoma** *Blood Medicine* Root used as a blood purifier. *Liver Aid* Root used for jaundice. (175:28, 78) **Iroquois** *Blood Medicine* Compound decoction of roots taken for blood disorders and as a blood purifier. (87:313) *Contraceptive* Used as a contraceptive. (87:312) *Gynecological Aid* Used to stop menses. (87:313) *Pediatric Aid* and *Pulmonary Aid* Decoction of root given to children for whooping cough. (87:312) *Tonic* Compound decoction taken as a blood medicine and tonic. (87:313)
- *Food*—**Saanich** *Unspecified* Young stems cooked and used for food. (182:85)

Rumex orbiculatus Gray, Greater Water Dock
- *Drug*—**Cree, Woodlands** *Antirheumatic (External)* Decoction of whole plant applied externally to painful joints. (109:58) **Meskwaki** *Antidote* Decoction of root taken as an antidote for poison. (as *R. britanica* 152:237) **Potawatomi** *Blood Medicine* Root used as a blood purifier. (as *R. britannica* 154:73)

Rumex patientia L., Patience Dock
- *Drug*—**Cherokee** *Antidiarrheal* Infusion taken for dysentery. *Blood Medicine* Infusion of root taken "for blood." *Dermatological Aid* Juice and infusion of root used as poultice and salve for various skin problems. *Kidney Aid* Infusion of root taken "to correct fluids." *Laxative* Infusion of root used for constipation. *Throat Aid* Leaves rubbed in mouth for sore throat. *Veterinary Aid* Beaten roots fed to horses for "sick stomach." (80:32)

Rumex paucifolius Nutt., Fewleaved Dock
- *Food*—**Klamath** *Unspecified* Leaves and stems eaten fresh. Ripe seeds used for food. (as *R. geyeri* 45:95) **Montana Indian** *Vegetable* Herbage eaten raw. (as *R. geyeri* 19:22)

Rumex salicifolius Weinm., Willow Dock
- *Drug*—**Blackfoot** *Dermatological Aid* Boiled root used for many complaints, generally for swellings. (114:274) **Gosiute** *Blood Medicine* Roots used as a blood medicine. *Cathartic* Decoction of roots used for severe constipation. (39:380) **Kawaiisu** *Analgesic* Mashed roots used as a salve on sore limbs. *Dermatological Aid* Dried, pounded roots used as powder on sores and plant salve used on cuts. *Gastrointestinal Aid* Infusion of roots taken for stomachaches. *Misc. Disease Remedy* Mashed roots used as a salve for chickenpox. *Orthopedic Aid* Mashed roots used as a salve on sore limbs. *Unspecified* Stems, seeds, and roots used as medicine. (206:60)
- *Food*—**Kawaiisu** *Porridge* Seeds parched with hot coals, pounded, and cooked to the consistency of "thick gravy." *Unspecified* Stems boiled with sugar, or roasted, inner pulp pushed out of the burned skin and eaten hot or cold. (206:60) **Klamath** *Unspecified* Seeds used for food. (45:95) **Montana Indian** *Unspecified* Seeds used for food. *Vegetable* Spring leaves used for "greens." (19:22)

Rumex salicifolius var. *mexicanus* (Meisn.) C. L. Hitchc., Mexican Dock
- *Drug*—**Apache, White Mountain** *Gynecological Aid* Infusion of leaves taken by childless women to become pregnant. *Throat Aid* Infusion of leaves used for sore throats. (as *R. mexicanus* 136:160) **Blackfoot** *Antirheumatic (External)* Decoction of plant used for swellings. *Panacea* Decoction of plant used for many complaints. (as *R. mexicanus* 97:34) **Cree, Woodlands** *Antirheumatic (External)* Decoction of whole plant applied externally to painful joints. (as *R. mexicanus* 109:58) **Houma** *Abortifacient* Decoction of white root used to regulate menstruation. *Febrifuge* Decoction of red root taken for fever. *Gastrointestinal Aid* Decoction of plant taken for intestinal disorders. *Liver Aid* Decoction of plant taken for liver trouble. Infusion of yellow root in gin taken for jaundice. (as *R. mexicana* 158:56, 57) **Keres, Western** *Burn Dressing* Poultice of crushed roots or paste of burned, ground roots and water used for burns. (as *R. mexicanus* 171:67) **Navajo, Ramah** *Ceremonial Medicine* and *Emetic* Whole plant used as a ceremonial emetic. *Panacea* Root used as a "life medicine." (as *R. mexicanus* 191:24) **Zuni** *Reproductive Aid* Strong infusion of root made and given to women by their husbands to help them to become pregnant. A strong infusion of root was made by the husband of a childless wife and given to her morning, noon, sunset, and bedtime for a month to help her to become pregnant. If the medicine did not work, it was because the wife's heart was not good. (as *R. mexicanus* 166:85) *Throat Aid* Ground root or infusion taken for sore throat, especially by sword-swallower. (as *R. mexicanus* 166:59)
- *Food*—**Cochiti** *Vegetable* Leaves used as greens. (as *R. mexicanus* 32:50)
- *Dye*—**Houma** *Unspecified* Roots used to make a dye for cane and palmetto splints in baskets. (as *R. mexicana* 158:56)
- *Other*—**Keres, Western** *Fuel* Crushed, dried roots used as tinder. (as *R. mexicanus* 171:67)

Rumex sp., Dock
- *Drug*—**Cowlitz** *Dermatological Aid* Decoction of stalks used as an antiseptic wash for leg sores. *Disinfectant* Decoction of stalks used as an antiseptic wash for leg sores. (79:29) **Eskimo, Kuskokwagmiut** *Dermatological Aid* Plant "widely recognized as an effective astringent." (122:31) **Iroquois** *Antidiarrheal* Infusion of grains with other plant segments used for diarrhea. (141:41) *Antihemorrhagic* Infusion of mature grains used for hemorrhages. (142:85) **Luiseño** *Unspecified* Decoction of roots used for medicinal purposes. (155:233)
- *Food*—**Algonquin, Quebec** *Substitution Food* Used as substitute for rhubarb in pies or added to water and the solution used as a salt substitute. (18:84) **Chehalis** *Unspecified* Green stalks of large plants cooked over maple and cedar limbs on hot rocks and used for food. (79:29) **Pima, Gila River** *Vegetable* Leaves boiled or boiled, strained, refried, and eaten as greens.[11] (133:5)

Rumex venosus Pursh, Veiny Dock
- *Drug*—**Arapaho** *Dermatological Aid* Stems and leaves used as a wash for sores. (121:47) **Lakota** *Gynecological Aid* Infusion of roots given to women to expel the afterbirth. (139:55) **Paiute** *Analgesic* Decoction of roots taken for stomachaches. *Antidiarrheal* Decoction of root taken for diarrhea. *Antirheumatic (Internal)* Decoction of root taken for rheumatism. *Blood Medicine* Decoction of roots taken as a blood purifier or tonic. (180:132, 133) *Burn Dressing* Powdered roots dusted on burns. (121:47) Poultice of dried or raw roots applied and decoction used as a wash for burns. *Cold Remedy* Decoction of root taken for colds. *Cough Medicine* Decoction of root taken for coughs. (180:132, 133) *Dermatological Aid* Powdered roots dusted on impetigo. (121:47) Poultice of root applied or decoction used as a wash for wounds, sores, and swellings. *Gastrointestinal Aid* Decoction of root taken for stomachaches, stomach trouble, and gallbladder trouble. *Kidney Aid* Decoction of root taken for kidney disorders. *Misc. Disease Remedy* Decoction of roots taken for influenza. *Pulmonary Aid* Decoction of root taken for pneumonia. *Tonic* Decoction of root taken as a blood purifier or tonic. *Venereal Aid* Simple or compound decoction of root taken for venereal disease. **Shoshoni** *Blood Medicine* Decoction of roots taken as a blood purifier or tonic. *Burn Dressing* Poultice of dried or raw roots applied and decoction used as a wash for burns. (180:132, 133) *Cathartic* Decoction of whole plant taken as a physic. (118:42) *Dermatological Aid* Poultice of root applied or decoction used as a wash for wounds, sores, and swellings. *Tonic* Decoction of root taken as a blood purifier or tonic. *Venereal Aid* Decoction of root taken for venereal disease. (180:132, 133)
- *Food*—**San Felipe** *Unspecified* Young stems eaten like rhubarb. (32:50)
- *Dye*—**Cheyenne** *Red* Roots and dried leaves boiled and used as a red dye. (76:172) *Unspecified* Used to make a dye. (83:46) *Yellow* Roots and dried leaves boiled and used as a yellow dye. (76:172) **Great Basin Indian** *Orange* Peeled root used to make a burnt orange dye. The procedure involved was described by children at the Wind River Community Day School as follows: "We break the roots into inch pieces. We then spread them out very thin on papers. We place them in the sun. We let it get very dry. After it is very dry we put it into water. We let it soak for a few days. We then boil it in the water it has soaked in. After it has boiled a long time we put some alum in it. This sets the color." (121:47)

Rumex verticillatus L., Swamp Dock
- *Drug*—**Choctaw** *Misc. Disease Remedy* Decoction of leaves used in bath to prevent smallpox. (25:23) Infusion of leaves used as a bath to prevent smallpox. (177:21)

Rumex violascens Rech. f., Violet Dock
- *Food*—**Pima, Gila River** *Unspecified* Leaves boiled and eaten. (133:7)

Ruppia, Potamogetonaceae
Ruppia sp., Sea Grass
- *Food*—**Seri** *Staple* Seeds made into a meal. (50:134)

Rutaceae, see *Citrus, Cneoridium, Fortunella, Melicope, Pelea, Ptelea, Ruta, Thamnosma, Zanthoxylum*

Ruta, Rutaceae

Ruta chalepensis L., Fringed Rue
- *Drug*—**Costanoan** *Cough Medicine* Decoction of plant used for coughs. *Ear Medicine* Heated leaves placed inside the ear for earaches. *Gastrointestinal Aid* Plant used for stomach pain. *Orthopedic Aid* Plant used for paralysis. (21:22) **Diegueño** *Ear Medicine* Mashed leaves wrapped in a piece of cotton and placed in the ear for earaches. (84:39)

Ruta graveolens L., Common Rue
- *Drug*—**Cherokee** *Anthelmintic* Decoction of top or leaves boiled into a syrup and taken for worms. *Orthopedic Aid* Added to whisky and taken for palsy. *Poultice* Used as poultice for gangrenous parts. *Sedative* Added to whisky and taken for hysterics. (80:53) **Diegueño** *Ear Medicine* Sprig put in the ear for an earache. *Gastrointestinal Aid* Infusion of leaves taken for stomachaches. (88:218)

Sabal, Arecaceae
Sabal minor (Jacq.) Pers., Dwarf Palmetto
- *Drug*—**Houma** *Eye Medicine* Crushed, small root juice rubbed into sore eyes as a counterirritant. *Hypotensive* Decoction of dried root taken for high blood pressure. *Kidney Aid* Decoction of root taken for kidney trouble. (as *S. adansonii* 158:55, 56) *Stimulant* Decoction of dried root taken for "swimming in head." (as *S. adansonii* 158:55)
- *Food*—**Houma** *Bread & Cake* Fresh root slices baked and eaten as bread. (as *S. adansonii* 158:55)

Sabal palmetto (Walt.) Lodd. ex J. A. & J. H. Schultes, Cabbage Palmetto
- *Drug*—**Seminole** *Analgesic, Dietary Aid,* and *Febrifuge* Berries or seeds used for grass sickness: low fever, headache, and weight loss. (169:242)
- *Food*—**Seminole** *Unspecified* Plant used for food. (169:506)
- *Fiber*—**Seminole** *Building Material* Plant used to make houses. (169:506)
- *Other*—**Seminole** *Cooking Tools* Plant used to make food paddles. Plant used to make skin drying frames and potato drying mats. *Hunting & Fishing Item* Plant used to make arrows, hunting dance staffs, fish drags, and fish poison. *Toys & Games* Plant used to make ball sticks. (169:506)

Sabatia, Gentianaceae
Sabatia angularis (L.) Pursh, Rose Pink
- *Drug*—**Cherokee** *Analgesic* and *Gynecological Aid* Infusion taken for periodic pains. (80:53)

Sabatia campanulata (L.) Torr., Slender Rosegentian
- *Drug*—**Seminole** *Analgesic, Antidiarrheal, Eye Medicine,* and *Febrifuge* Infusion of roots taken for sun sickness: eye disease, headache, high fever, and diarrhea. (169:206)

Saccharum, Poaceae
Saccharum officinarum L., Sugarcane
- *Food*—**Seminole** *Unspecified* Plant used for food. (169:471)

Sadleria, Blechnaceae, fern

Sadleria cyatheoides Kaulfuss, Amaumau Fern
- *Drug*—Hawaiian *Dermatological Aid* Plant and other ingredients pounded, squeezed, and the resulting juice applied to boils and pimples. *Pulmonary Aid* Shoots and other ingredients used for lung troubles. *Respiratory Aid* Infusion of powdered inner bark and other ingredients taken for asthma and kindred troubles. (2:16)
- *Food*—Hawaiian *Beverage* Plant powdered and used to make a beverage similar to coffee or tea. (2:16)

Sagittaria, Alismataceae

Sagittaria cuneata Sheldon, Arumleaf Arrowhead
- *Drug*—Cheyenne *Unspecified* Leaves used as an ingredient in a medicinal mixture. *Veterinary Aid* Dried leaves given to horses for urinary troubles or put into sore mouth. (83:6) Chippewa *Unspecified* Plant characterized as having some medicinal uses. (as *S. arifolia* 71:124) Navajo *Analgesic* Plant used for headaches. (as *S. arifolia* 55:24) Ojibwa *Gastrointestinal Aid* Corms eaten for indigestion. (as *S. arifolia* 153:353) *Unspecified* Used as a medicine for humans. *Veterinary Aid* Used as a medicine for horses. (as *S. arifolia* 153:396)
- *Food*—Klamath *Unspecified* Rootstocks used for food. (as *S. arifolia* 45:90) Menominee *Winter Use Food* Boiled, sliced potatoes strung on a string for winter use. (as *S. arifolia* 151:61) Montana Indian *Unspecified* Tubers eaten raw or boiled. (as *S. arifolia* 19:22) Ojibwa *Forage* Recognized as a favorite food of ducks and geese. *Staple* Corms, a most valued food, boiled fresh, dried, or candied with maple sugar. Muskrat and beavers store them in large caches, which the Indians learned to recognize and appropriate. (as *S. arifolia* 153:396) Paiute, Northern *Unspecified* Roots used for food. (59:44)

Sagittaria lancifolia L., Bulltongue Arrowhead
- *Drug*—Seminole *Dermatological Aid* Plant used for alligator bites. (169:298)

Sagittaria latifolia Willd., Broadleaf Arrowhead
- *Drug*—Cherokee *Febrifuge* and *Pediatric Aid* Infusion of leaves given, one sip, and used to bathe feverish baby. (80:23) Chippewa *Gastrointestinal Aid* Infusion of root taken for indigestion. (53:342) *Unspecified* Plant characterized as having some medicinal uses. (71:124) Iroquois *Antirheumatic (Internal)* Infusion of plant taken for rheumatism. *Dermatological Aid* Compound decoction taken for "boils around the abdomen of children." Compound decoction used as a wash on parts affected by "Italian itch." *Laxative* Compound decoction taken for constipation. *Pediatric Aid* Compound decoction taken for "boils around the abdomen of children." (87:273) Infusion of whole plant and rhizomes from another plant given to children who scream during the night. (141:65) Lakota *Unspecified* Roots used for food and eaten as medicine. (139:26) Potawatomi *Dermatological Aid* Poultice of pounded corms applied to wounds and sores. (154:37) Thompson *Love Medicine* Plant used as a love charm and for "witchcraft." (187:112)
- *Food*—Chippewa *Dried Food* "Potatoes" at the end of the roots dried, boiled, and used for food. (53:319) Cocopa *Unspecified* Tubers baked, peeled, and eaten whole or mashed. (37:207) Dakota *Unspecified* Roasted or boiled tubers used for food. (70:65) Klamath *Unspecified* Species used for food. (45:90) Lakota *Unspecified* Roots used for food and eaten as medicine. (139:26) Meskwaki *Forage* Muskrats gathered these corms for winter store of food and found to save the trouble of digging. *Winter Use Food* Boiled, sliced potatoes strung on a piece of basswood string and hung for winter supply. (152:254) Omaha *Unspecified* Tubers cooked as a farinaceous food. (68:325) Roasted or boiled tubers used for food. Pawnee *Unspecified* Roasted or boiled tubers used for food. (70:65) Pomo *Unspecified* Potato-like tubers eaten. (11:89) Potawatomi *Unspecified* Plant, growing along the streams and lakes, used as food by many tribes. Several days were required to cook the potatoes properly. The potatoes were cooked in a hole 6 feet deep. Thus, an article, unfit to eat raw, was made very nutritious and very palatable. (154:94) *Vegetable* Potatoes, deer meat, and maple sugar made a very tasty dish. *Winter Use Food* Boiled, sliced potatoes strung on a string and hung for storage and winter use. (154:95) Thompson *Dried Food* Cooked root, dried, soaked, and used with fish for food. *Unspecified* Cooked roots used for food. (187:112) Winnebago *Unspecified* Roasted or boiled tubers used for food. (70:65)
- *Other*—Cocopa *Toys & Games* Tubers used in gambling games. (37:207) Iroquois *Fertilizer* Decoction of root used as a corn medicine, when starting to plant corn. (87:273) Potawatomi *Hunting & Fishing Item* Favorite food with ducks and geese and planted by hunting clubs to attract these birds. (154:94)

Sagittaria latifolia Willd. var. latifolia, Broadleaf Arrowhead
- *Food*—Omaha *Vegetable* Bulbs boiled and eaten as vegetables. (as *S. variabilis* 58:341)

Sagittaria sp., Arrowhead
- *Drug*—Algonquin, Quebec *Tuberculosis Remedy* Root used for tuberculosis. (18:133)
- *Food*—Algonquin, Quebec *Unspecified* Species used for food. (18:73) Cheyenne *Unspecified* Stalk, below the blossom, peeled and eaten raw. (76:170) Dakota *Unspecified* Boiled tubers peeled and used for food. (69:358) Great Basin Indian *Unspecified* Tubers ground and eaten. (121:46)

Salicaceae, see *Populus, Salix*

Salicornia, Chenopodiaceae

Salicornia maritima Wolff & Jefferies, Slender Glasswort
- *Food*—Gosiute *Bread & Cake* Seeds ground into a meal and used to make a "sweet bread." (as *S. herbacea* 39:380)

Salicornia virginica L., Virginia Glasswort
- *Drug*—Heiltzuk *Analgesic* and *Antirheumatic (External)* Plant used for arthritic pain, rheumatism, aches, pain, and swelling. (43:91)
- *Food*—Salish, Coast *Unspecified* Fleshy stems used for food. (182:80)

Salix, Salicaceae

Salix alaxensis (Anderss.) Coville, Feltleaf Willow
- *Food*—Alaska Native *Dietary Aid* Young, tender leaves and shoots used as sources for vitamin C. *Snack Food* Inner bark eaten raw with seal oil and sugar as a winter tidbit. *Unspecified* Leaves used for food. Young, new shoots eaten raw or dipped in seal oil. (85:59) Eskimo, Alaska *Unspecified* Leaf tips eaten raw with seal oil in early spring. (1:34) Eskimo, Inupiat *Sweetener* Flowers sucked by children for the

sweet nectar. *Unspecified* Leaf buds used for food. Tender, new shoots peeled and eaten. Juice sucked from the stem. Juicy cambium, tasted like watermelon or cucumber, used for food. Tiny, green leaves used for food. (98:7)
- *Other*—**Eskimo, Alaska** *Smoke Plant* Plant gathered in late summer, burned to ashes, and added to chewing tobacco. *Snuff* Plant gathered in late summer, burned to ashes, and added to snuff. (1:34)

Salix alba L., White Willow
- *Drug*—**Cherokee** *Antidiarrheal* Infusion of bark taken to check bowels. *Dermatological Aid* Decoction or infusion of bark used as a wash to make the hair grow. Bark used as a poultice. *Febrifuge* Infusion taken for fever. *Respiratory Aid* Root chewed by ballplayers "for wind." *Throat Aid* Infusion of inner bark taken for lost voice and root chewed for hoarseness. (80:61) Decoction of inner bark taken for hoarseness. (177:12) *Tonic* Bark used as a tonic. (80:61)

Salix amygdaloides Anderss., Peachleaf Willow
- *Drug*—**Cheyenne** *Antidiarrheal* Infusion of bark shavings used for diarrhea. (82:67) Infusion of bark taken for diarrhea. *Ceremonial Medicine* Plant used in the Sun Dance ceremony. *Dermatological Aid* Poultice of bark applied to bleeding cuts. (83:37) *Gastrointestinal Aid* Infusion of bark shavings used for stomach ailments. (82:67) *Hemostat* Poultice of bark applied to bleeding cuts. *Panacea* Infusion of bark taken for diarrhea and other ailments. (83:37) **Okanagan-Colville** *Orthopedic Aid* Decoction of branch tips used for soaking the feet and legs for cramps. (188:135)
- *Fiber*—**Cheyenne** *Building Material* Branches used to build sweat lodges. *Furniture* Young twigs made into cages and used to carry children on travois. Slender shoots bound with sinew and used as backrests. *Mats, Rugs & Bedding* Wood made into mattresses and used to keep beds above the ground. (83:37) **Gosiute** *Basketry* Wood used to make baskets, fish weirs, and water jugs. (39:380) **Ute** *Basketry* Used in basketry. (38:36)
- *Other*—**Cheyenne** *Cooking Tools* Branches used to make meat drying racks. *Musical Instrument* Wood used to make drums. *Paint* Charcoal used as a black paint for Sun Dancers. *Tools* Sticks bent and used to remove hair from hides. (83:37) **Gosiute** *Containers* Wood used to make baskets and water jugs. *Hunting & Fishing Item* Wood used to make fish weirs. (39:380)

Salix arbusculoides Anderss., Littletree Willow
- *Drug*—**Eskimo, Inuktitut** *Dermatological Aid* Poultice of shredded, inner bark applied to skin sores. (202:186) **Eskimo, Kuskokwagmiut** *Dermatological Aid* Poultice of inner bark applied to sores. *Eye Medicine* Leaves placed in the corners of watery eyes. *Oral Aid* Leaves chewed for sore mouth. (122:30)

Salix babylonica L., Weeping Willow
- *Drug*—**Cherokee** *Antidiarrheal* Infusion of bark taken to check bowels. *Dermatological Aid* Decoction or infusion of bark used as a wash to make the hair grow. Bark used as a poultice. *Febrifuge* Infusion taken for fever. *Respiratory Aid* Root chewed by ballplayers "for wind." *Throat Aid* Infusion of inner bark taken for lost voice and root chewed for hoarseness. *Tonic* Bark used as a tonic. (80:61)

Salix bebbiana Sarg., Bebb Willow
- *Drug*—**Cree, Woodlands** *Dermatological Aid* Poultice of chewed root inner bark applied to a deep cut. (109:58) **Menominee** *Unspecified* Plant used medicinally. (as *S. rostrata* 151:52) **Okanagan-Colville** *Dermatological Aid* Poultice of inner cambium and powdered tree fungus applied to serious cuts. *Gynecological Aid* Shredded inner bark used for sanitary napkins to "heal a woman's insides." Decoction of branches taken by women for several months after childbirth to increase the blood flow. *Hemostat* Poultice of bark and sap applied as a wad to bleeding wounds. *Orthopedic Aid* Poultice of damp inner bark applied to the skin over a broken bone. *Pediatric Aid* Decoction of branches taken by women after childbirth and helped the baby through the breast milk. (188:136)
- *Fiber*—**Cree, Woodlands** *Basketry* Stems used to rim birch bark baskets. (109:58) **Okanagan-Colville** *Clothing* Bark twisted into cord and used to make bags and dresses. *Cordage* Branches or bark twisted into strong rope. *Sewing Material* Bark used for sewing birch bark onto basket frames. (188:136)
- *Other*—**Cree, Woodlands** *Containers* Stems used to make a stopper for a sturgeon skin jar. *Cooking Tools* Stems used to make a fish roasting stick. *Fasteners* Bark used to tie or fasten many things.[72] *Hunting & Fishing Item* Stems used to make bows and arrows. Bark made into netting and used to catch fish. *Tools* Stems used to make a bead weaving loom. Bark made into netting to clean pitch used in sealing birchbark canoes. *Toys & Games* Branches used to make whistles. (109:58) **Okanagan-Colville** *Containers* Bark twisted into cord and used to make bags and dresses. *Fasteners* Bark twisted into cord and used to tie things together. (188:136)

Salix bonplandiana Kunth, Red Willow
- *Drug*—**Costanoan** *Febrifuge* Bark used for fevers. (as *S. laevigata* 21:21) **Kawaiisu** *Antidiarrheal* Infusion of roots taken for diarrhea. (as *S. laevigata* 206:61)
- *Food*—**Kawaiisu** *Candy* Sticky, sweet substance relished like candy and honey. (as *S. laevigata* 206:61)
- *Fiber*—**Havasupai** *Basketry* Young shoots used for basketry. *Building Material* Wood used for fence posts and as fuel for fires. (197:214) **Karok** *Basketry* Roots used to make baskets. Twigs used as warp sticks. (as *S. laevigata* 148:381) **Kawaiisu** *Basketry* Young, green stems used to make baskets. *Building Material* Used in house construction. Used as the poles for the winter house and sweat house construction. *Furniture* Used to make the oval and Y-shaped cradles. (as *S. laevigata* 206:61)
- *Other*—**Havasupai** *Fuel* Wood used for fence posts and as fuel for fires. (197:214) **Karok** *Protection* Used as a protective charm by those ferrying turbulent waters. (as *S. laevigata* 148:381) **Kawaiisu** *Containers* Twigs with leaves used as "wrappers" to hold buckeye nuts and fish. *Hunting & Fishing Item* Used to make the bows and arrows for hunting small game and birds. *Smoking Tools* Twigs with leaves used as "wrappers" to hold tobacco. *Tools* Used to make long needles for sewing tule into mats. *Toys & Games* Split stems used to make clappers and whistles. (as *S. laevigata* 206:61)

Salix candida Fluegge ex Willd., Sageleaf Willow
- *Drug*—**Meskwaki** *Unspecified* Compound used as a medicine. (152:245) **Ojibwa** *Gastrointestinal Aid* Plant used for stomach troubles. *Sedative* Plant used for trembling. *Stimulant* Plant used for fainting. *Unspecified* Bark used for medicinal purposes. (135:243) **Ojibwa,**

South *Cough Medicine* Decoction of inner bark taken for coughs. (91:200)

Salix caroliniana Michx., Coastal Plain Willow
- *Drug*—**Houma** *Blood Medicine* Decoction of roots and bark taken for "feebleness" due to thin blood. *Febrifuge* Decoction of roots and bark taken for fever. (as *S. longipes* 158:60) **Seminole** *Analgesic* Infusion of bark taken as an emetic for rainbow sickness: fever, stiff neck, and backache. (as *S. amphibia* 169:210) Infusion of bark taken as an emetic for thunder sickness: fever, dizziness, headache, and diarrhea. (as *S. amphibia* 169:213) Infusion of plant taken by men for menstruation sickness: stomachache, headache, and body soreness. (as *S. amphibia* 169:248) *Antidiarrheal* Infusion of bark taken as an emetic for thunder sickness: fever, dizziness, headache, and diarrhea. (as *S. amphibia* 169:213) *Antirheumatic (External)* Plant used for fire sickness: fever and body aches. (as *S. amphibia* 169:203) Cold infusion of plant used as a bath for body aches. (as *S. amphibia* 169:215) *Antirheumatic (Internal)* Infusion of plant taken by men for menstruation sickness: stomachache, headache, and body soreness. (as *S. amphibia* 169:248) *Blood Medicine* Decoction of bark taken for menstruation sickness: yellow eyes and skin, weakness, and shaking head.[13] (as *S. amphibia* 169:247) *Ceremonial Medicine* Plant used as a ceremonial emetic. (as *S. amphibia* 169:163) Bark used as an emetic in purification after funerals, at doctor's school, and after death of patient. (as *S. amphibia* 169:167) Roots taken by students in medical training. (as *S. amphibia* 169:95) *Dermatological Aid* Plant used for gunshot wounds. (as *S. amphibia* 169:302) *Emetic* Decoction of plant and other plants taken as an emetic by doctors to strengthen his internal medicine. (as *S. amphibia* 169:145) Bark used as an emetic to "clean the insides." (as *S. amphibia* 169:167) Infusion of bark taken as an emetic for rainbow sickness: fever, stiff neck, and backache. (as *S. amphibia* 169:210) Infusion of bark taken as an emetic for thunder sickness: fever, dizziness, headache, and diarrhea. (as *S. amphibia* 169:213) Bark used as an emetic. (as *S. amphibia* 169:288) Infusion of plant taken as an emetic to vomit the object the witch "shot" into the body. (as *S. amphibia* 169:398) Infusion of roots taken as an emetic during religious ceremonies. (as *S. amphibia* 169:408) *Eye Medicine* Infusion of inner bark taken and used as a bath for mist sickness: eye disease, fever, and chills. (as *S. amphibia* 169:209) *Febrifuge* Plant used for fire sickness: fever and body aches. (as *S. amphibia* 169:203) Plant used for dance fire sickness: fever. (as *S. amphibia* 169:206) Infusion of inner bark taken and used as a bath for mist sickness: eye disease, fever, and chills. (as *S. amphibia* 169:209) Infusion of bark taken as an emetic for rainbow sickness: fever, stiff neck, and backache. (as *S. amphibia* 169:210) Infusion of bark taken as an emetic for thunder sickness: fever, dizziness, headache, and diarrhea. (as *S. amphibia* 169:213) Plant used as a fever medicine. (as *S. amphibia* 169:283) Bark used for fevers. (as *S. amphibia* 169:288) *Gastrointestinal Aid* Infusion of plant taken by men for menstruation sickness: stomachache, headache, and body soreness. (as *S. amphibia* 169:248) *Hunting Medicine* Infusion of roots used as a hunting medicine to increase hunting luck. (as *S. amphibia* 169:371) *Love Medicine* Bark used as a medicine to prevent adultery. (as *S. amphibia* 169:249) *Oral Aid* Plant used for lion sickness: panting, staring, and tongue hanging out. (as *S. amphibia* 169:232) *Orthopedic Aid* Infusion of bark used as a bath for hot feet. (as *S. amphibia* 169:288) *Other* Decoction of bark taken for menstruation sickness: yellow eyes and skin, weakness, and shaking head.[13] (as *S. amphibia* 169:247) Plant used for lightning sickness. (as *S. amphibia* 169:305) *Preventive Medicine* Plant made into medicine and used to prevent the new mother's condition from contaminating the camp. (as *S. amphibia* 169:325) *Respiratory Aid* Plant used for lion sickness: panting, staring, and tongue hanging out. *Stimulant* Plant used for lion sickness: panting, staring, and tongue hanging out. (as *S. amphibia* 169:232) Infusion of bark taken and used as a bath for menstruation sickness: lassitude, laziness, and weakness.[73] *Strengthener* Infusion of bark taken and used as a bath for menstruation sickness: lassitude, laziness, and weakness.[73] (as *S. amphibia* 169:244) Decoction of bark taken for menstruation sickness: yellow eyes and skin, weakness, and shaking head.[13] (as *S. amphibia* 169:247) *Unspecified* Plant used for medicinal purposes. (as *S. amphibia* 169:161) Plant used medicinally. (as *S. amphibia* 169:164) *Vertigo Medicine* Infusion of bark taken as an emetic for thunder sickness: fever, dizziness, headache, and diarrhea. (as *S. amphibia* 169:213)
- *Other*—**Seminole** *Toys & Games* Plant used to make ball sticks. (as *S. amphibia* 169:492)

Salix cordata Michx., Heartleaf Willow
- *Drug*—**Malecite** *Dermatological Aid* Bark placed in hot water and used for blisters. (116:251) *Dietary Aid* Infusion of bark used for stimulating the appetite. (116:253) **Micmac** *Cold Remedy* Bark used for colds and to stimulate the appetite. *Dermatological Aid* Bark used for blisters. *Dietary Aid* Bark used to stimulate the appetite. (40:61) **Thompson** *Dermatological Aid* Poultice of fresh bark applied to bruises and skin eruptions. (164:471)

Salix discolor Muhl., Pussy Willow
- *Drug*—**Algonquin, Tête-de-Boule** *Gynecological Aid* Infusion of young branches used to start lactation. *Throat Aid* Inner bark powdered, made into a paste, and applied to "sick" throats. (132:130) **Blackfoot** *Analgesic* Decoction of new twigs taken as a painkiller. *Febrifuge* Decoction of new twigs taken for fevers. (97:28) **Cree, Woodlands** *Antidiarrheal* Infusion of inner bark taken for diarrhea. (109:58) **Iroquois** *Emetic* Compound decoction taken to vomit during initial stages of consumption, and to reduce loneliness. *Hemorrhoid Remedy* Infusion of bark used for bleeding piles. *Psychological Aid* Compound decoction taken to vomit to reduce loneliness. *Tuberculosis Remedy* Compound decoction taken to vomit during initial stages of consumption. (87:294) **Ojibwa** *Gastrointestinal Aid* Plant used for stomach troubles. *Sedative* Plant used for trembling. *Stimulant* Plant used for fainting. *Unspecified* Bark used for medicinal purposes. (135:243) **Potawatomi** *Hemostat* Decoction of root bark used for hemorrhages. (154:81, 82) *Panacea* Bark used as a universal remedy. (154:81)
- *Fiber*—**Cree, Woodlands** *Basketry* Stems used to rim birch bark baskets. *Cordage* Bark used to make rope. (109:58)
- *Dye*—**Blackfoot** *Red* Spring buds used to make a red dye. (97:32)
- *Other*—**Cree, Woodlands** *Containers* Stems used to make a stopper for a sturgeon skin jar. *Cooking Tools* Stems used to make a fish roasting stick. *Fasteners* Bark used to tie or fasten many things.[72] *Hunting & Fishing Item* Stems used to make bows and arrows. Bark made into

netting and used to catch fish. *Tools* Stems used to make a bead weaving loom. Bark made into netting to clean pitch used in sealing birchbark canoes. *Toys & Games* Branches used to make whistles. (109:58)

Salix eriocephala Michx., Missouri River Willow
- *Fiber*—**Ute** *Basketry* Used in basketry. (as *S. cordata* 38:36)
- *Other*—**Lakota** *Walking Sticks* Wood used to make canes and staffs. (as *S. missouriensis* 106:34)

Salix exigua Nutt., Sandbar Willow
- *Drug*—**Montana Indian** *Adjuvant* Poles used for framework of "sweat tepee" for colds and rheumatism. *Antirheumatic (External)* Poles used for framework of "sweat tepee" for rheumatism. *Cold Remedy* Poles used for framework of "sweat tepee" for colds. *Febrifuge* Bark used for certain fevers. (19:22) **Navajo, Ramah** *Ceremonial Medicine* and *Emetic* Decoction of leaves used as ceremonial emetic. (191:22) **Paiute, Northern** *Venereal Aid* Decoction of dried roots taken for venereal diseases. (59:128) **Zuni** *Cough Medicine* Infusion of bark taken for coughs. *Throat Aid* Infusion of bark taken for sore throat. (27:378)
- *Food*—**Navajo** *Beverage* Leaves used to make a drink "like orange juice." *Fodder* and *Forage* Leaves and bark used as food for both wild and domesticated animals. (90:155)
- *Fiber*—**Blackfoot** *Building Material* Used to make the framework of the sweat lodges. (97:30) **Flathead** *Basketry* Willow made into baskets cemented with gum and used to cook fish. (19:22) **Havasupai** *Basketry* Young shoots used for basketry. (197:215) **Jemez** *Building Material* Straight branches used for the inside roofs. (44:27) **Keres, Western** *Basketry* Young branches used to make baskets. *Mats, Rugs & Bedding* Young branches used to make mats. (171:67) **Lakota** *Building Material* Branches used for building sweat lodges. (106:33) **Mandan** *Mats, Rugs & Bedding* Leaves woven into mats and used in the sweat tepees. **Montana Indian** *Cordage* Used extensively for cordage. (19:22) **Okanagan-Colville** *Clothing* Bark twisted into cord and used to make bags and dresses. *Cordage* Bark used to make excellent cord. *Sewing Material* Bark used for sewing birch bark onto basket frames. (188:136) **Paiute, Northern** *Basketry* Used to make baskets. (60:75) **Pomo** *Basketry* Used for basket body material. (as *S. argophylla* 117:296) **Tewa** *Basketry* Used for basketry. (as *S. argophylla* 138:49)
- *Other*—**Havasupai** *Tools* Used to make tongs for removing cactus fruit. (197:215) **Lakota** *Fasteners* Peeled bark used for tying together sweat lodge poles. (106:33) **Navajo, Ramah** *Ceremonial Items* Stem used to make Lightningway hoop. (191:22) **Okanagan-Colville** *Containers* Branches used under fish to keep them clean. Bark twisted into cord and used to make bags and dresses. *Fasteners* Bark twisted into cord and used to tie things together. (188:136) **Tubatulabal** *Cooking Tools* Used to wrap up fish before roasting. (193:11)

Salix fragilis L., Crack Willow
- *Drug*—**Ojibwa** *Dermatological Aid* Poultice of bark applied to sores as a styptic and healing aid. *Hemostat* Bark used as a styptic and poultice for sores. (153:388)

Salix fuscescens Anderss., Alaska Bog Willow
- *Drug*—**Eskimo, Western** *Eye Medicine* Cotton used by old men in inner corner of eye for watery sore eye. (108:60) *Oral Aid* Leaves chewed for mouth sores. (108:17, 60)

Salix gooddingii Ball, Goodding's Willow
- *Drug*—**Pima** *Febrifuge* Decoction of leaves and bark taken as a febrifuge. (47:108)
- *Food*—**Cocopa & Yuma** *Unspecified* Honeydew obtained from cut branches. (37:218) **Mohave** *Beverage* Young shoots used to make tea. (37:201) **Pima** *Unspecified* Catkins eaten raw. (47:108) **Yuma** *Beverage* Leaves and twig bark steeped to make tea. *Unspecified* Bark eaten raw or cooked in hot ashes. (37:201)
- *Fiber*—**Cahuilla** *Furniture* Wood used to make cradleboards. (15:135) **Pima** *Basketry* Used as foundations for outdoor storage baskets. *Mats, Rugs & Bedding* Bark used as padding in baby cradles. (47:108) *Sewing Material* Small, green branches split in two, peeled, twisted, dried, and used for sewing coiled baskets. (47:116)
- *Other*—**Pima** *Containers* Used to make bird cages. *Hunting & Fishing Item* Used to make bows. (47:108)

Salix hindsiana Benth., Hinds's Willow
- *Drug*—**Pomo, Kashaya** *Throat Aid* Decoction of bark or leaves used for sore throats. Infusion of leaves used for laryngitis. (72:118)
- *Food*—**Kawaiisu** *Candy* Sticky, sweet substance relished like candy and honey. (206:61)
- *Fiber*—**Costanoan** *Basketry* Shoots used in basketry. (21:249) **Karok** *Basketry* Stems used for the main ribs in baskets as an alternate for *Corylus* stems. (6:53) Twigs used to make the warp sticks for twined baskets. Roots scraped, dried, and used on the inside of the overlaid twined baskets. (as *S. sessilifolia* var. *hindsiana* 148:381) **Kawaiisu** *Building Material* Used in house construction. Used as the poles for the winter house and sweat house construction. (206:61) **Pomo, Kashaya** *Basketry* Root used in twined baskets. Switches used for twined baskets and foundations in coiled baskets. *Building Material* Large branches used as the framework for thatched summer homes, sudatories, and other construction. (72:118) **Tolowa** *Basketry* Roots used in basketry. (6:53) **Ukiah** *Basketry* Roots used to make baskets. (as *S. argyrophylla* 41:331)
- *Other*—**Costanoan** *Fuel* Twigs used for kindling. (21:249) **Kawaiisu** *Containers* Twigs with leaves used as "wrappers" to hold buckeye nuts and fish. *Hunting & Fishing Item* Used to make bows and arrows for hunting small game and birds. *Smoking Tools* Twigs with leaves used as "wrappers" to hold tobacco. *Tools* Used to make long needles for sewing tule into mats. *Toys & Games* Split stems used to make clappers and whistles. (206:61) **Tubatulabal** *Cooking Tools* Used to wrap up fish before roasting. (193:11) **Ukiah** *Hunting & Fishing Item* Straight wands made into arrows. Large limbs used to make weirs for catching fish. (as *S. argyrophylla* 41:331)

Salix hookeriana Barratt ex Hook., Dune Willow
- *Drug*—**Makah** *Antidote* Leaves used as an antidote for shellfish poisoning. *Dermatological Aid* Infusion of roots used as a hair wash. (79:27) **Quileute** *Sports Medicine* Roots rubbed on bodies of athletes in training. (as *S. piperi* 79:26)
- *Food*—**Quileute** *Spice* Leaves put in cooking baskets and used as a food flavoring. (as *S. piperi* 79:26)
- *Fiber*—**Makah** *Basketry* Used to make baskets. **Nitinaht** *Mats, Rugs & Bedding* Soft roots used as a towel to rub down after bathing. (as *S. piperi* 67:242) **Quileute** *Basketry* Bark used extensively for basketry. (as *S. piperi* 79:26)

- *Other*—**Quileute** *Hunting & Fishing Item* Young trees used for fish weir poles. (as *S. piperi* 79:26)

Salix humilis Marsh., Prairie Willow

- *Drug*—**Cherokee** *Antidiarrheal* Infusion of bark taken to check bowels. *Dermatological Aid* Decoction or infusion of bark used as a wash to make the hair grow. Bark used as a poultice. *Febrifuge* Infusion taken for fever. *Respiratory Aid* Root chewed by ballplayers "for wind." *Throat Aid* Infusion of inner bark taken for lost voice and root chewed for hoarseness. *Tonic* Bark used as a tonic. (80:61) **Delaware** *Venereal Aid* Infusion of roots and other plants used for scrofula and venereal disease. (176:34) **Delaware, Oklahoma** *Tuberculosis Remedy* Compound infusion of plant used for scrofula. *Venereal Aid* Compound infusion of plant used for venereal disease. (175:29, 78) **Menominee** *Antidiarrheal* Root taken from shrub bearing insect galls and used for dysentery and diarrhea. *Gastrointestinal Aid* Root taken only from shrub bearing insect galls and used for spasmodic colic. (151:52) *Tonic* Decoction of stalk taken as a general tonic. (54:133) **Meskwaki** *Antidiarrheal* Infusion of root used for flux and enemas. *Hemostat* Leaves used for stopping a hemorrhage. *Laxative* Infusion of root used for flux and giving enemas. (152:245)

Salix humilis var. tristis (Ait.) Griggs, Prairie Willow

- *Drug*—**Catawba** *Gynecological Aid* Plant used for sore nipples. *Oral Aid* and *Pediatric Aid* Infusion of roots used as a wash for children with sore mouths. (as *S. tristis* 177:13) **Delaware** *Reproductive Aid* Infusion of roots used by women for displacement of the womb. *Venereal Aid* Infusion of plant and roots of other plants used for scrofula and venereal disease. (as *S. tristis* 176:34) **Seminole** *Analgesic, Antidiarrheal*, and *Eye Medicine* Infusion of plant taken for sun sickness: eye disease, headache, high fever, and diarrhea. (as *S. tristis* 169:208) *Febrifuge* Infusion of plant taken and rubbed on the body for high fevers. (as *S. tristis* 169:202) Infusion of plant taken for sun sickness: eye disease, headache, high fever, and diarrhea. (as *S. tristis* 169:208) *Hunting Medicine* Infusion of plant used as a hunting medicine to increase hunting luck. (as *S. tristis* 169:371) *Other* Plant used to make a medicine and given to students in medical training. (as *S. tristis* 169:103)

Salix interior Rowlee, Sandbar Willow

- *Drug*—**Iroquois** *Analgesic* Infusion of stems and other plant parts used for side pains. (141:39) **Thompson** *Unspecified* Roots used medicinally for unspecified purpose. (164:465)
- *Fiber*—**Chippewa** *Basketry* Cut, peeled willows dipped in hot water to make them tough and pliable and made into baskets. The best time for weaving willow baskets of various forms and sizes was during the springtime. (as *S. longifolia* 71:126) **Cree, Woodlands** *Basketry* Stems used to rim birch bark baskets. (109:58) **Gosiute** *Basketry* Wood used to make baskets, fish weirs, and water jugs. (as *S. longifolia* 39:380) **Omaha** *Basketry* Peeled stems used in basketry. (70:73) **Ute** *Basketry* Used in basketry. (as *S. longifolia* 38:36)
- *Dye*—**Potawatomi** *Red* Willow and some other species of willow used for a scarlet dye. (as *S. longifolia* 154:123)
- *Other*—**Cree, Woodlands** *Containers* Stems used to make a stopper for a sturgeon skin jar. *Cooking Tools* Stems used to make a fish roasting stick. *Fasteners* Bark used to tie or fasten many things.[72] *Hunting & Fishing Item* Stems used to make bows and arrows. Bark made into netting and used to catch fish. *Tools* Stems used to make a bead weaving loom. Bark made into netting to clean pitch used in sealing birchbark canoes. *Toys & Games* Branches used to make whistles. (109:58) **Gosiute** *Containers* Wood used to make baskets and water jugs. *Hunting & Fishing Item* Wood used to make fish weirs. (as *S. longifolia* 39:380) **Thompson** *Fasteners* Slender and tough stems used as withes. *Stable Gear* Bark used to weave saddle blankets. (as *S. longifolia* 164:498)

Salix irrorata Anderss., Sandbar Willow

- *Fiber*—**Apache, White Mountain** *Basketry* Withes used to make baskets and water jugs. Withes used to make baskets to be sold. (136:150) Split withes used to make baskets. *Building Material* Split withes used as tepee and wickiup thatching. (136:160) **Tewa** *Basketry* Used for basketry. (138:49) **Zuni** *Basketry* Slender switches used to make baskets. *Building Material* Stems formerly used for filling between the house rafters. (166:81)
- *Other*—**Apache, White Mountain** *Containers* Withes used to make baskets and water jugs. (136:150) *Cooking Tools* Withes tied together and used to stir mush and other foods. *Toys & Games* Wood used to make the poles and hoops for the pole game. Split withes used to make the three dice and throwing sticks for the setdilth game. (136:160) **Keres, Western** *Ceremonial Items* Branches made into prayer sticks. (171:68) **Zuni** *Cooking Tools* Eight or 12 willows trimmed at the ends, tied together, and used for stirring fire toasted foods. The willows were used for stirring corn, popcorn, and any other food toasted over a fire. (166:70)

Salix lasiolepis Benth., Arroyo Willow

- *Drug*—**Costanoan** *Cold Remedy* Infusion of bark or young leaves or decoction of flowers used for colds. (21:21) **Mendocino Indian** *Antidiarrheal* Infusion of leaves taken for diarrhea. *Dermatological Aid* Decoction of bark used as a wash for the itch. *Diaphoretic* Infusion of bark taken to cause sweating for any disease. *Febrifuge* Infusion of bark taken for chills and fever. *Panacea* Infusion of bark taken to cause sweating for any disease. (41:331) **Mewuk** *Febrifuge* Decoction of bark used for fevers. *Misc. Disease Remedy* Decoction of bark used for measles. (as *S. dasiolipis* 117:366)
- *Fiber*—**California Indian** *Cordage* Inner bark used in spring to make rope. (118:60) **Costanoan** *Basketry* Shoots used in basketry. (21:248) **Kawaiisu** *Basketry* Stems split for coiled baskets and as weft in twined baskets; unsplit as warps in twined baskets. (206:61) **Mendocino Indian** *Cordage* Tough, inner fiber formerly used to make rope and garments. **Round Valley Indian** *Building Material* Branches used as thatching to provide shade around houses. (41:331) **Shoshoni** *Basketry* Strands used to make baskets. (117:445)
- *Other*—**Mendocino Indian** *Smoke Plant* Inner bark portions dried, powdered, and used as substitutes for chewing tobacco. **Round Valley Indian** *Fuel* Wood used for fuel. *Protection* Trees planted in circles and used to protect the dancers from the sun and wind. (41:331)

Salix lasiolepis Benth. var. lasiolepis, Tracy Willow

- *Fiber*—**Diegueño** *Basketry* Branches used to make acorn storage baskets. (84:39)

Salix lucida Muhl., Shining Willow

- *Drug*—**Micmac** *Hemostat* Bark used for bleeding. *Respiratory Aid*

Bark used for asthma. (40:61) **Montagnais** *Analgesic* Infusion of leaves taken and poultice of bark applied for headache. (156:315) **Ojibwa** *Dermatological Aid* Poultice of bark used for sores and applied to bleeding cuts. *Hemostat* Bark used on bleeding cuts. (153:388) **Penobscot** *Respiratory Aid* Bark smoked for asthma. (156:309)
- *Other*—**Montagnais** *Smoke Plant* Dried bark smoked as a substitute for tobacco. (156:315) **Ojibwa** *Smoke Plant* Peeled, toasted, and flaked bark used for "kinnikinnick" or smoking mixture. (153:422)

Salix lucida ssp. *lasiandra* (Benth.) E. Murr., Pacific Willow
- *Drug*—**Bella Coola** *Antidiarrheal* Cold infusion of charred, pulverized sticks taken for diarrhea. *Dermatological Aid* Folded inner bark inserted in knife cuts and used for incisions. (as *S. lasiandra* 150:53) **Navajo, Ramah** *Ceremonial Medicine* Decoction of leaves used as ceremonial emetic. *Disinfectant* Painted internode of stem held by baby for "lightning infection." *Emetic* Decoction of leaves used as ceremonial emetic. *Pediatric Aid* Painted internode of stem held by baby for "lightning infection." (as *S. lasiandra* 191:22) **Okanagan-Colville** *Orthopedic Aid* Decoction of branch tips used for soaking the feet and legs for cramps. (as *S. lasiandra* 188:135) **Pomo, Kashaya** *Cold Remedy* Decoction of leaves used for colds. *Throat Aid* Decoction of leaves used for sore throats. (as *S. lasiandra* 72:118)
- *Fiber*—**Chehalis** *Cordage* Inner bark twisted and made into two-ply strings. (as *S. lasiandra* 79:26) **Gosiute** *Basketry* Wood used to make baskets, fish weirs, and water jugs. (as *S. lasiandra* 39:380) **Navajo, Ramah** *Building Material* Used for hogan construction. (as *S. lasiandra* 191:22) **Panamint** *Basketry* Used as the primary basket material. Twigs used to make the withes for the three-rod foundation coils of baskets. (as *S. lasiandra* 105:77) **Pomo, Kashaya** *Basketry* Branches used as the warp for twined baskets and foundation in coiled baskets. (as *S. lasiandra* 72:118) **Shoshoni** *Basketry* Used for foundation in baskets. Bark used as thread in baskets. (as *S. lasiandra* 118:7) **Ute** *Basketry* Used in basketry. (as *S. lasiandra* 38:36)
- *Other*—**Cowlitz** *Tools* Wood used to make drills for fire drills. (as *S. lasiandra* 79:26) **Gosiute** *Containers* Wood used to make baskets and water jugs. *Hunting & Fishing Item* Wood used to make fish weirs. (as *S. lasiandra* 39:380) **Haisla & Hanaksiala** *Cooking Tools* Branches used to hang drying oolichans (candlefish). *Tools* Wood used to make mallet heads. (as *S. lasiandra* 43:287) **Navajo, Ramah** *Ceremonial Items* Stem used to make Lightningway hoop. *Tools* Stem used to make loom frames. *Toys & Games* Branches used to make hobbyhorses for children. (as *S. lasiandra* 191:22) **Nitinaht** *Cooking Tools* Wood used to make single-pronged barbecue sticks for roasting salmon. (as *S. lasiandra* 186:127) **Salish, Coast** *Hunting & Fishing Item* Wood used to make bows. (as *S. lasiandra* 182:88)

Salix melanopsis Nutt., Dusky Willow
- *Drug*—**Montana Indian** *Adjuvant* and *Antirheumatic (External)* Poles used for framework of "sweat tepee" for rheumatism. *Cold Remedy* Poles used for framework of "sweat tepee" for colds. *Febrifuge* Bark used for certain fevers. (as *S. fluviatilis* 19:22)
- *Fiber*—**Dakota** *Basketry* Peeled stems used in basketry. (as *S. fluviatilis* 69:361) **Flathead** *Basketry* Willow made into baskets cemented with gum and used to cook fish. **Mandan** *Mats, Rugs & Bedding* Leaves woven into mats and used in the sweat tepees. **Montana Indian** *Cordage* Used extensively for cordage. (as *S. fluviatilis* 19:22)

Salix myricoides Muhl. var. *myricoides*, Bayberry Willow
- *Drug*—**Iroquois** *Venereal Aid* Compound used for syphilis with chancres. (as *S. glaucophylloides* 87:294)

Salix nigra Marsh., Black Willow
- *Drug*—**Cherokee** *Antidiarrheal* Infusion of bark taken to check bowels. *Dermatological Aid* Decoction or infusion of bark used as a wash to make the hair grow. Bark used as a poultice. *Febrifuge* Infusion taken for fever. *Respiratory Aid* Root chewed by ballplayers "for wind." *Throat Aid* Infusion of inner bark taken for lost voice and root chewed for hoarseness. *Tonic* Bark used as a tonic. (80:61) **Houma** *Blood Medicine* Decoction of roots and bark taken for "feebleness" due to thin blood. *Febrifuge* Decoction of roots and bark taken for fever. (158:60) **Iroquois** *Carminative* Compound decoction taken for stomach gas. *Cough Medicine* Compound decoction taken for coughs. *Throat Aid* Used for mouth and throat abscesses. (87:294) **Koasati** *Analgesic* Infusion of roots taken for headaches. *Febrifuge* Cold infusion of roots taken for fevers. *Gastrointestinal Aid* Decoction of roots taken for dyspepsia. (177:13) **Micmac** *Dermatological Aid* Poultice of bruised leaves used on sprains and bruises. Poultice of scraped root and spirits applied to bruises and sprains. *Orthopedic Aid* Poultice of bruised leaves applied to sprains and bruises. Poultice of scraped root and spirits applied to sprains and broken bones. (as *S. vulgare* 156:317)
- *Fiber*—**Papago** *Basketry* Twigs split in half lengthwise, sun dried, and used as the foundation of coiled basketry. (34:56) Used for sewing coiled basketry. (34:58) *Other* Twigs used for curved structures in wrapped weaving. (34:53)

Salix pedicellaris Pursh, Bog Willow
- *Drug*—**Ojibwa** *Gastrointestinal Aid* Bark used for stomach troubles. (153:388, 389)

Salix planifolia ssp. *pulchra* (Cham.) Argus, Tealeaf Willow
- *Drug*—**Eskimo, Alaska** *Anesthetic* Bark and leaves chewed to numb the mouth and throat. *Eye Medicine* "Cotton" used to dry "moist eyes." *Oral Aid* Bark and leaves chewed for mouth sores. (1:34) **Eskimo, Inupiat** *Oral Aid* Leaves made the mouth smell good. (as *S. pulchra* 98:10) **Eskimo, Nunivak** *Analgesic* Infusion of leaves and bark used as an analgesic. *Oral Aid* Plant chewed for sore mouth. (as *S. pulchra* 149:325) **Eskimo, Western** *Oral Aid* Leaves chewed for mouth sores. (as *S. pulchra* 108:17)
- *Food*—**Alaska Native** *Dietary Aid* Shoots probably the first spring source of vitamin C. Leaves used as one of the richest sources of vitamin C. *Unspecified* Shoots peeled and eaten raw. Young, tender leaves mixed with seal oil and eaten raw. *Winter Use Food* Leaves mixed with seal oil and stored in barrels, kegs, or seal pokes for winter use. (as *S. pulchra* 85:61) **Eskimo, Alaska** *Unspecified* Young leaves gathered in the spring and eaten raw with seal oil. (1:34) Cambium layer scraped off and eaten. *Vegetable* Young shoots and catkins used fresh or in seal oil. (as *S. pulchra* 4:715) *Winter Use Food* Leaves soaked in seal oil and saved for future use. (1:34) Young shoots and catkins stored in oil for winter use. (as *S. pulchra* 4:715) **Eskimo, Arctic** *Unspecified* Leaves used for food. (as *S. pulchra* 128:29) **Eskimo, Inupiat** *Beverage* Dried leaves used to make tea. *Soup* Dried leaves used in soups. *Vegetable* Leaves used as greens in fresh salads. *Winter Use*

Food Leaves preserved in seal or fish oil or canned for winter use and eaten with meat or fish. (as *S. pulchra* 98:10)

Salix prolixa Anderss., Mackenzie's Willow
- *Fiber*—**Montana Indian** *Sporting Equipment* Younger stems used extensively for making walking sticks. (as *S. mackenziana* 19:23)
- *Other*—**Blackfoot** *Ceremonial Items* Wood used to make ceremonial sticks. (as *S. mackenzieana* 97:32)

Salix pyrifolia Anderss., Balsam Willow
- *Drug*—**Ojibwa** *Gastrointestinal Aid* Plant used for stomach troubles. *Sedative* Plant used for trembling. *Stimulant* Plant used for fainting. *Unspecified* Bark used for medicinal purposes. (as *S. balsamifera* 135:243)

Salix rotundifolia Trautv., Least Willow
- *Drug*—**Eskimo, Nunivak** *Analgesic* Infusion of leaves and bark used as an analgesic. *Oral Aid* Plant chewed for sore mouth. (149:325)

Salix scouleriana Barratt ex Hook., Scouler's Willow
- *Drug*—**Bella Coola** *Dermatological Aid* Folded inner bark inserted in knife cuts and used for incisions. (150:54) **Okanagan-Colville** *Dermatological Aid* Poultice of inner cambium and powdered tree fungus applied to serious cuts. *Gynecological Aid* Shredded inner bark used for sanitary napkins to "heal a woman's insides." Decoction of branches taken by women for several months after childbirth to increase the blood flow. *Hemostat* Poultice of bark and sap applied as a wad to bleeding wounds. *Orthopedic Aid* Poultice of damp inner bark applied to the skin over a broken bone. *Pediatric Aid* Decoction of branches taken by women after childbirth and helped the baby through the breast milk. (188:136) **Sanpoil** *Antidiarrheal* Decoction of roots taken to counteract diarrhea. (131:220)
- *Food*—**Shuswap** *Preservative* Wood used to smoke salmon. (123:68)
- *Fiber*—**Gosiute** *Basketry* Wood used to make baskets, fish weirs, and water jugs. (as *S. flavescens* 39:380) **Okanagan-Colville** *Clothing* Bark twisted into cord and used to make bags and dresses. *Cordage* Branches and bark twisted into strong rope. *Sewing Material* Bark used for sewing birch bark onto basket frames. (188:136) **Tolowa** *Basketry* Roots used to make baskets. (6:53) **Wet'suwet'en** *Cordage* Bark strips used for cord or rope. (73:154) **Yurok** *Basketry* Roots used to make baskets. (6:53)
- *Other*—**Gosiute** *Containers* Wood used to make baskets and water jugs. *Hunting & Fishing Item* Wood used to make fish weirs. (as *S. flavescens* 39:380) **Haisla** *Toys & Games* Whips used to lash opponents in the Haisla "hoop and pole game." Players attempted to spear a rolling hoop. If a player succeeded in spearing the hoop, he was lashed by his opponent with a willow whip. If, however, the spearman retrieved his spear, the hoop, and the whip, his team would get the next throw of the hoop. **Haisla & Hanaksiala** *Cleaning Agent* Leaves used to wipe slime from fish. *Containers* Sticks used to string discoidal basaltic seaweed cooking stones when not being used. *Cooking Tools* Leaves layered with salmon and western hemlock boughs to allow the fish to drain. Branches used as barbecue racks for salmon. *Hunting & Fishing Item* Withes used as fish stringers to transport fish downstream. *Walking Sticks* Wood used to make walking sticks or canes. (43:288) **Karok** *Containers* Roots used to make the fire hearth. *Tools* Roots used to make the fire drill. (148:381) **Okanagan-Colville** *Containers* Bark twisted into cord and used to make bags and dresses. *Fasteners* Bark twisted into cord and used to tie things together. (188:136) **Shuswap** *Ceremonial Items* Inner bark headbands used by pubescent girls and young men, in ritual isolation and training. *Fasteners* Tough, bark strips poked through the roots of dog tooth violet, to hang them for drying. (123:68) **Wet'suwet'en** *Cooking Tools* Bark strips used for hanging fish. *Hunting & Fishing Item* Bark strips used for twining into nets or fish line. (73:154)

Salix sericea Marsh., Silky Willow
- *Drug*—**Iroquois** *Oral Aid* Used for mouth and throat abscesses. (87:294)

Salix sessilifolia Nutt., Northwest Sandbar Willow
- *Fiber*—**Pomo** *Basketry* Used in basketry as foundation rods. (118:6)

Salix sitchensis Sanson ex Bong., Sitka Willow
- *Drug*—**Eskimo, Alaska** *Dermatological Aid* Poultice of pounded bark applied to wounds. (149:331) **Karok** *Ceremonial Medicine* Roots and branches used in the World Renewal ceremony fire. (148:381) **Klallam** *Tonic* Decoction of peeled bark taken as a tonic. (79:26) **Okanagan-Colville** *Gastrointestinal Aid* Infusion of stalks taken for stomach ailments. (188:136) **Skagit** *Tonic* Decoction of peeled bark taken as a tonic. (79:26)
- *Fiber*—**Clallam** *Cordage* Bark made into string. (57:203) **Hoh** *Basketry* Limbs split, pared, scraped, and used to make baskets. (137:60) **Klallam** *Cordage* Bark peeled, twisted, and used to make string. (79:26) **Quileute** *Basketry* Limbs split, pared, scraped, and used to make baskets. (137:60) **Quinault** *Cordage* Bark used to make lines for tumplines and slings. **Snohomish** *Cordage* Bark used to make a two-ply string. (79:26) **Tolowa** *Basketry* Roots used to make baskets. **Yurok** *Basketry* Roots used to make baskets. (6:53)
- *Other*—**Haisla** *Toys & Games* Whips used to lash opponents in the "hoop and pole game." **Haisla & Hanaksiala** *Cleaning Agent* Leaves used to wipe slime from fish. *Containers* Sticks used to string discoidal basaltic seaweed cooking stones when not being used. *Cooking Tools* Leaves layered with salmon and western hemlock boughs to allow the fish to drain. Branches used as barbecue racks for salmon. *Hunting & Fishing Item* Withes used as fish stringers to transport fish downstream. *Walking Sticks* Wood used to make walking sticks or canes. (43:288) **Quinault** *Hunting & Fishing Item* Bark used to make the harpoon lines in sea lion hunting gear. (79:26)

Salix sitchensis var. *coulteri* Jepson, Sitka Willow
- *Fiber*—**Karok** *Basketry* Roots used to make baskets. (148:381)
- *Other*—**Karok** *Containers* Roots used as a fire hearth. *Cooking Tools* Twigs used by fishermen to string salmon steaks while drying. *Protection* Branch tied to the bow of a boat as a charm against danger when crossing the river in high water. *Tools* Roots used as a fire drill. Whole plant beaten with a stick to make the wind blow on hot days. (148:381) **Pomo** *Fuel* Used for firewood. (66:12)

Salix sp., Willow
- *Drug*—**Abnaki** *Eye Medicine* Used for sore eyes. (144:155) Decoction of bark and bark from two other plants used for eye pain. (144:170) **Alabama** *Febrifuge* Cold infusion of roots taken or used as a bath for fevers. (177:12) **Blackfoot** *Antidiarrheal* Infusion of plant used to counteract the laxative effect of the chokecherry infusion. *Antihem-*

orrhagic Infusion of fresh, crushed roots used for internal hemorrhage. (86:68) *Dermatological Aid* Dried, crushed roots soaked in water and grease used as a tonic for dandruff and straightening the hair. (86:124) Infusion of roots mixed with kidney fat and applied to head sores. (86:78) *Eye Medicine* Infusion of roots used for bloodshot or troublesome eyes. (86:82) *Gastrointestinal Aid* Infusion of fresh, crushed roots used for "waist troubles." (86:68) *Throat Aid* Infusion of roots swallowed for throat constrictions. (86:74) *Veterinary Aid* Chewed roots spat into the horse's eye for cloudiness and bloodshot. (86:90) **Cheyenne** *Dermatological Aid* Bark peeled and used for cuts. (82:67) **Chickasaw** *Analgesic* Roots used for headache or poultice of branches applied for severe headache. *Antidiarrheal* Decoction of roots taken for dysentery. *Hemostat* Roots used for nosebleed. (177:12) **Chippewa** *Antidiarrheal* Root taken alone or in compounds for dysentery. *Gastrointestinal Aid* Compound decoction of inner bark taken for indigestion. (53:342) **Comanche** *Eye Medicine* Stem ashes used for sore eyes. (29:524) **Costanoan** *Dermatological Aid* Infusion of leaves used as a hair rinse. Leaves used as a paste rubbed into the scalp for falling hair. (21:21) **Cree** *Venereal Aid* Decoction of bark used as a wash for syphilitic sores. (16:495) **Creek** *Antiemetic* Compound infusion of root taken for fever with nausea and vomiting. *Antirheumatic (External)* Compound infusion of root used as a wash for rheumatism. (172:655) *Antirheumatic (Internal)* Plant used for rheumatism. (177:12) *Dermatological Aid* Compound infusion of root used as a wash for swellings. (172:655) *Emetic* Plant used as an emetic. (177:12) *Febrifuge* Compound infusion of root taken for fever with nausea and vomiting. (172:655) Roots used for fevers. (177:12) *Gastrointestinal Aid* Compound infusion of root taken for biliousness. (172:655) Roots used for biliousness. (177:12) *Kidney Aid* Compound infusion of root taken for dropsy. (172:655) Decoction of roots taken or used as a bath for dropsy. (177:12) *Misc. Disease Remedy* Compound infusion of root taken for malaria. (172:655) Roots used for malaria. (177:12) *Other* Compound infusion of root taken for dropsy and "deer sickness." *Venereal Aid* Compound infusion of root taken for "the clap." (172:655) **Crow** *Analgesic* Bark chewed for headache. *Emetic* Stem tips formerly chewed as an emetic. *Oral Aid* Bark chewed for tooth hygiene. (82:67) **Eskimo, Inuktitut** *Dermatological Aid* Poultice of wetted leaves applied to bee stings. (202:192) **Eskimo, Western** *Antihemorrhagic* Decoction of leaves or bark taken for lung hemorrhage. (108:24) *Oral Aid* Decoction of inner bark used as a wash for mouth sores. (108:5, 6) *Preventive Medicine* Leaves eaten in early summer to protect against disease. (108:43) *Pulmonary Aid* Decoction of leaves or bark taken for lung hemorrhage. (108:24) Strong decoction of leaves or bark taken regularly for lung hemorrhage. (108:5, 6) *Throat Aid* Decoction of inner and outer bark gargled for sore throat. (108:23) Decoction of bark gargled for sore throat. (108:5, 6) **Flathead** *Dermatological Aid* Bark chewed and used for cuts. *Eye Medicine* Leaves and young stem tips used to make an eyewash. (82:67) **Iroquois** *Emetic* Compound decoction taken to vomit during initial stages of consumption. *Tuberculosis Remedy* Compound decoction taken to vomit during initial stages of consumption. (87:294) **Isleta** *Dermatological Aid* Decoction of leaves used as a skin bath. (100:42) **Kiowa** *Antirheumatic (External)* Infusion of leaves used as a wash for rheumatic aches. *Pulmonary Aid* Infusion of leaves used as a wash for pneumonia. *Toothache Remedy* Bark chewed for toothaches. (192:19) **Klallam** *Throat Aid* Decoction of bark taken for sore throats. *Tuberculosis Remedy* Decoction of bark taken for tuberculosis. (79:26) **Makah** *Unspecified* Used for medicine. (67:242) **Nez Perce** *Emetic* Formerly used as an emetic. *Gastrointestinal Aid* Twigs used to "clean out" the insides. (82:67) **Ojibwa** *Respiratory Aid* Heated poultice of inner bark applied to the throat for diphtheria. (5:2302) **Okanagan-Colville** *Dermatological Aid* Poultice of inner cambium and powdered tree fungus applied to serious cuts. *Gynecological Aid* Shredded inner bark used for sanitary napkins to "heal a woman's insides." Decoction of branches taken by women for several months after childbirth to increase the blood flow. *Hemostat* Poultice of bark and sap applied as a wad to bleeding wounds. *Orthopedic Aid* Poultice of damp inner bark applied to the skin over a broken bone. *Pediatric Aid* Decoction of branches taken by women after childbirth and helped the baby through the breast milk. (188:136) **Omaha** *Ceremonial Medicine* Plant used for the ritual of mourning. (68:322) Stems thrust in gashes on forearms of grieving young men at funeral ceremony. (70:73, 74) **Paiute** *Antidiarrheal* Burned root taken as pills or infusion of burned stems taken for diarrhea. *Blood Medicine* Decoction of root taken as a blood purifier. *Cathartic* Decoction of woody stems taken as a physic. *Dermatological Aid* Powder from dried, pulverized roots applied to syphilitic or purulent sores. Powder of dried stem bark applied to infant's navel. *Diuretic* Infusion of burned stems taken by adults and children for "failure to urinate." *Laxative* Infusion of young twigs with salt taken as a laxative. *Misc. Disease Remedy* Infusion of burned stems taken by adults and children for intestinal influenza. *Orthopedic Aid* Compound decoction of roots taken for lumbago. *Pediatric Aid* Infusion of burned stems given to children for diarrhea, taken by adults and children for "failure to urinate," and taken by adults and children for intestinal influenza. Powder of dried, stem bark applied to infant's navel. *Tonic* Decoction of root bark taken as a spring tonic. *Venereal Aid* Several species used in various ways for venereal disease. (180:133–136) **Penobscot** *Cold Remedy* Infusion of bark taken in large quantities for colds. (156:309) **Shoshoni** *Analgesic* Decoction of root taken for stomachaches. *Dermatological Aid* Decoction of leaves and twigs rubbed into the scalp for dandruff. *Gastrointestinal Aid* Decoction of root taken for stomachaches. *Toothache Remedy* Poultice of mashed roots applied to the gums for toothaches. *Venereal Aid* Several species used in various ways for venereal disease. (180:133–136) **Sikani** *Dermatological Aid* Young willow chewed and saliva applied to sores. (150:54) **Tanana, Upper** *Oral Aid* Fresh leaves chewed for mouth sores. (102:7) **Thompson** *Analgesic* Decoction of plant used as a wash for pain and swelling, especially of feet. *Dermatological Aid* Decoction of plant used as a wash for pain and swelling, especially of feet. (164:471) Leaves used in shoes or moccasins as padding for tired or sore feet. Decoction of bark used as a wash for sores. (187:279) *Orthopedic Aid* Decoction of plant used as a wash for pain and swelling, especially of feet. (164:471) Branches used as splints for broken limbs and rubbed on compound fractures. Decoction of bark used for bathing broken bones. Poultice of boiled bark used for a period of time for broken bones. *Toothache Remedy* Bark chewed for toothache. (187:279)

- ***Food*—Blackfoot** *Unspecified* Scraped cambium eaten especially by children. Peeled galls used for food. (86:105) **Eskimo, Inuktitut** *Soup* Leaves added to stews and soups. *Unspecified* Leaves eaten raw in spring. (202:189) Early leaves used for food. (202:192) **Tanana, Upper Unspecified** Sap and leaves eaten raw. Young sprouts and sap used for food. (102:7) **Thompson** *Forage* Plant enjoyed by moose. (187:279)

- **Fiber**—**Abnaki** *Basketry* Used to make baskets and whistles. (144:166) **Blackfoot** *Building Material* Boughs used to make sweat lodge frames. (86:122) Wands, sharpened at both ends, used to construct the dome shape of the sweat lodge. Branches used to construct the many different sweat lodges. The Horn Society sweat lodge was constructed of 14 willows. The Holy Woman's sweat lodge in the Sun Dance was made of 100 willows, which had been gathered by members of the Pigeon Society. The Motokiks Society (a woman's society) constructed its lodge with 12 willows. Medicine Pipes had 14. The used frame of a sweat lodge was left on the prairie. (86:17) Sticks used as lodgepoles for small hunting tepees. *Canoe Material* Wood used to make the circular frames for bull boats. (97:32) *Furniture* Boughs used to make backrest slats. (86:122) Sticks used to make backrests, part of the furniture of the tepee. (97:32) **Costanoan** *Building Material* Wood made into poles used as basic element of house construction. *Cordage* Bark braided into rope. (21:249) **Dakota** *Building Material* Poles used to sustain the thatch of the earth lodges and to form the frame of the bath lodges. (70:73) **Hopi** *Building Material* Used in roof construction. (200:72) **Hualapai** *Basketry* Used as the frame in coiled basketry. *Building Material* Used to make shelters. (195:29) **Isleta** *Basketry* Twigs used to make baskets. *Building Material* Branches and leaves used for thatching houses. (100:42) **Kiowa** *Building Material* Branches used in construction of summer shelters. (192:19) **Klamath** *Snow Gear* Wood used to make frames for snowshoes. (45:94) **Maidu** *Basketry* Stalks used to make the coiled foundations and coarse twine in the manufacture of baskets. (173:71) **Montana Indian** *Basketry* Wood used to make baskets. *Cordage* Wood used to make ropes. *Furniture* Wood used to make backrests for tepees and cradleboards. *Mats, Rugs & Bedding* Wood used to make mattresses for tepees. *Snow Gear* Wood used to make snowshoes. (82:67) **Navajo** *Basketry* Branches used to make permanent carrying baskets. *Cordage* Branches used to make a braided strap worn across the forehead to support a water bottle. *Furniture* Branches used to make cradle canopies. (55:38) **Okanagan-Colville** *Clothing* Bark twisted into cord and used to make bags and dresses. (188:136) **Omaha** *Building Material* Young poles used for the framework of the vapor bath lodges. (68:324) **Paiute** *Basketry* Strong, flexible willows used to make baskets. *Building Material* Willow used to construct house and sweat house frames. Willow used to make sheds for wind drying fish. *Cordage* Woven willow bark made into string and used to make salmon traps. *Furniture* Willow used to make baby baskets and cradleboards. (111:61) **Pawnee** *Building Material* Poles used to sustain the thatch of the earth lodges and to form the frame of the bath lodges. (70:73) **Pomo** *Basketry* Stem used as foundation material in basketry. (9:138) **Tanana, Upper** *Basketry* Stems used to make baskets for storing dried fish and meat. Stems used to make rims for birch bark baskets. Stems used to make basket rims. *Building Material* Stems used to make the semispherical frame of the traditional style sweat house. Stems and branches used as the siding on a smokehouse. Stems used in the construction of various shelters. *Cordage* Split, outer bark twisted into twine. Stems used to make line. Bark used to make line. Stems used to make fish hangers and lashings. *Snow Gear* Stems and spruce bark used to make a temporary sled for transporting meat across ice. (102:7) **Thompson** *Basketry* Split withes used for weaving baskets. *Building Material* Branches with the bark and leaves twisted and used for tying and binding in construction.[74] Split withes heated and twisted to make cabling for suspension bridges in "the old days." *Canoe Material* Dry logs lashed together to make rafts. Branches used in making fish traps, weirs, and rafts. (187:279) *Clothing* Bark of dead trees used to make capes and aprons. (164:499) *Cordage* Long shoots made into rope and used in lashing together fish drying racks and fish weir stakes. Softened stems twisted to make rope and used to lash together fish drying racks. Split withes used to make string and rope. *Furniture* Branches used to make cradle hoops. (187:279) *Mats, Rugs & Bedding* Bark of dead trees used to make mats and fiber blankets. (164:499) **Winnebago** *Building Material* Poles used to sustain the thatch of the earth lodges and to form the frame of the bath lodges. (70:73) **Yuki** *Basketry* Used as basket material. (103:423)

- **Dye**—**Micmac** *Black* Roots used to make a black dye. **Montagnais** *Black* Roots used to make a black dye. (156:317)

- **Other**—**Abnaki** *Containers* Used to make containers. (144:156) *Toys & Games* Used to make baskets and whistles. (144:166) **Blackfoot** *Containers* Wood used to make the top and bottom hoop of buckets, basins, and other containers. *Fasteners* Wood used to make tepee pegs and pins. (97:32) *Incense & Fragrance* Gall pitch used for incense during the annual ceremonies of the Motokiks and Kaispa Societies. (86:17) *Toys & Games* Branch with loosened bark used as a buzzing whistle. (86:122) **Chippewa** *Smoke Plant* Used for smoking and general utility. (53:378) **Eskimo, Inuktitut** *Cooking Tools* Small branches used to string fish for drying. *Fuel* Wood used for firewood. (202:189) *Smoke Plant* Dried leaves added to tobacco in place of shelf fungus. (202:182) *Snuff* Ground galls used for snuff. (202:192) **Hopi** *Ceremonial Items* Used to make prayer sticks. Occasionally used in ceremonies. (200:72) **Hualapai** *Fuel* Used for firewood. (195:29) **Keresan** *Ceremonial Items* Used extensively in making prayer sticks. (198:564) **Kiowa** *Protection* Leafy stems used to make wreaths worn by the women and children as sunshades during long walks. (192:19) **Kwakiutl, Southern** *Hunting & Fishing Item* Bark used to make fishing line and reef nets. (183:292) **Luiseño** *Hunting & Fishing Item* Plant used to make bows. (155:233) **Micmac** *Smoke Plant* Leaves used as tobacco. (159:258) **Montana Indian** *Cooking Tools* Wood used to make meat racks. *Fasteners* Wood used to make pins and pegs for tepees. *Hunting & Fishing Item* Wood used to make fish and fox traps. *Musical Instrument* Wood used to make drums. *Stable Gear* Wood used to make stirrups and hoops for catching horses. *Tools* Wood used to make scrapers for removing hair from hides. *Toys & Games* Wood used to make gambling wheels. *Walking Sticks* Wood used to make walking sticks. (82:67) **Navajo** *Ceremonial Items* Sticks used for the Night Chant and Mountain Chant. For the first day's ceremony of the Mountain Chant, willow sticks were gathered to make the emblem of the concentration of the four winds. A square was made with these sticks, leaving the ends projecting at the corners. The square was then placed over the invalid's head. For the rite of charcoal painting in the Night chant, a quantity of willow sticks, together with several pieces of pine bark, were burned to charcoal. The ashes of two different kinds of weeds, together with the ashes of two small feathers, were then added to the fat of a goat, mountain sheep, or other animal, made into balls, and daubed on the usual parts of the body. Branches used to make prayer sticks, prayer stick foundations, and plumed wands. Peeled sticks made into the talisman used in the Night Chant. *Containers* Branches used to make or sew water bottles. *Hunting & Fishing Item* Branches hardened by pounding with a stone and used to make lances. Branches used to make arrow shafts. *Tools* Branches made into hoops and used inside the buckskin sack of a bellows. Branches made into heddle sticks and used in weaving. (55:38) **Nitinaht** *Ceremonial Items* Soft roots used by young boys and girls as pre-scrubbers in the first stage of adulthood training. (186:127) **Okanagan-Colville** *Containers* Bark twisted into cord and used to make bags and dresses. (188:136) **Omaha** *Ceremonial Items* Stems used in funeral customs. The burial ceremony occurred 4 days after the death. The young men and friends of the family of the deceased accompanied the funeral party to the grave where they made parallel gashes in the skin of the forearm and thrust the willow stems into the gashes. The stems were bathed with the young men's blood who attested their sympathy to the living and sang the tribal Song to the Spirit. The song was one of cheer to the departing spirit and one of sympathy to the bereaved. (70:73) **Paiute** *Containers* Willow covered with pine pitch used to make water jugs. *Cooking Tools* Willows used to make drinking vessels and cooking vessels. Willow woven into a tray and used for winnowing seeds. *Hunting & Fishing Item* Willow used to build fish weirs. *Musical Instrument* Sticks wrapped formerly with buckskin but now with cloth and used for drum sticks. (111:61) **Tanana, Upper** *Containers* Stems used to make fish hangers, basket rims, lashing, and in the construction of various shelters. *Fasteners* Stems used to fasten spruce poles into a fence for capturing caribou. *Fuel* Wood used for firewood and smoking fish. *Hunting & Fishing Item* Split, outer bark twisted into twine and used to make a dip net. *Protection* Leafless stems or branches waved in the air to scare wolves away. Wolves were said to dislike the noise this made and would leave the area. (102:7) **Tewa** *Ceremonial Items* Twigs, one for every household in the village, used in December ceremonies. (138:48) **Thompson** *Containers* Bark of dead trees used to make capes and aprons. (164:499) Peeled, cleaned bark braided and woven together with Indian hemp fiber to make storage bags. *Fasteners* Softened stems twisted to make rope and used to lash together fish drying racks. Long shoots made into ropes and used for lashing together fish drying racks and fish weir stakes. Branches with the bark and leaves twisted and used for tying and binding in construction.[74] *Hunting & Fishing Item* Branches used in making fish traps, weirs, and rafts. (187:279)

Salix washingtonia, American Willow
- *Drug*—**Mahuna** *Blood Medicine* Infusion of pounded leaves taken as a blood tonic. (140:32)

Salsola, Chenopodiaceae

Salsola australis R. Br., Prickly Russian Thistle
- *Drug*—**Navajo** *Dermatological Aid* Poultice of chewed plants applied to ant, bee, and wasp stings. (as *S. kali* var. *tragus* 55:44) **Navajo, Ramah** *Dermatological Aid* Infusion of plant ashes used internally and externally for smallpox. *Misc. Disease Remedy* Infusion of plant ashes taken and used externally for smallpox and influenza. (as *S. pestifer* 191:25)
- *Food*—**Havasupai** *Forage* Young plants eaten by horses. (as *S. kali* 197:218) **Navajo** *Unspecified* Roasted seeds used for food. (as *S. kali* var. *tenuifolia* 90:155) Sprouts boiled and eaten with butter or small pieces of mutton fat. *Vegetable* Very young, raw sprouts chopped into salads. (as *S. kali* 110:27) **Navajo, Ramah** *Fodder* Young plants used for sheep and horse feed. (as *S. pestifer* 191:25)
- *Other*—**Keres, Western** *Unspecified* Plant known and named but no use was specified. (as *S. pestifer* 171:68)

Salvia, Lamiaceae

Salvia apiana Jepson, White Sage
- *Drug*—**Cahuilla** *Cold Remedy* Leaves eaten, smoked, and used in the sweat house for colds. *Dermatological Aid* Crushed leaves and water used as a hair shampoo, dye, and hair straightener. Poultice of fresh, crushed leaves applied before retiring to the armpits for body odors. *Eye Medicine* Seeds used as eye cleansers. *Hunting Medicine* Leaves used to prevent bad luck if a menstruating woman accidentally touched hunting equipment. (15:136) **Diegueño** *Blood Medicine* Infusion of leaves taken as a tonic for the blood. (88:219) *Cold Remedy* Decoction of leaves taken for colds. (84:39) *Cough Medicine* Infusion of leaves taken as a cough medicine. (88:219) *Misc. Disease Remedy* Leaves burned in hot coals to fumigate the house after a case of sickness such as measles. *Other* Decoction of leaves taken for a serious case of poison oak that "has entered the blood." (84:39) **Mahuna** *Gynecological Aid* Infusion of roots taken to heal internally and remove particles of afterbirth. (as *Ramona polystachya* 140:14)
- *Food*—**Cahuilla** *Spice* Leaves used as flavoring for mush. *Staple* Parched seeds ground into a flour and used to make mush. (15:136) **Diegueño** *Porridge* Seeds mixed with wheat or wild oats, toasted, ground fine, and eaten as a dry cereal. *Unspecified* Young stalks eaten raw. (84:39) **Luiseño** *Unspecified* Ripe stem tops peeled and eaten uncooked. Seeds eaten for food. (as *Ramona polystachya* 155:229)
- *Other*—**Cahuilla** *Hunting & Fishing Item* Fresh, crushed leaves applied to armpits by hunters to eliminate body odors and detection from game. (15:136)

Salvia carduacea Benth., Thistle Sage
- *Food*—**Cahuilla** *Porridge* Parched seeds ground into flour, mixed with other seeds, and used to make mush. (15:136) **Diegueño** *Spice* Seeds added to wheat to improve the flavor. (84:41) **Luiseño** *Unspecified* Seeds used for food. (155:229) **Tubatulabal** *Unspecified* Seeds used extensively for food. (193:15)

Salvia columbariae Benth., Chia
- *Drug*—**Cahuilla** *Disinfectant* Poultice of seed mush applied to infections. *Eye Medicine* Seeds used to cleanse the eyes or remove foreign matter from the eyes. (15:136) **Costanoan** *Eye Medicine* Gelatinous seeds placed in the eye to remove foreign objects. *Febrifuge* Infusion of seeds taken for fevers. (21:16) **Diegueño** *Strengthener* Seeds kept in the mouth and chewed during long journeys on foot, to give strength. (84:41) **Kawaiisu** *Eye Medicine* Seeds placed in the eye for irritation and inflammation. (206:62) **Mahuna** *Eye Medicine* Seeds placed under the eyelids while sleeping to remove sand particles. (140:54)

• **Food**—**Cahuilla** *Beverage* Seeds used to make a beverage. *Dietary Aid* Seeds used to render water palatable by removing the alkalies. *Staple* Parched seeds ground into flour and used to make cakes or mush. (15:136) **Costanoan** *Staple* Seeds used for pinole. (21:253) **Diegueño** *Spice* Seeds added to wheat to improve the flavor. (84:41) **Kawaiisu** *Beverage* Seeds parched, pounded, mixed with water, and used as a beverage. (206:62) **Luiseño** *Unspecified* Seeds used for food. (155:229) **Mahuna** *Porridge* Seeds winnowed, ground into a fine meal, and made into porridge. (140:54) **Mohave** *Beverage* Seeds ground and mixed with water. *Staple* Seeds used to make pinole. (37:187) **Paiute** *Porridge* Seeds used to make mush. (167:243) **Papago** *Beverage* Seeds steeped and used as tea-like drinks for refreshment. (34:27) **Pima** *Beverage* Seeds made into a popular, mucilaginous beverage. (146:77) **Pima, Gila River** *Beverage* Seeds used to make a mucilaginous drink. *Porridge* Seeds used to make a mucilaginous mass and eaten. (133:5) *Unspecified* Seeds eaten raw and parched. (133:7) **Pomo** *Staple* Seeds used to make pinoles. (11:87) Ground seeds used for pinole. (118:28) **Tubatulabal** *Unspecified* Seeds used extensively for food. (193:15) **Yavapai** *Unspecified* Species used for food. (65:258)

• **Fiber**—**Mahuna** *Building Material* Formerly used to cover dwellings. (140:54)

Salvia dorrii (Kellogg) Abrams, Grayball Sage

• **Drug**—**Kawaiisu** *Analgesic* Decoction of leaves used as a wash for headaches. *Gastrointestinal Aid* Infusion of leaves taken for stomachaches. *Witchcraft Medicine* Plant thrown into the fire to keep away the ghosts. (206:62) **Paiute** *Pediatric Aid* Infusion of leaves taken by children for colds and sore throat. (as *Ramona incana* 118:38) **Paiute, Northern** *Analgesic* Decoction of leaves taken and used as a wash for headaches. *Cold Remedy* Decoction of leaves taken for colds. (59:129) *Venereal Aid* Decoction of leaves taken for gonorrhea. (59:125)

• **Other**—**Kawaiisu** *Protection* Plant thrown on the fire at night to keep away the spirits and ghosts. (206:62)

Salvia dorrii ssp. *dorrii* var. *incana* (Benth.) Strachan, Purple Sage

• **Drug**—**Hopi** *Anticonvulsive* Smoke blown in face or plant taken in a drink for epilepsy or faintness. (as *S. carnosa* 200:33, 91) *Other* Plant used as a "deer medicine." *Stimulant* Plant used as a medicine for an epileptic or faint person. (as *S. carnosa* 200:91) **Okanagan-Colville** *Cold Remedy* Decoction or infusion of leaves used for colds. *Panacea* Decoction or infusion of leaves used for any illness of a general nature. (188:110) **Paiute** *Analgesic* Decoction of leaf or stem taken, used as a wash, and fumes inhaled for headaches. Decoction of leaf and sometimes stem taken for stomachaches. *Cold Remedy* Compound of dried plant smoked for colds. Compound poultice of crushed leaves applied for chest congestion from colds. Decoction of leaf or stem taken and poultice applied for colds. Infusion or simple or compound decoction of leaf and sometimes stem used for colds. *Cough Medicine* Decoction of leaf or stem taken and poultice applied for coughs. *Ear Medicine* Decoction of leaf used as drops and poultice of leaf and stem used for earaches. *Eye Medicine* Decoction of leaves used as an eyewash. *Febrifuge* Decoction of leaf or stem taken and poultice applied for fever. *Gastrointestinal Aid* Decoction of leaf and sometimes stem taken for stomachaches or indigestion. *Herbal Steam* Decoction of leaf and sometimes stem used as herbal steam for headaches. *Misc. Disease Remedy* Decoction of leaf and sometimes stem taken for fevers and influenza. *Poultice* Poultice of boiled plant tops applied to swollen leg veins. *Pulmonary Aid* Compound poultice of crushed leaves applied for chest congestion from colds. Decoction of leaves and sometimes stems taken for pneumonia. *Venereal Aid* Decoction of leaf and sometimes stem taken for venereal disease. **Shoshoni** *Analgesic* Decoction of leaf and sometimes stem taken for stomachaches. *Cold Remedy* Infusion or decoction of leaves, and sometimes stems, taken for colds. *Gastrointestinal Aid* Decoction of leaf and sometimes stem taken for stomachaches or indigestion. *Other* Decoction of plant tops used as a wash for swollen leg veins. *Pediatric Aid* and *Throat Aid* Decoction of leaf or stem given and used as a wash for children's sore throat. **Washo** *Cold Remedy* Infusion or decoction of leaves, and sometimes stems, taken for colds. *Respiratory Aid* Dried leaves smoked in a pipe to clear congested nasal passages. (as *S. carnosa* 180:136, 137)

Salvia lyrata L., Lyreleaf Sage

• **Drug**—**Catawba** *Dermatological Aid* Roots used as a salve for sores. (157:191) Root salve applied to sores. (177:55) **Cherokee** *Antidiarrheal* Infusion taken to check bowels. *Cold Remedy* Infusion taken for colds. *Cough Medicine* Infusion taken for coughs. *Diaphoretic* Used as a mild diaphoretic. *Gynecological Aid* Taken by weakly females. *Laxative* Infusion taken as a laxative. *Respiratory Aid* Syrup of leaves and honey taken for asthma. *Sedative* Infusion taken for nervous debility. *Stimulant* Taken by persons of phlegmatic habits. (80:53)

Salvia mellifera Greene, Black Sage

• **Drug**—**Costanoan** *Analgesic* Green leaves chewed for gas pains. Poultice of heated leaves applied to the ear for earache pain. *Carminative* Green leaves chewed for gas pains. *Cough Medicine* Decoction of plant taken for coughs. *Ear Medicine* Poultice of heated leaves applied to the ear for earache pain. *Heart Medicine* Infusion of green leaves taken for heart disorders. *Orthopedic Aid* Decoction of plant used as a bath for paralysis. *Throat Aid* Poultice of heated leaves applied to the neck for sore throats. (21:16) **Mahuna** *Cough Medicine* Infusion of plant taken for chronic bronchial coughs. *Respiratory Aid* Infusion of plant taken for chronic bronchial coughs. (as *Audibertias stachyoides* 140:19)

• **Food**—**Cahuilla** *Spice* Leaves and stalks used as a food flavoring. *Staple* Parched seeds ground into a meal. (15:136) **Luiseño** *Unspecified* Seeds used for food. (as *Ramona stachyoides* 155:229)

Salvia officinalis L., Kitchen Sage

• **Drug**—**Cherokee** *Antidiarrheal* Infusion taken to check bowels. *Cold Remedy* Infusion taken for colds. *Cough Medicine* Infusion taken for coughs. *Diaphoretic* Used as a mild diaphoretic. *Gynecological Aid* Taken by weakly females. *Laxative* Infusion taken as a laxative. *Respiratory Aid* Syrup of leaves and honey taken for asthma. *Sedative* Infusion taken for nervous debility. *Stimulant* Taken by persons of phlegmatic habits. (80:53) **Mohegan** *Anthelmintic* "Sage tea" taken as a vermifuge. *Panacea* Fresh leaves chewed to benefit the entire body. (176:75, 132) *Tonic* Green or dried leaves used to make a tonic. (176:132)

Salvia sp., Sage

• **Drug**—**Jemez** *Kidney Aid* Decoction of plant taken or raw plant eaten for kidney troubles. (44:27) **Rappahannock** *Misc. Disease Remedy* Infusion of leaves given frequently to children for measles. (161:33)

Sambucus, Caprifoliaceae
Sambucus canadensis L., American Elder

- **Drug**—**Algonquin, Quebec** *Emetic* Infusion of bark scraped upward and used as an emetic. *Laxative* Infusion of bark scraped downward and used as a laxative. (18:236) **Cherokee** *Antirheumatic (Internal)* Infusion of berry used for rheumatism. *Burn Dressing* Salve used for burns. *Cathartic* Used as a cathartic. *Dermatological Aid* Salve used for skin eruptions and infusion taken as tonic for boils. *Diaphoretic* Infusion of flowers taken to "sweat out fever." *Disinfectant* Leaves used to wash sores to prevent infection. *Diuretic* Used as a diuretic. *Emetic* Used as an emetic. *Kidney Aid* Taken for "dropsy." *Other* Decoction taken for "summer complaint." *Pediatric Aid* Given for "light sickness among children" and taken for "dropsy." (80:33) **Chickasaw** *Analgesic* Infusion of branches applied to head for severe headaches. (177:58) **Chippewa** *Emetic* Infusion of roots taken as an emetic. (71:142) **Choctaw** *Liver Aid* Decoction of seeds and roots taken for liver troubles. **Creek** *Breast Treatment* Poultice of pounded roots applied to swollen breasts. (177:58) *Gynecological Aid* Poultice of pounded root or stalk skin applied to women for swollen breast. (172:661) **Delaware** *Blood Medicine* Leaves and stems used as a blood purifier. *Dermatological Aid* Poultice of bark scrapings applied to sores, swellings, and wounds. *Liver Aid* Leaves and stems used for jaundice. *Pediatric Aid* Infusion of flowers used for infants with colic. (176:31) **Delaware, Oklahoma** *Blood Medicine* Leaves and stems used as a blood purifier. *Dermatological Aid* Poultice or salve of bark scrapings applied to wounds, sores, and swellings. *Gastrointestinal Aid* Infusion of flowers given to infants for colic. (175:26, 78) *Liver Aid* Leaves and stems used for jaundice. *Pediatric Aid* Infusion of flower given to infants for colic. (175:26) **Houma** *Analgesic* Decoction of bark used as a wash for pain. *Dermatological Aid* Decoction of bark used as a wash for swelling. *Tonic* Wine made from berries taken as a tonic. (158:60) **Iroquois** *Analgesic* Poultice of bark applied for headaches. (87:448) *Cathartic* Decoction of bark taken as a physic. Infusion of blossoms given to babies as a physic. (87:449) *Ceremonial Medicine* Decoction of flowers with other plants used as medicine to soak corn seeds before planting. (196:19) *Dermatological Aid* Compound poultice of plants and leaves applied to swellings of all kinds. Compound poultice of powdered roots and bark applied to baby's unhealed navel. (87:448) Poultice of bark applied to cuts. *Emetic* Infusion of bark taken as a spring emetic and to vomit up gall. (87:449) *Febrifuge* Berries used for fevers. (124:96) *Gastrointestinal Aid* Plant used for stomach troubles. *Heart Medicine* Infusion of pith taken for heart disease. (87:450) *Kidney Aid* Decoction of bark taken for the kidneys. (87:449) *Laxative* Compound decoction of twigs given to children as a laxative. (87:448) Infusion of bark taken as a laxative. (87:449) *Liver Aid* Infusion of bark taken to vomit up gall. (87:450) *Misc. Disease Remedy* Infusion of bark taken for the measles. (87:448) Compound decoction of plants taken for diphtheria. Infusion of berries taken and poultice applied to swellings caused by mumps. (87:450) *Pediatric Aid* Compound decoction of twigs given to children as a laxative. Compound poultice of powdered roots and bark applied to baby's unhealed navel. (87:448) Infusion of blossoms given to babies as a physic. (87:449) *Unspecified* Berries used for convalescents. (124:96) *Venereal Aid* Compound infusion used as wash on parts affected by venereal disease. (87:448) Decoction of pith taken for gonorrhea. (87:449) **Menominee** *Febrifuge* Infusion of dried flowers used as a febrifuge. (151:27) **Meskwaki** *Cathartic* Inner bark of young shoots used as a purgative. *Diuretic* Inner bark of young shoots used as a diuretic. *Gynecological Aid* Infusion of bark used in extremely difficult cases of parturition. *Pulmonary Aid* Root bark used to free lungs of phlegm. (152:207) **Micmac** *Cathartic* Berries, bark, and flower used as a purgative and bark used as a physic. *Emetic* Berries, bark, and flower used as a purgative and bark used as an emetic. *Sedative* Berries, bark, and flower used as a soporific and purgative. (40:61) **Mohegan** *Cathartic* Infusion of bark scraped downward and used as a physic. *Emetic* Infusion of bark scraped upward and used as an emetic. (174:265) Inner bark, scraped upward, used as an emetic. (176:75, 132) *Gastrointestinal Aid* Infusion of flowers given to babies with colic. (174:265) Infusion of dried flowers given to infants for colic. *Laxative* Inner bark, scraped downward, used as a laxative. *Pediatric Aid* Infusion of dried flowers given to infants for colic. (176:75, 132) **Rappahannock** *Antirheumatic (Internal)* Fermented decoction of berries taken for "rheumatism" (neuritis). (161:33) *Dermatological Aid* Compound infusion with bark used for swellings and sores. (161:34) **Seminole** *Ceremonial Medicine* Root bark used as a purification emetic after funerals, at doctor's school, and after death of patient. (as *S. simpsonii* 169:167) *Emetic* Decoction of root bark taken as an emetic for stomachaches. (as *S. simpsoni* 169:276) Root bark used as an emetic to "clean the insides." (as *S. simpsonii* 169:167) *Gastrointestinal Aid* Decoction of root bark taken as an emetic for stomachaches. (as *S. simpsoni* 169:276) **Thompson** *Toothache Remedy* Fresh bark used in hollow tooth for toothaches. (164:474)

- **Food**—**Cherokee** *Beverage* Berries used to make wine. (126:32) *Fruit* Fruit used for food. (80:33) *Pie & Pudding* Berries used to make pie. (126:32) *Preserves* Berries used to make jellies. (80:33) Berries used to make jelly. (126:32) **Chippewa** *Dried Food* Fruit dried for winter use. *Fruit* Fruit eaten fresh. (71:142) **Dakota** *Beverage* Blossoms dipped in hot water to make a pleasant drink. *Fruit* Fresh fruit used for food. (70:115) **Iroquois** *Bread & Cake* Fruit mashed, made into small cakes, and dried for future use. *Dried Food* Raw or cooked fruit sun or fire dried and stored for future use. (196:128) *Fruit* Fruits eaten raw. (124:96) Dried fruit taken as a hunting food. (196:128) *Porridge* Berries dried, soaked in cold water, heated slowly, and mixed with bread meal or hominy in winter. *Sauce & Relish* Berries used as a sauce. Berries dried, soaked in cold water, heated slowly, and used as a winter sauce. (124:96) Dried fruit cakes soaked in warm water and cooked as a sauce or mixed with corn bread. (196:128) **Meskwaki** *Fruit* Berries eaten raw. *Preserves* Berries cooked without sugar into a conserve. (152:256) **Omaha** *Beverage* Blossoms dipped in hot water to make a pleasant drink. *Fruit* Fresh fruit used for food. **Pawnee** *Beverage* Blossoms dipped in hot water to make a pleasant drink. *Fruit* Fresh fruit used for food. **Ponca** *Beverage* Blossoms dipped in hot water to make a pleasant drink. *Fruit* Fresh fruit used for food. (70:115) **Seminole** *Starvation Food* Plant used as a scarcity food. (as *S. simpsoni* 169:505)

- **Other**—**Dakota** *Toys & Games* Larger stems used by small boys to make popguns. (70:115) **Houma** *Hunting & Fishing Item* Stalks formerly used for blowguns. (158:60) **Menominee** *Toys & Games* Stems, after punching out the pith, used by children to make popguns. (151: 74) **Meskwaki** *Insecticide* Inner bark of young shoots used as a repellent for flies and insects. (152:207) *Toys & Games* Branch joints used as water squirt guns for playing or popguns for shooting pith corks. (152:268) **Omaha** *Toys & Games* Larger stems used by small boys to make popguns. **Pawnee** *Toys & Games* Larger stems used by small boys to make popguns. **Ponca** *Toys & Games* Larger stems used by small boys to make popguns. (70:115) **Seminole** *Tools* Stems used to make medicine tubes. (as *S. simpsoni* 169:172) Plant used to make medicine blowing tubes. *Toys & Games* Plant used to make toy blowguns. (as *S. simpsoni* 169:505)

Sambucus cerulea Raf., Blue Elderberry

- **Drug**—**Choctaw** *Dermatological Aid* Poultice of beaten leaves applied to swollen hands. *Gastrointestinal Aid* Decoction of root taken for dyspepsia. *Urinary Aid* Decoction of root taken for bladder troubles. (as *S. intermedia* 177:59) **Clallam** *Antidiarrheal* Infusion of bark used for diarrhea. (57:198) **Costanoan** *Cathartic* Decoction of leaves used as a purgative. *Cold Remedy* Decoction of leaves used for new colds. (21:24) **Houma** *Analgesic* Decoction of bark used as a wash for pain. *Dermatological Aid* Decoction of bark used as a wash for swelling. *Tonic* Wine made from berries taken as a tonic. (as *S. intermedia* 158:60) **Kawaiisu** *Analgesic* Infusion of leaves and flowers used as steam bath for headaches. *Blood Medicine* Decoction of leaves used as a wash on limb affected by blood poisoning. *Cold Remedy* Infusion of leaves and flowers used as steam bath for colds. *Diaphoretic* Infusion of leaves and flowers used as steam bath to cause perspiration. *Febrifuge* Infusion of flowers taken for fevers. *Herbal Steam* Infusion of leaves and flowers used as steam bath for headaches and colds. Infusion of leaves and flowers used as steam bath to cause perspiration. *Misc. Disease Remedy* Infusion of flowers taken for measles. (206:62) **Okanagan-Colville** *Antirheumatic (External)* Dead stalks used to make a steam bath for arthritis or rheumatism. (188:94) **Pomo, Kashaya** *Febrifuge* Infusion of dried flowers taken to break a fever. (72:42) **Thompson** *Antirheumatic (Internal)* Decoction of finely chopped bark taken for arthritis. One informant cautioned that this decoction must be boiled in a nonmetal pot or it would become poisonous. (187:199) *Toothache Remedy* Fresh bark used in hollow tooth for toothaches. (164:474) *Venereal Aid* Decoction of dried flowers taken for syphilis. (187:199) **Yuki** *Febrifuge* Infusion of flowers taken for fevers. (49:46)
- **Food**—**Costanoan** *Fruit* Fruits eaten for food. (21:254) **Kawaiisu** *Preserves* Dried or fresh berries used to make jelly. (206:62) **Okanagan-Colville** *Frozen Food* Berries frozen for future use. *Fruit* Berries used for food. (188:94) **Pomo, Kashaya** *Fruit* Tart berries eaten fresh in small quantities, canned, or cooked for pie filling. (72:42) **Salish, Coast** *Fruit* Berries cooked and used for food. (182:80) **Sanpoil & Nespelem** *Fruit* Berries eaten fresh. (131:102) **Thompson** *Dried Food* Fruits dried for future use. (164:490) Dried fruit used for food. (187:199) *Fruit* Fruits eaten fresh in large quantities. (164:490) Fresh fruit used for food. *Preserves* Berries cooked to make jam. *Spice* Berry juice used for marinating fish.[75] (187:199) **Yuki** *Dried Food* Berries dried and used as a winter food. *Fruit* Berries eaten raw. *Soup* Berries formerly made into soup. (as *S. coerules* 49:86) **Yurok** *Fruit* Fresh berries used for food. (6:53)
- **Other**—**Costanoan** *Fuel* Hollow twigs used in fire making. *Hunting & Fishing Item* Hollow twigs used for arrow shafts. *Musical Instrument* Hollow twigs used for flutes. *Smoking Tools* Hollow twigs used for pipes. (21:254) **Houma** *Hunting & Fishing Item* Stalks formerly used for blowguns. (as *S. intermedia* 158:60) **Kawaiisu** *Musical Instrument* Wood used to make a flute by pushing out the soft core and burning a single row of six holes. *Smoking Tools* Wood section hollowed out and used as a tobacco container. (206:62) **Okanagan-Colville** *Musical Instrument* Stems used to make "flageolets," a musical instrument. (188:94) **Pomo, Kashaya** *Toys & Games* Branches used to make whistles and clappers. (72:42) **Yuki** *Musical Instrument* Branches used to make flutes, clappers, and small whistles. (49:94) **Yurok** *Containers* Leaves used to pack sturgeon eggs while cooking. (6:53)

Sambucus cerulea Raf. **var.** *cerulea*, Blue Elderberry

- **Drug**—**Karok** *Ceremonial Medicine*, *Panacea*, and *Pediatric Aid* Infusion of branches used as a wash for sick child in the Brush Dance. (as *S. glauca* 148:389) **Klallam** *Antidiarrheal* Infusion of bark taken for diarrhea. (as *S. glauca* 79:47) **Luiseño** *Gynecological Aid* Flowers used for female complaints. (as *S. glauca* 155:229) **Mendocino Indian** *Dermatological Aid* Decoction of plant rubbed on bruises. *Febrifuge* Decoction of plant rubbed on the body for fevers. *Orthopedic Aid* Decoction of plant rubbed on sprains. *Veterinary Aid* Decoction of plant used as an antiseptic wash for itch and sores on animals. (as *S. glauca* 41:388) **Mewuk** *Unspecified* Flowers and roots used for medicine. (as *S. glauca* 117:366) **Miwok** *Misc. Disease Remedy* Decoction of blossoms taken for ague. (as *S. glauca* 12:172) **Montana Indian** *Antirheumatic (External)* Decoction of dried flowers applied externally for sprains and bruises. *Dermatological Aid* Decoction of dried flowers used as an antiseptic wash for open sores and itch. *Emetic* Inner bark used as a strong emetic. *Febrifuge* Decoction of dried flowers applied externally for fevers. *Gastrointestinal Aid* Decoction of dried flowers used internally for stomach troubles. *Pulmonary Aid* Decoction of dried flowers used internally for lung troubles. (as *S. glauca* 19:23) **Paiute** *Antirheumatic (External)* Poultice of heated stems applied to the body that ached from rheumatism and similar disorders. *Gastrointestinal Aid* Decoction of root scrapings taken for stomachaches. (as *S. glauca* 111:111) **Pomo, Little Lakes** *Antihemorrhagic* and *Tuberculosis Remedy* Decoction of plant taken for bleeding lungs from consumption. **Pomo, Potter Valley** *Disinfectant* Decoction of leaves used as an antiseptic wash. *Emetic* Inner bark used as a strong emetic. *Gastrointestinal Aid* Decoction of plant taken for stomachaches. (as *S. glauca* 41:388) **Quinault** *Emetic* Infusion of bark taken as an emetic. (as *S. glauca* 79:47) **Yokia** *Disinfectant* Decoction of leaves used as an antiseptic wash. *Emetic* Inner bark used as a strong emetic. *Gastrointestinal Aid* Decoction of plant taken for stomachaches. (as *S. glauca* 41:388) **Yokut** *Burn Dressing* Poultice of bruised leaves applied to burns. *Cathartic* Infusion of pith used as a purge. *Emetic* Infusion of flowers used as an emetic. (as *S. glauca* 117:436)
- **Food**—**Chehalis** *Winter Use Food* Berries steamed on rocks, cooled, and eaten in the winter. **Green River Group** *Winter Use Food* Berries steamed on rocks, cooled, and eaten in the winter. (as *S. glauca* 79:47) **Karok** *Fruit* Mashed berries used for food. (as *S. glauca* 148:389) **Klallam** *Fruit* Berries used for food. (as *S. glauca* 78:197) *Winter Use*

Food Berries steamed on rocks, cooled, and eaten in the winter. (as *S. glauca* 79:47) **Klamath** *Fruit* Berries used for food. (as *S. glauca* 45:104) **Luiseño** *Dried Food* Fruit eaten dried. *Fruit* Fruit eaten fresh. (as *S. glauca* 155:229) **Lummi** *Winter Use Food* Berries steamed on rocks, cooled, and eaten in the winter. (as *S. glauca* 79:47) **Mendocino Indian** *Dried Food* Berries formerly dried for winter use. *Fruit* Berries formerly eaten raw. *Pie & Pudding* Berries made into pies and used for food. *Preserves* Berries made into jelly and used for food. (as *S. glauca* 41:388) **Montana Indian** *Beverage* Fruit used to make. *Fruit* Fruit eaten raw or cooked. *Pie & Pudding* Fruit used to make pies. *Preserves* Fruit used to make jelly. (as *S. glauca* 19:23) **Paiute** *Dried Food* Fruits dried and eaten. *Fruit* Fruits eaten fresh. (as *S. glauca* 104:100) Berries eaten fresh or boiled. (as *S. glauca* 111:111) **Quinault** *Winter Use Food* Berries steamed on rocks, cooled, and eaten in the winter. **Skagit** *Winter Use Food* Berries steamed on rocks, cooled, and eaten in the winter. (as *S. glauca* 79:47) **Skagit, Upper** *Dried Food* Berries steamed, pulped, and dried for winter use. (as *S. glauca* 179:38) *Fruit* Fruit eaten fresh. (as *S. glauca* 179:37) **Skokomish** *Fruit* Berries eaten fresh. **Squaxin** *Winter Use Food* Berries steamed on rocks, cooled, and eaten in the winter. **Swinomish** *Winter Use Food* Berries steamed on rocks, cooled, and eaten in the winter. (as *S. glauca* 79:47) **Wintoon** *Fruit* Berries used for food. (as *S. glauca* 117:264) **Yokut** *Dried Food* Berries eaten dried. *Winter Use Food* Berries stored for winter use and cooked. (as *S. glauca* 117:436)
- *Other*—**Karok** *Musical Instrument* Lower branches used to make flutes. (as *S. glauca* 148:389) **Luiseño** *Hunting & Fishing Item* Wood used to make bows. (as *S. glauca* 155:229) **Mendocino Indian** *Fuel* Pith formerly used as a combustible material for starting fires. Soft wood used as a twirling stick to make fire by friction. *Musical Instrument* Pithless wood used for flutes and other dance instruments. *Toys & Games* Pithless wood used to make "squirt guns" and whistles. (as *S. glauca* 41:388) **Montana Indian** *Tools* Wood used as a twirling stick in fire making. (as *S. glauca* 19:23) **Paiute** *Musical Instrument* Stems used to make four-holed flutes. (as *S. glauca* 111:111) **Quinault** *Hunting & Fishing Item* Plugs inserted into pithless stems and used as whistles for calling elk. (as *S. glauca* 79:47) **Wintoon** *Musical Instrument* Hollow stems used as music sticks. (as *S. glauca* 117:264) **Yokut** *Musical Instrument* Hollow wood used for flutes. *Toys & Games* Hollow wood used for popguns. Split branches used for making bows for small children. (as *S. glauca* 117:436)

Sambucus cerulea var. *mexicana* (K. Presl ex DC.) L. Benson, Blue Elder

- *Drug*—**Cahuilla** *Cold Remedy* Infusion of blossoms taken for colds. *Febrifuge* Infusion of blossoms taken for fevers. *Gastrointestinal Aid* Infusion of blossoms taken for upset stomachs. *Laxative* Decoction of roots used for constipation. *Misc. Disease Remedy* Infusion of blossoms taken for flu. *Pediatric Aid* Infusion of blossoms given to newborn babies. *Toothache Remedy* Infusion of blossoms used for the teeth. (as *S. mexicana* 15:138) **Diegueño** *Febrifuge* Infusion of fresh or dried blossoms given to babies with fever. *Laxative* Infusion of fresh or dried blossoms used as an enema. *Pediatric Aid* Infusion of fresh or dried blossoms given to babies with fever. (as *S. mexicana* 84:41) **Pima** *Cold Remedy* Decoction of flowers taken for colds. *Febrifuge* Infusion of dried flowers taken to break a fever. *Gastrointestinal Aid* Decoction of flowers taken for stomachaches. *Throat Aid* Decoction of flowers taken for sore throats. (as *S. mexicana* 47:75)
- *Food*—**Cahuilla** *Dried Food* Berries dried for future use. *Fruit* Berries eaten fresh. *Preserves* Berries used to make jams and jellies. *Sauce & Relish* Berries cooked into a rich sauce. (as *S. mexicana* 15:138) **Diegueño** *Fruit* Berries eaten fresh. *Winter Use Food* Berries dried for winter use and boiled like raisins. (as *S. mexicana* 84:41) **Paiute** *Dried Food* Fruit sun dried, stored in buckskin bags, and hung up for winter use. *Fruit* Fruit eaten fresh. (as *S. mexicana* 167:245) **Pima** *Preserves* Berries used to make jams and jellies. (as *S. mexicana* 47:75) **Pima, Gila River** *Beverage* Fruits brewed into a wine. *Fruit* Fruits eaten for food. (as *S. mexicana* 133:5) *Snack Food* Fruits eaten raw as a snack food. (as *S. mexicana* 133:7)
- *Dye*—**Cahuilla** *Black* Berry juice used as a black dye for basket materials. *Orange* Stems used to make an orange dye. *Purple* Berry juice used as a purple dye for basket materials. *Yellow* Stems used to make a yellow dye. (as *S. mexicana* 15:138)
- *Other*—**Cahuilla** *Toys & Games* Twigs used in making whistles. (as *S. mexicana* 15:138) **Diegueño** *Toys & Games* Berries crushed by children when playing "soda pop." (as *S. mexicana* 84:41)

Sambucus cerulea var. *neomexicana* (Woot.) Rehd., New Mexican Elderberry

- *Drug*—**Navajo, Kayenta** *Disinfectant* Plant used for lightning infection. *Veterinary Aid* Plant used for livestock with lightning infection. (as *S. neomexicana* 205:43)

Sambucus cerulea var. *velutina* (Dur. & Hilg.) Schwerin, Blue Elderberry

- *Drug*—**Paiute** *Antidiarrheal* Infusion of dried flowers taken for diarrhea. (as *S. velutina* 180:138)
- *Food*—**Atsugewi** *Bread & Cake* Mashed berries mixed with manzanita flour and stored in dried cakes. (as *S. velutinus* 61:139) **Tubatulabal** *Fruit* Berries used for food. (as *S. velutina* 193:15)

Sambucus nigra L., European Black Elder

- *Drug*—**Cherokee** *Antirheumatic (Internal)* Infusion of berry used for rheumatism. *Burn Dressing* Salve used for burns. *Cathartic* Used as a cathartic. *Dermatological Aid* Salve used for skin eruptions and infusion taken as tonic for boils. *Diaphoretic* Infusion of flowers taken to "sweat out fever." *Disinfectant* Leaves used to wash sores to prevent infection. *Diuretic* Used as a diuretic. *Emetic* Used as an emetic. *Kidney Aid* Taken for "dropsy." *Other* Decoction taken for "summer complaint." *Pediatric Aid* Given for "light sickness among children" and taken for "dropsy." (80:33)
- *Food*—**Cherokee** *Fruit* Fruit used for food. *Preserves* Berries used to make jellies. (80:33)

Sambucus racemosa L., Scarlet Elderberry

- *Drug*—**Bella Coola** *Analgesic* Infusion of roots used as an emetic and purgative for stomach pain. *Cathartic* Infusion of root bark used or root bark chewed as a purgative. *Emetic* Infusion of root bark used or root bark chewed as an emetic. *Gastrointestinal Aid* Infusion of roots used as an emetic and purgative for stomach pain. **Carrier, Northern** *Cathartic* Decoction of root, second brewing only, taken as a purgative. **Carrier, Southern** *Cathartic* Decoction of root taken twice a day as a purgative. **Gitksan** *Cathartic* Infusion of root bark taken as a purgative. (150:64) *Emetic* Bark used as an emetic. (73:152) Infusion of root bark taken as an emetic. (150:64) *Witchcraft Medicine* Bark, juniper roots, and cow parsnip roots used for evil witchcraft victims. (74:24) **Hesquiat** *Analgesic* Roots rubbed on the skin for aching, tired muscles. *Antirheumatic (External)* Roots rubbed on the skin for aching, tired muscles. *Emetic* Raw roots chewed as an emetic. *Gastrointestinal Aid* Raw roots chewed to clean out the stomach. *Laxative* Raw roots chewed as a laxative. *Poison* Berries should always be eaten cooked, as they are potentially poisonous when raw. (185:63) **Kwakiutl** *Emetic* Root extract taken to induce vomiting. *Gynecological Aid* and *Herbal Steam* Infusion of bark used as steam bath to relax body of woman after childbirth. *Orthopedic Aid* Compound infusion of bark used as a foot bath for aching legs and feet. (183:280) **Malecite** *Emetic* Infusion of plant strips used with round wood as an emetic. (116:254) **Menominee** *Antidote* Decoction of scraped inner bark used as a quick emetic in cases of poisoning. (54:131) *Cathartic* Decoction of peeled twigs, a drastic purgative, taken for severe constipation. (151:27, 28) *Emetic* Decoction of scraped inner bark used as a quick emetic in cases of poisoning. (54:131) Decoction of inner bark and rind taken as a powerful emetic. (151:27, 28) **Micmac** *Emetic* Herbs used as an "emetic (with round wood)." (40:61) **Nitinaht** *Emetic* Bark soaked in water and taken as an emetic and purge. *Laxative* Bark used as a very strong laxative. *Strengthener* Bark used by athletes to "draw out all the slime in the system," for better wind and endurance. (67:318) **Ojibwa** *Cathartic* Decoction of inner bark, considered dangerous, taken as a cathartic. *Emetic* Decoction of inner bark, considered dangerous, taken as an emetic. (153:360, 361) *Unspecified* Infusion of roots used as a medicine. (135:237) **Okanagon** *Antirheumatic (Internal)* Plant used for rheumatism. *Dermatological Aid* Plant used for erysipelas. *Toothache Remedy* Bark placed in the hollow of a tooth for toothaches. (125:42) **Pomo** *Dermatological Aid* Decoction of roots used as a lotion on open sores and cuts. (66:15) **Potawatomi** *Cathartic* Infusion of inner bark taken as a physic and emetic. *Emetic* Infusion of stem bark taken as a strong emetic. (154:46) **Sikani** *Cathartic* Decoction of bark taken as a purgative. (150:64) **Thompson** *Antirheumatic (Internal)* Plant used for rheumatism. *Dermatological Aid* Plant used for erysipelas. (125:42) *Liver Aid* Infusion of white roots and cascara bark taken for liver diseases. (187:199) *Toothache Remedy* Bark placed in the hollow of a tooth for toothaches. (125:42) **Wet'suwet'en** *Unspecified* Bark used for medicine. (73:152)

- **Food**—**Apache, White Mountain** *Fruit* Berries used for food. (136:160) **Bella Coola** *Beverage* Berries used to make wine. *Dried Food* Berries formerly boiled into a thick sauce, dried, and used for food. *Preserves* Berries used to make jelly. (184:203) **Gosiute** *Fruit* Fruit used in season for food. (39:380) **Hesquiat** *Fruit* Fruit cooked with sugar and eaten. Berries should always be eaten cooked, as they are potentially poisonous when raw. *Preserves* Cooked fruit made excellent jelly and jam. (185:63) **Kitasoo** *Bread & Cake* Fruit cooked, dried into cakes, stored, reconstituted, and eaten. (43:329) **Kwakiutl, Southern** *Bread & Cake* Berries pit steamed, dried over fire into cakes, and eaten at noon. (183:280) **Makah** *Dried Food* Fruit steamed, sun dried, and placed in bentwood cedar boxes for storage. *Fruit* Fruit eaten fresh. Fruit mixed with sugar, steamed, and eaten. *Winter Use Food* Fruit canned for winter use. Berry clusters placed in alder bark cones and submerged in cold creeks for storage. **Nitinaht** *Fruit* Fruit used for food. (67:318) **Ojibwa** *Unspecified* Species used for food. (135:237) **Okanagon** *Fruit* Fruits eaten for food. (125:39) **Quileute** *Fruit* Fruit eaten fresh. *Winter Use Food* Fruit canned for winter use. (67:318) **Thompson** *Fruit* Fruits eaten for food. (125:39) Berries stewed or eaten fresh with salmon egg "cheese."[76] *Soup* Mashed berries dried in cakes, broken off, and added to salmon head soup and other dishes.[76] *Spice* Berry juice used to marinate salmon.[75] (187:199)
- **Other**—**Bella Coola** *Smoking Tools* Stems hollowed out and used as pipe bowls. (184:203) **Kwakiutl, Southern** *Toys & Games* Stems hollowed and used as blowguns by children. (183:261) **Makah** *Waterproofing Agent* Fruit or flower glue used to waterproof cedar bark rain hats. (67:318) **Salish, Coast** *Toys & Games* Stems hollowed out and used as blowguns by children. (182:80)

Sambucus racemosa ssp. *pubens* (Michx.) House, Red Elderberry

- **Food**—**Paiute, Northern** *Dried Food* Berries dried for future use. *Fruit* Berries used for food. *Soup* Berries dried and boiled into a soup. (59:50)

Sambucus racemosa ssp. *pubens* var. *arborescens* (Torr. & Gray) Gray, Pacific Red Elder

- **Drug**—**Cowlitz** *Orthopedic Aid* Poultice of leaves or bark applied to sore joints for the swelling. (as *S. callicarpa* 79:47) **Haisla & Hanaksiala** *Abortifacient* Leaves boiled and used to shorten pregnancy. *Gastrointestinal Aid* Berries cooked and eaten for stomach problems. *Reproductive Aid* Leaves boiled and used in aiding childbirth. **Hanaksiala** *Analgesic* Poultice of cooked shoots applied for pain. Cooked shoots put in bath water and used as a soak for pain. *Antirheumatic (External)* Poultice of cooked shoots applied to sore, arthritic areas. Cooked shoots put in bath water and used as a soak for sore, arthritic areas. *Dermatological Aid* Root slivers inserted into boil eruptions. (43:229) **Hoh** *Cold Remedy* Infusion of roots or bark used for colds. *Cough Medicine* Infusion of roots or bark used for coughs. *Gynecological Aid* Infusion of roots or bark used by women during confinement. (as *S. calliocarpa* 137:69) **Makah** *Dermatological Aid* Poultice of pounded leaves applied to an abscess or boil. (as *S. callicarpa* 79:47) **Nitinaht** *Cathartic* Infusion of bark and roots taken by boys and girls as a purgative to cleanse the system. *Psychological Aid* Bark used with black twinberry bark for nervous breakdowns. (186:100) **Quileute** *Cold Remedy* Infusion of roots or bark used for colds. *Cough Medicine* Infusion of roots or bark used for coughs. *Gynecological Aid* Infusion of roots or bark used by women during confinement. (as *S. calliocarpa* 137:69) **Quinault** *Gynecological Aid* Decoction of bark applied to breast after childbirth to start milk flow. **Squaxin** *Blood Medicine* and *Disinfectant* Infusion of leaves used as a wash on area infected with blood poisoning. (as *S. callicarpa* 79:47)
- **Food**—**Chehalis** *Winter Use Food* Berries steamed on rocks, cooled, and eaten in the winter. **Cowlitz** *Winter Use Food* Berries steamed on rocks, cooled, and eaten in the winter. **Green River Group** *Winter Use Food* Berries steamed on rocks, cooled, and eaten in the winter. (as *S. callicarpa* 79:47) **Haisla** *Fruit* Berries used extensively, especially to mix with and extend other berries. **Haisla & Hanaksiala** *Beverage* Berries used to make wine. *Dried Food* Berries dried for future use. *Winter Use Food* Berries formerly an important winter food.

Hanaksiala *Fruit* Berries mixed with blueberries, Pacific crabapples, and oolichan (candlefish) grease and eaten. (43:229) **Hoh** *Fruit* Berries pit baked and used for food. *Sauce & Relish* Fruits used to make a sauce. *Winter Use Food* Berries cooked, wrapped in skunk cabbage leaves, and preserved for winter use. (as *S. calliocarpa* 137:69) **Klallam** *Fruit* Berries used for food. (as *S. callicarpa* 78:197) *Winter Use Food* Berries steamed on rocks, cooled, and eaten in the winter. **Makah** *Winter Use Food* Berries steamed on rocks, cooled, and eaten in the winter. (as *S. callicarpa* 79:47) **Nitinaht** *Fruit* Berries formerly used for food. *Preserves* Berries pounded, dried, soaked in water until jam-like, mixed with sugar, and used for food. (186:100) **Oweekeno** *Dried Food* Berries dried for storage. *Preserves* Berries used to make jam. (43:90) **Quileute** *Fruit* Berries pit baked and used for food. *Sauce & Relish* Fruits used to make a sauce. (as *S. calliocarpa* 137:69) *Winter Use Food* Berries steamed on rocks, cooled, and eaten in the winter. (as *S. callicarpa* 79:47) Berries cooked, wrapped in skunk cabbage leaves, and preserved for winter use. (as *S. calliocarpa* 137:69) **Quinault** *Winter Use Food* Berries steamed on rocks, cooled, and eaten in the winter. **Skagit** *Winter Use Food* Berries steamed on rocks, cooled, and eaten in the winter. (as *S. callicarpa* 79:47) **Skagit, Upper** *Dried Food* Berries steamed, pulped, and dried for winter use. (as *S. callicarpa* 179:38) *Fruit* Fruit eaten fresh. (as *S. callicarpa* 179:37) **Skokomish** *Winter Use Food* Berries steamed on rocks, cooled, and eaten in the winter. **Snohomish** *Winter Use Food* Berries steamed on rocks, cooled, and eaten in the winter. **Squaxin** *Winter Use Food* Berries steamed on rocks, cooled, and eaten in the winter. **Swinomish** *Winter Use Food* Berries steamed on rocks, cooled, and eaten in the winter. (as *S. callicarpa* 79:47) **Yurok** *Fruit* Fresh berries used for food. (as *S. callicarpa* 6:53)
- *Other*—**Haisla & Hanaksiala** *Cooking Tools* Leaves used to whip soapberries. *Hunting & Fishing Item* Wood used to make points for oolichan (candlefish) and squirrel spears. (43:229) **Nitinaht** *Sacred Items* Pithy branches hollowed out and used to make ceremonial and sacred wolf whistles. *Toys & Games* Used to make children's whistles and "pea shooters." (186:100) **Yurok** *Containers* Leaves used to pack sturgeon eggs while cooking. (as *S. callicarpa* 6:53)

Sambucus racemosa ssp. *pubens* var. *leucocarpa* (Torr. & Gray) Cronq., European Red Elderberry
- *Drug*—**Mahuna** *Misc. Disease Remedy* Blossoms used for measles. (as *S. pubens* 140:10)
- *Food*—**Mahuna** *Fruit* Berries eaten mainly to quench the thirst. (as *S. pubens* 140:70)

Sambucus racemosa ssp. *pubens* var. *melanocarpa* (Gray) McMinn, Black Elderberry
- *Drug*—**Paiute** *Antidiarrheal* Dried ripe berries eaten or decoction of root taken for diarrhea. *Cold Remedy* Decoction of flowers taken for colds. *Cough Medicine* Decoction of flowers taken for coughs. *Dermatological Aid* Poultice of boiled, mashed root applied to cuts and wounds. Poultice of leaves applied to bruises. *Gynecological Aid* Poultice of boiled, mashed root applied to caked breasts. *Hemostat* Poultice of bruised leaves applied to bleeding wounds. *Pediatric Aid* and *Tonic* Decoction of flowers given to children as a spring tonic. **Shoshoni** *Antidiarrheal* Decoction of root taken for dysentery. *Blood Medicine* Decoction of root taken as a blood tonic. *Cold Remedy* Decoction of blossoms taken for colds. *Cough Medicine* Decoction of blossoms taken for coughs. *Tuberculosis Remedy* Decoction of blossoms taken for tuberculosis. (as *S. melanocarpa* 180:137, 138)

Sambucus racemosa ssp. *pubens* var. *microbotrys* (Rydb.) Kearney & Peebles, European Red Elderberry
- *Food*—**Apache** *Fruit* Fruits eaten fresh or cooked. (as *S. microbotrys* 32:50) **Apache, Chiricahua & Mescalero** *Preserves* Fruit cooked with a sweet substance, strained, and eaten as jelly. (as *S. microbotrys* 33:46)

Sambucus sp., Elderberry
- *Drug*—**Iroquois** *Dermatological Aid* Poultice of crushed leaves and other leaves applied to blue swellings. (141:61)
- *Food*—**Coeur d'Alene** *Fruit* Berries eaten fresh. (178:89) **Kiowa** *Fruit* Berries used for food. (192:52) **Okanagon** *Staple* Berries used as a principal food. (178:238) **Poliklah** *Fruit* Berries used for food. (117:173) **Spokan** *Fruit* Berries used for food. (178:343)
- *Other*—**Iroquois** *Tools* Stems made into the shuttle used for weaving. (142:99)

Sambucus tridentata DC., Antelope Brush
- *Drug*—**Paiute** *Emetic* Infusion of leaves taken as an emetic. *Laxative* Infusion of leaves taken as a laxative. (167:317)

Sanguinaria, Papaveraceae
Sanguinaria canadensis L., Blood Root
- *Drug*—**Abnaki** *Abortifacient* Used as an abortifacient. (144:154) *Veterinary Aid* Used as an abortifacient for horses. (144:167) **Algonquin** *Love Medicine* Used as a love charm and red dye for skin, clothing, and weapons. (22:142) **Algonquin, Quebec** *Heart Medicine* Root chewed for heart trouble. *Tonic* Rhizomes used to make a medicinal tonic. (18:171) **Cherokee** *Cough Medicine* Decoction of root in small doses and infusion with broomsedge used for coughs. *Dermatological Aid* Used as wash for ulcers and sores and infusion with vinegar used for tetterworm. *Nose Medicine* Used as "snuff for polypus." *Pulmonary Aid* Decoction of root taken in small doses for lung inflammations and croup. *Respiratory Aid* Pulverized root sniffed for catarrh. (80:26) **Chippewa** *Analgesic* Compound decoction of root taken for cramps. *Gastrointestinal Aid* Compound decoction of root taken for stomach cramps. (53:344) *Unspecified* Plant used medicinally. (71:131) **Delaware** *Gastrointestinal Aid* Combined with other roots and used as a stomach remedy. *Strengthener* Pea-sized piece of roots taken every morning for 30 days for general debility. (176:38) **Delaware, Oklahoma** *Gastrointestinal Aid* Compound containing root used as a "stomach remedy." *Panacea* Piece of root eaten daily "for general debility." (175:32, 80) *Tonic* Root used in a tonic. (175:80) **Delaware, Ontario** *Antiemetic* Infusion of powdered root taken for vomiting. *Blood Medicine* Compound containing root taken as a blood purifier. (175:68, 82) **Iroquois** *Analgesic* Infusion of roots taken for inside pain. (87:335) *Anthelmintic* Compound infusion of roots and whisky taken as blood remedy and for tapeworms. (87:338) *Antidiarrheal* Compound infusion of plants taken for diarrhea. *Antiemetic* Compound infusion of plants taken for vomiting. (87:336) *Antihemorrhagic* Infusion of roots taken for stomach and lump hemorrhages. (87:335) *Blood Medicine* Compound infusion of roots taken to purify the blood and loosen the bowels. Infusion of branches tak-

516   Sanguinaria canadensis (Papaveraceae)

en as a blood tonic, "don't vomit." (87:336) Compound infusion of roots taken as a blood purifier. (87:337) *Carminative* Compound decoction of roots taken for stomach gas. (87:336) *Cold Remedy* Dried plant used as a snuff for head colds. (87:335) Decoction or infusion of roots taken for colds. (87:337) Plant chewed for colds. (87:338) *Cough Medicine* Compound infusion of roots and liquor taken as consumption cough medicine. (87:336) Decoction of powdered roots or infusion of roots taken for coughs. (87:337) *Dermatological Aid* Infusion of mashed roots taken or poultice applied to cuts or poison ivy. Plant juice taken as a wound medicine. Poultice of plants applied for drawing thorns and slivers or on leg sores. (87:336) Infusion of split root used as a wash for cuts and boils. Poultice of cooked roots applied to cuts and wounds. (87:337) Decoction of roots taken for swellings above the waist, wounds, and sores. Plant chewed for sores and cuts. (87:338) *Ear Medicine* Infusion of dried root fragments used as ear drops for earaches. (141:44) *Emetic* Decoction of branches or infusion of roots taken as a spring emetic. (87:336) *Eye Medicine* Decoction of powdered root used as a wash for sore eyes. *Febrifuge* Infusion of plants or decoction of roots or powdered roots taken for fevers. (87:337) *Gastrointestinal Aid* Infusion of roots taken for stomach hemorrhages. (87:335) Compound infusion of plants taken for upset stomach. Decoction of smashed roots taken for stomach cramps. (87:336) Compound decoction of bark taken to clean the stomach and for ulcers. Compound taken for intestinal trouble. Decoction of dried roots taken for ulcers or by women who are ugly. (87:338) Decoction of rhizomes taken for stomachaches after a big meal. (141:44) *Gynecological Aid* Infusion of plant taken for menses. (87:335) *Heart Medicine* Compound decoction of roots taken to regulate the heart and make blood redder. *Hemorrhoid Remedy* Decoction of roots used to push piles back into intestines. (87:337) *Hemostat* Decoction of roots applied to bleeding ax cuts on the foot. (87:336) *Kidney Aid* Compound decoction of roots taken for fevers or the kidneys. (87:337) *Laxative* Compound infusion or decoction of roots taken to loosen the bowels. (87:336) *Liver Aid* Compound infusion of roots taken as a gall medicine. (87:337) *Other* Cold infusion of roots taken for sickness caught from a menstruating girl. (87:335) Decoction of smashed roots taken for hiccups. (87:336) *Panacea* Compound decoction of roots taken as a panacea. *Pediatric Aid* Compound infusion of roots taken for prenatal strength or as blood purifier. (87:337) *Pulmonary Aid* Infusion of plant taken for bleeding lungs. *Respiratory Aid* Dried plant used as a snuff for catarrh. (87:335) Decoction of roots taken for asthma. *Throat Aid* Plant chewed or poultice applied for sore throats. (87:338) *Tuberculosis Remedy* Compound infusion of roots and liquor taken as consumption cough medicine. Infusion or decoction of mashed roots taken for tuberculosis. (87:336) *Venereal Aid* Cold infusion or decoction of smashed roots taken for gonorrhea and syphilis. (87:335) *Witchcraft Medicine* Smoke from plant used as a wash for a person who has seen a dead person. (87:336) **Malecite** *Antihemorrhagic* Infusion of plant used for hemorrhages in patients suffering from consumption. (116:252) *Dermatological Aid* Decoction of plant used for black, infected cuts. (116:246) Roots used for infected cuts. *Hemorrhoid Remedy* Roots boiled and used for bleeding piles. *Tuberculosis Remedy* Roots used for consumption. (116:250) **Menominee** *Abortifacient* Compound decoction of root used for irregular periods. (54:133) *Adjuvant* Root often added to medicines to strengthen their effect. (151:44) *Dermatological Aid* Fresh root used to paint the face of a warrior. (151:78) **Meskwaki** *Adjuvant* Added to other medicines to strengthen their effect. *Analgesic* Root chewed and spittle applied to burn pains. *Burn Dressing* Infusion of root used as a wash for burns and chewed root spittle applied to burn pain. (152:234) **Micmac** *Abortifacient* Used as an abortifacient. *Cold Remedy* Infusion of roots used for colds. (145:56) *Dermatological Aid* Roots used for infected cuts. *Hemostat* Roots used for hemorrhages and to prevent bleeding. (40:61) *Love Medicine* Used as an aphrodisiac. *Throat Aid* Infusion of roots used for sore throats. (145:56) *Tuberculosis Remedy* Roots used for consumption with hemorrhage. (40:61) **Mohegan** *Blood Medicine* Infusion of plant used as a blood medicine. (174:264) Infusion of inner bark of dried root taken as a blood purifier. (176:75, 132) *Emetic* Infusion of plant used as an emetic. (174:264) *Tonic* Leaves used to make a tonic. (176:132) **Ojibwa** *Analgesic* Plant used for stomach pain, fainting, and trembling in fits. *Anticonvulsive* Plant used for trembling in fits or infusion of leaves taken for fits. *Antirheumatic (External)* Infusion of pounded plants used as wash for general illnesses and rheumatism. *Blood Medicine* Leaf infusion taken as blood medicine and bark decoction used for blood disease. (135:231) *Ceremonial Medicine* Juice used as face paint for the medicine lodge ceremony or when on warpath. (153:377) *Dermatological Aid* Poultice of plant applied or root infusion taken and used as a wash for sores and cuts. *Gastrointestinal Aid* Decoction or infusion of plants taken for stomach or bowel troubles. *Hemostat* Infusion of roots taken and used as a wash for bleeding foot cuts. *Laxative* Decoction of plants taken for sick stomach, bowels, or for constipation. *Panacea* Infusion of pounded plants used as a wash for general illnesses. *Stimulant* Infusion of leaves taken for fainting, fits, and as a blood medicine. (135:231) *Throat Aid* Root juice on maple sugar used for sore throat. (153:377, 378) *Venereal Aid* Infusion of root bark taken for gonorrhea. (135:231) **Penobscot** *Preventive Medicine* Bits of dried root worn as a necklace to prevent bleeding. (156:311) **Ponca** *Love Medicine* Root rubbed on palm of bachelor as a love charm. (70:83) **Potawatomi** *Misc. Disease Remedy* Infusion of root used for diphtheria, considered a throat disease. *Throat Aid* Root juice squeezed on maple sugar as throat lozenge for mild sore throat. (154:68)

- **Dye—Cherokee** *Red* Used to make a red dye. (80:26) Roots used as a red dye in basket making. (203:74) **Chippewa** *Red* Roots boiled with the inner barks of other trees and used to make a red dye. (53:371) Roots dug in the fall and used to make a red dye. (71:131) *Yellow* Green or dried roots pounded and steeped to make a dark yellow dye. (53:373) Double handful of shredded roots boiled with wild plum roots to make a dark yellow dye. (53:374) **Iroquois** *Orange-Yellow* Rhizomes used as an orange or yellow dye for sheets. (141:44) **Menominee** *Orange-Red* Boiled root used to dye mats orange-red. *Red* Boiled root used to dye mats red. (151:78) **Meskwaki** *Red* Root cooked to make a red face paint and to dye baskets and mats red. (152:271) **Ojibwa** *Orange* Fresh or dried roots used as an orange dye to paint faces with clan marks.[77] (153:426) *Red* Roots boiled to obtain a red dye. (96:114) *Yellow* Fresh or dried roots used as a dark yellow dye to paint faces with clan marks.[77] (153:426) **Omaha** *Red* Root boiled with objects as a red dye. *Unspecified* Root used as a decorative skin stain. **Ponca** *Red* Root boiled with objects as a red dye. *Un-*

specified Root used as a decorative skin stain. (70:83) **Potawatomi** *Unspecified* Root used as facial paint to put on clan and identification marks. (154:121) **Winnebago** *Red* Root boiled with objects as a red dye. *Unspecified* Root used as a decorative skin stain. (70:83)
- *Other*—**Delaware** *Ceremonial Items* Roots used to make the face paint for the Big House Ceremony. (176:38) **Delaware, Oklahoma** *Ceremonial Items* Root used as a ceremonial face paint. (175:32, 80) **Omaha** *Paint* Plant used as a red skin stain. (68:324)

Sanicula, Apiaceae

Sanicula bipinnata Hook. & Arn., Poison Sanicle
- *Drug*—**Miwok** *Snakebite Remedy* Poultice of boiled plant applied to snakebites. (12:172)
- *Food*—**Karok** *Vegetable* Leaves eaten as greens. (148:386)
- *Other*—**Karok** *Water Indicator* Plant grew near good luck water. This plant grew naturally in swamps. It was believed that even if you found it in a dry place, if you searched you would find water nearby. The water bubbled up in a little hole and disappeared again soon. This was good luck water. When a woman was making baskets, she went to it, if she knew where it was, and washed her hands. Then she would have good luck in making her baskets and perhaps would sell them at a high price. When people were gambling, they would go and wash their hands in a "lucky water." If you found a lucky water, you would not tell anyone, but would keep it secret, so that no unclean person would go near it. (148:386)

Sanicula bipinnatifida Dougl. ex Hook., Purple Sanicle
- *Drug*—**Miwok** *Panacea* Decoction of root taken as a cure-all. *Snakebite Remedy* Infusion of leaves applied to snakebites. (12:172)

Sanicula canadensis L., Canadian Blacksnakeroot
- *Drug*—**Chippewa** *Abortifacient* Decoction of powdered root taken for "stoppage of periods." (53:358) *Gynecological Aid* Compound decoction of root taken during confinement. (53:360) **Houma** *Heart Medicine* Hot decoction of root taken for heart trouble. (158:64)

Sanicula crassicaulis Poepp. ex DC., Pacific Blacksnakeroot
- *Drug*—**Miwok** *Dermatological Aid* Poultice of leaves used for rattlesnake bites and other wounds. *Snakebite Remedy* Poultice of leaves used for rattlesnake bites and other wounds. (as *S. menziesii* 12:173)
- *Other*—**Mendocino Indian** *Good Luck Charm* Roots chewed, rubbed on the body, and used for good luck in gambling. (as *S. menziesii* 41:373)

Sanicula marilandica L., Maryland Sanicle
- *Drug*—**Iroquois** *Antidote* Compound decoction of plants taken to vomit to counteract a poison. (87:397) *Dermatological Aid* Decoction of roots used as a wash and given to children with sore navels. (87:396) *Emetic* Compound decoction of plants taken to vomit to counteract a poison. (87:397) *Kidney Aid* Decoction of plants taken for dropsy. *Laxative* Compound decoction of roots taken to loosen the bowels. *Pediatric Aid* Decoction of roots used as a wash and given to children with sore navels. *Venereal Aid* Compound decoction of roots taken for venereal disease. (87:396) **Malecite** *Abortifacient* Infusion of bulb roots used for irregular menstruation. (116:258) **Menominee** *Witchcraft Medicine* Root thought to be used by sorcerers for evil purposes. (151:55) **Micmac** *Abortifacient* Roots used for irregular menstruation. *Analgesic* Roots used for menstrual pain. *Antirheumatic (Internal)* Roots used for rheumatism. *Gynecological Aid* Roots used for menstrual pain and slow parturition. *Kidney Aid* Roots used for kidney trouble. *Snakebite Remedy* Roots used as a snakebite remedy and for rheumatism. (40:61) **Ojibwa** *Febrifuge* Infusion of root used for various fevers. *Snakebite Remedy* Poultice of pounded root applied to rattlesnake bite or any snakebite. (153:391)

Sanicula odorata (Raf.) K. M. Pryer & L. R. Phillippe, Clustered Blacksnakeroot
- *Drug*—**Malecite** *Analgesic* Infusion of plant and spikenard used by women with back and side pain. *Kidney Aid* Infusion of plant and spikenard used for kidney trouble. (as *S. gregaria* 116:257) **Menominee** *Witchcraft Medicine* Root thought to be used by sorcerers for evil purpose. (as *S. gregaria* 151:56) **Meskwaki** *Dermatological Aid* Plant used as an astringent and for nosebleeds. *Hemostat* Steam of burning plant on hot stones inhaled for nosebleed. (as *S. gregaria* 152:250)

Sanicula smallii Bickn., Small's Blacksnakeroot
- *Drug*—**Cherokee** *Analgesic* and *Gastrointestinal Aid* Infusion with pink lady's slipper taken for stomach cramps and colic. *Orthopedic Aid* Used as a liniment and infusion taken for colic. (80:55)

Sanicula sp., Sanicle
- *Drug*—**Pomo, Kashaya** *Unspecified* Plant used medicinally. (72:102)
- *Food*—**Neeshenam** *Unspecified* Roots used for food. *Vegetable* Leaves eaten as greens. (129:377)

Sanicula tuberosa Torr., Turkey Pea
- *Food*—**Mendocino Indian** *Unspecified* Bulbs eaten raw. (41:374) **Neeshenam** *Unspecified* Eaten raw, roasted, or boiled. (as *S. luberosa* 129:377) **Pomo** *Unspecified* Tuberous roots eaten raw. (11:89)

Santalaceae, see *Comandra, Geocaulon, Pyrularia*

Sanvitalia, Asteraceae

Sanvitalia abertii Gray, Albert's Creeping Zinnia
- *Drug*—**Navajo** *Diaphoretic* Plant used to increase perspiration. *Oral Aid* Plant chewed for mouth sores. (55:88) **Navajo, Ramah** *Analgesic* Cold infusion of leaves taken and used as lotion for headache. Compound decoction used for menstrual pain. *Cold Remedy* Chewed leaves swallowed for cold. *Dermatological Aid* Poultice of chewed leaves or infusion of leaves applied to skin sores. *Febrifuge* Cold infusion of leaves taken and used as lotion for fever. *Gynecological Aid* Compound decoction of plant used for menstrual pain. *Oral Aid* Warm infusion used as mouthwash and leaves chewed for canker sores. *Panacea* Plant used as "life medicine." *Snakebite Remedy* Compound decoction of plant used for snakebite. *Throat Aid* Chewed leaves swallowed for sore throat. *Toothache Remedy* Leaves chewed for toothache. (191:53)

Sapindaceae, see *Dodonaea, Sapindus*

Sapindus, Sapindaceae

Sapindus saponaria L., Wingleaf Soapberry
- *Other*—**Seminole** *Jewelry* Plant used to make beads. (169:468)

Sapindus saponaria var. *drummondii* (Hook. & Arn.) L. Benson, Western Soapberry

518 *Sapindus saponaria* (Sapindaceae)

- *Drug*—**Kiowa** *Dermatological Aid* Poultice of sap applied to wounds. (as *S. drummondii* 192:41)
- *Other*—**Comanche** *Toys & Games* Stems used to make arrows for aratsi game. (as *S. drummondii* 29:524) **Papago** *Hunting & Fishing Item* and *Weapon* Wood used to make stone-tipped hunting arrows. (as *S. drummondii* 34:71)

Sapium, Euphorbiaceae
Sapium biloculare (S. Wats.) Pax, Mexican Jumpingbean
- *Drug*—**Seri** *Poison* Juice used for arrow poison. (50:138)
- *Other*—**Seri** *Hunting & Fishing Item* Straight stems used to make arrow shafts. (50:138)

Saponaria, Caryophyllaceae
Saponaria officinalis L., Bouncing Bet
- *Drug*—**Cherokee** *Dermatological Aid* Used as a poultice for boils. (80:26) **Mahuna** *Analgesic* Poultice of leaves applied to spleen pain. *Dermatological Aid* Root juice used as a hair tonic. (140:40)
- *Other*—**Cherokee** *Soap* Used for soap. (80:26) **Mahuna** *Soap* Juice used as a hair shampoo. (140:40)

Sapotaceae, see *Chrysophyllum, Sideroxylon*

Sarcobatus, Chenopodiaceae
Sarcobatus vermiculatus (Hook.) Torr., Greasewood
- *Drug*—**Cheyenne** *Blood Medicine* Sharpened stick used to draw out bad blood. *Ceremonial Medicine* Sharpened stick used in acupuncture ceremony. *Veterinary Aid* Stick used to make holes in horse's shoulder for sprained or bruised legs. (83:17) **Hopi** *Ceremonial Medicine* Plant used for kiva fuel. (56:18) **Keres, Western** *Dermatological Aid* Crushed leaves used for insect bites. *Emetic* Infusion of leaves used as an emetic for lightning shock. (171:68) **Navajo** *Dermatological Aid* Plant used for insect bites. (55:97) **Navajo, Ramah** *Gastrointestinal Aid* Warm infusion of leaves taken to kill a swallowed red ant. (191:25) **Paiute** *Antidiarrheal* Infusion of burned plant taken for diarrhea. *Antihemorrhagic* Infusion of burned plant taken for rectal bleeding. (180:138, 139) **Paiute, Northern** *Toothache Remedy* Wood or roots heated until burned or blackened and used on aching and decayed teeth. (59:129)
- *Food*—**Keres, Western** *Forage* Shrub used as winter pasture for sheep. (171:68) **Montana Indian** *Vegetable* Young twigs used for greens. (19:23) **Navajo** *Forage* Used as forage by sheep and eaten for the salt. (55:44) *Fruit* "Seeds" (actually fruits) used for food. (90:155) **Navajo, Ramah** *Fodder* Used for sheep and horse feed in the spring. (191:25) **Pima** *Forage* Succulent, young leaves and branches eaten by cattle and sheep. *Starvation Food* Seeds roasted and eaten during "hard times." (47:71)
- *Fiber*—**Hopi** *Building Material* Wood used for construction. (42:358) Strong wood used in general construction. (200:74) *Furniture* Wood used for clothes hooks in houses. (42:358)
- *Other*—**Cheyenne** *Ceremonial Items* Sticks used to make man designs upon which Sun Dancers stood. (83:17) *Hunting & Fishing Item* Used to make arrow shafts. (83:46) Wood used for arrow shafts. *Smoking Tools* Small sticks wrapped with buffalo hair and used as a tamper for tobacco pipes. (83:17) **Hopi** *Cooking Tools* Wood used for stirring rods. (42:358) Strong wood used to make stirring rods. (200:74) *Fuel* Wood used for fuel. (42:358) Shrub used as one of the four prescribed fuels for the kivas. (56:18) Strong wood used as the chief kiva fuel. (200:74) *Hunting & Fishing Item* Wood used for rabbit sticks and arrows. (42:358) Strong wood used to make arrows. (200:74) *Musical Instrument* Wood used for musical rasps. (42:358) Strong wood used to make musical rasps. (200:74) *Tools* Wood used for planting and lease rods. (42:358) Used to make planting sticks and poorer boomerangs. Used to make planting sticks and poorer boomerangs. (190:162) Strong wood used for rabbit sticks, planting sticks, lease rods, and clothes hooks. (200:74) **Jemez** *Toys & Games* Plant part kicked to see who kicked it the farthest, in racing games. (44:27) **Navajo** *Ceremonial Items* Roots carved into an image of a snake for the Lightning Chant, Beauty Chant, and Mountain Chant. *Fuel* Used as firewood. *Tools* Wood used to make planting sticks, knitting needles, heddle sticks, and distaff handles used in weaving. *Toys & Games* Wood used to make dice. *Weapon* Wood used to make war bows. (55:44) **Navajo, Ramah** *Cooking Tools* Stems tied together with buckskin and used for mush stirring sticks. (191:25) **Paiute** *Tools* Twigs used as needles in the manufacture of tule and cattail mats. (111:69)

Sarcocornia, Chenopodiaceae
Sarcocornia pacifica (Standl.) A. J. Scott, Pacific Swampfire
- *Food*—**Alaska Native** *Vegetable* Young plants used in salads or for pickles. (as *Salicornia pacifica* 85:57)

Sarcostemma, Asclepiadaceae
Sarcostemma cynanchoides ssp. *hartwegii* (Vail) R. Holm, Hartweg's Twinevine
- *Food*—**Luiseño** *Unspecified* Plant eaten raw with salt. (as *Philibertia heterophylla* 155:230) **Papago** *Candy* Gum-like secretions heated over coals and chewed by children. (as *Philibertella heterophylla* 34:28) **Pima** *Candy* Milk extracted from main stem, baked or boiled, and used as chewing gum. (as *Funastrum heterophyllum* 47:82) **Pima, Gila River** *Baby Food* Sap roasted on coals and eaten primarily by children. (133:7)

Sarraceniaceae, see *Darlingtonia, Sarracenia*

Sarracenia, Sarraceniaceae
Sarracenia purpurea L., Purple Pitcherplant
- *Drug*—**Algonquin, Quebec** *Gynecological Aid* Infusion of leaves taken to make childbirth easier. *Urinary Aid* Decoction of root tops taken for urinary difficulties. (18:173) **Algonquin, Tête-de-Boule** *Diuretic* Roots used as a diuretic. *Urinary Aid* Roots mixed with beaver kidneys and used for urinary tract diseases. (132:131) **Cree, Woodlands** *Abortifacient* Decoction or infusion of leaves taken for sickness associated with absence of menstrual period. *Gynecological Aid* Decoction of root given to women to prevent sickness after childbirth. Decoction of root and other herbs taken to expel the afterbirth. *Orthopedic Aid* Decoction taken for lower back pain. *Venereal Aid* Decoction of roots taken for venereal disease. (109:59) **Iroquois** *Basket Medicine* Used as a "basket medicine." *Dietary Aid* Plant used as a medicine for thirst. *Febrifuge* Compound decoction of leaves taken for recurring chills followed by fever. Infusion of dried leaves taken

for high fever and shakiness. (87:342) Infusion of leaves and other plant fragments used for chills. (141:43) *Liver Aid* Compound infusion of whole roots taken for liver sickness. *Love Medicine* Powdered plant sprinkled on person for a love medicine. *Pulmonary Aid* Cold decoction of whole plant taken for whooping cough. Plant used for pneumonia. *Sports Medicine* Powdered plant sprinkled on person for a lacrosse medicine. (87:343) **Malecite** *Tuberculosis Remedy* Infusion of plants used for consumption. (116:251) **Menominee** *Witchcraft Medicine* Plant thought to be used by sorcerers. (151:52, 53) **Micmac** *Antihemorrhagic* Herbs used for spitting blood. (40:61) Strong decoction of root taken for "spitting blood" and pulmonary complaints. (156:316) *Kidney Aid* Herbs used for kidney trouble and consumption. *Misc. Disease Remedy* Roots used for smallpox and herbs used for consumption. (40:61) *Pulmonary Aid* Decoction of root taken for "spitting blood and other pulmonary complaints." *Throat Aid* Infusion of root taken for sore throat. (156:316) *Tuberculosis Remedy* Herbs used for consumption. (40:61) **Montagnais** *Misc. Disease Remedy* Infusion of leaves used as medicine for smallpox. (156:314) **Ojibwa** *Gynecological Aid* Infusion of root used "to help a woman accomplish parturition." (153:389) **Penobscot** *Antihemorrhagic* Infusion of plant taken for "spitting up blood." *Kidney Aid* Infusion of plant "supposedly" taken for kidney trouble. (156:310) **Potawatomi** *Gynecological Aid* Foliage used to make a "squaw remedy." (154:82)

- *Other*—**Chippewa** *Toys & Games* Used for toys. (53:378) **Cree, Woodlands** *Cooking Tools* Leaf used by children as a toy kettle to cook meat over an open fire. (109:59) **Potawatomi** *Cooking Tools* Leaves used for a drinking cup when out in the woods or swamp. (154:123)

Sassafras, Lauraceae
Sassafras albidum (Nutt.) Nees, Sassafras

- *Drug*—**Cherokee** *Anthelmintic* Compound taken for worms. (80:54) Infusion of bark used as a wash or given to children with worms. (177:24) *Antidiarrheal* Infusion of root bark taken for diarrhea. *Antirheumatic (Internal)* Infusion taken for rheumatism. *Blood Medicine* Infusion taken to purify blood. *Cold Remedy* Infusion of root bark taken for colds. *Dermatological Aid* Taken for skin diseases and used to poultice wounds and sores. *Dietary Aid* Infusion of bark taken for "overfatness." *Eye Medicine* Used as a wash for sore eyes. *Misc. Disease Remedy* Taken for ague. (80:54) *Oral Aid* Roots chewed to remove odor caused by eating ramps (*Allium tricoccum*?). (126:44) *Pediatric Aid* Infusion of bark used as a wash or given to children with worms. (177:24) *Venereal Aid* Taken for venereal diseases. (80:54) **Chippewa** *Blood Medicine* Infusion of root bark taken to thin the blood. (as *S. variifolium* 71:130) **Choctaw** *Blood Medicine* Decoction of roots taken to thin the blood. *Misc. Disease Remedy* Decoction of roots taken for measles. (177:24) **Creek** *Unspecified* Plant used for unspecified medicinal purpose. (as *S. variifolium* 172:661) **Delaware** *Blood Medicine* Root bark used as a blood purifier. (176:30) **Delaware, Oklahoma** *Blood Medicine* Compound containing root bark used as a blood purifier. (175:25, 80) *Tonic* Bark used in a tonic. (175:80) **Houma** *Misc. Disease Remedy* Decoction of fresh or dried root taken for measles and scarlet fever. (158:60) **Iroquois** *Anthelmintic* Compound infusion of roots and whisky taken as blood remedy and for tapeworms. (87:334) *Antirheumatic (Internal)* Compound infusion of plant with whisky taken for rheumatism. *Blood Medicine* Decoction of pith from new sprouts used as blood medicine. (87:333) Decoction or infusion of bark taken as a blood purifier and for watery blood. Decoction or infusion of roots taken for watery blood or to clear the blood. Plant taken to thin the blood. (87:334) *Cold Remedy* Infusion of roots taken by women for colds. *Dermatological Aid* Leaves used as a poultice for wounds, cuts, and bruises. (87:333) *Eye Medicine* Infusion or decoction of plant used as a wash for sore eyes or cataracts. (87:334) *Febrifuge* and *Gynecological Aid* Infusion of roots taken by women with fevers after childbirth. *Hemostat* Decoction of pith from new sprouts used for nosebleed. *Hypotensive* Decoction of pith from new sprouts or roots taken for high blood pressure. (87:333) Plant taken for blood pressure. *Orthopedic Aid* Compound decoction of roots taken for swellings on the shins and calves. *Tonic* Taken as a tonic. (87:334) **Koasati** *Dermatological Aid* Poultice of mashed leaves applied to bee stings. *Heart Medicine* Decoction of roots taken for heart troubles. (177:24) **Mohegan** *Eye Medicine* Infusion of young shoots used as a wash for sore eyes. (as *S. officinale* 176:75, 132) *Tonic* Complex compound infusion including sassafras root taken as spring tonic. (as *S. sassafras* 174:266) Root, leaves, and bark mixed with other herbs to make a tonic. (as *S. officinale* 176:75, 132) **Nanticoke** *Febrifuge* Infusion of root taken to ward off "fever and ague." (175:56, 84) *Misc. Disease Remedy* Infusion of plant taken to prevent fever and ague. (175:56) **Rappahannock** *Burn Dressing* Decoction of branch pith used as wash for burns. *Dermatological Aid* Infusion of roots taken for the rash of measles. *Eye Medicine* Decoction of branch pith used as wash for sore eyes. *Febrifuge* Infusion of roots taken for the fever of measles. *Misc. Disease Remedy* Infusion of roots taken for the rash and fever of measles. *Sedative* Infusion of root taken as a nerve medicine. *Stimulant* Raw buds chewed to "increase vigor in males." *Tonic* Infusion of root taken as a spring tonic. (as *S. molle* 161:26) **Seminole** *Analgesic* Bark used for cow sickness: lower chest pain, digestive disturbances, and diarrhea. (169:188) Infusion of plant taken for wolf sickness: vomiting, stomach pain, diarrhea, and frequent urination. (169:227) Plant used for gallstones and bladder pain. (169:275) *Antidiarrheal* Bark used for cow sickness: lower chest pain, digestive disturbances, and diarrhea. (169:188) Infusion of plant taken by small children for raccoon sickness: diarrhea. (169:218) Infusion of bark taken by babies and adults for otter sickness: diarrhea and vomiting. (169:222) Infusion of plant taken for wolf sickness: vomiting, stomach pain, diarrhea, and frequent urination. (169:227) Decoction of plant taken for wolf ghost sickness: diarrhea and painful defecation. (169:228) *Antiemetic* Decoction of bark used for horse sickness: nausea, constipation, and blocked urination. (169:188) Infusion of bark taken by babies and adults for otter sickness: diarrhea and vomiting. (169:222) Infusion of bark taken as an emetic and rubbed on the body for cat sickness: nausea. (169:224) Infusion of plant taken for wolf sickness: vomiting, stomach pain, diarrhea, and frequent urination. (169:227) Decoction of roots taken and rubbed on the stomach for continuous vomiting. (169:307) *Cathartic* Decoction of plant taken for wolf ghost sickness: diarrhea and painful defecation. (169:228) *Ceremonial Medicine* Bark used as an emetic in purification after funerals, at doctor's school, and after death of patient. (169:167) *Cold Remedy* Infusion of plant used as a mouthwash and gargle for colds. *Cough Medicine* Plant used as a cough medicine. (169:281) *Dermatological Aid* Infusion of bark taken and used as bath for babies with monkey sickness:

fever, itch, and enlarged eyes. (169:219) *Dietary Aid* Infusion of bark taken by babies for opossum sickness: appetite loss and drooling. (169:220) Infusion of bark taken as an emetic by children and adults for dog sickness: appetite loss and drooling. (169:225) *Emetic* Bark used as an emetic to "clean the insides." (169:167) Infusion of bark taken as an emetic and rubbed on the body for cat sickness: nausea. (169:224) *Eye Medicine* Infusion of bark taken and used as bath for babies with monkey sickness: fever, itch, and enlarged eyes. *Febrifuge* Infusion of bark taken and used as bath for babies with monkey sickness: fever, itch, and enlarged eyes. (169:219) *Gastrointestinal Aid* Bark used for cow sickness: lower chest pain, digestive disturbances, and diarrhea. (169:188) Infusion of plant taken for wolf sickness: vomiting, stomach pain, diarrhea, and frequent urination. (169:227) *Laxative* Decoction of bark used for horse sickness: nausea, constipation, and blocked urination. (169:188) *Oral Aid* Infusion of bark taken by babies for opossum sickness: appetite loss and drooling. (169:220) Infusion of bark taken as an emetic by children and adults for dog sickness: appetite loss and drooling. (169:225) *Other* Infusion of plant taken and rubbed on the body for "mythical wolf" sickness. (169:306) *Pediatric Aid* Infusion of plant taken by small children for raccoon sickness: diarrhea. (169:218) Infusion of bark taken and used as bath for babies with monkey sickness: fever, itch, and enlarged eyes. (169:219) Infusion of bark taken by babies for opossum sickness: appetite loss and drooling. (169:220) Infusion of bark taken by babies and adults for otter sickness: diarrhea and vomiting. (169:222) Infusion of bark taken as an emetic by children and adults for dog sickness: appetite loss and drooling. (169:225) *Throat Aid* Infusion of plant used as a mouthwash and gargle for sore throats. (169:281) *Unspecified* Plant used for medicinal purposes. (169:161) Plant used as medicine. (169:158) Plant used medicinally. (169:164) *Urinary Aid* Decoction of bark used for horse sickness: nausea, constipation, and blocked urination. (169:188) Infusion of plant taken for wolf sickness: vomiting, stomach pain, diarrhea, and frequent urination. (169:227) Plant used for gallstones and bladder pain. (169:275)

- *Food*—**Cherokee** *Beverage* Roots and barks used to make a beverage tea. (80:54) Red and white roots, red roots preferred, used to make tea. (126:44) **Chippewa** *Beverage* Root bark used to make a pleasant, tea-like beverage. *Spice* Leaves used in meat soups for the bay leaf-like flavor. (as *S. variifolium* 71:130) **Choctaw** *Spice* Pounded, dry leaves added to soup for flavor. (as *Laurus sassafras* 25:8)
- *Fiber*—**Cherokee** *Furniture* Wood used to make furniture. (80:54)
- *Other*—**Cherokee** *Fertilizer* Flowers mixed with beans for planting. *Incense & Fragrance* Used to scent soap. (80:54)

Satureja, Lamiaceae

Satureja douglasii (Benth.) Briq., Yerba Buena

- *Drug*—**Cahuilla** *Cold Remedy* Decoction of plant parts taken for colds. *Febrifuge* Decoction of plant parts taken for fevers. (15:139) **Costanoan** *Anthelmintic* Decoction of plant used for pinworms. *Toothache Remedy* Poultice of warm leaves applied to jaw or plant held in mouth for toothaches. (21:17) **Karok** *Kidney Aid* Infusion of leaves taken for the kidneys. *Love Medicine* Infusion of leaves taken as an aphrodisiac. (6:54) **Luiseño** *Cold Remedy* Decoction of plant parts taken for colds. *Febrifuge* Decoction of plant parts taken for fevers. (15:139) *Unspecified* Infusion of plant used for medicinal purposes. (as *Micromeria douglasii* 155:229) **Mahuna** *Sedative* Infusion of plant taken as a sedative for insomnia. (as *Micromeria douglasii* 140:23) **Mendocino Indian** *Blood Medicine* Infusion of dried, leafy vines taken to purify the blood. *Gastrointestinal Aid* Infusion of dried, leafy vines taken for colic. (as *Micromeria chamissonis* 41:383) **Pomo** *Blood Medicine* Infusion of plant taken to purify the blood. *Dietary Aid* Decoction of plant taken for becoming thin. *Gastrointestinal Aid* Decoction of plant taken for upset stomach. (as *Micromeria chamissonis* 66:15) **Pomo, Kashaya** *Blood Medicine* Decoction of crawling stems and leaves used to purify the blood. *Cold Remedy* Decoction of crawling stems and leaves used for chest colds. *Gastrointestinal Aid* Decoction of crawling stems and leaves used for an upset stomach and thinness. *Sedative* Decoction of crawling stems and leaves used to make you sleepy. (72:121) **Saanich** *Blood Medicine* Infusion of leaves taken for the blood. (182:84) **Yurok** *Blood Medicine* Infusion of leaves taken for the blood. (6:54)
- *Food*—**Diegueño** *Beverage* Leaves used to make mint tea. (84:41) **Luiseño** *Beverage* Plant used to make a tea. (as *Micromeria douglasii* 155:211) **Mendocino Indian** *Substitution Food* Slender, leafy vines made into rolls, dried, and used as a substitute for tea. (as *Micromeria douglasii* 41:383) **Pomo, Kashaya** *Beverage* Decoction of crawling stems and leaves used as a beverage tea. (72:121) **Saanich** *Beverage* Leaves used to make a refreshing tea. (182:84) **Tolowa** *Beverage* Fresh leaves used to make a refreshing tea. (6:54)
- *Other*—**Karok** *Incense & Fragrance* Leaves put in hats and clothes and vines hung around the neck as perfume. (as *Micromeria chamissonis* 148:389)

Satureja hortensis L., Summer Savory

- *Drug*—**Cherokee** *Analgesic* Snuff of leaves used for headache. (80:54)

Saururaceae, see *Anemopsis*, *Saururus*

Saururus, Saururaceae

Saururus cernuus L., Lizard's Tail

- *Drug*—**Cherokee** *Dermatological Aid* Roasted and mashed roots used as poultice. (80:43) *Poultice* Roasted and mashed roots used as poultices. (203:74) **Choctaw** *Dermatological Aid* Poultice of boiled, mashed roots applied to wounds. (25:23) **Ojibwa** *Antirheumatic (External)* Infusion of pounded plants used as wash for rheumatism. *Gastrointestinal Aid* Infusion of plant taken for stomach troubles and plant used as stomach medicine. *Panacea* Infusion of pounded plants used as wash for general illnesses. (135:231) **Seminole** *Antirheumatic (External)* Plant used for fire sickness: fever and body aches. (169:204) Roots used for rheumatism. (169:286) *Dermatological Aid* Poultice of plant applied to spider bites. (169:307) *Emetic* Plant used as an emetic during religious ceremonies. (169:409) *Febrifuge* Plant used for fire sickness: fever and body aches. (169:204) *Other* Complex infusion of whole plant taken for chronic conditions. (169:272) *Unspecified* Plant used for medicinal purposes. (169:161)

Saxifragaceae, see *Boykinia*, *Darmera*, *Heuchera*, *Leptarrhena*, *Lithophragma*, *Mitella*, *Parnassia*, *Penthorum*, *Saxifraga*, *Tellima*, *Tiarella*, *Tolmiea*

Saxifraga, Saxifragaceae

Saxifraga ferruginea Graham, Rustyhair Saxifrage

- **Drug**—Bella Coola *Urinary Aid* Decoction of root and leaf taken for "strangulation of the bladder." (as *S. bongardi* 150:58)

Saxifraga micranthidifolia (Haw.) Steud., Lettuceleaf Saxifrage
- **Food**—Cherokee *Unspecified* Young growth boiled, fried with ramps (*Allium tricoccum*?), and eaten. (204:252) *Vegetable* Leaves used in salads or wilted in boiling water with bacon grease dripped on the top. (126:54)

Saxifraga nelsoniana D. Don **ssp. *nelsoniana*,** Brook Saxifrage
- **Food**—Alaska Native *Dietary Aid* Leaves prepared and eaten soon after picking as a good source of vitamin C and provitamin A. *Unspecified* Leaves mixed with seal or walrus oil and eaten raw. (as *S. punctata* 85:63) **Eskimo, Alaska** *Unspecified* Leaves and stalks eaten raw with seal oil and fish. (as *S. punctata* ssp. *nelsoniana* 1:36) Leaves eaten fresh or in oil. Leaves preserved for long periods in oil. (as *Saxifrage punctata* 4:715) *Winter Use Food* Leaves preserved in seal oil for later use. (as *Saxifraga punctata* ssp. *nelsoniana* 1:36) **Eskimo, Arctic** *Vegetable* Leaves eaten raw with seal blubber or as "sauerkraut." (as *S. punctata* 128:29) **Eskimo, Inupiat** *Vegetable* Leaves preserved in seal oil and eaten with fish or meat or used fresh in salads. (as *S. punctata* 98:21)

Saxifraga pensylvanica L., Eastern Swamp Saxifrage
- **Drug**—Cherokee *Dermatological Aid* Poultice of root used for sore, swollen muscles. (80:26) **Iroquois** *Blood Medicine* Compound infusion of roots and leaves taken as a blood purifier. *Ceremonial Medicine* Ingredient in "Little Water Medicine" ritual. *Kidney Aid* Infusion of roots taken for weak kidneys or dropsy. Compound used for dropsy. *Panacea* Compound infusion taken or placed on injured part, a "Little Water Medicine." (87:344) **Menominee** *Unspecified* Remedy known as "the rabbit's ear, " less famous than *Valeriana uliginosa*. (151:53)
- **Food**—Cherokee *Vegetable* Leaves eaten raw as greens. (80:26)

Saxifraga spicata D. Don, Spiked Saxifrage
- **Food**—Alaska Native *Dietary Aid* Leaves used as a fair source of vitamin C. *Vegetable* Young, tender leaves used as a salad green. (85:65) **Eskimo, Alaska** *Unspecified* Young, tender leaves eaten raw with seal oil and often with fish. Young stems eaten raw. (1:36)

Saxifraga tricuspidata Rottb., Three Toothed Saxifrage
- **Food**—Eskimo, Inuktitut *Unspecified* Flowers eaten as food. (202:186)

Scaevola, Goodeniaceae
Scaevola sericea Vahl, Beach Naupaka
- **Drug**—Hawaiian *Dermatological Aid* Root bark pounded, mixed with salt, and used for cuts and skin diseases. (as *Scaevolo frutescens* 2:72)

Schedonnardus, Poaceae
Schedonnardus paniculatus (Nutt.) Trel., Tumblegrass
- **Other**—Apache, Chiricahua & Mescalero *Containers* Moist grass laid onto hot stones to prevent steam from escaping. (33:36)

Schizachyrium, Poaceae
Schizachyrium scoparium (Michx.) Nash **ssp. *scoparium*,** Little Bluestem
- **Drug**—Comanche *Venereal Aid* Stem ashes used for syphilitic sores. (as *Andropogon scoparius* 29:520)
- **Fiber**—Lakota *Clothing* Grass rubbed into softness and used as fur-like insulation in moccasins during the winter. (as *Andropogon scoparius* 139:28)
- **Other**—Comanche *Ceremonial Items* Bundles of stems used as switches in the sweat lodge. (as *Andropogon scoparius* 29:520)

Schkuhria, Asteraceae
Schkuhria multiflora Hook. & Arn., Manyflower False Threadleaf
- **Drug**—Navajo, Ramah *Oral Aid* Plant chewed 10 minutes for mouth sores. (as *Bahia neomexicana* 191:49)

Schoenocrambe, Brassicaceae
Schoenocrambe linearifolia (Gray) Rollins, Slimleaf Plainsmustard
- **Drug**—Navajo, Ramah *Ceremonial Medicine* and *Eye Medicine* Infusion of leaves used as a ceremonial eyewash. *Oral Aid* Cold infusion of leaves used as a mouthwash for sore gums. (as *Hesperidanthus linearifolius* 191:29)

Scirpus, Cyperaceae
Scirpus acutus Muhl. ex Bigelow, Hardstem Bulrush
- **Drug**—Clallam *Other* Used to suck out the cause of an illness. (57:201) **Cree, Woodlands** *Hemostat* Poultice of stem pith applied under the dressing to stop bleeding. (109:59) **Montana Indian** *Dietary Aid* Roots chewed "as a preventative to thirst." (as *S. lacustris occid.* 19:23) **Navajo, Ramah** *Ceremonial Medicine* and *Emetic* Plant used as a ceremonial emetic. (191:19) **Thompson** *Hemostat* and *Pediatric Aid* Burned stalk ashes used on baby's bleeding navel. (187:115)
- **Food**—Cheyenne *Unspecified* Inner part of stems eaten raw. (83:8) **Cree, Woodlands** *Unspecified* Stem base and tender leaf bases eaten fresh as collected by boat. Stem base and roots used for food. (109:59) **Dakota** *Unspecified* Tender, white stem base eaten fresh and uncooked. (as *S. lacustris* 69:359) **Gosiute** *Unspecified* Lower, tender stem portions formerly used for food. (as *S. lacustris* var. *occidentalis* 39:381) **Hesquiat** *Forage* Cows were said to eat it. (185:53) **Hopi** *Unspecified* Lower end of the stalk eaten raw. (as *S. lacustris* 190:159) **Klamath** *Unspecified* Seeds used for food. (as *S. lacustris occidentalis* 45:92) **Montana Indian** *Sauce & Relish* Roots boiled with water and made into a syrup. *Staple* Roots made into flour and used to make bread. *Unspecified* Roots eaten raw. Seeds used for food. (as *S. lacustris occidentalis* 19:23) **Paiute, Northern** *Unspecified* Roots peeled and eaten raw, boiled, or roasted. (59:49) Shoots eaten raw. Rhizomes peeled and chewed to extract the juices. Basal lengths of stalks eaten fresh. (60:72) **Pomo** *Unspecified* New sprouts eaten in the spring. (as *S. lacustris* 117:284) *Vegetable* Roots eaten as greens. Young shoots eaten as greens. (as *S. lacustris* var. *occidentalis* 11:92) **Sioux** *Unspecified* Young, spring shoots used for food. (as *S. lacustris occidentalis* 19:23) **Tubatulabal** *Unspecified* Roots used for food. (193:15)
- **Fiber**—Dakota *Mats, Rugs & Bedding* Stems pressed flat between the fingers and used to make household mats. (as *S. lacustris* 69:359) **Hesquiat** *Basketry* Round, pithy stems used to make baskets, shopping basket handles, and basket lids. *Mats, Rugs & Bedding* Dried stems used to make mats; excellent mattresses. (185:53) **Karok** *Mats,*

Rugs & Bedding Used for making matting. (148:380) **Klallam** *Mats, Rugs & Bedding* Used to make mats. (79:22) **Klamath** *Basketry* Stems used for baskets. (as *S. lacustris occidentalis* 45:92) Outer surface split and twisted into weft cords and warp used in making baskets. (as *S. lacustris* 46:737) *Clothing* Stems used for hats. *Mats, Rugs & Bedding* Stems used for mats. (as *S. lacustris occidentalis* 45:92) **Maidu** *Basketry* Stalks used as the overlay twine warp, upper portions only, in the manufacture of baskets. (173:71) **Makah** *Mats, Rugs & Bedding* Used to make mats. (79:22) **Montana Indian** *Mats, Rugs & Bedding* Stems used for making mats. (as *S. lacustris occidentalis* 19:23) **Nitinaht** *Mats, Rugs & Bedding* Tall, round stems sun dried and sewn together to make mats, sleeping compartments, and mattresses. (186:81) **Okanagan-Colville** *Building Material* Stems dried, sewn into large mats and used to make "teepees," temporary shelters, doors, and windows. (188:36) **Paiute, Northern** *Basketry* Used for wefts and binding tule items. (60:81) *Building Material* Used to build houses. (59:49) *Clothing* Used to secure the edges of skirts. (60:81) Used as the warps for sandals. (60:83) Used as the warps and the twining wefts for clothing. (60:86) *Mats, Rugs & Bedding* Used to secure the edges of mats. (60:81) Used to make twined mats for the insides of houses. Used in a simple pile for seating. (60:87) **Pomo** *Basketry* Stems used to make plain-twined baskets. *Canoe Material* Stems used to make boats. (as *S. lacustris* var. *occidentalis* 9:140) *Clothing* Skins used as swaddling clothes or soft padding. (as *S. lacustris* var. *occidentalis* 11:92) **Pomo, Kashaya** *Basketry* Root used to form the black part of the basket design. (72:28) **Shuswap** *Building Material* Used for weaving entrance covers, curtains, windbreaks, and for drying salmon. (123:54) **Snohomish** *Mats, Rugs & Bedding* Used to make mats. (79:22) **Thompson** *Building Material* Stems laid in an overlapping fashion and used for roofing. *Clothing* Stems woven or sewn with Indian hemp twine to make capes, bags, and Indian doctor headdresses. *Mats, Rugs & Bedding* Stems woven or sewn with Indian hemp twine to make mats. Food was placed on tule mats in order to dry it. The mats were also used as place mats for eating, as floor covering in lodges, as room partitions in pit houses, as walls for summer lodges and as mattresses and pillows. (187:115)
- *Other*—**Dakota** *Toys & Games* Long stems made into a ball and used as an instrument in children's games. The long stems were made into a ball by bending over the base of several together and the remaining length braided together to form a swinging handle. (as *S. lacustris* 69:359) **Hopi** *Ceremonial Items* Ceremonially associated with water. (as *S. lacustris* 200:70) **Kansa** *Ceremonial Items* Braided stems used as one of the five coverings of the sacred clamshell. The sacred clamshell was kept wrapped in five coverings. The innermost covering was the bladder of a buffalo bull; the second, a covering made of the spotted skin of a fawn; the third, a covering made of braided cattails; the fourth, a very broad piece of deerskin; and finally the fifth and outermost covering was made of braided hair from the head of a buffalo bull. (as *S. lacustris* 68:323) **Okanagan-Colville** *Ceremonial Items* Stems used to make headdresses for Indian doctors. *Containers* Stems woven into storage bags and used to store dried meat, fish, and berries. *Cooking Tools* Stems dried, sewn into large mats and used for drying berries and cutting and drying meat and fish. (188:36) **Paiute, Northern** *Containers* Used to make egg bags for transporting gathered eggs. (60:78) *Fasteners* Used for wefts and binding tule items. Used as ties for sandals. (60:81) **Thompson** *Containers* Stems woven or sewn with Indian hemp twine to make bags. *Decorations* Stems woven or sewn with Indian hemp twine to make Indian doctor headdresses. (187:115)

Scirpus americanus Pers., American Bulrush
- *Drug*—**Kwakiutl** *Dermatological Aid* and *Pediatric Aid* Grass and oil used on child's head to make the hair grow long and thick. (183:272)
- *Food*—**Keres, Western** *Unspecified* Roots and tender shoots used for food. (as *S. alneyi* 171:68)
- *Fiber*—**Hesquiat** *Basketry* Rushes used in weaving handles for baskets and shopping bags. *Clothing* Rushes used in weaving hats. (185:54) **Kwakiutl, Southern** *Basketry* Leaves used to make baskets. (183:296) Leaves used to make baskets. *Clothing* Leaves used to make hats. (183:272) **Makah** *Basketry* Shoots used to make baskets. (67:332) **Nitinaht** *Basketry* Leaves used for the bottoms and ribs of "grass" baskets. (186:79) **Pomo, Kashaya** *Basketry* Root used to form the black part of the basket design. Root used to form the black part of the basket design. (72:28) **Salish, Coast** *Basketry* Flat, fibrous leaves sun dried and used to make baskets. *Mats, Rugs & Bedding* Flat, fibrous leaves dried and used to make mats. (182:73)
- *Other*—**Hanaksiala** *Hunting & Fishing Item* Plant used by children to make tiny snares to catch small fish. (as *S. pungens* ssp. *monophyllus* var. *longisetis* 43:192)

Scirpus cyperinus (L.) Kunth, Wool Grass
- *Fiber*—**Ojibwa** *Basketry* Small rushes formerly used for woven storage bags. *Mats, Rugs & Bedding* Small rushes used for a certain kind of mat. (153:418) **Potawatomi** *Mats, Rugs & Bedding* Fruiting tops used as a resilient material for stuffing and making pillows. (154:118)

Scirpus fluviatilis (Torr.) Gray, River Bulrush
- *Fiber*—**Pomo, Kashaya** *Basketry* Root used to form the black part of the basket design. Roots were stained by burying them in the mud with ashes until a black color was obtained. (72:28)

Scirpus maritimus L., Saltmarsh Bulrush
- *Food*—**Paiute, Northern** *Porridge* Seeds parched, ground into flour, and made into mush. (60:74) *Unspecified* Seeds used for food. (59:48) *Winter Use Food* Seeds parched and stored for later use. (60:74) **Pima, Gila River** *Unspecified* Roots eaten raw. (133:7)
- *Fiber*—**Paiute, Northern** *Basketry* Used for wefts and binding tule items. *Clothing* Used to secure the edges of skirts. (60:81) Used as the warps for sandals. (60:83) Used as the warps and the twining wefts for clothing. (60:86) *Mats, Rugs & Bedding* Used to secure the edges of mats. (60:81) Used to make twined mats for the insides of houses. Used in a simple pile for seating. (60:87) **Pomo** *Basketry* Root stock fibers used as design material for baskets. (9:137) Grass used for basketry. This sedge has a central section, coal black, which color can be deepened by burying in manure or in blue mud. (118:6)
- *Other*—**Paiute, Northern** *Fasteners* Used for wefts and binding tule items. Used as ties for sandals. (60:81) **Panamint** *Decorations* Roots used as dark brown or black decorations for baskets. (as *S. palendosus* 105:78)

Scirpus microcarpus J. & K. Presl, Panicled Bulrush
- *Drug*—**Malecite** *Dermatological Aid* Poultice of pounded roots used for abscesses. (as *S. rubrotinctus* 116:247) *Throat Aid* Infusion of

plants and blue flag used as a gargle for sore throats. (as *S. rubrotinctus* 116:248) **Micmac** *Dermatological Aid* Roots used for abscesses. *Throat Aid* Herbs used for sore throats. (as *S. rubrotinctus* 40:61)
- *Fiber*—**Makah** *Basketry* Leaves used for the bottom portion of baskets. (67:332) **Okanagan-Colville** *Basketry* Used to make berry picking and root digging baskets. (188:37)
- *Other*—**Okanagan-Colville** *Containers* Used over and under food in steaming pits. *Decorations* Used to make fringe for buckskin dresses. (188:37) **Thompson** *Tools* Grass formerly used to cut a newborn baby's umbilical cord. (187:116)

Scirpus nevadensis S. Wats., Nevada Bulrush
- *Drug*—**Cheyenne** *Ceremonial Medicine* Plant used in the Sun Dance ceremony. (83:8)
- *Food*—**Cheyenne** *Unspecified* Peeled roots eaten raw. (76:170)
- *Fiber*—**Cheyenne** *Mats, Rugs & Bedding* Stems woven into mats and used like blankets on wooden mattresses. (76:170)

Scirpus pallidus (Britt.) Fern., Cloaked Bulrush
- *Drug*—**Navajo, Ramah** *Ceremonial Medicine* and *Emetic* Plant used as a ceremonial emetic. (191:19)

Scirpus pungens Vahl, Threesquare Bulrush
- *Food*—**Paiute, Northern** *Porridge* Seeds ground lightly into a flour and boiled into a mush. (59:49) Seeds parched, ground into flour, and made into mush. (60:74)

Scirpus robustus Pursh, Alkali Bulrush
- *Food*—**Montana Indian** *Unspecified* Autumn tubers used for food. (as *S. campestris* 19:23) **Pomo** *Vegetable* Young shoots eaten as greens. Roots eaten as greens. (11:92)
- *Fiber*—**Kawaiisu** *Basketry* Slender rootstock cores used as pattern material for coiled baskets. (206:63) **Pomo** *Basketry* Stems used to make plain-twined baskets. Plant made into the heavy hoop used at the opening of the conical burden basket. *Building Material* Stems used for thatch. Stems used as foundation material. (9:140) *Clothing* Skins used as swaddling clothes or soft padding. (11:92) *Mats, Rugs & Bedding* Stems used to make mats. *Sewing Material* Root used as a weft or sewing material. (9:140) **Pomo, Kashaya** *Basketry* Root used to form the black part of the basket design. (72:28)

Scirpus sp., Rat Grass
- *Drug*—**Houma** *Orthopedic Aid* Decoction of plant used as a wash for weak legs. *Pediatric Aid* and *Sedative* Decoction of plant given to nervous, fretful, crying children. (158:60)
- *Food*—**Cahuilla** *Bread & Cake* Pollen used to make cakes. *Staple* Roots ground into flour. Seeds eaten raw or ground into mush. (15:139) **Costanoan** *Unspecified* Roots eaten raw or ground into flour and cooked. (21:255) **Keres, Western** *Unspecified* Tender shoots used for food. (171:68) **Luiseño** *Unspecified* Tender, young shoots eaten raw. (155:233) **Pima, Gila River** *Snack Food* Tubers eaten as snack food primarily by children. (133:5) **Thompson** *Unspecified* Fleshy rootstocks baked and eaten. (164:481)
- *Fiber*—**Cahuilla** *Building Material* Stalks used for roofing. *Mats, Rugs & Bedding* Stalks used for bedding, mats, and weaving materials. (15:139) **Costanoan** *Basketry* Roots used in basketry. *Building Material* Stems used as thatch. *Canoe Material* Stems bundled and made into rafts. (21:255) **Costanoan (Olhonean)** *Building Material* Plant used to make houses. (117:373) **Houma** *Brushes & Brooms* Plants dried and used to make brooms. (158:60) **Paiute** *Mats, Rugs & Bedding* Used to make mats. (111:52) **Pomo** *Basketry* Rootstocks used to make fine baskets. (41:316) **Thompson** *Mats, Rugs & Bedding* Grass-like herbs used extensively for table mats. (164:497)
- *Other*—**Cahuilla** *Ceremonial Items* Plant used to make ceremonial bundles and images for image burning ceremony. *Hunting & Fishing Item* Plant areas used by nesting water fowl and used as indicator by hunters of game. (15:139) **Costanoan** *Smoking Tools* Hollow stems made into pipes. (21:255) **Havasupai** *Toys & Games* Plant braided by children to make a whip. (197:211) **Thompson** *Containers* Grass-like herbs used extensively to weave coarse bags and nets. (164:497)

Scirpus tabernaemontani K. C. Gmel., Softstem Bulrush
- *Drug*—**Cherokee** *Emetic* Decoction used as emetic. (as *S. validus* 80:27) Decoction of plant taken as an emetic. (as *S. validus* 177:6) *Oral Aid* Compound used as medicine for "spoiled saliva." (as *S. validus* 80:27) **Cree, Woodlands** *Hemostat* Poultice of stem pith applied under the dressing to stop bleeding. (as *S. validus* 109:60) **Iroquois** *Snakebite Remedy* Compound decoction of roots and stems used as a poultice for snakebite. *Tuberculosis Remedy* Compound taken for consumption caused by *Nuphar lutea*. (as *S. validus* 87:275) **Potawatomi** *Love Medicine* Flowers used by women as a love medicine. (as *S. validus* 154:118)
- *Food*—**Chippewa** *Unspecified* Sweet bulbs eaten raw in midsummer. (as *S. validus* 53:320) **Cree, Woodlands** *Unspecified* Stem base and roots used for food. Stem base and tender leaf bases eaten fresh as collected by boat. (as *S. validus* 109:60) **Dakota** *Unspecified* Fresh, raw stems used for food. (as *S. validus* 70:69) **Hopi** *Unspecified* Lower end of the stalk eaten raw. (as *S. validus* 190:159) **Kawaiisu** *Unspecified* Tender, lower portions of the plant eaten raw. (as *S. validus* 206:63) **Lakota** *Unspecified* Species used for food. (as *S. validus* 139:26) **Ute** *Unspecified* Lower, tender portions of the plant used for food. (as *S. validus* 38:36)
- *Fiber*—**Chippewa** *Mats, Rugs & Bedding* Used for mats. (as *S. validus* 53:378) Plant used for weaving floor and wall mats. (as *S. validus* 71:124) **Dakota** *Mats, Rugs & Bedding* Stems used to weave into matting. (as *S. validus* 70:69) **Kawaiisu** *Building Material* Stems sewn with cord, made into mats, and used as interior and exterior walls in house construction. Stems used as wall filler in house construction. *Mats, Rugs & Bedding* Stems sewn with cord, made into mats, and used as rugs, meat platters, cradle linings, and sleeping mats. (as *S. validus* 206:63) **Lakota** *Mats, Rugs & Bedding* Used to make mats. (as *S. validus* 139:26) **Menominee** *Building Material* Bleached, sun dried rushes sewn with basswood string used for covering and side walls of wigwams. Bleached, dried rushes sewn with basswood string used for covering and side walls of medicine lodges. *Mats, Rugs & Bedding* Bleached, sun dried rushes used to weave mats. (as *S. validus* 151:74) **Meskwaki** *Mats, Rugs & Bedding* Long, bleached and dyed rushes used to make mats. (as *S. validus* 152:268) **Ojibwa** *Mats, Rugs & Bedding* Rushes used for the best mats. The bleached rushes were immersed in water for a few days and then cleansed. They selected long rushes, with small diameters, so that the pith content was small. When the mat was in service, such fibers were not readily crushed. The rush, when gathered, is an intense green, white only at the base

where it stands in water. All rushes were first bleached pure white, and afterwards colored as desired. They were pulled, rather than cut, in order to obtain the maximum length. When thoroughly bleached and dried, they dyed them with white men's dyes. Formerly they used native dyes, which they really preferred. The bleached rushes predominated in any rug, and were ivory white in color. The finished rug or mat was 3 feet wide and from 4 to 8 feet long, and sold for from $8 to $30 in 1923. The edge was bound securely with nettle fiber cord. (as *S. validus* 153:418) **Omaha** *Mats, Rugs & Bedding* Stems used to weave into matting. **Pawnee** *Mats, Rugs & Bedding* Stems used to weave into matting. **Ponca** *Mats, Rugs & Bedding* Stems used to weave into matting. (as *S. validus* 70:69) **Potawatomi** *Basketry* Entire, dyed stem used to make baskets. *Mats, Rugs & Bedding* Entire, dyed stem used to make mats. (as *S. validus* 154:112)
- *Other*—**Chippewa** *Toys & Games* Used for toys. (as *S. validus* 53:378) **Kawaiisu** *Ceremonial Items* Used to make images representing the deceased in the ceremony for the dead. *Cooking Tools* Stems sewn with cord, made into mats and used as rugs, meat platters, cradle linings, and sleeping mats. *Stable Gear* Used to make saddles. (as *S. validus* 206:63)

Scorzonella, Asteraceae
Scorzonella sp.
- *Food*—**Paiute** *Unspecified* Roots roasted and used for food. (104:103)

Scrophulariaceae, see *Agalinis, Aureolaria, Bacopa, Besseya, Castilleja, Chelone, Collinsia, Cordylanthus, Digitalis, Diplacus, Epixiphium, Keckiella, Limosella, Linaria, Macranthera, Melampyrum, Mimulus, Nothochelone, Orthocarpus, Pedicularis, Penstemon, Scrophularia, Verbascum, Veronica, Veronicastrum*

Scrophularia, Scrophulariaceae
Scrophularia californica Cham. & Schlecht., California Figwort
- *Drug*—**Costanoan** *Dermatological Aid* Poultice of heated twigs applied to swollen sores. Poultice of leaves applied to boils or swellings. *Disinfectant* Decoction of twigs used as a wash for infections. *Eye Medicine* Plant juice used as an eyewash for poor vision. Poultice of leaves applied to sore eyes. (21:15) **Pomo** *Dermatological Aid* Poultice of heated leaves applied to boils. (66:15) **Pomo, Kashaya** *Dermatological Aid* Poultice of fresh, warm leaves used to draw a boil to a head. Leaves used to draw the pus out of a burst boil. (72:49)

Scrophularia californica ssp. *floribunda* (Greene) R. J. Shaw, California Figwort
- *Drug*—**Diegueño** *Febrifuge* Infusion of roots taken for fevers. (88:219)

Scrophularia lanceolata Pursh, Lanceleaf Figwort
- *Drug*—**Iroquois** *Antihemorrhagic* Decoction of roots taken for hemorrhage after childbirth. *Blood Medicine* Decoction of roots taken by women for the blood after childbirth. *Cold Remedy* Infusion of roots taken to prevent cramps and colds after birth. (87:434) *Dermatological Aid* Poultice applied for the soreness of sunburns, sunstroke, and frostbite. *Gynecological Aid* Compound decoction of leaves and roots taken for a "sick womb." (87:433) Infusion of roots taken to prevent cramps and colds after birth. *Kidney Aid* Compound taken for dropsy. (87:434)

Scrophularia marilandica L., Carpenter's Square
- *Drug*—**Iroquois** *Gynecological Aid* Infusion of roots taken by women who are weak due to irregular menses. (87:433) **Meskwaki** *Unspecified* Roots used as a medicine. (152:247)

Scrophularia sp.
- *Food*—**Yavapai** *Vegetable* Boiled leaves used for greens. (65:258)

Scutellaria, Lamiaceae
Scutellaria angustifolia Pursh, Narrowleaf Skullcap
- *Drug*—**Miwok** *Eye Medicine* Decoction used as wash for sore eyes. (12:173)

Scutellaria californica Gray, California Skullcap
- *Drug*—**Mendocino Indian** *Febrifuge* Plant used for chills and fevers. (41:385)

Scutellaria elliptica Muhl. ex Spreng., Hairy Skullcap
- *Drug*—**Cherokee** *Abortifacient* Infusion of root used for monthly period. *Antidiarrheal* Infusion of root taken for monthly period and diarrhea. *Breast Treatment* Decoction taken for nerves and compound used for breast pains. *Gynecological Aid* Compound used for expelling afterbirth. *Kidney Aid* Root compound used as a kidney medicine. (80:55)

Scutellaria galericulata L., Marsh Skullcap
- *Drug*—**Delaware** *Gastrointestinal Aid* Plant tops used as a stomach stimulant. *Laxative* Plant tops used as a laxative. (176:37) **Delaware, Oklahoma** *Laxative* Plant top used as a laxative and "stomach stimulant." (175:31, 80) **Ojibwa** *Heart Medicine* Plant used for heart trouble. (153:372)

Scutellaria incana Biehler, Hoary Skullcap
- *Drug*—**Cherokee** *Abortifacient* Infusion of root used for monthly period. *Antidiarrheal* Infusion of root taken for monthly period and diarrhea. *Breast Treatment* Decoction taken for nerves and compound used for breast pains. *Gynecological Aid* Compound used for expelling afterbirth. *Kidney Aid* Root compound used as a kidney medicine. (80:55)

Scutellaria lateriflora L., Blue Skullcap
- *Drug*—**Cherokee** *Abortifacient* Infusion of root used for monthly period. *Antidiarrheal* Infusion of root taken for monthly period and diarrhea. *Breast Treatment* Decoction taken for nerves and compound used for breast pains. (80:55) *Emetic* Decoction of roots taken as an emetic to expel the afterbirth. (177:56) *Gynecological Aid* Compound used for expelling afterbirth. (80:55) Decoction of roots taken as an emetic to expel the afterbirth. (177:56) *Kidney Aid* Root compound used as a kidney medicine. (80:55) **Iroquois** *Misc. Disease Remedy* Infusion of powdered roots taken to prevent smallpox. *Throat Aid* Infusion of powdered roots taken to keep throat clean. (87:422)

Scutellaria parvula Michx., Small Skullcap
- *Drug*—**Meskwaki** *Antidiarrheal* Plant used for flux. (152:227)

Scutellaria sp., Skullcap
- *Drug*—**Miwok** *Cold Remedy* Decoction taken for colds. *Cough Medicine* Decoction taken for coughs. (12:173)

Sebastiania, Euphorbiaceae
Sebastiania fruticosa (Bartr.) Fern., Gulf Sebastiana
- *Drug*—**Alabama** *Laxative* Roots chewed "to produce a movement of the bowels." (as *S. ligustrina* 172:665)

Sedum, Crassulaceae
Sedum debile S. Wats., Orpine Stonecrop
- *Other*—**Gosiute** *Smoke Plant* Leaves formerly used to smoke. (39:381)

Sedum divergens S. Wats., Pacific Stonecrop
- *Drug*—**Okanagon** *Hemorrhoid Remedy* Plant used for piles. *Laxative* and *Pediatric Aid* Plant given to children as a laxative. **Thompson** *Hemorrhoid Remedy* Plant used for piles. *Laxative* and *Pediatric Aid* Plant given to children as a laxative. (125:41) Decoction of whole plant given to children for constipation. (164:463)
- *Food*—**Gitksan** *Unspecified* Leaves used for food. **Haisla** *Unspecified* Leaves used for food. **Wet'suwet'en** *Unspecified* Leaves used for food. (73:154)

Sedum integrifolium (Raf.) A. Nels. **ssp. integrifolium**, Entireleaf Stonecrop
- *Drug*—**Eskimo, Alaska** *Oral Aid* Roots chewed and the juice spat out for sores in the mouth. (as *S. rosea* ssp. *integrifolium* 1:36)
- *Food*—**Eskimo, Alaska** *Beverage* Red plant tops used to make a tea-like beverage. (as *S. rosea* ssp. *integrifolium* 1:36) *Unspecified* Plant eaten fresh, soured, or in oil. Root used for food. (as *Rhodiola integrifolia* 4:715)

Sedum lanceolatum Torr., Spearleaf Stonecrop
- *Drug*—**Okanagan-Colville** *Gynecological Aid* Infusion of stems, leaves, and flowers taken to clean out the womb after childbirth. *Laxative* Infusion of stems, leaves, and flowers taken or raw leaves chewed as a laxative. (188:98)

Sedum laxum (Britt.) Berger, Roseflower Stonecrop
- *Food*—**Tubatulabal** *Unspecified* Rolled in palm of hand with salt grass leaves and stems and eaten. (as *Cotyledon laxa* 193:19) Leaves used for food. (as *Cotyledon laxa* 193:15)

Sedum rosea (L.) Scop., Roseroot Stonecrop
- *Drug*—**Eskimo, Nunivak** *Unspecified* Infusion of flowers used for its medicinal value. (149:325) **Eskimo, Western** *Analgesic* and *Gastrointestinal Aid* Compound decoction of flowers taken for stomachache and intestinal discomfort. (as *S. roseum* 108:5) *Tuberculosis Remedy* Raw flowers eaten for tuberculosis. (as *S. roseum* 108:24, 60)
- *Food*—**Alaska Native** *Unspecified* Roots used for food. *Vegetable* Leaves and succulent, fleshy stems used raw in mixed salads or cooked as a green vegetable. (as *S. roseum* 85:67) **Eskimo, Arctic** *Vegetable* Young leaves and flowering stems eaten raw as salad, cooked as a potherb, or made into a "sauerkraut." (as *Rhodiola rosea* 128:28) **Eskimo, Inupiat** *Frozen Food* Fermented stems, leaves, and young flower buds frozen through the winter. *Staple* Roots stored buried in the sand and grass and used in hard times when short of food. *Unspecified* Roots eaten with oil. *Vegetable* Fermented stems, leaves, and young flower buds eaten with walrus blubber, any kind of blubber, or oil. Stems, leaves, and young flower buds were preserved by fermenting in water. A barrel was filled with clean, unchopped roseroot plants which were covered with water. Plants were pressed under the water with a plate and stored in a medium warm to cool place to ferment. If it was too cold, the roseroot would not ferment. If it was too warm, it might spoil first. In 2 to 3 weeks, or, when the plants were squashed together, the lids were taken off and another batch of roseroot was added and covered again. More water was added, when necessary, to cover the plants. Roseroot was continuously added and let to ferment down until the barrel was full. When the taste was just right, the batches were taken out and put into plastic bags, plants and juice together, and frozen. (as *S. roseum* 98:54)

Sedum sp., Sedum
- *Drug*—**Costanoan** *Dermatological Aid* Decoction of plant or powdered plant used on wounds. *Febrifuge* Decoction of plant used as a gargle for fevers. *Throat Aid* Decoction of plant used as a gargle for sore throats. (21:18) **Makah** *Urinary Aid* Leaves used for bed-wetting. (67:256) **Thompson** *Hemorrhoid Remedy* Decoction of several species used as a wash or poultice of plant applied for piles. (164:463)
- *Food*—**Costanoan** *Unspecified* Raw leaves and stems used for food. (21:251) **Makah** *Substitution Food* Plant used on journeys in place of unsafe water. (79:31) *Unspecified* Leaves used for food. (67:256)

Sedum spathulifolium Hook., Broadleaf Stonecrop
- *Drug*—**Bella Coola** *Gynecological Aid* Leaves eaten or poultice of warmed leaves used to start maternal milk flow. (150:57) Decoction of stalks taken for easy deliveries. (184:204) **Kuper Island Indian** *Hemostat* Leaves and stems squeezed and juice rubbed over bleeding wounds. (182:81) **Okanagon** *Hemorrhoid Remedy* Plant used for piles. *Pediatric Aid* and *Sedative* Infusion of whole plant used as a wash for babies with extreme nervousness. (125:41) **Songish** *Gynecological Aid* Leaves chewed by women in their ninth month of pregnancy to facilitate childbearing. (182:81) **Thompson** *Hemorrhoid Remedy* Plant used for piles. (125:41) *Laxative* Decoction of whole plant given to children for constipation. (164:463) *Oral Aid* Plant used for sore gums. (187:205) *Pediatric Aid* Infusion of whole plant used as a wash for babies with extreme nervousness. (125:41) Decoction of whole plant used as a wash to soothe cross babies. (164:462) Decoction of whole plant given to children for constipation. (164:463) *Sedative* Infusion of whole plant used as a wash for babies with extreme nervousness. (125:41) Decoction of whole plant used as a wash to soothe cross babies. (164:462)

Sedum stenopetalum Pursh, Wormleaf Stonecrop
- *Drug*—**Okanagan-Colville** *Venereal Aid* Infusion of whole plant taken for venereal disease. (188:98)

Sedum telephioides Michx., Allegheny Stonecrop
- *Drug*—**Delaware, Ontario** *Dermatological Aid* Poultice of crushed leaves applied to wounds as a disinfectant. *Disinfectant* Poultice of crushed leaves applied to infected wounds. (175:66)

Sedum telephium L. **ssp. telephium**, Witch's Moneybags
- *Drug*—**Iroquois** *Dermatological Aid* Plant used when babies cry or for bruises. Poultice of stalks and leaves applied to injuries resulting from witching. *Liver Aid* Compound infusion of whole roots taken for liver sickness. *Other* Rubbed on paralyzed face. *Pediatric Aid* Used for crying babies or bruises. *Veterinary Aid* Infusion of plant

given to horses as a diuretic. *Witchcraft Medicine* Poultice of stalks and leaves applied to injuries resulting from witching. (87:343) **Malecite** *Dermatological Aid* Poultice of pounded leaves used for boils and carbuncles. (as *S. purpureum* 116:247) **Micmac** *Dermatological Aid* Leaves used for boils and carbuncles. (as *S. purpureum* 40:62)

Selaginellaceae, see *Selaginella*

Selaginella, Selaginellaceae, spike moss

Selaginella densa Rydb., Lesser Spikemoss
- *Drug*—**Blackfoot** *Antihemorrhagic* Infusion of plant taken for spitting up blood. (86:74) *Gynecological Aid* Decoction of plant used to induce labor and expel the afterbirth. (86:61) *Narcotic* Plant eaten by a starving person for its doping effect and to make one feel unusually strong. (86:105) *Veterinary Aid* Powdered roots applied to the mouth of racehorses to make them hyperactive. (86:90)
- *Food*—**Blackfoot** *Spice* Dried plant used to spice meat. *Starvation Food* Very bitter berries eaten only through necessity and never stored. (86:105)

Sempervivum, Crassulaceae

Sempervivum sp., Houseleek
- *Drug*—**Mahuna** *Oral Aid* Plant used for the prevention of salivary gland swelling. (140:47)

Sempervivum tectorum L., Common Houseleek
- *Drug*—**Cherokee** *Dermatological Aid* Poultice of wilted plant applied to remove corns. *Ear Medicine* Juice warmed and used for earache. (80:42)

Senecio, Asteraceae

Senecio anonymus Wood, Small's Ragwort
- *Drug*—**Catawba** *Tuberculosis Remedy* Plant used for consumption. (as *S. smallii* 157:191)

Senecio aureus L., Golden Ragwort
- *Drug*—**Cherokee** *Gynecological Aid* Infusion taken to prevent pregnancy. *Heart Medicine* Infusion taken for heart trouble. (80:52) **Iroquois** *Blood Medicine* Decoction of plants taken for the blood. *Diaphoretic* Roots used as a diaphoretic. *Febrifuge* Infusion of rosettes given to children with fevers. *Kidney Aid* Decoction of plants taken for the kidneys. *Orthopedic Aid* Decoction of plants taken for broken bones. *Pediatric Aid* Infusion of rosettes given to children with fevers. (87:473)

Senecio congestus (R. Br.) DC., Marsh Fleabane
- *Drug*—**Eskimo, Inuktitut** *Poison* Plant considered poisonous. (202:187) **Eskimo, Western** *Poison* Roots considered poisonous. (108:17)
- *Food*—**Eskimo, Arctic** *Vegetable* Young leaves and flowering stems eaten raw as salad, cooked as a potherb, or made into a "sauerkraut." (128:27)

Senecio fendleri Gray, Fendler's Ragwort
- *Drug*—**Keres, Western** *Psychological Aid* Infusion of plant used for homesickness. (171:68) **Navajo, Ramah** *Ceremonial Medicine* Plant used in ceremonial chant lotion. *Dermatological Aid* Plant used for frozen feet. Poultice of moistened, crushed leaves and flowers applied to pimples or sores. *Gastrointestinal Aid* Decoction taken after swallowing an ant. *Panacea* Plant used as "life medicine." *Pediatric Aid* Strong decoction used for "birth injury." (191:53)

Senecio flaccidus var. *douglasii* (DC.) B. L. Turner & T. M. Barkl., Douglas's Groundsel
- *Drug*—**Costanoan** *Dermatological Aid* Infusion of plant used externally for infected sores or cuts. *Disinfectant* Infusion of plant used externally for infected sores. *Gynecological Aid* Infusion of plant taken by women for "lockjaw" after childbirth. *Kidney Aid* Infusion of plant taken for a "cold in the kidneys." *Other* Infusion of plant taken by women for "lockjaw" after childbirth. (as *S. douglasii* 21:27) **Kawaiisu** *Laxative* Infusion of leaves taken as a strong laxative and could cause death. *Unspecified* Infusion of leaves used as a very strong medicine. (as *S. douglasii* 206:63)
- *Fiber*—**Hopi** *Brushes & Brooms* Plant top used as a brush to remove spines from the prickly pears. (as *S. douglasii* 56:19) **Navajo** *Brushes & Brooms* Tops used as brooms to brush the spines from cactus fruit. (as *S. douglasii* 55:88) Plant top used as a brush to remove spines from the prickly pears. (as *S. douglasii* 56:19)
- *Other*—**Navajo** *Ceremonial Items* Plant boiled and taken before person goes into sweat house, to get a good voice for the Night Chant. (as *S. douglasii* 55:88)

Senecio flaccidus Less. var. *flaccidus*, Threadleaf Groundsel
- *Drug*—**Hopi** *Antirheumatic (External)* Poultice of flowers and leaves used for sore muscles. *Dermatological Aid* Poultice of ground leaf used for pimples and skin diseases. (as *S. longilobus* 42:359) Ground leaves applied to pimples. *Orthopedic Aid* Pounded plant smeared over sore muscles. (as *S. longilobus* 200:32, 98) **Jemez** *Stimulant* Plant placed on hot coals and smoke stimulated faint and sick person. (as *S. filifolius* 44:27) **Keres, Western** *Dermatological Aid* Leaves used in shoes to prevent sweaty feet. Plant mixed with artemisia and deer marrow to make a salve. *Gastrointestinal Aid* Infusion of 6-inch piece of twig used for stomach trouble. (as *Seneci filifolius* 171:69) **Navajo, Kayenta** *Dermatological Aid* Poultice of plant applied to boils. (as *Senecio longilobus* 205:49)
- *Other*—**Keres, Western** *Incense & Fragrance* Plant parts used in bed for the good smell. *Insecticide* Plant parts used in bed as a bedbug repellent. (as *Senecio filifolius* 171:69)

Senecio jacobaea L., Stinking Willie
- *Drug*—**Makah** *Unspecified* Infusion of plant used as a medicinal tea. (67:327)

Senecio multicapitatus Greenm. ex Rydb., Ragwort Groundsel
- *Drug*—**Navajo** *Dermatological Aid* Decoction of plants used as a steam bath for sores. (90:156) **Navajo, Ramah** *Cathartic* Decoction of root taken as a cathartic. *Gynecological Aid* Decoction of root used to facilitate delivery of placenta. (191:53) **Zuni** *Analgesic* Cold infusion of pulverized root rubbed over limbs for "aching bones." (166:59, 60) *Ceremonial Medicine* Infusion of powdered root ceremonially rubbed on limbs for "aching bones." (166:59) *Eye Medicine* Infusion of blossoms used as drops for inflamed eyes. (166:59, 60) *Orthopedic Aid* Infusion of powdered root ceremonially rubbed on limbs for "aching bones." (166:59)
- *Food*—**Navajo, Ramah** *Candy* Root bark used for chewing gum. *Star-

vation *Food* Roasted, ground seeds, cornmeal, and goat's milk made into flat cakes or mush in times of famine. (191:53)

Senecio multilobatus Torr. & Gray ex Gray, Lobeleaf Groundsel
- **Drug**—Navajo, Ramah *Analgesic* Decoction of plant used for menstrual pain. *Ceremonial Medicine* Plant rubbed on body as ceremonial medicine. *Gynecological Aid* Decoction of plant taken for menstrual pain. (191:53) Yavapai *Cold Remedy* Decoction of leaf stem inhaled for colds. *Dermatological Aid* Boiled or dried and powdered leaves used for sores. *Gastrointestinal Aid* Decoction of root taken for stomachache. *Nose Medicine* Decoction of leaf steam inhaled for sore nose. *Venereal Aid* Decoction of root used as a wash for gonorrheal sores. (65:261)

Senecio neomexicanus Gray, New Mexico Groundsel
- **Drug**—Navajo, Kayenta *Antidote* Plant used as an antidote for narcotics. *Burn Dressing* Powdered plant, poultice of plant applied and plant used as lotion for burns. *Disinfectant* Plant used for bear infections. (205:49) Navajo, Ramah *Hunting Medicine* Cold infusion used as lotion for good luck in hunting. (191:53)

Senecio pseudoarnica Less., Seaside Ragwort
- **Drug**—Aleut *Dermatological Aid* Poultice of leaves applied to drain cuts and boils. (8:427)
- **Food**—Eskimo, Alaska *Unspecified* Fleshy stems and boiled leaves used for food. The leaves were edible only when boiled. The plant was eaten only when young and tender. (1:38)

Senecio sp., Groundsel
- **Drug**—Ute *Unspecified* Used as medicine. (38:36)
- **Food**—Aleut *Unspecified* Species used for food. (7:29) Gosiute *Candy* Latex used as a chewing gum. (39:381)

Senecio spartioides Torr. & Gray, Broom Groundsel
- **Drug**—Hopi *Antirheumatic (External)* Poultice of flowers and leaves used for sore muscles. *Dermatological Aid* Poultice of ground leaf used for pimples and skin diseases. (42:360) Keres, Western *Gynecological Aid* Infusion of leaves used as a tonic after childbirth. (171:68)

Senecio triangularis Hook., Arrowleaf Groundsel
- **Drug**—Cheyenne *Analgesic* Infusion of pulverized leaves or roots taken for chest pains. *Sedative* Infusion of pulverized leaves or roots taken as a sedative. (76:190)

Senna, Fabaceae
Senna hebecarpa (Fern.) Irwin & Barneby, American Wild Sensitive Plant
- **Drug**—Cherokee *Analgesic* Infusion taken for cramps. *Cathartic* Infusion given to children and adults as a purgative. *Dermatological Aid* Poultice of root used for sores. *Febrifuge* Infusion taken for fever and infusion of root given to children for fever. *Heart Medicine* Infusion of root taken for heart trouble. *Misc. Disease Remedy* Infusion taken for "blacks" (hands and eye sockets turn black). *Pediatric Aid* Infusion given to children as a purgative and for fever. *Pulmonary Aid* Compound taken for pneumonia. *Stimulant* Compound infusion given for fainting spells. (as *Cassia hebecarpa* 80:54) Iroquois *Anthelmintic* Plant used as a worm remedy. *Laxative* Compound decoction of flowers taken as a laxative. (as *Cassia hebecarpa* 87:362)

Senna marilandica (L.) Link, Maryland Wild Sensitive Plant
- **Drug**—Cherokee *Analgesic* Infusion taken for cramps. *Cathartic* Infusion given to children and adults as a purgative. *Dermatological Aid* Poultice of root used for sores. *Febrifuge* Infusion taken for fever and infusion of root given to children for fever. *Heart Medicine* Infusion of root taken for heart trouble. *Misc. Disease Remedy* Infusion taken for "blacks" (hands and eye sockets turn black). *Pediatric Aid* Infusion given to children as a purgative and for fever. *Pulmonary Aid* Compound taken for pneumonia. *Stimulant* Compound infusion given for fainting spells. (as *Cassia marilandica* 80:54) Iroquois *Anthelmintic* Compound infusion taken for tapeworms. *Blood Medicine* Compound infusion taken as a blood remedy. (as *Cassia marilandica* 87:362) Meskwaki *Throat Aid* Seeds soaked until mucilaginous, then eaten for sore throat. (as *Cassia marilandica* 152:228)

Senna occidentalis (L.) Link, Septicweed
- **Drug**—Hawaiian *Dermatological Aid* Plant and other ingredients pounded, mixed, squeezed, and resulting juice used as wash for skin diseases. (as *Cassis occidentalis* 2:4)
- **Food**—Kiowa *Beverage* Ground seeds boiled to make a coffee-like beverage. (as *Ditremexa occidentalis* 192:31)
- **Other**—Kiowa *Fuel* Wood used for fuel. Wood used for fuel. (as *Ditremexa occidentalis* 192:31)

Senna tora (L.) Roxb., Senna
- **Drug**—Houma *Misc. Disease Remedy* Compound decoction of root taken for typhoid. (as *Cassia tora* 158:65)

Sequoia, Taxodiaceae, conifer
Sequoia sempervirens (Lamb. ex D. Don) Endl., Redwood
- **Drug**—Houma *Blood Medicine* Scraped "knees" or inner bark mixed with whisky and taken as a blood purifier. *Liver Aid* Infusion of inner bark in whisky taken for jaundice and to purify the blood. (as *Taxodium sempervirens* 158:61) Pomo *Ear Medicine* Poultice of heated leaves applied for earaches. *Stimulant* Gummy sap taken for rundown conditions. *Tonic* Infusion of gummy sap taken as a tonic. (66:11) Pomo, Kashaya *Ear Medicine* Poultice of warmed, new foliage used for earaches. *Stimulant* Gummy sap and water taken as medicine for rundown condition. (72:97) Tlingit *Venereal Aid* Compound poultice of bark applied for syphilis. (as *Taxodium sempervirens* 107:283)
- **Fiber**—Costanoan *Basketry* Sprouts from burls used in basketry. (21:248) Mendocino Indian *Building Material* Lumber used to construct houses. *Canoe Material* Fallen logs formerly hollowed out by fire and used as canoes. (41:309) Pomo, Kashaya *Building Material* Bark leaned together, in a circle, to make a house. (72:97) Tolowa *Building Material* Wood used for building materials. *Canoe Material* Wood used to make canoes. Yurok *Building Material* Wood used for building materials. *Canoe Material* Wood used to make canoes. (6:55)
- **Other**—Pomo, Kashaya *Toys & Games* Pieces of bark used by young girls to play dolls. (72:97) Yurok *Cash Crop* Wood used to make canoes and traded to the Karok. (6:55)

Serenoa, Arecaceae
Serenoa repens (Bartr.) Small, Saw Palmetto
- **Food**—Seminole *Fruit* Fruit and trunk used for food. (169:504)
- **Fiber**—Choctaw *Basketry* Stems used in basketry. (as *Serrenoa serru-*

lata 25:13) **Seminole** *Basketry* Leaf stems used to make medicine baskets. (169:160) Plant used to make baskets. *Brushes & Brooms* Plant used to make brushes. *Cordage* Plant used to make rope. (169:504)
- **Other—Seminole** *Cash Crop* Plant used to make tourist goods. *Ceremonial Items* Plant used to make dance fans and rattles. *Fuel* Plant used to make flint and steel punk. *Hunting & Fishing Item* Plant used to make fish drags. *Tools* Plant used to make fire fans. *Toys & Games* Plant used to make dolls. (169:504)

Sesamum, Pedaliaceae

Sesamum orientale L., Sesame
- **Drug—Cherokee** *Antidiarrheal* Decoction of leaves and seeds given for dysentery. *Cathartic* Oil of seed used as a cathartic. *Gynecological Aid* Decoction of leaves and seeds given for flux. *Misc. Disease Remedy* and *Pediatric Aid* Decoction of leaves and seeds given for cholera infantum. (as *Sesamum indicum* 80:25)

Shepherdia, Elaeagnaceae

Shepherdia argentea (Pursh) Nutt., Silver Buffalo Berry
- **Drug—Blackfoot** *Gastrointestinal Aid* Berries eaten for stomach troubles. *Laxative* Berries eaten as a mild laxative. (86:68) **Cheyenne** *Unspecified* Dried, smashed berries used as an ingredient in medicinal mixtures. (83:24) **Dakota** *Ceremonial Medicine* Fruit used occasionally in ceremonial feasts at female puberty rites. (as *Lepargyrea argentea* 70:106) **Navajo** *Febrifuge* Berries taken for fevers. (90:156)
- **Food—Arapaho** *Preserves* Berries used to make jelly. *Winter Use Food* Berries dried for winter use. (121:49) **Blackfoot** *Forage* Berries eaten by buffalo. (97:48) **Cheyenne** *Pie & Pudding* Berries boiled, flour and sugar added, and eaten as a pudding. (83:24) **Dakota** *Dried Food* Fruit dried for winter use. *Fruit* Fruit eaten fresh in season. (as *Lepargyrea argentea* 70:106) **Gosiute** *Fruit* Berries used for food. (39:381) **Lakota** *Dried Food* Berries dried and used for food. *Fruit* Berries eaten fresh. (106:43) Fruits eaten for food. (139:44) **Montana Indian** *Dried Food* Fruits dried and stored for winter use. (82:57) *Fruit* Fruit eaten fresh. (19:23) Fruits eaten fresh. (82:57) *Preserves* Fruit used to make jelly. (19:23) *Sauce & Relish* Fruits used to make a meat flavoring sauce. (82:57) *Winter Use Food* Fruit dried for winter use. (19:23) **Navajo** *Dried Food* Dried fruit used for food. *Fruit* Fresh fruit used for food. (90:156) **Omaha** *Dried Food* Fruit dried for winter use. (as *Lepargyraea argentea* 68:326) Fruit dried for winter use. (as *Lepargyrea argentea* 70:106) *Fruit* Fruit eaten fresh. (as *Lepargyraea argentea* 68:326) Fruit eaten fresh in season. (as *Lepargyrea argentea* 70:106) **Paiute** *Dried Food* Berries cooked, dried, boiled, drained, crushed, and used for food. *Fruit* Fruits eaten fresh. *Pie & Pudding* Berries, flour, and sugar mixed and eaten as a pudding. (104:100) **Paiute, Northern** *Dried Food* Berries dried for winter use. *Porridge* Berries dried, mashed, and eaten like a mush. (59:50) **Pawnee** *Dried Food* Fruit dried for winter use. *Fruit* Fruit eaten fresh in season. **Ponca** *Dried Food* Fruit dried for winter use. *Fruit* Fruit eaten fresh in season. (as *Lepargyrea argentea* 70:106) **Thompson** *Fruit* Scarlet, sour fruits eaten fresh. *Winter Use Food* Scarlet, sour fruits preserved for winter use. (as *Lepargyrea argentea* 164:489) **Ute** *Fruit* Berries formerly used for food. (38:36) **Winnebago** *Dried Food* Fruit dried for winter use. *Fruit* Fruit eaten fresh in season. (as *Lepargyrea argentea* 70:106)
- **Dye—Blackfoot** *Red* Berries used to make a red dye. (97:48)

Shepherdia canadensis (L.) Nutt., Russet Buffalo Berry
- **Drug—Algonquin, Quebec** *Orthopedic Aid* Poultice of hot-water-softened bark and pin cherry bark used to make broken bone plaster or bandage. *Unspecified* Infusion of bark used for a medicinal tea. (18:203) **Carrier** *Dermatological Aid* Berries, froth, or jelly eaten to reduce injury from mosquito bites. The berries were ripe in June and were eaten at this time to reduce injury from mosquito bites: they seem apparently to feel that the occurrence of berries and mosquitoes simultaneously was a divine indication that one was an antidote for the other. Decoction of branches used as a hair tonic for dyeing and curling the hair. The branches were taken in July, broken up, and boiled for 2 to 3 hours in water, until the liquid looked like brown coffee. The liquid was decanted off and bottled without further treatment but did not deteriorate over a long period of time. To use, the decoction was rubbed into the hair, which was simultaneously curled and dyed a brownish color. *Gynecological Aid* Infusion of roots used in childbirth. (89:12) *Laxative* Decoction of stems taken for constipation. (31:76) *Tuberculosis Remedy* Roots used for tuberculosis. (89:12) **Carrier, Northern** *Cathartic* Decoction of root taken as a purgative. (150:60) **Cree, Woodlands** *Antihemorrhagic* Infusion of roots taken for coughing up blood. *Antirheumatic (External)* Decoction of plant applied externally for aching limbs and arthritis. *Dermatological Aid* Decoction of plant applied externally to head and face sores. *Laxative* Infusion of inner bark, scraped from the stem with a downward motion, used as a laxative. *Venereal Aid* Decoction of stems taken for venereal disease. (109:60) **Eskimo, Inupiat** *Poison* Berries poisonous in great quantities. (98:111) **Flathead** *Eye Medicine* Bark solution used for sore eyes. (82:53) *Tuberculosis Remedy* Roots used for tuberculosis. *Unspecified* Wood used for drug purposes. (89:12) **Gitksan** *Antirheumatic (Internal)* Compound decoction of root taken three times a day for rheumatism. *Cough Medicine* Decoction of bark, branches, and leaves taken for chronic cough. *Venereal Aid* Decoction of roots, stem, and branches used as a wash for gonorrhea. (150:60) **Haisla & Hanaksiala** *Reproductive Aid* Berries given to women in labor to ease the birth. (43:236) **Kutenai** *Eye Medicine* Bark solution used for sore eyes. (82:53) **Okanagan-Colville** *Dermatological Aid* Decoction of branches used as a shampoo. *Hunting Medicine* Leaves chewed and spit out by hunter to stop a wounded deer from running. (188:99) **Salish** *Cathartic* Decoction of twigs taken as a mild physic. *Gastrointestinal Aid* Decoction of twigs taken as a tonic for the stomach. *Tonic* Decoction of twigs taken as a tonic for the stomach. (178:294) **Shuswap** *Cathartic* Decoction of plants taken by young men in training to purge themselves. *Gastrointestinal Aid* Decoction of berries taken for stomach. (123:61) *Tuberculosis Remedy* Roots used for tuberculosis. *Unspecified* Wood used for drug purposes. (89:12) **Sioux** *Antidiarrheal* Bark used for diarrhea. *Cathartic* Root used as a cathartic. *Poison* Fruit very acrid and considered poisonous. (19:24) **Tanana, Upper** *Antirheumatic (External)* Decoction of stems and leaves used as a wash for swellings. *Dermatological Aid* Decoction of stems and leaves used as a wash for cuts. Decoction of whole plant above the ground used as a wash for sores. *Panacea* Decoction of berries taken for sickness. *Tuberculosis Remedy* Decoction of stems and leaves taken for tuberculosis. (102:13) **Thompson** *Cancer Treat-*

ment Decoction of branches and leaves taken in a 1-cupful dose for stomach cancer. (187:209) *Cathartic* Decoction of root or decoction of dried stem and leaves taken as a physic. (as *Lepargyrea canadensis* 164:472) *Dermatological Aid* Berry juice used for acne and boils. (187:209) *Disinfectant* Decoction of stem and leaf taken by hunters and warriors in sweat house to purify. (164:505) *Gastrointestinal Aid* Decoction of bark taken as a tonic for the stomach. (as *Lepargyrea canadensis* 164:472) Berry juice and whip taken in a 1-teaspoon dose for indigestion. Berry juice used for digestive problems. Infusion of fruits and leaves used for ulcers. Berries eaten for stomach cancer. *Heart Medicine* Berry juice and whip taken in a 1-teaspoon dose for heart attacks. *Hypotensive* Decoction of branches and leaves taken in a 1-cupful dose for high blood pressure. Berries eaten for high blood pressure. *Laxative* Decoction of twigs and sticks used as a laxative. *Liver Aid* Berry juice used for gallstones. *Sedative* Infusion of fruits and leaves considered a good sedative. Berry whip said to make one sleepy. (187:209) *Tonic* Decoction of bark taken as a tonic for the stomach. (as *Lepargyrea canadensis* 164:472) Berry jam taken as a tonic. (187:209) *Tuberculosis Remedy* Roots used for tuberculosis. **Umatilla** *Gynecological Aid* Infusion of roots used in childbirth. (89:12)

- **Food**—**Alaska Native** *Dessert* Berries mixed with sugar and water, beaten with hands into foam, and used on desserts like whipped cream. (85:146) **Bella Coola** *Ice Cream* Berries mixed with water, whipped, and eaten as "Indian ice-cream." (184:204) **Blackfoot** *Starvation Food* Bitter berries eaten in lean times. (86:105) **Carrier** *Dried Food* Berries dried for future use. (31:76) *Fruit* Berries used for food. (89:12) *Ice Cream* Berries beaten by hand in a birch basket into Indian ice cream. (31:76) Berries used to make a froth similar to ice cream. The berries were macerated. In this process, it was most essential that all grease be kept away and the utensils be kept perfectly clean. A smooth froth, almost like ice cream of light consistency, was formed, which was edible and to those accustomed to it of good taste. Sugar was added to sweeten. This froth appeared to be formed from the saponins which were admixed with the other components of the fruit. (89:12) *Preserves* Berries used to make jam. (31:76) Berries used to make jelly. (89:12) **Cheyenne** *Preserves* Fruit used to make excellent preserves. (as *Lepargyraea canadensis* 76:181) **Clallam** *Ice Cream* Berries whipped until foamy and eaten as "Indian ice cream." (57:199) **Coeur d'Alene** *Fruit* Berries eaten fresh. (178:90) *Ice Cream* Berries used to make a froth similar to ice cream. (89:12) **Eskimo, Inupiat** *Dessert* Berries, water, and sugar whipped into a foamy dessert. (98:111) **Flathead** *Beverage* Berries, water, sugar, and lemon or vanilla used to make a drink. (19:24) *Ice Cream* Berries used to make a froth similar to ice cream. (89:12) **Haisla & Hanaksiala** *Dried Food* Berries dried for future use. *Ice Cream* Berries whipped into a froth and eaten as "Indian ice cream." *Special Food* Berries served at large gatherings, special occasions, and feasts. (43:236) **Kitasoo** *Ice Cream* Berries whipped into "Indian ice cream." *Winter Use Food* Berries canned for future use. (43:331) **Kwakiutl, Southern** *Fruit* Berries whipped until white and frothy and used for food. (183:282) **Lillooet** *Ice Cream* Berries used to make a froth similar to ice cream. (89:12) **Makah** *Dessert* Berries used to make a frothy dessert. (67:288) Berries whipped into a froth and used as dessert at feasts. (79:41) *Dried Food* Purchased berries dried or canned for storage. *Winter Use Food* Purchased berries dried or canned for storage. (67:288) **Montana Indian** *Dried Food* Berries sun dried and stored for future use. *Fruit* Berries eaten fresh. *Ice Cream* Berries used to make a frothy or foamy "Indian Ice Cream." (82:53) **Nanaimo** *Dessert* Berries crushed in water, beaten to make a froth, and eaten as a favorite dessert. (182:82) **Nitinaht** *Ice Cream* Berries whipped in small amounts of water and eaten as "Indian ice cream" at large feasts. (186:103) **Northwest Indian** *Beverage* Berries made into a foaming drink. (as *Lepargyrea canadensis* 118:17) **Okanagan-Colville** *Ice Cream* Berries used to make "Indian ice cream." (188:99) **Okanagon** *Beverage* Fruits fermented to make an alcoholic drink. *Fruit* Fruits eaten fresh by children. (125:39) *Staple* Berries used as a principal food. (178:239) **Oregon Indian** *Beverage* Berries used to make a foaming drink. (as *Lepargyrea canadensis* 118:20) **Oweekeno** *Fruit* Berries whipped, mixed with sugar, and eaten. *Winter Use Food* Berries canned for future use. (43:93) **Salish, Coast** *Dessert* Berries crushed in water, beaten to make a froth, and eaten as a favorite dessert. (182:82) **Sanpoil** *Ice Cream* Berries used to make a froth similar to ice cream. (89:12) **Shuswap** *Beverage* Canned berry juice used as a beverage during haying time. *Dessert* Berries canned with sugar, mixed with equal amount of water, and whipped into a foam; whipped cream. *Fruit* Berries mixed with timber grass, dried, water added, and beaten to a foam. (123:61) *Ice Cream* Berries used to make a froth similar to ice cream. (89:12) **Spokan** *Fruit* Berries used for food. (178:343) **Tanana, Upper** *Fruit* Raw berries whipped with sugar until frothy. Berries mixed with sugar, fried, and eaten. (102:13) **Thompson** *Beverage* Fruits fermented to make an alcoholic drink. (125:39) Berries squeezed through a rice bag or some other straining cloth and made into juice. *Bread & Cake* Soapberries dried on mats and formed into cakes. The berries were gathered in the summer, but were not hand picked because they were too soft. A clean mat was placed underneath the bush, then a branch laden with fruit was held and hit with a stick until the fruit fell off. The ripe berries were then placed in a basket, heated with hot rocks, and spread out on mats or on a layer of "timbergrass" set on a scaffolding and allowed to dry. A small fire was lit beneath so that the smoke would drive away the flies. The dried soapberry cakes were then broken off, placed in a birch bark basket with water, and "swished" with a whisk of maple bark tied to a stick. The mixture was originally sweetened with the "white" variety of saskatoon berries that were dried and soaked in water to reconstitute them. More recently, sugar was added to the whip to sweeten it. The sweetened froth was served in small containers, first to the men and then to the women, as a sort of dessert or confection. It was said that the soapberries must never come into contact with grease or oil or the berries would not whip. One informant said that special containers were used for the preparation of soapberries, not for cooking or any other purpose, so that the berries could be kept free of grease. It was said that pregnant women should never eat the soapberry whip. *Candy* Berries made into juice and used to make a frothy confection. *Forage* Berries eaten by bears. (187:209) *Fruit* Fruits eaten fresh by children. (125:39) *Preserves* Berries made into jam. (187:209) **Thompson, Lower** *Ice Cream* Berries used to make a froth similar to ice cream. (89:12)

- **Other**—**Gitksan** *Cash Crop* Berries smoke dried and used for trade. (43:236) Berries used for trade. **Nisga** *Cash Crop* Berries used for trade. (43:331) **Nuxalkmc** *Cash Crop* Berries smoke dried and used for trade. (43:236) **Okanagan-Colville** *Good Luck Charm* Infusion of branches taken to bring good luck. (188:99) **Shuswap** *Ceremonial Items* Decoction or infusion of berries taken during purification rites. (89:12) **Thompson** *Cash Crop* Berries used as an important trade item. *Good Luck Charm* Decoction of leafless branches taken by hunters to clean out their insides for good luck. The decoction was said to clean the hunters' insides. The hunters also used the decoction to wash themselves and their gear. *Hunting & Fishing Item* Decoction of leafless branches used as a wash for hunting gear. (187:209) **Tsimshian** *Cash Crop* Berries smoke dried and traded. (43:331) **Tsimshian, Coast** *Cash Crop* Berries smoke dried and used for trade. (43:236)

Shepherdia rotundifolia Parry, Roundleaf Buffalo Berry
- **Drug**—**Havasupai** *Poison* Dust from the underside of the leaves said to make the eyes sore and to cause blindness. (197:234) **Navajo, Kayenta** *Analgesic* Ash used as lotion for headaches. *Ceremonial Medicine* Plant used as a Plumeway emetic. *Dermatological Aid* Ash used as lotion to heal navels. *Emetic* Plant used as a Plumeway emetic. *Pediatric Aid* Ash used as lotion to heal navels. *Throat Aid* Ash used as lotion for sore throat. *Toothache Remedy* Ash used as lotion for toothaches. (205:32)

***Shepherdia* sp.**, Soap Berry
- **Food**—**Thompson** *Preserves* Berries collected in large quantities and cured. (178:237)

Sherardia, Rubiaceae

Sherardia arvensis L., Blue Fieldmadder
- **Dye**—**Cherokee** *Red* Used to make a red or rose dye. (as *Sheradia arvensis* 80:44)

Shinnersoseris, Asteraceae

Shinnersoseris rostrata (Gray) S. Tomb, Beaked Skeletonweed
- **Drug**—**Navajo** *Sedative* Plant smoked as a sedative. (as *Lygodesmia rostrata* 55:88)
- **Other**—**Navajo** *Ceremonial Items* Plant offered to the antelope at the Corral Chant. (as *Lygodesmia rostrata* 55:88)

Sicyos, Cucurbitaceae

Sicyos angulatus L., Oneseed Burr Cucumber
- **Drug**—**Iroquois** *Venereal Aid* Decoction of vine taken for venereal disease. *Veterinary Aid* Compound plants mixed with cow's feed for difficult birth of a calf. (87:451)

Sida, Malvaceae

***Sida* sp.**, Ilima
- **Drug**—**Hawaiian** *Gynecological Aid* Flowers and other plants chewed, squeezed, and resulting liquid placed in the vagina for womb troubles. *Laxative* and *Pediatric Aid* Chewed flowers given to infants and children as a laxative. *Respiratory Aid* Flowers, shoots, root bark, and other plants pounded, squeezed, and resulting liquid taken for asthma. *Strengthener* Bark and other plants pounded, squeezed, and resulting liquid taken for general debility. (2:26)
- **Other**—**Hawaiian** *Decorations* Plant used to make wreaths. (2:26)

Sidalcea, Malvaceae

Sidalcea malviflora (DC.) Gray ex Benth., Dwarf Checkermallow
- **Food**—**Luiseño** *Vegetable* Plant used as greens. (155:231) **Yana** *Spice* Dried, mashed leaves used to flavor black manzanita berries. (147:251)

Sidalcea neomexicana Gray, New Mexico Checkermallow
- **Drug**—**Navajo, Ramah** *Internal Medicine* Cold infusion of plant taken for internal injury. (191:36)

Sideroxylon, Sapotaceae

Sideroxylon foetidissimum Jacq., False Mastic
- **Drug**—**Seminole** *Love Medicine* Decoction of wood ashes placed on the tongue to cleanse the body and strengthen the marriage. (169:250)
- **Food**—**Seminole** *Unspecified* Plant used for food. (169:494)
- **Fiber**—**Seminole** *Caulking Material* Plant used to make mortar. (169:494)
- **Other**—**Seminole** *Cooking Tools* Plant used to make food paddles. *Tools* Plant used to make pestles. (169:494)

Sideroxylon lanuginosum Michx., Gum Bully
- **Food**—**Kiowa** *Candy* Outer bark yields a mucilaginous substance used as chewing gum. (192:46)

Sideroxylon lanuginosum Michx. **ssp. *lanuginosum***, Gum Bumelia
- **Food**—**Kiowa** *Candy* Outer bark yields a mucilaginous substance used as chewing gum. (as *Bumelia lanuginosa* 192:46)

Sideroxylon salicifolium (L.) Lam., White Bully
- **Other**—**Seminole** *Cooking Tools* Plant used to make food paddles. *Hunting & Fishing Item* Plant used to make arrows and bows. (as *Dipholis salicifolia* 169:467)

***Sideroxylon* sp.**, A-laa
- **Drug**—**Hawaiian** *Adjuvant* Bark, leaves, and wood mixed with other herbs to produce strong remedies. *Febrifuge* Bark, leaves, other ingredients, and water used as a wash and taken for strong fevers. *Strengthener* Bark, leaves, other ingredients, and water used as a wash and taken for general debility. (2:12)

Silene, Caryophyllaceae

Silene acaulis (L.) Jacq., Moss Campion
- **Drug**—**Gosiute** *Gastrointestinal Aid* and *Pediatric Aid* Plant used for children with colic. (39:381)

Silene acaulis* var. *exscapa (All.) DC., Moss Campion
- **Food**—**Eskimo, Inuktitut** *Unspecified* Raw root skins used for food. (202:182)

Silene campanulata S. Wats., Red Mountain Catchfly
- **Drug**—**Karok** *Pediatric Aid* and *Unspecified* Plant used as medicine for babies. (148:383)

Silene douglasii Hook. **var. *douglasii***, Douglas's Campion
- **Drug**—**Gosiute** *Analgesic, Emetic,* and *Gastrointestinal Aid* Infusion of roots taken as an emetic for stomach trouble pain. (as *S. multicaulis* 39:350) Warm infusion of pounded plant used as an emetic for

stomach pain. *Veterinary Aid* Plant used as a horse medicine. (as *S. multicaulis* 39:381) **Navajo, Ramah** *Dermatological Aid* and *Veterinary Aid* Cold infusion of plant used as a lotion for coyote bite on humans, sheep, or horses. (as *S. pringlei* 191:27)

Silene drummondii Hook. **var.** *drummondii*, Drummond's Campion
- *Drug*—**Navajo, Ramah** *Panacea* Root used as a "life medicine." (as *Lychnis drummondii* 191:27)

Silene laciniata Cav., Mexican Campion
- *Drug*—**Keres, Western** *Dermatological Aid* Crushed plant rubbed on ant bites. *Other* Infusion of plant used as a reducing medicine. (171:69)

Silene laciniata ssp. *greggii* (Gray) C. L. Hitchc. & Maguire, Gregg's Campion
- *Drug*—**Navajo, Ramah** *Burn Dressing* Poultice of leaves applied to burns. *Dermatological Aid* Decoction of root taken for mad dog or mad coyote bite. (191:27)

Silene latifolia ssp. *alba* (P. Mill.) Greuter & Burdet, Bladder Campion
- *Drug*—**Ojibwa** *Cathartic* Infusion of root used as a physic. (as *Lychnis alba* 153:361)

Silene menziesii Hook., Menzies's Campion
- *Drug*—**Okanagan-Colville** *Eye Medicine* Infusion of pounded roots used as eye drops for cataracts. (188:95)
- *Other*—**Gosiute** *Smoke Plant* Dried leaves powdered and formerly used to smoke as a tobacco. (39:381)

Silene noctiflora L., Nightflowering Silene
- *Drug*—**Menominee** *Unspecified* Plant used as a medicine. (151:28) **Navajo, Ramah** *Dermatological Aid* Poultice of leaves applied to prairie dog bite. (191:27)

Silene scouleri Hook., Scouler's Campion
- *Drug*—**Gosiute** *Analgesic*, *Emetic*, and *Gastrointestinal Aid* Warm infusion of pounded plant used as an emetic for stomach pain. *Veterinary Aid* Plant used as a horse medicine. (39:381)

Silene stellata (L.) Ait. f., Widowsfrill
- *Drug*—**Meskwaki** *Dermatological Aid* Poultice of root applied to dry up swellings that discharge pus. (152:208)

Silphium, Asteraceae

Silphium compositum Michx., Kidneyleaf Rosinweed
- *Drug*—**Cherokee** *Gynecological Aid* and *Stimulant* Used as a strong stimulant for whites and taken by weakly females. (80:53)

Silphium integrifolium Michx., Wholeleaf Rosinweed
- *Drug*—**Meskwaki** *Analgesic* and *Herbal Steam* Root steam directed to crippled area and used for pain. *Kidney Aid* Root used for "one with kidney trouble or who was crippled." *Orthopedic Aid* Root steam directed to crippled area and used for pain. *Urinary Aid* Infusion of leaves used for bladder troubles. (152:216)

Silphium laciniatum L., Compass Plant
- *Drug*—**Dakota** *Veterinary Aid* Decoction of pounded root used as a vermifuge for horses. *Witchcraft Medicine* Dried root burned during storms to act as a charm against lightning. (70:132) **Meskwaki** *Emetic* Decoction of smaller roots taken as an emetic. (152:216, 217) **Omaha** *Veterinary Aid* Roots given to horses as a tonic. (68:335) Decoction of root given to horses as a tonic. *Witchcraft Medicine* Dried root burned during storms to act as a charm against lightning. **Pawnee** *Tonic* Decoction of root taken for general debility. *Witchcraft Medicine* Dried root burned during storms to act as a charm against lightning. **Ponca** *Veterinary Aid* Decoction of root given to horses as a tonic. *Witchcraft Medicine* Dried root burned during storms to act as a charm against lightning. **Winnebago** *Witchcraft Medicine* Dried root burned during storms to act as a charm against lightning. (70:132)
- *Food*—**Dakota** *Candy* Gum from upper part of stem used as chewing gum by children. **Omaha** *Candy* Gum from upper part of stem used as chewing gum by children. **Pawnee** *Candy* Gum from upper part of stem used as chewing gum by children. **Ponca** *Candy* Gum from upper part of stem used as chewing gum by children. **Winnebago** *Candy* Gum from upper part of stem used as chewing gum by children. (70:132)

Silphium perfoliatum L., Cup Plant
- *Drug*—**Chippewa** *Abortifacient* Simple or compound decoction of root taken for "stoppage of periods." (53:358) *Analgesic* Decoction of root taken for back and chest pain. *Antihemorrhagic* Decoction of root taken for lung hemorrhage. (53:340) *Hemostat* Poultice of moistened, dried root applied to wounds as a styptic. (53:356) *Orthopedic Aid* Decoction of root taken for back pain. *Pulmonary Aid* Decoction of root taken for chest pain. (53:340) **Iroquois** *Emetic* Decoction of roots taken as an emetic. *Orthopedic Aid* Decoction of roots used as face wash for paralysis. *Pediatric Aid* Burned root soot placed on child's cheek to prevent them from seeing ghosts. *Witchcraft Medicine* Burned root soot placed on cheek to prevent sickness caused by the dead. (87:468) **Meskwaki** *Antiemetic* Root used to "alleviate the vomiting of pregnancy." *Gynecological Aid* Infusion of root taken by women to prevent premature birth. Root used to reduce profuse menstruation and as an antiemetic during pregnancy. (152:217) **Ojibwa** *Antirheumatic (Internal)* Infusion of root taken for lumbago and other rheumatic back pains. *Gastrointestinal Aid* Plant used for stomach trouble. *Hemostat* Plant used for hemorrhage. (153:365) **Omaha** *Analgesic* Smoke from burning plant inhaled for pain. (68:334) Root used in smoke treatment for neuralgia. *Antirheumatic (Internal)* Root used in smoke treatment for rheumatism. (70:132) *Cold Remedy* Smoke from burning plant inhaled for head colds. (68:334) Root used in smoke treatment for head cold. *Herbal Steam* Rootstock used in the vapor bath. **Ponca** *Analgesic* Root used in smoke treatment for neuralgia. *Antirheumatic (Internal)* Root used in smoke treatment for rheumatism. *Cold Remedy* Root used in smoke treatment for head cold. *Herbal Steam* Rootstock used in the vapor bath. **Winnebago** *Analgesic* Root used in smoke treatment for neuralgia. *Antirheumatic (Internal)* Root used in smoke treatment for rheumatism. *Ceremonial Medicine* Decoction of root used as a ceremonial emetic. *Cold Remedy* Root used in smoke treatment for head cold. *Emetic* Decoction of root used as a ceremonial emetic. *Herbal Steam* Rootstock used in the vapor bath. (70:132)

Simaroubaceae, see *Castela*

Simmondsiaceae, see *Simmondsia*

Simmondsia, Simmondsiaceae
Simmondsia chinensis (Link) Schneid., Jojoba
- *Drug*—**Papago** *Dermatological Aid* Poultice of dried and pulverized nuts applied to sores. Poultice of parched, pulverized nuts applied to sores. (as *S. californica* 34:65) **Yavapai** *Cathartic* Plant yielded oily food with cathartic qualities. *Dermatological Aid* Parched, charred berry charcoal rubbed on sores. (as *S. californica* 63:211)
- *Food*—**Cahuilla** *Beverage* Seeds eaten fresh or ground into powder and used to make a coffee-like beverage. (15:139) **Coahuilla** *Beverage* Ground nut meal boiled into a "coffee." (as *S. californica* 13:74) **Cocopa** *Bread & Cake* Kernels molded into oily cake, boiled, and eaten. *Unspecified* Nuts cleaned, winnowed, shelled, and eaten. (37:188) **Papago** *Unspecified* Nuts eaten fresh from the shell. (as *S. californica* 34:19) **Yavapai** *Preserves* Berries parched and ground to consistency of peanut butter. (as *S. californica* 65:258)

Sinapis, Brassicaceae
Sinapis alba L., White Mustard
- *Drug*—**Cherokee** *Dietary Aid* Taken to increase appetite. *Febrifuge* Taken for fever and "nervous fever." *Kidney Aid* Taken for "dropsy." *Misc. Disease Remedy* Taken for "ague." *Orthopedic Aid* Taken for palsy. *Pulmonary Aid* Given for "phthisic" or asthma. *Respiratory Aid* Used as a poultice for croup. *Stimulant* Taken as a stimulant. *Tonic* Taken as a tonic. (as *Brassica hirta* 80:46) **Hoh** *Unspecified* Plants used for medicine. (as *Brassica alba* 137:62) **Micmac** *Tuberculosis Remedy* Parts of plant used for tuberculosis of lungs. (40:62) **Quileute** *Unspecified* Plants used for medicine. (as *Brassica alba* 137:62)
- *Food*—**Hoh** *Spice* Used for flavoring. *Vegetable* Plants eaten as greens. **Quileute** *Spice* Used for flavoring. *Vegetable* Plants eaten as greens. (as *Brassica alba* 137:62)

Sinapis arvensis L., Charlock Mustard
- *Drug*—**Navajo, Ramah** *Ceremonial Medicine, Disinfectant*, and *Emetic* Plant used as a ceremonial emetic for "deer infection." (as *Brassica kaber* 191:28)

Sisymbrium, Brassicaceae
Sisymbrium altissimum L., Tall Tumblemustard
- *Drug*—**Navajo, Ramah** Plant probably used in emetics. (191:29)
- *Food*—**Navajo** *Porridge* Seeds used, with goat's milk, to make a mush. (55:50)

Sisymbrium irio L., London Rocket
- *Drug*—**Pima** *Eye Medicine* Dried seeds placed under the lids of sore eyes to cause weeping. (47:84)
- *Food*—**Cahuilla** *Vegetable* Immature leaves boiled or fried and used for greens. (15:140) **Mohave** *Starvation Food* Young shoots roasted and eaten as a famine food. (37:201) **Pima** *Porridge* Seeds formerly parched, ground, water added, and eaten as a gruel. *Winter Use Food* Seeds stored and used as a winter food. (47:84) **Pima, Gila River** *Beverage* Seeds mixed with water to make a drink. *Porridge* Seeds used to make a mucilaginous mass and eaten. *Staple* Seeds ground, parched, and used to make pinole. (133:5) *Unspecified* Seeds mixed with water and eaten. (133:7)

Sisymbrium officinale (L.) Scop., Hedge Mustard
- *Drug*—**Cherokee** *Pulmonary Aid* Used as a poultice for croup. (80:46) **Iroquois** *Veterinary Aid* Compound decoction of plants mixed with feed for horses with cramps. (87:341)
- *Food*—**Cherokee** *Vegetable* Leaves cooked and eaten as salad greens. (80:46) **Navajo** *Forage* Plant used by horses for forage. (55:50) *Porridge* Seeds ground and eaten as a mush or gruel. (165:223) *Soup* Seeds parched, ground into meal, and made into soup or stew. (32:22) Parched, ground seeds used to make soup or stew. (55:50) **Tubatulabal** *Unspecified* Leaves fried in grease and eaten. (193:16)

Sisyrinchium, Iridaceae
Sisyrinchium albidum Raf., White Blueeyed Grass
- *Drug*—**Menominee** *Antidote* Root used for poisonous bite inflicted by horse which has eaten root. *Veterinary Aid* Roots mixed with horse feed to make horse sleek, vicious, and bite poisonous. *Witchcraft Medicine* Plant kept in house or pocket as a charm to ward off snakes. (151:38)

Sisyrinchium angustifolium P. Mill., Narrowleaf Blueeyed Grass
- *Drug*—**Cherokee** *Antidiarrheal* Infusion of root given to children for diarrhea. *Gastrointestinal Aid* Eaten as cooked greens for "regular bowels." *Pediatric Aid* Infusion of root given to children for diarrhea. (as *S. augustifolium* 80:26) **Iroquois** *Laxative* Decoction of roots and stalks taken before morning meals for constipation. *Other* Compound with plant used for "summer complaint." (as *S. graminoides* 87:288) **Mahuna** *Anthelmintic* Infusion of plant taken for stomach worms. *Gastrointestinal Aid* Infusion of plant taken for stomach troubles and stomach worms. (140:6)
- *Food*—**Cherokee** *Vegetable* Mixed into other greens and eaten. (as *S. augustifolium* 204:252)

Sisyrinchium atlanticum Bickn., Eastern Blueeyed Grass
- *Drug*—**Menominee** *Gynecological Aid* Compound decoction of plant taken to help expel afterbirth. (54:133)

Sisyrinchium bellum S. Wats., Western Blueeyed Grass
- *Drug*—**Costanoan** *Febrifuge* Decoction of plant taken for chills. *Gastrointestinal Aid* Decoction of plant taken for stomachaches. (21:29) **Luiseño** *Cathartic* Roots used as a purgative. (155:233) **Pomo, Kashaya** *Gastrointestinal Aid* Infusion of washed roots taken for upset stomach, heartburn, and ulcers. *Respiratory Aid* Infusion of washed roots taken for asthma. (72:24)

Sisyrinchium campestre Bickn., Prairie Blueeyed Grass
- *Drug*—**Meskwaki** *Analgesic* Plant used for cramps and decoction used for hay fever. *Gynecological Aid* Compound decoction of plant base taken by women for injured womb. *Respiratory Aid* Decoction of whole plant used for hay fever. (152:224)

Sisyrinchium montanum Greene, Mountain Blueeyed Grass
- *Drug*—**Iroquois** *Cathartic* Used as a physic for old people. *Misc. Disease Remedy* Decoction taken for fevers such as malaria and scarlet fever but not typhoid. *Poison* "Feared it was poison." (87:288)

Sisyrinchium mucronatum Michx., Needletip Blueeyed Grass
- **Drug**—Navajo *Nose Medicine* Plant used for nose troubles. *Throat Aid* Plant used for throat troubles. (55:37)

Sisyrinchium nashii Bickn., Nash's Blueeyed Grass
- **Drug**—Seminole *Analgesic* Roots used with a song or spell as an analgesic. (as *S. fibrosum* 169:167) Infusion of roots used for moving sickness: moving pain in the waist region. (as *S. fibrosum* 169:285)

Sitanion, Poaceae

***Sitanion* sp.**, Wild Rye
- **Drug**—Miwok *Witchcraft Medicine* Used dry or green to strike patient with, before and after the shaman sucks the cause of the illness out of the patient. (12:173)

Sium, Apiaceae

Sium suave Walt., Hemlock Waterparsnip
- **Drug**—Iroquois *Analgesic* Infusion of smashed roots applied as poultice for pain from broken limb. *Anticonvulsive* Compound decoction of roots taken by women for epilepsy. *Orthopedic Aid* Infusion of smashed roots applied as poultice for pain from broken limb. Poultice of fried turnips applied to sprained muscles or out of joint limbs. (87:399) **Lakota** *Gastrointestinal Aid* Roots used for the stomach. (139:33) **Ojibwa** *Hunting Medicine* Seeds smoked over a fire to drive away and blind evil spirit that steals away one's hunting luck. (as *S. cicutaefolium* 153:432) **Shuswap** *Poison* White flowers considered poisonous. (123:57)
- **Food**—Algonquin, Quebec *Unspecified* Root used for food. (18:101) **Bella Coola** *Unspecified* Tubers eaten. **Carrier** *Unspecified* Tubers eaten. (184:200) **Cree, Woodlands** *Unspecified* Roots collected in early spring or late fall, roasted, fried, or eaten raw. (109:61) **Klamath** *Sauce & Relish* Herbage eaten as a relish. (as *S. cicutaefolium* 45:102) **Montana Indian** *Sauce & Relish* Herbage has an aromatic flavor and eaten as a relish. (as *S. cicutaefolium* 19:24) **Okanagan-Colville** *Unspecified* Roots eaten raw in the early spring. (188:71) **Okanagon** *Staple* Roots used as a principal food. (as *S. lineare* 178:238) **Shuswap** *Unspecified* Crispy, delicious roots eaten raw or steamed. (123:57) **Thompson** *Dried Food* Roots dug in the spring and fall, washed, pit cooked, and dried for later use. (187:159) *Forage* Rootstocks or rhizomes eaten by cattle. *Unspecified* Rootstocks or rhizomes eaten. (as *S. laeve* 164:482) Roots sometimes eaten raw. (187:159)
- **Other**—Lakota *Toys & Games* Stems used by children for whistles. (139:33)

Smallanthus, Asteraceae

Smallanthus uvedalia (L.) Mackenzie ex Small, Hairy Leafcup
- **Drug**—Cherokee *Burn Dressing* Bruised root used on burns. *Dermatological Aid* Bruised root used as a salve for cuts. Poultice of root used for "inflammations," and bruised root in hog's lard used for itch. (as *Polymnia uvedalia* 80:42) *Emetic* Decoction of roots taken to vomit to expel the afterbirth. (as *Polymnia uvedalia* 177:63) *Gynecological Aid* Compound decoction used "for expelling afterbirth." (as *Polymnia uvedalia* 80:42) Decoction of roots taken to vomit to expel the afterbirth. (as *Polymnia uvedalia* 177:63) *Other* Plant used for rheumatism and "white swelling." (as *Polymnia uvedalia* 80:42) **Iroquois** *Analgesic* Compound infusion of stalks and roots taken for back pain. *Antiemetic* Compound infusion of stalks and roots taken for vomiting. (as *Polymnia uvedalia* 87:467) *Disinfectant* Dried roots used as fumigant for ghosts. (as *Polymnia uvedalia* 87:468) *Febrifuge* Infusion of roots taken for fevers. *Kidney Aid* Compound infusion of stalks and roots taken for kidney troubles. *Orthopedic Aid* Compound infusion of stalks and roots taken for back pain. *Sedative* Dried root smoke used as rub to sleep without being disturbed by ghosts. (as *Polymnia uvedalia* 87:467)

Smilacaceae, see *Smilax*

Smilacina, Liliaceae

***Smilacina* sp.**, False Solomon's Seal
- **Drug**—Yana *Dermatological Aid* Poultice of pounded roots applied to swellings or boils. (147:253)

Smilax, Smilacaceae

Smilax auriculata Walt., Earleaf Greenbrier
- **Drug**—Seminole *Other* Complex infusion of buds, or buds and leaves, taken for chronic conditions. (169:272)

Smilax bona-nox L., Saw Greenbrier
- **Drug**—Choctaw *Tonic* Decoction of stems taken as a general tonic. (177:8) **Creek** *Other* Plant moistened and rubbed on face to make one young. (172:667) **Houma** *Urinary Aid* Decoction of root taken for urinary disturbances. (158:58, 59)
- **Food**—Choctaw *Bread & Cake* Tuberous roots dried, ground into flour, and used to make bread. **Houma** *Bread & Cake* Tuberous roots dried, ground into flour, and used to make bread. (158:58)
- **Other**—Comanche *Smoking Tools* Leaves used as cigarette wrappers. (29:524)

Smilax californica (A. DC.) Gray, California Greenbrier
- **Fiber**—Round Valley Indian *Basketry* Long, trailing limbs used to make baskets. **Ukiah** *Basketry* Long, trailing limbs used to make baskets. (41:329)

Smilax glauca Walt., Cat Greenbrier
- **Drug**—Cherokee *Analgesic* Rubbed in brier scratches for local pains. *Antirheumatic (Internal)* Used for rheumatism. *Burn Dressing* Parched and powdered leaves used for "scalds." *Dermatological Aid* Powdered and beaten leaves put on "galled places" and wilted leaves put on boils. *Gastrointestinal Aid* Infusion taken for stomach trouble. *Gynecological Aid* Compound decoction used to aid in expelling afterbirth. (80:37) Decoction of roots taken to cause discharge of the afterbirth. (177:8) *Orthopedic Aid* Rubbed in brier scratches for muscular cramps and twitching. *Other* Compound infusion of bark taken for "bad disease." (80:37)
- **Food**—Cherokee *Unspecified* Roots used for food. (80:37)

Smilax herbacea L., Smooth Carrionflower
- **Drug**—Cherokee *Analgesic* Rubbed in brier scratches for local pains. *Antirheumatic (Internal)* Used for rheumatism. *Burn Dressing* Parched and powdered leaves used for "scalds." *Dermatological Aid* Powdered and beaten leaves put on "galled places" and wilted leaves put on boils. *Gastrointestinal Aid* Infusion taken for stomach trouble. *Gynecological Aid* Compound decoction used to aid in expelling afterbirth. *Orthopedic Aid* Rubbed in brier scratches for muscular

cramps and twitching. *Other* Compound infusion of bark taken for "bad disease." (80:37) **Chippewa** *Analgesic* Decoction of root taken for back pain. *Cathartic* Compound decoction of root taken as a physic. *Kidney Aid* and *Orthopedic Aid* Decoction of root taken for kidney trouble. (53:346) **Iroquois** *Antirheumatic (External)* Compound decoction used as a wash and steam bath for rheumatism. *Dermatological Aid* Powdered root used as a "deodorant." (87:287) *Gastrointestinal Aid* Decoction of roots taken by the elderly for stomach troubles. *Psychological Aid* Compound decoction of plant taken for "loss of senses during menses." (87:286) **Ojibwa** *Pulmonary Aid* Root used for lung troubles. (153:374) **Omaha** *Throat Aid* Fruits used for hoarseness. (70:71)
- **Food**—**Cherokee** *Unspecified* Roots used for food. (80:37) **Meskwaki** *Fruit* Fruit of the carrion flower used as food. (152:262) **Omaha** *Fruit* Fruits eaten for their pleasant taste. (70:71)

Smilax laurifolia L., Laurel Greenbrier
- **Drug**—**Cherokee** *Burn Dressing* Compound of root bark used as a wash for burns. *Dermatological Aid* Astringent and slightly tonic root bark used as a wash for burns and sores. Compound of root bark used for pox and as a wash for sores. *Tonic* Root bark astringent and slightly tonic. (80:24) **Houma** *Urinary Aid* Decoction of root taken for urinary disturbances. (158:58, 59)
- **Food**—**Choctaw** *Bread & Cake* Pounded roots made into cakes and fried in grease. (25:8) Tuberous roots dried, ground into flour, and used to make bread. **Houma** *Bread & Cake* Tuberous roots dried, ground into flour, and used to make bread. (158:58)
- **Dye**—**Seminole** *Unspecified* Plant used to make buckskin dye. (169:488)

Smilax pseudochina L., Bamboovine
- **Drug**—**Cherokee** *Analgesic* Rubbed in brier scratches for local pains. *Antirheumatic (Internal)* Used for rheumatism. *Burn Dressing* Parched and powdered leaves used for "scalds." *Dermatological Aid* Powdered and beaten leaves put on "galled places" and wilted leaves put on boils. *Gastrointestinal Aid* Infusion taken for stomach trouble. *Gynecological Aid* Compound decoction used to aid in expelling afterbirth. *Orthopedic Aid* Rubbed in brier scratches for muscular cramps and twitching. *Other* Compound infusion of bark taken for "bad disease." (as *S. tamnifolia* 80:37)
- **Food**—**Cherokee** *Unspecified* Roots used for food. (as *S. tamnifolia* 80:37)

Smilax rotundifolia L., Roundleaf Greenbrier
- **Drug**—**Cherokee** *Analgesic* Rubbed in brier scratches for local pains. *Antirheumatic (Internal)* Used for rheumatism. *Burn Dressing* Parched and powdered leaves used for "scalds." *Dermatological Aid* Powdered and beaten leaves put on "galled places" and wilted leaves put on boils. *Gastrointestinal Aid* Infusion taken for stomach trouble. *Gynecological Aid* Compound decoction used to aid in expelling afterbirth. *Orthopedic Aid* Rubbed in brier scratches for muscular cramps and twitching. *Other* Compound infusion of bark taken for "bad disease." (80:37) **Koasati** *Analgesic* Plant splints used to scratch the back for headaches. (177:8)
- **Food**—**Cherokee** *Unspecified* Roots used for food. (80:37)
- **Other**—**Kiowa** *Protection* Plant used to block the path of pursuers because of its sharp thorns and its scandent nature. *Toys & Games* Plant used in an old hide and seek game. (192:18)

Smilax sp., Catbriar
- **Drug**—**Cree** *Venereal Aid* Rhizomes, rootlets, and false indigo used for syphilis. (16:495) **Creek** *Dermatological Aid* Decoction of pounded root poured on ulcers, especially on the legs. (172:661) Decoction of pounded roots used as a wash for leg ulcers. (177:7) **Rappahannock** *Unspecified* "An ingredient of a medicine made after diagnosis." (161:31)

Smilax tamnoides L., Bristly Greenbrier
- **Drug**—**Choctaw** *Tonic* Decoction of stems taken as a general tonic. (25:23) **Iroquois** *Eye Medicine* Poultice of bark applied to eyes to "draw out foreign substances." *Witchcraft Medicine* Plant used to make doll to be used "to kill a woman who is using you badly." (as *S. hispida* 87:287)
- **Other**—**Chippewa** *Malicious Magic* Vine boiled with pisabik, a type of powdered stone, and used as malicious magic. The mischief maker sprinkled the mixture upon the bed of a couple, thereby causing them to quarrel and separate. It was thought that the prickly character of the stem was transferred to the bed and irritated the couple, causing them to become ill-disposed toward each other. (as *S. hispida* 71:126)

Solanaceae, see *Capsicum, Chamaesaracha, Datura, Lycium, Lycopersicon, Nicotiana, Physalis, Quincula, Solanum*

Solanum, Solanaceae

Solanum carolinense L., Carolina Horsenettle
- **Drug**—**Cherokee** *Anthelmintic* Infusion of leaf used for worms. *Dermatological Aid* Used "for ulcers and proud flesh" and wilted plant used on poison ivy. *Oral Aid* and *Pediatric Aid* Roots strung around baby's neck for teething. *Throat Aid* Infusion of seed gargled for sore throat and taken for goiter. *Veterinary Aid* Cut berries fried in grease and grease used for dogs with mange. (80:46)
- **Other**—**Cherokee** *Insecticide* Crushed leaves in sweet milk used to kill flies. (80:46)

Solanum donianum Walp., Mullein Nightshade
- **Drug**—**Seminole** *Analgesic* Decoction of root bark poured onto the head for headaches. (as *S. verbascifolium* 169:278)
- **Other**—**Seminole** *Hunting & Fishing Item* Plant used to make arrows. (as *S. blodgettii* 169:470)

Solanum douglasii Dunal, Greenspot Nightshade
- **Drug**—**Cahuilla** *Eye Medicine* Berry juice used for sore or infected eyes, pink eye, and eye strain, and to improve vision in older people. *Poison* Whole plant considered poisonous. (15:140) **Luiseño** *Eye Medicine* Berry juice used for inflamed eyes. (155:229)
- **Food**—**Luiseño** *Vegetable* Leaves used for greens. (155:229)
- **Dye**—**Cahuilla** *Unspecified* Dark berries used as a dye. (15:140)
- **Other**—**Luiseño** *Decorations* Berry juice used for tattooing. (155:229)

Solanum dulcamara L., Climbing Nightshade
- **Drug**—**Delaware, Oklahoma** *Dermatological Aid* Plant used to make a salve for unspecified use. *Febrifuge* Compound containing root used for fever. (175:80) **Iroquois** *Gastrointestinal Aid* Compound de-

coction of plants taken for biliousness. (87:431) **Malecite** *Antiemetic* Infusion of roots used for nausea. (116:253) **Micmac** *Antiemetic* Roots used for nausea. (40:62) **Nootka** *Gastrointestinal Aid* Plant used for "derangement" of stomach and bowels. (170:81)

Solanum elaeagnifolium Cav., Silverleaf Nightshade
- *Drug*—**Apache, White Mountain** *Unspecified* Plant used for medicinal purposes. (136:160) **Isleta** *Laxative* Raw seedpods eaten or boiled into a syrup and taken as a laxative. (100:43) **Keresan** *Gynecological Aid* Infusion of plant taken by nursing mothers to sustain milk flow. (198:562) **Navajo** *Eye Medicine* Plant used for sore eyes. (55:75) *Nose Medicine* Plant used for nose troubles. *Throat Aid* Plant used for throat troubles. (55:97) **Pima** *Cold Remedy* Crushed, dried berries used for colds. (47:88) **Zuni** *Snakebite Remedy* Fresh or dried root chewed by medicine man before sucking snakebite and poultice applied to wound. *Toothache Remedy* Fruit chewed over sore tooth. (27:378) Chewed root placed in cavity of aching tooth. (166:60)
- *Food*—**Cochiti** *Substitution Food* Fruits used as a substitute for rennet in curdling milk. (32:51) **Navajo** *Cooking Agent* Dried or fresh berries added to goat's milk to make it curdle for cheese. (165:222) **Pima** *Fruit* Berries powdered, placed in milk, a piece of rabbit or cow stomach added, and liquid eaten as cheese. (47:88) *Substitution Food* Berries used as a substitute for rennet. (146:78) **Spanish American** *Substitution Food* Fruits used as a substitute for rennet in curdling milk. (32:51) **Zuni** *Beverage* Berries mixed with curdled goat milk and considered a delicious beverage. (166:70)
- *Other*—**Hopi** *Jewelry* Yellow fruits made into necklaces for clowns. (200:90) **Keresan** *Jewelry* Fruit made into a necklace worn by women. (198:562)

Solanum fendleri Gray ex Torr., Fendler's Horsenettle
- *Drug*—**Navajo** *Carminative* and *Gastrointestinal Aid* Raw tubers taken for gastric distress from hyperacidity. (90:163)
- *Food*—**Apache, Chiricahua & Mescalero** *Bread & Cake* Plant dried, stored, ground into flour, and used to make bread. (33:42) **Keresan** *Starvation Food* Raw potatoes mixed with clay or boiled with clay and eaten only in times of extreme scarcity. (198:562) **Pueblo** *Unspecified* Tubers eaten. (32:51) **Sia** *Vegetable* Potatoes eaten raw or cooked with clay to counteract the astringency. (199:107) **Zuni** *Unspecified* Raw tubers used for food. After every mouthful of potato, a bite of white clay was taken to counteract the unpleasant astringent effect of the potato in the mouth. (166:71)
- *Fiber*—**Apache, White Mountain** *Building Material* Used for thatching. (136:160)

Solanum jamesii Torr., Wild Potato
- *Food*—**Apache, Chiricahua & Mescalero** *Vegetable* Unpeeled potatoes boiled and eaten. (33:42) **Hopi** *Cooking Agent* Small potatoes used to make yeast. (190:166) *Unspecified* Plant boiled and eaten. (56:19) Tubers boiled and eaten with magnesia clay. (120:20) **Isleta** *Vegetable* Small tubers cooked as potatoes. (100:43) **Keres, Western** *Vegetable* Small tubers used for food. (171:70) **Keresan** *Starvation Food* Raw potatoes mixed with clay or boiled with clay and eaten only in times of extreme scarcity. (198:562) **Navajo** *Vegetable* Tubers eaten raw, boiled, or baked. (55:75) Potatoes mixed with white clay to remove the astringent effect on the mouth and eaten like mush. (165:221) **Navajo, Ramah** *Vegetable* Potato boiled with clay. *Winter Use Food* Potato dug with a stick, halved, sun dried, and stored in a pit for winter. (191:43) **Sia** *Vegetable* Potatoes eaten raw or cooked with clay to counteract the astringency. (199:107) **Tewa** *Vegetable* Tubers eaten. (138:73)

Solanum nigrum L., Black Nightshade
- *Drug*—**Cherokee** *Emetic* Taken as an emetic and to relieve loneliness because of death in family. *Psychological Aid* Infusion of leaves and stem taken "if lonesome because of death in family." (80:51) **Costanoan** *Dermatological Aid* Poultice of heated leaves applied to boils. *Misc. Disease Remedy* Decoction of plant used for scarlet fever. *Toothache Remedy* Plant smoke inhaled for toothaches. (21:14) **Delaware, Oklahoma** *Venereal Aid* Compound containing root used for venereal disease. (175:29, 80) **Houma** *Anthelmintic* Decoction of root given to babies for worms. *Dermatological Aid* Poultice of crushed, green leaves, and grease applied to sores. *Pediatric Aid* Decoction of root given to babies for worms. (158:65) **Iroquois** *Other* Plant used for injured person who had relapse. (87:431) **Karok** *Poison* Plant considered poisonous. (148:389) **Mendocino Indian** *Poison* Berries considered poisonous. (41:387) **Miwok** *Eye Medicine* Decoction used as wash for sore eyes. (12:173) **Ojibwa** *Ceremonial Medicine* Plant used for medicinal purposes or medicine ceremonies. (135:239) **Rappahannock** *Poison* and *Sedative* Weak infusion of dried leaves, poisonous, taken for sleeplessness. (161:34)
- *Food*—**Cherokee** *Vegetable* Young leaves used as a potherb. (80:51) Used as the most relished potherb. (203:74) **Mendocino Indian** *Fruit* Fully ripe berries used for food. (41:387) **Tubatulabal** *Fruit* Berries used for food. (193:15)

Solanum physalifolium Rusby, Hoe Nightshade
- *Drug*—**Paiute** *Antidiarrheal* Ripe fruit eaten or decoction of dried fruit taken for diarrhea. (as *S. villosum* 180:140)
- *Other*—**Navajo, Ramah** *Fertilizer* Leaves and berries soaked in water used on watermelon seed to insure a good crop. (as *S. villosum* 191:43)

Solanum rostratum Dunal, Buffalobur Nightshade
- *Drug*—**Zuni** *Gastrointestinal Aid* Infusion of powdered root, not an emetic, taken for "sick stomach." (166:60)
- *Other*—**Keres, Western** *Unspecified* Plant known and named but no use was specified. (171:70)

***Solanum* sp.**, Nightshade
- *Drug*—**Comanche** *Tonic* Decoction of root taken as a general tonic. *Tuberculosis Remedy* Decoction of root taken for tuberculosis. (29:524) **Isleta** *Heart Medicine* Infusion of leaves taken for heart trouble. (100:43)
- *Dye*—**Costanoan** *Blue* Fruits used to prepare a permanent blue dye for tattooing. (21:253)

Solanum triflorum Nutt., Cutleaf Nightshade
- *Drug*—**Blackfoot** *Antidiarrheal* and *Pediatric Aid* Decoction of berries given to children for diarrhea. (97:53) Decoction of berries given to children for diarrhea. (114:275) **Lakota** *Gastrointestinal Aid* Berries used for stomachaches. (139:60) **Navajo, Ramah** *Veterinary Aid* Cold infusion used as lotion on horses' sores. (191:43)
- *Food*—**Acoma** *Starvation Food* Berries eaten in times of food shortages. (32:52) **Keres, Western** *Starvation Food* Berries eaten in times of

famine. (171:70) **Laguna** *Starvation Food* Berries eaten in times of food shortages. (32:52) **Zuni** *Sauce & Relish* Ripe fruit boiled, ground, mixed with ground chile and salt, and eaten as a condiment with mush or bread. (166:71)
- *Other*—**Keres, Western** *Fertilizer* Planted with watermelons to make them more prolific and ripen early. Infusion of plant sprinkled on watermelons to make them more prolific and ripen early. (171:70) **Navajo, Ramah** *Fertilizer* Dried berries soaked in water and planted with watermelon seed to increase productivity. (191:43)

Solanum tuberosum L., Potato
- *Drug*—**Cherokee** *Emetic* Taken as an emetic and to relieve loneliness because of death in family. *Psychological Aid* Infusion of leaves and stem taken "if lonesome because of death in family." (80:51) **Iroquois** *Eye Medicine* Poultice of scraped potato applied to eye for inflammation. (87:431) **Rappahannock** *Dermatological Aid* Bruised tuber rubbed on warts. (161:27)
- *Food*—**Abnaki** *Vegetable* Tubers eaten. (144:171) **Cherokee** *Vegetable* Roots used for food. (80:51) **Haisla & Hanaksiala** *Vegetable* Tubers eaten. (43:293) **Iroquois** *Unspecified* Tubers eaten. (196:120) **Kitasoo** *Vegetable* Tubers eaten. (43:350) **Makah** *Vegetable* Potatoes dipped in oil and eaten with smoked fish. (67:314) **Menominee** *Vegetable* Deep purple potatoes used for food. (151:72) **Meskwaki** *Vegetable* Potatoes used for food. (152:264) **Navajo, Ramah** *Pie & Pudding* Cut, dried potatoes boiled into a pudding, in the winter months. *Winter Use Food* Potatoes cultivated, harvested, and stored in a root cellar for winter use. (191:43) **Ojibwa** *Soup* Potato cultivated and prized for use in soups. *Vegetable* Potato cultivated and always firm and crisp when cooked. (153:410) **Oweekeno** *Vegetable* Tubers eaten. (43:119) **Seminole** *Vegetable* Tubers eaten. (169:466) **Sia** *Vegetable* Cultivated potatoes used for food. (199:106)
- *Other*—**Okanagan-Colville** *Cash Crop* Potatoes attained a high degree of importance in the economy. (188:140)

Solanum umbelliferum Eschsch., Bluewitch Nightshade
- *Food*—**Costanoan** *Fruit* Fruits eaten for food. (21:253)

Solanum xantii Gray, Purple Nightshade
- *Drug*—**Kawaiisu** *Dermatological Aid* Poultice of heated plant applied to sores. *Orthopedic Aid* Poultice of heated plant applied to swollen leg and shoulder. (206:64)
- *Food*—**Miwok** *Fruit* Raw berries used for food. (12:162)

Solidago, Asteraceae

Solidago californica Nutt., California Goldenrod
- *Drug*—**Cahuilla** *Dermatological Aid* Used in making a hair rinse. *Gynecological Aid* Used as a medicine for feminine hygiene. (15:140) **Costanoan** *Burn Dressing* Decoction of leaves used as a wash for burns. *Dermatological Aid* Decoction of leaves used as a wash for sores. Toasted, crumbled leaves applied to wounds. (21:27) **Diegueño** *Dermatological Aid* Decoction of leaves and stems used to wash the hair to prevent falling hair. (84:43) **Kawaiisu** *Dermatological Aid* Infusion of leaves and flowers used as wash for boils and "holes" in neck and limbs, and for open sores and skin irritations. (206:64) **Miwok** *Antirheumatic (External)* Leaf powder dusted on sores. (12:170) *Dermatological Aid* Leaf powder applied to open sores. *Toothache Remedy* Decoction held in mouth for toothaches. (12:173)

Solidago canadensis L., Canada Goldenrod
- *Drug*—**Iroquois** *Analgesic* Infusion of roots and flowers used for side pains. (141:65) *Emetic* Compound infusion of roots taken as an emetic. Infusion of flowers taken as an emetic for too much gall. *Gastrointestinal Aid* and *Liver Aid* Infusion of flowers taken as an emetic for too much gall. *Love Medicine* Compound infusion of roots taken to kill a love medicine. *Other* Plant used as a "gambling medicine." *Pediatric Aid* and *Sedative* Compound infusion of tubers given to babies that start suddenly during sleep. (87:461) **Meskwaki** *Psychological Aid* Compound decoction used as wash for child who does not talk or laugh. (152:217) **Okanagan-Colville** *Antidiarrheal* Infusion of flower heads taken for diarrhea. *Febrifuge* Infusion of shoots given to children with fevers. *Misc. Disease Remedy* Decoction of flower heads taken for the flu. *Pediatric Aid* Infusion of shoots given to children with fevers. (188:84) **Potawatomi** *Febrifuge* Infusion of blossoms used for special kinds of fevers. (154:53) **Shuswap** *Gynecological Aid* Infusion of plant used as a bath for the mother at childbirth. (123:59) **Thompson** *Antidiarrheal* Decoction of plant used as a bath for babies with diarrhea. Decoction of plant tops taken for diarrhea. *Orthopedic Aid* Plant used to make a steam bath for crippled, paralyzed people. *Pediatric Aid* Decoction of plant used as a bath for babies with diarrhea, sleeplessness, or excessive crying. *Sedative* Decoction of plant used as a bath for babies with sleeplessness or excessive crying. *Veterinary Aid* Decoction of plant and wild tarragon used as a wash for horses with cuts and sores. (187:184) **Zuni** *Analgesic* Crushed blossoms chewed for sore throat. Infusion of crushed blossoms taken for body pain. (166:60)
- *Food*—**Gosiute** *Unspecified* Seeds used for food. (39:382) **Navajo, Kayenta** *Unspecified* Roots steeped or eaten. (205:50)
- *Other*—**Navajo, Kayenta** *Good Luck Charm* Used as a charm for success in gambling. *Smoke Plant* Roots smoked with other tobaccos. (205:50) **Okanagan-Colville** *Toys & Games* Clustered flower heads used by children as play whips. (188:84)

Solidago canadensis var. *scabra* Torr. & Gray, Canada Goldenrod
- *Drug*—**Chippewa** *Burn Dressing* Compound poultice of flowers applied to burns. (as *S. altissima* 53:352) *Dermatological Aid* Poultice of moistened, pulverized root applied to boils. (as *S. altissima* 53:348) Poultice of moistened, dry flowers applied to ulcers. (as *S. altissima* 53:354)

Solidago confinis Gray, Southern California Goldenrod
- *Drug*—**Kawaiisu** *Dermatological Aid* Infusion of leaves and flowers used as wash for boils and "holes" in neck and limbs, and for open sores and skin irritations. (206:64)

Solidago flexicaulis L., Zigzag Goldenrod
- *Drug*—**Chippewa** *Throat Aid* Root chewed for sore throat. (53:342) **Iroquois** *Gastrointestinal Aid* Compound decoction of plants taken for biliousness. (87:459) **Menominee** *Analgesic* Snuff of dried, powdered leaves used for headache. (54:129) *Hemostat* Compound of powdered, dried leaves inserted in nostrils to check nosebleed. (54:132) **Potawatomi** *Febrifuge* Infusion of whole plant used for certain fevers. (as *S. latifolia* 154:53)

Solidago gigantea Ait., Giant Goldenrod
- *Drug*—**Keres, Western** *Cathartic* Infusion of leaves used as a strong

physic. (as *Salidago pitcheri* 171:67) **Menominee** *Unspecified* Plant used in medicine. (as *Solidago serotina* 151:31) **Potawatomi** *Febrifuge* Infusion of blossoms used for various fevers. (as *S. serotina* 154:53)
- **Fiber**—**Keres, Western** *Basketry* Stems made into rough baskets. (as *Salidago pitcheri* 171:67)

Solidago juncea Ait., Early Goldenrod
- **Drug**—**Chippewa** *Anticonvulsive* Decoction of root taken for convulsions. (53:336) **Delaware** *Antidiarrheal* Infusion of green leaves used or green leaves chewed for diarrhea. *Febrifuge* Infusion of green leaves used or green leaves chewed for fever. (176:33) **Delaware, Oklahoma** *Antidiarrheal* Leaves chewed or infusion taken for diarrhea. *Febrifuge* Leaves chewed or infusion taken for fever. (175:28, 80) **Iroquois** *Antiemetic* Infusion of plants taken for nausea. *Emetic* Decoction of flowers taken as an emetic. *Febrifuge* Decoction of roots taken for fevers. *Gastrointestinal Aid* Decoction or infusion of flowers taken for biliousness and upset stomach. *Liver Aid* Infusion of plants taken for jaundice. (87:460)

Solidago multiradiata Ait., Mountain Goldenrod
- **Drug**—**Cree, Hudson Bay** *Tonic* Plant used as a tonic. (as *S. virgaurea* 92:303)

Solidago nemoralis Ait., Dyersweed Goldenrod
- **Drug**—**Houma** *Liver Aid* Decoction of root taken for "yellow jaundice." (158:66) **Iroquois** *Kidney Aid* Decoction of roots taken for the kidneys. (87:460) **Mahuna** *Burn Dressing* Decoction of leaves used as a wash or poultice of leaves applied to burns. *Dermatological Aid* Decoction of leaves used as a wash or poultice of leaves applied to skin ulcers. Decoction of plant used as a bath for the 7-year itch. *Disinfectant* Decoction of leaves used as a disinfecting wash for burns or skin ulcers. (140:12)
- **Food**—**Gosiute** *Unspecified* Seeds used for food. (39:382)
- **Other**—**Navajo** *Incense & Fragrance* Plant used as an incense. (55:88)

Solidago odora Ait., Anisescented Goldenrod
- **Drug**—**Cherokee** *Abortifacient* Used for "female obstructions." *Antidiarrheal* Infusion taken for bloody discharge from bowels. *Cold Remedy* Infusion taken for cold. *Cough Medicine* Infusion taken for coughs. *Diaphoretic* Infusion taken as a diaphoretic. *Febrifuge* Infusion taken for fever. *Misc. Disease Remedy* Infusion taken for nerves and measles. *Oral Aid* Root chewed for sore mouth. *Sedative* Infusion taken for "nerves" and infusion of root held in mouth for neuralgia. *Stimulant* Used as a stimulant. *Tonic* Used as a tonic. *Tuberculosis Remedy* Infusion of leaf taken for tuberculosis. (80:36)

Solidago rigida L., Stiff Goldenrod
- **Drug**—**Chippewa** *Cathartic* Decoction of root used as an enema. (53:364) *Diuretic* Infusion of root taken for "stoppage of urine." (53:348) **Meskwaki** *Dermatological Aid* Flowers made into a lotion and used on bee stings and for swollen faces. (152:217, 218)
- **Other**—**Lakota** *Containers* Leaves used under meat. (139:39)

Solidago rugosa P. Mill., Wrinkleleaf Goldenrod
- **Drug**—**Iroquois** *Liver Aid* Whole plant used for biliousness and as liver medicine. *Other* Decoction of flowers and leaves taken for dizziness, weakness, or sunstroke. (87:461)

Solidago simplex ssp. *simplex* var. *nana* (Gray) Kartesz & Gandhi, Dwarf Goldenrod
- **Drug**—**Thompson** *Dietary Aid* and *Tonic* Decoction of plant taken as a tonic to restore appetite. (as *S. decumbens* 164:468) *Venereal Aid* Decoction of whole plant taken for syphilis. (as *S. decumbens* 164:472)

Solidago sp., Goldenrod
- **Drug**—**Alabama** *Cold Remedy* Infusion of root taken for bad colds. (172:664) Infusion of roots taken for colds. (177:63) *Toothache Remedy* Root placed in cavity of tooth for toothaches. (172:664) Roots placed in cavity for toothaches. (177:63) **Algonquin, Quebec** *Heart Medicine* Infusion of plant used for heart disease. (18:239) **Blackfoot** *Nose Medicine* Infusion of plant taken for nasal congestion. Root chewed for sore throats, throat constrictions, or nasal congestion. *Panacea* Infusion of plant used by the diviner with his supernatural powers for illnesses. *Throat Aid* Infusion of plant taken for sore throats and throat constrictions. (86:74) **Cherokee** *Febrifuge* Infusion used for fevers. (203:74) **Chippewa** *Analgesic* Decoction of root applied externally for cramps. (53:344) *Febrifuge* Decoction of dried leaves taken for fever. (53:354) *Gastrointestinal Aid* Hot decoction of root applied externally for stomach cramps. (53:344) **Iroquois** *Antihemorrhagic* Compound decoction of twigs and roots taken for internal hemorrhage. (87:459)
- **Other**—**Omaha** *Season Indicator* Blooms used as an indicator of corn ripening. The goldenrod served the Omaha as a mark or sign in their floral calendar. When they were on the summer buffalo hunt, the sight of the goldenrod indicated to them that their corn was beginning to ripen at home. (70:133)

Solidago spathulata DC., Coast Goldenrod
- **Drug**—**Thompson** *Misc. Disease Remedy* Poultice of toasted, powdered leaves mixed with grease and used for mumps. (187:184)

Solidago speciosa Nutt., Showy Goldenrod
- **Drug**—**Meskwaki** *Burn Dressing* Infusion of root used for burns or steam scalds. (152:218)

Solidago speciosa var. *rigidiuscula* Torr. & Gray, Showy Goldenrod
- **Drug**—**Chippewa** *Antihemorrhagic* Decoction of root taken for hemorrhaging from the mouth after being wounded. (as *S. rigidiuscula* 53:352) *Dermatological Aid* Root or stalk combined with bear grease used as an ointment for the hair. (as *S. rigidiuscula* 53:350) *Gynecological Aid* Infusion of root taken to ease difficult labor. (as *S. rigidiuscula* 53:358) *Orthopedic Aid* Warm poultice of boiled stalk or root applied to sprains or strained muscles. (as *S. rigidiuscula* 53:362) *Pulmonary Aid* Cold decoction of root taken for lung hemorrhages. Compound decoction of root taken for lung trouble. *Stimulant* Decoction of root and stalk taken as a stimulant. (as *S. rigidiuscula* 53:340) *Tonic* Decoction of root and stalk taken as a tonic. (as *S. rigidiuscula* 53:364)

Solidago spectabilis (D. C. Eat.) Gray, Nevada Goldenrod
- **Food**—**Gosiute** *Unspecified* Seeds used for food. (39:382)

Solidago squarrosa Nutt., Downy Ragged Goldenrod
- **Drug**—**Iroquois** *Burn Dressing* Compound infusion of dried leaves

and roots used as wash for scalds and burns. *Emetic* Compound infusion of dried leaves and roots taken as an emetic. *Gastrointestinal Aid* Compound infusion of dried leaves and roots used as wash for bad stomach. *Venereal Aid* Compound infusion of dried leaves and roots used as wash for venereal disease. (87:459)

Solidago uliginosa Nutt., Bog Goldenrod
- **Drug**—**Potawatomi** *Dermatological Aid* Poultice of root used to bring a boil to a head. (154:53, 54)

Solidago ulmifolia Muhl. ex Willd., Elmleaf Goldenrod
- **Drug**—**Meskwaki** *Stimulant* Smoke of smudged plant directed up nostrils to revive unconscious patient. (152:218)

Solidago velutina DC., Threenerve Goldenrod
- **Drug**—**Navajo, Kayenta** *Pediatric Aid* and *Psychological Aid* Plant used as a lotion to bathe an infant hermaphrodite to become sensible. (as *S. sparsiflora* 205:50) **Navajo, Ramah** *Witchcraft Medicine* Cold infusion taken and used as lotion in witchcraft. (as *S. sparsiflora* 191:53)

Sonchus, Asteraceae

Sonchus arvensis L., Field Sowthistle
- **Drug**—**Cherokee** *Sedative* Infusion taken to calm nerves. (80:59) **Potawatomi** *Gynecological Aid* Infusion of leaves used for caked breast. (154:54)
- **Other**—**Potawatomi** *Hunting & Fishing Item* Hunters sucked the milk from branches to imitate the sound of a fawn nursing, to draw the doe near. (154:117)

Sonchus asper (L.) Hill, Spiny Sowthistle
- **Drug**—**Iroquois** *Pediatric Aid* and *Sedative* Compound infusion given to babies "who cry until they hold their breath." (87:478) **Navajo, Kayenta** *Heart Medicine* Plant smoked or taken for palpitations. *Poison* Plant considered poisonous. (205:50)
- **Food**—**Luiseño** *Vegetable* Plant used for greens. (155:228) **Mohave** *Starvation Food* Young shoots roasted and eaten as a famine food. (37:201) **Pima** *Unspecified* Tender leaves rubbed between the palms and eaten raw. Stalks peeled and eaten raw like celery. *Vegetable* Tender leaves cooked as greens. (47:106) **Pima, Gila River** *Unspecified* Leaves eaten raw. (133:7)

Sonchus oleraceus L., Common Sowthistle
- **Drug**—**Houma** *Abortifacient* Infusion of plant taken to "make tardy menstruation come." *Antidiarrheal* Infusion of whole plant taken to "correct looseness of bowels." *Pediatric Aid* and *Toothache Remedy* Infusion of plant given to children for teething. (158:64) **Pima** *Cathartic* Gum used as a cathartic. *Other* Gum used as a cure for the opium habit. (47:106)
- **Food**—**Houma** *Fodder* Plants used for hog feed. (158:64) **Kamia** *Vegetable* Boiled leaves used for food as greens. (62:24) **Pima** *Unspecified* Leaves and stems rubbed between the palms of the hands and eaten raw. (47:106) **Pima, Gila River** *Unspecified* Leaves eaten raw and boiled. (133:7) **Yaqui** *Vegetable* Tender, young leaves boiled in salted water with chile and eaten as greens. (47:106)

Sophora, Fabaceae

Sophora nuttalliana B. L. Turner, Silky Sophora
- **Food**—**Acoma** *Special Food* Sweet roots chewed as a delicacy. (as *S. serecia* 32:33) **Keres, Western** *Sweetener* Roots chewed for the sweet taste. (as *S. sericea* 171:71) **Laguna** *Special Food* Sweet roots chewed as a delicacy. (as *S. serecia* 32:33) **Navajo** *Forage* Plant used by sheep for forage. (as *S. sericea* 55:58) **San Felipe** *Special Food* Sweet roots chewed as a delicacy. (as *S. serecia* 32:33)

Sophora secundiflora (Ortega) Lag. ex DC., Mescal Bean
- **Drug**—**Comanche** *Ear Medicine* Decoction of ground beans used for earaches and ear sores. (as *S. seundiflora* 99:3)
- **Other**—**Comanche** *Decorations* Beans used as ornaments, possibly for ceremonial purposes. (29:524) *Protection* Seeds carried by medicine men as protection against contamination from menstrual blood. (as *S. seundiflora* 99:3)

Sorbus, Rosaceae

Sorbus americana Marsh., American Mountainash
- **Drug**—**Algonquin, Quebec** *Cold Remedy* Infusion of inner bark taken for colds. Infusion of terminal buds and inner bark taken for colds. *Tonic* Infusion of inner bark and sweet flag used as a tonic. (as *Pyrus americana* 18:177) **Algonquin, Tête-de-Boule** *Psychological Aid* Buds and inner bark fibers boiled and used for moral depression. *Strengthener* Buds and inner bark fibers boiled and used for general weakness. (132:131) **Iroquois** *Gastrointestinal Aid* Fruit used to facilitate digestion. (142:91) **Malecite** *Analgesic* Infusion of bark used for pain after childbirth. (as *Pyrus americana* 116:258) *Dermatological Aid* Poultice of burned bark used for boils. (as *Pyrus americana* 116:247) **Micmac** *Analgesic* Bark used for "mother pains." *Dermatological Aid* Bark used for boils. *Emetic* Parts of plant used as an emetic. (as *Pyrus americana* 40:60) *Gastrointestinal Aid* Infusion of root taken for colic. (156:317) *Gynecological Aid* Bark used for "mother pains." (as *Pyrus americana* 40:60) *Unspecified* Infusion of bark taken for unspecified purpose. (156:317) **Montagnais** *Blood Medicine* Decoction of bark taken to purify the blood and to stimulate the appetite. *Dietary Aid* Decoction of bark taken to stimulate the appetite and to purify the blood. (156:313) **Ojibwa** *Unspecified* Root bark used for medicinal purposes. (as *Pyrus americana* 135:236) *Venereal Aid* Infusion of root bark taken for gonorrhea. (as *Pyrus americana* 135:231) **Penobscot** *Emetic* Plant used as an emetic. (156:309) **Potawatomi** *Unspecified* Inner bark used as a medicine. (as *Pyrus americana* 154:78) **Tlingit** *Pulmonary Aid* Plant used for pleurisy. (as *Pyrus sambucifolius* 107:283)
- **Food**—**Algonquin, Quebec** *Fruit* Fruit used for food. (as *Pyrus americana* 18:90) **Montagnais** *Forage* Berries eaten by bears. *Fruit* Berries used for food. (156:313) **Ojibwa** *Fruit* Fruit used for food. (as *Pyrus sambucifolia* 135:236)
- **Fiber**—**Ojibwa** *Canoe Material* Wood used to make ribs for canoes. *Snow Gear* Wood used to make snowshoe frames. *Sporting Equipment* Wood used to lacrosse clubs. (as *Pyrus americana* 135:236)

Sorbus aucuparia L., European Mountainash
- **Drug**—**Potawatomi** *Cold Remedy* Infusion of leaves taken as a cold medicine. *Emetic* Infusion of leaves taken as an emetic for pneumonia, diphtheria, and croup. *Misc. Disease Remedy* Infusion of leaves taken as an emetic for diphtheria. *Pulmonary Aid* Infusion of leaves taken as an emetic for pneumonia and croup. (as *Pyrus aucuparia* 154:78)

Sorbus decora (Sarg.) Schneid., Northern Mountainash
- *Drug*—**Cree, Woodlands** *Antirheumatic (External)* Decoction of inner bark from the stem base used for rheumatism. *Orthopedic Aid* Decoction of peeled sticks taken for back pain. Decoction of inner bark from the stem base used for backaches. (109:61)

Sorbus scopulina Greene, Greene Mountainash
- *Drug*—**Okanagan-Colville** *Pediatric Aid* and *Urinary Aid* Infusion of branches given to young children with bed-wetting problems. (188:133) **Wet'suwet'en** *Febrifuge* Bark used for fevers. *Tonic* Bark used as a tonic. *Unspecified* Bark used for general sickness. (73:152)

Sorbus sitchensis M. Roemer, Western Mountainash
- *Drug*—**Bella Coola** *Antirheumatic (Internal)* Infusion of root and branch bark taken for rheumatism. *Gastrointestinal Aid* Infusion of root and branch bark taken for stomach troubles. (184:210) **Carrier** *Unspecified* Scraped bark used for medicine. (31:70) **Okanagan-Colville** *Pediatric Aid* and *Urinary Aid* Infusion of branches given to young children with bed-wetting problems. (188:133) **Thompson** *Ear Medicine* Warmed stick used in the ear for earache. *Kidney Aid* Infusion of branches taken for weak kidneys, to stop frequent urination. (187:273)
- *Food*—**Thompson** *Dried Food* Berries sometimes dried for storage. *Fruit* Berries boiled and eaten alone. *Soup* Berries boiled and eaten in soups such as salmon head soup. *Spice* Berries cooked with marmot to flavor meat, and a cluster of berries was added to the top of a jar of blueberries as a flavor when canning. *Winter Use Food* Berries usually buried and kept fresh. (187:273)
- *Other*—**Bella Coola** *Insecticide* Berries rubbed on the scalp for lice. (184:210)

Sorbus sitchensis* var. *grayi (Wenzig) C. L. Hitchc., Gray's Mountainash
- *Drug*—**Heiltzuk** *Dermatological Aid* Berries mashed and rubbed on the head for lice. (43:116)
- *Food*—**Heiltzuk** *Forage* Considered a food for black bears. (43:116)

Sorbus sitchensis M. Roemer **var. *sitchensis***, Sitka Mountainash
- *Drug*—**Bella Coola** *Antirheumatic (Internal)* Decoction of root bark or stem bark taken or used as bath for rheumatism. *Eye Medicine* Decoction of root bark or inner bark of stem used as eyewash. *Gastrointestinal Aid* Decoction of root bark or inner bark of stem taken "for the stomach." **Carrier, Southern** *Cold Remedy* Bark chewed for colds. **Gitksan** *Cathartic* Crushed fresh fruit eaten raw as a strong purgative. (as *Pyrus sitchensis* 150:59)
- *Food*—**Okanagon** *Fruit* Fruits occasionally used for food. **Thompson** *Fruit* Fruits occasionally used for food. (as *Pyrus sitchensis* 125:38) Fruits eaten except by some of the Upper Thompsons. (as *Pyrus sitchensis* 164:488)

***Sorghum*, Poaceae**

Sorghum bicolor (L.) Moench **ssp. *bicolor***, Broomcorn
- *Food*—**Hopi** *Sauce & Relish* Used as a sweet syrup. (as *S. vulgare* 200:66)

Sorghum halepense (L.) Pers., Johnson Grass
- *Food*—**Kiowa** *Fodder* Used as a fodder for horses. (192:16)
- *Other*—**Kiowa** *Toys & Games* Stems and leaves used by children to make grass whistles. (192:16)

Sparganiaceae, see *Sparganium*

***Sparganium*, Sparganiaceae**

Sparganium eurycarpum Engelm. ex Gray, Broadfruit Burreed
- *Drug*—**Iroquois** *Febrifuge* Infusion of whole plant and other plant leaves used for chills. (141:71) *Veterinary Aid* Compound of chopped plants mixed with cows feed for difficult birth of calf. *Witchcraft Medicine* Compound poultice bound to "soreness all over in men from being witched." (87:272)
- *Food*—**Klamath** *Unspecified* Rootstocks used for food. Bulb at base of stem used for food. (45:90) **Okanagan-Colville** *Fodder* Used as hay for cattle. (188:57)
- *Other*—**Okanagan-Colville** *Containers* Used in pit cooking. (188:57)

***Spartina*, Poaceae**

Spartina alterniflora Loisel., Smooth Cordgrass
- *Food*—**Iroquois** *Forage* Used as forage. (142:106)

Spartina pectinata Link, Prairie Cordgrass
- *Fiber*—**Omaha** *Building Material* Plant used to support earth coverings of lodges in permanent villages. **Ponca** *Building Material* Plant used as thatching to support the earth coverings of lodges in permanent villages. (as *S. michauxiana* 70:66)

***Sphaeralcea*, Malvaceae**

Sphaeralcea ambigua Gray, Desert Globemallow
- *Drug*—**Shoshoni** *Antiemetic* Decoction of root taken for upset stomach. (180:141, 142) *Antirheumatic (External)* Poultice of mashed root used for swollen feet. (118:43) Poultice of wilted plants bound onto rheumatic or swollen areas. *Cathartic* Decoction of root or whole plant taken as a physic. *Cold Remedy* Decoction of leaves taken for colds. (180:141, 142) *Contraceptive* Infusion of root taken as birth control. (118:46) Decoction of root taken as a contraceptive, "uncertain report." *Dermatological Aid* Poultice of crushed, raw root applied to swellings. Poultice of wilted plants applied for rheumatism or swellings. *Emetic* Decoction of root or whole plant taken as an emetic. *Eye Medicine* Decoction of leaves used as an eyewash. (180:141, 142) *Unspecified* Poultices of cooked root applied medicinally. (118:43) *Venereal Aid* Decoction of root or whole plant taken for venereal disease. *Veterinary Aid* Poultice of boiled plant applied to wire cuts on horses. (180:141, 142)
- *Other*—**Shoshoni** *Containers* Whole plant boiled, added to red clay, and used to make cups. (118:56) **Yavapai** *Containers* Stems used as a litter for drying mescal. (65:259)

Sphaeralcea angustifolia (Cav.) G. Don, Copper Globemallow
- *Drug*—**Navajo** *Cold Remedy* Plant used for colds. *Cough Medicine* Plant used for coughs. *Misc. Disease Remedy* Plant used for influenza. (90:163) **Pima** *Antidiarrheal* Decoction of leaves used for diarrhea. *Gastrointestinal Aid* Decoction of root used for biliousness. (146:79)

Sphaeralcea angustifolia* ssp. *cuspidata (Gray) Kearney, Copper Globemallow
- *Other*—**Kiowa** *Decorations* Flowers appealed to the esthetic nature,

picked and used for bouquets in the home. (as *S. cuspidata* 192:43)

Sphaeralcea angustifolia* ssp. *lobata (Woot.) Kearney, Copper Globemallow
- *Drug*—**Keres, Western** *Gynecological Aid* Roots used for medicine by pregnant women. (as *S. lobata* 171:71) **Navajo** *Ceremonial Medicine* Roots used as a ceremonial medicine. (as *S. lobata* 55:63) **Tewa** *Dermatological Aid* and *Disinfectant* Poultice of pulverized roots applied to purulent sores. *Snakebite Remedy* Poultice of pulverized roots applied to snakebites. (as *S. lobata* 138:60, 61)
- *Food*—**Navajo** *Unspecified* Seeds used for food. (as *S. lobata* 55:63)
- *Other*—**Keres, Western** *Fasteners* Root sap rubbed on the skin and used as glue for feathers and cotton during dances. *Preservative* Leaves used while drying wild leafed yucca, to prevent spoiling. (as *S. lobata* 171:71) **Tewa** *Paint* Root skin pounded into powder, water added, and used as face paint for dances. (as *S. lobata* 138:60)

Sphaeralcea coccinea (Nutt.) Rydb., Scarlet Globemallow
- *Drug*—**Cheyenne** *Ceremonial Medicine* Plant used in ceremonies. (83:30) **Lakota** *Dermatological Aid* Poultice of chewed roots applied to sores. (139:51) **Navajo, Kayenta** *Ceremonial Medicine* Plant used as a ceremonial fumigant ingredient. *Dermatological Aid* Dried plant used as a dusting powder for sores. Plant used as a lotion for skin diseases. *Dietary Aid* Plant used as a tonic to improve appetite. *Disinfectant* Plant used as a ceremonial fumigant ingredient. *Other* Plant used for hydrophobia. *Strengthener* Plant used as a medicine to give singer strength. (205:31) **Navajo, Ramah** *Panacea* Plant used as "life medicine." (191:36)
- *Food*—**Navajo, Kayenta** *Beverage* Used to make a beverage. (205:31)
- *Other*—**Lakota** *Protection* Root juices used by medicine men for protection of hands from fire or boiling water. (106:32) Roots chewed and hands rubbed with the roots by heyoka men, a special sort of medicine men, to protect their hands in scalding water. (139:51)

Sphaeralcea coccinea (Nutt.) Rydb. **ssp. *coccinea***, Scarlet Globemallow
- *Drug*—**Cheyenne** *Adjuvant* Infusion of ground leaves, stems, and roots mixed with other medicine to render it more palatable. (as *Malvastrum coccineum* 76:180) **Comanche** *Dermatological Aid* Infusion of plant used for swellings. (as *Malvastrum coccineum* 29:522) **Dakota** *Analgesic* Plant used for pain. (as *Malvastrum coccineum* 69:362) *Burn Dressing* Chewed plant rubbed on skin to protect against boiling water in ceremony. *Dermatological Aid* Chewed plant applied as a cooling and healing salve to sores and wounds. (as *Malvastrum coccineum* 70:103) **Keres, Western** *Gynecological Aid* Roots used by women when they become pregnant. (as *Malvastrum coccineum* 171:53) **Navajo** *Hemostat* Infusion of plants used to stop bleeding. *Witchcraft Medicine* Infusion of plants taken for diseases produced by witchcraft. (as *Malvastrum coccineum* 55:62)
- *Food*—**Navajo** *Starvation Food* Roots chewed during food shortages. (as *Malvastrum coccineum* 55:62)
- *Other*—**Gosiute** *Paint* Plant pounded in water into a gummy paste and applied over the rough inner surfaces of earthen dishes. (as *Malvastrum coccineum* 39:374)

Sphaeralcea digitata (Greene) Rydb., Slippery Globemallow
- *Drug*—**Navajo, Ramah** *Gastrointestinal Aid* Infusion of whole plant taken for stomachache. *Panacea* Root used as a "life medicine." (191:36)

Sphaeralcea emoryi* ssp. *variabilis (Cockerell) Kearney, Emory's Globemallow
- *Drug*—**Pima** *Antidiarrheal* Decoction of root taken for diarrhea. (47:80)

Sphaeralcea fendleri Gray, Fendler's Globemallow
- *Drug*—**Navajo, Kayenta** *Dermatological Aid* Plant used for sand cricket bites. *Oral Aid* Infusion of plant taken for sore mouth. (205:32) **Navajo, Ramah** *Antihemorrhagic* Cold, compound infusion of plant taken for internal injury and hemorrhage. *Dermatological Aid* Infusion of plant used as lotion for external injury. (191:36)
- *Other*—**Havasupai** *Containers* Juice made into a paste and mixed with clay before molding it into a pot. (197:232)

Sphaeralcea grossulariifolia (Hook. & Arn.) Rydb., Gooseberry-leaf Globemallow
- *Drug*—**Hopi** *Antihemorrhagic* Roots chewed or boiled with cactus root and used for difficult defecation. *Diuretic* Plant used for babies with bowel trouble. *Orthopedic Aid* Root chewed or boiled for broken bones. *Pediatric Aid* Plant used for babies with bowel trouble. (42:362)
- *Other*—**Havasupai** *Containers* Juice made into a paste and mixed with clay before molding it into a pot. (197:232)

Sphaeralcea incana Torr. ex Gray, Gray Globemallow
- *Drug*—**Hopi** *Antidiarrheal* Plant used as a diarrhea medicine. (56:16)

Sphaeralcea munroana (Dougl. ex Lindl.) Spach ex Gray, Munro's Globemallow
- *Other*—**Gosiute** *Paint* Plant pounded in water into a gummy paste and applied over the rough inner surfaces of earthen dishes. (as *Malvastrum munroanum* 39:374)

Sphaeralcea parvifolia A. Nels., Smallflower Globemallow
- *Drug*—**Hopi** *Antihemorrhagic* Root chewed or boiled with cactus root and used for difficult defecation. *Dermatological Aid* Plant used for sores, cuts, and wounds. *Diuretic* Plant used for babies with bowel trouble. *Orthopedic Aid* Chewed or boiled root used for broken bones. *Pediatric Aid* Plant used for babies with bowel trouble. (42:363)
- *Other*—**Havasupai** *Containers* Juice made into a paste and mixed with clay before molding it into a pot. (197:232) **Hopi** *Ceremonial Items* Plant used for midwinter ceremonials. (42:363)

***Sphaeralcea* sp.**, Globemallow
- *Drug*—**Hopi** *Antidiarrheal* Root chewed or compound decoction taken for diarrhea. (200:34, 85) *Gastrointestinal Aid* Root used for babies with bowel trouble. (200:85) *Orthopedic Aid* Root chewed or decoction used for broken bones, as "there's gristle in the root." (200:31, 85) *Pediatric Aid* Root used for babies with bowel troubles. (200:34, 85) **Luiseño** *Emetic* Decoction of leaves taken as an emetic. (155:231)

Sphagnaceae, see *Sphagnum*

Sphagnum, Sphagnaceae, moss

Sphagnum dusenii C. Jens., Sphagnum

- *Fiber*—**Ojibwa** *Mats, Rugs & Bedding* Moss gathered and dried to make mattresses. (153:422)

Sphagnum sp.
- *Drug*—**Carrier** *Dermatological Aid* Berries rubbed on children's sores. *Pediatric Aid* Berries rubbed on children's sores. (31:87) **Nitinaht** *Disinfectant* Used as a good disinfectant. (186:59)
- *Food*—**Carrier** *Beverage* Leaves used to make tea. (31:87) **Haisla & Hanaksiala** *Forage* Plant eaten by grizzly bears. (43:148)
- *Fiber*—**Bella Coola** *Mats, Rugs & Bedding* Used for padding and bedding. (184:196) **Carrier** *Clothing* Whole plant used as a diaper for small children. (31:87) **Chinook** *Clothing* Used by women for sanitary napkins. *Mats, Rugs & Bedding* Used for bedding in camp. (79:50) **Chippewa** *Other* Used as an absorbent. (53:378) **Eskimo, Inuktitut** *Caulking Material* Used to chink log houses. (202:189) **Haisla** *Clothing* Plant used for diapering infants. (43:148) **Nitinaht** *Clothing* Used for baby diapers. (186:59) **Potawatomi** *Mats, Rugs & Bedding* Dried moss used for making pillows and mattresses. (154:124)

Sphenosciadium, Apiaceae
Sphenosciadium capitellatum Gray, Woollyhead Parsnip
- *Drug*—**Paiute** *Venereal Aid* Decoction of roots applied to venereal sores. (167:317)
- *Other*—**Paiute** *Insecticide* Infusion of root used for lice. (167:317)

Spigelia, Loganiaceae
Spigelia anthelmia L., West Indian Pinkroot
- *Drug*—**Creek** *Anthelmintic* Plant considered a "well-known remedy" for children with worms. (172:669) Plant used as a worm remedy. (177:51) *Pediatric Aid* Plant considered a "well-known remedy" for children with worms. (172:669)

Spigelia marilandica (L.) L., Woodland Pinkroot
- *Drug*—**Cherokee** *Anthelmintic* Infusion taken for worms. (80:40) Decoction of roots taken for worms. (177:51) Used as a general vermifuge. **Creek** *Anthelmintic* Used as a general vermifuge. (203:74)

Spiraea, Rosaceae
Spiraea alba Du Roi, White Meadowsweet
- *Drug*—**Algonquin, Quebec** *Unspecified* Infusion of leaves and stems used as a medicinal tea. (18:176) **Iroquois** *Analgesic* Compound decoction of mashed and powdered dried roots taken for side pain. (87:349)

Spiraea alba var. *latifolia* (Ait.) Dippel, White Meadowsweet
- *Drug*—**Algonquin, Quebec** *Unspecified* Infusion of leaves and stems used as a medicinal tea. (as *Spirea latifolia* 18:176) **Iroquois** *Antiemetic* Decoction of plant with other leaves and branches taken for nausea and vomiting. (as *Spiraea latifolia* 141:47)
- *Food*—**Abnaki** *Beverage* Used to make tea. (as *S. latifolia* 144:152) Leaves used to make tea. (as *S. latifolia* 144:168)

Spiraea betulifolia Pallas, White Spirea
- *Drug*—**Okanagan-Colville** *Analgesic* Decoction of branches used for menstrual pains. Infusion of branches taken for abdominal pains. *Cold Remedy* Infusion of branches taken for colds. *Gastrointestinal Aid* Infusion of branches taken for abdominal pains. *Gynecological Aid* Decoction of branches used for menstrual pains or heavy or prolonged menstruation. *Kidney Aid* Infusion of branches taken for poor kidneys. *Other* Infusion of branches taken for ruptures. (188:133) **Shuswap** *Antidiarrheal* Decoction of roots and leaves taken for diarrhea. *Gastrointestinal Aid* Decoction of roots and leaves taken for the stomach. (123:67) **Thompson** *Cold Remedy* Infusion of whole plant taken for colds. *Gastrointestinal Aid* Infusion of whole plant taken for internal stomach problems. *Venereal Aid* Decoction of leaves and branches taken or used as a bath for venereal diseases. (187:274)
- *Food*—**Thompson** *Beverage* Plant used to make a tea-like beverage. (187:274)

Spiraea douglasii Hook., Douglas's Spirea
- *Drug*—**Lummi** *Antidiarrheal* Infusion of seeds taken for diarrhea. (79:33)
- *Food*—**Thompson** *Forage* Dried flower spikes eaten by grouse. (187:274)
- *Fiber*—**Thompson** *Brushes & Brooms* Branches used to make brooms. (187:274)
- *Other*—**Bella Coola** *Cooking Tools* Branches used to hang salmon for drying and smoking. (184:210) **Lummi** *Cooking Tools* Used for spreading and cooking salmon. (79:33)

Spiraea douglasii var. *menziesii* (Hook.) K. Presl, Menzies's Spirea
- *Other*—**Quinault** *Cooking Tools* Stems peeled and used to string clams for roasting. (as *S. menziesii* 79:33)

Spiraea ×pyramidata Greene, Pyramid Spirea
- *Drug*—**Thompson** *Tonic* Decoction of plant taken as a tonic. (164:471)
- *Food*—**Thompson** *Beverage* Flowers, stems, and leaves boiled and drunk as a tea. (164:494)

Spiraea salicifolia L., Willowleaf Meadowsweet
- *Drug*—**Mahuna** *Cold Remedy* Roots used for chest colds. *Cough Medicine* Roots used for coughs. (140:18) **Meskwaki** *Antidiarrheal* Immature seeds used for bloody flux. (152:243) **Ojibwa** *Hunting Medicine* Root used as a trapping medicine. (153:386) **Potawatomi** *Unspecified* Bark used for unspecified ailments. (154:79, 80)

Spiraea sp.
- *Drug*—**Iroquois** *Antidiarrheal* Compound decoction of shoots or vines used as a wash for babies with diarrhea. *Emetic* Compound decoction of bark taken to vomit. *Love Medicine* Infusion of bark from young twigs taken as an antidote for love medicine. *Orthopedic Aid* Infusion of bark used as a body wash to make children stronger to walk. *Pediatric Aid* Compound decoction of shoots or vines used as a wash for babies with diarrhea. Infusion of bark used as a body wash to make children stronger to walk. *Tuberculosis Remedy* Compound decoction of bark taken during initial stages of consumption. (87:349)

Spiraea splendens Baumann ex K. Koch var. *splendens*, Mountain Spirea
- *Drug*—**Blackfoot** *Laxative* Infusion of root used as an enema. (as *S.*

densiflora 86:68) *Venereal Aid* Infusion of roots taken for venereal complaints. (as *S. densiflora* 86:70)
- *Fiber*—**Blackfoot** *Brushes & Brooms* Flowering stems used as paintbrushes, especially on the large spaces of tepees. (as *S. densiflora* 86:123)

Spiraea stevenii (Schneid.) Rydb., Beauverd Spirea
- *Food*—**Tanana, Upper** *Beverage* Fresh or dried leaves made into tea. (as *S. beauverdiana* 102:8)
- *Other*—**Eskimo, Alaska** *Fuel* Plant burned for smoking fish. (as *S. beauverdiana* 1:36)

Spiraea tomentosa L., Steeple Bush
- *Drug*—**Algonquin, Quebec** *Unspecified* Infusion of leaves and stems used as a medicinal tea. (18:177) **Mohegan** *Antidiarrheal* Infusion of leaves taken for dysentery. (174:266) **Ojibwa** *Antiemetic* Infusion of leaves and flowers taken for the sickness of pregnancy. *Gynecological Aid* Infusion of leaves and flowers taken for the sickness of pregnancy. Infusion of leaves and flowers used to ease childbirth. (153:386)

Spiranthes, Orchidaceae

Spiranthes lacera var. *gracilis* (Bigelow) Luer, Northern Slender Ladiestresses
- *Drug*—**Ojibwa** *Hunting Medicine* Roots used as an ingredient of the hunting charm to bring game to the hunter. (as *S. gracilis* 153:431)

Spiranthes lucida (H. H. Eat.) Ames, Shining Ladiestresses
- *Drug*—**Cherokee** *Pediatric Aid* Warm infusion used as a wash for infants "to insure fast, healthy growth." *Urinary Aid* Roots with "twayblade" used "for urinary trouble." (80:42)

Spiranthes romanzoffiana Cham., Hooded Ladiestresses
- *Drug*—**Gosiute** *Venereal Aid* Plant used for venereal disease. (39:383)

Spiranthes sp., Spiral Orchid
- *Drug*—**Seminole** *Blood Medicine* Decoction of roots used as a blood purifier. (169:304)
- *Food*—**Paiute** *Unspecified* Roots used for food. (167:244)

Spirogyra, Zygnemataceae, alga

Spirogyra sp., Pond Scum
- *Other*—**Keres, Western** *Unspecified* Alga known and named but no use was specified. (171:72) **Thompson** *Hunting & Fishing Item* Used as a bait in fishing. (164:515)

Sporobolus, Poaceae

Sporobolus airoides (Torr.) Torr., Alkali Sacaton
- *Food*—**Hopi** *Starvation Food* Grain occasionally used for food during famines. (200:66)
- *Other*—**Apache, Chiricahua & Mescalero** *Containers* Moist grass laid onto hot stones to prevent steam from escaping. (33:36)

Sporobolus contractus A. S. Hitchc., Spike Dropseed
- *Food*—**Apache, Western** *Porridge* Seeds ground, mixed with cornmeal and water, and made into a mush. (as *S. strictus* 26:189) **Apache, White Mountain** *Bread & Cake* Seeds ground and used to make bread and pones. *Porridge* Seeds ground, mixed with meal and water, and eaten as mush. (as *S. strictus* 136:149) **Navajo** *Bread & Cake* Seeds used to make bread. (165:223)
- *Fiber*—**Zuni** *Building Material* Grass bunches fastened together and used to construct shelters in or near the distant fields. *Mats, Rugs & Bedding* Grass bunches fastened together to make mats for covering hatchways and other openings in the house. (as *S. strictus* 166:81)
- *Other*—**Apache, White Mountain** *Cash Crop* Plant gathered and sold. (as *S. strictus* 136:149)

Sporobolus cryptandrus (Torr.) Gray, Sand Dropseed
- *Drug*—**Navajo, Ramah** *Veterinary Aid* Cold infusion of plant applied to sores or bruises on horse's leg. (191:17)
- *Food*—**Apache, Chiricahua & Mescalero** *Bread & Cake* Seeds threshed, winnowed, ground, and the flour used to make bread. *Porridge* Seeds boiled and eaten as porridge. (33:48) **Hopi** *Bread & Cake* Plant used to make bread. *Pie & Pudding* Plant used to make pudding. (42:364) **Keres, Western** *Fodder* Grass considered good pony feed. (171:72) **Kiowa** *Fodder* Foliage was a valuable fodder. (as *Vilfa cryptandra* 192:17) **Navajo** *Bread & Cake* Seeds ground to make dumplings, rolls, griddle cakes, and tortillas. (55:26) *Forage* Used as forage by animals. (90:163) **Navajo, Ramah** *Porridge* Ground seeds alone or with corn made into mush or bread. (191:17)
- *Fiber*—**Navajo, Ramah** *Scouring Material* Bunch about a foot long, tied with string or yucca fiber, used as a brush for cleaning metates. (191:17)

Sporobolus flexuosus (Thurb. ex Vasey) Rydb., Mesa Dropseed
- *Food*—**Hopi** *Starvation Food* Grain occasionally used for food during famines. (200:66)

Sporobolus giganteus Nash, Giant Dropseed
- *Food*—**Hopi** *Porridge* Seeds threshed, ground with corn into fine meal and used to make a mush. (120:20) *Spice* Seeds used as flavoring for cornmeal. (42:365)
- *Other*—**Hopi** *Ceremonial Items* Stems used to make pahos (prayer sticks). Bundles of plant used to cover kiva entrance during Bean Ceremonial. Pollen used in the hunting ceremony. (42:365)

Sporobolus heterolepis (Gray) Gray, Prairie Dropseed
- *Drug*—**Ojibwa, South** *Dermatological Aid* Poultice of crushed root applied to sores. *Emetic* and *Liver Aid* Decoction of root taken as an emetic "to remove bile." (91:200)

Sporobolus wrightii Munro ex Scribn., Giant Sacaton
- *Food*—**Navajo, Ramah** *Porridge* Ground seeds alone or with corn made into mush or bread. (191:17) **Papago** *Dried Food* Seeds basket winnowed, parched, sun dried, cooked, stored, and used for food. (34:24)
- *Fiber*—**Apache, Chiricahua & Mescalero** *Brushes & Brooms* Stiff stems made into a brush and used to clean spines of cacti. (33:40)

Stachys, Lamiaceae

Stachys albens Gray, Whitestem Hedgenettle
- *Food*—**Kawaiisu** *Spice* Bunched leaves used as "cork" for the basketry water bottle, "it gives a good taste to the water." (206:65)

Stachys bullata Benth., California Hedgenettle
- *Drug*—**Costanoan** *Dermatological Aid* Poultice of plant applied or decoction taken for swollen sores. *Disinfectant* Poultice of plant ap-

plied or decoction taken for infected sores. *Ear Medicine* Poultice of heated leaves applied for earaches. *Gastrointestinal Aid* Poultice of plant applied or decoction of plant taken for stomachaches. *Throat Aid* Poultice of leaves applied or decoction of roots used as a gargle for sore throats. (21:17) **Pomo, Kashaya** *Dermatological Aid* Poultice of heated leaves used on boils to bring them to a head. (72:77)

Stachys ciliata Epling, Great Hedgenettle
- **Drug**—**Saanich** *Tonic* Infusion of pounded roots used as a spring tonic. (as *S. cooleyae* 182:84)
- **Food**—**Saanich** *Forage* Roots eaten by wounded deer. (as *S. cooleyae* 182:84)
- **Other**—**Haisla & Hanaksiala** *Hunting & Fishing Item* Plant fiber used to make nets. (as *S. cooleyae* 43:256) **Hesquiat** *Cooking Tools* Plants laid under fish to keep them clean. (as *S. cooleyae* 185:69)

Stachys mexicana Benth., Emerson Betony
- **Drug**—**Green River Group** *Dermatological Aid* Plant used for boils. **Puyallup** *Dermatological Aid* Plant used for boils. (as *S. ciliata* 79:45)
- **Food**—**Quinault** *Unspecified* Honey sucked out of the blossoms and used for food. (as *S. ciliata* 79:45)
- **Other**—**Makah** *Cooking Tools* Plants used to cover steaming sprouts. **Quinault** *Cooking Tools* Plants used to cover steaming sprouts. (as *S. ciliata* 79:45)

Stachys palustris L., Marsh Hedgenettle
- **Drug**—**Chippewa** *Gastrointestinal Aid* Infusion of fresh or dried leaves taken for "sudden colic." (53:344) **Delaware** *Venereal Aid* Used with common nightshade, "pretty flower," and button snakeroot for venereal disease. (176:35) **Delaware, Oklahoma** *Venereal Aid* Compound containing root used for venereal disease. (175:29, 80)
- **Food**—**Gosiute** *Unspecified* Seeds used for food. (39:383)

Stachys rothrockii Gray, Rothrock's Hedgenettle
- **Drug**—**Navajo, Ramah** *Ceremonial Medicine* Dried leaves used as a ceremonial medicine and plant used in chant lotion. *Dermatological Aid* Decoction of plant used as foot deodorant. *Disinfectant* Dried leaves used for "deer infection." (191:42)

Stachys tenuifolia Willd. var. *tenuifolia*, Smooth Hedgenettle
- **Drug**—**Meskwaki** *Cold Remedy* Infusion of leaves taken for bad colds. *Emetic* Infusion of leaves given as an emetic. (152:227)

Stanleya, Brassicaceae
Stanleya albescens M. E. Jones, White Princesplume
- **Food**—**Hopi** *Unspecified* Leaves boiled and eaten. (56:15) Boiled and eaten. (190:163) *Vegetable* Eaten as greens in the spring. (200:77)

Stanleya pinnata (Pursh) Britt., Desert Princesplume
- **Drug**—**Havasupai** *Poison* Fresh leaves considered poisonous. (197:220) **Navajo** *Gland Medicine* Poultice of plants applied to glandular swellings. (55:50) **Paiute** *Analgesic* Poultice of mashed root applied for throat pain. *Misc. Disease Remedy* Poultice of mashed root applied for congestion of diphtheria. *Throat Aid* Poultice of mashed root applied for throat pain. *Tonic* Decoction of root taken as a tonic for general debility after an illness. **Shoshoni** *Analgesic* Poultice of pulped root applied for rheumatic pains. *Antirheumatic (External)* Poultice of hot, pulped root applied for rheumatic pains. *Ear Medicine* Poultice of hot, pulped root applied for an earache. *Toothache Remedy* Poultice of root applied to gums or placed in cavity for toothache. (180:142) **Zuni** *Dermatological Aid* Poultice of fresh, chewed pods used for itching. (27:375) *Venereal Aid* Powdered plant applied, as a specific, to scraped syphilitic sores. (166:60)
- **Food**—**Havasupai** *Vegetable* Leaves boiled two or three times to remove poisons and eaten. (197:220) Young, fresh, tender leaves boiled, drained, balled into individual portions, and served. (197:66) **Hopi** *Vegetable* Boiled plant used for greens in the spring. (42:366) Eaten as greens in the spring. (200:77) **Kawaiisu** *Vegetable* Leaves and stems boiled, squeezed out in cold water to remove the bitterness, fried in grease, and eaten. (206:65) **Tewa** *Vegetable* Boiled plant used for greens in the spring. (42:366)

Staphyleaceae, see *Staphylea*

Staphylea, Staphyleaceae
Staphylea trifolia L., American Bladdernut
- **Drug**—**Iroquois** *Antirheumatic (Internal)* Compound infusion of plants taken for rheumatism. *Dermatological Aid* Infusion of powdered bark used as a wash for sore faces. *Gynecological Aid* Compound poultice of bark applied when a woman swells after copulation. *Pediatric Aid* and *Sedative* Infusion of bark used as a wash to keep children from crying. (87:377) **Meskwaki** *Ceremonial Medicine* Seeds used in gourd rattles for dream and medicine dances. (152:248)
- **Other**—**Meskwaki** *Sacred Items* Seeds considered sacred and used in the rattles of the medicine dance. (152:274)

Stellaria, Caryophyllaceae
Stellaria media (L.) Vill., Common Chickweed
- **Drug**—**Chippewa** *Eye Medicine* Decoction of leaves strained and used as a wash for sore eyes. (53:360) **Iroquois** *Antirheumatic (External)* Poultice of plant fragments with other plants applied for swellings. (141:41) *Dermatological Aid* Raw, compound poultice applied to cuts and wounds. (87:317)
- **Food**—**Iroquois** *Forage* Plant eaten by chickens. (142:86)

Stenandrium, Acanthaceae
Stenandrium dulce (Cav.) Nees, Sweet Shaggytuft
- **Drug**—**Seminole** *Pediatric Aid* and *Sedative* Plant and other plants used as a baby's charm for fear from dreams about raccoons or opossums. (as *S. floridanum* 169:221) *Stimulant* Decoction of whole plant used as a bath for hog sickness: unconsciousness. (as *S. floridanum* 169:229)

Stenanthium, Liliaceae
Stenanthium occidentale Gray, Western Stenanthium
- **Drug**—**Thompson** *Unspecified* Plant used medicinally. (187:129)

Stenocereus, Cactaceae
Stenocereus thurberi (Engelm.) Buxbaum, Organ Pipe Cactus
- **Food**—**Apache, Chiricahua & Mescalero** *Fruit* Fruit used for food. (as *Lemaireocereus thurberi* 33:40) **Papago** *Beverage* Juice used as a ceremonial drink. (as *Lemaireocereus thurberi* 34:26) *Bread & Cake* Seeds parched, stored, and used to make meal cakes. *Cooking Agent*

Seeds parched, ground, water added, and oil extracted. (as *Lemaireocereus thurberi* 34:22) *Dried Food* Fruits dried, stored in jars, and used as sweets. (as *Lemaireocereus thurberi* 34:46) *Fodder* Seeds parched and used as a chicken feed. *Preserves* Pulp boiled to a sweet, sticky mass and used like raspberry jam. Juice made into cactus jam and used as the most important sweet in the diet. (as *Lemaireocereus thurberi* 34:22) Fruits made into jam. *Sauce & Relish* Fruits made into a syrup. (as *Lemaireocereus thurberi* 34:46) *Staple* Seeds made into flour and used for food. *Unspecified* Pulp eaten fresh. (as *Lemaireocereus thurberi* 34:22) Fruits and seeds used for food. (as *Lemaireocereus thurberi* 36:59) **Papago & Pima** *Beverage* Fruit used to make wine. (as *Lemaireocereus thurberi* 35:10) Juice used to make wine. (as *Lemaireocereus thurberi* 35:34) *Candy* Used to make candy. (as *Lemaireocereus thurberi* 35:17) *Fruit* Fruit used for food. (as *Lemaireocereus thurberi* 35:11) *Porridge* Seeds mixed with the pulp, formed into a paste, and eaten. (as *Lemaireocereus thurberi* 35:7) *Preserves* Fruit boiled, without sugar, to make preserves. (as *Lemaireocereus thurberi* 35:17) *Sauce & Relish* Fruit used to make syrup. (as *Lemaireocereus thurberi* 35:11) **Seri** *Fruit* Fruits eaten for food. *Porridge* Seeds ground to a powder and made into a meal or paste. (as *Lemaireocereus thurberi* 50:134)
- **Fiber**—**Seri** *Caulking Material* Dried plant skeletons and sea lion oil used as a caulking compound. (as *Lemaireocereus thurberi* 50:136)
- **Other**—**Papago** *Ceremonial Items* Pulp boiled with water, strained, boiled again, and used as a ceremonial drink. (as *Lemaireocereus thurberi* 34:22) **Papago & Pima** *Lighting* Used to make torches. (as *Lemaireocereus thurberi* 35:34) *Paint* Fruit used to make ink. (as *Lemaireocereus thurberi* 35:7) **Seri** *Tools* Dried plant skeletons used as a straight, slender pole for knocking off ripe fruit. (as *Lemaireocereus thurberi* 50:136)

Stenotaphrum, Poaceae
Stenotaphrum secundatum (Walt.) Kuntze, St. Augustine Grass
- **Drug**—**Hawaiian** *Dermatological Aid* Leaf ash used for sores and navel sores on babies. Leaf ash used for skin ulcers. *Gynecological Aid* Leaves and stems pounded, resulting juice mixed with other ingredients and taken for excessive menses. Leaf ash used on the vagina and neighboring parts after giving birth. *Oral Aid* Leaf ash used for excessive saliva from babies' mouths. *Pediatric Aid* Leaf ash used for sores and navel sores on babies. Leaf ash used for excessive saliva from babies' mouths. (as *S. americanum* 2:10)

Stenotus, Asteraceae
Stenotus lanuginosus (Gray) Greene **var. lanuginosus**, Cespitose Goldenweed
- **Drug**—**Navajo** *Gastrointestinal Aid* Plant used for indigestion. *Nose Medicine* Plant used for nose troubles. (as *Aplopappus lanuginosus* 55:96) *Oral Aid* Plant used for sore gums. (as *Aplopappus lanuginosus* 55:97) *Throat Aid* Plant used for throat troubles. (as *Aplopappus lanuginosus* 55:96) *Toothache Remedy* Plant used as a toothache medicine. (as *Aplopappus lanuginosus* 55:80)

Stephanomeria, Asteraceae
Stephanomeria exigua Nutt., Small Wirelettuce
- **Drug**—**Hopi** *Diuretic* and *Venereal Aid* Plant used as a diuretic for venereal disease. (as *Ptiloria exigua* 200:35, 97) **Navajo, Kayenta** *Misc. Disease Remedy* Plant used for measles. (205:50)

Stephanomeria pauciflora (Torr.) A. Nels., Brownplume Wirelettuce
- **Drug**—**Hopi** *Gynecological Aid* Root used in various ways to increase mother's milk supply. (as *Ptiloria pauciflora* 200:36, 98) **Navajo, Kayenta** *Narcotic* Roots used as a narcotic. (205:50) **Navajo, Ramah** *Gynecological Aid* Strong infusion of root used to hasten delivery of placenta. *Panacea* Root used as a "life medicine." (191:53)
- **Food**—**Kawaiisu** *Candy* Thick liquid used as chewing gum. (206:65) **Navajo, Kayenta** *Candy* Used as chewing gum. (205:50) **Navajo, Ramah** *Candy* Root used for chewing gum. (191:53)
- **Other**—**Navajo, Kayenta** *Ceremonial Items* Used as a paint ingredient for chant arrows used in various ceremonies. (205:50)

Stephanomeria runcinata Nutt., Desert Wirelettuce
- **Drug**—**Keres, Western** *Eye Medicine* Milky sap used for sore eyes. (as *S. ramosa* 171:72)

Stephanomeria sp.
- **Drug**—**Paiute** *Febrifuge* and *Panacea* Plant used for various ailments, especially fever. (167:317)

Stephanomeria spinosa (Nutt.) S. Tomb, Thorn Skeletonweed
- **Drug**—**Cheyenne** *Cold Remedy, Diaphoretic,* and *Herbal Steam* Decoction of smashed roots taken and used as a steam bath to cause sweating for colds. *Misc. Disease Remedy* Decoction of smashed roots used as a steam bath for mumps. *Panacea* Plant used for almost every ailment. *Tuberculosis Remedy* Decoction of smashed roots used as a steam bath for tuberculosis. (as *Lygodesmia spinosa* 83:22) **Paiute** *Antidiarrheal* Decoction of plant tops taken for diarrhea. *Cathartic* Decoction of plant tops taken as a physic. *Dermatological Aid* Compound decoction of root used as a wash for swellings. Poultice of cottony fuzz applied to boils or sores to promote healing. *Emetic* Decoction of plant tops taken as an emetic. *Toothache Remedy* Cottony fuzz placed in cavity of aching tooth. **Shoshoni** *Antiemetic* Decoction of plant tops taken for vomiting. *Eye Medicine* Decoction of plant tops used as an eyewash. *Tonic* Compound decoction of root taken as a tonic. (as *Lygodesmia spinosa* 180:102, 103)

Stephanomeria tenuifolia (Raf.) Hall, Narrowleaf Wirelettuce
- **Drug**—**Apache, White Mountain** *Snakebite Remedy* Poultice of powdered plants applied to rattlesnake bites. (as *Ptiloria tenuifolia* 136:160) **Shoshoni** *Venereal Aid* Decoction of plant taken for venereal diseases. (180:143) **Thompson** *Unspecified* Plant used medicinally for unspecified purpose. (as *Ptiloria tenuifolia* 164:468) **Zuni** *Snakebite Remedy* Poultice of pulverized plant applied and infusion taken for rattlesnake bite. (as *Ptiloria tenuifolia* 166:58)

Stephanomeria virgata Benth., Rod Wirelettuce
- **Drug**—**Kawaiisu** *Eye Medicine* Milky plant juice used as an eye medicine. (206:65)

Sterculiaceae, see *Fremontodendron, Theobroma, Waltheria*

Stereocaulaceae, see *Stereocaulon*

Stereocaulon, Stereocaulaceae, lichen

Stereocaulon paschale (L.) Hoffm., Flat Lichen
- *Fiber*—**Eskimo, Inuktitut** *Canoe Material* Used to stuff caribou skins for rafts. (202:190)

Stictaceae, see *Sticta*

Sticta, Stictaceae, lichen

Sticta glomulifera, Tree Lichen
- *Food*—**Ojibwa** *Unspecified* Boiled until the lichens coagulate like scrambled eggs. (153:406)

Stillingia, Euphorbiaceae

***Stillingia* sp.**
- *Drug*—**Creek** *Abortifacient* Compound decoction of root used as a wash for irregular periods. (172:662) *Cathartic* Plant used as a cathartic. (177:36) *Gynecological Aid* Decoction of root taken and used as a wash after childbirth. (172:662) Decoction of mashed roots used as a bath and taken by women after giving birth. *Reproductive Aid* Cold infusion of pounded roots taken by men to regain potency. (177:36)

Stillingia sylvatica Garden ex L., Queen's Delight
- *Drug*—**Cherokee** *Venereal Aid* Decoction or tincture of root used for the worst forms of venereal disease or "clap." (80:51)
- *Other*—**Kiowa** *Season Indicator* Plant considered a rain symbol because the dry land terrapins get under it for shade. (192:37)

Stillingia sylvatica Garden ex L. **ssp. *sylvatica***, Queen's Delight
- *Drug*—**Seminole** *Antidiarrheal* Plant used as an astringent for diarrhea. (as *S. angustifolia* 169:168) Decoction of roots taken by babies and adults for bird sickness: diarrhea, vomiting, and appetite loss. (as *S. angustifolia* 169:234) Plant used as a diarrhea medicine. (as *S. angustifolia* 169:275) *Antiemetic* Decoction of roots taken by babies and adults for bird sickness: diarrhea, vomiting, and appetite loss. (as *S. angustifolia* 169:234) *Blood Medicine* Decoction of roots taken for menstruation sickness: yellow eyes and skin, weakness, and shaking head.[13] (as *S. angustifolia* 169:247) *Dietary Aid* Decoction of roots taken by babies and adults for bird sickness: diarrhea, vomiting, and appetite loss. (as *S. angustifolia* 169:234) *Other* Decoction of roots taken for menstruation sickness: yellow eyes and skin, weakness, and shaking head.[13] (as *S. angustifolia* 169:247) *Pediatric Aid* Decoction of roots taken by babies and adults for bird sickness: diarrhea, vomiting, and appetite loss. (as *S. angustifolia* 169:234) *Strengthener* Decoction of roots taken for menstruation sickness: yellow eyes and skin, weakness, and shaking head.[13] (as *S. angustifolia* 169:247)

Stipa, Poaceae

Stipa comata Trin. & Rupr., Needleandthread
- *Other*—**Blackfoot** *Season Indicator* Spread out appearance of grass indicated the best condition for hunting buffalo cows in the fall. (97:22) **Navajo, Ramah** *Toys & Games* Sharp pointed fruits with long awns used as play arrows by adults and children. (191:17) **Okanagan-Colville** *Toys & Games* Seeds thrown like darts by playing children. (188:57)

Stipa neomexicana (Thurb. ex Coult.) Scribn., New Mexico Needlegrass
- *Other*—**Hopi** *Ceremonial Items* Plant used for necklaces on the first initiates in the Wu-chim ceremony. (42:367)

Stipa robusta (Vasey) Scribn., Sleepygrass
- *Food*—**Navajo, Ramah** *Fodder* Used for sheep and horse feed. (191:17)

Stipa spartea Trin., Porcupine Grass
- *Fiber*—**Omaha** *Brushes & Brooms* Grains bound with sinew and used to make combs or hairbrushes. The grains were firmly bound with sinew, the points being broken or burned off, the grains forming the teeth, the awns bent back, making the attachment for combs or hairbrushes. (68:324) Stiff awns firmly bound into a bundle and the pointed grains burned off to make a hairbrush. **Pawnee** *Brushes & Brooms* Stiff awns firmly bound into a bundle and the pointed grains burned off to make a hairbrush. **Ponca** *Brushes & Brooms* Stiff awns firmly bound into a bundle and the pointed grains burned off to make a hairbrush. (70:66)
- *Other*—**Lakota** *Toys & Games* Grass heads used in mischievous games in which boys throw a bunch at people to stick them. (139:31) **Omaha** *Ceremonial Items* Stiff awns firmly bound into a bundle and the pointed grains burned off to make a ceremonial brush. **Pawnee** *Ceremonial Items* Stiff awns firmly bound into a bundle and the pointed grains burned off to make a ceremonial brush. **Ponca** *Ceremonial Items* Stiff awns firmly bound into a bundle and the pointed grains burned off to make a ceremonial brush. (70:66)

Stipa speciosa Trin. & Rupr., Desert Needlegrass
- *Food*—**Kawaiisu** *Unspecified* Seeds used for food. When ripe in June, the grass was cut off in bunches, tied together with stems of the grass, and thrown over the shoulder into the carrying basket suspended on one's back. Two procedures were used in preparing the seeds for food. First, the grass was spread out on a flat rock, where it was allowed to dry a half day and then threshed by burning. If the fire burned too quickly, green spear grass was added to slow it down. The burned stalks were stirred and lifted with a green stick so that the seeds would fall out. The seeds were gathered and winnowed by being poured from one basket to another. Boiled, the seeds swelled "like rice." A cupful would fill a pot. Second, the grass was dried for a day or two and the seeds beaten out. They would be boiled whole or first pounded to a meal and then cooked. (206:66) **Paiute** *Porridge* Seeds used to make mush. (167:243)

Streptanthus, Brassicaceae

Streptanthus cordatus Nutt., Heartleaf Twistflower
- *Drug*—**Navajo, Kayenta** *Eye Medicine* Root juice used as eye drops for sore eyes. (205:25)
- *Food*—**Navajo, Kayenta** *Vegetable* Used for greens in foods. (205:25)

Streptopus, Liliaceae

Streptopus amplexifolius (L.) DC., Claspleaf Twistedstalk
- *Drug*—**Makah** *Gynecological Aid* Chewed roots taken to produce labor in case of protracted delay. (79:25) **Micmac** *Antihemorrhagic* Parts of plant used for spitting blood. *Kidney Aid* Parts of plant used for kidney trouble. *Venereal Aid* Parts of plant used for gonorrhea. (40:62) **Montagnais** *Panacea* Infusion of stems and berries taken "for sickness in general." (156:314) **Penobscot** *Antihemorrhagic* Compound

infusion of plant taken for "spitting up blood." *Kidney Aid* Compound infusion of plant taken for kidney trouble. *Tonic* Compound infusion of plant taken as a tonic. *Venereal Aid* Compound infusion of plant taken for gonorrhea. (156:311) **Thompson** *Analgesic* Compound decoction of roots taken for internal pains. (164:459) *Dietary Aid* Infusion of whole plant taken for loss of appetite. *Gastrointestinal Aid* Infusion of whole plant taken for stomachache. (187:130)
- *Food*—**Alaska Native** *Fruit* Berries used for food. *Vegetable* Young, tender shoots used in salads. (85:69) **Cherokee** *Vegetable* Leaves cooked and eaten as greens. (80:59) **Hesquiat** *Forage* Eaten by deer. (185:55) **Montagnais** *Forage* Berries and roots eaten by snakes. (156:314) **Nitinaht** *Forage* Berries eaten by wolves. (186:86) **Okanagon** *Fruit* Bright-colored berries used for food. (125:38) **Oweekeno** *Forage* Berries eaten by frogs. (43:79) **Thompson** *Fruit* Bright-colored berries used for food. (125:38) Berries eaten in large quantities. (164:486)
- *Other*—**Hesquiat** *Toys & Games* Children play with the berries. (185:55) **Thompson** *Incense & Fragrance* Plant tied to the body, clothes, or hair and used as a scent. (164:502)

Streptopus roseus Michx., Rosy Twistedstalk
- *Drug*—**Chippewa** *Eye Medicine* Poultice of steeped root applied to sties. (53:360) **Iroquois** *Gynecological Aid* Infusion of roots taken for fallen womb. (87:284) **Montagnais** *Diaphoretic* Infusion of blossoms taken to cause sweating. (156:314) **Ojibwa** *Cathartic* Plant used as a physic and infusion taken as a cough remedy. *Cough Medicine* Infusion of plant taken as a cough remedy and used as a physic. (153:374) **Okanagon** *Tonic* Plant used as a tonic medicine. (125:42) **Potawatomi** *Cough Medicine* Root used to make a cough syrup or tea. (154:63) **Thompson** *Tonic* Plant used as a tonic medicine. (125:42) *Unspecified* Decoction of root used medicinally. (164:467)
- *Food*—**Cherokee** *Unspecified* Young growth boiled, fried, and eaten. (204:251) *Vegetable* Leaves cooked and eaten as greens. (80:59) Leaves and stalks mixed with wanegedum (angelico) and sweet salad and cooked as greens. *Winter Use Food* Leaves and stalks mixed with wanegedum (angelico) and sweet salad and canned for future use. (as *S. roseus* 126:48)

Strophostyles, Fabaceae
Strophostyles helvula (L.) Ell., Trailing Fuzzybean
- *Drug*—**Houma** *Misc. Disease Remedy* Compound decoction of bean taken for typhoid. (158:65) **Iroquois** *Dermatological Aid* Leaves rubbed on parts affected by poison ivy and warts. (87:365)
- *Food*—**Choctaw** *Unspecified* Boiled, mashed roots used for food. (as *Phaseolus diversifolius* 25:8)

Stylosanthes, Fabaceae
Stylosanthes biflora (L.) B.S.P., Sidebeak Pencilflower
- *Drug*—**Cherokee** *Abortifacient* Compound used to promote menstruation. *Gynecological Aid* Infusion of root used for female complaint. (80:57)

Styphelia, Epacridaceae
Styphelia tameiameiae (Cham. & Schlecht.) F. Muell., Pukiawe
- *Drug*—**Hawaiian** *Dermatological Aid* and *Tuberculosis Remedy* Flowers and other plants dried, pounded into a powder, and applied to scrofulous sores. (2:49)

Styracaceae, see *Halesia*

Suaeda, Chenopodiaceae
Suaeda arborescens
- *Food*—**Pima** *Spice* Added as flavoring to greens or cactus fruits. (146:78)
- *Other*—**Pima** *Containers* Used moistened with cottonwood to line pits for roasting saltbush overnight. (32:23) Layer of plant used to cover the embers in a baking pit. (32:36) *Cooking Tools* Used to cover coals in cooking pits. (146:71) Used to line saltbush roasting pits. (146:78)

Suaeda calceoliformis (Hook.) Moq., Pursh Seepweed
- *Food*—**Gosiute** *Unspecified* Seeds used for food. (as *S. depressa* 39:383) **Paiute** *Staple* Seeds parched, ground, and eaten as meal. (as *S. depressa* var. *erecta* 104:98) **Paiute, Northern** *Unspecified* Seeds used for food. (as *S. depressa* 59:47)
- *Other*—**Keres, Western** *Unspecified* Plant known and named but no use was specified. (as *S. erecta* 171:72)

Suaeda moquinii (Torr.) Greene, Mojave Seablite
- *Drug*—**Hopi** *Analgesic* Poultice of dried leaves used on sore places. (as *S. intermedia* 190:161) *Ceremonial Medicine* Plant used to bathe the doctor before administering to patients. (as *Dondia fruticosa* 200:31, 74) **Navajo, Kayenta** *Gastrointestinal Aid* Plant used for bleeding bowels. (as *S. torreyana* 205:21) **Paiute** *Dermatological Aid* Crushed fresh plants rubbed on chickenpox to stop itching and to dry sores. *Kidney Aid* Decoction of plant taken for kidney trouble. *Misc. Disease Remedy* Crushed fresh plants rubbed on chickenpox to stop itching and to dry sores. *Urinary Aid* Decoction of plant taken for bladder trouble. **Shoshoni** *Kidney Aid* Decoction of plant taken for kidney trouble. *Urinary Aid* Decoction of plant taken for bladder trouble. (as *S. torreyana* var. *ramosissima* 180:143)
- *Food*—**Navajo** *Porridge* Seeds boiled into a gruel. (as *S. torreyana* 55:45) **Pima, Gila River** *Unspecified* Leaves pit roasted and eaten. (133:7)
- *Other*—**Papago** *Cooking Tools* Used to line baking pits for roasting chollas. (as *Dondia nigra* 34:15)

Suaeda sp., Seablite
- *Drug*—**Cahuilla** *Dermatological Aid* Leaves boiled, mixed with clay, and used as a hair dye. (15:141)
- *Food*—**Cahuilla** *Staple* Seeds ground into flour and used to make mush and cakes. *Vegetable* Leaves boiled and used as greens. (15:141)
- *Dye*—**Cahuilla** *Black* Plants boiled in water and used as a black dye for palm mats. (15:141)
- *Other*—**Cahuilla** *Soap* Plant used in making soap. (15:141)

Suaeda suffrutescens S. Wats., Desert Seepweed
- *Food*—**Papago** *Spice* Leaves and stalks lined inside cooking holes to give cactus fruits a salty flavor. **Pima** *Spice* Leaves and stalks lined inside cooking holes to give cactus fruits a salty flavor. (as *Dondia suffrutescens* 95:264) Added as flavoring to greens or cactus fruits. (146:78)
- *Other*—**Papago** *Cooking Tools* Leaves and stalks used to line cooking pits. **Pima** *Cooking Tools* Leaves and stalks used to line cooking pits. (as *Dondia suffrutescens* 95:262)

Symphoricarpos, Caprifoliaceae

Symphoricarpos albus (L.) Blake, Common Snowberry
- **Drug**—**Chehalis** *Dermatological Aid* Berries used for the hair. Poultice of chewed leaves applied or infusion of leaves used as a wash for injuries. *Venereal Aid* Decoction of root bark taken for venereal disease. (79:47) **Chippewa** *Diuretic* Compound decoction of root taken for "stoppage of urine." (53:348) **Cowichan** *Burn Dressing* Berries rubbed on skin for burns. *Dermatological Aid* Berries rubbed on skin for rashes and sores. (182:80) **Cree, Woodlands** *Dermatological Aid* Infusion of whole plant taken and applied externally for skin rash. *Eye Medicine* Infusion of fruit used for sore eyes. *Febrifuge* and *Pediatric Aid* Decoction of roots and stems taken for fever associated with teething sickness. *Venereal Aid* Decoction of roots and stems used for venereal disease. (109:62) **Crow** *Veterinary Aid* Decoction of crushed roots used for horses failing to void. **Flathead** *Burn Dressing* Poultice of crushed leaves, fruits, and bark used for burns. *Dermatological Aid* Poultice of crushed leaves, fruits, and bark used for sores, cuts, and chapped and injured skin. *Eye Medicine* Bark and wild rose used to make an eyewash. Fruit chewed and the juice used for injured eyes. (82:59) **Green River Group** *Dermatological Aid* and *Disinfectant* Plant used to disinfect a festering sore. (79:47) **Hesquiat** *Dermatological Aid* Berry juice rubbed on warts or sores. (185:63) **Klallam** *Cold Remedy* Decoction of leaves taken for colds. (79:47) **Kutenai** *Abortifacient* Infusion of cut branches taken for menstrual disorders. (82:59) **Kwakiutl** *Eye Medicine* Burned berries and oil rubbed on inflamed eyes. (183:280) **Makah** *Witchcraft Medicine* Leaves chewed and swallowed to counteract evil charms. (186:102) **Miwok** *Cold Remedy* Infusion of root taken for colds. *Gastrointestinal Aid* Infusion of root taken for stomachache. (12:173) **Nez Perce** *Febrifuge* Infusion of twigs used for fevers. Infusion of twigs used for young children with fevers. *Pediatric Aid* Infusion of twigs used for young children with fevers. (82:59) **Nitinaht** *Diuretic* Infusion of bark taken for inability to urinate. (186:102) **Okanagan-Colville** *Cathartic* Decoction of branches, leaves, and berries taken as a physic to clean out the system. *Dermatological Aid* Poultice of mashed berries applied to children's skin sores. Poultice of mashed berries applied to itchy skin. Berries mashed and rubbed in the armpits as an antiperspirant. *Eye Medicine* Berries mashed, mixed with small amount of warm water, and put into the eyes for sore, running eyes. Infusion of roots used as an eyewash. *Pediatric Aid* Poultice of mashed berries applied to children's skin sores. *Poison* Berries considered poisonous. (188:95) **Saanich** *Burn Dressing* Berries rubbed on skin for burns. *Dermatological Aid* Berries rubbed on skin for rashes and sores. (182:80) **Sanpoil** *Dermatological Aid* Crushed berries rubbed in the armpits as an antiperspirant. *Diuretic* Decoction of leaves used by men for urine retention. *Eye Medicine* Mashed berries mixed with water and used as a wash for sore eyes. *Unspecified* Decoction of roots used "for illness of an indefinite character." *Veterinary Aid* Decoction of leaves used for animals with urine retention. (131:220) **Sioux** *Diuretic* Decoction of berries used as a diuretic. *Eye Medicine* Infusion used for sore eyes. (82:59) **Skagit** *Antidote* Berries eaten as an antidote for poisoning. *Oral Aid* and *Pediatric Aid* Infusion of plant given to babies with coated tongues. *Tuberculosis Remedy* Bark used for tuberculosis. (79:47) **Thompson** *Antidiarrheal* Chewed berry juice swallowed, infusion of berries taken, or mashed berries eaten for diarrhea. (187:200) *Dermatological Aid* Strong decoction of wood used to cleanse sores. (164:455) Strong decoction of wood used as a wash for sores. (164:458) Decoction of berries, bark, or leaves, sometimes mixed with bear grease, used as a wash for sores. *Eye Medicine* Strained decoction of scraped bark or leaves used as a wash for sore eyes or impending blindness. Crushed berries or decoction of berries used as a wash for sore eyes or impending blindness. (187:200) *Gastrointestinal Aid* Decoction of stems or roots taken for stomach trouble. (164:458) Sap from young shoots used for stomachache. *Gynecological Aid* Decoction of berries, bark, or leaves used as an antiseptic wash for breasts before nursing. Decoction of berries, bark, or leaves taken after a birth "to clean you out." *Laxative* Sap from young shoots used as a laxative. (187:200) *Pediatric Aid* Mild decoction of wood used as a wash to keep babies healthy. (164:458) *Poison* Berries considered "deadly poisonous" if more than two or three eaten. (164:511) Berries considered poisonous, even fatal. (164:489) Berries considered very poisonous. An antidote for poisoning from the berries was to eat a large quantity of lard. (187:200) **Wet'suwet'en** *Eye Medicine* Bark used to make an eye medicine. (73:152)
- **Food**—**Okanagan-Colville** *Forage* Berries eaten by ruffed grouse and other birds. (188:95) **Squaxin** *Dried Food* Berries dried and used for food. (79:47)
- **Fiber**—**Okanagan-Colville** *Brushes & Brooms* Branches tied together in a bundle to make a broom. (188:95) **Paiute** *Furniture* Wood used in the construction of cradleboards. (111:112)
- **Other**—**Blackfoot** *Paint* Green twigs burned and smoke used to blacken newly made pipes. (82:59) **Cheyenne** *Ceremonial Items* Bushes used to make the Sun Dance altar. (83:17) **Kwakiutl, Southern** *Tools* Branch tips ignited and used for cauterizing. (183:280) **Nez Perce** *Protection* Branches looped around cradleboards to protect babies from ghosts. (82:59) **Nitinaht** *Protection* Bark soaked, rubbed in water, and used as a skin wash for protection and truthfulness. *Toys & Games* Berries rubbed on the face during a game called "lehal." (186:102) **Paiute** *Tools* Sharpened stem used as a root digging stick. *Toys & Games* Sticks marked and used in a dice game. (111:112) **Thompson** *Soap* Mild decoction of wood used to wash babies. (164:455)

Symphoricarpos albus (L.) Blake **var. *albus***, Common Snowberry
- **Drug**—**Bella Coola** *Diuretic* Decoction of branches taken as a diuretic and for gonorrhea. *Venereal Aid* Decoction of branches taken as "the best cure for gonorrhea." **Carrier, Southern** *Eye Medicine* Juice of ripe berries used in sore eyes. (as *Symphoricarpus racemosa* 150:64) **Kwakiutl** *Analgesic* Moxa of tips while still on plant used for headache. (as *Symphoricarpor racemosus* 20:386) **Ojibwa** *Gynecological Aid* Infusion of root taken "to clear up the afterbirth" and hasten convalescence. (as *Symphoricarpos racemosus* 153:361)
- **Fiber**—**Yokia** *Brushes & Brooms* Slender twigs bound together and used as brooms for sweeping. (as *S. racemosus* 41:389)
- **Other**—**Hoh** *Ceremonial Items* Used in ceremonies. (as *S. racemosus* 137:68) **Paiute** *Hunting & Fishing Item* Shoots used to make small bird arrows. (as *Symphoricarpus racemosus* 118:52) **Quileute** *Ceremonial Items* Used in ceremonies. (as *Symphoricarpos racemosus* 137:68) **Shoshoni** *Hunting & Fishing Item* Shoots used to make small

bird arrows. (as *Symphoricarpus racemosus* 118:52) **Yokia** *Hunting & Fishing Item* Medium-sized branches used to make first-rate arrows. (as *Symphoricarpus racemosus* 41:389)

Symphoricarpos albus var. *laevigatus* (Fern.) Blake, Common Snowberry

- **Drug**—**Yuki** *Dermatological Aid* Infusion of plant used as a wash for sores. (as *S. rivularis* 49:47)
- **Food**—**Haisla & Hanaksiala** *Forage* Berries eaten by crows and bears. (43:231)
- **Fiber**—**Costanoan** *Brushes & Brooms* Brushy stems used in brooms. (as *S. rivularis* 21:254)

Symphoricarpos longiflorus Gray, Desert Snowberry

- **Drug**—**Paiute** *Analgesic* Decoction of plant taken for stomach pains. *Gastrointestinal Aid* Decoction of plant taken for indigestion or stomach pains. (180:143)

Symphoricarpos mollis Nutt., Creeping Snowberry

- **Drug**—**Shuswap** *Eye Medicine* Infusion of berries used as a wash for sore and tired eyes. (123:61)
- **Other**—**Shuswap** *Smoking Tools* Stems used to make pipestems. (123:61)

Symphoricarpos occidentalis Hook., Western Snowberry

- **Drug**—**Blackfoot** *Veterinary Aid* Decoction of berries given to horses for water retention. (97:55) **Dakota** *Eye Medicine* Infusion of leaves used as wash for weak or inflamed eyes. (70:116) **Meskwaki** *Gynecological Aid* Infusion of root taken to cleanse the afterbirth and aid in convalescence. (152:207) **Omaha** *Eye Medicine* Infusion of leaves used as wash for weak or inflamed eyes. **Ponca** *Eye Medicine* Infusion of leaves used as wash for weak or inflamed eyes. (70:116)
- **Food**—**Blackfoot** *Starvation Food* Fruits eaten in times of scarcity. (97:55) **Sioux** *Fruit* Fruit used for food. (19:24)
- **Fiber**—**Montana Indian** *Brushes & Brooms* Shrubs made into brooms. (19:24)
- **Other**—**Blackfoot** *Fuel* Green twigs used to make a fire to blacken the surface of newly made pipes. (97:55) **Lakota** *Toys & Games* Stems made into arrows used to shoot at dogs in play. (139:43) **Montana Indian** *Hunting & Fishing Item* Slender twigs used for arrow shafts. (19:24)

Symphoricarpos orbiculatus Moench, Coralberry

- **Drug**—**Dakota** *Eye Medicine* Decoction of inner bark or leaves used for sore eyes. (as *S. symphoricarpos* 69:367) Infusion of leaves used as wash for weak or inflamed eyes. (as *S. symphoricarpos* 70:116) **Ojibwa, South** *Eye Medicine* Cold decoction of root bark applied to sore eyes. (as *Symphoricarpus vulgaris* 91:200) **Omaha** *Eye Medicine* Infusion of leaves used as wash for weak or inflamed eyes. **Ponca** *Eye Medicine* Infusion of leaves used as wash for weak or inflamed eyes. (as *Symphoricarpus symphoricarpos* 70:116)
- **Other**—**Dakota** *Decorations* Wood made into charcoal and used in tattooing. (as *S. symphoricarpos* 69:367)

Symphoricarpos oreophilus Gray, Whortleleaf Snowberry

- **Drug**—**Navajo, Ramah** *Ceremonial Medicine* and *Emetic* Leaves used as a ceremonial emetic. (191:45)

Symphoricarpos oreophilus var. *parishii* (Rydb.) Cronq., Parish's Snowberry

- **Drug**—**Navajo, Kayenta** *Throat Aid* Plant used for sore throat. (as *S. parishii* 205:44)

Symphoricarpos sp., Snowberry

- **Drug**—**Chippewa** *Cathartic* Decoction of root, a very strong remedy, taken as a physic. (53:346)
- **Fiber**—**Havasupai** *Furniture* Stems used to make the rim of the shade for cradleboards. (197:243)
- **Other**—**Great Basin Indian** *Hunting & Fishing Item* Lightweight, pithy stems used to make bird arrow shafts. (121:50)

Symphytum, Boraginaceae

Symphytum officinale L., Common Comfrey

- **Drug**—**Cherokee** *Antidiarrheal* Taken for "flux" or dysentery. *Gastrointestinal Aid* Infusion taken for heartburn. *Gynecological Aid* Infusion taken for heartburn in pregnancy and for flooding after birth. *Laxative* Infusion taken for "costiveness" in pregnancy. *Orthopedic Aid* Used for sprains and bruises. *Venereal Aid* Infusion of roots in water used for gonorrhea. (80:30)

Symplocaceae, see *Symplocos*

Symplocarpus, Araceae

Symplocarpus foetidus (L.) Salisb. ex Nutt., Skunk Cabbage

- **Drug**—**Abnaki** *Antirheumatic (External)* Used for swellings. (144:153) **Chippewa** *Cough Medicine* Infusion of roots taken as a cough medicine. (71:124) **Delaware** *Analgesic* Poultice of crushed leaves applied for pain. *Misc. Disease Remedy* Small portions of leaves chewed for epilepsy. *Pulmonary Aid* Infusion of roots used for whooping cough. (176:37) **Delaware, Oklahoma** *Analgesic* Poultice of crushed leaves applied for pain. *Anticonvulsive* Leaves chewed by epileptics. *Cough Medicine* Infusion of root taken for whooping cough. (175:31, 80) **Iroquois** *Anthelmintic* Plant used for children with worms. (87:278) *Antirheumatic (External)* Steam from compound decoction of roots used for rheumatism. *Dermatological Aid* Poultice used on bite from a fight or dog and caused the biter's teeth to fall out. (87:277) Plant used for bad wounds. (87:278) *Gynecological Aid* Compound decoction of upper parts and seeds taken for "falling of the womb." Decoction of crushed stalks used as a douche for displacement of womb. (87:277) "Pass seed over female genitals to bring about childbirth." *Pediatric Aid* Used for children with worms. (87:278) *Tuberculosis Remedy* Infusion of powdered root taken for consumption. *Witchcraft Medicine* Poultice used on bite from a fight or dog and caused the biter's teeth to fall out. (87:277) **Malecite** *Unspecified* Used for medicines. (160:6) **Menominee** *Adjuvant* Root used as a seasoner with medicines. *Analgesic* Root used for cramps. (as *Spatheyma foetida* 151:23, 24) *Anticonvulsive* Compound infusion of dried, powdered root used by children and adults for convulsions. (as *Spatheyma foetida* 54:128, 129) *Dermatological Aid* Poultice of dried root applied to wounds. (as *Spatheyma foetida* 151:23, 24) *Heart Medicine* Decoction of root used for "weak heart." (as *Spatheyma foetida* 54:128, 129) *Hemostat* Root hairs used for hemorrhages. (as *Spatheyma foetida* 151:23, 24) *Pediatric Aid* Compound infusion used by children or adults for convulsions. (as *Spathyema foetida* 54:128) *Witch-*

craft Medicine Root used in tattooing, as a talisman against the return of diseases. (as *Spatheyma foetida* 151:23, 24) **Meskwaki** *Dermatological Aid* Poultice of leaf bases applied to swellings. *Toothache Remedy* Fine rootlets or root hairs used for toothache. *Unspecified* Seeds used as medicine. (152:203) **Micmac** *Analgesic* Herbs used for headache. (40:62) **Mohegan** *Anticonvulsive* Raw leaves rolled and chewed for fits. (as *Spathyema foetida* 174:268) Small piece of leaf eaten for epileptic seizures. (as *Spathyema foetida* 176:75, 132) **Nanticoke** *Cold Remedy* Infusion of leaves taken as a cold medicine. (175:55, 84)
- Food—**Iroquois** *Vegetable* Young leaves and shoots cooked and seasoned with salt, pepper, or butter. (196:118)
- Other—**Iroquois** *Soap* Infusion of powdered root used as wash to "cure strong smell under your arm." (87:277)

Symplocos, Symplocaceae

Symplocos tinctoria (L.) L'Hér., Common Sweetleaf
- Drug—**Choctaw** *Febrifuge* Decoction of scraped roots taken for fevers. (177:50)

Syringa, Oleaceae

Syringa vulgaris L., Common Lilac
- Drug—**Iroquois** *Oral Aid* Bark or leaves chewed by children for sore mouths. Bark or leaves chewed for sore mouth caused by kissing a girl with menses, or caused by smoking someone else's pipe. *Pediatric Aid* Bark or leaves chewed by children for sore mouths. (87:413)

Syzygium, Myrtaceae

Syzygium malaccense (L.) Merr. & Perry, Malaysian Apple
- Drug—**Hawaiian** *Dermatological Aid* Bark pounded, mixed with salt and coconut fibers, squeezed, and resulting liquid used on deep cuts. *Emetic* Bark chewed to cause vomiting and bring out the tough phlegm from the throat and lungs. *Gastrointestinal Aid* Bark and other plants pounded, squeezed, and the resulting liquid given to children for stomach weakness. *Oral Aid* Bark and other plants pounded, squeezed, and the resulting liquid taken for bad breath and mouth sores. *Pediatric Aid* Bark and other plants pounded, squeezed, and the resulting liquid given to children for general debility or for stomach weakness. *Strengthener* Bark and other plants pounded, squeezed, and the resulting liquid given to children for general debility. (as *Jambosa malaecensis* 2:31)

Taccaceae, see *Tacca*

Tacca, Taccaceae

Tacca leontopetaloides (L.) Kuntze, Batflower
- Food—**Hawaiian** *Unspecified* Tubers grated, roasted, and eaten. (as *T. pinnatifida* 112:68)

Taenidia, Apiaceae

Taenidia integerrima (L.) Drude, Yellow Pimpernel
- Drug—**Menominee** *Adjuvant* Plant used as a seasoner to make various female remedies taste good. *Pulmonary Aid* Infusion of root taken for pulmonary troubles. *Respiratory Aid* Steeped root chewed for "bronchial affections." (151:56) **Meskwaki** *Adjuvant* Root used as a seasoner for other medicines because of the good smell. (152:250) **Ojibwa** *Hunting Medicine* Seeds smoked in a pipe when one goes hunting for they will bring him luck. (153:432)

Tagetes, Asteraceae

Tagetes erecta L., Aztec Marigold
- Drug—**Cherokee** *Dermatological Aid* Infusion used as wash for "eczema." (80:44)
- Dye—**Cherokee** *Yellow* Flowers used to make a yellow dye. (80:44)

Tagetes micrantha Cav., Licorice Marigold
- Drug—**Navajo** *Cold Remedy* Plants used for colds. *Febrifuge* Plants used for fevers. *Gastrointestinal Aid* Plants used for stomach troubles. *Other* Plants used for "summer complaint." (55:89)

Talinum, Portulacaceae

Talinum parviflorum Nutt., Sunbright
- Drug—**Navajo, Ramah** *Dermatological Aid* Poultice of root bark applied to sores and infusion of root bark used as lotion. *Veterinary Aid* Cold simple or compound infusion given to livestock as an aphrodisiac. (191:26)

Tamaricaceae, see *Tamarix*

Tamarix, Tamaricaceae

Tamarix aphylla (L.) Karst., Athel Tamarisk
- Other—**Pima** *Fuel* Wood used in winter for fuel. (47:107)

Tamarix gallica L., French Tamarisk
- Fiber—**Keres, Western** *Building Material* Plant considered an important source of wood. (171:72)

Tanacetum, Asteraceae

Tanacetum parthenium (L.) Schultz-Bip., Feverfew
- Drug—**Cherokee** *Orthopedic Aid* Infusion used to bathe swollen feet. (as *Chrysanthemum parthenium* 80:34) **Mahuna** *Antirheumatic (Internal)* Plant used for rheumatism. (as *Crysanthemum parthenium* 140:60)

Tanacetum vulgare L., Common Tansy
- Drug—**Cherokee** *Analgesic* Infusion used for backache. *Anthelmintic* Given to children for worms. *Gynecological Aid* Worn around waist and in shoes to prevent miscarriages and abortions. *Orthopedic Aid* Infusion used for backache. *Pediatric Aid* Given to children for worms. *Tonic* Taken as a tonic. (80:58) **Cheyenne** *Other* Infusion of leaves and blossoms taken for weakness and dizziness. (76:190) *Strengthener* Infusion of pulverized leaves and blossoms taken for weakness. *Vertigo Medicine* Infusion of pulverized leaves and blossoms taken for dizziness. (76:190) **Chippewa** *Abortifacient* Decoction of leaf taken, especially by young girls, for "stoppage of period." (53:358) *Diaphoretic* Compound infusion of leaves taken to "produce profuse perspiration." (53:354) *Ear Medicine* Weak decoction of root used as drops for sore ear. (53:362) *Febrifuge* Compound infusion of leaves taken to "break up a fever." (53:354) *Throat Aid* Decoction of root gargled or dried root chewed for sore throat. (53:342) **Delaware,**

Ontario *Gastrointestinal Aid* Whole plant used for stomach disorders. (175:66, 82) **Iroquois** *Analgesic* Poultice of leaves applied to the head for headaches. Poultice of smashed leaves applied to body pains caused by too much gall. (as *Chrysanthemum vulgare* 87:472) *Cold Remedy* Poultice of plants applied for colds. *Dermatological Aid* Plant used for bruises and cuts. (as *Chrysanthemum vulgare* 87:473) *Liver Aid* Poultice of smashed leaves applied to body pains caused by too much gall. (as *Chrysanthemum vulgare* 87:472) *Orthopedic Aid* Plant used for bone decay and headaches. *Panacea* Poultice of plants applied for any ailment. (as *Chrysanthemum vulgare* 87:473) **Malecite** *Contraceptive* Infusion of plant used to prevent pregnancy. (116:259) *Kidney Aid* Infusion of dried leaves used for kidney troubles. *Veterinary Aid* Infusion of dried leaves used for horses with colic. (116:243) **Micmac** *Contraceptive* Herbs used to prevent pregnancy. *Kidney Aid* Leaves used for kidney trouble. (40:62) **Mohegan** *Dietary Aid* Cold, compound infusion taken as an appetizer. *Gastrointestinal Aid* Cold, compound infusion taken for the stomach. (174:266) Compound infusion of leaves taken as a stomach aid and to improve the appetite. (176:75, 132) **Nanticoke** *Diaphoretic* Whole plant used as a sudorific. (175:58, 84) **Ojibwa** *Febrifuge* Plant used as a fever medicine. (153:366) **Paiute** *Emetic* Decoction of leaves taken as an emetic. (180:143, 144) **Shinnecock** *Analgesic* Infusion of leaves taken for "inside pains." (30:119) **Shoshoni** *Antidiarrheal* Decoction of leaves taken for bloody diarrhea. *Dermatological Aid* Decoction of leaves used as an antiseptic wash. *Disinfectant* Decoction of leaves and sometimes stems used as a warm antiseptic wash. (180:143, 144)

- *Other*—**Ojibwa** *Hunting & Fishing Item* Yellow flowers used in the odorous hunting mixture smoked to attract deer. (153:429)

Taraxacum, Asteraceae

Taraxacum californicum Munz & Johnston, California Dandelion
- *Food*—**Cahuilla** *Vegetable* Stems and leaves gathered and eaten in spring and early summer. (15:141)

Taraxacum officinale G. H. Weber ex Wiggers, Common Dandelion
- *Drug*—**Aleut** *Dermatological Aid* Poultice of steamed or wilted leaves applied to indolent ulcers. *Gastrointestinal Aid* Poultice of steamed or wilted leaves applied to stomachaches. *Throat Aid* Poultice of steamed or wilted leaves applied to sore throats. (149:327) **Algonquin, Quebec** *Blood Medicine* Greens eaten to purify the blood. *Poultice* Leaves used for plasters or poultices. (18:242) **Bella Coola** *Analgesic* and *Gastrointestinal Aid* Decoction of root taken for stomach pain. (150:65) **Cherokee** *Blood Medicine* Infusion of root used for blood. *Sedative* Infusion of herb used to "calm nerves." *Toothache Remedy* Chewed for toothache. (80:31) *Unspecified* Leaves and stems used as medicine. (126:35) **Chippewa** *Gynecological Aid* Compound infusion of root taken to produce postpartum milk flow. (53:360) **Delaware** *Laxative* Plant used to make a "laxative-tonic." (176:39) **Delaware, Oklahoma** *Laxative* and *Tonic* Plant used to make a "laxative-tonic." (175:32, 80) **Hoh** *Unspecified* Used for medicine. (137:69) **Iroquois** *Analgesic* Compound decoction of dried plants taken for pain. Compound infusion of roots and bark taken for back pain. (87:477) Infusion of flowers, roots, and roots from another plant taken for lower back pain. (141:61) *Blood Medicine* Decoction of plants taken for anemia. (87:478) *Dermatological Aid* Compound infusion of plants and roots taken and used as wash for liver spots. (87:476) Compound decoction of bark and roots taken for sores caused by bad blood. (87:478) *Emetic* Infusion of roots taken as an emetic. (87:476) *Eye Medicine* Compound infusion of roots and bark taken for dark circles and puffy eyes. *Kidney Aid* Compound infusion of roots and bark taken for kidney trouble and dropsy. *Laxative* Compound decoction of flowers and leaves taken as a laxative. *Love Medicine* Decoction of roots used as wash for a love medicine. *Orthopedic Aid* Compound infusion of roots and bark taken for back pain. *Pulmonary Aid* Compound decoction of dried plants taken for swollen lungs. (87:477) *Toothache Remedy* Flower stem chewed for worms in the teeth that cause decay. *Urinary Aid* Poultice of smashed flowers applied to swollen testicles. (87:476) Decoction of plants used as wash on parts affected by smashed testicles. *Witchcraft Medicine* Decoction of roots used as a wash for an anti-witch medicine. (87:477) **Kiowa** *Gynecological Aid* Decoction of young leaves taken by women for menstrual cramps. (192:62) **Meskwaki** *Pulmonary Aid* Infusion of root taken for chest pain when other remedies fail. (152:218) **Mohegan** *Cathartic* Infusion of plant taken as a physic. (174:266) Strong infusion of dried leaves taken as a physic. (176:76, 132) *Tonic* Dandelion and white daisy used to make wines and taken as tonics. (30:121) Compound decoction or infusion of plants taken as a spring tonic. (174:266) Compound infusion of root taken as a tonic. (176:76, 132) **Ojibwa** *Blood Medicine* Roots used as a blood medicine. (135:238) *Gastrointestinal Aid* Infusion of root taken for heartburn. (153:366) **Papago** *Analgesic* Infusion of blossoms taken for menstrual cramps. *Gynecological Aid* Infusion of blossoms taken for menstrual cramps. (34:65) **Potawatomi** *Tonic* Root used as a bitter tonic. (154:54) **Quileute** *Unspecified* Used for medicine. (137:69) **Rappahannock** *Blood Medicine* Infusion of root taken as a blood tonic. *Gastrointestinal Aid* Infusion of root taken for dyspepsia. (161:34) **Shinnecock** *Tonic* Dandelion and white daisy used to make wines and taken as tonics. (30:121)

- *Food*—**Apache, Chiricahua & Mescalero** *Spice* Flower used to flavor drinks and make them stronger. (33:51) **Cherokee** *Vegetable* Leaves and stems used for potherbs and salads. (126:35) **Iroquois** *Beverage* Used to make wine. (142:99) *Vegetable* Young plants boiled and eaten as greens. (124:93) Cooked and seasoned with salt, pepper, or butter. (196:118) **Kiowa** *Vegetable* Young leaves used as greens. (192:62) **Malecite** *Unspecified* Species used for food. (160:6) **Menominee** *Vegetable* Leaves cooked with maple sap vinegar for a dish of greens. (151:65) **Meskwaki** *Vegetable* Spring leaves used as greens and cooked with pork. (152:257) **Micmac** *Vegetable* Leaves used as greens in food. (159:258) **Mohegan** *Unspecified* Cooked and used for food. (176:83) **Ojibwa** *Vegetable* Young leaves gathered in spring and cooked as greens with pork or venison and maple sap vinegar. (153:399) **Okanagan-Colville** *Vegetable* Leaves eaten as greens. (188:85) **Papago** *Vegetable* Cooked or uncooked leaves eaten as greens. (34:14) **Potawatomi** *Unspecified* Leaves cooked with maple sap vinegar and often combined with pork or deer meat. (154:98) **Ute** *Unspecified* Leaves formerly used as food. (38:36)

- *Other*—**Hesquiat** *Toys & Games* Hollow stems made into whistles. (185:62)

Taraxacum officinale ssp. *vulgare* (Lam.) Schinz & R. Keller, Common Dandelion

- *Drug*—Navajo, Ramah *Dermatological Aid* Poultice of crushed plant applied to swellings. *Gynecological Aid* Cold infusion of plant used to speed delivery of baby. (as *T. palustre* var. *vulgare* 191:53) **Tewa** *Dermatological Aid* Poultice of pulverized leaves mixed with dough applied to a bad bruise. *Orthopedic Aid* Poultice of pulverized fresh leaves used to dress bone fractures. (as *T. taraxacum* 138:61)
- *Food*—**Tewa** *Vegetable* Young plants eaten as greens. (as *T. taraxacum* 138:61)

Taraxacum sp., Dandelion
- *Food*—**Alaska Native** *Dietary Aid* Raw, fresh leaves used as an excellent source for vitamin C and provitamin A. *Vegetable* Young, tender leaves used raw or cooked as a green vegetable. (85:71) **Algonquin, Quebec** *Vegetable* Leaves used for greens. (18:109) **Carrier** *Vegetable* Leaves boiled and eaten. (31:81) **Cree** *Starvation Food* Infusion of roots given in copious drafts at hourly intervals. (16:494) **Eskimo, Alaska** *Unspecified* Boiled leaves of young plants eaten, but not considered an important food source. (1:38) Scalded leaves used for food. (4:716) **Eskimo, Arctic** *Vegetable* Tender, young leaves eaten in salads and as a potherb. (128:29) **Eskimo, Inuktitut** *Spice* Used as a condiment in fish soup. (202:185)

Tauschia, Apiaceae

Tauschia arguta (Torr. & Gray) J. F. Macbr., Southern Umbrellawort
- *Drug*—**Luiseño** *Unspecified* Root used for medicinal purposes. (as *Deweya arguta* 155:230)

Tauschia parishii (Coult. & Rose) J. F. Macbr., Parish's Umbrellawort
- *Drug*—**Kawaiisu** *Analgesic* Dried root smoke inhaled for head pains. Dried root smoke used for eye pains. Ground root applied as a salve for aching limbs. Infusion of roots taken for inside pain. *Cold Remedy* Dried root smoke inhaled for head colds. *Dermatological Aid* Infusion of pounded roots used as a bath for swollen limbs. *Eye Medicine* Dried root smoke used for eye pains. *Orthopedic Aid* Ground root applied as a salve for aching limbs. Infusion of pounded roots used as a bath for swollen limbs. *Toothache Remedy* Mashed, ground root placed on hot rock, and cheek laid on rock for toothache. (206:66)

Taxaceae, see *Taxus, Torreya*

Taxodiaceae, see *Sequoia, Taxodium*

Taxodium, Taxodiaceae, conifer

Taxodium ascendens Brongn., Pond Cypress
- *Fiber*—**Seminole** *Building Material* Plant used to make houses. *Canoe Material* Wood used to make canoes. (169:471)
- *Other*—**Seminole** *Ceremonial Items* Plant used to make hunting dance posts. Plant used for burial purposes and made into coffin logs. (169:471) *Containers* Wood used to make medicine bowls. (169:95) *Cooking Tools* Plant used to make spoons and food paddles. *Hide Preparation* Plant used for tanning skins. *Hunting & Fishing Item* Plant used to make arrowheads. *Musical Instrument* Plant used to make drums. *Stable Gear* Plant used to make ox yokes and ox bows. *Tools* Plant used to make heddles, mortars, and pestles. *Toys & Games* Plant used to make ball poles, spoon ball sticks, and dolls. (169:471)

Taxodium distichum (L.) L. C. Rich., Bald Cypress
- *Fiber*—**Choctaw** *Cordage* Bark used to make cordage. (as *Cupressus disticha* 25:15)

Taxus, Taxaceae, conifer

Taxus baccata L., English Yew
- *Drug*—**Iroquois** *Abortifacient* Compound taken for menstruation when stopped by a cold. *Adjuvant* "Put in all medicines to give them strength." *Antirheumatic (External)* Steam from compound decoction used for rheumatism. *Antirheumatic (Internal)* Compound decoction taken for rheumatism. *Cold Remedy* Compound decoction taken for colds. Compound taken for menstruation when stopped by a cold. *Cough Medicine* Compound decoction taken for coughs. *Diaphoretic* Used for colds and sweating. *Orthopedic Aid* Decoction of twigs used for finger or leg numbness. *Respiratory Aid* Steam from decoction used for chest colds. *Tuberculosis Remedy* Compound decoction taken during the early stages of consumption. (87:264)

Taxus brevifolia Nutt., Pacific Yew
- *Drug*—**Bella Coola** *Pulmonary Aid* Decoction of branches with leaves taken "for the lungs." (150:48) **Chehalis** *Diaphoretic* Infusion of crushed leaves used as a wash to cause perspiring. *Panacea* Infusion of crushed leaves used as a wash to improve general health. **Cowlitz** *Dermatological Aid* Poultice of ground leaves applied to wounds. (79:16) **Haihais** *Gastrointestinal Aid* Decoction of wood and bark used for stomach pains. *Internal Medicine* Decoction of wood and bark used for internal ailments. (43:319) **Hanaksiala** *Urinary Aid* Cooled decoction of small wood pieces taken for bloody urine. (43:187) **Karok** *Blood Medicine* Decoction of bark taken as a "blood medicine." (6:57) *Gastrointestinal Aid* Decoction of twig bark taken for stomachaches. (148:379) **Kitasoo** *Gastrointestinal Aid* Decoction of wood and bark used for stomach pains. *Internal Medicine* Decoction of wood and bark used for internal ailments. (43:319) **Klallam** *Analgesic* Decoction of leaves taken for internal injury or pain. (79:16) **Mendocino Indian** *Poison* Seeds considered poisonous. (41:305) **Okanagan-Colville** *Burn Dressing* Wood scrapings and Vaseline used as a sunburn ointment. (188:35) **Quinault** *Dermatological Aid* Poultice of chewed leaves applied to wounds. *Pulmonary Aid* Decoction of dried bark taken as lung medicine. **Swinomish** *Strengthener* Smooth twigs of plant rubbed on the body to gain strength. (79:16) **Thompson** *Panacea* Decoction of bark taken for any illness. (187:111) **Tsimshian** *Cancer Treatment* Plant used for cancer. *Internal Medicine* Plant used for internal ailments. (43:187) **Yurok** *Blood Medicine* Decoction of bark taken to "purify the blood." (6:57)
- *Food*—**Karok** *Fruit* Berries eaten one at a time. (6:57) **Mendocino Indian** *Fruit* Red, fleshy berries used for food. (41:305)
- *Fiber*—**Clallam** *Canoe Material* Wood used to make canoe paddles. (57:195) **Cowlitz** *Brushes & Brooms* Used to make combs. (79:16) **Haisla** *Canoe Material* Wood used to make boat ribs. (43:187) **Hesquiat** *Canoe Material* Extremely strong and resilient wood used for implements requiring strength, such as paddles. (185:48) **Hoh** *Canoe Material* Wood used to make canoe paddles. (137:57) **Karok** *Sewing Material* Leaves used to sew sacks. (6:57) **Kwakiutl, Southern** *Brushes & Brooms* Wood used to make combs. (183:271) **Nitinaht** *Canoe Material* Wood used to make steering paddles and oarlock holders. *Sewing Material* Wood used to make mat making needles. (186:75)

Oweekeno *Brushes & Brooms* Wood used to make combs. *Canoe Material* Wood used to make canoe paddles. (43:75) **Pomo** *Basketry* Roots used in basketry. *Sewing Material* Roots twined as fabric. (66:11) **Pomo, Kashaya** *Basketry* Root used as weft in twined baskets and root very strong and especially good for hopper mortar baskets. (72:121) **Quileute** *Canoe Material* Wood used to make canoe paddles. (137:57) **Quinault** *Brushes & Brooms* Used to make combs. *Canoe Material* Used to make canoe bailers. (79:16) **Salish, Coast** *Brushes & Brooms* Wood used to make combs. *Canoe Material* Wood used to make paddles. (182:72) **Thompson** *Snow Gear* Twigs used to make snowshoe frames. (164:500)

- **Other**—**Bella Coola** *Hunting & Fishing Item* Wood used to make harpoon shafts. *Tools* Wood used to make adz handles, bark scrapers, and wedges. (184:198) **Chehalis** *Hunting & Fishing Item* Wood used to make bows and arrows. (79:16) **Clallam** *Cooking Tools* Wood used to make barbecue stakes. *Hunting & Fishing Item* Wood used to make bows and arrows. (57:195) *Smoke Plant* Needles mixed with tobacco and smoked. (57:199) *Tools* Wood used to make digging sticks. (57:195) **Costanoan** *Hunting & Fishing Item* Wood used to make bows. (21:248) **Cowlitz** *Musical Instrument* Used to make drum frames. *Tools* Used to make wedges for splitting logs and digging sticks for roots and clams. (79:16) **Flathead** *Hunting & Fishing Item* Branches used to make bows. (82:49) **Haihais** *Hunting & Fishing Item* Wood used to make bows, arrowheads, and halibut hooks. (43:319) **Haisla & Hanaksiala** *Weapon* Wood used to make war paddles and war clubs. **Hanaksiala** *Hunting & Fishing Item* Wood used to make bows. Wood used to make clubs to knock out fish, and halibut hooks. Sticks used to strike rods of cedar to make noise while driving animals to be killed. *Tools* Wood used to make digging sticks. (43:187) **Hesquiat** *Tools* Extremely strong and resilient wood used for implements requiring strength, such as needles, wedges and mat pressers, and sticks for prying open mussels and chitons. *Weapon* Extremely strong and resilient wood used for implements requiring strength: spear handles and war clubs. (185:48) **Hoh** *Ceremonial Items* Wood used in various ceremonies. *Hunting & Fishing Item* Wood used to make bows, arrows, and whaling harpoon stocks. *Toys & Games* Wood used to make game disks. (137:57) **Karok** *Containers* Bark used as the covering for stone knives. (148:379) *Hunting & Fishing Item* Wood used to make bows. (6:57) Wood used to make bows. *Smoking Tools* Wood used to make tobacco pipes. (148:379) *Tools* Wood used to make various tools. (6:57) Bark used as the handle for stone knives. (148:379) **Kitasoo** *Hunting & Fishing Item* Wood used to make bows, arrowheads, and halibut hooks. (43:319) **Klallam** *Hunting & Fishing Item* Wood used to make bows and arrows. *Smoke Plant* Needles dried, pulverized, and used in place of tobacco for smoking. (79:16) **Klamath** *Hunting & Fishing Item* Wood used for bows. (45:89) **Kwakiutl, Southern** *Cooking Tools* Wood used to make bowls. (183:271) *Tools* Wood used as a bark lifter to peel red cedar bark off the trees. (183:267) Used to make curved digging sticks for prying wild clover roots. (183:285) Wood used to make fire tongs. Trees used to measure strength. Men who could twist a yew tree from crown to butt were considered very strong. (183:271) **Makah** *Containers* Used to make trinket boxes. *Cooking Tools* Used to make spoons and dishes. *Hunting & Fishing Item* Wood used to make bows, arrows, and whale harpoon shafts. (79:16) **Mendocino Indian** *Hunting & Fishing Item* Formerly used to make the strongest bows. (41:305) **Montana Indian** *Hunting & Fishing Item* Wood used for making bows. (19:24) **Nitinaht** *Ceremonial Items* Branches used as scrubbers in the manhood training rituals of young boys. *Hunting & Fishing Item* Wood used to make bows, clubs, lances, whaling harpoon shafts, and other fishing gear. (186:75) *Tools* Wood used to make chest-high digging sticks. (67:226) Wood used to make digging sticks for harvesting pacific cinquefoil roots. (186:118) Wood used to make digging and prying sticks, mat pressers, and wedges. (186:75) **Okanagan-Colville** *Hunting & Fishing Item* Wood used to make bows. (188:35) **Oweekeno** *Cooking Tools* Wood used to make serving bowls for food. *Hunting & Fishing Item* Wood made into bows for use with arrows. *Tools* Wood used to make digging sticks. (43:75) **Paiute** *Hunting & Fishing Item* Wood used to make bows. (111:39) **Pit River** *Cash Crop* Formerly sold to the Ukiah Indians. (41:305) **Pomo** *Cooking Tools* Wood used for mush stirrers. *Hunting & Fishing Item* Wood used for bows. (66:11) **Pomo, Kashaya** *Cooking Tools* Wood used to make mush stirrers. *Hunting & Fishing Item* Wood used to make bows. *Tools* Wood used to make digging sticks. (72:121) **Quileute** *Ceremonial Items* Wood used in various ceremonies. *Hunting & Fishing Item* Wood used to make bows, arrows, and whaling harpoon stocks. *Toys & Games* Wood used to make game disks. (137:57) **Quinault** *Cooking Tools* Used to make spoons. *Hunting & Fishing Item* Wood used to make bows, arrows, all harpoon shafts, clubs, and dip net frameworks. *Tools* Used to make digging sticks for roots and clams. (79:16) **Salish, Coast** *Fasteners* Pegs dipped in boiling pitch and driven into holes in the corners of a box to seal it. *Hunting & Fishing Item* Wood used to make bows, harpoon shafts, and halibut hooks. *Toys & Games* Wood used to make gambling disks. *Weapon* Wood used to make weapons. (182:72) **Samish** *Hunting & Fishing Item* Wood used to make bows, arrows, all harpoon shafts, clubs, and dip net frameworks. *Smoke Plant* Needles dried, pulverized, and used in place of tobacco for smoking. *Tools* Used to make wedges for splitting logs. (79:16) **Shasta** *Hunting & Fishing Item* Wood used to make bows. (117:217) **Snohomish** *Hunting & Fishing Item* Wood used to make bows and arrows. **Swinomish** *Hunting & Fishing Item* Wood used to make bows, arrows, all harpoon shafts, clubs, and dip net frameworks. *Smoke Plant* Needles dried, pulverized, and used in place of tobacco for smoking. *Tools* Used to make wedges for splitting logs and digging sticks for roots and clams. *Weapon* Used to make war clubs. (79:16) **Thompson** *Ceremonial Items* Branches used by bereaved people to scrub and purify themselves. (187:111) *Hunting & Fishing Item* Wood used for making bows. (164:499) *Tools* Wood "hardened" in the fire and used to make wedges, ax handles, and digging sticks. (187:111) **Tolowa** *Smoking Tools* Wood used to make pipes. **Yurok** *Hunting & Fishing Item* Wood used to make bows. (6:57)

Taxus canadensis Marsh., Canada Yew

- **Drug**—**Abnaki** *Antirheumatic (External)* Leaves used for rheumatism. (144:155) *Antirheumatic (Internal)* Infusion of leaves taken for rheumatism. (144:163) **Algonquin, Quebec** *Antirheumatic (Internal)* Decoction of needles used for rheumatism. *Gynecological Aid* Used in a sudatory taken by women experiencing complications after childbirth and other complaints. *Poultice* Needles used for poultices. *Unspecified* Used in a sudatory taken by women experiencing com-

plications after childbirth and other complaints. (18:123) **Algonquin, Tête-de-Boule** *Abortifacient* Infusion of young branches, alone or with other plants, used for stomachaches and irregular menses. (132:132) **Chippewa** *Antirheumatic (External)* Compound decoction of twigs used as herbal steam for rheumatism. *Antirheumatic (Internal)* Compound decoction of twigs taken for rheumatism. *Herbal Steam* Compound decoction of twigs taken or used as herbal steam for rheumatism. (53:362) *Other* Plant used as one of the ingredients of "the thirty-two medicine." (71:122) **Malecite** *Unspecified* Bark used as medicine. (160:6) **Menominee** *Antirheumatic (External)* Herbal steam from branches used in sudatory for rheumatism and numbness. *Herbal Steam* Branches used in herbal steam for rheumatism, numbness, and paralysis. (151:54) **Micmac** *Analgesic* and *Blood Medicine* Parts of plant used for afterbirth pain and blood clots. *Febrifuge* Parts of plant used for fever. *Gastrointestinal Aid* Bark used for bowel and internal troubles. *Gynecological Aid* Parts of plant used for afterbirth pain and clots. *Misc. Disease Remedy* Parts of plant used for scurvy. (as *T. minor* 40:62) **Montagnais** *Febrifuge* "Brew" from plant used for weakness and fever. *Stimulant* Compound containing plant used for weakness and fever. (as *T. minor* 156:315) **Penobscot** *Cold Remedy* Infusion of twigs taken for colds. (as *T. minor* 156:309) **Potawatomi** *Diuretic* Infusion of leaves used as a diuretic. *Venereal Aid* Compound containing leaves used for gonorrhea. (154:84, 85)
- **Food**—**Iroquois** *Beverage* Fruits, leaves, cold water, and maple water fermented into a "little beer." (141:34) **Penobscot** *Beverage* Twigs used to make a beverage. (as *T. minor* 156:309)
- **Dye**—**Micmac** *Green* Leaves used to make a green dye. **Montagnais** *Green* Leaves used to make a green dye. (156:317)

Taxus cuspidata Sieb. & Zucc., Japanese Yew
- **Fiber**—**Aleut** *Canoe Material* Drift wood bent into ribs for boats. (7:29)

***Taxus* sp.**, Yew
- **Other**—**Wintoon** *Hunting & Fishing Item* Wood used to make bows. (117:264)

Tellima, Saxifragaceae

Tellima grandiflora (Pursh) Dougl. ex Lindl., Bigflower Tellima
- **Drug**—**Nitinaht** *Psychological Aid* Plants chewed as medicine to stop dreams of having sexual intercourse with the dead. *Unspecified* Used as a "special medicine." (186:127) **Skagit** *Dietary Aid* Decoction of pounded plants taken to restore the appetite. *Panacea* Decoction of pounded plants taken for any kind of sickness. (79:31)

Teloschistaceae, see *Xanthoria*

Tephrosia, Fabaceae

Tephrosia florida (F. G. Dietr.) C. E. Wood, Florida Hoarypea
- **Drug**—**Choctaw** *Dermatological Aid* Decoction of beaten roots applied to sores. **Koasati** *Snakebite Remedy* Infusion of roots applied to snakebites. (as *T. ambigua* 177:33)

Tephrosia hispidula (Michx.) Pers., Sprawling Hoarypea
- **Drug**—**Choctaw** *Cough Medicine* Root chewed and juice swallowed, too much would loosen the bowels, for bad coughs. (as *T. elegans* 28:287)

Tephrosia purpurea (L.) Pers., Fishpoison
- **Drug**—**Hawaiian** *Dermatological Aid* Leaves or buds, salt, baked coconut, water, and child's urine applied to skin diseases and cuts. *Poison* Poisonous herb. (as *T. piscatoria* 2:4) **Seminole** *Hemostat* Decoction of plant used for nosebleeds. (as *Cracca purpurea* 169:304)
- **Other**—**Hawaiian** *Hunting & Fishing Item* Leaves used to poison fish. (as *T. piscatoria* 2:4)

Tephrosia virginiana (L.) Pers., Virginia Tephrosia
- **Drug**—**Catawba** *Analgesic* Leaves put in shoes for pain in the flesh of the body (fever). (157:187) *Antirheumatic (External)* Leaves placed in shoe for rheumatism. (177:33) *Antirheumatic (Internal)* Plant used for rheumatism. *Febrifuge* Leaves put in shoes for pain in the flesh of the body (fever). (157:187) **Cherokee** *Anthelmintic* Infusion taken for worms. *Dermatological Aid* Decoction of roots used as shampoo by women to prevent hair loss. *Kidney Aid* Compound used for kidneys. *Orthopedic Aid* Compound rubbed on limbs of ballplayers to toughen them. (80:31) Decoction of roots given to children to make them strong and muscular. (177:33) *Pediatric Aid* Infusion of root given to children to make them strong and muscular. (80:31) Decoction of roots given to children to make them strong and muscular. (177:33) *Stimulant* Decoction taken for "lassitude." (80:31) **Creek** *Abortifacient* Compound decoction of plant taken and used as wash for irregular menstruation. *Reproductive Aid* Compound infusion of root used in "cases of loss of manhood." (as *Cracca virginiana* 172:658) Cold infusion of roots taken by men to regain potency. (177:33) *Tuberculosis Remedy* Roots, a very strong medicine, used in cases of "pulmonary consumption." (as *Cracca virginiana* 172:658) Plant used for pulmonary tuberculosis. (177:33) *Urinary Aid* Cold, compound infusion of root taken for bladder trouble. (as *Cracca virginiana* 172:658) Cold infusion of mashed roots taken for bladder troubles. (177:33) **Mahuna** *Gynecological Aid* Infusion of plant taken for women's diseases. (as *Cracca virginiana* 140:15) **Natchez** *Cough Medicine* Plant used as a cough medicine. (177:33)
- **Other**—**Seminole** *Hunting & Fishing Item* Plant used for fish poison. (as *Cracca virginiana* 169:485)

Tetraclea, Verbenaceae

Tetraclea coulteri Gray, Coulter's Wrinklefruit
- **Drug**—**Navajo, Ramah** *Ceremonial Medicine* Plant used in a ceremonial chant lotion. *Febrifuge* Plant used as a fever medicine. (191:42)

Tetradymia, Asteraceae

Tetradymia axillaris A. Nels., Longspine Horsebrush
- **Other**—**Kawaiisu** *Tools* Spines used as tattooing needles. (as *T. spinosa* var. *longispina* 206:66)

Tetradymia canescens DC., Spineless Horsebrush
- **Drug**—**Hopi** *Gynecological Aid* Decoction of leaf and root taken after birth to shrink uterus and stop discharge. (200:35, 98) *Tonic* Plant used as a tonic. (200:98) **Navajo** *Abortifacient* Infusion of plant used as bath for (inducing?) menstruation. (90:156) **Navajo, Ramah** *Analgesic* Cold simple or compound infusion of leaves taken for various aches and pain. *Ceremonial Medicine* Plant used as a ceremonial emetic. *Cold Remedy* Cold simple or compound infusion of leaf used for colds. *Cough Medicine* Cold simple or compound infusion of

leaves taken for cough. *Disinfectant* Cold simple or compound infusion of leaf used for "ghost infection." *Emetic* Plant used as a ceremonial emetic. *Febrifuge* Cold simple or compound infusion of leaves taken for fever. *Gastrointestinal Aid* Cold simple or compound infusion of leaves taken for stomachache. *Herbal Steam* Leaves used as sweat bath medicine. *Orthopedic Aid* Cold simple or compound infusion of leaves taken for backache. *Witchcraft Medicine* Leaves used as for protection from witches. (191:53) **Shoshoni** *Cathartic* Infusion or decoction of dried plant taken as a physic. *Venereal Aid* Decoction of plant taken for venereal diseases. (180:144)
- **Dye**—**Navajo, Ramah** *Yellow* Flowers with two other plants used as a yellow dye for wool. (191:53)
- **Other**—**Navajo, Ramah** *Ceremonial Items* Plant ash used for Evilway blackening. *Protection* Cold infusion of plant used as a bath by undertakers to prevent the ghost from following. Burning leaf smoke used by undertakers after a burial to prevent the ghost from following. (191:53)

Tetradymia comosa Gray, Hairy Horsebrush
- **Drug**—**Paiute** *Cold Remedy* Simple or compound decoction of stems and leaves taken for colds. *Cough Medicine* Simple or compound decoction of stems and leaves taken for coughs. *Gastrointestinal Aid* Decoction of stems and leaves used for stomachaches. *Misc. Disease Remedy* Compound decoction of stems taken for influenza. *Pulmonary Aid* Compound decoction of stems taken for pneumonia. **Shoshoni** *Antidiarrheal* Decoction of bark or root taken for diarrhea. *Cold Remedy* Decoction of stems and leaves taken for colds. *Cough Medicine* Decoction of stems and leaves taken for coughs. *Dermatological Aid* Decoction of stems and turpentine used as a wash for swellings from cuts or bruises. *Gastrointestinal Aid* Decoction of stems and leaves used for stomachaches. (180:144, 145)

Tetradymia stenolepis Greene, Mojave Cottonthorn
- **Drug**—**Kawaiisu** *Dermatological Aid* Spines used for warts. (206:66)
- **Other**—**Kawaiisu** *Tools* Spines used as tattooing needles. (206:66)

Tetraneuris, Asteraceae
Tetraneuris acaulis var. *arizonica* (Greene) Parker, Arizona Hymenoxys
- **Drug**—**Hopi** *Analgesic* Poultice of plant applied for hip and back pain, especially in pregnancy. (as *Actinea acaulis* 200:33, 94) *Antirheumatic (External)* Used for severe pains in hips and back. *Gynecological Aid* Used for severe pains in hips and back, especially in pregnant state. (as *Hymenoxys aculis* var. *arizonica* 42:327) Poultice of plant applied for hip and back pain, especially in pregnancy. (as *Actinea acaulis* 200:35, 94) *Orthopedic Aid* Poultice of plant applied to hip and back pain, especially during pregnancy. (as *Actinea acaulis* 200:94) *Stimulant* Used as a stimulant. (as *Hymenoxys aculis* var. *arizonica* 42:327) Plant used to make a stimulating drink. (as *Actinea acaulis* 200:31, 94)

Tetraneuris argentea (Gray) Greene, Perkysue
- **Drug**—**Navajo, Kayenta** *Ceremonial Medicine* Plant used in special ceremony for illness caused by lunar eclipse. *Dermatological Aid* Plant used as a lotion for eczema. *Psychological Aid* Plant used for dreaming of being bitten by an "alligator." (as *Actinea argentea* 205:44) **Navajo, Ramah** *Disinfectant* Plant used for "coyote infection." *Gastrointestinal Aid* Plant used for heartburn. *Other* Cold infusion taken and used as lotion for emergency treatment of injuries. *Panacea* Plant used as "life medicine." *Witchcraft Medicine* Plant used for protection from witches. (as *Actinea argentea* 191:47)

Tetraneuris scaposa (DC.) Greene, Stemmy Hymenoxys
- **Drug**—**Zuni** *Eye Medicine* Infusion of plant used as an eyewash, not for persons with "bad heart." (166:60, 61)

Thalia, Marantaceae
Thalia geniculata L., Bent Alligatorflag
- **Food**—**Seminole** *Unspecified* Plant used for food. (169:505)

Thalictrum, Ranunculaceae
Thalictrum dasycarpum Fisch. & Avé-Lall., Purple Meadowrue
- **Drug**—**Lakota** *Dermatological Aid* Seeds chewed and rubbed on the hands as a lotion. *Veterinary Aid* Seeds given to horses to make them lively. (139:56) **Meskwaki** *Love Medicine* Used as a love medicine to reconcile a quarrelsome couple. (152:240) **Ojibwa** *Febrifuge* Infusion of root used for fevers. (153:383) **Pawnee** *Veterinary Aid* Plant mixed with clay and rubbed on muzzle of horses as a stimulant. **Ponca** *Love Medicine* Plant tops rubbed in hands of bachelors as a love charm. (70:80) **Potawatomi** *Dermatological Aid* Seeds peppered on surface of poultices to make them more effective. *Gastrointestinal Aid* Compound containing leaves and seeds used for cramps. (154:75) *Hunting Medicine* Dried seeds smoked while hunting to bring good luck. *Love Medicine* Seeds mixed with tobacco and smoked when going to call upon a favorite lady friend. (154:123) Seeds used as a love medicine by the Prairie Potawatomi. (154:75)
- **Other**—**Dakota** *Incense & Fragrance* Mature fruits stored in bags for their agreeable odor and rubbed or scattered on clothing. The tops were broken off the mature fruits and stored in bags for their agreeable odor being rubbed and scattered on clothing at any time when the effect was desired. Although the smell was not considered to be very fragrant by Europeans, the Indians treasured it because of its suggestion of the fresh outdoors. (as *T. purpurascens* 69:360) Mature, lightly scented fruits stored or scattered over clothing for the pleasant odor. *Toys & Games* Hollow stems used by small boys to make toy flutes. **Omaha** *Toys & Games* Hollow stems used by small boys to make toy flutes. **Pawnee** *Toys & Games* Hollow stems used by small boys to make toy flutes. **Ponca** *Toys & Games* Hollow stems used by small boys to make toy flutes. (70:80)

Thalictrum dioicum L., Early Meadowrue
- **Drug**—**Cherokee** *Antidiarrheal* Infusion of root taken for diarrhea. *Antiemetic* Infusion of root taken for vomiting. (80:53) **Iroquois** *Eye Medicine* Decoction of roots used as a wash for sore eyes from a head cold. *Heart Medicine* Decoction of roots taken for heart palpitations. *Other* Plant used to "make you crazy." (87:327)

Thalictrum fendleri Engelm. ex Gray, Fendler's Meadowrue
- **Drug**—**Keres, Western** *Cold Remedy* Infusion of plant used for colds. (171:72) **Navajo, Ramah** *Ceremonial Medicine* Decoction of plant taken as ceremonial medicine. (191:28) **Shoshoni** *Venereal Aid* Decoction of root used for gonorrhea. **Washo** *Cold Remedy* Decoction of root taken as a cold remedy. (180:145)

- *Dye*—**Navajo, Ramah** *Black* Upper branch ash used an Enemyway blackening. (191:28)
- *Other*—**Navajo** *Ceremonial Items* Used to make tea to drink and bathe in on fifth night after blackening ceremony of War Dance. (55:48)

Thalictrum occidentale Gray, Western Meadowrue
- *Drug*—**Blackfoot** *Dermatological Aid* Powdered fruits mixed with water and used as cosmetic on the hair and body. (86:125) *Pulmonary Aid* Infusion of seeds used for chest pains. (86:74) **Gitksan** *Analgesic* Root chewed and juice swallowed for headache. *Blood Medicine* Root chewed and juice swallowed to improve blood circulation. *Eye Medicine* Root chewed and juice swallowed for eye trouble. *Orthopedic Aid* Root chewed and juice swallowed for sore legs. *Pulmonary Aid* Root chewed and juice swallowed to loosen phlegm. (150:57) **Thompson** *Dermatological Aid* Poultice of mashed roots applied to open wounds. (187:250)
- *Food*—**Blackfoot** *Spice* Fruit used to spice pemmican, dried meat, and broths. (86:105)
- *Other*—**Blackfoot** *Incense & Fragrance* Fruits kept as a household and clothing deodorant. (as *T. occidentalis* 86:123) Seeds and leaves kept for the pleasant smell. (97:35) Berries dried and placed in small buckskin bags for perfume. (114:278) *Insecticide* Seeds and leaves placed among clothing or other possessions as an insect repellent. (97:35) *Paint* Fruits often crushed and mixed with paint for a robe. (as *T. occidentalis* 86:123)

Thalictrum polycarpum (Torr.) S. Wats., Fendler's Meadowrue
- *Drug*—**Kawaiisu** *Poison* Root caused death when eaten by cows and horses. (206:67) **Mendocino Indian** *Panacea* Root used as a universal charm and panacea. *Poison* Stems considered poisonous. **Wailaki** *Analgesic* Crushed stem and leaf juice used as a wash for headaches. (41:348) **Yuki** *Orthopedic Aid* Poultice of pounded plant applied to sprains. (49:47)

Thalictrum pubescens Pursh, King of the Meadow
- *Drug*—**Iroquois** *Hemostat* Infusion of smashed plant used to wash the head and neck for nosebleeds. *Liver Aid* Compound infusion of roots taken as a gall medicine. (87:327)
- *Food*—**Montagnais** *Spice* Leaves used to flavor salmon. (as *T. polygamum* 156:315)

Thalictrum sp., Meadowrue
- *Drug*—**California Indian** *Dermatological Aid* Dried roots used for shampoo. (118:57) **Washo** *Cold Remedy* Infusion of roots taken for colds. (118:37)
- *Other*—**Blackfoot** *Incense & Fragrance* Dried, whole plant used for perfume. **Gros Ventre** *Incense & Fragrance* Dried, whole plant used for perfume. (118:57)

Thalictrum sparsiflorum Turcz. ex Fisch. & C. A. Mey., Fewflower Meadowrue
- *Drug*—**Blackfoot** *Veterinary Aid* Dried leaves ground into powder and given to horses to make them long-winded, spirited, and enduring. (97:35) **Cheyenne** *Veterinary Aid* Dried, powdered plant used to make a horse spirited, long-winded, and enduring. (76:173) Flower used for horses as perfume and medicine for long-windedness and endurance. (83:34) **Great Basin Indian** *Dermatological Aid* Dried seeds and roots used as a perfume. Powdered root used as a shampoo. (121:47)

Thalictrum thalictroides (L.) Eames & Boivin, Rue Anemone
- *Drug*—**Cherokee** *Antidiarrheal* Infusion of root taken for diarrhea. *Antiemetic* Infusion of root taken for vomiting. (80:53)

Thamnosma, Rutaceae

Thamnosma montana Torr. & Frém., Turpentine Broom
- *Drug*—**Havasupai** *Emetic* Decoction of leaves taken one to three times a day to cause vomiting. *Gastrointestinal Aid* Pounded leaves rubbed onto a hurting abdomen. *Laxative* Decoction of leaves taken one to three times a day to act as a laxative. (197:229) **Kawaiisu** *Analgesic* Decoction of stems taken for chest pains. *Cold Remedy* Decoction of stems taken for colds. *Dermatological Aid* Crushed stems rubbed into open wounds. *Diaphoretic* Powdered plant caused men to sweat. *Hallucinogen* Infusion of plant taken by medicine men "to go crazy like coyotes." *Hunting Medicine* Powdered plant used as an aid in hunting. Powdered plant put in deer tracks as an aid in hunting. This procedure would slow down the deer so that it could be overtaken. *Snakebite Remedy* Powdered plant used to keep snakes away. *Veterinary Aid* Powdered plant caused horses to sweat. (206:67) **Paiute** *Gynecological Aid* Decoction of stems used as a wash or douche for female complaints. *Misc. Disease Remedy* Decoction of stems taken for smallpox. (180:145, 146) **Pima** *Venereal Aid* Decoction of plant taken for gonorrhea. (146:80) **Shoshoni** *Cold Remedy* Decoction of stems taken or dried stems smoked with tobacco for colds. *Tonic* Decoction of stems taken as a tonic. (180:145, 146)
- *Other*—**Kawaiisu** *Protection* Powdered plant kept on one's person to keep snakes away. (206:67)

Thaspium, Apiaceae

Thaspium barbinode (Michx.) Nutt., Hairyjoint Meadowparsnip
- *Drug*—**Chippewa** *Gastrointestinal Aid* and *Pediatric Aid* Decoction of root given to children for colic. (53:344)

Theaceae, see *Camellia*

Thelesperma, Asteraceae

Thelesperma filifolium (Hook.) Gray **var. filifolium**, Stiff Greenthread
- *Drug*—**Keres, Western** *Pediatric Aid* and *Tuberculosis Remedy* Plant formerly used for children with tuberculosis. (as *Theleosperma trifidum* 171:72)
- *Food*—**Keres, Western** *Beverage* Infusion of plant used as a beverage. (as *Theleosperma trifidum* 171:72) **Tewa** *Beverage* Leaves steeped and the tea drunk as a beverage. (as *Thelesperma trifidum* 138:61)

Thelesperma longipes Gray, Longstalk Greenthread
- *Food*—**Isleta** *Beverage* Young plant leaves boiled to make a beverage resembling commercial tea. *Winter Use Food* Plants stored well for future use. (100:43) **Navajo** *Substitution Food* Leaves and stems used as a substitute for tea. (55:89)

Thelesperma megapotamicum (Spreng.) Kuntze, Hopi Tea Greenthread
- *Drug*—**Keres, Western** *Pediatric Aid* and *Tuberculosis Remedy* Plant

formerly used for children with tuberculosis. (as *Theleosperma gracile* 171:72) **Navajo** *Stimulant* Infusion of leaves and stems taken as a "nervous stimulant." *Toothache Remedy* Infusion of leaves and stems taken for the teeth. (as *Thelesperma gracile* 55:89)
- **Food**—**Apache, Chiricahua & Mescalero** *Beverage* Leaves and young stems boiled to make a nonintoxicating beverage. Fresh or stored portions boiled in water and liquid consumed with or without sugar. (as *T. gracile* 33:53) **Hopi** *Beverage* Flowers used to make a beverage. (as *T. gracile* 56:15) Used to make coffee. (as *T. gracile* 190:168) Flowers and tips of young leaves dried, boiled, and used to make tea. (as *T. gracile* 200:98) **Keres, Western** *Beverage* Infusion of plant used as a beverage. (as *Theleosperma gracile* 171:72) **Keresan** *Beverage* Leaves and roots boiled to make tea. (as *Thelesperma gracile* 198:563) **Navajo** *Substitution Food* Leaves and stems used as a substitute for tea. (as *T. gracile* 55:89) **Navajo, Ramah** *Beverage* Decoction of leaves and flowers, with lots of sugar, used as a tea. (191:53) **Tewa** *Beverage* Leaves steeped and the tea drunk as a beverage. (as *T. gracile* 138:61)
- **Dye**—**Hopi** *Red-Brown* Flowers used as a reddish brown dye for basket making yucca fibers. (as *T. gracile* 56:15) Used to make a fine reddish brown basketry and textile dye. (as *T. gracile* 200:98) **Navajo, Ramah** *Orange-Yellow* Boiled roots used as an orange-yellow dye for wool. (191:53)

Thelesperma subnudum Gray, Sand Fringedpod
- **Food**—**Hopi** *Beverage* Flowers and tips of young leaves dried, boiled, and used to make tea. (200:98)
- **Dye**—**Hopi** *Red-Brown* Used to make a fine reddish brown basketry and textile dye. (200:98) **Navajo** *Orange* Leaves, stems, and blossoms used as an orange dye for wool. (55:89)

Thelypodiopsis, Brassicaceae
Thelypodiopsis elegans (M. E. Jones) Rydb., Westwater Tumblemustard
- **Drug**—**Navajo, Kayenta** *Veterinary Aid* Plant used as a charm to make a horse run fast. (as *Sisymbrium elegans* 205:24)

Thelypodium, Brassicaceae
Thelypodium integrifolium (Nutt.) Endl. ex Walp. **ssp. *integrifolium*,** Entireleaved Thelypody
- **Food**—**Mohave** *Starvation Food* Young shoots roasted and eaten as a famine food. (as *T. lilacinum* 37:201)

Thelypodium wrightii Gray, Wright's Thelypody
- **Drug**—**Navajo** *Dermatological Aid* Plant used for swellings. (55:97)
- **Other**—**Zuni** *Fertilizer* Seeds crushed by women and planted with beans to ensure a proliferative crop. (166:85)

Thelypodium wrightii Gray **ssp. *wrightii*,** Wright's Thelypody
- **Drug**—**Navajo, Kayenta** *Eye Medicine* Ashes rubbed on lids for eye disease. *Pediatric Aid* and *Sedative* Plant tied to cradle bow to make baby sleep. (as *Stanleyella wrightii* 205:25)
- **Food**—**Pueblo** *Dried Food* Young plants boiled, pressed, rolled into balls, dried, and stored for winter use. *Soup* Plant made into a stew with wild onions, wild celery, tallow, or bits of meat. *Unspecified* Young plants boiled, pressed, rolled into balls, and eaten. *Vegetable* Young plants boiled with a pinch of salt and eaten as greens. (as *Stanleyella wrightii* 32:25) **Tewa** *Unspecified* Species used for food. (as *Stanleyella wrightii* 138:61)
- **Other**—**Tewa** *Paint* Used to make paint for pottery. (as *Stanleyella wrightii* 138:61)

Thelypteridaceae, see *Phegopteris*, *Thelypteris*

Thelypteris, Thelypteridaceae, fern
Thelypteris kunthii (Desv.) Morton, Kunth's Maiden Fern
- **Drug**—**Seminole** *Orthopedic Aid* and *Psychological Aid* Leaves used for old paint woman sickness: insanity and weakness of the limbs and neck. (as *Dryopteris normalis* 169:267)

Thelypteris palustris Schott, Eastern Marsh Fern
- **Drug**—**Iroquois** *Gynecological Aid* Roots used for woman's troubles. (87:256)

Theobroma, Sterculiaceae
Theobroma cacao L., Cacao
- **Food**—**Haisla & Hanaksiala** *Beverage* Used to make a beverage. (43:294)

Theophrastaceae, see *Jacquinia*

Thermopsis, Fabaceae
Thermopsis macrophylla Hook. & Arn., California Goldenbanner
- **Drug**—**Pomo** *Eye Medicine* Cold decoction of leaves used as a wash for sore eyes. (66:13) **Pomo, Kashaya** *Eye Medicine* Cooled decoction of leaves used as an eyewash for sore eyes and vision difficulties. *Gynecological Aid* Infusion of leaves or root and bark used to slow down menstrual flow. (72:66)

Thermopsis rhombifolia (Nutt. ex Pursh) Nutt. ex Richards., Prairie Thermopsis
- **Drug**—**Cheyenne** *Analgesic* Dried leaves burned and smoke inhaled for headaches. *Cold Remedy* Dried leaves burned and smoke inhaled for colds. (83:30)
- **Dye**—**Blackfoot** *Yellow* Yellow petals rubbed on arrow shafts for coloring. (86:123)
- **Other**—**Blackfoot** *Season Indicator* Flowering signaled the time to collect buffalo tongues in preparation for the Sun Dance. (86:123) Flowers indicated the prime buffalo hunting season. **Flathead** *Season Indicator* Flowers indicated the prime buffalo hunting season. (97:41)

Thermopsis rhombifolia var. *montana* (Nutt.) Isely, Mountain Thermopsis
- **Drug**—**Navajo, Ramah** *Analgesic* Compound containing plant used as fumigant for headache. *Cough Medicine* Decoction of plant taken as cough medicine. *Eye Medicine* Compound containing plant used as fumigant for sore eyes. *Hunting Medicine* Compound containing plant used as fumigant for sickness caused by hunting. *Witchcraft Medicine* Plant used as a lotion for protection from witches. (as *T. pinetorum* 191:34)

Thlaspi, Brassicaceae
Thlaspi arvense L., Field Pennycress
- **Drug**—**Iroquois** *Throat Aid* Infusion of plant taken for sore throats. (87:341)

- *Food*—Cherokee *Vegetable* Leaves used for food. (126:37) **Havasupai** *Unspecified* Seeds used in a variety of ways. (197:221)

Thlaspi montanum L., Alpine Pennycress
- *Food*—Havasupai *Unspecified* Seeds used in a variety of ways. (197:221)

Thlaspi montanum var. *fendleri* (Gray) P. Holmgren, Fendler's Pennycress
- *Drug*—Navajo, Ramah *Ceremonial Medicine* Plant used in ceremonial chant lotion. *Dermatological Aid* Cold infusion used internally and externally for itch. *Preventive Medicine* Cold infusion taken and used as lotion to prevent injury from deer. *Witchcraft Medicine* Cold infusion taken and used as lotion to protect from witches. (as *T. fendleri* 191:29)

Thuja, Cupressaceae, conifer

Thuja excelsa, Cedar
- *Drug*—Tlingit *Venereal Aid* Compound infusion of sprouts and bark taken for syphilis. (107:283)

Thuja occidentalis L., Eastern Arborvitae
- *Drug*—Abnaki *Antirheumatic (External)* Used for swellings. (144:155) Poultice of powdered leaves applied to swellings. (144:163) *Panacea* Leaves made into pillows and used as a panacea. (144:155) **Algonquin, Quebec** *Antirheumatic (Internal)* Decoction of branches taken for rheumatism. *Cold Remedy* Branches used in the steam bath for colds. *Dermatological Aid* Poultice of powdered, rotten wood used for rashes and skin irritations. *Febrifuge* Branches used in the steam bath for fevers. *Gastrointestinal Aid* Infusion of cones used for babies with colic. *Gynecological Aid* Branches used in the steam bath for women after childbirth. Infusion of plant taken for menstrual disorders. *Pediatric Aid* Infusion of cones used for babies with colic. *Toothache Remedy* Decoction of crushed branches used as a steam for toothache. (18:130) **Chippewa** *Analgesic* Compound containing charcoal pricked into temples with needles for headache. (53:338) *Ceremonial Medicine* Twigs burned for incense in religious ceremonies. (71:123) *Cough Medicine* Compound containing leaves taken as a cough syrup. (53:340) *Dermatological Aid* Plant used as a deodorant. *Disinfectant* and *Misc. Disease Remedy* Twigs burned as a disinfectant to fumigate a house for smallpox. *Unspecified* Leaves combined with ground hemlock for medicinal purposes. (71:123) **Cree, Woodlands** *Panacea* Powdered branches and many herbs used for various ailments. *Pulmonary Aid* Decoction of branches taken for pneumonia. *Urinary Aid* Decoction of needle covered branches or juice taken for urine retention or sore bladder. (109:62) **Iroquois** *Antirheumatic (External)* Decoction of plant tips used as a foot bath for rheumatism. Steam from compound decoction used as a bath for rheumatism. *Blood Medicine* Fermented compound decoction taken when "blood gets bad and cold." *Cold Remedy* Steam from decoction of leaves inhaled for colds. *Dermatological Aid* Decoction used as a wash or poultice applied to cuts, bruises, and sores. *Febrifuge* Fermented compound decoction taken for fever. *Gynecological Aid* Infusion used by "women during confinement." Steam from compound decoction used as a bath for parturition. (87:270) Infusion of leaves taken by women as a tonic and diaphoretic to increase the milk flow. *Hunting Medicine* Branches, without fruit, used in the vapor baths by hunters. (141:35) *Orthopedic Aid* Decoction used as a wash for weakness in the hips due to untreated broken coccyx. Decoction of plant used as a wash or poultice of leaves applied to sprains. Fermented compound decoction taken for soreness. (87:270) Poultice of bough and milk decoction mixed with grease and applied for paralysis. (141:35) *Stimulant* Fermented compound decoction taken when a "person is tired." (87:270) **Malecite** *Burn Dressing* Dried under bark pounded, mixed with grease, and used for burns. (as *Thuya occidentalis* 116:247) *Cough Medicine* Infusion of boughs used for coughs. (116:249) *Toothache Remedy* Gum used to fill cavities and for tooth pain. (as *Thuya occidentalis* 116:248) *Tuberculosis Remedy* Infusion of bark used for consumption. (116:251) **Menominee** *Abortifacient* Infusion of dried inner bark taken during a cold to treat suppressed menses. *Adjuvant* Inner bark used as a seasoner for enhancing medicines. (151:46) *Dermatological Aid* Compound poultice of dried, powdered leaves applied to swellings. (54:134) *Diaphoretic* Plant used in the sudatory. *Stimulant* Smudge of leaves used to revive "lost consciousness." (151:46) **Micmac** *Analgesic* Stems used for headaches. *Burn Dressing* Inner bark, bark, and stems used for burns. *Cough Medicine* Inner bark, bark, and stems used for cough. *Orthopedic Aid* Leaves used for swollen feet and hands and stems used for headaches. *Toothache Remedy* Gum used for toothache. *Tuberculosis Remedy* Inner bark, bark, and stems used for consumption. (40:62) **Montagnais** *Diaphoretic* Infusion of bruised twigs taken to cause sweating. (as *Arbor vitae* 156:315) **Ojibwa** *Analgesic* Infusion of leaves used for headache. *Blood Medicine* Decoction of leaves taken as a blood purifier. *Ceremonial Medicine* Smoke used to purify sacred objects, hands, and bodies of participants. *Cough Medicine* Decoction of leaves taken for coughs. *Diaphoretic* Compound containing leaves used in the sweat bath. (153:380) **Penobscot** *Analgesic* Compound poultice of bark used on cuts made in painful area to treat pain. (as *Arbor vitae* 156:311) *Dermatological Aid* Poultice of leaves applied to swollen hands or feet. (as *Arbor vitae* 156:309) *Panacea* Compound poultice of bark applied "for all kinds of trouble." (as *Arbor vitae* 156:311) **Potawatomi** *Adjuvant* Plant used as a seasoner for medicines. *Unspecified* Poultice of leaves used for unspecified ailments. *Witchcraft Medicine* Leaves burned on coals to purify patient and exorcise evil spirits. (154:70, 71)
- *Food*—Chippewa *Beverage* Leaves used to make a hot, tea-like beverage. (71:123) **Ojibwa** *Beverage* Leaves steeped for tea. (as *Thuga canadensis* 5:2234)
- *Fiber*—Algonquin, Tête-de-Boule *Canoe Material* Plant used to make canoe skeletons. (132:132) **Chippewa** *Canoe Material* Wood used to make canoe ribs. *Snow Gear* Wood used to make tobbogans. (71:123) **Iroquois** *Caulking Material* Bark pieces packed into joints during construction. (142:83) **Malecite** *Basketry* Used to make basket splints. *Canoe Material* Wood used for canoe slats. (160:6) **Menominee** *Basketry* Bark used to weave bags. (151:76) **Micmac** *Canoe Material* Used to make canoe slats. (159:258) **Ojibwa** *Basketry* Tough, stringy bark used in making fiber bags. *Canoe Material* Light, strong, straight-grained wood used for canoe frames and ribs. (153:422)
- *Other*—Chippewa *Hunting & Fishing Item* Wood used to make sturgeon spears. *Incense & Fragrance* Leaves used as perfume for clothing. (71:123) **Iroquois** *Insecticide* Branches used in closets to prevent moths. (142:83) **Malecite** *Hide Preparation* Bark used for tanning

hides. *Hunting & Fishing Item* Wood used to make arrows. (160:6) **Menominee** *Protection* Leaves used as mothballs and clothes stored away with layers of leaf sprays to keep out the moths. (151:81) **Micmac** *Fuel* Wood used for kindling and fuel. *Hunting & Fishing Item* Used to make arrow shafts. (159:258) **Ojibwa** *Incense & Fragrance* Pungent fragrance of leaves and wood always used as an acceptable incense to Winabojo. *Sacred Items* This tree and the white cedar were worshipped as the two most useful trees in the forest. (153:421) **Potawatomi** *Ceremonial Items* Preserved or fresh leaves used as a smudge to exorcise evil spirits and purify sacred objects. *Lighting* Rolled bark used for night hunting torches. *Protection* Preserved or fresh leaves used as a smudge to exorcise evil spirits and purify sacred objects. (154:122)

Thuja plicata Donn ex D. Don, Western Red Cedar

- **Drug**—**Bella Coola** *Analgesic* Decoction of powdered leaves used externally for various internal pains. *Antirheumatic (External)* Infusion of leaves used externally for rheumatism. (150:49) Poultice of pounded bough tips and eulachon (candlefish) grease applied to the back and chest for rheumatism. Poultice of pounded bough tips and eulachon grease applied to the back and chest for a swollen neck. (184:197) *Cough Medicine* Infusion of leaves used externally for coughs. *Dermatological Aid* Very soft bark used to bind wounds and cover poultices. *Gastrointestinal Aid* Simple decoction, compound decoction or infusion of leaf taken and used externally for stomach pain. (150:49) Poultice of pounded bough tips and eulachon (candlefish) grease applied to the back and chest for stomach pains. (184:197) *Heart Medicine* Infusion of leaves used externally for heart trouble. (150:49) Poultice of pounded bough tips and eulachon (candlefish) grease applied to the back and chest for heart trouble. (184:197) *Other* Infusion of leaves used externally for swollen neck. (150:49) *Respiratory Aid* Poultice of pounded bough tips and eulachon (candlefish) grease applied to the back and chest for bronchitis. (184:197) **Chehalis** *Abortifacient* Chewed bark or decoction of bark taken to induce menstruation. (79:19) **Clallam** *Tuberculosis Remedy* Decoction of small limbs used for tuberculosis. (57:195) **Cowlitz** *Cold Remedy* Decoction of plant tips and roots taken as a cold medicine. *Toothache Remedy* Buds chewed for toothaches. (79:19) **Haisla** *Dermatological Aid* Moxa of inner bark used as a counter irritant for the skin. **Hanaksiala** *Antidiarrheal* Infusion of pounded, bough tip leaves in cold water taken for diarrhea. (43:162) **Hoh** *Unspecified* Infusion of green bark juice used for medicine. (137:57) **Klallam** *Tuberculosis Remedy* Decoction of branches taken as a tuberculosis medicine. (79:19) **Kwakiutl** *Dermatological Aid* Poultice of inner bark applied to carbuncles. Shredded bark used to cauterize sores and swellings. *Eye Medicine* Sticks broken in front of the eye for sties. *Hemostat* Shredded bark used to cauterize sores and swellings. *Orthopedic Aid* Compound poultice of leaves applied to sore backs. (183:266) **Lummi** *Antiemetic* Tips chewed by men to avoid nausea while burying a corpse. *Pulmonary Aid* Chewed buds taken for sore lungs. (79:19) **Makah** *Cough Medicine* Infusion of boughs taken for coughs. *Dermatological Aid* Bark pounded until soft as cotton and used to rub the face. (67:228) **Nez Perce** *Antidiarrheal* Leaves used for diarrhea. *Cold Remedy* Infusion of boughs used for colds. *Cough Medicine* Infusion of boughs used for coughs. *Pediatric Aid* Infusion of boughs sweetened and used for children with colds or coughs. (82:54) **Okanagan-Colville** *Antirheumatic (External)* Infusion of boughs used to soak painful joints from arthritis and rheumatism. Weak infusion of boughs taken for painful joints from arthritis and rheumatism. *Dermatological Aid* Decoction of boughs and three plants used for washing the skin and hair during sweat bathing.[78] Infusion of boughs used as a hair wash for dandruff and scalp "germs." *Poison* Infusion of boughs considered toxic in large doses. *Tonic* Decoction of boughs and three plants taken as a sweat house tonic.[78] Weak infusion of boughs taken as a sweat house tonic. (188:20) **Oweekeno** *Ceremonial Medicine* Wood made into shamanistic soul catchers to use in ritual healing. (43:66) **Quileute** *Unspecified* Infusion of green bark juice used for medicine. (137:57) **Quinault** *Dermatological Aid* Infusion of twigs used as a wash for venereal disease sores. *Febrifuge* Infusion of seeds and twigs taken for fevers. *Kidney Aid* Infusion of bark and twigs taken for kidney trouble. *Venereal Aid* Infusion of twigs used as a wash for venereal disease sores. **Skagit** *Cough Medicine* Decoction of leaves taken for coughs. **Skokomish** *Oral Aid* Decoction of buds used as a gargle. (79:19) **Thompson** *Gynecological Aid* Compound decoction of twigs taken after childbirth. (164:461) *Misc. Disease Remedy* Decoction of old or green cones taken for leprosy. *Psychological Aid* Tree or spruce tree said to cause vivid dreams for those who slept under them. (187:94)

- **Food**—**Hesquiat** *Starvation Food* Branches used to make fish traps became "fish flavored" and the sticks were boiled for broth. (185:35) **Kwakiutl, Southern** *Candy* Pitch used as chewing gum. (183:293) **Montana Indian** *Dried Food* Spring cambium pressed into cakes and dried for storage. *Unspecified* Spring cambium eaten fresh. (19:25) **Salish, Coast** *Dried Food* Cambium dried and eaten in spring. *Unspecified* Cambium eaten fresh in spring. (182:71)

- **Fiber**—**Bella Coola** *Basketry* Inner bark used to make baskets. *Clothing* Inner bark used to make capes and shaman collars. *Mats, Rugs & Bedding* Inner bark used to make mats. *Scouring Material* Inner bark used for sponges. (184:197) **Chehalis** *Clothing* Bark coarsely shredded and plaited into skirts, capes, and dresses for women. *Mats, Rugs & Bedding* Bark finely shredded and used as padding for infants' cradles, sanitary pads, and towels. (79:19) **Clallam** *Basketry* Bark used for basketry. Roots used for coiled and imbricated baskets. *Building Material* Wood used for house planks, posts, and roof boards. *Canoe Material* Wood used for canoes. *Clothing* Bark used for clothing, sanitary napkins, and towels. *Cordage* Limbs used to make rope. *Mats, Rugs & Bedding* Bark used to pad cradles. (57:195) **Flathead** *Basketry* Bark strips used to make baskets. (82:54) **Gitksan** *Basketry* Inner bark strips used for basketry. (73:152) *Building Material* Bark sheets used for roofing. (73:153) *Cordage* Inner bark strips used for cordage. (73:152) *Mats, Rugs & Bedding* Bark sheets used for tarpaulins. (73:153) Inner bark strips used for mat making. **Haisla** *Basketry* Inner bark strips used for basketry. *Cordage* Inner bark strips used for cordage. *Mats, Rugs & Bedding* Inner bark strips used for mat making. (73:152) **Haisla & Hanaksiala** *Basketry* Bark used for basketry. Branches used in basketry. Roots used as the main structural elements in baskets. *Building Material* Bark used to partition longhouses and make temporary shelters. Bark used for roofing. Wood used to make dancing platforms. Wood used for house posts and planks in traditional house construction. *Canoe Material* Bark used to cover canoes to keep the rain out. Wood used to carve canoes. *Clothing* Bark used to make hats of various types. Inner bark used to make rain capes. Bark woven into hip length leggings to wear through deep snow. *Cordage* Branches twisted together to make a type of rope used to tie covers to storage boxes. *Furniture* Bark used to make baby cradles. *Mats, Rugs & Bedding* Bark used to make mats to sit on. Inner bark used to make blankets. (43:162) **Hesquiat** *Basketry* Very long, straight branches or withes used to make baskets. Cedar withes were also split and used to make a wide variety of strong openwork baskets for berry picking and other purposes. These were strengthened at the corners with twisted cedar ropes. The withes were divided with two vertical cuts. The two outer rounds were often used for finer construction and possibly the active weft; the inner piece, flattened on two sides, was usually used for the "framework" (warp and passive weft) of the baskets. Cleaned, finely split inner bark used to weave baskets. *Building Material* Wood prized as a material for house construction: shakes, shingles, and poles. *Canoe Material* Wood prized as a material for carving, especially canoes. *Clothing* Cleaned, finely split inner bark used to weave capes, skirts, and aprons. Shredded inner bark used for diapers. *Cordage* Very long, straight branches or withes used to make ropes. For large ropes, such as those used in whaling, the entire branch would be used. For smaller ropes, the withes were split off into three parts; the heartwood would be removed and the outer part twisted into rope. Branches used for such ropes would be 5 to 8 cm in diameter. Hesquiat cedar ropes were quite famous and were often traded to other tribes. The large ropes were used long ago as whale hunting lines, anchor lines, and binding lines for tying on house planks or tying together one's effects when moving. *Mats, Rugs & Bedding* Cleaned, finely split inner bark used to weave mats. *Scouring Material* Finely beaten bark used as a sponge. Young boys, when they were starting to walk, were rubbed with this bark dipped in cold octopus broth to make them tough. Girls were told to rub the same liquid on their hands to make them strong. (185:35) **Hoh** *Basketry* Roots split and used to make baskets. *Building Material* Used for house construction. Bark used to make wigwams, summer houses, and hunting lodges. *Canoe Material* Used to make dugout canoes. *Clothing* Bark inner fibers formerly used to make clothing. Inner bark shredded and used to make skirts. Bark used to make an overcoat for fishing in stormy weather. *Cordage* Twigs and roots twisted and used as ropes. *Mats, Rugs & Bedding* Bark used to line cradles. *Other* Twigs used for drying the body after a bath. (137:57) **Kutenai** *Basketry* Roots split, peeled, and used to make watertight baskets. *Canoe Material* Wood used to make canoe frames. (82:54) **Kwakiutl** *Building Material* Wood used for building homes. *Canoe Material* Wood used for making canoes. *Clothing* Bark used to make clothing. *Cordage* Bark used to make ropes. (as *T. gigantea* 20:8) **Kwakiutl, Southern** *Basketry* Roots and bark used to make baskets. (183:296) Fibrous bark used to make baskets. Long, straight branches used for weaving baskets. Roots washed, scorched over a fire, skinned, split in two, scraped, and used to make baskets. *Brushes & Brooms* Shredded bark used for paintbrushes. *Building Material* Long, straight branches used to make nets for tying wall boards onto house frames. *Canoe Material* Wood steamed for flexibility and malleability and used to make canoes. (183:266) *Clothing* Inner bark used to make clothing. Roots and bark used to make hats. (183:296) Fibrous bark used to make hats. Shredded bark used for diapers. Roots washed, scorched over a fire, skinned, split in two, scraped, and used to make hats. *Cordage* Fibrous bark used to make twine and ropes. (183:266) *Mats, Rugs & Bedding* Inner bark used to make mats and blankets. (183:296) Fibrous bark used to make mats. *Other* Shredded bark used for napkins, towels, and bandages. *Sewing Material* Long, straight branches used for "sewing wood." (183:266) **Makah** *Basketry* Inner bark used for basketry. *Canoe Material* Wood used to make canoes. *Clothing* Bark pounded until soft and made into clothes. (67:228) Used to make the lining and headbands of rain hats. (79:19) *Mats, Rugs & Bedding* Bark used to line baby cradles. (67:228) Bark cut into narrow strips and woven into mats. (79:19) **Montana Indian** *Basketry* Bark used to make frame work of baskets. Inner bark used to make sacks. Roots used for making baskets. *Building Material* Wood used for shingles, lumber, fence posts, and for paving. *Canoe Material* Bark used to make canoes. *Clothing* Inner bark woven into cloth. *Cordage* Inner bark used to make cords for fish nets. (19:25) **Nez Perce** *Basketry* Roots used to make baskets. *Building Material* Bark formerly used to make lean-to shelters while camping. Bark used to cover the roofs of semisubterranean houses. *Canoe Material* Wood hollowed out with fire, smoothed, trimmed, and used to make canoes. *Furniture* Wood used to make cradleboards. (82:54) **Nitinaht** *Basketry* Young limbs twisted and used to make basket frames and ribs. Bark used to make baskets. Inner bark strips used to make storage baskets and for basket bottoms, rims, and ribs. *Building Material* Wood used for house posts and wall and roofing house boards. *Canoe Material* Wood used to make dugout canoes. *Clothing* Bark used to make waterproof hats, capes, and other articles of clothing. (186:67) *Cordage* Used to make ropes. (67:228) Young limbs twisted and used for rope material. Bark used to make rope. Inner bark split into thin strips, spun, and used for rope and twine. *Mats, Rugs & Bedding* Bark used to make mats. Inner bark woven with dog hair and cottonwood bast fiber to make blankets. (186:67) **Okanagan-Colville** *Basketry* Bark used for weaving rough baskets. Split roots used to make a high-quality, coiled, watertight, and strong basket. *Brushes & Brooms* Boughs used to scrub the skin in the sweat house. *Building Material* Logs used to make A-frame shelters and to cover the rafters of an underground house. The logs were burned into 6-meter lengths and boards split off with horn wedges. Bark used to make a covering for sweat house frames and insulation for tule tepees. Bark used to make a raised storage cache. *Canoe Material* Wood used to make canoes, frames for birchbark canoes, and paddles. *Mats, Rugs & Bedding* Bark used, when tule not available, for weaving mats. (188:20) **Okanagon** *Clothing* Inner bark and stems used to make clothing. *Cordage* Inner bark and stems used to make rope and twine. *Mats, Rugs & Bedding* Inner bark and stems used to make mats and blankets. *Sewing Material* Inner bark and stems used to make thread. (125:39) **Oweekeno** *Basketry* Fibrous tissue used to make baskets. Inner bark used to make baskets. (43:63) Inner bark and roots used to make baskets. (43:66) *Building Material* Wood used for construction. *Canoe Material* Wood used to make dugout canoes. Fibrous tissue used to make canoe bailers. Inner bark used to make canoe bailers. Inner bark used to make sails for canoes. *Clothing* Fibrous tissue used to make napkins and clothing. Inner bark used to make women's aprons, conical rain capes, and napkins. (43:63) Inner bark used to make dance costumes. Wood used to make hoops worn by dancers as part of their costumes. (43:66) *Cordage* Inner bark made into cordage and used as lanyards or lines and to secure boxes. *Mats, Rugs & Bedding* Fibrous tissue used for weaving mats and screens. (43:63) **Quileute** *Basketry* Roots split and used to make baskets. *Building Material* Used for house construction. Bark used to make wigwams, summer houses, and hunting lodges. *Canoe Material* Used to make dugout canoes. (137:57) *Clothing* Used to make the lining and headbands of rain hats. (79:19) Bark inner fibers formerly used to make clothing. Inner bark shredded and used to make skirts. Bark used to make an overcoat for fishing in stormy weather. *Cordage* Twigs and roots twisted and used as ropes. *Mats, Rugs & Bedding* Bark used to line cradles. *Other* Twigs used for drying the body after a bath. (137:57) **Quinault** *Basketry* Limbs used for openwork baskets. *Clothing* Used to make the lining and headbands of rain hats. *Sewing Material* Roots used to sew the corners of wooden boxes. (79:19) **Salish, Coast** *Basketry* Inner bark beaten to separate the fibers and used to make baskets. *Brushes & Brooms* Wood used to make combs. *Building Material* Wood used to make houses. *Canoe Material* Wood used to make canoes, paddles, and canoe bailers. *Clothing* Inner bark beaten to separate the fibers and used to make hats, diapers, work aprons, and clothing. Inner bark beaten to separate the fibers and used for covering drummers' hands in winter dances. *Cordage* Young, slender branches used to make ropes. *Furniture* Wood used to make cradles. *Mats, Rugs & Bedding* Inner bark soaked, beaten to separate the fibers, and used to make mats and bedding. (182:71) **Shuswap** *Basketry* Root shaves woven into baskets. (123:50) **Snuqualmie** *Building Material* Limbs used to tie the poles of summer houses together. **Squaxin** *Basketry* Limbs used for openwork baskets. Roots used to make coiled and imbricated basketry. (79:19) **Thompson** *Basketry* Shredded roots used in making basketry. (164:496) Split roots made into coiled, watertight baskets and used for boiling food. The outer strips of the roots were used to make the bottom of the basket, the center core was used in the coils, and the bark of the roots was used for making the edges because of its toughness. The roots could not be harvested in the winter because it was too difficult to scrape the bark off of the stiff, brown roots. In the spring and early fall, the bark was more easily removed and the roots were white. Split roots made into watertight baskets and used for boiling food. Wood splints sometimes used for the inner foundation of baskets rather than root bundles. (187:94) *Canoe Material* Used to make dugout canoes. (164:496) *Clothing* Inner bark and stems used to make clothing. (125:39) Inner root softened, split into strips, and used to make rain cloaks, aprons, and capes. (164:496) *Cordage* Inner bark and stems used to make rope and twine. (125:39) *Furniture* Outer bark used to kneel and sit on in canoes. (164:496) *Mats, Rugs & Bedding* Inner bark and stems used to make mats and blankets. (125:39) Inner root softened, split into strips, and used to make mats. Bark piled up and used as a bed. (164:496) *Scouring Material* Inner bark, maple or cottonwood inner bark used to make scouring pads. (187:94) *Sewing Material* Inner bark and stems used to make thread. (125:39) **Tsimshian** *Building Material* Wood used for construction. (43:315) Bark sheets used for roofing and tarpaulins. (73:153) *Canoe Material* Wood used to make canoes. *Clothing* Inner bark used to make diapers for infants and rainproof capes. *Mats, Rugs & Bedding* Inner bark used to make mats. (43:315) Bark sheets used for tarpaulins. (73:153) **Wet'suwet'en** *Basketry* Inner bark strips used for basketry. (73:152) *Building Material* Bark sheets used for roofing. (73:153) *Cordage* Inner bark strips used for cordage. (73:152) *Mats, Rugs & Bedding* Bark sheets used for tarpaulins. (73:153) Inner bark strips used for mat making. (73:152)

- **Dye**—**Thompson** *Green* Leaves and twigs used as a green dye. (164:501)

- **Other**—**Bella Coola** *Decorations* Inner bark used for mask decoration. *Fuel* Inner bark used for tinder. (184:197) **Cowlitz** *Cooking Tools* Bark used to line cooking pits or cut in broad strips, plaited, and used for dishes. (79:19) **Flathead** *Containers* Whole bark sheets used to make storage containers. (82:54) **Haisla** *Hunting & Fishing Item* Boughs used to camouflage canoes, especially during duck hunting. *Lighting* Branches covered with Sitka spruce pitch used for night fishing. *Protection* Inner bark dyed and tied to pets to protect them from the dog eater. **Haisla & Hanaksiala** *Ceremonial Items* Wood used to make ceremonial whistles. Inner bark rings worn around the neck and on the legs by shamans. Bark used for many ceremonial purposes. Wood used to make a special type of "rattle" for ceremonial activities. Wood used to make coffins. *Cleaning Agent* Wood used as toilet sticks prior to the introduction of toilet paper. *Containers* Wood used to make boxes to store many types of food. *Cooking Tools* Inner bark used to whip soapberries. Wood boards used to separate oval-leaved blueberries from the leaves. Wood made into tongs and used to transfer boiling stones from the fire to cooking vessels. *Designs* Wood used to make totem poles and other carvings. *Fasteners* Withes used with pitch to fasten arrowheads to the shafts. *Fuel* Bark and branches used as kindling to start fires. *Hunting & Fishing Item* Inner bark used to make fishing line. Wood made into small sticks to prop open conical nets used to catch oolichans (candlefish). Wood used to make the shafts of oolichan spears and arrows. Wood used to make the hook used to lift oolichan nets to empty their contents into a canoe. Wood made into noisemakers and used to round up animals to be hunted. **Hanaksiala** *Ceremonial Items* Bark used in the fire dancer's headdress to produce sparks and fire. A baking powder can with a copper tube in front and a flexible hose in the back was filled with burning red cedar outer bark. An aide to the Fire Dancer blew on the hose in the back to produce sparks and smoke at the front of the headdress, which concealed the can and embers. (43:162) **Hesquiat** *Cooking Tools* Wood used to make serving dippers. *Fuel* Shredded inner bark twisted tightly, tied with cedar bark strips, and used as tinder to start a fire. A story involving cedar—called "nuhtume" by Hesquiat—tells of the time Deer stole fire. This version has Deer taking fire from the Wolves; another says that Deer stole it from Chief Red-winged Blackbird. In both versions Deer attached the soft, shredded cedar bark to his elbows, knees, and horns. He allowed this dry cedar bark to catch fire when he visited the fire's owner. Deer escaped by jumping out through the roof. Since, however, he had to jump into water while fleeing, the only fire that continued to burn was that in the cedar bark on his horns. Thus, Deer did bring back fire but suffered burnt knees and elbows. *Hunting & Fishing Item* Inner bark used to make a scabbard for a sword or spear. Young branches used to make fish traps. *Tools* Long, dried kindling tied with cedar bark and used as torch to burn out excess wood in making a canoe. There is a story about Deer and the Chickadees in which Deer was using such a bundle of cedar pieces while making a canoe. He became so absorbed in watching the dancing of the Chickadees that he burned right through the bottom of the canoe. Turner's informant Alice Paul pointed out that a deer is still like that—when he sees anything moving, he stares at it. (185:35) **Hoh** *Cooking Tools* Wood used to make dishes, ladles, and spoons. *Lighting* Inner bark used as lampwicks in oil lamps. (137:57) **Kutenai** *Containers* Bark used to make a storage trunk. *Cooking Tools* Wood used to make bowls. (82:54) **Kwakiutl** *Cooking Tools* Wood used for making utensils. *Tools* Wood used for making implements. (as *T. gigantea* 20:8) **Kwakiutl, Southern** *Containers* Wood steamed for flexibility and malleability and used to make boxes. *Cooking Tools* Boughs used in drying seaweed for eating. *Fuel* Wood used for fuel and making friction fires. *Hunting & Fishing Item* Fibrous bark used to make fish nets. Boughs weighted, placed underwater at estuaries, and used for catching herring spawn. Long, straight branches used to make fishing line. (183:266) **Lummi** *Ceremonial Items* Singed limbs used as brooms to sweep off the walls of a house after removal of a corpse. (79:19) **Makah** *Ceremonial Items* Wood used to make totem poles. *Containers* Wood used to make boxes. *Cooking Tools* Wood used to make fish barbecuing sticks and cross pieces. (67:228) *Hunting & Fishing Item* Used to make arrow shafts. Limbs used for towing dead whales out of the water. (79:19) **Montana Indian** *Hunting & Fishing Item* Wood made into floats for fish nets. (19:25) **Nitinaht** *Ceremonial Items* Wood used to make masks. Boughs used by boys in training for manhood. *Containers* Wood used to make storage boxes. Bark sheets used to make cases for storing whale harpoon heads. *Cooking Tools* Wood used to make bowls, cooking boxes, and spreading sticks to prevent dried fish from curling up. *Decorations* Bark used for mask adornments. *Hunting & Fishing Item* Wood used for the slats or poles of fish weirs, gaff sticks, and to make arrow shafts. Young limbs twisted and used as basket traps for fish. Green boughs used to catch herring spawn. (186:67) **Okanagan-Colville** *Ceremonial Items* Infusion of cedar, fir, and rose branches of equal size used as bathing water for sweat house ritual. Boughs used in church on Palm Sunday. *Hunting & Fishing Item* Wood used to make bows and arrows and dip net frames. *Musical Instrument* Wood used to make drum hoops. (188:20) **Okanagon** *Containers* Inner bark and stems used to make bags. (125:39) **Oweekeno** *Ceremonial Items* Fibrous tissue used to make ceremonial head, neck, ankle, and wrist rings. (43:63) Wood and inner bark used for shamanistic or other ceremonial activities. Inner bark used to make head rings worn by dancers. Bark used to make neck, wrist, and ankle rings worn by dancers. Inner bark woven into dance screens. (43:66) *Containers* Wood used to make containers. Inner bark used to make boxes. *Cooking Tools* Wood used to make sticks to dry oolichans (candlefish) and made into barbecue racks for salmon and other fish. *Decorations* Wood used for carving. Inner bark used to make masks. *Fuel* Wood used as kindling. *Hunting & Fishing Item* Boughs used to gather herring roe. Wood used to make arrows. (43:63) Branches used to gather herring roe. *Musical Instrument* Wood made into whistles or trumpets and used to announce the dances in the tsaika and glualaxa series. (43:66) *Tools* Wood used to make implements. (43:63) **Paiute** *Musical Instrument* Wood used for drum frames. (111:46) **Quileute** *Containers* Bark cut into strips and used for storage of string dried clams and smelts. (79:19) *Cooking Tools* Wood used to make dishes, ladles, and spoons. (137:57) *Hunting & Fishing Item* Limbs used for towing dead whales out of the water. (79:19) *Lighting* Inner bark used as lampwicks in oil lamps. (137:57) *Musical Instrument* Small limbs strung with pecten shells and used as dance rattles. *Tools* Used to make the spindle for spinning mountain goat wool. **Quinault** *Hunting & Fishing Item* Limbs used for towing dead whales out of the water. Limbs used for weaving with vine maple sticks to make fish weirs. (79:19) **Saanich** *Ceremonial Items* Boughs used to vigorously scrub the body for purity while bathing. *Hunting & Fishing Item* Roots woven with gooseberry and wild rose roots to make reef nets. **Salish, Coast** *Ceremonial Items* Wood used to make totems, masks, and ceremonial drums. *Containers* Wood used to make coffins and boxes. Wood used for the hearth to make friction fires. *Cooking Tools* Inner bark beaten to separate the fibers and used for threading clams. *Fuel* Outer bark used for fuel. Inner bark beaten to separate the fibers and used for tinder. Wood used as fuel and also as the drill or hearth in making friction fires. *Hunting & Fishing Item* Inner bark beaten to separate the fibers and used to make fishing line. Young, slender branches used to make nets. Wood used to make herring rakes. *Tools* Wood used for the drill to make friction fires. **Salish, Cowichan** *Hunting & Fishing Item* Roots woven with gooseberry and wild rose roots to make reef nets. (182:71) **Shuswap** *Cash Crop* Root sold at annual tribal gathering. *Ceremonial Items* Bark made into a headband and worn by pubescent girls and young men in ritual isolation. *Jewelry* Bark used to make necklaces and belts worn by dancers at tribal gatherings. (123:50) **Skagit** *Ceremonial Items* Limbs burned at night and waved through the houses after a death to scare off the ghosts. **Squaxin** *Hunting & Fishing Item* Used to make herring rakes. Limbs used for weaving with vine maple sticks to make fish weirs. (79:19) **Thompson** *Cash Crop* Roots made into coiled baskets and considered an important trading item. Widows and women who did not have ready access to meat or fish used to make baskets and trade for food. (187:94) *Containers* Inner bark and stems used to make bags. (125:39) Inner root softened, split into strips, and used to make bags. (164:496) *Good Luck Charm* Root "doll formation" considered to bring good luck to anyone who found it. (187:94) **Tsimshian** *Ceremonial Items* Wood made into horns used for ritual purposes. *Cleaning Agent* Wood made into toilet sticks. *Containers* Wood used to make storage boxes and chests. Inner bark used to make containers for collecting, storing, and cooking foods. *Decorations* Wood used to carve totem poles. *Fuel* Inner bark used to start fires. *Hunting & Fishing Item* Wood used to make hunting and fishing tools. *Tools* Wood used to make woodworking tools. (43:315)

Thymelaeaceae, see *Daphne, Dirca, Wikstroemia*

Thymophylla, Asteraceae

Thymophylla acerosa (DC.) Strother, Pricklyleaf Dogweed
- *Drug*—**Isleta** *Febrifuge* Decoction of leaves used as a body bath for fevers. (as *Aciphyllaea acerosa* 100:20)
- *Other*—**Keres, Western** *Smoke Plant* Leaves mixed with tobacco for the flavor. (as *Aciphyllea acerosa* 171:24)

Thymophylla pentachaeta* var. *belenidium (DC.) Strother, Fiveneedle Pricklyleaf

- *Drug*—Paiute *Cathartic* Decoction of root taken as a physic. (as *Dyssodia thurberi* 180:67)

Thymophylla pentachaeta (DC.) Small **var. pentachaeta**, Fiveneedle Pricklyleaf
- *Drug*—Navajo, Kayenta *Psychological Aid* Plant used for dreaming of being pursued by a deer. (as *Dyssodia pentachaeta* 205:46)
- *Other*—Navajo, Kayenta *Hunting & Fishing Item* Plant eaten or used as lotion by hunters so deer cannot smell them. (as *Dyssodia pentachaeta* 205:46)

Thymus, Lamiaceae
Thymus praecox ssp. *arcticus* (Dur.) Jalas, Creeping Thyme
- *Drug*—Delaware, Ontario *Febrifuge* Compound infusion of plant taken for chills and fever. (as *T. serpyllum* 175:56, 84)

Thysanocarpus, Brassicaceae
Thysanocarpus curvipes Hook., Sand Fringepod
- *Drug*—Mendocino Indian *Gastrointestinal Aid* Decoction of whole plant taken for stomachaches. (as *T. elegans* 41:352)
- *Food*—Mendocino Indian *Staple* Seeds used in pinole mixtures. (as *T. elegans* 41:352)

Tiarella, Saxifragaceae
Tiarella cordifolia L., Heartleaf Foamflower
- *Drug*—Cherokee *Oral Aid* Infusion held in mouth to "remove white coat from tongue." (80:34) **Iroquois** *Dermatological Aid* Poultice of smashed roots applied to wounds. *Dietary Aid* Infusion of roots and leaves given to fatten little children. (87:344) *Eye Medicine* Infusion of dried leaves used as drops for sore eyes. (87:345) *Hunting Medicine* Decoction of whole plant used as a wash for the rifle, a "hunting medicine." *Oral Aid* Infusion of smashed roots given to babies with sore mouths. *Orthopedic Aid* Compound poultice of smashed plants applied to sore backs of babies. (87:344) *Other* Compound decoction of dried roots given to children with "summer complaint." (87:345) *Pediatric Aid* Compound poultice of smashed plants applied to sore backs of babies. Infusion of roots and leaves given to fatten little children. (87:344) Compound decoction of dried roots given to children with "summer complaint." (87:345) *Tonic* Infusion of roots and leaves taken as a tonic. (87:344) **Malecite** *Antidiarrheal* and *Pediatric Aid* Infusion of roots used by children for diarrhea. (116:255) **Micmac** *Antidiarrheal* Roots used for diarrhea. (40:62)

Tiarella trifoliata L., Threeleaf Foamflower
- *Drug*—Quileute *Cough Medicine* Raw leaves chewed as a cough medicine. (79:31)

Tiliaceae, see *Tilia*

Tilia, Tiliaceae
Tilia americana L., American Basswood
- *Drug*—Algonquin, Quebec *Eye Medicine* Infusion of leaves used as an eyewash. *Unspecified* Poultice of leaves used for medicinal purposes. (18:200) **Cherokee** *Antidiarrheal* Compound of inner bark used for dysentery. (80:24) Infusion of inner bark taken for dysentery. (177:42) *Cough Medicine* Jelly used for coughs. *Dermatological Aid* Decoction of bark mixed with cornmeal and used as poultice for boils. *Gastrointestinal Aid* Inside bark and twigs used during pregnancy for heartburn, weak stomach, and bowels. Used "when stomach has been overheated by too free use of spirituous liquors." *Snakebite Remedy* Bark from tree struck by lightning chewed and spit on snakebite. *Tuberculosis Remedy* Jelly used for consumption. (80:24) **Iroquois** *Antihemorrhagic* Compound decoction of roots and bark taken for internal hemorrhage. *Burn Dressing* Compound decoction of leaves applied as poultice to burns or scalds. *Diuretic* Infusion of bark taken to increase urination. *Emetic* Compound decoction taken to vomit during initial stages of consumption. (87:384) *Gynecological Aid* Infusion of branches and bark and buds from another plant taken before giving birth. (141:51) *Orthopedic Aid* Decoction of branches used as wash for babies that do not walk but should. (87:383) Compound poultice of leaves applied to broken bones and swollen areas. *Other* Infusion of plant used for severe injuries. *Panacea* Compound infusion of twigs and roots taken as a panacea. (87:384) *Pediatric Aid* Decoction of branches used as wash for babies that do not walk but should. (87:383) *Stimulant* Infusion of shoots taken when feeling worn out. *Tuberculosis Remedy* Compound decoction taken to vomit during initial stages of consumption. (87:384) **Malecite** *Anthelmintic* Infusion of roots or bark used for worms. (116:255) **Meskwaki** *Dermatological Aid* Poultice of boiled inner bark applied to cause boils to open. *Pulmonary Aid* Decoction of twigs taken for lung trouble. (152:248) **Micmac** *Anthelmintic* Roots used for worms. *Dermatological Aid* Bark used for suppurating wounds. (40:62)
- *Food*—Chippewa *Vegetable* Young twigs and buds cooked as greens or eaten raw. (71:136)
- *Fiber*—Algonquin, Tête-de-Boule *Basketry* Used to make baskets stronger. (132:132) **Cherokee** *Building Material* Wood used for lumber. *Cordage* Boiled bark twisted into rope. *Furniture* Used to make chair bottoms. (80:24) **Chippewa** *Canoe Material* Wood used to make dugout canoes. (71:136) *Cordage* Used for twine and general utility. (53:378) Bast made into cordage of all sorts. The bast was boiled and rubbed on a stick to separate the fibers which were spun into thread for sewing, fine yarn for weaving bags, and made into cordage of all sorts. *Sewing Material* Bast made into thread for sewing and fine yarn for weaving bags. (71:136) **Lakota** *Cordage* Inner bark fibers used to make cordage. (139:60) **Malecite** *Clothing* Fiber used to make belts. *Cordage* Fiber used to make ropes. *Sewing Material* Fiber used for sewing birch bark. (160:6) **Menominee** *Basketry* Basswood fiber used for baskets and fish nets. *Cordage* Bast and bark fiber used for cordage. *Mats, Rugs & Bedding* Basswood fiber used for matting. *Snow Gear* Basswood fiber used for nets for snowshoes. (151:76) **Meskwaki** *Basketry* Inner bark two-ply cord used to make baskets and fish nets. *Clothing* Inner bark two-ply cord used to make shoes. *Cordage* Inner bark boiled in lye water, dried, seasoned, and twisted into two-ply cord. *Mats, Rugs & Bedding* Inner bark two-ply cord used to make mats. (152:269) **Ojibwa** *Cordage* Inner bark of young sprouts used to make twine and rope. (135:232) Tough, fibrous bark of young trees furnished ready cordage and string. The women stripped the bark and peeled the outer edge from the inner fiber with their teeth. The rolls were then kept in coils or were boiled and kept as coils until needed, being soaked again when used, to make them pliable. While there were countless uses for this cordage, perhaps the most important was in tying the poles together for the framework of the wigwam

or medicine lodge. When these crossings of poles were lashed together with wet bark fiber, it was easy to get a tight knot which shrank when dry and made an even tighter joint. The bark of an elm or a balsam, cut into broad strips, was then sewed into place on the framework with basswood string. An oak wood awl was used to punch holes in the bark, but Smith notes that, when they made his wigwam, they used an old file end for an awl. He reports that he lived in this new wigwam all the time he was among the Pillager Ojibwe and scarcely a night passed without a group of them visiting him and sitting around the campfire, telling old-time stories. (153:422) *Sewing Material* Inner bark of young sprouts used to make thread. (135:232) **Omaha** *Basketry* Inner bark used to make baskets. *Cordage* Inner bark used to make ropes and cordage. (68:324) Inner bark fiber used to make cordage and rope. **Pawnee** *Mats, Rugs & Bedding* Inner bark fiber used for spinning cordage and weaving matting. **Ponca** *Cordage* Inner bark fiber used to make cordage and rope. (70:102) **Potawatomi** *Basketry* Bark string used for fashioning bags. *Cordage* Bark string used for making cordage. *Sewing Material* Bark string used for sewing the edges of mats. (154:114)
- *Other*—**Cherokee** *Decorations* Wood used to carve. *Paper* Wood used for pulpwood. (80:24) **Chippewa** *Containers* Bast made into thread for sewing, fine yarn for weaving bags, and into other cordage of all sorts. *Fasteners* Bast strips used for tying the poles of the framework of houses. *Tools* Wood used to make spiles for drawing out maple sap from trees into buckets during sugar making time. (71:136) **Menominee** *Hunting & Fishing Item* Basswood fiber used for matting, baskets, fish nets, and nets for snowshoes. (151:76) **Meskwaki** *Hunting & Fishing Item* Inner bark two-ply cord used to make mats, baskets, fish nets, and shoes. (152:269)

Tilia americana L. **var. *americana***, American Basswood
- *Fiber*—**Abnaki** *Basketry* Inner bark used to make baskets. (as *T. glabra* 144:156)

Tilia americana* var. *heterophylla (Vent.) Loud., American Basswood
- *Drug*—**Cherokee** *Antidiarrheal* Compound of inner bark used for dysentery. *Cough Medicine* Jelly used for coughs. *Dermatological Aid* Decoction of bark mixed with cornmeal and used as poultice for boils. *Gastrointestinal Aid* Inside bark and twigs used during pregnancy for heartburn, weak stomach, and bowels. Used "when stomach has been overheated by too free use of spirituous liquors." *Snakebite Remedy* Bark from tree struck by lightning chewed and spit on snakebite. *Tuberculosis Remedy* Jelly used for consumption. (as *T. heterophylla* 80:24)
- *Fiber*—**Cherokee** *Building Material* Wood used for lumber. *Cordage* Boiled bark twisted into rope. *Furniture* Used to make chair bottoms. (as *T. heterophylla* 80:24)
- *Other*—**Cherokee** *Decorations* Wood used to carve. *Paper* Wood used for pulpwood. (as *T. heterophylla* 80:24)

Tillandsia, Bromeliaceae

***Tillandsia* sp.**, Airplant
- *Other*—**Seminole** *Cooking Tools* Plant used as a water supply for cooking during the dry season. (169:472)

Tillandsia usneoides (L.) L., Spanish Moss
- *Drug*—**Houma** *Febrifuge* Decoction of moss taken for chills and fever. (158:59)
- *Fiber*—**Houma** *Cordage* Dried fibers twisted and used for cordage. *Mats, Rugs & Bedding* Dried fibers twisted and used to make floor mats. (158:59)
- *Other*—**Seminole** *Cooking Tools* Used to remove scum in cooking. *Hide Preparation* Used for tanning. (as *Dendropogon usneoides* 169:472)

Tiquilia, Boraginaceae

Tiquilia latior (I. M. Johnston) A. Richards., Matted Crinklemat
- *Drug*—**Navajo, Kayenta** *Gastrointestinal Aid* Plant used for gastrointestinal disease. (as *Coldenia hispidissima* var. *latior* 205:39)

Tolmiea, Saxifragaceae

Tolmiea menziesii (Pursh) Torr. & Gray, Youth on Age
- *Drug*—**Cowlitz** *Dermatological Aid* Poultice of fresh leaves applied to boils. (79:31)
- *Food*—**Makah** *Unspecified* Sprouts eaten raw in early spring. (79:31)

Torreya, Taxaceae, conifer

Torreya californica Torr., California Nutmeg
- *Drug*—**Costanoan** *Analgesic* Smashed nuts and fat rubbed on temples for headaches. *Diaphoretic* Smashed nuts and fat rubbed on body to cause sweating. *Febrifuge* Smashed nuts and fat rubbed on body for chills. *Gastrointestinal Aid* Nuts chewed for indigestion. (as *Tumion californica* 21:6) **Pomo** *Tuberculosis Remedy* Decoction of nuts taken for tuberculosis. (66:11) **Pomo, Kashaya** *Tuberculosis Remedy* Decoction of cracked, soaked nut taken for tuberculosis. (72:78)
- *Food*—**Mendocino Indian** *Unspecified* Whole nuts roasted and used for food. (as *Tumion californicum* 41:305)
- *Fiber*—**Pomo** *Basketry* Roots used to make baskets. (as *Tumion californicum* 41:305) Roots used as splints in basketry. (as *Tumion californica* 118:6)
- *Other*—**Pomo** *Tools* Rigid, sharp-pointed leaves formerly used as needles to prick pitch soot into the skin in tattooing. (as *Tumion californicum* 41:305)

Touchardia, Urticaceae

Touchardia latifolia Gaud., Olona
- *Drug*—**Hawaiian** *Laxative* Slimy substance, water, plant milk, and watermelon juice mixed, strained, and taken as a laxative. *Pediatric Aid* Shoots chewed and given to infants for bodily ailments or weaknesses. *Strengthener* Shoots chewed and given to infants for bodily ailments or weaknesses. Shoots chewed for bodily ailments or weaknesses. (2:71)
- *Fiber*—**Hawaiian** *Cordage* Plant fiber used to make fishing lines or ropes. (2:71)

Townsendia, Asteraceae

Townsendia exscapa (Richards.) Porter, Stemless Townsendia
- *Drug*—**Blackfoot** *Veterinary Aid* Decoction of roots given to tired horses to relieve them. (as *T. sericea* 97:61) **Navajo** *Ceremonial Medicine* Chewed and spit upon ceremonial knots to unravel them, "un-

tying medicine." (55:89) **Navajo, Ramah** *Gynecological Aid* Plant chewed or infusion taken to ease delivery. (191:54)
- **Other**—**Navajo, Ramah** *Ceremonial Items* Pollen shaken from a live horned toad onto the unraveler string and used in the "unraveling ceremony." (191:54)

Townsendia incana Nutt., Hoary Townsendia
- **Drug**—**Hopi** *Reproductive Aid* Plant taken to induce pregnancy and insure male child. *Throat Aid* Plant used to clear the throat. (as *T. arizonica* 200:35, 99) **Navajo** *Gynecological Aid* Plant used in labor to facilitate delivery of the baby. (90:156) **Navajo, Kayenta** *Gynecological Aid* Plant used to expedite labor. *Unspecified* Plant used as a strong medicine. (as *T. arizonica* 205:50)

Townsendia sp.
- **Drug**—**Navajo** *Gynecological Aid* Plant used to accelerate deliverance. (55:89)

Townsendia strigosa Nutt., Hairy Townsendia
- **Drug**—**Hopi** *Throat Aid* Plant used to clear the throat. (42:368) **Keres, Western** *Gynecological Aid* Dried leaves ground into a powder and used on lacerations at childbirth. Infusion of plant used as a wash or douche. (171:73) **Navajo** *Gastrointestinal Aid* Decoction of crushed, dried leaves taken for stomach troubles. *Gynecological Aid* Decoction of crushed, dried leaves taken to accelerate deliverance. *Nose Medicine* Dried, pulverized plants used as a snuff for nose troubles. *Throat Aid* Dried, pulverized plants used as a snuff for throat troubles. (55:89)
- **Other**—**Navajo** *Ceremonial Items* Chewed by Night Chant chanter and spit upon prayer sticks to facilitate the disentanglement of string. Juice spat upon the wand used in the Night Chant. (55:89)

Toxicodendron, Anacardiaceae

Toxicodendron diversilobum (Torr. & Gray) Greene, Pacific Poison Oak
- **Drug**—**Diegueño** *Eye Medicine* Decoction of roots used in the eyes for tiny sores inside the lids and to improve vision. (84:43) **Karok** *Contraceptive* Leaf swallowed in the spring as a prophylactic. (as *Rhus diversiloba* 148:385) *Poison* Plant considered poisonous. (6:58) **Mahuna** *Preventive Medicine* Infusion of dried roots taken as an immunity against any further poisoning. (as *Rhus diversiloba* 140:11) **Mendocino Indian** *Dermatological Aid* Moxa of plant used for warts and ringworms. (as *Rhus diversiloba* 41:364) **Tolowa** *Antidote* Buds eaten in the spring to obtain immunity from the plant poisons. *Poison* Plant considered poisonous. (6:58) **Wailaki** *Snakebite Remedy* Poultice of fresh leaves applied to rattlesnake bites. (as *Rhus diversiloba* 41:364) **Yuki** *Dermatological Aid* Plant juice used on warts. (as *Rhus diversiloba* 49:46) **Yurok** *Poison* Plant considered poisonous. (6:58)
- **Food**—**Karok** *Candy* Plant chewed like tobacco, "just to raise heck." (6:58) **Mendocino Indian** *Forage* Fruits eaten by yellowhammers and squirrels as forage. Fruits and leaves eaten by hogs as forage. (as *Rhus diversiloba* 41:364)
- **Fiber**—**Costanoan** *Basketry* Shoots used in basketry. (21:251) **Mendocino Indian** *Basketry* Slender stems used for circular withes in basket making. (as *Rhus diversiloba* 41:364)
- **Dye**—**Pomo** *Black* Charcoal or soot used for tattoo pigment. Ashes rubbed on children to make skin color darker. The ashes were rubbed on those children who were fathered by a white man to make them look more "Indian" in color. (as *Rhus diversiloba* 66:14) Juice used as a black dye for blackroot sedge. (as *Rhus diversiloba* 118:6)
- **Other**—**Costanoan** *Containers* Leaves used to wrap bread. (21:251) **Karok** *Containers* Leaves used to cover soap plant while cooking in the earth oven. (as *Rhus diversiloba* 148:385) *Cooking Tools* Branches used to prop salmon filet open. (6:58) Twigs used to spit the salmon steaks during smoking. (as *Rhus diversiloba* 148:385) **Mendocino Indian** *Cooking Tools* Fresh leaves formerly used to wrap up acorn meal for baking. *Decorations* Black juice used to apply temporary tattoo marks onto the skin. (as *Rhus diversiloba* 41:364) **Pomo, Kashaya** *Designs* Burned wood ashes made into a paste for tattooing. (as *Rhus diversiloba* 72:81)

Toxicodendron pubescens P. Mill., Atlantic Poison Oak
- **Drug**—**Cherokee** *Emetic* Decoction given as an emetic. (as *Rhus toxicodendron* 80:41) Decoction of bark taken as an emetic. (as *Rhus toxicodendron* 177:37) **Delaware** *Unspecified* Poultice of roasted, crushed roots regarded as medicinally valuable. (as *Rhus toxicodendron* 176:33) **Delaware, Oklahoma** *Dermatological Aid* Poultice and salve containing root used on chronic sores and swollen glands. (as *Rhus toxicondendron* 175:27, 78) **Iroquois** *Oral Aid* Poultice of plant applied to infectious sores on lips. *Pediatric Aid* Given to nervous boys and girls. *Sedative* Given to nervous boys and girls. (as *Rhus toxicodendron* ssp. *radicans* 87:373) **Meskwaki** *Dermatological Aid* Poultice of pounded root applied to swelling "to make it open," dangerous. (as *Rhus toxicodendron* 152:201) **Omaha** *Poison* Plant considered poisonous. (as *T. toxicodendron* 70:100) **Paiute** *Poison* Plant considered poisonous. (as *Rhus toxicodendron* 111:88) **Ponca** *Poison* Plant considered poisonous. (as *T. toxicodendron* 70:100) **Potawatomi** *Dermatological Aid* Poultice of pounded root applied to swellings by skilled medicine men. *Poison* Plant considered poisonous. (as *Rhus toxicodendron* 154:38)

Toxicodendron radicans (L.) Kuntze, Eastern Poison Ivy
- **Drug**—**Houma** *Tonic* Decoction of leaves taken as a tonic and "rejuvenator." (158:59)

Toxicodendron radicans (L.) Kuntze ssp. *radicans*, Eastern Poison Ivy
- **Drug**—**Algonquin, Quebec** *Dermatological Aid* Leaves rubbed on the skin affected by a poison ivy reaction. (as *Rhus radicans* 18:194) **Cherokee** *Emetic* Decoction given as an emetic. (as *Rhus radicans* 80:41) **Kiowa** *Dermatological Aid* Plant used for running or nonhealing sores. Whole or broken leaves rubbed over boils or skin eruptions. (as *Rhus radicans* 192:38) **Navajo, Ramah** *Poison* Compound containing plant used to poison arrows. (as *Rhus radicans* 191:35) **Thompson** *Poison* Plant considered poisonous because it caused skin irritations. One informant said that it affected her eyes, causing temporary blindness. (as *Rhus radicans* 187:149)
- **Other**—**Navajo, Ramah** *Good Luck Charm* Leaf chewed and given to opponent to insure good luck in gambling. (as *Rhus radicans* 191:35)

Toxicodendron rydbergii (Small ex Rydb.) Greene, Western Poison Ivy
- **Drug**—**Iroquois** *Blood Medicine* Poultice of plant applied to the skin

as a vesicant for water in the blood. Poultice of plant applied to the skin as a vesicant for water in the blood. (as *Rhus radicans* var. *rydbergii* 141:52) **Lakota** *Poison* Poisonous plant caused a rash resembling venereal disease. (139:33)

Toxicodendron vernix (L.) Kuntze, Poison Sumac
- *Drug*—**Cherokee** *Dermatological Aid* Plant considered poison and used in some form as a wash for foul ulcers. *Febrifuge* Plant considered poison and taken in some form for fever. *Misc. Disease Remedy* Plant considered poison and taken in some form for ague. *Poison* Plant considered poison and taken in some form for clap and "gleet" or ulcerated bladder. *Respiratory Aid* Plant considered poison and taken in some form for asthma and phthisic. *Urinary Aid* Plant considered poison and taken in some form for ulcerated bladder. *Venereal Aid* Plant considered poison and taken in some form for clap and "gleet." (as *Rhus vernix* 80:57) **Chippewa** *Poison* Plant considered poisonous. (as *Rhus vernix* 71:135)

Tradescantia, Commelinaceae

Tradescantia bracteata Small ex Britt., Longbract Spiderwort
- *Other*—**Lakota** *Paint* Flowers made into a blue jelly-like paint used for painting moccasins. (139:26)

Tradescantia occidentalis (Britt.) Smyth, Prairie Spiderwort
- *Drug*—**Meskwaki** *Diuretic* Infusion of root used as a "urinary." *Psychological Aid* Root gum inserted in cut on head "to stop craziness." (152:209) **Navajo, Kayenta** *Love Medicine* Plant used as an aphrodisiac. (205:16) **Navajo, Ramah** *Disinfectant* Cold infusion of root used internally and externally for "deer infection." *Internal Medicine* Decoction of root taken for internal injury. *Veterinary Aid* Cold simple or compound infusion given to livestock as an aphrodisiac. (191:20)
- *Food*—**Acoma** *Unspecified* Tender shoots eaten without preparation. (32:53) **Hopi** *Vegetable* Plant used for greens. (42:369) **Keres, Western** *Unspecified* Tender shoots eaten for food. (171:73) **Laguna** *Unspecified* Tender shoots eaten without preparation. (32:53)

Tradescantia pinetorum Greene, Pinewoods Spiderwort
- *Drug*—**Navajo, Ramah** *Veterinary Aid* Cold simple or compound infusion given to livestock as an aphrodisiac. (191:20)

***Tradescantia* sp.**, Spiderwort
- *Food*—**Cherokee** *Vegetable* Leaves relished as greens. (203:75)

Tradescantia virginiana L., Virginia Spiderwort
- *Drug*—**Cherokee** *Analgesic* Infusion used for stomachache from overeating. *Antihemorrhagic* Compound infusion taken for "female ailments or rupture." *Cancer Treatment* Poultice of root used for cancer. *Dermatological Aid* Plant mashed and rubbed on insect bites. *Gastrointestinal Aid* Infusion used for stomachache from overeating. *Gynecological Aid* Compound infusion taken for "female ailments or rupture." *Kidney Aid* Compound used for kidney trouble. *Laxative* Infusion taken as a laxative and plant mashed and rubbed on insect bites. (80:56, 57)
- *Food*—**Cherokee** *Unspecified* Young growth parboiled, fried, frequently mixed with other greens, and eaten. (204:252) *Vegetable* Leaves and stems mixed with other greens or grease and parboiled until tender. (126:33)

Tragia, Euphorbiaceae

Tragia nepetifolia Cav., Catnip Noseburn
- *Drug*—**Navajo, Kayenta** *Snakebite Remedy* Plant used as a lotion to keep snakes away. (205:31) **Navajo, Ramah** *Panacea* Plant used as "life medicine." (191:35)
- *Other*—**Navajo, Kayenta** *Protection* Plant sprinkled on hogan during rain storm for protection from lightning. (205:31)

Tragia ramosa Torr., Branched Noseburn
- *Drug*—**Keres, Western** *Dermatological Aid* Infusion of plant used on ant bites. *Pediatric Aid* and *Strengthener* Nettle used to strike male infants, to increase pain threshold during battle. (171:73)

Tragopogon, Asteraceae

Tragopogon porrifolius L., Salsify
- *Drug*—**Navajo, Ramah** *Ceremonial Medicine* Plant used as a ceremonial emetic. *Dermatological Aid* Cold infusion taken or used as lotion for mad coyote bite on humans or livestock. *Emetic* Plant used as a ceremonial emetic. *Veterinary Aid* Cold infusion taken or used as lotion for mad coyote bite on humans or livestock. (191:54)
- *Food*—**Navajo** *Beverage* Latex used as milk. (90:156) **Okanagon** *Unspecified* Stems broken at the base and hardened juice chewed for food. **Thompson** *Unspecified* Stems broken at the base and hardened juice chewed for food. (125:38) **Thompson, Upper (Nicola Band)** *Unspecified* Latex exposed to air, chewed, and swallowed. **Thompson, Upper (Spences Bridge)** *Unspecified* Latex exposed to air, chewed, and swallowed. (164:484)

Tragopogon pratensis L., Meadow Salsify
- *Drug*—**Navajo, Ramah** *Ceremonial Medicine* Cold infusion gargled as ceremonial treatment for throat trouble. *Dermatological Aid* Cold infusion used as lotion for boils. *Throat Aid* Cold infusion gargled as ceremonial treatment for throat trouble. *Veterinary Aid* Cold infusion used in large amounts for internal injury to horses. (191:54)
- *Food*—**Okanagan-Colville** *Forage* Plant eaten by deer, horses, and cattle. (188:85)

***Tragopogon* sp.**, Salsify
- *Food*—**Cherokee** *Unspecified* Young plants boiled, fried, and eaten. (204:253) **Thompson** *Candy* Milky latex chewed for pleasure and later swallowed. The stems were broken, the sap squeezed out, allowed to harden, and then chewed for pleasure. (187:185) **Thompson, Upper (Nicola Band)** *Unspecified* Milky juice obtained from broken stem bases, dried, and chewed with great enjoyment. (164:493)
- *Fiber*—**Thompson** *Mats, Rugs & Bedding* Fluffy pappus of fruits used to make "cotton" for stuffing pillows. (187:185)

Trautvetteria, Ranunculaceae

Trautvetteria caroliniensis (Walt.) Vail, Carolina Bugbane
- *Drug*—**Bella Coola** *Dermatological Aid* Poultice of pounded roots, too strong for children, applied to boils. (as *T. grandis* 150:57)

Trema, Ulmaceae

Trema micranthum (L.) Blume, Trema
- *Drug*—**Seminole** *Reproductive Aid* Bark used for difficult deliveries. (as *T. floridanus* 169:323)

Trentepohliaceae, see *Trentepohlia*

Trentepohlia, Trentepohliaceae, alga

Trentepohlia aurea (L.) Martius
- *Other*—**Haisla & Hanaksiala** *Paint* Plant used for face paint. (43:132)

Triadenum, Clusiaceae

Triadenum virginicum (L.) Raf., Virginia Marsh St. Johnswort
- *Drug*—**Potawatomi** *Febrifuge* Infusion of leaves used for fevers. (as *Hypericum virginicum* 154:60)

Trianthema, Aizoaceae

Trianthema portulacastrum L., Desert Horsepurslane
- *Food*—**Pima** *Vegetable* Plants cooked and eaten as greens in summer. (47:64)

Tribulus, Zygophyllaceae

Tribulus terrestris L., Puncturevine
- *Drug*—**Navajo** *Ceremonial Medicine* Plant used as a traditional ceremonial medicine. (90:163)

Tricholomataceae, see *Armillaria*, *Pleurotus*

Trichostema, Lamiaceae

Trichostema lanatum Benth., Woolly Bluecurls
- *Drug*—**Cahuilla** *Gastrointestinal Aid* Decoction of leaves and flowers taken for stomach ailments. (15:141) **Mahuna** *Panacea* Plant used for many ailments. (140:48)

Trichostema lanceolatum Benth., Vinegar Weed
- *Drug*—**Concow** *Analgesic* Infusion of leaves used as a wash for feverish headaches. *Febrifuge* Infusion of leaves used as a wash for feverish headaches. *Misc. Disease Remedy* Infusion of leaves used as a wash for typhoid fever. (41:385) **Costanoan** *Analgesic* Ground leaves rubbed on the skin for pain. *Cold Remedy* Cold infusion of leaves used for colds. Ground leaves rubbed on the face and chest for colds. *Dermatological Aid* and *Disinfectant* Decoction of plant used for infected sores. *Gastrointestinal Aid* Decoction of plant used for stomachaches. (21:17) **Kawaiisu** *Cold Remedy* Infusion of plants taken for colds. *Gastrointestinal Aid* Infusion of plants taken for stomachaches. (206:67) **Miwok** *Analgesic* Decoction of leaves and flowers taken for headaches. *Cold Remedy* Decoction of leaves and flowers taken for colds. (12:173) *Dermatological Aid* Decoction of flat leaves used as bath for pustules and prevented skin eruptions caused by smallpox. *Gynecological Aid* Decoction steam used for uterine trouble. (12:173, 174) *Misc. Disease Remedy* Decoction of leaves and flowers taken for malaria. Decoction of leaves and flowers taken for ague. (12:173) Decoction of leaves and flowers used as a bath for ague, and smallpox. (12:173, 174) *Stimulant* Decoction of leaves and flowers taken for general debility. (12:173) *Toothache Remedy* Leaves chewed and placed in cavity or around aching tooth. (12:173, 174) *Urinary Aid* Decoction of leaves and flowers taken for stricture of the bladder. (12:173) **Tubatulabal** *Analgesic* Infusion of entire plant snuffed up the nose for headaches. *Hemostat* Infusion of entire plant snuffed up the nose for nosebleeds. (193:59)
- *Food*—**Kawaiisu** *Beverage* Leaves used to make a nonmedicinal beverage. (206:67)
- *Other*—**Concow** *Hunting & Fishing Item* Fresh leaves mashed and used as a fish poison. (41:385) **Karok** *Insecticide* Plant used in bedding to keep fleas away. (148:389) **Kawaiisu** *Insecticide* Crushed leaves used in bed rolls to discourage fleas. (206:67) **Numlaki** *Hunting & Fishing Item* Fresh leaves mashed and used as a fish poison. (41:385)

Trichostema sp., Blue Stinkweed
- *Drug*—**Karok** *Unspecified* Used for medicine. (117:209)

Tridens, Poaceae

Tridens muticus (Torr.) Nash **var. muticus**, Rough Tridens
- *Food*—**Navajo, Ramah** *Fodder* Used for sheep and horse feed. (as *Triodia mutica* 191:17)

Trientalis, Primulaceae

Trientalis borealis Raf. **ssp. borealis**, Maystar
- *Drug*—**Montagnais** *Panacea* Infusion of plant used for "general sickness." *Tuberculosis Remedy* Infusion of plant taken for consumption. (as *T. americana* 156:314)
- *Other*—**Ojibwa** *Hunting & Fishing Item* Root mixed with many others to make smoking scent that attracted the deer to the hunter. (as *T. americana* 153:431)

Trientalis borealis ssp. latifolia (Hook.) Hultén, Broadleaf Starflower
- *Drug*—**Cowlitz** *Eye Medicine* Infusion of plant juice used as an eyewash. (as *T. latifolia* 79:45) **Paiute** *Eye Medicine* Plant used as an eye medicine. (as *T. latifolia* 111:105)

Trifolium, Fabaceae

Trifolium albopurpureum Torr. & Gray, Rancheria Clover
- *Food*—**Pomo, Kashaya** *Unspecified* Leaves eaten alone or with salt or peppernut cakes. (72:36)

Trifolium bifidum Gray, Notchleaf Clover
- *Food*—**Mendocino Indian** *Staple* Seeds eaten as a pinole. *Unspecified* Eaten sparingly when young. (41:360)

Trifolium ciliolatum Benth., Foothill Clover
- *Food*—**Luiseño** *Unspecified* Plant eaten both cooked and raw. Seeds used for food. (155:231) **Miwok** *Dried Food* Steamed clover dried for later use. Dried clover stored for later use. *Unspecified* Raw or steamed clover used for food. (12:160) **Wailaki** *Forage* Eaten by horses with impunity. *Unspecified* Species used for food. **Yokia** *Unspecified* Species used for food. (41:360)

Trifolium cyathiferum Lindl., Cup Clover
- *Food*—**Mendocino Indian** *Unspecified* Flowers eaten as food. (41:361)

Trifolium dichotomum Hook. & Arn., Branched Indian Clover
- *Food*—**Mendocino Indian** *Staple* Seeds eaten as a pinole. *Unspecified* Young leaves sparingly eaten. (41:361)

Trifolium dubium Sibthorp, Suckling Clover
- *Drug*—**Navajo, Ramah** *Ceremonial Medicine* and *Emetic* Plant used as a ceremonial emetic. (191:34) *Hemostat* Poultice of chopped plant applied to cut as hemostat. (191:33)

Trifolium fucatum Lindl., Bull Clover
- *Food*—**Mendocino Indian** *Unspecified* Herbage used for food. Flowers and seedpods used for food. Seeds eaten raw. (as *T. virescens* 41:361) **Pomo, Kashaya** *Unspecified* Sweet flowers and leaves eaten alone, with salt or peppernut cakes. (72:35)

Trifolium gracilentum Torr. & Gray, Pinpoint Clover
- *Food*—**Luiseño** *Unspecified* Plant eaten both cooked and raw. (155:231)

Trifolium hybridum L., Alsike Clover
- *Drug*—**Iroquois** *Gynecological Aid* Cold infusion of plant used as a wash for breasts to increase milk flow. *Veterinary Aid* Cold infusion of plant used as a wash for cow's teats to increase milk flow. (87:364)

Trifolium microcephalum Pursh, Smallhead Clover
- *Food*—**Luiseño** *Unspecified* Plant cooked and eaten. (155:231)

Trifolium obtusiflorum Hook. f., Clammy Clover
- *Food*—**Luiseño** *Unspecified* Plant cooked and eaten. (155:231) **Mendocino Indian** *Unspecified* Leaves eaten after acid exudation washed away. (41:361)

Trifolium pratense L., Red Clover
- *Drug*—**Algonquin, Quebec** *Pulmonary Aid* Infusion of plant taken for whooping cough. (18:188) **Cherokee** *Febrifuge* Infusion taken for fevers. *Gynecological Aid* Infusion taken for fevers and leukorrhea. *Kidney Aid* Infusion taken for "Bright's disease." (80:29) **Iroquois** *Blood Medicine* Decoction of flowers taken as a blood medicine. *Gynecological Aid* Cold infusion of blossoms taken by women for the change of life. (87:363) **Rappahannock** *Blood Medicine* Infusion of stems and leaves used as an ingredient of a blood medicine. (161:31) **Shinnecock** *Cancer Treatment* Teaspoonful of powder mixed in boiling water and taken for cancer. (30:119) **Thompson** *Cancer Treatment* Infusion of heads taken for stomach cancer. (187:224)
- *Food*—**Shuswap** *Fodder* Used with timothy as a good feed for cows. (123:64) **Thompson** *Fodder* Plant used as food for livestock. (187:224)

Trifolium repens L., White Clover
- *Drug*—**Cherokee** *Febrifuge* Infusion taken for fevers. *Gynecological Aid* Infusion taken for fevers and leukorrhea. *Kidney Aid* Infusion taken for "Bright's disease." (80:29) **Iroquois** *Dermatological Aid* Compound infusion of whole plant used as a wash for liver spots. *Eye Medicine* Poultice of plant applied to eyes for paralysis. (87:363) Infusion of flowers used as an eyewash. *Respiratory Aid* Infusion of flowers, leaves, and roots of another plant used for asthma. (141:50) **Mohegan** *Cold Remedy* Infusion of plant taken for colds. (174:270) Infusion of dried leaves taken for colds. (176:76, 132) *Cough Medicine* Infusion of plant taken for coughs. (174:270) Infusion of dried leaves taken for coughs. (176:76, 132)

Trifolium sp., Clover
- *Drug*—**Apalachee** *Unspecified* Plant water used for medicinal purposes. (81:98) **Costanoan** *Cathartic* Decoction of foliage used or foliage eaten as a purgative. (21:19) **Pomo** *Antiemetic* Decoction of blossoms taken for vomiting. (66:13)
- *Food*—**Anticosti** *Beverage* Flowers used to make wine. (143:67) **Cahuilla** *Porridge* Seeds ground into a mush. *Vegetable* Leaves eaten raw or boiled as greens. (15:141) **Costanoan** *Unspecified* Foliage used for food. (21:250) **Diegueño** *Vegetable* Leaves cooked and eaten as greens. (84:43) **Mendocino Indian** *Forage* Eaten as forage by many kinds of animals. *Unspecified* Fresh green foliage eaten before flowering. (41:359) **Miwok** *Dried Food* Steamed clover used for food and dried for later use. *Unspecified* Clover eaten raw or steamed. (12:161) **Round Valley Indian** *Unspecified* Fresh green foliage eaten before flowering. (41:359) **Yuki** *Unspecified* Eaten raw or steamed. (49:88)

Trifolium variegatum Nutt., Whitetip Clover
- *Food*—**Mendocino Indian** *Vegetable* Eaten considerably as greens. (41:361) **Pomo, Kashaya** *Unspecified* Sweet flowers and leaves eaten alone, with salt or peppernut cakes. (72:36)

Trifolium willdenowii Spreng., Tomcat Clover
- *Food*—**Luiseño** *Unspecified* Plant eaten cooked or raw. Seeds used for food. (as *T. tridentatum* 155:231) **Miwok** *Unspecified* Raw or steamed leaves eaten before plant bloomed. Raw or steamed buds eaten before plant bloomed. (as *T. tridentatum* 12:160) *Winter Use Food* Steamed, dried leaves soaked in water or boiled before eating in winter. (as *T. tridentatum* 12:161) **Paiute** *Unspecified* Seeds used for food. *Vegetable* Whole plant used, without cooking, as greens. (as *T. tridentatum* 167:243) Young, tender plants eaten uncooked as greens. (as *T. tridentatum* 167:244)

Trifolium wormskioldii Lehm., Cow Clover
- *Drug*—**Makah** *Unspecified* Used for medicine. (67:281)
- *Food*—**Bella Coola** *Unspecified* White, brittle roots formerly pit steamed or boiled and eaten with eulachon (candlefish) grease and stink salmon. (184:205) **Haisla & Hanaksiala** *Unspecified* Rhizomes used for food. (43:251) **Hesquiat** *Unspecified* Steamed or boiled rhizomes eaten with oil or stink salmon eggs. Stink salmon eggs made by placing salmon roe in a codfish stomach, plugging it with a cedarwood cork, and allowing the eggs to ferment. (185:68) **Kawaiisu** *Vegetable* Green leaves eaten raw with salt. (206:68) **Kitasoo** *Unspecified* Rhizomes used for food. (43:337) **Kwakiutl, Southern** *Vegetable* Thin, wiry roots dried, steamed, or boiled, dipped in oil, and eaten as vegetables. (183:285) **Makah** *Unspecified* Roots steamed and eaten. (as *T. fimbriatum* 79:38) **Mendocino Indian** *Unspecified* Flowers and leaves used for food. (41:362) **Miwok** *Beverage* Wilted, dry leaves soaked and stirred in cold water to make a sour drink. *Unspecified* Leaves and flowers, never cooked or dried, eaten raw. (as *T. involucratum* 12:160) **Nitinaht** *Dried Food* Rhizomes steam cooked, dried, resteamed, dipped in oil, and eaten in winter. (186:110) *Unspecified* Roots eaten raw or cooked with fermented salmon eggs. (67:281) Rhizomes eaten as accompaniments to cooked duck. (186:131) Rhizomes steam cooked, cooled, and eaten immediately. (186:110) *Vegetable* Rhizomes formerly steamed, dried, and used as a vegetable food in winter. (186:63) **Nuxalkmc** *Unspecified* Rhizomes used for food. (43:251) **Oweekeno** *Unspecified* Rhizomes cooked with riceroot and eaten. (43:102) **Paiute** *Vegetable* Leaves eaten uncooked as greens. (as *T. involucratum* var. *fendlari* 167:244) **Pomo, Kashaya** *Unspecified* Flowers and leaves eaten alone or with salt or peppernut cakes. (72:38) **Tubatulabal** *Unspecified* Leaves and stems used extensively for food. (as *T. involucratum* 193:15)

Triglochin, Juncaginaceae

Triglochin maritimum L., Seaside Arrowgrass

- **Drug**—Blackfoot *Poison* Leaves known to be poisonous to stock. (97:19)
- **Food**—Gosiute *Unspecified* Seeds used for food. (39:383) **Klamath** *Beverage* Roasted and used as a substitute for coffee. *Unspecified* Parched plant used for food. (45:90) **Montana Indian** *Unspecified* Seeds used for food. (19:25) **Salish** *Vegetable* Grass-like plant eaten as a vegetable. (185:54)

Trillium, Liliaceae

Trillium chloropetalum (Torr.) T. J. Howell, Giant Wakerobin
- **Drug**—Costanoan *Analgesic* Poultice of heated plant applied to the chest for chest pains. (21:28) **Yurok** *Burn Dressing* Poultice of bulb scrapings applied to burns. (6:59)

Trillium erectum L., Red Trillium
- **Drug**—Abnaki *Panacea* Used by children for maladies. (144:155) Bulbs ground and given to sick children with unidentified illnesses. (144:174) *Pediatric Aid* Used by children for maladies. (144:155) Bulbs ground and given to sick children with unidentified illnesses. (144:174) **Cherokee** *Cancer Treatment* Poultice used for "putrid ulcers, tumors and inflamed parts." *Cough Medicine* Taken for coughs. *Dermatological Aid* Poultice used for "putrid ulcers, tumors and inflamed parts." *Gastrointestinal Aid* Taken for bowel complaints. *Gynecological Aid* Infusion used for profuse menstruation, hemorrhages, and the change of life. *Respiratory Aid* Taken for asthma. (80:59) **Iroquois** *Dermatological Aid* Infusion of rhizomes and flowers from another plant used for pimples and sunburn. (141:66)

Trillium grandiflorum (Michx.) Salisb., Snow Trillium
- **Drug**—Chippewa *Antirheumatic (External)* Decoction of root "pricked in with needles" near sore joints. *Ear Medicine* Decoction of root bark used as drops for sore ear. (53:362) **Menominee** *Abortifacient* Decoction of root taken for "irregularity of the menses." *Analgesic* Infusion of grated root taken for cramps. *Disinfectant* Decoction of root purified man after intercourse with menstruating woman. (151:41) *Diuretic* Infusion of root used as a diuretic. (54:131) *Eye Medicine* Poultice of grated, raw root applied to eye swellings. (151:41) **Potawatomi** *Gynecological Aid* Infusion of root taken for sore nipples and, to hasten the effect, the teats are pierced with a dog whisker. (154:63)

Trillium ovatum Pursh, Pacific Trillium
- **Drug**—Karok *Dermatological Aid* Plant juice applied to boils. (148:381) **Lummi** *Eye Medicine* Juice from smashed plants used as drops for sore eyes. **Makah** *Love Medicine* Poultice of pounded bulbs applied as a love medicine. (79:25) **Paiute** *Eye Medicine* Decoction of fresh or dried corms used as an eyewash. Fresh root juice dripped into an afflicted eye. (111:58) **Quileute** *Dermatological Aid* Poultice of scraped bulbs applied to boils. **Skagit** *Eye Medicine* Infusion of roots used as a wash for sore eyes. *Poison* Plant considered poisonous. (79:25) **Thompson** *Eye Medicine* Powdered root dropped or blown into sore eyes. (164:472) Infusion of roots placed in the eyes with an eye dropper. (187:130)

Trillium petiolatum Pursh, Idaho Trillium
- **Food**—Okanagan-Colville *Appetizer* Roots used to make a tea and taken as an appetizer. (188:50)

Trillium sessile L., Toadshade
- **Drug**—Concow *Poison* Plant considered poisonous. **Wailaki** *Dermatological Aid* Poultice of bruised leaves and crushed roots applied to boils. *Other* Decoction of plant taken to prevent deep and lasting sleep. *Panacea* Decoction of plants taken for any kind of sickness. **Yuki** *Dermatological Aid* Poultice of bruised leaves and crushed roots applied to boils. *Other* Decoction of plant taken to prevent deep and lasting sleep. *Panacea* Decoction of plants taken for any kind of sickness. (41:329)

Trillium sp., Trillium
- **Drug**—Iroquois *Analgesic* Compound of dried roots used as a snuff for headaches. (87:286) *Antirheumatic (External)* Infusion of powdered plant used on joints for rheumatism, then put on powder. (87:285) *Blood Medicine* Used for the blood. *Cold Remedy* Compound of dried roots used as a snuff for colds. *Dermatological Aid* Crushed plant used as a liniment for chapped hands and sun tan oil. Plant used as wash for the irritated parts from the itch. *Gynecological Aid* Decoction of root taken for "food for woman in the womb." *Hunting Medicine* Infusion of one smashed root used as a soak for fishing line "to catch fish." *Orthopedic Aid* Compound decoction of roots taken for stiff muscles. *Poison* Plant not used, "said to be poisonous." *Respiratory Aid* Compound of dried roots used as a snuff for catarrh. (87:286) *Witchcraft Medicine* Compound used to "detect bewitchment." (87:285) Dry root carried for luck and for the protection of teeth. (87:286) **Pomo, Kashaya** *Poison* Plant considered poisonous. (72:116)

Trillium undulatum Willd., Painted Trillium
- **Drug**—Algonquin, Tête-de-Boule *Gynecological Aid* Flowers, sepals, and leaves eaten to accelerate the delivery. (132:133)

Triodanis, Campanulaceae

Triodanis perfoliata (L.) Nieuwl. **var. perfoliata**, Clasping Venus's Lookingglass
- **Drug**—Cherokee *Gastrointestinal Aid* Liquid compound of root taken for dyspepsia from overeating. (as *Specularia perfoliata* 80:60) Infusion of roots taken and used as a bath for dyspepsia. (as *Specularia perfoliata* 177:60) **Meskwaki** *Emetic* Used as an emetic and "will make one sick all day long." (as *Specularia perfoliata* 152:206)
- **Other**—Meskwaki *Smoke Plant* Smoked at ceremonies. (as *Specularia perfoliata* 152:272)

Triosteum, Caprifoliaceae

Triosteum perfoliatum L., Feverwort
- **Drug**—Cherokee *Dermatological Aid* Infusion used for soaking sore feet and ooze used as a wash for leg swelling. *Emetic* Plant used as an emetic. *Febrifuge* Plant used as a febrifuge. *Orthopedic Aid* Infusion used for soaking sore feet and ooze used as a wash for leg swelling. (80:39) **Creek** *Unspecified* Plant used medicinally for unspecified purpose. (172:667) **Iroquois** *Abortifacient* Infusion of roots taken for irregular or profuse menses. (87:444) *Analgesic* Compound decoction of roots taken for urinating pain. *Cold Remedy* Cold infusion of roots taken for bad colds. *Diaphoretic* Root used to cause sweating. (87:445) *Dietary Aid* Infusion of roots given to babies and adults to fatten them. (87:444) *Diuretic* Root used as a diuretic. *Gastrointesti-*

nal Aid Decoction or poultice of roots used for stomach trouble caused by witchcraft. (87:445) *Gynecological Aid* Decoction or infusion of roots taken for irregular or profuse menses. (87:444) *Laxative* Compound decoction of roots taken as a laxative. (87:445) *Pediatric Aid* Infusion of roots given to babies to fatten them. (87:444) *Pulmonary Aid* Root used for pneumonia. *Throat Aid* Cold infusion of roots taken for dry throats. *Urinary Aid* Compound decoction of roots taken for urinating pain. *Venereal Aid* Root used for venereal disease. *Witchcraft Medicine* Decoction of roots taken for stomach trouble caused by witchcraft. Poultice of smashed roots applied for stomach trouble caused by witchcraft. (87:445) **Meskwaki** *Analgesic* Root used on a newborn infant with a sore head. *Cathartic* Decoction of root used as a "drink for cleansing the system." *Dermatological Aid* Poultice of root applied to old, raw sores and root used for snakebite. *Pediatric Aid* Root used on a newborn infant with a sore head. *Snakebite Remedy* Root used for snakebite and poultice used for old, raw sores. (152:207, 208)

Tripterocalyx, Nyctaginaceae

Tripterocalyx carnea var. *wootonii* (Standl.) L. A. Gal., Wooton's Sandpuffs
- *Drug*—**Hopi** *Pediatric Aid* and *Sedative* Plant placed on child's head to induce sleep. (as *T. wootoni* 200:36, 75) **Navajo, Ramah** *Other* Cold infusion of plant used for injury from falling off a horse. *Witchcraft Medicine* Cold infusion of plant taken to give protection from witches. (as *T. wootonii* 191:26) **Zuni** *Diaphoretic* Crushed leaves used in sweat bath for snakebite. *Snakebite Remedy* Infusion of powdered root taken for snakebite. *Throat Aid* Infusion of powdered root taken for swollen glands, especially in the throat. (as *T. wootonii* 166:61) *Witchcraft Medicine* Poultice of powdered seeds and water applied to swellings caused by being witched by a bullsnake. (as *T. wootonii* 27:377)

Trisetum, Poaceae

Trisetum spicatum (L.) Richter, Spike Trisetum
- *Food*—**Gosiute** *Unspecified* Seeds used for food. (as *T. subspicatum* 39:383)

Triteleia, Liliaceae

Triteleia grandiflora Lindl., Wild Hyacinth
- *Drug*—**Okanagan-Colville** *Poison* Plant considered poisonous. (as *Brodiaea douglasii* 188:41) **Thompson** *Adjuvant* Bulb used in medicine bag "to make the bag more potent." (164:508) *Unspecified* Bulbs eaten and used medicinally. (187:131)
- *Food*—**Okanagan-Colville** *Unspecified* Corms used for food. (as *Brodiaea douglasii* 188:41) **Okanagon** *Unspecified* Bulbs formerly steamed and used for food. (as *Hookera douglasii* 125:37) **Pomo** *Vegetable* Potatoes used for food. (as *Brodiaea grandiflora* 117:284) **Thompson** *Unspecified* Bulbs formerly steamed and used for food. (as *Hookera douglasii* 125:37) Bulbs eaten with yellowbell bulbs. (187:131) **Thompson, Upper (Lytton Band)** *Unspecified* Large bulbs eaten. **Thompson, Upper (Spences Bridge)** *Unspecified* Large bulbs eaten. (164:481) **Yana** *Unspecified* Roots roasted and eaten. (as *Brodiaea grandiflora* 147:251)

Triteleia hyacinthina (Lindl.) Greene, Hyacinth Brodiaea
- *Food*—**Pomo** *Unspecified* Bulbs eaten for food. (as *Hookera nyocintlima* var. *lactea* 11:90)

Triteleia hyacinthina (Lindl.) Greene var. *hyacinthina*, Fool's Onion
- *Food*—**Atsugewi** *Unspecified* Cooked in earth oven and used for food. (as *Brodiaea hyacinthina* 61:138) **Mendocino Indian** *Unspecified* Bulbs eaten raw or cooked. (as *Hesperoscordum lacteum* 41:326) **Miwok** *Unspecified* Bulbs steamed in earth oven and used as food. (as *Brodiaea hyacinthina* 12:156) **Neeshenam** *Unspecified* Eaten raw, roasted, or boiled. (as *Hesperoscordium lacteum* 129:377) **Paiute** *Unspecified* Bulbs roasted and used for food. (as *Brodiaea hyacinthina* 104:102) *Vegetable* Root eaten boiled and mashed, like potatoes. (as *Brodiaea hyacinthina* 111:55)

Triteleia ixioides (Ait. f.) Greene, Prettyface
- *Food*—**Miwok** *Unspecified* Bulbs used for food. (as *Brodiaea ixioides* 12:156)

Triteleia ixioides ssp. *scabra* (Greene) Lenz, Prettyface
- *Food*—**Neeshenam** *Unspecified* Eaten raw, roasted, or boiled. (as *Calliproa lutea* 129:377)

Triteleia laxa Benth., Grass Nut
- *Food*—**Karok** *Unspecified* Bulbs roasted and eaten. (as *Brodiaea laxa* 148:380) **Pomo** *Unspecified* Bulbs eaten for food. Bulbs eaten for food. Bulbs eaten as potatoes. (11:90) Corms cooked in hot ashes or boiled for food. (as *Brodiaea laxa* 66:12) **Pomo, Kashaya** *Vegetable* Baked or boiled corms eaten like baked or boiled potatoes. (as *Brodiaea laxa* 72:25) **Yuki** *Unspecified* Bulbs used for food. (41:327) Bulbs cooked and used for food. (as *Brodiaea laxa* 49:86)
- *Other*—**Pomo, Kashaya** *Ceremonial Items* Flowers used in dance wreaths at the Strawberry Festival. (as *Brodiaea laxa* 72:25)

Triteleia peduncularis Lindl., Longray Tripletlily
- *Food*—**Mendocino Indian** *Unspecified* Bulbs used for food. (41:329) **Pomo** *Unspecified* Bulbs eaten for food. (11:90)

Triticum, Poaceae

Triticum aestivum L., Common Wheat
- *Drug*—**Iroquois** *Veterinary Aid* Wheat flour, rhizomes from another plant, and raspberry leaves given to cows at birthing. (141:67)
- *Food*—**Apache, White Mountain** *Bread & Cake* Seeds used to make bread. (as *T. vulgare* 136:161) **Cahuilla** *Porridge* Parched seeds ground into flour and used to make mush. (15:142) **Haisla & Hanaksiala** *Bread & Cake* Grains used to make bread. *Staple* Grains used for food. (43:208) **Navajo, Ramah** *Fodder* Used for horse feed. *Unspecified* Species used for food. (191:17) **Okanagan-Colville** *Unspecified* Kernels boiled until opened and eaten. (188:57) **Papago** *Bread & Cake* Used for making native breads. (34:38) *Staple* Grains trampled, winnowed, softened with water, pounded, dried, and ground into flour. (34:37) **Pima** *Bread & Cake* Seeds ground into meal, water and salt added, and dough used to make tortillas and cakes. (as *T. sativum* 47:73) Ground into flour and used to make bread. *Porridge* Parched, ground, and eaten as a thin gruel. (as *T. sativum* 146:76) **Pomo, Kashaya** *Porridge* Seed used in mush and to make flour for bread. (72:54) **Sia** *Staple* Corn and wheat, the most important foods, used for food. (as *T. vulgare* 199:106) **Zuni** *Beverage* Dried, ground wheat

mixed with water to make a beverage. *Bread & Cake* Wheat made into flour and used to make doughnuts. (as *T. vulgare* 166:71)
- *Other*—**Cahuilla** *Cash Crop* Used to trade with travelers passing through their territory. (15:142)

Triticum durum Desf., Durum Wheat
- *Food*—**Pima** *Unspecified* Species used for food. (36:116)

Triticum polonicum L., Polish Wheat
- *Food*—**Pima** *Unspecified* Species used for food. (36:116)

Trollius, Ranunculaceae
Trollius laxus Salisb., American Globeflower
- *Drug*—**Cherokee** *Oral Aid* Infusion of leaves and stem used for "thrash." (80:36)

Tsuga, Pinaceae, conifer
Tsuga canadensis (L.) Carr., Eastern Hemlock
- *Drug*—**Abnaki** *Antirheumatic (External)* Used as a medicine for rheumatism. (144:155) *Antirheumatic (Internal)* Infusion of leaves taken for rheumatism. (144:163) Decoction of leaves taken for rheumatism. (144:164) *Dermatological Aid* Used for "slight" itches. (144:155) **Algonquin, Quebec** *Antirheumatic (External)* Decoction of branches boiled down to a thick syrup or paste and used as a poultice for arthritis. *Cold Remedy* Infusion of inner bark taken for colds. *Dermatological Aid* Decoction of inner bark applied externally to eczema and other similar skin conditions. *Disinfectant* Poultice of crushed branch tips applied to the infected navel of an infant. *Gynecological Aid* Used in the sudatory by women experiencing complications in childbirth and for other complaints. *Pediatric Aid* Poultice of crushed branch tips applied to the infected navel of an infant. *Unspecified* Used in the sudatory by women experiencing complications in childbirth and for other complaints. Decoction of branches used as a medicinal tea. (18:125) **Cherokee** *Antidiarrheal* Root chewed "to check bowels." *Dermatological Aid* Poultice of bark used for itching armpits. *Gynecological Aid* Compound decoction used to aid in expelling afterbirth. *Kidney Aid* Infusion of stem tips taken for kidneys. (80:38) **Chippewa** *Antidiarrheal* Infusion of twigs taken for dysentery. (71:123) *Hemostat* Pulverized inner bark applied to wounds as a styptic. (53:356) **Delaware** *Antirheumatic (External)* Hot infusion of roots or twigs used as a steam treatment for muscular swellings and stiff joints. (176:36) **Delaware, Oklahoma** *Antirheumatic (External)* Decoction of roots or twigs used as herbal steam for rheumatism. (175:30, 80) *Herbal Steam* Infusion of roots or twigs used as herbal steam for rheumatism. (175:30) **Iroquois** *Antirheumatic (External)* Steam from decoction of leaves and bark used for rheumatism and as a foot soak. (87:268) *Antirheumatic (Internal)* Compound decoction taken for rheumatism. (87:269) *Blood Medicine* Fermented compound decoction taken when "blood gets bad and cold." (87:268) *Cold Remedy* Cold infusion of twigs and bark taken for mild colds with fever. Decoction of smashed needles taken for colds. Infusion of twigs and bark taken for colds. *Cough Medicine* Compound decoction taken for coughs. *Diaphoretic* Infusion of twigs and bark taken to induce sweating. (87:269) *Febrifuge* Fermented compound decoction taken for fever. (87:268) Cold infusion of twigs and bark taken for mild colds with fever. *Misc. Disease Remedy* Compound decoction taken for cholera. (87:269) *Orthopedic Aid* Fermented compound decoction taken for soreness. *Stimulant* Fermented compound decoction taken when a "person is tired." (87:268) *Tuberculosis Remedy* Compound decoction taken during early stages of consumption. *Venereal Aid* Compound powder poultice "put in bag, place penis in bag and tie around waist." (87:269) **Malecite** *Antidiarrheal* Infusion of young bark and fir buds or cones used for diarrhea. (116:244) Infusion of inside bark used for diarrhea. (116:255) *Cold Remedy* Infusion of boughs used for colds. (116:249) *Dermatological Aid* Outer layer of dried trees used as a powder for prickly heat. (116:250) *Kidney Aid* Infusion of boughs used for cold in the kidneys. (116:249) *Pediatric Aid* Outer layer of dried trees used for chafed babies. (116:250) *Unspecified* Bark used as medicine. (160:6) **Menominee** *Analgesic* Infusion of inner bark taken for abdominal pains. *Cold Remedy* Infusion of inner bark taken, 1 quart required, for cold. (151:46) *Dermatological Aid* Compound containing branches steamed for swellings. (54:134) *Diaphoretic* Leaves used in the sudatory. *Gastrointestinal Aid* Infusion of inner bark used for abdominal pains. (151:46) *Herbal Steam* Herbal steam from compound containing branches used for swellings. (54:134) **Micmac** *Antidiarrheal* Inner bark used for diarrhea. (40:62, 63) *Cold Remedy* Bark and stems used for colds. (40:62) *Cough Medicine* Bark used as cough medicine and for grippe. *Dermatological Aid* Inner bark used for chapped skin. (40:62, 63) *Gastrointestinal Aid* Parts of plant used for bowel, stomach, and internal troubles. *Kidney Aid* Roots and stems used for "cold in kidney." (40:62) *Misc. Disease Remedy* Bark used for grippe and inner bark used for scurvy. (40:62, 63) *Urinary Aid* Roots and stems used for "cold in bladder." (40:62) **Ojibwa** *Adjuvant* Leaves used to flavor medicinal tea. (153:380) Leaves made into a tea and used as a beverage and to disguise medicine. (153:408) *Dermatological Aid* Bark used for cuts, wounds, and bleeding wounds. *Hemostat* Bark used for bleeding wounds. (153:380) **Potawatomi** *Antidiarrheal* Compound of inner bark used for flux. *Cold Remedy* and *Diaphoretic* Infusion of leaves taken to cause perspiration, which broke up a cold. (154:71)
- *Food*—**Chippewa** *Beverage* Leaves used to make a beverage. (53:317) **Iroquois** *Beverage* Branches and maple water used to make tea. (141:36) **Micmac** *Beverage* Bark used to make a beverage. (159:258) **Ojibwa** *Beverage* Leaves made into a tea and used as a beverage and to disguise medicine. (153:408)
- *Fiber*—**Cherokee** *Basketry* Inner bark used to make baskets. *Building Material* Wood used for lumber. (80:38) **Chippewa** *Building Material* Bark strips used for wigwam coverings. (71:123)
- *Dye*—**Cherokee** *Red-Brown* Bark used to make a rosy tan dye. (80:38) **Chippewa** *Red-Brown* Bark used to make a mahogany-colored dye. (53:371) **Malecite** *Unspecified* Bark used to make a dye and tanning material. (160:6) **Menominee** *Red* Boiled bark used for dark red coloring. (151:78) **Micmac** *Unspecified* Bark used to make a dye. (159:258) **Ojibwa** *Mordant* Bark used with a little rock dust to set the color. *Red-Brown* Bark used with a little rock dust to dye materials a dark red brown. (153:426)
- *Other*—**Cherokee** *Fuel* Wood used for firewood. *Paper* Wood used for pulpwood. *Preservative* Bark used as a source for tannic acid. (80:38) **Chippewa** *Cleaning Agent* Boiled bark used to make a wash to clean rust from steel traps and to prevent further rusting. (71:123)

Malecite *Hide Preparation* Bark used to make a dye and tanning material. (160:6) **Micmac** *Fuel* Wood used for kindling and fuel. (159:258) **Ojibwa** *Fuel* Bark used for fuel, when reboiling pitch, because the heat was easy to regulate. (153:422)

Tsuga caroliniana Engelm., Carolina Hemlock

- **Drug**—**Cherokee** *Antidiarrheal* Root chewed "to check bowels." *Dermatological Aid* Poultice of bark used for itching armpits. *Gynecological Aid* Compound decoction used to aid in expelling afterbirth. (80:38) Decoction of roots taken to cause discharge of the afterbirth. (177:5) *Kidney Aid* Infusion of stem tips taken for kidneys. (80:38)
- **Fiber**—**Cherokee** *Basketry* Inner bark used to make baskets. *Building Material* Wood used for lumber. (80:38)
- **Dye**—**Cherokee** *Red-Brown* Bark used to make a rosy tan dye. (80:38)
- **Other**—**Cherokee** *Fuel* Wood used for firewood. *Paper* Wood used for pulpwood. *Preservative* Bark used as a source for tannic acid. (80:38)

Tsuga heterophylla (Raf.) Sarg., Western Hemlock

- **Drug**—**Bella Coola** *Antirheumatic (External)* Poultice of compound containing gum applied to the arms for rheumatism. *Burn Dressing* Poultice of chewed leaves applied to burns. (150:51) Leaves chewed and used for burns. (184:198) *Dermatological Aid* Warm gum applied to cuts. *Heart Medicine* Poultice of compound containing gum applied to the chest for heart trouble. *Internal Medicine* Moxa of twigs applied to the skin for "various internal ailments." (150:51) Twigs burned and used to cauterize the skin for internal ailments. (184:198) **Chehalis** *Tuberculosis Remedy* Decoction of pounded bark taken for tuberculosis. *Venereal Aid* Decoction of pounded bark taken for syphilis. (79:17) **Clallam** *Antihemorrhagic* Bark boiled, added to licorice ferns, and used for hemorrhages. (57:195) **Cowlitz** *Antihemorrhagic* Infusion of bark taken for hemorrhage. *Dermatological Aid* Infusion of plants used as wash for skin sores. Pitch used to prevent chapping. *Eye Medicine* Infusion of plants used as wash for sore eyes. (79:17) **Gitksan** *Gastrointestinal Aid* Cambium used as a "cleanser." *Liver Aid* Cambium used for the gallbladder. *Other* Inner bark used for elimination of hard or sharp swallowed objects. (73:152) **Hesquiat** *Antirheumatic (Internal)* Decoction or infusion of bark, from inside of a crevice, taken for rheumatic fever. *Dermatological Aid* Gum and deer grease used on fur seal hunters faces to prevent skin from cracking and peeling in the sun. Gum and deer grease used for healing sores on the face. Poultice of chewed needles used for burns. Pitch and deer fat used on faces to heal abrasions cause by rubbing on hunting camouflage. Pitch and deer fat used as salve to prevent and soothe sunburn. *Eye Medicine* Boughs used by girls, at puberty ceremony, to prevent eye disease to herself and future children. *Other* Decoction or infusion of bark, from inside of a crevice, taken for phlebitis. *Tuberculosis Remedy* Decoction or infusion of bark, from inside of a crevice, taken for tuberculosis and rheumatic fever. (185:44) **Hoh** *Emetic* Bark used as an emetic. (137:58) **Klallam** *Antihemorrhagic* Compound infusion of bark taken for hemorrhage. *Dietary Aid* Infusion of plant tips taken to stimulate appetite. *Tuberculosis Remedy* Infusion of plant tips taken for tuberculosis. (79:17) **Kwakiutl** *Antidiarrheal* Compound decoction of plants or bark taken for diarrhea. (183:264) *Burn Dressing* Cold infusion of scraped, pounded bark applied to burns. *Dermatological Aid* Cold infusion of scraped, pounded bark applied to sores. Moxa of twigs used to cauterize warts and moles. *Eye Medicine* Infusion of boughs used as a wash for upper lids of inflamed eyes. *Gynecological Aid* Hemlock used as wash and tree tips prayed to by pregnant women to aid delivery. *Hunting Medicine* Branches rubbed on hunters and fishermen to purify them. (183:270) **Kwakwaka'wakw** *Ceremonial Medicine* Plant considered to have special powers to purify and cure. (43:71) **Makah** *Dermatological Aid* Pitch used to prevent sunburn. Poultice of bark applied to obstinate sores. (79:17) Pitch used as a sunburn preventative and rubbed on hair to remove lice. (186:74) *Hemostat* Poultice of plant applied to bleeding wounds. *Internal Medicine* Decoction of plant taken for internal injury. (79:17) **Nitinaht** *Dermatological Aid* Infusion of bark, grand fir and red alder barks taken for bruises. *Internal Medicine* Infusion of bark, grand fir and red alder barks taken for internal injuries. *Orthopedic Aid* Infusion of bark, grand fir and red alder barks taken for broken bones. (186:74) **Oweekeno** *Ceremonial Medicine* Plant considered to have special powers to purify and cure. (43:71) **Quileute** *Dermatological Aid* Poultice of chewed plants applied to swellings. (79:17) *Emetic* Bark used as an emetic. (137:58) **Quinault** *Cold Remedy* Poultice of pitch and ground bark applied to child's chest for a cold. *Laxative* Infusion of plants taken as a laxative. *Pediatric Aid* Poultice of pitch and ground bark applied to child's chest for a cold. (79:17) **Shuswap** *Tuberculosis Remedy* Decoction of bark taken for tuberculosis. (123:53) **Skagit** *Dermatological Aid* and *Eye Medicine* Infusion of plants used as wash for sore eyes. *Throat Aid* Infusion of bark taken for sore throats. (79:17) **Thompson** *Cold Remedy* Infusion of bark used for colds. *Disinfectant* Boughs considered an important disinfectant. One informant noted that those desiring to cleanse themselves would scrub their bodies with the branches. *Misc. Disease Remedy* Infusion of bark used for influenza. (187:111)
- **Food**—**Alaska Native** *Bread & Cake* Inner bark roasted in a pit oven, sometimes mixed with dried berries, and pressed into cakes. (85:146) **Bella Coola** *Dried Food* Inner bark steamed overnight, pounded, formed into balls, sun dried, and eaten in winter. (184:198) **Cowlitz** *Spice* Branch tips used to flavor cooking bear meat. (79:17) **Gitksan** *Staple* Cambium pit cooked, pounded, formed into cakes, dried, stored, and eaten as a staple food. *Sweetener* Cambium used as a sweetener for other foods. *Winter Use Food* Inner bark used as a survival food in winter. (73:150) **Haisla** *Dried Food* Cambium cooked in skunk cabbage leaves, pounded, dried, and stored for winter use. (43:180) *Staple* Cambium pit cooked, pounded, formed into cakes, dried, stored, and eaten as a staple food. *Sweetener* Cambium used as a sweetener for other foods. *Winter Use Food* Inner bark used as a survival food in winter. (73:150) **Haisla & Hanaksiala** *Winter Use Food* Cambium dried, pounded, served with oolichan (candlefish) grease and Pacific crabapples, and used as winter food. **Hanaksiala** *Bread & Cake* Cambium formed into cakes, cooked, dried, powdered, mixed with water, grease, and fruit, and eaten. (43:180) **Hesquiat** *Candy* Pitch, from the outside of a crevice, chewed like gum. (185:44) **Kitasoo** *Dried Food* Inner bark boiled and dried for storage. *Unspecified* Inner bark soaked in oolichan (candlefish) grease, drained, cooked, and eaten with sugar. (43:318) **Montana Indian** *Bread & Cake* Cambium made into a coarse bread. (19:25) **Nitinaht** *Starvation Food* Old leaves eaten sparingly to keep alive when hungry in the woods. (67:238) Light, green branch tips eaten to relieve hunger

when lost in the woods. (186:74) *Unspecified* Young growth used for food. (67:238) **Oweekeno** *Dried Food* Inner bark dried for future use. *Unspecified* Inner bark used for food. (43:71) **Salish, Coast** *Dried Food* Cambium dried in sheets and used as a winter food. (182:72) **Wet'suwet'en** *Sweetener* Cambium used as a sweetener for other foods. *Unspecified* Cambium used for food. *Winter Use Food* Inner bark used as a survival food in winter. (73:150)

- *Fiber*—**Haisla & Hanaksiala** *Building Material* Boughs used for making temporary shelters or huts by hunters. (43:180) **Hesquiat** *Mats, Rugs & Bedding* Boughs used as a mattress when camping. (185:44) **Kwakwaka'wakw** *Building Material* Boughs made into huts and used to house girls following their first menstruation. Boughs made into special shelters for hunters, fishers, shamans and their initiates. (43:71) **Makah** *Building Material* Wood used for lumber. (67:238) **Oweekeno** *Building Material* Boughs made into special shelters for hunters, fishers, shamans and their initiates. (43:71) **Quinault** *Building Material* Boughs used as shelters for hunters. *Canoe Material* Young trees used for poling canoes upstream. **Skagit** *Canoe Material* Young trees used for poling canoes upstream. **Snohomish** *Basketry* Inner bark used as a dye for baskets. *Canoe Material* Young trees used for poling canoes upstream. (79:17)

- *Dye*—**Bella Coola** *Brown* Bark boiled and used as a brown dye for fish nets. (184:198) **Chehalis** *Unspecified* Inner bark made into a dye and used on fish nets to make them invisible to fish and attract them. (79:17) **Clallam** *Red-Brown* Bark boiled and used to make a reddish brown dye. (57:195) **Hesquiat** *Red-Brown* Bark, especially from the inside of a crevice, used as a reddish brown dye. Bark was used to dye the rope used in halibut fishing to make it invisible to the fish. Canoes were often painted with a solution of this bark in water. This stain was made by steeping the bark in water for many days, until the liquid was bright red. (185:44) **Klallam** *Red-Brown* Bark boiled and used as a red-brown dye. **Makah** *Red-Brown* Inner bark pounded, boiled, and used as a red-brown dye. (79:17) **Nitinaht** *Black* Bark chopped into small pieces, pounded, crushed, and boiled to make a black dye. (186:74) *Brown* Bark used as a brown dye for basketry material and gill nets "so the fish won't see it." (67:238) Bark chopped into small pieces, pounded, crushed, and boiled to make different shades of brown dye. (186:74)

- *Other*—**Bella Coola** *Cleaning Agent* Bark boiled and used on traps to remove rust and give them a clean smell. **Bellabella** *Hunting & Fishing Item* Boughs used to collect herring spawn. (184:198) **Chehalis** *Fuel* Wood used extensively for firewood. (79:17) **Clallam** *Hunting & Fishing Item* Saplings used for fish trap stanchions. Limbs cut, placed around rocks in tidal areas and used to gather lingcod eggs. (57:195) **Haisla & Hanaksiala** *Ceremonial Items* Boughs used among several botanical materials in ritual purification practices. *Cooking Tools* Boughs used under cleaned salmon while draining. *Good Luck Charm* Boughs rubbed on the body for luck on hunting expeditions. *Hunting & Fishing Item* Boughs used to gather herring roe. Boughs used as camouflage by hunters. *Protection* Boughs made into hoops and used to combat witchcraft. *Weapon* Timber used as the "blade" of an executioner's device similar to the guillotine. **Hanaksiala** *Cooking Tools* Boughs used to wrap and cover foods in earth ovens. Boughs used in processing salmon. *Fasteners* Wood made into rings and used to fasten oolichan (candlefish) nets to the piles. *Hide Preparation* Bark used to smoke hides into a dark brown color. *Hunting & Fishing Item* Wood used to make piles for anchoring oolichan (candlefish) nets. *Jewelry* Wood knots used to make labrets worn by women of high social rank. *Tools* Wood used to make digging sticks. Wood used to make mallets. (43:180) **Hesquiat** *Ceremonial Items* Boughs used by girls at puberty for rubbing ceremony. Girls at puberty were brushed on the arms and face with boughs, which were bundled together with soft, fern fronds. The bundle was fist-sized, with needled hemlock twigs sticking out from both ends. Before the rubbing ceremony began, the girl would go down to the edge of the water at sunrise and, four separate times, would dip the branches in the water, suck the water from the branches, then blow it out in a fine spray. At the same time, she would dip her face in the water with her eyes open, four times, each time lifting her head and spraying out the water. This was said to prevent eye disease to herself and future children. All bundles used for this purpose were subsequently deposited in a special place. *Hunting & Fishing Item* Entire trees anchored upside down under the water to collect herring spawn. At herring spawning time, in spring, entire hemlock trees were cut and anchored upside down under the water. The spawn stuck on the branches. The branches were then broken off and the spawn peeled away and eaten fresh, usually after steam cooking. If for later use, the branches were hung outside to be wind dried. When one wanted to eat some, he would soak the branches and rub them to remove the spawn. While being cooked, the eggs would swell and float to the surface and could be scooped off and eaten. Also, hemlock and fern bundles rubbed on hunters' faces and arms to prevent sea mammals from noticing them. (185:44) **Hoh** *Hide Preparation* Bark used for tanning hides. (137:58) **Kitasoo** *Hunting & Fishing Item* Boughs submerged in herring spawning areas to collect the roe. (43:318) **Klallam** *Hunting & Fishing Item* Twigs laid in rivers and used to collect herring eggs. (78:202) Saplings used as stanchions of fish traps strung across a stream. (79:17) **Kwakiutl, Southern** *Ceremonial Items* Wood used to carve masks. *Cooking Tools* Wood used to carve dishes and spoons. *Musical Instrument* Wood used to carve rattles. (183:296) **Kwakwaka'wakw** *Protection* Four rings of boughs used to negate the effects of evil spirits. (43:71) **Lummi** *Hunting & Fishing Item* Young trees used as poles for large dip nets. **Makah** *Hunting & Fishing Item* Boughs immersed in the water and used to collect herring eggs. *Insecticide* Compound of powdered bark and oil or pitch used on the hair to remove vermin. (79:17) **Montana Indian** *Hide Preparation* Bark used for tanning. (19:25) **Nitinaht** *Ceremonial Items* Boughs used as scrubbers in the manhood training rituals of young boys. *Fuel* Slow burning wood used to bank up fires overnight. (186:74) *Hunting & Fishing Item* Wood knots used to make halibut hooks. (186:53) Boughs used to catch herring spawn. Hard, dense knots used to make molded halibut hooks. (186:74) **Oweekeno** *Ceremonial Items* Inner bark and grease made into realistic eyeballs and used in the corpse for the hamatsa ceremony. Boughs made into headdresses and worn by dancers in the Ghost dance. *Cooking Tools* Boughs used to roast fern roots. *Hunting & Fishing Item* Fresh boughs used to gather herring roe. Wood used to make octopus spears. *Protection* Four rings of boughs and ritual bathing used to negate the effects of evil spirits. *Toys & Games* Boughs used to make play houses. (43:71) **Quileute** *Hide Preparation* Bark used for tanning hides. (137:58)

Waterproofing Agent Bark finely chopped, boiled, and the juice applied to baskets to make them watertight. **Quinault** *Fuel* Wood used extensively for firewood. *Hunting & Fishing Item* Young trees used as walkways on top of fish weirs. *Paint* Bark mashed with salmon eggs to make a yellow-orange paint for dip nets and paddles. (79:17) **Salish, Coast** *Hide Preparation* Bark used to cure hides. (182:72) **Skagit** *Hunting & Fishing Item* Young trees used in fish trap construction. (79:17) **Thompson** *Incense & Fragrance* Boughs steamed or rubbed on furniture and used as a room deodorizer and disinfectant. *Tools* Broken boughs used by the handful like a bath brush. (187:111)

Tsuga mertensiana (Bong.) Carr., Mountain Hemlock
- **Drug**—**Bella Coola** *Burn Dressing* Poultice of chewed leaves applied to burns. *Dermatological Aid* Warm gum applied to cuts. *Internal Medicine* Moxa of twigs applied to the skin for "various internal ailments." (150:51) **Hoh** *Emetic* Bark used as an emetic. (137:58) **Kwakiutl** *Ceremonial Medicine, Dermatological Aid*, and *Pediatric Aid* Bark used in ritual to make children as light-skinned as the inner bark. (183:271) **Quileute** *Emetic* Bark used as an emetic. (137:58) **Thompson** *Cold Remedy* Infusion of bark used for colds. *Disinfectant* Boughs steamed or rubbed on furniture and used as a room deodorizer and disinfectant. *Misc. Disease Remedy* Infusion of bark used for influenza. (187:111) **Tlingit** *Toothache Remedy* Compound containing warmed seeds used for toothache. *Venereal Aid* Compound poultice of sap applied for syphilis. (as *Pinus mertensiana* 107:284)
- **Fiber**—**Paiute** *Mats, Rugs & Bedding* Boughs used for camping beds. (111:44) **Thompson** *Brushes & Brooms* Broken boughs used by the handful as a bath brush. (187:111) *Mats, Rugs & Bedding* Fragrant branches and leaves used for bedding. (164:500)
- **Other**—**Hoh** *Hide Preparation* Bark used for tanning hides. **Quileute** *Hide Preparation* Bark used for tanning hides. (137:58) **Thompson** *Incense & Fragrance* Boughs steamed or rubbed on furniture and used as a room deodorizer and disinfectant. (187:111)

Turricula, Hydrophyllaceae

Turricula parryi (Gray) J. F. Macbr., Common Turricula
- **Drug**—**Kawaiisu** *Antirheumatic (External)* Infusion of leaves used as a wash for rheumatism. *Dermatological Aid* Infusion of leaves used as a wash for swellings. (206:68) **Luiseño** *Unspecified* Plant used for medicinal purposes. (as *Eriodictyon parryi* 155:230)

Tussilago, Asteraceae

Tussilago farfara L., Coltsfoot
- **Drug**—**Iroquois** *Cough Medicine* and *Tuberculosis Remedy* Compound infusion of roots taken as a consumption cough medicine. (87:473)

Typhaceae, see *Typha*

Typha, Typhaceae

Typha angustifolia L., Narrowleaf Cattail
- **Drug**—**Malecite** *Kidney Aid* Infusion of one root used for gravel. (116:257) **Micmac** *Urinary Aid* Roots used for gravel. (40:63)
- **Food**—**Hopi** *Candy* Mature heads chewed with tallow as gum. (200:64) **Pima** *Bread & Cake* Pollen baked into brownish biscuits and used for food. *Porridge* Pollen mixed with ground wheat, stirred into boiling water, and eaten as a gruel. *Unspecified* Tender, white stalks eaten raw. (47:64)
- **Fiber**—**Pima** *Basketry* Flower stalks split, dried, and used for basket weaving. *Building Material* Green leaves woven into roofs. *Mats, Rugs & Bedding* Green leaves woven into mats. Silky down used as stuffing for pillows. (47:64)
- **Other**—**Hopi** *Ceremonial Items* Plant associated ceremoniously with water. (200:64) **Pima** *Decorations* Dry, yellow pollen used to decorate the face, chest, and back. (47:64)

Typha domingensis Pers., Southern Cattail
- **Food**—**Havasupai** *Fruit* Ripe, fruiting heads eaten "like corn." (197:208) **Kawaiisu** *Unspecified* White stem bases and brown flowers eaten raw. Green seeds used for food. (206:68) **Paiute, Northern** *Bread & Cake* Seeds gathered into a dough, kneaded, made into flat cakes, and roasted under hot coals. Seeds roasted, winnowed, ground into fine flour, boiled, made into round cakes, and sun dried. (59:48) Dried rhizomes ground into flour, made into mush, and the mush used to make cakes. Pollen mixed with water, kneaded, formed into cakes, and baked. (60:69) *Dried Food* Roots dried for future use. (59:49) Rhizomes peeled and dried for future use. (60:69) *Porridge* Roots dried, ground into flour, and made into a sweet mush. (59:49) Dried rhizomes ground into flour and made into mush. Seeds roasted, ground into a meal, and stone boiled into a mush. (60:69) *Soup* Seeds ground into meal and made into soup. (59:48) *Staple* Seeds roasted, ground into a meal, and eaten with a little water without boiling. (60:69) *Unspecified* Seeds eaten fresh and raw or cooked. Roots peeled, chewed, juice swallowed, and the stringy pulp spat out. Seeds roasted, ground, and eaten in powder form. (59:48) Roots peeled and eaten fresh. (59:49) Stalks used for food. Rhizomes peeled and eaten fresh. Green spikes eaten fresh.[79] Seeds roasted and eaten.[80] (60:69) **Pima, Gila River** *Unspecified* Roots eaten raw. (133:7)
- **Fiber**—**Havasupai** *Building Material* Stalks and leaves used in thatching houses. (197:208) **Kawaiisu** *Building Material* Long leaves used in house construction. (206:68) **Paiute, Northern** *Basketry* Used for wefts and binding tule items. (60:81) *Building Material* Used for house building materials. (60:69) Used to make shingles for house exteriors. (60:87) *Canoe Material* Used in the construction of duck decoys and boats. (60:69) *Clothing* Used to secure the edges of skirts. (60:81) Leaves twined as weft rows for sandals. (60:83) Used as the twining wefts for clothing. (60:86) *Mats, Rugs & Bedding* Used to secure the edges of mats. (60:81) Leaves used in a simple pile for seating. (60:87)
- **Other**—**Havasupai** *Paint* Pollen used as a face paint. *Toys & Games* Stalk used to make toy arrows. (197:208) **Paiute, Northern** *Fasteners* Used for wefts and binding tule items. Used as ties for sandals. (60:81) *Hunting & Fishing Item* Used in the construction of duck decoys and boats. (60:69)

Typha latifolia L., Broadleaf Cattail
- **Drug**—**Algonquin, Quebec** *Dermatological Aid* Poultice of crushed roots applied to wounds. *Disinfectant* Poultice of crushed roots applied to infections. (18:132) **Apache, Mescalero** *Unspecified* Pollen used as medicine. (14:46) **Cahuilla** *Dermatological Aid* Roots used for bleeding wounds. (15:142) **Cheyenne** *Ceremonial Medicine* Leaves used in the Sun Dance ceremony. (83:13) *Gastrointestinal Aid* Infu-

574 *Typha latifolia* (Typhaceae)

sion of dried, pulverized root and white base of the leaves taken for abdominal cramps. (76:170) **Dakota** *Burn Dressing* Down used as a dressing for burns and scalds. *Dermatological Aid* and *Pediatric Aid* Down used on infants to prevent chafing. (70:64, 65) **Delaware** *Kidney Aid* Roots used for kidney stones. (176:36) **Delaware, Oklahoma** *Kidney Aid* Root used to dissolve kidney stones. (175:30, 80) **Houma** *Pulmonary Aid* Decoction of stalks taken for whooping cough. (158:60) **Iroquois** *Gynecological Aid* "Patient sleeps on mattress made of plant" for cysts of breast. *Hemostat* Infusion of roots used as a wash for bleeding cuts. *Misc. Disease Remedy* "Patient sleeps on mattress made of plant" for cysts of yellow fever. *Orthopedic Aid* Decoction of smashed root applied as poultice for sprains. *Venereal Aid* Root chewed by women for gonorrhea. *Veterinary Aid* Infusion of roots used as a wash for horses with bleeding cuts. (87:271) **Mahuna** *Hemostat* and *Pediatric Aid* Plant used for newborns with bleeding navels. (140:56) **Malecite** *Dermatological Aid* Greased leaves used for sores. (116:246) **Meskwaki** *Dermatological Aid* Poultice of fuzz applied to old sores on neck. (152:248) **Micmac** *Dermatological Aid* Leaves used for sores. (40:63) **Montana Indian** *Burn Dressing* Poultice of fruit spikes' "down" applied to burns and scalds. (19:25) **Navajo, Ramah** *Ceremonial Medicine* and *Emetic* Whole plant used as a ceremonial emetic. (191:14, 15) **Ojibwa** *Dermatological Aid* Poultice of root inner skin applied to carbuncles and boils. (5:2306) *Other* Fruit fuzz used as a war medicine. (153:390) **Ojibwa, South** *Dermatological Aid* Poultice of crushed root applied to sores. (91:200) **Okanagan-Colville** *Dermatological Aid* Cottony fluff used as dressing for wounds. (188:57) **Omaha** *Burn Dressing* Down used as a dressing for burns and scalds. (70:64, 65) *Ceremonial Medicine* Plant used in various rituals. (68:322) *Dermatological Aid* and *Pediatric Aid* Down used on infants to prevent chafing. **Pawnee** *Burn Dressing* Down used as a dressing for burns and scalds. *Dermatological Aid* and *Pediatric Aid* Down used on infants to prevent chafing. (70:64, 65) **Plains Indian** *Dermatological Aid* and *Pediatric Aid* Down used on infants to prevent chafing. (82:60) **Ponca** *Burn Dressing* Down used as a dressing for burns and scalds. *Dermatological Aid* and *Pediatric Aid* Down used on infants to prevent chafing. (70:64, 65) **Potawatomi** *Dermatological Aid* Poultice of pounded root applied to inflammations. (154:85) **Sioux** *Burn Dressing* Poultice of fruit spikes' "down" applied to burns and scalds. *Dermatological Aid* and *Misc. Disease Remedy* Poultice of fruit spikes, down, and coyote fat applied to smallpox pustules. (19:25) **Washo** *Antidiarrheal* Young flowering heads eaten for diarrhea. (180:146) **Winnebago** *Burn Dressing* Down used as a dressing for burns and scalds. *Dermatological Aid* and *Pediatric Aid* Down used on infants to prevent chafing. (70:64, 65)

- **Food**—**Acoma** *Unspecified* Roots and tender shoots salted and eaten as food. (32:53) **Alaska Native** *Unspecified* Shoots eaten boiled or roasted. Green flower spikes boiled in salted water and eaten. (85:137) **Apache** *Unspecified* Roots used for food. (32:53) **Apache, Chiricahua & Mescalero** *Unspecified* Rootstocks cooked with meat. (33:47) **Apache, Mescalero** *Unspecified* Species used for food. Stem bases eaten raw or cooked with other foods in early spring. (14:46) **Blackfoot** *Unspecified* Rootstocks used for food. (97:19) **Cahuilla** *Porridge* Pollen used to make cakes and mush. *Staple* Dried roots ground into a meal. (15:142) **Carrier** *Unspecified* Stems peeled and eaten. (31:85) **Chehalis** *Unspecified* Roots and inner stalks baked in ashes and eaten. (79:21) **Clallam** *Unspecified* Fleshy interior eaten raw or pit cooked. (57:197) **Costanoan** *Unspecified* Roots used for food. Shoots used for food. Pollen used for food. (21:255) **Cree, Woodlands** *Dried Food* Peeled roots dried over a fire for winter storage. *Unspecified* Fresh stem bases and young shoots eaten in July. Raw roots used for food. Roots taken just before the plant bloomed and eaten raw or dipped in boiling water. (109:62) **Gosiute** *Unspecified* Seeds roasted and used for food. (39:383) **Keres, Western** *Unspecified* Roots and tender shoots eaten with salt for food. (171:73) **Klamath** *Unspecified* Rootstocks used for food. (45:90) **Laguna** *Unspecified* Roots and tender shoots salted and eaten as food. (32:53) **Lakota** *Staple* Pollen used as flour. (106:50) **Mendocino Indian** *Unspecified* Roots and stem bases used for food. (41:310) **Montana Indian** *Unspecified* Young roots and shoots eaten raw. (19:25) **Navajo, Ramah** *Unspecified* Rhizomes eaten raw in summer. Young stalks just appearing above the ground used for food. (191:14) **Ojibwa** *Dried Food* Green flower dried and used for food. *Staple* Pollen used for flour. *Unspecified* Green flower boiled and used for food. (5:2226) **Okanagan-Colville** *Fruit* Young, fruiting heads boiled or roasted and eaten. *Unspecified* Young sprouts used for food. Roots pit cooked and eaten. (188:57) **Paiute** *Unspecified* Seeds used for food. (32:53) **Paiute, Northern** *Bread & Cake* Seeds gathered into a dough, kneaded, made into flat cakes, and roasted under hot coals. Seeds roasted, winnowed, ground into fine flour, boiled, made into round cakes, and sun dried. (59:48) Dried rhizomes ground into flour, made into mush, and the mush used to make cakes. Pollen mixed with water, kneaded, formed into cakes, and baked. (60:69) *Dried Food* Roots dried for future use. (59:49) Rhizomes peeled and dried for future use. (60:69) *Porridge* Roots dried, ground into flour, and made into a sweet mush. (59:49) Dried rhizomes ground into flour and made into mush. Seeds roasted, ground into a meal and stone boiled into a mush. (60:69) *Soup* Seeds ground into meal and made into soup. (59:48) *Staple* Seeds roasted, ground into a meal, and eaten with a little water without boiling. (60:69) *Unspecified* Seeds eaten fresh and raw or cooked. Roots peeled, chewed, juice swallowed, and the stringy pulp spat out. Seeds roasted, ground and eaten in powder form. (59:48) Roots peeled and eaten fresh. (59:49) Stalks used for food. Rhizomes peeled and eaten fresh. Green spikes eaten fresh.[79] Seeds roasted and eaten.[80] (60:69) **Pomo** *Vegetable* Roots eaten as greens. Young shoots eaten as greens. (11:92) **San Felipe** *Unspecified* Shoots ground, mixed with cornmeal, and used as food. (32:53) **Sioux** *Unspecified* Young roots and shoots eaten raw. (19:25) **Tanana, Upper** *Unspecified* Lower part of stem used for food. (102:9) **Thompson** *Unspecified* Rootstocks used as an important food. (164:482) **Tubatulabal** *Unspecified* Roots used extensively for food. (193:15) **Yuma** *Bread & Cake* Pollen shaped into flat cakes and baked. *Dried Food* Rhizomes dried, stored temporarily, pounded, and boiled with fish. Pollen dried and stored for future use. *Porridge* Young shoots used in combination with corn or tepary meal to make mush. Pollen boiled in water into a thin gruel. *Spice* Pollen used as flavoring. *Unspecified* Fleshy rhizomes eaten without preparation. Young shoots eaten raw. Pollen gathered, sifted, and eaten raw. (37:207)

- **Fiber**—**Algonquin, Tête-de-Boule** *Mats, Rugs & Bedding* Wool used to stuff mattresses. (132:132) **Apache, Mescalero** *Mats, Rugs & Bedding* Leaves used for lodge floor covering. (14:46) **Apache, White**

Mountain *Building Material* Flags used for thatching the tepees and wickiups. (136:161) **Blackfoot** *Clothing* Down used as diapers. (97:19) **Cahuilla** *Mats, Rugs & Bedding* Stalks used as matting materials and bedding. (15:142) **Chehalis** *Basketry* Used to make lightweight baskets. *Clothing* Fruiting stalks made into mats and used for raincoats and capes. *Mats, Rugs & Bedding* Fruiting stalks used to make mats for hangings, screens, mattresses, and coverings. Fruiting stalks made into mats and used for kneeling pads in canoes. (79:21) **Cheyenne** *Basketry* Leaves formerly used to make baskets. (83:13) **Chippewa** *Basketry* Used for baskets. (53:378) *Building Material* Leaves used as mats for roofing wigwams. The leaves were laid parallel, with one overlapping another, and were then stitched together with basswood, dogbane, or swamp milkweed fiber. The leaves were said to be quite effective in shedding the rain. (71:124) *Mats, Rugs & Bedding* Used for mats. (53:378) **Clallam** *Basketry* Used to make baskets. *Mats, Rugs & Bedding* Used to make mats. (57:197) **Cowlitz** *Basketry* Used to make lightweight baskets. *Clothing* Fruiting stalks made into mats and used for raincoats and capes. *Mats, Rugs & Bedding* Fruiting stalks used to make mats for hangings, screens, mattresses, and coverings. Fruiting stalks made into mats and used for kneeling pads in canoes. (79:21) **Dakota** *Mats, Rugs & Bedding* Down used for filling pillows, padding cradles, and quilting baby wrappings. Leaves used by children to make mats while playing. The leaves became very brittle once dried and therefore were not put to serious use by adults. (69:359) **Hesquiat** *Basketry* Dried leaves used in weaving the bottoms of baskets and in making bags. *Mats, Rugs & Bedding* Dried leaves used in making mattresses. (185:58) **Hopi** *Mats, Rugs & Bedding* Used to make sleeping mats. (34:56) **Iroquois** *Mats, Rugs & Bedding* Flowers used to stuff pillows. (142:107) **Isleta** *Building Material* Cattails used to support the mud of thatched roofs. (100:44) **Klallam** *Basketry* Used to make lightweight baskets. *Clothing* Fruiting stalks made into mats and used for raincoats and capes. *Mats, Rugs & Bedding* Fruiting stalks used to make mats for hangings, screens, mattresses, and coverings. Fruiting stalks made into mats and used for kneeling pads in canoes. (79:21) **Klamath** *Mats, Rugs & Bedding* Leaves used in the manufacture of mats. Fruiting spike down used as a stuffing material for pillows. (45:90) **Kwakiutl, Southern** *Mats, Rugs & Bedding* Leaves used to make mats. (183:275) **Lakota** *Clothing* Fluffy tops used as padding for baby diapers. *Furniture* Fluffy tops used as padding for baby cradleboards. (106:50) *Mats, Rugs & Bedding* Fluffy fruits made into a waterproof quilt and placed over sheeting used for babies. (139:32) **Makah** *Basketry* Used to make lightweight baskets. *Clothing* Fruiting stalks made into mats and used for raincoats and capes. *Mats, Rugs & Bedding* Fruiting stalks used to make mats for hangings, screens, and mattresses. Fruiting stalks made into mats and used for kneeling pads in canoes. (79:21) **Mendocino Indian** *Mats, Rugs & Bedding* Down of the fruiting parts used for bedding. (41:310) **Menominee** *Caulking Material* Root used as a natural oakum for caulking leaks in boats. *Mats, Rugs & Bedding* Leaves used to make mats to cover the winter lodges. (151:77) **Meskwaki** *Building Material* Leaves used to make rainproof and windproof mats for the side walls of the wigwam. *Caulking Material* Root used as a natural oakum to caulk canoes. *Mats, Rugs & Bedding* Fuzz of the fruit used for pillows and comfort material. (152:269) **Montana Indian** *Mats, Rugs & Bedding* Leaves used for matting. Fruit spike "down" used for bedding. (19:25) Leaves used to make mats for tepees, sweat baths, and Sundance lodges. (82:60) **Navajo, Ramah** *Basketry* Leaves used to make storage baskets, medicine baskets, and water jugs. *Mats, Rugs & Bedding* Leaves used to make bed mats. (191:14) **Nimpkish** *Clothing* Inner bark used to make clothing. *Mats, Rugs & Bedding* Inner bark used to make mats and blankets. (183:296) **Nitinaht** *Basketry* Leaves used to make openwork baskets. *Clothing* Leaves used to make skirts. (186:88) **Ojibwa** *Mats, Rugs & Bedding* Blades used to weave mats. (135:245) Leaves used to make wind- and rainproof mats placed on the sides of the medicine lodge. Fuzz or seed used to make mattresses and sleeping bags. Fuzz or seed used to make a quilt and the quilt used to make a sleeping bag. (153:423) **Okanagan-Colville** *Building Material* Leaves woven into mats and used for door coverings, sweat houses, and "A-frame"-type shelters. *Clothing* Cottony fruiting heads used as "insoles" for moccasins. Cottony fluff used as diapers. (188:57) **Okanagon** *Mats, Rugs & Bedding* Leaves and stems used to make mats. (125:39) **Paiute** *Building Material* Leaves and fruiting stalks sewn and woven into mats wrapped around willow frames to form houses. (111:37) **Paiute, Northern** *Basketry* Used for wefts and binding tule items. (60:81) *Building Material* Used for house building materials. (60:69) Used to make shingles for house exteriors. (60:87) *Canoe Material* Used in the construction of duck decoys and boats. (60:69) *Clothing* Used to secure the edges of skirts. (60:81) Leaves twined as weft rows for sandals. (60:83) Used as the twining wefts for clothing. (60:86) *Mats, Rugs & Bedding* Used to secure the edges of mats. (60:81) Leaves used in a simple pile for seating. (60:87) **Pomo** *Clothing* Skins used as swaddling clothes or soft padding. (11:92) **Potawatomi** *Building Material* Leaves sewn together to make a windproof and waterproof side mat to be applied to the wigwam. *Mats, Rugs & Bedding* Catkins used to make an infant's quilt. (154:114) Fuzzy seeds used to make a soft comforter on which to place a newborn infant. (154:124) **Quinault** *Basketry* Used to make lightweight baskets. *Clothing* Fruiting stalks made into mats and used for raincoats and capes. *Mats, Rugs & Bedding* Fruiting stalks used to make mats for hangings, screens, mattresses, and coverings. Fruiting stalks made into mats and used for kneeling pads in canoes. (79:21) **Salish, Coast** *Basketry* Flat leaves sun dried, split, and spun to make baskets. *Mats, Rugs & Bedding* Flat leaves sun dried and woven into mats for insulating winter house walls and kneeling in canoes. (182:77) **Shuswap** *Mats, Rugs & Bedding* Used to make mats for bedding, drying berries, and for mealtimes. Cotton used for pillows and mattresses. (123:55) **Snohomish** *Basketry* Used to make lightweight baskets. *Clothing* Fruiting stalks made into mats and used for raincoats and capes. *Mats, Rugs & Bedding* Fruiting stalks used to make mats for hangings, screens, mattresses, and coverings. Fruiting stalks made into mats and used for kneeling pads in canoes. *Sewing Material* Leaves peeled and made into two-ply strings for sewing mats. **Squaxin** *Basketry* Used to make lightweight baskets. *Clothing* Fruiting stalks made into mats and used for raincoats and capes. *Mats, Rugs & Bedding* Fruiting stalks used to make mats for hangings, screens, mattresses, and coverings. Fruiting stalks made into mats and used for kneeling pads in canoes. (79:21) **Thompson** *Building Material* Leaves and stems made into mats used as wall insulators and in constructing temporary summer houses. The mats were twined with fiber of Indian hemp or silverberry. *Clothing* Downy seed fluff used as dia-

pers for infants. (187:144) *Mats, Rugs & Bedding* Leaves and stems used to make mats. (125:39) Stalks used to make bedding. Leaves and stems stripped and made into mats. (164:496) Fruiting heads substituted for down and used in stuffing pillows and making beds. (164:498) Leaves and stems made into twined mats and used as mattresses and place mats. Downy seed fluff used in quilts and as a substitute for feathers in stuffing pillows and mattresses. The seed fluff was considered to be "really warm." (187:144) **Tolowa** *Clothing* Leaves woven together to make raincoats. **Yurok** *Clothing* Leaves woven together to make raincoats. *Mats, Rugs & Bedding* Leaves woven together to make a mat. (6:59)
- **Other**—**Apache, Chiricahua & Mescalero** *Ceremonial Items* Pollen sprinkled as a cross onto largest mescal crown. (33:36) **Apache, Mescalero** *Ceremonial Items* Pollen used in many ceremonies. Leaves used as ground covering for ceremonial tepees during the puberty ritual. (14:46) **Apache, White Mountain** *Ceremonial Items* Pollen used in religious ceremonies. (136:151) **Cahuilla** *Ceremonial Items* Stalks used in constructing ceremonial bundles. (15:142) **Chehalis** *Decorations* Leaf edges used to make imbrications on coiled baskets. (79:21) **Keres, Western** *Ceremonial Items* Ripened cattails shaken in the rain dance to produce clouds. (171:73) **Keresan** *Ceremonial Items* Fuzz used ritually like prayer meal or pollen. (198:560) **Navajo** *Ceremonial Items* Pollen used in the ceremonies. Leaves used to make ceremonial necklaces and wristbands for the Male Shooting Chant. (55:24) **Navajo, Ramah** *Containers* Leaves used to make bed mats, storage baskets, medicine baskets, and water jugs. *Good Luck Charm* Leaves used to make mats hung up in the hogan to bring rain. *Protection* Leaves used to make mats hung up in the hogan to protect it, the people, and the sheep from lightning. (191:14) **Ojibwa** *Weapon* Fuzz of the fruit thrown into the eyes of their enemies, claiming that it blinded them. (153:432) **Okanagan-Colville** *Containers* Leaves used to make storage bags for dried roots and berries. *Protection* Leaves wrapped around the sweat house entrance frame to prevent from getting burned on the wood. (188:57) **Okanagon** *Containers* Leaves and stems used to make bags. (125:39) **Omaha** *Ceremonial Items* Leaf used as one of the required articles in dressing the sacred pipes. (68:323) **Paiute, Northern** *Fasteners* Used for wefts and binding tule items. Used as ties for sandals. (60:81) *Hunting & Fishing Item* Used in the construction of duck decoys and boats. (60:69) **Potawatomi** *Waterproofing Agent* Leaves sewn together to make a windproof and waterproof side mat to be applied to the wigwam. (154:114) **Quinault** *Containers* Used to make large wallet-like packsacks. (79:21) **Shuswap** *Cooking Tools* Used to make mats for bedding, drying berries, and for mealtimes. (123:55) **Thompson** *Containers* Leaves and stems used to make bags. (125:39) Leaves and stems stripped and made into bags. (164:496) *Cooking Tools* Leaves and stems used in making "trays" or plates. *Decorations* Leaves used to make headdresses for Indian doctors. (187:144)

Typha sp., Cattail
- **Drug**—**Omaha** *Burn Dressing* Pulverized root used as paste for burns and covered with ripe blossoms. (58:584)
- **Food**—**Hualapai** *Bread & Cake* Roots peeled, dried, ground into a flour, and used to make bread. *Unspecified* Young shoots eaten raw or boiled. (195:18) **Pomo, Kashaya** *Unspecified* Young shoots used for food. (72:32) **Yuki** *Unspecified* Young stem bases eaten raw. (49:87)
- **Other**—**Hualapai** *Toys & Games* Used to make the hoop for a game. (195:18)

Ulmaceae, see *Celtis, Trema, Ulmus*

Ulmus, Ulmaceae
Ulmus americana L., American Elm
- **Drug**—**Cheyenne** *Gynecological Aid* and *Pediatric Aid* Infusion of bark taken by pregnant women to insure stability of children. (83:39) **Choctaw** *Analgesic* and *Gynecological Aid* Decoction of inner bark taken for menstrual cramps. (177:18) **Delaware** *Cold Remedy* Infusion of inner bark used for colds. *Cough Medicine* Infusion of inner bark used for severe coughs. (176:31) **Delaware, Oklahoma** *Cold Remedy* Infusion of inner bark taken for colds. *Cough Medicine* Infusion of inner bark taken for severe coughs. (175:26, 80) **Houma** *Antidiarrheal* Compound decoction of bark taken for dysentery. (158:56) **Iroquois** *Antidiarrheal* and *Antiemetic* Compound decoction taken for "summer disease—vomiting, diarrhea and cramps." *Antihemorrhagic* Compound decoction with smashed twigs taken for internal hemorrhage. *Gastrointestinal Aid* Compound decoction taken for "summer disease": vomiting, diarrhea, and cramps. *Gynecological Aid* Compound decoction of bark taken to facilitate childbirth and for parturition. Infusion of root bark taken for excessive menstruation. *Hemorrhoid Remedy* Infusion taken for piles caused by contact with menstruating woman. *Orthopedic Aid* Compound decoction of bark taken for broken bones. (87:304) *Other* Infusion of bark used for ruptures caused by exaggerated efforts. (141:40) **Koasati** *Dermatological Aid* Decoction of bark taken and used as a wash for gun wounds. *Gastrointestinal Aid* Infusion of inner bark taken and used as a bath for appendicitis. (177:18) **Meskwaki** *Eye Medicine* Decoction of root bark applied to sore eyes as an eye lotion. (152:251) **Mohegan** *Cold Remedy* Infusion of bark taken for colds. (174:266) Infusion of inner bark taken for colds. (176:76, 132) *Cough Medicine* Infusion of bark taken for coughs. (174:266) Infusion of inner bark taken for coughs. (176:76, 132) **Montana Indian** *Cancer Treatment* Inner bark used as an emollient for tumors. (19:25) **Ojibwa** *Venereal Aid* Infusion of root bark taken for gonorrhea. (135:231) **Penobscot** *Antihemorrhagic* Infusion of bark taken for "bleeding at the lungs." *Pulmonary Aid* Infusion of bark taken for pulmonary hemorrhage. (156:311) **Potawatomi** *Analgesic* Bark used for cramps. *Antidiarrheal* Bark used for diarrhea. (154:86)
- **Food**—**Cheyenne** *Beverage* Red, inner bark used like coffee. (83:39)
- **Fiber**—**Chippewa** *Building Material* Bark peeled during raspberry ripening time and used in the winter for house roofing material. (71:129) **Dakota** *Building Material* Forked trees used for the posts in building the earth lodge. (70:75) **Lakota** *Building Material* Wood used for posts and other building materials. (106:31) **Omaha** *Building Material* Forked trees used for the posts in building the earth lodge. **Pawnee** *Building Material* Forked trees used for the posts in building the earth lodge. **Ponca** *Building Material* Forked trees used for the posts in building the earth lodge. (70:75)
- **Other**—**Dakota** *Cooking Tools* Log sections used to make corn mor-

tars and pestles. *Fuel* Wood used for fuel. *Stable Gear* Wood used for saddle trees. *Tools* Wood used to make small mortars and pestles for grinding medicines and perfumes. (70:75) *Toys & Games* Fibrous inner bark used for popgun wads. (70:116) **Iroquois** *Stable Gear* Bark used to make dog sled harnesses. (141:40) **Lakota** *Fuel* Wood used for fuel. (106:31) **Omaha** *Cooking Tools* Log sections used to make corn mortars and pestles. *Fuel* Wood used for fuel. *Stable Gear* Wood used for saddle trees. *Tools* Wood used to make small mortars and pestles for grinding medicines and perfumes. (70:75) *Toys & Games* Fibrous inner bark used for popgun wads. (70:116) **Pawnee** *Cooking Tools* Log sections used to make corn mortars and pestles. *Fuel* Wood used for fuel. *Stable Gear* Wood used for saddle trees. *Tools* Wood used to make small mortars and pestles for grinding medicines and perfumes. (70:75) *Toys & Games* Fibrous inner bark used for popgun wads. (70:116) **Ponca** *Cooking Tools* Log sections used to make corn mortars and pestles. *Fuel* Wood used for fuel. *Stable Gear* Wood used for saddle trees. *Tools* Wood used to make small mortars and pestles for grinding medicines and perfumes. (70:75) *Toys & Games* Fibrous inner bark used for popgun wads. **Winnebago** *Toys & Games* Fibrous inner bark used for popgun wads. (70:116)

Ulmus rubra Muhl., Slippery Elm
- **Drug**—**Alabama** *Gynecological Aid* Decoction of bark and gunpowder taken as sympathetic magic for delayed labor. (as *U. fulva* 172:665) Decoction of bark taken for prolonged labor. **Catawba** *Tuberculosis Remedy* Bark used for consumption. (as *U. fulva* 177:19) **Cherokee** *Antidiarrheal* Taken to soothe stomach and bowels and for dysentery and bowels of pregnant women. (80:33) Decoction of inner bark taken for dysentery. (as *U. fulva* 177:19) *Burn Dressing* Poultice of inside bark used for burns. *Cold Remedy* Used for colds. *Cough Medicine* Used for coughs. *Dermatological Aid* Poultice of inside bark used for old sores and wounds. *Eye Medicine* Decoction of bark used to wash eyes. *Gastrointestinal Aid* Taken to soothe stomach and bowels and for heartburn and bowels of pregnant women. *Gynecological Aid* Taken to soothe stomach and bowels and for heartburn and bowels of pregnant women. (80:33) Decoction of bark taken to ease labor. (as *U. fulva* 177:19) *Laxative* Used as a mild laxative and "soothes stomach and bowels." *Respiratory Aid* Used for catarrh. *Throat Aid* Used for "quinsies." *Tuberculosis Remedy* Used for "quinsies," coughs, "consumptions and breast complaints." (80:33) **Chippewa** *Throat Aid* Decoction of bark gargled or dried root chewed for sore throat. (as *U. fulva* 53:342) **Creek** *Witchcraft Medicine* Decoction of bark with gunpowder taken to speed delivery, sympathetic magic. (as *U. fulva* 172:665) **Dakota** *Laxative* Decoction of inner bark taken as a laxative. (as *U. fulva* 70:76) **Iroquois** *Blood Medicine* Complex compound used as a blood purifier. *Emetic* Compound decoction taken to vomit for sleepiness and weakness. *Eye Medicine* Infusion of bark used as drops and as a wash for sore eyes. (87:305) *Gastrointestinal Aid* Decoction of bark taken to clean stomach and used for biliousness. (87:306) *Gynecological Aid* Compound decoction of bark taken to facilitate childbirth and for parturition. Infusion of bark taken for dry birth. (87:305) *Kidney Aid* Decoction of bark taken for kidneys. (87:306) *Respiratory Aid* Compound of leaves smoked and exhaled through the nostrils for catarrh. *Stimulant* Compound decoction with black center of tree taken when feeling drowsy. (87:305) *Throat Aid* Raw bark chewed for sore throats. (87:306) *Tuberculosis Remedy* Compound decoction used as a poultice for infected and swollen tubercular glands. (87:304) **Kiowa** *Oral Aid* Fresh, inner bark used as a masticatory. (as *U. fulva* 192:23) **Mahuna** *Orthopedic Aid* Poultice of bark applied to broken and fractured arms or legs. (as *U. pubescens* 140:27) **Menominee** *Cathartic* Infusion of inner bark taken as a physic. *Dermatological Aid* Poultice of inner bark applied to draw pus from a wound. (as *U. fulva* 151:56, 57) **Meskwaki** *Dermatological Aid* Poultice of bark applied to old sores. *Gynecological Aid* Decoction of root taken by women to ease childbirth. (as *U. fulva* 152:251) **Micmac** *Dermatological Aid* Bark used for suppurating wounds. *Pulmonary Aid* Bark used for bleeding lungs. (as *U. fulva* 40:63) **Mohegan** *Cough Medicine* Bark used for coughs. (176:132) *Pulmonary Aid* Inner bark chewed to soothe the lungs. (176:76) *Throat Aid* Inner bark chewed for sore throat. (176:76, 132) **Ojibwa** *Dermatological Aid* Infusion of roots taken and used as a wash for bleeding foot cuts. *Gastrointestinal Aid* Infusion of plants taken for stomach troubles. *Hemostat* Infusion of roots taken and used as a wash for bleeding foot cuts. (as *U. fluva* 135:231) *Throat Aid* Inner bark used for dry, sore throat. (as *U. fulva* 153:392) *Venereal Aid* Plant used for gonorrhea. (as *U. fluva* 135:240) **Omaha** *Laxative* Decoction of inner bark taken as a laxative. **Pawnee** *Laxative* Decoction of inner bark taken as a laxative. **Ponca** *Laxative* Decoction of inner bark taken as a laxative. (as *U. fulva* 70:76) **Potawatomi** *Dermatological Aid* Inner bark used for boils. *Eye Medicine* Poultice of chewed inner bark applied to eye inflammations. *Throat Aid* Inner bark used to lubricate throat for removal of lodged bone. (as *U. fulva* 154:86, 87) **Winnebago** *Laxative* Decoction of inner bark taken as a laxative. (as *U. fulva* 70:76)
- **Food**—**Kiowa** *Beverage* Inner bark used to brew a "tea." *Winter Use Food* Dried, stored inner bark used to brew a "tea" during the winter. (as *U. fulva* 192:23) **Omaha** *Preservative* Bark cooked with rendering fat as a preservative. (as *U. fulva* 68:325) Inner bark cooked with buffalo fat as a preservative, to prevent it from becoming rancid. (as *U. fulva* 70:76) *Snack Food* Bark cooked with rendering fat and prized by children as special tidbits. *Spice* Bark cooked with rendering fat as a flavoring. (as *U. fulva* 68:325) Inner bark cooked with buffalo fat for its desirable flavor when rendering out the tallow. *Unspecified* Inner bark cooked with buffalo fat and prized by children as tidbits. (as *U. fulva* 70:76)
- **Fiber**—**Cherokee** *Sporting Equipment* Inner bark chewed and spat onto baseball glove to make the ball stick to the glove. (80:33) **Dakota** *Building Material* Forked trees used for the posts in building the earth lodge. *Cordage* Inner bark fibers used to make ropes and cords. (as *U. fulva* 70:76) **Menominee** *Basketry* Boiled bark used to make baskets and fish nets. *Mats, Rugs & Bedding* Boiled bark used to make matting. *Snow Gear* Boiled bark used to make nets for snowshoes. (as *U. fulva* 151:77) **Meskwaki** *Building Material* Bark strips used to make sides of the winter wigwam and roof rainproof. (as *U. fulva* 152:270) **Ojibwa** *Building Material* Stripped bark used as a wigwam cover, for the sides of the wigwam. (as *U. fulva* 153:423) **Omaha** *Basketry* Inner bark used to make baskets. (as *U. fulva* 68:324) *Building Material* Forked trees used for the posts in building the earth lodge. (as *U. fulva* 70:76) *Cordage* Inner bark used to make ropes and cordage. (as *U. fulva* 68:324) Inner bark fiber used to make cords and ropes. **Pawnee** *Building Material* Forked trees used for the posts in building the

earth lodge. *Cordage* Inner bark fiber used to make ropes and cords. **Ponca** *Building Material* Forked trees used for the posts in building the earth lodge. *Cordage* Inner bark fiber used to make cords and ropes. (as *U. fulva* 70:76) **Potawatomi** *Basketry* Bark used to make boxes and baskets. (as *U. fulva* 154:115) Bark used to make baskets. (as *U. fulva* 154:124) *Building Material* Bark used to make wigwam sides. (as *U. fulva* 154:115) **Winnebago** *Building Material* Forked trees used for the posts in building the earth lodge. *Cordage* Inner bark fiber used to make cords and ropes. (as *U. fulva* 70:76)
- *Other*—**Dakota** *Cooking Tools* Log sections used to make corn mortars and pestles. *Fuel* Weathered bark used to catch the spark in fire making. Wood used for fuel. *Tools* Wood used to make small mortars and pestles for grinding medicines and perfumes. (as *U. fulva* 70:76) *Toys & Games* Fibrous inner bark used for popgun wads. (as *U. fulva* 70:116) **Kiowa** *Stable Gear* Forked branches used for the saddle frames. (as *U. fulva* 192:23) **Menominee** *Hunting & Fishing Item* Boiled bark used to make matting, baskets, fish nets, and nets for snowshoes. (as *U. fulva* 151:77) **Omaha** *Cooking Tools* Log sections used to make corn mortars and pestles. *Fuel* Weathered bark used to catch the spark in fire making. Wood used for fuel. *Tools* Wood used to make small mortars and pestles for grinding medicines and perfumes. (as *U. fulva* 70:76) *Toys & Games* Fibrous inner bark used for popgun wads. (as *U. fulva* 70:116) **Pawnee** *Cooking Tools* Log sections used to make corn mortars and pestles. *Fuel* Weathered bark used to catch the spark in fire making. Wood used for fuel. *Tools* Wood used to make small mortars and pestles for grinding medicines and perfumes. (as *U. fulva* 70:76) *Toys & Games* Fibrous inner bark used for popgun wads. (as *U. fulva* 70:116) **Ponca** *Cooking Tools* Log sections used to make corn mortars and pestles. *Fuel* Weathered bark used to catch the spark in fire making. Wood used for fuel. *Tools* Wood used to make small mortars and pestles for grinding medicines and perfumes. (as *U. fulva* 70:76) *Toys & Games* Fibrous inner bark used for popgun wads. (as *U. fulva* 70:116) **Winnebago** *Cooking Tools* Log sections used to make corn mortars and pestles. *Fuel* Weathered bark used to catch the spark in fire making. Wood used for fuel. *Tools* Wood used to make small mortars and pestles for grinding medicines and perfumes. (as *U. fulva* 70:76) *Toys & Games* Fibrous inner bark used for popgun wads. (as *U. fulva* 70:116)

Ulmus sp., Elm
- *Drug*—**Creek** *Toothache Remedy* Unknown plant part, possibly branches, used for toothache. (172:660)

Ulmus thomasii Sarg., Rock Elm
- *Fiber*—**Dakota** *Building Material* Forked trees used for the posts in building the earth lodge. **Omaha** *Building Material* Forked trees used for the posts in building the earth lodge. **Ponca** *Building Material* Forked trees used for the posts in building the earth lodge. (70:75)
- *Other*—**Dakota** *Cooking Tools* Log sections used to make corn mortars and pestles. *Fuel* Wood used as fuel. *Stable Gear* Wood used for saddle trees. *Tools* Wood used to make small mortars and pestles for grinding medicines and perfumes. **Omaha** *Fuel* Wood used as fuel. *Stable Gear* Wood used for saddle trees. *Tools* Wood used to make small mortars and pestles for grinding medicines and perfumes. **Ponca** *Cooking Tools* Log sections used to make corn mortars and pestles. *Fuel* Wood used as fuel. *Stable Gear* Wood used for saddle trees. *Tools* Wood used to make small mortars and pestles for grinding medicines and perfumes. (70:75)

Ulvaceae, see *Enteromorpha*, *Ulva*

Ulva, Ulvaceae, alga
Ulva lactuca L., Sea Lettuce
- *Food*—**Pomo, Kashaya** *Spice* Peppery seaweed used as flavoring with other seaweeds. (72:127)

Umbellularia, Lauraceae
Umbellularia californica (Hook. & Arn.) Nutt., California Laurel
- *Drug*—**Cahuilla** *Analgesic* Leaves used for headaches. (15:143) **Karok** *Analgesic* Infusion of plant taken by women for the pains of afterbirth. *Cold Remedy* Plant used as steam bath or burning boughs used to fumigate house for colds. (148:383) *Dermatological Aid* Poultice of ground seeds applied to sores. (6:59) *Disinfectant* Boughs used in fire to fumigate the house for colds and other sicknesses. *Gynecological Aid* Infusion of plant taken by women for the pains of afterbirth. *Panacea* Plant used as steam bath or burning bough to fumigate house for any sickness. (148:383) **Mendocino Indian** *Analgesic* Decoction taken, used as wash, or poultice of leaves applied for headaches. *Antirheumatic (External)* Infusion of leaves used as a bath for rheumatism. *Dermatological Aid* Decoction of leaves used as wash for vermin on the head. *Gastrointestinal Aid* Decoction of plant taken for stomachaches and headaches. Poultice of leaves applied as counterirritant for chronic stomach complaints. *Panacea* Burning leaf vapor used for many diseases. *Stimulant* Nuts eaten as a stimulant. (41:349) **Miwok** *Analgesic* Poultice of leaves and twigs bound to forehead for headaches. (12:174) **Pomo** *Analgesic* Poultice of heated leaves applied for rheumatic and neuralgic pains. *Antirheumatic (External)* Poultice of heated leaves applied for rheumatic pains. (66:13) **Pomo, Kashaya** *Analgesic* Poultice of leaves used for rheumatic and neuralgic pains. Indian doctor would sing and hit you with little branches for pain or headache. *Antirheumatic (External)* Poultice of leaves used for rheumatic pains. *Cold Remedy* Decoction of leaves taken for colds. Indian doctor would sing and hit you with little branches for a cold. *Dermatological Aid* Decoction of leaves used to wash sores. Peppernut charcoal rubbed into a man's mustache to groom it. *Gynecological Aid* Decoction of leaves taken for menstrual cramps and clotting. *Respiratory Aid* Decoction of leaves taken to clear up "slime" in the chest. *Throat Aid* Decoction of leaves taken for sore throats. (72:90) **Yuki** *Analgesic* Crushed leaves inhaled for headaches. (49:47) Poultice of crushed leaves applied as compress for headaches. *Antirheumatic (External)* Decoction of leaves used as wash for rheumatism. (118:43) *Respiratory Aid* Crushed leaves inhaled to open the nasal passages. (49:47) **Yurok** *Unspecified* Plant used for medicinal purposes. (6:59)
- *Food*—**Concow** *Beverage* Root bark used to make a drink. (41:349) **Costanoan** *Bread & Cake* Kernels roasted or ground into flour for cakes. *Fruit* Fruits eaten raw or boiled. (21:249) **Karok** *Unspecified* Seeds shelled, roasted, and eaten. (6:59) Nuts hulled, stored, parched in ashes, cracked open, and eaten. *Winter Use Food* Nuts hulled and stored in big baskets for winter use. (148:383) **Mendocino Indian** *Bread & Cake* Nuts roasted, shelled, pounded into a small mass, and

molded into "bread." *Sauce & Relish* Nuts eaten as a relish. Nuts used as a condiment. *Unspecified* Nuts roasted and eaten. (41:349) **Pomo, Kashaya** *Dried Food* Fruit sun dried. *Fruit* Fleshy end of the husk eaten raw and the kernel roasted and eaten whole. *Unspecified* Roasted kernels or kernel meal cakes eaten with greens, buckeye meal, acorn meal, mush, or seaweed. *Winter Use Food* Pounded kernel meal used to make sun dried, flat cakes and stored for winter use. (72:90) **Tolowa** *Fruit* Fruit shelled, roasted, and eaten. (6:59) **Yuki** *Unspecified* Species used for food. (48:47) Nuts roasted in hot ashes, cracked, and eaten. (49:87) **Yurok** *Fruit* Fruit used for food. *Unspecified* Seeds baked in the sand under a fire and used for food. (6:59)

- **Other**—**Costanoan** *Hunting & Fishing Item* Burning leaf smoke used to drive ground squirrels from burrows. *Incense & Fragrance* Leaves hung in bunches to freshen air. (21:249) *Insecticide* Burning leaf smoke used to rid one of fleas. (21:7, 249) **Karok** *Ceremonial Items* Foliage placed on fire during the Brush Dance to drive evil spirits away. (6:59) *Toys & Games* Leaves thrown into the fire by children to hear them crack like firecrackers. (148:383) **Mendocino Indian** *Insecticide* Leaves used for fleas. (41:349) **Pomo, Kashaya** *Hunting & Fishing Item* Leaves rubbed on a man's body before he went hunting to take the body smell away. *Protection* Small, leafy branches hung in homes for protection against any harm that might come into the homes. *Sacred Items* Plant very sacred to the Kashaya Pomo because of its ceremonial uses. (72:90) **Yuki** *Insecticide* Leafy branches used for fleas. (49:91) **Yurok** *Good Luck Charm* Leaves burned to take bad luck away, in the house, or the smoke waved over people as they leave. *Insecticide* Plant put under the bed to rid it of fleas. (6:59)

Umbellularia sp., Pepperwood

- **Food**—**Midoo** *Unspecified* Nuts buried in mud for a long time to remove the bitterness and used for food. (117:314) **Pomo** *Unspecified* Nuts roasted in ashes and eaten with fresh clover. (117:284)
- **Fiber**—**Mewuk** *Building Material* Branches used to make arbors or shades. (117:349) **Yokut** *Building Material* Plant used to make shelters and canopies. (117:420)

Urticaceae, see *Laportea, Pilea, Pipturus, Touchardia, Urtica*

Urtica, Urticaceae

Urtica dioica L., Stinging Nettle

- **Drug**—**Cherokee** *Gastrointestinal Aid* Taken for upset stomach. *Misc. Disease Remedy* Infusion taken for "ague." (80:46) **Cree, Woodlands** *Gynecological Aid* Decoction of plant taken to keep blood flowing after childbirth. (109:63) **Hesquiat** *Analgesic* Nettles rubbed on body for aches, pains, and backaches. *Antirheumatic (External)* Poultice of steamed leaves and roots used on swollen, sore, arthritic legs, ankles, and joints. *Gastrointestinal Aid* Nettles rubbed on stomach when sore. (185:76) **Iroquois** *Witchcraft Medicine* Compound of plant and dried snake's blood used as a "witching medicine." (87:307) **Kwakiutl** *Analgesic* Plant rubbed on the skin for chest pains. *Dermatological Aid* Plant juice rubbed into the scalp to prevent hair from falling out. *Gynecological Aid* Plant juice taken by overdue, pregnant women. *Other* Moxa of plant fiber used to cauterize the skin for various ailments. *Venereal Aid* Plant used for locomotor ataxia. (183:292) **Kwakiutl, Southern** *Analgesic* Fiber used to cauterize headaches. *Antirheumatic (External)* Fiber used to cauterize swellings. (183:297) **Lakota** *Gastrointestinal Aid* Infusion of roots taken for stomach pains. (139:61) **Makah** *Hunting Medicine* Used to rub the whale hunters' bodies in order to be strong. Leaves rubbed on fishing line to give it a green color or used as medicine for good fishing.[81] *Oral Aid* Stems put under splints to hasten the healing process. *Reproductive Aid* Infusion of roots given to expectant mothers. *Stimulant* Used to rub down after the morning bath. (67:246) **Nitinaht** *Antirheumatic (External)* Plants whipped over body for arthritis and rheumatism. *Tonic* Young shoots chewed and swallowed as a tonic to prevent sickness. *Unspecified* Plants whipped over body by married persons for affection and faithfulness of spouses. (186:128) **Okanagan-Colville** *Antirheumatic (External)* Fresh plants used to beat the skin after "sweathousing" and for rheumatic and arthritic pain. (188:140) **Paiute, Northern** *Antirheumatic (External)* Leaves and stalks used to whip aching arms and legs for rheumatism. *Dermatological Aid* Decoction of roots taken for hives and itches. (59:126) **Shuswap** *Analgesic* and *Antirheumatic (External)* Decoction of stems and roots used as a sweat bath for rheumatism pain. *Dermatological Aid* Used for bathing and drinking. *Herbal Steam* Decoction of stems and roots used as a sweat bath for rheumatism pain. (123:70) **Thompson** *Antirheumatic (External)* Nettles used for arthritis. One informant said that a neighboring woman who had arthritis hit her skin all over with nettles and within 2 months, she was getting strong and healthy again. *Dermatological Aid* Decoction of roots used as a hair tonic for growing long, silky hair. Plant tops used for skin disease. *Hemorrhoid Remedy* Decoction of roots used as a soaking solution for bleeding hemorrhoids. *Orthopedic Aid* Poultice of leaves and stalks used for paralyzed limbs. *Unspecified* Infusion of leaves and tops used as medicine. (187:289)

- **Food**—**Iroquois** *Vegetable* Cooked and seasoned with salt, pepper, or butter. (196:118) **Makah** *Unspecified* Plant tops used for food. (67:246) **Mohegan** *Vegetable* Combined with pigweed, mustard, plantain, and dock and used as mixed greens. (176:83) **Okanagan-Colville** *Vegetable* New growths dipped in boiling water and eaten as greens. (188:140) **Shuswap** *Beverage* Used for bathing and drinking. (123:70) **Thompson** *Vegetable* Greens cooked as green vegetables. (187:288) Plant tops eaten as a potherb after the arrival of the Chinese. (187:289)
- **Fiber**—**Hesquiat** *Cordage* Dried, peeled stems used to make twine, ropes, and herring nets. (185:76) **Lakota** *Cordage* Stem fibers used to make cordage. (139:61) **Makah** *Basketry* Fibers used in weaving baskets. *Cordage* Fibers used to make string. **Nitinaht** *Cordage* Fibers, yellow cedar bark, or cottonwood fibers and dog hair used to make stronger ropes. (67:246) Stems dried, pounded, and spun to make twine for binding and sewing purposes. (186:128) **Thompson** *Cordage* Plant tops used to make twine and fine thread. The plant tops were made into twine in the same manner as Indian hemp. (187:289)
- **Dye**—**Makah** *Green* Leaves rubbed on fishing line to give it a green color or used as medicine for good fishing.[81] (67:246)
- **Other**—**Cherokee** *Hunting & Fishing Item* Stems twisted for bow strings. (80:46) **Hesquiat** *Hunting & Fishing Item* Dried, peeled stems used to make twine, ropes, and herring nets. (185:76) **Makah** *Ceremonial Items* Used in rituals for whaling and fishing. *Protection* Used to rub seal hunters bodies to protect themselves from the weather. **Nitinaht** *Hunting & Fishing Item* Fibers used to make fish and duck nets. (67:246) Plants rubbed on fishing lines to eliminate human

odor. (186:112) Used as the leader for fishing. (186:53) Stems dried, pounded, and spun to make fishing lines and duck nets. Plants rubbed in hands to eliminate human odor before touching fishing gear. (186:128) **Okanagan-Colville** *Ceremonial Items* Plant used to make a tea taken during "sweathousing" and used to "wash" the skin and hair. *Protection* Boiled plant used as a bath to combat witchcraft and a jinx by an evil person. (188:140) **Thompson** *Hunting & Fishing Item* Plant tops made into twine and fine thread and used for making scoop nets. (187:289)

Urtica dioica **ssp.** *gracilis* (Ait.) Seland., California Nettle
- *Drug*—**Abnaki** *Hemostat* Used for bloody noses. (as *U. procera* 144:154) Powdered leaves used as a snuff for nosebleeds. (as *U. procera* 144:166) **Bella Coola** *Analgesic* and *Antirheumatic (External)* Used in a sweat bath for pains similar to rheumatism all over the body. (as *U. lyallii* 150:55) *Hemostat* Burning stem fibers used to cauterize sores and swellings. (184:211) *Herbal Steam* Used in a sweat bath for pains similar to rheumatism all over the body. *Orthopedic Aid* Used to sting paralyzed limbs daily to cause sores and revive sensation. (as *U. lyallii* 150:55) **Carrier** *Analgesic* Used as whip for pain. *Dermatological Aid* Rubbed on the skin for rashes. (as *U. gracilis* 31:83) **Chehalis** *Antirheumatic (External)* Whole stalk used as a whip on a person with rheumatism. *Dermatological Aid* Decoction of roots used as a hair wash. (as *U. lyallii* 79:28) **Chippewa** *Antidiarrheal* Infusion of root taken for dysentery. (as *U. gracilis* 53:344) *Diuretic* Compound decoction of root taken for "stoppage of urine." (as *U. gracilis* 53:348) **Cowlitz** *Gynecological Aid* Infusion of nettles taken by women about to deliver a child. *Orthopedic Aid* Poultice of sprouts applied or person whipped with stalk for paralysis. (as *U. lyallii* 79:28) **Gitksan** *Antihemorrhagic* Decoction of plant taken for hemorrhage and many illnesses. *Unspecified* Decoction of entire plant taken for many illnesses. *Urinary Aid* Decoction of plant taken for bladder trouble and many illnesses. (as *U. lyallii* 150:55) **Haisla & Hanaksiala** *Antirheumatic (External)* Poultice of roots applied to aches and swollen joints. (43:294) **Klallam** *Analgesic* Infusion of stalks rubbed on the body for soreness or stiffness. *Orthopedic Aid* Infusion of stalks rubbed on the body for soreness or stiffness. *Panacea* Infusion of nettles taken for many ailments. **Lummi** *Gynecological Aid* Infusion of nettles taken by women to relax the muscles during childbirth. (as *U. lyallii* 79:28) **Miwok** *Antirheumatic (External)* Decoction of root used as a bath for rheumatism. Leaf powder rubbed on affected rheumatic body parts. (as *U. gracilis* 12:174) **Navajo, Ramah** *Poison* Plant considered poisonous. (as *U. gracilis* 191:23) **Ojibwa** *Dermatological Aid* Poultice of soaked leaves applied to heat rash. (as *U. lyallii* 153:392) **Paiute** *Antirheumatic (External)* Decoction of leaves used as a wash for rheumatism. *Cold Remedy* Decoction of leaves taken for colds. *Diaphoretic, Herbal Steam,* and *Misc. Disease Remedy* Plant used as an inhalant in the sweat bath for grippe or pneumonia. *Pulmonary Aid* Plant fumes inhaled in the sweat bath for pneumonia or grippe. *Unspecified* Plant used as switch on body as a counterirritant for unspecified purposes. (as *U. gracilis* 180:146) **Pomo, Kashaya** *Analgesic* Nettle used to strike the skin as a counterirritant for rheumatism and other such pains. *Antirheumatic (External)* Nettle used to strike the skin as a counterirritant for rheumatism. (as *U. lyallii* 72:77) **Potawatomi** *Febrifuge* Infusion of root used for intermittent fevers. *Unspecified* Infusion of leaves used medicinally for unspecified purposes. (as *U. lyallii* 154:87) **Quileute** *Antirheumatic (External)* Whole stalk used as a whip on a person with rheumatism. *Antirheumatic (Internal)* Infusion of pounded roots taken for rheumatism. **Quinault** *Analgesic* Decoction of peeled bark taken for headaches. *Gynecological Aid* Plant tips chewed by women during labor. *Hemostat* Decoction of peeled bark taken for nosebleeds. *Orthopedic Aid* Whole stalk used as a whip on a person with paralysis. **Samish** *Tonic* Decoction of plant taken as a general tonic. (as *U. lyallii* 79:28) **Shoshoni** *Antirheumatic (External)* Hot poultice of mashed leaves applied for rheumatism. *Blood Medicine* Compound decoction of roots taken as a blood tonic and for general debility. *Cold Remedy* Decoction of leaves taken for colds. *Tonic* Compound decoction of root taken as a blood tonic and for general debility. (as *U. gracilis* 180:146) **Sioux** *Diuretic* Root used for urine retention. (as *U. gracilis* 19:25) **Skagit** *Cold Remedy* Decoction of plant taken for colds. **Skokomish** *Dermatological Aid* Decoction of roots used as a hair wash. **Snohomish** *Cold Remedy* Infusion of nettles taken or nettles rubbed on the body for colds. **Squaxin** *Gynecological Aid* Infusion of crushed leaves taken by women having difficulties in childbirth. **Swinomish** *Tonic* Decoction of plant taken as a general tonic. (as *U. lyallii* 79:28) **Tanaina** *Antirheumatic (Internal)* Plant used for diseases from rheumatism to tuberculosis. *Misc. Disease Remedy* Plant used for diseases from rheumatism to tuberculosis. *Tuberculosis Remedy* Plant used for diseases from rheumatism to tuberculosis. (as *U. lyallii* 149:329) **Thompson** *Orthopedic Aid* Plant dipped in water and rubbed on stiff, sore joints and muscles. (as *U. lyallii* 164:471, 472) **Tolowa** *Dermatological Aid* Poultice of fresh, pounded leaves applied to skin inflammations. (as *U. lyallii* 6:60)
- *Food*—**Alaska Native** *Dietary Aid* Fresh, green leaves used as a good source of provitamin A, vitamin C, and some of the minerals. *Substitution Food* Leaves used as a good substitute for spinach. *Unspecified* Leaves boiled and eaten. (as *U. lyalli* 85:73) **Cowichan** *Vegetable* Young stems and leaves boiled and eaten like spinach. (182:90) **Haisla & Hanaksiala** *Preserves* Plant boiled with sugar to make jam. *Unspecified* Young shoots steamed and eaten with bear meat. (43:294) **Hoh** *Vegetable* Plant tops eaten as greens. (as *U. lyallii* 137:61) **Montana Indian** *Vegetable* Young shoots used as a potherb. (as *U. gracilis* 19:25) **Oweekeno** *Unspecified* Plant fried and eaten. (as *U. dioca* ssp. *gracilis* var. *lyallii* 43:119) **Quileute** *Vegetable* Plant tops eaten as greens. (as *U. lyallii* 137:61) **Saanich** *Vegetable* Young stems and leaves boiled and eaten like spinach. (182:90) **Skagit, Upper** *Unspecified* Tender shoots cooked and eaten. (as *U. lyallii* 179:42)
- *Fiber*—**Bella Coola** *Cordage* Stem fibers sun dried and used to make twine. (184:211) **Dakota** *Clothing* Dried stalk fiber used to make cloth. *Cordage* Dried stalk fiber used to make twine and cordage. (as *U. gracilis* 70:77) **Eskimo, Inuktitut** *Cordage* Dried stem fibers used for twine. (as *U. lyalli* 202:186) **Haisla & Hanaksiala** *Cordage* Fiber used to make cordage, bindings, and nets. (43:294) **Hoh** *Cordage* Roots formerly twisted and made into ropes. (as *U. lyallii* 137:61) **Kwakiutl, Southern** *Cordage* Plants split, dried, pounded, and used to make twine and rope. (183:292) **Menominee** *Basketry* Plant made into hemp twine and used to make fiber bags. (as *U. gracilis* 151:77) **Montana Indian** *Cordage* Bark used for cordage. (as *U. gracilis* 19:25) **Ojibwa** *Sewing Material* Bark or rind used as a fine, stout sewing fiber. (as *U. lyalli* 153:423) **Omaha** *Cordage* Dried stalk fiber used to

make twine and cordage. The fiber was separated from the nettle by either crumpling the dried stalks in the hands or gently pounding it with stones. *Sewing Material* Dried stalk fiber used to make cloth. (as *U. gracilis* 70:77) **Oweekeno** *Cordage* Fiber used to make bow strings, fishing line for jigging, ropes, and oolichan (candlefish) traps. (as *U. dioca* ssp. *gracilis* var. *lyallii* 43:119) **Pawnee** *Clothing* Dried stalk fiber used to make cloth. *Cordage* Dried stalk fiber used to make twine and cordage. **Ponca** *Clothing* Dried stalk fiber used to make cloth. *Cordage* Dried stalk fiber used to make twine and cordage. (as *U. gracilis* 70:77) **Potawatomi** *Cordage* Outer rind twisted into a two-strand cord and used for sewing cattail mats and baskets. (as *U. lyallii* 154:115) **Quileute** *Cordage* Roots formerly twisted and made into ropes. (as *U. lyallii* 137:61) **Skagit, Upper** *Cordage* Mature shoot fibers used to make cordage. (as *U. lyallii* 179:42) **Winnebago** *Clothing* Dried stalk fiber used to make cloth. *Cordage* Dried stalk fiber used to make twine and cordage. (as *U. gracilis* 70:77)

- *Other*—**Bella Coola** *Hunting & Fishing Item* Stem fibers sun dried and used to make fish nets. (184:211) **Dakota** *Ceremonial Items* Dried stalk fiber made into cloth and used in the Sacred Bundle of the Tent of War. *Stable Gear* Dried stalk fiber made into ropes and used to hobble horses. *Toys & Games* Plant fiber used by little boys as wadding for popguns. (as *U. gracilis* 70:77) **Haisla & Hanaksiala** *Cash Crop* Plant used for trade. *Fasteners* Fiber used to make cordage, bindings, and nets. *Hunting & Fishing Item* Fiber used to make nets and bow strings. (43:294) **Kwakiutl, Southern** *Hunting & Fishing Item* Plants split, dried, pounded, and used to make fishing nets. (183:292) **Lummi** *Hunting & Fishing Item* Bark peeled, dried, made into a two-ply string, and used for duck nets. **Makah** *Ceremonial Items* Rubbed on the body to purify it after handling a corpse. (as *U. lyallii* 79:28) **Omaha** *Ceremonial Items* Dried stalk fiber made into cloth and used in the Sacred Bundle of the Tent of War. *Stable Gear* Dried stalk fiber made into ropes and used to hobble horses. *Toys & Games* Plant fiber used by little boys as wadding for popguns. (as *U. gracilis* 70:77) **Oweekeno** *Hunting & Fishing Item* Fiber used to make bow strings, fishing line for jigging, ropes, and oolichan (candlefish) traps. Fiber used to make fishing nets. (as *U. dioca* ssp. *gracilis* var. *lyallii* 43:119) **Pawnee** *Ceremonial Items* Dried stalk fiber made into cloth and used in the Sacred Bundle of the Tent of War. *Stable Gear* Dried stalk fiber made into ropes and used to hobble horses. *Toys & Games* Plant fiber used by little boys as wadding for popguns. **Ponca** *Ceremonial Items* Dried stalk fiber made into cloth and used in the Sacred Bundle of the Tent of War. *Stable Gear* Dried stalk fiber made into ropes and used to hobble horses. *Toys & Games* Plant fiber used by little boys as wadding for popguns. (as *U. gracilis* 70:77) **Quileute** *Hunting & Fishing Item* Rubbed on the bodies of sealers to keep them awake during the night. **Skokomish** *Hunting & Fishing Item* Bark peeled, dried, made into a two-ply string, and used for duck nets. **Snohomish** *Hunting & Fishing Item* Bark peeled, dried, made into a two-ply string, and used for duck nets. (as *U. lyallii* 79:28) **Tsimshian** *Hunting & Fishing Item* Fiber used to make fishing nets. Fiber used to make fishing nets. (43:351) **Winnebago** *Ceremonial Items* Dried stalk fiber made into cloth and used in the Sacred Bundle of the Tent of War. *Stable Gear* Dried stalk fiber made into ropes and used to hobble horses. *Toys & Games* Plant fiber used by little boys as wadding for popguns. (as *U. gracilis* 70:77)

Urtica dioica ssp. *holosericea* (Nutt.) Thorne, Stinging Nettle

- *Drug*—**Cahuilla** *Analgesic* Used for headaches and sore backs. *Antirheumatic (External)* Used for rheumatism and muscular stiffness. (as *U. holosericea* 15:143) **Diegueño** *Antirheumatic (External)* Nettles used to whip rheumatic or arthritic joints. (as *U. holosericea* 84:43) **Kawaiisu** *Analgesic* Poultice of leaves applied for headaches. Poultice of mashed plant applied to the neck for pain. Stems and leaves used as a counterirritant for sore limbs and back. *Dermatological Aid* "Children walk through nettles to toughen their skin." Poultice of leaves applied for sores. *Orthopedic Aid* Stems and leaves used as a counterirritant for sore limbs and back. *Pediatric Aid* "Children walk through nettles to prepare them for practice of witchcraft." "Children walk through nettles to toughen their skin." "Younger people walk through nettles to procure dreams." *Psychological Aid* "Younger people walk through nettles to procure dreams." *Witchcraft Medicine* "Children walk through nettles to prepare them for practice of witchcraft." (as *U. holosericea* 206:68) **Mahuna** *Antirheumatic (External)* Leaves rubbed on parts ailing from inflammatory rheumatism. (as *U. holosericea* 140:59) **Miwok** *Analgesic* Branch used to strike affected parts for certain pains. (as *U. gracilis* var. *holosericea* 12:174) **Pomo** *Analgesic* and *Antirheumatic (External)* Nettle used to strike skin as a counterirritant for rheumatic pains. (as *U. gracilis* var. *holosericea* 66:13)
- *Food*—**Cahuilla** *Vegetable* Leaves eaten raw or boiled as greens. (as *U. holosericea* 15:143)
- *Fiber*—**Cahuilla** *Basketry* Fibers used in basket making. *Cordage* Fibers used to make bowstrings and cordage. (as *U. holosericea* 15:143) **Kawaiisu** *Cordage* Outer stem layers separated into long strands and two or three twisted into a cord. (as *U. holosericea* 206:68) **Klamath** *Cordage* Stems used in the manufacture of cords and nets. (as *U. breweri* 45:95) *Snow Gear* Stems used for mesh on snowshoes. (as *U. breweri* 45:94) **Luiseño** *Clothing* Plant fiber made into twine and used to make front aprons worn by women. *Unspecified* Plant sometimes used for fiber. (as *U. holosericea* 155:202)
- *Other*—**Diegueño** *Ceremonial Items* Nettles used to whip young men going through the ceremony to become dancers. (as *U. holosericea* 84:43) **Kawaiisu** *Containers* Outer stem layers made into cord and used to make carrying nets. *Hunting & Fishing Item* Outer stem layers made into cord and used to make rabbit nets. (as *U. holosericea* 206:68) **Klamath** *Hunting & Fishing Item* Stems used in the manufacture of cords and nets. (as *U. breweri* 45:95) **Luiseño** *Containers* Plant fibers made into twine and used to make large-meshed nets for carrying bulky or heavy articles. Plant made into twine and used to make network sacks for carrying acorns and other small seeds. *Hunting & Fishing Item* Plant fiber made into twine and sometimes used to make bowstrings. Plant fiber made into twine, occasionally used to make long nets and draw nets for catching rabbits. Plant fiber made into twine and used to make fishing nets.[14] Plant fiber made into twine and used to make slings. (as *U. holosericea* 155:202)

Urtica sp., Nettle

- *Drug*—**Costanoan** *Analgesic* Leaves used to strike aching joints "to reduce the pain." *Dermatological Aid* Decoction of plant used for hives. *Disinfectant* Decoction of plant used for infected sores. *Orthopedic Aid* Leaves used to strike aching joints "to reduce the pain."

582 *Urtica* sp. (Urticaceae)

(21:21) **Sanpoil** *Antirheumatic (External)* Sprig brushed against the affected part for rheumatism. *Dermatological Aid* Sprig brushed against the affected part for swelling. (131:219) **Thompson** *Disinfectant* Plant used to make a wash poured on the body as a purifier after sweat bath. (164:505)

Urtica urens L., Dwarf Nettle
- **Drug**—**Shuswap** *Analgesic* and *Antirheumatic (External)* Decoction of stems and roots used as a sweat bath for rheumatism pain. *Dermatological Aid* Used for bathing and drinking. *Herbal Steam* Decoction of stems and roots used as a sweat bath for rheumatism pain. (123:70)
- **Food**—**Shuswap** *Beverage* Used for bathing and drinking. (123:70)

Usneaceae, see *Evernia, Letharia, Ramalina, Usnea*

Usnea, Usneaceae, lichen
Usnea longissima
- **Drug**—**Nitinaht** *Dermatological Aid* Used for wound dressing material and as bandages. (maidenhair moss 186:55)
- **Fiber**—**Hanaksiala** *Mats, Rugs & Bedding* Plant used as mattresses at seasonal camps. (old man's beard lichen 43:144) **Nitinaht** *Clothing* Used for baby diapers and female sanitary napkins. (maidenhair moss 186:55)
- **Other**—**Nitinaht** *Cooking Tools* Used for wiping salmon. (maidenhair moss 186:55) **Oweekeno** *Cleaning Agent* Whole plant used by women to wipe slime when cleaning salmon. (old man's beard lichen 43:50)

Usnea sp., Tree Lichen
- **Drug**—**Makah** *Dermatological Aid* Used for boils. **Nitinaht** *Dermatological Aid* Used to bandage wounds. (67:211)
- **Food**—**Hesquiat** *Forage* Plant browsed by deer. (185:17)
- **Fiber**—**Hanaksiala** *Mats, Rugs & Bedding* Plant used as mattresses at seasonal camps. (43:144) **Makah** *Mats, Rugs & Bedding* Used in bags as pillows when feathers unavailable. **Nitinaht** *Clothing* Used to diaper babies and for women's sanitary napkins. (67:211)
- **Dye**—**Makah** *Yellow* Used as a source of yellow dye. (67:211)
- **Other**—**Bella Coola** *Decorations* Formerly used to decorate dance masks. (184:195)

Ustilaginaceae, see *Ustilago*

Ustilago, Ustilaginaceae, fungus
Ustilago zeae (Beckm.) Ung., Corn Smut
- **Food**—**Apache, White Mountain** *Unspecified* Smut boiled and eaten. (136:161) **Hopi** *Unspecified* Used with sweet corn as food. (200:100)
- **Other**—**Hopi** *Ceremonial Items* Black powder used as a ceremonial body paint. (200:100)

Uvularia, Liliaceae
Uvularia grandiflora Sm., Largeflower Bellwort
- **Drug**—**Menominee** *Dermatological Aid* Plant used for swellings. (151:41) **Ojibwa** *Analgesic* and *Gastrointestinal Aid* Root used for stomach pain, perhaps pleurisy. *Pulmonary Aid* Root used for "pain in the solar plexus, which may mean pleurisy." (153:374) **Potawatomi** *Analgesic* Infusion of root used for backache or with lard as a salve for sore muscles. (154:56, 57) *Orthopedic Aid* Infusion of root mixed with lard and used as salve to massage sore muscles and tendons. Infusion of root used for backaches. (154:64)

Uvularia perfoliata L., Perfoliate Bellwort
- **Drug**—**Iroquois** *Cough Medicine* Infusion of roots given to children as cough medicine. *Eye Medicine* Infusion of smashed roots used as a wash for sore eyes of any degree. *Orthopedic Aid* Plant used several ways, internally and externally, for broken bones. *Pediatric Aid* Infusion of roots given to children as cough medicine. (87:280)
- **Food**—**Cherokee** *Unspecified* Plants boiled, fried, and eaten with fat. (204:252)

Uvularia sessilifolia L., Sessileleaf Bellwort
- **Drug**—**Cherokee** *Antidiarrheal* Infusion of root taken for diarrhea. *Dermatological Aid* Used as poultice for boils. (80:25) **Iroquois** *Blood Medicine* Infusion of roots taken as a blood purifier. *Orthopedic Aid* Infusion of roots taken and roots used as a poultice for broken bones. (87:281) **Ojibwa** *Hunting Medicine* Root used as a part of the hunting medicine to bring a buck deer near the hunter. (as *Oakesia sessilifolia* 153:430)
- **Food**—**Cherokee** *Vegetable* Leaves eaten as cooked greens. (80:25)

Vaccinium, Ericaceae
Vaccinium alpinum Bigel., Blueberry
- **Food**—**Eskimo, Alaska** *Fruit* Fruit used for food. (4:715)

Vaccinium angustifolium Ait., Lowbush Blueberry
- **Drug**—**Algonquin, Quebec** *Gastrointestinal Aid* Infusion of leaves given to infants for colic. *Gynecological Aid* Infusion of leaves used by women after a miscarriage. Infusion of roots used by women to induce labor. *Pediatric Aid* Infusion of leaves given to infants for colic. (18:217) **Chippewa** *Psychological Aid* Dried flowers placed on hot stones as inhalant for "craziness." (53:338) **Iroquois** *Ceremonial Medicine* Berries used ceremonially by those desiring health and prosperity for the coming season. (as *V. pennsylvanicum* 196:142) **Ojibwa** *Blood Medicine* Infusion of leaves taken as a blood purifier. (as *V. pennsylvanicum* 153:369)
- **Food**—**Abnaki** *Fruit* Fruits eaten for food. (144:171) Fruit used for food. (as *V. augustifolium* 144:152) **Algonquin, Quebec** *Fruit* Berries used fresh. Berries canned or used to make fruit pemmican and pâté. *Pie & Pudding* Berries used to make pies, cobblers, and upside-down cakes. *Preserves* Berries made into preserves and butter. (18:104) **Chippewa** *Fruit* Berries boiled, seasoned, combined with moose fat and deer tallow, and used for food. (53:321) **Iroquois** *Bread & Cake* Fruit mashed, made into small cakes, and dried for future use. *Dried Food* Raw or cooked fruit sun or fire dried and stored for future use. *Fruit* Dried fruit taken as a hunting food. (as *V. pennsylvanicum* 196:128) *Preserves* Flowers used to make preserves. (142:96) *Sauce & Relish* Dried fruit cakes soaked in warm water and cooked as a sauce or mixed with corn bread. (as *V. pennsylvanicum* 196:128) *Unspecified* Flowers eaten fresh. (142:96) **Menominee** *Winter Use Food* Sun dried

berries and dried sweet corn sweetened with maple sugar and stored for winter use. (as *V. pennsylvanicum* 151:66) **Ojibwa** *Dried Food* Berries sun dried for winter use. (as *V. pennsylvanicum* 135:238) Berries dried like currants and cooked in winter with corn, rice, and venison. (as *V. pennsylvanicum* 153:401) *Fruit* Berries eaten fresh. *Winter Use Food* Berries canned for future use. (as *V. pennsylvanicum* 135:238)
- *Other*—**Algonquin, Quebec** *Cash Crop* Berries gathered and sold. (18:104) **Ojibwa** *Cash Crop* Berries gathered and sold to the nearby stores. (as *V. pennsylvanicum* 135:238)

Vaccinium cespitosum Michx., Dwarf Blueberry
- *Food*—**Alaska Native** *Bread & Cake* Berries cooked in muffins. Berries eaten raw or cooked in pies, puddings, and muffins. *Dietary Aid* Berries used as a fair source of vitamin C. *Frozen Food* Berries frozen or canned for winter use. *Fruit* Berries eaten raw or cooked in pies, puddings, and muffins. *Pie & Pudding* Berries eaten raw or cooked in pies, puddings, and muffins. *Winter Use Food* Berries frozen or canned for winter use. (as *V. paludicola* 85:107) **Bella Coola** *Fruit* Berries formerly used for food. (184:205) **Okanagan-Colville** *Dried Food* Berries dried for future use. *Forage* Berries eaten by domestic sheep. *Fruit* Berries eaten fresh. *Winter Use Food* Berries canned for future use. (188:102) **Paiute** *Fruit* Berries eaten fresh. *Winter Use Food* Berries sometimes canned. (111:102) **Shuswap** *Fruit* Berries used for food. (123:63) **Thompson** *Dried Food* Berries dried loose like raisins or canned and used for food. *Pie & Pudding* Berries used in pies. *Preserves* Berries used in jams. *Winter Use Food* Berries dried loose like raisins or canned and used for food in winter. (187:217)
- *Other*—**Gosiute** *Smoke Plant* Leaves formerly dried and used as a tobacco. (39:384)

Vaccinium cespitosum Michx. var. *cespitosum*, Tall Bilberry
- *Food*—**Hanaksiala** *Bread & Cake* Berries dried in the form of cakes and reconstituted during the winter. (43:244) **Kitasoo** *Bread & Cake* Berries dried into cakes and eaten. *Fruit* Berries eaten fresh. (43:335) **Oweekeno** *Fruit* Berries used for food. (43:98) *Special Food* Berries picked for feasts. (43:99)

Vaccinium cespitosum var. *paludicola* (Camp) Hultén, Dwarf Blueberry
- *Food*—**Hanaksiala** *Bread & Cake* Berries dried in the form of cakes and reconstituted during the winter. (43:244) **Kitasoo** *Bread & Cake* Berries dried into cakes and eaten. *Fruit* Berries eaten fresh. (43:335) **Oweekeno** *Fruit* Berries used for food. (43:98) *Special Food* Berries picked for feasts. (43:99)

Vaccinium corymbosum L., Highbush Blueberry
- *Food*—**Algonquin, Quebec** *Fruit* Berries canned or used to make fruit pemmican and pâté. Berries used fresh. *Pie & Pudding* Berries used to make pies, cobblers, and upside-down cakes. *Preserves* Berries made into preserves and butter. (18:104) **Iroquois** *Bread & Cake* Fruit mashed, made into small cakes, and dried for future use. *Dried Food* Raw or cooked fruit sun or fire dried and stored for future use. *Fruit* Dried fruit taken as hunting food. *Sauce & Relish* Dried fruit cakes soaked in warm water and cooked as a sauce or mixed with corn bread. (196:128)
- *Other*—**Algonquin, Quebec** *Cash Crop* Berries gathered and sold. (18:104)

Vaccinium deliciosum Piper, Blueleaved Huckleberry
- *Food*—**Hoh** *Fruit* Berries eaten raw. *Sauce & Relish* Berries stewed and made into a sauce. *Winter Use Food* Berries canned and used as a winter food. **Quileute** *Fruit* Berries eaten raw. *Sauce & Relish* Berries stewed and made into a sauce. *Winter Use Food* Berries canned and used as a winter food. (137:67) **Thompson** *Dried Food* Berries dried loose like raisins. *Pie & Pudding* Berries used in pies. *Preserves* Berries used in jams. *Winter Use Food* Berries canned and used for food. (187:217)

Vaccinium macrocarpon Ait., Cranberry
- *Drug*—**Montagnais** *Pulmonary Aid* Infusion of branches used as a medicine for pleurisy. (as *Oxycoccus macrocarpus* 156:316)
- *Food*—**Algonquin, Quebec** *Fruit* Berries used for food. (18:105) **Algonquin, Tête-de-Boule** *Fruit* Fruits eaten for food. (132:134) **Anticosti** *Winter Use Food* Fruits stored for winter use. (143:68) **Chippewa** *Fruit* Berries cooked and used for food. (as *Oxycoccus macrocarpus* 53:321) **Iroquois** *Bread & Cake* Fruit mashed, made into small cakes, and dried for future use. *Dried Food* Raw or cooked fruit sun or fire dried and stored for future use. *Fruit* Dried fruit taken as a hunting food. *Sauce & Relish* Dried fruit cakes soaked in warm water and cooked as a sauce or mixed with corn bread. (196:128) **Ojibwa** *Fruit* Fruit used for food. (135:238)
- *Other*—**Ojibwa** *Cash Crop* Fruit sold by the bushels. (135:238)

Vaccinium membranaceum Dougl. ex Torr., Blue Huckleberry
- *Drug*—**Flathead** *Antirheumatic (Internal)* Infusion of roots and stems taken for rheumatism and arthritis. *Heart Medicine* Infusion of roots and stems taken for heart trouble. (as *V. globulare* 82:63)
- *Food*—**Alaska Native** *Bread & Cake* Berries cooked in muffins. *Dietary Aid* Berries used as a fair source of vitamin C. *Frozen Food* Berries frozen for winter use. *Fruit* Berries eaten raw. *Pie & Pudding* Berries cooked in pies and puddings. *Winter Use Food* Berries frozen or canned for winter use. (85:107) **Bella Coola** *Fruit* Berries used for food. (184:205) **Coeur d'Alene** *Dried Food* Berries dried and used for food. *Fruit* Berries eaten fresh, boiled and eaten, or mashed and eaten. *Soup* Berries dried, boiled with roots, and eaten as soup. (178:90) **Hanaksiala** *Bread & Cake* Berries dried in the form of cakes and reconstituted during the winter. (43:244) **Kitasoo** *Dried Food* Berries dried for future use. *Fruit* Berries eaten fresh. (43:335) **Klamath** *Winter Use Food* Dried berries stored for winter use. (45:103) **Montana Indian** *Bread & Cake* Berries used to make pancakes and muffins. *Dried Food* Fruits sun dried and stored for winter use. (as *V. globulare* 82:63) *Fruit* Fruit eaten fresh. (19:25) Berries eaten fresh. (as *V. globulare* 82:63) *Pie & Pudding* Fruit used for making pies. *Preserves* Fruit used for making jelly. (19:25) Berries used to make jams and jellies. (as *V. globulare* 82:63) *Winter Use Food* Fruit dried for winter. (19:25) **Okanagan-Colville** *Dried Food* Berries dried, boiled, and eaten. *Fruit* Berries eaten fresh. *Winter Use Food* Berries canned for future use. (188:103) **Okanagon** *Staple* Berries used as a principal food. (178:239) **Oweekeno** *Special Food* Berries picked for feasts. (43:99) **Paiute** *Dried Food* Berries eaten dried. *Fruit* Berries eaten fresh. *Preserves* Berries canned or refrigerated for future use. (111:102) **Shuswap** *Fruit* Berries used for food. (123:63) **Skagit, Upper** *Dried Food* Berries pulped, dried, and stored for winter use. *Fruit* Berries eaten fresh. (179:38) **Spokan** *Fruit* Berries used for food. (178:343) **Thompson** *Dried Food*

Berries dried or canned for future use. *Frozen Food* Berries frozen for future use. (187:218) *Fruit* Sweet berries eaten as a favorite food. (164:490) Berries eaten fresh. *Preserves* Berries made into jam. (187:218)

Vaccinium myrsinites Lam., Shiny Blueberry
- *Drug*—**Seminole** *Analgesic* and *Antidiarrheal* Infusion of leaves taken for sun sickness: eye disease, headache, high fever and diarrhea. (169:208) *Ceremonial Medicine* Infusion of plant added to food after a recent death. (169:342) *Cold Remedy* Infusion of plant taken for colds. (169:283) *Emetic* Plant used as an emetic during religious ceremonies. (169:409) *Eye Medicine* and *Febrifuge* Infusion of leaves taken for sun sickness: eye disease, headache, high fever, and diarrhea. (169:208) Infusion of plant taken for fevers. (169:283) *Pediatric Aid* Plant used for chronically ill babies. (169:328) *Stimulant* Decoction of roots used as a bath for hog sickness: unconsciousness. (169:229) *Unspecified* Plant used for medicinal purposes. (169:162)
- *Food*—**Seminole** *Unspecified* Plant used for food. (169:494)

Vaccinium myrtilloides Michx., Velvetleaf Huckleberry
- *Drug*—**Cree, Woodlands** *Abortifacient* Decoction of leafy stems, or of plant, used to bring menstruation. *Contraceptive* Decoction of stems used to prevent pregnancy. *Diaphoretic* Decoction of leafy stem, or of plant, used to make a person sweat. *Gynecological Aid* Decoction of leafy stems, or of plant, used as a "woman's medicine," used to bring blood after childbirth, and used to slow excessive menstrual bleeding. *Reproductive Aid* Decoction of leafy stems taken to prevent miscarriage. Decoction of plant taken to prevent miscarriage. (109:63) **Potawatomi** *Unspecified* Root bark used for unspecified ailment. (as *V. canadense* 154:57)
- *Food*—**Abnaki** *Fruit* Fruit used for food. (*Vaccinium canadense* 144:152) Fruits eaten for food. (as *V. canadense* 144:171) **Algonquin, Quebec** *Fruit* Fruit gathered to eat and sell.[82] (18:103) **Algonquin, Tête-de-Boule** *Unspecified* Flowers eaten as food. (as *V. canadense* 132:133) **Cree, Woodlands** *Fruit* Berries eaten raw. Sun dried berries boiled or pounded into pemmican and eaten. *Preserves* Berries made into jam and eaten with fish and bannock. (109:63) **Hesquiat** *Pie & Pudding* Berries ordered and used to make pies. *Preserves* Berries ordered and used to make preserves. (185:67) **Hoh** *Fruit* Berries eaten raw. *Sauce & Relish* Berries stewed and made into a sauce. *Winter Use Food* Berries canned and used as a winter food. (as *V. macrophyllum* 137:67) **Iroquois** *Preserves* Flowers used to make preserves. *Unspecified* Flowers eaten fresh. (as *V. canadense* 142:96) **Ojibwa** *Dried Food* Berries sun dried for winter use. *Fruit* Berries eaten fresh. *Winter Use Food* Berries canned for future use. (as *V. canadense* 135:238) **Potawatomi** *Dried Food* Berries and low sweet blueberry were important items of food and used dried. *Fruit* Berries and low sweet blueberry were important items of food and used fresh or canned. (as *V. canadense* 154:99) **Quileute** *Fruit* Berries eaten raw. *Sauce & Relish* Berries stewed and made into a sauce. *Winter Use Food* Berries canned and used as a winter food. (as *V. macrophyllum* 137:67) **Thompson** *Pie & Pudding* Berries made into pies. (187:218)
- *Dye*—**Cree, Woodlands** *Unspecified* Berries used to dye porcupine quills. (109:63)
- *Other*—**Algonquin, Quebec** *Cash Crop* Fruit gathered to eat and sell.[82] (18:103) **Ojibwa** *Cash Crop* Berries gathered and sold to the nearby stores. (as *V. canadense* 135:238)

Vaccinium myrtillus L., Whortleberry
- *Food*—**Okanagon** *Staple* Berries used as a principal food. (178:239)

Vaccinium myrtillus var. *oreophilum* (Rydb.) Dorn, Whortleberry
- *Food*—**Thompson** *Fruit* Small, black berries eaten. (as *V. oreophilum* 164:486)

Vaccinium ovalifolium Sm., Ovalleaf Blueberry
- *Drug*—**Makah** *Gynecological Aid* Infusion of leaves and sugar given to mothers after childbirth. (67:305)
- *Food*—**Alaska Native** *Bread & Cake* Berries eaten raw or cooked in pies, puddings, and muffins. *Dietary Aid* Berries used as a fair source of vitamin C. *Frozen Food* Berries frozen or canned for winter use. *Fruit* Berries eaten raw or cooked in pies, puddings, and muffins. *Pie & Pudding* Berries eaten raw or cooked in pies, puddings, and muffins. *Winter Use Food* Berries frozen or canned for winter use. (85:107) **Bella Coola** *Bread & Cake* Berries formerly dried in cakes and used for food. *Fruit* Berries used for food. (as *V. alaskense* 184:205) **Chinook, Lower** *Dried Food* Berries dried and eaten. *Fruit* Berries eaten fresh. (79:44) **Clallam** *Dried Food* Berries eaten dried. *Fruit* Berries eaten fresh. (57:200) **Haisla & Hanaksiala** *Fruit* Berries used for food. (43:245) **Hanaksiala** *Dried Food* Berries dried for winter use. *Fruit* Berries eaten fresh. (as *V. alaskaense* 43:243) **Hesquiat** *Fruit* Berries eaten with oil of whale, dogfish, hair seal, or sea lion. (as *V. alaskaense* 185:65) Berries eaten with oil. *Preserves* Berries preserved or made into jam. (185:67) **Hoh** *Fruit* Berries eaten raw. *Sauce & Relish* Berries stewed and made into a sauce. *Winter Use Food* Berries canned and used as a winter food. (137:68) **Kitasoo** *Bread & Cake* Berries dried into cakes and eaten. (as *V. alaskaense* 43:334) Berries dried into cakes and eaten. (43:335) *Fruit* Berries eaten fresh. (as *V. alaskaense* 43:334) Berries eaten fresh. (43:335) **Klallam** *Dried Food* Berries dried and eaten. *Fruit* Berries eaten fresh. (79:44) **Kwakiutl, Southern** *Fruit* Berries used for food. (as *V. alaskaense* 183:283) Berries used for food. (183:284) **Makah** *Bread & Cake* Fruit formed into cakes, dried, and stored for future use. (as *V. alaskaense* 67:304) Fruit dried into cakes and stored for future use. (67:305) *Dried Food* Berries dried and eaten. (79:44) *Fruit* Fruit eaten fresh. (as *V. alaskaense* 67:304) Fruit eaten fresh. (67:305) Berries eaten fresh. (79:44) *Winter Use Food* Fruit canned for future use. (as *V. alaskaense* 67:304) Fruit canned for winter use. (67:305) **Nitinaht** *Dried Food* Fruits mashed, poured into rectangular frames to dry, soaked, boiled, and eaten in winter. (as *V. alaskaense* 186:107) Fruits formerly mashed, poured into rectangular frames to dry, soaked, boiled, and eaten in winter. (186:108) *Fruit* Fruits eaten fresh. (as *V. alaskaense* 186:107) Fruits eaten fresh. (186:108) *Special Food* Fruits eaten at impromptu village feasts. Fruits eaten at impromptu village feasts. (186:107) **Oweekeno** *Fruit* Berries used for food. *Winter Use Food* Berries preserved for future use. (as *V. alaskaense* 43:97) Berries preserved for winter use. (43:99) **Paiute** *Dried Food* Berries eaten dried. *Fruit* Berries eaten fresh. *Preserves* Berries canned or refrigerated for future use. (111:102) **Quileute** *Dried Food* Berries dried and eaten. *Fruit* Berries eaten fresh. (79:44) Berries eaten raw. *Sauce & Relish* Berries stewed and made into a sauce. *Winter Use Food* Berries canned and used as a winter food. (137:68) **Quinault** *Dried Food* Berries dried and eaten. *Fruit* Berries eaten fresh. (79:44) **Salish, Coast** *Dried Food*

Berries dried and used for food. *Fruit* Berries eaten fresh. (182:83) **Shuswap** *Fruit* Berries used for food. (123:63) **Tanana, Upper** *Fruit* Berries used for food. (115:36) **Thompson** *Bread & Cake* Berries scattered thinly on a mat and dried over a fire or mashed up and dried into a thin cake. (187:220) *Dried Food* Berries soaked, mashed, and dried for winter use. The berries were soaked, mashed, and then placed on drying racks with a small fire lit beneath them to keep away the flies. (as *V. alaskaense* 187:217) *Fruit* Sweet berries eaten as a favorite food. (164:490) Berries eaten fresh. (as *V. alaskaense* 187:217)
- **Dye**—**Hesquiat** *Purple* Berries and devil's club inner bark boiled to make a purple stain. (as *V. alaskaense* 185:65)
- **Other**—**Okanagon** *Smoke Plant* Leaves mixed with other plant leaves and smoked. **Thompson** *Smoke Plant* Leaves mixed with other plant leaves and smoked. (125:39) Leaves smoked as a "kinnikinnick." (164:495)

Vaccinium ovatum Pursh, Evergreen Huckleberry
- **Drug**—**Makah** *Gynecological Aid* Infusion of leaves and sugar given to mothers after childbirth to gain their strength. (67:306) **Pomo, Kashaya** *Misc. Disease Remedy* Decoction of leaves taken for diabetes. (72:60)
- **Food**—**Costanoan** *Fruit* Raw fruit used for food. (21:252) **Hesquiat** *Fruit* Berries eaten with oil. *Preserves* Berries cooked and made into jam or jelly. Berries stored with water in jars. (185:67) **Hoh** *Fruit* Berries eaten raw. *Sauce & Relish* Berries stewed and made into a sauce. *Winter Use Food* Berries canned and used as a winter food. (137:67) **Karok** *Fruit* Fresh berries used for food. (6:60) *Winter Use Food* Berries stored in baskets for future use. (148:388) **Makah** *Bread & Cake* Fruit dried into cakes and stored for future use. *Fruit* Fruit used for food. (67:306) **Nitinaht** *Fruit* Ripe berries used for food. *Pie & Pudding* Berries used in pies. (186:108) **Poliklah** *Fruit* Berries used extensively for food. (117:173) **Pomo** *Dried Food* Dried berries stored in large coiled baskets, boiled, and eaten. (66:15) *Fruit* Berries eaten fresh. (41:377) Raw or stone boiled berries used for food. (66:15) *Pie & Pudding* Berries made into pies and eaten. (41:377) **Pomo, Kashaya** *Dried Food* Berries eaten dried. *Fruit* Berries eaten fresh. *Pie & Pudding* Berries used for dumplings, pies, puddings, and toppings. (72:60) **Quileute** *Fruit* Berries eaten raw. *Sauce & Relish* Berries stewed and made into a sauce. *Winter Use Food* Berries canned and used as a winter food. (137:67) **Quinault** *Fruit* Berries eaten fresh. (79:44) **Salish** *Fruit* Berries used for food. (182:83) **Thompson** *Fruit* Fruit used for food. (187:220) **Tolowa** *Dried Food* Berries dried and stored. *Fruit* Fresh berries used for food. (6:60)
- **Other**—**Karok** *Containers* Leaves used to cover soap plant bulbs in the earth oven. (148:388)

Vaccinium oxycoccos L., Small Cranberry
- **Drug**—**Mohegan** *Unspecified* Plant used medicinally for unspecified purpose. (as *Oxycoccus microcarpus* 176:130) **Ojibwa** *Antiemetic* Infusion of plant taken by person with slight nausea. (153:369)
- **Food**—**Alaska Native** *Fruit* Fruit eaten raw. Fruit cooked and used for food. (85:103) **Algonquin, Quebec** *Fruit* Berries used for food. (18:105) **Algonquin, Tête-de-Boule** *Fruit* Fruits eaten for food. (132:134) **Anticosti** *Winter Use Food* Fruits stored for winter use. (143:68) **Clallam** *Beverage* Leaves used to make tea. (57:200) **Cree, Woodlands** *Fruit* Fresh berries used for food. Berries stewed and eaten with smoked fish. *Winter Use Food* Berries picked in the fall and stored outside in birch bark containers for winter use. (as *Oxycoccus quadripetalus* 109:47) **Eskimo, Alaska** *Fruit* Berries eaten occasionally, but not considered an important food source. (as *Oxycoccus microcarpus* 1:37) **Eskimo, Inupiat** *Dessert* Berries whipped with frozen fish eggs and eaten as a frozen dessert. Raw berries mashed with canned milk and seal oil into a dessert. *Fruit* Berries cooked with fish eggs, fish (whitefish, sheefish, or pike), blubber, and eaten. *Pie & Pudding* Berries boiled with sugar, water, and flour into a pudding. *Sauce & Relish* Berries boiled with dried fruit and eaten with meat or used as topping for ice cream, yogurt, or cake. Whole or mashed berries used cooked or raw, whipped with fat, and made into a sauce. Berries boiled with sugar, water, and flour into a topping for hot cakes or bread. Berries boiled with sugar, water, and flour and eaten with meats. *Winter Use Food* Berries boiled, cooled, blackberries or blueberries added, and stored for winter use. (98:104) **Haisla & Hanaksiala** *Winter Use Food* Berries boiled and stored in barrels of oolichan (candlefish) grease for winter use. (43:247) **Hesquiat** *Forage* Berries eaten by geese. *Fruit* Raw berries, without sugar, eaten with oil. *Preserves* Berries stored with water in jars. Berries made into jam. (185:67) **Iroquois** *Bread & Cake* Fruit mashed, made into small cakes, and dried for future use. *Dried Food* Raw or cooked fruit sun or fire dried and stored for future use. *Fruit* Dried fruit taken as a hunting food. *Sauce & Relish* Dried fruit cakes soaked in warm water and cooked as a sauce or mixed with corn bread. (196:128) **Kitasoo** *Fruit* Berries used for food. (43:336) **Klallam** *Fruit* Berries stored in boxes or baskets until soft and brown and used for food. (79:45) **Makah** *Fruit* Fruit eaten fresh. (67:307) Berries stored in boxes or baskets until soft and brown and used for food. (79:45) *Pie & Pudding* Fruit used to make pies. *Preserves* Fruit used to make jam and jellies. *Winter Use Food* Fruit canned for future use. (67:307) **Menominee** *Fruit* Berries sweetened with maple sugar and eaten. (151:65) **Nitinaht** *Fruit* Berries formerly eaten in fall. (186:109) **Ojibwa** *Fruit* Fruit used for food. (135:238) This was an important wild food. (153:401) **Oweekeno** *Fruit* Berries used for food. (43:100) **Potawatomi** *Fruit* Berries sweetened with maple sugar and always used as an article of food. (154:99) **Quinault** *Fruit* Berries stored in boxes or baskets until soft and brown and used for food. (79:45) **Salish, Coast** *Dried Food* Fruits dried and used for food. *Fruit* Berries eaten fresh. (182:83) **Tanana, Upper** *Frozen Food* Berries frozen for future use. Berries eaten raw, plain, or mixed raw with sugar, grease, or the combination of the two. Berries fried in grease with sugar or dried fish eggs. Berries boiled with sugar and flour to thicken. *Pie & Pudding* Berries used to make pies. *Preserves* Berries used to make jam and jelly. *Winter Use Food* Berries preserved alone or in grease and stored in a birch bark basket in an underground cache. (as *Oxycoccus microcarpus* 102:10) **Thompson** *Fruit* Fresh fruit used for food. This fruit was not dried because it remained fresh for a long time and could be picked any time until winter. (187:221)
- **Other**—**Ojibwa** *Cash Crop* Fruit sold by the bushels. (135:238)

Vaccinium parvifolium Sm., Red Huckleberry
- **Drug**—**Skagit** *Cold Remedy* Decoction of bark taken for colds. (79:44) **Skagit, Upper** *Cold Remedy* Decoction of bark taken for colds. (179:38)

- **Food**—**Alaska Native** *Preserves* Fruit used to make a very superior jelly. (85:105) **Bella Coola** *Bread & Cake* Berries formerly dried in cakes and used for food. *Frozen Food* Berries frozen and used for food. *Fruit* Berries used for food. (184:205) **Clallam** *Fruit* Berries eaten fresh. (57:200) **Haisla & Hanaksiala** *Fruit* Fruit eaten, sometimes at feasts. *Special Food* Fruit eaten, sometimes at feasts. (43:248) **Hesquiat** *Beverage* Berries made excellent wine. *Fruit* Raw berries eaten with oil. *Preserves* Berries used for jam. (185:67) **Hoh** *Fruit* Berries eaten raw. *Sauce & Relish* Berries stewed and made into a sauce. *Winter Use Food* Berries canned and used as a winter food. (137:68) **Karok** *Fruit* Berries eaten raw. (148:388) **Kitasoo** *Fruit* Berries used for food. (43:337) **Klallam** *Fruit* Berries used for food. (79:44) **Kwakiutl, Southern** *Special Food* Berries boiled, mixed with red salmon spawn and oil and eaten at feasts in winter ceremonies. (183:284) **Lummi** *Fruit* Berries used for food. (79:44) **Makah** *Frozen Food* Fruit frozen for future use. *Fruit* Fruit eaten fresh. (67:308) Berries used for food. (79:44) *Winter Use Food* Fruit canned for winter use. (67:308) **Nitinaht** *Dried Food* Berries mashed, poured into frames, dried into cakes, soaked, boiled, and eaten with oil or syrup. *Frozen Food* Berries frozen and used for food. *Fruit* Berries eaten fresh. (186:109) *Special Food* Fruits eaten at impromptu village feasts. (186:107) **Oweekeno** *Winter Use Food* Berries preserved for future use. (43:101) **Paiute** *Dried Food* Berries eaten dried. *Fruit* Berries eaten fresh. (111:103) **Poliklah** *Fruit* Berries used extensively for food. (117:173) **Pomo, Kashaya** *Fruit* Berries eaten fresh. (72:61) **Quileute** *Fruit* Berries used for food. (79:44) Berries eaten raw. *Sauce & Relish* Berries stewed and made into a sauce. *Winter Use Food* Berries canned and used as a winter food. (137:68) **Quinault** *Beverage* Leaves used to make tea. *Fruit* Berries used for food. (79:44) **Salish, Coast** *Dried Food* Fruits dried and used for food. *Fruit* Berries eaten fresh or cooked. (182:83) **Skagit** *Fruit* Berries used for food. (79:44) **Skagit, Upper** *Fruit* Berries eaten fresh. (179:38) **Skokomish** *Fruit* Berries used for food. **Snohomish** *Fruit* Berries used for food. **Swinomish** *Fruit* Berries used for food. (79:44) **Thompson** *Bread & Cake* Berries used in pancakes and muffins. (187:221) *Fruit* Bright red, acidic berries eaten in large quantities. (164:490) *Preserves* Berries used to make jam. (187:221) **Tolowa** *Dried Food* Berries dried and stored. *Fruit* Fresh berries used for food. **Yurok** *Fruit* Fresh berries used for food. (6:61)
- **Fiber**—**Karok** *Brushes & Brooms* Stems and twigs used to make brooms. (148:388) **Poliklah** *Brushes & Brooms* Long, straight, green branches used for brooms. (117:173) **Tolowa** *Brushes & Brooms* Branches used as a broom. **Yurok** *Brushes & Brooms* Branches used as brooms. (6:61)
- **Other**—**Kwakiutl, Southern** *Fasteners* Twigs used to fasten skunk cabbage leaves onto berry baskets. (183:286) **Thompson** *Smoke Plant* Leaves smoked as a "kinnikinnick." (164:495)

Vaccinium scoparium Leib. ex Coville, Grouse Whortleberry

- **Drug**—**Cheyenne** *Antiemetic* Infusion of dried, pulverized leaves and stems taken for nausea. *Dietary Aid* Dried, pulverized berries given to children with poor appetites. Infusion of dried, pulverized leaves and stems taken to increase appetite. *Pediatric Aid* Dried, pulverized berries given to children with poor appetites. (76:183)
- **Food**—**Klamath** *Dried Food* Dried berries used for food. *Fruit* Fresh berries used for food. (45:103) **Okanagan-Colville** *Fruit* Berries used for food. (188:105) **Thompson** *Dried Food* Berries dried loose like raisins. *Pie & Pudding* Berries used in pies. *Preserves* Berries used in jams. *Winter Use Food* Berries canned and used for food. (187:217) **Yurok** *Fruit* Fresh berries used for food. (6:61)
- **Fiber**—**Yurok** *Brushes & Brooms* Branches used as brooms. (6:61)

Vaccinium sp., Blueberry

- **Drug**—**Alabama** *Panacea* Plant used for many unspecified ailments. (172:664) **Eskimo, Western** *Antidiarrheal* Decoction of berries taken for diarrhea. (108:15) **Iroquois** *Dermatological Aid* Compound decoction used as wash for parts affected by "Italian itch." *Other* Compound decoction of plants taken for stricture. (87:411) **Makah** *Kidney Aid* Infusion of leaves used for kidney disorders. (67:310)
- **Food**—**Carrier** *Dried Food* Berries dried for future use. (31:73) **Chippewa** *Dried Food* Fruits of several different species dried for winter use. *Fruit* Fruits of several different species eaten fresh. (71:139) **Iroquois** *Bread & Cake* Fruits dried, soaked in water, and used in breads. *Pie & Pudding* Fruits dried, soaked in water, and used in pudding. *Porridge* Berries dried, soaked in cold water, heated slowly, and mixed with bread meal or hominy in winter. *Sauce & Relish* Fruits dried, soaked in water, and used as a sauce. Berries dried, soaked in cold water, heated slowly, and used as a winter sauce. *Soup* Fruits dried, soaked in water, and used in soups. (124:96) **Makah** *Bread & Cake* Berries formerly dried into cakes. *Winter Use Food* Berries canned for future use. (67:310) **Sanpoil & Nespelem** *Dried Food* Berries eaten sun dried. *Fruit* Berries eaten fresh. (131:102) **Thompson** *Fruit* Bluish black berries eaten in large quantities. (164:487) Berries eaten. (164:490) *Preserves* Berries collected in large quantities and cured. (178:237)
- **Other**—**Thompson** *Smoke Plant* Leaves of several species smoked as "kinnikinnick." (164:495)

Vaccinium uliginosum L., Bog Blueberry

- **Drug**—**Makah** *Gynecological Aid* Infusion of leaves and sugar given to mothers a few days after childbirth to gain their strength. (67:309)
- **Food**—**Alaska Native** *Bread & Cake* Berries cooked in muffins. *Dietary Aid* Berries used as a fair source of vitamin C. *Frozen Food* Berries frozen for winter use. *Fruit* Berries eaten raw. *Pie & Pudding* Berries cooked in pies and puddings. *Winter Use Food* Berries frozen or canned for winter use. (85:107) **Eskimo, Alaska** *Fruit* Fresh berries used for food. (1:37) Fruit used for food. (4:715) **Eskimo, Inuktitut** *Fruit* Berries used for food. (202:184) Berries used for food. (202:186) **Eskimo, Inupiat** *Cooking Agent* Berry juice made into a vinegar and used to pickle meats and greens. *Dessert* Fresh or frozen berries mixed with sugar and seal oil and eaten as a dessert. *Frozen Food* Berries frozen for future use. *Fruit* Berries eaten fresh. Berries mixed with raw, fresh fish eggs and eaten with seal oil and sugar. Berries mixed with sourdock and fermented. Berries mixed with potatoes, cabbage, or lettuce and pickled. Berries mixed with blubber and blackberries and eaten with or without sugar. *Ice Cream* Fresh or frozen berries used to make ice cream or yogurt. *Pie & Pudding* Stored berries used to make traditional desserts and pies. Berries and water boiled, flour paste, sugar, or honey added, and eaten hot or cold as a pudding. *Sauce & Relish* Berry pudding used as a topping for hot cakes, bread, or desserts. *Winter Use Food* Berries boiled with water, mixed with blackberries, and stored in a poke or barrel for winter

use. (98:78) **Haisla & Hanaksiala** *Fruit* Fruit used for food. (43:249) **Kitasoo** *Bread & Cake* Berries dried into cakes and eaten. *Fruit* Berries eaten fresh. (43:335) **Koyukon** *Frozen Food* Berries frozen for winter use. (119:54) **Kwakiutl, Southern** *Fruit* Berries used for food. (183:284) **Oweekeno** *Winter Use Food* Berries preserved for winter use. (43:99) **Salish, Coast** *Dried Food* Berries dried and used for food. *Fruit* Berries eaten fresh. (182:83) **Tanana, Upper** *Fruit* Berries used for food. (77:28) Berries used for food. (as *V. uglinosum* 115:36) *Preserves* Berries gathered and preserved in quantity. (102:9) *Winter Use Food* Berries mixed with grease and preserved in caches. (77:28)

Vaccinium vitis-idaea L., Lingonberry
- *Drug*—**Tanana, Upper** *Cold Remedy* Berries eaten raw or juice used for colds. *Cough Medicine* Berries eaten raw or juice used for coughs. *Throat Aid* Raw berries chewed or juice gargled for sore throat. (102:9)
- *Food*—**Alaska Native** *Preserves* Berries mixed with rose hip pulp and sugar to make jam. *Sauce & Relish* Berries cooked as a sauce. *Winter Use Food* Berries stored for future use. (85:109) **Carrier** *Preserves* Berries used to make jam. (31:76) **Eskimo, Arctic** *Beverage* Juice diluted and sweetened to make a refreshing beverage. *Frozen Food* Berries frozen and stored until the next spring. *Preserves* Berries used to make jams and jellies. (128:22) **Eskimo, Inupiat** *Dessert* Berries whipped with frozen fish eggs and eaten as a frozen dessert. Raw berries mashed with canned milk and seal oil into a dessert. *Fruit* Berries cooked with fish eggs, fish (whitefish, sheefish, or pike), blubber, and eaten. *Pie & Pudding* Berries boiled with sugar, water, and flour into a pudding. *Sauce & Relish* Berries boiled with dried fruit and eaten with meat or used as topping for ice cream, yogurt, or cake. Whole or mashed berries used cooked or raw, whipped with fat, and made into a sauce. Berries boiled with sugar, water, and flour into a topping for hot cakes or bread. Berries boiled with sugar, water, and flour and eaten with meats. *Winter Use Food* Berries boiled, cooled, blackberries or blueberries added, and stored for winter use. (98:86) **Koyukon** *Frozen Food* Berries frozen for winter use. (as *V. vitis* 119:55) **Tanana, Upper** *Frozen Food* Berries frozen for future use. (102:9) *Fruit* Berries used for food. (77:28) Berries eaten raw, plain, or mixed raw with sugar, grease, or the combination of the two. Berries fried in grease with sugar or dried fish eggs. Berries boiled with sugar and flour to thicken. (102:9) Berries used for food. (115:36) *Pie & Pudding* Berries used to make pies. *Preserves* Berries used to make jam and jelly. (102:9) *Winter Use Food* Berries preserved in caches. (77:28) Berries preserved alone or in grease and stored in a birch bark basket in an underground cache. (102:9)

Vaccinium vitis-idaea ssp. *minus* (Lodd.) Hultén, Northern Mountain Cranberry
- *Food*—**Anticosti** *Preserves* Fruit used to make jams and jellies. (143:68) **Cree, Woodlands** *Frozen Food* Berries stored during the winter by freezing outside. *Fruit* Berries mixed with boiled fish eggs, livers, air bladders, and fat, and eaten. *Snack Food* Berries eaten raw as a nibble. *Soup* Berries stewed and served with fish or meat. (109:64) **Eskimo, Alaska** *Fruit* Berries eaten occasionally, but not considered an important food source. (1:37) Fruit used for food. (4:715) **Eskimo, Inuktitut** *Fruit* Berries used for food. (202:183) **Haida** *Fruit* Berries used for food. **Hesquiat** *Fruit* Berries used for food. **Oweekeno** *Fruit* Berries used for food. **Tsimshian** *Fruit* Berries used for food. (43:101)
- *Dye*—**Cree, Woodlands** *Unspecified* Berries used to color porcupine quills. (109:64)
- *Other*—**Cree, Woodlands** *Jewelry* Firm, ripe berries strung on a string to make a necklace. (109:64) **Eskimo, Inuktitut** *Smoke Plant* Leaves used as a tobacco additive or substitute. (202:183)

Valerianaceae, see *Valeriana*, *Valerianella*

Valeriana, Valerianaceae

Valeriana acutiloba Rydb., Sharpleaf Valerian
- *Drug*—**Navajo, Ramah** *Cough Medicine* Plant used for cough. *Hunting Medicine* Plant rubbed on hunter's body for good luck in hunting. *Misc. Disease Remedy* Plant used for influenza. *Tuberculosis Remedy* Plant used for tuberculosis. *Witchcraft Medicine* Plant used for protection from witches. (191:45)

Valeriana capitata Pallas ex Link, Capitate Valerian
- *Drug*—**Eskimo** *Gastrointestinal Aid* Plant used for stomach troubles. (149:325) **Eskimo, Inuktitut** *Gastrointestinal Aid* Used for "stomach troubles." (202:182)
- *Other*—**Tanana, Upper** *Incense & Fragrance* Dried roots used as incense at potlatches. (102:18)

Valeriana dioica var. *sylvatica* S. Wats., Woods Valerian
- *Drug*—**Blackfoot** *Gastrointestinal Aid* Decoction of roots taken for stomach troubles. (as *V. septentrionalis* 97:56) Hot drink made from root taken for stomach trouble. (as *V. septentrionalis* 114:275) Infusion of roots taken for stomach trouble. (as *V. septentrionalis* 118:45) *Veterinary Aid* Decoction of roots given to horses for colic or distemper. (as *V. septentrionalis* 97:56) **Carrier, Northern** *Dermatological Aid* Oil of blossoms mixed with bear fat and used as a hair tonic to help growth. (as *V. septentrionalis* 150:64) **Cree, Woodlands** *Analgesic* Chewed roots rubbed on the head and temples for headaches. *Anticonvulsive* Poultice of roots applied externally, especially to babies, for seizures. *Cold Remedy* Roots used in a smoking mixture for colds. *Ear Medicine* Poultice of chewed roots applied to the ear for earaches. *Gynecological Aid* Powdered roots and many other herbs used for menstrual troubles. *Love Medicine* Roots used as a love medicine to attract a person whose affections were desired. It was a kind of bad medicine and viewed as a curse. The root of this plant was said by some to be an ingredient in love medicine; others denied this. *Panacea* Powdered roots and many other herbs used for various ailments. Infusion of roots used as an all purpose medicine. *Pediatric Aid* Poultice of roots applied externally, especially to babies, for seizures. *Pulmonary Aid* Decoction of roots used for pneumonia. (as *V. septentrionalis* 109:64) **Gosiute** *Poison* Roots used for arrow poison. (as *V. sylvatica* 39:384) **Thompson** *Adjuvant* Dried, powdered roots and leaves mixed with tobacco as a flavoring. (as *V. sylvatica* 164:495) *Analgesic* Decoction of roots taken for pains. (as *V. sylvatica* 164:460) *Antidiarrheal* Decoction of roots taken for diarrhea. (187:290) *Ceremonial Medicine* Root used as ceremonial medicine for unspecified purpose. *Cold Remedy* Decoction of roots taken for colds. (as *V. sylvatica* 164:460) Bitter infusion of roots taken for colds. Decoction of roots taken for colds. (187:290) *Dermatological Aid* Poultice of various plant parts used on cuts, wounds, bruises, or inflammations. (as *V. sylvatica* 164:460) Chewed leaves spat on cuts and

bruises. *Gastrointestinal Aid* Decoction of roots, leaves, stems, and flowers taken for ulcers and stomach trouble. (187:290) *Hunting Medicine* Decoction of whole plant taken and used as a wash by hunters for good luck. (as *V. sylvatica* 164:506) *Misc. Disease Remedy* Decoction of roots, leaves, stems, and flowers taken for influenza. *Tuberculosis Remedy* Decoction of roots taken for "shadow on your lung." (187:290) *Veterinary Aid* Decoction of roots used as a lotion or wash for swellings or sores on horses. (as *V. sylvatica* 164:513)
- *Food*—Thompson *Forage* Leaves eaten by deer. (187:290)

Valeriana edulis Nutt. ex Torr. & Gray, Edible Valerian
- *Drug*—Blackfoot *Poison* Raw roots considered poisonous. (97:56) Gosiute *Antirheumatic (External)* Pounded roots rubbed on parts affected by rheumatism. (39:350, 384) *Dermatological Aid* Pounded roots rubbed on skin for swollen bruises. (39:349) Pounded roots rubbed on swollen and bruised parts. (39:384) Menominee *Analgesic* Poultice of pulverized root applied to painful, bleeding cuts and wounds. *Anthelmintic* Root used as a tapeworm medicine. *Dermatological Aid* Poultice of pulverized root applied to draw out inflammation of boils. Poultice of pulverized root applied to painful, bleeding cuts and wounds. *Hemostat* Pulverized root placed on painful and bleeding cuts. (151:57) Meskwaki *Hemostat* Root used for hemorrhages. (152:251) Snake *Poison* Raw root poisonous and cooked root used for food. (19:26)
- *Food*—Klamath *Unspecified* Steamed, cooked roots used for food. (45:104) Montana Indian *Unspecified* Large quantities of roots cooked in a kiln until black and very sticky. (19:26) Okanagan-Colville *Winter Use Food* Roots stored in underground pits for about a year. (188:142) Paiute *Unspecified* Roots cooked overnight and eaten. (104:103) Boiled root used for food. (118:16) Sanpoil & Nespelem *Unspecified* Roots used for food. (188:142)

Valeriana sitchensis Bong., Sitka Valerian
- *Drug*—Okanagon *Analgesic* Decoction of roots taken for pains. *Cold Remedy* Decoction of roots taken for colds. *Dermatological Aid* Poultice of roots applied to cuts, wounds, bruises, and inflamed regions. (125:40) Thompson *Adjuvant* Dried, powdered roots and leaves mixed with tobacco as a flavoring. (164:495) *Analgesic* Decoction of roots taken for pains. (125:40) Decoction of root taken for pains. (164:460) *Antidiarrheal* Decoction of roots taken for diarrhea. (187:290) *Ceremonial Medicine* Root used as ceremonial medicine for unspecified purpose. (164:460) *Cold Remedy* Decoction of roots taken for colds. (125:40) Decoction of roots taken as a cold remedy. (164:460) Decoction of roots taken for colds. Bitter infusion of roots used for colds. (187:290) *Dermatological Aid* Poultice of roots applied to cuts, wounds, bruises, and inflamed regions. (125:40) Poultice of various plant parts used on cuts, wounds, bruises, or inflammations. (164:460) Chewed leaves spat on cuts and bruises. *Gastrointestinal Aid* Decoction of roots, leaves, stems, and flowers taken for ulcers or stomach trouble. (187:290) *Hunting Medicine* Decoction of whole plant taken and used as a wash by hunters for good luck. (164:506) *Misc. Disease Remedy* Decoction of roots, leaves, stems, and flowers taken for influenza. *Tuberculosis Remedy* Decoction of roots taken for "shadow on your lung." (187:290) *Veterinary Aid* Decoction of roots used as a lotion or wash for swellings or sores on horses. (164:513)
- *Food*—Thompson *Forage* Leaves eaten by deer. (187:290)
- *Other*—Blackfoot *Smoke Plant* Leaves sometimes used in the tobacco mixture. (97:56) Okanagon *Smoke Plant* Leaves mixed with other plant leaves and smoked. (125:39) Tanana, Upper *Incense & Fragrance* Dried roots used as incense at potlatches. (102:18) Thompson *Smoke Plant* Leaves mixed with other plant leaves and smoked. (125:39)

Valeriana sp.
- *Drug*—Gosiute *Other* Root salve applied to limbs for edema. (39:349)

Valeriana uliginosa (Torr. & Gray) Rydb., Mountain Valerian
- *Drug*—Menominee *Analgesic* Infusion of root taken for cramps and "disorders of the head." *Cathartic* Large doses of plant produced purging. *Dermatological Aid* Poultice of pulverized root applied to cuts and wounds. *Hunting Medicine* Root chewed and spit on fishhook to lure fish to Indian, but not to white man. *Psychological Aid* Large doses of plant produced mental stupor. *Pulmonary Aid* Infusion of root taken for "disorders of the throat and lungs." *Sedative* Plant used as a "feeble sedative to the nervous system." *Throat Aid* Infusion of root taken for "disorders of the throat and lungs." (151:57)
- *Other*—Menominee *Good Luck Charm* Root held in mouth while arguing to keep opponent from winning argument. (151:57)

Valerianella, Valerianaceae
Valerianella locusta (L.) Lat., Lewiston Cornsalad
- *Food*—Cherokee *Vegetable* Leaves cooked and eaten as greens. (as *V. olitoria* 80:30) Leaves used as a potherb. (as *V. olitoria* 126:59)

Vanclevea, Asteraceae
Vanclevea stylosa (Eastw.) Greene, Pillar False Gumweed
- *Drug*—Navajo, Kayenta *Dermatological Aid* Compound poultice of plants applied to tarantula or solpugid (wind scorpion) bites. (205:51)

Vancouveria, Berberidaceae
Vancouveria chrysantha Greene, Golden Insideout Flower
- *Drug*—Tolowa *Unspecified* Used for medicine. (6:61)

Vancouveria hexandra (Hook.) Morr. & Dcne., White Insideout Flower
- *Drug*—Yurok *Cough Medicine* Leaves eaten for coughs. (6:61)

Veratrum, Liliaceae
Veratrum californicum Dur., California False Hellebore
- *Drug*—Paiute *Antirheumatic (External)* Poultice of root applied for rheumatism. Pulped root or decoction of root applied with friction as a liniment. (180:147, 14) *Burn Dressing* Plant used for burns. (104:196) *Contraceptive* Decoction of root taken as a contraceptive "to insure permanent sterility." Decoction of root taken by both men and women as a contraceptive. (180:147, 148) *Dermatological Aid* Poultice of dried, pounded roots applied to bruises. (104:196) Dry, powdered root sprinkled on sores to promote healing. Poultice of mashed, raw root applied to boils, sores, cuts, or swellings. *Disinfectant* Poultice of root applied for infections. (180:147, 148) *Febrifuge* Infusion of plant used as a febrifuge. (104:197) *Gland Medicine* Poultice of root applied for enlarged throat glands and blood poisoning. *Gynecological Aid*

Poultice of root applied for sore nipples. (180:147, 148) *Panacea* Plant used for a variety of maladies. *Snakebite Remedy* Poultice of chewed roots applied to snakebites. (104:195) Poultice of pulped root applied to snakebites. *Throat Aid* Poultice of root applied for sore throats and enlarged glands from tonsillitis. (180:147, 148) *Toothache Remedy* Powdered roots rubbed on the face for toothaches. (104:197) *Venereal Aid* Decoction of roots taken for venereal disease. (104:198) Decoction of root taken for venereal disease. (180:147, 148) *Veterinary Aid* Poultice of mashed roots applied to saddle sores on horses. (111:54) **Paiute, Northern** *Antirheumatic (External)* Poultice of mashed roots applied to swollen arms and legs. *Cold Remedy* Roots grated, chewed, and the juice swallowed for colds. *Dermatological Aid* Poultice of mashed roots applied to rattlesnake bites to draw out the poison. Poultice of grated roots applied to cuts, sores, and snakebites. *Orthopedic Aid* Poultice of mashed plant applied to sprains and broken bones. (59:127) **Shoshoni** *Antirheumatic (External)* Poultice of root applied for rheumatism. Pulped root applied with friction as a liniment. *Cold Remedy* Raw root chewed and juice swallowed for "heavy colds." (180:147, 148) *Contraceptive* Infusion of fresh or cured root taken for birth control to ensure sterility for life. (118:46) Decoction of root taken as a contraceptive "to insure permanent sterility." Decoction of root taken by both men and women as a contraceptive. *Dermatological Aid* Poultice of mashed, raw root applied to boils, sores, cuts, or swellings. *Disinfectant* Poultice of root applied for infections. *Gland Medicine* Poultice of root applied for enlarged throat glands and blood poisoning. *Gynecological Aid* Poultice of root applied for sore nipples. *Snakebite Remedy* Poultice of pulped root applied to snakebites. *Throat Aid* Poultice of root applied for sore throats and enlarged glands from tonsillitis. Raw root chewed and juice swallowed for sore throats or inflamed tonsils. (180:147, 148) **Thompson** *Blood Medicine* Decoction of root ashes taken for blood disorders. (164:460) *Poison* Plant considered poisonous if eaten in large quantities. (164:512) *Respiratory Aid* Dried, powdered root used as snuff to cause sneezing to clear a head cold. *Venereal Aid* Decoction of root ashes taken for blood disorders, especially syphilis. (164:460) **Washo** *Antirheumatic (External)* Pulped root applied with friction as a liniment. (180:147, 148) *Contraceptive* Infusion of fresh or cured root taken for birth control to ensure sterility for life. (118:46) *Dermatological Aid* Dry, powdered root sprinkled on sores to promote healing. *Emetic* Decoction of root taken as an emetic. (180:147, 148)

- *Food*—**Miwok** *Unspecified* Roasted in hot ashes, peeled, and eaten. (12:158)
- *Other*—**Karok** *Decorations* Inner white stem torn into ribbons and braided into the girls' hair for ornaments. (148:380) **Paiute** *Preservative* Leaves used to cover huckleberries and keep the berries fresh. (111:54) **Paiute, Northern** *Protection* Roots tied on the ankle or calf to drive away rattlesnakes. (59:127)

Veratrum californicum Dur. **var.** *californicum*, California False Hellebore

- *Drug*—**Blackfoot** *Analgesic* Small piece of root snuffed up nose for headache. *Poison* Root poisonous to eat. (as *V. speciosum* 114:275)
- *Other*—**Blackfoot** *Snuff* Pounded, dry root used for snuff. (as *V. speciosum* 118:62)

Veratrum sp., False Hellebore

- *Drug*—**Shoshoni** *Snakebite Remedy* Poultice of mashed, raw root applied to snakebites. *Veterinary Aid* Poultice of mashed, raw root applied to horses for snakebites. (118:49)

Veratrum viride Ait., American False Hellebore

- *Drug*—**Alaska Native** *Poison* Plant considered poisonous. (as *V. eschscholtzii* 85:161) **Bella Coola** *Analgesic* Decoction of bulbs taken or pieces of root swallowed for stomach pain. *Antirheumatic (External)* Poultice of compound containing bulb applied to arms for rheumatism. *Cough Medicine* Decoction of bulb taken for chronic cough. (150:53) *Dermatological Aid* Roots used for skin washes and compresses for bruises. (74:26) *Emetic* Decoction of bulb taken or raw root eaten as an emetic for stomach pains. (150:53) Outer roots used as an emetic. (as *V. eschscholtzii* 184:199) *Gastrointestinal Aid* Decoction of bulb taken or raw root eaten for stomach pain. *Heart Medicine* Poultice of compound containing bulb applied to chest for heart trouble. *Laxative* Decoction of bulb taken for constipation and cough. (150:53) *Orthopedic Aid* Roots used as compresses for sprains and fractures. (74:26) *Poison* Overdose of raw root considered fatal. *Venereal Aid* Decoction of bulb taken for gonorrhea and chronic cough. (150:53) **Blackfoot** *Analgesic* Poisonous roots dried, crushed, and snuffed for headaches. (as *V. eschscholtzii* 97:25) *Gastrointestinal Aid* Infusion of plant taken for indigestion. (86:69) *Nose Medicine* Stems scraped and the powder snuffed to induce sneezing. (86:74) *Oral Aid* Leaves chewed by children for drooling. (86:105) *Poison* Poisonous roots ingested for suicide. (as *V. eschscholtzii* 97:25) **Carrier, Southern** *Emetic* Infusion of dried, powdered root taken as "an emetic for sickness." *Poison* Infusion of powdered root, strong infusion fatal, taken as an emetic. (150:53) **Cherokee** *Analgesic* Compound used as liniment for pains. (80:40) Infusion of leaves used as wash for aches and pains. *Antirheumatic (External)* Infusion of roots rubbed on "leg scratches" for rheumatism. (177:8) *Orthopedic Aid* Compound used as a liniment for pains or sore muscles. Compound used as liniment for sore muscles. (80:40) *Panacea* Infusion of roots rubbed on "leg scratches" for kindred ailments. *Stimulant* Infusion of roots rubbed on "leg scratches" for languor. (177:8) **Cowlitz** *Analgesic* Poultice of leaves applied for pain. *Poison* Plant considered poisonous. (as *V. eschscholtzii* 79:24) **Flathead** *Nose Medicine* Snuff of dried, powdered rootstocks used to clear the nasal passages. (82:73) **Gitksan** *Hunting Medicine* Roots used in purification rituals for hunting and trapping. *Other* Smoke used to assist the spirits of sleepwalkers to return to their bodies. (74:26) **Haisla** *Antirheumatic (External)* Decoction of cleaned, sliced, dried, and boiled roots used for rheumatism. *Cold Remedy* Plant used as a snuff for colds. *Emetic* Plant used as an emetic. *Hemostat* Poultice of roots applied to stop flow of blood from areas cut to release disease-causing objects. *Laxative* Decoction of cleaned, sliced, dried, and boiled roots used as a laxative. *Sedative* Decoction of cleaned, sliced, dried, and boiled roots used as a sedative. **Haisla & Hanaksiala** *Ceremonial Medicine* Roots put on the ends of arrows by the shaman to shoot towards "disease spirits." *Poison* Plant considered highly toxic. **Hanaksiala** *Antirheumatic (External)* Poultice of roots, rhizomes, and Sitka spruce pitch applied to sore areas. *Antirheumatic (Internal)* Infusion of roots taken for various types of swellings. *Blood Medicine* Plant used for any blood-related disorder. *Hypotensive* Plant used for high blood pressure. *Respiratory Aid* Poultice of roots, rhizomes, and Sitka spruce pitch paste applied to chest for respiratory afflictions. *Unspecified* Decoction of roots used as an internal medicine for unspecified purposes. (43:201) **Iroquois** *Analgesic* Compound of dried plants used as snuff for catarrh, headaches, and colds. *Cold Remedy* Dried plants or powdered roots used as snuff for catarrh, headaches, and colds. *Dermatological Aid* Compound infusion of roots used as a poultice to break open boils. *Respiratory Aid* Dried plants or dried roots used as snuff for catarrh, headaches, and colds. *Tuberculosis Remedy* Dried root used as snuff for tuberculosis. (87:280) **Kitasoo** *Antirheumatic (External)* Decoction of roots and rhizomes applied externally for arthritis. (43:323) **Kutenai** *Nose Medicine* Snuff of dried, powdered rootstocks used to clear the nasal passages. (82:73) **Kwakiutl** *Abortifacient* Juice taken by women to bring about an abortion. *Analgesic* Infusion of scraped roots taken to induce vomiting for internal pains. Poultice of roots or leaves applied to the chest for chest pains. Roots and oil rubbed on bloody back for back pains. *Cold Remedy* Powdered leaves used as snuff for bad colds. *Dermatological Aid* Cold infusion of roots rubbed on the scalp for dandruff. Poultice of scraped roots applied to swellings. *Emetic* Infusion of scraped roots taken to induce vomiting for internal pains. *Hunting Medicine* Leaves rubbed on body to purify hunter. *Laxative* Large, fleshy roots held in the mouth as a laxative. (183:273) *Orthopedic Aid* Compound poultice of roots applied to sore backs. (183:267) Roots and oil rubbed on bloody back for back pains. *Panacea* Braided leaves with the root suspended worn around the neck by sick people. *Pediatric Aid* Root hung around child's neck to ward off disease-causing spirits. *Poison* Secondary roots considered poisonous. (183:273) **Okanagan-Colville** *Antirheumatic (External)* Roots used for rheumatism or arthritis. (74:26) Poultice of raw roots applied carefully to the skin for rheumatic and arthritic pains. *Cold Remedy* Roots dried, powdered, and used as a snuff to cause sneezing and clear the sinuses for colds. *Hunting Medicine* Roots rubbed in each eye and the tip of the nose of hunting dogs during "training." *Orthopedic Aid* Poultice of leaves applied to the skin by hunters for backaches. *Poison* Plant considered poisonous. *Witchcraft Medicine* Plant used in "witchcraft" or "plhax" to jinx people. (188:50) **Okanogan** *Blood Medicine* Decoction of dried, burned roots taken for blood disorders. (125:40) **Oweekeno** *Antirheumatic (External)* Dried roots used in bath water for arthritis or rheumatism. *Poison* Plant poisonous to animals. (43:79) **Quinault** *Antirheumatic (Internal)* Decoction of whole plant taken for rheumatism. *Poison* Plant considered poisonous. (as *V. eschscholtzii* 79:24) **Salish** *Cold Remedy* Dried, powdered roots used as a snuff for colds. *Poison* Plant considered poisonous. (178:294) **Salish, Coast** *Analgesic* Roots used in small doses for internal pains. *Antirheumatic (External)* Roots used in small doses for rheumatism. *Diaphoretic* Roots used in medicinal sweat baths. (182:76) **Shuswap** *Dermatological Aid* Plant used to make hair grow on a bald head. *Poison* Plant considered poisonous. (123:55) **Thompson** *Antirheumatic (External)* Plant used externally for arthritis.[83] Washed, mashed roots rubbed on body parts with water or mixed with snowbrush for arthritis. The informants cautioned never to use this medicine internally or to let it get near the eyes, mouth, or open sores because of its extreme toxicity. Decoction of plant used for sore feet.[83] (187:131) *Blood Medicine* Decoction of dried, burned roots taken for blood disorders. (125:40) *Misc.*

Disease Remedy Poultice of root used for phlebitis.[83] *Orthopedic Aid* Decoction of plant considered good for broken bones.[83] *Pediatric Aid* Burned leaf ashes and Vaseline used as a diaper ointment to prevent babies from messing diapers. The ashes were probably originally mixed with deer or bear fat. The ointment was very strong and had to be used with extreme caution. One informant cautioned that it should not be taken internally because of its extreme toxicity. *Throat Aid* Poultice of boiled roots used for sore throat.[83] (187:131) **Tsimshian** *Antirheumatic (External)* Leaves used for arthritis. *Dermatological Aid* Roots used for scalp disease. *Psychological Aid* Roots used for insanity. *Respiratory Aid* Roots used as a snuff for sinus infections. (74:26)

- *Food*—**Blackfoot** *Soup* Leaves used to make soups. (86:105)
- *Fiber*—**Hanaksiala** *Mats, Rugs & Bedding* Leaves used to wipe the body off after bathing in water. (43:201)
- *Other*—**Bella Coola** *Good Luck Charm* Roots used for luck. (74:26) *Protection* Outer roots hung in homes or boiled for protection from supernatural powers. (as *V. eschscholtzii* 184:199) **Gitksan** *Ceremonial Items* Roots grated, added to laundry water, and used to purify clothing. Roots burned as a smudge or fumigant to purify houses. *Good Luck Charm* Root pieces carried as amulets for luck in hunting and gambling. *Soap* Roots grated, added to laundry water, and used to cleanse clothing. **Haisla** *Ceremonial Items* Roots burned and used as a fumigant to drive away evil spirits. *Good Luck Charm* Roots used as amulets for luck. (74:26) *Protection* Plants presence believed to repel ghosts, illness, and evil. **Hanaksiala** *Ceremonial Items* Plant used for ceremonial purification. Plant used for preparation for shamanistic activities, dancing, hunting, and fishing. *Good Luck Charm* Decoction of plant and devil's club used as a wash for areas occupied by corpse to give good luck. *Protection* Plant used to combat witchcraft. Decoction of plant and devil's club used as a wash for areas occupied by corpse to kill poison. (43:201) **Kitasoo** *Protection* Roots and rhizomes burned and smoke used as protection from ghosts and "demons." (43:323) **Kwakiutl** *Ceremonial Items* Dried roots used as a charm to call rain. *Protection* Root used as a charm for protection against all evils. (183:273) **Oweekeno** *Ceremonial Items* Roots burned to fumigate the houses of recently deceased persons. (43:79) **Salish, Coast** *Ceremonial Items* Roots carried as charms to ward off evil spirits or to kill sea monsters. (182:76)

Verbascum, Scrophulariaceae
Verbascum thapsus L., Common Mullein
- *Drug*—**Abnaki** *Pediatric Aid* Roots used to make a necklace worn by teething babies. (144:171) Used by children for teething. (144:155) *Toothache Remedy* Plant made into a magical necklace and worn by children for teething. (144:154) Roots used to make a necklace worn by teething babies. (144:171) Used by children for teething. (144:155) **Atsugewi** *Antirheumatic (Internal)* Decoction of leaves taken for rheumatism. *Cold Remedy* Decoction of leaves taken for colds. *Dermatological Aid* Poultice of pounded, raw leaves applied to cuts. Poultice of raw leaves applied to cuts. *Diaphoretic* and *Unspecified* Crushed leaves rubbed over the body during a sweat bath. (61:140) **Catawba** *Analgesic* Poultice of smashed leaves applied to pain. *Dermatological Aid* Poultice of smashed leaves applied to swellings, bruises, and wounds. *Orthopedic Aid* Poultice of smashed leaves ap-

plied to sprains. *Pediatric Aid* Decoction of roots given to children with croup. *Pulmonary Aid* Decoction of roots given to children with croup. (157:190) **Cherokee** *Analgesic* Compound taken for "pains." *Cough Medicine* Compound decoction of leaf taken with brown sugar or honey as a cough syrup. *Dermatological Aid* Leaves rubbed under armpits for "prickly rash" and flowers used on sores. *Gland Medicine* Scalded leaves used on swollen glands. *Gynecological Aid* Infusion of root or leaf taken for "female trouble." *Kidney Aid* Infusion of root taken "for kidneys" and used to bathe legs for "dropsy." *Misc. Disease Remedy* Leaves wrapped around neck for mumps and scalded leaves used on swollen glands. (80:45) Poultice of beaten leaves applied to the throat for diphtheria. (177:57) **Creek** *Cough Medicine* Compound decoction of root taken for coughs. (172:660, 661) Decoction of roots taken as a cough remedy. (177:57) **Delaware** *Antirheumatic (External)* Poultice of heated leaves applied to the joints and body for rheumatism pain and swelling. *Cough Medicine* Leaves combined with coltsfoot, plum root, and glycerin and used as a syrup for coughs. *Pulmonary Aid* Leaves combined with coltsfoot, plum root, and glycerin and used as a syrup for lung trouble. *Respiratory Aid* Leaves combined with coltsfoot, plum root, and glycerin and used as a syrup for catarrh. (176:36) **Delaware, Oklahoma** *Analgesic* Poultice of heated leaves applied for rheumatic pain. (175:30, 31) *Antirheumatic (External)* Poultice of heated leaves used for rheumatism. (175:30, 80) *Cold Remedy* Leaves used for colds. (175:80) *Cough Medicine* Compound containing leaves taken for catarrh, coughs, and lung trouble. (175:30, 31) *Pulmonary Aid* Compound decoction of leaves taken for lung trouble. *Respiratory Aid* Compound decoction of leaves taken for catarrh. (175:30, 80) **Delaware, Ontario** *Analgesic* Poultice of crushed leaves applied to bruises for pain and swelling. (175:66, 82) *Dermatological Aid* Poultice of crushed leaves applied to bruises for swelling and pain. (175:66) **Hopi** *Anticonvulsive* Leaves smoked with *Onosmodium* for "fits," craziness, and witchcraft. (200:33, 92) *Psychological Aid* Compound of plant smoked by persons not in their "right mind." *Witchcraft Medicine* Compound of plant smoked as cure for persons "with power to charm." (200:92) **Iroquois** *Anticonvulsive* Dried leaves smoked for bad hiccups. *Antidiarrheal* Compound decoction of roots and leaves taken for diarrhea with blood. *Blood Medicine* Complex compound taken as blood purifier. *Dermatological Aid* Decoction of whole plant given to babies with a rash. Poultice of leaves applied to swellings, abscesses, sores, and erysipelas. *Ear Medicine* Poultice of heated leaves applied for earaches. (87:432) *Febrifuge* Decoction of leaves taken for fevers. *Hemorrhoid Remedy* Poultice of leaves applied to piles. (87:433) *Laxative* Infusion of leaves given to babies to regulate their bowels. (87:431) *Misc. Disease Remedy* Poultice of leaves applied to face for mumps. *Orthopedic Aid* Compound decoction given and applied as poultice to baby's broken coccyx. (87:432) *Pediatric Aid* Infusion of leaves given to babies to regulate their bowels. (87:431) Compound decoction given and applied as poultice to baby's broken coccyx. Decoction of whole plant given to babies with a rash. (87:432) *Respiratory Aid* Dried leaves smoked for catarrh and asthma. (87:431) Infusion of leaves, roots, and flowers from another plant used for asthma. (141:57) *Toothache Remedy* Poultice of leaves applied to face for toothaches. (87:431) *Tuberculosis Remedy* Dried leaves smoked for consumption. (87:432) **Malecite** *Dermatological Aid* Leaves used for sores and cuts. (116:246) *Respiratory Aid* Plants smoked and used for catarrh. (116:248) **Menominee** *Pulmonary Aid* Root used for pulmonary disease. (151:53) **Micmac** *Dermatological Aid* Parts of plant used for sores and cuts. *Respiratory Aid* Parts of plant used for catarrh and leaves used for asthma. (40:63) **Mohegan** *Cold Remedy* Infusion of leaves taken with sugar for colds. (30:118) *Cough Medicine* Compound infusion of leaves taken for coughs. (174:270) Infusion of leaves mixed with molasses and taken for coughs. (176:77, 132) *Respiratory Aid* Leaves smoked for asthma. (174:265) Dried leaves smoked for asthma. (176:77, 132) *Throat Aid* Leaves smoked for sore throat. (174:265) Dried leaves smoked for sore throat. (176:77, 13) **Nanticoke** *Febrifuge* Poultice of leaves in vinegar bound to neck, head, wrists, and feet for fever. (175:56, 84) **Navajo** *Veterinary Aid* Plants "lighted and smoked for worms in sheep's nose." (90:156) **Navajo, Ramah** *Cough Medicine* Leaves smoked for cough. *Febrifuge* Leaves smoked for fever. *Psychological Aid* Dried leaves smoked in corn husk "to clear the mind if lost." *Strengthener* Cold infusion of leaves rubbed on bodies of hunters for strength. *Veterinary Aid* Cold infusion of leaves rubbed on bodies of horses for strength. (191:45) **Ojibwa** *Heart Medicine* Peeled roots used as a heart stimulant. (5:2304) **Penobscot** *Respiratory Aid* Dried, powdered leaves smoked for asthma. (156:310) **Potawatomi** *Respiratory Aid* Dried leaves smoked in a pipe for asthma and smudged leaves inhaled for catarrh. *Stimulant* Leaves smudged to revive unconscious person. (154:83, 84) **Rappahannock** *Dermatological Aid* Decoction of leaves rubbed on human or cattle swellings. *Orthopedic Aid* Decoction of leaves used as a hot poultice for human or animal sprains. *Veterinary Aid* Decoction of leaves rubbed or poultice applied to cattle swellings and sprains. (161:28) **Salish** *Tuberculosis Remedy* Decoction of plants taken for consumption. (178:293) **Shinnecock** *Cold Remedy* Infusion of leaves taken with sugar for colds. (30:118) **Thompson** *Cold Remedy* Decoction of leaves taken for colds. *Cough Medicine* Decoction of leaves taken for coughs. *Dermatological Aid* Juice rubbed on warts. (187:287) **Zuni** *Dermatological Aid* Poultice of powdered root applied to sores, rashes, and skin infections. Infusion of root used for athlete's foot infection. (27:378)

- *Other*—**Isleta** *Ceremonial Items* and *Smoke Plant* Leaves used as a ceremonial tobacco. (100:44) **Menominee** *Smoke Plant* Leaf gathered and smoked as an Indian tobacco. (151:53) **Thompson** *Ceremonial Items* Plant used in the sweat lodge. *Smoke Plant* Leaf used for smoking. It was cautioned that too much smoking of this plant was poisonous. (187:287)

Verbenaceae, see *Aloysia, Callicarpa, Glandularia, Phryma, Phyla, Tetraclea, Verbena, Vitex*

Verbena, Verbenaceae

Verbena bracteata Lag. & Rodr., Bigbract Verbena
- *Drug*—**Navajo, Ramah** *Ceremonial Medicine* Plant used in ceremonial chant lotion. *Dermatological Aid* Poultice of plant applied to centipede bite. (191:41)

Verbena hastata L., Swamp Verbena
- *Drug*—**Cherokee** *Abortifacient* Used for female obstructions. *Analgesic* Used for afterpains and taken as a tonic. *Antidiarrheal* Compound taken for flux, old bowel complaints, and dysentery. *Breast*

Treatment Taken as a tonic for breast complaints. *Cold Remedy* Taken for colds. *Cough Medicine* Taken for coughs. *Dermatological Aid* Astringent root compound taken for flux. *Diaphoretic* Used as a sudorific. *Emetic* Leaves, seeds, and roots used as an emetic. *Febrifuge* Leaves, seeds, and roots used for early stages of fever. *Gastrointestinal Aid* Compound used for flux and old bowel complaints. Taken to strengthen stomach. *Kidney Aid* One quart proof spirits and one handful of roots used for dropsy. *Tonic* Taken as a general or breast complaint tonic. (80:60) **Chippewa** *Hemostat* Snuff of dried flowers used for nosebleed. (53:356) **Dakota** *Analgesic* and *Gastrointestinal Aid* Decoction of leaves taken for stomachache. (70:111) **Delaware, Oklahoma** *Febrifuge* Compound containing root used for "chills." (175:80) **Iroquois** *Anthelmintic* Compound decoction of roots taken for worms. *Ear Medicine* Compound infusion of roots used as drops for earaches. *Gastrointestinal Aid* Compound decoction of roots taken for stomach cramps. *Other* Decoction of whole plant taken for "summer complaint." (87:422) *Poultice* Poultice of cut, wetted root applied to the head to cool off. (141:58) *Witchcraft Medicine* Cold infusion of smashed leaves used to make an obnoxious person leave. (87:422) **Mahuna** *Febrifuge* and *Gastrointestinal Aid* Roots used for complicated stomach fevers. (140:9) **Menominee** *Urinary Aid* Infusion of root taken to "clear up cloudy urine." (151:58)
- *Food*—**Concow** *Staple* Seeds used to make pinole. (41:383) **Omaha** *Beverage* Leaves steeped to make a tea-like beverage. (70:111)

Verbena lasiostachys Link, Western Vervain
- *Drug*—**Costanoan** *Febrifuge* Infusion of plant used for fevers. *Gastrointestinal Aid* Infusion of plant used for "fever of the stomach." *Misc. Disease Remedy* Infusion of plant used for typhoid fever. (21:15)

Verbena macdougalii Heller, Macdougal Verbena
- *Drug*—**Navajo, Ramah** *Ceremonial Medicine* Plant used in various ceremonial ways as a lotion and fumigant. *Febrifuge* Cold infusion taken and used as lotion for fever. (191:41)

Verbena officinalis L., Herb of the Cross
- *Drug*—**Houma** *Kidney Aid* and *Liver Aid* Decoction of root taken for kidney and liver trouble. (158:65)

Verbena sp.
- *Drug*—**Iroquois** *Cold Remedy* Decoction of roots taken for colds. (87:421)

Verbena stricta Vent., Hoary Verbena
- *Drug*—**Dakota** *Gastrointestinal Aid* Infusion of leaves taken for stomachaches. (69:363)
- *Food*—**Omaha** *Beverage* Leaves used to make a hot, aqueous, tea-like beverage. (68:329)

Verbena urticifolia L., White Vervain
- *Drug*—**Meskwaki** *Gynecological Aid* Infusion of root used for profuse menstruation. *Stimulant* Root eaten to revive patient and restore him to health. (152:251, 252)

Verbesina, Asteraceae
Verbesina encelioides (Cav.) Benth. & Hook. f. ex Gray, Golden Crownbeard
- *Drug*—**Hopi** *Dermatological Aid* Infusion of plant used as a wash for fever or itch from spider bites. (200:32, 99) Infusion of plant said to remove fever and itch from a spider bite. *Febrifuge* Infusion of plant said to remove fever and itch from a spider bite. (200:99) **Navajo, Kayenta** *Dermatological Aid* Infusion of plant taken and plant used as a lotion for spider bites. (205:51)

Verbesina encelioides ssp. *exauriculata* (Robins. & Greenm.) J. R. Coleman, Golden Crownbeard
- *Drug*—**Navajo** *Gastrointestinal Aid* Infusion of dried, crushed leaves taken for stomach troubles. (55:90) **Navajo, Ramah** *Hunting Medicine* Petals chewed for good luck in hunting. (191:54) **Zuni** *Analgesic, Emetic,* and *Gastrointestinal Aid* Blossoms chewed and swallowed with water as an emetic for stomach cramps. (as *Ximenesia exauriculata* 166:63) *Snakebite Remedy* Compound poultice of root applied with much ceremony to rattlesnake bite. (as *Ximinesia exauriculata* 166:53, 54)
- *Food*—**Navajo** *Unspecified* Seeds used for food. (55:90)
- *Other*—**Hopi** *Soap* Plant soaked in bath water. (200:99) **Navajo, Ramah** *Ceremonial Items* Used to make antelope prayer stick in Plumeway. *Insecticide* Hollow piece of dried stem used in procedure to rid corn of cutworms. Put four worms into a hollow piece of dried stem about 5 inches long, take it to an ancient Pueblo ruin, and stick it in the ground at the edge of the ruin. When the worms have disappeared, bury the stem, cover it with a piece of old pottery, and draw a line around it with an arrowhead. *Protection* Flowers hung in the hogan or worn in a hatband as protection from lightning. (191:54)

Verbesina sp.
- *Dye*—**Keres, Western** *Yellow* Petals mixed with white clay and used as a yellow dye for cotton. (as *Ximenesia* 171:74)

Verbesina virginica L., White Crownbeard
- *Drug*—**Chickasaw** *Abortifacient* Plant used as a "deobstruant." *Diuretic* Plant used as a diuretic. *Gynecological Aid* Infusion of root used for "Fluor Albus" and uterine weakness. *Stimulant* Plant used as a stimulant. *Venereal Aid* Plant used to treat venereal disease. (28:289) **Choctaw** *Febrifuge* Infusion of root taken "during attacks of fever." (25:23) Cold infusion of pounded roots taken for fevers. (177:64) **Seminole** *Analgesic* Infusion of leaf taken for bear sickness: fever, headache, thirst, constipation, and blocked urination. (169:198) *Antirheumatic (External)* Plant used for fire sickness: fever and body aches. (169:203) *Ceremonial Medicine* Root bark used as a purification emetic after funerals, at doctor's school, and after death of patient. *Emetic* Root bark used as an emetic to "clean the insides." (169:167) Decoction of root bark taken as an emetic for stomachaches. (169:276) *Eye Medicine* Infusion of plant taken and used as a bath for mist sickness: eye disease, fever, and chills. (169:209) *Febrifuge* Infusion of leaf taken for bear sickness: fever, headache, thirst, constipation, and blocked urination. (169:198) Plant used for fire sickness: fever and body aches. (169:203) Infusion of plant taken and used as a bath for mist sickness: eye disease, fever, and chills. (169:209) Plant used as a fever medicine. (169:283) *Gastrointestinal Aid* Decoction of root bark taken as an emetic for stomachaches. (169:276) *Laxative* and *Oral Aid* Infusion of leaf taken for bear sickness: fever, headache, thirst, constipation, and blocked urination. (169:198) *Unspecified* Plant used for medicinal purposes. (169:162) *Urinary Aid* Infusion of

leaf taken for bear sickness: fever, headache, thirst, constipation, and blocked urination. (169:198)

Vernonia, Asteraceae

Vernonia glauca (L.) Willd., Broadleaf Ironweed
- **Drug**—**Cherokee** *Analgesic* Infusion of root given for pains after childbirth. *Blood Medicine* Infusion taken for blood. *Gastrointestinal Aid* Infusion of root given "for stomach ulcers or hemorrhage." *Gynecological Aid* Infusion of root given for pains after childbirth. Various infusions given "for monthly period" and to "prevent menstruation." *Toothache Remedy* Infusion of root used for loose teeth. (80:41)

Vernonia missurica Raf., Missouri Ironweed
- **Drug**—**Kiowa** *Dermatological Aid* Decoction of plants used as a wash for dandruff. (192:62)
- **Food**—**Kiowa** *Candy* Pressed blossoms made into small wads for a short, sweet chew. (192:62)
- **Fiber**—**Kiowa** *Mats, Rugs & Bedding* Tall stems used as an overnight bed. Tall stems gathered, piled 6 inches thick on the ground, and arranged in a rectangular form. Used when traveling. (192:62)
- **Dye**—**Kiowa** *Purple* Flowers used as a purple dye. (192:62)

Vernonia noveboracensis (L.) Michx., New York Ironweed
- **Drug**—**Cherokee** *Analgesic* Infusion of root given for pains after childbirth. *Blood Medicine* Infusion taken for blood. *Gastrointestinal Aid* Infusion of root given "for stomach ulcers or hemorrhage." *Gynecological Aid* Infusion of root given for pains after childbirth. Various infusions given "for monthly period" and to "prevent menstruation." *Toothache Remedy* Infusion of root used for loose teeth. (80:41)

Vernonia sp., Ironweed
- **Drug**—**Natchez** *Antidiarrheal* Decoction of whole plant taken for dysentery. (177:64)

Veronica, Scrophulariaceae

Veronica americana Schwein. ex Benth., American Speedwell
- **Drug**—**Navajo, Ramah** *Ceremonial Medicine* and *Emetic* Plant used as a ceremonial emetic. (191:45)

Veronica officinalis L., Common Gypsyweed
- **Drug**—**Cherokee** *Cough Medicine* Taken with sugar for coughs. *Dermatological Aid* Used to poultice boils. *Ear Medicine* Warm juice used for earaches. *Febrifuge* Taken by thirsty patient for chills. (80:56) *Gynecological Aid* Decoction of roots taken to ease childbirth. (177:57) **Iroquois** *Emetic* Decoction of plants taken as an emetic to neutralize witchcraft and spoil hunting. *Veterinary Aid* Infusion of plants mixed with feed and used as wash for dried up cow's udders. *Witchcraft Medicine* Decoction of plants taken as an emetic to neutralize witchcraft and spoil hunting. (87:436)

Veronica peregrina ssp. *xalapensis* (Kunth) Pennell, Hairy Purslane Speedwell
- **Drug**—**Navajo, Ramah** *Ceremonial Medicine* Plant used as a ceremonial emetic and fumigant. *Disinfectant* Decoction taken for "deer infection" from overeating venison. *Emetic* Plant used as a ceremonial emetic. *Gastrointestinal Aid* Decoction taken for "deer infection" from overeating venison. *Hunting Medicine* Plant chewed and blown toward deer for good luck in hunting. (191:45)

Veronica serpyllifolia L., Thymeleaf Speedwell
- **Drug**—**Cherokee** *Cough Medicine* Taken with sugar for coughs. *Dermatological Aid* Used to poultice boils. *Ear Medicine* Warm juice used for earaches. *Febrifuge* Taken by thirsty patient for chills. (80:56)

Veronicastrum, Scrophulariaceae

Veronicastrum virginicum (L.) Farw., Culver's Root
- **Drug**—**Cherokee** *Analgesic* Infusion used for backache. *Cathartic* Used as a purgative. *Diaphoretic* Root used as a diaphoretic. *Disinfectant* Used as an antiseptic. *Febrifuge* "Good for typhus and bilious fevers." *Gastrointestinal Aid* Chewed for colic. *Liver Aid* Compound taken for "inactive liver." *Misc. Disease Remedy* "Good for typhus and bilious fevers." *Tonic* Used as a tonic. (80:31) **Chippewa** *Blood Medicine* Decoction of root taken "to cleanse the blood." (as *Leptandra virginica* 53:346) Compound decoction of root used as cathartic to cleanse blood in scrofula cases. (as *Leptandra virginica* 53:354) *Cathartic* Decoction of root taken as a physic. (as *Leptandra virginica* 53:346) Compound decoction of root used as a blood-cleansing cathartic for scrofula sores. *Tuberculosis Remedy* Compound decoction of root used as a blood-cleansing cathartic for scrofula sores, and to cleanse blood in scrofula cases. (as *Leptandra virginica* 53:354) **Iroquois** *Antidiarrheal* Infusion of roots taken for chills or diarrhea. *Antirheumatic (Internal)* Plant used for rheumatism. *Cathartic* Decoction or infusion of roots taken as a physic. *Cough Medicine* Infusion of roots taken for coughs and as a physic. *Emetic* Infusion of roots taken as an emetic. *Febrifuge* Infusion of roots taken for chills and fevers. *Gastrointestinal Aid* Infusion of smashed roots taken for biliousness and gallstones. (87:435) *Heart Medicine* Decoction of roots taken as a physic or for a bad heart. (87:436) *Panacea* Infusion of roots taken for all ailments and fevers. (87:435) *Poison* Plant considered poisonous. (87:436) *Pulmonary Aid* Infusion of smashed roots taken for chest lumps caused by a cold drink. *Witchcraft Medicine* Infusion of roots taken as a witch medicine. (87:435) **Menominee** *Cathartic* Strong decoction of root taken as powerful cathartic. (as *Leptandra virginica* 54:131) Root used as a strong physic. *Ceremonial Medicine* Used as a "reviver" to purify whoever had been defiled by touch of bereaved. (151:53, 54) *Emetic* Strong decoction of root taken as powerful emetic. *Laxative* Infusion of root taken as a mild laxative. (as *Leptandra virginica* 54:131) **Meskwaki** *Anticonvulsive* Root used for fits and constipation. *Gynecological Aid* Root used by women in labor. *Kidney Aid* Root used to dissolve gravel in the kidneys and for fits. *Laxative* Root used for constipation and fits. *Misc. Disease Remedy* Infusion of root taken for "ague of long standing." *Stimulant* Root used by women for weakness. (152:247) **Ojibwa, South** *Cathartic* Decoction of crushed root taken as a cathartic. (91:200)

Viburnum, Caprifoliaceae

Viburnum acerifolium L., Mapleleaf Viburnum
- **Drug**—**Cherokee** *Anticonvulsive* Infusion taken to prevent recurrent spasms. *Diaphoretic* Root bark taken as a diaphoretic. *Febrifuge* Compound infusion taken for fever. *Misc. Disease Remedy* Compound infusion taken for smallpox and ague. *Oral Aid* Infusion of bark used as a wash for sore tongue. *Tonic* Root bark taken as a tonic. (as *Virburnum acerifolium* 80:62) **Chippewa** *Analgesic* Decoction of inner bark taken for cramps. (53:344) *Emetic* Compound decoction

of scraped inner bark taken as an emetic. Cool infusion of bark taken as an emetic. (53:346) *Gastrointestinal Aid* Decoction of inner bark taken for stomach cramps. (53:344) **Iroquois** *Analgesic* Infusion of bark taken and applied as poultice for pain caused by witchcraft. *Gynecological Aid* Infusion of plants taken to suppress excessive menses. *Urinary Aid* Infusion of plants taken by men for stricture and painful urination. *Witchcraft Medicine* Infusion of bark taken and applied as poultice for pain caused by witchcraft. (87:447) **Menominee** *Analgesic* Infusion of inner bark taken for cramps. (151:29) *Gastrointestinal Aid* Infusion of inner bark taken for colic. (151:28)

Viburnum dentatum L. var. *dentatum*, Southern Arrowwood

- **Other**—**Ojibwa** *Smoke Plant* Bark furnished one of the ingredients of "kinnikinnick," a smoking mixture. (as *V. pubescens* 153:417)

Viburnum dentatum var. *lucidum* Ait., Southern Arrowwood

- **Drug**—**Iroquois** *Contraceptive* Decoction of twigs taken by women to prevent conception. (as *V. recognitum* 87:447) *Orthopedic Aid* Compound poultice applied to swollen legs of woman after birth of the baby. (as *V. recognitum* 87:448)

Viburnum edule (Michx.) Raf., Mooseberry Viburnum

- **Drug**—**Bella Coola** *Cold Remedy* Bark chewed and juice swallowed for "cold on the lungs." *Pulmonary Aid* Bark chewed and juice swallowed for whooping cough and "cold on the lungs." (as *V. pauciflorum* 150:64) **Carrier** *Cough Medicine* Decoction of stems taken for coughs. (31:77) **Carrier, Northern** *Antidiarrheal* Infusion of crushed inner bark taken for dysentery. *Cathartic* Infusion of crushed inner bark taken as a purgative. (as *V. pauciflorum* 150:64) **Cree, Woodlands** *Dermatological Aid* Poultice of chewed, unopened flower buds applied to lip sores. *Pediatric Aid* Decoction of roots taken for sickness associated with teething. *Throat Aid* Twig tips chewed and swallowed for sore throats. Plant used in a gargle taken for sore throats. Infusion of leaves and stems taken for sore throats. *Toothache Remedy* Decoction of roots taken for sickness associated with teething. (109:65) **Eskimo, Chugach** *Throat Aid* Decoction of leaves used as a gargle for sore throats. (as *V. pauciflorum* 149:326) **Gitksan** *Analgesic* Compound decoction taken for headache. *Cough Medicine* Compound decoction of twigs and bark taken for coughs. *Eye Medicine* Compound decoction taken for weak eyes. *Tuberculosis Remedy* Compound decoction of twigs and bark taken for consumption. (as *V. pauciflorum* 150:64) **Tanana, Upper** *Cold Remedy* Berries eaten for colds. *Gastrointestinal Aid* Decoction of bark taken for stomach troubles. *Throat Aid* Berries eaten for sore throats. (102:11)
- **Food**—**Alaska Native** *Preserves* Berries used to make jelly. (85:111) **Bella Coola** *Winter Use Food* Berries mixed with grease and other berries and used as a winter food. (184:203) **Carrier** *Preserves* Berries used to make jelly. (31:77) **Cree, Woodlands** *Frozen Food* Fruit collected in the fall and frozen or left to freeze on the bush and eaten as a nibble. *Preserves* Fruit used to make jam or jelly. *Snack Food* Fruit eaten raw as a nibble. Fruit left to freeze on the bush and eaten as a nibble. (109:65) **Eskimo, Inuktitut** *Fruit* Berries used for food. (202:188) **Eskimo, Inupiat** *Beverage* Berries used to make juice. *Dessert* Berries mixed with other berries and used to make traditional dessert. *Frozen Food* Berries frozen and stored for future use. *Fruit* Berries eaten fresh or cooked. *Ice Cream* Berries, oil, and water used to make ice cream. *Preserves* Berries used to make jam or jelly. *Sauce & Relish* Berries used to make catsup, syrup, juice, jam, or jelly. (98:106) **Haisla & Hanaksiala** *Fruit* Berries used for food. *Winter Use Food* Berries stored with oolichan (candlefish) grease in barrels for winter use. (43:232) **Hesquiat** *Fruit* Raw berries gathered to eat with oil "on the spot." (185:63) **Kitasoo** *Winter Use Food* Berries stored for future use. (43:329) **Koyukon** *Frozen Food* Berries frozen for winter use. (119:55) **Kwakiutl, Southern** *Preserves* Berries steamed, covered with oil, preserved, and used for food. *Special Food* Fresh, ripe berries eaten at feasts only. *Winter Use Food* Green berries steamed, covered with water, and used as a winter food. (183:281) **Okanagan-Colville** *Fruit* Berries used for food. (188:95) **Oweekeno** *Preserves* Berries used to make jam. (43:91) **Salish, Coast** *Fruit* Berries cooked and eaten with oil. Berries eaten raw. (182:80) **Tanana, Upper** *Frozen Food* Berries frozen for future use. (102:11) *Fruit* Fruit used for food. (77:28) Berries eaten raw, plain, or mixed raw with sugar, grease, or the combination of the two. Berries boiled with sugar and flour to thicken. (102:11) Berries mixed with fat and used for food. (as *V. pauciflorum* 115:37) *Pie & Pudding* Berries used to make pies. *Preserves* Berries used to make jam and jelly. *Winter Use Food* Berries preserved alone or in grease and stored in a birch bark basket in an underground cache. (102:11) **Thompson** *Dried Food* Fruit eaten dried. *Fruit* Fruit eaten fresh. *Preserves* Fruit made into jelly. *Sauce & Relish* Fruit made into a sauce with the seeds strained and cornstarch added. *Soup* Fruit cooked in soups. (187:201) **Unspecified** Small, acidic drupes eaten. (as *V. pauciflorum* 164:487)
- **Fiber**—**Koyukon** *Basketry* Stems used to reinforce birch bark basket rims. (119:55) **Tanana, Upper** *Basketry* Stems used for birch bark basket rims. (102:11)

Viburnum lantanoides Michx., Hobble Bush

- **Drug**—**Algonquin, Tête-de-Boule** *Analgesic* Leaves mashed and rubbed on the head for migraines. (132:134) **Iroquois** *Anthelmintic* Compound decoction of plants taken for worms caused by venereal disease. *Blood Medicine* Decoction of roots taken as a blood medicine. *Gynecological Aid* Decoction of roots taken as fertility drug by women. *Love Medicine* Plant used as a love medicine. *Pulmonary Aid* Decoction of plants taken for a sore chest and loss of breath. *Venereal Aid* Compound decoction of plants taken for worms caused by venereal disease. (87:446)

Viburnum lentago L., Nannyberry

- **Drug**—**Chippewa** *Urinary Aid* Infusion of leaves taken or poultice of leaves applied for dysuria. (71:142) **Delaware, Ontario** *Misc. Disease Remedy* Compound infusion of leaves taken for measles. (175:66, 82) **Iroquois** *Antihemorrhagic* Decoction of roots taken for spitting blood. *Emetic* and *Tuberculosis Remedy* Compound decoction taken to vomit during initial stages of consumption. (87:448) **Malecite** *Abortifacient* Infusion of roots used for irregular menstruation. (116:258) **Micmac** *Abortifacient* Roots used for irregular menstruation. (40:63) **Ojibwa** *Diuretic* Infusion of inner bark used as a diuretic. (153:361)
- **Food**—**Dakota** *Fruit* Fruit eaten from the hand, but not gathered in quantity. (70:115) **Iroquois** *Bread & Cake* Fruit mashed, made into small cakes, and dried for future use. *Dried Food* Raw or cooked fruit sun or fire dried and stored for future use. *Fruit* Dried fruit taken as a hunting food. *Sauce & Relish* Dried fruit cakes soaked in warm water

and cooked as a sauce or mixed with corn bread. (196:128) **Menominee** *Fruit* Berries used for food. (151:63) **Ojibwa** *Fruit* Berries eaten fresh from the bush. *Preserves* Berries used in jam with wild grapes. (153:398) **Omaha** *Fruit* Fruit eaten from the hand, but not gathered in quantity. **Pawnee** *Fruit* Fruit eaten from the hand, but not gathered in quantity. **Ponca** *Fruit* Fruit eaten from the hand, but not gathered in quantity. **Winnebago** *Fruit* Fruit eaten from the hand, but not gathered in quantity. (70:115)

Viburnum nudum var. *cassinoides* (L.) Torr. & Gray, Possumhaw

- *Drug*—**Cherokee** *Anticonvulsive* Infusion taken to prevent recurrent spasms. *Diaphoretic* Root bark used as a diaphoretic. *Febrifuge* Compound infusion taken for fever. *Misc. Disease Remedy* Compound infusion taken for smallpox and ague. *Oral Aid* Infusion of bark used as a wash for sore tongue. *Tonic* Root bark used as a tonic. (as *Virburnum cassinoides* 80:62)
- *Food*—**Abnaki** *Fruit* Fruit used for food. (as *Viburnum cassinoides* 144:152) *Unspecified* Grains used for food. (as *V. cassinoides* 144:173) **Algonquin, Quebec** *Fruit* Berries used for food. (as *V. cassinoides* 18:107)

Viburnum opulus L., European Cranberrybush Viburnum

- *Drug*—**Iroquois** *Blood Medicine* Berries considered "good" for the blood. *Liver Aid* Berries considered "good" for the liver. (124:96) **Meskwaki** *Analgesic* Decoction of root taken by "one who feels pain over his entire body." (152:208) **Micmac** *Misc. Disease Remedy* Bark used for swollen glands and mumps. (40:63) **Montagnais** *Eye Medicine* Decoction of plant used as a salve for sore eyes. (156:316) **Penobscot** *Misc. Disease Remedy* Infusion of berries taken for swollen glands and mumps. (156:310)
- *Food*—**Iroquois** *Fruit* Berries used as a favorite autumn food. (124:96) **Ojibwa** *Fruit* Fruit used for food. (135:237) **Shuswap** *Fruit* Berries eaten raw. *Preserves* Berries made into jelly. (123:61)
- *Other*—**Dakota** *Toys & Games* Stalks without the pith used to make popguns in the absence of elderberry. **Omaha** *Toys & Games* Stalks without the pith used to make popguns in the absence of elderberry. **Pawnee** *Toys & Games* Stalks without the pith used to make popguns in the absence of elderberry. **Ponca** *Toys & Games* Stalks without the pith used to make popguns in the absence of elderberry. **Winnebago** *Toys & Games* Stalks, without the pith, used to make popguns in the absence of elderberry. (70:115)

Viburnum opulus var. *americanum* Ait., American Cranberry Viburnum

- *Drug*—**Chippewa** *Gynecological Aid* Infusion of roots taken for prolapse of the uterus. (as *V. americanum* 71:141) **Iroquois** *Blood Medicine* Decoction of plants taken as a blood purifier and blood medicine. Infusion of bark taken to vomit for bad blood. (87:446) *Cold Remedy* Compound infusion of plants taken for colds and fevers. (87:447) *Emetic* Infusion of bark taken to vomit for bad blood and fever. *Febrifuge* Decoction of roots given to babies with fevers. Infusion of bark taken to vomit for fever. *Gynecological Aid* Decoction of branches taken for fallen womb after birth. (87:446) Compound decoction of stalks taken to prevent hemorrhage after childbirth. *Heart Medicine* Compound decoction of root taken to regulate the heart and as

blood medicine. *Kidney Aid* Compound decoction of roots taken for fevers and kidneys. (87:447) *Laxative* Infusion of bark taken as a laxative. *Pediatric Aid* Decoction of roots given to babies with fevers. (87:446) Compound infusion of roots taken as blood purifier or for prenatal strength. (87:447) *Pulmonary Aid* Compound decoction of bark taken by fat people who have difficulty breathing. (87:446) *Strengthener* Compound infusion of roots taken as blood purifier or for prenatal strength. (87:447) **Ojibwa** *Analgesic* Infusion of inner bark taken for stomach cramps. *Cathartic* Infusion of inner bark taken as a physic. *Gastrointestinal Aid* Infusion of inner bark taken for stomach cramps. (153:361)
- *Food*—**Algonquin, Quebec** *Fruit* Berries eaten fresh. *Preserves* Berries made into preserves. (as *V. trilobum* 18:107) **Chippewa** *Sauce & Relish* Fresh and dried fruits used as an acid sauce. (as *V. americanum* 71:141) **Iroquois** *Bread & Cake* Fruit mashed, made into small cakes, and dried for future use. *Dried Food* Raw or cooked fruit sun or fire dried and stored for future use. *Fruit* Dried fruit taken as a hunting food. *Sauce & Relish* Dried fruit cakes soaked in warm water and cooked as a sauce or mixed with corn bread. (196:128) **Menominee** *Fruit* Berries used for food. (151:63) **Okanagon** *Fruit* Fruits occasionally used for food. **Thompson** *Fruit* Fruits occasionally used for food. (125:38)
- *Other*—**Chippewa** *Hunting & Fishing Item* Fruit used as bait for snares set for snowshoe rabbits before guns had come into common use. (as *V. americanum* 71:141)

Viburnum prunifolium L., Black Haw

- *Drug*—**Cherokee** *Anticonvulsive* Infusion taken to prevent recurrent spasms. *Diaphoretic* Root bark taken as a diaphoretic. *Febrifuge* Compound infusion taken for fever. *Misc. Disease Remedy* Compound infusion taken for smallpox and ague. *Oral Aid* Infusion of bark used as a wash for sore tongue. *Tonic* Root bark taken as a tonic. (as *Virburnum prunifolium* 80:62) **Delaware** *Reproductive Aid* Root bark combined with leaves of other plants and used to strengthen female generative organs. (176:31) **Delaware, Oklahoma** *Gynecological Aid* Compound containing root bark used as a tonic for the "female generative organs." (175:26, 80) **Micmac** *Gynecological Aid* Infusion of plant taken before and during parturition. (as *Viburnum pomifolium* 194:28)
- *Food*—**Meskwaki** *Fruit* Berries eaten raw. *Preserves* Berries cooked into a jam. (152:256)

Viburnum sp.

- *Drug*—**Iroquois** *Contraceptive* Infusion of plant taken as a contraceptive. (87:445)

Vicia, Fabaceae

Vicia americana Muhl. ex Willd., American Vetch

- *Drug*—**Iroquois** *Love Medicine* Infusion of roots used by women as a love medicine. (87:365) **Keres, Western** *Dermatological Aid* Leaves rubbed in hands and applied to spider bites. (171:74) **Navajo, Kayenta** *Veterinary Aid* Plant smoked by horse to increase the horse's endurance. (205:29) **Navajo, Ramah** *Eye Medicine* Infusion of plant used as an eyewash. *Panacea* Plant used as "life medicine." (191:33) **Okanagan-Colville** *Ceremonial Medicine* Infusion of tops used as a bathing solution in the sweat house. (188:106) **Squaxin** *Analgesic*

Infusion of crushed leaves used as a bath for soreness. (79:39)
- *Food*—Acoma *Unspecified* Seeds used for food. (32:32) **Keres, Western** *Vegetable* Black peas used for food. (171:74) **Laguna** *Unspecified* Seeds used for food. Whole pods used for food. (32:32) **Mendocino Indian** *Fodder* Used for fodder. *Vegetable* Stems baked or boiled and eaten as greens. (41:362) **Montana Indian** *Unspecified* Cooked and eaten for greens. (19:26) **Thompson** *Fodder* Plant used as fodder for horses and cattle. (164:515)
- *Fiber*—Yuki *Cordage* Stout roots used for tying. (41:362)
- *Other*—Mendocino Indian *Good Luck Charm* Small bunch of roots kept in the pocket for good luck while gambling. (41:362)

Vicia caroliniana Walt., Carolina Vetch
- *Drug*—Cherokee *Analgesic* Used for back pains, local pains, and rubbed on stomach cramps. (80:60) Infusion of plant rubbed into scratches made over location of muscle pain. (177:34) *Antirheumatic (Internal)* Compound used for rheumatism. (80:60) *Emetic* Infusion of plant taken as an emetic. (177:34) *Gastrointestinal Aid* Decoction taken for dyspepsia. *Misc. Disease Remedy* Used for "blacks" and compound used for rheumatism. *Orthopedic Aid* Used for back pains, to toughen muscles, for muscular cramps and twitching. (80:60) Infusion of plant rubbed into scratches made over location of muscle pain. (177:34) *Respiratory Aid* Compound taken by ballplayers for wind during game. (80:60)

Vicia faba L., Horsebean
- *Drug*—Navajo *Ceremonial Medicine* Plant used in the Coyote Chant for medicine. *Poison* Plant considered poisonous. (55:59)
- *Food*—Sia *Vegetable* Cultivated beans used for food. (199:106)
- *Other*—Navajo *Smoke Plant* Crushed leaves mixed with tobacco and smoked. (55:59)

Vicia melilotoides, Vetch
- *Food*—Apache, Chiricahua & Mescalero *Dried Food* Ripe pods dried, stored, and soaked and boiled when needed. *Unspecified* Ripe pods cooked and eaten. (33:49)

Vicia nigricans ssp. *gigantea* (Hook.) Lassetter & Gunn., Giant Vetch
- *Drug*—Costanoan *Laxative* Decoction of roots used as a laxative. (as *V. gigantea* 21:19) **Makah** *Dermatological Aid* Infusion of roots used as a hair wash. (as *V. gigantea* 79:39) *Love Medicine* Used as a "love medicine." If you want your girlfriend to love you, take the plant and rub down with it after bathing, and she will love you forever. (as *V. gigantea* 67:283) **Quinault** *Love Medicine* Roots rubbed on woman and placed under pillow to bring back husband. (as *V. gigantea* 79:39) **Saanich** *Dermatological Aid* Infusion of pounded roots used as a hair tonic for falling hair and dandruff. (as *Vicea gigantea* 182:85)
- *Food*—Kwakiutl, Southern *Fruit* Pea-like fruits roasted and used for food. (as *Vicea gigantea* 183:285)
- *Other*—Costanoan *Hunting & Fishing Item* Foliage used as camouflage during deer hunting. (as *Vicia gigantea* 21:250) **Nitinaht** *Hunting & Fishing Item* Plants rubbed on hands and fishing lines to eliminate human odor. (as *V. gigantea* 186:112)

Vicia sativa ssp. *nigra* (L.) Ehrh., Common Vetch
- *Drug*—Iroquois *Abortifacient* Decoction of plant taken by women with suppressed menses. *Gynecological Aid* Decoction of plant taken by women with swollen external sexual organs. *Love Medicine* Cold infusion of plant used as a love medicine. (as *V. angustifolia* 87:365) **Snohomish** *Analgesic* Infusion of plant used on the hair for headaches. (as *V. angustifolia* 79:39)

Vicia sp., Vetch
- *Food*—Papago *Dried Food* Threshed, dried on the ground or roofs, stored, and used for food. (34:33) **Thompson** *Forage* Used as common forage plants. (164:515)

Vicia villosa Roth, Winter Vetch
- *Drug*—Rappahannock *Dermatological Aid* Compound infusion with dried leaves taken for sores. *Gastrointestinal Aid* Compound infusion with dried leaves taken for stomach pain. (161:35)

Vigna, Fabaceae

Vigna luteola (Jacq.) Benth., Hairypod Cowpea
- *Drug*—Hawaiian *Dermatological Aid* Whole plant and other plants pounded, squeezed, and resulting liquid applied to boils and ruptured skin. *Pediatric Aid* Flowers and buds chewed by the mother and given to infants for general body weakness. *Respiratory Aid* Whole plant and other plants pounded, squeezed, the resulting liquid heated and taken for asthma. *Strengthener* Flowers and buds chewed by the mother and given to infants for general body weakness. (as *V. luta* 2:33)

Vigna unguiculata (L.) Walp., Blackeyed Pea
- *Food*—Cocopa *Unspecified* Whole pods boiled, strings removed, and used for food; very hungry people ate strings. (as *V. sinensis* 64:264) **Cocopa, Maricopa, Mohave & Yuma** *Porridge* Ripe seeds parched, ground into flour, and boiled with corn to make mush. (as *V. sinensis* 37:129) **Kamia** *Unspecified* Species used for food. (as *V. sinensis* 62:21) **Mohave** *Dried Food* Ripe pods trampled, winnowed, and dried thoroughly. *Unspecified* Unripe seeds cooked in salted water and eaten. (as *V. sinensis* 37:129) **Papago** *Unspecified* Species used for food. **Pima** *Unspecified* Species used for food. (as *V. sinensis* 36:120)

Violaceae, see *Hybanthus, Viola*

Viola, Violaceae

Viola adunca Sm., Hookedspur Violet
- *Drug*—Blackfoot *Antirheumatic (External)* Infusion of roots and leaves applied to sore and swollen joints. (86:79) *Pediatric Aid* and *Respiratory Aid* Infusion of leaves and roots given to asthmatic children. (86:74) **Carrier, Southern** *Analgesic* and *Gastrointestinal Aid* Decoction of entire plant taken for stomach pain. (150:60) **Klallam** *Analgesic* Poultice of smashed flowers applied to the chest or side for pain. **Makah** *Gynecological Aid* Roots and leaves chewed by women during labor. (79:40) **Tolowa** *Eye Medicine* Poultice of chewed leaves applied to sore eyes. (6:62)
- *Dye*—Blackfoot *Blue* Plant used to dye arrows blue. (86:123)

Viola bicolor Pursh, Field Pansy
- *Drug*—Cherokee *Analgesic* Poultice of leaves used for headache. *Antidiarrheal* Infusion taken for dysentery. *Blood Medicine* Infusion taken for blood. *Cold Remedy* Infusion taken for colds. *Cough Medi-*

cine Infusion with sugar taken for cough. *Dermatological Aid* Poultice of crushed root applied to boils. *Respiratory Aid* Infusion sprayed up nose for catarrh. *Tonic* Infusion taken as spring tonic. (as *V. rafinesquil* 80:60)
• *Other*—**Cherokee** *Insecticide* Infusion of root used to soak corn before planting to keep off insects. (as *V. rafinesquil* 80:60)

Viola biflora L., Twoflower Violet
• *Other*—**Eskimo, Inuktitut** *Incense & Fragrance* Stems with blossoms placed among clothes. (202:191)

Viola blanda Willd., Sweet White Violet
• *Food*—**Cherokee** *Vegetable* Leaves and stems mixed with other greens, parboiled, rinsed, and fried with grease and salt until soft. (126:60)

Viola canadensis L., Canadian White Violet
• *Drug*—**Ojibwa, South** *Analgesic* Decoction of root used for pains near the bladder. (91:201)

Viola conspersa Reichenb., American Dog Violet
• *Drug*—**Ojibwa** *Heart Medicine* Infusion of whole plant taken for heart trouble. (153:392)

Viola cucullata Ait., Marsh Blue Violet
• *Drug*—**Cherokee** *Analgesic* Poultice of leaves used for headache. *Antidiarrheal* Infusion taken for dysentery. *Blood Medicine* Infusion taken for blood. *Cold Remedy* Infusion taken for colds. *Cough Medicine* Infusion with sugar taken for cough. *Dermatological Aid* Poultice of crushed root applied to boils. *Respiratory Aid* Infusion sprayed up nose for catarrh. *Tonic* Infusion taken as spring tonic. (80:60) **Ute** *Unspecified* Roots used as medicine. (38:37)
• *Other*—**Cherokee** *Insecticide* Infusion of root used to soak corn before planting to keep off insects. (80:60)

Viola epipsila Ledeb., Dwarf Marsh Violet
• *Other*—**Tanana, Upper** *Incense & Fragrance* Dried roots used as incense at potlatches. (102:18)

Viola nephrophylla Greene, Northern Bog Violet
• *Drug*—**Navajo, Ramah** *Ceremonial Medicine* and *Emetic* Plant used as a ceremonial emetic. (191:36)

Viola pedata L., Birdfoot Violet
• *Drug*—**Cherokee** *Analgesic* Poultice of leaves used for headache. *Antidiarrheal* Infusion taken for dysentery. *Blood Medicine* Infusion taken for blood. *Cold Remedy* Infusion taken for colds. *Cough Medicine* Infusion with sugar taken for cough. *Dermatological Aid* Poultice of crushed root applied to boils. *Respiratory Aid* Infusion sprayed up nose for catarrh. *Tonic* Infusion taken as spring tonic. (80:60)
• *Other*—**Cherokee** *Insecticide* Infusion of root used to soak corn before planting to keep off insects. (80:60)

Viola pedunculata Torr. & Gray, California Golden Violet
• *Food*—**Diegueño** *Vegetable* Young leaves, picked before the flowers appear in the spring, boiled once and eaten as greens. (84:43) **Luiseño** *Vegetable* Leaves used as greens. (155:230)

Viola pubescens Ait., Downy Yellow Violet
• *Drug*—**Cherokee** *Analgesic* Poultice of leaves used for headache. *Antidiarrheal* Infusion taken for dysentery. *Blood Medicine* Infusion taken for blood. *Cold Remedy* Infusion taken for colds. *Cough Medicine* Infusion with sugar taken for cough. *Dermatological Aid* Poultice of crushed root applied to boils. *Respiratory Aid* Infusion sprayed up nose for catarrh. *Tonic* Infusion taken as spring tonic. (80:60) **Iroquois** *Dermatological Aid* Decoction of plant taken and used as wash for facial eruptions. (87:387) **Ojibwa, South** *Throat Aid* Decoction of root taken in small doses for sore throat. (91:201) **Potawatomi** *Heart Medicine* Root used for various heart diseases. (154:87, 88)
• *Other*—**Cherokee** *Insecticide* Infusion of root used to soak corn before planting to keep off insects. (80:60)

Viola pubescens Ait. **var. *pubescens***, Smooth Yellow Violet
• *Drug*—**Iroquois** *Gastrointestinal Aid* Compound decoction of plants taken for indigestion. (as *V. eriocarpa* 87:387)
• *Food*—**Cherokee** *Vegetable* Leaves and stems mixed with other greens, parboiled, rinsed, and fried with grease and salt until soft. (as *V. pensylvanica* 126:60)

Viola rotundifolia Michx., Roundleaf Yellow Violet
• *Drug*—**Cherokee** *Analgesic* Poultice of leaves used for headache. *Antidiarrheal* Infusion taken for dysentery. *Blood Medicine* Infusion taken for blood. *Cold Remedy* Infusion taken for colds. *Cough Medicine* Infusion with sugar taken for cough. *Dermatological Aid* Poultice of crushed root applied to boils. *Respiratory Aid* Infusion sprayed up nose for catarrh. *Tonic* Infusion taken as spring tonic. (80:60)
• *Other*—**Cherokee** *Insecticide* Infusion of root used to soak corn before planting to keep off insects. (80:60)

Viola sagittata Ait., Arrowleaf Violet
• *Drug*—**Iroquois** *Witchcraft Medicine* Compound used to detect bewitchment. (87:386)

Viola sempervirens Greene, Evergreen Violet
• *Other*—**Karok** *Toys & Games* Flowers used by children during play. (as *V. sarmentosa* 148:386)

Viola sororia Willd., Common Blue Violet
• *Drug*—**Cherokee** *Analgesic* Poultice of leaves used for headache. *Antidiarrheal* Infusion taken for dysentery. *Blood Medicine* Infusion taken for blood. *Cold Remedy* Infusion taken for colds. *Cough Medicine* Infusion with sugar taken for cough. *Dermatological Aid* Poultice of crushed root applied to boils. *Respiratory Aid* Infusion sprayed up nose for catarrh. *Tonic* Infusion taken as spring tonic. (as *V. papilionacea* 80:60)
• *Other*—**Cherokee** *Insecticide* Infusion of root used to soak corn before planting to keep off insects. (as *V. papilionacea* 80:60)

Viola sp., Violet
• *Drug*—**Costanoan** *Dermatological Aid* Plant juice used as a salve and poultice of boiled plants applied to sores. (21:9) **Iroquois** *Dermatological Aid* Poultice of smashed plants applied to wounds. *Eye Medicine* Infusion of roots used as a wash for eyes. *Panacea* Compound decoction of roots taken as a panacea. *Pediatric Aid* Infusion of roots taken by mother and baby when baby gets sick. *Veterinary Aid* Poultice of smashed roots applied to boils on horse's head. (87:386) **Thompson** *Veterinary Aid* Infusion of mashed leaves and stalks used as a medicine for dogs. Juice used in dogs' noses to clear out

the nostrils and enable them to track deer better. (187:291)
- **Food**—**Cherokee** *Vegetable* Leaves and stems mixed with other greens, parboiled, rinsed, and fried with grease and salt until soft. (126:60) Leaves cooked with other potherbs and eaten. (204:253)
- **Other**—**Omaha** *Toys & Games* Violets used by children in playing games. The children separated into two teams, one team taking the name of their tribe and the other of another tribe such as the Dakota. Each team collected violets and the two parties sat down facing each other and snapped violets at each other until there were none remaining. The victorious team taunted the other as being poor fighters. (70:103)

Viola striata Ait., Striped Cream Violet
- **Drug**—**Iroquois** *Witchcraft Medicine* Plant used to make a girl sick and crazy by her rejected suitor after he has been refused by her parents. (87:387)

Viscaceae, see *Arceuthobium, Phoradendron*

Vitaceae, see *Ampelopsis, Cissus, Parthenocissus, Vitis*

Vitex, Verbenaceae

Vitex trifolia var. *unifoliolata*, Simpleleaf Chastetree
- **Drug**—**Hawaiian** *Unspecified* Leaves and wood thoroughly pounded and used to make a good bath. (2:72)
- **Food**—**Hawaiian** *Unspecified* Leaves eaten with dried coconut. (2:72)

Vitis, Vitaceae

Vitis aestivalis Michx., Summer Grape
- **Drug**—**Cherokee** *Antidiarrheal* Compound taken for diarrhea. *Blood Medicine* Infusion of leaf taken "for blood." *Gastrointestinal Aid* Taken as a "fall tonic" and infusion taken "for stomach." *Gynecological Aid* Wilted leaves used to draw soreness from breast after birth of a child. *Liver Aid* Infusion of leaf taken for liver. *Oral Aid* Compound decoction used to wash child's mouth for thrush. *Other* Compound infusion of bark taken for "bad disease." *Pediatric Aid* Compound decoction used to wash child's mouth for thrush. *Tonic* Taken as a "fall tonic." *Urinary Aid* Compound taken for "irregular urination." (80:37) Infusion of bark taken for urinary troubles. (177:41) **Choctaw** *Febrifuge* Used as a "refrigerant." *Gynecological Aid* "Water of the grape vine" taken and used as a wash to induce lactation. *Tonic* Used as a tonic. (28:287)
- **Food**—**Cherokee** *Beverage* Fruit mixed with sour grape, pokeberry juice, sugar, and cornmeal used as a juice. *Bread & Cake* Fruit used to make juice and dumplings. (126:60) *Fruit* Fruit used for food. (80:37) Raw fruit used for food. (126:60)

Vitis aestivalis Michx. var. *aestivalis*, Summer Grape
- **Drug**—**Seminole** *Analgesic* Decoction of leaves and stems taken for headaches. (as *V. rufotomentosa* 169:282) *Ceremonial Medicine* Infusion of plant added to food after a recent death. (as *V. rufotomentosa* 169:342) *Emetic* Plant used as an emetic during religious ceremonies. (as *V. rufotomentosa* 169:409) *Febrifuge* Decoction of leaves and stems taken for fevers. *Gastrointestinal Aid* Decoction of leaves and stems taken for stomachaches. (as *V. rufotomentosa* 169:282) *Pediatric Aid* Plant used for chronically ill babies. (as *V. rufotomentosa* 169:328)
- **Food**—**Seminole** *Unspecified* Plant used for food. (as *V. rufotomentosa* 169:475)
- **Fiber**—**Seminole** *Building Material* Plant used to make houses. *Cordage* Plant used for coffin lashing. (as *V. rufotomentosa* 169:475)

Vitis arizonica Engelm., Canyon Grape
- **Drug**—**Navajo** *Love Medicine* Vine used to make a cross and put on top of the basket of cornmeal and paper bread (flat bread), offered in courtship. (55:62)
- **Food**—**Apache, Chiricahua & Mescalero** *Dried Food* Fruit dried and eaten like raisins. *Fruit* Raw fruit eaten fresh. (33:44) **Apache, Mescalero** *Fruit* Fruits eaten fresh. (14:50) **Apache, Western** *Beverage* Juice boiled to make wine. *Dried Food* Berries pounded, dried, and stored in sacks. *Fruit* Ripe berries eaten raw. (26:190) **Havasupai** *Fruit* Fruit used for food. (197:231) **Isleta** *Fruit* Fruit considered an important part of the diet. (100:44) **Jemez** *Fruit* Grapes used for food. (44:28)
- **Other**—**Havasupai** *Toys & Games* Vines used to make the hoop of the hoop and pole game. (197:231) **Jemez** *Ceremonial Items* Berry juice mixed with white clay and used as a body paint for dancers. (44:28)

Vitis californica Benth., California Wild Grape
- **Food**—**Costanoan** *Fruit* Raw fruits used for food. (21:251) **Karok** *Fruit* Fruit used for food. (148:386) **Mendocino Indian** *Preserves* Fruits used to make jelly. (41:369) **Pomo** *Fruit* Raw berries used for food. (66:14) **Pomo, Kashaya** *Fruit* Berries eaten fresh. (72:51) **Tubatulabal** *Fruit* Berries used for food. (193:15) **Wintoon** *Fruit* Fruit used for food. (117:264) **Yurok** *Fruit* Fresh berries used for food. (6:62)
- **Fiber**—**Karok** *Basketry* Roots used as basketry material for the basket bottoms. (6:62) Roots used to make baskets. *Cordage* Vines used to moor a boat and smaller vines twisted to make ropes. (148:386) **Pomo** *Basketry* Sapwood used as binding material in basketry. (9:139) Woody parts of vines used for the rims of large, cone-shaped carrying baskets. (41:369) *Cordage* Vines used to make withes to tie things when hunting or traveling. Vines used to tie western serviceberry thatch in place on the winter house. (66:14) *Sewing Material* Vines soaked in water and hot ashes, bark removed, wood split into strands and used as thread. (41:369) **Pomo, Kashaya** *Cordage* Vine made a very strong cord used to tie bundles and for lashing. *Furniture* Vine used to make the hoop on a baby basket. (72:51)
- **Other**—**Karok** *Containers* Leaves put over bulbs while cooking in the earth oven. (148:386) *Tools* Stems used for lashings. (6:62) **Miwok** *Cooking Tools* Green leaves used as top layer, over the hot stones, in the earth oven. (12:139) **Pomo** *Fasteners* Sapwood used as a binding material for brush fences, rafters, stringers, and posts of earth lodges. (9:139)

Vitis cinerea (Engelm.) Engelm. ex Millard, Graybark Grape
- **Food**—**Dakota** *Dried Food* Fruit dried for winter use. *Fruit* Fruit eaten fresh. (70:102) **Kiowa** *Dried Food* Grapes gathered in large quantities and dried for later use. *Fruit* Grapes gathered in large quantities and eaten raw. *Preserves* Grapes gathered in large quantities and made into jams and jellies. (192:42) **Omaha** *Dried Food* Fruit dried for winter use. *Fruit* Fruit eaten fresh. **Pawnee** *Beverage* Fresh sap used as grape juice. *Dried Food* Fruit dried for winter use. *Fruit* Fruit eaten fresh. **Ponca** *Dried Food* Fruit dried for winter use. *Fruit* Fruit eaten fresh. **Winnebago** *Dried Food* Fruit dried for winter use. *Fruit* Fruit eaten fresh. (70:102)

Vitis cinerea* var. *baileyana (Munson) Comeaux, Graybark Grape
- **Food**—**Cherokee** *Beverage* Fruit mixed with sour grape, pokeberry juice, sugar, and cornmeal used as a juice. *Bread & Cake* Fruit used to make juice and dumplings. *Fruit* Raw fruit used for food. (as *V. baileyana* 126:60)

Vitis girdiana Munson, Valley Grape
- **Drug**—**Diegueño** *Dermatological Aid* Sap rubbed on falling or thin hair to keep it healthy and make it grow. (84:43)
- **Food**—**Cahuilla** *Beverage* Fruit used to make wine. *Dried Food* Fruit dried into raisins. *Fruit* Fruit eaten fresh and cooked in stews. *Porridge* Fruit used to make mush. (15:144) **Diegueño** *Dried Food* Fruit dried into raisins and cooked. *Fruit* Fruit eaten fresh. (84:43) **Luiseño** *Fruit* Cooked fruit used for food. (155:231)

Vitis labrusca L., Fox Grape
- **Drug**—**Cherokee** *Antidiarrheal* Compound taken for diarrhea. *Blood Medicine* Infusion of leaf taken "for blood." *Gastrointestinal Aid* Taken as a "fall tonic" and infusion taken "for stomach." *Gynecological Aid* Wilted leaves used to draw soreness from breast after birth of a child. *Liver Aid* Infusion of leaf taken for liver. *Oral Aid* Compound decoction used to wash child's mouth for thrush. *Other* Compound infusion of bark taken for "bad disease." *Pediatric Aid* Compound decoction used to wash child's mouth for thrush. *Tonic* Taken as a "fall tonic." *Urinary Aid* Compound taken for "irregular urination." (80:37) Infusion of bark taken for urinary troubles. (177:41) **Iroquois** *Veterinary Aid* Decoction of roots mixed with feed to assist horse conception. (87:383) **Mohegan** *Analgesic* Poultice of leaves bound to the head for headache. (174:264) Poultice of leaves applied to painful area. (176:77, 132) *Febrifuge* Poultice of leaves bound to the head for fever. (174:264)
- **Food**—**Cherokee** *Beverage* Fruit mixed with sour grape, pokeberry juice, sugar, and cornmeal used as a juice. *Bread & Cake* Fruit used to make juice and dumplings. (126:60) *Fruit* Fruit used for food. (80:37) Raw fruit used for food. (126:60)

Vitis munsoniana Simpson ex Munson, Munson's Grape
- **Drug**—**Seminole** *Ceremonial Medicine* Infusion of plant added to food after a recent death. (169:342) *Emetic* Plant used as an emetic during religious ceremonies. (169:409) *Pediatric Aid* Plant used for chronically ill babies. (169:328) *Snakebite Remedy* Infusion of plant used to wash snakebites. (169:297)
- **Food**—**Seminole** *Unspecified* Plant used for food. (169:479)

Vitis riparia Michx., Riverbank Grape
- **Food**—**Omaha** *Dried Food* Fruit dried for winter use. *Fruit* Fruit eaten fresh in season. (68:326)

Vitis rotundifolia Michx., Muscadine
- **Food**—**Cherokee** *Beverage* Fruit mixed with sour grape, pokeberry juice, sugar, and cornmeal used as a juice. *Bread & Cake* Fruit used to make juice and dumplings. *Fruit* Raw fruit used for food. (126:60)

Vitis rupestris Scheele, Sand Grape
- **Drug**—**Delaware** *Reproductive Aid* Vine mixed with other plants and used to increase fertility. *Strengthener* Vine mixed with other plants and used as a tonic for frail women. (176:31) **Delaware, Oklahoma** *Reproductive Aid* Compound containing vine used as a tonic by frail women and increases fertility. (175:26, 80) *Tonic* Compound containing vine used as a tonic by frail women and increases fertility. (175:26)

Vitis shuttleworthii House, Calloose Grape
- **Other**—**Seminole** *Hunting & Fishing Item* Plant used to make a deer snare. (169:478)

***Vitis* sp.**, Wild Grape
- **Drug**—**Chippewa** *Dermatological Aid* Vine sap used as a wash for hair. *Pulmonary Aid* Roots and branches used for pulmonary troubles. (71:136) **Mohegan** *Analgesic* Poultice of leaves applied to head for headache. **Shinnecock** *Analgesic* Poultice of leaves applied to head for headache. (30:121)
- **Food**—**Comanche** *Dried Food* Dried fruits stored for later use. *Fruit* Fresh fruits used for food. (29:524) **Hualapai** *Beverage* Fruit used to make juice. *Dried Food* Fruit sun dried and stored for later use. *Fruit* Fruit eaten raw from the vine. (195:23)

Vitis vinifera L., Domestic Grape
- **Food**—**Haisla & Hanaksiala** *Fruit* Fruit used for food. (43:296)

Vitis vulpina L., Frost Grape
- **Drug**—**Cherokee** *Antidiarrheal* Compound taken for diarrhea. *Blood Medicine* Infusion of leaf taken "for blood." *Gastrointestinal Aid* Taken as a "fall tonic" and infusion taken for stomach. *Gynecological Aid* Wilted leaves used to draw soreness from breast after birth of a child. *Liver Aid* Infusion of leaf taken for liver. *Oral Aid* Compound decoction used to wash child's mouth for thrush. *Other* Compound infusion of bark taken for "bad disease." *Pediatric Aid* Compound decoction used to wash child's mouth for thrush. *Tonic* Taken as a "fall tonic." *Urinary Aid* Compound taken for "irregular urination." (80:37) **Chippewa** *Antirheumatic (Internal)* Infusion of root taken for rheumatism. (as *V. cordifolia* 53:362) *Misc. Disease Remedy* Infusion of root taken for diabetes. (as *V. cordifolia* 53:364) **Delaware, Oklahoma** *Dermatological Aid* Sap considered beneficial to the hair. (175:26) *Gynecological Aid* Vine sap used for leukorrhea. (175:26, 80) **Iroquois** *Blood Medicine* Compound decoction of plants taken to make blood. (87:383) Infusion of plant with root bark from another plant and wine taken for anemia. (141:54) *Gastrointestinal Aid* Infusion of plant given to children with stomach troubles. *Kidney Aid* Decoction of plant taken for kidney trouble. *Other* Vines chewed for hiccups. *Pediatric Aid* Infusion of plant given to children with urinating or stomach troubles. *Urinary Aid* Decoction of plant taken for burning urination. Infusion of plant given to children with urinating troubles. (87:383) **Menominee** *Eye Medicine* Ripe grape squeezed into eye to remove rice husk at rice-making camp. (54:134) Seed or juice used to remove foreign matter from eyes. (as *V. cordifolia* 151:58) **Meskwaki** *Analgesic* Infusion of twigs held in child's mouth for pain of Indian turnip poisoning. *Antidote* Infusion of twigs held in mouth by children as antidote for poisoning. *Psychological Aid* Decoction of twigs taken for insanity. (as *V. cordifolia* 152:252) **Ojibwa** *Gastrointestinal Aid* Sap used for stomach and bowel troubles. *Gynecological Aid* Decoction of twigs taken to facilitate passing of afterbirth. (153:392)
- **Food**—**Cherokee** *Fruit* Fruit used for food. (80:37) **Cheyenne** *Fruit* Fruit eaten fresh and never dried. (76:180) **Chippewa** *Fruit* Fruits eaten raw. (as *V. cordifolia* 53:321) **Crow** *Dried Food* Fruit eaten dried.

Fruit Fruit eaten fresh. (19:26) **Dakota** *Dried Food* Fruit dried for winter use. *Fruit* Fruit eaten fresh. (70:102) **Iroquois** *Bread & Cake* Fruit mashed, made into small cakes, and dried for future use. *Dried Food* Raw or cooked fruit sun or fire dried and stored for future use. *Fruit* Dried fruit taken as a hunting food. *Sauce & Relish* Dried fruit cakes soaked in warm water and cooked as a sauce or mixed with corn bread. (196:128) *Unspecified* Fresh shoots eaten without peeling. (196:119) **Keres, Western** *Fruit* Fruit considered an important food. (171:74) **Lakota** *Fruit* Fruits eaten for food. (139:61) *Special Food* Vine pieces sucked or chewed for thirst during the Sun Dance. (106:44) **Mahuna** *Fruit* Berries eaten mainly to quench the thirst. (140:70) **Menominee** *Dried Food* Berries eaten dried. *Fruit* Berries eaten fresh. *Preserves* Berries eaten preserved or jellied. (as *V. cordifolia* 151:72) **Meskwaki** *Dessert* Grapes, touched by frost, prized as a dessert fruit. (as *V. cordifolia* 152:265) **Ojibwa** *Preserves* Frosted grapes made into jelly for winter use. (153:411) **Omaha** *Dried Food* Fruit dried for winter use. (68:326) Fruit dried for winter use. (70:102) *Fruit* Fruit eaten fresh in season. (68:326) Fruit eaten fresh. **Pawnee** *Beverage* Fresh sap used as grape juice. *Dried Food* Fruit dried for winter use. *Fruit* Fruit eaten fresh. **Ponca** *Dried Food* Fruit dried for winter use. *Fruit* Fruit eaten fresh. (70:102) **Sioux** *Dried Food* Fruit eaten dried. *Fruit* Fruit eaten fresh. (19:26) **Winnebago** *Dried Food* Fruit dried for winter use. *Fruit* Fruit eaten fresh. (70:102)

Vittariaceae, see *Vittaria*

Vittaria, Vittariaceae, fern

Vittaria lineata (L.) Sm., Appalachian Shoestring Fern
- *Drug*—**Seminole** *Other* Complex infusion of leaves taken for chronic conditions. (169:272) *Pediatric Aid* Plant used for chronically ill babies. (169:329) *Psychological Aid* Infusion of plant used to steam and bathe the body for insanity. (169:292) *Unspecified* Plant used for medicinal purposes. (169:162)

Vulpia, Poaceae

Vulpia octoflora* var. *glauca (Nutt.) Fern., Sixweeks Fescue
- *Food*—**Gosiute** *Unspecified* Seeds used for food. (as *Festuca tenella* 39:369)

Vulpia octoflora (Walt.) Rydb. **var. *octoflora***, Sixweeks Fescue
- *Food*—**Navajo, Kayenta** *Unspecified* Seeds roasted and used for food. (as *Festuca octoflora* 205:16)

Waldsteinia, Rosaceae

Waldsteinia fragarioides (Michx.) Tratt., Barren Strawberry
- *Drug*—**Iroquois** *Blood Medicine* Compound decoction of plants taken as a blood remedy. *Snakebite Remedy* Poultice of smashed plants applied to snakebites. (87:352)

Waltheria, Sterculiaceae

Waltheria indica L., Uhaloa
- *Drug*—**Hawaiian** *Dietary Aid* Root bark, buds, leaves, and other plants pounded and resulting liquid taken for losing weight. *Laxative* and *Pediatric Aid* Buds chewed by the mother and given to infants as a laxative. *Pulmonary Aid* Whole plant and other plants pounded, squeezed, and resulting liquid taken for pulmonary complications. (as *W. americana* 2:37) *Respiratory Aid* Buds and leaves pounded, mixed with water, and taken for asthma. (as *W. americana* 2:12) Whole plant and other plants pounded, squeezed, resulting liquid heated and taken for chronic asthma. Root bark, buds, leaves, and other plants pounded and resulting liquid taken for asthma. *Strengthener* Root bark, buds, leaves, and other plants pounded and resulting liquid taken for run-down condition. *Throat Aid* Root bark chewed and the juice swallowed for sore throats. (as *W. americana* 2:37)

Washingtonia, Arecaceae

Washingtonia filifera (L. Linden) H. Wendl., California Fan Palm
- *Food*—**Cahuilla** *Beverage* Fruit soaked in water to make a beverage. *Dried Food* Fruit sun dried for future use. *Fruit* Fruit eaten fresh. *Porridge* Fruit and seed ground into a flour and used to make mush. *Preserves* Fruit used to make jelly. (15:145) **Cocopa** *Beverage* Juice squeezed out of fruits, added to water, and used as a beverage. (37:204) **Pima, Gila River** *Baby Food* Fruits eaten raw primarily by children. (133:7) *Snack Food* Fruit eaten primarily by children as a snack food. (133:5)
- *Fiber*—**Cahuilla** *Building Material* Long fronds used to make the sides and roofs of waterproof and windproof houses. Long fronds used to make ramadas. (15:145)
- *Other*—**Cahuilla** *Ceremonial Items* Seeds used as filling material for gourd rattles. Leaves used to make images of the dead burned in the memorial rites. *Cooking Tools* Stems used to make cooking utensils, spoons, and stirring implements. *Hunting & Fishing Item* Stems used to make bows. *Tools* Leaves used for flailing and hulling dried seeds. *Toys & Games* Leaves used to make children's play hoops. (15:145) **Cocopa** *Musical Instrument* Seeds used in gourd rattles. (37:187)

Wikstroemia, Thymelaeaceae

***Wikstroemia* sp.**, Akia Launui
- *Drug*—**Hawaiian** *Laxative* Plant pounded, squeezed, the resulting liquid poured into a sweet potato and eaten as a laxative. *Respiratory Aid* Plant and other ingredients pounded, squeezed, and the resulting liquid taken for asthma. (2:8)

Wislizenia, Capparaceae

Wislizenia refracta Engelm. **ssp. *refracta***, Spectacle Fruit
- *Food*—**Hopi** *Unspecified* Young plants boiled for food. (as *W. melilotoides* 200:78)

Woodsia, Dryopteridaceae, fern

Woodsia neomexicana Windham, New Mexico Cliff Fern
- *Drug*—**Keres, Western** *Gynecological Aid* Infusion of plant used as a douche at childbirth. (as *W. mexicana* 171:74) **Navajo, Ramah** *Dermatological Aid* Cold, compound infusion taken and used as a lotion for injury, a "life medicine." (as *W. mexicana* 191:11)

Woodsia scopulina D. C. Eat., Rocky Mountain Woodsia
- *Other*—**Okanagan-Colville** *Water Indicator* Ferns considered to be a sign of water when traveling through the mountains. (188:18)

Woodwardia, Blechnaceae, fern

Woodwardia fimbriata Sm., Giant Chainfern
- *Fiber*—**Karok** *Basketry* Leaf fiber used to make baskets. Fibers were pounded from the rachis and dyed with *Alnus* bark for use in basketry. The fibers were then dried and coiled for storage. They were soaked to unroll and used. **Tolowa** *Unspecified* Leaves used to obtain fiber. The two fibers of each leaf were collected at the end of June or the first part of July. The rachis was pounded and fibers fell out. For designs, the fibers were dyed with *Alnus* bark. **Yurok** *Unspecified* Leaves used to obtain fiber. Fibers were harvested when leaves were fully grown and then dyed with alder bark. The strands could be dyed by chewing in *Alnus* bark and running the fibers through the mouth or by pounding the bark in a mortar and pestle. (6:63)
- *Other*—**Pomo, Kashaya** *Cooking Tools* Long leaves used to line the top and bottom of earth oven for baking acorn bread and other foods. (72:45)

Woodwardia radicans (L.) J. Sm., Rooting Chainfern
- *Drug*—**Luiseño** *Analgesic* Decoction of roots used externally and internally for pain from injuries. (155:234)
- *Fiber*—**Hupa** *Basketry* Dyed fronds used in basketry. (118:4) **Karok** *Basketry* Stems used in basket making. (148:378)
- *Other*—**Pomo** *Cooking Tools* Fronds used for lining the top and bottom of an earth oven in baking acorn bread. (66:11)

Woodwardia sp., Fern
- *Fiber*—**Poliklah** *Basketry* Stems used to make the designs on baskets. (117:170)

Wyethia, Asteraceae

Wyethia amplexicaulis (Nutt.) Nutt., Mulesears Wyethia
- *Drug*—**Gosiute** *Dermatological Aid* Infusion of roots used for bruise swellings. (39:349) Poultice of roots applied to bruised limbs. *Orthopedic Aid* Poultice of roots applied to swollen limbs. (39:384) **Okanagan-Colville** *Antirheumatic (External)* Poultice of warmed, pounded roots applied for arthritic or rheumatic pain. (188:85) **Paiute** *Emetic* Infusion or decoction of pulverized root taken as an emetic. **Shoshoni** *Dermatological Aid* Poultice of pulped root applied to swellings. *Emetic* Infusion or decoction of pulverized root taken as an emetic. *Misc. Disease Remedy* Decoction of root used as a wash for measles. *Venereal Aid* Compound decoction of root taken as an "unfailing cure for syphilis." (180:148)
- *Food*—**Gosiute** *Unspecified* Seeds used for food. (39:384) **Montana Indian** *Unspecified* Root heated, fermented, and eaten. (19:26)

Wyethia angustifolia (DC.) Nutt., California Compassplant
- *Drug*—**Costanoan** *Dermatological Aid* Poultice of pounded root lather used to draw blisters. *Pulmonary Aid* Poultice of pounded root lather used for lung problems. (21:27) **Miwok** *Diaphoretic* Decoction of leaves used as a bath for fever and produced perspiration. *Febrifuge* Decoction of leaves used as a bath for fever. (12:174) **Yuki** *Emetic* Decoction of roots taken as an emetic. (49:47)
- *Food*—**Costanoan** *Staple* Seeds eaten in pinole. *Unspecified* Raw stems used for food. (21:255) **Pomo** *Staple* Seeds used to make pinoles. (11:87) **Pomo, Kashaya** *Staple* Seeds ground to mix with pinole. *Winter Use Food* Seeds dried for winter use. (72:111) **Yuki** *Spice* Seeds used to flavor pinole. (49:85)

Wyethia glabra Gray, Coast Range Mulesears
- *Food*—**Pomo, Kashaya** *Unspecified* Seeds used in pinole or eaten fresh. (72:74)

Wyethia helenioides (DC.) Nutt., Whitehead Wyethia
- *Other*—**Miwok** *Cooking Tools* Green leaves used as top layer, over the hot stones, in the earth oven. (12:139)

Wyethia helianthoides Nutt., Sunflower Wyethia
- *Food*—**Montana Indian** *Unspecified* Plant heated, fermented, and eaten. (19:26)

Wyethia longicaulis Gray, Humboldt Mulesears
- *Drug*—**Mendocino Indian** *Analgesic* Decoction of roots used as a wash for headaches. *Antirheumatic (External)* Poultice of baked roots applied for rheumatism. *Burn Dressing* Poultice of dried, powdered roots applied to burns. *Dermatological Aid* Poultice of dried, powdered roots applied to running sores. *Emetic* Decoction of roots taken as an emetic. *Eye Medicine* Decoction of roots used as a wash for inflamed and sore eyes. *Gastrointestinal Aid* Decoction of roots taken for stomach complaints. (41:396)
- *Food*—**Mendocino Indian** *Staple* Seeds used with parched wheat for pinole. *Unspecified* Leaves and young stems used for food. (41:396) **Pomo** *Staple* Seeds used to make pinoles. (11:87)

Wyethia mollis Gray, Woolly Wyethia
- *Drug*—**Klamath** *Antirheumatic (External)* Poultice of root used for swellings. (45:106) *Dermatological Aid* Poultice of smashed roots applied to swellings. (163:131) **Paiute** *Blood Medicine* Decoction of root taken as a blood tonic. *Cathartic* Decoction of root taken as a physic or emetic. *Cold Remedy* Simple or compound decoction of chopped roots taken for colds. (180:148, 149) *Dermatological Aid* Poultice of crushed leaves applied to swellings. (104:196) *Febrifuge* Compound decoction of chopped roots taken for colds and fevers. (180:148, 149) *Orthopedic Aid* Poultice of crushed leaves applied to broken bones and sprains. (104:196) *Tonic* Decoction of root taken as a blood tonic and for colds. *Tuberculosis Remedy* Decoction of root taken for tuberculosis. *Venereal Aid* Decoction of root taken for venereal diseases. **Shoshoni** *Cathartic* Decoction of root taken as a physic or emetic. **Washo** *Cathartic* Decoction of root taken as a physic. *Emetic* Decoction of root taken as an emetic. (180:148, 149)
- *Food*—**Paiute** *Staple* Seeds parched, ground, and eaten as meal. *Unspecified* Stems and seeds eaten raw. (104:98) **Paiute, Northern** *Porridge* Seeds parched, winnowed, ground, and used to make mush. *Unspecified* Seeds roasted, ground, sometimes mixed with other seeds, and eaten. *Winter Use Food* Seeds stored for winter use. (59:47)

Wyethia ovata Torr. & Gray ex Torr., Southern Mulesears
- *Food*—**Paiute** *Unspecified* Species used for food. (167:242)

Wyethia scabra Hook., Badlands Wyethia
- *Drug*—**Hopi** *Emetic* Plant said to be a very strong, potentially lethal

emetic. (200:34, 99) *Poison* Plant sometimes used as an emetic, but if not vomited it would kill the person. (200:99) **Navajo, Kayenta** *Emetic* and *Gastrointestinal Aid* Plant used as an emetic for stomachaches. (205:51)

Wyethia sp., Narrow Leaf Wyethia
• *Food*—**Pomo** *Staple* Seeds used to make pinole. (117:284)

Xanthium, Asteraceae

Xanthium sp., Cockleburr
• *Drug*—**Jemez** *Urinary Aid* Decoction of plant taken for urinary disorders. (44:28)
• *Food*—**Yavapai** *Staple* Parched, ground seeds made into a meal and eaten dry or moistened and whole seeds stored. (65:258)

Xanthium spinosum L., Spiny Cockleburr
• *Drug*—**Cherokee** *Emetic* Infusion of root given to induce vomiting. *Gastrointestinal Aid* Taken for cramps. *Pulmonary Aid* Infusion given for croup. *Snakebite Remedy* Roots chewed for rattlesnake bite. *Throat Aid* Infusion of bur used "to unstick object in throat." (80:29, 30) **Mahuna** *Kidney Aid* Plant used for the kidneys. (140:69)

Xanthium strumarium L., Rough Cockleburr
• *Drug*—**Costanoan** *Urinary Aid* Decoction of seeds used for bladder ailments. (21:28) **Lakota** *Ceremonial Medicine* Used as a medicine in ceremonies. (139:40) **Paiute, Northern** *Oral Aid* Burs rubbed on sore gums to take the pain, poison, and blood out. (as *Zanthium strumarium* 59:130)

Xanthium strumarium var. canadense (P. Mill.) Torr. & Gray, Canada Cockleburr
• *Drug*—**Apache, White Mountain** *Blood Medicine* Roots and leaves used as a blood medicine. (as *X. commune* 136:161) **Houma** *Febrifuge* Decoction of root taken for high fever. (as *X. commune* 158:60) **Keres, Western** *Dermatological Aid* Poultice of ground seed powder used on open sores or saddle galls. (as *X. commune* 171:74) **Koasati** *Gynecological Aid* Decoction of roots taken to remove the afterbirth. (as *X. commune* 177:64) **Mahuna** *Antirheumatic (Internal)* Plant used for rheumatism. *Kidney Aid* Plant used for diseased kidneys. *Orthopedic Aid* Plant used for total paralysis. *Tuberculosis Remedy* Plant used for tuberculosis. *Venereal Aid* Plant used for gonorrhea. (as *X. canadense* 140:69) **Navajo** *Dermatological Aid* Plant used as a liniment for the armpit to remove excessive perspiration. (as *X. canadense* 55:90) Plant used to decrease perspiration. (as *X. canadense* 90:164) **Pima** *Antidiarrheal* Decoction of burs taken for diarrhea. *Eye Medicine* Pulp used for sore eyes. (as *X. saccharatum* 47:97) Pulp mixed with soot and used for sore eyes. (as *X. canadense* 146:80) *Laxative* Decoction of burs taken for constipation. *Veterinary Aid* Poultice of leaves applied to screwworm sores in livestock. (as *X. saccharatum* 47:97) **Rappahannock** *Dermatological Aid* Decoction of seeds used as salve for sores. (as *X. canadense* 161:31) *Panacea* Compound decoction used for complaint. (as *X. canadense* 161:30) **Tewa** *Antidiarrheal* Plant used for diarrhea. *Antiemetic* Plant used for vomiting. *Pediatric Aid* and *Urinary Aid* Plant used as fumigant for children with urinary disorders. (as *X. commune* 138:49) **Zuni** *Ceremonial Medicine* Chewed seeds rubbed on body prior to cactus ceremony to protect from spines. *Dermatological Aid* Compound poultice of seeds applied to wounds or used to remove splinters. (as *X. commune* 166:62, 63)
• *Food*—**Apache, White Mountain** *Bread & Cake* Seeds ground and used to make bread. (as *X. commune* 136:161) **Costanoan** *Staple* Seeds eaten in pinole. (21:255) **Zuni** *Bread & Cake* Seeds ground, mixed with cornmeal, made into pats, and steamed. (as *X. commune* 32:54) Seeds ground with cornmeal, made into cakes or balls, steamed, and used for food. (as *X. commune* 166:71)
• *Other*—**Keres, Western** *Paint* Ground, seed powder used as a blue paint for the mask dancers. (as *X. commune* 171:74) **Navajo, Ramah** *Ceremonial Items* Leaf ash used as ceremonial blackening. (as *X. italicum* 191:54) **Pima** *Cooking Tools* Leaves used in roasting pits as containers for beans. (as *X. commune* 32:45)

Xanthium strumarium var. glabratum (DC.) Cronq., Rough Cockleburr
• *Drug*—**Iroquois** *Witchcraft Medicine* Plant used in a witching medicine. (as *X. orientale* 87:469)

Xanthorhiza, Ranunculaceae

Xanthorhiza simplicissima Marsh., Yellow Root
• *Drug*—**Catawba** *Cold Remedy* Decoction of roots taken for colds. *Gastrointestinal Aid* Decoction of roots taken for ulcerated stomachs. (as *Xanthorrhiza apiifolia* 157:188) Decoction of plant taken for ulcerated stomach. (as *Zanthorhiza apiifolia* 177:22) *Liver Aid* Decoction of roots taken for jaundice. (as *Xanthorrhiza apiifolia* 157:188) Decoction of plant taken for jaundice. (as *Zanthorhiza apiifolia* 177:22) **Cherokee** *Analgesic* Infusion of root taken for cramps. *Blood Medicine* Compound decoction taken as a blood tonic. *Cancer Treatment* Ashes "burnt from greenswitch" used for cancer. *Dermatological Aid* Astringent and tonic infusion of root used for piles. *Eye Medicine* Poultice used for sore eyes. *Hemorrhoid Remedy* Infusion of root used for piles. *Oral Aid* Stem chewed for sore mouth. *Sedative* Infusion taken for nerves. *Throat Aid* Stem chewed for sore throat. *Tonic* Taken as a tonic. (80:62)
• *Dye*—**Cherokee** *Yellow* Entire plant crushed and used to make a yellow dye. (80:62)

Xanthoria, Teloschistaceae, lichen

Xanthoria elegans (Link) Th. Fr.
• *Other*—**Haisla & Hanaksiala** *Paint* Plant used as pigment for face paint. (43:144) **Kitasoo** *Paint* Used as a yellow pigment for paint. (43:311)

Xanthosoma, Araceae

Xanthosoma atrovirens K. Koch & Bouche, Yautia Amarilla
• *Food*—**Seminole** *Unspecified* Plant used for food. (169:465)

Xerophyllum, Liliaceae

Xerophyllum sp., Beargrass
• *Fiber*—**Hahwunkwut** *Basketry* Plant used to make cooking bowls, mush baskets, and other small baskets. (117:183) **Karok** *Basketry* Used to make baskets. *Clothing* Used to make basket hats for men

and women. (117:211) **Konomeho** *Basketry* Plant used to make the design on baskets. (117:234) **Poliklah** *Basketry* Used to make baskets. (117:170) **Wintoon** *Basketry* Plant purchased and used extensively in basket overlay and designs. (117:264)
- *Other*—**Hahwunkwut** *Cooking Tools* Plant used to make cooking bowls, mush baskets, and other small baskets. (117:183) **Poliklah** *Jewelry* Blades braided to make necklaces. (117:170)

Xerophyllum tenax (Pursh) Nutt., Common Beargrass
- *Drug*—**Blackfoot** *Dermatological Aid* Poultice of chewed roots applied to wounds. *Hemostat* Grated roots used for bleeding. *Orthopedic Aid* Decoction of grated roots used for breaks and sprains. (97:25) **Pomo, Kashaya** *Dermatological Aid* Roots washed, rubbed to make a lather and used to wash sores. (72:52)
- *Fiber*—**Chehalis** *Basketry* Plant used to decorate baskets. **Cowlitz** *Basketry* Plant used to decorate baskets. (79:23) **Hesquiat** *Basketry* Purchased grass shaved and dyed to make baskets and edging for mats and baskets. (185:56) **Hoh** *Basketry* Grass used to make baskets. (137:60) **Hupa** *Basketry* Grass used as a border pattern in baskets. (118:2) Leaves used to make baskets. (148:380) **Karok** *Basketry* New sprouts used to make baskets, especially for designs. (6:63) Leaves used to make baskets. (148:380) **Klallam** *Basketry* Plant used to decorate baskets. (79:23) **Maidu** *Basketry* Blades used as overlay twine in the manufacture of baskets. (173:71) **Makah** *Basketry* Leaves used for basketry. (67:343) **Montana Indian** *Basketry* Leaves woven into watertight baskets used for cooking. (as *X. douglasii* 19:27) **Nitinaht** *Basketry* Grass used as twining material for wrapped and twined baskets. (186:87) **Okanagan-Colville** *Basketry* Used to decorate baskets. (188:50) **Quileute** *Basketry* Grass used to make baskets. (137:60) **Quinault** *Basketry* Plant used to decorate baskets. (79:23) **Yana** *Clothing* Grass wrapped around the ties of the women's belts. (147:253) **Yurok** *Clothing* Small leaves used to make dresses. Plants were burned every year. Leaves were harvested in the spring when they first began to grow out from their charred rhizome. Prior to use, the leaves were soaked in water to make them pliable, but if left too long they turned green. (6:63)
- *Other*—**Hupa** *Decorations* Plant used as dress ornaments. **Karok** *Decorations* Plant used as dress ornaments. (148:380) **Makah** *Cash Crop* Leaves sold to the Nootkan and Hesquiat tribes. **Nitinaht** *Tools* Leaves used for trimming the edges of mats. (186:87)

Ximenia, Olacaceae
Ximenia americana L., Tallow Wood
- *Drug*—**Seminole** *Antirheumatic (External)* Decoction of inner bark or "beans" used as a body rub and steam for deer sickness: numb, painful limbs and joints. (169:192) Decoction of inner bark used as a body rub and steam for joint swellings. (169:193) *Oral Aid* Decoction of roots used as a mouthwash for sore mouth and gums. *Orthopedic Aid* Infusion of roots used as a steam and rubbed on the legs for soreness. (169:307) *Unspecified* Plant used medicinally. (169:164)
- *Food*—**Seminole** *Unspecified* Plant used for food. (169:488)

Xylorhiza, Asteraceae
Xylorhiza tortifolia (Torr. & Gray) Greene, Mojave Aster
- *Other*—**Havasupai** *Incense & Fragrance* Ground leaves carried in the clothes and used as perfume by men and women to counteract body odors. (197:250)

Xyridaceae, see *Xyris*

Xyris, Xyridaceae
Xyris ambigua Bey. ex Kunth, Coastalplain Yelloweyed Grass
- *Drug*—**Seminole** *Cold Remedy* Infusion of herbage rubbed on the chest for colds. *Pulmonary Aid* Infusion of herbage rubbed on the chest for pulmonary disorders. (169:281)

Xyris caroliniana Walt., Carolina Yelloweyed Grass
- *Drug*—**Cherokee** *Antidiarrheal* Infusion of root taken for diarrhea. *Pediatric Aid* Infusion of root taken for diarrhea and "good for children." (80:62)

Xyris sp., Yelloweyed Grass
- *Drug*—**Seminole** *Cold Remedy* Infusion of herbage rubbed on the chest for colds. *Pulmonary Aid* Infusion of herbage rubbed on the chest for pulmonary disorders. (169:281) *Witchcraft Medicine* Flowering stems twisted into a witchcraft medicine. (169:395)

Yucca, Agavaceae
Yucca aloifolia L., Aloe Yucca
- *Drug*—**Choctaw** *Dermatological Aid* Boiled mashed root with grease or tallow used as salve for various purposes. (25:24) *Unspecified* Decoction of roots used as a salve for various purposes. (177:9)

Yucca angustissima Engelm. ex Trel., Narrowleaf Yucca
- *Drug*—**Apache** *Snakebite Remedy* Emulsion used in cases of snake or insect bites. (17:51) **Hopi** *Ceremonial Medicine* Crushed root used in purification ceremony. *Dermatological Aid* Crushed root used as shampoo for baldness. *Disinfectant* Crushed root used in purification ceremony. (200:71) *Laxative* Root used as a strong laxative. (200:34, 71)
- *Food*—**Hopi** *Fruit* Fruit sometimes used for food. (17:64) Fruits pit baked with lamb's-quarter leaves and eaten with corn dumplings in salted water. (120:18) **Southwest Indians** *Starvation Food* Used when agricultural reserves dwindled. (17:10)
- *Fiber*—**Apache** *Cordage* Leaves split and used as string. (17:40) **Hopi** *Basketry* Leaves used in many types of basketry. (200:71) *Brushes & Brooms* Leaf splints used as brushes to apply color to pottery. (17:50) *Sewing Material* Narrow, split leaf strips used as sewing material for coiled plaques. (17:33) **Southwest Indians** *Cordage* Leaves used to make strings. (17:37)
- *Other*—**Apache** *Soap* Roots pounded and placed in water to form suds used in bathing and shampooing. (17:57) **Havasupai** *Fasteners* Leaves used to tie or repair holes in sacking. *Toys & Games* Leaves used as tally sticks to keep track of scores in the hidden-ball game. (197:213) **Hopi** *Ceremonial Items* Roots crushed to make soap, used ceremonially as a purification rite, and suds associated with clouds. Used as a whip during ceremonies. (200:71) *Decorations* Shredded leaves used to make the packing for the spirals of a plaque. (17:34) *Paint* Juice used as a varnish on certain kachinas. *Soap* Roots crushed

604 *Yucca angustissima* (Agavaceae)

with stones and used as soap. (200:71) **Southwest Indians** *Ceremonial Items* Roots made into suds and used during marriage ceremonies. As part of the marriage ceremony, the groom's head was washed by his future mother-in-law, while that of the bride was washed by the bridegroom's mother. (17:55)

Yucca baccata Torr., Banana Yucca

- **Drug**—**Keresan** *Dermatological Aid* Used for washing hair. (198:564) **Navajo** *Antiemetic* Infusion of pulverized leaves taken for vomiting. *Gastrointestinal Aid* Plant used for heartburn. (55:32) **Navajo, Ramah** *Ceremonial Medicine* Suds made from root used for ceremonial purification baths. *Gynecological Aid* Juice used to lubricate midwife's hand while removing retained placenta. (191:21) **Pima** *Cathartic* Fruits eaten raw as a cathartic. (146:72) **Tewa** *Ceremonial Medicine* Infusion of root used as a wash in adoption and name-giving ceremonies. *Emetic* Unspecified plant part chewed as ritual emetic. *Gynecological Aid* Fruit eaten to promote easy childbirth. (138:49, 50)
- **Food**—**Acoma** *Beverage* Dried fruits dissolved in water to make a drink. *Bread & Cake* Fruits baked, boiled, dried, rolled into loaves, and stored for winter use. (32:54) *Dried Food* Fruits sun dried and stored for winter use. *Fruit* Fruits eaten raw, boiled, or baked. (32:55) *Preserves* Dried fruits eaten as a paste. *Sauce & Relish* Dried fruits dissolved in water and used as a dip. (32:54) *Starvation Food* Tender crowns roasted and eaten in times of food shortages. (32:55) **Apache** *Beverage* Baked fruit pounded to a pulp, drained, and juice drunk. *Bread & Cake* Fruit roasted, pulp made into cakes, and stored. *Sauce & Relish* Baked fruit pounded to a pulp, drained, and juice poured over cakes. (17:18) *Soup* Young leaves cooked in soups or with meat. *Unspecified* Flowers eaten as food only if obtained at the proper time. (32:56) *Vegetable* Flowers eaten as a vegetable only if obtained before the summer rains. (17:19) **Apache, Chiricahua & Mescalero** *Bread & Cake* Fruit roasted, split, seeds removed, and pulp ground into large cakes. Fruit pulp ground, made into large cakes, and stored indefinitely. *Soup* Leaves cooked in soups. *Unspecified* Leaves boiled with meat. *Vegetable* Flowers eaten if obtained before the summer rain, otherwise they taste bitter. (33:39) **Apache, Mescalero** *Beverage* Fruits used to make a drink. *Dried Food* Ripe fruits cooked, split, cleaned of seeds, dried, and used for food. *Sauce & Relish* Fruits made into a syrup and placed on fruits before drying. (14:33) **Apache, Western** *Dried Food* Fruit roasted, dried, wrapped, and stored indefinitely. (26:181) *Sauce & Relish* Fruit pounded together to make gravy. (26:182) **Apache, White Mountain** *Dried Food* Pods dried for future use. *Unspecified* Pods roasted and used for food. (136:147) **Cochiti** *Unspecified* Used as a source of food. (32:14) **Havasupai** *Beverage* Plant used to make a drink. (197:66) *Dried Food* Sheet of fruit flesh dried and the bits eaten dry when needed. (17:17) Fruits split, sun dried, and prepared for storage in the shape of a mat. (197:212) **Hopi** *Fruit* Baked fruits used for food. (42:371) Fruit used for food. (56:17) Fruits eaten for food. (138:51) Large fruits oven baked. (200:71) *Preserves* Fruits sun dried, boiled into jam, and eaten with corn dumplings or boiled bread. (120:18) **Hualapai** *Beverage* Fruit used to make a fermented beverage. *Dried Food* Fruit baked, prepared, and dried for winter use. *Fruit* Fruit eaten raw. *Staple* Fruit cooked and ground into a meal. (195:39) **Isleta** *Forage* Fruit often eaten by deer which left few for the Isletans. *Fruit* Fruit baked, seasoned, and used for food. (100:45) *Unspecified* Used as a source of food. (32:14) *Winter Use Food* Sun dried fruit used for winter storage. (100:45) **Jemez** *Unspecified* Used as a source of food. (32:14) **Keres, Western** *Fruit* Fruit eaten when thoroughly ripe. *Sauce & Relish* Soaked, cooked fruit made into a syrup and used as hot chocolate. *Winter Use Food* Cooked, dried fruit stored for winter use. (171:74) **Keresan** *Fruit* Fruit used for food. (198:564) **Laguna** *Beverage* Dried fruits dissolved in water to make a drink. *Bread & Cake* Fruits baked, boiled, dried, rolled into loaves, and stored for winter use. (32:54) *Dried Food* Fruits sun dried and stored for winter use. *Fruit* Fruits eaten raw, boiled, or baked. (32:55) *Preserves* Dried fruits eaten as a paste. *Sauce & Relish* Dried fruits dissolved in water and used as a dip. (32:54) *Starvation Food* Tender crowns roasted and eaten in times of food shortages. (32:55) **Navajo** *Bread & Cake* Pulp made into cakes, dried, and stored for winter use. (17:20) Ripe fruits dried, ground, kneaded into small cakes, and slightly roasted. (32:54) Baked or dried fruits ground, made into small cakes, and roasted again. (55:32) *Dessert* Fruit boiled in water with or without sugar and eaten as a dessert. (165:221) *Dried Food* Fruits dried and stored for winter use. (32:54) Fruit dried and carried, when at war, with grass seeds and jerked venison. (55:32) Fruit dried for winter use. (110:31) Ripe fruit, with seeds removed, boiled down like jam, made into rolls, and dried for winter use. (165:221) *Fruit* Fruits eaten ripe or cooked. (17:20) Fruit eaten raw or cooked. (32:54) Fruit eaten when picked or cooked. (55:32) Fruit eaten raw or baked in hot coals. (110:31) Dried fruit rolls soaked in hot water and eaten with corn mush. (165:221) *Porridge* Ripe fruits dried, ground, kneaded into small cakes, and boiled with cornmeal into a mush. (32:54) Baked or dried fruits ground, made into cakes, roasted again, mixed with cornmeal, and made into gruel. (55:32) Dried fruit cakes boiled with cornmeal into a gruel. (110:31) *Preserves* Fruit used to make jelly. (55:32) *Sauce & Relish* Fruit pulp made into cakes and mixed with water to make a syrup eaten with meat or bread. (17:20) Dried fruit cakes mixed with water to make a syrup and eaten with meat and bread. (110:31) *Special Food* Dried fruit eaten by warriors at war. (17:20) *Winter Use Food* Fruit cut in half, dried, and stored for winter use. Baked or dried fruits ground, made into small cakes, roasted again, and stored for winter use. (55:32) **Navajo, Ramah** *Bread & Cake* Fruit molded into foot-long rolls. *Preserves* Fruit used to make preserves. (191:21) **Papago** *Dried Food* Seeds dried and used for food. *Fruit* Fruits eaten fresh. *Porridge* Fresh fruits made into a gruel and used for food. *Staple* Fruits used as an important staple crop. *Unspecified* Pulp ground, cooked, and used for food. (34:23) Species used for food. (36:61) **Pima** *Beverage* Dried fruit made into cakes, cooked in water, and liquid drunk. *Bread & Cake* Dried fruit made into cakes. (17:16) *Dried Food* Fruits boiled, dried, ground, boiled with flour, and used for food. (146:72) *Porridge* Dried fruit made into cakes, ground, and cooked with cornmeal to make gruel. (17:16) **Pima, Gila River** *Candy* Plant dried and used as sweets. (133:6) *Fruit* Fruits eaten raw and pit roasted. (133:7) **Pueblo** *Dried Food* Fruits sun dried and stored for winter use. *Fruit* Fruits eaten raw, boiled, or baked. (32:55) **San Felipe** *Dried Food* Partly matured fruits cooked to form a semiliquid substance, dried, and stored for winter use. *Fruit* Ripe fruits eaten without preparation. (32:54) *Unspecified* Used as a source of food. (32:14) **Sia** *Fruit* Fruit used for food. (199:107) **Southwest Indians** *Beverage* Dried fruit pulp boiled in water and drunk.

Bread & Cake Pulp mixed with chokecherries and made into cake. (17:11) Pulp patted into cakes and dried thoroughly. (17:15) *Dessert* Preserved fruit soaked in water, cooked to a thick syrup, and eaten as a sauce-like dessert. (17:12) *Dried Food* Pulp cooked to a paste, dried, and stored for winter use. (17:10) Rind dried and eaten without cooking. Partially ripened fruits gathered, cooked, the pulp dried and stored for winter use. Pulp cooked to a paste and dried for winter use. (17:11) Fruits dried between beds of grass, split into halves, dried further in the sun, and stored. (17:15) *Fruit* Fruit eaten raw. (17:10) Fruits eaten for food. (17:9) *Preserves* Fruit pared, pulp chewed, cooked, dried, and eaten as a conserve. (17:12) *Staple* Seeds dried, stored in baskets and ground into meal when needed. (17:15) *Sweetener* Fruit pared, pulp chewed, cooked, dried, and conserve dissolved in water to sweeten beverages. (17:12) *Unspecified* Greatly sought after and utilized as food. (17:10) Green pods gathered and allowed to mature in sun or dwellings. (17:9) **Tewa** *Fruit* Fruits formerly eaten. (138:50) **Walapai** *Beverage* Fruits mixed with water and liquid drunk. *Dried Food* Fruit dried, folded, and stored for winter use. (17:17) **Yavapai** *Beverage* Dried fruit soaked in water until pulp dissolved and liquid drunk. (17:16) Sun dried fruit boiled and used as a beverage. *Dried Food* Sun dried fruit boiled and used for food. *Fruit* Fruit cooked in coals and used for food. (65:258) *Unspecified* Flower stalks gathered before blossoming, roasted in fire, and prepared for use. (17:16) **Zuni** *Fruit* Fruits pared and eaten raw or boiled and skinned. (32:54) Fruit eaten fresh or boiled, cooled, and the skin peeled off with a knife. (166:72) *Preserves* Flesh cooked, made into pats, sun dried, and eaten as a conserve. (32:54) Fruit made into conserves and used for food. (166:72) *Sauce & Relish* Flesh cooked, made into pats, sun dried, and mixed with water to form a syrup. (32:54) *Sweetener* Fruit made into conserves and used as a sweetener before the introduction of coffee and sugar. (166:72)

- *Fiber*—**Apache** *Basketry* Leaves used for the main portion of the baskets. (17:35) *Cordage* Leaves split and used as string. (17:40) **Apache, Mescalero** *Basketry* Small roots used for basket work. *Cordage* Leaves used to make twine or rope. (14:33) **Apache, Western** *Cordage* Leaves split and sections tied together by square knots to make cordage. (26:182) **Apache, White Mountain** *Clothing* Leaves reduced to fiber and made into cloth. *Cordage* Leaves used to make string. (136:147) **Havasupai** *Cordage* Leaf fiber braided into ropes. The leaves contained a good fiber. The terminal spine and a section of the back of the leaf were removed and pounded to free this fiber from the fleshy portion of the leaf. The fiber was often braided into rope, three- to six-ply, from 12 to 30 feet long that was used for many purposes, including handling horses. (197:212) **Hopi** *Basketry* Leaves used in basketry. (42:371) Used for basketry. (56:17) Leaves occasionally used in basketry. (200:71) **Hualapai** *Clothing* Stems used to make shoes. *Cordage* Stems used to make rope. (195:39) **Isleta** *Basketry* Fibers used to make baskets. *Brushes & Brooms* Fibers used to make small brushes for pottery decoration. *Cordage* Fibers used to make cords and ropes. (100:45) **Jemez** *Basketry* Leaves used to make baskets. (17:34) **Keres, Western** *Cordage* Leaf fibers used to make ropes. Leaves were pounded between stones to separate the fibers which were used for ropes. (171:74) **Navajo, Ramah** *Basketry* Leaves used to make ceremonial and utilitarian baskets. *Brushes & Brooms* Leaves made into brushes used for cleaning baskets. Leaf slivers made into paintbrushes. *Cordage* Leaf fiber made into string or rope and used for temporary or emergency purposes. *Snow Gear* Fiber used to tie snowshoes to the feet. *Sporting Equipment* Leaves made into a ball thrown into the air for archery target practice. Roots made into ball for shinny game, played at night. (191:21) **Papago** *Basketry* Used as the foundation in coiled basketry. (17:34) Leaves used as foundation in coiled basketry. (34:57) *Building Material* Leaves split and used for the weft of wrapped weaving in house frames. (34:53) *Cordage* Leaves split and used as tying material. (17:40) **Pima** *Basketry* Used in basketry. (17:34) *Brushes & Brooms* Fiber used to make hairbrushes. (17:50) *Cordage* Leaf fiber used to make cords or ropes. (17:40) *Mats, Rugs & Bedding* Leaves woven into mats. (17:34) Split leaves used in weaving mats. (17:37) **Southwest Indians** *Cordage* Split leaves or fibers used as tying material. (17:39) **Tewa** *Cordage* Fleshy leaves boiled, chewed, and the fibers twisted into cord and rope. (138:50) **Yavapai** *Cordage* Leaf fiber used to tie grass stems of mescal to make a brush. (65:259) **Zuni** *Basketry* Split leaves used to make winnowing baskets, baskets for serving food and for transporting materials. (166:78) Interlaced leaves used to make baskets. (166:81) *Clothing* Leaves boiled, chewed, and fiber woven into skirts and kilts. (17:45) Leaf fibers used in weaving fabrics. The leaves were folded about 4 inches in length and tied with a fragment of the leaf. These folds were boiled with a small quantity of cedar ashes. When sufficiently cooked, the leaves were placed in a bowl or basket and cooled. When the leaves were cooled, youths and young women peeled off the epidermis and chewed the leaves, starting at one end and progressing to the other. After chewing the leaves, the fibers were separated, straightened out, tied with fiber strings, and hung in an inner room to dry. When required for weaving, the fibers were soaked in water to soften them. (166:78) *Cordage* Leaves boiled, chewed, and made into a double-stranded cord. (17:40) Leaf fibers made into cords used to tie prayer plume offerings together and for other ceremonial uses. Split leaves used in place of cords or rope. (166:78) *Mats, Rugs & Bedding* Leaves split and plaited into mats to cover various vessels. (17:36) Dried leaves split, plaited, and made into water-carrying head pads. (17:47) Split leaves plaited into mats for covering hatchways, grain vases, and other vessels. The leaves are also used for making pads for supporting water vases upon the head. (166:78)

- *Dye*—**Navajo, Ramah** *Unspecified* Leaf juice used as a medium for pigments of pottery paints and slips. (191:21)

- *Other*—**Apache** *Containers* Thick portion of stalk used as hearth when making fire. (17:51) *Designs* Roots used to produce a red pattern in baskets. (17:35) *Soap* Roots pounded and placed in water to form suds used in bathing and shampooing. (17:57) *Tools* Stalk used to make fire drills. (17:51) **Apache, Mescalero** *Soap* Large roots used to make soap. (14:33) **Apache, Western** *Soap* Roots used as soap. (26:182) **Apache, White Mountain** *Soap* Roots used for soap. (136:148) **Havasupai** *Soap* Roots used as a soap for washing the hair. *Tools* Terminal spines used as needles. *Toys & Games* Ring of leaves wrapped in buckskin used in the hoop and pole game. *Waterproofing Agent* Dried leaves boiled with gum, hardened, powdered, mixed with water, and used to waterproof baskets. (197:212) **Hopi** *Soap* Roots used for soap. (42:371) Root used for soap. (56:17) Roots used as soap. (200:71) **Hualapai** *Soap* Roots used for soap. (195:39) **Isleta** *Decorations* Fibers used to make plaques. (100:45) **Keres, Western** *Ceremo-*

nial Items Leaves used as whips during the initiations. Switches used as whips by the masked personage. *Soap* Crushed leaves mixed with water for soap. (171:74) **Navajo** *Ceremonial Items* Roots used ceremonially. (110:31) *Musical Instrument* Stout leaves used as drumsticks. (17:36) *Soap* Roots used to wash hair and garments. (110:31) **Navajo, Ramah** *Ceremonial Items* Leaves used to make ceremonial drumstick. Leaves stuck into snowballs, mixed with red clay, and used to stop the snow and rain. *Fasteners* Leaf juice mixed with pottery paste. *Soap* Root made into soap used for washing wool or clothing, shampooing the hair, and bathing the body. The root, pounded with rocks to remove the bark and to soften it, was stirred vigorously in warm water to whip up suds. (191:21) **Papago** *Soap* Plant macerated, placed in water to form suds, and used for washing hair. **Pima** *Soap* Plant macerated, placed in water to form suds, and used for washing hair. (17:56) Stems reduced to pulp and used as soap. (146:72) **Pima, Desert** *Cash Crop* Used for trade. (133:6) **Southwest Indians** *Ceremonial Items* Plant suspended from a ring carried on the back of a god impersonator. (17:53) *Soap* Leaves pounded and used in washing. (17:55) **Tewa** *Hunting & Fishing Item* Leaves baked in warm pit, chewed, and fiber woven into fishing nets. (17:45) *Soap* Roots bruised, placed in water, and used to wash woolens, cotton fabrics, feathers, and hair. (138:49) **Yavapai** *Soap* Root, stem, and leaves pounded and worked in water to form lather for washing hair and body. (17:56) **Zuni** *Ceremonial Items* Plant used ceremonially for a great variety of purposes. Narrow leaf bands worn around the head by personators of anthropic gods. The personators of anthropic gods adorned their wrists and ankles with yucca ribbons, and the novitiate into the medicine order of a secret fraternity had his or her wrists adorned with them also. (166:99) *Soap* Roots pounded, made into suds in cold water, and used for washing. (17:55) *Tools* Leaves used to make cincture pads for supporting water vases upon the head. (166:78)

Yucca baileyi var. *navajoa* (J. M. Webber) J. M. Webber, Navajo Yucca
- **Drug**—**Hopi** *Laxative* Plant used as a laxative. (as *Y. navaja* 42:370)
- **Fiber**—**Hopi** *Basketry* Leaves used in basketry. *Brushes & Brooms* Twigs used to make snow brooms. Plant used for paintbrushes. (as *Y. navaja* 42:370)
- **Other**—**Hopi** *Ceremonial Items* Twigs used to make the masks for the kachinas. Plant used as whips in ceremonies. *Hunting & Fishing Item* Plant used as an anchor for bird traps. *Paint* Juice used as a varnish for sacred kachinas. *Soap* Crushed roots used for soap. (as *Y. navaja* 42:370)

Yucca brevifolia Engelm., Joshua Tree
- **Food**—**Cahuilla** *Unspecified* Blossoms used for food. (15:150) **Kawaiisu** *Dried Food* Fruit pit roasted, mashed, dried, and stored for future use.84 *Fruit* Fruit pit roasted and eaten.84 (206:69) **Southwest Indians** *Fruit* Fruits eaten for food. (17:63) **Tubatulabal** *Unspecified* Immature pods used for food. (193:16)
- **Fiber**—**Cahuilla** *Clothing* Fibers used to make sandals. *Cordage* Fibers used to make nets. (15:150) **Kawaiisu** *Basketry* Dark red rootstock core used as pattern material in coiled basketry. The core was split into strands, soaked, and worked in with the coiling so that the color was always on the outside. (206:69) **Shoshoni** *Basketry* Roots used to make baskets. (as *Y. arborescens* 117:445) **Southwest Indians** *Clothing* Made into cords and used as base for fur robe garments. (17:43)
- **Dye**—**Shoshoni** *Black* Roots used as black dye in basketry. *Red* Roots used as red dye in basketry. (118:8)
- **Other**—**Panamint** *Designs* Red-brown inner roots used for basket designs. (105:78) **Southwest Indians** *Designs* Roots used to make brown designs. (17:35)

Yucca elata (Engelm.) Engelm., Soaptree Yucca
- **Drug**—**Apache, Western** *Ceremonial Medicine* Peeled stalk shaped like a short snake, heated, eaten by a practitioner and spit at the sick. (26:182)
- **Food**—**Apache** *Unspecified* Species used for food. (17:64) *Vegetable* Flowers boiled and eaten as a vegetable. (17:19) **Apache, Chiricahua & Mescalero** *Dried Food* Stems baked overnight, dried, broken into pieces, softened, and eaten. (33:38) *Vegetable* Flowers boiled and eaten as a vegetable. (33:39) **Apache, Mescalero** *Dried Food* Trunks pit cooked, dried, and stored for future food use. Flowers boiled, dried, and stored for future food use. *Soup* Flowers used as fresh vegetables in soups. *Staple* Trunks pit cooked, pounded, and made into flour. *Unspecified* Young stalks cooked, peeled, and eaten hot. (14:40) **Apache, Western** *Dried Food* Boiled blossoms dried and stored in a dry place. *Unspecified* Blossoms boiled with seeds, fat, or bones. Stalk charred and eaten like sugar cane. (26:182) Blossoms used for food. (26:193) **Papago** *Unspecified* Species used for food. (17:64) **Southwest Indians** *Dried Food* Stems baked, dried, softened in water, and eaten. (17:19)
- **Fiber**—**Apache, Chiricahua & Mescalero** *Basketry* Leaves woven into shallow or tray baskets to carry prepared mescal home. (33:37) **Apache, Western** *Cordage* Leaves used to make cordage. *Furniture* Leaves used for the head shade of cradleboards. (26:182) **Papago** *Basketry* Leaf used as the binding element in coarse coiled ware. (17:34) *Building Material* Used for the weft of wrapped weaving in house frames. (34:54) *Sewing Material* Species used to sew coils into tight baskets. (17:62) **Pima** *Mats, Rugs & Bedding* Leaves twilled into mats. (17:60)
- **Other**—**Apache, Western** *Decorations* Red roots used in basket decorations. *Soap* Roots used for soap. (26:183) **Navajo** *Ceremonial Items* Leaves made into scourges and used in the Night Chant. A leaf was taken from the east side of the plant and one from the west. The leaves were then split in two and the interchanged halves bound together to form the scourge. These scourges were carried by the different personators in the Night Chant. *Soap* Roots used for soap. *Toys & Games* Used to make the 102 counting sticks for the moccasin game. (55:33) **Pima** *Containers* Fibers made into carrying nets. (17:35) Cactus ribs bound together to form the frame for containers used to carry crops. (17:46) *Soap* Used as soap. (146:72) **Southwest Indians** *Fasteners* Leaves tied to make a fastening loop for sandals. (17:28)

Yucca filamentosa L., Adam's Needle
- **Drug**—**Catawba** *Dermatological Aid* Root rubbed on the body for skin disease. (157:188) Roots rubbed on body or decoction of roots taken for skin disease. (177:9) **Cherokee** *Ceremonial Medicine* Used as an ingredient with broom sedge and amaranth in green corn medicine. *Dermatological Aid* Beaten root used as salve for sores.

Misc. Disease Remedy Infusion taken for diabetes. (80:25) *Sedative* Used as a soporific. (203:75) **Nanticoke** *Orthopedic Aid* Poultice of roots applied to sprains. (175:56)
- *Other*—**Cherokee** *Hunting & Fishing Item* Used to intoxicate fish. (80:25) Pounded roots strewed on water to "intoxicate fishes." (203:75) *Soap* Roots pounded, boiled, and used instead of soap to wash blankets. (80:25)

Yucca glauca Nutt., Small Soapweed
- *Drug*—**Blackfoot** *Antirheumatic (External)* Decoction of grated roots used for sprains. (86:80) *Dermatological Aid* Decoction of root used as a tonic for falling hair. Poultice of roots applied to inflamed and bleeding cuts. (114:274) *Hemostat* Poultice of grated roots applied to bleeding cuts. (97:25) Poultice of roots applied to inflamed and bleeding cuts. *Herbal Steam* Decoction of grated roots used as herbal steam for breaks and sprains. (114:274) *Orthopedic Aid* Decoction of grated roots used for breaks. (86:80) Decoction of grated roots used as herbal steam for breaks and sprains. (114:274) *Veterinary Aid* Decoction of roots applied to saddle sores. (97:25) **Cheyenne** *Dermatological Aid* Decoction of dried root used as hair wash for dandruff and to prevent baldness. Smashed root applied as powder or used as a wash for sores, scabs, and skin outbreaks. (83:12) **Dakota** *Dermatological Aid* Root used to wash the scalp to make the hair grow. (69:358) **Isleta** *Dermatological Aid* Root mixed with ground stolons from vine mesquite grass and used as a hair wash to make the hair grow. (100:45) **Keres, Western** *Herbal Steam* Tender heart shoots used in the sweat bath. *Strengthener* Infusion of tender heart shoots used for weakness. (171:76) **Kiowa** *Dermatological Aid* Plant used for dandruff, baldness, and skin irritations. (192:18) **Lakota** *Abortifacient* Roots and prickly pear cactus roots used as "medicine for not give birth." *Dermatological Aid* Infusion of roots used to soak the hair as a vermin killer and to make the hair grow. *Gastrointestinal Aid* Infusion of pulverized roots taken for stomachaches. *Gynecological Aid* Roots and prickly pear cactus roots used by mothers when they cannot give birth. *Veterinary Aid* Burning root fumes used to allow a horse to be caught and haltered easily. (139:28) **Navajo** *Dermatological Aid* Roots used to wash hair. (55:33) *Laxative* Plant used as a laxative. *Stimulant* Plant used as a delirifacient, a drug that causes delirium. (90:164) **Navajo, Ramah** *Contraceptive* Rotten root used to make suds taken to induce menopause. *Gynecological Aid* Cold infusion of root used to expedite delivery of baby or placenta. *Poison* Compound containing leaf juice used to poison arrows. (191:21) **Omaha** *Unspecified* Root used in smoke treatment for unspecified illnesses. **Pawnee** *Herbal Steam* Root used in smoke treatment for unspecified illnesses. (70:71)
- *Food*—**Acoma** *Dried Food* Fruits sun dried and stored for winter use. *Fruit* Fruits eaten raw, boiled, or baked. (32:55) **Apache** *Unspecified* Stalks roasted, boiled, or eaten raw. (17:19) Flowers eaten as food. (32:56) **Apache, Chiricahua & Mescalero** *Unspecified* Stalks roasted, boiled, or eaten raw. *Vegetable* Stalks boiled, dried, and stored to be used as vegetables. (33:38) **Apache, White Mountain** *Dried Food* Pods dried for future use. *Unspecified* Pods roasted and used for food. (136:147) **Cochiti** *Unspecified* Used as a source of food. (32:14) **Isleta** *Forage* Fruit often eaten by deer, which left few for the Isletans. *Fruit* Fruit baked, seasoned, and used for food. (100:45) *Unspecified* Used as a source of food. (32:14) *Winter Use Food* Sun dried fruit stored for winter use. (100:45) **Jemez** *Unspecified* Used as a source of food. (32:14) **Keres, Western** *Beverage* Dried, fruit cakes used to make a beverage. *Bread & Cake* Boiled, dried fruit made into cakes. (171:76) *Fruit* Fruit eaten when thoroughly ripe. *Sauce & Relish* Soaked, cooked fruit make into a syrup and used as hot chocolate. (171:74) Dried, fruit cakes made into a syrup. *Starvation Food* Roasted hearts used for food in times of famine. *Unspecified* Tender heart shoots eaten for food. (171:76) *Winter Use Food* Cooked, dried fruit stored for winter use. (171:74) **Laguna** *Dried Food* Fruits sun dried and stored for winter use. *Fruit* Fruits eaten raw, boiled, or baked. (32:55) **Lakota** *Unspecified* Flowers and buds eaten fresh or cooked. (106:51) **Navajo** *Fruit* Fruit eaten raw or baked in ashes. *Winter Use Food* Fruit sliced and dried for winter use. (55:33) **Navajo, Ramah** *Fruit* Fruit roasted in ashes and eaten. *Unspecified* Flower buds roasted in ashes and leaves boiled with salt and used for food. (191:21) **Pueblo** *Dried Food* Fruits sun dried and stored for winter use. *Fruit* Fruits eaten raw, boiled, or baked. (32:55) **San Felipe** *Unspecified* Used as a source of food. (32:14) **Southwest Indians** *Sauce & Relish* Pods boiled in water and made into pickles. *Starvation Food* Crowns roasted and used in times of food shortage. (17:14) Used when agricultural reserves dwindled. (17:9) *Unspecified* Tender stalks boiled or baked. Seeds eaten with the pods. (17:14) **Tewa** *Fruit* Fruits eaten for food. (138:52) **Zuni** *Unspecified* Seedpods boiled and used for food. (166:73)
- *Fiber*—**Apache, Mescalero** *Basketry* Leaves split and used to make baskets. (14:40) **Apache, White Mountain** *Cordage* Leaves used to make "moccasin strings" and cords. (136:147) **Dakota** *Cordage* Leaves macerated to clear the fibers and with the sharp leaf points attached, twined into thread. *Sewing Material* Sharp points of leaves used as needles. (70:71) **Hopi** *Brushes & Brooms* Leaf splints used as brushes to apply color to pottery. (17:50) **Isleta** *Basketry* Fibers used to make baskets. *Brushes & Brooms* Fibers used to make brushes for pottery decoration. *Cordage* Fibers used to make cords and ropes. (100:45) **Keres, Western** *Basketry* Leaves used to make baskets. *Brushes & Brooms* Chewed leaves used to make brushes for pottery paint. *Cordage* Crushed leaf fibers twisted and used for ropes. (171:76) **Navajo, Ramah** *Cordage* Fiber made into string to tie hoops, prayer sticks, chant arrows, and other ceremonial equipment. (191:21) **Omaha** *Cordage* Leaves macerated to clear the fibers and with the sharp leaf points attached, twined into thread. *Sewing Material* Sharp leaf points used as needles. (70:71) **Papago** *Basketry* Leaves used as foundation in coiled basketry. (34:57) Inner leaves dried, split, and used for sewing nonwatertight trade baskets. (34:58) *Building Material* Leaves split into strands, made into cords, and used for lashing house frames together and thatching. (34:61) Strips of fiber used to tie mesquite pole tips to other poles in forming house skeletons. (34:66) *Cordage* Leaves split into strands, made into cords, and used for tying up bundles of material. (34:61) **Pawnee** *Cordage* Leaves macerated to clear the fibers and with the sharp leaf points attached, twined into thread. *Sewing Material* Sharp leaf points used as needles. **Ponca** *Cordage* Leaves macerated to clear the fibers and with the sharp leaf points attached, twined into thread. *Sewing Material* Sharp leaf points used as needles. (70:71) **Southwest Indians** *Basketry* Leaves used in the manufacture of baskets. (17:34) *Brushes & Brooms* Leaves used as brushes to place designs upon pottery. (17:49) *Cordage* Split leaves or fibers used as tying material. (17:39) **Tewa**

Basketry Used to make sifting baskets. *Brushes & Brooms* Narrow slips used as paintbrushes in decorating pottery. *Clothing* Used to make bandoleers and neckties for clowns and dancers. *Cordage* Fibrous leaves split into narrow strips and used for tying material. (138:52) **Zuni** *Brushes & Brooms* Leaves made into brushes and used for decorating pottery, ceremonial masks, alters, and other objects. (166:82) *Cordage* Leaves soaked in water to soften them and made into rope by knotting them together. The fibers of the leaves were separated and lengthened for making a coarse cord. (166:79) *Mats, Rugs & Bedding* Dried leaves split, plaited, and made into water-carrying head pads. (17:47) Leaves used for making mats, cincture pads, and other articles. (166:79)
- *Dye*—**Navajo, Ramah** *Black* Juice mixed with yellow soil for a black dye. *Red* Juice boiled alone for a red dye. (191:21)
- *Other*—**Apache** *Soap* Roots used for soap. (136:148) **Apache, White Mountain** *Toys & Games* Leaves used as counters in various games. (136:147) **Blackfoot** *Soap* Roots used as a soap substitute and hair wash. (97:25) **Dakota** *Containers* Dried stems peeled and used to make a hearth. (69:358) Dried, peeled stems used to make a hearth, to contain the fire. (70:71) *Fuel* Leaves bound in a slender bundle and used as a substitute for wood. The slender bundle of leaves formed the fire drill, which was placed in a hearth and twirled by the hands until it smoldered, upon which time it was blown upon to ignite the flame. *Hide Preparation* Boiled root used in tanning hides. The yucca roots were boiled and then cooled and sprinkled over the hides after they had been treated with the brain-liver-marrow dressing. *Soap* Root used like soap to wash the scalp. (69:358) Root used like soap, especially for washing the hair. *Tools* Hard, sharp-pointed blades bound with sinew and used in place of wood to make the fire drill. (70:71) **Isleta** *Decorations* Fibers used to make plaques. *Soap* Roots pounded until soft, soaked in water, and used as a soap for bathing, washing hair and blankets. (100:45) **Keres, Western** *Fasteners* Leaves used as strings to tie chili peppers. *Soap* Crushed roots rubbed on hair or clothes for soap or crushed, roasted roots soaked in water for soap. This soap caused more itching than that from wild leafed yucca, but was considered better to keep the natural color of the hair. *Tools* Flower stalks used as spindles to start fire by friction. (171:76) **Kiowa** *Soap* Root used to wash clothes and hair. (192:18) **Lakota** *Soap* Roots boiled and used as shampoo or as a substitute for soap. (106:51) Roots used to make soap. (139:28) **Navajo** *Ceremonial Items* Plant used to stir the water for the ceremonial baths. *Soap* Roots used for soap. *Toys & Games* Used to make the 102 counting sticks for the moccasin game. (55:33) **Navajo, Ramah** *Soap* Root used for soap. (191:21) **Omaha** *Soap* Root used like soap, especially for washing the hair. (70:71) **Papago** *Fasteners* Used to bind women's hair over their foreheads while racing. Fiber used to tie saguaro needles together. *Soap* Roots dried, pulverized, mixed with water, and the suds used for washing the hair. (34:51) **Pawnee** *Soap* Root used like soap, especially for washing the hair. **Ponca** *Soap* Root used like soap, especially for washing the hair. (70:71) **Southwest Indians** *Soap* Roots roasted, soaked in water, removed, and soapy liquid used to wash hair and cloth. (17:54) Leaves pounded and used in washing. (17:55) *Tools* Stalk used as a spindle in making fire by friction. (17:52) **Tewa** *Soap* Roots used to make lather. **Tewa of Hano** *Ceremonial Items* Used to make whips to beat novices during some initiation ceremonies. (138:52) **Zuni** *Soap* Peeled roots pounded, made into suds, and used for washing the head, wool garments, and blankets. (166:83)

Yucca glauca Nutt. **var.** *glauca*, Soapweed Yucca
- *Drug*—**Montana Indian** *Dermatological Aid* Root used as a substitute for soap and as a wash for the hair. (as *Y. angustifolia* 19:27)
- *Fiber*—**Hopi** *Basketry* Used for basketry. (as *Y. angustifolia* 56:17)
- *Other*—**Hopi** *Soap* Root used for soap. (as *Y. angustifolia* 56:17)

Yucca harrimaniae Trel., Spanish Bayonet
- *Food*—**Southwest Indians** *Starvation Food* Used when agricultural reserves dwindled. (17:9)

Yucca louisianensis Trel., Gulf Coast Yucca
- *Other*—**Comanche** *Soap* Roots used as soap. (29:524)

Yucca schidigera Roezl ex Ortgies, Mojave Yucca
- *Food*—**Cahuilla** *Unspecified* Fruit pods eaten raw or roasted. (15:150) **Hualapai** *Dried Food* Fruits baked, prepared, and dried for winter use. *Fruit* Fruit eaten raw. *Staple* Fruits cooked and ground into a meal. (as *Y. mohavensis* 195:40) **Luiseño** *Unspecified* Blossoms cooked in water and used for food. (as *Y. mohavensis* 155:195) Pods roasted on coals and used for food. (as *Y. mohavensis* 155:196) **Mohave** *Fruit* Fruit peeled and eaten without preparation. (37:204) **Southwest Indians** *Beverage* Fruit cooked and made into a drink. *Dried Food* Dried fruit used for food. (as *Y. mohavensis* 17:18) *Fruit* Fruit used for food. (17:22) Fruits eaten for food. (17:63)
- *Fiber*—**Cahuilla** *Basketry* Fiber used as starting material for baskets. *Brushes & Brooms* Fiber used to make brushes for body painting. *Cordage* Fiber used to make bowstrings, netting, and strings for shell money. (15:150) **Diegueño** *Clothing* Fibers used to make sandals. (84:45) **Hualapai** *Clothing* Strong fibers used to make shoes and sandals. (as *Y. mohavensis* 195:40) **Southwest Indians** *Clothing* Made into cords and used as base for fur robe garments. (17:43)
- *Other*—**Cahuilla** *Soap* Scraped, mashed roots and water used as soap. (15:150) **Luiseño** *Hunting & Fishing Item* Plant fiber used to make fishing lines. (as *Y. mohavensis* 155:200) Leaf fiber formerly used to make fishing lines. (as *Y. mohavensis* 155:203)

Yucca schottii Engelm., Schott's Yucca
- *Food*—**Apache, San Carlos** *Dried Food* Fruits dried and used for food. *Fruit* Fruits cooked, skins peeled off, and pulp used for food. (95:258) **Papago** *Fruit* Fruits cooked and eaten with white flour. Fruits eaten raw. **Pima** *Fruit* Fruits cooked and eaten with white flour. Fruits eaten raw. (95:262)
- *Other*—**Papago** *Cash Crop* Dried fruits sold to the Pimas. (95:262)

Yucca sp., Soap Root
- *Drug*—**Navajo** *Poison* Plant considered poisonous. *Throat Aid* Poultice of plants applied to the head for sore throats. (55:34)
- *Food*—**Navajo** *Forage* Buds eaten by sheep. (55:34) **Papago** *Staple* Fruits dried and used as a staple food. (34:45) **Yavapai** *Fruit* Boiled fruit used for food. *Unspecified* Flower stalk picked before blooming, roasted in fire, and used for food. (65:258)
- *Fiber*—**Navajo** *Basketry* Fiber used to secure the butts of the first twigs around a small stick at the bottom of the basket. Leaf pith braid woven into a basket. *Building Material* Pith twisted with mountain grass and used for roofing. *Clothing* Fiber used to make knitted leg-

gings. Yucca fiber and grass fiber used to make the earliest costume. Pith twisted with mountain grass and used to make leggings and shoes. Used to make moccasin uppers and dresses. *Cordage* Strands used to tie rolled skins into a rabbit skin blanket. Fiber used to tie butt and tip of corn husks filled with dough. *Mats, Rugs & Bedding* Pith twisted with mountain grass and used to make mats for bedding and blankets. Fiber and grass used to make sleeping mats. (55:34)

- *Other*—**Great Basin Indian** *Soap* Root suds used to wash wool. (121:47) **Jemez** *Soap* Roots boiled, fibrous part removed, water rubbed into lather, and used to wash hair and woolen blankets. (44:28) **Navajo** *Ceremonial Items* Leaves used for ceremonial purposes. On the sixth day of the Mountain Chant Ceremony, before the couriers were sent on their way, a basin of water containing soap root was brought in, and after the medicine man had daubed the couriers with a little of the suds, they washed themselves from head to foot and cleaned their hair as well. The Lashing God in the Night Chant carried a ring of yucca leaves on his back and suspended from this by its roots was a complete plant of soapweed. He held in his hand yucca scourges which were made from the leaves taken from the east and west sides of the plant. For the yucca that hangs at his back, a specimen was selected whose roots stuck well out of the ground and was kicked out with the foot. Masks made of the leaves were also used in the Night Chant. In one of the dances of the last night of the Mountain Chant, yucca was made to grow from the root, through buds and flowers to the ripe fruit. (55:32) Roots, pollen, and leaves used during many different ceremonies. Pitch used to cover bullroarers for some of the ceremonies. Fiber used to string cakes baked for Fire God and attached to his right arm on ninth day of Night Chant. Leaf strips intertwined with sprigs of fir and used to make necklaces and wristbands for ceremonies. *Containers* Wood tied to stalk with shallow holes and used at the hearth to hold a fireset. *Jewelry* Leaves used to make bracelets worn by scouts. *Musical Instrument* Folded leaves used as drumsticks to beat basket drums. *Protection* Leaf juice mixed with powders and applied to shields. Yucca leaves were heated over a fire and the juice wrung out of them into an earthen vessel. The juice was then mixed with powders and applied to the shield with a pointed stick to make it live in the power of the sun, the serpent, the bear, the lightning, and the rainbow. (55:34) *Soap* Used for cleansing purposes. (55:32) Roots used to wash wool and hides. Suds and ashes used to wash newborn babies. *Tools* Used to make a brush to apply colored clays to pottery. *Toys & Games* Leaves made into a ball and used to play "shooting the yucca," a Navajo game played with a ball made of bark and wound with yucca leaves which had been previously placed in hot ashes to make them flexible. A stick of scrub oak was attached to this by a yucca cord, to give momentum to the light ball. The ball was thrown into the air and the archers discharged their arrows at it as soon as it was drawn downward by the weight of the stick. Fiber used to make a ring for a game similar to "ring toss." *Waterproofing Agent* Leaf pitch used for waterproofing baskets. (55:34)

Yucca torreyi Shafer, Torrey's Yucca

- *Food*—**Apache** *Beverage* Baked fruit pounded to a pulp, drained, and juice drunk. *Bread & Cake* Fruit roasted, pulp made into cakes, and stored. *Sauce & Relish* Baked fruit pounded to a pulp, drained and juice poured over cakes. (17:18) **Apache, Chiricahua & Mescalero** *Bread & Cake* Fruit roasted, split, seeds removed, and pulp ground into large cakes. Fruit pulp ground, made into large cakes, and stored indefinitely. (as *Y. macrocarpa* 33:39) **Southwest Indians** *Fruit* Fruits eaten occasionally. (17:63)
- *Fiber*—**Apache** *Basketry* Leaves used for the main portion of the baskets. (17:35)
- *Other*—**Apache** *Designs* Roots used to produce a red pattern in baskets. (17:35)

Yucca whipplei Torr., Chaparral Yucca

- *Food*—**Cahuilla** *Bread & Cake* Roasted stalks dried, ground, and mixed with water to make cakes. *Dried Food* Flowers and stalks sun dried and preserved. *Unspecified* Less mature flowers parboiled and very mature flowers boiled three times with salt and eaten. *Vegetable* Sliced stalks parboiled and cooked like squash. (15:150) **Luiseño** *Unspecified* Roasted stalks used for food. Plant head roasted in an earth oven and formerly used for food. Blossoms cooked in water and used for food. (155:195) **Mahuna** *Sauce & Relish* Stalks pit roasted and used to make a syrup. *Unspecified* Flowers eaten as food. (140:58) **Tubatulabal** *Unspecified* Stalks used for food in late winter and early spring. (193:11) Stalks used extensively for food. (193:15)
- *Fiber*—**Mahuna** *Basketry* Pods used for basketry. (140:58)
- *Dye*—**Mahuna** *White* Pods used for bleaching buckskin fiber a pure white. (140:58)

Yucca whipplei var. *caespitosa* M. E. Jones, Chaparral Yucca

- *Food*—**Kawaiisu** *Dried Food* Apical meristems pit roasted, mashed, dried, and stored for future use.[84] *Unspecified* Apical meristems pit roasted and eaten.[84] Stalks roasted, peeled, and eaten. (206:69)
- *Fiber*—**Kawaiisu** *Basketry* Split rootstock cores used as pattern material in coiled basketry. *Building Material* Green leaves made into strong cord used to bind the vertical and horizontal poles of the winter house. (206:69)
- *Other*—**Kawaiisu** *Containers* Small stalk section used as a stopper for the basketry water bottle. (206:69)

Yucca whipplei Torr. ssp. *whipplei*, Chaparral Yucca

- *Food*—**Diegueño** *Unspecified* Young stalk peeled, roasted, and eaten in the spring. Blossoms picked before opening in the spring, boiled twice, and eaten. (84:45)
- *Fiber*—**Diegueño** *Basketry* Leaf fibers used as the foundation for the beginning of a coiled basket. (84:17) *Furniture* Leaf fibers used to make lattice work cradle on a frame of scrub oak. (84:45)

Zamiaceae, see *Zamia*

Zamia, Zamiaceae, conifer

Zamia pumila L., Coontie

- *Food*—**Seminole** *Unspecified* Plant used for food. (as *Z. floridana* 169:489)

Zanthoxylum, Rutaceae

Zanthoxylum americanum P. Mill., Common Prickly Ash

- **Drug**—**Alabama** *Dermatological Aid* Decoction of bark used as a wash for itching. (172:663) Infusion of inner bark rubbed on itchy area. (as *Xanthoxylum americanum* 177:35) *Toothache Remedy* Pounded inner bark used for toothache. (172:663) Inner bark put into cavity and packed around the tooth for toothaches. (as *Xanthoxylum americanum* 177:35) **Cherokee** *Antirheumatic (External)* Infusion used as a wash for swollen joints. (80:51) **Chippewa** *Cold Remedy* Infusion of bark taken for colds. *Cough Medicine* Infusion of bark taken for coughs. (as *Xanthoxylum americanum* 71:134) *Orthopedic Aid* Decoction of root used as a wash for paralysis and to strengthen child's legs and feet. *Pediatric Aid* Decoction of root used as a wash to strengthen legs and feet of weak children. (53:364) *Pulmonary Aid* Infusion of bark taken for all pulmonary troubles. (as *Xanthoxylum americanum* 71:134) *Throat Aid* Decoction of root gargled or taken for sore throat. (53:342) **Comanche** *Burn Dressing* Roots pulverized and powder used for burns. (99:6) *Febrifuge* Infusion of bark taken for fever. (29:524) Infusion of pulverized roots used for fever. (99:6) *Throat Aid* Inner bark placed in throat for sore throat. *Toothache Remedy* Root bark held against tooth for toothache. (29:524) Roots pulverized and used for toothache. (99:6) **Creek** *Veterinary Aid* Infusion of bark used to rub on dog's nose to improve his scent. (172:663) **Delaware** *Heart Medicine* Infusion of inner bark used for heart trouble. (as *Xanthoxylum americanum* 176:35) **Delaware, Oklahoma** *Heart Medicine* Infusion of inner bark taken sparingly for heart trouble. (as *Xanthoxylum americanum* 175:30, 80) *Tonic* Bark used alone and in compound as a tonic. (as *Xanthoxylum americanum* 175:80) **Iroquois** *Abortifacient* Decoction of bark taken to promote miscarriage. (87:368) *Analgesic* Bark smoked for toothaches or neuralgia. (87:367) Compound infusion taken for back pain. Infusion of bark taken for pain after confinement. *Anthelmintic* Decoction of bark taken for worms. *Antiemetic* Compound infusion taken for vomiting. *Diuretic* Infusion of roots taken when water stops because of gonorrhea. *Emetic* Compound decoction taken to vomit during initial stages of consumption. *Gastrointestinal Aid* Decoction of bark taken for cramps. *Gynecological Aid* Infusion of bark taken for pain after confinement. *Kidney Aid* Compound infusion taken for kidney trouble. *Orthopedic Aid* Compound infusion taken for back pain. *Toothache Remedy* Bark smoked, chewed, or placed into the tooth for toothaches. *Tuberculosis Remedy* Compound decoction taken to vomit during initial stages of consumption. *Venereal Aid* Infusion of roots taken when water stops because of gonorrhea. (87:368) **Menominee** *Adjuvant* Infusion of berries used as a seasoner in medicines. (151:51) *Analgesic* Poultice of pounded inner bark used for rheumatism and sharp pains. *Antirheumatic (External)* Poultice of pounded inner bark applied for rheumatism and sharp pains. (54:133) *Cold Remedy* Decoction of inner bark used for cold settled in the chest. (54:130) *Dermatological Aid* Infusion of berries sprayed from mouth onto sores. Poultice of root bark applied to swellings in special rite. *Other* Infusion of berries taken for minor maladies. *Pulmonary Aid* Infusion of berries sprayed on the chest or throat for bronchial diseases. (151:51, 52) *Sedative* Compound infusion taken and rubbed on body to quiet person near convulsions. (54:128) **Meskwaki** *Cough Medicine* Bark and berries used to make cough syrup. *Expectorant* Bark and berries used as an expectorant. *Hemostat* Bark and berry medicine used for hemorrhages. *Kidney Aid* Compound infusion of root used for kidney trouble. *Strengthener* Compound decoction of inner bark given to strengthen convalescent patient. *Toothache Remedy* Powdered inner bark used for toothache. *Tuberculosis Remedy* Bark and berries used for hemorrhages and tuberculosis. (152:244, 245) **Mohegan** *Heart Medicine* Infusion of bark taken for heart, 3 days on, 3 days off, before resuming. (as *Xanthoxylum americanum* 174:269) Infusion of inner bark taken in small doses for heart trouble. (as *Xanthoxylum americanum* 176:77, 132) **Ojibwa** *Respiratory Aid* Infusion of berries sprayed on chest for congestion from bronchitis. *Throat Aid* Bark or berries used for sore throat and tonsillitis. (153:387) **Pawnee** *Veterinary Aid* Fruits used as diuretic for horses. (70:98) **Potawatomi** *Venereal Aid* Root bark used for gonorrhea, an historic reference (1796) (154:80)
- **Other**—**Omaha** *Incense & Fragrance* Fruits used by young men as perfume. (as *Xanthoxylum americanum* 68:323) Fruits used by young men as perfume. (70:98)

Zanthoxylum clava-herculis L., Hercules's Club
- **Drug**—**Houma** *Orthopedic Aid* Salve of grated root mixed with whisky rubbed on swollen limbs. *Toothache Remedy* Poultice of grated root and bark applied to aching teeth. (as *Xanthoxylum clava-hercules* 158:61)

Zanthoxylum fagara (L.) Sarg., Lime Pricklyash
- **Other**—**Seminole** *Hunting & Fishing Item* Plant used to make bows and arrows. (169:475)

Zea, Poaceae

Zea mays L., Corn
- **Drug**—**Cherokee** *Dermatological Aid* "Smut" from plant used as salve. *Kidney Aid* Infusion taken for "gravel." *Pulmonary Aid* Parched grains eaten for "long wind." (80:30) **Keres, Western** *Panacea* Pollen eaten for almost any kind of medicine. (171:77) **Mohegan** *Dermatological Aid* Decoction of dried cobs used as a wash for poison ivy rash. (176:77) **Navajo** *Ceremonial Medicine* Poultice of plant applied as ceremonial medicine for sore throats. Leaves used in mixture for the Night Chant medicine. One part of the Night Chant medicine consists of a mixture of "blue pollen," wild plants, and tobacco to which was added the leaves from corn plants gathered in the east, south, west, and north corners of the field, squash from the southeast side, bean leaves from the southwest, watermelon leaves from the northwest, and muskmelon leaves from the northeast. *Throat Aid* Poultice of plant applied as ceremonial medicine for sore throats. (55:27) **Tewa** *Analgesic* Blue cornmeal and water used for "palpitations or pains." *Dermatological Aid* Warm ear of corn rubbed with foot for child's glandular swelling in neck. *Gynecological Aid* Black corn with red streaks good for menstruating woman. *Heart Medicine* Blue cornmeal and water used for "heart-sickness," "palpitations or pains." *Pediatric Aid* Warm ear of corn rubbed with foot for child's glandular swelling in neck. (138:97)
- **Food**—**Abnaki** *Soup* Seeds used to make soup. (144:175) **Cahuilla** *Porridge* Ground into a meal, boiled, and eaten. (15:153) **Cherokee** *Vegetable* Corn used for food. (80:30) **Chippewa** *Porridge* Used to make a "hominy." *Soup* Kernels pounded into a meal and used to make "parched corn soup." *Vegetable* Fresh ears roasted in the husks

and used for food. (53:319) **Choctaw** *Vegetable* Seeds parched and mixed with water or boiled with or without meat. (25:8–9) **Dakota** *Staple* Ripe, parched corn ground into a meal and used for food. Ripe corn hulled with lye from ashes and used to make hominy. *Sweetener* Sun dried corn silks ground with parched corn for sweetness. *Winter Use Food* Sun dried corn silks stored for future use. (70:67) **Delaware** *Bread & Cake* Dry, unparched corn made into flour and used to make bread. *Dried Food* Ears boiled, cooled, the grains dried and used for food. *Porridge* Ears sun dried, grains pounded into hominy grits and used for food. *Soup* Dried corn boiled in alkaline liquid and hulls combined with fresh or dried meat for stew. *Staple* Used as the staple vegetable food to provide nourishment for the soul and the body. *Unspecified* Ears roasted and used for food. Dried corn boiled in alkaline liquid and hulls eaten with milk and sugar or fried with potatoes. (176:55) **Havasupai** *Bread & Cake* Seeds used to make wafer bread. (as *Z. arnylacea* 162:103) Seeds used to make bread. Seeds parched, ground fine, mixed with salt water into thin gruel, and cooked in thin layer into piki. Seeds ground, added to boiling water, kneaded, rolled in corn husks, boiled, and eaten as tamales. Seeds parched, ground fine, boiled, thickened, made into balls, and eaten as dumplings. Seeds ground, kneaded into a thick paste, rolled into little balls, boiled, and eaten as marbles. (197:66) *Porridge* Seeds parched, ground, and used to make mush. *Soup* Seeds parched, ground, and used to make soup. *Staple* Seeds ground and eaten as a ground or parched meal. (197:67) *Vegetable* Seeds eaten fresh, baked on the cob, roasted, or boiled. *Winter Use Food* Seeds pit baked and stored for winter use. (197:66) **Hopi** *Bread & Cake* Seeds ground into meal and used to make wafer bread. (200:67) *Dried Food* Pit baked, husked, strung, and sun dried. (as *Z. mays saccharata* 200:69) *Porridge* Grains soaked in water with juniper ash, boiled, and washed to make hominy. Made into hominy and other dishes, plant constituted the main food supply. *Staple* Ground into meal. (as *Z. mays amylacea* 200:67) *Sweetener* Ears pit baked, husked, strung, sun dried, and used as a sweetener in the winter. *Unspecified* Pit baked and eaten immediately. (as *Z. mays saccharata* 200:69) **Iroquois** *Baby Food* Seeds boiled into a liquor and used in the preparation of food for infants. Seeds used to make a meal gruel for babies. (196:71) *Beverage* Stalks cut between the joints and chewed to quench the thirst. (196:119) Dried, roasted seeds boiled in water to make coffee. (196:145) Seeds boiled into a liquor and used as a beverage or made into soup. *Bread & Cake* Seeds ground into a meal or flour and used to make boiled bread. Seeds ground, mixed with hot water, molded, dropped into boiling water, and eaten as dumplings. *Pie & Pudding* Seeds, pumpkin mush, and maple sugar used to make pudding. *Porridge* Seeds used to make hominy. *Sauce & Relish* Seeds used to make succotash. *Snack Food* Seeds used to make popcorn. *Soup* Seeds used with beans, squash, and meats to make soups and broths. *Special Food* Seeds used to make wedding bread or bread placed in the coffin with the corpse. Seeds used for ceremonial occasions, such as False-Face Society functions. *Vegetable* Seeds eaten raw or cooked while traveling or hunting. Corn on the cob roasted and eaten. (196:71) **Isleta** *Beverage* Ground corn used to make a slightly intoxicating beverage. *Bread & Cake* Cornmeal used to make various breads. *Candy* Parched corn eaten as a confection. *Porridge* Cornmeal used to make a mush. *Staple* Parched corn eaten as a staple. *Sweetener* Evaporated liquid from crushed, soaked stalks used to make sugar. *Unspecified* Corn husks used to wrap tamales. *Winter Use Food* Cornmeal used to make mush, dried, and stored for winter use. (100:46) **Kamia** *Unspecified* Species used for food. (62:21) **Keres, Western** *Staple* Cornmeal used as one of the main foods. (171:76) *Vegetable* Roasted corn ears eaten warm for food. *Winter Use Food* Roasted corn ears dried and stored for winter use. (171:77) **Kiowa** *Fodder* Valued as a fodder for livestock. *Unspecified* Valued as a food. (192:17) **Menominee** *Beverage* Scorched or parched corn often used as a substitute for coffee. *Special Food* Parched, ground corn mixed with bear oil and used as trail ration. *Staple* Roasted popcorn pounded into a meal added to dried venison, maple sugar, or wild rice or all three. *Substitution Food* Scorched or parched corn often used as a substitute for coffee. *Vegetable* Ears roasted and made into hominy. *Winter Use Food* Ears parboiled and the kernels sun dried for winter use. (151:66) **Meskwaki** *Unspecified* Boiled or parched corn eaten or made into corn hominy grits. *Winter Use Food* Boiled or parched corn stored for winter use. (152:257) **Navajo** *Beverage* Cornmeal and juniper ash water used to make a beverage. *Bread & Cake* Corn and juniper ash used to make bread and dumplings. *Porridge* Corn and meat boiled all night into hominy. Cornmeal and juniper ash water used to make mush. *Special Food* Cornmeal porridge, served in wedding baskets, used as a nuptial dish. *Staple* Green corn roasted, shelled, ground, dried, and wrapped in corn husks, like tamales, for journeys. *Unspecified* Immature corn pounded, mixed with pumpkin, wrapped in a corn husk, and baked in ashes. *Vegetable* Leaves eaten like lettuce. (55:27) **Navajo, Ramah** *Fodder* Used as horse feed. *Special Food* Cornmeal used to make ceremonial cakes. *Vegetable* Young corn and cob eaten. *Winter Use Food* Roasted, dried corn on the cob stored for winter use. (191:18) **Ojibwa** *Dried Food* Kernels dried for winter use. *Vegetable* Several sorts of corn were grown, modern and ancient. Ears were roasted and made into hominy. (153:402) **Omaha** *Staple* Ripe, parched corn ground into a meal and used for food. Ripe corn hulled with lye from ashes and used to make hominy. (70:67) *Sweetener* Sun dried corn silks ground with parched corn for sweetness. *Winter Use Food* Sun dried corn silks stored for future use. (70:68) **Papago** *Dried Food* Grains parched, dried on mats on the roofs, and used for food. *Special Food* Cornmeal used ceremonially. *Staple* Whole ears roasted in open pits, dried, grains removed, winnowed, and ground into meal. *Unspecified* Whole ears roasted in open pits, dried, grains removed, winnowed, and cooked whole with meat. (34:34) **Pawnee** *Staple* Ripe, parched corn ground into a meal and used for food. Ripe corn hulled with lye from ashes and used to make hominy. *Sweetener* Sun dried corn silks ground with parched corn for sweetness. *Winter Use Food* Sun dried corn silks stored for future use. (70:67) **Pima** *Bread & Cake* Ground, baked in large cakes, and used for food. *Porridge* Boiled with ashes, dried, hulls washed off, dried, parched with coals, and made into gruel. (146:72) **Ponca** *Staple* Ripe, parched corn ground into a meal and used for food. Ripe corn hulled with lye from ashes and used to make hominy. *Sweetener* Sun dried corn silks ground with parched corn for sweetness. *Winter Use Food* Sun dried corn silks stored for future use. (70:67) **Potawatomi** *Winter Use Food* Elm bark bags, filled with corn or beans and peas, buried in the ground to keep for the winter. (154:101) **Pueblo** *Special Food* Cornmeal used ceremonially.

(34:34) **Seminole** *Unspecified* Seeds used for food. (169:473) **Sia** *Staple* Corn and wheat, the most important foods, used for food. (199:106) **Tewa** *Beverage* Corn ground and sifted into boiling water to make a gruel formerly drunk in the morning. Cornmeal mixed with cold water and drunk as a nourishing drink. *Bread & Cake* Corn ground on a metate, formed into cakes, rolled, and baked. *Forage* Husks, stalks, and leaves used for stock winter forage. *Staple* Used as a staple food. (138:78) **Zuni** *Beverage* Popped corn ground as fine as possible, mixed with cold water, strained, and used as a beverage.[85] *Bread & Cake* Toasted or untoasted corn ground into a flour and used to make bread. *Porridge* Corn used to make gruel. *Snack Food* Corn used to make popcorn. *Staple* Toasted or untoasted corn ground into a flour and used to make bread eaten as a staple on journeys. (166:73)

- *Fiber*—**Navajo** *Building Material* Stalks sometimes used for thatching. (55:27)
- *Other*—**Cahuilla** *Ceremonial Items* Sprinkled on images of the dead during mourning ceremonies. (15:153) **Cherokee** *Toys & Games* Shucks used to make dolls. (80:30) **Chippewa** *Fasteners* Husks turned back and used to suspend corn ears from the ceiling. (53:319) **Delaware** *Ceremonial Items* Used in the ceremonial diet of the participants of the Big House Ceremony. (176:55) **Hopi** *Ceremonial Items* Used in almost all ceremonies either as cornmeal, as an actual ear of corn, or as a painting. Ceremonially associated with the northeast direction. (as *Z. mays amylacea* 200:67) Whole ears boiled and given as presents during the winter ceremonies. Ceremonially associated with the nadir. (as *Z. mays saccharata* 200:69) **Iroquois** *Containers* Dried husks woven into small bottles or receptacles for salt. (196:154) *Cooking Tools* Husks used as wrappers for boiling the double wedding corn bread package. *Planting Seeds* Seeds selected for qualities such as size, flavor, color, and early maturity and used for planting. (196:71) **Isleta** *Ceremonial Items* Corn husks used as cigarette papers for the ceremonial cigarettes. Corn silks and pollen used in the "Corn Dances." Cornmeal smeared on the body in the burial ceremony. (100:46) **Keres, Western** *Ceremonial Items* Corn pollen and cornmeal used for many ceremonial purposes. Cornmeal sprinkled by everyone before eating and prayer repeated. (171:77) **Kiowa** *Smoke Plant* Shucks used for cigarette wrappings and used in the peyote ceremony. (192:17) **Navajo** *Containers* Husks used as casings for blood sausage. *Fuel* Cob pith used as punk (tinder). *Sacred Items* Sacred pollen used in innumerable ceremonies. Cornmeal, considered less sacred than corn pollen, used in innumerable ceremonies. *Smoke Plant* Husks used for cigarette papers. *Tools* Cobs used to beat leather while dyeing. (55:27) **Navajo, Ramah** *Ceremonial Items* Pollen used in all ceremonials and also for personal ceremonies. *Containers* Cobs used to close up pottery jars. *Smoke Plant* Dry husks of young corn used to roll cigarettes, when paper not available. *Tools* Cobs used by pottery makers to smooth the pottery. (191:18) **Omaha** *Protection* Chewed seeds scattered around the corn fields to protect the harvest from blackbirds. When the corn was approaching maturity, blackbirds attacked the fields for food. To prevent further damage, men chewed some grains and scattered them around the corn fields to deter the birds from the fields. (70:68) **Seminole** *Ceremonial Items* Plant used for religious scarification and at busk ceremonies. *Hunting & Fishing Item* Plant used to make arrowheads. (169:473) **Tewa** *Cash Crop* Formerly bartered with the Comanche for prepared buffalo hides. *Ceremonial Items* Husks twisted and used to make the framework and mounts for feathers in ceremonial ornaments. *Fuel* Cobs used as fuel in emergencies and as fire lighters. *Smoke Plant* Husks made into cigarettes. *Tools* Cobs used to make handles and holders. *Toys & Games* Cobs used to make feathered darts and to stuff kickballs. (138:78) **Zuni** *Ceremonial Items* Popped corn ground, made into a beverage, and used ceremonially.[85] (166:73) Corn ears carried or secretly worn in dances by personators of anthropic gods. Ribboned corn husks used as hair decorations in ceremonies. Cornmeal wrapped in husks given to theurgists visiting the sick. The packages were always presented with a prayer and the recipient prayed. White cornmeal made into a mush and used ceremonially during the death of a rain priest. Balls of husks covered with woven cotton used ceremonially to insure bountiful crops. The balls of corn husks covered with woven cotton were used with long fringes of white cotton ceremonial sashes symbolizing corn and a desire for bountiful crops. *Decorations* Ribboned corn husks used as hair decorations in ceremonies. *Toys & Games* Ribboned husks made into small, square pads and used by young people in games. Small plumes were attached to the small, square pads upright, in the center, forming the shuttlecocks for use in the game of battledore and shuttlecock. (166:99)

Zea mays L. var. *rugosa* Bonaf., Corn

- *Food*—**Navajo** *Bread & Cake* Sweet cornmeal and herb roots made into cakes and baked in a pit. (55:30)
- *Other*—**Navajo** *Ceremonial Items* Cornmeal mush used to make images for ceremonies. In the Bead Chant, an image of a wildcat was made of sweet corn; in the Mountain Chant, an image of the bear was made of sweet corn; and in the Coyote Chant, effigies of a coyote and a kit fox were made in sweet corn; other images such as the dog, chicken, cat, and pig were also reproduced in sweet corn. To make these animals, a stiff mush was made of corn, which was kneaded to the desired shape, omitting the extremities such as tail, ears, and feet. White shell, turquoise, and cannel coal was used for the eyes. (55:30)

Zephyranthes, Liliaceae

Zephyranthes sp., Zephyr Lily

- *Drug*—**Seminole** *Toothache Remedy* Infusion of bulbs used for toothaches. (169:304)

Zeuxine, Orchidaceae

Zeuxine strateumatica (L.) Schlechter, Soldier's Orchid

- *Drug*—**Seminole** *Gynecological Aid* Decoction of whole plant taken and used as a wash for barrenness. *Reproductive Aid* Decoction of whole plant used as a wash for impotency. (169:318)

Zigadenus, Liliaceae

Zigadenus elegans Pursh, Mountain Deathcamas

- *Drug*—**Alaska Native** *Poison* Plant considered poisonous. (85:163) **Eskimo, Inupiat** *Poison* Whole plant considered poisonous. (98:139) **Keres, Western** *Antirheumatic (External)* Infusion of 11 plants used as an athletic rubdown. *Diaphoretic* Eleven plants used in sweat bath. An infusion of 11 plants was sprinkled on hot rocks to form dense steam. Men stayed in the sweat bath for 3 to 5 minutes for sev-

eral successive evenings before the deer hunt or race. *Poison* Plant considered poisonous. *Strengthener* Root rubbed on muscles as a strengthener. Infusion of 11 plants used as an emetic before breakfast prior to athletic events or deer hunts. Infusion of 11 plants used during athletic training every morning prior to breakfast until the brew had been depleted, to give long endurance, a keen eye, and so that the deer could not smell you. (171:78) **Navajo, Ramah** *Dermatological Aid* Cold infusion of plant used as a lotion for mad coyote bite. (191:21) **Thompson** *Analgesic* Pulverized, baked root used as salve on painful areas, especially back and feet. (164:463) *Poison* Bulb caused "human poisoning" and leaves usually fatal to cattle. *Veterinary Aid* Leaves usually fatal to cattle. (164:512)
- *Other*—**Thompson** *Ceremonial Items* The father of a girl reaching puberty had to snare a grouse, cut off its head, remove its eyes, and in their places put two small roots of this plant and another in the mouth. Otherwise, the father was not able to snare any more grouse. (164:508)

Zigadenus nuttallii (Gray) S. Wats., Nuttall's Deathcamas
- *Drug*—**Gosiute** *Emetic* Plant used as an emetic. *Venereal Aid* Plant used for venereal affections. (39:384) **Ute** *Poison* Bulbs considered poisonous. (38:37)

Zigadenus paniculatus (Nutt.) S. Wats., Foothill Deathcamas
- *Drug*—**Navajo, Kayenta** *Veterinary Aid* Infusion of plant given to sheep with bloat. (205:17) **Paiute** *Analgesic* Poultice of bulb used for neuralgia. *Antirheumatic (External)* Poultice of root, sometimes mixed with tobacco, applied for rheumatism. *Dermatological Aid* Poultice of bulb used for swellings. *Emetic* Decoction of root taken as an emetic, in spite of poisonous nature of plant. *Orthopedic Aid* Poultice of bulb used for sprains and lameness. *Toothache Remedy* Poultice of bulb used for toothache. (180:149) **Paiute, Northern** *Antirheumatic (External)* Poultice of roasted, mashed bulbs applied to swollen parts or used for rheumatism. (59:128) **Shoshoni** *Analgesic* Poultice of bulb used for neuralgia. *Antirheumatic (External)* Poultice of crushed raw or roasted root applied for rheumatism. *Dermatological Aid* Poultice of bulb used for swellings. *Emetic* Decoction of root taken as an emetic, in spite of poisonous nature of plant. *Orthopedic Aid* Poultice of bulb used for sprains and lameness. *Toothache Remedy* Poultice of bulb used for toothache. **Washo** *Analgesic* Poultice of bulb used for neuralgia. *Antirheumatic (External)* Poultice of crushed, raw root applied for rheumatism. *Dermatological Aid* Poultice of bulb used for swellings. *Orthopedic Aid* Poultice of bulb used for sprains and lameness. *Toothache Remedy* Poultice of bulb used for toothache. (180:149)
- *Food*—**Navajo, Kayenta** *Unspecified* Bulbs cooked with meat and corn and used for food. *Vegetable* Plants used as greens. (205:17)

Zigadenus venenosus S. Wats., Meadow Deathcamas
- *Drug*—**Chehalis** *Emetic* Plant sometimes used as a violent emetic. (79:23) **Haisla & Hanaksiala** *Poison* Roots considered highly toxic. (43:203) **Klamath** *Emetic* Roots mixed with rootstocks of blue flag and used for vomiting. (45:93) **Lakota** *Poison* Plant poisonous to humans. (139:28) **Mendocino Indian** *Analgesic* Poultice of mashed bulbs applied to painful bruises and sprains. *Antirheumatic (External)* Poultice of mashed bulbs applied to rheumatism. *Dermatological Aid* Poultice of mashed bulbs applied for boils and painful bruises. *Orthopedic Aid* Poultice of mashed bulbs applied to painful bruises and sprains. *Poison* Root considered poisonous. (41:321) **Montana Indian** *Analgesic* Poultice of cooked, mashed bulbs applied to strain and bruise pains. *Antirheumatic (External)* Poultice of mashed bulbs applied for rheumatism. *Dermatological Aid* Poultice of cooked, mashed bulbs applied to boils. (19:27) **Okanagan-Colville** *Poison* Bulbs considered extremely poisonous. (188:50) **Okanagon** *Antirheumatic (External)* Bulbs mashed and used for rheumatism. (125:37) **Paiute** *Analgesic* Poultice of bulb used for rheumatic pains. *Antirheumatic (External)* Poultice of bulb used for rheumatic pains. *Burn Dressing* Poultice of bulb used for burns. *Dermatological Aid* Poultice of bulb used for swellings. (180:149, 150) *Poison* Seeds and roots considered a deadly poison if eaten by humans or horses. (111:54) *Snakebite Remedy* Poultice of bulb used for rattlesnake bites. (180:149, 150) **Paiute, Northern** *Antirheumatic (External)* Poultice of roasted, mashed bulbs applied to swollen parts or used for rheumatism. (59:128) **Pomo, Kashaya** *Poison* Plant considered poisonous. (72:30) **Shuswap** *Orthopedic Aid* Poultice of plant applied to sore legs. *Poison* Plant considered poisonous. (123:55) **Squaxin** *Emetic* Plant sometimes used as a violent emetic. (79:23) **Thompson** *Antirheumatic (External)* Bulbs mashed and used for rheumatism. (125:37) *Orthopedic Aid* Mashed bulbs rubbed on broken bones to help them heal more quickly. (187:133) **Yuki** *Poison* Bulbs considered poisonous. (49:94)
- *Food*—**Karok** *Unspecified* Bulbs used for food. (6:64)
- *Other*—**Okanagan-Colville** *Hunting & Fishing Item* Mashed bulbs used as an arrow poison. (188:50)

Zingiberaceae, see *Curcuma, Zingiber*

Zingiber, Zingiberaceae

Zingiber zerumbet (L.) Sm., Bitter Ginger
- *Drug*—**Hawaiian** *Analgesic* Roots or bulbs pounded with salt and the resulting juice used as a head wash for headaches. *Dermatological Aid* Juice used for hair dressing. Leaf ashes, other ashes, and nut juice used for cuts and skin sores. Roots and other roots pounded with salt, mixed with urine, and used for ringworm and white skin blotches. Roots and other roots pounded with salt and used for itch and kindred afflictions of the skin. Flowers and roots and other plants pounded, mixed with water, and rubbed on the body during massages. Roots with other plant parts mixed with water and used as a bath for bruises. *Orthopedic Aid* Roots with other plant parts mixed with water and used as a bath for slight sprains. *Toothache Remedy* Roots cooked and used in the tooth hollow for toothaches. (2:19)

Zinnia, Asteraceae

Zinnia acerosa (DC.) Gray, Desert Zinnia
- *Drug*—**Keres, Western** *Antirheumatic (External)* Crushed plant paste mixed with salt and used on swellings or aches. *Psychological Aid* Plant given to children to quickly learn to talk. (as *Crassina pulmila* 171:39)

Zinnia grandiflora Nutt., Rocky Mountain Zinnia
- *Drug*—**Keres, Western** *Kidney Aid* Hot infusion of plant drunk for kidney trouble. *Other* Infusion of plant used as a bath for excessive

sweating. (as *Crassina grandiflora* 171:38) **Navajo** *Nose Medicine* Plant used for nose troubles. *Throat Aid* Plant used for throat troubles. (55:97) **Navajo, Ramah** *Analgesic* Decoction of plant taken for stomachache, heartburn, and as a cathartic. *Cathartic* Decoction of plant taken as a cathartic. *Ceremonial Medicine* Plant used as a ceremonial emetic. *Emetic* Plant used as a ceremonial emetic. *Gastrointestinal Aid* Decoction of plant taken for stomachache or heartburn. (191:54) **Zuni** *Dermatological Aid* Poultice of powdered plant applied to bruises. *Diaphoretic* Plant used in a sweat bath for fever. *Eye Medicine* Cold infusion of blossoms used as an eyewash. *Febrifuge* Smoke from powdered plant inhaled in sweat bath for fever. (as *Crassina grandiflora* 166:45)
- *Dye*—**Keres, Western** *Yellow* Flowers, ground with white clay or mixed with warm water, used as yellow dye for wool. Flowers rubbed into buckskin as a yellow dye. (as *Crassina grandiflora* 171:38)
- *Other*—**Keres, Western** *Paint* Flowers ground into a paste and used as a dark red body paint. (as *Crassina grandiflora* 171:38)

Zizania, Poaceae

Zizania aquatica L., Annual Wildrice
- *Food*—**Dakota** *Staple* Grain used as an important and prized food item. (69:360) Rice considered an important dietary element. (70:67) **Menominee** *Staple* Rice cooked with deer broth, pork, or butter and seasoned with maple sugar. (151:67) **Meskwaki** *Unspecified* Rice used for food. (152:259) **Ojibwa** *Bread & Cake* Seeds used to make gem cakes, duck stuffing, and fowl stuffing. *Breakfast Food* Seeds steamed into puffed rice and eaten for breakfast with sugar and cream. *Special Food* Seeds boiled with rabbit excrements, eaten and esteemed as a luxury. (135:246) **Omaha** *Staple* Grains used as a staple food. (68:328) Rice considered an important dietary element. **Ponca** *Staple* Rice considered an important dietary element. (70:67) **Thompson** *Unspecified* Rice cooked with meat. (187:144) **Winnebago** *Staple* Rice considered an important dietary element. (70:67)
- *Other*—**Ojibwa** *Cash Crop* Seeds scorched, winnowed, and sold as breakfast food. (135:246) **Thompson** *Cash Crop* Rice used for trading. (187:144)

Zizania palustris L., Northern Wildrice
- *Food*—**Chippewa** *Unspecified* Cooked alone or with meat and used as the principal cereal food. (53:318) **Ojibwa** *Staple* Formed an important staple in the diet, cooked with deer broth and maple sugar and eaten. (153:403) **Potawatomi** *Dried Food* Rice gathered and dried for a winter supply of food. *Pie & Pudding* Wild rice sweetened with maple sugar and used to make pudding. *Staple* Rice valuable for cooking with wild fowl or game and maple sugar used to season the mixture. (154:101)

Zizia, Apiaceae

Zizia aurea (L.) W. D. J. Koch, Golden Zizia
- *Drug*—**Meskwaki** *Analgesic* Compound containing flower stalks used as snuff for sick headache. *Febrifuge* Root used for fevers and compound containing flower stalks used for headache. (152:250)

Ziziphus, Rhamnaceae

Ziziphus obtusifolia (Hook. ex Torr. & Gray) Gray, Lote Bush
- *Drug*—**Pima** *Eye Medicine* Decoction of pounded root used as a wash for sore eyes. (as *Zizyphus lycioides* 146:79)
- *Food*—**Papago** *Beverage* Fruits formerly fermented and used for a beverage. (as *Zizyphus lycioides* 34:26) *Sauce & Relish* Fruits boiled to a syrup and used for food. (as *Zizyphus lycioides* 34:19) **Pima** *Fruit* Black berries beaten with sticks and eaten raw. (as *Zizyphus lycioides* 146:76) **Yavapai** *Beverage* Mashed berries added to water and use as a drink. (as *Zizyphus cycioides* 65:258)

Ziziphus obtusifolia var. *canescens* (Gray) M. C. Johnston, Lote Bush
- *Drug*—**Pima** *Analgesic* and *Antirheumatic (External)* Thorns used to prick the skin over rheumatic pains. *Dermatological Aid* Infusion of roots used as a shampoo. *Eye Medicine* Decoction of roots used as a wash for sore eyes. (as *Condalia lycioides* var. *canescens* 47:50)
- *Food*—**Maricopa** *Dried Food* Berries dried and stored, to be soaked in hot water and used later. *Fruit* Fruits mashed into a concoction and eaten. **Mohave** *Dried Food* Berries dried and stored, to be soaked in hot water and used later. **Maricopa & Mohave** *Fruit* Fruits mashed into a concoction and eaten. (as *Condalia lycioides* var. *canescens* 37:204) **Pima** *Fodder* Seeds squeezed out from boiled berries and fed to chickens. *Fruit* Ripe, black berries eaten raw. *Sauce & Relish* Berries boiled and used to make a syrup. (as *Condalia lycioides* var. *canescens* 47:50) **Pima, Gila River** *Fruit* Fruits eaten raw and boiled. (133:7)
- *Other*—**Apache** *Soap* Root used for washing the hair. (as *Condalia lycioides* var. *canescens* 47:50) **Havasupai** *Tools* Branch used to make a planting stick. (as *Condalia lycioides* var. *canescens* 162:102)

Ziziphus parryi Torr., Parry's Jujube
- *Food*—**Cahuilla** *Fruit* Drupes eaten fresh. *Porridge* Drupes dried and ground into flour for mush. *Staple* Leached nutlet of the drupe ground into a flour. (as *Condalia parryi* 15:56)

Zosteraceae, see *Phyllospadix, Zostera*

Zostera, Zosteraceae

Zostera marina L., Seawrack
- *Food*—**Bellabella** *Unspecified* Plants eaten raw with eulachon (candlefish) grease. (184:200) **Cowichan** *Spice* Fleshy roots and leaf bases used to flavor seal, porpoise, and deer meat. (182:77) **Hesquiat** *Forage* Brownish "roots" (actually rhizomes) eaten by black brants, Canada geese, mallard ducks, and cattle. *Unspecified* Brownish "roots" (actually rhizomes) cleaned, washed, and eaten raw. Greenish "root" eaten raw. (185:59) **Kwakiutl, Southern** *Special Food* Stems and roots dipped in oil and eaten during feasts. (183:274) *Unspecified* Plants eaten raw with eulachon (candlefish) grease. (184:200) **Nitinaht** *Unspecified* Fleshy, whitish rhizomes formerly eaten raw. (186:89) **Oweekeno** *Unspecified* Leaves picked with attached herring spawn and eaten. (43:82) **Saanich** *Spice* Fleshy roots and leaf bases used to flavor seal, porpoise, and deer meat. (182:77)
- *Other*—**Hesquiat** *Hunting & Fishing Item* Leaves used to collect herring spawn. (185:59) **Nitinaht** *Hunting & Fishing Item* Leaves formerly used to trap herring spawn. (186:89)

Zygnemataceae, see *Spirogyra*

Zygophyllaceae, see *Guajacum, Kallstroemia, Larrea, Tribulus*

Notes

1. The Thompson Indians said that if too much balsam pitch were taken as medicine, it could make one stout (187:97). *Abies amabilis, A. grandis.*
2. The Thompson Indians mixed the boiled boughs of three types of fir with decoctions of leaves from a broad-leaved plant from the Okanagan River, sweet grass from the Thompson River, and deer grease, and then used it to perfume the hair (187:97). *Abies amabilis, A. grandis, A. lasiocarpa.*
3. The wood was always taken early in the morning from the sunrise side of the tree (187:145). Thompson Indians, *Acer circinatum.*
4. The twigs and bark of new growth were boiled with water. A clay that contained iron was mixed with grease and then roasted; then it was mixed with the boiled twig and bark water. Tanned hides were soaked in this solution for 2 or 3 days to get the right color; treatment for a shorter period of time resulted in a brownish color, and for a longer period in black (70:100). Omaha and Winnebago, *Acer saccharinum.*
5. The Hesquiat used dried kelp stipes to play a beach game, something like hockey. This game was played in winter on the beach in front of the village when large quantities of several kinds of kelp drifted ashore (185:24). *Costaria costata, Lessoniopsis littoralis, Postelsia palmaeformis.*
6. These strong, tough seaweeds grow in the subtidal and intertidal zones. Sometimes, herring spawn on the stipes and fronds of these short kelps, and then the plants are gathered by the Hesquiat and dipped briefly in hot water or dried for later use. The spawn were taken off the longer types and the alga discarded, or, in the case of the broad, leafy types, the alga was eaten along with the eggs. If the kelps with spawn were dried first, they were simply soaked in water before being eaten (185:24). *Alaria marginata, Costaria costata, Lessoniopsis littoralis, Postelsia palmaeformis.*
7. Thompson Indians cleaned onion bulbs and twined them together in mats before they were cooked. They were tied together by their leaves in big bunches, about 15 cm across. They were dipped in water, but not soaked, then laid in the cooking pit interspersed with layers of beard tongue and alder leaves. The bulbs were steam cooked overnight, and after being cooked they became extremely sweet and were considered a delicacy. The cooked bulbs were eaten after meat was eaten. It was very important as a refreshment. Sometimes, the bulbs were cooked with black tree lichen (187:117). *Allium cernuum.*
8. A dull reddish dye was made by the Navajo from the alder and several other plants. The woman first burned some of the twigs of the juniper or spruce, then crushed and boiled the root bark of the mountain mahogany. Only the bark was used because the roots themselves contain no color-bearing material. To this was added the powdered bark of the alder together with a ground lichen. This was put together and boiled until it was thought to be right, then it was strained, and the wool or yarn was soaked in it overnight. This produced a dull reddish color on wool and a fine tan color on buckskin (55:39). *Alnus incana.*
9. The bread was carried by personators of anthropic gods and thrown by them to the populace between the dances (166:87). Navajo, *Amaranthus cruentus.*
10. Blackfoot Indian ceremonial events often included the transfer of a tepee design or the opening of a medicine pipe bundle or a beaver bundle. Women prepared soup from berries, assorted roots, fat, and water. At an appointed time during the ceremony this soup was served to all participants. The soup was blessed, and an offering of one of the berries was put back into the ground before eating began. A few mouthfuls were taken; then the remainder of the soup was given to one or another of the women, who would take it home to her children (86:26). *Amelanchier alnifolia, Prunus virginiana.*
11. Star mallow, wild heliotrope, fiddlenecks, and wild sorrel were dropped entirely from the Pima diet in more recent times. Amadeo Rea, the author of the work cited on the Pima, suspects they were spring starvation season foods (133:5). *Amsinckia* sp., *Eremalche exilis, Phacelia* sp., *Rumex* sp.
12. Arikara and Mandan boys used thorn apple thorns for arrow points on their grass arrows, and practiced their archery skills by hunting frogs (70:68). *Andropogon gerardii.*
13. For the Seminole, if a man had sexual intercourse with a woman during her menstrual period, the results were more serious than other menstruation sicknesses. A doctor would never do this, as it would damage the "medicine" that he had in his body. Other men were, sometimes, willing to take the risk. Several plants were used as medicine to treat the resulting illness if it occurred (169:248). *Angadenia berteroi, Cephalanthus occidentalis, Salix caroliniana, Stillingia sylvatica.*
14. Luiseño made fine-meshed fishing nets for stream fishing and large-meshed nets for sea fishing from twine made from several fibrous plant species (155:202). *Apocynum cannabinum, Asclepias eriocarpa, Urtica dioica.*
15. The plant could also be boiled and the steam used for colds. The Thompson Indian patient would lie in the steam while he was covered with a blanket (187:169). *Artemisia dracunculus.*
16. Havasupai placed leafy stems of sagebrush on the fire and the resulting fumes were inhaled for a cold or runny nose. The fresh leaves were chewed for a cough. More often, the leaves were pounded a bit, boiled in water for 10 to 15 or 20 minutes and then cooled and drunk. The usual dose, a cupful, was taken before breakfast and again after supper. This treatment was continued until the cold was gone, usually in 2 or 3 days. Another method was to take 2 tablespoonfuls every half hour for 1 day. This method was good for a cold, cough, sore throat, or stomachache (197:246). *Artemisia tridentata.*
17. The Yurok took a large snail's body from the shell, smashed it, and put it inside a wild ginger leaf. The preparation was steamed and then held over the umbilical cord with twine. The umbilical cord would fall off and the hole would heal (6:19). *Asarum caudatum.*
18. Cahuilla pounded the stem of several kinds of milkweed to loosen the fiber,

which then was extracted by rubbing the stem between the palms of the hands. The fiber was rolled on the thigh to produce cordage; its many uses testified to its natural strength and durability (15:43). *Asclepias* sp.

19. After the Zuni decorated sticks with milkweed cords, they were used as offerings and were planted in their fields and in sacred springs. An excavation was made in the bed of the spring in which the offerings were deposited after stones had been attached to them (166:88). *Asclepias subverticillata*.
20. The fruits were not planted with the seed corn but were discarded before planting. Omaha and Ponca informants could not give a reason for this process as they said they had forgotten the origin of the old custom (70:91). *Astragalus crassicarpus*.
21. Thompson Indians reconstituted the dried roots by soaking in water overnight. After they were cooked, the roots tasted sweet and were eaten as a "sort of dessert" after a meal of dried fish or some other food as a main course (187:175). *Balsamorhiza sagittata*.
22. Birch bark was particularly important in the storage of food. It could be stripped off in the fall when it was quite papery and could be split into thin sheets. These were weighted down with rocks to flatten them and then used to line the bottoms of berry baskets to keep the baskets from getting stained. The bark was also placed between layers of dried salmon in storage and used in the storage of cooked roots such as lily corms (187:189). Thompson Indians, *Betula papyrifera*.
23. Thistle roots were pit cooked after which they usually turned dark brown. One or two bags of dried roots were stored each year by a family and were said to be "full of vitamins." One informant said that the roots caused "gas" if too many were eaten (187:178). Thompson Indians, *Cirsium edule, C. hookerianum, C. undulatum, C. vulgare*.
24. The stem was stripped of its outer layers, leaving only the terminal leaf or inflorescence hanging loosely on a flexible strand. Opponents held firmly to the base of the strand and attempted to hook the opponent's leaf or inflorescence. After engaging, the opponents pulled and the opponent who retained his leaf or inflorescence won (6:25). Karok and Yurok, *Claytonia parviflora, C. perfoliata*.
25. Beeplant was so important economically to the Hopi and Tewa that it was listed in songs along with corn, pumpkins, and cotton, the three main cultivated plants (32:24). *Cleome serrulata*.
26. The Zuni boiled beeplant for a long time after which the concoction was allowed to evaporate. The precipitated paste was then used with black mineral paint to color sticks of plume offerings to anthropic gods (166:96). *Cleome serrulata*.
27. A decoction of red osier dogwood was used by the Thompson Indians to bathe 4- to 6-month-old babies four to six times over a period of several weeks. It was said to make the baby strong, bright, and good-natured (187:204). *Cornus sericea*.
28. Yokut placed the salt grass, when dry, on a dry hide or a large piece of canvas or cloth and then beat it for a long time until the tiny black, salty specks on the stem and narrow blades fell off and collected on the cloth. This material was kept in bottles or jars (formerly in baskets). When needed for medicine, it was put in hot water and boiled until it formed a dark reddish brown gum. The informant remarked that it should be "cooked like gravy until the gum comes" (117:423). *Distichlis spicata*.
29. Thompson Indians peeled silverberry bark off in strips as long as possible in the spring or fall when it was "kind of dry" and split it with a knife (originally of stone). The grayish outer bark was removed and the inner bark scraped, cleaned, and cut into desired widths. At this stage, it could be dried for future use. The long, even strands of fresh or dried inner bark, after it had been soaked, could be spun on the bare leg into a strong two-ply twine used for many different purposes (187:207). *Elaeagnus commutata*.
30. Woodland Cree ground the leafy branches of black crowberry and mixed them with grease, which was then applied externally for fevers; or, a branch was chewed as a diuretic (109:38). *Empetrum nigrum*.
31. Thompson Indians used a decoction or infusion of horsetail immediately after childbirth to expel the afterbirth more quickly. The stems were usually gathered ahead of time, being pulled up, the roots cut off, and the stems cut into short segments and dried. Then, just before childbirth, a small handful of the cut stems was steeped in boiling water for 5 or 10 minutes. The woman could drink this tea for several days (187:86). *Equisetum arvense, E. hyemale, E. laevigatum*.
32. Thompson Indians used the stem liquid of horsetail, which was collected in the springtime and could be stored in the refrigerator and used whenever needed. The stem decoction was used as an eyewash for sore or itchy eyes or for impending blindness such as that caused by cataracts (187:86). *Equisetum hyemale, E. laevigatum*.
33. The stems were used by Thompson Indians to do the final polishing of wooden spoons and to polish the soft rock used for pipe bowls (187:86). *Equisetum hyemale, E. laevigatum*.
34. The stem segments of horsetail were pulled apart and the stem liquid was extracted by the Thompson Indians; the liquid was splashed over weeds as an herbicide (187:86). *Equisetum hyemale, E. laevigatum*.
35. Okanagan-Colville children played a game with the stems of several kinds of buckwheat. They broke off the main stem, leaving one side branch attached to make a "hook." Each child took one of these. They hooked them together and pulled. The first one to break his or her stick lost the game (188:112). *Eriogonum compositum, E. heracleoides, E. niveum*.
36. The symptoms of the Seminole disease "dead people's sickness" are numb and painful legs, neck, shoulders, and perhaps the backbone. The sufferer is sleepy during the day, has "short breath," and loses appetite. Fever and headaches also accompany this illness (169:257). *Eryngium yuccifolium, Persea borbonia*.
37. This emetic was taken by the Seminole doctor after one of his patients had died. It prevented the next patient from getting worse instead of improving under the doctor's care (169:184). *Eryngium yuccifolium, Persea borbonia*.
38. The wild strawberries could be eaten only after the strawberries were danced over and blessed; they were eaten on picnic day by Kasha Pomo at the Strawberry Festival (72:110). *Fragaria chiloensis, F. vesca*.
39. Western Keres used juniper twigs in basketry. The twigs were bent into shape by allowing them to remain in the sun or, for quicker bending, were pushed into the hot ashes under a fire (171:48). *Juniperus monosperma*.
40. Juniper berries were gathered by Western Keres, mixed with chopped meat, put into a clean deer stomach, and roasted as a spicy sausage (171:48). *Juniperus monosperma*.
41. Juniper was chewed or an infusion of leaves was taken by Hopi and Tewa women immediately after giving birth. During the lying-in period, all of the mother's food was prepared with a decoction of leaves. Her head was washed with yucca suds and her body bathed with a hot infusion of leaves while her clothes were washed in water in which leaves of juniper had been placed (42:330). *Juniperus monosperma*.
42. If upon growing up, the Hopi or Tewa child misbehaves, he is taken at the request of the mother and held by some other woman in a blanket over a smoldering fire of juniper. He soon escapes, half-suffocated, supposedly a better and (probably) wiser youngster (42:330). *Juniperus monosperma*.
43. A handle was attached to the bottle gourd and its contents were removed and replaced with small gravel or seeds from another plant. The rattle was used in rituals by the Dakota, Omaha, and Ponca (70:117). *Lagenaria siceraria*.
44. The Karok would camp in tan oak groves when harvesting the acorns. Certain villages had certain fruit crops. Fruits were gathered after they had fallen from the trees but before insects invaded them. While younger men hunted, the remainder of the people played games centered around removing the shells from the seed. When the seeds were ground, a basket with a hole

in the bottom large enough to include the stone mortar was placed over the mortar to keep the acorn flour in place. It was then leached in sand with cold water. The finished flour was mixed with water to make a paste, which could be cooked in several ways. A gruel was most often made by cooking the paste in cooking baskets. Hot rocks were placed into the paste to bring it to boiling. The rocks were kept from burning the basket with "acorn paddles." The rocks were placed in and out of the gruel with twigs bent into a U shape. Males ate gruel with wooden spoons; the females used mussel shells. The cake of acorn meal that formed around the hot rocks was given to children as sort of a treat. Gruel was flavored with venison, herbs, etc. The paste was occasionally baked as patties in hot coals. Flour was stored in large storage baskets (6:35). *Lithocarpus densiflorus*.

45. Kashaya Pomo Indians dried acorns in the sun before storing them. The acorns were cracked open and the inner nuts put in a winnowing basket and rubbed to remove the chaff. They were then put into a mortar basket and pounded with a pestle to the consistency of flour. This flour was sifted with a basket and placed in a basin of clean sand, and water was poured over it many times to remove the bitter flavor. The water was poured over a bundle of leaves or branches that served to break the fall of the water and not splash sand into the food. The ground and leached meal was then cooked into mush or thinned with water to make soup. If pancakes or bread were to be made, the flour was ground coarser and was left soaking longer in the water. For bread, the dough was shaped into cakes that were wrapped in large leaves and baked in the coals. Red earth could be added to the dough to make a dark sweet bread. Another method produced moldy acorns that were made into mush. The acorns were not dried in the sun but were left in the house until they turned greenish with mold. The mold was rubbed off. These nuts were pounded together with whitened dry acorns and made into mush. Another method was to leave cracked acorns in a pool for 4 or 5 months. They were then removed from the shell and cooked without pulverizing. They could be used for soup or mush, or eaten whole (72:83). *Lithocarpus densiflorus, Quercus agrifolia, Q. garryana, Q. kelloggii*.

46. The woody stems of hollyleaved barberry or Cascade Oregongrape were broken, the bark scraped off, and boiling water was poured over it to make an infusion that the Thompson Indians used to make an eyewash (187:187). *Mahonia aquifolium, M. nervosa*.

47. Before a Cahuilla ritual was conducted, tobacco was smoked by the ritual leaders and shamans and the smoke was blown in the sacred directions: north, east, west, south, and up or center. This helped to clear the area of any malevolent force that might interfere with the ritual. Throughout ceremonies, especially those honoring the recent dead, everyone was obliged to smoke tobacco, as they are even in more recent times. At funerals, smoking served to concentrate power that would aid the dead in their arduous journey to the other world (15:90). *Nicotiana clevelandii, N. glauca, N. trigonophylla*.

48. The infusion of devil's club sticks was taken by Thompson Indians in doses of about ½ cup before meals, to replace milk and other beverages (187:164). *Oplopanax horridus*.

49. The stems of pricklypear were used for food during times of famine by the Thompson Indians because they could be harvested even during the winter and early spring when few other foods were available (187:194). *Opuntia fragilis, O. polyacantha*.

50. Thompson Indians baked pricklypear stems in hot coals until the spines were singed off and then squeezed them until the fleshy centers popped out. The inside part was eaten and considered quite tasty. In more recent years, some people mixed cactus stems with fruit cakes, but traditionally, cactus stems were eaten with northern black currants or other types of berries (187:194). *Opuntia fragilis, O. polyacantha*.

51. Freshly peeled stems of pricklypear were rubbed over painted objects to fix the color by Dakota and Pawnee Indians (70:104). *Opuntia humifusa*.

52. Omaha and Ponca knew that horses were very fond of the roots of sweetroot and would come running if people whistled for them while holding the roots in their hands; then the horses could easily be caught (70:107). *Osmorhiza longistylis*.

53. The blossoms of fetid marigold were chewed by Zuni of both sexes—but especially by women—then spit into the hands and rubbed on the neck, limbs, and clothing as perfume (166:83). *Pectis papposa*.

54. A rabbit stick treated with Torrey's penstemon in this manner was sure to kill any rabbit at which it was aimed, provided the Zuni thrower had a good heart (166:95). *Penstemon barbatus* ssp. *torreyi*.

55. When a Zuni infant boy showed signs of timidity, his father carried a small quantity of cornmeal wrapped in a bit of corn husk to the warrior of his choice, presented it, and asked that the warrior apply a medicine made of slimleaf bean on his child's body so that he might have a brave heart and never be afraid of the enemy. The warrior then chewed the crushed leaves and blossoms and the powdered root, spit the mass into his hands, and rubbed it all over the child's body (166:85). *Phaseolus angustissimus*.

56. Thompson Indians said that this treatment should be three times over a period of days (187:230). *Philadelphus lewisii*.

57. Upper Tanana Indians dug spruce roots by hand or with an ax, preferably from a tree that was not crowded by other trees. The roots of a tree growing in an open place were less likely to be entangled with the roots of other trees and were therefore easier to dig. Spruce roots in moist ground where moss grows were also easier to gather than those found in dry soil. Before using spruce roots, the Upper Tanana peeled the bark off by hand or with a knife. After peeling them, they sometimes dyed them by boiling berries and soaking the roots in the juice. Spruce roots could be dried for future use but had to be soaked in water to make them pliable before being used. They could be dug anytime during the year when the ground was not frozen (102:2). *Picea glauca*.

58. Karok placed pine cones in a trench and covered them with dirt. A fire was built on top. After roasting, the cones were broken open to release the seeds. Some were stored over winter (6:44). *Pinus lambertiana*.

59. The game, pile pile, was played frequently in Neah Bay by the Makah earlier in the twentieth century and went as follows: One would take a deep breath and pluck the pinnae one by one from a swordfern, saying "pile" for each, all in one breath. The winner was the one who reached the top (apex) of the frond, pulled the most pinnae off, before running out of breath. In Neah Bay anyone could play this game, but among the Nitinaht, where adults also played pile pile, the game was a family privilege exercised at big potlatches (67:221). *Polystichum munitum*.

60. Dried chokecherry cakes were used in winter with dried corn or cooked alone with sugar by the Omaha Indians. (68:326). *Prunus virginiana*.

61. Pemmican was made by the Sioux and other tribes by mixing certain berries, such as buffalo berry, choke cherry, and serviceberry, with the fat of the buffalo, pounding up the whole, which was then packed away in skins. Sometimes jerked buffalo was put into an oven to render it brittle, beaten up on a skin with these berries, some marrow fat being added to give consistency, and finally packed in skin bags. This was a regular article of commerce and highly prized by the old trappers and hunters for its portability as a condensed food and for its keeping qualities. Later the flesh and tallow of the ox was substituted for that of the buffalo and was still used to some extent until the turn of the twentieth century (19:19). *Prunus virginiana*.

62. The blossoms of the woolly paperflower were used by the Zuni for painting the masks and for coloring their limbs and bodies yellow when they impersonated anthropic gods. The flowers were ground into a meal and mixed with yellow ocher and urine (166:97). *Psilostrophe tagetina*.

63. The leaves of the pallid hoptree, which had a strong odor, were pounded up together with jimson weed, scorpions, and other noxious items by the Havasupai for arrow poison. It was said that even a slight scratch from an

arrow so treated would insure death (197:229). *Ptelea trifoliata*.

64. Luiseño Indians used several methods to remove the bitterness from acorn meal. The meal was either leached with hot water, placed in a rush basket and warm water poured over it, or placed in a sand hole and warm water poured over it to soak away the bitterness (155:194). *Quercus agrifolia, Q. chrysolepis, Q. dumosa, Q. engelmannii, Q. kelloggii, Q. wislizeni*.

65. Kawaiisu Indians stored acorns in several different ways. The granaries, elevated about a foot above the ground to keep out rodents, were made of hardwood poles, usually oak, with sides, top, and bottom covered with bark and lined with gray California buckwheat leaves. Stone-lined pits were covered with brush, and acorns were piled on a large flat stone and covered with bark (206:56). *Quercus chrysolepis, Q. douglasii, Q. dumosa, Q. garryana, Q. kelloggii, Q. lobata, Q. wislizeni*.

66. A decoction of the branches of wax currant along with squaw currant, red osier dogwood branches, and the boughs of Douglas fir or tamarack was used by Thompson Indians to bathe babies four to six times over a period of several weeks to make them strong, independent, and good-natured (187:226). *Ribes cereum*.

67. Thompson Indians stored and dried rose hips to make tea (187:267). *Rosa acicularis, R. woodsii*.

68. Rose hips were eaten by Thompson Indians with a kind of whitish berry or mixed with wild mushrooms (187:267). *Rosa acicularis, R. woodsii*.

69. Okanagan-Colville doctors swept out new graves with rose branches to prevent someone else's spirit from being buried with the dead person (188:131). *Rosa acicularis, R. gymnocarpa, R. nutkana, R. woodsii*.

70. Among the Thompson Indians, after a death, an orphan, widower, or widow placed rose branches under the mattress to protect against sickness and to "keep ghosts away." Widows and widowers could fish only if rose branches were first swept around their gill nets (187:267). *Rosa acicularis, R. nutkana, R. pisocarpa, R. woodsii*.

71. Navajo sometimes dried dock roots and stored them indefinitely. When ready for use in dyes, the dried roots were ground. By this aging process various shades were obtained, from a grayed yellow to a dull red. Several handfuls of the fresh roots boiled in water yielded a lemon yellow, and when more of the root was used and boiled longer, a soft orange or orange-brown was obtained. If the mixture was boiled in an iron vessel, the reaction formed a red-brown or mahogany dye. When mixed with indigo, a green dye was produced (55:43). *Rumex hymenosepalus*.

72. Woodland Cree used willow bark to tie fish roasting sticks, rockfish net weights, birch bark moose calls, and snowshoe frames, which were tied together before they were bent. They also used the bark to tie bundles to be carried by hand (109:58). *Salix bebbiana, S. discolor, S. interior*.

73. According to the Seminole, if a woman does not eat by herself during her menstrual period, the other inhabitants of her camp will get sick. Both sexes are thus affected, but men are particularly vulnerable (169:244). *Salix caroliniana*.

74. The Thompson Indians used the branches of willows for tying and binding, especially in constructing their underground winter pit houses and in making fish traps, weirs, and rafts (187:279). *Salix* sp.

75. Thompson Indians mashed elderberries in birch bark baskets, and when a fish had been split open along the backbone in preparation for barbecuing, it was soaked for a while in elderberry juice to flavor it. Then the fish was taken out, the berry seeds scraped off, and the fish barbecued and eaten (187:199). *Sambucus cerulea, S. racemosa*.

76. The Thompson Indians said elderberries tasted like sulfur but nonetheless ate them in a variety of ways (187:199). *Sambucus racemosa*.

77. The roots of bloodroot were used by the Ojibwa in four or five combinations in dyeing various materials (153:426). *Sanguinaria canadensis*.

78. Western red cedar was used along with Douglas fir boughs, rose branches, and often a small quantity of stinging nettle in several medical compounds by the Okanagan-Colville Indians (188:20). *Thuja plicata*.

79. The fresh green spikes of cattail eaten by the Paiute later formed the brown pollen and seed-bearing heads (60:69). *Typha domingensis, T. latifolia*.

80. Brown seed heads of cattail were cut from the stalks by the Paiute and taken to an area where the earth had been dampened and packed into a hard, crusted surface. After warming the spikes in the sun, the cattail fluff was removed from the spikes and placed on the ground in a layer about 2 inches deep. This was set on fire and the mixture stirred until all the fluff was burned. Remaining on the hard, packed earth were thousands of tiny black cattail seeds. These were gathered into a finely woven basketry tray and the seed was winnowed by gently tossing it in the air during a slight breeze (60:69). *Typha domingensis, T. latifolia*.

81. A Makah informant said, "As a child I saw my father when he'd take this halibut line, fish line, and he'd tighten it from one end of the yard to the other, while it was being stretched like that, otherwise they coil and tangle you know. He'd take a handful of those [stinging nettle] leaves and he'd rub it along the line and it gave it kind of a green color. I don't know if that was just for the color or if he thought there was some medicine in it or something, for good fishing or something. Might have been just to tint the line" (67:246). *Urtica dioica*.

82. Blueberries were probably the most highly regarded wild plant food of the various Algonquin groups in the study area in Quebec. They were locally abundant and individuals and families from bands not in the vicinity of good blueberry fields traveled considerable distances in order to pick berries for themselves and to sell. During the course of field studies, members of the River Desert band traveled northwest of their area 90 miles and more. Members of the Weymontaching Cree band were met near Clova in the Obedjiwan band area. They had traveled to this location by train in order to pick blueberries and meet with relatives (18:103). *Vaccinium myrtilloides*.

83. One Thompson Indian informant cautioned that, while American false hellebore had a number of uses as an external medicine, it should not be taken internally because of its extreme toxicity (187:131). *Veratrum viride*.

84. Yucca was a valued food supply for the Kawaiisu. In the early spring the apical meristem, the "heart" or "cabbage," was removed with the aid of an oak or a juniper shaft having a shovel-like cutting edge at one end. The "cabbage" was roasted in a pit about 3 feet deep and perhaps 5 feet in diameter. A large stone was placed in the center of the pit, with smaller stones radiating out from the center. Firewood, preferably Douglas oak, was thrown in and the hot fire burned for a half day. From time to time more stones, and probably wood, were added. As the fire died down, it was covered with sand or dirt to about ground level, leaving a small aperture in the center. The sand was tapped down, causing flames to shoot out of the hole. Then a layer of dry pine needles was put on and the "cabbages," having been skinned, were placed on top. Two or three families shared the same oven. The direction in which "cabbages" were laid identified the owner. Another layer of dry pine needles apparently mixed with silky California broom covered the "cabbages." More sand or dirt, patted down and smoothed with a basketry tray, completed the mound, which would now reach a height of 5 feet but which settled in the roasting process. The roasting continued for two nights, during which no sexual intercourse was permitted. Otherwise it was said that "it won't cook." When the "cabbages" were pulled out, they were so hot they burned the hands. They were cooled before they were eaten. To be stored, they were pulled apart, mashed a little, and dried. They could not be dried or stored uncooked. After storage they were soaked and eaten (206:69). *Yucca brevifolia*.

85. Although this corn beverage could be consumed by Zuni at any time, it was used especially by the rain priests and personators of anthropic gods during ceremonies. Another native beverage was also made by the Zuni. Water was poured over sprouted corn, allowed to stand for some days, and then used as a beverage (166:73). *Zea mays*.

Bibliography

1. Ager, Thomas A., and Lynn Price Ager. 1980. Ethnobotany of the Eskimos of Nelson Island, Alaska. Arctic Anthropology 27: 26–48.
2. Akana, Akaiko. 1922. Hawaiian Herbs of Medicinal Value. Honolulu: Pacific Book House.
3. Aller, Wilma F. 1954. Aboriginal Food Utilization of Vegetation by the Indians of the Great Lake Region as Recorded in the Jesuit Relations. Wisconsin Archeologist 35: 59–73.
4. Anderson, J. P. 1939. Plants Used by the Eskimo of the Northern Bering Sea and Arctic Regions of Alaska. American Journal of Botany 26: 714–716.
5. Arnason, Thor, Richard J. Hebda and Timothy Johns. 1981. Use of Plants for Food and Medicine by Native Peoples of Eastern Canada. Canadian Journal of Botany 59(11): 2189–2325.
6. Baker, Marc A. 1981. The Ethnobotany of the Yurok, Tolowa and Karok Indians of Northwest California. M.A. Thesis, Humboldt State University, Arcata, California.
7. Bank, Theodore P., II. 1951. Botanical and Ethnobotanical Studies in the Aleutian Islands I. Aleutian Vegetation and Aleut Culture. Botanical and Ethnobotanical Studies Papers, Michigan Academy of Science, Arts and Letters, 37: 13–30.
8. Bank, Theodore P., II. 1953. Botanical and Ethnobotanical Studies in the Aleutian Islands II. Health and Medical Lore of the Aleuts. Botanical and Ethnobotanical Studies Papers, Michigan Academy of Science, Arts and Letters, 38: 415–431.
9. Barrett, S. A. 1908. Pomo Indian Basketry. University of California Publications in American Archaeology and Ethnology 7: 134–308.
10. Barrett, S. A. 1917. The Washoe Indians. Bulletin of the Public Museum of the City of Milwaukee 2(1): 1–52.
11. Barrett, S. A. 1952. Material Aspects of Pomo Culture. Bulletin of the Public Museum of the City of Milwaukee, Number 20.
12. Barrett, S. A., and E. W. Gifford. 1933. Miwok Material Culture. Bulletin of the Public Museum of the City of Milwaukee 2(4): 11.
13. Barrows, David Prescott. 1967. The Ethno-Botany of the Coahuilla Indians of Southern California. Banning, California: Malki Museum Press. Originally published in 1900.
14. Basehart, Harry W. 1974. Apache Indians XII. Mescalero Apache Subsistence Patterns and Socio-Political Organization. New York: Garland Publishing.
15. Bean, Lowell John, and Katherine Siva Saubel. 1972. Temalpakh (From the Earth); Cahuilla Indian Knowledge and Usage of Plants. Banning, California: Malki Museum Press.
16. Beardsley, Gretchen 1941. Notes on Cree Medicines, Based on Collections Made by I. Cowie in 1892. Papers of the Michigan Academy of Science, Arts and Letters 28: 483–496.
17. Bell, Willis H., and Edward F. Castetter. 1941. Ethnobiological Studies in the Southwest VII. The Utilization of Yucca, Sotol and Beargrass by the Aborigines in the American Southwest. University of New Mexico Bulletin 5(5): 1–74.
18. Black, Meredith Jean. 1980. Algonquin Ethnobotany: An Interpretation of Aboriginal Adaptation in South Western Quebec. Ottawa: National Museums of Canada. Mercury Series, Number 65.
19. Blankinship, J. W. 1905. Native Economic Plants of Montana. Bozeman: Montana Agricultural College Experimental Station, Bulletin 56.
20. Boas, Franz 1966. Kwakiutl Ethnography. Chicago: University of Chicago Press.
21. Bocek, Barbara R. 1984. Ethnobotany of Costanoan Indians, California, Based on Collections by John P. Harrington. Economic Botany 38(2): 240–255.
22. Bradley, Will T. 1936. Medical Practices of the New England Aborigines. Journal of the American Pharmaceutical Association 25(2): 138–147.
23. Brugge, David M. 1965. Navajo Use of *Agave*. Kiva 31(2): 88–98.
24. Burgesse, J. Allen 1944. The Woman and the Child among the Lac-St.-Jean Montagnais. Primitive Man 17: 1–18.
25. Bushnell, David I., Jr. 1909. The Choctaw of Bayou Lacomb, St. Tammany Parish, Louisiana. Smithsonian Institution—Bureau of American Ethnology Bulletin, Number 48.
26. Buskirk, Winfred. 1986. The Western Apache: Living with the Land before 1950. Norman: University of Oklahoma Press.
27. Camazine, Scott, and Robert A. Bye 1980. A Study of the Medical Ethnobotany of the Zuni Indians of New Mexico. Journal of Ethnopharmacology 2: 365–388.
28. Campbell, T. N. 1951. Medicinal Plants Used by Choctaw, Chickasaw, and Creek Indians in the Early Nineteenth Century. Journal of the Washington Academy of Sciences 41(9): 285–290.
29. Carlson, Gustav G., and Volney H. Jones. 1940. Some Notes on Uses of Plants by the Comanche Indians. Papers of the Michigan Academy of Science, Arts and Letters 25: 517–542.
30. Carr, Lloyd G., and Carlos Westey. 1945. Surviving Folktales and Herbal Lore among the Shinnecock Indians. Journal of American Folklore 58: 113–123.
31. Carrier Linguistic Committee. 1973. Plants of Carrier Country. Fort St. James, British Columbia: Carrier Linguistic Committee.
32. Castetter, Edward F. 1935. Ethnobiological Studies in the American Southwest I. Uncultivated Native Plants Used as Sources of Food. University of New Mexico Bulletin 4(1): 1–44.
33. Castetter, Edward F., and M. E. Opler. 1936. Ethnobiological Studies in the American Southwest III. The Ethnobiology of the Chiricahua and Mescalero Apache. University of New Mexico Bulletin 4(5): 1–63.
34. Castetter, Edward F., and Ruth M. Underhill. 1935. Ethnobiological Studies

34. ...in the American Southwest II. The Ethnobiology of the Papago Indians. University of New Mexico Bulletin 4(3): 1–84.
35. Castetter, Edward F., and Willis H. Bell. 1937. Ethnobiological Studies in the American Southwest IV. The Aboriginal Utilization of the Tall Cacti in the American South. University of New Mexico Bulletin 5: 1–48.
36. Castetter, Edward F., and Willis H. Bell. 1942. Pima and Papago Indian Agriculture. Albuquerque: University of New Mexico Press.
37. Castetter, Edward F., and Willis H. Bell. 1951. Yuman Indian Agriculture. Albuquerque: University of New Mexico Press.
38. Chamberlin, Ralph V. 1909. Some Plant Names of the Ute Indians. American Anthropologist 11: 27–40.
39. Chamberlin, Ralph V. 1911. The Ethno-Botany of the Gosiute Indians of Utah. Memoirs of the American Anthropological Association 2(5): 331–405.
40. Chandler, R. Frank, Lois Freeman, and Shirley N. Hooper. 1979. Herbal Remedies of the Maritime Indians. Journal of Ethnopharmacology 1: 49–68.
41. Chesnut, V. K. 1902. Plants Used by the Indians of Mendocino County, California. Contributions from the U.S. National Herbarium 7: 295–408.
42. Colton, Harold S. 1974. Hopi History and Ethnobotany. Pages 279–373 in D. A. Horr (editor), Hopi Indians. New York: Garland Publishing.
43. Compton, Brian Douglas. 1993. Upper North Wakashan and Southern Tsimshian Ethnobotany: The Knowledge and Usage of Plants . . . Ph.D. Dissertation, University of British Columbia, Vancouver.
44. Cook, Sarah Louise. 1930. The Ethnobotany of Jemez Indians. M.A. Thesis, University of New Mexico, Albuquerque.
45. Coville, Frederick V. 1897. Notes on the Plants Used by the Klamath Indians of Oregon. Contributions from the U.S. National Herbarium 5(2): 87–110.
46. Coville, Frederick V. 1904. Wokas, a Primitive Food of the Klamath Indians. Washington, D.C.: Smithsonian Institution, U.S. National Museum.
47. Curtin, L. S. M. 1949. By the Prophet of the Earth. Sante Fe, New Mexico: San Vicente Foundation.
48. Curtin, L. S. M. 1957. Some Plants Used by the Yuki Indians . . . I. Historical Review and Medicinal Plants. Masterkey 31: 40–48.
49. Curtin, L. S. M. 1957. Some Plants Used by the Yuki Indians . . . II. Food Plants. Masterkey 31: 85–94.
50. Dawson, E. Yale. 1944. Some Ethnobotanical Notes on the Seri Indians. Desert Plant Life 9: 133–138.
51. Densmore, Frances. 1913. Chippewa Music—II. Smithsonian Institution—Bureau of American Ethnology Bulletin, Number 53.
52. Densmore, Frances. 1918. Teton Sioux Music. Smithsonian Institution—Bureau of American Ethnology Bulletin, Number 61.
53. Densmore, Frances. 1928. Uses of Plants by the Chippewa Indians. Smithsonian Institution—Bureau of American Ethnology Annual Report 44: 273–379.
54. Densmore, Francis. 1932. Menominee Music. Smithsonian Institution—Bureau of American Ethnology Bulletin, Number 102.
55. Elmore, Francis H. 1944. Ethnobotany of the Navajo. Sante Fe, New Mexico: School of American Research.
56. Fewkes, J. Walter. 1896. A Contribution to Ethnobotany. American Anthropologist 9: 14–21.
57. Fleisher, Mark S. 1980. The Ethnobotany of the Clallam Indians of Western Washington. Northwest Anthropological Research Notes 14(2): 192–210.
58. Fletcher, Alice C., and Francis la Flesche. 1911. The Omaha Tribe. Smithsonian Institution—Bureau of American Ethnology Annual Report, Number 27.
59. Fowler, Catherine S. 1989. Willard Z. Park's Ethnographic Notes on the Northern Paiute of Western Nevada 1933–1940. Salt Lake City: University of Utah Press.
60. Fowler, Catherine S. 1990. Tule Technology: Northern Paiute Uses of Marsh Resources in Western Nevada. Washington, D.C.: Smithsonian Institution Press.
61. Garth, Thomas R. 1953. Atsugewi Ethnography. Anthropological Records 14(2): 140–141.
62. Gifford, E. W. 1931. The Kamia of Imperial Valley. Washington, D.C.: U.S. Government Printing Office.
63. Gifford, E. W. 1932. The Southeastern Yavapai. University of California Publications in American Archaeology and Ethnology 29: 177–252.
64. Gifford, E. W. 1933. The Cocopa. University of California Publications in American Archaeology and Ethnology 31: 263–270.
65. Gifford, E. W. 1936. Northeastern and Western Yavapai. University of California Publications in American Archaeology and Ethnology 34: 247–345.
66. Gifford, E. W. 1967. Ethnographic Notes on the Southwestern Pomo. Anthropological Records 25: 10–15.
67. Gill, Steven J. 1983. Ethnobotany of the Makah and Ozette People, Olympic Peninsula, Washington (USA). Ph.D. Thesis, Washington State University, Pullman.
68. Gilmore, Melvin R. 1913. A Study in the Ethnobotany of the Omaha Indians. Nebraska State Historical Society Collections 17: 314–357.
69. Gilmore, Melvin R. 1913. Some Native Nebraska Plants with Their Uses by the Dakota. Collections of the Nebraska State Historical Society 17: 358–370.
70. Gilmore, Melvin R. 1919. Uses of Plants by the Indians of the Missouri River Region. Smithsonian Institution—Bureau of American Ethnology Annual Report, Number 33.
71. Gilmore, Melvin R. 1933. Some Chippewa Uses of Plants. Ann Arbor: University of Michigan Press.
72. Goodrich, Jennie, and Claudia Lawson. 1980. Kashaya Pomo Plants. Los Angeles: American Indian Studies Center, University of California, Los Angeles.
73. Gottesfeld, Leslie M. J. 1992. The Importance of Bark Products in the Aboriginal Economies of Northwestern British Columbia, Canada. Economic Botany 46(2): 148–157.
74. Gottesfeld, Leslie M. J., and Beverley Anderson. 1988. Gitksan Traditional Medicine: Herbs And Healing. Journal of Ethnobiology 8(1): 13–33.
75. Grinnell, George Bird. 1905. Some Cheyenne Plant Medicines. American Anthropologist 7: 37–43.
76. Grinnell, George Bird. 1972. The Cheyenne Indians—Their History and Ways of Life, Volume 2. Lincoln: University of Nebraska Press.
77. Guédon, Marie-Francoise. 1974. People of Tetlin, Why Are You Singing? Ottawa: National Museums of Canada. Mercury Series, Number 9.
78. Gunther, Erna. 1927. Klallam Ethnography. Seattle: University of Washington Press.
79. Gunther, Erna. 1973. Ethnobotany of Western Washington. Revised edition. Seattle: University of Washington Press.
80. Hamel, Paul B., and Mary U. Chiltoskey. 1975. Cherokee Plants and Their Uses—A 400 Year History. Sylva, North Carolina: Herald Publishing.
81. Hann, John H. 1986. The Use and Processing of Plants by Indians of Spanish Florida. Southeastern Archaeology 5(2): 1–102.
82. Hart, Jeff. 1992. Montana Native Plants and Early Peoples. Helena: Montana Historical Society Press.
83. Hart, Jeffrey A. 1981. The Ethnobotany of the Northern Cheyenne Indians of Montana. Journal of Ethnopharmacology 4: 1–55.
84. Hedges, Ken. 1986. Santa Ysabel Ethnobotany. San Diego Museum of Man Ethnic Technology Notes, Number 20.
85. Heller, Christine A. 1953. Edible and Poisonous Plants of Alaska. College, Alaska: Cooperative Agricultural Extension Service.
86. Hellson, John C. 1974. Ethnobotany of the Blackfoot Indians. Ottawa: National Museums of Canada. Mercury Series, Number 19.

87. Herrick, James William. 1977. Iroquois Medical Botany. Ph.D. Thesis, State University of New York, Albany.
88. Hinton, Leanne. 1975. Notes on La Huerta Diegueno Ethnobotany. Journal of California Anthropology 2: 214–222.
89. Hocking, George M. 1949. From Pokeroot to Penicillin. Rocky Mountain Druggist, November 1949, pages 12, 38.
90. Hocking, George M. 1956. Some Plant Materials Used Medicinally and Otherwise by the Navaho Indians in the Chaco Canyon, New Mexico. Palacio 56: 146–165.
91. Hoffman, W. J. 1891. The Midewiwin or "Grand Medicine Society" of the Ojibwa. Smithsonian Institution—Bureau of American Ethnology Annual Report, Number 7.
92. Holmes, E. M. 1884. Medicinal Plants Used by Cree Indians, Hudson's Bay Territory. Pharmaceutical Journal and Transactions 15: 302–304.
93. Holt, Catharine. 1946. Shasta Ethnography. Anthropological Records 3(4): 308.
94. Howard, James. 1965. The Ponca Tribe. Smithsonian Institution—Bureau of American Ethnology Bulletin, Number 195.
95. Hrdlicka, Ales. 1908. Physiological and Medical Observations among the Indians of Southwestern United States and Northern Mexico. Smithsonian Institution—Bureau of American Ethnology Bulletin 34: 1–427.
96. Jenness, Diamond. 1935. The Ojibwa Indians of Parry Island, Their Social and Religious Life. National Museums of Canada Bulletin, Number 78, Anthropological Series, Number 17.
97. Johnston, Alex. 1987. Plants and the Blackfoot. Lethbridge, Alberta: Lethbridge Historical Society.
98. Jones, Anore. 1983. Nauriat Niginaqtuat = Plants That We Eat. Kotzebue, Alaska: Maniilaq Association Traditional Nutrition Program.
99. Jones, David E. 1968. Comanche Plant Medicine. Papers in Anthropology 9: 1–13.
100. Jones, Volney H. 1931. The Ethnobotany of the Isleta Indians. M.A. Thesis, University of New Mexico, Albuquerque.
101. Jones, Volney H. 1938. An Ancient Food Plant of the Southwest and Plateau Regions. Palacio 44: 41–53.
102. Kari, Priscilla Russe. 1985. Upper Tanana Ethnobotany. Anchorage: Alaska Historical Commission.
103. Kelly, Isabel T. 1930. Yuki Basketry. University of California Publications in American Archaeology and Ethnology 24: 421–444.
104. Kelly, Isabel T. 1932. Ethnography of the Surprise Valley Paiute. University of California Publications in American Archaeology and Ethnology 31(3): 67–210.
105. Kirk, Ruth E. 1952. Panamint Basketry: A Dying Art. Masterkey 26: 76–86.
106. Kraft, Shelly Katherene. 1990. Recent Changes in the Ethnobotany of Standing Rock Indian Reservation. M.A. Thesis, University of North Dakota, Grand Forks.
107. Krause, Aurel. 1956. The Tlingit Indians. Translated by Erna Gunther. Seattle: University of Washington Press.
108. Lantis, Margaret. 1959. Folk Medicine and Hygiene. Anthropological Papers of the University of Alaska 8: 1–75.
109. Leighton, Anna L. 1985. Wild Plant Use by the Woods Cree (Nihithawak) of East-Central Saskatchewan. Ottawa. National Museums of Canada. Mercury Series, Number 101.
110. Lynch, Regina H. 1986. Cookbook. Chinle, Arizona: Navajo Curriculum Center, Rough Rock Demonstration School.
111. Mahar, James Michael. 1953. Ethnobotany of the Oregon Paiutes of the Warm Springs Indian Reservation. B.A. Thesis, Reed College, Portland, Oregon.
112. Malo, David. 1903. Hawaiian Antiquities. Honolulu: Hawaiian Gazette Co., Ltd.
113. Mandelbaum, David G. 1940. The Plains Cree. Anthropological Papers of the American Museum of Natural History 37: 202–203.
114. McClintock, Walter. 1909. Medizinal- und Nutzpflanzen der Schwarzfuss Indianer. Zeitschrift für Ethnologie 41: 273–279.
115. McKennan, Robert A. 1959. The Upper Tanana Indians. Yale University Publications in Anthropology, Number 55.
116. Mechling, W. H. 1959. The Malecite Indians with Notes on the Micmacs. Anthropologica 8: 239–263.
117. Merriam, C. Hart. 1966. Ethnographic Notes on California Indian Tribes. Berkeley: University of California Archaeological Research Facility.
118. Murphey, Edith Van Allen. 1990. Indian Uses of Native Plants. Glenwood, Illinois: Meyerbooks. Originally published in 1959.
119. Nelson, Richard K. 1983. Make Prayers to the Raven—A Koyukon View of the Northern Forest. Chicago: University of Chicago Press.
120. Nequatewa, Edmund. 1943. Some Hopi Recipes for the Preparation of Wild Plant Foods. Plateau 18: 18–20.
121. Nickerson, Gifford S. 1966. Some Data on Plains and Great Basin Indian Uses of Certain Native Plants. Tebiwa 9(1): 45–51.
122. Oswalt, W. H. 1957. A Western Eskimo Ethnobotany. Anthropological Papers of the University of Alaska 6: 17–36.
123. Palmer, Gary. 1975. Shuswap Indian Ethnobotany. Syesis 8: 29–51.
124. Parker, Arthur Caswell. 1910. Iroquois Uses of Maize and Other Food Plants. Albany, New York: University of the State of New York.
125. Perry, F. 1952. Ethno-Botany of the Indians in the Interior of British Columbia. Museum and Art Notes 2(2): 36–43.
126. Perry, Myra Jean. 1975. Food Use of "Wild" Plants by Cherokee Indians. M.S. Thesis, University of Tennessee, Knoxville.
127. Porsild, A. E. 1937. Edible Roots and Berries of Northern Canada. Toronto: Canada Department of Mines and Resources, National Museum of Canada.
128. Porsild, A. E. 1953. Edible Plants of the Arctic. Arctic 6: 15–34.
129. Powers, Stephen. 1874. Aboriginal Botany. Proceedings of the California Academy of Science 5: 373–379.
130. Radin, Paul. 1923. The Winnebago Tribe. Smithsonian Institution—Bureau of American Ethnology Annual Report, Number 37.
131. Ray, Verne F. 1932. The Sanpoil and Nespelem: Salishan Peoples of Northeastern Washington. University of Washington Publications in Anthropology, Volume 5.
132. Raymond, Marcel. 1945. Notes Ethnobotaniques sur les Tête-de-Boule de Manouan. Contributions de l'Institut Botanique de l'Université de Montréal 55: 113–134.
133. Rea, Amadeo M. 1991. Gila River Pima Dietary Reconstruction. Arid Lands Newsletter 31: 3–10.
134. Reagan, Albert. 1934. Various Uses of Plants by West Coast Indians. Washington Historical Quarterly 25: 133–137.
135. Reagan, Albert B. 1928. Plants Used by the Bois Fort Chippewa (Ojibwa) Indians of Minnesota. Wisconsin Archeologist 7(4): 230–248.
136. Reagan, Albert B. 1929. Plants Used by the White Mountain Apache Indians of Arizona. Wisconsin Archeologist 8: 143–161.
137. Reagan, Albert B. 1936. Plants Used by the Hoh and Quileute Indians. Kansas Academy of Science 37: 55–70.
138. Robbins, W. W., J. P. Harrington, and B. Freire-Marreco 1916. Ethnobotany of the Tewa Indians. Smithsonian Institution—Bureau of American Ethnology Bulletin, Number 55.
139. Rogers, Dilwyn J. 1980. Lakota Names and Traditional Uses of Native Plants by Sicangu (Brule) People in the Rosebud Area, South Dakota. St. Francis, South Dakota: Rosebud Educational Society.
140. Romero, John Bruno. 1954. The Botanical Lore of the California Indians. New York: Vantage Press.
141. Rousseau, Jacques. 1945. Le Folklore Botanique de Caughnawaga. Contributions de l'Institut Botanique de l'Université de Montréal 55: 7–72.

142. Rousseau, Jacques. 1945. Le Folklore Botanique de l'Ile aux Coudres. Contributions de l'Institut Botanique de l'Université de Montréal 55: 75–111.
143. Rousseau, Jacques. 1946. Notes Sur l'Ethnobotanique d'Anticosti. Archives de Folklore 1: 60–71.
144. Rousseau, Jacques. 1947. Ethnobotanique Abénakise. Archives de Folklore 11: 145–182.
145. Rousseau, Jacques. 1948. Ethnobotanique et Ethnozoologie Gaspésiennes. Archives de Folklore 3: 51–64.
146. Russell, Frank. 1908. The Pima Indians. Smithsonian Institution—Bureau of American Ethnology Annual Report 26: 1–390.
147. Sapir, Edward, and Leslie Spier. 1943. Notes on the Culture of the Yana. Anthropological Records 3(3): 252–253.
148. Schenck, Sara M., and E. W. Gifford. 1952. Karok Ethnobotany. Anthropological Records 13(6): 377–392.
149. Smith, G. Warren. 1973. Arctic Pharmacognosia. Arctic 26: 324–333.
150. Smith, Harlan I. 1929. Materia Medica of the Bella Coola and Neighboring Tribes of British Columbia. National Museum of Canada Bulletin 56: 47–68.
151. Smith, Huron H. 1923. Ethnobotany of the Menomini Indians. Bulletin of the Public Museum of the City of Milwaukee 4: 1–174.
152. Smith, Huron H. 1928. Ethnobotany of the Meskwaki Indians. Bulletin of the Public Museum of the City of Milwaukee 4: 175–326.
153. Smith, Huron H. 1932. Ethnobotany of the Ojibwe Indians. Bulletin of the Public Museum of Milwaukee 4: 327–525.
154. Smith, Huron H. 1933. Ethnobotany of the Forest Potawatomi Indians. Bulletin of the Public Museum of the City of Milwaukee 7: 1–230.
155. Sparkman, Philip S. 1908. The Culture of the Luiseño Indians. University of California Publications in American Archaeology and Ethnology 8(4): 187–234.
156. Speck, Frank G. 1917. Medicine Practices of the Northeastern Algonquians. Proceedings of the 19th International Congress of Americanists, pages 303–321.
157. Speck, Frank G. 1937. Catawba Medicines and Curative Practices. Publications of the Philadelphia Anthropological Society 1: 179–197.
158. Speck, Frank G. 1941. A List of Plant Curatives Obtained from the Houma Indians of Louisiana. Primitive Man 14: 49–75.
159. Speck, Frank G., and R. W. Dexter. 1951. Utilization of Animals and Plants by the Micmac Indians of New Brunswick. Journal of the Washington Academy of Sciences 41: 250–259.
160. Speck, Frank G., and R. W. Dexter. 1952. Utilization of Animals and Plants by the Malecite Indians of New Brunswick. Journal of the Washington Academy of Sciences 42: 1–7.
161. Speck, Frank G., R. B. Hassrick, and E. S. Carpenter. 1942. Rappahannock Herbals, Folk-Lore and Science of Cures. Proceedings of the Delaware County Institute of Science 10: 7–55.
162. Spier, Leslie. 1928. Havasupai Ethnography. Anthropological Papers of the American Museum of Natural History 29(3): 101–123, 284–285.
163. Spier, Leslie. 1930. Klamath Ethnography. University of California Publications in American Archaeology and Ethnology 30: 1–338.
164. Steedman, E. V. 1928. The Ethnobotany of the Thompson Indians of British Columbia. Smithsonian Institution—Bureau of American Ethnology Annual Report 45: 441–522.
165. Steggerda, Morris. 1941. Navajo Foods and Their Preparation. Journal of the American Dietetic Association 17(3): 217–225.
166. Stevenson, Matilda Coxe. 1915. Ethnobotany of the Zuni Indians. Smithsonian Institution—Bureau of American Ethnology Annual Report, Number 30.
167. Steward, Julian H. 1933. Ethnography of the Owens Valley Paiute. University of California Publications in American Archaeology and Ethnology 33(3): 233–250.
168. Stewart, Kenneth M. 1965. Mohave Indian Gathering of Wild Plants. Kiva 31(1): 46–53.
169. Sturtevant, William Curtis. 1955. The Mikasuki Seminole: Medical Beliefs and Practices. Ph.D. Thesis, Yale University, New Haven, Connecticut. Ann Arbor: University Microfilms.
170. Swan, James Gilchrist. 1869. The Indians of Cape Flattery . . . Washington Territory. Washington, D.C.: Smithsonian Institution.
171. Swank, George R. 1932. The Ethnobotany of the Acoma and Laguna Indians. M.A. Thesis, University of New Mexico, Albuquerque.
172. Swanton, John R 1928. Religious Beliefs and Medical Practices of the Creek Indians. Smithsonian Institution—Bureau of American Ethnology Annual Report 42: 473–672.
173. Swartz, B. K., Jr. 1958. A Study of Material Aspects of Northeastern Maidu Basketry. Kroeber Anthropological Society Publications 19: 67–84.
174. Tantaquidgeon, Gladys. 1928. Mohegan Medicinal Practices, Weather-Lore and Superstitions. Smithsonian Institution—Bureau of American Ethnology Annual Report 43: 264–270.
175. Tantaquidgeon, Gladys. 1942. A Study of Delaware Indian Medicine Practice and Folk Beliefs. Harrisburg: Pennsylvania Historical Commission.
176. Tantaquidgeon, Gladys. 1972. Folk Medicine of the Delaware and Related Algonkian Indians. Harrisburg: Pennsylvania Historical Commission Anthropological Papers, Number 3.
177. Taylor, Linda Averill. 1940. Plants Used as Curatives by Certain Southeastern Tribes. Cambridge, Massachusetts: Botanical Museum of Harvard University.
178. Teit, James A. 1928. The Salishan Tribes of the Western Plateaus. Smithsonian Institution—Bureau of American Ethnology Annual Report, Number 45.
179. Theodoratus, Robert J. 1989. Loss, Transfer, and Reintroduction in the Use of Wild Plant Foods in the Upper Skagit Valley. Northwest Anthropological Research Notes 23(1): 35–52.
180. Train, Percy, James R. Henrichs, and W. Andrew Archer. 1941. Medicinal Uses of Plants by Indian Tribes of Nevada. Washington, D.C.: U.S. Department of Agriculture.
181. Turner, Lucien M. 1890. Ethnology of the Ungava District, Hudson Bay Territory. Smithsonian Institution—Bureau of American Ethnology Annual Report 11: 159–350.
182. Turner, Nancy Chapman, and Marcus A. M. Bell. 1971. The Ethnobotany of the Coast Salish Indians of Vancouver Island, I and II. Economic Botany 25(1): 63–104, 335–339.
183. Turner, Nancy Chapman, and Marcus A. M. Bell. 1973. The Ethnobotany of the Southern Kwakiutl Indians of British Columbia. Economic Botany 27: 257–310.
184. Turner, Nancy J. 1973. The Ethnobotany of the Bella Coola Indians of British Columbia. Syesis 6: 193–220.
185. Turner, Nancy J., and Barbara S. Efrat 1982. Ethnobotany of the Hesquiat Indians of Vancouver Island. Victoria: British Columbia Provincial Museum.
186. Turner, Nancy J., John Thomas, Barry F. Carlson, and Robert T. Ogilvie. 1983. Ethnobotany of the Nitinaht Indians of Vancouver Island. Victoria: British Columbia Provincial Museum.
187. Turner, Nancy J., Laurence C. Thompson, M. Terry Thompson, and Annie Z. York. 1990. Thompson Ethnobotany: Knowledge and Usage of Plants by the Thompson Indians of British Columbia. Victoria: Royal British Columbia Museum.
188. Turner, Nancy J., R. Bouchard, and Dorothy I. D. Kennedy. 1980. Ethnobotany of the Okanagan-Colville Indians of British Columbia and Washington. Victoria: British Columbia Provincial Museum.
189. Veniamenov, I. 1840. Notes on the Islands in the Unalaska District. Translated by Human Relations Area Files, New Haven, Connecticut.
190. Vestal, Paul A. 1940. Notes on a Collection of Plants from the Hopi Indian

Region of Arizona Made by J. G. Owens in 1891. Botanical Museum Leaflets (Harvard University) 8(8): 153–168.
191. Vestal, Paul A. 1952. The Ethnobotany of the Ramah Navaho. Papers of the Peabody Museum of American Archaeology and Ethnology 40(4): 1–94.
192. Vestal, Paul A., and Richard Evans Schultes. 1939. The Economic Botany of the Kiowa Indians. Cambridge, Massachusetts: Botanical Museum of Harvard University.
193. Voegelin, Ermine W. 1938. Tubatulabal Ethnography. Anthropological Records 2(1): 1–84.
194. Wallis, Wilson D. 1922. Medicines Used by the Micmac Indians. American Anthropologist 24: 24–30.
195. Watahomigie, Lucille J. 1982. Hualapai Ethnobotany. Peach Springs, Arizona: Hualapai Bilingual Program, Peach Springs School District Number 8.
196. Waugh, F. W. 1916. Iroquis Foods and Food Preparation. Ottawa: Canada Department of Mines.
197. Weber, Steven A., and P. David Seaman. 1985. Havasupai Habitat: A. F. Whiting's Ethnography of a Traditional Indian Culture. Tucson: University of Arizona Press.
198. White, Leslie A. 1945. Notes on the Ethnobotany of the Keres. Papers of the Michigan Academy of Arts, Sciences and Letters 30: 557-568.
199. White, Leslie A. 1962. The Pueblo of Sia, New Mexico. Smithsonian Institution—Bureau of American Ethnology Bulletin, Number 184.
200. Whiting, Alfred F. 1939. Ethnobotany of the Hopi. Museum of Northern Arizona Bulletin, Number 15.
201. Willoughby, C. 1889. Indians of the Quinaielt Agency, Washington Territory. Smithsonian Institution Annual Report for 1886.
202. Wilson, Michael R. 1978. Notes on Ethnobotany in Inuktitut. Western Canadian Journal of Anthropology 8: 180–196.
203. Witthoft, John. 1947. An Early Cherokee Ethnobotanical Note. Journal of the Washington Academy of Sciences 37(3): 73–75.
204. Witthoft, John. 1977. Cherokee Indian Use of Potherbs. Journal of Cherokee Studies 2(2): 250–255.
205. Wyman, Leland C., and Stuart K. Harris. 1951. The Ethnobotany of the Kayenta Navaho. Albuquerque: University of New Mexico Press.
206. Zigmond, Maurice L. 1981. Kawaiisu Ethnobotany. Salt Lake City: University of Utah Press.

Index of Tribes

Plant usages are listed under the names of Native American groups, which are arranged alphabetically. Usages are categorized under the five main categories—Drug, Food, Fiber, Dye, Other—and then particular usages are listed alphabetically. Plants are identified below to the level of species. If subspecies or varieties appear in the Catalog of Plants, check under those names, too, for all usages given below. For example, one may find Comanche below and under Drug, Cold Remedy, see that *Rhus trilobata* was used. The specific ethnobotanical information and the sources from which the information was obtained may be found by turning to *Rhus trilobata*, *R. trilobata* var. *pilosissima*, and *R. trilobata* var. *trilobata* in the Catalog of Plants.

Abnaki
Drug
Abortifacient: *Fraxinus americana, Sanguinaria canadensis*
Analgesic: *Arctium minus, Carum carvi, Cornus canadensis, Plantago major, Ranunculus acris*
Anthelmintic: *Cirsium arvense, Populus tremuloides*
Antihemorrhagic: *Maianthemum racemosum, Polygonatum pubescens*
Antirheumatic (External): *Arctium minus, Mitchella repens, Plantago major, Symplocarpus foetidus, Taxus canadensis, Thuja occidentalis, Tsuga canadensis*
Antirheumatic (Internal): *Arctium minus, Taxus canadensis, Tsuga canadensis*
Blood Medicine: *Aralia nudicaulis*
Carminative: *Acorus calamus*
Cold Remedy: *Achillea millefolium, Asarum canadense, Chimaphila umbellata, Coptis trifolia, Kalmia angustifolia, Ledum groenlandicum*
Cough Medicine: *Asarum canadense, Coptis trifolia, Epilobium angustifolium, E.* sp., *Larix laricina, Pinus strobus*
Dermatological Aid: *Abies balsamea, Alnus incana, Tsuga canadensis*
Disinfectant: *Abies balsamea*
Eye Medicine: *Cornus sericea, Corylus cornuta, Salix* sp.
Febrifuge: *Achillea millefolium, Arctium minus*
Gastrointestinal Aid: *Polypodium virginianum*
Hemostat: *Urtica dioica*
Love Medicine: *Nuphar* sp.
Misc. Disease Remedy: *Achillea millefolium, Arctium minus*
Nose Medicine: *Chimaphila umbellata, Kalmia angustifolia, Ledum groenlandicum*
Orthopedic Aid: *Eupatorium perfoliatum*
Other: *Botrychium virginianum, Gymnocarpium disjunctum*
Panacea: *Abies balsamea, Mentha canadensis, Thuja occidentalis, Trillium erectum*
Pediatric Aid: *Achillea millefolium, Botrychium virginianum, Cirsium arvense, Mentha canadensis, Trillium erectum, Verbascum thapsus*
Poison: *Caltha palustris, Iris versicolor*
Psychological Aid: *Nuphar* sp., *Nymphaea* sp.
Respiratory Aid: *Acer pensylvanicum*
Sedative: *Mentha canadensis*
Toothache Remedy: *Verbascum thapsus*
Unspecified: *Abies balsamea, Botrychium virginianum*
Urinary Aid: *Picea glauca*
Veterinary Aid: *Sanguinaria canadensis*
Food
Beverage: *Gaultheria procumbens, Spiraea alba*
Forage: *Cladonia rangiferina*
Fruit: *Amelanchier* sp., *Aronia melanocarpa, Cornus canadensis, Crataegus* sp., *Fragaria virginiana, Prunus virginiana, Rubus idaeus, Vaccinium angustifolium, V. myrtilloides, Viburnum nudum*
Preserves: *Rubus* sp.
Sauce & Relish: *Cardamine diphylla*
Snack Food: *Osmunda cinnamomea*
Soup: *Zea mays*
Sweetener: *Acer rubrum*
Unspecified: *Acorus calamus, Aronia melanocarpa, Caltha palustris, Cardamine diphylla, Osmunda cinnamomea, Viburnum nudum*
Vegetable: *Caltha palustris, Phaseolus vulgaris, Solanum tuberosum*
Fiber
Basketry: *Anthoxanthum odoratum, Apocynum* sp., *Betula papyrifera, Fraxinus nigra, Laportea canadensis, Salix* sp., *Tilia americana*
Canoe Material: *Betula papyrifera*
Snow Gear: *Fraxinus americana*
Other
Containers: *Anthoxanthum odoratum, Betula papyrifera, Salix* sp.
Smoke Plant: *Cornus sericea*
Toys & Games: *Salix* sp.

Acoma
Food
Appetizer: *Rhus trilobata*
Beverage: *Yucca baccata*
Bread & Cake: *Yucca baccata*
Candy: *Asclepias speciosa, Populus deltoides*
Dried Food: *Amaranthus blitoides, A. hybridus, A. retroflexus, Opuntia imbricata, Prunus virginiana, Yucca baccata, Y. glauca*
Fruit: *Atriplex argentea, Ceanothus fendleri, Celtis laevigata, Juniperus monosperma, Opuntia engelmannii, Prunus virginiana, Ribes cereum, Yucca baccata, Y. glauca*
Porridge: *Cleome serrulata, Opuntia engelmannii, Prosopis glandulosa*
Preserves: *Ribes cereum, Yucca baccata*
Sauce & Relish: *Lycium pallidum, Yucca baccata*
Soup: *Juniperus osteosperma*
Special Food: *Sophora nuttalliana*
Spice: *Agastache pallidiflora, Berlandiera lyrata, Juniperus monosperma, Monarda fistulosa, M. pectinata, Pectis angustifolia, Rhus trilobata*
Staple: *Amaranthus blitoides, Dalea candida, Quercus gambelii*
Starvation Food: *Juniperus monosperma, Opuntia clavata, Phoradendron juniperinum, Solanum triflorum, Yucca baccata*
Unspecified: *Abronia fragrans, Allium cernuum, Astragalus lentiginosus, Cyperus squarrosus, Lathyrus polymorphus, Opuntia imbricata, Plantago major, Prosopis glandulosa, Quercus gambelii, Ranunculus inamoenus, Tradescantia occidentalis, Typha latifolia, Vicia americana*
Vegetable: *Amaranthus blitoides, A. hybridus, A. retroflexus, Cymopterus bulbosus, Lactuca sativa, Portulaca oleracea*

Alabama
Drug
Antidiarrheal: *Cornus* sp., *Hypericum hypericoides, H. multicaule, Pinus* sp.
Antirheumatic (External): *Callicarpa americana*
Ceremonial Medicine: *Ilex vomitoria*
Cold Remedy: *Desmodium* sp., *Solidago* sp.
Dermatological Aid: *Quercus rubra, Q.* sp., *Zanthoxylum americanum*
Diaphoretic: *Callicarpa americana*
Emetic: *Desmodium* sp., *Eryngium aquaticum, Ilex vomitoria, Quercus* sp.
Eye Medicine: *Hypericum hypericoides, H. multicaule, Ilex opaca, I.* sp.
Febrifuge: *Callicarpa americana, Cercis canadensis, Salix* sp.
Gastrointestinal Aid: *Erythrina herbacea*
Gynecological Aid: *Ulmus rubra*
Herbal Steam: *Callicarpa americana*
Laxative: *Sebastiania fruticosa*
Orthopedic Aid: *Ceanothus americanus, Hypericum hypericoides, Quercus rubra*
Panacea: *Vaccinium* sp.
Pediatric Aid: *Hypericum hypericoides, Quercus rubra*
Pulmonary Aid: *Cercis canadensis, Desmodium* sp., *Quercus rubra, Q.* sp.
Respiratory Aid: *Cercis canadensis*
Sedative: *Gnaphalium obtusifolium*
Throat Aid: *Quercus rubra, Q.* sp.
Toothache Remedy: *Solidago* sp., *Zanthoxylum americanum*
Urinary Aid: *Morus rubra*

Alaska Native
Drug
Poison: *Actaea rubra, Astragalus* sp., *Caltha palustris, Cicuta douglasii, C. maculata, C. virosa, Hedysarum boreale, Lupinus nootkatensis, Oxytropis* sp., *Pteridium aquilinum, Veratrum viride, Zigadenus elegans*
Food
Beverage: *Ledum groenlandicum, Rosa acicularis*
Bread & Cake: *Amelanchier alnifolia, Menyanthes trifoliata, Tsuga heterophylla, Vaccinium cespitosum, V. membranaceum, V. ovalifolium, V. uliginosum*
Cooking Agent: *Malus fusca*
Dessert: *Shepherdia canadensis*
Dietary Aid: *Chenopodium album, C. capitatum, Claytonia sibirica, C. tuberosa, Epilobium angustifolium, Fragaria chiloensis, Honckenya peploides, Ligusticum scothicum, Oxyria digyna, Polygonum bistorta, Rosa acicularis, Rubus chamaemorus, Rumex arcticus, Salix alaxensis, S. planifolia, Saxifraga nelsoniana, S. spicata, Taraxacum* sp., *Urtica dioica, Vaccinium cespitosum, V. membranaceum, V. ovalifolium, V. uliginosum*
Dried Food: *Amelanchier alnifolia, Fritillaria camschatcensis, Menyanthes trifoliata, Porphyra laciniata, Rhodymenia palmata*
Frozen Food: *Rubus chamaemorus, Vaccinium cespitosum, V. membranaceum, V. ovalifolium, V. uliginosum*
Fruit: *Amelanchier alnifolia, Arctostaphylos alpina, Elaeagnus commutata, Fragaria chiloensis, Gaultheria shallon, Geocaulon lividum, Ribes triste, Rubus chamaemorus, R. idaeus, R. parviflorus, R. spectabilis, Streptopus amplexifolius, Vaccinium cespitosum, V. membranaceum, V. ovalifolium, V. oxycoccos, V. uliginosum*
Ice Cream: *Honckenya peploides*
Pie & Pudding: *Amelanchier alnifolia, Empetrum nigrum, Polygonum alpinum, Rubus chamaemorus, Vaccinium cespitosum, V. membranaceum, V. ovalifolium, V. uliginosum*
Preserves: *Empetrum nigrum, Fragaria chiloensis, Ribes triste, Rosa acicularis, Rubus arcticus, R. idaeus, R. pedatus, R. spectabilis, Vaccinium parvifolium, V. vitis-idaea, Viburnum edule*
Sauce & Relish: *Rosa acicularis, Vaccinium vitis-idaea*
Snack Food: *Porphyra laciniata, Salix alaxensis*
Soup: *Claytonia tuberosa, Fritillaria camschatcensis, Hippuris vulgaris, Parrya nudicaulis, Polygonum bistorta, Porphyra laciniata, Rhodymenia palmata*
Staple: *Fritillaria camschatcensis*
Starvation Food: *Menyanthes trifoliata*
Substitution Food: *Chenopodium album, Ligusticum scothicum, Pteridium aquilinum, Urtica dioica*
Unspecified: *Allium schoenoprasum, Anemone narcissiflora, Angelica lucida, Caltha palustris, Carex aquatilis, Claytonia acutifolia, Dryopteris carthusiana, Epilobium angustifolium, Eriophorum angustifolium, Hedysarum alpinum, Heracleum maximum, Honckenya peploides, Ligusticum scothicum, Lupinus nootkatensis, Oxyria digyna, Oxytropis nigrescens, Parrya nudicaulis, Pedicularis lanata, Plantago maritima, Polygonum bistorta, Ranunculus pallasii, Rhodymenia palmata, Salix alaxensis, S. planifolia, Saxifraga nelsoniana, Sedum rosea, Typha latifolia, Urtica dioica*
Vegetable: *Angelica lucida, Arabis lyrata, Argentina egedii, Barbarea orthoceras, Chenopodium album, C. capitatum, Claytonia sibirica, C. tuberosa, Cochlearia officinalis, Dryopteris carthusiana, Epilobium angustifolium, E. latifolium, Honckenya peploides, Ligusticum scothicum, Nuphar lutea, Petasites frigidus, Plantago macrocarpa, Polygonum alpinum, P. bistorta, Rumex arcticus, Sarcocornia pacifica, Saxifraga spicata, Sedum rosea, Streptopus amplexifolius, Taraxacum* sp.
Winter Use Food: *Hippuris vulgaris, Ligusticum scothicum, Parrya nudicaulis, Plantago maritima, Pteridium aquilinum, Rubus chamaemorus, Rumex arcticus, Salix planifolia, Vaccinium cespitosum, V. membranaceum, V. ovalifolium, V. uliginosum, V. vitis-idaea*
Other
Containers: *Petasites frigidus*
Hunting & Fishing Item: *Nereocystis luetkeana*

Aleut
Drug
Analgesic: *Achillea millefolium, Angelica lucida, Lycopodium clavatum, Matricaria discoidea, Menyanthes trifoliata*

Antihemorrhagic: *Anemone narcissiflora*
Antirheumatic (External): *Artemisia vulgaris*
Antirheumatic (Internal): *Menyanthes trifoliata*
Carminative: *Matricaria discoidea, Menyanthes trifoliata*
Cold Remedy: *Achillea millefolium, Angelica lucida, Conioselinum gmelinii, Geum calthifolium, Heracleum maximum*
Dermatological Aid: *Artemisia vulgaris, Geum calthifolium, Heracleum maximum, Rumex acetosella, Senecio pseudoarnica, Taraxacum officinale*
Gastrointestinal Aid: *Achillea millefolium, Matricaria discoidea, Taraxacum officinale*
Gynecological Aid: *Lycopodium clavatum*
Hemostat: *Achillea millefolium*
Laxative: *Iris setosa, Matricaria discoidea, Menyanthes trifoliata*
Misc. Disease Remedy: *Leptarrhena pyrolifolia*
Orthopedic Aid: *Artemisia vulgaris, Heracleum maximum*
Panacea: *Matricaria discoidea*
Poison: *Aconitum maximum, A.* sp., *Equisetum* sp., *Ranunculus occidentalis*
Throat Aid: *Achillea millefolium, Angelica lucida, Conioselinum gmelinii, Geranium erianthum, Geum calthifolium, Heracleum maximum, Taraxacum officinale*
Tonic: *Angelica lucida, Artemisia vulgaris, Geum calthifolium, Matricaria discoidea, Menyanthes trifoliata, Plantago macrocarpa, Polygonum bistorta*
Tuberculosis Remedy: *Achillea millefolium*
Unspecified: *Caltha* sp., *Matricaria* sp., *Menyanthes* sp.

Food
Unspecified: *Achillea* sp., *Angelica* sp., *Claytonia* sp., *Cochlearia* sp., *Conioselinum* sp., *Heracleum* sp., *Senecio* sp.

Fiber
Canoe Material: *Taxus cuspidata*
Other: *Elymus* sp., *Lathyrus* sp.

Algonquin
Drug
Analgesic: *Actaea rubra*
Love Medicine: *Sanguinaria canadensis*
Other: *Juniperus communis*

Algonquin, Quebec
Drug
Analgesic: *Achillea millefolium, Comptonia peregrina, Gaultheria procumbens, Ledum groenlandicum*
Anticonvulsive: *Asarum canadense*
Antidiarrheal: *Coptis trifolia, Rubus idaeus*
Antirheumatic (External): *Tsuga canadensis*
Antirheumatic (Internal): *Maianthemum racemosum, Rhus hirta, Taxus canadensis, Thuja occidentalis*
Blood Medicine: *Prunus pensylvanica, Taraxacum officinale*
Cathartic: *Cornus canadensis, Lonicera dioica*
Ceremonial Medicine: *Ledum groenlandicum*
Cold Remedy: *Achillea millefolium, Acorus calamus, Cornus sericea, Gaultheria procumbens, Kalmia angustifolia, Sorbus americana, Thuja occidentalis, Tsuga canadensis*
Cough Medicine: *Acorus calamus, Larix laricina, Picea glauca, Prunus nigra, P. pensylvanica, P. virginiana*
Dermatological Aid: *Abies balsamea, Betula papyrifera, Clintonia borealis, Nuphar lutea, Picea glauca, P. mariana, Populus balsamifera, Thuja occidentalis, Toxicodendron radicans, Tsuga canadensis, Typha latifolia*
Dietary Aid: *Rhus hirta*
Disinfectant: *Clintonia borealis, Larix laricina, Nuphar lutea, Populus balsamifera, Prunus pensylvanica, Tsuga canadensis, Typha latifolia*
Ear Medicine: *Fraxinus* sp.
Emetic: *Alnus incana, Sambucus canadensis*
Eye Medicine: *Coptis trifolia, Tilia americana*
Febrifuge: *Asarum canadense, Cardamine diphylla, Prunella vulgaris, Thuja occidentalis*
Gastrointestinal Aid: *Corylus cornuta, Gaultheria hispidula, Lilium canadense, L. philadelphicum, Thuja occidentalis, Vaccinium angustifolium*
Gynecological Aid: *Abies balsamea, Acorus calamus, Cypripedium acaule, Eupatorium maculatum, Linnaea borealis, Lonicera dioica, Lycopodium* sp., *Picea glauca, Sarracenia purpurea, Taxus canadensis, Thuja occidentalis, Tsuga canadensis, Vaccinium angustifolium*
Heart Medicine: *Abies balsamea, Acorus calamus, Aralia hispida, Cardamine diphylla, Coptis trifolia, Corylus cornuta, Polypodium virginianum, Sanguinaria canadensis, Solidago* sp.
Hemorrhoid Remedy: *Cornus sericea*
Hemostat: *Polygonum* sp.
Internal Medicine: *Picea glauca, P. mariana*
Kidney Aid: *Aralia nudicaulis, Epigaea repens, Lonicera dioica*
Laxative: *Abies balsamea, Alnus incana, Dirca palustris, Picea glauca, Sambucus canadensis*
Misc. Disease Remedy: *Aralia racemosa, Euphorbia* sp.
Orthopedic Aid: *Shepherdia canadensis*
Pediatric Aid: *Aralia nudicaulis, Asarum canadense, Betula papyrifera, Cardamine diphylla, Lycopodium* sp., *Thuja occidentalis, Tsuga canadensis, Vaccinium angustifolium*
Poison: *Kalmia angustifolia*
Poultice: *Abies balsamea, Achillea millefolium, Plantago major, Taraxacum officinale, Taxus canadensis*
Preventive Medicine: *Acorus calamus*
Pulmonary Aid: *Prunus pensylvanica, Trifolium pratense*
Respiratory Aid: *Achillea millefolium*
Tonic: *Ledum groenlandicum, Sanguinaria canadensis, Sorbus americana*
Toothache Remedy: *Alnus incana, Coptis trifolia, Thuja occidentalis*
Tuberculosis Remedy: *Aralia racemosa, Sagittaria* sp.
Unspecified: *Abies balsamea, Acer pensylvanicum, Aralia hispida, Betula lenta, Chelone glabra, Cornus* sp., *Gaultheria procumbens, Larix laricina, Ledum groenlandicum, Picea glauca, P. mariana, Prunus nigra, Rhus hirta, Rubus idaeus, Shepherdia canadensis, Spiraea alba, S. tomentosa, Taxus canadensis, Tilia americana, Tsuga canadensis*
Urinary Aid: *Lycopodium* sp., *Sarracenia purpurea*
Venereal Aid: *Cypripedium acaule, Eupatorium maculatum*
Veterinary Aid: *Acer pensylvanicum*

Food
Beverage: *Aralia nudicaulis, Gaultheria procumbens, Ledum groenlandicum, Prunus virginiana, Rhus hirta*
Bread & Cake: *Humulus lupulus*
Candy: *Picea glauca*
Fruit: *Amelanchier* sp., *Crataegus* sp., *Fragaria virginiana, Gaultheria hispidula, Prunus nigra, P. pensylvanica, P. virginiana, Ribes cynosbati, R. glandulosum, Rubus idaeus, R. odoratus, Sorbus americana, Vaccinium angustifolium, V. corymbosum, V. macrocarpon, V. myrtilloides, V. oxycoccos, Viburnum nudum, V. opulus*
Pie & Pudding: *Vaccinium angustifolium, V. corymbosum*
Preserves: *Fragaria virginiana, Prunus nigra, P. pensylvanica, P. virginiana, Ribes cynosbati, Rubus idaeus, R.* sp., *Vaccinium angustifolium, V. corymbosum, Viburnum opulus*
Sauce & Relish: *Acer rubrum, A. saccharum, Betula papyrifera, Cardamine diphylla*
Snack Food: *Cornus canadensis, Gaultheria procumbens*
Substitution Food: *Betula alleghaniensis, Rumex* sp.
Sweetener: *Acer rubrum, A. saccharum*

Unspecified: *Corylus cornuta, Fagus grandifolia, Juglans cinerea, Sagittaria* sp., *Sium suave*
Vegetable: *Aster macrophyllus, Claytonia virginica, Rorippa nasturtium-aquaticum, Taraxacum* sp.
Other
Cash Crop: *Vaccinium angustifolium, V. corymbosum, V. myrtilloides*

Algonquin, Tête-de-Boule
Drug
Abortifacient: *Taxus canadensis*
Analgesic: *Achillea millefolium, Kalmia angustifolia, Viburnum lantanoides*
Antihemorrhagic: *Geum rivale*
Antirheumatic (External): *Populus tremuloides*
Burn Dressing: *Anaphalis margaritacea, Iris versicolor, Plantago major*
Cold Remedy: *Abies balsamea, Coptis trifolia, Cornus canadensis, Gaultheria procumbens, Pinus resinosa, P. strobus*
Dermatological Aid: *Acer spicatum, Clintonia borealis, Epilobium angustifolium, Iris versicolor, Plantago major*
Diuretic: *Diervilla lonicera, Sarracenia purpurea*
Ear Medicine: *Aralia nudicaulis*
Febrifuge: *Mentha canadensis*
Gastrointestinal Aid: *Cypripedium acaule, Gaultheria procumbens*
Gynecological Aid: *Cornus canadensis, Salix discolor, Trillium undulatum*
Heart Medicine: *Corylus cornuta*
Hemostat: *Prunus pensylvanica*
Kidney Aid: *Cypripedium acaule*
Laxative: *Larix laricina*
Misc. Disease Remedy: *Gaultheria procumbens*
Pediatric Aid: *Cypripedium acaule, Prunus pensylvanica*
Poison: *Kalmia angustifolia*
Psychological Aid: *Fraxinus pennsylvanica, Sorbus americana*
Respiratory Aid: *Coptis trifolia, Rubus* sp.
Stimulant: *Fraxinus pennsylvanica*
Strengthener: *Sorbus americana*
Throat Aid: *Salix discolor*
Urinary Aid: *Cypripedium acaule, Rubus idaeus, Sarracenia purpurea*
Food
Beverage: *Nuphar lutea*
Fruit: *Fragaria virginiana, Rubus idaeus, Vaccinium macrocarpon, V. oxycoccos*
Spice: *Allium sativum*
Unspecified: *Nuphar lutea, Vaccinium myrtilloides*
Fiber
Basketry: *Betula papyrifera, Picea glauca, P. mariana, Tilia americana*
Building Material: *Betula papyrifera*
Canoe Material: *Betula papyrifera, Picea glauca, P. mariana, Thuja occidentalis*
Caulking Material: *Pinus banksiana*
Mats, Rugs & Bedding: *Abies balsamea, Typha latifolia*
Snow Gear: *Picea glauca, P. mariana*
Dye
Yellow: *Alnus incana*
Other
Containers: *Betula papyrifera*
Designs: *Betula papyrifera*
Insecticide: *Clintonia borealis*

Anticosti
Drug
Dermatological Aid: *Populus balsamifera*
Kidney Aid: *Abies balsamea, Larix laricina*
Sedative: *Gaultheria hispidula*
Throat Aid: *Abies balsamea*
Unspecified: *Ledum groenlandicum*
Food
Beverage: *Anaphalis margaritacea, Beta vulgaris, Juniperus communis, Larix laricina, Ledum groenlandicum, Picea mariana, Trifolium* sp.
Forage: *Heracleum maximum*
Preserves: *Ribes* sp., *Rubus chamaemorus, Vaccinium vitis-idaea*
Soup: *Allium schoenoprasum*
Spice: *Ligusticum scothicum*
Unspecified: *Rumex acetosella*
Winter Use Food: *Rubus chamaemorus, Vaccinium macrocarpon, V. oxycoccos*
Fiber
Snow Gear: *Betula papyrifera*
Other
Ceremonial Items: *Anaphalis margaritacea*

Apache
Drug
Snakebite Remedy: *Yucca angustissima*
Unspecified: *Frasera speciosa*
Food
Beverage: *Yucca baccata, Y. torreyi*
Bread & Cake: *Prosopis glandulosa, Prunus virginiana, Yucca baccata, Y. torreyi*
Dried Food: *Agave americana, A. parryi*
Fruit: *Juniperus deppeana, Mahonia haematocarpa, Oenothera albicaulis, Rhus microphylla, R. trilobata, Ribes inerme, Sambucus racemosa*
Preserves: *Prosopis glandulosa*
Sauce & Relish: *Yucca baccata, Y. torreyi*
Soup: *Yucca baccata*
Special Food: *Fragaria vesca*
Staple: *Agastache pallidiflora, Agave americana, A. parryi, Rhus trilobata*
Unspecified: *Agave parryi, Allium geyeri, Frasera speciosa, Hoffmannseggia glauca, Typha latifolia, Yucca baccata, Y. elata, Y. glauca*
Vegetable: *Chenopodium album, C. leptophyllum, Yucca baccata, Y. elata*
Fiber
Basketry: *Yucca baccata, Y. torreyi*
Cordage: *Yucca angustissima, Y. baccata*
Other
Containers: *Yucca baccata*
Designs: *Yucca baccata, Y. torreyi*
Hunting & Fishing Item: *Frasera speciosa*
Soap: *Yucca angustissima, Y. baccata, Y. glauca, Ziziphus obtusifolia*
Tools: *Yucca baccata*

Apache, Chiricahua & Mescalero
Drug
Narcotic: *Broussonetia papyrifera, Opuntia leptocaulis*
Food
Beverage: *Artemisia dracunculus, Cheilanthes fendleri, Dasylirion wheeleri, Hedeoma nana, Monarda fistulosa, Parthenium incanum, Prosopis glandulosa, P. pubescens, Thelesperma megapotamicum*
Bread & Cake: *Amaranthus albus, A. retroflexus, Camelina microcarpa, Capsella bursa-pastoris, Celtis laevigata, Cirsium pallidum, Crataegus erythropoda, Dasylirion wheeleri, Descurainia incana, Dyssodia papposa, Helianthus annuus, Morus microphylla, Muhlenbergia* sp., *Panicum bulbosum, Pinus ponderosa, Populus tremuloides, Prosopis glandulosa, P. pubescens, Ribes leptanthum, R. pinetorum, R. wolfii, Rubus arizonensis, Solanum fendleri, Sporobolus cryptandrus, Yucca baccata, Y. torreyi*
Candy: *Asclepias speciosa, Populus deltoides*
Dried Food: *Acer negundo, Lathyrus lanszwertii, Mammillaria grahamii, Prosopis pubescens, Rhus trilobata, Vicia melilotoides, Vitis arizonica, Yucca elata*
Fodder: *Cyperus fendlerianus*

Fruit: *Carnegia gigantea, Celtis laevigata, Crataegus erythropoda, Echinocereus coccineus, E. engelmannii, E. fendleri, E. polyacanthus, E. rigidissimus, Forestiera pubescens, Fragaria vesca, Juniperus deppeana, J. scopulorum, Mammillaria grahamii, M. mainiae, Morus microphylla, Prunus virginiana, Quercus gambelii, Q. grisea, Ribes leptanthum, R. mescalerium, R. wolfii, Rosa woodsii, Rubus arizonensis, Stenocereus thurberi, Vitis arizonica*
Pie & Pudding: *Pinus edulis, Prosopis glandulosa*
Porridge: *Sporobolus cryptandrus*
Preserves: *Celtis laevigata, Juniperus deppeana, Mahonia haematocarpa, Prunus virginiana, Rhus microphylla, R. trilobata, Ribes wolfii, Rosa woodsii, Sambucus racemosa*
Sauce & Relish: *Helianthus annuus, Juniperus monosperma, Oenothera albicaulis, Panicum bulbosum, P. obtusum*
Soup: *Oenothera albicaulis, Yucca baccata*
Special Food: *Oenothera albicaulis, Physalis subulata, Pinus edulis, Prosopis pubescens, Rhus glabra*
Spice: *Allium cernuum, A. geyeri, Artemisia ludoviciana, Cymopterus acaulis, Dracocephalum parviflorum, Hedeoma nana, Humulus lupulus, Mentha canadensis, Monarda fistulosa, Prosopis glandulosa, Quercus grisea, Taraxacum officinale*
Starvation Food: *Pinus ponderosa*
Substitution Food: *Carnegia gigantea, Prosopis glandulosa*
Sweetener: *Acer glabrum, A. negundo*
Unspecified: *Agave parryi, Amaranthus albus, A. retroflexus, Calylophus lavandulifolius, Camelina microcarpa, Capsella bursa-pastoris, Chenopodium album, Cirsium pallidum, Cymopterus acaulis, Cyperus fendlerianus, Dasylirion wheeleri, Descurainia incana, Dyssodia papposa, Epixiphium wislizeni, Hoffmannseggia glauca, Jamesia americana, Juglans major, Lathyrus lanszwertii, Ligusticum porteri, Matelea producta, Nolina microcarpa, Oxalis violacea, Pinus edulis, P. flexilis, P. ponderosa, Populus tremuloides, Portulaca oleracea, Proboscidea louisianica, Prosopis glandulosa, Rumex aquaticus, Typha latifolia, Vicia melilotoides, Yucca baccata, Y. glauca*
Vegetable: *Agave parryi, Allium cernuum, A. geyeri, Capsella bursa-pastoris, Dasylirion wheeleri, Dyssodia papposa, Nolina microcarpa, Robinia neomexicana, Solanum jamesii, Yucca baccata, Y. elata, Y. glauca*
Winter Use Food: *Crataegus erythropoda, Juglans major, Morus microphylla, Prunus virginiana, Quercus gambelii, Q. grisea, Robinia neomexicana, Rubus arizonensis*

Fiber
Basketry: *Yucca elata*
Brushes & Brooms: *Sporobolus wrightii*

Other
Ceremonial Items: *Artemisia* sp., *Bouteloua* sp., *Broussonetia papyrifera, Gutierrezia* sp., *Typha latifolia*
Containers: *Alopecurus aequalis, Andropogon gerardii, Bouteloua curtipendula, Muhlenbergia pauciflora, M. wrightii, Nolina microcarpa, Schedonnardus paniculatus, Sporobolus airoides*
Fuel: *Juniperus* sp., *Quercus* sp.
Tools: *Quercus* sp.

Apache, Mescalero
Drug
Antidiarrheal: *Prunus* sp.
Burn Dressing: *Prunus* sp.
Cold Remedy: *Pinus edulis*
Dermatological Aid: *Opuntia* sp.
Disinfectant: *Opuntia* sp.
Ear Medicine: *Prosopis pubescens*
Eye Medicine: *Mahonia haematocarpa, Opuntia* sp., *Prosopis glandulosa*
Pediatric Aid: *Prosopis glandulosa*
Unspecified: *Typha latifolia*
Urinary Aid: *Prosopis glandulosa*

Food
Beverage: *Dasylirion wheeleri, Prosopis glandulosa, Yucca baccata*
Bread & Cake: *Agave parryi, Dasylirion wheeleri, Prunus* sp.
Dried Food: *Dasylirion wheeleri, Opuntia* sp., *Pinus edulis, Prunus americana, Robinia neomexicana, Yucca baccata, Y. elata*
Fruit: *Echinocereus* sp., *Juniperus* sp., *Mahonia haematocarpa, Morus microphylla, Opuntia* sp., *Prunus* sp., *Vitis arizonica*
Sauce & Relish: *Morus microphylla, Yucca baccata*
Soup: *Dasylirion wheeleri, Yucca elata*
Special Food: *Pinus edulis*
Spice: *Populus tremuloides*
Staple: *Prosopis glandulosa, Yucca elata*
Unspecified: *Dasylirion wheeleri, Juglans major, Quercus* sp., *Robinia neomexicana, Typha latifolia, Yucca elata*

Fiber
Basketry: *Yucca baccata, Y. glauca*
Building Material: *Juglans major, Juniperus* sp., *Quercus* sp.
Cordage: *Yucca baccata*
Furniture: *Dasylirion wheeleri, Pinus edulis, Quercus* sp.
Mats, Rugs & Bedding: *Nolina microcarpa, Typha latifolia*

Dye
Yellow: *Mahonia haematocarpa*

Other
Ceremonial Items: *Dasylirion wheeleri, Pinus edulis, Typha latifolia*
Cooking Tools: *Nolina microcarpa, Quercus* sp.
Fuel: *Juniperus* sp.
Hunting & Fishing Item: *Juniperus* sp., *Prosopis glandulosa, Prunus* sp., *Robinia neomexicana*
Paint: *Juglans major*
Smoking Tools: *Dasylirion wheeleri*
Soap: *Yucca baccata*
Stable Gear: *Quercus* sp.
Tools: *Dasylirion wheeleri, Juniperus* sp., *Quercus* sp.
Toys & Games: *Quercus* sp.
Waterproofing Agent: *Pinus edulis*
Weapon: *Morus microphylla*

Apache, San Carlos
Food
Beverage: *Ferocactus wislizeni*
Bread & Cake: *Carnegia gigantea*
Dried Food: *Yucca schottii*
Fruit: *Canotia holacantha, Carnegia gigantea, Mammillaria grahamii, Yucca schottii*
Porridge: *Ferocactus wislizeni, Opuntia* sp.
Staple: *Opuntia* sp.
Unspecified: *Agave* sp., *Dichelostemma pulchellum, Peniocereus greggii*

Apache, Western
Drug
Antidiarrheal: *Opuntia* sp.
Burn Dressing: *Opuntia* sp.
Ceremonial Medicine: *Yucca elata*
Dermatological Aid: *Juglans* sp., *Pinus edulis, P. monophylla*
Ear Medicine: *Prosopis pubescens*
Laxative: *Opuntia* sp.
Pediatric Aid: *Opuntia* sp.
Pulmonary Aid: *Juniperus* sp.
Veterinary Aid: *Cucurbita foetidissima, Juglans* sp.

Apache, Western

Food
Beverage: *Agave palmeri, A. parryi, Carnegia gigantea, Juniperus* sp., *Martynia* sp., *Prosopis glandulosa, Rhus trilobata, Vitis arizonica*
Bread & Cake: *Carnegia gigantea, Helianthus* sp., *Prosopis glandulosa*
Candy: *Agave palmeri, A. parryi, Helianthus* sp., *Pinus edulis, P. monophylla, Prosopis glandulosa*
Dried Food: *Agave palmeri, A. parryi, Carnegia gigantea, Juniperus* sp., *Prosopis glandulosa, Vitis arizonica, Yucca baccata, Y. elata*
Fruit: *Canotia holacantha, Carnegia gigantea, Prunus virginiana, Rhus trilobata, Vitis arizonica*
Pie & Pudding: *Carnegia gigantea*
Porridge: *Bouteloua gracilis, Carnegia gigantea, Helianthus* sp., *Muhlenbergia rigens, Opuntia* sp., *Oryzopsis hymenoides, Pinus edulis, P. monophylla, Prosopis glandulosa, Sporobolus contractus*
Sauce & Relish: *Juglans* sp., *Juniperus* sp., *Yucca baccata*
Soup: *Opuntia* sp.
Special Food: *Helianthus* sp.
Spice: *Juniperus* sp.
Staple: *Pinus edulis, P. monophylla, Prosopis glandulosa*
Substitution Food: *Agave palmeri, A. parryi, Prosopis glandulosa*
Unspecified: *Agave palmeri, A. parryi, Chenopodium incanum, C. leptophyllum, Juglans* sp., *Nolina microcarpa, Pinus* sp., *Quercus emoryi, Q. gambelii, Q. ×pauciloba, Yucca elata*
Vegetable: *Cleome serrulata*
Winter Use Food: *Martynia* sp., *Pinus edulis, P. monophylla*

Fiber
Basketry: *Rhus trilobata*
Brushes & Brooms: *Bouteloua gracilis*
Building Material: *Nolina microcarpa*
Cordage: *Yucca baccata, Y. elata*
Furniture: *Yucca elata*
Sewing Material: *Agave palmeri, A. parryi*

Other
Ceremonial Items: *Carnegia gigantea*
Containers: *Carnegia gigantea*
Cooking Tools: *Carnegia gigantea, Nolina microcarpa*
Decorations: *Agave palmeri, A. parryi, Yucca elata*
Fasteners: *Prosopis glandulosa*
Fuel: *Prosopis glandulosa*
Lighting: *Juniperus* sp.
Paint: *Agave palmeri, A. parryi*
Protection: *Nolina microcarpa*
Soap: *Nolina microcarpa, N.* sp., *Yucca baccata, Y. elata*
Tools: *Agave palmeri, A. parryi*
Waterproofing Agent: *Pinus edulis, P. monophylla*
Weapon: *Agave palmeri, A. parryi*

Apache, White Mountain
Drug
Anticonvulsive: *Juniperus californica, J. monosperma, J. occidentalis*
Antidiarrheal: *Phragmites australis*
Blood Medicine: *Malacothrix glabrata, Xanthium strumarium*
Burn Dressing: *Cercocarpus montanus*
Cathartic: *Croton texensis*
Ceremonial Medicine: *Cornus sericea, Datura wrightii, Dimorphocarpa wislizeni, Eriogonum jamesii, Mahonia fremontii, Nicotiana attenuata*
Cold Remedy: *Juniperus californica, J. monosperma, J. occidentalis*
Cough Medicine: *Juniperus californica, J. monosperma, J. occidentalis, Pseudotsuga menziesii*
Dermatological Aid: *Dimorphocarpa wislizeni*
Disinfectant: *Datura wrightii*
Eye Medicine: *Linum puberulum*
Gastrointestinal Aid: *Croton texensis, Phragmites australis*
Gynecological Aid: *Juniperus californica, J. monosperma, J. occidentalis, Rumex salicifolius*
Laxative: *Mentzelia pumila*
Narcotic: *Datura wrightii*
Oral Aid: *Chamaesyce serpyllifolia, Eriogonum jamesii*
Snakebite Remedy: *Helianthus annuus, Stephanomeria tenuifolia*
Throat Aid: *Dimorphocarpa wislizeni, Rumex salicifolius*
Unspecified: *Berula erecta, Eriogonum jamesii, Polygonum lapathifolium, Purshia mexicana, Solanum elaeagnifolium*
Venereal Aid: *Ephedra nevadensis, Pinus edulis*
Witchcraft Medicine: *Penstemon barbatus*

Food
Beverage: *Agave americana, A. decipiens, Artemisia tridentata, Chamaesyce serpyllifolia, Datura wrightii, Ephedra nevadensis, Quercus ×pauciloba*
Bread & Cake: *Bouteloua gracilis, Cucurbita pepo, Muhlenbergia rigens, Oryzopsis hymenoides, Quercus ×pauciloba, Sporobolus contractus, Triticum aestivum, Xanthium strumarium*
Candy: *Lactuca tatarica, Populus angustifolia, P. deltoides, Pseudotsuga menziesii*
Cooking Agent: *Chamaesyce serpyllifolia*
Dried Food: *Chamaesyce serpyllifolia, Opuntia imbricata, O. whipplei, Yucca baccata, Y. glauca*
Fodder: *Elytrigia repens, Muhlenbergia rigens, Oryzopsis hymenoides*
Fruit: *Astragalus lentiginosus, Carnegia gigantea, Cereus* sp., *Juniperus monosperma, J. occidentalis, J. osteosperma, Opuntia imbricata, O. whipplei, Physalis hederifolia, Rhus trilobata, Ribes cereum, Sambucus racemosa*
Porridge: *Bouteloua gracilis, Muhlenbergia rigens, Oryzopsis hymenoides, Sporobolus contractus*
Preserves: *Carnegia gigantea, Cereus* sp.
Spice: *Artemisia tridentata*
Staple: *Cycloloma atriplicifolium, Helianthus annuus*
Unspecified: *Agave americana, A. decipiens, Allium bisceptrum, Amaranthus albus, A. blitoides, Artemisia carruthii, Asclepias subverticillata, Berula erecta, Chenopodium incanum, C. leptophyllum, Chrysothamnus nauseosus, Cucurbita pepo, Elytrigia repens, Mammillaria* sp., *Muhlenbergia rigens, Oryzopsis hymenoides, Pinus edulis, Populus angustifolia, P. deltoides, Quercus gambelii, Q. ×pauciloba, Ustilago zeae, Yucca baccata, Y. glauca*
Vegetable: *Phaseolus vulgaris, Robinia neomexicana*
Winter Use Food: *Agave americana*

Fiber
Basketry: *Rhus trilobata, Salix irrorata*
Brushes & Brooms: *Bouteloua gracilis*
Building Material: *Salix irrorata, Solanum fendleri, Typha latifolia*
Clothing: *Yucca baccata*
Cordage: *Yucca baccata, Y. glauca*

Dye
Red: *Coreopsis tinctoria*
Red-Brown: *Alnus incana*
Unspecified: *Castilleja integra, C. miniata, C. minor, C. parviflora*
Yellow: *Chrysothamnus nauseosus, Psilostrophe tagetina*

Other
Cash Crop: *Bouteloua gracilis, Muhlenbergia rigens, Oryzopsis hymenoides, Sporobolus contractus*
Ceremonial Items: *Cucurbita foetidissima, Typha latifolia*
Containers: *Salix irrorata*
Cooking Tools: *Salix irrorata*
Hide Preparation: *Quercus gambelii, Q. ×pauciloba*

Hunting & Fishing Item: *Cercocarpus montanus, Phragmites australis*
Paint: *Amaranthus cruentus*
Smoke Plant: *Phragmites australis*
Smoking Tools: *Phragmites australis*
Soap: *Yucca baccata*
Toys & Games: *Salix irrorata, Yucca glauca*
Waterproofing Agent: *Pinus edulis*

Apalachee
Drug
Unspecified: *Adiantum* sp., *Cuminum cyminum, Rumex acetosa, Trifolium* sp.
Food
Fruit: *Citrullus lanatus*
Other
Cash Crop: *Nicotiana rustica*
Smoke Plant: *Nicotiana rustica*

Arapaho
Drug
Analgesic: *Psoralidium lanceolatum*
Antirheumatic (External): *Rosa woodsii*
Cough Medicine: *Artemisia frigida*
Dermatological Aid: *Psoralidium lanceolatum, Rumex hymenosepalus, R. venosus*
Disinfectant: *Juniperus communis*
Gastrointestinal Aid: *Juniperus communis*
Misc. Disease Remedy: *Juniperus communis*
Throat Aid: *Psoralidium lanceolatum*
Tonic: *Ivesia gordonii*
Unspecified: *Heuchera cylindrica*
Food
Beverage: *Rosa woodsii*
Preserves: *Shepherdia argentea*
Starvation Food: *Ipomoea leptophylla*
Sweetener: *Frasera speciosa*
Winter Use Food: *Shepherdia argentea*
Dye
Blue: *Comandra umbellata*
Orange: *Rosa woodsii*
Other
Ceremonial Items: *Artemisia frigida*
Incense & Fragrance: *Juniperus communis*
Jewelry: *Elaeagnus commutata*

Arikara
Other
Hunting & Fishing Item: *Andropogon gerardii*

Atsugewi
Drug
Analgesic: *Ligusticum grayi*
Antirheumatic (External): *Eriodictyon californicum*
Antirheumatic (Internal): *Verbascum thapsus*
Blood Medicine: *Quercus* sp.
Burn Dressing: *Arctostaphylos patula*
Cold Remedy: *Eriodictyon californicum, Ligusticum grayi, Quercus* sp., *Verbascum thapsus*
Cough Medicine: *Ligusticum grayi*
Dermatological Aid: *Arctostaphylos patula, Prunus virginiana, Verbascum thapsus*

Diaphoretic: *Verbascum thapsus*
Gastrointestinal Aid: *Ligusticum grayi*
Herbal Steam: *Eriodictyon californicum*
Panacea: *Ligusticum grayi*
Pediatric Aid: *Ligusticum grayi*
Pulmonary Aid: *Eriodictyon californicum*
Unspecified: *Verbascum thapsus*
Food
Beverage: *Arctostaphylos patula, Rhus trilobata*
Bread & Cake: *Amsinckia lycopsoides, Arctostaphylos patula, Balsamorhiza deltoidea, B. hookeri, B. sagittata, Chenopodium carinatum, Descurainia pinnata, Perideridia bolanderi, Prunus subcordata, Sambucus cerulea*
Dried Food: *Amelanchier alnifolia, Juniperus occidentalis, Perideridia bolanderi, Rhus trilobata*
Fruit: *Frangula rubra, Juniperus occidentalis, Ribes roezlii*
Porridge: *Amelanchier alnifolia, Prunus virginiana*
Preserves: *Rhus trilobata*
Soup: *Perideridia bolanderi*
Substitution Food: *Ligusticum grayi*
Unspecified: *Brodiaea coronaria, Cirsium drummondii, Dichelostemma multiflorum, Lilium pardalinum, Lomatium nudicaule, L. triternatum, L. utriculatum, Perideridia bolanderi, Pteridium aquilinum, Triteleia hyacinthina*
Vegetable: *Ligusticum grayi*
Winter Use Food: *Arctostaphylos patula, Ligusticum grayi*
Other
Hunting & Fishing Item: *Ligusticum grayi*

Bannock
Drug
Cold Remedy: *Monarda* sp.
Throat Aid: *Glycyrrhiza lepidota*
Food
Appetizer: *Monarda* sp.

Bellabella
Food
Unspecified: *Zostera marina*
Other
Hunting & Fishing Item: *Tsuga heterophylla*

Bella Coola
Drug
Adjuvant: *Chamaecyparis nootkatensis*
Analgesic: *Alnus incana, Antennaria howellii, Aralia nudicaulis, Aruncus dioicus, Asarum caudatum, Geum macrophyllum, Heracleum maximum, Juniperus communis, Lactuca biennis, Ledum groenlandicum, Mentha canadensis, Nuphar lutea, Physocarpus opulifolius, Picea sitchensis, Polypodium virginianum, Populus balsamifera, Prenanthes alata, Pseudotsuga menziesii, Rosa* sp., *Rumex aquaticus, Sambucus racemosa, Taraxacum officinale, Thuja plicata, Urtica dioica, Veratrum viride*
Antidiarrheal: *Aruncus dioicus, Lactuca biennis, Pseudotsuga menziesii, Salix lucida*
Antidote: *Dryopteris carthusiana, D. filix-mas*
Antiemetic: *Lactuca biennis*
Antihemorrhagic: *Arceuthobium americanum, Lactuca biennis*
Antirheumatic (External): *Alnus incana, Asarum caudatum, Heracleum maximum, Picea sitchensis, Pinus contorta, Plagiomnium insigne, Populus balsamifera, Rhizomnium glabrescens, Rumex aquaticus, Thuja plicata, Tsuga heterophylla, Urtica dioica, Veratrum viride*
Antirheumatic (Internal): *Nuphar lutea, Oplopanax horridus, Pseudotsuga*

[Antirheumatic (Internal)]
 menziesii, Sorbus sitchensis
Blood Medicine: *Nuphar lutea, Plagiomnium insigne, Rhizomnium glabrescens*
Breast Treatment: *Achillea millefolium, Plagiomnium insigne, Rhizomnium glabrescens*
Burn Dressing: *Achillea millefolium, Picea sitchensis, Prenanthes alata, Tsuga heterophylla, T. mertensiana*
Cathartic: *Alnus rubra, Angelica genuflexa, Cicuta douglasii, Empetrum nigrum, Oplopanax horridus, Osmorhiza berteroi, Pseudotsuga menziesii, Rosa* sp., *Sambucus racemosa*
Ceremonial Medicine: *Picea sitchensis*
Cold Remedy: *Prenanthes alata, Pseudotsuga menziesii, Ribes divaricatum, Viburnum edule*
Cough Medicine: *Juniperus communis, Lonicera involucrata, Thuja plicata, Veratrum viride*
Dermatological Aid: *Achillea millefolium, Clintonia uniflora, Epilobium angustifolium, Gaultheria shallon, Geum macrophyllum, Heracleum maximum, Lonicera involucrata, Picea sitchensis, Pinus contorta, Plagiomnium insigne, Populus balsamifera, Ranunculus acris, Rhizomnium glabrescens, Ribes lacustre, Rumex aquaticus, Salix lucida, S. scouleriana, Thuja plicata, Trautvetteria caroliniensis, Tsuga heterophylla, T. mertensiana, Veratrum viride*
Diaphoretic: *Populus balsamifera, Rosa nutkana*
Disinfectant: *Picea sitchensis*
Diuretic: *Aruncus dioicus, Myrica gale, Picea sitchensis, Pseudotsuga menziesii, Symphoricarpos albus*
Emetic: *Cicuta douglasii, Oplopanax horridus, Osmorhiza berteroi, O.* sp., *Physocarpus capitatus, P. opulifolius, Pseudotsuga menziesii, Sambucus racemosa, Veratrum viride*
Eye Medicine: *Abies amabilis, A. grandis, Athyrium filix-femina, Clintonia uniflora, Cornus sericea, Malus fusca, Ribes divaricatum, R. laxiflorum, Rosa nutkana, Sorbus sitchensis*
Gastrointestinal Aid: *Abies amabilis, A. grandis, Aralia nudicaulis, Aruncus dioicus, Asarum caudatum, Geum macrophyllum, Juniperus communis, Ledum groenlandicum, Lysichiton americanus, Mentha canadensis, Picea sitchensis, Polypodium virginianum, Pseudotsuga menziesii, Rosa* sp., *Rubus spectabilis, Sambucus racemosa, Sorbus sitchensis, Taraxacum officinale, Thuja plicata, Veratrum viride*
Gynecological Aid: *Sedum spathulifolium*
Heart Medicine: *Lactuca biennis, Nuphar lutea, Picea sitchensis, Pinus contorta, Prunella vulgaris, Prunus emarginata, Thuja plicata, Tsuga heterophylla, Veratrum viride*
Hemostat: *Urtica dioica*
Herbal Steam: *Rumex aquaticus, Urtica dioica*
Internal Medicine: *Tsuga heterophylla, T. mertensiana*
Laxative: *Frangula purshiana, Physocarpus opulifolius, Picea sitchensis, Pseudotsuga menziesii, Ribes lacustre, Veratrum viride*
Misc. Disease Remedy: *Aruncus dioicus*
Oral Aid: *Polypodium glycyrrhiza*
Orthopedic Aid: *Heracleum maximum, Populus balsamifera, Urtica dioica, Veratrum viride*
Other: *Thuja plicata*
Pediatric Aid: *Achillea millefolium, Prenanthes alata*
Poison: *Veratrum viride*
Pulmonary Aid: *Alnus incana, Arceuthobium americanum, Heracleum maximum, Juniperus communis, Osmorhiza* sp., *Populus balsamifera, Taxus brevifolia, Viburnum edule*
Respiratory Aid: *Achillea millefolium, Thuja plicata*
Throat Aid: *Abies amabilis, A. grandis, Polypodium glycyrrhiza, P. virginianum, Populus balsamifera, Ribes divaricatum*
Tuberculosis Remedy: *Abies amabilis, A. grandis, Anaphalis margaritacea, Nuphar lutea, Physocarpus opulifolius, Pinus contorta, Populus balsamifera, Prunus emarginata*
Unspecified: *Alnus incana, A. viridis, Artemisia ludoviciana, Ligusticum scothicum, Paxistima myrsinites, Ribes bracteosum*
Urinary Aid: *Saxifraga ferruginea*
Venereal Aid: *Amelanchier alnifolia, Aruncus dioicus, Lonicera involucrata, Myrica gale, Nuphar lutea, Physocarpus opulifolius, Picea sitchensis, Populus tremuloides, Pseudotsuga menziesii, Ribes bracteosum, Symphoricarpos albus, Veratrum viride*

Food
Beverage: *Aralia nudicaulis, Ledum groenlandicum, Sambucus racemosa*
Bread & Cake: *Gaultheria shallon, Rubus leucodermis, R. spectabilis, Vaccinium ovalifolium, V. parvifolium*
Dietary Aid: *Dryopteris filix-mas*
Dried Food: *Egregia menziesii, Populus balsamifera, Porphyra* sp., *Sambucus racemosa, Tsuga heterophylla*
Forage: *Clintonia uniflora, Lonicera involucrata*
Frozen Food: *Vaccinium parvifolium*
Fruit: *Amelanchier alnifolia, Cornus unalaschkensis, Crataegus douglasii, Fragaria vesca, F. virginiana, Maianthemum dilatatum, M. stellatum, Malus fusca, Ribes bracteosum, R. divaricatum, R. lacustre, R. laxiflorum, Rosa nutkana, Rubus idaeus, R. leucodermis, R. spectabilis, Vaccinium cespitosum, V. membranaceum, V. ovalifolium, V. parvifolium*
Ice Cream: *Shepherdia canadensis*
Preserves: *Rubus idaeus, R. leucodermis, R. parviflorus, Sambucus racemosa*
Sauce & Relish: *Ribes divaricatum*
Special Food: *Arctostaphylos uva-ursi*
Unspecified: *Allium cernuum, Angelica lucida, Argentina egedii, Dryopteris filix-mas, Egregia menziesii, Fritillaria camschatcensis, Heracleum maximum, Populus balsamifera, Pteridium aquilinum, Rubus parviflorus, R. spectabilis, Rumex acetosella, Sium suave, Trifolium wormskioldii*
Vegetable: *Epilobium angustifolium, Rumex aquaticus*
Winter Use Food: *Viburnum edule*

Fiber
Basketry: *Betula papyrifera, Picea sitchensis, Prunus emarginata, Thuja plicata*
Canoe Material: *Betula papyrifera*
Clothing: *Chamaecyparis nootkatensis, Thuja plicata*
Cordage: *Apocynum androsaemifolium, Urtica dioica*
Mats, Rugs & Bedding: *Chamaecyparis nootkatensis, Hylocomium splendens, Rhytidiadelphus loreus, Sphagnum* sp., *Thuja plicata*
Scouring Material: *Equisetum hyemale, Thuja plicata*

Dye
Brown: *Tsuga heterophylla*
Red: *Alnus rubra*

Other
Cleaning Agent: *Tsuga heterophylla*
Containers: *Lysichiton americanus, Nereocystis luetkeana, Prunus emarginata*
Cooking Tools: *Acer glabrum, Angelica genuflexa, Cornus sericea, Lysichiton americanus, Spiraea douglasii*
Decorations: *Alectoria sarmentosa, Chamaecyparis nootkatensis, Lycopodium clavatum, Thuja plicata, Usnea* sp.
Fuel: *Alnus rubra, Fucus* sp., *Pseudotsuga menziesii, Thuja plicata*
Good Luck Charm: *Veratrum viride*
Hunting & Fishing Item: *Apocynum androsaemifolium, Egregia menziesii, Taxus brevifolia, Urtica dioica*
Insecticide: *Sorbus sitchensis*
Protection: *Oplopanax horridus, Ribes lacustre, Veratrum viride*
Smoking Tools: *Ribes divaricatum, Sambucus racemosa*
Tools: *Acer glabrum, Angelica genuflexa, Taxus brevifolia*
Toys & Games: *Castilleja miniata, C. unalaschcensis*

Blackfoot
Drug

Abortifacient: *Anemone multifida, Artemisia campestris, Betula occidentalis, Draba incerta, Physaria didymocarpa*

Analgesic: *Abies lasiocarpa, Achillea millefolium, Acorus calamus, Allium* sp., *Anemone multifida, Glycyrrhiza lepidota, Liatris punctata, Mentha canadensis, Penstemon acuminatus, Physaria didymocarpa, Salix discolor, Veratrum californicum, V. viride*

Antidiarrheal: *Argentina anserina, Conyza canadensis, Delphinium bicolor, Erigeron philadelphicus, Escobaria vivipara, Fragaria* sp., *F. virginiana, Heracleum maximum, Heuchera cylindrica, H. richardsonii, Matricaria discoidea, Perideridia gairdneri, Polyporus* sp., *Prunus virginiana, Rosa acicularis, R.* sp., *Salix* sp., *Solanum triflorum*

Antiemetic: *Allium* sp., *Juniperus scopulorum, Penstemon acuminatus, Perideridia gairdneri*

Antihemorrhagic: *Angelica dawsonii, Astragalus canadensis, Castilleja* sp., *Conyza canadensis, Erigeron philadelphicus, Lycoperdon* sp., *Mahonia aquifolium, M. repens, Salix* sp., *Selaginella densa*

Antirheumatic (External): *Achillea millefolium, Allium* sp., *Angelica dawsonii, Artemisia campestris, Asclepias viridiflora, Glycyrrhiza lepidota, Heuchera parvifolia, Juniperus scopulorum, J.* sp., *Pediomelum esculentum, Physaria didymocarpa, Populus* sp., *Pulsatilla* sp., *Rumex crispus, R. salicifolius, Viola adunca, Yucca glauca*

Antirheumatic (Internal): *Balsamorhiza sagittata*

Blood Medicine: *Geum triflorum*

Breast Treatment: *Gaillardia aristata, Osmorhiza occidentalis, Perideridia gairdneri*

Cathartic: *Acer glabrum, Amelanchier alnifolia, Lonicera involucrata, Prunus virginiana*

Ceremonial Medicine: *Abies lasiocarpa, Angelica dawsonii, Hierochloe odorata, Lithospermum incisum, Lomatium dissectum*

Cold Remedy: *Abies lasiocarpa, Actaea pachypoda, A. rubra, Allium* sp., *Artemisia frigida, Cornus sericea, Geranium viscosissimum, Hierochloe odorata, Lycopus* sp., *Osmorhiza berteroi*

Cough Medicine: *Actaea pachypoda, A. rubra, Allium* sp., *Artemisia campestris, A. frigida, Geum triflorum, Glycyrrhiza lepidota, Hierochloe odorata, Lygodesmia juncea, Monarda fistulosa, Osmorhiza occidentalis, Perideridia gairdneri, Pyrola* sp.

Dermatological Aid: *Abies lasiocarpa, Achillea millefolium, Allium* sp., *Angelica dawsonii, Apocynum cannabinum, Arctostaphylos uva-ursi, Argentina anserina, Artemisia campestris, A. frigida, A. ludoviciana, Asclepias viridiflora, Astragalus canadensis, Balsamorhiza sagittata, Delphinium bicolor, Elaeagnus commutata, Epilobium angustifolium, Equisetum arvense, Evernia vulpina, Gaillardia aristata, Geum triflorum, Heracleum maximum, Heuchera cylindrica, H. parviflora, H. parvifolia, Hierochloe odorata, Juniperus* sp., *Liatris punctata, Lycopodium complanatum, Lygodesmia juncea, Mahonia repens, Monarda fistulosa, Oenothera cespitosa, Orobanche fasciculata, O. ludoviciana, Osmorhiza occidentalis, Oxytropis lagopus, O. sericea, Perideridia gairdneri, Physaria didymocarpa, Prunella vulgaris, Pulsatilla patens, Pyrola* sp., *Rumex crispus, R. salicifolius, Salix* sp., *Thalictrum occidentale, Xerophyllum tenax, Yucca glauca*

Dietary Aid: *Angelica dawsonii, Lomatium dissectum, Physaria didymocarpa*

Disinfectant: *Mahonia repens*

Diuretic: *Achillea millefolium, Castilleja* sp., *Equisetum arvense, Lygodesmia juncea, Perideridia gairdneri, Pyrola* sp.

Ear Medicine: *Allium* sp., *Amelanchier alnifolia, Oxytropis sericea, Pediomelum esculentum, Physaria didymocarpa, Pyrola* sp.

Emetic: *Abies lasiocarpa, Allium* sp., *Argentina anserina, Lonicera involucrata, Monarda fistulosa, Phragmites australis*

Eye Medicine: *Allium* sp., *Amelanchier alnifolia, Artemisia campestris, Asclepias viridiflora, Disporum trachycarpum, Dodecatheon pulchellum, Escobaria vivipara, Gaillardia aristata, Geranium viscosissimum, Geum triflorum, Heuchera cylindrica, H. parvifolia, Hierochloe odorata, Monarda fistulosa, Osmorhiza occidentalis, Pediomelum esculentum, Physaria didymocarpa, Prunella vulgaris, Pyrola* sp., *Salix* sp.

Febrifuge: *Abies lasiocarpa, Aconitum* sp., *Artemisia frigida, Salix discolor*

Gastrointestinal Aid: *Acorus calamus, Amelanchier alnifolia, Angelica dawsonii, Artemisia campestris, A. frigida, Evernia vulpina, Gaillardia aristata, Heuchera flabellifolia, Liatris punctata, Lonicera involucrata, Lupinus* sp., *Lygodesmia juncea, Mahonia aquifolium, M. repens, Pediomelum esculentum, Penstemon acuminatus, Physaria didymocarpa, Populus tremuloides, Rosa* sp., *Salix* sp., *Shepherdia argentea, Valeriana dioica, Veratrum viride*

Gynecological Aid: *Achillea millefolium, Artemisia frigida, Camassia quamash, Lygodesmia juncea, Osmorhiza occidentalis, Populus tremuloides, Pulsatilla patens, Pyrola* sp., *Ribes americanum, Selaginella densa*

Heart Medicine: *Mentha canadensis*

Hemostat: *Artemisia frigida, Heuchera parvifolia, Lycoperdon* sp., *Lycopodium complanatum, Xerophyllum tenax, Yucca glauca*

Herbal Steam: *Gutierrezia sarothrae, Yucca glauca*

Kidney Aid: *Juniperus horizontalis, Lygodesmia juncea, Mahonia repens, Monarda fistulosa, Ribes americanum*

Laxative: *Amelanchier alnifolia, Apocynum cannabinum, Aster* sp., *Crataegus chrysocarpa, Epilobium angustifolium, Perideridia gairdneri, Phlox hoodii, Pyrola* sp., *Ribes* sp., *Shepherdia argentea, Spiraea splendens*

Liver Aid: *Achillea millefolium, Cornus sericea, Grindelia squarrosa, Populus* sp.

Misc. Disease Remedy: *Allium* sp., *Angelica dawsonii, Artemisia frigida, Lupinus* sp., *Rhus trilobata*

Narcotic: *Selaginella densa*

Nose Medicine: *Draba incerta, Gaillardia aristata, Osmorhiza occidentalis, Solidago* sp., *Veratrum viride*

Oral Aid: *Abies lasiocarpa, Arctostaphylos uva-ursi, Artemisia campestris, Asclepias viridiflora, Delphinium bicolor, Dodecatheon pulchellum, Geum triflorum, Glycyrrhiza lepidota, Heuchera parvifolia, Veratrum viride*

Orthopedic Aid: *Equisetum arvense, Gaillardia aristata, Juniperus* sp., *Lygodesmia juncea, Pediomelum esculentum, Physaria didymocarpa, Xerophyllum tenax, Yucca glauca*

Other: *Angelica dawsonii, Geranium viscosissimum*

Panacea: *Achillea millefolium, Lomatium triternatum, Perideridia gairdneri, Populus tremuloides, Rumex salicifolius, Solidago* sp.

Pediatric Aid: *Allium* sp., *Amelanchier alnifolia, Angelica dawsonii, Arctostaphylos uva-ursi, Artemisia campestris, A. ludoviciana, Asclepias viridiflora, Aster* sp., *Astragalus canadensis, Delphinium bicolor, Dodecatheon pulchellum, Elaeagnus commutata, Epilobium angustifolium, Heuchera parvifolia, Lycopus* sp., *Lygodesmia juncea, Pediomelum esculentum, Phlox hoodii, Physaria didymocarpa, Prunus virginiana, Pyrola* sp., *Rosa acicularis, Solanum triflorum, Viola adunca*

Poison: *Aconitum* sp., *Cornus sericea, Delphinium* sp., *Oxytropis* sp., *Triglochin maritimum, Valeriana edulis, Veratrum californicum, V. viride*

Pulmonary Aid: *Abies lasiocarpa, Acorus calamus, Artemisia ludoviciana, Astragalus canadensis, Juniperus communis, Lomatium triternatum, Lonicera involucrata, Lycopodium complanatum, Pediomelum esculentum, Phlox hoodii, Rubus parviflorus, Thalictrum occidentale*

Respiratory Aid: *Aconitum* sp., *Allium* sp., *Artemisia ludoviciana, Geum triflorum, Gutierrezia sarothrae, Lupinus* sp., *Perideridia gairdneri, Polygala senega, Viola adunca*

Snakebite Remedy: *Heuchera cylindrica*

Stimulant: *Abies lasiocarpa, Artemisia frigida, Delphinium bicolor, Lomatium dissectum*

Strengthener: *Hierochloe odorata, Lomatium macrocarpum, L. triternatum, Perideridia gairdneri*

Throat Aid: *Achillea millefolium, Aconitum* sp., *Acorus calamus, Artemisia ludoviciana, Asclepias viridiflora, Geum triflorum, Glycyrrhiza lepidota, Hierochloe odorata, Lewisia rediviva, Monarda fistulosa, Osmorhiza berteroi, Oxytropis lagopus, O.* sp., *Pediomelum esculentum, Perideridia gairdneri, Physaria didymocarpa, Prunus virginiana, Pyrola* sp., *Salix* sp., *Solidago* sp.

Tonic: *Geum triflorum, Juniperus* sp., *Lomatium dissectum, Lygodesmia juncea*

Toothache Remedy: *Acorus calamus, Echinacea angustifolia, Pediomelum esculentum, Physaria didymocarpa*

Tuberculosis Remedy: *Abies lasiocarpa, Alnus incana, Pinus contorta*

Unspecified: *Angelica dawsonii, Delphinium* sp., *Pediomelum esculentum, Prunus virginiana*

Venereal Aid: *Abies lasiocarpa, Hierochloe odorata, Juniperus communis, Lycopodium complanatum, Spiraea splendens*

Veterinary Aid: *Abies lasiocarpa, Achillea millefolium, Actaea rubra, Allium* sp., *Angelica dawsonii, Artemisia campestris, A. frigida, A. ludoviciana, A.* sp., *Aster* sp., *Clematis occidentalis, Equisetum arvense, E. hyemale, Gaillardia aristata, Geum triflorum, Glycyrrhiza lepidota, Heuchera cylindrica, H. parvifolia, H.* sp., *Hierochloe odorata, Juniperus horizontalis, Lomatium dissectum, L. macrocarpum, Lupinus* sp., *Lycoperdon* sp., *Lygodesmia juncea, Mahonia repens, Monarda* sp., *Muhlenbergia richardsonis, Osmorhiza berteroi, Perideridia gairdneri, Physaria didymocarpa, Prunella vulgaris, Salix* sp., *Selaginella densa, Symphoricarpos occidentalis, Thalictrum sparsiflorum, Townsendia exscapa, Valeriana dioica, Yucca glauca*

Food

Beverage: *Achillea millefolium, Arctostaphylos uva-ursi, Equisetum hyemale, Eriogonum umbellatum, Fragaria virginiana, Lithospermum incisum, Mentha arvensis, M. canadensis, Prunus virginiana*

Bread & Cake: *Camassia quamash*

Candy: *Abies lasiocarpa, Antennaria rosea, Artemisia ludoviciana, Elaeagnus commutata, Escobaria vivipara, Osmorhiza occidentalis, Pinus contorta*

Dessert: *Amelanchier alnifolia*

Dried Food: *Amelanchier alnifolia, A. arborea, Arctostaphylos uva-ursi, Lewisia pygmaea, Pediomelum esculentum, Prunus virginiana*

Fodder: *Populus* sp., *P. tremuloides*

Forage: *Artemisia cana, Buchloe dactyloides, Carex nebrascensis, Chrysothamnus nauseosus, Leymus cinereus, Muhlenbergia richardsonis, Shepherdia argentea*

Fruit: *Amelanchier alnifolia, Arctostaphylos uva-ursi, Cornus sericea, Crataegus chrysocarpa, Disporum trachycarpum, Elaeagnus commutata, Escobaria vivipara, Fragaria virginiana, Mahonia repens, Prunus virginiana, Ribes aureum, R. oxyacanthoides, Rosa acicularis, R. nutkana, R.* sp., *Rubus parviflorus*

Preserves: *Amelanchier alnifolia, Rosa nutkana*

Snack Food: *Amelanchier alnifolia, Cornus sericea, Perideridia gairdneri, Populus tremuloides*

Soup: *Amelanchier alnifolia, A. arborea, Asclepias viridiflora, Elaeagnus commutata, Erythronium grandiflorum, Fritillaria pudica, Gaillardia aristata, Heracleum maximum, Lilium philadelphicum, Perideridia gairdneri, Polygonum bistortoides, Prunus virginiana, Ribes oxyacanthoides, Veratrum viride*

Special Food: *Amelanchier alnifolia, Camassia quamash, Populus tremuloides, Prunus virginiana*

Spice: *Acer glabrum, Allium cernuum, A.* sp., *Artemisia frigida, Asclepias viridiflora, Geranium viscosissimum, Mentha arvensis, M. canadensis, Pentaphylloides floribunda, Perideridia gairdneri, Prunus virginiana, Selaginella densa, Thalictrum occidentale*

Staple: *Amelanchier alnifolia, Astragalus canadensis, Perideridia gairdneri, Prunus virginiana*

Starvation Food: *Rosa* sp., *Selaginella densa, Shepherdia canadensis, Symphoricarpos occidentalis*

Unspecified: *Amelanchier arborea, Asclepias viridiflora, Astragalus canadensis, Camassia quamash, C. scilloides, Cirsium* sp., *Epilobium angustifolium, Eriogonum flavum, Erythronium grandiflorum, Fritillaria pudica, Heracleum maximum, Lewisia rediviva, Liatris punctata, Lilium philadelphicum, Lithospermum incisum, Lomatium simplex, L. triternatum, Musineon divaricatum, Pediomelum esculentum, Perideridia gairdneri, Pinus ponderosa, Populus balsamifera, P. deltoides, P. tremuloides, Salix* sp., *Typha latifolia*

Vegetable: *Allium cernuum, A.* sp., *Claytonia lanceolata, Heracleum maximum, Perideridia gairdneri*

Winter Use Food: *Allium* sp., *Amelanchier alnifolia, Arctostaphylos uva-ursi, Camassia quamash*

Fiber

Brushes & Brooms: *Spiraea splendens*

Building Material: *Betula occidentalis, Pinus contorta, Populus* sp., *Salix exigua, S.* sp.

Canoe Material: *Salix* sp.

Clothing: *Carex* sp., *Typha latifolia*

Cordage: *Argentina anserina, Elaeagnus commutata*

Furniture: *Pinus contorta, Prunus virginiana, Salix* sp.

Mats, Rugs & Bedding: *Artemisia ludoviciana, Juniperus horizontalis, Leymus cinereus, Pentaphylloides floribunda*

Dye

Blue: *Delphinium bicolor, Viola adunca*

Mordant: *Heuchera cylindrica, Lycopodium complanatum*

Orange: *Alnus incana*

Red: *Equisetum arvense, Orthocarpus luteus, Salix discolor, Shepherdia argentea*

Red-Brown: *Alnus incana, Orthocarpus luteus*

Unspecified: *Aster* sp., *Castilleja* sp., *Osmorhiza occidentalis, Rubus parviflorus*

Yellow: *Castilleja* sp., *Evernia vulpina, Mahonia repens, Phlox hoodii, Thermopsis rhombifolia*

Other

Cash Crop: *Acorus calamus, Amelanchier alnifolia*

Ceremonial Items: *Abies lasiocarpa, Amelanchier alnifolia, Angelica dawsonii, Arctostaphylos uva-ursi, Artemisia frigida, A. ludoviciana, Carex nebrascensis, Clematis occidentalis, Cornus sericea, Heracleum maximum, Hierochloe odorata, Juniperus horizontalis, J. scopulorum, Lithospermum incisum, Lophophora williamsii, Lupinus* sp., *Lycoperdon* sp., *Nicotiana quadrivalvis, Polyporus* sp., *Populus* sp., *Salix prolixa*

Containers: *Acer glabrum, Prunus virginiana, Salix* sp.

Cooking Tools: *Balsamorhiza sagittata, Betula occidentalis, B. papyrifera, Gaillardia aristata, Heracleum maximum, Monarda fistulosa, Pinus contorta, Prunus virginiana*

Decorations: *Gentiana affinis, Hierochloe odorata, Juniperus horizontalis, Pediomelum esculentum, Populus* sp.

Designs: *Lycoperdon* sp.

Fasteners: *Pinus contorta, Salix* sp.

Fuel: *Pentaphylloides floribunda, Populus* sp., *Symphoricarpos occidentalis*

Good Luck Charm: *Angelica dawsonii, Artemisia ludoviciana, Lomatium triternatum*

Hide Preparation: *Artemisia campestris, A. frigida, Juniperus* sp., *Lomatium dissectum, L. triternatum*

Hunting & Fishing Item: *Amelanchier alnifolia, Mentha canadensis, Pseudotsuga menziesii*

Incense & Fragrance: *Abies lasiocarpa, Allium* sp., *Artemisia frigida, A. ludoviciana, Balsamorhiza sagittata, Geum triflorum, Hierochloe odorata, Lupinus* sp., *Lycoperdon* sp., *Matricaria discoidea, Monarda fistulosa, Osmorhiza occidentalis, Polyporus* sp., *Populus* sp., *Salix* sp., *Thalictrum occidentale, T.* sp.

Insecticide: *Artemisia frigida, A. ludoviciana, Matricaria discoidea, Thalictrum occidentale*

Jewelry: *Arctostaphylos uva-ursi, Aster* sp., *Elaeagnus commutata, Juniperus* sp., *Oxytropis sericea, Rosa* sp.

Malicious Charm: *Angelica dawsonii, Artemisia ludoviciana*

Musical Instrument: *Heracleum maximum, Pinus contorta*
Paint: *Abies lasiocarpa, Symphoricarpos albus, Thalictrum occidentale*
Paper: *Artemisia frigida, A. ludoviciana*
Preservative: *Populus* sp.
Protection: *Abies lasiocarpa, Artemisia ludoviciana, Clematis occidentalis, Populus balsamifera, P. deltoides, Rubus parviflorus*
Season Indicator: *Arctostaphylos uva-ursi, Stipa comata, Thermopsis rhombifolia*
Smoke Plant: *Antennaria rosea, Arctostaphylos uva-ursi, Chimaphila umbellata, Cornus sericea, Nicotiana attenuata, N. quadrivalvis, Valeriana sitchensis*
Smoking Tools: *Cornus sericea*
Snuff: *Veratrum californicum*
Soap: *Allium* sp., *Artemisia campestris, A. frigida, Elaeagnus commutata, Equisetum arvense, Hierochloe odorata, Yucca glauca*
Stable Gear: *Alnus incana*
Tools: *Artemisia frigida, Monarda fistulosa, Pinus ponderosa, Prunus virginiana*
Toys & Games: *Amelanchier alnifolia, Cornus sericea, Elaeagnus commutata, Escobaria vivipara, Heracleum maximum, Lithospermum ruderale, Pinus contorta, Populus tremuloides, Pulsatilla patens, Ribes oxyacanthoides, Salix* sp.
Waterproofing Agent: *Castilleja* sp., *Epilobium angustifolium, Gaillardia aristata, Lygodesmia juncea, Perideridia gairdneri, Pinus contorta*
Weapon: *Glycyrrhiza lepidota*

Cahuilla
Drug
Analgesic: *Asclepias* sp., *Datura wrightii, Populus fremontii, Rosa californica, Umbellularia californica, Urtica dioica*
Antidiarrheal: *Arctostaphylos glandulosa, A. glauca, A. pungens, Conyza canadensis, Matricaria discoidea*
Antidote: *Datura wrightii*
Antihemorrhagic: *Hyptis emoryi*
Antirheumatic (External): *Adenostoma fasciculatum, A. sparsifolium, Datura wrightii, Eriodictyon trichocalyx, Larrea tridentata, Populus fremontii, Urtica dioica*
Antirheumatic (Internal): *Eriodictyon trichocalyx*
Blood Medicine: *Ephedra nevadensis, Eriodictyon trichocalyx*
Burn Dressing: *Opuntia acanthocarpa*
Cancer Treatment: *Larrea tridentata*
Cathartic: *Cucurbita foetidissima, Eriogonum* sp., *Opuntia ficus-indica*
Cold Remedy: *Adenostoma sparsifolium, Anemopsis californica, Artemisia californica, Atriplex lentiformis, Eriodictyon trichocalyx, Eucalyptus* sp., *Larrea tridentata, Rhus ovata, Salvia apiana, Sambucus cerulea, Satureja douglasii*
Cough Medicine: *Eriodictyon trichocalyx, Rhus ovata*
Dermatological Aid: *Anemopsis californica, Arctostaphylos glandulosa, A. glauca, A. pungens, Baccharis salicifolia, Bursera microphylla, Chamaesyce melanadenia, Chlorogalum pomeridianum, Cucurbita foetidissima, Ericameria palmeri, Isocoma acradenia, Larrea tridentata, Lepidium nitidum, Nicotiana clevelandii, N. glauca, N. trigonophylla, Opuntia acanthocarpa, O. ficus-indica, Pinus monophylla, P. quadrifolia, Populus fremontii, Prosopis glandulosa, Ricinus communis, Salvia apiana, Solidago californica, Suaeda* sp., *Typha latifolia*
Disinfectant: *Adenostoma fasciculatum, Artemisia tridentata, Larrea tridentata, Phoradendron* sp., *Salvia columbariae*
Ear Medicine: *Chamaesyce melanadenia, Croton californicus, Nicotiana clevelandii, N. glauca, N. trigonophylla*
Emetic: *Adenostoma sparsifolium, Cucurbita foetidissima, Larrea tridentata, Nicotiana clevelandii, N. glauca, N. trigonophylla*
Eye Medicine: *Eriogonum* sp., *Phoradendron* sp., *Prosopis glandulosa, Salvia apiana, S. columbariae, Solanum douglasii*
Febrifuge: *Eriodictyon trichocalyx, Euphorbia* sp., *Sambucus cerulea, Satureja douglasii*
Gastrointestinal Aid: *Adenostoma sparsifolium, Anemopsis californica, Arbutus menziesii, Chenopodium californicum, Descurainia pinnata, Eriogonum* sp., *Larrea tridentata, Matricaria discoidea, Monardella villosa, Sambucus cerulea, Trichostema lanatum*
Gynecological Aid: *Artemisia californica, Baccharis salicifolia, Eriogonum* sp., *Larrea tridentata, Solidago californica*
Hallucinogen: *Datura wrightii*
Hunting Medicine: *Datura wrightii, Nicotiana clevelandii, N. glauca, N. trigonophylla, Salvia apiana*
Kidney Aid: *Marrubium vulgare*
Laxative: *Adenostoma sparsifolium, Frangula californica, Opuntia ficus-indica, Sambucus cerulea*
Misc. Disease Remedy: *Euphorbia* sp., *Sambucus cerulea*
Nose Medicine: *Atriplex lentiformis*
Oral Aid: *Eriodictyon trichocalyx, Euphorbia* sp.
Orthopedic Aid: *Arundo donax, Phragmites australis*
Other: *Cucurbita foetidissima, Datura wrightii, Ephedra nevadensis*
Panacea: *Bursera microphylla*
Pediatric Aid: *Artemisia californica, Croton californicus, Eschscholzia* sp., *Rosa californica, Sambucus cerulea*
Poison: *Astragalus* sp., *Croton californicus, Datura wrightii, Ricinus communis, Solanum douglasii*
Pulmonary Aid: *Adenostoma sparsifolium, Anemopsis californica, Larrea tridentata*
Respiratory Aid: *Anemopsis californica, Artemisia tridentata, Croton californicus, Datura wrightii, Eriodictyon trichocalyx, Larrea tridentata*
Sedative: *Eschscholzia* sp.
Snakebite Remedy: *Datura wrightii*
Sports Medicine: *Datura wrightii*
Throat Aid: *Ericameria palmeri, Eriodictyon trichocalyx, Isocoma acradenia*
Tonic: *Frangula californica, Larrea tridentata*
Toothache Remedy: *Chrysothamnus nauseosus, Datura wrightii, Encelia farinosa, Gutierrezia microcephala, Sambucus cerulea*
Tuberculosis Remedy: *Eriodictyon trichocalyx*
Unspecified: *Apocynum cannabinum, Artemisia californica, Datura wrightii, Phoradendron* sp., *Prosopis pubescens*
Veterinary Aid: *Adenostoma sparsifolium, Anemopsis californica, Cucurbita foetidissima, Datura wrightii, Populus fremontii*

Food
Baby Food: *Pinus monophylla, P. quadrifolia*
Beverage: *Allenrolfea occidentalis, Arctostaphylos glandulosa, A. glauca, A. pungens, Ephedra nevadensis, Eriodictyon trichocalyx, Ferocactus cylindraceus, Fouquieria splendens, Nicotiana clevelandii, N. glauca, N. trigonophylla, Phoradendron* sp., *Pinus monophylla, P. quadrifolia, Prosopis glandulosa, P. pubescens, Rhus integrifolia, R. trilobata, Romneya coulteri, Rosa californica, Rubus leucodermis, R. parviflorus, R. vitifolius, Salvia columbariae, Simmondsia chinensis, Vitis girdiana, Washingtonia filifera*
Bread & Cake: *Allenrolfea occidentalis, Prosopis glandulosa, P. pubescens, Quercus agrifolia, Q. chrysolepis, Q. dumosa, Q. kelloggii, Scirpus* sp., *Yucca whipplei*
Candy: *Chenopodium californicum*
Dietary Aid: *Salvia columbariae*
Dried Food: *Agave deserti, Amelanchier pallida, Arctostaphylos glandulosa, A. glauca, A. pungens, Cucurbita moschata, Ferocactus cylindraceus, Juniperus californica, Lasthenia glabrata, Lycium andersonii, L. fremontii, Opuntia acanthocarpa, O. basilaris, O. bigelovii, O. ficus-indica, O. ramosissima, Pinus monophylla, P. quadrifolia, Prosopis pubescens, Prunus virginiana, Quercus agrifolia, Q. chrysolepis, Q. dumosa, Q. kelloggii, Rhus ovata, Rubus leucodermis, R. parviflorus, R. vitifolius, Sambucus cerulea, Vitis girdiana, Washingtonia filifera, Yucca whipplei*
Fruit: *Amelanchier pallida, Arctostaphylos glandulosa, A. pungens, Atriplex*

[Fruit] *semibaccata, Citrullus lanatus, Fragaria vesca, Heteromeles arbutifolia, Juniperus californica, Lycium andersonii, L. fremontii, Opuntia acanthocarpa, O. engelmannii, O. ficus-indica, O. ×occidentalis, O. ramosissima, Phoradendron* sp., *Prunus andersonii, P. emarginata, P. fasciculata, P. fremontii, P. ilicifolia, P. virginiana, Rhamnus crocea, Rhus ovata, R. trilobata, Ribes malvaceum, R. montigenum, Rubus leucodermis, R. parviflorus, R. vitifolius, Sambucus cerulea, Vitis girdiana, Washingtonia filifera, Ziziphus parryi*

Porridge: *Acacia greggii, Allenrolfea occidentalis, Amaranthus fimbriatus, Arctostaphylos glandulosa, A. glauca, A. pungens, Atriplex lentiformis, Avena fatua, Chaenactis glabriuscula, Cucurbita foetidissima, Fouquieria splendens, Hirschfeldia incana, Juniperus californica, Lasthenia californica, L. glabrata, Layia glandulosa, L. platyglossa, Medicago polymorpha, Mentzelia albicaulis, M. involucrata, M. puberula, Opuntia basilaris, Panicum urvilleanum, Parkinsonia florida, Pinus monophylla, P. quadrifolia, Prosopis glandulosa, Quercus agrifolia, Q. chrysolepis, Q. dumosa, Q. kelloggii, Rhus ovata, Salvia carduacea, Trifolium* sp., *Triticum aestivum, Typha latifolia, Vitis girdiana, Washingtonia filifera, Zea mays, Ziziphus parryi*

Preserves: *Prunus andersonii, Rhamnus* sp., *Sambucus cerulea, Washingtonia filifera*

Sauce & Relish: *Arctostaphylos glandulosa, A. glauca, A. pungens, Sambucus cerulea*

Soup: *Opuntia ramosissima, Rhus trilobata*

Special Food: *Quercus agrifolia, Q. chrysolepis, Q. dumosa, Q. kelloggii*

Spice: *Allium validum, Astragalus* sp., *Descurainia pinnata, Distichlis spicata, Salvia apiana, S. mellifera*

Staple: *Arctostaphylos glandulosa, A. glauca, A. pungens, Arthrocnemum subterminale, Asclepias* sp., *Chenopodium californicum, Echinocactus polycephalus, Eriophyllum confertiflorum, Ferocactus cylindraceus, Helianthus annuus, Olneya tesota, Opuntia acanthocarpa, O. bigelovii, O. engelmannii, O. ficus-indica, O. parryi, O. ramosissima, Prosopis glandulosa, P. pubescens, Prunus virginiana, Salvia apiana, S. columbariae, S. mellifera, Scirpus* sp., *Suaeda* sp., *Typha latifolia, Ziziphus parryi*

Starvation Food: *Bromus tectorum, Hemizonia fasciculata*

Sweetener: *Castilleja foliolosa, Rhus ovata*

Unspecified: *Adenostoma sparsifolium, Agave deserti, Apiastrum angustifolium, Avena barbata, Bloomeria crocea, Calochortus catalinae, C. concolor, C. flexuosus, C. palmeri, Capsella bursa-pastoris, Chilopsis linearis, Cirsium drummondii, Cucurbita moschata, Dichelostemma pulchellum, Dudleya* sp., *Eriogonum* sp., *Ferocactus cylindraceus, Fouquieria splendens, Hesperocallis undulata, Hordeum murinum, H. vulgare, Malva* sp., *Nolina bigelovii, Opuntia basilaris, O. bigelovii, O. engelmannii, O. ficus-indica, O. ×occidentalis, Orobanche cooperi, Pluchea sericea, Proboscidea althaeifolia, Prosopis glandulosa, Rosa californica, Yucca brevifolia, Y. schidigera, Y. whipplei*

Vegetable: *Acacia greggii, Allium validum, Amaranthus fimbriatus, Apium graveolens, Asclepias* sp., *Camissonia claviformis, Capsella bursa-pastoris, Chenopodium californicum, C. fremontii, C. humile, C. murale, Chlorogalum pomeridianum, Claytonia perfoliata, C. spathulata, Descurainia pinnata, Erodium* sp., *Hirschfeldia incana, Hydrocotyle* sp., *Opuntia basilaris, Rorippa nasturtium-aquaticum, Rumex hymenosepalus, Sisymbrium irio, Suaeda* sp., *Taraxacum californicum, Trifolium* sp., *Urtica dioica, Yucca whipplei*

Winter Use Food: *Citrullus lanatus, Hirschfeldia incana*

Fiber

Basketry: *Agave deserti, Juncus acutus, J. effusus, J. lesueurii, J. textilis, Muhlenbergia rigens, Pinus monophylla, P. quadrifolia, Rhus trilobata, Urtica dioica, Yucca schidigera*

Brushes & Brooms: *Chlorogalum pomeridianum, Yucca schidigera*

Building Material: *Acacia greggii, Acer macrophyllum, Adenostoma fasciculatum, A. sparsifolium, Arctostaphylos glandulosa, A. glauca, A. pungens, Artemisia tridentata, Baccharis salicifolia, Calocedrus decurrens, Chilopsis linearis, Ericameria palmeri, Fouquieria splendens, Isocoma acradenia, Leymus condensatus, Lotus scoparius, Pinus monophylla, P. quadrifolia, Platanus racemosa, Pluchea sericea, Prosopis glandulosa, P. pubescens, Scirpus* sp., *Washingtonia filifera*

Clothing: *Adenostoma sparsifolium, Agave deserti, Chilopsis linearis, Prosopis glandulosa, Yucca brevifolia*

Cordage: *Agave deserti, Asclepias* sp., *Chilopsis linearis, Prosopis glandulosa, Urtica dioica, Yucca brevifolia, Y. schidigera*

Furniture: *Salix gooddingii*

Mats, Rugs & Bedding: *Scirpus* sp., *Typha latifolia*

Scouring Material: *Agave deserti, Cuscuta californica, Distichlis spicata, Equisetum* sp.

Unspecified: *Apocynum cannabinum*

Dye

Black: *Phoradendron* sp., *Sambucus cerulea, Suaeda* sp.

Orange: *Sambucus cerulea*

Purple: *Sambucus cerulea*

Unspecified: *Agave deserti, Cucurbita foetidissima, Solanum douglasii*

Yellow: *Hoita macrostachya, Palafoxia arida, Sambucus cerulea*

Other

Cash Crop: *Pinus monophylla, P. quadrifolia, Quercus agrifolia, Q. chrysolepis, Q. dumosa, Q. kelloggii, Triticum aestivum*

Ceremonial Items: *Agave deserti, Cucurbita foetidissima, Datura wrightii, Leymus condensatus, Nicotiana clevelandii, N. glauca, N. trigonophylla, Scirpus* sp., *Typha latifolia, Washingtonia filifera, Zea mays*

Cooking Tools: *Cucurbita foetidissima, Ferocactus cylindraceus, Populus fremontii, Prosopis glandulosa, Washingtonia filifera*

Decorations: *Eschscholzia* sp., *Penstemon centranthifolius, Pentachaeta aurea*

Fasteners: *Pinus monophylla, P. quadrifolia, Prosopis glandulosa*

Fuel: *Acacia greggii, Acer macrophyllum, Adenostoma fasciculatum, A. sparsifolium, Agave deserti, Arctostaphylos glandulosa, A. glauca, A. pungens, Ceanothus* sp., *Fouquieria splendens, Olneya tesota, Pinus monophylla, P. quadrifolia, Prosopis glandulosa, Quercus agrifolia, Q. chrysolepis, Q. dumosa, Q. kelloggii*

Good Luck Charm: *Bursera microphylla, Datura wrightii*

Hide Preparation: *Rumex hymenosepalus*

Hunting & Fishing Item: *Adenostoma fasciculatum, A. sparsifolium, Agave deserti, Arctostaphylos glandulosa, A. glauca, A. pungens, Asclepias* sp., *Chilopsis linearis, Chlorogalum pomeridianum, Leymus condensatus, Pluchea sericea, Prosopis glandulosa, P. pubescens, Quercus agrifolia, Q. chrysolepis, Q. dumosa, Q. kelloggii, Salvia apiana, Scirpus* sp., *Washingtonia filifera*

Incense & Fragrance: *Pinus monophylla, P. quadrifolia*

Jewelry: *Quercus agrifolia, Q. chrysolepis, Q. dumosa, Q. kelloggii*

Lighting: *Adenostoma fasciculatum*

Musical Instrument: *Arundo donax, Phragmites australis, Quercus agrifolia, Q. chrysolepis, Q. dumosa, Q. kelloggii*

Paper: *Prosopis glandulosa*

Protection: *Chilopsis linearis, Nicotiana clevelandii, N. glauca, N. trigonophylla, Parkinsonia florida, Prosopis glandulosa*

Smoke Plant: *Arctostaphylos glandulosa, A. glauca, A. pungens, Artemisia californica, Datura wrightii, Nicotiana clevelandii, N. glauca, N. trigonophylla*

Smoking Tools: *Arctostaphylos glandulosa, A. glauca, A. pungens*

Soap: *Atriplex lentiformis, Chlorogalum pomeridianum, Cucurbita foetidissima, Suaeda* sp., *Yucca schidigera*

Stable Gear: *Agave deserti*

Tools: *Agave deserti, Arctostaphylos glandulosa, A. glauca, A. pungens, Chilopsis linearis, Olneya tesota, Proboscidea althaeifolia, Prosopis glandulosa, P. pubescens, Washingtonia filifera*

Toys & Games: *Quercus agrifolia, Q. chrysolepis, Q. dumosa, Q. kelloggii, Sambucus cerulea, Washingtonia filifera*

Index of Tribes

California Indian
Drug
Antirheumatic (Internal): *Heracleum maximum*
Dermatological Aid: *Thalictrum* sp.
Kidney Aid: *Echinocystis* sp., *Mentha arvensis*
Snakebite Remedy: *Asclepias fascicularis*
Toothache Remedy: *Eschscholzia californica*
Food
Vegetable: *Heracleum maximum*
Fiber
Cordage: *Apocynum cannabinum*, *Hoita macrostachya*, *Salix lasiolepis*
Sewing Material: *Hoita macrostachya*
Other
Containers: *Hoita macrostachya*
Hunting & Fishing Item: *Philadelphus lewisii*
Paint: *Plagiobothrys fulvus*
Smoking Tools: *Cercocarpus montanus*
Soap: *Ceanothus* sp.
Tools: *Cercocarpus montanus*
Weapon: *Cercocarpus montanus*

Canadian Indian
Food
Unspecified: *Astragalus australis*

Carrier
Drug
Abortifacient: *Rubus* sp.
Analgesic: *Cornus sericea*, *Oplopanax horridus*, *Urtica dioica*
Anthelmintic: *Populus* sp.
Antihemorrhagic: *Alnus* sp., *Fragaria* sp.
Antirheumatic (External): *Achillea millefolium*, *Heracleum maximum*
Blood Medicine: *Chenopodium album*
Cold Remedy: *Picea* sp.
Cough Medicine: *Plantago major*, *Viburnum edule*
Dermatological Aid: *Abies* sp., *Arctostaphylos uva-ursi*, *Lonicera involucrata*, *Populus* sp., *Shepherdia canadensis*, *Sphagnum* sp., *Urtica dioica*
Eye Medicine: *Abies* sp., *Lonicera involucrata*, *Rosa nutkana*
Gastrointestinal Aid: *Plantago major*
Gynecological Aid: *Shepherdia canadensis*
Heart Medicine: *Alnus* sp.
Kidney Aid: *Equisetum hyemale*
Laxative: *Plantago major*, *Shepherdia canadensis*
Pediatric Aid: *Populus* sp., *Sphagnum* sp.
Pulmonary Aid: *Cornus sericea*
Toothache Remedy: *Achillea millefolium*
Tuberculosis Remedy: *Abies* sp., *Juniperus communis*, *Picea* sp., *Shepherdia canadensis*
Unspecified: *Sorbus sitchensis*
Urinary Aid: *Equisetum hyemale*
Food
Beverage: *Sphagnum* sp.
Candy: *Picea mariana*
Dried Food: *Amelanchier* sp., *Shepherdia canadensis*, *Vaccinium* sp.
Forage: *Lathyrus* sp.
Fruit: *Arctostaphylos uva-ursi*, *Fragaria* sp., *Rubus spectabilis*, *Shepherdia canadensis*
Ice Cream: *Shepherdia canadensis*
Preserves: *Ribes* sp., *Rosa nutkana*, *Rubus* sp., *Shepherdia canadensis*, *Vaccinium vitis-idaea*, *Viburnum edule*
Soup: *Arctostaphylos uva-ursi*
Unspecified: *Heracleum maximum*, *Linnaea borealis*, *Sium suave*, *Typha latifolia*
Vegetable: *Taraxacum* sp.
Fiber
Basketry: *Betula papyrifera*
Building Material: *Abies* sp.
Canoe Material: *Betula papyrifera*, *Populus balsamifera*
Clothing: *Populus* sp., *Sphagnum* sp.
Mats, Rugs & Bedding: *Picea* sp.
Sewing Material: *Betula papyrifera*
Snow Gear: *Betula papyrifera*, *Picea mariana*
Dye
Black: *Alnus* sp.
Other
Cooking Tools: *Betula papyrifera*, *Picea mariana*, *Rubus parviflorus*
Hide Preparation: *Abies* sp.
Hunting & Fishing Item: *Alnus* sp.
Paint: *Chenopodium capitatum*
Smoke Plant: *Arctostaphylos uva-ursi*

Carrier, Northern
Drug
Analgesic: *Artemisia ludoviciana*, *Oplopanax horridus*
Antidiarrheal: *Viburnum edule*
Burn Dressing: *Abies* sp.
Cathartic: *Abies* sp., *Juniperus communis*, *Oplopanax horridus*, *Sambucus racemosa*, *Shepherdia canadensis*, *Viburnum edule*
Cough Medicine: *Juniperus communis*
Dermatological Aid: *Abies* sp., *Cornus sericea*, *Heracleum maximum*, *Lonicera involucrata*, *Pinus contorta*, *Populus balsamifera*, *Ribes* sp., *Rosa* sp., *Rubus* sp., *Valeriana dioica*
Eye Medicine: *Pinus contorta*, *Populus balsamifera*
Gastrointestinal Aid: *Alnus rubra*, *Oplopanax horridus*
Orthopedic Aid: *Cornus sericea*, *Lonicera involucrata*, *Pinus contorta*, *Ribes* sp., *Rosa* sp., *Rubus* sp.
Stimulant: *Cornus sericea*, *Lonicera involucrata*, *Pinus contorta*, *Ribes* sp., *Rosa* sp., *Rubus* sp.
Unspecified: *Cornus canadensis*

Carrier, Southern
Drug
Analgesic: *Juniperus communis*, *Lycopodium clavatum*, *Picea sitchensis*, *Pinus contorta*, *Populus tremuloides*, *Viola adunca*
Antihemorrhagic: *Arceuthobium americanum*
Antirheumatic (External): *Mnium affine*
Cathartic: *Oplopanax horridus*, *Physocarpus opulifolius*, *Sambucus racemosa*
Cold Remedy: *Achillea millefolium*, *Anemone multifida*, *Mentha canadensis*, *Sorbus sitchensis*
Cough Medicine: *Populus balsamifera*
Dermatological Aid: *Achillea millefolium*, *Alnus rubra*, *Artemisia ludoviciana*, *Geum macrophyllum*
Dietary Aid: *Arceuthobium americanum*
Emetic: *Physocarpus opulifolius*, *Veratrum viride*
Eye Medicine: *Cornus canadensis*, *Lonicera involucrata*, *Orthilia secunda*, *Picea sitchensis*, *Pyrola asarifolia*, *Symphoricarpos albus*
Gastrointestinal Aid: *Mentha canadensis*, *Picea sitchensis*, *Pinus contorta*, *Populus tremuloides*, *Viola adunca*
Gynecological Aid: *Oplopanax horridus*
Orthopedic Aid: *Achillea millefolium*, *Artemisia ludoviciana*
Panacea: *Anemone multifida*, *Geum macrophyllum*
Poison: *Veratrum viride*

Pulmonary Aid: *Mentha canadensis, Populus balsamifera*
Tuberculosis Remedy: *Arceuthobium americanum*
Unspecified: *Abies grandis*
Other
Insecticide: *Heracleum maximum, Populus balsamifera*

Catawba
Drug
Analgesic: *Andropogon glomeratus, Aplectrum hyemale, Arnica acaulis, Chimaphila umbellata, Glyceria obtusa, Hexastylis arifolia, Pedicularis canadensis, Tephrosia virginiana, Verbascum thapsus*
Antidiarrheal: *Aletris farinosa*
Antirheumatic (External): *Tephrosia virginiana*
Antirheumatic (Internal): *Tephrosia virginiana*
Burn Dressing: *Parthenium integrifolium*
Cold Remedy: *Hedeoma pulegioides, Xanthorhiza simplicissima*
Dermatological Aid: *Aplectrum hyemale, Ilex opaca, Lechea* sp., *Orbexilum pedunculatum, Salvia lyrata, Verbascum thapsus, Yucca filamentosa*
Febrifuge: *Tephrosia virginiana*
Gastrointestinal Aid: *Aletris farinosa, Hexastylis arifolia, Pedicularis canadensis, Xanthorhiza simplicissima*
Gynecological Aid: *Oxydendrum arboreum, Salix humilis*
Heart Medicine: *Erigeron strigosus, Hexastylis arifolia*
Kidney Aid: *Manfreda virginica*
Liver Aid: *Xanthorhiza simplicissima*
Misc. Disease Remedy: *Ilex opaca, Marshallia obovata, Prunella vulgaris*
Oral Aid: *Salix humilis*
Orthopedic Aid: *Andropogon glomeratus, Arnica acaulis, Chimaphila umbellata, Glyceria obtusa, Hexastylis arifolia, Orbexilum pedunculatum, Verbascum thapsus*
Pediatric Aid: *Salix humilis, Verbascum thapsus*
Pulmonary Aid: *Verbascum thapsus*
Snakebite Remedy: *Manfreda virginica*
Tuberculosis Remedy: *Senecio anonymus, Ulmus rubra*
Veterinary Aid: *Lachnanthes caroliana, Parthenium integrifolium*

Chehalis
Drug
Abortifacient: *Thuja plicata*
Analgesic: *Rosa nutkana*
Antidiarrheal: *Achillea millefolium*
Antirheumatic (External): *Urtica dioica*
Cold Remedy: *Abies grandis, Mentha* sp.
Contraceptive: *Geum macrophyllum, Lonicera ciliosa*
Dermatological Aid: *Delphinium menziesii, Lonicera ciliosa, Marah oreganus, Symphoricarpos albus, Urtica dioica*
Diaphoretic: *Taxus brevifolia*
Emetic: *Zigadenus venenosus*
Gynecological Aid: *Rosa nutkana*
Love Medicine: *Eriophyllum lanatum*
Misc. Disease Remedy: *Holodiscus discolor*
Panacea: *Taxus brevifolia*
Poison: *Delphinium menziesii*
Tuberculosis Remedy: *Marah oreganus, Tsuga heterophylla*
Venereal Aid: *Symphoricarpos albus, Tsuga heterophylla*
Food
Fruit: *Amelanchier alnifolia, Fragaria ×ananassa, Ribes sanguineum, Rubus parviflorus, R. spectabilis*
Soup: *Camassia quamash*
Spice: *Amelanchier alnifolia*
Unspecified: *Corylus cornuta, Pteridium aquilinum, Quercus garryana, Rubus spectabilis, Rumex acetosella, R.* sp., *Typha latifolia*
Winter Use Food: *Corylus cornuta, Sambucus cerulea, S. racemosa*
Fiber
Basketry: *Typha latifolia, Xerophyllum tenax*
Clothing: *Thuja plicata, Typha latifolia*
Cordage: *Corylus cornuta, Salix lucida*
Mats, Rugs & Bedding: *Thuja plicata, Typha latifolia*
Dye
Unspecified: *Tsuga heterophylla*
Yellow: *Mahonia aquifolium*
Other
Cooking Tools: *Acer macrophyllum, Holodiscus discolor, Polystichum munitum*
Decorations: *Typha latifolia*
Fuel: *Abies grandis, Pseudotsuga menziesii, Tsuga heterophylla*
Good Luck Charm: *Pseudotsuga menziesii*
Hunting & Fishing Item: *Acer circinatum, Holodiscus discolor, Pseudotsuga menziesii, Taxus brevifolia*
Toys & Games: *Fomes* sp.

Cherokee
Drug
Abortifacient: *Angelica atropurpurea, Apocynum cannabinum, Aristolochia serpentaria, Armoracia rusticana, Asarum canadense, Asplenium trichomanes, Carya alba, C. laciniosa, C. pallida, Chamaemelum nobile, Cimicifuga racemosa, Echinocystis lobata, Erigeron philadelphicus, E. pulchellus, Hedeoma pulegioides, Hydrangea arborescens, Hypericum gentianoides, H. hypericoides, H. perforatum, Hyssopus officinalis, Juniperus virginiana, Lindera benzoin, Monarda didyma, M. fistulosa, Nepeta cataria, Orbexilum pedunculatum, Petroselinum crispum, Platanus occidentalis, Polygala senega, Scutellaria elliptica, S. incana, S. lateriflora, Solidago odora, Stylosanthes biflora, Verbena hastata*
Adjuvant: *Eupatorium maculatum, E. purpureum, Gleditsia triacanthos, Mentha ×piperita, M. spicata, Prunella vulgaris*
Analgesic: *Acer rubrum, A. saccharinum, Acorus calamus, Alnus serrulata, Anaphalis margaritacea, Anthemis cotula, Arisaema triphyllum, Aristolochia serpentaria, Artemisia biennis, Asarum canadense, Asclepias perennis, A. quadrifolia, A. syriaca, A. tuberosa, Aster linariifolius, A. novae-angliae, Blephilia ciliata, Cardamine diphylla, Carya alba, C. laciniosa, C. pallida, Castanea pumila, Celastrus scandens, Chenopodium botrys, Chimaphila maculata, Cimicifuga racemosa, Cirsium altissimum, C. vulgare, Clematis virginiana, Cornus alternifolia, C. florida, Cunila marina, Cypripedium acaule, C. parviflorum, Daphne mezereum, Desmodium nudiflorum, D. perplexum, Epigaea repens, Erigeron philadelphicus, E. pulchellus, Euonymus americana, Gnaphalium obtusifolium, Hamamelis virginiana, Hedeoma pulegioides, Hepatica nobilis, Humulus lupulus, Kalmia latifolia, Lactuca canadensis, Leucothoe axillaris, Liatris spicata, Lobelia cardinalis, L. inflata, L. siphilitica, Magnolia acuminata, M. macrophylla, Mentha ×piperita, M. spicata, Mitchella repens, Monarda didyma, M. fistulosa, Nicotiana rustica, N. tabacum, Panax quinquefolius, P. trifolius, Papaver somniferum, Pastinaca sativa, Phoradendron leucarpum, Plantago aristata, P. lanceolata, P. major, Platanthera ciliaris, Polygonum aviculare, P. hydropiper, P. persicaria, Prenanthes serpentaria, P. trifoliolata, Pycnanthemum flexuosum, P. incanum, Rhododendron maximum, Rubus idaeus, R. occidentalis, R. odoratus, Sabatia angularis, Sanicula smallii, Satureja hortensis, Senna hebecarpa, S. marilandica, Smilax glauca, S. herbacea, S. pseudochina, S. rotundifolia, Tanacetum vulgare, Tradescantia virginiana, Veratrum viride, Verbascum thapsus, Verbena hastata, Vernonia glauca, V. noveboracensis, Veronicastrum virginicum, Vicia caroliniana, Viola bicolor, V. cucullata, V. pedata, V. pubescens, V. rotundifolia, V. sororia, Xanthorhiza simplicissima*

Index of Tribes 639

Anthelmintic: *Acorus calamus, Amelanchier arborea, A. canadensis, Artemisia biennis, Asarum canadense, Bidens bipinnata, Cerastium fontanum, Chelone glabra, Chenopodium botrys, Comptonia peregrina, Cornus alternifolia, C. florida, Cucurbita pepo, Cypripedium acaule, C. parviflorum, C. pubescens, Dryopteris filix-mas, Fagus grandifolia, Gleditsia triacanthos, Juniperus virginiana, Liriodendron tulipifera, Lobelia cardinalis, L. siphilitica, Manfreda virginica, Melia azedarach, Morus alba, M. rubra, Nepeta cataria, Nicotiana rustica, N. tabacum, Nyssa sylvatica, Oxalis corniculata, O. violacea, Pinus glabra, P. virginiana, Podophyllum peltatum, Portulaca oleracea, Prunus persica, Rosa palustris, R. virginiana, Rudbeckia fulgida, R. hirta, Ruta graveolens, Sassafras albidum, Solanum carolinense, Spigelia marilandica, Tanacetum vulgare, Tephrosia virginiana*

Anticonvulsive: *Acorus calamus, Anthemis cotula, Caulophyllum thalictroides, Cypripedium acaule, C. parviflorum, Erigeron philadelphicus, E. pulchellus, Monotropa uniflora, Nepeta cataria, Nicotiana rustica, N. tabacum, Panax quinquefolius, Papaver somniferum, Phoradendron leucarpum, Viburnum acerifolium, V. nudum, V. prunifolium*

Antidiarrheal: *Acer rubrum, A. saccharinum, Acorus calamus, Ageratina altissima, Agrimonia gryposepala, A. parviflora, Amelanchier arborea, Amphicarpaea bracteata, Andropogon virginicus, Asarum canadense, Asclepias tuberosa, Aster linariifolius, A. novae-angliae, Aureolaria flava, A. laevigata, A. pedicularia, A. virginica, Berberis canadensis, Betula lenta, B. nigra, Campanula divaricata, Carex sp., Carpinus caroliniana, Coreopsis tinctoria, Cornus alternifolia, C. florida, Diospyros virginiana, Epigaea repens, Fragaria virginiana, Frasera caroliniensis, Gaultheria procumbens, Gaylussacia baccata, Hedeoma pulegioides, Heuchera americana, Hypericum gentianoides, H. hypericoides, H. perforatum, Juglans cinerea, Lilium canadense, Liquidambar styraciflua, Liriodendron tulipifera, Magnolia acuminata, M. macrophylla, Manfreda virginica, Menispermum canadense, Mitchella repens, Morus alba, M. rubra, Nyssa sylvatica, Oxydendrum arboreum, Pedicularis canadensis, Pinus glabra, P. virginiana, Plantago aristata, P. lanceolata, P. major, Platanthera ciliaris, Platanus occidentalis, Polygonatum biflorum, Polygonum aviculare, P. hydropiper, Potentilla simplex, Pycnanthemum flexuosum, P. incanum, Quercus alba, Q. falcata, Q. imbricaria, Q. rubra, Q. stellata, Q. velutina, Rheum rhaponticum, Ribes rotundifolium, Rosa palustris, R. sp., Rubus alleghaniensis, R. argutus, R. flagellaris, R. trivialis, Rumex crispus, R. patientia, Salix alba, S. babylonica, S. humilis, S. nigra, Salvia lyrata, S. officinalis, Sassafras albidum, Scutellaria elliptica, S. incana, S. lateriflora, Sesamum orientale, Sisyrinchium angustifolium, Solidago odora, Symphytum officinale, Thalictrum dioicum, T. thalictroides, Tilia americana, Tsuga canadensis, T. caroliniana, Ulmus rubra, Uvularia sessilifolia, Verbena hastata, Viola bicolor, V. cucullata, V. pedata, V. pubescens, V. rotundifolia, V. sororia, Vitis aestivalis, V. labrusca, V. vulpina, Xyris caroliniana*

Antidote: *Cornus alternifolia, C. florida, Plantago aristata, P. lanceolata, P. major*

Antiemetic: *Baptisia australis, B. tinctoria, Chamaemelum nobile, Clethra acuminata, Frasera caroliniensis, Hydrangea arborescens, H. cinerea, Lupinus perennis, Mentha ×piperita, M. spicata, Oxalis corniculata, O. violacea, Prunus persica, Pteridium aquilinum, Rhus copallinum, R. glabra, R. hirta, Thalictrum dioicum, T. thalictroides*

Antihemorrhagic: *Achillea millefolium, Allium tricoccum, Erigeron philadelphicus, E. pulchellus, Euonymus americana, Hosta lancifolia, Lachnanthes caroliana, Lupinus perennis, Tradescantia virginiana*

Antirheumatic (External): *Adiantum pedatum, Aesculus pavia, Celastrus scandens, Coronilla varia, Ipomoea pandurata, Kalmia latifolia, Leucothoe axillaris, Osmunda cinnamomea, Phacelia purshii, Pinus glabra, P. virginiana, Polystichum acrostichoides, Porteranthus stipulatus, P. trifoliatus, Rhododendron calendulaceum, Rubus idaeus, R. occidentalis, R. odoratus, Veratrum viride, Zanthoxylum americanum*

Antirheumatic (Internal): *Adiantum pedatum, Aletris farinosa, Anthemis cotula, Apocynum cannabinum, Aralia spinosa, Arctium lappa, Aristolochia serpentaria, Armoracia rusticana, Caulophyllum thalictroides, Chimaphila maculata, Cimicifuga racemosa, Dryopteris marginalis, Echinocystis lobata, Eupatorium maculatum, E. purpureum, Gnaphalium obtusifolium, Humulus lupulus, Juniperus virginiana, Leucothoe axillaris, Lilium canadense, Liriodendron tulipifera, Lobelia cardinalis, L. siphilitica, Panax trifolius, Phytolacca americana, Pinus glabra, P. virginiana, Podophyllum peltatum, Polygala senega, Polystichum acrostichoides, Populus balsamifera, P. nigra, Rhododendron maximum, Rubus allegheniensis, R. argutus, R. flagellaris, R. trivialis, Sambucus canadensis, S. nigra, Sassafras albidum, Smilax glauca, S. herbacea, S. pseudochina, S. rotundifolia, Vicia caroliniana*

Blood Medicine: *Agrimonia gryposepala, A. parviflora, Alnus serrulata, Aralia nudicaulis, Arctium lappa, Asarum canadense, Barbarea vulgaris, Bignonia capreolata, Cornus alternifolia, C. florida, Goodyera pubescens, G. repens, Lindera benzoin, Ostrya virginiana, Oxalis corniculata, O. violacea, Phytolacca americana, Prunus cerasus, P. pensylvanica, P. serotina, P. virginiana, Pyrrhopappus carolinianus, Rumex crispus, R. patientia, Sassafras albidum, Taraxacum officinale, Vernonia glauca, V. noveboracensis, Viola bicolor, V. cucullata, V. pedata, V. pubescens, V. rotundifolia, V. sororia, Vitis aestivalis, V. labrusca, V. vulpina, Xanthorhiza simplicissima*

Breast Treatment: *Asarum canadense, Asplenium rhizophyllum, A. trichomanes, Collinsonia canadensis, Eupatorium pilosum, Hepatica nobilis, Humulus lupulus, Marrubium vulgare, Panax trifolius, Polygonatum biflorum, Scutellaria elliptica, S. incana, S. lateriflora, Verbena hastata*

Burn Dressing: *Aralia racemosa, Goodyera pubescens, G. repens, Hydrangea arborescens, Plantago aristata, P. lanceolata, P. major, Prunella vulgaris, Rhus copallinum, R. glabra, R. hirta, Sambucus canadensis, S. nigra, Smallanthus uvedalia, Smilax glauca, S. herbacea, S. laurifolia, S. pseudochina, S. rotundifolia, Ulmus rubra*

Cancer Treatment: *Aesculus pavia, Arnoglossum atriplicifolium, Chamaesyce maculata, Chimaphila maculata, Cynoglossum virginianum, Euphorbia corollata, Hackelia virginiana, Hydrangea arborescens, Hydrastis canadensis, Lachnanthes caroliana, Oxalis corniculata, O. violacea, Tradescantia virginiana, Trillium erectum, Xanthorhiza simplicissima*

Carminative: *Acorus calamus, Aletris farinosa, Allium canadense, A. sativum, A. vineale, Angelica atropurpurea, Aralia spinosa, Arisaema triphyllum, Foeniculum vulgare, Liatris spicata, Mentha ×piperita, M. spicata, Monarda didyma, M. fistulosa*

Cathartic: *Allium canadense, A. sativum, A. vineale, Alnus incana, A. serrulata, Baptisia australis, B. tinctoria, Chamaesyce maculata, Eupatorium perfoliatum, Euphorbia corollata, Gentianella quinquefolia, Hydrangea arborescens, H. cinerea, Juglans cinerea, Morus alba, M. rubra, Nicotiana rustica, N. tabacum, Platanus occidentalis, Podophyllum peltatum, Polygala senega, Prunus persica, Rhamnus cathartica, Rheum rhaponticum, Ricinus communis, Rubus idaeus, R. occidentalis, R. odoratus, Sambucus canadensis, S. nigra, Senna hebecarpa, S. marilandica, Sesamum orientale, Veronicastrum virginicum*

Ceremonial Medicine: *Amaranthus hybridus, A. retroflexus, A. spinosus, Ambrosia artemisiifolia, A. trifida, Andropogon virginicus, Cicuta maculata, Clematis virginiana, Cucurbita pepo, Cynoglossum virginianum, Impatiens capensis, I. pallida, Lactuca canadensis, Lycopus virginicus, Nicotiana rustica, N. tabacum, Pinus virginiana, Rhododendron maximum, Yucca filamentosa*

Cold Remedy: *Acorus calamus, Allium cernuum, A. tricoccum, Anaphalis margaritacea, Angelica atropurpurea, Arisaema triphyllum, Aristolochia serpentaria, Armoracia rusticana, Asarum canadense, Betula lenta, B. nigra, Capsicum annuum, Cardamine diphylla, Carya alba, C. laciniosa, C. pallida, Chenopodium botrys, Chimaphila maculata, Cimicifuga racemosa, Cunila marina, Cypripedium acaule, C. parviflorum, Erigeron philadelphicus, E. pulchellus, Eupatorium perfoliatum, E. pilosum, Foeniculum vulgare, Gaultheria procumbens, Glechoma hederacea, Gnaphalium obtusifolium, Goodyera pubescens, G. repens, Hamamelis virginiana, Hedeoma pulegioides, Hyssopus officinalis, Juniperus virginiana, Lindera benzoin, Linum usitatissimum,*

[Cold Remedy]
Lobelia cardinalis, L. siphilitica, Marrubium vulgare, Melissa officinalis, Mentha ×piperita, M. spicata, Monarda didyma, M. fistulosa, Nepeta cataria, Obolaria virginica, Picea rubens, Pinus glabra, P. virginiana, Polygala senega, Porteranthus stipulatus, P. trifoliatus, Prunus cerasus, P. pensylvanica, P. serotina, P. virginiana, Pycnanthemum flexuosum, P. incanum, Salvia lyrata, S. officinalis, Sassafras albidum, Solidago odora, Ulmus rubra, Verbena hastata, Viola bicolor, V. cucullata, V. pedata, V. pubescens, V. rotundifolia, V. sororia

Contraceptive: *Cicuta maculata*

Cough Medicine: *Aletris farinosa, Alnus serrulata, Anaphalis margaritacea, Apocynum cannabinum, Aralia racemosa, Arisaema triphyllum, Aristolochia serpentaria, Asarum canadense, Asplenium trichomanes, Castanea dentata, Celastrus scandens, Cimicifuga racemosa, Erigeron philadelphicus, E. pulchellus, Galium circaezans, Glycyrrhiza glabra, Gnaphalium obtusifolium, Hedeoma pulegioides, Hosta lancifolia, Hyssopus officinalis, Inula helenium, Ipomoea pandurata, Lindera benzoin, Linum usitatissimum, Liriodendron tulipifera, Marrubium vulgare, Nepeta cataria, Obolaria virginica, Pedicularis canadensis, Pinus glabra, P. virginiana, Platanus occidentalis, Prunus americana, P. cerasus, P. pensylvanica, P. serotina, P. virginiana, Rubus idaeus, R. occidentalis, R. odoratus, R. sp., Salvia lyrata, S. officinalis, Sanguinaria canadensis, Solidago odora, Tilia americana, Trillium erectum, Ulmus rubra, Verbascum thapsus, Verbena hastata, Veronica officinalis, V. serpyllifolia, Viola bicolor, V. cucullata, V. pedata, V. pubescens, V. rotundifolia, V. sororia*

Dermatological Aid: *Abies fraseri, Acer rubrum, A. saccharinum, Achillea millefolium, Acorus calamus, Actaea pachypoda, Aesculus pavia, Agrimonia gryposepala, A. parviflora, Alisma subcordatum, Allium cernuum, Alnus serrulata, Amaranthus hybridus, A. retroflexus, A. spinosus, Ambrosia artemisiifolia, A. trifida, Amianthium muscitoxicum, Andropogon virginicus, Anthemis cotula, Apocynum cannabinum, Aralia racemosa, A. spinosa, Arctium minus, Arisaema triphyllum, Aristolochia macrophylla, Arnoglossum atriplicifolium, Artemisia biennis, Aruncus dioicus, Asarum canadense, A. sp., Asclepias perennis, A. quadrifolia, A. syriaca, Brassica oleracea, Calycanthus floridus, Carpinus caroliniana, Carya alba, C. laciniosa, C. pallida, Castanea dentata, C. pumila, Caulophyllum thalictroides, Celastrus scandens, Chamaemelum nobile, Chamaesyce maculata, Chelone glabra, Chimaphila maculata, Cimicifuga racemosa, Collinsonia canadensis, Comandra umbellata, Cornus alternifolia, C. florida, Corylus americana, Cuscuta gronovii, Cynoglossum virginianum, Datura stramonium, Daucus carota, Diospyros virginiana, Erigeron philadelphicus, E. pulchellus, Erythronium americanum, Euonymus americana, Euphorbia corollata, Geranium maculatum, Glechoma hederacea, Hackelia virginiana, Hamamelis virginiana, Heteranthera reniformis, Heuchera americana, Hosta lancifolia, Hydrangea arborescens, Hydrastis canadensis, Hypericum gentianoides, H. hypericoides, H. perforatum, Impatiens capensis, I. pallida, Iris cristata, I. verna, I. virginica, Jeffersonia diphylla, Juglans nigra, Juniperus virginiana, Kalmia latifolia, Lachnanthes caroliana, Lagenaria siceraria, Lepidium virginicum, Leucothoe axillaris, Lindera benzoin, Liquidambar styraciflua, Liriodendron tulipifera, Lobelia cardinalis, L. inflata, L. siphilitica, Lyonia mariana, Malva neglecta, Melia azedarach, Menispermum canadense, Mirabilis nyctaginea, Mitchella repens, Monotropa uniflora, Nepeta cataria, Nicotiana rustica, N. tabacum, Oxalis corniculata, O. violacea, Oxydendrum arboreum, Panax trifolius, Passiflora incarnata, Pedicularis canadensis, Phytolacca americana, Pilea pumila, Pinus glabra, P. virginiana, Piper nigrum, Plantago aristata, P. lanceolata, P. major, Platanus occidentalis, Podophyllum peltatum, Polygonatum biflorum, Polygonum aviculare, P. hydropiper, P. persicaria, Polypodium virginianum, Populus balsamifera, P. nigra, Porteranthus stipulatus, P. trifoliatus, Potentilla simplex, Prunella vulgaris, Prunus cerasus, P. pensylvanica, P. persica, P. serotina, P. virginiana, Pycnanthemum flexuosum, P. incanum, Pyrola americana, P. elliptica, Pyrularia pubera, Quercus alba, Q. falcata, Q. imbricaria, Q. rubra, Q. stellata, Q. velutina, Ranunculus abortivus, R. acris, R. recurvatus, Rhamnus cathartica, Rheum rhaponticum, Rhus copallinum, R. glabra, Ricinus communis, Rubus allegheniensis, R. argutus, R. flagellaris, R. idaeus, R. occidentalis, R. odoratus, R. trivialis, Rudbeckia fulgida, R. hirta, Rumex acetosella, R. crispus, R. patientia, Salix alba, S. babylonica, S. humilis, S. nigra, Sambucus canadensis, S. nigra, Sanguinaria canadensis, Saponaria officinalis, Sassafras albidum, Saururus cernuus, Saxifraga pensylvanica, Sempervivum tectorum, Senna hebecarpa, S. marilandica, Smallanthus uvedalia, Smilax glauca, S. herbacea, S. laurifolia, S. pseudochina, S. rotundifolia, Solanum carolinense, Tagetes erecta, Tephrosia virginiana, Tilia americana, Toxicodendron vernix, Tradescantia virginiana, Trillium erectum, Triosteum perfoliatum, Tsuga canadensis, T. caroliniana, Ulmus rubra, Uvularia sessilifolia, Verbascum thapsus, Verbena hastata, Veronica officinalis, V. serpyllifolia, Viola bicolor, V. cucullata, V. pedata, V. pubescens, V. rotundifolia, V. sororia, Xanthorhiza simplicissima, Yucca filamentosa, Zea mays*

Diaphoretic: *Acorus calamus, Anthemis cotula, Aralia racemosa, A. spinosa, Arisaema triphyllum, Carya alba, C. laciniosa, C. pallida, Cornus alternifolia, C. florida, Cunila marina, Daphne mezereum, Diphylleia cymosa, Erigeron philadelphicus, E. pulchellus, Eupatorium perfoliatum, Euphorbia ipecacuanhae, Hedeoma pulegioides, Ilex cassine, Juniperus virginiana, Liatris spicata, Lindera benzoin, Mitchella repens, Monarda didyma, M. fistulosa, Nicotiana rustica, N. tabacum, Obolaria virginica, Orbexilum pedunculatum, Polygala senega, Salvia lyrata, S. officinalis, Sambucus canadensis, S. nigra, Solidago odora, Verbena hastata, Veronicastrum virginicum, Viburnum acerifolium, V. nudum, V. prunifolium*

Dietary Aid: *Agrimonia gryposepala, A. parviflora, Aplectrum hyemale, Arctium lappa, Armoracia rusticana, Asparagus officinalis, Brassica napus, B. nigra, Chelone glabra, Chenopodium album, Crataegus spathulata, Frasera caroliniensis, Goodyera pubescens, G. repens, Hydrastis canadensis, Lilium canadense, Mitchella repens, Oenothera biennis, Passiflora incarnata, Phlox maculata, Pilea pumila, Rudbeckia laciniata, Sassafras albidum, Sinapis alba*

Disinfectant: *Ambrosia artemisiifolia, A. trifida, Aralia racemosa, Aristolochia serpentaria, Cornus alternifolia, C. florida, Diphylleia cymosa, Euonymus americana, Eupatorium perfoliatum, Frasera caroliniensis, Hydrangea arborescens, Kalmia latifolia, Pteridium aquilinum, Quercus alba, Q. falcata, Q. imbricaria, Q. rubra, Q. stellata, Q. velutina, Sambucus canadensis, S. nigra, Veronicastrum virginicum*

Diuretic: *Acorus calamus, Ageratina altissima, Allium canadense, A. sativum, A. vineale, Aristolochia serpentaria, Armoracia rusticana, Cimicifuga racemosa, Cucurbita pepo, Diphylleia cymosa, Erigeron philadelphicus, E. pulchellus, Eupatorium maculatum, E. perfoliatum, E. purpureum, Ipomoea pandurata, Liatris spicata, Mitchella repens, Monarda didyma, M. fistulosa, Nicotiana rustica, N. tabacum, Polygala senega, Sambucus canadensis, S. nigra*

Ear Medicine: *Allium canadense, A. sativum, A. tricoccum, A. vineale, Passiflora incarnata, Podophyllum peltatum, Portulaca oleracea, Rudbeckia fulgida, R. hirta, Sempervivum tectorum, Veronica officinalis, V. serpyllifolia*

Emetic: *Adiantum pedatum, Alnus incana, A. serrulata, Anthemis cotula, Aralia spinosa, Asarum canadense, Asplenium rhizophyllum, Baptisia australis, B. tinctoria, Botrychium virginianum, Calycanthus floridus, Carya alba, C. laciniosa, C. pallida, Chimaphila maculata, Clethra acuminata, Collinsonia canadensis, Coronilla varia, Corylus americana, Dryopteris marginalis, Epigaea repens, Eryngium aquaticum, Eupatorium perfoliatum, Euphorbia ipecacuanhae, Goodyera pubescens, G. repens, Hepatica nobilis, Hydrangea arborescens, H. cinerea, Ilex cassine, I. vomitoria, Juncus effusus, Lobelia inflata, Nicotiana rustica, N. tabacum, Nyssa sylvatica, Platanus occidentalis, Polystichum acrostichoides, Porteranthus stipulatus, P. trifoliatus, Pyrularia pubera, Quercus alba, Q. falcata, Q. imbricaria, Q. rubra, Q. stellata, Q. velutina, Robinia hispida, R. pseudoacacia, Rubus idaeus, R. occidentalis, R. odoratus, Sambucus canadensis, S. nigra, Scirpus tabernaemontani, Scutellaria lateriflora, Smallanthus uvedalia, Solanum nigrum, S. tuberosum, Toxicodendron pubescens, T. radicans, Triosteum perfoliatum, Verbena hastata, Vicia caroliniana, Xanthium spinosum*

Expectant: *Allium canadense, A. sativum, A. vineale, Aralia racemosa, Arisaema triphyllum, Asclepias tuberosa, Euonymus americana, Euphorbia ipecacuanhae, Galium circaezans, Glycyrrhiza glabra, Hedeoma pulegioides, Ipomoea pandurata, Liatris spicata, Nicotiana rustica, N. tabacum, Panax quinquefolius, Polygala senega*

Eye Medicine: *Acer rubrum, A. saccharinum, Alnus incana, A. serrulata, Anaphalis margaritacea, Aruncus dioicus, Asarum canadense, Calycanthus floridus, Erigeron philadelphicus, E. pulchellus, Goodyera pubescens, G. repens, Maianthemum racemosum, Monotropa uniflora, Nyssa sylvatica, Plantago aristata, P. lanceolata, P. major, Rhamnus cathartica, Sassafras albidum, Ulmus rubra, Xanthorhiza simplicissima*

Febrifuge: *Achillea millefolium, Adiantum pedatum, Ageratina altissima, Agrimonia gryposepala, A. parviflora, Aletris farinosa, Allium cernuum, Alnus serrulata, Ambrosia artemisiifolia, A. trifida, Angelica atropurpurea, Anthemis cotula, Aristolochia serpentaria, Asarum canadense, Aster linariifolius, A. novae-angliae, Brassica napus, B. nigra, Capsicum annuum, Castanea pumila, Chelone glabra, Chimaphila maculata, Clethra acuminata, Cornus alternifolia, C. florida, Cunila marina, Cystopteris protrusa, Dennstaedtia punctilobula, Echinocystis lobata, Erythronium americanum, Eupatorium perfoliatum, Hamamelis virginiana, Hedeoma pulegioides, Hypericum gentianoides, H. hypericoides, H. perforatum, Hyssopus officinalis, Linum usitatissimum, Liriodendron tulipifera, Lobelia cardinalis, L. siphilitica, Melissa officinalis, Mentha arvensis, M. ×piperita, M. spicata, Monarda didyma, M. fistulosa, Nepeta cataria, Osmunda cinnamomea, Phytolacca americana, Pinus glabra, P. virginiana, Polystichum acrostichoides, Potentilla simplex, P. sp., Prunus cerasus, P. pensylvanica, P. persica, P. serotina, P. virginiana, Pycnanthemum flexuosum, P. incanum, Quercus alba, Q. falcata, Q. imbricaria, Q. rubra, Q. stellata, Q. velutina, Sagittaria latifolia, Salix alba, S. babylonica, S. humilis, S. nigra, Senna hebecarpa, S. marilandica, Sinapis alba, Solidago odora, S. sp., Toxicodendron vernix, Trifolium pratense, T. repens, Triosteum perfoliatum, Verbena hastata, Veronica officinalis, V. serpyllifolia, Veronicastrum virginicum, Viburnum acerifolium, V. nudum, V. prunifolium*

Gastrointestinal Aid: *Abies fraseri, Achillea millefolium, Acorus calamus, Aesculus pavia, Agrimonia gryposepala, A. parviflora, Alisma subcordatum, Allium cernuum, Alnus incana, A. serrulata, Antennaria plantaginifolia, Aristolochia serpentaria, Armoracia rusticana, Artemisia biennis, Asarum canadense, Betula lenta, B. nigra, Capsicum annuum, Carya alba, C. laciniosa, C. pallida, Castanea dentata, Caulophyllum thalictroides, Ceanothus americanus, Celastrus scandens, Chamaemelum nobile, Cirsium altissimum, C. vulgare, Clematis virginiana, Clethra acuminata, Cornus alternifolia, C. florida, Cypripedium acaule, C. parviflorum, Diospyros virginiana, Epigaea repens, Eryngium aquaticum, Eupatorium perfoliatum, Foeniculum vulgare, Fragaria virginiana, Frasera caroliniensis, Fraxinus americana, F. nigra, Galium triflorum, Gaultheria procumbens, Gentianella quinquefolia, Gleditsia triacanthos, Hepatica nobilis, Heuchera americana, Hexastylis virginica, Hieracium venosum, Hydrangea arborescens, Hydrastis canadensis, Hypericum gentianoides, H. hypericoides, H. perforatum, Ilex opaca, Impatiens capensis, I. pallida, Lachnanthes caroliana, Leonurus cardiaca, Liatris spicata, Ligusticum canadense, Liriodendron tulipifera, Lobelia cardinalis, L. inflata, L. siphilitica, Lysimachia quadrifolia, Magnolia acuminata, M. macrophylla, Malus coronaria, M. pumila, Matricaria discoidea, Menispermum canadense, Mentha ×piperita, M. spicata, Mitchella repens, Monarda didyma, M. fistulosa, Nepeta cataria, Nicotiana rustica, N. tabacum, Nyssa sylvatica, Obolaria virginica, Orbexilum pedunculatum, Oxydendrum arboreum, Panax quinquefolius, P. trifolius, Pedicularis canadensis, Penstemon laevigatus, Pinus glabra, P. virginiana, Plantago aristata, P. lanceolata, P. major, Platanus occidentalis, Polygonatum biflorum, Polystichum acrostichoides, Populus balsamifera, P. nigra, Prunus cerasus, P. pensylvanica, P. persica, P. serotina, P. virginiana, Pycnanthemum flexuosum, P. incanum, Quercus alba, Q. falcata, Q. imbricaria, Q. rubra, Q. stellata, Q. velutina, Rubus idaeus, R. occidentalis, R. odoratus, Sanicula smallii, Sisyrinchium angustifolium, Smilax glauca, S. herbacea, S. pseudochina, S. rotundifolia, Symphytum officinale, Tilia americana, Tradescantia virginiana, Trillium erectum, Triodanis perfoliata, Ulmus rubra, Urtica dioica, Verbena hastata, Vernonia glauca, V. noveboracensis, Veronicastrum virginicum, Vicia caroliniana, Vitis aestivalis, V. labrusca, V. vulpina, Xanthium spinosum*

Gland Medicine: *Verbascum thapsus*

Gynecological Aid: *Abies fraseri, Acer rubrum, A. saccharinum, Achillea millefolium, Aesculus pavia, Agrimonia gryposepala, A. parviflora, Aletris farinosa, Alnus incana, A. serrulata, Amaranthus hybridus, A. retroflexus, A. spinosus, Antennaria plantaginifolia, Aquilegia canadensis, Aralia racemosa, Arctium lappa, Artemisia biennis, Aruncus dioicus, Asarum canadense, Asclepias tuberosa, Baptisia australis, B. tinctoria, Castanea dentata, Caulophyllum thalictroides, Celastrus scandens, Chamaesyce maculata, Cornus alternifolia, C. florida, Cunila marina, Cypripedium acaule, C. parviflorum, Euonymus americana, Eupatorium maculatum, E. purpureum, Euphorbia corollata, E. ipecacuanhae, Foeniculum vulgare, Fraxinus americana, F. nigra, Hamamelis virginiana, Helenium autumnale, Heuchera americana, Humulus lupulus, Hydrangea cinerea, Impatiens capensis, I. pallida, Inula helenium, Lachnanthes caroliana, Liquidambar styraciflua, Lysimachia quadrifolia, Menispermum canadense, Mitchella repens, Nyssa sylvatica, Orbexilum pedunculatum, Panax quinquefolius, Petroselinum crispum, Phoradendron leucarpum, Pinus glabra, P. virginiana, Plantago aristata, P. lanceolata, P. major, Platanus occidentalis, Polygonatum biflorum, Prunus cerasus, P. pensylvanica, P. serotina, P. virginiana, Quercus velutina, Rhododendron calendulaceum, Rhus copallinum, R. glabra, R. hirta, Rubus idaeus, R. occidentalis, R. odoratus, Rudbeckia fulgida, R. hirta, Sabatia angularis, Salvia lyrata, S. officinalis, Scutellaria elliptica, S. incana, S. lateriflora, Senecio aureus, Sesamum orientale, Silphium compositum, Smallanthus uvedalia, Smilax glauca, S. herbacea, S. pseudochina, S. rotundifolia, Stylosanthes biflora, Symphytum officinale, Tanacetum vulgare, Tradescantia virginiana, Trifolium pratense, T. repens, Trillium erectum, Tsuga canadensis, T. caroliniana, Ulmus rubra, Verbascum thapsus, Vernonia glauca, V. noveboracensis, Veronica officinalis, Vitis aestivalis, V. labrusca, V. vulpina*

Hallucinogen: *Ilex vomitoria*

Heart Medicine: *Adiantum pedatum, Alnus serrulata, Aquilegia canadensis, Asarum canadense, Asclepias tuberosa, Castanea dentata, Consolida ajacis, Crataegus spathulata, Delphinium tricorne, Hypoxis hirsuta, Monarda didyma, M. fistulosa, Pycnanthemum flexuosum, P. incanum, Rhododendron maximum, Senecio aureus, Senna hebecarpa, S. marilandica*

Hemorrhoid Remedy: *Achillea millefolium, Aesculus pavia, Alnus serrulata, Diospyros virginiana, Heuchera americana, Lachnanthes caroliana, Malus coronaria, M. pumila, Mentha ×piperita, M. spicata, Mitchella repens, Oenothera biennis, Pinus glabra, P. virginiana, Rubus allegheniensis, R. argutus, R. flagellaris, R. trivialis, Xanthorhiza simplicissima*

Hemostat: *Geranium maculatum, Hypericum gentianoides, H. hypericoides, H. perforatum, Lobelia cardinalis, L. siphilitica, Monarda didyma, M. fistulosa*

Hunting Medicine: *Erythronium americanum*

Hypotensive: *Alnus serrulata, Hydrangea arborescens, Phoradendron leucarpum*

Kidney Aid: *Abies fraseri, Acalypha virginica, Acorus calamus, Allium canadense, A. cernuum, A. sativum, A. vineale, Alnus serrulata, Anthemis cotula, Apocynum cannabinum, Arctostaphylos uva-ursi, Asclepias perennis, A. quadrifolia, A. syriaca, Brassica napus, B. nigra, Citrullus lanatus, Clematis virginiana, Comandra umbellata, Cucurbita pepo, Cypripedium acaule, C. parviflorum, Echinocystis lobata, Epigaea repens, Equisetum arvense, E. hyemale, Erigeron philadelphicus, E. pulchellus, Eupatorium maculatum, E. purpureum, Fragaria virginiana, Galax urceolata, Gaylussacia baccata, Goodyera pubescens, G. repens, Hackelia virginiana, Helianthemum canadense, Humulus lupulus, Ilex cassine, Ipomoea pandurata, Jeffersonia diphylla, Liatris spicata, Lysimachia quadrifolia, Lythrum alatum, Nicotiana rustica, N. tabacum, Panax trifolius, Petroselinum crispum, Phytolacca americana, Pinus glabra, P. virgini-*

[Kidney Aid] *ana, Polygala senega, Porteranthus stipulatus, P. trifoliatus, Prunus americana, Rudbeckia fulgida, R. hirta, Rumex crispus, R. patientia, Sambucus canadensis, S. nigra, Scutellaria elliptica, S. incana, S. lateriflora, Sinapis alba, Tephrosia virginiana, Tradescantia virginiana, Trifolium pratense, T. repens, Tsuga canadensis, T. caroliniana, Verbascum thapsus, Verbena hastata, Zea mays*

Laxative: *Abies fraseri, Asclepias perennis, A. quadrifolia, A. syriaca, A. tuberosa, Chelone glabra, Cimicifuga racemosa, Equisetum arvense, E. hyemale, Eupatorium pilosum, Galium aparine, Gentianella quinquefolia, Hepatica nobilis, Ipomoea pandurata, Menispermum canadense, Morus alba, M. rubra, Phytolacca americana, Pinus glabra, P. virginiana, Podophyllum peltatum, Rheum rhaponticum, Rumex crispus, R. patientia, Salvia lyrata, S. officinalis, Symphytum officinale, Tradescantia virginiana, Ulmus rubra*

Liver Aid: *Aletris farinosa, Allium cernuum, Asplenium trichomanes, Carya alba, C. laciniosa, C. pallida, Clethra acuminata, Diospyros virginiana, Fragaria virginiana, Hepatica nobilis, Hydrangea arborescens, H. cinerea, Iris cristata, I. verna, I. virginica, Manfreda virginica, Panax trifolius, Parthenocissus quinquefolia, Passiflora incarnata, Porteranthus stipulatus, P. trifoliatus, Veronicastrum virginicum, Vitis aestivalis, V. labrusca, V. vulpina*

Love Medicine: *Hackelia virginiana, Phoradendron leucarpum*

Misc. Disease Remedy: *Acalypha virginica, Acer rubrum, A. saccharinum, Adiantum pedatum, Ageratina altissima, Allium canadense, A. sativum, A. vineale, Angelica atropurpurea, Aristolochia serpentaria, Asarum canadense, Brassica napus, B. nigra, Carya alba, C. laciniosa, C. pallida, Castanea dentata, Chenopodium botrys, Cornus alternifolia, C. florida, Cypripedium acaule, C. parviflorum, Diphylleia cymosa, Eupatorium maculatum, E. perfoliatum, E. purpureum, Fragaria virginiana, Glechoma hederacea, Gleditsia triacanthos, Gnaphalium obtusifolium, Impatiens capensis, I. pallida, Ipomoea pandurata, Juglans nigra, Juniperus virginiana, Lindera benzoin, Liriodendron tulipifera, Melissa officinalis, Mentha ×piperita, M. spicata, Monarda didyma, M. fistulosa, Nicotiana rustica, N. tabacum, Panax trifolius, Picea rubens, Pinus glabra, P. virginiana, Platanus occidentalis, Prunus cerasus, P. pensylvanica, P. serotina, P. virginiana, Pteridium aquilinum, Ribes rotundifolium, Sassafras albidum, Senna hebecarpa, S. marilandica, Sesamum orientale, Sinapis alba, Solidago odora, Toxicodendron vernix, Urtica dioica, Verbascum thapsus, Veronicastrum virginicum, Viburnum acerifolium, V. nudum, V. prunifolium, Vicia caroliniana, Yucca filamentosa*

Nose Medicine: *Aristolochia serpentaria, Helenium autumnale, Helianthus giganteus, Sanguinaria canadensis*

Oral Aid: *Alnus serrulata, Angelica atropurpurea, Armoracia rusticana, Carya alba, C. laciniosa, C. pallida, Clitoria mariana, Desmodium nudiflorum, D. perplexum, Diospyros virginiana, Gaultheria procumbens, Geranium maculatum, Gnaphalium obtusifolium, Heuchera americana, Juncus effusus, J. tenuis, Lachnanthes caroliana, Malus coronaria, Oxalis corniculata, O. violacea, Oxydendrum arboreum, Panax quinquefolius, Potentilla simplex, Prunus cerasus, P. pensylvanica, P. serotina, P. virginiana, Quercus alba, Q. falcata, Q. imbricaria, Q. rubra, Q. stellata, Q. velutina, Ranunculus abortivus, R. acris, R. recurvatus, Rubus allegheniensis, R. argutus, R. flagellaris, R. trivialis, Sassafras albidum, Scirpus tabernaemontani, Solanum carolinense, Solidago odora, Tiarella cordifolia, Trollius laxus, Viburnum acerifolium, V. nudum, V. prunifolium, Vitis aestivalis, V. labrusca, V. vulpina, Xanthorhiza simplicissima*

Orthopedic Aid: *Aesculus pavia, Alnus serrulata, Aralia racemosa, A. spinosa, Arisaema triphyllum, Aruncus dioicus, Brassica napus, B. nigra, Carya alba, C. laciniosa, C. pallida, Clematis virginiana, Coronilla varia, Elytrigia repens, Euonymus americana, Gnaphalium obtusifolium, Hydrangea arborescens, Ilex opaca, Juncus effusus, J. tenuis, Kalmia latifolia, Liriodendron tulipifera, Lobelia inflata, L. spicata, Ostrya virginiana, Pinus glabra, P. virginiana, Plantago aristata, P. lanceolata, P. major, Porteranthus stipulatus, P. trifoliatus, Ruta graveolens, Sanicula smallii, Sinapis alba, Smilax glauca, S. herbacea, S. pseudochina, S. rotundifolia, Symphytum officinale, Tanacetum parthenium, T. vulgare, Tephrosia virginiana, Triosteum perfoliatum, Veratrum viride, Vicia caroliniana*

Other: *Adiantum pedatum, Aureolaria flava, A. laevigata, A. pedicularia, Euonymus americana, Eupatorium maculatum, E. purpureum, Leucothoe axillaris, Liquidambar styraciflua, Lobelia inflata, Lycopus virginicus, Nicotiana rustica, N. tabacum, Nyssa sylvatica, Panax quinquefolius, P. trifolius, Phytolacca americana, Pinus glabra, P. sp., P. virginiana, Platanus occidentalis, Polygala senega, P. verticillata, Rhododendron maximum, Sambucus canadensis, S. nigra, Smallanthus uvedalia, Smilax glauca, S. herbacea, S. pseudochina, S. rotundifolia, Vitis aestivalis, V. labrusca, V. vulpina*

Panacea: *Kalmia latifolia, Veratrum viride*

Pediatric Aid: *Agrimonia gryposepala, A. parviflora, Allium canadense, A. cernuum, A. sativum, A. vineale, Alnus serrulata, Amelanchier canadensis, Antennaria plantaginifolia, Aplectrum hyemale, Calycanthus floridus, Castanea dentata, Cerastium fontanum, Chamaesyce maculata, Chimaphila maculata, Cimicifuga racemosa, Citrullus lanatus, Coix lacryma-jobi, Cornus alternifolia, C. florida, Epigaea repens, Euphorbia corollata, Foeniculum vulgare, Geranium maculatum, Glechoma hederacea, Hydrangea arborescens, Impatiens capensis, I. pallida, Juncus effusus, J. tenuis, Lilium canadense, Liriodendron tulipifera, Lycopus virginicus, Marrubium vulgare, Mentha ×piperita, M. spicata, Mitchella repens, Monotropa uniflora, Nepeta cataria, Nyssa sylvatica, Oxalis corniculata, O. violacea, Panax trifolius, Passiflora incarnata, Phlox maculata, Pilea pumila, Plantago aristata, P. lanceolata, P. major, Platanus occidentalis, Polygonum aviculare, P. hydropiper, Rosa virginiana, Sagittaria latifolia, Sambucus canadensis, S. nigra, Sassafras albidum, Senna hebecarpa, S. marilandica, Sesamum orientale, Sisyrinchium angustifolium, Solanum carolinense, Spiranthes lucida, Tanacetum vulgare, Tephrosia virginiana, Vitis aestivalis, V. labrusca, V. vulpina, Xyris caroliniana*

Poison: *Amianthium muscitoxicum, Aralia spinosa, Calycanthus floridus, Castilleja coccinea, Chimaphila maculata, Consolida ajacis, Delphinium tricorne, Juglans nigra, Phytolacca americana, Podophyllum peltatum, Polygonum aviculare, P. hydropiper, Rheum rhaponticum, Toxicodendron vernix*

Poultice: *Aesculus sp., Capsicum annuum, Cirsium altissimum, C. sp., C. vulgare, Cornus florida, Liriodendron tulipifera, Polygonum aviculare, P. hydropiper, Ruta graveolens, Saururus cernuus*

Preventive Medicine: *Crataegus spathulata*

Psychological Aid: *Cynoglossum virginianum, Fragaria virginiana, Hackelia virginiana, Solanum nigrum, S. tuberosum*

Pulmonary Aid: *Abies fraseri, Aletris farinosa, Allium canadense, A. cernuum, A. sativum, A. tricoccum, A. vineale, Ambrosia artemisiifolia, A. trifida, Anemone virginiana, Apocynum cannabinum, Aralia racemosa, Aristolochia serpentaria, Asclepias tuberosa, Brassica napus, B. nigra, Cercis canadensis, Epigaea repens, Eryngium yuccifolium, Euphorbia ipecacuanhae, Gleditsia triacanthos, Hedeoma pulegioides, Hyssopus officinalis, Inula helenium, Lepidium virginicum, Lindera benzoin, Linum usitatissimum, Lobelia cardinalis, L. inflata, L. siphilitica, Malus pumila, Mertensia virginica, Oxydendrum arboreum, Panax trifolius, Polygala senega, Polygonatum biflorum, Polygonum virginianum, Polystichum acrostichoides, Potentilla simplex, Sanguinaria canadensis, Senna hebecarpa, S. marilandica, Sinapis alba, Sisymbrium officinale, Xanthium spinosum, Zea mays*

Reproductive Aid: *Cinchona calisaya*

Respiratory Aid: *Achillea millefolium, Adiantum pedatum, Allium canadense, A. cernuum, A. sativum, A. vineale, Anaphalis margaritacea, Anthemis cotula, Apocynum cannabinum, Aralia racemosa, Armoracia rusticana, Aster linariifolius, A. novae-angliae, Brassica napus, B. nigra, Datura stramonium, Euonymus americana, Eupatorium pilosum, Galium circaezans, Glycyrrhiza glabra, Gnaphalium obtusifolium, Hyssopus officinalis, Inula helenium, Ipomoea pandurata, Lindera benzoin, Lobelia inflata, Magnolia acuminata, M. macrophylla, Oxydendrum arboreum, Pimpinella anisum, Pinus glabra, P. virginiana, Porteranthus stipulatus, P. trifoliatus, Quercus alba, Q. falcata, Q. imbri-*

caria, Q. rubra, Q. stellata, Q. velutina, Salix alba, S. babylonica, S. humilis, S. nigra, Salvia lyrata, S. officinalis, Sanguinaria canadensis, Sinapis alba, Toxicodendron vernix, Trillium erectum, Ulmus rubra, Vicia caroliniana, Viola bicolor, V. cucullata, V. pedata, V. pubescens, V. rotundifolia, V. sororia

Sedative: Achillea millefolium, Angelica atropurpurea, Anthemis cotula, Asarum canadense, Caulophyllum thalictroides, Chamaemelum nobile, Cimicifuga racemosa, Cypripedium acaule, C. parviflorum, Fragaria virginiana, Galax urceolata, Gentianella quinquefolia, Humulus lupulus, Lactuca canadensis, Leonurus cardiaca, Liquidambar styraciflua, Liriodendron tulipifera, Mentha ×piperita, M. spicata, Monarda didyma, M. fistulosa, Nepeta cataria, Oxydendrum arboreum, Papaver somniferum, Pinus glabra, P. virginiana, Ranunculus abortivus, R. acris, R. recurvatus, Ribes rotundifolium, Ruta graveolens, Salvia lyrata, S. officinalis, Solidago odora, Sonchus arvensis, Taraxacum officinale, Xanthorhiza simplicissima, Yucca filamentosa

Snakebite Remedy: Amphicarpaea bracteata, Aristolochia serpentaria, Botrychium virginianum, Cunila marina, Eryngium yuccifolium, Hypericum gentianoides, H. hypericoides, H. perforatum, Liriodendron tulipifera, Lycopus virginicus, Nicotiana rustica, N. tabacum, Osmunda cinnamomea, Plantago aristata, P. lanceolata, P. major, Polygala senega, Rudbeckia fulgida, R. hirta, Tilia americana, Xanthium spinosum

Sports Medicine: Chamaecrista fasciculata, C. nictitans

Stimulant: Acorus calamus, Actaea pachypoda, Aesculus pavia, Ageratina altissima, Allium canadense, A. sativum, A. vineale, Arisaema triphyllum, Aristolochia serpentaria, Asarum canadense, Brassica napus, B. nigra, Capsicum annuum, Chamaecrista fasciculata, C. nictitans, Cimicifuga racemosa, Cornus alternifolia, C. florida, Cunila marina, Daphne mezereum, Erythronium americanum, Eupatorium perfoliatum, Gentianella quinquefolia, Hedeoma pulegioides, Hydrangea arborescens, Hydrastis canadensis, Lactuca canadensis, Leonurus cardiaca, Leucothoe axillaris, Liatris spicata, Liriodendron tulipifera, Melissa officinalis, Menispermum canadense, Mentha ×piperita, M. spicata, Nepeta cataria, Panax trifolius, Papaver somniferum, Pinus glabra, P. virginiana, Piper nigrum, Populus balsamifera, P. nigra, Rubus allegheniensis, R. argutus, R. flagellaris, R. trivialis, Salvia lyrata, S. officinalis, Senna hebecarpa, S. marilandica, Silphium compositum, Sinapis alba, Solidago odora, Tephrosia virginiana, Veratrum viride

Strengthener: Hypericum gentianoides, H. hypericoides, H. perforatum, Juncus effusus, J. tenuis, Rheum rhaponticum

Throat Aid: Acorus calamus, Actaea pachypoda, Allium cernuum, Anaphalis margaritacea, Angelica atropurpurea, Arisaema triphyllum, Aristolochia serpentaria, Armoracia rusticana, Bidens bipinnata, Cardamine diphylla, Cornus alternifolia, C. florida, C. foemina, Diospyros virginiana, Eupatorium perfoliatum, Galium circaezans, Glycyrrhiza glabra, Gnaphalium obtusifolium, Hamamelis virginiana, Lachnanthes caroliana, Lobelia inflata, Malus pumila, M. sylvestris, Marrubium vulgare, Oxalis corniculata, O. violacea, Prunus cerasus, P. pensylvanica, P. serotina, P. virginiana, Quercus alba, Q. falcata, Q. imbricaria, Q. rubra, Q. stellata, Q. velutina, Ranunculus abortivus, R. acris, R. recurvatus, Rubus allegheniensis, R. argutus, R. flagellaris, R. trivialis, Rumex crispus, R. patientia, Salix alba, S. babylonica, S. humilis, S. nigra, Solanum carolinense, Ulmus rubra, Xanthium spinosum, Xanthorhiza simplicissima

Tonic: Ageratina altissima, Aletris farinosa, Amelanchier arborea, Anthemis cotula, Aralia racemosa, A. spinosa, Aristolochia serpentaria, Armoracia rusticana, Brassica napus, B. nigra, Cichorium intybus, Cimicifuga racemosa, Cinchona calisaya, Cornus alternifolia, C. florida, Cunila marina, Euonymus americana, Eupatorium maculatum, E. perfoliatum, E. pilosum, E. purpureum, Foeniculum vulgare, Frasera caroliniensis, Gentianella quinquefolia, Hydrastis canadensis, Lindera benzoin, Melissa officinalis, Nepeta cataria, Orbexilum pedunculatum, Osmunda cinnamomea, Panax quinquefolius, Polygonatum biflorum, Pteridium aquilinum, Quercus alba, Q. falcata, Q. imbricaria, Q. rubra, Q. stellata, Q. velutina, Rubus allegheniensis, R. argutus, R. flagellaris, R. idaeus, R. occidentalis, R. odoratus, R. trivialis, Salix alba, S. babylonica, S. humilis, S. nigra, Sinapis alba, Smilax laurifolia, Solidago odora, Tanacetum vulgare, Verbena hastata, Veronicastrum virginicum, Viburnum acerifolium, V. nudum, V. prunifolium, Viola bicolor, V. cucullata, V. pedata, V. pubescens, V. rotundifolia, V. sororia, Vitis aestivalis, V. labrusca, V. vulpina, Xanthorhiza simplicissima

Toothache Remedy: Actaea pachypoda, Alnus serrulata, Aralia spinosa, Aristolochia serpentaria, Baptisia australis, B. tinctoria, Caulophyllum thalictroides, Ceanothus americanus, Chamaesyce maculata, Coix lacryma-jobi, Diospyros virginiana, Dryopteris marginalis, Erigenia bulbosa, Eryngium yuccifolium, Euphorbia corollata, Fragaria virginiana, Goodyera pubescens, G. repens, Hedeoma pulegioides, Juglans cinerea, J. nigra, Magnolia acuminata, M. macrophylla, Nicotiana rustica, N. tabacum, Ostrya virginiana, Polystichum acrostichoides, Populus balsamifera, P. nigra, Porteranthus stipulatus, P. trifoliatus, Robinia hispida, R. pseudoacacia, Rubus idaeus, R. occidentalis, R. odoratus, Taraxacum officinale, Vernonia glauca, V. noveboracensis

Tuberculosis Remedy: Aletris farinosa, Arisaema triphyllum, Chimaphila maculata, Cimicifuga racemosa, Hamamelis virginiana, Inula helenium, Ipomoea pandurata, Mertensia virginica, Panax trifolius, Pinus glabra, P. virginiana, Solidago odora, Tilia americana, Ulmus rubra

Unspecified: Coix lacryma-jobi, Phytolacca americana, Taraxacum officinale

Urinary Aid: Abies fraseri, Acalypha virginica, Achillea millefolium, Acorus calamus, Ageratina altissima, Aletris farinosa, Allium cernuum, Alnus serrulata, Ampelopsis cordata, Arctium lappa, Arctostaphylos uva-ursi, Aristolochia macrophylla, Armoracia rusticana, Aruncus dioicus, Asclepias perennis, A. quadrifolia, A. syriaca, Betula lenta, B. nigra, Calycanthus floridus, Carpinus caroliniana, Chamaesyce hypericifolia, C. maculata, Chimaphila maculata, Citrullus lanatus, Cucurbita pepo, Cynoglossum virginianum, Echium vulgare, Elytrigia repens, Euonymus americana, Eupatorium maculatum, E. pilosum, E. purpureum, Euphorbia corollata, Fragaria virginiana, Houstonia caerulea, Humulus lupulus, Ilex cassine, Ipomoea pandurata, Iris cristata, I. verna, I. virginica, Jeffersonia diphylla, Linum usitatissimum, Liparis loeselii, Lysimachia quadrifolia, Mentha ×piperita, M. spicata, Nyssa sylvatica, Petroselinum crispum, Pinus glabra, P. virginiana, Plantago aristata, P. lanceolata, P. major, Platanus occidentalis, Polygonum aviculare, P. hydropiper, P. persicaria, Prunus americana, Quercus alba, Q. falcata, Q. imbricaria, Q. rubra, Q. stellata, Q. velutina, Rhus copallinum, R. glabra, R. hirta, Rubus allegheniensis, R. argutus, R. flagellaris, R. trivialis, Spiranthes lucida, Toxicodendron vernix, Vitis aestivalis, V. labrusca, V. vulpina

Venereal Aid: Abies fraseri, Aralia spinosa, Arctium lappa, Asclepias perennis, A. quadrifolia, A. syriaca, Chamaesyce maculata, Daphne mezereum, Diospyros virginiana, Euonymus americana, Euphorbia corollata, Hypericum gentianoides, H. hypericoides, H. perforatum, Lachnanthes caroliana, Lobelia cardinalis, L. siphilitica, Menispermum canadense, Panax trifolius, Pinus glabra, P. virginiana, Populus balsamifera, P. nigra, Rubus allegheniensis, R. argutus, R. flagellaris, R. trivialis, Rudbeckia fulgida, R. hirta, Sassafras albidum, Stillingia sylvatica, Symphytum officinale, Toxicodendron vernix

Vertigo Medicine: Nicotiana rustica, N. tabacum

Veterinary Aid: Apocynum androsaemifolium, Asclepias perennis, A. quadrifolia, A. syriaca, Chimaphila maculata, Collinsonia canadensis, Lactuca canadensis, Lepidium virginicum, Leucothoe axillaris, Lycopus virginicus, Mitchella repens, Pedicularis canadensis, Porteranthus stipulatus, P. trifoliatus, Robinia hispida, R. pseudoacacia, Rumex crispus, R. patientia, Solanum carolinense

Food

Beverage: Gaultheria procumbens, Gleditsia triacanthos, Hamamelis virginiana, Hydrangea arborescens, Lindera benzoin, Liquidambar styraciflua, Morus rubra, Passiflora incarnata, Phytolacca americana, Prunus americana, Robinia pseudoacacia, Rubus allegheniensis, R. argutus, R. flagellaris, R. trivialis, Sambucus canadensis, Sassafras albidum, Vitis aestivalis, V. cinerea, V. labrusca, V. rotundifolia

Bread & Cake: *Amphicarpaea bracteata, Castanea dentata, Coix lacryma-jobi, Gaylussacia baccata, Morus rubra, Phaseolus lunatus, P. vulgaris, Polygonatum biflorum, Vitis aestivalis, V. cinerea, V. labrusca, V. rotundifolia*

Candy: *Liquidambar styraciflua*

Cooking Agent: *Carya* sp., *Phytolacca americana, Rubus* sp.

Dried Food: *Carya* sp., *Juglans nigra, Ligusticum canadense, Malus angustifolia, Phytolacca americana, Rudbeckia laciniata*

Fodder: *Aplectrum hyemale*

Frozen Food: *Gaylussacia baccata, Rudbeckia laciniata*

Fruit: *Amelanchier arborea, A. laevis, Arctostaphylos uva-ursi, Asimina triloba, Crataegus macrosperma, Diospyros virginiana, Fragaria virginiana, Gaultheria procumbens, Gaylussacia baccata, Malus coronaria, Mitchella repens, Morus alba, M. rubra, Passiflora incarnata, Physalis heterophylla, P.* sp.*, Podophyllum peltatum, Prunus americana, P. cerasus, P. pensylvanica, P. persica, P. serotina, P. virginiana, Pyrus communis, Rhus copallinum, R. glabra, R. hirta, Rubus allegheniensis, R. argutus, R. flagellaris, R. idaeus, R. occidentalis, R. odoratus, R. trivialis, Sambucus canadensis, S. nigra, Vitis aestivalis, V. cinerea, V. labrusca, V. rotundifolia, V. vulpina*

Pie & Pudding: *Diospyros virginiana, Gaylussacia baccata, Prunus pensylvanica, Rubus occidentalis, R. odoratus, R.* sp., *Sambucus canadensis*

Porridge: *Juglans nigra*

Preserves: *Fragaria virginiana, Gaylussacia baccata, G. ursina, Malus angustifolia, Morus rubra, Prunus americana, P. pensylvanica, Rubus occidentalis, R. odoratus, R.* sp., *Sambucus canadensis, S. nigra*

Sauce & Relish: *Armoracia rusticana, Fragaria virginiana, Liriodendron tulipifera, Oxydendrum arboreum*

Snack Food: *Chimaphila maculata*

Soup: *Carya* sp., *Phaseolus lunatus, P. vulgaris*

Spice: *Arnoglossum atriplicifolium, Capsella bursa-pastoris, Carya* sp., *Chenopodium album, Eupatorium purpureum, Lindera benzoin, Mentha ×piperita, M. spicata, Piper nigrum, Polygonatum biflorum*

Starvation Food: *Lilium canadense*

Substitution Food: *Apios americana, Castanea dentata*

Sweetener: *Acer saccharum*

Unspecified: *Allium cernuum, A. tricoccum, Amphicarpaea bracteata, Barbarea verna, B. vulgaris, Beta vulgaris, Carya alba, C. laciniosa, C. pallida, Castanea dentata, Cercis canadensis, Chelone glabra, Chenopodium album, Citrullus lanatus, Corylus americana, Cucurbita pepo, Dentaria* sp., *Gaultheria procumbens, Gaylussacia baccata, Gleditsia triacanthos, Hydrangea arborescens, Ipomoea pandurata, Juglans cinerea, J. nigra, Lagenaria siceraria, Lepidium campestre, L. virginicum, Ligusticum canadense, Mentha arvensis, M. ×piperita, M. spicata, Monarda didyma, M. fistulosa, Oenothera* sp., *Oxalis corniculata, O. violacea, Oxypolis rigidior, Passiflora incarnata, Phacelia dubia, Phytolacca americana, Polygonatum biflorum, Polygonum hydropiper, Polystichum acrostichoides, Pycnanthemum flexuosum, P. incanum, Ranunculus* sp., *Rheum rhaponticum, Rhododendron calendulaceum, Rubus* sp., *Rudbeckia laciniata, Rumex acetosella, Saxifraga micranthidifolia, Smilax glauca, S. herbacea, S. pseudochina, S. rotundifolia, Streptopus roseus, Tradescantia virginiana, Tragopogon* sp., *Uvularia perfoliata*

Vegetable: *Allium canadense, A. tricoccum, Amphicarpaea bracteata, Apios americana, Asparagus officinalis, Barbarea verna, B. vulgaris, Brassica napus, B. oleracea, B. rapa, Capsella bursa-pastoris, Cardamine diphylla, Chelone* sp., *Chenopodium album, Cucurbita pepo, Helianthus tuberosus, Hydrangea arborescens, Ipomoea batatas, I. pandurata, Lactuca canadensis, Lepidium* sp., *Ligusticum canadense, Oenothera biennis, O. fruticosa, Oxalis stricta, Passiflora incarnata, Pedicularis canadensis, Penthorum sedoides, Phacelia dubia, Phaseolus lunatus, P. vulgaris, Phytolacca americana, Pisum sativum, Plantago major, Polygonatum biflorum, Polygonum cuspidatum, Prenanthes serpentaria, P. trifoliolata, Prunella vulgaris, Ranunculus abortivus, R. acris, R. recurvatus, Rorippa nasturtium-aquaticum, Rubus* sp., *Rudbeckia laciniata, Rumex acetosa, R. acetosella, R. crispus, Saxifraga micranthidifolia, S. pensylvanica, Sisymbrium officinale, Sisyrinchium angustifolium, Solanum nigrum, S. tuberosum, Streptopus amplexifolius, S. roseus, Taraxacum officinale, Thlaspi arvense, Tradescantia* sp., *T. virginiana, Uvularia sessilifolia, Valerianella locusta, Viola blanda, V. pubescens, V.* sp., *Zea mays*

Winter Use Food: *Ligusticum canadense, Morus rubra, Polygonatum biflorum, Ribes cynosbati, R. rotundifolium, Rubus occidentalis, R. odoratus, R.* sp., *Rudbeckia laciniata, Streptopus roseus*

Fiber

Basketry: *Acer rubrum, A. saccharinum, Arundinaria gigantea, Carya alba, C. laciniosa, C. pallida, Lonicera japonica, Picea rubens, Quercus alba, Q. falcata, Q. imbricaria, Q. rubra, Q. stellata, Q. velutina, Tsuga canadensis, T. caroliniana*

Building Material: *Acer rubrum, A. saccharinum, A. saccharum, Aesculus flava, Arundinaria gigantea, Betula alleghaniensis, Castanea dentata, Cladrastis kentukea, Fagus grandifolia, Fraxinus pennsylvanica, Gleditsia triacanthos, Halesia carolina, Juglans cinerea, Juniperus virginiana, Liriodendron tulipifera, Magnolia acuminata, M. macrophylla, Picea rubens, Pinus echinata, P. pungens, P. rigida, P. strobus, P. taeda, Platanus occidentalis, Prunus cerasus, P. pensylvanica, P. serotina, P. virginiana, Quercus alba, Q. falcata, Q. imbricaria, Q. prinus, Q. rubra, Q. stellata, Q. velutina, Robinia hispida, R. pseudoacacia, Tilia americana, Tsuga canadensis, T. caroliniana*

Canoe Material: *Liriodendron tulipifera, Pinus echinata, P. pungens, P. rigida, P. strobus, P. taeda*

Clothing: *Apocynum cannabinum, Asclepias tuberosa, Panicum* sp.

Cordage: *Apocynum cannabinum, Asimina triloba, Juncus effusus, J. tenuis, Tilia americana*

Furniture: *Acer rubrum, A. saccharinum, A. saccharum, Aesculus flava, Carya alba, C. laciniosa, C. pallida, Juglans nigra, Juniperus virginiana, Liriodendron tulipifera, Magnolia acuminata, M. macrophylla, Prunus cerasus, P. pensylvanica, P. serotina, P. virginiana, Quercus alba, Q. falcata, Q. imbricaria, Q. rubra, Q. stellata, Q. velutina, Sassafras albidum, Tilia americana*

Snow Gear: *Oxydendrum arboreum*

Sporting Equipment: *Crataegus spathulata, Fraxinus pennsylvanica, Ulmus rubra*

Dye

Black: *Juglans cinerea, Rhus copallinum, R. glabra, R. hirta*

Blue: *Baptisia australis, B. tinctoria*

Brown: *Castanea dentata, Juglans cinerea, J. nigra, Quercus prinus*

Green: *Juglans nigra*

Red: *Coreopsis* sp., *C. tinctoria, Rhus copallinum, R. glabra, R. hirta, Sanguinaria canadensis, Sherardia arvensis*

Red-Brown: *Tsuga canadensis, T. caroliniana*

Unspecified: *Hydrastis canadensis, Ilex opaca*

Yellow: *Andropogon virginicus, Malus pumila, Tagetes erecta, Xanthorhiza simplicissima*

Other

Cash Crop: *Panax quinquefolius*

Ceremonial Items: *Aesculus flava, Lagenaria siceraria*

Containers: *Aesculus flava, Carya alba, C. laciniosa, C. pallida*

Cooking Tools: *Carya alba, C. laciniosa, C. pallida, C.* sp., *Eupatorium maculatum, E. purpureum, Fraxinus pennsylvanica, Ilex opaca, Lagenaria siceraria, Oxydendrum arboreum, Quercus alba, Q. falcata, Q. imbricaria, Q. rubra, Q. stellata, Q. velutina, Rhododendron maximum*

Decorations: *Acer rubrum, A. saccharinum, A. saccharum, Aesculus flava, Cladrastis kentukea, Cornus alternifolia, C. florida, Ilex opaca, Juglans nigra, Juniperus virginiana, Kalmia latifolia, Oxydendrum arboreum, Pinus echinata, P. pungens, P. rigida, P. strobus, P. taeda, Prunus cerasus, P. pensylvanica, P. serotina, P. virginiana, Rhododendron calendulaceum, R. maximum, Tilia americana*

Fasteners: *Fagus grandifolia, Platanus occidentalis, Quercus palustris, Robinia*

hispida, R. pseudoacacia
Fertilizer: *Sassafras albidum*
Fuel: *Acer pensylvanicum, Arundinaria gigantea, Carya alba, C. laciniosa, C. pallida, Castanea dentata, Fraxinus pennsylvanica, Oxydendrum arboreum, Quercus alba, Q. falcata, Q. imbricaria, Q. rubra, Q. stellata, Q. velutina, Tsuga canadensis, T. caroliniana*
Good Luck Charm: *Aesculus pavia*
Hunting & Fishing Item: *Aesculus* sp., *Arundinaria gigantea, Asclepias perennis, A. quadrifolia, A. syriaca, Carya alba, C. laciniosa, C. pallida, Cirsium altissimum, C. vulgare, Dionaea muscipula, Juglans nigra, Oxydendrum arboreum, Platanthera ciliaris, Quercus* sp., *Robinia hispida, R. pseudoacacia, Urtica dioica, Yucca filamentosa*
Incense & Fragrance: *Calycanthus floridus, Pinus glabra, P. virginiana, Sassafras albidum*
Insecticide: *Cicuta maculata, Hedeoma pulegioides, Ipomoea pandurata, Juniperus virginiana, Lobelia inflata, Melia azedarach, Mirabilis nyctaginea, Podophyllum peltatum, Solanum carolinense, Viola bicolor, V. cucullata, V. pedata, V. pubescens, V. rotundifolia, V. sororia*
Jewelry: *Coix lacryma-jobi, Echium vulgare*
Lighting: *Arundinaria gigantea*
Musical Instrument: *Arundinaria gigantea*
Paper: *Aesculus flava, Liriodendron tulipifera, Magnolia acuminata, M. macrophylla, Tilia americana, Tsuga canadensis, T. caroliniana*
Preservative: *Tsuga canadensis, T. caroliniana*
Protection: *Crataegus spathulata, Gleditsia triacanthos*
Smoke Plant: *Anaphalis margaritacea, Cardamine diphylla, Gaultheria procumbens, Nicotiana rustica, N. tabacum*
Smoking Tools: *Oxydendrum arboreum, Rhododendron maximum*
Snuff: *Asarum canadense, A.* sp.
Soap: *Saponaria officinalis, Yucca filamentosa*
Tools: *Carya alba, C. laciniosa, C. pallida, Cornus alternifolia, C. florida, Fraxinus pennsylvanica*
Toys & Games: *Zea mays*
Weapon: *Arundinaria gigantea*

Cheyenne
Drug
Adjuvant: *Abies* sp., *Osmorhiza berteroi, Sphaeralcea coccinea*
Analgesic: *Achillea millefolium, Acorus calamus, Agastache foeniculum, Ambrosia psilostachya, Arctostaphylos uva-ursi, Artemisia ludoviciana, Balsamorhiza incana, B. sagittata, Capsella bursa-pastoris, Echinacea angustifolia, Erigeron peregrinus, Lithospermum ruderale, Lomatium dissectum, L. orientale, Oplopanax horridus, Ratibida columnifera, Senecio triangularis, Thermopsis rhombifolia*
Antidiarrheal: *Ambrosia psilostachya, Frasera speciosa, Glycyrrhiza lepidota, Lomatium orientale, Pediomelum esculentum, Prunus virginiana, Rosa* sp., *Salix amygdaloides*
Antiemetic: *Achillea millefolium, Mentha arvensis, M. canadensis, Vaccinium scoparium*
Antihemorrhagic: *Ambrosia psilostachya, Boykinia jamesii, Epilobium angustifolium, Pterospora andromedea, Rumex crispus*
Antirheumatic (External): *Heuchera cylindrica, Lithospermum ruderale, Lomatium* sp., *Onosmodium molle*
Antirheumatic (Internal): *Cucurbita* sp., *Echinacea pallida, Heuchera cylindrica, Mentzelia laevicaulis*
Blood Medicine: *Actaea rubra, Sarcobatus vermiculatus*
Breast Treatment: *Lygodesmia juncea, Mertensia ciliata*
Burn Dressing: *Echinacea pallida, Pediomelum esculentum, Rhus trilobata*
Ceremonial Medicine: *Abies lasiocarpa, A.* sp., *Acer negundo, Acorus calamus, Actaea rubra, Anaphalis margaritacea, Artemisia frigida, A. ludoviciana, Carex nebrascensis, Glycyrrhiza lepidota, Helianthus* sp., *Hierochloe odorata, Juniperus communis, J. horizontalis, J. scopulorum, Koeleria macrantha, Madia glomerata, Matricaria discoidea, Mentha arvensis, Monarda fistulosa, Pentaphylloides floribunda, Prunus americana, Psoralidium lanceolatum, Salix amygdaloides, Sarcobatus vermiculatus, Scirpus nevadensis, Sphaeralcea coccinea, Typha latifolia*
Cold Remedy: *Achillea millefolium, Acorus calamus, Agastache foeniculum, Ambrosia psilostachya, Arabis glabra, Arctostaphylos uva-ursi, Balsamorhiza incana, B. sagittata, Chrysothamnus nauseosus, Echinacea pallida, Juniperus communis, J. horizontalis, J. scopulorum, Osmorhiza berteroi, Rhus trilobata, Stephanomeria spinosa, Thermopsis rhombifolia*
Cough Medicine: *Achillea millefolium, Arctostaphylos uva-ursi, Chrysothamnus nauseosus, Juniperus communis, J. horizontalis, J. scopulorum, Pedicularis groenlandica*
Dermatological Aid: *Actaea rubra, Allium brevistylum, Astragalus adsurgens, Chrysothamnus nauseosus, Echinacea pallida, Grindelia squarrosa, Heuchera cylindrica, Koeleria macrantha, Lomatium* sp., *Matricaria discoidea, Mentha arvensis, Mertensia ciliata, Monarda fistulosa, Nicotiana* sp., *Onosmodium molle, Pinus ponderosa, Pterospora andromedea, Ratibida columnifera, Rumex crispus, Salix amygdaloides, S.* sp., *Yucca glauca*
Diaphoretic: *Achillea millefolium, Agastache foeniculum, Stephanomeria spinosa*
Dietary Aid: *Actaea rubra, Amelanchier alnifolia, Echinacea pallida, Mentzelia laevicaulis, Prunus virginiana, Vaccinium scoparium*
Disinfectant: *Anaphalis margaritacea, Heterotheca villosa, Pterospora andromedea*
Diuretic: *Acorus calamus, Citrullus lanatus, Cucurbita* sp., *Rhus trilobata*
Ear Medicine: *Aster cusickii, Cucurbita* sp., *Mentzelia laevicaulis*
Eye Medicine: *Asclepias speciosa, Grindelia squarrosa, Rosa* sp.
Febrifuge: *Achillea millefolium, Agastache foeniculum, Balsamorhiza sagittata, Cucurbita* sp., *Echinacea pallida, Juniperus communis, J. horizontalis, J. scopulorum, Mentzelia laevicaulis, Pediomelum argophyllum*
Gastrointestinal Aid: *Acorus calamus, Ambrosia psilostachya, Balsamorhiza incana, B. sagittata, Epilobium angustifolium, Glycyrrhiza lepidota, Lomatium dissectum, L. orientale, Osmorhiza longistylis, Parnassia fimbriata, Rosa* sp., *Salix amygdaloides, Typha latifolia*
Gynecological Aid: *Actaea rubra, Artemisia frigida, Balsamorhiza sagittata, Eriogonum umbellatum, Juniperus communis, J. horizontalis, J. scopulorum, Lygodesmia juncea, Mertensia ciliata, Oxytropis* sp., *Ulmus americana*
Heart Medicine: *Achillea millefolium, Agastache foeniculum, Cucurbita* sp., *Mentha arvensis*
Hemostat: *Achillea millefolium, Artemisia frigida, A. ludoviciana, Geranium richardsonii, Pterospora andromedea, Rhus trilobata, Salix amygdaloides*
Herbal Steam: *Agastache foeniculum, Juniperus communis, J. horizontalis, J. scopulorum, Madia glomerata, Stephanomeria spinosa*
Kidney Aid: *Cucurbita* sp., *Osmorhiza longistylis*
Laxative: *Acorus calamus, Ambrosia psilostachya, Cucurbita* sp.
Love Medicine: *Juniperus communis, J. horizontalis, J. scopulorum, Madia glomerata, Mentha arvensis*
Misc. Disease Remedy: *Chrysothamnus nauseosus, Echinacea pallida, Lygodesmia juncea, Mentzelia laevicaulis, Mertensia ciliata, Stephanomeria spinosa*
Nose Medicine: *Artemisia ludoviciana, Geranium richardsonii, Pterospora andromedea*
Oral Aid: *Balsamorhiza sagittata, Echinacea angustifolia, E. pallida, Prunus americana*
Orthopedic Aid: *Arctostaphylos uva-ursi, Echinacea angustifolia, Erigeron peregrinus, Heuchera cylindrica, Lithospermum incisum, Onosmodium molle*
Other: *Erigeron peregrinus, Pentaphylloides floribunda, Tanacetum vulgare*
Panacea: *Acorus calamus, Arabis glabra, Balsamorhiza sagittata, Salix amygdaloides, Stephanomeria spinosa*

Pediatric Aid: *Acorus calamus, Amelanchier alnifolia, Arabis glabra, Lomatium orientale, Lygodesmia juncea, Oxytropis* sp., *Parnassia fimbriata, Prunus virginiana, Ulmus americana, Vaccinium scoparium*
Poison: *Astragalus* sp., *Pentaphylloides floribunda, Pulsatilla patens*
Psychological Aid: *Arctostaphylos uva-ursi, Artemisia ludoviciana, Chrysothamnus nauseosus, Lithospermum incisum, Lygodesmia juncea, Madia glomerata*
Pulmonary Aid: *Agastache foeniculum, Boykinia jamesii, Juniperus scopulorum, Pterospora andromedea, Rumex crispus*
Reproductive Aid: *Rhus trilobata*
Respiratory Aid: *Achillea millefolium, Artemisia ludoviciana*
Sedative: *Heterotheca villosa, Juniperus communis, J. horizontalis, J. scopulorum, Lithospermum incisum, Senecio triangularis*
Snakebite Remedy: *Ratibida columnifera*
Stimulant: *Abies lasiocarpa, Erigeron peregrinus, Koeleria macrantha, Lithospermum incisum, Mentha arvensis, Osmorhiza berteroi, Parnassia fimbriata, Phlox multiflora, Pulsatilla patens*
Strengthener: *Tanacetum vulgare*
Throat Aid: *Achillea millefolium, Balsamorhiza sagittata, Echinacea angustifolia, E. pallida, Juniperus communis, J. horizontalis, J. scopulorum*
Tonic: *Lomatium dissectum*
Toothache Remedy: *Balsamorhiza sagittata, Echinacea angustifolia, E. pallida, Rhus trilobata*
Tuberculosis Remedy: *Achillea millefolium, Chrysothamnus nauseosus, Cucurbita* sp., *Stephanomeria spinosa*
Unspecified: *Agastache foeniculum, Amelanchier alnifolia, Anaphalis margaritacea, Arctostaphylos uva-ursi, Artemisia ludoviciana, Calochortus gunnisonii, Mahonia repens, Matricaria discoidea, Mentzelia laevicaulis, Pediomelum esculentum, Perideridia gairdneri, Prunus virginiana, Pulsatilla patens, Rosa* sp., *Sagittaria cuneata, Shepherdia argentea*
Venereal Aid: *Cucurbita* sp., *Madia glomerata*
Vertigo Medicine: *Tanacetum vulgare*
Veterinary Aid: *Anaphalis margaritacea, Astragalus* sp., *Calochortus gunnisonii, Equisetum arvense, E. hyemale, Monarda fistulosa, Rhus trilobata, Sagittaria cuneata, Sarcobatus vermiculatus, Thalictrum sparsiflorum*
Witchcraft Medicine: *Abies lasiocarpa, Acorus calamus, Hierochloe odorata*

Food
Beverage: *Agastache foeniculum, Amelanchier alnifolia, Arabis glabra, Mentha arvensis, Ulmus americana*
Bread & Cake: *Prunus virginiana*
Candy: *Acer negundo, Asclepias speciosa, Pinus ponderosa*
Cooking Agent: *Opuntia polyacantha, Pediomelum esculentum*
Dried Food: *Calochortus gunnisonii, Crataegus douglasii, Escobaria vivipara, Opuntia polyacantha, Pediomelum esculentum, P. hypogaeum, Perideridia gairdneri, Ribes cereum, R. lacustre*
Fodder: *Populus balsamifera, P. deltoides*
Fruit: *Asclepias speciosa, Crataegus douglasii, Escobaria vivipara, Fragaria virginiana, Mahonia repens, Opuntia polyacantha, Physalis heterophylla, Prunus virginiana, Ribes lacustre, R. oxyacanthoides, Rosa woodsii, Rubus idaeus, Vitis vulpina*
Pie & Pudding: *Amelanchier alnifolia, Pediomelum esculentum, Prunus americana, P. virginiana, Rosa* sp., *Shepherdia argentea*
Porridge: *Calochortus gunnisonii, Perideridia gairdneri*
Preserves: *Shepherdia canadensis*
Sauce & Relish: *Asclepias speciosa*
Soup: *Asclepias speciosa, Opuntia polyacantha*
Special Food: *Amelanchier alnifolia, Cirsium edule, Prunus americana*
Spice: *Allium drummondii, A. schoenoprasum*
Starvation Food: *Ipomoea leptophylla*
Unspecified: *Allium drummondii, A. schoenoprasum, Apios tuberosum, Asclepias speciosa, Calochortus gunnisonii, Castilleja sessiliflora, Cirsium edule, Glycyrrhiza lepidota, Helianthus tuberosus, Nuphar lutea, Pediomelum esculentum, P. hypogaeum, Perideridia gairdneri, Pinus ponderosa, Polygonum bistortoides, Populus deltoides, Quercus macrocarpa, Rosa* sp., *Rumex crispus, Sagittaria* sp., *Scirpus acutus, S. nevadensis*
Vegetable: *Apios tuberosum*
Winter Use Food: *Amelanchier alnifolia, Calochortus gunnisonii, Crataegus douglasii, Opuntia polyacantha, Pediomelum hypogaeum, Perideridia gairdneri, Prunus americana, P. virginiana, Ribes aureum, R. oxyacanthoides*

Fiber
Basketry: *Eleocharis* sp., *Juncus balticus, Typha latifolia*
Brushes & Brooms: *Koeleria macrantha*
Building Material: *Fraxinus pennsylvanica, Pinus contorta, Populus balsamifera, P. deltoides, P. tremuloides, Salix amygdaloides*
Furniture: *Salix amygdaloides*
Mats, Rugs & Bedding: *Monarda fistulosa, Salix amygdaloides, Scirpus nevadensis*
Other: *Eleocharis* sp., *Juncus balticus*

Dye
Black: *Leymus cinereus*
Blue: *Pinus ponderosa*
Green: *Populus deltoides*
Purple: *Populus deltoides*
Red: *Populus deltoides, Rumex venosus*
Unspecified: *Leymus cinereus, Pinus ponderosa, Populus deltoides, Rumex crispus, R. venosus*
White: *Populus deltoides*
Yellow: *Letharia vulpina, Rumex crispus, R. venosus*

Other
Ceremonial Items: *Abies lasiocarpa, Acer negundo, Anaphalis margaritacea, Artemisia ludoviciana, Cornus sericea, Hierochloe odorata, Leymus cinereus, Psoralidium lanceolatum, Sarcobatus vermiculatus, Symphoricarpos albus*
Cooking Tools: *Acer negundo, Artemisia* sp., *Eleocharis* sp., *Fraxinus pennsylvanica, Juniperus scopulorum, Salix amygdaloides*
Decorations: *Juncus balticus*
Fuel: *Acer negundo*
Hunting & Fishing Item: *Cornus sericea, Fraxinus pennsylvanica, Juniperus scopulorum, Prunus virginiana, Ribes* sp., *Rosa* sp., *Sarcobatus vermiculatus*
Incense & Fragrance: *Hierochloe odorata, Juniperus scopulorum, Matricaria discoidea, Mentha arvensis, Monarda fistulosa*
Musical Instrument: *Heracleum maximum, Juniperus scopulorum, Pinus ponderosa, Salix amygdaloides*
Paint: *Hierochloe odorata, Populus balsamifera, P. deltoides, Salix amygdaloides*
Protection: *Acorus calamus, Anaphalis margaritacea, Eleocharis* sp., *Hierochloe odorata, Juniperus scopulorum, Pentaphylloides floribunda*
Smoke Plant: *Acorus calamus, Arctostaphylos uva-ursi, Cornus sericea, Rhus glabra, R. trilobata*
Smoking Tools: *Fraxinus pennsylvanica, Sarcobatus vermiculatus*
Tools: *Artemisia ludoviciana, Salix amygdaloides*
Toys & Games: *Eleocharis* sp.

Cheyenne, Northern
Drug
Disinfectant: *Grindelia squarrosa*
Eye Medicine: *Grindelia squarrosa*

Chickasaw
Drug
Abortifacient: *Verbesina virginica*
Analgesic: *Robinia* sp., *Salix* sp., *Sambucus canadensis*
Antidiarrheal: *Populus* sp., *Salix* sp.
Dermatological Aid: *Heuchera americana*

Diaphoretic: *Botrychium virginianum*
Diuretic: *Verbesina virginica*
Emetic: *Aureolaria pedicularia, Botrychium virginianum, Chaerophyllum procumbens*
Expectorant: *Botrychium virginianum*
Eye Medicine: *Cephalanthus occidentalis, Hedeoma pulegioides*
Gynecological Aid: *Verbesina virginica*
Hemostat: *Salix* sp.
Misc. Disease Remedy: *Aureolaria pedicularia*
Poison: *Chaerophyllum procumbens*
Psychological Aid: *Gaylussacia* sp.
Sedative: *Gaylussacia* sp.
Stimulant: *Verbesina virginica*
Tonic: *Heuchera americana*
Toothache Remedy: *Ageratina altissima*
Urinary Aid: *Citrullus lanatus*
Venereal Aid: *Verbesina virginica*

Chinook
Fiber
Clothing: *Sphagnum* sp.
Mats, Rugs & Bedding: *Sphagnum* sp.

Chinook, Lower
Food
Dried Food: *Arctostaphylos uva-ursi, Vaccinium ovalifolium*
Fruit: *Arctostaphylos uva-ursi, Fragaria* sp., *Malus fusca, Rubus spectabilis, Vaccinium ovalifolium*
Unspecified: *Equisetum arvense, Rubus spectabilis*

Chippewa
Drug
Abortifacient: *Aralia nudicaulis, A. racemosa, Artemisia dracunculus, Eupatorium perfoliatum, Hepatica nobilis, Pycnanthemum virginianum, Ribes triste, Rubus frondosus, Sanicula canadensis, Silphium perfoliatum, Tanacetum vulgare*
Adjuvant: *Asarum canadense*
Analgesic: *Abies balsamea, Achillea millefolium, Agastache foeniculum, Andropogon gerardii, Apocynum androsaemifolium, A.* sp., *Arctostaphylos uva-ursi, Betula nigra, Capsella bursa-pastoris, Carya ovata, Caulophyllum thalictroides, Cirsium* sp., *Conyza canadensis, Corylus americana, Crataegus* sp., *Diervilla lonicera, Euthamia graminifolia, Heuchera americana, Maianthemum racemosum, Monarda fistulosa, Polygonatum biflorum, Polygonum persicaria, P. punctatum, Populus balsamifera, Potentilla arguta, Prunus virginiana, Pulsatilla patens, Quercus macrocarpa, Rhus hirta, Ribes glandulosum, R. oxyacanthoides, Rubus occidentalis, Sanguinaria canadensis, Silphium perfoliatum, Smilax herbacea, Solidago* sp., *Thuja occidentalis, Viburnum acerifolium*
Anthelmintic: *Monarda fistulosa, Prunus americana, P. serotina*
Anticonvulsive: *Actaea pachypoda, Apocynum androsaemifolium, Artemisia frigida, Astragalus crassicarpus, Hepatica nobilis, Lathyrus venosus, Polygala senega, Polygonum pensylvanicum, Rosa arkansana, Solidago juncea*
Antidiarrheal: *Amelanchier canadensis, Artemisia dracunculus, Betula lenta, Capsella bursa-pastoris, Comarum palustre, Cornus sericea, Geranium maculatum, Heuchera* sp., *Potentilla arguta, Rhus glabra, Rubus allegheniensis, R. idaeus, Salix* sp., *Tsuga canadensis, Urtica dioica*
Antidote: *Artemisia ludoviciana*
Antihemorrhagic: *Caulophyllum thalictroides, Corylus* sp., *Ostrya virginiana, Prunus virginiana, Silphium perfoliatum, Solidago speciosa*
Antirheumatic (External): *Abies balsamea, Anaphalis margaritacea, Eupatorium maculatum, E. perfoliatum, Juniperus virginiana, Lycopodium obscurum, Ostrya virginiana, Picea glauca, Plantago major, Taxus canadensis, Trillium grandiflorum*
Antirheumatic (Internal): *Juniperus virginiana, Taxus canadensis, Vitis vulpina*
Blood Medicine: *Alnus incana, Aralia nudicaulis, Gaultheria procumbens, Larix laricina, Pedicularis canadensis, Prunus virginiana, Sassafras albidum, Veronicastrum virginicum*
Burn Dressing: *Agastache foeniculum, Clintonia borealis, Larix laricina, Ledum groenlandicum, Monarda fistulosa, Prunus serotina, Rudbeckia laciniata, Solidago canadensis*
Cancer Treatment: *Celastrus scandens*
Carminative: *Mentha canadensis*
Cathartic: *Acorus calamus, Amphicarpaea bracteata, Betula papyrifera, Celastrus scandens, Dirca palustris, Fraxinus* sp., *Juglans cinerea, Prunella vulgaris, Prunus virginiana, Smilax herbacea, Solidago rigida, Symphoricarpos* sp., *Veronicastrum virginicum*
Ceremonial Medicine: *Comptonia peregrina, Thuja occidentalis*
Cold Remedy: *Acorus calamus, Allium stellatum, Apocynum androsaemifolium, Caltha palustris, Castilleja coccinea, Eupatorium purpureum, Gaultheria procumbens, Monarda fistulosa, Rhus glabra, Zanthoxylum americanum*
Cough Medicine: *Acorus calamus, Agastache foeniculum, Aralia racemosa, Arctium minus, Ceanothus herbaceus, Cornus alternifolia, Ostrya virginiana, Symplocarpus foetidus, Thuja occidentalis, Zanthoxylum americanum*
Dermatological Aid: *Abies balsamea, Acer saccharinum, Achillea millefolium, Anemone canadensis, Aralia nudicaulis, A. racemosa, Artemisia dracunculus, Asarum canadense, Caltha palustris, Celastrus scandens, Clintonia borealis, Cornus sericea, Cypripedium pubescens, Dirca palustris, Epilobium angustifolium, Erysimum cheiranthoides, Galium aparine, Hamamelis virginiana, Hepatica nobilis, Heracleum maximum, Impatiens capensis, Iris versicolor, Lactuca canadensis, Ledum groenlandicum, Lilium philadelphicum, Monarda fistulosa, Osmorhiza claytonii, Pinus strobus, Plantago major, Populus balsamifera, P. tremuloides, Potentilla arguta, Prunus americana, P. serotina, P.* sp., *P. virginiana, Rumex crispus, R. obtusifolius, Solidago canadensis, S. speciosa, Thuja occidentalis, Vitis* sp.
Diaphoretic: *Caltha palustris, Tanacetum vulgare*
Disinfectant: *Amelanchier canadensis, Artemisia frigida, Equisetum hyemale, Prunus americana, P. serotina, P. virginiana, Thuja occidentalis*
Diuretic: *Andropogon gerardii, Athyrium filix-femina, Caltha palustris, Celastrus scandens, Lonicera dioica, Ribes* sp., *Solidago rigida, Symphoricarpos albus, Urtica dioica*
Ear Medicine: *Apocynum androsaemifolium, Aster nemoralis, Campanula rotundifolia, Tanacetum vulgare, Trillium grandiflorum*
Emetic: *Allium tricoccum, Alnus incana, Caltha palustris, Caulophyllum thalictroides, Hamamelis virginiana, Lathyrus venosus, Physocarpus opulifolius, Rhus glabra, Sambucus canadensis, Viburnum acerifolium*
Eye Medicine: *Alnus incana, Arisaema triphyllum, Chimaphila umbellata, Cornus alternifolia, C. sericea, Diervilla lonicera, Hamamelis virginiana, Heuchera americana, Hordeum jubatum, Rosa* sp., *Rubus idaeus, R. occidentalis, Stellaria media, Streptopus roseus*
Febrifuge: *Comptonia peregrina, Nepeta cataria, Pycnanthemum virginianum, Solidago* sp., *Tanacetum vulgare*
Gastrointestinal Aid: *Andropogon gerardii, Artemisia frigida, Asarum canadense, Betula nigra, Capsella bursa-pastoris, Caulophyllum thalictroides, Ceanothus americanus, Conyza canadensis, Cypripedium pubescens, Diervilla lonicera, Geum triflorum, Heuchera americana, H.* sp., *Polygonum persicaria, P. punctatum, Prunus virginiana, Quercus macrocarpa, Rhus hirta, Rudbeckia laciniata, Sagittaria latifolia, Salix* sp., *Sanguinaria canadensis, Solidago* sp., *Stachys palustris, Thaspium barbinode, Viburnum acerifolium*
Gynecological Aid: *Actaea rubra, Alnus incana, Amelanchier canadensis, Artemisia dracunculus, Asclepias syriaca, Caltha palustris, Cirsium* sp., *Conyza canadensis, Crataegus* sp., *Eupatorium purpureum, Geum canadense, Mai-*

[Gynecological Aid] anthemum racemosum, Osmorhiza longistylis, Populus balsamifera, P. tremuloides, Prenanthes alba, Prunus sp., Ribes glandulosum, R. oxyacanthoides, Rubus allegheniensis, R. occidentalis, Sanicula canadensis, Solidago speciosa, Taraxacum officinale, Viburnum opulus

Heart Medicine: *Apocynum androsaemifolium, Artemisia dracunculus, Dalea purpurea, Polygala senega, Populus balsamifera, P. tremuloides, Quercus macrocarpa, Q. rubra*

Hemostat: *Anemone canadensis, Apocynum androsaemifolium, Aralia nudicaulis, Artemisia frigida, Astragalus crassicarpus, Lathyrus venosus, Polygala senega, Quercus sp., Rosa arkansana, Silphium perfoliatum, Tsuga canadensis, Verbena hastata*

Herbal Steam: *Abies balsamea, Achillea millefolium, Anaphalis margaritacea, Artemisia dracunculus, Carya ovata, Juniperus virginiana, Polygonatum biflorum, Taxus canadensis*

Hunting Medicine: *Acorus calamus, Arctostaphylos uva-ursi, Aster novae-angliae, A. puniceus, Cornus alternifolia, Eupatorium perfoliatum, Hepatica nobilis*

Kidney Aid: *Ostrya virginiana, Smilax herbacea*

Laxative: *Ceanothus americanus, Diervilla lonicera*

Liver Aid: *Hepatica nobilis*

Misc. Disease Remedy: *Fragaria virginiana, Polygonatum biflorum, Prunus serotina, Rubus idaeus, Thuja occidentalis, Vitis vulpina*

Oral Aid: *Geranium maculatum, Heuchera* sp., *Nymphaea odorata, Rhus glabra*

Orthopedic Aid: *Anaphalis margaritacea, Aralia racemosa, Artemisia absinthium, Asarum canadense, Castilleja coccinea, Mirabilis nyctaginea, Populus balsamifera, Silphium perfoliatum, Smilax herbacea, Solidago speciosa, Zanthoxylum americanum*

Other: *Apocynum* sp., *Taxus canadensis*

Pediatric Aid: *Acorus calamus, Actaea pachypoda, Allium stellatum, Apocynum androsaemifolium, Artemisia dracunculus, Asclepias incarnata, Celastrus scandens, Eupatorium maculatum, Fragaria virginiana, Geranium maculatum, Hepatica nobilis, Heuchera* sp., *Prunus serotina, Rhus glabra, Rudbeckia hirta, Rumex obtusifolius, Thaspium barbinode, Zanthoxylum americanum*

Poison: *Toxicodendron vernix*

Psychological Aid: *Apocynum androsaemifolium, Vaccinium angustifolium*

Pulmonary Aid: *Betula lenta, Caulophyllum thalictroides, Ceanothus americanus, Corylus* sp., *Dirca palustris, Euthamia graminifolia, Fagus grandifolia, Lonicera* sp., *Ostrya virginiana, Prunus virginiana, Pulsatilla patens, Quercus macrocarpa, Rubus frondosus, Silphium perfoliatum, Solidago speciosa, Vitis* sp., *Zanthoxylum americanum*

Respiratory Aid: *Acorus calamus, Ceanothus americanus, Juniperus communis, Rhus glabra*

Sedative: *Eupatorium maculatum, Polygonatum biflorum*

Snakebite Remedy: *Botrychium virginianum, Eupatorium perfoliatum, Lilium canadense, Plantago major*

Stimulant: *Achillea millefolium, Artemisia frigida, Astragalus crassicarpus, Fraxinus* sp., *Geum triflorum, Heliopsis helianthoides, Lathyrus venosus, Polygala senega, Rosa arkansana, Solidago speciosa*

Strengthener: *Artemisia dracunculus, Asclepias incarnata*

Throat Aid: *Acorus calamus, Heracleum maximum, Osmorhiza claytonii, Phryma leptostachya, Potentilla norvegica, Prunus virginiana, Solidago flexicaulis, Tanacetum vulgare, Ulmus rubra, Zanthoxylum americanum*

Tonic: *Artemisia frigida, Astragalus crassicarpus, Fraxinus* sp., *Gaultheria procumbens, Lathyrus venosus, Polygala senega, Rosa arkansana, Solidago speciosa*

Toothache Remedy: *Acorus calamus, Cypripedium pubescens*

Tuberculosis Remedy: *Caltha palustris, Crataegus* sp., *Iris versicolor, Prunus serotina, P. virginiana, Veronicastrum virginicum*

Unspecified: *Comptonia peregrina, Lindera benzoin, Mitchella repens, Polygala senega, Pyrola elliptica, Sagittaria cuneata, S. latifolia, Sanguinaria canadensis, Thuja occidentalis*

Urinary Aid: *Equisetum arvense, Lonicera dioica, Ribes triste, Viburnum lentago*

Venereal Aid: *Chimaphila umbellata*

Vertigo Medicine: *Apocynum androsaemifolium*

Veterinary Aid: *Achillea millefolium, Aralia nudicaulis, Geum triflorum, Liatris scariosa, Osmorhiza longistylis, Pediomelum argophyllum, Rudbeckia laciniata*

Witchcraft Medicine: *Lilium philadelphicum*

Food

Appetizer: *Asclepias syriaca*

Beverage: *Comptonia peregrina, Gaultheria hispidula, G. procumbens, Ledum groenlandicum, Lindera benzoin, Mentha canadensis, Picea rubens, Prunus serotina, P. virginiana, Rubus idaeus, Sassafras albidum, Thuja occidentalis, Tsuga canadensis*

Bread & Cake: *Crataegus* sp., *Prunus americana, P. serotina, Ribes triste, Rubus frondosus, R. idaeus*

Dried Food: *Amelanchier canadensis, Lycopus asper, Prunus virginiana, Ribes americanum, Rubus allegheniensis, R. idaeus, R. occidentalis, R. odoratus, Sagittaria latifolia, Sambucus canadensis, Vaccinium* sp.

Fruit: *Amelanchier canadensis, Amphicarpaea bracteata, Cornus canadensis, Fragaria vesca, F. virginiana, Podophyllum peltatum, Prunus americana, P. serotina, Ribes americanum, R. cynosbati, R. missouriense, R. rubrum, R. triste, Rubus allegheniensis, R. canadensis, R. frondosus, R. idaeus, R. occidentalis, R. odoratus, R. pubescens, Sambucus canadensis, Vaccinium angustifolium, V. macrocarpon, V.* sp., *Vitis vulpina*

Porridge: *Zea mays*

Preserves: *Asclepias syriaca*

Sauce & Relish: *Viburnum opulus*

Soup: *Zea mays*

Spice: *Arctostaphylos uva-ursi, Asarum canadense, Gaultheria procumbens, Lindera benzoin, Mentha canadensis, Pycnanthemum virginianum, Sassafras albidum*

Sweetener: *Acer saccharinum*

Unspecified: *Amphicarpaea bracteata, Aster* sp., *Corylus americana, Fagus grandifolia, Helianthus tuberosus, Lathyrus palustris, Parthenocissus quinquefolia, Quercus macrocarpa, Q.* sp., *Scirpus tabernaemontani, Zizania palustris*

Vegetable: *Apios americana, Asclepias syriaca, Caltha palustris, Quercus macrocarpa, Tilia americana, Zea mays*

Winter Use Food: *Corylus americana*

Fiber

Basketry: *Fraxinus nigra, Salix interior, Typha latifolia*

Building Material: *Betula papyrifera, Carpinus caroliniana, Fraxinus nigra, Ostrya virginiana, Tsuga canadensis, Typha latifolia, Ulmus americana*

Canoe Material: *Betula pubescens, Picea rubens, Thuja occidentalis, Tilia americana*

Caulking Material: *Picea rubens*

Clothing: *Arctium minus*

Cordage: *Apocynum cannabinum, Asclepias incarnata, Laportea canadensis, Tilia americana*

Mats, Rugs & Bedding: *Juncus effusus, Juniperus virginiana, Scirpus tabernaemontani, Typha latifolia*

Other: *Dicranum bonjeanii, Sphagnum* sp.

Scouring Material: *Equisetum hyemale*

Sewing Material: *Tilia americana*

Snow Gear: *Fraxinus* sp., *Thuja occidentalis*

Dye

Black: *Alnus incana, Cornus sericea, Corylus americana, Juglans cinerea, J. nigra, Quercus macrocarpa, Q.* sp.

Blue: *Fraxinus nigra*

Brown: *Juglans cinerea, J. nigra*
Mordant: *Acorus calamus, Prunus americana*
Red: *Alnus incana, Betula papyrifera, Cornus sericea, Lithospermum caroliniense, Prunus americana, Quercus* sp., *Rhus glabra, Sanguinaria canadensis*
Red-Brown: *Juniperus virginiana, Tsuga canadensis*
Yellow: *Alnus incana, Coptis trifolia, Cornus sericea, Prunus americana, Rhus glabra, Sanguinaria canadensis*
Other
Cash Crop: *Acer saccharum, Panax quinquefolius*
Ceremonial Items: *Hierochloe odorata*
Cleaning Agent: *Acer saccharinum, Quercus bicolor, Tsuga canadensis*
Containers: *Juncus effusus, Larix laricina, Tilia americana*
Cooking Tools: *Acer saccharum, Betula papyrifera, Phragmites australis*
Designs: *Betula papyrifera*
Fasteners: *Dirca palustris, Tilia americana, Zea mays*
Fuel: *Fraxinus nigra*
Good Luck Charm: *Onosmodium molle, Panax quinquefolius*
Hunting & Fishing Item: *Asclepias syriaca, Carya ovata, Fraxinus americana, Thuja occidentalis, Viburnum opulus*
Incense & Fragrance: *Asarum canadense, Comptonia peregrina, Polygonatum biflorum, Thuja occidentalis*
Malicious Charm: *Equisetum arvense*
Malicious Magic: *Smilax tamnoides*
Musical Instrument: *Corylus americana*
Paint: *Bovista pila, Lithospermum caroliniense*
Protection: *Acorus calamus, Agastache foeniculum, Apocynum androsaemifolium, Aralia racemosa, Artemisia frigida, Lathyrus venosus, Plantago major*
Smoke Plant: *Arctostaphylos uva-ursi, Cicuta maculata, Cornus rugosa, Rhus glabra, Salix* sp.
Tools: *Comptonia peregrina, Crataegus* sp., *Nyssa sylvatica, Quercus* sp., *Tilia americana*
Toys & Games: *Allium stellatum, Clintonia borealis, Pinus resinosa, Sarracenia purpurea, Scirpus tabernaemontani*
Weapon: *Cornus alternifolia*

Choctaw
Drug
Abortifacient: *Galium boreale*
Analgesic: *Aralia racemosa, Aristolochia serpentaria, Arundinaria gigantea, Gnaphalium obtusifolium, Hypericum crux-andreae, Monarda fistulosa, Prenanthes aspera, Pycnanthemum incanum, Quercus marilandica, Ulmus americana*
Anthelmintic: *Pinus echinata, P. virginiana*
Antidiarrheal: *Callicarpa americana, Cephalanthus occidentalis, Malvella leprosa, Rubus* sp.
Antidote: *Eryngium aquaticum*
Antihemorrhagic: *Ceanothus* sp.
Blood Medicine: *Arisaema triphyllum, Berchemia scandens, Sassafras albidum*
Burn Dressing: *Malvella leprosa*
Cathartic: *Monarda fistulosa*
Cold Remedy: *Gnaphalium obtusifolium, Pycnanthemum albescens*
Contraceptive: *Galium boreale*
Cough Medicine: *Echinacea purpurea, Tephrosia hispidula*
Dermatological Aid: *Aralia spinosa, Baptisia alba, Chionanthus virginicus, Galium uniflorum, Geranium maculatum, Heuchera americana, Liquidambar styraciflua, Magnolia grandiflora, Obolaria virginica, Polygala lutea, Populus deltoides, Sambucus cerulea, Saururus cernuus, Tephrosia florida, Yucca aloifolia*
Diaphoretic: *Asclepias verticillata, Galium asprellum, G. boreale, G. uniflorum, Pycnanthemum albescens*
Disinfectant: *Chionanthus virginicus*
Diuretic: *Eryngium aquaticum, Galium asprellum, G. boreale, G. uniflorum, Prenanthes aspera*
Expectorant: *Aralia racemosa, Eryngium aquaticum*
Eye Medicine: *Aralia racemosa, Cephalanthus occidentalis, Hypericum crux-andreae, H. hypericoides, Ilex opaca*
Febrifuge: *Cephalanthus occidentalis, Myrica cerifera, Pluchea foetida, Symplocos tinctoria, Verbesina virginica, Vitis aestivalis*
Gastrointestinal Aid: *Aralia racemosa, Aristolochia serpentaria, Callicarpa americana, Echinacea purpurea, Hypericum hypericoides, Malvella leprosa, Quercus stellata, Sambucus cerulea*
Gynecological Aid: *Polygonum aviculare, Quercus marilandica, Ulmus americana, Vitis aestivalis*
Herbal Steam: *Populus deltoides*
Kidney Aid: *Bignonia capreolata, Magnolia grandiflora*
Liver Aid: *Sambucus canadensis*
Misc. Disease Remedy: *Galium asprellum, Rumex verticillatus, Sassafras albidum*
Oral Aid: *Pityopsis graminifolia*
Other: *Callicarpa americana, Prenanthes aspera*
Panacea: *Pycnanthemum incanum*
Pediatric Aid: *Aralia racemosa, Monarda fistulosa*
Poultice: *Aralia spinosa*
Pulmonary Aid: *Ceanothus* sp., *Gnaphalium obtusifolium*
Snakebite Remedy: *Asclepias verticillata, Eryngium aquaticum, Populus deltoides*
Stimulant: *Ageratina altissima, Aralia racemosa, Asclepias verticillata, Eryngium aquaticum, Prenanthes aspera*
Throat Aid: *Myrica cerifera*
Tonic: *Ageratina altissima, Cephalanthus occidentalis, Erythrina herbacea, Heuchera americana, Rubus* sp., *Smilax bona-nox, S. tamnoides, Vitis aestivalis*
Toothache Remedy: *Ageratina altissima, Cephalanthus occidentalis*
Unspecified: *Yucca aloifolia*
Urinary Aid: *Sambucus cerulea*
Venereal Aid: *Eryngium aquaticum, Geranium maculatum*
Food
Bread & Cake: *Smilax bona-nox, S. laurifolia*
Soup: *Carya alba*
Spice: *Sassafras albidum*
Staple: *Quercus nigra*
Unspecified: *Strophostyles helvula*
Vegetable: *Zea mays*
Fiber
Basketry: *Arundinaria gigantea, Serenoa repens*
Cordage: *Taxodium distichum*
Dye
Red: *Nyssa aquatica, Quercus texana*
Yellow: *Rumex crispus*
Other
Hunting & Fishing Item: *Arundinaria gigantea*
Insecticide: *Carya alba*
Paint: *Quercus laurifolia, Q. texana, Q. virginiana*

Chumash
Drug
Hallucinogen: *Datura wrightii*
Unspecified: *Datura wrightii*

Clallam
Drug
Antidiarrheal: *Sambucus cerulea*

Antihemorrhagic: *Tsuga heterophylla*
Cold Remedy: *Achillea millefolium*
Dermatological Aid: *Alnus rubra, Frangula purshiana, Geum macrophyllum, Lysichiton americanus*
Gastrointestinal Aid: *Alnus rubra*
Gynecological Aid: *Achillea millefolium*
Other: *Scirpus acutus*
Pulmonary Aid: *Alnus rubra*
Tuberculosis Remedy: *Thuja plicata*
Food
Beverage: *Epilobium angustifolium, Vaccinium oxycoccos*
Bread & Cake: *Gaultheria shallon*
Dried Food: *Acer circinatum, A. macrophyllum, Populus balsamifera, Vaccinium ovalifolium*
Fruit: *Fragaria chiloensis, F. vesca, F. virginiana, Mahonia nervosa, Malus fusca, Ribes divaricatum, Rubus parviflorus, R. spectabilis, R. ursinus, Vaccinium ovalifolium, V. parvifolium*
Ice Cream: *Shepherdia canadensis*
Staple: *Pteridium aquilinum*
Sweetener: *Alnus rubra*
Unspecified: *Abronia latifolia, Acer circinatum, A. macrophyllum, Allium cernuum, Camassia quamash, Daucus pusillus, Dryopteris expansa, Equisetum telmateia, Lilium columbianum, Populus balsamifera, Typha latifolia*
Fiber
Basketry: *Thuja plicata, Typha latifolia*
Building Material: *Thuja plicata*
Canoe Material: *Acer macrophyllum, Taxus brevifolia, Thuja plicata*
Clothing: *Epilobium angustifolium, Thuja plicata*
Cordage: *Salix sitchensis, Thuja plicata*
Mats, Rugs & Bedding: *Thuja plicata, Typha latifolia*
Dye
Red-Brown: *Tsuga heterophylla*
Other
Cooking Tools: *Alnus rubra, Pteridium aquilinum, Taxus brevifolia*
Fuel: *Alnus rubra, Pseudotsuga menziesii*
Hunting & Fishing Item: *Oplopanax horridus, Prunus emarginata, Pseudotsuga menziesii, Taxus brevifolia, Tsuga heterophylla*
Incense & Fragrance: *Achillea millefolium*
Smoke Plant: *Arctostaphylos uva-ursi, Taxus brevifolia*
Tools: *Taxus brevifolia*
Toys & Games: *Polystichum munitum*

Coahuilla
Drug
Analgesic: *Adenostoma sparsifolium, Chrysothamnus nauseosus, Ericameria palmeri, Eriodictyon californicum, Eriogonum fasciculatum, Rhus ovata*
Cathartic: *Acourtia microcephala, Adenostoma sparsifolium*
Cough Medicine: *Chrysothamnus nauseosus, Rhus ovata*
Dermatological Aid: *Adenostoma sparsifolium, Eriodictyon californicum*
Emetic: *Adenostoma sparsifolium*
Eye Medicine: *Baccharis salicifolia, Eriogonum fasciculatum*
Gastrointestinal Aid: *Adenostoma sparsifolium, Artemisia tridentata, Eriogonum fasciculatum, Larrea tridentata*
Hallucinogen: *Datura wrightii*
Orthopedic Aid: *Ericameria palmeri, Eriodictyon californicum*
Poison: *Datura wrightii*
Tuberculosis Remedy: *Larrea tridentata*
Unspecified: *Ephedra nevadensis*
Veterinary Aid: *Adenostoma fasciculatum, Cucurbita foetidissima, Datura wrightii, Eriodictyon californicum, Larrea tridentata*

Food
Beverage: *Ephedra nevadensis, Simmondsia chinensis*
Candy: *Asclepias erosa*
Unspecified: *Adenostoma sparsifolium*
Fiber
Building Material: *Adenostoma sparsifolium*
Other
Fuel: *Adenostoma sparsifolium*
Smoke Plant: *Nicotiana attenuata*

Cochiti
Food
Dried Food: *Prunus virginiana*
Fruit: *Juniperus monosperma, Prunus virginiana*
Special Food: *Fragaria vesca*
Staple: *Quercus gambelii*
Substitution Food: *Solanum elaeagnifolium*
Unspecified: *Echinocereus fendleri, E. triglochidiatus, Lathyrus polymorphus, Quercus gambelii, Yucca baccata, Y. glauca*
Vegetable: *Amaranthus albus, A. retroflexus, Atriplex powellii, Cymopterus bulbosus, Rumex salicifolius*

Cocopa
Drug
Dermatological Aid: *Ephedra trifurca*
Food
Beverage: *Washingtonia filifera*
Bread & Cake: *Cucumis* sp., *Panicum hirticaule, P. sonorum, Simmondsia chinensis*
Dried Food: *Citrullus lanatus, Cucumis* sp., *Cucurbita pepo, Pholisma sonorae, Polygonum argyrocoleon*
Fruit: *Citrullus lanatus, Cucumis* sp., *Opuntia echinocarpa, O. engelmannii*
Porridge: *Chasmanthium latifolium, Echinochloa colona, Eragrostis mexicana, Eriochloa aristata, Olneya tesota, Parkinsonia florida, P. microphylla*
Sauce & Relish: *Panicum hirticaule*
Staple: *Descurainia pinnata, Echinochloa colona, Eragrostis mexicana, Eriochloa aristata, Phaseolus acutifolius*
Unspecified: *Agave deserti, Amaranthus caudatus, A. palmeri, Chasmanthium latifolium, Cucumis* sp., *Cucurbita pepo, Cyperus odoratus, Echinochloa crus-galli, Hoffmannseggia glauca, Pholisma sonorae, Pinus monophylla, Prosopis glandulosa, P. velutina, Rumex crispus, Sagittaria latifolia, Simmondsia chinensis, Vigna unguiculata*
Vegetable: *Amaranthus caudatus, A. palmeri, Chenopodium fremontii, Cucurbita pepo, Descurainia obtusa*
Winter Use Food: *Amaranthus caudatus, A. palmeri, Chasmanthium latifolium, Citrullus lanatus, Echinochloa crus-galli, Panicum hirticaule, P. sonorum, Phaseolus acutifolius, Prosopis glandulosa, P. velutina*
Other
Cash Crop: *Agave deserti, Quercus turbinella*
Hide Preparation: *Cucurbita moschata*
Musical Instrument: *Glaucothea armata, Lagenaria siceraria, Washingtonia filifera*
Protection: *Cucurbita pepo*
Smoking Tools: *Phragmites australis*
Toys & Games: *Sagittaria latifolia*

Cocopa & Yuma
Food
Unspecified: *Salix gooddingii*

Cocopa, Maricopa, Mohave & Yuma
Food
Porridge: *Vigna unguiculata*

Coeur d'Alene
Food
Bread & Cake: *Crataegus* sp.
Dried Food: *Arctostaphylos uva-ursi, Prunus virginiana, Rubus* sp., *Vaccinium membranaceum*
Fruit: *Amelanchier* sp., *Arctostaphylos uva-ursi, Crataegus* sp., *Fragaria vesca, Mahonia* sp., *Prunus emarginata, P.* sp., *P. virginiana, Ribes* sp., *Rosa* sp., *Rubus leucodermis, R.* sp., *Sambucus* sp., *Shepherdia canadensis, Vaccinium membranaceum*
Ice Cream: *Shepherdia canadensis*
Soup: *Arctostaphylos uva-ursi, Prunus virginiana, Rubus* sp., *Vaccinium membranaceum*
Unspecified: *Alectoria jubata, Balsamorhiza* sp., *Heracleum maximum, Peucedanum* sp., *Pinus albicaulis, P. contorta, P. ponderosa, Populus* sp.
Vegetable: *Allium* sp., *Camassia scilloides, Claytonia* sp., *Lewisia rediviva*
Other
Tools: *Amelanchier* sp., *Crataegus* sp., *Philadelphus lewisii*

Comanche
Drug
Adjuvant: *Poliomintha incana*
Analgesic: *Juniperus pinchotii*
Burn Dressing: *Zanthoxylum americanum*
Ceremonial Medicine: *Juniperus pinchotii, Lophophora williamsii, Matelea biflora, M. cynanchoides*
Cold Remedy: *Rhus trilobata*
Dermatological Aid: *Amphiachyris dracunculoides, Artemisia ludoviciana, Carya illinoinensis, Juglans nigra, Matelea biflora, M. cynanchoides, Sphaeralcea coccinea*
Disinfectant: *Juniperus virginiana*
Ear Medicine: *Sophora secundiflora*
Eye Medicine: *Argemone polyanthemos, Elymus* sp., *Maclura pomifera, Salix* sp.
Febrifuge: *Helenium autumnale, Zanthoxylum americanum*
Gastrointestinal Aid: *Eriogonum longifolium, Matelea biflora, M. cynanchoides, Prosopis glandulosa*
Gynecological Aid: *Artemisia filifolia, Helenium microcephalum, Juniperus pinchotii, Matelea biflora, M. cynanchoides*
Heart Medicine: *Helenium microcephalum*
Hypotensive: *Helenium microcephalum*
Misc. Disease Remedy: *Matelea biflora, M. cynanchoides*
Narcotic: *Lophophora williamsii*
Orthopedic Aid: *Matelea biflora, M. cynanchoides*
Other: *Juniperus pinchotii*
Pediatric Aid: *Matelea biflora, M. cynanchoides*
Pulmonary Aid: *Gutierrezia sarothrae*
Respiratory Aid: *Helenium microcephalum*
Throat Aid: *Echinacea* sp., *Zanthoxylum americanum*
Tonic: *Solanum* sp.
Toothache Remedy: *Echinacea* sp., *Zanthoxylum americanum*
Tuberculosis Remedy: *Solanum* sp.
Urinary Aid: *Liatris punctata*
Venereal Aid: *Cirsium undulatum, Schizachyrium scoparium*
Food
Beverage: *Ilex* sp., *Lespedeza capitata*
Candy: *Crataegus* sp., *Dalea purpurea*
Dried Food: *Prunus* sp., *Vitis* sp.
Fruit: *Celtis laevigata, Crataegus* sp., *Diospyros texana, D. virginiana, Juniperus virginiana, Morus rubra, Opuntia* sp., *Prunus angustifolia, Rhus glabra, Ribes aureum, Vitis* sp.
Staple: *Agastache pallidiflora, Agave americana, A. parryi, Prosopis glandulosa*
Starvation Food: *Quercus marilandica*
Unspecified: *Allium* sp., *Caesalpinia jamesii, Camassia scilloides, Carya illinoinensis, Cirsium undulatum, Cymopterus acaulis, Juglans nigra, Nelumbo lutea, Nuphar lutea, Pediomelum hypogaeum, Quercus* sp.
Winter Use Food: *Carya illinoinensis, Juglans nigra, Prunus angustifolia*
Fiber
Brushes & Brooms: *Gutierrezia sarothrae*
Building Material: *Quercus* sp.
Other
Ceremonial Items: *Schizachyrium scoparium*
Decorations: *Sophora secundiflora*
Hide Preparation: *Croton monanthogynus*
Hunting & Fishing Item: *Cornus asperifolia, Maclura pomifera*
Protection: *Sophora secundiflora*
Smoke Plant: *Rhus glabra*
Smoking Tools: *Quercus marilandica, Smilax bona-nox*
Soap: *Yucca louisianensis*
Toys & Games: *Cephalanthus occidentalis, Sapindus saponaria*
Weapon: *Coryphantha* sp.

Concow
Drug
Analgesic: *Croton setigerus, Pogogyne douglasii, Trichostema lanceolatum*
Burn Dressing: *Cynoglossum grande*
Dermatological Aid: *Arctostaphylos manzanita, Petasites frigidus*
Emetic: *Arbutus menziesii*
Febrifuge: *Croton setigerus, Trichostema lanceolatum*
Gastrointestinal Aid: *Pogogyne douglasii*
Misc. Disease Remedy: *Croton setigerus, Petasites frigidus, Trichostema lanceolatum*
Poison: *Trillium sessile*
Tuberculosis Remedy: *Petasites frigidus*
Veterinary Aid: *Arctostaphylos manzanita*
Food
Beverage: *Umbellularia californica*
Bread & Cake: *Quercus* sp.
Porridge: *Quercus* sp.
Staple: *Ceanothus integerrimus, Verbena hastata*
Substitution Food: *Pogogyne douglasii*
Unspecified: *Petasites frigidus*
Fiber
Basketry: *Acer macrophyllum, Ceanothus integerrimus*
Clothing: *Acer macrophyllum*
Cordage: *Asclepias eriocarpa*
Sewing Material: *Hoita macrostachya*
Dye
Black: *Quercus lobata*
Other
Hunting & Fishing Item: *Trichostema lanceolatum*

Costanoan
Drug
Abortifacient: *Equisetum laevigatum, E.* sp.
Analgesic: *Adiantum jordanii, Anemopsis californica, Angelica* sp., *Artemisia californica, A. douglasiana, Cirsium* sp., *Clematis ligusticifolia, Datura wrightii, Eriodictyon californicum, Gnaphalium californicum, Juniperus*

[Analgesic]
californica, Malva nicaeensis, Matricaria discoidea, Salvia mellifera, Torreya californica, Trichostema lanceolatum, Trillium chloropetalum, Urtica sp.

Anthelmintic: *Satureja douglasii*

Anticonvulsive: *Matricaria discoidea*

Antidiarrheal: *Artemisia dracunculus, Capsella bursa-pastoris, Croton setigerus, Galium* sp., *Quercus* sp., *Rubus vitifolius*

Antihemorrhagic: *Pellaea mucronata*

Antirheumatic (External): *Artemisia californica, A. douglasiana, Eriodictyon californicum, Galium* sp., *Opuntia* sp.

Antirheumatic (Internal): *Cupressus macrocarpa, Pinus sabiniana, Rosa californica*

Blood Medicine: *Adiantum jordanii, A. pedatum, Chamaesyce maculata, Daucus pusillus, Eriodictyon californicum, Hoita orbicularis, Juglans californica, Limonium californicum, Monardella villosa, Pellaea mucronata*

Burn Dressing: *Navarretia atractyloides, Solidago californica*

Carminative: *Salvia mellifera*

Cathartic: *Datura wrightii, Frangula californica, Nicotiana quadrivalvis, Sambucus cerulea, Trifolium* sp.

Ceremonial Medicine: *Nicotiana quadrivalvis*

Cold Remedy: *Artemisia californica, Asclepias eriocarpa, Daucus pusillus, Eriodictyon californicum, Eriogonum latifolium, Gnaphalium californicum, Helenium puberulum, Rosa californica, Salix lasiolepis, Sambucus cerulea, Trichostema lanceolatum*

Contraceptive: *Equisetum laevigatum, E.* sp., *Maianthemum racemosum*

Cough Medicine: *Artemisia californica, Eriogonum latifolium, Lonicera* sp., *Lotus scoparius, Marrubium vulgare, Orthocarpus* sp., *Ruta chalepensis, Salvia mellifera*

Dermatological Aid: *Achillea millefolium, Anemopsis californica, Angelica* sp., *Artemisia californica, A. douglasiana, Asclepias eriocarpa, Baccharis douglasii, B. salicifolia, Castilleja affinis, Ceanothus* sp., *Chamaesyce maculata, Chlorogalum pomeridianum, Cirsium* sp., *Datura wrightii, Daucus pusillus, Dryopteris arguta, Epilobium canum, Equisetum laevigatum, Eriodictyon californicum, Frangula californica, Galium* sp., *Grindelia camporum, Helenium puberulum, Lithocarpus densiflorus, Lonicera* sp., *Malva nicaeensis, Marah macrocarpus, Marrubium vulgare, Matricaria discoidea, Pellaea mucronata, Penstemon centranthifolius, Pteridium aquilinum, Ranunculus* sp., *Rosa californica, Rubus vitifolius, Salix* sp., *Scrophularia californica, Sedum* sp., *Senecio flaccidus, Solanum nigrum, Solidago californica, Stachys bullata, Trichostema lanceolatum, Urtica* sp., *Viola* sp., *Wyethia angustifolia*

Diaphoretic: *Juniperus californica, Torreya californica*

Disinfectant: *Anemopsis californica, Baccharis douglasii, Castilleja affinis, Cirsium* sp., *Epilobium canum, Eriodictyon californicum, Lonicera* sp., *Matricaria discoidea, Penstemon centranthifolius, Rubus vitifolius, Scrophularia californica, Senecio flaccidus, Stachys bullata, Trichostema lanceolatum, Urtica* sp.

Ear Medicine: *Artemisia douglasiana, Medicago sativa, Nicotiana quadrivalvis, Ruta chalepensis, Salvia mellifera, Stachys bullata*

Emetic: *Lathyrus vestitus, Malva nicaeensis, Nicotiana quadrivalvis, Pellaea mucronata*

Eye Medicine: *Chamaesyce maculata, Datura wrightii, Eriodictyon californicum, Salvia columbariae, Scrophularia californica*

Febrifuge: *Cornus sericea, Daucus pusillus, Epilobium canum, Fraxinus latifolia, Hoita orbicularis, Malva nicaeensis, Matricaria discoidea, Pellaea mucronata, Phacelia californica, Plantago major, Rorippa nasturtium-aquaticum, Rosa californica, Salix bonplandiana, Salvia columbariae, Sedum* sp., *Sisyrinchium bellum, Torreya californica, Verbena lasiostachys*

Gastrointestinal Aid: *Achillea millefolium, Adiantum jordanii, A. pedatum, Angelica* sp., *Artemisia dracunculus, Cirsium* sp., *Gnaphalium californicum, Malva nicaeensis, Matricaria discoidea, Melissa officinalis, Paeonia brownii, Rosa californica, Ruta chalepensis, Sisyrinchium bellum, Stachys bullata, Torreya californica, Trichostema lanceolatum, Verbena lasiostachys*

Gynecological Aid: *Adiantum jordanii, Anemopsis californica, Senecio flaccidus*

Hallucinogen: *Datura wrightii*

Heart Medicine: *Salvia mellifera*

Hemorrhoid Remedy: *Aesculus californica*

Kidney Aid: *Baccharis douglasii, Diplacus aurantiacus, Disporum hookeri, Rorippa nasturtium-aquaticum, Rosa californica, Senecio flaccidus*

Laxative: *Frangula californica, Paeonia brownii, Plantago major, Vicia nigricans*

Liver Aid: *Rorippa nasturtium-aquaticum*

Love Medicine: *Datura wrightii*

Misc. Disease Remedy: *Erodium cicutarium, Solanum nigrum, Verbena lasiostachys*

Orthopedic Aid: *Ambrosia psilostachya, Angelica* sp., *Chenopodium californicum, Lonicera* sp., *Populus* sp., *Ruta chalepensis, Salvia mellifera, Urtica* sp.

Other: *Asclepias eriocarpa, Malva nicaeensis, Senecio flaccidus*

Panacea: *Baccharis pilularis, Epilobium canum, Lathyrus vestitus, Platanus racemosa*

Pediatric Aid: *Artemisia dracunculus, Epilobium canum, Eschscholzia californica, Malva nicaeensis, Matricaria discoidea, Melissa officinalis*

Poison: *Aesculus californica, Chlorogalum pomeridianum, Eschscholzia californica*

Pulmonary Aid: *Marrubium vulgare, Monardella villosa, Paeonia brownii, Wyethia angustifolia*

Respiratory Aid: *Artemisia californica, A. douglasiana, Asclepias eriocarpa, Cirsium* sp., *Datura wrightii, Eriodictyon californicum, Limonium californicum, Monardella villosa*

Sedative: *Eschscholzia californica*

Snakebite Remedy: *Daucus pusillus, Fraxinus latifolia*

Throat Aid: *Datisca glomerata, Rosa californica, Salvia mellifera, Sedum* sp., *Stachys bullata*

Toothache Remedy: *Achillea millefolium, Aesculus californica, Artemisia californica, Lithocarpus densiflorus, Quercus* sp., *Satureja douglasii, Solanum nigrum*

Tuberculosis Remedy: *Eriodictyon californicum*

Unspecified: *Amsinckia douglasiana*

Urinary Aid: *Arctostaphylos* sp., *Artemisia douglasiana, A. dracunculus, Diplacus aurantiacus, Epilobium canum, Equisetum laevigatum, E.* sp., *Eriogonum fasciculatum, Hordeum murinum, Limonium californicum, Rumex crispus, Xanthium strumarium*

Venereal Aid: *Limonium californicum*

Food

Beverage: *Arctostaphylos* sp.

Bread & Cake: *Umbellularia californica*

Dried Food: *Arctostaphylos* sp., *Heteromeles arbutifolia*

Fruit: *Aesculus californica, Amelanchier pallida, Arbutus menziesii, Arctostaphylos* sp., *Fragaria* sp., *Frangula californica, Heteromeles arbutifolia, Juniperus californica, Maianthemum racemosum, Opuntia* sp., *Prunus ilicifolia, P. virginiana, Ribes* sp., *Rubus leucodermis, R. vitifolius, Sambucus cerulea, Solanum umbelliferum, Umbellularia californica, Vaccinium ovatum, Vitis californica*

Staple: *Calandrinia ciliata, Elymus glaucus, Festuca* sp., *Hemizonia corymbosa, Hordeum murinum, Layia platyglossa, Lupinus* sp., *Rumex crispus, Salvia columbariae, Wyethia angustifolia, Xanthium strumarium*

Unspecified: *Acer macrophyllum, Alnus rhombifolia, Brodiaea* sp., *Calandrinia ciliata, Camissonia ovata, Chlorogalum pomeridianum, Cirsium* sp., *Claytonia perfoliata, Corylus cornuta, Cyperus esculentus, Dryopteris arguta, Erodium cicutarium, Helianthus annuus, Heracleum maximum, Juglans californica, Lithocarpus densiflorus, Lobularia maritima, Oenanthe sarmentosa, Pinus sabiniana, Platanus racemosa, Polystichum munitum, Populus* sp., *Prunus ilicifolia, Pteridium aquilinum, Quercus agrifolia, Q.* sp., *Raphanus sativus,*

Scirpus sp., *Sedum* sp., *Trifolium* sp., *Typha latifolia*, *Wyethia angustifolia*
Vegetable: *Rumex crispus*
Winter Use Food: *Allium* sp., *Asyneuma prenanthoides*
Fiber
Basketry: *Adenostoma fasciculatum*, *Carex* sp., *Cornus sericea*, *Corylus cornuta*, *Equisetum arvense*, *E. hyemale*, *Juncus* sp., *Pteridium aquilinum*, *Salix hindsiana*, *S. lasiolepis*, *Scirpus* sp., *Sequoia sempervirens*, *Toxicodendron diversilobum*
Brushes & Brooms: *Chlorogalum pomeridianum*, *Symphoricarpos albus*
Building Material: *Lotus scoparius*, *Salix* sp., *Scirpus* sp.
Canoe Material: *Scirpus* sp.
Clothing: *Eriodictyon californicum*, *Juncus* sp.
Cordage: *Asclepias fascicularis*, *Juncus* sp., *Salix* sp.
Mats, Rugs & Bedding: *Juncus* sp.
Dye
Blue: *Solanum* sp.
Red: *Datisca glomerata*
Unspecified: *Alnus rhombifolia*, *Cornus sericea*, *Lithocarpus densiflorus*
Yellow: *Datisca glomerata*
Other
Containers: *Pteridium aquilinum*, *Toxicodendron diversilobum*
Cooking Tools: *Bouteloua* sp., *Platanus racemosa*, *Polystichum munitum*, *Quercus* sp.
Fuel: *Quercus* sp., *Salix hindsiana*, *Sambucus cerulea*
Hunting & Fishing Item: *Adenostoma fasciculatum*, *Chlorogalum pomeridianum*, *Corylus cornuta*, *Croton setigerus*, *Hemizonia corymbosa*, *Prunus ilicifolia*, *Sambucus cerulea*, *Taxus brevifolia*, *Umbellularia californica*, *Vicia nigricans*
Incense & Fragrance: *Umbellularia californica*
Insecticide: *Eschscholzia californica*, *Umbellularia californica*
Lighting: *Artemisia douglasiana*, *A. dracunculus*
Musical Instrument: *Sambucus cerulea*
Protection: *Fraxinus latifolia*, *Pteridium aquilinum*
Smoking Tools: *Sambucus cerulea*, *Scirpus* sp.
Snuff: *Helenium puberulum*
Soap: *Ceanothus* sp., *Chenopodium californicum*, *Chlorogalum pomeridianum*, *Marah macrocarpus*
Tools: *Artemisia douglasiana*

Costanoan (Olhonean)
Drug
Abortifacient: *Heteromeles arbutifolia*
Antihemorrhagic: *Arctostaphylos tomentosa*
Cathartic: *Rhamnus* sp.
Dermatological Aid: *Rhamnus* sp.
Food
Beverage: *Arctostaphylos pumila*
Fiber
Building Material: *Scirpus* sp.

Cowichan
Drug
Burn Dressing: *Arbutus menziesii*, *Symphoricarpos albus*
Cold Remedy: *Lomatium nudicaule*, *Polypodium virginianum*, *Prunus emarginata*
Dermatological Aid: *Arbutus menziesii*, *Moneses uniflora*, *Symphoricarpos albus*
Gastrointestinal Aid: *Polypodium virginianum*
Misc. Disease Remedy: *Arbutus menziesii*
Panacea: *Malus fusca*, *Prunus emarginata*
Throat Aid: *Lomatium nudicaule*, *Polypodium virginianum*
Unspecified: *Lonicera ciliosa*

Food
Fruit: *Rosa nutkana*
Special Food: *Camassia leichtlinii*, *C. quamash*
Spice: *Acer macrophyllum*, *Zostera marina*
Unspecified: *Cirsium brevistylum*, *Daucus pusillus*
Vegetable: *Urtica dioica*
Fiber
Cordage: *Ribes divaricatum*, *R. lacustre*, *R. lobbii*
Other
Ceremonial Items: *Lomatium nudicaule*
Cleaning Agent: *Galium* sp.
Cooking Tools: *Acer macrophyllum*
Fuel: *Galium* sp.
Hunting & Fishing Item: *Ribes divaricatum*, *R. lacustre*, *R. lobbii*, *Rosa nutkana*
Preservative: *Arbutus menziesii*

Cowlitz
Drug
Analgesic: *Alnus rubra*, *Athyrium filix-femina*, *Veratrum viride*
Anthelmintic: *Fraxinus latifolia*
Antihemorrhagic: *Tsuga heterophylla*
Antirheumatic (External): *Lysichiton americanus*, *Oplopanax horridus*
Burn Dressing: *Rubus parviflorus*
Cold Remedy: *Mentha* sp., *Oplopanax horridus*, *Pseudotsuga menziesii*, *Thuja plicata*
Dermatological Aid: *Achillea millefolium*, *Adenocaulon bicolor*, *Claytonia sibirica*, *Clintonia uniflora*, *Mahonia* sp., *Polystichum munitum*, *Pseudotsuga menziesii*, *Ribes divaricatum*, *Rumex* sp., *Taxus brevifolia*, *Tolmiea menziesii*, *Tsuga heterophylla*
Disinfectant: *Rumex* sp.
Eye Medicine: *Clintonia uniflora*, *Oxalis oregana*, *Trientalis borealis*, *Tsuga heterophylla*
Gastrointestinal Aid: *Achillea millefolium*
Gynecological Aid: *Urtica dioica*
Laxative: *Frangula purshiana*
Love Medicine: *Galium aparine*
Misc. Disease Remedy: *Polypodium virginianum*
Oral Aid: *Mahonia* sp.
Orthopedic Aid: *Alnus rubra*, *Sambucus racemosa*, *Urtica dioica*
Pediatric Aid: *Rosa nutkana*
Poison: *Galium aparine*, *Oplopanax horridus*, *Veratrum viride*
Pulmonary Aid: *Arctium minus*
Strengthener: *Rosa nutkana*
Tonic: *Goodyera oblongifolia*
Toothache Remedy: *Thuja plicata*
Tuberculosis Remedy: *Achlys triphylla*, *Anemone* sp., *Quercus garryana*
Food
Beverage: *Fragaria vesca*
Candy: *Pseudotsuga menziesii*
Dried Food: *Equisetum hyemale*, *Fragaria vesca*, *Oemleria cerasiformis*, *Ribes divaricatum*, *Rubus leucodermis*, *R. ursinus*
Fruit: *Fragaria vesca*, *Mahonia* sp., *Malus fusca*, *Oemleria cerasiformis*, *Ribes divaricatum*, *Rubus leucodermis*, *R. parviflorus*, *R. spectabilis*, *R. ursinus*
Spice: *Tsuga heterophylla*
Unspecified: *Dryopteris expansa*, *Equisetum telmateia*, *Hydrophyllum tenuipes*, *Lysichiton americanus*, *Oenanthe sarmentosa*, *Oxalis oregana*, *Pteridium aquilinum*, *Quercus garryana*, *Rubus spectabilis*
Winter Use Food: *Corylus cornuta*, *Sambucus racemosa*
Fiber
Basketry: *Equisetum telmateia*, *Typha latifolia*, *Xerophyllum tenax*

Brushes & Brooms: *Philadelphus lewisii, Quercus garryana, Taxus brevifolia*
Canoe Material: *Fraxinus latifolia*
Clothing: *Typha latifolia*
Cordage: *Acer macrophyllum*
Mats, Rugs & Bedding: *Polystichum munitum, Typha latifolia*
Scouring Material: *Equisetum hyemale*
Unspecified: *Fomes* sp.
Dye
Unspecified: *Mahonia* sp.
Other
Cooking Tools: *Athyrium filix-femina, Polystichum munitum, Thuja plicata*
Fuel: *Pseudotsuga menziesii, Quercus garryana*
Good Luck Charm: *Pseudotsuga menziesii*
Hunting & Fishing Item: *Pseudotsuga menziesii*
Insecticide: *Equisetum hyemale*
Musical Instrument: *Taxus brevifolia*
Soap: *Rubus parviflorus*
Tools: *Fraxinus latifolia, Quercus garryana, Salix lucida, Taxus brevifolia*

Cree
Drug
Abortifacient: *Equisetum hyemale, Grindelia squarrosa, Populus grandidentata*
Burn Dressing: *Ledum groenlandicum, Plantago* sp.
Ceremonial Medicine: *Agastache foeniculum*
Dermatological Aid: *Cirsium discolor, Heracleum maximum*
Diuretic: *Ledum groenlandicum*
Emetic: *Ledum groenlandicum*
Gastrointestinal Aid: *Acorus calamus*
Gynecological Aid: *Grindelia squarrosa, Populus grandidentata*
Heart Medicine: *Ligusticum canbyi*
Kidney Aid: *Anthemis* sp., *Grindelia squarrosa*
Poison: *Heracleum maximum*
Throat Aid: *Acorus calamus, Picea* sp.
Toothache Remedy: *Heracleum maximum*
Venereal Aid: *Baptisia* sp., *Grindelia squarrosa, Heracleum maximum, Picea* sp., *Salix* sp., *Smilax* sp.
Food
Beverage: *Elaeagnus commutata, Ledum groenlandicum*
Fruit: *Elaeagnus commutata*
Starvation Food: *Taraxacum* sp.
Sweetener: *Acer negundo*
Vegetable: *Allium cernuum*
Fiber
Cordage: *Elaeagnus commutata*
Dye
Red: *Galium boreale*
Other
Jewelry: *Elaeagnus commutata*
Smoke Plant: *Arctostaphylos uva-ursi, Cornus sericea*

Cree, Alberta
Drug
Hallucinogen: *Acorus calamus*
Stimulant: *Acorus calamus*
Unspecified: *Acorus calamus*

Cree, Hudson Bay
Drug
Analgesic: *Ledum groenlandicum*
Antidiarrheal: *Conyza canadensis, Kalmia latifolia, Prunus virginiana*
Antirheumatic (External): *Ledum groenlandicum*
Cathartic: *Actaea rubra, Apocynum cannabinum, Geocaulon lividum, Iris versicolor, Populus* sp.
Cold Remedy: *Cornus sericea*
Cough Medicine: *Cornus sericea, Populus* sp.
Dermatological Aid: *Alnus viridis, Apocynum cannabinum, Betula pubescens, Geocaulon lividum, Juniperus communis, Ledum groenlandicum, Populus* sp., *Pyrus* sp.
Disinfectant: *Juniperus communis*
Diuretic: *Galium boreale, Juniperus virginiana*
Emetic: *Apocynum cannabinum, Cornus sericea, Geocaulon lividum, Lobelia kalmii*
Febrifuge: *Cornus sericea*
Gastrointestinal Aid: *Kalmia angustifolia, Mentha canadensis*
Kidney Aid: *Alnus viridis*
Liver Aid: *Iris versicolor*
Misc. Disease Remedy: *Pyrus* sp.
Orthopedic Aid: *Ledum groenlandicum*
Poison: *Aconitum heterophyllum, Kalmia latifolia*
Pulmonary Aid: *Pyrus* sp.
Throat Aid: *Prunella vulgaris*
Tonic: *Kalmia angustifolia, Solidago multiradiata*
Unspecified: *Apocynum cannabinum, Geocaulon lividum*

Cree, Plains
Food
Dried Food: *Amelanchier alnifolia*

Cree, Woodlands
Drug
Abortifacient: *Abies balsamea, Alnus viridis, Arctostaphylos uva-ursi, Aster puniceus, Sarracenia purpurea, Vaccinium myrtilloides*
Adjuvant: *Acorus calamus*
Analgesic: *Achillea millefolium, A.* sp., *Acorus calamus, Artemisia frigida, Heracleum maximum, Mentha canadensis, Nuphar lutea, Valeriana dioica*
Anticonvulsive: *Valeriana dioica*
Antidiarrheal: *Arctostaphylos uva-ursi, Heuchera richardsonii, Juniperus communis, Picea mariana, Prunus virginiana, Salix discolor*
Antiemetic: *Larix laricina*
Antihemorrhagic: *Achillea millefolium, Acorus calamus, Agastache foeniculum, Chimaphila umbellata, Mentha canadensis, Pyrola asarifolia, Shepherdia canadensis*
Antirheumatic (External): *Acorus calamus, Chenopodium album, Cicuta maculata, Heracleum maximum, Nuphar lutea, Rumex orbiculatus, R. salicifolius, Shepherdia canadensis, Sorbus decora*
Antirheumatic (Internal): *Chenopodium album, Picea glauca*
Blood Medicine: *Picea glauca, Polygala senega*
Breast Treatment: *Ledum groenlandicum*
Burn Dressing: *Achillea* sp., *Betula papyrifera, Ledum groenlandicum, Picea mariana*
Cold Remedy: *Abies balsamea, Acorus calamus, Amelanchier alnifolia, Mentha canadensis, Valeriana dioica*
Contraceptive: *Vaccinium myrtilloides*
Cough Medicine: *Abies balsamea, Acorus calamus, Amelanchier alnifolia, Juniperus communis, Rosa acicularis*
Dermatological Aid: *Abies balsamea, Acorus calamus, Aralia nudicaulis, Betula papyrifera, Epilobium angustifolium, Heracleum maximum, Larix laricina, Ledum groenlandicum, Nuphar lutea, Petasites sagittatus, Picea glauca, P. mariana, Pinus banksiana, Populus tremuloides, Salix bebbiana, Shepherdia canadensis, Symphoricarpos albus, Viburnum edule*
Diaphoretic: *Amelanchier alnifolia, Aster puniceus, Betula papyrifera, Geum*

Index of Tribes

aleppicum, *Vaccinium myrtilloides*
Disinfectant: *Populus balsamifera*
Diuretic: *Empetrum nigrum*, *Ledum groenlandicum*, *Lonicera dioica*
Ear Medicine: *Acorus calamus*, *Mitella nuda*, *Valeriana dioica*
Eye Medicine: *Alnus incana*, *Apocynum androsaemifolium*, *Cornus sericea*, *Diervilla lonicera*, *Heuchera richardsonii*, *Prunus pensylvanica*, *Pyrola asarifolia*, *Rosa acicularis*, *Symphoricarpos albus*
Febrifuge: *Achillea millefolium*, *Acorus calamus*, *Amelanchier alnifolia*, *Artemisia frigida*, *Aster puniceus*, *Juniperus communis*, *Mentha canadensis*, *Symphoricarpos albus*
Gastrointestinal Aid: *Acorus calamus*, *Alisma plantago-aquatica*, *Astragalus americanus*
Gynecological Aid: *Actaea rubra*, *Apocynum androsaemifolium*, *Aralia nudicaulis*, *Arctostaphylos uva-ursi*, *Aster puniceus*, *Betula papyrifera*, *Diervilla lonicera*, *Juniperus communis*, *Lonicera dioica*, *Matteuccia struthiopteris*, *Ribes glandulosum*, *R. hudsonianum*, *R. oxyacanthoides*, *Rubus chamaemorus*, *R. idaeus*, *Sarracenia purpurea*, *Urtica dioica*, *Vaccinium myrtilloides*, *Valeriana dioica*
Heart Medicine: *Alisma plantago-aquatica*, *Campanula rotundifolia*, *Chimaphila umbellata*, *Rubus idaeus*
Hemostat: *Acorus calamus*, *Mentha canadensis*, *Populus balsamifera*, *P. tremuloides*, *Scirpus acutus*, *S. tabernaemontani*
Kidney Aid: *Juniperus communis*
Laxative: *Alisma plantago-aquatica*, *Alnus incana*, *Shepherdia canadensis*
Love Medicine: *Valeriana dioica*
Misc. Disease Remedy: *Alisma plantago-aquatica*, *Amelanchier alnifolia*, *Astragalus americanus*, *Lonicera dioica*
Oral Aid: *Achillea sibirica*, *Aralia nudicaulis*, *Mentha canadensis*, *Picea mariana*, *Polygala senega*, *Polygonum amphibium*
Orthopedic Aid: *Acorus calamus*, *Aster puniceus*, *Betula papyrifera*, *Calla palustris*, *Chimaphila umbellata*, *Matteuccia struthiopteris*, *Sarracenia purpurea*, *Sorbus decora*
Panacea: *Acorus calamus*, *Alisma plantago-aquatica*, *Aralia nudicaulis*, *Geum aleppicum*, *Nuphar lutea*, *Polygala senega*, *Polygonum amphibium*, *Thuja occidentalis*, *Valeriana dioica*
Pediatric Aid: *Achillea millefolium*, *A. sibirica*, *Acorus calamus*, *Amelanchier alnifolia*, *Aralia nudicaulis*, *Arctostaphylos uva-ursi*, *Aster puniceus*, *Betula papyrifera*, *Carum carvi*, *Empetrum nigrum*, *Geum aleppicum*, *G. macrophyllum*, *Juniperus communis*, *Ledum groenlandicum*, *Picea glauca*, *Rubus idaeus*, *Symphoricarpos albus*, *Valeriana dioica*, *Viburnum edule*
Poison: *Calla palustris*
Pulmonary Aid: *Acorus calamus*, *Amelanchier alnifolia*, *Aralia nudicaulis*, *Chimaphila umbellata*, *Juniperus communis*, *Ledum groenlandicum*, *Thuja occidentalis*, *Valeriana dioica*
Reproductive Aid: *Rubus chamaemorus*, *Vaccinium myrtilloides*
Respiratory Aid: *Juniperus communis*
Sedative: *Carum carvi*
Stimulant: *Alisma plantago-aquatica*
Throat Aid: *Acorus calamus*, *Geum aleppicum*, *Picea mariana*, *Polygala senega*, *Viburnum edule*
Toothache Remedy: *Achillea millefolium*, *A. sibirica*, *A. sp.*, *Acorus calamus*, *Amelanchier alnifolia*, *Aralia nudicaulis*, *Aster puniceus*, *Betula papyrifera*, *Geum aleppicum*, *G. macrophyllum*, *Juniperus communis*, *Mentha canadensis*, *Monotropa uniflora*, *Picea mariana*, *Polygala senega*, *Rubus idaeus*, *Viburnum edule*
Tuberculosis Remedy: *Abies balsamea*, *Polypodium virginianum*
Urinary Aid: *Thuja occidentalis*
Venereal Aid: *Acorus calamus*, *Betula papyrifera*, *Lonicera dioica*, *Picea mariana*, *Populus tremuloides*, *Sarracenia purpurea*, *Shepherdia canadensis*, *Symphoricarpos albus*

Food
Beverage: *Agastache foeniculum*, *Ledum groenlandicum*, *Mentha canadensis*, *Ribes glandulosum*
Candy: *Picea glauca*, *P. mariana*
Dried Food: *Amelanchier alnifolia*, *Nuphar lutea*, *Typha latifolia*
Frozen Food: *Vaccinium vitis-idaea*, *Viburnum edule*
Fruit: *Amelanchier alnifolia*, *Arctostaphylos uva-ursi*, *Empetrum nigrum*, *Prunus virginiana*, *Ribes glandulosum*, *R. oxyacanthoides*, *Rubus arcticus*, *R. chamaemorus*, *R. idaeus*, *R. pubescens*, *Vaccinium myrtilloides*, *V. oxycoccos*, *V. vitis-idaea*
Preservative: *Amelanchier alnifolia*, *Betula papyrifera*, *Populus tremuloides*
Preserves: *Prunus pensylvanica*, *Ribes hudsonianum*, *Vaccinium myrtilloides*, *Viburnum edule*
Sauce & Relish: *Betula papyrifera*, *Prunus virginiana*
Snack Food: *Amelanchier alnifolia*, *Cornus canadensis*, *Fragaria virginiana*, *Lilium philadelphicum*, *Rosa acicularis*, *Vaccinium vitis-idaea*, *Viburnum edule*
Soup: *Vaccinium vitis-idaea*
Spice: *Allium schoenoprasum*, *Carum carvi*, *Mentha canadensis*
Staple: *Carum carvi*
Substitution Food: *Betula papyrifera*
Unspecified: *Allium schoenoprasum*, *Betula papyrifera*, *Corylus cornuta*, *Heracleum maximum*, *Lilium philadelphicum*, *Pinus banksiana*, *Populus tremuloides*, *Rubus idaeus*, *Scirpus acutus*, *S. tabernaemontani*, *Sium suave*, *Typha latifolia*
Winter Use Food: *Corylus cornuta*, *Vaccinium oxycoccos*

Fiber
Basketry: *Amelanchier alnifolia*, *Betula papyrifera*, *Cornus sericea*, *Picea glauca*, *Salix bebbiana*, *S. discolor*, *S. interior*
Building Material: *Abies balsamea*, *Betula papyrifera*, *Picea glauca*, *P. mariana*, *Populus tremuloides*
Canoe Material: *Abies balsamea*, *Betula papyrifera*, *Picea glauca*, *P. mariana*
Caulking Material: *Alnus incana*, *Picea glauca*, *P. mariana*
Cordage: *Picea mariana*, *Salix discolor*
Mats, Rugs & Bedding: *Calamagrostis canadensis*, *Picea glauca*, *P. mariana*
Sewing Material: *Epilobium angustifolium*, *Picea glauca*, *P. mariana*
Snow Gear: *Alnus incana*, *Betula papyrifera*, *Larix laricina*

Dye
Brown: *Alnus incana*, *Cornus sericea*
Orange-Red: *Alnus incana*
Red-Brown: *Alnus incana*
Unspecified: *Alnus incana*, *Vaccinium myrtilloides*, *V. vitis-idaea*
Yellow: *Alnus incana*
Yellow-Brown: *Picea glauca*

Other
Containers: *Betula papyrifera*, *Calamagrostis canadensis*, *Salix bebbiana*, *S. discolor*, *S. interior*
Cooking Tools: *Betula papyrifera*, *Lonicera dioica*, *Lycopodium annotinum*, *L. obscurum*, *Picea glauca*, *Salix bebbiana*, *S. discolor*, *S. interior*, *Sarracenia purpurea*
Designs: *Picea mariana*
Fasteners: *Picea glauca*, *Salix bebbiana*, *S. discolor*, *S. interior*
Fuel: *Betula papyrifera*
Hide Preparation: *Betula papyrifera*, *Larix laricina*, *Picea glauca*, *P. mariana*, *Pinus banksiana*
Hunting & Fishing Item: *Achillea millefolium*, *A. sp.*, *Artemisia frigida*, *Betula papyrifera*, *Myrica gale*, *Picea glauca*, *Populus balsamifera*, *P. tremuloides*, *Salix bebbiana*, *S. discolor*, *S. interior*
Jewelry: *Rosa acicularis*, *Vaccinium vitis-idaea*
Season Indicator: *Epilobium angustifolium*
Smoke Plant: *Arctostaphylos uva-ursi*, *Cornus sericea*

Smoking Tools: *Lonicera dioica*
Tools: *Picea glauca, Salix bebbiana, S. discolor, S. interior*
Toys & Games: *Picea mariana, Populus tremuloides, Rosa acicularis, Salix bebbiana, S. discolor, S. interior*

Creek
Drug
Abortifacient: *Stillingia* sp., *Tephrosia virginiana*
Adjuvant: *Gnaphalium obtusifolium, Panax* sp.
Alterative: *Persea palustris*
Analgesic: *Angelica* sp., *Celastrus scandens, Eryngium yuccifolium, Erythrina herbacea, Juniperus* sp., *Lindera benzoin*
Anthelmintic: *Angelica* sp., *Spigelia anthelmia, S. marilandica*
Antidiarrheal: *Prunus* sp., *Quercus stellata, Rhus copallinum, R. glabra*
Antiemetic: *Gnaphalium obtusifolium, Salix* sp.
Antihemorrhagic: *Aralia* sp.
Antirheumatic (External): *Juniperus* sp., *Liatris* sp., *Monarda* sp., *Salix* sp.
Antirheumatic (Internal): *Eryngium yuccifolium, Liatris* sp., *Lindera benzoin, Monarda* sp., *Salix* sp.
Blood Medicine: *Eryngium yuccifolium, Juniperus* sp., *Lindera benzoin*
Breast Treatment: *Sambucus canadensis*
Carminative: *Angelica* sp.
Cathartic: *Eryngium yuccifolium, Euphorbia* sp., *Ilex vomitoria, Iris verna, I. versicolor, Stillingia* sp.
Ceremonial Medicine: *Panax* sp.
Cold Remedy: *Gnaphalium obtusifolium*
Cough Medicine: *Verbascum thapsus*
Dermatological Aid: *Heuchera americana, Monarda* sp., *Panax quinquefolius, Salix* sp., *Smilax* sp.
Diaphoretic: *Lindera benzoin, Monarda* sp., *Panax quinquefolius, P.* sp., *Persea palustris*
Diuretic: *Ipomoea pandurata*
Ear Medicine: *Monarda* sp.
Emetic: *Ilex vomitoria, Lindera benzoin, Morus rubra, Myrica* sp., *Salix* sp.
Febrifuge: *Chenopodium ambrosioides, Panax quinquefolius, P.* sp., *Persea palustris, Salix* sp.
Gastrointestinal Aid: *Angelica* sp., *Eryngium yuccifolium, Ligusticum canadense, Salix* sp.
Gynecological Aid: *Celastrus scandens, Porteranthus* sp., *Sambucus canadensis, Stillingia* sp.
Hemostat: *Panax quinquefolius, P.* sp.
Herbal Steam: *Lindera benzoin, Malus* sp.
Kidney Aid: *Eryngium yuccifolium, Impatiens* sp., *Ipomoea pandurata, Monarda* sp., *Persea palustris, Populus* sp., *Salix* sp.
Liver Aid: *Frangula caroliniana*
Misc. Disease Remedy: *Gleditsia triacanthos, Gnaphalium obtusifolium, Malus* sp., *Panicum* sp., *Salix* sp.
Orthopedic Aid: *Angelica* sp., *Celastrus scandens, Populus* sp., *Quercus* sp.
Other: *Rhus glabra, Salix* sp., *Smilax bona-nox*
Panacea: *Chenopodium ambrosioides, Eryngium yuccifolium, Gleditsia triacanthos*
Pediatric Aid: *Angelica* sp., *Baptisia* sp., *Gleditsia triacanthos, Panax* sp., *Quercus* sp., *Spigelia anthelmia*
Poison: *Lomatium nuttallii*
Psychological Aid: *Gnaphalium obtusifolium, Monarda* sp.
Pulmonary Aid: *Panax quinquefolius, P.* sp., *Phoradendron leucarpum*
Reproductive Aid: *Stillingia* sp., *Tephrosia virginiana*
Sedative: *Angelica* sp., *Eryngium yuccifolium, Gnaphalium obtusifolium, Monarda* sp.
Snakebite Remedy: *Eryngium yuccifolium, Manfreda virginica*
Stimulant: *Baptisia* sp., *Morus rubra*
Tonic: *Chenopodium ambrosioides, Heuchera americana, Juniperus* sp.
Toothache Remedy: *Achillea millefolium, Ulmus* sp.
Tuberculosis Remedy: *Aesculus* sp., *Betula* sp., *Collinsia violacea, Nyssa* sp., *N. sylvatica, Penstemon* sp., *Phoradendron leucarpum, Platanus occidentalis, Tephrosia virginiana*
Unspecified: *Bignonia capreolata, Camassia scilloides, Lomatium nuttallii, Sassafras albidum, Triosteum perfoliatum*
Urinary Aid: *Celastrus scandens, Morus rubra, Tephrosia virginiana*
Venereal Aid: *Eryngium yuccifolium, Parthenocissus* sp., *Salix* sp.
Veterinary Aid: *Zanthoxylum americanum*
Witchcraft Medicine: *Gnaphalium obtusifolium, Monarda* sp., *Ulmus rubra*

Crow
Drug
Analgesic: *Equisetum hyemale, Oplopanax horridus, Salix* sp.
Antidiarrheal: *Juniperus scopulorum, Prunus virginiana, Rosa* sp.
Antihemorrhagic: *Juniperus scopulorum*
Antirheumatic (External): *Lomatium macrocarpum, Rosa* sp.
Burn Dressing: *Achillea millefolium, Prunus virginiana*
Ceremonial Medicine: *Abies lasiocarpa, Catabrosa aquatica, Glyceria fluitans, Juniperus chinense, Lobelia inflata, Madia glomerata*
Cold Remedy: *Abies lasiocarpa, Echinacea pallida, Grindelia squarrosa, Ligusticum canbyi, Lomatium macrocarpum*
Cough Medicine: *Abies lasiocarpa, Grindelia squarrosa, Ligusticum canbyi*
Dermatological Aid: *Achillea millefolium, Artemisia ludoviciana, Prunus virginiana*
Dietary Aid: *Juniperus scopulorum*
Diuretic: *Equisetum hyemale*
Ear Medicine: *Ligusticum canbyi*
Emetic: *Salix* sp.
Eye Medicine: *Artemisia dracunculus*
Gastrointestinal Aid: *Echinacea pallida, Juniperus scopulorum, Rosa* sp.
Gynecological Aid: *Juniperus scopulorum*
Laxative: *Abies lasiocarpa*
Nose Medicine: *Rosa* sp.
Oral Aid: *Arctostaphylos uva-ursi, Rosa* sp., *Salix* sp.
Pulmonary Aid: *Grindelia squarrosa*
Respiratory Aid: *Grindelia squarrosa, Ligusticum canbyi, Monarda fistulosa*
Throat Aid: *Lomatium macrocarpum, Rosa* sp.
Toothache Remedy: *Echinacea pallida*
Veterinary Aid: *Symphoricarpos albus*
Food
Dried Food: *Vitis vulpina*
Fruit: *Escobaria missouriensis, Prunus americana, Vitis vulpina*
Sauce & Relish: *Asclepias speciosa*
Unspecified: *Asclepias speciosa, Catabrosa aquatica, Glyceria fluitans, Leucocrinum montanum, Madia glomerata, Musineon divaricatum*
Winter Use Food: *Prunus americana*
Fiber
Building Material: *Populus tremuloides*
Mats, Rugs & Bedding: *Matricaria discoidea*
Dye
Mordant: *Opuntia polyacantha*
Other
Ceremonial Items: *Artemisia ludoviciana, Lomatium macrocarpum*
Fasteners: *Amelanchier alnifolia, Cornus sericea, Pinus ponderosa, Prunus virginiana*
Incense & Fragrance: *Ligusticum canbyi, Lomatium macrocarpum, Monarda fistulosa*

Musical Instrument: *Cornus sericea*
Paint: *Asclepias speciosa*
Smoke Plant: *Ligusticum canbyi*
Soap: *Morchella* sp.
Tools: *Cornus sericea*

Dakota
Drug
Abortifacient: *Artemisia frigida*
Analgesic: *Callirhoe involucrata, Echinacea angustifolia, Helianthus annuus, Humulus lupulus, Monarda fistulosa, Penstemon grandiflorus, Ratibida columnifera, Sphaeralcea coccinea, Verbena hastata*
Anthelmintic: *Echinacea pallida, Mirabilis nyctaginea*
Antidiarrheal: *Ambrosia artemisiifolia, Dalea aurea*
Antidote: *Echinacea angustifolia*
Antiemetic: *Ambrosia artemisiifolia*
Burn Dressing: *Echinacea angustifolia, Sphaeralcea coccinea, Typha latifolia*
Carminative: *Acorus calamus, Mentha canadensis*
Ceremonial Medicine: *Acorus calamus, Artemisia* sp., *Hierochloe odorata, Prunus virginiana, Shepherdia argentea*
Cold Remedy: *Acorus calamus, Callirhoe involucrata, Hedeoma hispida, H.* sp., *Juniperus virginiana*
Cough Medicine: *Acorus calamus, Juniperus virginiana*
Dermatological Aid: *Echinacea pallida, Humulus lupulus, Lilium philadelphicum, Mirabilis nyctaginea, Opuntia humifusa, Ratibida columnifera, Rumex altissimus, R. crispus, Sphaeralcea coccinea, Typha latifolia, Yucca glauca*
Dietary Aid: *Hedeoma hispida, H.* sp.
Disinfectant: *Artemisia* sp.
Ear Medicine: *Glycyrrhiza lepidota*
Eye Medicine: *Echinacea pallida, Symphoricarpos occidentalis, S. orbiculatus*
Febrifuge: *Acorus calamus, Astragalus canadensis, Glycyrrhiza lepidota, Humulus lupulus, Mentzelia nuda, Mirabilis nyctaginea*
Gastrointestinal Aid: *Acorus calamus, Artemisia* sp., *Dalea aurea, Grindelia squarrosa, Humulus lupulus, Monarda fistulosa, Quercus* sp., *Verbena hastata, V. stricta*
Laxative: *Gymnocladus dioicus, Ulmus rubra*
Love Medicine: *Lomatium foeniculaceum*
Misc. Disease Remedy: *Echinacea angustifolia, Juniperus virginiana*
Other: *Echinacea angustifolia*
Panacea: *Acorus calamus, Cucurbita foetidissima, Ratibida columnifera*
Pediatric Aid: *Astragalus canadensis, Glycyrrhiza lepidota, Grindelia squarrosa, Quercus* sp., *Typha latifolia*
Poison: *Dalea enneandra*
Psychological Aid: *Acorus calamus*
Pulmonary Aid: *Helianthus annuus*
Snakebite Remedy: *Echinacea angustifolia, E. pallida*
Stimulant: *Gymnocladus dioicus*
Tonic: *Gentiana saponaria, Hedeoma* sp.
Toothache Remedy: *Acorus calamus, Echinacea angustifolia, Glycyrrhiza lepidota*
Tuberculosis Remedy: *Psoralidium tenuiflorum*
Veterinary Aid: *Clematis ligusticifolia, Dyssodia papposa, Echinacea angustifolia, Glycyrrhiza lepidota, Gutierrezia sarothrae, Juniperus virginiana, Silphium laciniatum*
Witchcraft Medicine: *Artemisia* sp., *Silphium laciniatum*

Food
Beverage: *Agastache foeniculum, Ceanothus americanus, Mentha canadensis, Ratibida columnifera, Rubus idaeus, R. occidentalis, Sambucus canadensis*
Bread & Cake: *Prunus virginiana*
Candy: *Populus deltoides, P.* sp., *Silphium laciniatum*
Dried Food: *Opuntia humifusa, Pediomelum esculentum, Physalis heterophylla, Prunus americana, P. pumila, P. virginiana, Rubus idaeus, R. occidentalis, Shepherdia argentea, Vitis cinerea, V. vulpina*
Fodder: *Linum lewisii, Populus* sp.
Forage: *Dyssodia papposa, Populus deltoides*
Fruit: *Amelanchier alnifolia, Fragaria vesca, F. virginiana, Opuntia humifusa, Prunus americana, P. pumila, P. virginiana, Ribes missouriense, Rubus idaeus, R. occidentalis, Sambucus canadensis, Shepherdia argentea, Viburnum lentago, Vitis cinerea, V. vulpina*
Sauce & Relish: *Allium canadense, Physalis heterophylla, Prunus americana, P. pumila*
Soup: *Carya ovata, Chenopodium album, Corylus americana, Juglans nigra, Nelumbo lutea*
Spice: *Allium canadense, Celtis occidentalis, Mentha canadensis*
Staple: *Zea mays, Zizania aquatica*
Starvation Food: *Opuntia humifusa, Rosa arkansana*
Sweetener: *Acer negundo, A. saccharinum, A. saccharum, Agastache foeniculum, Carya ovata, Populus* sp., *Zea mays*
Unspecified: *Acorus calamus, Allium canadense, Amphicarpaea bracteata, Apios americana, Asclepias syriaca, Astragalus crassicarpus, Carya ovata, Chenopodium album, Corylus americana, Helianthus tuberosus, Juglans nigra, Linum lewisii, Mentha canadensis, Nelumbo lutea, Pediomelum esculentum, Physalis lanceolata, Populus deltoides, Quercus macrocarpa, Q. rubra, Sagittaria latifolia, S.* sp., *Scirpus acutus, S. tabernaemontani*
Winter Use Food: *Physalis heterophylla, Prunus virginiana, Zea mays*

Fiber
Basketry: *Salix melanopsis*
Brushes & Brooms: *Prunus americana*
Building Material: *Pinus contorta, Salix* sp., *Ulmus americana, U. rubra, U. thomasii*
Clothing: *Psoralidium tenuiflorum, Urtica dioica*
Cordage: *Ulmus rubra, Urtica dioica, Yucca glauca*
Mats, Rugs & Bedding: *Artemisia dracunculus, A. frigida, A. ludoviciana, Scirpus acutus, S. tabernaemontani, Typha latifolia*
Sewing Material: *Yucca glauca*

Dye
Black: *Gymnocladus dioicus, Juglans nigra*
Mordant: *Opuntia humifusa, Rhus trilobata*
Red: *Rhus trilobata*
Yellow: *Populus deltoides*

Other
Ceremonial Items: *Acer negundo, Hierochloe odorata, Lagenaria siceraria, Nelumbo lutea, Populus* sp., *Prunus americana, Urtica dioica*
Containers: *Betula papyrifera, Yucca glauca*
Cooking Tools: *Betula papyrifera, Ulmus americana, U. rubra, U. thomasii*
Decorations: *Acer negundo, Symphoricarpos orbiculatus*
Fuel: *Ceanothus americanus, Populus* sp., *Ulmus americana, U. rubra, U. thomasii, Yucca glauca*
Hide Preparation: *Prunus americana, Yucca glauca*
Hunting & Fishing Item: *Amelanchier alnifolia, Fraxinus pennsylvanica*
Incense & Fragrance: *Melilotus officinalis, Monarda fistulosa, Thalictrum dasycarpum*
Lighting: *Betula papyrifera*
Musical Instrument: *Populus deltoides*
Protection: *Juniperus virginiana, Psoralidium tenuiflorum*
Season Indicator: *Prunus virginiana*
Smoke Plant: *Cornus amomum, C. sericea, Nicotiana quadrivalvis, Rhus glabra, Rosa arkansana*
Smoking Tools: *Fraxinus pennsylvanica*
Soap: *Yucca glauca*

Stable Gear: *Ulmus americana, U. thomasii, Urtica dioica*
Tools: *Ulmus americana, U. rubra, U. thomasii, Yucca glauca*
Toys & Games: *Amelanchier alnifolia, Artemisia* sp., *Betula papyrifera, Opuntia humifusa, Physalis lanceolata, Populus deltoides, Prunus americana, Quercus macrocarpa, Sambucus canadensis, Scirpus acutus, Thalictrum dasycarpum, Ulmus americana, U. rubra, Urtica dioica, Viburnum opulus*

Delaware
Drug
Abortifacient: *Acorus calamus, Mitchella repens*
Analgesic: *Cornus canadensis, Symplocarpus foetidus*
Anthelmintic: *Eryngium aquaticum, Juglans nigra, Prunus persica*
Antidiarrheal: *Prunus serotina, Rubus allegheniensis, Solidago juncea*
Antiemetic: *Cercis canadensis, Prunus persica*
Antirheumatic (External): *Aesculus glabra, Asclepias tuberosa, Cirsium vulgare, Gaultheria procumbens, Goodyera pubescens, Iris versicolor, Mitchella repens, Phytolacca americana, Tsuga canadensis, Verbascum thapsus*
Antirheumatic (Internal): *Arctium minus*
Blood Medicine: *Ambrosia artemisiifolia, Arctium minus, Chimaphila umbellata, Comptonia peregrina, Gelsemium sempervirens, Gentianopsis crinita, Gleditsia triacanthos, Myrica* sp., *Phytolacca americana, Rumex crispus, R. obtusifolius, Sambucus canadensis, Sassafras albidum*
Cathartic: *Maianthemum stellatum, Pimpinella anisum*
Cold Remedy: *Acorus calamus, Platanus occidentalis, Quercus velutina, Ulmus americana*
Cough Medicine: *Acorus calamus, Gleditsia triacanthos, Petasites frigidus, Prunus serotina, Quercus alba, Q. rubra, Ulmus americana, Verbascum thapsus*
Dermatological Aid: *Baptisia tinctoria, Celastrus scandens, Chimaphila umbellata, Comptonia peregrina, Datura stramonium, Juglans nigra, Monarda punctata, Phytolacca americana, Rhus copallinum, Sambucus canadensis*
Disinfectant: *Quercus alba*
Ear Medicine: *Aesculus glabra, Humulus lupulus*
Febrifuge: *Cercis canadensis, Eupatorium perfoliatum, Monarda punctata, Prunella vulgaris, Solidago juncea*
Gastrointestinal Aid: *Acorus calamus, Angelica atropurpurea, Cardamine diphylla, Gentianopsis crinita, Hedeoma pulegioides, Inula helenium, Juglans nigra, Maianthemum stellatum, Pimpinella anisum, Quercus palustris, Sanguinaria canadensis, Scutellaria galericulata*
Gland Medicine: *Phytolacca americana*
Gynecological Aid: *Asclepias tuberosa, Baptisia tinctoria, Goodyera pubescens, Leonurus cardiaca, Maianthemum stellatum, Plantago major, Quercus alba*
Heart Medicine: *Zanthoxylum americanum*
Hemorrhoid Remedy: *Datura stramonium*
Kidney Aid: *Achillea millefolium, Gaultheria procumbens, Hydrangea arborescens, Iris versicolor, Myrica* sp., *Typha latifolia*
Laxative: *Podophyllum peltatum, Scutellaria galericulata, Taraxacum officinale*
Liver Aid: *Achillea millefolium, Celastrus scandens, Iris versicolor, Rumex crispus, R. obtusifolius, Sambucus canadensis*
Misc. Disease Remedy: *Asclepias* sp., *Daucus carota, Lobelia cardinalis, Symplocarpus foetidus*
Narcotic: *Crotalaria sagittalis*
Oral Aid: *Rhus copallinum*
Panacea: *Lophophora williamsii*
Pediatric Aid: *Chrysopsis mariana, Nepeta cataria, Prunus persica, Sambucus canadensis*
Poison: *Aesculus glabra*
Pulmonary Aid: *Asclepias tuberosa, Chimaphila umbellata, Comptonia peregrina, Goodyera pubescens, Petasites frigidus, Symplocarpus foetidus, Verbascum thapsus*
Reproductive Aid: *Salix humilis, Viburnum prunifolium, Vitis rupestris*
Respiratory Aid: *Petasites frigidus, Verbascum thapsus*
Sedative: *Chrysopsis mariana, Humulus lupulus*
Stimulant: *Arctium minus, Humulus lupulus, Phytolacca americana*
Strengthener: *Populus deltoides, Sanguinaria canadensis, Vitis rupestris*
Throat Aid: *Helianthemum canadense, Platanus occidentalis, Quercus alba, Q. rubra, Q. velutina*
Tonic: *Aristolochia serpentaria, Chrysopsis mariana, Cimicifuga racemosa, Cornus florida, Gaultheria procumbens, Inula helenium, Panax quinquefolius, Podophyllum peltatum, Prunus serotina*
Toothache Remedy: *Humulus lupulus*
Tuberculosis Remedy: *Lophophora williamsii*
Unspecified: *Panax quinquefolius, Plantago major, Toxicodendron pubescens*
Urinary Aid: *Chimaphila umbellata, Comptonia peregrina*
Venereal Aid: *Cardamine diphylla, Chimaphila umbellata, Comptonia peregrina, Crotalaria sagittalis, Echinacea purpurea, Iris versicolor, Rhus copallinum, R. hirta, Salix humilis, Stachys palustris*
Food
Bread & Cake: *Apios americana, Zea mays*
Dried Food: *Zea mays*
Porridge: *Zea mays*
Soup: *Zea mays*
Staple: *Zea mays*
Unspecified: *Agaricus campestris, Apios americana, Rumex acetosella, Zea mays*
Winter Use Food: *Apios americana*
Other
Ceremonial Items: *Sanguinaria canadensis, Zea mays*
Insecticide: *Juglans nigra*

Delaware, Oklahoma
Drug
Abortifacient: *Acorus calamus, Mitchella repens*
Analgesic: *Acorus calamus, Cornus canadensis, Hedeoma pulegioides, Helianthemum canadense, Quercus palustris, Symplocarpus foetidus, Verbascum thapsus*
Anthelmintic: *Eryngium aquaticum*
Anticonvulsive: *Asclepias* sp., *Symplocarpus foetidus*
Antidiarrheal: *Prunus serotina, Rubus allegheniensis, R. canadensis, Solidago juncea*
Antiemetic: *Cercis canadensis*
Antirheumatic (External): *Cirsium vulgare, Juniperus virginiana, Mitchella repens, Phytolacca americana, Tsuga canadensis, Verbascum thapsus*
Antirheumatic (Internal): *Arctium minus, Asclepias tuberosa, Gaultheria procumbens, Goodyera pubescens, Iris versicolor, Phytolacca americana*
Blood Medicine: *Ambrosia artemisiifolia, Arctium minus, Chimaphila umbellata, Comptonia peregrina, Gelsemium sempervirens, Gentianopsis crinita, Gleditsia triacanthos, Myrica* sp., *Phytolacca americana, Rumex crispus, R. obtusifolius, Sambucus canadensis, Sassafras albidum*
Cathartic: *Betula alleghaniensis, Frangula caroliniana, Fraxinus americana, Juglans nigra, Maianthemum stellatum, Morus nigra, Pimpinella anisum*
Ceremonial Medicine: *Artemisia frigida, Rhus copallinum*
Cold Remedy: *Acorus calamus, Platanus occidentalis, Quercus velutina, Ulmus americana, Verbascum thapsus*
Cough Medicine: *Acorus calamus, Gleditsia triacanthos, Petasites frigidus, Prunus serotina, Quercus alba, Q. rubra, Symplocarpus foetidus, Ulmus americana, Verbascum thapsus*
Dermatological Aid: *Ambrosia artemisiifolia, Baptisia tinctoria, Chimaphila umbellata, Comptonia peregrina, Datura stramonium, Gelsemium sempervirens, Juglans nigra, Menispermum canadense, Quercus alba, Rhus copallinum, Sambucus canadensis, Solanum dulcamara, Toxicodendron pubescens, Vitis vulpina*

Disinfectant: *Quercus alba*
Ear Medicine: *Aesculus glabra, Humulus lupulus*
Emetic: *Betula alleghaniensis, Frangula caroliniana, Fraxinus americana, Juglans nigra, Morus nigra*
Expectorant: *Chimaphila umbellata, Comptonia peregrina*
Febrifuge: *Cercis canadensis, Eupatorium perfoliatum, Monarda punctata, Prunella vulgaris, Solanum dulcamara, Solidago juncea, Verbena hastata*
Gastrointestinal Aid: *Angelica atropurpurea, Betula alleghaniensis, Cardamine diphylla, Frangula caroliniana, Fraxinus americana, Gentianopsis crinita, Hedeoma pulegioides, Inula helenium, Juglans nigra, Maianthemum stellatum, Morus nigra, Pimpinella anisum, Quercus palustris, Sambucus canadensis, Sanguinaria canadensis*
Gynecological Aid: *Asclepias tuberosa, Baptisia tinctoria, Goodyera pubescens, Leonurus cardiaca, Maianthemum stellatum, Mitchella repens, Myrica* sp., *Plantago major, Populus deltoides, Quercus alba, Viburnum prunifolium, Vitis vulpina*
Heart Medicine: *Zanthoxylum americanum*
Hemorrhoid Remedy: *Datura stramonium*
Herbal Steam: *Cirsium vulgare, Juniperus virginiana, Mitchella repens, Phytolacca americana, Tsuga canadensis*
Kidney Aid: *Achillea millefolium, Iris versicolor, Myrica* sp., *Typha latifolia*
Laxative: *Inula helenium, Podophyllum peltatum, Scutellaria galericulata, Taraxacum officinale*
Liver Aid: *Achillea millefolium, Betula alleghaniensis, Frangula caroliniana, Fraxinus americana, Hydrangea arborescens, Iris versicolor, Juglans nigra, Morus nigra, Rumex crispus, R. obtusifolius, Sambucus canadensis*
Love Medicine: *Podophyllum peltatum*
Misc. Disease Remedy: *Daucus carota, Lobelia cardinalis*
Narcotic: *Crotalaria sagittalis*
Oral Aid: *Rhus copallinum*
Panacea: *Panax quinquefolius, Quercus alba, Sanguinaria canadensis*
Pediatric Aid: *Nepeta cataria, Sambucus canadensis*
Poison: *Aesculus glabra*
Pulmonary Aid: *Asclepias tuberosa, Goodyera pubescens, Petasites frigidus, Verbascum thapsus*
Reproductive Aid: *Vitis rupestris*
Respiratory Aid: *Petasites frigidus, Verbascum thapsus*
Sedative: *Chrysopsis mariana, Humulus lupulus*
Stimulant: *Arctium minus, Humulus lupulus, Maianthemum stellatum, Phytolacca americana*
Throat Aid: *Platanus occidentalis, Quercus alba, Q. rubra, Q. velutina*
Tonic: *Anaphalis margaritacea, Aralia nudicaulis, Aristolochia serpentaria, Chrysopsis mariana, Cimicifuga racemosa, Cornus florida, Gaultheria procumbens, Gleditsia triacanthos, Helianthemum canadense, Humulus lupulus, Inula helenium, Maianthemum racemosum, Mentha ×piperita, Nepeta cataria, Panax quinquefolius, Pimpinella anisum, Podophyllum peltatum, Prunus serotina, Sanguinaria canadensis, Sassafras albidum, Taraxacum officinale, Vitis rupestris, Zanthoxylum americanum*
Toothache Remedy: *Humulus lupulus*
Tuberculosis Remedy: *Cardamine diphylla, Chimaphila umbellata, Comptonia peregrina, Iris versicolor, Maianthemum stellatum, Salix humilis*
Unspecified: *Goodyera pubescens, Plantago major*
Urinary Aid: *Chimaphila umbellata, Comptonia peregrina*
Venereal Aid: *Cardamine diphylla, Crotalaria sagittalis, Echinacea purpurea, Eryngium aquaticum, Maianthemum stellatum, Rhus copallinum, R. hirta, Salix humilis, Solanum nigrum, Stachys palustris*

Other
Ceremonial Items: *Rhus copallinum, Sanguinaria canadensis*
Insecticide: *Juglans nigra*
Smoke Plant: *Angelica atropurpurea*

Delaware, Ontario
Drug
Analgesic: *Arctium minus, Armoracia rusticana, Pinus strobus, Verbascum thapsus*
Antidiarrheal: *Rhus hirta*
Antiemetic: *Quercus muehlenbergii, Sanguinaria canadensis*
Blood Medicine: *Arctium minus, Sanguinaria canadensis*
Cold Remedy: *Acorus calamus, Inula helenium, Populus tremuloides*
Dermatological Aid: *Pinus strobus, Plantago major, Sedum telephioides, Verbascum thapsus*
Disinfectant: *Sedum telephioides*
Febrifuge: *Thymus praecox*
Gastrointestinal Aid: *Eupatorium perfoliatum, Tanacetum vulgare*
Gynecological Aid: *Carpinus caroliniana, Carya alba, C. ovata, Juniperus communis, Ostrya virginiana, Prunus serotina, Quercus alba*
Kidney Aid: *Pinus strobus*
Misc. Disease Remedy: *Viburnum lentago*
Pediatric Aid: *Nepeta cataria, Pinus strobus*
Pulmonary Aid: *Pinus strobus*
Sedative: *Nepeta cataria*
Tonic: *Carpinus caroliniana, Carya alba, C. ovata, Juniperus communis, Ostrya virginiana, Prunus serotina, Quercus alba*
Tuberculosis Remedy: *Celastrus scandens*
Unspecified: *Pinus strobus*

Diegueño
Drug
Analgesic: *Juniperus californica, Lamarckia aurea*
Antidiarrheal: *Eriogonum fasciculatum, Fragaria vesca, Gutierrezia sarothrae, Rubus ursinus*
Antidote: *Cuscuta californica*
Antihemorrhagic: *Pellaea mucronata*
Antirheumatic (External): *Hazardia squarrosa, Larrea tridentata, Urtica dioica*
Blood Medicine: *Ephedra californica, Grindelia hallii, Platanus racemosa, Salvia apiana*
Cathartic: *Frangula californica*
Cold Remedy: *Artemisia tridentata, Eriodictyon lanatum, Marrubium vulgare, Salvia apiana*
Cough Medicine: *Artemisia tridentata, Baccharis sarothroides, Croton californicus, Eriodictyon lanatum, Prunus ilicifolia, Salvia apiana*
Dermatological Aid: *Ambrosia psilostachya, Baccharis salicifolia, Ceanothus leucodermis, Chamaesyce albomarginata, Dudleya pulverulenta, Ericameria brachylepis, Frangula californica, Heteromeles arbutifolia, Malva parviflora, Phoradendron macrophyllum, Populus fremontii, Quercus agrifolia, Ricinus communis, Solidago californica, Vitis girdiana*
Dietary Aid: *Ephedra californica*
Ear Medicine: *Ruta chalepensis, R. graveolens*
Emetic: *Eriogonum fasciculatum*
Eye Medicine: *Physalis philadelphica, Prosopis glandulosa, Quercus dumosa, Rhus trilobata, Toxicodendron diversilobum*
Febrifuge: *Brickellia californica, Malva parviflora, Matricaria discoidea, Prosopis glandulosa, Rosa californica, Sambucus cerulea, Scrophularia californica*
Gastrointestinal Aid: *Adenostoma sparsifolium, Ambrosia psilostachya, Baccharis sarothroides, Ephedra californica, Lepidium nitidum, Paeonia californica, Ruta graveolens*
Gynecological Aid: *Matricaria discoidea, Rhus ovata*
Hallucinogen: *Datura wrightii*
Heart Medicine: *Eriogonum fasciculatum*
Hypotensive: *Equisetum laevigatum, Juniperus californica*
Kidney Aid: *Arctostaphylos* sp., *Ephedra californica*

Laxative: *Sambucus cerulea*
Misc. Disease Remedy: *Ericameria brachylepis, Salvia apiana*
Orthopedic Aid: *Larrea tridentata, Populus fremontii*
Other: *Salvia apiana*
Pediatric Aid: *Eriogonum fasciculatum, Marrubium vulgare, Matricaria discoidea, Rhus trilobata, Rosa californica, Sambucus cerulea*
Poison: *Datura wrightii*
Pulmonary Aid: *Marrubium vulgare*
Respiratory Aid: *Artemisia tridentata, Dudleya pulverulenta, Platanus racemosa*
Strengthener: *Salvia columbariae*
Toothache Remedy: *Adenostoma sparsifolium*
Unspecified: *Anemopsis californica, Datura wrightii, Monardella lanceolata*
Veterinary Aid: *Croton setigerus, Lonicera subspicata*

Food
Beverage: *Monardella lanceolata, Pellaea mucronata, Rhus integrifolia, Satureja douglasii*
Bread & Cake: *Prosopis glandulosa, Prunus ilicifolia*
Candy: *Eriodictyon lanatum, Marrubium vulgare, Quercus engelmannii*
Dried Food: *Eriodictyon lanatum, Opuntia basilaris, Rubus ursinus, Vitis girdiana*
Fodder: *Acacia greggii, Lotus scoparius*
Fruit: *Arctostaphylos glauca, Fragaria vesca, Heteromeles arbutifolia, Juniperus californica, Mammillaria dioica, Opuntia engelmannii, Prunus ilicifolia, Quercus dunnii, Rubus ursinus, Sambucus cerulea, Vitis girdiana*
Porridge: *Avena fatua, Prunus ilicifolia, Quercus agrifolia, Q. chrysolepis, Q. dumosa, Q. engelmannii, Q. kelloggii, Q. wislizeni, Salvia apiana*
Preserves: *Arctostaphylos* sp.
Spice: *Salvia carduacea, S. columbariae*
Staple: *Quercus peninsularis*
Starvation Food: *Juniperus californica*
Sweetener: *Justicia californica*
Unspecified: *Agave deserti, Cleome isomeris, Dudleya pulverulenta, Penstemon centranthifolius, Pinus monophylla, P. quadrifolia, Salvia apiana, Yucca whipplei*
Vegetable: *Brassica nigra, Chenopodium album, Claytonia perfoliata, Erodium cicutarium, Lepidium nitidum, Opuntia engelmannii, Paeonia californica, Rorippa nasturtium-aquaticum, Trifolium sp., Viola pedunculata*
Winter Use Food: *Sambucus cerulea*

Fiber
Basketry: *Juncus textilis, Muhlenbergia rigens, Pinus coulteri, P. jeffreyi, P. ponderosa, Quercus dumosa, Rhus trilobata, Salix lasiolepis, Yucca whipplei*
Brushes & Brooms: *Arctostaphylos glauca*
Building Material: *Pinus jeffreyi, P. ponderosa*
Clothing: *Yucca schidigera*
Furniture: *Quercus dumosa, Yucca whipplei*

Dye
Black: *Quercus chrysolepis*
Red: *Plagiobothrys arizonicus*

Other
Cash Crop: *Opuntia basilaris, Phaseolus vulgaris*
Ceremonial Items: *Datura wrightii, Urtica dioica*
Cooking Tools: *Prosopis glandulosa*
Fuel: *Adenostoma fasciculatum, A. sparsifolium, Arctostaphylos glauca, Populus fremontii, Prosopis glandulosa, Quercus agrifolia*
Good Luck Charm: *Pellaea mucronata*
Incense & Fragrance: *Eriodictyon lanatum*
Lubricant: *Opuntia engelmannii*
Smoke Plant: *Nicotiana glauca*
Soap: *Atriplex canescens, Chenopodium californicum, Cucurbita foetidissima, Lotus scoparius*
Tools: *Prosopis glandulosa*
Toys & Games: *Sambucus cerulea*

Eskimo
Drug
Cancer Treatment: *Artemisia tilesii*
Disinfectant: *Artemisia tilesii*
Febrifuge: *Artemisia tilesii*
Gastrointestinal Aid: *Valeriana capitata*
Panacea: *Angelica lucida*
Preventive Medicine: *Angelica lucida*
Psychological Aid: *Angelica lucida*
Unspecified: *Potentilla nana*

Food
Fruit: *Rubus chamaemorus*

Eskimo, Alaska
Drug
Anesthetic: *Salix planifolia*
Antihemorrhagic: *Equisetum sylvaticum, Ledum palustre, Matricaria discoidea*
Antirheumatic (External): *Artemisia* sp., *A. tilesii*
Cold Remedy: *Artemisia* sp., *Moneses uniflora, Pinus contorta*
Cough Medicine: *Moneses uniflora, Pinus contorta*
Dermatological Aid: *Alnus viridis, Artemisia* sp., *Picea glauca, Salix sitchensis*
Eye Medicine: *Salix planifolia*
Gastrointestinal Aid: *Ledum palustre*
Hemostat: *Artemisia tilesii*
Laxative: *Artemisia tilesii, Epilobium angustifolium*
Oral Aid: *Salix planifolia, Sedum integrifolium*
Respiratory Aid: *Artemisia tilesii*
Tonic: *Artemisia tilesii*
Unspecified: *Achillea millefolium, Ledum palustre, Picea glauca, Pinus contorta*

Food
Beverage: *Comarum palustre, Iris setosa, Lathyrus japonicus, Ledum palustre, Pentaphylloides floribunda, Sedum integrifolium*
Candy: *Matricaria discoidea, Picea glauca*
Fruit: *Arctostaphylos alpina, Cornus canadensis, C. suecica, Empetrum nigrum, Rubus arcticus, Vaccinium alpinum, V. oxycoccos, V. uliginosum, V. vitis-idaea*
Ice Cream: *Anemone narcissiflora, Empetrum nigrum*
Soup: *Claytonia* sp., *Dryopteris expansa, Hippuris vulgaris*
Sour: *Pedicularis* sp.
Spice: *Ledum palustre*
Unspecified: *Anemone narcissiflora, Angelica lucida, Artemisia tilesii, Astragalus polaris, Caltha palustris, Claytonia acutifolia, C.* sp., *Dryopteris expansa, Epilobium angustifolium, Equisetum arvense, Fucus* sp., *Honckenya peploides, Mertensia maritima, Oxyria digyna, Pedicularis lanata, P.* sp., *Ranunculus pallasii, Ribes triste, Rumex arcticus, Salix alaxensis, S. planifolia, Saxifraga nelsoniana, S. spicata, Sedum integrifolium, Senecio pseudoarnica, Taraxacum* sp.
Vegetable: *Angelica lucida, Hedysarum* sp., *Hippuris tetraphylla, Ligusticum scothicum, Petasites frigidus, Salix planifolia*
Winter Use Food: *Empetrum nigrum, Ligusticum scothicum, Rubus chamaemorus, Rumex arcticus, Salix planifolia, Saxifraga nelsoniana*

Fiber
Basketry: *Leymus mollis*
Building Material: *Cassiope tetragona, Picea* sp.
Canoe Material: *Picea* sp.
Clothing: *Poa* sp.
Cordage: *Leymus mollis*
Mats, Rugs & Bedding: *Eriophorum angustifolium, Leymus mollis, Petasites frigidus*

Dye
Orange: *Alnus viridis*
Unspecified: *Alnus viridis, Iris setosa*
Other
Cash Crop: *Leymus mollis*
Ceremonial Items: *Angelica lucida*
Cleaning Agent: *Artemisia tilesii*
Containers: *Petasites frigidus*
Cooking Tools: *Leymus mollis, Petasites frigidus, Picea* sp.
Fuel: *Alnus viridis, Betula nana, Picea* sp., *Spiraea stevenii*
Smoke Plant: *Alnus viridis, Petasites frigidus, Salix alaxensis*
Smoking Tools: *Angelica lucida*
Snuff: *Petasites frigidus, Salix alaxensis*
Tools: *Picea* sp.
Weapon: *Picea* sp.

Eskimo, Arctic
Drug
Poison: *Actaea rubra, Cicuta* sp.
Food
Beverage: *Ledum groenlandicum, L. palustre, Pentaphylloides floribunda, Polygonum alpinum, Vaccinium vitis-idaea*
Forage: *Arctostaphylos alpina, A. rubra, Hedysarum alpinum, H. boreale*
Frozen Food: *Empetrum nigrum, Vaccinium vitis-idaea*
Fruit: *Arctostaphylos uva-ursi, Empetrum nigrum*
Ice Cream: *Rubus chamaemorus*
Pie & Pudding: *Polygonum alpinum*
Preserves: *Vaccinium vitis-idaea*
Unspecified: *Oxyria digyna, Pedicularis lanata, Polygonum alpinum, Salix planifolia*
Vegetable: *Claytonia tuberosa, Epilobium latifolium, Hedysarum alpinum, H. boreale, Honckenya peploides, Pedicularis lanata, Petasites frigidus, Rumex arcticus, Saxifraga nelsoniana, Sedum rosea, Senecio congestus, Taraxacum* sp.
Other
Smoke Plant: *Arctostaphylos uva-ursi*

Eskimo, Chugach
Drug
Throat Aid: *Viburnum edule*

Eskimo, Greenland
Food
Unspecified: *Oxyria digyna*
Vegetable: *Angelica archangelica, Epilobium latifolium*

Eskimo, Inuktitut
Drug
Antidiarrheal: *Rumex arcticus*
Antirheumatic (Internal): *Alnus viridis*
Dermatological Aid: *Artemisia tilesii, Picea glauca, P. mariana, Salix arbusculoides, S.* sp.
Gastrointestinal Aid: *Valeriana capitata*
Laxative: *Fomes igniarius*
Poison: *Senecio congestus*
Respiratory Aid: *Picea glauca, P. mariana*
Strengthener: *Nephroma arcticum*
Unspecified: *Angelica lucida, Eriophorum angustifolium, Matricaria discoidea*
Veterinary Aid: *Nicotiana* sp.
Food
Fodder: *Alectoria nigricans, A. nitidula, A. ochroleuca, Cornicularia divergens*
Frozen Food: *Rubus chamaemorus*
Fruit: *Rubus chamaemorus, Vaccinium uliginosum, V. vitis-idaea, Viburnum edule*
Ice Cream: *Dryopteris campyloptera, Hippuris vulgaris, Rumex arcticus*
Sauce & Relish: *Cetraria cucullata*
Soup: *Hippuris vulgaris, Ranunculus lapponicus, Salix* sp.
Spice: *Allium schoenoprasum, Cetraria crispa, Rorippa islandica, Taraxacum* sp.
Unspecified: *Angelica lucida, Cicuta virosa, Epilobium angustifolium, E. latifolium, Eriophorum angustifolium, Hippuris vulgaris, Nephroma arcticum, Oxyria digyna, Picea glauca, P. mariana, Rumex arcticus, Salix* sp., *Saxifraga tricuspidata, Silene acaulis*
Winter Use Food: *Rubus arcticus*
Fiber
Building Material: *Picea glauca, P. mariana*
Canoe Material: *Parmelia saxatilis, Peltigera aphthosa, Picea glauca, P. mariana, Stereocaulon paschale*
Caulking Material: *Sphagnum* sp.
Clothing: *Eriophorum angustifolium, E. scheuchzeri, Poa* sp.
Cordage: *Picea glauca, P. mariana, Urtica dioica*
Mats, Rugs & Bedding: *Betula nana, Dryopteris fragrans, Eriophorum angustifolium, Picea glauca, P. mariana, Poa* sp.
Snow Gear: *Betula* sp.
Dye
Red-Brown: *Alnus viridis*
Other
Cash Crop: *Fomes igniarius*
Ceremonial Items: *Picea glauca, P. mariana*
Containers: *Betula* sp., *Picea glauca, P. mariana*
Cooking Tools: *Picea glauca, P. mariana, Populus* sp., *Salix* sp.
Fuel: *Alectoria nigricans, A. nitidula, A. ochroleuca, Betula nana, B.* sp., *Cassiope tetragona, Cornicularia divergens, Ledum palustre, Picea glauca, P. mariana, Populus* sp., *Salix* sp.
Hunting & Fishing Item: *Betula* sp., *Leymus mollis, Picea glauca, P. mariana, Populus balsamifera*
Incense & Fragrance: *Artemisia tilesii, Matricaria discoidea, Viola biflora*
Insecticide: *Alnus viridis, Fomes igniarius, Populus balsamifera*
Lighting: *Eriophorum* sp.
Paper: *Poa* sp.
Preservative: *Alnus viridis*
Season Indicator: *Matricaria discoidea*
Smoke Plant: *Arctostaphylos uva-ursi, Artemisia tilesii, Nicotiana* sp., *Petasites frigidus, Populus balsamifera, Salix* sp., *Vaccinium vitis-idaea*
Snuff: *Salix* sp.

Eskimo, Inupiat
Drug
Carminative: *Chenopodium album*
Cold Remedy: *Juniperus communis, Petasites frigidus*
Cough Medicine: *Juniperus communis*
Dermatological Aid: *Epilobium latifolium*
Eye Medicine: *Epilobium latifolium*
Misc. Disease Remedy: *Juniperus communis*
Oral Aid: *Salix planifolia*
Poison: *Aconitum delphiniifolium, Caltha palustris, Cicuta virosa, Hedysarum boreale, Iris setosa, Lathyrus japonicus, Ledum palustre, Lupinus arcticus, Ranunculus pallasii, Shepherdia canadensis, Zigadenus elegans*
Respiratory Aid: *Juniperus communis, Petasites frigidus*
Unspecified: *Ledum palustre*
Food
Beverage: *Ledum palustre, Polygonum alpinum, Rosa acicularis, Salix planifolia, Viburnum edule*

662 Eskimo, Inupiat

Cooking Agent: *Vaccinium uliginosum*
Dessert: *Empetrum nigrum, Pedicularis lanata, Polygonum alpinum, Ribes triste, Rubus arcticus, R. chamaemorus, R. idaeus, Rumex arcticus, Shepherdia canadensis, Vaccinium oxycoccos, V. uliginosum, V. vitis-idaea, Viburnum edule*
Dried Food: *Chenopodium album*
Frozen Food: *Arctostaphylos uva-ursi, Chenopodium album, Hedysarum alpinum, Oxytropis maydelliana, Rosa acicularis, Rubus chamaemorus, Sedum rosea, Vaccinium uliginosum, Viburnum edule*
Fruit: *Arctostaphylos alpina, A. uva-ursi, Empetrum nigrum, Ribes triste, Rubus arcticus, R. chamaemorus, Vaccinium oxycoccos, V. uliginosum, V. vitis-idaea, Viburnum edule*
Ice Cream: *Arctostaphylos uva-ursi, Rosa acicularis, Rubus chamaemorus, Vaccinium uliginosum, Viburnum edule*
Pie & Pudding: *Empetrum nigrum, Vaccinium oxycoccos, V. uliginosum, V. vitis-idaea*
Preserves: *Rosa acicularis, Viburnum edule*
Sauce & Relish: *Polygonum alpinum, Ribes triste, Rosa acicularis, Vaccinium oxycoccos, V. uliginosum, V. vitis-idaea, Viburnum edule*
Soup: *Allium schoenoprasum, Salix planifolia*
Spice: *Ligusticum scothicum*
Staple: *Sedum rosea*
Sweetener: *Epilobium angustifolium, Salix alaxensis*
Unspecified: *Angelica lucida, Epilobium latifolium, Equisetum pratense, Eriophorum angustifolium, Pedicularis lanata, Polygonum alpinum, P. bistorta, Rosa acicularis, Salix alaxensis, Sedum rosea*
Vegetable: *Allium schoenoprasum, Chenopodium album, Epilobium angustifolium, Hedysarum alpinum, Honckenya peploides, Ligusticum scothicum, Oxyria digyna, Oxytropis maydelliana, Pedicularis lanata, Polygonum alpinum, P. bistorta, Rumex arcticus, Salix planifolia, Saxifraga nelsoniana, Sedum rosea*
Winter Use Food: *Arctostaphylos rubra, A. uva-ursi, Empetrum nigrum, Epilobium angustifolium, Equisetum pratense, Eriophorum angustifolium, Hedysarum alpinum, Honckenya peploides, Oxytropis maydelliana, Polygonum alpinum, Rubus arcticus, R. chamaemorus, Rumex arcticus, Salix planifolia, Vaccinium oxycoccos, V. uliginosum, V. vitis-idaea*

Other

Containers: *Petasites frigidus*

Eskimo, Kuskokwagmiut
Drug

Adjuvant: *Matricaria discoidea*
Antidiarrheal: *Rumex acetosa*
Cold Remedy: *Artemisia* sp., *Matricaria discoidea*
Cough Medicine: *Picea glauca*
Dermatological Aid: *Artemisia tilesii, Rumex* sp., *Salix arbusculoides*
Dietary Aid: *Ranunculus lapponicus*
Disinfectant: *Angelica lucida*
Eye Medicine: *Salix arbusculoides*
Gastrointestinal Aid: *Matricaria discoidea*
Oral Aid: *Salix arbusculoides*
Orthopedic Aid: *Artemisia tilesii*
Panacea: *Eriophorum angustifolium*
Pediatric Aid: *Ledum palustre*
Poison: *Cicuta virosa*
Unspecified: *Artemisia* sp.
Veterinary Aid: *Nicotiana* sp.

Eskimo, Nunivak
Drug

Analgesic: *Salix planifolia, S. rotundifolia*
Cold Remedy: *Artemisia* sp.
Dermatological Aid: *Artemisia* sp., *Picea glauca*
Oral Aid: *Salix planifolia, S. rotundifolia*
Panacea: *Picea glauca*
Unspecified: *Achillea millefolium, Artemisia* sp., *Ledum palustre, Sedum rosea*

Eskimo, Western
Drug

Adjuvant: *Matricaria discoidea*
Analgesic: *Betula nana, Dryopteris campyloptera, Epilobium angustifolium, Ledum palustre, Sedum rosea*
Antidiarrheal: *Rubus* sp., *Vaccinium* sp.
Antihemorrhagic: *Salix* sp.
Cold Remedy: *Matricaria discoidea*
Cough Medicine: *Picea glauca*
Dermatological Aid: *Eriophorum russeolum, E. scheuchzeri*
Eye Medicine: *Eriophorum russeolum, Salix fuscescens*
Gastrointestinal Aid: *Artemisia tilesii, Betula nana, Dryopteris campyloptera, Epilobium angustifolium, Ledum palustre, Matricaria discoidea, Sedum rosea*
Laxative: *Caltha palustris*
Oral Aid: *Salix fuscescens, S. planifolia, S.* sp.
Orthopedic Aid: *Artemisia tilesii*
Poison: *Cicuta virosa, Ligusticum scothicum, Senecio congestus*
Preventive Medicine: *Salix* sp.
Pulmonary Aid: *Salix* sp.
Throat Aid: *Salix* sp.
Tuberculosis Remedy: *Sedum rosea*

Other

Ceremonial Items: *Artemisia tilesii*

Flathead
Drug

Analgesic: *Hierochloe odorata, Opuntia polyacantha, Pinus ponderosa*
Anthelmintic: *Prunus virginiana*
Anticonvulsive: *Ligusticum canbyi*
Antidiarrheal: *Heuchera cylindrica, Matricaria discoidea, Opuntia polyacantha, Prunus virginiana*
Antirheumatic (External): *Achillea millefolium, Artemisia dracunculus, Mahonia repens, Nuphar lutea, Pinus ponderosa*
Antirheumatic (Internal): *Vaccinium membranaceum*
Breast Treatment: *Lewisia rediviva*
Burn Dressing: *Arctostaphylos uva-ursi, Balsamorhiza sagittata, Pinus contorta, Symphoricarpos albus*
Cathartic: *Balsamorhiza sagittata, Frangula purshiana, Rhus glabra*
Ceremonial Medicine: *Juniperus scopulorum*
Cold Remedy: *Achillea millefolium, Artemisia ludoviciana, A. tridentata, Grindelia squarrosa, Hierochloe odorata, Juniperus scopulorum, Matricaria discoidea, Mentha arvensis, Monarda fistulosa, Populus balsamifera, P. deltoides*
Contraceptive: *Mahonia repens*
Cough Medicine: *Grindelia squarrosa, Mahonia repens, Mentha arvensis, Monarda fistulosa*
Dermatological Aid: *Abies lasiocarpa, Achillea millefolium, Artemisia dracunculus, A. ludoviciana, Mahonia repens, Nuphar lutea, Pinus contorta, P. ponderosa, Populus balsamifera, P. deltoides, Salix* sp., *Symphoricarpos albus*
Disinfectant: *Achillea millefolium*
Diuretic: *Equisetum hyemale*
Ear Medicine: *Arctostaphylos uva-ursi*
Eye Medicine: *Chimaphila umbellata, Monarda fistulosa, Prunus virginiana, Rosa* sp., *Salix* sp., *Shepherdia canadensis, Symphoricarpos albus*
Febrifuge: *Achillea millefolium, Hierochloe odorata, Juniperus scopulorum, Mentha arvensis, Monarda fistulosa*
Gastrointestinal Aid: *Asclepias speciosa, Heuchera cylindrica, Matricaria discoidea*

Gynecological Aid: *Lewisia rediviva, Mahonia repens, Monarda fistulosa, Pinus ponderosa*
Heart Medicine: *Lewisia rediviva, Vaccinium membranaceum*
Misc. Disease Remedy: *Monarda fistulosa*
Nose Medicine: *Veratrum viride*
Oral Aid: *Abies lasiocarpa*
Other: *Populus tremuloides*
Pediatric Aid: *Abies lasiocarpa, Matricaria discoidea*
Poison: *Frangula purshiana*
Pulmonary Aid: *Artemisia tridentata, Balsamorhiza sagittata, Grindelia squarrosa, Juniperus scopulorum, Lewisia rediviva, Monarda fistulosa*
Respiratory Aid: *Grindelia squarrosa, Hierochloe odorata*
Tonic: *Mentha arvensis*
Toothache Remedy: *Mentha arvensis, Monarda fistulosa*
Tuberculosis Remedy: *Balsamorhiza sagittata, Grindelia squarrosa, Rhus glabra, Shepherdia canadensis*
Unspecified: *Ligusticum canbyi, Shepherdia canadensis*
Urinary Aid: *Balsamorhiza sagittata*
Venereal Aid: *Mahonia repens, Nuphar lutea, Populus balsamifera, P. deltoides*
Veterinary Aid: *Amelanchier alnifolia, Grindelia squarrosa, Juniperus scopulorum, Nuphar lutea*

Food
Beverage: *Camassia quamash, Shepherdia canadensis*
Bread & Cake: *Perideridia gairdneri*
Candy: *Larix occidentalis, Pinus contorta*
Dessert: *Mahonia repens*
Dried Food: *Lomatium macrocarpum*
Fruit: *Cornus sericea, Mahonia repens*
Ice Cream: *Shepherdia canadensis*
Pie & Pudding: *Amelanchier alnifolia*
Preservative: *Matricaria discoidea, Monarda fistulosa*
Sauce & Relish: *Allium cernuum, Arctostaphylos uva-ursi, Camassia quamash, Larix occidentalis*
Soup: *Camassia quamash*
Staple: *Allium cernuum*
Unspecified: *Balsamorhiza sagittata, Camassia quamash, Cirsium scariosum, Fritillaria pudica, Larix occidentalis, Lomatium macrocarpum, Pinus contorta, Populus balsamifera, P. deltoides*

Fiber
Basketry: *Salix exigua, S. melanopsis, Thuja plicata*

Dye
Orange: *Alnus incana*
Red: *Alnus incana*
Red-Brown: *Alnus incana*

Other
Containers: *Betula papyrifera, Thuja plicata*
Decorations: *Hierochloe odorata*
Fuel: *Artemisia tridentata*
Hunting & Fishing Item: *Amelanchier alnifolia, Cornus sericea, Taxus brevifolia*
Incense & Fragrance: *Abies lasiocarpa*
Insecticide: *Hierochloe odorata, Mentha arvensis, Monarda fistulosa*
Protection: *Rosa sp.*
Season Indicator: *Thermopsis rhombifolia*
Soap: *Ligusticum canbyi*

Gabrielino
Drug
Hallucinogen: *Datura wrightii*
Unspecified: *Datura wrightii*

Gitksan
Drug
Analgesic: *Alnus rubra, Angelica genuflexa, Oplopanax horridus, Thalictrum occidentale, Viburnum edule*
Antihemorrhagic: *Alnus viridis, Calla palustris, Castilleja miniata, Juniperus sp., Lysichiton americanus, Nuphar lutea, Oplopanax horridus, Urtica dioica*
Antirheumatic (External): *Abies grandis, Heracleum maximum, Lysichiton americanus, Oplopanax horridus*
Antirheumatic (Internal): *Anemone multifida, Lysichiton americanus, Maianthemum racemosum, Malus fusca, Picea sitchensis, Shepherdia canadensis*
Blood Medicine: *Pinus contorta, Thalictrum occidentale*
Cancer Treatment: *Oplopanax horridus*
Cathartic: *Abies sp., Alnus rubra, A. viridis, Castilleja miniata, Juniperus sp., Maianthemum racemosum, Oplopanax horridus, Pinus contorta, Populus tremuloides, Sambucus racemosa, Sorbus sitchensis*
Cold Remedy: *Abies lasiocarpa, Oplopanax horridus, Picea glauca*
Cough Medicine: *Abies lasiocarpa, Alnus rubra, Castilleja miniata, Oplopanax horridus, Picea glauca, Shepherdia canadensis, Viburnum edule*
Dermatological Aid: *Abies grandis, A. sp., Descurainia incana, Heracleum maximum, Juniperus sp., Lysichiton americanus, Maianthemum racemosum, Oplopanax horridus, Populus tremuloides*
Diaphoretic: *Anemone multifida*
Dietary Aid: *Malus fusca*
Diuretic: *Abies sp., Alnus incana, Castilleja miniata, Juniperus sp., Ledum groenlandicum, Malus fusca, Oplopanax horridus, Pinus contorta*
Emetic: *Alnus rubra, Sambucus racemosa*
Eye Medicine: *Angelica genuflexa, Calla palustris, Castilleja miniata, Lonicera involucrata, Malus fusca, Thalictrum occidentale, Viburnum edule*
Gastrointestinal Aid: *Oplopanax horridus, Tsuga heterophylla*
Gynecological Aid: *Nuphar lutea*
Hunting Medicine: *Veratrum viride*
Kidney Aid: *Castilleja miniata, Juniperus sp., Maianthemum racemosum*
Laxative: *Abies sp., Alnus incana, Malus fusca*
Liver Aid: *Tsuga heterophylla*
Misc. Disease Remedy: *Abies lasiocarpa, Calla palustris, Lysichiton americanus, Oplopanax horridus, Picea glauca*
Orthopedic Aid: *Castilleja miniata, Maianthemum racemosum, Oplopanax horridus, Thalictrum occidentale*
Other: *Abies sp., Oplopanax horridus, Tsuga heterophylla, Veratrum viride*
Poison: *Lysichiton americanus*
Pulmonary Aid: *Abies grandis, Castilleja miniata, Lysichiton americanus, Nuphar lutea, Oplopanax horridus, Thalictrum occidentale*
Respiratory Aid: *Calla palustris, Oplopanax horridus*
Sedative: *Lysichiton americanus*
Strengthener: *Juniperus sp.*
Throat Aid: *Achillea millefolium*
Tonic: *Abies lasiocarpa, Alnus viridis, Calla palustris, Oplopanax horridus, Picea glauca, Pinus contorta*
Tuberculosis Remedy: *Abies sp., Malus fusca, Oplopanax horridus, Picea sitchensis, Pinus contorta, Viburnum edule*
Unspecified: *Alnus incana, A. rubra, Juniperus communis, Pinus contorta, Prunus pensylvanica, Ribes lacustre, Urtica dioica*
Urinary Aid: *Urtica dioica*
Venereal Aid: *Abies sp., Alnus incana, A. viridis, Oplopanax horridus, Pinus contorta, Shepherdia canadensis*
Witchcraft Medicine: *Heracleum maximum, Juniperus communis, Sambucus racemosa*

Food
Staple: *Tsuga heterophylla*

Sweetener: *Tsuga heterophylla*
Unspecified: *Epilobium angustifolium, Heracleum maximum, Picea glauca, Pinus contorta, Sedum divergens*
Winter Use Food: *Tsuga heterophylla*
Fiber
Basketry: *Acer glabrum, Betula papyrifera, Thuja plicata*
Building Material: *Thuja plicata*
Cordage: *Thuja plicata*
Mats, Rugs & Bedding: *Acer glabrum, Thuja plicata*
Other
Cash Crop: *Shepherdia canadensis*
Ceremonial Items: *Juniperus communis, Oplopanax horridus, Veratrum viride*
Containers: *Betula papyrifera*
Good Luck Charm: *Veratrum viride*
Lighting: *Betula papyrifera*
Soap: *Veratrum viride*

Gosiute
Drug
Analgesic: *Achillea millefolium, Aquilegia coerulea, Chaenactis douglasii, Mentha canadensis, Silene douglasii, S. scouleri*
Anthelmintic: *Pinus monophylla*
Antidiarrheal: *Geranium caespitosum*
Antirheumatic (External): *Achillea millefolium, Artemisia tridentata, Maianthemum stellatum, Valeriana edulis*
Antirheumatic (Internal): *Artemisia tridentata*
Blood Medicine: *Prunus virginiana, Rumex salicifolius*
Burn Dressing: *Cercocarpus ledifolius, Mentzelia albicaulis, Petrophyton caespitosum*
Cathartic: *Arenaria triflora, Heuchera* sp., *Mitella* sp., *Rumex salicifolius*
Cold Remedy: *Artemisia tridentata, Juniperus californica, Lomatium graveolens, Mentha canadensis*
Cough Medicine: *Artemisia tridentata, Grindelia squarrosa, Juniperus californica, Mentha canadensis*
Dermatological Aid: *Achillea millefolium, Balsamorhiza sagittata, Cirsium eatonii, Collomia linearis, Geranium caespitosum, Gilia* sp., *Heuchera rubescens, Linum lewisii, Lomatium dissectum, Mentzelia laevicaulis, Phlox gracilis, Potentilla glandulosa, Valeriana edulis, Wyethia amplexicaulis*
Disinfectant: *Lomatium dissectum*
Diuretic: *Lithospermum ruderale*
Emetic: *Silene douglasii, S. scouleri, Zigadenus nuttallii*
Eye Medicine: *Ambrosia psilostachya, Antennaria dioica, Cleome serrulata, Eriogonum ovalifolium*
Febrifuge: *Artemisia tridentata, Krascheninnikovia lanata*
Gastrointestinal Aid: *Achillea millefolium, Aquilegia coerulea, Arenaria congesta, A. triflora, Eriogonum ovalifolium, Heuchera rubescens, H.* sp., *Lomatium graveolens, Mitella* sp., *M. trifida, Petrophyton caespitosum, Prunus virginiana, Silene acaulis, S. douglasii, S. scouleri*
Heart Medicine: *Aquilegia coerulea*
Hemostat: *Balsamorhiza sagittata, Prunus virginiana*
Kidney Aid: *Lithospermum ruderale*
Laxative: *Mitella* sp.
Narcotic: *Cornus sericea*
Orthopedic Aid: *Chaenactis douglasii, Lomatium dissectum, Wyethia amplexicaulis*
Other: *Valeriana* sp.
Panacea: *Aquilegia coerulea, Artemisia tridentata*
Pediatric Aid: *Arenaria triflora, Heuchera rubescens, H.* sp., *Mitella* sp., *M. trifida, Prunus virginiana, Silene acaulis*
Poison: *Aconitum fischeri, Brickellia grandiflora, Delphinium bicolor, Erigeron grandiflorus, Valeriana dioica*
Throat Aid: *Lomatium graveolens*
Unspecified: *Angelica pinnata, Brickellia grandiflora, Carex* sp., *Clematis ligusticifolia, Geum macrophyllum, Kalmia polifolia, Lomatium graveolens, Oenothera cespitosa, Potentilla glandulosa, Purshia mexicana*
Venereal Aid: *Chamaebatiaria millefolium, Eriogonum ovalifolium, Parnassia fimbriata, Spiranthes romanzoffiana, Zigadenus nuttallii*
Veterinary Aid: *Astragalus convallarius, Galium aparine, Lomatium dissectum, Lygodesmia grandiflora, Silene douglasii, S. scouleri*
Food
Beverage: *Mentha canadensis*
Bread & Cake: *Salicornia maritima*
Candy: *Asclepias asperula, Chrysothamnus viscidiflorus, Pseudotsuga menziesii, Senecio* sp.
Cooking Agent: *Balsamorhiza sagittata, Brickellia grandiflora, Helianthus annuus*
Dried Food: *Amelanchier alnifolia, Calochortus nuttallii, Prunus virginiana, Ribes aureum, R. divaricatum, R. lacustre, R. oxyacanthoides, R. velutinum*
Fruit: *Amelanchier alnifolia, Fragaria vesca, Juniperus osteosperma, Prunus virginiana, Rhus glabra, R. trilobata, Ribes aureum, R. divaricatum, R. lacustre, R. oxyacanthoides, R. velutinum, Rosa californica, R. woodsii, Rubus leucodermis, R. parviflorus, Sambucus racemosa, Shepherdia argentea*
Porridge: *Descurainia pinnata, Prunus virginiana*
Unspecified: *Agastache urticifolia, Agoseris aurantiaca, Allium acuminatum, A. bisceptrum, Amaranthus* sp., *Amsinckia tessellata, Artemisia biennis, A. dracunculus, A. ludoviciana, A. tripartita, Atriplex canescens, A. confertifolia, A. truncata, Balsamorhiza hookeri, B. sagittata, Bromus marginatus, Calochortus nuttallii, Camassia scilloides, Carex utriculata, Catabrosa aquatica, Chenopodium capitatum, C. leptophyllum, C. rubrum, Cinna arundinacea, Cirsium drummondii, C. eatonii, C. undulatum, Claytonia caroliniana, Crepis runcinata, Cymopterus longipes, C. montanus, Deschampsia cespitosa, Dracocephalum parviflorum, Elymus canadensis, E. sibiricus, Elytrigia repens, Festuca brachyphylla, Fritillaria pudica, Helianthus annuus, Heliomeris multiflora, Lactuca ludoviciana, Lithospermum multiflorum, L. ruderale, Lomatium dissectum, Mammillaria* sp., *Oenothera biennis, Opuntia polyacantha, Orobanche fasciculata, Oryzopsis hymenoides, Perideridia gairdneri, Pinus edulis, P. monophylla, Poa arida, P. secunda, Puccinellia distans, Quercus ×pauciloba, Ranunculus aquatilis, Rorippa nasturtium-aquaticum, Scirpus acutus, Solidago canadensis, S. nemoralis, S. spectabilis, Stachys palustris, Suaeda calceoliformis, Triglochin maritimum, Trisetum spicatum, Typha latifolia, Vulpia octoflora, Wyethia amplexicaulis*
Winter Use Food: *Camassia scilloides, Perideridia gairdneri*
Fiber
Basketry: *Amelanchier alnifolia, Populus angustifolia, Salix amygdaloides, S. interior, S. lucida, S. scouleriana*
Building Material: *Juniperus osteosperma*
Furniture: *Amelanchier alnifolia*
Snow Gear: *Cornus sericea*
Other
Containers: *Artemisia tridentata, Juniperus osteosperma, Salix amygdaloides, S. interior, S. lucida, S. scouleriana*
Fuel: *Artemisia tridentata*
Hunting & Fishing Item: *Amelanchier alnifolia, Cercocarpus ledifolius, Salix amygdaloides, S. interior, S. lucida, S. scouleriana*
Paint: *Sphaeralcea coccinea, S. munroana*
Smoke Plant: *Cornus sericea, Nicotiana attenuata, Rhus glabra, Sedum debile, Silene menziesii, Vaccinium cespitosum*
Toys & Games: *Equisetum hyemale*

Great Basin Indian
Drug
Analgesic: *Dugaldia hoopesii, Ipomopsis congesta*
Antidiarrheal: *Mahonia repens*
Antihemorrhagic: *Artemisia frigida*
Antirheumatic (External): *Pulsatilla patens*
Blood Medicine: *Ipomopsis aggregata, Mahonia repens*
Cold Remedy: *Lomatium dissectum*
Cough Medicine: *Lomatium dissectum*
Dermatological Aid: *Achillea millefolium, Clematis ligusticifolia, Lomatium dissectum, Thalictrum sparsiflorum*
Eye Medicine: *Linum lewisii*
Gastrointestinal Aid: *Mentha canadensis*
Heart Medicine: *Chaenactis douglasii*
Laxative: *Achillea millefolium*
Misc. Disease Remedy: *Lomatium dissectum*
Other: *Ceanothus velutinus, Lomatium dissectum*
Pediatric Aid: *Chaenactis douglasii*
Respiratory Aid: *Dugaldia hoopesii, Lomatium dissectum*
Throat Aid: *Glycyrrhiza lepidota*
Tonic: *Lomatium dissectum*
Toothache Remedy: *Iris missouriensis*
Unspecified: *Lithospermum incisum, Lomatium dissectum*
Food
Beverage: *Lomatium dissectum, Rhus trilobata*
Dried Food: *Amelanchier alnifolia, Prunus virginiana*
Fruit: *Amelanchier alnifolia*
Unspecified: *Allium schoenoprasum, Perideridia gairdneri, Sagittaria* sp.
Vegetable: *Lomatium dissectum*
Winter Use Food: *Calochortus nuttallii*
Fiber
Basketry: *Rhus trilobata*
Cordage: *Apocynum androsaemifolium, Clematis ligusticifolia, Linum lewisii, Psoralidium lanceolatum*
Dye
Black: *Rhus trilobata*
Blue: *Delphinium* sp., *Lithospermum incisum*
Brown: *Allium schoenoprasum, Arctostaphylos uva-ursi*
Green: *Achillea millefolium, Artemisia frigida*
Mordant: *Juniperus monosperma*
Orange: *Alnus* sp., *Mahonia repens, Rumex venosus*
Purple: *Purshia tridentata*
Red: *Galium boreale, Prunus virginiana*
Red-Brown: *Castilleja* sp., *Prunus virginiana, Rhus trilobata*
Yellow: *Atriplex garrettii, Juniperus monosperma, Orthocarpus luteus*
Other
Ceremonial Items: *Artemisia frigida*
Hunting & Fishing Item: *Symphoricarpos* sp.
Smoke Plant: *Antennaria rosea, Arctostaphylos uva-ursi, Cornus sericea, Lomatium dissectum*
Soap: *Yucca* sp.

Green River Group
Drug
Cathartic: *Cornus nuttallii*
Cold Remedy: *Abies grandis, Oplopanax horridus*
Cough Medicine: *Polypodium virginianum*
Dermatological Aid: *Oplopanax horridus, Stachys mexicana, Symphoricarpos albus*
Disinfectant: *Symphoricarpos albus*
Emetic: *Cornus nuttallii, Physocarpus capitatus*
Laxative: *Frangula purshiana*
Food
Fruit: *Rubus leucodermis, R. spectabilis*
Unspecified: *Pteridium aquilinum, Rubus spectabilis*
Winter Use Food: *Sambucus cerulea, S. racemosa*
Other
Fuel: *Pseudotsuga menziesii*
Hunting & Fishing Item: *Pseudotsuga menziesii*
Toys & Games: *Cornus nuttallii*

Gros Ventre
Drug
Analgesic: *Mentha arvensis*
Antidiarrheal: *Prunus virginiana, Rosa* sp.
Ceremonial Medicine: *Helianthus annuus*
Febrifuge: *Artemisia ludoviciana*
Gastrointestinal Aid: *Rosa* sp.
Stimulant: *Helianthus annuus*
Food
Staple: *Helianthus annuus*
Unspecified: *Helianthus annuus*
Other
Incense & Fragrance: *Thalictrum* sp.
Soap: *Hierochloe odorata*

Hahwunkwut
Food
Bread & Cake: *Lithocarpus densiflorus*
Porridge: *Lithocarpus densiflorus*
Staple: *Lithocarpus densiflorus*
Unspecified: *Pteridium aquilinum*
Fiber
Basketry: *Adiantum* sp., *Corylus* sp., *Picea sitchensis, Xerophyllum* sp.
Other
Cooking Tools: *Adiantum* sp., *Picea sitchensis, Xerophyllum* sp.

Haida
Food
Fruit: *Vaccinium vitis-idaea*
Other
Decorations: *Acer macrophyllum*

Haihais
Drug
Gastrointestinal Aid: *Taxus brevifolia*
Internal Medicine: *Taxus brevifolia*
Unspecified: *Aruncus dioicus*
Food
Unspecified: *Conioselinum gmelinii*
Other
Hunting & Fishing Item: *Taxus brevifolia*

Haisla
Drug
Analgesic: *Polypodium glycyrrhiza*
Antirheumatic (External): *Oplopanax horridus, Veratrum viride*
Cancer Treatment: *Oplopanax horridus*
Cold Remedy: *Oplopanax horridus, Veratrum viride*
Cough Medicine: *Oplopanax horridus*
Dermatological Aid: *Alnus rubra, Heracleum maximum, Oplopanax horridus, Ribes bracteosum, Thuja plicata*

Emetic: *Veratrum viride*
Gastrointestinal Aid: *Oplopanax horridus*
Hemostat: *Veratrum viride*
Laxative: *Populus tremuloides, Rumex aquaticus, Veratrum viride*
Misc. Disease Remedy: *Oplopanax horridus*
Reproductive Aid: *Ribes bracteosum*
Respiratory Aid: *Oplopanax horridus, Polypodium glycyrrhiza*
Sedative: *Veratrum viride*
Tonic: *Abies amabilis, Alnus rubra, Oplopanax horridus*
Tuberculosis Remedy: *Oplopanax horridus*
Unspecified: *Abies amabilis, Pseudotsuga menziesii*

Food
Bread & Cake: *Rubus parviflorus*
Dried Food: *Tsuga heterophylla*
Fruit: *Amelanchier alnifolia, Sambucus racemosa*
Staple: *Tsuga heterophylla*
Sweetener: *Tsuga heterophylla*
Unspecified: *Abies amabilis, Conioselinum gmelinii, Epilobium angustifolium, Heracleum maximum, Populus balsamifera, Sedum divergens*
Vegetable: *Camassia* sp.
Winter Use Food: *Tsuga heterophylla*

Fiber
Basketry: *Thuja plicata*
Canoe Material: *Taxus brevifolia*
Clothing: *Chamaecyparis nootkatensis, Sphagnum* sp.
Cordage: *Epilobium angustifolium, Thuja plicata*
Mats, Rugs & Bedding: *Thuja plicata*

Other
Ceremonial Items: *Abies amabilis, Acer glabrum, Juniperus communis, Oplopanax horridus, Veratrum viride*
Cleaning Agent: *Epilobium angustifolium*
Fuel: *Alnus rubra*
Good Luck Charm: *Veratrum viride*
Hunting & Fishing Item: *Pseudotsuga menziesii, Thuja plicata*
Lighting: *Picea sitchensis, Thuja plicata*
Paint: *Oplopanax horridus*
Protection: *Thuja plicata, Veratrum viride*
Tools: *Pinus contorta*
Toys & Games: *Salix scouleriana, S. sitchensis*

Haisla & Hanaksiala
Drug
Abortifacient: *Sambucus racemosa*
Antirheumatic (External): *Nicotiana tabacum, Oplopanax horridus, Picea sitchensis, Pinus contorta, Urtica dioica*
Antirheumatic (Internal): *Malus fusca*
Burn Dressing: *Conocephalum conicum, Lysichiton americanus, Populus balsamifera*
Ceremonial Medicine: *Malus fusca, Veratrum viride*
Cold Remedy: *Ledum groenlandicum, Picea sitchensis, Polypodium glycyrrhiza*
Cough Medicine: *Picea sitchensis, Polypodium glycyrrhiza*
Dermatological Aid: *Clintonia uniflora, Oplopanax horridus, Picea sitchensis, Populus balsamifera*
Dietary Aid: *Ledum groenlandicum*
Emetic: *Oplopanax horridus*
Eye Medicine: *Clintonia uniflora, Oplopanax horridus*
Gastrointestinal Aid: *Oplopanax horridus, Sambucus racemosa*
Laxative: *Frangula purshiana, Oplopanax horridus*
Oral Aid: *Picea sitchensis, Populus tremuloides, Prunus virginiana*
Poison: *Bovista pila, B. plumbea, Bovistella* sp., *Calvatia* sp., *Cicuta douglasii, C. virosa, Lycoperdon* sp., *Oenanthe sarmentosa, Veratrum viride, Zigadenus venenosus*
Reproductive Aid: *Sambucus racemosa, Shepherdia canadensis*
Throat Aid: *Moneses uniflora, Polypodium glycyrrhiza*
Tuberculosis Remedy: *Fomitopsis officinalis, Ledum groenlandicum, Nuphar lutea, Picea sitchensis*
Unspecified: *Achillea millefolium, Nuphar lutea*
Urinary Aid: *Lysichiton americanus*

Food
Beverage: *Camellia sinensis, Ledum groenlandicum, Rosa nutkana, Rubus idaeus, R. spectabilis, Sambucus racemosa, Theobroma cacao*
Bread & Cake: *Triticum aestivum*
Candy: *Picea sitchensis*
Dessert: *Cornus canadensis, C. unalaschkensis*
Dried Food: *Cornus canadensis, C. unalaschkensis, Egregia menziesii, Fucus gardneri, Porphyra abbottae, Rubus spectabilis, Sambucus racemosa, Shepherdia canadensis*
Forage: *Achillea millefolium, Blechnum spicant, Cornus sericea, Dryopteris* sp., *Equisetum arvense, E. scirpoides, Lysichiton americanus, Prunus virginiana, Rubus pedatus, Sphagnum* sp., *Symphoricarpos albus*
Fruit: *Citrus limon, C. sinensis, Crataegus douglasii, Fragaria chiloensis, F. vesca, F. virginiana, Gaultheria shallon, Lycopersicon esculentum, Maianthemum dilatatum, Malus fusca, M. sylvestris, Prunus virginiana, Ribes divaricatum, R. lacustre, R. laxiflorum, Rubus discolor, R. idaeus, R. leucodermis, R. pedatus, R. spectabilis, R. ursinus, Vaccinium ovalifolium, V. parvifolium, V. uliginosum, Viburnum edule, Vitis vinifera*
Ice Cream: *Shepherdia canadensis*
Preservative: *Alnus rubra*
Preserves: *Fragaria chiloensis, F. vesca, F. virginiana, Urtica dioica*
Sauce & Relish: *Porphyra abbottae*
Special Food: *Rubus spectabilis, Shepherdia canadensis, Vaccinium parvifolium*
Spice: *Piper nigrum*
Staple: *Oryza sativa, Triticum aestivum*
Unspecified: *Argentina egedii, Avena sativa, Conioselinum gmelinii, Epilobium angustifolium, Exobasidium* sp., *Fritillaria camschatcensis, Heracleum maximum, Lupinus littoralis, L. nootkatensis, Macrocystis integrifolia, Populus balsamifera, Rheum rhabarbarum, Rosa nutkana, Trifolium wormskioldii, Urtica dioica*
Vegetable: *Allium cepa, A. cernuum, Brassica oleracea, B. rapa, Daucus carota, Rumex aquaticus, Solanum tuberosum*
Winter Use Food: *Malus fusca, Sambucus racemosa, Tsuga heterophylla, Vaccinium oxycoccos, Viburnum edule*

Fiber
Basketry: *Hierochloe alpina, H. hirta, Picea sitchensis, Thuja plicata*
Building Material: *Thuja plicata, Tsuga heterophylla*
Canoe Material: *Chamaecyparis nootkatensis, Thuja plicata*
Clothing: *Chamaecyparis nootkatensis, Thuja plicata*
Cordage: *Epilobium angustifolium, Thuja plicata, Urtica dioica*
Furniture: *Thuja plicata*
Mats, Rugs & Bedding: *Chamaecyparis nootkatensis, Thuja plicata*
Snow Gear: *Abies amabilis, Acer glabrum*

Other
Cash Crop: *Urtica dioica*
Ceremonial Items: *Acer glabrum, Alnus rubra, Fomes* sp., *Fomitopsis* sp., *Ganoderma* sp., *Oplopanax horridus, Polyporus* sp., *Rosa nutkana, Rubus spectabilis, Thuja plicata, Tsuga heterophylla*
Cleaning Agent: *Cornus sericea, Salix scouleriana, S. sitchensis, Thuja plicata*
Containers: *Angelica genuflexa, Dryopteris* sp., *Leymus mollis, Picea sitchensis, Salix scouleriana, S. sitchensis, Thuja plicata*
Cooking Tools: *Acer glabrum, Alnus rubra, A. viridis, Lysichiton americanus,*

Rubus spectabilis, Salix lucida, S. scouleriana, S. sitchensis, Sambucus racemosa, Thuja plicata, Tsuga heterophylla
Decorations: *Betula papyrifera, Blechnum spicant, Chamaecyparis nootkatensis, Prunus emarginata*
Designs: *Nereocystis luetkeana, Thuja plicata*
Fasteners: *Chamaecyparis nootkatensis, Leymus mollis, Picea sitchensis, Pseudotsuga menziesii, Thuja plicata, Urtica dioica*
Fuel: *Thuja plicata*
Good Luck Charm: *Oplopanax horridus, Tsuga heterophylla*
Hunting & Fishing Item: *Chamaecyparis nootkatensis, Epilobium angustifolium, Malus fusca, Picea sitchensis, Salix scouleriana, S. sitchensis, Sambucus racemosa, Stachys ciliata, Thuja plicata, Tsuga heterophylla, Urtica dioica*
Incense & Fragrance: *Angelica genuflexa*
Paint: *Bryoria capillaris, B. glabra, B. trichodes, Candelaria concolor, Candelariella vitellina, Conocephalum conicum, Oplopanax horridus, Pinus contorta, Pseudephebe pubescens, Trentepohlia aurea, Xanthoria elegans*
Preservative: *Populus balsamifera*
Protection: *Fomes* sp., *Fomitopsis* sp., *Ganoderma* sp., *Oplopanax horridus, Polyporus* sp., *Tsuga heterophylla*
Season Indicator: *Rubus spectabilis*
Smoke Plant: *Nicotiana tabacum*
Soap: *Angelica genuflexa, Oplopanax horridus*
Tools: *Acer glabrum, Fucus gardneri, Malus fusca, Picea sitchensis, Pinus contorta, Salix lucida*
Toys & Games: *Fomes* sp., *Fomitopsis* sp., *Ganoderma* sp., *Heracleum maximum, Polyporus* sp.
Walking Sticks: *Salix scouleriana, S. sitchensis*
Weapon: *Taxus brevifolia, Tsuga heterophylla*

Hanaksiala
Drug
Analgesic: *Sambucus racemosa*
Antidiarrheal: *Thuja plicata*
Antirheumatic (External): *Rumex aquaticus, Sambucus racemosa, Veratrum viride*
Antirheumatic (Internal): *Veratrum viride*
Blood Medicine: *Veratrum viride*
Cold Remedy: *Oplopanax horridus*
Dermatological Aid: *Juniperus communis, Rumex aquaticus, Sambucus racemosa*
Gastrointestinal Aid: *Abies amabilis, Porphyra abbottae, Pseudotsuga menziesii*
Gynecological Aid: *Pseudotsuga menziesii*
Hemorrhoid Remedy: *Abies amabilis*
Hypotensive: *Veratrum viride*
Laxative: *Picea sitchensis*
Orthopedic Aid: *Porphyra abbottae*
Panacea: *Porphyra abbottae*
Respiratory Aid: *Veratrum viride*
Throat Aid: *Pseudotsuga menziesii*
Unspecified: *Ribes bracteosum, Veratrum viride*
Urinary Aid: *Taxus brevifolia*
Food
Beverage: *Kalmia microphylla, Maianthemum racemosum, Rumex aquaticus*
Bread & Cake: *Tsuga heterophylla, Vaccinium cespitosum, V. membranaceum*
Candy: *Aquilegia formosa, Gentiana douglasiana*
Dried Food: *Amelanchier alnifolia, Vaccinium ovalifolium*
Fruit: *Arctostaphylos uva-ursi, Sambucus racemosa, Vaccinium ovalifolium*
Special Food: *Arctostaphylos uva-ursi*
Unspecified: *Angelica genuflexa, Conioselinum gmelinii, Dryopteris* sp., *Rumex acetosella*

Winter Use Food: *Ribes bracteosum, Rumex aquaticus*
Fiber
Basketry: *Prunus emarginata*
Clothing: *Lycopodium clavatum*
Cordage: *Epilobium angustifolium*
Furniture: *Abies lasiocarpa*
Mats, Rugs & Bedding: *Alectoria sarmentosa, Populus balsamifera, Usnea longissima, U.* sp., *Veratrum viride*
Other
Ceremonial Items: *Acer glabrum, Fritillaria camschatcensis, Picea sitchensis, Thuja plicata, Veratrum viride*
Containers: *Abies amabilis, A. lasiocarpa*
Cooking Tools: *Antitrichia curtipendula, Malus fusca, Rubus parviflorus, Tsuga heterophylla*
Designs: *Abies lasiocarpa*
Fasteners: *Tsuga heterophylla*
Good Luck Charm: *Veratrum viride*
Hide Preparation: *Tsuga heterophylla*
Hunting & Fishing Item: *Cornus sericea, Scirpus americanus, Taxus brevifolia, Tsuga heterophylla*
Jewelry: *Tsuga heterophylla*
Protection: *Veratrum viride*
Season Indicator: *Fritillaria camschatcensis*
Tools: *Taxus brevifolia, Tsuga heterophylla*
Toys & Games: *Angelica genuflexa*

Havasupai
Drug
Analgesic: *Aloysia wrightii, Eriogonum corymbosum, Mahonia repens, Porophyllum gracile*
Antirheumatic (External): *Mahonia repens, Phlox austromontana, P. longifolia, Porophyllum gracile*
Antirheumatic (Internal): *Aloysia wrightii, Porophyllum gracile*
Cold Remedy: *Artemisia tridentata, Frasera speciosa, Juniperus osteosperma, Mahonia repens, Phlox austromontana, P. longifolia, Purshia mexicana*
Cough Medicine: *Artemisia tridentata*
Dermatological Aid: *Artemisia tridentata, Atriplex canescens, Datura wrightii, Pinus edulis, P. monophylla, Porophyllum gracile*
Emetic: *Ephedra viridis, Thamnosma montana*
Gastrointestinal Aid: *Artemisia tridentata, Chaetopappa ericoides, Frasera speciosa, Mahonia repens, Phlox austromontana, P. longifolia, Porophyllum gracile, Ptelea trifoliata, Thamnosma montana*
Laxative: *Ephedra viridis, Erioneuron pulchellum, Mahonia repens, Purshia mexicana, Thamnosma montana*
Misc. Disease Remedy: *Atriplex canescens*
Narcotic: *Datura wrightii*
Nose Medicine: *Artemisia tridentata*
Pediatric Aid: *Chaetopappa ericoides, Mahonia repens, Phlox austromontana, P. longifolia, Ptelea trifoliata*
Poison: *Ptelea trifoliata, Shepherdia rotundifolia, Stanleya pinnata*
Psychological Aid: *Aloysia wrightii*
Throat Aid: *Artemisia tridentata, Pluchea sericea*
Unspecified: *Artemisia campestris, A. ludoviciana, Mahonia repens, Pseudotsuga menziesii*
Venereal Aid: *Frasera speciosa*
Veterinary Aid: *Pinus edulis, P. monophylla*
Food
Beverage: *Agave* sp., *A. utahensis, Aloysia wrightii, Descurainia* sp., *Ephedra fasciculata, E. nevadensis, E. torreyana, E. viridis, Eriogonum microthecum, Ficus carica, Juniperus osteosperma, Lycium pallidum, Opuntia phaeacantha,*

[Beverage]
 Prosopis glandulosa, Prunus persica, Rhus trilobata, Yucca baccata
Bread & Cake: *Acacia greggii, Amaranthus hybridus, Chenopodium fremontii, C.* sp., *Helianthus* sp., *Koeleria macrantha, Lepidium lasiocarpum, Opuntia phaeacantha, Oryzopsis hymenoides, Poa fendleriana, Zea mays*
Candy: *Populus fremontii, Prosopis glandulosa*
Dried Food: *Cucurbita moschata, Echinocactus* sp., *Ficus carica, Helianthus annuus, H. petiolaris, Juniperus osteosperma, Lycium pallidum, Opuntia phaeacantha, Proboscidea parviflora, Prunus persica, Rhus trilobata, Yucca baccata*
Fodder: *Coleogyne ramosissima, Krascheninnikovia lanata*
Forage: *Amelanchier utahensis, Koeleria macrantha, Salsola australis*
Fruit: *Citrullus lanatus, Ficus carica, Opuntia phaeacantha, Typha domingensis, Vitis arizonica*
Porridge: *Amaranthus hybridus, Citrullus lanatus, Cucurbita moschata, Echinocactus* sp., *Pectis papposa, Phaseolus* sp., *Plantago patagonica, Quercus gambelii, Zea mays*
Preserves: *Artemisia* sp., *Descurainia* sp., *Gaillardia pinnatifida, Gilia sinuata, Helianthus annuus, H. petiolaris, Lepidium lasiocarpum, Mentzelia albicaulis, Pinus edulis*
Sauce & Relish: *Pectis angustifolia, P. papposa*
Soup: *Amaranthus hybridus, Cucurbita moschata, Mentzelia albicaulis, Oryzopsis hymenoides, Pectis papposa, Phaseolus acutifolius, P. lunatus, P.* sp., *P. vulgaris, Pinus edulis, Quercus gambelii, Zea mays*
Spice: *Pinus edulis, P. monophylla, Quercus gambelii*
Staple: *Chenopodium* sp., *Descurainia* sp., *Helianthus annuus, H. petiolaris, Lepidium lasiocarpum, Oryzopsis hymenoides, Poa fendleriana, Zea mays*
Starvation Food: *Allium bisceptrum, Cirsium* sp.
Unspecified: *Allium cepa, Amaranthus hybridus, Astragalus* sp., *Calochortus nuttallii, Chenopodium* sp., *Cleome serrulata, Koeleria macrantha, Lepidium lasiocarpum, L. montanum, Lotus mearnsii, Mentzelia albicaulis, Phoradendron juniperinum, Pinus edulis, P. monophylla, P. ponderosa, Poa fendleriana, Proboscidea parviflora, Quercus gambelii, Rorippa nasturtium-aquaticum, Thlaspi arvense, T. montanum*
Vegetable: *Amaranthus hybridus, Cucurbita moschata, Eriogonum inflatum, Phaseolus acutifolius, P. lunatus, P. vulgaris, Rumex crispus, Stanleya pinnata, Zea mays*
Winter Use Food: *Ficus carica, Koeleria macrantha, Phaseolus acutifolius, P. lunatus, P. vulgaris, Zea mays*

Fiber
Basketry: *Acacia greggii, Amelanchier utahensis, Baccharis emoryi, Chilopsis linearis, Fallugia paradoxa, Populus fremontii, Proboscidea parviflora, Rhus trilobata, Salix bonplandiana, S. exigua*
Brushes & Brooms: *Acacia greggii, Agave utahensis*
Building Material: *Apocynum ×floribundum, Artemisia tridentata, Baccharis emoryi, Cercis canadensis, Fraxinus pennsylvanica, Juniperus osteosperma, Nolina microcarpa, Pinus edulis, P. monophylla, Pluchea sericea, Populus fremontii, Salix bonplandiana, Typha domingensis*
Caulking Material: *Pinus edulis, P. monophylla*
Clothing: *Apocynum ×floribundum, Purshia mexicana*
Cordage: *Apocynum ×floribundum, Yucca baccata*
Furniture: *Amelanchier utahensis, Fallugia paradoxa, Fraxinus pennsylvanica, Pluchea sericea, Prosopis glandulosa, Symphoricarpos* sp.
Mats, Rugs & Bedding: *Ephedra* sp., *Nolina microcarpa, Phragmites australis, Purshia mexicana*

Dye
Red: *Cercocarpus ledifolius*
Yellow: *Mahonia fremontii*

Other
Cash Crop: *Agave* sp., *Pinus edulis, P. monophylla*
Ceremonial Items: *Agave* sp., *Pseudotsuga menziesii*
Containers: *Artemisia tridentata, Cucurbita moschata, Echinocactus* sp., *Lagenaria siceraria, Opuntia phaeacantha, O.* sp., *Sphaeralcea fendleri, S. grossulariifolia, S. parvifolia*
Cooking Tools: *Agave utahensis, Amelanchier utahensis, Echinocactus* sp., *Eriogonum inflatum, Nolina microcarpa, Populus fremontii*
Fasteners: *Yucca angustissima*
Fuel: *Baccharis emoryi, Celtis laevigata, Fraxinus pennsylvanica, Gossypium* sp., *Juniperus osteosperma, Pinus edulis, P. monophylla, Populus fremontii, Prosopis glandulosa, Ptelea trifoliata, Purshia mexicana, Salix bonplandiana*
Hunting & Fishing Item: *Amelanchier utahensis, Cercis canadensis, Fendlera rupicola, Fraxinus pennsylvanica, Mortonia sempervirens, Penstemon pachyphyllus, Phragmites australis, Pluchea sericea, Ptelea trifoliata, Ribes cereum*
Incense & Fragrance: *Melilotus* sp., *Xylorhiza tortifolia*
Jewelry: *Echinocactus* sp.
Musical Instrument: *Lagenaria siceraria, Populus fremontii*
Paint: *Pinus edulis, P. monophylla, Typha domingensis*
Planting Seeds: *Citrullus lanatus, Cucurbita moschata*
Season Indicator: *Populus fremontii*
Smoke Plant: *Nicotiana attenuata, N. trigonophylla*
Smoking Tools: *Phragmites australis*
Soap: *Yucca baccata*
Tools: *Amelanchier utahensis, Baccharis emoryi, Cercis canadensis, Equisetum hyemale, Fraxinus pennsylvanica, Gossypium* sp., *Opuntia phaeacantha, Pinus edulis, P. monophylla, Pluchea sericea, Quercus gambelii, Q. turbinella, Salix exigua, Yucca baccata, Ziziphus obtusifolia*
Toys & Games: *Apocynum ×floribundum, Baccharis emoryi, Cucurbita foetidissima, Fraxinus pennsylvanica, Garrya* sp., *Juniperus osteosperma, Scirpus* sp., *Typha domingensis, Vitis arizonica, Yucca angustissima, Y. baccata*
Waterproofing Agent: *Pinus edulis, P. monophylla, Yucca baccata*
Weapon: *Ptelea trifoliata, Ribes cereum*

Hawaiian
Drug
Abortifacient: *Aleurites moluccana, Cocos nucifera*
Adjuvant: *Sideroxylon* sp.
Analgesic: *Alocasia macrorrhizos, Cibotium chamissoi, Eucalyptus* sp., *Freycinetia arborea, Ipomoea indica, Merremia dissecta, Metrosideros polymorpha, Pandanus tectorius, Piper methysticum, Ricinus communis, Zingiber zerumbet*
Antidiarrheal: *Psidium guajava*
Antiemetic: *Antidesma pulvinatum, Rubus hawaiensis, R. macraei*
Antihemorrhagic: *Digitaria setigera, Psidium guajava*
Antirheumatic (External): *Cibotium chamissoi, Conyza canadensis, Cyperus* sp., *Eucalyptus* sp., *Plumbago zeylanica*
Blood Medicine: *Asplenium horridum, Bobea* sp., *Caesalpinia bonduc, C. kavaiensis, Catharanthus roseus, Cibotium chamissoi, Curcuma longa, Hibiscus* sp., *Ludwigia bonariensis, Murdannia nudiflora, Pelea* sp., *Polygonum densiflorum, Rumex giganteus*
Breast Treatment: *Carica papaya, Chamaesyce multiformis, Clermontia arborescens*
Burn Dressing: *Alocasia macrorrhizos*
Ceremonial Medicine: *Dodonaea viscosa, Pelea* sp.
Cold Remedy: *Cyperus laevigatus, Desmodium sandwicense, Piper methysticum*
Dermatological Aid: *Aleurites moluccana, Alyxia oliviformis, Antidesma pulvinatum, Artocarpus altilis, Asplenium nidus, A. pseudofalcatum, Bobea* sp., *Canavalia galeata, Carica papaya, Cenchrus calyculatus, Chenopodium oahuense, Clermontia arborescens, Cocos nucifera, Coprosma* sp., *Cucurbita maxima, Cyperus laevigatus, Diplazium meyenianum, Dodonaea viscosa, Eucalyptus* sp., *Ipomoea indica, Jacquemontia ovalifolia, Ludwigia bonariensis, Merremia dissecta, Morinda citrifolia, Nicotiana glauca, N. tabacum, Osteomeles anthyllidifolia, Pelea* sp., *Piper methysticum, Plumbago zeylanica, Psidium*

guajava, Rubus hawaiensis, R. macraei, Sadleria cyatheoides, Scaevola sericea, Senna occidentalis, Stenotaphrum secundatum, Styphelia tameiameiae, Syzygium malaccense, Tephrosia purpurea, Vigna luteola, Zingiber zerumbet
Diaphoretic: *Acacia koa*
Dietary Aid: *Alocasia macrorrhizos, Bidens* sp., *Chamaesyce multiformis, Chenopodium oahuense, Cibotium chamissoi, Phegopteris* sp., *Waltheria indica*
Emetic: *Ipomoea* sp., *Syzygium malaccense*
Eye Medicine: *Digitaria setigera, Piper methysticum*
Febrifuge: *Artemisia australis, Cordyline fruticosa, Dioscorea* sp., *Eucalyptus* sp., *Pleomele aurea, Ricinus communis, Sideroxylon* sp.
Gastrointestinal Aid: *Abutilon incanum, Aleurites moluccana, Bidens* sp., *Piper methysticum, Rubus hawaiensis, R. macraei, Syzygium malaccense*
Gynecological Aid: *Cassytha filiformis, Cucurbita maxima, Freycinetia arborea, Hibiscus tiliaceus, Ipomoea* sp., *Peperomia* sp., *Phegopteris* sp., *Piper methysticum, Sida* sp., *Stenotaphrum secundatum*
Heart Medicine: *Rumex giganteus*
Herbal Steam: *Ochrosia compta*
Laxative: *Aleurites moluccana, Alocasia macrorrhizos, Caesalpinia bonduc, Chamaesyce multiformis, Colocasia esculenta, Cucurbita maxima, Digitaria setigera, Hibiscus* sp., *H. tiliaceus, Ipomoea indica, I.* sp., *I. tiliacea, Merremia dissecta, Mucuna gigantea, Odontoglossum chinensis, Opuntia tunicata, Osteomeles anthyllidifolia, Pandanus tectorius, Peperomia* sp., *Peucedanum sandwicense, Sida* sp., *Touchardia latifolia, Waltheria indica, Wikstroemia* sp.
Love Medicine: *Alocasia macrorrhizos*
Misc. Disease Remedy: *Dodonaea viscosa, Piper methysticum, Rumex giganteus*
Nose Medicine: *Cordyline fruticosa, Curcuma longa*
Oral Aid: *Artocarpus altilis, Asplenium horridum, A. nidus, A. pseudofalcatum, Curcuma longa, Piper methysticum, Stenotaphrum secundatum, Syzygium malaccense*
Orthopedic Aid: *Conyza canadensis, Eucalyptus* sp., *Ipomoea indica, Merremia dissecta, Morinda citrifolia, Psidium guajava, Zingiber zerumbet*
Other: *Cocos nucifera, Peperomia* sp.
Pediatric Aid: *Acacia koa, Aleurites moluccana, Asplenium nidus, Bidens* sp., *Broussaisia arguta, Caesalpinia bonduc, Chamaesyce multiformis, Chenopodium oahuense, Cyrtandra* sp., *Digitaria setigera, Freycinetia arborea, Hibiscus* sp., *H. tiliaceus, Ipomoea indica, I.* sp., *Jacquemontia ovalifolia, Ludwigia bonariensis, Merremia dissecta, Ochrosia compta, Pandanus tectorius, Pelea* sp., *Peperomia* sp., *Peucedanum sandwicense, Piper methysticum, Pipturus* sp., *Psidium guajava, Ricinus communis, Sida* sp., *Stenotaphrum secundatum, Syzygium malaccense, Touchardia latifolia, Vigna luteola, Waltheria indica*
Poison: *Tephrosia purpurea*
Psychological Aid: *Cucurbita maxima, Microlepia setosa*
Pulmonary Aid: *Artemisia australis, Caesalpinia bonduc, Cordyline fruticosa, Hibiscus tiliaceus, Hydrocotyle poltata, Peperomia* sp., *Piper methysticum, Pleomele aurea, Sadleria cyatheoides, Waltheria indica*
Reproductive Aid: *Artemisia australis, Broussaisia arguta, Chamaesyce multiformis, Cyrtandra* sp., *Ipomoea pes-caprae, Opuntia tunicata, Peucedanum sandwicense, Rumex giganteus*
Respiratory Aid: *Aleurites moluccana, Bidens* sp., *Cassytha filiformis, Cheirodendron gaudicchaudii, Clermontia arborescens, Cordyline fruticosa, Desmodium sandwicense, Ipomoea* sp., *Peperomia* sp., *Pleomele aurea, Sadleria cyatheoides, Sida* sp., *Vigna luteola, Waltheria indica, Wikstroemia* sp.
Sedative: *Cibotium chamissoi, Cordyline fruticosa, Ipomoea* sp., *Piper methysticum*
Stimulant: *Asplenium horridum, Piper methysticum*
Strengthener: *Acacia koa, Aleurites moluccana, Asplenium nidus, Bidens* sp., *Broussaisia arguta, Chamaesyce multiformis, Chenopodium oahuense, Cocos nucifera, Cyperus laevigatus, Cyrtandra* sp., *Desmodium sandwicense, Digitaria setigera, Freycinetia arborea, Hibiscus* sp., *Hydrocotyle poltata, Ipomoea indica, I.* sp., *Jacquemontia ovalifolia, Merremia dissecta, Musa ×paradisiaca, Ochrosia compta, Pandanus tectorius, Pelea* sp., *Peperomia* sp., *Piper methysticum, Pipturus* sp., *Portulaca oleracea, Rumex giganteus, Sida* sp., *Sideroxylon* sp., *Syzygium malaccense, Touchardia latifolia, Vigna luteola, Waltheria indica*
Throat Aid: *Bidens* sp., *Hibiscus tiliaceus, Waltheria indica*
Toothache Remedy: *Zingiber zerumbet*
Tuberculosis Remedy: *Aleurites moluccana, Antidesma pulvinatum, Asplenium nidus, Cenchrus calyculatus, Chamaesyce multiformis, Coprosma* sp., *Desmodium sandwicense, Nicotiana glauca, N. tabacum, Pittosporum* sp., *Rumex giganteus, Styphelia tameiameiae*
Unspecified: *Chamaesyce multiformis, Cocos nucifera, Colocasia esculenta, Coprosma* sp., *Dioscorea* sp., *Morinda citrifolia, Vitex trifolia*
Urinary Aid: *Piper methysticum*
Venereal Aid: *Cyperus laevigatus, Erythrina sandwicensis, Hydrocotyle poltata, Melicope cinerea*
Food
Beverage: *Cordyline fruticosa, Jacquemontia ovalifolia, Sadleria cyatheoides*
Fruit: *Artocarpus altilis, Carica papaya, Clermontia arborescens, Dioscorea bulbifera*
Unspecified: *Colocasia esculenta, Cordyline fruticosa, Dioscorea pentaphylla, D.* sp., *Diplazium meyenianum, Ipomoea cairica, Jacquemontia ovalifolia, Marattia* sp., *Ochrosia compta, Phegopteris* sp., *Tacca leontopetaloides, Vitex trifolia*
Vegetable: *Colocasia esculenta, Dioscorea* sp., *Ipomoea* sp.
Fiber
Canoe Material: *Acacia koa, Bobea* sp.
Clothing: *Pandanus tectorius*
Cordage: *Cyperus laevigatus, Touchardia latifolia*
Mats, Rugs & Bedding: *Cibotium chamissoi, Pandanus tectorius*
Dye
Red: *Asplenium horridum*
Other
Containers: *Cordyline fruticosa, Murdannia nudiflora*
Decorations: *Artemisia australis, Sida* sp.
Fuel: *Aleurites moluccana, Cheirodendron gaudicchaudii*
Hunting & Fishing Item: *Cordyline fruticosa, Tephrosia purpurea*
Lighting: *Aleurites moluccana*
Tools: *Cyperus laevigatus, Pipturus* sp.

Heiltzuk
Drug
Analgesic: *Salicornia virginica*
Antirheumatic (External): *Plagiomnium juniperinum, Salicornia virginica*
Dermatological Aid: *Sorbus sitchensis*
Food
Forage: *Sorbus sitchensis*
Unspecified: *Conioselinum gmelinii, Rumex aquaticus*
Fiber
Cordage: *Epilobium angustifolium*
Other
Cash Crop: *Macrocystis integrifolia*
Ceremonial Items: *Juniperus communis*
Smoke Plant: *Arctostaphylos uva-ursi*

Hesquiat
Drug
Analgesic: *Achillea millefolium, Picea sitchensis, Ranunculus repens, Sambucus racemosa, Urtica dioica*
Anthelmintic: *Frangula purshiana*
Antidote: *Physocarpus capitatus*

Antirheumatic (External): *Physocarpus capitatus, Ranunculus repens, Sambucus racemosa, Urtica dioica*
Antirheumatic (Internal): *Physocarpus capitatus, Tsuga heterophylla*
Burn Dressing: *Lysichiton americanus*
Cancer Treatment: *Athyrium filix-femina, Blechnum spicant, Dryopteris campyloptera, Polystichum munitum, Pteridium aquilinum*
Carminative: *Polypodium glycyrrhiza*
Cough Medicine: *Achillea millefolium, Boschniakia hookeri, Polypodium glycyrrhiza*
Dermatological Aid: *Abies grandis, Blechnum spicant, Claytonia sibirica, Lysichiton americanus, Maianthemum dilatatum, Nicotiana tabacum, Picea sitchensis, Plantago major, Populus balsamifera, Ranunculus repens, Symphoricarpos albus, Tsuga heterophylla*
Emetic: *Physocarpus capitatus, Sambucus racemosa*
Eye Medicine: *Claytonia sibirica, Tsuga heterophylla*
Gastrointestinal Aid: *Achillea millefolium, Frangula purshiana, Geum macrophyllum, Rubus ursinus, Sambucus racemosa, Urtica dioica*
Gynecological Aid: *Geum macrophyllum, Ranunculus repens*
Internal Medicine: *Achillea millefolium*
Laxative: *Frangula purshiana, Physocarpus capitatus, Sambucus racemosa*
Misc. Disease Remedy: *Alnus rubra*
Oral Aid: *Menziesia ferruginea, Polypodium glycyrrhiza*
Other: *Ranunculus repens, Rubus ursinus, Tsuga heterophylla*
Poison: *Bromus carinatus, Kalmia polifolia, Sambucus racemosa*
Respiratory Aid: *Adiantum pedatum*
Strengthener: *Adiantum pedatum, Postelsia palmaeformis*
Throat Aid: *Polypodium glycyrrhiza*
Tuberculosis Remedy: *Alnus rubra, Maianthemum dilatatum, Tsuga heterophylla*
Unspecified: *Halosaccion glandiforme, Leathesia difformis, Lysichiton americanus, Nuphar lutea*

Food
Beverage: *Ledum groenlandicum, Vaccinium parvifolium*
Candy: *Picea sitchensis, Pinus contorta, Polypodium scouleri, Tsuga heterophylla*
Dessert: *Cornus sericea*
Dried Food: *Alaria marginata, Costaria costata, Fritillaria camschatcensis, Gaultheria shallon, Lessoniopsis littoralis, Malus fusca, Phyllospadix torreyi, Postelsia palmaeformis, Rubus parviflorus*
Forage: *Alectoria* sp., *Amelanchier alnifolia, Anthoxanthum odoratum, Enteromorpha intestinalis, Fritillaria camschatcensis, Heracleum maximum, Lonicera involucrata, Lysichiton americanus, Menyanthes trifoliata, Potamogeton* sp., *Ranunculus repens, Rosa nutkana, Scirpus acutus, Streptopus amplexifolius, Usnea* sp., *Vaccinium oxycoccos, Zostera marina*
Fruit: *Amelanchier alnifolia, Fragaria chiloensis, F. vesca, F. virginiana, Maianthemum dilatatum, Ribes bracteosum, R. divaricatum, R. laxiflorum, Rosa nutkana, Rubus leucodermis, R. parviflorus, R. ursinus, Sambucus racemosa, Vaccinium ovalifolium, V. ovatum, V. oxycoccos, V. parvifolium, V. vitis-idaea, Viburnum edule*
Pie & Pudding: *Vaccinium myrtilloides*
Preserves: *Ribes bracteosum, Rubus parviflorus, Sambucus racemosa, Vaccinium myrtilloides, V. ovalifolium, V. ovatum, V. oxycoccos, V. parvifolium*
Sour: *Rumex acetosella*
Special Food: *Cornus canadensis*
Spice: *Gaultheria shallon, Rubus parviflorus*
Starvation Food: *Blechnum spicant, Thuja plicata*
Unspecified: *Argentina egedii, Boschniakia hookeri, Cirsium brevistylum, C. vulgare, Heracleum maximum, Oenanthe sarmentosa, Phyllospadix scouleri, Porphyra perforata, Rubus spectabilis, Trifolium wormskioldii, Zostera marina*
Vegetable: *Camassia quamash, Equisetum arvense, Fritillaria camschatcensis, Polypodium glycyrrhiza, Pteridium aquilinum*

Fiber
Basketry: *Carex aquatilis, C. obnupta, Leymus mollis, Phyllospadix torreyi, Prunus emarginata, Scirpus acutus, S. americanus, Thuja plicata, Typha latifolia, Xerophyllum tenax*
Building Material: *Picea sitchensis, Thuja plicata*
Canoe Material: *Alnus rubra, Chamaecyparis nootkatensis, Taxus brevifolia, Thuja plicata*
Clothing: *Abies grandis, Carex obnupta, Chamaecyparis nootkatensis, Polystichum munitum, Scirpus americanus, Thuja plicata*
Cordage: *Juncus effusus, Nereocystis luetkeana, Thuja plicata, Urtica dioica*
Mats, Rugs & Bedding: *Chamaecyparis nootkatensis, Polystichum munitum, Scirpus acutus, Thuja plicata, Tsuga heterophylla, Typha latifolia*
Scouring Material: *Thuja plicata*
Sporting Equipment: *Alaria marginata, Costaria costata, Lessoniopsis littoralis, Postelsia palmaeformis, P.* sp.

Dye
Brown: *Physocarpus capitatus*
Purple: *Lonicera involucrata, Vaccinium ovalifolium*
Red: *Alnus rubra*
Red-Brown: *Tsuga heterophylla*
Unspecified: *Oplopanax horridus*

Other
Ceremonial Items: *Picea sitchensis, Tsuga heterophylla*
Containers: *Lysichiton americanus, Nereocystis luetkeana, Polystichum munitum*
Cooking Tools: *Alnus rubra, Chamaecyparis nootkatensis, Gaultheria shallon, Rubus spectabilis, Stachys ciliata, Thuja plicata*
Decorations: *Lycopodium clavatum*
Fasteners: *Picea sitchensis, Pinus contorta, Prunus emarginata*
Fertilizer: *Egregia menziesii, Fucus gardneri*
Fuel: *Alnus rubra, Picea sitchensis, Pseudotsuga menziesii, Thuja plicata*
Hunting & Fishing Item: *Oplopanax horridus, Thuja plicata, Tsuga heterophylla, Urtica dioica, Zostera marina*
Incense & Fragrance: *Abies grandis*
Preservative: *Pinus contorta*
Smoke Plant: *Arctostaphylos uva-ursi, Nicotiana tabacum*
Smoking Tools: *Ribes laxiflorum, Rubus spectabilis*
Snuff: *Nicotiana tabacum*
Tools: *Carex obnupta, Frangula purshiana, Holodiscus discolor, Malus fusca, Taxus brevifolia, Thuja plicata*
Toys & Games: *Acer macrophyllum, Ambrosia chamissonis, Macrocystis integrifolia, Phyllospadix torreyi, Physocarpus capitatus, Streptopus amplexifolius, Taraxacum officinale*
Weapon: *Picea sitchensis, Taxus brevifolia*

Hidatsa
Other
Hunting & Fishing Item: *Andropogon gerardii*

Hoh
Drug
Blood Medicine: *Mahonia nervosa, Prunus emarginata*
Ceremonial Medicine: *Equisetum hyemale, E. laevigatum*
Cold Remedy: *Sambucus racemosa*
Cough Medicine: *Pinus monticola, Sambucus racemosa*
Emetic: *Tsuga heterophylla, T. mertensiana*
Gynecological Aid: *Sambucus racemosa*
Pediatric Aid: *Nepeta cataria*
Tonic: *Cornus canadensis, C. nuttallii, C. sericea*
Unspecified: *Alnus rubra, Arctium minus, Arctostaphylos uva-ursi, Brassica nigra, Juncus ensifolius, Mentha canadensis, M. ×piperita, Moricandia arvensis,*

Oplopanax horridus, Populus balsamifera, Sinapis alba, Taraxacum officinale, Thuja plicata
Venereal Aid: *Malus fusca*

Food
Dried Food: *Equisetum hyemale, E. laevigatum*
Forage: *Lysichiton americanus*
Fruit: *Amelanchier alnifolia, Fragaria chiloensis, Gaultheria ovatifolia, Malus fusca, Ribes sanguineum, Rosa pisocarpa, Rubus idaeus, R. laciniatus, R. lasiococcus, R. leucodermis, R. nivalis, R. parviflorus, R. spectabilis, R. ursinus, Sambucus racemosa, Vaccinium deliciosum, V. myrtilloides, V. ovalifolium, V. ovatum, V. parvifolium*
Preserves: *Gaultheria ovatifolia, Mahonia nervosa, Prunus emarginata*
Sauce & Relish: *Gaultheria ovatifolia, Sambucus racemosa, Vaccinium deliciosum, V. myrtilloides, V. ovalifolium, V. ovatum, V. parvifolium*
Special Food: *Equisetum hyemale, E. laevigatum*
Spice: *Brassica nigra, Lysichiton americanus, Moricandia arvensis, Sinapis alba*
Unspecified: *Allium acuminatum, A. cernuum, Camassia quamash, Equisetum hyemale, Lepidium virginicum*
Vegetable: *Brassica nigra, Cirsium edule, Heracleum maximum, Lepidium virginicum, Moricandia arvensis, Sinapis alba, Urtica dioica*
Winter Use Food: *Ribes sanguineum, Rosa pisocarpa, Rubus idaeus, R. laciniatus, R. lasiococcus, R. leucodermis, R. nivalis, R. parviflorus, R. spectabilis, R. ursinus, Sambucus racemosa, Vaccinium deliciosum, V. myrtilloides, V. ovalifolium, V. ovatum, V. parvifolium*

Fiber
Basketry: *Picea* sp., *Salix sitchensis, Thuja plicata, Xerophyllum tenax*
Building Material: *Picea engelmannii, P. sitchensis, Pseudotsuga menziesii, Thuja plicata*
Canoe Material: *Chamaecyparis nootkatensis, Taxus brevifolia, Thuja plicata*
Clothing: *Thuja plicata*
Cordage: *Picea engelmannii, P. sitchensis, Thuja plicata, Urtica dioica*
Mats, Rugs & Bedding: *Thuja plicata*
Other: *Thuja plicata*

Other
Ceremonial Items: *Arctostaphylos uva-ursi, Cornus canadensis, C. nuttallii, C. sericea, Digitalis purpurea, Juniperus scopulorum, Rubus spectabilis, Symphoricarpos albus, Taxus brevifolia*
Containers: *Lysichiton americanus*
Cooking Tools: *Lysichiton americanus, Thuja plicata*
Hide Preparation: *Tsuga heterophylla, T. mertensiana*
Hunting & Fishing Item: *Taxus brevifolia*
Lighting: *Thuja plicata*
Smoke Plant: *Arbutus menziesii, Arctostaphylos tomentosa, A. uva-ursi, Cornus canadensis, C. nuttallii, C. sericea*
Toys & Games: *Picea engelmannii, P. sitchensis, Taxus brevifolia*

Hopi
Drug
Adjuvant: *Psilostrophe sparsiflora*
Analgesic: *Conyza canadensis, Cryptantha cinerea, Cycloloma atriplicifolium, Epilobium ciliatum, Eriogonum* sp., *Gaillardia pinnatifida, Heterotheca villosa, Ipomopsis longiflora, Petradoria pumila, Suaeda moquinii, Tetraneuris acaulis*
Anthelmintic: *Cirsium calcareum*
Anticonvulsive: *Atriplex confertifolia, A. obovata, Macromeria viridiflora, Salvia dorrii, Verbascum thapsus*
Antidiarrheal: *Opuntia whipplei, Sphaeralcea incana, S.* sp.
Antidote: *Physaria newberryi*
Antihemorrhagic: *Eriogonum* sp., *Lithospermum incisum, Sphaeralcea grossulariifolia, S. parvifolia*
Antirheumatic (External): *Hymenoxys bigelovii, Juniperus monosperma, Poliomintha incana, Senecio flaccidus, S. spartioides, Tetraneuris acaulis*
Antirheumatic (Internal): *Cycloloma atriplicifolium*
Blood Medicine: *Lithospermum incisum*
Breast Treatment: *Petradoria pumila*
Burn Dressing: *Krascheninnikovia lanata*
Cancer Treatment: *Cryptantha flava*
Carminative: *Gutierrezia microcephala*
Cathartic: *Hymenoxys bigelovii*
Ceremonial Medicine: *Astragalus* sp., *Atriplex canescens, Calochortus aureus, Chrysothamnus parryi, Delphinium geraniifolium, D. scaposum, Equisetum laevigatum, Hymenopappus filifolius, Lesquerella intermedia, Lycium pallidum, Nicotiana attenuata, Oenothera albicaulis, Physaria newberryi, Poa fendleriana, Rhus trilobata, Sarcobatus vermiculatus, Suaeda moquinii, Yucca angustissima*
Cold Remedy: *Rumex hymenosepalus*
Contraceptive: *Castilleja linariifolia*
Cough Medicine: *Machaeranthera grindelioides*
Dermatological Aid: *Artemisia filifolia, Chrysothamnus viscidiflorus, Cirsium calcareum, Cryptantha crassisepala, Dimorphocarpa wislizeni, Helianthus annuus, H. anomalus, H. petiolaris, Juniperus* sp., *Mirabilis coccineus, Pinus edulis, P. monophylla, Purshia stansburiana, Rhus trilobata, Rumex hymenosepalus, Senecio flaccidus, S. spartioides, Sphaeralcea parvifolia, Verbesina encelioides, Yucca angustissima*
Dietary Aid: *Chamaesyce fendleri*
Disinfectant: *Juniperus* sp., *Pinus edulis, P. monophylla, Yucca angustissima*
Diuretic: *Gaillardia pinnatifida, Sphaeralcea grossulariifolia, S. parvifolia, Stephanomeria exigua*
Ear Medicine: *Lupinus pusillus, Poliomintha incana*
Emetic: *Arenaria eastwoodiae, Astragalus* sp., *Croton texensis, Dalea candida, Delphinium scaposum, Hymenopappus filifolius, Lesquerella intermedia, Purshia stansburiana, Robinia neomexicana, Wyethia scabra*
Eye Medicine: *Croton texensis, Lupinus kingii, L. pusillus, Oenothera cespitosa*
Febrifuge: *Cycloloma atriplicifolium, Krascheninnikovia lanata, Verbesina encelioides*
Gastrointestinal Aid: *Artemisia filifolia, A. tridentata, Gutierrezia* sp., *Ipomopsis longiflora, Juniperus monosperma, J.* sp., *Linum australe, Phoradendron juniperinum, Ribes cereum, Sphaeralcea* sp.
Gynecological Aid: *Asclepias subverticillata, A. verticillata, Castilleja linariifolia, Delphinium geraniifolium, D. scaposum, Eriogonum* sp., *Hymenoxys bigelovii, Ipomopsis aggregata, Juniperus monosperma, J. osteosperma, J.* sp., *Lesquerella intermedia, Linum australe, Lygodesmia grandiflora, Machaeranthera canescens, M. tanacetifolia, Mirabilis multiflora, Petradoria pumila, Reverchonia arenaria, Stephanomeria pauciflora, Tetradymia canescens, Tetraneuris acaulis*
Hallucinogen: *Datura wrightii, Mirabilis multiflora*
Laxative: *Cirsium calcareum, Juniperus monosperma, Yucca angustissima, Y. baileyi*
Narcotic: *Datura wrightii*
Nose Medicine: *Chaetopappa ericoides*
Oral Aid: *Chamaesyce fendleri, Mahonia fremontii*
Orthopedic Aid: *Artemisia tridentata, Eriogonum* sp., *Juniperus* sp., *Krascheninnikovia lanata, Senecio flaccidus, Sphaeralcea grossulariifolia, S. parvifolia, S.* sp., *Tetraneuris acaulis*
Other: *Datura wrightii, Helianthus petiolaris, Juniperus osteosperma, Salvia dorrii*
Panacea: *Chaetopappa ericoides*
Pediatric Aid: *Abronia elliptica, Chaetopappa ericoides, Chamaesyce fendleri, Juniperus monosperma, J.* sp., *Sphaeralcea grossulariifolia, S. parvifolia, S.* sp., *Tripterocalyx carnea*

Poison: *Datura wrightii, Oxytropis lambertii, Wyethia scabra*

Psychological Aid: *Datura wrightii, Macromeria viridiflora, Plantago patagonica, Verbascum thapsus*

Reproductive Aid: *Chaetopappa ericoides, Juniperus monosperma, Townsendia incana*

Sedative: *Abronia elliptica, Chaetopappa ericoides, Tripterocalyx carnea*

Snakebite Remedy: *Gaura parviflora, Lesquerella intermedia*

Stimulant: *Chaetopappa ericoides, Datura wrightii, Hymenoxys bigelovii, Machaeranthera canescens, M. tanacetifolia, Salvia dorrii, Tetraneuris acaulis*

Throat Aid: *Cirsium calcareum, Townsendia incana, T. strigosa*

Tonic: *Ephedra viridis, Tetradymia canescens*

Toothache Remedy: *Hymenopappus filifolius, Mentzelia albicaulis, M. pumila, Oenothera cespitosa, Parryella filifolia*

Tuberculosis Remedy: *Erysimum capitatum, E. inconspicuum, Pinus edulis, Rhus trilobata*

Unspecified: *Hymenoxys bigelovii, Lithospermum incisum, Marrubium vulgare, Phoradendron juniperinum, P. sp., Rhus trilobata*

Venereal Aid: *Ephedra torreyana, E. viridis, Stephanomeria exigua*

Veterinary Aid: *Mirabilis multiflora, Phacelia crenulata*

Witchcraft Medicine: *Macromeria viridiflora, Phoradendron juniperinum, Pinus edulis, P. monophylla, Verbascum thapsus*

Food

Beverage: *Bidens amplectens, Hymenopappus filifolius, Hymenoxys cooperi, Ipomopsis aggregata, Rhus trilobata, Thelesperma megapotamicum, T. subnudum*

Bread & Cake: *Chenopodium graveolens, Eriogonum corymbosum, Hymenopappus filifolius, Muhlenbergia rigens, Oryzopsis hymenoides, Panicum capillare, Sporobolus cryptandrus, Zea mays*

Candy: *Astragalus ceramicus, Calochortus nuttallii, Dalea lanata, Erodium cicutarium, Populus sp., Typha angustifolia*

Cooking Agent: *Amaranthus cruentus, Carthamus tinctorius, Citrullus lanatus, Cucurbita moschata, Solanum jamesii*

Dried Food: *Capsicum annuum, Cucumis melo, Cucurbita moschata, Eriogonum corymbosum, Helianthus sp., Monarda fistulosa, Pectis angustifolia, Poliomintha incana, Prunus persica, Zea mays*

Fodder: *Croton texensis, Helianthus annuus, H. anomalus, H. petiolaris*

Forage: *Bouteloua gracilis*

Fruit: *Juniperus monosperma, J. osteosperma, Lycium pallidum, Opuntia erinacea, O. polyacantha, Prunus persica, Rhus trilobata, Ribes cereum, Rosa woodsii, Yucca angustissima, Y. baccata*

Pie & Pudding: *Atriplex confertifolia, Sporobolus cryptandrus*

Porridge: *Amaranthus blitoides, Chenopodium album, C. fremontii, C. leptophyllum, Cycloloma atriplicifolium, Lycium pallidum, Monolepis nuttalliana, Sporobolus giganteus, Zea mays*

Preserves: *Lycium pallidum, Yucca baccata*

Sauce & Relish: *Coriandrum sativum, Mentha canadensis, Sorghum bicolor*

Special Food: *Cucurbita moschata, Pinus edulis*

Spice: *Allium cernuum, A. geyeri, Artemisia frigida, Atriplex confertifolia, A. obovata, Capsicum annuum, Chenopodium fremontii, Chrysothamnus viscidiflorus, Coriandrum sativum, Descurainia pinnata, Eriogonum hookeri, Lygodesmia grandiflora, Mentha canadensis, Pectis angustifolia, Poliomintha incana, Sporobolus giganteus*

Staple: *Citrullus lanatus, Digitaria cognata, Mentzelia albicaulis, Oryzopsis hymenoides, Panicum capillare, Zea mays*

Starvation Food: *Amaranthus acanthochiton, Lycium pallidum, Oryzopsis hymenoides, Sporobolus airoides, S. flexuosus*

Substitution Food: *Atriplex canescens*

Sweetener: *Dalea lanata, Echinocereus fendleri, Zea mays*

Unspecified: *Allium cernuum, A. geyeri, A. vineale, Amaranthus arenicola, A. blitoides, A. powellii, Artemisia dracunculus, Asclepias speciosa, A. verticillata, Astragalus ceramicus, Atriplex argentea, A. confertifolia, A. powellii, Calochortus aureus, C. nuttallii, Castilleja linariifolia, Chenopodium album, Citrullus lanatus, Cleome serrulata, Coriandrum sativum, Cucumis melo, Cucurbita maxima, C. moschata, Cycloloma cornutum, Cymopterus multinervatus, C. newberryi, Descurainia pinnata, Helianthus tuberosus, Lycium pallidum, Lygodesmia grandiflora, Malus sylvestris, Mentzelia albicaulis, Monarda citriodora, Opuntia erinacea, O. polyacantha, O. sp., O. whipplei, Pectis angustifolia, Pinus edulis, P. monophylla, Poliomintha incana, Portulaca oleracea, Prunus armeniaca, P. dulcis, Pyrus communis, Scirpus acutus, S. tabernaemontani, Solanum jamesii, Stanleya albescens, Ustilago zeae, Wislizenia refracta, Zea mays*

Vegetable: *Amaranthus acanthochiton, A. blitoides, A. powellii, Atriplex argentea, A. confertifolia, A. obovata, A. powellii, A. saccaria, Chenopodium fremontii, C. incanum, Cleome serrulata, Descurainia obtusa, D. pinnata, Pseudocymopterus montanus, Stanleya albescens, S. pinnata, Tradescantia occidentalis*

Fiber

Basketry: *Bouteloua gracilis, Hilaria jamesii, Parryella filifolia, Rhus trilobata, Yucca angustissima, Y. baccata, Y. baileyi, Y. glauca*

Brushes & Brooms: *Aristida purpurea, Hilaria jamesii, Muhlenbergia pungens, Senecio flaccidus, Yucca angustissima, Y. baileyi, Y. glauca*

Building Material: *Chrysothamnus nauseosus, C. parryi, C. viscidiflorus, Juniperus monosperma, Phragmites australis, Pinus ponderosa, Populus sp., Salix sp., Sarcobatus vermiculatus*

Clothing: *Purshia stansburiana*

Furniture: *Rhus trilobata, Sarcobatus vermiculatus*

Mats, Rugs & Bedding: *Purshia stansburiana, Typha latifolia*

Sewing Material: *Yucca angustissima*

Dye

Black: *Helianthus sp.*

Blue: *Atriplex canescens*

Green: *Chrysothamnus nauseosus, C. parryi*

Mordant: *Atriplex canescens, Rhus trilobata*

Purple: *Helianthus sp.*

Red: *Amaranthus cruentus, Cycloloma atriplicifolium*

Red-Brown: *Thelesperma megapotamicum, T. subnudum*

Unspecified: *Amaranthus cruentus, Hymenopappus filifolius, Hymenoxys cooperi, Ipomopsis aggregata, Pectis angustifolia, Pinus edulis, P. monophylla, Rumex hymenosepalus*

Yellow: *Carthamus tinctorius, Chrysothamnus nauseosus, C. parryi, C. viscidiflorus*

Other

Ceremonial Items: *Amelanchier utahensis, Argemone polyanthemos, Aristida purpurea, Artemisia carruthii, A. filifolia, A. frigida, Atriplex canescens, Calamovilfa gigantea, Calochortus aureus, C. nuttallii, Castilleja linariifolia, Cercocarpus intricatus, C. montanus, Chrysothamnus nauseosus, C. viscidiflorus, Cucumis melo, Delphinium geraniifolium, D. scaposum, Equisetum laevigatum, Fendlera rupicola, Forestiera pubescens, Fraxinus anomala, Gutierrezia sarothrae, Helianthus petiolaris, H. sp., Hilaria jamesii, Hymenoxys cooperi, Juncus balticus, J. torreyi, Juniperus monosperma, Krascheninnikovia lanata, Lagenaria siceraria, Lupinus pusillus, Lycium pallidum, Mahonia repens, Nicotiana attenuata, N. trigonophylla, Oenothera albicaulis, O. cespitosa, O. pallida, Parryella filifolia, Penstemon ambiguus, P. eatonii, P. utahensis, Phragmites australis, Pinus edulis, P. ponderosa, Plantago patagonica, Populus sp., P. tremuloides, Proboscidea louisianica, Pseudotsuga menziesii, Psilostrophe sparsiflora, Purshia stansburiana, Quercus gambelii, Rhus trilobata, Salix sp., Scirpus acutus, Sphaeralcea parvifolia, Sporobolus giganteus, Stipa neomexicana, Typha angustifolia, Ustilago zeae, Yucca angustissima, Y. baileyi, Zea mays*

Cleaning Agent: *Juniperus monosperma*

Containers: *Chenopodium fremontii, Cucurbita moschata, Lagenaria siceraria, Populus* sp.
Cooking Tools: *Chrysothamnus viscidiflorus, Gutierrezia microcephala, Juniperus monosperma, Lagenaria siceraria, Sarcobatus vermiculatus*
Decorations: *Aristida purpurea, Calamovilfa gigantea, Castilleja affinis, C. linariifolia, Chrysothamnus depressus, C. greenei, C. nauseosus, C. parryi, C. viscidiflorus, Gutierrezia microcephala, G. sarothrae, Helianthus annuus, H. anomalus, H. petiolaris, Hilaria jamesii, Ipomopsis aggregata, Juniperus monosperma, Lagenaria siceraria, Oenothera albicaulis, Penstemon utahensis, Petradoria pumila, Yucca angustissima*
Fasteners: *Pinus edulis, P. monophylla*
Fuel: *Chrysothamnus parryi, Juniperus monosperma, J. osteosperma, Rhus trilobata, Sarcobatus vermiculatus*
Hunting & Fishing Item: *Amelanchier pallida, A. utahensis, Calamovilfa gigantea, Chrysothamnus nauseosus, Fallugia paradoxa, Lagenaria siceraria, Mirabilis multiflora, Purshia stansburiana, Ribes cereum, Sarcobatus vermiculatus, Yucca baileyi*
Incense & Fragrance: *Rhus trilobata*
Insecticide: *Boerhavia erecta*
Jewelry: *Juniperus osteosperma, Solanum elaeagnifolium*
Malicious Charm: *Juniperus monosperma*
Musical Instrument: *Cucurbita moschata, Lagenaria siceraria, Populus* sp., *Sarcobatus vermiculatus*
Paint: *Castilleja linariifolia, Chrysothamnus parryi, Descurainia pinnata, Yucca angustissima, Y. baileyi*
Protection: *Chrysothamnus nauseosus, Phlox stansburyi, Pinus edulis*
Season Indicator: *Helianthus petiolaris, Penstemon ambiguus, P. eatonii, P. utahensis, Pinus ponderosa*
Smoke Plant: *Foeniculum vulgare, Mentzelia albicaulis, M. pumila, Nicotiana attenuata, N. trigonophylla, Oenothera albicaulis, O. cespitosa*
Soap: *Verbesina encelioides, Yucca angustissima, Y. baccata, Y. baileyi, Y. glauca*
Stable Gear: *Parryella filifolia*
Tools: *Asclepias verticillata, Cercocarpus montanus, Forestiera pubescens, Juniperus monosperma, Lagenaria siceraria, Mahonia fremontii, Pinus ponderosa, Populus* sp., *Prunus persica, Sarcobatus vermiculatus*
Toys & Games: *Calochortus aureus, Delphinium scaposum, Dimorphocarpa wislizeni, Juniperus monosperma, Populus* sp.
Waterproofing Agent: *Pinus edulis, P. monophylla*

Houma
Drug
Abortifacient: *Rumex salicifolius, Sonchus oleraceus*
Analgesic: *Chenopodium ambrosioides, Desmodium paniculatum, Hypericum hypericoides, Lagenaria siceraria, Pleopeltis polypodioides, Polygonum punctatum, Sambucus canadensis, S. cerulea*
Anthelmintic: *Chenopodium ambrosioides, Myrica cerifera, Nyssa sylvatica, Solanum nigrum*
Antidiarrheal: *Quercus pagoda, Q. virginiana, Sonchus oleraceus, Ulmus americana*
Antiemetic: *Gonolobus* sp., *Panax quinquefolius*
Antirheumatic (External): *Quercus alba*
Antirheumatic (Internal): *Panax quinquefolius*
Blood Medicine: *Cocculus carolinus, Ipomoea sagittata, Magnolia virginiana, Passiflora incarnata, Salix caroliniana, S. nigra, Sequoia sempervirens*
Burn Dressing: *Plantago cordata*
Cold Remedy: *Gamochaeta purpurea, Magnolia virginiana*
Dermatological Aid: *Chamaesyce nutans, Cirsium horridulum, Ipomoea sagittata, Juglans nigra, Liquidambar styraciflua, Parthenocissus quinquefolia, Plantago cordata, Polymnia canadensis, Sambucus canadensis, S. cerulea, Solanum nigrum*
Diaphoretic: *Liquidambar styraciflua*
Expectorant: *Cirsium horridulum*
Eye Medicine: *Sabal minor*
Febrifuge: *Cornus florida, C. foemina, Eupatorium serotinum, Hypericum hypericoides, Laportea canadensis, Liquidambar styraciflua, Magnolia virginiana, Pluchea* sp., *Rumex salicifolius, Salix caroliniana, S. nigra, Tillandsia usneoides, Xanthium strumarium*
Gastrointestinal Aid: *Chamaesyce nutans, Rumex salicifolius*
Gynecological Aid: *Ambrosia artemisiifolia, Andropogon* sp., *Conyza canadensis, Erigeron philadelphicus, Hypericum hypericoides*
Heart Medicine: *Ipomoea sagittata, Sanicula canadensis*
Hemorrhoid Remedy: *Pluchea* sp.
Hypotensive: *Juglans nigra, Sabal minor*
Kidney Aid: *Arundinaria gigantea, Sabal minor, Verbena officinalis*
Liver Aid: *Rumex salicifolius, Sequoia sempervirens, Solidago nemoralis, Verbena officinalis*
Misc. Disease Remedy: *Bignonia capreolata, Cornus florida, C. foemina, Eupatorium serotinum, Gamochaeta purpurea, Modiola caroliniana, Parthenocissus quinquefolia, Sassafras albidum, Senna tora, Strophostyles helvula*
Oral Aid: *Pleopeltis polypodioides*
Orthopedic Aid: *Phoradendron leucarpum, Phyla nodiflora, Polygonum punctatum, Quercus pagoda, Scirpus* sp., *Zanthoxylum clava-herculis*
Panacea: *Forestiera acuminata, Phoradendron leucarpum*
Pediatric Aid: *Andropogon* sp., *Chamaesyce nutans, Chenopodium ambrosioides, Phyla nodiflora, Pleopeltis polypodioides, Scirpus* sp., *Solanum nigrum, Sonchus oleraceus*
Pulmonary Aid: *Iresine diffusa, Typha latifolia*
Reproductive Aid: *Berchemia scandens*
Sedative: *Scirpus* sp.
Snakebite Remedy: *Ipomoea sagittata, Melothria pendula*
Stimulant: *Arundinaria gigantea, Desmodium paniculatum, Sabal minor*
Throat Aid: *Celtis laevigata, C. occidentalis, Cirsium horridulum, Modiola caroliniana, Quercus pagoda*
Tonic: *Quercus pagoda, Sambucus canadensis, S. cerulea, Toxicodendron radicans*
Toothache Remedy: *Hypericum hypericoides, Sonchus oleraceus, Zanthoxylum clava-herculis*
Tuberculosis Remedy: *Apium graveolens, Lepidium virginicum*
Urinary Aid: *Smilax bona-nox, S. laurifolia*
Venereal Aid: *Celtis laevigata, C. occidentalis*
Vertigo Medicine: *Pleopeltis polypodioides*
Food
Bread & Cake: *Sabal minor, Smilax bona-nox, S. laurifolia*
Fodder: *Sonchus oleraceus*
Unspecified: *Cirsium horridulum*
Fiber
Brushes & Brooms: *Scirpus* sp.
Cordage: *Tillandsia usneoides*
Mats, Rugs & Bedding: *Tillandsia usneoides*
Dye
Red: *Quercus virginiana*
Unspecified: *Rumex salicifolius*
Other
Cooking Tools: *Lagenaria siceraria*
Fasteners: *Berchemia scandens*
Hunting & Fishing Item: *Arundinaria gigantea, Sambucus canadensis, S. cerulea*
Lighting: *Myrica cerifera, M. heterophylla*
Musical Instrument: *Lagenaria siceraria*

Hualapai
Drug
Analgesic: *Artemisia* sp.
Antirheumatic (External): *Atriplex* sp., *Baccharis* sp.
Antirheumatic (Internal): *Purshia mexicana*
Burn Dressing: *Opuntia* sp.
Cold Remedy: *Artemisia* sp., *Larrea tridentata*
Dermatological Aid: *Acourtia wrightii*, *Agave* sp., *Eriodictyon angustifolium*, *Juniperus* sp., *Opuntia* sp., *Purshia mexicana*
Disinfectant: *Larrea tridentata*
Expectorant: *Pinus edulis*
Gastrointestinal Aid: *Artemisia* sp., *Eriodictyon angustifolium*, *Erodium* sp., *Mahonia fremontii*
Laxative: *Eriodictyon angustifolium*, *Mahonia fremontii*
Liver Aid: *Mahonia fremontii*
Orthopedic Aid: *Eriodictyon angustifolium*, *Fouquieria* sp.
Other: *Pinus edulis*
Panacea: *Juniperus* sp.
Pulmonary Aid: *Artemisia* sp.
Respiratory Aid: *Larrea tridentata*
Unspecified: *Ephedra* sp.
Food
Beverage: *Arctostaphylos* sp., *Mahonia fremontii*, *Opuntia* sp., *Pinus edulis*, *Rhus trilobata*, *Rumex hymenosepalus*, *Vitis* sp., *Yucca baccata*
Bread & Cake: *Cymopterus* sp., *Pinus edulis*, *Quercus turbinella*, *Typha* sp.
Candy: *Pinus edulis*
Dried Food: *Arctostaphylos* sp., *Celtis laevigata*, *Ferocactus* sp., *Mentzelia* sp., *Opuntia* sp., *Prosopis pubescens*, *P.* sp., *Vitis* sp., *Yucca baccata*, *Y. schidigera*
Fruit: *Arctostaphylos* sp., *Celtis laevigata*, *Ferocactus* sp., *Mahonia fremontii*, *Morus* sp., *Nolina bigelovii*, *Opuntia* sp., *Physalis* sp., *Rhus trilobata*, *Vitis* sp., *Yucca baccata*, *Y. schidigera*
Porridge: *Mentzelia* sp., *Pinus edulis*
Preserves: *Physalis* sp.
Sauce & Relish: *Physalis* sp.
Soup: *Cymopterus* sp., *Pinus edulis*, *Quercus gambelii*, *Q. turbinella*
Staple: *Agave* sp., *Mentzelia* sp., *Yucca baccata*, *Y. schidigera*
Starvation Food: *Juniperus* sp.
Sweetener: *Agave* sp.
Unspecified: *Cymopterus* sp., *Helianthus* sp., *Juglans major*, *Pinus edulis*, *Proboscidea parviflora*, *Prosopis pubescens*, *Quercus gambelii*, *Q. turbinella*, *Typha* sp.
Vegetable: *Allium* sp.
Winter Use Food: *Agave* sp., *Allium* sp., *Helianthus* sp.
Fiber
Basketry: *Acacia* sp., *Morus* sp., *Nolina bigelovii*, *Populus* sp., *Proboscidea parviflora*, *Rhus trilobata*, *Salix* sp.
Brushes & Brooms: *Gutierrezia* sp., *Hymenoclea* sp.
Building Material: *Baccharis* sp., *Fouquieria* sp., *Nolina* sp., *Pluchea sericea*, *Purshia mexicana*, *Salix* sp.
Clothing: *Agave* sp., *Purshia mexicana*, *Yucca baccata*, *Y. schidigera*
Cordage: *Agave* sp., *Yucca baccata*
Furniture: *Acacia* sp., *Chilopsis linearis*, *Fallugia paradoxa*, *Pluchea sericea*, *Prosopis pubescens*, *P.* sp., *Robinia neomexicana*
Mats, Rugs & Bedding: *Agave* sp.
Sewing Material: *Chilopsis linearis*
Dye
Black: *Helianthus* sp.
Purple: *Helianthus* sp.
Unspecified: *Juglans major*, *Prosopis* sp., *Rhus trilobata*, *Rumex hymenosepalus*
Yellow: *Mahonia fremontii*
Other
Cash Crop: *Pinus ponderosa*
Ceremonial Items: *Gutierrezia* sp.
Containers: *Agave* sp.
Fasteners: *Pinus edulis*
Fuel: *Baccharis* sp., *Hymenoclea* sp., *Salix* sp.
Hide Preparation: *Rumex hymenosepalus*
Hunting & Fishing Item: *Fraxinus velutina*, *Hymenoclea* sp., *Morus* sp., *Nolina* sp., *Phragmites australis*, *Pluchea sericea*, *Robinia neomexicana*
Insecticide: *Rhus trilobata*
Musical Instrument: *Phragmites australis*, *Populus* sp.
Protection: *Rhus trilobata*
Smoke Plant: *Nicotiana trigonophylla*
Soap: *Yucca baccata*
Tools: *Fraxinus velutina*
Toys & Games: *Typha* sp.
Walking Sticks: *Fraxinus velutina*, *Morus* sp.
Waterproofing Agent: *Pinus edulis*

Hupa
Food
Bread & Cake: *Lithocarpus densiflorus*
Porridge: *Lithocarpus densiflorus*
Staple: *Lithocarpus densiflorus*, *Madia elegans*
Unspecified: *Lithocarpus densiflorus*
Fiber
Basketry: *Woodwardia radicans*, *Xerophyllum tenax*
Other
Decorations: *Xerophyllum tenax*

Huron
Food
Starvation Food: *Apios americana*, *Arachis hypogaea*, *Helianthus tuberosus*, *Lilium canadense*, *Nelumbo lutea*

Iroquois
Drug
Abortifacient: *Aralia racemosa*, *Ceanothus americanus*, *Celastrus scandens*, *Celtis occidentalis*, *Euonymus americana*, *Fagus grandifolia*, *Fragaria* sp., *Hypericum ellipticum*, *Malus coronaria*, *Rhus* sp., *Taxus baccata*, *Triosteum perfoliatum*, *Vicia sativa*, *Zanthoxylum americanum*
Adjuvant: *Asarum canadense*, *Chimaphila umbellata*, *Lobelia cardinalis*, *Taxus baccata*
Alterative: *Oxalis* sp., *Rhus glabra*
Analgesic: *Achillea millefolium*, *Adiantum pedatum*, *Aesculus hippocastanum*, *Alnus incana*, *Angelica atropurpurea*, *Arctium* sp., *Arisaema triphyllum*, *Asarum canadense*, *Cardamine concatenata*, *Chimaphila umbellata*, *Collinsonia canadensis*, *Cornus amomum*, *C. sericea*, *Cypripedium acaule*, *C. arietinum*, *C. pubescens*, *Dirca palustris*, *Epigaea repens*, *Epilobium angustifolium*, *Equisetum arvense*, *Eupatorium perfoliatum*, *Fraxinus nigra*, *Gentiana andrewsii*, *Hedeoma pulegioides*, *Hepatica nobilis*, *Heracleum maximum*, *Inula helenium*, *Ipomoea pandurata*, *Juglans cinerea*, *J. nigra*, *Lactuca canadensis*, *Larix laricina*, *Lobelia cardinalis*, *Lonicera oblongifolia*, *Mentha spicata*, *Mitchella repens*, *Monarda* sp., *Nepeta cataria*, *Nuphar lutea*, *N.* sp., *Osmorhiza* sp., *Osmunda cinnamomea*, *Panax trifolius*, *Plantago* sp., *Platanthera psycodes*, *Polygonum hydropiper*, *Polystichum acrostichoides*, *Prunella vulgaris*, *Prunus serotina*, *Ranunculus acris*, *Rhus* sp., *Rubus allegheniensis*, *R. idaeus*, *Salix interior*, *Sambucus canadensis*, *Sanguinaria canadensis*, *Sium suave*, *Smallanthus uvedalia*, *Solidago canadensis*, *Spiraea alba*, *Tanacetum vulgare*, *Taraxacum officinale*, *Trillium* sp., *Triosteum perfoliatum*, *Veratrum viride*, *Viburnum acerifolium*, *Zanthoxylum americanum*

Index of Tribes   675

Anthelmintic: *Achillea millefolium, Acorus calamus, Allium tricoccum, Anemone canadensis, Apocynum androsaemifolium, Aralia racemosa, Carya ovata, Chimaphila umbellata, Coptis trifolia, Elytrigia repens, Euonymus europaea, Gaultheria procumbens, Gentianella quinquefolia, Helianthus strumosus, Juglans cinerea, Magnolia acuminata, Maianthemum racemosum, Panax quinquefolius, Populus balsamifera, P. deltoides, P. tremuloides, Rhus* sp., *Rudbeckia hirta, Sanguinaria canadensis, Sassafras albidum, Senna hebecarpa, S. marilandica, Symplocarpus foetidus, Verbena hastata, Viburnum lantanoides, Zanthoxylum americanum*

Anticonvulsive: *Achillea millefolium, Asarum canadense, Claytonia virginica, Conyza canadensis, Geum aleppicum, Lobelia cardinalis, Lycopodium* sp., *Medeola virginiana, Mimulus ringens, Mitchella repens, Nuphar lutea, Osmunda regalis, Polystichum acrostichoides, Pyrola elliptica, Ranunculus abortivus, Rhus* sp., *Sium suave, Verbascum thapsus*

Antidiarrheal: *Achillea millefolium, Agrimonia gryposepala, Ambrosia artemisiifolia, A. trifida, Anaphalis margaritacea, Anemone virginiana, Anthemis cotula, Apocynum cannabinum, Aralia racemosa, Argentina anserina, Arisaema triphyllum, Carpinus caroliniana, Ceanothus americanus, Chenopodium album, Collinsonia canadensis, Corylus americana, Epifagus virginiana, Eupatorium maculatum, Fragaria* sp., *Gentianella quinquefolia, Geranium maculatum, Geum aleppicum, G. rivale, Hamamelis virginiana, Hieracium pilosella, Hydrastis canadensis, Jeffersonia diphylla, Linaria vulgaris, Lonicera* sp., *Nepeta cataria, Osmorhiza* sp., *Plantago* sp., *Polygonum aviculare, Polystichum acrostichoides, Porteranthus trifoliatus, Potentilla canadensis, Prunella vulgaris, Prunus virginiana, Pteridium aquilinum, Quercus macrocarpa, Ranunculus acris, Rubus allegheniensis, R. occidentalis, R. odoratus, Rumex crispus, R.* sp., *Sanguinaria canadensis, Spiraea* sp., *Ulmus americana, Verbascum thapsus, Veronicastrum virginicum*

Antidote: *Cardamine douglassii, Hydrophyllum canadense, Laportea canadensis, Maianthemum racemosum, Mentha arvensis, Mertensia virginica, Mimulus ringens, Nicotiana rustica, Parthenocissus quinquefolia, Portulaca oleracea, Quercus macrocarpa, Ranunculus abortivus, Rhamnus alnifolia, Ribes americanum, Sanicula marilandica*

Antiemetic: *Achillea millefolium, Agrimonia gryposepala, Anthemis cotula, Asarum canadense, Coptis trifolia, Corylus americana, Hamamelis virginiana, Mitchella repens, Nepeta cataria, Panax quinquefolius, Plantago* sp., *Prunella vulgaris, Ribes americanum, Sanguinaria canadensis, Smallanthus uvedalia, Solidago juncea, Spiraea alba, Ulmus americana, Zanthoxylum americanum*

Antihemorrhagic: *Acer spicatum, Actaea rubra, Alnus incana, Corylus americana, Cynoglossum officinale, Prunus virginiana, Ranunculus acris, Rumex* sp., *Sanguinaria canadensis, Scrophularia lanceolata, Solidago* sp., *Tilia americana, Ulmus americana, Viburnum lentago*

Antirheumatic (External): *Abies balsamea, Actaea rubra, Adiantum pedatum, Angelica atropurpurea, Arctium minus, A.* sp., *Asclepias syriaca, Asparagus officinalis, Carya ovata, Caulophyllum thalictroides, Cimicifuga racemosa, Collinsonia canadensis, Corylus cornuta, Eupatorium maculatum, Fraxinus nigra, Heracleum maximum, Inula helenium, Lathyrus japonicus, Lemna trisulca, Maianthemum racemosum, Osmunda cinnamomea, Phytolacca americana, Picea* sp., *Plantago* sp., *Platanus occidentalis, Polygonum persicaria, Polystichum acrostichoides, Rumex crispus, Smilax herbacea, Stellaria media, Symplocarpus foetidus, Taxus baccata, Thuja occidentalis, Trillium* sp., *Tsuga canadensis*

Antirheumatic (Internal): *Abies balsamea, Achillea millefolium, Adiantum pedatum, Aralia racemosa, Baptisia tinctoria, Carya ovata, Caulophyllum thalictroides, Chimaphila umbellata, Collinsonia canadensis, Epigaea repens, Equisetum arvense, Gaultheria procumbens, Hamamelis virginiana, Juniperus virginiana, Larix laricina, Maianthemum racemosum, Onoclea sensibilis, Picea* sp., *Pinus rigida, P. strobus, Populus balsamifera, Pteridium aquilinum, Pyrola elliptica, Sagittaria latifolia, Sassafras albidum, Staphylea trifolia, Taxus baccata, Tsuga canadensis, Veronicastrum virginicum*

Basket Medicine: *Agrimonia gryposepala, Clintonia umbellulata, Corallorrhiza maculata, Desmodium glutinosum, Epilobium angustifolium, Galium* sp., *Lobelia cardinalis, Sarracenia purpurea*

Blood Medicine: *Acer rubrum, A. saccharum, Achillea millefolium, Acorus calamus, Ageratina altissima, Ambrosia trifida, Amelanchier arborea, A. canadensis, Angelica atropurpurea, Anthemis cotula, Apocynum cannabinum, Aralia nudicaulis, A. racemosa, Arctium minus, A.* sp., *Arisaema triphyllum, Armoracia rusticana, Asarum canadense, Asparagus officinalis, Aster macrophyllus, A.* sp., *Betula alleghaniensis, B. lenta, Ceanothus americanus, Celastrus scandens, Chimaphila umbellata, Cimicifuga racemosa, Collinsonia canadensis, Coptis trifolia, Cornus florida, Corylus americana, Cypripedium parviflorum, Dalibarda repens, Daucus carota, Diervilla lonicera, Dirca palustris, Fagus grandifolia, Fragaria* sp., *Fraxinus americana, Gaultheria procumbens, Gaylussacia baccata, Hamamelis virginiana, Hepatica nobilis, Huperzia lucidula, Ipomoea pandurata, Iris versicolor, Juglans cinerea, J. nigra, Larix laricina, Lobelia spicata, Lonicera canadensis, Lycopodium obscurum, Maianthemum racemosum, Mitchella repens, Myriophyllum sibiricum, Nuphar lutea, Onoclea sensibilis, Osmorhiza* sp., *Osmunda claytoniana, O. regalis, Oxalis stricta, Panax quinquefolius, Pinus strobus, Plantago major, P.* sp., *Polystichum acrostichoides, Populus balsamifera, Prunella vulgaris, Prunus serotina, P. virginiana, Pteridium aquilinum, Pyrola elliptica, Ranunculus abortivus, R. acris, Rhamnus alnifolia, Rubus allegheniensis, R. idaeus, R. odoratus, R.* sp., *Rumex crispus, R. obtusifolius, Sanguinaria canadensis, Sassafras albidum, Saxifraga pensylvanica, Scrophularia lanceolata, Senecio aureus, Senna marilandica, Taraxacum officinale, Thuja occidentalis, Toxicodendron rydbergii, Trifolium pratense, Trillium* sp., *Tsuga canadensis, Ulmus rubra, Uvularia sessilifolia, Verbascum thapsus, Viburnum lantanoides, V. opulus, Vitis vulpina, Waldsteinia fragarioides*

Breast Treatment: *Cardamine diphylla*

Burn Dressing: *Chenopodium album, Fagus grandifolia, Physalis heterophylla, Pinus rigida, Plantago* sp., *Portulaca oleracea, Prunus pensylvanica, P. serotina, Solidago squarrosa, Tilia americana*

Cancer Treatment: *Abies balsamea, Aralia nudicaulis, Chimaphila umbellata, Cirsium vulgare, Cynoglossum officinale, Ostrya virginiana*

Carminative: *Cardamine diphylla, Eupatorium maculatum, Hydrastis canadensis, Inula helenium, Mitchella repens, Polygonatum pubescens, Rhus* sp., *Salix nigra, Sanguinaria canadensis*

Cathartic: *Ageratina altissima, Alnus incana, Asarum canadense, Cornus rugosa, Dirca palustris, Euonymus europaea, Fraxinus americana, Ilex verticillata, Inula helenium, Iris versicolor, Juglans cinerea, Lobelia inflata, Mitchella repens, Osmorhiza* sp., *Phytolacca americana, Podophyllum peltatum, Porteranthus trifoliatus, Rhamnus alnifolia, Rhus* sp., *Rubus idaeus, R. occidentalis, Rumex crispus, Sambucus canadensis, Sisyrinchium montanum, Veronicastrum virginicum*

Ceremonial Medicine: *Elymus canadensis, E. hystrix, Gaylussacia baccata, Phragmites australis, Podophyllum peltatum, Sambucus canadensis, Saxifraga pensylvanica, Vaccinium angustifolium*

Cold Remedy: *Abies balsamea, Acorus calamus, Angelica atropurpurea, Aralia nudicaulis, Asarum canadense, Aster prenanthoides, A. puniceus, Betula lenta, Cardamine concatenata, Ceanothus americanus, Celastrus scandens, Celtis occidentalis, Cornus alternifolia, C. canadensis, C. sericea, Erigeron pulchellus, Eupatorium maculatum, E. perfoliatum, Gaultheria procumbens, Hamamelis virginiana, Huperzia lucidula, Inula helenium, Juniperus communis, J. virginiana, Larix laricina, Lindera benzoin, Mentha ×piperita, M. spicata, Nepeta cataria, Osmunda cinnamomea, Phytolacca americana, Picea* sp., *Pinus strobus, Populus alba, Porteranthus trifoliatus, Prunella vulgaris, Prunus serotina, Rhus* sp., *Rubus allegheniensis, R. odoratus, Sanguinaria canadensis, Sassafras albidum, Scrophularia lanceolata, Tanacetum vulgare, Taxus baccata, Thuja occidentalis, Trillium* sp., *Triosteum perfoliatum, Tsuga canadensis, Veratrum viride, Verbena* sp., *Viburnum opulus*

Contraceptive: *Arisaema triphyllum, Asclepias syriaca, Claytonia virginica, Erythronium americanum, Hepatica nobilis, Rumex obtusifolius, Viburnum dentatum, V.* sp.

Cough Medicine: *Abies balsamea, Aralia nudicaulis, A. racemosa, Asarum canadense, Botrychium virginianum, Cornus alternifolia, C. sericea, Erigeron pulchellus, Hamamelis virginiana, Inula helenium, Ipomoea pandurata, Juniperus communis, J. virginiana, Larix laricina, Lobelia siphilitica, Nepeta cataria, Ostrya virginiana, Picea* sp., *Pinus strobus, Prunella vulgaris, Prunus pensylvanica, P. serotina, P. virginiana, Rhus* sp., *Rubus allegheniensis, Salix nigra, Sanguinaria canadensis, Taxus baccata, Tsuga canadensis, Tussilago farfara, Uvularia perfoliata, Veronicastrum virginicum*

Dermatological Aid: *Abies balsamea, Acer saccharum, Acorus calamus, Agastache nepetoides, Apocynum androsaemifolium, Aquilegia canadensis, Aralia nudicaulis, A. racemosa, Arctium minus, A.* sp., *Arisaema triphyllum, Asarum canadense, Asclepias incarnata, A. syriaca, Aster novae-angliae, Betula alleghaniensis, Brassica napus, Cardamine diphylla, Carpinus caroliniana, Carya cordiformis, C. ovata, Castanea dentata, Ceanothus americanus, Cerastium arvense, Chimaphila umbellata, Cichorium intybus, Cicuta maculata, Circaea lutetiana, Cirsium discolor, Clematis virginiana, Collinsonia canadensis, Cornus alternifolia, C. amomum, C. racemosa, Corylus americana, Cynoglossum officinale, Cypripedium acaule, Daucus carota, Dipsacus fullonum, Dirca palustris, Erigeron philadelphicus, Erythronium americanum, Eupatorium perfoliatum, Fagus grandifolia, Fraxinus americana, Galium aparine, Geranium maculatum, Hamamelis virginiana, Heracleum maximum, Hieracium* sp., *Huperzia lucidula, Impatiens capensis, I. pallida, Inula helenium, Juglans cinerea, J. nigra, Juncus bufonius, Lobelia cardinalis, L. inflata, L. kalmii, L. spicata, Lonicera canadensis, Maianthemum racemosum, Malus coronaria, Malva neglecta, Melilotus officinalis, Mitchella repens, Nicotiana rustica, Oenothera biennis, Onoclea sensibilis, Ostrya virginiana, Panax quinquefolius, Pastinaca sativa, Phytolacca americana, Picea* sp., *Pinus rigida, P. strobus, Plantago major, P.* sp., *Platanthera orbiculata, Platanus occidentalis, Podophyllum peltatum, Polygala paucifolia, Polygonum aviculare, Polystichum acrostichoides, Populus balsamifera, P. grandidentata, Portulaca oleracea, Prenanthes alba, P. trifoliolata, Prunus serotina, P. virginiana, Pyrola elliptica, Quercus macrocarpa, Q. rubra, Rhamnus alnifolia, Rhus* sp., *Ribes americanum, Rubus allegheniensis, R. idaeus, R. odoratus, R.* sp., *Rumex crispus, Sagittaria latifolia, Sambucus canadensis, S.* sp., *Sanguinaria canadensis, Sanicula marilandica, Sassafras albidum, Scrophularia lanceolata, Sedum telephium, Smilax herbacea, Staphylea trifolia, Stellaria media, Strophostyles helvula, Symplocarpus foetidus, Tanacetum vulgare, Taraxacum officinale, Thuja occidentalis, Tiarella cordifolia, Trifolium repens, Trillium erectum, T.* sp., *Vaccinium* sp., *Veratrum viride, Verbascum thapsus, Viola pubescens, V.* sp.

Diaphoretic: *Ageratina altissima, Angelica atropurpurea, Arctium* sp., *Asarum canadense, Heracleum maximum, Porteranthus trifoliatus, Senecio aureus, Taxus baccata, Triosteum perfoliatum, Tsuga canadensis*

Dietary Aid: *Asarum canadense, Cardamine concatenata, Daucus carota, Euonymus europaea, Hamamelis virginiana, Panax quinquefolius, Pinus strobus, Rhus hirta, R.* sp., *Rumex crispus, Sarracenia purpurea, Tiarella cordifolia, Triosteum perfoliatum*

Disinfectant: *Asarum canadense, Cicuta maculata, Smallanthus uvedalia*

Diuretic: *Adiantum pedatum, Arctium* sp., *Argentina anserina, Asclepias incarnata, Celastrus scandens, Chimaphila umbellata, Cucurbita pepo, Daucus carota, Equisetum hyemale, Impatiens capensis, Inula helenium, Juniperus virginiana, Lithospermum officinale, Lycopodium* sp., *Rubus odoratus, Tilia americana, Triosteum perfoliatum, Zanthoxylum americanum*

Ear Medicine: *Acorus calamus, Arctium* sp., *Balsamita major, Coptis trifolia, Fraxinus americana, F. nigra, Hydrastis canadensis, Lobelia kalmii, Malus sylvestris, Panax quinquefolius, Sanguinaria canadensis, Verbascum thapsus, Verbena hastata*

Emetic: *Acer pensylvanicum, Achillea millefolium, Acorus calamus, Adiantum pedatum, Agrimonia gryposepala, Alnus incana, Anemone virginiana, Anthemis cotula, Apocynum cannabinum, Arctium* sp., *Caltha palustris, Carex oligosperma, C. prasina, Caulophyllum thalictroides, Coptis trifolia, Cornus alternifolia, C. amomum, C. rugosa, C. sericea, Corylus cornuta, Dirca palustris, Elodea canadensis, Fraxinus americana, Galeopsis tetrahit, Geranium maculatum, Geum aleppicum, Hamamelis virginiana, Hydrastis canadensis, Ilex verticillata, Juglans cinerea, Juncus bufonius, J. tenuis, Laportea canadensis, Linaria vulgaris, Lobelia inflata, L. kalmii, L. spicata, Lonicera dioica, Lysimachia quadrifolia, Malva neglecta, Mentha canadensis, M. spicata, Mitella diphylla, Monarda* sp., *Myriophyllum sibiricum, Onopordum acanthium, Pedicularis canadensis, Penstemon fruticosus, Physalis heterophylla, Phytolacca americana, Picea* sp., *Pinus strobus, Polystichum acrostichoides, Prunella vulgaris, Prunus serotina, Ranunculus abortivus, Rhamnus* sp., *Rhus* sp., *Rubus idaeus, R. occidentalis, Rumex crispus, Salix discolor, S.* sp., *Sambucus canadensis, Sanguinaria canadensis, Sanicula marilandica, Silphium perfoliatum, Solidago canadensis, S. juncea, S. squarrosa, Spiraea* sp., *Taraxacum officinale, Tilia americana, Ulmus rubra, Veronica officinalis, Veronicastrum virginicum, Viburnum lentago, V. opulus, Zanthoxylum americanum*

Expectorant: *Phytolacca americana, Rhus* sp.

Eye Medicine: *Acer rubrum, A. saccharum, Anaphalis margaritacea, Apocynum cannabinum, Aralia nudicaulis, Arisaema triphyllum, Coptis trifolia, Cornus alternifolia, C. sericea, Dirca palustris, Equisetum hyemale, Fragaria* sp., *Galium* sp., *Gentiana andrewsii, Hydrastis canadensis, Impatiens capensis, Lactuca canadensis, Leucanthemum vulgare, Malus coronaria, M. sylvestris, Mitella diphylla, Panax quinquefolius, Polygonatum pubescens, Prenanthes trifoliolata, Pyrola elliptica, Ranunculus abortivus, Rhus* sp., *Rosa acicularis, Sanguinaria canadensis, Sassafras albidum, Smilax tamnoides, Solanum tuberosum, Taraxacum officinale, Thalictrum dioicum, Tiarella cordifolia, Trifolium repens, Ulmus rubra, Uvularia perfoliata, Viola* sp.

Febrifuge: *Achillea millefolium, Angelica atropurpurea, Anthemis cotula, Arctium* sp., *Arisaema triphyllum, Asarum canadense, Aster lanceolatus, A. novae-angliae, A. prenanthoides, A. puniceus, Athyrium filix-femina, Betula lenta, Cardamine diphylla, Caulophyllum thalictroides, Celastrus scandens, Chimaphila umbellata, Clintonia umbellulata, Conyza canadensis, Cornus canadensis, Cypripedium parviflorum, Eupatorium maculatum, E. perfoliatum, Gentiana andrewsii, Geum aleppicum, G. rivale, Hydrastis canadensis, Hypericum perforatum, Impatiens capensis, I. pallida, Inula helenium, Larix laricina, Lindera benzoin, Linnaea borealis, Lobelia cardinalis, Lonicera dioica, Lythrum salicaria, Malva moschata, Melilotus officinalis, Mentha canadensis, M. ×piperita, M. spicata, Mitchella repens, Monarda* sp., *Nepeta cataria, Nuphar lutea, Osmorhiza* sp., *Oxalis stricta, Panax quinquefolius, Plantago* sp., *Polygonum hydropiper, Polystichum acrostichoides, Porteranthus trifoliatus, Prunella vulgaris, Prunus serotina, Rhus* sp., *Rorippa sylvestris, Sambucus canadensis, Sanguinaria canadensis, Sarracenia purpurea, Sassafras albidum, Senecio aureus, Smallanthus uvedalia, Solidago juncea, Sparganium eurycarpum, Thuja occidentalis, Tsuga canadensis, Verbascum thapsus, Veronicastrum virginicum, Viburnum opulus*

Gastrointestinal Aid: *Acer spicatum, Achillea millefolium, Acorus calamus, Amphicarpaea bracteata, Anaphalis margaritacea, Anthemis cotula, Apocynum androsaemifolium, A. cannabinum, Aralia nudicaulis, Arctium* sp., *Asclepias syriaca, Baptisia tinctoria, Bellis perennis, Cardamine concatenata, Carex prasina, Coptis trifolia, Cornus amomum, C. racemosa, Corylus americana, Crataegus punctata, C.* sp., *Cypripedium arietinum, Desmodium canadense, Epigaea repens, Eupatorium maculatum, E. perfoliatum, Euphorbia helioscopia, Fragaria* sp., *Fraxinus americana, Gentianella quinquefolia, Heracleum maximum, Hydrastis canadensis, Ilex verticillata, Inula helenium, Ipomoea pandurata, Leonurus cardiaca, Linnaea borealis, Lobelia cardinalis, Malva neglecta, Mentha spicata, Mitchella repens, Nepeta cataria, Nuphar lutea, Onoclea sensibilis, Oxalis stricta, Panax quinquefolius, Pedicularis canadensis, Physalis heterophylla, Picea glauca, P. mariana, Pinus strobus, Plantago major,*

P. sp., *Polygonum hydropiper, Populus tremuloides, Prunella vulgaris, Pyrola elliptica, Ranunculus abortivus, Rhus* sp., *Rosa eglanteria, Rubus odoratus, Rumex crispus, Sambucus canadensis, Sanguinaria canadensis, Smilax herbacea, Solanum dulcamara, Solidago canadensis, S. flexicaulis, S. juncea, S. squarrosa, Sorbus americana, Triosteum perfoliatum, Ulmus americana, U. rubra, Verbena hastata, Veronicastrum virginicum, Viola pubescens, Vitis vulpina, Zanthoxylum americanum*

Gland Medicine: *Chamaesyce glyptosperma*

Gynecological Aid: *Abies balsamea, Adiantum pedatum, Ageratina altissima, Alisma plantago-aquatica, Amelanchier arborea, A. canadensis, Angelica atropurpurea, Antennaria plantaginifolia, Apocynum androsaemifolium, Aralia racemosa, Arctium* sp., *Arisaema triphyllum, Asclepias syriaca, Aster* sp., *Betula alleghaniensis, B. lenta, B. papyrifera, Campanula aparinoides, Carex brevior, Carpinus caroliniana, Celastrus scandens, Celtis occidentalis, Cerastium arvense, Chamaesyce glyptosperma, Chenopodium album, Chimaphila umbellata, Cimicifuga racemosa, Coeloglossum viride, Cornus alternifolia, Corylus americana, Crataegus punctata, Daucus carota, Diervilla lonicera, Dirca palustris, Echium vulgare, Epigaea repens, Euonymus americana, Eupatorium maculatum, Fragaria* sp., *Hamamelis virginiana, Hepatica nobilis, Impatiens pallida, Inula helenium, Iris versicolor, Juglans cinerea, Laportea canadensis, Lilium philadelphicum, Lobelia cardinalis, Lonicera dioica, L. oblongifolia, Lycopodium obscurum, Lysimachia thyrsiflora, Maianthemum racemosum, M. stellatum, Malus coronaria, Mitchella repens, Onoclea sensibilis, Osmunda cinnamomea, O. regalis, Panax quinquefolius, Penstemon fruticosus, Physocarpus opulifolius, Plantago major, P.* sp., *Platanthera psycodes, Polygonum arenastrum, Polystichum acrostichoides, Prunella vulgaris, Prunus serotina, P. virginiana, Pteridium aquilinum, Quercus ilicifolia, Rhus hirta, R.* sp., *Rosa acicularis, Rubus idaeus, R. odoratus, Rumex obtusifolius, Sanguinaria canadensis, Sassafras albidum, Scrophularia lanceolata, S. marilandica, Staphylea trifolia, Streptopus roseus, Symplocarpus foetidus, Thelypteris palustris, Thuja occidentalis, Tilia americana, Trifolium hybridum, T. pratense, Trillium* sp., *Triosteum perfoliatum, Typha latifolia, Ulmus americana, U. rubra, Viburnum acerifolium, V. lantanoides, V. opulus, Vicia sativa, Zanthoxylum americanum*

Hallucinogen: *Cardamine concatenata, Clematis virginiana*

Heart Medicine: *Ambrosia artemisiifolia, Cardamine concatenata, Clintonia borealis, Collinsonia canadensis, Festuca subverticillata, Fragaria* sp., *Geranium maculatum, Hamamelis virginiana, Hydrastis canadensis, Inula helenium, Nuphar lutea, Pedicularis canadensis, Plantago* sp., *Polygonum persicaria, Rudbeckia hirta, Sambucus canadensis, Sanguinaria canadensis, Thalictrum dioicum, Veronicastrum virginicum, Viburnum opulus*

Hemorrhoid Remedy: *Betula populifolia, Cirsium discolor, C. vulgare, Corydalis sempervirens, Eupatorium perfoliatum, Lactuca tatarica, Mentha canadensis, Oenothera biennis, Prunella vulgaris, Rhus* sp., *Rumex crispus, Salix discolor, Sanguinaria canadensis, Ulmus americana, Verbascum thapsus*

Hemostat: *Cirsium vulgare, Cornus sericea, Juglans cinerea, Lactuca canadensis, Lycopodium* sp., *Mitchella repens, Plantago* sp., *Rumex crispus, Sanguinaria canadensis, Sassafras albidum, Thalictrum pubescens, Typha latifolia*

Hunting Medicine: *Acer rubrum, Aster* sp., *Cardamine concatenata, Corallorhiza maculata, Fraxinus americana, Heracleum maximum, Maianthemum racemosum, Osmorhiza* sp., *Panax trifolius, Polygonatum pubescens, Prenanthes trifoliolata, Thuja occidentalis, Tiarella cordifolia, Trillium* sp.

Hypotensive: *Rubus idaeus, Sassafras albidum*

Internal Medicine: *Dirca palustris, Epilobium angustifolium*

Kidney Aid: *Alisma plantago-aquatica, Aquilegia canadensis, Aralia racemosa, Arctium* sp., *Asclepias incarnata, A. syriaca, Aster prenanthoides, Celastrus scandens, Chimaphila umbellata, Clematis virginiana, Collinsonia canadensis, Cornus rugosa, Cynoglossum officinale, Dirca palustris, Elymus canadensis, Epigaea repens, Epilobium angustifolium, Equisetum hyemale, Euonymus* sp., *Eupatorium maculatum, E. perfoliatum, Gaultheria procumbens, Hamamelis virginiana, Impatiens capensis, Juniperus communis, Lactuca canadensis, Maianthemum canadense, Mitchella repens, Osmunda regalis, Ribes americanum, Rubus idaeus, R. odoratus, Rumex crispus, Sambucus canadensis, Sanguinaria canadensis, Sanicula marilandica, Saxifraga pensylvanica, Scrophularia lanceolata, Senecio aureus, Smallanthus uvedalia, Solidago nemoralis, Taraxacum officinale, Ulmus rubra, Viburnum opulus, Vitis vulpina, Zanthoxylum americanum*

Laxative: *Acer pensylvanicum, Apocynum cannabinum, Aster macrophyllus, Chimaphila umbellata, Cornus amomum, Dirca palustris, Eupatorium perfoliatum, Fraxinus americana, F. nigra, Geranium maculatum, Juglans cinerea, J. nigra, Lycopus asper, Monarda* sp., *Nepeta cataria, Osmorhiza* sp., *Pinus rigida, Podophyllum peltatum, Populus balsamifera, Ranunculus recurvatus, Rhus* sp., *Rubus odoratus, Sagittaria latifolia, Sambucus canadensis, Sanguinaria canadensis, Sanicula marilandica, Senna hebecarpa, Sisyrinchium angustifolium, Taraxacum officinale, Triosteum perfoliatum, Verbascum thapsus, Viburnum opulus*

Liver Aid: *Adiantum pedatum, Agrimonia* sp., *Apocynum androsaemifolium, Aralia racemosa, Baptisia tinctoria, Caulophyllum thalictroides, Chelone glabra, Eupatorium maculatum, Fagus grandifolia, Gaylussacia baccata, Gentiana andrewsii, Hydrastis canadensis, Impatiens capensis, Ipomoea pandurata, Jeffersonia diphylla, Juglans cinerea, Panax quinquefolius, Phytolacca americana, Pinus strobus, Prunus serotina, Pteridium aquilinum, Rhus* sp., *Rubus idaeus, R. occidentalis, Sambucus canadensis, Sanguinaria canadensis, Sarracenia purpurea, Sedum telephium, Solidago canadensis, S. juncea, S. rugosa, Tanacetum vulgare, Thalictrum pubescens, Viburnum opulus*

Love Medicine: *Anemone virginiana, Aster novae-angliae, A.* sp., *Caltha palustris, Cardamine concatenata, C. diphylla, Corallorrhiza maculata, Dirca palustris, Eupatorium maculatum, Galium triflorum, Geranium maculatum, Geum aleppicum, G. canadense, Lilium philadelphicum, Linaria vulgaris, Lobelia cardinalis, L. inflata, L. kalmii, L. spicata, Lonicera dioica, Malva neglecta, Mitchella repens, Osmorhiza* sp., *Penstemon fruticosus, Phytolacca americana, Polygonum arenastrum, Populus alba, Prenanthes trifoliolata, Rumex crispus, Sarracenia purpurea, Solidago canadensis, Spiraea* sp., *Taraxacum officinale, Viburnum lantanoides, Vicia americana, V. sativa*

Misc. Disease Remedy: *Achillea millefolium, Acorus calamus, Angelica atropurpurea, Aralia nudicaulis, Armoracia rusticana, Asarum canadense, Aster puniceus, Ceanothus americanus, Cinna arundinacea, Clintonia borealis, Dirca palustris, Eupatorium perfoliatum, Hamamelis virginiana, Heracleum maximum, Inula helenium, Ipomoea pandurata, Lindera benzoin, Mentha spicata, Mitchella repens, Nuphar lutea, Panax quinquefolius, Pinus strobus, Polypodium virginianum, Populus tremuloides, Porteranthus trifoliatus, Prunella vulgaris, Prunus virginiana, Quercus bicolor, Ranunculus abortivus, Rhus* sp., *Rumex crispus, Sambucus canadensis, Scutellaria lateriflora, Sisyrinchium montanum, Tsuga canadensis, Typha latifolia, Verbascum thapsus*

Nose Medicine: *Arisaema triphyllum*

Oral Aid: *Ceanothus americanus, Cirsium arvense, Coptis trifolia, Cornus sericea, Fragaria* sp., *Geranium maculatum, Hydrophyllum virginianum, Juglans cinerea, Nepeta cataria, Oxalis stricta, Populus ×jackii, Rhus* sp., *Salix sericea, Syringa vulgaris, Tiarella cordifolia, Toxicodendron pubescens*

Orthopedic Aid: *Abies balsamea, Acer pensylvanicum, Adiantum pedatum, Alisma plantago-aquatica, Ambrosia artemisiifolia, Angelica atropurpurea, A. venenosa, Aralia racemosa, Arctium* sp., *Arisaema triphyllum, Asclepias incarnata, Baptisia tinctoria, Betula lenta, Cicuta maculata, Cimicifuga racemosa, Cornus racemosa, Cypripedium parviflorum, Dirca palustris, Epilobium angustifolium, Equisetum arvense, Eupatorium perfoliatum, Galium triflorum, Gentiana andrewsii, Gnaphalium uliginosum, Hamamelis virginiana, Hepatica nobilis, Iris versicolor, Lactuca canadensis, Malus sylvestris, Malva neglecta, Mitchella repens, Oenothera perennis, Osmunda cinnamomea, Parthenocissus quinquefolia, Pedicularis canadensis, Phytolacca americana, Pinus strobus, Plantago* sp., *Polygala paucifolia, Polygonum arenastrum, P. aviculare, Polystichum acrostichoides, Prunella vulgaris, Pyrola elliptica, Quercus*

[Orthopedic Aid]
bicolor, Ranunculus abortivus, Rhamnus alnifolia, Rhus sp., Ribes americanum, Rumex crispus, Sassafras albidum, Senecio aureus, Silphium perfoliatum, Sium suave, Smallanthus uvedalia, Spiraea sp., Tanacetum vulgare, Taraxacum officinale, Taxus baccata, Thuja occidentalis, Tiarella cordifolia, Tilia americana, Trillium sp., Tsuga canadensis, Typha latifolia, Ulmus americana, Uvularia perfoliata, U. sessilifolia, Verbascum thapsus, Viburnum dentatum, Zanthoxylum americanum

Other: Agrimonia gryposepala, Alisma plantago-aquatica, Angelica atropurpurea, Anthemis cotula, Asclepias incarnata, A. syriaca, A. tuberosa, Cardamine diphylla, Carex platyphylla, C. vulpinoidea, Carpinus caroliniana, Ceanothus americanus, Celastrus scandens, Circaea lutetiana, Elymus canadensis, Equisetum hyemale, Euonymus obovata, E. sp., Eupatorium perfoliatum, E. purpureum, Festuca subverticillata, Fraxinus nigra, Hepatica nobilis, Ilex verticillata, Ipomoea pandurata, Lespedeza sp., Lobelia cardinalis, L. inflata, L. sp., Luzula sp., Maianthemum racemosum, Mentha ×piperita, M. spicata, Monarda sp., Myriophyllum verticillatum, Nepeta cataria, Nuphar lutea, Oxalis stricta, Parthenocissus quinquefolia, Phragmites australis, Pinus strobus, Polygala verticillata, Quercus ilicifolia, Q. sp., Ribes americanum, R. rotundifolium, Rubus allegheniensis, Sanguinaria canadensis, Sedum telephium, Sisyrinchium angustifolium, Solanum nigrum, Solidago canadensis, S. rugosa, Thalictrum dioicum, Tiarella cordifolia, Tilia americana, Ulmus americana, Vaccinium sp., Verbena hastata, Vitis vulpina

Panacea: Achillea millefolium, Ageratina altissima, Cardamine concatenata, Collinsonia canadensis, Cornus sericea, Epilobium angustifolium, Geum canadense, Hamamelis virginiana, Inula helenium, Lindera benzoin, Lobelia cardinalis, Medeola virginiana, Osmunda cinnamomea, Panax quinquefolius, Pinus strobus, Platanthera psycodes, Prunella vulgaris, Rumex crispus, Sanguinaria canadensis, Saxifraga pensylvanica, Tanacetum vulgare, Tilia americana, Veronicastrum virginicum, Viola sp.

Pediatric Aid: Achillea millefolium, Acorus calamus, Adiantum pedatum, Agrimonia gryposepala, Allium sp., A. tricoccum, Anthemis cotula, Apocynum cannabinum, Argentina anserina, Arisaema triphyllum, Asarum canadense, Asclepias incarnata, Aster prenanthoides, Carpinus caroliniana, Castanea dentata, Ceanothus americanus, Celastrus scandens, Chimaphila umbellata, Cimicifuga racemosa, Claytonia virginica, Collinsonia canadensis, Conyza canadensis, Coptis trifolia, Cornus alternifolia, C. amomum, Corylus americana, C. cornuta, Cucurbita pepo, Diervilla lonicera, Equisetum arvense, Euonymus europaea, Euphorbia helioscopia, Fagopyrum esculentum, Fragaria sp., Galium triflorum, Geranium maculatum, Helianthus strumosus, Hepatica nobilis, Inula helenium, Jeffersonia diphylla, Linaria vulgaris, Linnaea borealis, Lithospermum officinale, Lonicera canadensis, L. dioica, L. sp., Lycopus asper, Malva neglecta, Medeola virginiana, Mentha canadensis, M. spicata, Mitchella repens, Monarda sp., Myriophyllum sibiricum, M. verticillatum, Nepeta cataria, Nuphar lutea, Onoclea sensibilis, Osmorhiza sp., Osmunda regalis, Panax quinquefolius, Pinus strobus, Platanthera psycodes, Polygala paucifolia, P. verticillata, Polygonum aviculare, P. hydropiper, Polystichum acrostichoides, Populus tremuloides, Prunella vulgaris, Prunus serotina, P. virginiana, Pyrola elliptica, Rhamnus alnifolia, Rhus sp., Rorippa sylvestris, Rubus allegheniensis, R. occidentalis, R. odoratus, Rudbeckia hirta, Rumex obtusifolius, Sagittaria latifolia, Sambucus canadensis, Sanguinaria canadensis, Sanicula marilandica, Sedum telephium, Senecio aureus, Silphium perfoliatum, Solidago canadensis, Sonchus asper, Spiraea sp., Staphylea trifolia, Symplocarpus foetidus, Syringa vulgaris, Tiarella cordifolia, Tilia americana, Toxicodendron pubescens, Triosteum perfoliatum, Uvularia perfoliata, Verbascum thapsus, Viburnum opulus, Viola sp., Vitis vulpina

Poison: Angelica atropurpurea, A. venenosa, Cardamine rhomboidea, Celastrus scandens, Cicuta maculata, Datura stramonium, Dipsacus fullonum, Eupatorium perfoliatum, Lycopus asper, L. virginicus, Onopordum acanthium, Parthenocissus quinquefolia, Podophyllum peltatum, Sisyrinchium montanum, Trillium sp., Veronicastrum virginicum

Poultice: Cornus amomum, Verbena hastata

Psychological Aid: Actaea rubra, Asarum canadense, Cannabis sativa, Cardamine diphylla, Cornus sericea, Corylus cornuta, Eupatorium perfoliatum, Gentiana andrewsii, Ilex verticillata, Juglans cinerea, J. nigra, Laportea canadensis, Lobelia cardinalis, L. inflata, Lonicera canadensis, L. oblongifolia, Mitchella repens, Nicotiana rustica, Pinus strobus, Plantago sp., Polygonum punctatum, Prunella vulgaris, Quercus alba, Ranunculus abortivus, Salix discolor, Smilax herbacea

Pulmonary Aid: Acer saccharum, Aesculus hippocastanum, Angelica atropurpurea, Anthemis cotula, Aralia racemosa, Asarum canadense, Aster puniceus, Campanulastrum americanum, Cardamine diphylla, Cornus amomum, C. sericea, Cypripedium pubescens, Erigeron philadelphicus, Eupatorium perfoliatum, Gentianella quinquefolia, Hamamelis virginiana, Hepatica nobilis, Hydrastis canadensis, Inula helenium, Nuphar lutea, Panax trifolius, Pinus strobus, Plantago sp., Prunella vulgaris, Prunus serotina, Rhus sp., Rubus occidentalis, Rumex obtusifolius, Sanguinaria canadensis, Sarracenia purpurea, Taraxacum officinale, Triosteum perfoliatum, Veronicastrum virginicum, Viburnum lantanoides, V. opulus

Reproductive Aid: Athyrium filix-femina, Fraxinus americana, F. nigra, Hypericum perforatum, Lycopodium complanatum, Rhus hirta, Rumex crispus

Respiratory Aid: Acorus calamus, Anaphalis margaritacea, Arisaema triphyllum, Asarum canadense, Cornus alternifolia, Corylus americana, Gnaphalium uliginosum, Hamamelis virginiana, Ilex verticillata, Inula helenium, Iris versicolor, Mentha spicata, Panax quinquefolius, Pilea pumila, Plantago major, Prunella vulgaris, Prunus serotina, Quercus bicolor, Rubus allegheniensis, Sanguinaria canadensis, Taxus baccata, Trifolium repens, Trillium sp., Ulmus rubra, Veratrum viride, Verbascum thapsus

Sedative: Allium sp., Anthemis cotula, Cornus amomum, Cypripedium parviflorum, Leonurus cardiaca, Linaria vulgaris, Linnaea borealis, Lonicera canadensis, L. oblongifolia, Nepeta cataria, Prunella vulgaris, Rhamnus alnifolia, Smallanthus uvedalia, Solidago canadensis, Sonchus asper, Staphylea trifolia, Toxicodendron pubescens

Snakebite Remedy: Adiantum pedatum, Calla palustris, Fraxinus americana, Maianthemum racemosum, Prenanthes alba, P. altissima, P. trifoliolata, Ranunculus abortivus, Scirpus tabernaemontani, Waldsteinia fragarioides

Sports Medicine: Asclepias tuberosa, Cornus sericea, Dicentra cucullaria, Juncus tenuis, Panax trifolius, Sarracenia purpurea

Stimulant: Achillea millefolium, Arisaema triphyllum, Asarum canadense, Betula lenta, Cannabis sativa, Chimaphila umbellata, Collinsonia canadensis, Cypripedium parviflorum, Hydrastis canadensis, Larix laricina, Malva moschata, Monarda sp., Myriophyllum verticillatum, Oenothera biennis, Panax quinquefolius, Polystichum acrostichoides, Prenanthes alba, Prunus serotina, Rubus idaeus, Thuja occidentalis, Tilia americana, Tsuga canadensis, Ulmus rubra

Strengthener: Aralia racemosa, Asclepias incarnata, Collinsonia canadensis, Dirca palustris, Juncus bufonius, Oenothera biennis, Plantago sp., Podophyllum peltatum, Rumex crispus, Viburnum opulus

Throat Aid: Acorus calamus, Aralia nudicaulis, Asarum canadense, Coptis trifolia, Geranium maculatum, Nepeta cataria, Pinus strobus, Polystichum acrostichoides, Porteranthus trifoliatus, Prunus serotina, Quercus sp., Rhus sp., Salix nigra, Sanguinaria canadensis, Scutellaria lateriflora, Thlaspi arvense, Triosteum perfoliatum, Ulmus rubra

Tonic: Allium tricoccum, Aralia racemosa, Asarum canadense, Caulophyllum thalictroides, Chimaphila umbellata, Dirca palustris, Juniperus communis, Leonurus cardiaca, Osmorhiza sp., Panax quinquefolius, Populus alba, Rhamnus alnifolia, Rubus idaeus, Rumex crispus, R. obtusifolius, Sassafras albidum, Tiarella cordifolia

Toothache Remedy: Acorus calamus, Antennaria plantaginifolia, Anthemis cotula, Asclepias incarnata, Celastrus scandens, Corylus americana, C. cornuta,

Equisetum arvense, Fragaria sp., *Hamamelis virginiana, Juglans cinerea, Magnolia acuminata, Polymnia canadensis, Ranunculus abortivus, R. bulbosus, R. hispidus, R. recurvatus, Taraxacum officinale, Verbascum thapsus, Zanthoxylum americanum*

Tuberculosis Remedy: *Abies balsamea, Alisma plantago-aquatica, Amphicarpaea bracteata, Anemone virginiana, Aralia nudicaulis, A. racemosa, Arisaema triphyllum, Asarum canadense, Aster puniceus, Botrychium virginianum, Cardamine diphylla, Carpinus caroliniana, Corallorrhiza maculata, Cornus alternifolia, C. canadensis, C. rugosa, C. sericea, Cynoglossum officinale, Cypripedium parviflorum, C. pubescens, Dirca palustris, Epilobium angustifolium, Erigeron pulchellus, Eupatorium maculatum, Fagus grandifolia, Hamamelis virginiana, Hieracium* sp., *Hydrastis canadensis, Inula helenium, Ipomoea pandurata, Juglans cinerea, Laportea canadensis, Lobelia cardinalis, Lonicera dioica, Malus coronaria, Nicotiana rustica, Onoclea sensibilis, Ostrya virginiana, Panax quinquefolius, Pedicularis canadensis, Picea* sp., *Pinus strobus, Platanthera orbiculata, Polystichum acrostichoides, Prunella vulgaris, Prunus serotina, P. virginiana, Pteridium aquilinum, Quercus alba, Q. bicolor, Q.* sp., *Rhus* sp., *Rubus allegheniensis, Salix discolor, S.* sp., *Sanguinaria canadensis, Scirpus tabernaemontani, Spiraea* sp., *Symplocarpus foetidus, Taxus baccata, Tilia americana, Tsuga canadensis, Tussilago farfara, Ulmus rubra, Veratrum viride, Verbascum thapsus, Viburnum lentago, Zanthoxylum americanum*

Unspecified: *Acer saccharinum, A. saccharum, Asclepias syriaca, Betula lenta, Coptis trifolia, Fragaria virginiana, Rubus canadensis, Sambucus canadensis*

Urinary Aid: *Abies balsamea, Actaea pachypoda, Alnus incana, Aralia racemosa, Asarum canadense, Asclepias incarnata, Ceanothus americanus, Celastrus scandens, Chimaphila umbellata, Citrullus lanatus, Clematis virginiana, Cornus amomum, Diervilla lonicera, Dirca palustris, Elytrigia repens, Epilobium angustifolium, Equisetum hyemale, Euonymus americana, E. europaea, E. obovata, Fraxinus nigra, Galium* sp., *G. triflorum, Impatiens capensis, Juglans cinerea, Lonicera oblongifolia, Mitchella repens, Parthenocissus quinquefolia, P. vitacea, Plantago* sp., *Populus tremuloides, Pteridium aquilinum, Rhus* sp., *Rosa eglanteria, Rubus idaeus, Taraxacum officinale, Triosteum perfoliatum, Viburnum acerifolium, Vitis vulpina*

Venereal Aid: *Abies balsamea, Achillea millefolium, Adiantum pedatum, Ageratina altissima, Alnus incana, Amelanchier arborea, Anthemis cotula, Aralia nudicaulis, A. racemosa, Asarum canadense, Aster macrophyllus, Athyrium filix-femina, Ceanothus americanus, Chimaphila umbellata, Clematis virginiana, Coptis trifolia, Cornus alternifolia, C. amomum, C. drummondii, C. racemosa, Cynoglossum officinale, Dalibarda repens, Diervilla lonicera, Dirca palustris, Equisetum hyemale, Euonymus* sp., *Eupatorium maculatum, E. perfoliatum, Fraxinus americana, Galium* sp., *Gaultheria procumbens, Geranium maculatum, Hamamelis virginiana, Juglans cinerea, Larix laricina, Lindera benzoin, Lobelia inflata, Lonicera canadensis, L. dioica, Lycopodium sabinifolium, Magnolia acuminata, Mertensia virginica, Mitchella repens, Onoclea sensibilis, Osmorhiza* sp., *Osmunda cinnamomea, O. claytoniana, Panax quinquefolius, Physalis heterophylla, Pinus strobus, Polygala paucifolia, Polystichum acrostichoides, Populus tremuloides, Prunella vulgaris, Prunus serotina, Pteridium aquilinum, Ranunculus bulbosus, R. recurvatus, Rhamnus alnifolia, Rhus* sp., *Rubus idaeus, R. occidentalis, R. odoratus, Rumex crispus, Salix myricoides, Sambucus canadensis, Sanguinaria canadensis, Sanicula marilandica, Sicyos angulatus, Solidago squarrosa, Triosteum perfoliatum, Tsuga canadensis, Typha latifolia, Viburnum lantanoides, Zanthoxylum americanum*

Veterinary Aid: *Actaea rubra, Ageratina altissima, Apocynum androsaemifolium, Arctium* sp., *Arisaema triphyllum, Asarum canadense, Butomus umbellatus, Carex prasina, Carum carvi, Castanea dentata, Chelidonium majus, Cicuta maculata, Corallorrhiza maculata, Cornus racemosa, Eupatorium perfoliatum, Fraxinus americana, F. nigra, Geum aleppicum, Hybanthus concolor, Inula helenium, Juncus tenuis, Myosotis* sp., *Nuphar lutea, Osmunda cinnamomea, Pinus strobus, Podophyllum peltatum, Polygonum arenastrum, P. pensylvanicum, P. persicaria, Populus balsamifera, P. deltoides, P. tremuloides,* *Prunus virginiana, Pteridium aquilinum, Quercus alba, Ranunculus* sp., *Rhus* sp., *Rubus idaeus, Sedum telephium, Sicyos angulatus, Sisymbrium officinale, Sparganium eurycarpum, Trifolium hybridum, Triticum aestivum, Typha latifolia, Veronica officinalis, Viola* sp., *Vitis labrusca*

Witchcraft Medicine: *Acorus calamus, Ageratina altissima, Alnus incana, Amaranthus retroflexus, Anemone canadensis, A. virginiana, Angelica atropurpurea, Aquilegia canadensis, Arctium* sp., *Asarum canadense, Asclepias incarnata, Cardamine douglassii, Chelone glabra, Corallorrhiza maculata, Cornus sericea, Crataegus punctata, C. submollis, Epilobium angustifolium, Euonymus obovata, Eupatorium perfoliatum, Fagopyrum esculentum, Galeopsis tetrahit, Gentiana andrewsii, Hepatica nobilis, Ipomoea pandurata, Juglans nigra, Laportea canadensis, Linaria vulgaris, Lobelia cardinalis, L. inflata, L. siphilitica, Lythrum salicaria, Maianthemum racemosum, Medeola virginiana, Mentha ×piperita, Nuphar lutea, Onopordum acanthium, Oxalis stricta, Phytolacca americana, Pinus strobus, Potamogeton* sp., *Pteridium aquilinum, Quercus alba, Q. bicolor, Rosa acicularis, Rubus allegheniensis, R. occidentalis, Sanguinaria canadensis, Sedum telephium, Silphium perfoliatum, Smilax tamnoides, Sparganium eurycarpum, Symplocarpus foetidus, Taraxacum officinale, Trillium* sp., *Triosteum perfoliatum, Urtica dioica, Verbena hastata, Veronica officinalis, Veronicastrum virginicum, Viburnum acerifolium, Viola sagittata, V. striata, Xanthium strumarium*

Food

Baby Food: *Carya ovata, Juglans cinerea, Zea mays*

Beverage: *Acer saccharinum, A. saccharum, Alisma plantago-aquatica, Aralia nudicaulis, Betula lenta, Carya cordiformis, C. ovata, Castanea dentata, Corylus americana, Fagus grandifolia, Juglans cinerea, J. nigra, Monarda fistulosa, Prunus americana, Quercus* sp., *Rhus glabra, Rubus canadensis, Taraxacum officinale, Taxus canadensis, Tsuga canadensis, Zea mays*

Bread & Cake: *Acer rubrum, A. saccharinum, A. saccharum, Amelanchier canadensis, Asimina triloba, Carya cordiformis, C. ovata, Castanea dentata, Citrullus lanatus, Corylus americana, Crataegus pruinosa, C. submollis, Cucumis melo, C. sativus, Cucurbita maxima, C. moschata, C. pepo, Fagus grandifolia, Fragaria vesca, F. virginiana, Gaultheria procumbens, Gaylussacia baccata, Juglans cinerea, J. nigra, Malus coronaria, M. sylvestris, Mitchella repens, Morus rubra, Phaseolus coccineus, P. lunatus, P. vulgaris, Physalis* sp., *Podophyllum peltatum, Prunus americana, P. nigra, P. pensylvanica, P. persica, P. serotina, P. virginiana, Pyrus communis, Quercus* sp., *Ribes americanum, R. triste, Rubus canadensis, R. idaeus, R. occidentalis, R. odoratus, R. pubescens, Sambucus canadensis, Vaccinium angustifolium, V. corymbosum, V. macrocarpon, V. oxycoccos, V.* sp., *Viburnum lentago, V. opulus, Vitis vulpina, Zea mays*

Dried Food: *Amelanchier canadensis, Arctium lappa, Asimina triloba, Citrullus lanatus, Crataegus pruinosa, C. submollis, Cucumis melo, C. sativus, Cucurbita maxima, C. moschata, C. pepo, Fragaria vesca, F. virginiana, Gaultheria procumbens, Gaylussacia baccata, Malus coronaria, M. sylvestris, Mitchella repens, Morus rubra, Phaseolus coccineus, P. lunatus, P. vulgaris, Physalis* sp., *Podophyllum peltatum, Prunus americana, P. nigra, P. pensylvanica, P. persica, P. serotina, P. virginiana, Pyrus communis, P.* sp., *Ribes americanum, R. triste, Rubus canadensis, R. idaeus, R. occidentalis, R. odoratus, R. pubescens, Sambucus canadensis, Vaccinium angustifolium, V. corymbosum, V. macrocarpon, V. oxycoccos, Viburnum lentago, V. opulus, Vitis vulpina*

Forage: *Artemisia biennis, Spartina alterniflora, Stellaria media*

Fruit: *Amelanchier canadensis, A.* sp., *Asimina triloba, Corylus cornuta, Crataegus pruinosa, C. submollis, Fragaria vesca, F. virginiana, Gaultheria procumbens, Gaylussacia baccata, Malus coronaria, M. sylvestris, Mitchella repens, Morus rubra, Physalis* sp., *Podophyllum peltatum, Prunus americana, P. nigra, P. pensylvanica, P. persica, P. serotina, P. virginiana, Pyrus communis, P.* sp., *Ribes americanum, R. triste, Rubus canadensis, R. idaeus, R. occidentalis, R. odoratus, R. pubescens, R.* sp., *Sambucus canadensis, Vaccinium angustifolium, V. corymbosum, V. macrocarpon, V. oxycoccos, Viburnum lentago, V. opulus, Vitis vulpina*

Pie & Pudding: *Carya cordiformis, C. ovata, Castanea dentata, Corylus americana, Cucurbita pepo, Fagus grandifolia, Gaylussacia baccata, Juglans cinerea, J. nigra, Quercus* sp., *Rubus occidentalis, Vaccinium* sp., *Zea mays*

Porridge: *Gaylussacia baccata, Rubus occidentalis, Sambucus canadensis, Vaccinium* sp., *Zea mays*

Preserves: *Vaccinium angustifolium, V. myrtilloides*

Sauce & Relish: *Amelanchier canadensis, Asimina triloba, Carya cordiformis, C. ovata, Castanea dentata, Corylus americana, Crataegus pruinosa, C. submollis, Cucurbita pepo, Fagus grandifolia, Fragaria vesca, F. virginiana, Gaultheria procumbens, Gaylussacia baccata, Juglans cinerea, J. nigra, Malus coronaria, M. sylvestris, Mitchella repens, Morus rubra, Physalis* sp., *Podophyllum peltatum, Prunus americana, P. nigra, P. pensylvanica, P. persica, P. serotina, P. virginiana, Pyrus communis, P.* sp., *Ribes americanum, R. triste, Rubus canadensis, R. idaeus, R. occidentalis, R. odoratus, R. pubescens, R.* sp., *Sambucus canadensis, Vaccinium angustifolium, V. corymbosum, V. macrocarpon, V. oxycoccos, V.* sp., *Viburnum lentago, V. opulus, Vitis vulpina, Zea mays*

Snack Food: *Zea mays*

Soup: *Arctium lappa, Carya cordiformis, C. ovata, Castanea dentata, Corylus americana, Fagus grandifolia, Gaylussacia baccata, Juglans cinerea, J. nigra, Phaseolus coccineus, P. lunatus, P. vulgaris, Prunus virginiana, Quercus* sp., *Rubus occidentalis, Vaccinium* sp., *Zea mays*

Special Food: *Carya cordiformis, C. ovata, Castanea dentata, Citrullus lanatus, Corylus americana, Cucumis melo, C. sativus, Cucurbita maxima, C. moschata, C. pepo, Fagus grandifolia, Juglans cinerea, J. nigra, Quercus* sp., *Rubus* sp., *Zea mays*

Spice: *Castanea dentata, Polygonum hydropiper, Rumex acetosella*

Staple: *Carya cordiformis, C. ovata, Castanea dentata, Corylus americana, Fagus grandifolia, Juglans cinerea, J. nigra*

Sweetener: *Acer saccharinum, A. saccharum*

Unspecified: *Apios americana, Cardamine concatenata, C. diphylla, Carya cordiformis, C. ovata, Castanea dentata, Claytonia virginica, Corylus americana, Fagus grandifolia, Helianthus tuberosus, Juglans cinerea, J. nigra, Oxalis* sp., *Pinus strobus, Quercus alba, Q. bicolor, Q. prinus, Q. rubra, Q.* sp., *Ranunculus cymbalaria, Rhus glabra, Rubus aculeatissimus, Solanum tuberosum, Vaccinium angustifolium, V. myrtilloides, Vitis vulpina*

Vegetable: *Allium canadense, A. tricoccum, Amaranthus retroflexus, Arctium lappa, Asclepias syriaca, Asparagus officinalis, Brassica nigra, B.* sp., *Caltha palustris, Chenopodium album, Citrullus lanatus, Cucumis melo, C. sativus, Cucurbita maxima, C. moschata, C. pepo, Hydrophyllum virginianum, Lathyrus japonicus, Onoclea sensibilis, Oxalis corniculata, Pedicularis canadensis, P. lanceolata, Phaseolus coccineus, P. lunatus, P. vulgaris, Phytolacca americana, Portulaca oleracea, Rorippa nasturtium-aquaticum, Rumex acetosella, R. crispus, Symplocarpus foetidus, Taraxacum officinale, Urtica dioica, Zea mays*

Fiber

Basketry: *Cornus sericea, Fraxinus americana, Hierochloe odorata*

Brushes & Brooms: *Picea glauca, P. mariana*

Caulking Material: *Thuja occidentalis*

Cordage: *Dirca palustris*

Furniture: *Carya cordiformis, Fraxinus americana, F.* sp.

Mats, Rugs & Bedding: *Dryopteris* sp., *Onoclea* sp., *Osmunda* sp., *Typha latifolia*

Dye

Brown: *Alnus viridis, Ledum groenlandicum*

Green: *Allium cepa*

Orange-Yellow: *Sanguinaria canadensis*

Yellow: *Allium cepa*

Other

Ceremonial Items: *Cardamine concatenata, Eupatorium perfoliatum, Fragaria virginiana, Lagenaria siceraria, Nicotiana rustica, Phaseolus coccineus, P. lunatus, P. vulgaris*

Containers: *Zea mays*

Cooking Tools: *Acer rubrum, Zea mays*

Fertilizer: *Bromus ciliatus, Elymus canadensis, Osmorhiza* sp., *Podophyllum peltatum, Sagittaria latifolia*

Fuel: *Picea glauca, P. mariana*

Good Luck Charm: *Mitella diphylla*

Incense & Fragrance: *Melilotus officinalis*

Insecticide: *Carya cordiformis, C. ovata, Castanea dentata, Corylus americana, Fagus grandifolia, Hackelia virginiana, Juglans cinerea, J. nigra, Pinus rigida, Thuja occidentalis*

Planting Seeds: *Zea mays*

Protection: *Angelica atropurpurea, Platanthera grandiflora, Rubus odoratus*

Season Indicator: *Amelanchier canadensis*

Smoke Plant: *Nicotiana rustica, N. tabacum, Picea* sp.

Soap: *Symplocarpus foetidus*

Stable Gear: *Ulmus americana*

Tools: *Cornus amomum, Sambucus* sp.

Toys & Games: *Hordeum jubatum*

Isleta

Drug

Analgesic: *Frasera speciosa, Lepidium densiflorum*

Antirheumatic (External): *Echinocereus* sp., *Juniperus monosperma, Larrea tridentata, Marrubium vulgare, Pseudotsuga menziesii*

Antirheumatic (Internal): *Nolina microcarpa*

Blood Medicine: *Anemopsis californica*

Dermatological Aid: *Allium cernuum, Anemopsis californica, Ephedra torreyana, Gaura parviflora, Geastrum* sp., *Glycyrrhiza lepidota, Gutierrezia sarothrae, Larrea tridentata, Oenothera cespitosa, Panicum obtusum, Pinus edulis, Salix* sp., *Yucca glauca*

Disinfectant: *Anemopsis californica, Larrea tridentata*

Ear Medicine: *Croton texensis*

Emetic: *Juniperus monosperma*

Eye Medicine: *Mentha canadensis, Prosopis glandulosa*

Febrifuge: *Chrysothamnus* sp., *Gutierrezia sarothrae, Thymophylla acerosa*

Gastrointestinal Aid: *Artemisia frigida, Hymenopappus newberryi, Lactuca sativa, Plantago major, Rumex crispus*

Gynecological Aid: *Juniperus monosperma*

Heart Medicine: *Solanum* sp.

Laxative: *Artemisia* sp., *Croton texensis, Solanum elaeagnifolium*

Orthopedic Aid: *Populus tremuloides, Pseudotsuga menziesii*

Pediatric Aid: *Hymenopappus newberryi, Rosa woodsii*

Poison: *Atriplex canescens, Chrysothamnus nauseosus, Hackelia floribunda, Heterotheca villosa*

Psychological Aid: *Hymenoxys richardsonii*

Pulmonary Aid: *Anemopsis californica, Cucurbita foetidissima, Frasera speciosa, Nolina microcarpa*

Reproductive Aid: *Quercus gambelii*

Respiratory Aid: *Asclepias latifolia*

Throat Aid: *Allium cernuum, Frasera speciosa*

Toothache Remedy: *Chrysothamnus* sp.

Venereal Aid: *Chrysothamnus* sp., *Gutierrezia sarothrae, Hymenoxys* sp.

Food

Beverage: *Androsace* sp., *Echinocereus triglochidiatus, Holodiscus dumosus, Hymenopappus* sp., *Osmorhiza depauperata, Pseudocymopterus* sp., *Thelesperma longipes, Zea mays*

Bread & Cake: *Cleome serrulata, Echinocereus triglochidiatus, Koeleria macrantha, Nolina microcarpa, Prosopis glandulosa, Zea mays*

Candy: *Apocynum cannabinum, Echinocereus triglochidiatus, Hymenoxys richardsonii, Populus deltoides, Prosopis glandulosa, Zea mays*

Fodder: *Equisetum laevigatum*
Forage: *Artemisia frigida, Erodium cicutarium, Lotus wrightii, Yucca baccata, Y. glauca*
Fruit: *Amelanchier utahensis, Cucurbita foetidissima, Echinocereus triglochidiatus, Fragaria vesca, Juniperus deppeana, Lycium pallidum, Nolina microcarpa, Opuntia* sp., *Philadelphus microphyllus, Physalis* sp., *Prunus americana, Ribes cereum, R. leptanthum, Rubus parviflorus, Vitis arizonica, Yucca baccata, Y. glauca*
Porridge: *Koeleria macrantha, Nolina microcarpa, Zea mays*
Preserves: *Echinocereus triglochidiatus, Nolina microcarpa, Opuntia* sp., *Ribes cereum*
Sauce & Relish: *Aletes anisatus, Echinocereus triglochidiatus, Rhus trilobata*
Special Food: *Fragaria vesca, Rubus parviflorus*
Spice: *Hedeoma nana, Monarda fistulosa*
Staple: *Koeleria macrantha, Nolina microcarpa, Pinus edulis, Quercus gambelii, Zea mays*
Sweetener: *Zea mays*
Unspecified: *Allium cernuum, Asparagus officinalis, Atriplex* sp., *Echinocereus triglochidiatus, Hedeoma nana, Pinus edulis, Polyporus harlowii, Populus deltoides, Prosopis pubescens, Yucca baccata, Y. glauca, Zea mays*
Vegetable: *Aletes anisatus, Allium cernuum, Amaranthus retroflexus, Asparagus officinalis, Atriplex argentea, A.* sp., *Chenopodium* sp., *Cleome serrulata, Echinocereus triglochidiatus, Mimulus glabratus, Portulaca oleracea, Rumex crispus, Solanum jamesii*
Winter Use Food: *Allium cernuum, Pinus edulis, Polyporus harlowii, Portulaca oleracea, Thelesperma longipes, Yucca baccata, Y. glauca, Zea mays*

Fiber
Basketry: *Nolina microcarpa, Salix* sp., *Yucca baccata, Y. glauca*
Brushes & Brooms: *Bromus* sp., *Nolina microcarpa, Yucca baccata, Y. glauca*
Building Material: *Juncus* sp., *Pinus ponderosa, Populus deltoides, Salix* sp., *Typha latifolia*
Canoe Material: *Populus deltoides*
Clothing: *Gossypium* sp.
Cordage: *Nolina microcarpa, Yucca baccata, Y. glauca*

Dye
Red: *Alnus incana, Cercocarpus montanus, Prunus americana*
Unspecified: *Cleome serrulata*

Other
Cash Crop: *Populus deltoides*
Ceremonial Items: *Helianthus annuus, Nolina microcarpa, Polanisia dodecandra, Pseudotsuga menziesii, Verbascum thapsus, Zea mays*
Decorations: *Clematis columbiana, Yucca baccata, Y. glauca*
Fasteners: *Koeleria macrantha*
Fuel: *Juniperus monosperma, Pinus ponderosa*
Hunting & Fishing Item: *Acer glabrum, Fallugia paradoxa, Prosopis glandulosa, Prunus virginiana*
Insecticide: *Melilotus indicus*
Lighting: *Chrysothamnus nauseosus, Echinocereus triglochidiatus*
Smoke Plant: *Polanisia dodecandra, Verbascum thapsus*
Soap: *Gutierrezia sarothrae, Panicum obtusum, Yucca glauca*
Stable Gear: *Nolina microcarpa*
Tools: *Quercus gambelii*
Water Indicator: *Forestiera pubescens*
Weapon: *Atriplex canescens, Chrysothamnus nauseosus*

Jemez
Drug
Analgesic: *Croton texensis*
Antirheumatic (Internal): *Croton texensis*
Cathartic: *Astragalus* sp., *Dalea formosa*
Cold Remedy: *Chrysothamnus* sp.
Dermatological Aid: *Atriplex canescens, Grindelia* sp., *Gutierrezia sarothrae, Helianthus annuus, Opuntia* sp.
Gastrointestinal Aid: *Artemisia* sp., *Juniperus monosperma*
Gynecological Aid: *Erodium cicutarium, Gutierrezia sarothrae, Juniperus monosperma*
Kidney Aid: *Salvia* sp.
Misc. Disease Remedy: *Croton texensis*
Oral Aid: *Chrysothamnus* sp., *Rhus trilobata*
Stimulant: *Atriplex canescens, Senecio flaccidus*
Throat Aid: *Chrysothamnus* sp.
Urinary Aid: *Xanthium* sp.
Veterinary Aid: *Grindelia* sp., *Macranthera* sp.

Food
Beverage: *Hymenopappus* sp., *Juniperus communis*
Bread & Cake: *Cleome serrulata*
Forage: *Melilotus officinalis*
Fruit: *Berberis fendleri, Juniperus monosperma, J. scopulorum, Lycium pallidum, Opuntia* sp., *Rhus trilobata, Ribes leptanthum, Vitis arizonica*
Special Food: *Lycium pallidum*
Unspecified: *Amaranthus retroflexus, Asclepias* sp., *A. subverticillata, Astragalus lentiginosus, Pinus edulis, Populus deltoides, Robinia neomexicana, Yucca baccata, Y. glauca*
Vegetable: *Cleome serrulata*

Fiber
Basketry: *Nolina microcarpa, Rhus trilobata, Yucca baccata*
Brushes & Brooms: *Fallugia paradoxa, Koeleria macrantha*
Building Material: *Baileya multiradiata, Helianthus annuus, Juniperus monosperma, Pinus ponderosa, Salix exigua*
Cordage: *Geranium atropurpureum*

Dye
Red: *Alnus incana, Betula occidentalis, Cercocarpus montanus, Pinus edulis*

Other
Cash Crop: *Pinus edulis*
Ceremonial Items: *Carex* sp., *Forestiera pubescens, Juniperus monosperma, Lobelia cardinalis, Parthenocissus vitacea, Pentaphylloides floribunda, Pseudotsuga menziesii, Pterospora andromedea, Vitis arizonica*
Decorations: *Calamovilfa gigantea, Clematis ligusticifolia, Helianthus annuus, Iris missouriensis*
Good Luck Charm: *Conopholis alpina, Oenothera elata*
Hunting & Fishing Item: *Cornus sericea, Quercus gambelii, Robinia neomexicana*
Insecticide: *Gutierrezia sarothrae*
Preservative: *Castilleja integra*
Protection: *Erodium cicutarium*
Sacred Items: *Carex* sp.
Smoke Plant: *Arctostaphylos uva-ursi*
Soap: *Helianthus annuus, Yucca* sp.
Tools: *Rhus trilobata*
Toys & Games: *Sarcobatus vermiculatus*

Kamia
Food
Bread & Cake: *Anemopsis californica*
Porridge: *Anemopsis californica, Cyperus erythrorhizos*
Staple: *Atriplex torreyi*
Unspecified: *Citrullus lanatus, Cucumis* sp., *Cucurbita pepo, Phaseolus acutifolius, Prosopis glandulosa, P. pubescens, Vigna unguiculata, Zea mays*
Vegetable: *Sonchus oleraceus*

Kansa
Other
Ceremonial Items: *Scirpus acutus*

Karok
Drug
Analgesic: *Artemisia vulgaris, Osmorhiza berteroi, Pentagrama triangularis, Umbellularia californica*
Antidiarrheal: *Arctostaphylos nevadensis*
Antidote: *Arctostaphylos nevadensis*
Antirheumatic (External): *Aralia californica, Artemisia douglasiana*
Blood Medicine: *Taxus brevifolia*
Cathartic: *Frangula purshiana*
Ceremonial Medicine: *Arbutus menziesii, Equisetum hyemale, Pyrola asarifolia, Rhododendron macrophyllum, Salix sitchensis, Sambucus cerulea*
Cold Remedy: *Artemisia vulgaris, Eriodictyon californicum, Prunus virginiana, Pseudotsuga menziesii, Umbellularia californica*
Contraceptive: *Toxicodendron diversilobum*
Dermatological Aid: *Achillea millefolium, Ceanothus velutinus, Maianthemum racemosum, Marah oreganus, Trillium ovatum, Umbellularia californica*
Diaphoretic: *Monardella odoratissima*
Dietary Aid: *Ligusticum apiifolium, Lomatium californicum, Oxalis oregana, Rubus parviflorus*
Disinfectant: *Pseudotsuga menziesii, Umbellularia californica*
Eye Medicine: *Equisetum hyemale, Gnaphalium microcephalum*
Gastrointestinal Aid: *Eriogonum nudum, Lupinus albifrons, Taxus brevifolia*
Gynecological Aid: *Anthemis cotula, Artemisia vulgaris, Ceanothus integerrimus, Chimaphila menziesii, Darmera peltata, Hierochloe occidentalis, Lotus humistratus, Pentagrama triangularis, Quercus garryana, Umbellularia californica*
Herbal Steam: *Cornus nuttallii, Pseudotsuga menziesii, Rumex conglomeratus*
Kidney Aid: *Chimaphila menziesii, Satureja douglasii*
Love Medicine: *Acer circinatum, Calystegia occidentalis, Galium triflorum, Monardella odoratissima, Populus balsamifera, Satureja douglasii*
Misc. Disease Remedy: *Mahonia aquifolium*
Orthopedic Aid: *Chimaphila umbellata*
Other: *Elymus glaucus, Mahonia aquifolium*
Panacea: *Artemisia vulgaris, Mahonia aquifolium, Osmorhiza berteroi, Petasites frigidus, Pyrola picta, Sambucus cerulea, Umbellularia californica*
Pediatric Aid: *Maianthemum racemosum, Mimulus cardinalis, Mirabilis greenei, Petasites frigidus, Prunus virginiana, Pyrola asarifolia, P. picta, Sambucus cerulea, Silene campanulata*
Poison: *Heracleum maximum, Mahonia aquifolium, Marah oreganus, Solanum nigrum, Toxicodendron diversilobum*
Preventive Medicine: *Fraxinus latifolia, Osmorhiza berteroi*
Psychological Aid: *Osmorhiza berteroi, Penstemon laetus*
Pulmonary Aid: *Eriodictyon californicum*
Sedative: *Pyrola asarifolia*
Tonic: *Abies grandis, Mahonia pumila, Rubus parviflorus*
Tuberculosis Remedy: *Eriodictyon californicum*
Unspecified: *Asarum* sp., *Gnaphalium* sp., *Juncus effusus, Monardella* sp., *Nicotiana quadrivalvis, Silene campanulata, Trichostema* sp.
Urinary Aid: *Chimaphila menziesii*
Veterinary Aid: *Hierochloe occidentalis*

Food
Beverage: *Arctostaphylos canescens, A. manzanita, A. nevadensis, Eriodictyon californicum, Pseudotsuga menziesii*
Bread & Cake: *Lithocarpus densiflorus*
Candy: *Agoseris aurantiaca, Asclepias cordifolia, A. eriocarpa, A.* sp., *Toxicodendron diversilobum*
Dried Food: *Amelanchier alnifolia, Arbutus menziesii, Arctostaphylos canescens, A. manzanita, A. nevadensis, A. patula*
Forage: *Ceanothus integerrimus, Disporum smithii, Oemleria cerasiformis*
Frozen Food: *Arbutus menziesii*
Fruit: *Amelanchier alnifolia, Arbutus menziesii, Arctostaphylos canescens, A. manzanita, A. nevadensis, Fragaria vesca, Gaultheria shallon, Heteromeles arbutifolia, Paxistima myrsinites, Prunus virginiana, Quercus chrysolepis, Q. kelloggii, Ribes cruentum, R. divaricatum, R. roezlii, Rubus leucodermis, R. parviflorus, R. vitifolius, Sambucus cerulea, Taxus brevifolia, Vaccinium ovatum, V. parvifolium, Vitis californica*
Porridge: *Bromus diandrus, B. hordeaceus, Elymus glaucus, Lithocarpus densiflorus*
Preservative: *Alnus rhombifolia*
Spice: *Pseudotsuga menziesii*
Staple: *Lithocarpus densiflorus*
Unspecified: *Allium acuminatum, A. bolanderi, Apocynum cannabinum, Armillaria ponderosa, Avena sativa, Balsamorhiza deltoidea, Boschniakia strobilacea, Calochortus pulchellus, Castanopsis chrysophylla, Chlorogalum pomeridianum, Corylus cornuta, Dichelostemma multiflorum, D. pulchellum, Epilobium canum, Eriogonum nudum, Heracleum maximum, Lilium occidentale, L. pardalinum, Lomatium californicum, Perideridia gairdneri, Pinus lambertiana, Quercus chrysolepis, Q. garryana, Q. kelloggii, Q. sadleriana, Rorippa nasturtium-aquaticum, Triteleia laxa, Umbellularia californica, Zigadenus venenosus*
Vegetable: *Angelica tomentosa, Camassia quamash, Crepis acuminata, Darmera peltata, Eriogonum nudum, Grindelia robusta, Lathyrus graminifolius, Osmorhiza berteroi, Sanicula bipinnata*
Winter Use Food: *Castanopsis chrysophylla, Corylus cornuta, Lithocarpus densiflorus, Pinus lambertiana, Umbellularia californica, Vaccinium ovatum*

Fiber
Basketry: *Adiantum aleuticum, A. pedatum, Alnus rhombifolia, A. rubra, Amelanchier alnifolia, Ceanothus integerrimus, Corylus cornuta, Fraxinus latifolia, Juncus effusus, J. ensifolius, Pinus ponderosa, Populus balsamifera, Salix bonplandiana, S. hindsiana, S. sitchensis, Vitis californica, Woodwardia fimbriata, W. radicans, Xerophyllum* sp., *X. tenax*
Brushes & Brooms: *Calocedrus decurrens, Chamaecyparis lawsoniana, Chlorogalum pomeridianum, Vaccinium parvifolium*
Building Material: *Calocedrus decurrens, Chamaecyparis lawsoniana, Pinus lambertiana, Pseudotsuga menziesii*
Clothing: *Xerophyllum* sp.
Cordage: *Corylus cornuta, Iris macrosiphon, Vitis californica*
Furniture: *Chamaecyparis lawsoniana*
Mats, Rugs & Bedding: *Acer macrophyllum, Chamaecyparis lawsoniana, Scirpus acutus*
Scouring Material: *Equisetum hyemale, E. laevigatum*
Sewing Material: *Taxus brevifolia*
Snow Gear: *Arbutus menziesii, Corylus cornuta*
Unspecified: *Arbutus menziesii*

Dye
Black: *Gaultheria shallon*
Unspecified: *Alnus rhombifolia, Letharia vulpina*
Yellow: *Datisca glomerata, Heracleum maximum, Mahonia pumila*

Other
Ceremonial Items: *Angelica tomentosa, Umbellularia californica*
Containers: *Acer macrophyllum, Arbutus menziesii, Arctostaphylos canescens, A. manzanita, A. nevadensis, Pteridium aquilinum, Salix scouleriana, S. sitchensis, Taxus brevifolia, Toxicodendron diversilobum, Vaccinium ovatum, Vitis californica*
Cooking Tools: *Acer circinatum, A. macrophyllum, Arctostaphylos canescens, A. manzanita, A. nevadensis, Athyrium filix-femina, Dryopteris arguta, Pteridi-*

um aquilinum, Salix sitchensis, Toxicodendron diversilobum
Decorations: *Adiantum pedatum, Chrysothamnus nauseosus, Dichelostemma ida-maia, Epipactis gigantea, Lilium rubescens, Pinus attenuata, P. sabiniana, Pseudotsuga menziesii, Veratrum californicum, Xerophyllum tenax*
Fasteners: *Pinus lambertiana, Populus balsamifera, Prunus virginiana*
Good Luck Charm: *Cornus nuttallii, Osmorhiza berteroi, Pseudotsuga menziesii*
Hunting & Fishing Item: *Amelanchier alnifolia, Arbutus menziesii, Cornus sericea, Corylus cornuta, Iris macrosiphon, Philadelphus lewisii, Physocarpus capitatus, Pseudotsuga menziesii, Taxus brevifolia*
Incense & Fragrance: *Boykinia occidentalis, Monardella odoratissima, Satureja douglasii*
Insecticide: *Artemisia douglasiana, Grindelia robusta, Trichostema lanceolatum*
Jewelry: *Pinus lambertiana*
Musical Instrument: *Ranunculus californicus, Sambucus cerulea*
Paint: *Delphinium decorum, Mahonia aquifolium, Prunus virginiana*
Protection: *Salix bonplandiana, S. sitchensis*
Smoke Plant: *Lomatium californicum, Nicotiana quadrivalvis*
Smoking Tools: *Philadelphus lewisii, Taxus brevifolia*
Soap: *Chlorogalum angustifolium, C. pomeridianum*
Tools: *Alnus rubra, Arbutus menziesii, Cercocarpus montanus, Pseudotsuga menziesii, Salix scouleriana, S. sitchensis, Taxus brevifolia, Vitis californica*
Toys & Games: *Castilleja parviflora, Claytonia parviflora, C. sibirica, Eriogonum nudum, Hastingsia alba, Heteromeles arbutifolia, Holodiscus discolor, Polystichum munitum, Umbellularia californica, Viola sempervirens*
Walking Sticks: *Arctostaphylos canescens, A. manzanita, A. nevadensis*
Water Indicator: *Sanicula bipinnata*

Kawaiisu
Drug
Abortifacient: *Artemisia douglasiana, Phoradendron villosum*
Analgesic: *Acamptopappus sphaerocephalus, Arenaria macradenia, Artemisia tridentata, Aster subulatus, Astragalus pachypus, A. purshii, Chaenactis santolinoides, Cicuta douglasii, Croton setigerus, Cupressus nevadensis, Datura wrightii, Dicentra chrysantha, Encelia virginensis, Erigeron foliosus, Gnaphalium stramineum, Grindelia camporum, Larrea tridentata, Lessingia glandulifera, Lomatium dissectum, Mentha arvensis, Mimulus guttatus, Nicotiana quadrivalvis, Purshia glandulosa, Rhamnus crocea, Rumex salicifolius, Salvia dorrii, Sambucus cerulea, Tauschia parishii, Thamnosma montana, Urtica dioica*
Antidiarrheal: *Eriogonum sp., Salix bonplandiana*
Antidote: *Frangula californica*
Antirheumatic (External): *Arenaria macradenia, Artemisia dracunculus, Datura wrightii, Encelia virginensis, Ericameria linearifolia, Phoradendron villosum, Turricula parryi*
Antirheumatic (Internal): *Quercus wislizeni, Rhamnus crocea*
Blood Medicine: *Ephedra viridis, Rhamnus crocea, Sambucus cerulea*
Burn Dressing: *Argemone munita, Frangula californica, Mentzelia albicaulis, Quercus douglasii, Q. lobata, Q. wislizeni*
Carminative: *Pinus lambertiana*
Cathartic: *Fremontodendron californicum*
Ceremonial Medicine: *Datura wrightii*
Cold Remedy: *Anemopsis californica, Artemisia tridentata, Balsamorhiza deltoidea, Corethrogyne filaginifolia, Cupressus nevadensis, Eriodictyon californicum, Eriogonum nudum, Garrya flavescens, Lomatium californicum, Marrubium vulgare, Osmorhiza brachypoda, Phacelia californica, Rhamnus crocea, Sambucus cerulea, Tauschia parishii, Thamnosma montana, Trichostema lanceolatum*
Contraceptive: *Pinus monophylla*
Cough Medicine: *Anemopsis californica, Artemisia tridentata, Balsamorhiza deltoidea, Cercocarpus montanus, Cupressus nevadensis, Eriogonum nudum, Marrubium vulgare, Osmorhiza brachypoda, Phacelia californica, Rhamnus crocea*
Dermatological Aid: *Anemopsis californica, Arenaria macradenia, Artemisia douglasiana, Asclepias californica, Castilleja stenantha, Chenopodium californicum, Datura wrightii, Distichlis spicata, Eriastrum densifolium, Ericameria linearifolia, Eriogonum umbellatum, Eschscholzia parishii, Frangula californica, Lomatium dissectum, L. utriculatum, Lythrum californicum, Marah horridus, Mentha arvensis, Nicotiana quadrivalvis, Osmorhiza brachypoda, Pinus monophylla, Quercus douglasii, Q. lobata, Ranunculus cymbalaria, Rhamnus crocea, Rumex salicifolius, Solanum xantii, Solidago californica, S. confinis, Tauschia parishii, Tetradymia stenolepis, Thamnosma montana, Turricula parryi, Urtica dioica*
Diaphoretic: *Corethrogyne filaginifolia, Sambucus cerulea, Thamnosma montana*
Disinfectant: *Frangula californica, Larrea tridentata*
Diuretic: *Rhamnus crocea*
Ear Medicine: *Cercocarpus ledifolius, Marah horridus, Nicotiana quadrivalvis, Plantago lanceolata, P. major*
Emetic: *Chenopodium californicum, Lomatium californicum, Nicotiana quadrivalvis, Phacelia ramosissima, Purshia glandulosa*
Eye Medicine: *Pinus lambertiana, Quercus douglasii, Q. lobata, Salvia columbariae, Stephanomeria virgata, Tauschia parishii*
Febrifuge: *Sambucus cerulea*
Gastrointestinal Aid: *Eriodictyon californicum, Garrya flavescens, Lomatium californicum, Nicotiana quadrivalvis, Phacelia californica, P. ramosissima, Rhamnus crocea, Rumex salicifolius, Salvia dorrii, Trichostema lanceolatum*
Gynecological Aid: *Astragalus pachypus, A. purshii, Cercocarpus ledifolius, Cupressus nevadensis, Nicotiana quadrivalvis, Pinus monophylla, Purshia glandulosa, Rhamnus crocea*
Hallucinogen: *Datura wrightii, Nicotiana quadrivalvis, Thamnosma montana*
Heart Medicine: *Croton setigerus, Dicentra chrysantha, Distichlis spicata, Eriogonum sp.*
Hemorrhoid Remedy: *Aesculus californica*
Hemostat: *Cuscuta californica, Frangula californica, Nicotiana quadrivalvis*
Herbal Steam: *Artemisia tridentata, Corethrogyne filaginifolia, Mimulus guttatus, Sambucus cerulea*
Hunting Medicine: *Thamnosma montana*
Internal Medicine: *Cercocarpus montanus*
Kidney Aid: *Cupressus nevadensis, Rhamnus crocea*
Laxative: *Distichlis spicata, Frangula californica, Garrya flavescens, Pinus lambertiana, Prunus virginiana, Purshia glandulosa, Rhamnus crocea, Senecio flaccidus*
Liver Aid: *Rhamnus crocea*
Misc. Disease Remedy: *Anemopsis californica, Artemisia tridentata, Rumex salicifolius, Sambucus cerulea*
Orthopedic Aid: *Chaenactis santolinoides, Cicuta douglasii, Cupressus nevadensis, Datura wrightii, Delphinium parryi, Ephedra viridis, Ericameria linearifolia, Grindelia camporum, Gutierrezia californica, Larrea tridentata, Lomatium dissectum, L. utriculatum, Mimulus guttatus, Penstemon rostriflorus, Populus fremontii, Rumex salicifolius, Solanum xantii, Tauschia parishii, Urtica dioica*
Other: *Artemisia douglasiana, Nicotiana quadrivalvis*
Pediatric Aid: *Artemisia douglasiana, Datura wrightii, Pinus lambertiana, P. monophylla, Urtica dioica*
Poison: *Aesculus californica, Chenopodium californicum, Cicuta douglasii, Datura wrightii, Nicotiana quadrivalvis, Thalictrum polycarpum*
Poultice: *Populus fremontii*
Psychological Aid: *Nicotiana quadrivalvis, Urtica dioica*
Respiratory Aid: *Arenaria macradenia, Marrubium vulgare, Nicotiana quadrivalvis*
Sedative: *Nicotiana quadrivalvis*

Snakebite Remedy: *Achillea millefolium, Chamaesyce albomarginata, Thamnosma montana*
Stimulant: *Nicotiana quadrivalvis, Phacelia californica, Rhamnus crocea*
Throat Aid: *Lomatium californicum*
Toothache Remedy: *Aster subulatus, Erigeron foliosus, Melica imperfecta, Nicotiana quadrivalvis, Tauschia parishii*
Unspecified: *Alnus rhombifolia, Frangula californica, Larrea tridentata, Lythrum californicum, Platanus racemosa, Rumex salicifolius, Senecio flaccidus*
Venereal Aid: *Calystegia longipes, Cercocarpus ledifolius, Coleogyne ramosissima, Distichlis spicata, Eriastrum densifolium, Eriodictyon californicum, Eriogonum umbellatum, Eschscholzia parishii, Garrya flavescens, Mahonia dictyota, Phacelia ramosissima, Purshia glandulosa, Rhamnus crocea*
Veterinary Aid: *Anemopsis californica, Chamaesyce albomarginata, Encelia virginensis, Ericameria linearifolia, Larrea tridentata, Thamnosma montana*
Witchcraft Medicine: *Salvia dorrii, Urtica dioica*

Food

Beverage: *Arctostaphylos glauca, Cheilanthes covillei, Descurainia pinnata, D. sophia, Distichlis spicata, Ephedra californica, E. nevadensis, E. viridis, Eriogonum baileyi, E. roseum, E. wrightii, Lepidium fremontii, Lycium andersonii, Mentha arvensis, M. spicata, Monardella linoides, M. odoratissima, M. viridis, Pellaea mucronata, Platanus racemosa, Salvia columbariae, Trichostema lanceolatum*
Bread & Cake: *Aesculus californica, Juniperus californica, Prosopis glandulosa, Quercus chrysolepis, Q. douglasii, Q. dumosa, Q. garryana, Q. kelloggii, Q. lobata, Q. wislizeni*
Candy: *Asclepias californica, Salix bonplandiana, S. hindsiana, Stephanomeria pauciflora*
Dried Food: *Eriophyllum ambiguum, Juniperus californica, Lycium andersonii, Ribes californicum, R. quercetorum, R. roezlii, R. velutinum, Yucca brevifolia, Y. whipplei*
Forage: *Erodium cicutarium, Leymus triticoides, Nemophila menziesii*
Fruit: *Amelanchier pallida, Arctostaphylos glauca, Frangula californica, Juniperus californica, Lycium andersonii, Prunus virginiana, Ribes californicum, R. quercetorum, R. roezlii, R. velutinum, Rosa californica, R. woodsii, Yucca brevifolia*
Porridge: *Deschampsia danthonioides, Elymus multisetus, Eriogonum inflatum, E. plumatella, Leymus triticoides, Melica imperfecta, Nama demissum, Pinus monophylla, P. sabiniana, Prosopis glandulosa, Rumex crispus, R. hymenosepalus, R. salicifolius*
Preserves: *Mentzelia affinis, M. albicaulis, M. congesta, M. dispersa, Prunus virginiana, Ribes californicum, R. quercetorum, R. roezlii, R. velutinum, Sambucus cerulea*
Spice: *Chrysothamnus nauseosus, Lotus procumbens, L. unifoliolatus, Stachys albens*
Staple: *Eriogonum baileyi, E. davidsonii, E. inflatum, E. pusillum, E. roseum, E. wrightii, Helianthus annuus, Juniperus californica, Oryzopsis hymenoides, Quercus chrysolepis, Q. douglasii, Q. dumosa, Q. garryana, Q. kelloggii, Q. lobata, Q. wislizeni*
Sweetener: *Coreopsis bigelovii, Phragmites australis, Pinus lambertiana*
Unspecified: *Avena fatua, Carex douglasii, Castanopsis sempervirens, Caulanthus inflatus, Cirsium californicum, C. congdonii, C. occidentale, Cistanthe monandra, Cleome isomeris, Dendromecon rigida, Ephedra californica, E. nevadensis, E. viridis, Eriogonum angulosum, E. nudum, E. pusillum, Hordeum jubatum, Mimulus cardinalis, Opuntia basilaris, Pholisma arenarium, Pinus lambertiana, P. monophylla, P. ponderosa, P. sabiniana, Pterospora andromedea, Rorippa nasturtium-aquaticum, Rumex crispus, R. hymenosepalus, R. salicifolius, Scirpus tabernaemontani, Stipa speciosa, Typha domingensis, Yucca whipplei*
Vegetable: *Agoseris retrorsa, Allium* sp., *Amsinckia tessellata, Atriplex serenana, Caulanthus coulteri, Chenopodium album, Claytonia perfoliata, Coreopsis bigelovii, Lomatium californicum, L. utriculatum, Perideridia pringlei, Phacelia distans, P. ramosissima, Stanleya pinnata, Trifolium wormskioldii*
Winter Use Food: *Descurainia pinnata, D. sophia, Mentzelia affinis, M. albicaulis, M. congesta, M. dispersa, Pinus monophylla, Quercus chrysolepis, Q. douglasii, Q. dumosa, Q. garryana, Q. kelloggii, Q. lobata, Q. wislizeni*

Fiber

Basketry: *Juncus balticus, Muhlenbergia rigens, Proboscidea louisianica, Pteridium aquilinum, Quercus douglasii, Q. wislizeni, Rhus trilobata, Rosa californica, R. woodsii, Salix bonplandiana, S. lasiolepis, Scirpus robustus, Yucca brevifolia, Y. whipplei*
Brushes & Brooms: *Chlorogalum pomeridianum*
Building Material: *Fremontodendron californicum, Gutierrezia californica, Juniperus californica, Lotus procumbens, Pinus ponderosa, P. sabiniana, Quercus chrysolepis, Q. douglasii, Q. dumosa, Q. garryana, Q. kelloggii, Q. lobata, Q. wislizeni, Salix bonplandiana, S. hindsiana, Scirpus tabernaemontani, Typha domingensis, Yucca whipplei*
Clothing: *Artemisia tridentata*
Cordage: *Asclepias fascicularis, Fremontodendron californicum, Potamogeton diversifolius, Urtica dioica*
Furniture: *Fremontodendron californicum, Salix bonplandiana*
Mats, Rugs & Bedding: *Scirpus tabernaemontani*

Other

Ceremonial Items: *Amelanchier pallida, Artemisia douglasiana, A. tridentata, Pinus ponderosa, Scirpus tabernaemontani*
Containers: *Artemisia tridentata, Eriogonum fasciculatum, Fremontodendron californicum, Pinus ponderosa, P. sabiniana, Potamogeton diversifolius, Salix bonplandiana, S. hindsiana, Urtica dioica, Yucca whipplei*
Cooking Tools: *Aesculus californica, Eriogonum nudum, Juniperus californica, Quercus douglasii, Scirpus tabernaemontani*
Fasteners: *Bloomeria crocea, Chlorogalum pomeridianum, Dichelostemma pulchellum, Quercus chrysolepis, Q. douglasii, Q. dumosa, Q. garryana, Q. kelloggii, Q. lobata, Q. wislizeni*
Fuel: *Artemisia tridentata, Ceanothus cuneatus, Quercus douglasii*
Hide Preparation: *Alnus rhombifolia*
Hunting & Fishing Item: *Amelanchier pallida, Atriplex canescens, A. confertifolia, Baccharis salicifolia, Ceanothus cuneatus, Ericameria palmeri, Fremontodendron californicum, Juniperus californica, Maianthemum stellatum, Phragmites australis, Potamogeton diversifolius, Prunus fasciculata, P. virginiana, Salix bonplandiana, S. hindsiana, Urtica dioica*
Incense & Fragrance: *Artemisia douglasiana*
Insecticide: *Osmorhiza brachypoda, Trichostema lanceolatum*
Jewelry: *Pinus monophylla*
Musical Instrument: *Phragmites australis, Sambucus cerulea*
Paint: *Plagiobothrys jonesii, P. nothofulvus*
Protection: *Nicotiana attenuata, N. quadrivalvis, Salvia dorrii, Thamnosma montana*
Smoke Plant: *Dendromecon rigida, Nicotiana attenuata, N. quadrivalvis, Pinus sabiniana*
Smoking Tools: *Cercocarpus montanus, Eriogonum insigne, E. nudum, Lonicera interrupta, Phragmites australis, Salix bonplandiana, S. hindsiana, Sambucus cerulea*
Soap: *Ceanothus leucodermis, Chenopodium californicum, Cucurbita foetidissima*
Stable Gear: *Scirpus tabernaemontani*
Tools: *Artemisia tridentata, Echinocactus polycephalus, Ephedra californica, E. nevadensis, E. viridis, Eriogonum fasciculatum, Fraxinus latifolia, Hordeum jubatum, Larrea tridentata, Muhlenbergia rigens, Phragmites australis, Prunus fasciculata, P. virginiana, Salix bonplandiana, S. hindsiana, Tetradymia axillaris, T. stenolepis*
Toys & Games: *Mentzelia involucrata, Phragmites australis, Quercus chrysole-*

pis, Q. douglasii, Q. dumosa, Q. garryana, Q. kelloggii, Q. lobata, Q. wislizeni, Salix bonplandiana, S. hindsiana
Waterproofing Agent: *Pinus monophylla*

Keresan
Drug
Ceremonial Medicine: *Gutierrezia* sp.
Dermatological Aid: *Yucca baccata*
Emetic: *Chrysothamnus* sp., *Gutierrezia* sp.
Febrifuge: *Glycyrrhiza lepidota, Mentha canadensis*
Gastrointestinal Aid: *Chrysothamnus* sp.
Gynecological Aid: *Solanum elaeagnifolium*
Veterinary Aid: *Astragalus* sp., *Ipomoea leptophylla*
Food
Beverage: *Thelesperma megapotamicum*
Dried Food: *Prunus armeniaca, P. persica*
Fruit: *Juniperus monosperma, J. scopulorum, Prunus armeniaca, P. persica, Rhus trilobata, Yucca baccata*
Soup: *Capsicum annuum*
Spice: *Coriandrum sativum, Cucumis melo*
Starvation Food: *Solanum fendleri, S. jamesii*
Unspecified: *Cleome serrulata, Dalea candida, Opuntia engelmannii*
Vegetable: *Amaranthus cruentus, Chenopodium* sp., *Cleome serrulata*
Winter Use Food: *Pinus edulis*
Fiber
Basketry: *Nolina microcarpa*
Brushes & Brooms: *Fallugia paradoxa*
Dye
Red: *Cercocarpus* sp.
Other
Ceremonial Items: *Ipomoea leptophylla, Phaseolus acutifolius, Pseudotsuga menziesii, Salix* sp., *Typha latifolia*
Cooking Tools: *Lagenaria siceraria*
Hunting & Fishing Item: *Amaranthus cruentus, Fallugia paradoxa*
Jewelry: *Solanum elaeagnifolium*
Paint: *Cleome serrulata*
Smoke Plant: *Arctostaphylos uva-ursi, Rhus trilobata*
Toys & Games: *Lagenaria siceraria*

Keres, Western
Drug
Analgesic: *Artemisia* sp., *Baccharis* sp., *Mentha canadensis, Plantago patagonica*
Antidiarrheal: *Juniperus monosperma, Phoradendron juniperinum, Plantago patagonica, Portulaca oleracea*
Antirheumatic (External): *Abies concolor, Artemisia ludoviciana, A.* sp., *Baccharis* sp., *Brickellia ambigens, B. grandiflora, Chaetopappa ericoides, Epixiphium wislizeni, Erysimum capitatum, Gilia rigidula, Gutierrezia sarothrae, Lesquerella fendleri, Lygodesmia juncea, Oenothera albicaulis, Phacelia crenulata, Phoradendron juniperinum, Picea parryana, Zigadenus elegans, Zinnia acerosa*
Antirheumatic (Internal): *Abies concolor, Calochortus gunnisonii*
Blood Medicine: *Aquilegia elegantula, Atriplex* sp., *Plantago major, Portulaca oleracea*
Burn Dressing: *Anemopsis californica, Aristida divaricata, Conyza canadensis, Rumex salicifolius*
Carminative: *Brickellia ambigens, B. grandiflora*
Cathartic: *Aletes acaulis, Bahia dissecta, Croton texensis, Gutierrezia sarothrae, Solidago gigantea*
Cold Remedy: *Gnaphalium canescens, Picea parryana, Thalictrum fendleri*
Cough Medicine: *Ephedra torreyana, Glycyrrhiza lepidota, Prunus virginiana*
Dermatological Aid: *Alnus incana, Anemopsis californica, Baileya multiradiata, Buchloe dactyloides, Conyza canadensis, Croton texensis, Cucurbita foetidissima, Dalea candida, Datura wrightii, Geranium caespitosum, Gnaphalium canescens, Juniperus monosperma, Opuntia clavata, O. imbricata, Pinus edulis, Psorothamnus scoparius, Sarcobatus vermiculatus, Senecio flaccidus, Silene laciniata, Tragia ramosa, Vicia americana, Xanthium strumarium*
Diaphoretic: *Artemisia ludoviciana, A.* sp., *Ephedra torreyana, Gutierrezia sarothrae, Juniperus monosperma, Zigadenus elegans*
Dietary Aid: *Abronia fragrans, Brickellia ambigens, B. grandiflora*
Diuretic: *Cirsium pallidum, Mentzelia multiflora*
Ear Medicine: *Geastrum* sp., *Juniperus monosperma, Opuntia imbricata*
Emetic: *Aletes acaulis, Bahia dissecta, Chenopodium graveolens, Dalea formosa, Eriogonum rotundifolium, Gutierrezia sarothrae, Ipomopsis laxiflora, I. longiflora, Juniperus monosperma, Lesquerella fendleri, Panicum capillare, Penstemon ambiguus, Psorothamnus scoparius, Pterospora andromedea, Quercus gambelii, Rhus trilobata, Sarcobatus vermiculatus*
Eye Medicine: *Chamaesyce albomarginata, Gutierrezia sarothrae, Houstonia rubra, Prosopis glandulosa, Stephanomeria runcinata*
Febrifuge: *Artemisia* sp., *Dyssodia papposa, Eriogonum tenellum, Gaura parviflora, Mentha canadensis*
Gastrointestinal Aid: *Amaranthus hybridus, A. retroflexus, Arabis fendleri, Asclepias involucrata, Cymopterus bulbosus, Grindelia decumbens, G. fastigiata, Houstonia rubra, Ipomoea leptophylla, Juniperus monosperma, Pectis angustifolia, Picea parryana, Pinus edulis, Polygonum lapathifolium, Psorothamnus scoparius, Rhus trilobata, Senecio flaccidus*
Gynecological Aid: *Ambrosia psilostachya, Apocynum cannabinum, Asclepias subverticillata, Chamaesyce albomarginata, Cheilanthes fendleri, Eriogonum tenellum, Euphorbia dentata, Gaillardia pinnatifida, G. pulchella, Ratibida columnifera, Rhus trilobata, Senecio spartioides, Sphaeralcea angustifolia, S. coccinea, Townsendia strigosa, Woodsia neomexicana*
Heart Medicine: *Eriogonum jamesii*
Hemorrhoid Remedy: *Croton texensis, Equisetum laevigatum*
Hemostat: *Helianthus niveus*
Herbal Steam: *Yucca glauca*
Hunting Medicine: *Ratibida tagetes*
Kidney Aid: *Artemisia* sp., *Ephedra torreyana, Zinnia grandiflora*
Laxative: *Juniperus monosperma*
Liver Aid: *Brickellia ambigens, B. grandiflora*
Misc. Disease Remedy: *Rosa nutkana*
Nose Medicine: *Dimorphocarpa wislizeni*
Oral Aid: *Ceanothus fendleri, Portulaca oleracea, Quercus gambelii, Rhus trilobata*
Other: *Artemisia ludoviciana, Draba helleriana, Dyssodia papposa, Mahonia aquifolium, Silene laciniata, Zinnia grandiflora*
Panacea: *Zea mays*
Pediatric Aid: *Dalea nana, Phoradendron juniperinum, Phragmites australis, Quercus gambelii, Rosa nutkana, Thelesperma filifolium, T. megapotamicum, Tragia ramosa*
Poison: *Cryptantha crassisepala, Datura wrightii, Hymenoxys richardsonii, Ranunculus sceleratus, Zigadenus elegans*
Preventive Medicine: *Mahonia aquifolium*
Psychological Aid: *Abronia fragrans, Berlandiera lyrata, Chrysothamnus* sp., *Crepis runcinata, Datura ferox, Eriogonum jamesii, E. rotundifolium, Gaillardia pinnatifida, G. pulchella, Linum rigidum, Mentzelia multiflora, Pectis angustifolia, Senecio fendleri, Zinnia acerosa*
Pulmonary Aid: *Orobanche fasciculata*
Sedative: *Berlandiera lyrata, Eriogonum rotundifolium, Gaura parviflora, Ratibida tagetes*
Snakebite Remedy: *Epixiphium wislizeni, Glandularia bipinnatifida, Gutierrezia sarothrae*

Stimulant: *Nepeta cataria*
Strengthener: *Artemisia* sp., *Cercocarpus montanus*, *Dalea formosa*, *D. nana*, *Opuntia imbricata*, *Tragia ramosa*, *Yucca glauca*, *Zigadenus elegans*
Throat Aid: *Glandularia bipinnatifida*, *Phacelia crenulata*
Tuberculosis Remedy: *Commelina dianthifolia*, *Conopholis alpina*, *Mentzelia multiflora*, *Thelesperma filifolium*, *T. megapotamicum*
Unspecified: *Ratibida tagetes*
Urinary Aid: *Ephedra torreyana*
Veterinary Aid: *Artemisia* sp., *Cynodon dactylon*, *Gutierrezia sarothrae*, *Opuntia clavata*

Food
Beverage: *Rhus trilobata*, *Thelesperma filifolium*, *T. megapotamicum*, *Yucca glauca*
Bread & Cake: *Yucca glauca*
Candy: *Asclepias speciosa*, *A. subverticillata*, *Hymenoxys richardsonii*, *Populus deltoides*
Cooking Agent: *Opuntia engelmannii*
Fodder: *Geranium caespitosum*, *Phoradendron juniperinum*, *Sporobolus cryptandrus*
Forage: *Atriplex argentea*, *Bouteloua gracilis*, *Krascheninnikovia lanata*, *Sarcobatus vermiculatus*
Fruit: *Ceanothus fendleri*, *Celtis occidentalis*, *Echinocereus triglochidiatus*, *Opuntia engelmannii*, *O. phaeacantha*, *Physalis longifolia*, *P. subulata*, *Prunus persica*, *P. virginiana*, *Rhus trilobata*, *Ribes cereum*, *R. inerme*, *Vitis vulpina*, *Yucca baccata*, *Y. glauca*
Porridge: *Cleome serrulata*, *Prosopis glandulosa*
Sauce & Relish: *Lycium pallidum*, *Yucca baccata*, *Y. glauca*
Spice: *Agastache pallidiflora*, *Allium geyeri*, *Berlandiera lyrata*, *Juniperus monosperma*, *Monarda pectinata*, *Pectis angustifolia*, *Rhus trilobata*
Staple: *Dalea candida*, *Quercus gambelii*, *Zea mays*
Starvation Food: *Juniperus monosperma*, *Opuntia clavata*, *O. imbricata*, *Phoradendron juniperinum*, *Solanum triflorum*, *Yucca glauca*
Sweetener: *Dalea candida*, *Sophora nuttalliana*
Unspecified: *Amaranthus hybridus*, *A. retroflexus*, *Aster laevis*, *Astragalus cyaneus*, *Atriplex argentea*, *Pinus edulis*, *Plantago major*, *Quercus gambelii*, *Ranunculus inamoenus*, *Scirpus americanus*, *S.* sp., *Tradescantia occidentalis*, *Typha latifolia*, *Yucca glauca*
Vegetable: *Allium cernuum*, *Amaranthus hybridus*, *A. retroflexus*, *Atriplex powellii*, *Cleome serrulata*, *Cymopterus acaulis*, *Cyperus squarrosus*, *Lactuca sativa*, *Lathyrus polymorphus*, *Portulaca oleracea*, *Prosopis glandulosa*, *Solanum jamesii*, *Vicia americana*, *Zea mays*
Winter Use Food: *Amaranthus hybridus*, *A. retroflexus*, *Opuntia engelmannii*, *O. imbricata*, *O. polyacantha*, *Prunus persica*, *P. virginiana*, *Yucca baccata*, *Y. glauca*, *Zea mays*

Fiber
Basketry: *Juniperus monosperma*, *Nolina microcarpa*, *Rhus trilobata*, *Salix exigua*, *Solidago gigantea*, *Yucca glauca*
Brushes & Brooms: *Brickellia ambigens*, *B. grandiflora*, *Bromus marginatus*, *Cercocarpus montanus*, *Chrysothamnus nauseosus*, *Dalea candida*, *Erigeron asper*, *E. flagellaris*, *Fallugia paradoxa*, *Yucca glauca*
Building Material: *Juniperus monosperma*, *Pinus ponderosa*, *Tamarix gallica*
Clothing: *Juniperus monosperma*
Cordage: *Yucca baccata*, *Y. glauca*
Mats, Rugs & Bedding: *Nolina microcarpa*, *Salix exigua*
Sewing Material: *Opuntia imbricata*

Dye
Green: *Juniperus monosperma*
Red: *Alnus incana*, *Cercocarpus montanus*
Unspecified: *Aster laevis*
Yellow: *Psilostrophe tagetina*, *Verbesina* sp., *Zinnia grandiflora*

Other
Cash Crop: *Arundinaria* sp., *Pinus edulis*
Ceremonial Items: *Abronia fragrans*, *Acer negundo*, *Castilleja integra*, *C. miniata*, *Cucurbita foetidissima*, *Juniperus monosperma*, *Picea parryana*, *Populus deltoides*, *Pseudotsuga menziesii*, *Salix irrorata*, *Typha latifolia*, *Yucca baccata*, *Zea mays*
Containers: *Cercocarpus montanus*
Cooking Tools: *Lagenaria siceraria*, *Pinus edulis*
Decorations: *Abies concolor*, *Clematis bigelovii*, *Lathyrus polymorphus*, *Lygodesmia juncea*, *Penstemon barbatus*
Fasteners: *Sphaeralcea angustifolia*, *Yucca glauca*
Fertilizer: *Descurainia pinnata*, *D. sophia*, *Solanum triflorum*
Fuel: *Dalea formosa*, *Juniperus monosperma*, *Populus deltoides*, *Rumex salicifolius*
Good Luck Charm: *Picea parryana*
Hide Preparation: *Cercocarpus montanus*
Hunting & Fishing Item: *Cercocarpus montanus*, *Chrysothamnus nauseosus*, *Fallugia paradoxa*, *Juniperus monosperma*, *Monarda fistulosa*, *Prunus virginiana*, *Purshia stansburiana*, *Rhus trilobata*, *Robinia neomexicana*
Incense & Fragrance: *Senecio flaccidus*
Insecticide: *Artemisia dracunculus*, *Gutierrezia sarothrae*, *Melilotus officinalis*, *Pascopyrum smithii*, *Senecio flaccidus*
Lighting: *Opuntia imbricata*
Musical Instrument: *Arundinaria* sp.
Paint: *Aster laevis*, *Erysimum capitatum*, *Linum puberulum*, *Lithospermum* sp., *Opuntia engelmannii*, *Pinus edulis*, *Psilostrophe tagetina*, *Rhus trilobata*, *Xanthium strumarium*, *Zinnia grandiflora*
Preservative: *Descurainia pinnata*, *D. sophia*, *Juniperus monosperma*, *Sphaeralcea angustifolia*
Protection: *Chenopodium graveolens*
Season Indicator: *Opuntia imbricata*
Smoke Plant: *Datura wrightii*, *Mirabilis multiflora*, *M. nyctaginea*, *Thymophylla acerosa*
Snuff: *Dimorphocarpa wislizeni*
Soap: *Hedeoma nana*, *Yucca baccata*, *Y. glauca*
Tools: *Cercocarpus montanus*, *Opuntia engelmannii*, *O. imbricata*, *Quercus pungens*, *Yucca glauca*
Toys & Games: *Arundinaria* sp., *Cucurbita foetidissima*, *Lagenaria siceraria*
Unspecified: *Agoseris aurantiaca*, *Cirsium parryi*, *Clematis columbiana*, *C. ligusticifolia*, *Corydalis aurea*, *Desmanthus illinoensis*, *Echinocereus coccineus*, *Elymus glaucus*, *Fomes* sp., *Gleditsia triacanthos*, *Hymenopappus filifolius*, *Ipomoea leptophylla*, *Ipomopsis aggregata*, *I. pumila*, *Juniperus monosperma*, *Lappula occidentalis*, *Lepidium densiflorum*, *Lotus wrightii*, *Medicago sativa*, *Moricandia arvensis*, *Panicum capillare*, *Parthenocissus vitacea*, *Phragmites australis*, *Psoralidium tenuiflorum*, *Pyrrhopappus pauciflorus*, *Ribes montigenum*, *Salsola australis*, *Solanum rostratum*, *Spirogyra* sp., *Suaeda calceoliformis*
Water Indicator: *Mimulus glabratus*

Kiliwa
Other
Cash Crop: *Phaseolus vulgaris*

Kimsquit
Food
Dried Food: *Arctostaphylos uva-ursi*
Unspecified: *Lupinus nootkatensis*

Kiowa
Drug
Analgesic: *Lophophora williamsii*

Anthelmintic: *Artemisia vulgaris, Juglans nigra*
Antihemorrhagic: *Cephalanthus occidentalis*
Antirheumatic (External): *Lophophora williamsii, Salix* sp.
Burn Dressing: *Cirsium ochrocentrum*
Cold Remedy: *Lophophora williamsii*
Cough Medicine: *Echinacea angustifolia*
Dermatological Aid: *Ambrosia psilostachya, Artemisia ludoviciana, Centaurea americana, Cirsium ochrocentrum, Hierochloe odorata, Lophophora williamsii, Monarda pectinata, Opuntia* sp.*, Sapindus saponaria, Toxicodendron radicans, Vernonia missurica, Yucca glauca*
Emetic: *Aesculus glabra, Cucurbita foetidissima*
Febrifuge: *Lophophora williamsii*
Gastrointestinal Aid: *Artemisia ludoviciana, Lophophora williamsii, Mentha* sp.*, Penstemon grandiflorus, Rhus trilobata*
Gynecological Aid: *Taraxacum officinale*
Hemostat: *Opuntia* sp.
Herbal Steam: *Artemisia ludoviciana*
Misc. Disease Remedy: *Lophophora williamsii, Quincula lobata, Rhus trilobata*
Narcotic: *Lophophora williamsii*
Oral Aid: *Bothriochloa saccharoides, Helianthus annuus, Juniperus virginiana, Oxalis stricta, Ulmus rubra*
Orthopedic Aid: *Lophophora williamsii*
Other: *Rhus glabra*
Panacea: *Lophophora williamsii*
Poison: *Citrullus lanatus*
Pulmonary Aid: *Artemisia ludoviciana, Lophophora williamsii, Salix* sp.
Snakebite Remedy: *Ribes aureum*
Throat Aid: *Artemisia ludoviciana, Echinacea angustifolia*
Toothache Remedy: *Salix* sp.
Tuberculosis Remedy: *Carya illinoinensis, Lophophora williamsii, Rhus glabra*
Venereal Aid: *Lophophora williamsii*
Veterinary Aid: *Ambrosia psilostachya*
Food
Beverage: *Paronychia jamesii, Quercus nigra, Q. stellata, Rhus trilobata, Senna occidentalis, Ulmus rubra*
Candy: *Apocynum cannabinum, Euphorbia marginata, Mentha* sp.*, Opuntia* sp.*, Prunus gracilis, Sideroxylon lanuginosum, Vernonia missurica*
Dried Food: *Quercus stellata, Vitis cinerea*
Fodder: *Bouteloua curtipendula, B. hirsuta, Bromus catharticus, Cyperus schweinitzii, Dichanthelium oligosanthes, Elymus canadensis, Paspalum setaceum, Prosopis chilensis, P. glandulosa, Sorghum halepense, Sporobolus cryptandrus, Zea mays*
Forage: *Paspalum setaceum*
Fruit: *Asclepias* sp.*, Celtis occidentalis, Opuntia* sp.*, Prunus americana, P. virginiana, Ribes aureum, Rhus trilobata, Ribes aureum, Sambucus* sp.*, Vitis cinerea*
Preserves: *Opuntia* sp.*, Quincula lobata, Ribes aureum, Vitis cinerea*
Starvation Food: *Ipomoea leptophylla*
Unspecified: *Cirsium ochrocentrum, Dalea candida, Eriogonum longifolium, Helianthus annuus, Juglans nigra, Liatris punctata, Pyrrhopappus carolinianus, Quercus nigra, Zea mays*
Vegetable: *Nymphaea* sp.*, Prosopis chilensis, P. glandulosa, Taraxacum officinale*
Winter Use Food: *Prunus americana, P. gracilis, P. virginiana, Ulmus rubra*
Fiber
Building Material: *Populus deltoides, Salix* sp.
Mats, Rugs & Bedding: *Amorpha fruticosa, Artemisia vulgaris, Hierochloe odorata, Vernonia missurica*
Dye
Blue-Black: *Juglans nigra*
Orange-Yellow: *Rhus glabra*
Pink: *Parthenocissus quinquefolia*
Purple: *Vernonia missurica*
Unspecified: *Argemone polyanthemos, Eriogonum annuum*
Yellow: *Maclura pomifera*
Other
Ceremonial Items: *Acer negundo, Delphinium carolinianum, Maclura pomifera, Plantago virginica, Populus deltoides*
Cleaning Agent: *Artemisia vulgaris*
Cooking Tools: *Asclepias* sp.*, Psoralidium tenuiflorum*
Decorations: *Asclepias* sp.*, Bouteloua curtipendula, Callirhoe involucrata, Gaillardia pulchella, Oenothera rhombipetala, Sphaeralcea angustifolia*
Fasteners: *Amaranthus blitoides*
Fuel: *Ambrosia psilostachya, Celtis occidentalis, Cercis canadensis, Populus deltoides, Quercus nigra, Q. rubra, Q. stellata, Senna occidentalis*
Good Luck Charm: *Erigeron divergens, Gaillardia pulchella*
Hide Preparation: *Eriogonum annuum*
Hunting & Fishing Item: *Dalea enneandra, Maclura pomifera, Opuntia* sp.
Incense & Fragrance: *Hierochloe odorata, Juniperus virginiana*
Jewelry: *Phytolacca americana*
Musical Instrument: *Juniperus virginiana*
Paper: *Artemisia filifolia*
Protection: *Cirsium ochrocentrum, Opuntia* sp.*, Salix* sp.*, Smilax rotundifolia*
Season Indicator: *Cercis canadensis, Stillingia sylvatica*
Smoke Plant: *Quercus nigra, Q. stellata, Rhus glabra, R. trilobata, Zea mays*
Soap: *Cucurbita foetidissima, Yucca glauca*
Stable Gear: *Ulmus rubra*
Toys & Games: *Dalea enneandra, Quincula lobata, Smilax rotundifolia, Sorghum halepense*

Kitasoo
Drug
Antidote: *Dryopteris* sp.
Antirheumatic (External): *Veratrum viride*
Cathartic: *Oenanthe sarmentosa*
Cold Remedy: *Ledum groenlandicum*
Cough Medicine: *Polypodium glycyrrhiza*
Dietary Aid: *Elliottia pyroliflorus*
Emetic: *Oenanthe sarmentosa*
Gastrointestinal Aid: *Taxus brevifolia*
Gynecological Aid: *Nuphar lutea*
Internal Medicine: *Taxus brevifolia*
Respiratory Aid: *Ledum groenlandicum*
Throat Aid: *Polypodium glycyrrhiza*
Unspecified: *Abies amabilis, Aruncus dioicus, Maianthemum racemosum*
Food
Beverage: *Ledum groenlandicum*
Bread & Cake: *Porphyra abbottae, Sambucus racemosa, Vaccinium cespitosum, V. ovalifolium, V. uliginosum*
Dried Food: *Picea sitchensis, Tsuga heterophylla, Vaccinium membranaceum*
Fruit: *Amelanchier alnifolia, Cornus unalaschkensis, Fragaria chiloensis, F. vesca, F. virginiana, Gaultheria shallon, Maianthemum dilatatum, Oemleria cerasiformis, Ribes bracteosum, Rubus discolor, R. idaeus, R. leucodermis, R. pedatus, R. spectabilis, Vaccinium cespitosum, V. membranaceum, V. ovalifolium, V. oxycoccos, V. parvifolium, V. uliginosum*
Ice Cream: *Shepherdia canadensis*
Pie & Pudding: *Rheum rhabarbarum*
Preserves: *Rheum rhabarbarum*
Special Food: *Malus fusca*
Unspecified: *Abies amabilis, Argentina egedii, Egregia menziesii, Fritillaria*

[Unspecified] camschatcensis, Lupinus sp., Macrocystis integrifolia, Oryza sativa, Porphyra abbottae, Rubus spectabilis, Trifolium wormskioldii, Tsuga heterophylla
Vegetable: Brassica oleracea, B. rapa, Conioselinum gmelinii, Daucus carota, Dryopteris sp., Heracleum maximum, Rumex aquaticus, Solanum tuberosum
Winter Use Food: Malus fusca, Shepherdia canadensis, Viburnum edule
Fiber
Basketry: Chamaecyparis nootkatensis
Building Material: Chamaecyparis nootkatensis, Picea sitchensis
Canoe Material: Chamaecyparis nootkatensis
Clothing: Chamaecyparis nootkatensis
Cordage: Chamaecyparis nootkatensis
Mats, Rugs & Bedding: Chamaecyparis nootkatensis
Other
Containers: Chamaecyparis nootkatensis
Fuel: Alnus rubra, Pseudotsuga menziesii
Good Luck Charm: Juniperus communis
Hunting & Fishing Item: Chamaecyparis nootkatensis, Epilobium angustifolium, Nereocystis luetkeana, Oplopanax horridus, Taxus brevifolia, Tsuga heterophylla
Paint: Xanthoria elegans
Preservative: Alnus rubra
Protection: Picea sitchensis, Veratrum viride
Tools: Chamaecyparis nootkatensis
Toys & Games: Lysichiton americanus

Kitkatla
Other
Cash Crop: Macrocystis integrifolia

Klallam
Drug
Analgesic: Taxus brevifolia, Urtica dioica, Viola adunca
Antidiarrheal: Alnus rubra, Sambucus cerulea
Antihemorrhagic: Tsuga heterophylla
Burn Dressing: Gaultheria shallon
Cold Remedy: Achillea millefolium, Rubus sp., Symphoricarpos albus
Cough Medicine: Polypodium virginianum
Dermatological Aid: Achillea millefolium, Aruncus dioicus, Dryopteris expansa, Galium triflorum, Lonicera ciliosa, Lysichiton americanus
Dietary Aid: Tsuga heterophylla
Eye Medicine: Malus fusca, Populus balsamifera, Ribes divaricatum
Gynecological Aid: Achillea millefolium, Geum macrophyllum
Laxative: Frangula purshiana
Love Medicine: Conium maculatum
Orthopedic Aid: Urtica dioica
Panacea: Urtica dioica
Poison: Conium maculatum, Disporum hookeri
Throat Aid: Salix sp.
Tonic: Salix sitchensis
Tuberculosis Remedy: Acer macrophyllum, Lysichiton americanus, Salix sp., Thuja plicata, Tsuga heterophylla
Food
Bread & Cake: Gaultheria shallon
Candy: Pseudotsuga menziesii
Dried Food: Rubus leucodermis, R. parviflorus, Vaccinium ovalifolium
Fruit: Fragaria ×ananassa, Mahonia aquifolium, M. nervosa, Ribes lobbii, R. sanguineum, Rubus leucodermis, Sambucus cerulea, S. racemosa, Vaccinium ovalifolium, V. oxycoccos, V. parvifolium
Unspecified: Abronia latifolia, Allium cernuum, Equisetum telmateia, Lilium columbianum, Polystichum munitum, Pteridium aquilinum, Rubus leucodermis, R. parviflorus, Rumex aquaticus
Winter Use Food: Sambucus cerulea, S. racemosa
Fiber
Basketry: Typha latifolia, Xerophyllum tenax
Canoe Material: Acer macrophyllum
Clothing: Typha latifolia
Cordage: Salix sitchensis
Mats, Rugs & Bedding: Scirpus acutus, Typha latifolia
Dye
Red-Brown: Tsuga heterophylla
Unspecified: Mahonia nervosa
Other
Cooking Tools: Holodiscus discolor
Fuel: Pseudotsuga menziesii
Hunting & Fishing Item: Holodiscus discolor, Oplopanax horridus, Pseudotsuga menziesii, Taxus brevifolia, Tsuga heterophylla
Smoke Plant: Taxus brevifolia
Toys & Games: Cornus nuttallii, Polystichum munitum

Klamath
Drug
Antidiarrheal: Artemisia tridentata
Antirheumatic (External): Artemisia tridentata, Wyethia mollis
Burn Dressing: Eriogonum umbellatum
Cough Medicine: Purshia tridentata
Dermatological Aid: Chrysothamnus nauseosus, Ericameria bloomeri, Wyethia mollis
Emetic: Frangula purshiana, Iris missouriensis, Purshia tridentata, Zigadenus venenosus
Eye Medicine: Artemisia tridentata, Pinus contorta
Herbal Steam: Calocedrus decurrens
Orthopedic Aid: Artemisia tridentata
Poison: Cicuta maculata
Pulmonary Aid: Purshia tridentata
Respiratory Aid: Purshia tridentata
Unspecified: Heracleum maximum
Food
Beverage: Carex sp., Mentha canadensis, Triglochin maritimum
Bread & Cake: Nuphar lutea
Dried Food: Camassia quamash, Lomatium canbyi, Nuphar lutea, Perideridia oregana, Prunus subcordata, P. virginiana, Ribes hirtellum, Rubus leucodermis, Vaccinium scoparium
Fruit: Amelanchier alnifolia, Arctostaphylos patula, Fragaria virginiana, Lonicera conjugialis, Prunus emarginata, P. subcordata, Ribes aureum, R. cereum, R. hirtellum, Rosa woodsii, Rubus leucodermis, Sambucus cerulea, Vaccinium scoparium
Pie & Pudding: Camassia leichtlinii
Porridge: Lomatium canbyi, Nuphar lutea, Polygonum douglasii
Preservative: Achillea millefolium
Sauce & Relish: Pinus ponderosa, Sium suave
Snack Food: Ipomopsis aggregata
Special Food: Nuphar lutea
Staple: Nuphar lutea
Starvation Food: Pinus ponderosa
Unspecified: Agrostis perennans, Amaranthus blitoides, Amelanchier alnifolia, Balsamorhiza deltoidea, B. sagittata, Beckmannia syzigachne, Calochortus macrocarpus, Camassia quamash, Carex sp., Castanopsis chrysophylla, Chenopodium fremontii, Corylus cornuta, Descurainia incana, Glyceria fluitans, Heracleum maximum, Leymus condensatus, Madia glomerata, Ment-

zelia albicaulis, Nuphar lutea, Nymphaea sp., Perideridia gairdneri, Phragmites australis, Pinus lambertiana, P. ponderosa, Rubus vitifolius, Rumex paucifolius, R. salicifolius, Sagittaria cuneata, S. latifolia, Scirpus acutus, Sparganium eurycarpum, Triglochin maritimum, Typha latifolia, Valeriana edulis
Winter Use Food: *Amelanchier alnifolia, Vaccinium membranaceum*
Fiber
Basketry: *Calocedrus decurrens, Juncus balticus, Linum lewisii, Phragmites australis, Scirpus acutus*
Canoe Material: *Pinus contorta, P. ponderosa, Pseudotsuga menziesii*
Clothing: *Populus balsamifera, P. tremuloides, Scirpus acutus*
Cordage: *Linum lewisii, Urtica dioica*
Mats, Rugs & Bedding: *Carex* sp., *Juncus balticus, Linum lewisii, Scirpus acutus, Typha latifolia*
Scouring Material: *Equisetum hyemale*
Snow Gear: *Linum lewisii, Salix* sp., *Urtica dioica*
Dye
Brown: *Abies concolor*
Orange: *Alnus incana*
Purple: *Purshia tridentata*
Other
Containers: *Pinus contorta*
Fuel: *Artemisia tridentata, Calocedrus decurrens, Pinus ponderosa*
Hide Preparation: *Abies concolor*
Hunting & Fishing Item: *Cercocarpus ledifolius, Juniperus occidentalis, Linum lewisii, Philadelphus lewisii, Phragmites australis, Rosa woodsii, Taxus brevifolia, Urtica dioica*
Smoke Plant: *Arctostaphylos nevadensis, A. patula, Nicotiana attenuata*
Smoking Tools: *Rosa woodsii*
Tools: *Artemisia tridentata, Cercocarpus ledifolius*
Weapon: *Prunus emarginata*

Koasati
Drug
Analgesic: *Bignonia capreolata, Pteridium aquilinum, Pycnanthemum incanum, Salix nigra, Smilax rotundifolia*
Anthelmintic: *Chenopodium ambrosioides*
Antidiarrheal: *Parthenium hysterophorus*
Antirheumatic (Internal): *Baptisia alba, Bignonia capreolata, Cephalanthus occidentalis, Hypericum hypericoides, Liatris acidota*
Cough Medicine: *Berchemia scandens*
Dermatological Aid: *Acer rubrum, Chionanthus virginicus, Helenium amarum, Ilex opaca, Magnolia grandiflora, Nyssa sylvatica, Sassafras albidum, Ulmus americana*
Emetic: *Eryngium aquaticum, Eupatorium perfoliatum*
Eye Medicine: *Aralia spinosa*
Febrifuge: *Gnaphalium obtusifolium, Monarda fistulosa, Salix nigra*
Gastrointestinal Aid: *Callicarpa americana, Castanea pumila, Myrica cerifera, Prunus* sp., *Salix nigra, Ulmus americana*
Gynecological Aid: *Gossypium herbaceum, Xanthium strumarium*
Heart Medicine: *Sassafras albidum*
Hemostat: *Pycnanthemum incanum*
Herbal Steam: *Helenium amarum*
Kidney Aid: *Helenium amarum*
Orthopedic Aid: *Cephalanthus occidentalis, Prunus persica, Rhus copallinum*
Other: *Liquidambar styraciflua*
Pediatric Aid: *Gnaphalium obtusifolium, Myrica cerifera, Rhus copallinum*
Snakebite Remedy: *Tephrosia florida*
Stimulant: *Pycnanthemum incanum*
Throat Aid: *Aesculus* sp.
Urinary Aid: *Eupatorium perfoliatum*

Konomeho
Fiber
Basketry: *Xerophyllum* sp.

Koyukon
Drug
Ceremonial Medicine: *Picea glauca*
Dermatological Aid: *Picea glauca, P. mariana*
Hunting Medicine: *Picea glauca, P. mariana*
Kidney Aid: *Picea glauca, P. mariana*
Panacea: *Picea glauca, P. mariana*
Unspecified: *Betula papyrifera, Picea glauca, P. mariana*
Food
Beverage: *Empetrum nigrum*
Candy: *Picea glauca*
Frozen Food: *Vaccinium uliginosum, V. vitis-idaea, Viburnum edule*
Fruit: *Rosa acicularis, Rubus arcticus, R. chamaemorus, R. idaeus*
Unspecified: *Allium schoenoprasum, Polygonum alpinum, Rumex arcticus*
Winter Use Food: *Arctostaphylos alpina, A. uva-ursi*
Fiber
Basketry: *Betula papyrifera, Viburnum edule*
Building Material: *Picea glauca, P. mariana*
Canoe Material: *Betula papyrifera, Picea glauca, P. mariana*
Caulking Material: *Picea glauca*
Mats, Rugs & Bedding: *Equisetum* sp.
Snow Gear: *Betula papyrifera, Picea glauca, P. mariana*
Other
Ceremonial Items: *Picea glauca*
Containers: *Betula papyrifera*
Cooking Tools: *Picea glauca, P. mariana*
Fasteners: *Picea glauca, P. mariana*
Fuel: *Betula papyrifera, Equisetum* sp., *Picea glauca, P. mariana*
Good Luck Charm: *Picea glauca*
Hide Preparation: *Picea glauca, P. mariana*
Hunting & Fishing Item: *Betula papyrifera*
Protection: *Picea glauca*
Tools: *Picea mariana*

Kuper Island Indian
Drug
Hemostat: *Sedum spathulifolium*

Kutenai
Drug
Abortifacient: *Alnus incana, Symphoricarpos albus*
Antidiarrheal: *Prunus virginiana, Rosa* sp.
Antirheumatic (External): *Heuchera cylindrica, Mentha arvensis*
Blood Medicine: *Mahonia repens*
Cathartic: *Frangula purshiana*
Cold Remedy: *Juniperus scopulorum, Mentha arvensis*
Cough Medicine: *Mentha arvensis*
Dermatological Aid: *Abies lasiocarpa, Achillea millefolium, Artemisia ludoviciana, Balsamorhiza sagittata, Cicuta douglasii, Larix occidentalis, Nuphar lutea, Populus balsamifera, P. deltoides*
Disinfectant: *Achillea millefolium*
Emetic: *Cicuta douglasii*
Eye Medicine: *Chimaphila umbellata, Shepherdia canadensis*
Febrifuge: *Juniperus scopulorum, Mentha arvensis*
Gastrointestinal Aid: *Rosa* sp.
Kidney Aid: *Chimaphila umbellata, Mahonia repens, Mentha arvensis, Monarda fistulosa*

Misc. Disease Remedy: *Juniperus scopulorum*
Nose Medicine: *Veratrum viride*
Pulmonary Aid: *Juniperus scopulorum*
Respiratory Aid: *Populus balsamifera, P. deltoides*
Throat Aid: *Rhus glabra*
Tonic: *Mentha arvensis*
Tuberculosis Remedy: *Heuchera cylindrica, Larix occidentalis, Pinus contorta, Populus deltoides*
Unspecified: *Ligusticum canbyi*
Veterinary Aid: *Apocynum cannabinum*

Food
Appetizer: *Mahonia repens*
Cooking Agent: *Lewisia rediviva*
Dessert: *Lewisia rediviva, Mahonia repens*
Dried Food: *Lewisia rediviva*
Fruit: *Cornus sericea*
Sauce & Relish: *Allium cernuum, Larix occidentalis*
Staple: *Allium cernuum*
Unspecified: *Balsamorhiza sagittata, Camassia quamash, Cirsium scariosum, Lewisia rediviva, Matricaria discoidea, Pinus contorta, Populus balsamifera, P. deltoides*

Fiber
Basketry: *Thuja plicata*
Canoe Material: *Thuja plicata*
Cordage: *Apocynum cannabinum*
Sewing Material: *Apocynum cannabinum*

Dye
Orange: *Alnus incana*
Red-Brown: *Alnus incana*

Other
Cash Crop: *Camassia quamash*
Ceremonial Items: *Artemisia frigida, Larix occidentalis*
Containers: *Thuja plicata*
Cooking Tools: *Thuja plicata*
Hide Preparation: *Larix occidentalis*
Incense & Fragrance: *Achillea millefolium, Matricaria discoidea, Monarda fistulosa*
Insecticide: *Artemisia frigida, Mentha arvensis*
Jewelry: *Matricaria discoidea*
Tools: *Cornus sericea*

Kwakiutl
Drug
Abortifacient: *Veratrum viride*
Analgesic: *Alnus rubra, Angelica lucida, Argentina anserina, Galium triflorum, Lomatium nudicaule, Lonicera involucrata, Menziesia ferruginea, Moneses uniflora, Nuphar lutea, Oemleria cerasiformis, Oplopanax horridus, Picea sitchensis, Symphoricarpos albus, Urtica dioica, Veratrum viride*
Antidiarrheal: *Blechnum spicant, Cicuta douglasii, Juniperus communis, Picea sitchensis, Polypodium glycyrrhiza, Pseudotsuga menziesii, Ribes lobbii, R.* sp.*, Rubus ursinus, Tsuga heterophylla*
Antiemetic: *Menyanthes trifoliata, Polypodium glycyrrhiza, Rubus parviflorus, R. ursinus*
Antihemorrhagic: *Alnus rubra, Aralia nudicaulis, Kalmia polifolia, Malus fusca, Menyanthes trifoliata, Polypodium glycyrrhiza, Rubus parviflorus, R. ursinus*
Antirheumatic (External): *Achillea millefolium, Chamaecyparis nootkatensis, Conioselinum gmelinii, Lomatium nudicaule, Lonicera involucrata*
Blood Medicine: *Juniperus communis*
Burn Dressing: *Rubus spectabilis, Tsuga heterophylla*
Cancer Treatment: *Epilobium angustifolium, Prunus emarginata*
Cathartic: *Cicuta douglasii, Physocarpus capitatus*
Ceremonial Medicine: *Abies grandis, Tsuga mertensiana*
Cold Remedy: *Achillea millefolium, Lomatium nudicaule, Picea sitchensis, Veratrum viride*
Cough Medicine: *Abies grandis, Aralia nudicaulis, Aruncus dioicus, Lomatium nudicaule, Picea sitchensis, Pinus contorta, P. monticola*
Dermatological Aid: *Abies grandis, Acer macrophyllum, Achillea millefolium, Allium cernuum, Alnus rubra, Anaphalis margaritacea, Argentina anserina, A. egedii, Chamaecyparis nootkatensis, Cicuta douglasii, Crataegus douglasii, Drosera rotundifolia, Epilobium angustifolium, Equisetum arvense, E. telmateia, Heracleum maximum, Kalmia polifolia, Lomatium nudicaule, Lonicera involucrata, Lysichiton americanus, Malus fusca, Menziesia ferruginea, Moneses uniflora, Nicotiana* sp.*, Oemleria cerasiformis, Oplopanax horridus, Picea sitchensis, Pinus monticola, Plantago major, Populus balsamifera, Prunus emarginata, Pseudotsuga menziesii, Ribes lobbii, R.* sp.*, Rubus parviflorus, R. spectabilis, Rumex aquaticus, Scirpus americanus, Thuja plicata, Tsuga heterophylla, T. mertensiana, Urtica dioica, Veratrum viride*
Dietary Aid: *Menyanthes trifoliata, Prunus emarginata*
Disinfectant: *Picea sitchensis*
Emetic: *Cicuta douglasii, Oenanthe sarmentosa, Osmorhiza berteroi, Physocarpus capitatus, Sambucus racemosa, Veratrum viride*
Eye Medicine: *Argentina egedii, Symphoricarpos albus, Thuja plicata, Tsuga heterophylla*
Gastrointestinal Aid: *Frangula purshiana, Lomatium nudicaule, Menyanthes trifoliata, Menziesia ferruginea, Oplopanax horridus, Pinus contorta, P. monticola, Rumex aquaticus*
Gynecological Aid: *Achillea millefolium, Heracleum maximum, Lomatium nudicaule, Lonicera involucrata, Polystichum munitum, Prunus emarginata, Rubus parviflorus, Sambucus racemosa, Tsuga heterophylla, Urtica dioica*
Heart Medicine: *Menziesia ferruginea, Prunus emarginata*
Hemostat: *Prunus emarginata, Thuja plicata*
Herbal Steam: *Achillea millefolium, Angelica lucida, Chamaecyparis nootkatensis, Conioselinum gmelinii, Lomatium nudicaule, Lonicera involucrata, Lysichiton americanus, Nicotiana* sp.*, Oplopanax horridus, Sambucus racemosa*
Hunting Medicine: *Angelica lucida, Lomatium nudicaule, Tsuga heterophylla, Veratrum viride*
Internal Medicine: *Anaphalis margaritacea, Rubus parviflorus*
Kidney Aid: *Chamaecyparis nootkatensis, Picea sitchensis*
Laxative: *Abies grandis, Frangula purshiana, Lomatium nudicaule, Oplopanax horridus, Physocarpus capitatus, Veratrum viride*
Love Medicine: *Aruncus dioicus, Drosera rotundifolia, Monotropa hypopithys, Platanthera stricta*
Misc. Disease Remedy: *Cicuta douglasii, Menyanthes trifoliata*
Narcotic: *Arctostaphylos uva-ursi, Ledum groenlandicum*
Oral Aid: *Abies grandis, Cirsium remotifolium, C.* sp.*, Prunus emarginata, Ribes lobbii, R.* sp.
Orthopedic Aid: *Lomatium nudicaule, Lonicera involucrata, Nuphar lutea, Sambucus racemosa, Thuja plicata, Veratrum viride*
Other: *Lysichiton americanus, Picea sitchensis, Urtica dioica*
Panacea: *Achillea millefolium, Chamaecyparis nootkatensis, Lomatium nudicaule, Veratrum viride*
Pediatric Aid: *Cirsium remotifolium, Heracleum maximum, Lupinus littoralis, Lysichiton americanus, Phyllospadix torreyi, Prunus emarginata, Ribes* sp.*, Rubus spectabilis, Scirpus americanus, Tsuga mertensiana, Veratrum viride*
Poison: *Cicuta douglasii, Oplopanax horridus, Osmorhiza berteroi, Veratrum viride*
Preventive Medicine: *Prunus emarginata*
Reproductive Aid: *Pinus monticola*
Respiratory Aid: *Alnus rubra, Juniperus communis, Nuphar lutea*

Sedative: *Glaux maritima, Lupinus littoralis*
Stimulant: *Angelica lucida, Conioselinum gmelinii, Lysichiton americanus*
Strengthener: *Chamaecyparis nootkatensis, Phyllospadix torreyi*
Throat Aid: *Lomatium nudicaule*
Tonic: *Abies grandis*
Tuberculosis Remedy: *Abies grandis, Alnus rubra, Oplopanax horridus, Prunus emarginata*
Unspecified: *Angelica lucida, Pseudotsuga menziesii*
Venereal Aid: *Physocarpus capitatus, Urtica dioica*
Witchcraft Medicine: *Oplopanax horridus*
Food
Dried Food: *Erythronium oregonum*
Unspecified: *Erythronium oregonum, Lupinus polyphyllus*
Fiber
Building Material: *Thuja plicata*
Canoe Material: *Thuja plicata*
Clothing: *Thuja plicata*
Cordage: *Thuja plicata*
Other
Ceremonial Items: *Polystichum munitum, Veratrum viride*
Cooking Tools: *Thuja plicata*
Protection: *Oplopanax horridus, Veratrum viride*
Tools: *Thuja plicata*

Kwakiutl, Southern
Drug
Analgesic: *Alnus rubra, Fucus gardneri, Lonicera involucrata, Nicotiana* sp., *Urtica dioica*
Antirheumatic (External): *Chamaecyparis nootkatensis, Fucus gardneri, Urtica dioica*
Burn Dressing: *Nereocystis luetkeana*
Dermatological Aid: *Nereocystis luetkeana*
Orthopedic Aid: *Nereocystis luetkeana*
Pediatric Aid: *Nereocystis luetkeana*
Strengthener: *Chamaecyparis nootkatensis, Fucus gardneri*
Venereal Aid: *Fucus gardneri*
Food
Beverage: *Ledum groenlandicum*
Bread & Cake: *Porphyra* sp., *Sambucus racemosa*
Candy: *Picea sitchensis, Thuja plicata*
Dessert: *Porphyra* sp.
Dietary Aid: *Polypodium glycyrrhiza*
Dried Food: *Argentina egedii, Erythronium revolutum, Fritillaria camschatcensis, Gaultheria shallon, Porphyra* sp., *Rubus parviflorus, R. spectabilis*
Fruit: *Amelanchier alnifolia, Arctostaphylos uva-ursi, Cornus canadensis, Crataegus douglasii, Mahonia aquifolium, M. nervosa, Maianthemum dilatatum, Oemleria cerasiformis, Ribes bracteosum, R. laxiflorum, R. lobbii, Rubus parviflorus, R. spectabilis, Shepherdia canadensis, Vaccinium ovalifolium, V. uliginosum, Vicia nigricans*
Preserves: *Viburnum edule*
Snack Food: *Porphyra* sp.
Special Food: *Argentina egedii, Gaultheria shallon, Malus fusca, Oemleria cerasiformis, Ribes lobbii, Vaccinium parvifolium, Viburnum edule, Zostera marina*
Starvation Food: *Polypodium glycyrrhiza*
Unspecified: *Allium cernuum, Aralia nudicaulis, Camassia leichtlinii, C. quamash, Castilleja miniata, Dryopteris campyloptera, Erythronium revolutum, Glaux maritima, Heracleum maximum, Lupinus littoralis, Polypodium glycyrrhiza, Polystichum munitum, Populus balsamifera, Porphyra* sp., *Pteridium aquilinum, Rubus spectabilis, Zostera marina*

Vegetable: *Trifolium wormskioldii*
Winter Use Food: *Ribes bracteosum, Viburnum edule*
Fiber
Basketry: *Leymus mollis, Picea sitchensis, Scirpus americanus, Thuja plicata*
Brushes & Brooms: *Taxus brevifolia, Thuja plicata*
Building Material: *Thuja plicata*
Canoe Material: *Chamaecyparis nootkatensis, Thuja plicata*
Clothing: *Chamaecyparis nootkatensis, Leymus mollis, Picea sitchensis, Scirpus americanus, Thuja plicata*
Cordage: *Picea sitchensis, Thuja plicata, Urtica dioica*
Furniture: *Chamaecyparis nootkatensis*
Mats, Rugs & Bedding: *Chamaecyparis nootkatensis, Picea sitchensis, Thuja plicata, Typha latifolia*
Other: *Thuja plicata*
Scouring Material: *Equisetum arvense, E. telmateia*
Sewing Material: *Picea sitchensis, Thuja plicata*
Dye
Black: *Alnus rubra*
Brown: *Alnus rubra*
Orange: *Alnus rubra*
Red: *Alnus rubra*
Other
Cash Crop: *Malus fusca*
Ceremonial Items: *Acer macrophyllum, Alnus rubra, Nereocystis luetkeana, Tsuga heterophylla*
Containers: *Adiantum pedatum, Athyrium filix-femina, Lysichiton americanus, Nereocystis luetkeana, Polystichum munitum, Thuja plicata*
Cooking Tools: *Acer macrophyllum, Adiantum pedatum, Alnus rubra, Chamaecyparis nootkatensis, Leymus mollis, Lysichiton americanus, Nereocystis luetkeana, Pinus contorta, Polystichum munitum, Rubus parviflorus, R. spectabilis, Taxus brevifolia, Thuja plicata, Tsuga heterophylla*
Fasteners: *Picea sitchensis, Vaccinium parvifolium*
Fuel: *Alnus rubra, Dryopteris campyloptera, Thuja plicata*
Hunting & Fishing Item: *Chamaecyparis nootkatensis, Holodiscus discolor, Macrocystis integrifolia, Nereocystis luetkeana, Picea sitchensis, Salix* sp., *Thuja plicata, Urtica dioica*
Musical Instrument: *Acer macrophyllum, Alnus rubra, Tsuga heterophylla*
Sacred Items: *Picea sitchensis*
Tools: *Holodiscus discolor, Pinus contorta, Symphoricarpos albus, Taxus brevifolia*
Toys & Games: *Sambucus racemosa*

Kwakwaka'wakw
Drug
Antiemetic: *Kalmia microphylla*
Antihemorrhagic: *Kalmia microphylla*
Ceremonial Medicine: *Tsuga heterophylla*
Dermatological Aid: *Moneses uniflora*
Emetic: *Oenanthe sarmentosa*
Unspecified: *Pinus contorta*
Food
Unspecified: *Conioselinum gmelinii*
Fiber
Building Material: *Tsuga heterophylla*
Other
Fuel: *Pinus contorta*
Hunting & Fishing Item: *Pinus contorta*
Protection: *Tsuga heterophylla*
Tools: *Pinus contorta*

Laguna
Food
Appetizer: *Rhus trilobata*
Beverage: *Yucca baccata*
Bread & Cake: *Yucca baccata*
Candy: *Asclepias speciosa, Populus deltoides*
Dried Food: *Amaranthus blitoides, A. hybridus, A. retroflexus, Opuntia imbricata, Prunus virginiana, Yucca baccata, Y. glauca*
Fruit: *Atriplex argentea, Ceanothus fendleri, Celtis laevigata, Juniperus monosperma, Opuntia engelmannii, Prunus virginiana, Ribes cereum, Yucca baccata, Y. glauca*
Porridge: *Cleome serrulata, Opuntia engelmannii, Prosopis glandulosa*
Preserves: *Ribes cereum, Yucca baccata*
Sauce & Relish: *Lycium pallidum, Yucca baccata*
Special Food: *Sophora nuttalliana*
Spice: *Agastache pallidiflora, Berlandiera lyrata, Juniperus monosperma, Monarda fistulosa, M. pectinata, Pectis angustifolia, Rhus trilobata*
Staple: *Amaranthus blitoides, Dalea candida, Quercus gambelii*
Starvation Food: *Juniperus monosperma, Opuntia clavata, Phoradendron juniperinum, Solanum triflorum, Yucca baccata*
Unspecified: *Abronia fragrans, Allium cernuum, Astragalus lentiginosus, Cyperus squarrosus, Lathyrus polymorphus, Opuntia imbricata, Plantago major, Prosopis glandulosa, Quercus gambelii, Ranunculus inamoenus, Tradescantia occidentalis, Typha latifolia, Vicia americana*
Vegetable: *Amaranthus blitoides, A. hybridus, A. retroflexus, Cymopterus bulbosus, Lactuca sativa, Portulaca oleracea*

Lakota
Drug
Abortifacient: *Opuntia sp., Yucca glauca*
Analgesic: *Astragalus canadensis, Clematis ligusticifolia, Dyssodia papposa, Psoralidium tenuiflorum, Ratibida columnifera*
Antidiarrheal: *Artemisia ludoviciana, Asclepias pumila, A. viridiflora, Heuchera richardsonii, Lygodesmia juncea, Rumex altissimus*
Antihemorrhagic: *Astragalus canadensis, Dyssodia papposa, Glycyrrhiza lepidota, Grindelia squarrosa, Rumex altissimus*
Antirheumatic (External): *Ambrosia artemisiifolia, Amphicarpaea bracteata, Antennaria parvifolia, Echinacea angustifolia, Euphorbia marginata, Onosmodium molle*
Cancer Treatment: *Fritillaria atropurpurea*
Cathartic: *Dalea villosa*
Ceremonial Medicine: *Xanthium strumarium*
Cold Remedy: *Artemisia ludoviciana, Gutierrezia sarothrae, Juniperus virginiana*
Cough Medicine: *Acorus calamus, Astragalus canadensis, Gutierrezia sarothrae, Monarda fistulosa, Pycnanthemum virginianum*
Dermatological Aid: *Achillea millefolium, Echinacea angustifolia, Heuchera richardsonii, Sphaeralcea coccinea, Thalictrum dasycarpum, Yucca glauca*
Dietary Aid: *Asclepias stenophylla, Physalis heterophylla*
Diuretic: *Artemisia campestris, Eriogonum annuum, Mirabilis linearis, Opuntia sp.*
Eye Medicine: *Monarda fistulosa*
Gastrointestinal Aid: *Artemisia campestris, Croton texensis, Echinacea angustifolia, Ipomoea leptophylla, Ratibida columnifera, Rumex altissimus, Sium suave, Solanum triflorum, Urtica dioica, Yucca glauca*
Gynecological Aid: *Artemisia campestris, Asclepias verticillata, A. viridiflora, Astragalus gracilis, Euphorbia marginata, Opuntia sp., Rumex venosus, Yucca glauca*
Hemostat: *Monarda fistulosa*
Hypotensive: *Acorus calamus*
Kidney Aid: *Lepidium densiflorum*
Misc. Disease Remedy: *Acorus calamus, Glycyrrhiza lepidota*
Oral Aid: *Echinacea angustifolia, Eriogonum annuum*
Orthopedic Aid: *Acorus calamus*
Other: *Echinacea angustifolia*
Pediatric Aid: *Asclepias pumila, A. viridiflora, Eriogonum annuum, Lygodesmia juncea*
Poison: *Astragalus racemosus, Cicuta maculata, Conium maculatum, Delphinium carolinianum, Dichanthelium oligosanthes, Oxytropis lambertii, Toxicodendron rydbergii, Zigadenus venenosus*
Preventive Medicine: *Chamaesyce geyeri*
Pulmonary Aid: *Astragalus canadensis, Lithospermum caroliniense, Monarda fistulosa*
Reproductive Aid: *Dyssodia papposa*
Sedative: *Artemisia campestris*
Snakebite Remedy: *Opuntia humifusa*
Stimulant: *Monarda fistulosa*
Throat Aid: *Acorus calamus, Artemisia ludoviciana, Dalea villosa, Echinacea angustifolia*
Toothache Remedy: *Acorus calamus, Echinacea angustifolia, Glycyrrhiza lepidota*
Unspecified: *Ambrosia trifida, Artemisia cana, A. tridentata, Asclepias speciosa, Glycyrrhiza lepidota, Pediomelum argophyllum, P. cuspidatum, Sagittaria latifolia*
Vertigo Medicine: *Gutierrezia sarothrae*
Veterinary Aid: *Astragalus crassicarpus, Hymenopappus tenuifolius, Onosmodium molle, Pediomelum argophyllum, Ratibida columnifera, Thalictrum dasycarpum, Yucca glauca*

Food
Beverage: *Agastache foeniculum, Amelanchier alnifolia, Ceanothus herbaceus, Coreopsis tinctoria, Juniperus virginiana, Mentha arvensis, Opuntia humifusa, Prunus virginiana, Rosa arkansana*
Candy: *Dalea purpurea*
Cooking Agent: *Asclepias speciosa, Humulus lupulus*
Dietary Aid: *Liatris punctata*
Dried Food: *Opuntia humifusa, Pediomelum esculentum, Rosa arkansana, Shepherdia argentea*
Fodder: *Astragalus canadensis, Equisetum hyemale, Populus ×acuminata*
Forage: *Artemisia cana, Clematis ligusticifolia, Oxytropis lambertii, Pascopyrum smithii, Populus deltoides*
Fruit: *Amelanchier alnifolia, Astragalus crassicarpus, Crataegus chrysocarpa, Fragaria vesca, Morus sp., Opuntia humifusa, Prunus americana, P. pumila, P. virginiana, Ribes americanum, R. aureum, R. missouriense, Rosa woodsii, Rubus occidentalis, Shepherdia argentea, Vitis vulpina*
Pie & Pudding: *Prunus virginiana*
Preserves: *Asclepias speciosa, Rosa arkansana*
Soup: *Allium textile, Hedeoma drummondii, Pediomelum esculentum, Quercus macrocarpa, Rosa arkansana*
Special Food: *Monarda fistulosa, Prunus virginiana, Vitis vulpina*
Spice: *Juniperus virginiana*
Staple: *Quercus velutina, Typha latifolia*
Starvation Food: *Amelanchier alnifolia, Helianthus tuberosus, Prunus americana, Ribes americanum, R. aureum, R. missouriense, Rosa arkansana*
Unspecified: *Acorus calamus, Allium drummondii, A. textile, Asclepias speciosa, Carya ovata, Cypripedium sp., Helianthus tuberosus, Juglans nigra, Lomatium orientale, Lygodesmia juncea, Nuphar lutea, Nymphaea sp., Pediomelum esculentum, Polygonum amphibium, Quercus macrocarpa, Sagittaria latifolia, Scirpus tabernaemontani, Yucca glauca*
Vegetable: *Brassica oleracea, Chenopodium album*

Fiber
Basketry: *Pediomelum argophyllum*

Building Material: *Populus ×acuminata, Salix exigua, Ulmus americana*
Clothing: *Schizachyrium scoparium, Typha latifolia*
Cordage: *Tilia americana, Urtica dioica*
Furniture: *Typha latifolia*
Mats, Rugs & Bedding: *Scirpus tabernaemontani, Typha latifolia*
Dye
White: *Eriogonum annuum*
Yellow: *Argemone polyanthemos*
Other
Ceremonial Items: *Artemisia ludoviciana, Hierochloe odorata, Prunus virginiana*
Containers: *Solidago rigida*
Cooking Tools: *Ratibida columnifera*
Decorations: *Calamovilfa longifolia*
Fasteners: *Salix exigua*
Fuel: *Ipomoea leptophylla, Populus ×acuminata, Ulmus americana*
Hunting & Fishing Item: *Amelanchier alnifolia, Amorpha fruticosa, Fraxinus pennsylvanica, Ostrya virginiana, Prunus virginiana, Ribes aureum*
Incense & Fragrance: *Artemisia campestris, A. ludoviciana, Hierochloe odorata, Juniperus virginiana, Oenothera biennis*
Insecticide: *Psoralidium tenuiflorum*
Paint: *Celastrus scandens, Ostrya virginiana, Penstemon angustifolius, Prunus pumila, Tradescantia bracteata*
Paper: *Ambrosia artemisiifolia*
Protection: *Artemisia ludoviciana, Penstemon gracilis, Sphaeralcea coccinea*
Smoke Plant: *Arctostaphylos uva-ursi, Cornus sericea, Rhus aromatica, R. glabra*
Smoking Tools: *Calamovilfa longifolia, Fraxinus pennsylvanica*
Soap: *Yucca glauca*
Stable Gear: *Gaura coccinea*
Toys & Games: *Amelanchier alnifolia, Bouteloua gracilis, Sium suave, Stipa spartea, Symphoricarpos occidentalis*
Walking Sticks: *Salix eriocephala*

Lillooet
Food
Ice Cream: *Shepherdia canadensis*

Lolahnkok
Drug
Emetic: *Asarum* sp.
Love Medicine: *Angelica* sp.

Luiseño
Drug
Abortifacient: *Croton californicus*
Analgesic: *Woodwardia radicans*
Cathartic: *Marah macrocarpus, Mirabilis californica, Sisyrinchium bellum*
Ceremonial Medicine: *Datura wrightii*
Cold Remedy: *Satureja douglasii*
Dermatological Aid: *Baccharis douglasii, Hoita macrostachya, Quercus dumosa*
Emetic: *Ambrosia artemisiifolia, Cordylanthus* sp., *Sphaeralcea* sp.
Eye Medicine: *Solanum douglasii*
Febrifuge: *Centaurium venustum, Satureja douglasii*
Gynecological Aid: *Sambucus cerulea*
Hallucinogen: *Datura wrightii*
Narcotic: *Datura wrightii*
Snakebite Remedy: *Chamaesyce polycarpa*
Toothache Remedy: *Ribes indecorum, R. malvaceum*
Unspecified: *Artemisia dracunculus, A. furcata, Cneoridium dumosum, Datura wrightii, Ericameria parishii, Eriodictyon crassifolium, E. tomentosum, Monardella lanceolata, Pellaea mucronata, Rumex* sp., *Satureja douglasii, Tauschia arguta, Turricula parryi*
Food
Beverage: *Monardella lanceolata, Pellaea mucronata, Satureja douglasii*
Candy: *Asclepias eriocarpa, Eschscholzia californica*
Dried Food: *Heteromeles arbutifolia, Opuntia* sp., *Sambucus cerulea*
Fruit: *Arctostaphylos parryana, Carpobrotus aequilateralus, Opuntia* sp., *Prunus ilicifolia, P. virginiana, Rubus parviflorus, R. vitifolius, Sambucus cerulea, Vitis girdiana*
Porridge: *Prunus ilicifolia, Quercus agrifolia, Q. chrysolepis, Q. dumosa, Q. engelmannii, Q. kelloggii, Q. wislizeni, Rhus trilobata*
Staple: *Avena fatua, Opuntia* sp., *Prosopis glandulosa, Prunus ilicifolia, Quercus agrifolia, Q. chrysolepis, Q. dumosa, Q. engelmannii, Q. kelloggii, Q. wislizeni*
Substitution Food: *Quercus chrysolepis, Q. dumosa, Q. engelmannii, Q. wislizeni*
Unspecified: *Artemisia dracunculus, Bloomeria crocea, Boschniakia hookeri, Bromus diandrus, Calandrinia ciliata, Carduus* sp., *Chenopodium californicum, Chlorogalum parviflorum, Cucurbita foetidissima, Dichelostemma pulchellum, Ericameria parishii, Gilia capitata, Helianthus annuus, Layia glandulosa, Lepidium nitidum, Lupinus* sp., *Malacothrix californica, Prunus ilicifolia, Salvia apiana, S. carduacea, S. columbariae, S. mellifera, Sarcostemma cynanchoides, Scirpus* sp., *Trifolium ciliolatum, T. gracilentum, T. microcephalum, T. obtusiflorum, T. willldenowii, Yucca schidigera, Y. whipplei*
Vegetable: *Apium graveolens, Brassica nigra, Calandrinia ciliata, Carduus* sp., *Chenopodium album, Claytonia perfoliata, Eschscholzia californica, Hoita orbicularis, Lepidium nitidum, Lotus strigosus, Lupinus* sp., *Phacelia ramosissima, Portulaca oleracea, Rorippa nasturtium-aquaticum, Sidalcea malviflora, Solanum douglasii, Sonchus asper, Viola pedunculata*
Winter Use Food: *Quercus agrifolia, Q. chrysolepis, Q. dumosa, Q. engelmannii, Q. kelloggii, Q. wislizeni*
Fiber
Basketry: *Juncus mertensianus, Muhlenbergia rigens, Rhus trilobata*
Brushes & Brooms: *Chlorogalum pomeridianum*
Building Material: *Pluchea sericea*
Clothing: *Apocynum cannabinum, Asclepias eriocarpa, Muhlenbergia rigens, Urtica dioica*
Cordage: *Apocynum cannabinum, Asclepias eriocarpa*
Unspecified: *Urtica dioica*
Dye
Unspecified: *Rubus vitifolius*
Yellow: *Hoita macrostachya*
Other
Ceremonial Items: *Artemisia californica, Datura wrightii*
Containers: *Apocynum cannabinum, Asclepias eriocarpa, Juncus mertensianus, Muhlenbergia rigens, Urtica dioica*
Cooking Tools: *Juncus mertensianus, Muhlenbergia rigens*
Decorations: *Solanum douglasii*
Fasteners: *Adenostoma fasciculatum*
Hunting & Fishing Item: *Adenostoma fasciculatum, Apocynum cannabinum, Artemisia furcata, Asclepias eriocarpa, Heterotheca grandiflora, Leymus condensatus, Pluchea sericea, Salix* sp., *Sambucus cerulea, Urtica dioica, Yucca schidigera*
Paint: *Marah macrocarpus*
Smoke Plant: *Nicotiana* sp.
Soap: *Chenopodium californicum, Chlorogalum pomeridianum, Cucurbita foetidissima*
Tools: *Baccharis douglasii, Rhus trilobata*

Lummi
Drug
Analgesic: *Achillea millefolium, Ribes lacustre*
Antidiarrheal: *Holodiscus discolor, Spiraea douglasii*

Antiemetic: *Thuja plicata*
Dermatological Aid: *Achlys triphylla, Adiantum aleuticum, Aruncus dioicus*
Diaphoretic: *Achillea millefolium*
Emetic: *Achlys triphylla, Petasites frigidus*
Eye Medicine: *Holodiscus discolor, Trillium ovatum*
Gynecological Aid: *Cirsium* sp., *Oplopanax horridus, Opuntia* sp., *Polystichum munitum, Prunus emarginata, Urtica dioica*
Laxative: *Cornus nuttallii, Frangula purshiana*
Misc. Disease Remedy: *Achillea millefolium, Aruncus dioicus*
Oral Aid: *Holodiscus discolor*
Orthopedic Aid: *Holodiscus discolor*
Pulmonary Aid: *Thuja plicata*
Tonic: *Ribes laxiflorum*
Tuberculosis Remedy: *Lonicera ciliosa, Pinus monticola*
Food
Beverage: *Rosa nutkana*
Dried Food: *Amelanchier alnifolia, Rosa nutkana*
Fruit: *Mahonia* sp., *Oemleria cerasiformis, Ribes laxiflorum, Rubus spectabilis, Vaccinium parvifolium*
Unspecified: *Corylus cornuta, Lilium columbianum, Pteridium aquilinum, Rubus spectabilis*
Winter Use Food: *Sambucus cerulea*
Fiber
Brushes & Brooms: *Philadelphus lewisii*
Furniture: *Acer macrophyllum*
Dye
Red-Brown: *Alnus rubra*
Other
Ceremonial Items: *Thuja plicata*
Cooking Tools: *Acer macrophyllum, Holodiscus discolor, Spiraea douglasii*
Fuel: *Pseudotsuga menziesii*
Hunting & Fishing Item: *Acer circinatum, Holodiscus discolor, Pseudotsuga menziesii, Tsuga heterophylla, Urtica dioica*
Paint: *Oplopanax horridus*
Tools: *Philadelphus lewisii*

Lummi, Washington
Other
Paint: *Oplopanax horridus*

Mahuna
Drug
Abortifacient: *Chenopodium ambrosioides, Iva axillaris*
Analgesic: *Ephedra* sp., *Malva neglecta, Phytolacca americana, Rosa californica, Saponaria officinalis*
Anthelmintic: *Sisyrinchium angustifolium*
Antidiarrheal: *Capsella bursa-pastoris, Diplacus aurantiacus, Hedeoma pulegioides, Rubus procumbens*
Antihemorrhagic: *Dennstaedtia punctilobula*
Antirheumatic (External): *Nicotiana glauca, Urtica dioica*
Antirheumatic (Internal): *Adiantum capillus-veneris, Eriodictyon californicum, Phlox subulata, Phyla lanceolata, Pinus monticola, Tanacetum parthenium, Xanthium strumarium*
Blood Medicine: *Ephedra* sp., *Eriogonum elongatum, Fouquieria splendens, Lotus scoparius, Pellaea atropurpurea, Salix washingtonia*
Burn Dressing: *Solidago nemoralis*
Cathartic: *Eriogonum elatum, Frangula californica*
Cold Remedy: *Spiraea salicifolia*
Contraceptive: *Iva axillaris*
Cough Medicine: *Eriodictyon californicum, Marrubium vulgare, Prunus ilicifolia, P. serotina, Rumex hymenosepalus, Salvia mellifera, Spiraea salicifolia*
Dermatological Aid: *Allium canadense, A. vineale, Ambrosia artemisiifolia, Anaphalis margaritacea, Anemopsis californica, Brodiaea* sp., *Clematis ligusticifolia, Datura wrightii, Euphorbia* sp., *Frasera* sp., *Grindelia humilis, G. squarrosa, Kalmia latifolia, Keckiella cordifolia, Larrea tridentata, Marah macrocarpus, Opuntia* sp., *Phytolacca americana, Piperia* sp., *Plantago major, Saponaria officinalis, Solidago nemoralis*
Dietary Aid: *Allium bisceptrum, Claytonia perfoliata, Lepidium densiflorum, Panicum capillare, Rhus trilobata*
Disinfectant: *Anemopsis californica, Grindelia squarrosa, Larrea tridentata, Solidago nemoralis*
Eye Medicine: *Piperia* sp., *Salvia columbariae*
Febrifuge: *Centaurium muehlenbergii, Juniperus californica, Mirabilis californica, Paeonia brownii, Rosa californica, R. gallica, Verbena hastata*
Gastrointestinal Aid: *Centaurium muehlenbergii, Chamaemelum nobile, Ephedra* sp., *Eriogonum umbellatum, Larrea tridentata, Malva neglecta, Monardella villosa, Platanus occidentalis, Rhus trilobata, Rosa californica, Sisyrinchium angustifolium, Verbena hastata*
Gynecological Aid: *Artemisia californica, Salvia apiana, Tephrosia virginiana*
Hemostat: *Quercus agrifolia, Typha latifolia*
Hypotensive: *Eriogonum elongatum*
Kidney Aid: *Apium* sp., *Centaurea melitensis, Croton pottsii, Ephedra* sp., *Malva* sp., *Pellaea atropurpurea, Xanthium spinosum, X. strumarium*
Laxative: *Centaurium muehlenbergii, Eupatorium purpureum*
Liver Aid: *Rorippa nasturtium-aquaticum*
Misc. Disease Remedy: *Juniperus californica, Sambucus racemosa*
Narcotic: *Datura wrightii*
Oral Aid: *Echinocactus* sp., *Eriogonum* sp., *Persea planifolia, Sempervivum* sp.
Orthopedic Aid: *Ulmus rubra, Xanthium strumarium*
Panacea: *Trichostema lanatum*
Pediatric Aid: *Chamaemelum nobile, Quercus agrifolia, Typha latifolia*
Poison: *Astragalus mollissimus, Datura wrightii, Eschscholzia californica, Kalmia latifolia, Phytolacca americana*
Preventive Medicine: *Pellaea atropurpurea, Toxicodendron diversilobum*
Pulmonary Aid: *Eriodictyon californicum, Paeonia brownii*
Respiratory Aid: *Andromeda polifolia, Eriodictyon californicum, Helenium autumnale, Platanus occidentalis, Salvia mellifera*
Sedative: *Mentha canadensis, M. spicata, Satureja douglasii*
Snakebite Remedy: *Allium canadense, A. vineale, Datura wrightii, Piperia* sp.
Throat Aid: *Marrubium vulgare, Nicotiana glauca, Rumex hymenosepalus*
Toothache Remedy: *Achillea millefolium, Persea planifolia, Quercus rubra*
Tuberculosis Remedy: *Nicotiana glauca, Xanthium strumarium*
Unspecified: *Quercus virginiana*
Urinary Aid: *Equisetum hyemale*
Venereal Aid: *Anagallis* sp., *Cercocarpus montanus, Ephedra* sp., *Xanthium strumarium*
Food
Bread & Cake: *Prosopis glandulosa*
Dessert: *Quercus virginiana*
Dried Food: *Prosopis glandulosa*
Forage: *Kalmia latifolia*
Fruit: *Arctostaphylos* sp., *Heteromeles arbutifolia, Prunus ilicifolia, P. serotina, Rhus integrifolia, R. trilobata, Ribes amarum, R. sanguineum, Rubus procumbens, Sambucus racemosa, Vitis vulpina*
Porridge: *Prosopis glandulosa, Salvia columbariae*
Sauce & Relish: *Yucca whipplei*
Unspecified: *Pteridium aquilinum, Quercus virginiana, Yucca whipplei*
Fiber
Basketry: *Rhus trilobata, Yucca whipplei*
Building Material: *Salvia columbariae*

Mats, Rugs & Bedding: *Quercus virginiana*
Dye
Black: *Quercus virginiana*
Brown: *Quercus virginiana*
Gray: *Quercus virginiana*
Red: *Quercus virginiana*
Unspecified: *Phytolacca americana*
White: *Quercus virginiana, Yucca whipplei*
Yellow: *Quercus virginiana*
Other
Fasteners: *Asclepias syriaca*
Hunting & Fishing Item: *Croton setigerus, Echinocactus* sp.
Protection: *Ceanothus oliganthus, Pinus sabiniana*
Soap: *Chlorogalum pomeridianum, Cucurbita foetidissima, Saponaria officinalis*

Maidu
Fiber
Basketry: *Acer macrophyllum, Amelanchier alnifolia, Calocedrus decurrens, Carex barbarae, Ceanothus integerrimus, Cercis canadensis, Pinus ponderosa, Prunus virginiana, Pseudotsuga menziesii, Quercus kelloggii, Salix* sp., *Scirpus acutus, Xerophyllum tenax*
Brushes & Brooms: *Chlorogalum pomeridianum*
Sewing Material: *Acer macrophyllum*
Other
Decorations: *Adiantum pedatum, Cercis canadensis, Pteridium aquilinum*

Makah
Drug
Abortifacient: *Lysichiton americanus*
Adjuvant: *Frangula purshiana*
Analgesic: *Allium cernuum, Lysichiton americanus, Rubus spectabilis*
Antidiarrheal: *Equisetum hyemale, Malus fusca*
Antidote: *Salix hookeriana*
Antihemorrhagic: *Adiantum aleuticum*
Antirheumatic (External): *Aruncus dioicus, Lysichiton americanus, Mnium punctatum, Oplopanax horridus*
Blood Medicine: *Achillea millefolium, Ledum groenlandicum, Lysichiton americanus, Malus fusca, Picea sitchensis, Prunus emarginata, Rubus parviflorus*
Cathartic: *Achillea millefolium*
Cough Medicine: *Polypodium virginianum, Thuja plicata*
Dermatological Aid: *Adiantum aleuticum, Aruncus dioicus, Galium triflorum, Lonicera involucrata, Malus fusca, Myosotis laxa, Picea sitchensis, Rubus spectabilis, Salix hookeriana, Sambucus racemosa, Thuja plicata, Tsuga heterophylla, Usnea* sp., *Vicia nigricans*
Diaphoretic: *Achillea millefolium*
Emetic: *Lonicera involucrata*
Eye Medicine: *Heracleum maximum, Ribes divaricatum*
Gastrointestinal Aid: *Adiantum aleuticum, A. pedatum, Blechnum spicant, Lysichiton americanus, Malus fusca, Picea sitchensis*
Gynecological Aid: *Achillea millefolium, Athyrium filix-femina, Ledum groenlandicum, Lonicera involucrata, Streptopus amplexifolius, Vaccinium ovalifolium, V. ovatum, V. uliginosum, Viola adunca*
Heart Medicine: *Malus fusca*
Hemostat: *Camellia sinensis, Tsuga heterophylla*
Hunting Medicine: *Urtica dioica*
Internal Medicine: *Aruncus dioicus, Malus fusca, Tsuga heterophylla*
Kidney Aid: *Aruncus dioicus, Ledum groenlandicum, Vaccinium* sp.
Laxative: *Frangula purshiana, Malus fusca, Oemleria cerasiformis, Oenanthe sarmentosa, Prunus emarginata*
Love Medicine: *Disporum hookeri, D. smithii, Trillium ovatum, Vicia nigricans*
Oral Aid: *Gaultheria shallon, Urtica dioica*
Orthopedic Aid: *Malus fusca*
Other: *Achillea millefolium, Ambrosia chamissonis, Oxalis oregana*
Panacea: *Malus fusca*
Pediatric Aid: *Picea sitchensis*
Poison: *Lonicera involucrata*
Pulmonary Aid: *Allium cernuum, Blechnum spicant, Lysichiton americanus, Malus fusca*
Reproductive Aid: *Maianthemum dilatatum, Urtica dioica*
Stimulant: *Urtica dioica*
Strengthener: *Ambrosia chamissonis, Picea sitchensis*
Throat Aid: *Achillea millefolium*
Tonic: *Heracleum maximum, Holodiscus discolor, Malus fusca, Prunus emarginata*
Toothache Remedy: *Pteridium aquilinum, Rubus spectabilis*
Tuberculosis Remedy: *Malus fusca, Oemleria cerasiformis*
Unspecified: *Aruncus dioicus, Heracleum maximum, Holodiscus discolor, Ledum groenlandicum, Leymus mollis, Lysichiton americanus, Malus fusca, Oemleria cerasiformis, Oplopanax horridus, Polypodium glycyrrhiza, Salix* sp., *Senecio jacobaea, Trifolium wormskioldii*
Urinary Aid: *Sedum* sp.
Venereal Aid: *Aruncus dioicus*
Witchcraft Medicine: *Symphoricarpos albus*
Food
Beverage: *Gaultheria shallon, Ledum groenlandicum, Rosa nutkana*
Bread & Cake: *Gaultheria shallon, Vaccinium ovalifolium, V. ovatum, V.* sp.
Candy: *Picea sitchensis*
Dessert: *Shepherdia canadensis*
Dietary Aid: *Polypodium glycyrrhiza*
Dried Food: *Gaultheria shallon, Sambucus racemosa, Shepherdia canadensis, Vaccinium ovalifolium*
Frozen Food: *Vaccinium parvifolium*
Fruit: *Arctostaphylos uva-ursi, Cornus canadensis, Fragaria chiloensis, Frangula purshiana, Gaultheria shallon, Malus fusca, Ribes bracteosum, R. divaricatum, R. laxiflorum, Rubus discolor, R. laciniatus, R. leucodermis, R. parviflorus, R. spectabilis, R. ursinus, Sambucus racemosa, Vaccinium ovalifolium, V. ovatum, V. oxycoccos, V. parvifolium*
Pie & Pudding: *Gaultheria shallon, Rubus laciniatus, R. leucodermis, R. ursinus, Vaccinium oxycoccos*
Preservative: *Rubus parviflorus*
Preserves: *Fragaria chiloensis, Gaultheria shallon, Mahonia aquifolium, M. nervosa, Malus fusca, Ribes laxiflorum, Rubus laciniatus, R. leucodermis, R. ursinus, Vaccinium oxycoccos*
Special Food: *Rubus spectabilis*
Spice: *Blechnum spicant, Gaultheria shallon, Polystichum munitum*
Substitution Food: *Sedum* sp.
Unspecified: *Abronia latifolia, Allium cernuum, Angelica lucida, Argentina egedii, Camassia quamash, Equisetum telmateia, Heracleum maximum, Oxalis oregana, Phyllospadix scouleri, P. serrulatus, P.* sp., *P. torreyi, Picea sitchensis, Polypodium glycyrrhiza, P. scouleri, Polystichum munitum, Pteridium aquilinum, Ranunculus reptans, Rosa nutkana, Rubus parviflorus, R. spectabilis, Sedum* sp., *Tolmiea menziesii, Trifolium wormskioldii, Urtica dioica*
Vegetable: *Lathyrus japonicus, Solanum tuberosum*
Winter Use Food: *Rubus spectabilis, Sambucus racemosa, Shepherdia canadensis, Vaccinium ovalifolium, V. oxycoccos, V. parvifolium, V.* sp.
Fiber
Basketry: *Acer circinatum, A. macrophyllum, Adiantum aleuticum, A. pedatum, Carex aquatilis, C. obnupta, Heracleum maximum, Phyllospadix torreyi, Salix hookeriana, Scirpus americanus, S. microcarpus, Thuja plicata, Typha latifolia, Urtica dioica, Xerophyllum tenax*

Building Material: *Tsuga heterophylla*
Canoe Material: *Thuja plicata*
Clothing: *Thuja plicata, Typha latifolia*
Cordage: *Nereocystis luetkeana, Urtica dioica*
Furniture: *Alnus rubra*
Mats, Rugs & Bedding: *Scirpus acutus, Thuja plicata, Typha latifolia, Usnea* sp.
Dye
Green: *Urtica dioica*
Red-Brown: *Tsuga heterophylla*
Unspecified: *Aruncus dioicus, Lonicera involucrata, Mahonia* sp.
Yellow: *Mahonia aquifolium, M. nervosa, Usnea* sp.
Other
Cash Crop: *Arctostaphylos uva-ursi, Xerophyllum tenax*
Ceremonial Items: *Thuja plicata, Urtica dioica*
Containers: *Alnus rubra, Lysichiton americanus, Nereocystis luetkeana, Taxus brevifolia, Thuja plicata*
Cooking Tools: *Alnus rubra, Gaultheria shallon, Holodiscus discolor, Lysichiton americanus, Polystichum munitum, Pteridium aquilinum, Stachys mexicana, Taxus brevifolia, Thuja plicata*
Decorations: *Boykinia occidentalis*
Fasteners: *Oemleria cerasiformis, Picea sitchensis*
Hunting & Fishing Item: *Holodiscus discolor, Nereocystis luetkeana, Oplopanax horridus, Picea sitchensis, Taxus brevifolia, Thuja plicata, Tsuga heterophylla*
Incense & Fragrance: *Galium triflorum*
Insecticide: *Tsuga heterophylla*
Jewelry: *Rosa nutkana*
Preservative: *Alnus rubra*
Protection: *Urtica dioica*
Smoke Plant: *Alnus rubra, Arctostaphylos uva-ursi, Gaultheria shallon*
Toys & Games: *Nereocystis luetkeana, Oenanthe sarmentosa, Polystichum munitum*
Waterproofing Agent: *Sambucus racemosa*

Malecite
Drug
Abortifacient: *Fragaria virginiana, Lilium canadense, Rubus pubescens, Sanicula marilandica, Viburnum lentago*
Adjuvant: *Lilium philadelphicum*
Analgesic: *Aralia racemosa, Cornus sericea, Inula helenium, Sanicula odorata, Sorbus americana*
Anthelmintic: *Tilia americana*
Anticonvulsive: *Cornus canadensis*
Antidiarrheal: *Geum rivale, Quercus rubra, Rubus* sp., *Tiarella cordifolia, Tsuga canadensis*
Antiemetic: *Solanum dulcamara*
Antihemorrhagic: *Sanguinaria canadensis*
Blood Medicine: *Chimaphila umbellata, Rhus hirta*
Burn Dressing: *Thuja occidentalis*
Cathartic: *Juglans cinerea*
Cold Remedy: *Acorus calamus, Aralia racemosa, Inula helenium, Larix laricina, Polygala senega, Prunus serotina, Tsuga canadensis*
Contraceptive: *Chelone glabra, Pontederia cordata, Tanacetum vulgare*
Cough Medicine: *Geum aleppicum, Lilium philadelphicum, Nemopanthus mucronatus, Prunus serotina, Rhus hirta, Thuja occidentalis*
Dermatological Aid: *Achillea millefolium, Arctium lappa, Arisaema triphyllum, Betula populifolia, Comptonia peregrina, Fagus grandifolia, Juniperus communis, Kalmia angustifolia, Lilium philadelphicum, Maianthemum racemosum, Populus balsamifera, Prunus pensylvanica, Salix cordata, Sanguinaria canadensis, Scirpus microcarpus, Sedum telephium, Sorbus americana, Tsuga canadensis, Typha latifolia, Verbascum thapsus*
Dietary Aid: *Populus grandidentata, Salix cordata*
Emetic: *Sambucus racemosa*
Eye Medicine: *Acer spicatum, Cornus sericea*
Febrifuge: *Lilium philadelphicum, Nemopanthus mucronatus, Rhus hirta*
Gastrointestinal Aid: *Asarum canadense, Mentha canadensis*
Heart Medicine: *Inula helenium*
Hemorrhoid Remedy: *Sanguinaria canadensis*
Kidney Aid: *Aralia racemosa, Ledum groenlandicum, Nemopanthus mucronatus, Polygonum hydropiper, Sanicula odorata, Tanacetum vulgare, Tsuga canadensis, Typha angustifolia*
Laxative: *Abies balsamea*
Liver Aid: *Impatiens capensis*
Misc. Disease Remedy: *Heracleum maximum*
Oral Aid: *Alnus incana, Coptis trifolia*
Orthopedic Aid: *Achillea millefolium, Kalmia angustifolia*
Pediatric Aid: *Asarum canadense, Cardamine diphylla, Coptis trifolia, Geum aleppicum, G. rivale, Mentha canadensis, Prunus pensylvanica, Rubus* sp., *Tiarella cordifolia, Tsuga canadensis*
Preventive Medicine: *Acorus calamus*
Pulmonary Aid: *Geum aleppicum, Mentha canadensis, Polypodium virginianum*
Respiratory Aid: *Comptonia peregrina, Cornus sericea, Verbascum thapsus*
Sedative: *Mentha canadensis*
Strengthener: *Aralia racemosa, Larix laricina*
Throat Aid: *Cardamine diphylla, Cornus sericea, Iris versicolor, Polystichum acrostichoides, Scirpus microcarpus*
Tonic: *Cardamine diphylla, Juniperus communis*
Toothache Remedy: *Thuja occidentalis*
Tuberculosis Remedy: *Aralia racemosa, Chimaphila umbellata, Juniperus communis, Larix laricina, Lilium philadelphicum, Nemopanthus mucronatus, Prunus serotina, Rhus hirta, Sanguinaria canadensis, Sarracenia purpurea, Thuja occidentalis*
Unspecified: *Abies balsamea, Acorus calamus, Alnus* sp., *Betula* sp., *Coptis trifolia, Larix laricina, Picea glauca, P. mariana, Populus ×jackii, Prunus* sp., *Quercus* sp., *Symplocarpus foetidus, Taxus canadensis, Tsuga canadensis*
Venereal Aid: *Abies balsamea, Aralia racemosa, Arctium lappa, Larix laricina*
Veterinary Aid: *Tanacetum vulgare*
Food
Beverage: *Betula* sp., *Ledum groenlandicum, Pyrola* sp.
Fruit: *Prunus* sp.
Sauce & Relish: *Acer saccharum*
Spice: *Mentha canadensis*
Sweetener: *Acer saccharum*
Unspecified: *Allium* sp., *Brassica* sp., *Chenopodium* sp., *Helianthus tuberosus, Phytolacca americana, Quercus* sp., *Taraxacum officinale*
Fiber
Basketry: *Acer rubrum, Fraxinus nigra, Hierochloe odorata, Thuja occidentalis*
Brushes & Brooms: *Betula* sp.
Building Material: *Picea glauca*
Canoe Material: *Acer saccharum, Betula papyrifera, Fraxinus americana, Picea glauca, Thuja occidentalis*
Clothing: *Tilia americana*
Cordage: *Tilia americana*
Mats, Rugs & Bedding: *Abies balsamea, Picea glauca, P. mariana*
Sewing Material: *Abies balsamea, Picea glauca, P. mariana, Tilia americana*
Snow Gear: *Betula* sp., *Fraxinus americana*
Dye
Unspecified: *Tsuga canadensis*
Other
Containers: *Betula papyrifera*

Hide Preparation: *Thuja occidentalis, Tsuga canadensis*
Hunting & Fishing Item: *Larix laricina, Ostrya virginiana, Populus ×jackii, Thuja occidentalis*
Lighting: *Acer saccharum*
Smoking Tools: *Prunus* sp.
Tools: *Ostrya virginiana*
Waterproofing Agent: *Abies balsamea, Picea glauca, P. mariana*

Mandan
Drug
Ceremonial Medicine: *Helianthus annuus*
Stimulant: *Helianthus annuus*
Food
Staple: *Helianthus annuus*
Unspecified: *Helianthus annuus*
Fiber
Mats, Rugs & Bedding: *Salix exigua, S. melanopsis*
Other
Hunting & Fishing Item: *Andropogon gerardii*

Maricopa
Drug
Antirheumatic (External): *Atriplex polycarpa*
Food
Beverage: *Carnegia gigantea, Prosopis glandulosa*
Dried Food: *Cucurbita moschata, Ziziphus obtusifolia*
Fruit: *Condalia hookeri, Cucurbita moschata, Lycium fremontii, Opuntia echinocarpa, O. engelmannii, O.* sp., *Ziziphus obtusifolia*
Porridge: *Phoradendron californicum*
Staple: *Allenrolfea occidentalis*
Unspecified: *Opuntia acanthocarpa, Prosopis velutina*
Vegetable: *Cucurbita moschata*
Other
Smoking Tools: *Phragmites australis*

Maricopa & Mohave
Food
Fruit: *Ziziphus obtusifolia*

Mendocino Indian
Drug
Abortifacient: *Phoradendron leucarpum*
Analgesic: *Achillea millefolium, Angelica* sp., *Artemisia furcata, Eschscholzia californica, Heteromeles arbutifolia, Umbellularia californica, Wyethia longicaulis, Zigadenus venenosus*
Antidiarrheal: *Alnus rhombifolia, Artemisia furcata, Prunus virginiana, Rubus vitifolius, Salix lasiolepis*
Antidote: *Artemisia furcata*
Antihemorrhagic: *Alnus rhombifolia*
Antirheumatic (External): *Anthemis cotula, Artemisia furcata, Marah oreganus, Umbellularia californica, Wyethia longicaulis, Zigadenus venenosus*
Blood Medicine: *Alnus rhombifolia, Grindelia* sp., *Mahonia repens, Monardella villosa, Satureja douglasii*
Burn Dressing: *Alnus rhombifolia, Pinus sabiniana, Wyethia longicaulis*
Cathartic: *Frangula californica, Pinus lambertiana*
Cold Remedy: *Angelica* sp., *Anthemis cotula, Aralia californica, Artemisia furcata, Clematis ligusticifolia, Eriodictyon californicum, Grindelia* sp., *Lithophragma affine*
Dermatological Aid: *Achillea millefolium, Asclepias eriocarpa, Chlorogalum pomeridianum, Eschscholzia californica, Marah oreganus, Mentzelia laevicaulis, Pinus sabiniana, Polygonum aviculare, Populus fremontii, Salix lasiolepis, Sambucus cerulea, Toxicodendron diversilobum, Umbellularia californica, Wyethia longicaulis, Zigadenus venenosus*
Diaphoretic: *Alnus rhombifolia, Salix lasiolepis*
Dietary Aid: *Quercus* sp.
Emetic: *Alnus rhombifolia, Eschscholzia californica, Wyethia longicaulis*
Eye Medicine: *Achillea millefolium, Angelica* sp., *Anthemis cotula, Artemisia furcata, Clarkia purpurea, Lonicera interrupta, Polypodium californicum, Wyethia longicaulis*
Febrifuge: *Angelica* sp., *Aralia californica, Artemisia furcata, Cercis canadensis, Salix lasiolepis, Sambucus cerulea, Scutellaria californica*
Gastrointestinal Aid: *Achillea millefolium, Alnus rhombifolia, Angelica* sp., *Aralia californica, Arceuthobium occidentale, Artemisia furcata, Calocedrus decurrens, Chenopodium album, Eschscholzia californica, Grindelia* sp., *Heteromeles arbutifolia, Lithophragma affine, Mahonia repens, Mentzelia laevicaulis, Monardella villosa, Satureja douglasii, Thysanocarpus curvipes, Umbellularia californica, Wyethia longicaulis*
Gynecological Aid: *Alnus rhombifolia, Artemisia furcata, Eschscholzia californica*
Herbal Steam: *Artemisia furcata*
Kidney Aid: *Frangula californica*
Laxative: *Grindelia* sp.
Misc. Disease Remedy: *Frangula californica*
Narcotic: *Delphinium nudicaule, Eschscholzia californica*
Orthopedic Aid: *Achillea millefolium, Lathyrus jepsonii, Sambucus cerulea, Zigadenus venenosus*
Panacea: *Salix lasiolepis, Thalictrum polycarpum, Umbellularia californica*
Poison: *Aesculus californica, Allium unifolium, Arctostaphylos manzanita, Asclepias fascicularis, Chlorogalum pomeridianum, Cicuta* sp., *Crataegus rivularis, Delphinium hesperium, Marah oreganus, Phoradendron leucarpum, Quercus chrysolepis, Solanum nigrum, Taxus brevifolia, Thalictrum polycarpum, Zigadenus venenosus*
Psychological Aid: *Angelica* sp., *Frangula californica*
Pulmonary Aid: *Aralia californica*
Respiratory Aid: *Angelica* sp., *Artemisia furcata, Eriodictyon californicum*
Sedative: *Prunus virginiana*
Snakebite Remedy: *Angelica* sp.
Stimulant: *Umbellularia californica*
Throat Aid: *Clematis ligusticifolia*
Tonic: *Prunus virginiana*
Toothache Remedy: *Aesculus californica, Eschscholzia californica, Phoradendron leucarpum*
Tuberculosis Remedy: *Achillea millefolium, Alnus rhombifolia, Aralia californica, Eschscholzia californica*
Urinary Aid: *Marah oreganus*
Venereal Aid: *Helenium puberulum, Marah oreganus*
Veterinary Aid: *Aesculus californica, Populus fremontii, Pteridium aquilinum, Sambucus cerulea*
Food
Beverage: *Arctostaphylos manzanita, A. tomentosa*
Bread & Cake: *Boisduvalia densiflora, Quercus douglasii, Q. garryana, Q. kelloggii, Q. lobata, Umbellularia californica*
Cooking Agent: *Madia sativa*
Dried Food: *Juniperus californica, Prunus subcordata, Rubus leucodermis, R. vitifolius, Sambucus cerulea*
Fodder: *Hordeum marinum, Lathyrus jepsonii, Leymus triticoides, Plantago lanceolata, Vicia americana*
Forage: *Aesculus californica, Arbutus menziesii, Arctostaphylos manzanita, Asclepias eriocarpa, Brodiaea coronaria, Carex vicaria, Ceanothus cuneatus, Croton setigerus, Equisetum variegatum, Juncus effusus, Lupinus luteolus, Medicago polymorpha, Nuphar lutea, Orthocarpus lithospermoides, Pedicularis densiflora, Robinia pseudoacacia, Toxicodendron diversilobum, Trifolium* sp.

Fruit: *Aesculus californica, Amelanchier alnifolia, Arctostaphylos manzanita, Fragaria vesca, Heteromeles arbutifolia, Prunus subcordata, Ribes californicum, R. divaricatum, Rubus leucodermis, R. parviflorus, R. vitifolius, Sambucus cerulea, Solanum nigrum, Taxus brevifolia*
Pie & Pudding: *Sambucus cerulea*
Porridge: *Rumex crispus*
Preserves: *Prunus virginiana, Sambucus cerulea, Vitis californica*
Sauce & Relish: *Armoracia rusticana, Umbellularia californica*
Soup: *Quercus douglasii, Q. garryana, Q. kelloggii*
Spice: *Perideridia kelloggii*
Staple: *Amaranthus retroflexus, Avena fatua, Boisduvalia densiflora, Bromus marginatus, Capsella bursa-pastoris, Clarkia purpurea, Hemizonia luzulifolia, Hordeum murinum, Layia platyglossa, Leymus triticoides, Madia gracilis, Perideridia kelloggii, Plagiobothrys fulvus, Ranunculus occidentalis, Thysanocarpus curvipes, Trifolium bifidum, T. dichotomum, Wyethia longicaulis*
Starvation Food: *Pinus sabiniana*
Substitution Food: *Grindelia sp., Microseris laciniata, Mimulus guttatus, Monardella villosa, Satureja douglasii*
Unspecified: *Allium bolanderi, A. unifolium, Amsinckia lycopsoides, Angelica sp., Arctostaphylos manzanita, Brodiaea coronaria, Calochortus pulchellus, C. tolmiei, Castanopsis chrysophylla, Chlorogalum pomeridianum, Claytonia perfoliata, Dichelostemma pulchellum, Eriogonum latifolium, Helenium puberulum, Lithocarpus densiflorus, Lomatium utriculatum, Lonicera interrupta, Microseris laciniata, Nuphar lutea, Pedicularis densiflora, Perideridia kelloggii, Pinus lambertiana, Plagiobothrys fulvus, Sanicula tuberosa, Torreya californica, Trifolium bifidum, T. cyathiferum, T. dichotomum, T. fucatum, T. obtusiflorum, T. sp., T. wormskioldii, Triteleia hyacinthina, T. peduncularis, Typha latifolia, Umbellularia californica, Wyethia longicaulis*
Vegetable: *Brassica rapa, Chenopodium album, Claytonia perfoliata, Eschscholzia californica, Heracleum maximum, Lupinus affinis, L. luteolus, Platystemon californicus, Rumex crispus, Trifolium variegatum, Vicia americana*
Winter Use Food: *Arctostaphylos manzanita, Rubus leucodermis*

Fiber
Basketry: *Calycanthus occidentalis, Carex sp., Cercis canadensis, Juncus effusus, Lonicera interrupta, Pteridium aquilinum, Toxicodendron diversilobum*
Brushes & Brooms: *Perideridia kelloggii*
Building Material: *Arbutus menziesii, Pinus ponderosa, Sequoia sempervirens*
Canoe Material: *Sequoia sempervirens*
Clothing: *Apocynum cannabinum*
Cordage: *Apocynum cannabinum, Carex sp., Cercis canadensis, Hoita macrostachya, Juncus effusus, Salix lasiolepis*
Furniture: *Philadelphus lewisii*
Mats, Rugs & Bedding: *Chlorogalum pomeridianum, Typha latifolia*
Scouring Material: *Equisetum variegatum*
Sewing Material: *Apocynum cannabinum*

Dye
Red: *Plagiobothrys fulvus*
Unspecified: *Alnus rhombifolia*

Other
Cooking Tools: *Chlorogalum pomeridianum, Toxicodendron diversilobum*
Decorations: *Arbutus menziesii, Chlorogalum pomeridianum, Dodecatheon hendersonii, Pinus sabiniana, Toxicodendron diversilobum*
Fasteners: *Chlorogalum pomeridianum, Pinus ponderosa*
Fuel: *Alnus rhombifolia, Arctostaphylos manzanita, Crataegus rivularis, Pinus ponderosa, Populus fremontii, Sambucus cerulea*
Good Luck Charm: *Angelica sp., Daucus pusillus, Sanicula crassicaulis, Vicia americana*
Hunting & Fishing Item: *Alnus rhombifolia, Amelanchier alnifolia, Apocynum cannabinum, Baccharis pilularis, Ceanothus cuneatus, Cercocarpus montanus, Chlorogalum pomeridianum, Croton setigerus, Datisca glomerata, Hoita macrostachya, Iris douglasiana, Marah oreganus, Taxus brevifolia*
Insecticide: *Pogogyne douglasii, Umbellularia californica*
Musical Instrument: *Pinus sabiniana, Sambucus cerulea*
Paint: *Chlorogalum pomeridianum*
Smoke Plant: *Nicotiana quadrivalvis, Salix lasiolepis*
Stable Gear: *Arbutus menziesii*
Tools: *Aesculus californica, Arbutus menziesii, Cercocarpus montanus, Fraxinus latifolia*
Toys & Games: *Sambucus cerulea*
Walking Sticks: *Fraxinus latifolia*
Weapon: *Cercocarpus montanus*

Menominee
Drug
Abortifacient: *Acorus calamus, Artemisia campestris, Quercus ellipsoidalis, Q. macrocarpa, Sanguinaria canadensis, Thuja occidentalis, Trillium grandiflorum*
Adjuvant: *Abies balsamea, Arctostaphylos uva-ursi, Chimaphila umbellata, Comptonia peregrina, Corylus americana, Fraxinus nigra, Ptelea trifoliata, Robinia pseudoacacia, Rubus idaeus, Sanguinaria canadensis, Symplocarpus foetidus, Taenidia integerrima, Thuja occidentalis, Zanthoxylum americanum*
Alterative: *Alnus incana, Helenium autumnale*
Analgesic: *Abies balsamea, Acorus calamus, Angelica atropurpurea, Aralia racemosa, Echinocystis lobata, Gnaphalium obtusifolium, Helenium autumnale, Heuchera americana, Hydrophyllum virginianum, Pinus strobus, Polygonatum biflorum, Solidago flexicaulis, Symplocarpus foetidus, Trillium grandiflorum, Tsuga canadensis, Valeriana edulis, V. uliginosa, Viburnum acerifolium, Zanthoxylum americanum*
Anthelmintic: *Apocynum cannabinum, Valeriana edulis*
Anticonvulsive: *Symplocarpus foetidus*
Antidiarrheal: *Adiantum pedatum, Betula papyrifera, Cornus alternifolia, C. amomum, Geranium maculatum, Hepatica nobilis, Heuchera americana, Hydrophyllum virginianum, Prunella vulgaris, Prunus virginiana, Rubus canadensis, Salix humilis*
Antidote: *Clintonia borealis, Sambucus racemosa, Sisyrinchium albidum*
Antihemorrhagic: *Polygonum pensylvanicum*
Antirheumatic (External): *Taxus canadensis, Zanthoxylum americanum*
Antirheumatic (Internal): *Gaultheria procumbens*
Blood Medicine: *Aralia racemosa, Chimaphila umbellata, Diervilla lonicera*
Burn Dressing: *Plantago rugelii*
Cancer Treatment: *Cornus alternifolia*
Cathartic: *Acorus calamus, Juglans cinerea, Sambucus racemosa, Ulmus rubra, Valeriana uliginosa, Veronicastrum virginicum*
Ceremonial Medicine: *Cornus amomum, Hamamelis virginiana, Veronicastrum virginicum*
Cold Remedy: *Abies balsamea, Acorus calamus, Alnus incana, Helenium autumnale, Populus balsamifera, Tsuga canadensis, Zanthoxylum americanum*
Cough Medicine: *Ceanothus americanus, Rhus hirta*
Dermatological Aid: *Abies balsamea, Achillea millefolium, Alnus incana, Anemone virginiana, Angelica atropurpurea, Aralia nudicaulis, A. racemosa, Arctium lappa, Asclepias tuberosa, Capsella bursa-pastoris, Castilleja sessiliflora, Epilobium angustifolium, Hierochloe odorata, Lactuca canadensis, Larix laricina, Lepidium virginicum, Lilium philadelphicum, Nuphar lutea, Petasites frigidus, Picea glauca, Pinus strobus, Plantago rugelii, Populus balsamifera, Prunus virginiana, Rhus hirta, Sanguinaria canadensis, Symplocarpus foetidus, Thuja occidentalis, Tsuga canadensis, Ulmus rubra, Uvularia grandiflora, Valeriana edulis, V. uliginosa, Zanthoxylum americanum*
Diaphoretic: *Nepeta cataria, Thuja occidentalis, Tsuga canadensis*
Dietary Aid: *Osmorhiza claytonii*
Disinfectant: *Gnaphalium obtusifolium, Trillium grandiflorum*

Diuretic: *Cucurbita pepo, Diervilla lonicera, Dirca palustris, Trillium grandiflorum*

Emetic: *Sambucus racemosa, Veronicastrum virginicum*

Eye Medicine: *Arisaema triphyllum, Osmorhiza claytonii, Quercus velutina, Rubus allegheniensis, Trillium grandiflorum, Vitis vulpina*

Febrifuge: *Achillea millefolium, Eupatorium perfoliatum, Leucanthemum vulgare, Mentha canadensis, Sambucus canadensis*

Gastrointestinal Aid: *Acorus calamus, Aralia racemosa, Asarum canadense, Cardamine maxima, Ceanothus americanus, Heuchera americana, Rhus hirta, Rosa carolina, Salix humilis, Tsuga canadensis, Viburnum acerifolium*

Gynecological Aid: *Adiantum pedatum, Arisaema dracontium, Caulophyllum thalictroides, Chimaphila umbellata, Comptonia peregrina, Cypripedium pubescens, Equisetum hyemale, Eupatorium purpureum, Hepatica nobilis, Mitchella repens, Osmunda cinnamomea, Physocarpus opulifolius, Polygonum pensylvanicum, Pteridium aquilinum, Rhus hirta, Sisyrinchium atlanticum*

Hallucinogen: *Cypripedium pubescens*

Heart Medicine: *Liatris spicata, Symplocarpus foetidus*

Hemorrhoid Remedy: *Cornus alternifolia, Rhus hirta*

Hemostat: *Equisetum sylvaticum, Solidago flexicaulis, Symplocarpus foetidus, Valeriana edulis*

Herbal Steam: *Maianthemum racemosum, Taxus canadensis, Tsuga canadensis*

Hunting Medicine: *Heracleum maximum, Panax quinquefolius, Valeriana uliginosa*

Internal Medicine: *Picea glauca*

Kidney Aid: *Dirca palustris, Equisetum hyemale, E. sylvaticum, Galium triflorum, Hypericum ascyron*

Laxative: *Veronicastrum virginicum*

Love Medicine: *Castilleja coccinea, Dicentra cucullaria, Echinocystis lobata, Pedicularis canadensis*

Oral Aid: *Coptis trifolia*

Orthopedic Aid: *Asclepias tuberosa, Hamamelis virginiana*

Other: *Hamamelis virginiana, Larix laricina, Zanthoxylum americanum*

Panacea: *Echinocystis lobata, Ligusticum filicinum, Ptelea trifoliata*

Pediatric Aid: *Achillea millefolium, Coptis trifolia, Monarda fistulosa, Prunella vulgaris, Prunus virginiana, Symplocarpus foetidus*

Poison: *Comptonia peregrina*

Psychological Aid: *Diervilla lonicera, Gnaphalium obtusifolium, Panax quinquefolius, Valeriana uliginosa*

Pulmonary Aid: *Abies balsamea, Aralia nudicaulis, Asclepias syriaca, Hypericum ascyron, Mentha arvensis, M. ×piperita, Nepeta cataria, Prunus virginiana, Rhus hirta, Taenidia integerrima, Valeriana uliginosa, Verbascum thapsus, Zanthoxylum americanum*

Respiratory Aid: *Maianthemum racemosum, Monarda fistulosa, Taenidia integerrima*

Sedative: *Lithospermum canescens, Mitchella repens, Nepeta cataria, Valeriana uliginosa, Zanthoxylum americanum*

Snakebite Remedy: *Carex plantaginea*

Stimulant: *Gnaphalium obtusifolium, Polygonatum biflorum, Thuja occidentalis*

Throat Aid: *Coptis trifolia, Valeriana uliginosa*

Tonic: *Asclepias tuberosa, Betula papyrifera, Comptonia peregrina, Echinocystis lobata, Panax quinquefolius, Salix humilis*

Toothache Remedy: *Coptis trifolia*

Tuberculosis Remedy: *Arctium lappa, Hypericum ascyron, Rhus hirta, Rubus occidentalis*

Unspecified: *Abies balsamea, Fagus grandifolia, Menyanthes trifoliata, Myriophyllum spicatum, Osmunda regalis, Pinus banksiana, Quercus alba, Rubus allegheniensis, Salix bebbiana, Saxifraga pensylvanica, Silene noctiflora, Solidago gigantea*

Urinary Aid: *Cypripedium acaule, Eupatorium purpureum, Lonicera canadensis, Verbena hastata*

Venereal Aid: *Lonicera canadensis*

Veterinary Aid: *Alnus incana, Larix laricina, Lupinus perennis, Pedicularis canadensis, Prunus virginiana, Sisyrinchium albidum*

Witchcraft Medicine: *Arisaema triphyllum, Gnaphalium obtusifolium, Heracleum maximum, Lupinus perennis, Sanicula marilandica, S. odorata, Sarracenia purpurea, Sisyrinchium albidum, Symplocarpus foetidus*

Food

Beverage: *Ceanothus americanus, Prunus virginiana, Quercus ellipsoidalis, Rhus hirta, Zea mays*

Dried Food: *Vitis vulpina*

Fruit: *Fragaria virginiana, Podophyllum peltatum, Prunus pumila, P. serotina, P. virginiana, Ribes cynosbati, Rubus allegheniensis, R. idaeus, R. occidentalis, Vaccinium oxycoccos, Viburnum lentago, V. opulus, Vitis vulpina*

Pie & Pudding: *Quercus alba, Rubus allegheniensis*

Porridge: *Quercus alba*

Preserves: *Podophyllum peltatum, Prunus pumila, Ribes cynosbati, Vitis vulpina*

Soup: *Asclepias incarnata, Osmunda cinnamomea*

Special Foods: *Zea mays*

Staple: *Phaseolus vulgaris, Quercus alba, Zea mays, Zizania aquatica*

Starvation Food: *Celastrus scandens*

Substitution Food: *Ceanothus americanus, Zea mays*

Sweetener: *Acer saccharum*

Unspecified: *Allium canadense, Aralia racemosa, Asclepias incarnata, Corylus americana, Juglans cinerea*

Vegetable: *Apios americana, Caltha palustris, Cardamine maxima, Hydrophyllum virginianum, Nuphar lutea, Solanum tuberosum, Taraxacum officinale, Zea mays*

Winter Use Food: *Allium tricoccum, Amelanchier canadensis, A. laevis, A. sanguinea, Apios americana, Asclepias incarnata, Corylus americana, Cucurbita pepo, Fagus grandifolia, Rhus hirta, Ribes cynosbati, Rubus allegheniensis, Sagittaria cuneata, Vaccinium angustifolium, Zea mays*

Fiber

Basketry: *Hierochloe odorata, Laportea canadensis, Thuja occidentalis, Tilia americana, Ulmus rubra, Urtica dioica*

Building Material: *Scirpus tabernaemontani*

Caulking Material: *Typha latifolia*

Cordage: *Apocynum androsaemifolium, Asclepias syriaca, Dirca palustris, Tilia americana*

Mats, Rugs & Bedding: *Scirpus tabernaemontani, Tilia americana, Typha latifolia, Ulmus rubra*

Scouring Material: *Equisetum hyemale*

Sewing Material: *Apocynum androsaemifolium, Asclepias syriaca, Hierochloe odorata, Pinus banksiana*

Snow Gear: *Tilia americana, Ulmus rubra*

Dye

Black: *Juglans cinerea*

Brown: *Juglans cinerea*

Orange-Red: *Sanguinaria canadensis*

Orange-Yellow: *Impatiens capensis*

Red: *Ranunculus recurvatus, Sanguinaria canadensis, Tsuga canadensis*

Red-Brown: *Alnus incana*

Yellow: *Oxalis corniculata, O. montana, O. stricta, Rhus hirta*

Other

Designs: *Acer spicatum*

Good Luck Charm: *Valeriana uliginosa*

Hunting & Fishing Item: *Apocynum androsaemifolium, Asclepias syriaca, Fraxinus nigra, Krigia biflora, Osmunda cinnamomea, Panax quinquefolius, Tilia americana, Ulmus rubra*

Insecticide: *Podophyllum peltatum*

Protection: *Heracleum maximum, Panax quinquefolius, Thuja occidentalis*

Sacred Items: *Arisaema dracontium, Hierochloe odorata, Lithospermum canescens, Mitella diphylla*
Smoke Plant: *Cornus alternifolia, C. amomum, Verbascum thapsus*
Toys & Games: *Sambucus canadensis*

Meskwaki
Drug
Adjuvant: *Aralia racemosa, Asarum canadense, Cirsium vulgare, Desmodium illinoense, Juniperus virginiana, Panax quinquefolius, Ptelea trifoliata, Rubus idaeus, Sanguinaria canadensis, Taenidia integerrima*
Alterative: *Pycnanthemum virginianum*
Analgesic: *Acorus calamus, Actaea pachypoda, Anemone cylindrica, Arctium minus, Asarum canadense, Athyrium filix-femina, Celastrus scandens, Cirsium discolor, Comandra umbellata, Coreopsis tripteris, Cornus racemosa, Crataegus calpodendron, Dioscorea villosa, Echinacea angustifolia, Echinocystis lobata, Erigeron philadelphicus, Geranium maculatum, Heracleum maximum, Lycopus americanus, Monarda punctata, Platanus occidentalis, Sanguinaria canadensis, Silphium integrifolium, Sisyrinchium campestre, Triosteum perfoliatum, Viburnum opulus, Vitis vulpina, Zizia aurea*
Anthelmintic: *Amorpha canescens, Asclepias incarnata, Eupatorium perfoliatum, Euphorbia corollata, Lonicera dioica, Quercus macrocarpa, Rhus hirta, Ribes americanum*
Anticonvulsive: *Echinacea angustifolia, Maianthemum racemosum, Veronicastrum virginicum*
Antidiarrheal: *Agalinis tenuifolia, Aquilegia canadensis, Ceanothus americanus, Cornus racemosa, Dalea purpurea, Geranium maculatum, Parthenocissus quinquefolia, Polygonum amphibium, P. pensylvanicum, Polytaenia nuttallii, Quercus alba, Salix humilis, Scutellaria parvula, Spiraea salicifolia*
Antidote: *Asarum canadense, Eryngium yuccifolium, Lespedeza capitata, Rubus allegheniensis, Rumex orbiculatus, Vitis vulpina*
Antiemetic: *Prunus nigra, Silphium perfoliatum*
Antihemorrhagic: *Alnus incana, Coreopsis tripteris*
Antirheumatic (Internal): *Euphorbia corollata, Podophyllum peltatum*
Blood Medicine: *Phlox pilosa, Platanus occidentalis*
Burn Dressing: *Acorus calamus, Anemone cylindrica, Aralia nudicaulis, Artemisia campestris, Geranium maculatum, Helianthus grosseserratus, Iris versicolor, Mirabilis nyctaginea, Plantago major, Sanguinaria canadensis, Solidago speciosa*
Cancer Treatment: *Crepis runcinata, Pedicularis canadensis*
Carminative: *Asclepias incarnata*
Cathartic: *Acorus calamus, Asclepias incarnata, Euphorbia corollata, Juglans cinerea, J. nigra, Podophyllum peltatum, Polemonium reptans, Sambucus canadensis, Triosteum perfoliatum*
Ceremonial Medicine: *Aquilegia canadensis, Arisaema triphyllum, Eryngium yuccifolium, Maianthemum racemosum, Staphylea trifolia*
Cold Remedy: *Brassica nigra, Comandra umbellata, Erigeron philadelphicus, Gleditsia triacanthos, Helenium autumnale, Iris versicolor, Monarda fistulosa, M. punctata, Physostegia parviflora, Platanus occidentalis, Populus tremuloides, Ranunculus flabellaris, Stachys tenuifolia*
Cough Medicine: *Acorus calamus, Campanulastrum americanum, Penthorum sedoides, Populus tremuloides, Zanthoxylum americanum*
Dermatological Aid: *Achillea millefolium, Amorpha canescens, Aralia nudicaulis, Artemisia ludoviciana, Baptisia alba, Ceanothus americanus, Chenopodium album, Crepis runcinata, Euonymus atropurpurea, Fraxinus americana, Helianthus decapetalus, Heracleum maximum, Heuchera americana, Impatiens capensis, Iris versicolor, Liatris punctata, Napaea dioica, Pedicularis canadensis, Phlox pilosa, Plantago major, Platanus occidentalis, Prunus virginiana, Rhus glabra, Rosa blanda, Sanicula odorata, Silene stellata, Solidago rigida, Symplocarpus foetidus, Tilia americana, Toxicodendron pubescens, Triosteum perfoliatum, Typha latifolia, Ulmus rubra*
Diaphoretic: *Ageratina altissima, Conyza canadensis*
Dietary Aid: *Osmorhiza longistylis, Platanus occidentalis, Rhus glabra*
Diuretic: *Agastache scrophulariifolia, Asclepias incarnata, Carya cordiformis, Laportea canadensis, Plantago major, Polemonium reptans, Sambucus canadensis, Tradescantia occidentalis*
Ear Medicine: *Asarum canadense*
Emetic: *Acer negundo, Asclepias incarnata, Cephalanthus occidentalis, Galium aparine, Podophyllum peltatum, Silphium laciniatum, Stachys tenuifolia, Triodanis perfoliata*
Expectorant: *Zanthoxylum americanum*
Eye Medicine: *Anemone canadensis, A. cylindrica, Euonymus atropurpurea, Hepatica nobilis, Osmorhiza longistylis, Rubus allegheniensis, Ulmus americana*
Febrifuge: *Achillea millefolium, Gleditsia triacanthos, Pycnanthemum virginianum, Zizia aurea*
Gastrointestinal Aid: *Acorus calamus, Aquilegia canadensis, Asarum canadense, Ceanothus americanus, Cirsium discolor, Echinacea angustifolia, Helenium autumnale, Heracleum maximum, Lycopus americanus, Monarda punctata, Plantago major, Prunus virginiana, Rosa blanda, Rubus allegheniensis*
Gynecological Aid: *Actaea pachypoda, Antennaria plantaginifolia, Apocynum androsaemifolium, Aralia racemosa, Arctium minus, Athyrium filix-femina, Caulophyllum thalictroides, Celastrus scandens, Cucurbita pepo, Dioscorea villosa, Gentiana andrewsii, Glycyrrhiza glabra, Lactuca sativa, Napaea dioica, Polygonum amphibium, Polytaenia nuttallii, Ribes cynosbati, Sambucus canadensis, Silphium perfoliatum, Sisyrinchium campestre, Symphoricarpos occidentalis, Ulmus rubra, Verbena urticifolia, Veronicastrum virginicum*
Heart Medicine: *Filipendula rubra, Polygala senega*
Hemorrhoid Remedy: *Baptisia alba, Geranium maculatum, Napaea dioica, Polygonum pensylvanicum, Prunus virginiana, Rosa blanda*
Hemostat: *Agrimonia gryposepala, Gentianella quinquefolia, Platanus occidentalis, Ranunculus abortivus, Salix humilis, Sanicula odorata, Valeriana edulis, Zanthoxylum americanum*
Herbal Steam: *Aster ericoides, A. laevis, A. lateriflorus, Silphium integrifolium*
Hunting Medicine: *Asarum canadense, Napaea dioica*
Internal Medicine: *Pedicularis canadensis*
Kidney Aid: *Apocynum androsaemifolium, A. cannabinum, Baptisia alba, Galium concinnum, Liatris scariosa, Silphium integrifolium, Veronicastrum virginicum, Zanthoxylum americanum*
Laxative: *Carya cordiformis, Fraxinus nigra, Maianthemum racemosum, Pediomelum argophyllum, Rhamnus alnifolia, Salix humilis, Veronicastrum virginicum*
Love Medicine: *Aquilegia canadensis, Cypripedium acaule, Eupatorium purpureum, Filipendula rubra, Iodanthus pinnatifidus, Lobelia cardinalis, L. siphilitica, Panax quinquefolius, Pedicularis canadensis, Phlox pilosa, Thalictrum dasycarpum*
Misc. Disease Remedy: *Achillea millefolium, Apocynum cannabinum, Dalea purpurea, Galium concinnum, Gleditsia triacanthos, Heracleum maximum, Maianthemum racemosum, Malus ioensis, Platanus occidentalis, Pycnanthemum virginianum, Veronicastrum virginicum*
Oral Aid: *Cornus racemosa, Geranium maculatum, Polygonum amphibium, Prunus americana*
Orthopedic Aid: *Coreopsis palmata, Silphium integrifolium*
Other: *Hepatica nobilis, Juniperus virginiana*
Panacea: *Apocynum cannabinum, Carya cordiformis, Clematis viorna, Echinocystis lobata, Fraxinus nigra, Heuchera americana, Morus rubra, Osmorhiza longistylis, Panax quinquefolius*
Pediatric Aid: *Adiantum pedatum, Alnus incana, Cornus racemosa, Maianthemum racemosum, Panax quinquefolius, Polygonum amphibium, Populus tremuloides, Triosteum perfoliatum*
Poison: *Arisaema triphyllum, Helenium autumnale*

Poultice: *Iodanthus pinnatifidus*
Psychological Aid: *Ambrosia trifida, Anemone cylindrica, Aster lateriflorus, Gnaphalium obtusifolium, Gymnocladus dioicus, Maianthemum racemosum, Solidago canadensis, Tradescantia occidentalis, Vitis vulpina*
Pulmonary Aid: *Asarum canadense, Comandra umbellata, Helianthus strumosus, Heliopsis helianthoides, Iris versicolor, Platanus occidentalis, Ptelea trifoliata, Quercus alba, Q. velutina, Sambucus canadensis, Taraxacum officinale, Tilia americana*
Respiratory Aid: *Anemone virginiana, Baptisia alba, Comandra umbellata, Erigeron philadelphicus, Helenium autumnale, Monarda punctata, Ranunculus flabellaris, Sisyrinchium campestre*
Sedative: *Arisaema triphyllum, Humulus lupulus, Maianthemum racemosum, Prunus virginiana*
Snakebite Remedy: *Arisaema triphyllum, Baptisia alba, Ceanothus americanus, Eryngium yuccifolium, Eupatorium perfoliatum, Fraxinus americana, Gentiana andrewsii, Hypericum ascyron, Juglans nigra, Triosteum perfoliatum*
Stimulant: *Actaea pachypoda, Ageratina altissima, Anemone cylindrica, A. virginiana, Aralia nudicaulis, Aster ericoides, A. laevis, A. novae-angliae, A. praealtus, Cornus racemosa, Crataegus calpodendron, Gnaphalium obtusifolium, Juniperus virginiana, Maianthemum racemosum, Monarda punctata, Physalis virginiana, Polygonatum biflorum, Pycnanthemum virginianum, Solidago ulmifolia, Verbena urticifolia, Veronicastrum virginicum*
Strengthener: *Zanthoxylum americanum*
Throat Aid: *Artemisia ludoviciana, Asarum canadense, Senna marilandica*
Tonic: *Gleditsia triacanthos*
Toothache Remedy: *Geranium maculatum, Symplocarpus foetidus, Zanthoxylum americanum*
Tuberculosis Remedy: *Acorus calamus, Artemisia ludoviciana, Campanulastrum americanum, Cornus racemosa, Hypericum ascyron, Zanthoxylum americanum*
Unspecified: *Agastache scrophulariifolia, Capsella bursa-pastoris, Hypericum punctatum, Physalis heterophylla, Salix candida, Scrophularia marilandica, Symplocarpus foetidus*
Urinary Aid: *Actaea pachypoda, Aquilegia canadensis, Caulophyllum thalictroides, Crataegus calpodendron, Diervilla lonicera, Eryngium yuccifolium, Galium concinnum, Laportea canadensis, Liatris punctata, L. scariosa, Mirabilis nyctaginea, Silphium integrifolium*
Venereal Aid: *Diervilla lonicera, Equisetum hyemale, Liatris punctata*
Veterinary Aid: *Artemisia ludoviciana, Celtis occidentalis, Liatris punctata, Maianthemum racemosum, Osmorhiza longistylis*

Food
Beverage: *Ceanothus americanus, Gymnocladus dioicus, Prunus virginiana, Quercus alba, Rhus glabra, Rubus occidentalis*
Dessert: *Ribes cynosbati, Vitis vulpina*
Fodder: *Equisetum arvense, E. hyemale*
Forage: *Sagittaria latifolia*
Fruit: *Crataegus calpodendron, Physalis heterophylla, P. virginiana, Podophyllum peltatum, Prunus americana, P. nigra, P. virginiana, Ribes americanum, Rubus alleghemiensis, R. occidentalis, Sambucus canadensis, Smilax herbacea, Viburnum prunifolium*
Pie & Pudding: *Rubus alleghemiensis*
Porridge: *Celtis occidentalis, Quercus alba*
Preserves: *Fragaria virginiana, Malus ioensis, Podophyllum peltatum, Prunus americana, P. nigra, Rubus alleghemiensis, Sambucus canadensis, Viburnum prunifolium*
Soup: *Asclepias syriaca*
Sour: *Oxalis stricta*
Spice: *Allium canadense, Asarum canadense*
Sweetener: *Acer saccharum*
Unspecified: *Amphicarpaea bracteata, Citrullus lanatus, Corylus americana, Gymnocladus dioicus, Juglans nigra, Nelumbo lutea, Zea mays, Zizania aquatica*
Vegetable: *Apios americana, Asclepias syriaca, Heracleum maximum, Lilium philadelphicum, Solanum tuberosum, Taraxacum officinale*
Winter Use Food: *Allium canadense, Apios americana, Asclepias syriaca, Carya ovata, Corylus americana, Cucurbita pepo, Juglans cinerea, Malus ioensis, Nelumbo lutea, Rubus alleghemiensis, R. occidentalis, Sagittaria latifolia, Zea mays*

Fiber
Basketry: *Corylus americana, Fraxinus americana, F. nigra, Tilia americana*
Brushes & Brooms: *Corylus americana*
Building Material: *Betula papyrifera, Typha latifolia, Ulmus rubra*
Canoe Material: *Betula papyrifera*
Caulking Material: *Typha latifolia*
Clothing: *Tilia americana*
Cordage: *Apocynum cannabinum, Asclepias syriaca, Laportea canadensis, Tilia americana*
Mats, Rugs & Bedding: *Scirpus tabernaemontani, Tilia americana, Typha latifolia*
Scouring Material: *Equisetum hyemale*
Sewing Material: *Apocynum androsaemifolium, A. cannabinum*

Dye
Black: *Juglans nigra*
Orange: *Oxalis stricta*
Red: *Sanguinaria canadensis*
Yellow: *Rhus glabra*

Other
Ceremonial Items: *Lobelia cardinalis*
Designs: *Acer saccharum*
Hunting & Fishing Item: *Fraxinus nigra, Lathyrus palustris, Pycnanthemum virginianum, Tilia americana*
Incense & Fragrance: *Aquilegia canadensis*
Insecticide: *Artemisia ludoviciana, Sambucus canadensis*
Sacred Items: *Staphylea trifolia*
Smoke Plant: *Cornus racemosa, Triodanis perfoliata*
Toys & Games: *Sambucus canadensis*

Mewuk
Drug
Antidote: *Angelica tomentosa*
Antiemetic: *Dryopteris arguta*
Antihemorrhagic: *Dryopteris arguta*
Antirheumatic (External): *Heracleum maximum*
Blood Medicine: *Grindelia camporum*
Cathartic: *Artemisia ludoviciana, Frangula californica*
Dermatological Aid: *Achillea sp., Artemisia ludoviciana*
Disinfectant: *Artemisia ludoviciana*
Febrifuge: *Salix lasiolepis*
Gastrointestinal Aid: *Achillea sp., Artemisia ludoviciana*
Misc. Disease Remedy: *Heracleum maximum, Salix lasiolepis*
Other: *Artemisia ludoviciana*
Panacea: *Artemisia ludoviciana*
Poison: *Chlorogalum pomeridianum*
Pulmonary Aid: *Achillea sp.*
Unspecified: *Artemisia ludoviciana, Sambucus cerulea*
Veterinary Aid: *Artemisia ludoviciana*

Food
Beverage: *Arctostaphylos viscida*
Bread & Cake: *Quercus kelloggii*
Fruit: *Arctostaphylos viscida*
Porridge: *Quercus kelloggii*

Staple: *Madia elegans*
Unspecified: *Angelica tomentosa, Heracleum maximum, Pinus sabiniana, Quercus kelloggii*

Fiber
Basketry: *Ceanothus cuneatus, C. integerrimus, Cercis canadensis, Cladium mariscus, Philadelphus lewisii, Pinus sabiniana, Pteridium aquilinum, Quercus kelloggii, Rhus trilobata*
Building Material: *Abies concolor, Epilobium* sp., *Libocedrus* sp., *Pinus ponderosa, Umbellularia* sp.
Caulking Material: *Chlorogalum pomeridianum*

Other
Cooking Tools: *Ceanothus* sp., *Libocedrus* sp.
Decorations: *Artemisia ludoviciana*
Fasteners: *Chlorogalum* sp.
Fuel: *Ceanothus cuneatus*
Smoke Plant: *Nicotiana attenuata, N. quadrivalvis*
Tools: *Aesculus* sp., *Ceanothus cuneatus, C. integerrimus, Quercus kelloggii*

Micmac
Drug
Abortifacient: *Aletris farinosa, Fragaria virginiana, Lilium canadense, Rubus pubescens, Sanguinaria canadensis, Sanicula marilandica, Viburnum lentago*
Analgesic: *Alnus* sp., *Aralia racemosa, Asarum canadense, Comptonia peregrina, Cornus sericea, Inula helenium, Kalmia angustifolia, Myrica cerifera, Ranunculus acris, Sanicula marilandica, Sorbus americana, Symplocarpus foetidus, Taxus canadensis, Thuja occidentalis*
Anthelmintic: *Apocynum cannabinum, Tilia americana*
Anticonvulsive: *Aristolochia serpentaria, Cornus canadensis*
Antidiarrheal: *Abies balsamea, Geum rivale, Prunus virginiana, Quercus rubra, Rubus fruticosus, Tiarella cordifolia, Tsuga canadensis*
Antiemetic: *Alnus* sp., *Mentha canadensis, Solanum dulcamara*
Antihemorrhagic: *Acer pensylvanicum, Aralia racemosa, Baptisia tinctoria, Galium aparine, Pyrola asarifolia, Sarracenia purpurea, Streptopus amplexifolius*
Antirheumatic (External): *Achillea millefolium, Myrica cerifera*
Antirheumatic (Internal): *Alnus* sp., *Chimaphila umbellata, Juniperus communis, Sanicula marilandica*
Blood Medicine: *Chimaphila umbellata, Panax quinquefolius, Taxus canadensis*
Burn Dressing: *Abies balsamea, Thuja occidentalis*
Cathartic: *Alnus* sp., *Daucus carota, Juglans cinerea, Ribes uva-crispa, Rumex crispus, Sambucus canadensis*
Cold Remedy: *Abies balsamea, Acer pensylvanicum, Achillea millefolium, Acorus calamus, Angelica sylvestris, Aralia racemosa, Brassica napus, Geum rivale, Inula helenium, Larix laricina, Ledum groenlandicum, Nymphaea odorata, Pinus strobus, Polygala senega, Populus tremuloides, Prunus serotina, Salix cordata, Sanguinaria canadensis, Tsuga canadensis*
Contraceptive: *Chelone glabra, Pontederia cordata, Tanacetum vulgare*
Cough Medicine: *Acer alba, A. pensylvanicum, Acorus calamus, Angelica sylvestris, Aralia nudicaulis, A. racemosa, Brassica napus, Geum aleppicum, G. rivale, Ilex aquifolium, Lilium philadelphicum, Nymphaea odorata, Picea glauca, Pinus strobus, Prunus serotina, Rubus chamaemorus, R. hispidus, Thuja occidentalis, Tsuga canadensis*
Dermatological Aid: *Abies balsamea, Achillea millefolium, Alnus* sp., *Aralia racemosa, Arctium lappa, A. minus, Arisaema triphyllum, Betula populifolia, Chimaphila umbellata, Comptonia peregrina, Hydrastis canadensis, Iris versicolor, Juniperus communis, Kalmia angustifolia, Larix americana, L. laricina, Lilium philadelphicum, Maianthemum racemosum, Myrica cerifera, Nuphar lutea, Nymphaea odorata, Picea glauca, Pinus strobus, Populus balsamifera, Prunus pensylvanica, Salix cordata, S. nigra, Sanguinaria canadensis, Scirpus microcarpus, Sedum telephium, Sorbus americana, Tilia americana, Tsuga canadensis, Typha latifolia, Ulmus rubra, Verbascum thapsus*
Diaphoretic: *Achillea millefolium*
Dietary Aid: *Populus tremuloides, Quercus alba, Rhus hirta, Salix cordata*
Diuretic: *Larix americana, Ledum groenlandicum, Polypodium virginianum*
Ear Medicine: *Nicotiana tabacum, Rhus glabra*
Emetic: *Betula populifolia, Euphorbia corollata, Sambucus canadensis, S. racemosa, Sorbus americana*
Eye Medicine: *Acer spicatum, Aralia racemosa, Cornus sericea*
Febrifuge: *Alnus* sp., *Ilex aquifolium, Lilium philadelphicum, Lycopodium* sp., *Rubus chamaemorus, R. hispidus, Taxus canadensis*
Gastrointestinal Aid: *Abies balsamea, Aletris farinosa, Asarum canadense, Chimaphila umbellata, Picea glauca, Sorbus americana, Taxus canadensis, Tsuga canadensis*
Gland Medicine: *Nymphaea odorata*
Gynecological Aid: *Aralia racemosa, Fraxinus americana, Heracleum sphondylium, Leonurus cardiaca, Sanicula marilandica, Sorbus americana, Taxus canadensis, Viburnum prunifolium*
Heart Medicine: *Inula helenium*
Hemorrhoid Remedy: *Quercus alba*
Hemostat: *Alnus* sp., *Nicotiana tabacum, Phytolacca americana, Pinus strobus, Salix lucida, Sanguinaria canadensis*
Kidney Aid: *Acer pensylvanicum, Aralia racemosa, Baptisia tinctoria, Chimaphila umbellata, Cimicifuga racemosa, Eupatorium perfoliatum, Galium aparine, Ledum groenlandicum, Pinus strobus, Pyrola asarifolia, Sanicula marilandica, Sarracenia purpurea, Streptopus amplexifolius, Tanacetum vulgare, Tsuga canadensis*
Laxative: *Abies balsamea*
Liver Aid: *Impatiens capensis*
Love Medicine: *Sanguinaria canadensis*
Misc. Disease Remedy: *Acer pensylvanicum, Acorus calamus, Alnus* sp., *Brassica napus, Chimaphila umbellata, Heracleum maximum, Iris versicolor, Ledum groenlandicum, Nymphaea odorata, Picea glauca, Pinus strobus, Prunus pensylvanica, P. serotina, Sarracenia purpurea, Taxus canadensis, Tsuga canadensis, Viburnum opulus*
Oral Aid: *Alnus incana, Berberis vulgaris, Coptis trifolia*
Orthopedic Aid: *Abies balsamea, Acer pensylvanicum, Achillea millefolium, Alnus* sp., *Arisaema triphyllum, Comptonia peregrina, Juniperus communis, Kalmia angustifolia, Nuphar lutea, Nymphaea odorata, Salix nigra, Thuja occidentalis*
Other: *Adiantum pedatum, Betula alleghaniensis*
Panacea: *Acorus calamus, Iris versicolor, Kalmia angustifolia*
Pediatric Aid: *Geum rivale, Mentha canadensis, Pteridium aquilinum, Rubus fruticosus*
Poison: *Actaea* sp., *Kalmia angustifolia*
Pulmonary Aid: *Acorus calamus, Alnus* sp., *Geum aleppicum, Mentha canadensis, Polypodium virginianum, Sarracenia purpurea, Ulmus rubra*
Respiratory Aid: *Comptonia peregrina, Cornus sericea, Ledum groenlandicum, Salix lucida, Verbascum thapsus*
Sedative: *Cardamine diphylla, Cypripedium acaule, Sambucus canadensis*
Snakebite Remedy: *Sanicula marilandica*
Stimulant: *Aralia racemosa, Comptonia peregrina, Larix laricina, Myrica cerifera, Pteridium aquilinum*
Throat Aid: *Angelica sylvestris, Aralia racemosa, Berberis vulgaris, Cardamine diphylla, Cornus sericea, Iris versicolor, Polystichum acrostichoides, Rhexia virginica, Rhus hirta, Sanguinaria canadensis, Sarracenia purpurea, Scirpus microcarpus*
Tonic: *Aletris farinosa, Asarum canadense, Cardamine diphylla, Juniperus communis, Ledum groenlandicum, Prunus serotina*
Toothache Remedy: *Thuja occidentalis*
Tuberculosis Remedy: *Aralia racemosa, Chimaphila umbellata, Ilex aquifolium, Juniperus communis, Larix laricina, Lilium philadelphicum, Limonium caro-

linianum, *L. vulgare, Prunus serotina, Rubus chamaemorus, R. hispidus, Sanguinaria canadensis, Sarracenia purpurea, Sinapis alba, Thuja occidentalis*
Unspecified: *Acorus calamus, Coptis trifolia, Gaultheria hispidula, Heracleum sphondylium, Menyanthes trifoliata, Sorbus americana*
Urinary Aid: *Chimaphila umbellata, Clintonia uniflora, Conioselinum chinense, Heracleum sphondylium, Ilex aquifolium, Petroselinum crispum, Platanthera dilatata, Rumex crispus, Tsuga canadensis, Typha angustifolia*
Venereal Aid: *Abies balsamea, Acer pensylvanicum, Aralia racemosa, Arctium lappa, Baptisia tinctoria, Eupatorium perfoliatum, Fagus grandifolia, Galium aparine, Larix laricina, Populus balsamifera, Pyrola asarifolia, Streptopus amplexifolius*

Food
Beverage: *Abies balsamea, Acer pensylvanicum, A. saccharum, Acorus calamus, Aralia nudicaulis, Ledum groenlandicum, Mitchella repens, Picea glauca, P. mariana, Pinus strobus, Prunus* sp., *Tsuga canadensis*
Sauce & Relish: *Acer saccharum*
Unspecified: *Helianthus tuberosus*
Vegetable: *Taraxacum officinale*

Fiber
Basketry: *Acer rubrum, Betula papyrifera, Fraxinus nigra, Hierochloe odorata*
Building Material: *Betula papyrifera*
Canoe Material: *Betula papyrifera, Thuja occidentalis*
Mats, Rugs & Bedding: *Abies balsamea, Hierochloe odorata, Picea glauca, P. mariana*
Sewing Material: *Picea mariana*
Snow Gear: *Fagus grandifolia*

Dye
Black: *Salix* sp.
Green: *Taxus canadensis*
Red: *Galium tinctorium*
Unspecified: *Alnus* sp., *Tsuga canadensis*

Other
Containers: *Betula papyrifera*
Cooking Tools: *Betula papyrifera*
Fuel: *Abies balsamea, Juniperus* sp., *Larix laricina, Picea glauca, P. mariana, Pinus strobus, Thuja occidentalis, Tsuga canadensis*
Hunting & Fishing Item: *Acer saccharum, Betula* sp., *Thuja occidentalis*
Lighting: *Betula* sp.
Smoke Plant: *Cornus foemina, Salix* sp.
Tools: *Fraxinus americana*

Midoo
Drug
Throat Aid: *Quercus douglasii*
Food
Fruit: *Arctostaphylos patula, A. viscida, Rhus aromatica*
Unspecified: *Umbellularia* sp.
Other
Hunting & Fishing Item: *Philadelphus* sp.
Smoke Plant: *Nicotiana attenuata, N. quadrivalvis*

Missouri River Indian
Fiber
Mats, Rugs & Bedding: *Equisetum hyemale*
Dye
Yellow: *Populus balsamifera, P. deltoides*

Miwok
Drug
Analgesic: *Achillea millefolium, Angelica breweri, Artemisia douglasiana, A. vulgaris, Balsamorhiza sagittata, Centaurium exaltatum, Lepechinia calycina, Monardella lanceolata, Polygala cornuta, Rosa californica, Trichostema lanceolatum, Umbellularia californica, Urtica dioica*
Antidiarrheal: *Mentha spicata, Mimulus* sp.
Antihemorrhagic: *Epilobium canum, Pellaea mucronata*
Antirheumatic (External): *Chenopodium ambrosioides, Datisca glomerata, Ericameria arborescens, Eriodictyon californicum, Eriophyllum lanatum, Geranium oreganum, Gnaphalium viscosum, Malva parviflora, Navarretia cotulifolia, Solidago californica, Urtica dioica*
Antirheumatic (Internal): *Agastache urticifolia, Artemisia douglasiana, A. vulgaris, Balsamorhiza sagittata, Chamaebatia foliolosa, Cupressus* sp., *Eriodictyon californicum, Mahonia pinnata*
Blood Medicine: *Chamaesyce ocellata, Pellaea mucronata*
Burn Dressing: *Clematis lasiantha, Pinus sabiniana*
Cathartic: *Epilobium canum, Frangula rubra*
Ceremonial Medicine: *Artemisia douglasiana, A. vulgaris*
Cold Remedy: *Achillea millefolium, Angelica breweri, Aristolochia californica, Chamaebatia foliolosa, Cupressus* sp., *Ericameria cuneata, Eriodictyon californicum, Gnaphalium viscosum, Keckiella breviflora, Monardella lanceolata, M. odoratissima, Polygala cornuta, Pycnanthemum californicum, Scutellaria* sp., *Symphoricarpos albus, Trichostema lanceolatum*
Cough Medicine: *Chamaebatia foliolosa, Eriodictyon californicum, Polygala cornuta, Quercus lobata, Q. wislizeni, Scutellaria* sp.
Dermatological Aid: *Asclepias speciosa, Chamaesyce serpyllifolia, Chenopodium ambrosioides, Clematis lasiantha, Datisca glomerata, Ericameria arborescens, Erigeron foliosus, Eriodictyon californicum, Grindelia robusta, Hypericum concinnum, Mahonia pinnata, Malva parviflora, Phacelia heterophylla, Pinus sabiniana, Polygonum bistortoides, Quercus lobata, Q. wislizeni, Rumex conglomeratus, Sanicula crassicaulis, Solidago californica, Trichostema lanceolatum*
Diaphoretic: *Balsamorhiza sagittata, Wyethia angustifolia*
Dietary Aid: *Arbutus menziesii, Arctostaphylos manzanita, A. tomentosa, A. viscida*
Emetic: *Polygala cornuta*
Eye Medicine: *Pinus lambertiana, Scutellaria angustifolia, Solanum nigrum*
Febrifuge: *Centaurium venustum, Erigeron foliosus, Holocarpha virgata, Lepechinia calycina, Monardella odoratissima, Wyethia angustifolia*
Gastrointestinal Aid: *Arbutus menziesii, Arctostaphylos manzanita, A. tomentosa, A. viscida, Centaurium exaltatum, Ericameria arborescens, Eriodictyon californicum, Gnaphalium viscosum, Mahonia pinnata, Mentha spicata, Rosa californica, Symphoricarpos albus*
Gynecological Aid: *Epilobium canum, Ericameria arborescens, Trichostema lanceolatum*
Hallucinogen: *Datura wrightii*
Kidney Aid: *Epilobium canum, Galium triflorum*
Misc. Disease Remedy: *Achillea millefolium, Agastache urticifolia, Centaurium venustum, Chamaebatia foliolosa, Erigeron foliosus, Holocarpha virgata, Lepechinia calycina, Mahonia pinnata, Sambucus cerulea, Trichostema lanceolatum*
Orthopedic Aid: *Ericameria arborescens, Eriodictyon californicum*
Other: *Pellaea mucronata*
Panacea: *Sanicula bipinnatifida*
Pediatric Aid: *Quercus lobata, Q. wislizeni*
Poultice: *Mentzelia* sp.
Pulmonary Aid: *Centaurium venustum*
Snakebite Remedy: *Chamaesyce ocellata, C. serpyllifolia, Daucus pusillus, Osmorhiza* sp., *Sanicula bipinnata, S. bipinnatifida, S. crassicaulis*
Stimulant: *Trichostema lanceolatum*
Toothache Remedy: *Centaurium exaltatum, Chenopodium ambrosioides, Erigeron foliosus, Pentagrama triangularis, Solidago californica, Trichostema lanceolatum*

Tuberculosis Remedy: *Centaurium exaltatum, Epilobium canum, Mahonia pinnata*
Unspecified: *Asclepias cordifolia, Equisetum* sp.
Urinary Aid: *Epilobium canum, Trichostema lanceolatum*
Venereal Aid: *Asclepias speciosa, Chamaebatia foliolosa, Chenopodium ambrosioides, Epilobium canum*
Witchcraft Medicine: *Artemisia douglasiana, A. vulgaris, Sitanion* sp.
Food
Beverage: *Arbutus menziesii, Arctostaphylos manzanita, A. tomentosa, A. viscida, Castilleja applegatei, C. parviflora, Mentha spicata, Monardella odoratissima, Trifolium wormskioldii*
Bread & Cake: *Quercus douglasii, Q. kelloggii, Q. lobata, Q.* sp., *Q. wislizeni*
Cooking Agent: *Asclepias fascicularis, Darmera peltata, Lotus unifoliolatus*
Dried Food: *Chenopodium album, Heuchera micrantha, Hypericum scouleri, Pinus ponderosa, Trifolium ciliolatum, T.* sp.
Fruit: *Physocarpus capitatus, Ribes nevadense, R. roezlii, Solanum xantii*
Porridge: *Avena barbata, Quercus douglasii, Q. kelloggii, Q. wislizeni*
Sauce & Relish: *Lupinus latifolius*
Soup: *Aesculus californica, Avena barbata, Quercus douglasii, Q. kelloggii, Q. lobata, Q.* sp., *Q. wislizeni*
Staple: *Hypericum scouleri, Madia elegans, M. gracilis, Quercus lobata*
Substitution Food: *Perideridia bolanderi*
Sweetener: *Pinus lambertiana*
Unspecified: *Balsamorhiza sagittata, Boisduvalia densiflora, B. stricta, Brodiaea coronaria, B.* sp., *Bromus diandrus, Calandrinia ciliata, Calochortus luteus, C. venustus, Chlorogalum pomeridianum, Clarkia amoena, C. biloba, C. purpurea, C. unguiculata, Claytonia perfoliata, Corylus cornuta, Dichelostemma pulchellum, Hemizonia fitchii, Juniperus occidentalis, Lathyrus vestitus, Lupinus densiflorus, Madia sativa, Mentzelia* sp., *Navarretia* sp., *Oenanthe* sp., *Orthocarpus attenuatus, Perideridia bolanderi, P. gairdneri, P. kelloggii, Pinus lambertiana, P. sabiniana, Ranunculus californicus, Trifolium ciliolatum, T.* sp., *T. willdenowii, T. wormskioldii, Triteleia hyacinthina, T. ixioides, Veratrum californicum*
Vegetable: *Aquilegia formosa, Asclepias fascicularis, Chenopodium album, Clarkia purpurea, Conyza canadensis, Delphinium hesperium, D.* sp., *Eriogonum nudum, Heuchera micrantha, Lathyrus vestitus, Mimulus guttatus, M. moschatus, Osmorhiza berteroi, Rumex acetosella, R. conglomeratus*
Winter Use Food: *Aesculus californica, Arbutus menziesii, Arctostaphylos manzanita, A. tomentosa, A. viscida, Castilleja* sp., *Chlorogalum pomeridianum, Lupinus latifolius, Quercus douglasii, Q. kelloggii, Q. lobata, Q. wislizeni, Trifolium willdenowii*
Fiber
Basketry: *Pinus sabiniana*
Brushes & Brooms: *Eriogonum nudum*
Building Material: *Pinus sabiniana*
Mats, Rugs & Bedding: *Pinus sabiniana*
Other
Cash Crop: *Quercus douglasii*
Containers: *Lupinus latifolius*
Cooking Tools: *Arctostaphylos* sp., *Quercus kelloggii, Q.* sp., *Vitis californica, Wyethia helenioides*
Protection: *Angelica breweri*

Modesse
Drug
Antidote: *Rhododendron occidentale*
Antirheumatic (External): *Frangula californica*
Blood Medicine: *Mahonia* sp.
Cathartic: *Frangula californica*
Cough Medicine: *Ceanothus velutinus, Equisetum* sp.
Dermatological Aid: *Asarum* sp.
Febrifuge: *Ceanothus velutinus*
Kidney Aid: *Croton setigerus*
Poison: *Evernia* sp.
Urinary Aid: *Equisetum* sp.
Food
Fruit: *Amelanchier alnifolia, Prunus subcordata, P. virginiana*
Preserves: *Mahonia* sp.
Staple: *Quercus kelloggii*
Starvation Food: *Aesculus californica*
Fiber
Basketry: *Cercis canadensis, Cornus glabrata*
Cordage: *Asclepias* sp.
Scouring Material: *Equisetum* sp.
Snow Gear: *Acer circinatum*
Other
Hunting & Fishing Item: *Cercocarpus montanus, Philadelphus* sp.
Tools: *Ceanothus cuneatus, Cercocarpus montanus, Rhus trilobata*

Modoc
Food
Unspecified: *Lomatium canbyi*
Fiber
Unspecified: *Apocynum cannabinum*

Mohave
Food
Beverage: *Lycium andersonii, Prosopis pubescens, Salix gooddingii, Salvia columbariae*
Bread & Cake: *Olneya tesota, Prosopis glandulosa*
Dried Food: *Lycium andersonii, Olneya tesota, Prosopis glandulosa, Vigna unguiculata, Ziziphus obtusifolia*
Fruit: *Opuntia echinocarpa, O. engelmannii, Physalis hederifolia, P. pubescens, Yucca schidigera*
Porridge: *Quercus turbinella*
Staple: *Agastache pallidiflora, Agave americana, A. parryi, Allenrolfea occidentalis, Helianthus annuus, Salvia columbariae*
Starvation Food: *Baccharis salicifolia, Chloracantha spinosa, Parkinsonia florida, P. microphylla, Sisymbrium irio, Sonchus asper, Thelypodium integrifolium*
Unspecified: *Amaranthus caudatus, A. palmeri, Ammannia coccinea, Camissonia brevipes, Cyperus odoratus, Vigna unguiculata*
Vegetable: *Amaranthus caudatus, A. palmeri, Chenopodium fremontii, C. murale, Prosopis glandulosa, P. pubescens, Rumex crispus*
Winter Use Food: *Amaranthus caudatus, A. palmeri, Helianthus annuus*
Other
Musical Instrument: *Lagenaria siceraria*
Smoke Plant: *Nicotiana trigonophylla*

Mohegan
Drug
Abortifacient: *Acorus calamus, Pedicularis canadensis*
Analgesic: *Acorus calamus, Alnus incana, A.* sp., *Arctium minus, Arisaema triphyllum, Brassica nigra, Capsella bursa-pastoris, Humulus lupulus, Iris versicolor, Monotropa uniflora, Picea abies, Pinus strobus, Quercus alba, Vitis labrusca, V.* sp.
Anthelmintic: *Artemisia absinthium, Capsella bursa-pastoris, Lindera* sp., *Mentha ×piperita, M. spicata, Rubus hispidus, Salvia officinalis*
Anticonvulsive: *Symplocarpus foetidus*
Antidiarrheal: *Prunus serotina, Rubus hispidus, Spiraea tomentosa*
Antirheumatic (External): *Aesculus glabra, A. hippocastanum, Arctium minus, Arisaema triphyllum, Quercus alba*

Antirheumatic (Internal): *Acorus calamus, Castanea dentata*
Blood Medicine: *Crotalaria sagittalis, Rumex crispus, Sanguinaria canadensis*
Burn Dressing: *Impatiens capensis, Plantago major*
Cathartic: *Euonymus atropurpurea, Sambucus canadensis, Taraxacum officinale*
Cold Remedy: *Acorus calamus, Allium cepa, Anaphalis margaritacea, Arctium minus, Castanea dentata, Eupatorium perfoliatum, Monotropa uniflora, Pinus strobus, Prunus serotina, Quercus alba, Trifolium repens, Ulmus americana, Verbascum thapsus*
Cough Medicine: *Acer saccharinum, A. saccharum, Barbarea vulgaris, Maianthemum racemosum, Pinus strobus, Trifolium repens, Ulmus americana, U. rubra, Verbascum thapsus*
Dermatological Aid: *Baptisia tinctoria, Chimaphila umbellata, Comptonia peregrina, Datura stramonium, Hamamelis virginiana, Impatiens capensis, Picea abies, Pinus strobus, Plantago major, Zea mays*
Dietary Aid: *Achillea millefolium, Tanacetum vulgare*
Ear Medicine: *Humulus lupulus, Nicotiana tabacum*
Emetic: *Sambucus canadensis, Sanguinaria canadensis*
Eye Medicine: *Sassafras albidum*
Febrifuge: *Anthemis cotula, Berberis vulgaris, Eupatorium perfoliatum, Monarda punctata, Monotropa uniflora, Prunella vulgaris, Vitis labrusca*
Gastrointestinal Aid: *Achillea millefolium, Aster umbellatus, Capsella bursa-pastoris, Eupatorium perfoliatum, Hedeoma pulegioides, Maianthemum racemosum, Mentha canadensis, Nepeta cataria, Prunus serotina, Rumex acetosella, Sambucus canadensis, Tanacetum vulgare*
Gynecological Aid: *Leonurus cardiaca, Phytolacca americana*
Heart Medicine: *Zanthoxylum americanum*
Kidney Aid: *Achillea millefolium, Caulophyllum thalictroides, Echium vulgare, Gaultheria procumbens, Myrica* sp.
Laxative: *Sambucus canadensis*
Liver Aid: *Achillea millefolium*
Misc. Disease Remedy: *Daucus carota*
Oral Aid: *Cirsium arvense, Coptis trifolia, Goodyera pubescens, Pyrola elliptica*
Orthopedic Aid: *Alnus incana, A.* sp., *Arisaema triphyllum, Impatiens capensis, Quercus alba*
Panacea: *Acorus calamus, Anthemis cotula, Eupatorium perfoliatum, Panax quinquefolius, Salvia officinalis*
Pediatric Aid: *Cirsium arvense, Coptis trifolia, Goodyera pubescens, Lindera* sp., *Mentha ×piperita, Nepeta cataria, Sambucus canadensis*
Poison: *Arisaema triphyllum, Phytolacca americana*
Pulmonary Aid: *Asclepias tuberosa, Castanea dentata, Cirsium arvense, Inula helenium, Ulmus rubra*
Respiratory Aid: *Prunus americana, Verbascum thapsus*
Sedative: *Humulus lupulus*
Snakebite Remedy: *Aristolochia serpentaria, Plantago major*
Throat Aid: *Amaranthus retroflexus, Arisaema triphyllum, Berberis vulgaris, Pyrola elliptica, Rhus hirta, Ulmus rubra, Verbascum thapsus*
Tonic: *Acorus calamus, Aralia nudicaulis, Betula lenta, Eupatorium perfoliatum, Leonurus cardiaca, Leucanthemum vulgare, Maianthemum racemosum, Panax quinquefolius, Prunus serotina, Rumex crispus, Salvia officinalis, Sanguinaria canadensis, Sassafras albidum, Taraxacum officinale*
Toothache Remedy: *Armoracia rusticana, Brassica nigra, Humulus lupulus, Rorippa nasturtium-aquaticum*
Tuberculosis Remedy: *Cirsium arvense, Inula helenium*
Unspecified: *Vaccinium oxycoccos*
Veterinary Aid: *Inula helenium, Quercus alba*
Food
Cooking Agent: *Apios americana*
Sweetener: *Acer saccharum*
Unspecified: *Apios americana, Asclepias syriaca, Caltha palustris, Chenopodium album, Phytolacca americana, Taraxacum officinale*
Vegetable: *Amaranthus retroflexus, Brassica nigra, Plantago major, Rumex crispus, Urtica dioica*
Other
Water Indicator: *Hamamelis virginiana, Malus sylvestris, Prunus americana*

Montagnais
Drug
Analgesic: *Iris versicolor, Kalmia angustifolia, Maianthemum canadense, Ranunculus acris, Salix lucida*
Anthelmintic: *Populus tremuloides*
Blood Medicine: *Alnus* sp., *Hudsonia tomentosa, Ledum groenlandicum, Sorbus americana*
Cathartic: *Lycopodium dendroideum*
Cold Remedy: *Kalmia angustifolia, Pinus strobus*
Cough Medicine: *Anaphalis margaritacea, Gnaphalium obtusifolium, Hypericum perforatum, Picea mariana, Polygala polygama*
Dermatological Aid: *Platanthera orbiculata*
Diaphoretic: *Chimaphila umbellata, Streptopus roseus, Thuja occidentalis*
Dietary Aid: *Abies balsamea, Sorbus americana*
Diuretic: *Lonicera canadensis*
Expectorant: *Larix laricina*
Eye Medicine: *Viburnum opulus*
Febrifuge: *Achillea millefolium, Ledum groenlandicum, Lycopodium clavatum, Mitchella repens, Taxus canadensis*
Gastrointestinal Aid: *Kalmia angustifolia, Lycopodium dendroideum*
Liver Aid: *Ledum groenlandicum*
Misc. Disease Remedy: *Sarracenia purpurea*
Orthopedic Aid: *Cornus canadensis, Linnaea borealis, Moneses uniflora, Pteridium aquilinum*
Panacea: *Asarum canadense, Pontederia cordata, Pyrola asarifolia, Streptopus amplexifolius, Trientalis borealis*
Pediatric Aid: *Ledum groenlandicum, Pteridium aquilinum*
Poison: *Kalmia angustifolia*
Pulmonary Aid: *Picea rubens, Vaccinium macrocarpon*
Stimulant: *Aralia nudicaulis, Lycopodium clavatum, Pyrola elliptica, Taxus canadensis*
Throat Aid: *Picea rubens, Pinus strobus, Rhexia virginica*
Tonic: *Aralia nudicaulis, Picea glauca*
Tuberculosis Remedy: *Anaphalis margaritacea, Cirsium arvense, Gnaphalium obtusifolium, Pinus strobus, Trientalis borealis*
Food
Beverage: *Aralia nudicaulis, Rhexia virginica*
Dietary Aid: *Betula papyrifera*
Forage: *Aralia nudicaulis, Sorbus americana, Streptopus amplexifolius*
Fruit: *Sorbus americana*
Spice: *Thalictrum pubescens*
Dye
Black: *Salix* sp.
Green: *Taxus canadensis*
Other
Smoke Plant: *Salix lucida*

Montana Indian
Drug
Adjuvant: *Salix exigua, S. melanopsis*
Alterative: *Rubus parviflorus*
Analgesic: *Cicuta douglasii, Clematis hirsutissima, Zigadenus venenosus*
Antidiarrheal: *Artemisia tridentata, Geranium oreganum, Matricaria discoidea*
Antirheumatic (External): *Juniperus scopulorum, Pseudotsuga menziesii, Salix exigua, S. melanopsis, Sambucus cerulea, Zigadenus venenosus*

Burn Dressing: *Typha latifolia*
Cancer Treatment: *Orobanche fasciculata, Ulmus americana*
Cathartic: *Achillea millefolium, Apocynum androsaemifolium, Aralia nudicaulis*
Ceremonial Medicine: *Catabrosa aquatica, Glyceria fluitans, Juniperus scopulorum, Oplopanax horridus*
Cold Remedy: *Abies lasiocarpa, Clematis ligusticifolia, Salix exigua, S. melanopsis*
Dermatological Aid: *Abies lasiocarpa, Artemisia cana, Cornus sericea, Dalea purpurea, Erythronium grandiflorum, Geranium oreganum, Mentzelia laevicaulis, Pediomelum argophyllum, Sambucus cerulea, Yucca glauca, Zigadenus venenosus*
Dietary Aid: *Artemisia cana, Scirpus acutus*
Emetic: *Frangula purshiana, Iris missouriensis, Purshia tridentata, Sambucus cerulea*
Eye Medicine: *Artemisia tridentata*
Febrifuge: *Abies lasiocarpa, Apocynum androsaemifolium, Chimaphila umbellata, Mahonia repens, Salix exigua, S. melanopsis, Sambucus cerulea*
Gastrointestinal Aid: *Mahonia repens, Mentzelia laevicaulis, Sambucus cerulea*
Gynecological Aid: *Matricaria discoidea, Monarda fistulosa*
Kidney Aid: *Juniperus chinense, Mahonia repens*
Misc. Disease Remedy: *Rhus trilobata, Rubus parviflorus*
Oral Aid: *Echinacea angustifolia*
Orthopedic Aid: *Artemisia tridentata*
Poison: *Apocynum androsaemifolium, Cicuta maculata, Lomatium dissectum, Orobanche fasciculata*
Pulmonary Aid: *Abies lasiocarpa, Artemisia frigida, Purshia tridentata, Sambucus cerulea*
Snakebite Remedy: *Cicuta douglasii, Echinacea angustifolia*
Throat Aid: *Clematis ligusticifolia, Glycyrrhiza lepidota*
Tonic: *Apocynum androsaemifolium, Aralia nudicaulis, Artemisia cana, Glycyrrhiza lepidota, Mahonia repens*
Tuberculosis Remedy: *Artemisia frigida*
Unspecified: *Artemisia cana, Frangula purshiana, Mentha canadensis*
Venereal Aid: *Grindelia squarrosa, Pseudotsuga menziesii*
Veterinary Aid: *Clematis hirsutissima*

Food

Beverage: *Amelanchier alnifolia, Mahonia repens, Prunus virginiana, Sambucus cerulea*
Bread & Cake: *Amelanchier alnifolia, Camassia quamash, Helianthus annuus, Lomatium cous, Pediomelum esculentum, Prunus virginiana, Tsuga heterophylla, Vaccinium membranaceum*
Cooking Agent: *Nuphar lutea, Pediomelum esculentum*
Dried Food: *Arctostaphylos uva-ursi, Lomatium cous, Opuntia polyacantha, Pediomelum esculentum, Shepherdia argentea, S. canadensis, Thuja plicata, Vaccinium membranaceum*
Fodder: *Claytonia lanceolata, C. multicaulis, Opuntia polyacantha, Pascopyrum smithii, Populus angustifolia, P. balsamifera, P. deltoides, P. tremuloides*
Forage: *Amelanchier alnifolia, Bouteloua gracilis, Claytonia lanceolata, Erythronium grandiflorum, Fritillaria pudica, Pascopyrum smithii*
Fruit: *Amelanchier alnifolia, Arctostaphylos uva-ursi, Crataegus columbiana, Elaeagnus commutata, Fragaria sp., Mahonia repens, Moneses uniflora, Opuntia polyacantha, Parthenocissus quinquefolia, Prunus virginiana, Rhus trilobata, Ribes americanum, R. aureum, R. cereum, R. hudsonianum, R. lacustre, R. oxyacanthoides, R. viscosissimum, Rosa acicularis, R. nutkana, R. woodsii, Rubus idaeus, R. leucodermis, Sambucus cerulea, Shepherdia argentea, S. canadensis, Vaccinium membranaceum*
Ice Cream: *Shepherdia canadensis*
Pie & Pudding: *Amelanchier alnifolia, Prunus virginiana, Sambucus cerulea, Vaccinium membranaceum*
Porridge: *Helianthus annuus, Lomatium cous, Nuphar lutea, Pediomelum esculentum*
Preserves: *Amelanchier alnifolia, Mahonia repens, Opuntia polyacantha, Prunus virginiana, Sambucus cerulea, Shepherdia argentea, Vaccinium membranaceum*
Sauce & Relish: *Acer negundo, Claytonia cordifolia, C. parviflora, C. perfoliata, Scirpus acutus, Shepherdia argentea, Sium suave*
Snack Food: *Nuphar lutea*
Soup: *Amelanchier alnifolia, Arctostaphylos uva-ursi, Lomatium cous, Nuphar lutea, Prunus virginiana*
Staple: *Balsamorhiza sagittata, Camassia quamash, Chenopodium album, Lomatium ambiguum, L. cous, L. simplex, L. triternatum, Polygonum douglasii, Prunus virginiana, Scirpus acutus*
Starvation Food: *Alectoria fremontii, Arctostaphylos uva-ursi, Pinus contorta, Rosa sp.*
Substitution Food: *Pseudotsuga menziesii*
Sweetener: *Camassia quamash*
Unspecified: *Amaranthus blitoides, Astragalus crassicarpus, Balsamorhiza sagittata, Beckmannia syzigachne, Calochortus sp., Camassia scilloides, Carex sp., Claytonia lanceolata, C. multicaulis, Descurainia incana, Erythronium grandiflorum, Fritillaria pudica, Glycyrrhiza lepidota, Lewisia rediviva, Leymus condensatus, Lomatium ambiguum, L. cous, L. dissectum, L. simplex, L. triternatum, Matricaria discoidea, Mentzelia albicaulis, Microseris nutans, Nuphar lutea, Opuntia polyacantha, Oryzopsis hymenoides, Pediomelum esculentum, Perideridia gairdneri, Phragmites australis, Pinus albicaulis, P. flexilis, P. ponderosa, Populus angustifolia, P. balsamifera, P. deltoides, P. tremuloides, Pteridium aquilinum, Rosa sp., Rubus parviflorus, Rumex aquaticus, R. crispus, R. salicifolius, Sagittaria cuneata, Scirpus acutus, S. robustus, Thuja plicata, Triglochin maritimum, Typha latifolia, Valeriana edulis, Vicia americana, Wyethia amplexicaulis, W. helianthoides*
Vegetable: *Allium sp., Amaranthus blitoides, Argentina anserina, Balsamorhiza sagittata, Chenopodium album, Cirsium undulatum, Heracleum maximum, Lomatium simplex, L. triternatum, Oxyria digyna, Pediomelum esculentum, Perideridia gairdneri, Rumex aquaticus, R. crispus, R. paucifolius, R. salicifolius, Sarcobatus vermiculatus, Urtica dioica*
Winter Use Food: *Amelanchier alnifolia, Camassia scilloides, Crataegus columbiana, Fragaria sp., Lonicera involucrata, Pediomelum esculentum, Prunus virginiana, Shepherdia argentea, Vaccinium membranaceum*

Fiber

Basketry: *Juncus balticus, Linum lewisii, Phragmites australis, Pinus contorta, Salix sp., Thuja plicata, Xerophyllum tenax*
Brushes & Brooms: *Symphoricarpos occidentalis*
Building Material: *Juniperus scopulorum, Pinus contorta, P. ponderosa, Populus angustifolia, P. balsamifera, P. deltoides, P. tremuloides, Pseudotsuga menziesii, Thuja plicata*
Canoe Material: *Betula papyrifera, Pinus ponderosa, Thuja plicata*
Clothing: *Evernia vulpina, Thuja plicata*
Cordage: *Apocynum androsaemifolium, Linum lewisii, Populus tremuloides, Salix exigua, S. melanopsis, S. sp., Thuja plicata, Urtica dioica*
Furniture: *Salix sp.*
Mats, Rugs & Bedding: *Carex sp., Evernia vulpina, Juncus balticus, Linum lewisii, Salix sp., Scirpus acutus, Typha latifolia*
Scouring Material: *Equisetum hyemale*
Snow Gear: *Linum lewisii, Salix sp.*
Sporting Equipment: *Salix prolixa*

Dye

Orange: *Alnus incana*
Purple: *Lithospermum incisum, Purshia tridentata*
Yellow: *Bazzania trilobata, Evernia vulpina, Mahonia repens*

Other

Ceremonial Items: *Cornus sericea, Hierochloe odorata*

Cooking Tools: *Acer negundo, Salix* sp.
Decorations: *Clematis ligusticifolia*
Fasteners: *Salix* sp.
Fuel: *Populus angustifolia, P. balsamifera, P. deltoides, P. tremuloides*
Hide Preparation: *Alnus incana, Tsuga heterophylla*
Hunting & Fishing Item: *Acer glabrum, Cercocarpus ledifolius, Cornus sericea, Juniperus scopulorum, Linum lewisii, Philadelphus lewisii, Phragmites australis, Prunus virginiana, Salix* sp., *Symphoricarpos occidentalis, Taxus brevifolia, Thuja plicata*
Incense & Fragrance: *Hierochloe odorata, Matricaria discoidea, Mentha arvensis*
Insecticide: *Matricaria discoidea*
Musical Instrument: *Acer negundo, Salix* sp.
Paint: *Evernia vulpina, Prunus virginiana*
Season Indicator: *Bouteloua gracilis*
Smoke Plant: *Arctostaphylos uva-ursi, Chimaphila umbellata, Cornus sericea, Nicotiana quadrivalvis, Populus angustifolia, P. balsamifera, P. deltoides, P. tremuloides*
Smoking Tools: *Acer negundo, Phragmites australis*
Stable Gear: *Salix* sp.
Tools: *Artemisia tridentata, Asclepias speciosa, Cercocarpus ledifolius, Crataegus columbiana, Pinus ponderosa, Salix* sp., *Sambucus cerulea*
Toys & Games: *Salix* sp.
Walking Sticks: *Salix* sp.
Weapon: *Amelanchier alnifolia*

Montauk
Drug
Toothache Remedy: *Nicotiana tabacum*

Nanaimo
Food
Dessert: *Shepherdia canadensis*

Nanticoke
Drug
Analgesic: *Pinus echinata*
Anthelmintic: *Aristolochia serpentaria*
Burn Dressing: *Impatiens capensis*
Cathartic: *Pinus echinata*
Cold Remedy: *Acorus calamus, Monarda punctata, Symplocarpus foetidus*
Dermatological Aid: *Arctium minus, Impatiens capensis, Opuntia humifusa*
Diaphoretic: *Hedeoma pulegioides, Tanacetum vulgare*
Febrifuge: *Eupatorium perfoliatum, Hepatica nobilis, Sassafras albidum, Verbascum thapsus*
Gastrointestinal Aid: *Acorus calamus*
Kidney Aid: *Hedeoma pulegioides*
Liver Aid: *Hedeoma pulegioides*
Misc. Disease Remedy: *Chimaphila maculata, C. umbellata, Sassafras albidum*
Orthopedic Aid: *Baptisia tinctoria, Impatiens capensis, Pinus echinata, Populus deltoides, Yucca filamentosa*
Pediatric Aid: *Acorus calamus, Peltandra virginica*
Tonic: *Leiophyllum buxifolium*
Unspecified: *Peltandra virginica*

Narraganset
Drug
Cold Remedy: *Prunus serotina*

Natchez
Drug
Anthelmintic: *Chenopodium ambrosioides*
Antidiarrheal: *Eryngium yuccifolium, Vernonia* sp.
Cold Remedy: *Antennaria* sp., *Collinsia violacea, Penstemon* sp.
Cough Medicine: *Antennaria* sp., *Collinsia violacea, Penstemon* sp., *Tephrosia virginiana*
Dermatological Aid: *Rhus aromatica, R. hirta*
Diuretic: *Hypericum* sp.
Emetic: *Ilex vomitoria*
Febrifuge: *Aristolochia serpentaria, Chenopodium ambrosioides, Panicum* sp.
Hemostat: *Eryngium yuccifolium*
Kidney Aid: *Asclepias* sp.
Misc. Disease Remedy: *Panicum* sp.
Pediatric Aid: *Chenopodium ambrosioides, Hypericum hypericoides, H.* sp.
Pulmonary Aid: *Collinsia violacea, Penstemon* sp.
Tuberculosis Remedy: *Collinsia violacea*
Urinary Aid: *Hypericum hypericoides*
Venereal Aid: *Asclepias* sp.
Witchcraft Medicine: *Potentilla canadensis*

Navajo
Drug
Abortifacient: *Tetradymia canescens*
Analgesic: *Artemisia tridentata, A. tripartita, Clematis ligusticifolia, Cryptantha fulvocanescens, Dalea candida, Datura* sp., *Eriogonum alatum, Gentiana affinis, Gutierrezia sarothrae, Hedeoma drummondii, Juniperus osteosperma, Machaeranthera pinnatifida, Monarda fistulosa, M. pectinata, Portulaca oleracea, Sagittaria cuneata*
Antidiarrheal: *Eriogonum* sp., *Lactuca virosa*
Antidote: *Eupatorium purpureum*
Antiemetic: *Dugaldia hoopesii, Lactuca virosa, Yucca baccata*
Antihemorrhagic: *Krascheninnikovia lanata*
Antirheumatic (Internal): *Corydalis aurea, Mahonia repens, Mirabilis multiflora*
Blood Medicine: *Hymenopappus filifolius*
Burn Dressing: *Castilleja integra, Cosmos* sp., *Gaura parviflora, Phlox caespitosa*
Carminative: *Pectis angustifolia, Solanum fendleri*
Cathartic: *Chamaesyce lata, Euphorbia robusta, Machaeranthera parviflora, Phlox caespitosa, Prunus persica*
Ceremonial Medicine: *Artemisia tridentata, A. tripartita, Datura* sp., *Gutierrezia sarothrae, Hedeoma nana, Helianthus annuus, Ipomopsis longiflora, Juniperus scopulorum, Kochia scoparia, Lomatium dissectum, Mentha* sp., *Nicotiana* sp., *Oreoxis alpina, Parthenocissus vitacea, Pectis angustifolia, Phlox caespitosa, Picea* sp., *Pinus edulis, P. ponderosa, P.* sp., *Sphaeralcea angustifolia, Townsendia exscapa, Tribulus terrestris, Vicia faba, Zea mays*
Cold Remedy: *Artemisia tridentata, Cryptantha fulvocanescens, Lithospermum incisum, Sphaeralcea angustifolia, Tagetes micrantha*
Contraceptive: *Lithospermum incisum, L. ruderale, Ricinus communis*
Cough Medicine: *Cryptantha fulvocanescens, Ephedra viridis, Lithospermum incisum, Sphaeralcea angustifolia*
Dermatological Aid: *Abronia fragrans, Achillea millefolium, Artemisia tripartita, Atriplex canescens, A.* sp., *Camissonia tanacetifolia, Chrysothamnus greenei, Comandra umbellata, Eupatorium purpureum, Euphorbia robusta, Gaura parviflora, Gutierrezia sarothrae, Juniperus osteosperma, J. scopulorum, Kochia scoparia, Krascheninnikovia lanata, Lesquerella fendleri, Lithospermum incisum, Lupinus brevicaulis, L. lyallii, Machaeranthera alta, Mirabilis multiflora, Nicotiana* sp., *Oenothera* sp., *Opuntia* sp., *Orobanche fasciculata, Phoradendron juniperinum, Pinus edulis, Salsola australis, Sarcobatus vermiculatus, Senecio multicapitatus, Thelypodium wrightii, Xanthium strumarium, Yucca glauca*
Diaphoretic: *Sanvitalia abertii*
Dietary Aid: *Chenopodium album, Cryptantha* sp., *Helianthus annuus*

Disinfectant: *Lepidium lasiocarpum*
Diuretic: *Draba rectifructa, Penstemon barbatus, Phlox caespitosa*
Emetic: *Amelanchier* sp., *Chrysothamnus viscidiflorus, Cirsium vulgare, Cordylanthus ramosus, Eriogonum rotundifolium, Ipomopsis longiflora, Mentzelia multiflora, Pinus edulis*
Eye Medicine: *Datura* sp., *Solanum elaeagnifolium*
Febrifuge: *Artemisia tridentata, Cirsium neomexicanum, Datura* sp., *Shepherdia argentea, Tagetes micrantha*
Gastrointestinal Aid: *Abronia fragrans, Artemisia tridentata, Asclepias* sp., *Castilleja angustifolia, C. integra, C. lineata, Cercocarpus montanus, Chamaesyce fendleri, C. lata, Cucurbita pepo, Dalea candida, Ephedra trifurca, Fendlera rupicola, Helianthus nuttallii, Heuchera bracteata, Ipomopsis aggregata, I. longiflora, Lactuca virosa, Pectis angustifolia, Plantago patagonica, Portulaca oleracea, Solanum fendleri, Stenotus lanuginosus, Tagetes micrantha, Townsendia strigosa, Verbesina encelioides, Yucca baccata*
Gland Medicine: *Stanleya pinnata*
Gynecological Aid: *Ambrosia tenuifolia, Amelanchier utahensis, Artemisia tridentata, Asclepias hallii, Clematis ligusticifolia, Cordylanthus ramosus, Cryptantha fulvocanescens, Erigeron divergens, Eriogonum* sp., *Euphorbia robusta, Lappula occidentalis, Phlox caespitosa, Purshia tridentata, Rorippa alpina, Townsendia incana, T.* sp., *T. strigosa*
Heart Medicine: *Echinocereus coccineus*
Hemostat: *Cordylanthus ramosus, Lappula occidentalis, Sphaeralcea coccinea*
Kidney Aid: *Ephedra nevadensis, E. trifurca*
Laxative: *Plantago patagonica, Yucca glauca*
Love Medicine: *Vitis arizonica*
Misc. Disease Remedy: *Astragalus kentrophyta, Chrysothamnus greenei, Gaillardia pinnatifida, Juniperus* sp., *Krascheninnikovia lanata, Oxytropis* sp., *Sphaeralcea angustifolia*
Nose Medicine: *Asclepias verticillata, Machaeranthera canescens, Sisyrinchium mucronatum, Solanum elaeagnifolium, Stenotus lanuginosus, Townsendia strigosa, Zinnia grandiflora*
Oral Aid: *Heuchera bracteata, Mirabilis multiflora, Sanvitalia abertii, Stenotus lanuginosus*
Orthopedic Aid: *Cordylanthus ramosus*
Other: *Orobanche* sp., *Tagetes micrantha*
Panacea: *Argythamnia cyanophylla, Portulaca oleracea*
Pediatric Aid: *Lithospermum incisum, Plantago patagonica*
Poison: *Astragalus* sp., *Echinocereus coccineus. Opuntia polyacantha, Vicia faba, Yucca* sp.
Pulmonary Aid: *Dalea purpurea*
Reproductive Aid: *Lupinus brevicaulis*
Respiratory Aid: *Amorpha nana, Oxytropis* sp., *Physaria newberryi*
Sedative: *Ceanothus fendleri, Frasera speciosa, Gutierrezia sarothrae, Shinnersoseris rostrata*
Snakebite Remedy: *Gutierrezia sarothrae, Penstemon* sp.
Sports Medicine: *Artemisia tridentata*
Stimulant: *Achillea millefolium, Gentiana affinis, Picea* sp., *Rumex crispus, Thelesperma megapotamicum, Yucca glauca*
Strengthener: *Cupressus* sp.
Throat Aid: *Asclepias verticillata, Eriogonum rotundifolium, Machaeranthera canescens, Marrubium vulgare, Sisyrinchium mucronatum, Solanum elaeagnifolium, Stenotus lanuginosus, Townsendia strigosa, Yucca* sp., *Zea mays, Zinnia grandiflora*
Tonic: *Achillea millefolium, Clematis ligusticifolia*
Toothache Remedy: *Dalea candida, Datura* sp., *Heuchera bracteata, Phlox caespitosa, Stenotus lanuginosus, Thelesperma megapotamicum*
Unspecified: *Artemisia ludoviciana, Delphinium menziesii, D. scaposum, Eriogonum rotundifolium, Juniperus scopulorum, Prunus virginiana, Rumex hymenosepalus*
Venereal Aid: *Cordylanthus ramosus, C. wrightii, Ephedra nevadensis, E. trifurca, E. viridis, Ranunculus cymbalaria*
Veterinary Aid: *Atriplex confertifolia, Datura* sp., *D. wrightii, Gutierrezia microcephala, G. sarothrae, Ipomopsis longiflora, Muhlenbergia dubia, Verbascum thapsus*
Witchcraft Medicine: *Eriogonum fasciculatum, Gentiana affinis, Sphaeralcea coccinea*

Food

Baby Food: *Calochortus nuttallii*
Beverage: *Agave* sp., *Anthemis* sp., *Arctostaphylos pringlei, Dalea purpurea, Ephedra torreyana, E. viridis, Lycium pallidum, Opuntia phaeacantha, Penstemon* sp., *Phoradendron juniperinum, Rhus trilobata, Salix exigua, Tragopogon porrifolius, Zea mays*
Bread & Cake: *Amaranthus retroflexus, Artemisia carruthii, Chenopodium fremontii, C.* sp., *Cleome serrulata, Cupressus* sp., *Helianthus annuus, Opuntia phaeacantha, Oryzopsis hymenoides, Pinus edulis, Rhus trilobata, Sporobolus contractus, S. cryptandrus, Yucca baccata, Zea mays*
Candy: *Chloracantha spinosa, Dugaldia hoopesii, Hymenoxys richardsonii, Lactuca tatarica, Opuntia phaeacantha, Pinus edulis, Populus angustifolia, P. deltoides, Quercus pungens*
Cooking Agent: *Opuntia phaeacantha, Solanum elaeagnifolium*
Dessert: *Yucca baccata*
Dried Food: *Agave* sp., *Allium macropetalum, Amelanchier utahensis, Chenopodium album, Citrullus lanatus, Cleome serrulata, Cucumis melo, Cucurbita moschata, Cymopterus acaulis, Datura wrightii, Daucus pusillus, Lycium pallidum, Opuntia phaeacantha, O.* sp., *Rhus trilobata, Shepherdia argentea, Yucca baccata*
Fodder: *Artemisia filifolia, Atriplex argentea, A. canescens, Avena sativa, Bouteloua* sp., *Juniperus monosperma, Oryzopsis hymenoides, Salix exigua*
Forage: *Amaranthus blitoides, Atriplex argentea, A. canescens, Cercocarpus montanus, Chrysothamnus nauseosus, Dimorphocarpa wislizeni, Ipomopsis aggregata, Juniperus* sp., *Kochia scoparia, Krascheninnikovia lanata, Oryzopsis hymenoides, Oxytropis* sp., *Penstemon* sp., *Portulaca oleracea, Purshia stansburiana, P. tridentata, Salix exigua, Sarcobatus vermiculatus, Sisymbrium officinale, Sophora nuttalliana, Sporobolus cryptandrus, Yucca* sp.
Fruit: *Amelanchier alnifolia, A. utahensis, Arctostaphylos pringlei, Celtis laevigata, Cupressus* sp., *Datura wrightii, Echinocereus* sp., *Juniperus monosperma, Lycium pallidum, Opuntia phaeacantha, O.* sp., *Phoradendron juniperinum, Physalis lanceolata, P. pubescens, P. subulata, Prunus* sp., *Rhus trilobata, Ribes cereum, R. inerme, Rosa woodsii, Rubus arizonensis, Sarcobatus vermiculatus, Shepherdia argentea, Yucca baccata, Y. glauca*
Pie & Pudding: *Atriplex canescens*
Porridge: *Agave* sp., *Amaranthus blitoides, A. retroflexus, Arctostaphylos pringlei, Artemisia carruthii, Chenopodium* sp., *Cupressus* sp., *Descurainia pinnata, Helianthus annuus, Oryzopsis hymenoides, Pinus edulis, Prunus virginiana, Rhus trilobata, Rumex hymenosepalus, Sisymbrium altissimum, S. officinale, Suaeda moquinii, Yucca baccata, Zea mays*
Preserves: *Arctostaphylos pringlei, Cucurbita moschata, Opuntia phaeacantha, Physalis pubescens, Pinus edulis, Yucca baccata*
Sauce & Relish: *Allium cernuum, Cleome multicaulis, Cucumis melo, Cucurbita moschata, Opuntia* sp., *Yucca baccata*
Soup: *Agave* sp., *Allium cernuum, Cleome multicaulis, C. serrulata, Cymopterus acaulis, Lycium pallidum, Phaseolus vulgaris, Sisymbrium officinale*
Special Food: *Castilleja lineata, Cleome multicaulis, Dalea candida, Datura wrightii, Fragaria vesca, Lycium pallidum, Pinus edulis, Yucca baccata, Zea mays*
Spice: *Allium cernuum, Cleome serrulata, Cymopterus acaulis, C. purpureus*
Staple: *Amaranthus blitoides, A. palmeri, A.* sp., *Chenopodium album, C.* sp., *Opuntia phaeacantha, Oryzopsis hymenoides, Physalis pubescens, Pinus edulis, Quercus* sp., *Rhus trilobata, Zea mays*

Starvation Food: *Calochortus nuttallii, Juniperus monosperma, Sphaeralcea coccinea*

Substitution Food: *Cymopterus montanus, Thelesperma longipes, T. megapotamicum*

Sweetener: *Amaranthus palmeri, A.* sp.

Unspecified: *Agave* sp., *Allium cepa, A. cernuum, A. macropetalum, Amaranthus retroflexus, Artemisia carruthii, Asclepias* sp., *Calochortus aureus, C. luteus, C. nuttallii, Capsicum annuum, Cercis canadensis, Cleome serrulata, Cuscuta* sp., *Cymopterus montanus, Daucus pusillus, Eriogonum alatum, E. rotundifolium, Humulus lupulus, Juglans major, Lomatium orientale, Mammillaria* sp., *Mentzelia multiflora, Neomammillaria* sp., *Opuntia phaeacantha, O.* sp., *Oryzopsis hymenoides, Panicum capillare, Pinus edulis, Populus angustifolia, P. deltoides, Portulaca oleracea, Quercus gambelii, Q.* sp., *Rumex hymenosepalus, Salsola australis, Sphaeralcea angustifolia, Verbesina encelioides, Zea mays*

Vegetable: *Allium cernuum, A. macropetalum, Amaranthus blitoides, A. retroflexus, Chenopodium album, Cleome multicaulis, C. serrulata, Cucurbita pepo, Cymopterus acaulis, Phaseolus vulgaris, Rumex hymenosepalus, Salsola australis, Solanum jamesii, Zea mays*

Winter Use Food: *Allium cepa, A. cernuum, A. macropetalum, Amaranthus retroflexus, Amelanchier utahensis, Cleome serrulata, Daucus pusillus, Lycium pallidum, Yucca baccata, Y. glauca*

Fiber

Basketry: *Quercus* sp., *Rhus trilobata, Salix* sp., *Yucca* sp.

Brushes & Brooms: *Muhlenbergia cuspidata, M. pungens, M.* sp., *Senecio flaccidus*

Building Material: *Artemisia tridentata, Cercocarpus montanus, Chrysothamnus viscidiflorus, Juniperus monosperma, J. osteosperma, J.* sp., *Parthenocissus vitacea, Pinus edulis, P.* sp., *Populus deltoides, P.* sp., *Yucca* sp., *Zea mays*

Clothing: *Juniperus* sp., *Purshia tridentata, Yucca* sp.

Cordage: *Agave* sp., *Gossypium* sp., *Salix* sp., *Yucca* sp.

Furniture: *Pinus edulis, Populus angustifolia, P. deltoides, Quercus ×pauciloba, Salix* sp.

Mats, Rugs & Bedding: *Agave utahensis, Juniperus* sp., *Purshia stansburiana, Yucca* sp.

Sewing Material: *Gossypium* sp., *Rhus trilobata*

Dye

Black: *Gaultheria humifusa, Pinus edulis, Rhus trilobata*

Blue: *Delphinium scaposum, Lupinus lyallii, Rhus trilobata*

Brown: *Alnus incana, Ephedra viridis, Juglans major, J. regia, Mirabilis* sp., *Purshia stansburiana, Quercus pungens, Rumex hymenosepalus*

Green: *Allium* sp., *Chrysothamnus nauseosus, Iris* sp., *Juniperus monosperma, Lupinus lyallii, Prunus* sp., *Rumex hymenosepalus*

Mordant: *Rhus trilobata*

Orange: *Chrysothamnus viscidiflorus, Rumex hymenosepalus, Thelesperma subnudum*

Orange-Yellow: *Dahlia pinnata*

Purple: *Cerasus crenulata, Mirabilis* sp., *Prunus* sp.

Red: *Alnus incana, Helianthus annuus, Juniperus occidentalis, Mirabilis* sp., *Opuntia polyacantha, Prunus americana, Rumex hymenosepalus*

Red-Brown: *Celtis laevigata, Heuchera bracteata*

Red-Yellow: *Malus sylvestris*

Unspecified: *Juniperus* sp., *Nolina microcarpa*

Yellow: *Artemisia* sp., *Atriplex canescens, Chrysothamnus nauseosus, C.* sp., *C. viscidiflorus, Dugaldia hoopesii, Gutierrezia sarothrae, Hymenoxys richardsonii, Mahonia fremontii, Mirabilis* sp., *Petradoria pumila, Prunus persica, Rumex hymenosepalus*

Yellow-Brown: *Purshia stansburiana*

Other

Cash Crop: *Chrysothamnus nauseosus, Pinus edulis, Pseudotsuga menziesii*

Ceremonial Items: *Amaranthus albus, Arabis holboellii, Artemisia* sp., *Arundo donax, Aster* sp., *Astragalus allochrous, A. mollissimus, Bambusa* sp., *Bouteloua gracilis, Brickellia grandiflora, Campanula uniflora, Cercocarpus montanus, Chenopodium graveolens, C.* sp., *Chrysothamnus nauseosus, Dalea candida, Delphinium scaposum, Eriogonum alatum, Fendlera rupicola, Forestiera pubescens, Gutierrezia sarothrae, Helianthus annuus, Ipomopsis longiflora, Juniperus monosperma, J. occidentalis, J.* sp., *J. virginiana, Krascheninnikovia lanata, Lagenaria siceraria, Lithospermum incisum, Lupinus brevicaulis, L.* sp., *Mentzelia multiflora, Nicotiana attenuata, N. trigonophylla, Oxytropis lambertii, Phoradendron* sp., *Phragmites australis, Picea pungens, P.* sp., *Pinus edulis, P. flexilis, P.* sp., *Populus deltoides, P.* sp., *P. tremuloides, Prunus* sp., *P. virginiana, Pseudotsuga menziesii, Purshia stansburiana, Quercus pungens, Q.* sp., *Rhus trilobata, Rosa woodsii, Salix* sp., *Sarcobatus vermiculatus, Senecio flaccidus, Shinnersoseris rostrata, Thalictrum fendleri, Townsendia strigosa, Typha latifolia, Yucca baccata, Y. elata, Y. glauca, Y.* sp., *Zea mays*

Containers: *Artemisia tripartita, Juniperus* sp., *Lagenaria siceraria, Pinus* sp., *Populus deltoides, P.* sp., *Quercus* sp., *Rhus trilobata, Salix* sp., *Yucca* sp., *Zea mays*

Cooking Tools: *Agave* sp., *Lagenaria siceraria*

Decorations: *Ipomopsis aggregata, Juniperus monosperma, Rhus trilobata*

Designs: *Opuntia* sp.

Fasteners: *Opuntia polyacantha, Pinus edulis*

Fuel: *Chrysothamnus nauseosus, Juniperus monosperma, J.* sp., *Pinus edulis, P.* sp., *Populus deltoides, P.* sp., *Sarcobatus vermiculatus, Zea mays*

Good Luck Charm: *Juniperus monosperma*

Hunting & Fishing Item: *Fendlera rupicola, Fraxinus cuspidata, Helianthus annuus, Phragmites australis, Prosopis glandulosa, Purshia tridentata, Quercus* sp., *Rhus trilobata, Ribes cereum, Salix* sp.

Incense & Fragrance: *Cercis canadensis, Monarda punctata, Pinus edulis, Solidago nemoralis*

Insecticide: *Chenopodium* sp., *Fendlera rupicola*

Jewelry: *Cupressus* sp., *Juniperus* sp., *Pinus edulis, Punica granatum, Yucca* sp.

Musical Instrument: *Yucca baccata, Y.* sp.

Paint: *Quercus pungens*

Paper: *Artemisia filifolia, A. frigida*

Protection: *Gaura parviflora, Phoradendron* sp., *Ricinus communis, Yucca* sp.

Sacred Items: *Bouteloua hirsuta, Lycium pallidum, Nicotiana tabacum, Prunus virginiana, Rhus trilobata, Zea mays*

Smoke Plant: *Ambrosia tenuifolia, Psoralidium tenuiflorum, Vicia faba, Zea mays*

Soap: *Yucca baccata, Y. elata, Y. glauca, Y.* sp.

Stable Gear: *Fraxinus cuspidata*

Tools: *Acer negundo, Agave* sp., *Artemisia tridentata, Celtis laevigata, Cercocarpus montanus, Cleome multicaulis, Dipsacus fullonum, Fendlera rupicola, Gossypium* sp., *Gutierrezia sarothrae, Juncus* sp., *Picea* sp., *Pinus edulis, Populus deltoides, P.* sp., *Quercus ×pauciloba, Q. pungens, Q.* sp., *Ribes cereum, Rosa woodsii, Salix* sp., *Sarcobatus vermiculatus, Yucca* sp., *Zea mays*

Toys & Games: *Betula occidentalis, Cercocarpus montanus, Juniperus* sp., *Pinus* sp., *Populus* sp., *Purshia stansburiana, Quercus* sp., *Sarcobatus vermiculatus, Yucca elata, Y. glauca, Y.* sp.

Waterproofing Agent: *Pinus edulis, P.* sp., *Yucca* sp.

Weapon: *Juniperus monosperma, Quercus* sp., *Sarcobatus vermiculatus*

Navajo, Kayenta

Drug

Analgesic: *Achillea millefolium, Androsace septentrionalis, Aquilegia triternata, Cymopterus purpurascens, Equisetum laevigatum, Erigeron concinnus, E. divergens, Eriogonum racemosum, Euphorbia robusta, Juniperus scopulorum, Linum lewisii, Pseudotsuga menziesii, Shepherdia rotundifolia*

Anticonvulsive: *Arabis perennans, Mimulus eastwoodiae*

Antidiarrheal: *Asclepias nyctaginifolia, Corydalis aurea, Dracocephalum parviflorum, Gutierrezia sarothrae, Psilostrophe sparsiflora, Pyrola chlorantha*

Antidote: *Senecio neomexicanus*

Antiemetic: *Cymopterus purpurascens, Malacothrix sonchoides*

Blood Medicine: *Antennaria parvifolia, Conioselinum scopulorum, Ipomopsis gunnisonii, I. longiflora, Psilostrophe sparsiflora*

Burn Dressing: *Chenopodium album, Potentilla hippiana, Senecio neomexicanus*

Cathartic: *Abronia fragrans, Fendlera rupicola, Ipomopsis aggregata*

Ceremonial Medicine: *Apocynum cannabinum, Aquilegia triternata, Arctostaphylos patula, Artemisia campestris, Asclepias speciosa, Astragalus sesquiflorus, Brickellia californica, Bromus tectorum, Ceanothus fendleri, Chrysothamnus nauseosus, Cleome lutea, Cordylanthus wrightii, Cornus sericea, Epipactis gigantea, Eriogonum divaricatum, Fendlera rupicola, Frangula betulifolia, Gutierrezia sarothrae, Helianthus annuus, Heterotheca villosa, Lesquerella intermedia, Maianthemum stellatum, Mentzelia multiflora, Oenothera cespitosa, O. elata, O. pallida, Petrophyton caespitosum, Psathyrotes pilifera, Pseudocymopterus montanus, Pseudostellaria jamesiana, Ribes cereum, Shepherdia rotundifolia, Sphaeralcea coccinea, Tetraneuris argentea*

Cold Remedy: *Oenothera elata*

Dermatological Aid: *Abronia fragrans, Adiantum capillus-veneris, Astragalus calycosus, A. sesquiflorus, Brickellia californica, B. oblongifolia, Castilleja angustifolia, Chamaesaracha coronopus, Chamaesyce revoluta, Chenopodium capitatum, Clematis ligusticifolia, Cleome lutea, Conyza canadensis, Corallorrhiza maculata, Corydalis aurea, Cryptantha crassisepala, Cymopterus newberryi, Dalea candida, D. lanata, Descurainia incana, Dimorphocarpa wislizeni, Encelia frutescens, Erigeron neomexicanus, Eriogonum alatum, E. cernuum, E. inflatum, Erodium cicutarium, Frasera albomarginata, Gayophytum ramosissimum, Gilia leptomeria, Heuchera parvifolia, Hymenopappus filifolius, Ipomopsis aggregata, I. gunnisonii, Isocoma pluriflora, Iva xanthifolia, Lappula occidentalis, Leptodactylon pungens, Lygodesmia grandiflora, Mentha arvensis, Mentzelia multiflora, Mirabilis oxybaphoides, M. pumila, Nama hispidum, Oenothera brachycarpa, O. cespitosa, O. elata, O. pallida, Orobanche ludoviciana, Penstemon ambiguus, P. eatonii, Petradoria pumila, Phyla cuneifolia, Poliomintha incana, Potentilla hippiana, Psathyrotes pilifera, Pseudostellaria jamesiana, Psilostrophe sparsiflora, Psoralidium lanceolatum, Rhus trilobata, Ribes cereum, Senecio flaccidus, Shepherdia rotundifolia, Sphaeralcea coccinea, S. fendleri, Tetraneuris argentea, Vanclevea stylosa, Verbesina encelioides*

Diaphoretic: *Abronia fragrans*

Dietary Aid: *Sphaeralcea coccinea*

Disinfectant: *Artemisia campestris, Conyza canadensis, Corydalis aurea, Cryptantha cinerea, C.* sp., *Eriogonum umbellatum, Erodium cicutarium, Gaura parviflora, Gutierrezia sarothrae, Helianthus annuus, Linum lewisii, Lupinus pusillus, Mentha arvensis, Mentzelia multiflora, Penstemon ambiguus, P. eatonii, Pseudotsuga menziesii, Sambucus cerulea, Senecio neomexicanus, Sphaeralcea coccinea*

Ear Medicine: *Astragalus pattersonii, Conyza canadensis, Lupinus pusillus*

Emetic: *Abronia fragrans, Apocynum cannabinum, Arctostaphylos patula, Asclepias speciosa, Astragalus lonchocarpus, A. pattersonii, A. sesquiflorus, Atriplex canescens, Brickellia californica, Ceanothus fendleri, Chrysothamnus nauseosus, Cornus sericea, Eriogonum umbellatum, E. wrightii, Frangula betulifolia, Ipomopsis aggregata, Oenothera elata, O. pallida, Penstemon eatonii, P. jamesii, Psathyrotes pilifera, Pseudocymopterus montanus, Ribes cereum, Shepherdia rotundifolia, Wyethia scabra*

Eye Medicine: *Achillea millefolium, Astragalus pattersonii, Chenopodium capitatum, Comandra umbellata, Cryptantha flava, Lesquerella intermedia, Streptanthus cordatus, Thelypodium wrightii*

Febrifuge: *Achillea millefolium, Baccharis salicifolia, Cirsium rothrockii*

Gastrointestinal Aid: *Abronia fragrans, Astragalus* sp., *Atriplex canescens, Celtis laevigata, Conyza canadensis, Cryptantha flava, Cymopterus purpurascens, Epilobium angustifolium, Erigeron neomexicanus, Eriogonum* sp., *E. umbellatum, Gutierrezia sarothrae, Ipomopsis aggregata, Lepidium densiflorum, L. montanum, Mentzelia multiflora, Mirabilis linearis, Monarda pectinata, Penstemon eatonii, Pseudotsuga menziesii, Rhus trilobata, Rumex maritimus, Suaeda moquinii, Tiquilia latior, Wyethia scabra*

Gynecological Aid: *Acourtia wrightii, Aquilegia micrantha, Campanula parryi, Conioselinum scopulorum, Cordylanthus wrightii, Corydalis aurea, Cryptantha flava, Delphinium scaposum, Erigeron concinnus, Euphorbia incisa, Gaura parviflora, Gilia subnuda, Houstonia rubra, Ipomopsis longiflora, Lathyrus eucosmus, Mirabilis linearis, Oenothera cespitosa, Potentilla hippiana, Psilostrophe sparsiflora, Pyrola chlorantha, Townsendia incana*

Heart Medicine: *Polypogon monspeliensis, Sonchus asper*

Hemorrhoid Remedy: *Dimorphocarpa wislizeni*

Hemostat: *Aquilegia micrantha, Lupinus pusillus, Nicotiana attenuata, Penstemon eatonii, Pyrola chlorantha*

Kidney Aid: *Chaetopappa ericoides, Eriogonum cernuum, Leptodactylon pungens, Linum australe, Oenothera pallida*

Laxative: *Artemisia tridentata, Oxytropis lambertii*

Love Medicine: *Tradescantia occidentalis*

Misc. Disease Remedy: *Astragalus pattersonii, Cirsium rothrockii, Oenothera elata, Portulaca oleracea, Stephanomeria exigua*

Narcotic: *Comandra umbellata, Datura wrightii, Nicotiana attenuata, Petrophyton caespitosum, Stephanomeria pauciflora*

Nose Medicine: *Evolvulus nuttallianus*

Oral Aid: *Comandra umbellata, Sphaeralcea fendleri*

Orthopedic Aid: *Astragalus* sp., *Cymopterus purpurascens, Datura wrightii, Epilobium ciliatum, Equisetum laevigatum, Eriogonum divaricatum, E. racemosum, Penstemon eatonii*

Other: *Apocynum cannabinum, Astragalus calycosus, A. pattersonii, Camissonia multijuga, Chamaesaracha coronopus, Epipactis gigantea, Gaillardia pinnatifida, Geranium atropurpureum, Hymenopappus filifolius, Lepidium montanum, Sphaeralcea coccinea*

Panacea: *Abronia fragrans, Cirsium rothrockii, Dalea candida, Dracocephalum parviflorum, Eriogonum alatum, Frasera speciosa, Mahonia repens, Mirabilis linearis, Psilostrophe sparsiflora, Ranunculus cymbalaria, Rumex hymenosepalus*

Pediatric Aid: *Asclepias nyctaginifolia, Brickellia californica, B. oblongifolia, Conyza canadensis, Cryptantha cinerea, Dimorphocarpa wislizeni, Dracocephalum parviflorum, Epipactis gigantea, Helianthus annuus, Isocoma pluriflora, Lepidium densiflorum, Mentha arvensis, Penstemon jamesii, Pyrola chlorantha, Ribes cereum, Shepherdia rotundifolia, Solidago velutina, Thelypodium wrightii*

Poison: *Sonchus asper*

Poultice: *Astragalus calycosus, A. lonchocarpus*

Psychological Aid: *Adiantum capillus-veneris, Arabis perennans, Eriogonum jamesii, Gayophytum ramosissimum, Solidago velutina, Tetraneuris argentea, Thymophylla pentachaeta*

Respiratory Aid: *Asclepias asperula, A. auriculata, Conioselinum scopulorum*

Sedative: *Gilia leptomeria, Ipomopsis polycladon, Lepidium densiflorum, Quercus×pauciloba, Thelypodium wrightii*

Snakebite Remedy: *Artemisia tridentata, Conioselinum scopulorum, Cryptantha cinerea, Eremocrinum albomarginatum, Eriogonum divaricatum, Penstemon eatonii, P. palmeri, Tragia nepetifolia*

Strengthener: *Sphaeralcea coccinea*

Throat Aid: *Astragalus pattersonii, A.* sp., *Descurainia incana, Shepherdia rotundifolia, Symphoricarpos oreophilus*

Tonic: *Gilia leptomeria, Ipomopsis polycladon*

Toothache Remedy: *Asclepias involucrata, Dimorphocarpa wislizeni, Lycium pallidum, Phlox austromontana, Shepherdia rotundifolia*

Unspecified: *Townsendia incana*
Urinary Aid: *Chaetopappa ericoides*
Venereal Aid: *Kochia americana, Oenothera pallida*
Veterinary Aid: *Asclepias hallii, Collinsia parviflora, Corydalis aurea, Cryptantha cinerea, Dalea candida, Delphinium scaposum, Euphorbia incisa, Frasera speciosa, Iva xanthifolia, Oenothera pallida, Penstemon ambiguus, P. eatonii, Reverchonia arenaria, Sambucus cerulea, Thelypodiopsis elegans, Vicia americana, Zigadenus paniculatus*
Witchcraft Medicine: *Androsace septentrionalis, Euphorbia robusta, Fallugia paradoxa, Gaillardia pinnatifida*

Food
Beverage: *Sphaeralcea coccinea*
Candy: *Stephanomeria pauciflora*
Fruit: *Arctostaphylos patula, Mimulus eastwoodiae, Mirabilis linearis*
Porridge: *Carex sp., Eriogonum cernuum, Heliotropium convolvulaceum, Oxytropis lambertii, Plantago patagonica, Rumex maritimus*
Spice: *Encelia frutescens, Mentha arvensis*
Substitution Food: *Atriplex powellii*
Unspecified: *Artemisia campestris, Cercis canadensis, Chrysothamnus nauseosus, Comandra umbellata, Conioselinum scopulorum, Eriogonum racemosum, Gaura parviflora, Mirabilis linearis, Solidago canadensis, Vulpia octoflora, Zigadenus paniculatus*
Vegetable: *Antennaria parvifolia, Cymopterus newberryi, Lygodesmia grandiflora, Mirabilis oxybaphoides, Phacelia heterophylla, Phlox stansburyi, Streptanthus cordatus, Zigadenus paniculatus*

Fiber
Basketry: *Rhus trilobata*

Other
Ceremonial Items: *Astragalus lentiginosus, Cymopterus purpurascens, Dimorphocarpa wislizeni, Erodium cicutarium, Fendlera rupicola, Fraxinus anomala, Geranium atropurpureum, Gutierrezia sarothrae, Mahonia repens, Pyrola chlorantha, Rumex hymenosepalus, Stephanomeria pauciflora*
Decorations: *Monroa squarrosa, Oenothera cespitosa*
Fasteners: *Chaenactis stevioides*
Good Luck Charm: *Solidago canadensis*
Hunting & Fishing Item: *Thymophylla pentachaeta*
Protection: *Dalea flavescens, Phlox stansburyi, Tragia nepetifolia*
Smoke Plant: *Solidago canadensis*
Soap: *Polypogon monspeliensis*

Navajo, Ramah
Drug
Adjuvant: *Oenothera coronopifolia*
Analgesic: *Ageratina herbacea, Androsace septentrionalis, Apocynum cannabinum, Arabis perennans, Arenaria lanuginosa, Artemisia tridentata, Atriplex argentea, A. canescens, Bahia dissecta, Brickellia grandiflora, Campanula rotundifolia, Castilleja linariifolia, Chamaesyce albomarginata, C. fendleri, C. serpyllifolia, Chrysothamnus nauseosus, Clematis hirsutissima, C. ligusticifolia, Cordylanthus wrightii, Corydalis aurea, Dalea candida, Datura wrightii, Dracocephalum parviflorum, Erigeron concinnus, E. speciosus, Eriogonum jamesii, E. leptophyllum, Erysimum capitatum, Galium fendleri, Gutierrezia sarothrae, Heuchera novomexicana, H. parvifolia, Ipomopsis longiflora, Juniperus monosperma, J. scopulorum, Lotus wrightii, Marrubium vulgare, Menodora scabra, Monarda pectinata, M. punctata, Nicotiana attenuata, Oenothera coronopifolia, Oxalis drummondii, Penstemon barbatus, P. jamesii, Pericome caudata, Pinus edulis, Polygonum aviculare, P. ramosissimum, Potentilla norvegica, Prunus virginiana, Psilostrophe tagetina, Psoralidium lanceolatum, Quercus gambelii, Q. ×pauciloba, Ratibida tagetes, Rhus trilobata, Sanvitalia abertii, Senecio multilobatus, Tetradymia canescens, Thermopsis rhombifolia, Zinnia grandiflora*
Antidiarrheal: *Chamaesyce fendleri, C. serpyllifolia, Eriogonum alatum*
Antidote: *Amaranthus retroflexus, Bouteloua gracilis, Chenopodium album, C. incanum, Gutierrezia sarothrae, Krascheninnikovia lanata*
Antiemetic: *Gaillardia pinnatifida, Gaura coccinea*
Antihemorrhagic: *Sphaeralcea fendleri*
Antirheumatic (Internal): *Bahia dissecta, Ipomopsis longiflora*
Blood Medicine: *Castilleja integra, Cleome serrulata, Eriogonum racemosum*
Burn Dressing: *Castilleja integra, Mirabilis linearis, M. oblongifolia, Oenothera flava, Penstemon barbatus, Pinus edulis, Silene laciniata*
Cathartic: *Chrysothamnus nauseosus, Glycyrrhiza lepidota, Lotus wrightii, Orthocarpus purpureoalbus, Petradoria pumila, Psilostrophe tagetina, Quercus gambelii, Ratibida tagetes, Senecio multicapitatus, Zinnia grandiflora*
Ceremonial Medicine: *Achillea millefolium, Agastache pallidiflora, Agoseris aurantiaca, Amelanchier utahensis, Antennaria parvifolia, Apocynum cannabinum, A. ×floribundum, Arceuthobium campylopodum, A. vaginatum, Arctostaphylos pungens, A. uva-ursi, Asclepias asperula, A. tuberosa, Aster praealtus, Astragalus allochrous, A. bisulcatus, A. humistratus, A. kentrophyta, A. mollissimus, A. praelongus, Atriplex canescens, Baccharis wrightii, Besseya plantaginea, Brickellia grandiflora, Calochortus gunnisonii, Campanula rotundifolia, Carex microptera, Ceanothus fendleri, Chamaesyce fendleri, C. serpyllifolia, Chrysothamnus nauseosus, Cleome serrulata, Coreopsis tinctoria, Cornus sericea, Cosmos parviflorus, Cryptantha cinerea, Cuscuta megalocarpa, Cyperus esculentus, Datura wrightii, Dichanthelium oligosanthes, Draba helleriana, Echinochloa crus-pavonis, Eleocharis montevidensis, E. rostellata, Erigeron bellidiastrum, E. canus, E. divergens, E. eximius, E. flagellaris, Eriogonum alatum, E. annuum, Erysimum capitatum, Fallugia paradoxa, Forestiera pubescens, Galium fendleri, Gnaphalium stramineum, Gutierrezia sarothrae, Heterotheca villosa, Houstonia wrightii, Hymenoxys richardsonii, Ipomopsis longiflora, I. multiflora, Iris missouriensis, Juniperus communis, J. monosperma, J. scopulorum, Lactuca serriola, Lesquerella rectipes, Limosella aquatica, Lonicera arizonica, L. involucrata, Lupinus caudatus, Luzula multiflora, Lycium pallidum, L. torreyi, Mahonia repens, Maianthemum stellatum, Monarda pectinata, Monolepis nuttalliana, Nicotiana attenuata, Oenothera albicaulis, O. pallida, O. primiveris, Orthocarpus purpureoalbus, Parryella filifolia, Paxistima myrsinites, Penstemon jamesii, Pericome caudata, Peteria scoparia, Petradoria pumila, Picea engelmannii, Pinus edulis, P. flexilis, P. ponderosa, Polygonum lapathifolium, Potamogeton natans, Prunus virginiana, Pseudocymopterus montanus, Pseudotsuga menziesii, Psilostrophe tagetina, Psoralidium lanceolatum, Purshia stansburiana, P. tridentata, Pyrrhopappus pauciflorus, Quercus gambelii, Ranunculus cymbalaria, Ratibida tagetes, Ribes pinetorum, Rorippa palustris, Rosa woodsii, Rumex crispus, R. hymenosepalus, R. salicifolius, Salix exigua, S. lucida, Schoenocrambe linearifolia, Scirpus acutus, S. pallidus, Senecio fendleri, S. multilobatus, Sinapis arvensis, Stachys rothrockii, Symphoricarpos oreophilus, Tetraclea coulteri, Tetradymia canescens, Thalictrum fendleri, Thlaspi montanum, Tragopogon porrifolius, T. pratensis, Trifolium dubium, Typha latifolia, Verbena bracteata, V. macdougalii, Veronica americana, V. peregrina, Viola nephrophylla, Yucca baccata, Zinnia grandiflora*
Cold Remedy: *Chrysothamnus nauseosus, Juniperus scopulorum, Melilotus officinalis, Pinus edulis, Sanvitalia abertii, Tetradymia canescens*
Contraceptive: *Bahia dissecta, Erigeron speciosus, Eriogonum jamesii, Phlox stansburyi, Rhus trilobata, Yucca glauca*
Cough Medicine: *Agastache pallidiflora, Artemisia carruthii, A. frigida, A. tridentata, Atriplex canescens, Brickellia californica, B. eupatorioides, Chrysothamnus nauseosus, Cryptantha fendleri, Draba helleriana, Ephedra torreyana, Erigeron eximius, Eriogonum alatum, Humulus lupulus, Hymenopappus filifolius, Iva xanthifolia, Juniperus communis, J. monosperma, Mirabilis linearis, Monarda pectinata, M. punctata, Nicotiana attenuata, Penstemon barbatus, Pericome caudata, Pinus edulis, P. flexilis, P. ponderosa, Purshia stansburiana, Ratibida tagetes, Tetradymia canescens, Thermopsis rhombifolia, Valeriana acutiloba, Verbascum thapsus*

Dermatological Aid: *Abronia fragrans, Agastache pallidiflora, Agoseris aurantiaca, Allionia incarnata, Arenaria lanuginosa, Artemisia carruthii, A. dracunculus, A. tridentata, Asclepias tuberosa, Astragalus humistratus, Atriplex argentea, A. canescens, Besseya plantaginea, Bouteloua gracilis, B. simplex, Calochortus gunnisonii, Campanula parryi, Chamaesyce fendleri, C. serpyllifolia, Cheilanthes wootonii, Cleome serrulata, Convolvulus arvensis, Conyza canadensis, Cryptantha cinerea, Cystopteris fragilis, Dimorphocarpa wislizeni, Draba reptans, Drymaria glandulosa, Dyssodia papposa, Erigeron flagellaris, Eriogonum abertianum, E. alatum, E. annuum, E. cernuum, Geranium lentum, Grindelia nuda, Gutierrezia sarothrae, Helianthella parryi, Helianthus annuus, Heterotheca villosa, Heuchera novomexicana, Houstonia wrightii, Hymenopappus filifolius, Hymenoxys richardsonii, Ipomopsis longiflora, Juniperus monosperma, Krascheninnikovia lanata, Lappula occidentalis, Limosella aquatica, Lupinus argenteus, L. caudatus, L. kingii, Machaeranthera gracilis, Mahonia repens, Malacothrix fendleri, Mirabilis multiflora, Monarda fistulosa, Monolepis nuttalliana, Myosurus cupulatus, M. minimus, Oenothera primiveris, Opuntia macrorhiza, Oxalis drummondii, Parthenocissus vitacea, Pennisetum glaucum, Penstemon barbatus, P. fendleri, Pericome caudata, Peteria scoparia, Petradoria pumila, Phlox gracilis, P. stansburyi, Picradeniopsis oppositifolia, Potentilla hippiana, Psilostrophe tagetina, Rhus trilobata, Rumex crispus, Salsola australis, Sanvitalia abertii, Senecio fendleri, Silene douglasii, S. laciniata, S. noctiflora, Sphaeralcea fendleri, Stachys rothrockii, Talinum parviflorum, Taraxacum officinale, Thlaspi montanum, Tragopogon porrifolius, T. pratensis, Verbena bracteata, Woodsia neomexicana, Zigadenus elegans*

Diaphoretic: *Artemisia carruthii, A. tridentata, Juniperus monosperma, Pericome caudata*

Dietary Aid: *Hilaria jamesii, Plantago patagonica*

Disinfectant: *Agastache pallidiflora, Agoseris aurantiaca, Apocynum ×floribundum, Brickellia brachyphylla, Campanula rotundifolia, Carex inops, Coreopsis tinctoria, Dalea candida, Dichanthelium oligosanthes, Equisetum laevigatum, Erigeron canus, E. concinnus, E. divergens, E. flagellaris, Eriogonum annuum, Forestiera pubescens, Gutierrezia sarothrae, Heuchera novomexicana, Houstonia wrightii, Ipomopsis longiflora, Lathyrus eucosmus, Leptodactylon pungens, Lotus wrightii, Marrubium vulgare, Oenothera villosa, Phlox stansburyi, Plantago major, Purshia tridentata, Salix lucida, Sinapis arvensis, Stachys rothrockii, Tetradymia canescens, Tetraneuris argentea, Tradescantia occidentalis, Veronica peregrina*

Diuretic: *Besseya plantaginea, Hieracium fendleri*

Ear Medicine: *Pinus edulis*

Emetic: *Achillea millefolium, Agoseris aurantiaca, Amelanchier utahensis, Apocynum cannabinum, A. ×floribundum, Arctostaphylos pungens, A. uva-ursi, Asclepias asperula, Astragalus allochrous, A. bisulcatus, A. mollissimus, A. praelongus, Atriplex canescens, Baccharis wrightii, Besseya plantaginea, Brickellia grandiflora, Carex microptera, Ceanothus fendleri, Chrysothamnus nauseosus, Cornus sericea, Cuscuta megalocarpa, Cyperus esculentus, Draba helleriana, Echinochloa crus-pavonis, Eleocharis montevidensis, E. rostellata, Erysimum capitatum, Fallugia paradoxa, Forestiera pubescens, Galium fendleri, Gnaphalium stramineum, Grindelia nuda, Heterotheca villosa, Hymenoxys richardsonii, Iris missouriensis, Juniperus communis, J. monosperma, Lactuca serriola, Lonicera arizonica, L. involucrata, Lupinus caudatus, Luzula multiflora, Lycium pallidum, L. torreyi, Mahonia repens, Maianthemum stellatum, Monolepis nuttalliana, Oenothera pallida, Parryella filifolia, Paxistima myrsinites, Pericome caudata, Petradoria pumila, Picea engelmannii, Pinus edulis, P. flexilis, P. ponderosa, Potamogeton natans, Prunus virginiana, Pseudocymopterus montanus, Pseudotsuga menziesii, Purshia stansburiana, P. tridentata, Pyrrhopappus pauciflorus, Quercus gambelii, Ranunculus cymbalaria, Ribes pinetorum, Rosa woodsii, Rumex crispus, R. salicifolius, Salix exigua, S. lucida, Scirpus acutus, S. pallidus, Sinapis arvensis, Sisymbrium altissimum, Symphoricarpos oreophilus, Tetradymia canescens, Tragopogon porrifolius, Trifolium dubium, Typha latifolia, Veronica americana, V. peregrina, Viola nephrophylla, Zinnia grandiflora*

Eye Medicine: *Arenaria lanuginosa, Aster praealtus, Astragalus bisulcatus, Campanula rotundifolia, Cirsium calcareum, C. neomexicanum, C. undulatum, Draba helleriana, Dracocephalum parviflorum, Erigeron divergens, Ipomopsis longiflora, Lesquerella rectipes, Lithospermum incisum, Machaeranthera gracilis, Malacothrix fendleri, Mentzelia laciniata, M. multiflora, Psilostrophe tagetina, Rorippa palustris, Schoenocrambe linearifolia, Thermopsis rhombifolia, Vicia americana*

Febrifuge: *Agastache pallidiflora, Ageratina herbacea, Arenaria lanuginosa, Artemisia carruthii, Brickellia californica, Chrysothamnus nauseosus, Dalea candida, Dracocephalum parviflorum, Erigeron eximius, Gilia inconspicua, Gutierrezia sarothrae, Juniperus communis, J. monosperma, J. scopulorum, Mentha arvensis, Monarda pectinata, M. punctata, Pericome caudata, Pinus edulis, P. flexilis, P. ponderosa, Purshia tridentata, Ratibida columnifera, R. tagetes, Sanvitalia abertii, Tetraclea coulteri, Tetradymia canescens, Verbascum thapsus, Verbena macdougalii*

Gastrointestinal Aid: *Apocynum cannabinum, Aster praealtus, Atriplex canescens, Brassica juncea, Carex inops, Castilleja linariifolia, Cercocarpus montanus, Chamaesyce albomarginata, C. fendleri, C. serpyllifolia, Convolvulus arvensis, Corydalis aurea, Dalea candida, Dyssodia papposa, Ephedra torreyana, Eriogonum jamesii, Gaillardia pinnatifida, Gaura hexandra, Grindelia nuda, Gutierrezia sarothrae, Heterotheca villosa, Heuchera parvifolia, Ipomopsis longiflora, Juniperus monosperma, J. scopulorum, Linum lewisii, L. puberulum, Lotus wrightii, Machaeranthera tanacetifolia, Marrubium vulgare, Menodora scabra, Myosurus cupulatus, Oenothera coronopifolia, Orthocarpus purpureoalbus, Penstemon barbatus, Phoradendron juniperinum, Picradeniopsis oppositifolia, Polygonum aviculare, P. ramosissimum, Prunus virginiana, Pseudocymopterus montanus, Psilostrophe tagetina, Psoralidium lanceolatum, Ratibida tagetes, Rhus trilobata, Sarcobatus vermiculatus, Senecio fendleri, Sphaeralcea digitata, Tetradymia canescens, Tetraneuris argentea, Veronica peregrina, Zinnia grandiflora*

Gynecological Aid: *Androsace occidentalis, Artemisia tridentata, Bahia dissecta, Bouteloua gracilis, Calochortus gunnisonii, Capsicum annuum, Castilleja integra, C. linariifolia, Cercocarpus montanus, Chamaesyce fendleri, C. serpyllifolia, Chrysothamnus depressus, C. nauseosus, Cordylanthus wrightii, Corydalis aurea, Echeandia flavescens, Erigeron speciosus, Eriogonum jamesii, E. leptophyllum, Erysimum capitatum, Euphorbia lurida, Gutierrezia sarothrae, Heuchera parvifolia, Ipomopsis longiflora, Juniperus monosperma, Leptodactylon pungens, Marrubium vulgare, Menodora scabra, Opuntia macrorhiza, Pennellia micrantha, Penstemon barbatus, P. linarioides, Pericome caudata, Phlox stansburyi, Psoralidium lanceolatum, Purshia tridentata, Quercus gambelii, Rhus trilobata, Rumex hymenosepalus, Sanvitalia abertii, Senecio multicapitatus, S. multilobatus, Stephanomeria pauciflora, Taraxacum officinale, Townsendia exscapa, Yucca baccata, Y. glauca*

Hallucinogen: *Datura wrightii*

Heart Medicine: *Heterotheca villosa*

Hemostat: *Chamaesyce albomarginata, C. fendleri, C. serpyllifolia, Erigeron flagellaris, Gayophytum ramosissimum, Trifolium dubium*

Herbal Steam: *Picradeniopsis oppositifolia, Tetradymia canescens*

Hunting Medicine: *Antennaria rosulata, Aster praealtus, Besseya plantaginea, Campanula rotundifolia, Castilleja miniata, Cercocarpus montanus, Chamaebatiaria millefolium, Datura wrightii, Erigeron eximius, E. formosissimus, Helianthus petiolaris, Hieracium fendleri, Humulus lupulus, Ipomopsis aggregata, Limosella aquatica, Lonicera utahensis, Mirabilis linearis, Monolepis nuttalliana, Oenothera villosa, Pinus flexilis, Purshia tridentata, Ranunculus inamoenus, Senecio neomexicanus, Thermopsis rhombifolia, Valeriana acutiloba, Verbesina encelioides, Veronica peregrina*

Internal Medicine: *Aster praealtus, Machaeranthera gracilis, Sidalcea neomexicana, Tradescantia occidentalis*

Kidney Aid: *Draba helleriana, Juniperus scopulorum*

Laxative: *Mahonia repens*

Misc. Disease Remedy: *Artemisia carruthii, Asclepias tuberosa, Brickellia grandiflora, Erigeron eximius, Galium fendleri, Hedeoma drummondii, Holodiscus discolor, Humulus lupulus, Iva xanthifolia, Lycium pallidum, L. torreyi, Marrubium vulgare, Mentha arvensis, Monarda pectinata, Pericome caudata, Peteria scoparia, Pinus edulis, Psoralidium tenuiflorum, Salsola australis, Valeriana acutiloba*

Narcotic: *Datura wrightii*

Nose Medicine: *Atriplex canescens, Chaetopappa ericoides, Heterotheca villosa, Polygonum douglasii*

Oral Aid: *Abronia fragrans, Eriogonum alatum, Phlox gracilis, Rumex crispus, Sanvitalia abertii, Schkuhria multiflora, Schoenocrambe linearifolia*

Orthopedic Aid: *Agoseris aurantiaca, Clematis columbiana, C. ligusticifolia, Corydalis aurea, Hackelia floribunda, Heuchera novomexicana, Menodora scabra, Mirabilis oxybaphoides, Oenothera albicaulis, Tetradymia canescens*

Other: *Atriplex argentea, Brickellia eupatorioides, Corydalis aurea, Eriogonum annuum, Helianthus annuus, Malva neglecta, Pinus edulis, Tetraneuris argentea, Tripterocalyx carnea*

Panacea: *Acer glabrum, Agoseris aurantiaca, Amelanchier utahensis, Androsace septentrionalis, Arabis fendleri, Arenaria fendleri, Argythamnia cyanophylla, Artemisia carruthii, A. frigida, A. ludoviciana, Astragalus humistratus, A. kentrophyta, Besseya plantaginea, Bouteloua gracilis, Calliandra humilis, Calochortus aureus, C. gunnisonii, Calylophus hartwegii, Cercocarpus montanus, Cheilanthes wootonii, Cirsium neomexicanum, C. undulatum, Coreopsis tinctoria, Cryptantha cinerea, Cymopterus bulbosus, Dalea candida, Draba helleriana, Erigeron divergens, Eriogonum alatum, E. annuum, E. jamesii, E. leptophyllum, Fragaria vesca, Gaura coccinea, Geranium atropurpureum, G. lentum, G. richardsonii, Glandularia wrightii, Gnaphalium stramineum, Gutierrezia sarothrae, Helianthella parryi, Helianthus petiolaris, Heliomeris longifolia, Heterotheca villosa, Heuchera novomexicana, H. parvifolia, Houstonia wrightii, Hymenopappus filifolius, Ipomopsis longiflora, Krascheninnikovia lanata, Linanthus nuttallii, Linum puberulum, Lithospermum incisum, L. multiflorum, Lotus wrightii, Lupinus kingii, Lycium pallidum, L. torreyi, Menodora scabra, Myosurus aristatus, Oenothera albicaulis, O. cespitosa, O. coronopifolia, O. elata, O. flava, Orobanche fasciculata, Penstemon barbatus, P. virgatus, Physalis pubescens, Picradeniopsis oppositifolia, Plantago major, Polygonum ramosissimum, Potentilla crinita, P. hippiana, P. pensylvanica, Prunus virginiana, Quercus gambelii, Ratibida tagetes, Rumex crispus, R. salicifolius, Sanvitalia abertii, Senecio fendleri, Silene drummondii, Sphaeralcea coccinea, S. digitata, Stephanomeria pauciflora, Tetraneuris argentea, Tragia nepetifolia, Vicia americana*

Pediatric Aid: *Androsace occidentalis, Antennaria rosulata, Besseya plantaginea, Brickellia brachyphylla, Capsicum annuum, Cryptantha cinerea, Eriogonum alatum, Gaura coccinea, Grindelia nuda, Hilaria jamesii, Juniperus monosperma, Orobanche fasciculata, Ratibida tagetes, Salix lucida, Senecio fendleri*

Poison: *Hackelia floribunda, Hordeum jubatum, Hymenoxys richardsonii, Toxicodendron radicans, Urtica dioica, Yucca glauca*

Preventive Medicine: *Thlaspi montanum*

Psychological Aid: *Frasera speciosa, Verbascum thapsus*

Respiratory Aid: *Arenaria fendleri, A. lanuginosa, Clematis hirsutissima, Erysimum capitatum, Gaillardia pinnatifida, Lesquerella rectipes, Machaeranthera gracilis, M. tanacetifolia*

Snakebite Remedy: *Artemisia filifolia, Aster falcatus, A. praealtus, Chaetopappa ericoides, Conyza canadensis, Cryptantha fulvocanescens, Erigeron divergens, E. flagellaris, Eriogonum leptophyllum, Gutierrezia sarothrae, Mentzelia albicaulis, Oenothera pallida, Sanvitalia abertii*

Stimulant: *Juniperus monosperma, Mentha arvensis*

Strengthener: *Frasera speciosa, Verbascum thapsus*

Throat Aid: *Bouteloua simplex, Cleome serrulata, Corydalis aurea, Oenothera albicaulis, O. flava, O. pallida, Penstemon jamesii, Psilostrophe tagetina, Sanvitalia abertii, Tragopogon pratensis*

Toothache Remedy: *Artemisia frigida, Astragalus bisulcatus, Atriplex canescens, Chaetopappa ericoides, Chamaesyce fendleri, C. serpyllifolia, Cryptantha fulvocanescens, Descurainia pinnata, D. sophia, Erysimum capitatum, Heterotheca villosa, Lesquerella rectipes, Lycium pallidum, L. torreyi, Mentzelia albicaulis, Pennellia micrantha, Pericome caudata, Sanvitalia abertii*

Tuberculosis Remedy: *Valeriana acutiloba*

Urinary Aid: *Gutierrezia sarothrae*

Venereal Aid: *Androsace septentrionalis, Arenaria lanuginosa, Baccharis wrightii, Coreopsis tinctoria, Draba helleriana, Heterotheca villosa, Heuchera parvifolia, Potentilla norvegica, Psoralidium lanceolatum, Ratibida tagetes*

Veterinary Aid: *Artemisia carruthii, A. tridentata, Asclepias asperula, Atriplex canescens, Bouteloua gracilis, Cerastium beeringianum, Chamaesyce fendleri, C. serpyllifolia, Cirsium neomexicanum, C. undulatum, Commelina dianthifolia, Dalea candida, Datura wrightii, Echeandia flavescens, Elymus trachycaulus, Erigeron flagellaris, Eriogonum abertianum, Frasera speciosa, Grindelia nuda, Gutierrezia sarothrae, Ipomopsis longiflora, Juniperus monosperma, Lathyrus eucosmus, Mirabilis linearis, Nicotiana attenuata, Penstemon barbatus, Peteria scoparia, Psoralidium tenuiflorum, Ratibida columnifera, Rumex hymenosepalus, Silene douglasii, Solanum triflorum, Sporobolus cryptandrus, Talinum parviflorum, Tradescantia occidentalis, T. pinetorum, Tragopogon porrifolius, T. pratensis, Verbascum thapsus*

Witchcraft Medicine: *Agastache pallidiflora, Agoseris aurantiaca, Androsace septentrionalis, Antennaria parvifolia, A. rosulata, Aster oblongifolius, Besseya plantaginea, Campanula rotundifolia, Castilleja miniata, Clematis hirsutissima, C. ligusticifolia, Datura wrightii, Draba helleriana, Erigeron eximius, Eriogonum annuum, Heliomeris multiflora, Humulus lupulus, Iva xanthifolia, Juniperus scopulorum, Limosella aquatica, Myosurus aristatus, Pericome caudata, Peteria scoparia, Psoralidium lanceolatum, Solidago velutina, Tetradymia canescens, Tetraneuris argentea, Thermopsis rhombifolia, Thlaspi montanum, Tripterocalyx carnea, Valeriana acutiloba*

Food

Beverage: *Mirabilis multiflora, M. oxybaphoides, Thelesperma megapotamicum*

Bread & Cake: *Amaranthus retroflexus, Chenopodium album, C. incanum, Descurainia pinnata, D. sophia, Prunus virginiana, Yucca baccata*

Candy: *Lygodesmia juncea, Pinus edulis, Senecio multicapitatus, Stephanomeria pauciflora*

Dried Food: *Allium cernuum, Cleome serrulata, Cymopterus bulbosus, Lycium pallidum, L. torreyi, Machaeranthera gracilis, Opuntia macrorhiza, Rhus trilobata*

Fodder: *Aristida purpurea, Atriplex canescens, A. rosea, Avena sativa, Beckmannia syzigachne, Bromus anomalus, B. tectorum, Chamaebatiaria millefolium, Cleome serrulata, Corydalis aurea, Cryptantha cinerea, C. fendleri, Descurainia pinnata, D. sophia, Echinochloa crus-pavonis, Elymus elymoides, E. trachycaulus, Erodium cicutarium, Helianthus annuus, Heliomeris longifolia, H. multiflora, Heterotheca villosa, Lappula occidentalis, Lepidium montanum, Marrubium vulgare, Medicago sativa, Monolepis nuttalliana, Muhlenbergia filiformis, M. mexicana, Oryzopsis hymenoides, Panicum capillare, P. obtusum, Phaseolus vulgaris, Poa fendleriana, Rorippa teres, Salsola australis, Sarcobatus vermiculatus, Stipa robusta, Tridens muticus, Triticum aestivum, Zea mays*

Forage: *Astragalus mollissimus, Bouteloua gracilis, B. simplex, Carex microptera, Hilaria jamesii, Panicum obtusum*

Fruit: *Amelanchier utahensis, Echinocereus coccineus, Juniperus deppeana, Lycium pallidum, L. torreyi, Opuntia macrorhiza, Parthenocissus vitacea, Physalis pubescens, Prunus persica, P. virginiana, Rhus trilobata, Ribes cereum, R. pinetorum, Rosa woodsii, Yucca glauca*

Pie & Pudding: *Solanum tuberosum*

Porridge: *Atriplex rosea, Eriogonum alatum, Oryzopsis hymenoides, Sporobolus cryptandrus, S. wrightii*

Preserves: *Pinus edulis, Yucca baccata*
Special Food: *Amaranthus retroflexus, Chenopodium album, C. incanum, Zea mays*
Spice: *Atriplex canescens, Capsicum annuum, Cucurbita maxima, C. pepo, Portulaca oleracea*
Staple: *Amaranthus albus, A. cruentus, Mentzelia multiflora, Quercus gambelii*
Starvation Food: *Pinus edulis, Populus tremuloides, Senecio multicapitatus*
Unspecified: *Allium cepa, A. cernuum, A. drummondii, Calochortus aureus, C. gunnisonii, Ceanothus fendleri, Chenopodium leptophyllum, C. watsonii, Citrullus lanatus, Cleome serrulata, Cucumis melo, Cucurbita maxima, C. pepo, Cymopterus bulbosus, Eriogonum alatum, Fragaria vesca, Helianthus annuus, Juniperus monosperma, Mammillaria wrightii, Orobanche fasciculata, Phaseolus lunatus, Pinus ponderosa, Pseudocymopterus montanus, Quercus grisea, Q. ×pauciloba, Rhus trilobata, Triticum aestivum, Typha latifolia, Yucca glauca*
Vegetable: *Amaranthus retroflexus, Pisum sativum, Portulaca oleracea, Solanum jamesii, Zea mays*
Winter Use Food: *Amaranthus retroflexus, Chenopodium album, C. incanum, Citrullus lanatus, Cucurbita maxima, C. pepo, Cymopterus bulbosus, Juniperus deppeana, Opuntia macrorhiza, Phaseolus vulgaris, Pinus edulis, Prunus virginiana, Solanum jamesii, S. tuberosum, Zea mays*

Fiber
Basketry: *Parryella filifolia, Pinus edulis, Rhus trilobata, Typha latifolia, Yucca baccata*
Brushes & Brooms: *Aristida purpurea, Artemisia tridentata, Muhlenbergia andina, M. mexicana, Phleum pratense, Yucca baccata*
Building Material: *Juniperus monosperma, Pinus edulis, P. ponderosa, Populus ×acuminata, Quercus gambelii, Salix lucida*
Clothing: *Juniperus monosperma, Purshia stansburiana, Rhus trilobata*
Cordage: *Asclepias subverticillata, Yucca baccata, Y. glauca*
Furniture: *Juniperus monosperma, Pinus ponderosa, Quercus gambelii*
Mats, Rugs & Bedding: *Chrysothamnus nauseosus, Purshia stansburiana, P. tridentata, Typha latifolia*
Scouring Material: *Festuca idahoensis, Koeleria macrantha, Sporobolus cryptandrus*
Snow Gear: *Pinus ponderosa, Yucca baccata*
Sporting Equipment: *Yucca baccata*

Dye
Black: *Atriplex rosea, Pinus edulis, Quercus ×pauciloba, Rhus trilobata, Thalictrum fendleri, Yucca glauca*
Brown: *Cercocarpus montanus*
Orange-Yellow: *Thelesperma megapotamicum*
Red: *Atriplex canescens, Cercocarpus montanus, Yucca glauca*
Red-Brown: *Alnus incana*
Unspecified: *Juniperus monosperma, Quercus gambelii, Yucca baccata*
Yellow: *Atriplex canescens, Chrysothamnus nauseosus, Tetradymia canescens*
Yellow-Brown: *Rumex hymenosepalus*

Other
Cash Crop: *Phaseolus vulgaris, Pinus edulis*
Ceremonial Items: *Ambrosia acanthicarpa, Amelanchier utahensis, Artemisia biennis, A. campestris, A. carruthii, A. frigida, A. ludoviciana, Cercocarpus montanus, Chrysothamnus nauseosus, Delphinium nuttallianum, D. scaposum, D. tenuisectum, Forestiera pubescens, Gutierrezia sarothrae, Helianthus annuus, Lagenaria siceraria, Lesquerella fendleri, Lycium pallidum, L. torreyi, Mentzelia multiflora, Opuntia whipplei, Pedicularis procera, Pinus edulis, Plantago patagonica, Populus ×acuminata, P. tremuloides, Prunus virginiana, Pseudotsuga menziesii, Purshia tridentata, Quercus gambelii, Rhus trilobata, Rosa woodsii, Salix exigua, S. lucida, Tetradymia canescens, Townsendia exscapa, Verbesina encelioides, Xanthium strumarium, Yucca baccata, Zea mays*
Containers: *Juniperus monosperma, Lagenaria siceraria, Pinus edulis, P. ponderosa, Rhus trilobata, Typha latifolia, Zea mays*
Cooking Tools: *Populus tremuloides, Sarcobatus vermiculatus*
Fasteners: *Yucca baccata*
Fertilizer: *Corydalis aurea, Rorippa sinuata, Solanum physalifolium, S. triflorum*
Fuel: *Juniperus monosperma, Pinus edulis, P. ponderosa*
Good Luck Charm: *Arctostaphylos pungens, A. uva-ursi, Euphorbia lurida, Hackelia floribunda, Juniperus communis, Toxicodendron radicans, Typha latifolia*
Hunting & Fishing Item: *Delphinium scopulorum, Fraxinus cuspidata, Juniperus monosperma, Rhus trilobata, Ribes cereum, R. pinetorum*
Incense & Fragrance: *Croton texensis*
Insecticide: *Grindelia nuda, Heterotheca villosa, Verbesina encelioides*
Lighting: *Juniperus monosperma*
Protection: *Chenopodium graveolens, Quercus grisea, Q. ×pauciloba, Tetradymia canescens, Typha latifolia, Verbesina encelioides*
Season Indicator: *Ribes cereum*
Smoke Plant: *Juniperus communis, Nicotiana attenuata, Zea mays*
Soap: *Yucca baccata, Y. glauca*
Stable Gear: *Pinus edulis, P. ponderosa*
Tools: *Cercocarpus montanus, Fraxinus cuspidata, Lagenaria siceraria, Pinus edulis, Populus ×acuminata, Quercus gambelii, Salix lucida, Zea mays*
Toys & Games: *Pinus edulis, Salix lucida, Stipa comata*
Weapon: *Rhus trilobata*

Neeshenam
Drug
Antirheumatic (External): *Galium* sp.
Antirheumatic (Internal): *Quercus* sp.
Burn Dressing: *Quercus* sp.
Cold Remedy: *Quercus* sp.
Cough Medicine: *Quercus* sp.
Dermatological Aid: *Quercus* sp.
Misc. Disease Remedy: *Croton setigerus*
Psychological Aid: *Quercus* sp.
Toothache Remedy: *Frangula californica*

Food
Bread & Cake: *Blennosperma nanum, Bromus carinatus, Madia elegans, Quercus gambelii, Ranunculus californicus*
Fruit: *Heteromeles arbutifolia*
Porridge: *Blennosperma nanum, Bromus carinatus, Madia elegans, Quercus gambelii, Ranunculus californicus*
Staple: *Blennosperma nanum, Bromus carinatus, Madia elegans, Ranunculus californicus*
Unspecified: *Allium cepa, Chlorogalum pomeridianum, Dichelostemma pulchellum, D. volubile, Quercus kelloggii, Q. wislizeni, Sanicula sp., S. tuberosa, Triteleia hyacinthina, T. ixioides*
Vegetable: *Angelica* sp., *Claytonia perfoliata, Dudleya lanceolata, Eschscholzia californica, Mimulus tilingii, Sanicula* sp.

Fiber
Basketry: *Cercis canadensis*
Clothing: *Juncus* sp.
Cordage: *Asclepias* sp.

Other
Hunting & Fishing Item: *Asclepias* sp.
Smoke Plant: *Nicotiana plumbaginifolia*

Nevada Indian
Drug
Cold Remedy: *Corallorrhiza maculata, Eriogonum umbellatum*
Dermatological Aid: *Clematis ligusticifolia*
Heart Medicine: *Chaenactis stevioides, Peniocereus greggii*
Kidney Aid: *Iris missouriensis*

Love Medicine: *Castilleja* sp.
Pediatric Aid: *Chaenactis stevioides*
Urinary Aid: *Iris missouriensis*
Veterinary Aid: *Lomatium dissectum, Rumex crispus*
Food
Pie & Pudding: *Perideridia gairdneri*
Unspecified: *Perideridia gairdneri*
Winter Use Food: *Perideridia gairdneri*

Nez Perce
Drug
Antidiarrheal: *Thuja plicata*
Antirheumatic (External): *Populus balsamifera, P. deltoides*
Blood Medicine: *Lewisia rediviva*
Cold Remedy: *Juniperus scopulorum, Larix occidentalis, Thuja plicata*
Cough Medicine: *Larix occidentalis, Thuja plicata*
Dermatological Aid: *Lomatium dissectum, Rhus glabra*
Dietary Aid: *Lomatium dissectum*
Emetic: *Salix* sp.
Eye Medicine: *Lomatium dissectum*
Febrifuge: *Juniperus scopulorum, Symphoricarpos albus*
Gastrointestinal Aid: *Salix* sp.
Gynecological Aid: *Lewisia rediviva*
Pediatric Aid: *Symphoricarpos albus, Thuja plicata*
Pulmonary Aid: *Juniperus scopulorum*
Respiratory Aid: *Lomatium dissectum*
Throat Aid: *Larix occidentalis*
Tuberculosis Remedy: *Lomatium dissectum*
Veterinary Aid: *Clematis ligusticifolia, Populus balsamifera, P. deltoides*
Food
Starvation Food: *Rosa* sp.
Unspecified: *Balsamorhiza incana, B. sagittata, Cirsium scariosum, Lomatium dissectum*
Fiber
Basketry: *Thuja plicata*
Building Material: *Thuja plicata*
Canoe Material: *Thuja plicata*
Cordage: *Apocynum cannabinum*
Furniture: *Thuja plicata*
Dye
Orange: *Alnus incana*
Red-Brown: *Alnus incana*
Other
Cash Crop: *Camassia quamash*
Ceremonial Items: *Abies lasiocarpa*
Containers: *Apocynum cannabinum*
Fasteners: *Pinus ponderosa*
Hunting & Fishing Item: *Larix occidentalis*
Incense & Fragrance: *Abies lasiocarpa*
Lighting: *Pinus ponderosa*
Protection: *Rosa* sp., *Symphoricarpos albus*

Nimpkish
Fiber
Clothing: *Typha latifolia*
Mats, Rugs & Bedding: *Typha latifolia*

Nisga
Fiber
Canoe Material: *Populus balsamifera*

Other
Cash Crop: *Shepherdia canadensis*

Nisqually
Food
Dried Food: *Camassia quamash*
Fruit: *Fragaria* sp., *Rubus leucodermis*
Unspecified: *Quercus garryana*
Fiber
Building Material: *Acer macrophyllum*

Nitinaht
Drug
Anticonvulsive: *Postelsia palmaeformis*
Antirheumatic (External): *Urtica dioica*
Antirheumatic (Internal): *Oplopanax horridus*
Burn Dressing: *Lysichiton americanus, Maianthemum dilatatum*
Cathartic: *Huperzia selago, Sambucus racemosa*
Ceremonial Medicine: *Adiantum pedatum, Lomatium nudicaule*
Cold Remedy: *Achillea millefolium, Lomatium nudicaule*
Cough Medicine: *Malus fusca, Polypodium glycyrrhiza*
Dermatological Aid: *Alectoria sarmentosa, Alnus rubra, Galium aparine, Maianthemum dilatatum, Nereocystis luetkeana, Pinus contorta, P. monticola, Plantago major, Populus balsamifera, Tsuga heterophylla, Usnea longissima, U.* sp.
Dietary Aid: *Ledum groenlandicum, Malus fusca*
Disinfectant: *Frangula purshiana, Plantago major, Sphagnum* sp.
Diuretic: *Symphoricarpos albus*
Emetic: *Huperzia selago, Sambucus racemosa*
Eye Medicine: *Conocephalum conicum*
Febrifuge: *Aruncus dioicus*
Gastrointestinal Aid: *Frangula purshiana, Gaultheria shallon, Huperzia selago, Plantago major*
Gynecological Aid: *Oenanthe sarmentosa, Polytrichum commune*
Internal Medicine: *Abies amabilis, A. grandis, Alnus rubra, Tsuga heterophylla*
Kidney Aid: *Conocephalum conicum*
Laxative: *Frangula purshiana, Mahonia aquifolium, M. nervosa, Ribes bracteosum, Sambucus racemosa*
Misc. Disease Remedy: *Aruncus dioicus*
Orthopedic Aid: *Alnus rubra, Oplopanax horridus, Tsuga heterophylla*
Other: *Anaphalis margaritacea, Nuphar lutea*
Panacea: *Achillea millefolium, Malus fusca, Prunus emarginata*
Preventive Medicine: *Abies amabilis*
Psychological Aid: *Conocephalum conicum, Lonicera involucrata, Postelsia palmaeformis, Sambucus racemosa, Tellima grandiflora*
Pulmonary Aid: *Alnus rubra*
Reproductive Aid: *Gaultheria shallon, Halosaccion glandiforme*
Respiratory Aid: *Polypodium glycyrrhiza*
Strengthener: *Ledum groenlandicum, Lessoniopsis littoralis, Leymus mollis, Postelsia palmaeformis, Sambucus racemosa*
Throat Aid: *Achillea millefolium*
Tonic: *Malus fusca, Urtica dioica*
Tuberculosis Remedy: *Alnus rubra, Mahonia aquifolium, M. nervosa, Peltigera aphthosa*
Unspecified: *Abies grandis, Alnus rubra, Mahonia aquifolium, M. nervosa, Malus fusca, Nuphar lutea, Oplopanax horridus, Rosa nutkana, Tellima grandiflora, Urtica dioica*
Urinary Aid: *Peltigera aphthosa, P. canina*
Witchcraft Medicine: *Menziesia ferruginea*
Food
Beverage: *Equisetum telmateia, Ledum groenlandicum*

Candy: *Abies amabilis, A. grandis, Castilleja hispida, C. miniata, Lonicera ciliosa*
Dessert: *Argentina egedii, Ribes bracteosum, Rubus spectabilis*
Dried Food: *Argentina egedii, Camassia leichtlinii, C. quamash, Gaultheria shallon, Trifolium wormskioldii, Vaccinium ovalifolium, V. parvifolium*
Forage: *Arctostaphylos uva-ursi, Disporum hookeri, Malus fusca, Petasites frigidus, Streptopus amplexifolius*
Frozen Food: *Gaultheria shallon, Vaccinium parvifolium*
Fruit: *Arctostaphylos uva-ursi, Cornus canadensis, C. sericea, Fragaria chiloensis, F. vesca, F. virginiana, Gaultheria shallon, Malus fusca, Oemleria cerasiformis, Ribes divaricatum, Rosa nutkana, Rubus leucodermis, R. parviflorus, R. spectabilis, R. ursinus, Sambucus racemosa, Vaccinium ovalifolium, V. ovatum, V. oxycoccos, V. parvifolium*
Ice Cream: *Shepherdia canadensis*
Pie & Pudding: *Vaccinium ovatum*
Preserves: *Gaultheria shallon, Sambucus racemosa*
Special Food: *Vaccinium ovalifolium, V. parvifolium*
Spice: *Gaultheria shallon*
Starvation Food: *Blechnum spicant, Gaultheria shallon, Tsuga heterophylla*
Substitution Food: *Equisetum telmateia*
Unspecified: *Allium cernuum, Argentina egedii, Camassia* sp., *Equisetum telmateia, Hedophyllum sessile, Heracleum maximum, Laminaria groenlandica, Lilium columbianum, Polystichum munitum, Porphyra* sp., *Pteridium aquilinum, Rosa nutkana, Rubus parviflorus, Trifolium wormskioldii, Tsuga heterophylla, Zostera marina*
Vegetable: *Camassia leichtlinii, C. quamash, Pteridium aquilinum, Trifolium wormskioldii*
Fiber
Basketry: *Carex obnupta, Phyllospadix torreyi, Picea sitchensis, Prunus emarginata, Scirpus americanus, Thuja plicata, Typha latifolia, Xerophyllum tenax*
Building Material: *Thuja plicata*
Canoe Material: *Acer macrophyllum, Alnus rubra, Chamaecyparis nootkatensis, Taxus brevifolia, Thuja plicata*
Caulking Material: *Picea sitchensis*
Clothing: *Alectoria sarmentosa, Chamaecyparis nootkatensis, Cirsium brevistylum, Sphagnum* sp., *Thuja plicata, Typha latifolia, Usnea longissima, U.* sp.
Cordage: *Chamaecyparis nootkatensis, Nereocystis luetkeana, Populus balsamifera, Thuja plicata, Urtica dioica*
Mats, Rugs & Bedding: *Chamaecyparis nootkatensis, Polystichum munitum, Pteridium aquilinum, Salix hookeriana, Scirpus acutus, Thuja plicata*
Sewing Material: *Holodiscus discolor, Leymus mollis, Taxus brevifolia*
Dye
Black: *Tsuga heterophylla*
Brown: *Alnus rubra, Tsuga heterophylla*
Red: *Alnus rubra*
Yellow: *Gaultheria shallon, Mahonia aquifolium, M. nervosa, M.* sp.
Other
Cash Crop: *Porphyra perforata*
Ceremonial Items: *Acer macrophyllum, Alnus rubra, Fucus gardneri, Oplopanax horridus, Picea sitchensis, Pinus contorta, P. monticola, Prunus emarginata, Salix* sp., *Taxus brevifolia, Thuja plicata, Tsuga heterophylla*
Containers: *Acer circinatum, Chamaecyparis nootkatensis, Lysichiton americanus, Nereocystis luetkeana, Thuja plicata*
Cooking Tools: *Acer circinatum, A. macrophyllum, Alectoria sarmentosa, Alnus rubra, Athyrium filix-femina, Blechnum spicant, Dryopteris campyloptera, Gaultheria shallon, Holodiscus discolor, Lysichiton americanus, Nereocystis luetkeana, Polystichum munitum, Salix lucida, Thuja plicata, Usnea longissima*
Decorations: *Thuja plicata*
Fasteners: *Leymus mollis, Picea sitchensis, Prunus emarginata*
Fuel: *Alnus rubra, Chamaecyparis nootkatensis, Pseudotsuga menziesii, Tsuga heterophylla*
Good Luck Charm: *Fucus gardneri*
Hunting & Fishing Item: *Abies amabilis, A. grandis, Acer circinatum, Castilleja hispida, C. miniata, Hedophyllum sessile, Holodiscus discolor, Laminaria groenlandica, Lomatium nudicaule, Malus fusca, Nereocystis luetkeana, Oplopanax horridus, Phyllospadix torreyi, Picea sitchensis, Pinus contorta, P. monticola, Populus balsamifera, Pseudotsuga menziesii, Taxus brevifolia, Thuja plicata, Tsuga heterophylla, Urtica dioica, Vicia nigricans, Zostera marina*
Incense & Fragrance: *Abies amabilis, A. grandis, Artemisia suksdorfii, Lomatium nudicaule, Populus balsamifera*
Paint: *Gaultheria shallon, Oplopanax horridus, Picea sitchensis, Populus balsamifera*
Preservative: *Alnus rubra*
Protection: *Cirsium brevistylum, Fomitopsis pinicola, Ganoderma applanatum, Symphoricarpos albus*
Sacred Items: *Chamaecyparis nootkatensis, Sambucus racemosa*
Smoke Plant: *Arctostaphylos uva-ursi*
Soap: *Galium aparine, G. triflorum*
Tools: *Frangula purshiana, Holodiscus discolor, Malus fusca, Nereocystis luetkeana, Taxus brevifolia, Xerophyllum tenax*
Toys & Games: *Angelica genuflexa, Heracleum maximum, Holodiscus discolor, Lessoniopsis littoralis, Menziesia ferruginea, Polystichum munitum, Postelsia palmaeformis, Rubus spectabilis, Sambucus racemosa, Symphoricarpos albus*
Water Indicator: *Alnus rubra*
Waterproofing Agent: *Picea sitchensis*

Nootka
Drug
Alterative: *Polypodium glycyrrhiza*
Cancer Treatment: *Pyrola elliptica*
Gastrointestinal Aid: *Pyrola* sp., *Solanum dulcamara*
Strengthener: *Postelsia palmaeformis*
Unspecified: *Ledum groenlandicum*
Venereal Aid: *Polypodium glycyrrhiza*
Fiber
Basketry: *Carex obnupta*
Other
Smoke Plant: *Arctostaphylos uva-ursi*
Tools: *Holodiscus discolor*

Nootka, Manhousat
Drug
Love Medicine: *Lonicera involucrata*

Northwest Indian
Food
Beverage: *Shepherdia canadensis*
Unspecified: *Glycyrrhiza lepidota*

Numlaki
Food
Bread & Cake: *Arctostaphylos manzanita*
Porridge: *Arctostaphylos manzanita*
Staple: *Arctostaphylos manzanita, Calandrinia ciliata, Pogogyne douglasii*
Other
Hunting & Fishing Item: *Trichostema lanceolatum*

Nuxalkmc
Drug
Cough Medicine: *Lonicera involucrata*

Dermatological Aid: *Lonicera involucrata*
Emetic: *Oenanthe sarmentosa*
Venereal Aid: *Lonicera involucrata*
Food
Fruit: *Arctostaphylos uva-ursi*
Unspecified: *Conioselinum gmelinii, Trifolium wormskioldii*
Other
Cash Crop: *Arctostaphylos uva-ursi, Shepherdia canadensis*
Ceremonial Items: *Rubus spectabilis*

Oglala
Drug
Poison: *Celastrus scandens*
Food
Beverage: *Amorpha canescens, Dalea purpurea, Ratibida columnifera*
Other
Jewelry: *Echinocystis lobata*
Smoke Plant: *Amorpha canescens*
Toys & Games: *Croton texensis*

Ojibwa
Drug
Adjuvant: *Betula papyrifera, Rubus idaeus, Tsuga canadensis*
Analgesic: *Acorus calamus, Apocynum androsaemifolium, Arctium minus, Aster macrophyllus, Betula papyrifera, Caulophyllum thalictroides, Cirsium vulgare, Comarum palustre, Comptonia peregrina, Cynoglossum virginianum, Erigeron strigosus, Euthamia graminifolia, Fragaria virginiana, Impatiens capensis, Maianthemum canadense, M. racemosum, Picea mariana, Polygonum amphibium, Pteridium aquilinum, Sanguinaria canadensis, Thuja occidentalis, Uvularia grandiflora, Viburnum opulus*
Anthelmintic: *Corylus cornuta*
Anticonvulsive: *Aralia nudicaulis, Monarda fistulosa, Picea mariana, Pinus banksiana, Sanguinaria canadensis*
Antidiarrheal: *Comptonia peregrina, Cornus racemosa, Geranium maculatum, Hydrophyllum virginianum, Ilex verticillata, Rubus allegheniensis, Rumex crispus*
Antidote: *Clintonia borealis*
Antiemetic: *Spiraea tomentosa, Vaccinium oxycoccos*
Antirheumatic (External): *Arctostaphylos alpina, A. uva-ursi, Populus alba, P. balsamifera, Sanguinaria canadensis, Saururus cernuus*
Antirheumatic (Internal): *Gaultheria procumbens, Silphium perfoliatum*
Blood Medicine: *Aralia nudicaulis, Arctium lappa, Arctostaphylos alpina, A. uva-ursi, Betula alleghaniensis, B. papyrifera, Mentha canadensis, Nepeta cataria, Populus alba, P. balsamifera, Quercus rubra, Sanguinaria canadensis, Taraxacum officinale, Thuja occidentalis, Vaccinium angustifolium*
Burn Dressing: *Plantago major*
Cathartic: *Acorus calamus, Euphorbia corollata, Iris versicolor, Polygonatum biflorum, Potentilla norvegica, Sambucus racemosa, Silene latifolia, Streptopus roseus, Viburnum opulus*
Ceremonial Medicine: *Abies balsamea, Achillea millefolium, Anemone canadensis, Apocynum androsaemifolium, Arctostaphylos alpina, A. uva-ursi, Cornus sericea, Mitchella repens, Rhus aromatica, R. copallinum, R. glabra, Sanguinaria canadensis, Solanum nigrum, Thuja occidentalis*
Cold Remedy: *Abies balsamea, Acorus calamus, Erigeron philadelphicus, Populus balsamifera, Prunus serotina*
Cough Medicine: *Abies balsamea, Nymphaea odorata, Polygonatum biflorum, Prunus pensylvanica, P. serotina, Streptopus roseus, Thuja occidentalis*
Dermatological Aid: *Abies balsamea, Achillea millefolium, Aralia nudicaulis, Athyrium filix-femina, Clintonia borealis, Corylus americana, C. cornuta, Epilobium angustifolium, Galium trifidum, Heracleum maximum, Impatiens pallida, Nuphar lutea, Oenothera biennis, Plantago major, Populus balsamifera, P. tremuloides, Quercus macrocarpa, Rhus glabra, Rosa virginiana, Rumex crispus, Salix fragilis, S. lucida, Sanguinaria canadensis, Tsuga canadensis, Typha latifolia, Ulmus rubra, Urtica dioica*
Diaphoretic: *Abies balsamea, Mentha canadensis, Thuja occidentalis*
Dietary Aid: *Asarum canadense*
Disinfectant: *Larix laricina, Picea glauca*
Diuretic: *Apocynum androsaemifolium, Cucurbita maxima, Dirca palustris, Galium aparine, Humulus lupulus, Laportea canadensis, Lycopodium obscurum, Malaxis unifolia, Prenanthes alba, Rubus allegheniensis, Viburnum lentago*
Emetic: *Acer negundo, Caulophyllum thalictroides, Cornus alternifolia, Iris versicolor, Sambucus racemosa*
Eye Medicine: *Abies balsamea, Acer rubrum, A. spicatum, Arisaema triphyllum, Melampyrum lineare, Rhus glabra, Rubus idaeus*
Febrifuge: *Achillea millefolium, Erigeron philadelphicus, Hedeoma pulegioides, Mentha canadensis, Monarda fistulosa, Sanicula marilandica, Tanacetum vulgare, Thalictrum dasycarpum*
Gastrointestinal Aid: *Acorus calamus, Actaea rubra, Alnus incana, Aquilegia canadensis, Arctium minus, Asarum canadense, Betula papyrifera, Cardamine maxima, Celastrus scandens, Chimaphila umbellata, Cirsium arvense, C. vulgare, Comarum palustre, Comptonia peregrina, Cornus canadensis, Dryopteris cristata, Echinocystis lobata, Equisetum palustre, Fragaria virginiana, Hedeoma pulegioides, Lathyrus ochroleucus, Mentha canadensis, Monarda punctata, Pedicularis canadensis, Picea mariana, Polygonum amphibium, Rosa blanda, Sagittaria cuneata, Salix candida, S. discolor, S. pedicellaris, S. pyrifolia, Sanguinaria canadensis, Saururus cernuus, Silphium perfoliatum, Taraxacum officinale, Ulmus rubra, Uvularia grandiflora, Viburnum opulus, Vitis vulpina*
Gynecological Aid: *Actaea rubra, Amelanchier laevis, Antennaria howellii, Apocynum androsaemifolium, Asclepias syriaca, Athyrium filix-femina, Betula pumila, Caulophyllum thalictroides, Clintonia borealis, Crataegus* sp., *Cypripedium pubescens, Geum macrophyllum, Glyceria canadensis, Lactuca biennis, Maianthemum canadense, M. racemosum, Onoclea sensibilis, Osmorhiza claytonii, O. longistylis, Pastinaca sativa, Prenanthes alba, Prunella vulgaris, Pteridium aquilinum, Ribes triste, Sarracenia purpurea, Spiraea tomentosa, Symphoricarpos albus, Vitis vulpina*
Heart Medicine: *Acorus calamus, Quercus rubra, Scutellaria galericulata, Verbascum thapsus, Viola conspersa*
Hemorrhoid Remedy: *Cornus racemosa*
Hemostat: *Eriophorum callitrix, Populus grandidentata, Rhus glabra, R. hirta, Rosa virginiana, Salix fragilis, S. lucida, Sanguinaria canadensis, Silphium perfoliatum, Tsuga canadensis, Ulmus rubra*
Herbal Steam: *Linaria vulgaris*
Hunting Medicine: *Acorus calamus, Aralia nudicaulis, Aster cordifolius, A. macrophyllus, Cicuta maculata, Conyza canadensis, Euthamia graminifolia, Heracleum maximum, Hieracium canadense, Polygonum amphibium, Prunella vulgaris, Pyrola americana, Ranunculus pensylvanicus, Rumex crispus, Sium suave, Spiraea salicifolia, Spiranthes lacera, Taenidia integerrima, Uvularia sessilifolia*
Kidney Aid: *Equisetum arvense, E. sylvaticum, Galium aparine, Maianthemum canadense, M. racemosum*
Laxative: *Equisetum palustre, Monarda punctata, Sanguinaria canadensis*
Love Medicine: *Coeloglossum viride, Pedicularis canadensis*
Narcotic: *Arctostaphylos alpina, A. uva-ursi*
Oral Aid: *Apocynum androsaemifolium, Coptis trifolia, Geranium maculatum, Rhus hirta*
Orthopedic Aid: *Mirabilis nyctaginea, Plantago major, Populus tremuloides, Quercus macrocarpa*
Other: *Nepeta cataria, Typha latifolia*

Panacea: *Arctostaphylos alpina, A. uva-ursi, Populus alba, P. balsamifera, Sanguinaria canadensis, Saururus cernuus*
Pediatric Aid: *Coptis trifolia, Cornus canadensis, Eupatorium purpureum, Fragaria virginiana, Hydrophyllum virginianum, Monarda fistulosa*
Poison: *Clintonia borealis, Pastinaca sativa*
Pulmonary Aid: *Anemone cylindrica, Botrychium virginianum, Campanula rotundifolia, Galium tinctorium, Prunus virginiana, Smilax herbacea, Uvularia grandiflora*
Respiratory Aid: *Betula pumila, Linaria vulgaris, Monarda fistulosa, Populus balsamifera, Quercus rubra, Salix* sp., *Zanthoxylum americanum*
Sedative: *Salix candida, S. discolor, S. pyrifolia*
Snakebite Remedy: *Plantago major, Sanicula marilandica*
Stimulant: *Abies balsamea, Anaphalis margaritacea, Aralia nudicaulis, Corydalis aurea, Lycopodium complanatum, Maianthemum racemosum, Picea mariana, Pinus banksiana, P. resinosa, P. strobus, Salix candida, S. discolor, S. pyrifolia, Sanguinaria canadensis*
Throat Aid: *Acorus calamus, Anemone canadensis, Apocynum androsaemifolium, Maianthemum canadense, M. racemosum, Osmorhiza claytonii, O. longistylis, Pedicularis canadensis, Rhus hirta, Sanguinaria canadensis, Ulmus rubra, Zanthoxylum americanum*
Tonic: *Arctium minus, Echinocystis lobata, Fraxinus pennsylvanica*
Tuberculosis Remedy: *Anemone cylindrica, Botrychium virginianum, Galium trifidum, Nymphaea odorata*
Unspecified: *Aralia racemosa, Arctostaphylos alpina, A. uva-ursi, Arisaema triphyllum, Aster macrophyllus, Baptisia tinctoria, Brassica rapa, Celastrus scandens, Cicuta maculata, Cornus sericea, Equisetum hyemale, Fraxinus americana, Impatiens capensis, Larix laricina, Maianthemum canadense, Monarda punctata, Nemopanthus mucronatus, Picea mariana, Pinus resinosa, P. strobus, Plantago major, Polygala senega, Populus alba, Quercus rubra, Rhus aromatica, R. copallinum, R. glabra, Ribes americanum, R. hudsonianum, R. oxyacanthoides, R. rubrum, Rosa virginiana, Rumex altissimus, Sagittaria cuneata, Salix candida, S. discolor, S. pyrifolia, Sambucus racemosa, Sorbus americana*
Urinary Aid: *Agrimonia gryposepala, Diervilla lonicera, Dirca palustris, Galium aparine, Laportea canadensis*
Venereal Aid: *Acer saccharinum, Quercus rubra, Sanguinaria canadensis, Sorbus americana, Ulmus americana, U. rubra*
Veterinary Aid: *Artemisia ludoviciana, Lathyrus ochroleucus, L. palustris, Sagittaria cuneata*

Food
Beverage: *Acer negundo, A. saccharum, Andromeda polifolia, Betula alleghaniensis, Chamaedaphne calyculata, Gaultheria procumbens, Juniperus horizontalis, Ledum groenlandicum, Mentha canadensis, Nepeta cataria, Prunus serotina, Rhus glabra, R. hirta, Thuja occidentalis, Tsuga canadensis*
Bread & Cake: *Zizania aquatica*
Breakfast Food: *Zizania aquatica*
Cooking Agent: *Humulus lupulus*
Dietary Aid: *Asarum canadense*
Dried Food: *Allium tricoccum, Amelanchier laevis, Cucurbita maxima, C. pepo, Prunus americana, P. pensylvanica, P. pumila, P. serotina, P. virginiana, Ribes americanum, R. hudsonianum, R. missouriense, R. oxyacanthoides, R. rubrum, R. triste, Rubus idaeus, Typha latifolia, Vaccinium angustifolium, V. myrtilloides, Zea mays*
Fodder: *Equisetum arvense, Hydrophyllum virginianum, Lathyrus ochroleucus, L. palustris, Maianthemum racemosum*
Forage: *Erigeron philadelphicus, Sagittaria cuneata*
Fruit: *Amelanchier canadensis, Crataegus chrysocarpa, C.* sp., *Empetrum nigrum, Fragaria virginiana, Gaultheria procumbens, Malus coronaria, Prunus americana, P. nigra, P. pensylvanica, P. pumila, P. serotina, P. virginiana, Ribes americanum, R. cynosbati, R. hudsonianum, R. missouriense, R. oxyacanthoides, R. rubrum, R. triste, Rubus idaeus, R. occidentalis, Sorbus americana, Vaccinium angustifolium, V. macrocarpon, V. myrtilloides, V. oxycoccos, Viburnum lentago, V. opulus*
Preservative: *Betula papyrifera, Comptonia peregrina*
Preserves: *Fragaria virginiana, Prunus nigra, Ribes americanum, R. cynosbati, R. oxyacanthoides, R. triste, Rubus allegheniensis, R. idaeus, Viburnum lentago, Vitis vulpina*
Sauce & Relish: *Cardamine diphylla*
Soup: *Aster macrophyllus, Celastrus scandens, Prunus americana, P. pensylvanica, P. serotina, P. virginiana, Pteridium aquilinum, Quercus alba, Solanum tuberosum*
Sour: *Acer saccharum*
Special Food: *Parthenocissus quinquefolia, Zizania aquatica*
Spice: *Asarum canadense*
Staple: *Quercus rubra, Sagittaria cuneata, Typha latifolia, Zizania palustris*
Sweetener: *Acer nigrum, A. saccharinum, A. saccharum*
Unspecified: *Amphicarpaea bracteata, Asclepias syriaca, Aster macrophyllus, Caltha palustris, Carya ovata, Corylus americana, C. cornuta, Fagus grandifolia, Fraxinus pennsylvanica, Gaylussacia baccata, Juglans cinerea, Nelumbo lutea, Nymphaea odorata, Parthenocissus quinquefolia, Pinus strobus, Populus deltoides, P. grandidentata, Pteridium aquilinum, Quercus macrocarpa, Q. velutina, Rosa acicularis, R. virginiana, Sambucus racemosa, Sticta glomulifera, Typha latifolia*
Vegetable: *Allium cernuum, Amphicarpaea bracteata, Asclepias syriaca, Cardamine maxima, Chenopodium album, Cucumis sativus, Heracleum maximum, Lagenaria siceraria, Lathyrus ochroleucus, L. palustris, Maianthemum racemosum, Phaseolus lunatus, P. vulgaris, Solanum tuberosum, Taraxacum officinale, Zea mays*
Winter Use Food: *Asclepias syriaca, Fragaria virginiana, Rhus hirta, Rubus canadensis, R. occidentalis, Vaccinium angustifolium, V. myrtilloides*

Fiber
Basketry: *Anthoxanthum odoratum, Betula papyrifera, B. pumila, Corylus americana, Fraxinus nigra, F. pennsylvanica, Larix laricina, Scirpus cyperinus, Thuja occidentalis*
Brushes & Brooms: *Corylus americana*
Building Material: *Betula alleghaniensis, B. lenta, B. papyrifera, Cornus alternifolia, Juniperus chinense, J. communis, J. virginiana, Pinus resinosa, Quercus alba, Ulmus rubra*
Canoe Material: *Abies balsamea, Betula alleghaniensis, B. lenta, B. papyrifera, Fraxinus americana, Larix laricina, Picea mariana, Sorbus americana, Thuja occidentalis*
Caulking Material: *Picea mariana, Pinus resinosa, P. strobus*
Cordage: *Tilia americana*
Furniture: *Fraxinus pennsylvanica, Juniperus chinense, J. communis, J. virginiana*
Mats, Rugs & Bedding: *Juncus dudleyi, J. stygius, Juniperus chinense, J. communis, J. virginiana, Pinus banksiana, P. strobus, Scirpus cyperinus, S. tabernaemontani, Sphagnum dusenii, Typha latifolia*
Scouring Material: *Equisetum hyemale*
Sewing Material: *Apocynum androsaemifolium, Laportea canadensis, Larix laricina, Pinus banksiana, Tilia americana, Urtica dioica*
Snow Gear: *Fraxinus americana, F. pennsylvanica, Sorbus americana*
Sporting Equipment: *Sorbus americana*

Dye
Brown: *Juglans cinerea, Myrica gale*
Mordant: *Corylus americana, Prunus nigra, Quercus macrocarpa, Q. velutina, Tsuga canadensis*
Orange: *Rhus glabra, R. hirta, Sanguinaria canadensis*
Red: *Betula papyrifera, Ranunculus pensylvanicus, Sanguinaria canadensis*
Red-Brown: *Tsuga canadensis*
Red-Yellow: *Quercus velutina*

Unspecified: *Baptisia tinctoria, Quercus rubra*
Yellow: *Alnus incana, Coptis trifolia, Impatiens capensis, Myrica gale, Sanguinaria canadensis*
Other
Cash Crop: *Juniperus chinense, J. communis, J. virginiana, Vaccinium angustifolium, V. macrocarpon, V. myrtilloides, V. oxycoccos, Zizania aquatica*
Ceremonial Items: *Achillea millefolium, Anthoxanthum odoratum, Betula alleghaniensis, B. lenta, B. papyrifera, Juniperus chinense, J. communis, J. virginiana, Lagenaria siceraria*
Containers: *Betula alleghaniensis, B. lenta, B. papyrifera*
Cooking Tools: *Acer saccharum, Betula alleghaniensis, B. lenta, B. papyrifera, Lagenaria siceraria*
Designs: *Acer rubrum, A. spicatum*
Fasteners: *Dirca palustris*
Fuel: *Betula papyrifera, Tsuga canadensis*
Hide Preparation: *Quercus rubra*
Hunting & Fishing Item: *Acer nigrum, A. pensylvanicum, A. saccharinum, A. spicatum, Asclepias syriaca, Carya ovata, Cornus alternifolia, Crataegus sp., Erigeron philadelphicus, Fraxinus pennsylvanica, Heracleum maximum, Lactuca biennis, Tanacetum vulgare, Trientalis borealis*
Incense & Fragrance: *Thuja occidentalis*
Lighting: *Betula papyrifera*
Musical Instrument: *Corylus americana*
Paper: *Betula papyrifera, Populus alba, P. balsamifera, P. deltoides, P. grandidentata, P. tremuloides*
Preservative: *Betula papyrifera*
Protection: *Betula papyrifera, Iris versicolor, Plantago major*
Sacred Items: *Betula papyrifera, Thuja occidentalis*
Smoke Plant: *Achillea millefolium, Cornus alternifolia, C. racemosa, Erigeron philadelphicus, Salix lucida, Viburnum dentatum*
Tools: *Crataegus sp., Quercus alba*
Toys & Games: *Acer nigrum, A. pensylvanicum, A. saccharinum, A. spicatum*
Waterproofing Agent: *Pinus resinosa, P. strobus*
Weapon: *Typha latifolia*

Ojibwa, South
Drug
Analgesic: *Actaea rubra, Amorpha canescens, Anemone canadensis, A. sp., Chamaecyparis thyoides, Geum aleppicum, Juniperus virginiana, Lappula squarrosa, Larix laricina, Maianthemum racemosum, Monarda fistulosa, Pinus resinosa, P. strobus, Prunus pensylvanica, P. serotina, Rubus idaeus, R. occidentalis, Viola canadensis*
Antidiarrheal: *Acer nigrum, A. saccharinum, Crataegus chrysocarpa, Picea glauca, Prunus americana, Quercus alba, Q. rubra, Rhus aromatica*
Antirheumatic (Internal): *Phryma leptostachya*
Cathartic: *Veronicastrum virginicum*
Cold Remedy: *Abies balsamea, Polygala senega*
Cough Medicine: *Geum aleppicum, Polygala senega, Salix candida*
Dermatological Aid: *Abies balsamea, Botrychium virginianum, Helianthus occidentalis, Populus deltoides, Prunus serotina, Rumex crispus, Sporobolus heterolepis, Typha latifolia*
Diaphoretic: *Abies balsamea*
Diuretic: *Acer saccharinum, Betula alleghaniensis*
Emetic: *Acer pensylvanicum, Sporobolus heterolepis*
Eye Medicine: *Fraxinus nigra, Rosa blanda, R. virginiana, Symphoricarpos orbiculatus*
Gastrointestinal Aid: *Actaea rubra, Amorpha canescens, Monarda fistulosa, Polygala senega, Prunus pensylvanica, Rubus idaeus, R. occidentalis*
Gynecological Aid: *Maianthemum racemosum, Prunus virginiana*
Hemostat: *Maianthemum racemosum, Panax trifolius*
Herbal Steam: *Larix laricina, Pinus strobus*
Liver Aid: *Sporobolus heterolepis*
Orthopedic Aid: *Anemone canadensis*
Pulmonary Aid: *Geum aleppicum, Prunus serotina*
Throat Aid: *Polygala senega, Viola pubescens*
Unspecified: *Picea mariana*
Venereal Aid: *Abies balsamea*

Okanagan-Colville
Drug
Abortifacient: *Pinus ponderosa*
Analgesic: *Achillea millefolium, Artemisia dracunculus, Erigeron philadelphicus, Gaillardia aristata, Mentha arvensis, Penstemon fruticosus, Potentilla gracilis, Ranunculus glaberrimus, Rorippa nasturtium-aquaticum, Spiraea betulifolia*
Antidiarrheal: *Achillea millefolium, Arabis sparsiflora, Crataegus douglasii, Epilobium minutum, Lactuca tatarica, Lesquerella douglasii, Potentilla gracilis, Prunus virginiana, Ribes lacustre, Rubus idaeus, Solidago canadensis*
Antihemorrhagic: *Arctostaphylos uva-ursi, Juniperus scopulorum, Leymus cinereus, Lithospermum ruderale, Pinus ponderosa*
Antirheumatic (External): *Achillea millefolium, Artemisia dracunculus, Asclepias speciosa, Crataegus douglasii, Juniperus scopulorum, Larix occidentalis, Pseudoroegneria spicata, Ranunculus glaberrimus, Sambucus cerulea, Thuja plicata, Urtica dioica, Veratrum viride, Wyethia amplexicaulis*
Antirheumatic (Internal): *Equisetum arvense, E. hyemale, E. laevigatum, Frangula purshiana, Larix occidentalis, Lomatium dissectum, Mentha arvensis, Potentilla gracilis*
Blood Medicine: *Arctostaphylos uva-ursi, Chimaphila umbellata, Cornus sericea, Frangula purshiana, Heuchera cylindrica, Larix occidentalis, Lomatium dissectum, Lonicera utahensis, Mahonia aquifolium, Phlox longifolia, Potentilla gracilis, Pseudotsuga menziesii*
Burn Dressing: *Balsamorhiza sagittata, Ceanothus sanguineus, Holodiscus discolor, Taxus brevifolia*
Cancer Treatment: *Larix occidentalis*
Carminative: *Artemisia ludoviciana*
Cathartic: *Achillea millefolium, Philadelphus lewisii, Rubus idaeus, Symphoricarpos albus*
Ceremonial Medicine: *Antennaria rosea, Ceanothus velutinus, Ligusticum canbyi, Rosa acicularis, R. gymnocarpa, R. nutkana, R. woodsii, Vicia americana*
Cold Remedy: *Achillea millefolium, Agastache urticifolia, Amelanchier alnifolia, Artemisia absinthium, A. frigida, A. tridentata, A. tripartita, Asarum caudatum, Chimaphila umbellata, Cornus sericea, Equisetum laevigatum, Eriogonum compositum, E. heracleoides, E. niveum, Erythronium grandiflorum, Geum triflorum, Juniperus communis, Ligusticum canbyi, Lomatium ambiguum, L. macrocarpum, L. triternatum, Maianthemum racemosum, Mentha arvensis, Monardella odoratissima, Nepeta cataria, Paxistima myrsinites, Penstemon fruticosus, Polygonum amphibium, Prunus virginiana, Ribes lacustre, Salvia dorrii, Spiraea betulifolia, Veratrum viride*
Contraceptive: *Amelanchier alnifolia, Apocynum cannabinum, Arabis sparsiflora, Clematis ligusticifolia, Cornus sericea*
Cough Medicine: *Abies grandis, A. lasiocarpa, Oplopanax horridus, Prunus virginiana*
Dermatological Aid: *Abies grandis, A. lasiocarpa, Achillea millefolium, Agoseris glauca, Artemisia dracunculus, Asclepias speciosa, Aster conspicuus, Calochortus macrocarpus, Castilleja hispida, C. thompsonii, Ceanothus velutinus, Clematis columbiana, C. ligusticifolia, Cornus sericea, Disporum trachycarpum, Epilobium brachycarpum, Equisetum arvense, E. hyemale, E. laevigatum, Eriogonum compositum, E. heracleoides, E. niveum, Erysimum asperum, Fragaria vesca, F. virginiana, Goodyera oblongifolia, Heracleum maximum, Heuchera cylindrica, Juniperus scopulorum, Larix occidentalis, Lewisia rediviva,*

[Dermatological Aid]
Leymus cinereus, Lomatium dissectum, Lonicera utahensis, Opuntia fragilis, O. polyacantha, Penstemon fruticosus, Pinus ponderosa, Plantago major, P. patagonica, Populus tremuloides, Potentilla gracilis, P. recta, Prunus virginiana, Pseudotsuga menziesii, Pteryxia terebinthina, Rhus glabra, Rosa acicularis, R. gymnocarpa, R. nutkana, R. woodsii, Rubus parviflorus, Salix bebbiana, S. scouleriana, S. sp., Shepherdia canadensis, Symphoricarpos albus, Thuja plicata, Typha latifolia

Diaphoretic: *Artemisia tridentata, A. tripartita, Balsamorhiza sagittata*

Dietary Aid: *Abies grandis, A. lasiocarpa, Alnus incana, A. viridis, Chimaphila umbellata, Geum triflorum, Lomatium dissectum, Maianthemum racemosum*

Disinfectant: *Fragaria vesca, F. virginiana*

Diuretic: *Equisetum arvense, E. hyemale, E. laevigatum, Opuntia fragilis, O. polyacantha*

Emetic: *Pseudotsuga menziesii*

Eye Medicine: *Arabis sparsiflora, Arctostaphylos uva-ursi, Chaenactis douglasii, Dodecatheon pulchellum, Equisetum hyemale, Erigeron pumilus, Lupinus sericeus, L. sulphureus, L. wyethii, Mahonia aquifolium, Pinus ponderosa, Populus tremuloides, Ribes cereum, Silene menziesii, Symphoricarpos albus*

Febrifuge: *Agastache urticifolia, Collomia grandiflora, Geum triflorum, Ipomopsis aggregata, Mentha arvensis, Osmorhiza occidentalis, Pinus ponderosa, Pseudotsuga menziesii, Solidago canadensis*

Gastrointestinal Aid: *Abies grandis, A. lasiocarpa, Achillea millefolium, Anaphalis margaritacea, Arabis sparsiflora, Artemisia absinthium, Cornus sericea, Lesquerella douglasii, Mentha arvensis, Penstemon fruticosus, Pinus contorta, P. ponderosa, Prunus emarginata, P. virginiana, Rubus idaeus, R. parviflorus, Salix sitchensis, Spiraea betulifolia*

Gland Medicine: *Abies grandis, A. lasiocarpa*

Gynecological Aid: *Alnus incana, A. viridis, Artemisia absinthium, A. dracunculus, Chrysothamnus nauseosus, Cornus sericea, Geum macrophyllum, G. triflorum, Lonicera involucrata, Potentilla arguta, Rhus glabra, Salix bebbiana, S. scouleriana, S. sp., Sedum lanceolatum, Spiraea betulifolia*

Heart Medicine: *Cornus sericea, Rhus glabra*

Hemorrhoid Remedy: *Aster conspicuus*

Hemostat: *Disporum trachycarpum, Salix bebbiana, S. scouleriana, S. sp.*

Hunting Medicine: *Acer glabrum, Physocarpus malvaceus, Shepherdia canadensis, Veratrum viride*

Internal Medicine: *Ligusticum canbyi, Potentilla recta*

Kidney Aid: *Arctostaphylos uva-ursi, Chimaphila umbellata, Gaillardia aristata, Ledum groenlandicum, Mahonia aquifolium, Paxistima myrsinites, Spiraea betulifolia*

Laxative: *Achillea millefolium, Agoseris glauca, Asarum caudatum, Collomia grandiflora, Frangula purshiana, Ipomopsis aggregata, Lonicera utahensis, Rubus idaeus, Sedum lanceolatum*

Love Medicine: *Apocynum androsaemifolium, Arnica cordifolia, A. latifolia, Geum triflorum, Matricaria discoidea*

Misc. Disease Remedy: *Artemisia absinthium, A. frigida, Geum triflorum, Juniperus scopulorum, Lewisia rediviva, Lomatium macrocarpum, Penstemon fruticosus, P. richardsonii, Solidago canadensis*

Oral Aid: *Artemisia tridentata, A. tripartita, Crataegus douglasii, Fragaria vesca, F. virginiana, Lomatium macrocarpum*

Orthopedic Aid: *Artemisia absinthium, Ceanothus velutinus, Crepis atribarba, Equisetum arvense, E. hyemale, E. laevigatum, Gaillardia aristata, Heracleum maximum, Lomatium dissectum, Salix amygdaloides, S. bebbiana, S. lucida, S. scouleriana, S. sp., Veratrum viride*

Other: *Abies grandis, A. lasiocarpa, Cornus sericea, Juniperus scopulorum, Ligusticum canbyi, Lomatium dissectum, Rhus glabra, Spiraea betulifolia*

Panacea: *Cornus sericea, Salvia dorrii*

Pediatric Aid: *Agastache urticifolia, Alnus incana, A. viridis, Artemisia dracunculus, Ceanothus velutinus, Cornus sericea, Crataegus douglasii, Epilobium minutum, Equisetum hyemale, E. laevigatum, Fragaria vesca, F. virginiana, Heuchera cylindrica, Lactuca tatarica, Ligusticum canbyi, Lomatium macrocarpum, Mentha arvensis, Monardella odoratissima, Phlox longifolia, Salix bebbiana, S. scouleriana, S. sp., Solidago canadensis, Sorbus scopulina, S. sitchensis, Symphoricarpos albus*

Poison: *Aconitum columbianum, Cicuta douglasii, Juniperus scopulorum, Lomatium dissectum, Lonicera involucrata, Nuphar lutea, Nymphaea odorata, Pteridium aquilinum, Ranunculus glaberrimus, Symphoricarpos albus, Thuja plicata, Triteleia grandiflora, Veratrum viride, Zigadenus venenosus*

Reproductive Aid: *Antennaria rosea, Cypripedium montanum, Goodyera oblongifolia*

Respiratory Aid: *Lomatium macrocarpum, Picea engelmannii, P. glauca*

Stimulant: *Equisetum arvense, E. hyemale, E. laevigatum*

Strengthener: *Abies grandis, A. lasiocarpa. Artemisia ludoviciana*

Throat Aid: *Artemisia tridentata, A. tripartita, Heuchera cylindrica, Lomatium ambiguum, L. triternatum, Pinus contorta*

Tonic: *Amelanchier alnifolia, Artemisia dracunculus, Hieracium cynoglossoides, H. scouleri, Juniperus communis, Prunus virginiana, Pteryxia terebinthina, Thuja plicata*

Toothache Remedy: *Achillea millefolium, Alnus incana, A. viridis, Aster conspicuus, Nuphar lutea, Nymphaea odorata, Osmorhiza occidentalis, Penstemon fruticosus*

Tuberculosis Remedy: *Abies grandis, A. lasiocarpa, Artemisia absinthium, Chimaphila umbellata, Erigeron linearis, Frasera montana, Juniperus communis, Lomatium dissectum, Oplopanax horridus, Paxistima myrsinites, Picea engelmannii, P. glauca*

Unspecified: *Aster conspicuus, Cornus sericea, Disporum trachycarpum, Geranium viscosissimum*

Urinary Aid: *Arctostaphylos uva-ursi, Sorbus scopulina, S. sitchensis*

Venereal Aid: *Artemisia absinthium, Aster conspicuus, Equisetum arvense, E. hyemale, E. laevigatum, Gaillardia aristata, Leymus cinereus, Populus balsamifera, Potentilla gracilis, Pterospora andromedea, Rhus glabra, Sedum stenopetalum*

Vertigo Medicine: *Rorippa nasturtium-aquaticum*

Veterinary Aid: *Aster conspicuus, A. foliaceus, Equisetum arvense, E. hyemale, E. laevigatum, Lomatium dissectum, Matricaria discoidea, Penstemon fruticosus*

Witchcraft Medicine: *Aconitum columbianum, Juncus mertensianus, Lewisia rediviva, Lilium columbianum, Machaeranthera canescens, Matricaria discoidea, Pinus ponderosa, Veratrum viride*

Food

Appetizer: *Trillium petiolatum*

Beverage: *Juniperus scopulorum, Ledum groenlandicum, Mentha arvensis, Monardella odoratissima, Picea engelmannii, P. glauca, Rhododendron albiflorum, Rhus glabra*

Bread & Cake: *Crataegus columbiana, C. douglasii, Lilium columbianum, Prunus virginiana, Ribes aureum, R. oxyacanthoides*

Candy: *Agoseris glauca, Larix occidentalis, Pinus ponderosa*

Dessert: *Rubus leucodermis*

Dried Food: *Allium cernuum, A. douglasii, A. geyeri, Amelanchier alnifolia, Balsamorhiza sagittata, Camassia quamash, Erythronium grandiflorum, Fritillaria pudica, Lewisia rediviva, Lomatium ambiguum, L. canbyi, L. cous, L. triternatum, Perideridia gairdneri, Ribes aureum, Rubus idaeus, R. leucodermis, Vaccinium cespitosum, V. membranaceum*

Fodder: *Equisetum arvense, E. hyemale, E. laevigatum, Juncus effusus, Koeleria macrantha, Leymus cinereus, Sparganium eurycarpum*

Forage: *Ceanothus sanguineus, C. velutinus, Cornus sericea, Crataegus douglasii, Epilobium angustifolium, Larix occidentalis, Lonicera ciliosa, L. involucrata, Lupinus sericeus, L. sulphureus, L. wyethii, Lysichiton americanus, Nepeta cataria, Paxistima myrsinites, Pinus contorta, Pseudoroegneria spicata,*

Purshia tridentata, Ribes cereum, R. oxyacanthoides, Rosa acicularis, R. gymnocarpa, R. nutkana, R. woodsii, Symphoricarpos albus, Tragopogon pratensis, Vaccinium cespitosum

Frozen Food: *Amelanchier alnifolia, Pinus ponderosa, Rubus idaeus, Sambucus cerulea*

Fruit: *Amelanchier alnifolia, Arctostaphylos uva-ursi, Cornus sericea, Crataegus columbiana, Elaeagnus commutata, Fragaria vesca, F. virginiana, Lonicera utahensis, Mahonia aquifolium, Prunus virginiana, Ribes aureum, R. cereum, R. oxyacanthoides, Rubus idaeus, R. leucodermis, R. parviflorus, Sambucus cerulea, Typha latifolia, Vaccinium cespitosum, V. membranaceum, V. scoparium, Viburnum edule*

Ice Cream: *Shepherdia canadensis*

Pie & Pudding: *Amelanchier alnifolia, Perideridia gairdneri*

Sauce & Relish: *Camassia quamash, Lomatium dissectum*

Soup: *Opuntia fragilis, O. polyacantha*

Spice: *Lilium columbianum, Lomatium ambiguum, L. triternatum, Medicago sativa, Rosa acicularis, R. gymnocarpa, R. nutkana, R. woodsii*

Starvation Food: *Rorippa nasturtium-aquaticum*

Substitution Food: *Lomatium ambiguum, L. triternatum*

Sweetener: *Amelanchier alnifolia, Comandra umbellata*

Unspecified: *Astragalus miser, Balsamorhiza hookeri, B. sagittata, Calochortus macrocarpus, Camassia quamash, Citrullus lanatus, Claytonia lanceolata, Cucumis melo, Cucurbita pepo, Equisetum laevigatum, Erythronium grandiflorum, Fritillaria pudica, Lewisia rediviva, Lilium columbianum, Lomatium canbyi, L. cous, L. farinosum, L. geyeri, L. macrocarpum, Maianthemum racemosum, Matricaria discoidea, Opuntia fragilis, O. polyacantha, Perideridia gairdneri, Phaseolus sp., Physocarpus malvaceus, Pinus albicaulis, P. contorta, P. ponderosa, Pisum sativum, Pseudotsuga menziesii, Rosa acicularis, R. gymnocarpa, R. nutkana, R. woodsii, Sium suave, Triteleia grandiflora, Triticum aestivum, Typha latifolia*

Vegetable: *Allium cernuum, A. douglasii, A. geyeri, Brassica oleracea, B. rapa, Heracleum maximum, Rorippa nasturtium-aquaticum, Rumex acetosella, Taraxacum officinale, Urtica dioica*

Winter Use Food: *Amelanchier alnifolia, Claytonia lanceolata, Fragaria vesca, F. virginiana, Perideridia gairdneri, Pinus albicaulis, P. ponderosa, Prunus virginiana, Pseudotsuga menziesii, Ribes oxyacanthoides, Rubus idaeus, Vaccinium cespitosum, V. membranaceum, Valeriana edulis*

Fiber

Basketry: *Betula occidentalis, B. papyrifera, Scirpus microcarpus, Thuja plicata, Xerophyllum tenax*

Brushes & Brooms: *Symphoricarpos albus, Thuja plicata*

Building Material: *Acer glabrum, Cornus sericea, Crataegus douglasii, Pinus ponderosa, Pseudotsuga menziesii, Scirpus acutus, Thuja plicata, Typha latifolia*

Canoe Material: *Betula occidentalis, B. papyrifera, Pinus monticola, P. ponderosa, Populus balsamifera, Thuja plicata*

Caulking Material: *Calamagrostis rubescens*

Clothing: *Calamagrostis rubescens, Elaeagnus commutata, Penstemon fruticosus, Salix bebbiana, S. exigua, S. scouleriana, S. sp., Typha latifolia*

Cordage: *Amelanchier alnifolia, Apocynum androsaemifolium, A. cannabinum, Asclepias speciosa, Convolvulus arvensis, Cornus sericea, Elaeagnus commutata, Salix bebbiana, S. exigua, S. scouleriana*

Furniture: *Betula occidentalis, B. papyrifera, Holodiscus discolor, Philadelphus lewisii*

Mats, Rugs & Bedding: *Abies grandis, A. lasiocarpa, Artemisia dracunculus, Beckmannia syzigachne, Elaeagnus commutata, Eleocharis palustris, Leymus cinereus, Lupinus sericeus, L. sulphureus, L. wyethii, Thuja plicata*

Scouring Material: *Equisetum arvense, E. hyemale, E. laevigatum*

Sewing Material: *Salix bebbiana, S. exigua, S. scouleriana*

Snow Gear: *Acer glabrum, Philadelphus lewisii*

Dye

Blue: *Delphinium nuttallianum, Penstemon confertus, P. pruinosus*

Brown: *Betula occidentalis, B. papyrifera*

Red: *Dodecatheon pulchellum*

Red-Brown: *Alnus incana, A. viridis*

Yellow: *Mahonia aquifolium*

Other

Cash Crop: *Corylus cornuta, Lewisia rediviva, Ribes oxyacanthoides, Solanum tuberosum*

Ceremonial Items: *Juniperus communis, Phalaris arundinacea, Pinus contorta, P. ponderosa, Pseudotsuga menziesii, Rosa acicularis, R. gymnocarpa, R. nutkana, R. woodsii, Scirpus acutus, Thuja plicata, Urtica dioica*

Containers: *Calamagrostis rubescens, Carex concinnoides, Elaeagnus commutata, Elytrigia repens, Equisetum arvense, E. hyemale, E. laevigatum, Leymus cinereus, Lysichiton americanus, Phleum pratense, Pinus contorta, Populus balsamifera, Pseudotsuga menziesii, Pteridium aquilinum, Rosa acicularis, R. gymnocarpa, R. nutkana, R. woodsii, Rubus parviflorus, Salix bebbiana, S. exigua, S. scouleriana, S. sp., Scirpus acutus, S. microcarpus, Sparganium eurycarpum, Typha latifolia*

Cooking Tools: *Acer glabrum, Artemisia dracunculus, Calamagrostis rubescens, Cornus sericea, Elaeagnus commutata, Phalaris arundinacea, Scirpus acutus*

Decorations: *Prunus emarginata, Scirpus microcarpus*

Fasteners: *Acer glabrum, Holodiscus discolor, Pinus ponderosa, Populus balsamifera, Salix bebbiana, S. exigua, S. scouleriana*

Fuel: *Acer glabrum, Artemisia tridentata, A. tripartita, Betula occidentalis, Ceanothus sanguineus, Purshia tridentata*

Good Luck Charm: *Aquilegia formosa, Lithospermum ruderale, Rosa acicularis, R. gymnocarpa, R. nutkana, R. woodsii, Shepherdia canadensis*

Hide Preparation: *Artemisia tridentata, A. tripartita, Pinus ponderosa*

Hunting & Fishing Item: *Amelanchier alnifolia, Cornus sericea, Holodiscus discolor, Juniperus scopulorum, Leymus cinereus, Lomatium dissectum, Monardella odoratissima, Opuntia fragilis, O. polyacantha, Osmorhiza occidentalis, Phalaris arundinacea, Philadelphus lewisii, Pseudotsuga menziesii, Rosa acicularis, R. gymnocarpa, R. nutkana, R. woodsii, Taxus brevifolia, Thuja plicata, Zigadenus venenosus*

Incense & Fragrance: *Anaphalis margaritacea, Hierochloe odorata*

Insecticide: *Achillea millefolium, Artemisia absinthium, A. dracunculus*

Malicious Charm: *Physocarpus malvaceus*

Musical Instrument: *Acer glabrum, Holodiscus discolor, Sambucus cerulea, Thuja plicata*

Paint: *Larix occidentalis*

Preservative: *Ceanothus sanguineus, Cornus sericea*

Protection: *Juniperus communis, J. scopulorum, Philadelphus lewisii, Rosa acicularis, R. gymnocarpa, R. nutkana, R. woodsii, Typha latifolia, Urtica dioica*

Season Indicator: *Astragalus miser, Crataegus douglasii, Fritillaria pudica, Larix occidentalis, Lupinus sericeus, L. sulphureus, L. wyethii, Opuntia fragilis, O. polyacantha, Philadelphus lewisii, Pinus contorta, Populus balsamifera, Prunus virginiana, Pseudotsuga menziesii, Rhus glabra, Ribes cereum*

Smoke Plant: *Arctostaphylos uva-ursi, Clematis ligusticifolia, Cornus sericea, Ligusticum canbyi, Nicotiana attenuata*

Smoking Tools: *Philadelphus lewisii*

Soap: *Philadelphus lewisii, Pseudotsuga menziesii*

Stable Gear: *Juniperus scopulorum*

Tools: *Acer glabrum, Amelanchier alnifolia, Astragalus miser, Balsamorhiza sagittata, Holodiscus discolor, Philadelphus lewisii, Populus balsamifera, P. tremuloides*

Toys & Games: *Acer glabrum, Eriogonum compositum, E. heracleoides, E. niveum, Holodiscus discolor, Juniperus scopulorum, Leymus cinereus, Solidago canadensis, Stipa comata*

Water Indicator: *Athyrium filix-femina, Gymnocarpium dryopteris, Lithospermum ruderale, Pteridium aquilinum, Woodsia scopulina*
Weapon: *Cicuta douglasii, Juniperus scopulorum*

Okanagon
Drug
Analgesic: *Arabis drummondii, Artemisia dracunculus, Asclepias speciosa, Ceanothus velutinus, Luetkea pectinata, Mentha canadensis, Valeriana sitchensis*
Antidiarrheal: *Artemisia campestris*
Antiemetic: *Rubus pubescens*
Antihemorrhagic: *Arctostaphylos uva-ursi, Rubus pubescens*
Antirheumatic (External): *Artemisia dracunculus, Populus tremuloides, Pseudotsuga menziesii, Zigadenus venenosus*
Antirheumatic (Internal): *Actaea rubra, Populus tremuloides, Sambucus racemosa*
Blood Medicine: *Aralia nudicaulis, Oplopanax horridus, Veratrum viride*
Burn Dressing: *Equisetum laevigatum*
Cathartic: *Abies grandis, Heracleum maximum*
Cold Remedy: *Mentha canadensis, Pinus contorta, Valeriana sitchensis*
Cough Medicine: *Pinus contorta*
Dermatological Aid: *Achillea millefolium, Aralia nudicaulis, Caltha leptosepala, Chaenactis douglasii, Chimaphila umbellata, Crataegus douglasii, Linum lewisii, Luetkea pectinata, Mentha canadensis, Oplopanax horridus, Rhododendron albiflorum, Sambucus racemosa, Valeriana sitchensis*
Diaphoretic: *Artemisia dracunculus*
Dietary Aid: *Actaea rubra, Aster foliaceus*
Eye Medicine: *Abies grandis, Achillea millefolium, Arctostaphylos uva-ursi, Juniperus communis, Penstemon fruticosus, Pinus ponderosa*
Gastrointestinal Aid: *Asarum caudatum, Aster foliaceus, Crataegus douglasii, Luetkea pectinata, Mentha canadensis, Oplopanax horridus, Populus tremuloides, Pulsatilla occidentalis, Rhododendron albiflorum, Rubus pubescens*
Gynecological Aid: *Artemisia campestris, A. dracunculus, Chimaphila umbellata, Cornus sericea, Goodyera oblongifolia, Luetkea pectinata*
Hemorrhoid Remedy: *Sedum divergens, S. spathulifolium*
Hemostat: *Anemone multifida*
Kidney Aid: *Arabis drummondii, Arctostaphylos uva-ursi, Juniperus communis, Penstemon fruticosus, Pseudotsuga menziesii*
Laxative: *Sedum divergens*
Oral Aid: *Rhus glabra*
Orthopedic Aid: *Arabis drummondii, Artemisia dracunculus, Penstemon fruticosus, Pinus contorta*
Pediatric Aid: *Linum lewisii, Mentha canadensis, Sedum divergens, S. spathulifolium*
Sedative: *Sedum spathulifolium*
Snakebite Remedy: *Achillea millefolium, Chaenactis douglasii*
Stimulant: *Potentilla glandulosa*
Throat Aid: *Pinus contorta*
Tonic: *Achillea millefolium, Aralia nudicaulis, Arctostaphylos uva-ursi, Clematis ligusticifolia, Heracleum maximum, Juniperus communis, Oplopanax horridus, Potentilla glandulosa, Rubus pubescens, Streptopus roseus*
Toothache Remedy: *Sambucus racemosa*
Urinary Aid: *Arabis drummondii, Arctostaphylos uva-ursi, Juniperus scopulorum, Pseudotsuga menziesii*

Food
Beverage: *Shepherdia canadensis*
Bread & Cake: *Amelanchier alnifolia, Gaultheria shallon*
Dried Food: *Lomatium dissectum*
Forage: *Hydrophyllum fendleri*
Fruit: *Arctostaphylos uva-ursi, Crataegus columbiana, C. douglasii, Gaultheria shallon, Lonicera involucrata, Maianthemum racemosum, M. stellatum, Ribes cereum, Rubus spectabilis, Sambucus racemosa, Shepherdia canadensis, Sorbus sitchensis, Streptopus amplexifolius, Viburnum opulus*
Soup: *Arctostaphylos uva-ursi*
Staple: *Allium cernuum, Amelanchier alnifolia, A.* sp., *Arctostaphylos uva-ursi, Argentina anserina, Balsamorhiza sagittata, Calochortus macrocarpus, Camassia scilloides, Cirsium undulatum, Claytonia lanceolata, Comandra umbellata, Cornus sericea, Crataegus rivularis, Elaeagnus commutata, Erythronium grandiflorum, Ferula dissoluta, Fragaria vesca, Fritillaria lanceolata, F. pudica, Heracleum maximum, Hydrophyllum occidentale, Lewisia rediviva, Lilium columbianum, Lomatium macrocarpum, Lycopus uniflorus, Mahonia* sp., *Opuntia* sp., *Peucedanum* sp., *Pinus contorta, P. ponderosa, Prunus virginiana, Ribes cereum, R. lacustre, R.* sp., *Rosa* sp., *Rubus leucodermis, R.* sp., *Sambucus* sp., *Shepherdia canadensis, Sium suave, Vaccinium membranaceum, V. myrtillus*
Unspecified: *Allium cernuum, Argentina anserina, Balsamorhiza sagittata, Calochortus macrocarpus, Camassia quamash, C. scilloides, Cirsium edule, C. hookerianum, Claytonia lanceolata, Corylus cornuta, Epilobium angustifolium, Erythronium grandiflorum, Fritillaria lanceolata, F. pudica, Heracleum maximum, Hydrophyllum fendleri, Lewisia rediviva, Lilium columbianum, Lithospermum incisum, Lomatium macrocarpum, Opuntia polyacantha, Osmorhiza berteroi, Pinus contorta, P. ponderosa, Pteridium aquilinum, Rubus leucodermis, R. spectabilis, R. ursinus, Tragopogon porrifolius, Triteleia grandiflora*
Vegetable: *Lomatium nudicaule*
Winter Use Food: *Lewisia columbiana, L. rediviva*

Fiber
Basketry: *Phragmites australis, Prunus emarginata*
Clothing: *Apocynum cannabinum, Thuja plicata*
Cordage: *Apocynum cannabinum, Thuja plicata*
Mats, Rugs & Bedding: *Asarum caudatum, Phragmites australis, Prunus emarginata, Thuja plicata, Typha latifolia*
Scouring Material: *Equisetum laevigatum*
Sewing Material: *Apocynum cannabinum, Thuja plicata*

Other
Cash Crop: *Amelanchier alnifolia, Apocynum cannabinum, Corylus cornuta, Lewisia columbiana, L. rediviva*
Containers: *Artemisia tridentata, Prunus emarginata, Thuja plicata, Typha latifolia*
Decorations: *Phragmites australis*
Fasteners: *Apocynum androsaemifolium*
Hide Preparation: *Artemisia tridentata*
Hunting & Fishing Item: *Apocynum cannabinum*
Smoke Plant: *Arctostaphylos uva-ursi, Cornus sericea, Nicotiana attenuata, Rosa gymnocarpa, Vaccinium ovalifolium, Valeriana sitchensis*
Stable Gear: *Artemisia tridentata*
Tools: *Amelanchier alnifolia*
Toys & Games: *Lithospermum incisum, Rosa gymnocarpa*

Omaha
Drug
Abortifacient: *Artemisia dracunculus, A. frigida*
Analgesic: *Amorpha canescens, Aquilegia canadensis, Cucurbita foetidissima, Dyssodia papposa, Echinacea angustifolia, Heracleum maximum, Lespedeza capitata, Physalis lanceolata, Pulsatilla patens, Rhus glabra, Silphium perfoliatum*
Anesthetic: *Echinacea angustifolia*
Antidiarrheal: *Chamaesyce serpyllifolia*
Antidote: *Echinacea angustifolia, Rhus glabra*
Antirheumatic (External): *Amorpha canescens, Lespedeza capitata, Pulsatilla patens*

Antirheumatic (Internal): *Silphium perfoliatum*
Blood Medicine: *Echinacea angustifolia*
Burn Dressing: *Echinacea angustifolia, Typha latifolia, T.* sp.
Carminative: *Acorus calamus, Mentha canadensis*
Cathartic: *Heracleum maximum*
Ceremonial Medicine: *Acorus calamus, Artemisia ludoviciana, A.* sp., *Asclepias tuberosa, Fraxinus pennsylvanica, Hierochloe odorata, Juniperus virginiana, Lophophora williamsii, Populus deltoides, Prunus virginiana, Salix* sp., *Typha latifolia*
Cold Remedy: *Acorus calamus, Juniperus virginiana, Silphium perfoliatum*
Cough Medicine: *Acorus calamus, Juniperus virginiana*
Dermatological Aid: *Amorpha canescens, Asclepias tuberosa, Echinacea angustifolia, Eriogonum fasciculatum, Humulus lupulus, Impatiens capensis, I. pallida, Iris versicolor, Liatris scariosa, Osmorhiza longistylis, Oxalis stricta, Physalis lanceolata, P. viscosa, Prunus americana, Rhus glabra, Typha latifolia*
Diaphoretic: *Juniperus virginiana*
Dietary Aid: *Cucurbita foetidissima, Gymnocladus dioicus, Liatris scariosa*
Disinfectant: *Artemisia* sp.
Diuretic: *Rhus glabra*
Ear Medicine: *Iris versicolor*
Eye Medicine: *Echinacea angustifolia, Iris versicolor, Lygodesmia juncea, Rosa arkansana, Symphoricarpos occidentalis, S. orbiculatus*
Febrifuge: *Acorus calamus, Andropogon gerardii, Aquilegia canadensis, Artemisia ludoviciana, Caulophyllum thalictroides*
Gastrointestinal Aid: *Acorus calamus, Artemisia* sp., *Asclepias exaltata, Chamaesyce serpyllifolia, Heracleum maximum, Liatris scariosa, Physalis lanceolata, Quercus* sp., *Rubus idaeus, R. occidentalis*
Gynecological Aid: *Cucurbita foetidissima, Gymnocladus dioicus, Lygodesmia juncea, Rhus glabra*
Hemostat: *Artemisia ludoviciana, Gymnocladus dioicus, Rhus glabra*
Herbal Steam: *Juniperus virginiana, Silphium perfoliatum*
Kidney Aid: *Gymnocladus dioicus, Ribes americanum*
Laxative: *Gymnocladus dioicus, Ulmus rubra*
Love Medicine: *Aquilegia canadensis, Artemisia dracunculus, Lomatium foeniculaceum*
Misc. Disease Remedy: *Echinacea angustifolia*
Other: *Echinacea angustifolia, Echinocactus williamsii*
Panacea: *Acorus calamus, Anemone canadensis, Cucurbita foetidissima*
Pediatric Aid: *Quercus* sp., *Rubus idaeus, R. occidentalis, Typha latifolia*
Poison: *Toxicodendron pubescens*
Pulmonary Aid: *Asclepias tuberosa*
Respiratory Aid: *Asclepias tuberosa*
Snakebite Remedy: *Echinacea angustifolia*
Stimulant: *Andropogon gerardii, Gymnocladus dioicus*
Throat Aid: *Smilax herbacea*
Tonic: *Acorus calamus, Cucurbita foetidissima, Gymnocladus dioicus, Liatris scariosa*
Toothache Remedy: *Acorus calamus, Echinacea angustifolia*
Unspecified: *Artemisia dracunculus, Physalis lanceolata, Yucca glauca*
Urinary Aid: *Rhus glabra*
Veterinary Aid: *Acorus calamus, Echinacea angustifolia, Juniperus virginiana, Liatris scariosa, Silphium laciniatum*
Witchcraft Medicine: *Artemisia* sp., *Silphium laciniatum*

Food

Beverage: *Agastache foeniculum, Ceanothus americanus, Crataegus chrysocarpa, C. mollis, Mentha canadensis, Rubus idaeus, R. occidentalis, Sambucus canadensis, Verbena hastata, V. stricta*
Bread & Cake: *Prunus virginiana*
Candy: *Populus deltoides, Silphium laciniatum*
Cooking Agent: *Lithospermum canescens*
Dried Food: *Fragaria virginiana, Morus rubra, Pediomelum esculentum, Physalis heterophylla, Prunus americana, P. pumila, P. virginiana, Ribes missouriense, Rubus idaeus, R. occidentalis, Shepherdia argentea, Vitis cinerea, V. riparia, V. vulpina*
Fodder: *Osmorhiza longistylis, Oxalis stricta, O. violacea*
Fruit: *Amelanchier alnifolia, Celtis occidentalis, Crataegus chrysocarpa, C. mollis, Fragaria vesca, F. virginiana, Helianthus tuberosus, Malus ioensis, Morus rubra, Prunus americana, P. pumila, P. virginiana, Ribes missouriense, Rubus idaeus, R. occidentalis, Sambucus canadensis, Shepherdia argentea, Smilax herbacea, Viburnum lentago, Vitis cinerea, V. riparia, V. vulpina*
Preservative: *Ulmus rubra*
Sauce & Relish: *Allium canadense, Physalis heterophylla, Prunus americana, P. pumila*
Snack Food: *Ulmus rubra*
Soup: *Carya ovata, Chenopodium album, Corylus americana, Juglans nigra, Nelumbo lutea, Pediomelum esculentum*
Spice: *Allium canadense, Ulmus rubra*
Staple: *Zea mays, Zizania aquatica*
Starvation Food: *Crataegus chrysocarpa, C. mollis, Rosa arkansana*
Sweetener: *Acer negundo, A. saccharinum, Agastache foeniculum, Carya ovata, Zea mays*
Unspecified: *Allium canadense, A.* sp., *Amphicarpaea bracteata, Apios americana, Carya ovata, Corylus americana, Helianthus tuberosus, Juglans nigra, Lathyrus brachycalyx, Linum lewisii, Nelumbo lutea, Oxalis stricta, O. violacea, Pediomelum esculentum, Quercus macrocarpa, Q. rubra, Rumex crispus, Sagittaria latifolia, Ulmus rubra*
Vegetable: *Apios americana, Asclepias syriaca, Nelumbo lutea, Sagittaria latifolia*
Winter Use Food: *Amphicarpaea bracteata, Zea mays*

Fiber

Basketry: *Salix interior, Tilia americana, Ulmus rubra*
Brushes & Brooms: *Prunus americana, Stipa spartea*
Building Material: *Andropogon gerardii, Fraxinus pennsylvanica, Populus deltoides, Salix* sp., *Spartina pectinata, Ulmus americana, U. rubra, U. thomasii*
Cordage: *Tilia americana, Ulmus rubra, Urtica dioica, Yucca glauca*
Mats, Rugs & Bedding: *Artemisia dracunculus, A. frigida, A. ludoviciana, Scirpus tabernaemontani*
Scouring Material: *Equisetum* sp.
Sewing Material: *Urtica dioica, Yucca glauca*
Snow Gear: *Carya cordiformis, C. glabra, C. laciniosa, C. ovata*

Dye

Black: *Acer saccharinum, Juglans nigra, Quercus rubra*
Red: *Sanguinaria canadensis*
Unspecified: *Sanguinaria canadensis*
Yellow: *Populus deltoides, Rhus glabra*

Other

Ceremonial Items: *Acer negundo, Astragalus crassicarpus, Fraxinus pennsylvanica, Heracleum maximum, Hierochloe odorata, Lagenaria siceraria, Nelumbo lutea, Populus deltoides, Salix* sp., *Stipa spartea, Typha latifolia, Urtica dioica*
Cooking Tools: *Astragalus canadensis, Ulmus americana, U. rubra*
Decorations: *Acer negundo*
Fuel: *Ceanothus americanus, Populus deltoides, Ulmus americana, U. rubra, U. thomasii*
Good Luck Charm: *Melia azedarach*
Hunting & Fishing Item: *Amelanchier alnifolia, Cornus asperifolia, Fraxinus pennsylvanica, Maclura pomifera*
Incense & Fragrance: *Acorus calamus, Aquilegia canadensis, Galium triflorum, Hierochloe odorata, Juniperus virginiana, Monarda fistulosa, Rosa arkansana, Zanthoxylum americanum*
Jewelry: *Melia azedarach*

Musical Instrument: *Populus deltoides*
Paint: *Sanguinaria canadensis*
Protection: *Juniperus virginiana, Zea mays*
Sacred Items: *Populus deltoides*
Season Indicator: *Prunus americana, P. virginiana, Solidago* sp.
Smoke Plant: *Cornus amomum, C. sericea, Nicotiana quadrivalvis, Rhus glabra, Rosa arkansana*
Smoking Tools: *Fraxinus pennsylvanica*
Soap: *Yucca glauca*
Stable Gear: *Ulmus americana, U. thomasii, Urtica dioica*
Tools: *Ulmus americana, U. rubra, U. thomasii*
Toys & Games: *Amelanchier alnifolia, Andropogon gerardii, Artemisia* sp., *Asclepias syriaca, Astragalus canadensis, Betula papyrifera, Desmanthus illinoensis, Populus deltoides, Quercus macrocarpa, Ribes missouriense, Sambucus canadensis, Thalictrum dasycarpum, Ulmus americana, U. rubra, Urtica dioica, Viburnum opulus, Viola* sp.

Oregon Indian
Drug
Antidiarrheal: *Prunus virginiana*
Dermatological Aid: *Clematis ligusticifolia*
Febrifuge: *Clematis ligusticifolia, Cleome serrulata*
Love Medicine: *Lomatium* sp.
Veterinary Aid: *Lomatium dissectum*
Food
Beverage: *Shepherdia canadensis*
Fruit: *Crataegus columbiana*
Soup: *Lomatium cous*
Other
Hide Preparation: *Lomatium dissectum*

Oregon Indian, Warm Springs
Drug
Unspecified: *Ceanothus velutinus*
Food
Unspecified: *Lewisia rediviva*

Oto
Drug
Antiemetic: *Ambrosia artemisiifolia*
Laxative: *Gymnocladus dioicus*
Pulmonary Aid: *Arctium minus*

Oweekeno
Drug
Analgesic: *Oplopanax horridus*
Antirheumatic (External): *Oplopanax horridus, Picea sitchensis, Plagiomnium insigne, Veratrum viride*
Ceremonial Medicine: *Thuja plicata, Tsuga heterophylla*
Cold Remedy: *Abies amabilis, Ledum groenlandicum, Oplopanax horridus*
Cough Medicine: *Polypodium glycyrrhiza*
Dermatological Aid: *Maianthemum dilatatum, Oplopanax horridus, Peltigera* sp., *Picea sitchensis, Plagiomnium insigne, Populus balsamifera*
Gastrointestinal Aid: *Picea sitchensis*
Internal Medicine: *Plagiomnium insigne*
Oral Aid: *Malus fusca*
Panacea: *Oplopanax horridus*
Poison: *Oplopanax horridus, Ribes lacustre, Veratrum viride*
Throat Aid: *Ledum groenlandicum, Polypodium glycyrrhiza*
Tonic: *Oplopanax horridus*
Unspecified: *Picea sitchensis*
Veterinary Aid: *Peltigera* sp.
Food
Beverage: *Camellia sinensis, Ledum groenlandicum, Rubus parviflorus*
Bread & Cake: *Porphyra abbottae*
Candy: *Picea sitchensis*
Dried Food: *Egregia menziesii, Macrocystis integrifolia, Porphyra abbottae, Sambucus racemosa, Tsuga heterophylla*
Forage: *Cornus unalaschkensis, Lysichiton americanus, Maianthemum dilatatum, Streptopus amplexifolius*
Fruit: *Amelanchier alnifolia, Arctostaphylos uva-ursi, Cornus unalaschkensis, Fragaria chiloensis, F. vesca, F. virginiana, Gaultheria shallon, Lonicera involucrata, Malus fusca, M. sylvestris, Prunus* sp., *Ribes bracteosum, R. divaricatum, R. laxiflorum, Rubus idaeus, R. leucodermis, R. parviflorus, R. spectabilis, Shepherdia canadensis, Vaccinium cespitosum, V. ovalifolium, V. oxycoccos, V. vitis-idaea*
Preserves: *Fragaria chiloensis, F. vesca, F. virginiana, Gaultheria shallon, Rubus parviflorus, R. spectabilis, Sambucus racemosa, Viburnum edule*
Special Food: *Vaccinium cespitosum, V. membranaceum*
Unspecified: *Allium cepa, A. cernuum, Argentina egedii, Brassica rapa, Conioselinum gmelinii, Daucus carota, Dryopteris* sp., *Egregia menziesii, Epilobium angustifolium, Fritillaria camschatcensis, Heracleum maximum, Macrocystis integrifolia, Nereocystis luetkeana, Oplopanax horridus, Oryza sativa, Populus balsamifera, Porphyra abbottae, Rubus parviflorus, R. spectabilis, Rumex aquaticus, Trifolium wormskioldii, Tsuga heterophylla, Urtica dioica, Zostera marina*
Vegetable: *Solanum tuberosum*
Winter Use Food: *Egregia menziesii, Macrocystis integrifolia, Nereocystis luetkeana, Rubus spectabilis, Shepherdia canadensis, Vaccinium ovalifolium, V. parvifolium, V. uliginosum*
Fiber
Basketry: *Picea sitchensis, Thuja plicata*
Brushes & Brooms: *Taxus brevifolia*
Building Material: *Thuja plicata, Tsuga heterophylla*
Canoe Material: *Alnus rubra, Chamaecyparis nootkatensis, Taxus brevifolia, Thuja plicata*
Clothing: *Chamaecyparis nootkatensis, Polystichum munitum, Thuja plicata*
Cordage: *Chamaecyparis nootkatensis, Epilobium angustifolium, Thuja plicata, Urtica dioica*
Mats, Rugs & Bedding: *Chamaecyparis nootkatensis, Polystichum munitum, Thuja plicata*
Dye
Red: *Alnus rubra*
Yellow: *Letharia vulpina*
Other
Cash Crop: *Arctostaphylos uva-ursi*
Ceremonial Items: *Alnus rubra, Dryopteris* sp., *Juniperus communis, Rosa nutkana, Rubus spectabilis, Thuja plicata, Tsuga heterophylla, Veratrum viride*
Cleaning Agent: *Usnea longissima*
Containers: *Chamaecyparis nootkatensis, Lysichiton americanus, Polystichum munitum, Rubus spectabilis, Thuja plicata*
Cooking Tools: *Acer glabrum, Gaultheria shallon, Lysichiton americanus, Malus fusca, Taxus brevifolia, Thuja plicata, Tsuga heterophylla*
Decorations: *Chamaecyparis nootkatensis, Thuja plicata*
Fasteners: *Populus balsamifera*
Fuel: *Alnus rubra, Pseudotsuga menziesii, Pteridium aquilinum, Thuja plicata*
Good Luck Charm: *Pinguicula vulgaris*
Hunting & Fishing Item: *Cornus sericea, Nereocystis luetkeana, Taxus brevifolia, Thuja plicata, Tsuga heterophylla, Urtica dioica*
Jewelry: *Lycopodium clavatum*
Musical Instrument: *Alnus rubra, Thuja plicata*
Paint: *Fomes* sp.

Protection: *Tsuga heterophylla*
Smoke Plant: *Nicotiana tabacum*
Tools: *Alnus rubra, Chamaecyparis nootkatensis, Cornus sericea, Taxus brevifolia, Thuja plicata*
Toys & Games: *Fucus gardneri, Tsuga heterophylla*

Paipai
Other
Cash Crop: *Phaseolus vulgaris*

Paiute
Drug
Abortifacient: *Porophyllum gracile*
Adjuvant: *Ephedra nevadensis*
Analgesic: *Achillea millefolium, Agastache urticifolia, Aquilegia formosa, Artemisia douglasiana, A. ludoviciana, A. spinescens, A. tridentata, Asclepias cryptoceras, Balsamorhiza sagittata, Cercocarpus ledifolius, Chaenactis douglasii, Cicuta maculata, Erigeron aphanactis, Eriogonum umbellatum, Hypericum scouleri, Iris missouriensis, Juniperus occidentalis, J. osteosperma, Larrea tridentata, Marrubium vulgare, Mentha canadensis, Mirabilis alipes, Monardella odoratissima, Osmorhiza occidentalis, Phragmites australis, Pinus monophylla, Prunus virginiana, Psathyrotes ramosissima, Psorothamnus polydenius, Purshia tridentata, Rumex crispus, R. venosus, Salvia dorrii, Stanleya pinnata, Symphoricarpos longiflorus, Zigadenus paniculatus, Z. venenosus*
Anthelmintic: *Nicotiana attenuata*
Antidiarrheal: *Artemisia ludoviciana, Cercocarpus ledifolius, Ephedra viridis, Eriastrum eremicum, Ericameria nana, Erigeron caespitosus, Eriodictyon angustifolium, Eriogonum sphaerocephalum, Eryngium alismifolium, Haplopappus* sp.*, Heliotropium curassavicum, Holodiscus dumosus, Ipomopsis congesta, Mahonia repens, Pinus monophylla, Pluchea sericea, Prunus andersonii, Psathyrotes ramosissima, Psorothamnus polydenius, Rosa woodsii, Rumex crispus, R. venosus, Salix* sp.*, Sambucus cerulea, S. racemosa, Sarcobatus vermiculatus, Solanum physalifolium, Stephanomeria spinosa*
Antiemetic: *Eriodictyon angustifolium, Mirabilis alipes, Pinus monophylla*
Antihemorrhagic: *Juniperus occidentalis, J. osteosperma, Psorothamnus fremontii, Sarcobatus vermiculatus*
Antirheumatic (External): *Achillea millefolium, Aquilegia formosa, Artemisia dracunculus, A. ludoviciana, A. spinescens, Asclepias speciosa, Brachyactis frondosa, Cicuta douglasii, C. maculata, Eriogonum umbellatum, Gutierrezia sarothrae, Helianthus annuus, Heracleum maximum, Juniperus occidentalis, J. osteosperma, Larrea tridentata, Lomatium dissectum, Nicotiana attenuata, Psorothamnus polydenius, Rumex crispus, Sambucus cerulea, Urtica dioica, Veratrum californicum, Zigadenus paniculatus, Z. venenosus*
Antirheumatic (Internal): *Ephedra viridis, Pinus monophylla, Prunus andersonii, Rumex venosus*
Blood Medicine: *Achillea millefolium, Brachyactis frondosa, Cercocarpus ledifolius, Corallorrhiza maculata, Datura wrightii, Ephedra viridis, Juniperus communis, J. occidentalis, J. osteosperma, Mahonia nervosa, M. repens, Rumex crispus, R. venosus, Salix* sp.*, Wyethia mollis*
Burn Dressing: *Argemone polyanthemos, Cercocarpus ledifolius, Ephedra nevadensis, Larrea tridentata, Mirabilis alipes, Rosa woodsii, Rumex crispus, R. hymenosepalus, R. venosus, Veratrum californicum, Zigadenus venenosus*
Carminative: *Linum lewisii, Mentha canadensis*
Cathartic: *Artemisia* sp.*, Cucurbita foetidissima, Eriastrum filifolium, Erigeron aphanactis, Ipomopsis aggregata, I. congesta, Mirabilis alipes, Nicotiana attenuata, Osmorhiza occidentalis, Phlox longifolia, Psathyrotes ramosissima, Purshia mexicana, P. tridentata, Salix* sp.*, Stephanomeria spinosa, Thymophylla pentachaeta, Wyethia mollis*
Ceremonial Medicine: *Artemisia tridentata*
Cold Remedy: *Abies procera, Achillea millefolium, Agastache urticifolia, Angelica breweri, Aquilegia formosa, Artemisia spinescens, A. tridentata, A. vulgaris, Calocedrus decurrens, Cercocarpus ledifolius, Chaenactis douglasii, Chrysothamnus viscidiflorus, Ephedra viridis, Ericameria nana, Eriodictyon angustifolium, Eriogonum ovalifolium, E. sphaerocephalum, E. umbellatum, Heracleum maximum, Holodiscus dumosus, Ipomopsis aggregata, I. congesta, Juniperus monosperma, J. occidentalis, J. osteosperma, Larrea tridentata, Lomatium dissectum, Mentha canadensis, Monardella odoratissima, Nicotiana attenuata, Orobanche californica, Osmorhiza occidentalis, Pinus monophylla, Plantago major, Prunus andersonii, P. virginiana, Psorothamnus polydenius, Purshia mexicana, P. tridentata, Rosa woodsii, Rumex venosus, Salvia dorrii, Sambucus racemosa, Tetradymia comosa, Urtica dioica, Wyethia mollis*
Contraceptive: *Cuscuta* sp.*, Veratrum californicum*
Cough Medicine: *Abies procera, Achillea millefolium, Angelica breweri, Aquilegia formosa, Artemisia spinescens, Asclepias speciosa, Cercocarpus ledifolius, Chaenactis douglasii, Chrysothamnus viscidiflorus, Ericameria nana, Eriodictyon angustifolium, Grindelia nana, G. squarrosa, Juniperus occidentalis, J. osteosperma, Ligusticum filicinum, Lomatium dissectum, Mahonia repens, Maianthemum stellatum, Paeonia brownii, Populus tremuloides, Prunus virginiana, Psorothamnus polydenius, Rumex venosus, Salvia dorrii, Sambucus racemosa, Tetradymia comosa*
Dermatological Aid: *Abies concolor, A. procera, Achillea millefolium, Agastache urticifolia, Angelica breweri, A.* sp.*, Antennaria anaphaloides, Aquilegia formosa, Argemone hispida, A. polyanthemos, Artemisia dracunculus, A. ludoviciana, A. spinescens, A. tridentata, A. vulgaris, Asclepias cryptoceras, A. speciosa, Balsamorhiza sagittata, Cercocarpus ledifolius, Chaenactis douglasii, Chamaesyce ocellata, Cicuta maculata, Crepis modocensis, Cucurbita foetidissima, Descurainia sophia, Ephedra viridis, Helianthella uniflora, Heracleum maximum, Hypericum scouleri, Iris missouriensis, Iva axillaris, Juniperus monosperma, J. occidentalis, J. osteosperma, Keckiella breviflora, Krameria grayi, Krascheninnikovia lanata, Larrea tridentata, Leucocrinum montanum, Linum lewisii, Lomatium dissectum, Maianthemum stellatum, Mentha canadensis, Mirabilis alipes, M. bigelovii, Nicotiana attenuata, Nothochelone nemorosa, Osmorhiza occidentalis, Paeonia brownii, Penstemon deustus, P. richardsonii, Pinus monophylla, P. ponderosa, Plantago major, Prunus virginiana, Psathyrotes ramosissima, Psorothamnus polydenius, Purshia tridentata, Rhus trilobata, Ribes aureum, Rosa* sp.*, R. woodsii, Rumex crispus, R. hymenosepalus, R. venosus, Salix* sp.*, Sambucus racemosa, Stephanomeria spinosa, Suaeda moquinii, Veratrum californicum, Wyethia mollis, Zigadenus paniculatus, Z. venenosus*
Diaphoretic: *Artemisia tridentata, Chrysothamnus viscidiflorus, Urtica dioica*
Dietary Aid: *Paeonia brownii*
Disinfectant: *Artemisia ludoviciana, A. tridentata, Balsamorhiza sagittata, Juniperus occidentalis, J. osteosperma, Krameria grayi, Lomatium dissectum, Osmorhiza occidentalis, Psorothamnus polydenius, Veratrum californicum*
Diuretic: *Abronia villosa, Artemisia spinescens, Heliotropium curassavicum, Juniperus occidentalis, J. osteosperma, Lupinus* sp.*, Psorothamnus polydenius, Salix* sp.*
Ear Medicine: *Iris missouriensis, Maianthemum stellatum, Salvia dorrii*
Emetic: *Achillea millefolium, Artemisia tridentata, Chenopodium album, Cucurbita foetidissima, Eriastrum filifolium, Erigeron aphanactis, Heliotropium curassavicum, Ipomopsis aggregata, I. congesta, Nicotiana attenuata, Psathyrotes ramosissima, Purshia tridentata, Sambucus tridentata, Stephanomeria spinosa, Tanacetum vulgare, Wyethia amplexicaulis, Zigadenus paniculatus*
Expectorant: *Eriodictyon angustifolium, Grindelia nana, G. squarrosa, Phragmites australis*
Eye Medicine: *Achillea millefolium, Achlys triphylla, Artemisia ludoviciana, Chamaesyce ocellata, Cornus canadensis, Desmanthus illinoensis, Ericameria nana, Erigeron caespitosus, Heuchera rubescens, Krameria grayi, Krascheninnikovia lanata, Leymus condensatus, Linum lewisii, Maianthemum stellatum, Marah oreganus, Monardella odoratissima, Osmorhiza occidentalis, Paeonia brownii, Penstemon deustus, Phlox longifolia, Physaria chambersii,*

[Eye Medicine]
: *Prosopis pubescens, Prunus emarginata, P. virginiana, Psathyrotes annua, Salvia dorrii, Trientalis borealis, Trillium ovatum*

Febrifuge: *Achillea millefolium, Artemisia furcata, A. ludoviciana, A. tridentata, Ericameria nana, Juniperus occidentalis, J. osteosperma, Mentha canadensis, Osmorhiza occidentalis, Pinus monophylla, Populus tremuloides, Salvia dorrii, Stephanomeria sp., Veratrum californicum, Wyethia mollis*

Gastrointestinal Aid: *Achillea millefolium, Agastache urticifolia, Aquilegia formosa, Artemisia ludoviciana, A. spinescens, A. tridentata, Balsamorhiza hookeri, B. sagittata, Cercocarpus ledifolius, Ephedra viridis, Eriastrum eremicum, Ericameria nana, Erigeron aphanactis, Eriogonum umbellatum, Frangula purshiana, Haplopappus sp., Holodiscus dumosus, Ipomopsis congesta, Iris missouriensis, Juniperus occidentalis, J. osteosperma, Larrea tridentata, Linum lewisii, Mahonia repens, Maianthemum stellatum, Mentha canadensis, Monardella odoratissima, Osmorhiza occidentalis, Penstemon deustus, Phlox longifolia, Pinus monophylla, Pluchea sericea, Populus balsamifera, Psathyrotes ramosissima, Psorothamnus polydenius, Rosa sp., Rumex crispus, R. venosus, Salvia dorrii, Sambucus cerulea, Symphoricarpos longiflorus, Tetradymia comosa*

Gland Medicine: *Veratrum californicum*

Gynecological Aid: *Achillea millefolium, Artemisia dracunculus, A. ludoviciana, A. tridentata, Juniperus occidentalis, J. osteosperma, Maianthemum stellatum, Phoenicaulis cheiranthoides, Pinus monophylla, Sambucus racemosa, Thamnosma montana, Veratrum californicum*

Hallucinogen: *Datura wrightii*

Heart Medicine: *Cercocarpus ledifolius, Chaenactis douglasii, Helianthus cusickii, Paeonia brownii*

Hemostat: *Gutierrezia sarothrae, Mahonia nervosa, Sambucus racemosa*

Herbal Steam: *Artemisia ludoviciana, Lomatium dissectum, Prunus virginiana, Salvia dorrii, Urtica dioica*

Kidney Aid: *Achillea millefolium, Angelica breweri, Clematis ligusticifolia, Ephedra viridis, Juniperus occidentalis, J. osteosperma, Nicotiana attenuata, Paeonia brownii, Psorothamnus polydenius, Rumex venosus, Suaeda moquinii*

Laxative: *Anemopsis californica, Potentilla sp., Psathyrotes ramosissima, Salix sp., Sambucus tridentata*

Liver Aid: *Psathyrotes ramosissima, Purshia tridentata*

Misc. Disease Remedy: *Artemisia douglasiana, A. ludoviciana, A. tridentata, Asclepias speciosa, Ericameria nana, Juniperus monosperma, J. occidentalis, J. osteosperma, Larrea tridentata, Lomatium dissectum, Nicotiana attenuata, Osmorhiza occidentalis, Pinus monophylla, Prunus andersonii, Psorothamnus polydenius, Purshia tridentata, Rosa woodsii, Rumex venosus, Salix sp., Salvia dorrii, Stanleya pinnata, Suaeda moquinii, Tetradymia comosa, Thamnosma montana, Urtica dioica*

Narcotic: *Datura wrightii*

Orthopedic Aid: *Achillea millefolium, Anemopsis californica, Artemisia dracunculus, A. ludoviciana, A. tridentata, A. vulgaris, Chaenactis douglasii, Chamaebatiaria millefolium, Cicuta maculata, Eriogonum umbellatum, Gutierrezia sarothrae, Helianthella uniflora, Hypericum scouleri, Juniperus communis, Lomatium dissectum, Maianthemum stellatum, Pinus monophylla, Psorothamnus polydenius, Ribes aureum, Salix sp., Wyethia mollis, Zigadenus paniculatus*

Other: *Datura wrightii, Glycyrrhiza lepidota, Mentha arvensis*

Panacea: *Aquilegia formosa, Larrea tridentata, Lomatium dissectum, Stephanomeria sp., Veratrum californicum*

Pediatric Aid: *Artemisia furcata, A. ludoviciana, A. tridentata, Cercocarpus ledifolius, Ephedra viridis, Eriastrum eremicum, Mentha canadensis, Penstemon deustus, Phlox longifolia, Rosa woodsii, Salix sp., Salvia dorrii, Sambucus racemosa*

Poison: *Toxicodendron pubescens, Zigadenus venenosus*

Poultice: *Linum lewisii, Salvia dorrii*

Psychological Aid: *Mirabilis alipes*

Pulmonary Aid: *Abies concolor, Artemisia spinescens, A. tridentata, Cercocarpus ledifolius, Corallorrhiza maculata, Eriodictyon angustifolium, Grindelia nana, G. squarrosa, Juniperus occidentalis, J. osteosperma, Lomatium dissectum, Orobanche californica, Osmorhiza occidentalis, Phragmites australis, Pinus monophylla, Plantago major, Psorothamnus polydenius, Purshia tridentata, Rumex venosus, Salvia dorrii, Tetradymia comosa, Urtica dioica*

Respiratory Aid: *Achillea millefolium, Artemisia tridentata, Lomatium dissectum, Nicotiana attenuata*

Snakebite Remedy: *Asclepias speciosa, Chaenactis douglasii, Cicuta maculata, Nicotiana attenuata, Osmorhiza occidentalis, Psathyrotes ramosissima, Veratrum californicum, Zigadenus venenosus*

Stimulant: *Mirabilis alipes, Rosa sp.*

Throat Aid: *Angelica breweri, Aquilegia formosa, Eriastrum virgatum, Heliotropium curassavicum, Lomatium dissectum, Machaeranthera canescens, Mentha canadensis, Osmorhiza occidentalis, Pinus monophylla, Stanleya pinnata, Veratrum californicum*

Tonic: *Artemisia dracunculus, A. ludoviciana, A. tridentata, Brachyactis frondosa, Ephedra viridis, Juniperus communis, J. occidentalis, J. osteosperma, Phoenicaulis cheiranthoides, Pinus monophylla, Prunus andersonii, Purshia tridentata, Rosa woodsii, Rumex crispus, R. venosus, Salix sp., Sambucus racemosa, Stanleya pinnata, Wyethia mollis*

Toothache Remedy: *Achillea millefolium, Iris missouriensis, Penstemon sp., Psathyrotes annua, P. ramosissima, Psorothamnus polydenius, Stephanomeria spinosa, Veratrum californicum, Zigadenus paniculatus*

Tuberculosis Remedy: *Abies concolor, Achillea millefolium, Asclepias speciosa, Balsamorhiza sagittata, Cercocarpus ledifolius, Eriastrum eremicum, Eriodictyon angustifolium, Eriogonum microthecum, Helianthus cusickii, Lomatium dissectum, Nicotiana attenuata, Paeonia brownii, Pastinaca sativa, Pinus monophylla, Prunus andersonii, P. virginiana, Psorothamnus polydenius, Purshia tridentata, Wyethia mollis*

Unspecified: *Linum lewisii, Lophophora williamsii, Urtica dioica*

Urinary Aid: *Achillea millefolium, Artemisia spinescens, Balsamorhiza hookeri, Ephedra viridis, Eriogonum microthecum, Grindelia nana, G. squarrosa, Iris missouriensis, Lupinus sp., Mahonia repens, Suaeda moquinii*

Venereal Aid: *Abies concolor, Achillea millefolium, Anemopsis californica, Angelica sp., Artemisia ludoviciana, A. vulgaris, Balsamorhiza sagittata, Cercocarpus ledifolius, Cucurbita foetidissima, Ephedra nevadensis, E. viridis, Heuchera rubescens, Ipomopsis congesta, Iris missouriensis, Juniperus communis, J. occidentalis, J. osteosperma, Krameria grayi, Larrea tridentata, Lomatium dissectum, Mahonia repens, Osmorhiza occidentalis, Phlox longifolia, Pinus monophylla, Psathyrotes ramosissima, Psorothamnus polydenius, Purshia mexicana, P. tridentata, Rumex crispus, R. venosus, Salix sp., Salvia dorrii, Sphenosciadium capitellatum, Veratrum californicum, Wyethia mollis*

Veterinary Aid: *Achillea millefolium, Descurainia sophia, Geum triflorum, Juniperus occidentalis, Lomatium dissectum, Paeonia brownii, Veratrum californicum*

Food

Beverage: *Balsamorhiza sagittata, Castanopsis chrysophylla, Descurainia sophia, Ephedra sp., E. viridis, Mentha arvensis, M. canadensis, Prunus virginiana*

Bread & Cake: *Lomatium canbyi*

Candy: *Amelanchier alnifolia, Asclepias speciosa, Balsamorhiza sagittata, Chrysothamnus nauseosus, C. viscidiflorus, Juncus balticus, Larix occidentalis, Pinus ponderosa*

Dried Food: *Amelanchier alnifolia, A. utahensis, Arctostaphylos sp., Balsamorhiza ×terebinthacea, Camassia quamash, Crataegus douglasii, Dichelostemma pulchellum, Frangula californica, Lewisia rediviva, Lomatium bicolor, L. canbyi, L. macrocarpum, Perideridia gairdneri, P. oregana, Pinus monophylla, P. ponderosa, P. sp., Prunus virginiana, Ribes aureum, Sambucus cerulea, Shep-*

herdia argentea, Vaccinium membranaceum, V. ovalifolium, V. parvifolium
Fodder: *Juncus ensifolius*
Forage: *Arctostaphylos patula, Ceanothus velutinus, Lupinus* sp., *Ribes bracteosum*
Fruit: *Amelanchier alnifolia, A. utahensis, Arctostaphylos nevadensis, Crataegus douglasii, Fragaria* sp., *Frangula californica, Juniperus occidentalis, Prunus virginiana, Ribes aureum, R. cereum, R. sanguineum, R.* sp., *Rosa* sp., *Rubus parviflorus, R.* sp., *R. spectabilis, Sambucus cerulea, Shepherdia argentea, Vaccinium cespitosum, V. membranaceum, V. ovalifolium, V. parvifolium*
Ice Cream: *Descurainia sophia*
Pie & Pudding: *Balsamorhiza sagittata, Camassia quamash, Shepherdia argentea*
Porridge: *Balsamorhiza sagittata, Cyperus rotundus, Helianthus annuus, Oryzopsis hymenoides, Pinus monophylla, Quercus dunnii, Q. kelloggii, Salvia columbariae, Stipa speciosa*
Preserves: *Vaccinium membranaceum, V. ovalifolium*
Sauce & Relish: *Allium acuminatum, A. bisceptrum, A. platycaule, A. pleianthum, Mentzelia albicaulis, M. laevicaulis, Oryzopsis hymenoides, Ribes lacustre*
Soup: *Pinus monophylla*
Spice: *Alnus* sp., *Pseudotsuga menziesii*
Staple: *Agastache pallidiflora, Agave americana, A. parryi, Artemisia tridentata, Chenopodium album, C. nevadense, Descurainia sophia, Dichelostemma pulchellum, Helianthus annuus, Mentzelia albicaulis, Oryzopsis hymenoides, Pinus monophylla, Piptatherum miliaceum, Suaeda calceoliformis, Wyethia mollis*
Starvation Food: *Artemisia tridentata, Lomatium grayi*
Sweetener: *Castilleja* sp., *Phragmites australis*
Unspecified: *Agropyron* sp., *Allium acuminatum, A. bisceptrum, A. platycaule, Alnus* sp., *Asclepias fascicularis, A. speciosa, Atriplex* sp., *Balsamorhiza sagittata, B. ×terebinthacea, Bidens laevis, Calochortus macrocarpus, Camassia quamash, Chenopodium album, C. fremontii, Cirsium pastoris, C. tioganum, Comandra umbellata, Corylus cornuta, Crepis occidentalis, Cyperus rotundus, Dalea* sp., *Dichelostemma pulchellum, Echinochloa crus-galli, Elaeagnus commutata, Eleocharis* sp., *Eragrostis secundiflora, Fritillaria pudica, Haplopappus* sp., *Helianthus bolanderi, Juncus balticus, Lewisia rediviva, Leymus condensatus, Lilium parvum, Lomatium bicolor, L. canbyi, L. grayi, L. macrocarpum, L. piperi, Oenothera elata, Perideridia gairdneri, P. oregana, Pinus monophylla, P. monticola, P. ponderosa, Polygonum* sp., *Prosopis glandulosa, P. pubescens, Quercus garryana, Rorippa curvisiliqua, Rosa pisocarpa, Scorzonella* sp., *Spiranthes* sp., *Trifolium willdenowii, Triteleia hyacinthina, Typha latifolia, Valeriana edulis, Wyethia mollis, W. ovata*
Vegetable: *Allium acuminatum, Calochortus macrocarpus, Glyptopleura marginata, Lomatium nevadense, L. nudicaule, Trifolium willdenowii, T. wormskioldii, Triteleia hyacinthina*
Winter Use Food: *Amelanchier alnifolia, Balsamorhiza sagittata, Camassia quamash, Corylus cornuta, Helianthus annuus, Juniperus occidentalis, Lewisia rediviva, Lomatium watsonii, Perideridia gairdneri, Pinus monophylla, Prunus virginiana, Quercus dunnii, Q. kelloggii, Vaccinium cespitosum*

Fiber
Basketry: *Calocedrus decurrens, Salix* sp.
Brushes & Brooms: *Leymus condensatus, Philadelphus lewisii*
Building Material: *Artemisia tridentata, Calocedrus decurrens, Juniperus occidentalis, Pinus contorta, P. ponderosa, Populus tremuloides, Salix* sp., *Typha latifolia*
Clothing: *Artemisia tridentata, Juniperus occidentalis, Purshia tridentata*
Cordage: *Artemisia tridentata, Salix* sp.
Furniture: *Cornus sericea, Salix* sp., *Symphoricarpos albus*
Mats, Rugs & Bedding: *Abies grandis, Bromus tectorum, Picea engelmannii, Pseudotsuga menziesii, Scirpus* sp., *Tsuga mertensiana*

Dye
Orange: *Alnus* sp.

Other
Ceremonial Items: *Artemisia dracunculus, A. ludoviciana, Pinus monticola, P.* sp.
Containers: *Anaphalis margaritacea, Collomia grandiflora, Salix* sp.
Cooking Tools: *Carex* sp., *Quercus garryana, Salix* sp.
Decorations: *Artemisia tridentata, Paxistima myrsinites, Polystichum munitum*
Designs: *Prunus virginiana*
Fasteners: *Pinus ponderosa*
Fuel: *Artemisia tridentata, Juniperus occidentalis, Populus balsamifera, P. tremuloides, Purshia tridentata*
Good Luck Charm: *Datura wrightii*
Hunting & Fishing Item: *Acer* sp., *Juniperus occidentalis, Oenothera elata, Prunus virginiana, Salix* sp., *Symphoricarpos albus, Taxus brevifolia*
Insecticide: *Cymopterus globosus, Osmorhiza occidentalis, Sphenosciadium capitellatum*
Jewelry: *Paeonia brownii, Rosa* sp.
Lighting: *Artemisia tridentata, Juniperus occidentalis*
Malicious Magic: *Castilleja miniata, Lomatium triternatum*
Musical Instrument: *Juniperus occidentalis, Salix* sp., *Sambucus cerulea, Thuja plicata*
Preservative: *Pinus ponderosa, Pteridium aquilinum, Veratrum californicum*
Protection: *Oenothera elata, Rosa* sp.
Smoke Plant: *Arctostaphylos nevadensis, A. patula, A. uva-ursi, Ceanothus* sp., *Nicotiana attenuata*
Stable Gear: *Betula occidentalis*
Tools: *Artemisia tridentata, Ceanothus cuneatus, Chrysothamnus nauseosus, C. viscidiflorus, Juniperus occidentalis, Philadelphus lewisii, Sarcobatus vermiculatus, Symphoricarpos albus*
Toys & Games: *Equisetum* sp., *Eriogonum elatum, Juniperus occidentalis, Polystichum munitum, Symphoricarpos albus*
Waterproofing Agent: *Pinus ponderosa*

Paiute, Nevada
Food
Unspecified: *Calochortus nuttallii*

Paiute, Northern
Drug
Analgesic: *Artemisia tridentata, Juniperus osteosperma, Lomatium dissectum, Osmorhiza occidentalis, Salvia dorrii*
Anthelmintic: *Purshia tridentata*
Antidiarrheal: *Artemisia tridentata*
Antihemorrhagic: *Angelica lineariloba, Cercocarpus ledifolius*
Antirheumatic (External): *Artemisia tridentata, Atriplex confertifolia, Heracleum maximum, Juniperus osteosperma, Lomatium dissectum, Osmorhiza occidentalis, Urtica dioica, Veratrum californicum, Zigadenus paniculatus, Z. venenosus*
Cathartic: *Psorothamnus polydenius*
Cold Remedy: *Achillea millefolium, Artemisia douglasiana, A. tridentata, Atriplex confertifolia, Juniperus osteosperma, Lomatium dissectum, Mentha canadensis, Osmorhiza occidentalis, Psorothamnus polydenius, Salvia dorrii, Veratrum californicum*
Cough Medicine: *Achillea millefolium*
Dermatological Aid: *Achillea millefolium, Asclepias cryptoceras, Ephedra viridis, Lomatium dissectum, Mirabilis alipes, Osmorhiza occidentalis, Penstemon deustus, Rumex crispus, Urtica dioica, Veratrum californicum*
Diaphoretic: *Achillea millefolium*
Diuretic: *Lomatium dissectum*
Emetic: *Artemisia tridentata*
Eye Medicine: *Osmorhiza occidentalis*
Febrifuge: *Artemisia douglasiana, A. tridentata, Mentha canadensis*

Gastrointestinal Aid: *Eriastrum sparsiflorum, Purshia tridentata*
Hallucinogen: *Datura wrightii*
Kidney Aid: *Achillea millefolium*
Laxative: *Purshia tridentata*
Misc. Disease Remedy: *Achillea millefolium, Psorothamnus polydenius*
Oral Aid: *Xanthium strumarium*
Orthopedic Aid: *Veratrum californicum*
Panacea: *Lomatium dissectum*
Pediatric Aid: *Artemisia tridentata*
Poison: *Datura wrightii*
Pulmonary Aid: *Angelica lineariloba, Psorothamnus polydenius*
Stimulant: *Artemisia tridentata*
Throat Aid: *Achillea millefolium, Lomatium dissectum, Psorothamnus polydenius*
Toothache Remedy: *Sarcobatus vermiculatus*
Tuberculosis Remedy: *Cercocarpus ledifolius*
Venereal Aid: *Salix exigua, Salvia dorrii*
Vertigo Medicine: *Lomatium dissectum*

Food

Beverage: *Descurainia incana, D. pinnata, D. sophia, Ephedra viridis, Juncus balticus*
Bread & Cake: *Allium anceps, Pinus monophylla, Prunus virginiana, Rumex crispus, Typha domingensis, T. latifolia*
Candy: *Artemisia tridentata, Chrysothamnus nauseosus, Phragmites australis, Pinus jeffreyi, P. monophylla*
Dried Food: *Allium parvum, Calochortus leichtlinii, C. nuttallii, Cyperus esculentus, Lycium andersonii, Mentzelia albicaulis, Pinus monophylla, Prunus virginiana, Ribes aureum, Sambucus racemosa, Shepherdia argentea, Typha domingensis, T. latifolia*
Fruit: *Juniperus occidentalis, Lycium andersonii, Prunus virginiana, Ribes aureum, Sambucus racemosa*
Ice Cream: *Pinus monophylla*
Porridge: *Atriplex argentea, Lycium andersonii, Mentzelia albicaulis, Oryzopsis hymenoides, Perideridia gairdneri, Prunus virginiana, Ribes aureum, Scirpus maritimus, S. pungens, Shepherdia argentea, Typha domingensis, T. latifolia, Wyethia mollis*
Soup: *Allium parvum, Calochortus leichtlinii, C. nuttallii, Oryzopsis hymenoides, Perideridia gairdneri, Pinus monophylla, Sambucus racemosa, Typha domingensis, T. latifolia*
Special Food: *Oryzopsis hymenoides*
Staple: *Chenopodium fremontii, Helianthus annuus, Oryzopsis hymenoides, Perideridia gairdneri, Quercus kelloggii, Typha domingensis, T. latifolia*
Starvation Food: *Rumex crispus*
Unspecified: *Allium anceps, A. nevadense, A.* sp., *Balsamorhiza hookeri, Chenopodium nevadense, Claytonia umbellata, Cyperus esculentus, Descurainia incana, D. pinnata, D. sophia, Eleocharis palustris, Helianthus cusickii, Lewisia rediviva, Lomatium macrocarpum, L. nevadense, Opuntia polyacantha, Orobanche fasciculata, Perideridia gairdneri, Sagittaria cuneata, Scirpus acutus, S. maritimus, Suaeda calceoliformis, Typha domingensis, T. latifolia, Wyethia mollis*
Vegetable: *Allium* sp., *Calochortus leichtlinii, C. nuttallii, Claytonia perfoliata, Glyptopleura marginata, Lewisia rediviva*
Winter Use Food: *Descurainia incana, D. pinnata, D. sophia, Oryzopsis hymenoides, Pinus jeffreyi, Prunus virginiana, Scirpus maritimus, Wyethia mollis*

Fiber

Basketry: *Salix exigua, Scirpus acutus, S. maritimus, Typha domingensis, T. latifolia*
Building Material: *Scirpus acutus, Typha domingensis, T. latifolia*
Canoe Material: *Typha domingensis, T. latifolia*
Clothing: *Scirpus acutus, S. maritimus, Typha domingensis, T. latifolia*
Cordage: *Apocynum cannabinum*
Mats, Rugs & Bedding: *Apocynum cannabinum, Scirpus acutus, S. maritimus, Typha domingensis, T. latifolia*

Other

Cash Crop: *Pinus jeffreyi*
Containers: *Scirpus acutus*
Fasteners: *Scirpus acutus, S. maritimus, Typha domingensis, T. latifolia*
Hunting & Fishing Item: *Apocynum cannabinum, Lomatium dissectum, Phragmites australis, Typha domingensis, T. latifolia*
Protection: *Veratrum californicum*
Smoke Plant: *Lomatium dissectum, Prunus virginiana*

Paiute, Southern

Food
Unspecified: *Leymus condensatus*

Panamint

Fiber
Basketry: *Rhus trilobata, Salix lucida*

Other
Decorations: *Juncus balticus, Proboscidea louisianica, Scirpus maritimus*
Designs: *Yucca brevifolia*
Tools: *Echinocactus polycephalus, Larrea tridentata*

Papago

Drug

Analgesic: *Larrea tridentata, Phoradendron* sp., *Taraxacum officinale*
Antirheumatic (External): *Larrea tridentata*
Dermatological Aid: *Larrea tridentata, Peniocereus greggii, Prosopis velutina, Rumex hymenosepalus, Simmondsia chinensis*
Emetic: *Anemopsis californica, Larrea tridentata*
Gastrointestinal Aid: *Phoradendron* sp.
Gynecological Aid: *Larrea tridentata, Phoradendron* sp., *Taraxacum officinale*
Orthopedic Aid: *Larrea tridentata*
Pediatric Aid: *Larrea tridentata*
Snakebite Remedy: *Larrea tridentata*
Throat Aid: *Rumex hymenosepalus*
Toothache Remedy: *Phaseolus acutifolius*

Food

Beverage: *Baccharis sarothroides, Carnegia gigantea, Datura wrightii, Descurainia pinnata, Ephedra nevadensis, Ferocactus wislizeni, Opuntia engelmannii, Peniocereus greggii, Salvia columbariae, Stenocereus thurberi, Ziziphus obtusifolia*
Bread & Cake: *Carnegia gigantea, Gossypium* sp., *Stenocereus thurberi, Triticum aestivum*
Candy: *Encelia farinosa, Prosopis velutina, Quercus emoryi, Sarcostemma cynanchoides*
Cooking Agent: *Carnegia gigantea, Stenocereus thurberi*
Dried Food: *Amaranthus palmeri, Ambrosia tenuifolia, Carnegia gigantea, Cicer arietinum, Cucurbita moschata, C. pepo, Descurainia pinnata, Lens culinaris, Lepidium thurberi, Lycium fremontii, Monolepis nuttalliana, Olneya tesota, Opuntia* sp., *Parkinsonia aculeata, P. microphylla, Phaseolus acutifolius, P. vulgaris, Pholisma sonorae, Phoradendron californicum, Proboscidea louisianica, Prosopis velutina, Sporobolus wrightii, Stenocereus thurberi, Vicia* sp., *Yucca baccata, Zea mays*
Fodder: *Carnegia gigantea, Stenocereus thurberi*
Fruit: *Carnegia gigantea, Celtis laevigata, Condalia globosa, Cucurbita maxima, C. moschata, C. pepo, Lycium fremontii, Opuntia* sp., *Yucca baccata, Y. schottii*
Porridge: *Yucca baccata*
Preserves: *Carnegia gigantea, Prosopis velutina, Stenocereus thurberi*

Sauce & Relish: *Carnegia gigantea, Opuntia engelmannii, O.* sp.*, Stenocereus thurberi, Ziziphus obtusifolia*
Soup: *Atriplex wrightii, Chenopodium album*
Special Food: *Capsicum annuum, Fouquieria splendens, Zea mays*
Spice: *Atriplex wrightii, Capsicum annuum, Suaeda suffrutescens*
Staple: *Agastache pallidiflora, Agave americana, A. parryi, Amaranthus palmeri, Ambrosia tenuifolia, Carnegia gigantea, Cucurbita pepo, Olneya tesota, Opuntia echinocarpa, O. engelmannii, O. fulgida, O. spinosior, O. versicolor, Parkinsonia microphylla, Phaseolus acutifolius, Pholisma sonorae, Prosopis glandulosa, P. velutina, Stenocereus thurberi, Triticum aestivum, Yucca baccata, Y.* sp.*, Zea mays*
Starvation Food: *Malva* sp.
Unspecified: *Agave americana, A. deserti, A. palmeri, A. schottii, Allium unifolium, Amaranthus palmeri, Ambrosia tenuifolia, Atriplex lentiformis, Capsicum annuum, Carnegia gigantea, Chenopodium murale, Cicer arietinum, Descurainia pinnata, Dichelostemma pulchellum, Hordeum vulgare, Lens culinaris, Monolepis nuttalliana, Olneya tesota, Opuntia* sp.*, Parkinsonia aculeata, P. microphylla, Peniocereus greggii, Phaseolus vulgaris, Pholisma sonorae, Pisum sativum, Proboscidea althaeifolia, P. louisianica, P. parviflora, Prosopis velutina, Quercus emoryi, Q. oblongifolia, Simmondsia chinensis, Stenocereus thurberi, Vigna unguiculata, Yucca baccata, Y. elata, Zea mays*
Vegetable: *Agave americana, Amaranthus palmeri, Ambrosia tenuifolia, Atriplex wrightii, Chenopodium murale, C.* sp.*, Dasylirion wheeleri, Ferocactus wislizeni, Opuntia echinocarpa, O. engelmannii, O. fulgida, O. imbricata, O.* sp.*, O. versicolor, Peniocereus greggii, Rumex hymenosepalus, Taraxacum officinale*

Fiber
Basketry: *Acacia greggii, Carnegia gigantea, Dasylirion wheeleri, Jatropha cardiophylla, Morus microphylla, Nolina bigelovii, N. erumpens, N. microcarpa, Proboscidea louisianica, Prosopis velutina, Salix nigra, Yucca baccata, Y. elata, Y. glauca*
Brushes & Brooms: *Agave americana, Larrea tridentata*
Building Material: *Agave americana, Carnegia gigantea, Fouquieria splendens, Larrea tridentata, Olneya tesota, Prosopis velutina, Yucca baccata, Y. elata, Y. glauca*
Clothing: *Celtis laevigata, Dasylirion wheeleri*
Cordage: *Agave lechuguilla, Yucca baccata, Y. glauca*
Mats, Rugs & Bedding: *Dasylirion wheeleri*
Other: *Acacia greggii, Prosopis velutina, Salix nigra*
Sewing Material: *Yucca elata*
Unspecified: *Gossypium thurberi*

Dye
Red: *Krameria erecta*
Yellow: *Arundo donax*

Other
Cash Crop: *Agave americana, A. lechuguilla, Allium unifolium, Yucca schottii*
Ceremonial Items: *Fouquieria splendens, Stenocereus thurberi*
Containers: *Agave americana*
Cooking Tools: *Carnegia gigantea, Lagenaria siceraria, Prosopis velutina, Suaeda moquinii, S. suffrutescens*
Decorations: *Larrea tridentata*
Fasteners: *Coursetia glandulosa, Encelia farinosa, Prosopis velutina, Yucca glauca*
Fuel: *Prosopis velutina*
Hunting & Fishing Item: *Acacia greggii, Baccharis sarothroides, Carnegia gigantea, Larrea tridentata, Lycium fremontii, Morus microphylla, Sapindus saponaria*
Incense & Fragrance: *Acacia greggii*
Musical Instrument: *Carnegia gigantea, Lagenaria siceraria, Larrea tridentata, Olneya tesota*
Protection: *Larrea tridentata, Opuntia* sp.*, Prosopis velutina*
Season Indicator: *Carnegia gigantea*
Smoke Plant: *Nicotiana attenuata, N. tabacum, N. trigonophylla*
Smoking Tools: *Phragmites australis*
Soap: *Cucurbita foetidissima, Yucca baccata, Y. glauca*
Tools: *Acacia greggii, Carnegia gigantea, Fouquieria splendens, Larrea tridentata, Olneya tesota, Pluchea sericea, Prosopis velutina*
Waterproofing Agent: *Encelia farinosa*
Weapon: *Baccharis sarothroides, Larrea tridentata, Sapindus saponaria*

Papago & Pima
Food
Beverage: *Stenocereus thurberi*
Candy: *Carnegia gigantea, Stenocereus thurberi*
Cooking Agent: *Pachycereus pringlei*
Fruit: *Carnegia gigantea, Machaerocereus eruca, M. gummosus, Myrtillocactus cochal, Peniocereus striatus, Stenocereus thurberi*
Porridge: *Stenocereus thurberi*
Preserves: *Carnegia gigantea, Stenocereus thurberi*
Sauce & Relish: *Carnegia gigantea, Stenocereus thurberi*

Fiber
Building Material: *Pachycereus pringlei*
Furniture: *Pachycereus pringlei*

Other
Hunting & Fishing Item: *Pachycereus pringlei*
Lighting: *Stenocereus thurberi*
Paint: *Stenocereus thurberi*
Tools: *Carnegia gigantea*

Pawnee
Drug
Abortifacient: *Artemisia frigida*
Analgesic: *Aquilegia canadensis, Arisaema triphyllum, Echinacea angustifolia, Gymnocladus dioicus, Ipomoea leptophylla*
Antidiarrheal: *Liatris scariosa, Rhus glabra, Rumex hymenosepalus*
Antidote: *Echinacea angustifolia*
Antirheumatic (External): *Arisaema triphyllum, Artemisia dracunculus*
Burn Dressing: *Echinacea angustifolia, Rosa arkansana, Typha latifolia*
Carminative: *Acorus calamus, Mentha canadensis*
Ceremonial Medicine: *Acorus calamus, Artemisia* sp.*, Hierochloe odorata, Prunus virginiana*
Cold Remedy: *Acorus calamus, Juniperus virginiana*
Cough Medicine: *Acorus calamus, Juniperus virginiana*
Dermatological Aid: *Desmanthus illinoensis, Heracleum maximum, Opuntia humifusa, Typha latifolia*
Disinfectant: *Artemisia* sp.
Ear Medicine: *Glycyrrhiza lepidota*
Febrifuge: *Acorus calamus, Aquilegia canadensis, Glycyrrhiza lepidota, Penstemon grandiflorus*
Gastrointestinal Aid: *Acorus calamus, Artemisia* sp.*, Baptisia bracteata, Quercus* sp.
Gynecological Aid: *Helianthus annuus, Mirabilis nyctaginea, Rhus glabra*
Herbal Steam: *Yucca glauca*
Laxative: *Ulmus rubra*
Love Medicine: *Aquilegia canadensis, Cuscuta compacta, Lobelia cardinalis, Lomatium foeniculaceum, Panax quinquefolius*
Misc. Disease Remedy: *Echinacea angustifolia*
Oral Aid: *Mirabilis nyctaginea*
Other: *Aster* sp.*, Echinacea angustifolia*
Panacea: *Acorus calamus, Cucurbita foetidissima, Dalea candida, D. purpurea*

Pediatric Aid: *Croton texensis, Glycyrrhiza lepidota, Liatris scariosa, Mirabilis nyctaginea, Quercus* sp., *Typha latifolia*
Poison: *Euphorbia marginata*
Sedative: *Ipomoea leptophylla, Juniperus virginiana*
Snakebite Remedy: *Echinacea angustifolia*
Stimulant: *Gymnocladus dioicus, Ipomoea leptophylla, Osmorhiza longistylis*
Tonic: *Silphium laciniatum*
Toothache Remedy: *Acorus calamus, Echinacea angustifolia, Glycyrrhiza lepidota*
Unspecified: *Artemisia dracunculus*
Veterinary Aid: *Echinacea angustifolia, Grindelia squarrosa, Juniperus virginiana, Oxalis violacea, Thalictrum dasycarpum, Zanthoxylum americanum*
Witchcraft Medicine: *Artemisia* sp., *Silphium laciniatum*

Food
Beverage: *Agastache foeniculum, Ceanothus americanus, Mentha canadensis, Rubus idaeus, R. occidentalis, Sambucus canadensis, Vitis cinerea, V. vulpina*
Bread & Cake: *Prunus virginiana*
Candy: *Populus deltoides, Silphium laciniatum*
Dried Food: *Opuntia humifusa, Pediomelum esculentum, Physalis heterophylla, Prunus americana, P. pumila, P. virginiana, Rubus idaeus, R. occidentalis, Shepherdia argentea, Vitis cinerea, V. vulpina*
Fodder: *Oxalis stricta, O. violacea*
Forage: *Oxalis stricta*
Fruit: *Celtis occidentalis, Fragaria vesca, F. virginiana, Opuntia humifusa, Prunus americana, P. virginiana, Rubus idaeus, R. occidentalis, Sambucus canadensis, Shepherdia argentea, Viburnum lentago, Vitis cinerea, V. vulpina*
Sauce & Relish: *Allium canadense, Physalis heterophylla, Prunus americana, P. pumila*
Soup: *Carya ovata, Chenopodium album, Juglans nigra, Nelumbo lutea*
Spice: *Allium canadense*
Staple: *Zea mays*
Starvation Food: *Opuntia humifusa, Rosa arkansana*
Sweetener: *Acer negundo, Agastache foeniculum, Carya ovata, Zea mays*
Unspecified: *Allium canadense, Amphicarpaea bracteata, Apios americana, Carya ovata, Gymnocladus dioicus, Helianthus tuberosus, Juglans nigra, Linum lewisii, Nelumbo lutea, Nuphar lutea, Oxalis stricta, O. violacea, Pediomelum esculentum, Quercus macrocarpa, Q. rubra, Sagittaria latifolia*
Vegetable: *Asclepias syriaca*
Winter Use Food: *Zea mays*

Fiber
Brushes & Brooms: *Artemisia dracunculus, Dalea purpurea, Prunus americana, Stipa spartea*
Building Material: *Salix* sp., *Ulmus americana, U. rubra*
Clothing: *Urtica dioica*
Cordage: *Ulmus rubra, Urtica dioica, Yucca glauca*
Mats, Rugs & Bedding: *Artemisia dracunculus, A. frigida, A. ludoviciana, Scirpus tabernaemontani, Tilia americana*
Scouring Material: *Equisetum* sp.
Sewing Material: *Yucca glauca*

Dye
Black: *Juglans nigra*
Mordant: *Opuntia humifusa*
Orange: *Cuscuta compacta*
Yellow: *Populus deltoides*

Other
Ceremonial Items: *Nelumbo lutea, Stipa spartea, Urtica dioica*
Cooking Tools: *Amorpha fruticosa, Ulmus americana, U. rubra*
Fuel: *Ceanothus americanus, Ulmus americana, U. rubra*
Hunting & Fishing Item: *Cornus asperifolia, Fraxinus pennsylvanica, Maclura pomifera*
Incense & Fragrance: *Aquilegia canadensis*
Musical Instrument: *Populus deltoides*
Paint: *Chenopodium album, Phytolacca americana*
Protection: *Juniperus virginiana*
Season Indicator: *Prunus virginiana*
Smoke Plant: *Arctostaphylos uva-ursi, Cornus amomum, Nicotiana quadrivalvis, Rhus glabra, Rosa arkansana*
Smoking Tools: *Fraxinus pennsylvanica*
Soap: *Yucca glauca*
Stable Gear: *Ulmus americana, Urtica dioica*
Tools: *Ulmus americana, U. rubra*
Toys & Games: *Amelanchier alnifolia, Arisaema triphyllum, Artemisia* sp., *Asclepias syriaca, Betula papyrifera, Desmanthus illinoensis, Populus deltoides, Quercus macrocarpa, Sambucus canadensis, Thalictrum dasycarpum, Ulmus americana, U. rubra, Urtica dioica, Viburnum opulus*

Pend d'Oreille, Upper
Other
Cash Crop: *Camassia quamash*

Penobscot
Drug
Analgesic: *Alnus* sp., *Kalmia angustifolia, Thuja occidentalis*
Anthelmintic: *Apocynum cannabinum*
Anticonvulsive: *Aristolochia serpentaria*
Antidiarrheal: *Prunus virginiana*
Antiemetic: *Alnus* sp.
Antihemorrhagic: *Acer pensylvanicum, Aralia racemosa, Baptisia tinctoria, Eupatorium perfoliatum, Galium aparine, Pyrola asarifolia, Sarracenia purpurea, Streptopus amplexifolius, Ulmus americana*
Burn Dressing: *Abies balsamea, Impatiens capensis*
Cold Remedy: *Populus tremuloides, Salix* sp., *Taxus canadensis*
Cough Medicine: *Aralia nudicaulis, Prunus serotina*
Dermatological Aid: *Abies balsamea, Acer pensylvanicum, Arctium minus, Chimaphila umbellata, Comptonia peregrina, Impatiens capensis, Nuphar lutea, Nymphaea odorata, Picea* sp., *Thuja occidentalis*
Diaphoretic: *Populus tremuloides*
Dietary Aid: *Quercus alba*
Emetic: *Sorbus americana*
Gastrointestinal Aid: *Alnus* sp.
Gynecological Aid: *Fraxinus americana*
Hemorrhoid Remedy: *Quercus alba*
Herbal Steam: *Iris versicolor*
Kidney Aid: *Acer pensylvanicum, Aralia racemosa, Baptisia tinctoria, Cimicifuga racemosa, Eupatorium perfoliatum, Galium aparine, Pyrola asarifolia, Sarracenia purpurea, Streptopus amplexifolius*
Misc. Disease Remedy: *Iris versicolor, Viburnum opulus*
Oral Aid: *Berberis vulgaris, Coptis trifolia*
Orthopedic Aid: *Arisaema triphyllum, Impatiens capensis*
Panacea: *Kalmia angustifolia, Thuja occidentalis*
Poison: *Arisaema triphyllum*
Preventive Medicine: *Iris versicolor, Sanguinaria canadensis*
Pulmonary Aid: *Ulmus americana*
Reproductive Aid: *Panax quinquefolius*
Respiratory Aid: *Salix lucida, Verbascum thapsus*
Sedative: *Cypripedium acaule*
Throat Aid: *Berberis vulgaris*
Tonic: *Acer pensylvanicum, Aralia racemosa, Baptisia tinctoria, Eupatorium perfoliatum, Galium aparine, Prunus serotina, Pyrola asarifolia, Streptopus amplexifolius*
Unspecified: *Lycopodium dendroideum, Mitchella repens*

Venereal Aid: *Acer pensylvanicum, Aralia racemosa, Baptisia tinctoria, Eupatorium perfoliatum, Galium aparine, Pyrola asarifolia, Streptopus amplexifolius*
Food
Beverage: *Taxus canadensis*
Candy: *Picea* sp.

Pima
Drug
Analgesic: *Ambrosia ambrosioides, Datura discolor, Encelia farinosa, Isocoma pluriflora, Krameria grayi, Larrea tridentata, Proboscidea althaeifolia, P. parviflora, Prosopis velutina, Ricinus communis, Ziziphus obtusifolia*
Anthelmintic: *Helianthus annuus*
Antidiarrheal: *Larrea tridentata, Plantago ovata, Pluchea sericea, Prosopis velutina, Sphaeralcea angustifolia, S. emoryi, Xanthium strumarium*
Antihemorrhagic: *Ambrosia ambrosioides*
Antirheumatic (External): *Atriplex polycarpa, Larrea tridentata, Proboscidea althaeifolia, P. parviflora, Ziziphus obtusifolia*
Antirheumatic (Internal): *Larrea tridentata*
Burn Dressing: *Prosopis velutina*
Carminative: *Larrea tridentata*
Cathartic: *Asclepias subulata, Phoradendron californicum, Prosopis velutina, Ricinus communis, Sonchus oleraceus, Yucca baccata*
Cold Remedy: *Anemopsis californica, Cyperus esculentus, Larrea tridentata, Rumex hymenosepalus, Sambucus cerulea, Solanum elaeagnifolium*
Cough Medicine: *Ambrosia ambrosioides, Anemopsis californica, Cyperus esculentus, Isocoma pluriflora, Krameria grayi, Rumex hymenosepalus*
Dermatological Aid: *Anemopsis californica, Atriplex lentiformis, Chamaesyce polycarpa, Cucurbita pepo, Datura discolor, Descurainia pinnata, Ephedra fasciculata, E. trifurca, Heliotropium curassavicum, Krameria erecta, K. grayi, Larrea tridentata, Malva parviflora, Orobanche ludoviciana, Phoradendron californicum, Pluchea sericea, Populus fremontii, Prosopis pubescens, P. velutina, Ricinus communis, Rumex hymenosepalus, Ziziphus obtusifolia*
Diaphoretic: *Anemopsis californica, Chamaesyce polycarpa*
Disinfectant: *Krameria grayi, Prosopis velutina*
Ear Medicine: *Datura discolor, Mammillaria grahamii*
Emetic: *Anemopsis californica, Asclepias subulata, Chamaesyce polycarpa, Larrea tridentata, Prosopis velutina*
Eye Medicine: *Asclepias subulata, Datura discolor, Krameria grayi, Pluchea sericea, Prosopis velutina, Sisymbrium irio, Xanthium strumarium, Ziziphus obtusifolia*
Febrifuge: *Helianthus annuus, Krameria grayi, Larrea tridentata, Salix gooddingii, Sambucus cerulea*
Gastrointestinal Aid: *Anemopsis californica, Asclepias subulata, Chamaesyce polycarpa, Datura discolor, Larrea tridentata, Opuntia acanthocarpa, Phoradendron californicum, Pluchea sericea, Prosopis velutina, Sambucus cerulea, Sphaeralcea angustifolia*
Gynecological Aid: *Ambrosia ambrosioides, Carnegia gigantea, Datura discolor, Opuntia engelmannii, O. phaeacantha, Prosopis pubescens*
Hemorrhoid Remedy: *Datura discolor*
Hemostat: *Acourtia wrightii*
Laxative: *Chamaesyce polycarpa, Pectis papposa, Ricinus communis, Xanthium strumarium*
Misc. Disease Remedy: *Peniocereus greggii*
Oral Aid: *Larrea tridentata, Prosopis velutina, Rumex hymenosepalus*
Orthopedic Aid: *Carnegia gigantea, Isocoma pluriflora*
Other: *Anemopsis californica, Datura discolor, Ephedra trifurca, Prosopis velutina, Sonchus oleraceus*
Panacea: *Asclepias subulata, Larrea tridentata*
Pediatric Aid: *Krameria grayi, Pluchea sericea, Prosopis velutina*
Poison: *Asclepias subulata, Chamaesyce polycarpa, Cuscuta* sp., *Ricinus communis*
Sedative: *Pluchea sericea*
Snakebite Remedy: *Chamaesyce polycarpa, Cyperus esculentus*
Strengthener: *Larrea tridentata*
Throat Aid: *Anemopsis californica, Krameria grayi, Rumex hymenosepalus, Sambucus cerulea*
Toothache Remedy: *Larrea tridentata*
Tuberculosis Remedy: *Anemopsis californica, Larrea tridentata*
Unspecified: *Larrea tridentata, Rumex hymenosepalus*
Urinary Aid: *Larrea tridentata*
Venereal Aid: *Anemopsis californica, Ephedra antisyphilitica, E. fasciculata, E. trifurca, Thamnosma montana*
Veterinary Aid: *Cyperus esculentus, Helianthus annuus, Pluchea sericea, Xanthium strumarium*
Food
Beverage: *Carnegia gigantea, Datura discolor, Descurainia* sp., *Ephedra fasciculata, Ferocactus cylindraceus, F. wislizeni, Lycium fremontii, Prosopis glandulosa, P. pubescens, P. velutina, Salvia columbariae, Yucca baccata*
Bread & Cake: *Atriplex nuttallii, A. polycarpa, Carnegia gigantea, Prosopis velutina, Rumex hymenosepalus, Triticum aestivum, Typha angustifolia, Yucca baccata, Zea mays*
Candy: *Agave deserti, Encelia farinosa, Ferocactus cylindraceus, F. wislizeni, Helianthus annuus, Populus deltoides, P. fremontii, Prosopis glandulosa, P. pubescens, P. velutina, Rumex hymenosepalus, Sarcostemma cynanchoides*
Cooking Agent: *Atriplex coronata, A. elegans, A. serenana*
Dessert: *Carnegia gigantea*
Dried Food: *Agave americana, Atriplex lentiformis, A.* sp., *Carnegia gigantea, Cucurbita moschata, Descurainia pinnata, Gossypium* sp., *Olneya tesota, Opuntia imbricata, Yucca baccata*
Fodder: *Carnegia gigantea, Plantago ovata, Ziziphus obtusifolia*
Forage: *Atriplex polycarpa, A.* sp., *Malva parviflora, Pluchea sericea, Prosopis pubescens, Sarcobatus vermiculatus*
Fruit: *Carnegia gigantea, Citrullus lanatus, Condalia hookeri, Cucurbita maxima, C. moschata, C. pepo, Echinocereus engelmannii, Lycium fremontii, Opuntia arbuscula, O. engelmannii, O. imbricata, O. leptocaulis, O. versicolor, Phoradendron californicum, Solanum elaeagnifolium, Yucca schottii, Ziziphus obtusifolia*
Pie & Pudding: *Rumex hymenosepalus*
Porridge: *Atriplex lentiformis, Carnegia gigantea, Prosopis velutina, Sisymbrium irio, Triticum aestivum, Typha angustifolia, Yucca baccata, Zea mays*
Preserves: *Sambucus cerulea*
Sauce & Relish: *Agave americana, Carnegia gigantea, Condalia hookeri, Ziziphus obtusifolia*
Spice: *Atriplex coronata, A. elegans, A. nuttallii, A. serenana, Capsicum annuum, Suaeda arborescens, S. suffrutescens*
Staple: *Atriplex* sp., *Carnegia gigantea, Chenopodium murale, Descurainia pinnata, Helianthus annuus, Monolepis nuttalliana, Olneya tesota, Prosopis pubescens, P. velutina, Quercus oblongifolia*
Starvation Food: *Agave americana, Atriplex lentiformis, A. polycarpa, Sarcobatus vermiculatus*
Substitution Food: *Carnegia gigantea, Ferocactus wislizeni, Prosopis velutina, Solanum elaeagnifolium*
Sweetener: *Prosopis velutina*
Unspecified: *Acacia greggii, Agave deserti, Allenrolfea occidentalis, Amaranthus palmeri, Amsinckia spectabilis, A. tessellata, Atriplex coronata, A. elegans, A. nuttallii, A. serenana, Capsicum annuum, Carnegia gigantea, Cicer arietinum, Cucurbita foetidissima, C. pepo, Cyperus odoratus, Ferocactus cylindraceus, F. wislizeni, Helianthus annuus, Hoffmannseggia glauca, Hordeum vulgare, Lens culinaris, Lithospermum* sp., *Malva nicaeensis, Monolepis nuttalliana,*

[Unspecified]
   Opuntia engelmannii, O. phaeacantha, O. sp., *Orobanche ludoviciana, Parkinsonia florida, P. microphylla, Pisum sativum, Populus deltoides, Proboscidea parviflora, Prosopis velutina, Rumex hymenosepalus, Salix gooddingii, Sonchus asper, S. oleraceus, Triticum durum, T. polonicum, Typha angustifolia, Vigna unguiculata*
Vegetable: *Amaranthus palmeri, Atriplex wrightii, Chenopodium* sp., *Hoffmannseggia glauca, Monolepis nuttalliana, Rumex crispus, R. hymenosepalus, Sonchus asper, Trianthema portulacastrum*
Winter Use Food: *Sisymbrium irio*

Fiber
Basketry: *Nolina microcarpa, Populus fremontii, Proboscidea althaeifolia, P. parviflora, Salix gooddingii, Typha angustifolia, Yucca baccata*
Brushes & Brooms: *Agave lechuguilla, Baccharis sarothroides, Yucca baccata*
Building Material: *Acacia greggii, Fouquieria splendens, Populus fremontii, Prosopis pubescens, P. velutina, Typha angustifolia*
Cordage: *Agave deserti, Yucca baccata*
Furniture: *Acacia greggii, Prosopis velutina*
Mats, Rugs & Bedding: *Dasylirion wheeleri, Phragmites australis, Salix gooddingii, Typha angustifolia, Yucca baccata, Y. elata*
Sewing Material: *Salix gooddingii*

Dye
Black: *Prosopis velutina*
Brown: *Krameria grayi, Rumex hymenosepalus*
Red-Brown: *Rumex hymenosepalus*
Unspecified: *Maclura pomifera*
Yellow: *Rumex crispus, R. hymenosepalus*

Other
Cash Crop: *Agastache pallidiflora, Agave americana, A. parryi, Opuntia acanthocarpa*
Ceremonial Items: *Lagenaria siceraria*
Containers: *Populus deltoides, Salix gooddingii, Suaeda arborescens, Yucca elata*
Cooking Tools: *Parkinsonia florida, Prosopis velutina, Suaeda arborescens, S. suffrutescens, Xanthium strumarium*
Decorations: *Fouquieria splendens, Typha angustifolia*
Designs: *Proboscidea parviflora*
Fuel: *Acacia greggii, Atriplex* sp., *Encelia farinosa, Isocoma pluriflora, Olneya tesota, Populus fremontii, Prosopis pubescens, P. velutina, Tamarix aphylla*
Hunting & Fishing Item: *Acacia greggii, Ferocactus wislizeni, Maclura pomifera, Salix gooddingii*
Lighting: *Helianthus annuus*
Musical Instrument: *Phragmites australis*
Paint: *Encelia farinosa, Prosopis velutina*
Season Indicator: *Carnegia gigantea*
Smoke Plant: *Nicotiana attenuata, N. rustica, N. tabacum, N. trigonophylla*
Soap: *Atriplex lentiformis, Yucca baccata, Y. elata*
Tools: *Olneya tesota*
Toys & Games: *Melilotus indicus, Prosopis velutina*

Pima, Desert
Other
Cash Crop: *Amoreuxia palmatifida, Capsicum annuum, Quercus* sp., *Yucca baccata*

Pima, Gila River
Food
Baby Food: *Dichelostemma pulchellum, Mammillaria grahamii, Physalis acutifolia, Sarcostemma cynanchoides, Washingtonia filifera*
Beverage: *Carnegia gigantea, Descurainia pinnata, Plantago* sp., *Prosopis velutina, Salvia columbariae, Sambucus cerulea, Sisymbrium irio*
Bread & Cake: *Prosopis velutina*
Candy: *Agave deserti, Carnegia gigantea, Yucca baccata*
Dried Food: *Agave deserti, Amaranthus palmeri, Carnegia gigantea, Opuntia acanthocarpa, O. arbuscula*
Fruit: *Capsicum annuum, Lycium* sp., *Opuntia engelmannii, O. leptocaulis, Phoradendron californicum, Sambucus cerulea, Yucca baccata, Ziziphus obtusifolia*
Porridge: *Carnegia gigantea, Descurainia pinnata, Plantago* sp., *Salvia columbariae, Sisymbrium irio*
Preserves: *Carnegia gigantea*
Sauce & Relish: *Carnegia gigantea*
Snack Food: *Cucurbita digitata, Cyperus* sp., *Dichelostemma pulchellum, Echinocereus engelmannii, Mammillaria grahamii, Phoradendron californicum, Physalis acutifolia, Populus fremontii, Proboscidea* sp., *Prosopis pubescens, P. velutina, Quercus turbinella, Sambucus cerulea, Scirpus* sp., *Washingtonia filifera*
Special Food: *Ferocactus* sp.
Staple: *Agave deserti, Capsicum annuum, Carnegia gigantea, Descurainia pinnata, Opuntia acanthocarpa, Prosopis pubescens, P. velutina, Sisymbrium irio*
Starvation Food: *Acacia greggii, Allenrolfea occidentalis, Atriplex lentiformis*
Unspecified: *Allenrolfea occidentalis, Amaranthus palmeri, Amoreuxia palmatifida, Atriplex elegans, A. lentiformis, A. wrightii, Chenopodium album, C. pratericola, Cyperus* sp., *Descurainia pinnata, Eremalche exilis, Ferocactus* sp., *Gossypium hirsutum, Hoffmannseggia glauca, Olneya tesota, Opuntia acanthocarpa, O. arbuscula, O. ×kelvinensis, O.* sp., *Orobanche cooperi, Parkinsonia microphylla, Phacelia* sp., *Phalaris caroliniana, P. minor, Plantago ovata, P. patagonica, Portulaca oleracea, Proboscidea* sp., *Rumex crispus, R. violascens, Salvia columbariae, Scirpus maritimus, Sisymbrium irio, Sonchus asper, S. oleraceus, Suaeda moquinii, Typha domingensis*
Vegetable: *Amaranthus palmeri, Amsinckia* sp., *Eremalche exilis, Monolepis nuttalliana, Phacelia* sp., *Portulaca* sp., *Rumex* sp.
Winter Use Food: *Prosopis velutina*

Other
Cash Crop: *Prosopis velutina*
Season Indicator: *Prosopis pubescens, P. velutina*

Pima, Lehi
Food
Beverage: *Ephedra fasciculata*

Pit River
Other
Cash Crop: *Taxus brevifolia*
Protection: *Amelanchier* sp.

Plains Indian
Drug
Dermatological Aid: *Typha latifolia*
Pediatric Aid: *Typha latifolia*
Veterinary Aid: *Hierochloe odorata*

Dye
Black: *Rhus glabra*
Gray: *Rhus glabra*
Yellow: *Rhus glabra*

Other
Smoke Plant: *Rhus glabra*

Poliklah
Drug
Cold Remedy: *Garrya* sp.
Eye Medicine: *Artemisia ludoviciana*
Pediatric Aid: *Ceanothus thyrsiflorus*
Poison: *Lonicera involucrata*

Food
Bread & Cake: *Lithocarpus densiflorus*
Fruit: *Sambucus* sp., *Vaccinium ovatum*, *V. parvifolium*
Porridge: *Lithocarpus densiflorus*
Staple: *Lithocarpus densiflorus*
Fiber
Basketry: *Adiantum* sp., *Corylus* sp., *Picea sitchensis*, *Woodwardia* sp., *Xerophyllum* sp.
Brushes & Brooms: *Vaccinium parvifolium*
Other
Containers: *Lysichiton* sp.
Hunting & Fishing Item: *Holodiscus discolor*, *Philadelphus lewisii*
Jewelry: *Xerophyllum* sp.
Sacred Items: *Lomatium californicum*

Pomo
Drug
Abortifacient: *Phoradendron villosum*
Analgesic: *Umbellularia californica*, *Urtica dioica*
Antidiarrheal: *Arctostaphylos columbiana*, *Croton setigerus*
Antiemetic: *Perideridia kelloggii*, *Trifolium* sp.
Antihemorrhagic: *Ligusticum apiifolium*
Antirheumatic (External): *Heracleum maximum*, *Oxalis oregana*, *Umbellularia californica*, *Urtica dioica*
Blood Medicine: *Satureja douglasii*
Cold Remedy: *Calycanthus occidentalis*
Dermatological Aid: *Alnus rhombifolia*, *Aralia californica*, *Arbutus menziesii*, *Artemisia vulgaris*, *Asarum caudatum*, *Chlorogalum pomeridianum*, *Gnaphalium* sp., *G. stramineum*, *Heracleum maximum*, *Marah fabaceus*, *Sambucus racemosa*, *Scrophularia californica*
Dietary Aid: *Satureja douglasii*
Ear Medicine: *Sequoia sempervirens*
Expectorant: *Eriodictyon californicum*
Eye Medicine: *Thermopsis macrophylla*
Gastrointestinal Aid: *Satureja douglasii*
Gynecological Aid: *Amelanchier alnifolia*, *Artemisia vulgaris*, *Convolvulus arvensis*, *Iris macrosiphon*, *I. tenuissima*
Laxative: *Frangula californica*
Panacea: *Aralia californica*
Pediatric Aid: *Alnus rhombifolia*, *Artemisia vulgaris*, *Gnaphalium* sp.
Poison: *Aesculus californica*, *Clintonia andrewsiana*, *Croton setigerus*, *Frangula californica*
Pulmonary Aid: *Ligusticum apiifolium*
Stimulant: *Sequoia sempervirens*
Tonic: *Sequoia sempervirens*
Tuberculosis Remedy: *Torreya californica*
Unspecified: *Arctostaphylos glandulosa*, *Pinus lambertiana*
Food
Bread & Cake: *Lithocarpus densiflorus*, *Porphyra perforata*, *Quercus chrysolepis*, *Q. garryana*, *Q. kelloggii*, *Q. lobata*
Dried Food: *Arctostaphylos tomentosa*, *Castanopsis chrysophylla*, *Porphyra perforata*, *Vaccinium ovatum*
Fruit: *Arbutus menziesii*, *Carpobrotus aequilateralus*, *Fragaria vesca*, *Gaultheria shallon*, *Heteromeles arbutifolia*, *Rubus leucodermis*, *R. parviflorus*, *R. spectabilis*, *Vaccinium ovatum*, *Vitis californica*
Pie & Pudding: *Vaccinium ovatum*
Porridge: *Arctostaphylos tomentosa*, *Lithocarpus densiflorus*, *Madia sativa*, *Melica bulbosa*, *Quercus chrysolepis*, *Q. garryana*, *Q. kelloggii*, *Q. lobata*
Soup: *Lithocarpus densiflorus*, *Quercus kelloggii*, *Q. lobata*
Spice: *Allium unifolium*
Staple: *Avena fatua*, *Boisduvalia densiflora*, *Hemizonia clevelandii*, *H. luzulifolia*, *Lolium temulentum*, *Madia capitata*, *M. elegans*, *M. gracilis*, *M. sativa*, *Perideridia gairdneri*, *P. kelloggii*, *Ranunculus occidentalis*, *Salvia columbariae*, *Wyethia angustifolia*, *W. longicaulis*, *W.* sp.
Unspecified: *Allium unifolium*, *Avena fatua*, *A. sativa*, *Brodiaea coronaria*, *Calochortus pulchellus*, *C.* sp., *C. vestae*, *Castanopsis chrysophylla*, *Dichelostemma pulchellum*, *D. volubile*, *Lithocarpus densiflorus*, *Macrocystis luetkeana*, *Melica bulbosa*, *Pinus lambertiana*, *P. sabiniana*, *Postelsia palmaeformis*, *Quercus agrifolia*, *Q. garryana*, *Q. kelloggii*, *Sagittaria latifolia*, *Sanicula tuberosa*, *Scirpus acutus*, *Triteleia hyacinthina*, *T. laxa*, *T. peduncularis*, *Umbellularia* sp.
Vegetable: *Perideridia gairdneri*, *Scirpus acutus*, *S. robustus*, *Triteleia grandiflora*, *Typha latifolia*
Winter Use Food: *Avena fatua*, *Madia sativa*, *Porphyra perforata*, *Rubus leucodermis*
Fiber
Basketry: *Carex barbarae*, *C. mendocinensis*, *C.* sp., *Cercis canadensis*, *Cladium* sp., *Corylus cornuta*, *Juncus balticus*, *Juniperus occidentalis*, *Pinus muricata*, *P. sabiniana*, *Pteridium aquilinum*, *Rhus trilobata*, *Salix exigua*, *S. sessilifolia*, *S.* sp., *Scirpus acutus*, *S. maritimus*, *S. robustus*, *S.* sp., *Taxus brevifolia*, *Torreya californica*, *Vitis californica*
Building Material: *Artemisia ludoviciana*, *Scirpus robustus*
Canoe Material: *Scirpus acutus*
Clothing: *Scirpus acutus*, *S. robustus*, *Typha latifolia*
Cordage: *Carex* sp., *Lupinus arboreus*, *Vitis californica*
Mats, Rugs & Bedding: *Scirpus robustus*
Sewing Material: *Scirpus robustus*, *Taxus brevifolia*, *Vitis californica*
Sporting Equipment: *Juncus effusus*
Dye
Black: *Toxicodendron diversilobum*
Other
Cooking Tools: *Chlorogalum* sp., *Corylus cornuta*, *Polystichum munitum*, *Rubus parviflorus*, *Taxus brevifolia*, *Woodwardia radicans*
Decorations: *Rhododendron macrophyllum*, *R. occidentale*
Fasteners: *Pinus sabiniana*, *Vitis californica*
Fuel: *Pinus muricata*, *Pseudotsuga menziesii*, *Quercus kelloggii*, *Salix sitchensis*
Hunting & Fishing Item: *Carex* sp., *Corylus cornuta*, *Holodiscus discolor*, *Iris macrosiphon*, *I. tenuissima*, *Marah fabaceus*, *Pinus sabiniana*, *Quercus garryana*, *Taxus brevifolia*
Lighting: *Carex* sp.
Smoking Tools: *Lonicera hispidula*
Soap: *Arctostaphylos manzanita*
Tools: *Adiantum pedatum*, *Ceanothus oliganthus*, *Juncus balticus*, *Torreya californica*
Toys & Games: *Acer macrophyllum*

Pomo, Calpella
Drug
Blood Medicine: *Linanthus ciliatus*
Cold Remedy: *Arctostaphylos manzanita*, *Linanthus ciliatus*
Cough Medicine: *Linanthus ciliatus*
Pediatric Aid: *Linanthus ciliatus*

Pomo, Kashaya
Drug
Abortifacient: *Garrya elliptica*, *Phoradendron villosum*
Analgesic: *Artemisia douglasiana*, *Umbellularia californica*, *Urtica dioica*
Antidiarrheal: *Arctostaphylos columbiana*, *A. glandulosa*, *Croton setigerus*, *Rubus leucodermis*, *R. ursinus*
Antiemetic: *Perideridia kelloggii*

Antirheumatic (External): *Heracleum maximum, Oxalis oregana, Umbellularia californica, Urtica dioica*
Blood Medicine: *Eriodictyon californicum, Horkelia californica, Ligusticum apiifolium, Satureja douglasii*
Cold Remedy: *Angelica tomentosa, Calycanthus occidentalis, Salix lucida, Satureja douglasii, Umbellularia californica*
Cough Medicine: *Eriodictyon californicum, Lithocarpus densiflorus*
Dermatological Aid: *Achillea millefolium, Alnus rhombifolia, A. rubra, Angelica tomentosa, Aralia californica, Arbutus menziesii, Artemisia douglasiana, Asarum caudatum, Chlorogalum pomeridianum, Equisetum arvense, Eriodictyon californicum, Marah fabaceus, Phacelia californica, Pteridium aquilinum, Scrophularia californica, Stachys bullata, Umbellularia californica, Xerophyllum tenax*
Ear Medicine: *Sequoia sempervirens*
Expectorant: *Calycanthus occidentalis, Nereocystis luetkeana*
Eye Medicine: *Diplacus aurantiacus, Foeniculum vulgare, Thermopsis macrophylla*
Febrifuge: *Eriodictyon californicum, Sambucus cerulea*
Gastrointestinal Aid: *Angelica tomentosa, Artemisia douglasiana, Calycanthus occidentalis, Foeniculum vulgare, Rubus leucodermis, Satureja douglasii, Sisyrinchium bellum*
Gynecological Aid: *Amelanchier pallida, Angelica tomentosa, Artemisia douglasiana, Convolvulus arvensis, Equisetum telmateia, Eschscholzia californica, Rubus ursinus, Thermopsis macrophylla, Umbellularia californica*
Kidney Aid: *Equisetum laevigatum*
Laxative: *Frangula californica, Melilotus indicus*
Love Medicine: *Arbutus menziesii*
Misc. Disease Remedy: *Vaccinium ovatum*
Oral Aid: *Angelica tomentosa*
Other: *Angelica tomentosa, Rubus leucodermis*
Poison: *Amanita muscaria, Clintonia andrewsiana, Lycoperdon* sp., *Trillium* sp., *Zigadenus venenosus*
Pulmonary Aid: *Ligusticum apiifolium*
Respiratory Aid: *Sisyrinchium bellum, Umbellularia californica*
Sedative: *Dodecatheon hendersonii, Satureja douglasii*
Stimulant: *Sequoia sempervirens*
Throat Aid: *Angelica tomentosa, Arbutus menziesii, Calycanthus occidentalis, Nereocystis luetkeana, Salix hindsiana, S. lucida, Umbellularia californica*
Toothache Remedy: *Asarum caudatum*
Tuberculosis Remedy: *Ligusticum apiifolium, Torreya californica*
Unspecified: *Sanicula* sp.

Food
Beverage: *Ledum ×columbianum, Satureja douglasii*
Candy: *Pinus lambertiana*
Dried Food: *Arctostaphylos glandulosa, Juglans hindsii, Lithocarpus densiflorus, Porphyra lanceolata, P. perforata, Postelsia palmaeformis, Quercus agrifolia, Q. garryana, Q. kelloggii, Umbellularia californica, Vaccinium ovatum*
Forage: *Lithocarpus densiflorus, Quercus dumosa*
Fruit: *Arbutus menziesii, Carpobrotus aequilateralus, Fragaria chiloensis, F. vesca, Gaultheria shallon, Heteromeles arbutifolia, Ribes californicum, Rosa californica, R. gymnocarpa, Rubus leucodermis, R. parviflorus, R. spectabilis, R. ursinus, Sambucus cerulea, Umbellularia californica, Vaccinium ovatum, V. parvifolium, Vitis californica*
Pie & Pudding: *Gaultheria shallon, Rubus ursinus, Vaccinium ovatum*
Porridge: *Lithocarpus densiflorus, Quercus agrifolia, Q. garryana, Q. kelloggii, Q. lobata, Triticum aestivum*
Sauce & Relish: *Rubus ursinus*
Sour: *Oxalis oregana*
Spice: *Lomatium macrocarpum, Ulva lactuca*
Staple: *Avena* sp., *Elymus* sp., *Madia elegans, M. sativa, Pinus sabiniana, Wyethia angustifolia*
Sweetener: *Pinus lambertiana*
Unspecified: *Aesculus californica, Angelica tomentosa, Brassica* sp., *Castanopsis chrysophylla, Heracleum maximum, Hordeum* sp., *Juglans hindsii, Lomatium macrocarpum, Nereocystis luetkeana, Perideridia kelloggii, Pinus lambertiana, P. muricata, P. sabiniana, Porphyra lanceolata, P. perforata, Postelsia palmaeformis, Trifolium albopurpureum, T. fucatum, T. variegatum, T. wormskioldii, Typha* sp., *Umbellularia californica, Wyethia glabra*
Vegetable: *Agaricus campestris, A. silvicola, Allium dichlamydeum, Boletus edulis, Brodiaea coronaria, Calochortus amabilis, C. luteus, C. tolmiei, C. vestae, Cantharellus cibarius, Cyperus esculentus, Dentinum repandum, Dichelostemma pulchellum, Hericium coralloides, Peziza aurantia, Piperia elegans, P. unalascensis, Pleurotus ostreatus, Triteleia laxa*
Winter Use Food: *Arbutus menziesii, Castanopsis chrysophylla, Nereocystis luetkeana, Pinus lambertiana, P. muricata, P. sabiniana, Rubus leucodermis, Umbellularia californica, Wyethia angustifolia*

Fiber
Basketry: *Adiantum jordanii, Cercis canadensis, Corylus cornuta, Pteridium aquilinum, Salix hindsiana, S. lucida, Scirpus acutus, S. americanus, S. fluviatilis, S. robustus, Taxus brevifolia*
Brushes & Brooms: *Chlorogalum pomeridianum, Perideridia kelloggii*
Building Material: *Amelanchier pallida, Salix hindsiana, Sequoia sempervirens*
Clothing: *Asclepias eriocarpa, A. speciosa, Ramalina menziesii*
Cordage: *Asclepias eriocarpa, A. speciosa, Carex barbarae, Lupinus arboreus, Nereocystis luetkeana, Vitis californica*
Furniture: *Cornus nuttallii, Holodiscus discolor, Vitis californica*
Sewing Material: *Carex barbarae*

Dye
Black: *Juglans hindsii, Quercus kelloggii*
Yellow: *Oenothera elata*

Other
Ceremonial Items: *Aquilegia eximia, Ceanothus griseus, C. thyrsiflorus, Fragaria chiloensis, F. vesca, Iris douglasiana, Lithocarpus densiflorus, Lupinus affinis, L. albifrons, L. arboreus, L. densiflorus, L. luteolus, L. nanus, L. polyphyllus, L. succulentus, L. versicolor, Orthocarpus densiflorus, O. faucibarbatus, O. purpurascens, Rhododendron macrophyllum, R. occidentale, Triteleia laxa*
Containers: *Heracleum maximum*
Cooking Tools: *Polystichum munitum, Rubus parviflorus, Taxus brevifolia, Woodwardia fimbriata*
Designs: *Lonicera hispidula, Toxicodendron diversilobum*
Fasteners: *Equisetum telmateia, Grindelia integrifolia, Pinus muricata, Pseudotsuga menziesii*
Fuel: *Arbutus menziesii, Pseudotsuga menziesii*
Hunting & Fishing Item: *Aesculus californica, Chlorogalum pomeridianum, Corylus cornuta, Gnaphalium stramineum, Holodiscus discolor, Marah fabaceus, Nereocystis luetkeana, Pinus muricata, Taxus brevifolia, Umbellularia californica*
Jewelry: *Adiantum aleuticum, A. jordanii, Juncus balticus*
Lighting: *Arctostaphylos columbiana*
Musical Instrument: *Lithocarpus densiflorus*
Protection: *Angelica tomentosa, Umbellularia californica*
Sacred Items: *Umbellularia californica*
Smoke Plant: *Artemisia douglasiana, Nicotiana quadrivalvis*
Smoking Tools: *Lonicera hispidula*
Soap: *Ceanothus griseus, C. thyrsiflorus, Chlorogalum pomeridianum*
Tools: *Aesculus californica, Arctostaphylos columbiana, A. glandulosa, Equisetum telmateia, Taxus brevifolia*
Toys & Games: *Acer macrophyllum, Arbutus menziesii, Heracleum maximum, Juncus balticus, Oxalis oregana, Pinus lambertiana, Ranunculus californicus, Sambucus cerulea, Sequoia sempervirens*

Pomo, Little Lakes
Drug
Analgesic: *Arctostaphylos manzanita*
Antidiarrheal: *Arctostaphylos manzanita*
Antihemorrhagic: *Sambucus cerulea*
Antirheumatic (External): *Pseudotsuga menziesii*
Cold Remedy: *Arbutus menziesii*
Herbal Steam: *Pseudotsuga menziesii*
Tuberculosis Remedy: *Sambucus cerulea*
Venereal Aid: *Pseudotsuga menziesii*

Pomo, Potter Valley
Drug
Disinfectant: *Sambucus cerulea*
Emetic: *Sambucus cerulea*
Gastrointestinal Aid: *Cynoglossum grande, Sambucus cerulea*
Venereal Aid: *Cynoglossum grande*

Ponca
Drug
Abortifacient: *Artemisia frigida*
Analgesic: *Aquilegia canadensis, Echinacea angustifolia, Lespedeza capitata, Physalis lanceolata, Silphium perfoliatum*
Antidiarrheal: *Lygodesmia juncea, Prunus virginiana*
Antidote: *Echinacea angustifolia*
Antirheumatic (External): *Lespedeza capitata*
Antirheumatic (Internal): *Silphium perfoliatum*
Burn Dressing: *Artemisia dracunculus, Echinacea angustifolia, Typha latifolia*
Carminative: *Acorus calamus, Mentha canadensis*
Ceremonial Medicine: *Acorus calamus, Artemisia* sp., *Hierochloe odorata, Prunus virginiana*
Cold Remedy: *Acorus calamus, Juniperus virginiana, Silphium perfoliatum*
Cough Medicine: *Acorus calamus, Juniperus virginiana*
Dermatological Aid: *Iris versicolor, Mirabilis nyctaginea, Physalis lanceolata, Plantago major, Typha latifolia*
Disinfectant: *Artemisia* sp.
Ear Medicine: *Iris versicolor*
Eye Medicine: *Iris versicolor, Lygodesmia juncea, Symphoricarpos occidentalis, S. orbiculatus*
Febrifuge: *Acorus calamus, Aquilegia canadensis, Caulophyllum thalictroides*
Gastrointestinal Aid: *Acorus calamus, Artemisia* sp., *Asclepias exaltata, Physalis lanceolata, Quercus* sp.
Gynecological Aid: *Chamaesyce serpyllifolia, Lygodesmia juncea*
Hallucinogen: *Lophophora williamsii*
Herbal Steam: *Juniperus virginiana, Silphium perfoliatum*
Laxative: *Gymnocladus dioicus, Ulmus rubra*
Love Medicine: *Aquilegia canadensis, Lomatium foeniculaceum, Sanguinaria canadensis, Thalictrum dasycarpum*
Misc. Disease Remedy: *Echinacea angustifolia*
Other: *Echinacea angustifolia*
Panacea: *Acorus calamus, Anemone canadensis, Cucurbita foetidissima*
Pediatric Aid: *Quercus* sp., *Typha latifolia*
Poison: *Toxicodendron pubescens*
Pulmonary Aid: *Asclepias tuberosa*
Respiratory Aid: *Asclepias tuberosa*
Snakebite Remedy: *Echinacea angustifolia*
Stimulant: *Gymnocladus dioicus*
Toothache Remedy: *Acorus calamus, Echinacea angustifolia*
Tuberculosis Remedy: *Grindelia squarrosa*
Unspecified: *Artemisia dracunculus, Lophophora williamsii, Physalis lanceolata*
Veterinary Aid: *Echinacea angustifolia, Juniperus virginiana, Silphium laciniatum*
Witchcraft Medicine: *Artemisia* sp., *Silphium laciniatum*

Food
Beverage: *Agastache foeniculum, Ceanothus americanus, Mentha canadensis, Rubus idaeus, R. occidentalis, Sambucus canadensis*
Bread & Cake: *Prunus virginiana*
Candy: *Dalea purpurea, Populus deltoides, Silphium laciniatum*
Cooking Agent: *Lithospermum canescens*
Dried Food: *Pediomelum esculentum, Physalis heterophylla, Prunus americana, P. pumila, P. virginiana, Rubus idaeus, R. occidentalis, Shepherdia argentea, Vitis cinerea, V. vulpina*
Fodder: *Osmorhiza longistylis, Oxalis stricta, O. violacea*
Fruit: *Amelanchier alnifolia, Crataegus chrysocarpa, Fragaria vesca, F. virginiana, Malus ioensis, Prunus americana, P. virginiana, Ribes missouriense, Rubus idaeus, R. occidentalis, Sambucus canadensis, Shepherdia argentea, Viburnum lentago, Vitis cinerea, V. vulpina*
Sauce & Relish: *Allium canadense, Physalis heterophylla, Prunus americana, P. pumila*
Soup: *Carya ovata, Corylus americana, Juglans nigra, Nelumbo lutea*
Spice: *Allium canadense*
Staple: *Zea mays, Zizania aquatica*
Starvation Food: *Crataegus chrysocarpa, Rosa arkansana*
Sweetener: *Acer negundo, A. saccharinum, Agastache foeniculum, Carya ovata, Zea mays*
Unspecified: *Allium canadense, Amphicarpaea bracteata, Apios americana, Carya ovata, Corylus americana, Helianthus tuberosus, Juglans nigra, Lathyrus brachycalyx, Linum lewisii, Nelumbo lutea, Oxalis stricta, O. violacea, Pediomelum esculentum, Quercus macrocarpa, Q. rubra*
Vegetable: *Asclepias syriaca*
Winter Use Food: *Zea mays*

Fiber
Brushes & Brooms: *Prunus americana, Stipa spartea*
Building Material: *Andropogon gerardii, Fraxinus pennsylvanica, Spartina pectinata, Ulmus americana, U. rubra, U. thomasii*
Clothing: *Urtica dioica*
Cordage: *Tilia americana, Ulmus rubra, Urtica dioica, Yucca glauca*
Mats, Rugs & Bedding: *Artemisia dracunculus, A. frigida, A. ludoviciana, Scirpus tabernaemontani*
Scouring Material: *Equisetum* sp.
Sewing Material: *Yucca glauca*

Dye
Black: *Juglans nigra*
Red: *Sanguinaria canadensis*
Unspecified: *Sanguinaria canadensis*
Yellow: *Populus deltoides*

Other
Ceremonial Items: *Astragalus crassicarpus, Fraxinus pennsylvanica, Lagenaria siceraria, Lophophora williamsii, Nelumbo lutea, Stipa spartea, Urtica dioica*
Cooking Tools: *Astragalus canadensis, Ulmus americana, U. rubra, U. thomasii*
Fuel: *Ceanothus americanus, Ulmus americana, U. rubra, U. thomasii*
Good Luck Charm: *Anemone cylindrica, Melia azedarach*
Hunting & Fishing Item: *Amelanchier alnifolia, Cornus asperifolia, Fraxinus pennsylvanica, Maclura pomifera, Prunus virginiana*
Incense & Fragrance: *Aquilegia canadensis, Galium triflorum*
Jewelry: *Melia azedarach*
Musical Instrument: *Populus deltoides*
Protection: *Juniperus virginiana*
Season Indicator: *Prunus virginiana*

Smoke Plant: *Cornus amomum, C. sericea, Nicotiana quadrivalvis, Rhus glabra, Rosa arkansana*
Smoking Tools: *Fraxinus pennsylvanica*
Soap: *Yucca glauca*
Stable Gear: *Ulmus americana, U. thomasii, Urtica dioica*
Tools: *Ulmus americana, U. rubra, U. thomasii*
Toys & Games: *Amelanchier alnifolia, Andropogon gerardii, Artemisia* sp., *Asclepias syriaca, Astragalus canadensis, Betula papyrifera, Desmanthus illinoensis, Populus deltoides, Quercus macrocarpa, Sambucus canadensis, Thalictrum dasycarpum, Ulmus americana, U. rubra, Urtica dioica, Viburnum opulus*

Potawatomi
Drug
Adjuvant: *Betula alleghaniensis, B. papyrifera, Cirsium vulgare, Panax quinquefolius, Prunus serotina, Thuja occidentalis*
Alterative: *Aralia hispida, Gentiana alba*
Analgesic: *Aster furcatus, Coptis trifolia, Equisetum arvense, Gaultheria procumbens, Impatiens capensis, Prunus pensylvanica, Rosa blanda, Ulmus americana, Uvularia grandiflora*
Anthelmintic: *Rhus hirta*
Antidiarrheal: *Alnus incana, Cornus sericea, Epilobium ciliatum, Ostrya virginiana, Quercus rubra, Tsuga canadensis, Ulmus americana*
Antiemetic: *Asarum canadense*
Antihemorrhagic: *Acorus calamus, Ostrya virginiana*
Antirheumatic (Internal): *Gaultheria procumbens*
Blood Medicine: *Arctium minus, Rumex orbiculatus*
Burn Dressing: *Eupatorium purpureum, Fagus grandifolia*
Cathartic: *Juglans cinerea, Pedicularis canadensis, Rhamnus alnifolia, Sambucus racemosa*
Cold Remedy: *Abies balsamea, Aronia melanocarpa, Polygonum careyi, Rudbeckia hirta, Sorbus aucuparia, Tsuga canadensis*
Cough Medicine: *Acer spicatum, Prunus pensylvanica, Streptopus roseus*
Dermatological Aid: *Abies balsamea, Alnus incana, Aralia nudicaulis, A. racemosa, Calla palustris, Chamaedaphne calyculata, Comptonia peregrina, Corylus cornuta, Fagus grandifolia, Impatiens capensis, Iris versicolor, Larix laricina, Nuphar lutea, Pastinaca sativa, Picea mariana, Pinus banksiana, P. strobus, Plantago major, Populus balsamifera, Ranunculus pensylvanicus, Sagittaria latifolia, Solidago uliginosa, Thalictrum dasycarpum, Toxicodendron pubescens, Typha latifolia, Ulmus rubra*
Diaphoretic: *Tsuga canadensis*
Disinfectant: *Aralia nudicaulis, Picea mariana*
Diuretic: *Apocynum androsaemifolium, Diervilla lonicera, Dirca palustris, Lonicera canadensis, Taxus canadensis*
Ear Medicine: *Panax quinquefolius*
Emetic: *Sambucus racemosa, Sorbus aucuparia*
Expectorant: *Acer saccharum*
Eye Medicine: *Acer rubrum, Cornus alternifolia, Osmorhiza longistylis, Panax quinquefolius, Prunus virginiana, Ribes cynosbati, Rubus allegheniensis, R. idaeus, Ulmus rubra*
Febrifuge: *Chamaedaphne calyculata, Euthamia graminifolia, Gaultheria procumbens, Mentha canadensis, Polygonum careyi, P. lapathifolium, Solidago canadensis, S. flexicaulis, S. gigantea, Triadenum virginicum, Urtica dioica*
Gastrointestinal Aid: *Crataegus chrysocarpa, Fragaria vesca, Goodyera repens, Impatiens capensis, Osmorhiza longistylis, Thalictrum dasycarpum*
Gynecological Aid: *Actaea rubra, Adiantum pedatum, Alnus incana, Athyrium filix-femina, Caulophyllum thalictroides, Eupatorium purpureum, Goodyera repens, Linnaea borealis, Monotropa uniflora, Ribes cynosbati, Sarracenia purpurea, Sonchus arvensis, Trillium grandiflorum*
Heart Medicine: *Apocynum androsaemifolium, Viola pubescens*
Hemorrhoid Remedy: *Alnus incana*
Hemostat: *Agrimonia gryposepala, Lycopodium clavatum, L. obscurum, Rhus hirta, Salix discolor*
Hunting Medicine: *Thalictrum dasycarpum*
Internal Medicine: *Pedicularis canadensis*
Kidney Aid: *Apocynum androsaemifolium, Equisetum arvense*
Love Medicine: *Platanthera ×media, Scirpus tabernaemontani, Thalictrum dasycarpum*
Misc. Disease Remedy: *Chenopodium album, Rhus hirta, Sanguinaria canadensis, Sorbus aucuparia*
Oral Aid: *Coptis trifolia*
Orthopedic Aid: *Equisetum arvense, Hamamelis virginiana, Impatiens capensis, Rosa blanda, Uvularia grandiflora*
Other: *Diervilla lonicera, Hepatica nobilis, Larix laricina*
Panacea: *Nemopanthus mucronatus, Salix discolor*
Pediatric Aid: *Coptis trifolia*
Poison: *Pastinaca sativa, Toxicodendron pubescens*
Pulmonary Aid: *Chenopodium capitatum, Galeopsis tetrahit, Impatiens capensis, Mentha canadensis, Pinus banksiana, Sorbus aucuparia*
Respiratory Aid: *Acorus calamus, Verbascum thapsus*
Snakebite Remedy: *Goodyera repens*
Stimulant: *Achillea millefolium, Artemisia frigida, Aster novae-angliae, Maianthemum racemosum, Pinus banksiana, P. resinosa, Verbascum thapsus*
Throat Aid: *Maianthemum canadense, Plantago major, Rhus hirta, Sanguinaria canadensis, Ulmus rubra*
Tonic: *Amelanchier stolonifera, Aralia hispida, Arctium minus, Juglans cinerea, Nemopanthus mucronatus, Prunus virginiana, Taraxacum officinale*
Tuberculosis Remedy: *Abies balsamea*
Unspecified: *Asclepias syriaca, Aster shortii, Botrychium virginianum, Clintonia borealis, Epilobium angustifolium, Hordeum jubatum, Lactuca biennis, Ledum groenlandicum, Mimulus glabratus, Nymphaea odorata, Oenothera biennis, Polygonum amphibium, Potentilla norvegica, Rosa blanda, Sorbus americana, Spiraea salicifolia, Thuja occidentalis, Urtica dioica, Vaccinium myrtilloides*
Urinary Aid: *Apocynum androsaemifolium, Equisetum arvense, Goodyera repens, Juniperus communis*
Venereal Aid: *Diervilla lonicera, Taxus canadensis, Zanthoxylum americanum*
Veterinary Aid: *Alnus incana, Conyza canadensis, Larix laricina, Populus tremuloides*
Witchcraft Medicine: *Achillea millefolium, Anaphalis margaritacea, Aster umbellatus, Thuja occidentalis*

Food
Beverage: *Acer saccharum, Ledum groenlandicum, Prunus pumila, P. serotina*
Candy: *Acer saccharum*
Dessert: *Oxalis montana*
Dried Food: *Amelanchier stolonifera, Fragaria vesca, Vaccinium myrtilloides, Zizania palustris*
Fodder: *Larix laricina, Osmorhiza longistylis, Pedicularis canadensis*
Fruit: *Amelanchier stolonifera, Aronia melanocarpa, Cornus canadensis, Crataegus chrysocarpa, Maianthemum canadense, Prunus pensylvanica, P. serotina, P. virginiana, Ribes cynosbati, Rubus allegheniensis, R. idaeus, Vaccinium myrtilloides, V. oxycoccos*
Pie & Pudding: *Zizania palustris*
Porridge: *Quercus rubra*
Preservative: *Myrica gale*
Preserves: *Ribes cynosbati, Rubus idaeus*
Soup: *Allium canadense, Aralia racemosa, Asclepias syriaca, Parmelia physodes*
Sour: *Acer saccharum, Nemopanthus mucronatus, Rhus hirta*
Spice: *Asarum canadense*
Staple: *Zizania palustris*

Starvation Food: *Celastrus scandens*
Sweetener: *Acer saccharum*
Unspecified: *Arisaema triphyllum, Fagus grandifolia, Helianthus tuberosus, Nelumbo lutea, Sagittaria latifolia, Taraxacum officinale*
Vegetable: *Allium tricoccum, Apios americana, Chenopodium album, Parmelia physodes, Phaseolus vulgaris, Sagittaria latifolia*
Winter Use Food: *Amelanchier stolonifera, Carya ovata, Corylus cornuta, Fragaria vesca, Juglans cinerea, Nelumbo lutea, Sagittaria latifolia, Zea mays*
Fiber
Basketry: *Anthoxanthum odoratum, Fraxinus pennsylvanica, Iris versicolor, Scirpus tabernaemontani, Tilia americana, Ulmus rubra*
Brushes & Brooms: *Corylus cornuta*
Building Material: *Betula alleghaniensis, B. papyrifera, Typha latifolia, Ulmus rubra*
Canoe Material: *Betula papyrifera*
Caulking Material: *Pinus strobus*
Cordage: *Dirca palustris, Tilia americana, Urtica dioica*
Mats, Rugs & Bedding: *Abies balsamea, Iris versicolor, Scirpus cyperinus, S. tabernaemontani, Sphagnum* sp., *Typha latifolia*
Sewing Material: *Anthoxanthum odoratum, Apocynum androsaemifolium, Asclepias syriaca, Pinus banksiana, Tilia americana*
Dye
Brown: *Alnus incana, Ledum groenlandicum*
Orange: *Impatiens capensis*
Red: *Alnus incana, Chenopodium capitatum, Salix interior*
Red-Brown: *Quercus rubra*
Unspecified: *Hepatica nobilis, Sanguinaria canadensis*
Yellow: *Coptis trifolia, Impatiens capensis, Ranunculus pensylvanicus, Rudbeckia hirta*
Other
Ceremonial Items: *Thuja occidentalis*
Cooking Tools: *Betula papyrifera, Fagus grandifolia, Fraxinus pennsylvanica, Sarracenia purpurea*
Designs: *Acer spicatum, Quercus rubra*
Good Luck Charm: *Adiantum pedatum, Eupatorium purpureum, Maianthemum canadense*
Hunting & Fishing Item: *Acer rubrum, Carya ovata, Sagittaria latifolia, Sonchus arvensis*
Insecticide: *Comptonia peregrina, Myrica gale*
Lighting: *Pinus banksiana, Thuja occidentalis*
Protection: *Achillea millefolium, Anaphalis margaritacea, Thuja occidentalis*
Sacred Items: *Epigaea repens*
Smoke Plant: *Arctostaphylos uva-ursi, Cornus sericea, Rhus hirta*
Waterproofing Agent: *Pinus banksiana, Typha latifolia*

Pueblo
Food
Dried Food: *Descurainia pinnata, D. sophia, Monarda fistulosa, Polanisia dodecandra, Thelypodium wrightii, Yucca baccata, Y. glauca*
Fruit: *Celtis laevigata, Physalis longifolia, P. subulata, Prunus virginiana, Ribes inerme, Yucca baccata, Y. glauca*
Preserves: *Mahonia haematocarpa*
Soup: *Descurainia pinnata, D. sophia, Polanisia dodecandra, Thelypodium wrightii*
Special Food: *Polyporus harlowii, Zea mays*
Spice: *Allium geyeri, Atriplex argentea, Monarda fistulosa, Pectis angustifolia, P. papposa*
Staple: *Cleome serrulata*
Unspecified: *Descurainia pinnata, D. sophia, Helianthus annuus, Hoffmannseggia glauca, Pinus edulis, Polanisia dodecandra, Quercus gambelii, Solanum fendleri, Thelypodium wrightii*
Vegetable: *Amaranthus blitoides, A. retroflexus, Atriplex powellii, Chenopodium album, C. leptophyllum, Descurainia pinnata, D. sophia, Polanisia dodecandra, Thelypodium wrightii*
Other
Cash Crop: *Pinus edulis*

Puyallup
Drug
Dermatological Aid: *Stachys mexicana*
Food
Fruit: *Fragaria* sp., *Rubus leucodermis*

Quileute
Drug
Abortifacient: *Castilleja angustifolia*
Analgesic: *Lysichiton americanus*
Antidiarrheal: *Alnus rubra*
Antidote: *Lonicera involucrata*
Antirheumatic (External): *Achillea millefolium, Urtica dioica*
Antirheumatic (Internal): *Anaphalis margaritacea, Urtica dioica*
Blood Medicine: *Mahonia nervosa, Prunus emarginata*
Burn Dressing: *Fragaria chiloensis, Rubus spectabilis*
Ceremonial Medicine: *Equisetum hyemale, E. laevigatum*
Cold Remedy: *Sambucus racemosa*
Cough Medicine: *Petasites frigidus, Pinus monticola, Sambucus racemosa, Tiarella trifoliata*
Dermatological Aid: *Actaea rubra, Aquilegia formosa, Aruncus dioicus, Claytonia sibirica, Gaultheria shallon, Geum macrophyllum, Leucanthemum vulgare, Lysichiton americanus, Oxalis oregana, Polystichum munitum, Prunella vulgaris, Rosa nutkana, Trillium ovatum, Tsuga heterophylla*
Emetic: *Lonicera involucrata, Tsuga heterophylla, T. mertensiana*
Eye Medicine: *Claytonia sibirica*
Febrifuge: *Achillea millefolium, Lysichiton americanus*
Gynecological Aid: *Lysichiton americanus, Sambucus racemosa*
Laxative: *Frangula purshiana*
Love Medicine: *Galium triflorum*
Orthopedic Aid: *Blechnum spicant*
Panacea: *Achillea millefolium, Blechnum spicant, Frangula purshiana*
Pediatric Aid: *Achillea millefolium, Nepeta cataria*
Pulmonary Aid: *Malus fusca*
Sports Medicine: *Equisetum hyemale, Salix hookeriana*
Tonic: *Aruncus dioicus, Cornus canadensis, C. nuttallii, C. sericea*
Tuberculosis Remedy: *Boykinia occidentalis*
Unspecified: *Alnus rubra, Arctium minus, Arctostaphylos uva-ursi, Brassica nigra, Juncus ensifolius, Leymus mollis, Mentha canadensis, M. ×piperita, Moricandia arvensis, Oplopanax horridus, Populus balsamifera, Sinapis alba, Taraxacum officinale, Thuja plicata*
Urinary Aid: *Claytonia sibirica*
Venereal Aid: *Frangula purshiana, Malus fusca*
Food
Beverage: *Rubus ursinus*
Bread & Cake: *Gaultheria shallon*
Dried Food: *Equisetum hyemale, E. laevigatum, Rubus ursinus, Vaccinium ovalifolium*
Fodder: *Equisetum telmateia*
Forage: *Lysichiton americanus*
Fruit: *Amelanchier alnifolia, Fragaria chiloensis, Gaultheria ovatifolia, G. shallon, Malus fusca, Ribes sanguineum, Rosa pisocarpa, Rubus idaeus, R. laciniatus, R. lasiococcus, R. leucodermis, R. nivalis, R. parviflorus, R. spectabilis, R. ursinus, Sambucus racemosa, Vaccinium deliciosum, V. myrtilloides, V. ovali-

[Fruit]
 folium, V. ovatum, V. parvifolium
Preserves: *Gaultheria ovatifolia, Mahonia nervosa, Prunus emarginata*
Sauce & Relish: *Gaultheria ovatifolia, Sambucus racemosa, Vaccinium deliciosum, V. myrtilloides, V. ovalifolium, V. ovatum, V. parvifolium*
Special Food: *Equisetum hyemale, E. laevigatum*
Spice: *Brassica nigra, Lysichiton americanus, Moricandia arvensis, Salix hookeriana, Sinapis alba*
Unspecified: *Allium acuminatum, A. cernuum, Argentina egedii, Athyrium filix-femina, Camassia quamash, Equisetum hyemale, E. telmateia, Heracleum maximum, Lepidium virginicum, Lilium columbianum, Lysichiton americanus, Oxalis oregana, Polystichum munitum, Pteridium aquilinum, Ranunculus reptans, Rubus spectabilis*
Vegetable: *Brassica nigra, Cirsium edule, Heracleum maximum, Lepidium virginicum, Moricandia arvensis, Sinapis alba, Urtica dioica*
Winter Use Food: *Ribes sanguineum, Rosa pisocarpa, Rubus idaeus, R. laciniatus, R. lasiococcus, R. leucodermis, R. nivalis, R. parviflorus, R. spectabilis, R. ursinus, Sambucus racemosa, Vaccinium deliciosum, V. myrtilloides, V. ovalifolium, V. ovatum, V. parvifolium*

Fiber
Basketry: *Equisetum telmateia, Heracleum maximum, Picea sitchensis, P. sp., Salix hookeriana, S. sitchensis, Thuja plicata, Xerophyllum tenax*
Building Material: *Picea engelmannii, P. sitchensis, Pseudotsuga menziesii, Thuja plicata*
Canoe Material: *Alnus rubra, Chamaecyparis nootkatensis, Menziesia ferruginea, Taxus brevifolia, Thuja plicata*
Caulking Material: *Picea sitchensis*
Clothing: *Picea sitchensis, Thuja plicata*
Cordage: *Picea engelmannii, P. sitchensis, Thuja plicata, Urtica dioica*
Mats, Rugs & Bedding: *Polystichum munitum, Thuja plicata*
Other: *Thuja plicata*

Dye
Red-Brown: *Alnus rubra*

Other
Ceremonial Items: *Arctostaphylos uva-ursi, Cornus canadensis, C. nuttallii, C. sericea, Digitalis purpurea, Juniperus scopulorum, Rubus spectabilis, Symphoricarpos albus, Taxus brevifolia*
Cleaning Agent: *Athyrium filix-femina*
Containers: *Lysichiton americanus, Nereocystis luetkeana, Ribes bracteosum, Rubus parviflorus, Thuja plicata*
Cooking Tools: *Lysichiton americanus, Polystichum munitum, Pteridium aquilinum, Thuja plicata*
Good Luck Charm: *Lysichiton americanus*
Hide Preparation: *Tsuga heterophylla, T. mertensiana*
Hunting & Fishing Item: *Acer circinatum, Malus fusca, Nereocystis luetkeana, Picea sitchensis, Rubus spectabilis, Salix hookeriana, Taxus brevifolia, Thuja plicata, Urtica dioica*
Lighting: *Thuja plicata*
Musical Instrument: *Thuja plicata*
Paint: *Lonicera involucrata*
Smoke Plant: *Arbutus menziesii, Arctostaphylos tomentosa, A. uva-ursi, Cornus canadensis, C. nuttallii, C. sericea*
Tools: *Malus fusca, Ribes bracteosum, Thuja plicata*
Toys & Games: *Oenanthe sarmentosa, Phyllospadix scouleri, Picea engelmannii, P. sitchensis, Taxus brevifolia*
Waterproofing Agent: *Tsuga heterophylla*

Quinault
Drug
Abortifacient: *Equisetum hyemale*
Analgesic: *Allium cernuum, Heracleum maximum, Malus fusca, Nuphar lutea, Rubus spectabilis, Urtica dioica*
Antidiarrheal: *Gaultheria shallon*
Antirheumatic (External): *Nuphar lutea*
Antirheumatic (Internal): *Ledum groenlandicum, Veratrum viride*
Blood Medicine: *Malus fusca, Pinus monticola*
Burn Dressing: *Polystichum munitum, Rubus spectabilis*
Cold Remedy: *Tsuga heterophylla*
Cough Medicine: *Mahonia sp., Polypodium virginianum*
Dermatological Aid: *Actaea rubra, Aruncus dioicus, Galium triflorum, Geum macrophyllum, Lonicera involucrata, Petasites frigidus, Picea sitchensis, Pinus contorta, Polystichum munitum, Populus balsamifera, Prunella vulgaris, Pseudotsuga menziesii, Rosa nutkana, Rubus spectabilis, Taxus brevifolia, Thuja plicata*
Disinfectant: *Populus balsamifera, Rubus spectabilis*
Emetic: *Sambucus cerulea*
Eye Medicine: *Achillea millefolium, Equisetum hyemale, Maianthemum dilatatum, Malus fusca, Oxalis oregana, Petasites frigidus*
Febrifuge: *Thuja plicata*
Gastrointestinal Aid: *Blechnum spicant, Gaultheria shallon, Mahonia sp., Pinus monticola*
Gynecological Aid: *Claytonia sibirica, Geum macrophyllum, Lonicera involucrata, Prunus emarginata, Rubus spectabilis, Sambucus racemosa, Urtica dioica*
Hemostat: *Urtica dioica*
Kidney Aid: *Thuja plicata*
Laxative: *Frangula purshiana, Prunus emarginata, Tsuga heterophylla*
Love Medicine: *Menziesia ferruginea, Vicia nigricans*
Oral Aid: *Lonicera involucrata*
Orthopedic Aid: *Heracleum maximum, Urtica dioica*
Panacea: *Geum macrophyllum, Lysichiton americanus*
Pediatric Aid: *Tsuga heterophylla*
Poison: *Veratrum viride*
Pulmonary Aid: *Allium cernuum, Taxus brevifolia*
Throat Aid: *Picea sitchensis, Pinus contorta*
Tonic: *Achillea millefolium*
Tuberculosis Remedy: *Achillea millefolium, Populus balsamifera*
Urinary Aid: *Lysichiton americanus*
Venereal Aid: *Rosa nutkana, Thuja plicata*

Food
Beverage: *Vaccinium parvifolium*
Bread & Cake: *Gaultheria shallon*
Candy: *Picea sitchensis, Pseudotsuga menziesii*
Dried Food: *Vaccinium ovalifolium*
Fodder: *Equisetum telmateia*
Fruit: *Malus fusca, Oemleria cerasiformis, Rubus parviflorus, R. spectabilis, Vaccinium ovalifolium, V. ovatum, V. oxycoccos, V. parvifolium*
Preservative: *Ribes divaricatum*
Special Food: *Fragaria chiloensis*
Unspecified: *Allium cernuum, Athyrium filix-femina, Equisetum telmateia, Heracleum maximum, Lilium columbianum, Oxalis oregana, Polystichum munitum, Pteridium aquilinum, Rosa nutkana, Rubus spectabilis, Stachys mexicana*
Winter Use Food: *Sambucus cerulea, S. racemosa*

Fiber
Basketry: *Acer circinatum, Adiantum aleuticum, Thuja plicata, Typha latifolia, Xerophyllum tenax*
Brushes & Brooms: *Taxus brevifolia*
Building Material: *Acer circinatum, Populus balsamifera, Tsuga heterophylla*
Canoe Material: *Taxus brevifolia, Tsuga heterophylla*

Caulking Material: *Picea sitchensis*
Clothing: *Thuja plicata, Typha latifolia*
Cordage: *Salix sitchensis*
Mats, Rugs & Bedding: *Epilobium angustifolium, Typha latifolia*
Scouring Material: *Equisetum hyemale*
Sewing Material: *Thuja plicata*
Dye
Red-Brown: *Alnus rubra*
Other
Containers: *Alnus rubra, Leymus mollis, Nereocystis luetkeana, Rubus parviflorus, Typha latifolia*
Cooking Tools: *Blechnum spicant, Petasites frigidus, Polystichum munitum, Spiraea douglasii, Stachys mexicana, Taxus brevifolia*
Decorations: *Cornus nuttallii*
Fasteners: *Prunus emarginata*
Fuel: *Acer macrophyllum, Pseudotsuga menziesii, Tsuga heterophylla*
Hunting & Fishing Item: *Nereocystis luetkeana, Picea sitchensis, Pseudotsuga menziesii, Salix sitchensis, Sambucus cerulea, Taxus brevifolia, Thuja plicata, Tsuga heterophylla*
Paint: *Acer circinatum, Tsuga heterophylla*
Tools: *Taxus brevifolia*

Rappahannock
Drug
Abortifacient: *Lindera benzoin*
Analgesic: *Brassica oleracea, Liriodendron tulipifera, Nepeta cataria*
Anesthetic: *Rubus* sp.
Anthelmintic: *Chenopodium ambrosioides*
Antidiarrheal: *Cornus florida, Hieracium scabrum, Juglans nigra, Liquidambar styraciflua, Phytolacca americana, Rubus hispidus, R.* sp.
Antirheumatic (Internal): *Nepeta cataria, Phytolacca americana, Sambucus canadensis*
Blood Medicine: *Chimaphila umbellata, Cornus florida, Eupatorium purpureum, Gentianopsis crinita, Perilla frutescens, Rumex crispus, Taraxacum officinale, Trifolium pratense*
Burn Dressing: *Sassafras albidum*
Cold Remedy: *Gleditsia triacanthos, Marrubium vulgare, Prunus serotina*
Cough Medicine: *Gleditsia triacanthos, Marrubium vulgare, Prunus serotina*
Dermatological Aid: *Andropogon glomeratus, Aralia spinosa, Arisaema triphyllum, Aristolochia serpentaria, Asclepias syriaca, Centaurea* sp., *Datura stramonium, Fagus grandifolia, Morus rubra, Nuphar lutea, Phytolacca americana, Pinus echinata, Polygonatum biflorum, Portulaca oleracea, Rubus hispidus, Sambucus canadensis, Sassafras albidum, Solanum tuberosum, Verbascum thapsus, Vicia villosa, Xanthium strumarium*
Dietary Aid: *Chimaphila umbellata, Prunus serotina, Quercus rubra*
Ear Medicine: *Nicotiana tabacum*
Emetic: *Pinus echinata*
Eye Medicine: *Sassafras albidum*
Febrifuge: *Allium* sp., *Aralia spinosa, Aristolochia serpentaria, Datura stramonium, Gnaphalium obtusifolium, Hexastylis arifolia, Juglans nigra, Nuphar lutea, Plantago major, Sassafras albidum*
Gastrointestinal Aid: *Acorus calamus, Chimaphila umbellata, Gaylussacia* sp., *Juglans nigra, Rubus* sp., *Taraxacum officinale, Vicia villosa*
Gynecological Aid: *Aletris farinosa, Hedeoma pulegioides, Lindera benzoin*
Hallucinogen: *Magnolia virginiana*
Hemorrhoid Remedy: *Andropogon glomeratus, Phytolacca americana*
Hypotensive: *Allium vineale*
Kidney Aid: *Citrullus lanatus, Pinus virginiana, Prunus persica*
Love Medicine: *Liriodendron tulipifera*
Misc. Disease Remedy: *Salvia* sp., *Sassafras albidum*
Oral Aid: *Diospyros virginiana*
Orthopedic Aid: *Polygonatum biflorum, Verbascum thapsus*
Other: *Juniperus virginiana*
Panacea: *Alnus glutinosa, Cypripedium acaule, Iris versicolor, Rhus hirta, Xanthium strumarium*
Pediatric Aid: *Acorus calamus, Nepeta cataria*
Poison: *Datura stramonium, Prunus serotina, Solanum nigrum*
Poultice: *Brassica* sp.
Pulmonary Aid: *Allium vineale, Datura stramonium, Hexastylis arifolia, Juniperus virginiana*
Respiratory Aid: *Gnaphalium obtusifolium, Hexastylis arifolia, Juniperus virginiana*
Sedative: *Acorus calamus, Sassafras albidum, Solanum nigrum*
Snakebite Remedy: *Aristolochia serpentaria, Asclepias tuberosa*
Stimulant: *Liriodendron tulipifera, Sassafras albidum*
Throat Aid: *Datura stramonium, Diospyros virginiana*
Tonic: *Acorus calamus, Chenopodium ambrosioides, Chimaphila umbellata, Cornus florida, Eupatorium perfoliatum, Nepeta cataria, Prunus serotina, Quercus rubra, Rubus hispidus, Sassafras albidum*
Toothache Remedy: *Nicotiana tabacum*
Unspecified: *Prunus americana, Smilax* sp.
Veterinary Aid: *Chimaphila umbellata, Citrullus lanatus, Liquidambar styraciflua, Pinus echinata, Verbascum thapsus*
Food
Beverage: *Diospyros virginiana*
Candy: *Gnaphalium obtusifolium*
Snack Food: *Cucurbita pepo*
Dye
Unspecified: *Rhus* sp.
Other
Insecticide: *Hedeoma pulegioides*
Protection: *Brassica oleracea*

Ree
Drug
Ceremonial Medicine: *Helianthus annuus*
Stimulant: *Helianthus annuus*
Food
Staple: *Helianthus annuus*
Unspecified: *Helianthus annuus*

Rocky Boy
Food
Fruit: *Rosa nutkana*
Preserves: *Rosa nutkana*

Round Valley Indian
Drug
Analgesic: *Eriogonum latifolium*
Antidiarrheal: *Marrubium vulgare*
Antirheumatic (Internal): *Eriodictyon californicum*
Blood Medicine: *Eriodictyon californicum*
Cold Remedy: *Marrubium vulgare*
Dermatological Aid: *Humulus lupulus, Rhus trilobata*
Eye Medicine: *Eriogonum latifolium*
Febrifuge: *Eriodictyon californicum*
Gastrointestinal Aid: *Eriogonum latifolium*
Gynecological Aid: *Eriogonum latifolium*
Misc. Disease Remedy: *Eriodictyon californicum, Rhus trilobata*
Respiratory Aid: *Eriodictyon californicum*
Tuberculosis Remedy: *Eriodictyon californicum*

Food
Bread & Cake: *Quercus* sp.
Dried Food: *Prunus virginiana*
Fruit: *Prunus virginiana*
Porridge: *Quercus* sp.
Spice: *Calocedrus decurrens*
Substitution Food: *Pseudotsuga menziesii*
Unspecified: *Camassia leichtlinii, Trifolium* sp.
Fiber
Basketry: *Corylus cornuta, Pseudotsuga menziesii, Pteridium aquilinum, Smilax californica*
Brushes & Brooms: *Chlorogalum pomeridianum*
Building Material: *Salix lasiolepis*
Other
Cooking Tools: *Calocedrus decurrens*
Fuel: *Salix lasiolepis*
Hunting & Fishing Item: *Calocedrus decurrens, Corylus cornuta*
Lighting: *Pseudotsuga menziesii*
Protection: *Salix lasiolepis*

Saanich
Drug
Antidiarrheal: *Rubus parviflorus*
Antirheumatic (External): *Chimaphila umbellata, Goodyera oblongifolia*
Blood Medicine: *Equisetum arvense, E. telmateia, Satureja douglasii*
Burn Dressing: *Symphoricarpos albus*
Cold Remedy: *Achillea millefolium, Arbutus menziesii, Lomatium nudicaule, Polypodium virginianum, Prunus emarginata*
Dermatological Aid: *Abies grandis, Symphoricarpos albus, Vicia nigricans*
Gastrointestinal Aid: *Cornus sericea, Polypodium virginianum, Rubus parviflorus*
Laxative: *Physocarpus capitatus*
Misc. Disease Remedy: *Juniperus scopulorum*
Panacea: *Malus fusca, Prunus emarginata*
Psychological Aid: *Prunus emarginata, Ribes divaricatum, R. lacustre, R. lobbii*
Respiratory Aid: *Cornus sericea*
Throat Aid: *Achillea millefolium, Lomatium nudicaule, Polypodium virginianum*
Tonic: *Alnus rubra, Stachys ciliata*
Toothache Remedy: *Achillea millefolium*
Food
Beverage: *Epilobium angustifolium, Ledum groenlandicum, Rubus ursinus, Satureja douglasii*
Candy: *Lonicera ciliosa*
Forage: *Stachys ciliata*
Fruit: *Amelanchier alnifolia, Oemleria cerasiformis, Rosa nutkana*
Spice: *Acer macrophyllum, Mentha arvensis, Zostera marina*
Unspecified: *Cirsium brevistylum, Daucus pusillus, Equisetum arvense, E. telmateia, Fritillaria lanceolata, Rorippa nasturtium-aquaticum, Rumex obtusifolius*
Vegetable: *Rumex acetosella, Urtica dioica*
Fiber
Cordage: *Ribes divaricatum, R. lacustre, R. lobbii*
Other
Cash Crop: *Paxistima myrsinites*
Ceremonial Items: *Lomatium nudicaule, Polystichum munitum, Rubus ursinus, Thuja plicata*
Cooking Tools: *Holodiscus discolor*
Hunting & Fishing Item: *Holodiscus discolor, Philadelphus lewisii, Ribes divaricatum, R. lacustre, R. lobbii, Thuja plicata*
Insecticide: *Achlys triphylla*
Preservative: *Arbutus menziesii, Cornus nuttallii*
Protection: *Cirsium brevistylum*
Tools: *Holodiscus discolor*

Salish
Drug
Cathartic: *Shepherdia canadensis*
Cold Remedy: *Artemisia tridentata, Veratrum viride*
Dermatological Aid: *Ipomopsis aggregata, Nicotiana attenuata*
Disinfectant: *Juniperus virginiana*
Eye Medicine: *Achillea millefolium, Ipomopsis aggregata*
Gastrointestinal Aid: *Shepherdia canadensis*
Oral Aid: *Geranium oreganum*
Pediatric Aid: *Ipomopsis aggregata*
Poison: *Veratrum viride*
Tonic: *Ledum groenlandicum, Lupinus polyphyllus, Shepherdia canadensis*
Tuberculosis Remedy: *Verbascum thapsus*
Unspecified: *Aconitum delphiniifolium, Apocynum androsaemifolium, Mentha canadensis, Penstemon fruticosus*
Venereal Aid: *Arabis drummondii, Populus tremuloides*
Food
Fruit: *Cornus canadensis, Vaccinium ovatum*
Unspecified: *Rosa nutkana*
Vegetable: *Camassia leichtlinii, C. quamash, Triglochin maritimum*
Other
Hunting & Fishing Item: *Holodiscus discolor*
Tools: *Holodiscus discolor*

Salish, Coast
Drug
Analgesic: *Lomatium utriculatum, Oplopanax horridus, Veratrum viride*
Antirheumatic (External): *Oplopanax horridus, Veratrum viride*
Cathartic: *Cicuta douglasii*
Dermatological Aid: *Abies grandis, Heracleum maximum, Moneses uniflora, Pinus contorta, Prunella vulgaris*
Diaphoretic: *Veratrum viride*
Emetic: *Cicuta douglasii*
Gastrointestinal Aid: *Lomatium utriculatum*
Internal Medicine: *Lomatium nudicaule*
Misc. Disease Remedy: *Pinus contorta*
Other: *Ribes divaricatum, R. lacustre, R. lobbii*
Tonic: *Frangula purshiana*
Food
Beverage: *Fragaria chiloensis, F. vesca, F. virginiana, Ledum groenlandicum*
Bread & Cake: *Gaultheria shallon, Pinus contorta, Pteridium aquilinum, Ribes bracteosum, R. divaricatum, R. hudsonianum, R. lacustre, R. lobbii, R. sanguineum, Rubus leucodermis, R. parviflorus, R. ursinus*
Dessert: *Shepherdia canadensis*
Dried Food: *Camassia leichtlinii, C. quamash, Pinus monticola, Polypodium virginianum, Thuja plicata, Tsuga heterophylla, Vaccinium ovalifolium, V. oxycoccos, V. parvifolium, V. uliginosum*
Fruit: *Amelanchier alnifolia, Arctostaphylos uva-ursi, Crataegus douglasii, Fragaria chiloensis, F. vesca, F. virginiana, Gaultheria shallon, Maianthemum dilatatum, Malus fusca, Rubus leucodermis, R. parviflorus, R. spectabilis, R. ursinus, Sambucus cerulea, Vaccinium ovalifolium, V. oxycoccos, V. parvifolium, V. uliginosum, Viburnum edule*
Preserves: *Mahonia aquifolium, M. nervosa*
Substitution Food: *Polypodium virginianum*
Unspecified: *Acer macrophyllum, Allium acuminatum, A. cernuum, Alnus*

rubra, Argentina egedii, Athyrium filix-femina, Corylus cornuta, Daucus pusillus, Dryopteris campyloptera, Fritillaria lanceolata, Glaux maritima, Heracleum maximum, Pinus contorta, P. monticola, Polypodium virginianum, Pteridium aquilinum, Quercus garryana, Rosa nutkana, Rubus parviflorus, R. spectabilis, Salicornia virginica, Thuja plicata
Vegetable: *Camassia leichtlinii, C. quamash*
Fiber
Basketry: *Carex* sp., *Equisetum telmateia, Prunus emarginata, Scirpus americanus, Thuja plicata, Typha latifolia*
Brushes & Brooms: *Taxus brevifolia, Thuja plicata*
Building Material: *Thuja plicata*
Canoe Material: *Populus balsamifera, Taxus brevifolia, Thuja plicata*
Caulking Material: *Pseudotsuga menziesii*
Clothing: *Thuja plicata*
Cordage: *Carex* sp., *Thuja plicata*
Furniture: *Thuja plicata*
Mats, Rugs & Bedding: *Polystichum munitum, Scirpus americanus, Thuja plicata, Typha latifolia*
Dye
Red-Brown: *Alnus rubra*
Yellow: *Mahonia aquifolium*
Other
Cash Crop: *Corylus cornuta*
Ceremonial Items: *Thuja plicata, Veratrum viride*
Containers: *Prunus emarginata, Thuja plicata*
Cooking Tools: *Alnus rubra, Arbutus menziesii, Holodiscus discolor, Lysichiton americanus, Polystichum munitum, Thuja plicata*
Fasteners: *Pinus contorta, P. monticola, Taxus brevifolia*
Fuel: *Acer macrophyllum, Prunus emarginata, Pseudotsuga menziesii, Thuja plicata*
Hide Preparation: *Tsuga heterophylla*
Hunting & Fishing Item: *Alnus rubra, Amelanchier alnifolia, Cornus nuttallii, Corylus cornuta, Holodiscus discolor, Malus fusca, Prunus emarginata, Pseudotsuga menziesii, Salix lucida, Taxus brevifolia, Thuja plicata*
Insecticide: *Allium acuminatum, A. cernuum*
Protection: *Geum macrophyllum*
Smoke Plant: *Arctostaphylos uva-ursi, Cornus nuttallii*
Tools: *Holodiscus discolor, Malus fusca, Prunus emarginata, Ribes divaricatum, R. lacustre, R. lobbii, Thuja plicata*
Toys & Games: *Arbutus menziesii, Sambucus racemosa, Taxus brevifolia*
Weapon: *Taxus brevifolia*

Salish, Cowichan
Drug
Throat Aid: *Arbutus menziesii*
Fiber
Sewing Material: *Holodiscus discolor*
Other
Cooking Tools: *Holodiscus discolor*
Hunting & Fishing Item: *Thuja plicata*

Salish, Halkomelem
Other
Cash Crop: *Camassia leichtlinii, C. quamash*

Salish, Straits
Food
Unspecified: *Fritillaria camschatcensis*
Other
Cash Crop: *Camassia leichtlinii, C. quamash*

Samish
Drug
Cough Medicine: *Gaultheria shallon*
Dermatological Aid: *Malus fusca*
Gastrointestinal Aid: *Malus fusca*
Other: *Lysichiton americanus*
Tonic: *Mahonia aquifolium, Urtica dioica*
Tuberculosis Remedy: *Gaultheria shallon*
Food
Bread & Cake: *Gaultheria shallon*
Fruit: *Mahonia aquifolium, Malus fusca, Oemleria cerasiformis, Rubus parviflorus*
Unspecified: *Lilium columbianum, Rubus parviflorus*
Other
Containers: *Lysichiton americanus*
Hunting & Fishing Item: *Amelanchier alnifolia, Taxus brevifolia*
Smoke Plant: *Taxus brevifolia*
Tools: *Taxus brevifolia*

Sanel
Food
Substitution Food: *Brickellia californica*

San Felipe
Food
Dried Food: *Prunus virginiana, Yucca baccata*
Fruit: *Juniperus deppeana, Opuntia engelmannii, Physalis longifolia, Prunus virginiana, Ptelea trifoliata, Yucca baccata*
Porridge: *Equisetum laevigatum, Opuntia engelmannii*
Special Food: *Sophora nuttalliana*
Staple: *Dalea candida, Quercus gambelii*
Starvation Food: *Platanthera sparsiflora*
Unspecified: *Cleome serrulata, Opuntia polyacantha, Portulaca oleracea, Quercus gambelii, Rumex venosus, Typha latifolia, Yucca baccata, Y. glauca*
Vegetable: *Portulaca oleracea, Rudbeckia laciniata*
Winter Use Food: *Opuntia polyacantha*

San Ildefonso
Food
Fruit: *Juniperus monosperma*
Spice: *Monarda fistulosa*
Unspecified: *Dalea candida*

Sanpoil
Drug
Abortifacient: *Achillea millefolium*
Adjuvant: *Equisetum hyemale*
Analgesic: *Artemisia ludoviciana, Balsamorhiza sagittata, Heracleum maximum*
Antidiarrheal: *Eriogonum compositum, E. heracleoides, Prunus virginiana, Salix scouleriana*
Antiemetic: *Mahonia aquifolium*
Antihemorrhagic: *Purshia tridentata*
Antirheumatic (External): *Urtica* sp.
Cathartic: *Chaenactis douglasii, Frangula purshiana*
Cold Remedy: *Achillea millefolium, Artemisia dracunculus, A. ludoviciana, A. tridentata, Mentha canadensis, Monardella odoratissima, Oplopanax horridus, Ribes bracteosum*
Dermatological Aid: *Alnus incana, Arctostaphylos uva-ursi, Artemisia ludoviciana, Balsamorhiza sagittata, Ceanothus sanguineus, Clematis ligusticifolia, Geranium viscosissimum, Heracleum maximum, Holodiscus discolor, Juniperus scopulorum, Osmorhiza* sp., *Pinus* sp., *Rhus glabra, Symphoricarpos albus, Urtica* sp.

Diaphoretic: *Artemisia tridentata*
Diuretic: *Symphoricarpos albus*
Eye Medicine: *Geranium viscosissimum, Heracleum maximum, Mahonia aquifolium, Symphoricarpos albus*
Gastrointestinal Aid: *Artemisia tridentata, Mahonia aquifolium*
Gynecological Aid: *Apocynum androsaemifolium*
Laxative: *Artemisia tridentata, Purshia tridentata*
Misc. Disease Remedy: *Artemisia tridentata*
Oral Aid: *Rhus glabra*
Panacea: *Mentha canadensis*
Pediatric Aid: *Arctostaphylos uva-ursi, Equisetum hyemale, Mentha canadensis, Monardella odoratissima, Ribes bracteosum*
Tuberculosis Remedy: *Artemisia tridentata, Chaenactis douglasii, Grindelia nana, Juniperus scopulorum, Mahonia aquifolium*
Unspecified: *Juniperus scopulorum, Osmorhiza sp., Symphoricarpos albus*
Veterinary Aid: *Chrysothamnus nauseosus, Clematis ligusticifolia, Osmorhiza sp., Symphoricarpos albus*
Food
Beverage: *Mentha arvensis*
Fruit: *Mahonia aquifolium*
Ice Cream: *Shepherdia canadensis*
Preserves: *Mahonia aquifolium*
Special Food: *Balsamorhiza sagittata, Lomatium dissectum*
Unspecified: *Lomatium macrocarpum*
Other
Containers: *Balsamorhiza sagittata*
Insecticide: *Chrysothamnus nauseosus*

Sanpoil & Nespelem
Drug
Unspecified: *Prunus virginiana*
Food
Beverage: *Mentha canadensis, Monardella odoratissima, Prunus virginiana*
Bread & Cake: *Amelanchier alnifolia, Camassia sp., Prunus virginiana*
Dried Food: *Arctostaphylos uva-ursi, Crataegus douglasii, Daucus carota, Helianthus annuus, Opuntia polyacantha, Rubus sp., Vaccinium sp.*
Fruit: *Amelanchier alnifolia, Cornus sericea, Crataegus columbiana, C. douglasii, Fragaria virginiana, Mahonia aquifolium, Prunus virginiana, Ribes cereum, R. oxyacanthoides, R. sp., Rubus parviflorus, R. sp., Sambucus cerulea, Vaccinium sp.*
Porridge: *Lewisia rediviva*
Soup: *Arctostaphylos uva-ursi*
Starvation Food: *Rosa woodsii*
Unspecified: *Allium sp., Camassia sp., Corylus cornuta, Helianthus annuus, Larix occidentalis, Petasites frigidus, Pinus ponderosa, Valeriana edulis*
Vegetable: *Daucus carota*
Winter Use Food: *Helianthus annuus*
Fiber
Basketry: *Equisetum hyemale*

Santa Clara
Food
Special Food: *Dalea candida*
Other
Ceremonial Items: *Gossypium sp., Martynia sp.*
Hunting & Fishing Item: *Rhus trilobata*

Seminole
Drug
Abortifacient: *Persea borbonia, Pterocaulon virgatum*
Analgesic: *Andropogon floridanus, Bidens coronata, Cephalanthus occidentalis, Desmodium incanum, Dichanthelium laxiflorum, D. strigosum, Eleocharis geniculata, Eryngium yuccifolium, Galactia volubilis, Juniperus virginiana, Lagenaria siceraria, Lechea minor, Liatris laxa, Licania michauxii, Myrica cerifera, Pediomelum canescens, Persea borbonia, Phytolacca americana, Pinguicula lutea, P. pumila, Pinus elliottii, Quercus phellos, Q. virginiana, Rudbeckia hirta, Sabal palmetto, Sabatia campanulata, Salix caroliniana, S. humilis, Sassafras albidum, Sisyrinchium nashii, Solanum donianum, Vaccinium myrsinites, Verbesina virginica, Vitis aestivalis*
Antidiarrheal: *Andropogon floridanus, Bidens coronata, Cephalanthus occidentalis, Eleocharis geniculata, Eryngium yuccifolium, Galactia volubilis, Juniperus virginiana, Lechea minor, Liatris laxa, Licania michauxii, Persea borbonia, Pterocaulon virgatum, Sabatia campanulata, Salix caroliniana, S. humilis, Sassafras albidum, Stillingia sylvatica, Vaccinium myrsinites*
Antiemetic: *Andropogon floridanus, Cephalanthus occidentalis, Chamaecrista fasciculata, Erythrina herbacea, Galactia volubilis, Lechea minor, Liatris laxa, Licania michauxii, Persea borbonia, Sassafras albidum, Stillingia sylvatica*
Antihemorrhagic: *Eryngium yuccifolium, Pterocaulon virgatum*
Antirheumatic (External): *Barbula unguiculata, Bidens coronata, Bryum capillare, Chaptalia tomentosa, Dichanthelium laxiflorum, D. strigosum, Eleocharis geniculata, Eryngium yuccifolium, Erythrina herbacea, Juniperus virginiana, Liatris laxa, Octoblephorum albidum, Panax quinquefolius, Panicum sp., Pediomelum canescens, Persea borbonia, Phoradendron leucarpum, Pinus elliottii, Polygala lutea, P. rugelii, Quercus phellos, Q. virginiana, Salix caroliniana, Saururus cernuus, Verbesina virginica, Ximenia americana*
Antirheumatic (Internal): *Eryngium yuccifolium, Phytolacca americana, Salix caroliniana*
Blood Medicine: *Cephalanthus occidentalis, Chenopodium ambrosioides, Polygala lutea, P. rugelii, Salix caroliniana, Spiranthes sp., Stillingia sylvatica*
Cathartic: *Arundinaria gigantea, Cephalanthus occidentalis, Hypericum brachyphyllum, H. fasciculatum, Ricinus communis, Sassafras albidum*
Ceremonial Medicine: *Eryngium yuccifolium, Persea borbonia, Piloblephis rigida, Salix caroliniana, Sambucus canadensis, Sassafras albidum, Vaccinium myrsinites, Verbesina virginica, Vitis aestivalis, V. munsoniana*
Cold Remedy: *Conyza canadensis, Juniperus virginiana, Pediomelum canescens, Piloblephis rigida, Pityopsis graminifolia, Pterocaulon virgatum, Sassafras albidum, Vaccinium myrsinites, Xyris ambigua, X. sp.*
Cough Medicine: *Andropogon floridanus, Bacopa caroliniana, Conyza canadensis, Dichanthelium laxiflorum, D. strigosum, Hydrocotyle umbellata, Juniperus virginiana, Nymphaea sp., Nymphoides cordata, Panicum sp., Pediomelum canescens, Sassafras albidum*
Dermatological Aid: *Acer rubrum, Angadenia berteroi, Aster carolinianus, Callicarpa americana, Drosera capillaris, Eryngium yuccifolium, Ficus aurea, Hyptis pectinata, Iris sp., Ludwigia virgata, Mikania batatifolia, Panax quinquefolius, Paspalidium geminatum, Phragmites australis, Piloblephis rigida, Pinus elliottii, Quercus phellos, Q. virginiana, Rhus copallinum, Sagittaria lancifolia, Salix caroliniana, Sassafras albidum, Saururus cernuus*
Dietary Aid: *Eryngium yuccifolium, Galactia volubilis, Lechea minor, Liatris laxa, Persea borbonia, Sabal palmetto, Sassafras albidum, Stillingia sylvatica*
Emetic: *Eleocharis geniculata, Eryngium yuccifolium, Eupatorium perfoliatum, Juniperus virginiana, Persea borbonia, Phoradendron leucarpum, Piloblephis rigida, Salix caroliniana, Sambucus canadensis, Sassafras albidum, Saururus cernuus, Vaccinium myrsinites, Verbesina virginica, Vitis aestivalis, V. munsoniana*
Eye Medicine: *Bidens coronata, Juniperus virginiana, Persea borbonia, Sabatia campanulata, Salix caroliniana, S. humilis, Sassafras albidum, Vaccinium myrsinites, Verbesina virginica*
Febrifuge: *Acrostichum danaeifolium, Barbula unguiculata, Bidens coronata, Bryum capillare, Cephalanthus occidentalis, Cicuta maculata, Desmodium incanum, Eleocharis geniculata, Eryngium yuccifolium, Eupatorium perfoliatum, Galactia volubilis, Juniperus virginiana, Lechea minor, Myrica cerifera,*

Octoblephorum albidum, Persea borbonia, Piloblephis rigida, Pityopsis graminifolia, Pterocaulon virgatum, Rudbeckia hirta, Sabal palmetto, Sabatia campanulata, Salix caroliniana, S. humilis, Sassafras albidum, Saururus cernuus, Vaccinium myrsinites, Verbesina virginiana, Vitis aestivalis

Gastrointestinal Aid: *Andropogon floridanus, Cephalanthus occidentalis, Chenopodium ambrosioides, Desmodium incanum, Eryngium yuccifolium, Lechea minor, Liatris laxa, Licania michauxii, Myrica cerifera, Persea borbonia, Pinguicula lutea, P. pumila, Pterocaulon virgatum, Rubus trivialis, Salix caroliniana, Sambucus canadensis, Sassafras albidum, Verbesina virginiana, Vitis aestivalis*

Gynecological Aid: *Pterocaulon virgatum, Zeuxine strateumatica*

Heart Medicine: *Eryngium yuccifolium, Polygala lutea, P. rugelii*

Hemorrhoid Remedy: *Acer rubrum, Pinus elliottii, Quercus phellos, Q. virginiana*

Hemostat: *Tephrosia purpurea*

Hunting Medicine: *Salix caroliniana, S. humilis*

Kidney Aid: *Annona reticulata, Mitchella repens*

Laxative: *Cephalanthus occidentalis, Erythrina herbacea, Persea borbonia, Sassafras albidum, Verbesina virginiana*

Love Medicine: *Chrysobalanus icaco, Chrysophyllum oliviforme, Conyza canadensis, Myrica cerifera, Panax quinquefolius, Persea borbonia, Quercus phellos, Q. virginiana, Salix caroliniana, Sideroxylon foetidissimum*

Oral Aid: *Persea borbonia, Salix caroliniana, Sassafras albidum, Verbesina virginiana, Ximenia americana*

Orthopedic Aid: *Acer rubrum, Eryngium yuccifolium, Juniperus virginiana, Persea borbonia, Pinus elliottii, Pteridium caudatum, Pterocaulon virgatum, Quercus phellos, Q. virginiana, Salix caroliniana, Thelypteris kunthii, Ximenia americana*

Other: *Angadenia berteroi, Aster simmondsii, Bacopa caroliniana, Berchemia scandens, Cephalanthus occidentalis, Clematis baldwinii, Commelina erecta, Coreopsis leavenworthii, Juniperus virginiana, Licania michauxii, Nymphoides cordata, Osmunda regalis, Persea borbonia, Phlebodium aureum, Polygala lutea, Polypodium incanum, Pterocaulon virgatum, Rubus cuneifolius, Salix caroliniana, S. humilis, Sassafras albidum, Saururus cernuus, Smilax auriculata, Stillingia sylvatica, Vittaria lineata*

Panacea: *Eryngium yuccifolium, Persea borbonia*

Pediatric Aid: *Galactia volubilis, Juniperus virginiana, Lechea minor, Liatris laxa, Osmunda regalis, Panax quinquefolius, Persea borbonia, Phlebodium aureum, Phoradendron leucarpum, Piloblephis rigida, Sassafras albidum, Stenandrium dulce, Stillingia sylvatica, Vaccinium myrsinites, Vitis aestivalis, V. munsoniana, Vittaria lineata*

Preventive Medicine: *Salix caroliniana*

Psychological Aid: *Hyptis pectinata, Ilex vomitoria, Juniperus virginiana, Lagenaria siceraria, Licania michauxii, Osmunda regalis, Persea borbonia, Phlebodium aureum, Polypodium incanum, Thelypteris kunthii, Vittaria lineata*

Pulmonary Aid: *Andropogon floridanus, Chenopodium ambrosioides, Dichanthelium laxiflorum, D. strigosum, Panicum sp., Persea borbonia, Pterocaulon virgatum, Xyris ambigua, X. sp.*

Reproductive Aid: *Galactia volubilis, Justicia crassifolia, Licania michauxii, Persea borbonia, Trema micranthum, Zeuxine strateumatica*

Respiratory Aid: *Bacopa caroliniana, Conyza canadensis, Eryngium yuccifolium, Hydrocotyle umbellata, Nymphaea sp., Nymphoides cordata, Panax quinquefolius, Persea borbonia, Polygala lutea, P. rugelii, Salix caroliniana*

Sedative: *Bacopa caroliniana, Chenopodium ambrosioides, Hydrocotyle umbellata, Juniperus virginiana, Nymphaea sp., Nymphoides cordata, Panax quinquefolius, Persea borbonia, Stenandrium dulce*

Snakebite Remedy: *Eryngium yuccifolium, Manfreda virginica, Platanthera ciliaris, Polygala rugelii, Vitis munsoniana*

Stimulant: *Chenopodium ambrosioides, Eryngium yuccifolium, Galactia volubilis, Juniperus virginiana, Persea borbonia, Piloblephis rigida, Salix caroliniana, Stenandrium dulce, Vaccinium myrsinites*

Strengthener: *Cephalanthus occidentalis, Habenaria odontopetala, Salix caroliniana, Stillingia sylvatica*

Throat Aid: *Andropogon floridanus, Crotalaria rotundifolia, Dichanthelium laxiflorum, D. strigosum, Panicum sp., Sassafras albidum*

Tonic: *Panax quinquefolius*

Toothache Remedy: *Zephyranthes sp.*

Unspecified: *Cephalanthus occidentalis, Eleocharis geniculata, Eryngium yuccifolium, Galactia volubilis, Juniperus virginiana, Licania michauxii, Osmunda regalis, Panax quinquefolius, Pediomelum canescens, Persea borbonia, Phlebodium aureum, Pterocaulon virgatum, Salix caroliniana, Sassafras albidum, Saururus cernuus, Vaccinium myrsinites, Verbesina virginiana, Vittaria lineata, Ximenia americana*

Urinary Aid: *Andropogon floridanus, Callicarpa americana, Cephalanthus occidentalis, Chaptalia tomentosa, Eleocharis geniculata, Erythrina herbacea, Hypericum fasciculatum, Licania michauxii, Persea borbonia, Rhus copallinum, Sassafras albidum, Verbesina virginiana*

Venereal Aid: *Rhus copallinum*

Vertigo Medicine: *Eleocharis geniculata, Juniperus virginiana, Persea borbonia, Salix caroliniana*

Witchcraft Medicine: *Juniperus virginiana, Panax quinquefolius, Xyris sp.*

Food

Candy: *Ficus aurea*

Fodder: *Quercus virginiana*

Forage: *Licania michauxii*

Fruit: *Serenoa repens*

Spice: *Piloblephis rigida*

Starvation Food: *Sambucus canadensis*

Unspecified: *Allium cepa, Ananas comosus, Annona glabra, A. reticulata, Apios americana, Arachis hypogaea, Ardisia escallonoides, Brassica oleracea, Carica papaya, Carya sp., Celtis laevigata, Chrysobalanus icaco, Chrysophyllum oliviforme, Citrullus lanatus, Citrus aurantifolia, C. aurantium, C. limon, C. ×paradisi, C. reticulata, C. sinensis, Coccoloba diversifolia, Cocos nucifera, Colocasia esculenta, Cucumis melo, C. sativus, Cucurbita moschata, C. sp., Diospyros virginiana, Ficus aurea, Fortunella sp., Lycopersicon esculentum, Mangifera indica, Manihot esculenta, Morus rubra, Musa sp., Oryza sativa, Peltandra virginica, Prunus persica, Psidium guajava, Pyrus communis, Quercus phellos, Q. pumila, Sabal palmetto, Saccharum officinarum, Sideroxylon foetidissimum, Thalia geniculata, Vaccinium myrsinites, Vitis aestivalis, V. munsoniana, Xanthosoma atrovirens, Ximenia americana, Zamia pumila, Zea mays*

Vegetable: *Ipomoea batatas, Solanum tuberosum*

Fiber

Basketry: *Pinus elliottii, Serenoa repens*

Brushes & Brooms: *Serenoa repens*

Building Material: *Pinus elliottii, Sabal palmetto, Taxodium ascendens, Vitis aestivalis*

Canoe Material: *Taxodium ascendens*

Caulking Material: *Quercus virginiana, Sideroxylon foetidissimum*

Cordage: *Serenoa repens, Vitis aestivalis*

Furniture: *Pinus elliottii*

Dye

Unspecified: *Conocarpus erectus, Rhizophora mangle, Smilax laurifolia*

Other

Cash Crop: *Acrostichum danaeifolium, Serenoa repens*

Ceremonial Items: *Cissus verticillata, Cocos nucifera, Persea borbonia, Pinus elliottii, Serenoa repens, Taxodium ascendens, Zea mays*

Cleaning Agent: *Annona glabra, Myrica cerifera, Quercus phellos, Q. virginiana*

Containers: *Bursera simaruba, Taxodium ascendens*

Cooking Tools: *Acer rubrum, Annona glabra, Ardisia escallonoides, Catopsis sp., Cocos nucifera, Ficus aurea, Guzmania sp., Lagenaria siceraria, Myrcianthes*

[Cooking Tools]
fragrans, Persea borbonia, Sabal palmetto, Sideroxylon foetidissimum, S. salicifolium, Taxodium ascendens, Tillandsia sp., *T. usneoides*
Fasteners: *Ficus aurea, Pinus elliottii*
Fuel: *Conocarpus erectus, Quercus virginiana, Serenoa repens*
Hide Preparation: *Pinus elliottii, Quercus virginiana, Taxodium ascendens, Tillandsia usneoides*
Hunting & Fishing Item: *Acer rubrum, Ardisia escallonoides, Arundinaria gigantea, Carya* sp., *Chrysobalanus icaco, Cirsium horridulum, Cissus verticillata, Citrus aurantium, C. limon, Eugenia axillaris, Ficus aurea, Forestiera segregata, Ilex vomitoria, Maclura pomifera, Morus rubra, Myrcianthes fragrans, Ocotea coriacea, Psidium guajava, Sabal palmetto, Serenoa repens, Sideroxylon salicifolium, Solanum donianum, Taxodium ascendens, Tephrosia virginiana, Vitis shuttleworthii, Zanthoxylum fagara, Zea mays*
Jewelry: *Eleocharis equisetoides, Sapindus saponaria*
Lighting: *Pinus elliottii*
Musical Instrument: *Arundinaria gigantea, Canna flaccida, Taxodium ascendens*
Protection: *Juniperus virginiana*
Smoke Plant: *Myrica cerifera*
Smoking Tools: *Lyonia fruticosa*
Soap: *Ilex cassine*
Stable Gear: *Acer rubrum, Taxodium ascendens*
Tools: *Arundinaria gigantea, Celtis laevigata, Chamaecrista fasciculata, Cladium mariscus, Myrcianthes fragrans, Phragmites australis, Quercus virginiana, Sambucus canadensis, Serenoa repens, Sideroxylon foetidissimum, Taxodium ascendens*
Toys & Games: *Pinus elliottii, Quercus phellos, Sabal palmetto, Salix caroliniana, Sambucus canadensis, Serenoa repens, Taxodium ascendens*
Weapon: *Arundinaria gigantea*

Seri
Drug
Poison: *Sapium biloculare*
Unspecified: *Guajacum coulteri*
Food
Beverage: *Ferocactus coulteri, F. wislizeni*
Fruit: *Carnegia gigantea, Pachycereus pringlei, Stenocereus thurberi*
Porridge: *Acacia greggii, Carnegia gigantea, Olneya tesota, Pachycereus pringlei, Prosopis glandulosa, Stenocereus thurberi*
Staple: *Ruppia* sp.
Unspecified: *Hymenoclea monogyra, Jacquinia pungens*
Fiber
Basketry: *Acacia willardiana, Bursera laxiflora, B. microphylla, Caesalpinia gracilis, Jatropha cardiophylla, J. spatulata*
Building Material: *Fouquieria splendens*
Canoe Material: *Phragmites australis*
Caulking Material: *Carnegia gigantea, Pachycereus pringlei, Stenocereus thurberi*
Cordage: *Prosopis glandulosa*
Mats, Rugs & Bedding: *Prosopis glandulosa*
Sewing Material: *Prosopis glandulosa*
Other
Ceremonial Items: *Jacquinia pungens*
Hunting & Fishing Item: *Acacia willardiana, Prosopis glandulosa, Sapium biloculare*
Tools: *Carnegia gigantea, Pachycereus pringlei, Stenocereus thurberi*

Shasta
Drug
Analgesic: *Helianthus cusickii*
Carminative: *Helianthus cusickii*
Cold Remedy: *Clematis lasiantha*
Dermatological Aid: *Helianthus cusickii*
Disinfectant: *Helianthus cusickii*
Febrifuge: *Helianthus cusickii*
Herbal Steam: *Clematis lasiantha, Helianthus cusickii*
Preventive Medicine: *Helianthus cusickii*
Food
Bread & Cake: *Lithocarpus densiflorus, Pinus lambertiana, P. ponderosa, P. sabiniana, Quercus chrysolepis, Q. garryana, Q. kelloggii*
Dried Food: *Pinus lambertiana, P. ponderosa, P. sabiniana*
Fruit: *Oemleria cerasiformis*
Porridge: *Lithocarpus densiflorus, Quercus chrysolepis, Q. garryana, Q. kelloggii*
Soup: *Lithocarpus densiflorus, Quercus chrysolepis, Q. garryana, Q. kelloggii*
Staple: *Lithocarpus densiflorus, Quercus chrysolepis, Q. garryana, Q. kelloggii*
Unspecified: *Fritillaria recurva, Pinus lambertiana, P. ponderosa, P. sabiniana*
Other
Hunting & Fishing Item: *Amelanchier* sp., *Taxus brevifolia*
Season Indicator: *Ranunculus occidentalis*

Shinnecock
Drug
Analgesic: *Brassica nigra, Hedeoma pulegioides, Quercus alba, Tanacetum vulgare, Vitis* sp.
Antidiarrheal: *Rubus hispidus*
Antirheumatic (External): *Aesculus hippocastanum*
Antirheumatic (Internal): *Nepeta cataria*
Blood Medicine: *Acorus calamus*
Cancer Treatment: *Trifolium pratense*
Cold Remedy: *Allium cepa, Eupatorium perfoliatum, Prunus serotina, Verbascum thapsus*
Cough Medicine: *Barbarea vulgaris, Pinus strobus*
Dermatological Aid: *Alcea rosea, Comptonia peregrina, Impatiens capensis, Pinus rigida, Plantago major*
Diaphoretic: *Eupatorium perfoliatum*
Disinfectant: *Allium cepa*
Ear Medicine: *Allium cepa, Nicotiana tabacum*
Emetic: *Brassica nigra*
Eye Medicine: *Plantago major*
Febrifuge: *Allium cepa, Eupatorium perfoliatum*
Gastrointestinal Aid: *Prunus serotina*
Gynecological Aid: *Leonurus cardiaca*
Kidney Aid: *Gaultheria procumbens*
Liver Aid: *Berberis vulgaris*
Misc. Disease Remedy: *Allium cepa*
Oral Aid: *Acorus calamus*
Orthopedic Aid: *Quercus alba*
Pulmonary Aid: *Humulus lupulus*
Sedative: *Humulus lupulus*
Tonic: *Leucanthemum vulgare, Taraxacum officinale*
Toothache Remedy: *Brassica nigra, Nicotiana tabacum*
Urinary Aid: *Hibiscus moscheutos*

Shoshoni
Drug
Abortifacient: *Porophyllum gracile*
Adjuvant: *Angelica breweri*
Analgesic: *Achillea millefolium, Angelica breweri, Aquilegia formosa, Artemisia ludoviciana, A. tridentata, Chamaebatiaria millefolium, Claytonia perfoliata, Clematis ligusticifolia, Crepis acuminata, Eriastrum filifolium, Ericameria nana, Erigeron aphanactis, Eriodictyon angustifolium, Grindelia nana, G. squarrosa, Helianthella uniflora, Hypericum scouleri, Ipomopsis aggregata,*

Iris missouriensis, Iva axillaris, Machaeranthera canescens, Mahonia repens, Monardella odoratissima, Opuntia basilaris, Osmorhiza occidentalis, Penstemon deustus, P. eatonii, Pinus monophylla, Populus balsamifera, Psathyrotes ramosissima, Psorothamnus polydenius, Rumex crispus, Salix sp., Salvia dorrii, Stanleya pinnata, Zigadenus paniculatus

Anesthetic: *Achillea millefolium*

Anthelmintic: *Juniperus occidentalis, J. osteosperma, Nicotiana attenuata*

Anticonvulsive: *Anemopsis californica*

Antidiarrheal: *Achillea millefolium, Aquilegia formosa, Artemisia ludoviciana, Asclepias speciosa, Cercocarpus ledifolius, Chrysothamnus nauseosus, Enceliopsis nudicaulis, Ephedra viridis, Eriogonum sphaerocephalum, Heuchera rubescens, Ipomopsis congesta, Iva axillaris, Lithospermum ruderale, Mahonia repens, Osmorhiza occidentalis, Paeonia brownii, Phlox longifolia, Psathyrotes ramosissima, Psorothamnus polydenius, Rosa woodsii, Sambucus racemosa, Tanacetum vulgare, Tetradymia comosa*

Antidote: *Artemisia tridentata*

Antiemetic: *Achillea millefolium, Pinus monophylla, Prunus virginiana, Sphaeralcea ambigua, Stephanomeria spinosa*

Antihemorrhagic: *Purshia tridentata*

Antirheumatic (External): *Achillea millefolium, Angelica breweri, Arabis puberula, Arenaria congesta, Artemisia tridentata, Chrysothamnus viscidiflorus, Claytonia perfoliata, Clematis ligusticifolia, Eriastrum filifolium, Eriodictyon angustifolium, Eriogonum microthecum, E. umbellatum, Helianthella uniflora, Ipomopsis aggregata, Iris missouriensis, Lomatium dissectum, Lonicera interrupta, Nicotiana attenuata, Plantago major, Rumex crispus, Sphaeralcea ambigua, Stanleya pinnata, Urtica dioica, Veratrum californicum, Zigadenus paniculatus*

Antirheumatic (Internal): *Mahonia repens, Penstemon deustus*

Blood Medicine: *Arenaria congesta, Castilleja linariifolia, Caulanthus crassicaulis, Cercocarpus ledifolius, Corallorrhiza maculata, Ephedra viridis, Ipomopsis aggregata, Juniperus communis, Machaeranthera canescens, Mahonia repens, Monardella odoratissima, Populus balsamifera, Rosa woodsii, Rumex crispus, R. venosus, Sambucus racemosa, Urtica dioica*

Burn Dressing: *Abronia villosa, Argemone polyanthemos, Brassica sp., Cercocarpus ledifolius, Clematis ligusticifolia, Ephedra viridis, Iris missouriensis, Juniperus occidentalis, J. osteosperma, Paeonia brownii, Penstemon eatonii, Rosa woodsii, Rumex crispus, R. venosus*

Carminative: *Achillea millefolium, Mentha canadensis*

Cathartic: *Agastache urticifolia, Argemone hispida, A. polyanthemos, Artemisia dracunculus, A. ludoviciana, Atriplex canescens, Castilleja linariifolia, Cucurbita foetidissima, Ephedra viridis, Eriastrum filifolium, Hedeoma nana, Ipomopsis aggregata, I. congesta, Machaeranthera canescens, Monardella odoratissima, Nicotiana attenuata, Osmorhiza occidentalis, Psathyrotes ramosissima, Purshia mexicana, P. tridentata, Rumex crispus, R. venosus, Sphaeralcea ambigua, Tetradymia canescens, Wyethia mollis*

Cold Remedy: *Abies lasiocarpa, Achillea millefolium, Anemopsis californica, Angelica breweri, Artemisia dracunculus, A. ludoviciana, A. nova, A. tridentata, Cercocarpus ledifolius, Chrysothamnus nauseosus, Ephedra viridis, Ericameria nana, Eriodictyon angustifolium, Eriogonum ovalifolium, E. umbellatum, Gutierrezia sarothrae, Heracleum maximum, Ipomopsis congesta, Iva axillaris, Juniperus monosperma, J. occidentalis, J. osteosperma, Larrea tridentata, Lomatium dissectum, Mentha canadensis, Monardella odoratissima, Osmorhiza occidentalis, Penstemon deustus, Pinus monophylla, Psorothamnus polydenius, Rosa woodsii, Salvia dorrii, Sambucus racemosa, Sphaeralcea ambigua, Tetradymia comosa, Thamnosma montana, Urtica dioica, Veratrum californicum*

Contraceptive: *Juniperus sp., Lithospermum ruderale, Sphaeralcea ambigua, Veratrum californicum*

Cough Medicine: *Artemisia ludoviciana, A. nova, A. tridentata, Cercocarpus ledifolius, Chrysothamnus nauseosus, Enceliopsis nudicaulis, Ericameria nana, Eriodictyon angustifolium, Eriogonum microthecum, Grindelia nana, G. squarrosa, Heracleum maximum, Juniperus occidentalis, J. osteosperma, Lomatium dissectum, Mahonia repens, Osmorhiza occidentalis, Paeonia brownii, Pinus monophylla, Psathyrotes ramosissima, Psorothamnus polydenius, Sambucus racemosa, Tetradymia comosa*

Dermatological Aid: *Abies concolor, Abronia turbinata, Achillea millefolium, Anemopsis californica, Aquilegia formosa, Arenaria congesta, Argemone polyanthemos, Artemisia dracunculus, A. ludoviciana, A. spinescens, A. tridentata, Asclepias speciosa, Astragalus sp., Balsamorhiza sagittata, Cercocarpus ledifolius, Chaenactis douglasii, Clematis ligusticifolia, Ephedra nevadensis, E. viridis, Grindelia nana, G. squarrosa, Gutierrezia sarothrae, Hypericum scouleri, Ipomopsis aggregata, I. congesta, Iris missouriensis, Juniperus monosperma, J. occidentalis, J. osteosperma, Krameria grayi, Krascheninnikovia lanata, Leucocrinum montanum, Linum lewisii, Lomatium dissectum, Mimulus guttatus, Nicotiana attenuata, Opuntia basilaris, Osmorhiza occidentalis, Paeonia brownii, Penstemon deustus, Phlox longifolia, Pinus aristata, P. monophylla, Plantago major, Psathyrotes ramosissima, Purshia mexicana, P. tridentata, Rosa woodsii, Rubus leucodermis, Rumex crispus, R. venosus, Salix sp., Sphaeralcea ambigua, Tanacetum vulgare, Tetradymia comosa, Veratrum californicum, Wyethia amplexicaulis, Zigadenus paniculatus*

Diaphoretic: *Artemisia tridentata*

Disinfectant: *Anemopsis californica, Artemisia ludoviciana, A. tridentata, Grindelia nana, G. squarrosa, Gutierrezia sarothrae, Holodiscus dumosus, Ipomopsis aggregata, I. congesta, Juniperus occidentalis, J. osteosperma, Lomatium dissectum, Osmorhiza occidentalis, Penstemon deustus, Pinus monophylla, Plantago major, Psorothamnus polydenius, Purshia mexicana, Tanacetum vulgare, Veratrum californicum*

Diuretic: *Ephedra nevadensis, Heliotropium curassavicum, Juniperus occidentalis, J. osteosperma, Larrea tridentata, Lupinus sp., Rosa woodsii*

Ear Medicine: *Iris missouriensis, Penstemon deustus, Stanleya pinnata*

Emetic: *Aquilegia formosa, Argemone polyanthemos, Artemisia tridentata, Castilleja linariifolia, Chaenactis douglasii, Cucurbita foetidissima, Eriastrum filifolium, Grindelia nana, G. squarrosa, Heliotropium curassavicum, Holodiscus dumosus, Ipomopsis aggregata, I. congesta, Nicotiana attenuata, Psathyrotes ramosissima, Purshia tridentata, Ribes cereum, Sphaeralcea ambigua, Wyethia amplexicaulis, Zigadenus paniculatus*

Expectorant: *Eriodictyon angustifolium, Grindelia nana, G. squarrosa*

Eye Medicine: *Amelanchier sp., Arenaria aculeata, Argemone polyanthemos, Artemisia dracunculus, A. ludoviciana, A. tridentata, Astragalus sp., Balsamorhiza sagittata, Cercocarpus ledifolius, Chamaesyce polycarpa, Cicuta maculata, Crepis acuminata, C. modocensis, Ericameria nana, Erigeron aphanactis, Frasera albomarginata, Ipomopsis congesta, Krascheninnikovia lanata, Leptodactylon pungens, Leymus condensatus, Linum lewisii, Lomatium dissectum, Machaeranthera canescens, Maianthemum stellatum, Osmorhiza occidentalis, Paeonia brownii, Penstemon deustus, Phlox longifolia, Physaria chambersii, Prunus virginiana, Sphaeralcea ambigua, Stephanomeria spinosa*

Febrifuge: *Artemisia ludoviciana, A. tridentata, Heuchera rubescens, Mentha canadensis, Osmorhiza occidentalis, Pinus monophylla*

Gastrointestinal Aid: *Achillea millefolium, Anemopsis californica, Aquilegia formosa, Artemisia ludoviciana, A. tridentata, Brickellia oblongifolia, Chaenactis douglasii, Chamaebatiaria millefolium, Chrysothamnus nauseosus, Clematis ligusticifolia, Ephedra viridis, Ericameria nana, Erigeron aphanactis, Eriodictyon angustifolium, Grindelia nana, G. squarrosa, Hedeoma nana, Holodiscus dumosus, Ipomopsis congesta, Iris missouriensis, Iva axillaris, Maianthemum stellatum, Mentha canadensis, Monardella odoratissima, Osmorhiza occidentalis, Pedicularis centranthera, Penstemon deustus, Phlox longifolia, Pinus monophylla, Plantago major, Prunus virginiana, Psathyrotes annua, P. ramosissima, Psorothamnus fremontii, P. polydenius, Salix sp., Salvia dorrii, Tetradymia comosa*

Gland Medicine: *Veratrum californicum*
Gynecological Aid: *Artemisia ludoviciana, A. tridentata, Crepis acuminata, C. modocensis, Maianthemum stellatum, Osmorhiza occidentalis, Veratrum californicum*
Hallucinogen: *Datura wrightii*
Heart Medicine: *Cercocarpus ledifolius, Heuchera rubescens, Juniperus occidentalis, J. osteosperma*
Hemostat: *Artemisia spinescens*
Herbal Steam: *Artemisia dracunculus, Lomatium dissectum, Prunus virginiana*
Kidney Aid: *Chaenactis douglasii, Clematis ligusticifolia, Ephedra viridis, Equisetum* sp., *Grindelia squarrosa, Ipomopsis congesta, Juniperus occidentalis, J. osteosperma, Mahonia repens, Paeonia brownii, Pinus monophylla, Plantago major, Psorothamnus polydenius, Purshia mexicana, Suaeda moquinii*
Laxative: *Psathyrotes ramosissima*
Liver Aid: *Heuchera rubescens, Ipomopsis congesta, Linum lewisii, Psathyrotes ramosissima, Pyrola asarifolia, Rumex crispus*
Misc. Disease Remedy: *Artemisia ludoviciana, A. spinescens, Cercocarpus ledifolius, Chrysothamnus viscidiflorus, Grindelia nana, G. squarrosa, Gutierrezia sarothrae, Heliotropium curassavicum, Ipomopsis congesta, Juniperus monosperma, J. occidentalis, J. osteosperma, Lomatium dissectum, Osmorhiza occidentalis, Pinus monophylla, Psorothamnus polydenius, Purshia mexicana, P. tridentata, Wyethia amplexicaulis*
Narcotic: *Datura wrightii*
Oral Aid: *Juniperus occidentalis, J. osteosperma*
Orthopedic Aid: *Achillea millefolium, Artemisia tridentata, Cicuta maculata, Clematis ligusticifolia, Eriogonum microthecum, E. umbellatum, Grindelia nana, G. squarrosa, Hypericum scouleri, Lomatium dissectum, Penstemon deustus, Pinus monophylla, Ribes aureum, Zigadenus paniculatus*
Other: *Purshia tridentata, Salvia dorrii*
Panacea: *Frasera speciosa, Lomatium dissectum*
Pediatric Aid: *Angelica breweri, Artemisia ludoviciana, A. tridentata, Cercocarpus ledifolius, Ephedra viridis, Gilia* sp., *Iva axillaris, Mentha canadensis, Pedicularis centranthera, Penstemon deustus, Phlox longifolia, Psathyrotes annua, Salvia dorrii*
Poison: *Cicuta maculata*
Poultice: *Pinus monophylla*
Pulmonary Aid: *Abies concolor, Angelica breweri, Artemisia spinescens, A. tridentata, Corallorrhiza maculata, Eriodictyon angustifolium, Lomatium dissectum, Osmorhiza occidentalis, Pinus monophylla, Psorothamnus polydenius*
Respiratory Aid: *Lomatium dissectum*
Snakebite Remedy: *Chamaesyce albomarginata, Osmorhiza occidentalis, Psathyrotes ramosissima, Veratrum californicum, V.* sp.
Stimulant: *Aquilegia formosa, Catabrosa aquatica*
Throat Aid: *Artemisia dracunculus, A. tridentata, Heliotropium curassavicum, Heracleum maximum, Juniperus occidentalis, J. osteosperma, Lomatium dissectum, Osmorhiza occidentalis, Paeonia brownii, Salvia dorrii, Veratrum californicum*
Tonic: *Anemopsis californica, Angelica breweri, Artemisia tridentata, Catabrosa aquatica, Chamaesyce albomarginata, C. polycarpa, Chrysothamnus nauseosus, Ephedra viridis, Frasera speciosa, Heuchera rubescens, Ipomopsis aggregata, Juniperus communis, J. occidentalis, J. osteosperma, Machaeranthera canescens, Monardella odoratissima, Osmorhiza occidentalis, Populus balsamifera, Purshia tridentata, Rosa woodsii, Rumex crispus, R. venosus, Stephanomeria spinosa, Thamnosma montana, Urtica dioica*
Toothache Remedy: *Achillea millefolium, Argemone polyanthemos, Artemisia tridentata, Astragalus* sp., *Chrysothamnus viscidiflorus, Heracleum maximum, Hypericum scouleri, Iris missouriensis, Juniperus occidentalis, J. osteosperma, Nicotiana attenuata, Osmorhiza occidentalis, Salix* sp., *Stanleya pinnata, Zigadenus paniculatus*
Tuberculosis Remedy: *Angelica breweri, Cercocarpus ledifolius, Eriodictyon angustifolium, Eriogonum microthecum, Glossopetalon spinescens, Heracleum maximum, Lomatium dissectum, Nicotiana attenuata, Paeonia brownii, Populus balsamifera, Psathyrotes ramosissima, Psorothamnus polydenius, Sambucus racemosa*
Unspecified: *Cercocarpus ledifolius, Holodiscus dumosus, Ribes aureum, Sphaeralcea ambigua*
Urinary Aid: *Ephedra viridis, Grindelia nana, G. squarrosa, Psathyrotes annua, Psorothamnus polydenius, Suaeda moquinii*
Venereal Aid: *Anemopsis californica, Angelica breweri, Aquilegia formosa, Arctostaphylos patula, Arenaria congesta, Artemisia dracunculus, Asclepias speciosa, Astragalus calycosus, Balsamorhiza sagittata, Castilleja linariifolia, Cercocarpus ledifolius, Clematis ligusticifolia, Cordylanthus ramosus, Cucurbita foetidissima, Enceliopsis nudicaulis, Ephedra nevadensis, E. viridis, Eriastrum filifolium, Erigeron concinnus, Eriodictyon angustifolium, Grindelia nana, G. squarrosa, Heliotropium curassavicum, Heuchera rubescens, Holodiscus dumosus, Hypericum scouleri, Ipomopsis aggregata, I. congesta, Iris missouriensis, Juniperus occidentalis, J. osteosperma, J. scopulorum, Larrea tridentata, Lomatium dissectum, Mahonia repens, Maianthemum stellatum, Osmorhiza occidentalis, Paeonia brownii, Penstemon deustus, Pinus monophylla, Populus balsamifera, P. tremuloides, Psathyrotes ramosissima, Psorothamnus polydenius, Purshia mexicana, P. tridentata, Rumex crispus, R. venosus, Salix* sp., *Sphaeralcea ambigua, Stephanomeria tenuifolia, Tetradymia canescens, Thalictrum fendleri, Wyethia amplexicaulis*
Veterinary Aid: *Angelica breweri, Asclepias cryptoceras, Cucurbita foetidissima, Heuchera rubescens, Ipomopsis congesta, Lomatium dissectum, Pinus monophylla, Sphaeralcea ambigua, Veratrum* sp.

Food
Beverage: *Ephedra* sp., *Lithospermum incisum, Rhus trilobata*
Candy: *Asclepias speciosa*
Spice: *Allium falcifolium, Apium* sp., *Artemisia dracunculus, Astragalus* sp., *Osmorhiza occidentalis*
Starvation Food: *Leymus condensatus*
Unspecified: *Madia elegans, Pinus monophylla*
Winter Use Food: *Rhus trilobata*

Fiber
Basketry: *Epicampes* sp., *Martynia* sp., *Salix lasiolepis, S. lucida, Yucca brevifolia*
Cordage: *Apocynum* sp., *Fremontodendron* sp.
Sewing Material: *Muhlenbergia rigens*

Dye
Black: *Carex exsiccata, Proboscidea louisianica, Yucca brevifolia*
Green: *Juncus mexicanus*
Red: *Yucca brevifolia*

Other
Cash Crop: *Camassia quamash, Nicotiana attenuata*
Containers: *Clematis ligusticifolia, Sphaeralcea ambigua*
Fasteners: *Fremontodendron californicum*
Hunting & Fishing Item: *Cercocarpus ledifolius, Clematis ligusticifolia, Symphoricarpos albus*
Musical Instrument: *Equisetum arvense*
Smoke Plant: *Arenaria* sp.

Shuswap
Drug
Analgesic: *Aster* sp., *Betula papyrifera, Lysichiton americanus, Nuphar lutea, Urtica dioica, U. urens*
Antidiarrheal: *Heuchera cylindrica, Rubus arcticus, Spiraea betulifolia*
Antirheumatic (External): *Nuphar lutea, Urtica dioica, U. urens*
Blood Medicine: *Achillea millefolium, Mahonia repens, Maianthemum racemosum*
Cathartic: *Shepherdia canadensis*

Cold Remedy: *Artemisia campestris, A. tridentata, Matricaria discoidea, Phacelia linearis*
Cough Medicine: *Artemisia campestris, Pinus contorta*
Dermatological Aid: *Abies grandis, Achillea millefolium, Alnus incana, Artemisia campestris, Balsamorhiza sagittata, Heracleum maximum, Heuchera cylindrica, Ledum groenlandicum, Lesquerella douglasii, Lithospermum ruderale, Lomatium dissectum, Lysichiton americanus, Nuphar lutea, Opuntia fragilis, Picea glauca, Pinus ponderosa, Plantago major, Populus balsamifera, Urtica dioica, U. urens, Veratrum viride*
Diaphoretic: *Alnus incana, Juniperus communis, J. scopulorum, Lesquerella douglasii, Plantago major*
Disinfectant: *Artemisia tridentata, Lithospermum ruderale*
Eye Medicine: *Arnica cordifolia, Castilleja angustifolia, Ledum groenlandicum, Rudbeckia hirta, Symphoricarpos mollis*
Gastrointestinal Aid: *Cirsium undulatum, Shepherdia canadensis, Spiraea betulifolia*
Gynecological Aid: *Artemisia dracunculus, Solidago canadensis*
Heart Medicine: *Matricaria discoidea*
Herbal Steam: *Urtica dioica, U. urens*
Internal Medicine: *Heracleum maximum*
Kidney Aid: *Cornus sericea*
Laxative: *Frangula purshiana*
Misc. Disease Remedy: *Ceanothus velutinus, Juniperus scopulorum*
Orthopedic Aid: *Lysichiton americanus, Nuphar lutea, Zigadenus venenosus*
Panacea: *Abies grandis, Artemisia campestris, Juniperus communis, J. scopulorum, Lysichiton americanus, Picea glauca, Pinus ponderosa, Ribes lacustre*
Pediatric Aid: *Cornus sericea, Pinus ponderosa*
Poison: *Cicuta douglasii, Galium boreale, Platanthera dilatata, Sium suave, Veratrum viride, Zigadenus venenosus*
Stimulant: *Artemisia dracunculus, Pinus ponderosa*
Throat Aid: *Opuntia fragilis*
Toothache Remedy: *Abies grandis, Picea glauca*
Tuberculosis Remedy: *Abies grandis, Artemisia campestris, Picea glauca, Pinus contorta, P. monticola, Shepherdia canadensis, Tsuga heterophylla*
Unspecified: *Alnus incana, Shepherdia canadensis*
Urinary Aid: *Cornus sericea, Heracleum maximum, Nicotiana attenuata, Penstemon fruticosus*
Witchcraft Medicine: *Artemisia dracunculus, A. tridentata*
Food
Beverage: *Abies grandis, Ledum groenlandicum, Mentha arvensis, Prunus virginiana, Rosa nutkana, Shepherdia canadensis, Urtica dioica, U. urens*
Candy: *Abies grandis*
Dessert: *Shepherdia canadensis*
Dried Food: *Fragaria virginiana*
Fodder: *Medicago sativa, Phleum pratense, Trifolium pratense*
Forage: *Allium cernuum, Calochortus macrocarpus*
Fruit: *Amelanchier alnifolia, Cornus sericea, Disporum trachycarpum, Mahonia repens, Prunus emarginata, Ribes hudsonianum, R. lacustre, Rubus idaeus, R. leucodermis, R. parviflorus, Shepherdia canadensis, Vaccinium cespitosum, V. membranaceum, V. ovalifolium, Viburnum opulus*
Ice Cream: *Shepherdia canadensis*
Preservative: *Cornus sericea, Salix scouleriana*
Preserves: *Crataegus douglasii, Viburnum opulus*
Spice: *Allium cernuum, Angelica arguta, Cornus sericea, Lomatium macrocarpum*
Sweetener: *Pseudotsuga menziesii*
Unspecified: *Abies lasiocarpa, Argentina anserina, Balsamorhiza sagittata, Calochortus macrocarpus, Cirsium undulatum, Corylus cornuta, Fritillaria lanceolata, F. pudica, Heracleum maximum, Lewisia rediviva, Lilium columbianum, Lomatium dissectum, L. macrocarpum, Opuntia fragilis, Pinus contorta, P. monticola, Sium suave*
Vegetable: *Angelica arguta, Chenopodium album*
Winter Use Food: *Erythronium grandiflorum, Prunus virginiana*
Fiber
Basketry: *Betula papyrifera, Prunus emarginata, Thuja plicata*
Building Material: *Aulacomnium* sp., *Dicranum* sp., *Lycopodium annotinum, Scirpus acutus*
Canoe Material: *Populus balsamifera*
Cordage: *Apocynum cannabinum*
Mats, Rugs & Bedding: *Abies lasiocarpa, Pteridium aquilinum, Typha latifolia*
Snow Gear: *Acer circinatum, Philadelphus lewisii*
Dye
Unspecified: *Alnus incana*
Other
Cash Crop: *Abies lasiocarpa, Thuja plicata*
Ceremonial Items: *Delphinium nuttallianum, Salix scouleriana, Shepherdia canadensis, Thuja plicata*
Containers: *Heracleum maximum, Rubus parviflorus*
Cooking Tools: *Cornus sericea, Typha latifolia*
Decorations: *Epilobium angustifolium, Fritillaria pudica*
Fasteners: *Salix scouleriana*
Fertilizer: *Aulacomnium* sp., *Dicranum* sp., *Lycopodium annotinum*
Fuel: *Abies lasiocarpa, Pinus ponderosa*
Hide Preparation: *Pinus ponderosa, Populus balsamifera*
Hunting & Fishing Item: *Amelanchier alnifolia, Juniperus scopulorum, Philadelphus lewisii*
Insecticide: *Artemisia dracunculus, A. frigida, Ceanothus velutinus, Juniperus scopulorum*
Jewelry: *Thuja plicata*
Paint: *Prunus virginiana*
Protection: *Athyrium filix-femina, Pteridium aquilinum, Rosa nutkana*
Smoke Plant: *Arctostaphylos uva-ursi, Cornus sericea, Nicotiana attenuata*
Smoking Tools: *Symphoricarpos mollis*
Soap: *Philadelphus lewisii*
Stable Gear: *Apocynum cannabinum*
Tools: *Equisetum arvense*
Toys & Games: *Populus tremuloides*

Sia
Drug
Febrifuge: *Mentha canadensis*
Veterinary Aid: *Ipomoea leptophylla*
Food
Fruit: *Yucca baccata*
Staple: *Triticum aestivum, Zea mays*
Unspecified: *Amaranthus cruentus, Capsicum annuum, Citrullus lanatus, Cleome serrulata, Cucumis melo, Cucurbita maxima, C. moschata, Opuntia engelmannii, Pinus edulis*
Vegetable: *Amaranthus cruentus, Cleome serrulata, Phaseolus acutifolius, P. vulgaris, Solanum fendleri, S. jamesii, S. tuberosum, Vicia faba*

Sierra
Food
Staple: *Pteridium aquilinum*

Sikani
Drug
Analgesic: *Heracleum maximum*
Anthelmintic: *Populus tremuloides*
Antirheumatic (External): *Heracleum maximum*
Cathartic: *Sambucus racemosa*

Cough Medicine: *Picea sitchensis, Pinus contorta*
Dermatological Aid: *Abies* sp., *Heracleum maximum, Populus tremuloides, Salix* sp.
Eye Medicine: *Rosa* sp.

Similkameen
Food
Unspecified: *Fritillaria* sp.

Sioux
Drug
Abortifacient: *Artemisia frigida*
Adjuvant: *Prunus virginiana*
Analgesic: *Echinacea pallida*
Antidiarrheal: *Prunus virginiana, Shepherdia canadensis*
Antidote: *Echinacea pallida*
Antihemorrhagic: *Rhus glabra*
Burn Dressing: *Echinacea pallida, Typha latifolia*
Cathartic: *Shepherdia canadensis*
Ceremonial Medicine: *Prunus virginiana*
Cold Remedy: *Juniperus scopulorum*
Dermatological Aid: *Opuntia polyacantha, Rhus glabra, Typha latifolia*
Diuretic: *Symphoricarpos albus, Urtica dioica*
Ear Medicine: *Glycyrrhiza lepidota, Polygala alba*
Eye Medicine: *Lygodesmia juncea, Symphoricarpos albus*
Febrifuge: *Glycyrrhiza lepidota, Juniperus scopulorum*
Gastrointestinal Aid: *Grindelia squarrosa, Monarda fistulosa*
Gynecological Aid: *Lygodesmia juncea, Monarda fistulosa*
Hemostat: *Nuphar lutea, Prunus virginiana*
Kidney Aid: *Grindelia squarrosa*
Misc. Disease Remedy: *Juniperus scopulorum, Typha latifolia*
Pediatric Aid: *Glycyrrhiza lepidota*
Poison: *Shepherdia canadensis*
Pulmonary Aid: *Juniperus scopulorum*
Toothache Remedy: *Echinacea pallida, Glycyrrhiza lepidota*
Urinary Aid: *Rhus glabra*
Veterinary Aid: *Echinacea pallida, Glycyrrhiza lepidota*
Food
Dried Food: *Vitis vulpina*
Fruit: *Symphoricarpos occidentalis, Vitis vulpina*
Sauce & Relish: *Polygonum amphibium*
Soup: *Pediomelum esculentum*
Staple: *Acer negundo*
Unspecified: *Helianthus maximiliani, Lygodesmia juncea, Scirpus acutus, Typha latifolia*
Winter Use Food: *Pediomelum esculentum*
Dye
Mordant: *Opuntia polyacantha*
Other
Ceremonial Items: *Acer negundo, Fraxinus pennsylvanica, Hierochloe odorata*
Hunting & Fishing Item: *Fraxinus pennsylvanica, Mentha arvensis*
Protection: *Echinacea pallida*
Smoke Plant: *Artemisia ludoviciana*
Soap: *Artemisia frigida*
Toys & Games: *Equisetum hyemale*

Sioux, Fort Peck
Drug
Abortifacient: *Acorus calamus*
Panacea: *Acorus calamus*

Sioux, Teton
Drug
Analgesic: *Echinacea angustifolia, Erysimum asperum*
Antidiarrheal: *Heuchera americana*
Cold Remedy: *Monarda fistulosa*
Dermatological Aid: *Mirabilis nyctaginea*
Febrifuge: *Monarda fistulosa*
Gastrointestinal Aid: *Echinacea angustifolia, Erysimum asperum*
Orthopedic Aid: *Mirabilis nyctaginea*
Pulmonary Aid: *Lithospermum incisum*
Throat Aid: *Echinacea angustifolia*
Toothache Remedy: *Echinacea angustifolia*

Skagit
Drug
Anthelmintic: *Dicentra formosa*
Antidiarrheal: *Achillea millefolium, Frangula purshiana*
Antidote: *Symphoricarpos albus*
Antirheumatic (External): *Petasites frigidus, Pinus monticola*
Cold Remedy: *Aruncus dioicus, Prunus emarginata, Ribes laxiflorum, Urtica dioica, Vaccinium parvifolium*
Cough Medicine: *Thuja plicata*
Dermatological Aid: *Achlys triphylla, Aruncus dioicus, Dicentra formosa, Eriophyllum lanatum, Frangula purshiana, Heuchera micrantha, Pinus monticola, Pseudotsuga menziesii, Rubus parviflorus, Tsuga heterophylla*
Dietary Aid: *Asarum caudatum, Tellima grandiflora*
Disinfectant: *Pseudotsuga menziesii*
Eye Medicine: *Ribes lacustre, Rosa nutkana, Trillium ovatum, Tsuga heterophylla*
Gastrointestinal Aid: *Rubus ursinus*
Gynecological Aid: *Oplopanax horridus, Prunus emarginata, Ribes lacustre*
Laxative: *Frangula purshiana*
Oral Aid: *Symphoricarpos albus*
Panacea: *Tellima grandiflora*
Pediatric Aid: *Symphoricarpos albus*
Poison: *Trillium ovatum*
Throat Aid: *Aruncus dioicus, Claytonia sibirica, Rosa nutkana, Tsuga heterophylla*
Tonic: *Asarum caudatum, Claytonia sibirica, Gaultheria shallon, Lonicera ciliosa, Salix sitchensis*
Toothache Remedy: *Dicentra formosa*
Tuberculosis Remedy: *Achlys triphylla, Asarum caudatum, Oplopanax horridus, Petasites frigidus, Pinus monticola, Symphoricarpos albus*
Venereal Aid: *Mahonia nervosa*
Food
Beverage: *Rosa nutkana*
Bread & Cake: *Gaultheria shallon*
Dried Food: *Rubus ursinus*
Fruit: *Amelanchier alnifolia, Mahonia nervosa, Oemleria cerasiformis, Ribes laxiflorum, Rubus parviflorus, R. ursinus, Vaccinium parvifolium*
Preserves: *Mahonia nervosa*
Unspecified: *Corylus cornuta, Lilium columbianum, Pteridium aquilinum, Rosa nutkana, Rubus parviflorus*
Winter Use Food: *Sambucus cerulea, S. racemosa*
Fiber
Canoe Material: *Acer macrophyllum, Pinus monticola, Tsuga heterophylla*
Furniture: *Acer circinatum*
Dye
Green: *Frangula purshiana*
Unspecified: *Mahonia nervosa*
Yellow: *Mahonia aquifolium*

Other
Ceremonial Items: *Thuja plicata*
Cooking Tools: *Acer macrophyllum, Holodiscus discolor*
Fuel: *Pseudotsuga menziesii*
Hunting & Fishing Item: *Cornus nuttallii, Holodiscus discolor, Philadelphus lewisii, Pseudotsuga menziesii, Tsuga heterophylla*
Tools: *Acer circinatum*
Toys & Games: *Cornus nuttallii*

Skagit, Upper
Drug
Cold Remedy: *Ribes laxiflorum, Vaccinium parvifolium*
Dermatological Aid: *Polypodium virginianum*
Expectorant: *Polypodium virginianum*
Gastrointestinal Aid: *Rubus ursinus*
Gynecological Aid: *Ribes lacustre*
Laxative: *Frangula purshiana, Polypodium virginianum*
Other: *Gaultheria shallon*
Throat Aid: *Claytonia sibirica, Ribes divaricatum, Rosa nutkana*
Food
Beverage: *Rosa nutkana*
Dried Food: *Amelanchier alnifolia, Fragaria* sp., *Gaultheria shallon, Mahonia aquifolium, M. nervosa, Rubus leucodermis, R. ursinus, Sambucus cerulea, S. racemosa, Vaccinium membranaceum*
Fruit: *Amelanchier alnifolia, Fragaria* sp., *Gaultheria shallon, Mahonia aquifolium, M. nervosa, Maianthemum racemosum, Malus fusca, Oemleria cerasiformis, Ribes divaricatum, R. lacustre, R. laxiflorum, R. sanguineum, Rubus leucodermis, R. parviflorus, R. spectabilis, R. ursinus, Sambucus cerulea, S. racemosa, Vaccinium membranaceum, V. parvifolium*
Spice: *Rosa nutkana*
Unspecified: *Alnus rubra, Camassia quamash, Corylus cornuta, Equisetum* sp., *Lilium columbianum, Perideridia gairdneri, Pteridium aquilinum, Rubus parviflorus, R. spectabilis, Urtica dioica*
Winter Use Food: *Corylus cornuta*
Fiber
Cordage: *Urtica dioica*
Other
Containers: *Acer macrophyllum*
Cooking Tools: *Alnus rubra*
Fuel: *Alnus rubra*
Smoke Plant: *Arctostaphylos uva-ursi*

Skokomish
Drug
Analgesic: *Lysichiton americanus*
Antidiarrheal: *Fragaria* sp.
Cathartic: *Lysichiton americanus*
Cold Remedy: *Arbutus menziesii, Prunus emarginata, Rhododendron albiflorum*
Dermatological Aid: *Adiantum aleuticum, Claytonia sibirica, Lysichiton americanus, Rhododendron albiflorum, Urtica dioica*
Febrifuge: *Lysichiton americanus*
Gastrointestinal Aid: *Arbutus menziesii, Rhododendron albiflorum*
Gynecological Aid: *Prunus emarginata*
Oral Aid: *Thuja plicata*
Throat Aid: *Arbutus menziesii, Rhododendron albiflorum*
Tuberculosis Remedy: *Epilobium angustifolium, Ribes laxiflorum*
Food
Bread & Cake: *Gaultheria shallon*
Fruit: *Arctostaphylos uva-ursi, Fragaria* sp., *Sambucus cerulea, Vaccinium parvifolium*
Unspecified: *Lilium columbianum, Lysichiton americanus, Oenanthe sarmentosa, Pteridium aquilinum, Rosa nutkana*
Winter Use Food: *Ranunculus* sp., *Sambucus racemosa*
Fiber
Cordage: *Corylus cornuta*
Mats, Rugs & Bedding: *Epilobium angustifolium*
Scouring Material: *Equisetum telmateia*
Other
Hunting & Fishing Item: *Urtica dioica*

Snake
Drug
Poison: *Valeriana edulis*

Snohomish
Drug
Analgesic: *Vicia sativa*
Antidiarrheal: *Achillea millefolium*
Cold Remedy: *Linnaea borealis, Urtica dioica*
Dermatological Aid: *Claytonia sibirica, Dryopteris expansa, Geum macrophyllum, Philadelphus lewisii*
Eye Medicine: *Cornus sericea*
Poison: *Conium maculatum*
Throat Aid: *Epilobium angustifolium, Rosa pisocarpa*
Food
Bread & Cake: *Gaultheria shallon*
Fruit: *Mahonia aquifolium, Oemleria cerasiformis, Rubus parviflorus, Vaccinium parvifolium*
Unspecified: *Corylus cornuta, Pteridium aquilinum, Rosa nutkana*
Winter Use Food: *Sambucus racemosa*
Fiber
Basketry: *Prunus emarginata, Tsuga heterophylla, Typha latifolia*
Canoe Material: *Acer macrophyllum, Tsuga heterophylla*
Clothing: *Typha latifolia*
Cordage: *Salix sitchensis*
Mats, Rugs & Bedding: *Scirpus acutus, Typha latifolia*
Sewing Material: *Typha latifolia*
Dye
Red-Brown: *Alnus rubra*
Unspecified: *Mahonia nervosa*
Yellow: *Mahonia aquifolium*
Other
Cooking Tools: *Acer macrophyllum, Cornus nuttallii, Holodiscus discolor*
Hunting & Fishing Item: *Holodiscus discolor, Taxus brevifolia, Urtica dioica*
Toys & Games: *Amelanchier alnifolia, Fomes* sp.

Snuqualmie
Food
Unspecified: *Juncus effusus, Oenanthe sarmentosa*
Fiber
Building Material: *Thuja plicata*
Cordage: *Juncus effusus*

Songish
Drug
Abortifacient: *Carex* sp.
Cold Remedy: *Lomatium nudicaule*
Gynecological Aid: *Sedum spathulifolium*
Love Medicine: *Osmorhiza purpurea*
Throat Aid: *Lomatium nudicaule*
Other
Ceremonial Items: *Lomatium nudicaule*
Hunting & Fishing Item: *Apocynum cannabinum*

Southwest Indians
Food
Beverage: *Yucca baccata, Y. schidigera*
Bread & Cake: *Dasylirion texanum, D. wheeleri, Yucca baccata*
Dessert: *Yucca baccata*
Dried Food: *Yucca baccata, Y. elata, Y. schidigera*
Fruit: *Carnegia gigantea, Yucca baccata, Y. brevifolia, Y. schidigera, Y. torreyi*
Preserves: *Yucca baccata*
Sauce & Relish: *Yucca glauca*
Staple: *Yucca baccata*
Starvation Food: *Yucca angustissima, Y. glauca, Y. harrimaniae*
Sweetener: *Yucca baccata*
Unspecified: *Opuntia* sp., *Prosopis glandulosa, Yucca baccata, Y. glauca*
Fiber
Basketry: *Nolina microcarpa, N.* sp., *Rhus trilobata, Yucca glauca*
Brushes & Brooms: *Yucca glauca*
Clothing: *Cowania* sp., *Yucca brevifolia, Y. schidigera*
Cordage: *Nolina microcarpa, Yucca angustissima, Y. baccata, Y. glauca*
Mats, Rugs & Bedding: *Nolina microcarpa, N.* sp.
Other
Ceremonial Items: *Yucca angustissima, Y. baccata*
Containers: *Carnegia gigantea*
Designs: *Yucca brevifolia*
Fasteners: *Yucca elata*
Soap: *Yucca baccata, Y. glauca*
Tools: *Yucca glauca*

Spanish American
Food
Beverage: *Ribes leptanthum*
Candy: *Hymenoxys richardsonii*
Dried Food: *Monarda fistulosa*
Preserves: *Mahonia haematocarpa, Prunus virginiana, Ribes leptanthum*
Spice: *Monarda fistulosa*
Substitution Food: *Solanum elaeagnifolium*
Vegetable: *Amaranthus blitoides, A. retroflexus, Chenopodium album, C. leptophyllum*
Other
Cash Crop: *Pinus edulis*

Spokan
Food
Fruit: *Amelanchier* sp., *Arctostaphylos uva-ursi, Cornus sericea, Crataegus* sp., *Fragaria vesca, Mahonia* sp., *Prunus virginiana, Ribes* sp., *Rosa* sp., *Rubus leucodermis, R.* sp., *Sambucus* sp., *Shepherdia canadensis, Vaccinium membranaceum*
Unspecified: *Alectoria jubata, Allium* sp., *Balsamorhiza* sp., *Camassia scilloides, Cirsium undulatum, Claytonia* sp., *Fritillaria pudica, Heracleum maximum, Lewisia rediviva, Opuntia* sp., *Pinus albicaulis, P. contorta, P. ponderosa*

Squaxin
Drug
Analgesic: *Marah oreganus, Vicia americana*
Blood Medicine: *Holodiscus discolor, Mahonia aquifolium, Sambucus racemosa*
Cold Remedy: *Pseudotsuga menziesii*
Dermatological Aid: *Achillea millefolium, Adenocaulon bicolor, Frangula purshiana, Populus balsamifera*
Disinfectant: *Populus balsamifera, Sambucus racemosa*
Emetic: *Zigadenus venenosus*
Gastrointestinal Aid: *Achillea millefolium*
Gynecological Aid: *Lonicera ciliosa, Rosa pisocarpa, Urtica dioica*
Laxative: *Frangula purshiana*
Orthopedic Aid: *Marah oreganus*
Poison: *Marah oreganus*
Throat Aid: *Mahonia aquifolium, Populus balsamifera*
Tuberculosis Remedy: *Adenocaulon bicolor, Rumex acetosella*
Food
Dried Food: *Rubus parviflorus, Symphoricarpos albus*
Fruit: *Arctostaphylos uva-ursi, Fragaria ×ananassa, Mahonia aquifolium, Oemleria cerasiformis, Ribes sanguineum, Rubus parviflorus, R. spectabilis*
Unspecified: *Corylus cornuta, Pteridium aquilinum, Quercus garryana, Rosa pisocarpa, Rubus spectabilis*
Winter Use Food: *Sambucus cerulea, S. racemosa*
Fiber
Basketry: *Thuja plicata, Typha latifolia*
Building Material: *Populus balsamifera*
Canoe Material: *Holodiscus discolor*
Clothing: *Typha latifolia*
Cordage: *Populus balsamifera*
Mats, Rugs & Bedding: *Pteridium aquilinum, Typha latifolia*
Other
Containers: *Acer macrophyllum, Polystichum munitum*
Cooking Tools: *Holodiscus discolor, Pteridium aquilinum*
Hunting & Fishing Item: *Holodiscus discolor, Thuja plicata*

Stony Indian
Drug
Antihemorrhagic: *Juniperus scopulorum*

Swinomish
Drug
Analgesic: *Pseudotsuga menziesii*
Antirheumatic (External): *Juniperus scopulorum*
Cold Remedy: *Alnus rubra, Lonicera ciliosa, Pseudotsuga menziesii*
Cough Medicine: *Gaultheria shallon*
Dermatological Aid: *Alnus rubra, Malus fusca*
Disinfectant: *Juniperus scopulorum*
Gastrointestinal Aid: *Alnus rubra, Malus fusca*
Gynecological Aid: *Lonicera ciliosa*
Laxative: *Frangula purshiana*
Love Medicine: *Osmorhiza berteroi*
Oral Aid: *Pseudotsuga menziesii*
Other: *Achillea millefolium, Epilobium angustifolium, Lysichiton americanus*
Panacea: *Juniperus scopulorum*
Poison: *Epilobium angustifolium, Ribes lacustre*
Strengthener: *Taxus brevifolia*
Throat Aid: *Lonicera ciliosa, Polystichum munitum, Pseudotsuga menziesii, Ribes divaricatum*
Tonic: *Juniperus scopulorum, Mahonia aquifolium, Pseudotsuga menziesii, Urtica dioica*
Tuberculosis Remedy: *Alnus rubra, Gaultheria shallon, Ribes divaricatum*
Venereal Aid: *Ribes divaricatum*
Food
Bread & Cake: *Gaultheria shallon*
Dried Food: *Amelanchier alnifolia*
Fruit: *Amelanchier alnifolia, Fragaria vesca, Mahonia aquifolium, Malus fusca, Oemleria cerasiformis, Ribes divaricatum, Rubus parviflorus, R. spectabilis, Vaccinium parvifolium*
Unspecified: *Alnus rubra, Corylus cornuta, Equisetum telmateia, Juncus ensifolius, Lilium columbianum, Pteridium aquilinum, Rosa nutkana, Rubus parviflorus, R. spectabilis*
Winter Use Food: *Sambucus cerulea, S. racemosa*

Fiber
Basketry: *Equisetum telmateia*
Furniture: *Acer macrophyllum*
Scouring Material: *Equisetum telmateia*
Dye
Brown: *Pseudotsuga menziesii*
Other
Containers: *Alnus rubra, Lysichiton americanus*
Cooking Tools: *Acer macrophyllum, Holodiscus discolor, Pteridium aquilinum*
Fuel: *Pseudotsuga menziesii*
Hunting & Fishing Item: *Amelanchier alnifolia, Holodiscus discolor, Pseudotsuga menziesii, Taxus brevifolia*
Smoke Plant: *Taxus brevifolia*
Tools: *Taxus brevifolia*
Weapon: *Taxus brevifolia*

Tanaina
Drug
Antirheumatic (Internal): *Artemisia tilesii, Petasites frigidus, Urtica dioica*
Misc. Disease Remedy: *Artemisia tilesii, Petasites frigidus, Urtica dioica*
Tuberculosis Remedy: *Artemisia tilesii, Petasites frigidus, Urtica dioica*
Unspecified: *Heracleum maximum*

Tanana, Upper
Drug
Analgesic: *Huperzia selago, Linnaea borealis, Nuphar lutea*
Antidiarrheal: *Empetrum nigrum*
Antirheumatic (External): *Artemisia tilesii, Juniperus communis, Ledum palustre, Shepherdia canadensis*
Antirheumatic (Internal): *Ledum palustre, Picea glauca*
Blood Medicine: *Artemisia tilesii, Ledum palustre, Rosa acicularis*
Cancer Treatment: *Artemisia alaskana, A. arctica, A. frigida*
Carminative: *Alnus viridis*
Cold Remedy: *Artemisia alaskana, A. arctica, A. frigida, Cyperus sp., Empetrum nigrum, Gentianella propinqua, Juniperus communis, Ledum palustre, Picea glauca, Polygonum alpinum, Populus balsamifera, P. tremuloides, Ribes hudsonianum, Rosa acicularis, Vaccinium vitis-idaea, Viburnum edule*
Cough Medicine: *Artemisia alaskana, A. arctica, A. frigida, A. tilesii, Gentianella propinqua, Juniperus communis, Ledum palustre, Picea glauca, Polygonum alpinum, Populus balsamifera, P. tremuloides, Vaccinium vitis-idaea*
Dermatological Aid: *Artemisia tilesii, Ledum palustre, Picea glauca, Populus balsamifera, Shepherdia canadensis*
Disinfectant: *Artemisia tilesii, Ledum palustre, Picea glauca*
Emetic: *Rosa acicularis*
Eye Medicine: *Artemisia alaskana, A. arctica, A. frigida, A. tilesii, Empetrum nigrum, Ribes triste*
Febrifuge: *Alnus viridis, Rosa acicularis*
Gastrointestinal Aid: *Ledum palustre, Rosa acicularis, Viburnum edule*
Gynecological Aid: *Pentaphylloides floribunda*
Hemorrhoid Remedy: *Picea glauca*
Kidney Aid: *Empetrum nigrum, Juniperus communis*
Laxative: *Arctostaphylos uva-ursi*
Misc. Disease Remedy: *Artemisia alaskana, A. arctica, A. frigida, Ledum palustre*
Oral Aid: *Artemisia tilesii, Picea glauca, Salix sp.*
Orthopedic Aid: *Betula papyrifera*
Panacea: *Juniperus communis, Ledum palustre, Populus balsamifera, Ribes hudsonianum, Shepherdia canadensis*
Pediatric Aid: *Linnaea borealis*
Poison: *Hedysarum boreale*
Psychological Aid: *Linnaea borealis*
Pulmonary Aid: *Picea glauca*
Respiratory Aid: *Ledum palustre, Picea glauca*
Stimulant: *Alnus viridis*
Throat Aid: *Juniperus communis, Ledum palustre, Picea glauca, Vaccinium vitis-idaea, Viburnum edule*
Tuberculosis Remedy: *Juniperus communis, Picea glauca, Shepherdia canadensis*
Unspecified: *Ribes triste*
Vertigo Medicine: *Ledum palustre*
Food
Beverage: *Andromeda polifolia, Hedysarum alpinum, Ledum palustre, Rosa acicularis, Spiraea stevenii*
Candy: *Picea glauca*
Fodder: *Boschniakia rossica, Epilobium angustifolium, Picea glauca*
Frozen Food: *Allium schoenoprasum, Empetrum nigrum, Myriophyllum spicatum, Polygonum alpinum, Rubus arcticus, R. chamaemorus, R. idaeus, Rumex aquaticus, R. arcticus, Vaccinium oxycoccos, V. vitis-idaea, Viburnum edule*
Fruit: *Arctostaphylos rubra, A. uva-ursi, Elaeagnus commutata, Empetrum nigrum, Ribes hudsonianum, R. laxiflorum, R. triste, Rosa acicularis, Rubus arcticus, R. chamaemorus, R. idaeus, Shepherdia canadensis, Vaccinium ovalifolium, V. oxycoccos, V. uliginosum, V. vitis-idaea, Viburnum edule*
Pie & Pudding: *Empetrum nigrum, Rubus arcticus, R. chamaemorus, R. idaeus, Vaccinium oxycoccos, V. vitis-idaea, Viburnum edule*
Preservative: *Alnus viridis, Epilobium angustifolium, Populus tremuloides*
Preserves: *Empetrum nigrum, Ribes hudsonianum, Rosa acicularis, Rubus arcticus, R. chamaemorus, R. idaeus, Vaccinium oxycoccos, V. uliginosum, V. vitis-idaea, Viburnum edule*
Soup: *Elaeagnus commutata*
Spice: *Ledum palustre*
Starvation Food: *Picea glauca*
Unspecified: *Allium schoenoprasum, Betula papyrifera, Epilobium angustifolium, Equisetum arvense, Hedysarum boreale, Myriophyllum spicatum, Picea glauca, Populus balsamifera, P. tremuloides, Rosa acicularis, Salix sp., Typha latifolia*
Vegetable: *Hedysarum alpinum, Polygonum alpinum, Rumex aquaticus, R. arcticus*
Winter Use Food: *Arctostaphylos uva-ursi, Empetrum nigrum, Hedysarum alpinum, H. boreale, Rosa acicularis, Rubus arcticus, R. chamaemorus, R. idaeus, Vaccinium oxycoccos, V. uliginosum, V. vitis-idaea, Viburnum edule*
Fiber
Basketry: *Betula papyrifera, Picea glauca, Salix sp., Viburnum edule*
Brushes & Brooms: *Picea glauca*
Building Material: *Artemisia tilesii, Betula papyrifera, Picea glauca, Salix sp.*
Canoe Material: *Betula papyrifera, Picea glauca*
Clothing: *Betula papyrifera, Poaceae sp.*
Cordage: *Picea glauca, Salix sp.*
Furniture: *Betula papyrifera*
Mats, Rugs & Bedding: *Alnus viridis, Picea glauca, Poaceae sp.*
Snow Gear: *Betula papyrifera, Picea glauca, Salix sp.*
Dye
Unspecified: *Alnus viridis*
Other
Containers: *Artemisia tilesii, Betula papyrifera, Picea glauca, Salix sp.*
Cooking Tools: *Betula papyrifera, Picea glauca, Populus balsamifera*
Fasteners: *Picea glauca, Salix sp.*
Fuel: *Alnus viridis, Betula papyrifera, Picea glauca, Populus balsamifera, P. tremuloides, Salix sp.*
Hide Preparation: *Picea glauca*
Hunting & Fishing Item: *Alnus viridis, Betula papyrifera, Eriophorum sp., Picea glauca, Poaceae sp., Salix sp.*

Incense & Fragrance: *Juniperus communis, Valeriana capitata, V. sitchensis, Viola epipsila*
Insecticide: *Epilobium angustifolium, Picea glauca*
Jewelry: *Elaeagnus commutata*
Musical Instrument: *Betula papyrifera*
Paint: *Chenopodium capitatum*
Preservative: *Populus balsamifera*
Protection: *Betula papyrifera*, Poaceae sp., *Salix* sp.
Season Indicator: *Arctostaphylos rubra*
Smoke Plant: *Populus balsamifera, P. tremuloides*
Tools: *Artemisia tilesii, Betula papyrifera, Epilobium angustifolium, Ledum palustre, Picea glauca*, Poaceae sp.
Toys & Games: *Populus balsamifera*

Tarahumara
Drug
Ceremonial Medicine: *Dasylirion durangensis*

Tewa
Drug
Analgesic: *Angelica* sp., *Gutierrezia sarothrae, Ipomopsis longiflora, Juniperus monosperma, Malva* sp., *Monarda fistulosa, Zea mays*
Antidiarrheal: *Angelica* sp., *Ephedra antisyphilitica, Kallstroemia californica, Xanthium strumarium*
Antiemetic: *Angelica* sp., *Artemisia campestris, Xanthium strumarium*
Antirheumatic (External): *Juniperus monosperma, Poliomintha incana*
Burn Dressing: *Krascheninnikovia lanata*
Carminative: *Artemisia filifolia, A. frigida, A. tridentata, Gutierrezia microcephala*
Ceremonial Medicine: *Nicotiana attenuata, Yucca baccata*
Contraceptive: *Castilleja linariifolia*
Cough Medicine: *Artemisia tridentata, Nicotiana attenuata*
Dermatological Aid: *Abies concolor, Argyrochosma fendleri, Artemisia filifolia, Fallugia paradoxa, Gossypium* sp., *Ipomopsis longiflora, Juniperus monosperma, Kallstroemia californica, Penstemon barbatus, Pinus edulis, Sphaeralcea angustifolia, Taraxacum officinale, Zea mays*
Disinfectant: *Gutierrezia sarothrae, Juniperus monosperma, Sphaeralcea angustifolia*
Diuretic: *Juniperus monosperma*
Ear Medicine: *Gutierrezia sarothrae, Poliomintha incana, Prosopis pubescens*
Emetic: *Yucca baccata*
Expectorant: *Artemisia tridentata*
Eye Medicine: *Monarda fistulosa*
Febrifuge: *Artemisia campestris, Krascheninnikovia lanata, Monarda fistulosa*
Gastrointestinal Aid: *Angelica* sp., *Artemisia filifolia, A. frigida, A. tridentata, Cleome serrulata, Gutierrezia sarothrae, Juniperus monosperma, Phoradendron juniperinum*
Gynecological Aid: *Angelica* sp., *Asclepias* sp., *Castilleja linariifolia, Gutierrezia sarothrae, Juniperus monosperma, Nicotiana attenuata, Yucca baccata, Zea mays*
Heart Medicine: *Zea mays*
Herbal Steam: *Juniperus monosperma*
Internal Medicine: *Juniperus monosperma*
Kidney Aid: *Mirabilis multiflora*
Laxative: *Cercocarpus montanus, Cucurbita foetidissima, Juniperus monosperma*
Misc. Disease Remedy: *Gutierrezia sarothrae*
Nose Medicine: *Nicotiana attenuata*
Oral Aid: *Chrysothamnus nauseosus*
Orthopedic Aid: *Juniperus monosperma, Taraxacum officinale*
Pediatric Aid: *Chrysothamnus nauseosus, Gossypium* sp., *Gutierrezia sarothrae, Juniperus monosperma, Xanthium strumarium, Zea mays*
Reproductive Aid: *Juniperus monosperma*
Snakebite Remedy: *Sphaeralcea angustifolia*
Throat Aid: *Monarda fistulosa*
Tonic: *Ephedra viridis*
Toothache Remedy: *Juniperus monosperma, Nicotiana attenuata, Parryella filifolia*
Urinary Aid: *Populus tremuloides, Xanthium strumarium*
Venereal Aid: *Ephedra viridis*

Food
Beverage: *Thelesperma filifolium, T. megapotamicum, Zea mays*
Bread & Cake: *Zea mays*
Candy: *Hymenoxys richardsonii*
Dried Food: *Poliomintha incana*
Forage: *Equisetum arvense, Gutierrezia sarothrae, Zea mays*
Fruit: *Aster dumosus, Celtis laevigata, Juniperus monosperma, J. scopulorum, Maianthemum racemosum, Opuntia phaeacantha, Physalis subulata, Prunus virginiana, Rhus trilobata, Ribes cereum, Yucca baccata, Y. glauca*
Spice: *Poliomintha incana*
Staple: *Phaseolus vulgaris, Zea mays*
Unspecified: *Allium cernuum, Amaranthus blitoides, A. retroflexus, Asclepias* sp., *Cleome serrulata, Liatris punctata, Mammillaria* sp., *Pinus edulis, Poliomintha incana, Portulaca oleracea, Quercus gambelii, Thelypodium wrightii*
Vegetable: *Cleome serrulata, Solanum jamesii, Stanleya pinnata, Taraxacum officinale*

Fiber
Basketry: *Parryella filifolia, Rhus trilobata, Salix exigua, S. irrorata, Yucca glauca*
Brushes & Brooms: *Bouteloua curtipendula, Fallugia paradoxa, Hilaria jamesii, Panicum capillare, Yucca glauca*
Building Material: *Juniperus monosperma*
Clothing: *Yucca glauca*
Cordage: *Asclepias* sp., *Yucca baccata, Y. glauca*

Dye
Mordant: *Atriplex canescens*
Red: *Alnus incana*
Unspecified: *Pinus edulis*
Yellow: *Chrysothamnus nauseosus*

Other
Cash Crop: *Zea mays*
Ceremonial Items: *Artemisia filifolia, Atriplex canescens, Castilleja linariifolia, Cercocarpus intricatus, Juniperus monosperma, Krascheninnikovia lanata, Parryella filifolia, Penstemon ambiguus, Pinus edulis, Pseudotsuga menziesii, Salix* sp., *Zea mays*
Cleaning Agent: *Juniperus monosperma*
Cooking Tools: *Gutierrezia microcephala, Juniperus monosperma*
Decorations: *Gutierrezia microcephala, Hilaria jamesii, Juniperus monosperma, Populus deltoides*
Fasteners: *Pinus edulis*
Fuel: *Artemisia tridentata, Juniperus monosperma, Pinus edulis, Zea mays*
Hunting & Fishing Item: *Fallugia paradoxa, Juniperus monosperma, Maclura pomifera, Phragmites australis, Prunus virginiana, Ribes cereum, Robinia neomexicana, Yucca baccata*
Incense & Fragrance: *Rosa* sp.
Lighting: *Juniperus monosperma*
Malicious Charm: *Juniperus monosperma*
Paint: *Castilleja linariifolia, Cleome serrulata, Sphaeralcea angustifolia, Thelypodium wrightii*
Protection: *Chrysothamnus nauseosus, Pinus edulis*
Season Indicator: *Penstemon ambiguus*
Smoke Plant: *Nicotiana attenuata, Rhus glabra, Zea mays*

Smoking Tools: *Abies concolor, Acer negundo, Helianthus annuus*
Soap: *Yucca baccata, Y. glauca*
Stable Gear: *Parryella filifolia*
Tools: *Celtis laevigata, Cercocarpus montanus, Juniperus monosperma, Quercus gambelii, Zea mays*
Toys & Games: *Juniperus monosperma, Phragmites australis, Zea mays*
Waterproofing Agent: *Pinus edulis*
Weapon: *Quercus gambelii*

Tewa of Hano
Food
Cooking Agent: *Atriplex canescens*
Fruit: *Opuntia imbricata*
Special Food: *Juniperus monosperma*
Unspecified: *Descurainia* sp., *Monarda fistulosa*
Fiber
Building Material: *Juniperus monosperma*
Sewing Material: *Quercus gambelii*
Other
Ceremonial Items: *Gossypium* sp., *Gutierrezia sarothrae, Martynia* sp., *Pinus ponderosa, Pseudotsuga menziesii, Yucca glauca*
Decorations: *Castilleja linariifolia*
Fuel: *Juniperus monosperma*
Paint: *Descurainia* sp.
Tools: *Quercus gambelii*
Waterproofing Agent: *Pinus edulis*

Thompson
Drug
Adjuvant: *Triteleia grandiflora, Valeriana dioica, V. sitchensis*
Analgesic: *Arabis drummondii, Artemisia dracunculus, Asclepias speciosa, Ceanothus velutinus, Cicuta douglasii, Disporum* sp., *Erigeron* sp., *Eriogonum androsaceum, E. heracleoides, E.* sp., *Frangula purshiana, Gaillardia aristata, Geum* sp., *G. triflorum, Luetkea pectinata, Maianthemum racemosum, M. stellatum, Malus fusca, Mentha canadensis, Nuphar lutea, Oxytropis* sp., *Paxistima myrsinites, Pinus contorta, Platanthera leucostachys, Salix* sp., *Streptopus amplexifolius, Valeriana dioica, V. sitchensis, Zigadenus elegans*
Anthelmintic: *Cornus sericea*
Anticonvulsive: *Calypso bulbosa, Lonicera ciliosa*
Antidiarrheal: *Acer circinatum, Achillea millefolium, Artemisia campestris, Balsamorhiza sagittata, Ceanothus velutinus, Chrysothamnus nauseosus, Cornus sericea, Crataegus douglasii, Fragaria vesca, F. virginiana, Picea sitchensis, Polygonum aviculare, Prunus virginiana, Ribes cereum, Rosa acicularis, R. nutkana, R. pisocarpa, R. woodsii, Solidago canadensis, Symphoricarpos albus, Valeriana dioica, V. sitchensis*
Antiemetic: *Acer glabrum, Cornus sericea, Rosa acicularis, R. nutkana, R. pisocarpa, R. woodsii, Rubus pubescens*
Antihemorrhagic: *Arctostaphylos uva-ursi, Athyrium filix-femina, Pteridium aquilinum, Rubus idaeus, R. pubescens*
Antirheumatic (External): *Achillea millefolium, Arnica cordifolia, Artemisia dracunculus, A. ludoviciana, A. tridentata, Ceanothus velutinus, Eriogonum androsaceum, E. heracleoides, E.* sp., *Geum* sp., *G. triflorum, Juniperus scopulorum, Mahonia aquifolium, M. nervosa, Mentha canadensis, Nuphar lutea, Oxytropis* sp., *Penstemon fruticosus, Philadelphus lewisii, Pinus contorta, P. ponderosa, Platanthera leucostachys, Populus tremuloides, Pseudotsuga menziesii, Pteridium aquilinum, Pulsatilla occidentalis, Ranunculus uncinatus, Urtica dioica, Veratrum viride, Zigadenus venenosus*
Antirheumatic (Internal): *Actaea rubra, Ceanothus velutinus, Juniperus communis, Mahonia aquifolium, M. nervosa, Maianthemum racemosum, M. stellatum, Populus tremuloides, Sambucus cerulea, S. racemosa*
Blood Medicine: *Aralia nudicaulis, Cornus nuttallii, Mahonia aquifolium, M. nervosa, Oplopanax horridus, Veratrum californicum, V. viride*
Breast Treatment: *Philadelphus lewisii*
Burn Dressing: *Equisetum laevigatum, E.* sp., *Larix occidentalis, Lomatium dissectum*
Cancer Treatment: *Ceanothus velutinus, Gaillardia aristata, Larix occidentalis, Maianthemum racemosum, Picea engelmannii, Shepherdia canadensis, Trifolium pratense*
Cathartic: *Abies grandis, Frangula purshiana, Heracleum maximum, Juniperus communis, Penstemon confertus, Shepherdia canadensis*
Ceremonial Medicine: *Eriogonum heracleoides, Heracleum maximum, Populus balsamifera, Valeriana dioica, V. sitchensis*
Cold Remedy: *Abies amabilis, A. grandis, A. lasiocarpa, Achillea millefolium, Antennaria* sp., *Artemisia dracunculus, A. ludoviciana, A. tridentata, Aruncus dioicus, Betula papyrifera, Chrysothamnus nauseosus, Cornus sericea, Juniperus communis, J. scopulorum, Lomatium dissectum, L. nudicaule, Maianthemum stellatum, Mentha arvensis, M. canadensis, Osmorhiza occidentalis, Pinus contorta, Polypodium glycyrrhiza, P. hesperium, Prunus virginiana, Pseudotsuga menziesii, Pteridium aquilinum, Ribes hudsonianum, Spiraea betulifolia, Tsuga heterophylla, T. mertensiana, Valeriana dioica, V. sitchensis, Verbascum thapsus*
Contraceptive: *Amelanchier alnifolia, Betula papyrifera*
Cough Medicine: *Abies lasiocarpa, Antennaria* sp., *Betula papyrifera, Larix occidentalis, Picea engelmannii, Pinus contorta, Prunus virginiana, Rosa woodsii, Rumex crispus, Verbascum thapsus*
Dermatological Aid: *Abies lasiocarpa, Achillea millefolium, Agoseris glauca, Alnus rubra, Aquilegia formosa, Arabis drummondii, Aralia nudicaulis, Arnica cordifolia, Artemisia dracunculus, A.* sp., *Aruncus dioicus, Asarum caudatum, Asclepias speciosa, Astragalus purshii, A.* sp., *Caltha leptosepala, Chaenactis douglasii, Chimaphila umbellata, Clematis columbiana, C. ligusticifolia, Comandra umbellata, Cornus canadensis, Crataegus douglasii, Epilobium angustifolium, Equisetum arvense, Erigeron compositus, E. filifolius, E.* sp., *Eriogonum androsaceum, E. heracleoides, E.* sp., *Fragaria virginiana, Helianthus annuus, H. petiolaris, Heuchera cylindrica, H. micrantha, Hierochloe odorata, Juniperus scopulorum, Larix occidentalis, Leptarrhena pyrolifolia, Linum lewisii, Lomatium dissectum, Lonicera involucrata, Luetkea pectinata, Lysichiton americanus, Mentha canadensis, Monotropa uniflora, Nicotiana attenuata, N. tabacum, Nuphar lutea, Oplopanax horridus, Oxytropis monticola, O.* sp., *Paxistima myrsinites, Penstemon confertus, Philadelphus lewisii, Picea engelmannii, Pinus contorta, P. ponderosa, Plantago major, Polemonium elegans, P. pulcherrimum, Populus balsamifera, P. tremuloides, Potentilla gracilis, Pseudotsuga menziesii, Pteridium aquilinum, Ranunculus glaberrimus, Rhododendron albiflorum, Rosa acicularis, R. nutkana, R. pisocarpa, R. woodsii, Rubus parviflorus, Salix cordata, S.* sp., *Sambucus racemosa, Shepherdia canadensis, Symphoricarpos albus, Thalictrum occidentale, Urtica dioica, Valeriana dioica, V. sitchensis, Verbascum thapsus*
Diaphoretic: *Artemisia dracunculus, Lomatium nudicaule, Oxytropis* sp.
Dietary Aid: *Actaea rubra, Arctostaphylos uva-ursi, Asclepias speciosa, Aster foliaceus, Balsamorhiza sagittata, Ceanothus velutinus, Larix occidentalis, Lonicera involucrata, Oplopanax horridus, Pteridium aquilinum, Solidago simplex, Streptopus amplexifolius*
Disinfectant: *Artemisia ludoviciana, Astragalus purshii, Eriogonum heracleoides, Geum triflorum, Heracleum maximum, Juniperus scopulorum, Oxytropis campestris, Pinus contorta, Platanthera leucostachys, Populus tremuloides, Pseudotsuga menziesii, Ranunculus uncinatus, Shepherdia canadensis, Tsuga heterophylla, T. mertensiana, Urtica* sp.
Diuretic: *Arabis drummondii, Arctostaphylos uva-ursi, Juniperus scopulorum, Pseudotsuga menziesii*
Ear Medicine: *Pinus ponderosa, Sorbus sitchensis*
Eye Medicine: *Abies grandis, Achillea millefolium, Arctostaphylos uva-ursi,*

[Eye Medicine]
Calochortus macrocarpus, Campanula rotundifolia, Claytonia perfoliata, Comandra umbellata, Cryptogramma sitchensis, Equisetum hyemale, E. laevigatum, E. sp., *Eriogonum heracleoides, Juniperus communis, Mahonia aquifolium, M. nervosa, Penstemon fruticosus, P.* sp., *Picea sitchensis, Pinus ponderosa, Pulsatilla occidentalis, Ribes lacustre, Rosa gymnocarpa, Symphoricarpos albus, Trillium ovatum*

Febrifuge: *Lomatium nudicaule*

Gastrointestinal Aid: *Abies* sp., *Achillea millefolium, Amelanchier alnifolia, Artemisia ludoviciana, Aruncus dioicus, Asarum caudatum, Aster foliaceus, Chaenactis douglasii, Chrysothamnus nauseosus, Cornus nuttallii, Crataegus douglasii, Erigeron* sp., *Eriogonum androsaceum, E. heracleoides, E.* sp., *Frangula purshiana, Juniperus communis, J. scopulorum, Larix occidentalis, Luetkea pectinata, Maianthemum racemosum, Mentha canadensis, Nuphar lutea, Oplopanax horridus, Penstemon confertus, P. fruticosus, Populus tremuloides, Pulsatilla occidentalis, Rhododendron albiflorum, Rhus glabra, Ribes hudsonianum, R. lacustre, R. oxyacanthoides, Rubus idaeus, R. pubescens, Shepherdia canadensis, Spiraea betulifolia, Streptopus amplexifolius, Symphoricarpos albus, Valeriana dioica, V. sitchensis*

Gland Medicine: *Erigeron* sp.

Gynecological Aid: *Acer glabrum, Amelanchier alnifolia, Artemisia campestris, A. dracunculus, Calamagrostis rubescens, Chimaphila umbellata, Cornus sericea, Equisetum hyemale, E. laevigatum, E.* sp., *Erigeron* sp., *Geranium viscosissimum, Goodyera oblongifolia, Juniperus scopulorum, Larix occidentalis, Luetkea pectinata, Maianthemum racemosum, Phacelia hastata, Populus balsamifera, Prunus virginiana, Pyrola* sp., *Ribes lacustre, Rosa acicularis, R. nutkana, R. pisocarpa, R. woodsii, Symphoricarpos albus, Thuja plicata*

Heart Medicine: *Juniperus communis, J. scopulorum, Maianthemum racemosum, Shepherdia canadensis*

Hemorrhoid Remedy: *Philadelphus lewisii, Plantago major, Sedum divergens, S.* sp., *S. spathulifolium, Urtica dioica*

Hemostat: *Anemone multifida, Scirpus acutus*

Herbal Steam: *Eriogonum androsaceum, E. heracleoides, E.* sp., *Geum* sp., *G. triflorum, Mentha canadensis, Oxytropis* sp., *Platanthera leucostachys, Ranunculus uncinatus*

Hunting Medicine: *Astragalus purshii, Platanthera leucostachys, Polystichum munitum, Rosa gymnocarpa, Valeriana dioica, V. sitchensis*

Hypotensive: *Juniperus communis, Shepherdia canadensis*

Internal Medicine: *Paxistima myrsinites, Rhus glabra*

Kidney Aid: *Arabis drummondii, Arctostaphylos uva-ursi, Juniperus communis, J. scopulorum, Penstemon fruticosus, P.* sp., *Pseudotsuga menziesii, Sorbus sitchensis*

Laxative: *Frangula purshiana, Mahonia aquifolium, M. nervosa, Oplopanax horridus, Prunus virginiana, Sedum divergens, S. spathulifolium, Shepherdia canadensis, Symphoricarpos albus*

Liver Aid: *Cryptogramma sitchensis, Frangula purshiana, Heuchera cylindrica, H. micrantha, Sambucus racemosa, Shepherdia canadensis*

Love Medicine: *Aquilegia formosa, Delphinium menziesii, Dodecatheon jeffreyi, Geranium viscosissimum, Geum* sp., *Platanthera leucostachys, Sagittaria latifolia*

Misc. Disease Remedy: *Acer circinatum, Achillea millefolium, Anaphalis margaritacea, Artemisia ludoviciana, Aruncus dioicus, Gaillardia aristata, Geum* sp., *Juniperus scopulorum, Mentha arvensis, Oplopanax horridus, Pinus contorta, Prunus virginiana, Rubus leucodermis, Solidago spathulata, Thuja plicata, Tsuga heterophylla, T. mertensiana, Valeriana dioica, V. sitchensis, Veratrum viride*

Oral Aid: *Arctostaphylos uva-ursi, Heuchera cylindrica, H. micrantha, Hieracium* sp., *Polypodium glycyrrhiza, P. hesperium, Pseudotsuga menziesii, Rhus glabra, Sedum spathulifolium*

Orthopedic Aid: *Abies lasiocarpa, Achillea millefolium, Arabis drummondii, Arctostaphylos uva-ursi, Artemisia dracunculus, A. ludoviciana, Aruncus dioicus, Ceanothus velutinus, Chimaphila umbellata, Cicuta douglasii, Erigeron compositus, Eriogonum androsaceum, E. heracleoides, E.* sp., *Geum* sp., *G. triflorum, Larix occidentalis, Lomatium dissectum, Lonicera involucrata, Oxytropis* sp., *Paxistima myrsinites, Penstemon fruticosus, P.* sp., *Pinus contorta, Platanthera leucostachys, Populus balsamifera, Prunus emarginata, Pseudotsuga menziesii, Pteridium aquilinum, Ranunculus uncinatus, Salix* sp., *Solidago canadensis, Urtica dioica, Veratrum viride, Zigadenus venenosus*

Other: *Chrysothamnus nauseosus, Clematis ligusticifolia, Eriogonum androsaceum, E. heracleoides, E.* sp., *Juniperus scopulorum, Ribes lacustre*

Panacea: *Abies amabilis, A. grandis, A. lasiocarpa, Achillea millefolium, Artemisia tridentata, Ceanothus velutinus, Chrysothamnus nauseosus, Cicuta douglasii, Cornus sericea, Crataegus douglasii, Larix occidentalis, Oplopanax horridus, Picea sitchensis, Pinus monticola, Ribes hudsonianum, Taxus brevifolia*

Pediatric Aid: *Abies grandis, Achillea millefolium, Artemisia dracunculus, Asarum caudatum, Clematis ligusticifolia, Cornus sericea, Fragaria vesca, F. virginiana, Larix occidentalis, Linum lewisii, Lomatium macrocarpum, Mentha canadensis, Pinus contorta, P. ponderosa, Polygonum aviculare, Populus tremuloides, Ribes cereum, R. hudsonianum, Rosa woodsii, Rubus parviflorus, Scirpus acutus, Sedum divergens, S. spathulifolium, Solidago canadensis, Symphoricarpos albus, Veratrum viride*

Poison: *Actaea rubra, Anemone multifida, Artemisia dracunculus, Asclepias speciosa, Cicuta douglasii, Cornus sericea, Lonicera involucrata, Lupinus polyphyllus, L. sericeus, L.* sp., *Ranunculus glaberrimus, R. repens, R. sceleratus, R.* sp., *R. uncinatus, Rhus glabra, Rosa gymnocarpa, Symphoricarpos albus, Toxicodendron radicans, Veratrum californicum, Zigadenus elegans*

Psychological Aid: *Cornus nuttallii, Lewisia pygmaea, Lysichiton americanus, Mahonia nervosa, Picea engelmannii, Populus tremuloides, Thuja plicata*

Pulmonary Aid: *Actaea rubra, Cornus nuttallii, Eriogonum heracleoides, Philadelphus lewisii, Pinus contorta*

Reproductive Aid: *Lomatium macrocarpum, Lonicera ciliosa*

Respiratory Aid: *Larix occidentalis, Veratrum californicum*

Sedative: *Asarum caudatum, Balsamorhiza sagittata, Lomatium macrocarpum, Lonicera ciliosa, Pinus ponderosa, Ribes hudsonianum, Sedum spathulifolium, Shepherdia canadensis, Solidago canadensis*

Snakebite Remedy: *Acer glabrum, Achillea millefolium, Chaenactis douglasii, Chamaesyce glyptosperma*

Stimulant: *Aralia nudicaulis, Artemisia tridentata, Chaenactis douglasii, Potentilla glandulosa*

Strengthener: *Aquilegia formosa, A.* sp., *Cornus sericea, Larix occidentalis, Ribes cereum*

Throat Aid: *Artemisia tridentata, Corylus cornuta, Erigeron* sp., *Heuchera cylindrica, H. micrantha, Lonicera involucrata, Maianthemum racemosum, Pinus contorta, Polypodium glycyrrhiza, P. hesperium, Ribes hudsonianum, Rosa woodsii, Veratrum viride*

Tonic: *Abies* sp., *Acer macrophyllum, Achillea millefolium, Amelanchier alnifolia, Aralia nudicaulis, Arctostaphylos uva-ursi, Asarum caudatum, Asclepias speciosa, Chaenactis douglasii, Chimaphila umbellata, Clematis ligusticifolia, Geum triflorum, Heracleum maximum, Juniperus communis, Lonicera ciliosa, Mahonia aquifolium, Oplopanax horridus, Potentilla glandulosa, Prunella vulgaris, Prunus virginiana, Pseudotsuga menziesii, Ribes lacustre, R. oxyacanthoides, Rosa gymnocarpa, Rubus idaeus, R. pubescens, Shepherdia canadensis, Solidago simplex, Spiraea ×pyramidata, Streptopus roseus*

Toothache Remedy: *Achillea millefolium, Alnus rubra, A. viridis, Arabis holboellii, Clematis ligusticifolia, Salix* sp., *Sambucus canadensis, S. cerulea, S. racemosa*

Tuberculosis Remedy: *Abies amabilis, A. grandis, A. lasiocarpa, Arnica cordifolia, Artemisia tridentata, Cassiope mertensiana, C.* sp., *Chrysothamnus nauseosus, Eriogonum heracleoides, Gaillardia aristata, Juniperus communis,*

J. scopulorum, Larix occidentalis, Paxistima myrsinites, Phyllodoce empetriformis, P. sp., *Ranunculus* sp., *Ribes hudsonianum, Shepherdia canadensis, Valeriana dioica, V. sitchensis*

Unspecified: *Abies amabilis, A. grandis, A. lasiocarpa, Achillea millefolium, Anemone* sp., *Apocynum cannabinum, Arnica latifolia, Artemisia campestris, A. frigida, A.* sp., *Asarum caudatum, Calypso bulbosa, Castilleja miniata, C.* sp., *Ceanothus velutinus, Chaenactis douglasii, Dicentra formosa, Epilobium* sp., *Erigeron filifolius, Eriogonum heracleoides, Geranium richardsonii, G. viscosissimum, Geum* sp., *Hackelia hispida, Heracleum maximum, Heuchera cylindrica, H. micrantha, Hieracium* sp., *Kalmia polifolia, Ligusticum canbyi, Linnaea borealis, Lupinus polyphyllus, L. rivularis, Mentha canadensis, Mentzelia laevicaulis, Pedicularis bracteosa, P. racemosa, Phacelia linearis, Picea sitchensis, Pinus monticola, Populus balsamifera, Prunus emarginata, P. virginiana, Pulsatilla occidentalis, Salix interior, Stenanthium occidentale, Stephanomeria tenuifolia, Streptopus roseus, Triteleia grandiflora, Urtica dioica*

Urinary Aid: *Achillea millefolium, Arabis drummondii, Arctostaphylos uva-ursi, Chrysothamnus nauseosus, Clematis ligusticifolia, Equisetum arvense, E. hyemale, E. laevigatum, E.* sp., *E. telmateia, Juniperus scopulorum, Lonicera involucrata, Pseudotsuga menziesii*

Venereal Aid: *Abies grandis, Achillea millefolium, Actaea rubra, Apocynum cannabinum, Arabis drummondii, Artemisia frigida, Aster foliaceus, Ceanothus velutinus, Chrysothamnus nauseosus, Elaeagnus commutata, Eriogonum androsaceum, E. heracleoides, E.* sp., *Heracleum maximum, Mahonia aquifolium, M. nervosa, Populus tremuloides, Rhus glabra, Rosa acicularis, R. nutkana, R. pisocarpa, R. woodsii, Sambucus cerulea, Solidago simplex, Spiraea betulifolia, Veratrum californicum*

Veterinary Aid: *Achlys triphylla, Aquilegia formosa, A.* sp., *Artemisia dracunculus, Cicuta douglasii, Juniperus scopulorum, Leymus cinereus, Lomatium dissectum, Lupinus polyphyllus, L. sericeus, L.* sp., *Penstemon fruticosus, Pinus ponderosa, Solidago canadensis, Valeriana dioica, V. sitchensis, Viola* sp., *Zigadenus elegans*

Witchcraft Medicine: *Abies grandis, Geranium viscosissimum, Lithospermum ruderale*

Food

Beverage: *Abies grandis, Amelanchier alnifolia, Arctostaphylos uva-ursi, Chimaphila umbellata, Geum triflorum, Juniperus communis, Ledum groenlandicum, Lomatium nudicaule, Penstemon confertus, Pinus contorta, Prunella vulgaris, Prunus virginiana, Ribes divaricatum, R. inerme, R. oxyacanthoides, Rosa acicularis, R. gymnocarpa, R. nutkana, R. pisocarpa, R. woodsii, Shepherdia canadensis, Spiraea betulifolia, S. ×pyramidata*

Bread & Cake: *Amelanchier alnifolia, Balsamorhiza sagittata, Claytonia lanceolata, Gaultheria shallon, Lewisia rediviva, Oemleria cerasiformis, Rubus idaeus, Shepherdia canadensis, Vaccinium ovalifolium, V. parvifolium*

Candy: *Agoseris glauca, Erythronium grandiflorum, Larix occidentalis, Lonicera ciliosa, Pinus contorta, Polypodium glycyrrhiza, P. hesperium, Shepherdia canadensis, Tragopogon* sp.

Dessert: *Balsamorhiza sagittata, Cornus sericea, Lycopus uniflorus, Opuntia fragilis, O. polyacantha, Prunus emarginata*

Dried Food: *Agaricus* sp., *Allium cernuum, Amelanchier alnifolia, Balsamorhiza sagittata, Cirsium edule, C. hookerianum, C. undulatum, C. vulgare, Cornus sericea, Crataegus douglasii, Erythronium grandiflorum, Fragaria vesca, F. virginiana, Fritillaria lanceolata, Heracleum maximum, Lewisia rediviva, Lilium columbianum, Lomatium dissectum, L. macrocarpum, L. nudicaule, Mahonia aquifolium, Nuphar lutea, Pinus albicaulis, P. contorta, Prunus virginiana, Ribes lacustre, R. sanguineum, Rubus idaeus, R. leucodermis, R. spectabilis, R. ursinus, Sagittaria latifolia, Sambucus cerulea, Sium suave, Sorbus sitchensis, Vaccinium cespitosum, V. deliciosum, V. membranaceum, V. ovalifolium, V. scoparium, Viburnum edule*

Fodder: *Astragalus giganteus, A. miser, Carex atherodes, C. obnupta, C. rostrata, Castilleja miniata, Elymus trachycaulus, Epilobium angustifolium, Lupinus* sp., *Trifolium pratense, Vicia americana*

Forage: *Aquilegia formosa, Arctostaphylos uva-ursi, Astragalus purshii, Bromus* sp., *Carex atherodes, C. obnupta, C. rostrata, C.* sp., *Ceanothus velutinus, Hackelia diffusa, Heracleum maximum, Hydrophyllum fendleri, H. occidentale, Lathyrus nevadensis, Lonicera ciliosa, L. involucrata, Maianthemum racemosum, Oxytropis campestris, Paxistima myrsinites, Penstemon fruticosus, Phragmites australis, Populus balsamifera, P. tremuloides, Ribes hudsonianum, Rosa acicularis, R. woodsii, Salix* sp., *Shepherdia canadensis, Sium suave, Spiraea douglasii, Valeriana dioica, V. sitchensis, Vicia* sp.

Frozen Food: *Amelanchier alnifolia, Heracleum maximum, Lomatium nudicaule, Ribes lacustre, Rubus idaeus, Vaccinium membranaceum*

Fruit: *Amelanchier alnifolia, Arctostaphylos uva-ursi, Citrus limon, C. medica, C. sinensis, Cornus sericea, Crataegus columbiana, C. douglasii, Cucumis melo, Disporum hookeri, Fragaria vesca, F. virginiana, Gaultheria shallon, Lomatium nudicaule, Lonicera involucrata, Mahonia aquifolium, Maianthemum racemosum, M. stellatum, Malus fusca, Oemleria cerasiformis, Prunus emarginata, P. virginiana, Ribes cereum, R. divaricatum, R. hudsonianum, R. inerme, R. oxyacanthoides, R. sanguineum, Rosa gymnocarpa, Rubus idaeus, R. leucodermis, R. parviflorus, R. pedatus, R. spectabilis, R. ursinus, Sambucus cerulea, S. racemosa, Shepherdia argentea, S. canadensis, Sorbus sitchensis, Streptopus amplexifolius, Vaccinium membranaceum, V. myrtillus, V. ovalifolium, V. ovatum, V. oxycoccos, V. parvifolium, V.* sp., *Viburnum edule, V. opulus*

Pie & Pudding: *Amelanchier alnifolia, Erythronium grandiflorum, Gaultheria shallon, Lewisia rediviva, Lomatium macrocarpum, Ribes divaricatum, R. inerme, R. oxyacanthoides, Vaccinium cespitosum, V. deliciosum, V. myrtilloides, V. scoparium*

Porridge: *Pinus albicaulis, P. ponderosa*

Preservative: *Rhododendron albiflorum*

Preserves: *Amelanchier alnifolia, A.* sp., *Crataegus douglasii, Gaultheria shallon, Mahonia aquifolium, M. nervosa, Prunus* sp., *Ribes lacustre, Rosa acicularis, R. woodsii, Rubus idaeus, Sambucus cerulea, Shepherdia canadensis, S.* sp., *Vaccinium cespitosum, V. deliciosum, V. membranaceum, V. parvifolium, V. scoparium, V.* sp., *Viburnum edule*

Sauce & Relish: *Acer macrophyllum, Prunus virginiana, Rosa acicularis, R. woodsii, Viburnum edule*

Snack Food: *Corylus cornuta*

Soup: *Arctostaphylos uva-ursi, Cirsium edule, Erythronium grandiflorum, Lilium columbianum, Sambucus racemosa, Sorbus sitchensis, Viburnum edule*

Special Food: *Allium cernuum, Lewisia rediviva*

Spice: *Amelanchier alnifolia, Fragaria vesca, Fritillaria lanceolata, Lilium columbianum, Lomatium macrocarpum, L. nudicaule, Maianthemum racemosum, Penstemon fruticosus, Ribes sanguineum, Sambucus cerulea, S. racemosa, Sorbus sitchensis*

Staple: *Balsamorhiza sagittata, Pteridium aquilinum*

Starvation Food: *Balsamorhiza sagittata, Opuntia fragilis, O. polyacantha*

Sweetener: *Amelanchier alnifolia, Pseudotsuga menziesii, Rubus parviflorus*

Unspecified: *Abies lasiocarpa, A.* sp., *Acer macrophyllum, Alectoria jubata, Allium acuminatum, A. cernuum, Amelanchier alnifolia, Argentina anserina, Astragalus miser, Balsamorhiza sagittata, Calochortus macrocarpus, Camassia quamash, C. scilloides, Carex rostrata, Cirsium edule, C. hookerianum, C. undulatum, C. vulgare, Claytonia lanceolata, Cornus sericea, Corylus cornuta, Crataegus columbiana, C. douglasii, Cucumis* sp., *Dryopteris arguta, D. expansa, D.* sp., *Epilobium angustifolium, Erythronium grandiflorum, Fritillaria lanceolata, F. pudica, Helianthus* sp., *Heracleum maximum, Hieracium* sp., *Hydrophyllum fendleri, H. occidentale, Larix occidentalis, Lewisia columbiana, L. pygmaea, L. rediviva, Lilium columbianum, Lithospermum incisum, Lomatium dissectum, L. macrocarpum, L. nudicaule, Lycopus uniflorus, Maianthemum racemosum, Mentha arvensis, Opuntia fragilis, O. polyacantha, Osmorhiza berteroi, Peucedanum* sp., *Picea engelmannii, Pinus albicaulis,*

[Unspecified]
P. contorta, P. monticola, P. ponderosa, Polystichum munitum, Prunus virginiana, Pteridium aquilinum, Rosa acicularis, R. nutkana, R. pisocarpa, R. woodsii, Rubus leucodermis, R. occidentalis, R. parviflorus, R. spectabilis, R. ursinus, Rumex acetosella, Sagittaria latifolia, Scirpus sp., Sium suave, Tragopogon porrifolius, Triteleia grandiflora, Typha latifolia, Viburnum edule, Zizania aquatica

Vegetable: *Acer macrophyllum, Capsella bursa-pastoris, Chenopodium album, Heracleum maximum, Lomatium nudicaule, Maianthemum racemosum, Urtica dioica*

Winter Use Food: *Amelanchier alnifolia, Claytonia lanceolata, Heracleum maximum, Lewisia columbiana, L. rediviva, Lomatium nudicaule, Opuntia fragilis, O. polyacantha, Pinus albicaulis, Prunus virginiana, Shepherdia argentea, Sorbus sitchensis, Vaccinium cespitosum, V. deliciosum, V. scoparium*

Fiber

Basketry: *Acer macrophyllum, Betula papyrifera, Erigeron peregrinus, Leymus cinereus, L. triticoides, Philadelphus lewisii, Phragmites australis, Picea engelmannii, Prunus emarginata, Rhus glabra, Salix sp., Thuja plicata*

Brushes & Brooms: *Carex sp., Philadelphus lewisii, Spiraea douglasii, Tsuga mertensiana*

Building Material: *Abies grandis, Betula papyrifera, Lonicera ciliosa, Picea engelmannii, Pinus contorta, P. ponderosa, Prunus emarginata, Pseudotsuga menziesii, Salix sp., Scirpus acutus, Typha latifolia*

Canoe Material: *Abies grandis, Betula papyrifera, Picea engelmannii, Populus balsamifera, Salix sp., Thuja plicata*

Caulking Material: *Pseudotsuga menziesii*

Clothing: *Apocynum cannabinum, Artemisia tridentata, Asclepias speciosa, Calamagrostis rubescens, Carex atherodes, C. obnupta, Clematis ligusticifolia, Elaeagnus commutata, Salix sp., Scirpus acutus, Thuja plicata, Typha latifolia*

Cordage: *Acer glabrum, Apocynum androsaemifolium, A. cannabinum, Asclepias speciosa, Elaeagnus commutata, Lonicera ciliosa, Prunus emarginata, Salix sp., Thuja plicata, Urtica dioica*

Furniture: *Acer circinatum, A. glabrum, Betula papyrifera, Cornus sericea, Philadelphus lewisii, Rosa acicularis, R. gymnocarpa, R. nutkana, R. pisocarpa, R. woodsii, Salix sp., Thuja plicata*

Mats, Rugs & Bedding: *Abies grandis, A. lasiocarpa, Artemisia tridentata, Asarum caudatum, Elaeagnus commutata, Lomatium macrocarpum, Phragmites australis, Populus balsamifera, Prunus emarginata, Pseudotsuga menziesii, Rosa gymnocarpa, Salix sp., Scirpus acutus, S. sp., Thuja plicata, Tragopogon sp., Tsuga mertensiana, Typha latifolia*

Scouring Material: *Acer macrophyllum, Equisetum hyemale, E. laevigatum, E. sp., Pinus contorta, Populus balsamifera, Thuja plicata*

Sewing Material: *Apocynum cannabinum, Lonicera ciliosa, Thuja plicata*

Snow Gear: *Acer circinatum, A. glabrum, Alnus incana, Pseudotsuga menziesii, Taxus brevifolia*

Dye

Black: *Cornus nuttallii, Lonicera ciliosa*

Blue: *Corylus cornuta, Delphinium menziesii*

Brown: *Alnus rubra, Cornus nuttallii*

Green: *Thuja plicata*

Red: *Alnus rubra, Chenopodium capitatum*

Unspecified: *Clintonia uniflora, Penstemon fruticosus, Rhus glabra, Rubus leucodermis*

Yellow: *Evernia vulpina, Mahonia aquifolium, M. nervosa*

Other

Cash Crop: *Amelanchier alnifolia, Balsamorhiza sagittata, Corylus cornuta, Erythronium grandiflorum, Lewisia columbiana, L. rediviva, Shepherdia canadensis, Thuja plicata, Zizania aquatica*

Ceremonial Items: *Elaeagnus commutata, Equisetum sp., Leymus cinereus, Nicotiana tabacum, Pseudotsuga menziesii, Rosa gymnocarpa, Taxus brevifolia, Verbascum thapsus, Zigadenus elegans*

Containers: *Acer macrophyllum, Apocynum cannabinum, Artemisia tridentata, Betula papyrifera, Elaeagnus commutata, Epilobium angustifolium, Mentha arvensis, Pinus ponderosa, Polystichum munitum, Prunus emarginata, Salix sp., Scirpus acutus, S. sp., Thuja plicata, Typha latifolia*

Cooking Tools: *Acer macrophyllum, Amelanchier alnifolia, Betula papyrifera, Calamagrostis rubescens, Corylus cornuta, Elaeagnus commutata, Leymus cinereus, Phragmites australis, Picea engelmannii, Pinus ponderosa, Pseudotsuga menziesii, Typha latifolia*

Decorations: *Betula papyrifera, Castilleja miniata, C. sp., Elaeagnus commutata, Lithospermum ruderale, Lycopodium clavatum, Ornithogalum umbellatum, Penstemon fruticosus, Phragmites australis, Prunus emarginata, Scirpus acutus, Typha latifolia*

Designs: *Dodecatheon pulchellum, Pedicularis bracteosa, Phragmites australis, Prunus virginiana*

Fasteners: *Apocynum androsaemifolium, Asclepias speciosa, Opuntia fragilis, O. polyacantha, Prunus emarginata, Salix interior, S. sp.*

Fertilizer: *Equisetum hyemale, E. laevigatum, E. sp.*

Fuel: *Alnus rubra, Artemisia ludoviciana, Juniperus virginiana, Populus balsamifera, Pseudotsuga menziesii*

Good Luck Charm: *Aquilegia formosa, Dodecatheon jeffreyi, Epilobium sp., Juniperus scopulorum, Lewisia pygmaea, Picea engelmannii, Platanthera leucostachys, Pseudotsuga menziesii, Rosa acicularis, R. nutkana, R. pisocarpa, R. woodsii, Shepherdia canadensis, Thuja plicata*

Hide Preparation: *Artemisia tridentata*

Hunting & Fishing Item: *Acer circinatum, A. glabrum, Alnus incana, Amelanchier alnifolia, Apocynum cannabinum, Cornus nuttallii, C. sericea, Corylus cornuta, Crataegus columbiana, C. douglasii, Epilobium angustifolium, Holodiscus discolor, Juniperus scopulorum, Pinus contorta, Polygonum amphibium, Populus tremuloides, Pseudotsuga menziesii, Ribes hudsonianum, Rosa acicularis, R. gymnocarpa, R. nutkana, R. pisocarpa, R. woodsii, Salix sp., Shepherdia canadensis, Spirogyra sp., Taxus brevifolia, Urtica dioica*

Incense & Fragrance: *Abies amabilis, A. grandis, A. lasiocarpa, Alnus rhombifolia, Chenopodium botrys, Fragaria vesca, Hierochloe odorata, Lomatium nudicaule, Mentha arvensis, M. canadensis, Pinus ponderosa, Rhododendron albiflorum, Streptopus amplexifolius, Tsuga heterophylla, T. mertensiana*

Insecticide: *Achlys triphylla, Anemone multifida, Artemisia ludoviciana, Juniperus scopulorum, Mentha arvensis*

Jewelry: *Elaeagnus commutata, Opuntia polyacantha, Phragmites australis, Rosa gymnocarpa*

Musical Instrument: *Juniperus scopulorum*

Paint: *Chenopodium capitatum, Delphinium menziesii, Evernia vulpina, Larix occidentalis, Lithospermum incisum, Rosa acicularis, R. nutkana, R. pisocarpa, R. woodsii*

Paper: *Betula papyrifera*

Plant Indicator: *Lewisia rediviva, Monotropa uniflora*

Protection: *Abies grandis, Arctostaphylos uva-ursi, Artemisia ludoviciana, Holodiscus discolor, Juniperus scopulorum, Lithospermum ruderale, Lysichiton americanus, Picea sitchensis, Populus tremuloides, Rosa acicularis, R. nutkana, R. pisocarpa, R. woodsii*

Smoke Plant: *Arctostaphylos uva-ursi, Cornus sericea, Nicotiana attenuata, Rosa gymnocarpa, Vaccinium ovalifolium, V. parvifolium, V. sp., Valeriana sitchensis, Verbascum thapsus*

Smoking Tools: *Pinus ponderosa*

Soap: *Abies lasiocarpa, Ceanothus velutinus, Juniperus communis, Linum lewisii, Mentha arvensis, Picea engelmannii, Populus balsamifera, Symphoricarpos albus*

Stable Gear: *Artemisia sp., A. tridentata, Populus balsamifera, Salix interior*

Tools: *Acer circinatum, Amelanchier alnifolia, Cornus nuttallii, Crataegus columbiana, C. douglasii, Equisetum sp., Philadelphus lewisii, Populus bal-*

Index of Tribes 757

samifera, *Prunus virginiana, Rosa gymnocarpa, Scirpus microcarpus, Taxus brevifolia, Tsuga heterophylla*
Toys & Games: *Erythronium grandiflorum, Hierochloe odorata, Lithospermum incisum, Rosa gymnocarpa*
Water Indicator: *Alnus viridis*
Waterproofing Agent: *Arctostaphylos uva-ursi, Pinus ponderosa*
Weapon: *Holodiscus discolor*

Thompson, Lower
Food
Ice Cream: *Shepherdia canadensis*

Thompson, Upper (Fraser Band)
Fiber
Canoe Material: *Pinus ponderosa, Populus tremuloides*
Other
Fuel: *Pinus ponderosa*

Thompson, Upper (Lytton Band)
Food
Fruit: *Ribes sanguineum*
Unspecified: *Triteleia grandiflora*
Fiber
Canoe Material: *Pinus ponderosa*
Other
Cash Crop: *Lewisia rediviva, Lomatium macrocarpum*
Fuel: *Pinus ponderosa*

Thompson, Upper (Nicola Band)
Food
Unspecified: *Lewisia columbiana, Peucedanum* sp., *Tragopogon porrifolius, T.* sp.
Other
Fuel: *Artemisia frigida, Juniperus virginiana*

Thompson, Upper (Spences Bridge)
Food
Unspecified: *Opuntia* sp., *Tragopogon porrifolius, Triteleia grandiflora*

Tlingit
Drug
Antidiarrheal: *Picea glauca*
Blood Medicine: *Aruncus dioicus*
Cough Medicine: *Osmorhiza claytonii*
Dermatological Aid: *Boschniakia glabra, Kalmia polifolia, Oplopanax horridus, Petasites frigidus*
Gastrointestinal Aid: *Oxytropis* sp.
Herbal Steam: *Artemisia vulgaris*
Pulmonary Aid: *Artemisia vulgaris, Asplenium* sp., *Coptis macrosepala, Sorbus americana*
Toothache Remedy: *Picea sitchensis, Tsuga mertensiana*
Unspecified: *Menyanthes trifoliata*
Venereal Aid: *Claytonia sibirica, Heuchera glabra, Ledum palustre, Picea sitchensis, Pinus contorta, Sequoia sempervirens, Thuja excelsa, Tsuga mertensiana*
Other
Decorations: *Acer macrophyllum*

Tolowa
Drug
Anthelmintic: *Artemisia douglasiana*
Anticonvulsive: *Pteridium aquilinum*
Antidote: *Toxicodendron diversilobum*
Antirheumatic (External): *Artemisia douglasiana, Lysichiton americanus, Oxalis oregana, Petasites frigidus*
Blood Medicine: *Mahonia pumila*
Cough Medicine: *Mahonia pumila*
Dermatological Aid: *Oxalis oregana, Plantago australis, Urtica dioica*
Disinfectant: *Asarum caudatum, Oxalis oregana*
Eye Medicine: *Viola adunca*
Laxative: *Frangula purshiana*
Oral Aid: *Equisetum telmateia*
Orthopedic Aid: *Artemisia douglasiana*
Other: *Lysichiton americanus*
Pediatric Aid: *Artemisia douglasiana, Equisetum telmateia*
Poison: *Lonicera involucrata, Toxicodendron diversilobum*
Unspecified: *Vancouveria chrysantha*
Food
Beverage: *Ledum glandulosum, Satureja douglasii*
Bread & Cake: *Arctostaphylos $cinerea, A. nevadensis, A. uva-ursi*
Dried Food: *Corylus cornuta, Vaccinium ovatum, V. parvifolium*
Fruit: *Fragaria chiloensis, Gaultheria shallon, Oemleria cerasiformis, Quercus kelloggii, Ribes lobbii, Rubus spectabilis, Umbellularia californica, Vaccinium ovatum, V. parvifolium*
Spice: *Ribes nevadense*
Staple: *Lithocarpus densiflorus*
Unspecified: *Castanopsis chrysophylla, Corylus cornuta, Heracleum maximum, Ligusticum californicum, Lysichiton americanus, Nuphar lutea, Oxalis oregana, Porphyra lanceolata, Rubus spectabilis*
Fiber
Basketry: *Adiantum aleuticum, Alnus rubra, Corylus cornuta, Juncus effusus, Salix hindsiana, S. scouleriana, S. sitchensis*
Brushes & Brooms: *Vaccinium parvifolium*
Building Material: *Sequoia sempervirens*
Canoe Material: *Sequoia sempervirens*
Clothing: *Acer macrophyllum, Arbutus menziesii, Typha latifolia*
Cordage: *Iris innominata, I. tenax*
Snow Gear: *Arbutus menziesii*
Unspecified: *Woodwardia fimbriata*
Dye
Unspecified: *Alnus rubra, Lithocarpus densiflorus*
Other
Containers: *Lysichiton americanus*
Cooking Tools: *Lysichiton americanus*
Decorations: *Juniperus communis*
Hunting & Fishing Item: *Pinus contorta*
Jewelry: *Arbutus menziesii, Juniperus communis*
Smoke Plant: *Nicotiana quadrivalvis*
Smoking Tools: *Taxus brevifolia*
Toys & Games: *Physocarpus capitatus*

Tsimshian
Drug
Antirheumatic (External): *Veratrum viride*
Cancer Treatment: *Taxus brevifolia*
Ceremonial Medicine: *Oenanthe sarmentosa*
Dermatological Aid: *Veratrum viride*
Hunting Medicine: *Picea sitchensis*
Internal Medicine: *Taxus brevifolia*
Psychological Aid: *Veratrum viride*
Respiratory Aid: *Veratrum viride*
Unspecified: *Argentina egedii*

Food
Fruit: *Empetrum nigrum, Rubus parviflorus, Vaccinium vitis-idaea*
Fiber
Building Material: *Pinus contorta, Thuja plicata*
Canoe Material: *Thuja plicata*
Clothing: *Thuja plicata*
Mats, Rugs & Bedding: *Thuja plicata*
Other
Cash Crop: *Porphyra abbottae, Shepherdia canadensis*
Ceremonial Items: *Oplopanax horridus, Picea sitchensis, Thuja plicata*
Cleaning Agent: *Rubus parviflorus, Thuja plicata*
Containers: *Lysichiton americanus, Thuja plicata*
Cooking Tools: *Lysichiton americanus*
Decorations: *Acer macrophyllum, Thuja plicata*
Fuel: *Thuja plicata*
Hunting & Fishing Item: *Heracleum maximum, Nereocystis luetkeana, Thuja plicata, Urtica dioica*
Protection: *Pinus contorta*
Smoke Plant: *Nicotiana quadrivalvis, N. tabacum*
Tools: *Thuja plicata*

Tsimshian, Coast
Other
Cash Crop: *Shepherdia canadensis*

Tubatulabal
Drug
Analgesic: *Trichostema lanceolatum*
Antidiarrheal: *Eriogonum fasciculatum, Heliotropium curassavicum*
Antirheumatic (External): *Encelia virginensis, Ericameria cooperi, E. linearifolia*
Antirheumatic (Internal): *Datura wrightii*
Blood Medicine: *Ephedra viridis*
Cold Remedy: *Anemopsis californica, Loeseliastrum matthewsii*
Dermatological Aid: *Argemone polyanthemos, Chorizanthe staticoides, Datura wrightii, Echinocystis brandegei, Eriogonum baileyi, E. gracillimum, E. roseum, Marah horridus*
Gastrointestinal Aid: *Datura wrightii, Diplacus aurantiacus, D. longiflorus, Eriogonum fasciculatum*
Hemorrhoid Remedy: *Argemone polyanthemos*
Hemostat: *Trichostema lanceolatum*
Laxative: *Datura wrightii*
Pediatric Aid: *Echinocystis brandegei, Eriogonum fasciculatum, Marah horridus*
Sedative: *Datura wrightii*
Venereal Aid: *Ephedra viridis*
Food
Beverage: *Monardella candicans*
Candy: *Asclepias erosa*
Fruit: *Arctostaphylos* sp., *Juniperus osteosperma, Lycium torreyi, Ribes quercetorum, Sambucus cerulea, Solanum nigrum, Vitis californica*
Unspecified: *Aesculus californica, Allium hyalinum, A. lacunosum, A. peninsulare, Calochortus palmeri, C. venustus, Cirsium occidentale, Coreopsis bigelovii, Distichlis spicata, Echinochloa crus-galli, Ephedra viridis, Eriogonum* sp., *Heliotropium curassavicum, Lotus scoparius, Mentzelia albicaulis, M. gracilenta, Monardella candicans, Nolina bigelovii, Opuntia basilaris, O. caseyi, Pellaea mucronata, Pholistoma membranaceum, Pinus monophylla, P. sabiniana, Polypogon monspeliensis, Quercus chrysolepis, Q. douglasii, Q. dumosa, Q. kelloggii, Q. lobata, Q. wislizeni, Salvia carduacea, S. columbariae, Scirpus acutus, Sedum laxum, Sisymbrium officinale, Trifolium wormskioldii, Typha latifolia, Yucca brevifolia, Y. whipplei*
Vegetable: *Rorippa nasturtium-aquaticum*

Other
Cooking Tools: *Artemisia cana, A. tridentata, Eriogonum elongatum, Salix exigua, S. hindsiana*
Fuel: *Artemisia cana, A. tridentata*
Tools: *Ceanothus cuneatus*

Ukiah
Fiber
Basketry: *Pteridium aquilinum, Salix hindsiana, Smilax californica*
Other
Hunting & Fishing Item: *Salix hindsiana*

Umatilla
Drug
Gynecological Aid: *Shepherdia canadensis*
Food
Unspecified: *Perideridia gairdneri*

Ute
Drug
Analgesic: *Comandra umbellata*
Cold Remedy: *Grindelia squarrosa*
Dermatological Aid: *Achillea millefolium, Collinsia parviflora, Lomatium dissectum, Phlox gracilis*
Diuretic: *Lithospermum ruderale*
Gastrointestinal Aid: *Abronia fragrans, Castilleja parviflora, Cryptantha sericea*
Narcotic: *Datura wrightii*
Panacea: *Achillea millefolium*
Poison: *Fritillaria atropurpurea, Zigadenus nuttallii*
Unspecified: *Artemisia tridentata, Descurainia pinnata, Eriogonum ovalifolium, Fritillaria atropurpurea, Hedysarum boreale, Iva axillaris, Matricaria discoidea, Senecio* sp., *Viola cucullata*
Veterinary Aid: *Lomatium dissectum*
Food
Dried Food: *Amelanchier alnifolia*
Fruit: *Amelanchier alnifolia, Rhus trilobata, Ribes aureum, Rosa woodsii, Shepherdia argentea*
Staple: *Agastache pallidiflora, Agave americana, A. parryi*
Starvation Food: *Calochortus nuttallii*
Unspecified: *Agoseris* sp., *Allium acuminatum, A. bisceptrum, Balsamorhiza sagittata, Calochortus nuttallii, Claytonia caroliniana, Cymopterus longipes, Elymus canadensis, Fritillaria pudica, Perideridia gairdneri, Scirpus tabernaemontani, Taraxacum officinale*
Fiber
Basketry: *Salix amygdaloides, S. eriocephala, S. interior, S. lucida*
Other
Fasteners: *Ipomopsis aggregata*
Toys & Games: *Equisetum laevigatum*

Wailaki
Drug
Analgesic: *Chlorogalum pomeridianum, Thalictrum polycarpum*
Antirheumatic (External): *Chlorogalum pomeridianum, Polypodium californicum*
Antirheumatic (Internal): *Pinus sabiniana*
Carminative: *Chlorogalum pomeridianum*
Dermatological Aid: *Chlorogalum pomeridianum, Erythronium oregonum, Evernia vulpina, Polypodium californicum, Trillium sessile*
Disinfectant: *Chlorogalum pomeridianum*
Diuretic: *Chlorogalum pomeridianum*
Gastrointestinal Aid: *Chlorogalum pomeridianum*
Laxative: *Chlorogalum pomeridianum*

Other: *Trillium sessile*
Panacea: *Trillium sessile*
Snakebite Remedy: *Toxicodendron diversilobum*
Food
Candy: *Pinus sabiniana*
Forage: *Robinia pseudoacacia, Trifolium ciliolatum*
Fruit: *Arbutus menziesii*
Unspecified: *Trifolium ciliolatum*
Fiber
Basketry: *Carex* sp.
Clothing: *Carex* sp.
Dye
Red: *Alnus rhombifolia*
Other
Hunting & Fishing Item: *Acer macrophyllum, Philadelphus lewisii*

Walapai
Drug
Venereal Aid: *Aloysia wrightii*
Food
Beverage: *Yucca baccata*
Dried Food: *Yucca baccata*
Fiber
Furniture: *Acacia* sp.
Dye
Yellow: *Mahonia fremontii*
Other
Toys & Games: *Cowania* sp.

Warihio
Food
Beverage: *Panicum sonorum*
Staple: *Panicum sonorum*

Washo
Drug
Adjuvant: *Angelica breweri*
Analgesic: *Artemisia douglasiana, A. ludoviciana, Juniperus occidentalis, J. osteosperma, Monardella odoratissima, Osmorhiza occidentalis, Paeonia brownii, Zigadenus paniculatus*
Antidiarrheal: *Heracleum maximum, Ipomopsis congesta, Mentha canadensis, Typha latifolia*
Antirheumatic (External): *Arenaria congesta, Artemisia douglasiana, Veratrum californicum, Zigadenus paniculatus*
Blood Medicine: *Maianthemum stellatum*
Burn Dressing: *Argemone polyanthemos*
Cathartic: *Ipomopsis congesta, Osmorhiza occidentalis, Purshia tridentata, Wyethia mollis*
Cold Remedy: *Artemisia ludoviciana, A. tridentata, Ipomopsis congesta, Juniperus occidentalis, J. osteosperma, Lomatium dissectum, Mentha canadensis, Monardella odoratissima, Osmorhiza occidentalis, Pinus monophylla, Rosa woodsii, Salvia dorrii, Thalictrum fendleri, T.* sp.
Contraceptive: *Veratrum californicum*
Cough Medicine: *Angelica breweri, Artemisia ludoviciana, Lomatium dissectum*
Dermatological Aid: *Achillea millefolium, Argemone polyanthemos, Lomatium dissectum, Maianthemum stellatum, Pedicularis attollens, Veratrum californicum, Zigadenus paniculatus*
Disinfectant: *Artemisia tridentata, Balsamorhiza sagittata, Juniperus occidentalis, J. osteosperma*
Emetic: *Ipomopsis congesta, Veratrum californicum, Wyethia mollis*
Eye Medicine: *Phlox longifolia*
Febrifuge: *Mentha canadensis*
Gastrointestinal Aid: *Ipomopsis congesta, Mentha canadensis, Monardella odoratissima, Osmorhiza occidentalis*
Gynecological Aid: *Balsamorhiza hookeri, Ephedra viridis*
Hemostat: *Maianthemum stellatum*
Herbal Steam: *Lomatium dissectum*
Kidney Aid: *Ipomopsis congesta*
Misc. Disease Remedy: *Angelica breweri, Lomatium dissectum, Osmorhiza occidentalis*
Orthopedic Aid: *Zigadenus paniculatus*
Panacea: *Lomatium dissectum*
Pediatric Aid: *Lomatium dissectum, Mentha canadensis*
Pulmonary Aid: *Lomatium dissectum, Osmorhiza occidentalis*
Respiratory Aid: *Angelica breweri, Lomatium dissectum, Salvia dorrii*
Throat Aid: *Angelica breweri, Lomatium dissectum*
Tonic: *Artemisia tridentata, Maianthemum stellatum, Pedicularis attollens*
Toothache Remedy: *Heracleum maximum, Zigadenus paniculatus*
Tuberculosis Remedy: *Abies concolor, Lomatium dissectum, Paeonia brownii*
Venereal Aid: *Pinus monophylla*
Food
Beverage: *Rosa* sp.
Candy: *Lygodesmia* sp.
Dried Food: *Pinus monophylla*
Fruit: *Rosa nutkana*
Porridge: *Pinus monophylla*
Fiber
Basketry: *Cornus sericea*
Other
Ceremonial Items: *Artemisia tridentata*
Designs: *Cornus sericea*
Hunting & Fishing Item: *Calocedrus decurrens, Clematis ligusticifolia, Philadelphus* sp.

West Coast Indian
Drug
Cathartic: *Frangula purshiana*
Panacea: *Frangula purshiana*
Poison: *Frangula purshiana*
Venereal Aid: *Frangula purshiana*

Wet'suwet'en
Drug
Antirheumatic (External): *Oplopanax horridus*
Burn Dressing: *Lonicera involucrata*
Cancer Treatment: *Oplopanax horridus*
Cold Remedy: *Abies lasiocarpa, Oplopanax horridus, Picea glauca*
Cough Medicine: *Abies lasiocarpa, Oplopanax horridus, Picea glauca, Prunus pensylvanica*
Dermatological Aid: *Cornus sericea, Lonicera involucrata, Oplopanax horridus*
Disinfectant: *Lonicera involucrata*
Eye Medicine: *Symphoricarpos albus*
Febrifuge: *Cornus sericea, Sorbus scopulina*
Gastrointestinal Aid: *Oplopanax horridus*
Hemorrhoid Remedy: *Cornus sericea*
Misc. Disease Remedy: *Abies lasiocarpa, Oplopanax horridus, Picea glauca*
Respiratory Aid: *Oplopanax horridus*
Tonic: *Abies lasiocarpa, Oplopanax horridus, Picea glauca, Sorbus scopulina*
Tuberculosis Remedy: *Oplopanax horridus*
Unspecified: *Sambucus racemosa, Sorbus scopulina*

Food
Dried Food: *Pinus contorta*
Sweetener: *Tsuga heterophylla*
Unspecified: *Epilobium angustifolium, Heracleum maximum, Picea glauca, Pinus contorta, Sedum divergens, Tsuga heterophylla*
Winter Use Food: *Tsuga heterophylla*
Fiber
Basketry: *Thuja plicata*
Building Material: *Thuja plicata*
Cordage: *Salix scouleriana, Thuja plicata*
Mats, Rugs & Bedding: *Thuja plicata*
Other
Ceremonial Items: *Oplopanax horridus*
Containers: *Betula papyrifera*
Cooking Tools: *Salix scouleriana*
Hunting & Fishing Item: *Salix scouleriana*
Lighting: *Betula papyrifera*

Weyot
Drug
Antidiarrheal: *Lathyrus* sp.
Unspecified: *Gnaphalium* sp.

Winnebago
Drug
Analgesic: *Echinacea angustifolia, Physalis lanceolata, Silphium perfoliatum*
Anticonvulsive: *Heracleum maximum*
Antidiarrheal: *Rhus* sp.
Antidote: *Echinacea angustifolia*
Antirheumatic (Internal): *Silphium perfoliatum*
Burn Dressing: *Echinacea angustifolia, Typha latifolia*
Carminative: *Acorus calamus, Mentha canadensis*
Ceremonial Medicine: *Acorus calamus, Artemisia* sp., *Hierochloe odorata, Lophophora williamsii, Prunus virginiana, Silphium perfoliatum*
Cold Remedy: *Acorus calamus, Silphium perfoliatum*
Cough Medicine: *Acorus calamus*
Dermatological Aid: *Achillea millefolium, Monarda fistulosa, Osmorhiza longistylis, Physalis lanceolata, Typha latifolia*
Disinfectant: *Artemisia* sp.
Ear Medicine: *Achillea millefolium*
Emetic: *Silphium perfoliatum*
Febrifuge: *Acorus calamus, Artemisia dracunculus*
Gastrointestinal Aid: *Acorus calamus, Artemisia* sp., *Physalis lanceolata, Quercus* sp.
Gynecological Aid: *Euonymus atropurpurea, Ribes americanum*
Herbal Steam: *Silphium perfoliatum*
Laxative: *Gymnocladus dioicus, Ulmus rubra*
Love Medicine: *Artemisia dracunculus, Lomatium foeniculaceum*
Misc. Disease Remedy: *Echinacea angustifolia*
Other: *Echinacea angustifolia*
Panacea: *Acorus calamus, Cucurbita foetidissima*
Pediatric Aid: *Quercus* sp., *Typha latifolia*
Snakebite Remedy: *Echinacea angustifolia*
Stimulant: *Gymnocladus dioicus, Heracleum maximum*
Tonic: *Acorus calamus, Gentiana saponaria*
Toothache Remedy: *Acorus calamus, Echinacea angustifolia*
Unspecified: *Artemisia dracunculus, Physalis lanceolata*
Veterinary Aid: *Echinacea angustifolia*
Witchcraft Medicine: *Artemisia* sp., *Silphium laciniatum*
Food
Beverage: *Agastache foeniculum, Ceanothus americanus, Fragaria vesca, F. virginiana, Mentha canadensis*
Candy: *Silphium laciniatum*
Dried Food: *Pediomelum esculentum, Prunus americana, Shepherdia argentea, Vitis cinerea, V. vulpina*
Fruit: *Amelanchier alnifolia, Crataegus chrysocarpa, Fragaria vesca, F. virginiana, Prunus americana, Ribes missouriense, Shepherdia argentea, Viburnum lentago, Vitis cinerea, V. vulpina*
Sauce & Relish: *Allium canadense, Prunus americana*
Soup: *Carya ovata, Corylus americana, Juglans nigra, Nelumbo lutea*
Spice: *Allium canadense*
Staple: *Zizania aquatica*
Starvation Food: *Crataegus chrysocarpa*
Sweetener: *Acer negundo, A. saccharinum, Agastache foeniculum, Carya ovata*
Unspecified: *Allium canadense, Amphicarpaea bracteata, Apios americana, Carya ovata, Corylus americana, Erythronium mesochoreum, Gymnocladus dioicus, Helianthus tuberosus, Juglans nigra, Linum lewisii, Nelumbo lutea, Pediomelum esculentum, Quercus macrocarpa, Sagittaria latifolia*
Vegetable: *Asclepias syriaca*
Fiber
Brushes & Brooms: *Prunus americana*
Building Material: *Salix* sp., *Ulmus rubra*
Clothing: *Urtica dioica*
Cordage: *Ulmus rubra, Urtica dioica*
Mats, Rugs & Bedding: *Artemisia dracunculus, A. frigida, A. ludoviciana*
Dye
Black: *Acer saccharinum, Juglans nigra*
Red: *Sanguinaria canadensis*
Unspecified: *Sanguinaria canadensis*
Yellow: *Rhus glabra*
Other
Ceremonial Items: *Nelumbo lutea, Urtica dioica*
Cooking Tools: *Ulmus rubra*
Fuel: *Ceanothus americanus, Ulmus rubra*
Hunting & Fishing Item: *Amelanchier alnifolia, Cornus asperifolia, Fraxinus pennsylvanica, Mentha canadensis*
Incense & Fragrance: *Mentha canadensis*
Smoke Plant: *Cornus amomum, Nicotiana quadrivalvis, Rhus glabra*
Smoking Tools: *Fraxinus pennsylvanica*
Stable Gear: *Urtica dioica*
Tools: *Ulmus rubra*
Toys & Games: *Amelanchier alnifolia, Artemisia* sp., *Asclepias syriaca, Betula papyrifera, Equisetum* sp., *Gymnocladus dioicus, Quercus macrocarpa, Ulmus americana, U. rubra, Urtica dioica, Viburnum opulus*

Wintoon
Drug
Unspecified: *Garrya* sp.
Food
Dried Food: *Quercus chrysolepis*
Fruit: *Amelanchier* sp., *Arctostaphylos patula, A. viscida, Prunus subcordata, P. virginiana, Rhus trilobata, Ribes* sp., *Rubus parviflorus, R. vitifolius, Sambucus cerulea, Vitis californica*
Staple: *Hemizonia* sp., *Madia* sp.
Unspecified: *Quercus chrysolepis, Q. lobata*
Fiber
Basketry: *Adiantum* sp., *Cercis* sp., *Cornus glabrata, Corylus cornuta, Pinus ponderosa, Rhus trilobata, Xerophyllum* sp.
Building Material: *Pinus sabiniana*
Cordage: *Apocynum* sp., *Asclepias* sp., *Iris* sp.

Dye
Red: *Alnus rubra*
Yellow: *Datisca glomerata, Mahonia* sp.
Other
Hunting & Fishing Item: *Physocarpus capitatus, Taxus* sp.
Lighting: *Rhamnus crocea*
Musical Instrument: *Sambucus cerulea*
Tools: *Aristolochia* sp., *Cercocarpus montanus*

Yana
Drug
Analgesic: *Angelica tomentosa*
Antidiarrheal: *Angelica tomentosa*
Burn Dressing: *Pteridium aquilinum*
Cold Remedy: *Angelica tomentosa*
Cough Medicine: *Iris* sp.
Dermatological Aid: *Smilacina* sp.
Febrifuge: *Calycadenia fremontii*
Panacea: *Angelica tomentosa*
Food
Bread & Cake: *Quercus douglasii*
Dried Food: *Quercus douglasii*
Porridge: *Berberis* sp., *Quercus douglasii*
Spice: *Sidalcea malviflora*
Staple: *Aesculus californica, Quercus douglasii*
Unspecified: *Angelica tomentosa*, Apiaceae sp., *Brodiaea minor, Clarkia rhomboidea, Fritillaria* sp., *Helianthella californica, Lilium pardalinum, Perideridia gairdneri, P. pringlei, Triteleia grandiflora*
Fiber
Clothing: *Xerophyllum tenax*

Yaqui
Food
Vegetable: *Chenopodium* sp., *Sonchus oleraceus*

Yavapai
Drug
Antirheumatic (External): *Baccharis pteronioides, Larrea tridentata*
Cathartic: *Arenaria* sp., *Iris missouriensis, Simmondsia chinensis*
Cold Remedy: *Cyperus* sp., *Senecio multilobatus*
Cough Medicine: *Rumex crispus*
Dermatological Aid: *Castela emoryi, Cyperus* sp., *Euphorbia* sp., *Larrea tridentata, Pellaea mucronata, Perezia* sp., *Rumex crispus, Senecio multilobatus, Simmondsia chinensis*
Gastrointestinal Aid: *Arenaria* sp., *Clematis ligusticifolia, Cyperus* sp., *Mimulus guttatus, Rumex crispus, Senecio multilobatus*
Gynecological Aid: *Calliandra eriophylla, Pellaea mucronata*
Hunting Medicine: *Datura wrightii*
Nose Medicine: *Senecio multilobatus*
Pediatric Aid: *Perezia* sp., *Rumex crispus*
Throat Aid: *Larrea tridentata, Rumex crispus*
Toothache Remedy: *Rumex crispus*
Venereal Aid: *Baccharis pteronioides, Euphorbia* sp., *Larrea tridentata, Mirabilis coccineus, Rumex crispus, Senecio multilobatus*
Food
Beverage: *Agave* sp., *Arctostaphylos pungens, Carnegia gigantea, Juglans major, Juniperus deppeana, J. osteosperma, Psoralidium tenuiflorum, Rhus trilobata, Yucca baccata, Ziziphus obtusifolia*
Bread & Cake: *Carnegia gigantea, Juniperus deppeana, J. osteosperma, Olneya tesota, Opuntia* sp.
Cooking Agent: *Quercus emoryi, Q. gambelii*
Dried Food: *Agave* sp., *Carnegia gigantea, Opuntia* sp., *Yucca baccata*
Fruit: *Arctostaphylos pungens, Carnegia gigantea, Echinocereus engelmannii, Mahonia fremontii, Morus microphylla, Opuntia chlorotica, O. echinocarpa, O. erinacea, O.* sp., *Rhus ovata, Ribes aureum, Yucca baccata, Y.* sp.
Preserves: *Simmondsia chinensis*
Snack Food: *Fouquieria splendens*
Staple: *Juniperus deppeana, J. osteosperma, Olneya tesota, Prosopis glandulosa, Xanthium* sp.
Starvation Food: *Rumex crispus*
Unspecified: *Agave* sp., *Celtis laevigata, Chenopodium* sp., *Cirsium neomexicanum, Cordylanthus* sp., *Echinocereus engelmannii, Frasera* sp., *Juglans major, Prosopis glandulosa, Quercus gambelii, Rhus trilobata, Salvia columbariae, Yucca baccata, Y.* sp.
Vegetable: *Amaranthus* sp., *Chenopodium* sp., *Lupinus* sp., *Scrophularia* sp.
Winter Use Food: *Juglans major, Quercus emoryi*
Fiber
Brushes & Brooms: *Agave* sp., *Chrysothamnus viscidiflorus, Eriogonum* sp.
Building Material: *Nolina microcarpa*
Cordage: *Yucca baccata*
Other
Containers: *Sphaeralcea ambigua*
Fuel: *Ceanothus greggii, Juniperus deppeana, J. osteosperma*
Lighting: *Juniperus osteosperma*
Smoke Plant: *Nicotiana attenuata*
Smoking Tools: *Eriogonum inflatum*
Soap: *Yucca baccata*

Yokia
Drug
Analgesic: *Pinus sabiniana*
Antidiarrheal: *Matricaria discoidea*
Antirheumatic (External): *Pinus sabiniana*
Dermatological Aid: *Fraxinus latifolia, Pinus sabiniana, Rhus trilobata*
Diaphoretic: *Pinus sabiniana*
Disinfectant: *Sambucus cerulea*
Emetic: *Sambucus cerulea*
Gastrointestinal Aid: *Sambucus cerulea*
Misc. Disease Remedy: *Frangula californica, Rhus trilobata*
Oral Aid: *Iris douglasiana*
Pediatric Aid: *Iris douglasiana*
Tuberculosis Remedy: *Pinus sabiniana*
Food
Unspecified: *Asclepias fascicularis, Trifolium ciliolatum*
Vegetable: *Lathyrus jepsonii*
Fiber
Brushes & Brooms: *Symphoricarpos albus*
Sewing Material: *Hoita macrostachya*
Other
Containers: *Arctostaphylos manzanita*
Hunting & Fishing Item: *Calycanthus occidentalis, Symphoricarpos albus*

Yokut
Drug
Burn Dressing: *Sambucus cerulea*
Cathartic: *Atriplex* sp., *Sambucus cerulea*
Ceremonial Medicine: *Datura wrightii*
Cold Remedy: *Distichlis spicata, Lobelia* sp.
Diaphoretic: *Eriodictyon californicum*
Dietary Aid: *Distichlis spicata*
Emetic: *Lobelia* sp., *Sambucus cerulea*
Gastrointestinal Aid: *Datura wrightii*

Other: *Datura wrightii*
Unspecified: *Artemisia ludoviciana*
Food
Dried Food: *Sambucus cerulea*
Fruit: *Rhus trilobata*
Unspecified: *Quercus douglasii, Q. kelloggii, Q. lobata*
Winter Use Food: *Sambucus cerulea*
Fiber
Basketry: *Cladium* sp., *Pteridium* sp.
Building Material: *Umbellularia* sp.
Cordage: *Fremontodendron californicum*
Other
Musical Instrument: *Sambucus cerulea*
Toys & Games: *Sambucus cerulea*

Yuki
Drug
Analgesic: *Artemisia douglasiana, Lomatium californicum, Umbellularia californica*
Antidiarrheal: *Artemisia douglasiana, Quercus lobata, Rumex crispus*
Antirheumatic (External): *Artemisia douglasiana, Lomatium californicum, Umbellularia californica*
Cold Remedy: *Achillea millefolium, Lomatium californicum, Populus fremontii*
Cough Medicine: *Eriodictyon californicum, Marrubium vulgare*
Dermatological Aid: *Arbutus menziesii, Artemisia douglasiana, Eriodictyon californicum, Evernia vulpina, Lonicera interrupta, Populus fremontii, Rumex crispus, Symphoricarpos albus, Toxicodendron diversilobum, Trillium sessile*
Diuretic: *Equisetum telmateia*
Emetic: *Arbutus menziesii, Wyethia angustifolia*
Eye Medicine: *Equisetum variegatum*
Febrifuge: *Sambucus cerulea*
Gastrointestinal Aid: *Arbutus menziesii*
Gynecological Aid: *Artemisia douglasiana*
Orthopedic Aid: *Artemisia douglasiana, Thalictrum polycarpum*
Other: *Lomatium californicum, Trillium sessile*
Panacea: *Trillium sessile*
Pediatric Aid: *Rumex crispus*
Poison: *Anthemis cotula, Zigadenus venenosus*
Respiratory Aid: *Achillea millefolium, Umbellularia californica*
Throat Aid: *Populus fremontii*
Unspecified: *Boykinia occidentalis, Rhamnus crocea*
Veterinary Aid: *Arbutus menziesii, Artemisia douglasiana*
Food
Beverage: *Arctostaphylos manzanita, Mentha spicata*
Bread & Cake: *Hordeum vulgare, Lithocarpus densiflorus, Quercus kelloggii, Q. lobata*
Candy: *Pinus lambertiana*
Dried Food: *Rubus leucodermis, Sambucus cerulea*
Forage: *Chlorogalum pomeridianum*
Fruit: *Arbutus menziesii, Arctostaphylos manzanita, Prunus virginiana, Rubus leucodermis, Sambucus cerulea*
Porridge: *Lithocarpus densiflorus, Quercus kelloggii, Q. lobata*
Soup: *Lithocarpus densiflorus, Sambucus cerulea*
Spice: *Wyethia angustifolia*
Staple: *Arctostaphylos manzanita, Avena* sp., *Lolium temulentum, Pogogyne douglasii*
Substitution Food: *Hordeum vulgare, Linanthus ciliatus, Pseudotsuga menziesii*
Unspecified: *Aesculus californica, Allium unifolium, Calochortus tolmiei, Camassia leichtlinii, C. quamash, Corylus cornuta, Cynoglossum grande, Dichelostemma pulchellum, Dodecatheon hendersonii, Heracleum maximum, Lomatium californicum, Perideridia kelloggii, Trifolium* sp., *Triteleia laxa, Typha* sp., *Umbellularia californica*
Vegetable: *Lupinus* sp., *Plagiobothrys nothofulvus*
Fiber
Basketry: *Carex* sp., *Cercis canadensis, Cornus glabrata, C.* sp., *Corylus cornuta, Salix* sp.
Brushes & Brooms: *Perideridia kelloggii*
Building Material: *Pseudotsuga menziesii*
Cordage: *Vicia americana*
Scouring Material: *Equisetum telmateia*
Sewing Material: *Cercis canadensis*
Other
Fasteners: *Pinus sabiniana*
Fuel: *Fraxinus latifolia*
Good Luck Charm: *Lomatium californicum*
Hunting & Fishing Item: *Apocynum cannabinum, Cercocarpus montanus, Chlorogalum pomeridianum, Lomatium californicum, Philadelphus lewisii*
Incense & Fragrance: *Mentha spicata*
Insecticide: *Umbellularia californica*
Musical Instrument: *Sambucus cerulea*
Protection: *Ligusticum porteri, Lomatium californicum*
Smoke Plant: *Nicotiana* sp.
Smoking Tools: *Fraxinus latifolia*
Soap: *Chlorogalum pomeridianum*
Walking Sticks: *Cercocarpus montanus*

Yuma
Drug
Narcotic: *Datura wrightii*
Food
Beverage: *Lycium exsertum, L. fremontii, Prosopis glandulosa, Salix gooddingii*
Bread & Cake: *Olneya tesota, Prosopis glandulosa, Typha latifolia*
Dried Food: *Cucurbita moschata, Lycium exsertum, L. fremontii, Olneya tesota, Prosopis glandulosa, Typha latifolia*
Fruit: *Physalis hederifolia, P. pubescens*
Porridge: *Atriplex lentiformis, Echinochloa crus-galli, Lycium exsertum, L. fremontii, Typha latifolia*
Spice: *Typha latifolia*
Staple: *Agastache pallidiflora, Agave americana, A. parryi, Allenrolfea occidentalis, Amaranthus palmeri, Echinochloa crus-galli, Panicum hirticaule, Prosopis glandulosa*
Starvation Food: *Baccharis salicifolia, Parkinsonia florida, P. microphylla*
Unspecified: *Amaranthus palmeri, Ammannia coccinea, Atriplex lentiformis, A. polycarpa, Echinochloa crus-galli, Hesperocallis undulata, Phragmites australis, Prosopis glandulosa, Salix gooddingii, Typha latifolia*
Vegetable: *Amaranthus palmeri*
Winter Use Food: *Amaranthus palmeri*
Other
Containers: *Lagenaria siceraria*
Hunting & Fishing Item: *Ferocactus wislizeni*
Musical Instrument: *Lagenaria siceraria*
Smoke Plant: *Nicotiana trigonophylla*
Smoking Tools: *Phragmites australis*

Yurok
Drug
Anthelmintic: *Artemisia douglasiana*
Antirheumatic (External): *Artemisia douglasiana, Chimaphila umbellata, Lysichiton americanus*
Blood Medicine: *Satureja douglasii, Taxus brevifolia*

Burn Dressing: *Trillium chloropetalum*
Cold Remedy: *Eriodictyon californicum*
Cough Medicine: *Eriodictyon californicum, Vancouveria hexandra*
Dermatological Aid: *Asarum caudatum, Plantago australis, P. major*
Disinfectant: *Polypodium californicum*
Eye Medicine: *Achillea millefolium*
Kidney Aid: *Chimaphila umbellata*
Laxative: *Frangula purshiana*
Misc. Disease Remedy: *Lysichiton americanus*
Orthopedic Aid: *Artemisia douglasiana*
Pediatric Aid: *Artemisia douglasiana, Asarum caudatum*
Poison: *Toxicodendron diversilobum*
Strengthener: *Lithocarpus densiflorus*
Unspecified: *Populus balsamifera, Rosa pisocarpa, Umbellularia californica*
Food
Beverage: *Ledum glandulosum, Marah oreganus, Pseudotsuga menziesii, Rubus vitifolius*
Bread & Cake: *Lithocarpus densiflorus*
Candy: *Asclepias cordifolia, Pseudotsuga menziesii*
Cooking Agent: *Dryopteris arguta*
Fruit: *Arbutus menziesii, Arctostaphylos ×cinerea, A. uva-ursi, Fragaria chiloensis, F. vesca, Gaultheria shallon, Heteromeles arbutifolia, Prunus virginiana, Quercus kelloggii, Ribes roezlii, Rubus leucodermis, R. parviflorus, R. spectabilis, R. vitifolius, Sambucus cerulea, S. racemosa, Umbellularia californica, Vaccinium parvifolium, V. scoparium, Vitis californica*
Soup: *Lithocarpus densiflorus*
Staple: *Lithocarpus densiflorus*
Unspecified: *Aquilegia formosa, Armillaria ponderosa, Boschniakia strobilacea, Castanopsis chrysophylla, Claytonia sibirica, Corylus cornuta, Equisetum telmateia, Heracleum maximum, Lysichiton americanus, Oxalis oregana, Porphyra lanceolata, Umbellularia californica*
Vegetable: *Brodiaea elegans*
Fiber
Basketry: *Adiantum aleuticum, Alnus rubra, Corylus cornuta, Juncus effusus, Picea sitchensis, Salix scouleriana, S. sitchensis*
Brushes & Brooms: *Vaccinium parvifolium, V. scoparium*
Building Material: *Sequoia sempervirens*
Canoe Material: *Sequoia sempervirens*
Clothing: *Typha latifolia, Xerophyllum tenax*
Mats, Rugs & Bedding: *Blechnum spicant, Polystichum munitum, Typha latifolia*
Snow Gear: *Arbutus menziesii*
Unspecified: *Woodwardia fimbriata*
Dye
Unspecified: *Alnus rubra, Letharia vulpina*
Yellow: *Mahonia aquifolium*
Other
Cash Crop: *Sequoia sempervirens*
Ceremonial Items: *Lomatium californicum*
Containers: *Dryopteris arguta, Lysichiton americanus, Pteridium aquilinum, Sambucus cerulea, S. racemosa*
Cooking Tools: *Lysichiton americanus, Pteridium aquilinum*
Decorations: *Juniperus communis*
Designs: *Dryopteris arguta, Pentagrama triangularis*
Good Luck Charm: *Umbellularia californica*
Hunting & Fishing Item: *Taxus brevifolia*
Insecticide: *Darlingtonia californica, Umbellularia californica*
Season Indicator: *Cardamine californica*
Tools: *Populus balsamifera*
Toys & Games: *Acer glabrum, Chamaecyparis lawsoniana, Claytonia perfoliata, Marah oreganus*

Yurok, South Coast (Nererner)
Drug
Cathartic: *Frangula purshiana*
Dermatological Aid: *Artemisia ludoviciana*
Eye Medicine: *Artemisia ludoviciana*
Food
Dried Food: *Fucus* sp.
Staple: *Lithocarpus densiflorus*
Other
Tools: *Garrya elliptica*

Zuni
Drug
Abortifacient: *Ambrosia acanthicarpa*
Analgesic: *Artemisia carruthii, Bahia dissecta, Chaetopappa ericoides, Chenopodium graveolens, Erysimum* sp., *Gilia* sp., *Ipomopsis multiflora, Senecio multicapitatus, Solidago canadensis, Verbesina encelioides*
Anesthetic: *Datura wrightii*
Antidiarrheal: *Plantago patagonica*
Antiemetic: *Croton texensis*
Antirheumatic (External): *Ageratina occidentalis, Artemisia tridentata, Bahia dissecta, Berula erecta, Chaetopappa ericoides, Cucurbita foetidissima, C. pepo, Erysimum capitatum, Gutierrezia sarothrae, Juniperus monosperma, Ligusticum porteri, Lobelia cardinalis, Oenothera coronopifolia, O. elata, O. triloba*
Burn Dressing: *Achillea millefolium, Krascheninnikovia lanata*
Carminative: *Pectis papposa*
Cathartic: *Chamaesyce serpyllifolia, Croton texensis, Polygonum lapathifolium*
Ceremonial Medicine: *Ligusticum porteri, Lithospermum incisum, Senecio multicapitatus, Xanthium strumarium*
Cold Remedy: *Artemisia frigida, A. tridentata, Chaetopappa ericoides*
Contraceptive: *Cirsium ochrocentrum, Juniperus monosperma*
Cough Medicine: *Salix exigua*
Dermatological Aid: *Ageratina occidentalis, Artemisia tridentata, Aster falcatus, A. lanceolatus, Atriplex argentea, A. canescens, Berula erecta, Calliandra humilis, Campanula parryi, Chaetopappa ericoides, Cryptantha cinerea, Cucurbita pepo, Dalea compacta, Datura wrightii, Dimorphocarpa wislizeni, Eriogonum fasciculatum, Erodium cicutarium, Erysimum* sp., *Hymenopappus filifolius, Hymenoxys richardsonii, Ipomopsis longiflora, I. multiflora, Lithospermum incisum, Lobelia cardinalis, Oenothera triloba, Phacelia neomexicana, Picradeniopsis woodhousei, Pinus edulis, Rumex crispus, Stanleya pinnata, Verbascum thapsus, Xanthium strumarium, Zinnia grandiflora*
Diaphoretic: *Cirsium ochrocentrum, Gutierrezia sarothrae, Pinus edulis, Tripterocalyx carnea, Zinnia grandiflora*
Dietary Aid: *Mirabilis multiflora*
Disinfectant: *Pinus edulis, Psoralidium tenuiflorum*
Diuretic: *Cirsium ochrocentrum, Croton texensis, Gilia* sp., *Gutierrezia sarothrae, Mirabilis linearis, Pinus edulis*
Emetic: *Chamaesyce serpyllifolia, Cirsium ochrocentrum, Dimorphocarpa wislizeni, Eriogonum alatum, Erysimum capitatum, Gilia* sp., *Hymenopappus filifolius, Machaeranthera canescens, Mirabilis linearis, Phoradendron juniperinum, Picradeniopsis woodhousei, Polygonum lapathifolium, Ratibida columnifera, Verbesina encelioides*
Eye Medicine: *Eriogonum jamesii, Linum puberulum, Pectis papposa, Rorippa sinuata, Senecio multicapitatus, Tetraneuris scaposa, Zinnia grandiflora*
Febrifuge: *Gilia* sp., *Zinnia grandiflora*
Gastrointestinal Aid: *Abronia fragrans, Atriplex argentea, Croton texensis, Dalea compacta, Eriogonum jamesii, Erodium cicutarium, Hymenoxys richardsonii, Lithospermum incisum, Mirabilis linearis, M. multiflora, Picradeniopsis woodhousei, Psoralidium lanceolatum, Solanum rostratum, Verbesina encelioides*
Gynecological Aid: *Chaetopappa ericoides, Chamaesyce albomarginata,*

[Gynecological Aid]
C. polycarpa, C. serpyllifolia, Eriogonum fasciculatum, Juniperus monosperma, Phoradendron juniperinum
Hemorrhoid Remedy: *Orobanche fasciculata*
Hemostat: *Aster lanceolatus, Juniperus monosperma, Phoradendron juniperinum*
Herbal Steam: *Chenopodium graveolens*
Hunting Medicine: *Atriplex canescens, Penstemon barbatus*
Kidney Aid: *Lithospermum incisum*
Laxative: *Gilia sp., Mentzelia pumila*
Misc. Disease Remedy: *Cirsium ochrocentrum*
Narcotic: *Datura wrightii*
Oral Aid: *Eriogonum jamesii, Glycyrrhiza lepidota*
Orthopedic Aid: *Senecio multicapitatus*
Other: *Eriogonum alatum*
Pediatric Aid: *Aster falcatus, Ipomopsis longiflora, Iris missouriensis, Mentzelia pumila, Mirabilis multiflora, Phaseolus angustissimus*
Poultice: *Grindelia nuda*
Psychological Aid: *Dimorphocarpa wislizeni*
Pulmonary Aid: *Ipomopsis multiflora*
Reproductive Aid: *Coreopsis tinctoria, Rumex salicifolius*
Respiratory Aid: *Conyza canadensis*
Snakebite Remedy: *Amsonia tomentosa, Astragalus amphioxys, Croton texensis, Gaura parviflora, Grindelia nuda, Helianthus annuus, Nicotiana attenuata, Psilostrophe tagetina, Solanum elaeagnifolium, Stephanomeria tenuifolia, Tripterocalyx carnea, Verbesina encelioides*
Stimulant: *Cryptantha crassisepala*
Strengthener: *Aster falcatus, Gutierrezia sarothrae, Iris missouriensis, Mentzelia pumila, Phaseolus angustissimus*
Throat Aid: *Eriogonum fasciculatum, Gilia sp., Ligusticum porteri, Lithospermum incisum, Rumex salicifolius, Salix exigua, Tripterocalyx carnea*
Toothache Remedy: *Ambrosia acanthicarpa, Solanum elaeagnifolium*
Unspecified: *Asclepias involucrata, Machaeranthera tanacetifolia*
Urinary Aid: *Gutierrezia sarothrae*
Venereal Aid: *Cirsium ochrocentrum, Croton texensis, Ephedra nevadensis, Pinus edulis, Stanleya pinnata*
Veterinary Aid: *Caesalpinia jamesii*
Witchcraft Medicine: *Lotus wrightii, Tripterocalyx carnea*

Food
Beverage: *Coreopsis tinctoria, Ephedra nevadensis, Solanum elaeagnifolium, Triticum aestivum, Zea mays*
Bread & Cake: *Amaranthus blitoides, Artemisia carruthii, Chenopodium leptophyllum, Triticum aestivum, Xanthium strumarium, Zea mays*
Candy: *Chamaesyce serpyllifolia, Dalea lasiathera, Hymenopappus filifolius, Lactuca tatarica, Populus angustifolia, P. deltoides*
Cooking Agent: *Amaranthus cruentus*
Dried Food: *Astragalus lentiginosus, Cleome serrulata, Cucurbita pepo, Opuntia whipplei*
Forage: *Asclepias involucrata*
Fruit: *Lycium pallidum, Opuntia whipplei, Physalis longifolia, Ribes cereum, Yucca baccata*
Porridge: *Atriplex powellii, Cycloloma atriplicifolium, Opuntia whipplei, Zea mays*
Preserves: *Yucca baccata*
Sauce & Relish: *Coriandrum sativum, Physalis hederifolia, Solanum triflorum, Yucca baccata*
Snack Food: *Zea mays*
Special Food: *Cucurbita pepo*
Spice: *Dalea lasiathera*
Staple: *Cycloloma atriplicifolium, Oryzopsis hymenoides, Zea mays*
Sweetener: *Chamaesyce serpyllifolia, Yucca baccata*
Unspecified: *Artemisia carruthii, Asclepias subverticillata, Astragalus lentiginosus, Chenopodium leptophyllum, Cleome serrulata, Cucurbita pepo, Oenothera triloba, Oryzopsis hymenoides, Populus angustifolia, P. deltoides, Ribes cereum, Solanum fendleri, Yucca glauca*
Vegetable: *Chenopodium album, C. leptophyllum, Coriandrum sativum, Phaseolus vulgaris*
Winter Use Food: *Pinus edulis*

Fiber
Basketry: *Chrysothamnus nauseosus, Parryella filifolia, Rhus trilobata, Salix irrorata, Yucca baccata*
Brushes & Brooms: *Bouteloua gracilis, Yucca glauca*
Building Material: *Salix irrorata, Sporobolus contractus*
Clothing: *Asclepias subverticillata, Gossypium hirsutum, Yucca baccata*
Cordage: *Asclepias subverticillata, Gossypium hirsutum, Yucca baccata, Y. glauca*
Mats, Rugs & Bedding: *Sporobolus contractus, Yucca baccata, Y. glauca*

Dye
Black: *Castilleja integra*
Red: *Coreopsis tinctoria*
Red-Brown: *Alnus incana*
Yellow: *Chrysothamnus nauseosus, Psilostrophe tagetina*

Other
Ceremonial Items: *Amaranthus cruentus, Asclepias subverticillata, Atriplex canescens, Cleome serrulata, Cucurbita pepo, Datura wrightii, Eriogonum jamesii, Erysimum sp., Gossypium hirsutum, Helianthus annuus, Mahonia fremontii, Muhlenbergia rigens, Nicotiana attenuata, Oenothera albicaulis, Opuntia imbricata, Pectis papposa, Penstemon barbatus, Polanisia dodecandra, Psilostrophe tagetina, Yucca baccata, Zea mays*
Containers: *Cucurbita pepo*
Cooking Tools: *Bouteloua gracilis, Cucurbita pepo, Salix irrorata*
Decorations: *Artemisia frigida, Cleome serrulata, Zea mays*
Fertilizer: *Artemisia frigida, Thelypodium wrightii*
Fuel: *Juniperus monosperma*
Incense & Fragrance: *Pectis papposa*
Paint: *Amaranthus cruentus, Cleome serrulata, Mahonia fremontii, Psilostrophe tagetina*
Protection: *Cycloloma atriplicifolium, Lycium pallidum*
Smoke Plant: *Nicotiana attenuata*
Soap: *Yucca baccata, Y. glauca*
Tools: *Juniperus monosperma, Yucca baccata*
Toys & Games: *Zea mays*
Weapon: *Lithospermum incisum*

Index of Usages

Plants are listed under the categories of usage, which are arranged first by the five main categories—Drug, Food, Fiber, Dye, Other—and then alphabetically by the particular usage. Plants are identified below to the level of genus. For all usages given below, check all entries under that genus in the Catalog of Plants. For example, one may find Drug, Cold Remedy, below and see that *Rhus* was used by the Cahuilla, Cheyenne, Chippewa, Comanche, and Iroquois. The specific ethnobotanical information and the sources from which the information was obtained may be found by turning to *Rhus* in the Catalog of Plants and examining Drug, Cold Remedy, under the entries for that genus.

Drug
Abortifacient
Abies: Cree, Woodlands
Achillea: Sanpoil
Acorus: Delaware; Delaware, Oklahoma; Menominee; Mohegan; Sioux, Fort Peck
Aletris: Micmac
Aleurites: Hawaiian
Alnus: Cree, Woodlands; Kutenai
Ambrosia: Zuni
Anemone: Blackfoot
Angelica: Cherokee
Apocynum: Cherokee
Aralia: Chippewa; Iroquois
Arctostaphylos: Cree, Woodlands
Aristolochia: Cherokee
Armoracia: Cherokee
Artemisia: Blackfoot; Chippewa; Dakota; Kawaiisu; Menominee; Omaha; Pawnee; Ponca; Sioux
Asarum: Cherokee
Asplenium: Cherokee
Aster: Cree, Woodlands
Betula: Blackfoot
Carex: Songish
Carya: Cherokee
Castilleja: Quileute
Ceanothus: Iroquois
Celastrus: Iroquois
Celtis: Iroquois
Chamaemelum: Cherokee
Chenopodium: Mahuna
Cimicifuga: Cherokee
Cocos: Hawaiian
Croton: Luiseño

Draba: Blackfoot
Echinocystis: Cherokee
Equisetum: Costanoan; Cree; Quinault
Erigeron: Cherokee
Euonymus: Iroquois
Eupatorium: Chippewa
Fagus: Iroquois
Fragaria: Iroquois; Malecite; Micmac
Fraxinus: Abnaki
Galium: Choctaw
Garrya: Pomo, Kashaya
Grindelia: Cree
Hedeoma: Cherokee
Hepatica: Chippewa
Heteromeles: Costanoan (Olhonean)
Hydrangea: Cherokee
Hypericum: Cherokee; Iroquois
Hyssopus: Cherokee
Iva: Mahuna
Juniperus: Cherokee
Lilium: Malecite; Micmac
Lindera: Cherokee; Rappahannock
Lysichiton: Makah
Malus: Iroquois
Mitchella: Delaware; Delaware, Oklahoma
Monarda: Cherokee
Nepeta: Cherokee
Opuntia: Lakota
Orbexilum: Cherokee
Pedicularis: Mohegan
Persea: Seminole
Petroselinum: Cherokee
Phoradendron: Kawaiisu; Mendocino Indian; Pomo; Pomo, Kashaya

Physaria: Blackfoot
Pinus: Okanagan-Colville
Platanus: Cherokee
Polygala: Cherokee
Populus: Cree
Porophyllum: Paiute; Shoshoni
Pterocaulon: Seminole
Pycnanthemum: Chippewa
Quercus: Menominee
Rhus: Iroquois
Ribes: Chippewa
Rubus: Carrier; Chippewa; Malecite; Micmac
Rumex: Houma
Sambucus: Haisla & Hanaksiala
Sanguinaria: Abnaki; Menominee; Micmac
Sanicula: Chippewa; Malecite; Micmac
Sarracenia: Cree, Woodlands
Scutellaria: Cherokee
Silphium: Chippewa
Solidago: Cherokee
Sonchus: Houma
Stillingia: Creek
Stylosanthes: Cherokee
Symphoricarpos: Kutenai
Tanacetum: Chippewa
Taxus: Algonquin, Tête-de-Boule; Iroquois
Tephrosia: Creek
Tetradymia: Navajo
Thuja: Chehalis; Menominee
Trillium: Menominee
Triosteum: Iroquois
Vaccinium: Cree, Woodlands
Veratrum: Kwakiutl
Verbena: Cherokee

Verbesina: Chickasaw
Viburnum: Malecite; Micmac
Vicia: Iroquois
Yucca: Lakota
Zanthoxylum: Iroquois

Adjuvant
Abies: Cheyenne; Menominee
Acorus: Cree, Woodlands
Angelica: Shoshoni; Washo
Aralia: Meskwaki
Arctostaphylos: Menominee
Asarum: Chippewa; Iroquois; Meskwaki
Betula: Ojibwa; Potawatomi
Chamaecyparis: Bella Coola
Chimaphila: Iroquois; Menominee
Cirsium: Meskwaki; Potawatomi
Comptonia: Menominee
Corylus: Menominee
Desmodium: Meskwaki
Ephedra: Paiute
Equisetum: Sanpoil
Eupatorium: Cherokee
Frangula: Makah
Fraxinus: Menominee
Gleditsia: Cherokee
Gnaphalium: Creek
Juniperus: Meskwaki
Lilium: Malecite
Lobelia: Iroquois
Matricaria: Eskimo, Kuskokwagmiut; Eskimo, Western
Mentha: Cherokee
Oenothera: Navajo, Ramah
Osmorhiza: Cheyenne
Panax: Creek; Meskwaki; Potawatomi
Poliomintha: Comanche
Prunella: Cherokee
Prunus: Potawatomi; Sioux
Psilostrophe: Hopi
Ptelea: Menominee; Meskwaki
Robinia: Menominee
Rubus: Menominee; Meskwaki; Ojibwa
Salix: Montana Indian
Sanguinaria: Menominee; Meskwaki
Sideroxylon: Hawaiian
Sphaeralcea: Cheyenne
Symplocarpus: Menominee
Taenidia: Menominee; Meskwaki
Taxus: Iroquois
Thuja: Menominee; Potawatomi
Triteleia: Thompson
Tsuga: Ojibwa
Valeriana: Thompson
Zanthoxylum: Menominee

Alterative
Alnus: Menominee
Aralia: Potawatomi
Gentiana: Potawatomi
Helenium: Menominee
Oxalis: Iroquois

Persea: Creek
Polypodium: Nootka
Pycnanthemum: Meskwaki
Rhus: Iroquois
Rubus: Montana Indian

Analgesic
Abies: Blackfoot; Chippewa; Menominee
Acamptopappus: Kawaiisu
Acer: Cherokee
Achillea: Aleut; Algonquin, Quebec; Algonquin, Tête-de-Boule; Blackfoot; Cheyenne; Chippewa; Cree, Woodlands; Gosiute; Hesquiat; Iroquois; Lummi; Mendocino Indian; Miwok; Navajo, Kayenta; Okanagan-Colville; Paiute; Shoshoni
Acorus: Blackfoot; Cherokee; Cheyenne; Cree, Woodlands; Delaware, Oklahoma; Menominee; Meskwaki; Mohegan; Ojibwa
Actaea: Algonquin; Meskwaki; Ojibwa, South
Adenostoma: Coahuilla
Adiantum: Costanoan; Iroquois
Aesculus: Iroquois
Agastache: Cheyenne; Chippewa; Paiute
Ageratina: Navajo, Ramah
Allium: Blackfoot; Makah; Quinault
Alnus: Bella Coola; Cherokee; Cowlitz; Gitksan; Iroquois; Kwakiutl; Kwakiutl, Southern; Micmac; Mohegan; Penobscot
Alocasia: Hawaiian
Aloysia: Havasupai
Ambrosia: Cheyenne; Pima
Amorpha: Ojibwa, South; Omaha
Anaphalis: Cherokee
Andropogon: Catawba; Chippewa; Seminole
Androsace: Navajo, Kayenta; Navajo, Ramah
Anemone: Blackfoot; Meskwaki; Ojibwa, South
Anemopsis: Costanoan
Angelica: Aleut; Costanoan; Creek; Gitksan; Iroquois; Kwakiutl; Mendocino Indian; Menominee; Miwok; Shoshoni; Tewa; Yana
Antennaria: Bella Coola
Anthemis: Cherokee
Aplectrum: Catawba
Apocynum: Chippewa; Navajo, Ramah; Ojibwa
Aquilegia: Gosiute; Navajo, Kayenta; Omaha; Paiute; Pawnee; Ponca; Shoshoni
Arabis: Navajo, Ramah; Okanagon; Thompson
Aralia: Bella Coola; Choctaw; Malecite; Menominee; Micmac
Arctium: Abnaki; Delaware, Ontario; Iroquois; Meskwaki; Mohegan; Ojibwa
Arctostaphylos: Cheyenne; Chippewa; Pomo, Little Lakes
Arenaria: Kawaiisu; Navajo, Ramah
Argentina: Kwakiutl
Arisaema: Cherokee; Iroquois; Mohegan; Pawnee
Aristolochia: Cherokee; Choctaw
Armoracia: Delaware, Ontario
Arnica: Catawba
Artemisia: Carrier, Northern; Cherokee; Cheyenne; Costanoan; Cree, Woodlands; Hualapai; Karok; Kawaiisu; Keres, Western; Mendocino Indian; Miwok; Navajo; Navajo, Ramah; Okanagan-Colville; Okanagon; Paiute; Paiute, Northern; Pomo, Kashaya; Sanpoil; Shoshoni; Thompson; Washo; Yuki; Zuni
Aruncus: Bella Coola
Arundinaria: Choctaw
Asarum: Bella Coola; Cherokee; Iroquois; Meskwaki; Micmac
Asclepias: Cahuilla; Cherokee; Okanagon; Paiute; Thompson
Aster: Cherokee; Kawaiisu; Ojibwa; Potawatomi; Shuswap
Astragalus: Kawaiisu; Lakota
Athyrium: Cowlitz; Meskwaki
Atriplex: Navajo, Ramah
Baccharis: Keres, Western
Bahia: Navajo, Ramah; Zuni
Balsamorhiza: Cheyenne; Miwok; Paiute; Sanpoil
Betula: Chippewa; Eskimo, Western; Ojibwa; Shuswap
Bidens: Seminole
Bignonia: Koasati
Blephilia: Cherokee
Brassica: Mohegan; Rappahannock; Shinnecock
Brickellia: Navajo, Ramah
Callirhoe: Dakota
Campanula: Navajo, Ramah
Capsella: Cheyenne; Chippewa; Mohegan
Cardamine: Cherokee; Iroquois
Carum: Abnaki
Carya: Cherokee; Chippewa
Castanea: Cherokee
Castilleja: Navajo, Ramah
Caulophyllum: Chippewa; Ojibwa
Ceanothus: Okanagon; Thompson
Celastrus: Cherokee; Creek; Meskwaki
Centaurium: Miwok
Cephalanthus: Seminole
Cercocarpus: Paiute
Chaenactis: Gosiute; Kawaiisu; Paiute
Chaetopappa: Zuni
Chamaebatiaria: Shoshoni
Chamaecyparis: Ojibwa, South
Chamaesyce: Navajo, Ramah
Chenopodium: Cherokee; Houma; Zuni
Chimaphila: Catawba; Cherokee; Iroquois
Chlorogalum: Wailaki
Chrysothamnus: Coahuilla; Navajo, Ramah
Cibotium: Hawaiian
Cicuta: Kawaiisu; Montana Indian; Paiute; Thompson
Cimicifuga: Cherokee
Cirsium: Cherokee; Chippewa; Costanoan; Meskwaki; Ojibwa
Claytonia: Shoshoni
Clematis: Cherokee; Costanoan; Lakota; Montana Indian; Navajo; Navajo, Ramah; Shoshoni
Collinsonia: Iroquois
Comandra: Meskwaki; Ute

Comarum: Ojibwa
Comptonia: Algonquin, Quebec; Micmac; Ojibwa
Conyza: Chippewa; Hopi
Coptis: Potawatomi
Cordylanthus: Navajo, Ramah
Coreopsis: Meskwaki
Cornus: Abnaki; Carrier; Cherokee; Delaware; Delaware, Oklahoma; Iroquois; Malecite; Meskwaki; Micmac
Corydalis: Navajo, Ramah
Corylus: Chippewa
Crataegus: Chippewa; Meskwaki
Crepis: Shoshoni
Croton: Concow; Jemez; Kawaiisu
Cryptantha: Hopi; Navajo
Cucurbita: Omaha
Cunila: Cherokee
Cupressus: Kawaiisu
Cycloloma: Hopi
Cymopterus: Navajo, Kayenta
Cynoglossum: Ojibwa
Cypripedium: Cherokee; Iroquois
Dalea: Navajo; Navajo, Ramah
Daphne: Cherokee
Datura: Cahuilla; Costanoan; Kawaiisu; Navajo; Navajo, Ramah; Pima
Desmodium: Cherokee; Houma; Seminole
Dicentra: Kawaiisu
Dichanthelium: Seminole
Diervilla: Chippewa
Dioscorea: Meskwaki
Dirca: Iroquois
Disporum: Thompson
Dracocephalum: Navajo, Ramah
Dryopteris: Eskimo, Western
Dugaldia: Great Basin Indian
Dyssodia: Lakota; Omaha
Echinacea: Cheyenne; Dakota; Meskwaki; Omaha; Pawnee; Ponca; Sioux; Sioux, Teton; Winnebago
Echinocystis: Menominee; Meskwaki
Eleocharis: Seminole
Encelia: Kawaiisu; Pima
Ephedra: Mahuna
Epigaea: Cherokee; Iroquois
Epilobium: Eskimo, Western; Hopi; Iroquois
Equisetum: Crow; Iroquois; Navajo, Kayenta; Potawatomi
Eriastrum: Shoshoni
Ericameria: Coahuilla; Shoshoni
Erigeron: Cherokee; Cheyenne; Kawaiisu; Meskwaki; Navajo, Kayenta; Navajo, Ramah; Ojibwa; Okanagan-Colville; Paiute; Shoshoni; Thompson
Eriodictyon: Coahuilla; Costanoan; Shoshoni
Eriogonum: Coahuilla; Havasupai; Hopi; Navajo; Navajo, Kayenta; Navajo, Ramah; Paiute; Round Valley Indian; Thompson
Eryngium: Creek; Seminole
Erysimum: Navajo, Ramah; Sioux, Teton; Zuni
Erythrina: Creek
Eschscholzia: Mendocino Indian

Eucalyptus: Hawaiian
Euonymus: Cherokee
Eupatorium: Iroquois
Euphorbia: Navajo, Kayenta
Euthamia: Chippewa; Ojibwa
Fragaria: Ojibwa
Frangula: Thompson
Frasera: Isleta
Fraxinus: Iroquois
Freycinetia: Hawaiian
Fucus: Kwakiutl, Southern
Gaillardia: Hopi; Okanagan-Colville; Thompson
Galactia: Seminole
Galium: Kwakiutl; Navajo, Ramah
Gaultheria: Algonquin, Quebec; Potawatomi
Gentiana: Iroquois; Navajo
Geranium: Meskwaki
Geum: Bella Coola; Ojibwa, South; Thompson
Gilia: Zuni
Glyceria: Catawba
Glycyrrhiza: Blackfoot
Gnaphalium: Cherokee; Choctaw; Costanoan; Kawaiisu; Menominee
Grindelia: Kawaiisu; Shoshoni
Gutierrezia: Navajo; Navajo, Ramah; Tewa
Gymnocladus: Pawnee
Hamamelis: Cherokee
Hedeoma: Cherokee; Delaware, Oklahoma; Iroquois; Navajo; Shinnecock
Helenium: Menominee
Helianthella: Shoshoni
Helianthemum: Delaware, Oklahoma
Helianthus: Dakota; Shasta
Hepatica: Cherokee; Iroquois
Heracleum: Bella Coola; Cree, Woodlands; Iroquois; Meskwaki; Omaha; Quinault; Sanpoil; Sikani
Heteromeles: Mendocino Indian
Heterotheca: Hopi
Heuchera: Chippewa; Menominee; Navajo, Ramah
Hexastylis: Catawba
Hierochloe: Flathead
Humulus: Cherokee; Dakota; Mohegan
Huperzia: Tanana, Upper
Hydrophyllum: Menominee
Hypericum: Choctaw; Houma; Paiute; Shoshoni
Impatiens: Ojibwa; Potawatomi
Inula: Iroquois; Malecite; Micmac
Ipomoea: Hawaiian; Iroquois; Pawnee
Ipomopsis: Great Basin Indian; Hopi; Navajo, Ramah; Shoshoni; Tewa; Zuni
Iris: Mohegan; Montagnais; Paiute; Shoshoni
Isocoma: Pima
Iva: Shoshoni
Juglans: Iroquois
Juniperus: Bella Coola; Carrier, Southern; Comanche; Costanoan; Creek; Diegueño; Navajo; Navajo, Kayenta; Navajo, Ramah; Ojibwa, South; Paiute; Paiute, Northern; Seminole; Tewa; Washo
Kalmia: Algonquin, Tête-de-Boule; Cherokee; Micmac; Montagnais; Penobscot

Krameria: Pima
Lactuca: Bella Coola; Cherokee; Iroquois
Lagenaria: Houma; Seminole
Lamarckia: Diegueño
Lappula: Ojibwa, South
Larix: Iroquois; Ojibwa, South
Larrea: Kawaiisu; Paiute; Papago; Pima
Lechea: Seminole
Ledum: Algonquin, Quebec; Bella Coola; Cree, Hudson Bay; Eskimo, Western
Lepechinia: Miwok
Lepidium: Isleta
Lespedeza: Omaha; Ponca
Lessingia: Kawaiisu
Leucothoe: Cherokee
Liatris: Blackfoot; Cherokee; Seminole
Licania: Seminole
Ligusticum: Atsugewi
Lindera: Creek
Linnaea: Tanana, Upper
Linum: Navajo, Kayenta
Liriodendron: Rappahannock
Lithospermum: Cheyenne
Lobelia: Cherokee; Iroquois
Lomatium: Cheyenne; Kawaiisu; Kwakiutl; Paiute, Northern; Salish, Coast; Yuki
Lonicera: Iroquois; Kwakiutl; Kwakiutl, Southern
Lophophora: Kiowa
Lotus: Navajo, Ramah
Luetkea: Okanagon; Thompson
Lycopodium: Aleut; Carrier, Southern
Lycopus: Meskwaki
Lysichiton: Makah; Quileute; Shuswap; Skokomish
Machaeranthera: Navajo; Shoshoni
Magnolia: Cherokee
Mahonia: Havasupai; Shoshoni
Maianthemum: Chippewa; Montagnais; Ojibwa; Ojibwa, South; Thompson
Malus: Quinault; Thompson
Malva: Costanoan; Mahuna; Tewa
Marah: Squaxin
Marrubium: Navajo, Ramah; Paiute
Matricaria: Aleut; Costanoan
Menodora: Navajo, Ramah
Mentha: Bella Coola; Blackfoot; Cherokee; Cree, Woodlands; Gosiute; Gros Ventre; Iroquois; Kawaiisu; Keres, Western; Okanagan-Colville; Okanagon; Paiute; Thompson
Menyanthes: Aleut
Menziesia: Kwakiutl
Merremia: Hawaiian
Metrosideros: Hawaiian
Mimulus: Kawaiisu
Mirabilis: Paiute
Mitchella: Cherokee; Iroquois
Monarda: Cherokee; Chippewa; Choctaw; Dakota; Iroquois; Meskwaki; Navajo; Navajo, Ramah; Ojibwa, South; Tewa
Monardella: Miwok; Paiute; Shoshoni; Washo
Moneses: Kwakiutl

Monotropa: Mohegan
Myrica: Micmac; Seminole
Nepeta: Iroquois; Rappahannock
Nicotiana: Cherokee; Kawaiisu; Kwakiutl, Southern; Navajo, Ramah
Nuphar: Bella Coola; Cree, Woodlands; Iroquois; Kwakiutl; Quinault; Shuswap; Tanana, Upper; Thompson
Oemleria: Kwakiutl
Oenothera: Navajo, Ramah
Oplopanax: Carrier; Carrier, Northern; Cheyenne; Crow; Gitksan; Kwakiutl; Oweekeno; Salish, Coast
Opuntia: Flathead; Shoshoni
Osmorhiza: Iroquois; Karok; Paiute; Paiute, Northern; Shoshoni; Washo
Osmunda: Iroquois
Oxalis: Navajo, Ramah
Oxytropis: Thompson
Paeonia: Washo
Panax: Cherokee; Iroquois
Pandanus: Hawaiian
Papaver: Cherokee
Pastinaca: Cherokee
Paxistima: Thompson
Pedicularis: Catawba
Pediomelum: Seminole
Penstemon: Blackfoot; Dakota; Navajo, Ramah; Okanagan-Colville; Shoshoni
Pentagrama: Karok
Pericome: Navajo, Ramah
Persea: Seminole
Petradoria: Hopi
Phoradendron: Cherokee; Papago
Phragmites: Paiute
Physalis: Omaha; Ponca; Winnebago
Physaria: Blackfoot
Physocarpus: Bella Coola
Phytolacca: Mahuna; Seminole
Picea: Bella Coola; Carrier, Southern; Hesquiat; Kwakiutl; Mohegan; Ojibwa
Pinguicula: Seminole
Pinus: Carrier, Southern; Delaware, Ontario; Flathead; Menominee; Mohegan; Nanticoke; Navajo, Ramah; Ojibwa, South; Paiute; Seminole; Shoshoni; Thompson; Yokia
Piper: Hawaiian
Plantago: Abnaki; Cherokee; Iroquois; Keres, Western
Platanthera: Cherokee; Iroquois; Thompson
Platanus: Meskwaki
Pleopeltis: Houma
Pogogyne: Concow
Polygala: Miwok
Polygonatum: Chippewa; Menominee
Polygonum: Cherokee; Chippewa; Houma; Iroquois; Navajo, Ramah; Ojibwa
Polypodium: Bella Coola; Haisla
Polystichum: Iroquois
Populus: Bella Coola; Cahuilla; Carrier, Southern; Chippewa; Shoshoni

Porophyllum: Havasupai
Portulaca: Navajo
Potentilla: Chippewa; Navajo, Ramah; Okanagan-Colville
Prenanthes: Bella Coola; Cherokee; Choctaw
Proboscidea: Pima
Prosopis: Pima
Prunella: Iroquois
Prunus: Chippewa; Iroquois; Navajo, Ramah; Ojibwa, South; Paiute; Potawatomi
Psathyrotes: Paiute; Shoshoni
Pseudotsuga: Bella Coola; Navajo, Kayenta; Swinomish
Psilostrophe: Navajo, Ramah
Psoralidium: Arapaho; Lakota; Navajo, Ramah
Psorothamnus: Paiute; Shoshoni
Pteridium: Koasati; Ojibwa
Pulsatilla: Chippewa; Omaha
Purshia: Kawaiisu; Paiute
Pycnanthemum: Cherokee; Choctaw; Koasati
Quercus: Chippewa; Choctaw; Delaware, Oklahoma; Mohegan; Navajo, Ramah; Seminole; Shinnecock
Ranunculus: Abnaki; Hesquiat; Iroquois; Micmac; Montagnais; Okanagan-Colville
Ratibida: Cheyenne; Dakota; Lakota; Navajo, Ramah
Rhamnus: Kawaiisu
Rhododendron: Cherokee
Rhus: Chippewa; Coahuilla; Iroquois; Navajo, Ramah; Omaha
Ribes: Chippewa; Lummi
Ricinus: Hawaiian; Pima
Robinia: Chickasaw
Rorippa: Okanagan-Colville
Rosa: Bella Coola; Cahuilla; Chehalis; Mahuna; Miwok; Potawatomi
Rubus: Cherokee; Chippewa; Iroquois; Makah; Ojibwa, South; Quinault
Rudbeckia: Seminole
Rumex: Bella Coola; Kawaiisu; Paiute; Shoshoni
Sabal: Seminole
Sabatia: Cherokee; Seminole
Sagittaria: Navajo
Salicornia: Heiltzuk
Salix: Blackfoot; Chickasaw; Crow; Eskimo, Nunivak; Iroquois; Koasati; Montagnais; Seminole; Shoshoni; Thompson
Salvia: Costanoan; Kawaiisu; Paiute; Paiute, Northern; Shoshoni
Sambucus: Bella Coola; Chickasaw; Hanaksiala; Hesquiat; Houma; Iroquois; Kawaiisu
Sanguinaria: Chippewa; Iroquois; Meskwaki; Ojibwa
Sanicula: Cherokee; Malecite; Micmac
Sanvitalia: Navajo, Ramah
Saponaria: Mahuna
Sassafras: Seminole
Satureja: Cherokee
Sedum: Eskimo, Western

Senecio: Cheyenne; Navajo, Ramah; Zuni
Senna: Cherokee
Shepherdia: Navajo, Kayenta
Silene: Gosiute
Silphium: Chippewa; Meskwaki; Omaha; Ponca; Winnebago
Sisyrinchium: Meskwaki; Seminole
Sium: Iroquois
Smallanthus: Iroquois
Smilax: Cherokee; Chippewa; Koasati
Solanum: Seminole
Solidago: Chippewa; Iroquois; Menominee; Zuni
Sorbus: Malecite; Micmac
Sphaeralcea: Dakota
Spiraea: Iroquois; Okanagan-Colville
Stanleya: Paiute; Shoshoni
Streptopus: Thompson
Suaeda: Hopi
Symphoricarpos: Kwakiutl; Paiute
Symplocarpus: Delaware; Delaware, Oklahoma; Menominee; Micmac
Tanacetum: Cherokee; Iroquois; Shinnecock
Taraxacum: Bella Coola; Iroquois; Papago
Tauschia: Kawaiisu
Taxus: Klallam; Micmac
Tephrosia: Catawba
Tetradymia: Navajo, Ramah
Tetraneuris: Hopi
Thalictrum: Gitksan; Wailaki
Thamnosma: Kawaiisu
Thermopsis: Cheyenne; Navajo, Ramah
Thuja: Bella Coola; Chippewa; Micmac; Ojibwa; Penobscot
Torreya: Costanoan
Tradescantia: Cherokee
Trichostema: Concow; Costanoan; Miwok; Tubatulabal
Trillium: Costanoan; Iroquois; Menominee
Triosteum: Iroquois; Meskwaki
Tsuga: Menominee
Ulmus: Choctaw; Potawatomi
Umbellularia: Cahuilla; Karok; Mendocino Indian; Miwok; Pomo; Pomo, Kashaya; Yuki
Urtica: Bella Coola; Cahuilla; Carrier; Costanoan; Hesquiat; Kawaiisu; Klallam; Kwakiutl; Kwakiutl, Southern; Miwok; Pomo; Pomo, Kashaya; Quinault; Shuswap
Uvularia: Ojibwa; Potawatomi
Vaccinium: Seminole
Valeriana: Cree, Woodlands; Menominee; Okanagon; Thompson
Veratrum: Bella Coola; Blackfoot; Cherokee; Cowlitz; Iroquois; Kwakiutl; Salish, Coast
Verbascum: Catawba; Cherokee; Delaware, Oklahoma; Delaware, Ontario
Verbena: Cherokee; Dakota
Verbesina: Seminole; Zuni
Vernonia: Cherokee
Veronicastrum: Cherokee
Viburnum: Algonquin, Tête-de-Boule; Chippewa;

Gitksan; Iroquois; Menominee; Meskwaki; Ojibwa
Vicia: Cherokee; Snohomish; Squaxin
Viola: Carrier, Southern; Cherokee; Klallam; Ojibwa, South
Vitis: Meskwaki; Mohegan; Seminole; Shinnecock
Woodwardia: Luiseño
Wyethia: Mendocino Indian
Xanthorhiza: Cherokee
Zanthoxylum: Iroquois; Menominee
Zea: Tewa
Zigadenus: Mendocino Indian; Montana Indian; Paiute; Shoshoni; Thompson; Washo
Zingiber: Hawaiian
Zinnia: Navajo, Ramah
Zizia: Meskwaki
Ziziphus: Pima

Anesthetic
Achillea: Shoshoni
Datura: Zuni
Echinacea: Omaha
Rubus: Rappahannock
Salix: Eskimo, Alaska

Anthelmintic
Achillea: Iroquois
Acorus: Cherokee; Iroquois
Allium: Iroquois
Amelanchier: Cherokee
Amorpha: Meskwaki
Anemone: Iroquois
Angelica: Creek
Apocynum: Iroquois; Menominee; Micmac; Penobscot
Aralia: Iroquois
Aristolochia: Nanticoke
Artemisia: Cherokee; Kiowa; Mohegan; Tolowa; Yurok
Asarum: Cherokee
Asclepias: Meskwaki
Bidens: Cherokee
Capsella: Mohegan
Carya: Iroquois
Cerastium: Cherokee
Chelone: Cherokee
Chenopodium: Cherokee; Houma; Koasati; Natchez; Rappahannock
Chimaphila: Iroquois
Cirsium: Abnaki; Hopi
Comptonia: Cherokee
Coptis: Iroquois
Cornus: Cherokee; Thompson
Corylus: Ojibwa
Cucurbita: Cherokee
Cypripedium: Cherokee
Dicentra: Skagit
Dryopteris: Cherokee
Echinacea: Dakota
Elytrigia: Iroquois
Eryngium: Delaware; Delaware, Oklahoma
Euonymus: Iroquois
Eupatorium: Meskwaki

Euphorbia: Meskwaki
Fagus: Cherokee
Frangula: Hesquiat
Fraxinus: Cowlitz
Gaultheria: Iroquois
Gentianella: Iroquois
Gleditsia: Cherokee
Helianthus: Iroquois; Pima
Juglans: Delaware; Iroquois; Kiowa
Juniperus: Cherokee; Shoshoni
Lindera: Mohegan
Liriodendron: Cherokee
Lobelia: Cherokee
Lonicera: Meskwaki
Magnolia: Iroquois
Maianthemum: Iroquois
Manfreda: Cherokee
Melia: Cherokee
Mentha: Mohegan
Mirabilis: Dakota
Monarda: Chippewa
Morus: Cherokee
Myrica: Houma
Nepeta: Cherokee
Nicotiana: Cherokee; Paiute; Shoshoni
Nyssa: Cherokee; Houma
Oxalis: Cherokee
Panax: Iroquois
Pinus: Cherokee; Choctaw; Gosiute
Podophyllum: Cherokee
Populus: Abnaki; Carrier; Iroquois; Montagnais; Sikani
Portulaca: Cherokee
Prunus: Cherokee; Chippewa; Delaware; Flathead
Purshia: Paiute, Northern
Quercus: Meskwaki
Rhus: Iroquois; Meskwaki; Potawatomi
Ribes: Meskwaki
Rosa: Cherokee
Rubus: Mohegan
Rudbeckia: Cherokee; Iroquois
Ruta: Cherokee
Salvia: Mohegan
Sanguinaria: Iroquois
Sassafras: Cherokee; Iroquois
Satureja: Costanoan
Senna: Iroquois
Sisyrinchium: Mahuna
Solanum: Cherokee; Houma
Spigelia: Cherokee; Creek
Symplocarpus: Iroquois
Tanacetum: Cherokee
Tephrosia: Cherokee
Tilia: Malecite; Micmac
Valeriana: Menominee
Verbena: Iroquois
Viburnum: Iroquois
Zanthoxylum: Iroquois

Anticonvulsive
Achillea: Iroquois

Acorus: Cherokee
Actaea: Chippewa
Anemopsis: Shoshoni
Anthemis: Cherokee
Apocynum: Chippewa
Arabis: Navajo, Kayenta
Aralia: Ojibwa
Aristolochia: Micmac; Penobscot
Artemisia: Chippewa
Asarum: Algonquin, Quebec; Iroquois
Asclepias: Delaware, Oklahoma
Astragalus: Chippewa
Atriplex: Hopi
Calypso: Thompson
Caulophyllum: Cherokee
Claytonia: Iroquois
Conyza: Iroquois
Cornus: Malecite; Micmac
Cypripedium: Cherokee
Echinacea: Meskwaki
Erigeron: Cherokee
Geum: Iroquois
Hepatica: Chippewa
Heracleum: Winnebago
Juniperus: Apache, White Mountain
Lathyrus: Chippewa
Ligusticum: Flathead
Lobelia: Iroquois
Lonicera: Thompson
Lycopodium: Iroquois
Macromeria: Hopi
Maianthemum: Meskwaki
Matricaria: Costanoan
Medeola: Iroquois
Mimulus: Iroquois; Navajo, Kayenta
Mitchella: Iroquois
Monarda: Ojibwa
Monotropa: Cherokee
Nepeta: Cherokee
Nicotiana: Cherokee
Nuphar: Iroquois
Osmunda: Iroquois
Panax: Cherokee
Papaver: Cherokee
Phoradendron: Cherokee
Picea: Ojibwa
Pinus: Ojibwa
Polygala: Chippewa
Polygonum: Chippewa
Polystichum: Iroquois
Postelsia: Nitinaht
Pteridium: Tolowa
Pyrola: Iroquois
Ranunculus: Iroquois
Rhus: Iroquois
Rosa: Chippewa
Salvia: Hopi
Sanguinaria: Ojibwa
Sium: Iroquois
Solidago: Chippewa

Symplocarpus: Delaware, Oklahoma; Menominee; Mohegan
Valeriana: Cree, Woodlands
Verbascum: Hopi; Iroquois
Veronicastrum: Meskwaki
Viburnum: Cherokee

Antidiarrheal
Abies: Micmac
Acer: Cherokee; Ojibwa, South; Thompson
Achillea: Chehalis; Iroquois; Okanagan-Colville; Shoshoni; Skagit; Snohomish; Thompson
Acorus: Cherokee
Adiantum: Menominee
Agalinis: Meskwaki
Ageratina: Cherokee
Agrimonia: Cherokee; Iroquois
Aletris: Catawba
Alnus: Klallam; Mendocino Indian; Potawatomi; Quileute
Ambrosia: Cheyenne; Dakota; Iroquois
Amelanchier: Cherokee; Chippewa
Amphicarpaea: Cherokee
Anaphalis: Iroquois
Andropogon: Cherokee; Seminole
Anemone: Iroquois
Angelica: Tewa; Yana
Anthemis: Iroquois
Apocynum: Iroquois
Aquilegia: Meskwaki; Shoshoni
Arabis: Okanagan-Colville
Aralia: Iroquois
Arctostaphylos: Cahuilla; Cree, Woodlands; Karok; Pomo; Pomo, Kashaya; Pomo, Little Lakes
Argentina: Blackfoot; Iroquois
Arisaema: Iroquois
Artemisia: Chippewa; Costanoan; Klamath; Lakota; Mendocino Indian; Montana Indian; Okanagon; Paiute; Paiute, Northern; Shoshoni; Thompson; Yuki
Aruncus: Bella Coola
Asarum: Cherokee
Asclepias: Cherokee; Lakota; Navajo, Kayenta; Shoshoni
Aster: Cherokee
Aureolaria: Cherokee
Balsamorhiza: Thompson
Berberis: Cherokee
Betula: Cherokee; Chippewa; Menominee
Bidens: Seminole
Blechnum: Kwakiutl
Callicarpa: Choctaw
Campanula: Cherokee
Capsella: Chippewa; Costanoan; Mahuna
Carex: Cherokee
Carpinus: Cherokee; Iroquois
Ceanothus: Iroquois; Meskwaki; Thompson
Cephalanthus: Choctaw; Seminole
Cercocarpus: Paiute; Shoshoni
Chamaesyce: Navajo, Ramah; Omaha
Chenopodium: Iroquois

Chrysothamnus: Shoshoni; Thompson
Cicuta: Kwakiutl
Collinsonia: Iroquois
Comarum: Chippewa
Comptonia: Ojibwa
Conyza: Blackfoot; Cahuilla; Cree, Hudson Bay
Coptis: Algonquin, Quebec
Coreopsis: Cherokee
Cornus: Alabama; Cherokee; Chippewa; Menominee; Meskwaki; Ojibwa; Potawatomi; Rappahannock; Thompson
Corydalis: Navajo, Kayenta
Corylus: Iroquois
Crataegus: Ojibwa, South; Okanagan-Colville; Thompson
Croton: Costanoan; Pomo; Pomo, Kashaya
Dalea: Dakota; Meskwaki
Delphinium: Blackfoot
Diospyros: Cherokee
Diplacus: Mahuna
Dracocephalum: Navajo, Kayenta
Eleocharis: Seminole
Empetrum: Tanana, Upper
Enceliopsis: Shoshoni
Ephedra: Paiute; Shoshoni; Tewa
Epifagus: Iroquois
Epigaea: Cherokee
Epilobium: Okanagan-Colville; Potawatomi
Equisetum: Makah
Eriastrum: Paiute
Ericameria: Paiute
Erigeron: Blackfoot; Paiute
Eriodictyon: Paiute
Eriogonum: Diegueño; Kawaiisu; Navajo; Navajo, Ramah; Paiute; Sanpoil; Shoshoni; Tubatulabal
Eryngium: Natchez; Paiute; Seminole
Escobaria: Blackfoot
Eupatorium: Iroquois
Fragaria: Blackfoot; Cherokee; Diegueño; Iroquois; Skokomish; Thompson
Frangula: Skagit
Frasera: Cherokee; Cheyenne
Galactia: Seminole
Galium: Costanoan
Gaultheria: Cherokee; Quinault
Gaylussacia: Cherokee
Gentianella: Iroquois
Geranium: Chippewa; Gosiute; Iroquois; Menominee; Meskwaki; Montana Indian; Ojibwa
Geum: Iroquois; Malecite; Micmac
Glycyrrhiza: Cheyenne
Gutierrezia: Diegueño; Navajo, Kayenta
Hamamelis: Iroquois
Haplopappus: Paiute
Hedeoma: Cherokee; Mahuna
Heliotropium: Paiute; Tubatulabal
Hepatica: Menominee
Heracleum: Blackfoot; Washo
Heuchera: Blackfoot; Cherokee; Chippewa; Cree, Woodlands; Flathead; Lakota; Menominee; Shoshoni; Shuswap; Sioux, Teton

Hieracium: Iroquois; Rappahannock
Holodiscus: Lummi; Paiute
Hydrastis: Iroquois
Hydrophyllum: Menominee; Ojibwa
Hypericum: Alabama; Cherokee
Ilex: Ojibwa
Ipomopsis: Paiute; Shoshoni; Washo
Iva: Shoshoni
Jeffersonia: Iroquois
Juglans: Cherokee; Rappahannock
Juniperus: Cree, Woodlands; Crow; Keres, Western; Kwakiutl; Seminole
Kallstroemia: Tewa
Kalmia: Cree, Hudson Bay
Lactuca: Bella Coola; Navajo; Okanagan-Colville
Larrea: Pima
Lathyrus: Weyot
Lechea: Seminole
Lesquerella: Okanagan-Colville
Liatris: Pawnee; Seminole
Licania: Seminole
Lilium: Cherokee
Linaria: Iroquois
Liquidambar: Cherokee; Rappahannock
Liriodendron: Cherokee
Lithospermum: Shoshoni
Lomatium: Cheyenne
Lonicera: Iroquois
Lygodesmia: Lakota; Ponca
Magnolia: Cherokee
Mahonia: Great Basin Indian; Paiute; Shoshoni
Malus: Makah
Malvella: Choctaw
Manfreda: Cherokee
Marrubium: Round Valley Indian
Matricaria: Blackfoot; Cahuilla; Flathead; Montana Indian; Yokia
Menispermum: Cherokee
Mentha: Miwok; Washo
Mimulus: Miwok
Mitchella: Cherokee
Morus: Cherokee
Nepeta: Iroquois
Nyssa: Cherokee
Opuntia: Apache, Western; Flathead; Hopi
Osmorhiza: Iroquois; Shoshoni
Ostrya: Potawatomi
Oxydendrum: Cherokee
Paeonia: Shoshoni
Parthenium: Koasati
Parthenocissus: Meskwaki
Pedicularis: Cherokee
Pediomelum: Cheyenne
Perideridia: Blackfoot
Persea: Seminole
Phlox: Shoshoni
Phoradendron: Keres, Western
Phragmites: Apache, White Mountain
Phytolacca: Rappahannock

Picea: Cree, Woodlands; Kwakiutl; Ojibwa, South; Thompson; Tlingit
Pinus: Alabama; Cherokee; Paiute
Plantago: Cherokee; Iroquois; Keres, Western; Pima; Zuni
Platanthera: Cherokee
Platanus: Cherokee
Pluchea: Paiute; Pima
Polygonatum: Cherokee
Polygonum: Cherokee; Iroquois; Meskwaki; Thompson
Polypodium: Kwakiutl
Polyporus: Blackfoot
Polystichum: Iroquois
Polytaenia: Meskwaki
Populus: Chickasaw
Porteranthus: Iroquois
Portulaca: Keres, Western
Potentilla: Cherokee; Chippewa; Iroquois; Okanagan-Colville
Prosopis: Pima
Prunella: Iroquois; Menominee
Prunus: Apache, Mescalero; Blackfoot; Cheyenne; Cree, Hudson Bay; Cree, Woodlands; Creek; Crow; Delaware; Delaware, Oklahoma; Flathead; Gros Ventre; Iroquois; Kutenai; Mendocino Indian; Menominee; Micmac; Mohegan; Ojibwa, South; Okanagan-Colville; Oregon Indian; Paiute; Penobscot; Ponca; Sanpoil; Sioux; Thompson
Psathyrotes: Paiute; Shoshoni
Pseudotsuga: Bella Coola; Kwakiutl
Psidium: Hawaiian
Psilostrophe: Navajo, Kayenta
Psorothamnus: Paiute; Shoshoni
Pteridium: Iroquois
Pterocaulon: Seminole
Pycnanthemum: Cherokee
Pyrola: Navajo, Kayenta
Quercus: Cherokee; Costanoan; Creek; Houma; Iroquois; Malecite; Meskwaki; Micmac; Ojibwa, South; Potawatomi; Yuki
Ranunculus: Iroquois
Rheum: Cherokee
Rhus: Chippewa; Creek; Delaware, Ontario; Ojibwa, South; Pawnee; Winnebago
Ribes: Cherokee; Kwakiutl; Okanagan-Colville; Thompson
Rosa: Blackfoot; Cherokee; Cheyenne; Crow; Gros Ventre; Kutenai; Paiute; Shoshoni; Thompson
Rubus: Algonquin, Quebec; Cherokee; Chippewa; Choctaw; Costanoan; Delaware; Delaware, Oklahoma; Diegueño; Eskimo, Western; Iroquois; Kwakiutl; Mahuna; Malecite; Mendocino Indian; Menominee; Micmac; Mohegan; Ojibwa; Okanagan-Colville; Pomo, Kashaya; Rappahannock; Saanich; Shinnecock; Shuswap
Rumex: Cherokee; Eskimo, Inuktitut; Eskimo, Kuskokwagmiut; Iroquois; Lakota; Ojibwa; Paiute; Pawnee; Yuki
Sabatia: Seminole

Salix: Bella Coola; Blackfoot; Cherokee; Cheyenne; Chickasaw; Chippewa; Cree, Woodlands; Kawaiisu; Mendocino Indian; Menominee; Meskwaki; Paiute; Sanpoil; Seminole
Salvia: Cherokee
Sambucus: Clallam; Klallam; Paiute; Shoshoni
Sanguinaria: Iroquois
Sarcobatus: Paiute
Sassafras: Cherokee; Seminole
Scutellaria: Cherokee; Meskwaki
Sesamum: Cherokee
Shepherdia: Sioux
Sisyrinchium: Cherokee
Solanum: Blackfoot; Paiute
Solidago: Cherokee; Delaware; Delaware, Oklahoma; Okanagan-Colville; Thompson
Sonchus: Houma
Sphaeralcea: Hopi; Pima
Spiraea: Iroquois; Lummi; Meskwaki; Mohegan; Shuswap
Stephanomeria: Paiute
Stillingia: Seminole
Symphoricarpos: Thompson
Symphytum: Cherokee
Tanacetum: Shoshoni
Tetradymia: Shoshoni
Thalictrum: Cherokee
Thuja: Hanaksiala; Nez Perce
Tiarella: Malecite; Micmac
Tilia: Cherokee
Tsuga: Cherokee; Chippewa; Kwakiutl; Malecite; Micmac; Potawatomi
Typha: Washo
Ulmus: Cherokee; Houma; Iroquois; Potawatomi
Urtica: Chippewa
Uvularia: Cherokee
Vaccinium: Eskimo, Western; Seminole
Valeriana: Thompson
Verbascum: Iroquois
Verbena: Cherokee
Vernonia: Natchez
Veronicastrum: Iroquois
Viburnum: Carrier, Northern
Viola: Cherokee
Vitis: Cherokee
Xanthium: Pima; Tewa
Xyris: Cherokee

Antidote
Amaranthus: Navajo, Ramah
Angelica: Mewuk
Arctostaphylos: Karok
Artemisia: Chippewa; Mendocino Indian; Shoshoni
Asarum: Meskwaki
Bouteloua: Navajo, Ramah
Cardamine: Iroquois
Chenopodium: Navajo, Ramah
Clintonia: Menominee; Ojibwa
Cornus: Cherokee
Cuscuta: Diegueño
Datura: Cahuilla

Dryopteris: Bella Coola; Kitasoo
Echinacea: Dakota; Omaha; Pawnee; Ponca; Sioux; Winnebago
Eryngium: Choctaw; Meskwaki
Eupatorium: Navajo
Frangula: Kawaiisu
Gutierrezia: Navajo, Ramah
Hydrophyllum: Iroquois
Krascheninnikovia: Navajo, Ramah
Laportea: Iroquois
Lespedeza: Meskwaki
Lonicera: Quileute
Maianthemum: Iroquois
Mentha: Iroquois
Mertensia: Iroquois
Mimulus: Iroquois
Nicotiana: Iroquois
Parthenocissus: Iroquois
Physaria: Hopi
Physocarpus: Hesquiat
Plantago: Cherokee
Portulaca: Iroquois
Quercus: Iroquois
Ranunculus: Iroquois
Rhamnus: Iroquois
Rhododendron: Modesse
Rhus: Omaha
Ribes: Iroquois
Rubus: Meskwaki
Rumex: Meskwaki
Salix: Makah
Sambucus: Menominee
Sanicula: Iroquois
Senecio: Navajo, Kayenta
Sisyrinchium: Menominee
Symphoricarpos: Skagit
Toxicodendron: Tolowa
Vitis: Meskwaki

Antiemetic
Acer: Thompson
Achillea: Cheyenne; Iroquois; Shoshoni
Agrimonia: Iroquois
Allium: Blackfoot
Alnus: Micmac; Penobscot
Ambrosia: Dakota; Oto
Andropogon: Seminole
Angelica: Tewa
Anthemis: Iroquois
Antidesma: Hawaiian
Artemisia: Tewa
Asarum: Iroquois; Potawatomi
Baptisia: Cherokee
Cephalanthus: Seminole
Cercis: Delaware; Delaware, Oklahoma
Chamaecrista: Seminole
Chamaemelum: Cherokee
Clethra: Cherokee
Coptis: Iroquois
Cornus: Thompson
Corylus: Iroquois

Croton: Zuni
Cymopterus: Navajo, Kayenta
Dryopteris: Mewuk
Dugaldia: Navajo
Eriodictyon: Paiute
Erythrina: Seminole
Frasera: Cherokee
Gaillardia: Navajo, Ramah
Galactia: Seminole
Gaura: Navajo, Ramah
Gnaphalium: Creek
Gonolobus: Houma
Hamamelis: Iroquois
Hydrangea: Cherokee
Juniperus: Blackfoot
Kalmia: Kwakwaka'wakw
Lactuca: Bella Coola; Navajo
Larix: Cree, Woodlands
Lechea: Seminole
Liatris: Seminole
Licania: Seminole
Lupinus: Cherokee
Mahonia: Sanpoil
Malacothrix: Navajo, Kayenta
Mentha: Cherokee; Cheyenne; Micmac
Menyanthes: Kwakiutl
Mirabilis: Paiute
Mitchella: Iroquois
Nepeta: Iroquois
Oxalis: Cherokee
Panax: Houma; Iroquois
Penstemon: Blackfoot
Perideridia: Blackfoot; Pomo; Pomo, Kashaya
Persea: Seminole
Pinus: Paiute; Shoshoni
Plantago: Iroquois
Polypodium: Kwakiutl
Prunella: Iroquois
Prunus: Cherokee; Delaware; Meskwaki; Shoshoni
Pteridium: Cherokee
Quercus: Delaware, Ontario
Rhus: Cherokee
Ribes: Iroquois
Rosa: Thompson
Rubus: Hawaiian; Kwakiutl; Okanagon; Thompson
Salix: Creek
Sanguinaria: Delaware, Ontario; Iroquois
Sassafras: Seminole
Silphium: Meskwaki
Smallanthus: Iroquois
Solanum: Malecite; Micmac
Solidago: Iroquois
Sphaeralcea: Shoshoni
Spiraea: Iroquois; Ojibwa
Stephanomeria: Shoshoni
Stillingia: Seminole
Thalictrum: Cherokee
Thuja: Lummi
Trifolium: Pomo
Ulmus: Iroquois
Vaccinium: Cheyenne; Ojibwa
Xanthium: Tewa
Yucca: Navajo
Zanthoxylum: Iroquois

Antihemorrhagic
Acer: Iroquois; Micmac; Penobscot
Achillea: Cherokee; Cree, Woodlands
Acorus: Cree, Woodlands; Potawatomi
Actaea: Iroquois
Adiantum: Makah
Agastache: Cree, Woodlands
Allium: Cherokee
Alnus: Carrier; Gitksan; Iroquois; Kwakiutl; Mendocino Indian; Meskwaki
Ambrosia: Cheyenne; Pima
Anemone: Aleut
Angelica: Blackfoot; Paiute, Northern
Aralia: Creek; Kwakiutl; Micmac; Penobscot
Arceuthobium: Bella Coola; Carrier, Southern
Arctostaphylos: Costanoan (Olhonean); Okanagan-Colville; Okanagon; Thompson
Artemisia: Great Basin Indian
Astragalus: Blackfoot; Lakota
Athyrium: Thompson
Baptisia: Micmac; Penobscot
Boykinia: Cheyenne
Calla: Gitksan
Castilleja: Blackfoot; Gitksan
Caulophyllum: Chippewa
Ceanothus: Choctaw
Cephalanthus: Kiowa
Cercocarpus: Paiute, Northern
Chimaphila: Cree, Woodlands
Conyza: Blackfoot
Coreopsis: Meskwaki
Corylus: Chippewa; Iroquois
Cynoglossum: Iroquois
Dennstaedtia: Mahuna
Digitaria: Hawaiian
Dryopteris: Mewuk
Dyssodia: Lakota
Epilobium: Cheyenne; Miwok
Equisetum: Eskimo, Alaska
Erigeron: Blackfoot; Cherokee
Eriogonum: Hopi
Eryngium: Seminole
Euonymus: Cherokee
Eupatorium: Penobscot
Fragaria: Carrier
Galium: Micmac; Penobscot
Geum: Algonquin, Tête-de-Boule
Glycyrrhiza: Lakota
Grindelia: Lakota
Hosta: Cherokee
Hyptis: Cahuilla
Juniperus: Crow; Gitksan; Okanagan-Colville; Paiute; Stony Indian
Kalmia: Kwakiutl; Kwakwaka'wakw
Krascheninnikovia: Navajo
Lachnanthes: Cherokee
Lactuca: Bella Coola
Ledum: Eskimo, Alaska
Leymus: Okanagan-Colville
Ligusticum: Pomo
Lithospermum: Hopi; Okanagan-Colville
Lupinus: Cherokee
Lycoperdon: Blackfoot
Lysichiton: Gitksan
Mahonia: Blackfoot
Maianthemum: Abnaki
Malus: Kwakiutl
Matricaria: Eskimo, Alaska
Mentha: Cree, Woodlands
Menyanthes: Kwakiutl
Nuphar: Gitksan
Oplopanax: Gitksan
Ostrya: Chippewa; Potawatomi
Pellaea: Costanoan; Diegueño; Miwok
Pinus: Okanagan-Colville
Polygonatum: Abnaki
Polygonum: Menominee
Polypodium: Kwakiutl
Prunus: Chippewa; Iroquois
Psidium: Hawaiian
Psorothamnus: Paiute
Pteridium: Thompson
Pterocaulon: Seminole
Pterospora: Cheyenne
Purshia: Sanpoil; Shoshoni
Pyrola: Cree, Woodlands; Micmac; Penobscot
Ranunculus: Iroquois
Rhus: Sioux
Rubus: Kwakiutl; Okanagon; Thompson
Rumex: Cheyenne; Iroquois; Lakota
Salix: Blackfoot; Eskimo, Western
Sambucus: Pomo, Little Lakes
Sanguinaria: Iroquois; Malecite
Sarcobatus: Paiute
Sarracenia: Micmac; Penobscot
Scrophularia: Iroquois
Selaginella: Blackfoot
Shepherdia: Cree, Woodlands
Silphium: Chippewa
Solidago: Chippewa; Iroquois
Sphaeralcea: Hopi; Navajo, Ramah
Streptopus: Micmac; Penobscot
Tilia: Iroquois
Tradescantia: Cherokee
Tsuga: Clallam; Cowlitz; Klallam
Ulmus: Iroquois; Penobscot
Urtica: Gitksan
Viburnum: Iroquois

Antirheumatic (External)
Abies: Chippewa; Gitksan; Iroquois; Keres, Western
Achillea: Blackfoot; Carrier; Flathead; Gosiute; Kwakiutl; Micmac; Okanagan-Colville; Paiute; Quileute; Shoshoni; Thompson
Acorus: Cree, Woodlands
Actaea: Iroquois
Adenostoma: Cahuilla

Index of Usages

Adiantum: Cherokee; Iroquois
Aesculus: Cherokee; Delaware; Mohegan; Shinnecock
Ageratina: Zuni
Allium: Blackfoot
Alnus: Bella Coola
Ambrosia: Lakota
Amorpha: Omaha
Amphicarpaea: Lakota
Anaphalis: Chippewa
Angelica: Blackfoot; Iroquois; Shoshoni
Antennaria: Lakota
Anthemis: Mendocino Indian
Aquilegia: Paiute
Arabis: Shoshoni
Aralia: Karok
Arctium: Abnaki; Iroquois; Mohegan
Arctostaphylos: Ojibwa
Arenaria: Kawaiisu; Shoshoni; Washo
Arisaema: Mohegan; Pawnee
Arnica: Thompson
Artemisia: Aleut; Blackfoot; Costanoan; Eskimo, Alaska; Flathead; Gosiute; Karok; Kawaiisu; Keres, Western; Klamath; Mendocino Indian; Okanagan-Colville; Okanagon; Paiute; Paiute, Northern; Pawnee; Shoshoni; Tanana, Upper; Thompson; Tolowa; Washo; Yuki; Yurok; Zuni
Aruncus: Makah
Asarum: Bella Coola
Asclepias: Blackfoot; Delaware; Iroquois; Okanagan-Colville; Paiute
Asparagus: Iroquois
Atriplex: Hualapai; Maricopa; Paiute, Northern; Pima
Baccharis: Hualapai; Keres, Western; Yavapai
Bahia: Zuni
Barbula: Seminole
Berula: Zuni
Bidens: Seminole
Brachyactis: Paiute
Brickellia: Keres, Western
Bryum: Seminole
Callicarpa: Alabama
Carya: Iroquois
Caulophyllum: Iroquois
Ceanothus: Thompson
Celastrus: Cherokee
Chaetopappa: Keres, Western; Zuni
Chamaecyparis: Kwakiutl; Kwakiutl, Southern
Chaptalia: Seminole
Chenopodium: Cree, Woodlands; Miwok
Chimaphila: Saanich; Yurok
Chlorogalum: Wailaki
Chrysothamnus: Shoshoni
Cibotium: Hawaiian
Cicuta: Cree, Woodlands; Paiute
Cimicifuga: Iroquois
Cirsium: Delaware; Delaware, Oklahoma
Claytonia: Shoshoni
Clematis: Shoshoni

Collinsonia: Iroquois
Conioselinum: Kwakiutl
Conyza: Hawaiian
Coronilla: Cherokee
Corylus: Iroquois
Crataegus: Okanagan-Colville
Cucurbita: Zuni
Cyperus: Hawaiian
Datisca: Miwok
Datura: Cahuilla; Kawaiisu
Dichanthelium: Seminole
Echinacea: Lakota
Echinocereus: Isleta
Eleocharis: Seminole
Encelia: Kawaiisu; Tubatulabal
Epixiphium: Keres, Western
Eriastrum: Shoshoni
Ericameria: Kawaiisu; Miwok; Tubatulabal
Eriodictyon: Atsugewi; Cahuilla; Costanoan; Miwok; Shoshoni
Eriogonum: Paiute; Shoshoni; Thompson
Eriophyllum: Miwok
Eryngium: Seminole
Erysimum: Keres, Western; Zuni
Erythrina: Seminole
Eucalyptus: Hawaiian
Eupatorium: Chippewa; Iroquois
Euphorbia: Lakota
Frangula: Modesse
Fraxinus: Iroquois
Fucus: Kwakiutl, Southern
Galium: Costanoan; Neeshenam
Gaultheria: Delaware
Geranium: Miwok
Geum: Thompson
Gilia: Keres, Western
Glycyrrhiza: Blackfoot
Gnaphalium: Miwok
Goodyera: Delaware; Saanich
Gutierrezia: Keres, Western; Paiute; Zuni
Hazardia: Diegueño
Helianthella: Shoshoni
Helianthus: Paiute
Heracleum: Bella Coola; Carrier; Cree, Woodlands; Gitksan; Iroquois; Mewuk; Paiute; Paiute, Northern; Pomo; Pomo, Kashaya; Sikani
Heuchera: Blackfoot; Cheyenne; Kutenai
Hymenoxys: Hopi
Inula: Iroquois
Ipomoea: Cherokee
Ipomopsis: Shoshoni
Iris: Delaware; Shoshoni
Juniperus: Blackfoot; Chippewa; Creek; Delaware, Oklahoma; Hopi; Isleta; Montana Indian; Okanagan-Colville; Paiute; Paiute, Northern; Seminole; Swinomish; Tanana, Upper; Tewa; Thompson; Zuni
Kalmia: Cherokee
Larix: Okanagan-Colville
Larrea: Cahuilla; Diegueño; Isleta; Paiute; Papago; Pima; Yavapai

Lathyrus: Iroquois
Ledum: Cree, Hudson Bay; Tanana, Upper
Lemna: Iroquois
Lespedeza: Omaha; Ponca
Lesquerella: Keres, Western
Leucothoe: Cherokee
Liatris: Creek; Seminole
Ligusticum: Zuni
Lithospermum: Cheyenne
Lobelia: Zuni
Lomatium: Cheyenne; Crow; Kwakiutl; Paiute; Paiute, Northern; Shoshoni; Yuki
Lonicera: Kwakiutl; Shoshoni
Lophophora: Kiowa
Lycopodium: Chippewa
Lygodesmia: Keres, Western
Lysichiton: Cowlitz; Gitksan; Makah; Tolowa; Yurok
Mahonia: Flathead; Havasupai; Thompson
Maianthemum: Gosiute; Iroquois
Malva: Miwok
Marah: Mendocino Indian
Marrubium: Isleta
Mentha: Kutenai; Thompson
Mitchella: Abnaki; Delaware; Delaware, Oklahoma
Mnium: Carrier, Southern; Makah
Monarda: Creek
Myrica: Micmac
Navarretia: Miwok
Nicotiana: Haisla & Hanaksiala; Mahuna; Paiute; Shoshoni
Nuphar: Cree, Woodlands; Flathead; Quinault; Shuswap; Thompson
Octoblepharum: Seminole
Oenothera: Keres, Western; Zuni
Onosmodium: Cheyenne; Lakota
Oplopanax: Cowlitz; Gitksan; Haisla; Haisla & Hanaksiala; Makah; Oweekeno; Salish, Coast; Wet'suwet'en
Opuntia: Costanoan
Osmorhiza: Paiute, Northern
Osmunda: Cherokee; Iroquois
Ostrya: Chippewa
Oxalis: Pomo; Pomo, Kashaya; Tolowa
Oxytropis: Thompson
Panax: Seminole
Panicum: Seminole
Pediomelum: Blackfoot; Seminole
Penstemon: Thompson
Persea: Seminole
Petasites: Skagit; Tolowa
Phacelia: Cherokee; Keres, Western
Philadelphus: Thompson
Phlox: Havasupai
Phoradendron: Kawaiisu; Keres, Western; Seminole
Physaria: Blackfoot
Physocarpus: Hesquiat
Phytolacca: Delaware; Delaware, Oklahoma; Iroquois

Picea: Bella Coola; Chippewa; Haisla & Hanaksiala; Iroquois; Oweekeno
Pinus: Bella Coola; Cherokee; Flathead; Haisla & Hanaksiala; Keres, Western; Seminole; Skagit; Thompson; Yokia
Plagiomnium: Bella Coola; Heiltzuk; Oweekeno
Plantago: Abnaki; Chippewa; Iroquois; Shoshoni
Platanthera: Thompson
Platanus: Iroquois
Plumbago: Hawaiian
Poliomintha: Hopi; Tewa
Polygala: Seminole
Polygonum: Iroquois
Polypodium: Wailaki
Polystichum: Cherokee; Iroquois
Populus: Algonquin, Tête-de-Boule; Bella Coola; Blackfoot; Cahuilla; Nez Perce; Ojibwa; Okanagon; Thompson
Porophyllum: Havasupai
Porteranthus: Cherokee
Proboscidea: Pima
Pseudoroegneria: Okanagan-Colville
Pseudotsuga: Isleta; Montana Indian; Okanagon; Pomo, Little Lakes; Thompson
Psorothamnus: Paiute
Pteridium: Thompson
Pulsatilla: Blackfoot; Great Basin Indian; Omaha; Thompson
Quercus: Houma; Mohegan; Seminole
Ranunculus: Hesquiat; Okanagan-Colville; Thompson
Rhizomnium: Bella Coola
Rhododendron: Cherokee
Rosa: Arapaho; Crow
Rubus: Cherokee
Rumex: Bella Coola; Blackfoot; Cree, Woodlands; Hanaksiala; Iroquois; Paiute; Shoshoni
Salicornia: Heiltzuk
Salix: Creek; Kiowa; Montana Indian; Seminole
Sambucus: Hanaksiala; Hesquiat; Montana Indian; Okanagan-Colville; Paiute
Sanguinaria: Ojibwa
Saururus: Ojibwa; Seminole
Senecio: Hopi
Shepherdia: Cree, Woodlands; Tanana, Upper
Smilax: Iroquois
Solidago: Miwok
Sorbus: Cree, Woodlands
Sphaeralcea: Shoshoni
Stanleya: Shoshoni
Stellaria: Iroquois
Symplocarpus: Abnaki; Iroquois
Taxus: Abnaki; Chippewa; Iroquois; Menominee
Tephrosia: Catawba
Tetraneuris: Hopi
Thuja: Abnaki; Bella Coola; Iroquois; Okanagan-Colville
Trillium: Chippewa; Iroquois
Tsuga: Abnaki; Algonquin, Quebec; Bella Coola; Delaware; Delaware, Oklahoma; Iroquois
Turricula: Kawaiisu
Umbellularia: Mendocino Indian; Pomo; Pomo, Kashaya; Yuki
Urtica: Bella Coola; Cahuilla; Chehalis; Diegueño; Haisla & Hanaksiala; Hesquiat; Kwakiutl, Southern; Mahuna; Miwok; Nitinaht; Okanagan-Colville; Paiute; Paiute, Northern; Pomo; Pomo, Kashaya; Quileute; Sanpoil; Shoshoni; Shuswap; Thompson
Valeriana: Gosiute
Veratrum: Bella Coola; Cherokee; Haisla; Hanaksiala; Kitasoo; Okanagan-Colville; Oweekeno; Paiute; Paiute, Northern; Salish, Coast; Shoshoni; Thompson; Tsimshian; Washo
Verbascum: Delaware; Delaware, Oklahoma
Verbesina: Seminole
Viola: Blackfoot
Wyethia: Klamath; Mendocino Indian; Okanagan-Colville
Ximenia: Seminole
Yucca: Blackfoot
Zanthoxylum: Cherokee; Menominee
Zigadenus: Keres, Western; Mendocino Indian; Montana Indian; Okanagon; Paiute; Paiute, Northern; Shoshoni; Thompson; Washo
Zinnia: Keres, Western
Ziziphus: Pima

Antirheumatic (Internal)

Abies: Iroquois; Keres, Western
Achillea: Iroquois
Acorus: Mohegan
Actaea: Okanagon; Thompson
Adiantum: Cherokee; Iroquois; Mahuna
Agastache: Miwok
Aletris: Cherokee
Alnus: Eskimo, Inuktitut; Micmac
Aloysia: Havasupai
Anaphalis: Quileute
Anemone: Gitksan
Anthemis: Cherokee
Apocynum: Cherokee
Aralia: Cherokee; Iroquois
Arctium: Abnaki; Cherokee; Delaware; Delaware, Oklahoma
Aristolochia: Cherokee
Armoracia: Cherokee
Artemisia: Gosiute; Miwok; Tanaina
Asclepias: Delaware, Oklahoma
Bahia: Navajo, Ramah
Balsamorhiza: Blackfoot; Miwok
Baptisia: Iroquois; Koasati
Bignonia: Koasati
Calochortus: Keres, Western
Carya: Iroquois
Castanea: Mohegan
Caulophyllum: Cherokee; Iroquois
Ceanothus: Thompson
Cephalanthus: Koasati
Chamaebatia: Miwok
Chenopodium: Cree, Woodlands
Chimaphila: Cherokee; Iroquois; Micmac
Cimicifuga: Cherokee
Collinsonia: Iroquois
Corydalis: Navajo
Croton: Jemez
Cucurbita: Cheyenne
Cupressus: Costanoan; Miwok
Cycloloma: Hopi
Datura: Tubatulabal
Dryopteris: Cherokee
Echinacea: Cheyenne
Echinocystis: Cherokee
Ephedra: Paiute
Epigaea: Iroquois
Equisetum: Iroquois; Okanagan-Colville
Eriodictyon: Cahuilla; Mahuna; Miwok; Round Valley Indian
Eryngium: Creek; Seminole
Eupatorium: Cherokee
Euphorbia: Meskwaki
Frangula: Okanagan-Colville
Gaultheria: Delaware, Oklahoma; Iroquois; Menominee; Ojibwa; Potawatomi
Gnaphalium: Cherokee
Goodyera: Delaware, Oklahoma
Hamamelis: Iroquois
Heracleum: California Indian
Heuchera: Cheyenne
Humulus: Cherokee
Hypericum: Koasati
Ipomopsis: Navajo, Ramah
Iris: Delaware, Oklahoma
Juniperus: Cherokee; Chippewa; Iroquois; Micmac; Thompson
Larix: Iroquois; Okanagan-Colville
Larrea: Pima
Ledum: Quinault; Tanana, Upper
Leucothoe: Cherokee
Liatris: Creek; Koasati
Lilium: Cherokee
Lindera: Creek
Liriodendron: Cherokee
Lobelia: Cherokee
Lomatium: Okanagan-Colville
Lysichiton: Gitksan
Mahonia: Miwok; Navajo; Shoshoni; Thompson
Maianthemum: Algonquin, Quebec; Gitksan; Iroquois; Thompson
Malus: Gitksan; Haisla & Hanaksiala
Mentha: Okanagan-Colville
Mentzelia: Cheyenne
Menyanthes: Aleut
Mirabilis: Navajo
Monarda: Creek
Nepeta: Rappahannock; Shinnecock
Nolina: Isleta
Nuphar: Bella Coola
Onoclea: Iroquois
Oplopanax: Bella Coola; Nitinaht
Panax: Cherokee; Houma

Penstemon: Shoshoni
Petasites: Tanaina
Phlox: Mahuna
Phryma: Ojibwa, South
Phyla: Mahuna
Physocarpus: Hesquiat
Phytolacca: Cherokee; Delaware, Oklahoma; Rappahannock; Seminole
Picea: Cree, Woodlands; Gitksan; Iroquois; Tanana, Upper
Pinus: Cherokee; Costanoan; Iroquois; Mahuna; Paiute; Wailaki
Podophyllum: Cherokee; Meskwaki
Polygala: Cherokee
Polystichum: Cherokee
Populus: Cherokee; Iroquois; Okanagon; Thompson
Porophyllum: Havasupai
Potentilla: Okanagan-Colville
Prunus: Paiute
Pseudotsuga: Bella Coola
Pteridium: Iroquois
Purshia: Hualapai
Pyrola: Iroquois
Quercus: Kawaiisu; Neeshenam
Rhamnus: Kawaiisu
Rhododendron: Cherokee
Rhus: Algonquin, Quebec
Rosa: Costanoan
Rubus: Cherokee
Rumex: Paiute
Sagittaria: Iroquois
Salix: Creek; Seminole
Sambucus: Cherokee; Okanagon; Rappahannock; Thompson
Sanicula: Micmac
Sassafras: Cherokee; Iroquois
Shepherdia: Gitksan
Silphium: Ojibwa; Omaha; Ponca; Winnebago
Smilax: Cherokee
Sorbus: Bella Coola
Staphylea: Iroquois
Tanacetum: Mahuna
Taxus: Abnaki; Algonquin, Quebec; Chippewa; Iroquois
Tephrosia: Catawba
Thuja: Algonquin, Quebec
Tsuga: Abnaki; Hesquiat; Iroquois
Urtica: Quileute; Tanaina
Vaccinium: Flathead
Veratrum: Hanaksiala; Quinault
Verbascum: Atsugewi
Veronicastrum: Iroquois
Vicia: Cherokee
Vitis: Chippewa
Xanthium: Mahuna

Basket Medicine
Agrimonia: Iroquois
Clintonia: Iroquois
Corallorrhiza: Iroquois
Desmodium: Iroquois

Epilobium: Iroquois
Galium: Iroquois
Lobelia: Iroquois
Sarracenia: Iroquois

Blood Medicine
Acer: Iroquois
Achillea: Iroquois; Makah; Paiute; Shuswap
Acorus: Iroquois; Shinnecock
Actaea: Cheyenne
Adiantum: Costanoan
Ageratina: Iroquois
Agrimonia: Cherokee
Alnus: Cherokee; Chippewa; Mendocino Indian; Montagnais
Ambrosia: Delaware; Delaware, Oklahoma; Iroquois
Amelanchier: Iroquois
Anemopsis: Isleta
Angelica: Iroquois
Antennaria: Navajo, Kayenta
Anthemis: Iroquois
Apocynum: Iroquois
Aquilegia: Keres, Western
Aralia: Abnaki; Cherokee; Chippewa; Iroquois; Menominee; Ojibwa; Okanagon; Thompson
Arctium: Cherokee; Delaware; Delaware, Oklahoma; Delaware, Ontario; Iroquois; Ojibwa; Potawatomi
Arctostaphylos: Ojibwa; Okanagan-Colville
Arenaria: Shoshoni
Arisaema: Choctaw; Iroquois
Armoracia: Iroquois
Artemisia: Tanana, Upper
Aruncus: Tlingit
Asarum: Cherokee; Iroquois
Asparagus: Iroquois
Asplenium: Hawaiian
Aster: Iroquois
Atriplex: Keres, Western
Barbarea: Cherokee
Berchemia: Choctaw
Betula: Iroquois; Ojibwa
Bignonia: Cherokee
Bobea: Hawaiian
Brachyactis: Paiute
Caesalpinia: Hawaiian
Castilleja: Navajo, Ramah; Shoshoni
Catharanthus: Hawaiian
Caulanthus: Shoshoni
Ceanothus: Iroquois
Celastrus: Iroquois
Cephalanthus: Seminole
Cercocarpus: Paiute; Shoshoni
Chamaesyce: Costanoan; Miwok
Chenopodium: Carrier; Seminole
Chimaphila: Delaware; Delaware, Oklahoma; Iroquois; Malecite; Menominee; Micmac; Okanagan-Colville; Rappahannock
Cibotium: Hawaiian
Cimicifuga: Iroquois

Cleome: Navajo, Ramah
Cocculus: Houma
Collinsonia: Iroquois
Comptonia: Delaware; Delaware, Oklahoma
Conioselinum: Navajo, Kayenta
Coptis: Iroquois
Corallorrhiza: Paiute; Shoshoni
Cornus: Cherokee; Iroquois; Okanagan-Colville; Rappahannock; Thompson
Corylus: Iroquois
Crotalaria: Mohegan
Curcuma: Hawaiian
Cypripedium: Iroquois
Dalibarda: Iroquois
Datura: Paiute
Daucus: Costanoan; Iroquois
Diervilla: Iroquois; Menominee
Dirca: Iroquois
Echinacea: Omaha
Ephedra: Cahuilla; Diegueño; Kawaiisu; Mahuna; Paiute; Shoshoni; Tubatulabal
Equisetum: Saanich
Eriodictyon: Cahuilla; Costanoan; Pomo, Kashaya; Round Valley Indian
Eriogonum: Mahuna; Navajo, Ramah
Eryngium: Creek
Eupatorium: Rappahannock
Fagus: Iroquois
Fouquieria: Mahuna
Fragaria: Iroquois
Frangula: Okanagan-Colville
Fraxinus: Iroquois
Gaultheria: Chippewa; Iroquois
Gaylussacia: Iroquois
Gelsemium: Delaware; Delaware, Oklahoma
Gentianopsis: Delaware; Delaware, Oklahoma; Rappahannock
Geum: Blackfoot
Gleditsia: Delaware; Delaware, Oklahoma
Goodyera: Cherokee
Grindelia: Diegueño; Mendocino Indian; Mewuk
Hamamelis: Iroquois
Hepatica: Iroquois
Heuchera: Okanagan-Colville
Hibiscus: Hawaiian
Hoita: Costanoan
Holodiscus: Squaxin
Horkelia: Pomo, Kashaya
Hudsonia: Montagnais
Huperzia: Iroquois
Hymenopappus: Navajo
Ipomoea: Houma; Iroquois
Ipomopsis: Great Basin Indian; Navajo, Kayenta; Shoshoni
Iris: Iroquois
Juglans: Costanoan; Iroquois
Juniperus: Creek; Kwakiutl; Paiute; Shoshoni
Larix: Chippewa; Iroquois; Okanagan-Colville
Ledum: Makah; Montagnais; Tanana, Upper
Lewisia: Nez Perce

Ligusticum: Pomo, Kashaya
Limonium: Costanoan
Linanthus: Pomo, Calpella
Lindera: Cherokee; Creek
Lithospermum: Hopi
Lobelia: Iroquois
Lomatium: Okanagan-Colville
Lonicera: Iroquois; Okanagan-Colville
Lotus: Mahuna
Ludwigia: Hawaiian
Lycopodium: Iroquois
Lysichiton: Makah
Machaeranthera: Shoshoni
Magnolia: Houma
Mahonia: Great Basin Indian; Hoh; Kutenai; Mendocino Indian; Modesse; Okanagan-Colville; Paiute; Quileute; Shoshoni; Shuswap; Squaxin; Thompson; Tolowa
Maianthemum: Iroquois; Shuswap; Washo
Malacothrix: Apache, White Mountain
Malus: Makah; Quinault
Mentha: Ojibwa
Mitchella: Iroquois
Monardella: Costanoan; Mendocino Indian; Shoshoni
Murdannia: Hawaiian
Myrica: Delaware; Delaware, Oklahoma
Myriophyllum: Iroquois
Nepeta: Ojibwa
Nuphar: Bella Coola; Iroquois
Onoclea: Iroquois
Oplopanax: Okanagon; Thompson
Osmorhiza: Iroquois
Osmunda: Iroquois
Ostrya: Cherokee
Oxalis: Cherokee; Iroquois
Panax: Iroquois; Micmac
Passiflora: Houma
Pedicularis: Chippewa
Pelea: Hawaiian
Pellaea: Costanoan; Mahuna; Miwok
Perilla: Rappahannock
Phlox: Meskwaki; Okanagan-Colville
Phytolacca: Cherokee; Delaware; Delaware, Oklahoma
Picea: Cree, Woodlands; Makah
Pinus: Gitksan; Iroquois; Quinault
Plagiomnium: Bella Coola
Plantago: Iroquois; Keres, Western
Platanus: Diegueño; Meskwaki
Polygala: Cree, Woodlands; Seminole
Polygonum: Hawaiian
Polystichum: Iroquois
Populus: Iroquois; Ojibwa; Shoshoni
Portulaca: Keres, Western
Potentilla: Okanagan-Colville
Prunella: Iroquois
Prunus: Algonquin, Quebec; Cherokee; Chippewa; Gosiute; Hoh; Iroquois; Makah; Quileute
Pseudotsuga: Okanagan-Colville

Psilostrophe: Navajo, Kayenta
Pteridium: Iroquois
Pyrola: Iroquois
Pyrrhopappus: Cherokee
Quercus: Atsugewi; Ojibwa
Ranunculus: Iroquois
Rhamnus: Iroquois; Kawaiisu
Rhizomnium: Bella Coola
Rhus: Malecite
Rosa: Shoshoni; Tanana, Upper
Rubus: Iroquois; Makah
Rumex: Cherokee; Delaware; Delaware, Oklahoma; Gosiute; Hawaiian; Iroquois; Mohegan; Paiute; Potawatomi; Rappahannock; Shoshoni
Salix: Houma; Mahuna; Paiute; Seminole
Salvia: Diegueño
Sambucus: Delaware; Delaware, Oklahoma; Kawaiisu; Shoshoni; Squaxin
Sanguinaria: Delaware, Ontario; Iroquois; Mohegan; Ojibwa
Sarcobatus: Cheyenne
Sassafras: Cherokee; Chippewa; Choctaw; Delaware; Delaware, Oklahoma; Iroquois
Satureja: Mendocino Indian; Pomo; Pomo, Kashaya; Saanich; Yurok
Saxifraga: Iroquois
Scrophularia: Iroquois
Senecio: Iroquois
Senna: Iroquois
Sequoia: Houma
Sorbus: Montagnais
Spiranthes: Seminole
Stillingia: Seminole
Taraxacum: Algonquin, Quebec; Cherokee; Iroquois; Ojibwa; Rappahannock
Taxus: Karok; Micmac; Yurok
Thalictrum: Gitksan
Thuja: Iroquois; Ojibwa
Toxicodendron: Iroquois
Trifolium: Iroquois; Rappahannock
Trillium: Iroquois
Tsuga: Iroquois
Ulmus: Iroquois
Urtica: Shoshoni
Uvularia: Iroquois
Vaccinium: Ojibwa
Veratrum: Hanaksiala; Okanagon; Thompson
Verbascum: Iroquois
Vernonia: Cherokee
Veronicastrum: Chippewa
Viburnum: Iroquois
Viola: Cherokee
Vitis: Cherokee; Iroquois
Waldsteinia: Iroquois
Wyethia: Paiute
Xanthium: Apache, White Mountain
Xanthorhiza: Cherokee

Breast Treatment
Achillea: Bella Coola
Asarum: Cherokee

Asplenium: Cherokee
Cardamine: Iroquois
Carica: Hawaiian
Chamaesyce: Hawaiian
Clermontia: Hawaiian
Collinsonia: Cherokee
Eupatorium: Cherokee
Gaillardia: Blackfoot
Hepatica: Cherokee
Humulus: Cherokee
Ledum: Cree, Woodlands
Lewisia: Flathead
Lygodesmia: Cheyenne
Marrubium: Cherokee
Mertensia: Cheyenne
Osmorhiza: Blackfoot
Panax: Cherokee
Perideridia: Blackfoot
Petradoria: Hopi
Philadelphus: Thompson
Plagiomnium: Bella Coola
Polygonatum: Cherokee
Rhizomnium: Bella Coola
Sambucus: Creek
Scutellaria: Cherokee
Verbena: Cherokee

Burn Dressing
Abies: Carrier, Northern; Micmac; Penobscot
Abronia: Shoshoni
Achillea: Bella Coola; Cree, Woodlands; Crow; Zuni
Acorus: Meskwaki
Agastache: Chippewa
Alnus: Mendocino Indian
Alocasia: Hawaiian
Anaphalis: Algonquin, Tête-de-Boule
Anemone: Meskwaki
Anemopsis: Keres, Western
Aralia: Cherokee; Meskwaki
Arbutus: Cowichan
Arctostaphylos: Atsugewi; Flathead
Argemone: Kawaiisu; Paiute; Shoshoni; Washo
Aristida: Keres, Western
Artemisia: Meskwaki; Ponca
Balsamorhiza: Flathead; Okanagan-Colville
Betula: Cree, Woodlands
Brassica: Shoshoni
Castilleja: Navajo; Navajo, Ramah
Ceanothus: Okanagan-Colville
Cercocarpus: Apache, White Mountain; Gosiute; Paiute; Shoshoni
Chenopodium: Iroquois; Navajo, Kayenta
Cirsium: Kiowa
Clematis: Miwok; Shoshoni
Clintonia: Chippewa
Conocephalum: Haisla & Hanaksiala
Conyza: Keres, Western
Cosmos: Navajo
Cynoglossum: Concow
Echinacea: Cheyenne; Dakota; Omaha; Pawnee; Ponca; Sioux; Winnebago

Ephedra: Paiute; Shoshoni
Equisetum: Okanagon; Thompson
Eriogonum: Klamath
Eupatorium: Potawatomi
Fagus: Iroquois; Potawatomi
Fragaria: Quileute
Frangula: Kawaiisu
Gaultheria: Klallam
Gaura: Navajo
Geranium: Meskwaki
Goodyera: Cherokee
Helianthus: Meskwaki
Holodiscus: Okanagan-Colville
Hydrangea: Cherokee
Impatiens: Mohegan; Nanticoke; Penobscot
Iris: Algonquin, Tête-de-Boule; Meskwaki; Shoshoni
Juniperus: Shoshoni
Krascheninnikovia: Hopi; Tewa; Zuni
Larix: Chippewa; Thompson
Larrea: Paiute
Ledum: Chippewa; Cree; Cree, Woodlands
Lomatium: Thompson
Lonicera: Wet'suwet'en
Lysichiton: Haisla & Hanaksiala; Hesquiat; Nitinaht
Maianthemum: Nitinaht
Malvella: Choctaw
Mentzelia: Gosiute; Kawaiisu
Mirabilis: Meskwaki; Navajo, Ramah; Paiute
Monarda: Chippewa
Navarretia: Costanoan
Nereocystis: Kwakiutl, Southern
Oenothera: Navajo, Ramah
Opuntia: Apache, Western; Cahuilla; Hualapai
Paeonia: Shoshoni
Parthenium: Catawba
Pediomelum: Cheyenne
Penstemon: Navajo, Ramah; Shoshoni
Petrophyton: Gosiute
Phlox: Navajo
Physalis: Iroquois
Picea: Bella Coola; Cree, Woodlands
Pinus: Flathead; Iroquois; Mendocino Indian; Miwok; Navajo, Ramah
Plantago: Algonquin, Tête-de-Boule; Cherokee; Cree; Houma; Iroquois; Menominee; Meskwaki; Mohegan; Ojibwa
Polystichum: Quinault
Populus: Haisla & Hanaksiala
Portulaca: Iroquois
Potentilla: Navajo, Kayenta
Prenanthes: Bella Coola
Prosopis: Pima
Prunella: Cherokee
Prunus: Apache, Mescalero; Chippewa; Crow; Iroquois
Pteridium: Yana
Quercus: Kawaiisu; Neeshenam
Rhus: Cherokee; Cheyenne
Rosa: Paiute; Pawnee; Shoshoni

Rubus: Cowlitz; Kwakiutl; Quileute; Quinault
Rudbeckia: Chippewa
Rumex: Keres, Western; Paiute; Shoshoni
Sambucus: Cherokee; Yokut
Sanguinaria: Meskwaki
Sassafras: Rappahannock
Senecio: Navajo, Kayenta
Silene: Navajo, Ramah
Smallanthus: Cherokee
Smilax: Cherokee
Solidago: Chippewa; Costanoan; Iroquois; Mahuna; Meskwaki
Sphaeralcea: Dakota
Symphoricarpos: Cowichan; Flathead; Saanich
Taxus: Okanagan-Colville
Thuja: Malecite; Micmac
Tilia: Iroquois
Trillium: Yurok
Tsuga: Bella Coola; Kwakiutl
Typha: Dakota; Montana Indian; Omaha; Pawnee; Ponca; Sioux; Winnebago
Ulmus: Cherokee
Veratrum: Paiute
Wyethia: Mendocino Indian
Zanthoxylum: Comanche
Zigadenus: Paiute

Cancer Treatment
Abies: Iroquois
Aesculus: Cherokee
Aralia: Iroquois
Arnoglossum: Cherokee
Artemisia: Eskimo; Tanana, Upper
Athyrium: Hesquiat
Blechnum: Hesquiat
Ceanothus: Thompson
Celastrus: Chippewa
Chamaesyce: Cherokee
Chimaphila: Cherokee; Iroquois
Cirsium: Iroquois
Cornus: Menominee
Crepis: Meskwaki
Cryptantha: Hopi
Cynoglossum: Cherokee; Iroquois
Dryopteris: Hesquiat
Epilobium: Kwakiutl
Euphorbia: Cherokee
Fritillaria: Lakota
Gaillardia: Thompson
Hackelia: Cherokee
Hydrangea: Cherokee
Hydrastis: Cherokee
Lachnanthes: Cherokee
Larix: Okanagan-Colville; Thompson
Larrea: Cahuilla
Maianthemum: Thompson
Oplopanax: Gitksan; Haisla; Wet'suwet'en
Orobanche: Montana Indian
Ostrya: Iroquois
Oxalis: Cherokee
Pedicularis: Meskwaki

Picea: Thompson
Polystichum: Hesquiat
Prunus: Kwakiutl
Pteridium: Hesquiat
Pyrola: Nootka
Shepherdia: Thompson
Taxus: Tsimshian
Tradescantia: Cherokee
Trifolium: Shinnecock; Thompson
Trillium: Cherokee
Ulmus: Montana Indian
Xanthorhiza: Cherokee

Carminative
Achillea: Shoshoni
Acorus: Abnaki; Cherokee; Dakota; Omaha; Pawnee; Ponca; Winnebago
Aletris: Cherokee
Allium: Cherokee
Alnus: Tanana, Upper
Angelica: Cherokee; Creek
Aralia: Cherokee
Arisaema: Cherokee
Artemisia: Okanagan-Colville; Tewa
Asclepias: Meskwaki
Brickellia: Keres, Western
Cardamine: Iroquois
Chenopodium: Eskimo, Inupiat
Chlorogalum: Wailaki
Eupatorium: Iroquois
Foeniculum: Cherokee
Gutierrezia: Hopi; Tewa
Helianthus: Shasta
Hydrastis: Iroquois
Inula: Iroquois
Larrea: Pima
Liatris: Cherokee
Linum: Paiute
Matricaria: Aleut
Mentha: Cherokee; Chippewa; Dakota; Omaha; Paiute; Pawnee; Ponca; Shoshoni; Winnebago
Menyanthes: Aleut
Mitchella: Iroquois
Monarda: Cherokee
Pectis: Navajo; Zuni
Pinus: Kawaiisu
Polygonatum: Iroquois
Polypodium: Hesquiat
Rhus: Iroquois
Salix: Iroquois
Salvia: Costanoan
Sanguinaria: Iroquois
Solanum: Navajo

Cathartic
Abies: Carrier, Northern; Gitksan; Okanagon; Thompson
Abronia: Navajo, Kayenta
Acer: Blackfoot
Achillea: Makah; Montana Indian; Okanagan-Colville
Acorus: Chippewa; Menominee; Meskwaki; Ojibwa

Acourtia: Coahuilla
Actaea: Cree, Hudson Bay
Adenostoma: Coahuilla
Agastache: Shoshoni
Ageratina: Iroquois
Aletes: Keres, Western
Allium: Cherokee
Alnus: Bella Coola; Cherokee; Gitksan; Iroquois; Micmac
Amelanchier: Blackfoot
Amphicarpaea: Chippewa
Angelica: Bella Coola
Apocynum: Cree, Hudson Bay; Montana Indian
Aralia: Montana Indian
Arenaria: Gosiute; Yavapai
Argemone: Shoshoni
Artemisia: Mewuk; Paiute; Shoshoni
Arundinaria: Seminole
Asarum: Iroquois
Asclepias: Meskwaki; Pima
Astragalus: Jemez
Atriplex: Shoshoni; Yokut
Bahia: Keres, Western
Balsamorhiza: Flathead
Baptisia: Cherokee
Betula: Chippewa; Delaware, Oklahoma
Castilleja: Gitksan; Shoshoni
Celastrus: Chippewa
Cephalanthus: Seminole
Chaenactis: Sanpoil
Chamaesyce: Cherokee; Navajo; Zuni
Chrysothamnus: Navajo, Ramah
Cicuta: Bella Coola; Kwakiutl; Salish, Coast
Cornus: Algonquin, Quebec; Green River Group; Iroquois
Croton: Apache, White Mountain; Keres, Western; Zuni
Cucurbita: Cahuilla; Paiute; Shoshoni
Dalea: Jemez; Lakota
Datura: Costanoan
Daucus: Micmac
Dirca: Chippewa; Iroquois
Empetrum: Bella Coola
Ephedra: Shoshoni
Epilobium: Miwok
Eriastrum: Paiute; Shoshoni
Erigeron: Paiute
Eriogonum: Cahuilla; Mahuna
Eryngium: Creek
Euonymus: Iroquois; Mohegan
Eupatorium: Cherokee
Euphorbia: Cherokee; Creek; Meskwaki; Navajo; Ojibwa
Fendlera: Navajo, Kayenta
Frangula: Costanoan; Delaware, Oklahoma; Diegueño; Flathead; Karok; Kutenai; Mahuna; Mendocino Indian; Mewuk; Miwok; Modesse; Sanpoil; Thompson; West Coast Indian; Yurok, South Coast (Nererner)
Fraxinus: Chippewa; Delaware, Oklahoma; Iroquois
Fremontodendron: Kawaiisu
Gentianella: Cherokee
Geocaulon: Cree, Hudson Bay
Glycyrrhiza: Navajo, Ramah
Gutierrezia: Keres, Western
Hedeoma: Shoshoni
Heracleum: Okanagon; Omaha; Thompson
Heuchera: Gosiute
Huperzia: Nitinaht
Hydrangea: Cherokee
Hymenoxys: Hopi
Hypericum: Seminole
Ilex: Creek; Iroquois
Inula: Iroquois
Ipomopsis: Navajo, Kayenta; Paiute; Shoshoni; Washo
Iris: Cree, Hudson Bay; Creek; Iroquois; Ojibwa; Yavapai
Juglans: Cherokee; Chippewa; Delaware, Oklahoma; Iroquois; Malecite; Menominee; Meskwaki; Micmac; Potawatomi
Juniperus: Carrier, Northern; Gitksan; Thompson
Lobelia: Iroquois
Lonicera: Algonquin, Quebec; Blackfoot
Lotus: Navajo, Ramah
Lycopodium: Montagnais
Lysichiton: Skokomish
Machaeranthera: Navajo; Shoshoni
Maianthemum: Delaware; Delaware, Oklahoma; Gitksan
Marah: Luiseño
Mirabilis: Luiseño; Paiute
Mitchella: Iroquois
Mitella: Gosiute
Monarda: Choctaw
Monardella: Shoshoni
Morus: Cherokee; Delaware, Oklahoma
Nicotiana: Cherokee; Costanoan; Paiute; Shoshoni
Oenanthe: Kitasoo
Oplopanax: Bella Coola; Carrier, Northern; Carrier, Southern; Gitksan
Opuntia: Cahuilla
Orthocarpus: Navajo, Ramah
Osmorhiza: Bella Coola; Iroquois; Paiute; Shoshoni; Washo
Pedicularis: Potawatomi
Penstemon: Thompson
Petradoria: Navajo, Ramah
Philadelphus: Okanagan-Colville
Phlox: Navajo; Paiute
Phoradendron: Pima
Physocarpus: Carrier, Southern; Kwakiutl
Phytolacca: Iroquois
Pimpinella: Delaware; Delaware, Oklahoma
Pinus: Gitksan; Mendocino Indian; Nanticoke
Platanus: Cherokee
Podophyllum: Cherokee; Iroquois; Meskwaki
Polemonium: Meskwaki
Polygala: Cherokee
Polygonatum: Ojibwa
Polygonum: Zuni
Populus: Cree, Hudson Bay; Gitksan
Porteranthus: Iroquois
Potentilla: Ojibwa
Prosopis: Pima
Prunella: Chippewa
Prunus: Blackfoot; Cherokee; Chippewa; Navajo
Psathyrotes: Paiute; Shoshoni
Pseudotsuga: Bella Coola
Psilostrophe: Navajo, Ramah
Psorothamnus: Paiute, Northern
Purshia: Paiute; Shoshoni; Washo
Quercus: Navajo, Ramah
Ratibida: Navajo, Ramah
Rhamnus: Cherokee; Costanoan (Olhonean); Iroquois; Potawatomi
Rheum: Cherokee
Rhus: Flathead; Iroquois
Ribes: Micmac
Ricinus: Cherokee; Pima; Seminole
Rosa: Bella Coola
Rubus: Cherokee; Iroquois; Okanagan-Colville
Rumex: Gosiute; Iroquois; Micmac; Shoshoni
Salix: Paiute
Sambucus: Bella Coola; Carrier, Northern; Carrier, Southern; Cherokee; Costanoan; Gitksan; Iroquois; Menominee; Meskwaki; Micmac; Mohegan; Nitinaht; Ojibwa; Potawatomi; Sikani; Yokut
Sassafras: Seminole
Senecio: Navajo, Ramah
Senna: Cherokee
Sesamum: Cherokee
Shepherdia: Carrier, Northern; Salish; Shuswap; Sioux; Thompson
Silene: Ojibwa
Simmondsia: Yavapai
Sisyrinchium: Iroquois; Luiseño
Smilax: Chippewa
Solidago: Chippewa; Keres, Western
Sonchus: Pima
Sorbus: Gitksan
Sphaeralcea: Shoshoni
Stephanomeria: Paiute
Stillingia: Creek
Streptopus: Ojibwa
Symphoricarpos: Chippewa; Okanagan-Colville
Taraxacum: Mohegan
Tetradymia: Shoshoni
Thymophylla: Paiute
Trifolium: Costanoan
Triosteum: Meskwaki
Ulmus: Menominee
Valeriana: Menominee
Veronicastrum: Cherokee; Chippewa; Iroquois; Menominee; Ojibwa, South
Viburnum: Carrier, Northern; Ojibwa
Wyethia: Paiute; Shoshoni; Washo
Yucca: Pima

Zinnia: Navajo, Ramah

Ceremonial Medicine

Abies: Blackfoot; Cheyenne; Crow; Kwakiutl; Ojibwa
Acer: Cheyenne
Achillea: Navajo, Ramah; Ojibwa
Acorus: Cheyenne; Dakota; Omaha; Pawnee; Ponca; Winnebago
Actaea: Cheyenne
Adiantum: Nitinaht
Agastache: Cree; Navajo, Ramah
Agoseris: Navajo, Ramah
Amaranthus: Cherokee
Ambrosia: Cherokee
Amelanchier: Navajo, Ramah
Anaphalis: Cheyenne
Andropogon: Cherokee
Anemone: Ojibwa
Angelica: Blackfoot
Antennaria: Navajo, Ramah; Okanagan-Colville
Apocynum: Navajo, Kayenta; Navajo, Ramah; Ojibwa
Aquilegia: Meskwaki; Navajo, Kayenta
Arbutus: Karok
Arceuthobium: Navajo, Ramah
Arctostaphylos: Navajo, Kayenta; Navajo, Ramah; Ojibwa
Arisaema: Meskwaki
Artemisia: Cheyenne; Dakota; Delaware, Oklahoma; Miwok; Navajo; Navajo, Kayenta; Omaha; Paiute; Pawnee; Ponca; Winnebago
Asclepias: Navajo, Kayenta; Navajo, Ramah; Omaha
Aster: Navajo, Ramah
Astragalus: Hopi; Navajo, Kayenta; Navajo, Ramah
Atriplex: Hopi; Navajo, Ramah
Baccharis: Navajo, Ramah
Besseya: Navajo, Ramah
Brickellia: Navajo, Kayenta; Navajo, Ramah
Bromus: Navajo, Kayenta
Calochortus: Hopi; Navajo, Ramah
Campanula: Navajo, Ramah
Carex: Cheyenne; Navajo, Ramah
Catabrosa: Crow; Montana Indian
Ceanothus: Navajo, Kayenta; Navajo, Ramah; Okanagan-Colville
Chamaesyce: Navajo, Ramah
Chrysothamnus: Hopi; Navajo, Kayenta; Navajo, Ramah
Cicuta: Cherokee
Clematis: Cherokee
Cleome: Navajo, Kayenta; Navajo, Ramah
Comptonia: Chippewa
Cordylanthus: Navajo, Kayenta
Coreopsis: Navajo, Ramah
Cornus: Apache, White Mountain; Menominee; Navajo, Kayenta; Navajo, Ramah; Ojibwa
Cosmos: Navajo, Ramah
Cryptantha: Navajo, Ramah
Cucurbita: Cherokee
Cuscuta: Navajo, Ramah
Cynoglossum: Cherokee

Cyperus: Navajo, Ramah
Dasylirion: Tarahumara
Datura: Apache, White Mountain; Kawaiisu; Luiseño; Navajo; Navajo, Ramah; Yokut
Delphinium: Hopi
Dichanthelium: Navajo, Ramah
Dimorphocarpa: Apache, White Mountain
Dodonaea: Hawaiian
Draba: Navajo, Ramah
Echinochloa: Navajo, Ramah
Eleocharis: Navajo, Ramah
Elymus: Iroquois
Epipactis: Navajo, Kayenta
Equisetum: Hoh; Hopi; Karok; Quileute
Erigeron: Navajo, Ramah
Eriogonum: Apache, White Mountain; Navajo, Kayenta; Navajo, Ramah; Thompson
Eryngium: Meskwaki; Seminole
Erysimum: Navajo, Ramah
Fallugia: Navajo, Ramah
Fendlera: Navajo, Kayenta
Forestiera: Navajo, Ramah
Frangula: Navajo, Kayenta
Fraxinus: Omaha
Galium: Navajo, Ramah
Gaylussacia: Iroquois
Glyceria: Crow; Montana Indian
Glycyrrhiza: Cheyenne
Gnaphalium: Navajo, Ramah
Gutierrezia: Keresan; Navajo; Navajo, Kayenta; Navajo, Ramah
Hamamelis: Menominee
Hedeoma: Navajo
Helianthus: Cheyenne; Gros Ventre; Mandan; Navajo; Navajo, Kayenta; Ree
Heracleum: Thompson
Heterotheca: Navajo, Kayenta; Navajo, Ramah
Hierochloe: Blackfoot; Cheyenne; Dakota; Omaha; Pawnee; Ponca; Winnebago
Houstonia: Navajo, Ramah
Hymenopappus: Hopi
Hymenoxys: Navajo, Ramah
Ilex: Alabama
Impatiens: Cherokee
Ipomopsis: Navajo; Navajo, Ramah
Iris: Navajo, Ramah
Juniperus: Cheyenne; Comanche; Crow; Flathead; Montana Indian; Navajo; Navajo, Ramah; Omaha
Kochia: Navajo
Koeleria: Cheyenne
Lactuca: Cherokee; Navajo, Ramah
Ledum: Algonquin, Quebec
Lesquerella: Hopi; Navajo, Kayenta; Navajo, Ramah
Ligusticum: Okanagan-Colville; Zuni
Limosella: Navajo, Ramah
Lithospermum: Blackfoot; Zuni
Lobelia: Crow
Lomatium: Blackfoot; Navajo; Nitinaht
Lonicera: Navajo, Ramah
Lophophora: Comanche; Omaha; Winnebago

Lupinus: Navajo, Ramah
Luzula: Navajo, Ramah
Lycium: Hopi; Navajo, Ramah
Lycopus: Cherokee
Madia: Cheyenne; Crow
Mahonia: Apache, White Mountain; Navajo, Ramah
Maianthemum: Meskwaki; Navajo, Kayenta; Navajo, Ramah
Malus: Haisla & Hanaksiala
Matelea: Comanche
Matricaria: Cheyenne
Mentha: Cheyenne; Navajo
Mentzelia: Navajo, Kayenta
Mitchella: Ojibwa
Monarda: Cheyenne; Navajo, Ramah
Monolepis: Navajo, Ramah
Nicotiana: Apache, White Mountain; Cherokee; Costanoan; Hopi; Navajo; Navajo, Ramah; Tewa
Oenanthe: Tsimshian
Oenothera: Hopi; Navajo, Kayenta; Navajo, Ramah
Oplopanax: Montana Indian
Oreoxis: Navajo
Orthocarpus: Navajo, Ramah
Panax: Creek
Parryella: Navajo, Ramah
Parthenocissus: Navajo
Paxistima: Navajo, Ramah
Pectis: Navajo
Pelea: Hawaiian
Penstemon: Navajo, Ramah
Pentaphylloides: Cheyenne
Pericome: Navajo, Ramah
Persea: Seminole
Peteria: Navajo, Ramah
Petradoria: Navajo, Ramah
Petrophyton: Navajo, Kayenta
Phlox: Navajo
Phragmites: Iroquois
Physaria: Hopi
Picea: Bella Coola; Koyukon; Navajo; Navajo, Ramah
Piloblephis: Seminole
Pinus: Cherokee; Navajo; Navajo, Ramah
Poa: Hopi
Podophyllum: Iroquois
Polygonum: Navajo, Ramah
Populus: Omaha; Thompson
Potamogeton: Navajo, Ramah
Prunus: Cheyenne; Dakota; Navajo, Ramah; Omaha; Pawnee; Ponca; Sioux; Winnebago
Psathyrotes: Navajo, Kayenta
Pseudocymopterus: Navajo, Kayenta; Navajo, Ramah
Pseudostellaria: Navajo, Kayenta
Pseudotsuga: Navajo, Ramah
Psilostrophe: Navajo, Ramah
Psoralidium: Cheyenne; Navajo, Ramah
Purshia: Navajo, Ramah
Pyrola: Karok
Pyrrhopappus: Navajo, Ramah
Quercus: Navajo, Ramah

Ranunculus: Navajo, Ramah
Ratibida: Navajo, Ramah
Rhododendron: Cherokee; Karok
Rhus: Delaware, Oklahoma; Hopi; Ojibwa
Ribes: Navajo, Kayenta; Navajo, Ramah
Rorippa: Navajo, Ramah
Rosa: Navajo, Ramah; Okanagan-Colville
Rumex: Navajo, Ramah
Salix: Cheyenne; Karok; Navajo, Ramah; Omaha; Seminole
Sambucus: Iroquois; Karok; Seminole
Sanguinaria: Ojibwa
Sarcobatus: Cheyenne; Hopi
Sassafras: Seminole
Saxifraga: Iroquois
Schoenocrambe: Navajo, Ramah
Scirpus: Cheyenne; Navajo, Ramah
Senecio: Navajo, Ramah; Zuni
Shepherdia: Dakota; Navajo, Kayenta
Silphium: Winnebago
Sinapis: Navajo, Ramah
Solanum: Ojibwa
Sphaeralcea: Cheyenne; Navajo; Navajo, Kayenta
Stachys: Navajo, Ramah
Staphylea: Meskwaki
Suaeda: Hopi
Symphoricarpos: Navajo, Ramah
Tetraclea: Navajo, Ramah
Tetradymia: Navajo, Ramah
Tetraneuris: Navajo, Kayenta
Thalictrum: Navajo, Ramah
Thlaspi: Navajo, Ramah
Thuja: Chippewa; Ojibwa; Oweekeno
Townsendia: Navajo
Tragopogon: Navajo, Ramah
Tribulus: Navajo
Trifolium: Navajo, Ramah
Tsuga: Kwakiutl; Kwakwaka'wakw; Oweekeno
Typha: Cheyenne; Navajo, Ramah; Omaha
Vaccinium: Iroquois; Seminole
Valeriana: Thompson
Veratrum: Haisla & Hanaksiala
Verbena: Navajo, Ramah
Verbesina: Seminole
Veronica: Navajo, Ramah
Veronicastrum: Menominee
Vicia: Navajo; Okanagan-Colville
Viola: Navajo, Ramah
Vitis: Seminole
Xanthium: Lakota; Zuni
Yucca: Apache, Western; Cherokee; Hopi; Navajo, Ramah; Tewa
Zea: Navajo
Zinnia: Navajo, Ramah

Cold Remedy
Abies: Algonquin, Tête-de-Boule; Blackfoot; Chehalis; Cree, Woodlands; Crow; Gitksan; Green River Group; Iroquois; Menominee; Micmac; Montana Indian; Ojibwa; Ojibwa, South; Oweekeno; Paiute; Potawatomi; Shoshoni; Thompson; Wet'suwet'en
Acer: Micmac
Achillea: Abnaki; Aleut; Algonquin, Quebec; Carrier, Southern; Cheyenne; Clallam; Flathead; Klallam; Kwakiutl; Micmac; Miwok; Nitinaht; Okanagan-Colville; Paiute; Paiute, Northern; Saanich; Sanpoil; Shoshoni; Thompson; Yuki
Acorus: Algonquin, Quebec; Cherokee; Cheyenne; Chippewa; Cree, Woodlands; Dakota; Delaware; Delaware, Oklahoma; Delaware, Ontario; Iroquois; Malecite; Menominee; Micmac; Mohegan; Nanticoke; Ojibwa; Omaha; Pawnee; Ponca; Winnebago
Actaea: Blackfoot
Adenostoma: Cahuilla
Agastache: Cheyenne; Okanagan-Colville; Paiute
Allium: Blackfoot; Cherokee; Chippewa; Mohegan; Shinnecock
Alnus: Menominee; Swinomish
Ambrosia: Cheyenne
Amelanchier: Cree, Woodlands; Okanagan-Colville
Anaphalis: Cherokee; Mohegan
Anemone: Carrier, Southern
Anemopsis: Cahuilla; Kawaiisu; Pima; Shoshoni; Tubatulabal
Angelica: Aleut; Cherokee; Iroquois; Mendocino Indian; Micmac; Miwok; Paiute; Pomo, Kashaya; Shoshoni; Yana
Antennaria: Natchez; Thompson
Anthemis: Mendocino Indian
Apocynum: Chippewa
Aquilegia: Paiute
Arabis: Cheyenne
Aralia: Iroquois; Malecite; Mendocino Indian; Micmac
Arbutus: Pomo, Little Lakes; Saanich; Skokomish
Arctium: Mohegan
Arctostaphylos: Cheyenne; Pomo, Calpella
Arisaema: Cherokee
Aristolochia: Cherokee; Miwok
Armoracia: Cherokee
Aronia: Potawatomi
Artemisia: Blackfoot; Cahuilla; Costanoan; Diegueño; Eskimo, Alaska; Eskimo, Kuskokwagmiut; Eskimo, Nunivak; Flathead; Gosiute; Havasupai; Hualapai; Karok; Kawaiisu; Lakota; Mendocino Indian; Navajo; Okanagan-Colville; Paiute; Paiute, Northern; Salish; Sanpoil; Shoshoni; Shuswap; Tanana, Upper; Thompson; Washo; Zuni
Aruncus: Skagit; Thompson
Asarum: Abnaki; Cherokee; Iroquois; Okanagan-Colville
Asclepias: Costanoan
Aster: Iroquois
Atriplex: Cahuilla; Paiute, Northern
Balsamorhiza: Cheyenne; Kawaiisu
Betula: Cherokee; Iroquois; Thompson
Brassica: Meskwaki; Micmac
Callirhoe: Dakota

Calocedrus: Paiute
Caltha: Chippewa
Calycanthus: Pomo; Pomo, Kashaya
Capsicum: Cherokee
Cardamine: Cherokee; Iroquois
Carya: Cherokee
Castanea: Mohegan
Castilleja: Chippewa
Ceanothus: Iroquois
Celastrus: Iroquois
Celtis: Iroquois
Cercocarpus: Paiute; Shoshoni
Chaenactis: Paiute
Chaetopappa: Zuni
Chamaebatia: Miwok
Chenopodium: Cherokee
Chimaphila: Abnaki; Cherokee; Okanagan-Colville
Chrysothamnus: Cheyenne; Jemez; Navajo, Ramah; Paiute; Shoshoni; Thompson
Cimicifuga: Cherokee
Clematis: Mendocino Indian; Montana Indian; Shasta
Collinsia: Natchez
Comandra: Meskwaki
Conioselinum: Aleut
Conyza: Seminole
Coptis: Abnaki; Algonquin, Tête-de-Boule
Corallorrhiza: Nevada Indian
Corethrogyne: Kawaiisu
Cornus: Algonquin, Quebec; Algonquin, Tête-de-Boule; Blackfoot; Cree, Hudson Bay; Iroquois; Okanagan-Colville; Thompson
Cryptantha: Navajo
Cunila: Cherokee
Cupressus: Kawaiisu; Miwok
Cyperus: Hawaiian; Pima; Tanana, Upper; Yavapai
Cypripedium: Cherokee
Daucus: Costanoan
Desmodium: Alabama; Hawaiian
Distichlis: Yokut
Echinacea: Cheyenne; Crow
Empetrum: Tanana, Upper
Ephedra: Paiute; Shoshoni
Equisetum: Okanagan-Colville
Ericameria: Miwok; Paiute; Shoshoni
Erigeron: Cherokee; Iroquois; Meskwaki; Ojibwa
Eriodictyon: Atsugewi; Cahuilla; Costanoan; Diegueño; Karok; Kawaiisu; Mendocino Indian; Miwok; Paiute; Shoshoni; Yurok
Eriogonum: Costanoan; Kawaiisu; Nevada Indian; Okanagan-Colville; Paiute; Shoshoni
Erythronium: Okanagan-Colville
Eucalyptus: Cahuilla
Eupatorium: Cherokee; Chippewa; Iroquois; Mohegan; Shinnecock
Foeniculum: Cherokee
Frasera: Havasupai
Gamochaeta: Houma
Garrya: Kawaiisu; Poliklah
Gaultheria: Algonquin, Quebec; Algonquin, Tête-

de-Boule; Cherokee; Chippewa; Iroquois
Gentianella: Tanana, Upper
Geranium: Blackfoot
Geum: Aleut; Micmac; Okanagan-Colville
Glechoma: Cherokee
Gleditsia: Meskwaki; Rappahannock
Gnaphalium: Cherokee; Choctaw; Costanoan; Creek; Keres, Western; Miwok
Goodyera: Cherokee
Grindelia: Crow; Flathead; Mendocino Indian; Ute
Gutierrezia: Lakota; Shoshoni
Hamamelis: Cherokee; Iroquois
Hedeoma: Catawba; Cherokee; Dakota
Helenium: Costanoan; Menominee; Meskwaki
Heracleum: Aleut; Paiute; Shoshoni
Hierochloe: Blackfoot; Flathead
Holodiscus: Paiute
Huperzia: Iroquois
Hyssopus: Cherokee
Inula: Delaware, Ontario; Iroquois; Malecite; Micmac
Ipomopsis: Paiute; Shoshoni; Washo
Iris: Meskwaki
Iva: Shoshoni
Juniperus: Apache, White Mountain; Cherokee; Cheyenne; Dakota; Eskimo, Inupiat; Flathead; Gosiute; Havasupai; Iroquois; Kutenai; Lakota; Navajo, Ramah; Nez Perce; Okanagan-Colville; Omaha; Paiute; Paiute, Northern; Pawnee; Ponca; Seminole; Shoshoni; Sioux; Tanana, Upper; Thompson; Washo
Kalmia: Abnaki; Algonquin, Quebec; Montagnais
Keckiella: Miwok
Larix: Iroquois; Malecite; Micmac; Nez Perce
Larrea: Cahuilla; Hualapai; Paiute; Pima; Shoshoni
Ledum: Abnaki; Haisla & Hanaksiala; Kitasoo; Micmac; Oweekeno; Tanana, Upper
Ligusticum: Atsugewi; Crow; Okanagan-Colville
Linanthus: Pomo, Calpella
Lindera: Cherokee; Iroquois
Linnaea: Snohomish
Linum: Cherokee
Lithophragma: Mendocino Indian
Lithospermum: Navajo
Lobelia: Cherokee; Yokut
Loeseliastrum: Tubatulabal
Lomatium: Cowichan; Crow; Gosiute; Great Basin Indian; Kawaiisu; Kwakiutl; Nitinaht; Okanagan-Colville; Paiute; Paiute, Northern; Saanich; Shoshoni; Songish; Thompson; Washo; Yuki
Lonicera: Swinomish
Lophophora: Kiowa
Lycopus: Blackfoot
Magnolia: Houma
Mahonia: Havasupai
Maianthemum: Okanagan-Colville; Thompson
Marrubium: Cherokee; Diegueño; Kawaiisu; Rappahannock; Round Valley Indian
Matricaria: Eskimo, Kuskokwagmiut; Eskimo, Western; Flathead; Shuswap

Melilotus: Navajo, Ramah
Melissa: Cherokee
Mentha: Carrier, Southern; Chehalis; Cherokee; Cowlitz; Cree, Woodlands; Flathead; Gosiute; Iroquois; Kutenai; Okanagan-Colville; Okanagon; Paiute; Paiute, Northern; Sanpoil; Shoshoni; Thompson; Washo
Monarda: Bannock; Cherokee; Chippewa; Flathead; Meskwaki; Nanticoke; Sioux, Teton
Monardella: Miwok; Okanagan-Colville; Paiute; Sanpoil; Shoshoni; Washo
Moneses: Eskimo, Alaska
Monotropa: Mohegan
Nepeta: Cherokee; Iroquois; Okanagan-Colville
Nicotiana: Paiute
Nymphaea: Micmac
Obolaria: Cherokee
Oenothera: Navajo, Kayenta
Oplopanax: Cowlitz; Gitksan; Green River Group; Haisla; Hanaksiala; Oweekeno; Sanpoil; Wet'suwet'en
Orobanche: Paiute
Osmorhiza: Blackfoot; Cheyenne; Kawaiisu; Paiute; Paiute, Northern; Shoshoni; Thompson; Washo
Osmunda: Iroquois
Paxistima: Okanagan-Colville
Pediomelum: Seminole
Penstemon: Natchez; Okanagan-Colville; Shoshoni
Petasites: Eskimo, Inupiat
Phacelia: Kawaiisu; Shuswap
Phlox: Havasupai
Physostegia: Meskwaki
Phytolacca: Iroquois
Picea: Carrier; Cherokee; Gitksan; Haisla & Hanaksiala; Iroquois; Kwakiutl; Tanana, Upper; Wet'suwet'en
Piloblephis: Seminole
Pinus: Algonquin, Tête-de-Boule; Apache, Mescalero; Cherokee; Eskimo, Alaska; Iroquois; Keres, Western; Micmac; Mohegan; Montagnais; Navajo, Ramah; Okanagon; Paiute; Shoshoni; Thompson; Washo
Piper: Hawaiian
Pityopsis: Seminole
Plantago: Paiute
Platanus: Delaware; Delaware, Oklahoma; Meskwaki
Polygala: Cherokee; Malecite; Micmac; Miwok; Ojibwa, South
Polygonum: Okanagan-Colville; Potawatomi; Tanana, Upper
Polypodium: Cowichan; Haisla & Hanaksiala; Saanich; Thompson
Populus: Delaware, Ontario; Flathead; Iroquois; Menominee; Meskwaki; Micmac; Ojibwa; Penobscot; Tanana, Upper; Yuki
Porteranthus: Cherokee; Iroquois
Prenanthes: Bella Coola
Prunella: Iroquois
Prunus: Cherokee; Cowichan; Iroquois; Karok;

Malecite; Micmac; Mohegan; Narraganset; Ojibwa; Okanagan-Colville; Paiute; Rappahannock; Saanich; Shinnecock; Skagit; Skokomish; Thompson
Pseudotsuga: Bella Coola; Cowlitz; Karok; Squaxin; Swinomish; Thompson
Psorothamnus: Paiute; Paiute, Northern; Shoshoni
Pteridium: Thompson
Pterocaulon: Seminole
Purshia: Havasupai; Paiute
Pycnanthemum: Cherokee; Choctaw; Miwok
Quercus: Atsugewi; Delaware; Delaware, Oklahoma; Mohegan; Neeshenam
Ranunculus: Meskwaki
Rhamnus: Kawaiisu
Rhododendron: Skokomish
Rhus: Cahuilla; Cheyenne; Chippewa; Comanche; Iroquois
Ribes: Bella Coola; Okanagan-Colville; Sanpoil; Skagit; Skagit, Upper; Tanana, Upper; Thompson
Rosa: Costanoan; Paiute; Shoshoni; Tanana, Upper; Washo
Rubus: Iroquois; Klallam
Rudbeckia: Potawatomi
Rumex: Hopi; Paiute; Pima
Salix: Costanoan; Micmac; Montana Indian; Penobscot; Pomo, Kashaya
Salvia: Cahuilla; Cherokee; Diegueño; Okanagan-Colville; Paiute; Paiute, Northern; Shoshoni; Washo
Sambucus: Cahuilla; Costanoan; Hoh; Kawaiisu; Paiute; Pima; Quileute; Shoshoni
Sanguinaria: Iroquois; Micmac
Sanvitalia: Navajo, Ramah
Sassafras: Cherokee; Iroquois; Seminole
Satureja: Cahuilla; Luiseño; Pomo, Kashaya
Scrophularia: Iroquois
Scutellaria: Miwok
Senecio: Yavapai
Silphium: Omaha; Ponca; Winnebago
Solanum: Pima
Solidago: Alabama; Cherokee
Sorbus: Algonquin, Quebec; Carrier, Southern; Potawatomi
Sphaeralcea: Navajo; Shoshoni
Spiraea: Mahuna; Okanagan-Colville; Thompson
Stachys: Meskwaki
Stephanomeria: Cheyenne
Symphoricarpos: Klallam; Miwok
Symplocarpus: Nanticoke
Tagetes: Navajo
Tanacetum: Iroquois
Tauschia: Kawaiisu
Taxus: Iroquois; Penobscot
Tetradymia: Navajo, Ramah; Paiute; Shoshoni
Thalictrum: Keres, Western; Washo
Thamnosma: Kawaiisu; Shoshoni
Thermopsis: Cheyenne
Thuja: Algonquin, Quebec; Cowlitz; Iroquois; Nez Perce

Trichostema: Costanoan; Kawaiisu; Miwok
Trifolium: Mohegan
Trillium: Iroquois
Triosteum: Iroquois
Tsuga: Algonquin, Quebec; Iroquois; Malecite; Menominee; Micmac; Potawatomi; Quinault; Thompson
Ulmus: Cherokee; Delaware; Delaware, Oklahoma; Mohegan
Umbellularia: Karok; Pomo, Kashaya
Urtica: Paiute; Shoshoni; Skagit; Snohomish
Vaccinium: Seminole; Skagit; Skagit, Upper; Tanana, Upper
Valeriana: Cree, Woodlands; Okanagon; Thompson
Veratrum: Haisla; Iroquois; Kwakiutl; Okanagan-Colville; Paiute, Northern; Salish; Shoshoni
Verbascum: Atsugewi; Delaware, Oklahoma; Mohegan; Shinnecock; Thompson
Verbena: Cherokee; Iroquois
Viburnum: Bella Coola; Iroquois; Tanana, Upper
Viola: Cherokee
Wyethia: Paiute
Xanthorhiza: Catawba
Xyris: Seminole
Zanthoxylum: Chippewa; Menominee

Contraceptive
Amelanchier: Okanagan-Colville; Thompson
Apocynum: Okanagan-Colville
Arabis: Okanagan-Colville
Arisaema: Iroquois
Asclepias: Iroquois
Bahia: Navajo, Ramah
Betula: Thompson
Castilleja: Hopi; Tewa
Chelone: Malecite; Micmac
Cicuta: Cherokee
Cirsium: Zuni
Claytonia: Iroquois
Clematis: Okanagan-Colville
Cornus: Okanagan-Colville
Cuscuta: Paiute
Equisetum: Costanoan
Erigeron: Navajo, Ramah
Eriogonum: Navajo, Ramah
Erythronium: Iroquois
Galium: Choctaw
Geum: Chehalis
Hepatica: Iroquois
Iva: Mahuna
Juniperus: Shoshoni; Zuni
Lithospermum: Navajo; Shoshoni
Lonicera: Chehalis
Mahonia: Flathead
Maianthemum: Costanoan
Phlox: Navajo, Ramah
Pinus: Kawaiisu
Pontederia: Malecite; Micmac
Rhus: Navajo, Ramah
Ricinus: Navajo
Rumex: Iroquois

Sphaeralcea: Shoshoni
Tanacetum: Malecite; Micmac
Toxicodendron: Karok
Vaccinium: Cree, Woodlands
Veratrum: Paiute; Shoshoni; Washo
Viburnum: Iroquois
Yucca: Navajo, Ramah

Cough Medicine
Abies: Cree, Woodlands; Crow; Gitksan; Iroquois; Kwakiutl; Ojibwa; Okanagan-Colville; Paiute; Thompson; Wet'suwet'en
Acer: Micmac; Mohegan; Potawatomi
Achillea: Cheyenne; Hesquiat; Paiute; Paiute, Northern
Acorus: Algonquin, Quebec; Chippewa; Cree, Woodlands; Dakota; Delaware; Delaware, Oklahoma; Lakota; Meskwaki; Micmac; Omaha; Pawnee; Ponca; Winnebago
Actaea: Blackfoot
Agastache: Chippewa; Navajo, Ramah
Aletris: Cherokee
Allium: Blackfoot
Alnus: Cherokee; Gitksan
Ambrosia: Pima
Amelanchier: Cree, Woodlands
Anaphalis: Cherokee; Montagnais
Andropogon: Seminole
Anemopsis: Kawaiisu; Pima
Angelica: Micmac; Paiute; Washo
Antennaria: Natchez; Thompson
Apocynum: Cherokee
Aquilegia: Paiute
Aralia: Cherokee; Chippewa; Iroquois; Kwakiutl; Micmac; Penobscot
Arctium: Chippewa
Arctostaphylos: Cheyenne
Arisaema: Cherokee
Aristolochia: Cherokee
Artemisia: Arapaho; Blackfoot; Costanoan; Diegueño; Gosiute; Havasupai; Kawaiisu; Navajo, Ramah; Paiute; Shoshoni; Shuswap; Tanana, Upper; Tewa; Washo
Aruncus: Kwakiutl
Asarum: Abnaki; Cherokee; Iroquois
Asclepias: Paiute
Asplenium: Cherokee
Astragalus: Lakota
Atriplex: Navajo, Ramah
Baccharis: Diegueño
Bacopa: Seminole
Balsamorhiza: Kawaiisu
Barbarea: Mohegan; Shinnecock
Berchemia: Koasati
Betula: Thompson
Boschniakia: Hesquiat
Botrychium: Iroquois
Brassica: Micmac
Brickellia: Navajo, Ramah
Campanulastrum: Meskwaki
Castanea: Cherokee

Castilleja: Gitksan
Ceanothus: Chippewa; Menominee; Modesse
Celastrus: Cherokee
Cercocarpus: Kawaiisu; Paiute; Shoshoni
Chaenactis: Paiute
Chamaebatia: Miwok
Chrysothamnus: Cheyenne; Coahuilla; Navajo, Ramah; Paiute; Shoshoni
Cimicifuga: Cherokee
Collinsia: Natchez
Conyza: Seminole
Coptis: Abnaki
Cornus: Chippewa; Cree, Hudson Bay; Iroquois
Croton: Diegueño
Cryptantha: Navajo; Navajo, Ramah
Cupressus: Kawaiisu
Cyperus: Pima
Dichanthelium: Seminole
Draba: Navajo, Ramah
Echinacea: Choctaw; Kiowa
Enceliopsis: Shoshoni
Ephedra: Keres, Western; Navajo; Navajo, Ramah
Epilobium: Abnaki
Equisetum: Modesse
Ericameria: Paiute; Shoshoni
Erigeron: Cherokee; Iroquois; Navajo, Ramah
Eriodictyon: Cahuilla; Diegueño; Mahuna; Miwok; Paiute; Pomo, Kashaya; Shoshoni; Yuki; Yurok
Eriogonum: Costanoan; Kawaiisu; Navajo, Ramah; Shoshoni
Galium: Cherokee
Gaultheria: Samish; Swinomish
Gentianella: Tanana, Upper
Geum: Blackfoot; Malecite; Micmac; Ojibwa, South
Gleditsia: Delaware; Delaware, Oklahoma; Rappahannock
Glycyrrhiza: Blackfoot; Cherokee; Keres, Western
Gnaphalium: Cherokee; Montagnais
Grindelia: Crow; Flathead; Gosiute; Paiute; Shoshoni
Gutierrezia: Lakota
Hamamelis: Iroquois
Hedeoma: Cherokee
Heracleum: Shoshoni
Hierochloe: Blackfoot
Hosta: Cherokee
Humulus: Navajo, Ramah
Hydrocotyle: Seminole
Hymenopappus: Navajo, Ramah
Hypericum: Montagnais
Hyssopus: Cherokee
Ilex: Micmac
Inula: Cherokee; Iroquois
Ipomoea: Cherokee; Iroquois
Iris: Yana
Isocoma: Pima
Iva: Navajo, Ramah
Juniperus: Apache, White Mountain; Bella Coola; Carrier, Northern; Cheyenne; Cree, Woodlands; Dakota; Eskimo, Inupiat; Gosiute; Iroquois;

Navajo, Ramah; Omaha; Paiute; Pawnee; Ponca; Seminole; Shoshoni; Tanana, Upper
Krameria: Pima
Larix: Abnaki; Algonquin, Quebec; Iroquois; Nez Perce; Thompson
Ledum: Tanana, Upper
Ligusticum: Atsugewi; Crow; Paiute
Lilium: Malecite; Micmac
Linanthus: Pomo, Calpella
Lindera: Cherokee
Linum: Cherokee
Liriodendron: Cherokee
Lithocarpus: Pomo, Kashaya
Lithospermum: Navajo
Lobelia: Iroquois
Lomatium: Great Basin Indian; Kwakiutl; Paiute; Shoshoni; Washo
Lonicera: Bella Coola; Costanoan; Nuxalkmc
Lotus: Costanoan
Lygodesmia: Blackfoot
Machaeranthera: Hopi
Mahonia: Flathead; Paiute; Quinault; Shoshoni; Tolowa
Maianthemum: Mohegan; Paiute
Malus: Nitinaht
Marrubium: Cherokee; Costanoan; Kawaiisu; Mahuna; Rappahannock; Yuki
Mentha: Flathead; Gosiute; Kutenai
Mirabilis: Navajo, Ramah
Monarda: Blackfoot; Flathead; Lakota; Navajo, Ramah
Moneses: Eskimo, Alaska
Nemopanthus: Malecite
Nepeta: Cherokee; Iroquois
Nicotiana: Navajo, Ramah; Tewa
Nymphaea: Micmac; Ojibwa; Seminole
Nymphoides: Seminole
Obolaria: Cherokee
Oplopanax: Gitksan; Haisla; Okanagan-Colville; Wet'suwet'en
Orthocarpus: Costanoan
Osmorhiza: Blackfoot; Kawaiisu; Shoshoni; Tlingit
Ostrya: Chippewa; Iroquois
Paeonia: Paiute; Shoshoni
Panicum: Seminole
Pedicularis: Cherokee; Cheyenne
Pediomelum: Seminole
Penstemon: Natchez; Navajo, Ramah
Penthorum: Meskwaki
Pericome: Navajo, Ramah
Perideridia: Blackfoot
Petasites: Delaware; Delaware, Oklahoma; Quileute
Phacelia: Kawaiisu
Picea: Algonquin, Quebec; Eskimo, Kuskokwagmiut; Eskimo, Western; Gitksan; Haisla & Hanaksiala; Iroquois; Kwakiutl; Micmac; Montagnais; Sikani; Tanana, Upper; Thompson; Wet'suwet'en
Pinus: Abnaki; Cherokee; Eskimo, Alaska; Hoh; Iroquois; Kwakiutl; Micmac; Mohegan; Navajo, Ramah; Okanagon; Quileute; Shinnecock; Shoshoni; Shuswap; Sikani; Thompson
Plantago: Carrier
Platanus: Cherokee
Polygala: Miwok; Montagnais; Ojibwa, South
Polygonatum: Ojibwa
Polygonum: Tanana, Upper
Polypodium: Green River Group; Haisla & Hanaksiala; Hesquiat; Kitasoo; Klallam; Makah; Nitinaht; Oweekeno; Quinault
Populus: Carrier, Southern; Cree, Hudson Bay; Meskwaki; Paiute; Tanana, Upper
Prunella: Iroquois
Prunus: Algonquin, Quebec; Cherokee; Delaware; Delaware, Oklahoma; Diegueño; Iroquois; Keres, Western; Mahuna; Malecite; Micmac; Ojibwa; Okanagan-Colville; Paiute; Penobscot; Potawatomi; Rappahannock; Thompson; Wet'suwet'en
Psathyrotes: Shoshoni
Pseudotsuga: Apache, White Mountain
Psorothamnus: Paiute; Shoshoni
Purshia: Klamath; Navajo, Ramah
Pycnanthemum: Lakota
Pyrola: Blackfoot
Quercus: Delaware; Delaware, Oklahoma; Miwok; Neeshenam
Ratibida: Navajo, Ramah
Rhamnus: Kawaiisu
Rhus: Cahuilla; Coahuilla; Iroquois; Malecite; Menominee
Rosa: Cree, Woodlands; Thompson
Rubus: Cherokee; Iroquois; Micmac
Rumex: Mahuna; Paiute; Pima; Thompson; Yavapai
Ruta: Costanoan
Salix: Iroquois; Ojibwa, South; Zuni
Salvia: Cherokee; Costanoan; Diegueño; Mahuna; Paiute
Sambucus: Hoh; Paiute; Quileute; Shoshoni
Sanguinaria: Cherokee; Iroquois
Sassafras: Seminole
Scutellaria: Miwok
Shepherdia: Gitksan
Solidago: Cherokee
Sphaeralcea: Navajo
Spiraea: Mahuna
Streptopus: Ojibwa; Potawatomi
Symplocarpus: Chippewa; Delaware, Oklahoma
Taxus: Iroquois
Tephrosia: Choctaw; Natchez
Tetradymia: Navajo, Ramah; Paiute; Shoshoni
Thermopsis: Navajo, Ramah
Thuja: Bella Coola; Chippewa; Makah; Malecite; Micmac; Nez Perce; Ojibwa; Skagit
Tiarella: Quileute
Tilia: Cherokee
Trifolium: Mohegan
Trillium: Cherokee
Tsuga: Iroquois; Micmac
Tussilago: Iroquois
Ulmus: Cherokee; Delaware; Delaware, Oklahoma; Mohegan
Uvularia: Iroquois
Vaccinium: Tanana, Upper
Valeriana: Navajo, Ramah
Vancouveria: Yurok
Veratrum: Bella Coola
Verbascum: Cherokee; Creek; Delaware; Delaware, Oklahoma; Mohegan; Navajo, Ramah; Thompson
Verbena: Cherokee
Veronica: Cherokee
Veronicastrum: Iroquois
Viburnum: Carrier; Gitksan
Viola: Cherokee
Zanthoxylum: Chippewa; Meskwaki

Dermatological Aid

Abies: Abnaki; Algonquin, Quebec; Blackfoot; Carrier; Carrier, Northern; Cherokee; Chippewa; Cree, Woodlands; Flathead; Gitksan; Hesquiat; Iroquois; Kutenai; Kwakiutl; Menominee; Micmac; Montana Indian; Ojibwa; Ojibwa, South; Okanagan-Colville; Paiute; Penobscot; Potawatomi; Saanich; Salish, Coast; Shoshoni; Shuswap; Sikani; Tewa; Thompson
Abronia: Navajo; Navajo, Kayenta; Navajo, Ramah; Shoshoni
Acer: Algonquin, Tête-de-Boule; Cherokee; Chippewa; Iroquois; Koasati; Kwakiutl; Penobscot; Seminole
Achillea: Bella Coola; Blackfoot; Carrier, Southern; Cherokee; Chippewa; Costanoan; Cowlitz; Crow; Flathead; Gosiute; Great Basin Indian; Karok; Klallam; Kutenai; Kwakiutl; Lakota; Malecite; Mendocino Indian; Menominee; Meskwaki; Mewuk; Micmac; Navajo; Ojibwa; Okanagan-Colville; Okanagon; Paiute; Paiute, Northern; Pomo, Kashaya; Shoshoni; Shuswap; Squaxin; Thompson; Ute; Washo; Winnebago
Achlys: Lummi; Skagit
Acorus: Cherokee; Cree, Woodlands; Iroquois
Acourtia: Hualapai
Actaea: Cherokee; Cheyenne; Quileute; Quinault
Adenocaulon: Cowlitz; Squaxin
Adenostoma: Coahuilla
Adiantum: Lummi; Makah; Navajo, Kayenta; Skokomish
Aesculus: Cherokee
Agastache: Iroquois; Navajo, Ramah; Paiute
Agave: Hualapai
Ageratina: Zuni
Agoseris: Navajo, Ramah; Okanagan-Colville; Thompson
Agrimonia: Cherokee
Alcea: Shinnecock
Alectoria: Nitinaht
Aleurites: Hawaiian
Alisma: Cherokee
Allionia: Navajo, Ramah
Allium: Blackfoot; Cherokee; Cheyenne; Isleta; Kwakiutl; Mahuna
Alnus: Abnaki; Carrier, Southern; Cherokee; Clallam; Cree, Hudson Bay; Eskimo, Alaska; Haisla;

[*Alnus*]
　Keres, Western; Kwakiutl; Menominee; Micmac; Nitinaht; Pomo; Pomo, Kashaya; Potawatomi; Sanpoil; Shuswap; Swinomish; Thompson
Alyxia: Hawaiian
Amaranthus: Cherokee
Ambrosia: Cherokee; Delaware, Oklahoma; Diegueño; Kiowa; Mahuna
Amianthium: Cherokee
Amorpha: Meskwaki; Omaha
Amphiachyris: Comanche
Anaphalis: Kwakiutl; Mahuna
Andropogon: Cherokee; Rappahannock
Anemone: Chippewa; Menominee
Anemopsis: Cahuilla; Costanoan; Isleta; Kawaiisu; Keres, Western; Mahuna; Pima; Shoshoni
Angadenia: Seminole
Angelica: Blackfoot; Costanoan; Menominee; Paiute; Pomo, Kashaya
Antennaria: Paiute
Anthemis: Cherokee
Antidesma: Hawaiian
Aplectrum: Catawba
Apocynum: Blackfoot; Cherokee; Cree, Hudson Bay; Iroquois
Aquilegia: Iroquois; Paiute; Quileute; Shoshoni; Thompson
Arabis: Thompson
Aralia: Cherokee; Chippewa; Choctaw; Cree, Woodlands; Iroquois; Menominee; Meskwaki; Micmac; Ojibwa; Okanagon; Pomo; Pomo, Kashaya; Potawatomi; Rappahannock; Thompson
Arbutus: Cowichan; Pomo; Pomo, Kashaya; Yuki
Arctium: Cherokee; Iroquois; Malecite; Menominee; Micmac; Nanticoke; Penobscot
Arctostaphylos: Atsugewi; Blackfoot; Cahuilla; Carrier; Concow; Sanpoil
Arenaria: Kawaiisu; Navajo, Ramah; Shoshoni
Argemone: Paiute; Shoshoni; Tubatulabal; Washo
Argentina: Blackfoot; Kwakiutl
Argyrochosma: Tewa
Arisaema: Cherokee; Iroquois; Malecite; Micmac; Rappahannock
Aristolochia: Cherokee; Rappahannock
Arnica: Thompson
Arnoglossum: Cherokee
Artemisia: Aleut; Blackfoot; Carrier, Southern; Cherokee; Chippewa; Comanche; Costanoan; Crow; Eskimo, Alaska; Eskimo, Inuktitut; Eskimo, Kuskokwagmiut; Eskimo, Nunivak; Flathead; Havasupai; Hopi; Kawaiisu; Kiowa; Kutenai; Meskwaki; Mewuk; Montana Indian; Navajo; Navajo, Ramah; Okanagan-Colville; Paiute; Pomo; Pomo, Kashaya; Sanpoil; Shoshoni; Shuswap; Tanana, Upper; Tewa; Thompson; Yuki; Yurok, South Coast (Nererner); Zuni
Artocarpus: Hawaiian
Aruncus: Cherokee; Klallam; Lummi; Makah; Quileute; Quinault; Skagit; Thompson
Asarum: Cherokee; Chippewa; Iroquois; Modesse; Pomo; Pomo, Kashaya; Thompson; Yurok
Asclepias: Blackfoot; Cherokee; Costanoan; Iroquois; Kawaiisu; Mendocino Indian; Menominee; Miwok; Navajo, Ramah; Okanagan-Colville; Omaha; Paiute; Paiute, Northern; Rappahannock; Shoshoni; Thompson
Asplenium: Hawaiian
Aster: Iroquois; Okanagan-Colville; Seminole; Zuni
Astragalus: Blackfoot; Cheyenne; Navajo, Kayenta; Navajo, Ramah; Shoshoni; Thompson
Athyrium: Ojibwa
Atriplex: Havasupai; Jemez; Navajo; Navajo, Ramah; Pima; Zuni
Baccharis: Cahuilla; Costanoan; Diegueño; Luiseño
Baileya: Keres, Western
Balsamorhiza: Blackfoot; Gosiute; Kutenai; Paiute; Sanpoil; Shoshoni; Shuswap
Baptisia: Choctaw; Delaware; Delaware, Oklahoma; Meskwaki; Mohegan
Berula: Zuni
Besseya: Navajo, Ramah
Betula: Algonquin, Quebec; Cree, Hudson Bay; Cree, Woodlands; Iroquois; Malecite; Micmac
Blechnum: Hesquiat
Bobea: Hawaiian
Boschniakia: Tlingit
Botrychium: Ojibwa, South
Bouteloua: Navajo, Ramah
Brassica: Cherokee; Iroquois
Brickellia: Navajo, Kayenta
Brodiaea: Mahuna
Buchloe: Keres, Western
Bursera: Cahuilla
Calla: Potawatomi
Calliandra: Zuni
Callicarpa: Seminole
Calochortus: Navajo, Ramah; Okanagan-Colville
Caltha: Chippewa; Okanagon; Thompson
Calycanthus: Cherokee
Camissonia: Navajo
Campanula: Navajo, Ramah; Zuni
Canavalia: Hawaiian
Capsella: Menominee
Cardamine: Iroquois
Carica: Hawaiian
Carpinus: Cherokee; Iroquois
Carya: Cherokee; Comanche; Iroquois
Castanea: Cherokee; Iroquois
Castela: Yavapai
Castilleja: Costanoan; Kawaiisu; Menominee; Navajo, Kayenta; Okanagan-Colville
Caulophyllum: Cherokee
Ceanothus: Costanoan; Diegueño; Iroquois; Karok; Meskwaki; Okanagan-Colville; Sanpoil
Celastrus: Cherokee; Chippewa; Delaware
Cenchrus: Hawaiian
Centaurea: Kiowa; Rappahannock
Cerastium: Iroquois
Cercocarpus: Paiute; Shoshoni
Chaenactis: Okanagon; Paiute; Shoshoni; Thompson
Chaetopappa: Zuni
Chamaecyparis: Kwakiutl
Chamaedaphne: Potawatomi
Chamaemelum: Cherokee
Chamaesaracha: Navajo, Kayenta
Chamaesyce: Cahuilla; Cherokee; Costanoan; Diegueño; Houma; Miwok; Navajo, Kayenta; Navajo, Ramah; Paiute; Pima
Cheilanthes: Navajo, Ramah
Chelone: Cherokee
Chenopodium: Hawaiian; Kawaiisu; Meskwaki; Miwok; Navajo, Kayenta
Chimaphila: Cherokee; Delaware; Delaware, Oklahoma; Iroquois; Micmac; Mohegan; Okanagon; Penobscot; Thompson
Chionanthus: Choctaw; Koasati
Chlorogalum: Cahuilla; Costanoan; Mendocino Indian; Pomo; Pomo, Kashaya; Wailaki
Chorizanthe: Tubatulabal
Chrysothamnus: Cheyenne; Hopi; Klamath; Navajo
Cichorium: Iroquois
Cicuta: Iroquois; Kutenai; Kwakiutl; Paiute
Cimicifuga: Cherokee
Circaea: Iroquois
Cirsium: Costanoan; Cree; Gosiute; Hopi; Houma; Iroquois; Kiowa
Claytonia: Cowlitz; Hesquiat; Quileute; Skokomish; Snohomish
Clematis: Great Basin Indian; Iroquois; Mahuna; Miwok; Navajo, Kayenta; Nevada Indian; Okanagan-Colville; Oregon Indian; Sanpoil; Shoshoni; Thompson
Cleome: Navajo, Kayenta; Navajo, Ramah
Clermontia: Hawaiian
Clintonia: Algonquin, Quebec; Algonquin, Tête-de-Boule; Bella Coola; Chippewa; Cowlitz; Haisla & Hanaksiala; Ojibwa
Cocos: Hawaiian
Collinsia: Ute
Collinsonia: Cherokee; Iroquois
Collomia: Gosiute
Comandra: Cherokee; Navajo; Thompson
Comptonia: Delaware; Delaware, Oklahoma; Malecite; Micmac; Mohegan; Penobscot; Potawatomi; Shinnecock
Convolvulus: Navajo, Ramah
Conyza: Keres, Western; Navajo, Kayenta; Navajo, Ramah
Coprosma: Hawaiian
Corallorrhiza: Navajo, Kayenta
Cornus: Carrier, Northern; Cherokee; Chippewa; Iroquois; Montana Indian; Okanagan-Colville; Thompson; Wet'suwet'en
Corydalis: Navajo, Kayenta
Corylus: Cherokee; Iroquois; Ojibwa; Potawatomi
Crataegus: Kwakiutl; Okanagon; Thompson
Crepis: Meskwaki; Paiute
Croton: Keres, Western
Cryptantha: Hopi; Navajo, Kayenta; Navajo,

Ramah; Zuni
Cucurbita: Cahuilla; Hawaiian; Keres, Western; Paiute; Pima; Zuni
Cuscuta: Cherokee
Cymopterus: Navajo, Kayenta
Cynoglossum: Cherokee; Iroquois
Cyperus: Hawaiian; Yavapai
Cypripedium: Chippewa; Iroquois
Cystopteris: Navajo, Ramah
Dalea: Keres, Western; Montana Indian; Navajo, Kayenta; Zuni
Datisca: Miwok
Datura: Cherokee; Costanoan; Delaware; Delaware, Oklahoma; Havasupai; Kawaiisu; Keres, Western; Mahuna; Mohegan; Pima; Rappahannock; Tubatulabal; Zuni
Daucus: Cherokee; Costanoan; Iroquois
Delphinium: Blackfoot; Chehalis
Descurainia: Gitksan; Navajo, Kayenta; Paiute; Pima
Desmanthus: Pawnee
Dicentra: Skagit
Dimorphocarpa: Apache, White Mountain; Hopi; Navajo, Kayenta; Navajo, Ramah; Zuni
Diospyros: Cherokee
Diplazium: Hawaiian
Dipsacus: Iroquois
Dirca: Chippewa; Iroquois
Disporum: Okanagan-Colville
Distichlis: Kawaiisu
Dodonaea: Hawaiian
Draba: Navajo, Ramah
Drosera: Kwakiutl; Seminole
Drymaria: Navajo, Ramah
Dryopteris: Costanoan; Klallam; Snohomish
Dudleya: Diegueño
Dyssodia: Navajo, Ramah
Echinacea: Cheyenne; Dakota; Lakota; Omaha
Echinocystis: Tubatulabal
Elaeagnus: Blackfoot
Encelia: Navajo, Kayenta
Ephedra: Cocopa; Isleta; Paiute; Paiute, Northern; Pima; Shoshoni
Epilobium: Algonquin, Tête-de-Boule; Bella Coola; Blackfoot; Chippewa; Costanoan; Cree, Woodlands; Eskimo, Inupiat; Kwakiutl; Menominee; Ojibwa; Okanagan-Colville; Thompson
Equisetum: Blackfoot; Costanoan; Kwakiutl; Okanagan-Colville; Pomo, Kashaya; Thompson
Eriastrum: Kawaiisu
Ericameria: Cahuilla; Diegueño; Kawaiisu; Klamath; Miwok
Erigeron: Cherokee; Iroquois; Miwok; Navajo, Kayenta; Navajo, Ramah; Thompson
Eriodictyon: Coahuilla; Costanoan; Hualapai; Miwok; Pomo, Kashaya; Yuki
Eriogonum: Kawaiisu; Navajo, Kayenta; Navajo, Ramah; Okanagan-Colville; Omaha; Thompson; Tubatulabal; Zuni
Eriophorum: Eskimo, Western
Eriophyllum: Skagit

Erodium: Navajo, Kayenta; Zuni
Eryngium: Seminole
Erysimum: Chippewa; Okanagan-Colville; Zuni
Erythronium: Cherokee; Iroquois; Montana Indian; Wailaki
Eschscholzia: Kawaiisu; Mendocino Indian
Eucalyptus: Hawaiian
Euonymus: Cherokee; Meskwaki
Eupatorium: Iroquois; Navajo
Euphorbia: Cherokee; Mahuna; Navajo; Yavapai
Evernia: Blackfoot; Wailaki; Yuki
Fagus: Iroquois; Malecite; Potawatomi; Rappahannock
Fallugia: Tewa
Ficus: Seminole
Fragaria: Okanagan-Colville; Thompson
Frangula: Clallam; Costanoan; Diegueño; Kawaiisu; Skagit; Squaxin
Frasera: Mahuna; Navajo, Kayenta
Fraxinus: Iroquois; Meskwaki; Yokia
Gaillardia: Blackfoot
Galium: Chippewa; Choctaw; Costanoan; Iroquois; Klallam; Makah; Nitinaht; Ojibwa; Quinault
Gaultheria: Bella Coola; Quileute
Gaura: Isleta; Navajo
Gayophytum: Navajo, Kayenta
Geastrum: Isleta
Gelsemium: Delaware, Oklahoma
Geocaulon: Cree, Hudson Bay
Geranium: Cherokee; Choctaw; Gosiute; Iroquois; Keres, Western; Montana Indian; Navajo, Ramah; Sanpoil
Geum: Aleut; Bella Coola; Blackfoot; Carrier, Southern; Clallam; Quileute; Quinault; Snohomish
Gilia: Gosiute; Navajo, Kayenta
Glechoma: Cherokee
Glycyrrhiza: Isleta
Gnaphalium: Keres, Western; Pomo
Goodyera: Okanagan-Colville
Gossypium: Tewa
Grindelia: Cheyenne; Costanoan; Jemez; Mahuna; Miwok; Navajo, Ramah; Shoshoni
Gutierrezia: Isleta; Jemez; Navajo; Navajo, Ramah; Shoshoni
Hackelia: Cherokee
Hamamelis: Cherokee; Chippewa; Iroquois; Mohegan
Helenium: Costanoan; Koasati
Helianthella: Navajo, Ramah; Paiute
Helianthus: Hopi; Jemez; Meskwaki; Navajo, Ramah; Ojibwa, South; Shasta; Thompson
Heliotropium: Pima
Hepatica: Chippewa
Heracleum: Aleut; Bella Coola; Blackfoot; Carrier, Northern; Chippewa; Cree; Cree, Woodlands; Gitksan; Haisla; Iroquois; Kwakiutl; Meskwaki; Ojibwa; Okanagan-Colville; Paiute; Pawnee; Pomo; Salish, Coast; Sanpoil; Shuswap; Sikani
Heteranthera: Cherokee
Heteromeles: Diegueño

Heterotheca: Navajo, Ramah
Heuchera: Blackfoot; Cherokee; Cheyenne; Chickasaw; Choctaw; Creek; Gosiute; Lakota; Meskwaki; Navajo, Kayenta; Navajo, Ramah; Okanagan-Colville; Shuswap; Skagit; Thompson
Hieracium: Iroquois
Hierochloe: Blackfoot; Kiowa; Menominee; Thompson
Hoita: Luiseño
Holodiscus: Sanpoil
Hosta: Cherokee
Houstonia: Navajo, Ramah
Humulus: Dakota; Omaha; Round Valley Indian
Huperzia: Iroquois
Hydrangea: Cherokee
Hydrastis: Cherokee; Micmac
Hymenopappus: Navajo, Kayenta; Navajo, Ramah; Zuni
Hymenoxys: Navajo, Ramah; Zuni
Hypericum: Cherokee; Miwok; Paiute; Shoshoni
Hyptis: Seminole
Ilex: Catawba; Koasati
Impatiens: Cherokee; Chippewa; Iroquois; Meskwaki; Mohegan; Nanticoke; Ojibwa; Omaha; Penobscot; Potawatomi; Shinnecock
Inula: Iroquois
Ipomoea: Hawaiian; Houma
Ipomopsis: Navajo, Kayenta; Navajo, Ramah; Salish; Shoshoni; Tewa; Zuni
Iris: Algonquin, Tête-de-Boule; Cherokee; Chippewa; Meskwaki; Micmac; Omaha; Paiute; Ponca; Potawatomi; Seminole; Shoshoni
Isocoma: Cahuilla; Navajo, Kayenta
Iva: Navajo, Kayenta; Paiute
Jacquemontia: Hawaiian
Jeffersonia: Cherokee
Juglans: Apache, Western; Cherokee; Comanche; Delaware; Delaware, Oklahoma; Houma; Iroquois
Juncus: Iroquois
Juniperus: Blackfoot; Cherokee; Cree, Hudson Bay; Gitksan; Hanaksiala; Hopi; Hualapai; Keres, Western; Malecite; Micmac; Navajo; Navajo, Ramah; Okanagan-Colville; Paiute; Sanpoil; Shoshoni; Tewa; Thompson
Kallstroemia: Tewa
Kalmia: Cherokee; Kwakiutl; Mahuna; Malecite; Micmac; Tlingit
Keckiella: Mahuna; Paiute
Kochia: Navajo
Koeleria: Cheyenne
Krameria: Paiute; Pima; Shoshoni
Krascheninnikovia: Navajo; Navajo, Ramah; Paiute; Shoshoni
Lachnanthes: Cherokee
Lactuca: Chippewa; Menominee
Lagenaria: Cherokee
Lappula: Navajo, Kayenta; Navajo, Ramah
Larix: Cree, Woodlands; Kutenai; Menominee; Micmac; Okanagan-Colville; Potawatomi; Thompson

Larrea: Cahuilla; Isleta; Mahuna; Paiute; Papago; Pima; Yavapai
Lechea: Catawba
Ledum: Chippewa; Cree, Hudson Bay; Cree, Woodlands; Shuswap; Tanana, Upper
Lepidium: Cahuilla; Cherokee; Menominee
Leptarrhena: Thompson
Leptodactylon: Navajo, Kayenta
Lesquerella: Navajo; Shuswap
Leucanthemum: Quileute
Leucocrinum: Paiute; Shoshoni
Leucothoe: Cherokee
Lewisia: Okanagan-Colville
Leymus: Okanagan-Colville
Liatris: Blackfoot; Meskwaki; Omaha
Lilium: Chippewa; Dakota; Malecite; Menominee; Micmac
Limosella: Navajo, Ramah
Lindera: Cherokee
Linum: Gosiute; Okanagon; Paiute; Shoshoni; Thompson
Liquidambar: Cherokee; Choctaw; Houma
Liriodendron: Cherokee
Lithocarpus: Costanoan
Lithospermum: Navajo; Shuswap; Zuni
Lobelia: Cherokee; Iroquois; Zuni
Lomatium: Cheyenne; Gosiute; Great Basin Indian; Kawaiisu; Kwakiutl; Nez Perce; Okanagan-Colville; Paiute; Paiute, Northern; Shoshoni; Shuswap; Thompson; Ute; Washo
Lonicera: Bella Coola; Carrier; Carrier, Northern; Chehalis; Costanoan; Iroquois; Klallam; Kwakiutl; Makah; Nuxalkmc; Okanagan-Colville; Quinault; Thompson; Wet'suwet'en; Yuki
Lophophora: Kiowa
Ludwigia: Hawaiian; Seminole
Luetkea: Okanagon; Thompson
Lupinus: Navajo; Navajo, Ramah
Lycopodium: Blackfoot
Lygodesmia: Blackfoot; Navajo, Kayenta
Lyonia: Cherokee
Lysichiton: Clallam; Gitksan; Hesquiat; Klallam; Kwakiutl; Quileute; Shuswap; Skokomish; Thompson
Lythrum: Kawaiisu
Machaeranthera: Navajo; Navajo, Ramah
Magnolia: Choctaw; Koasati
Mahonia: Blackfoot; Cowlitz; Flathead; Miwok; Navajo, Ramah
Maianthemum: Gitksan; Hesquiat; Iroquois; Karok; Malecite; Micmac; Nitinaht; Oweekeno; Paiute; Washo
Malacothrix: Navajo, Ramah
Malus: Iroquois; Kwakiutl; Makah; Samish; Swinomish
Malva: Cherokee; Costanoan; Diegueño; Iroquois; Miwok; Pima
Marah: Chehalis; Costanoan; Karok; Kawaiisu; Mahuna; Mendocino Indian; Pomo; Pomo, Kashaya; Tubatulabal

Marrubium: Costanoan
Matelea: Comanche
Matricaria: Cheyenne; Costanoan
Melia: Cherokee
Melilotus: Iroquois
Menispermum: Cherokee; Delaware, Oklahoma
Mentha: Cheyenne; Kawaiisu; Navajo, Kayenta; Okanagon; Paiute; Thompson
Mentzelia: Gosiute; Mendocino Indian; Montana Indian; Navajo, Kayenta
Menziesia: Kwakiutl
Merremia: Hawaiian
Mertensia: Cheyenne
Mikania: Seminole
Mimulus: Shoshoni
Mirabilis: Cherokee; Dakota; Hopi; Navajo; Navajo, Kayenta; Navajo, Ramah; Paiute; Paiute, Northern; Ponca; Sioux, Teton
Mitchella: Cherokee; Iroquois
Monarda: Blackfoot; Cheyenne; Chippewa; Creek; Delaware; Kiowa; Navajo, Ramah; Winnebago
Moneses: Cowichan; Kwakiutl; Kwakwaka'wakw; Salish, Coast
Monolepis: Navajo, Ramah
Monotropa: Cherokee; Thompson
Morinda: Hawaiian
Morus: Rappahannock
Myosotis: Makah
Myosurus: Navajo, Ramah
Myrica: Micmac
Nama: Navajo, Kayenta
Napaea: Meskwaki
Nepeta: Cherokee
Nereocystis: Kwakiutl, Southern; Nitinaht
Nicotiana: Cahuilla; Cherokee; Cheyenne; Hawaiian; Hesquiat; Iroquois; Kawaiisu; Kwakiutl; Navajo; Paiute; Salish; Shoshoni; Thompson
Nothochelone: Paiute
Nuphar: Algonquin, Quebec; Cree, Woodlands; Flathead; Kutenai; Menominee; Micmac; Ojibwa; Penobscot; Potawatomi; Rappahannock; Shuswap; Thompson
Nymphaea: Micmac; Penobscot
Nyssa: Koasati
Obolaria: Choctaw
Oemleria: Kwakiutl
Oenothera: Blackfoot; Iroquois; Isleta; Navajo; Navajo, Kayenta; Navajo, Ramah; Ojibwa; Zuni
Onoclea: Iroquois
Onosmodium: Cheyenne
Oplopanax: Gitksan; Green River Group; Haisla; Haisla & Hanaksiala; Kwakiutl; Okanagon; Oweekeno; Thompson; Tlingit; Wet'suwet'en
Opuntia: Apache, Mescalero; Cahuilla; Dakota; Hualapai; Jemez; Keres, Western; Kiowa; Mahuna; Nanticoke; Navajo; Navajo, Ramah; Okanagan-Colville; Pawnee; Shoshoni; Shuswap; Sioux
Orbexilum: Catawba
Orobanche: Blackfoot; Navajo; Navajo, Kayenta; Pima

Osmorhiza: Blackfoot; Chippewa; Kawaiisu; Omaha; Paiute; Paiute, Northern; Sanpoil; Shoshoni; Winnebago
Osteomeles: Hawaiian
Ostrya: Iroquois
Oxalis: Cherokee; Navajo, Ramah; Omaha; Quileute; Tolowa
Oxydendrum: Cherokee
Oxytropis: Blackfoot; Thompson
Paeonia: Paiute; Shoshoni
Panax: Cherokee; Creek; Iroquois; Seminole
Panicum: Isleta
Parthenocissus: Houma; Navajo, Ramah
Paspalidium: Seminole
Passiflora: Cherokee
Pastinaca: Iroquois; Potawatomi
Paxistima: Thompson
Pedicularis: Cherokee; Meskwaki; Washo
Pediomelum: Montana Indian
Pelea: Hawaiian
Pellaea: Costanoan; Yavapai
Peltigera: Oweekeno
Peniocereus: Papago
Pennisetum: Navajo, Ramah
Penstemon: Costanoan; Navajo, Kayenta; Navajo, Ramah; Okanagan-Colville; Paiute; Paiute, Northern; Shoshoni; Tewa; Thompson
Perezia: Yavapai
Pericome: Navajo, Ramah
Perideridia: Blackfoot
Petasites: Concow; Cree, Woodlands; Menominee; Quinault; Tlingit
Peteria: Navajo, Ramah
Petradoria: Navajo, Kayenta; Navajo, Ramah
Phacelia: Miwok; Pomo, Kashaya; Zuni
Philadelphus: Snohomish; Thompson
Phlox: Gosiute; Meskwaki; Navajo, Ramah; Shoshoni; Ute
Phoradendron: Diegueño; Navajo; Pima
Phragmites: Seminole
Phyla: Navajo, Kayenta
Physalis: Omaha; Ponca; Winnebago
Physaria: Blackfoot
Phytolacca: Cherokee; Delaware; Iroquois; Mahuna; Rappahannock
Picea: Algonquin, Quebec; Bella Coola; Cree, Woodlands; Eskimo, Alaska; Eskimo, Inuktitut; Eskimo, Nunivak; Haisla & Hanaksiala; Hesquiat; Iroquois; Koyukon; Kwakiutl; Makah; Menominee; Micmac; Mohegan; Oweekeno; Penobscot; Potawatomi; Quinault; Shuswap; Tanana, Upper; Thompson
Picradeniopsis: Navajo, Ramah; Zuni
Pilea: Cherokee
Piloblephis: Seminole
Pinus: Apache, Western; Bella Coola; Cahuilla; Carrier, Northern; Cherokee; Cheyenne; Chippewa; Cree, Woodlands; Delaware, Ontario; Flathead; Havasupai; Hopi; Iroquois; Isleta; Kawaiisu; Keres, Western; Kwakiutl; Mendocino Indian; Menominee; Micmac; Miwok; Mohegan; Navajo; Nitinaht;

Index of Usages

Okanagan-Colville; Paiute; Potawatomi; Quinault; Rappahannock; Salish, Coast; Sanpoil; Seminole; Shinnecock; Shoshoni; Shuswap; Skagit; Tewa; Thompson; Yokia; Zuni
Piper: Cherokee; Hawaiian
Piperacea: Mahuna
Plagiomnium: Bella Coola; Oweekeno
Plantago: Algonquin, Tête-de-Boule; Cherokee; Chippewa; Delaware, Ontario; Hesquiat; Houma; Iroquois; Kwakiutl; Mahuna; Menominee; Meskwaki; Mohegan; Nitinaht; Ojibwa; Okanagan-Colville; Paiute; Ponca; Potawatomi; Shinnecock; Shoshoni; Shuswap; Thompson; Tolowa; Yurok
Platanthera: Iroquois; Montagnais
Platanus: Cherokee; Iroquois; Meskwaki
Pluchea: Pima
Plumbago: Hawaiian
Podophyllum: Cherokee; Iroquois
Polemonium: Thompson
Poliomintha: Navajo, Kayenta
Polygala: Choctaw; Iroquois
Polygonatum: Cherokee; Rappahannock
Polygonum: Cherokee; Iroquois; Mendocino Indian; Miwok
Polymnia: Houma
Polypodium: Cherokee; Skagit, Upper; Wailaki
Polystichum: Cowlitz; Iroquois; Quileute; Quinault
Populus: Algonquin, Quebec; Anticosti; Bella Coola; Cahuilla; Carrier; Carrier, Northern; Cherokee; Chippewa; Choctaw; Cree, Hudson Bay; Cree, Woodlands; Diegueño; Flathead; Gitksan; Haisla & Hanaksiala; Hesquiat; Iroquois; Kutenai; Kwakiutl; Malecite; Mendocino Indian; Menominee; Micmac; Nitinaht; Ojibwa; Ojibwa, South; Okanagan-Colville; Oweekeno; Pima; Potawatomi; Quinault; Shuswap; Sikani; Squaxin; Tanana, Upper; Thompson; Yuki
Porophyllum: Havasupai
Porteranthus: Cherokee
Portulaca: Iroquois; Rappahannock
Potentilla: Cherokee; Chippewa; Gosiute; Navajo, Kayenta; Navajo, Ramah; Okanagan-Colville; Thompson
Prenanthes: Iroquois
Prosopis: Cahuilla; Papago; Pima
Prunella: Blackfoot; Cherokee; Quileute; Quinault; Salish, Coast
Prunus: Atsugewi; Cherokee; Chippewa; Crow; Iroquois; Kwakiutl; Malecite; Menominee; Meskwaki; Micmac; Ojibwa, South; Okanagan-Colville; Omaha; Paiute
Psathyrotes: Navajo, Kayenta; Paiute; Shoshoni
Pseudostellaria: Navajo, Kayenta
Pseudotsuga: Cowlitz; Kwakiutl; Okanagan-Colville; Quinault; Skagit; Thompson
Psidium: Hawaiian
Psilostrophe: Navajo, Kayenta; Navajo, Ramah
Psoralidium: Arapaho; Navajo, Kayenta
Psorothamnus: Keres, Western; Paiute
Pteridium: Costanoan; Pomo, Kashaya; Thompson

Pterospora: Cheyenne
Pteryxia: Okanagan-Colville
Pulsatilla: Blackfoot
Purshia: Hopi; Hualapai; Paiute; Shoshoni
Pycnanthemum: Cherokee
Pyrola: Blackfoot; Cherokee; Iroquois
Pyrularia: Cherokee
Pyrus: Cree, Hudson Bay
Quercus: Alabama; Cherokee; Delaware, Oklahoma; Diegueño; Iroquois; Kawaiisu; Luiseño; Miwok; Neeshenam; Ojibwa; Seminole
Ranunculus: Bella Coola; Cherokee; Costanoan; Hesquiat; Kawaiisu; Potawatomi; Thompson
Ratibida: Cheyenne; Dakota
Rhamnus: Cherokee; Costanoan (Olhonean); Iroquois; Kawaiisu
Rheum: Cherokee
Rhizomnium: Bella Coola
Rhododendron: Okanagon; Skokomish; Thompson
Rhus: Cherokee; Delaware; Delaware, Oklahoma; Hopi; Iroquois; Menominee; Meskwaki; Natchez; Navajo, Kayenta; Navajo, Ramah; Nez Perce; Ojibwa; Okanagan-Colville; Omaha; Paiute; Round Valley Indian; Sanpoil; Seminole; Sioux; Yokia
Ribes: Bella Coola; Carrier, Northern; Cowlitz; Haisla; Iroquois; Kwakiutl; Navajo, Kayenta; Paiute
Ricinus: Cahuilla; Cherokee; Diegueño; Pima
Rosa: Carrier, Northern; Costanoan; Meskwaki; Ojibwa; Okanagan-Colville; Paiute; Quileute; Quinault; Shoshoni; Thompson
Rubus: Carrier, Northern; Cherokee; Costanoan; Hawaiian; Iroquois; Kwakiutl; Makah; Okanagan-Colville; Quinault; Rappahannock; Shoshoni; Skagit; Thompson
Rudbeckia: Cherokee
Rumex: Aleut; Arapaho; Bella Coola; Blackfoot; Cherokee; Cheyenne; Chippewa; Cowlitz; Dakota; Eskimo, Kuskokwagmiut; Hanaksiala; Hopi; Iroquois; Kawaiisu; Kwakiutl; Miwok; Navajo, Ramah; Ojibwa; Ojibwa, South; Paiute; Paiute, Northern; Papago; Pima; Shoshoni; Yavapai; Yuki; Zuni
Sadleria: Hawaiian
Sagittaria: Iroquois; Potawatomi; Seminole
Salix: Bella Coola; Blackfoot; Cherokee; Cheyenne; Costanoan; Cree, Woodlands; Creek; Eskimo, Alaska; Eskimo, Inuktitut; Eskimo, Kuskokwagmiut; Flathead; Isleta; Makah; Malecite; Mendocino Indian; Micmac; Ojibwa; Okanagan-Colville; Paiute; Seminole; Shoshoni; Sikani; Thompson
Salsola: Navajo; Navajo, Ramah
Salvia: Cahuilla; Catawba
Sambucus: Cherokee; Choctaw; Delaware; Delaware, Oklahoma; Hanaksiala; Houma; Iroquois; Makah; Mendocino Indian; Montana Indian; Okanagon; Paiute; Pomo; Rappahannock; Thompson
Sanguinaria: Cherokee; Iroquois; Malecite; Menominee; Micmac; Ojibwa
Sanicula: Iroquois; Meskwaki; Miwok

Sanvitalia: Navajo, Ramah
Sapindus: Kiowa
Saponaria: Cherokee; Mahuna
Sarcobatus: Keres, Western; Navajo
Sassafras: Cherokee; Iroquois; Koasati; Rappahannock; Seminole
Saururus: Cherokee; Choctaw; Seminole
Saxifraga: Cherokee
Scaevola: Hawaiian
Scirpus: Kwakiutl; Malecite; Micmac
Scrophularia: Costanoan; Iroquois; Pomo; Pomo, Kashaya
Sedum: Costanoan; Delaware, Ontario; Iroquois; Malecite; Micmac
Sempervivum: Cherokee
Senecio: Aleut; Costanoan; Hopi; Keres, Western; Navajo; Navajo, Kayenta; Navajo, Ramah; Yavapai
Senna: Cherokee; Hawaiian
Shepherdia: Carrier; Cree, Woodlands; Navajo, Kayenta; Okanagan-Colville; Tanana, Upper; Thompson
Silene: Keres, Western; Meskwaki; Navajo, Ramah
Simmondsia: Papago; Yavapai
Smallanthus: Cherokee
Smilacina: Yana
Smilax: Cherokee; Creek; Iroquois
Solanum: Cherokee; Costanoan; Delaware, Oklahoma; Houma; Kawaiisu; Rappahannock
Solidago: Cahuilla; Chippewa; Costanoan; Diegueño; Kawaiisu; Mahuna; Meskwaki; Miwok; Potawatomi
Sorbus: Heiltzuk; Malecite; Micmac
Sphaeralcea: Comanche; Dakota; Hopi; Lakota; Navajo, Kayenta; Navajo, Ramah; Shoshoni; Tewa
Sphagnum: Carrier
Sporobolus: Ojibwa, South
Stachys: Costanoan; Green River Group; Navajo, Ramah; Pomo, Kashaya; Puyallup
Stanleya: Zuni
Staphylea: Iroquois
Stellaria: Iroquois
Stenotaphrum: Hawaiian
Stephanomeria: Paiute
Strophostyles: Iroquois
Styphelia: Hawaiian
Suaeda: Cahuilla; Paiute
Symphoricarpos: Chehalis; Cowichan; Cree, Woodlands; Flathead; Green River Group; Hesquiat; Okanagan-Colville; Saanich; Sanpoil; Thompson; Yuki
Symplocarpus: Iroquois; Menominee; Meskwaki
Syzygium: Hawaiian
Tagetes: Cherokee
Talinum: Navajo, Ramah
Tanacetum: Iroquois; Shoshoni
Taraxacum: Aleut; Iroquois; Navajo, Ramah; Tewa
Tauschia: Kawaiisu
Taxus: Cowlitz; Quinault
Tephrosia: Cherokee; Choctaw; Hawaiian
Tetradymia: Kawaiisu; Shoshoni

Tetraneuris: Navajo, Kayenta
Thalictrum: Blackfoot; California Indian; Great Basin Indian; Lakota; Potawatomi; Thompson
Thamnosma: Kawaiisu
Thelypodium: Navajo
Thlaspi: Navajo, Ramah
Thuja: Algonquin, Quebec; Bella Coola; Chippewa; Haisla; Iroquois; Kwakiutl; Makah; Menominee; Okanagan-Colville; Penobscot; Quinault
Tiarella: Iroquois
Tilia: Cherokee; Meskwaki; Micmac
Tolmiea: Cowlitz
Toxicodendron: Algonquin, Quebec; Cherokee; Delaware, Oklahoma; Kiowa; Mendocino Indian; Meskwaki; Potawatomi; Yuki
Tradescantia: Cherokee
Tragia: Keres, Western
Tragopogon: Navajo, Ramah
Trautvetteria: Bella Coola
Trichostema: Costanoan; Miwok
Trifolium: Iroquois
Trillium: Cherokee; Iroquois; Karok; Quileute; Wailaki; Yuki
Triosteum: Cherokee; Meskwaki
Tsuga: Abnaki; Algonquin, Quebec; Bella Coola; Cherokee; Cowlitz; Hesquiat; Kwakiutl; Makah; Malecite; Menominee; Micmac; Nitinaht; Ojibwa; Quileute; Skagit
Turricula: Kawaiisu
Typha: Algonquin, Quebec; Cahuilla; Dakota; Malecite; Meskwaki; Micmac; Ojibwa; Ojibwa, South; Okanagan-Colville; Omaha; Pawnee; Plains Indian; Ponca; Potawatomi; Sioux; Winnebago
Ulmus: Cherokee; Koasati; Menominee; Meskwaki; Micmac; Ojibwa; Potawatomi
Umbellularia: Karok; Mendocino Indian; Pomo, Kashaya
Urtica: Carrier; Chehalis; Costanoan; Kawaiisu; Kwakiutl; Ojibwa; Paiute, Northern; Sanpoil; Shuswap; Skokomish; Thompson; Tolowa
Usnea: Makah; Nitinaht
Uvularia: Cherokee; Menominee
Vaccinium: Iroquois
Valeriana: Carrier, Northern; Gosiute; Menominee; Okanagon; Thompson
Vanclevea: Navajo, Kayenta
Veratrum: Bella Coola; Iroquois; Kwakiutl; Paiute; Paiute, Northern; Shoshoni; Shuswap; Tsimshian; Washo
Verbascum: Atsugewi; Catawba; Cherokee; Delaware, Ontario; Iroquois; Malecite; Micmac; Rappahannock; Thompson; Zuni
Verbena: Cherokee; Navajo, Ramah
Verbesina: Hopi; Navajo, Kayenta
Vernonia: Kiowa
Veronica: Cherokee
Viburnum: Cree, Woodlands
Vicia: Keres, Western; Makah; Rappahannock; Saanich
Vigna: Hawaiian
Viola: Cherokee; Costanoan; Iroquois
Vitis: Chippewa; Delaware, Oklahoma; Diegueño
Woodsia: Navajo, Ramah
Wyethia: Costanoan; Gosiute; Klamath; Mendocino Indian; Paiute; Shoshoni
Xanthium: Keres, Western; Navajo; Rappahannock; Zuni
Xanthorhiza: Cherokee
Xerophyllum: Blackfoot; Pomo, Kashaya
Yucca: Blackfoot; Catawba; Cherokee; Cheyenne; Choctaw; Dakota; Hopi; Isleta; Keresan; Kiowa; Lakota; Montana Indian; Navajo
Zanthoxylum: Alabama; Menominee
Zea: Cherokee; Mohegan; Tewa
Zigadenus: Mendocino Indian; Montana Indian; Navajo, Ramah; Paiute; Shoshoni; Washo
Zingiber: Hawaiian
Zinnia: Zuni
Ziziphus: Pima

Diaphoretic
Abies: Ojibwa; Ojibwa, South
Abronia: Navajo, Kayenta
Acacia: Hawaiian
Achillea: Cheyenne; Lummi; Makah; Micmac; Paiute, Northern
Acorus: Cherokee
Agastache: Cheyenne
Ageratina: Iroquois; Meskwaki
Alnus: Mendocino Indian; Shuswap
Amelanchier: Cree, Woodlands
Anemone: Gitksan
Anemopsis: Pima
Angelica: Iroquois
Anthemis: Cherokee
Aralia: Cherokee
Arctium: Iroquois
Arisaema: Cherokee
Artemisia: Keres, Western; Navajo, Ramah; Okanagan-Colville; Okanagon; Paiute; Sanpoil; Shoshoni; Thompson
Asarum: Iroquois
Asclepias: Choctaw
Aster: Cree, Woodlands
Balsamorhiza: Miwok; Okanagan-Colville
Betula: Cree, Woodlands
Botrychium: Chickasaw
Callicarpa: Alabama
Caltha: Chippewa
Carya: Cherokee
Chamaesyce: Pima
Chimaphila: Montagnais
Chrysothamnus: Paiute
Cirsium: Zuni
Conyza: Meskwaki
Corethrogyne: Kawaiisu
Cornus: Cherokee
Cunila: Cherokee
Daphne: Cherokee
Diphylleia: Cherokee
Ephedra: Keres, Western
Erigeron: Cherokee
Eriodictyon: Yokut
Eupatorium: Cherokee; Shinnecock
Euphorbia: Cherokee
Galium: Choctaw
Geum: Cree, Woodlands
Gutierrezia: Keres, Western; Zuni
Hedeoma: Cherokee; Nanticoke
Heracleum: Iroquois
Ilex: Cherokee
Juniperus: Cherokee; Costanoan; Keres, Western; Navajo, Ramah; Omaha; Shuswap
Lesquerella: Shuswap
Liatris: Cherokee
Lindera: Cherokee; Creek
Liquidambar: Houma
Lomatium: Thompson
Mentha: Ojibwa
Mitchella: Cherokee
Monarda: Cherokee; Creek
Monardella: Karok
Nepeta: Menominee
Nicotiana: Cherokee
Obolaria: Cherokee
Orbexilum: Cherokee
Oxytropis: Thompson
Panax: Creek
Pericome: Navajo, Ramah
Persea: Creek
Pinus: Yokia; Zuni
Plantago: Shuswap
Polygala: Cherokee
Populus: Bella Coola; Penobscot
Porteranthus: Iroquois
Pycnanthemum: Choctaw
Rosa: Bella Coola
Salix: Mendocino Indian
Salvia: Cherokee
Sambucus: Cherokee; Kawaiisu
Sanvitalia: Navajo
Senecio: Iroquois
Solidago: Cherokee
Stephanomeria: Cheyenne
Streptopus: Montagnais
Tanacetum: Chippewa; Nanticoke
Taxus: Chehalis; Iroquois
Thamnosma: Kawaiisu
Thuja: Menominee; Montagnais; Ojibwa
Torreya: Costanoan
Triosteum: Iroquois
Tripterocalyx: Zuni
Tsuga: Iroquois; Menominee; Potawatomi
Urtica: Paiute
Vaccinium: Cree, Woodlands
Veratrum: Salish, Coast
Verbascum: Atsugewi
Verbena: Cherokee
Veronicastrum: Cherokee
Viburnum: Cherokee
Wyethia: Miwok

Zigadenus: Keres, Western
Zinnia: Zuni

Dietary Aid
Abies: Montagnais; Okanagan-Colville
Abronia: Keres, Western
Achillea: Mohegan
Actaea: Cheyenne; Okanogan; Thompson
Agrimonia: Cherokee
Allium: Mahuna
Alnus: Okanagan-Colville
Alocasia: Hawaiian
Amelanchier: Cheyenne
Angelica: Blackfoot
Aplectrum: Cherokee
Arbutus: Miwok
Arceuthobium: Carrier, Southern
Arctium: Cherokee
Arctostaphylos: Miwok; Thompson
Armoracia: Cherokee
Artemisia: Montana Indian
Asarum: Iroquois; Ojibwa; Skagit
Asclepias: Lakota; Thompson
Asparagus: Cherokee
Aster: Okanagon; Thompson
Balsamorhiza: Thompson
Bidens: Hawaiian
Brassica: Cherokee
Brickellia: Keres, Western
Cardamine: Iroquois
Ceanothus: Thompson
Chamaesyce: Hawaiian; Hopi
Chelone: Cherokee
Chenopodium: Cherokee; Hawaiian; Navajo
Chimaphila: Okanagan-Colville; Rappahannock
Cibotium: Hawaiian
Claytonia: Mahuna
Crataegus: Cherokee
Cryptantha: Navajo
Cucurbita: Omaha
Daucus: Iroquois
Distichlis: Yokut
Echinacea: Cheyenne
Elliottia: Kitasoo
Ephedra: Diegueño
Eryngium: Seminole
Euonymus: Iroquois
Frasera: Cherokee
Galactia: Seminole
Geum: Okanagan-Colville
Goodyera: Cherokee
Gymnocladus: Omaha
Hamamelis: Iroquois
Hedeoma: Dakota
Helianthus: Navajo
Hilaria: Navajo, Ramah
Hydrastis: Cherokee
Juniperus: Crow
Larix: Thompson
Lechea: Seminole
Ledum: Haisla & Hanaksiala; Nitinaht

Lepidium: Mahuna
Liatris: Omaha; Seminole
Ligusticum: Karok
Lilium: Cherokee
Lomatium: Blackfoot; Karok; Nez Perce; Okanagan-Colville
Lonicera: Thompson
Maianthemum: Okanagan-Colville
Malus: Gitksan; Nitinaht
Mentzelia: Cheyenne
Menyanthes: Kwakiutl
Mirabilis: Zuni
Mitchella: Cherokee
Oenothera: Cherokee
Oplopanax: Thompson
Osmorhiza: Menominee; Meskwaki
Oxalis: Karok
Paeonia: Paiute
Panax: Iroquois
Panicum: Mahuna
Passiflora: Cherokee
Persea: Seminole
Phegopteris: Hawaiian
Phlox: Cherokee
Physalis: Lakota
Physaria: Blackfoot
Pilea: Cherokee
Pinus: Iroquois
Plantago: Navajo, Ramah
Platanus: Meskwaki
Populus: Malecite; Micmac
Prunus: Cheyenne; Kwakiutl; Rappahannock
Pteridium: Thompson
Quercus: Mendocino Indian; Micmac; Penobscot; Rappahannock
Ranunculus: Eskimo, Kuskokwagmiut
Rhus: Algonquin, Quebec; Iroquois; Mahuna; Meskwaki; Micmac
Rubus: Karok
Rudbeckia: Cherokee
Rumex: Iroquois
Sabal: Seminole
Salix: Malecite; Micmac
Sarracenia: Iroquois
Sassafras: Cherokee; Seminole
Satureja: Pomo
Scirpus: Montana Indian
Sinapis: Cherokee
Solidago: Thompson
Sorbus: Montagnais
Sphaeralcea: Navajo, Kayenta
Stillingia: Seminole
Streptopus: Thompson
Tanacetum: Mohegan
Tellima: Skagit
Tiarella: Iroquois
Triosteum: Iroquois
Tsuga: Klallam
Vaccinium: Cheyenne
Waltheria: Hawaiian

Disinfectant
Abies: Abnaki
Achillea: Flathead; Kutenai
Adenostoma: Cahuilla
Agastache: Navajo, Ramah
Agoseris: Navajo, Ramah
Allium: Shinnecock
Ambrosia: Cherokee
Amelanchier: Chippewa
Anaphalis: Cheyenne
Anemopsis: Costanoan; Isleta; Mahuna; Shoshoni
Angelica: Eskimo, Kuskokwagmiut
Apocynum: Navajo, Ramah
Aralia: Cherokee; Potawatomi
Aristolochia: Cherokee
Artemisia: Cahuilla; Chippewa; Dakota; Eskimo; Mewuk; Navajo, Kayenta; Omaha; Paiute; Pawnee; Ponca; Shoshoni; Shuswap; Tanana, Upper; Thompson; Washo; Winnebago
Asarum: Iroquois; Tolowa
Astragalus: Thompson
Baccharis: Costanoan
Balsamorhiza: Paiute; Washo
Brickellia: Navajo, Ramah
Campanula: Navajo, Ramah
Carex: Navajo, Ramah
Castilleja: Costanoan
Chionanthus: Choctaw
Chlorogalum: Wailaki
Cicuta: Iroquois
Cirsium: Costanoan
Clintonia: Algonquin, Quebec
Conyza: Navajo, Kayenta
Coreopsis: Navajo, Ramah
Cornus: Cherokee
Corydalis: Navajo, Kayenta
Cryptantha: Navajo, Kayenta
Dalea: Navajo, Ramah
Datura: Apache, White Mountain
Dichanthelium: Navajo, Ramah
Diphylleia: Cherokee
Epilobium: Costanoan
Equisetum: Chippewa; Navajo, Ramah
Erigeron: Navajo, Ramah
Eriodictyon: Costanoan
Eriogonum: Navajo, Kayenta; Navajo, Ramah; Thompson
Erodium: Navajo, Kayenta
Euonymus: Cherokee
Eupatorium: Cherokee
Forestiera: Navajo, Ramah
Fragaria: Okanagan-Colville
Frangula: Kawaiisu; Nitinaht
Frasera: Cherokee
Gaura: Navajo, Kayenta
Geum: Thompson
Gnaphalium: Menominee
Grindelia: Cheyenne, Northern; Mahuna; Shoshoni
Gutierrezia: Navajo, Kayenta; Navajo, Ramah; Shoshoni; Tewa

Helianthus: Navajo, Kayenta; Shasta
Heracleum: Thompson
Heterotheca: Cheyenne
Heuchera: Navajo, Ramah
Holodiscus: Shoshoni
Houstonia: Navajo, Ramah
Hydrangea: Cherokee
Ipomopsis: Navajo, Ramah; Shoshoni
Juniperus: Arapaho; Comanche; Cree, Hudson Bay; Hopi; Paiute; Salish; Shoshoni; Swinomish; Tewa; Thompson; Washo
Kalmia: Cherokee
Krameria: Paiute; Pima
Larix: Algonquin, Quebec; Ojibwa
Larrea: Cahuilla; Hualapai; Isleta; Kawaiisu; Mahuna
Lathyrus: Navajo, Ramah
Ledum: Tanana, Upper
Lepidium: Navajo
Leptodactylon: Navajo, Ramah
Linum: Navajo, Kayenta
Lithospermum: Shuswap
Lomatium: Gosiute; Paiute; Shoshoni
Lonicera: Costanoan; Wet'suwet'en
Lotus: Navajo, Ramah
Lupinus: Navajo, Kayenta
Mahonia: Blackfoot
Marrubium: Navajo, Ramah
Matricaria: Costanoan
Mentha: Navajo, Kayenta
Mentzelia: Navajo, Kayenta
Nuphar: Algonquin, Quebec
Oenothera: Navajo, Ramah
Opuntia: Apache, Mescalero
Osmorhiza: Paiute; Shoshoni
Oxalis: Tolowa
Oxytropis: Thompson
Penstemon: Costanoan; Navajo, Kayenta; Shoshoni
Phlox: Navajo, Ramah
Phoradendron: Cahuilla
Picea: Bella Coola; Kwakiutl; Ojibwa; Potawatomi; Tanana, Upper
Pinus: Hopi; Shoshoni; Thompson; Zuni
Plantago: Navajo, Ramah; Nitinaht; Shoshoni
Platanthera: Thompson
Polypodium: Yurok
Populus: Algonquin, Quebec; Cree, Woodlands; Quinault; Squaxin; Thompson
Prosopis: Pima
Prunus: Algonquin, Quebec; Chippewa
Pseudotsuga: Karok; Navajo, Kayenta; Skagit; Thompson
Psoralidium: Zuni
Psorothamnus: Paiute; Shoshoni
Pteridium: Cherokee
Pterospora: Cheyenne
Purshia: Navajo, Ramah; Shoshoni
Quercus: Cherokee; Delaware; Delaware, Oklahoma
Ranunculus: Thompson
Rubus: Costanoan; Quinault

Rumex: Cowlitz
Salix: Navajo, Ramah
Salvia: Cahuilla
Sambucus: Cherokee; Navajo, Kayenta; Pomo, Potter Valley; Squaxin; Yokia
Scrophularia: Costanoan
Sedum: Delaware, Ontario
Senecio: Costanoan; Navajo, Kayenta
Shepherdia: Thompson
Sinapis: Navajo, Ramah
Smallanthus: Iroquois
Solidago: Mahuna
Sphaeralcea: Navajo, Kayenta; Tewa
Sphagnum: Nitinaht
Stachys: Costanoan; Navajo, Ramah
Symphoricarpos: Green River Group
Tanacetum: Shoshoni
Tetradymia: Navajo, Ramah
Tetraneuris: Navajo, Ramah
Thuja: Chippewa
Tradescantia: Navajo, Ramah
Trichostema: Costanoan
Trillium: Menominee
Tsuga: Algonquin, Quebec; Thompson
Typha: Algonquin, Quebec
Umbellularia: Karok
Urtica: Costanoan; Thompson
Veratrum: Paiute; Shoshoni
Veronica: Navajo, Ramah
Veronicastrum: Cherokee
Yucca: Hopi

Diuretic
Abies: Gitksan
Abronia: Paiute
Achillea: Blackfoot
Acorus: Cherokee; Cheyenne
Adiantum: Iroquois
Agastache: Meskwaki
Ageratina: Cherokee
Allium: Cherokee
Alnus: Gitksan
Andropogon: Chippewa
Apocynum: Ojibwa; Potawatomi
Arabis: Thompson
Arctium: Iroquois
Arctostaphylos: Thompson
Argentina: Iroquois
Aristolochia: Cherokee
Armoracia: Cherokee
Artemisia: Lakota; Paiute
Aruncus: Bella Coola
Asclepias: Iroquois; Meskwaki
Athyrium: Chippewa
Besseya: Navajo, Ramah
Betula: Ojibwa, South
Caltha: Chippewa
Carya: Meskwaki
Castilleja: Blackfoot; Gitksan
Celastrus: Chippewa; Iroquois
Chimaphila: Iroquois

Chlorogalum: Wailaki
Cimicifuga: Cherokee
Cirsium: Keres, Western; Zuni
Citrullus: Cheyenne
Croton: Zuni
Cucurbita: Cherokee; Cheyenne; Iroquois; Menominee; Ojibwa
Daucus: Iroquois
Diervilla: Algonquin, Tête-de-Boule; Menominee; Potawatomi
Diphylleia: Cherokee
Dirca: Menominee; Ojibwa; Potawatomi
Draba: Navajo
Empetrum: Cree, Woodlands
Ephedra: Shoshoni
Equisetum: Blackfoot; Crow; Flathead; Iroquois; Okanagan-Colville; Yuki
Erigeron: Cherokee
Eriogonum: Lakota
Eryngium: Choctaw
Eupatorium: Cherokee
Gaillardia: Hopi
Galium: Choctaw; Cree, Hudson Bay; Ojibwa
Gilia: Zuni
Gutierrezia: Zuni
Heliotropium: Paiute; Shoshoni
Hieracium: Navajo, Ramah
Humulus: Ojibwa
Hypericum: Natchez
Impatiens: Iroquois
Inula: Iroquois
Ipomoea: Cherokee; Creek
Juniperus: Cree, Hudson Bay; Gitksan; Iroquois; Paiute; Shoshoni; Tewa; Thompson
Laportea: Meskwaki; Ojibwa
Larix: Micmac
Larrea: Shoshoni
Ledum: Cree; Cree, Woodlands; Gitksan; Micmac
Liatris: Cherokee
Lithospermum: Gosiute; Iroquois; Ute
Lomatium: Paiute, Northern
Lonicera: Chippewa; Cree, Woodlands; Montagnais; Potawatomi
Lupinus: Paiute; Shoshoni
Lycopodium: Iroquois; Ojibwa
Lygodesmia: Blackfoot
Malaxis: Ojibwa
Malus: Gitksan
Mentzelia: Keres, Western
Mirabilis: Lakota; Zuni
Mitchella: Cherokee
Monarda: Cherokee
Myrica: Bella Coola
Nicotiana: Cherokee
Oplopanax: Gitksan
Opuntia: Lakota; Okanagan-Colville
Penstemon: Navajo
Perideridia: Blackfoot
Phlox: Navajo
Picea: Bella Coola

Pinus: Gitksan; Zuni
Plantago: Meskwaki
Polemonium: Meskwaki
Polygala: Cherokee
Polypodium: Micmac
Prenanthes: Choctaw; Ojibwa
Pseudotsuga: Bella Coola; Thompson
Psorothamnus: Paiute
Pyrola: Blackfoot
Rhamnus: Kawaiisu
Rhus: Cheyenne; Omaha
Ribes: Chippewa
Rosa: Shoshoni
Rubus: Iroquois; Ojibwa
Salix: Paiute
Sambucus: Cherokee; Meskwaki
Sarracenia: Algonquin, Tête-de-Boule
Solidago: Chippewa
Sphaeralcea: Hopi
Stephanomeria: Hopi
Symphoricarpos: Bella Coola; Chippewa; Nitinaht; Sanpoil; Sioux
Taxus: Potawatomi
Tilia: Iroquois
Tradescantia: Meskwaki
Trillium: Menominee
Triosteum: Iroquois
Urtica: Chippewa; Sioux
Verbesina: Chickasaw
Viburnum: Ojibwa
Zanthoxylum: Iroquois

Ear Medicine
Achillea: Winnebago
Acorus: Cree, Woodlands; Iroquois
Aesculus: Delaware; Delaware, Oklahoma
Allium: Blackfoot; Cherokee; Shinnecock
Amelanchier: Blackfoot
Apocynum: Chippewa
Aralia: Algonquin, Tête-de-Boule
Arctium: Iroquois
Arctostaphylos: Flathead
Artemisia: Costanoan
Asarum: Meskwaki
Aster: Cheyenne; Chippewa
Astragalus: Navajo, Kayenta
Balsamita: Iroquois
Campanula: Chippewa
Cercocarpus: Kawaiisu
Chamaesyce: Cahuilla
Conyza: Navajo, Kayenta
Coptis: Iroquois
Croton: Cahuilla; Isleta
Cucurbita: Cheyenne
Datura: Pima
Fraxinus: Algonquin, Quebec; Iroquois
Geastrum: Keres, Western
Glycyrrhiza: Dakota; Pawnee; Sioux
Gutierrezia: Tewa
Humulus: Delaware; Delaware, Oklahoma; Mohegan

Hydrastis: Iroquois
Iris: Omaha; Paiute; Ponca; Shoshoni
Juniperus: Keres, Western
Ligusticum: Crow
Lobelia: Iroquois
Lupinus: Hopi; Navajo, Kayenta
Maianthemum: Paiute
Malus: Iroquois
Mammillaria: Pima
Marah: Kawaiisu
Medicago: Costanoan
Mentzelia: Cheyenne
Mitella: Cree, Woodlands
Monarda: Creek
Nicotiana: Cahuilla; Costanoan; Kawaiisu; Micmac; Mohegan; Rappahannock; Shinnecock
Opuntia: Keres, Western
Oxytropis: Blackfoot
Panax: Iroquois; Potawatomi
Passiflora: Cherokee
Pediomelum: Blackfoot
Penstemon: Shoshoni
Physaria: Blackfoot
Pinus: Navajo, Ramah; Thompson
Plantago: Kawaiisu
Podophyllum: Cherokee
Poliomintha: Hopi; Tewa
Polygala: Sioux
Portulaca: Cherokee
Prosopis: Apache, Mescalero; Apache, Western; Tewa
Pyrola: Blackfoot
Rhus: Micmac
Rudbeckia: Cherokee
Ruta: Costanoan; Diegueño
Salvia: Costanoan; Paiute
Sanguinaria: Iroquois
Sempervivum: Cherokee
Sequoia: Pomo; Pomo, Kashaya
Sophora: Comanche
Sorbus: Thompson
Stachys: Costanoan
Stanleya: Shoshoni
Tanacetum: Chippewa
Trillium: Chippewa
Valeriana: Cree, Woodlands
Verbascum: Iroquois
Verbena: Iroquois
Veronica: Cherokee

Emetic
Abies: Blackfoot
Abronia: Navajo, Kayenta
Acer: Iroquois; Meskwaki; Ojibwa; Ojibwa, South
Achillea: Iroquois; Navajo, Ramah; Paiute
Achlys: Lummi
Acorus: Iroquois
Adenostoma: Cahuilla; Coahuilla
Adiantum: Cherokee; Iroquois
Aesculus: Kiowa
Agoseris: Navajo, Ramah

Agrimonia: Iroquois
Aletes: Keres, Western
Allium: Blackfoot; Chippewa
Alnus: Algonquin, Quebec; Cherokee; Chippewa; Gitksan; Iroquois; Mendocino Indian
Ambrosia: Luiseño
Amelanchier: Navajo; Navajo, Ramah
Anemone: Iroquois
Anemopsis: Papago; Pima
Anthemis: Cherokee; Iroquois
Apocynum: Cree, Hudson Bay; Iroquois; Navajo, Kayenta; Navajo, Ramah
Aquilegia: Shoshoni
Aralia: Cherokee
Arbutus: Concow; Yuki
Arctium: Iroquois
Arctostaphylos: Navajo, Kayenta; Navajo, Ramah
Arenaria: Hopi
Argemone: Shoshoni
Argentina: Blackfoot
Artemisia: Paiute; Paiute, Northern; Shoshoni
Asarum: Cherokee; Lolahnkok
Asclepias: Meskwaki; Navajo, Kayenta; Navajo, Ramah; Pima
Asplenium: Cherokee
Astragalus: Hopi; Navajo, Kayenta; Navajo, Ramah
Atriplex: Navajo, Kayenta; Navajo, Ramah
Aureolaria: Chickasaw
Baccharis: Navajo, Ramah
Bahia: Keres, Western
Baptisia: Cherokee
Besseya: Navajo, Ramah
Betula: Delaware, Oklahoma; Micmac
Botrychium: Cherokee; Chickasaw
Brassica: Shinnecock
Brickellia: Navajo, Kayenta; Navajo, Ramah
Caltha: Chippewa; Iroquois
Calycanthus: Cherokee
Carex: Iroquois; Navajo, Ramah
Carya: Cherokee
Castilleja: Shoshoni
Caulophyllum: Chippewa; Iroquois; Ojibwa
Ceanothus: Navajo, Kayenta; Navajo, Ramah
Cephalanthus: Meskwaki
Chaenactis: Shoshoni
Chaerophyllum: Chickasaw
Chamaesyce: Pima; Zuni
Chenopodium: Kawaiisu; Keres, Western; Paiute
Chimaphila: Cherokee
Chrysothamnus: Keresan; Navajo; Navajo, Kayenta; Navajo, Ramah
Cicuta: Bella Coola; Kutenai; Kwakiutl; Salish, Coast
Cirsium: Navajo; Zuni
Clethra: Cherokee
Collinsonia: Cherokee
Coptis: Iroquois
Cordylanthus: Luiseño; Navajo
Cornus: Cree, Hudson Bay; Green River Group; Iroquois; Navajo, Kayenta; Navajo, Ramah; Ojibwa
Coronilla: Cherokee

Corylus: Cherokee; Iroquois
Croton: Hopi
Cucurbita: Cahuilla; Kiowa; Paiute; Shoshoni
Cuscuta: Navajo, Ramah
Cyperus: Navajo, Ramah
Dalea: Hopi; Keres, Western
Delphinium: Hopi
Desmodium: Alabama
Dimorphocarpa: Zuni
Dirca: Iroquois
Draba: Navajo, Ramah
Dryopteris: Cherokee
Echinochloa: Navajo, Ramah
Eleocharis: Navajo, Ramah; Seminole
Elodea: Iroquois
Ephedra: Havasupai
Epigaea: Cherokee
Eriastrum: Paiute; Shoshoni
Erigeron: Paiute
Eriogonum: Diegueño; Keres, Western; Navajo; Navajo, Kayenta; Zuni
Eryngium: Alabama; Cherokee; Koasati; Seminole
Erysimum: Navajo, Ramah; Zuni
Eschscholzia: Mendocino Indian
Eupatorium: Cherokee; Koasati; Seminole
Euphorbia: Cherokee; Micmac
Fallugia: Navajo, Ramah
Forestiera: Navajo, Ramah
Frangula: Delaware, Oklahoma; Klamath; Montana Indian; Navajo, Kayenta
Fraxinus: Delaware, Oklahoma; Iroquois
Galeopsis: Iroquois
Galium: Meskwaki; Navajo, Ramah
Geocaulon: Cree, Hudson Bay
Geranium: Iroquois
Geum: Iroquois
Gilia: Zuni
Gnaphalium: Navajo, Ramah
Goodyera: Cherokee
Grindelia: Navajo, Ramah; Shoshoni
Gutierrezia: Keres, Western; Keresan
Hamamelis: Chippewa; Iroquois
Heliotropium: Paiute; Shoshoni
Hepatica: Cherokee
Heterotheca: Navajo, Ramah
Holodiscus: Shoshoni
Huperzia: Nitinaht
Hydrangea: Cherokee
Hydrastis: Iroquois
Hymenopappus: Hopi; Zuni
Hymenoxys: Navajo, Ramah
Ilex: Alabama; Cherokee; Creek; Iroquois; Natchez
Ipomoea: Hawaiian
Ipomopsis: Keres, Western; Navajo; Navajo, Kayenta; Paiute; Shoshoni; Washo
Iris: Klamath; Montana Indian; Navajo, Ramah; Ojibwa
Juglans: Delaware, Oklahoma; Iroquois
Juncus: Cherokee; Iroquois
Juniperus: Isleta; Keres, Western; Navajo, Ramah; Seminole
Lactuca: Navajo, Ramah
Laportea: Iroquois
Larrea: Cahuilla; Papago; Pima
Lathyrus: Chippewa; Costanoan
Ledum: Cree
Lesquerella: Hopi; Keres, Western
Linaria: Iroquois
Lindera: Creek
Lobelia: Cherokee; Cree, Hudson Bay; Iroquois; Yokut
Lomatium: Kawaiisu
Lonicera: Blackfoot; Iroquois; Makah; Navajo, Ramah; Quileute
Lupinus: Navajo, Ramah
Luzula: Navajo, Ramah
Lycium: Navajo, Ramah
Lysimachia: Iroquois
Machaeranthera: Zuni
Mahonia: Navajo, Ramah
Maianthemum: Navajo, Ramah
Malva: Costanoan; Iroquois
Mentha: Iroquois
Mentzelia: Navajo
Mirabilis: Zuni
Mitella: Iroquois
Monarda: Blackfoot; Iroquois
Monolepis: Navajo, Ramah
Morus: Creek; Delaware, Oklahoma
Myrica: Creek
Myriophyllum: Iroquois
Nicotiana: Cahuilla; Cherokee; Costanoan; Kawaiisu; Paiute; Shoshoni
Nyssa: Cherokee
Oenanthe: Kitasoo; Kwakiutl; Kwakwaka'wakw; Nuxalkmc
Oenothera: Navajo, Kayenta; Navajo, Ramah
Onopordum: Iroquois
Oplopanax: Bella Coola; Haisla & Hanaksiala
Osmorhiza: Bella Coola; Kwakiutl
Panicum: Keres, Western
Parryella: Navajo, Ramah
Paxistima: Navajo, Ramah
Pedicularis: Iroquois
Pellaea: Costanoan
Penstemon: Iroquois; Keres, Western; Navajo, Kayenta
Pericome: Navajo, Ramah
Persea: Seminole
Petasites: Lummi
Petradoria: Navajo, Ramah
Phacelia: Kawaiisu
Phoradendron: Seminole; Zuni
Phragmites: Blackfoot
Physalis: Iroquois
Physocarpus: Bella Coola; Carrier, Southern; Chippewa; Green River Group; Hesquiat; Kwakiutl
Phytolacca: Iroquois
Picea: Iroquois; Navajo, Ramah
Picradeniopsis: Zuni
Piloblephis: Seminole
Pinus: Iroquois; Navajo; Navajo, Ramah; Rappahannock
Platanus: Cherokee
Podophyllum: Meskwaki
Polygala: Miwok
Polygonum: Zuni
Polystichum: Cherokee; Iroquois
Porteranthus: Cherokee
Potamogeton: Navajo, Ramah
Prosopis: Pima
Prunella: Iroquois
Prunus: Iroquois; Navajo, Ramah
Psathyrotes: Navajo, Kayenta; Paiute; Shoshoni
Pseudocymopterus: Navajo, Kayenta; Navajo, Ramah
Pseudotsuga: Bella Coola; Navajo, Ramah; Okanagan-Colville
Psorothamnus: Keres, Western
Pterospora: Keres, Western
Purshia: Hopi; Kawaiisu; Klamath; Montana Indian; Navajo, Ramah; Paiute; Shoshoni
Pyrrhopappus: Navajo, Ramah
Pyrularia: Cherokee
Quercus: Alabama; Cherokee; Keres, Western; Navajo, Ramah
Ranunculus: Iroquois; Navajo, Ramah
Ratibida: Zuni
Rhamnus: Iroquois
Rhus: Chippewa; Iroquois; Keres, Western
Ribes: Navajo, Kayenta; Navajo, Ramah; Shoshoni
Robinia: Cherokee; Hopi
Rosa: Navajo, Ramah; Tanana, Upper
Rubus: Cherokee; Iroquois
Rumex: Iroquois; Navajo, Ramah
Salix: Creek; Crow; Iroquois; Navajo, Ramah; Nez Perce; Seminole
Sambucus: Algonquin, Quebec; Bella Coola; Cherokee; Chippewa; Gitksan; Hesquiat; Iroquois; Kwakiutl; Malecite; Menominee; Micmac; Mohegan; Montana Indian; Nitinaht; Ojibwa; Paiute; Pomo, Potter Valley; Potawatomi; Quinault; Seminole; Yokia; Yokut
Sanguinaria: Iroquois; Mohegan
Sanicula: Iroquois
Sarcobatus: Keres, Western
Sassafras: Seminole
Saururus: Seminole
Scirpus: Cherokee; Navajo, Ramah
Scutellaria: Cherokee
Shepherdia: Navajo, Kayenta
Silene: Gosiute
Silphium: Iroquois; Meskwaki; Winnebago
Sinapis: Navajo, Ramah
Sisymbrium: Navajo, Ramah
Smallanthus: Cherokee
Solanum: Cherokee
Solidago: Iroquois
Sorbus: Micmac; Penobscot; Potawatomi
Sphaeralcea: Luiseño; Shoshoni

Spiraea: Iroquois
Sporobolus: Ojibwa, South
Stachys: Meskwaki
Stephanomeria: Paiute
Symphoricarpos: Navajo, Ramah
Syzygium: Hawaiian
Tanacetum: Paiute
Taraxacum: Iroquois
Tetradymia: Navajo, Ramah
Thamnosma: Havasupai
Tilia: Iroquois
Toxicodendron: Cherokee
Tragopogon: Navajo, Ramah
Trifolium: Navajo, Ramah
Triodanis: Meskwaki
Triosteum: Cherokee
Tsuga: Hoh; Quileute
Typha: Navajo, Ramah
Ulmus: Iroquois
Vaccinium: Seminole
Veratrum: Bella Coola; Carrier, Southern; Haisla; Kwakiutl; Washo
Verbena: Cherokee
Verbesina: Seminole; Zuni
Veronica: Iroquois; Navajo, Ramah
Veronicastrum: Iroquois; Menominee
Viburnum: Chippewa; Iroquois
Vicia: Cherokee
Viola: Navajo, Ramah
Vitis: Seminole
Wyethia: Hopi; Mendocino Indian; Navajo, Kayenta; Paiute; Shoshoni; Washo; Yuki
Xanthium: Cherokee
Yucca: Tewa
Zanthoxylum: Iroquois
Zigadenus: Chehalis; Gosiute; Klamath; Paiute; Shoshoni; Squaxin
Zinnia: Navajo, Ramah

Expectorant

Acer: Potawatomi
Allium: Cherokee
Aralia: Cherokee; Choctaw
Arisaema: Cherokee
Artemisia: Tewa
Asclepias: Cherokee
Botrychium: Chickasaw
Calycanthus: Pomo, Kashaya
Chimaphila: Delaware, Oklahoma
Cirsium: Houma
Comptonia: Delaware, Oklahoma
Eriodictyon: Paiute; Pomo; Shoshoni
Eryngium: Choctaw
Euonymus: Cherokee
Euphorbia: Cherokee
Galium: Cherokee
Glycyrrhiza: Cherokee
Grindelia: Paiute; Shoshoni
Hedeoma: Cherokee
Ipomoea: Cherokee
Larix: Montagnais
Liatris: Cherokee
Nereocystis: Pomo, Kashaya
Nicotiana: Cherokee
Panax: Cherokee
Phragmites: Paiute
Phytolacca: Iroquois
Pinus: Hualapai
Polygala: Cherokee
Polypodium: Skagit, Upper
Rhus: Iroquois
Zanthoxylum: Meskwaki

Eye Medicine

Abies: Bella Coola; Carrier; Ojibwa; Okanagon; Thompson
Acer: Cherokee; Iroquois; Malecite; Micmac; Ojibwa; Potawatomi
Achillea: Mendocino Indian; Navajo, Kayenta; Okanagon; Paiute; Quinault; Salish; Thompson; Yurok
Achlys: Paiute
Allium: Blackfoot
Alnus: Cherokee; Chippewa; Cree, Woodlands
Ambrosia: Gosiute
Amelanchier: Blackfoot; Shoshoni
Anaphalis: Cherokee; Iroquois
Anemone: Meskwaki
Angelica: Gitksan; Mendocino Indian
Antennaria: Gosiute
Anthemis: Mendocino Indian
Apocynum: Cree, Woodlands; Iroquois
Arabis: Okanagan-Colville
Aralia: Choctaw; Iroquois; Koasati; Micmac
Arctostaphylos: Okanagan-Colville; Okanagon; Thompson
Arenaria: Navajo, Ramah; Shoshoni
Argemone: Comanche; Shoshoni
Argentina: Kwakiutl
Arisaema: Chippewa; Iroquois; Menominee; Ojibwa
Arnica: Shuswap
Artemisia: Blackfoot; Crow; Klamath; Mendocino Indian; Montana Indian; Paiute; Poliklah; Shoshoni; Tanana, Upper; Yurok, South Coast (Nererner)
Aruncus: Cherokee
Asarum: Cherokee
Asclepias: Blackfoot; Cheyenne; Pima
Aster: Navajo, Ramah
Astragalus: Navajo, Kayenta; Navajo, Ramah; Shoshoni
Athyrium: Bella Coola
Baccharis: Coahuilla
Balsamorhiza: Shoshoni
Bidens: Seminole
Calla: Gitksan
Calochortus: Thompson
Calycanthus: Cherokee
Campanula: Navajo, Ramah; Thompson
Castilleja: Gitksan; Shuswap
Cephalanthus: Chickasaw; Choctaw
Cercocarpus: Shoshoni
Chaenactis: Okanagan-Colville
Chamaesyce: Costanoan; Keres, Western; Paiute; Shoshoni
Chenopodium: Navajo, Kayenta
Chimaphila: Chippewa; Flathead; Kutenai
Cicuta: Shoshoni
Cirsium: Navajo, Ramah
Clarkia: Mendocino Indian
Claytonia: Hesquiat; Quileute; Thompson
Cleome: Gosiute
Clintonia: Bella Coola; Cowlitz; Haisla & Hanaksiala
Comandra: Navajo, Kayenta; Thompson
Conocephalum: Nitinaht
Coptis: Algonquin, Quebec; Iroquois
Cornus: Abnaki; Bella Coola; Carrier, Southern; Chippewa; Cree, Woodlands; Iroquois; Malecite; Micmac; Paiute; Potawatomi; Snohomish
Corylus: Abnaki
Crepis: Shoshoni
Croton: Hopi
Cryptantha: Navajo, Kayenta
Cryptogramma: Thompson
Datura: Costanoan; Navajo; Pima
Desmanthus: Paiute
Diervilla: Chippewa; Cree, Woodlands
Digitaria: Hawaiian
Diplacus: Pomo, Kashaya
Dirca: Iroquois
Disporum: Blackfoot
Dodecatheon: Blackfoot; Okanagan-Colville
Draba: Navajo, Ramah
Dracocephalum: Navajo, Ramah
Echinacea: Dakota; Omaha
Elymus: Comanche
Empetrum: Tanana, Upper
Epilobium: Eskimo, Inupiat
Equisetum: Iroquois; Karok; Okanagan-Colville; Quinault; Thompson; Yuki
Ericameria: Paiute; Shoshoni
Erigeron: Cherokee; Navajo, Ramah; Okanagan-Colville; Paiute; Shoshoni
Eriodictyon: Costanoan
Eriogonum: Cahuilla; Coahuilla; Gosiute; Round Valley Indian; Thompson; Zuni
Eriophorum: Eskimo, Western
Escobaria: Blackfoot
Euonymus: Meskwaki
Foeniculum: Pomo, Kashaya
Fragaria: Iroquois
Frasera: Shoshoni
Fraxinus: Ojibwa, South
Gaillardia: Blackfoot
Galium: Iroquois
Gentiana: Iroquois
Geranium: Blackfoot; Sanpoil
Geum: Blackfoot
Gnaphalium: Karok
Goodyera: Cherokee
Grindelia: Cheyenne; Cheyenne, Northern
Gutierrezia: Keres, Western

Hamamelis: Chippewa
Hedeoma: Chickasaw
Hepatica: Meskwaki
Heracleum: Makah; Sanpoil
Heuchera: Blackfoot; Chippewa; Cree, Woodlands; Paiute
Hierochloe: Blackfoot
Holodiscus: Lummi
Hordeum: Chippewa
Houstonia: Keres, Western
Hydrastis: Iroquois
Hypericum: Alabama; Choctaw
Ilex: Alabama; Choctaw
Impatiens: Iroquois
Ipomopsis: Navajo, Ramah; Salish; Shoshoni
Iris: Omaha; Ponca
Juniperus: Okanagon; Seminole; Thompson
Krameria: Paiute; Pima
Krascheninnikovia: Paiute; Shoshoni
Lactuca: Iroquois
Ledum: Shuswap
Leptodactylon: Shoshoni
Lesquerella: Navajo, Kayenta; Navajo, Ramah
Leucanthemum: Iroquois
Leymus: Paiute; Shoshoni
Linum: Apache, White Mountain; Great Basin Indian; Paiute; Shoshoni; Zuni
Lithospermum: Navajo, Ramah
Lomatium: Nez Perce; Shoshoni
Lonicera: Carrier; Carrier, Southern; Gitksan; Mendocino Indian
Lupinus: Hopi; Okanagan-Colville
Lygodesmia: Omaha; Ponca; Sioux
Machaeranthera: Navajo, Ramah; Shoshoni
Maclura: Comanche
Mahonia: Apache, Mescalero; Okanagan-Colville; Sanpoil; Thompson
Maianthemum: Cherokee; Paiute; Quinault; Shoshoni
Malacothrix: Navajo, Ramah
Malus: Bella Coola; Gitksan; Iroquois; Klallam; Quinault
Marah: Paiute
Melampyrum: Ojibwa
Mentha: Isleta
Mentzelia: Navajo, Ramah
Mitella: Iroquois
Monarda: Blackfoot; Flathead; Lakota; Tewa
Monardella: Paiute
Monotropa: Cherokee
Nyssa: Cherokee
Oenothera: Hopi
Oplopanax: Haisla & Hanaksiala
Opuntia: Apache, Mescalero
Orthilia: Carrier, Southern
Osmorhiza: Blackfoot; Menominee; Meskwaki; Paiute; Paiute, Northern; Potawatomi; Shoshoni
Oxalis: Cowlitz; Quinault
Paeonia: Paiute; Shoshoni
Panax: Iroquois; Potawatomi
Pectis: Zuni
Pediomelum: Blackfoot
Penstemon: Okanagon; Paiute; Shoshoni; Thompson
Persea: Seminole
Petasites: Quinault
Phlox: Paiute; Shoshoni; Washo
Phoradendron: Cahuilla
Physalis: Diegueño
Physaria: Blackfoot; Paiute; Shoshoni
Picea: Carrier, Southern; Thompson
Pinus: Carrier, Northern; Kawaiisu; Klamath; Miwok; Okanagan-Colville; Okanagon; Thompson
Piper: Hawaiian
Piperacea: Mahuna
Plantago: Cherokee; Shinnecock
Pluchea: Pima
Polygonatum: Iroquois
Polypodium: Mendocino Indian
Populus: Carrier, Northern; Klallam; Okanagan-Colville
Prenanthes: Iroquois
Prosopis: Apache, Mescalero; Cahuilla; Diegueño; Isleta; Keres, Western; Paiute; Pima
Prunella: Blackfoot
Prunus: Cree, Woodlands; Flathead; Paiute; Potawatomi; Shoshoni
Psathyrotes: Paiute
Psilostrophe: Navajo, Ramah
Pulsatilla: Thompson
Pyrola: Blackfoot; Carrier, Southern; Cree, Woodlands; Iroquois
Quercus: Diegueño; Kawaiisu; Menominee
Ranunculus: Iroquois
Rhamnus: Cherokee
Rhus: Diegueño; Iroquois; Ojibwa
Ribes: Bella Coola; Klallam; Makah; Okanagan-Colville; Potawatomi; Skagit; Tanana, Upper; Thompson
Rorippa: Navajo, Ramah; Zuni
Rosa: Bella Coola; Carrier; Cheyenne; Chippewa; Cree, Woodlands; Flathead; Iroquois; Ojibwa, South; Omaha; Sikani; Skagit; Thompson
Rubus: Chippewa; Menominee; Meskwaki; Ojibwa; Potawatomi
Rudbeckia: Shuswap
Sabal: Houma
Sabatia: Seminole
Salix: Abnaki; Blackfoot; Comanche; Eskimo, Alaska; Eskimo, Kuskokwagmiut; Eskimo, Western; Flathead; Seminole
Salvia: Cahuilla; Costanoan; Kawaiisu; Mahuna; Paiute
Sanguinaria: Iroquois
Sassafras: Cherokee; Iroquois; Mohegan; Rappahannock; Seminole
Schoencrambe: Navajo, Ramah
Scrophularia: Costanoan
Scutellaria: Miwok
Senecio: Zuni
Shepherdia: Flathead; Kutenai
Silene: Okanagan-Colville
Sisymbrium: Pima
Smilax: Iroquois
Solanum: Cahuilla; Iroquois; Luiseño; Miwok; Navajo
Sorbus: Bella Coola
Sphaeralcea: Shoshoni
Stellaria: Chippewa
Stephanomeria: Kawaiisu; Keres, Western; Shoshoni
Streptanthus: Navajo, Kayenta
Streptopus: Chippewa
Symphoricarpos: Carrier, Southern; Cree, Woodlands; Dakota; Flathead; Kwakiutl; Ojibwa, South; Okanagan-Colville; Omaha; Ponca; Sanpoil; Shuswap; Sioux; Thompson; Wet'suwet'en
Taraxacum: Iroquois
Tauschia: Kawaiisu
Tetraneuris: Zuni
Thalictrum: Gitksan; Iroquois
Thelypodium: Navajo, Kayenta
Thermopsis: Navajo, Ramah; Pomo; Pomo, Kashaya
Thuja: Kwakiutl
Tiarella: Iroquois
Tilia: Algonquin, Quebec
Toxicodendron: Diegueño
Trientalis: Cowlitz; Paiute
Trifolium: Iroquois
Trillium: Lummi; Menominee; Paiute; Skagit; Thompson
Tsuga: Cowlitz; Hesquiat; Kwakiutl; Skagit
Ulmus: Cherokee; Iroquois; Meskwaki; Potawatomi
Uvularia: Iroquois
Vaccinium: Seminole
Verbesina: Seminole
Viburnum: Gitksan; Montagnais
Vicia: Navajo, Ramah
Viola: Iroquois; Tolowa
Vitis: Menominee
Wyethia: Mendocino Indian
Xanthium: Pima
Xanthorhiza: Cherokee
Zinnia: Zuni
Ziziphus: Pima

Febrifuge

Abies: Blackfoot; Montana Indian
Achillea: Abnaki; Cherokee; Cheyenne; Cree, Woodlands; Flathead; Iroquois; Menominee; Meskwaki; Montagnais; Navajo, Kayenta; Ojibwa; Paiute; Quileute
Aconitum: Blackfoot
Acorus: Cree, Woodlands; Dakota; Omaha; Pawnee; Ponca; Winnebago
Acrostichum: Seminole
Adiantum: Cherokee
Agastache: Cheyenne; Navajo, Ramah; Okanagan-Colville
Ageratina: Cherokee; Navajo, Ramah
Agrimonia: Cherokee
Aletris: Cherokee

Index of Usages

Allium: Cherokee; Rappahannock; Shinnecock
Alnus: Cherokee; Micmac; Tanana, Upper
Ambrosia: Cherokee
Amelanchier: Cree, Woodlands
Andropogon: Omaha
Angelica: Cherokee; Iroquois; Mendocino Indian
Anthemis: Cherokee; Iroquois; Mohegan
Apocynum: Montana Indian
Aquilegia: Omaha; Pawnee; Ponca
Aralia: Mendocino Indian; Rappahannock
Arctium: Abnaki; Iroquois
Arenaria: Navajo, Ramah
Arisaema: Iroquois
Aristolochia: Cherokee; Natchez; Rappahannock
Artemisia: Blackfoot; Cree, Woodlands; Eskimo; Gosiute; Gros Ventre; Hawaiian; Keres, Western; Mendocino Indian; Navajo; Navajo, Ramah; Omaha; Paiute; Paiute, Northern; Shoshoni; Tewa; Winnebago
Aruncus: Nitinaht
Asarum: Algonquin, Quebec; Cherokee; Iroquois
Aster: Cherokee; Cree, Woodlands; Iroquois
Astragalus: Dakota
Athyrium: Iroquois
Baccharis: Navajo, Kayenta
Balsamorhiza: Cheyenne
Barbula: Seminole
Berberis: Mohegan
Betula: Iroquois
Bidens: Seminole
Boebera: Keres, Western
Brassica: Cherokee
Brickellia: Diegueño; Navajo, Ramah
Bryum: Seminole
Callicarpa: Alabama
Calycadenia: Yana
Capsicum: Cherokee
Cardamine: Algonquin, Quebec; Iroquois
Castanea: Cherokee
Caulophyllum: Iroquois; Omaha; Ponca
Ceanothus: Modesse
Celastrus: Iroquois
Centaurium: Luiseño; Mahuna; Miwok
Cephalanthus: Choctaw; Seminole
Cercis: Alabama; Delaware; Delaware, Oklahoma; Mendocino Indian
Chamaedaphne: Potawatomi
Chelone: Cherokee
Chenopodium: Creek; Natchez
Chimaphila: Cherokee; Iroquois; Montana Indian
Chrysothamnus: Isleta; Navajo, Ramah
Cicuta: Seminole
Cirsium: Navajo; Navajo, Kayenta
Clematis: Oregon Indian
Cleome: Oregon Indian
Clethra: Cherokee
Clintonia: Iroquois
Collomia: Okanagan-Colville
Comptonia: Chippewa
Conyza: Iroquois

Cordyline: Hawaiian
Cornus: Cherokee; Costanoan; Cree, Hudson Bay; Houma; Iroquois; Wet'suwet'en
Croton: Concow
Cucurbita: Cheyenne
Cunila: Cherokee
Cycloloma: Hopi
Cypripedium: Iroquois
Cystopteris: Cherokee
Dalea: Navajo, Ramah
Datura: Navajo; Rappahannock
Daucus: Costanoan
Dennstaedtia: Cherokee
Desmodium: Seminole
Dioscorea: Hawaiian
Dracocephalum: Navajo, Ramah
Echinacea: Cheyenne
Echinocystis: Cherokee
Eleocharis: Seminole
Epilobium: Costanoan
Ericameria: Paiute
Erigeron: Miwok; Navajo, Ramah; Ojibwa
Eriodictyon: Cahuilla; Pomo, Kashaya; Round Valley Indian
Eriogonum: Keres, Western
Eryngium: Seminole
Erythronium: Cherokee
Eucalyptus: Hawaiian
Eupatorium: Cherokee; Delaware; Delaware, Oklahoma; Houma; Iroquois; Menominee; Mohegan; Nanticoke; Seminole; Shinnecock
Euphorbia: Cahuilla
Euthamia: Potawatomi
Fraxinus: Costanoan
Galactia: Seminole
Gaultheria: Potawatomi
Gaura: Keres, Western
Gentiana: Iroquois
Geum: Iroquois; Okanagan-Colville
Gilia: Navajo, Ramah; Zuni
Gleditsia: Meskwaki
Glycyrrhiza: Dakota; Keresan; Pawnee; Sioux
Gnaphalium: Koasati; Rappahannock
Gutierrezia: Isleta; Navajo, Ramah
Hamamelis: Cherokee
Hedeoma: Cherokee; Ojibwa
Helenium: Comanche
Helianthus: Pima; Shasta
Hepatica: Nanticoke
Heuchera: Shoshoni
Hexastylis: Rappahannock
Hierochloe: Flathead
Hoita: Costanoan
Holocarpha: Miwok
Humulus: Dakota
Hydrastis: Iroquois
Hypericum: Cherokee; Houma; Iroquois
Hyssopus: Cherokee
Ilex: Micmac
Impatiens: Iroquois

Inula: Iroquois
Ipomopsis: Okanagan-Colville
Juglans: Rappahannock
Juniperus: Cheyenne; Cree, Woodlands; Flathead; Kutenai; Mahuna; Navajo, Ramah; Nez Perce; Paiute; Seminole; Sioux
Krameria: Pima
Krascheninnikovia: Gosiute; Hopi; Tewa
Laportea: Houma
Larix: Iroquois
Larrea: Pima
Lechea: Seminole
Ledum: Montagnais
Lepechinia: Miwok
Leucanthemum: Menominee
Lilium: Malecite; Micmac
Lindera: Iroquois
Linnaea: Iroquois
Linum: Cherokee
Liquidambar: Houma
Liriodendron: Cherokee
Lobelia: Cherokee; Iroquois
Lomatium: Thompson
Lonicera: Iroquois
Lophophora: Kiowa
Lycopodium: Micmac; Montagnais
Lysichiton: Quileute; Skokomish
Lythrum: Iroquois
Magnolia: Houma
Mahonia: Montana Indian
Malva: Costanoan; Diegueño; Iroquois
Matricaria: Costanoan; Diegueño
Melilotus: Iroquois
Melissa: Cherokee
Mentha: Algonquin, Tête-de-Boule; Cherokee; Cree, Woodlands; Flathead; Iroquois; Keres, Western; Keresan; Kutenai; Menominee; Navajo, Ramah; Ojibwa; Okanagan-Colville; Paiute; Paiute, Northern; Potawatomi; Shoshoni; Sia; Washo
Mentzelia: Cheyenne; Dakota
Mirabilis: Dakota; Mahuna
Mitchella: Iroquois; Montagnais
Monarda: Cherokee; Delaware; Delaware, Oklahoma; Flathead; Iroquois; Koasati; Mohegan; Navajo, Ramah; Ojibwa; Sioux, Teton; Tewa
Monardella: Miwok
Monotropa: Mohegan
Myrica: Choctaw; Seminole
Nemopanthus: Malecite
Nepeta: Cherokee; Chippewa; Iroquois
Nuphar: Iroquois; Rappahannock
Octoblephorum: Seminole
Osmorhiza: Iroquois; Okanagan-Colville; Paiute; Shoshoni
Osmunda: Cherokee
Oxalis: Iroquois
Paeonia: Mahuna
Panax: Creek; Iroquois
Panicum: Natchez
Pediomelum: Cheyenne

Pellaea: Costanoan
Penstemon: Pawnee
Pericome: Navajo, Ramah
Persea: Creek; Seminole
Phacelia: Costanoan
Phytolacca: Cherokee
Piloblephis: Seminole
Pinus: Cherokee; Navajo, Ramah; Okanagan-Colville; Paiute; Shoshoni
Pityopsis: Seminole
Plantago: Costanoan; Iroquois; Rappahannock
Pleomele: Hawaiian
Pluchea: Choctaw; Houma
Polygonum: Iroquois; Potawatomi
Polystichum: Cherokee; Iroquois
Populus: Paiute
Porteranthus: Iroquois
Potentilla: Cherokee
Prosopis: Diegueño
Prunella: Algonquin, Quebec; Delaware; Delaware, Oklahoma; Iroquois; Mohegan
Prunus: Cherokee; Iroquois
Pseudotsuga: Okanagan-Colville
Pterocaulon: Seminole
Purshia: Navajo, Ramah
Pycnanthemum: Cherokee; Chippewa; Meskwaki
Quercus: Cherokee
Ratibida: Navajo, Ramah
Rhus: Iroquois; Malecite
Ricinus: Hawaiian
Rorippa: Costanoan; Iroquois
Rosa: Costanoan; Diegueño; Mahuna; Tanana, Upper
Rubus: Micmac
Rudbeckia: Seminole
Rumex: Houma
Sabal: Seminole
Sabatia: Seminole
Sagittaria: Cherokee
Salix: Alabama; Blackfoot; Cherokee; Costanoan; Creek; Houma; Koasati; Mendocino Indian; Mewuk; Montana Indian; Pima; Seminole
Salvia: Costanoan; Paiute
Sambucus: Cahuilla; Diegueño; Iroquois; Kawaiisu; Mendocino Indian; Menominee; Montana Indian; Pima; Pomo, Kashaya; Yuki
Sanguinaria: Iroquois
Sanicula: Ojibwa
Sanvitalia: Navajo, Ramah
Sarracenia: Iroquois
Sassafras: Iroquois; Nanticoke; Rappahannock; Seminole
Satureja: Cahuilla; Luiseño
Saururus: Seminole
Scrophularia: Diegueño
Scutellaria: Mendocino Indian
Sedum: Costanoan
Senecio: Iroquois
Senna: Cherokee
Shepherdia: Navajo

Sideroxylon: Hawaiian
Sinapis: Cherokee
Sisyrinchium: Costanoan
Smallanthus: Iroquois
Solanum: Delaware, Oklahoma
Solidago: Cherokee; Chippewa; Delaware; Delaware, Oklahoma; Iroquois; Okanagan-Colville; Potawatomi
Sorbus: Wet'suwet'en
Sparganium: Iroquois
Stephanomeria: Paiute
Symphoricarpos: Cree, Woodlands; Nez Perce
Symplocos: Choctaw
Tagetes: Navajo
Tanacetum: Chippewa; Ojibwa
Taxus: Micmac; Montagnais
Tephrosia: Catawba
Tetraclea: Navajo, Ramah
Tetradymia: Navajo, Ramah
Thalictrum: Ojibwa
Thuja: Algonquin, Quebec; Iroquois; Quinault
Thymophylla: Isleta
Thymus: Delaware, Ontario
Tillandsia: Houma
Torreya: Costanoan
Toxicodendron: Cherokee
Triadenum: Potawatomi
Trichostema: Concow
Trifolium: Cherokee
Triosteum: Cherokee
Tsuga: Iroquois
Urtica: Potawatomi
Vaccinium: Seminole
Veratrum: Paiute
Verbascum: Iroquois; Nanticoke; Navajo, Ramah
Verbena: Cherokee; Costanoan; Delaware, Oklahoma; Mahuna; Navajo, Ramah
Verbesina: Choctaw; Hopi; Seminole
Veronica: Cherokee
Veronicastrum: Cherokee; Iroquois
Viburnum: Cherokee; Iroquois
Vitis: Choctaw; Mohegan; Seminole
Wyethia: Miwok; Paiute
Xanthium: Houma
Zanthoxylum: Comanche
Zinnia: Zuni
Zizia: Meskwaki

Gastrointestinal Aid

Abies: Bella Coola; Cherokee; Hanaksiala; Micmac; Okanagan-Colville; Thompson
Abronia: Navajo; Navajo, Kayenta; Ute; Zuni
Abutilon: Hawaiian
Acer: Iroquois
Achillea: Aleut; Cherokee; Costanoan; Cowlitz; Gosiute; Hesquiat; Iroquois; Mendocino Indian; Mewuk; Mohegan; Okanagan-Colville; Paiute; Shoshoni; Squaxin; Thompson
Acorus: Blackfoot; Cherokee; Cheyenne; Cree; Cree, Woodlands; Dakota; Delaware; Iroquois; Menominee; Meskwaki; Nanticoke; Ojibwa; Omaha; Pawnee; Ponca; Rappahannock; Winnebago
Actaea: Ojibwa; Ojibwa, South
Adenostoma: Cahuilla; Coahuilla; Diegueño
Adiantum: Costanoan; Makah
Aesculus: Cherokee
Agastache: Paiute
Agrimonia: Cherokee
Aletris: Catawba; Micmac
Aleurites: Hawaiian
Alisma: Cherokee; Cree, Woodlands
Allium: Cherokee
Alnus: Carrier, Northern; Cherokee; Clallam; Mendocino Indian; Ojibwa; Penobscot; Swinomish
Amaranthus: Keres, Western
Ambrosia: Cheyenne; Diegueño
Amelanchier: Blackfoot; Thompson
Amorpha: Ojibwa, South
Amphicarpaea: Iroquois
Anaphalis: Iroquois; Okanagan-Colville
Andropogon: Chippewa; Seminole
Anemopsis: Cahuilla; Pima; Shoshoni
Angelica: Blackfoot; Costanoan; Creek; Delaware; Delaware, Oklahoma; Mendocino Indian; Pomo, Kashaya; Tewa
Antennaria: Cherokee
Anthemis: Iroquois
Apocynum: Iroquois; Navajo, Ramah
Aquilegia: Gosiute; Meskwaki; Ojibwa; Paiute; Shoshoni
Arabis: Keres, Western; Okanagan-Colville
Aralia: Bella Coola; Choctaw; Iroquois; Mendocino Indian; Menominee
Arbutus: Cahuilla; Miwok; Skokomish; Yuki
Arceuthobium: Mendocino Indian
Arctium: Iroquois; Ojibwa
Arctostaphylos: Miwok
Arenaria: Gosiute; Yavapai
Aristolochia: Cherokee; Choctaw
Armoracia: Cherokee
Artemisia: Blackfoot; Cherokee; Chippewa; Coahuilla; Costanoan; Dakota; Eskimo, Western; Havasupai; Hopi; Hualapai; Isleta; Jemez; Kiowa; Lakota; Mendocino Indian; Mewuk; Navajo; Okanagan-Colville; Omaha; Paiute; Pawnee; Pomo, Kashaya; Ponca; Sanpoil; Shoshoni; Tewa; Thompson; Winnebago
Aruncus: Bella Coola; Thompson
Asarum: Bella Coola; Cherokee; Chippewa; Malecite; Menominee; Meskwaki; Micmac; Ojibwa; Okanagon; Thompson
Asclepias: Flathead; Iroquois; Keres, Western; Navajo; Omaha; Pima; Ponca
Aster: Mohegan; Navajo, Ramah; Okanagon; Thompson
Astragalus: Cree, Woodlands; Navajo, Kayenta
Atriplex: Navajo, Kayenta; Navajo, Ramah; Zuni
Baccharis: Diegueño
Balsamorhiza: Cheyenne; Paiute
Baptisia: Iroquois; Pawnee
Bellis: Iroquois

Betula: Cherokee; Chippewa; Delaware, Oklahoma; Eskimo, Western; Ojibwa
Bidens: Hawaiian
Blechnum: Makah; Quinault
Brassica: Navajo, Ramah
Brickellia: Shoshoni
Callicarpa: Choctaw; Koasati
Calocedrus: Mendocino Indian
Calycanthus: Pomo, Kashaya
Capsella: Chippewa; Mohegan
Capsicum: Cherokee
Cardamine: Delaware; Delaware, Oklahoma; Iroquois; Menominee; Ojibwa
Carex: Iroquois; Navajo, Ramah
Carya: Cherokee
Castanea: Cherokee; Koasati
Castilleja: Navajo; Navajo, Ramah; Ute
Caulophyllum: Cherokee; Chippewa
Ceanothus: Cherokee; Chippewa; Menominee; Meskwaki
Celastrus: Cherokee; Ojibwa
Celtis: Navajo, Kayenta
Centaurium: Mahuna; Miwok
Cephalanthus: Seminole
Cercocarpus: Navajo; Navajo, Ramah; Paiute
Chaenactis: Shoshoni; Thompson
Chaetopappa: Havasupai
Chamaebatiaria: Shoshoni
Chamaemelum: Cherokee; Mahuna
Chamaesyce: Houma; Navajo; Navajo, Ramah; Omaha; Pima
Chenopodium: Cahuilla; Mendocino Indian; Seminole
Chimaphila: Micmac; Ojibwa; Rappahannock
Chlorogalum: Wailaki
Chrysothamnus: Keresan; Shoshoni; Thompson
Cirsium: Cherokee; Costanoan; Meskwaki; Ojibwa; Shuswap
Clematis: Cherokee; Shoshoni; Yavapai
Cleome: Tewa
Clethra: Cherokee
Comarum: Ojibwa
Comptonia: Ojibwa
Convolvulus: Navajo, Ramah
Conyza: Chippewa; Navajo, Kayenta
Coptis: Iroquois
Cornus: Cherokee; Iroquois; Ojibwa; Okanagan-Colville; Saanich; Thompson
Corydalis: Navajo, Ramah
Corylus: Algonquin, Quebec; Iroquois
Crataegus: Iroquois; Okanagon; Potawatomi; Thompson
Croton: Apache, White Mountain; Lakota; Zuni
Cryptantha: Navajo, Kayenta; Ute
Cucurbita: Navajo
Cymopterus: Keres, Western; Navajo, Kayenta
Cynoglossum: Pomo, Potter Valley
Cyperus: Yavapai
Cypripedium: Algonquin, Tête-de-Boule; Cherokee; Chippewa; Iroquois

Dalea: Dakota; Navajo; Navajo, Ramah; Zuni
Datura: Pima; Tubatulabal; Yokut
Descurainia: Cahuilla
Desmodium: Iroquois; Seminole
Diervilla: Chippewa
Diospyros: Cherokee
Diplacus: Tubatulabal
Dryopteris: Eskimo, Western; Ojibwa
Dyssodia: Navajo, Ramah
Echinacea: Choctaw; Crow; Lakota; Meskwaki; Sioux, Teton
Echinocystis: Ojibwa
Ephedra: Diegueño; Mahuna; Navajo; Navajo, Ramah; Paiute; Shoshoni
Epigaea: Cherokee; Iroquois
Epilobium: Cheyenne; Eskimo, Western; Navajo, Kayenta
Equisetum: Ojibwa
Eriastrum: Paiute; Paiute, Northern
Ericameria: Miwok; Paiute; Shoshoni
Erigeron: Navajo, Kayenta; Paiute; Shoshoni; Thompson
Eriodictyon: Hualapai; Kawaiisu; Miwok; Shoshoni
Eriogonum: Cahuilla; Coahuilla; Comanche; Gosiute; Karok; Mahuna; Navajo, Kayenta; Navajo, Ramah; Paiute; Round Valley Indian; Thompson; Tubatulabal; Zuni
Erodium: Hualapai; Zuni
Eryngium: Cherokee; Creek; Seminole
Erysimum: Sioux, Teton
Erythrina: Alabama
Eschscholzia: Mendocino Indian
Eupatorium: Cherokee; Delaware, Ontario; Iroquois; Mohegan
Euphorbia: Iroquois
Evernia: Blackfoot
Fendlera: Navajo
Foeniculum: Cherokee; Pomo, Kashaya
Fragaria: Cherokee; Iroquois; Ojibwa; Potawatomi
Frangula: Delaware, Oklahoma; Hesquiat; Kwakiutl; Nitinaht; Paiute; Thompson
Frasera: Cherokee; Havasupai
Fraxinus: Cherokee; Delaware, Oklahoma; Iroquois
Gaillardia: Blackfoot; Navajo, Ramah
Galium: Cherokee
Garrya: Kawaiisu
Gaultheria: Algonquin, Quebec; Algonquin, Tête-de-Boule; Cherokee; Nitinaht; Quinault
Gaura: Navajo, Ramah
Gaylussacia: Rappahannock
Gentianella: Cherokee; Iroquois
Gentianopsis: Delaware; Delaware, Oklahoma
Geum: Bella Coola; Chippewa; Hesquiat
Gleditsia: Cherokee
Glycyrrhiza: Cheyenne
Gnaphalium: Costanoan; Miwok
Goodyera: Potawatomi
Grindelia: Dakota; Keres, Western; Mendocino Indian; Navajo, Ramah; Shoshoni; Sioux

Gutierrezia: Hopi; Navajo, Kayenta; Navajo, Ramah; Tewa
Haplopappus: Paiute
Hedeoma: Delaware; Delaware, Oklahoma; Mohegan; Ojibwa; Shoshoni
Helenium: Meskwaki
Helianthus: Navajo
Hepatica: Cherokee
Heracleum: Iroquois; Meskwaki; Omaha
Heteromeles: Mendocino Indian
Heterotheca: Navajo, Ramah
Heuchera: Blackfoot; Cherokee; Chippewa; Flathead; Gosiute; Menominee; Navajo; Navajo, Ramah
Hexastylis: Catawba; Cherokee
Hieracium: Cherokee
Holodiscus: Paiute; Shoshoni
Houstonia: Keres, Western
Humulus: Dakota
Huperzia: Nitinaht
Hydrangea: Cherokee
Hydrastis: Cherokee; Iroquois
Hymenopappus: Isleta
Hymenoxys: Zuni
Hypericum: Cherokee; Choctaw
Ilex: Cherokee; Iroquois
Impatiens: Cherokee; Potawatomi
Inula: Delaware; Delaware, Oklahoma; Iroquois
Ipomoea: Iroquois; Keres, Western; Lakota
Ipomopsis: Hopi; Navajo; Navajo, Kayenta; Navajo, Ramah; Paiute; Shoshoni; Washo
Iris: Paiute; Shoshoni
Iva: Shoshoni
Juglans: Delaware; Delaware, Oklahoma; Rappahannock
Juniperus: Arapaho; Bella Coola; Crow; Hopi; Jemez; Keres, Western; Navajo, Ramah; Paiute; Tewa; Thompson
Kalmia: Cree, Hudson Bay; Montagnais
Lachnanthes: Cherokee
Lactuca: Isleta; Navajo
Larix: Thompson
Larrea: Cahuilla; Coahuilla; Mahuna; Paiute; Pima
Lathyrus: Ojibwa
Lechea: Seminole
Ledum: Bella Coola; Eskimo, Alaska; Eskimo, Western; Tanana, Upper
Leonurus: Cherokee; Iroquois
Lepidium: Diegueño; Navajo, Kayenta
Lesquerella: Okanagan-Colville
Liatris: Blackfoot; Cherokee; Omaha; Seminole
Licania: Seminole
Ligusticum: Atsugewi; Cherokee; Creek
Lilium: Algonquin, Quebec
Linnaea: Iroquois
Linum: Hopi; Navajo, Ramah; Paiute
Liriodendron: Cherokee
Lithophragma: Mendocino Indian
Lithospermum: Zuni
Lobelia: Cherokee; Iroquois

Lomatium: Cheyenne; Gosiute; Kawaiisu; Kwakiutl; Salish, Coast
Lonicera: Blackfoot
Lophophora: Kiowa
Lotus: Navajo, Ramah
Luetkea: Okanagon; Thompson
Lupinus: Blackfoot; Karok
Lycopodium: Montagnais
Lycopus: Meskwaki
Lygodesmia: Blackfoot
Lysichiton: Bella Coola; Makah
Lysimachia: Cherokee
Machaeranthera: Navajo, Ramah
Magnolia: Cherokee
Mahonia: Blackfoot; Havasupai; Hualapai; Mendocino Indian; Miwok; Montana Indian; Paiute; Quinault; Sanpoil
Maianthemum: Delaware; Delaware, Oklahoma; Mohegan; Paiute; Shoshoni; Thompson
Malus: Cherokee; Makah; Samish; Swinomish
Malva: Costanoan; Iroquois; Mahuna
Malvella: Choctaw
Marrubium: Navajo, Ramah
Matelea: Comanche
Matricaria: Aleut; Cahuilla; Cherokee; Costanoan; Eskimo, Kuskokwagmiut; Eskimo, Western; Flathead
Melissa: Costanoan
Menispermum: Cherokee
Menodora: Navajo, Ramah
Mentha: Bella Coola; Carrier, Southern; Cherokee; Cree, Hudson Bay; Great Basin Indian; Iroquois; Kiowa; Malecite; Miwok; Mohegan; Ojibwa; Okanagan-Colville; Okanagon; Paiute; Shoshoni; Thompson; Washo
Mentzelia: Mendocino Indian; Montana Indian; Navajo, Kayenta
Menyanthes: Kwakiutl
Menziesia: Kwakiutl
Mimulus: Yavapai
Mirabilis: Navajo, Kayenta; Zuni
Mitchella: Cherokee; Iroquois
Mitella: Gosiute
Monarda: Cherokee; Dakota; Meskwaki; Navajo, Kayenta; Ojibwa; Ojibwa, South; Sioux
Monardella: Cahuilla; Mahuna; Mendocino Indian; Paiute; Shoshoni; Washo
Morus: Delaware, Oklahoma
Myosurus: Navajo, Ramah
Myrica: Koasati; Seminole
Nepeta: Cherokee; Iroquois; Mohegan
Nicotiana: Cherokee; Kawaiisu
Nuphar: Iroquois; Thompson
Nyssa: Cherokee
Obolaria: Cherokee
Oenothera: Navajo, Ramah
Onoclea: Iroquois
Oplopanax: Carrier, Northern; Gitksan; Haisla; Haisla & Hanaksiala; Kwakiutl; Okanagon; Thompson; Wet'suwet'en
Opuntia: Pima
Orbexilum: Cherokee
Orthocarpus: Navajo, Ramah
Osmorhiza: Cheyenne; Paiute; Potawatomi; Shoshoni; Washo
Oxalis: Iroquois
Oxydendrum: Cherokee
Oxytropis: Tlingit
Paeonia: Costanoan; Diegueño
Panax: Cherokee; Iroquois
Parnassia: Cheyenne
Pectis: Keres, Western; Navajo
Pedicularis: Catawba; Cherokee; Iroquois; Ojibwa; Shoshoni
Pediomelum: Blackfoot
Penstemon: Blackfoot; Cherokee; Kiowa; Navajo, Kayenta; Navajo, Ramah; Okanagan-Colville; Paiute; Shoshoni; Thompson
Persea: Seminole
Petrophyton: Gosiute
Phacelia: Kawaiisu
Phlox: Havasupai; Paiute; Shoshoni
Phoradendron: Hopi; Navajo, Ramah; Papago; Pima; Tewa
Phragmites: Apache, White Mountain
Physalis: Iroquois; Omaha; Ponca; Winnebago
Physaria: Blackfoot
Picea: Bella Coola; Carrier, Southern; Iroquois; Makah; Micmac; Ojibwa; Oweekeno
Picradeniopsis: Navajo, Ramah; Zuni
Pimpinella: Delaware; Delaware, Oklahoma
Pinguicula: Seminole
Pinus: Carrier, Southern; Cherokee; Iroquois; Keres, Western; Kwakiutl; Okanagan-Colville; Paiute; Quinault; Shoshoni
Piper: Hawaiian
Plantago: Carrier; Cherokee; Iroquois; Isleta; Meskwaki; Navajo; Nitinaht; Shoshoni
Platanus: Cherokee; Mahuna
Pluchea: Paiute; Pima
Pogogyne: Concow
Polygala: Ojibwa, South
Polygonatum: Cherokee
Polygonum: Chippewa; Iroquois; Keres, Western; Navajo, Ramah; Ojibwa
Polypodium: Abnaki; Bella Coola; Cowichan; Saanich
Polystichum: Cherokee
Populus: Blackfoot; Carrier, Southern; Cherokee; Iroquois; Okanagon; Paiute; Thompson
Porophyllum: Havasupai
Porphyra: Hanaksiala
Portulaca: Navajo
Prosopis: Comanche; Pima
Prunella: Iroquois
Prunus: Cherokee; Chippewa; Gosiute; Koasati; Meskwaki; Mohegan; Navajo, Ramah; Ojibwa, South; Okanagan-Colville; Shinnecock; Shoshoni
Psathyrotes: Paiute; Shoshoni
Pseudocymopterus: Navajo, Ramah
Pseudotsuga: Bella Coola; Hanaksiala; Navajo, Kayenta
Psilostrophe: Navajo, Ramah
Psoralidium: Navajo, Ramah; Zuni
Psorothamnus: Keres, Western; Paiute; Shoshoni
Ptelea: Havasupai
Pterocaulon: Seminole
Pulsatilla: Okanagon; Thompson
Purshia: Paiute, Northern
Pycnanthemum: Cherokee
Pyrola: Iroquois; Nootka
Quercus: Cherokee; Chippewa; Choctaw; Dakota; Delaware; Delaware, Oklahoma; Omaha; Pawnee; Ponca; Winnebago
Ranunculus: Iroquois
Ratibida: Lakota; Navajo, Ramah
Rhamnus: Kawaiisu
Rhododendron: Okanagon; Skokomish; Thompson
Rhus: Chippewa; Iroquois; Keres, Western; Kiowa; Mahuna; Menominee; Navajo, Kayenta; Navajo, Ramah; Thompson
Ribes: Hopi; Thompson
Rosa: Bella Coola; Blackfoot; Cheyenne; Costanoan; Crow; Gros Ventre; Iroquois; Kutenai; Mahuna; Menominee; Meskwaki; Miwok; Ojibwa; Paiute; Tanana, Upper
Rubus: Bella Coola; Cherokee; Hawaiian; Hesquiat; Iroquois; Meskwaki; Ojibwa, South; Okanagan-Colville; Okanagon; Omaha; Pomo, Kashaya; Rappahannock; Saanich; Seminole; Skagit; Skagit, Upper; Thompson
Rudbeckia: Chippewa
Rumex: Houma; Iroquois; Isleta; Kawaiisu; Kwakiutl; Lakota; Mohegan; Navajo, Kayenta; Paiute; Yavapai
Ruta: Costanoan; Diegueño
Sagittaria: Chippewa; Ojibwa
Salix: Blackfoot; Cheyenne; Chippewa; Creek; Koasati; Menominee; Nez Perce; Ojibwa; Okanagan-Colville; Seminole; Shoshoni
Salvia: Kawaiisu; Paiute; Shoshoni
Sambucus: Bella Coola; Cahuilla; Choctaw; Delaware, Oklahoma; Haisla & Hanaksiala; Hesquiat; Iroquois; Mohegan; Montana Indian; Paiute; Pima; Pomo, Potter Valley; Seminole; Yokia
Sanguinaria: Chippewa; Delaware; Delaware, Oklahoma; Iroquois; Ojibwa
Sanicula: Cherokee
Sarcobatus: Navajo, Ramah
Sassafras: Seminole
Satureja: Mendocino Indian; Pomo; Pomo, Kashaya
Saururus: Ojibwa
Scutellaria: Delaware
Sedum: Eskimo, Western
Senecio: Keres, Western; Navajo, Ramah; Yavapai
Shepherdia: Blackfoot; Salish; Shuswap; Thompson
Silene: Gosiute
Silphium: Ojibwa
Sisyrinchium: Cherokee; Costanoan; Mahuna; Pomo, Kashaya

Sium: Lakota
Smilax: Cherokee; Iroquois
Solanum: Iroquois; Lakota; Navajo; Nootka; Zuni
Solidago: Chippewa; Iroquois
Sorbus: Bella Coola; Iroquois; Micmac
Sphaeralcea: Hopi; Navajo, Ramah; Pima
Spiraea: Okanagan-Colville; Shuswap; Thompson
Stachys: Chippewa; Costanoan
Stenotus: Navajo
Streptopus: Thompson
Suaeda: Navajo, Kayenta
Symphoricarpos: Miwok; Paiute; Thompson
Symphytum: Cherokee
Syzygium: Hawaiian
Tagetes: Navajo
Tanacetum: Delaware, Ontario; Mohegan
Taraxacum: Aleut; Bella Coola; Ojibwa; Rappahannock
Taxus: Haihais; Karok; Kitasoo; Micmac
Tetradymia: Navajo, Ramah; Paiute; Shoshoni
Tetraneuris: Navajo, Ramah
Thalictrum: Potawatomi
Thamnosma: Havasupai
Thaspium: Chippewa
Thuja: Algonquin, Quebec; Bella Coola
Thysanocarpus: Mendocino Indian
Tilia: Cherokee
Tiquilia: Navajo, Kayenta
Torreya: Costanoan
Townsendia: Navajo
Tradescantia: Cherokee
Trichostema: Cahuilla; Costanoan; Kawaiisu
Trillium: Cherokee
Triodanis: Cherokee
Triosteum: Iroquois
Tsuga: Gitksan; Menominee; Micmac
Typha: Cheyenne
Ulmus: Cherokee; Iroquois; Koasati; Ojibwa
Umbellularia: Mendocino Indian
Urtica: Cherokee; Hesquiat; Lakota
Uvularia: Ojibwa
Vaccinium: Algonquin, Quebec
Valeriana: Blackfoot; Eskimo; Eskimo, Inuktitut; Thompson
Veratrum: Bella Coola; Blackfoot
Verbena: Cherokee; Costanoan; Dakota; Iroquois; Mahuna
Verbesina: Navajo; Seminole; Zuni
Vernonia: Cherokee
Veronica: Navajo, Ramah
Veronicastrum: Cherokee; Iroquois
Viburnum: Chippewa; Menominee; Ojibwa; Tanana, Upper
Vicia: Cherokee; Rappahannock
Viola: Carrier, Southern; Iroquois
Vitis: Cherokee; Iroquois; Ojibwa; Seminole
Wyethia: Mendocino Indian; Navajo, Kayenta
Xanthium: Cherokee
Xanthorhiza: Catawba
Yucca: Lakota; Navajo
Zanthoxylum: Iroquois
Zinnia: Navajo, Ramah

Gland Medicine
Abies: Okanagan-Colville
Chamaesyce: Iroquois
Erigeron: Thompson
Nymphaea: Micmac
Phytolacca: Delaware
Stanleya: Navajo
Veratrum: Paiute; Shoshoni
Verbascum: Cherokee

Gynecological Aid
Abies: Algonquin, Quebec; Cherokee; Iroquois
Acer: Cherokee; Thompson
Achillea: Blackfoot; Cherokee; Clallam; Klallam; Kwakiutl; Makah; Paiute
Acorus: Algonquin, Quebec
Acourtia: Navajo, Kayenta
Actaea: Cheyenne; Chippewa; Cree, Woodlands; Meskwaki; Ojibwa; Potawatomi
Adiantum: Costanoan; Iroquois; Menominee; Potawatomi
Aesculus: Cherokee
Ageratina: Iroquois
Agrimonia: Cherokee
Aletris: Cherokee; Rappahannock
Alisma: Iroquois
Alnus: Cherokee; Chippewa; Mendocino Indian; Okanagan-Colville; Potawatomi
Amaranthus: Cherokee
Ambrosia: Houma; Keres, Western; Navajo; Pima
Amelanchier: Chippewa; Iroquois; Navajo; Ojibwa; Pomo; Pomo, Kashaya; Thompson
Andropogon: Houma
Androsace: Navajo, Ramah
Anemopsis: Costanoan
Angelica: Iroquois; Pomo, Kashaya; Tewa
Antennaria: Cherokee; Iroquois; Meskwaki; Ojibwa
Anthemis: Karok
Apocynum: Cree, Woodlands; Iroquois; Keres, Western; Meskwaki; Ojibwa; Sanpoil
Aquilegia: Cherokee; Navajo, Kayenta
Aralia: Cherokee; Cree, Woodlands; Iroquois; Meskwaki; Micmac
Arctium: Cherokee; Iroquois; Meskwaki
Arctostaphylos: Cree, Woodlands
Arisaema: Iroquois; Menominee
Artemisia: Blackfoot; Cahuilla; Cherokee; Cheyenne; Chippewa; Comanche; Karok; Lakota; Mahuna; Mendocino Indian; Navajo; Navajo, Ramah; Okanagan-Colville; Okanagon; Paiute; Pomo; Pomo, Kashaya; Shoshoni; Shuswap; Thompson; Yuki
Aruncus: Cherokee
Asarum: Cherokee
Asclepias: Cherokee; Chippewa; Delaware; Delaware, Oklahoma; Hopi; Iroquois; Keres, Western; Lakota; Navajo; Ojibwa; Tewa
Aster: Cree, Woodlands; Iroquois
Astragalus: Kawaiisu; Lakota

Athyrium: Makah; Meskwaki; Ojibwa; Potawatomi
Baccharis: Cahuilla
Bahia: Navajo, Ramah
Balsamorhiza: Cheyenne; Washo
Baptisia: Cherokee; Delaware; Delaware, Oklahoma
Betula: Cree, Woodlands; Iroquois; Ojibwa
Bouteloua: Navajo, Ramah
Calamagrostis: Thompson
Calliandra: Yavapai
Calochortus: Navajo, Ramah
Caltha: Chippewa
Camassia: Blackfoot
Campanula: Iroquois; Navajo, Kayenta
Capsicum: Navajo, Ramah
Carex: Iroquois
Carnegia: Pima
Carpinus: Delaware, Ontario; Iroquois
Carya: Delaware, Ontario
Cassytha: Hawaiian
Castanea: Cherokee
Castilleja: Hopi; Navajo, Ramah; Tewa
Caulophyllum: Cherokee; Menominee; Meskwaki; Ojibwa; Potawatomi
Ceanothus: Karok
Celastrus: Cherokee; Creek; Iroquois; Meskwaki
Celtis: Iroquois
Cerastium: Iroquois
Cercocarpus: Kawaiisu; Navajo, Ramah
Chaetopappa: Zuni
Chamaesyce: Cherokee; Iroquois; Keres, Western; Navajo, Ramah; Ponca; Zuni
Cheilanthes: Keres, Western
Chenopodium: Iroquois
Chimaphila: Iroquois; Karok; Menominee; Okanagon; Thompson
Chrysothamnus: Navajo, Ramah; Okanagan-Colville
Cimicifuga: Iroquois
Cirsium: Chippewa; Lummi
Claytonia: Quinault
Clematis: Navajo
Clintonia: Ojibwa
Coeloglossum: Iroquois
Comptonia: Menominee
Conioselinum: Navajo, Kayenta
Convolvulus: Pomo; Pomo, Kashaya
Conyza: Chippewa; Houma
Cordylanthus: Navajo; Navajo, Kayenta; Navajo, Ramah
Cornus: Algonquin, Tête-de-Boule; Cherokee; Iroquois; Okanagan-Colville; Okanagon; Thompson
Corydalis: Navajo, Kayenta; Navajo, Ramah
Corylus: Iroquois
Crataegus: Chippewa; Iroquois; Ojibwa
Crepis: Shoshoni
Cryptantha: Navajo; Navajo, Kayenta
Cucurbita: Hawaiian; Meskwaki; Omaha
Cunila: Cherokee
Cupressus: Kawaiisu
Cypripedium: Algonquin, Quebec; Cherokee; Menominee; Ojibwa

Darmera: Karok
Datura: Pima
Daucus: Iroquois
Delphinium: Hopi; Navajo, Kayenta
Diervilla: Cree, Woodlands; Iroquois
Dioscorea: Meskwaki
Dirca: Iroquois
Echeandia: Navajo, Ramah
Echium: Iroquois
Ephedra: Washo
Epigaea: Iroquois
Epilobium: Miwok
Equisetum: Menominee; Pomo, Kashaya; Thompson
Ericameria: Miwok
Erigeron: Houma; Navajo; Navajo, Kayenta; Navajo, Ramah; Thompson
Eriogonum: Cahuilla; Cheyenne; Hopi; Keres, Western; Navajo; Navajo, Ramah; Round Valley Indian; Zuni
Erodium: Jemez
Erysimum: Navajo, Ramah
Eschscholzia: Mendocino Indian; Pomo, Kashaya
Euonymus: Cherokee; Iroquois; Winnebago
Eupatorium: Algonquin, Quebec; Cherokee; Chippewa; Iroquois; Menominee; Potawatomi
Euphorbia: Cherokee; Keres, Western; Lakota; Navajo; Navajo, Kayenta; Navajo, Ramah
Foeniculum: Cherokee
Fragaria: Iroquois
Fraxinus: Cherokee; Micmac; Penobscot
Freycinetia: Hawaiian
Gaillardia: Keres, Western
Gaura: Navajo, Kayenta
Gentiana: Meskwaki
Geranium: Thompson
Geum: Chippewa; Hesquiat; Klallam; Ojibwa; Okanagan-Colville; Quinault
Gilia: Navajo, Kayenta
Glyceria: Ojibwa
Glycyrrhiza: Meskwaki
Goodyera: Delaware; Delaware, Oklahoma; Okanagon; Potawatomi; Thompson
Gossypium: Koasati
Grindelia: Cree
Gutierrezia: Jemez; Navajo, Ramah; Tewa
Gymnocladus: Omaha
Hamamelis: Cherokee; Iroquois
Hedeoma: Rappahannock
Helenium: Cherokee; Comanche
Helianthus: Pawnee
Hepatica: Iroquois; Menominee
Heracleum: Kwakiutl; Micmac
Heuchera: Cherokee; Navajo, Ramah
Hibiscus: Hawaiian
Hierochloe: Karok
Houstonia: Navajo, Kayenta
Humulus: Cherokee
Hydrangea: Cherokee
Hymenoxys: Hopi
Hypericum: Houma
Impatiens: Cherokee; Iroquois
Inula: Cherokee; Iroquois
Ipomoea: Hawaiian
Ipomopsis: Hopi; Navajo, Kayenta; Navajo, Ramah
Iris: Iroquois; Pomo
Juglans: Iroquois
Juniperus: Apache, White Mountain; Cheyenne; Comanche; Cree, Woodlands; Crow; Delaware, Ontario; Hopi; Isleta; Jemez; Navajo, Ramah; Paiute; Tewa; Thompson; Zuni
Lachnanthes: Cherokee
Lactuca: Meskwaki; Ojibwa
Laportea: Iroquois
Lappula: Navajo
Larix: Thompson
Larrea: Cahuilla; Papago
Lathyrus: Navajo, Kayenta
Ledum: Makah
Leonurus: Delaware; Delaware, Oklahoma; Micmac; Mohegan; Shinnecock
Leptodactylon: Navajo, Ramah
Lesquerella: Hopi
Lewisia: Flathead; Nez Perce
Lilium: Iroquois
Lindera: Rappahannock
Linnaea: Algonquin, Quebec; Potawatomi
Linum: Hopi
Liquidambar: Cherokee
Lobelia: Iroquois
Lomatium: Kwakiutl
Lonicera: Algonquin, Quebec; Cree, Woodlands; Iroquois; Kwakiutl; Makah; Okanagan-Colville; Quinault; Squaxin; Swinomish
Lotus: Karok
Luetkea: Okanagon; Thompson
Lycopodium: Aleut; Algonquin, Quebec; Iroquois
Lygodesmia: Blackfoot; Cheyenne; Hopi; Omaha; Ponca; Sioux
Lysichiton: Quileute
Lysimachia: Cherokee; Iroquois
Machaeranthera: Hopi
Mahonia: Flathead
Maianthemum: Chippewa; Delaware; Delaware, Oklahoma; Iroquois; Ojibwa; Ojibwa, South; Paiute; Shoshoni; Thompson
Malus: Iroquois
Marrubium: Navajo, Ramah
Matelea: Comanche
Matricaria: Diegueño; Montana Indian
Matteuccia: Cree, Woodlands
Menispermum: Cherokee
Menodora: Navajo, Ramah
Mertensia: Cheyenne
Mirabilis: Hopi; Navajo, Kayenta; Pawnee
Mitchella: Cherokee; Delaware, Oklahoma; Iroquois; Menominee
Monarda: Flathead; Montana Indian; Sioux
Monotropa: Potawatomi
Myrica: Delaware, Oklahoma
Napaea: Meskwaki
Nicotiana: Kawaiisu; Tewa
Nuphar: Gitksan; Kitasoo
Nyssa: Cherokee
Oenanthe: Nitinaht
Oenothera: Navajo, Kayenta
Onoclea: Iroquois; Ojibwa
Oplopanax: Carrier, Southern; Lummi; Skagit
Opuntia: Lakota; Lummi; Navajo, Ramah; Pima
Orbexilum: Cherokee
Osmorhiza: Blackfoot; Chippewa; Ojibwa; Shoshoni
Osmunda: Iroquois; Menominee
Ostrya: Delaware, Ontario
Oxydendrum: Catawba
Oxytropis: Cheyenne
Panax: Cherokee; Iroquois
Pastinaca: Ojibwa
Pellaea: Yavapai
Pennellia: Navajo, Ramah
Penstemon: Iroquois; Navajo, Ramah
Pentagrama: Karok
Pentaphylloides: Tanana, Upper
Peperomia: Hawaiian
Pericome: Navajo, Ramah
Petradoria: Hopi
Petroselinum: Cherokee
Phacelia: Thompson
Phegopteris: Hawaiian
Phlox: Navajo; Navajo, Ramah
Phoenicaulis: Paiute
Phoradendron: Cherokee; Papago; Zuni
Physocarpus: Iroquois; Menominee
Phytolacca: Mohegan
Picea: Algonquin, Quebec
Pinus: Cherokee; Flathead; Kawaiisu; Paiute
Piper: Hawaiian
Plantago: Cherokee; Delaware; Delaware, Oklahoma; Iroquois
Platanthera: Iroquois
Platanus: Cherokee
Polygonatum: Cherokee
Polygonum: Choctaw; Iroquois; Menominee; Meskwaki
Polystichum: Iroquois; Kwakiutl; Lummi
Polytaenia: Meskwaki
Polytrichum: Nitinaht
Populus: Blackfoot; Chippewa; Cree; Delaware, Oklahoma; Thompson
Porteranthus: Creek
Potentilla: Navajo, Kayenta; Okanagan-Colville
Prenanthes: Chippewa; Ojibwa
Prosopis: Pima
Prunella: Iroquois; Ojibwa
Prunus: Cherokee; Chippewa; Delaware, Ontario; Iroquois; Kwakiutl; Lummi; Ojibwa, South; Quinault; Skagit; Skokomish; Thompson
Pseudotsuga: Hanaksiala
Psilostrophe: Navajo, Kayenta
Psoralidium: Navajo, Ramah
Pteridium: Iroquois; Menominee; Ojibwa

Pterocaulon: Seminole
Pulsatilla: Blackfoot
Purshia: Kawaiisu; Navajo; Navajo, Ramah
Pyrola: Blackfoot; Navajo, Kayenta; Thompson
Quercus: Cherokee; Choctaw; Delaware; Delaware, Oklahoma; Delaware, Ontario; Iroquois; Karok; Navajo, Ramah
Ranunculus: Hesquiat
Ratibida: Keres, Western
Reverchonia: Hopi
Rhamnus: Kawaiisu
Rhododendron: Cherokee
Rhus: Cherokee; Diegueño; Iroquois; Keres, Western; Menominee; Navajo, Ramah; Okanagan-Colville; Omaha; Pawnee
Ribes: Blackfoot; Chippewa; Cree, Woodlands; Meskwaki; Ojibwa; Potawatomi; Skagit; Skagit, Upper; Thompson; Winnebago
Rorippa: Navajo
Rosa: Chehalis; Iroquois; Squaxin; Thompson
Rubus: Cherokee; Chippewa; Cree, Woodlands; Iroquois; Kwakiutl; Pomo, Kashaya; Quinault
Rudbeckia: Cherokee
Rumex: Apache, White Mountain; Iroquois; Lakota; Navajo, Ramah
Sabatia: Cherokee
Salix: Algonquin, Tête-de-Boule; Catawba; Okanagan-Colville
Salvia: Cherokee; Mahuna
Sambucus: Creek; Hoh; Kwakiutl; Luiseño; Meskwaki; Paiute; Quileute; Quinault
Sanguinaria: Iroquois
Sanicula: Chippewa; Micmac
Sanvitalia: Navajo, Ramah
Sarracenia: Algonquin, Quebec; Cree, Woodlands; Ojibwa; Potawatomi
Sassafras: Iroquois
Scrophularia: Iroquois
Scutellaria: Cherokee
Sedum: Bella Coola; Okanagan-Colville; Songish
Selaginella: Blackfoot
Senecio: Cherokee; Costanoan; Keres, Western; Navajo, Ramah
Sesamum: Cherokee
Shepherdia: Carrier; Umatilla
Sida: Hawaiian
Silphium: Cherokee; Meskwaki
Sisyrinchium: Menominee; Meskwaki
Smallanthus: Cherokee
Smilax: Cherokee
Solanum: Keresan
Solidago: Cahuilla; Chippewa; Shuswap
Sonchus: Potawatomi
Sorbus: Micmac
Sphaeralcea: Keres, Western
Spiraea: Ojibwa; Okanagan-Colville
Staphylea: Iroquois
Stenotaphrum: Hawaiian
Stephanomeria: Hopi; Navajo, Ramah
Stillingia: Creek

Streptopus: Iroquois; Makah
Stylosanthes: Cherokee
Symphoricarpos: Meskwaki; Ojibwa; Thompson
Symphytum: Cherokee
Symplocarpus: Iroquois
Tanacetum: Cherokee
Taraxacum: Chippewa; Kiowa; Navajo, Ramah; Papago
Taxus: Algonquin, Quebec; Micmac
Tephrosia: Mahuna
Tetradymia: Hopi
Tetraneuris: Hopi
Thamnosma: Paiute
Thelypteris: Iroquois
Thermopsis: Pomo, Kashaya
Thuja: Algonquin, Quebec; Iroquois; Thompson
Tilia: Iroquois
Townsendia: Keres, Western; Navajo; Navajo, Kayenta; Navajo, Ramah
Tradescantia: Cherokee
Trichostema: Miwok
Trifolium: Cherokee; Iroquois
Trillium: Algonquin, Tête-de-Boule; Cherokee; Iroquois; Potawatomi
Triosteum: Iroquois
Tsuga: Algonquin, Quebec; Cherokee; Kwakiutl
Typha: Iroquois
Ulmus: Alabama; Cherokee; Cheyenne; Choctaw; Iroquois; Meskwaki
Umbellularia: Karok; Pomo, Kashaya
Urtica: Cowlitz; Cree, Woodlands; Kwakiutl; Lummi; Quinault; Squaxin
Vaccinium: Algonquin, Quebec; Cree, Woodlands; Makah
Valeriana: Cree, Woodlands
Veratrum: Paiute; Shoshoni
Verbascum: Cherokee
Verbena: Meskwaki
Verbesina: Chickasaw
Vernonia: Cherokee
Veronica: Cherokee
Veronicastrum: Meskwaki
Viburnum: Chippewa; Delaware, Oklahoma; Iroquois; Micmac
Vicia: Iroquois
Viola: Makah
Vitis: Cherokee; Choctaw; Delaware, Oklahoma; Ojibwa
Woodsia: Keres, Western
Xanthium: Koasati
Yucca: Lakota; Navajo, Ramah; Tewa
Zanthoxylum: Iroquois
Zea: Tewa
Zeuxine: Seminole

Hallucinogen
Acorus: Cree, Alberta
Cardamine: Iroquois
Clematis: Iroquois
Cypripedium: Menominee
Datura: Cahuilla; Chumash; Coahuilla; Costanoan; Diegueño; Gabrielino; Hopi; Kawaiisu; Luiseño; Miwok; Navajo, Ramah; Paiute; Paiute, Northern; Shoshoni
Ilex: Cherokee
Lophophora: Ponca
Magnolia: Rappahannock
Mirabilis: Hopi
Nicotiana: Kawaiisu
Thamnosma: Kawaiisu

Heart Medicine
Abies: Algonquin, Quebec
Achillea: Cheyenne
Acorus: Algonquin, Quebec; Ojibwa
Adiantum: Cherokee
Agastache: Cheyenne
Alisma: Cree, Woodlands
Alnus: Carrier; Cherokee
Ambrosia: Iroquois
Apocynum: Chippewa; Potawatomi
Aquilegia: Cherokee; Gosiute
Aralia: Algonquin, Quebec
Artemisia: Chippewa
Asarum: Cherokee
Asclepias: Cherokee
Campanula: Cree, Woodlands
Cardamine: Algonquin, Quebec; Iroquois
Castanea: Cherokee
Cercocarpus: Paiute; Shoshoni
Chaenactis: Great Basin Indian; Nevada Indian; Paiute
Chimaphila: Cree, Woodlands
Clintonia: Iroquois
Collinsonia: Iroquois
Consolida: Cherokee
Coptis: Algonquin, Quebec
Cornus: Okanagan-Colville
Corylus: Algonquin, Quebec; Algonquin, Tête-de-Boule
Crataegus: Cherokee
Croton: Kawaiisu
Cucurbita: Cheyenne
Dalea: Chippewa
Delphinium: Cherokee
Dicentra: Kawaiisu
Distichlis: Kawaiisu
Echinocereus: Navajo
Erigeron: Catawba
Eriogonum: Diegueño; Kawaiisu; Keres, Western
Eryngium: Seminole
Festuca: Iroquois
Filipendula: Meskwaki
Fragaria: Iroquois
Geranium: Iroquois
Hamamelis: Iroquois
Helenium: Comanche
Helianthus: Paiute
Heterotheca: Navajo, Ramah
Heuchera: Shoshoni
Hexastylis: Catawba
Hydrastis: Iroquois

Hypoxis: Cherokee
Inula: Iroquois; Malecite; Micmac
Ipomoea: Houma
Juniperus: Shoshoni; Thompson
Lactuca: Bella Coola
Lewisia: Flathead
Liatris: Menominee
Ligusticum: Cree
Maianthemum: Thompson
Malus: Makah
Matricaria: Shuswap
Mentha: Blackfoot; Cheyenne
Menziesia: Kwakiutl
Monarda: Cherokee
Nuphar: Bella Coola; Iroquois
Paeonia: Paiute
Pedicularis: Iroquois
Peniocereus: Nevada Indian
Picea: Bella Coola
Pinus: Bella Coola
Plantago: Iroquois
Polygala: Chippewa; Meskwaki; Seminole
Polygonum: Iroquois
Polypodium: Algonquin, Quebec
Polypogon: Navajo, Kayenta
Populus: Chippewa
Prunella: Bella Coola
Prunus: Bella Coola; Kwakiutl
Pycnanthemum: Cherokee
Quercus: Chippewa; Ojibwa
Rhododendron: Cherokee
Rhus: Okanagan-Colville
Rubus: Cree, Woodlands
Rudbeckia: Iroquois
Rumex: Hawaiian
Salvia: Costanoan
Sambucus: Iroquois
Sanguinaria: Algonquin, Quebec; Iroquois
Sanicula: Houma
Sassafras: Koasati
Scutellaria: Ojibwa
Senecio: Cherokee
Senna: Cherokee
Shepherdia: Thompson
Solanum: Isleta
Solidago: Algonquin, Quebec
Sonchus: Navajo, Kayenta
Symplocarpus: Menominee
Thalictrum: Iroquois
Thuja: Bella Coola
Tsuga: Bella Coola
Vaccinium: Flathead
Veratrum: Bella Coola
Verbascum: Ojibwa
Veronicastrum: Iroquois
Viburnum: Iroquois
Viola: Ojibwa; Potawatomi
Zanthoxylum: Delaware; Delaware, Oklahoma; Mohegan
Zea: Tewa

Hemorrhoid Remedy
Abies: Hanaksiala
Acer: Seminole
Achillea: Cherokee
Aesculus: Cherokee; Costanoan; Kawaiisu
Alnus: Cherokee; Potawatomi
Andropogon: Rappahannock
Argemone: Tubatulabal
Aster: Okanagan-Colville
Baptisia: Meskwaki
Betula: Iroquois
Cirsium: Iroquois
Cornus: Algonquin, Quebec; Menominee; Ojibwa; Wet'suwet'en
Corydalis: Iroquois
Croton: Keres, Western
Datura: Delaware; Delaware, Oklahoma; Pima
Dimorphocarpa: Navajo, Kayenta
Diospyros: Cherokee
Equisetum: Keres, Western
Eupatorium: Iroquois
Geranium: Meskwaki
Heuchera: Cherokee
Lachnanthes: Cherokee
Lactuca: Iroquois
Malus: Cherokee
Mentha: Cherokee; Iroquois
Mitchella: Cherokee
Napaea: Meskwaki
Oenothera: Cherokee; Iroquois
Orobanche: Zuni
Philadelphus: Thompson
Phytolacca: Rappahannock
Picea: Tanana, Upper
Pinus: Cherokee; Seminole
Plantago: Thompson
Pluchea: Houma
Polygonum: Meskwaki
Prunella: Iroquois
Prunus: Meskwaki
Quercus: Micmac; Penobscot; Seminole
Rhus: Iroquois; Menominee
Rosa: Meskwaki
Rubus: Cherokee
Rumex: Iroquois
Salix: Iroquois
Sanguinaria: Iroquois; Malecite
Sedum: Okanagon; Thompson
Ulmus: Iroquois
Urtica: Thompson
Verbascum: Iroquois
Xanthorhiza: Cherokee

Hemostat
Achillea: Aleut; Cheyenne
Acorus: Cree, Woodlands
Acourtia: Pima
Agrimonia: Meskwaki; Potawatomi
Alnus: Micmac
Anemone: Chippewa; Okanagon; Thompson
Apocynum: Chippewa
Aquilegia: Navajo, Kayenta
Aralia: Chippewa
Artemisia: Blackfoot; Cheyenne; Chippewa; Eskimo, Alaska; Omaha; Shoshoni
Aster: Zuni
Astragalus: Chippewa
Balsamorhiza: Gosiute
Camellia: Makah
Chamaesyce: Navajo, Ramah
Cirsium: Iroquois
Cordylanthus: Navajo
Cornus: Iroquois
Cuscuta: Kawaiisu
Disporum: Okanagan-Colville
Equisetum: Menominee
Erigeron: Navajo, Ramah
Eriophorum: Ojibwa
Eryngium: Natchez
Frangula: Kawaiisu
Gayophytum: Navajo, Ramah
Gentianella: Meskwaki
Geranium: Cherokee; Cheyenne
Gutierrezia: Paiute
Gymnocladus: Omaha
Helianthus: Keres, Western
Heuchera: Blackfoot
Hypericum: Cherokee
Juglans: Iroquois
Juniperus: Zuni
Lactuca: Iroquois
Lappula: Navajo
Lathyrus: Chippewa
Lobelia: Cherokee
Lupinus: Navajo, Kayenta
Lycoperdon: Blackfoot
Lycopodium: Blackfoot; Iroquois; Potawatomi
Mahonia: Paiute
Maianthemum: Ojibwa, South; Washo
Mentha: Cree, Woodlands
Mitchella: Iroquois
Monarda: Cherokee; Lakota
Nicotiana: Kawaiisu; Micmac; Navajo, Kayenta
Nuphar: Sioux
Opuntia: Kiowa
Panax: Creek; Ojibwa, South
Penstemon: Navajo, Kayenta
Phoradendron: Zuni
Phytolacca: Micmac
Pinus: Micmac
Plantago: Iroquois
Platanus: Meskwaki
Polygala: Chippewa
Polygonum: Algonquin, Quebec
Populus: Cree, Woodlands; Ojibwa
Prunus: Algonquin, Tête-de-Boule; Gosiute; Kwakiutl; Sioux
Pterospora: Cheyenne
Pycnanthemum: Koasati
Pyrola: Navajo, Kayenta
Quercus: Chippewa; Mahuna

Index of Usages 803

Ranunculus: Meskwaki
Rhus: Cheyenne; Ojibwa; Omaha; Potawatomi
Rosa: Chippewa; Ojibwa
Rumex: Iroquois
Salix: Cheyenne; Chickasaw; Meskwaki; Micmac; Ojibwa; Okanagan-Colville; Potawatomi
Sambucus: Paiute
Sanguinaria: Iroquois; Micmac; Ojibwa
Sanicula: Meskwaki
Sassafras: Iroquois
Scirpus: Cree, Woodlands; Thompson
Sedum: Kuper Island Indian
Silphium: Chippewa; Ojibwa
Solidago: Menominee
Sphaeralcea: Navajo
Symplocarpus: Menominee
Tephrosia: Seminole
Thalictrum: Iroquois
Thuja: Kwakiutl
Trichostema: Tubatulabal
Trifolium: Navajo, Ramah
Tsuga: Chippewa; Makah; Ojibwa
Typha: Iroquois; Mahuna
Ulmus: Ojibwa
Urtica: Abnaki; Bella Coola; Quinault
Valeriana: Menominee; Meskwaki
Veratrum: Haisla
Verbena: Chippewa
Xerophyllum: Blackfoot
Yucca: Blackfoot
Zanthoxylum: Meskwaki

Herbal Steam
Abies: Chippewa
Achillea: Chippewa; Kwakiutl
Agastache: Cheyenne
Anaphalis: Chippewa
Angelica: Kwakiutl
Artemisia: Chippewa; Kawaiisu; Kiowa; Mendocino Indian; Paiute; Shoshoni; Tlingit
Aster: Meskwaki
Callicarpa: Alabama
Calocedrus: Klamath
Carya: Chippewa
Chamaecyparis: Kwakiutl
Chenopodium: Zuni
Cirsium: Delaware, Oklahoma
Clematis: Shasta
Conioselinum: Kwakiutl
Corethrogyne: Kawaiisu
Cornus: Karok
Eriodictyon: Atsugewi
Eriogonum: Thompson
Geum: Thompson
Gutierrezia: Blackfoot
Helenium: Koasati
Helianthus: Shasta
Iris: Penobscot
Juniperus: Cheyenne; Chippewa; Delaware, Oklahoma; Omaha; Ponca; Tewa
Larix: Ojibwa, South

Linaria: Ojibwa
Lindera: Creek
Lomatium: Kwakiutl; Paiute; Shoshoni; Washo
Lonicera: Kwakiutl
Lysichiton: Kwakiutl
Madia: Cheyenne
Maianthemum: Menominee
Malus: Creek
Mentha: Thompson
Mimulus: Kawaiisu
Mitchella: Delaware, Oklahoma
Nicotiana: Kwakiutl
Ochrosia: Hawaiian
Oplopanax: Kwakiutl
Oxytropis: Thompson
Phytolacca: Delaware, Oklahoma
Picradeniopsis: Navajo, Ramah
Pinus: Ojibwa, South
Platanthera: Thompson
Polygonatum: Chippewa
Populus: Choctaw
Prunus: Paiute; Shoshoni
Pseudotsuga: Karok; Pomo, Little Lakes
Ranunculus: Thompson
Rumex: Bella Coola; Karok
Salvia: Paiute
Sambucus: Kawaiisu; Kwakiutl
Silphium: Meskwaki; Omaha; Ponca; Winnebago
Stephanomeria: Cheyenne
Taxus: Chippewa; Menominee
Tetradymia: Navajo, Ramah
Tsuga: Delaware, Oklahoma; Menominee
Urtica: Bella Coola; Paiute; Shuswap
Yucca: Blackfoot; Keres, Western; Pawnee

Hunting Medicine
Acer: Iroquois; Okanagan-Colville
Acorus: Chippewa; Ojibwa
Angelica: Kwakiutl
Antennaria: Navajo, Ramah
Aralia: Ojibwa
Arctostaphylos: Chippewa
Asarum: Meskwaki
Aster: Chippewa; Iroquois; Navajo, Ramah; Ojibwa
Astragalus: Thompson
Atriplex: Zuni
Besseya: Navajo, Ramah
Campanula: Navajo, Ramah
Cardamine: Iroquois
Castilleja: Navajo, Ramah
Cercocarpus: Navajo, Ramah
Chamaebatiaria: Navajo, Ramah
Cicuta: Ojibwa
Conyza: Ojibwa
Corallorrhiza: Iroquois
Cornus: Chippewa
Datura: Cahuilla; Navajo, Ramah; Yavapai
Erigeron: Navajo, Ramah
Erythronium: Cherokee
Eupatorium: Chippewa
Euthamia: Ojibwa

Fraxinus: Iroquois
Helianthus: Navajo, Ramah
Hepatica: Chippewa
Heracleum: Iroquois; Menominee; Ojibwa
Hieracium: Navajo, Ramah; Ojibwa
Humulus: Navajo, Ramah
Ipomopsis: Navajo, Ramah
Limosella: Navajo, Ramah
Lomatium: Kwakiutl
Lonicera: Navajo, Ramah
Maianthemum: Iroquois
Mirabilis: Navajo, Ramah
Monolepis: Navajo, Ramah
Napaea: Meskwaki
Nicotiana: Cahuilla
Oenothera: Navajo, Ramah
Osmorhiza: Iroquois
Panax: Iroquois; Menominee
Penstemon: Zuni
Physocarpus: Okanagan-Colville
Picea: Koyukon; Tsimshian
Pinus: Navajo, Ramah
Platanthera: Thompson
Polygonatum: Iroquois
Polygonum: Ojibwa
Polystichum: Thompson
Prenanthes: Iroquois
Prunella: Ojibwa
Purshia: Navajo, Ramah
Pyrola: Ojibwa
Ranunculus: Navajo, Ramah; Ojibwa
Ratibida: Keres, Western
Rosa: Thompson
Rumex: Ojibwa
Salix: Seminole
Salvia: Cahuilla
Senecio: Navajo, Ramah
Shepherdia: Okanagan-Colville
Sium: Ojibwa
Spiraea: Ojibwa
Spiranthes: Ojibwa
Taenidia: Ojibwa
Thalictrum: Potawatomi
Thamnosma: Kawaiisu
Thermopsis: Navajo, Ramah
Thuja: Iroquois
Tiarella: Iroquois
Trillium: Iroquois
Tsuga: Kwakiutl
Urtica: Makah
Uvularia: Ojibwa
Valeriana: Menominee; Navajo, Ramah; Thompson
Veratrum: Gitksan; Kwakiutl; Okanagan-Colville
Verbesina: Navajo, Ramah
Veronica: Navajo, Ramah

Hypotensive
Acorus: Lakota
Allium: Rappahannock
Alnus: Cherokee
Equisetum: Diegueño

Eriogonum: Mahuna
Helenium: Comanche
Hydrangea: Cherokee
Juglans: Houma
Juniperus: Diegueño; Thompson
Phoradendron: Cherokee
Rubus: Iroquois
Sabal: Houma
Sassafras: Iroquois
Shepherdia: Thompson
Veratrum: Hanaksiala

Internal Medicine
Abies: Nitinaht
Achillea: Hesquiat
Alnus: Nitinaht
Anaphalis: Kwakiutl
Aruncus: Makah
Aster: Navajo, Ramah
Cercocarpus: Kawaiisu
Dirca: Iroquois
Epilobium: Iroquois
Heracleum: Shuswap
Juniperus: Tewa
Ligusticum: Okanagan-Colville
Lomatium: Salish, Coast
Machaeranthera: Navajo, Ramah
Malus: Makah
Paxistima: Thompson
Pedicularis: Meskwaki; Potawatomi
Picea: Algonquin, Quebec; Menominee
Plagiomnium: Oweekeno
Potentilla: Okanagan-Colville
Rhus: Thompson
Rubus: Kwakiutl
Sidalcea: Navajo, Ramah
Taxus: Haihais; Kitasoo; Tsimshian
Tradescantia: Navajo, Ramah
Tsuga: Bella Coola; Makah; Nitinaht

Kidney Aid
Abies: Anticosti; Cherokee
Acalypha: Cherokee
Acer: Micmac; Penobscot
Achillea: Delaware; Delaware, Oklahoma; Mohegan; Paiute; Paiute, Northern
Acorus: Cherokee
Alisma: Iroquois
Allium: Cherokee
Alnus: Cherokee; Cree, Hudson Bay
Angelica: Paiute
Annona: Seminole
Anthemis: Cherokee; Cree
Apium: Mahuna
Apocynum: Cherokee; Meskwaki; Potawatomi
Aquilegia: Iroquois
Arabis: Okanagon; Thompson
Aralia: Algonquin, Quebec; Iroquois; Malecite; Micmac; Penobscot
Arctium: Iroquois
Arctostaphylos: Cherokee; Diegueño; Okanagan-Colville; Okanagon; Thompson
Artemisia: Keres, Western
Aruncus: Makah
Arundinaria: Houma
Asclepias: Cherokee; Iroquois; Natchez
Aster: Iroquois
Baccharis: Costanoan
Baptisia: Meskwaki; Micmac; Penobscot
Bignonia: Choctaw
Brassica: Cherokee
Castilleja: Gitksan
Caulophyllum: Mohegan
Celastrus: Iroquois
Centaurea: Mahuna
Chaenactis: Shoshoni
Chaetopappa: Navajo, Kayenta
Chamaecyparis: Kwakiutl
Chimaphila: Iroquois; Karok; Kutenai; Micmac; Okanagan-Colville; Yurok
Cimicifuga: Micmac; Penobscot
Citrullus: Cherokee; Rappahannock
Clematis: Cherokee; Iroquois; Paiute; Shoshoni
Collinsonia: Iroquois
Comandra: Cherokee
Conocephalum: Nitinaht
Cornus: Iroquois; Shuswap
Croton: Mahuna; Modesse
Cucurbita: Cherokee; Cheyenne
Cupressus: Kawaiisu
Cynoglossum: Iroquois
Cypripedium: Algonquin, Tête-de-Boule; Cherokee
Diplacus: Costanoan
Dirca: Iroquois; Menominee
Disporum: Costanoan
Draba: Navajo, Ramah
Echinocystis: California Indian; Cherokee
Echium: Mohegan
Elymus: Iroquois
Empetrum: Tanana, Upper
Ephedra: Diegueño; Keres, Western; Mahuna; Navajo; Paiute; Shoshoni
Epigaea: Algonquin, Quebec; Cherokee; Iroquois
Epilobium: Iroquois; Miwok
Equisetum: Carrier; Cherokee; Iroquois; Menominee; Ojibwa; Pomo, Kashaya; Potawatomi; Shoshoni
Erigeron: Cherokee
Eriogonum: Navajo, Kayenta
Eryngium: Creek
Euonymus: Iroquois
Eupatorium: Cherokee; Iroquois; Micmac; Penobscot
Fragaria: Cherokee
Frangula: Mendocino Indian
Gaillardia: Okanagan-Colville
Galax: Cherokee
Galium: Menominee; Meskwaki; Micmac; Miwok; Ojibwa; Penobscot
Gaultheria: Delaware; Iroquois; Mohegan; Shinnecock
Gaylussacia: Cherokee
Goodyera: Cherokee
Grindelia: Cree; Shoshoni; Sioux
Gymnocladus: Omaha
Hackelia: Cherokee
Hamamelis: Iroquois
Hedeoma: Nanticoke
Helenium: Koasati
Helianthemum: Cherokee
Humulus: Cherokee
Hydrangea: Delaware
Hypericum: Menominee
Ilex: Cherokee
Impatiens: Creek; Iroquois
Ipomoea: Cherokee; Creek
Ipomopsis: Shoshoni; Washo
Iris: Delaware; Delaware, Oklahoma; Nevada Indian
Jeffersonia: Cherokee
Juniperus: Blackfoot; Cree, Woodlands; Gitksan; Iroquois; Montana Indian; Navajo, Ramah; Okanagon; Paiute; Shoshoni; Tanana, Upper; Thompson
Lactuca: Iroquois
Larix: Anticosti
Ledum: Makah; Malecite; Micmac; Okanagan-Colville
Lepidium: Lakota
Leptodactylon: Navajo, Kayenta
Liatris: Cherokee; Meskwaki
Linum: Navajo, Kayenta
Lithospermum: Gosiute; Zuni
Lonicera: Algonquin, Quebec
Lygodesmia: Blackfoot
Lysimachia: Cherokee
Lythrum: Cherokee
Magnolia: Choctaw
Mahonia: Blackfoot; Kutenai; Montana Indian; Okanagan-Colville; Shoshoni
Maianthemum: Gitksan; Iroquois; Ojibwa
Malva: Mahuna
Manfreda: Catawba
Marrubium: Cahuilla
Mentha: California Indian; Kutenai
Mirabilis: Tewa
Mitchella: Iroquois; Seminole
Monarda: Blackfoot; Creek; Kutenai
Myrica: Delaware; Delaware, Oklahoma; Mohegan
Nemopanthus: Malecite
Nicotiana: Cherokee; Paiute
Oenothera: Navajo, Kayenta
Osmorhiza: Cheyenne
Osmunda: Iroquois
Ostrya: Chippewa
Paeonia: Paiute; Shoshoni
Panax: Cherokee
Paxistima: Okanagan-Colville
Pellaea: Mahuna
Penstemon: Okanagon; Thompson
Persea: Creek
Petroselinum: Cherokee
Phytolacca: Cherokee

Picea: Koyukon; Kwakiutl
Pinus: Cherokee; Delaware, Ontario; Micmac; Rappahannock; Shoshoni
Plantago: Shoshoni
Polygala: Cherokee
Polygonum: Malecite
Populus: Creek
Porteranthus: Cherokee
Prunus: Cherokee; Rappahannock
Pseudotsuga: Okanagon; Thompson
Psorothamnus: Paiute; Shoshoni
Purshia: Shoshoni
Pyrola: Micmac; Penobscot
Rhamnus: Kawaiisu
Ribes: Blackfoot; Iroquois; Omaha
Rorippa: Costanoan
Rosa: Costanoan
Rubus: Iroquois
Rudbeckia: Cherokee
Rumex: Cherokee; Iroquois; Paiute
Sabal: Houma
Salix: Creek
Salvia: Jemez
Sambucus: Cherokee; Iroquois
Sanguinaria: Iroquois
Sanicula: Iroquois; Malecite; Micmac
Sarracenia: Micmac; Penobscot
Satureja: Karok
Saxifraga: Iroquois
Scrophularia: Iroquois
Scutellaria: Cherokee
Senecio: Costanoan; Iroquois
Silphium: Meskwaki
Sinapis: Cherokee
Smallanthus: Iroquois
Smilax: Chippewa
Solidago: Iroquois
Sorbus: Thompson
Spiraea: Okanagan-Colville
Streptopus: Micmac; Penobscot
Suaeda: Paiute; Shoshoni
Tanacetum: Malecite; Micmac
Taraxacum: Iroquois
Tephrosia: Cherokee
Thuja: Quinault
Tradescantia: Cherokee
Trifolium: Cherokee
Tsuga: Cherokee; Malecite; Micmac
Typha: Delaware; Delaware, Oklahoma; Malecite
Ulmus: Iroquois
Vaccinium: Makah
Verbascum: Cherokee
Verbena: Cherokee; Houma
Veronicastrum: Meskwaki
Viburnum: Iroquois
Vitis: Iroquois
Xanthium: Mahuna
Zanthoxylum: Iroquois; Meskwaki
Zea: Cherokee
Zinnia: Keres, Western

Laxative

Abies: Algonquin, Quebec; Cherokee; Crow; Gitksan; Kwakiutl; Malecite; Micmac
Acer: Iroquois
Achillea: Great Basin Indian; Okanagan-Colville
Acorus: Cheyenne
Adenostoma: Cahuilla
Agoseris: Okanagan-Colville
Aleurites: Hawaiian
Alisma: Cree, Woodlands
Alnus: Algonquin, Quebec; Cree, Woodlands; Gitksan
Alocasia: Hawaiian
Ambrosia: Cheyenne
Amelanchier: Blackfoot
Anemopsis: Paiute
Apocynum: Blackfoot; Iroquois
Arctostaphylos: Tanana, Upper
Artemisia: Eskimo, Alaska; Isleta; Navajo, Kayenta; Sanpoil
Asarum: Okanagan-Colville
Asclepias: Cherokee
Aster: Blackfoot; Iroquois
Caesalpinia: Hawaiian
Caltha: Eskimo, Western
Carya: Meskwaki
Ceanothus: Chippewa
Centaurium: Mahuna
Cephalanthus: Seminole
Cercocarpus: Tewa
Chamaesyce: Hawaiian; Pima
Chelone: Cherokee
Chimaphila: Iroquois
Chlorogalum: Wailaki
Cimicifuga: Cherokee
Cirsium: Hopi
Collomia: Okanagan-Colville
Colocasia: Hawaiian
Cornus: Iroquois; Lummi
Crataegus: Blackfoot
Croton: Isleta
Cucurbita: Cheyenne; Hawaiian; Tewa
Datura: Tubatulabal
Diervilla: Chippewa
Digitaria: Hawaiian
Dirca: Algonquin, Quebec; Iroquois
Distichlis: Kawaiisu
Ephedra: Havasupai
Epilobium: Blackfoot; Eskimo, Alaska
Equisetum: Cherokee; Ojibwa
Eriodictyon: Hualapai
Erioneuron: Havasupai
Erythrina: Seminole
Eupatorium: Cherokee; Iroquois; Mahuna
Fomes: Eskimo, Inuktitut
Frangula: Bella Coola; Cahuilla; Costanoan; Cowlitz; Green River Group; Haisla & Hanaksiala; Hesquiat; Kawaiisu; Klallam; Kwakiutl; Lummi; Makah; Nitinaht; Okanagan-Colville; Pomo; Pomo, Kashaya; Quileute; Quinault; Shuswap; Skagit; Skagit, Upper; Squaxin; Swinomish; Thompson; Tolowa; Yurok
Fraxinus: Iroquois; Meskwaki
Galium: Cherokee
Garrya: Kawaiisu
Gentianella: Cherokee
Geranium: Iroquois
Gilia: Zuni
Grindelia: Mendocino Indian
Gymnocladus: Dakota; Omaha; Oto; Ponca; Winnebago
Hepatica: Cherokee
Hibiscus: Hawaiian
Inula: Delaware, Oklahoma
Ipomoea: Cherokee; Hawaiian
Ipomopsis: Okanagan-Colville
Iris: Aleut
Juglans: Iroquois
Juniperus: Hopi; Keres, Western; Tewa
Larix: Algonquin, Tête-de-Boule
Lomatium: Kwakiutl
Lonicera: Okanagan-Colville
Lycopus: Iroquois
Mahonia: Havasupai; Hualapai; Navajo, Ramah; Nitinaht; Thompson
Maianthemum: Meskwaki
Malus: Gitksan; Makah
Matricaria: Aleut
Melilotus: Pomo, Kashaya
Menispermum: Cherokee
Mentzelia: Apache, White Mountain; Zuni
Menyanthes: Aleut
Merremia: Hawaiian
Mitella: Gosiute
Monarda: Iroquois; Ojibwa
Morus: Cherokee
Mucuna: Hawaiian
Nepeta: Iroquois
Odontoglossum: Hawaiian
Oemleria: Makah
Oenanthe: Makah
Oplopanax: Haisla & Hanaksiala; Kwakiutl; Thompson
Opuntia: Apache, Western; Cahuilla; Hawaiian
Osmorhiza: Iroquois
Osteomeles: Hawaiian
Oxytropis: Navajo, Kayenta
Paeonia: Costanoan
Pandanus: Hawaiian
Pectis: Pima
Pediomelum: Meskwaki
Peperomia: Hawaiian
Perideridia: Blackfoot
Persea: Seminole
Peucedanum: Hawaiian
Phlox: Blackfoot
Physocarpus: Bella Coola; Hesquiat; Kwakiutl; Saanich
Phytolacca: Cherokee
Picea: Algonquin, Quebec; Bella Coola; Hanaksiala

Pinus: Cherokee; Iroquois; Kawaiisu
Plantago: Carrier; Costanoan; Navajo
Podophyllum: Cherokee; Delaware; Delaware, Oklahoma; Iroquois
Polypodium: Skagit, Upper
Populus: Haisla; Iroquois
Potentilla: Paiute
Prunus: Kawaiisu; Makah; Quinault; Thompson
Psathyrotes: Paiute; Shoshoni
Pseudotsuga: Bella Coola
Purshia: Havasupai; Kawaiisu; Paiute, Northern; Sanpoil
Pyrola: Blackfoot
Ranunculus: Iroquois
Rhamnus: Kawaiisu; Meskwaki
Rheum: Cherokee
Rhus: Iroquois
Ribes: Bella Coola; Blackfoot; Nitinaht
Ricinus: Pima
Rubus: Iroquois; Okanagan-Colville
Rumex: Cherokee; Haisla
Sagittaria: Iroquois
Salix: Meskwaki; Paiute
Salvia: Cherokee
Sambucus: Algonquin, Quebec; Cahuilla; Diegueño; Hesquiat; Iroquois; Mohegan; Nitinaht; Paiute
Sanguinaria: Iroquois; Ojibwa
Sanicula: Iroquois
Sassafras: Seminole
Scutellaria: Delaware; Delaware, Oklahoma
Sebastiania: Alabama
Sedum: Okanagan-Colville; Okanagon; Thompson
Senecio: Kawaiisu
Senna: Iroquois
Shepherdia: Blackfoot; Carrier; Cree, Woodlands; Thompson
Sida: Hawaiian
Sisyrinchium: Iroquois
Solanum: Isleta
Spiraea: Blackfoot
Symphoricarpos: Thompson
Symphytum: Cherokee
Taraxacum: Delaware; Delaware, Oklahoma; Iroquois
Thamnosma: Havasupai
Touchardia: Hawaiian
Tradescantia: Cherokee
Triosteum: Iroquois
Tsuga: Quinault
Ulmus: Cherokee; Dakota; Omaha; Pawnee; Ponca; Winnebago
Veratrum: Bella Coola; Haisla; Kwakiutl
Verbascum: Iroquois
Verbesina: Seminole
Veronicastrum: Menominee; Meskwaki
Viburnum: Iroquois
Vicia: Costanoan
Waltheria: Hawaiian
Wikstroemia: Hawaiian

Xanthium: Pima
Yucca: Hopi; Navajo

Liver Aid
Achillea: Blackfoot; Delaware; Delaware, Oklahoma; Mohegan
Adiantum: Iroquois
Agrimonia: Iroquois
Aletris: Cherokee
Allium: Cherokee
Apocynum: Iroquois
Aralia: Iroquois
Asplenium: Cherokee
Baptisia: Iroquois
Berberis: Shinnecock
Betula: Delaware, Oklahoma
Brickellia: Keres, Western
Carya: Cherokee
Caulophyllum: Iroquois
Celastrus: Delaware
Chelone: Iroquois
Clethra: Cherokee
Cornus: Blackfoot
Cryptogramma: Thompson
Diospyros: Cherokee
Eupatorium: Iroquois
Fagus: Iroquois
Fragaria: Cherokee
Frangula: Creek; Delaware, Oklahoma; Thompson
Fraxinus: Delaware, Oklahoma
Gaylussacia: Iroquois
Gentiana: Iroquois
Grindelia: Blackfoot
Hedeoma: Nanticoke
Hepatica: Cherokee; Chippewa
Heuchera: Shoshoni; Thompson
Hydrangea: Cherokee; Delaware, Oklahoma
Hydrastis: Iroquois
Impatiens: Iroquois; Malecite; Micmac
Ipomoea: Iroquois
Ipomopsis: Shoshoni
Iris: Cherokee; Cree, Hudson Bay; Delaware; Delaware, Oklahoma
Jeffersonia: Iroquois
Juglans: Delaware, Oklahoma; Iroquois
Ledum: Montagnais
Linum: Shoshoni
Mahonia: Hualapai
Manfreda: Cherokee
Morus: Delaware, Oklahoma
Panax: Cherokee; Iroquois
Parthenocissus: Cherokee
Passiflora: Cherokee
Phytolacca: Iroquois
Pinus: Iroquois
Populus: Blackfoot
Porteranthus: Cherokee
Prunus: Iroquois
Psathyrotes: Paiute; Shoshoni
Pteridium: Iroquois
Purshia: Paiute

Pyrola: Shoshoni
Rhamnus: Kawaiisu
Rhus: Iroquois
Rorippa: Costanoan; Mahuna
Rubus: Iroquois
Rumex: Delaware; Delaware, Oklahoma; Houma; Shoshoni
Sambucus: Choctaw; Delaware; Delaware, Oklahoma; Iroquois; Thompson
Sanguinaria: Iroquois
Sarracenia: Iroquois
Sedum: Iroquois
Sequoia: Houma
Shepherdia: Thompson
Solidago: Houma; Iroquois
Sporobolus: Ojibwa, South
Tanacetum: Iroquois
Thalictrum: Iroquois
Tsuga: Gitksan
Verbena: Houma
Veronicastrum: Cherokee
Viburnum: Iroquois
Vitis: Cherokee
Xanthorhiza: Catawba

Love Medicine
Acer: Karok
Alocasia: Hawaiian
Anemone: Iroquois
Angelica: Lolahnkok
Apocynum: Okanagan-Colville
Aquilegia: Meskwaki; Omaha; Pawnee; Ponca; Thompson
Arbutus: Pomo, Kashaya
Arnica: Okanagan-Colville
Artemisia: Omaha; Winnebago
Aruncus: Kwakiutl
Aster: Iroquois
Caltha: Iroquois
Calystegia: Karok
Cardamine: Iroquois
Castilleja: Menominee; Nevada Indian
Chrysobalanus: Seminole
Chrysophyllum: Seminole
Coeloglossum: Ojibwa
Conium: Klallam
Conyza: Seminole
Corallorhiza: Iroquois
Cuscuta: Pawnee
Cypripedium: Meskwaki
Datura: Costanoan
Delphinium: Thompson
Dicentra: Menominee
Dirca: Iroquois
Disporum: Makah
Dodecatheon: Thompson
Drosera: Kwakiutl
Echinocystis: Menominee
Eriophyllum: Chehalis
Eupatorium: Iroquois; Meskwaki
Filipendula: Meskwaki

Galium: Cowlitz; Iroquois; Karok; Quileute
Geranium: Iroquois; Thompson
Geum: Iroquois; Okanagan-Colville; Thompson
Hackelia: Cherokee
Iodanthus: Meskwaki
Juniperus: Cheyenne
Lilium: Iroquois
Linaria: Iroquois
Liriodendron: Rappahannock
Lobelia: Iroquois; Meskwaki; Pawnee
Lomatium: Dakota; Omaha; Oregon Indian; Pawnee; Ponca; Winnebago
Lonicera: Iroquois; Nootka, Manhousat
Madia: Cheyenne
Malva: Iroquois
Matricaria: Okanagan-Colville
Mentha: Cheyenne
Menziesia: Quinault
Mitchella: Iroquois
Monardella: Karok
Monotropa: Kwakiutl
Myrica: Seminole
Nuphar: Abnaki
Osmorhiza: Iroquois; Songish; Swinomish
Panax: Meskwaki; Pawnee; Seminole
Pedicularis: Menominee; Meskwaki; Ojibwa
Penstemon: Iroquois
Persea: Seminole
Phlox: Meskwaki
Phoradendron: Cherokee
Phytolacca: Iroquois
Platanthera: Kwakiutl; Potawatomi; Thompson
Podophyllum: Delaware, Oklahoma
Polygonum: Iroquois
Populus: Iroquois; Karok
Prenanthes: Iroquois
Quercus: Seminole
Rumex: Iroquois
Sagittaria: Thompson
Salix: Seminole
Sanguinaria: Algonquin; Micmac; Ponca
Sarracenia: Iroquois
Satureja: Karok
Scirpus: Potawatomi
Sideroxylon: Seminole
Solidago: Iroquois
Spiraea: Iroquois
Taraxacum: Iroquois
Thalictrum: Meskwaki; Ponca; Potawatomi
Tradescantia: Navajo, Kayenta
Trillium: Makah
Valeriana: Cree, Woodlands
Viburnum: Iroquois
Vicia: Iroquois; Makah; Quinault
Vitis: Navajo

Miscellaneous Disease Remedy

Abies: Gitksan; Wet'suwet'en
Acalypha: Cherokee
Acer: Cherokee; Micmac; Thompson
Achillea: Abnaki; Iroquois; Lummi; Meskwaki; Miwok; Paiute, Northern; Thompson
Acorus: Iroquois; Lakota; Micmac
Adiantum: Cherokee
Agastache: Miwok
Ageratina: Cherokee
Alisma: Cree, Woodlands
Allium: Blackfoot; Cherokee; Shinnecock
Alnus: Hesquiat; Micmac
Amelanchier: Cree, Woodlands
Anaphalis: Thompson
Anemopsis: Kawaiisu
Angelica: Blackfoot; Cherokee; Iroquois; Washo
Apocynum: Meskwaki
Aralia: Algonquin, Quebec; Iroquois
Arbutus: Cowichan
Arctium: Abnaki
Aristolochia: Cherokee
Armoracia: Iroquois
Artemisia: Blackfoot; Kawaiisu; Navajo, Ramah; Okanagan-Colville; Paiute; Sanpoil; Shoshoni; Tanaina; Tanana, Upper; Thompson
Aruncus: Bella Coola; Lummi; Nitinaht; Thompson
Asarum: Cherokee; Iroquois
Asclepias: Delaware; Navajo, Ramah; Paiute
Aster: Iroquois
Astragalus: Cree, Woodlands; Navajo; Navajo, Kayenta
Atriplex: Havasupai
Aureolaria: Chickasaw
Bignonia: Houma
Brassica: Cherokee; Micmac
Brickellia: Navajo, Ramah
Calla: Gitksan
Carya: Cherokee
Castanea: Cherokee
Ceanothus: Iroquois; Shuswap
Centaurium: Miwok
Cercocarpus: Shoshoni
Chamaebatia: Miwok
Chenopodium: Cherokee; Potawatomi
Chimaphila: Micmac; Nanticoke
Chrysothamnus: Cheyenne; Navajo; Shoshoni
Cicuta: Kwakiutl
Cinna: Iroquois
Cirsium: Navajo, Kayenta; Zuni
Clintonia: Iroquois
Cornus: Cherokee; Houma
Croton: Concow; Jemez; Neeshenam
Cypripedium: Cherokee
Dalea: Meskwaki
Daucus: Delaware; Delaware, Oklahoma; Mohegan
Diphylleia: Cherokee
Dirca: Iroquois
Dodonaea: Hawaiian
Echinacea: Cheyenne; Dakota; Omaha; Pawnee; Ponca; Winnebago
Ericameria: Diegueño; Paiute
Erigeron: Miwok; Navajo, Ramah
Eriodictyon: Round Valley Indian

Erodium: Costanoan
Eupatorium: Cherokee; Houma; Iroquois
Euphorbia: Algonquin, Quebec; Cahuilla
Fragaria: Cherokee; Chippewa
Frangula: Mendocino Indian; Yokia
Gaillardia: Navajo; Thompson
Galium: Choctaw; Meskwaki; Navajo, Ramah
Gamochaeta: Houma
Gaultheria: Algonquin, Tête-de-Boule
Geum: Okanagan-Colville; Thompson
Glechoma: Cherokee
Gleditsia: Cherokee; Creek; Meskwaki
Glycyrrhiza: Lakota
Gnaphalium: Cherokee; Creek
Grindelia: Shoshoni
Gutierrezia: Shoshoni; Tewa
Hamamelis: Iroquois
Hedeoma: Navajo, Ramah
Heliotropium: Shoshoni
Heracleum: Iroquois; Malecite; Meskwaki; Mewuk; Micmac
Holocarpha: Miwok
Holodiscus: Chehalis; Navajo, Ramah
Humulus: Navajo, Ramah
Ilex: Catawba
Impatiens: Cherokee
Inula: Iroquois
Ipomoea: Cherokee; Iroquois
Ipomopsis: Shoshoni
Iris: Micmac; Penobscot
Iva: Navajo, Ramah
Juglans: Cherokee
Juniperus: Arapaho; Cherokee; Dakota; Eskimo, Inupiat; Kutenai; Mahuna; Navajo; Okanagan-Colville; Paiute; Saanich; Shoshoni; Shuswap; Sioux; Thompson
Krascheninnikovia: Navajo
Larrea: Paiute
Ledum: Micmac; Tanana, Upper
Lepechinia: Miwok
Leptarrhena: Aleut
Lewisia: Okanagan-Colville
Lindera: Cherokee; Iroquois
Liriodendron: Cherokee
Lobelia: Delaware; Delaware, Oklahoma
Lomatium: Great Basin Indian; Okanagan-Colville; Paiute; Shoshoni; Washo
Lonicera: Cree, Woodlands
Lophophora: Kiowa
Lupinus: Blackfoot
Lycium: Navajo, Ramah
Lygodesmia: Cheyenne
Lysichiton: Gitksan; Yurok
Mahonia: Karok; Miwok
Maianthemum: Meskwaki
Malus: Creek; Meskwaki
Marrubium: Navajo, Ramah
Marshallia: Catawba
Matelea: Comanche
Melissa: Cherokee

Mentha: Cherokee; Iroquois; Navajo, Ramah; Thompson
Mentzelia: Cheyenne
Menyanthes: Kwakiutl
Mertensia: Cheyenne
Mitchella: Iroquois
Modiola: Houma
Monarda: Cherokee; Flathead; Navajo, Ramah
Nicotiana: Cherokee; Paiute
Nuphar: Iroquois
Nymphaea: Micmac
Oenothera: Navajo, Kayenta
Oplopanax: Gitksan; Haisla; Thompson; Wet'suwet'en
Osmorhiza: Paiute; Shoshoni; Washo
Oxytropis: Navajo
Panax: Cherokee; Iroquois
Panicum: Creek; Natchez
Parthenocissus: Houma
Peniocereus: Pima
Penstemon: Okanagan-Colville
Pericome: Navajo, Ramah
Petasites: Concow; Tanaina
Peteria: Navajo, Ramah
Picea: Cherokee; Gitksan; Micmac; Wet'suwet'en
Pinus: Cherokee; Iroquois; Micmac; Navajo, Ramah; Paiute; Salish, Coast; Shoshoni; Thompson
Piper: Hawaiian
Platanus: Cherokee; Meskwaki
Polygonatum: Chippewa
Polypodium: Cowlitz; Iroquois
Populus: Iroquois
Porteranthus: Iroquois
Portulaca: Navajo, Kayenta
Prunella: Catawba; Iroquois
Prunus: Cherokee; Chippewa; Iroquois; Micmac; Paiute; Thompson
Psoralidium: Navajo, Ramah
Psorothamnus: Paiute; Paiute, Northern; Shoshoni
Pteridium: Cherokee
Purshia: Paiute; Shoshoni
Pycnanthemum: Meskwaki
Pyrus: Cree, Hudson Bay
Quercus: Iroquois
Quincula: Kiowa
Ranunculus: Iroquois
Rhus: Blackfoot; Iroquois; Kiowa; Montana Indian; Potawatomi; Round Valley Indian; Yokia
Ribes: Cherokee
Rosa: Keres, Western; Paiute
Rubus: Chippewa; Montana Indian; Thompson
Rumex: Choctaw; Hawaiian; Iroquois; Kawaiisu; Paiute
Salix: Creek; Mewuk; Paiute
Salsola: Navajo, Ramah
Salvia: Diegueño; Paiute; Rappahannock
Sambucus: Cahuilla; Iroquois; Kawaiisu; Mahuna; Miwok
Sanguinaria: Potawatomi
Sarracenia: Micmac; Montagnais

Sassafras: Cherokee; Choctaw; Houma; Nanticoke; Rappahannock
Scutellaria: Iroquois
Senna: Cherokee; Houma
Sesamum: Cherokee
Sinapis: Cherokee
Sisyrinchium: Iroquois
Solanum: Costanoan
Solidago: Cherokee; Okanagan-Colville; Thompson
Sorbus: Potawatomi
Sphaeralcea: Navajo
Stanleya: Paiute
Stephanomeria: Cheyenne; Navajo, Kayenta
Strophostyles: Houma
Suaeda: Paiute
Symplocarpus: Delaware
Taxus: Micmac
Tetradymia: Paiute
Thamnosma: Paiute
Thuja: Chippewa; Thompson
Toxicodendron: Cherokee
Trichostema: Concow; Miwok
Tsuga: Iroquois; Micmac; Thompson
Typha: Iroquois; Sioux
Urtica: Cherokee; Paiute; Tanaina
Vaccinium: Pomo, Kashaya
Valeriana: Navajo, Ramah; Thompson
Veratrum: Thompson
Verbascum: Cherokee; Iroquois
Verbena: Costanoan
Veronicastrum: Cherokee; Meskwaki
Viburnum: Cherokee; Delaware, Ontario; Micmac; Penobscot
Vicia: Cherokee
Vitis: Chippewa
Wyethia: Shoshoni
Yucca: Cherokee

Narcotic

Arctostaphylos: Kwakiutl; Ojibwa
Broussonetia: Apache, Chiricahua & Mescalero
Comandra: Navajo, Kayenta
Cornus: Gosiute
Crotalaria: Delaware; Delaware, Oklahoma
Datura: Apache, White Mountain; Havasupai; Hopi; Luiseño; Mahuna; Navajo, Kayenta; Navajo, Ramah; Paiute; Shoshoni; Ute; Yuma; Zuni
Delphinium: Mendocino Indian
Eschscholzia: Mendocino Indian
Ledum: Kwakiutl
Lophophora: Comanche; Kiowa
Nicotiana: Navajo, Kayenta
Opuntia: Apache, Chiricahua & Mescalero
Petrophyton: Navajo, Kayenta
Selaginella: Blackfoot
Stephanomeria: Navajo, Kayenta

Nose Medicine

Arisaema: Iroquois
Aristolochia: Cherokee
Artemisia: Cheyenne; Havasupai
Asclepias: Navajo

Atriplex: Cahuilla; Navajo, Ramah
Chaetopappa: Hopi; Navajo, Ramah
Chimaphila: Abnaki
Cordyline: Hawaiian
Curcuma: Hawaiian
Dimorphocarpa: Keres, Western
Draba: Blackfoot
Evolvulus: Navajo, Kayenta
Gaillardia: Blackfoot
Geranium: Cheyenne
Helenium: Cherokee
Helianthus: Cherokee
Heterotheca: Navajo, Ramah
Kalmia: Abnaki
Ledum: Abnaki
Machaeranthera: Navajo
Nicotiana: Tewa
Osmorhiza: Blackfoot
Polygonum: Navajo, Ramah
Pterospora: Cheyenne
Rosa: Crow
Sanguinaria: Cherokee
Senecio: Yavapai
Sisyrinchium: Navajo
Solanum: Navajo
Solidago: Blackfoot
Stenotus: Navajo
Townsendia: Navajo
Veratrum: Blackfoot; Flathead; Kutenai
Zinnia: Navajo

Oral Aid

Abies: Blackfoot; Flathead; Kwakiutl
Abronia: Navajo, Ramah
Achillea: Cree, Woodlands
Acorus: Shinnecock
Alnus: Cherokee; Malecite; Micmac
Angelica: Cherokee; Pomo, Kashaya
Apocynum: Ojibwa
Aralia: Cree, Woodlands
Arctostaphylos: Blackfoot; Crow; Thompson
Armoracia: Cherokee
Artemisia: Blackfoot; Okanagan-Colville; Tanana, Upper
Artocarpus: Hawaiian
Asclepias: Blackfoot
Asplenium: Hawaiian
Balsamorhiza: Cheyenne
Berberis: Micmac; Penobscot
Bothriochloa: Kiowa
Carya: Cherokee
Ceanothus: Iroquois; Keres, Western
Chamaesyce: Apache, White Mountain; Hopi
Chrysothamnus: Jemez; Tewa
Cirsium: Iroquois; Kwakiutl; Mohegan
Clitoria: Cherokee
Comandra: Navajo, Kayenta
Coptis: Iroquois; Malecite; Menominee; Micmac; Mohegan; Ojibwa; Penobscot; Potawatomi
Cornus: Iroquois; Meskwaki
Crataegus: Okanagan-Colville

Curcuma: Hawaiian
Delphinium: Blackfoot
Desmodium: Cherokee
Diospyros: Cherokee; Rappahannock
Dodecatheon: Blackfoot
Echinacea: Cheyenne; Lakota; Montana Indian
Echinocactus: Mahuna
Equisetum: Tolowa
Eriodictyon: Cahuilla
Eriogonum: Apache, White Mountain; Lakota; Mahuna; Navajo, Ramah; Zuni
Euphorbia: Cahuilla
Fragaria: Iroquois; Okanagan-Colville
Gaultheria: Cherokee; Makah
Geranium: Cherokee; Chippewa; Iroquois; Meskwaki; Ojibwa; Salish
Geum: Blackfoot
Glycyrrhiza: Blackfoot; Zuni
Gnaphalium: Cherokee
Goodyera: Mohegan
Helianthus: Kiowa
Heuchera: Blackfoot; Cherokee; Chippewa; Navajo; Thompson
Hieracium: Thompson
Holodiscus: Lummi
Hydrophyllum: Iroquois
Iris: Yokia
Juglans: Iroquois
Juncus: Cherokee
Juniperus: Kiowa; Shoshoni
Lachnanthes: Cherokee
Larrea: Pima
Lomatium: Okanagan-Colville
Lonicera: Quinault
Mahonia: Cowlitz; Hopi
Malus: Cherokee; Oweekeno
Mentha: Cree, Woodlands
Menziesia: Hesquiat
Mirabilis: Navajo; Pawnee
Nepeta: Iroquois
Nymphaea: Chippewa
Oxalis: Cherokee; Iroquois; Kiowa
Oxydendrum: Cherokee
Panax: Cherokee
Persea: Mahuna; Seminole
Phlox: Navajo, Ramah
Picea: Cree, Woodlands; Haisla & Hanaksiala; Tanana, Upper
Piper: Hawaiian
Pityopsis: Choctaw
Pleopeltis: Houma
Polygala: Cree, Woodlands
Polygonum: Cree, Woodlands; Meskwaki
Polypodium: Bella Coola; Hesquiat; Thompson
Populus: Haisla & Hanaksiala; Iroquois
Portulaca: Keres, Western
Potentilla: Cherokee
Prosopis: Pima
Prunus: Cherokee; Cheyenne; Haisla & Hanaksiala; Kwakiutl; Meskwaki

Pseudotsuga: Swinomish; Thompson
Pyrola: Mohegan
Quercus: Cherokee; Keres, Western
Ranunculus: Cherokee
Rhus: Chippewa; Delaware; Delaware, Oklahoma; Iroquois; Jemez; Keres, Western; Ojibwa; Okanagon; Sanpoil; Thompson
Ribes: Kwakiutl
Rosa: Crow
Rubus: Cherokee
Rumex: Navajo, Ramah; Pima
Salix: Catawba; Crow; Eskimo, Alaska; Eskimo, Inupiat; Eskimo, Kuskokwagmiut; Eskimo, Nunivak; Eskimo, Western; Iroquois; Seminole; Tanana, Upper
Sanvitalia: Navajo; Navajo, Ramah
Sassafras: Cherokee; Seminole
Schkuhria: Navajo, Ramah
Schoenocrambe: Navajo, Ramah
Scirpus: Cherokee
Sedum: Eskimo, Alaska; Thompson
Sempervivum: Mahuna
Solanum: Cherokee
Solidago: Cherokee
Sphaeralcea: Navajo, Kayenta
Stenotaphrum: Hawaiian
Stenotus: Navajo
Symphoricarpos: Skagit
Syringa: Iroquois
Syzygium: Hawaiian
Thuja: Skokomish
Tiarella: Cherokee; Iroquois
Toxicodendron: Iroquois
Trollius: Cherokee
Ulmus: Kiowa
Urtica: Makah
Veratrum: Blackfoot
Verbesina: Seminole
Viburnum: Cherokee
Vitis: Cherokee
Xanthium: Paiute, Northern
Xanthorhiza: Cherokee
Ximenia: Seminole

Orthopedic Aid
Abies: Iroquois; Micmac; Thompson
Acer: Iroquois; Micmac; Seminole
Achillea: Carrier, Southern; Malecite; Mendocino Indian; Micmac; Paiute; Shoshoni; Thompson
Acorus: Cree, Woodlands; Lakota
Adiantum: Iroquois
Aesculus: Cherokee
Agoseris: Navajo, Ramah
Alisma: Iroquois
Alnus: Cherokee; Cowlitz; Micmac; Mohegan; Nitinaht
Ambrosia: Costanoan; Iroquois
Anaphalis: Chippewa
Andropogon: Catawba
Anemone: Ojibwa, South
Anemopsis: Paiute

Angelica: Costanoan; Creek; Iroquois
Arabis: Okanagon; Thompson
Aralia: Cherokee; Chippewa; Iroquois
Arctium: Iroquois
Arctostaphylos: Cheyenne; Thompson
Arisaema: Cherokee; Iroquois; Micmac; Mohegan; Penobscot
Arnica: Catawba
Artemisia: Aleut; Carrier, Southern; Chippewa; Eskimo, Kuskokwagmiut; Eskimo, Western; Hopi; Klamath; Montana Indian; Okanagan-Colville; Okanagon; Paiute; Shoshoni; Thompson; Tolowa; Yuki; Yurok
Aruncus: Cherokee; Thompson
Arundo: Cahuilla
Asarum: Chippewa
Asclepias: Iroquois; Menominee
Aster: Cree, Woodlands
Astragalus: Navajo, Kayenta
Baptisia: Iroquois; Nanticoke
Betula: Cree, Woodlands; Iroquois; Tanana, Upper
Blechnum: Quileute
Brassica: Cherokee
Calla: Cree, Woodlands
Carnegia: Pima
Carya: Cherokee
Castilleja: Chippewa; Gitksan
Ceanothus: Alabama; Okanagan-Colville; Thompson
Celastrus: Creek
Cephalanthus: Koasati
Chaenactis: Gosiute; Kawaiisu; Paiute
Chamaebatiaria: Paiute
Chenopodium: Costanoan
Chimaphila: Catawba; Cree, Woodlands; Karok; Thompson
Cicuta: Iroquois; Kawaiisu; Paiute; Shoshoni; Thompson
Cimicifuga: Iroquois
Clematis: Cherokee; Navajo, Ramah; Shoshoni
Comptonia: Micmac
Conyza: Hawaiian
Cordylanthus: Navajo
Coreopsis: Meskwaki
Cornus: Carrier, Northern; Iroquois; Montagnais
Coronilla: Cherokee
Corydalis: Navajo, Ramah
Crepis: Okanagan-Colville
Cupressus: Kawaiisu
Cymopterus: Navajo, Kayenta
Cypripedium: Iroquois
Datura: Kawaiisu; Navajo, Kayenta
Delphinium: Kawaiisu
Dirca: Iroquois
Echinacea: Cheyenne
Elytrigia: Cherokee
Ephedra: Kawaiisu
Epilobium: Iroquois; Navajo, Kayenta
Equisetum: Blackfoot; Iroquois; Navajo, Kayenta; Okanagan-Colville; Potawatomi

Ericameria: Coahuilla; Kawaiisu; Miwok
Erigeron: Cheyenne; Thompson
Eriodictyon: Coahuilla; Hualapai; Miwok
Eriogonum: Hopi; Navajo, Kayenta; Paiute; Shoshoni; Thompson
Eryngium: Seminole
Eucalyptus: Hawaiian
Euonymus: Cherokee
Eupatorium: Abnaki; Iroquois
Fouquieria: Hualapai
Gaillardia: Blackfoot; Okanagan-Colville
Galium: Iroquois
Gentiana: Iroquois
Geum: Thompson
Glyceria: Catawba
Gnaphalium: Cherokee; Iroquois
Grindelia: Kawaiisu; Shoshoni
Gutierrezia: Kawaiisu; Paiute
Hackelia: Navajo, Ramah
Hamamelis: Iroquois; Menominee; Potawatomi
Helianthella: Paiute
Hepatica: Iroquois
Heracleum: Aleut; Bella Coola; Okanagan-Colville; Quinault
Heuchera: Cheyenne; Navajo, Ramah
Hexastylis: Catawba
Holodiscus: Lummi
Hydrangea: Cherokee
Hypericum: Alabama; Paiute; Shoshoni
Ilex: Cherokee
Impatiens: Mohegan; Nanticoke; Penobscot; Potawatomi
Ipomoea: Hawaiian
Iris: Iroquois
Isocoma: Pima
Juncus: Cherokee
Juniperus: Blackfoot; Hopi; Micmac; Paiute; Seminole; Tewa
Kalmia: Cherokee; Malecite; Micmac
Krascheninnikovia: Hopi
Lactuca: Iroquois
Larix: Thompson
Larrea: Diegueño; Kawaiisu; Papago
Lathyrus: Mendocino Indian
Ledum: Cree, Hudson Bay
Linnaea: Montagnais
Liriodendron: Cherokee
Lithospermum: Cheyenne
Lobelia: Cherokee
Lomatium: Gosiute; Kawaiisu; Kwakiutl; Okanagan-Colville; Paiute; Shoshoni; Thompson
Lonicera: Carrier, Northern; Costanoan; Kwakiutl; Thompson
Lophophora: Kiowa
Lygodesmia: Blackfoot
Lysichiton: Shuswap
Maianthemum: Gitksan; Paiute
Malus: Iroquois; Makah
Malva: Iroquois
Marah: Squaxin

Matelea: Comanche
Matteuccia: Cree, Woodlands
Menodora: Navajo, Ramah
Merremia: Hawaiian
Mimulus: Kawaiisu
Mirabilis: Chippewa; Navajo, Ramah; Ojibwa; Sioux, Teton
Mitchella: Iroquois
Moneses: Montagnais
Morinda: Hawaiian
Nereocystis: Kwakiutl, Southern
Nuphar: Kwakiutl; Micmac; Shuswap
Nymphaea: Micmac
Oenothera: Iroquois; Navajo, Ramah
Onosmodium: Cheyenne
Oplopanax: Gitksan; Nitinaht
Orbexilum: Catawba
Osmunda: Iroquois
Ostrya: Cherokee
Oxytropis: Thompson
Parthenocissus: Iroquois
Paxistima: Thompson
Pedicularis: Iroquois
Pediomelum: Blackfoot
Penstemon: Kawaiisu; Navajo, Kayenta; Okanogon; Shoshoni; Thompson
Persea: Seminole
Phoradendron: Houma
Phragmites: Cahuilla
Phyla: Houma
Physaria: Blackfoot
Phytolacca: Iroquois
Pinus: Carrier, Northern; Cherokee; Iroquois; Nanticoke; Okanogon; Paiute; Seminole; Shoshoni; Thompson
Plantago: Cherokee; Iroquois; Ojibwa
Platanthera: Thompson
Polygala: Iroquois
Polygonatum: Rappahannock
Polygonum: Houma; Iroquois
Polystichum: Iroquois
Populus: Bella Coola; Chippewa; Costanoan; Creek; Diegueño; Isleta; Kawaiisu; Nanticoke; Ojibwa; Thompson
Porphyra: Hanaksiala
Porteranthus: Cherokee
Prunella: Iroquois
Prunus: Koasati; Thompson
Pseudotsuga: Isleta; Thompson
Psidium: Hawaiian
Psorothamnus: Paiute
Pteridium: Montagnais; Seminole; Thompson
Pterocaulon: Seminole
Pyrola: Iroquois
Quercus: Alabama; Creek; Houma; Iroquois; Mohegan; Ojibwa; Seminole; Shinnecock
Ranunculus: Iroquois; Thompson
Rhamnus: Iroquois
Rhus: Iroquois; Koasati
Ribes: Carrier, Northern; Iroquois; Paiute; Shoshoni

Rosa: Carrier, Northern; Potawatomi
Rubus: Carrier, Northern
Rumex: Iroquois; Kawaiisu
Ruta: Cherokee; Costanoan
Salix: Micmac; Okanagan-Colville; Paiute; Seminole; Thompson
Salvia: Costanoan
Sambucus: Cowlitz; Kwakiutl; Mendocino Indian
Sanicula: Cherokee
Sarracenia: Cree, Woodlands
Sassafras: Iroquois
Scirpus: Houma
Senecio: Hopi; Iroquois; Zuni
Shepherdia: Algonquin, Quebec
Silphium: Chippewa; Iroquois; Meskwaki
Sinapis: Cherokee
Sium: Iroquois
Smallanthus: Iroquois
Smilax: Cherokee; Chippewa
Solanum: Kawaiisu
Solidago: Chippewa; Thompson
Sorbus: Cree, Woodlands
Sphaeralcea: Hopi
Spiraea: Iroquois
Symphytum: Cherokee
Tanacetum: Cherokee; Iroquois
Taraxacum: Iroquois; Tewa
Tauschia: Kawaiisu
Taxus: Iroquois
Tephrosia: Cherokee
Tetradymia: Navajo, Ramah
Tetraneuris: Hopi
Thalictrum: Gitksan; Yuki
Thelypteris: Seminole
Thuja: Iroquois; Kwakiutl; Micmac
Tiarella: Iroquois
Tilia: Iroquois
Trillium: Iroquois
Triosteum: Cherokee
Tsuga: Iroquois; Nitinaht
Typha: Iroquois
Ulmus: Iroquois; Mahuna
Urtica: Bella Coola; Costanoan; Cowlitz; Kawaiisu; Klallam; Quinault; Thompson
Uvularia: Iroquois; Potawatomi
Veratrum: Bella Coola; Cherokee; Kwakiutl; Okanagan-Colville; Paiute, Northern; Thompson
Verbascum: Catawba; Iroquois; Rappahannock
Viburnum: Iroquois
Vicia: Cherokee
Wyethia: Gosiute; Paiute
Xanthium: Mahuna
Xerophyllum: Blackfoot
Ximenia: Seminole
Yucca: Blackfoot; Nanticoke
Zanthoxylum: Chippewa; Houma; Iroquois
Zigadenus: Mendocino Indian; Paiute; Shoshoni; Shuswap; Thompson; Washo
Zingiber: Hawaiian

Other

Abies: Gitksan; Okanagan-Colville
Achillea: Makah; Swinomish
Adiantum: Cherokee; Micmac
Agrimonia: Iroquois
Alisma: Iroquois
Ambrosia: Makah
Anaphalis: Nitinaht
Anemopsis: Pima
Angadenia: Seminole
Angelica: Blackfoot; Iroquois; Pomo, Kashaya
Anthemis: Iroquois
Apocynum: Chippewa; Navajo, Kayenta
Artemisia: Kawaiisu; Keres, Western; Mewuk
Asclepias: Costanoan; Iroquois
Aster: Pawnee; Seminole
Astragalus: Navajo, Kayenta
Atriplex: Navajo, Ramah
Aureolaria: Cherokee
Bacopa: Seminole
Berchemia: Seminole
Betula: Micmac
Botrychium: Abnaki
Brickellia: Navajo, Ramah
Callicarpa: Choctaw
Camissonia: Navajo, Kayenta
Cardamine: Iroquois
Carex: Iroquois
Carpinus: Iroquois
Ceanothus: Great Basin Indian; Iroquois
Celastrus: Iroquois
Cephalanthus: Seminole
Chamaesaracha: Navajo, Kayenta
Chrysothamnus: Thompson
Circaea: Iroquois
Clematis: Seminole; Thompson
Cocos: Hawaiian
Commelina: Seminole
Coreopsis: Seminole
Cornus: Okanagan-Colville
Corydalis: Navajo, Ramah
Cucurbita: Cahuilla
Datura: Cahuilla; Hopi; Paiute; Pima; Yokut
Diervilla: Potawatomi
Draba: Keres, Western
Dyssodia: Keres, Western
Echinacea: Dakota; Lakota; Omaha; Pawnee; Ponca; Winnebago
Echinocactus: Omaha
Elymus: Iroquois; Karok
Ephedra: Cahuilla; Pima
Epilobium: Swinomish
Epipactis: Navajo, Kayenta
Equisetum: Iroquois
Erigeron: Cheyenne
Eriogonum: Navajo, Ramah; Thompson; Zuni
Euonymus: Cherokee; Iroquois
Eupatorium: Cherokee; Iroquois
Festuca: Iroquois
Fraxinus: Iroquois
Gaillardia: Navajo, Kayenta
Gaultheria: Skagit, Upper
Geranium: Blackfoot; Navajo, Kayenta
Glycyrrhiza: Paiute
Gymnocarpium: Abnaki
Hamamelis: Menominee
Helianthus: Hopi; Navajo, Ramah
Hepatica: Iroquois; Meskwaki; Potawatomi
Hymenopappus: Navajo, Kayenta
Ilex: Iroquois
Ipomoea: Iroquois
Juniperus: Algonquin; Comanche; Hopi; Meskwaki; Okanagan-Colville; Rappahannock; Seminole; Thompson
Larix: Menominee; Potawatomi
Lepidium: Navajo, Kayenta
Lespedeza: Iroquois
Leucothoe: Cherokee
Licania: Seminole
Ligusticum: Okanagan-Colville
Liquidambar: Cherokee; Koasati
Lobelia: Cherokee; Iroquois
Lomatium: Great Basin Indian; Okanagan-Colville; Yuki
Luzula: Iroquois
Lycopus: Cherokee
Lysichiton: Kwakiutl; Samish; Swinomish; Tolowa
Mahonia: Karok; Keres, Western
Maianthemum: Iroquois
Malva: Costanoan; Navajo, Ramah
Mentha: Iroquois; Paiute
Monarda: Iroquois
Myriophyllum: Iroquois
Nepeta: Iroquois; Ojibwa
Nicotiana: Cherokee; Kawaiisu
Nuphar: Iroquois; Nitinaht
Nymphoides: Seminole
Nyssa: Cherokee
Oplopanax: Gitksan
Orobanche: Navajo
Osmunda: Seminole
Oxalis: Iroquois; Makah
Panax: Cherokee
Parthenocissus: Iroquois
Pellaea: Miwok
Pentaphylloides: Cheyenne
Peperomia: Hawaiian
Persea: Seminole
Phlebodium: Seminole
Phragmites: Iroquois
Phytolacca: Cherokee
Picea: Kwakiutl
Pinus: Cherokee; Hualapai; Iroquois; Navajo, Ramah
Platanus: Cherokee
Polygala: Cherokee; Iroquois; Seminole
Polypodium: Seminole
Populus: Flathead
Prenanthes: Choctaw
Prosopis: Pima
Pterocaulon: Seminole
Purshia: Shoshoni
Quercus: Iroquois
Ranunculus: Hesquiat
Rhododendron: Cherokee
Rhus: Creek; Kiowa; Okanagan-Colville
Ribes: Iroquois; Salish, Coast; Thompson
Rubus: Hesquiat; Iroquois; Pomo, Kashaya; Seminole
Salix: Creek; Seminole
Salvia: Diegueño; Hopi; Shoshoni
Sambucus: Cherokee
Sanguinaria: Iroquois
Sassafras: Seminole
Saururus: Seminole
Scirpus: Clallam
Sedum: Iroquois
Senecio: Costanoan
Silene: Keres, Western
Sisyrinchium: Iroquois
Smallanthus: Cherokee
Smilax: Cherokee; Creek; Seminole
Solanum: Iroquois
Solidago: Iroquois
Sonchus: Pima
Sphaeralcea: Navajo, Kayenta
Spiraea: Okanagan-Colville
Stillingia: Seminole
Tagetes: Navajo
Tanacetum: Cheyenne
Taxus: Chippewa
Tetraneuris: Navajo, Ramah
Thalictrum: Iroquois
Thuja: Bella Coola
Tiarella: Iroquois
Tilia: Iroquois
Trillium: Wailaki; Yuki
Tripterocalyx: Navajo, Ramah
Tsuga: Gitksan; Hesquiat
Typha: Ojibwa
Ulmus: Iroquois
Urtica: Kwakiutl
Vaccinium: Iroquois
Valeriana: Gosiute
Veratrum: Gitksan
Verbena: Iroquois
Vitis: Cherokee; Iroquois
Vittaria: Seminole
Zanthoxylum: Menominee
Zinnia: Keres, Western

Panacea

Abies: Abnaki; Shuswap; Thompson
Abronia: Navajo, Kayenta
Acer: Navajo, Ramah
Achillea: Blackfoot; Iroquois; Kwakiutl; Nitinaht; Quileute; Thompson; Ute
Acorus: Cheyenne; Cree, Woodlands; Dakota; Micmac; Mohegan; Omaha; Pawnee; Ponca; Sioux, Fort Peck; Winnebago
Ageratina: Iroquois

Agoseris: Navajo, Ramah
Alisma: Cree, Woodlands
Alnus: Rappahannock
Amelanchier: Navajo, Ramah
Androsace: Navajo, Ramah
Anemone: Carrier, Southern; Omaha; Ponca
Angelica: Eskimo; Yana
Anthemis: Mohegan
Apocynum: Meskwaki
Aquilegia: Gosiute; Paiute
Arabis: Cheyenne; Navajo, Ramah
Aralia: Cree, Woodlands; Pomo
Arctostaphylos: Ojibwa
Arenaria: Navajo, Ramah
Argythamnia: Navajo; Navajo, Ramah
Artemisia: Gosiute; Karok; Mewuk; Navajo, Ramah; Shuswap; Thompson
Asarum: Montagnais
Asclepias: Pima
Astragalus: Navajo, Ramah
Baccharis: Costanoan
Balsamorhiza: Cheyenne
Besseya: Navajo, Ramah
Blechnum: Quileute
Bouteloua: Navajo, Ramah
Bursera: Cahuilla
Calliandra: Navajo, Ramah
Calochortus: Navajo, Ramah
Calylophus: Navajo, Ramah
Cardamine: Iroquois
Carya: Meskwaki
Ceanothus: Thompson
Cercocarpus: Navajo, Ramah
Chaetopappa: Hopi
Chamaecyparis: Kwakiutl
Cheilanthes: Navajo, Ramah
Chenopodium: Creek
Chrysothamnus: Thompson
Cicuta: Thompson
Cirsium: Navajo, Kayenta; Navajo, Ramah
Clematis: Meskwaki
Collinsonia: Iroquois
Coreopsis: Navajo, Ramah
Cornus: Iroquois; Okanagan-Colville; Thompson
Crataegus: Thompson
Cryptantha: Navajo, Ramah
Cucurbita: Dakota; Omaha; Pawnee; Ponca; Winnebago
Cymopterus: Navajo, Ramah
Cypripedium: Rappahannock
Dalea: Navajo, Kayenta; Navajo, Ramah; Pawnee
Draba: Navajo, Ramah
Dracocephalum: Navajo, Kayenta
Echinocystis: Menominee; Meskwaki
Epilobium: Costanoan; Iroquois
Erigeron: Navajo, Ramah
Eriogonum: Navajo, Kayenta; Navajo, Ramah
Eriophorum: Eskimo, Kuskokwagmiut
Eryngium: Creek; Seminole
Eupatorium: Mohegan
Forestiera: Houma
Fragaria: Navajo, Ramah
Frangula: Quileute; West Coast Indian
Frasera: Navajo, Kayenta; Shoshoni
Fraxinus: Meskwaki
Gaura: Navajo, Ramah
Geranium: Navajo, Ramah
Geum: Carrier, Southern; Cree, Woodlands; Iroquois; Quinault
Glandularia: Navajo, Ramah
Gleditsia: Creek
Gnaphalium: Navajo, Ramah
Gutierrezia: Navajo, Ramah
Hamamelis: Iroquois
Helianthella: Navajo, Ramah
Helianthus: Navajo, Ramah
Heliomeris: Navajo, Ramah
Heterotheca: Navajo, Ramah
Heuchera: Meskwaki; Navajo, Ramah
Houstonia: Navajo, Ramah
Hymenopappus: Navajo, Ramah
Inula: Iroquois
Ipomopsis: Navajo, Ramah
Iris: Micmac; Rappahannock
Juniperus: Hualapai; Shuswap; Swinomish; Tanana, Upper
Kalmia: Cherokee; Micmac; Penobscot
Krascheninnikovia: Navajo, Ramah
Larix: Thompson
Larrea: Paiute; Pima
Lathyrus: Costanoan
Ledum: Tanana, Upper
Ligusticum: Atsugewi; Menominee
Linanthus: Navajo, Ramah
Lindera: Iroquois
Linum: Navajo, Ramah
Lithospermum: Navajo, Ramah
Lobelia: Iroquois
Lomatium: Blackfoot; Kwakiutl; Paiute; Paiute, Northern; Shoshoni; Washo
Lophophora: Delaware; Kiowa
Lotus: Navajo, Ramah
Lupinus: Navajo, Ramah
Lycium: Navajo, Ramah
Lysichiton: Quinault; Shuswap
Mahonia: Karok; Navajo, Kayenta
Malus: Cowichan; Makah; Nitinaht; Saanich
Matricaria: Aleut
Medeola: Iroquois
Menodora: Navajo, Ramah
Mentha: Abnaki; Sanpoil
Mirabilis: Navajo, Kayenta
Morus: Meskwaki
Myosurus: Navajo, Ramah
Nemopanthus: Potawatomi
Nuphar: Cree, Woodlands
Oenothera: Navajo, Ramah
Oplopanax: Oweekeno; Thompson
Orobanche: Navajo, Ramah
Osmorhiza: Karok; Meskwaki
Osmunda: Iroquois
Panax: Delaware, Oklahoma; Iroquois; Meskwaki; Mohegan
Penstemon: Navajo, Ramah
Perideridia: Blackfoot
Persea: Seminole
Petasites: Karok
Phoradendron: Houma
Physalis: Navajo, Ramah
Picea: Eskimo, Nunivak; Koyukon; Shuswap; Thompson
Picradeniopsis: Navajo, Ramah
Pinus: Iroquois; Shuswap; Thompson
Plantago: Navajo, Ramah
Platanthera: Iroquois
Platanus: Costanoan
Polygala: Cree, Woodlands
Polygonum: Cree, Woodlands; Navajo, Ramah
Pontederia: Montagnais
Populus: Blackfoot; Ojibwa; Tanana, Upper
Porphyra: Hanaksiala
Portulaca: Navajo
Potentilla: Navajo, Ramah
Prunella: Iroquois
Prunus: Cowichan; Navajo, Ramah; Nitinaht; Saanich
Psilostrophe: Navajo, Kayenta
Ptelea: Menominee
Pycnanthemum: Choctaw
Pyrola: Karok; Montagnais
Quercus: Delaware, Oklahoma; Navajo, Ramah
Ranunculus: Navajo, Kayenta
Ratibida: Dakota; Navajo, Ramah
Rhus: Rappahannock
Ribes: Shuswap; Tanana, Upper; Thompson
Rumex: Blackfoot; Iroquois; Navajo, Kayenta; Navajo, Ramah
Salix: Cheyenne; Mendocino Indian; Potawatomi
Salvia: Mohegan; Okanagan-Colville
Sambucus: Karok
Sanguinaria: Delaware, Oklahoma; Iroquois; Ojibwa
Sanicula: Miwok
Sanvitalia: Navajo, Ramah
Saururus: Ojibwa
Saxifraga: Iroquois
Senecio: Navajo, Ramah
Shepherdia: Tanana, Upper
Silene: Navajo, Ramah
Solidago: Blackfoot
Sphaeralcea: Navajo, Ramah
Stephanomeria: Cheyenne; Navajo, Ramah; Paiute
Streptopus: Montagnais
Tanacetum: Iroquois
Taxus: Chehalis; Thompson
Tellima: Skagit
Tetraneuris: Navajo, Ramah
Thalictrum: Mendocino Indian
Thuja: Abnaki; Cree, Woodlands; Penobscot
Tilia: Iroquois

Tragia: Navajo, Ramah
Trichostema: Mahuna
Trientalis: Montagnais
Trillium: Abnaki; Wailaki; Yuki
Umbellularia: Karok; Mendocino Indian
Urtica: Klallam
Vaccinium: Alabama
Valeriana: Cree, Woodlands
Veratrum: Cherokee; Kwakiutl; Paiute
Veronicastrum: Iroquois
Vicia: Navajo, Ramah
Viola: Iroquois
Xanthium: Rappahannock
Zea: Keres, Western

Pediatric Aid
Abies: Flathead; Thompson
Abronia: Hopi
Acacia: Hawaiian
Achillea: Abnaki; Bella Coola; Cree, Woodlands; Iroquois; Menominee; Quileute; Thompson
Acorus: Cheyenne; Chippewa; Cree, Woodlands; Iroquois; Nanticoke; Rappahannock
Actaea: Chippewa
Adiantum: Iroquois; Meskwaki
Agastache: Okanagan-Colville
Agrimonia: Cherokee; Iroquois
Aleurites: Hawaiian
Allium: Blackfoot; Cherokee; Chippewa; Iroquois
Alnus: Cherokee; Meskwaki; Okanagan-Colville; Pomo
Amelanchier: Blackfoot; Cherokee; Cheyenne; Cree, Woodlands
Andropogon: Houma
Androsace: Navajo, Ramah
Angelica: Blackfoot; Creek; Shoshoni
Antennaria: Cherokee; Navajo, Ramah
Anthemis: Iroquois
Aplectrum: Cherokee
Apocynum: Chippewa; Iroquois
Arabis: Cheyenne
Aralia: Algonquin, Quebec; Choctaw; Cree, Woodlands
Arctostaphylos: Blackfoot; Cree, Woodlands; Sanpoil
Arenaria: Gosiute
Argentina: Iroquois
Arisaema: Iroquois
Artemisia: Blackfoot; Cahuilla; Chippewa; Costanoan; Kawaiisu; Okanagan-Colville; Paiute; Paiute, Northern; Pomo; Shoshoni; Thompson; Tolowa; Yurok
Asarum: Algonquin, Quebec; Iroquois; Malecite; Thompson; Yurok
Asclepias: Blackfoot; Chippewa; Iroquois; Lakota; Navajo, Kayenta
Asplenium: Hawaiian
Aster: Blackfoot; Cree, Woodlands; Iroquois; Zuni
Astragalus: Blackfoot; Dakota
Baptisia: Creek
Besseya: Navajo, Ramah
Betula: Algonquin, Quebec; Cree, Woodlands

Bidens: Hawaiian
Botrychium: Abnaki
Brickellia: Navajo, Kayenta; Navajo, Ramah
Broussaisia: Hawaiian
Caesalpinia: Hawaiian
Calycanthus: Cherokee
Capsicum: Navajo, Ramah
Cardamine: Algonquin, Quebec; Malecite
Carpinus: Iroquois
Carum: Cree, Woodlands
Castanea: Cherokee; Iroquois
Ceanothus: Iroquois; Okanagan-Colville; Poliklah
Celastrus: Chippewa; Iroquois
Cerastium: Cherokee
Cercocarpus: Paiute; Shoshoni
Chaenactis: Great Basin Indian; Nevada Indian
Chaetopappa: Havasupai; Hopi
Chamaemelum: Mahuna
Chamaesyce: Cherokee; Hawaiian; Hopi; Houma
Chenopodium: Hawaiian; Houma; Natchez
Chimaphila: Cherokee; Iroquois
Chrysopsis: Delaware
Chrysothamnus: Tewa
Cimicifuga: Cherokee; Iroquois
Cirsium: Abnaki; Kwakiutl; Mohegan
Citrullus: Cherokee
Claytonia: Iroquois
Clematis: Thompson
Coix: Cherokee
Collinsonia: Iroquois
Conyza: Iroquois; Navajo, Kayenta
Coptis: Iroquois; Malecite; Menominee; Mohegan; Ojibwa; Potawatomi
Cornus: Cherokee; Iroquois; Meskwaki; Ojibwa; Okanagan-Colville; Shuswap; Thompson
Corylus: Iroquois
Crataegus: Okanagan-Colville
Croton: Cahuilla; Pawnee
Cryptantha: Navajo, Kayenta; Navajo, Ramah
Cucurbita: Iroquois
Cypripedium: Algonquin, Tête-de-Boule
Cyrtandra: Hawaiian
Dalea: Keres, Western
Datura: Kawaiisu
Delphinium: Blackfoot
Diervilla: Iroquois
Digitaria: Hawaiian
Dimorphocarpa: Navajo, Kayenta
Dodecatheon: Blackfoot
Dracocephalum: Navajo, Kayenta
Echinocystis: Tubatulabal
Elaeagnus: Blackfoot
Empetrum: Cree, Woodlands
Ephedra: Paiute; Shoshoni
Epigaea: Cherokee
Epilobium: Blackfoot; Costanoan; Okanagan-Colville
Epipactis: Navajo, Kayenta
Equisetum: Iroquois; Okanagan-Colville; Sanpoil; Tolowa

Eriastrum: Paiute
Eriogonum: Diegueño; Lakota; Navajo, Ramah; Tubatulabal
Eschscholzia: Cahuilla; Costanoan
Euonymus: Iroquois
Eupatorium: Chippewa; Ojibwa
Euphorbia: Cherokee; Iroquois
Fagopyrum: Iroquois
Foeniculum: Cherokee
Fragaria: Chippewa; Iroquois; Ojibwa; Okanagan-Colville; Thompson
Freycinetia: Hawaiian
Galactia: Seminole
Galium: Iroquois
Gaura: Navajo, Ramah
Geranium: Cherokee; Chippewa; Iroquois
Geum: Cree, Woodlands; Malecite; Micmac
Gilia: Shoshoni
Glechoma: Cherokee
Gleditsia: Creek
Glycyrrhiza: Dakota; Pawnee; Sioux
Gnaphalium: Koasati; Pomo
Goodyera: Mohegan
Gossypium: Tewa
Grindelia: Dakota; Navajo, Ramah
Gutierrezia: Tewa
Helianthus: Iroquois; Navajo, Kayenta
Hepatica: Chippewa; Iroquois
Heracleum: Kwakiutl
Heuchera: Blackfoot; Chippewa; Gosiute; Okanagan-Colville
Hibiscus: Hawaiian
Hilaria: Navajo, Ramah
Hydrangea: Cherokee
Hydrophyllum: Ojibwa
Hymenopappus: Isleta
Hypericum: Alabama; Natchez
Impatiens: Cherokee
Inula: Iroquois
Ipomoea: Hawaiian
Ipomopsis: Salish; Zuni
Iris: Yokia; Zuni
Isocoma: Navajo, Kayenta
Iva: Shoshoni
Jacquemontia: Hawaiian
Jeffersonia: Iroquois
Juncus: Cherokee
Juniperus: Cree, Woodlands; Hopi; Navajo, Ramah; Seminole; Tewa
Krameria: Pima
Lactuca: Okanagan-Colville
Larix: Thompson
Larrea: Papago
Lechea: Seminole
Ledum: Cree, Woodlands; Eskimo, Kuskokwagmiut; Montagnais
Lepidium: Navajo, Kayenta
Liatris: Pawnee; Seminole
Ligusticum: Atsugewi; Okanagan-Colville
Lilium: Cherokee

Linanthus: Pomo, Calpella
Linaria: Iroquois
Lindera: Mohegan
Linnaea: Iroquois; Tanana, Upper
Linum: Okanagon; Thompson
Liriodendron: Cherokee
Lithospermum: Iroquois; Navajo
Lomatium: Cheyenne; Okanagan-Colville; Thompson; Washo
Lonicera: Iroquois
Ludwigia: Hawaiian
Lupinus: Kwakiutl
Lycopodium: Algonquin, Quebec
Lycopus: Blackfoot; Cherokee; Iroquois
Lygodesmia: Blackfoot; Cheyenne; Lakota
Lysichiton: Kwakiutl
Mahonia: Havasupai
Maianthemum: Karok; Meskwaki
Malva: Costanoan; Iroquois
Marah: Tubatulabal
Marrubium: Cherokee; Diegueño
Matelea: Comanche
Matricaria: Costanoan; Diegueño; Flathead
Medeola: Iroquois
Melissa: Costanoan
Mentha: Abnaki; Cherokee; Iroquois; Malecite; Micmac; Mohegan; Navajo, Kayenta; Okanagan-Colville; Okanagon; Paiute; Sanpoil; Shoshoni; Thompson; Washo
Mentzelia: Zuni
Merremia: Hawaiian
Mimulus: Karok
Mirabilis: Karok; Pawnee; Zuni
Mitchella: Cherokee; Iroquois
Mitella: Gosiute
Monarda: Choctaw; Iroquois; Menominee; Ojibwa
Monardella: Okanagan-Colville; Sanpoil
Monotropa: Cherokee
Myrica: Koasati
Myriophyllum: Iroquois
Nepeta: Cherokee; Delaware; Delaware, Oklahoma; Delaware, Ontario; Hoh; Iroquois; Mohegan; Quileute; Rappahannock
Nereocystis: Kwakiutl, Southern
Nuphar: Iroquois
Nyssa: Cherokee
Ochrosia: Hawaiian
Onoclea: Iroquois
Opuntia: Apache, Western
Orobanche: Navajo, Ramah
Osmorhiza: Iroquois
Osmunda: Iroquois; Seminole
Oxalis: Cherokee
Oxytropis: Cheyenne
Panax: Cherokee; Creek; Iroquois; Meskwaki; Seminole
Pandanus: Hawaiian
Parnassia: Cheyenne
Passiflora: Cherokee
Pedicularis: Shoshoni
Pediomelum: Blackfoot
Pelea: Hawaiian
Peltandra: Nanticoke
Penstemon: Navajo, Kayenta; Paiute; Shoshoni
Peperomia: Hawaiian
Perezia: Yavapai
Persea: Seminole
Petasites: Karok
Peucedanum: Hawaiian
Phaseolus: Zuni
Phlebodium: Seminole
Phlox: Blackfoot; Cherokee; Havasupai; Okanagan-Colville; Paiute; Shoshoni
Phoradendron: Keres, Western; Seminole
Phragmites: Keres, Western
Phyla: Houma
Phyllospadix: Kwakiutl
Physaria: Blackfoot
Picea: Cree, Woodlands; Makah
Pilea: Cherokee
Piloblephis: Seminole
Pinus: Delaware, Ontario; Iroquois; Kawaiisu; Shuswap; Thompson
Piper: Hawaiian
Pipturus: Hawaiian
Plantago: Cherokee; Navajo
Platanthera: Iroquois
Platanus: Cherokee
Pleopeltis: Houma
Pluchea: Pima
Polygala: Iroquois
Polygonum: Cherokee; Iroquois; Meskwaki; Thompson
Polystichum: Iroquois
Populus: Carrier; Iroquois; Meskwaki; Thompson
Prenanthes: Bella Coola
Prosopis: Apache, Mescalero; Pima
Prunella: Iroquois; Menominee
Prunus: Algonquin, Tête-de-Boule; Blackfoot; Cheyenne; Chippewa; Delaware; Gosiute; Iroquois; Karok; Kwakiutl; Malecite; Menominee
Psathyrotes: Shoshoni
Psidium: Hawaiian
Ptelea: Havasupai
Pteridium: Micmac; Montagnais
Pyrola: Blackfoot; Iroquois; Karok; Navajo, Kayenta
Quercus: Alabama; Creek; Dakota; Keres, Western; Mahuna; Miwok; Omaha; Pawnee; Ponca; Winnebago
Ratibida: Navajo, Ramah
Rhamnus: Iroquois
Rhus: Chippewa; Diegueño; Iroquois; Koasati
Ribes: Kwakiutl; Navajo, Kayenta; Sanpoil; Thompson
Ricinus: Hawaiian
Rorippa: Iroquois
Rosa: Blackfoot; Cahuilla; Cherokee; Cowlitz; Diegueño; Isleta; Keres, Western; Paiute; Thompson
Rubus: Cree, Woodlands; Iroquois; Kwakiutl; Malecite; Micmac; Omaha; Thompson
Rudbeckia: Chippewa; Iroquois
Rumex: Chippewa; Iroquois; Yavapai; Yuki
Sagittaria: Cherokee; Iroquois
Salix: Catawba; Navajo, Ramah; Okanagan-Colville; Paiute
Salvia: Paiute; Shoshoni
Sambucus: Cahuilla; Cherokee; Delaware; Delaware, Oklahoma; Diegueño; Iroquois; Karok; Mohegan; Paiute
Sanguinaria: Iroquois
Sanicula: Iroquois
Sassafras: Cherokee; Seminole
Scirpus: Houma; Kwakiutl; Thompson
Sedum: Iroquois; Okanagon; Thompson
Senecio: Iroquois; Navajo, Ramah
Senna: Cherokee
Sesamum: Cherokee
Shepherdia: Navajo, Kayenta
Sida: Hawaiian
Silene: Gosiute; Karok
Silphium: Iroquois
Sisyrinchium: Cherokee
Solanum: Blackfoot; Cherokee; Houma
Solidago: Iroquois; Navajo, Kayenta; Okanagan-Colville; Thompson
Sonchus: Houma; Iroquois
Sorbus: Okanagan-Colville
Sphaeralcea: Hopi
Sphagnum: Carrier
Spigelia: Creek
Spiraea: Iroquois
Spiranthes: Cherokee
Staphylea: Iroquois
Stenandrium: Seminole
Stenotaphrum: Hawaiian
Stillingia: Seminole
Symphoricarpos: Cree, Woodlands; Nez Perce; Okanagan-Colville; Skagit; Thompson
Symplocarpus: Iroquois; Menominee
Syringa: Iroquois
Syzygium: Hawaiian
Tanacetum: Cherokee
Tephrosia: Cherokee
Thaspium: Chippewa
Thelesperma: Keres, Western
Thelypodium: Navajo, Kayenta
Thuja: Algonquin, Quebec; Nez Perce
Tiarella: Iroquois; Malecite
Tilia: Iroquois
Touchardia: Hawaiian
Toxicodendron: Iroquois
Tragia: Keres, Western
Trillium: Abnaki
Triosteum: Iroquois; Meskwaki
Tripterocalyx: Hopi
Tsuga: Algonquin, Quebec; Kwakiutl; Malecite; Quinault
Typha: Dakota; Mahuna; Omaha; Pawnee; Plains Indian; Ponca; Winnebago
Ulmus: Cheyenne

Urtica: Kawaiisu
Uvularia: Iroquois
Vaccinium: Algonquin, Quebec; Cheyenne; Seminole
Valeriana: Cree, Woodlands
Veratrum: Kwakiutl; Thompson
Verbascum: Abnaki; Catawba; Iroquois
Viburnum: Cree, Woodlands; Iroquois
Vigna: Hawaiian
Viola: Blackfoot; Iroquois
Vitis: Cherokee; Iroquois; Seminole
Vittaria: Seminole
Waltheria: Hawaiian
Xanthium: Tewa
Xyris: Cherokee
Zanthoxylum: Chippewa
Zea: Tewa

Poison

Aconitum: Aleut; Blackfoot; Cree, Hudson Bay; Eskimo, Inupiat; Gosiute; Okanagan-Colville
Actaea: Alaska Native; Eskimo, Arctic; Micmac; Thompson
Aesculus: Costanoan; Delaware; Delaware, Oklahoma; Kawaiisu; Mendocino Indian; Pomo
Allium: Mendocino Indian
Amanita: Pomo, Kashaya
Amianthium: Cherokee
Anemone: Thompson
Angelica: Iroquois
Anthemis: Yuki
Apocynum: Montana Indian
Aralia: Cherokee
Arctostaphylos: Mendocino Indian
Arisaema: Meskwaki; Mohegan; Penobscot
Artemisia: Thompson
Asclepias: Mendocino Indian; Pima; Thompson
Astragalus: Alaska Native; Cahuilla; Cheyenne; Lakota; Mahuna; Navajo
Atriplex: Isleta
Bovista: Haisla & Hanaksiala
Bovistella: Haisla & Hanaksiala
Brickellia: Gosiute
Bromus: Hesquiat
Calla: Cree, Woodlands
Caltha: Abnaki; Alaska Native; Eskimo, Inupiat
Calvatia: Haisla & Hanaksiala
Calycanthus: Cherokee
Cardamine: Iroquois
Castilleja: Cherokee
Celastrus: Iroquois; Oglala
Chaerophyllum: Chickasaw
Chamaesyce: Pima
Chenopodium: Kawaiisu
Chimaphila: Cherokee
Chlorogalum: Costanoan; Mendocino Indian; Mewuk
Chrysothamnus: Isleta
Cicuta: Alaska Native; Eskimo, Arctic; Eskimo, Inupiat; Eskimo, Kuskokwagmiut; Eskimo, Western; Haisla & Hanaksiala; Iroquois; Kawaiisu; Klamath; Kwakiutl; Lakota; Mendocino Indian; Montana Indian; Okanagan-Colville; Shoshoni; Shuswap; Thompson
Citrullus: Kiowa
Clintonia: Ojibwa; Pomo; Pomo, Kashaya
Comptonia: Menominee
Conium: Klallam; Lakota; Snohomish
Consolida: Cherokee
Cornus: Blackfoot; Thompson
Crataegus: Mendocino Indian
Croton: Cahuilla; Pomo
Cryptantha: Keres, Western
Cuscuta: Pima
Dalea: Dakota
Datura: Cahuilla; Coahuilla; Diegueño; Hopi; Iroquois; Kawaiisu; Keres, Western; Mahuna; Paiute, Northern; Rappahannock
Delphinium: Blackfoot; Chehalis; Cherokee; Gosiute; Lakota; Mendocino Indian
Dichanthelium: Lakota
Dipsacus: Iroquois
Disporum: Klallam
Echinocereus: Navajo
Epilobium: Swinomish
Equisetum: Aleut
Erigeron: Gosiute
Eschscholzia: Costanoan; Mahuna
Eupatorium: Iroquois
Euphorbia: Pawnee
Evernia: Modesse
Frangula: Flathead; Pomo; West Coast Indian
Fritillaria: Ute
Galium: Cowlitz; Shuswap
Hackelia: Isleta; Navajo, Ramah
Hedysarum: Alaska Native; Eskimo, Inupiat; Tanana, Upper
Helenium: Meskwaki
Heracleum: Cree; Karok
Heterotheca: Isleta
Hordeum: Navajo, Ramah
Hymenoxys: Keres, Western; Navajo, Ramah
Iris: Abnaki; Eskimo, Inupiat
Juglans: Cherokee
Juniperus: Okanagan-Colville
Kalmia: Algonquin, Quebec; Algonquin, Tête-de-Boule; Cree, Hudson Bay; Hesquiat; Mahuna; Micmac; Montagnais
Lathyrus: Eskimo, Inupiat
Ledum: Eskimo, Inupiat
Ligusticum: Eskimo, Western
Lomatium: Creek; Montana Indian; Okanagan-Colville
Lonicera: Makah; Okanagan-Colville; Poliklah; Thompson; Tolowa
Lupinus: Alaska Native; Eskimo, Inupiat; Thompson
Lycoperdon: Haisla & Hanaksiala; Pomo, Kashaya
Lycopus: Iroquois
Lysichiton: Gitksan
Mahonia: Karok
Marah: Karok; Mendocino Indian; Squaxin
Nicotiana: Kawaiisu
Nuphar: Okanagan-Colville
Nymphaea: Okanagan-Colville
Oenanthe: Haisla & Hanaksiala
Onopordum: Iroquois
Oplopanax: Cowlitz; Kwakiutl; Oweekeno
Opuntia: Navajo
Orobanche: Montana Indian
Osmorhiza: Kwakiutl
Oxytropis: Alaska Native; Blackfoot; Hopi; Lakota
Parthenocissus: Iroquois
Pastinaca: Ojibwa; Potawatomi
Pentaphylloides: Cheyenne
Phoradendron: Mendocino Indian
Phytolacca: Cherokee; Mahuna; Mohegan
Platanthera: Shuswap
Podophyllum: Cherokee; Iroquois
Polygonum: Cherokee
Prunus: Rappahannock
Ptelea: Havasupai
Pteridium: Alaska Native; Okanagan-Colville
Pulsatilla: Cheyenne
Quercus: Mendocino Indian
Ranunculus: Aleut; Eskimo, Inupiat; Keres, Western; Okanagan-Colville; Thompson
Rheum: Cherokee
Rhus: Thompson
Ribes: Oweekeno; Swinomish
Ricinus: Cahuilla; Pima
Rosa: Thompson
Sambucus: Hesquiat
Sapium: Seri
Senecio: Eskimo, Inuktitut; Eskimo, Western
Shepherdia: Eskimo, Inupiat; Havasupai; Sioux
Sisyrinchium: Iroquois
Sium: Shuswap
Solanum: Cahuilla; Karok; Mendocino Indian; Rappahannock
Sonchus: Navajo, Kayenta
Stanleya: Havasupai
Symphoricarpos: Okanagan-Colville; Thompson
Taxus: Mendocino Indian
Tephrosia: Hawaiian
Thalictrum: Kawaiisu; Mendocino Indian
Thuja: Okanagan-Colville
Toxicodendron: Cherokee; Chippewa; Karok; Lakota; Navajo, Ramah; Omaha; Paiute; Ponca; Potawatomi; Thompson; Tolowa; Yurok
Triglochin: Blackfoot
Trillium: Concow; Iroquois; Pomo, Kashaya; Skagit
Triteleia: Okanagan-Colville
Urtica: Navajo, Ramah
Valeriana: Blackfoot; Gosiute; Snake
Veratrum: Alaska Native; Bella Coola; Blackfoot; Carrier, Southern; Cowlitz; Haisla & Hanaksiala; Kwakiutl; Okanagan-Colville; Oweekeno; Quinault; Salish; Shuswap; Thompson
Veronicastrum: Iroquois
Vicia: Navajo
Wyethia: Hopi

Yucca: Navajo; Navajo, Ramah
Zigadenus: Alaska Native; Eskimo, Inupiat; Haisla & Hanaksiala; Keres, Western; Lakota; Mendocino Indian; Okanagan-Colville; Paiute; Pomo, Kashaya; Shuswap; Thompson; Ute; Yuki

Poultice
Abies: Algonquin, Quebec
Achillea: Algonquin, Quebec
Aesculus: Cherokee
Aralia: Choctaw
Astragalus: Navajo, Kayenta
Brassica: Rappahannock
Capsicum: Cherokee
Cirsium: Cherokee
Cornus: Cherokee; Iroquois
Grindelia: Zuni
Iodanthus: Meskwaki
Linum: Paiute
Liriodendron: Cherokee
Mentzelia: Miwok
Pinus: Shoshoni
Plantago: Algonquin, Quebec
Polygonum: Cherokee
Populus: Kawaiisu
Ruta: Cherokee
Salvia: Paiute
Saururus: Cherokee
Taraxacum: Algonquin, Quebec
Taxus: Algonquin, Quebec
Verbena: Iroquois

Preventive Medicine
Abies: Nitinaht
Acorus: Algonquin, Quebec; Malecite
Angelica: Eskimo
Chamaesyce: Lakota
Crataegus: Cherokee
Fraxinus: Karok
Helianthus: Shasta
Iris: Penobscot
Mahonia: Keres, Western
Osmorhiza: Karok
Pellaea: Mahuna
Prunus: Kwakiutl
Salix: Eskimo, Western; Seminole
Sanguinaria: Penobscot
Thlaspi: Navajo, Ramah
Toxicodendron: Mahuna

Psychological Aid
Abronia: Keres, Western
Acorus: Dakota
Actaea: Iroquois
Adiantum: Navajo, Kayenta
Aloysia: Havasupai
Ambrosia: Meskwaki
Anemone: Meskwaki
Angelica: Eskimo; Mendocino Indian
Apocynum: Chippewa
Arabis: Navajo, Kayenta
Arctostaphylos: Cheyenne
Artemisia: Cheyenne

Asarum: Iroquois
Aster: Meskwaki
Berlandiera: Keres, Western
Cannabis: Iroquois
Cardamine: Iroquois
Chrysothamnus: Cheyenne; Keres, Western
Conocephalum: Nitinaht
Cornus: Iroquois; Thompson
Corylus: Iroquois
Crepis: Keres, Western
Cucurbita: Hawaiian
Cynoglossum: Cherokee
Datura: Hopi; Keres, Western
Diervilla: Menominee
Dimorphocarpa: Zuni
Eriogonum: Keres, Western; Navajo, Kayenta
Eupatorium: Iroquois
Fragaria: Cherokee
Frangula: Mendocino Indian
Frasera: Navajo, Ramah
Fraxinus: Algonquin, Tête-de-Boule
Gaillardia: Keres, Western
Gaylussacia: Chickasaw
Gayophytum: Navajo, Kayenta
Gentiana: Iroquois
Gnaphalium: Creek; Menominee; Meskwaki
Gymnocladus: Meskwaki
Hackelia: Cherokee
Hymenoxys: Isleta
Hyptis: Seminole
Ilex: Iroquois; Seminole
Juglans: Iroquois
Juniperus: Seminole
Lagenaria: Seminole
Laportea: Iroquois
Lewisia: Thompson
Licania: Seminole
Linnaea: Tanana, Upper
Linum: Keres, Western
Lithospermum: Cheyenne
Lobelia: Iroquois
Lonicera: Iroquois; Nitinaht
Lygodesmia: Cheyenne
Lysichiton: Thompson
Macromeria: Hopi
Madia: Cheyenne
Mahonia: Thompson
Maianthemum: Meskwaki
Mentzelia: Keres, Western
Microlepia: Hawaiian
Mirabilis: Paiute
Mitchella: Iroquois
Monarda: Creek
Nicotiana: Iroquois; Kawaiisu
Nuphar: Abnaki
Nymphaea: Abnaki
Osmorhiza: Karok
Osmunda: Seminole
Panax: Menominee
Pectis: Keres, Western

Penstemon: Karok
Persea: Seminole
Phlebodium: Seminole
Picea: Thompson
Pinus: Iroquois
Plantago: Hopi; Iroquois
Polygonum: Iroquois
Polypodium: Seminole
Populus: Thompson
Postelsia: Nitinaht
Prunella: Iroquois
Prunus: Saanich
Quercus: Iroquois; Neeshenam
Ranunculus: Iroquois
Ribes: Saanich
Salix: Iroquois
Sambucus: Nitinaht
Senecio: Keres, Western
Smilax: Iroquois
Solanum: Cherokee
Solidago: Meskwaki; Navajo, Kayenta
Sorbus: Algonquin, Tête-de-Boule
Tellima: Nitinaht
Tetraneuris: Navajo, Kayenta
Thelypteris: Seminole
Thuja: Thompson
Thymophylla: Navajo, Kayenta
Tradescantia: Meskwaki
Urtica: Kawaiisu
Vaccinium: Chippewa
Valeriana: Menominee
Veratrum: Tsimshian
Verbascum: Hopi; Navajo, Ramah
Vitis: Meskwaki
Vittaria: Seminole
Zinnia: Keres, Western

Pulmonary Aid
Abies: Blackfoot; Cherokee; Gitksan; Menominee; Montana Indian; Paiute; Shoshoni
Acer: Iroquois
Achillea: Mewuk
Acorus: Blackfoot; Cree, Woodlands; Micmac
Actaea: Thompson
Adenostoma: Cahuilla
Aesculus: Iroquois
Agastache: Cheyenne
Aletris: Cherokee
Allium: Cherokee; Makah; Quinault; Rappahannock
Alnus: Bella Coola; Clallam; Micmac; Nitinaht
Ambrosia: Cherokee
Amelanchier: Cree, Woodlands
Andropogon: Seminole
Anemone: Cherokee; Ojibwa
Anemopsis: Cahuilla; Isleta
Angelica: Iroquois; Paiute, Northern; Shoshoni
Anthemis: Iroquois
Apocynum: Cherokee
Aralia: Cherokee; Cree, Woodlands; Iroquois; Mendocino Indian; Menominee
Arceuthobium: Bella Coola

Arctium: Cowlitz; Oto
Aristolochia: Cherokee
Artemisia: Blackfoot; Flathead; Hawaiian; Hualapai; Kiowa; Montana Indian; Paiute; Shoshoni; Tlingit
Asarum: Iroquois; Meskwaki
Asclepias: Cherokee; Delaware; Delaware, Oklahoma; Menominee; Mohegan; Omaha; Ponca
Asplenium: Tlingit
Aster: Iroquois
Astragalus: Blackfoot; Lakota
Balsamorhiza: Flathead
Betula: Chippewa
Blechnum: Makah
Botrychium: Ojibwa
Boykinia: Cheyenne
Brassica: Cherokee
Caesalpinia: Hawaiian
Campanula: Ojibwa
Campanulastrum: Iroquois
Cardamine: Iroquois
Castanea: Mohegan
Castilleja: Gitksan
Caulophyllum: Chippewa
Ceanothus: Chippewa; Choctaw
Centaurium: Miwok
Cercis: Alabama; Cherokee
Cercocarpus: Paiute
Chenopodium: Potawatomi; Seminole
Chimaphila: Cree, Woodlands; Delaware
Cirsium: Mohegan
Collinsia: Natchez
Comandra: Meskwaki
Comptonia: Delaware
Coptis: Tlingit
Corallorrhiza: Paiute; Shoshoni
Cordyline: Hawaiian
Cornus: Carrier; Iroquois; Thompson
Corylus: Chippewa
Cucurbita: Isleta
Cypripedium: Iroquois
Dalea: Navajo
Datura: Rappahannock
Desmodium: Alabama
Dichanthelium: Seminole
Dirca: Chippewa
Epigaea: Cherokee
Erigeron: Iroquois
Eriodictyon: Atsugewi; Karok; Mahuna; Paiute; Shoshoni
Eriogonum: Thompson
Eryngium: Cherokee
Eupatorium: Iroquois
Euphorbia: Cherokee
Euthamia: Chippewa
Fagus: Chippewa
Frasera: Isleta
Galeopsis: Potawatomi
Galium: Ojibwa
Gentianella: Iroquois
Geum: Malecite; Micmac; Ojibwa, South

Gleditsia: Cherokee
Gnaphalium: Choctaw
Goodyera: Delaware; Delaware, Oklahoma
Grindelia: Crow; Flathead; Paiute
Gutierrezia: Comanche
Hamamelis: Iroquois
Hedeoma: Cherokee
Helianthus: Dakota; Meskwaki
Heliopsis: Meskwaki
Hepatica: Iroquois
Heracleum: Bella Coola
Hexastylis: Rappahannock
Hibiscus: Hawaiian
Humulus: Shinnecock
Hydrastis: Iroquois
Hydrocotyle: Hawaiian
Hypericum: Menominee
Hyssopus: Cherokee
Impatiens: Potawatomi
Inula: Cherokee; Iroquois; Mohegan
Ipomopsis: Zuni
Iresine: Houma
Iris: Meskwaki
Juniperus: Apache, Western; Bella Coola; Blackfoot; Cheyenne; Cree, Woodlands; Flathead; Kutenai; Nez Perce; Paiute; Rappahannock; Sioux
Larrea: Cahuilla
Ledum: Cree, Woodlands
Lepidium: Cherokee
Lewisia: Flathead
Ligusticum: Pomo; Pomo, Kashaya
Lindera: Cherokee
Linum: Cherokee
Lithospermum: Lakota; Sioux, Teton
Lobelia: Cherokee
Lomatium: Blackfoot; Paiute; Shoshoni; Washo
Lonicera: Blackfoot; Chippewa
Lophophora: Kiowa
Lycopodium: Blackfoot
Lysichiton: Gitksan; Makah
Malus: Cherokee; Makah; Quileute
Marrubium: Costanoan; Diegueño
Mentha: Carrier, Southern; Malecite; Menominee; Micmac; Potawatomi
Mertensia: Cherokee
Monarda: Flathead; Lakota
Monardella: Costanoan
Nepeta: Menominee
Nolina: Isleta
Nuphar: Gitksan; Iroquois
Oplopanax: Gitksan
Orobanche: Keres, Western; Paiute
Osmorhiza: Bella Coola; Paiute; Shoshoni; Washo
Ostrya: Chippewa
Oxydendrum: Cherokee
Paeonia: Costanoan; Mahuna
Panax: Cherokee; Creek; Iroquois
Panicum: Seminole
Pediomelum: Blackfoot
Penstemon: Natchez

Peperomia: Hawaiian
Persea: Seminole
Petasites: Delaware; Delaware, Oklahoma
Philadelphus: Thompson
Phlox: Blackfoot
Phoradendron: Creek
Phragmites: Paiute
Picea: Montagnais; Tanana, Upper
Pinus: Delaware, Ontario; Iroquois; Paiute; Potawatomi; Shoshoni; Thompson
Piper: Hawaiian
Plantago: Iroquois; Paiute
Platanus: Meskwaki
Pleomele: Hawaiian
Polygala: Cherokee
Polygonatum: Cherokee
Polygonum: Cherokee
Polypodium: Malecite; Micmac
Polystichum: Cherokee
Populus: Bella Coola; Carrier, Southern
Potentilla: Cherokee
Prunella: Iroquois
Prunus: Algonquin, Quebec; Chippewa; Iroquois; Menominee; Ojibwa; Ojibwa, South
Psorothamnus: Paiute; Paiute, Northern; Shoshoni
Ptelea: Meskwaki
Pterocaulon: Seminole
Pterospora: Cheyenne
Pulsatilla: Chippewa
Purshia: Klamath; Montana Indian; Paiute
Pyrus: Cree, Hudson Bay
Quercus: Alabama; Chippewa; Meskwaki
Rhus: Iroquois; Menominee
Rubus: Blackfoot; Chippewa; Iroquois
Rumex: Cheyenne; Iroquois; Paiute
Sadleria: Hawaiian
Salix: Eskimo, Western; Kiowa
Salvia: Paiute
Sambucus: Meskwaki; Montana Indian
Sanguinaria: Cherokee; Iroquois
Sarracenia: Iroquois; Micmac
Senna: Cherokee
Silphium: Chippewa
Sinapis: Cherokee
Sisymbrium: Cherokee
Smilax: Ojibwa
Solidago: Chippewa
Sorbus: Potawatomi; Tlingit
Symplocarpus: Delaware
Taenidia: Menominee
Taraxacum: Iroquois; Meskwaki
Taxus: Bella Coola; Quinault
Tetradymia: Paiute
Thalictrum: Blackfoot; Gitksan
Thuja: Cree, Woodlands; Lummi
Tilia: Meskwaki
Trifolium: Algonquin, Quebec
Triosteum: Iroquois
Typha: Houma
Ulmus: Micmac; Mohegan; Penobscot

Urtica: Paiute
Uvularia: Ojibwa
Vaccinium: Montagnais
Valeriana: Cree, Woodlands; Menominee
Verbascum: Catawba; Delaware; Delaware, Oklahoma; Menominee
Veronicastrum: Iroquois
Viburnum: Bella Coola; Iroquois
Vitis: Chippewa
Waltheria: Hawaiian
Wyethia: Costanoan
Xanthium: Cherokee
Xyris: Seminole
Zanthoxylum: Chippewa; Menominee
Zea: Cherokee

Reproductive Aid
Antennaria: Okanagan-Colville
Artemisia: Hawaiian
Athyrium: Iroquois
Berchemia: Houma
Broussaisia: Hawaiian
Chaetopappa: Hopi
Chamaesyce: Hawaiian
Cinchona: Cherokee
Coreopsis: Zuni
Cypripedium: Okanagan-Colville
Cyrtandra: Hawaiian
Dyssodia: Lakota
Fraxinus: Iroquois
Galactia: Seminole
Gaultheria: Nitinaht
Goodyera: Okanagan-Colville
Halosaccion: Nitinaht
Hypericum: Iroquois
Ipomoea: Hawaiian
Juniperus: Hopi; Tewa
Justicia: Seminole
Licania: Seminole
Lomatium: Thompson
Lonicera: Thompson
Lupinus: Navajo
Lycopodium: Iroquois
Maianthemum: Makah
Opuntia: Hawaiian
Panax: Penobscot
Persea: Seminole
Peucedanum: Hawaiian
Pinus: Kwakiutl
Quercus: Isleta
Rhus: Cheyenne; Iroquois
Ribes: Haisla
Rubus: Cree, Woodlands
Rumex: Hawaiian; Iroquois; Zuni
Salix: Delaware
Sambucus: Haisla & Hanaksiala
Shepherdia: Haisla & Hanaksiala
Stillingia: Creek
Tephrosia: Creek
Townsendia: Hopi
Trema: Seminole

Urtica: Makah
Vaccinium: Cree, Woodlands
Viburnum: Delaware
Vitis: Delaware; Delaware, Oklahoma
Zeuxine: Seminole

Respiratory Aid
Acer: Abnaki
Achillea: Algonquin, Quebec; Bella Coola; Cherokee; Cheyenne; Paiute; Yuki
Aconitum: Blackfoot
Acorus: Chippewa; Iroquois; Potawatomi
Adiantum: Cherokee; Hesquiat
Aleurites: Hawaiian
Allium: Blackfoot; Cherokee
Alnus: Kwakiutl
Amorpha: Navajo
Anaphalis: Cherokee; Iroquois
Andromeda: Mahuna
Anemone: Meskwaki
Anemopsis: Cahuilla
Angelica: Mendocino Indian; Washo
Anthemis: Cherokee
Apocynum: Cherokee
Aralia: Cherokee
Arenaria: Kawaiisu; Navajo, Ramah
Arisaema: Iroquois
Armoracia: Cherokee
Artemisia: Blackfoot; Cahuilla; Cheyenne; Costanoan; Diegueño; Eskimo, Alaska; Mendocino Indian; Paiute
Asarum: Iroquois
Asclepias: Costanoan; Isleta; Navajo, Kayenta; Omaha; Ponca
Aster: Cherokee
Bacopa: Seminole
Baptisia: Meskwaki
Betula: Ojibwa
Bidens: Hawaiian
Brassica: Cherokee
Calla: Gitksan
Cassytha: Hawaiian
Ceanothus: Chippewa
Cercis: Alabama
Cheirodendron: Hawaiian
Cirsium: Costanoan
Clematis: Navajo, Ramah
Clermontia: Hawaiian
Comandra: Meskwaki
Comptonia: Malecite; Micmac
Conioselinum: Navajo, Kayenta
Conyza: Seminole; Zuni
Coptis: Algonquin, Tête-de-Boule
Cordyline: Hawaiian
Cornus: Iroquois; Malecite; Micmac; Saanich
Corylus: Iroquois
Croton: Cahuilla
Datura: Cahuilla; Cherokee; Costanoan
Desmodium: Hawaiian
Dudleya: Diegueño
Dugaldia: Great Basin Indian

Erigeron: Meskwaki
Eriodictyon: Cahuilla; Costanoan; Mahuna; Mendocino Indian; Round Valley Indian
Eryngium: Seminole
Erysimum: Navajo, Ramah
Euonymus: Cherokee
Eupatorium: Cherokee
Gaillardia: Navajo, Ramah
Galium: Cherokee
Geum: Blackfoot
Glycyrrhiza: Cherokee
Gnaphalium: Cherokee; Iroquois; Rappahannock
Grindelia: Crow; Flathead
Gutierrezia: Blackfoot
Hamamelis: Iroquois
Helenium: Comanche; Mahuna; Meskwaki
Hexastylis: Rappahannock
Hierochloe: Flathead
Hydrocotyle: Seminole
Hyssopus: Cherokee
Ilex: Iroquois
Inula: Cherokee; Iroquois
Ipomoea: Cherokee; Hawaiian
Iris: Iroquois
Juniperus: Chippewa; Cree, Woodlands; Eskimo, Inupiat; Kwakiutl; Rappahannock
Larix: Thompson
Larrea: Cahuilla; Hualapai
Ledum: Kitasoo; Micmac; Tanana, Upper
Lesquerella: Navajo, Ramah
Ligusticum: Crow
Limonium: Costanoan
Linaria: Ojibwa
Lindera: Cherokee
Lobelia: Cherokee
Lomatium: Great Basin Indian; Nez Perce; Okanagan-Colville; Paiute; Shoshoni; Washo
Lupinus: Blackfoot
Machaeranthera: Navajo, Ramah
Magnolia: Cherokee
Maianthemum: Menominee
Marrubium: Kawaiisu
Mentha: Iroquois
Monarda: Crow; Menominee; Meskwaki; Ojibwa
Monardella: Costanoan
Nicotiana: Kawaiisu; Paiute
Nuphar: Kwakiutl
Nymphaea: Seminole
Nymphoides: Seminole
Oplopanax: Gitksan; Haisla; Wet'suwet'en
Oxydendrum: Cherokee
Oxytropis: Navajo
Panax: Iroquois; Seminole
Peperomia: Hawaiian
Perideridia: Blackfoot
Persea: Seminole
Petasites: Delaware; Delaware, Oklahoma; Eskimo, Inupiat
Physaria: Navajo
Picea: Eskimo, Inuktitut; Okanagan-Colville;

Tanana, Upper
Pilea: Iroquois
Pimpinella: Cherokee
Pinus: Cherokee
Plantago: Iroquois
Platanus: Diegueño; Mahuna
Pleomele: Hawaiian
Polygala: Blackfoot; Seminole
Polypodium: Haisla; Nitinaht
Populus: Kutenai; Ojibwa
Porteranthus: Cherokee
Prunella: Iroquois
Prunus: Iroquois; Mohegan
Purshia: Klamath
Quercus: Cherokee; Iroquois; Ojibwa
Ranunculus: Meskwaki
Rhus: Chippewa
Rubus: Algonquin, Tête-de-Boule; Iroquois
Sadleria: Hawaiian
Salix: Cherokee; Micmac; Ojibwa; Penobscot; Seminole
Salvia: Cherokee; Mahuna; Washo
Sanguinaria: Cherokee; Iroquois
Sida: Hawaiian
Sinapis: Cherokee
Sisyrinchium: Meskwaki; Pomo, Kashaya
Taenidia: Menominee
Taxus: Iroquois
Thuja: Bella Coola
Toxicodendron: Cherokee
Trifolium: Iroquois
Trillium: Cherokee; Iroquois
Ulmus: Cherokee; Iroquois
Umbellularia: Pomo, Kashaya; Yuki
Veratrum: Hanaksiala; Iroquois; Thompson; Tsimshian
Verbascum: Delaware; Delaware, Oklahoma; Iroquois; Malecite; Micmac; Mohegan; Penobscot; Potawatomi
Vicia: Cherokee
Vigna: Hawaiian
Viola: Blackfoot; Cherokee
Waltheria: Hawaiian
Wikstroemia: Hawaiian
Zanthoxylum: Ojibwa

Sedative
Abronia: Hopi
Achillea: Cherokee
Acorus: Rappahannock
Allium: Iroquois
Angelica: Cherokee; Creek
Anthemis: Cherokee; Iroquois
Arisaema: Meskwaki
Artemisia: Lakota
Asarum: Cherokee; Thompson
Bacopa: Seminole
Balsamorhiza: Thompson
Berlandiera: Keres, Western
Cardamine: Micmac
Carum: Cree, Woodlands

Caulophyllum: Cherokee
Ceanothus: Navajo
Chaetopappa: Hopi
Chamaemelum: Cherokee
Chenopodium: Seminole
Chrysopsis: Delaware; Delaware, Oklahoma
Cibotium: Hawaiian
Cimicifuga: Cherokee
Cordyline: Hawaiian
Cornus: Iroquois
Cypripedium: Cherokee; Iroquois; Micmac; Penobscot
Datura: Tubatulabal
Dodecatheon: Pomo, Kashaya
Eriogonum: Keres, Western
Eryngium: Creek
Eschscholzia: Cahuilla; Costanoan
Eupatorium: Chippewa
Fragaria: Cherokee
Frasera: Navajo
Galax: Cherokee
Gaultheria: Anticosti
Gaura: Keres, Western
Gaylussacia: Chickasaw
Gentianella: Cherokee
Gilia: Navajo, Kayenta
Glaux: Kwakiutl
Gnaphalium: Alabama; Creek
Gutierrezia: Navajo
Heterotheca: Cheyenne
Humulus: Cherokee; Delaware; Delaware, Oklahoma; Meskwaki; Mohegan; Shinnecock
Hydrocotyle: Seminole
Ipomoea: Hawaiian; Pawnee
Ipomopsis: Navajo, Kayenta
Juniperus: Cheyenne; Pawnee; Seminole
Lactuca: Cherokee
Leonurus: Cherokee; Iroquois
Lepidium: Navajo, Kayenta
Linaria: Iroquois
Linnaea: Iroquois
Liquidambar: Cherokee
Liriodendron: Cherokee
Lithospermum: Cheyenne; Menominee
Lomatium: Thompson
Lonicera: Iroquois; Thompson
Lupinus: Kwakiutl
Lysichiton: Gitksan
Maianthemum: Meskwaki
Mentha: Abnaki; Cherokee; Mahuna; Malecite
Mitchella: Menominee
Monarda: Cherokee; Creek
Nepeta: Cherokee; Delaware, Ontario; Iroquois; Menominee
Nicotiana: Kawaiisu
Nymphaea: Seminole
Nymphoides: Seminole
Oxydendrum: Cherokee
Panax: Seminole
Papaver: Cherokee

Persea: Seminole
Pinus: Cherokee; Thompson
Piper: Hawaiian
Pluchea: Pima
Polygonatum: Chippewa
Prunella: Iroquois
Prunus: Mendocino Indian; Meskwaki
Pyrola: Karok
Quercus: Navajo, Kayenta
Ranunculus: Cherokee
Ratibida: Keres, Western
Rhamnus: Iroquois
Ribes: Cherokee; Thompson
Ruta: Cherokee
Salix: Ojibwa
Salvia: Cherokee
Sambucus: Micmac
Sassafras: Rappahannock
Satureja: Mahuna; Pomo, Kashaya
Scirpus: Houma
Sedum: Okanagon; Thompson
Senecio: Cheyenne
Shepherdia: Thompson
Shinnersoseris: Navajo
Smallanthus: Iroquois
Solanum: Rappahannock
Solidago: Cherokee; Iroquois; Thompson
Sonchus: Cherokee; Iroquois
Staphylea: Iroquois
Stenandrium: Seminole
Taraxacum: Cherokee
Thelypodium: Navajo, Kayenta
Toxicodendron: Iroquois
Tripterocalyx: Hopi
Valeriana: Menominee
Veratrum: Haisla
Xanthorhiza: Cherokee
Yucca: Cherokee
Zanthoxylum: Menominee

Snakebite Remedy
Acer: Thompson
Achillea: Kawaiisu; Okanagon; Thompson
Adiantum: Iroquois
Allium: Mahuna
Amphicarpaea: Cherokee
Amsonia: Zuni
Angelica: Mendocino Indian
Arisaema: Meskwaki
Aristolochia: Cherokee; Mohegan; Rappahannock
Artemisia: Navajo, Kayenta; Navajo, Ramah
Asclepias: California Indian; Choctaw; Paiute; Rappahannock
Aster: Navajo, Ramah
Astragalus: Zuni
Baptisia: Meskwaki
Botrychium: Cherokee; Chippewa
Calla: Iroquois
Carex: Menominee
Ceanothus: Meskwaki
Chaenactis: Okanagon; Paiute; Thompson

Chaetopappa: Navajo, Ramah
Chamaesyce: Kawaiisu; Luiseño; Miwok; Pima; Shoshoni; Thompson
Cicuta: Montana Indian; Paiute
Conioselinum: Navajo, Kayenta
Conyza: Navajo, Ramah
Croton: Zuni
Cryptantha: Navajo, Kayenta; Navajo, Ramah
Cunila: Cherokee
Cyperus: Pima
Datura: Cahuilla; Mahuna
Daucus: Costanoan; Miwok
Echinacea: Dakota; Montana Indian; Omaha; Pawnee; Ponca; Winnebago
Epixiphium: Keres, Western
Eremocrinum: Navajo, Kayenta
Erigeron: Navajo, Ramah
Eriogonum: Navajo, Kayenta; Navajo, Ramah
Eryngium: Cherokee; Choctaw; Creek; Meskwaki; Seminole
Eupatorium: Chippewa; Meskwaki
Fraxinus: Costanoan; Iroquois; Meskwaki
Gaura: Hopi; Zuni
Gentiana: Meskwaki
Glandularia: Keres, Western
Goodyera: Potawatomi
Grindelia: Zuni
Gutierrezia: Keres, Western; Navajo; Navajo, Ramah
Helianthus: Apache, White Mountain; Zuni
Heuchera: Blackfoot
Hypericum: Cherokee; Meskwaki
Ipomoea: Houma
Juglans: Meskwaki
Larrea: Papago
Lesquerella: Hopi
Lilium: Chippewa
Liriodendron: Cherokee
Lycopus: Cherokee
Maianthemum: Iroquois
Manfreda: Catawba; Creek; Seminole
Melothria: Houma
Mentzelia: Navajo, Ramah
Nicotiana: Cherokee; Paiute; Zuni
Oenothera: Navajo, Ramah
Opuntia: Lakota
Osmorhiza: Miwok; Paiute; Shoshoni
Osmunda: Cherokee
Penstemon: Navajo; Navajo, Kayenta
Piperacea: Mahuna
Plantago: Cherokee; Chippewa; Mohegan; Ojibwa
Platanthera: Seminole
Polygala: Cherokee; Seminole
Populus: Choctaw
Prenanthes: Iroquois
Psathyrotes: Paiute; Shoshoni
Psilostrophe: Zuni
Ranunculus: Iroquois
Ratibida: Cheyenne
Ribes: Kiowa
Rudbeckia: Cherokee
Sanicula: Micmac; Miwok; Ojibwa
Sanvitalia: Navajo, Ramah
Scirpus: Iroquois
Solanum: Zuni
Sphaeralcea: Tewa
Stephanomeria: Apache, White Mountain; Zuni
Tephrosia: Koasati
Thamnosma: Kawaiisu
Tilia: Cherokee
Toxicodendron: Wailaki
Tragia: Navajo, Kayenta
Triosteum: Meskwaki
Tripterocalyx: Zuni
Veratrum: Paiute; Shoshoni
Verbesina: Zuni
Vitis: Seminole
Waldsteinia: Iroquois
Xanthium: Cherokee
Yucca: Apache
Zigadenus: Paiute

Sports Medicine
Artemisia: Navajo
Asclepias: Iroquois
Chamaecrista: Cherokee
Cornus: Iroquois
Datura: Cahuilla
Dicentra: Iroquois
Equisetum: Quileute
Juncus: Iroquois
Panax: Iroquois
Salix: Quileute
Sarracenia: Iroquois

Stimulant
Abies: Blackfoot; Cheyenne; Ojibwa
Achillea: Chippewa; Iroquois; Navajo; Potawatomi
Acorus: Cherokee; Cree, Alberta
Actaea: Cherokee; Meskwaki
Aesculus: Cherokee
Ageratina: Cherokee; Choctaw; Meskwaki
Alisma: Cree, Woodlands
Allium: Cherokee
Alnus: Tanana, Upper
Anaphalis: Ojibwa
Andropogon: Omaha
Anemone: Meskwaki
Angelica: Kwakiutl
Aquilegia: Shoshoni
Aralia: Choctaw; Meskwaki; Micmac; Montagnais; Ojibwa; Thompson
Arctium: Delaware; Delaware, Oklahoma
Arisaema: Cherokee; Iroquois
Aristolochia: Cherokee
Artemisia: Blackfoot; Chippewa; Paiute, Northern; Potawatomi; Shuswap; Thompson
Arundinaria: Houma
Asarum: Cherokee; Iroquois
Asclepias: Choctaw
Asplenium: Hawaiian
Aster: Meskwaki; Potawatomi
Astragalus: Chippewa
Atriplex: Jemez
Baptisia: Creek
Betula: Iroquois
Brassica: Cherokee
Cannabis: Iroquois
Capsicum: Cherokee
Catabrosa: Shoshoni
Chaenactis: Thompson
Chaetopappa: Hopi
Chamaecrista: Cherokee
Chenopodium: Seminole
Chimaphila: Iroquois
Cimicifuga: Cherokee
Collinsonia: Iroquois
Comptonia: Micmac
Conioselinum: Kwakiutl
Cornus: Carrier, Northern; Cherokee; Meskwaki
Corydalis: Ojibwa
Crataegus: Meskwaki
Cryptantha: Zuni
Cunila: Cherokee
Cypripedium: Iroquois
Daphne: Cherokee
Datura: Hopi
Delphinium: Blackfoot
Desmodium: Houma
Equisetum: Okanagan-Colville
Erigeron: Cheyenne
Eryngium: Choctaw; Seminole
Erythronium: Cherokee
Eupatorium: Cherokee
Fraxinus: Algonquin, Tête-de-Boule; Chippewa
Galactia: Seminole
Gentiana: Navajo
Gentianella: Cherokee
Geum: Chippewa
Gnaphalium: Menominee; Meskwaki
Gymnocladus: Dakota; Omaha; Pawnee; Ponca; Winnebago
Hedeoma: Cherokee
Helianthus: Gros Ventre; Mandan; Ree
Heliopsis: Chippewa
Heracleum: Winnebago
Humulus: Delaware; Delaware, Oklahoma
Hydrangea: Cherokee
Hydrastis: Cherokee; Iroquois
Hymenoxys: Hopi
Ipomoea: Pawnee
Juniperus: Meskwaki; Navajo, Ramah; Seminole
Koeleria: Cheyenne
Lactuca: Cherokee
Larix: Iroquois; Micmac
Lathyrus: Chippewa
Leonurus: Cherokee
Leucothoe: Cherokee
Liatris: Cherokee
Liriodendron: Cherokee; Rappahannock
Lithospermum: Cheyenne
Lomatium: Blackfoot
Lonicera: Carrier, Northern

Lycopodium: Montagnais; Ojibwa
Lysichiton: Kwakiutl
Machaeranthera: Hopi
Maianthemum: Delaware, Oklahoma; Meskwaki; Ojibwa; Potawatomi
Malva: Iroquois
Melissa: Cherokee
Menispermum: Cherokee
Mentha: Cherokee; Cheyenne; Navajo, Ramah
Mirabilis: Paiute
Monarda: Iroquois; Lakota; Meskwaki
Morus: Creek
Myrica: Micmac
Myriophyllum: Iroquois
Nepeta: Cherokee; Keres, Western
Nicotiana: Kawaiisu
Oenothera: Iroquois
Osmorhiza: Cheyenne; Pawnee
Panax: Cherokee; Iroquois
Papaver: Cherokee
Parnassia: Cheyenne
Persea: Seminole
Phacelia: Kawaiisu
Phlox: Cheyenne
Physalis: Meskwaki
Phytolacca: Delaware; Delaware, Oklahoma
Picea: Navajo; Ojibwa
Piloblephis: Seminole
Pinus: Carrier, Northern; Cherokee; Ojibwa; Potawatomi; Shuswap
Piper: Cherokee; Hawaiian
Polygala: Chippewa
Polygonatum: Menominee; Meskwaki
Polystichum: Iroquois
Populus: Cherokee
Potentilla: Okanagon; Thompson
Prenanthes: Choctaw; Iroquois
Prunus: Iroquois
Pteridium: Micmac
Pulsatilla: Cheyenne
Pycnanthemum: Koasati; Meskwaki
Pyrola: Montagnais
Rhamnus: Kawaiisu
Ribes: Carrier, Northern
Rosa: Carrier, Northern; Chippewa; Paiute
Rubus: Carrier, Northern; Cherokee; Iroquois
Rumex: Navajo
Sabal: Houma
Salix: Ojibwa; Seminole
Salvia: Cherokee; Hopi
Sanguinaria: Ojibwa
Sassafras: Rappahannock
Senecio: Jemez
Senna: Cherokee
Sequoia: Pomo; Pomo, Kashaya
Silphium: Cherokee
Sinapis: Cherokee
Solidago: Cherokee; Chippewa; Meskwaki
Stenandrium: Seminole
Taxus: Montagnais

Tephrosia: Cherokee
Tetraneuris: Hopi
Thelesperma: Navajo
Thuja: Iroquois; Menominee
Tilia: Iroquois
Trichostema: Miwok
Tsuga: Iroquois
Ulmus: Iroquois
Umbellularia: Mendocino Indian
Urtica: Makah
Vaccinium: Seminole
Veratrum: Cherokee
Verbascum: Potawatomi
Verbena: Meskwaki
Verbesina: Chickasaw
Veronicastrum: Meskwaki
Yucca: Navajo

Strengthener
Abies: Okanagan-Colville
Acacia: Hawaiian
Adiantum: Hesquiat
Aleurites: Hawaiian
Ambrosia: Makah
Aquilegia: Thompson
Aralia: Iroquois; Malecite
Artemisia: Chippewa; Keres, Western; Okanagan-Colville
Asclepias: Chippewa; Iroquois
Asplenium: Hawaiian
Aster: Zuni
Bidens: Hawaiian
Broussaisia: Hawaiian
Cephalanthus: Seminole
Cercocarpus: Keres, Western
Chamaecyparis: Kwakiutl; Kwakiutl, Southern
Chamaesyce: Hawaiian
Chenopodium: Hawaiian
Cocos: Hawaiian
Collinsonia: Iroquois
Cornus: Thompson
Cupressus: Navajo
Cyperus: Hawaiian
Cyrtandra: Hawaiian
Dalea: Keres, Western
Desmodium: Hawaiian
Digitaria: Hawaiian
Dirca: Iroquois
Frasera: Navajo, Ramah
Freycinetia: Hawaiian
Fucus: Kwakiutl, Southern
Gutierrezia: Zuni
Habenaria: Seminole
Hibiscus: Hawaiian
Hierochloe: Blackfoot
Hydrocotyle: Hawaiian
Hypericum: Cherokee
Ipomoea: Hawaiian
Iris: Zuni
Jacquemontia: Hawaiian
Juncus: Cherokee; Iroquois

Juniperus: Gitksan
Larix: Malecite; Thompson
Larrea: Pima
Ledum: Nitinaht
Lessoniopsis: Nitinaht
Leymus: Nitinaht
Lithocarpus: Yurok
Lomatium: Blackfoot
Mentzelia: Zuni
Merremia: Hawaiian
Musa: Hawaiian
Nephroma: Eskimo, Inuktitut
Ochrosia: Hawaiian
Oenothera: Iroquois
Opuntia: Keres, Western
Pandanus: Hawaiian
Pelea: Hawaiian
Peperomia: Hawaiian
Perideridia: Blackfoot
Phaseolus: Zuni
Phyllospadix: Kwakiutl
Picea: Makah
Piper: Hawaiian
Pipturus: Hawaiian
Plantago: Iroquois
Podophyllum: Iroquois
Populus: Delaware
Portulaca: Hawaiian
Postelsia: Hesquiat; Nitinaht; Nootka
Rheum: Cherokee
Ribes: Thompson
Rosa: Cowlitz
Rumex: Hawaiian; Iroquois
Salix: Seminole
Salvia: Diegueño
Sambucus: Nitinaht
Sanguinaria: Delaware
Sida: Hawaiian
Sideroxylon: Hawaiian
Sorbus: Algonquin, Tête-de-Boule
Sphaeralcea: Navajo, Kayenta
Stillingia: Seminole
Syzygium: Hawaiian
Tanacetum: Cheyenne
Taxus: Swinomish
Touchardia: Hawaiian
Tragia: Keres, Western
Verbascum: Navajo, Ramah
Viburnum: Iroquois
Vigna: Hawaiian
Vitis: Delaware
Waltheria: Hawaiian
Yucca: Keres, Western
Zanthoxylum: Meskwaki
Zigadenus: Keres, Western

Throat Aid
Abies: Anticosti; Bella Coola
Achillea: Aleut; Blackfoot; Cheyenne; Gitksan; Makah; Nitinaht; Paiute, Northern; Saanich
Aconitum: Blackfoot

Acorus: Blackfoot; Cherokee; Chippewa; Cree; Cree, Woodlands; Iroquois; Lakota; Ojibwa
Actaea: Cherokee
Aesculus: Koasati
Allium: Cherokee; Isleta
Amaranthus: Mohegan
Anaphalis: Cherokee
Andropogon: Seminole
Anemone: Ojibwa
Anemopsis: Pima
Angelica: Aleut; Cherokee; Micmac; Paiute; Pomo, Kashaya; Washo
Apocynum: Ojibwa
Aquilegia: Paiute
Aralia: Iroquois; Micmac
Arbutus: Pomo, Kashaya; Salish, Cowichan; Skokomish
Arisaema: Cherokee; Mohegan
Aristolochia: Cherokee
Armoracia: Cherokee
Artemisia: Blackfoot; Havasupai; Kiowa; Lakota; Meskwaki; Okanagan-Colville; Shoshoni; Thompson
Aruncus: Skagit
Asarum: Iroquois; Meskwaki
Asclepias: Blackfoot; Navajo
Astragalus: Navajo, Kayenta
Balsamorhiza: Cheyenne
Berberis: Micmac; Mohegan; Penobscot
Bidens: Cherokee; Hawaiian
Bouteloua: Navajo, Ramah
Calycanthus: Pomo, Kashaya
Cardamine: Cherokee; Malecite; Micmac
Celtis: Houma
Chrysothamnus: Jemez
Cirsium: Hopi; Houma
Claytonia: Skagit; Skagit, Upper
Clematis: Mendocino Indian; Montana Indian
Cleome: Navajo, Ramah
Conioselinum: Aleut
Coptis: Iroquois; Menominee
Cornus: Cherokee; Malecite; Micmac
Corydalis: Navajo, Ramah
Corylus: Thompson
Crotalaria: Seminole
Dalea: Lakota
Datisca: Costanoan
Datura: Rappahannock
Descurainia: Navajo, Kayenta
Dichanthelium: Seminole
Dimorphocarpa: Apache, White Mountain
Diospyros: Cherokee; Rappahannock
Echinacea: Cheyenne; Comanche; Kiowa; Lakota; Sioux, Teton
Epilobium: Snohomish
Eriastrum: Paiute
Ericameria: Cahuilla
Erigeron: Thompson
Eriodictyon: Cahuilla
Eriogonum: Navajo; Zuni
Eupatorium: Cherokee
Frasera: Isleta
Galium: Cherokee
Geranium: Aleut; Iroquois
Geum: Aleut; Blackfoot; Cree, Woodlands
Gilia: Zuni
Glandularia: Keres, Western
Glycyrrhiza: Bannock; Blackfoot; Cherokee; Great Basin Indian; Montana Indian
Gnaphalium: Cherokee
Hamamelis: Cherokee
Helianthemum: Delaware
Heliotropium: Paiute; Shoshoni
Heracleum: Aleut; Chippewa; Shoshoni
Heuchera: Okanagan-Colville; Thompson
Hibiscus: Hawaiian
Hierochloe: Blackfoot
Iris: Malecite; Micmac
Isocoma: Cahuilla
Juniperus: Cheyenne; Shoshoni; Tanana, Upper
Krameria: Pima
Lachnanthes: Cherokee
Larix: Nez Perce
Larrea: Yavapai
Ledum: Oweekeno; Tanana, Upper
Lewisia: Blackfoot
Ligusticum: Zuni
Lithospermum: Zuni
Lobelia: Cherokee
Lomatium: Cowichan; Crow; Gosiute; Kawaiisu; Kwakiutl; Okanagan-Colville; Paiute; Paiute, Northern; Saanich; Shoshoni; Songish; Washo
Lonicera: Swinomish; Thompson
Machaeranthera: Navajo; Paiute
Mahonia: Squaxin
Maianthemum: Ojibwa; Potawatomi; Thompson
Malus: Cherokee
Marrubium: Cherokee; Mahuna; Navajo
Mentha: Paiute
Modiola: Houma
Monarda: Blackfoot; Tewa
Moneses: Haisla & Hanaksiala
Myrica: Choctaw
Nepeta: Iroquois
Nereocystis: Pomo, Kashaya
Nicotiana: Mahuna
Oenothera: Navajo, Ramah
Opuntia: Shuswap
Osmorhiza: Blackfoot; Chippewa; Ojibwa; Paiute; Shoshoni
Oxalis: Cherokee
Oxytropis: Blackfoot
Paeonia: Shoshoni
Panicum: Seminole
Pedicularis: Ojibwa
Pediomelum: Blackfoot
Penstemon: Navajo, Ramah
Perideridia: Blackfoot
Phacelia: Keres, Western
Phryma: Chippewa
Physaria: Blackfoot
Picea: Cree; Cree, Woodlands; Montagnais; Quinault; Tanana, Upper
Pinus: Iroquois; Montagnais; Okanagan-Colville; Okanagon; Paiute; Quinault; Thompson
Plantago: Potawatomi
Platanus: Delaware; Delaware, Oklahoma
Pluchea: Havasupai
Polygala: Cree, Woodlands; Ojibwa, South
Polypodium: Bella Coola; Cowichan; Haisla & Hanaksiala; Hesquiat; Kitasoo; Oweekeno; Saanich; Thompson
Polystichum: Iroquois; Malecite; Micmac; Swinomish
Populus: Bella Coola; Squaxin; Yuki
Porteranthus: Iroquois
Potentilla: Chippewa
Prunella: Cree, Hudson Bay
Prunus: Blackfoot; Cherokee; Chippewa; Iroquois
Pseudotsuga: Hanaksiala; Swinomish
Psilostrophe: Navajo, Ramah
Psoralidium: Arapaho
Psorothamnus: Paiute, Northern
Pyrola: Blackfoot; Mohegan
Quercus: Alabama; Cherokee; Delaware; Delaware, Oklahoma; Houma; Iroquois; Midoo
Ranunculus: Cherokee
Rhexia: Micmac; Montagnais
Rhododendron: Skokomish
Rhus: Iroquois; Kutenai; Micmac; Mohegan; Ojibwa; Potawatomi
Ribes: Bella Coola; Skagit, Upper; Swinomish; Thompson
Rosa: Costanoan; Crow; Skagit; Skagit, Upper; Snohomish; Thompson
Rubus: Cherokee
Rumex: Apache, White Mountain; Cherokee; Mahuna; Papago; Pima; Yavapai; Zuni
Salix: Algonquin, Tête-de-Boule; Blackfoot; Cherokee; Eskimo, Western; Iroquois; Klallam; Pomo, Kashaya; Zuni
Salvia: Costanoan; Shoshoni
Sambucus: Pima
Sanguinaria: Iroquois; Micmac; Ojibwa; Potawatomi
Sanvitalia: Navajo, Ramah
Sarracenia: Micmac
Sassafras: Seminole
Scirpus: Malecite; Micmac
Scutellaria: Iroquois
Sedum: Costanoan
Senna: Meskwaki
Shepherdia: Navajo, Kayenta
Sisyrinchium: Navajo
Smilax: Omaha
Solanum: Cherokee; Navajo
Solidago: Blackfoot; Chippewa
Stachys: Costanoan
Stanleya: Paiute
Stenotus: Navajo

Symphoricarpos: Navajo, Kayenta
Tanacetum: Chippewa
Taraxacum: Aleut
Thlaspi: Iroquois
Townsendia: Hopi; Navajo
Tragopogon: Navajo, Ramah
Triosteum: Iroquois
Tripterocalyx: Zuni
Tsuga: Skagit
Ulmus: Cherokee; Chippewa; Iroquois; Mohegan; Ojibwa; Potawatomi
Umbellularia: Pomo, Kashaya
Vaccinium: Tanana, Upper
Valeriana: Menominee
Veratrum: Paiute; Shoshoni; Thompson
Verbascum: Mohegan
Viburnum: Cree, Woodlands; Eskimo, Chugach; Tanana, Upper
Viola: Ojibwa, South
Waltheria: Hawaiian
Xanthium: Cherokee
Xanthorhiza: Cherokee
Yucca: Navajo
Zanthoxylum: Chippewa; Comanche; Ojibwa
Zea: Navajo
Zinnia: Navajo

Tonic

Abies: Gitksan; Haisla; Karok; Kwakiutl; Thompson; Wet'suwet'en
Acer: Penobscot; Thompson
Achillea: Navajo; Okanagon; Quinault; Thompson
Acorus: Mohegan; Omaha; Rappahannock; Winnebago
Ageratina: Cherokee; Choctaw
Aletris: Cherokee; Micmac
Allium: Iroquois
Alnus: Gitksan; Haisla; Saanich
Amelanchier: Cherokee; Okanagan-Colville; Potawatomi; Thompson
Anaphalis: Delaware, Oklahoma
Anemopsis: Shoshoni
Angelica: Aleut; Shoshoni
Anthemis: Cherokee
Apocynum: Montana Indian
Aralia: Cherokee; Delaware, Oklahoma; Iroquois; Mohegan; Montagnais; Montana Indian; Okanagon; Penobscot; Potawatomi; Thompson
Arctium: Ojibwa; Potawatomi
Arctostaphylos: Okanagon; Thompson
Aristolochia: Cherokee; Delaware; Delaware, Oklahoma
Armoracia: Cherokee
Artemisia: Aleut; Chippewa; Eskimo, Alaska; Montana Indian; Okanagan-Colville; Paiute; Shoshoni; Washo
Aruncus: Quileute
Asarum: Iroquois; Micmac; Skagit; Thompson
Asclepias: Menominee; Thompson
Astragalus: Chippewa
Baptisia: Penobscot

Betula: Menominee; Mohegan
Brachyactis: Paiute
Brassica: Cherokee
Calla: Gitksan
Cardamine: Malecite; Micmac
Carpinus: Delaware, Ontario
Carya: Delaware, Ontario
Catabrosa: Shoshoni
Caulophyllum: Iroquois
Cephalanthus: Choctaw
Chaenactis: Thompson
Chamaesyce: Shoshoni
Chenopodium: Creek; Rappahannock
Chimaphila: Iroquois; Rappahannock; Thompson
Chrysopsis: Delaware; Delaware, Oklahoma
Chrysothamnus: Shoshoni
Cichorium: Cherokee
Cimicifuga: Cherokee; Delaware; Delaware, Oklahoma
Cinchona: Cherokee
Claytonia: Skagit
Clematis: Navajo; Okanagon; Thompson
Comptonia: Menominee
Cornus: Cherokee; Delaware; Delaware, Oklahoma; Hoh; Quileute; Rappahannock
Cucurbita: Omaha
Cunila: Cherokee
Dirca: Iroquois
Echinocystis: Menominee; Ojibwa
Ephedra: Hopi; Paiute; Shoshoni; Tewa
Erythrina: Choctaw
Euonymus: Cherokee
Eupatorium: Cherokee; Mohegan; Penobscot; Rappahannock
Foeniculum: Cherokee
Frangula: Cahuilla; Salish, Coast
Frasera: Cherokee; Shoshoni
Fraxinus: Chippewa; Ojibwa
Galium: Penobscot
Gaultheria: Chippewa; Delaware; Delaware, Oklahoma; Skagit
Gentiana: Dakota; Winnebago
Gentianella: Cherokee
Geum: Aleut; Blackfoot; Thompson
Gilia: Navajo, Kayenta
Gleditsia: Delaware, Oklahoma; Meskwaki
Glycyrrhiza: Montana Indian
Goodyera: Cowlitz
Gymnocladus: Omaha
Hedeoma: Dakota
Helianthemum: Delaware, Oklahoma
Heracleum: Makah; Okanagon; Thompson
Heuchera: Chickasaw; Choctaw; Creek; Shoshoni
Hieracium: Okanagan-Colville
Holodiscus: Makah
Humulus: Delaware, Oklahoma
Hydrastis: Cherokee
Inula: Delaware; Delaware, Oklahoma
Ipomopsis: Navajo, Kayenta; Shoshoni
Ivesia: Arapaho

Juglans: Potawatomi
Juniperus: Blackfoot; Creek; Delaware, Ontario; Iroquois; Malecite; Micmac; Okanagan-Colville; Okanagon; Paiute; Shoshoni; Swinomish; Thompson
Kalmia: Cree, Hudson Bay
Larrea: Cahuilla
Lathyrus: Chippewa
Ledum: Algonquin, Quebec; Micmac; Salish
Leiophyllum: Nanticoke
Leonurus: Iroquois; Mohegan
Leucanthemum: Mohegan; Shinnecock
Liatris: Omaha
Lindera: Cherokee
Lomatium: Blackfoot; Cheyenne; Great Basin Indian
Lonicera: Skagit; Thompson
Lupinus: Salish
Lygodesmia: Blackfoot
Machaeranthera: Shoshoni
Mahonia: Karok; Montana Indian; Samish; Swinomish; Thompson
Maianthemum: Delaware, Oklahoma; Mohegan; Washo
Malus: Makah; Nitinaht
Matricaria: Aleut
Melissa: Cherokee
Mentha: Delaware, Oklahoma; Flathead; Kutenai
Menyanthes: Aleut
Monardella: Shoshoni
Nemopanthus: Potawatomi
Nepeta: Cherokee; Delaware, Oklahoma; Rappahannock
Oplopanax: Gitksan; Haisla; Okanagon; Oweekeno; Thompson; Wet'suwet'en
Orbexilum: Cherokee
Osmorhiza: Iroquois; Shoshoni
Osmunda: Cherokee
Ostrya: Delaware, Ontario
Panax: Cherokee; Delaware; Delaware, Oklahoma; Iroquois; Menominee; Mohegan; Seminole
Pedicularis: Washo
Phoenicaulis: Paiute
Picea: Gitksan; Montagnais; Wet'suwet'en
Pimpinella: Delaware, Oklahoma
Pinus: Gitksan; Paiute
Plantago: Aleut
Podophyllum: Delaware; Delaware, Oklahoma
Polygala: Chippewa
Polygonatum: Cherokee
Polygonum: Aleut
Populus: Iroquois; Shoshoni
Potentilla: Okanagon; Thompson
Prunella: Thompson
Prunus: Delaware; Delaware, Oklahoma; Delaware, [*Prunus*]
  Ontario; Makah; Mendocino Indian; Micmac; Mohegan; Okanagan-Colville; Paiute; Penobscot; Potawatomi; Rappahannock; Thompson
Pseudotsuga: Swinomish; Thompson
Pteridium: Cherokee

Pteryxia: Okanagan-Colville
Purshia: Paiute; Shoshoni
Pyrola: Penobscot
Quercus: Cherokee; Delaware, Ontario; Houma; Rappahannock
Rhamnus: Iroquois
Ribes: Lummi; Thompson
Rosa: Chippewa; Paiute; Shoshoni; Thompson
Rubus: Cherokee; Choctaw; Iroquois; Karok; Okanagon; Rappahannock; Thompson
Rumex: Iroquois; Mohegan; Paiute; Shoshoni
Salix: Cherokee; Klallam; Menominee; Paiute; Skagit
Salvia: Mohegan
Sambucus: Houma; Paiute
Sanguinaria: Algonquin, Quebec; Delaware, Oklahoma; Mohegan
Sassafras: Delaware, Oklahoma; Iroquois; Mohegan; Rappahannock
Sequoia: Pomo
Shepherdia: Salish; Thompson
Silphium: Pawnee
Sinapis: Cherokee
Smilax: Cherokee; Choctaw
Solanum: Comanche
Solidago: Cherokee; Chippewa; Cree, Hudson Bay; Thompson
Sorbus: Algonquin, Quebec; Wet'suwet'en
Spiraea: Thompson
Stachys: Saanich
Stanleya: Paiute
Stephanomeria: Shoshoni
Streptopus: Okanagon; Penobscot; Thompson
Tanacetum: Cherokee
Taraxacum: Delaware, Oklahoma; Mohegan; Potawatomi; Shinnecock
Tetradymia: Hopi
Thamnosma: Shoshoni
Thuja: Okanagan-Colville
Tiarella: Iroquois
Toxicodendron: Houma
Urtica: Nitinaht; Samish; Shoshoni; Swinomish
Verbena: Cherokee
Veronicastrum: Cherokee
Viburnum: Cherokee
Viola: Cherokee
Vitis: Cherokee; Choctaw; Delaware, Oklahoma
Wyethia: Paiute
Xanthorhiza: Cherokee
Zanthoxylum: Delaware, Oklahoma

Toothache Remedy

Abies: Shuswap
Achillea: Carrier; Costanoan; Cree, Woodlands; Creek; Mahuna; Okanagan-Colville; Paiute; Saanich; Shoshoni; Thompson
Acorus: Blackfoot; Chippewa; Cree, Woodlands; Dakota; Iroquois; Lakota; Omaha; Pawnee; Ponca; Winnebago
Actaea: Cherokee
Adenostoma: Diegueño
Aesculus: Costanoan; Mendocino Indian
Ageratina: Chickasaw; Choctaw
Alnus: Algonquin, Quebec; Cherokee; Okanagan-Colville; Thompson
Ambrosia: Zuni
Amelanchier: Cree, Woodlands
Antennaria: Iroquois
Anthemis: Iroquois
Arabis: Thompson
Aralia: Cherokee; Cree, Woodlands
Argemone: Shoshoni
Aristolochia: Cherokee
Armoracia: Mohegan
Artemisia: Costanoan; Navajo, Ramah; Shoshoni
Asarum: Pomo, Kashaya
Asclepias: Iroquois; Navajo, Kayenta
Aster: Cree, Woodlands; Kawaiisu; Okanagan-Colville
Astragalus: Navajo, Ramah; Shoshoni
Atriplex: Navajo, Ramah
Balsamorhiza: Cheyenne
Baptisia: Cherokee
Betula: Cree, Woodlands
Brassica: Mohegan; Shinnecock
Caulophyllum: Cherokee
Ceanothus: Cherokee
Celastrus: Iroquois
Centaurium: Miwok
Cephalanthus: Choctaw
Chaetopappa: Navajo, Ramah
Chamaesyce: Cherokee; Navajo, Ramah
Chenopodium: Miwok
Chrysothamnus: Cahuilla; Isleta; Shoshoni
Clematis: Thompson
Coix: Cherokee
Coptis: Algonquin, Quebec; Menominee
Corylus: Iroquois
Cryptantha: Navajo, Ramah
Cypripedium: Chippewa
Dalea: Navajo
Datura: Cahuilla; Navajo
Descurainia: Navajo, Ramah
Dicentra: Skagit
Dimorphocarpa: Navajo, Kayenta
Diospyros: Cherokee
Dryopteris: Cherokee
Echinacea: Blackfoot; Cheyenne; Comanche; Crow; Dakota; Lakota; Omaha; Pawnee; Ponca; Sioux; Sioux, Teton; Winnebago
Encelia: Cahuilla
Equisetum: Iroquois
Erigenia: Cherokee
Erigeron: Kawaiisu; Miwok
Eryngium: Cherokee
Erysimum: Navajo, Ramah
Eschscholzia: California Indian; Mendocino Indian
Euphorbia: Cherokee
Fragaria: Cherokee; Iroquois
Frangula: Neeshenam
Geranium: Meskwaki
Geum: Cree, Woodlands
Glycyrrhiza: Dakota; Lakota; Pawnee; Sioux
Goodyera: Cherokee
Gutierrezia: Cahuilla
Hamamelis: Iroquois
Hedeoma: Cherokee
Heracleum: Cree; Shoshoni; Washo
Heterotheca: Navajo, Ramah
Heuchera: Navajo
Humulus: Delaware; Delaware, Oklahoma; Mohegan
Hymenopappus: Hopi
Hypericum: Houma; Shoshoni
Iris: Great Basin Indian; Paiute; Shoshoni
Juglans: Cherokee; Iroquois
Juniperus: Cree, Woodlands; Shoshoni; Tewa
Larrea: Pima
Lesquerella: Navajo, Ramah
Lithocarpus: Costanoan
Lycium: Navajo, Kayenta; Navajo, Ramah
Magnolia: Cherokee; Iroquois
Melica: Kawaiisu
Mentha: Cree, Woodlands; Flathead
Mentzelia: Hopi; Navajo, Ramah
Monarda: Flathead
Monotropa: Cree, Woodlands
Nicotiana: Cherokee; Kawaiisu; Montauk; Rappahannock; Shinnecock; Shoshoni; Tewa
Nuphar: Okanagan-Colville
Nymphaea: Okanagan-Colville
Oenothera: Hopi
Osmorhiza: Okanagan-Colville; Shoshoni
Ostrya: Cherokee
Parryella: Hopi; Tewa
Pediomelum: Blackfoot
Pennellia: Navajo, Ramah
Penstemon: Okanagan-Colville; Paiute
Pentagrama: Miwok
Pericome: Navajo, Ramah
Persea: Mahuna
Phaseolus: Papago
Phlox: Navajo; Navajo, Kayenta
Phoradendron: Mendocino Indian
Physaria: Blackfoot
Picea: Cree, Woodlands; Shuswap; Tlingit
Polygala: Cree, Woodlands
Polymnia: Iroquois
Polystichum: Cherokee
Populus: Cherokee
Porteranthus: Cherokee
Psathyrotes: Paiute
Psorothamnus: Paiute
Pteridium: Makah
Quercus: Costanoan; Mahuna
Ranunculus: Iroquois
Rhus: Cheyenne
Ribes: Luiseño
Robinia: Cherokee
Rorippa: Mohegan
Rubus: Cherokee; Cree, Woodlands; Makah

Rumex: Yavapai
Salix: Kiowa; Shoshoni; Thompson
Sambucus: Cahuilla; Okanagon; Thompson
Sanvitalia: Navajo, Ramah
Sarcobatus: Paiute, Northern
Satureja: Costanoan
Shepherdia: Navajo, Kayenta
Solanum: Costanoan; Zuni
Solidago: Alabama; Miwok
Sonchus: Houma
Stanleya: Shoshoni
Stenotus: Navajo
Stephanomeria: Paiute
Symplocarpus: Meskwaki
Taraxacum: Cherokee; Iroquois
Tauschia: Kawaiisu
Thelesperma: Navajo
Thuja: Algonquin, Quebec; Cowlitz; Malecite; Micmac
Trichostema: Miwok
Tsuga: Tlingit
Ulmus: Creek
Veratrum: Paiute
Verbascum: Abnaki; Iroquois
Vernonia: Cherokee
Viburnum: Cree, Woodlands
Zanthoxylum: Alabama; Comanche; Houma; Iroquois; Meskwaki
Zephyranthes: Seminole
Zigadenus: Paiute; Shoshoni; Washo
Zingiber: Hawaiian

Tuberculosis Remedy
Abies: Bella Coola; Blackfoot; Carrier; Cree, Woodlands; Gitksan; Iroquois; Kwakiutl; Okanagan-Colville; Paiute; Potawatomi; Shuswap; Thompson; Washo
Acer: Klallam
Achillea: Aleut; Cheyenne; Mendocino Indian; Paiute; Quinault
Achlys: Cowlitz; Skagit
Acorus: Meskwaki
Adenocaulon: Squaxin
Aesculus: Creek
Aletris: Cherokee
Aleurites: Hawaiian
Alisma: Iroquois
Alnus: Blackfoot; Hesquiat; Kwakiutl; Mendocino Indian; Nitinaht; Swinomish
Amphicarpaea: Iroquois
Anaphalis: Bella Coola; Montagnais
Anemone: Cowlitz; Iroquois; Ojibwa
Anemopsis: Pima
Angelica: Shoshoni
Antidesma: Hawaiian
Apium: Houma
Aralia: Algonquin, Quebec; Iroquois; Malecite; Mendocino Indian; Micmac
Arceuthobium: Carrier, Southern
Arctium: Menominee
Arisaema: Cherokee; Iroquois

Arnica: Thompson
Artemisia: Meskwaki; Montana Indian; Okanagan-Colville; Sanpoil; Shuswap; Tanaina; Thompson
Asarum: Iroquois; Skagit
Asclepias: Paiute
Asplenium: Hawaiian
Aster: Iroquois
Balsamorhiza: Flathead; Paiute
Betula: Creek
Botrychium: Iroquois; Ojibwa
Boykinia: Quileute
Caltha: Chippewa
Campanulastrum: Meskwaki
Cardamine: Delaware, Oklahoma; Iroquois
Carpinus: Iroquois
Carya: Kiowa
Cassiope: Thompson
Celastrus: Delaware, Ontario
Cenchrus: Hawaiian
Centaurium: Miwok
Cercocarpus: Paiute; Paiute, Northern; Shoshoni
Chaenactis: Sanpoil
Chamaesyce: Hawaiian
Chimaphila: Cherokee; Delaware, Oklahoma; Malecite; Micmac; Okanagan-Colville
Chrysothamnus: Cheyenne; Thompson
Cimicifuga: Cherokee
Cirsium: Mohegan; Montagnais
Collinsia: Creek; Natchez
Commelina: Keres, Western
Comptonia: Delaware, Oklahoma
Conopholis: Keres, Western
Coprosma: Hawaiian
Corallorrhiza: Iroquois
Cornus: Iroquois; Meskwaki
Crataegus: Chippewa
Cucurbita: Cheyenne
Cynoglossum: Iroquois
Cypripedium: Iroquois
Desmodium: Hawaiian
Dirca: Iroquois
Epilobium: Iroquois; Miwok; Skokomish
Eriastrum: Paiute
Erigeron: Iroquois; Okanagan-Colville
Eriodictyon: Cahuilla; Costanoan; Karok; Paiute; Round Valley Indian; Shoshoni
Eriogonum: Paiute; Shoshoni; Thompson
Erysimum: Hopi
Eschscholzia: Mendocino Indian
Eupatorium: Iroquois
Fagus: Iroquois
Fomitopsis: Haisla & Hanaksiala
Frasera: Okanagan-Colville
Gaillardia: Thompson
Galium: Ojibwa
Gaultheria: Samish; Swinomish
Glossopetalon: Shoshoni
Gnaphalium: Montagnais
Grindelia: Flathead; Ponca; Sanpoil
Hamamelis: Cherokee; Iroquois

Helianthus: Paiute
Heracleum: Shoshoni
Heuchera: Kutenai
Hieracium: Iroquois
Hydrastis: Iroquois
Hypericum: Menominee; Meskwaki
Ilex: Micmac
Inula: Cherokee; Iroquois; Mohegan
Ipomoea: Cherokee; Iroquois
Iris: Chippewa; Delaware, Oklahoma
Juglans: Iroquois
Juniperus: Carrier; Malecite; Micmac; Okanagan-Colville; Sanpoil; Tanana, Upper; Thompson
Laportea: Iroquois
Larix: Kutenai; Malecite; Micmac; Thompson
Larrea: Coahuilla; Pima
Ledum: Haisla & Hanaksiala
Lepidium: Houma
Ligusticum: Pomo, Kashaya
Lilium: Malecite; Micmac
Limonium: Micmac
Lobelia: Iroquois
Lomatium: Nez Perce; Okanagan-Colville; Paiute; Shoshoni; Washo
Lonicera: Iroquois; Lummi
Lophophora: Delaware; Kiowa
Lysichiton: Klallam
Mahonia: Miwok; Nitinaht; Sanpoil
Maianthemum: Delaware, Oklahoma; Hesquiat
Malus: Gitksan; Iroquois; Makah
Marah: Chehalis
Mentzelia: Keres, Western
Mertensia: Cherokee
Nemopanthus: Malecite
Nicotiana: Hawaiian; Iroquois; Mahuna; Paiute; Shoshoni
Nuphar: Bella Coola; Haisla & Hanaksiala
Nymphaea: Ojibwa
Nyssa: Creek
Oemleria: Makah
Onoclea: Iroquois
Oplopanax: Gitksan; Haisla; Kwakiutl; Okanagan-Colville; Skagit; Wet'suwet'en
Ostrya: Iroquois
Paeonia: Paiute; Shoshoni; Washo
Panax: Cherokee; Iroquois
Pastinaca: Paiute
Paxistima: Okanagan-Colville; Thompson
Pedicularis: Iroquois
Peltigera: Nitinaht
Penstemon: Creek
Petasites: Concow; Skagit; Tanaina
Phoradendron: Creek
Phyllodoce: Thompson
Physocarpus: Bella Coola
Picea: Carrier; Gitksan; Haisla & Hanaksiala; Iroquois; Okanagan-Colville; Shuswap; Tanana, Upper
Pinus: Bella Coola; Blackfoot; Cherokee; Gitksan; Hopi; Iroquois; Kutenai; Lummi; Montagnais;

[*Pinus*]
  Paiute; Shuswap; Skagit; Yokia
Pittosporum: Hawaiian
Platanthera: Iroquois
Platanus: Creek
Polypodium: Cree, Woodlands
Polystichum: Iroquois
Populus: Bella Coola; Kutenai; Quinault; Shoshoni
Prunella: Iroquois
Prunus: Bella Coola; Chippewa; Iroquois; Kwakiutl; Malecite; Micmac; Paiute
Psathyrotes: Shoshoni
Psoralidium: Dakota
Psorothamnus: Paiute; Shoshoni
Pteridium: Iroquois
Purshia: Paiute
Quercus: Cowlitz; Iroquois
Ranunculus: Thompson
Rhus: Flathead; Hopi; Iroquois; Kiowa; Malecite; Menominee
Ribes: Skokomish; Swinomish; Thompson
Rubus: Iroquois; Menominee; Micmac
Rumex: Hawaiian; Squaxin
Sagittaria: Algonquin, Quebec
Salix: Delaware, Oklahoma; Iroquois; Klallam
Sambucus: Pomo, Little Lakes; Shoshoni
Sanguinaria: Iroquois; Malecite; Micmac
Sarracenia: Malecite; Micmac
Scirpus: Iroquois
Sedum: Eskimo, Western
Senecio: Catawba
Shepherdia: Carrier; Flathead; Shuswap; Tanana, Upper; Thompson
Sinapis: Micmac
Solanum: Comanche
Solidago: Cherokee
Spiraea: Iroquois
Stephanomeria: Cheyenne
Styphelia: Hawaiian
Symphoricarpos: Skagit
Symplocarpus: Iroquois
Taxus: Iroquois
Tephrosia: Creek
Thelesperma: Keres, Western
Thuja: Clallam; Klallam; Malecite; Micmac
Tilia: Cherokee; Iroquois
Torreya: Pomo; Pomo, Kashaya
Trientalis: Montagnais
Tsuga: Chehalis; Hesquiat; Iroquois; Klallam; Shuswap
Tussilago: Iroquois
Ulmus: Catawba; Cherokee; Iroquois
Urtica: Tanaina
Valeriana: Navajo, Ramah; Thompson
Veratrum: Iroquois
Verbascum: Iroquois; Salish
Veronicastrum: Chippewa
Viburnum: Gitksan; Iroquois
Wyethia: Paiute
Xanthium: Mahuna
Zanthoxylum: Iroquois; Meskwaki

Unspecified
Abies: Abnaki; Algonquin, Quebec; Carrier, Southern; Haisla; Kitasoo; Malecite; Menominee; Nitinaht; Thompson
Acer: Algonquin, Quebec; Iroquois
Achillea: Eskimo, Alaska; Eskimo, Nunivak; Haisla & Hanaksiala; Thompson
Aconitum: Salish
Acorus: Cree, Alberta; Malecite; Micmac
Adiantum: Apalachee
Agastache: Cheyenne; Meskwaki
Alnus: Bella Coola; Gitksan; Hoh; Kawaiisu; Malecite; Nitinaht; Quileute; Shuswap
Ambrosia: Lakota
Amelanchier: Cheyenne
Amsinckia: Costanoan
Anaphalis: Cheyenne
Anemone: Thompson
Anemopsis: Diegueño
Angelica: Blackfoot; Eskimo, Inuktitut; Gosiute; Kwakiutl
Apocynum: Cahuilla; Cree, Hudson Bay; Salish; Thompson
Aralia: Algonquin, Quebec; Ojibwa
Arctium: Hoh; Quileute
Arctostaphylos: Cheyenne; Hoh; Ojibwa; Pomo; Quileute
Argentina: Tsimshian
Arisaema: Ojibwa
Arnica: Thompson
Artemisia: Bella Coola; Cahuilla; Cheyenne; Eskimo, Kuskokwagmiut; Eskimo, Nunivak; Havasupai; Lakota; Luiseño; Mewuk; Montana Indian; Navajo; Omaha; Pawnee; Ponca; Thompson; Ute; Winnebago; Yokut
Aruncus: Haihais; Kitasoo; Makah
Asarum: Karok; Thompson
Asclepias: Iroquois; Lakota; Miwok; Potawatomi; Zuni
Aster: Ojibwa; Okanagan-Colville; Potawatomi
Baptisia: Ojibwa
Berula: Apache, White Mountain
Betula: Algonquin, Quebec; Iroquois; Koyukon; Malecite
Bignonia: Creek
Botrychium: Abnaki; Potawatomi
Boykinia: Yuki
Brassica: Hoh; Ojibwa; Quileute
Brickellia: Gosiute
Calochortus: Cheyenne
Caltha: Aleut
Calypso: Thompson
Camassia: Creek
Capsella: Meskwaki
Carex: Gosiute
Castilleja: Thompson
Ceanothus: Oregon Indian, Warm Springs; Thompson
Celastrus: Ojibwa
Cephalanthus: Seminole
Cercocarpus: Shoshoni
Chaenactis: Thompson
Chamaesyce: Hawaiian
Chelone: Algonquin, Quebec
Cicuta: Ojibwa
Clematis: Gosiute
Clintonia: Potawatomi
Cneoridium: Luiseño
Cocos: Hawaiian
Coix: Cherokee
Colocasia: Hawaiian
Comptonia: Chippewa
Coprosma: Hawaiian
Coptis: Iroquois; Malecite; Micmac
Cornus: Algonquin, Quebec; Carrier, Northern; Ojibwa; Okanagan-Colville
Cuminum: Apalachee
Datura: Cahuilla; Chumash; Diegueño; Gabrielino; Luiseño
Delphinium: Blackfoot; Navajo
Descurainia: Ute
Dicentra: Thompson
Dioscorea: Hawaiian
Disporum: Okanagan-Colville
Eleocharis: Seminole
Ephedra: Coahuilla; Hualapai
Epilobium: Potawatomi; Thompson
Equisetum: Miwok; Ojibwa
Ericameria: Luiseño
Erigeron: Thompson
Eriodictyon: Luiseño
Eriogonum: Apache, White Mountain; Navajo; Thompson; Ute
Eriophorum: Eskimo, Inuktitut
Eryngium: Seminole
Fagus: Menominee
Fragaria: Iroquois
Frangula: Kawaiisu; Montana Indian
Frasera: Apache
Fraxinus: Ojibwa
Fritillaria: Ute
Galactia: Seminole
Garrya: Wintoon
Gaultheria: Algonquin, Quebec; Micmac
Geocaulon: Cree, Hudson Bay
Geranium: Okanagan-Colville; Thompson
Geum: Gosiute; Thompson
Glycyrrhiza: Lakota
Gnaphalium: Karok; Weyot
Goodyera: Delaware, Oklahoma
Guajacum: Seri
Hackelia: Thompson
Halosaccion: Hesquiat
Hedysarum: Ute
Heracleum: Klamath; Makah; Micmac; Tanaina; Thompson
Heuchera: Arapaho; Thompson
Hieracium: Thompson
Holodiscus: Makah; Shoshoni

Hordeum: Potawatomi
Hymenoxys: Hopi
Hypericum: Meskwaki
Impatiens: Ojibwa
Iva: Ute
Juncus: Hoh; Karok; Quileute
Juniperus: Gitksan; Navajo; Sanpoil; Seminole
Kalmia: Gosiute; Thompson
Lactuca: Potawatomi
Larix: Algonquin, Quebec; Malecite; Ojibwa
Larrea: Kawaiisu; Pima
Leathesia: Hesquiat
Ledum: Algonquin, Quebec; Anticosti; Eskimo, Alaska; Eskimo, Inupiat; Eskimo, Nunivak; Makah; Nootka; Potawatomi
Leymus: Makah; Quileute
Licania: Seminole
Ligusticum: Bella Coola; Flathead; Kutenai; Thompson
Lindera: Chippewa
Linnaea: Thompson
Linum: Paiute
Lithospermum: Great Basin Indian; Hopi
Lomatium: Creek; Gosiute; Great Basin Indian
Lonicera: Cowichan
Lophophora: Paiute; Ponca
Lupinus: Thompson
Lycopodium: Penobscot
Lysichiton: Hesquiat; Makah
Lythrum: Kawaiisu
Machaeranthera: Zuni
Mahonia: Cheyenne; Havasupai; Nitinaht
Maianthemum: Kitasoo; Ojibwa
Malus: Makah; Nitinaht
Marrubium: Hopi
Matricaria: Aleut; Cheyenne; Eskimo, Inuktitut; Ute
Mentha: Hoh; Montana Indian; Quileute; Salish; Thompson
Mentzelia: Cheyenne; Thompson
Menyanthes: Aleut; Menominee; Micmac; Tlingit
Mimulus: Potawatomi
Mitchella: Chippewa; Penobscot
Monarda: Ojibwa
Monardella: Diegueño; Karok; Luiseño
Moricandia: Hoh; Quileute
Morinda: Hawaiian
Myriophyllum: Menominee
Nemopanthus: Ojibwa
Nicotiana: Karok
Nuphar: Haisla & Hanaksiala; Hesquiat; Nitinaht
Nymphaea: Potawatomi
Oemleria: Makah
Oenothera: Gosiute; Potawatomi
Oplopanax: Hoh; Makah; Nitinaht; Quileute
Osmorhiza: Sanpoil
Osmunda: Menominee; Seminole
Panax: Delaware; Seminole
Paxistima: Bella Coola
Pedicularis: Thompson
Pediomelum: Blackfoot; Cheyenne; Lakota; Seminole
Pellaea: Luiseño
Peltandra: Nanticoke
Penstemon: Salish
Perideridia: Cheyenne
Persea: Seminole
Phacelia: Thompson
Phlebodium: Seminole
Phoradendron: Cahuilla; Hopi
Physalis: Meskwaki; Omaha; Ponca; Winnebago
Phytolacca: Cherokee
Picea: Algonquin, Quebec; Eskimo, Alaska; Koyukon; Malecite; Ojibwa; Ojibwa, South; Oweekeno; Thompson
Pinus: Delaware, Ontario; Eskimo, Alaska; Gitksan; Kwakwaka'wakw; Menominee; Ojibwa; Pomo; Thompson
Plantago: Delaware; Delaware, Oklahoma; Ojibwa
Platanus: Kawaiisu
Polygala: Chippewa; Ojibwa
Polygonum: Apache, White Mountain; Potawatomi
Polypodium: Makah
Populus: Hoh; Malecite; Ojibwa; Quileute; Thompson; Yurok
Potentilla: Eskimo; Gosiute; Potawatomi
Prosopis: Cahuilla
Prunus: Algonquin, Quebec; Blackfoot; Cheyenne; Gitksan; Malecite; Navajo; Rappahannock; Sanpoil & Nespelem; Thompson
Pseudotsuga: Haisla; Havasupai; Kwakiutl
Pterocaulon: Seminole
Pulsatilla: Cheyenne; Thompson
Purshia: Apache, White Mountain; Gosiute
Pyrola: Chippewa
Quercus: Mahuna; Malecite; Menominee; Ojibwa
Ratibida: Keres, Western
Rhamnus: Yuki
Rhus: Algonquin, Quebec; Hopi; Ojibwa
Ribes: Bella Coola; Gitksan; Hanaksiala; Ojibwa; Shoshoni; Tanana, Upper
Rosa: Cheyenne; Nitinaht; Ojibwa; Potawatomi; Yurok
Rubus: Algonquin, Quebec; Iroquois; Menominee
Rumex: Apalachee; Kawaiisu; Luiseño; Navajo; Ojibwa; Pima
Sagittaria: Cheyenne; Chippewa; Lakota; Ojibwa
Salix: Makah; Menominee; Meskwaki; Ojibwa; Seminole; Thompson
Sambucus: Iroquois; Mewuk; Ojibwa; Wet'suwet'en
Sanguinaria: Chippewa
Sanicula: Pomo, Kashaya
Sassafras: Creek; Seminole
Satureja: Luiseño
Saururus: Seminole
Saxifraga: Menominee
Scrophularia: Meskwaki
Sedum: Eskimo, Nunivak
Senecio: Kawaiisu; Makah; Ute
Shepherdia: Algonquin, Quebec; Cheyenne; Flathead; Shuswap
Silene: Karok; Menominee
Sinapis: Hoh; Quileute
Smilax: Rappahannock
Solanum: Apache, White Mountain
Solidago: Menominee
Sorbus: Carrier; Micmac; Ojibwa; Potawatomi; Wet'suwet'en
Sphaeralcea: Shoshoni
Spiraea: Algonquin, Quebec; Potawatomi
Stenanthium: Thompson
Stephanomeria: Thompson
Streptopus: Thompson
Symphoricarpos: Sanpoil
Symplocarpus: Malecite; Meskwaki
Taraxacum: Cherokee; Hoh; Quileute
Tauschia: Luiseño
Taxus: Algonquin, Quebec; Malecite
Tellima: Nitinaht
Thuja: Chippewa; Hoh; Potawatomi; Quileute
Tilia: Algonquin, Quebec
Townsendia: Navajo, Kayenta
Toxicodendron: Delaware
Trichostema: Karok
Trifolium: Apalachee; Makah
Triosteum: Creek
Triteleia: Thompson
Tsuga: Algonquin, Quebec; Malecite
Turricula: Luiseño
Typha: Apache, Mescalero
Umbellularia: Yurok
Urtica: Gitksan; Nitinaht; Paiute; Potawatomi; Thompson
Vaccinium: Mohegan; Potawatomi; Seminole
Vancouveria: Tolowa
Veratrum: Hanaksiala
Verbascum: Atsugewi
Verbesina: Seminole
Viola: Ute
Vitex: Hawaiian
Vittaria: Seminole
Ximenia: Seminole
Yucca: Choctaw; Omaha

Urinary Aid
Abies: Cherokee; Iroquois
Acalypha: Cherokee
Achillea: Cherokee; Paiute; Thompson
Acorus: Cherokee
Actaea: Iroquois; Meskwaki
Ageratina: Cherokee
Agrimonia: Ojibwa
Aletris: Cherokee
Allium: Cherokee
Alnus: Cherokee; Iroquois
Ampelopsis: Cherokee
Andropogon: Seminole
Apocynum: Potawatomi
Aquilegia: Meskwaki
Arabis: Okanagon; Thompson
Aralia: Iroquois
Arctium: Cherokee

Arctostaphylos: Cherokee; Costanoan; Okanagan-Colville; Okanagon; Thompson
Aristolochia: Cherokee
Armoracia: Cherokee
Artemisia: Costanoan; Paiute
Aruncus: Cherokee
Asarum: Iroquois
Asclepias: Cherokee; Iroquois
Balsamorhiza: Flathead; Paiute
Betula: Cherokee
Callicarpa: Seminole
Calycanthus: Cherokee
Carpinus: Cherokee
Caulophyllum: Meskwaki
Ceanothus: Iroquois
Celastrus: Creek; Iroquois
Cephalanthus: Seminole
Chaetopappa: Navajo, Kayenta
Chamaesyce: Cherokee
Chaptalia: Seminole
Chimaphila: Cherokee; Delaware; Delaware, Oklahoma; Iroquois; Karok; Micmac
Chrysothamnus: Thompson
Citrullus: Cherokee; Chickasaw; Iroquois
Claytonia: Quileute
Clematis: Iroquois; Thompson
Clintonia: Micmac
Comptonia: Delaware; Delaware, Oklahoma
Conioselinum: Micmac
Cornus: Iroquois; Shuswap
Crataegus: Meskwaki
Cucurbita: Cherokee
Cynoglossum: Cherokee
Cypripedium: Algonquin, Tête-de-Boule; Menominee
Diervilla: Iroquois; Meskwaki; Ojibwa
Diplacus: Costanoan
Dirca: Iroquois; Ojibwa
Echium: Cherokee
Eleocharis: Seminole
Elytrigia: Cherokee; Iroquois
Ephedra: Keres, Western; Paiute; Shoshoni
Epilobium: Costanoan; Iroquois; Miwok
Equisetum: Carrier; Chippewa; Costanoan; Iroquois; Mahuna; Modesse; Potawatomi; Thompson
Eriogonum: Costanoan; Paiute
Eryngium: Meskwaki
Erythrina: Seminole
Euonymus: Cherokee; Iroquois
Eupatorium: Cherokee; Koasati; Menominee
Euphorbia: Cherokee
Fragaria: Cherokee
Fraxinus: Iroquois
Galium: Iroquois; Meskwaki; Ojibwa
Goodyera: Potawatomi
Grindelia: Paiute; Shoshoni
Gutierrezia: Navajo, Ramah; Zuni
Heracleum: Micmac; Shuswap
Hibiscus: Shinnecock

Hordeum: Costanoan
Houstonia: Cherokee
Humulus: Cherokee
Hypericum: Natchez; Seminole
Ilex: Cherokee; Micmac
Impatiens: Iroquois
Ipomoea: Cherokee
Iris: Cherokee; Nevada Indian; Paiute
Jeffersonia: Cherokee
Juglans: Iroquois
Juniperus: Okanagon; Potawatomi; Thompson
Laportea: Meskwaki; Ojibwa
Larrea: Pima
Liatris: Comanche; Meskwaki
Licania: Seminole
Limonium: Costanoan
Linum: Cherokee
Liparis: Cherokee
Lonicera: Chippewa; Iroquois; Menominee; Thompson
Lupinus: Paiute
Lycopodium: Algonquin, Quebec
Lysichiton: Haisla & Hanaksiala; Quinault
Lysimachia: Cherokee
Mahonia: Paiute
Marah: Mendocino Indian
Mentha: Cherokee
Mirabilis: Meskwaki
Mitchella: Iroquois
Morus: Alabama; Creek
Nicotiana: Shuswap
Nyssa: Cherokee
Parthenocissus: Iroquois
Peltigera: Nitinaht
Penstemon: Shuswap
Persea: Seminole
Petroselinum: Cherokee; Micmac
Picea: Abnaki
Pinus: Cherokee
Piper: Hawaiian
Plantago: Cherokee; Iroquois
Platanthera: Micmac
Platanus: Cherokee
Polygonum: Cherokee
Populus: Iroquois; Tewa
Prosopis: Apache, Mescalero
Prunus: Cherokee
Psathyrotes: Shoshoni
Pseudotsuga: Okanagon; Thompson
Psorothamnus: Shoshoni
Pteridium: Iroquois
Quercus: Cherokee
Rhus: Cherokee; Iroquois; Omaha; Seminole; Sioux
Ribes: Chippewa
Rosa: Iroquois
Rubus: Algonquin, Tête-de-Boule; Cherokee; Iroquois
Rumex: Costanoan; Micmac
Sambucus: Choctaw
Sarracenia: Algonquin, Quebec; Algonquin, Tête-de-Boule

Sassafras: Seminole
Saxifraga: Bella Coola
Sedum: Makah
Silphium: Meskwaki
Smilax: Houma
Sorbus: Okanagan-Colville
Spiranthes: Cherokee
Suaeda: Paiute; Shoshoni
Taraxacum: Iroquois
Taxus: Hanaksiala
Tephrosia: Creek
Thuja: Cree, Woodlands
Toxicodendron: Cherokee
Trichostema: Miwok
Triosteum: Iroquois
Tsuga: Micmac
Typha: Micmac
Urtica: Gitksan
Verbena: Menominee
Verbesina: Seminole
Viburnum: Chippewa; Iroquois
Vitis: Cherokee; Iroquois
Xanthium: Costanoan; Jemez; Tewa

Venereal Aid

Abies: Blackfoot; Cherokee; Gitksan; Iroquois; Malecite; Micmac; Ojibwa, South; Paiute; Thompson
Acer: Micmac; Ojibwa; Penobscot
Achillea: Iroquois; Paiute; Thompson
Acorus: Cree, Woodlands
Actaea: Thompson
Adiantum: Iroquois
Ageratina: Iroquois
Alnus: Gitksan; Iroquois
Aloysia: Walapai
Amelanchier: Bella Coola; Iroquois
Anagallis: Mahuna
Androsace: Navajo, Ramah
Anemopsis: Paiute; Pima; Shoshoni
Angelica: Paiute; Shoshoni
Anthemis: Iroquois
Apocynum: Thompson
Aquilegia: Shoshoni
Arabis: Salish; Thompson
Aralia: Cherokee; Iroquois; Malecite; Micmac; Penobscot
Arctium: Cherokee; Malecite; Micmac
Arctostaphylos: Shoshoni
Arenaria: Navajo, Ramah; Shoshoni
Artemisia: Okanagan-Colville; Paiute; Shoshoni; Thompson
Aruncus: Bella Coola; Makah
Asarum: Iroquois
Asclepias: Cherokee; Miwok; Natchez; Shoshoni
Aster: Iroquois; Okanagan-Colville; Thompson
Astragalus: Shoshoni
Athyrium: Iroquois
Baccharis: Navajo, Ramah; Yavapai
Balsamorhiza: Paiute; Shoshoni
Baptisia: Cree; Micmac; Penobscot

Index of Usages

Betula: Cree, Woodlands
Calystegia: Kawaiisu
Cardamine: Delaware; Delaware, Oklahoma
Castilleja: Shoshoni
Ceanothus: Iroquois; Thompson
Celtis: Houma
Cercocarpus: Kawaiisu; Mahuna; Paiute; Shoshoni
Chamaebatia: Miwok
Chamaebatiaria: Gosiute
Chamaesyce: Cherokee
Chenopodium: Miwok
Chimaphila: Chippewa; Delaware; Iroquois
Chrysothamnus: Isleta; Thompson
Cirsium: Comanche; Zuni
Claytonia: Tlingit
Clematis: Iroquois; Shoshoni
Coleogyne: Kawaiisu
Comptonia: Delaware
Coptis: Iroquois
Cordylanthus: Navajo; Shoshoni
Coreopsis: Navajo, Ramah
Cornus: Iroquois
Crotalaria: Delaware; Delaware, Oklahoma
Croton: Zuni
Cucurbita: Cheyenne; Paiute; Shoshoni
Cynoglossum: Iroquois; Pomo, Potter Valley
Cyperus: Hawaiian
Cypripedium: Algonquin, Quebec
Dalibarda: Iroquois
Daphne: Cherokee
Diervilla: Iroquois; Meskwaki; Potawatomi
Diospyros: Cherokee
Dirca: Iroquois
Distichlis: Kawaiisu
Draba: Navajo, Ramah
Echinacea: Delaware; Delaware, Oklahoma
Elaeagnus: Thompson
Enceliopsis: Shoshoni
Ephedra: Apache, White Mountain; Hopi; Mahuna; Navajo; Paiute; Pima; Shoshoni; Tewa; Tubatulabal; Zuni
Epilobium: Miwok
Equisetum: Iroquois; Meskwaki; Okanagan-Colville
Eriastrum: Kawaiisu; Shoshoni
Erigeron: Shoshoni
Eriodictyon: Kawaiisu; Shoshoni
Eriogonum: Gosiute; Kawaiisu; Thompson
Eryngium: Choctaw; Creek; Delaware, Oklahoma
Erythrina: Hawaiian
Eschscholzia: Kawaiisu
Euonymus: Cherokee; Iroquois
Eupatorium: Algonquin, Quebec; Iroquois; Micmac; Penobscot
Euphorbia: Cherokee; Yavapai
Fagus: Micmac
Frangula: Quileute; West Coast Indian
Frasera: Havasupai
Fraxinus: Iroquois
Fucus: Kwakiutl, Southern
Gaillardia: Okanagan-Colville

Galium: Iroquois; Micmac; Penobscot
Garrya: Kawaiisu
Gaultheria: Iroquois
Geranium: Choctaw; Iroquois
Grindelia: Cree; Montana Indian; Shoshoni
Gutierrezia: Isleta
Hamamelis: Iroquois
Helenium: Mendocino Indian
Heliotropium: Shoshoni
Heracleum: Cree; Thompson
Heterotheca: Navajo, Ramah
Heuchera: Navajo, Ramah; Paiute; Shoshoni; Tlingit
Hierochloe: Blackfoot
Holodiscus: Shoshoni
Hydrocotyle: Hawaiian
Hymenoxys: Isleta
Hypericum: Cherokee; Shoshoni
Ipomopsis: Paiute; Shoshoni
Iris: Delaware; Paiute; Shoshoni
Juglans: Iroquois
Juniperus: Blackfoot; Paiute; Shoshoni
Kochia: Navajo, Kayenta
Krameria: Paiute
Lachnanthes: Cherokee
Larix: Iroquois; Malecite; Micmac
Larrea: Paiute; Shoshoni; Yavapai
Ledum: Tlingit
Leymus: Okanagan-Colville
Liatris: Meskwaki
Limonium: Costanoan
Lindera: Iroquois
Lobelia: Cherokee; Iroquois
Lomatium: Paiute; Shoshoni
Lonicera: Bella Coola; Cree, Woodlands; Iroquois; Menominee; Nuxalkmc
Lophophora: Kiowa
Lycopodium: Blackfoot; Iroquois
Madia: Cheyenne
Magnolia: Iroquois
Mahonia: Flathead; Kawaiisu; Paiute; Shoshoni; Skagit; Thompson
Maianthemum: Delaware, Oklahoma; Shoshoni
Malus: Hoh; Quileute
Marah: Mendocino Indian
Melicope: Hawaiian
Menispermum: Cherokee
Mertensia: Iroquois
Mirabilis: Yavapai
Mitchella: Iroquois
Myrica: Bella Coola
Nuphar: Bella Coola; Flathead
Oenothera: Navajo, Kayenta
Onoclea: Iroquois
Oplopanax: Gitksan
Osmorhiza: Iroquois; Paiute; Shoshoni
Osmunda: Iroquois
Paeonia: Shoshoni
Panax: Cherokee; Iroquois
Parnassia: Gosiute
Parthenocissus: Creek

Penstemon: Shoshoni
Phacelia: Kawaiisu
Phlox: Paiute
Physalis: Iroquois
Physocarpus: Bella Coola; Kwakiutl
Picea: Bella Coola; Cree; Cree, Woodlands; Tlingit
Pinus: Apache, White Mountain; Cherokee; Gitksan; Iroquois; Paiute; Shoshoni; Tlingit; Washo; Zuni
Polygala: Iroquois
Polypodium: Nootka
Polystichum: Iroquois
Populus: Bella Coola; Cherokee; Cree, Woodlands; Flathead; Iroquois; Micmac; Okanagan-Colville; Salish; Shoshoni; Thompson
Potentilla: Navajo, Ramah; Okanagan-Colville
Prunella: Iroquois
Prunus: Iroquois
Psathyrotes: Paiute; Shoshoni
Pseudotsuga: Bella Coola; Montana Indian; Pomo, Little Lakes
Psoralidium: Navajo, Ramah
Psorothamnus: Paiute; Shoshoni
Pteridium: Iroquois
Pterospora: Okanagan-Colville
Purshia: Kawaiisu; Paiute; Shoshoni
Pyrola: Micmac; Penobscot
Quercus: Ojibwa
Ranunculus: Iroquois; Navajo
Ratibida: Navajo, Ramah
Rhamnus: Iroquois; Kawaiisu
Rhus: Delaware; Delaware, Oklahoma; Iroquois; Okanagan-Colville; Seminole; Thompson
Ribes: Bella Coola; Swinomish
Rosa: Quinault; Thompson
Rubus: Cherokee; Iroquois
Rudbeckia: Cherokee
Rumex: Iroquois; Paiute; Shoshoni; Yavapai
Salix: Cree; Creek; Delaware; Delaware, Oklahoma; Iroquois; Paiute; Paiute, Northern; Shoshoni
Salvia: Paiute; Paiute, Northern
Sambucus: Iroquois; Thompson
Sanguinaria: Iroquois; Ojibwa
Sanicula: Iroquois
Sarracenia: Cree, Woodlands
Sassafras: Cherokee
Schizachyrium: Comanche
Sedum: Okanagan-Colville
Senecio: Yavapai
Sequoia: Tlingit
Shepherdia: Cree, Woodlands; Gitksan
Sicyos: Iroquois
Smilax: Cree
Solanum: Delaware, Oklahoma
Solidago: Iroquois; Thompson
Sorbus: Ojibwa
Sphaeralcea: Shoshoni
Sphenosciadium: Paiute
Spiraea: Blackfoot; Thompson
Spiranthes: Gosiute
Stachys: Delaware; Delaware, Oklahoma

Stanleya: Zuni
Stephanomeria: Hopi; Shoshoni
Stillingia: Cherokee
Streptopus: Micmac; Penobscot
Symphoricarpos: Bella Coola; Chehalis; Cree, Woodlands
Symphytum: Cherokee
Taxus: Potawatomi
Tetradymia: Shoshoni
Thalictrum: Shoshoni
Thamnosma: Pima
Thuja: Quinault; Tlingit
Toxicodendron: Cherokee
Triosteum: Iroquois
Tsuga: Chehalis; Iroquois; Tlingit
Typha: Iroquois
Ulmus: Ojibwa
Urtica: Kwakiutl
Veratrum: Bella Coola; Paiute; Thompson
Verbesina: Chickasaw
Viburnum: Iroquois
Wyethia: Paiute; Shoshoni
Xanthium: Mahuna
Zanthoxylum: Iroquois; Potawatomi
Zigadenus: Gosiute

Vertigo Medicine
Apocynum: Chippewa
Eleocharis: Seminole
Gutierrezia: Lakota
Juniperus: Seminole
Ledum: Tanana, Upper
Lomatium: Paiute, Northern
Nicotiana: Cherokee
Persea: Seminole
Pleopeltis: Houma
Rorippa: Okanagan-Colville
Salix: Seminole
Tanacetum: Cheyenne

Veterinary Aid
Abies: Blackfoot
Acer: Algonquin, Quebec
Achillea: Blackfoot; Chippewa; Paiute
Achlys: Thompson
Acorus: Omaha
Actaea: Blackfoot; Iroquois
Adenostoma: Cahuilla; Coahuilla
Aesculus: Mendocino Indian
Ageratina: Iroquois
Allium: Blackfoot
Alnus: Menominee; Potawatomi
Ambrosia: Kiowa
Amelanchier: Flathead
Anaphalis: Cheyenne
Anemopsis: Cahuilla; Kawaiisu
Angelica: Blackfoot; Shoshoni
Apocynum: Cherokee; Iroquois; Kutenai
Aquilegia: Thompson
Aralia: Chippewa
Arbutus: Yuki
Arctium: Iroquois
Arctostaphylos: Concow
Arisaema: Iroquois
Artemisia: Blackfoot; Keres, Western; Meskwaki; Mewuk; Navajo, Ramah; Ojibwa; Thompson; Yuki
Asarum: Iroquois
Asclepias: Cherokee; Navajo, Kayenta; Navajo, Ramah; Shoshoni
Aster: Blackfoot; Okanagan-Colville
Astragalus: Cheyenne; Gosiute; Keresan; Lakota
Atriplex: Navajo; Navajo, Ramah
Bouteloua: Navajo, Ramah
Butomus: Iroquois
Caesalpinia: Zuni
Calochortus: Cheyenne
Carex: Iroquois
Carum: Iroquois
Castanea: Iroquois
Celtis: Meskwaki
Cerastium: Navajo, Ramah
Chamaesyce: Kawaiisu; Navajo, Ramah
Chelidonium: Iroquois
Chimaphila: Cherokee; Rappahannock
Chrysothamnus: Sanpoil
Cicuta: Iroquois; Thompson
Cirsium: Navajo, Ramah
Citrullus: Rappahannock
Clematis: Blackfoot; Dakota; Montana Indian; Nez Perce; Sanpoil
Collinsia: Navajo, Kayenta
Collinsonia: Cherokee
Commelina: Navajo, Ramah
Conyza: Potawatomi
Corallorrhiza: Iroquois
Cornus: Iroquois
Corydalis: Navajo, Kayenta
Croton: Diegueño
Cryptantha: Navajo, Kayenta
Cucurbita: Apache, Western; Cahuilla; Coahuilla; Shoshoni
Cynodon: Keres, Western
Cyperus: Pima
Dalea: Navajo, Kayenta; Navajo, Ramah
Datura: Cahuilla; Coahuilla; Navajo; Navajo, Ramah
Delphinium: Navajo, Kayenta
Descurainia: Paiute
Dyssodia: Dakota
Echeandia: Navajo, Ramah
Echinacea: Dakota; Omaha; Pawnee; Ponca; Sioux; Winnebago
Elymus: Navajo, Ramah
Encelia: Kawaiisu
Equisetum: Blackfoot; Cheyenne; Okanagan-Colville
Ericameria: Kawaiisu
Erigeron: Navajo, Ramah
Eriodictyon: Coahuilla
Eriogonum: Navajo, Ramah
Eupatorium: Iroquois
Euphorbia: Navajo, Kayenta
Frasera: Navajo, Kayenta; Navajo, Ramah
Fraxinus: Iroquois
Gaillardia: Blackfoot
Galium: Gosiute
Geum: Blackfoot; Chippewa; Iroquois; Paiute
Glycyrrhiza: Blackfoot; Dakota; Sioux
Grindelia: Flathead; Jemez; Navajo, Ramah; Pawnee
Gutierrezia: Dakota; Keres, Western; Navajo; Navajo, Ramah
Helianthus: Pima
Heuchera: Blackfoot; Shoshoni
Hierochloe: Blackfoot; Karok; Plains Indian
Hybanthus: Iroquois
Hymenopappus: Lakota
Inula: Iroquois; Mohegan
Ipomoea: Keresan; Sia
Ipomopsis: Navajo; Navajo, Ramah; Shoshoni
Iva: Navajo, Kayenta
Juglans: Apache, Western
Juncus: Iroquois
Juniperus: Blackfoot; Dakota; Flathead; Navajo, Ramah; Omaha; Paiute; Pawnee; Ponca; Thompson
Lachnanthes: Catawba
Lactuca: Cherokee
Larix: Menominee; Potawatomi
Larrea: Coahuilla; Kawaiisu
Lathyrus: Navajo, Ramah; Ojibwa
Lepidium: Cherokee
Leucothoe: Cherokee
Leymus: Thompson
Liatris: Chippewa; Meskwaki; Omaha
Liquidambar: Rappahannock
Lomatium: Blackfoot; Gosiute; Nevada Indian; Okanagan-Colville; Oregon Indian; Paiute; Shoshoni; Thompson; Ute
Lonicera: Diegueño
Lupinus: Blackfoot; Menominee; Thompson
Lycoperdon: Blackfoot
Lycopus: Cherokee
Lygodesmia: Blackfoot; Gosiute
Macranthera: Jemez
Mahonia: Blackfoot
Maianthemum: Meskwaki
Matricaria: Okanagan-Colville
Mirabilis: Hopi; Navajo, Ramah
Mitchella: Cherokee
Monarda: Blackfoot; Cheyenne
Muhlenbergia: Blackfoot; Navajo
Myosotis: Iroquois
Nicotiana: Eskimo, Inuktitut; Eskimo, Kuskokwagmiut; Navajo, Ramah
Nuphar: Flathead; Iroquois
Oenothera: Navajo, Kayenta
Onosmodium: Lakota
Opuntia: Keres, Western
Osmorhiza: Blackfoot; Chippewa; Meskwaki; Sanpoil
Osmunda: Iroquois
Oxalis: Pawnee

Paeonia: Paiute
Parthenium: Catawba
Pedicularis: Cherokee; Menominee
Pediomelum: Chippewa; Lakota
Peltigera: Oweekeno
Penstemon: Navajo, Kayenta; Navajo, Ramah; Okanagan-Colville; Thompson
Perideridia: Blackfoot
Peteria: Navajo, Ramah
Phacelia: Hopi
Physaria: Blackfoot
Pinus: Havasupai; Iroquois; Rappahannock; Shoshoni; Thompson
Pluchea: Pima
Podophyllum: Iroquois
Polygonum: Iroquois
Populus: Cahuilla; Iroquois; Mendocino Indian; Nez Perce; Potawatomi
Porteranthus: Cherokee
Prunella: Blackfoot
Prunus: Iroquois; Menominee
Psoralidium: Navajo, Ramah
Pteridium: Iroquois; Mendocino Indian
Quercus: Iroquois; Mohegan
Ranunculus: Iroquois
Ratibida: Lakota; Navajo, Ramah
Reverchonia: Navajo, Kayenta
Rhus: Cheyenne; Iroquois
Robinia: Cherokee
Rubus: Iroquois
Rudbeckia: Chippewa
Rumex: Cherokee; Navajo, Ramah; Nevada Indian
Sagittaria: Cheyenne; Ojibwa
Salix: Blackfoot
Sambucus: Mendocino Indian; Navajo, Kayenta
Sanguinaria: Abnaki
Sarcobatus: Cheyenne
Sedum: Iroquois
Selaginella: Blackfoot
Sicyos: Iroquois
Silene: Gosiute; Navajo, Ramah
Silphium: Dakota; Omaha; Ponca
Sisymbrium: Iroquois
Sisyrinchium: Menominee
Solanum: Cherokee; Navajo, Ramah
Solidago: Thompson
Sparganium: Iroquois
Sphaeralcea: Shoshoni
Sporobolus: Navajo, Ramah
Symphoricarpos: Blackfoot; Crow; Sanpoil
Talinum: Navajo, Ramah
Tanacetum: Malecite
Thalictrum: Blackfoot; Cheyenne; Lakota; Pawnee
Thamnosma: Kawaiisu
Thelypodiopsis: Navajo, Kayenta
Townsendia: Blackfoot
Tradescantia: Navajo, Ramah
Tragopogon: Navajo, Ramah
Trifolium: Iroquois
Triticum: Iroquois
Typha: Iroquois
Valeriana: Blackfoot; Thompson
Veratrum: Paiute; Shoshoni
Verbascum: Navajo; Navajo, Ramah; Rappahannock
Veronica: Iroquois
Vicia: Navajo, Kayenta
Viola: Iroquois; Thompson
Vitis: Iroquois
Xanthium: Pima
Yucca: Blackfoot; Lakota
Zanthoxylum: Creek; Pawnee
Zigadenus: Navajo, Kayenta; Thompson

Witchcraft Medicine

Abies: Cheyenne; Thompson
Achillea: Potawatomi
Aconitum: Okanagan-Colville
Acorus: Cheyenne; Iroquois
Agastache: Navajo, Ramah
Ageratina: Iroquois
Agoseris: Navajo, Ramah
Alnus: Iroquois
Amaranthus: Iroquois
Anaphalis: Potawatomi
Androsace: Navajo, Kayenta; Navajo, Ramah
Anemone: Iroquois
Angelica: Iroquois
Antennaria: Navajo, Ramah
Aquilegia: Iroquois
Arctium: Iroquois
Arisaema: Menominee
Artemisia: Dakota; Miwok; Omaha; Pawnee; Ponca; Shuswap; Winnebago
Asarum: Iroquois
Asclepias: Iroquois
Aster: Navajo, Ramah; Potawatomi
Besseya: Navajo, Ramah
Campanula: Navajo, Ramah
Cardamine: Iroquois
Castilleja: Navajo, Ramah
Chelone: Iroquois
Clematis: Navajo, Ramah
Corallorrhiza: Iroquois
Cornus: Iroquois
Crataegus: Iroquois
Datura: Navajo, Ramah
Draba: Navajo, Ramah
Epilobium: Iroquois
Erigeron: Navajo, Ramah
Eriogonum: Navajo; Navajo, Ramah
Euonymus: Iroquois
Eupatorium: Iroquois
Euphorbia: Navajo, Kayenta
Fagopyrum: Iroquois
Fallugia: Navajo, Kayenta
Gaillardia: Navajo, Kayenta
Galeopsis: Iroquois
Gentiana: Iroquois; Navajo
Geranium: Thompson
Gnaphalium: Creek; Menominee
Heliomeris: Navajo, Ramah
Hepatica: Iroquois
Heracleum: Gitksan; Menominee
Hierochloe: Cheyenne
Humulus: Navajo, Ramah
Ipomoea: Iroquois
Iva: Navajo, Ramah
Juglans: Iroquois
Juncus: Okanagan-Colville
Juniperus: Gitksan; Navajo, Ramah; Seminole
Laportea: Iroquois
Lewisia: Okanagan-Colville
Lilium: Chippewa; Okanagan-Colville
Limosella: Navajo, Ramah
Linaria: Iroquois
Lithospermum: Thompson
Lobelia: Iroquois
Lotus: Zuni
Lupinus: Menominee
Lythrum: Iroquois
Machaeranthera: Okanagan-Colville
Macromeria: Hopi
Maianthemum: Iroquois
Matricaria: Okanagan-Colville
Medeola: Iroquois
Mentha: Iroquois
Menziesia: Nitinaht
Monarda: Creek
Myosurus: Navajo, Ramah
Nuphar: Iroquois
Onopordum: Iroquois
Oplopanax: Kwakiutl
Oxalis: Iroquois
Panax: Seminole
Penstemon: Apache, White Mountain
Pericome: Navajo, Ramah
Peteria: Navajo, Ramah
Phoradendron: Hopi
Phytolacca: Iroquois
Pinus: Hopi; Iroquois; Okanagan-Colville
Potamogeton: Iroquois
Potentilla: Natchez
Psoralidium: Navajo, Ramah
Pteridium: Iroquois
Quercus: Iroquois
Rosa: Iroquois
Rubus: Iroquois
Salvia: Kawaiisu
Sambucus: Gitksan
Sanguinaria: Iroquois
Sanicula: Menominee
Sarracenia: Menominee
Sedum: Iroquois
Silphium: Dakota; Iroquois; Omaha; Pawnee; Ponca; Winnebago
Sisyrinchium: Menominee
Sitanion: Miwok
Smilax: Iroquois
Solidago: Navajo, Ramah
Sparganium: Iroquois
Sphaeralcea: Navajo

Symphoricarpos: Makah
Symplocarpus: Iroquois; Menominee
Taraxacum: Iroquois
Tetradymia: Navajo, Ramah
Tetraneuris: Navajo, Ramah
Thermopsis: Navajo, Ramah
Thlaspi: Navajo, Ramah
Thuja: Potawatomi
Trillium: Iroquois
Triosteum: Iroquois
Tripterocalyx: Navajo, Ramah; Zuni
Ulmus: Creek
Urtica: Iroquois; Kawaiisu
Valeriana: Navajo, Ramah
Veratrum: Okanagan-Colville
Verbascum: Hopi
Verbena: Iroquois
Veronica: Iroquois
Veronicastrum: Iroquois
Viburnum: Iroquois
Viola: Iroquois
Xanthium: Iroquois
Xyris: Seminole

Food

Appetizer

Asclepias: Chippewa
Mahonia: Kutenai
Monarda: Bannock
Rhus: Acoma; Laguna
Trillium: Okanagan-Colville

Baby Food

Calochortus: Navajo
Carya: Iroquois
Dichelostemma: Pima, Gila River
Juglans: Iroquois
Mammillaria: Pima, Gila River
Physalis: Pima, Gila River
Pinus: Cahuilla
Sarcostemma: Pima, Gila River
Washingtonia: Pima, Gila River
Zea: Iroquois

Beverage

Abies: Micmac; Shuswap; Thompson
Acer: Iroquois; Micmac; Ojibwa; Potawatomi
Achillea: Blackfoot
Acorus: Micmac
Agastache: Cheyenne; Cree, Woodlands; Dakota; Lakota; Omaha; Pawnee; Ponca; Winnebago
Agave: Apache, Western; Apache, White Mountain; Havasupai; Navajo; Yavapai
Alisma: Iroquois
Allenrolfea: Cahuilla
Aloysia: Havasupai
Amelanchier: Cheyenne; Lakota; Montana Indian; Thompson
Amorpha: Oglala
Anaphalis: Anticosti
Andromeda: Ojibwa; Tanana, Upper
Androsace: Isleta
Anthemis: Navajo
Arabis: Cheyenne
Aralia: Algonquin, Quebec; Bella Coola; Iroquois; Micmac; Montagnais
Arbutus: Miwok
Arctostaphylos: Atsugewi; Blackfoot; Cahuilla; Costanoan; Costanoan (Olhonean); Hualapai; Karok; Kawaiisu; Mendocino Indian; Mewuk; Miwok; Navajo; Thompson; Yavapai; Yuki
Artemisia: Apache, Chiricahua & Mescalero; Apache, White Mountain
Baccharis: Papago
Balsamorhiza: Paiute
Beta: Anticosti
Betula: Iroquois; Malecite; Ojibwa
Bidens: Hopi
Camassia: Flathead
Camellia: Haisla & Hanaksiala; Oweekeno
Carex: Klamath
Carnegia: Apache, Western; Maricopa; Papago; Pima; Pima, Gila River; Yavapai
Carya: Iroquois
Castanea: Iroquois
Castanopsis: Paiute
Castilleja: Miwok
Ceanothus: Dakota; Lakota; Menominee; Meskwaki; Omaha; Pawnee; Ponca; Winnebago
Chamaedaphne: Ojibwa
Chamaesyce: Apache, White Mountain
Cheilanthes: Apache, Chiricahua & Mescalero; Kawaiisu
Chimaphila: Thompson
Comarum: Eskimo, Alaska
Comptonia: Chippewa
Cordyline: Hawaiian
Coreopsis: Lakota; Zuni
Corylus: Iroquois
Crataegus: Omaha
Dalea: Navajo; Oglala
Dasylirion: Apache, Chiricahua & Mescalero; Apache, Mescalero
Datura: Apache, White Mountain; Papago; Pima
Descurainia: Havasupai; Kawaiisu; Paiute; Paiute, Northern; Papago; Pima; Pima, Gila River
Diospyros: Rappahannock
Distichlis: Kawaiisu
Echinocereus: Isleta
Elaeagnus: Cree
Empetrum: Koyukon
Ephedra: Apache, White Mountain; Cahuilla; Coahuilla; Havasupai; Kawaiisu; Navajo; Paiute; Paiute, Northern; Papago; Pima; Pima, Lehi; Shoshoni; Zuni
Epilobium: Clallam; Saanich
Equisetum: Blackfoot; Nitinaht
Eriodictyon: Cahuilla; Karok
Eriogonum: Blackfoot; Havasupai; Kawaiisu
Fagus: Iroquois
Ferocactus: Apache, San Carlos; Cahuilla; Papago; Pima; Seri
Ficus: Havasupai
Fouquieria: Cahuilla
Fragaria: Blackfoot; Cowlitz; Salish, Coast; Winnebago
Gaultheria: Abnaki; Algonquin, Quebec; Cherokee; Chippewa; Makah; Ojibwa
Geum: Thompson
Gleditsia: Cherokee
Gymnocladus: Meskwaki
Hamamelis: Cherokee
Hedeoma: Apache, Chiricahua & Mescalero
Hedysarum: Tanana, Upper
Holodiscus: Isleta
Hydrangea: Cherokee
Hymenopappus: Hopi; Isleta; Jemez
Hymenoxys: Hopi
Ilex: Comanche
Ipomopsis: Hopi
Iris: Eskimo, Alaska
Jacquemontia: Hawaiian
Juglans: Iroquois; Yavapai
Juncus: Paiute, Northern
Juniperus: Anticosti; Apache, Western; Havasupai; Jemez; Lakota; Ojibwa; Okanagan-Colville; Thompson; Yavapai
Kalmia: Hanaksiala
Larix: Anticosti
Lathyrus: Eskimo, Alaska
Ledum: Alaska Native; Algonquin, Quebec; Anticosti; Bella Coola; Chippewa; Cree; Cree, Woodlands; Eskimo, Alaska; Eskimo, Arctic; Eskimo, Inupiat; Haisla & Hanaksiala; Hesquiat; Kitasoo; Kwakiutl, Southern; Makah; Malecite; Micmac; Nitinaht; Ojibwa; Okanagan-Colville; Oweekeno; Pomo, Kashaya; Potawatomi; Saanich; Salish, Coast; Shuswap; Tanana, Upper; Thompson; Tolowa; Yurok
Lepidium: Kawaiisu
Lespedeza: Comanche
Lindera: Cherokee; Chippewa
Liquidambar: Cherokee
Lithospermum: Blackfoot; Shoshoni
Lomatium: Great Basin Indian; Thompson
Lycium: Havasupai; Kawaiisu; Mohave; Navajo; Pima; Yuma
Mahonia: Hualapai; Montana Indian
Maianthemum: Hanaksiala
Marah: Yurok
Martynia: Apache, Western
Mentha: Blackfoot; Cheyenne; Chippewa; Cree, Woodlands; Dakota; Gosiute; Kawaiisu; Klamath; Lakota; Miwok; Ojibwa; Okanagan-Colville; Omaha; Paiute; Pawnee; Ponca; Sanpoil; Sanpoil & Nespelem; Shuswap; Winnebago; Yuki
Mirabilis: Navajo, Ramah
Mitchella: Micmac
Monarda: Apache, Chiricahua & Mescalero; Iroquois
Monardella: Diegueño; Kawaiisu; Luiseño; Miwok; Okanagan-Colville; Sanpoil & Nespelem;

Tubatulabal
Morus: Cherokee
Nepeta: Ojibwa
Nicotiana: Cahuilla
Nuphar: Algonquin, Tête-de-Boule
Opuntia: Havasupai; Hualapai; Lakota; Navajo; Papago
Osmorhiza: Isleta
Panicum: Warihio
Paronychia: Kiowa
Parthenium: Apache, Chiricahua & Mescalero
Passiflora: Cherokee
Pellaea: Diegueño; Kawaiisu; Luiseño
Peniocereus: Papago
Penstemon: Navajo; Thompson
Pentaphylloides: Eskimo, Alaska; Eskimo, Arctic
Phoradendron: Cahuilla; Navajo
Phytolacca: Cherokee
Picea: Anticosti; Chippewa; Micmac; Okanagan-Colville
Pinus: Cahuilla; Hualapai; Micmac; Thompson
Plantago: Pima, Gila River
Platanus: Kawaiisu
Polygonum: Eskimo, Arctic; Eskimo, Inupiat
Prosopis: Apache, Chiricahua & Mescalero; Apache, Mescalero; Apache, Western; Cahuilla; Havasupai; Maricopa; Mohave; Pima; Pima, Gila River; Yuma
Prunella: Thompson
Prunus: Algonquin, Quebec; Blackfoot; Cherokee; Chippewa; Havasupai; Iroquois; Lakota; Menominee; Meskwaki; Micmac; Montana Indian; Ojibwa; Paiute; Potawatomi; Sanpoil & Nespelem; Shuswap; Thompson
Pseudocymopterus: Isleta
Pseudotsuga: Karok; Yurok
Psoralidium: Yavapai
Pyrola: Malecite
Quercus: Apache, White Mountain; Iroquois; Kiowa; Menominee; Meskwaki
Ratibida: Dakota; Oglala
Rhexia: Montagnais
Rhododendron: Okanagan-Colville
Rhus: Algonquin, Quebec; Apache, Western; Atsugewi; Cahuilla; Diegueño; Great Basin Indian; Havasupai; Hopi; Hualapai; Iroquois; Keres, Western; Kiowa; Menominee; Meskwaki; Navajo; Ojibwa; Okanagan-Colville; Shoshoni; Yavapai
Ribes: Cree, Woodlands; Spanish American; Thompson
Robinia: Cherokee
Romneya: Cahuilla
Rosa: Alaska Native; Arapaho; Cahuilla; Eskimo, Inupiat; Haisla & Hanaksiala; Lakota; Lummi; Makah; Shuswap; Skagit; Skagit, Upper; Tanana, Upper; Thompson; Washo
Rubus: Cahuilla; Cherokee; Chippewa; Dakota; Haisla & Hanaksiala; Iroquois; Meskwaki; Omaha; Oweekeno; Pawnee; Ponca; Quileute; Saanich; Yurok
Rumex: Hanaksiala; Hualapai
Sadleria: Hawaiian
Salix: Eskimo, Inupiat; Mohave; Navajo; Yuma
Salvia: Cahuilla; Kawaiisu; Mohave; Papago; Pima; Pima, Gila River
Sambucus: Bella Coola; Cherokee; Dakota; Haisla & Hanaksiala; Montana Indian; Omaha; Pawnee; Pima, Gila River; Ponca
Sassafras: Cherokee; Chippewa
Satureja: Diegueño; Luiseño; Pomo, Kashaya; Saanich; Tolowa
Sedum: Eskimo, Alaska
Senna: Kiowa
Shepherdia: Flathead; Northwest Indian; Okanagon; Oregon Indian; Shuswap; Thompson
Simmondsia: Cahuilla; Coahuilla
Sisymbrium: Pima, Gila River
Solanum: Zuni
Sphaeralcea: Navajo, Kayenta
Sphagnum: Carrier
Spiraea: Abnaki; Tanana, Upper; Thompson
Stenocereus: Papago; Papago & Pima
Taraxacum: Iroquois
Taxus: Iroquois; Penobscot
Thelesperma: Apache, Chiricahua & Mescalero; Hopi; Isleta; Keres, Western; Keresan; Navajo, Ramah; Tewa
Theobroma: Haisla & Hanaksiala
Thuja: Chippewa; Ojibwa
Tragopogon: Navajo
Trichostema: Kawaiisu
Trifolium: Anticosti; Miwok
Triglochin: Klamath
Triticum: Zuni
Tsuga: Chippewa; Iroquois; Micmac; Ojibwa
Ulmus: Cheyenne; Kiowa
Umbellularia: Concow
Urtica: Shuswap
Vaccinium: Clallam; Eskimo, Arctic; Hesquiat; Quinault
Verbena: Omaha
Viburnum: Eskimo, Inupiat
Vitis: Apache, Western; Cahuilla; Cherokee; Hualapai; Pawnee
Washingtonia: Cahuilla; Cocopa
Yucca: Acoma; Apache; Apache, Mescalero; Havasupai; Hualapai; Keres, Western; Laguna; Pima; Southwest Indians; Walapai; Yavapai
Zea: Iroquois; Isleta; Menominee; Navajo; Tewa; Zuni
Ziziphus: Papago; Yavapai

Bread & Cake
Acacia: Havasupai
Acer: Iroquois
Aesculus: Kawaiisu
Agave: Apache, Mescalero
Allenrolfea: Cahuilla
Allium: Paiute, Northern
Amaranthus: Apache, Chiricahua & Mescalero; Havasupai; Navajo; Navajo, Ramah; Zuni
Amelanchier: Alaska Native; Iroquois; Montana Indian; Okanagon; Sanpoil & Nespelem; Thompson
Amphicarpaea: Cherokee
Amsinckia: Atsugewi
Anemopsis: Kamia
Apios: Delaware
Arctostaphylos: Atsugewi; Numlaki; Tolowa
Artemisia: Navajo; Zuni
Asimina: Iroquois
Atriplex: Pima
Balsamorhiza: Atsugewi; Thompson
Blennosperma: Neeshenam
Boisduvalia: Mendocino Indian
Bouteloua: Apache, White Mountain
Bromus: Neeshenam
Camassia: Blackfoot; Montana Indian; Sanpoil & Nespelem
Camelina: Apache, Chiricahua & Mescalero
Capsella: Apache, Chiricahua & Mescalero
Carnegia: Apache, San Carlos; Apache, Western; Papago; Pima; Yavapai
Carya: Iroquois
Castanea: Cherokee; Iroquois
Celtis: Apache, Chiricahua & Mescalero
Chenopodium: Atsugewi; Havasupai; Hopi; Navajo; Navajo, Ramah; Zuni
Cirsium: Apache, Chiricahua & Mescalero
Citrullus: Iroquois
Claytonia: Thompson
Cleome: Isleta; Jemez; Navajo
Coix: Cherokee
Corylus: Iroquois
Crataegus: Apache, Chiricahua & Mescalero; Chippewa; Coeur d'Alene; Iroquois; Okanagan-Colville
Cucumis: Cocopa; Iroquois
Cucurbita: Apache, White Mountain; Iroquois
Cupressus: Navajo
Cymopterus: Hualapai
Dasylirion: Apache, Chiricahua & Mescalero; Apache, Mescalero; Southwest Indians
Descurainia: Apache, Chiricahua & Mescalero; Atsugewi; Navajo, Ramah
Dyssodia: Apache, Chiricahua & Mescalero
Echinocereus: Isleta
Eriogonum: Hopi
Fagus: Iroquois
Fragaria: Iroquois
Gaultheria: Bella Coola; Clallam; Iroquois; Klallam; Makah; Okanagon; Quileute; Quinault; Salish, Coast; Samish; Skagit; Skokomish; Snohomish; Swinomish; Thompson
Gaylussacia: Cherokee; Iroquois
Gossypium: Papago
Helianthus: Apache, Chiricahua & Mescalero; Apache, Western; Havasupai; Montana Indian; Navajo
Hordeum: Yuki
Humulus: Algonquin, Quebec
Hymenopappus: Hopi
Juglans: Iroquois

Juniperus: Kawaiisu; Yavapai
Koeleria: Havasupai; Isleta
Lepidium: Havasupai
Lewisia: Thompson
Lilium: Okanagan-Colville
Lithocarpus: Hahwunkwut; Hupa; Karok; Poliklah; Pomo; Shasta; Yuki; Yurok
Lomatium: Montana Indian; Paiute
Madia: Neeshenam
Malus: Iroquois
Menyanthes: Alaska Native
Mitchella: Iroquois
Morus: Apache, Chiricahua & Mescalero; Cherokee; Iroquois
Muhlenbergia: Apache, Chiricahua & Mescalero; Apache, White Mountain; Hopi
Nolina: Isleta
Nuphar: Klamath
Oemleria: Thompson
Olneya: Mohave; Yavapai; Yuma
Opuntia: Havasupai; Navajo; Yavapai
Oryzopsis: Apache, White Mountain; Havasupai; Hopi; Navajo
Panicum: Apache, Chiricahua & Mescalero; Cocopa; Hopi
Pediomelum: Montana Indian
Perideridia: Atsugewi; Flathead
Phaseolus: Cherokee; Iroquois
Physalis: Iroquois
Pinus: Apache, Chiricahua & Mescalero; Hualapai; Navajo; Paiute, Northern; Salish, Coast; Shasta
Poa: Havasupai
Podophyllum: Iroquois
Polygonatum: Cherokee
Populus: Apache, Chiricahua & Mescalero
Porphyra: Kitasoo; Kwakiutl, Southern; Oweekeno; Pomo
Prosopis: Apache; Apache, Chiricahua & Mescalero; Apache, Western; Cahuilla; Diegueño; Isleta; Kawaiisu; Mahuna; Mohave; Pima; Pima, Gila River; Yuma
Prunus: Apache; Apache, Mescalero; Atsugewi; Cheyenne; Chippewa; Dakota; Diegueño; Iroquois; Montana Indian; Navajo, Ramah; Okanagan-Colville; Omaha; Paiute, Northern; Pawnee; Ponca; Sanpoil & Nespelem
Pteridium: Salish, Coast
Pyrus: Iroquois
Quercus: Apache, White Mountain; Cahuilla; Concow; Hualapai; Iroquois; Kawaiisu; Mendocino Indian; Mewuk; Miwok; Neeshenam; Pomo; Round Valley Indian; Shasta; Yana; Yuki
Ranunculus: Neeshenam
Rhus: Navajo
Ribes: Apache, Chiricahua & Mescalero; Chippewa; Iroquois; Okanagan-Colville; Salish, Coast
Rubus: Apache, Chiricahua & Mescalero; Bella Coola; Chippewa; Haisla; Iroquois; Salish, Coast; Thompson
Rumex: Paiute, Northern; Pima

Sabal: Houma
Salicornia: Gosiute
Sambucus: Atsugewi; Iroquois; Kitasoo; Kwakiutl, Southern
Scirpus: Cahuilla
Shepherdia: Thompson
Simmondsia: Cocopa
Smilax: Choctaw; Houma
Solanum: Apache, Chiricahua & Mescalero
Sporobolus: Apache, Chiricahua & Mescalero; Apache, White Mountain; Hopi; Navajo
Stenocereus: Papago
Triticum: Apache, White Mountain; Haisla & Hanaksiala; Papago; Pima; Zuni
Tsuga: Alaska Native; Hanaksiala; Montana Indian
Typha: Hualapai; Paiute, Northern; Pima; Yuma
Umbellularia: Costanoan; Mendocino Indian
Vaccinium: Alaska Native; Bella Coola; Hanaksiala; Iroquois; Kitasoo; Makah; Montana Indian; Thompson
Viburnum: Iroquois
Vitis: Cherokee; Iroquois
Xanthium: Apache, White Mountain; Zuni
Yucca: Acoma; Apache; Apache, Chiricahua & Mescalero; Cahuilla; Keres, Western; Laguna; Navajo; Navajo, Ramah; Pima; Southwest Indians
Zea: Delaware; Havasupai; Hopi; Iroquois; Isleta; Navajo; Pima; Tewa; Zuni
Zizania: Ojibwa

Breakfast Food
Zizania: Ojibwa

Candy
Abies: Blackfoot; Nitinaht; Shuswap
Acer: Cheyenne; Potawatomi
Agave: Apache, Western; Pima; Pima, Gila River
Agoseris: Karok; Okanagan-Colville; Thompson
Amelanchier: Paiute
Antennaria: Blackfoot
Apocynum: Isleta; Kiowa
Aquilegia: Hanaksiala
Artemisia: Blackfoot; Paiute, Northern
Asclepias: Acoma; Apache, Chiricahua & Mescalero; Cheyenne; Coahuilla; Gosiute; Karok; Kawaiisu; Keres, Western; Laguna; Luiseño; Paiute; Shoshoni; Tubatulabal; Yurok
Astragalus: Hopi
Balsamorhiza: Paiute
Calochortus: Hopi
Carnegia: Papago & Pima; Pima, Gila River
Castilleja: Nitinaht
Chamaesyce: Zuni
Chenopodium: Cahuilla
Chloracantha: Navajo
Chrysothamnus: Gosiute; Paiute; Paiute, Northern
Crataegus: Comanche
Dalea: Comanche; Hopi; Lakota; Ponca; Zuni
Dugaldia: Navajo
Echinocereus: Isleta
Elaeagnus: Blackfoot
Encelia: Papago; Pima

Eriodictyon: Diegueño
Erodium: Hopi
Erythronium: Thompson
Eschscholzia: Luiseño
Escobaria: Blackfoot
Euphorbia: Kiowa
Ferocactus: Pima
Ficus: Seminole
Gentiana: Hanaksiala
Gnaphalium: Rappahannock
Helianthus: Apache, Western; Pima
Hymenopappus: Zuni
Hymenoxys: Isleta; Keres, Western; Navajo; Spanish American; Tewa
Juncus: Paiute
Lactuca: Apache, White Mountain; Navajo; Zuni
Larix: Flathead; Okanagan-Colville; Paiute; Thompson
Liquidambar: Cherokee
Lonicera: Nitinaht; Saanich; Thompson
Lygodesmia: Navajo, Ramah; Washo
Marrubium: Diegueño
Matricaria: Eskimo, Alaska
Mentha: Kiowa
Opuntia: Kiowa; Navajo
Osmorhiza: Blackfoot
Phragmites: Paiute, Northern
Picea: Algonquin, Quebec; Carrier; Cree, Woodlands; Eskimo, Alaska; Haisla & Hanaksiala; Hesquiat; Koyukon; Kwakiutl, Southern; Makah; Oweekeno; Penobscot; Quinault; Tanana, Upper
Pinus: Apache, Western; Blackfoot; Cheyenne; Flathead; Hesquiat; Hualapai; Navajo; Navajo, Ramah; Okanagan-Colville; Paiute; Paiute, Northern; Pomo, Kashaya; Thompson; Wailaki; Yuki
Polypodium: Hesquiat; Thompson
Populus: Acoma; Apache, Chiricahua & Mescalero; Apache, White Mountain; Dakota; Havasupai; Hopi; Isleta; Keres, Western; Laguna; Navajo; Omaha; Pawnee; Pima; Ponca; Zuni
Prosopis: Apache, Western; Havasupai; Isleta; Papago; Pima
Prunus: Kiowa
Pseudotsuga: Apache, White Mountain; Cowlitz; Gosiute; Klallam; Quinault; Yurok
Quercus: Diegueño; Navajo; Papago
Rumex: Pima
Salix: Kawaiisu
Sarcostemma: Papago; Pima
Senecio: Gosiute; Navajo, Ramah
Shepherdia: Thompson
Sideroxylon: Kiowa
Silphium: Dakota; Omaha; Pawnee; Ponca; Winnebago
Stenocereus: Papago & Pima
Stephanomeria: Kawaiisu; Navajo, Kayenta; Navajo, Ramah
Thuja: Kwakiutl, Southern
Toxicodendron: Karok
Tragopogon: Thompson

Tsuga: Hesquiat
Typha: Hopi
Vernonia: Kiowa
Yucca: Pima, Gila River
Zea: Isleta

Cooking Agent
Amaranthus: Hopi; Zuni
Apios: Mohegan
Asclepias: Lakota; Miwok
Atriplex: Pima; Tewa of Hano
Balsamorhiza: Gosiute
Brickellia: Gosiute
Carnegia: Papago
Carthamus: Hopi
Carya: Cherokee
Chamaesyce: Apache, White Mountain
Citrullus: Hopi
Cucurbita: Hopi
Darmera: Miwok
Dryopteris: Yurok
Helianthus: Gosiute
Humulus: Lakota; Ojibwa
Lewisia: Kutenai
Lithospermum: Omaha; Ponca
Lotus: Miwok
Madia: Mendocino Indian
Malus: Alaska Native
Nuphar: Montana Indian
Opuntia: Cheyenne; Keres, Western; Navajo
Pachycereus: Papago & Pima
Pediomelum: Cheyenne; Montana Indian
Phytolacca: Cherokee
Quercus: Yavapai
Rubus: Cherokee
Solanum: Hopi; Navajo
Stenocereus: Papago
Vaccinium: Eskimo, Inupiat

Dessert
Amelanchier: Blackfoot
Argentina: Nitinaht
Balsamorhiza: Thompson
Carnegia: Pima
Cornus: Haisla & Hanaksiala; Hesquiat; Thompson
Empetrum: Eskimo, Inupiat
Lewisia: Kutenai
Lycopus: Thompson
Mahonia: Flathead; Kutenai
Opuntia: Thompson
Oxalis: Potawatomi
Pedicularis: Eskimo, Inupiat
Polygonum: Eskimo, Inupiat
Porphyra: Kwakiutl, Southern
Prunus: Thompson
Quercus: Mahuna
Ribes: Eskimo, Inupiat; Meskwaki; Nitinaht
Rubus: Eskimo, Inupiat; Nitinaht; Okanagan-Colville
Rumex: Eskimo, Inupiat
Shepherdia: Alaska Native; Eskimo, Inupiat; Makah; Nanaimo; Salish, Coast; Shuswap

Vaccinium: Eskimo, Inupiat
Viburnum: Eskimo, Inupiat
Vitis: Meskwaki
Yucca: Navajo; Southwest Indians

Dietary Aid
Asarum: Ojibwa
Betula: Montagnais
Chenopodium: Alaska Native
Claytonia: Alaska Native
Dryopteris: Bella Coola
Epilobium: Alaska Native
Fragaria: Alaska Native
Honckenya: Alaska Native
Liatris: Lakota
Ligusticum: Alaska Native
Oxyria: Alaska Native
Polygonum: Alaska Native
Polypodium: Kwakiutl, Southern; Makah
Rosa: Alaska Native
Rubus: Alaska Native
Rumex: Alaska Native
Salix: Alaska Native
Salvia: Cahuilla
Saxifraga: Alaska Native
Taraxacum: Alaska Native
Urtica: Alaska Native
Vaccinium: Alaska Native

Dried Food
Acer: Apache, Chiricahua & Mescalero; Clallam
Agaricus: Thompson
Agave: Apache; Apache, Western; Cahuilla; Navajo; Pima; Pima, Gila River; Yavapai
Alaria: Hesquiat
Allium: Navajo; Navajo, Ramah; Ojibwa; Okanagan-Colville; Paiute, Northern; Thompson
Amaranthus: Acoma; Laguna; Papago; Pima, Gila River
Ambrosia: Papago
Amelanchier: Alaska Native; Atsugewi; Blackfoot; Cahuilla; Carrier; Chippewa; Cree, Plains; Cree, Woodlands; Gosiute; Great Basin Indian; Hanaksiala; Iroquois; Karok; Lummi; Navajo; Ojibwa; Okanagan-Colville; Paiute; Potawatomi; Skagit, Upper; Swinomish; Thompson; Ute
Arbutus: Karok
Arctium: Iroquois
Arctostaphylos: Blackfoot; Cahuilla; Chinook, Lower; Coeur d'Alene; Costanoan; Hualapai; Karok; Kimsquit; Montana Indian; Paiute; Pomo; Pomo, Kashaya; Sanpoil & Nespelem
Argentina: Kwakiutl, Southern; Nitinaht
Asimina: Iroquois
Astragalus: Zuni
Atriplex: Pima
Balsamorhiza: Okanagan-Colville; Paiute; Thompson
Calochortus: Cheyenne; Gosiute; Paiute, Northern
Camassia: Klamath; Nisqually; Nitinaht; Okanagan-Colville; Paiute; Salish, Coast
Capsicum: Hopi

Carnegia: Apache, Western; Papago; Pima; Pima, Gila River; Yavapai
Carya: Cherokee
Castanopsis: Pomo
Celtis: Hualapai
Chamaesyce: Apache, White Mountain
Chenopodium: Eskimo, Inupiat; Miwok; Navajo
Cicer: Papago
Cirsium: Thompson
Citrullus: Cocopa; Iroquois; Navajo
Cleome: Navajo; Navajo, Ramah; Zuni
Cornus: Haisla & Hanaksiala; Thompson
Corylus: Tolowa
Costaria: Hesquiat
Crataegus: Cheyenne; Iroquois; Paiute; Sanpoil & Nespelem; Thompson
Cucumis: Cocopa; Hopi; Iroquois; Navajo
Cucurbita: Cahuilla; Cocopa; Havasupai; Hopi; Iroquois; Maricopa; Navajo; Ojibwa; Papago; Pima; Yuma; Zuni
Cymopterus: Navajo; Navajo, Ramah
Cyperus: Paiute, Northern
Dasylirion: Apache, Mescalero
Datura: Navajo
Daucus: Navajo; Sanpoil & Nespelem
Descurainia: Papago; Pima; Pueblo
Dichelostemma: Paiute
Echinocactus: Havasupai
Egregia: Bella Coola; Haisla & Hanaksiala; Oweekeno
Equisetum: Cowlitz; Hoh; Quileute
Eriodictyon: Diegueño
Eriogonum: Hopi
Eriophyllum: Kawaiisu
Erythronium: Kwakiutl; Kwakiutl, Southern; Okanagan-Colville; Thompson
Escobaria: Cheyenne
Ferocactus: Cahuilla; Hualapai
Ficus: Havasupai
Fragaria: Cowlitz; Iroquois; Omaha; Potawatomi; Shuswap; Skagit, Upper; Thompson
Frangula: Paiute
Fritillaria: Alaska Native; Hesquiat; Kwakiutl, Southern; Okanagan-Colville; Thompson
Fucus: Haisla & Hanaksiala; Yurok, South Coast (Nererner)
Gaultheria: Hesquiat; Iroquois; Kwakiutl, Southern; Makah; Nitinaht; Skagit, Upper
Gaylussacia: Iroquois
Gossypium: Pima
Helianthus: Havasupai; Hopi; Sanpoil & Nespelem
Heracleum: Thompson
Heteromeles: Costanoan; Luiseño
Heuchera: Miwok
Hypericum: Miwok
Juglans: Cherokee; Pomo, Kashaya
Juniperus: Apache, Western; Atsugewi; Cahuilla; Havasupai; Kawaiisu; Mendocino Indian
Lasthenia: Cahuilla
Lathyrus: Apache, Chiricahua & Mescalero

Lens: Papago
Lepidium: Papago
Lessoniopsis: Hesquiat
Lewisia: Blackfoot; Kutenai; Okanagan-Colville; Paiute; Thompson
Ligusticum: Cherokee
Lilium: Thompson
Lithocarpus: Pomo, Kashaya
Lomatium: Flathead; Klamath; Montana Indian; Okanagan-Colville; Okanogan; Paiute; Thompson
Lycium: Cahuilla; Havasupai; Kawaiisu; Mohave; Navajo; Navajo, Ramah; Paiute, Northern; Papago; Yuma
Lycopus: Chippewa
Machaeranthera: Navajo, Ramah
Macrocystis: Oweekeno
Mahonia: Skagit, Upper; Thompson
Malus: Cherokee; Hesquiat; Iroquois
Mammillaria: Apache, Chiricahua & Mescalero
Mentzelia: Hualapai; Paiute, Northern
Menyanthes: Alaska Native
Mitchella: Iroquois
Monarda: Hopi; Pueblo; Spanish American
Monolepis: Papago
Morus: Iroquois; Omaha
Nuphar: Cree, Woodlands; Klamath; Thompson
Oemleria: Cowlitz
Olneya: Mohave; Papago; Pima; Yuma
Opuntia: Acoma; Apache, Mescalero; Apache, White Mountain; Cahuilla; Cheyenne; Dakota; Diegueño; Havasupai; Hualapai; Laguna; Lakota; Luiseño; Montana Indian; Navajo; Navajo, Ramah; Papago; Pawnee; Pima; Pima, Gila River; Sanpoil & Nespelem; Yavapai; Zuni
Parkinsonia: Papago
Pectis: Hopi
Pediomelum: Blackfoot; Cheyenne; Dakota; Lakota; Montana Indian; Omaha; Pawnee; Ponca; Winnebago
Perideridia: Atsugewi; Cheyenne; Klamath; Okanagan-Colville; Paiute
Phaseolus: Iroquois; Papago
Pholisma: Cocopa; Papago
Phoradendron: Papago
Phyllospadix: Hesquiat
Physalis: Dakota; Iroquois; Omaha; Pawnee; Ponca
Phytolacca: Cherokee
Picea: Kitasoo
Pinus: Apache, Mescalero; Cahuilla; Miwok; Paiute; Paiute, Northern; Salish, Coast; Shasta; Thompson; Washo; Wet'suwet'en
Podophyllum: Iroquois
Polanisia: Pueblo
Poliomintha: Hopi; Tewa
Polygonum: Cocopa
Polypodium: Salish, Coast
Populus: Bella Coola; Clallam
Porphyra: Alaska Native; Bella Coola; Haisla & Hanaksiala; Kwakiutl, Southern; Oweekeno; Pomo; Pomo, Kashaya

Postelsia: Hesquiat; Pomo, Kashaya
Proboscidea: Havasupai; Papago
Prosopis: Apache, Chiricahua & Mescalero; Apache, Western; Cahuilla; Hualapai; Mahuna; Mohave; Papago; Yuma
Prunus: Acoma; Apache, Mescalero; Blackfoot; Cahuilla; Chippewa; Cochiti; Coeur d'Alene; Comanche; Dakota; Gosiute; Great Basin Indian; Havasupai; Hopi; Iroquois; Keresan; Klamath; Laguna; Mendocino Indian; Ojibwa; Omaha; Paiute; Paiute, Northern; Pawnee; Ponca; Round Valley Indian; San Felipe; Thompson; Winnebago
Pyrus: Iroquois
Quercus: Cahuilla; Kiowa; Pomo, Kashaya; Wintoon; Yana
Rhodymenia: Alaska Native
Rhus: Apache, Chiricahua & Mescalero; Atsugewi; Cahuilla; Havasupai; Navajo; Navajo, Ramah
Ribes: Cheyenne; Chippewa; Cowlitz; Gosiute; Iroquois; Kawaiisu; Klamath; Ojibwa; Okanagan-Colville; Omaha; Paiute; Paiute, Northern; Thompson
Robinia: Apache, Mescalero
Rosa: Lakota; Lummi
Rubus: Cahuilla; Chippewa; Coeur d'Alene; Cowlitz; Dakota; Diegueño; Haisla & Hanaksiala; Hesquiat; Iroquois; Klallam; Klamath; Kwakiutl, Southern; Mendocino Indian; Ojibwa; Okanagan-Colville; Omaha; Pawnee; Ponca; Quileute; Sanpoil & Nespelem; Skagit; Skagit, Upper; Squaxin; Thompson; Yuki
Rudbeckia: Cherokee
Sagittaria: Chippewa; Thompson
Sambucus: Bella Coola; Cahuilla; Chippewa; Haisla & Hanaksiala; Iroquois; Luiseño; Makah; Mendocino Indian; Oweekeno; Paiute; Paiute, Northern; Skagit, Upper; Thompson; Yokut; Yuki
Shepherdia: Carrier; Dakota; Haisla & Hanaksiala; Lakota; Makah; Montana Indian; Navajo; Omaha; Paiute; Paiute, Northern; Pawnee; Ponca; Winnebago
Sium: Thompson
Sorbus: Thompson
Sporobolus: Papago
Stenocereus: Papago
Symphoricarpos: Squaxin
Thelypodium: Pueblo
Thuja: Montana Indian; Salish, Coast
Trifolium: Miwok; Nitinaht
Tsuga: Bella Coola; Haisla; Kitasoo; Oweekeno; Salish, Coast
Typha: Cree, Woodlands; Ojibwa; Paiute, Northern; Yuma
Umbellularia: Pomo, Kashaya
Vaccinium: Carrier; Chinook, Lower; Chippewa; Clallam; Coeur d'Alene; Hanaksiala; Iroquois; Kitasoo; Klallam; Klamath; Makah; Montana Indian; Nitinaht; Ojibwa; Okanagan-Colville; Paiute; Pomo; Pomo, Kashaya; Potawatomi; Quileute; Quinault; Salish, Coast; Sanpoil &

Nespelem; Skagit, Upper; Thompson; Tolowa
Viburnum: Iroquois; Thompson
Vicia: Apache, Chiricahua & Mescalero; Papago
Vigna: Mohave
Vitis: Apache, Chiricahua & Mescalero; Apache, Western; Cahuilla; Comanche; Crow; Dakota; Diegueño; Hualapai; Iroquois; Kiowa; Menominee; Omaha; Pawnee; Ponca; Sioux; Winnebago
Washingtonia: Cahuilla
Yucca: Acoma; Apache, Chiricahua & Mescalero; Apache, Mescalero; Apache, San Carlos; Apache, Western; Apache, White Mountain; Cahuilla; Havasupai; Hualapai; Kawaiisu; Laguna; Navajo; Papago; Pima; Pueblo; San Felipe; Southwest Indians; Walapai; Yavapai
Zea: Delaware; Hopi; Ojibwa; Papago
Zizania: Potawatomi
Ziziphus: Maricopa; Mohave

Fodder
Acacia: Diegueño
Alectoria: Eskimo, Inuktitut
Aplectrum: Cherokee
Aristida: Navajo, Ramah
Artemisia: Navajo
Astragalus: Lakota; Thompson
Atriplex: Navajo; Navajo, Ramah
Avena: Navajo; Navajo, Ramah
Beckmannia: Navajo, Ramah
Boschniakia: Tanana, Upper
Bouteloua: Kiowa; Navajo
Bromus: Kiowa; Navajo, Ramah
Carex: Thompson
Carnegia: Papago; Pima
Castilleja: Thompson
Chamaebatiaria: Navajo, Ramah
Claytonia: Montana Indian
Cleome: Navajo, Ramah
Coleogyne: Havasupai
Cornicularia: Eskimo, Inuktitut
Corydalis: Navajo, Ramah
Croton: Hopi
Cryptantha: Navajo, Ramah
Cyperus: Apache, Chiricahua & Mescalero; Kiowa
Descurainia: Navajo, Ramah
Dichanthelium: Kiowa
Echinochloa: Navajo, Ramah
Elymus: Kiowa; Navajo, Ramah; Thompson
Elytrigia: Apache, White Mountain
Epilobium: Tanana, Upper; Thompson
Equisetum: Isleta; Lakota; Meskwaki; Ojibwa; Okanagan-Colville; Quileute; Quinault
Erodium: Navajo, Ramah
Geranium: Keres, Western
Helianthus: Hopi; Navajo, Ramah
Heliomeris: Navajo, Ramah
Heterotheca: Navajo, Ramah
Hordeum: Mendocino Indian
Hydrophyllum: Ojibwa
Juncus: Okanagan-Colville; Paiute
Juniperus: Navajo

Koeleria: Okanagan-Colville
Krascheninnikovia: Havasupai
Lappula: Navajo, Ramah
Larix: Potawatomi
Lathyrus: Mendocino Indian; Ojibwa
Lepidium: Navajo, Ramah
Leymus: Mendocino Indian; Okanagan-Colville
Linum: Dakota
Lotus: Diegueño
Lupinus: Thompson
Maianthemum: Ojibwa
Marrubium: Navajo, Ramah
Medicago: Navajo, Ramah; Shuswap
Monolepis: Navajo, Ramah
Muhlenbergia: Apache, White Mountain; Navajo, Ramah
Opuntia: Montana Indian
Oryzopsis: Apache, White Mountain; Navajo; Navajo, Ramah
Osmorhiza: Omaha; Ponca; Potawatomi
Oxalis: Omaha; Pawnee; Ponca
Panicum: Navajo, Ramah
Pascopyrum: Montana Indian
Paspalum: Kiowa
Pedicularis: Potawatomi
Phaseolus: Navajo, Ramah
Phleum: Shuswap
Phoradendron: Keres, Western
Picea: Tanana, Upper
Plantago: Mendocino Indian; Pima
Poa: Navajo, Ramah
Populus: Blackfoot; Cheyenne; Dakota; Lakota; Montana Indian
Prosopis: Kiowa
Quercus: Seminole
Rorippa: Navajo, Ramah
Salix: Navajo
Salsola: Navajo, Ramah
Sarcobatus: Navajo, Ramah
Sonchus: Houma
Sorghum: Kiowa
Sparganium: Okanagan-Colville
Sporobolus: Keres, Western; Kiowa
Stenocereus: Papago
Stipa: Navajo, Ramah
Tridens: Navajo, Ramah
Trifolium: Shuswap; Thompson
Triticum: Navajo, Ramah
Vicia: Mendocino Indian; Thompson
Zea: Kiowa; Navajo, Ramah
Ziziphus: Pima

Forage

Achillea: Haisla & Hanaksiala
Aesculus: Mendocino Indian
Alectoria: Hesquiat
Allium: Shuswap
Amaranthus: Navajo
Amelanchier: Havasupai; Hesquiat; Montana Indian
Anthoxanthum: Hesquiat
Aquilegia: Thompson
Aralia: Montagnais
Arbutus: Mendocino Indian
Arctostaphylos: Eskimo, Arctic; Mendocino Indian; Nitinaht; Paiute; Thompson
Artemisia: Blackfoot; Iroquois; Isleta; Lakota
Asclepias: Mendocino Indian; Zuni
Astragalus: Navajo, Ramah; Thompson
Atriplex: Keres, Western; Navajo; Pima
Blechnum: Haisla & Hanaksiala
Bouteloua: Hopi; Keres, Western; Montana Indian; Navajo, Ramah
Brodiaea: Mendocino Indian
Bromus: Thompson
Buchloe: Blackfoot
Calochortus: Shuswap
Carex: Blackfoot; Mendocino Indian; Navajo, Ramah; Thompson
Ceanothus: Karok; Mendocino Indian; Okanagan-Colville; Paiute; Thompson
Cercocarpus: Navajo
Chlorogalum: Yuki
Chrysothamnus: Blackfoot; Navajo
Cladonia: Abnaki
Claytonia: Montana Indian
Clematis: Lakota
Clintonia: Bella Coola
Cornus: Haisla & Hanaksiala; Okanagan-Colville; Oweekeno
Crataegus: Okanagan-Colville
Croton: Mendocino Indian
Dimorphocarpa: Navajo
Disporum: Karok; Nitinaht
Dryopteris: Haisla & Hanaksiala
Dyssodia: Dakota
Enteromorpha: Hesquiat
Epilobium: Okanagan-Colville
Equisetum: Haisla & Hanaksiala; Mendocino Indian; Tewa
Erigeron: Ojibwa
Erodium: Isleta; Kawaiisu
Erythronium: Montana Indian
Fritillaria: Hesquiat; Montana Indian
Gutierrezia: Tewa
Hackelia: Thompson
Hedysarum: Eskimo, Arctic
Heracleum: Anticosti; Hesquiat; Thompson
Hilaria: Navajo, Ramah
Hydrophyllum: Okanagon; Thompson
Ipomopsis: Navajo
Juncus: Mendocino Indian
Juniperus: Navajo
Kalmia: Mahuna
Kochia: Navajo
Koeleria: Havasupai
Krascheninnikovia: Keres, Western; Navajo
Larix: Okanagan-Colville
Lathyrus: Carrier; Thompson
Leymus: Blackfoot; Kawaiisu
Licania: Seminole
Lithocarpus: Pomo, Kashaya
Lonicera: Bella Coola; Hesquiat; Okanagan-Colville; Thompson
Lotus: Isleta
Lupinus: Mendocino Indian; Okanagan-Colville; Paiute
Lysichiton: Haisla & Hanaksiala; Hesquiat; Hoh; Okanagan-Colville; Oweekeno; Quileute
Maianthemum: Oweekeno; Thompson
Malus: Nitinaht
Malva: Pima
Medicago: Mendocino Indian
Melilotus: Jemez
Menyanthes: Hesquiat
Muhlenbergia: Blackfoot
Nemophila: Kawaiisu
Nepeta: Okanagan-Colville
Nuphar: Mendocino Indian
Oemleria: Karok
Orthocarpus: Mendocino Indian
Oryzopsis: Navajo
Oxalis: Pawnee
Oxytropis: Lakota; Navajo; Thompson
Panicum: Navajo, Ramah
Pascopyrum: Lakota; Montana Indian
Paspalum: Kiowa
Paxistima: Okanagan-Colville; Thompson
Pedicularis: Mendocino Indian
Penstemon: Navajo; Thompson
Petasites: Nitinaht
Phragmites: Thompson
Pinus: Okanagan-Colville
Pluchea: Pima
Populus: Dakota; Lakota; Thompson
Portulaca: Navajo
Potamogeton: Hesquiat
Prosopis: Pima
Prunus: Haisla & Hanaksiala
Pseudoroegneria: Okanagan-Colville
Purshia: Navajo; Okanagan-Colville
Quercus: Pomo, Kashaya
Ranunculus: Hesquiat
Ribes: Okanagan-Colville; Paiute; Thompson
Robinia: Mendocino Indian; Wailaki
Rosa: Hesquiat; Okanagan-Colville; Thompson
Rubus: Haisla & Hanaksiala
Sagittaria: Meskwaki; Ojibwa
Salix: Navajo; Thompson
Salsola: Havasupai
Sarcobatus: Keres, Western; Navajo; Pima
Scirpus: Hesquiat
Shepherdia: Blackfoot; Thompson
Sisymbrium: Navajo
Sium: Thompson
Sophora: Navajo
Sorbus: Heiltzuk; Montagnais
Spartina: Iroquois
Sphagnum: Haisla & Hanaksiala
Spiraea: Thompson
Sporobolus: Navajo

Stachys: Saanich
Stellaria: Iroquois
Streptopus: Hesquiat; Montagnais; Nitinaht; Oweekeno
Symphoricarpos: Haisla & Hanaksiala; Okanagan-Colville
Toxicodendron: Mendocino Indian
Tragopogon: Okanagan-Colville
Trifolium: Mendocino Indian; Wailaki
Usnea: Hesquiat
Vaccinium: Hesquiat; Okanagan-Colville
Valeriana: Thompson
Vicia: Thompson
Yucca: Isleta; Navajo
Zea: Tewa
Zostera: Hesquiat

Frozen Food
Allium: Tanana, Upper
Amelanchier: Okanagan-Colville; Thompson
Arbutus: Karok
Arctostaphylos: Eskimo, Inupiat
Chenopodium: Eskimo, Inupiat
Empetrum: Eskimo, Arctic; Tanana, Upper
Gaultheria: Nitinaht
Gaylussacia: Cherokee
Hedysarum: Eskimo, Inupiat
Heracleum: Thompson
Lomatium: Thompson
Myriophyllum: Tanana, Upper
Oxytropis: Eskimo, Inupiat
Pinus: Okanagan-Colville
Polygonum: Tanana, Upper
Ribes: Thompson
Rosa: Eskimo, Inupiat
Rubus: Alaska Native; Eskimo, Inuktitut; Eskimo, Inupiat; Okanagan-Colville; Tanana, Upper; Thompson
Rudbeckia: Cherokee
Rumex: Tanana, Upper
Sambucus: Okanagan-Colville
Sedum: Eskimo, Inupiat
Vaccinium: Alaska Native; Bella Coola; Cree, Woodlands; Eskimo, Arctic; Eskimo, Inupiat; Koyukon; Makah; Nitinaht; Tanana, Upper; Thompson
Viburnum: Cree, Woodlands; Eskimo, Inupiat; Koyukon; Tanana, Upper

Fruit
Aesculus: Costanoan; Mendocino Indian
Amelanchier: Abnaki; Alaska Native; Algonquin, Quebec; Bella Coola; Blackfoot; Cahuilla; Chehalis; Cherokee; Chippewa; Coeur d'Alene; Costanoan; Cree, Woodlands; Dakota; Gosiute; Great Basin Indian; Haisla; Hesquiat; Hoh; Iroquois; Isleta; Karok; Kawaiisu; Kitasoo; Klamath; Kwakiutl, Southern; Lakota; Mendocino Indian; Modesse; Montana Indian; Navajo; Navajo, Ramah; Ojibwa; Okanagan-Colville; Omaha; Oweekeno; Paiute; Ponca; Potawatomi; Quileute; Saanich; Salish, Coast; Sanpoil & Nespelem; Shuswap; Skagit; Skagit, Upper; Spokan; Swinomish; Thompson; Ute; Winnebago; Wintoon
Amphicarpaea: Chippewa
Arbutus: Costanoan; Karok; Pomo; Pomo, Kashaya; Wailaki; Yuki; Yurok
Arctostaphylos: Alaska Native; Blackfoot; Cahuilla; Carrier; Cherokee; Chinook, Lower; Coeur d'Alene; Costanoan; Cree, Woodlands; Diegueño; Eskimo, Alaska; Eskimo, Arctic; Eskimo, Inupiat; Hanaksiala; Hualapai; Karok; Kawaiisu; Klamath; Kwakiutl, Southern; Luiseño; Mahuna; Makah; Mendocino Indian; Mewuk; Midoo; Montana Indian; Navajo; Navajo, Kayenta; Nitinaht; Nux-alkmc; Okanagan-Colville; Okanagon; Oweekeno; Paiute; Salish, Coast; Skokomish; Spokan; Squaxin; Tanana, Upper; Thompson; Tubatulabal; Wintoon; Yavapai; Yuki; Yurok
Aronia: Abnaki; Potawatomi
Artocarpus: Hawaiian
Asclepias: Cheyenne; Kiowa
Asimina: Cherokee; Iroquois
Aster: Tewa
Astragalus: Apache, White Mountain; Lakota
Atriplex: Acoma; Cahuilla; Laguna
Berberis: Jemez
Canotia: Apache, San Carlos; Apache, Western
Capsicum: Pima, Gila River
Carica: Hawaiian
Carnegia: Apache, Chiricahua & Mescalero; Apache, San Carlos; Apache, Western; Apache, White Mountain; Papago; Papago & Pima; Pima; Seri; Southwest Indians; Yavapai
Carpobrotus: Luiseño; Pomo; Pomo, Kashaya
Ceanothus: Acoma; Keres, Western; Laguna
Celtis: Acoma; Apache, Chiricahua & Mescalero; Comanche; Hualapai; Keres, Western; Kiowa; Laguna; Navajo; Omaha; Papago; Pawnee; Pueblo; Tewa
Cereus: Apache, White Mountain
Citrullus: Apalachee; Cahuilla; Cocopa; Havasupai; Pima
Citrus: Haisla & Hanaksiala; Thompson
Clermontia: Hawaiian
Condalia: Maricopa; Papago; Pima
Cornus: Abnaki; Bella Coola; Blackfoot; Chippewa; Eskimo, Alaska; Flathead; Kitasoo; Kutenai; Kwakiutl, Southern; Makah; Nitinaht; Okanagan-Colville; Oweekeno; Potawatomi; Salish; Sanpoil & Nespelem; Shuswap; Spokan; Thompson
Corylus: Iroquois
Crataegus: Abnaki; Algonquin, Quebec; Apache, Chiricahua & Mescalero; Bella Coola; Blackfoot; Cherokee; Cheyenne; Coeur d'Alene; Comanche; Haisla & Hanaksiala; Iroquois; Kwakiutl, Southern; Lakota; Meskwaki; Montana Indian; Ojibwa; Okanagan-Colville; Okanagon; Omaha; Oregon Indian; Paiute; Ponca; Potawatomi; Salish, Coast; Sanpoil & Nespelem; Spokan; Thompson; Winnebago
Cucumis: Cocopa; Thompson
Cucurbita: Isleta; Maricopa; Papago; Pima
Cupressus: Navajo
Datura: Navajo
Dioscorea: Hawaiian
Diospyros: Cherokee; Comanche
Disporum: Blackfoot; Shuswap; Thompson
Echinocereus: Apache, Chiricahua & Mescalero; Apache, Mescalero; Isleta; Keres, Western; Navajo; Navajo, Ramah; Pima; Yavapai
Elaeagnus: Alaska Native; Blackfoot; Cree; Montana Indian; Okanagan-Colville; Tanana, Upper
Empetrum: Cree, Woodlands; Eskimo, Alaska; Eskimo, Arctic; Eskimo, Inupiat; Ojibwa; Tanana, Upper; Tsimshian
Escobaria: Blackfoot; Cheyenne; Crow
Ferocactus: Hualapai
Ficus: Havasupai
Forestiera: Apache, Chiricahua & Mescalero
Fragaria: Abnaki; Alaska Native; Algonquin, Quebec; Algonquin, Tête-de-Boule; Apache, Chiricahua & Mescalero; Bella Coola; Blackfoot; Cahuilla; Carrier; Chehalis; Cherokee; Cheyenne; Chinook, Lower; Chippewa; Clallam; Coeur d'Alene; Costanoan; Cowlitz; Dakota; Diegueño; Gosiute; Haisla & Hanaksiala; Hesquiat; Hoh; Iroquois; Isleta; Karok; Kitasoo; Klallam; Klamath; Lakota; Makah; Mendocino Indian; Menominee; Montana Indian; Nisqually; Nitinaht; Ojibwa; Okanagan-Colville; Omaha; Oweekeno; Paiute; Pawnee; Pomo; Pomo, Kashaya; Ponca; Puyallup; Quileute; Salish, Coast; Sanpoil & Nespelem; Skagit, Upper; Skokomish; Spokan; Squaxin; Swinomish; Thompson; Tolowa; Winnebago; Yurok
Frangula: Atsugewi; Costanoan; Kawaiisu; Makah; Paiute
Gaultheria: Alaska Native; Algonquin, Quebec; Cherokee; Haisla & Hanaksiala; Hoh; Iroquois; Karok; Kitasoo; Makah; Nitinaht; Ojibwa; Okanagon; Oweekeno; Pomo; Pomo, Kashaya; Quileute; Salish, Coast; Skagit, Upper; Thompson; Tolowa; Yurok
Gaylussacia: Cherokee; Iroquois
Geocaulon: Alaska Native
Helianthus: Omaha
Heteromeles: Cahuilla; Costanoan; Diegueño; Karok; Mahuna; Mendocino Indian; Neeshenam; Pomo; Pomo, Kashaya; Yurok
Juniperus: Acoma; Apache; Apache, Chiricahua & Mescalero; Apache, Mescalero; Apache, White Mountain; Atsugewi; Cahuilla; Cochiti; Comanche; Costanoan; Diegueño; Gosiute; Hopi; Isleta; Jemez; Kawaiisu; Keresan; Laguna; Navajo; Navajo, Ramah; Paiute; Paiute, Northern; San Felipe; San Ildefonso; Tewa; Tubatulabal
Lomatium: Thompson
Lonicera: Klamath; Okanagan-Colville; Okanagon; Oweekeno; Thompson
Lycium: Cahuilla; Hopi; Isleta; Jemez; Kawaiisu; Maricopa; Navajo; Navajo, Ramah; Paiute, Northern; Papago; Pima; Pima, Gila River; Tubatulabal; Zuni

Lycopersicon: Haisla & Hanaksiala
Machaerocereus: Papago & Pima
Mahonia: Apache; Apache, Mescalero; Blackfoot; Cheyenne; Clallam; Coeur d'Alene; Cowlitz; Flathead; Hualapai; Klallam; Kwakiutl, Southern; Lummi; Montana Indian; Okanagan-Colville; Samish; Sanpoil; Sanpoil & Nespelem; Shuswap; Skagit; Skagit, Upper; Snohomish; Spokan; Squaxin; Swinomish; Thompson; Yavapai
Maianthemum: Bella Coola; Costanoan; Haisla & Hanaksiala; Hesquiat; Kitasoo; Kwakiutl, Southern; Okanogon; Potawatomi; Salish, Coast; Skagit, Upper; Tewa; Thompson
Malus: Bella Coola; Cherokee; Chinook, Lower; Clallam; Cowlitz; Haisla & Hanaksiala; Hoh; Iroquois; Makah; Nitinaht; Ojibwa; Omaha; Oweekeno; Ponca; Quileute; Quinault; Salish, Coast; Samish; Skagit, Upper; Swinomish; Thompson
Mammillaria: Apache, Chiricahua & Mescalero; Apache, San Carlos; Diegueño
Mimulus: Navajo, Kayenta
Mirabilis: Navajo, Kayenta
Mitchella: Cherokee; Iroquois
Moneses: Montana Indian
Morus: Apache, Chiricahua & Mescalero; Apache, Mescalero; Cherokee; Comanche; Hualapai; Iroquois; Lakota; Omaha; Yavapai
Myrtillocactus: Papago & Pima
Nolina: Hualapai; Isleta
Oemleria: Cowlitz; Kitasoo; Kwakiutl, Southern; Lummi; Nitinaht; Quinault; Saanich; Samish; Shasta; Skagit; Skagit, Upper; Snohomish; Squaxin; Swinomish; Thompson; Tolowa
Oenothera: Apache
Opuntia: Acoma; Apache, Mescalero; Apache, White Mountain; Cahuilla; Cheyenne; Cocopa; Comanche; Costanoan; Dakota; Diegueño; Havasupai; Hopi; Hualapai; Isleta; Jemez; Keres, Western; Kiowa; Laguna; Lakota; Luiseño; Maricopa; Mohave; Montana Indian; Navajo; Navajo, Ramah; Papago; Pawnee; Pima; Pima, Gila River; San Felipe; Tewa; Tewa of Hano; Yavapai; Zuni
Pachycereus: Seri
Parthenocissus: Montana Indian; Navajo; Navajo, Ramah
Passiflora: Cherokee
Paxistima: Karok
Peniocereus: Papago & Pima
Philadelphus: Isleta
Phoradendron: Cahuilla; Navajo; Pima; Pima, Gila River
Physalis: Apache, White Mountain; Cherokee; Cheyenne; Hualapai; Iroquois; Isleta; Keres, Western; Meskwaki; Mohave; Navajo; Navajo, Ramah; Pueblo; San Felipe; Tewa; Yuma; Zuni
Physocarpus: Miwok
Podophyllum: Cherokee; Chippewa; Iroquois; Menominee; Meskwaki
Prunus: Abnaki; Acoma; Algonquin, Quebec; Apache, Chiricahua & Mescalero; Apache, Mescalero; Apache, Western; Blackfoot; Cahuilla; Cherokee; Cheyenne; Chippewa; Cochiti; Coeur d'Alene; Comanche; Costanoan; Cree, Woodlands; Crow; Dakota; Diegueño; Gosiute; Haisla & Hanaksiala; Hopi; Iroquois; Isleta; Karok; Kawaiisu; Keres, Western; Keresan; Kiowa; Klamath; Laguna; Lakota; Luiseño; Mahuna; Malecite; Mendocino Indian; Menominee; Meskwaki; Modesse; Montana Indian; Navajo; Navajo, Ramah; Ojibwa; Okanagan-Colville; Omaha; Oweekeno; Paiute; Paiute, Northern; Pawnee; Ponca; Potawatomi; Pueblo; Round Valley Indian; San Felipe; Sanpoil & Nespelem; Shuswap; Spokan; Tewa; Thompson; Winnebago; Wintoon; Yuki; Yurok
Ptelea: San Felipe
Pyrus: Cherokee; Iroquois
Quercus: Apache, Chiricahua & Mescalero; Diegueño; Karok; Tolowa; Yurok
Rhamnus: Cahuilla
Rhus: Apache; Apache, Western; Apache, White Mountain; Cahuilla; Cherokee; Comanche; Gosiute; Hopi; Hualapai; Jemez; Keres, Western; Keresan; Kiowa; Mahuna; Midoo; Montana Indian; Navajo; Navajo, Ramah; Tewa; Ute; Wintoon; Yavapai; Yokut
Ribes: Acoma; Alaska Native; Algonquin, Quebec; Apache; Apache, Chiricahua & Mescalero; Apache, White Mountain; Atsugewi; Bella Coola; Blackfoot; Cahuilla; Chehalis; Cheyenne; Chippewa; Clallam; Coeur d'Alene; Comanche; Costanoan; Cowlitz; Cree, Woodlands; Dakota; Eskimo, Inupiat; Gosiute; Haisla & Hanaksiala; Hesquiat; Hoh; Hopi; Iroquois; Isleta; Jemez; Karok; Kawaiisu; Keres, Western; Kiowa; Kitasoo; Klallam; Klamath; Kwakiutl, Southern; Laguna; Lakota; Lummi; Mahuna; Makah; Mendocino Indian; Menominee; Meskwaki; Miwok; Montana Indian; Navajo; Navajo, Ramah; Nitinaht; Ojibwa; Okanagan-Colville; Okanogon; Omaha; Oweekeno; Paiute; Paiute, Northern; Pomo, Kashaya; Ponca; Potawatomi; Pueblo; Quileute; Sanpoil & Nespelem; Shuswap; Skagit; Skagit, Upper; Spokan; Squaxin; Swinomish; Tanana, Upper; Tewa; Thompson; Thompson, Upper (Lytton Band); Tolowa; Tubatulabal; Ute; Winnebago; Wintoon; Yavapai; Yurok; Zuni
Rosa: Apache, Chiricahua & Mescalero; Bella Coola; Blackfoot; Cheyenne; Coeur d'Alene; Cowichan; Gosiute; Hesquiat; Hoh; Hopi; Kawaiisu; Klamath; Koyukon; Lakota; Montana Indian; Navajo; Navajo, Ramah; Nitinaht; Paiute; Pomo, Kashaya; Quileute; Rocky Boy; Saanich; Spokan; Tanana, Upper; Thompson; Ute; Washo
Rubus: Abnaki; Alaska Native; Algonquin, Quebec; Algonquin, Tête-de-Boule; Apache, Chiricahua & Mescalero; Bella Coola; Blackfoot; Cahuilla; Carrier; Chehalis; Cherokee; Cheyenne; Chinook, Lower; Chippewa; Clallam; Coeur d'Alene; Costanoan; Cowlitz; Cree, Woodlands; Dakota; Diegueño; Eskimo; Eskimo, Alaska; Eskimo, Inuktitut; Eskimo, Inupiat; Gosiute; Green River Group; Haisla & Hanaksiala; Hesquiat; Hoh; Iroquois; Isleta; Karok; Kitasoo; Klallam; Klamath; Koyukon; Kwakiutl, Southern; Lakota; Luiseño; Lummi; Mahuna; Makah; Mendocino Indian; Menominee; Meskwaki; Montana Indian; Navajo; Nisqually; Nitinaht; Ojibwa; Okanagan-Colville; Okanogon; Omaha; Oweekeno; Paiute; Pawnee; Pomo; Pomo, Kashaya; Ponca; Potawatomi; Puyallup; Quileute; Quinault; Salish, Coast; Samish; Sanpoil & Nespelem; Shuswap; Skagit; Skagit, Upper; Snohomish; Spokan; Squaxin; Swinomish; Tanana, Upper; Thompson; Tolowa; Tsimshian; Wintoon; Yuki; Yurok
Sambucus: Apache; Apache, White Mountain; Cahuilla; Cherokee; Chippewa; Coeur d'Alene; Costanoan; Dakota; Diegueño; Gosiute; Haisla & Hanaksiala; Hesquiat; Hoh; Iroquois; Karok; Kiowa; Klallam; Klamath; Luiseño; Mahuna; Makah; Mendocino Indian; Meskwaki; Montana Indian; Nitinaht; Okanagan-Colville; Okanogon; Omaha; Paiute; Paiute, Northern; Pawnee; Pima, Gila River; Poliklah; Pomo, Kashaya; Ponca; Quileute; Salish, Coast; Sanpoil & Nespelem; Skagit, Upper; Skokomish; Spokan; Thompson; Tubatulabal; Wintoon; Yuki; Yurok
Sarcobatus: Navajo
Serenoa: Seminole
Shepherdia: Carrier; Coeur d'Alene; Dakota; Gosiute; Kwakiutl, Southern; Lakota; Montana Indian; Navajo; Okanagan; Omaha; Oweekeno; Paiute; Pawnee; Ponca; Shuswap; Spokan; Tanana, Upper; Thompson; Ute; Winnebago
Smilax: Meskwaki; Omaha
Solanum: Costanoan; Mendocino Indian; Miwok; Pima; Tubatulabal
Sorbus: Algonquin, Quebec; Montagnais; Ojibwa; Okanogon; Thompson
Stenocereus: Apache, Chiricahua & Mescalero; Papago & Pima; Seri
Streptopus: Alaska Native; Okanogon; Thompson
Symphoricarpos: Sioux
Taxus: Karok; Mendocino Indian
Typha: Havasupai; Okanagan-Colville
Umbellularia: Costanoan; Pomo, Kashaya; Tolowa; Yurok
Vaccinium: Abnaki; Alaska Native; Algonquin, Quebec; Algonquin, Tête-de-Boule; Bella Coola; Chinook, Lower; Chippewa; Clallam; Coeur d'Alene; Costanoan; Cree, Woodlands; Eskimo, Alaska; Eskimo, Inuktitut; Eskimo, Inupiat; Haida; Haisla & Hanaksiala; Hanaksiala; Hesquiat; Hoh; Iroquois; Karok; Kitasoo; Klallam; Klamath; Kwakiutl, Southern; Lummi; Makah; Menominee; Montana Indian; Nitinaht; Ojibwa; Okanagan-Colville; Oweekeno; Paiute; Poliklah; Pomo; Pomo, Kashaya; Potawatomi; Quileute; Quinault; Salish; Salish, Coast; Sanpoil & Nespelem; Shuswap; Skagit; Skagit, Upper; Skokomish; Snohomish; Spokan; Swinomish; Tanana, Upper; Thompson; Tolowa; Tsimshian; Yurok

Viburnum: Abnaki; Algonquin, Quebec; Dakota; Eskimo, Inuktitut; Eskimo, Inupiat; Haisla & Hanaksiala; Hesquiat; Iroquois; Menominee; Meskwaki; Ojibwa; Okanagan-Colville; Okanagon; Omaha; Pawnee; Ponca; Salish, Coast; Shuswap; Tanana, Upper; Thompson; Winnebago
Vicia: Kwakiutl, Southern
Vitis: Apache, Chiricahua & Mescalero; Apache, Mescalero; Apache, Western; Cahuilla; Cherokee; Cheyenne; Chippewa; Comanche; Costanoan; Crow; Dakota; Diegueño; Haisla & Hanaksiala; Havasupai; Hualapai; Iroquois; Isleta; Jemez; Karok; Keres, Western; Kiowa; Lakota; Luiseño; Mahuna; Menominee; Omaha; Pawnee; Pomo; Pomo, Kashaya; Ponca; Sioux; Tubatulabal; Winnebago; Wintoon; Yurok
Washingtonia: Cahuilla
Yucca: Acoma; Apache, San Carlos; Hopi; Hualapai; Isleta; Kawaiisu; Keres, Western; Keresan; Laguna; Mohave; Navajo; Navajo, Ramah; Papago; Pima; Pima, Gila River; Pueblo; San Felipe; Sia; Southwest Indians; Tewa; Yavapai; Zuni
Ziziphus: Cahuilla; Maricopa; Maricopa & Mohave; Pima; Pima, Gila River

Ice Cream
Anemone: Eskimo, Alaska
Arctostaphylos: Eskimo, Inupiat
Descurainia: Paiute
Dryopteris: Eskimo, Inuktitut
Empetrum: Eskimo, Alaska
Hippuris: Eskimo, Inuktitut
Honckenya: Alaska Native
Pinus: Paiute, Northern
Rosa: Eskimo, Inupiat
Rubus: Eskimo, Arctic; Eskimo, Inupiat
Rumex: Eskimo, Inuktitut
Shepherdia: Bella Coola; Carrier; Clallam; Coeur d'Alene; Flathead; Haisla & Hanaksiala; Kitasoo; Lillooet; Montana Indian; Nitinaht; Okanagan-Colville; Sanpoil; Shuswap; Thompson, Lower
Vaccinium: Eskimo, Inupiat
Viburnum: Eskimo, Inupiat

Pie & Pudding
Amelanchier: Alaska Native; Cheyenne; Flathead; Montana Indian; Okanagan-Colville; Thompson
Atriplex: Hopi; Navajo
Balsamorhiza: Paiute
Camassia: Klamath; Paiute
Carnegia: Apache, Western
Carya: Iroquois
Castanea: Iroquois
Corylus: Iroquois
Cucurbita: Iroquois
Diospyros: Cherokee
Empetrum: Alaska Native; Eskimo, Inupiat; Tanana, Upper
Erythronium: Thompson
Fagus: Iroquois
Gaultheria: Makah; Pomo, Kashaya; Thompson
Gaylussacia: Cherokee; Iroquois

Juglans: Iroquois
Lewisia: Thompson
Lomatium: Thompson
Pediomelum: Cheyenne
Perideridia: Nevada Indian; Okanagan-Colville
Pinus: Apache, Chiricahua & Mescalero
Polygonum: Alaska Native; Eskimo, Arctic
Prosopis: Apache, Chiricahua & Mescalero
Prunus: Cherokee; Cheyenne; Lakota; Montana Indian
Quercus: Iroquois; Menominee
Rheum: Kitasoo
Ribes: Thompson
Rosa: Cheyenne
Rubus: Alaska Native; Cherokee; Iroquois; Makah; Menominee; Meskwaki; Pomo, Kashaya; Tanana, Upper
Rumex: Pima
Sambucus: Cherokee; Mendocino Indian; Montana Indian
Shepherdia: Cheyenne; Paiute
Solanum: Navajo, Ramah
Sporobolus: Hopi
Vaccinium: Alaska Native; Algonquin, Quebec; Eskimo, Inupiat; Hesquiat; Iroquois; Makah; Montana Indian; Nitinaht; Pomo; Pomo, Kashaya; Tanana, Upper; Thompson
Viburnum: Tanana, Upper
Zea: Iroquois
Zizania: Potawatomi

Porridge
Acacia: Cahuilla; Seri
Agave: Navajo
Allenrolfea: Cahuilla
Amaranthus: Cahuilla; Havasupai; Hopi; Navajo
Amelanchier: Atsugewi
Anemopsis: Kamia
Arctostaphylos: Cahuilla; Navajo; Numlaki; Pomo
Artemisia: Navajo
Atriplex: Cahuilla; Navajo, Ramah; Paiute, Northern; Pima; Yuma; Zuni
Avena: Cahuilla; Diegueño; Miwok
Balsamorhiza: Paiute
Berberis: Yana
Blennosperma: Neeshenam
Bouteloua: Apache, Western; Apache, White Mountain
Bromus: Karok; Neeshenam
Calochortus: Cheyenne
Carex: Navajo, Kayenta
Carnegia: Apache, Western; Pima; Pima, Gila River; Seri
Celtis: Meskwaki
Chaenactis: Cahuilla
Chasmanthium: Cocopa
Chenopodium: Hopi; Navajo
Citrullus: Havasupai
Cleome: Acoma; Keres, Western; Laguna
Cucurbita: Cahuilla; Havasupai
Cupressus: Navajo

Cycloloma: Hopi; Zuni
Cyperus: Kamia; Paiute
Deschampsia: Kawaiisu
Descurainia: Gosiute; Navajo; Pima, Gila River
Echinocactus: Havasupai
Echinochloa: Cocopa; Yuma
Elymus: Karok; Kawaiisu
Equisetum: San Felipe
Eragrostis: Cocopa
Eriochloa: Cocopa
Eriogonum: Kawaiisu; Navajo, Kayenta; Navajo, Ramah
Ferocactus: Apache, San Carlos
Fouquieria: Cahuilla
Gaylussacia: Iroquois
Helianthus: Apache, Western; Montana Indian; Navajo; Paiute
Heliotropium: Navajo, Kayenta
Hirschfeldia: Cahuilla
Juglans: Cherokee
Juniperus: Cahuilla
Koeleria: Isleta
Lasthenia: Cahuilla
Layia: Cahuilla
Lewisia: Sanpoil & Nespelem
Leymus: Kawaiisu
Lithocarpus: Hahwunkwut; Hupa; Karok; Poliklah; Pomo; Pomo, Kashaya; Shasta; Yuki
Lomatium: Klamath; Montana Indian
Lycium: Hopi; Paiute, Northern; Yuma
Madia: Neeshenam; Pomo
Medicago: Cahuilla
Melica: Kawaiisu; Pomo
Mentzelia: Cahuilla; Hualapai; Paiute, Northern
Monolepis: Hopi
Muhlenbergia: Apache, Western; Apache, White Mountain
Nama: Kawaiisu
Nolina: Isleta
Nuphar: Klamath; Montana Indian
Olneya: Cocopa; Seri
Opuntia: Acoma; Apache, San Carlos; Apache, Western; Cahuilla; Laguna; San Felipe; Zuni
Oryzopsis: Apache, Western; Apache, White Mountain; Navajo; Navajo, Ramah; Paiute; Paiute, Northern
Oxytropis: Navajo, Kayenta
Pachycereus: Seri
Panicum: Cahuilla
Parkinsonia: Cahuilla; Cocopa
Pectis: Havasupai
Pediomelum: Montana Indian
Perideridia: Cheyenne; Paiute, Northern
Phaseolus: Havasupai
Phoradendron: Maricopa
Pinus: Apache, Western; Cahuilla; Hualapai; Kawaiisu; Navajo; Paiute; Thompson; Washo
Plantago: Havasupai; Navajo, Kayenta; Pima, Gila River
Polygonum: Klamath

Prosopis: Acoma; Apache, Western; Cahuilla; Kawaiisu; Keres, Western; Laguna; Mahuna; Pima; Seri
Prunus: Atsugewi; Diegueño; Gosiute; Luiseño; Navajo; Paiute, Northern
Quercus: Cahuilla; Concow; Diegueño; Havasupai; Luiseño; Menominee; Meskwaki; Mewuk; Miwok; Mohave; Neeshenam; Paiute; Pomo; Pomo, Kashaya; Potawatomi; Round Valley Indian; Shasta; Yana; Yuki
Ranunculus: Neeshenam
Rhus: Cahuilla; Luiseño; Navajo
Ribes: Paiute, Northern
Rubus: Iroquois
Rumex: Kawaiisu; Mendocino Indian; Navajo; Navajo, Kayenta
Salvia: Cahuilla; Diegueño; Mahuna; Paiute; Pima, Gila River
Sambucus: Iroquois
Scirpus: Paiute, Northern
Shepherdia: Paiute, Northern
Sisymbrium: Navajo; Pima; Pima, Gila River
Sporobolus: Apache, Chiricahua & Mescalero; Apache, Western; Apache, White Mountain; Hopi; Navajo, Ramah
Stenocereus: Papago & Pima; Seri
Stipa: Paiute
Suaeda: Navajo
Trifolium: Cahuilla
Triticum: Cahuilla; Pima; Pomo, Kashaya
Typha: Cahuilla; Paiute, Northern; Pima; Yuma
Vaccinium: Iroquois
Vigna: Cocopa, Maricopa, Mohave & Yuma
Vitis: Cahuilla
Washingtonia: Cahuilla
Wyethia: Paiute, Northern
Yucca: Navajo; Papago; Pima
Zea: Cahuilla; Chippewa; Delaware; Havasupai; Hopi; Iroquois; Isleta; Navajo; Pima; Zuni
Ziziphus: Cahuilla

Preservative
Achillea: Klamath
Alnus: Haisla & Hanaksiala; Karok; Tanana, Upper
Amelanchier: Cree, Woodlands
Betula: Cree, Woodlands; Ojibwa
Comptonia: Ojibwa
Cornus: Shuswap
Epilobium: Tanana, Upper
Matricaria: Flathead
Monarda: Flathead
Myrica: Potawatomi
Populus: Cree, Woodlands; Tanana, Upper
Rhododendron: Thompson
Ribes: Quinault
Rubus: Makah
Salix: Shuswap
Ulmus: Omaha

Preserves
Amelanchier: Blackfoot; Montana Indian; Thompson

Arctostaphylos: Diegueño; Navajo
Artemisia: Havasupai
Asclepias: Chippewa; Lakota
Carnegia: Apache, White Mountain; Papago; Papago & Pima; Pima, Gila River
Celtis: Apache, Chiricahua & Mescalero
Cereus: Apache, White Mountain
Crataegus: Shuswap; Thompson
Cucurbita: Navajo
Descurainia: Havasupai
Echinocereus: Isleta
Empetrum: Alaska Native; Tanana, Upper
Fragaria: Alaska Native; Algonquin, Quebec; Cherokee; Haisla & Hanaksiala; Makah; Meskwaki; Ojibwa; Oweekeno
Gaillardia: Havasupai
Gaultheria: Hoh; Makah; Nitinaht; Oweekeno; Quileute; Thompson
Gaylussacia: Cherokee
Gilia: Havasupai
Helianthus: Havasupai
Juniperus: Apache, Chiricahua & Mescalero
Lepidium: Havasupai
Lycium: Hopi
Mahonia: Apache, Chiricahua & Mescalero; Hoh; Makah; Modesse; Montana Indian; Pueblo; Quileute; Salish, Coast; Sanpoil; Skagit; Spanish American; Thompson
Malus: Cherokee; Makah; Meskwaki
Mentzelia: Havasupai; Kawaiisu
Morus: Cherokee
Nolina: Isleta
Opuntia: Isleta; Kiowa; Montana Indian; Navajo
Physalis: Hualapai; Navajo
Pinus: Havasupai; Navajo; Navajo, Ramah
Podophyllum: Menominee; Meskwaki
Prosopis: Apache; Papago
Prunus: Algonquin, Quebec; Apache, Chiricahua & Mescalero; Cahuilla; Cherokee; Cree, Woodlands; Hoh; Kawaiisu; Mendocino Indian; Menominee; Meskwaki; Montana Indian; Ojibwa; Quileute; Spanish American; Thompson
Quincula: Kiowa
Rhamnus: Cahuilla
Rheum: Kitasoo
Rhus: Apache, Chiricahua & Mescalero; Atsugewi
Ribes: Acoma; Alaska Native; Algonquin, Quebec; Anticosti; Apache, Chiricahua & Mescalero; Carrier; Cree, Woodlands; Hesquiat; Isleta; Kawaiisu; Kiowa; Laguna; Makah; Menominee; Ojibwa; Potawatomi; Spanish American; Tanana, Upper; Thompson
Rosa: Alaska Native; Apache, Chiricahua & Mescalero; Blackfoot; Carrier; Eskimo, Inupiat; Lakota; Rocky Boy; Tanana, Upper; Thompson
Rubus: Abnaki; Alaska Native; Algonquin, Quebec; Anticosti; Bella Coola; Carrier; Cherokee; Hesquiat; Makah; Meskwaki; Ojibwa; Oweekeno; Potawatomi; Tanana, Upper; Thompson
Sambucus: Apache, Chiricahua & Mescalero; Bella Coola; Cahuilla; Cherokee; Hesquiat; Kawaiisu; Mendocino Indian; Meskwaki; Montana Indian; Nitinaht; Oweekeno; Pima; Thompson
Shepherdia: Arapaho; Carrier; Cheyenne; Montana Indian; Thompson
Simmondsia: Yavapai
Stenocereus: Papago; Papago & Pima
Urtica: Haisla & Hanaksiala
Vaccinium: Alaska Native; Algonquin, Quebec; Anticosti; Carrier; Cree, Woodlands; Eskimo, Arctic; Hesquiat; Iroquois; Makah; Montana Indian; Paiute; Tanana, Upper; Thompson
Viburnum: Alaska Native; Algonquin, Quebec; Carrier; Cree, Woodlands; Eskimo, Inupiat; Kwakiutl, Southern; Meskwaki; Ojibwa; Oweekeno; Shuswap; Tanana, Upper; Thompson
Vitis: Kiowa; Mendocino Indian; Menominee; Ojibwa
Washingtonia: Cahuilla
Yucca: Acoma; Hopi; Laguna; Navajo; Navajo, Ramah; Southwest Indians; Zuni

Sauce & Relish
Acer: Algonquin, Quebec; Malecite; Micmac; Montana Indian; Thompson
Agave: Pima
Aletes: Isleta
Allium: Dakota; Flathead; Kutenai; Navajo; Omaha; Paiute; Pawnee; Ponca; Winnebago
Amelanchier: Iroquois
Arctostaphylos: Cahuilla; Flathead
Armoracia: Cherokee; Mendocino Indian
Asclepias: Cheyenne; Crow
Asimina: Iroquois
Betula: Algonquin, Quebec; Cree, Woodlands
Camassia: Flathead; Okanagan-Colville
Cardamine: Abnaki; Algonquin, Quebec; Ojibwa
Carnegia: Papago; Papago & Pima; Pima; Pima, Gila River
Carya: Iroquois
Castanea: Iroquois
Cetraria: Eskimo, Inuktitut
Claytonia: Montana Indian
Cleome: Navajo
Condalia: Pima
Coriandrum: Hopi; Zuni
Corylus: Iroquois
Crataegus: Iroquois
Cucumis: Navajo
Cucurbita: Iroquois; Navajo
Echinocereus: Isleta
Fagus: Iroquois
Fragaria: Cherokee; Iroquois
Gaultheria: Hoh; Iroquois; Quileute
Gaylussacia: Iroquois
Helianthus: Apache, Chiricahua & Mescalero
Juglans: Apache, Western; Iroquois
Juniperus: Apache, Chiricahua & Mescalero; Apache, Western
Larix: Flathead; Kutenai
Liriodendron: Cherokee

Lomatium: Okanagan-Colville
Lupinus: Miwok
Lycium: Acoma; Keres, Western; Laguna
Malus: Iroquois
Mentha: Hopi
Mentzelia: Paiute
Mitchella: Iroquois
Morus: Apache, Mescalero; Iroquois
Oenothera: Apache, Chiricahua & Mescalero
Opuntia: Navajo; Papago
Oryzopsis: Paiute
Oxydendrum: Cherokee
Panicum: Apache, Chiricahua & Mescalero; Cocopa
Pectis: Havasupai
Physalis: Dakota; Hualapai; Iroquois; Omaha; Pawnee; Ponca; Zuni
Pinus: Klamath
Podophyllum: Iroquois
Polygonum: Eskimo, Inupiat; Sioux
Porphyra: Haisla & Hanaksiala
Prunus: Cree, Woodlands; Dakota; Iroquois; Omaha; Pawnee; Ponca; Thompson; Winnebago
Pyrus: Iroquois
Rhus: Isleta
Ribes: Bella Coola; Eskimo, Inupiat; Iroquois; Paiute
Rosa: Alaska Native; Eskimo, Inupiat; Thompson
Rubus: Iroquois; Pomo, Kashaya
Sambucus: Cahuilla; Hoh; Iroquois; Quileute
Scirpus: Montana Indian
Shepherdia: Montana Indian
Sium: Klamath; Montana Indian
Solanum: Zuni
Sorghum: Hopi
Stenocereus: Papago; Papago & Pima
Umbellularia: Mendocino Indian
Vaccinium: Alaska Native; Eskimo, Inupiat; Hoh; Iroquois; Quileute
Viburnum: Chippewa; Eskimo, Inupiat; Iroquois; Thompson
Vitis: Iroquois
Yucca: Acoma; Apache; Apache, Mescalero; Apache, Western; Keres, Western; Laguna; Mahuna; Navajo; Southwest Indians; Zuni
Zea: Iroquois
Ziziphus: Papago; Pima

Snack Food
Amelanchier: Blackfoot; Cree, Woodlands
Chimaphila: Cherokee
Cornus: Algonquin, Quebec; Blackfoot; Cree, Woodlands
Corylus: Thompson
Cucurbita: Pima, Gila River; Rappahannock
Cyperus: Pima, Gila River
Dichelostemma: Pima, Gila River
Echinocereus: Pima, Gila River
Fouquieria: Yavapai
Fragaria: Cree, Woodlands
Gaultheria: Algonquin, Quebec
Ipomopsis: Klamath
Lilium: Cree, Woodlands

Mammillaria: Pima, Gila River
Nuphar: Montana Indian
Osmunda: Abnaki
Perideridia: Blackfoot
Phoradendron: Pima, Gila River
Physalis: Pima, Gila River
Populus: Blackfoot; Pima, Gila River
Porphyra: Alaska Native; Kwakiutl, Southern
Proboscidea: Pima, Gila River
Prosopis: Pima, Gila River
Quercus: Pima, Gila River
Rosa: Cree, Woodlands
Salix: Alaska Native
Sambucus: Pima, Gila River
Scirpus: Pima, Gila River
Ulmus: Omaha
Vaccinium: Cree, Woodlands
Viburnum: Cree, Woodlands
Washingtonia: Pima, Gila River
Zea: Iroquois; Zuni

Soup
Aesculus: Miwok
Agave: Navajo
Allium: Anticosti; Eskimo, Inupiat; Lakota; Navajo; Paiute, Northern; Potawatomi
Amaranthus: Havasupai
Amelanchier: Blackfoot; Montana Indian
Aralia: Potawatomi
Arctium: Iroquois
Arctostaphylos: Carrier; Coeur d'Alene; Montana Indian; Okanagon; Sanpoil & Nespelem; Thompson
Asclepias: Blackfoot; Cheyenne; Menominee; Meskwaki; Potawatomi
Aster: Ojibwa
Atriplex: Papago
Avena: Miwok
Calochortus: Paiute, Northern
Camassia: Chehalis; Flathead
Capsicum: Keresan
Carya: Cherokee; Choctaw; Dakota; Iroquois; Omaha; Pawnee; Ponca; Winnebago
Castanea: Iroquois
Celastrus: Ojibwa
Chenopodium: Dakota; Omaha; Papago; Pawnee
Cirsium: Thompson
Claytonia: Alaska Native; Eskimo, Alaska
Cleome: Navajo
Corylus: Dakota; Iroquois; Omaha; Ponca; Winnebago
Cucurbita: Havasupai
Cymopterus: Hualapai; Navajo
Dasylirion: Apache, Mescalero
Descurainia: Pueblo
Dryopteris: Eskimo, Alaska
Elaeagnus: Blackfoot; Tanana, Upper
Erythronium: Blackfoot; Thompson
Fagus: Iroquois
Fritillaria: Alaska Native; Blackfoot
Gaillardia: Blackfoot

Gaylussacia: Iroquois
Hedeoma: Lakota
Heracleum: Blackfoot
Hippuris: Alaska Native; Eskimo, Alaska; Eskimo, Inuktitut
Juglans: Dakota; Iroquois; Omaha; Pawnee; Ponca; Winnebago
Juniperus: Acoma
Lilium: Blackfoot; Thompson
Lithocarpus: Pomo; Shasta; Yuki; Yurok
Lomatium: Montana Indian; Oregon Indian
Lycium: Navajo
Mentzelia: Havasupai
Nelumbo: Dakota; Omaha; Pawnee; Ponca; Winnebago
Nuphar: Montana Indian
Oenothera: Apache, Chiricahua & Mescalero
Opuntia: Apache, Western; Cahuilla; Cheyenne; Okanagan-Colville
Oryzopsis: Havasupai; Paiute, Northern
Osmunda: Menominee
Parmelia: Potawatomi
Parrya: Alaska Native
Pectis: Havasupai
Pediomelum: Lakota; Omaha; Sioux
Perideridia: Atsugewi; Blackfoot; Paiute, Northern
Phaseolus: Cherokee; Havasupai; Iroquois; Navajo
Pinus: Havasupai; Hualapai; Paiute; Paiute, Northern
Polanisia: Pueblo
Polygonum: Alaska Native; Blackfoot
Porphyra: Alaska Native
Prunus: Blackfoot; Coeur d'Alene; Iroquois; Montana Indian; Ojibwa
Pteridium: Ojibwa
Quercus: Havasupai; Hualapai; Iroquois; Lakota; Mendocino Indian; Miwok; Ojibwa; Pomo; Shasta
Ranunculus: Eskimo, Inuktitut
Rhodymenia: Alaska Native
Rhus: Cahuilla
Ribes: Blackfoot
Rosa: Lakota
Rubus: Coeur d'Alene; Iroquois
Salix: Eskimo, Inuktitut; Eskimo, Inupiat
Sambucus: Paiute, Northern; Thompson; Yuki
Sisymbrium: Navajo
Solanum: Ojibwa
Sorbus: Thompson
Thelypodium: Pueblo
Typha: Paiute, Northern
Vaccinium: Coeur d'Alene; Cree, Woodlands; Iroquois
Veratrum: Blackfoot
Viburnum: Thompson
Yucca: Apache; Apache, Chiricahua & Mescalero; Apache, Mescalero
Zea: Abnaki; Chippewa; Delaware; Havasupai; Iroquois

Sour
Acer: Ojibwa; Potawatomi

Nemopanthus: Potawatomi
Oxalis: Meskwaki; Pomo, Kashaya
Pedicularis: Eskimo, Alaska
Rhus: Potawatomi
Rumex: Hesquiat

Special Food

Allium: Thompson
Amaranthus: Navajo, Ramah
Amelanchier: Blackfoot; Cheyenne
Arctostaphylos: Bella Coola; Hanaksiala
Argentina: Kwakiutl, Southern
Balsamorhiza: Sanpoil
Camassia: Blackfoot; Cowichan
Capsicum: Papago
Carya: Iroquois
Castanea: Iroquois
Castilleja: Navajo
Chenopodium: Navajo, Ramah
Cirsium: Cheyenne
Citrullus: Iroquois
Cleome: Navajo
Cornus: Hesquiat
Corylus: Iroquois
Cucumis: Iroquois
Cucurbita: Hopi; Iroquois; Zuni
Dalea: Navajo; Santa Clara
Datura: Navajo
Equisetum: Hoh; Quileute
Fagus: Iroquois
Ferocactus: Pima, Gila River
Fouquieria: Papago
Fragaria: Apache; Cochiti; Isleta; Navajo; Quinault
Gaultheria: Kwakiutl, Southern
Helianthus: Apache, Western
Juglans: Iroquois
Juniperus: Tewa of Hano
Lewisia: Thompson
Lomatium: Sanpoil
Lycium: Jemez; Navajo
Malus: Kitasoo; Kwakiutl, Southern
Monarda: Lakota
Nuphar: Klamath
Oemleria: Kwakiutl, Southern
Oenothera: Apache, Chiricahua & Mescalero
Oryzopsis: Paiute, Northern
Parthenocissus: Ojibwa
Physalis: Apache, Chiricahua & Mescalero
Pinus: Apache, Chiricahua & Mescalero; Apache, Mescalero; Hopi; Navajo
Polyporus: Pueblo
Populus: Blackfoot
Prosopis: Apache, Chiricahua & Mescalero
Prunus: Blackfoot; Cheyenne; Lakota
Quercus: Cahuilla; Iroquois
Rhus: Apache, Chiricahua & Mescalero
Ribes: Kwakiutl, Southern
Rubus: Haisla & Hanaksiala; Iroquois; Isleta; Makah
Shepherdia: Haisla & Hanaksiala
Sophora: Acoma; Laguna; San Felipe
Vaccinium: Haisla & Hanaksiala; Kwakiutl, Southern; Nitinaht; Oweekeno
Viburnum: Kwakiutl, Southern
Vitis: Lakota
Yucca: Navajo
Zea: Iroquois; Menominee; Navajo; Navajo, Ramah; Papago; Pueblo
Zizania: Ojibwa
Zostera: Kwakiutl, Southern

Spice

Acer: Blackfoot; Cowichan; Saanich
Agastache: Acoma; Keres, Western; Laguna
Allium: Algonquin, Tête-de-Boule; Apache, Chiricahua & Mescalero; Blackfoot; Cahuilla; Cheyenne; Cree, Woodlands; Dakota; Eskimo, Inuktitut; Hopi; Keres, Western; Meskwaki; Navajo; Omaha; Pawnee; Pomo; Ponca; Pueblo; Shoshoni; Shuswap; Winnebago
Alnus: Paiute
Amelanchier: Chehalis; Thompson
Angelica: Shuswap
Apium: Shoshoni
Arctostaphylos: Chippewa
Arnoglossum: Cherokee
Artemisia: Apache, Chiricahua & Mescalero; Apache, White Mountain; Blackfoot; Hopi; Shoshoni
Asarum: Chippewa; Meskwaki; Ojibwa; Potawatomi
Asclepias: Blackfoot
Astragalus: Cahuilla; Shoshoni
Atriplex: Hopi; Navajo, Ramah; Papago; Pima; Pueblo
Berlandiera: Acoma; Keres, Western; Laguna
Blechnum: Makah
Brassica: Hoh; Quileute
Calocedrus: Round Valley Indian
Capsella: Cherokee
Capsicum: Hopi; Navajo, Ramah; Papago; Pima
Carum: Cree, Woodlands
Carya: Cherokee
Castanea: Iroquois
Celtis: Dakota
Cetraria: Eskimo, Inuktitut
Chenopodium: Cherokee; Hopi
Chrysothamnus: Hopi; Kawaiisu
Cleome: Navajo
Coriandrum: Hopi; Keresan
Cornus: Shuswap
Cucumis: Keresan
Cucurbita: Navajo, Ramah
Cymopterus: Apache, Chiricahua & Mescalero; Navajo
Dalea: Zuni
Descurainia: Cahuilla; Hopi
Distichlis: Cahuilla
Dracocephalum: Apache, Chiricahua & Mescalero
Encelia: Navajo, Kayenta
Eriogonum: Hopi
Eupatorium: Cherokee
Fragaria: Thompson
Fritillaria: Thompson
Gaultheria: Chippewa; Hesquiat; Makah; Nitinaht
Geranium: Blackfoot
Hedeoma: Apache, Chiricahua & Mescalero; Isleta
Humulus: Apache, Chiricahua & Mescalero
Juniperus: Acoma; Apache, Western; Keres, Western; Laguna; Lakota
Ledum: Eskimo, Alaska; Tanana, Upper
Ligusticum: Anticosti; Eskimo, Inupiat
Lilium: Okanagan-Colville; Thompson
Lindera: Cherokee; Chippewa
Lomatium: Okanagan-Colville; Pomo, Kashaya; Shuswap; Thompson
Lotus: Kawaiisu
Lygodesmia: Hopi
Lysichiton: Hoh; Quileute
Maianthemum: Thompson
Medicago: Okanagan-Colville
Mentha: Apache, Chiricahua & Mescalero; Blackfoot; Cherokee; Chippewa; Cree, Woodlands; Dakota; Hopi; Malecite; Navajo, Kayenta; Saanich
Monarda: Acoma; Apache, Chiricahua & Mescalero; Isleta; Keres, Western; Laguna; Pueblo; San Ildefonso; Spanish American
Moricandia: Hoh; Quileute
Osmorhiza: Shoshoni
Pectis: Acoma; Hopi; Keres, Western; Laguna; Pueblo
Penstemon: Thompson
Pentaphylloides: Blackfoot
Perideridia: Blackfoot; Mendocino Indian
Piloblephis: Seminole
Pinus: Havasupai
Piper: Cherokee; Haisla & Hanaksiala
Poliomintha: Hopi; Tewa
Polygonatum: Cherokee
Polygonum: Iroquois
Polystichum: Makah
Populus: Apache, Mescalero
Portulaca: Navajo, Ramah
Prosopis: Apache, Chiricahua & Mescalero
Prunus: Blackfoot
Pseudotsuga: Karok; Paiute
Pycnanthemum: Chippewa
Quercus: Apache, Chiricahua & Mescalero; Havasupai
Rhus: Acoma; Keres, Western; Laguna
Ribes: Thompson; Tolowa
Rorippa: Eskimo, Inuktitut
Rosa: Okanagan-Colville; Skagit, Upper
Rubus: Hesquiat
Rumex: Iroquois
Salix: Quileute
Salvia: Cahuilla; Diegueño
Sambucus: Thompson
Sassafras: Chippewa; Choctaw
Selaginella: Blackfoot
Sidalcea: Yana
Sinapis: Hoh; Quileute
Sorbus: Thompson
Sporobolus: Hopi
Stachys: Kawaiisu

Suaeda: Papago; Pima
Taraxacum: Apache, Chiricahua & Mescalero; Eskimo, Inuktitut
Thalictrum: Blackfoot; Montagnais
Tsuga: Cowlitz
Typha: Yuma
Ulmus: Omaha
Ulva: Pomo, Kashaya
Wyethia: Yuki
Zostera: Cowichan; Saanich

Staple

Acer: Sioux
Aesculus: Yana
Agastache: Apache; Comanche; Mohave; Paiute; Papago; Ute; Yuma
Agave: Apache; Comanche; Hualapai; Mohave; Paiute; Papago; Pima, Gila River; Ute; Yuma
Allenrolfea: Maricopa; Mohave; Yuma
Allium: Flathead; Kutenai; Okanagon
Amaranthus: Acoma; Laguna; Mendocino Indian; Navajo; Navajo, Ramah; Papago; Yuma
Ambrosia: Papago
Amelanchier: Blackfoot; Okanagon
Arctostaphylos: Cahuilla; Numlaki; Okanagon; Yuki
Argentina: Okanagon
Artemisia: Paiute
Arthrocnemum: Cahuilla
Asclepias: Cahuilla
Astragalus: Blackfoot
Atriplex: Kamia; Pima
Avena: Luiseño; Mendocino Indian; Pomo; Pomo, Kashaya; Yuki
Balsamorhiza: Montana Indian; Okanagon; Thompson
Blennosperma: Neeshenam
Boisduvalia: Mendocino Indian; Pomo
Bromus: Mendocino Indian; Neeshenam
Calandrinia: Costanoan; Numlaki
Calochortus: Okanagon
Camassia: Montana Indian; Okanagon
Capsella: Mendocino Indian
Capsicum: Pima, Gila River
Carnegia: Papago; Pima; Pima, Gila River
Carum: Cree, Woodlands
Carya: Iroquois
Castanea: Iroquois
Ceanothus: Concow
Chenopodium: Cahuilla; Havasupai; Montana Indian; Navajo; Paiute; Paiute, Northern; Pima
Cirsium: Okanagon
Citrullus: Hopi
Clarkia: Mendocino Indian
Claytonia: Okanagon
Cleome: Pueblo
Comandra: Okanagon
Cornus: Okanagon
Corylus: Iroquois
Crataegus: Okanagon
Cucurbita: Papago
Cycloloma: Apache, White Mountain; Zuni
Dalea: Acoma; Keres, Western; Laguna; San Felipe
Descurainia: Cocopa; Havasupai; Paiute; Pima; Pima, Gila River
Dichelostemma: Paiute
Digitaria: Hopi
Echinocactus: Cahuilla
Echinochloa: Cocopa; Yuma
Elaeagnus: Okanagon
Elymus: Costanoan; Pomo, Kashaya
Eragrostis: Cocopa
Eriochloa: Cocopa
Eriogonum: Kawaiisu
Eriophyllum: Cahuilla
Erythronium: Okanagon
Fagus: Iroquois
Ferocactus: Cahuilla
Festuca: Costanoan
Fragaria: Okanagon
Fritillaria: Alaska Native; Okanagon
Helianthus: Apache, White Mountain; Cahuilla; Gros Ventre; Havasupai; Kawaiisu; Mandan; Mohave; Paiute; Paiute, Northern; Pima; Ree
Hemizonia: Costanoan; Mendocino Indian; Pomo; Wintoon
Heracleum: Okanagon
Hordeum: Costanoan; Mendocino Indian
Hydrophyllum: Okanagon
Hypericum: Miwok
Juglans: Iroquois
Juniperus: Kawaiisu; Yavapai
Koeleria: Isleta
Layia: Costanoan; Mendocino Indian
Lepidium: Havasupai
Lewisia: Okanagon
Leymus: Mendocino Indian
Lilium: Okanagon
Lithocarpus: Hahwunkwut; Hupa; Karok; Polikiah; Shasta; Tolowa; Yurok; Yurok, South Coast (Nererner)
Lolium: Pomo; Yuki
Lomatium: Montana Indian; Okanagon
Lupinus: Costanoan
Lycopus: Okanagon
Madia: Hupa; Mendocino Indian; Mewuk; Miwok; Neeshenam; Pomo; Pomo, Kashaya; Wintoon
Mahonia: Okanagon
Mentzelia: Hopi; Hualapai; Navajo, Ramah; Paiute
Monolepis: Pima
Nolina: Isleta
Nuphar: Klamath
Olneya: Cahuilla; Papago; Pima; Yavapai
Opuntia: Apache, San Carlos; Cahuilla; Luiseño; Navajo; Okanagon; Papago; Pima, Gila River
Oryza: Haisla & Hanaksiala
Oryzopsis: Havasupai; Hopi; Kawaiisu; Navajo; Paiute; Paiute, Northern; Zuni
Panicum: Hopi; Warihio; Yuma
Parkinsonia: Papago
Perideridia: Blackfoot; Mendocino Indian; Paiute, Northern; Pomo
Peucedanum: Okanagon
Phaseolus: Cocopa; Menominee; Papago; Tewa
Pholisma: Papago
Physalis: Navajo
Pinus: Apache, Western; Isleta; Navajo; Okanagon; Paiute; Pomo, Kashaya
Piptatherum: Paiute
Plagiobothrys: Mendocino Indian
Poa: Havasupai
Pogogyne: Numlaki; Yuki
Polygonum: Montana Indian
Prosopis: Apache, Mescalero; Apache, Western; Cahuilla; Comanche; Luiseño; Papago; Pima; Pima, Gila River; Yavapai; Yuma
Prunus: Blackfoot; Cahuilla; Luiseño; Montana Indian; Okanagon
Pteridium: Clallam; Sierra; Thompson
Quercus: Acoma; Choctaw; Cochiti; Diegueño; Isleta; Kawaiisu; Keres, Western; Laguna; Lakota; Luiseño; Menominee; Miwok; Modesse; Navajo; Navajo, Ramah; Ojibwa; Paiute, Northern; Pima; San Felipe; Shasta; Yana
Ranunculus: Mendocino Indian; Neeshenam; Pomo
Rhus: Apache; Navajo
Ribes: Okanagon
Rosa: Okanagon
Rubus: Okanagon
Rumex: Costanoan
Ruppia: Seri
Sagittaria: Ojibwa
Salvia: Cahuilla; Costanoan; Mohave; Pomo
Sambucus: Okanagon
Scirpus: Cahuilla; Montana Indian
Sedum: Eskimo, Inupiat
Shepherdia: Okanagon
Sisymbrium: Pima, Gila River
Sium: Okanagon
Stenocereus: Papago
Suaeda: Cahuilla; Paiute
Thysanocarpus: Mendocino Indian
Trifolium: Mendocino Indian
Triticum: Haisla & Hanaksiala; Papago; Sia
Tsuga: Gitksan; Haisla
Typha: Cahuilla; Lakota; Ojibwa; Paiute, Northern
Vaccinium: Okanagon
Verbena: Concow
Wyethia: Costanoan; Mendocino Indian; Paiute; Pomo; Pomo, Kashaya
Xanthium: Costanoan; Yavapai
Yucca: Apache, Mescalero; Hualapai; Papago; Southwest Indians
Zea: Dakota; Delaware; Havasupai; Hopi; Isleta; Keres, Western; Menominee; Navajo; Omaha; Papago; Pawnee; Ponca; Sia; Tewa; Zuni
Zizania: Dakota; Menominee; Ojibwa; Omaha; Ponca; Potawatomi; Winnebago
Ziziphus: Cahuilla

Starvation Food

Acacia: Pima, Gila River
Aesculus: Modesse

Index of Usages

Agave: Pima
Alectoria: Montana Indian
Allenrolfea: Pima, Gila River
Allium: Havasupai
Amaranthus: Hopi
Amelanchier: Lakota
Apios: Huron
Arachis: Huron
Arctostaphylos: Montana Indian
Artemisia: Paiute
Atriplex: Pima; Pima, Gila River
Baccharis: Mohave; Yuma
Balsamorhiza: Thompson
Blechnum: Hesquiat; Nitinaht
Bromus: Cahuilla
Calochortus: Navajo; Ute
Celastrus: Menominee; Potawatomi
Chlorantha: Mohave
Cirsium: Havasupai
Crataegus: Omaha; Ponca; Winnebago
Gaultheria: Nitinaht
Helianthus: Huron; Lakota
Hemizonia: Cahuilla
Ipomoea: Arapaho; Cheyenne; Kiowa
Juniperus: Acoma; Diegueño; Hualapai; Keres, Western; Laguna; Navajo
Leymus: Shoshoni
Lilium: Cherokee; Huron
Lomatium: Paiute
Lycium: Hopi
Malva: Papago
Menyanthes: Alaska Native
Nelumbo: Huron
Opuntia: Acoma; Dakota; Keres, Western; Laguna; Pawnee; Thompson
Oryzopsis: Hopi
Parkinsonia: Mohave; Yuma
Phoradendron: Acoma; Keres, Western; Laguna
Picea: Tanana, Upper
Pinus: Apache, Chiricahua & Mescalero; Klamath; Mendocino Indian; Montana Indian; Navajo, Ramah
Platanthera: San Felipe
Polypodium: Kwakiutl, Southern
Populus: Navajo, Ramah
Prunus: Lakota
Quercus: Comanche
Ribes: Lakota
Rorippa: Okanagan-Colville
Rosa: Blackfoot; Dakota; Lakota; Montana Indian; Nez Perce; Omaha; Pawnee; Ponca; Sanpoil & Nespelem
Rumex: Paiute, Northern; Yavapai
Sambucus: Seminole
Sarcobatus: Pima
Selaginella: Blackfoot
Senecio: Navajo, Ramah
Shepherdia: Blackfoot
Sisymbrium: Mohave
Solanum: Acoma; Keres, Western; Keresan; Laguna

Sonchus: Mohave
Sphaeralcea: Navajo
Sporobolus: Hopi
Symphoricarpos: Blackfoot
Taraxacum: Cree
Thelypodium: Mohave
Thuja: Hesquiat
Tsuga: Nitinaht
Yucca: Acoma; Keres, Western; Laguna; Southwest Indians

Substitution Food
Agave: Apache, Western
Apios: Cherokee
Atriplex: Hopi; Navajo, Kayenta
Betula: Algonquin, Quebec; Cree, Woodlands
Brickellia: Sanel
Carnegia: Apache, Chiricahua & Mescalero; Pima
Castanea: Cherokee
Ceanothus: Menominee
Chenopodium: Alaska Native
Cymopterus: Navajo
Equisetum: Nitinaht
Ferocactus: Pima
Grindelia: Mendocino Indian
Hordeum: Yuki
Ligusticum: Alaska Native; Atsugewi
Linanthus: Yuki
Lomatium: Okanagan-Colville
Microseris: Mendocino Indian
Mimulus: Mendocino Indian
Monardella: Mendocino Indian
Perideridia: Miwok
Pogogyne: Concow
Polypodium: Salish, Coast
Prosopis: Apache, Chiricahua & Mescalero; Apache, Western; Pima
Pseudotsuga: Montana Indian; Round Valley Indian; Yuki
Pteridium: Alaska Native
Quercus: Luiseño
Rumex: Algonquin, Quebec
Satureja: Mendocino Indian
Sedum: Makah
Solanum: Cochiti; Pima; Spanish American
Thelesperma: Navajo
Urtica: Alaska Native
Zea: Menominee

Sweetener
Acer: Abnaki; Algonquin, Quebec; Apache, Chiricahua & Mescalero; Cherokee; Chippewa; Cree; Dakota; Iroquois; Malecite; Menominee; Meskwaki; Mohegan; Ojibwa; Omaha; Pawnee; Ponca; Potawatomi; Winnebago
Agastache: Dakota; Omaha; Pawnee; Ponca; Winnebago
Agave: Hualapai
Alnus: Clallam
Amaranthus: Navajo
Amelanchier: Okanagan-Colville; Thompson
Camassia: Montana Indian

Carya: Dakota; Omaha; Pawnee; Ponca; Winnebago
Castilleja: Cahuilla; Paiute
Chamaesyce: Zuni
Comandra: Okanagan-Colville
Coreopsis: Kawaiisu
Dalea: Hopi; Keres, Western
Echinocereus: Hopi
Epilobium: Eskimo, Inupiat
Frasera: Arapaho
Justicia: Diegueño
Phragmites: Kawaiisu; Paiute
Pinus: Kawaiisu; Miwok; Pomo, Kashaya
Populus: Dakota
Prosopis: Pima
Pseudotsuga: Shuswap; Thompson
Rhus: Cahuilla
Rubus: Thompson
Salix: Eskimo, Inupiat
Sophora: Keres, Western
Tsuga: Gitksan; Haisla; Wet'suwet'en
Yucca: Southwest Indians; Zuni
Zea: Dakota; Hopi; Isleta; Omaha; Pawnee; Ponca

Unspecified
Abies: Haisla; Kitasoo; Shuswap; Thompson
Abronia: Acoma; Clallam; Klallam; Laguna; Makah
Acacia: Pima
Acer: Clallam; Costanoan; Salish, Coast; Thompson
Achillea: Aleut
Acorus: Abnaki; Dakota; Lakota
Adenostoma: Cahuilla; Coahuilla
Aesculus: Pomo, Kashaya; Tubatulabal; Yuki
Agaricus: Delaware
Agastache: Gosiute
Agave: Apache; Apache, Chiricahua & Mescalero; Apache, San Carlos; Apache, Western; Apache, White Mountain; Cahuilla; Cocopa; Diegueño; Navajo; Papago; Pima; Yavapai
Agoseris: Gosiute; Ute
Agropyron: Paiute
Agrostis: Klamath
Alectoria: Coeur d'Alene; Spokan; Thompson
Allenrolfea: Pima; Pima, Gila River
Allium: Acoma; Alaska Native; Apache; Apache, White Mountain; Bella Coola; Cherokee; Cheyenne; Clallam; Comanche; Cree, Woodlands; Dakota; Gosiute; Great Basin Indian; Havasupai; Hoh; Hopi; Isleta; Karok; Klallam; Koyukon; Kwakiutl, Southern; Laguna; Lakota; Makah; Malecite; Mendocino Indian; Menominee; Navajo; Navajo, Ramah; Neeshenam; Nitinaht; Okanagon; Omaha; Oweekeno; Paiute; Paiute, Northern; Papago; Pawnee; Pomo; Ponca; Quileute; Quinault; Salish, Coast; Sanpoil & Nespelem; Seminole; Spokan; Tanana, Upper; Tewa; Thompson; Tubatulabal; Ute; Winnebago; Yuki
Alnus: Costanoan; Paiute; Salish, Coast; Skagit, Upper; Swinomish
Amaranthus: Apache, Chiricahua & Mescalero; Apache, White Mountain; Cocopa; Gosiute; Havasupai; Hopi; Jemez; Keres, Western; Klamath;

[Amaranthus]
  Mohave; Montana Indian; Navajo; Papago; Pima; Pima, Gila River; Sia; Tewa; Yuma
Ambrosia: Papago
Amelanchier: Blackfoot; Klamath; Thompson
Ammannia: Mohave; Yuma
Amoreuxia: Pima, Gila River
Amphicarpaea: Cherokee; Chippewa; Dakota; Meskwaki; Ojibwa; Omaha; Pawnee; Ponca; Winnebago
Amsinckia: Gosiute; Mendocino Indian; Pima
Ananas: Seminole
Anemone: Alaska Native; Eskimo, Alaska
Angelica: Alaska Native; Aleut; Bella Coola; Eskimo, Alaska; Eskimo, Inuktitut; Eskimo, Inupiat; Hanaksiala; Makah; Mendocino Indian; Mewuk; Pomo, Kashaya; Yana
Annona: Seminole
Apiaceae sp.: Yana
Apiastrum: Cahuilla
Apios: Cheyenne; Dakota; Delaware; Iroquois; Mohegan; Omaha; Pawnee; Ponca; Seminole; Winnebago
Apocynum: Karok
Aquilegia: Yurok
Arachis: Seminole
Aralia: Kwakiutl, Southern; Menominee
Arctostaphylos: Mendocino Indian
Ardisia: Seminole
Argentina: Bella Coola; Haisla & Hanaksiala; Hesquiat; Kitasoo; Makah; Nitinaht; Okanagon; Oweekeno; Quileute; Salish, Coast; Shuswap; Thompson
Arisaema: Potawatomi
Armillaria: Karok; Yurok
Aronia: Abnaki
Artemisia: Apache, White Mountain; Eskimo, Alaska; Gosiute; Hopi; Luiseño; Navajo; Navajo, Kayenta; Zuni
Asclepias: Apache, White Mountain; Blackfoot; Cheyenne; Crow; Dakota; Hopi; Jemez; Lakota; Menominee; Mohegan; Navajo; Ojibwa; Paiute; Tewa; Yokia; Zuni
Asparagus: Isleta
Aster: Chippewa; Keres, Western; Ojibwa
Astragalus: Acoma; Blackfoot; Canadian Indian; Dakota; Eskimo, Alaska; Havasupai; Hopi; Jemez; Keres, Western; Laguna; Montana Indian; Okanagan-Colville; Thompson; Zuni
Athyrium: Quileute; Quinault; Salish, Coast
Atriplex: Gosiute; Hopi; Isleta; Keres, Western; Paiute; Papago; Pima; Pima, Gila River; Yuma
Avena: Cahuilla; Haisla & Hanaksiala; Karok; Kawaiisu; Pomo
Balsamorhiza: Coeur d'Alene; Flathead; Gosiute; Karok; Klamath; Kutenai; Miwok; Montana Indian; Nez Perce; Okanagan-Colville; Okanagon; Paiute; Paiute, Northern; Shuswap; Spokan; Thompson; Ute
Barbarea: Cherokee

Beckmannia: Klamath; Montana Indian
Berula: Apache, White Mountain
Beta: Cherokee
Betula: Cree, Woodlands; Tanana, Upper
Bidens: Paiute
Bloomeria: Cahuilla; Luiseño
Boisduvalia: Miwok
Boschniakia: Hesquiat; Karok; Luiseño; Yurok
Brassica: Malecite; Oweekeno; Pomo, Kashaya; Seminole
Brodiaea: Atsugewi; Costanoan; Mendocino Indian; Miwok; Pomo; Yana
Bromus: Gosiute; Luiseño; Miwok
Caesalpinia: Comanche
Calandrinia: Costanoan; Luiseño; Miwok
Calochortus: Cahuilla; Cheyenne; Gosiute; Havasupai; Hopi; Karok; Klamath; Mendocino Indian; Miwok; Montana Indian; Navajo; Navajo, Ramah; Okanagan-Colville; Okanagon; Paiute; Paiute, Nevada; Pomo; Shuswap; Thompson; Tubatulabal; Ute; Yuki
Caltha: Abnaki; Alaska Native; Eskimo, Alaska; Mohegan; Ojibwa
Calylophus: Apache, Chiricahua & Mescalero
Camassia: Blackfoot; Clallam; Comanche; Flathead; Gosiute; Hoh; Klamath; Kutenai; Kwakiutl, Southern; Makah; Montana Indian; Nitinaht; Okanagan-Colville; Okanagon; Paiute; Quileute; Round Valley Indian; Sanpoil & Nespelem; Skagit, Upper; Spokan; Thompson; Yuki
Camelina: Apache, Chiricahua & Mescalero
Camissonia: Costanoan; Mohave
Capsella: Apache, Chiricahua & Mescalero; Cahuilla
Capsicum: Navajo; Papago; Pima; Sia
Cardamine: Abnaki; Iroquois
Carduus: Luiseño
Carex: Alaska Native; Gosiute; Kawaiisu; Klamath; Montana Indian; Thompson
Carica: Seminole
Carnegia: Papago; Pima
Carya: Cherokee; Comanche; Dakota; Iroquois; Lakota; Ojibwa; Omaha; Pawnee; Ponca; Seminole; Winnebago
Castanea: Cherokee; Iroquois
Castanopsis: Karok; Kawaiisu; Klamath; Mendocino Indian; Pomo; Pomo, Kashaya; Tolowa; Yurok
Castilleja: Cheyenne; Hopi; Kwakiutl, Southern
Catabrosa: Crow; Gosiute
Caulanthus: Kawaiisu
Ceanothus: Navajo, Ramah
Celtis: Seminole; Yavapai
Cercis: Cherokee; Navajo; Navajo, Kayenta
Chasmanthium: Cocopa
Chelone: Cherokee
Chenopodium: Apache, Chiricahua & Mescalero; Apache, Western; Apache, White Mountain; Cherokee; Dakota; Gosiute; Havasupai; Hopi; Klamath; Luiseño; Malecite; Mohegan; Navajo, Ramah; Paiute; Paiute, Northern; Papago; Pima, Gila River; Yavapai; Zuni

Chilopsis: Cahuilla
Chlorogalum: Costanoan; Karok; Luiseño; Mendocino Indian; Miwok; Neeshenam
Chrysobalanus: Seminole
Chrysophyllum: Seminole
Chrysothamnus: Apache, White Mountain; Navajo, Kayenta
Cicer: Papago; Pima
Cicuta: Eskimo, Inuktitut
Cinna: Gosiute
Cirsium: Apache, Chiricahua & Mescalero; Atsugewi; Blackfoot; Cahuilla; Cheyenne; Comanche; Costanoan; Cowichan; Flathead; Gosiute; Hesquiat; Houma; Kawaiisu; Kiowa; Kutenai; Nez Perce; Okanagon; Paiute; Saanich; Shuswap; Spokan; Thompson; Tubatulabal; Yavapai
Cistanthe: Kawaiisu
Citrullus: Cherokee; Hopi; Kamia; Meskwaki; Navajo, Ramah; Okanagan-Colville; Seminole; Sia
Citrus: Seminole
Clarkia: Miwok; Yana
Claytonia: Alaska Native; Aleut; Costanoan; Eskimo, Alaska; Gosiute; Iroquois; Mendocino Indian; Miwok; Montana Indian; Okanagan-Colville; Okanagon; Paiute, Northern; Spokan; Thompson; Ute; Yurok
Cleome: Diegueño; Havasupai; Hopi; Kawaiisu; Keresan; Navajo; Navajo, Ramah; San Felipe; Sia; Tewa; Zuni
Coccoloba: Seminole
Cochlearia: Aleut
Cocos: Seminole
Colocasia: Hawaiian; Seminole
Comandra: Navajo, Kayenta; Paiute
Conioselinum: Aleut; Haihais; Haisla; Haisla & Hanaksiala; Hanaksiala; Heiltzuk; Kwakwaka'wakw; Navajo, Kayenta; Nuxalkmc; Oweekeno
Cordylanthus: Yavapai
Cordyline: Hawaiian
Coreopsis: Tubatulabal
Coriandrum: Hopi
Cornus: Thompson
Corylus: Algonquin, Quebec; Chehalis; Cherokee; Chippewa; Costanoan; Cree, Woodlands; Dakota; Iroquois; Karok; Klamath; Lummi; Menominee; Meskwaki; Miwok; Ojibwa; Okanagon; Omaha; Paiute; Ponca; Salish, Coast; Sanpoil & Nespelem; Shuswap; Skagit; Skagit, Upper; Snohomish; Squaxin; Swinomish; Thompson; Tolowa; Winnebago; Yuki; Yurok
Crataegus: Thompson
Crepis: Gosiute; Paiute
Cucumis: Cocopa; Hopi; Kamia; Navajo, Ramah; Okanagan-Colville; Seminole; Sia; Thompson
Cucurbita: Apache, White Mountain; Cahuilla; Cherokee; Cocopa; Hopi; Kamia; Luiseño; Navajo, Ramah; Okanagan-Colville; Pima; Seminole; Sia; Zuni
Cuscuta: Navajo
Cycloloma: Hopi

Cymopterus: Apache, Chiricahua & Mescalero; Comanche; Gosiute; Hopi; Hualapai; Navajo; Navajo, Ramah; Ute
Cynoglossum: Yuki
Cyperus: Acoma; Apache, Chiricahua & Mescalero; Cocopa; Costanoan; Laguna; Mohave; Paiute; Paiute, Northern; Pima; Pima, Gila River
Cypripedium: Lakota
Dalea: Keresan; Kiowa; Paiute; San Ildefonso
Dasylirion: Apache, Chiricahua & Mescalero; Apache, Mescalero
Daucus: Clallam; Cowichan; Navajo; Oweekeno; Saanich; Salish, Coast
Dendromecon: Kawaiisu
Dentaria: Cherokee
Deschampsia: Gosiute
Descurainia: Apache, Chiricahua & Mescalero; Hopi; Klamath; Montana Indian; Paiute, Northern; Papago; Pima, Gila River; Pueblo; Tewa of Hano
Dichelostemma: Apache, San Carlos; Atsugewi; Cahuilla; Karok; Luiseño; Mendocino Indian; Miwok; Neeshenam; Paiute; Papago; Pomo; Yuki
Dioscorea: Hawaiian
Diospyros: Seminole
Diplazium: Hawaiian
Distichlis: Tubatulabal
Dodecatheon: Yuki
Dracocephalum: Gosiute
Dryopteris: Alaska Native; Bella Coola; Clallam; Costanoan; Cowlitz; Eskimo, Alaska; Hanaksiala; Kwakiutl, Southern; Oweekeno; Salish, Coast; Thompson
Dudleya: Cahuilla; Diegueño
Dyssodia: Apache, Chiricahua & Mescalero
Echinocereus: Cochiti; Isleta; Yavapai
Echinochloa: Cocopa; Paiute; Tubatulabal; Yuma
Egregia: Bella Coola; Kitasoo; Oweekeno
Elaeagnus: Paiute
Eleocharis: Paiute; Paiute, Northern
Elymus: Gosiute; Ute
Elytrigia: Apache, White Mountain; Gosiute
Ephedra: Kawaiisu; Tubatulabal
Epilobium: Alaska Native; Blackfoot; Eskimo, Alaska; Eskimo, Inuktitut; Eskimo, Inupiat; Gitksan; Haisla; Haisla & Hanaksiala; Karok; Okanagon; Oweekeno; Tanana, Upper; Thompson; Wet'suwet'en
Epixiphium: Apache, Chiricahua & Mescalero
Equisetum: Chinook, Lower; Clallam; Cowlitz; Eskimo, Alaska; Eskimo, Inupiat; Hoh; Klallam; Makah; Nitinaht; Okanagan-Colville; Quileute; Quinault; Saanich; Skagit, Upper; Swinomish; Tanana, Upper; Yurok
Eragrostis: Paiute
Eremalche: Pima, Gila River
Ericameria: Luiseño
Eriogonum: Blackfoot; Cahuilla; Karok; Kawaiisu; Kiowa; Mendocino Indian; Navajo; Navajo, Kayenta; Navajo, Ramah; Tubatulabal

Eriophorum: Alaska Native; Eskimo, Inuktitut; Eskimo, Inupiat
Erodium: Costanoan
Erythronium: Blackfoot; Kwakiutl; Kwakiutl, Southern; Montana Indian; Okanagan-Colville; Okanagon; Thompson; Winnebago
Exobasidium: Haisla & Hanaksiala
Fagus: Algonquin, Quebec; Chippewa; Iroquois; Ojibwa; Potawatomi
Ferocactus: Cahuilla; Pima; Pima, Gila River
Festuca: Gosiute
Ficus: Seminole
Fortunella: Seminole
Fouquieria: Cahuilla
Fragaria: Navajo, Ramah
Frasera: Apache; Yavapai
Fraxinus: Ojibwa
Fritillaria: Bella Coola; Blackfoot; Flathead; Gosiute; Haisla & Hanaksiala; Kitasoo; Montana Indian; Okanagan-Colville; Okanagon; Oweekeno; Paiute; Saanich; Salish, Coast; Salish, Straits; Shasta; Shuswap; Similkameen; Spokan; Thompson; Ute; Yana
Fucus: Eskimo, Alaska
Gaultheria: Cherokee
Gaura: Navajo, Kayenta
Gaylussacia: Cherokee; Ojibwa
Gilia: Luiseño
Glaux: Kwakiutl, Southern; Salish, Coast
Gleditsia: Cherokee
Glyceria: Crow; Klamath
Glycyrrhiza: Cheyenne; Montana Indian; Northwest Indian
Gossypium: Pima, Gila River
Gymnocladus: Meskwaki; Pawnee; Winnebago
Haplopappus: Paiute
Hedeoma: Isleta
Hedophyllum: Nitinaht
Hedysarum: Alaska Native; Tanana, Upper
Helenium: Mendocino Indian
Helianthella: Yana
Helianthus: Cheyenne; Chippewa; Costanoan; Dakota; Gosiute; Gros Ventre; Hopi; Hualapai; Iroquois; Kiowa; Lakota; Luiseño; Malecite; Mandan; Micmac; Navajo, Ramah; Omaha; Paiute; Paiute, Northern; Pawnee; Pima; Ponca; Potawatomi; Pueblo; Ree; Sanpoil & Nespelem; Sioux; Thompson; Winnebago
Heliomeris: Gosiute
Heliotropium: Tubatulabal
Hemizonia: Miwok
Heracleum: Alaska Native; Aleut; Bella Coola; Blackfoot; Carrier; Coeur d'Alene; Costanoan; Cree, Woodlands; Gitksan; Haisla; Haisla & Hanaksiala; Hesquiat; Karok; Klamath; Kwakiutl, Southern; Makah; Mewuk; Nitinaht; Okanagon; Oweekeno; Pomo, Kashaya; Quileute; Quinault; Salish, Coast; Shuswap; Spokan; Thompson; Tolowa; Wet'suwet'en; Yuki; Yurok
Hesperocallis: Cahuilla; Yuma

Hieracium: Thompson
Hippuris: Eskimo, Inuktitut
Hoffmannseggia: Apache; Apache, Chiricahua & Mescalero; Cocopa; Pima; Pima, Gila River; Pueblo
Honckenya: Alaska Native; Eskimo, Alaska
Hordeum: Cahuilla; Kawaiisu; Papago; Pima; Pomo, Kashaya
Humulus: Navajo
Hydrangea: Cherokee
Hydrophyllum: Cowlitz; Okanagon; Thompson
Hymenoclea: Seri
Ipomoea: Cherokee; Hawaiian
Jacquemontia: Hawaiian
Jacquinia: Seri
Jamesia: Apache, Chiricahua & Mescalero
Juglans: Algonquin, Quebec; Apache, Chiricahua & Mescalero; Apache, Mescalero; Apache, Western; Cherokee; Comanche; Costanoan; Dakota; Hualapai; Iroquois; Kiowa; Lakota; Menominee; Meskwaki; Navajo; Ojibwa; Omaha; Pawnee; Pomo, Kashaya; Ponca; Winnebago; Yavapai
Juncus: Paiute; Snuqualmie; Swinomish
Juniperus: Miwok; Navajo, Ramah
Koeleria: Havasupai
Lactuca: Gosiute
Lagenaria: Cherokee
Laminaria: Nitinaht
Larix: Flathead; Sanpoil & Nespelem; Thompson
Lathyrus: Acoma; Apache, Chiricahua & Mescalero; Chippewa; Cochiti; Laguna; Miwok; Omaha; Ponca
Layia: Luiseño
Lens: Papago; Pima
Lepidium: Cherokee; Havasupai; Hoh; Luiseño; Quileute
Leucocrinum: Crow
Lewisia: Blackfoot; Kutenai; Montana Indian; Okanagan-Colville; Okanagon; Oregon Indian, Warm Springs; Paiute; Paiute, Northern; Shuswap; Spokan; Thompson; Thompson, Upper (Nicola Band)
Leymus: Klamath; Montana Indian; Paiute; Paiute, Southern
Liatris: Blackfoot; Kiowa; Tewa
Ligusticum: Alaska Native; Apache, Chiricahua & Mescalero; Cherokee; Tolowa
Lilium: Atsugewi; Blackfoot; Clallam; Cree, Woodlands; Karok; Klallam; Lummi; Nitinaht; Okanagan-Colville; Okanagon; Paiute; Quileute; Quinault; Samish; Shuswap; Skagit; Skagit, Upper; Skokomish; Swinomish; Thompson; Yana
Linnaea: Carrier
Linum: Dakota; Omaha; Pawnee; Ponca; Winnebago
Lithocarpus: Costanoan; Hupa; Mendocino Indian; Pomo
Lithospermum: Blackfoot; Gosiute; Okanagon; Pima; Thompson
Lobularia: Costanoan
Lomatium: Atsugewi; Blackfoot; Flathead; Gosiute; Karok; Lakota; Mendocino Indian; Modoc;

[*Lomatium*] Montana Indian; Navajo; Nez Perce; Okanagan-Colville; Okanogon; Paiute; Paiute, Northern; Pomo, Kashaya; Sanpoil; Shuswap; Thompson; Yuki
Lonicera: Mendocino Indian
Lotus: Havasupai; Tubatulabal
Lupinus: Alaska Native; Haisla & Hanaksiala; Kimsquit; Kitasoo; Kwakiutl; Kwakiutl, Southern; Luiseño; Miwok
Lycium: Hopi
Lycopersicon: Seminole
Lycopus: Thompson
Lygodesmia: Hopi; Lakota; Sioux
Lysichiton: Cowlitz; Quileute; Skokomish; Tolowa; Yurok
Macrocystis: Haisla & Hanaksiala; Kitasoo; Oweekeno; Pomo
Madia: Crow; Klamath; Miwok; Shoshoni
Maianthemum: Okanagan-Colville; Thompson
Malacothrix: Luiseño
Malus: Hopi
Malva: Cahuilla; Pima
Mammillaria: Apache, White Mountain; Gosiute; Navajo; Navajo, Ramah; Tewa
Mangifera: Seminole
Manihot: Seminole
Marattia: Hawaiian
Matelea: Apache, Chiricahua & Mescalero
Matricaria: Kutenai; Montana Indian; Okanagan-Colville
Melica: Pomo
Mentha: Cherokee; Dakota; Thompson
Mentzelia: Havasupai; Hopi; Klamath; Miwok; Montana Indian; Navajo; Tubatulabal
Mertensia: Eskimo, Alaska
Microseris: Mendocino Indian; Montana Indian
Mimulus: Kawaiisu
Mirabilis: Navajo, Kayenta
Monarda: Cherokee; Hopi; Tewa of Hano
Monardella: Tubatulabal
Monolepis: Papago; Pima
Morus: Seminole
Muhlenbergia: Apache, White Mountain
Musa: Seminole
Musineon: Blackfoot; Crow
Myriophyllum: Tanana, Upper
Navarretia: Miwok
Nelumbo: Comanche; Dakota; Meskwaki; Ojibwa; Omaha; Pawnee; Ponca; Potawatomi; Winnebago
Neomammillaria: Navajo
Nephroma: Eskimo, Inuktitut
Nereocystis: Oweekeno; Pomo, Kashaya
Nolina: Apache, Chiricahua & Mescalero; Apache, Western; Cahuilla; Tubatulabal
Nuphar: Algonquin, Tête-de-Boule; Cheyenne; Comanche; Klamath; Lakota; Mendocino Indian; Montana Indian; Pawnee; Tolowa
Nymphaea: Klamath; Lakota; Ojibwa
Ochrosia: Hawaiian
Oenanthe: Costanoan; Cowlitz; Hesquiat; Miwok; Skokomish; Snuqualmie
Oenothera: Cherokee; Gosiute; Paiute; Zuni
Olneya: Papago; Pima, Gila River
Oplopanax: Oweekeno
Opuntia: Acoma; Cahuilla; Gosiute; Hopi; Kawaiisu; Keresan; Laguna; Maricopa; Montana Indian; Navajo; Okanagan-Colville; Okanogon; Paiute, Northern; Papago; Pima; Pima, Gila River; San Felipe; Shuswap; Sia; Southwest Indians; Spokan; Thompson; Thompson, Upper (Spences Bridge); Tubatulabal
Orobanche: Cahuilla; Gosiute; Navajo, Ramah; Paiute, Northern; Pima; Pima, Gila River
Orthocarpus: Miwok
Oryza: Kitasoo; Oweekeno; Seminole
Oryzopsis: Apache, White Mountain; Gosiute; Montana Indian; Navajo; Zuni
Osmorhiza: Okanogon; Thompson
Osmunda: Abnaki
Oxalis: Apache, Chiricahua & Mescalero; Cherokee; Cowlitz; Iroquois; Makah; Omaha; Pawnee; Ponca; Quileute; Quinault; Tolowa; Yurok
Oxypolis: Cherokee
Oxyria: Alaska Native; Eskimo, Alaska; Eskimo, Arctic; Eskimo, Greenland; Eskimo, Inuktitut
Oxytropis: Alaska Native
Panicum: Navajo
Parkinsonia: Papago; Pima; Pima, Gila River
Parrya: Alaska Native
Parthenocissus: Chippewa; Ojibwa
Passiflora: Cherokee
Pectis: Hopi
Pedicularis: Alaska Native; Eskimo, Alaska; Eskimo, Arctic; Eskimo, Inupiat; Mendocino Indian
Pediomelum: Blackfoot; Cheyenne; Comanche; Dakota; Lakota; Montana Indian; Omaha; Pawnee; Ponca; Winnebago
Pellaea: Tubatulabal
Peltandra: Seminole
Peniocereus: Apache, San Carlos; Papago
Penstemon: Diegueño
Perideridia: Atsugewi; Blackfoot; Cheyenne; Gosiute; Great Basin Indian; Karok; Klamath; Mendocino Indian; Miwok; Montana Indian; Nevada Indian; Okanagan-Colville; Paiute; Paiute, Northern; Pomo, Kashaya; Skagit, Upper; Umatilla; Ute; Yana; Yuki
Petasites: Concow; Sanpoil & Nespelem
Peucedanum: Coeur d'Alene; Thompson; Thompson, Upper (Nicola Band)
Phacelia: Cherokee; Pima, Gila River
Phalaris: Pima, Gila River
Phaseolus: Kamia; Navajo, Ramah; Okanagan-Colville; Papago
Phegopteris: Hawaiian
Pholisma: Cocopa; Kawaiisu; Papago
Pholistoma: Tubatulabal
Phoradendron: Havasupai
Phragmites: Klamath; Montana Indian; Yuma
Phyllospadix: Hesquiat; Makah
Physalis: Dakota
Physocarpus: Okanagan-Colville
Phytolacca: Cherokee; Malecite; Mohegan
Picea: Eskimo, Inuktitut; Gitksan; Makah; Tanana, Upper; Thompson; Wet'suwet'en
Pinus: Apache, Chiricahua & Mescalero; Apache, Western; Apache, White Mountain; Blackfoot; Cheyenne; Cocopa; Coeur d'Alene; Costanoan; Cree, Woodlands; Diegueño; Flathead; Gitksan; Gosiute; Havasupai; Hopi; Hualapai; Iroquois; Isleta; Jemez; Karok; Kawaiisu; Keres, Western; Klamath; Kutenai; Mendocino Indian; Mewuk; Miwok; Montana Indian; Navajo; Navajo, Ramah; Ojibwa; Okanagan-Colville; Okanogon; Paiute; Pomo; Pomo, Kashaya; Pueblo; Salish, Coast; Sanpoil & Nespelem; Shasta; Shoshoni; Shuswap; Sia; Spokan; Tewa; Thompson; Tubatulabal; Wet'suwet'en
Pisum: Okanagan-Colville; Papago; Pima
Plagiobothrys: Mendocino Indian
Plantago: Acoma; Alaska Native; Keres, Western; Laguna; Pima, Gila River
Platanus: Costanoan
Pluchea: Cahuilla
Poa: Gosiute; Havasupai
Polanisia: Pueblo
Poliomintha: Hopi; Tewa
Polygonatum: Cherokee
Polygonum: Alaska Native; Cherokee; Cheyenne; Eskimo, Arctic; Eskimo, Inupiat; Koyukon; Lakota; Paiute
Polypodium: Kwakiutl, Southern; Makah; Salish, Coast
Polypogon: Tubatulabal
Polyporus: Isleta
Polystichum: Cherokee; Costanoan; Klallam; Kwakiutl, Southern; Makah; Nitinaht; Quileute; Quinault; Thompson
Populus: Apache, Chiricahua & Mescalero; Apache, White Mountain; Bella Coola; Blackfoot; Cheyenne; Clallam; Coeur d'Alene; Costanoan; Cree, Woodlands; Dakota; Flathead; Haisla; Haisla & Hanaksiala; Isleta; Jemez; Kutenai; Kwakiutl, Southern; Montana Indian; Navajo; Ojibwa; Oweekeno; Pima; Tanana, Upper; Zuni
Porphyra: Hesquiat; Kitasoo; Kwakiutl, Southern; Nitinaht; Oweekeno; Pomo, Kashaya; Tolowa; Yurok
Portulaca: Apache, Chiricahua & Mescalero; Hopi; Navajo; Pima, Gila River; San Felipe; Tewa
Postelsia: Pomo; Pomo, Kashaya
Proboscidea: Apache, Chiricahua & Mescalero; Cahuilla; Havasupai; Hualapai; Papago; Pima; Pima, Gila River
Prosopis: Acoma; Apache, Chiricahua & Mescalero; Cahuilla; Cocopa; Hualapai; Isleta; Kamia; Laguna; Maricopa; Paiute; Papago; Pima; Southwest Indians; Yavapai; Yuma
Prunus: Costanoan; Hopi; Luiseño; Seminole; Thompson

Pseudocymopterus: Navajo, Ramah
Pseudotsuga: Okanagan-Colville
Psidium: Seminole
Pteridium: Atsugewi; Bella Coola; Chehalis; Costanoan; Cowlitz; Green River Group; Hahwunkwut; Klallam; Kwakiutl, Southern; Lummi; Mahuna; Makah; Montana Indian; Nitinaht; Ojibwa; Okanagon; Quileute; Quinault; Salish, Coast; Skagit; Skagit, Upper; Skokomish; Snohomish; Squaxin; Swinomish; Thompson
Pterospora: Kawaiisu
Puccinellia: Gosiute
Pycnanthemum: Cherokee
Pyrrhopappus: Kiowa
Pyrus: Hopi; Seminole
Quercus: Acoma; Apache, Mescalero; Apache, Western; Apache, White Mountain; Chehalis; Cheyenne; Chippewa; Cochiti; Comanche; Costanoan; Cowlitz; Dakota; Gosiute; Havasupai; Hualapai; Iroquois; Karok; Keres, Western; Kiowa; Laguna; Lakota; Mahuna; Malecite; Mewuk; Navajo; Navajo, Ramah; Neeshenam; Nisqually; Ojibwa; Omaha; Paiute; Papago; Pawnee; Pomo; Ponca; Pueblo; Salish, Coast; San Felipe; Seminole; Squaxin; Tewa; Tubatulabal; Winnebago; Wintoon; Yavapai; Yokut
Ranunculus: Acoma; Alaska Native; Cherokee; Eskimo, Alaska; Gosiute; Iroquois; Keres, Western; Laguna; Makah; Miwok; Quileute
Raphanus: Costanoan
Rheum: Cherokee; Haisla & Hanaksiala
Rhododendron: Cherokee
Rhodymenia: Alaska Native
Rhus: Iroquois; Navajo, Ramah; Yavapai
Ribes: Eskimo, Alaska; Zuni
Robinia: Apache, Mescalero; Jemez
Rorippa: Gosiute; Havasupai; Karok; Kawaiisu; Paiute; Saanich
Rosa: Cahuilla; Cheyenne; Eskimo, Inupiat; Haisla & Hanaksiala; Makah; Montana Indian; Nitinaht; Ojibwa; Okanagan-Colville; Paiute; Quinault; Salish; Salish, Coast; Skagit; Skokomish; Snohomish; Squaxin; Swinomish; Tanana, Upper; Thompson
Rubus: Bella Coola; Chehalis; Cherokee; Chinook, Lower; Cowlitz; Cree, Woodlands; Green River Group; Hesquiat; Iroquois; Kitasoo; Klallam; Klamath; Kwakiutl, Southern; Lummi; Makah; Montana Indian; Nitinaht; Okanagon; Oweekeno; Quileute; Quinault; Salish, Coast; Samish; Skagit; Skagit, Upper; Squaxin; Swinomish; Thompson; Tolowa
Rudbeckia: Cherokee
Rumex: Anticosti; Apache, Chiricahua & Mescalero; Bella Coola; Chehalis; Cherokee; Cheyenne; Cocopa; Delaware; Eskimo, Alaska; Eskimo, Inuktitut; Hanaksiala; Heiltzuk; Kawaiisu; Klallam; Klamath; Koyukon; Montana Indian; Navajo; Omaha; Oweekeno; Pima; Pima, Gila River; Saanich; San Felipe; Thompson
Sabal: Seminole

Saccharum: Seminole
Sagittaria: Algonquin, Quebec; Cheyenne; Cocopa; Dakota; Great Basin Indian; Klamath; Lakota; Montana Indian; Omaha; Paiute, Northern; Pawnee; Pomo; Potawatomi; Thompson; Winnebago
Salicornia: Salish, Coast
Salix: Alaska Native; Blackfoot; Cocopa & Yuma; Eskimo, Alaska; Eskimo, Arctic; Eskimo, Inuktitut; Eskimo, Inupiat; Pima; Tanana, Upper; Yuma
Salsola: Navajo
Salvia: Diegueño; Luiseño; Pima, Gila River; Tubatulabal; Yavapai
Sambucus: Ojibwa
Sanicula: Mendocino Indian; Neeshenam; Pomo
Sarcostemma: Luiseño
Saxifraga: Alaska Native; Cherokee; Eskimo, Alaska; Eskimo, Inuktitut
Scirpus: Cheyenne; Chippewa; Costanoan; Cree, Woodlands; Dakota; Gosiute; Hopi; Kawaiisu; Keres, Western; Klamath; Lakota; Luiseño; Montana Indian; Paiute, Northern; Pima, Gila River; Pomo; Sioux; Thompson; Tubatulabal; Ute
Scorzonella: Paiute
Sedum: Alaska Native; Costanoan; Eskimo, Alaska; Eskimo, Inupiat; Gitksan; Haisla; Makah; Tubatulabal; Wet'suwet'en
Senecio: Aleut; Eskimo, Alaska
Sideroxylon: Seminole
Silene: Eskimo, Inuktitut
Simmondsia: Cocopa; Papago
Sisymbrium: Pima, Gila River; Tubatulabal
Sium: Algonquin, Quebec; Bella Coola; Carrier; Cree, Woodlands; Okanagan-Colville; Shuswap; Thompson
Smilax: Cherokee
Solanum: Hopi; Iroquois; Pueblo; Zuni
Solidago: Gosiute; Navajo, Kayenta
Sonchus: Pima; Pima, Gila River
Sparganium: Klamath
Sphaeralcea: Navajo
Spiranthes: Paiute
Stachys: Gosiute; Quinault
Stanleya: Hopi
Stenocereus: Papago
Sticta: Ojibwa
Stipa: Kawaiisu
Streptopus: Cherokee
Strophostyles: Choctaw
Suaeda: Gosiute; Paiute, Northern; Pima, Gila River
Tacca: Hawaiian
Taraxacum: Eskimo, Alaska; Malecite; Mohegan; Potawatomi; Ute
Thalia: Seminole
Thelypodium: Pueblo; Tewa
Thlaspi: Havasupai
Thuja: Montana Indian; Salish, Coast
Tolmiea: Makah
Torreya: Mendocino Indian
Tradescantia: Acoma; Cherokee; Keres, Western; Laguna

Tragopogon: Cherokee; Okanagon; Thompson; Thompson, Upper (Nicola Band); Thompson, Upper (Spences Bridge)
Trifolium: Bella Coola; Costanoan; Haisla & Hanaksiala; Hesquiat; Kitasoo; Luiseño; Makah; Mendocino Indian; Miwok; Nitinaht; Nuxalkmc; Oweekeno; Paiute; Pomo, Kashaya; Round Valley Indian; Tubatulabal; Wailaki; Yokia; Yuki
Triglochin: Gosiute; Klamath; Montana Indian
Trisetum: Gosiute
Triteleia: Atsugewi; Karok; Mendocino Indian; Miwok; Neeshenam; Okanagan-Colville; Okanagon; Paiute; Pomo; Thompson; Thompson, Upper (Lytton Band); Thompson, Upper (Spences Bridge); Yana; Yuki
Triticum: Navajo, Ramah; Okanagan-Colville; Pima
Tsuga: Kitasoo; Nitinaht; Oweekeno; Wet'suwet'en
Typha: Acoma; Alaska Native; Apache; Apache, Chiricahua & Mescalero; Apache, Mescalero; Blackfoot; Carrier; Chehalis; Clallam; Costanoan; Cree, Woodlands; Gosiute; Hualapai; Kawaiisu; Keres, Western; Klamath; Laguna; Mendocino Indian; Montana Indian; Navajo, Ramah; Ojibwa; Okanagan-Colville; Paiute; Paiute, Northern; Pima; Pima, Gila River; Pomo, Kashaya; San Felipe; Sioux; Tanana, Upper; Thompson; Tubatulabal; Yuki; Yuma
Ulmus: Omaha
Umbellularia: Karok; Mendocino Indian; Midoo; Pomo; Pomo, Kashaya; Yuki; Yurok
Urtica: Alaska Native; Haisla & Hanaksiala; Makah; Oweekeno; Skagit, Upper
Ustilago: Apache, White Mountain; Hopi
Uvularia: Cherokee
Vaccinium: Algonquin, Tête-de-Boule; Iroquois; Seminole
Valeriana: Klamath; Montana Indian; Paiute; Sanpoil & Nespelem
Veratrum: Miwok
Verbesina: Navajo
Viburnum: Abnaki; Thompson
Vicia: Acoma; Apache, Chiricahua & Mescalero; Laguna; Montana Indian
Vigna: Cocopa; Kamia; Mohave; Papago; Pima
Vitex: Hawaiian
Vitis: Iroquois; Seminole
Vulpia: Gosiute; Navajo, Kayenta
Wislizenia: Hopi
Wyethia: Costanoan; Gosiute; Mendocino Indian; Montana Indian; Paiute; Paiute, Northern; Pomo, Kashaya
Xanthosoma: Seminole
Ximenia: Seminole
Yucca: Apache; Apache, Chiricahua & Mescalero; Apache, Mescalero; Apache, Western; Apache, White Mountain; Cahuilla; Cochiti; Diegueño; Isleta; Jemez; Kawaiisu; Keres, Western; Lakota; Luiseño; Mahuna; Navajo, Ramah; Papago; San Felipe; Southwest Indians; Tubatulabal; Yavapai; Zuni

Zamia: Seminole
Zea: Delaware; Hopi; Isleta; Kamia; Kiowa; Meskwaki; Navajo; Papago; Seminole
Zigadenus: Karok; Navajo, Kayenta
Zizania: Chippewa; Meskwaki; Thompson
Zostera: Bellabella; Hesquiat; Kwakiutl, Southern; Nitinaht; Oweekeno

Vegetable

Acacia: Cahuilla
Acer: Thompson
Agaricus: Pomo, Kashaya
Agave: Apache, Chiricahua & Mescalero; Papago
Agoseris: Kawaiisu
Aletes: Isleta
Allium: Apache, Chiricahua & Mescalero; Blackfoot; Cahuilla; Cherokee; Coeur d'Alene; Cree; Eskimo, Inupiat; Haisla & Hanaksiala; Hualapai; Iroquois; Isleta; Kawaiisu; Keres, Western; Montana Indian; Navajo; Ojibwa; Okanagan-Colville; Paiute; Paiute, Northern; Pomo, Kashaya; Potawatomi
Amaranthus: Acoma; Cahuilla; Cochiti; Cocopa; Havasupai; Hopi; Iroquois; Isleta; Keres, Western; Keresan; Laguna; Mohave; Mohegan; Montana Indian; Navajo; Navajo, Ramah; Papago; Pima; Pima, Gila River; Pueblo; Sia; Spanish American; Yavapai; Yuma
Ambrosia: Papago
Amphicarpaea: Cherokee; Ojibwa
Amsinckia: Kawaiisu; Pima, Gila River
Angelica: Alaska Native; Eskimo, Alaska; Eskimo, Greenland; Karok; Neeshenam; Shuswap
Antennaria: Navajo, Kayenta
Apios: Cherokee; Cheyenne; Chippewa; Menominee; Meskwaki; Omaha; Potawatomi
Apium: Cahuilla; Luiseño
Aquilegia: Miwok
Arabis: Alaska Native
Arctium: Iroquois
Argentina: Alaska Native; Montana Indian
Asclepias: Cahuilla; Chippewa; Iroquois; Meskwaki; Miwok; Ojibwa; Omaha; Pawnee; Ponca; Winnebago
Asparagus: Cherokee; Iroquois; Isleta
Aster: Algonquin, Quebec
Atriplex: Cochiti; Hopi; Isleta; Kawaiisu; Keres, Western; Papago; Pima; Pueblo
Balsamorhiza: Montana Indian
Barbarea: Alaska Native; Cherokee
Boletus: Pomo, Kashaya
Brassica: Cherokee; Diegueño; Haisla & Hanaksiala; Hoh; Iroquois; Kitasoo; Lakota; Luiseño; Mendocino Indian; Mohegan; Okanagan-Colville; Quileute
Brodiaea: Pomo, Kashaya; Yurok
Calandrinia: Luiseño
Calochortus: Paiute; Paiute, Northern; Pomo, Kashaya
Caltha: Abnaki; Chippewa; Iroquois; Menominee
Camassia: Coeur d'Alene; Haisla; Hesquiat; Karok; Nitinaht; Salish; Salish, Coast
Camissonia: Cahuilla

Cantharellus: Pomo, Kashaya
Capsella: Apache, Chiricahua & Mescalero; Cahuilla; Cherokee; Thompson
Cardamine: Cherokee; Menominee; Ojibwa
Carduus: Luiseño
Caulanthus: Kawaiisu
Chelone: Cherokee
Chenopodium: Alaska Native; Apache; Cahuilla; Cherokee; Cocopa; Diegueño; Eskimo, Inupiat; Hopi; Iroquois; Isleta; Kawaiisu; Keresan; Lakota; Luiseño; Mendocino Indian; Miwok; Mohave; Montana Indian; Navajo; Ojibwa; Papago; Pima; Potawatomi; Pueblo; Shuswap; Spanish American; Thompson; Yaqui; Yavapai; Zuni
Chlorogalum: Cahuilla
Cirsium: Hoh; Montana Indian; Quileute
Citrullus: Iroquois
Clarkia: Miwok
Claytonia: Alaska Native; Algonquin, Quebec; Blackfoot; Cahuilla; Coeur d'Alene; Diegueño; Eskimo, Arctic; Kawaiisu; Luiseño; Mendocino Indian; Neeshenam; Paiute, Northern
Cleome: Apache, Western; Hopi; Isleta; Jemez; Keres, Western; Keresan; Navajo; Sia; Tewa
Cochlearia: Alaska Native
Colocasia: Hawaiian
Conioselinum: Kitasoo
Conyza: Miwok
Coreopsis: Kawaiisu
Coriandrum: Zuni
Crepis: Karok
Cucumis: Iroquois; Ojibwa
Cucurbita: Cherokee; Cocopa; Havasupai; Iroquois; Maricopa; Navajo
Cymopterus: Acoma; Cochiti; Keres, Western; Laguna; Navajo; Navajo, Kayenta
Cyperus: Keres, Western; Pomo, Kashaya
Darmera: Karok
Dasylirion: Apache, Chiricahua & Mescalero; Papago
Daucus: Haisla & Hanaksiala; Kitasoo; Sanpoil & Nespelem
Delphinium: Miwok
Dentinum: Pomo, Kashaya
Descurainia: Cahuilla; Cocopa; Hopi; Pueblo
Dichelostemma: Pomo, Kashaya
Dioscorea: Hawaiian
Dryopteris: Alaska Native; Kitasoo
Dudleya: Neeshenam
Dyssodia: Apache, Chiricahua & Mescalero
Echinocereus: Isleta
Epilobium: Alaska Native; Bella Coola; Eskimo, Arctic; Eskimo, Greenland; Eskimo, Inupiat
Equisetum: Hesquiat
Eremalche: Pima, Gila River
Eriogonum: Havasupai; Karok; Miwok
Erodium: Cahuilla; Diegueño
Eschscholzia: Luiseño; Mendocino Indian; Neeshenam
Ferocactus: Papago

Fritillaria: Hesquiat
Glyptopleura: Paiute; Paiute, Northern
Grindelia: Karok
Hedysarum: Eskimo, Alaska; Eskimo, Arctic; Eskimo, Inupiat; Tanana, Upper
Helianthus: Cherokee
Heracleum: Blackfoot; California Indian; Hoh; Kitasoo; Mendocino Indian; Meskwaki; Montana Indian; Ojibwa; Okanagan-Colville; Quileute; Thompson
Hericium: Pomo, Kashaya
Heuchera: Miwok
Hippuris: Eskimo, Alaska
Hirschfeldia: Cahuilla
Hoffmannseggia: Pima
Hoita: Luiseño
Honckenya: Alaska Native; Eskimo, Arctic; Eskimo, Inupiat
Hydrangea: Cherokee
Hydrocotyle: Cahuilla
Hydrophyllum: Iroquois; Menominee
Ipomoea: Cherokee; Hawaiian; Seminole
Lactuca: Acoma; Cherokee; Keres, Western; Laguna
Lagenaria: Ojibwa
Lathyrus: Iroquois; Karok; Keres, Western; Makah; Miwok; Ojibwa; Yokia
Lepidium: Cherokee; Diegueño; Hoh; Luiseño; Quileute
Lewisia: Coeur d'Alene; Paiute, Northern
Ligusticum: Alaska Native; Atsugewi; Cherokee; Eskimo, Alaska; Eskimo, Inupiat
Lilium: Meskwaki
Lomatium: Great Basin Indian; Kawaiisu; Montana Indian; Okanagon; Paiute; Thompson
Lotus: Luiseño
Lupinus: Luiseño; Mendocino Indian; Yavapai; Yuki
Lygodesmia: Navajo, Kayenta
Maianthemum: Ojibwa; Thompson
Mimulus: Isleta; Miwok; Neeshenam
Mirabilis: Navajo, Kayenta
Monolepis: Pima; Pima, Gila River
Moricandia: Hoh; Quileute
Nelumbo: Omaha
Nolina: Apache, Chiricahua & Mescalero
Nuphar: Alaska Native; Menominee
Nymphaea: Kiowa
Oenothera: Cherokee
Onoclea: Iroquois
Opuntia: Cahuilla; Diegueño; Papago
Osmorhiza: Karok; Miwok
Oxalis: Cherokee; Iroquois
Oxyria: Eskimo, Inupiat; Montana Indian
Oxytropis: Eskimo, Inupiat
Paeonia: Diegueño
Parmelia: Potawatomi
Passiflora: Cherokee
Pedicularis: Cherokee; Eskimo, Arctic; Eskimo, Inupiat; Iroquois
Pediomelum: Montana Indian
Peniocereus: Papago

Penthorum: Cherokee
Perideridia: Blackfoot; Kawaiisu; Montana Indian; Pomo
Petasites: Alaska Native; Eskimo, Alaska; Eskimo, Arctic
Peziza: Pomo, Kashaya
Phacelia: Cherokee; Kawaiisu; Luiseño; Navajo, Kayenta; Pima, Gila River
Phaseolus: Abnaki; Apache, White Mountain; Cherokee; Havasupai; Iroquois; Navajo; Ojibwa; Potawatomi; Sia; Zuni
Phlox: Navajo, Kayenta
Phytolacca: Cherokee; Iroquois
Piperia: Pomo, Kashaya
Pisum: Cherokee; Navajo, Ramah
Plagiobothrys: Yuki
Plantago: Alaska Native; Cherokee; Mohegan
Platystemon: Mendocino Indian
Pleurotus: Pomo, Kashaya
Polanisia: Pueblo
Polygonatum: Cherokee
Polygonum: Alaska Native; Cherokee; Eskimo, Inupiat; Tanana, Upper
Polypodium: Hesquiat
Portulaca: Acoma; Iroquois; Isleta; Keres, Western; Laguna; Luiseño; Navajo, Ramah; Pima, Gila River; San Felipe
Prenanthes: Cherokee
Prosopis: Keres, Western; Kiowa; Mohave
Prunella: Cherokee
Pseudocymopterus: Hopi
Pteridium: Hesquiat; Nitinaht
Quercus: Chippewa
Ranunculus: Cherokee
Robinia: Apache, Chiricahua & Mescalero; Apache, White Mountain
Rorippa: Algonquin, Quebec; Cahuilla; Cherokee; Diegueño; Iroquois; Luiseño; Okanagan-Colville; Tubatulabal
Rubus: Cherokee
Rudbeckia: Cherokee; San Felipe
Rumex: Alaska Native; Bella Coola; Cahuilla; Cherokee; Cochiti; Costanoan; Eskimo, Arctic; Eskimo, Inupiat; Haisla & Hanaksiala; Havasupai; Iroquois; Isleta; Kitasoo; Mendocino Indian; Miwok; Mohave; Mohegan; Montana Indian; Navajo; Okanagan-Colville; Papago; Pima; Pima, Gila River; Saanich; Tanana, Upper
Sagittaria: Omaha; Potawatomi
Salix: Eskimo, Alaska; Eskimo, Inupiat
Salsola: Navajo
Sanicula: Karok; Neeshenam
Sarcobatus: Montana Indian
Sarcocornia: Alaska Native
Saxifraga: Alaska Native; Cherokee; Eskimo, Arctic; Eskimo, Inupiat
Scirpus: Pomo
Scrophularia: Yavapai
Sedum: Alaska Native; Eskimo, Arctic; Eskimo, Inupiat

Senecio: Eskimo, Arctic
Sidalcea: Luiseño
Sinapis: Hoh; Quileute
Sisymbrium: Cahuilla; Cherokee
Sisyrinchium: Cherokee
Solanum: Abnaki; Apache, Chiricahua & Mescalero; Cherokee; Haisla & Hanaksiala; Isleta; Keres, Western; Kitasoo; Luiseño; Makah; Menominee; Meskwaki; Navajo; Navajo, Ramah; Ojibwa; Oweekeno; Seminole; Sia; Tewa
Sonchus: Kamia; Luiseño; Pima; Yaqui
Stanleya: Havasupai; Hopi; Kawaiisu; Tewa
Streptanthus: Navajo, Kayenta
Streptopus: Alaska Native; Cherokee
Suaeda: Cahuilla
Symplocarpus: Iroquois
Taraxacum: Alaska Native; Algonquin, Quebec; Cahuilla; Carrier; Cherokee; Eskimo, Arctic; Iroquois; Kiowa; Menominee; Meskwaki; Micmac; Ojibwa; Okanagan-Colville; Papago; Tewa
Thelypodium: Pueblo
Thlaspi: Cherokee
Tilia: Chippewa
Tradescantia: Cherokee; Hopi
Trianthema: Pima
Trifolium: Cahuilla; Diegueño; Kawaiisu; Kwakiutl, Southern; Mendocino Indian; Nitinaht; Paiute
Triglochin: Salish
Triteleia: Paiute; Pomo; Pomo, Kashaya
Typha: Pomo
Urtica: Cahuilla; Cowichan; Hoh; Iroquois; Mohegan; Montana Indian; Okanagan-Colville; Quileute; Saanich; Thompson
Uvularia: Cherokee
Valerianella: Cherokee
Vicia: Keres, Western; Mendocino Indian; Sia
Viola: Cherokee; Diegueño; Luiseño
Yucca: Apache; Apache, Chiricahua & Mescalero; Cahuilla
Zea: Cherokee; Chippewa; Choctaw; Havasupai; Iroquois; Keres, Western; Menominee; Navajo; Navajo, Ramah; Ojibwa
Zigadenus: Navajo, Kayenta

Winter Use Food

Aesculus: Miwok
Agave: Apache, White Mountain; Hualapai
Allium: Blackfoot; Costanoan; Hualapai; Isleta; Menominee; Meskwaki; Navajo
Amaranthus: Cocopa; Keres, Western; Mohave; Navajo; Navajo, Ramah; Yuma
Amelanchier: Blackfoot; Cheyenne; Klamath; Menominee; Montana Indian; Navajo; Okanagan-Colville; Paiute; Potawatomi; Thompson
Amphicarpaea: Omaha
Apios: Delaware; Menominee; Meskwaki
Arbutus: Miwok; Pomo, Kashaya
Arctostaphylos: Atsugewi; Blackfoot; Eskimo, Inupiat; Koyukon; Mendocino Indian; Miwok; Tanana, Upper

Asclepias: Menominee; Meskwaki; Ojibwa
Asyneuma: Costanoan
Avena: Pomo
Balsamorhiza: Paiute
Calochortus: Cheyenne; Great Basin Indian
Camassia: Blackfoot; Gosiute; Montana Indian; Paiute
Carya: Comanche; Meskwaki; Potawatomi
Castanopsis: Karok; Pomo, Kashaya
Castilleja: Miwok
Chasmanthium: Cocopa
Chenopodium: Navajo, Ramah
Chlorogalum: Miwok
Citrullus: Cahuilla; Cocopa; Navajo, Ramah
Claytonia: Okanagan-Colville; Thompson
Cleome: Navajo
Corylus: Chehalis; Chippewa; Cowlitz; Cree, Woodlands; Karok; Menominee; Meskwaki; Paiute; Potawatomi; Skagit, Upper
Crataegus: Apache, Chiricahua & Mescalero; Cheyenne; Montana Indian
Cucurbita: Menominee; Meskwaki; Navajo, Ramah
Cymopterus: Navajo, Ramah
Daucus: Navajo
Descurainia: Kawaiisu; Paiute, Northern
Echinochloa: Cocopa
Egregia: Oweekeno
Empetrum: Eskimo, Alaska; Eskimo, Inupiat; Tanana, Upper
Epilobium: Eskimo, Inupiat
Equisetum: Eskimo, Inupiat
Eriophorum: Eskimo, Inupiat
Erythronium: Shuswap
Fagus: Menominee
Ficus: Havasupai
Fragaria: Montana Indian; Ojibwa; Okanagan-Colville; Potawatomi
Hedysarum: Eskimo, Inupiat; Tanana, Upper
Helianthus: Hualapai; Mohave; Paiute; Sanpoil & Nespelem
Heracleum: Thompson
Hippuris: Alaska Native
Hirschfeldia: Cahuilla
Honckenya: Eskimo, Inupiat
Juglans: Apache, Chiricahua & Mescalero; Comanche; Meskwaki; Potawatomi; Yavapai
Juniperus: Navajo, Ramah; Paiute
Koeleria: Havasupai
Lewisia: Okanogan; Paiute; Thompson
Ligusticum: Alaska Native; Atsugewi; Cherokee; Eskimo, Alaska
Lithocarpus: Karok
Lomatium: Paiute; Thompson
Lonicera: Montana Indian
Lupinus: Miwok
Lycium: Navajo
Macrocystis: Oweekeno
Madia: Pomo
Malus: Haisla & Hanaksiala; Kitasoo; Meskwaki
Martynia: Apache, Western

Mentzelia: Kawaiisu
Morus: Apache, Chiricahua & Mescalero; Cherokee
Nelumbo: Meskwaki; Potawatomi
Nereocystis: Oweekeno; Pomo, Kashaya
Opuntia: Cheyenne; Keres, Western; Navajo, Ramah; San Felipe; Thompson
Oryzopsis: Paiute, Northern
Oxytropis: Eskimo, Inupiat
Panicum: Cocopa
Parrya: Alaska Native
Pediomelum: Cheyenne; Montana Indian; Sioux
Perideridia: Cheyenne; Gosiute; Nevada Indian; Okanagan-Colville; Paiute
Phaseolus: Cocopa; Havasupai; Navajo, Ramah
Physalis: Dakota
Pinus: Apache, Western; Isleta; Karok; Kawaiisu; Keresan; Navajo, Ramah; Okanagan-Colville; Paiute; Paiute, Northern; Pomo, Kashaya; Thompson; Zuni
Plantago: Alaska Native
Polygonatum: Cherokee
Polygonum: Eskimo, Inupiat
Polyporus: Isleta
Porphyra: Pomo
Portulaca: Isleta
Prosopis: Cocopa; Pima, Gila River
Prunus: Apache, Chiricahua & Mescalero; Cheyenne; Comanche; Crow; Dakota; Keres, Western; Kiowa; Montana Indian; Navajo, Ramah; Okanagan-Colville; Paiute; Paiute, Northern; Shuswap; Thompson
Pseudotsuga: Okanagan-Colville
Pteridium: Alaska Native
Quercus: Apache, Chiricahua & Mescalero; Kawaiisu; Luiseño; Miwok; Paiute; Yavapai
Ranunculus: Skokomish
Rhus: Menominee; Ojibwa; Shoshoni
Ribes: Cherokee; Cheyenne; Hanaksiala; Hoh; Kwakiutl, Southern; Menominee; Okanagan-Colville; Quileute
Robinia: Apache, Chiricahua & Mescalero
Rosa: Hoh; Quileute; Tanana, Upper
Rubus: Alaska Native; Anticosti; Apache, Chiricahua & Mescalero; Cherokee; Eskimo, Alaska; Eskimo, Inuktitut; Eskimo, Inupiat; Hoh; Makah; Mendocino Indian; Menominee; Meskwaki; Ojibwa; Okanagan-Colville; Oweekeno; Pomo; Pomo, Kashaya; Quileute; Tanana, Upper
Rudbeckia: Cherokee
Rumex: Alaska Native; Eskimo, Alaska; Eskimo, Inupiat; Hanaksiala
Sagittaria: Menominee; Meskwaki; Potawatomi
Salix: Alaska Native; Eskimo, Alaska; Eskimo, Inupiat
Sambucus: Chehalis; Cowlitz; Diegueño; Green River Group; Haisla & Hanaksiala; Hoh; Klallam; Lummi; Makah; Quileute; Quinault; Skagit; Skokomish; Snohomish; Squaxin; Swinomish; Yokut
Saxifraga: Eskimo, Alaska
Scirpus: Paiute, Northern
Shepherdia: Arapaho; Kitasoo; Makah; Montana Indian; Oweekeno; Thompson
Sisymbrium: Pima
Solanum: Navajo, Ramah
Sorbus: Thompson
Streptopus: Cherokee
Thelesperma: Isleta
Trifolium: Miwok
Tsuga: Gitksan; Haisla; Haisla & Hanaksiala; Wet'suwet'en
Ulmus: Kiowa
Umbellularia: Karok; Pomo, Kashaya
Vaccinium: Alaska Native; Anticosti; Cree, Woodlands; Eskimo, Inupiat; Haisla & Hanaksiala; Hoh; Karok; Klamath; Makah; Menominee; Montana Indian; Ojibwa; Okanagan-Colville; Oweekeno; Paiute; Quileute; Tanana, Upper; Thompson
Valeriana: Okanagan-Colville
Viburnum: Bella Coola; Haisla & Hanaksiala; Kitasoo; Kwakiutl, Southern; Tanana, Upper
Wyethia: Paiute, Northern; Pomo, Kashaya
Yucca: Isleta; Keres, Western; Navajo
Zea: Dakota; Havasupai; Isleta; Keres, Western; Menominee; Meskwaki; Navajo, Ramah; Omaha; Pawnee; Ponca; Potawatomi

Fiber
Basketry
Acacia: Havasupai; Hualapai; Papago; Seri
Acer: Cherokee; Concow; Gitksan; Maidu; Makah; Malecite; Micmac; Quinault; Thompson
Adenostoma: Costanoan
Adiantum: Hahwunkwut; Karok; Makah; Poliklah; Pomo, Kashaya; Quinault; Tolowa; Wintoon; Yurok
Agave: Cahuilla
Alnus: Karok; Tolowa; Yurok
Amelanchier: Cree, Woodlands; Gosiute; Havasupai; Karok; Maidu
Anthoxanthum: Abnaki; Ojibwa; Potawatomi
Apocynum: Abnaki
Arundinaria: Cherokee; Choctaw
Baccharis: Havasupai
Betula: Abnaki; Algonquin, Tête-de-Boule; Bella Coola; Carrier; Cree, Woodlands; Gitksan; Koyukon; Micmac; Ojibwa; Okanagan-Colville; Shuswap; Tanana, Upper; Thompson
Bouteloua: Hopi
Bursera: Seri
Caesalpinia: Seri
Calocedrus: Klamath; Maidu; Paiute
Calycanthus: Mendocino Indian
Carex: Costanoan; Hesquiat; Maidu; Makah; Mendocino Indian; Nitinaht; Nootka; Pomo; Salish, Coast; Wailaki; Yuki
Carnegia: Papago
Carya: Cherokee
Ceanothus: Concow; Karok; Maidu; Mewuk
Cercis: Maidu; Mendocino Indian; Mewuk; Modesse; Neeshenam; Pomo; Pomo, Kashaya; Wintoon; Yuki
Chamaecyparis: Kitasoo
Chilopsis: Havasupai
Chrysothamnus: Zuni
Cladium: Mewuk; Pomo; Yokut
Cornus: Costanoan; Cree, Woodlands; Iroquois; Modesse; Washo; Wintoon; Yuki
Corylus: Costanoan; Hahwunkwut; Karok; Meskwaki; Ojibwa; Poliklah; Pomo; Pomo, Kashaya; Round Valley Indian; Tolowa; Wintoon; Yuki; Yurok
Dasylirion: Papago
Eleocharis: Cheyenne
Epicampes: Shoshoni
Equisetum: Costanoan; Cowlitz; Quileute; Salish, Coast; Sanpoil & Nespelem; Swinomish
Erigeron: Thompson
Fallugia: Havasupai
Fraxinus: Abnaki; Chippewa; Iroquois; Karok; Malecite; Meskwaki; Micmac; Ojibwa; Potawatomi
Heracleum: Makah; Quileute
Hierochloe: Haisla & Hanaksiala; Iroquois; Malecite; Menominee; Micmac
Hilaria: Hopi
Iris: Potawatomi
Jatropha: Papago; Seri
Juncus: Cahuilla; Cheyenne; Costanoan; Diegueño; Karok; Kawaiisu; Klamath; Luiseño; Mendocino Indian; Montana Indian; Pomo; Tolowa; Yurok
Juniperus: Keres, Western; Pomo
Laportea: Abnaki; Menominee
Larix: Ojibwa
Leymus: Eskimo, Alaska; Hesquiat; Kwakiutl, Southern; Thompson
Linum: Klamath; Montana Indian
Lonicera: Cherokee; Mendocino Indian
Martynia: Shoshoni
Morus: Hualapai; Papago
Muhlenbergia: Cahuilla; Diegueño; Kawaiisu; Luiseño
Nolina: Hualapai; Isleta; Jemez; Keres, Western; Keresan; Papago; Pima; Southwest Indians
Parryella: Hopi; Navajo, Ramah; Tewa; Zuni
Pediomelum: Lakota
Philadelphus: Mewuk; Thompson
Phragmites: Klamath; Montana Indian; Okanagon; Thompson
Phyllospadix: Hesquiat; Makah; Nitinaht
Picea: Algonquin, Tête-de-Boule; Bella Coola; Cherokee; Cree, Woodlands; Hahwunkwut; Haisla & Hanaksiala; Hoh; Kwakiutl, Southern; Nitinaht; Oweekeno; Poliklah; Quileute; Tanana, Upper; Thompson; Yurok
Pinus: Cahuilla; Diegueño; Karok; Maidu; Mewuk; Miwok; Montana Indian; Navajo, Ramah; Pomo; Seminole; Wintoon
Populus: Gosiute; Havasupai; Hualapai; Karok; Pima
Proboscidea: Havasupai; Hualapai; Kawaiisu; Papago; Pima

Index of Usages 853

Prosopis: Papago
Prunus: Bella Coola; Hanaksiala; Hesquiat; Maidu; Nitinaht; Okanogon; Salish, Coast; Shuswap; Snohomish; Thompson
Pseudotsuga: Maidu; Round Valley Indian
Pteridium: Costanoan; Kawaiisu; Mendocino Indian; Mewuk; Pomo; Pomo, Kashaya; Round Valley Indian; Ukiah; Yokut
Quercus: Cherokee; Diegueño; Kawaiisu; Maidu; Mewuk; Navajo
Rhus: Apache, Western; Apache, White Mountain; Cahuilla; Diegueño; Great Basin Indian; Havasupai; Hopi; Hualapai; Jemez; Kawaiisu; Keres, Western; Luiseño; Mahuna; Mewuk; Navajo; Navajo, Kayenta; Navajo, Ramah; Panamint; Pomo; Southwest Indians; Tewa; Thompson; Wintoon; Zuni
Rosa: Kawaiisu
Salix: Abnaki; Apache, White Mountain; Chippewa; Costanoan; Cree, Woodlands; Dakota; Diegueño; Flathead; Gosiute; Havasupai; Hoh; Hualapai; Isleta; Karok; Kawaiisu; Keres, Western; Maidu; Makah; Montana Indian; Navajo; Omaha; Paiute; Paiute, Northern; Panamint; Papago; Pima; Pomo; Pomo, Kashaya; Quileute; Shoshoni; Tanana, Upper; Tewa; Thompson; Tolowa; Ukiah; Ute; Yuki; Yurok; Zuni
Scirpus: Costanoan; Hesquiat; Kawaiisu; Klamath; Kwakiutl, Southern; Maidu; Makah; Nitinaht; Ojibwa; Okanagan-Colville; Paiute, Northern; Pomo; Pomo, Kashaya; Potawatomi; Salish, Coast
Sequoia: Costanoan
Serenoa: Choctaw; Seminole
Smilax: Round Valley Indian; Ukiah
Solidago: Keres, Western
Taxus: Pomo; Pomo, Kashaya
Thuja: Bella Coola; Clallam; Flathead; Gitksan; Haisla; Haisla & Hanaksiala; Hesquiat; Hoh; Kutenai; Kwakiutl, Southern; Makah; Malecite; Menominee; Montana Indian; Nez Perce; Nitinaht; Ojibwa; Okanagan-Colville; Oweekeno; Quileute; Quinault; Salish, Coast; Shuswap; Squaxin; Thompson; Wet'suwet'en
Tilia: Abnaki; Algonquin, Tête-de-Boule; Menominee; Meskwaki; Omaha; Potawatomi
Torreya: Pomo
Toxicodendron: Costanoan; Mendocino Indian
Tsuga: Cherokee; Snohomish
Typha: Chehalis; Cheyenne; Chippewa; Clallam; Cowlitz; Hesquiat; Klallam; Makah; Navajo, Ramah; Nitinaht; Paiute, Northern; Pima; Quinault; Salish, Coast; Snohomish; Squaxin
Ulmus: Menominee; Omaha; Potawatomi
Urtica: Cahuilla; Makah; Menominee
Viburnum: Koyukon; Tanana, Upper
Vitis: Karok; Pomo
Woodwardia: Hupa; Karok; Poliklah
Xerophyllum: Chehalis; Cowlitz; Hahwunkwut; Hesquiat; Hoh; Hupa; Karok; Klallam; Konomeho; Maidu; Makah; Montana Indian; Nitinaht;

Okanagan-Colville; Poliklah; Quileute; Quinault; Wintoon
Yucca: Apache; Apache, Chiricahua & Mescalero; Apache, Mescalero; Cahuilla; Diegueño; Hopi; Isleta; Jemez; Kawaiisu; Keres, Western; Mahuna; Navajo; Navajo, Ramah; Papago; Pima; Shoshoni; Southwest Indians; Tewa; Zuni

Brushes & Brooms
Acacia: Havasupai
Agave: Havasupai; Papago; Pima; Yavapai
Arctostaphylos: Diegueño
Aristida: Hopi; Navajo, Ramah
Artemisia: Navajo, Ramah; Pawnee
Baccharis: Pima
Betula: Malecite
Bouteloua: Apache, Western; Apache, White Mountain; Tewa; Zuni
Brickellia: Keres, Western
Bromus: Isleta; Keres, Western
Calocedrus: Karok
Carex: Thompson
Cercocarpus: Keres, Western
Chamaecyparis: Karok
Chlorogalum: Cahuilla; Costanoan; Karok; Kawaiisu; Luiseño; Maidu; Pomo; Pomo, Kashaya; Round Valley Indian
Chrysothamnus: Keres, Western; Yavapai
Corylus: Meskwaki; Ojibwa; Potawatomi
Dalea: Keres, Western; Pawnee
Erigeron: Keres, Western
Eriogonum: Miwok; Yavapai
Fallugia: Jemez; Keres, Western; Keresan; Tewa
Gutierrezia: Comanche; Hualapai
Hilaria: Hopi; Tewa
Hymenoclea: Hualapai
Koeleria: Cheyenne; Jemez
Larrea: Papago
Leymus: Paiute
Muhlenbergia: Hopi; Navajo; Navajo, Ramah
Nolina: Isleta
Panicum: Tewa
Perideridia: Mendocino Indian; Pomo, Kashaya; Yuki
Philadelphus: Cowlitz; Lummi; Paiute; Thompson
Phleum: Navajo, Ramah
Picea: Iroquois; Tanana, Upper
Prunus: Dakota; Omaha; Pawnee; Ponca; Winnebago
Quercus: Cowlitz
Scirpus: Houma
Senecio: Hopi; Navajo
Serenoa: Seminole
Spiraea: Blackfoot; Thompson
Sporobolus: Apache, Chiricahua & Mescalero
Stipa: Omaha; Pawnee; Ponca
Symphoricarpos: Costanoan; Montana Indian; Okanagan-Colville; Yokia
Taxus: Cowlitz; Kwakiutl, Southern; Oweekeno; Quinault; Salish, Coast
Thuja: Kwakiutl, Southern; Okanagan-Colville;

Salish, Coast
Tsuga: Thompson
Vaccinium: Karok; Poliklah; Tolowa; Yurok
Yucca: Cahuilla; Hopi; Isleta; Keres, Western; Navajo, Ramah; Pima; Southwest Indians; Tewa; Zuni

Building Material
Abies: Carrier; Cree, Woodlands; Mewuk; Thompson
Acacia: Cahuilla; Pima
Acer: Cahuilla; Cherokee; Nisqually; Okanagan-Colville; Quinault
Adenostoma: Cahuilla; Coahuilla
Aesculus: Cherokee
Agave: Papago
Amelanchier: Pomo, Kashaya
Andropogon: Omaha; Ponca
Apocynum: Havasupai
Arbutus: Mendocino Indian
Arctostaphylos: Cahuilla
Artemisia: Cahuilla; Havasupai; Navajo; Paiute; Pomo; Tanana, Upper
Arundinaria: Cherokee
Aulacomnium: Shuswap
Baccharis: Cahuilla; Havasupai; Hualapai
Baileya: Jemez
Betula: Algonquin, Tête-de-Boule; Blackfoot; Cherokee; Chippewa; Cree, Woodlands; Meskwaki; Micmac; Ojibwa; Potawatomi; Tanana, Upper; Thompson
Calocedrus: Cahuilla; Karok; Paiute
Carnegia: Papago
Carpinus: Chippewa
Cassiope: Eskimo, Alaska
Castanea: Cherokee
Cercis: Havasupai
Cercocarpus: Navajo
Chamaecyparis: Karok; Kitasoo
Chilopsis: Cahuilla
Chrysothamnus: Hopi; Navajo
Cladrastis: Cherokee
Cornus: Ojibwa; Okanagan-Colville
Crataegus: Okanagan-Colville
Dicranum: Shuswap
Epilobium: Mewuk
Ericameria: Cahuilla
Fagus: Cherokee
Fouquieria: Cahuilla; Hualapai; Papago; Pima; Seri
Fraxinus: Cherokee; Cheyenne; Chippewa; Havasupai; Omaha; Ponca
Fremontodendron: Kawaiisu
Gleditsia: Cherokee
Gutierrezia: Kawaiisu
Halesia: Cherokee
Helianthus: Jemez
Isocoma: Cahuilla
Juglans: Apache, Mescalero; Cherokee
Juncus: Isleta
Juniperus: Apache, Mescalero; Cherokee; Gosiute; Havasupai; Hopi; Jemez; Kawaiisu; Keres, Western; Montana Indian; Navajo; Navajo, Ramah; Ojibwa; Paiute; Tewa; Tewa of Hano

Larrea: Papago
Leymus: Cahuilla
Libocedrus: Mewuk
Liriodendron: Cherokee
Lonicera: Thompson
Lotus: Cahuilla; Costanoan; Kawaiisu
Lycopodium: Shuswap
Magnolia: Cherokee
Nolina: Apache, Western; Havasupai; Hualapai; Yavapai
Olneya: Papago
Ostrya: Chippewa
Pachycereus: Papago & Pima
Parthenocissus: Navajo
Phragmites: Hopi
Picea: Cherokee; Cree, Woodlands; Eskimo, Alaska; Eskimo, Inuktitut; Hesquiat; Hoh; Kitasoo; Koyukon; Malecite; Quileute; Tanana, Upper; Thompson
Pinus: Blackfoot; Cahuilla; Cherokee; Cheyenne; Dakota; Diegueño; Havasupai; Hopi; Isleta; Jemez; Karok; Kawaiisu; Keres, Western; Mendocino Indian; Mewuk; Miwok; Montana Indian; Navajo; Navajo, Ramah; Ojibwa; Okanagan-Colville; Paiute; Seminole; Thompson; Tsimshian; Wintoon
Platanus: Cahuilla; Cherokee
Pluchea: Cahuilla; Havasupai; Hualapai; Luiseño
Populus: Blackfoot; Cheyenne; Cree, Woodlands; Crow; Havasupai; Hopi; Isleta; Kiowa; Lakota; Montana Indian; Navajo; Navajo, Ramah; Omaha; Paiute; Pima; Quinault; Squaxin
Prosopis: Cahuilla; Papago; Pima
Prunus: Cherokee; Thompson
Pseudotsuga: Hoh; Karok; Montana Indian; Okanagan-Colville; Quileute; Thompson; Yuki
Purshia: Hualapai
Quercus: Apache, Mescalero; Cherokee; Comanche; Kawaiisu; Navajo, Ramah; Ojibwa
Robinia: Cherokee
Sabal: Seminole
Salix: Apache, White Mountain; Blackfoot; Cheyenne; Costanoan; Dakota; Havasupai; Hopi; Hualapai; Isleta; Jemez; Kawaiisu; Kiowa; Lakota; Navajo, Ramah; Omaha; Paiute; Pawnee; Pomo, Kashaya; Round Valley Indian; Tanana, Upper; Thompson; Winnebago; Zuni
Salvia: Mahuna
Sarcobatus: Hopi
Scirpus: Cahuilla; Costanoan; Costanoan (Olhonean); Kawaiisu; Menominee; Okanagan-Colville; Paiute, Northern; Pomo; Shuswap; Thompson
Sequoia: Mendocino Indian; Pomo, Kashaya; Tolowa; Yurok
Solanum: Apache, White Mountain
Spartina: Omaha; Ponca
Sporobolus: Zuni
Tamarix: Keres, Western
Taxodium: Seminole
Thuja: Clallam; Gitksan; Haisla & Hanaksiala; Hesquiat; Hoh; Kwakiutl; Kwakiutl, Southern; Montana Indian; Nez Perce; Nitinaht; Okanagan-Colville; Oweekeno; Quileute; Salish, Coast; Snuqualmie; Tsimshian; Wet'suwet'en
Tilia: Cherokee
Tsuga: Cherokee; Chippewa; Haisla & Hanaksiala; Kwakwaka'wakw; Makah; Oweekeno; Quinault
Typha: Apache, White Mountain; Chippewa; Havasupai; Isleta; Kawaiisu; Meskwaki; Okanagan-Colville; Paiute; Paiute, Northern; Pima; Potawatomi; Thompson
Ulmus: Chippewa; Dakota; Lakota; Meskwaki; Ojibwa; Omaha; Pawnee; Ponca; Potawatomi; Winnebago
Umbellularia: Mewuk; Yokut
Vitis: Seminole
Washingtonia: Cahuilla
Yucca: Kawaiisu; Navajo; Papago
Zea: Navajo

Canoe Material
Abies: Cree, Woodlands; Ojibwa; Thompson
Acacia: Hawaiian
Acer: Clallam; Klallam; Malecite; Nitinaht; Skagit; Snohomish
Alnus: Hesquiat; Nitinaht; Oweekeno; Quileute
Betula: Abnaki; Algonquin, Tête-de-Boule; Bella Coola; Carrier; Chippewa; Cree, Woodlands; Koyukon; Malecite; Meskwaki; Micmac; Montana Indian; Ojibwa; Okanagan-Colville; Potawatomi; Tanana, Upper; Thompson
Bobea: Hawaiian
Chamaecyparis: Haisla & Hanaksiala; Hesquiat; Hoh; Kitasoo; Kwakiutl, Southern; Nitinaht; Oweekeno; Quileute
Fraxinus: Cowlitz; Malecite; Ojibwa
Holodiscus: Squaxin
Larix: Ojibwa
Liriodendron: Cherokee
Menziesia: Quileute
Parmelia: Eskimo, Inuktitut
Peltigera: Eskimo, Inuktitut
Phragmites: Seri
Picea: Algonquin, Tête-de-Boule; Chippewa; Cree, Woodlands; Eskimo, Alaska; Eskimo, Inuktitut; Koyukon; Malecite; Ojibwa; Tanana, Upper; Thompson
Pinus: Cherokee; Klamath; Montana Indian; Okanagan-Colville; Skagit; Thompson, Upper (Fraser Band); Thompson, Upper (Lytton Band)
Populus: Carrier; Isleta; Nisga; Okanagan-Colville; Salish, Coast; Shuswap; Thompson; Thompson, Upper (Fraser Band)
Pseudotsuga: Klamath
Salix: Blackfoot; Thompson
Scirpus: Costanoan; Pomo
Sequoia: Mendocino Indian; Tolowa; Yurok
Sorbus: Ojibwa
Stereocaulon: Eskimo, Inuktitut
Taxodium: Seminole
Taxus: Aleut; Clallam; Haisla; Hesquiat; Hoh; Nitinaht; Oweekeno; Quileute; Quinault; Salish, Coast
Thuja: Algonquin, Tête-de-Boule; Chippewa; Clallam; Haisla & Hanaksiala; Hesquiat; Hoh; Kutenai; Kwakiutl; Kwakiutl, Southern; Makah; Malecite; Micmac; Montana Indian; Nez Perce; Nitinaht; Ojibwa; Okanagan-Colville; Oweekeno; Quileute; Salish, Coast; Thompson; Tsimshian
Tilia: Chippewa
Tsuga: Quinault; Skagit; Snohomish
Typha: Paiute, Northern

Caulking Material
Alnus: Cree, Woodlands
Calamagrostis: Okanagan-Colville
Carnegia: Seri
Chlorogalum: Mewuk
Pachycereus: Seri
Picea: Chippewa; Cree, Woodlands; Koyukon; Nitinaht; Ojibwa; Quileute; Quinault
Pinus: Algonquin, Tête-de-Boule; Havasupai; Ojibwa; Potawatomi
Pseudotsuga: Salish, Coast; Thompson
Quercus: Seminole
Sideroxylon: Seminole
Sphagnum: Eskimo, Inuktitut
Stenocereus: Seri
Thuja: Iroquois
Typha: Menominee; Meskwaki

Clothing
Abies: Hesquiat
Acer: Concow; Tolowa
Adenostoma: Cahuilla
Agave: Cahuilla; Hualapai
Alectoria: Nitinaht
Apocynum: Cherokee; Havasupai; Luiseño; Mendocino Indian; Okanogan; Thompson
Arbutus: Tolowa
Arctium: Chippewa
Artemisia: Kawaiisu; Paiute; Thompson
Asclepias: Cherokee; Luiseño; Pomo, Kashaya; Thompson; Zuni
Betula: Tanana, Upper
Calamagrostis: Okanagan-Colville; Thompson
Carex: Blackfoot; Hesquiat; Thompson; Wailaki
Celtis: Papago
Chamaecyparis: Bella Coola; Haisla; Haisla & Hanaksiala; Hesquiat; Kitasoo; Kwakiutl, Southern; Nitinaht; Oweekeno
Chilopsis: Cahuilla
Cirsium: Nitinaht
Clematis: Thompson
Cowania: Southwest Indians
Dasylirion: Papago
Elaeagnus: Okanagan-Colville; Thompson
Epilobium: Clallam
Eriodictyon: Costanoan
Eriophorum: Eskimo, Inuktitut
Evernia: Montana Indian
Gossypium: Isleta; Zuni

Index of Usages 855

Juncus: Costanoan; Neeshenam
Juniperus: Keres, Western; Navajo; Navajo, Ramah; Paiute
Leymus: Kwakiutl, Southern
Lycopodium: Hanaksiala
Muhlenbergia: Luiseño
Pandanus: Hawaiian
Panicum: Cherokee
Penstemon: Okanagan-Colville
Picea: Kwakiutl, Southern; Quileute
Poa: Eskimo, Alaska; Eskimo, Inuktitut
Poaceae sp.: Tanana, Upper
Polystichum: Hesquiat; Oweekeno
Populus: Carrier; Klamath
Prosopis: Cahuilla
Psoralidium: Dakota
Purshia: Havasupai; Hopi; Hualapai; Navajo; Navajo, Ramah; Paiute
Ramalina: Pomo, Kashaya
Rhus: Navajo, Ramah
Salix: Okanagan-Colville; Thompson
Schizachyrium: Lakota
Scirpus: Hesquiat; Klamath; Kwakiutl, Southern; Paiute, Northern; Pomo; Thompson
Sphagnum: Carrier; Chinook; Haisla; Nitinaht
Thuja: Bella Coola; Chehalis; Clallam; Haisla & Hanaksiala; Hesquiat; Hoh; Kwakiutl; Kwakiutl, Southern; Makah; Montana Indian; Nitinaht; Okanagon; Oweekeno; Quileute; Quinault; Salish, Coast; Thompson; Tsimshian
Tilia: Malecite; Meskwaki
Typha: Blackfoot; Chehalis; Cowlitz; Klallam; Lakota; Makah; Nimpkish; Nitinaht; Okanagan-Colville; Paiute, Northern; Pomo; Quinault; Snohomish; Squaxin; Thompson; Tolowa; Yurok
Urtica: Dakota; Luiseño; Pawnee; Ponca; Winnebago
Usnea: Nitinaht
Xerophyllum: Karok; Yana; Yurok
Yucca: Apache, White Mountain; Cahuilla; Diegueño; Hualapai; Navajo; Southwest Indians; Tewa; Zuni

Cordage
Acer: Cowlitz; Thompson
Agave: Cahuilla; Hualapai; Navajo; Papago; Pima
Amelanchier: Okanagan-Colville
Apocynum: Bella Coola; California Indian; Cherokee; Chippewa; Great Basin Indian; Havasupai; Kutenai; Luiseño; Mendocino Indian; Menominee; Meskwaki; Montana Indian; Nez Perce; Okanagan-Colville; Okanagon; Paiute, Northern; Shoshoni; Shuswap; Thompson; Wintoon
Argentina: Blackfoot
Artemisia: Paiute
Asclepias: Cahuilla; Chippewa; Concow; Costanoan; Kawaiisu; Luiseño; Menominee; Meskwaki; Modesse; Navajo, Ramah; Neeshenam; Okanagan-Colville; Pomo, Kashaya; Tewa; Thompson; Wintoon; Zuni
Asimina: Cherokee

Carex: Mendocino Indian; Pomo; Pomo, Kashaya; Salish, Coast
Cercis: Mendocino Indian
Chamaecyparis: Kitasoo; Nitinaht; Oweekeno
Chilopsis: Cahuilla
Clematis: Great Basin Indian
Convolvulus: Okanagan-Colville
Cornus: Okanagan-Colville
Corylus: Chehalis; Karok; Skokomish
Cyperus: Hawaiian
Dirca: Iroquois; Menominee; Potawatomi
Elaeagnus: Blackfoot; Cree; Okanagan-Colville; Thompson
Epilobium: Haisla; Haisla & Hanaksiala; Hanaksiala; Heiltzuk; Oweekeno
Fremontodendron: Kawaiisu; Shoshoni; Yokut
Geranium: Jemez
Gossypium: Navajo; Zuni
Hoita: California Indian; Mendocino Indian
Iris: Karok; Tolowa; Wintoon
Juncus: Cherokee; Costanoan; Hesquiat; Mendocino Indian; Snuqualmie
Laportea: Chippewa; Meskwaki
Leymus: Eskimo, Alaska
Linum: Great Basin Indian; Klamath; Montana Indian
Lonicera: Thompson
Lupinus: Pomo; Pomo, Kashaya
Nereocystis: Hesquiat; Makah; Nitinaht; Pomo, Kashaya
Nolina: Isleta; Southwest Indians
Picea: Cree, Woodlands; Eskimo, Inuktitut; Hoh; Kwakiutl, Southern; Quileute; Tanana, Upper
Populus: Montana Indian; Nitinaht; Squaxin
Potamogeton: Kawaiisu
Prosopis: Cahuilla; Seri
Prunus: Thompson
Psoralidium: Great Basin Indian
Ribes: Cowichan; Saanich
Salix: California Indian; Chehalis; Clallam; Costanoan; Cree, Woodlands; Klallam; Mendocino Indian; Montana Indian; Navajo; Okanagan-Colville; Paiute; Quinault; Snohomish; Tanana, Upper; Thompson; Wet'suwet'en
Serenoa: Seminole
Taxodium: Choctaw
Thuja: Clallam; Gitksan; Haisla; Haisla & Hanaksiala; Hesquiat; Hoh; Kwakiutl; Kwakiutl, Southern; Montana Indian; Nitinaht; Okanagon; Oweekeno; Quileute; Salish, Coast; Thompson; Wet'suwet'en
Tilia: Cherokee; Chippewa; Lakota; Malecite; Menominee; Meskwaki; Ojibwa; Omaha; Ponca; Potawatomi
Tillandsia: Houma
Touchardia: Hawaiian
Ulmus: Dakota; Omaha; Pawnee; Ponca; Winnebago
Urtica: Bella Coola; Cahuilla; Dakota; Eskimo, Inuktitut; Haisla & Hanaksiala; Hesquiat; Hoh;

Kawaiisu; Klamath; Kwakiutl, Southern; Lakota; Makah; Montana Indian; Nitinaht; Omaha; Oweekeno; Pawnee; Ponca; Potawatomi; Quileute; Skagit, Upper; Thompson; Winnebago
Vicia: Yuki
Vitis: Karok; Pomo; Pomo, Kashaya; Seminole
Yucca: Apache; Apache, Mescalero; Apache, Western; Apache, White Mountain; Cahuilla; Dakota; Havasupai; Hualapai; Isleta; Keres, Western; Navajo; Navajo, Ramah; Omaha; Papago; Pawnee; Pima; Ponca; Southwest Indians; Tewa; Yavapai; Zuni

Furniture
Abies: Hanaksiala
Acacia: Hualapai; Pima; Walapai
Acer: Cherokee; Lummi; Skagit; Swinomish; Thompson
Aesculus: Cherokee
Alnus: Makah
Amelanchier: Gosiute; Havasupai
Betula: Okanagan-Colville; Tanana, Upper; Thompson
Carya: Cherokee; Iroquois
Chamaecyparis: Karok; Kwakiutl, Southern
Chilopsis: Hualapai
Cornus: Paiute; Pomo, Kashaya; Thompson
Dasylirion: Apache, Mescalero
Fallugia: Havasupai; Hualapai
Fraxinus: Havasupai; Iroquois; Ojibwa
Fremontodendron: Kawaiisu
Holodiscus: Okanagan-Colville; Pomo, Kashaya
Juglans: Cherokee
Juniperus: Cherokee; Navajo, Ramah; Ojibwa
Liriodendron: Cherokee
Magnolia: Cherokee
Pachycereus: Papago & Pima
Philadelphus: Mendocino Indian; Okanagan-Colville; Thompson
Pinus: Apache, Mescalero; Blackfoot; Navajo; Navajo, Ramah; Seminole
Pluchea: Havasupai; Hualapai
Populus: Navajo
Prosopis: Havasupai; Hualapai; Pima
Prunus: Blackfoot; Cherokee
Quercus: Apache, Mescalero; Cherokee; Diegueño; Navajo; Navajo, Ramah
Rhus: Hopi
Robinia: Hualapai
Rosa: Thompson
Salix: Blackfoot; Cahuilla; Cheyenne; Kawaiisu; Montana Indian; Navajo; Paiute; Thompson
Sarcobatus: Hopi
Sassafras: Cherokee
Symphoricarpos: Havasupai; Paiute
Thuja: Haisla & Hanaksiala; Nez Perce; Salish, Coast; Thompson
Tilia: Cherokee
Typha: Lakota
Vitis: Pomo, Kashaya
Yucca: Apache, Western; Diegueño

Mats, Rugs & Bedding

Abies: Algonquin, Tête-de-Boule; Malecite; Micmac; Okanagan-Colville; Paiute; Potawatomi; Shuswap; Thompson
Acer: Gitksan; Karok
Agave: Hualapai; Navajo
Alectoria: Hanaksiala
Alnus: Tanana, Upper
Amorpha: Kiowa
Apocynum: Paiute, Northern
Artemisia: Blackfoot; Dakota; Kiowa; Okanagan-Colville; Omaha; Pawnee; Ponca; Thompson; Winnebago
Asarum: Okanagan; Thompson
Beckmannia: Okanagan-Colville
Betula: Eskimo, Inuktitut
Blechnum: Yurok
Bromus: Paiute
Calamagrostis: Cree, Woodlands
Carex: Klamath; Montana Indian
Chamaecyparis: Bella Coola; Haisla & Hanaksiala; Hesquiat; Karok; Kitasoo; Kwakiutl, Southern; Nitinaht; Oweekeno
Chlorogalum: Mendocino Indian
Chrysothamnus: Navajo, Ramah
Cibotium: Hawaiian
Dasylirion: Papago; Pima
Dryopteris: Eskimo, Inuktitut; Iroquois
Elaeagnus: Okanagan-Colville; Thompson
Eleocharis: Okanagan-Colville
Ephedra: Havasupai
Epilobium: Quinault; Skokomish
Equisetum: Koyukon; Missouri River Indian
Eriophorum: Eskimo, Alaska; Eskimo, Inuktitut
Evernia: Montana Indian
Hierochloe: Kiowa; Micmac
Hylocomium: Bella Coola
Iris: Potawatomi
Juncus: Chippewa; Costanoan; Klamath; Montana Indian; Ojibwa
Juniperus: Blackfoot; Chippewa; Navajo; Ojibwa
Leymus: Blackfoot; Eskimo, Alaska; Okanagan-Colville
Linum: Klamath; Montana Indian
Lomatium: Thompson
Lupinus: Okanagan-Colville
Matricaria: Crow
Monarda: Cheyenne
Nolina: Apache, Mescalero; Havasupai; Keres, Western; Southwest Indians
Onoclea: Iroquois
Osmunda: Iroquois
Pandanus: Hawaiian
Pentaphylloides: Blackfoot
Petasites: Eskimo, Alaska
Phragmites: Havasupai; Okanagon; Pima; Thompson
Picea: Carrier; Cree, Woodlands; Eskimo, Inuktitut; Kwakiutl, Southern; Malecite; Micmac; Paiute; Tanana, Upper
Pinus: Miwok; Ojibwa
Poa: Eskimo, Inuktitut
Poaceae sp.: Tanana, Upper
Polystichum: Cowlitz; Hesquiat; Nitinaht; Oweekeno; Quileute; Salish, Coast; Yurok
Populus: Hanaksiala; Thompson
Prosopis: Seri
Prunus: Okanagon; Thompson
Pseudotsuga: Paiute; Thompson
Pteridium: Nitinaht; Shuswap; Squaxin
Purshia: Havasupai; Hopi; Navajo; Navajo, Ramah
Quercus: Mahuna
Rhytidiadelphus: Bella Coola
Rosa: Thompson
Salix: Cheyenne; Keres, Western; Mandan; Montana Indian; Nitinaht; Pima; Thompson
Scirpus: Cahuilla; Cheyenne; Chippewa; Dakota; Hesquiat; Karok; Kawaiisu; Klallam; Klamath; Lakota; Makah; Menominee; Meskwaki; Montana Indian; Nitinaht; Ojibwa; Omaha; Paiute; Paiute, Northern; Pawnee; Pomo; Ponca; Potawatomi; Salish, Coast; Snohomish; Thompson
Sphagnum: Bella Coola; Chinook; Ojibwa; Potawatomi
Sporobolus: Zuni
Thuja: Bella Coola; Chehalis; Clallam; Gitksan; Haisla; Haisla & Hanaksiala; Hesquiat; Hoh; Kwakiutl, Southern; Makah; Nitinaht; Okanagan-Colville; Okanagon; Oweekeno; Quileute; Salish, Coast; Thompson; Tsimshian; Wet'suwet'en
Tilia: Menominee; Meskwaki; Pawnee
Tillandsia: Houma
Tragopogon: Thompson
Tsuga: Hesquiat; Paiute; Thompson
Typha: Algonquin, Tête-de-Boule; Apache, Mescalero; Cahuilla; Chehalis; Chippewa; Clallam; Cowlitz; Dakota; Hesquiat; Hopi; Iroquois; Klallam; Klamath; Kwakiutl, Southern; Lakota; Makah; Mendocino Indian; Menominee; Meskwaki; Montana Indian; Navajo, Ramah; Nimpkish; Ojibwa; Okanagon; Paiute, Northern; Pima; Potawatomi; Quinault; Salish, Coast; Shuswap; Snohomish; Squaxin; Thompson; Yurok
Ulmus: Menominee
Usnea: Hanaksiala; Makah
Veratrum: Hanaksiala
Vernonia: Kiowa
Yucca: Navajo; Pima; Zuni

Other

Acacia: Papago
Dicranum: Chippewa
Eleocharis: Cheyenne
Elymus: Aleut
Juncus: Cheyenne
Lathyrus: Aleut
Prosopis: Papago
Salix: Papago
Sphagnum: Chippewa
Thuja: Hoh; Kwakiutl, Southern; Quileute

Scouring Material

Acer: Thompson
Agave: Cahuilla
Cuscuta: Cahuilla
Distichlis: Cahuilla
Equisetum: Bella Coola; Cahuilla; Chippewa; Cowlitz; Karok; Klamath; Kwakiutl, Southern; Mendocino Indian; Menominee; Meskwaki; Modesse; Montana Indian; Ojibwa; Okanagan-Colville; Okanagon; Omaha; Pawnee; Ponca; Quinault; Skokomish; Swinomish; Thompson; Yuki
Festuca: Navajo, Ramah
Koeleria: Navajo, Ramah
Pinus: Thompson
Populus: Thompson
Sporobolus: Navajo, Ramah
Thuja: Bella Coola; Hesquiat; Thompson

Sewing Material

Abies: Malecite
Acer: Maidu
Agave: Apache, Western
Anthoxanthum: Potawatomi
Apocynum: Kutenai; Mendocino Indian; Menominee; Meskwaki; Ojibwa; Okanagon; Potawatomi; Thompson
Asclepias: Menominee; Potawatomi
Betula: Carrier
Carex: Pomo, Kashaya
Cercis: Yuki
Chilopsis: Hualapai
Epilobium: Cree, Woodlands
Gossypium: Navajo
Hierochloe: Menominee
Hoita: California Indian; Concow; Yokia
Holodiscus: Nitinaht; Salish, Cowichan
Laportea: Ojibwa
Larix: Ojibwa
Leymus: Nitinaht
Lonicera: Thompson
Muhlenbergia: Shoshoni
Opuntia: Keres, Western
Picea: Cree, Woodlands; Kwakiutl, Southern; Malecite; Micmac
Pinus: Menominee; Ojibwa; Potawatomi
Prosopis: Seri
Quercus: Tewa of Hano
Rhus: Navajo
Salix: Okanagan-Colville; Pima
Scirpus: Pomo
Taxus: Karok; Nitinaht; Pomo
Thuja: Kwakiutl, Southern; Okanagon; Quinault; Thompson
Tilia: Chippewa; Malecite; Ojibwa; Potawatomi
Typha: Snohomish
Urtica: Ojibwa; Omaha
Vitis: Pomo
Yucca: Dakota; Hopi; Omaha; Papago; Pawnee; Ponca

Snow Gear

Abies: Haisla & Hanaksiala

Acer: Haisla & Hanaksiala; Modesse; Okanagan-Colville; Shuswap; Thompson
Alnus: Cree, Woodlands; Thompson
Arbutus: Karok; Tolowa; Yurok
Betula: Anticosti; Carrier; Cree, Woodlands; Eskimo, Inuktitut; Koyukon; Malecite; Tanana, Upper
Carya: Omaha
Cornus: Gosiute
Corylus: Karok
Fagus: Micmac
Fraxinus: Abnaki; Chippewa; Malecite; Ojibwa
Larix: Cree, Woodlands
Linum: Klamath; Montana Indian
Oxydendrum: Cherokee
Philadelphus: Okanagan-Colville; Shuswap
Picea: Algonquin, Tête-de-Boule; Carrier; Koyukon; Tanana, Upper
Pinus: Navajo, Ramah
Pseudotsuga: Thompson
Salix: Klamath; Montana Indian; Tanana, Upper
Sorbus: Ojibwa
Taxus: Thompson
Thuja: Chippewa
Tilia: Menominee
Ulmus: Menominee
Urtica: Klamath
Yucca: Navajo, Ramah

Sporting Equipment
Alaria: Hesquiat
Costaria: Hesquiat
Crataegus: Cherokee
Fraxinus: Cherokee
Juncus: Pomo
Lessoniopsis: Hesquiat
Postelsia: Hesquiat
Salix: Montana Indian
Sorbus: Ojibwa
Ulmus: Cherokee
Yucca: Navajo, Ramah

Unspecified
Apocynum: Cahuilla; Modoc
Arbutus: Karok
Fomes: Cowlitz
Gossypium: Papago
Urtica: Luiseño
Woodwardia: Tolowa; Yurok

Dye
Black
Acer: Omaha; Winnebago
Alnus: Carrier; Chippewa; Kwakiutl, Southern
Atriplex: Navajo, Ramah
Carex: Shoshoni
Castilleja: Zuni
Cornus: Chippewa; Thompson
Corylus: Chippewa
Gaultheria: Karok; Navajo
Gymnocladus: Dakota
Helianthus: Hopi; Hualapai
Juglans: Cherokee; Chippewa; Dakota; Menominee; Meskwaki; Omaha; Pawnee; Pomo, Kashaya; Ponca; Winnebago
Leymus: Cheyenne
Lonicera: Thompson
Phoradendron: Cahuilla
Pinus: Navajo; Navajo, Ramah
Proboscidea: Shoshoni
Prosopis: Pima
Quercus: Chippewa; Concow; Diegueño; Mahuna; Navajo, Ramah; Omaha; Pomo, Kashaya
Rhus: Cherokee; Great Basin Indian; Navajo; Navajo, Ramah; Plains Indian
Salix: Micmac; Montagnais
Sambucus: Cahuilla
Suaeda: Cahuilla
Thalictrum: Navajo, Ramah
Toxicodendron: Pomo
Tsuga: Nitinaht
Yucca: Navajo, Ramah; Shoshoni

Blue
Atriplex: Hopi
Baptisia: Cherokee
Comandra: Arapaho
Corylus: Thompson
Delphinium: Blackfoot; Great Basin Indian; Navajo; Okanagan-Colville; Thompson
Fraxinus: Chippewa
Lithospermum: Great Basin Indian
Lupinus: Navajo
Penstemon: Okanagan-Colville
Pinus: Cheyenne
Rhus: Navajo
Solanum: Costanoan
Viola: Blackfoot

Blue-Black
Juglans: Kiowa

Brown
Abies: Klamath
Allium: Great Basin Indian
Alnus: Cree, Woodlands; Iroquois; Kwakiutl, Southern; Navajo; Nitinaht; Potawatomi; Thompson
Arctostaphylos: Great Basin Indian
Betula: Okanagan-Colville
Castanea: Cherokee
Cercocarpus: Navajo, Ramah
Cornus: Cree, Woodlands; Thompson
Ephedra: Navajo
Juglans: Cherokee; Chippewa; Menominee; Navajo; Ojibwa
Krameria: Pima
Ledum: Iroquois; Potawatomi
Mirabilis: Navajo
Myrica: Ojibwa
Physocarpus: Hesquiat
Pseudotsuga: Swinomish
Purshia: Navajo
Quercus: Cherokee; Mahuna; Navajo
Rumex: Navajo; Pima
Tsuga: Bella Coola; Nitinaht

Gray
Quercus: Mahuna
Rhus: Plains Indian

Green
Achillea: Great Basin Indian
Allium: Iroquois; Navajo
Artemisia: Great Basin Indian
Chrysothamnus: Hopi; Navajo
Frangula: Skagit
Iris: Navajo
Juglans: Cherokee
Juncus: Shoshoni
Juniperus: Keres, Western; Navajo
Lupinus: Navajo
Populus: Cheyenne
Prunus: Navajo
Rumex: Navajo
Taxus: Micmac; Montagnais
Thuja: Thompson
Urtica: Makah

Mordant
Acorus: Chippewa
Atriplex: Hopi; Tewa
Corylus: Ojibwa
Heuchera: Blackfoot
Juniperus: Great Basin Indian
Lycopodium: Blackfoot
Opuntia: Crow; Dakota; Pawnee; Sioux
Prunus: Chippewa; Ojibwa
Quercus: Ojibwa
Rhus: Dakota; Hopi; Navajo
Tsuga: Ojibwa

Orange
Alnus: Blackfoot; Eskimo, Alaska; Flathead; Great Basin Indian; Klamath; Kutenai; Kwakiutl, Southern; Montana Indian; Nez Perce; Paiute
Chrysothamnus: Navajo
Cuscuta: Pawnee
Impatiens: Potawatomi
Mahonia: Great Basin Indian
Oxalis: Meskwaki
Rhus: Ojibwa
Rosa: Arapaho
Rumex: Great Basin Indian; Navajo
Sambucus: Cahuilla
Sanguinaria: Ojibwa
Thelesperma: Navajo

Orange-Red
Alnus: Cree, Woodlands
Sanguinaria: Menominee

Orange-Yellow
Dahlia: Navajo
Impatiens: Menominee
Rhus: Kiowa
Sanguinaria: Iroquois
Thelesperma: Navajo, Ramah

Pink
Parthenocissus: Kiowa

Purple
Cerasus: Navajo

Helianthus: Hopi; Hualapai
Lithospermum: Montana Indian
Lonicera: Hesquiat
Mirabilis: Navajo
Populus: Cheyenne
Prunus: Navajo
Purshia: Great Basin Indian; Klamath; Montana Indian
Sambucus: Cahuilla
Vaccinium: Hesquiat
Vernonia: Kiowa

Red
Alnus: Bella Coola; Chippewa; Flathead; Hesquiat; Isleta; Jemez; Keres, Western; Kwakiutl, Southern; Navajo; Nitinaht; Oweekeno; Potawatomi; Tewa; Thompson; Wailaki; Wintoon
Amaranthus: Hopi
Asplenium: Hawaiian
Atriplex: Navajo, Ramah
Betula: Chippewa; Jemez; Ojibwa
Cercocarpus: Havasupai; Isleta; Jemez; Keres, Western; Keresan; Navajo, Ramah
Chenopodium: Potawatomi; Thompson
Coreopsis: Apache, White Mountain; Cherokee; Zuni
Cornus: Chippewa
Cycloloma: Hopi
Datisca: Costanoan
Dodecatheon: Okanagan-Colville
Equisetum: Blackfoot
Galium: Cree; Great Basin Indian; Micmac
Helianthus: Navajo
Juniperus: Navajo
Krameria: Papago
Lithospermum: Chippewa
Mirabilis: Navajo
Nyssa: Choctaw
Opuntia: Navajo
Orthocarpus: Blackfoot
Pinus: Jemez
Plagiobothrys: Diegueño; Mendocino Indian
Populus: Cheyenne
Prunus: Chippewa; Great Basin Indian; Isleta; Navajo
Quercus: Chippewa; Choctaw; Houma; Mahuna
Ranunculus: Menominee; Ojibwa
Rhus: Cherokee; Chippewa; Dakota
Rumex: Cheyenne; Navajo
Salix: Blackfoot; Potawatomi
Sanguinaria: Cherokee; Chippewa; Menominee; Meskwaki; Ojibwa; Omaha; Ponca; Winnebago
Shepherdia: Blackfoot
Sherardia: Cherokee
Tsuga: Menominee
Yucca: Navajo, Ramah; Shoshoni

Red-Brown
Alnus: Apache, White Mountain; Blackfoot; Cree, Woodlands; Eskimo, Inuktitut; Flathead; Kutenai; Lummi; Menominee; Navajo, Ramah; Nez Perce; Okanagan-Colville; Quileute; Quinault; Salish, Coast; Snohomish; Zuni

Castilleja: Great Basin Indian
Celtis: Navajo
Heuchera: Navajo
Juniperus: Chippewa
Orthocarpus: Blackfoot
Prunus: Great Basin Indian
Quercus: Potawatomi
Rhus: Great Basin Indian
Rumex: Pima
Thelesperma: Hopi
Tsuga: Cherokee; Chippewa; Clallam; Hesquiat; Klallam; Makah; Ojibwa

Red-Yellow
Malus: Navajo
Quercus: Ojibwa

Unspecified
Agave: Cahuilla
Alnus: Costanoan; Cree, Woodlands; Eskimo, Alaska; Karok; Mendocino Indian; Micmac; Shuswap; Tanana, Upper; Tolowa; Yurok
Amaranthus: Hopi
Argemone: Kiowa
Aruncus: Makah
Aster: Blackfoot; Keres, Western
Baptisia: Ojibwa
Castilleja: Apache, White Mountain; Blackfoot
Cleome: Isleta
Clintonia: Thompson
Conocarpus: Seminole
Cornus: Costanoan
Cucurbita: Cahuilla
Eriogonum: Kiowa
Hepatica: Potawatomi
Hydrastis: Cherokee
Hymenopappus: Hopi
Hymenoxys: Hopi
Ilex: Cherokee
Ipomopsis: Hopi
Iris: Eskimo, Alaska
Juglans: Hualapai
Juniperus: Navajo; Navajo, Ramah
Letharia: Karok; Yurok
Leymus: Cheyenne
Lithocarpus: Costanoan; Tolowa
Lonicera: Makah
Maclura: Pima
Mahonia: Cowlitz; Klallam; Makah; Skagit; Snohomish
Nolina: Navajo
Oplopanax: Hesquiat
Osmorhiza: Blackfoot
Pectis: Hopi
Penstemon: Thompson
Phytolacca: Mahuna
Pinus: Cheyenne; Hopi; Tewa
Populus: Cheyenne
Prosopis: Hualapai
Quercus: Navajo, Ramah; Ojibwa
Rhizophora: Seminole
Rhus: Hualapai; Rappahannock; Thompson

Rubus: Blackfoot; Luiseño; Thompson
Rumex: Cheyenne; Hopi; Houma; Hualapai
Sanguinaria: Omaha; Pónca; Potawatomi; Winnebago
Smilax: Seminole
Solanum: Cahuilla
Tsuga: Chehalis; Malecite; Micmac
Vaccinium: Cree, Woodlands
Yucca: Navajo, Ramah

White
Eriogonum: Lakota
Populus: Cheyenne
Quercus: Mahuna
Yucca: Mahuna

Yellow
Allium: Iroquois
Alnus: Algonquin, Tête-de-Boule; Chippewa; Cree, Woodlands; Ojibwa
Andropogon: Cherokee
Argemone: Lakota
Artemisia: Navajo
Arundo: Papago
Atriplex: Great Basin Indian; Navajo; Navajo, Ramah
Bazzania: Montana Indian
Carthamus: Hopi
Castilleja: Blackfoot
Chrysothamnus: Apache, White Mountain; Hopi; Navajo; Navajo, Ramah; Tewa; Zuni
Coptis: Chippewa; Ojibwa; Potawatomi
Cornus: Chippewa
Datisca: Costanoan; Karok; Wintoon
Dugaldia: Navajo
Evernia: Blackfoot; Montana Indian; Thompson
Gaultheria: Nitinaht
Gutierrezia: Navajo
Heracleum: Karok
Hoita: Cahuilla; Luiseño
Hymenoxys: Navajo
Impatiens: Ojibwa; Potawatomi
Juniperus: Great Basin Indian
Letharia: Cheyenne; Oweekeno
Maclura: Kiowa
Mahonia: Apache, Mescalero; Blackfoot; Chehalis; Havasupai; Hualapai; Karok; Makah; Montana Indian; Navajo; Nitinaht; Okanagan-Colville; Salish, Coast; Skagit; Snohomish; Thompson; Walapai; Wintoon; Yurok
Malus: Cherokee
Mirabilis: Navajo
Myrica: Ojibwa
Oenothera: Pomo, Kashaya
Orthocarpus: Great Basin Indian
Oxalis: Menominee
Palafoxia: Cahuilla
Petradoria: Navajo
Phlox: Blackfoot
Populus: Dakota; Missouri River Indian; Omaha; Pawnee; Ponca
Prunus: Chippewa; Navajo

Psilostrophe: Apache, White Mountain; Keres, Western; Zuni
Quercus: Mahuna
Ranunculus: Potawatomi
Rhus: Chippewa; Menominee; Meskwaki; Omaha; Plains Indian; Winnebago
Rudbeckia: Potawatomi
Rumex: Cheyenne; Choctaw; Navajo; Pima
Sambucus: Cahuilla
Sanguinaria: Chippewa; Ojibwa
Tagetes: Cherokee
Tetradymia: Navajo, Ramah
Thermopsis: Blackfoot
Usnea: Makah
Verbesina: Keres, Western
Xanthorhiza: Cherokee
Zinnia: Keres, Western

Yellow-Brown

Picea: Cree, Woodlands
Purshia: Navajo
Rumex: Navajo, Ramah

Other

Cash Crop

Abies: Shuswap
Acer: Chippewa
Acorus: Blackfoot
Acrostichum: Seminole
Agastache: Pima
Agave: Cocopa; Havasupai; Papago; Pima
Allium: Papago
Amelanchier: Blackfoot; Okanagon; Thompson
Amoreuxia: Pima, Desert
Apocynum: Okanagon
Arctostaphylos: Makah; Nuxalkmc; Oweekeno
Arundinaria: Keres, Western
Balsamorhiza: Thompson
Bouteloua: Apache, White Mountain
Camassia: Kutenai; Nez Perce; Pend d'Oreille, Upper; Salish, Halkomelem; Salish, Straits; Shoshoni
Capsicum: Pima, Desert
Chrysothamnus: Navajo
Corylus: Okanagan-Colville; Okanagon; Salish, Coast; Thompson
Erythronium: Thompson
Fomes: Eskimo, Inuktitut
Juniperus: Ojibwa
Lewisia: Okanagan-Colville; Okanagon; Thompson; Thompson, Upper (Lytton Band)
Leymus: Eskimo, Alaska
Lomatium: Thompson, Upper (Lytton Band)
Macrocystis: Heiltzuk; Kitkatla
Malus: Kwakiutl, Southern
Muhlenbergia: Apache, White Mountain
Nicotiana: Apalachee; Shoshoni
Opuntia: Diegueño; Pima
Oryzopsis: Apache, White Mountain
Panax: Cherokee; Chippewa
Paxistima: Saanich
Phaseolus: Diegueño; Kiliwa; Navajo, Ramah; Paipai
Pinus: Cahuilla; Havasupai; Hualapai; Jemez; Keres, Western; Navajo; Navajo, Ramah; Paiute, Northern; Pueblo; Spanish American
Populus: Isleta
Porphyra: Nitinaht; Tsimshian
Prosopis: Pima, Gila River
Pseudotsuga: Navajo
Quercus: Cahuilla; Cocopa; Miwok; Pima, Desert
Ribes: Okanagan-Colville
Sequoia: Yurok
Serenoa: Seminole
Shepherdia: Gitksan; Nisga; Nuxalkmc; Thompson; Tsimshian; Tsimshian, Coast
Solanum: Okanagan-Colville
Sporobolus: Apache, White Mountain
Taxus: Pit River
Thuja: Shuswap; Thompson
Triticum: Cahuilla
Urtica: Haisla & Hanaksiala
Vaccinium: Algonquin, Quebec; Ojibwa
Xerophyllum: Makah
Yucca: Papago; Pima, Desert
Zea: Tewa
Zizania: Ojibwa; Thompson

Ceremonial Items

Abies: Blackfoot; Cheyenne; Haisla; Nez Perce
Abronia: Keres, Western
Acer: Cheyenne; Dakota; Haisla; Haisla & Hanaksiala; Hanaksiala; Keres, Western; Kiowa; Kwakiutl, Southern; Nitinaht; Omaha; Sioux
Achillea: Ojibwa
Aesculus: Cherokee
Agave: Cahuilla; Havasupai
Alnus: Haisla & Hanaksiala; Kwakiutl, Southern; Nitinaht; Oweekeno
Amaranthus: Navajo; Zuni
Ambrosia: Navajo, Ramah
Amelanchier: Blackfoot; Hopi; Kawaiisu; Navajo, Ramah
Anaphalis: Anticosti; Cheyenne
Angelica: Blackfoot; Eskimo, Alaska; Karok
Anthoxanthum: Ojibwa
Aquilegia: Pomo, Kashaya
Arabis: Navajo
Arctostaphylos: Blackfoot; Hoh; Quileute
Argemone: Hopi
Aristida: Hopi
Artemisia: Apache, Chiricahua & Mescalero; Arapaho; Blackfoot; Cheyenne; Crow; Eskimo, Western; Great Basin Indian; Hopi; Kawaiisu; Kutenai; Lakota; Luiseño; Navajo; Navajo, Ramah; Paiute; Tewa; Washo
Arundo: Navajo
Asclepias: Zuni
Aster: Navajo
Astragalus: Navajo; Navajo, Kayenta; Omaha; Ponca
Atriplex: Hopi; Tewa; Zuni
Bambusa: Navajo
Betula: Ojibwa
Bouteloua: Apache, Chiricahua & Mescalero; Navajo
Brickellia: Navajo
Broussonetia: Apache, Chiricahua & Mescalero
Calamovilfa: Hopi
Calochortus: Hopi
Campanula: Navajo
Cardamine: Iroquois
Carex: Blackfoot; Jemez
Carnegia: Apache, Western
Castilleja: Hopi; Keres, Western; Tewa
Ceanothus: Pomo, Kashaya
Cercocarpus: Hopi; Navajo; Navajo, Ramah; Tewa
Chenopodium: Navajo
Chrysothamnus: Hopi; Navajo; Navajo, Ramah
Cissus: Seminole
Clematis: Blackfoot
Cleome: Zuni
Cocos: Seminole
Cornus: Blackfoot; Cheyenne; Hoh; Montana Indian; Quileute
Cucumis: Hopi
Cucurbita: Apache, White Mountain; Cahuilla; Keres, Western; Zuni
Cymopterus: Navajo, Kayenta
Dalea: Navajo
Dasylirion: Apache, Mescalero
Datura: Cahuilla; Diegueño; Luiseño; Zuni
Delphinium: Hopi; Kiowa; Navajo; Navajo, Ramah; Shuswap
Digitalis: Hoh; Quileute
Dimorphocarpa: Navajo, Kayenta
Dryopteris: Oweekeno
Elaeagnus: Thompson
Equisetum: Hopi; Thompson
Eriogonum: Navajo; Zuni
Erodium: Navajo, Kayenta
Erysimum: Zuni
Eupatorium: Iroquois
Fendlera: Hopi; Navajo; Navajo, Kayenta
Fomes: Haisla & Hanaksiala
Fomitopsis: Haisla & Hanaksiala
Forestiera: Hopi; Jemez; Navajo; Navajo, Ramah
Fouquieria: Papago
Fragaria: Iroquois; Pomo, Kashaya
Fraxinus: Hopi; Navajo, Kayenta; Omaha; Ponca; Sioux
Fritillaria: Hanaksiala
Fucus: Nitinaht
Ganoderma: Haisla & Hanaksiala
Geranium: Navajo, Kayenta
Gossypium: Santa Clara; Tewa of Hano; Zuni
Gutierrezia: Apache, Chiricahua & Mescalero; Hopi; Hualapai; Navajo; Navajo, Kayenta; Navajo, Ramah; Tewa of Hano
Helianthus: Hopi; Isleta; Navajo; Navajo, Ramah; Zuni
Heracleum: Blackfoot; Omaha
Hierochloe: Blackfoot; Cheyenne; Chippewa; Dakota; Lakota; Montana Indian; Omaha; Sioux

Hilaria: Hopi
Hymenoxys: Hopi
Ipomoea: Keresan
Ipomopsis: Navajo
Iris: Pomo, Kashaya
Jacquinia: Seri
Juncus: Hopi
Juniperus: Blackfoot; Gitksan; Haisla; Heiltzuk; Hoh; Hopi; Jemez; Keres, Western; Navajo; Ojibwa; Okanagan-Colville; Oweekeno; Quileute; Tewa
Krascheninnikovia: Hopi; Navajo; Tewa
Lagenaria: Cherokee; Dakota; Hopi; Iroquois; Navajo; Navajo, Ramah; Ojibwa; Omaha; Pima; Ponca
Larix: Kutenai
Lesquerella: Navajo, Ramah
Leymus: Cahuilla; Cheyenne; Thompson
Lithocarpus: Pomo, Kashaya
Lithospermum: Blackfoot; Navajo
Lobelia: Jemez; Meskwaki
Lomatium: Cowichan; Crow; Saanich; Songish; Yurok
Lophophora: Blackfoot; Ponca
Lupinus: Blackfoot; Hopi; Navajo; Pomo, Kashaya
Lycium: Hopi; Navajo, Ramah
Lycoperdon: Blackfoot
Maclura: Kiowa
Mahonia: Hopi; Navajo, Kayenta; Zuni
Martynia: Santa Clara; Tewa of Hano
Mentzelia: Navajo; Navajo, Ramah
Muhlenbergia: Zuni
Nelumbo: Dakota; Omaha; Pawnee; Ponca; Winnebago
Nereocystis: Kwakiutl, Southern
Nicotiana: Blackfoot; Cahuilla; Hopi; Iroquois; Navajo; Thompson; Zuni
Nolina: Isleta
Oenothera: Hopi; Zuni
Oplopanax: Gitksan; Haisla; Haisla & Hanaksiala; Nitinaht; Tsimshian; Wet'suwet'en
Opuntia: Navajo, Ramah; Zuni
Orthocarpus: Pomo, Kashaya
Oxytropis: Navajo
Parryella: Hopi; Tewa
Parthenocissus: Jemez
Pectis: Zuni
Pedicularis: Navajo, Ramah
Penstemon: Hopi; Tewa; Zuni
Pentaphylloides: Jemez
Persea: Seminole
Phalaris: Okanagan-Colville
Phaseolus: Iroquois; Keresan
Phoradendron: Navajo
Phragmites: Hopi; Navajo
Picea: Eskimo, Inuktitut; Hanaksiala; Hesquiat; Koyukon; Navajo; Nitinaht; Tsimshian
Pinus: Apache, Mescalero; Hopi; Kawaiisu; Keres, Western; Navajo; Navajo, Ramah; Nitinaht; Okanagan-Colville; Paiute; Seminole; Tewa; Tewa of Hano
Plantago: Hopi; Kiowa; Navajo, Ramah

Polanisia: Isleta; Zuni
Polyporus: Blackfoot; Haisla & Hanaksiala
Polystichum: Kwakiutl; Saanich
Populus: Blackfoot; Dakota; Hopi; Keres, Western; Kiowa; Navajo; Navajo, Ramah; Omaha
Proboscidea: Hopi
Prunus: Dakota; Lakota; Navajo; Navajo, Ramah; Nitinaht
Pseudotsuga: Havasupai; Hopi; Isleta; Jemez; Keres, Western; Keresan; Navajo; Navajo, Ramah; Okanagan-Colville; Tewa; Tewa of Hano; Thompson
Psilostrophe: Hopi; Zuni
Psoralidium: Cheyenne
Pterospora: Jemez
Purshia: Hopi; Navajo; Navajo, Ramah
Pyrola: Navajo, Kayenta
Quercus: Hopi; Navajo; Navajo, Ramah
Rhododendron: Pomo, Kashaya
Rhus: Delaware, Oklahoma; Hopi; Navajo; Navajo, Ramah
Rosa: Haisla & Hanaksiala; Navajo; Navajo, Ramah; Okanagan-Colville; Oweekeno; Thompson
Rubus: Haisla & Hanaksiala; Hoh; Nuxalkmc; Oweekeno; Quileute; Saanich
Rumex: Navajo, Kayenta
Salix: Blackfoot; Hopi; Keres, Western; Keresan; Navajo; Navajo, Ramah; Nitinaht; Omaha; Shuswap; Tewa
Sanguinaria: Delaware; Delaware, Oklahoma
Sarcobatus: Cheyenne; Navajo
Schizachyrium: Comanche
Scirpus: Cahuilla; Hopi; Kansa; Kawaiisu; Okanagan-Colville
Senecio: Navajo
Serenoa: Seminole
Shepherdia: Shuswap
Shinnersoseris: Navajo
Sphaeralcea: Hopi
Sporobolus: Hopi
Stenocereus: Papago
Stephanomeria: Navajo, Kayenta
Stipa: Hopi; Omaha; Pawnee; Ponca
Symphoricarpos: Cheyenne; Hoh; Quileute
Taxodium: Seminole
Taxus: Hoh; Nitinaht; Quileute; Thompson
Tetradymia: Navajo, Ramah
Thalictrum: Navajo
Thuja: Haisla & Hanaksiala; Hanaksiala; Lummi; Makah; Nitinaht; Okanagan-Colville; Oweekeno; Potawatomi; Saanich; Salish, Coast; Shuswap; Skagit; Tsimshian
Townsendia: Navajo; Navajo, Ramah
Triteleia: Pomo, Kashaya
Tsuga: Haisla & Hanaksiala; Hesquiat; Kwakiutl, Southern; Nitinaht; Oweekeno
Typha: Apache, Chiricahua & Mescalero; Apache, Mescalero; Apache, White Mountain; Cahuilla; Hopi; Keres, Western; Keresan; Navajo; Omaha
Umbellularia: Karok

Urtica: Dakota; Diegueño; Makah; Okanagan-Colville; Omaha; Pawnee; Ponca; Winnebago
Ustilago: Hopi
Veratrum: Gitksan; Haisla; Hanaksiala; Kwakiutl; Oweekeno; Salish, Coast
Verbascum: Isleta; Thompson
Verbesina: Navajo, Ramah
Vitis: Jemez
Washingtonia: Cahuilla
Xanthium: Navajo, Ramah
Yucca: Hopi; Keres, Western; Navajo; Navajo, Ramah; Southwest Indians; Tewa of Hano; Zuni
Zea: Cahuilla; Delaware; Hopi; Isleta; Keres, Western; Navajo; Navajo, Ramah; Seminole; Tewa; Zuni
Zigadenus: Thompson

Cleaning Agent
Acer: Chippewa
Annona: Seminole
Artemisia: Eskimo, Alaska; Kiowa
Athyrium: Quileute
Cornus: Haisla & Hanaksiala
Epilobium: Haisla
Galium: Cowichan
Juniperus: Hopi; Tewa
Myrica: Seminole
Quercus: Chippewa; Seminole
Rubus: Tsimshian
Salix: Haisla & Hanaksiala
Thuja: Haisla & Hanaksiala; Tsimshian
Tsuga: Bella Coola; Chippewa
Usnea: Oweekeno

Containers
Abies: Hanaksiala
Acer: Blackfoot; Karok; Nitinaht; Skagit; Upper; Squaxin; Thompson
Adiantum: Kwakiutl, Southern
Aesculus: Cherokee
Agave: Hualapai; Papago
Alnus: Makah; Quinault; Swinomish
Alopecurus: Apache, Chiricahua & Mescalero
Anaphalis: Paiute
Andropogon: Apache, Chiricahua & Mescalero
Angelica: Haisla & Hanaksiala
Anthoxanthum: Abnaki
Apocynum: Luiseño; Nez Perce; Thompson
Arbutus: Karok
Arctostaphylos: Karok; Yokia
Artemisia: Gosiute; Havasupai; Kawaiisu; Navajo; Okanagon; Tanana, Upper; Thompson
Asclepias: Luiseño
Athyrium: Kwakiutl, Southern
Balsamorhiza: Sanpoil
Betula: Abnaki; Algonquin, Tête-de-Boule; Cree, Woodlands; Dakota; Eskimo, Inuktitut; Flathead; Gitksan; Koyukon; Malecite; Micmac; Ojibwa; Tanana, Upper; Thompson; Wet'suwet'en
Bouteloua: Apache, Chiricahua & Mescalero
Bursera: Seminole
Calamagrostis: Cree, Woodlands; Okanagan-Colville

Carex: Okanagan-Colville
Carnegia: Apache, Western; Southwest Indians
Carya: Cherokee
Cercocarpus: Keres, Western
Chamaecyparis: Kitasoo; Nitinaht; Oweekeno
Chenopodium: Hopi
Clematis: Shoshoni
Collomia: Paiute
Cordyline: Hawaiian
Cucurbita: Havasupai; Hopi; Zuni
Dryopteris: Haisla & Hanaksiala; Yurok
Echinocactus: Havasupai
Elaeagnus: Okanagan-Colville; Thompson
Elytrigia: Okanagan-Colville
Epilobium: Thompson
Equisetum: Okanagan-Colville
Eriogonum: Kawaiisu
Fremontodendron: Kawaiisu
Heracleum: Pomo, Kashaya; Shuswap
Hoita: California Indian
Juncus: Chippewa; Luiseño
Juniperus: Gosiute; Navajo; Navajo, Ramah
Lagenaria: Havasupai; Hopi; Navajo; Navajo, Ramah; Yuma
Larix: Chippewa
Leymus: Haisla & Hanaksiala; Okanagan-Colville; Quinault
Lupinus: Miwok
Lysichiton: Bella Coola; Hesquiat; Hoh; Kwakiutl, Southern; Makah; Nitinaht; Okanagan-Colville; Oweekeno; Poliklah; Quileute; Samish; Swinomish; Tolowa; Tsimshian; Yurok
Mentha: Thompson
Muhlenbergia: Apache, Chiricahua & Mescalero; Luiseño
Murdannia: Hawaiian
Nereocystis: Bella Coola; Hesquiat; Kwakiutl, Southern; Makah; Nitinaht; Quileute; Quinault
Nolina: Apache, Chiricahua & Mescalero
Opuntia: Havasupai
Petasites: Alaska Native; Eskimo, Alaska; Eskimo, Inupiat
Phleum: Okanagan-Colville
Picea: Eskimo, Inuktitut; Haisla & Hanaksiala; Tanana, Upper
Pinus: Kawaiisu; Klamath; Navajo; Navajo, Ramah; Okanagan-Colville; Thompson
Polystichum: Hesquiat; Kwakiutl, Southern; Oweekeno; Squaxin; Thompson
Populus: Hopi; Navajo; Okanagan-Colville; Pima
Potamogeton: Kawaiisu
Prunus: Bella Coola; Blackfoot; Okanagon; Salish, Coast; Thompson
Pseudotsuga: Okanagan-Colville
Pteridium: Costanoan; Karok; Okanagan-Colville; Yurok
Quercus: Navajo
Rhus: Navajo; Navajo, Ramah
Ribes: Quileute
Rosa: Okanagan-Colville
Rubus: Okanagan-Colville; Oweekeno; Quileute; Quinault; Shuswap
Salix: Abnaki; Apache, White Mountain; Blackfoot; Cree, Woodlands; Gosiute; Haisla & Hanaksiala; Karok; Kawaiisu; Navajo; Okanagan-Colville; Paiute; Pima; Tanana, Upper; Thompson
Sambucus: Yurok
Schedonnardus: Apache, Chiricahua & Mescalero
Scirpus: Okanagan-Colville; Paiute, Northern; Thompson
Solidago: Lakota
Sparganium: Okanagan-Colville
Sphaeralcea: Havasupai; Shoshoni; Yavapai
Sporobolus: Apache, Chiricahua & Mescalero
Suaeda: Pima
Taxodium: Seminole
Taxus: Karok; Makah
Thuja: Flathead; Haisla & Hanaksiala; Kutenai; Kwakiutl, Southern; Makah; Nitinaht; Okanagon; Oweekeno; Quileute; Salish, Coast; Thompson; Tsimshian
Tilia: Chippewa
Toxicodendron: Costanoan; Karok
Typha: Navajo, Ramah; Okanagan-Colville; Okanagon; Quinault; Thompson
Urtica: Kawaiisu; Luiseño
Vaccinium: Karok
Vitis: Karok
Yucca: Apache; Dakota; Kawaiisu; Navajo; Pima
Zea: Iroquois; Navajo; Navajo, Ramah

Cooking Tools

Acer: Bella Coola; Chehalis; Cheyenne; Chippewa; Cowichan; Haisla & Hanaksiala; Iroquois; Karok; Kwakiutl, Southern; Lummi; Montana Indian; Nitinaht; Ojibwa; Okanagan-Colville; Oweekeno; Seminole; Skagit; Snohomish; Swinomish; Thompson
Adiantum: Hahwunkwut; Kwakiutl, Southern
Aesculus: Kawaiisu
Agave: Havasupai; Navajo
Alectoria: Nitinaht
Alnus: Clallam; Haisla & Hanaksiala; Hesquiat; Kwakiutl, Southern; Makah; Nitinaht; Salish, Coast; Skagit, Upper
Amelanchier: Havasupai; Thompson
Amorpha: Pawnee
Angelica: Bella Coola
Annona: Seminole
Antitrichia: Hanaksiala
Arbutus: Salish, Coast
Arctostaphylos: Karok; Miwok
Ardisia: Seminole
Artemisia: Cheyenne; Okanagan-Colville; Tubatulabal
Asclepias: Kiowa
Astragalus: Omaha; Ponca
Athyrium: Cowlitz; Karok; Nitinaht
Balsamorhiza: Blackfoot
Betula: Blackfoot; Carrier; Chippewa; Cree, Woodlands; Dakota; Micmac; Ojibwa; Potawatomi; Tanana, Upper; Thompson
Blechnum: Nitinaht; Quinault
Bouteloua: Costanoan; Zuni
Calamagrostis: Okanagan-Colville; Thompson
Calocedrus: Round Valley Indian
Carex: Paiute
Carnegia: Apache, Western; Papago
Carya: Cherokee
Catopsis: Seminole
Ceanothus: Mewuk
Chamaecyparis: Hesquiat; Kwakiutl, Southern
Chlorogalum: Mendocino Indian; Pomo
Chrysothamnus: Hopi
Cocos: Seminole
Cornus: Bella Coola; Okanagan-Colville; Shuswap; Snohomish
Corylus: Pomo; Thompson
Cucurbita: Cahuilla; Zuni
Dryopteris: Karok; Nitinaht
Echinocactus: Havasupai
Elaeagnus: Okanagan-Colville; Thompson
Eleocharis: Cheyenne
Eriogonum: Havasupai; Kawaiisu; Tubatulabal
Eupatorium: Cherokee
Fagus: Potawatomi
Ferocactus: Cahuilla
Ficus: Seminole
Fraxinus: Cherokee; Cheyenne; Potawatomi
Gaillardia: Blackfoot
Gaultheria: Hesquiat; Makah; Nitinaht; Oweekeno
Gutierrezia: Hopi; Tewa
Guzmania: Seminole
Heracleum: Blackfoot
Holodiscus: Chehalis; Klallam; Lummi; Makah; Nitinaht; Saanich; Salish, Coast; Salish, Cowichan; Skagit; Snohomish; Squaxin; Swinomish
Ilex: Cherokee
Juncus: Luiseño
Juniperus: Cheyenne; Hopi; Kawaiisu; Tewa
Lagenaria: Cherokee; Hopi; Houma; Keres, Western; Keresan; Navajo; Ojibwa; Papago; Seminole
Leymus: Eskimo, Alaska; Kwakiutl, Southern; Thompson
Libocedrus: Mewuk
Lonicera: Cree, Woodlands
Lycopodium: Cree, Woodlands
Lysichiton: Bella Coola; Haisla & Hanaksiala; Hoh; Kwakiutl, Southern; Makah; Nitinaht; Oweekeno; Quileute; Salish, Coast; Tolowa; Tsimshian; Yurok
Malus: Hanaksiala; Oweekeno
Monarda: Blackfoot
Muhlenbergia: Luiseño
Myrcianthes: Seminole
Nereocystis: Kwakiutl, Southern; Nitinaht
Nolina: Apache, Mescalero; Apache, Western; Havasupai
Oxydendrum: Cherokee
Parkinsonia: Pima
Persea: Seminole

Petasites: Eskimo, Alaska; Quinault
Phalaris: Okanagan-Colville
Phragmites: Chippewa; Thompson
Picea: Carrier; Cree, Woodlands; Eskimo, Alaska; Eskimo, Inuktitut; Hahwunkwut; Koyukon; Tanana, Upper; Thompson
Pinus: Blackfoot; Keres, Western; Kwakiutl, Southern; Thompson
Platanus: Costanoan
Polystichum: Chehalis; Costanoan; Cowlitz; Kwakiutl, Southern; Makah; Nitinaht; Pomo; Pomo, Kashaya; Quileute; Quinault; Salish, Coast
Populus: Cahuilla; Eskimo, Inuktitut; Havasupai; Navajo, Ramah; Tanana, Upper
Prosopis: Cahuilla; Diegueño; Papago; Pima
Prunus: Blackfoot
Pseudotsuga: Thompson
Psoralidium: Kiowa
Pteridium: Clallam; Karok; Makah; Quileute; Squaxin; Swinomish; Yurok
Quercus: Apache, Mescalero; Cherokee; Costanoan; Kawaiisu; Miwok; Paiute
Ratibida: Lakota
Rhododendron: Cherokee
Rubus: Carrier; Haisla & Hanaksiala; Hanaksiala; Hesquiat; Kwakiutl, Southern; Pomo; Pomo, Kashaya
Sabal: Seminole
Salix: Apache, White Mountain; Cheyenne; Cree, Woodlands; Eskimo, Inuktitut; Haisla & Hanaksiala; Karok; Montana Indian; Nitinaht; Paiute; Tubatulabal; Wet'suwet'en; Zuni
Sambucus: Haisla & Hanaksiala
Sarcobatus: Hopi; Navajo, Ramah
Sarracenia: Cree, Woodlands; Potawatomi
Scirpus: Kawaiisu; Okanagan-Colville
Sideroxylon: Seminole
Spiraea: Bella Coola; Lummi; Quinault
Stachys: Hesquiat; Makah; Quinault
Suaeda: Papago; Pima
Taxodium: Seminole
Taxus: Clallam; Kwakiutl, Southern; Makah; Oweekeno; Pomo; Pomo, Kashaya; Quinault
Thuja: Cowlitz; Haisla & Hanaksiala; Hesquiat; Hoh; Kutenai; Kwakiutl; Kwakiutl, Southern; Makah; Nitinaht; Oweekeno; Quileute; Salish, Coast
Tillandsia: Seminole
Toxicodendron: Karok; Mendocino Indian
Tsuga: Haisla & Hanaksiala; Hanaksiala; Kwakiutl, Southern; Oweekeno
Typha: Shuswap; Thompson
Ulmus: Dakota; Omaha; Pawnee; Ponca; Winnebago
Usnea: Nitinaht
Vitis: Miwok
Washingtonia: Cahuilla
Woodwardia: Pomo; Pomo, Kashaya
Wyethia: Miwok
Xanthium: Pima

Xerophyllum: Hahwunkwut
Zea: Iroquois

Decorations

Abies: Keres, Western
Acer: Cherokee; Dakota; Haida; Omaha; Tlingit; Tsimshian
Adiantum: Karok; Maidu
Aesculus: Cherokee
Agave: Apache, Western
Alectoria: Bella Coola
Arbutus: Mendocino Indian
Aristida: Hopi
Artemisia: Hawaiian; Mewuk; Paiute; Zuni
Asclepias: Kiowa
Betula: Haisla & Hanaksiala; Thompson
Blechnum: Haisla & Hanaksiala
Bouteloua: Kiowa
Boykinia: Makah
Calamovilfa: Hopi; Jemez; Lakota
Callirhoe: Kiowa
Castilleja: Hopi; Tewa of Hano; Thompson
Cercis: Maidu
Chamaecyparis: Bella Coola; Haisla & Hanaksiala; Oweekeno
Chlorogalum: Mendocino Indian
Chrysothamnus: Hopi; Karok
Cladrastis: Cherokee
Clematis: Isleta; Jemez; Keres, Western; Montana Indian
Cleome: Zuni
Cornus: Cherokee; Quinault
Dichelostemma: Karok
Dodecatheon: Mendocino Indian
Elaeagnus: Thompson
Epilobium: Shuswap
Epipactis: Karok
Eschscholzia: Cahuilla
Fouquieria: Pima
Fritillaria: Shuswap
Gaillardia: Kiowa
Gentiana: Blackfoot
Gutierrezia: Hopi; Tewa
Helianthus: Hopi; Jemez
Hierochloe: Blackfoot; Flathead
Hilaria: Hopi; Tewa
Ilex: Cherokee
Ipomopsis: Hopi; Navajo
Iris: Jemez
Juglans: Cherokee
Juncus: Cheyenne; Panamint
Juniperus: Blackfoot; Cherokee; Hopi; Navajo; Tewa; Tolowa; Yurok
Kalmia: Cherokee
Lagenaria: Hopi
Larrea: Papago
Lathyrus: Keres, Western
Lilium: Karok
Lithospermum: Thompson
Lycopodium: Bella Coola; Hesquiat; Thompson
Lygodesmia: Keres, Western

Monroa: Navajo, Kayenta
Oenothera: Hopi; Kiowa; Navajo, Kayenta
Ornithogalum: Thompson
Oxydendrum: Cherokee
Paxistima: Paiute
Pediomelum: Blackfoot
Penstemon: Cahuilla; Hopi; Keres, Western; Thompson
Pentachaeta: Cahuilla
Petradoria: Hopi
Phragmites: Okanagon; Thompson
Pinus: Cherokee; Karok; Mendocino Indian
Polystichum: Paiute
Populus: Blackfoot; Tewa
Proboscidea: Panamint
Prunus: Cherokee; Haisla & Hanaksiala; Okanagan-Colville; Thompson
Pseudotsuga: Karok
Pteridium: Maidu
Rhododendron: Cherokee; Pomo
Rhus: Navajo
Scirpus: Okanagan-Colville; Panamint; Thompson
Sida: Hawaiian
Solanum: Luiseño
Sophora: Comanche
Sphaeralcea: Kiowa
Symphoricarpos: Dakota
Thuja: Bella Coola; Nitinaht; Oweekeno; Tsimshian
Tilia: Cherokee
Toxicodendron: Mendocino Indian
Typha: Chehalis; Pima; Thompson
Usnea: Bella Coola
Veratrum: Karok
Xerophyllum: Hupa; Karok
Yucca: Apache, Western; Hopi; Isleta
Zea: Zuni

Designs

Abies: Hanaksiala
Acer: Menominee; Meskwaki; Ojibwa; Potawatomi
Betula: Algonquin, Tête-de-Boule; Chippewa
Cornus: Washo
Dodecatheon: Thompson
Dryopteris: Yurok
Lonicera: Pomo, Kashaya
Lycoperdon: Blackfoot
Nereocystis: Haisla & Hanaksiala
Opuntia: Navajo
Pedicularis: Thompson
Pentagrama: Yurok
Phragmites: Thompson
Picea: Cree, Woodlands
Proboscidea: Pima
Prunus: Paiute; Thompson
Quercus: Potawatomi
Thuja: Haisla & Hanaksiala
Toxicodendron: Pomo, Kashaya
Yucca: Apache; Panamint; Southwest Indians

Fasteners

Acer: Okanagan-Colville
Adenostoma: Luiseño

Amaranthus: Kiowa
Amelanchier: Crow
Apocynum: Okanagon; Thompson
Asclepias: Mahuna; Thompson
Berchemia: Houma
Bloomeria: Kawaiisu
Chaenactis: Navajo, Kayenta
Chamaecyparis: Haisla & Hanaksiala
Chlorogalum: Kawaiisu; Mendocino Indian; Mewuk
Cornus: Crow
Coursetia: Papago
Dichelostemma: Kawaiisu
Dirca: Chippewa; Ojibwa
Encelia: Papago
Equisetum: Pomo, Kashaya
Fagus: Cherokee
Ficus: Seminole
Fremontodendron: Shoshoni
Grindelia: Pomo, Kashaya
Holodiscus: Okanagan-Colville
Ipomopsis: Ute
Koeleria: Isleta
Leymus: Haisla & Hanaksiala; Nitinaht
Oemleria: Makah
Opuntia: Navajo; Thompson
Picea: Cree, Woodlands; Haisla & Hanaksiala; Hesquiat; Koyukon; Kwakiutl, Southern; Makah; Nitinaht; Tanana, Upper
Pinus: Blackfoot; Cahuilla; Crow; Hesquiat; Hopi; Hualapai; Karok; Mendocino Indian; Navajo; Nez Perce; Okanagan-Colville; Paiute; Pomo; Pomo, Kashaya; Salish, Coast; Seminole; Tewa; Yuki
Platanus: Cherokee
Populus: Karok; Okanagan-Colville; Oweekeno
Prosopis: Apache, Western; Cahuilla; Papago
Prunus: Crow; Hesquiat; Karok; Nitinaht; Quinault; Thompson
Pseudotsuga: Haisla & Hanaksiala; Pomo, Kashaya
Quercus: Cherokee; Kawaiisu
Robinia: Cherokee
Salix: Blackfoot; Cree, Woodlands; Lakota; Montana Indian; Okanagan-Colville; Shuswap; Tanana, Upper; Thompson
Scirpus: Paiute, Northern
Sphaeralcea: Keres, Western
Taxus: Salish, Coast
Thuja: Haisla & Hanaksiala
Tilia: Chippewa
Tsuga: Hanaksiala
Typha: Paiute, Northern
Urtica: Haisla & Hanaksiala
Vaccinium: Kwakiutl, Southern
Vitis: Pomo
Yucca: Havasupai; Keres, Western; Navajo, Ramah; Papago; Southwest Indians
Zea: Chippewa

Fertilizer
Artemisia: Zuni
Aulacomnium: Shuswap
Bromus: Iroquois
Corydalis: Navajo, Ramah
Descurainia: Keres, Western
Dicranum: Shuswap
Egregia: Hesquiat
Elymus: Iroquois
Equisetum: Thompson
Fucus: Hesquiat
Lycopodium: Shuswap
Osmorhiza: Iroquois
Podophyllum: Iroquois
Rorippa: Navajo, Ramah
Sagittaria: Iroquois
Sassafras: Cherokee
Solanum: Keres, Western; Navajo, Ramah
Thelypodium: Zuni

Fuel
Abies: Chehalis; Micmac; Shuswap
Acacia: Cahuilla; Pima
Acer: Cahuilla; Cherokee; Cheyenne; Okanagan-Colville; Quinault; Salish, Coast
Adenostoma: Cahuilla; Coahuilla; Diegueño
Agave: Cahuilla
Alectoria: Eskimo, Inuktitut
Aleurites: Hawaiian
Alnus: Bella Coola; Clallam; Eskimo, Alaska; Haisla; Hesquiat; Kitasoo; Kwakiutl, Southern; Mendocino Indian; Nitinaht; Oweekeno; Skagit, Upper; Tanana, Upper; Thompson
Ambrosia: Kiowa
Arbutus: Pomo, Kashaya
Arctostaphylos: Cahuilla; Diegueño; Mendocino Indian
Artemisia: Flathead; Gosiute; Kawaiisu; Klamath; Okanagan-Colville; Paiute; Tewa; Thompson; Thompson, Upper (Nicola Band); Tubatulabal
Arundinaria: Cherokee
Atriplex: Pima
Baccharis: Havasupai; Hualapai
Betula: Cree, Woodlands; Eskimo, Alaska; Eskimo, Inuktitut; Koyukon; Ojibwa; Okanagan-Colville; Tanana, Upper
Calocedrus: Klamath
Carya: Cherokee
Cassiope: Eskimo, Inuktitut
Castanea: Cherokee
Ceanothus: Cahuilla; Dakota; Kawaiisu; Mewuk; Okanagan-Colville; Omaha; Pawnee; Ponca; Winnebago; Yavapai
Celtis: Havasupai; Kiowa
Cercis: Kiowa
Chamaecyparis: Nitinaht
Cheirodendron: Hawaiian
Chrysothamnus: Hopi; Navajo
Conocarpus: Seminole
Cornicularia: Eskimo, Inuktitut
Crataegus: Mendocino Indian
Dalea: Keres, Western
Dryopteris: Kwakiutl, Southern
Encelia: Pima
Equisetum: Koyukon
Fouquieria: Cahuilla
Fraxinus: Cherokee; Chippewa; Havasupai; Yuki
Fucus: Bella Coola
Galium: Cowichan
Gossypium: Havasupai
Hymenoclea: Hualapai
Ipomoea: Lakota
Isocoma: Pima
Juniperus: Apache, Chiricahua & Mescalero; Apache, Mescalero; Havasupai; Hopi; Isleta; Keres, Western; Micmac; Navajo; Navajo, Ramah; Paiute; Tewa; Tewa of Hano; Thompson; Thompson, Upper (Nicola Band); Yavapai; Zuni
Larix: Micmac
Ledum: Eskimo, Inuktitut
Olneya: Cahuilla; Pima
Oxydendrum: Cherokee
Pentaphylloides: Blackfoot
Picea: Eskimo, Alaska; Eskimo, Inuktitut; Hesquiat; Iroquois; Koyukon; Micmac; Tanana, Upper
Pinus: Cahuilla; Havasupai; Isleta; Klamath; Kwakwaka'wakw; Mendocino Indian; Micmac; Navajo; Navajo, Ramah; Pomo; Shuswap; Tewa; Thompson, Upper (Fraser Band); Thompson, Upper (Lytton Band)
Populus: Blackfoot; Dakota; Diegueño; Eskimo, Inuktitut; Havasupai; Keres, Western; Kiowa; Lakota; Mendocino Indian; Montana Indian; Navajo; Omaha; Paiute; Pima; Tanana, Upper; Thompson
Prosopis: Apache, Western; Cahuilla; Diegueño; Havasupai; Papago; Pima
Prunus: Salish, Coast
Pseudotsuga: Bella Coola; Chehalis; Clallam; Cowlitz; Green River Group; Hesquiat; Kitasoo; Klallam; Lummi; Nitinaht; Oweekeno; Pomo; Pomo, Kashaya; Quinault; Salish, Coast; Skagit; Swinomish; Thompson
Ptelea: Havasupai
Pteridium: Oweekeno
Purshia: Havasupai; Okanagan-Colville; Paiute
Quercus: Apache, Chiricahua & Mescalero; Cahuilla; Cherokee; Costanoan; Cowlitz; Diegueño; Kawaiisu; Kiowa; Pomo; Seminole
Rhus: Hopi
Rumex: Keres, Western
Salix: Costanoan; Eskimo, Inuktitut; Havasupai; Hualapai; Pomo; Round Valley Indian; Tanana, Upper
Sambucus: Costanoan; Mendocino Indian
Sarcobatus: Hopi; Navajo
Senna: Kiowa
Serenoa: Seminole
Spiraea: Eskimo, Alaska
Symphoricarpos: Blackfoot
Tamarix: Pima
Thuja: Bella Coola; Haisla & Hanaksiala; Hesquiat; Kwakiutl, Southern; Micmac; Oweekeno; Salish, Coast; Tsimshian

Tsuga: Chehalis; Cherokee; Micmac; Nitinaht; Ojibwa; Quinault
Ulmus: Dakota; Lakota; Omaha; Pawnee; Ponca; Winnebago
Yucca: Dakota
Zea: Navajo; Tewa

Good Luck Charm
Adiantum: Potawatomi
Aesculus: Cherokee
Anemone: Ponca
Angelica: Blackfoot; Mendocino Indian
Aquilegia: Okanagan-Colville; Thompson
Arctostaphylos: Navajo, Ramah
Artemisia: Blackfoot
Bursera: Cahuilla
Conopholis: Jemez
Cornus: Karok
Datura: Cahuilla; Paiute
Daucus: Mendocino Indian
Dodecatheon: Thompson
Epilobium: Thompson
Erigeron: Kiowa
Eupatorium: Potawatomi
Euphorbia: Navajo, Ramah
Fucus: Nitinaht
Gaillardia: Kiowa
Hackelia: Navajo, Ramah
Juniperus: Kitasoo; Navajo; Navajo, Ramah; Thompson
Lewisia: Thompson
Lithospermum: Okanagan-Colville
Lomatium: Blackfoot; Yuki
Lysichiton: Quileute
Maianthemum: Potawatomi
Melia: Omaha; Ponca
Mitella: Iroquois
Oenothera: Jemez
Onosmodium: Chippewa
Oplopanax: Haisla & Hanaksiala
Osmorhiza: Karok
Panax: Chippewa
Pellaea: Diegueño
Picea: Koyukon; Thompson
Pinguicula: Oweekeno
Pinus: Keres, Western
Platanthera: Thompson
Pseudotsuga: Chehalis; Cowlitz; Karok; Thompson
Rosa: Okanagan-Colville; Thompson
Sanicula: Mendocino Indian
Shepherdia: Okanagan-Colville; Thompson
Solidago: Navajo, Kayenta
Thuja: Thompson
Toxicodendron: Navajo, Ramah
Tsuga: Haisla & Hanaksiala
Typha: Navajo, Ramah
Umbellularia: Yurok
Valeriana: Menominee
Veratrum: Bella Coola; Gitksan; Haisla, Hanaksiala
Vicia: Mendocino Indian

Hide Preparation
Abies: Carrier; Klamath
Alnus: Kawaiisu; Montana Indian
Artemisia: Blackfoot; Okanagan-Colville; Okanagon; Thompson
Betula: Cree, Woodlands
Cercocarpus: Keres, Western
Croton: Comanche
Cucurbita: Cocopa
Eriogonum: Kiowa
Juniperus: Blackfoot
Larix: Cree, Woodlands; Kutenai
Lomatium: Blackfoot; Oregon Indian
Picea: Cree, Woodlands; Koyukon; Tanana, Upper
Pinus: Cree, Woodlands; Okanagan-Colville; Seminole; Shuswap
Populus: Shuswap
Prunus: Dakota
Quercus: Apache, White Mountain; Ojibwa; Seminole
Rumex: Cahuilla; Hualapai
Taxodium: Seminole
Thuja: Malecite
Tillandsia: Seminole
Tsuga: Hanaksiala; Hoh; Malecite; Montana Indian; Quileute; Salish, Coast
Yucca: Dakota

Hunting & Fishing Item
Abies: Nitinaht
Acacia: Papago; Pima; Seri
Acer: Chehalis; Isleta; Lummi; Micmac; Montana Indian; Nitinaht; Ojibwa; Paiute; Potawatomi; Quileute; Seminole; Thompson; Wailaki
Achillea: Cree, Woodlands
Adenostoma: Cahuilla; Costanoan; Luiseño
Aesculus: Cherokee; Pomo, Kashaya
Agave: Cahuilla
Alnus: Carrier; Mendocino Indian; Salish, Coast; Tanana, Upper; Thompson
Amaranthus: Keresan
Amelanchier: Blackfoot; Dakota; Flathead; Gosiute; Havasupai; Hopi; Karok; Kawaiisu; Lakota; Mendocino Indian; Okanagan-Colville; Omaha; Ponca; Salish, Coast; Samish; Shasta; Shuswap; Swinomish; Thompson; Winnebago
Amorpha: Lakota
Andropogon: Arikara; Hidatsa; Mandan
Apocynum: Bella Coola; Luiseño; Mendocino Indian; Menominee; Okanagon; Paiute, Northern; Songish; Thompson; Yuki
Arbutus: Karok
Arctostaphylos: Cahuilla
Ardisia: Seminole
Artemisia: Cree, Woodlands; Luiseño
Arundinaria: Cherokee; Choctaw; Houma; Seminole
Asclepias: Cahuilla; Cherokee; Chippewa; Luiseño; Menominee; Neeshenam; Ojibwa
Atriplex: Kawaiisu
Baccharis: Kawaiisu; Mendocino Indian; Papago

Betula: Cree, Woodlands; Eskimo, Inuktitut; Koyukon; Micmac; Tanana, Upper
Calamovilfa: Hopi
Calocedrus: Round Valley Indian; Washo
Calycanthus: Yokia
Carex: Pomo
Carnegia: Papago
Carya: Cherokee; Chippewa; Ojibwa; Potawatomi; Seminole
Castilleja: Nitinaht
Ceanothus: Kawaiisu; Mendocino Indian
Cercis: Havasupai
Cercocarpus: Apache, White Mountain; Gosiute; Keres, Western; Klamath; Mendocino Indian; Modesse; Montana Indian; Shoshoni; Yuki
Chamaecyparis: Haisla & Hanaksiala; Kitasoo; Kwakiutl, Southern
Chilopsis: Cahuilla
Chlorogalum: Cahuilla; Costanoan; Mendocino Indian; Pomo, Kashaya; Yuki
Chrysobalanus: Seminole
Chrysothamnus: Hopi; Keres, Western
Cirsium: Cherokee; Seminole
Cissus: Seminole
Citrus: Seminole
Clematis: Shoshoni; Washo
Cordyline: Hawaiian
Cornus: Cheyenne; Comanche; Flathead; Hanaksiala; Jemez; Karok; Montana Indian; Ojibwa; Okanagan-Colville; Omaha; Oweekeno; Pawnee; Ponca; Salish, Coast; Skagit; Thompson; Winnebago
Corylus: Costanoan; Karok; Pomo; Pomo, Kashaya; Round Valley Indian; Salish, Coast; Thompson
Crataegus: Ojibwa; Thompson
Croton: Costanoan; Mahuna; Mendocino Indian
Dalea: Kiowa
Datisca: Mendocino Indian
Delphinium: Navajo, Ramah
Dionaea: Cherokee
Echinocactus: Mahuna
Egregia: Bella Coola
Epilobium: Haisla & Hanaksiala; Kitasoo; Thompson
Ericameria: Kawaiisu
Erigeron: Ojibwa
Eriophorum: Tanana, Upper
Eugenia: Seminole
Fallugia: Hopi; Isleta; Keres, Western; Keresan; Tewa
Fendlera: Havasupai; Navajo
Ferocactus: Pima; Yuma
Ficus: Seminole
Forestiera: Seminole
Frasera: Apache
Fraxinus: Cheyenne; Chippewa; Dakota; Havasupai; Hualapai; Lakota; Menominee; Meskwaki; Navajo; Navajo, Ramah; Ojibwa; Omaha; Pawnee; Ponca; Sioux; Winnebago
Fremontodendron: Kawaiisu
Gnaphalium: Pomo, Kashaya

Hedophyllum: Nitinaht
Helianthus: Navajo
Hemizonia: Costanoan
Heracleum: Ojibwa; Tsimshian
Heterotheca: Luiseño
Hoita: Mendocino Indian
Holodiscus: Chehalis; Klallam; Kwakiutl, Southern; Lummi; Makah; Nitinaht; Okanagan-Colville; Poliklah; Pomo; Pomo, Kashaya; Saanich; Salish; Salish, Coast; Skagit; Snohomish; Squaxin; Swinomish; Thompson
Hymenoclea: Hualapai
Ilex: Seminole
Iris: Karok; Mendocino Indian; Pomo
Juglans: Cherokee
Juniperus: Apache, Mescalero; Cheyenne; Kawaiisu; Keres, Western; Klamath; Montana Indian; Navajo, Ramah; Okanagan-Colville; Paiute; Shuswap; Tewa; Thompson
Krigia: Menominee
Lactuca: Ojibwa
Lagenaria: Hopi
Laminaria: Nitinaht
Larix: Malecite; Nez Perce
Larrea: Papago
Lathyrus: Meskwaki
Leymus: Cahuilla; Eskimo, Inuktitut; Luiseño; Okanagan-Colville
Ligusticum: Atsugewi
Linum: Klamath; Montana Indian
Lomatium: Nitinaht; Okanagan-Colville; Paiute, Northern; Yuki
Lycium: Papago
Maclura: Comanche; Kiowa; Omaha; Pawnee; Pima; Ponca; Seminole; Tewa
Macrocystis: Kwakiutl, Southern
Maianthemum: Kawaiisu
Malus: Haisla & Hanaksiala; Nitinaht; Quileute; Salish, Coast
Marah: Mendocino Indian; Pomo; Pomo, Kashaya
Mentha: Blackfoot; Sioux; Winnebago
Mirabilis: Hopi
Monarda: Keres, Western
Monardella: Okanagan-Colville
Mortonia: Havasupai
Morus: Hualapai; Papago; Seminole
Myrcianthes: Seminole
Myrica: Cree, Woodlands
Nereocystis: Alaska Native; Kitasoo; Kwakiutl, Southern; Makah; Nitinaht; Oweekeno; Pomo, Kashaya; Quileute; Quinault; Tsimshian
Nolina: Hualapai
Ocotea: Seminole
Oenothera: Paiute
Oplopanax: Clallam; Hesquiat; Kitasoo; Klallam; Makah; Nitinaht
Opuntia: Kiowa; Okanagan-Colville
Osmorhiza: Okanagan-Colville
Osmunda: Menominee
Ostrya: Lakota; Malecite

Oxydendrum: Cherokee
Pachycereus: Papago & Pima
Panax: Menominee
Penstemon: Havasupai
Phalaris: Okanagan-Colville
Philadelphus: California Indian; Karok; Klamath; Midoo; Modesse; Montana Indian; Okanagan-Colville; Poliklah; Saanich; Shuswap; Skagit; Wailaki; Washo; Yuki
Phragmites: Apache, White Mountain; Havasupai; Hualapai; Kawaiisu; Klamath; Montana Indian; Navajo; Paiute, Northern; Tewa
Phyllospadix: Nitinaht
Physocarpus: Karok; Wintoon
Picea: Cree, Woodlands; Eskimo, Inuktitut; Haisla & Hanaksiala; Kwakiutl, Southern; Makah; Nitinaht; Quileute; Quinault; Tanana, Upper
Pinus: Kwakwaka'wakw; Nitinaht; Pomo; Pomo, Kashaya; Thompson; Tolowa
Platanthera: Cherokee
Pluchea: Cahuilla; Havasupai; Hualapai; Luiseño
Poaceae sp.: Tanana, Upper
Polygonum: Thompson
Populus: Cree, Woodlands; Eskimo, Inuktitut; Malecite; Nitinaht; Thompson
Potamogeton: Kawaiisu
Prosopis: Apache, Mescalero; Cahuilla; Isleta; Navajo; Seri
Prunus: Apache, Mescalero; Cheyenne; Clallam; Costanoan; Isleta; Kawaiisu; Keres, Western; Lakota; Montana Indian; Paiute; Ponca; Salish, Coast; Tewa
Pseudotsuga: Blackfoot; Chehalis; Clallam; Cowlitz; Green River Group; Haisla; Karok; Klallam; Lummi; Nitinaht; Okanagan-Colville; Quinault; Salish, Coast; Skagit; Swinomish; Thompson
Psidium: Seminole
Ptelea: Havasupai
Purshia: Hopi; Keres, Western; Navajo
Pycnanthemum: Meskwaki
Quercus: Cahuilla; Cherokee; Jemez; Navajo; Pomo
Rhus: Keres, Western; Navajo; Navajo, Ramah; Santa Clara
Ribes: Cheyenne; Cowichan; Havasupai; Hopi; Lakota; Navajo; Navajo, Ramah; Saanich; Tewa; Thompson
Robinia: Apache, Mescalero; Cherokee; Hualapai; Jemez; Keres, Western; Tewa
Rosa: Cheyenne; Cowichan; Klamath; Okanagan-Colville; Thompson
Rubus: Quileute
Sabal: Seminole
Sagittaria: Potawatomi
Salix: Cree, Woodlands; Gosiute; Haisla & Hanaksiala; Kawaiisu; Kwakiutl, Southern; Luiseño; Montana Indian; Navajo; Paiute; Pima; Quileute; Quinault; Salish, Coast; Tanana, Upper; Thompson; Ukiah; Wet'suwet'en
Salvia: Cahuilla
Sambucus: Costanoan; Haisla & Hanaksiala;

Houma; Luiseño; Quinault
Sapindus: Papago
Sapium: Seri
Sarcobatus: Cheyenne; Hopi
Scirpus: Cahuilla; Hanaksiala
Serenoa: Seminole
Shepherdia: Thompson
Sideroxylon: Seminole
Solanum: Seminole
Sonchus: Potawatomi
Spirogyra: Thompson
Stachys: Haisla & Hanaksiala
Symphoricarpos: Great Basin Indian; Montana Indian; Paiute; Shoshoni; Yokia
Tanacetum: Ojibwa
Taxodium: Seminole
Taxus: Bella Coola; Chehalis; Clallam; Costanoan; Flathead; Haihais; Hanaksiala; Hoh; Karok; Kitasoo; Klallam; Klamath; Makah; Mendocino Indian; Montana Indian; Nitinaht; Okanagan-Colville; Oweekeno; Paiute; Pomo; Pomo, Kashaya; Quileute; Quinault; Salish, Coast; Samish; Shasta; Snohomish; Swinomish; Thompson; Wintoon; Yurok
Tephrosia: Hawaiian; Seminole
Thuja: Chippewa; Haisla; Haisla & Hanaksiala; Hesquiat; Kwakiutl, Southern; Makah; Malecite; Micmac; Montana Indian; Nitinaht; Okanagan-Colville; Oweekeno; Quileute; Quinault; Saanich; Salish, Coast; Salish, Cowichan; Squaxin; Tsimshian
Thymophylla: Navajo, Kayenta
Tilia: Menominee; Meskwaki
Trichostema: Concow; Numlaki
Trientalis: Ojibwa
Tsuga: Bellabella; Clallam; Haisla & Hanaksiala; Hanaksiala; Hesquiat; Kitasoo; Klallam; Lummi; Makah; Nitinaht; Oweekeno; Quinault; Skagit
Typha: Paiute, Northern
Ulmus: Menominee
Umbellularia: Costanoan; Pomo, Kashaya
Urtica: Bella Coola; Cherokee; Haisla & Hanaksiala; Hesquiat; Kawaiisu; Klamath; Kwakiutl, Southern; Luiseño; Lummi; Nitinaht; Oweekeno; Quileute; Skokomish; Snohomish; Thompson; Tsimshian
Viburnum: Chippewa
Vicia: Costanoan; Nitinaht
Vitis: Seminole
Washingtonia: Cahuilla
Yucca: Cherokee; Hopi; Luiseño; Tewa
Zanthoxylum: Seminole
Zea: Seminole
Zigadenus: Okanagan-Colville
Zostera: Hesquiat; Nitinaht

Incense & Fragrance
Abies: Blackfoot; Flathead; Hesquiat; Nez Perce; Nitinaht; Thompson
Acacia: Papago
Achillea: Clallam; Kutenai
Acorus: Omaha

Allium: Blackfoot
Alnus: Thompson
Anaphalis: Okanagan-Colville
Angelica: Haisla & Hanaksiala
Aquilegia: Meskwaki; Omaha; Pawnee; Ponca
Artemisia: Blackfoot; Eskimo, Inuktitut; Kawaiisu; Lakota; Nitinaht
Asarum: Chippewa
Balsamorhiza: Blackfoot
Boykinia: Karok
Calycanthus: Cherokee
Cercis: Navajo
Chenopodium: Thompson
Comptonia: Chippewa
Croton: Navajo, Ramah
Eriodictyon: Diegueño
Fragaria: Thompson
Galium: Makah; Omaha; Ponca
Geum: Blackfoot
Hierochloe: Blackfoot; Cheyenne; Kiowa; Lakota; Montana Indian; Okanagan-Colville; Omaha; Thompson
Juniperus: Arapaho; Cheyenne; Kiowa; Lakota; Omaha; Tanana, Upper
Ligusticum: Crow
Lomatium: Crow; Nitinaht; Thompson
Lupinus: Blackfoot
Lycoperdon: Blackfoot
Matricaria: Blackfoot; Cheyenne; Eskimo, Inuktitut; Kutenai; Montana Indian
Melilotus: Dakota; Havasupai; Iroquois
Mentha: Cheyenne; Montana Indian; Thompson; Winnebago; Yuki
Monarda: Blackfoot; Cheyenne; Crow; Dakota; Kutenai; Navajo; Omaha
Monardella: Karok
Oenothera: Lakota
Osmorhiza: Blackfoot
Pectis: Zuni
Pinus: Cahuilla; Cherokee; Navajo; Thompson
Polygonatum: Chippewa
Polyporus: Blackfoot
Populus: Blackfoot; Nitinaht
Rhododendron: Thompson
Rhus: Hopi
Rosa: Omaha; Tewa
Salix: Blackfoot
Sassafras: Cherokee
Satureja: Karok
Senecio: Keres, Western
Solidago: Navajo
Streptopus: Thompson
Thalictrum: Blackfoot; Dakota; Gros Ventre
Thuja: Chippewa; Ojibwa
Tsuga: Thompson
Umbellularia: Costanoan
Valeriana: Tanana, Upper
Viola: Eskimo, Inuktitut; Tanana, Upper
Xylorhiza: Havasupai
Zanthoxylum: Omaha

Insecticide
Achillea: Okanagan-Colville
Achlys: Saanich; Thompson
Allium: Salish, Coast
Alnus: Eskimo, Inuktitut
Anemone: Thompson
Artemisia: Blackfoot; Karok; Keres, Western; Kutenai; Meskwaki; Okanagan-Colville; Shuswap; Thompson
Boerhavia: Hopi
Carya: Choctaw; Iroquois
Castanea: Iroquois
Ceanothus: Shuswap
Chenopodium: Navajo
Chrysothamnus: Sanpoil
Cicuta: Cherokee
Clintonia: Algonquin, Tête-de-Boule
Comptonia: Potawatomi
Corylus: Iroquois
Cymopterus: Paiute
Darlingtonia: Yurok
Epilobium: Tanana, Upper
Equisetum: Cowlitz
Eschscholzia: Costanoan
Fagus: Iroquois
Fendlera: Navajo
Fomes: Eskimo, Inuktitut
Grindelia: Karok; Navajo, Ramah
Gutierrezia: Jemez; Keres, Western
Hackelia: Iroquois
Hedeoma: Cherokee; Rappahannock
Heracleum: Carrier, Southern
Heterotheca: Navajo, Ramah
Hierochloe: Flathead
Ipomoea: Cherokee
Juglans: Delaware; Delaware, Oklahoma; Iroquois
Juniperus: Cherokee; Shuswap; Thompson
Lobelia: Cherokee
Matricaria: Blackfoot; Montana Indian
Melia: Cherokee
Melilotus: Isleta; Keres, Western
Mentha: Flathead; Kutenai; Thompson
Mirabilis: Cherokee
Monarda: Flathead
Myrica: Potawatomi
Osmorhiza: Kawaiisu; Paiute
Pascopyrum: Keres, Western
Picea: Tanana, Upper
Pinus: Iroquois
Podophyllum: Cherokee; Menominee
Pogogyne: Mendocino Indian
Populus: Carrier, Southern; Eskimo, Inuktitut
Psoralidium: Lakota
Rhus: Hualapai
Sambucus: Meskwaki
Senecio: Keres, Western
Solanum: Cherokee
Sorbus: Bella Coola
Sphenosciadium: Paiute
Thalictrum: Blackfoot

Thuja: Iroquois
Trichostema: Karok; Kawaiisu
Tsuga: Makah
Umbellularia: Costanoan; Mendocino Indian; Yuki; Yurok
Verbesina: Navajo, Ramah
Viola: Cherokee

Jewelry
Adiantum: Pomo, Kashaya
Arbutus: Tolowa
Arctostaphylos: Blackfoot
Aster: Blackfoot
Coix: Cherokee
Cupressus: Navajo
Echinocactus: Havasupai
Echinocystis: Oglala
Echium: Cherokee
Elaeagnus: Arapaho; Blackfoot; Cree; Tanana, Upper; Thompson
Eleocharis: Seminole
Juncus: Pomo, Kashaya
Juniperus: Blackfoot; Hopi; Navajo; Tolowa
Lycopodium: Oweekeno
Matricaria: Kutenai
Melia: Omaha; Ponca
Opuntia: Thompson
Oxytropis: Blackfoot
Paeonia: Paiute
Phragmites: Thompson
Phytolacca: Kiowa
Pinus: Karok; Kawaiisu; Navajo
Punica: Navajo
Quercus: Cahuilla
Rosa: Blackfoot; Cree, Woodlands; Makah; Paiute; Thompson
Sapindus: Seminole
Solanum: Hopi; Keresan
Thuja: Shuswap
Tsuga: Hanaksiala
Vaccinium: Cree, Woodlands
Xerophyllum: Poliklah
Yucca: Navajo

Lighting
Acer: Malecite
Adenostoma: Cahuilla
Aleurites: Hawaiian
Arctostaphylos: Pomo, Kashaya
Artemisia: Costanoan; Paiute
Arundinaria: Cherokee
Betula: Dakota; Gitksan; Micmac; Ojibwa; Wet'suwet'en
Carex: Pomo
Chrysothamnus: Isleta
Echinocereus: Isleta
Eriophorum: Eskimo, Inuktitut
Helianthus: Pima
Juniperus: Apache, Western; Navajo, Ramah; Paiute; Tewa; Yavapai
Myrica: Houma
Opuntia: Keres, Western

Picea: Haisla
Pinus: Nez Perce; Potawatomi; Seminole
Pseudotsuga: Round Valley Indian
Rhamnus: Wintoon
Stenocereus: Papago & Pima
Thuja: Haisla; Hoh; Potawatomi; Quileute

Lubricant
Opuntia: Diegueño

Malicious Charm
Angelica: Blackfoot
Artemisia: Blackfoot
Equisetum: Chippewa
Juniperus: Hopi; Tewa
Physocarpus: Okanagan-Colville

Malicious Magic
Castilleja: Paiute
Lomatium: Paiute
Smilax: Chippewa

Musical Instrument
Acer: Kwakiutl, Southern; Montana Indian; Okanagan-Colville
Alnus: Kwakiutl, Southern; Oweekeno
Arundinaria: Cherokee; Keres, Western; Seminole
Arundo: Cahuilla
Betula: Tanana, Upper
Canna: Seminole
Carnegia: Papago
Cornus: Crow
Corylus: Chippewa; Ojibwa
Cucurbita: Hopi
Equisetum: Shoshoni
Glaucothea: Cocopa
Heracleum: Blackfoot; Cheyenne
Holodiscus: Okanagan-Colville
Juniperus: Cheyenne; Kiowa; Paiute; Thompson
Lagenaria: Cocopa; Havasupai; Hopi; Houma; Mohave; Papago; Yuma
Larrea: Papago
Lithocarpus: Pomo, Kashaya
Olneya: Papago
Phragmites: Cahuilla; Hualapai; Kawaiisu; Pima
Pinus: Blackfoot; Cheyenne; Mendocino Indian
Populus: Dakota; Havasupai; Hopi; Hualapai; Omaha; Pawnee; Ponca
Quercus: Cahuilla
Ranunculus: Karok
Salix: Cheyenne; Montana Indian; Paiute
Sambucus: Costanoan; Karok; Kawaiisu; Mendocino Indian; Okanagan-Colville; Paiute; Wintoon; Yokut; Yuki
Sarcobatus: Hopi
Taxodium: Seminole
Taxus: Cowlitz
Thuja: Okanagan-Colville; Oweekeno; Paiute; Quileute
Tsuga: Kwakiutl, Southern
Washingtonia: Cocopa
Yucca: Navajo

Paint
Abies: Blackfoot

Acer: Quinault
Agave: Apache, Western
Amaranthus: Apache, White Mountain; Zuni
Asclepias: Crow
Aster: Keres, Western
Bovista: Chippewa
Bryoria: Haisla & Hanaksiala
Candelaria: Haisla & Hanaksiala
Candelariella: Haisla & Hanaksiala
Castilleja: Hopi; Tewa
Celastrus: Lakota
Chenopodium: Carrier; Pawnee; Tanana, Upper; Thompson
Chlorogalum: Mendocino Indian
Chrysothamnus: Hopi
Cleome: Keresan; Tewa; Zuni
Conocephalum: Haisla & Hanaksiala
Delphinium: Karok; Thompson
Descurainia: Hopi; Tewa of Hano
Encelia: Pima
Erysimum: Keres, Western
Evernia: Montana Indian; Thompson
Fomes: Oweekeno
Gaultheria: Nitinaht
Hierochloe: Cheyenne
Juglans: Apache, Mescalero
Larix: Okanagan-Colville; Thompson
Linum: Keres, Western
Lithospermum: Chippewa; Keres, Western; Thompson
Lonicera: Quileute
Mahonia: Karok; Zuni
Marah: Luiseño
Oplopanax: Haisla; Haisla & Hanaksiala; Lummi; Lummi, Washington; Nitinaht
Opuntia: Keres, Western
Ostrya: Lakota
Penstemon: Lakota
Phytolacca: Pawnee
Picea: Nitinaht
Pinus: Haisla & Hanaksiala; Havasupai; Keres, Western
Plagiobothrys: California Indian; Kawaiisu
Populus: Cheyenne; Nitinaht
Prosopis: Pima
Prunus: Karok; Lakota; Montana Indian; Shuswap
Pseudephebe: Haisla & Hanaksiala
Psilostrophe: Keres, Western; Zuni
Quercus: Choctaw; Navajo
Rhus: Keres, Western
Rosa: Thompson
Salix: Cheyenne
Sanguinaria: Omaha
Sphaeralcea: Gosiute; Tewa
Stenocereus: Papago & Pima
Symphoricarpos: Blackfoot
Thalictrum: Blackfoot
Thelypodium: Tewa
Tradescantia: Lakota
Trentepohlia: Haisla & Hanaksiala

Tsuga: Quinault
Typha: Havasupai
Xanthium: Keres, Western
Xanthoria: Haisla & Hanaksiala; Kitasoo
Yucca: Hopi
Zinnia: Keres, Western

Paper
Aesculus: Cherokee
Ambrosia: Lakota
Artemisia: Blackfoot; Kiowa; Navajo
Betula: Ojibwa; Thompson
Liriodendron: Cherokee
Magnolia: Cherokee
Poa: Eskimo, Inuktitut
Populus: Ojibwa
Prosopis: Cahuilla
Tilia: Cherokee
Tsuga: Cherokee

Plant Indicator
Lewisia: Thompson
Monotropa: Thompson

Planting Seeds
Citrullus: Havasupai
Cucurbita: Havasupai
Zea: Iroquois

Preservative
Alnus: Eskimo, Inuktitut; Kitasoo; Makah; Nitinaht
Arbutus: Cowichan; Saanich
Betula: Ojibwa
Castilleja: Jemez
Ceanothus: Okanagan-Colville
Cornus: Okanagan-Colville; Saanich
Descurainia: Keres, Western
Juniperus: Keres, Western
Pinus: Hesquiat; Paiute
Populus: Blackfoot; Haisla & Hanaksiala; Tanana, Upper
Pteridium: Paiute
Sphaeralcea: Keres, Western
Tsuga: Cherokee
Veratrum: Paiute

Protection
Abies: Blackfoot; Thompson
Achillea: Potawatomi
Acorus: Cheyenne; Chippewa
Agastache: Chippewa
Amelanchier: Pit River
Anaphalis: Cheyenne; Potawatomi
Angelica: Iroquois; Miwok; Pomo, Kashaya
Apocynum: Chippewa
Aralia: Chippewa
Arctostaphylos: Thompson
Artemisia: Blackfoot; Chippewa; Lakota; Thompson
Athyrium: Shuswap
Betula: Ojibwa; Tanana, Upper
Brassica: Rappahannock
Ceanothus: Mahuna
Chenopodium: Keres, Western; Navajo, Ramah
Chilopsis: Cahuilla
Chrysothamnus: Hopi; Tewa

Cirsium: Kiowa; Nitinaht; Saanich
Clematis: Blackfoot
Crataegus: Cherokee
Cucurbita: Cocopa
Cycloloma: Zuni
Dalea: Navajo, Kayenta
Echinacea: Sioux
Eleocharis: Cheyenne
Erodium: Jemez
Fomes: Haisla & Hanaksiala
Fomitopsis: Haisla & Hanaksiala; Nitinaht
Fraxinus: Costanoan
Ganoderma: Haisla & Hanaksiala; Nitinaht
Gaura: Navajo
Geum: Salish, Coast
Gleditsia: Cherokee
Heracleum: Menominee
Hierochloe: Cheyenne
Holodiscus: Thompson
Iris: Ojibwa
Juniperus: Cheyenne; Dakota; Okanagan-Colville; Omaha; Pawnee; Ponca; Seminole; Thompson
Larrea: Papago
Lathyrus: Chippewa
Ligusticum: Yuki
Lithospermum: Thompson
Lomatium: Yuki
Lycium: Zuni
Lysichiton: Thompson
Nicotiana: Cahuilla; Kawaiisu
Nolina: Apache, Western
Oenothera: Paiute
Oplopanax: Bella Coola; Haisla & Hanaksiala; Kwakiutl
Opuntia: Kiowa; Papago
Panax: Menominee
Parkinsonia: Cahuilla
Penstemon: Lakota
Pentaphylloides: Cheyenne
Philadelphus: Okanagan-Colville
Phlox: Hopi; Navajo, Kayenta
Phoradendron: Navajo
Picea: Kitasoo; Koyukon; Thompson
Pinus: Hopi; Mahuna; Tewa; Tsimshian
Plantago: Chippewa; Ojibwa
Platanthera: Iroquois
Poaceae sp.: Tanana, Upper
Polyporus: Haisla & Hanaksiala
Populus: Blackfoot; Thompson
Prosopis: Cahuilla; Papago
Psoralidium: Dakota
Pteridium: Costanoan; Shuswap
Quercus: Navajo, Ramah
Rhus: Hualapai
Ribes: Bella Coola
Ricinus: Navajo
Rosa: Flathead; Nez Perce; Okanagan-Colville; Paiute; Shuswap; Thompson
Rubus: Blackfoot; Iroquois
Salix: Karok; Kiowa; Round Valley Indian; Tanana, Upper
Salvia: Kawaiisu
Smilax: Kiowa
Sophora: Comanche
Sphaeralcea: Lakota
Symphoricarpos: Nez Perce; Nitinaht
Tetradymia: Navajo, Ramah
Thamnosma: Kawaiisu
Thuja: Haisla; Menominee; Potawatomi
Tragia: Navajo, Kayenta
Tsuga: Haisla & Hanaksiala; Kwakwaka'wakw; Oweekeno
Typha: Navajo, Ramah; Okanagan-Colville
Umbellularia: Pomo, Kashaya
Urtica: Makah; Okanagan-Colville
Veratrum: Bella Coola; Haisla; Hanaksiala; Kitasoo; Kwakiutl; Paiute, Northern
Verbesina: Navajo, Ramah
Yucca: Navajo
Zea: Omaha

Sacred Items
Arisaema: Menominee
Betula: Ojibwa
Bouteloua: Navajo
Carex: Jemez
Chamaecyparis: Nitinaht
Epigaea: Potawatomi
Hierochloe: Menominee
Lithospermum: Menominee
Lomatium: Poliklah
Lycium: Navajo
Mitella: Menominee
Nicotiana: Navajo
Picea: Kwakiutl, Southern
Populus: Omaha
Prunus: Navajo
Rhus: Navajo
Sambucus: Nitinaht
Staphylea: Meskwaki
Thuja: Ojibwa
Umbellularia: Pomo, Kashaya
Zea: Navajo

Season Indicator
Amelanchier: Iroquois
Arctostaphylos: Blackfoot; Tanana, Upper
Astragalus: Okanagan-Colville
Bouteloua: Montana Indian
Cardamine: Yurok
Carnegia: Papago; Pima
Cercis: Kiowa
Crataegus: Okanagan-Colville
Epilobium: Cree, Woodlands
Fritillaria: Hanaksiala; Okanagan-Colville
Helianthus: Hopi
Larix: Okanagan-Colville
Lupinus: Okanagan-Colville
Matricaria: Eskimo, Inuktitut
Opuntia: Keres, Western; Okanagan-Colville
Penstemon: Hopi; Tewa
Philadelphus: Okanagan-Colville
Pinus: Hopi; Okanagan-Colville
Populus: Havasupai; Okanagan-Colville
Prosopis: Pima, Gila River
Prunus: Dakota; Okanagan-Colville; Omaha; Pawnee; Ponca
Pseudotsuga: Okanagan-Colville
Ranunculus: Shasta
Rhus: Okanagan-Colville
Ribes: Navajo, Ramah; Okanagan-Colville
Rubus: Haisla & Hanaksiala
Solidago: Omaha
Stillingia: Kiowa
Stipa: Blackfoot
Thermopsis: Blackfoot; Flathead

Smoke Plant
Achillea: Ojibwa
Acorus: Cheyenne
Alnus: Eskimo, Alaska; Makah
Ambrosia: Navajo
Amorpha: Oglala
Anaphalis: Cherokee
Angelica: Delaware, Oklahoma
Antennaria: Blackfoot; Great Basin Indian
Arbutus: Hoh; Quileute
Arctostaphylos: Blackfoot; Cahuilla; Carrier; Cheyenne; Chippewa; Clallam; Cree; Cree, Woodlands; Eskimo, Arctic; Eskimo, Inuktitut; Great Basin Indian; Heiltzuk; Hesquiat; Hoh; Jemez; Keresan; Klamath; Lakota; Makah; Montana Indian; Nitinaht; Nootka; Okanagan-Colville; Okanagon; Paiute; Pawnee; Potawatomi; Quileute; Salish, Coast; Shuswap; Skagit, Upper; Thompson
Arenaria: Shoshoni
Artemisia: Cahuilla; Eskimo, Inuktitut; Pomo, Kashaya; Sioux
Cardamine: Cherokee
Ceanothus: Paiute
Chimaphila: Blackfoot; Montana Indian
Cicuta: Chippewa
Clematis: Okanagan-Colville
Cornus: Abnaki; Blackfoot; Cheyenne; Chippewa; Cree; Cree, Woodlands; Dakota; Gosiute; Great Basin Indian; Hoh; Lakota; Menominee; Meskwaki; Micmac; Montana Indian; Ojibwa; Okanagan-Colville; Okanagon; Omaha; Pawnee; Ponca; Potawatomi; Quileute; Salish, Coast; Shuswap; Thompson; Winnebago
Datura: Cahuilla; Keres, Western
Dendromecon: Kawaiisu
Erigeron: Ojibwa
Foeniculum: Hopi
Gaultheria: Cherokee; Makah
Juniperus: Navajo, Ramah
Ligusticum: Crow; Okanagan-Colville
Lomatium: Great Basin Indian; Karok; Paiute, Northern
Mentzelia: Hopi
Mirabilis: Keres, Western
Myrica: Seminole
Nicotiana: Apalachee; Blackfoot; Cahuilla; Chero-

Index of Usages 869

kee; Coahuilla; Dakota; Diegueño; Eskimo, Inuktitut; Gosiute; Haisla & Hanaksiala; Havasupai; Hesquiat; Hopi; Hualapai; Iroquois; Karok; Kawaiisu; Klamath; Luiseño; Mendocino Indian; Mewuk; Midoo; Mohave; Montana Indian; Navajo, Ramah; Neeshenam; Okanagan-Colville; Okanagon; Omaha; Oweekeno; Paiute; Papago; Pawnee; Pima; Pomo, Kashaya; Ponca; Shuswap; Tewa; Thompson; Tolowa; Tsimshian; Winnebago; Yavapai; Yuki; Yuma; Zuni
Oenothera: Hopi
Petasites: Eskimo, Alaska; Eskimo, Inuktitut
Phragmites: Apache, White Mountain
Picea: Iroquois
Pinus: Kawaiisu
Polanisia: Isleta
Populus: Eskimo, Inuktitut; Montana Indian; Tanana, Upper
Prunus: Paiute, Northern
Psoralidium: Navajo
Quercus: Kiowa
Rhus: Cheyenne; Chippewa; Comanche; Dakota; Gosiute; Keresan; Kiowa; Lakota; Omaha; Pawnee; Plains Indian; Ponca; Potawatomi; Tewa; Winnebago
Rosa: Dakota; Okanagon; Omaha; Pawnee; Ponca; Thompson
Salix: Chippewa; Eskimo, Alaska; Eskimo, Inuktitut; Mendocino Indian; Micmac; Montagnais; Ojibwa
Sedum: Gosiute
Silene: Gosiute
Solidago: Navajo, Kayenta
Taxus: Clallam; Klallam; Samish; Swinomish
Thymophylla: Keres, Western
Triodanis: Meskwaki
Vaccinium: Eskimo, Inuktitut; Gosiute; Okanagon; Thompson
Valeriana: Blackfoot; Okanagon; Thompson
Verbascum: Isleta; Menominee; Thompson
Viburnum: Ojibwa
Vicia: Navajo
Zea: Kiowa; Navajo; Navajo, Ramah; Tewa

Smoking Tools
Abies: Tewa
Acer: Montana Indian; Tewa
Angelica: Eskimo, Alaska
Arctostaphylos: Cahuilla
Calamovilfa: Lakota
Cercocarpus: California Indian; Kawaiisu
Cornus: Blackfoot
Dasylirion: Apache, Mescalero
Eriogonum: Kawaiisu; Yavapai
Fraxinus: Cheyenne; Dakota; Lakota; Omaha; Pawnee; Ponca; Winnebago; Yuki
Helianthus: Tewa
Lonicera: Cree, Woodlands; Kawaiisu; Pomo; Pomo, Kashaya
Lyonia: Seminole
Oxydendrum: Cherokee
Philadelphus: Karok; Okanagan-Colville

Phragmites: Apache, White Mountain; Cocopa; Havasupai; Kawaiisu; Maricopa; Montana Indian; Papago; Yuma
Pinus: Thompson
Prunus: Malecite
Quercus: Comanche
Rhododendron: Cherokee
Ribes: Bella Coola; Hesquiat
Rosa: Klamath
Rubus: Hesquiat
Salix: Kawaiisu
Sambucus: Bella Coola; Costanoan; Kawaiisu
Sarcobatus: Cheyenne
Scirpus: Costanoan
Smilax: Comanche
Symphoricarpos: Shuswap
Taxus: Karok; Tolowa

Snuff
Asarum: Cherokee
Dimorphocarpa: Keres, Western
Helenium: Costanoan
Nicotiana: Hesquiat
Petasites: Eskimo, Alaska
Salix: Eskimo, Alaska; Eskimo, Inuktitut
Veratrum: Blackfoot

Soap
Abies: Thompson
Allium: Blackfoot
Angelica: Haisla & Hanaksiala
Arctostaphylos: Pomo
Artemisia: Blackfoot; Sioux
Atriplex: Cahuilla; Diegueño; Pima
Ceanothus: California Indian; Costanoan; Kawaiisu; Pomo, Kashaya; Thompson
Chenopodium: Costanoan; Diegueño; Kawaiisu; Luiseño
Chlorogalum: Cahuilla; Costanoan; Karok; Luiseño; Mahuna; Pomo, Kashaya; Yuki
Cucurbita: Cahuilla; Diegueño; Kawaiisu; Kiowa; Luiseño; Mahuna; Papago
Elaeagnus: Blackfoot
Equisetum: Blackfoot
Galium: Nitinaht
Gutierrezia: Isleta
Hedeoma: Keres, Western
Helianthus: Jemez
Hierochloe: Blackfoot; Gros Ventre
Ilex: Seminole
Juniperus: Thompson
Ligusticum: Flathead
Linum: Thompson
Lotus: Diegueño
Marah: Costanoan
Mentha: Thompson
Morchella: Crow
Nolina: Apache, Western
Oplopanax: Haisla & Hanaksiala
Panicum: Isleta
Philadelphus: Okanagan-Colville; Shuswap
Picea: Thompson

Polypogon: Navajo, Kayenta
Populus: Thompson
Pseudotsuga: Okanagan-Colville
Rubus: Cowlitz
Saponaria: Cherokee; Mahuna
Suaeda: Cahuilla
Symphoricarpos: Thompson
Symplocarpus: Iroquois
Veratrum: Gitksan
Verbesina: Hopi
Yucca: Apache; Apache, Mescalero; Apache, Western; Apache, White Mountain; Blackfoot; Cahuilla; Cherokee; Comanche; Dakota; Great Basin Indian; Havasupai; Hopi; Hualapai; Isleta; Jemez; Keres, Western; Kiowa; Lakota; Navajo; Navajo, Ramah; Omaha; Papago; Pawnee; Pima; Ponca; Southwest Indians; Tewa; Yavapai; Zuni
Ziziphus: Apache

Stable Gear
Acer: Seminole
Agave: Cahuilla
Alnus: Blackfoot
Apocynum: Shuswap
Arbutus: Mendocino Indian
Artemisia: Okanagon; Thompson
Betula: Paiute
Fraxinus: Navajo
Gaura: Lakota
Juniperus: Okanagan-Colville
Nolina: Isleta
Parryella: Hopi; Tewa
Pinus: Navajo, Ramah
Populus: Thompson
Quercus: Apache, Mescalero
Salix: Montana Indian; Thompson
Scirpus: Kawaiisu
Taxodium: Seminole
Ulmus: Dakota; Iroquois; Kiowa; Omaha; Pawnee; Ponca
Urtica: Dakota; Omaha; Pawnee; Ponca; Winnebago

Tools
Acacia: Papago
Acer: Bella Coola; Haisla & Hanaksiala; Navajo; Okanagan-Colville; Skagit; Thompson
Adiantum: Pomo
Aesculus: Mendocino Indian; Mewuk; Pomo, Kashaya
Agave: Apache, Western; Cahuilla; Navajo
Alnus: Karok; Oweekeno
Amelanchier: Coeur d'Alene; Havasupai; Okanagan-Colville; Okanagon; Thompson
Angelica: Bella Coola
Arbutus: Karok; Mendocino Indian
Arctostaphylos: Cahuilla; Pomo, Kashaya
Aristolochia: Wintoon
Artemisia: Blackfoot; Cheyenne; Costanoan; Kawaiisu; Klamath; Montana Indian; Navajo; Paiute; Tanana, Upper
Arundinaria: Seminole
Asclepias: Hopi; Montana Indian

Astragalus: Okanagan-Colville
Baccharis: Havasupai; Luiseño
Balsamorhiza: Okanagan-Colville
Betula: Tanana, Upper
Carex: Hesquiat
Carnegia: Papago; Papago & Pima; Seri
Carya: Cherokee
Ceanothus: Mewuk; Modesse; Paiute; Pomo; Tubatulabal
Celtis: Navajo; Seminole; Tewa
Cercis: Havasupai
Cercocarpus: California Indian; Hopi; Karok; Keres, Western; Klamath; Mendocino Indian; Modesse; Montana Indian; Navajo; Navajo, Ramah; Tewa; Wintoon
Chamaecrista: Seminole
Chamaecyparis: Kitasoo; Oweekeno
Chilopsis: Cahuilla
Chrysothamnus: Paiute
Cladium: Seminole
Cleome: Navajo
Comptonia: Chippewa
Cornus: Cherokee; Crow; Iroquois; Kutenai; Oweekeno; Thompson
Crataegus: Chippewa; Coeur d'Alene; Montana Indian; Ojibwa; Thompson
Cyperus: Hawaiian
Dasylirion: Apache, Mescalero
Dipsacus: Navajo
Echinocactus: Kawaiisu; Panamint
Ephedra: Kawaiisu
Epilobium: Tanana, Upper
Equisetum: Havasupai; Pomo, Kashaya; Shuswap; Thompson
Eriogonum: Kawaiisu
Fendlera: Navajo
Forestiera: Hopi
Fouquieria: Papago
Frangula: Hesquiat; Nitinaht
Fraxinus: Cherokee; Cowlitz; Havasupai; Hualapai; Kawaiisu; Mendocino Indian; Micmac; Navajo, Ramah
Fucus: Haisla & Hanaksiala
Garrya: Yurok, South Coast (Nererner)
Gossypium: Havasupai; Navajo
Gutierrezia: Navajo
Holodiscus: Hesquiat; Kwakiutl, Southern; Nitinaht; Nootka; Okanagan-Colville; Saanich; Salish; Salish, Coast
Hordeum: Kawaiisu
Juncus: Navajo; Pomo
Juniperus: Apache, Mescalero; Hopi; Paiute; Tewa; Zuni
Lagenaria: Hopi; Navajo, Ramah
Larrea: Kawaiisu; Panamint; Papago
Ledum: Tanana, Upper
Mahonia: Hopi
Malus: Haisla & Hanaksiala; Hesquiat; Nitinaht; Quileute; Salish, Coast
Monarda: Blackfoot
Muhlenbergia: Kawaiisu
Myrcianthes: Seminole
Nereocystis: Nitinaht
Nyssa: Chippewa
Olneya: Cahuilla; Papago; Pima
Opuntia: Havasupai; Keres, Western
Ostrya: Malecite
Pachycereus: Seri
Philadelphus: Coeur d'Alene; Lummi; Okanagan-Colville; Paiute; Thompson
Phragmites: Kawaiisu; Seminole
Picea: Cree, Woodlands; Eskimo, Alaska; Haisla & Hanaksiala; Koyukon; Navajo; Tanana, Upper
Pinus: Blackfoot; Haisla; Haisla & Hanaksiala; Havasupai; Hopi; Kwakiutl, Southern; Kwakwaka'wakw; Montana Indian; Navajo; Navajo, Ramah
Pipturus: Hawaiian
Pluchea: Havasupai; Papago
Poaceae sp.: Tanana, Upper
Populus: Hopi; Navajo; Navajo, Ramah; Okanagan-Colville; Thompson; Yurok
Proboscidea: Cahuilla
Prosopis: Cahuilla; Diegueño; Papago
Prunus: Blackfoot; Hopi; Kawaiisu; Salish, Coast; Thompson
Pseudotsuga: Karok
Quercus: Apache, Chiricahua & Mescalero; Apache, Mescalero; Chippewa; Cowlitz; Havasupai; Isleta; Keres, Western; Mewuk; Navajo; Navajo, Ramah; Ojibwa; Seminole; Tewa; Tewa of Hano
Rhus: Jemez; Luiseño; Modesse
Ribes: Navajo; Quileute; Salish, Coast
Rosa: Navajo; Thompson
Salix: Cheyenne; Cowlitz; Cree, Woodlands; Haisla & Hanaksiala; Havasupai; Karok; Kawaiisu; Montana Indian; Navajo; Navajo, Ramah
Sambucus: Iroquois; Montana Indian; Seminole
Sarcobatus: Hopi; Navajo; Paiute
Scirpus: Thompson
Serenoa: Seminole
Sideroxylon: Seminole
Stenocereus: Seri
Symphoricarpos: Kwakiutl, Southern; Paiute
Taxodium: Seminole
Taxus: Bella Coola; Clallam; Cowlitz; Hanaksiala; Hesquiat; Karok; Kwakiutl, Southern; Nitinaht; Oweekeno; Pomo, Kashaya; Quinault; Samish; Swinomish; Thompson
Tetradymia: Kawaiisu
Thuja: Hesquiat; Kwakiutl; Oweekeno; Quileute; Salish, Coast; Tsimshian
Tilia: Chippewa
Torreya: Pomo
Tsuga: Hanaksiala; Thompson
Ulmus: Dakota; Omaha; Pawnee; Ponca; Winnebago
Vitis: Karok
Washingtonia: Cahuilla
Xerophyllum: Nitinaht
Yucca: Apache; Dakota; Havasupai; Keres, Western; Navajo; Southwest Indians; Zuni
Zea: Navajo; Navajo, Ramah; Tewa
Ziziphus: Havasupai

Toys & Games
Acer: Hesquiat; Ojibwa; Okanagan-Colville; Pomo; Pomo, Kashaya; Yurok
Allium: Chippewa
Ambrosia: Hesquiat
Amelanchier: Blackfoot; Dakota; Lakota; Omaha; Pawnee; Ponca; Snohomish; Winnebago
Andropogon: Omaha; Ponca
Angelica: Hanaksiala; Nitinaht
Apocynum: Havasupai
Arbutus: Pomo, Kashaya; Salish, Coast
Arisaema: Pawnee
Artemisia: Dakota; Omaha; Pawnee; Ponca; Winnebago
Arundinaria: Keres, Western
Asclepias: Omaha; Pawnee; Ponca; Winnebago
Astragalus: Omaha; Ponca
Baccharis: Havasupai
Betula: Dakota; Navajo; Omaha; Pawnee; Ponca; Winnebago
Bouteloua: Lakota
Calochortus: Hopi
Castilleja: Bella Coola; Karok
Cephalanthus: Comanche
Cercocarpus: Navajo
Chamaecyparis: Yurok
Claytonia: Karok; Yurok
Clintonia: Chippewa
Cornus: Blackfoot; Green River Group; Klallam; Skagit
Cowania: Walapai
Croton: Oglala
Cucurbita: Havasupai; Keres, Western
Dalea: Kiowa
Delphinium: Hopi
Desmanthus: Omaha; Pawnee; Ponca
Dimorphocarpa: Hopi
Elaeagnus: Blackfoot
Eleocharis: Cheyenne
Equisetum: Gosiute; Paiute; Sioux; Ute; Winnebago
Eriogonum: Karok; Okanagan-Colville; Paiute
Erythronium: Thompson
Escobaria: Blackfoot
Fomes: Chehalis; Haisla & Hanaksiala; Snohomish
Fomitopsis: Haisla & Hanaksiala
Fraxinus: Havasupai
Fucus: Oweekeno
Ganoderma: Haisla & Hanaksiala
Garrya: Havasupai
Gymnocladus: Winnebago
Hastingsia: Karok
Heracleum: Blackfoot; Haisla & Hanaksiala; Nitinaht; Pomo, Kashaya
Heteromeles: Karok
Hierochloe: Thompson
Holodiscus: Karok; Nitinaht; Okanagan-Colville
Hordeum: Iroquois

Juncus: Pomo, Kashaya
Juniperus: Havasupai; Hopi; Navajo; Okanagan-Colville; Paiute; Tewa
Lagenaria: Keres, Western; Keresan
Lessoniopsis: Nitinaht
Leymus: Okanagan-Colville
Lithospermum: Blackfoot; Okanogon; Thompson
Lysichiton: Kitasoo
Macrocystis: Hesquiat
Marah: Yurok
Melilotus: Pima
Mentzelia: Kawaiisu
Menziesia: Nitinaht
Nereocystis: Makah
Oenanthe: Makah; Quileute
Opuntia: Dakota
Oxalis: Pomo, Kashaya
Phragmites: Kawaiisu; Tewa
Phyllospadix: Hesquiat; Quileute
Physalis: Dakota
Physocarpus: Hesquiat; Tolowa
Picea: Cree, Woodlands; Hoh; Quileute
Pinus: Blackfoot; Chippewa; Navajo; Navajo, Ramah; Pomo, Kashaya; Seminole
Polyporus: Haisla & Hanaksiala
Polystichum: Clallam; Karok; Klallam; Makah; Nitinaht; Paiute
Populus: Blackfoot; Cree, Woodlands; Dakota; Hopi; Navajo; Omaha; Pawnee; Ponca; Shuswap; Tanana, Upper
Postelsia: Nitinaht
Prosopis: Pima
Prunus: Dakota
Pulsatilla: Blackfoot
Purshia: Navajo
Quercus: Apache, Mescalero; Cahuilla; Dakota; Kawaiisu; Navajo; Omaha; Pawnee; Ponca; Seminole; Winnebago
Quincula: Kiowa
Ranunculus: Pomo, Kashaya
Ribes: Blackfoot; Omaha
Rosa: Cree, Woodlands; Okanogon; Thompson
Rubus: Nitinaht
Sabal: Seminole
Sagittaria: Cocopa
Salix: Abnaki; Apache, White Mountain; Blackfoot; Cree, Woodlands; Haisla; Kawaiisu; Montana Indian; Navajo, Ramah; Seminole
Sambucus: Cahuilla; Dakota; Diegueño; Kwakiutl, Southern; Mendocino Indian; Menominee; Meskwaki; Nitinaht; Omaha; Pawnee; Pomo, Kashaya; Ponca; Salish, Coast; Seminole; Yokut
Sapindus: Comanche
Sarcobatus: Jemez; Navajo
Sarracenia: Chippewa
Scirpus: Chippewa; Dakota; Havasupai
Sequoia: Pomo, Kashaya
Serenoa: Seminole
Sium: Lakota
Smilax: Kiowa

Solidago: Okanagan-Colville
Sorghum: Kiowa
Stipa: Lakota; Navajo, Ramah; Okanagan-Colville
Streptopus: Hesquiat
Symphoricarpos: Lakota; Nitinaht; Paiute
Taraxacum: Hesquiat
Taxodium: Seminole
Taxus: Hoh; Quileute; Salish, Coast
Thalictrum: Dakota; Omaha; Pawnee; Ponca
Tsuga: Oweekeno
Typha: Havasupai; Hualapai
Ulmus: Dakota; Omaha; Pawnee; Ponca; Winnebago
Umbellularia: Karok
Urtica: Dakota; Omaha; Pawnee; Ponca; Winnebago
Viburnum: Dakota; Omaha; Pawnee; Ponca; Winnebago
Viola: Karok; Omaha
Vitis: Havasupai
Washingtonia: Cahuilla
Yucca: Apache, White Mountain; Havasupai; Navajo
Zea: Cherokee; Tewa; Zuni

Unspecified
Agoseris: Keres, Western
Cirsium: Keres, Western
Clematis: Keres, Western
Corydalis: Keres, Western
Desmanthus: Keres, Western
Echinocereus: Keres, Western
Elymus: Keres, Western
Fomes: Keres, Western
Gleditsia: Keres, Western
Hymenopappus: Keres, Western
Ipomoea: Keres, Western
Ipomopsis: Keres, Western
Juniperus: Keres, Western
Lappula: Keres, Western
Lepidium: Keres, Western
Lotus: Keres, Western
Medicago: Keres, Western
Moricandia: Keres, Western
Panicum: Keres, Western
Parthenocissus: Keres, Western
Phragmites: Keres, Western
Psoralidium: Keres, Western
Pyrrhopappus: Keres, Western
Ribes: Keres, Western
Salsola: Keres, Western
Solanum: Keres, Western
Spirogyra: Keres, Western
Suaeda: Keres, Western

Walking Sticks
Arctostaphylos: Karok
Cercocarpus: Yuki
Fraxinus: Hualapai; Mendocino Indian
Morus: Hualapai
Salix: Haisla & Hanaksiala; Lakota; Montana Indian

Water Indicator
Alnus: Nitinaht; Thompson
Athyrium: Okanagan-Colville
Forestiera: Isleta

Gymnocarpium: Okanagan-Colville
Hamamelis: Mohegan
Lithospermum: Okanagan-Colville
Malus: Mohegan
Mimulus: Keres, Western
Prunus: Mohegan
Pteridium: Okanagan-Colville
Sanicula: Karok
Woodsia: Okanagan-Colville

Waterproofing Agent
Abies: Malecite
Arctostaphylos: Thompson
Castilleja: Blackfoot
Encelia: Papago
Epilobium: Blackfoot
Gaillardia: Blackfoot
Lygodesmia: Blackfoot
Perideridia: Blackfoot
Picea: Malecite; Nitinaht
Pinus: Apache, Mescalero; Apache, Western; Apache, White Mountain; Blackfoot; Havasupai; Hopi; Hualapai; Kawaiisu; Navajo; Ojibwa; Paiute; Potawatomi; Tewa; Tewa of Hano; Thompson
Sambucus: Makah
Tsuga: Quileute
Typha: Potawatomi
Yucca: Havasupai; Navajo

Weapon
Agave: Apache, Western
Amelanchier: Montana Indian
Arundinaria: Cherokee; Seminole
Atriplex: Isleta
Baccharis: Papago
Cercocarpus: California Indian; Mendocino Indian
Chrysothamnus: Isleta
Cicuta: Okanagan-Colville
Cornus: Chippewa
Coryphantha: Comanche
Glycyrrhiza: Blackfoot
Holodiscus: Thompson
Juniperus: Navajo; Okanagan-Colville
Larrea: Papago
Lithospermum: Zuni
Morus: Apache, Mescalero
Picea: Eskimo, Alaska; Hesquiat
Prunus: Klamath
Ptelea: Havasupai
Quercus: Navajo; Tewa
Rhus: Navajo, Ramah
Ribes: Havasupai
Sapindus: Papago
Sarcobatus: Navajo
Taxus: Haisla & Hanaksiala; Hesquiat; Salish, Coast; Swinomish
Tsuga: Haisla & Hanaksiala
Typha: Ojibwa

Index of Synonyms

Synonymous plant names used in the original sources are listed alphabetically and are cross-referenced to the names used in the Catalog of Plants. Names for species, subspecies, and varieties are all in lowercase, although some were capitalized in the original sources. Variations in names appearing in some sources are indicated in parentheses.

Abies canadensis, see *Picea glauca*
Abies nigra, see *Picea mariana*
Abies nobilis, see *A. procera*
Acanthochiton wrightii, see *Amaranthus acanthochiton*
Acer neomexicana, see *A. glabrum* var. *neomexicanum*
Acer pennsylvanicum, see *A. pensylvanicum*
Acer rubra, see *A. rubrum*
Acer saccharinum var. *nigrum,* see *A. nigrum*
Acer saccharophorum, see *A. saccharum*
Acerates auriculata, see *Asclepias auriculata*
Achillaea, see *Achillea*
Achillea borealis, see *A. millefolium* var. *borealis*
Achillea borealis ssp. *arenicola,* see *A. millefolium* var. *arenicola*
Achillea lanulosa, see *A. millefolium* var. *occidentalis*
Achillea millifolium, see *A. millefolium*
Achilles, see *Achillea*
Achlys triphylla, see *Achlys triphylla*
Aciphyllaea acerosa, see *Thymophylla acerosa*
Aciphyllea acerosa, see *Thymophylla acerosa*
Aconitum delphinifolium, see *A. delphiniifolium*
Acrostichum danaeaefolium, see *A. danaeifolium*
Actaea alba, see *A. pachypoda*
Actaea arguta, see *A. rubra* ssp. *arguta*
Actaea eburnea, see *A. pachypoda*
Actaea spicata, see *A. rubra*
Actaea spicata ssp. *rubra,* see *A. rubra* ssp. *rubra*
Actinea acaulis, see *Tetraneuris acaulis* var. *arizonica*
Actinea argentea, see *Tetraneuris argentea*
Actinea richardsoni, see *Hymenoxys richardsonii* var. *richardsonii*
Actinea scaposa, see *Tetraneuris argentea*
Actinella richardsoni, see *Hymenoxys richardsonii*

Acuan illinoensis, see *Desmanthus illinoensis*
Adenostoma fasiculatum, see *A. fasciculatum*
Adiantum jordani, see *A. jordanii*
Adiantum pedatum var. *aleuticum,* see *A. aleuticum*
Aesculus arguta, see *A. glabra* var. *arguta*
Aesculus californicus, see *A. californica*
Aesculus octandra, see *A. flava*
Agastache anethiodora, see *A. foeniculum*
Agastache neo(-)mexicana, see *A. pallidiflora* ssp. *neomexicana* var. *neomexicana*
Agastache scrophulariaefolia, see *A. scrophulariifolia*
Agastrache, see *Agastache*
Agave couesii, see *A. parryi*
Agave virginica, see *Manfreda virginica*
Agoseris gracilens, see *A. aurantiaca*
Agoseris purpurea, see *A. aurantiaca* var. *purpurea*
Agoseris villosa, see *A. glauca*
Agropyron occidentale, see *Pascopyrum smithii*
Agropyron repens, see *Elytrigia repens* var. *repens*
Agropyron smithii, see *Pascopyrum smithii*
Agropyron spicatum, see *Pseudoroegneria spicata* ssp. *spicata*
Agropyron tenerum, see *Elymus trachycaulus* ssp. *trachycaulus*
Agropyron trachycaulum, see *Elymus trachycaulus* ssp. *trachycaulus*
Agrostis cryptandra, see *Sporobolus cryptandrus*
Alium acuminatum, see *Allium acuminatum*
Alium bolanderi, see *Allium bolanderi*
Allionia coccinea, see *Mirabilis coccineus*
Allionia nyctaginea, see *Mirabilis nyctaginea*
Allium schoenoprosum var. *sibiricum,* see *A. schoenoprasum* var. *sibiricum*
Allium cernum, see *A. cernuum*
Allium deserticola, see *A. macropetalum*

Allium deserticolum, see *A. macropetalum*
Allium drummondi, see *A. drummondii*
Allium mutabile, see *A. canadense* var. *mobilense*
Allium nuttalli(i), see *A. drummondii*
Allium palmeri, see *A. bisceptrum* var. *palmeri*
Allium recurvatum, see *A. cernuum* var. *obtusum*
Allium sabulicola, see *A. geyeri* var. *tenerum*
Allium sibiricum, see *A. schoenoprasum* var. *sibiricum*
Alnus crispa, see *A. viridis* ssp. *crispa*
Alnus crispa ssp. *sinuata,* see *A. viridis* ssp. *sinuata*
Alnus frutica, see *A. viridis* ssp. *sinuata*
Alnus mollis, see *A. viridis* ssp. *crispa*
Alnus oregana, see *A. rubra*
Alnus oregona, see *A. rubra*
Alnus rugosa (var. *americana*), see *A. incana* ssp. *rugosa*
Alnus sinuata, see *A. viridis* ssp. *sinuata*
Alnus sitchensis, see *A. viridis* ssp. *sinuata*
Alnus tenuifolia, see *A. incana* ssp. *tenuifolia*
Alocasia macrorrhiza, see *A. macrorrhizos*
Alopecurus aristulatus, see *A. aequalis* var. *aequalis*
Althaea rosea, see *Alcea rosea*
Alycine apios, see *Apios tuberosum*
Alyxia clivaeformis, see *A. oliviformis*
Amaranth, see *Amaranthus*
Amaranthus graecizans, see *A. albus*
Amaranthus hybridus paniculatus, see *A. cruentus*
Amaranthus paniculatus, see *A. cruentus*
Amaranthus torreyi, see *A. arenicola*
Amarantus, see *Amaranthus*
Amblogyne torreyi, see *Amaranthus arenicola*
Ambrosia artemisiaefolia, see *A. artemisiifolia*
Ambrosia artemisifolia, see *A. artemisiifolia*
Ambrosia elatior, see *A. artemisiifolia* var. *elatior*
Amelancher, see *Amelanchier*
Amelanchier cusickii, see *A. alnifolia* var. *cusickii*

Amelanchier florida, see *A. alnifolia* var. *semiintegrifolia*
Amelanchier huronensis, see *A. sanguinea* var. *sanguinea*
Amelanchier mormonica, see *A. utahensis* ssp. *utahensis*
Amelanchier oblongifolia, see *A. arborea* var. *arborea*
Amelanchier prunifolia, see *A. utahensis*
Amelanchier spicata, see *A. stolonifera*
Amelanchier utahenis, see *A. utahensis*
Amelanchier venulosa, see *A. utahensis*
Amianthium muscaetoxicum, see *A. muscitoxicum*
Ammobroma sonorae, see *Pholisma sonorae*
Ammodenia peploides major, see *Honckenya peploides* ssp. *major*
Amoracia lapathifolia, see *Armoracia rusticana*
Amorpha angustifolia, see *A. fruticosa*
Amorpha fragrans, see *A. fruticosa*
Amorpha microphyllus, see *A. nana*
Ampelopsis quinquefolia, see *Parthenocissus quinquefolia* var. *quinquefolia*
Amphicarpa bracteata, see *Amphicarpaea bracteata*
Amphicarpa monoica, see *Amphicarpaea bracteata*
Amphicarpa pitcheri, see *Amphicarpaea bracteata*
Amphicarpaea monoica, see *A. bracteata*
Amsinckia parviflora, see *A. lycopsoides*
Amsinckia tesselata, see *A. tessellata*
Amsinkia, see *Amsinckia*
Amsonia brevifolia, see *A. tomentosa* var. *tomentosa*
Anacharis canadensis, see *Elodea canadensis*
Andromeda glaucophylla, see *A. polifolia* var. *glaucophylla*
Andropogon furcatus, see *A. gerardii*
Andropogon saccharoides, see *Bothriochloa saccharoides*
Andropogon scoparius, see *Schizachyrium scoparium* ssp. *scoparium*
Anemone globosa, see *A. multifida* var. *globosa*
Anemone nuttalliana, see *Pulsatilla patens* ssp. *multifida*
Anemone occidentalis, see *Pulsatilla occidentalis*
Anemone patens (var. *wolfgangiana*), see *Pulsatilla patens* ssp. *multifida*
Anemone pennsylvanicum, see *A. canadensis*
Angelica hendersoni, see *A. tomentosa* var. *hendersonii*
Angelica linearloba, see *A. lineariloba*
Anisostichus capreolata, see *Bignonia capreolata*
Anogra albicaulis, see *Oenothera albicaulis*
Anogra pallida, see *Oenothera pallida* ssp. *pallida*
Anogra runcinata, see *Oenothera pallida* ssp. *runcinata*
Antennaria aprica, see *A. parvifolia*
Antennaria margaritacea, see *Anaphalis margaritacea*
Antennaria neodioica, see *A. howellii* ssp. *neodioica*
Antennaria parviflora, see *A. parvifolia*
Anthemis nobilis, see *Chamaemelum nobile*
Anthemus, see *Anthemis*

Anthericum torreyi, see *Echeandia flavescens*
Antidesma palvinatum, see *A. pulvinatum*
Aphyllon fasciculatum, see *Orobanche fasciculata*
Apios apios, see *A. americana*
Apios (*t*)*uberosa,* see *A. americana*
Apium graveolum, see *A. graveolens*
Aplopappus gracilis, see *Machaeranthera gracilis*
Aplopappus heterophyllus, see *Isocoma pluriflora*
Aplopappus lanuginosus, see *Stenotus lanuginosus* var. *lanuginosus*
Aplopappus nanus, see *Ericameria nana*
Aplopappus nuttallii, see *Machaeranthera grindelioides* var. *grindelioides*
Aplopappus palmeri, see *Ericameria palmeri* var. *palmeri*
Aplopappus propinquus, see *Ericameria brachylepis*
Aplopappus squarrosus ssp. *grindelioides,* see *Hazardia squarrosa* var. *squarrosa*
Apocynum androsaemifolium var. *nevadense,* see *A. cannabinum*
Apocynum androsaemilifolium, see *A. androsaemifolium*
Apocynum angustifolium, see *A. cannabinum*
Apocynum cannabium, see *A. cannabinum*
Apocynum hypericifolium, see *A. cannabinum*
Apocynum medium, see *A.* ×*floribundum*
Apocynum pubescens, see *A. cannabinum*
Apocynum sibiricum var. *salignum,* see *A. cannabinum*
Apocynum suksdorfii var. *angustifolium,* see *A. cannabinum*
Apocynum viride, see *A. cannabinum*
Aquilegia truncata, see *A. formosa* var. *formosa*
Arabis drummondi, see *A. drummondii*
Arabis hoelboellii, see *A. holboellii*
Arachis hypogae, see *A. hypogaea*
Aragallus lagopus, see *Oxytropis lagopus*
Aralia medicalis, see *A. nudicaulis*
Aralia naudicalis, see *A. nudicaulis*
Aralia nudicalis, see *A. nudicaulis*
Aralia trifolia, see *Panax trifolius*
Arbor vitae, see *Thuja occidentalis*
Arcotostaphylos, see *Arctostaphylos*
Arctachylos, see *Arctostaphylos*
Arctaphulos, see *Arctostaphylos*
Arctos(-)*taphyl*(*l*)*os,* see *Arctostaphylos*
Arctostaphylos cinerea, see *A.* ×*cinerea*
Arctostaphylos mariposa, see *A. viscida* ssp. *mariposa*
Arctostaphylos parryi, see *A. parryana*
Arctostaphylus, see *Arctostaphylos*
Arctrostaphylos, see *Arctostaphylos*
Ardisia escallonioides, see *A. escallonoides*
Arenaria confusa, see *A. lanuginosa* ssp. *saxosa*
Arenaria peploides, see *Honckenya peploides* ssp. *peploides*
Arenaria saxosa, see *A. lanuginosa* ssp. *saxosa*
Argemone intermedia, see *A. polyanthemos*
Argemone platyceras, see *A. polyanthemos*

Argyreia tiliaefolia, see *Ipomoea tiliacea*
Arisaema atrorubens, see *A. triphyllum* ssp. *triphyllum*
Arisaema quinatum, see *A. triphyllum* ssp. *quinatum*
Aristida divircata, see *A. divaricata*
Aristida fendleriana, see *A. purpurea* var. *longiseta*
Aristida longiseta, see *A. purpurea* var. *longiseta*
Aristolchia, see *Aristolochia*
Armoracia lapathifolia, see *A. rusticana*
Artemesia, *Artemisia*
Artemisia aromatica, see *A. dracunculus*
Artemisia australis var. *escholsiana,* see *Artemisia australis*
Artemisia canadensis, see *A. campestris* ssp. *borealis* var. *borealis*
Artemisia discolor, see *A. ludoviciana* ssp. *incompta*
Artemisia forwoodii, see *A. campestris* ssp. *caudata*
Artemisia glauca, see *Artemisia dracunculus* ssp. *glauca*
Artemisia gnaphalo(*i*)*des,* see *A. ludoviciana* ssp. *ludoviciana*
Artemisia heterophylla, see *A. furcata* var. *heterophylla*
Artemisia lucoviciana, see *A. ludoviciana*
Artemisia ludiviciana, see *A. ludoviciana*
Artemisia mexicana, see *A. ludoviciana* ssp. *mexicana*
Artemisia pacifica, see *A. campestris* ssp. *pacifica*
Artemisia telesii, see *A. tilesii*
Artemisia trifida, see *A. tripartita* ssp. *tripartita*
Artemisia unalaskensis, see *A. vulgaris* var. *kamtschatica*
Artemisia vulgaris var. *heterophylla,* see *A. douglasiana*
Artemisia vulgaris var. *ludoviciana,* see *A. ludoviciana* ssp. *ludoviciana*
Artemisia wrightii, see *A. carruthii*
Artemisis, see *Artemisia*
Artocarpus incisa, see *A. altilis*
Arum esculentum, see *Colocasia esculenta*
Aruncus acuminatus, see *A. dioicus* var. *acuminatus*
Aruncus aruncus, see *A. dioicus* var. *vulgaris*
Aruncus sylvester, see *A. dioicus* var. *vulgaris*
Arundinaria macrosperma, see *A. gigantea*
Arundinaria tecta, see *A. gigantea* ssp. *tecta*
Arvtostaphylos, see *Arctostaphylos*
Asarum apiifolia, see *Hexastylis arifolia*
Asarum arifolia, see *Hexastylis arifolia* var. *arifolia*
Asarum arifolium, see *Hexastylis arifolia* var. *arifolia*
Ascepias incarnata, see *Asclepias incarnata*
Asclepia syriaca, see *Asclepias syriaca*
Asclepias galioides, see *A. subverticillata*
Asclepias mexicana, see *A. fascicularis*
Asclepias viridis flora, see *A. viridiflora*
Asclepiodora decumbens, see *Asclepias asperula* ssp. *capricornu*
Asclepisa syriaca, see *Asclepias syriaca*
Ascyrum, see *Hypericum*

Index of Synonyms

Ascyrum crux andreae, see *Hypericum crux-andreae*
Ascyrum hypericoides, see *Hypericum hypericoides* ssp. *hypericoides*
Ascyrum linifolium, see *Hypericum hypericoides* ssp. *hypericoides*
Aseyrum, see *Hypericum*
Aspidium acrostichoides, see *Polystichum acrostichoides*
Aspidium cristatum, see *Dryopteris cristata*
Aspidium spinulosum, see *Dryopteris carthusiana*
Asplenium cyclosorum, see *Athyrium filix-femina* ssp. *cyclosorum*
Asplenium filis-femina, see *Athyrium filix-femina*
Asplenium filix-femina, see *Athyrium filix-femina*
Aster arenosus, see *Chaetopappa ericoides*
Aster canescens, see *Machaeranthera canescens* ssp. *canescens* var. *canescens*
Aster cichoriaceus, see *Machaeranthera canescens* ssp. *glabra* var. *aristata*
Aster coerulescens, see *A. praealtus* var. *coerulescens*
Aster commutatus var. *crassulus*, see *A. falcatus* var. *crassulus*
Aster ericaefolius, see *Chaetopappa ericoides*
Aster exilis, see *A. subulatus* var. *ligulatus*
Aster frondosus, see *Brachyactis frondosa*
Aster geyeri, see *A. laevis* var. *geyeri*
Aster hesperius, see *A. lanceolatus* ssp. *hesperius*
Aster incanopilosus, see *A. falcatus* var. *commutatus*
Aster leucanthemifolius, see *Machaeranthera canescens* ssp. *canescens* var. *leucanthemifolia*
Aster leucelene, see *Chaetopappa ericoides*
Aster multiflorus, see *A. ericoides* var. *ericoides*
Aster paniculatus, see *A. lanceolatus* ssp. *lanceolatus* var. *lanceolatus*
Aster parviflorus, see *Machaeranthera parviflora*
Aster salicifolius, see *A. praealtus* var. *praealtus*
Aster spinosus, see *Chloracantha spinosa*
Aster tanacetifolius, see *Machaeranthera tanacetifolia*
Astragalus aboriginum, see *A. australis*
Astragalus caroliniana, see *A. canadensis* var. *canadensis*
Astragalus carolinianus, see *A. canadensis* var. *canadensis*
Astragalus caryocarpus, see *A. crassicarpus* var. *crassicarpus*
Astragalus ceramicus imperfectus, see *A. ceramicus* var. *filifolius*
Astragalus decumbens, see *A. miser* var. *decumbens*
Astragalus diphysus, see *A. lentiginosus* var. *diphysus*
Astragalus haydenianus, see *A. bisulcatus* var. *haydenianus*
Astragalus impensus, see *A. kentrophyta* var. *elatus*
Astragalus jemensis, see *A. cyaneus*
Astragalus junceus, see *A. convallarius* var. *convallarius*
Astragalus matthewsii, see *A. mollissimus* var. *matthewsii*
Astragalus nitidus, see *A. adsurgens* var. *robustior*
Astragalus palans, see *A. lentiginosus* var. *palans*
Astragalus pictus filifolius, see *A. ceramicus* var. *ceramicus*
Astragalus scaposus, see *A. calycosus* var. *scaposus*
Astragalus sonorae, see *A. humistratus* var. *sonorae*
Astralagus canadensis, see *Astragalus canadensis*
Atheropogon curtipendula, see *Bouteloua curtipendula* var. *curtipendula*
Athyrium angustum, see *A. filix-femina* ssp. *angustum*
Athyrium filix-foemina, see *A. filix-femina*
Atragene pseudalpina, see *Clematis columbiana* var. *columbiana*
Atriplex argenta, see *A. argentea*
Atriplex bracteosa, see *A. serenana*
Atriplex cornuta, see *A. argentea*
Atriplex expansa, see *A. argentea* ssp. *expansa*
Atriplex jonesii, see *A. confertifolia*
Atriplex philonitra, see *A. powellii*
Audibertias stachyoides, see *Salvia mellifera*
Azalea occidentalis, see *Rhododendron occidentale* var. *occidentale*
Baccharis consanguinea, see *B. pilularis*
Baccharis glutinosa, see *B. salicifolia*
Baccharis viminea, see *B. salicifolia*
Baeria chrysostoma, see *Lasthenia californica*
Bahia neomexicana, see *Schkuhria multiflora*
Bahia oppositifolia, see *Picradeniopsis oppositifolia*
Bahia woodhousei, see *Picradeniopsis woodhousei*
Balsamorhiza hirsuta, see *B. hookeri* var. *hirsuta*
Balsamorhiza terebinthacea, see *B.* ×*terebinthacea*
Balsamorr(i)hiza, see *Balsamorhiza*
Baptisia leucantha, see *B. alba* var. *macrophylla*
Batis vermiculata, see *Sarcobatus vermiculatus*
Beckmannia erucaeformis, see *B. syzigachne*
Beloperone californica, see *Justicia californica*
Benzoin aestivale, see *Lindera benzoin* var. *benzoin*
Benzoin, see *Lindera*
Berbaris aquifolium, see *Mahonia aquifolium*
Berbaris nervosa, see *Mahonia nervosa* var. *nervosa*
Berberis aquifolium, see *Mahonia aquifolium*
Berberis dictyota, see *Mahonia dictyota*
Berberis fermontii, see *Mahonia fremontii*
Berberis fremontii, see *Mahonia fremontii*
Berberis haematocarpa, see *Mahonia haematocarpa*
Berberis nervosa, see *Mahonia nervosa* var. *nervosa*
Berberis pinnata, see *Mahonia pinnata* ssp. *pinnata*
Berberis repens, see *Mahonia repens*
Berberis sp., see *Berberis* sp., and *Mahonia* sp.
Betula alba, see *B. pubescens* ssp. *pubescens*
Betula alba (var. *papyrifera*), see *B. papyrifera*
Betula excelsa, see *B. alleghaniensis*
Betula exilis, see *B. nana*
Betula fontinalis, see *B. occidentalis*
Betula glandulosa, see *B. nana*
Betula lutea, see *B. alleghaniensis* var. *alleghaniensis*
Betula occidentalis × *papyrifera*, see *B. occidentalis*
Bidens gracilis, see *B. amplectens*
Bidens levis, see *B. laevis*
Bigelovia douglasii (var. *stenophylla*), see *Chrysothamnus viscidiflorus*
Bigelovia graveolens, see *Chrysothamnus nauseosus*
Bigelovia howardii, see *Chrysothamnus parryi* ssp. *howardii*
Bigelovia parishii, see *Ericameria parishii*
Bigilovia graveolens, see *Chrysothamnus nauseosus*
Bignonia crucigera, see *B. capreolata*
Biscutella wislizeni, see *Dimorphocarpa wislizeni*
Bistorta bistortoides, see *Polygonum bistortoides*
Blennosperma californicum, see *B. nanum*
Blepharipappus platyglossus, see *Layia platyglossa*
Blitum capitatum, see *Chenopodium capitatum* var. *capitatum*
Blitum nuttallianum, see *Monolepis nuttalliana*
Bloomeria aurea, see *B. crocea* var. *aurea*
Boebera papposa, see *Dyssodia papposa*
Boerhaavia erecta, see *Boerhavia erecta*
Bossekia parviflora, see *Rubus parviflorus*
Bossekia parvifolia, see *Rubus parviflorus*
Botriochloa saccharoides, see *Bothriochloa saccharoides*
Botrychium virginicum, see *B. virginianum*
Bouteloua cligostachya, see *B. gracilis*
Bouteloua oligostachya, see *B. gracilis*
Boykinia elat(e)a, see *B. occidentalis*
Brassica alba, see *Sinapis alba*
Brassica arvensis, see *Moricandia arvensis*
Brassica campestris, see *B. rapa* var. *rapa*
Brassica geniculata, see *Hirschfeldia incana*
Brassica hirta, see *Sinapis alba*
Brassica kaber, see *Sinapis arvensis*
Brassica napa, see *B. napus*
Brassica oleraceae, see *B. oleracea*
Brauneria angustifolia, see *Echinacea angustifolia* var. *angustifolia*
Brauneria pallida, see *Echinacea pallida*
Brauneria purpurea, see *Echinacea purpurea*
Brayodendron texanum, see *Diospyros texana*
Brickelia grandiflora var. *minor*, see *Brickellia grandiflora*
Brodiaea capitata (var. *pauciflora*), see *Dichelostemma pulchellum*
Brodiaea congesta, see *Dichelostemma pulchellum*
Brodiaea douglasii, see *Triteleia grandiflora*
Brodiaea grandiflora, see *Triteleia grandiflora*
Brodiaea hyacinthina, see *Triteleia hyacinthina* var. *hyacinthina*
Brodiaea ida-maia, see *Dichelostemma ida-maia*
Brodiaea ixioides, see *Triteleia ixioides*
Brodiaea laxa, see *Triteleia laxa*
Brodiaea multiflora, see *Dichelostemma multiflorum*
Brodiaea pulchella, see *Dichelostemma pulchellum*
Brodiaea volubilis, see *Dichelostemma volubile*
Bromus breviaristatus, see *B. marginatus* var. *breviaristatus*
Bromus maximus, see *B. diandrus*

Index of Synonyms

Bromus rigidus (var. *gussonei*), see *B. diandrus*
Bromus unioloides, see *B. catharticus*
Broussaisa, see *Broussaisia*
Broussaisia pelluoida, see *B. arguta*
Broussonetia secundiflora, see *B. papyrifera*
Bulbilis dactyloides, see *Buchloe dactyloides*
Bumelia lanuginosa, see *Sideroxylon lanuginosum* ssp. *lanuginosum*
Bursa bursa-pastoris, see *Capsella bursa-pastoris*
Butneria occidentalis, see *Calycanthus occidentalis*
Cacalia atriplicifolia, see *Arnoglossum atriplicifolium*
Caesalpinia bonducella, see *C. bonduc*
Calandrinia caulescens (var. *mensiesii*), see *C. ciliata*
Calandrinia elegans, see *C. ciliata*
Calligonum canescens, see *Atriplex canescens*
Calliproa lutea, see *Triteleia ixioides* ssp. *scabra*
Callirrhoe involucrata, see *Callirhoe involucrata*
Calochortus caeruleus var. *maweanus*, see *C. tolmiei*
Calochortus gunnisoni, see *C. gunnisonii*
Calochortus luteus var. *oculatus*, see *C. vestae*
Calochortus maweanus, see *C. tolmiei*
Calochortus nuttall(i), see *C. nuttallii*
Calochortus nuttallii var. *aureus*, see *C. aureus*
Calohortus, see *Calochortus*
Calomovilfa gigantea, see *Calamovilfa gigantea*
Caltha asarifolia, see *C. palustris* var. *palustris*
Caltha rotundifolia, see *C. leptosepala* ssp. *leptosepala* var. *leptosepala*
Calycanthus fertilis, see *C. floridus* var. *glaucus*
Calyptridium monandrum, see *Cistanthe monandra*
Camassia esculenta, see *C. scilloides*
Cambrium sp., see *Quercus* sp.
Cammassia, see *Camassia*
Campanula americana, see *Campanulastrum americanum*
Campanula prenanthoides, see *Asyneuma prenanthoides*
Camptosorus rhizophyllus, see *Asplenium rhizophyllum*
Campylotheca sp., see *Bidens* sp.
Canotia holocantha, see *C. holacantha*
Capriola dactylon, see *Cynodon dactylon* var. *dactylon*
Capsicum annum, see *C. annuum*
Capsicum anuum, see *C. annuum*
Capsicum frutescens, see *C. annuum* var. *frutescens*
Capsicum frutescens var. *baccatum*, see *C. annuum* var. *frutescens*
Capsicum frutescens var. *longum*, see *C. annuum* var. *frutescens*
Cardamine bulbosa, see *C. rhomboidea*
Carduus altissimus, see *Cirsium altissimum*
Carduus edulis, see *Cirsium edule*
Carduus hookerianus, see *Cirsium hookerianum*
Carduus lanceolatus, see *Cirsium vulgare*
Carduus ochrocentrus, see *Cirsium ochrocentrum*

Carduus remaliflorus, see *Cirsium remotifolium*
Carex ex-siccata, see *C. exsiccata*
Carex festivella, see *C. microptera*
Carex mendocinoensis, see *C. mendocinensis*
Carex nebrascensis praevia, see *C. nebrascensis*
Carex nebraskensis, see *C. nebrascensis*
Carex pensylvanica var. *digyna*, see *C. inops* ssp. *heliophila*
Carex sitchensis, see *C. aquatilis* var. *dives*
Carnegiea gigantea, see *Carnegia gigantea*
Carum gairdneri, see *Perideridia gairdneri* ssp. *gairdneri*
Carum kelloggii, see *Perideridia kelloggii*
Carum oreganum, see *Perideridia oregana*
Carun kelloggii, see *Perideridia kelloggii*
Carya illinoensis, see *C. illinoinensis*
Carya pecan, see *C. illinoinensis*
Carya tomentosa, see *C. alba*
Cassia fasciculata, see *Chamaecrista fasciculata*
Cassia hebecarpa, see *Senna hebecarpa*
Cassia marilandica, see *Senna marilandica*
Cassia nictitans, see *Chamaecrista nictitans* ssp. *nictitans* var. *nictitans*
Cassia occidentalis, see *Senna occidentalis*
Cassia tora, see *Senna tora*
Cassis occidentalis, see *Senna occidentalis*
Cassitha filiformis, see *Cassytha filiformis*
Castalia odorata, see *Nymphaea odorata*
Castalia sp., see *Nymphaea* sp.
Castanopsis chrysophylla minor, see *C. chrysophylla*
Castilleia, see *Castilleja*
Castilleja chromosa, see *C. angustifolia* var. *dubia*
Castilleja confusa, see *C. miniata* var. *miniata*
Castilleja linariaefolia, see *C. linariifolia*
Castilleja parriflora, see *C. parviflora*
Castilleja pinetorum, see *C. applegatei* var. *pinetorum*
Castilleja unalaschensis, see *C. unalaschcensis*
Cathartolinum puberulum, see *Linum puberulum*
Ceanothus americana, see *C. americanus*
Ceanothus divaricatus, see *C. oliganthus* ssp. *oliganthus*
Ceanothus integerr(i)umus, see *C. integerrimus*
Ceanothus ovatus, see *C. herbaceus*
Celtis douglasii, see *C. laevigata* var. *reticulata*
Celtis laevigata brevipes, see *C. laevigata* var. *brevipes*
Celtis reticulata, see *C. laevigata* var. *reticulata*
Centromadia fitchii, see *Hemizonia fitchii*
Centunculus sp., see *Anagallis* sp.
Cephaelis ipecacuanha, see *Euphorbia ipecacuanhae*
Cerastium holosteoides, see *C. fontanum* ssp. *vulgare*
Cerasus demissa, see *Prunus virginiana* var. *demissa*
Cerasus ilicifolia, see *Prunus ilicifolia*
Ceratoides lanata, see *Krascheninnikovia lanata*
Ceratonia chilensis, see *Prosopis chilensis*
Cercidium floridum, see *Parkinsonia florida*
Cercidium microphyllum, see *Parkinsonia microphylla*

Cercidium torreyana, see *Prosopis glandulosa* var. *torreyana*
Cercidium torreyanum, see *Prosopis glandulosa* var. *torreyana*
Cercis occidentalis, see *C. canadensis* var. *texensis*
Cercocarpus betulaefolius, see *C. montanus* var. *glaber*
Cercocarpus betuloides, see *C. montanus* var. *glaber*
Cercocarpus parvifolius, see *C. montanus* var. *glaber*
Cercus gigantea, see *Carnegia gigantea*
Cercus occidentalis, see *Cercis canadensis* var. *texensis*
Cereus engelmanni, see *Echinocereus engelmannii*
Cereus gigantea, see *Carnegia gigantea*
Cereus giganteus, see *Carnegia gigantea*
Cereus greggii, see *Peniocereus greggii* var. *greggii*
Chaetopappa aurea, see *Pentachaeta aurea*
Chamaecrista brachiata, see *C. fasciculata*
Chamaecrista brachista, see *C. fasciculata*
Chamaecyperis nootkatensis, see *Chamaecyparis nootkatensis*
Chamaenerion angustifolium, see *Epilobium angustifolium* ssp. *angustifolium*
Chamoenerion angustifolium, see *Epilobium angustifolium* ssp. *angustifolium*
Cheilanthes wootoni, see *C. wootonii*
Cheirinia aspera, see *Erysimum asperum*
Cheirodendron caudicchaudii, see *C. gaudicchaudii*
Chenepodium, see *Chenopodium*
Chenopodium alba, see *C. album*
Chenopodium albidum, see *C. album*
Chenopodium ambrosiodes, see *C. ambrosioides*
Chenopodium cornutum, see *C. graveolens*
Chenopodium desiccatum var. *leptophylloides*, see *C. pratericola*
Chenopodium fremonti, see *C. fremontii*
Chenopodium fruticosum, see *Suaeda moquinii*
Chenopodium sandwicheum, see *C. oahuense*
Chenopodium watsoni, see *C. watsonii*
Chimaphila umbellatta, see *C. umbellata*
Chimaphila umbellata occidentalis, see *C. umbellata* ssp. *occidentalis*
Chimaphila umbellata var. *occidentalis*, see *C. umbellata* ssp. *occidentalis*
Chimaphylla, see *Chimaphila*
Chimapila, see *Chimaphila*
Chimphila, see *Chimaphila*
Chiogenes hispidula, see *Gaultheria hispidula*
Chionanthus virginica, see *C. virginicus*
Chloris curtipendula, see *Bouteloua curtipendula* var. *curtipendula*
Chlorogalum divaricatum, see *C. pomeridianum* var. *divaricatum*
Chondrophora nauseosa, see *Chrysothamnus nauseosus*
Chorogalum, see *Chlorogalum*
Chrysanthemum balsamita var. *tanacetoides*, see *Balsamita major*

Chrysanthemum leucanthemum (var. *pinnatifidum*), see *Leucanthemum vulgare*
Chrysanthemum parthenium, see *Tanacetum parthenium*
Chrysanthemum vulgare, see *Tanacetum vulgare*
Chrysobalanus oblongifolia, see *Licania michauxii*
Chrysobalanus oblongifolius, see *Licania michauxii*
Chrysobotrya odorata, see *Ribes aureum* var. *villosum*
Chrysolepis chrysophila, see *Castanopsis chrysophylla*
Chrysolepis sempervirens, see *Castanopsis sempervirens*
Chrysophylium oliviforme, see *Chrysophyllum oliviforme*
Chrysopsis foliosa, see *Heterotheca villosa* var. *villosa*
Chrysopsis graminea, see *Pityopsis graminifolia* var. *graminifolia*
Chrysopsis graminifolia, see *Pityopsis graminifolia* var. *graminifolia*
Chrysopsis hirsutissima, see *Heterotheca villosa* var. *hispida*
Chrysopsis hispida, see *Heterotheca villosa* var. *hispida*
Chrysopsis villosa, see *Heterotheca villosa* var. *villosa*
Chrysothamnus bigelovii, see *C. nauseosus* ssp. *bigelovii*
Chrysothamnus bloomeri, see *Ericameria bloomeri*
Chrysothamnus confinis, see *C. nauseosus*
Chrysothamnus graveolens, see *C. nauseosus* ssp. *graveolens*
Chrysothamnus howardi(i), see *C. parryi* ssp. *howardii*
Chrysothamnus latisquameus, see *C. nauseosus* ssp. *latisquameus*
Chrysothamnus nauseosa, see *C. nauseosus*
Chrysothamnus pinifolius, see *C. nauseosus* ssp. *consimilis*
Chrysothamnus speciosus gnaphalodes, see *C. nauseosus* ssp. *hololeucus*
Chrysothamnus stenophyllus, see *C. viscidiflorus* ssp. *viscidiflorus* var. *stenophyllus*
Chunbago zelanica, see *Plumbago zeylanica*
Cibatium whamissoi, see *Cibotium chamissoi*
Cicuta curtissii, see *C. maculata* var. *maculata*
Cicuta mackenzi(e)ana, see *C. virosa*
Cicuta occidentalis, see *C. maculata* var. *angustifolia*
Cicuta vagans, see *C. douglasii*
Cimaphila, see *Chimaphila*
Cinchona ledgeriana, see *C. calisaya*
Cinna arundinacea (var. *pendula*), see *C. arundinacea*
Circium, see *Cirsium*
Cirsium acaulescens, see *C. tioganum* var. *tioganum*
Cirsium lanceolatum, see *C. vulgare*
Cirsium occidentale var. *candidissimum*, see *C. pastoris*
Cirsium pulchellum, see *C. calcareum*
Cissus sicyoides, see *C. verticillata*
Citrullus citrullus, see *C. lanatus* var. *lanatus*
Citrullus vulgaris, see *C. lanatus* var. *lanatus*
Citrus paradisi, see *C. ×paradisi*
Cladothamnus pyroliflorus, see *Elliottia pyroliflorus*
Cladrastis lutea, see *C. kentukea*
Clarkia elegans, see *C. unguiculata*
Claytonia alsinoides, see *C. sibirica*
Claytonia sessilifolia, see *C. lanceolata* var. *sessilifolia*
Clematis douglasii, see *C. hirsutissima*
Clematis eriophora, see *C. hirsutissima*
Clematis lingusticifolia, see *C. ligusticifolia*
Clematis lugusticifolia, see *C. ligusticifolia*
Clematis pseudoalpina, see *C. columbiana* var. *columbiana*
Clematis verticellaris, see *C. occidentalis* var. *occidentalis*
Clematis virginica, see *C. virginiana*
Clematis virona, see *C. viorna*
Cleome integrifolia, see *C. serrulata*
Cleome serrulatum, see *C. serrulata*
Cleome sonorae, see *C. multicaulis*
Cleome surrulata, see *C. serrulata*
Clermontia arborescene, see *C. arborescens*
Cnicus drummondi, see *Cirsium drummondii*
Cnicus eatoni, see *Cirsium eatonii*
Cnicus eriocephalus, see *Cirsium undulatum*
Cnicus undulatus, see *Cirsium undulatum*
Coccolobis laurifolia, see *Coccoloba diversifolia*
Coelapleurum gmelini, see *Angelica lucida*
Coelopleurum gmelini, see *Angelica lucida*
Cogswellia cous, see *Lomatium cous*
Cogswellia daucifolia, see *Lomatium foeniculaceum* ssp. *daucifolium*
Cogswellia orientalis, see *Lomatium orientale*
Cogswellia sp., see *Lomatium* sp.
Coldenia hispidissima var. *latior*, see *Tiquilia latior*
Coleosanthus californicus, see *Brickellia californica*
Coloccasia antiquorum, see *Colocasia esculenta*
Comandra livida, see *Geocaulon lividum*
Comandra pallida, see *C. umbellata* ssp. *pallida*
Commandra, see *Comandra*
Commelina angustifolia, see *C. erecta* var. *angustifolia*
Commelina dianthifoli, see *C. dianthifolia*
Commelina nudiflora, see *Murdannia nudiflora*
Compositae, see Asteraceae
Condalia lycioides var. *canescens*, see *Ziziphus obtusifolia* var. *canescens*
Condalia obovata, see *C. hookeri* var. *hookeri*
Condalia parryi, see *Ziziphus parryi*
Condolis spathulata, see *Condalia globosa*
Conioselinum gmelini, see *C. gmelinii*
Conioselinum pacificum, see *C. gmelinii*
Conopholis mexicana, see *C. alpina* var. *mexicana*
Convolvulus fulcratus, see *Calystegia occidentalis* ssp. *fulcrata*
Coptis groenlandica, see *C. trifolia* ssp. *groenlandica*
Corallorhiza, see *Corallrrhiza*
Cordyline terminalis, see *C. fruticosa*
Coreopsis cardaminefolia, see *C. tinctoria* var. *tinctoria*
Cornus alba ssp. *stolonifera*, see *C. sericea* ssp. *sericea*
Cornus californica, see *C. sericea* ssp. *sericea*
Cornus instolonea, see *C. sericea* ssp. *sericea*
Cornus obliqua, see *C. amomum* ssp. *obliqua*
Cornus occidentalis, see *C. sericea* ssp. *occidentalis*
Cornus paniculata, see *C. racemosa*
Cornus pubescens, see *C. sericea* ssp. *occidentalis*
Cornus stolonifera, see *C. sericea* ssp. *sericea*
Cornus stolonifera var. *occidentalis*, see *C. sericea* ssp. *occidentalis*
Cornus stolonifora, see *C. sericea* ssp. *sericea*
Cornus stricta, see *C. foemina*
Cornus unalaschensis, see *C. unalaschkensis*
Corydalis montana, see *C. aurea* ssp. *occidentalis*
Corydalis montanum, see *C. aurea* ssp. *occidentalis*
Corylus americanus, see *C. americana*
Corylus californica, see *C. cornuta* var. *californica*
Corylus cornuta california, see *C. cornuta* var. *californica*
Corylus rostrata (var. *californica*), see *C. cornuta* var. *cornuta*
Coryphantha vivipara, see *Escobaria vivipara* var. *vivipara*
Cota sp., see *Anthemis* sp.
Cotyledon laxa, see *Sedum laxum*
Covillea glutinosa, see *Larrea tridentata* var. *tridentata*
Cowania mexicana, see *Purshia mexicana*
Cowania mexicana var. *stansburiana*, see *Purshia stansburiana*
Cowania stansburiana, see *Purshia stansburiana*
Cracca purpurea, see *Tephrosia purpurea*
Cracca virginiana, see *Tephrosia virginiana*
Crassina grandiflora, see *Zinnia grandiflora*
Crassina pulmila, see *Zinnia acerosa*
Crataegus brevispina, see *C. douglasii*
Crataegus cerronis, see *C. erythropoda*
Crataegus coccinea, see *C. chrysocarpa*
Crataegus rivalaris, see *C. rivularis*
Crataegus rotundifolia var. *bicknellii*, see *C. chrysocarpa*
Crataegus tomentosa, see *C. calpodendron*
Crepis atrabarba, see *C. atribarba*
Crepis glauca, see *C. runcinata* ssp. *glauca*
Crepis scopulorum, see *C. modocensis*
Crotalaria maritima, see *C. rotundifolia*
Crotolaria saggitalis, see *Crotalaria sagittalis*
Croton corymbosus, see *C. pottsii* var. *pottsii*
Cruciferae, see Brassicaceae
Cryptantha jamesii, see *C. cinerea* var. *jamesii*
Cryptantha jamesii var. *multicaulis*, see *C. cinerea* var. *cinerea*
Cryptanthe, see *Cryptantha*

Crypthantha, see *Cryptantha*
Cryptogramma crispa, see *C. sitchensis*
Crysanthemum parthenium, see *Tanacetum parthenium*
Cucerbita, see *Cucurbita*
Cucurbita citrullus, see *Citrullus lanatus* var. *lanatus*
Cucurbita lagenaria, see *Lagenaria siceraria*
Cucurbita melo, see *Cucumis melo*
Cucurbita perennis, see *C. foetidissima*
Cucurbita sativus, see *Cucumis sativus*
Cuminum cymium, see *C. cyminum*
Cunila origanoides, see *C. marina*
Cupressaceae sp., see *Cupressus* sp.
Cupressus disticha, see *Taxodium distichum*
Cupressus thyoides, see *Chamaecyparis thyoides*
Curcuma louza, see *C. longa*
Cuscuta curta, see *C. megalocarpa*
Cuscuta paradoxa, see *C. compacta*
Cyamus luteus, see *Nelumbo lutea*
Cycloloma artriplicifolium, see *C. atriplicifolium*
Cycopus uniflorus, see *Lycopus uniflorus*
Cymopteris, see *Cymopterus*
Cymopterus alpinus, see *Oreoxis alpina* ssp. *alpina*
Cymopterus fendleri, see *C. acaulis* var. *fendleri*
Cymopterus glomeratus, see *C. acaulis*
Cymopterus terebinthinus var. *terebinthinus*, see *Pteryxia terebinthina* var. *terebinthina*
Cympoterus, see *Cymopterus*
Cynoglossum boreale, see *C. virginianum* var. *boreale*
Cyperaceae sp., see *Cyperus* sp.
Cyperus erythrorrhizos, see *C. erythrorhizos*
Cyperus ferax, see *C. odoratus*
Cyperus inflexus, see *C. squarrosus*
Cyperus laevigata, see *C. laevigatus*
Cyprepedium, see *Cypripedium*
Cypripedium calceolus, see *C. parviflorum*
Cypripedium calceolus var. *pubescens*, see *C. pubescens*
Cypripedium calceolus ssp. *parviflorum*, see *C. parviflorum*
Cypripedium hirsutum, see *C. pubescens*
Cypripedium parviflorum (var. *pubescens*), see *C. pubescens*
Cytherea bulbosa, see *Calypso bulbosa* var. *americana*
Dalea fremontii, see *Psorothamnus fremontii* var. *fremontii*
Dalea polyadenia, see *Psorothamnus polydenius* var. *polydenius*
Dasi(o)phora fruticosa, see *Pentaphylloides floribunda*
Dasystephana puberula, see *Gentiana saponaria*
Dasystoma pedicularia, see *Aureolaria pedicularia* var. *pedicularia*
Datura metefoides, see *D. wrightii*
Datura meteloides, see *D. wrightii*
Datura quercifolia, see *D. ferox*
Delphinium ajacis, see *Consolida ajacis*

Delphinium nelsonii, see *D. nuttallianum* var. *nuttallianum*
Delphinium virescens, see *D. carolinianum* ssp. *virescens*
Dendropogon usneoides, see *Tillandsia usneoides*
Dentaria californica, see *Cardamine californica* var. *californica*
Dentaria diphylla, see *Cardamine diphylla*
Dentaria laciniata, see *Cardamine concatenata*
Dentaria maxima, see *Cardamine maxima*
Deschampsia caespitosa, see *D. cespitosa*
Descurainia incisa, see *D. incana* ssp. *incisa*
Descurainia richardsonii, see *D. incana* ssp. *incana*
Desmodium supina, see *D. incanum* var. *incanum*
Desmodium supinum, see *D. incanum* var. *incanum*
Desmodium uncinatum, see *D. sandwicense*
Deweya arguta, see *Tauschia arguta*
Dichelostemma, var. *brodiaca, capitata pauciflora*, see *D. pulchellum*
Dichelostemma capitatum, see *D. pulchellum*
Dichelostemma pulchella, see *D. pulchellum*
Dichrophyllum marginatum, see *Euphorbia marginata*
Dilazium arnottii, see *Diplazium meyenianum*
Dipholis salicifolia, see *Sideroxylon salicifolium*
Dipsacus glutinosus, see *Diplacus aurantiacus* ssp. *aurantiacus*
Dipsacus sylvestris, see *D. fullonum*
Disporum oreganum, see *D. hookeri* var. *oreganum*
Ditaxis cyanophylla, see *Argythamnia cyanophylla*
Dithyr(a)ea wislizeni(i), see *Dimorphocarpa wislizeni*
Dithyrea wizlizenii, see *Dimorphocarpa wislizeni*
Ditremexa occidentalis, see *Senna occidentalis*
Dodecatheon henersoni(i), see *D. hendersonii*
Dodecatheon pauciflorum, see *D. pulchellum* ssp. *pauciflorum*
Dodecatheon radicatum, see *D. pulchellum* ssp. *pulchellum*
Dondia fruticosa, see *Suaeda moquinii*
Dondia nigra, see *Suaeda moquinii*
Dondia suffrutescens, see *Suaeda suffrutescens*
Draba montana, see *D. rectifructa*
Dracaena aurea, see *Pleomele aurea*
Drymaria fendleri, see *D. glandulosa*
Drymocallis arguta, see *Potentilla arguta* ssp. *arguta*
Dryopteris assimilis, see *D. expansa*
Dryopteris austriaca, see *D. campyloptera*
Dryopteris dilatata, see *D. expansa*
Dryopteris dilatata ssp. *americana*, see *D. campyloptera*
Dryopteris disjuncta, see *Gymnocarpium disjunctum*
Dryopteris normalis, see *Thelypteris kunthii*
Dryopteris rigida arguta, see *D. arguta*
Dryopteris spinulosa, see *D. carthusiana*
Dysodia papposa, see *Dyssodia papposa*
Dyssodia pentachaeta, see *Thymophylla pentachaeta* var. *pentachaeta*

Dyssodia thurberi, see *Thymophylla pentachaeta* var. *belenidium*
Echeveris lanceolata, see *Dudleya lanceolata*
Echinocactus acanthodes, see *Ferocactus cylindraceus* var. *cylindraceus*
Echinocactus lecontei, see *Ferocactus cylindraceus* var. *lecontei*
Echinocactus wislizeni, see *Ferocactus wislizeni*
Echinocereus chrysocentrus, see *E. engelmannii* var. *chrysocentrus*
Echinocereus gonacanthus, see *E. triglochidiatus* var. *triglochidiatus*
Echinocereus gonocanthus, see *E. triglochidiatus* var. *triglochidiatus*
Echinocereus leeanus, see *E. polyacanthus*
Echinocereus wislizeni, see *Ferocactus wislizeni*
Echinochloa colonum, see *E. colona*
Echinochloa crusgalli, see *E. crus-galli*
Echinochloa crusgalli var. *zelayensis*, see *E. crus-pavonis* var. *macera*
Echinocystis fabacea, see *Marah fabaceus*
Echinocystis horrida, see *Marah horridus*
Echinocystis macrocarpa, see *Marah macrocarpus*
Echinocystis oregana(s), see *Marah oreganus*
Echinopanax horridum, see *Oplopanax horridus*
Echinospermum lappula, see *Lappula squarrosa*
Elaeagnus argentea, see *E. commutata*
Elaeagnus asgentae, see *E. commutata*
Eleagnus, see *Elaeagnus*
Eleocharis caribaea, see *E. geniculata*
Ellisea membranacea, see *Pholistoma membranaceum*
Ellisia membranacea, see *Pholistoma membranaceum*
Elymus arenarius ssp. *mollis*, see *Leymus mollis* ssp. *mollis*
Elymus ciner(e)us, see *Leymus cinereus*
Elymus condensatus, see *Leymus condensatus*
Elymus mollis (var. *mollis*), see *Leymus mollis* ssp. *mollis*
Elymus robustus, see *E. canadensis* var. *canadensis*
Elymus triticoides, see *Leymus triticoides*
Empertrum nigrum, see *Empetrum nigrum*
Encelia actoni, see *E. virginensis* var. *actonii*
Ephedra viridus, see *E. viridis*
Ephredra, see *Ephedra*
Epicampes rigens (*californica*), see *Muhlenbergia rigens*
Epicompes rigens, see *Muhlenbergia rigens*
Epilobium adenocaulon, see *E. ciliatum* ssp. *ciliatum*
Epilobium angustofolium, see *E. angustifolium*
Epilobium miniatum, see *E. minutum*
Epilobium paniculatum, see *E. brachycarpum*
Epipactis pubescens, see *Goodyera pubescens*
Epipactis repens var. *ophioides*, see *Goodyera repens*
Equisetum funstoni, see *E. laevigatum*
Equisetum hiemale, see *E. hyemale*

Index of Synonyms

Equisetum kansanum, see *E. laevigatum*
Equisetum loevigatum, see *E. laevigatum*
Equisetum silvaticum, see *E. sylvaticum*
Eremocarpus setigerus, see *Croton setigerus*
Ericameria monactis, see *E. cooperi*
Ericooma cuspidata, see *Oryzopsis hymenoides*
Erigen canadense, see *Conyza canadensis* var. *canadensis*
Erigeron canadense, see *Conyza canadensis* var. *canadensis*
Erigeron canadensis, see *Conyza canadensis* var. *canadensis*
Erigeron concinnus var. *aphanactis*, see *E. aphanactis* var. *aphanactis*
Erigeron formossissimus, see *E. formosissimus*
Erigeron macranthus, see *E. speciosus* var. *macranthus*
Erigeron nudiflorus, see *E. flagellaris*
Erigeron ramosus, see *E. strigosus* var. *strigosus*
Erigeron salsuginosus, see *E. peregrinus* ssp. *callianthemus*
Erigeron superbus, see *E. eximius*
Erigonum flavum, see *Eriogonum flavum*
Erigonum subalpinum, see *Eriogonum umbellatum* var. *majus*
Eriocoma cuspidata, see *Oryzopsis hymenoides*
Eriocoma membranacea, see *Oryzopsis hymenoides*
Eriodictyon californica, see *E. californicum*
Eriodictyon parryi, see *Turricula parryi*
Eriodictyon trichocalyx ssp. *lanatum*, see *E. lanatum*
Eriodyction, see *Eriodictyon*
Eriogonum androseceum, see *E. androsaceum*
Eriogonum stellatum, see *E. umbellatum* var. *stellatum*
Eriogonum subalpinum, see *E. umbellatum* var. *majus*
Eriogonum virgatum, see *E. roseum*
Eriophyllum caespitosum, see *E. lanatum* var. *leucophyllum*
Eryngium alismaefolium, see *E. alismifolium*
Eryngium synchaetum, see *E. yuccifolium* var. *synchaetum*
Eryngium yuccaefolium, see *E. yuccifolium*
Erysimum elatum, see *E. capitatum* var. *capitatum*
Erysimum wheeleri, see *E. capitatum* var. *capitatum*
Erythaea muehlenbergii, see *Centaurium muehlenbergii*
Erythraea venusta, see *Centaurium venustum*
Erythremia grandiflora, see *Lygodesmia grandiflora*
Erythrina monosperma, see *E. sandwicensis*
Erythronium giganteum, see *E. oregonum* ssp. *oregonum*
Erythronium grandiflorum parviflorum, see *E. grandiflorum* ssp. *grandiflorum*
Erythronium oreganum, see *E. oregonum*
Escholtzia, see *Eschscholzia*
Eschscholtzia, see *Eschscholzia*
Eschscholzia douglasii, see *E. californica*
Eugenia simpsoni, see *Myrcianthes fragrans*

Eulophus bolanderi, see *Perideridia bolanderi*
Eulophus pringlei, see *Perideridia pringlei*
Euonymus americanus, see *E. americana*
Euonymus atropurpureus, see *E. atropurpurea*
Euonymus europaeus, see *E. europaea*
Euonymus obovatus, see *E. obovata*
Eupatorium ageratoides, see *Ageratina altissima* var. *roanensis*
Eupatorium falcatum, see *E. purpureum* var. *purpureum*
Eupatorium herbaceum, see *Ageratina herbacea*
Eupatorium occidentale, see *Ageratina occidentalis*
Eupatorium purpurem, see *E. purpureum*
Eupatorium rugosum, see *Ageratina altissima* var. *altissima*
Eupatorium urticaefolium, see *Ageratina altissima* var. *roanensis*
Euphorbia alb(in)omarginata, see *Chamaesyce albomarginata*
Euphorbia arenicola, see *Chamaesyce ocellata* ssp. *arenicola*
Euphorbia fendleri, see *Chamaesyce fendleri* var. *fendleri*
Euphorbia geyeri, see *Chamaesyce geyeri*
Euphorbia glyptosperma, see *Chamaesyce glyptosperma*
Euphorbia hypericifolia, see *Chamaesyce hypericifolia*
Euphorbia lata, see *Chamaesyce lata*
Euphorbia maculata, see *Chamaesyce maculata*
Euphorbia melanadenia, see *Chamaesyce melanadenia*
Euphorbia montana, see *E. robusta*
Euphorbia multiformis, see *Chamaesyce multiformis* var. *multiformis*
Euphorbia nutans, see *Chamaesyce nutans*
Euphorbia ocellata, see *Chamaesyce ocellata* ssp. *ocellata*
Euphorbia polycarpa, see *Chamaesyce polycarpa* var. *polycarpa*
Euphorbia revoluta, see *Chamaesyce revoluta*
Euphorbia serpyllifolia, see *Chamaesyce serpyllifolia* ssp. *serpyllifolia*
Euphorhia corollata, see *Euphorbia corollata*
Eurotia lanata, see *Krascheninnikovia lanata*
Evolvulus pilosus, see *E. nuttallianus*
Evonymus, see *Euonymus*
Falcata comosa, see *Amphicarpaea bracteata*
Fatsia horrida, see *Oplopanax horridus*
Ferocactus wislizenii, see *F. wislizeni*
Ferula multifida, see *Lomatium dissectum* var. *multifidum*
Festuca obtusa, see *F. subverticillata*
Festuca octoflora, see *Vulpia octoflora* var. *octoflora*
Festuca ovina var. *brevifolia*, see *F. brachyphylla* ssp. *brachyphylla*
Festuca tenella, see *Vulpia octoflora* var. *glauca*
Flaxinus velutina, see *Fraxinus velutina*
Foeniculum officinale, see *F. vulgare*
Forestiera neomexicana, see *F. pubescens* var. *pubescens*

Forestiera porulosa, see *F. segregata* var. *segregata*
Forsellesia nevadensis, see *Glossopetalon spinescens* var. *aridum*
Fourquieria splendens, see *Fouquieria splendens*
Fragaria americana, see *F. vesca* ssp. *americana*
Fragaria bracteata, see *F. vesca* ssp. *bracteata*
Fragaria californica, see *F. vesca* ssp. *californica*
Fragaria cuneifolia, see *F. ×ananassa* var. *cuneifolia*
Fragaria vesca americana, see *F. vesca* ssp. *americana*
Fragaria virginana, see *F. virginiana*
Fragaria virginiana illinoensis, see *F. virginiana* ssp. *platypetala*
Frangula california, see *F. californica*
Franseria acanthicarpa, see *Ambrosia acanthicarpa*
Franseria ambrosioides, see *Ambrosia ambrosioides*
Franseria tenuifolia, see *Ambrosia tenuifolia*
Fraxinum pennsylvannica, see *Fraxinus pennsylvanica*
Fraxinus oregana, see *F. latifolia*
Fraxinus oregona, see *F. latifolia*
Fraxinus sambucifolia, see *F. nigra*
Fraxinus viridis, see *F. pennsylvanica*
Fremontia sp., see *Fremontodendron* sp.
Fremontodendron californicus, see *F. californicum*
Freycinetia arnotti, see *F. arborea*
Fritillaria camchatcensis, see *F. camschatcensis*
Fritillaria lancelota, see *F. lanceolata*
Frunella vulgaris, see *Prunella vulgaris*
Funastrum heterophyllum, see *Sarcostemma cynanchoides* ssp. *hartwegii*
Gaertneria acanthicarpa, see *Ambrosia acanthicarpa*
Gaillardia neo-mexicana, see *G. pulchella* var. *pulchella*
Galax aphylla, see *G. urceolata*
Galpinsia lavandulaefolia, see *Calylophus lavandulifolius*
Gaura gracilis, see *G. hexandra* ssp. *gracilis*
Gaura parvifolia, see *G. parviflora*
Gaylussacia resinosa, see *G. baccata*
Gentiana crinita, see *Gentianopsis crinita*
Gentiana flavida, see *G. alba*
Gentiana propinqua, see *Gentianella propinqua* ssp. *propinqua*
Gentiana quinquefolia, see *Gentianella quinquefolia* ssp. *quinquefolia*
Geoaster sp., see *Geastrum* sp.
Geoprumnon crassicarpum, see *Astragalus crassicarpus* var. *crassicarpus*
Geraneum maculatum, see *Geranium maculatum*
Geranium fremontii, see *G. caespitosum* var. *fremontii*
Geranium incisum, see *G. oreganum*
Gerardia tenuifolia, see *Agalinis tenuifolia* var. *tenuifolia*
Gerardia virginica, see *Aureolaria virginica*
Geum ciliatum, see *G. triflorum* var. *ciliatum*
Geum nivale, see *G. rivale*

Geum strictum, see *G. aleppicum*
Gilia acerosa, see *G. rigidula* ssp. *acerosa*
Gilia aggregata, see *Ipomopsis aggregata* ssp. *aggregata*
Gilia appregata, see *Ipomopsis aggregata*
Gilia attenuata, see *Ipomopsis aggregata* ssp. *attenuata*
Gilia congesta, see *Ipomopsis congesta* ssp. *congesta*
Gilia eremica var. *arizonica*, see *Eriastrum eremicum*
Gilia filifolia var. *sparsiflora*, see *Eriastrum filifolium*
Gilia gracilis, see *Phlox gracilis* ssp. *gracilis*
Gilia greeneana, see *Ipomopsis aggregata* ssp. *candida*
Gilia gunnisoni, see *Ipomopsis gunnisonii*
Gilia laxiflora, see *Ipomopsis laxiflora*
Gilia linearis, see *Collomia linearis*
Gilia longiflora, see *Ipomopsis longiflora* ssp. *longiflora*
Gilia multiflora, see *Ipomopsis multiflora*
Gilia polycladon, see *Ipomopsis polycladon*
Gilia pulmila, see *Ipomopsis pumila*
Gilia pungens, see *Leptodactylon pungens*
Gilia staminea, see *G. capitata* ssp. *staminea*
Gillenia stipulata, see *Porteranthus stipulatus*
Gillenia trifoliata, see *Porteranthus trifoliatus*
Glaucotheca armata, see *Glaucothea armata*
Glcyrrhiza, see *Glycyrrhiza*
Glecoma hederaceae, see *Glechoma hederacea*
Gleditsia triacanthas, see *G. triacanthos*
Glycerhize, see *Glycyrrhiza*
Glyceria aquatica, see *Catabrosa aquatica*
Glyceria distans, see *Puccinellia distans*
Glycerrhiza, see *Glycyrrhiza*
Glyceyrrhiza, see *Glycyrrhiza*
Glycine apios, see *Apios americana*
Glycyrhiza, see *Glycyrrhiza*
Glycyrrhiza lepiodata, see *G. lepidota*
Glycyrrihiza, see *Glycyrrhiza*
Gnaphalium chilense, see *G. stramineum*
Gnaphalium decurrens var. *californicum*, see *G. viscosum*
Gnaphalium polycephalum, see *G. obtusifolium*
Gnaphalium popycephalum, see *G. obtusifolium*
Gnaphalium purpureum, see *Gamochaeta purpurea*
Gnaphalium wrightii, see *G. canescens*
Godetia albescens, see *Clarkia purpurea* ssp. *quadrivulnera*
Godetia amoena, see *Clarkia amoena* ssp. *amoena*
Godetia biloba, see *Clarkia biloba* ssp. *biloba*
Godetia viminea, see *Clarkia purpurea* ssp. *viminea*
Gosypium, see *Gossypium*
Gramineae, see Poaceae
Grindelia aphanactis, see *G. nuda* var. *aphanactis*
Grindelia squarrosa serrulata, see *G. squarrosa* var. *serrulata*
Grindelia stricta venulosa, see *G. integrifolia* var. *macrophylla*
Grossularia irrigua, see *Ribes oxyacanthoides* ssp. *irriguum*
Grossularia leptantha, see *Ribes leptanthum*
Grossularia missouriensis, see *Ribes missouriense*
Grossularia oxyacanthoides, see *Ribes oxyacanthoides* ssp. *oxyacanthoides*
Grossularia setosa, see *Ribes oxyacanthoides* ssp. *setosum*
Guaiacum, see *Guajacum*
Guitierrezia dracunculoides, see *Amphiachyris dracunculoides*
Gutierrezia diversifolia, see *G. sarothrae*
Gutierrezia dracunculoides, see *Amphiachyris dracunculoides*
Gutierrezia euthamiae, see *G. sarothrae*
Gutierrezia filifolia, see *G. sarothrae*
Gutierrezia furfuracea, see *G. sarothrae*
Gutierrezia longifolia, see *G. sarothrae*
Gutierrezia lucida, see *G. microcephala*
Gutierrezia microcephalia, see *G. microcephala*
Gutierrezia cf. *sarothrae*, see *G. sarothrae*
Guttiferae, see Clusiaceae
Guzmannia, see *Guzmania*
Gymnocladus canadensis, see *G. dioicus*
Gymnocladus dioica, see *G. dioicus*
Gymnogramma triangularis, see *Pentagrama triangularis* ssp. *triangularis*
Gymnogramme triangularis, see *Pentagrama triangularis* ssp. *triangularis*
Gymnolomia multiflora, see *Heliomeris multiflora*
Gynoglossom virginianum, see *Cynoglossum virginianum*
Gyrotheca capitata, see *Lachnanthes caroliana*
Habenaria bracteata, see *Coeloglossum viride* var. *virescens*
Habenaria ciliaris, see *Platanthera ciliaris*
Habenaria dilatata, see *Platanthera dilatata* var. *dilatata*
Habenaria dilatata var. *media*, see *Platanthera* ×*media*
Habenaria elegans, see *Piperia elegans*
Habenaria fimbriata, see *Platanthera grandiflora*
Habenaria leucostachys, see *Platanthera leucostachys*
Habenaria orbiculata, see *Platanthera orbiculata* var. *orbiculata*
Habenaria psycodes, see *Platanthera psycodes*
Habenaria saccata, see *Platanthera stricta*
Habenaria sparsiflora, see *Platanthera sparsiflora* var. *sparsiflora*
Habenaria strictissima var. *odontopetala*, see *H. odontopetala*
Habenaria unalascensis, see *Piperia unalascensis*
Habenaria viridis ssp. *bracteata*, see *Coeloglossum viride* var. *virescens*
Hackelia virginia, see *H. virginiana*
Haplopappus aborescens, see *Ericameria arborescens*
Haplopappus acradenius, see *Isocoma acradenia* var. *acradenia*
Haplopappus cuneatus, see *Ericameria cuneata* var. *cuneata*
Haplopappus linearifolius var. *interior*, see *Ericameria linearifolia*
Haplopappus palmeri, see *Ericameria palmeri* var. *palmeri*
Haplopappus palmeri ssp. *pachylepus*, see *Ericameria palmeri* var. *pachylepis*
Haplopappus parishii, see *Ericameria parishii*
Haplopappus spinulosus ssp. *typicus*, see *Machaeranthera pinnatifida* ssp. *pinnatifida*
Haploppapus, see *Haplopappus*
Havernaria dilatata, see *Platanthera dilatata* var. *dilatata*
Hedeoma incana, see *Poliomintha incana*
Hedeoma pulegoides, see *H. pulegioides*
Hedera quinquefolia, see *Parthenocissus quinquefolia* var. *quinquefolia*
Hedysarum mackenzii, see *H. boreale* ssp. *mackenziei*
Heirchloe, see *Hierochloe*
Heirochloe, see *Hierochloe*
Helenium hoopesii, see *Dugaldia hoopesii*
Helenium tenuifolium, see *H. amarum* var. *amarum*
Helianthellia uniflora, see *Helianthella uniflora*
Helianthemum canadense, see *Helianthemum canadense*
Helianthus annus, see *H. annuus*
Helianthus canus, see *H. niveus* ssp. *canescens*
Helianthus grosse-serratus, see *H. grosseserratus*
Helianthus lenticularis, see *H. annuus*
Helianthus tuberosa, see *H. tuberosus*
Helinium microcephalum, see *Helenium microcephalum*
Heliopsis scabra, see *H. helianthoides* var. *scabra*
Heliotropum curassavicum, see *Heliotropium curassavicum*
Helmia bulbifera, see *Dioscorea bulbifera*
Hemizonia clevelandi, see *H. clevelandii*
Hemizonia luzulaefolia, see *H. luzulifolia*
Hemizonia multiglandulosa, see *Calycadenia fremontii*
Hemizonia virgata, see *Holocarpha virgata*
Hepatica acutiloba, see *H. nobilis* var. *acuta*
Hepatica americana, see *H. nobilis* var. *obtusa*
Hepatica triloba, see *H. nobilis* var. *obtusa*
Heracleum lan(a)tum, see *H. maximum*
Heracleum spondylium ssp. *montanum*, see *H. maximum*
Hermidium alipes, see *Mirabilis alipes*
Hesperidanthus linearifolius, see *Schoenocrambe linearifolia*
Hesperonia retrorsa, see *Mirabilis bigelovii* var. *retrorsa*
Hesperoscord(i)um lacteum, see *Triteleia hyacinthina* var. *hyacinthina*
Heterotheca grandifolia, see *H. grandiflora*
Heuchera devaricata, see *H. glabra*
Heuchera glabella, see *H. cylindrica* var. *glabella*

Index of Synonyms

Heuchera hispida, see *H. americana* var. *hispida*
Heuchera ovalifolia, see *H. cylindrica* var. *alpina*
Heydysarum mackenzii, see *Hedysarum boreale* ssp. *mackenziei*
Hibiscus palustris, see *H. moscheutos* ssp. *moscheutos*
Hicoria alba, see *Carya ovata*
Hicoria glabra, see *Carya glabra* var. *glabra*
Hicoria laciniosa, see *Carya laciniosa*
Hicoria minima, see *Carya cordiformis*
Hicoria ovata, see *Carya ovata*
Hierchloe, see *Hierochloe*
Hierechloe, see *Hierochloe*
Hierochloe odorata ssp. *hirta*, see *H. hirta* ssp. *hirta*
Hoffmannseggia densiflora, see *H. glauca*
Hoffmannseggia jamesii, see *Caesalpinia jamesii*
Hoffmannseggia stricta, see *H. glauca*
Hoffmanseggia, see *Hoffmannseggia*
Hoffmansegia, see *Hoffmannseggia*
Holcus halepensis, see *Sorghum halepense*
Holcus odoratus, see *Hierochloe odorata* ssp. *odorata*
Holocantha emoryi, see *Castela emoryi*
Holodiscus ariaefolius, see *H. discolor*
Holodiscus discolor var. *dumosus*, see *H. dumosus*
Hookera coronaria, see *Brodiaea coronaria*
Hookera douglasii, see *Triteleia grandiflora*
Hookera laxa, see *Triteleia laxa*
Hookera nyocintlima var. *lactea*, see *Triteleia hyacinthina*
Hookera volubilis, see *Dichelostemma volubile*
Hordeum glaucum, see *H. murinum* ssp. *glaucum*
Hordeum gussoneanum, see *H. marinum* ssp. *gussonianum*
Hordeum stebbinsi, see *H. murinum* ssp. *glaucum*
Hordeum vulgare pallidum typica, see *H. vulgare*
Horkelia gordonii, see *Ivesia gordonii*
Hosackia glabra, see *Lotus scoparius* var. *scoparius*
Hosta japonica, see *H. lancifolia*
Houttuynia californica, see *Anemopsis californica*
Hugelia virgata, see *Eriastrum virgatum*
Humulus americana, see *H. lupulus* var. *lupuloides*
Humulus americanus, see *H. lupulus* var. *lupuloides*
Humulus lupulus neomexicanus, see *H. lupulus* var. *neomexicanus*
Hydranga aborescens, see *Hydrangea arborescens*
Hydrogotyle, see *Hydrocotyle*
Hydrophyllum albifrons, see *H. fendleri* var. *albifrons*
Hydrotrida caroliniana, see *Bacopa caroliniana*
Hymenopappus arenosus, see *H. filifolius* var. *cinereus*
Hymenopappus lugens, see *H. filifolius* var. *lugens*
Hymenopappus nudatus, see *H. filifolius* var. *lugens*
Hymenopappus pauciflorus, see *H. filifolius* var. *pauciflorus*
Hymenoxys aculis var. *arizonica*, see *Tetraneuris acaulis* var. *arizonica*
Hymenoxys floribunda, see *H. richardsonii* var. *floribunda*

Hymenoxys metcalfei, see *H. richardsonii* var. *floribunda*
Hymenoxys richarsonii, see *H. richardsonii*
Hypericium, see *Hypericum*
Hypericum aspalathoides, see *H. brachyphyllum*
Hypericum formosum var. *scouleri*, see *H. scouleri* ssp. *scouleri*
Hypericum virginicum, see *Triadenum virginicum*
Hypopites monotropa, see *Monotropa hypopithys*
Hystrix patula, see *Elymus hystrix* var. *hystrix*
Impatiens biflora, see *I. capensis*
Impomea, see *Ipomoea*
Impomoea, see *Ipomoea*
Ionoxalis violacea, see *Oxalis violacea*
Ipomaea, see *Ipomoea*
Ipomea, see *Ipomoea*
Ipomoea dissecta, see *Merremia dissecta*
Ipomoea insularis, see *I. indica*
Ipomoea tuberculata, see *I. cairica*
Iresine paniculata, see *I. diffusa*
Isomeris arborea (var. *angustata*), see *Cleome isomeris*
Ivesia gordoni, see *I. gordonii*
Jacqemontia sandwicensis, see *Jacquemontia ovalifolia* ssp. *sandwicensis*
Jambosa malaecensis, see *Syzygium malaccense*
Juglan cinerea, see *Juglans cinerea*
Juglans squamosa, see *Carya alba*
Juncus xiphioides var. *triandrus*, see *J. ensifolius*
Juniperus bilicicola, see *J. virginiana* var. *silicicola*
Juniperus californica var. *utahensis*, see *J. osteosperma*
Juniperus cilicicola, see *J. virginiana* var. *silicicola*
Juniperus monosperma, see *Datura wrightii*
Juniperus pachyphlaea, see *J. deppeana*
Juniperus pachyphloea, see *J. deppeana*
Juniperus prostrata, see *J. horizontalis*
Juniperus sabina procumbens, see *J. horizontalis*
Juniperus sabina var. *procumbens*, see *J. horizontalis*
Juniperus sibirica, see *J. communis* var. *montana*
Juniperus sili(ci)cola, see *J. virginiana* var. *silicicola*
Juniperus utahensis, see *J. osteosperma*
Juniperus utaliensis, see *J. osteosperma*
Jussiaea villosa, see *Ludwigia bonariensis*
Kallstroemia brachystylis, see *K. californica*
Kalmia glauca, see *K. polifolia*
Kentrophyta montana, see *Astragalus kentrophyta* var. *kentrophyta*
Kochia trichophylla, see *K. scoparia*
Koeleria cristata, see *K. macrantha*
Koellia virginiana, see *Pycnanthemum virginianum*
Krameria glandulosa, see *K. erecta*
Krameria parvifolia, see *K. erecta*
Krigia amplexicaulis, see *K. biflora*
Krynitzkia sericea, see *Cryptantha sericea*
Kuhnia rosmarinifolia, see *Brickellia eupatorioides* var. *eupatorioides*
Kunzia tridentata, see *Purshia tridentata*

Labiatae, see Lamiaceae
Lachnanthes caroliniana, see *L. caroliana*
Lacinaria punctata, see *Liatris punctata* var. *punctata*
Lacinaria scariosa, see *Liatris scariosa* var. *scariosa*
Lacinaria sp., see *Liatris* sp.
Lacinaria spicata, see *Liatris spicata*
Laciniaria punctata, see *Liatris punctata* var. *punctata*
Laciniaria scariosa, see *Liatris scariosa* var. *scariosa*
Lactuca integrata, see *L. sativa*
Lactuca pulchella, see *L. tatarica* var. *pulchella*
Lactuca scariola var. *integrata*, see *L. sativa*
Lactuca spicata, see *L. biennis*
Lagenaria vulgaris, see *L. siceraria*
Langloisia matthewsii, see *Loeseliastrum matthewsii*
Lappula floribunda, see *Hackelia floribunda*
Lappula hispida, see *Hackelia hispida* var. *hispida*
Lappula redowskii, see *L. occidentalis* var. *occidentalis*
Lappula texana, see *L. occidentalis* var. *cupulata*
Lappula virginiana, see *Hackelia virginiana*
Larix americana, see *L. laricina*
Larix lar(a)cina, see *L. laricina*
Larrea divaricata, see *L. tridentata* var. *tridentata*
Larrea mexicana, see *L. tridentata*
Larus laricina, see *Larix laricina*
Lathyrus decaphyllus, see *L. polymorphus* ssp. *polymorphus* var. *polymorphus*
Lathyrus leucanthus, see *L. lanszwertii* var. *leucanthus*
Lathyrus maritimus, see *L. japonicus* var. *maritimus*
Lathyrus nuttallii, see *L. nevadensis* ssp. *lanceolatus* var. *nuttallii*
Lathyrus ornatus, see *L. brachycalyx* ssp. *brachycalyx*
Lathyrus watsoni, see *L. jepsonii* ssp. *californicus*
Laurus sassafras, see *Sassafras albidum*
Lavauxia triloba, see *Oenothera triloba*
Ledum decumbens, see *L. palustre* ssp. *decumbens*
Ledum glandulosum ssp. *columbianum*, see *L. ×columbianum*
Ledum latifolium, see *L. groenlandicum*
Ledum palustre ssp. *groenlandicum*, see *L. groenlandicum*
Leguminosae, see Fabaceae
Lemaireocereus thurberi, see *Stenocereus thurberi*
Lens esculenta, see *L. culinaris*
Leontodon carolinianum, see *Pyrrhopappus carolinianus*
Lepadena marginata, see *Euphorbia marginata*
Lepargyraea argentea, see *Shepherdia argentea*
Lepargyraea canadensis, see *Shepherdia canadensis*
Lepargyrea argentea, see *Shepherdia argentea*
Lepargyrea canadensis, see *Shepherdia canadensis*
Lepidium apetalum, see *L. densiflorum*
Lepidium epetalum, see *L. densiflorum*
Lepidium epitalum, see *L. densiflorum*

Lepidium menziesii, see *L. virginicum* var. *menziesii*
Leptandra virginica, see *Veronicastrum virginicum*
Leptarrhenia amplexifolia, see *Leptarrhena pyrolifolia*
Leptilon canadense, see *Conyza canadensis* var. *canadensis*
Leptoloma cognatum, see *Digitaria cognata* var. *cognata*
Leptotaenia californica, see *Lomatium californicum*
Leptotaenia dissecta, see *Lomatium dissectum* var. *dissectum*
Leptotaenia dissecta var. *multifida*, see *Lomatium dissectum* var. *multifidum*
Leptotaenia multifida, see *Lomatium dissectum* var. *multifidum*
Leptotoenia multifida, see *Lomatium dissectum* var. *multifidum*
Lessingia germanorum var. *vallicola*, see *L. glandulifera* var. *glandulifera*
Leucampyx newberri, see *Hymenopappus newberryi*
Leucelene ericoides, see *Chaetopappa ericoides*
Leucothoe catesbaei, see *L. axillaris*
Lib(r)ocedrus decurrens, see *Calocedrus decurrens*
Licium, see *Lycium*
Ligusticum apiodorum, see *L. apiifolium*
Ligusticum hultenii, see *L. scothicum* ssp. *hultenii*
Ligusticum scoticum, see *L. scothicum*
Ligusticum scoticum ssp. *hultenii*, see *L. scothicum* ssp. *hultenii*
Lilium parviflorum, see *L. ×columbianum*
Lilium umbellatum, see *L. philadelphicum* var. *andinum*
Lillium, see *Lilium*
Lilum columbianum, see *Lilium ×columbianum*
Limonium caro linianum, see *L. carolinianum*
Linanthastrum nuttallii, see *Linanthus nuttallii* ssp. *nuttallii*
Lingusticum, see *Ligusticum*
Linosyris howardi, see *Chrysothamnus parryi*
Linum aristatum var. *australe*, see *L. australe* var. *australe*
Linum lewisi, see *L. lewisii*
Lippia cuneifolia, see *Phyla cuneifolia*
Lippia lanceolata, see *Phyla lanceolata*
Lippia nodiflora, see *Phyla nodiflora*
Lippia wrightii, see *Aloysia wrightii*
Lithocarpus densiflora (var. *echinoides*), see *L. densiflorus*
Lithospermum angustifolium, see *L. incisum*
Lithospermum carolinense, see *L. caroliniense*
Lithospermum linearifolium, see *L. incisum*
Lithospermum pilosum, see *L. ruderale*
Lithospermun canescens, see *Lithospermum canescens*
Lithospernum ruderale, see *Lithospermum ruderale*
Lithosperum angustofolium, see *Lithospermum incisum*
Lobelia splendens, see *L. cardinalis* ssp. *graminea* var. *propinqua*
Lobelia syphilitica, see *L. siphilitica*
Lochnera rosea, see *Catharanthus roseus*
Lomatium leptocarpum, see *L. bicolor* var. *leptocarpum*
Lomatium montanum, see *L. cous*
Lomatium nudicaules, see *L. nudicaule*
Lomatium orientalis, see *L. orientale*
Lomatium platycarpum, see *L. simplex* var. *simplex*
Lomatium watsoni, see *L. watsonii*
Lonicera divica, see *L. dioica*
Lophanthus urticifolius, see *Agastache urticifolia*
Lophophora williamsi, see *L. williamsii*
Lotus americanus, see *L. unifoliolatus* var. *unifoliolatus*
Lotus purshianus, see *L. unifoliolatus* var. *unifoliolatus*
Lupinus aduncus, see *L. caudatus* ssp. *argophyllus*
Lupinus carnosulus, see *L. affinis*
Lupinus ingratus, see *L. argenteus* ssp. *ingratus*
Lupinus luteolous, see *L. luteolus*
Lupinus variicolor, see *L. versicolor*
Lychnis alba, see *Silene latifolia* ssp. *alba*
Lychnis drummondii, see *Silene drummondii* var. *drummondii*
Lycopodia, see *Lycopodium*
Lycopodium lucidulum, see *Huperzia lucidula*
Lycopodium sabinaefolium ssp. *sabinaefolium*, see *L. sabinifolium*
Lycopodium selago, see *Huperzia selago* var. *selago*
Lycopus lucidus ssp. *americanus*, see *L. asper*
Lygodesmia rostrata, see *Shinnersoseris rostrata*
Lygodesmia spinosa, see *Stephanomeria spinosa*
Lysichiton americanum, see *L. americanus*
Lysichiton camtschatcense, see *L. americanus*
Lysichiton kamtschatcense, see *L. americanus*
Lysichitum, see *Lysichiton*
Lythrum lanceolatum, see *L. alatum* var. *lanceolatum*
Machaeranthea sp., see *Macranthera* sp.
Machaeranthera glabella, see *M. canescens* ssp. *canescens* var. *canescens*
Maclura aurantiaca, see *M. pomifera*
Macrocystis l̦tkeana, see *M. luetkeana*
Madaria, see *Madia elegans*
Madia densifolia, see *M. elegans* ssp. *densifolia*
Madia dissitiflora, see *M. gracilis* ssp. *gracilis*
Magnolia glauca, see *M. virginiana*
Mahonia fremonti, see *M. fremontii*
Maianthemum bifolium, see *M. dilatatum*
Mallow sp., see *Malva* sp.
Malva borealis, see *M. nicaeensis*
Malva involucrata, see *Callirhoe involucrata*
Malva rotundifolia, see *M. neglecta*
Malvacea rubra, see *Malva* sp.
Malvastrum coccineum, see *Sphaeralcea coccinea* ssp. *coccinea*
Malvastrum munroanum, see *Sphaeralcea munroana*
Mamillaria, see *Mammillaria*
Mammillaria microcarpa, see *M. grahamii* var. *grahamii*
Mammillaria missouriensis, see *Escobaria missouriensis* var. *missouriensis*
Mammillaria viv(i)par(i)a, see *Escobaria vivipara* var. *vivipara*
Marah macrocarpa, see *M. macrocarpus*
Marginaria polypodioides, see *Pleopeltis polypodioides* ssp. *polypodioides*
Mariscus cladium, see *Cladium mariscus* ssp. *jamaicense*
Mariscus jamaicensis, see *Cladium mariscus* ssp. *jamaicense*
Martynia arenaria, see *Proboscidea althaeifolia*
Martynia fragrans, see *Proboscidea louisianica* ssp. *fragrans*
Martynia louisiana, see *Proboscidea louisianica* ssp. *louisianica*
Martynia parviflora, see *Proboscidea parviflora* ssp. *parviflora*
Martynia proboscidea, see *Proboscidea louisianica*
Martynia proboscides, see *Proboscidea louisianica*
Matricaria matricarioides, see *M. discoidea*
Matricaria suaveolens, see *M. discoidea*
Maurandia wislizeni, see *Epixiphium wislizeni*
Medicago denticulata, see *M. polymorpha*
Medicago hispida, see *M. polymorpha*
Meibomia sp., see *Desmodium* sp.
Melia azederach, see *M. azedarach*
Melilotus alba, see *M. officinalis*
Melilotus indica, see *M. indicus*
Melitotus alba, see *Melilotus albus*
Mentha aquatica, see *M. ×piperita*
Mentha arvenis, see *M. arvensis*
Mentha arvensis var. *canadensis*, see *M. canadensis*
Mentha arvensis var. *glabrata*, see *M. canadensis*
Mentha arvensis var. *lanata*, see *M. canadensis*
Mentha arvensis var. *villosa*, see *M. canadensis*
Mentha borealis, see *M. canadensis*
Mentha canadensis glabrata, see *M. canadensis*
Mentha penardi, see *M. canadensis*
Mentha piperita, see *M. ×piperita*
Mentzelia pumila var. *multiflora*, see *M. multiflora* var. *multiflora*
Mentzelia veatchiana, see *M. albicaulis* var. *veatchiana*
Meracleum lanatum, see *Heracleum maximum*
Mergantnes trifolia, see *Menyanthes trifoliata*
Mesembryanthemum aequilaterale, see *Carpobrotus aequilateralus*
Mesembryanthemum chilense, see *Carpobrotus aequilateralus*
Mestensia ciliata, see *Mertensia ciliata*
Metrosideros collins polym, see *M. polymorpha* var. *polymorpha*
Mezoneurum kauaiense, see *Caesalpinia kavaiensis*
Micrampelis lobata, see *Echinocystis lobata*
Micrampelis marah, see *Marah oreganus*
Micrapelis micracarpa, see *Marah macrocarpus* var. *major*
Microlepia strigosa, see *M. setosa*

Index of Synonyms

Micromeria chamissonis, see *Satureja douglasii*
Micromeria douglasii, see *Satureja douglasii*
Microstylis unifolia, see *Malaxis unifolia*
Mimulus aurantiacus, see *Diplacus aurantiacus* ssp. *aurantiacus*
Mimulus geyeri, see *M. glabratus* var. *jamesii*
Mimulus luteus, see *M. tilingii* var. *caespitosus*
Mimulus nasutus, see *M. guttatus*
Mirabilis linoides, see *Monardella linoides*
Mirabilis odoratissima, see *Monardella odoratissima*
Mirabilis viridis, see *Monardella viridis*
Moldavica parviflora, see *Dracocephalum parviflorum*
Monarda menthaefolia, see *M. fistulosa* ssp. *fistulosa* var. *menthifolia*
Monarda menthoefolia, see *M. fistulosa* ssp. *fistulosa* var. *menthifolia*
Monarda mollis, see *M. fistulosa* ssp. *fistulosa* var. *mollis*
Monarda nuttallii, see *M. pectinata*
Monarda scabra, see *M. fistulosa* ssp. *fistulosa* var. *mollis*
Monardella adoratissima, see *M. odoratissima*
Monardella sheltonii, see *M. villosa* ssp. *sheltonii*
Monardo fisiulosa var. *menthaefolia,* see *Monarda fistulosa* ssp. *fistulosa* var. *menthifolia*
Monensis, see *Moneses*
Monesus, see *Moneses*
Monolepis chenopodioides, see *M. nuttalliana*
Monolepis chenopoides, see *M. nuttalliana*
Monseses, see *Moneses*
Montia asarifolia, see *Claytonia cordifolia*
Montia parviflora, see *Claytonia parviflora*
Montia perfoliata, see *Claytonia perfoliata* ssp. *perfoliata* var. *perfoliata*
Montia sibirica, see *Claytonia sibirica* var. *sibirica*
Montia spathulata, see *Claytonia spathulata* var. *spathulata*
Mortonia scabrella, see *M. sempervirens* ssp. *scabrella*
Morus ruba, see *M. rubra*
Mucuna gigantia, see *M. gigantea*
Muhlenbergia foliosa, see *M. mexicana* var. *mexicana*
Muhlenbergia neomexicana, see *M. pauciflora*
Munroa squarrosa, see *Monroa squarrosa*
Musa sapientum, see *M.* ×*paradisiaca*
Museneon hookeri, see *Musineon divaricatum* var. *hookeri*
Musenium divaricatum, see *Musineon divaricatum* var. *divaricatum*
Myraca cirifera, see *Myrica cerifera*
Myrica asplenifolia, see *Comptonia peregrina*
Myrica caroliniensis, see *M. heterophylla*
Myriophyllum exalbescens, see *M. sibiricum*
Nabalus asper, see *Prenanthes aspera*
Nardosmia palmata, see *Petasites frigidus* var. *palmatus*
Nasturtium officinale, see *Rorippa nasturtium-aquaticum*
Nasturtium palustre, see *Rorippa nasturtium-aquaticum*
Navarretia cotulaefolia, see *N. cotulifolia*
Negundo aceroides, see *Acer negundo* var. *negundo*
Negundo interius, see *Acer negundo* var. *interius*
Negundo negundo, see *Acer negundo* var. *negundo*
Nelumbium luteim, see *Nelumbo lutea*
Nelumbium luteum, see *Nelumbo lutea*
Nemopanthus mucronata, see *N. mucronatus*
Neomammillaria grahami, see *Mammillaria grahamii*
Neomammillaria mainae, see *Mammillaria mainiae*
Neomammillaria olivia, see *Mammillaria grahamii*
Neomammillaria wrightii, see *Mammillaria wrightii*
Nereocystis luetkeama, see *N. luetkeana*
Nicotiana bigelovi(i) (var. *exaltata*), see *N. quadrivalvis* var. *bigelovii*
Nicotiana multivalvis, see *N. quadrivalvis*
Nicotiana palmeri, see *N. trigonophylla* var. *palmeri*
Nicotiana tobacum, see *N. tabacum*
Nicotiana trigonphylla, see *N. trigonophylla*
Nicotina, see *Nicotiana*
Nirabilis, see *Mirabilis*
Nolina parryi, see *N. bigelovii* var. *parryi*
Notholaena fendleri, see *Argyrochosma fendleri*
Nuphar advena, see *N. lutea* ssp. *advena*
Nuphar luteum, see *N. lutea*
Nuphar luteum ssp. *macrophyllum,* see *N. lutea* ssp. *advena*
Nuphar polysepala, see *N. lutea* ssp. *polysepala*
Nuphar polysepalum, see *N. lutea* ssp. *polysepala*
Nuphar variegatum, see *N. lutea* ssp. *variegata*
Nuttallia nuda, see *Mentzelia nuda* var. *nuda*
Nymphaea advena, see *Nuphar lutea* ssp. *advena*
Nymphaea polysepala, see *Nuphar lutea* ssp. *polysepala*
Nymphea polysepala, see *Nuphar lutea* ssp. *polysepala*
Nymphoides lacunosum, see *N. cordata*
Nymphozanthus polysepalus, see *Nuphar lutea* ssp. *polysepala*
Oakesia sessilifolia, see *Uvularia sessilifolia*
Obione confertifolia, see *Atriplex confertifolia*
Ochorosia sandwicensis, see *Ochrosia compta*
Octoblepharum albidum, see *Octoblephorum albidum*
Odonto-glossum chineusis, see *Odontoglossum chinensis*
Odostemon fremontii, *Mahonia fremontii*
Oemlaria cerasiformis, see *Oemleria cerasiformis*
Oenethera albicaulis, see *Oenothera albicaulis*
Oenothera brevipes, see *Camissonia brevipes* ssp. *brevipes*
Oenothera caespitosa, see *O. cespitosa*
Oenothera caespitosa var. *marginata,* see *O. cespitosa* ssp. *marginata*
Oenothera clavaeformis, see *Camissonia claviformis* ssp. *claviformis*
Oenothera ctenophylla, see *O. albicaulis*
Oenothera hartwegii var. *fendleri,* see *Calylophus hartwegii* ssp. *fendleri*
Oenothera hookeri (var. *hirsutissima*), see *O. elata* ssp. *hookeri*
Oenothera multijuga, see *Camissonia multijuga*
Oenothera ovata, see *Camissonia ovata*
Oenothera pinnatifida, see *O. albicaulis*
Oenothera procera, see *O. villosa* ssp. *strigosa*
Oenothera runcinata, see *O. pallida* ssp. *runcinata*
Oenothera tanacetifolia, see *Camissonia tanacetifolia* ssp. *tanacetifolia*
Onosmodium hispidissimum, see *O. molle* ssp. *hispidissimum*
Onosmodium occidentale, see *O. molle* ssp. *occidentale*
Onosmodium thurberi, see *Macromeria viridiflora*
Oplopanax horridum, see *O. horridus*
Opuntia arborescens, see *O. imbricata* var. *imbricata*
Opuntia basilans, see *O. basilaris*
Opuntia camanchica, see *O. phaeacantha* var. *camanchica*
Opuntia engelmanni, see *O. engelmannii*
Opuntia hystricina, see *O. erinacea* var. *hystricina*
Opuntia lindheimeri, see *O. engelmannii* var. *lindheimeri*
Opuntia megacantha, see *O. ficus-indica*
Opuntia megacarpa, see *O. engelmannii* var. *engelmannii*
Opuntia occidentalis, see *O.* ×*occidentalis*
Opuntia occidentalis var. *megacarpa,* see *O. engelmannii* var. *engelmannii*
Opuntia phaeacantha var. *discata,* see *O. engelmannii* var. *engelmannii*
Opuntia ploycantha, see *O. polyacantha*
Opuntia plumbea, see *O. macrorhiza* var. *macrorhiza*
Opuntia polycantha, see *O. polyacantha*
Opuntia rutila, see *O. polyacantha* var. *rufispina*
Opuntia tuna, see *O. tunicata*
Oreocarya fulvocanescens, see *Cryptantha fulvocanescens* var. *fulvocanescens*
Orobanche ludoviciana var. *Cooperi,* see *O. cooperi* ssp. *cooperi*
Orobanche multiflora var. *arenosa,* see *O. ludoviciana* ssp. *ludoviciana*
Orobanche tuberosa, see *Boschniakia hookeri*
Orobranche multiflora, see *Orobanche ludoviciana* ssp. *multiflora*
Orthocarpus purpureo-albus, see *O. purpureoalbus*
Oryzopsis cuspidata, see *O. hymenoides*
Oryzopsis hymenioides, see *O. hymenoides*
Oryzopsis membranacea, see *O. hymenoides*
Oryzopsis miliacea, see *Piptatherum miliaceum*
Osmaronia cerasiformis, see *Oemleria cerasiformis*
Osmorhiza brevistyla, see *O. claytonii*
Osmorhiza chilensis, see *O. berteroi*

Osmorhiza claytoni, see *O. claytonii*
Osmorhiza nuda (var. *brevipes*), see *O. berteroi*
Osmorhize, see *Osmorhiza*
Osmorhyza, see *Osmorhiza*
Osmorrhiza, see *Osmorhiza*
Oxalis acetosella, see *O. montana*
Oxalis amplifolia, see *O. drummondii*
Oxalis europaea, see *O. stricta*
Oxybaphus coccinea, see *Mirabilis coccineus*
Oxybaphus comatus, see *Mirabilis oblongifolia*
Oxybaphus linearis, see *Mirabilis linearis*
Oxybaphus nyctaginea, see *Mirabilis nyctaginea*
Oxybaphus nyctagineus, see *Mirabilis nyctaginea*
Oxybaphus pumilus, see *Mirabilis pumila*
Oxycoccus macrocarpus, see *Vaccinium macrocarpon*
Oxycoccus microcarpus, see *Vaccinium oxycoccos*
Oxycoccus quadripetalus, see *Vaccinium oxycoccos*
Oxydendron arboreum, see *Oxydendrum arboreum*
Oxytropis campestris spicats, see *O. campestris*
Oxytropis nigresceus, see *O. nigrescens*
Oxytropus sericea, see *Oxytropis sericea*
Pachistima myrsinites, see *Paxistima myrsinites*
Pachylobus caespitosus, see *Oenothera cespitosa* ssp. *cespitosa*
Pachylophus caespitosus, see *Oenothera cespitosa* ssp. *cespitosa*
Pachylophus hirsutus, see *Oenothera cespitosa*
Pachystima myrsinites, see *Paxistima myrsinites*
Padus melanocarpa, see *Prunus virginiana* var. *melanocarpa*
Padus nana, see *Prunus virginiana* var. *virginiana*
Palafoxia linearis, see *P. arida*
Palmae, see Arecaceae
Panax horridum, see *Oplopanax horridus*
Panax quinquefolia, see *P. quinquefolius*
Panax quinquefolium, see *P. quinquefolius*
Panax trifolium, see *P. trifolius*
Pandanus odoratissimus, see *P. tectorius*
Panicularia fluitans, see *Glyceria fluitans*
Panicum autumnale, see *Digitaria cognata*
Panicum barbipulvinatum, see *P. capillare*
Panicum bulbosum minor, see *P. bulbosum*
Panicum cognatum, see *Digitaria cognata*
Panicum oligosanthes, see *Dichanthelium oligosanthes* var. *oligosanthes*
Panicum paludivagum, see *Paspalidium geminatum* var. *paludivagum*
Panicum polycaulon, see *Dichanthelium strigosum* var. *glabrescens*
Panicum pruricus, see *Digitaria setigera*
Panicum scribnerianum, see *Dichanthelium oligosanthes* var. *scribnerianum*
Panicum seribnerianum, see *Dichanthelium oligosanthes* var. *scribnerianum*
Panicum xalapense, see *Dichanthelium laxiflorum*
Parkinsonia torreyana, see *P. florida*
Parosela aurea, see *Dalea aurea*
Parosela enneandra, see *Dalea enneandra*
Parosela formosa, see *Dalea formosa*
Parosela lanata, see *Dalea lanata* var. *lanata*
Parosela lasianthera, see *Dalea lasiathera*
Parosela nana, see *Dalea nana* var. *nana*
Parosela scoparia, see *Psorothamnus scoparius*
Parosela sp., see *Dalea* sp.
Parrya menziesii, see *Phoenicaulis cheiranthoides*
Parthencissus quinquefolia, see *Parthenocissus quinquefolia*
Paspalum stramineum, see *P. setaceum*
Pedasites speciosa, see *Petasites frigidus* var. *palmatus*
Pedicularis attolens, see *P. attollens*
Pedicularis grayi, see *P. procera*
Pedicularis kanei, see *P. lanata* ssp. *lanata*
Pelea cinerea, see *Melicope cinerea*
Pellaea compacta, see *P. mucronata* ssp. *californica*
Pellaea mucronata var. *mucronata*, see *P. mucronata* ssp. *mucronata*
Pellaea ornithopus, see *P. mucronata* ssp. *mucronata*
Peltigera apthosa, see *P. aphthosa*
Peltiphyllum peltatum, see *Darmera peltata*
Pennellia micrantha, see *P. micrantha*
Penoedanum leiocarpum, see *Lomatium nudicaule*
Penstemon berviflorus, see *Keckiella breviflora* ssp. *breviflora*
Penstemon breviflorus, see *Keckiella breviflora* ssp. *breviflora*
Penstemon bridgesii, see *P. rostriflorus*
Penstemon cordifolius, see *Keckiella cordifolia*
Penstemon douglas(s)ii, see *P. fruticosus*
Penstemon eatoni, see *P. eatonii*
Penstemon nemorosus, see *Nothochelone nemorosa*
Penstemon pubescens, see *P. fruticosus*
Penstemon scouleri, see *P. fruticosus* var. *scouleri*
Penstemon torreyi, see *P. barbatus* ssp. *torreyi*
Pentalostemum candidus, see *Dalea candida* var. *candida*
Pentalostemum oligophyllum, see *Dalea candida* var. *oligophylla*
Penthorum sediodes, see *P. sedoides*
Pentstemon, see *Penstemon*
Pepo foetidissima, see *Cucurbita foetidissima*
Pepo foetidissimus, see *Cucurbita foetidissima*
Peramium decipiens, see *Goodyera oblongifolia*
Perezia microcephala, see *Acourtia microcephala*
Perezia wrightii, see *Acourtia wrightii*
Perilla fructescens, see *P. frutescens*
Peritoma serrulatum, see *Cleome serrulata*
Persea pubescens, see *P. palustris*
Petalostemon candidus, see *Dalea candida* var. *candida*
Petalostemon cliogophyllus, see *Dalea candida* var. *oligophylla*
Petalostemon gracile var. *oligophyllum*, see *Dalea candida* var. *oligophylla*
Petalostemon oligophyllum, see *Dalea candida* var. *oligophylla*
Petalostemon oligophyllus, see *Dalea candida* var. *oligophylla*
Petalostemon purpureum, see *Dalea purpurea* var. *purpurea*
Petalostemon purpureus, see *Dalea purpurea* var. *purpurea*
Petalostemon villosum, see *Dalea villosa* var. *villosa*
Petalostemon violaceus, see *Dalea purpurea*
Petalostemum candidum, see *Dalea candida* var. *candida*
Petalostemum flavescens, see *Dalea flavescens*
Petalostemum oligophyllum, see *Dalea candida* var. *oligophylla*
Petalostemum purpureum, see *Dalea purpurea* var. *purpurea*
Petasite, see *Petasites*
Petasites frigida, see *P. frigidus*
Petasites hyperboreus, see *P. frigidus* var. *nivalis*
Petasites palmata, see *P. frigidus* var. *palmatus*
Petasites palmatum, see *P. frigidus* var. *palmatus*
Petasites palmatus, see *P. frigidus* var. *palmatus*
Petasites speciosus, see *P. frigidus* var. *palmatus*
Petroselinum sativum, see *P. crispum*
Peucedanum canbyi, see *Lomatium canbyi*
Peucedanum graveolens, see *Lomatium graveolens* var. *graveolens*
Peucedanum lecocarpum, see *Lomatium nudicaule*
Peucedanum leiocarpum, see *Lomatium nudicaule*
Peucedanum macrocarpum, see *Lomatium macrocarpum*
Peucedanum nudicaulis, see *Lomatium nudicaule*
Phacelia corrugata, see *P. crenulata* var. *corrugata*
Phacelia leucophylla, see *P. hastata* var. *hastata*
Phacelia menziesii, see *P. linearis*
Phaseolus diversifolius, see *Strophostyles helvula*
Phaseolus limensis, see *P. lunatus*
Phaseolus lunatus macrocarpus, see *P. lunatus*
Phaseolus multiflorus, see *P. coccineus*
Phellopterus bulbosus, see *Cymopterus bulbosus*
Phellopterus montanus, see *Cymopterus montanus*
Philadelphus gordonianus, see *P. lewisii* var. *gordonianus*
Philadelphus lewisi, see *P. lewisii*
Philibertella heterophylla, see *Sarcostemma cynanchoides* ssp. *hartwegii*
Philibertia heterophylla, see *Sarcostemma cynanchoides* ssp. *hartwegii*
Phlox douglasii var. *diffusa*, see *P. caespitosa*
Phoradendron flavescens, see *P. leucarpum*
Phoradendron flavescens var. *villosum*, see *P. villosum* ssp. *villosum*
Phoradendron ligatum, see *P. juniperinum* ssp. *juniperinum*
Phoradendron serotinum, see *P. leucarpum*
Phoradendron tomentosum ssp. *macrophyllum*, see *P. macrophyllum* ssp. *macrophyllum*
Photinea arbutifolia, see *Heteromeles arbutifolia* var. *arbutifolia*
Photinia arbutifolia, see *Heteromeles arbutifolia* var. *arbutifolia*
Phragmites communis (var. *berlandieri*), see *P. australis*

Index of Synonyms

Phragmites phragmites, see *P. australis*
Physalis acutifolia wrightii, see *P. acutifolia*
Physalis fendleri, see *P. hederifolia* var. *fendleri*
Physalis lobata, see *Quincula lobata*
Physalis neo(-)mexicana, see *P. subulata* var. *neomexicana*
Physalis viscora, see *P. viscosa*
Physaria didymocarpa lanata, see *P. didymocarpa* var. *lanata*
Phytolacca decandra, see *P. americana*
Picea alba, see *P. glauca*
Picea canadensis, see *P. glauca*
Picea edulis, see *Pinus edulis*
Picea engelmanni, see *P. engelmannii*
Picea glauca × engelmannii, see *P. glauca*
Picea nigra, see *P. mariana*
Picea rubra, see *P. rubens*
Piceae, see *Picea*
Pinus baksiana, see *P. banksiana*
Pinus brachyptera, see *P. ponderosa* var. *ponderosa*
Pinus canadensis, see *Picea glauca*
Pinus caribaea, see *P. elliottii*
Pinus inops, see *P. contorta*
Pinus mertensiana, see *Tsuga mertensiana*
Pinus microcarpa, see *Larix americana*
Pinus mitis, see *P. virginiana*
Pinus murrayana, see *P. contorta* var. *murrayana*
Pinus poderosa, see *P. ponderosa*
Pinus scopulorum, see *P. ponderosa* var. *scopulorum*
Pinus tuberculata, see *P. attenuata*
Piperacea sp., see *Piperia* sp.
Pirola asarifolia var. *incarnata*, see *Pyrola asarifolia*
Pirola picta, see *Pyrola picta*
Pittostorum sp., see *Pittosporum* sp.
Pityrogramma triangularis, see *Pentagrama triangularis* ssp. *triangularis*
Plagiobothrys campestris, see *P. fulvus* var. *campestris*
Plantago fastigiata, see *P. ovata*
Plantago hirtella, see *P. australis* ssp. *hirtella*
Plantago insularis, see *P. ovata*
Plantago purshii, see *P. patagonica*
Plantanus occidentalis, see *Platanus occidentalis*
Pluchea borealis, see *P. sericea*
Poa californica, see *P. arida*
Poa longiligula, see *P. fendleriana* ssp. *fendleriana*
Poa tenuifolia, see *P. secunda*
Podyphyllum, see *Podophyllum*
Pogogyne parviflora, see *P. douglasii* ssp. *parviflora*
Poinsettia dentata, see *Euphorbia dentata*
Polanisia trachycarpa, see *P. dodecandra* ssp. *trachysperma*
Polanisia trachysperma, see *P. dodecandra* ssp. *trachysperma*
Polemonium humile, see *P. pulcherrimum* ssp. *lindleyi*
Poliomentha ineana, see *Poliomintha incana*
Polygala polygamma, see *P. polygama*
Polygala rugellii, see *P. rugelii*
Polygala senaga, see *P. senega*
Polygonatum commutatum, see *P. biflorum* var. *commutatum*
Polygonum alaskanum, see *P. alpinum*
Polygonum coccineum, see *P. amphibium* var. *emersum*
Polygonum emersum, see *P. amphibium* var. *emersum*
Polygonum glabrum, see *P. densiflorum*
Polygonum hydopiper, see *P. hydropiper*
Polygonum muhlenbergii, see *P. amphibium* var. *emersum*
Polygonum pennsylvanicum (var. *laevigatum*), see *P. pensylvanicum*
Polygonum sawatchense, see *P. douglasii* ssp. *johnstonii*
Polymnia uvedalia, see *Smallanthus uvedalia*
Polypodium falcatum, see *P. glycyrrhiza*
Polypodium vulgare, see *P. virginianum*
Pontentilla anserina, see *Argentina anserina*
Pontentilla simplex, see *Potentilla simplex*
Potentilla anserina ssp. *pacifica*, see *Argentina egedii* ssp. *egedii*
Potentilla pacifica, see *Argentina egedii* ssp. *egedii*
Populus acuminata, see *P. ×acuminata*
Populus angulata, see *P. deltoides* ssp. *deltoides*
Populus aurea, see *P. tremuloides*
Populus candicans, see *P. balsamifera* ssp. *balsamifera*
Populus fremontii wislizeni, see *P. deltoides* ssp. *wislizeni*
Populus gileadensis, see *P. ×jackii*
Populus monilifera, see *P. deltoides* ssp. *monilifera*
Populus sargentii, see *P. deltoides* ssp. *monilifera*
Populus tacamahacca, see *P. balsamifera* ssp. *balsamifera*
Populus trichocarpa, see *P. balsamifera* ssp. *trichocarpa*
Populus wislizeni(i), see *P. deltoides* ssp. *wislizeni*
Porophyllum leucospermum, see *P. gracile*
Portualaca oleraces, see *Portulaca oleracea*
Portulaca retusa, see *P. oleracea* ssp. *oleracea*
Postelsia palmiformis, see *P. palmaeformis*
Potentilla anserina (ssp. *anserina*), see *Argentina anserina*
Potentilla anserina ssp. *pacifica*, see *Argentina egedii* ssp. *egedii*
Potentilla fruticosa, see *Pentaphylloides floribunda*
Potentilla hyparctica, see *P. nana*
Potentilla monspeliensis, see *P. norvegica* ssp. *monspeliensis*
Potentilla pacifica, see *Argentina egedii* ssp. *egedii*
Potentilla palustris, see *Comarum palustre*
Potentilla pennsylvanica, see *P. pensylvanica*
Prenanthus alba, see *Prenanthes alba*
Proboscidea altheaefolia, see *P. althaeifolia*
Promus virens, see *Bromus carinatus*
Prosopis chilensis (var. *glandulosa*), see *P. glandulosa* var. *glandulosa*
Prosopis juliflora (*glandulosa*), see *P. glandulosa* var. *glandulosa*
Prosopis juliflora var. *torreyana*, see *P. glandulosa* var. *torreyana*
Prosopis odorata, see *P. glandulosa* var. *torreyana*
Prosopis prosopis, see *P. glandulosa*
Prosopis pubescene, see *P. pubescens*
Prosopis veluntina, see *P. velutina*
Prosopis velutinea, see *P. velutina*
Prosopsis glandulosa, see *Prosopis glandulosa*
Prunus amygdalus, see *P. dulcis*
Prunus andersoni, see *P. andersonii*
Prunus besseyi, see *P. pumila* var. *besseyi*
Prunus cuneata, see *P. pumila* var. *susquehanae*
Prunus demissa, see *P. virginiana* var. *demissa*
Prunus emarginata var. *villosa*, see *P. emarginata*
Prunus fasiculata, see *P. fasciculata*
Prunus melanocarpa, see *P. virginiana* var. *melanocarpa*
Prunus oregana, see *P. subcordata* var. *oregana*
Prunus pennsylvanica (var. *nettlakea*), see *P. pensylvanica*
Prunus pennsylvanicus, see *P. pensylvanica*
Prunus pensylvania, see *P. pensylvanica*
Prunus scrotina, see *P. serotina*
Psedera quinquefokia, see *Parthenocissus quinquefolia* var. *quinquefolia*
Psedera quinquefolia, see *Parthenocissus quinquefolia* var. *quinquefolia*
Pseudocymopterus aletifolius, see *Aletes anisatus*
Pseudotsuga douglas(s)ii, see *P. menziesii*
Pseudotsuga mucronata, see *P. menziesii*
Pseudotsuga taxifolia, see *P. menziesii* var. *menziesii*
Psidium guayava, see *P. guajava*
Psilostrophe sparisflora, see *P. sparsiflora*
Psilostrophe tagetinae, see *P. tagetina*
Psoralea argophylla, see *Pediomelum argophyllum*
Psoralea canescens, see *Pediomelum canescens*
Psoralea cuspidata, see *Pediomelum cuspidatum*
Psoralea esculenta, see *Pediomelum esculentum*
Psoralea floribunda, see *Psoralidium tenuiflorum*
Psoralea hypog(a)ea(e), see *Pediomelum hypogaeum* var. *hypogaeum*
Psoralea lanceolata, see *Psoralidium lanceolatum*
Psoralea macrostachya, see *Hoita macrostachya*
Psoralea orbicularis, see *Hoita orbicularis*
Psoralea pedunculata, see *Orbexilum pedunculatum* var. *pedunculatum*
Psoralea psoralioides, see *Orbexilum pedunculatum* var. *psoralioides*
Psoralea tenuiflora, see *Psoralidium tenuiflorum*
Psoralea tenuifolia, see *Psoralidium tenuiflorum*
Psorglea pedunculata, see *Orbexilum pedunculatum* var. *pedunculatum*
Pteidium, see *Pteridium*
Ptelea pallida, see *P. trifoliata* ssp. *pallida* var. *pallida*
Ptelea tomentosa, see *P. trifoliata* ssp. *trifoliata* var. *mollis*
Pteridendia bolanden, see *Perideridia bolanderi*

Pteridium latiusculum, see *P. aquilinum* var. *latiusculum*
Pteridum quilina var. *lanuginosa*, see *Pteridium aquilinum* var. *pubescens*
Pteris aquilina, see *Pteridium aquilinum*
Pteris aquilina var. *lanuginosa*, see *Pteridium aquilinum* var. *pubescens*
Pteris caudata, see *Pteridium caudatum*
Pterocaulon undulatum, see *P. virgatum*
Pterospora andromeda, see *P. andromedea*
Ptiloria exigua, see *Stephanomeria exigua*
Ptiloria pauciflora, see *Stephanomeria pauciflora*
Ptiloria tenuifolia, see *Stephanomeria tenuifolia*
Pulsatilla hirsutissima, see *P. patens* ssp. *multifida*
Punex giguateus, see *Rumex giganteus*
Pycnanthemum virginanum, see *P. virginianum*
Pycnothymus rigidus, see *Piloblephis rigida*
Pyrola asarifolia var. *incarnata*, see *P. asarifolia*
Pyrola rotundifolia, see *P. americana*
Pyrola secunda, see *Orthilia secunda*
Pyrola uliginosa, see *P. asarifolia* ssp. *asarifolia*
Pyrrhopappus multicaulis, see *P. pauciflorus*
Pyrus americana, see *Sorbus americana*
Pyrus aucuparia, see *Sorbus aucuparia*
Pyrus coronaria, see *Malus coronaria* var. *coronaria*
Pyrus diversifolia, see *Malus fusca*
Pyrus fusca, see *Malus fusca*
Pyrus ioensis, see *Malus ioensis* var. *ioensis*
Pyrus malus, see *Malus sylvestris*
Pyrus melanocarpa, see *Aronia melanocarpa*
Pyrus sambucifolia, see *Sorbus americana*
Pyrus sambucifolius, see *Sorbus americana*
Pyrus sitchensis, see *Sorbus sitchensis* var. *sitchensis*
Quamasia, see *Camassia*
Quamoclidion multiflorum, see *Mirabilis multiflora* var. *multiflora*
Quercus aquatica, see *Q. nigra*
Quercus borealis var. *maxima*, see *Q. rubra* var. *rubra*
Quercus californica, see *Q. kelloggii*
Quercus californicum, see *Q. kelloggii*
Quercus californicus, see *Q. kelloggii*
Quercus chrysolepis, see *Q. chrysolepis*
Quercus densiflora, see *Lithocarpus densiflorus*
Quercus douglasi, see *Q. douglasii*
Quercus engelmanni, see *Q. engelmannii*
Quercus gambellii, see *Q. gambelii*
Quercus kellogii, see *Q. kelloggii*
Quercus obtusiloba, see *Q. laurifolia*
Quercus pagodaefolia, see *Q. pagoda*
Quercus palmeri, see *Q. dunnii*
Quercus ruba, see *Q. rubra*
Quercus sonomensis, see *Q. kelloggii*
Quercus undulata, see *Q.* ×*pauciloba*
Quercus utahensis, see *Q. gambelii* var. *gambelii*
Quercus virens, see *Q. virginiana*
Quercus wislizeni, see *Q. wislizeni*
Quercus wislizenii, see *Q. wislizeni*
Quercus wislizenii var. *frutescens*, see *Q. wislizeni* var. *frutescens*

Radicula alpina, see *Rorippa alpina*
Radicula curvisiliqua, see *Rorippa curvisiliqua* var. *curvisiliqua*
Radicula nasturtium-aquaticum, see *Rorippa nasturtium-aquaticum*
Radicula sinuata, see *Rorippa sinuata*
Raimannia rhombipetala, see *Oenothera rhombipetala*
Ramona incana, see *Salvia dorrii*
Ramona polystachya, see *Salvia apiana*
Ramona stachyoides, see *Salvia mellifera*
Ranunculus arcis, see *R. acris*
Ranunculus californicum, see *R. californicus*
Ranunculus delphinifolius, see *R. flabellaris*
Ranunculus douglasii, see *R. uncinatus*
Ranunculus eisenii, see *R. occidentalis* var. *eisenii*
Ranunculus eremogenes, see *R. sceleratus* var. *multifidus*
Ranunculus pallasu, see *R. pallasii*
Ranunculus pennsylvanicus, see *R. pensylvanicus*
Ranunculus septentrionalis, see *R. hispidus* var. *nitidus*
Raphanus sativum, see *R. sativus*
Ratibida columnaris (var. *pulcherrima*), see *R. columnifera*
Ratibida columnifera-pulcherrima, see *R. columnifera*
Razoumofskya occidentalis, see *Arceuthobium occidentale*
Rhabdadenia corallicola, see *Angadenia berteroi*
Rhamnus betulaefolia, see *Frangula betulifolia* ssp. *betulifolia*
Rhamnus californica, see *Frangula californica* ssp. *californica*
Rhamnus californica ssp. *occidentalis*, see *Frangula californica* ssp. *occidentalis*
Rhamnus californica ssp. *tomentella*, see *Frangula californica* ssp. *tomentella*
Rhamnus caroliniana, see *Frangula caroliniana*
Rhamnus catharticus, see *R. cathartica*
Rhamnus crocea ilicifolia, see *R. crocea* ssp. *ilicifolia*
Rhamnus ilicifolia, see *R. crocea* ssp. *ilicifolia*
Rhamnus purshiana, see *Frangula purshiana*
Rhamnus purshianus, see *Frangula purshiana*
Rhamnus rubra, see *Frangula rubra* ssp. *rubra*
Rhamus purshiana, see *Frangula purshiana*
Rhamus, see *Rhamnus*
Rhodiola integrifolia, see *Sedum integrifolium* ssp. *integrifolium*
Rhodiola rosea, see *Sedum rosea*
Rhododendron californicum, see *R. macrophyllum*
Rhus aromatica trilobata, see *R. trilobata* var. *trilobata*
Rhus cismontana, see *R. glabra*
Rhus copallina, see *R. copallinum*
Rhus diversiloba, see *Toxicodendron diversilobum*
Rhus emoryi, see *R. trilobata* var. *pilosissima*
Rhus leucantha, see *R. copallinum* var. *leucantha*
Rhus microphyllus, see *R. microphylla*

Rhus radicans, see *Toxicodendron radicans* ssp. *radicans*
Rhus radicans var. *rydbergii*, see *Toxicodendron rydbergii*
Rhus toxico(n)dendron (ssp. *radicans*), see *Toxicodendron pubescens*
Rhus typbira, see *R. hirta*
Rhus typhina, see *R. hirta*
Rhus vernix, see *Toxicodendron vernix*
Rhytideadelphus, see *Rhytidiadelphus*
Ribes floridum, see *R. americanum*
Ribes glutinosum, see *R. sanguineum* var. *glutinosum*
Ribes gracile, see *R. missouriense*
Ribes grossularia, see *R. uva-crispa* var. *sativum*
Ribes inebrians, see *R. cereum* var. *pedicellare*
Ribes inverme, see *R. inerme*
Ribes irriguum, see *R. oxyacanthoides* ssp. *irriguum*
Ribes laxifolium, see *R. laxiflorum*
Ribes leptanthum var. *brachyanthum*, see *R. velutinum*
Ribes missouriensis, see *R. missouriense*
Ribes odoratum, see *R. aureum* var. *villosum*
Ribes oxyacanthoides saxosum, see *R. hirtellum*
Ribes petiolare, see *R. hudsonianum* var. *petiolare*
Ribes roezlii var. *cruentum*, see *R. cruentum* var. *cruentum*
Ribes sanguinem, see *R. sanguineum*
Ribes setosum, see *R. oxyacanthoides* ssp. *setosum*
Ribes viscosisimum, see *R. viscosissimum*
Ribies, see *Ribes*
Robinia neo-mexicana, see *R. neomexicana*
Robinia pseud(o-)acacia, see *R. pseudoacacia*
Roripa, see *Rorippa*
Rorippa amoracia, see *Armoracia rusticana*
Rorippa armoracia, see *Armoracia rusticana*
Rorippa hispida, see *R. palustris* ssp. *hispida*
Rorippa nasturtium, see *R. nasturtium-aquaticum*
Rorippa obtusa, see *R. teres*
Rosa acidularis, see *R. acicularis*
Rosa arizonica, see *R. woodsii* var. *ultramontana*
Rosa californica var. *ultramontana*, see *R. woodsii* var. *ultramontana*
Rosa fendleri, see *R. woodsii* var. *woodsii*
Rosa humilis, see *R. carolina*
Rosa lucida, see *R. virginiana*
Rosa nelina, see *R. nutkana*
Rosa neomexicana, see *R. woodsii* var. *woodsii*
Rosa pratincola, see *R. arkansana* var. *suffulta*
Rosa rubiginosa, see *R. eglanteria*
Rosa sayi, see *R. acicularis* ssp. *sayi*
Rosa spaldingii, see *R. nutkana* var. *hispida*
Rubis, see *Rubus*
Rubus acaulis, see *R. arcticus* ssp. *acaulis*
Rubus alleghenienis, see *R. alleghaniensis*
Rubus arizonicus, see *R. arizonensis*
Rubus choemomorus, see *R. chamaemorus*
Rubus glabra, see *Rhus glabra*
Rubus hawaiiansis, see *R. hawaiensis*
Rubus hispidis, see *R. hispidus*

Index of Synonyms

Rubus idaeus(-)*aculeatissi*(*mus*), see *R. idaeus* ssp. *strigosus*
Rubus leucooderme, see *R. leucodermis*
Rubus macropetalus, see *R. ursinus* ssp. *macropetalus*
Rubus melanolasius, see *R. idaeus* ssp. *strigosus*
Rubus nigrobaccus, see *R. allegheniensis* var. *allegheniensis*
Rubus nutkanus, see *R. parviflorus*
Rubus parviflor(*a*), see *R. parviflorus*
Rubus parviflorus velutinus, see *R. parviflorus* var. *velutinus*
Rubus parvifolia, see *R. parviflorus*
Rubus pebescens, see *R. pubescens*
Rubus spectablilis, see *R. spectabilis*
Rubus strigosus, see *R. idaeus* ssp. *strigosus*
Rubus triflorus, see *R. pubescens* var. *pubescens*
Rubus villosus, see *R. procumbens*
Rudbeckia divergens, see *R. hirta* var. *angustifolia*
Rudbeckia laciniate, see *R. laciniata*
Rumex britan(*n*)*ica*, see *R. orbiculatus*
Rumex crispis, see *R. crispus*
Rumex fenestratus, see *R. aquaticus* var. *fenestratus*
Rumex fueginus, see *R. maritimus*
Rumex geyeri, see *R. paucifolius*
Rumex hymenocallis, see *R. hymenosepalus*
Rumex mexicana, see *R. salicifolius* var. *mexicanus*
Rumex mexicanus, see *R. salicifolius* var. *mexicanus*
Rumex obtusifolium, see *R. obtusifolius*
Rumex occidentalis (var. *procerus*), see *R. aquaticus* var. *fenestratus*
Ruta chalapensis, see *R. chalepensis*
Sabal adansonii, see *S. minor*
Sabbatia campanulata, see *Sabatia campanulata*
Sabina virginiana, see *Juniperus virginiana* var. *virginiana*
Sadleria cyatheoedes, see *S. cyatheoides*
Sagittaria arifolia, see *S. cuneata*
Sagittaria variabilis, see *S. latifolia* var. *latifolia*
Salicornia herbacea, see *S. maritima*
Salicornia pacifica, see *Sarcocornia pacifica*
Salicornia subterminalis, see *Arthrocnemum subterminale*
Salidago pitcheri, see *Solidago gigantea*
Salix alexensis, see *S. alaxensis*
Salix amphibia, see *S. caroliniana*
Salix argophylla, see *S. exigua*
Salix argyrophylla, see *S. hindsiana*
Salix balsamifera, see *S. pyrifolia*
Salix cordata, see *S. eriocephala*
Salix dasiolipis, see *S. lasiolepis*
Salix flavescens, see *S. scouleriana*
Salix fluviatilis, see *S. melanopsis*
Salix glaucophylloides, see *S. myricoides* var. *myricoides*
Salix laevigata, see *S. bonplandiana*
Salix lasiandra, see *S. lucida* ssp. *lasiandra*
Salix longifolia, see *S. interior*
Salix longipes, see *S. caroliniana*
Salix mackenzi(*e*)*ana*, see *S. prolixa*

Salix missouriensis, see *S. eriocephala*
Salix piperi, see *S. hookeriana*
Salix pulchra, see *S. planifolia* ssp. *pulchra*
Salix rostrata, see *S. bebbiana*
Salix sessilifolia var. *hindsiana*, see *S. hindsiana*
Salix tristis, see *S. humilis* var. *tristis*
Salix vulgare, see *S. nigra*
Salsola atriplicifolia, see *Cycloloma atriplicifolium*
Salsola kali, see *S. australis*
Salsola kali var. *tenuifolia*, see *S. australis*
Salsola kali var. *tragus*, see *S. australis*
Salsola pestifer, see *S. australis*
Salvia carduaceae, see *S. carduacea*
Salvia carnosa, see *S. dorrii* ssp. *dorrii* var. *incana*
Salvia columbaria, see *S. columbariae*
Salvia columbriae, see *S. columbariae*
Salvia dorii, see *S. dorrii*
Sambucus caerulea, see *S. cerulea*
Sambucus calli(*o*)*carpa*, see *S. racemosa* ssp. *pubens* var. *arborescens*
Sambucus coerulea, see *S. cerulea*
Sambucus coerules, see *S. cerulea*
Sambucus glauca, see *S. cerulea* var. *cerulea*
Sambucus intermedia, see *S. cerulea*
Sambucus melanocarpa, see *S. racemosa* ssp. *pubens* var. *melanocarpa*
Sambucus mexicana, see *S. cerulea* var. *mexicana*
Sambucus microbotrys, see *S. racemosa* ssp. *pubens* var. *microbotrys*
Sambucus neomexicana, see *S. cerulea* var. *neomexicana*
Sambucus pubens, see *S. racemosa* ssp. *pubens* var. *leucocarpa*
Sambucus simpsoni(*i*), see *S. canadensis*
Sambucus velutina, see *S. cerulea* var. *velutina*
Sambucus velutinus, see *S. cerulea* var. *velutina*
Sanguinaria candensis, see *S. canadensis*
Sanicula gregaria, see *S. odorata*
Sanicula luberosa, see *S. tuberosa*
Sanicula menziesii, see *S. crassicaulis*
Sanquinaria, see *Sanguinaria*
Sanvitalia aberti, see *S. abertii*
Sapindus drummondii, see *S. saponaria* var. *drummondii*
Saponaria officianalis, see *S. officinalis*
Sarcobatus vermicularis, see *S. vermiculatus*
Sarracena, see *Sarracenia*
Sassafras molle, see *S. albidum*
Sassafras officinale, see *S. albidum*
Sassafras sassafras, see *S. albidum*
Sassafras variifolium, see *S. albidum*
Savastana odorata, see *Hierochloe odorata* ssp. *odorata*
Saxifraga bongardi, see *S. ferruginea*
Saxifraga jamesi, see *Boykinia jamesii*
Saxifraga pennsylvanica, see *S. pensylvanica*
Saxifraga punctata (ssp. *nelsoniana*), see *S. nelsoniana* ssp. *nelsoniana*
Saxifrage, see *Saxifraga*
Scaevolo frutescens, see *Scaevola sericea*

Schedonardus paniculatus, see *Schedonnardus paniculatus*
Schmaltzia bakeri, see *Rhus trilobata*
Schmaltzia trilobata, see *Rhus trilobata* var. *trilobata*
Schoenolirion album, see *Hastingsia alba*
Scirpus alneyi, see *S. americanus*
Scirpus campestris, see *S. robustus*
Scirpus lacustris, see *S. acutus*
Scirpus lacustris (var.) *occid*(*entalis*), see *S. acutus*
Scirpus olneyi, see *S. americanus*
Scirpus palendosus, see *S. maritimus*
Scirpus pungens ssp. *monophyllus* var. *longisetis*, see *S. americanus*
Scirpus rubrotinctus, see *S. microcarpus*
Scirpus validus, see *S. tabernaemontani*
Scorzonella maxima, see *Microseris laciniata*
Sebastiana ligustrina, see *Sebastiania fruticosa*
Sedum purpureum, see *S. telephium* ssp. *telephium*
Sedum rosea ssp. *integrifolium*, see *S. integrifolium* ssp. *integrifolium*
Sedum roseum, see *S. rosea*
Seneci, see *Senecio*
Senecio douglasii, see *S. flaccidus* var. *douglasii*
Senecio filifolius, see *S. flaccidus* var. *flaccidus*
Senecio jacobea, see *S. jacobaea*
Senecio longilobus, see *S. flaccidus* var. *flaccidus*
Senecio pseudo-arnica, see *S. pseudoarnica*
Senecio smallii, see *S. anonymus*
Senecio spartoides, see *S. spartioides*
Sericotheca discolor, see *Holodiscus discolor*
Sericotheca dumosa, see *Aster dumosus*
Serrenoa serrulata, see *Serenoa repens*
Sesamum indicum, see *S. orientale*
Setaria lutescens, see *Pennisetum glaucum*
Sevastana odorata, see *Hierochloe odorata* ssp. *odorata*
Shepherdia candensis, see *S. canadensis*
Shepherdia stolonifera, see *Cornus sericea* ssp. *sericea*
Sheradia arvensis, see *Sherardia arvensis*
Sida hederacea, see *Malvella leprosa*
Sidalcea malvaeflora, see *S. malviflora*
Sieversia ciliata, see *Geum triflorum* var. *ciliatum*
Silene multicaulis, see *S. douglasii* var. *douglasii*
Silene pringlei, see *S. douglasii* var. *douglasii*
Simmondsia californica, see *S. chinensis*
Sismybrium officinale leiocarpum, see *Sisymbrium officinale*
Sisymbrium canescens, see *Descurainia pinnata* ssp. *pinnata*
Sisymbrium elegans, see *Thelypodiopsis elegans*
Sisymbrium incisum, see *Descurainia incana* ssp. *incisa*
Sisymbrium officinale leiocarpum, see *S. officinale*
Sisymbrium pinnatum, see *Descurainia pinnata* ssp. *pinnata*
Sisymbrium sophia, see *Descurainia sophia*
Sisyrinchium albidium, see *S. albidum*
Sisyrinchium augustifolium, see *S. angustifolium*

888 Index of Synonyms

Sisyrinchium fibrosum, see *S. nashii*
Sisyrinchium graminoides, see *S. angustifolium*
Sitanion hystrix, see *Elymus elymoides*
Sitanion jubatum, see *Elymus multisetus*
Sitilias caroliniana, see *Pyrrhopappus carolinianus*
Sitilias multicaulis, see *Pyrrhopappus pauciflorus*
Sium cicutaefolium, see *S. suave*
Sium laeve, see *S. suave*
Sium lineare, see *S. suave*
Sium sauve, see *S. suave*
Smilacina amplexicaulis, see *Maianthemum racemosum* ssp. *amplexicaule*
Smilacina racemosa, see *Maianthemum racemosum* ssp. *racemosum*
Smilacina stellata, see *Maianthemum stellatum*
Smilacrina borealis, see *Clintonia uniflora*
Smilax hispida, see *S. tamnoides*
Smilax tamnifolia, see *S. pseudochina*
Solanum blodgettii, see *S. donianum*
Solanum dulcamera, see *S. dulcamara*
Solanum verbascifolium, see *S. donianum*
Solanum villosum, see *S. physalifolium*
Solanum xanti, see *S. xantii*
Solidaga, see *Solidago*
Solidago altissima, see *S. canadensis* var. *scabra*
Solidago decumbens, see *S. simplex* ssp. *simplex* var. *nana*
Solidago graminifolia, see *Euthamia graminifolia* var. *graminifolia*
Solidago latifolia, see *S. flexicaulis*
Solidago petradoria, see *Petradoria pumila* ssp. *pumila*
Solidago pitcheri, see *S. gigantea*
Solidago pumila, see *Petradoria pumila*
Solidago rigidiuscula, see *S. speciosa* var. *rigidiuscula*
Solidago sarothrae, see *Gutierrezia sarothrae*
Solidago serotina, see *S. gigantea*
Solidago sparsiflora, see *S. velutina*
Solidago virgaurea, see *S. multiradiata*
Sophia halictorum, see *Descurainia pinnata* ssp. *halictorum*
Sophia incisa, see *Descurainia incana* ssp. *incisa*
Sophia pinnata, see *Descurainia pinnata* ssp. *pinnata*
Sophia sophia, see *Descurainia sophia*
Sophia sp., see *Descurainia* sp.
Sophora serecia, see *S. nuttalliana*
Sophora sericea, see *S. nuttalliana*
Sophora seundiflora, see *S. secundiflora*
Sorghum vulgare, see *S. bicolor* ssp. *bicolor*
Spartina michauxiana, see *S. pectinata*
Spatheyma foetida, see *Symplocarpus foetidus*
Spathyema foetida, see *Symplocarpus foetidus*
Specularia perfoliata, see *Triodanis perfoliata* var. *perfoliata*
Sphacele calycina, see *Lepechinia calycina*
Sphaeralcea cuspidata, see *S. angustifolia* ssp. *cuspidata*
Sphaeralcea grossulariaefolia, see *S. grossulariifolia*

Sphaeralcea lobata, see *S. angustifolia* ssp. *lobata*
Sphaeralcea parviflora, see *S. parvifolia*
Spigelia anthelmintica, see *S. anthelmia*
Spiraea aruncus, see *Aruncus dioicus* var. *vulgaris*
Spiraea beauverdiana, see *S. stevenii*
Spiraea caespitosa, see *Petrophyton caespitosum* var. *caespitosum*
Spiraea densiflora, see *S. splendens* var. *splendens*
Spiraea latifolia, see *S. alba* var. *latifolia*
Spiraea menziesii, see *S. douglasii* var. *menziesii*
Spiraea millefolium, see *Chamaebatiaria millefolium*
Spiraea pyramidata, see *S.* ×*pyramidata*
Spiragyra sp., see *Spirogyra* sp.
Spiranthes gracilis, see *S. lacera* var. *gracilis*
Spirea alba, see *Spiraea alba*
Spirea latifolia, see *Spiraea alba* var. *latifolia*
Spirea tomentosa, see *Spiraea tomentosa*
Sporobolus strictus, see *S. contractus*
Stachys ciliata, see *S. mexicana*
Stachys cooleyae, see *S. ciliata*
Stanleyella wrightii, see *Thelypodium wrightii* ssp. *wrightii*
Statice limonium, see *Limonium vulgare*
Stellaria jamesiana, see *Pseudostellaria jamesiana*
Stenandrium floridanum, see *S. dulce*
Stenotaphrum americanum, see *S. secundatum*
Stenotopsis linearifolius, see *Ericameria linearifolia*
Stephanomeria ramosa, see *S. runcinata*
Stephanomeria tenuiflora, see *S. tenuifolia*
Steptopus roseus, see *Streptopus roseus*
Stillingia angustifolia, see *S. sylvatica* ssp. *sylvatica*
Stipa hymenoides, see *Oryzopsis hymenoides*
Streptopus amplexifolium, see *S. amplexifolius*
Strombocarpa pubescens, see *Prosopis pubescens*
Strophostyles helvola, see *S. helvula*
Struthiopteris spicant, see *Blechnum spicant*
Styphelia tameiameia, see *S. tameiameiae*
Suaeda depressa (var. *erecta*), see *S. calceoliformis*
Suaeda erecta, see *S. calceoliformis*
Suaeda fruticosa, see *S. moquinii*
Suaeda intermedia, see *S. moquinii*
Suaeda torreyana (var. *ramosissima*), see *S. moquinii*
Svida stolonifera, see *Cornus sericea* ssp. *sericea*
Swertia caroliniensis, see *Frasera caroliniensis*
Swertia radiata, see *Frasera speciosa*
Symphoricarpor, see *Symphoricarpos*
Symphoricarpos parishii, see *S. oreophilus* var. *parishii*
Symphoricarpos racemosa, see *S. albus* var. *albus*
Symphoricarpos racemosus, see *S. albus* var. *albus*
Symphoricarpos rivularis, see *S. albus* var. *laevigatus*
Symphoricarpos symphoricarpos, see *S. orbiculatus*
Symphoricarpos vulgaris, see *S. orbiculatus*
Symphoricarpus, see *Symphoricarpos*
Tacca pinnatifida, see *T. leontopetaloides*
Taraxacum officinal, see *T. officinale*
Taraxacum palustre var. *vulgare*, see *T. officinale*

ssp. *vulgare*
Taraxacum taraxacum, see *T. officinale* ssp. *vulgare*
Taraxicum, see *Taraxacum*
Tatropolia cordiophylla, see *Jatropha cardiophylla*
Taxodium sempervirens, see *Sequoia sempervirens*
Taxus minor, see *T. canadensis*
Tellima affinis, see *Lithophragma affine*
Tephrosia ambigua, see *T. florida*
Tephrosia elegans, see *T. hispidula*
Tephrosia piscatoria, see *T. purpurea*
Tessaria sericea, see *Pluchea sericea*
Tetradymia spinosa var. *longispina*, see *T. axillaris*
Thacictrum occidentalis, see *Thalictrum occidentale*
Thalaspi, see *Thlaspi*
Thalesia fasciculata, see *Orobanche fasciculata*
Thalictrium polygamum, see *Thalictrum pubescens*
Thalictrum occidentalis, see *T. occidentale*
Thalictrum purpurascens, see *T. dasycarpum*
Thamnosma montanum, see *T. montana*
Thea sinensis, see *Camellia sinensis*
Theleosperma, see *Thelesperma*
Thelesperma gracile, see *T. megapotamicum*
Thelesperma trifidum, see *T. filifolium* var. *filifolium*
Thelypodium lilacinum, see *T. integrifolium* ssp. *integrifolium*
Thermopsis pinetorum, see *T. rhombifolia* var. *montana*
Therofon elatum, see *Boykinia occidentalis*
Thlaspi fendleri, see *T. montanum* var. *fendleri*
Thuga canadensis, see *Thuja occidentalis*
Thuja gigantea, see *T. plicata*
Thurberia thespesioides, see *Gossypium thurberi*
Thuya, see *Thuja*
Thymus serpyllum, see *T. praecox* ssp. *arcticus*
Thysanocarpus elegans, see *T. curvipes*
Tilia glabra, see *T. americana* var. *americana*
Tilia heterophylla, see *T. americana* var. *heterophylla*
Torresia macrophylla, see *Hierochloe occidentalis*
Torresia odorata, see *Hierochloe odorata* ssp. *odorata*
Tovara virginiana, see *Polygonum virginianum*
Townsendia arizonica, see *T. incana*
Townsendia sericea, see *T. exscapa*
Toxicodendron toxicodendron, see *T. pubescens*
Toxylon pomiferum, *Maclura pomifera*
Tradescentia virginiana, see *Tradescantia virginiana*
Tragia nepetaefolia, see *T. nepetifolia*
Trautvetteria grandis, see *T. caroliniensis*
Trema floridanus, see *T. micranthum*
Tridens pulchellus, see *Erioneuron pulchellum*
Trientalis americana, see *T. borealis* ssp. *borealis*
Trientalis latifolia, see *T. borealis* ssp. *latifolia*
Trifolium ciliatum, see *T. ciliolatum*
Trifolium fimbriatum, see *T. wormskioldii*
Trifolium involucratum (var. *fendlari*), see *T. wormskioldii*

Trifolium tridentatum, see *T. willdenowii*
Trifolium virescens, see *T. fucatum*
Trifolium wormskjoldii, see *T. wormskioldii*
Triglochin maritima, see *T. maritimum*
Triodia mutica, see *Tridens muticus* var. *muticus*
Tripterocalyx wootoni(i), see *T. carnea* var. *wootonii*
Trisetum subspicatum, see *T. spicatum*
Triticum sativum, see *T. aestivum*
Triticum vulgare, see *T. aestivum*
Troximon aurantiacum, see *Agoseris aurantiaca* var. *aurantiaca*
Troximon sp., see *Agoseris* sp.
Tsuga canadense, see *T. canadensis*
Tumion californica, see *Torreya californica*
Tumion californicum, see *Torreya californica*
Typa, see *Typha*
Ulmus americanus, see *U. americana*
Ulmus falva, see *U. rubra*
Ulmus fluva, see *U. rubra*
Ulmus fulva, see *U. rubra*
Ulmus pubescens, see *U. rubra*
Ulmus thomasi, see *U. thomasii*
Umbelliferae, see Apiaceae
Uniola palmeri, see *Chasmanthium latifolium*
Urtica breweri, see *U. dioica* ssp. *holosericea*
Urtica lyallii, see *U. dioica* ssp. *gracilis*
Urtica dioica var. *gracilis*, see *U. dioica* ssp. *gracilis*
Urtica dioica var. *lyallii*, see *U. dioica* ssp. *gracilis*
Urtica gracilis var. *holosericea*, see *U. dioica* ssp. *holosericea*
Urtica holosericea, see *U. dioica* ssp. *holosericea*
Urtica lyalli(i), see *U. dioica* ssp. *gracilis*
Urtica procera, see *U. dioica* ssp. *gracilis*
Urticastrum divaricatum, see *Laportea canadensis*
Uva-ursi uva-ursi, see *Arctostaphylos uva-ursi*
Vaccinium alask(a)ense, see *V. ovalifolium*
Vaccinium alaskensis, see *V. ovalifolium*
Vaccinium augustifolium, see *V. angustifolium*
Vaccinium caespitosa, see *V. cespitosum*
Vaccinium caespitosum (var. *caespitosum*), see *V. cespitosum* var. *cespitosum*
Vaccinium caespitosum var. *paludicola*, see *V. cespitosum* var. *paludicola*
Vaccinium canadense, see *V. myrtilloides*
Vaccinium globulare, see *V. membranaceum*
Vaccinium hispidotum, see *Gaultheria hispidula*
Vaccinium macrophyllum, see *V. myrtilloides*
Vaccinium membranaceun, see *V. membranaceum*
Vaccinium membranceum, see *V. membranaceum*
Vaccinium oreophilum, see *V. myrtillus* var. *oreophilum*
Vaccinium oxycoccus (var. *ovalifolium*), see *V. oxycoccos*
Vaccinium paludicola, see *V. cespitosum*
Vaccinium pennsylvanicum, see *V. angustifolium*
Vaccinium uglinosum, see *V. uliginosum*
Vaccinium vitis(idaea), see *V. vitis-idaea*
Vaccinium vitis-idea (ssp.) *minus*, see *V. vitis-idaea* ssp. *minus*
Vaccinum, see *Vaccinium*
Vagnera amplexicaulis, see *Maianthemum racemosum* ssp. *amplexicaule*
Vagnera racemosa, see *Maianthemum racemosum* ssp. *racemosum*
Vagnera stellata, see *Maianthemum stellatum*
Valerian capitata, see *Valeriana capitata*
Valerian sitchensis, see *Valeriana sitchensis*
Valeriana septentrionalis, see *V. dioica* var. *sylvatica*
Valeriana sylvatica, see *V. dioica* var. *sylvatica*
Valerianella olitoria, see *V. locusta*
Veratrum californica, see *V. californicum*
Veratrum eschscholtzii, see *V. viride*
Veratrum speciosum, see *V. californicum* var. *californicum*
Verbena lupinnatifida, see *Glandularia bipinnatifida*
Verbena urticaefolia, see *V. urticifolia*
Verbena wrightii, see *Glandularia wrightii*
Verbesina encelioides exauriculata, see *V. encelioides* ssp. *exauriculata*
Veronica serpylliafolia, see *V. serpyllifolia*
Veronica virginica, see *Veronicastrum virginicum*
Viburnum americanum, see *V. opulus* var. *americanum*
Viburnum cassinoides, see *V. nudum* var. *cassinoides*
Viburnum pauciflorum, see *V. edule*
Viburnum pomifolium, see *V. prunifolium*
Viburnum pubescens, see *V. dentatum* var. *dentatum*
Viburnum recognitum, see *V. dentatum* var. *lucidum*
Viburnum trilobum, see *V. opulus* var. *americanum*
Vicea, see *Vicia*
Vicia angustifolia, see *V. sativa* ssp. *nigra*
Vicia gigantea, see *V. nigricans* ssp. *gigantea*
Vigna luta, see *V. luteola*
Vigna sinensis, see *V. unguiculata*
Viguiera annua, see *Heliomeris longifolia* var. *annua*
Viguiera longifolia, see *Heliomeris longifolia* var. *longifolia*
Viguiera multiflora, see *Heliomeris multiflora* var. *multiflora*
Vilfa cryptandra, see *Sporobolus cryptandrus*
Villanova dissecta, see *Bahia dissecta*
Vincetoxicum productum, see *Matelea producta*
Viola eriocarpa, see *V. pubescens* var. *pubescens*
Viola papilionacea, see *V. sororia*
Viola pensylvanica, see *V. pubescens* var. *pubescens*
Viola rafinesquil, see *V. bicolor*
Viola sarmentosa, see *V. sempervirens*
Viorna baldwinii, see *Clematis baldwinii* var. *baldwinii*
Virburnum, see *Viburnum*
Vitis arizonicus, see *V. arizonica*
Vitis baileyana, see *V. cinerea* var. *baileyana*
Vitis cordifolia, see *V. vulpina*
Vitis rufotomentosa, see *V. aestivalis* var. *aestivalis*
Vitus, see *Vitis*
Waltheria americana, see *W. indica*
Washingtonia divaricata, see *Osmorhiza berteroi*
Washingtonia longistylis, see *Osmorhiza longistylis*
Washingtonia nuda, see *Osmorhiza berteroi*
Washingtonia obtusa, see *Osmorhiza depauperata*
Wilcoxia striata, see *Peniocereus striatus*
Winstroemia sp., see *Wikstroemia* sp.
Wislizenia melilotoides, see *W. refracta* ssp. *refracta*
Woodsia mexicana, see *W. neomexicana*
Xanthium canadense, see *X. strumarium* var. *canadense*
Xanthium commune, see *X. strumarium* var. *canadense*
Xanthium italicum, see *X. strumarium* var. *canadense*
Xanthium orientale, see *X. strumarium* var. *glabratum*
Xanthium saccharatum, see *X. strumarium* var. *canadense*
Xanthorrhiza apiifolia, see *Xanthorhiza simplicissima*
Xanthoxalis stricta, see *Oxalis stricta*
Xanthoxylum americanum, see *Zanthoxylum americanum*
Xanthoxylum clava-hercules, see *Zanthoxylum clava-herculis*
Xerophyllum douglasii, see *X. tenax*
Xerophyllum tenex, see *X. tenax*
Ximenesia exauriculata, see *Verbesina encelioides* ssp. *exauriculata*
Ximenesia sp., see *Verbesina* sp.
Ximinesia exauriculata, see *Verbesina encelioides* ssp. *exauriculata*
Xiris ambigua, see *Xyris ambigua*
Xiris, see *Xyris*
Xolisma fruticosa, see *Lyonia fruticosa*
Yucca angustifolia, see *Y. glauca* var. *glauca*
Yucca arborescens, see *Y. brevifolia*
Yucca bacatta, see *Y. baccata*
Yucca bacchata, see *Y. baccata*
Yucca macrocarpa, see *Y. torreyi*
Yucca mohavensis, see *Y. schidigera*
Yucca navaja, see *Y. baileyi* var. *navajoa*
Zamia floridana, see *Z. pumila*
Zanthium strumarium, see *Xanthium strumarium*
Zanthorhiza apiifolia, see *Xanthorhiza simplicissima*
Zanthoxylum americana, see *Z. americanum*
Zauschneria californica, see *Epilobium canum* ssp. *angustifolium*
Zauschneria latifolia, see *Epilobium canum* ssp. *latifolium*
Zea arnylacea, see *Z. mays*
Zea maize, see *Z. mays*
Zea mays amylacea, see *Z. mays*
Zea mays saccharata, see *Z. mays*
Ziziphus cycioides, see *Ziziphus obtusifolia*
Ziziphus lycioides, see *Ziziphus obtusifolia*
Zygadenus, see *Zigadenus*
Zigadenus nuttalli, see *Z. nuttallii*

Index of Common Names

Common plant names appearing in the original sources are listed alphabetically and are cross-referenced to the scientific names used in the Catalog of Plants. For brevity, not all common names have been able to be listed. For example, common names that include the genus name used in the Catalog of Plants are not usually listed in this index, so remember that some "common" names are in fact genus names, which are listed alphabetically in the Catalog of Plants. A few examples of such names are asparagus (*Asparagus*), citrus (*Citrus*), dahlia (*Dahlia*), iris (*Iris*), phlox (*Phlox*), and spiraea (*Spiraea*). But when the common name includes a different genus name (e.g., rock spiraea for *Holodiscus discolor, H. dumosus*, and *Petrophyton caespitosum*) it is listed. Plants are identified below to the level of species and a particular common name may apply only to a single subspecies or variety of that species. Word division as used in the original sources has been followed when possible, but indexing often required other word divisions. Some misspellings in the original sources have been corrected and some editorial conventions have been adopted to facilitate finding names in the index. Unlike scientific names, common plant names do not follow rules of nomenclature. This index is intended as a handy finding tool rather than as a consistently uniform catalog of common names.

A-a-li-i: *Dodonaea viscosa*
Acid Berry: *Rhus trilobata*
Aconite, Kamchatka: *Aconitum maximum*
Adam and Eve: *Aplectrum hyemale*
Adam's Needle: *Yucca filamentosa*
Adder's Mouth: *Malaxis unifolia*
Adder's Tongue, Yellow: *Erythronium americanum*
Adelia: *Forestiera pubescens*
Agrimony: *Agrimonia gryposepala, A. parviflora*
Agueweed: *Gentianella quinquefolia*
Ahakea: *Bobea* sp.
Ahinahina: *Artemisia australis*
Ahuawa: *Cyperus laevigatus*
Airplant: *Catopsis* sp., *Guzmania* sp., *Tillandsia* sp.
Akala: *Rubus hawaiensis, R. macraei*
Aki-aki: *Stenotaphrum secundatum*
Akia Launui: *Wikstroemia* sp.
A-koko: *Chamaesyce multiformis*
Ako-lea: *Phegopteris* sp.
A-laa: *Sideroxylon* sp.
Ala-ala-pu-loa: *Waltheria indica*
Ala-ala-waionui-pehu: *Peperomia* sp.
Alabama Supplejack: *Berchemia scandens*
Alaea: *Asplenium horridum*
Alani-kuahiwi: *Pelea* sp.
Alaweo: *Chenopodium oahuense*

Alder: *Alnus* spp., *Cornus sericea*
Alder, Black: *Alnus incana, Ilex verticillata*
Alder, Green: *Alnus viridis*
Alder, Hazel: *Alnus serrulata*
Alder, Hoary: *Alnus incana*
Alder, Meadow: *Alnus glutinosa*
Alder, Mountain: *Alnus incana, A. rhombifolia, A. viridis*
Alder, Paper Leaf: *Alnus incana*
Alder, Red: *Alnus rubra*
Alder, River: *Alnus incana*
Alder, Sitka: *Alnus viridis*
Alder, Smooth: *Alnus incana*
Alder, Speckled: *Alnus incana, A.* sp.
Alder, Thinleaf: *Alnus incana*
Alder, White: *Alnus rhombifolia, Clethra acuminata*
Alderleaf Buckthorn: *Rhamnus alnifolia*
Alderleaf Mountain Mahogany: *Cercocarpus montanus*
Alexander, Golden: *Zizia aurea*
Alfalfa: *Medicago sativa*
Alfilaria (or Alfileria): *Erodium cicutarium*
Alga, Tubular Green: *Enteromorpha intestinalis*
Algarrobo: *Prosopis chilensis*
Algerita: *Mahonia fremontii, M. haematocarpa*
Alkali Sacaton: *Sporobolus airoides*

Alligatorflag: *Thalia geniculata*
Alligator Tree: *Forestiera acuminata*
Allspice, Carolina: *Calycanthus floridus*
Almond, Desert: *Prunus fasciculata*
Almond, Sweet: *Prunus dulcis*
Aloe: *Agave americana, Manfreda virginica*
Alsike Clover: *Trifolium hybridum*
Alumroot, Alpine: *Heuchera cylindrica*
Alumroot, American: *Heuchera americana*
Alumroot, Beautiful: *Heuchera cylindrica*
Alumroot, Bracted: *Heuchera bracteata*
Alumroot, Bridger Mountain: *Heuchera flabellifolia*
Alumroot, Crevice: *Heuchera micrantha*
Alumroot, Littleflower: *Heuchera parviflora*
Alumroot, Littleleaf: *Heuchera parvifolia*
Alumroot, New Mexico: *Heuchera novomexicana*
Alumroot, Pink: *Heuchera rubescens*
Alumroot, Richardson's: *Heuchera richardsonii*
Alumroot, Rough Leaved: *Heuchera cylindrica*
Alumroot, Round Leaved: *Heuchera cylindrica*
Alumroot, Small Flowered: *Heuchera micrantha*
Alumroot, Small Leaved: *Heuchera flabellifolia*
Alyssum, Desert: *Lepidium fremontii*
Amaranth, Bloodroot: *Amaranthus palmeri*
Amaranth, Fringed: *Amaranthus fimbriatus*
Amaranth, Green: *Amaranthus retroflexus*

Amaranth, Mat: *Amaranthus blitoides*
Amaranth, Palmer: *Amaranthus palmeri*
Amaranth, Powell's: *Amaranthus powellii*
Amaranth, Prostrate: *Amaranthus blitoides*
Amaranth, Purple: *Amaranthus cruentus*
Amaranth, Red: *Amaranthus cruentus*
Amaranth, Redroot: *Amaranthus retroflexus*
Amaranth, Sandhill: *Amaranthus arenicola*
Amaranth, Slim: *Amaranthus hybridus*
Amaranth, Spiny: *Amaranthus spinosus*
A-mau-mau: *Sadleria cyatheoides*
Amole: *Yucca glauca*
Anemone, Mountain: *Pulsatilla occidentalis*
Anemone, Star: *Trientalis borealis*
Angelica: *Angelica* spp., *Ligusticum canadense*
Angelica Tree: *Aralia spinosa*
Angelico: *Ligusticum canadense*
Anglepod: *Matelea producta*
Anglestem Buckwheat: *Eriogonum angulosum*
Anise: *Osmorhiza occidentalis*, *Perideridia gairdneri*, *P. kelloggii*, *Pimpinella anisum*
Anise, Sweet: *Osmorhiza occidentalis*, *Perideridia kelloggii*
Annis: *Perideridia gairdneri*
Antelope Brush: *Purshia glandulosa*, *P. tridentata*, *Sambucus tridentata*
Antelope Horns: *Asclepias asperula*
Apache Plume: *Fallugia paradoxa*
Ape Keokeo: *Alocasia macrorrhizos*
Apple: *Malus sylvestris*, *Pyrus* sp.
Apple, Bake: *Rubus chamaemorus*
Apple, Cactus: *Opuntia engelmannii*
Apple, Gopher: *Licania michauxii*
Apple, Malaysian: *Syzygium malaccense*
Apple, May: *Matricaria discoidea*, *Podophyllum peltatum*
Apple, Pond: *Annona glabra*
Apple, Sweet: *Malus pumila*
Apple, Thorn: see Thorn Apple
Apple, Wild: *Malus sylvestris*
Apricot: *Prunus armeniaca*
Apricot, Desert: *Prunus fremontii*
Apricot, Old Field: *Passiflora incarnata*
Arbor Vitae, American (or Eastern): *Thuja occidentalis*
Arbor Vitae, Giant: *Thuja plicata*
Arbutus: *Arbutus menziesii*
Arbutus, Trailing: *Epigaea repens*
Argentea: *Atriplex* sp.
Armoglossum: *Arnoglossum atriplicifolium*
Arrow Bush: *Pluchea sericea*
Arrowhead, Arum Leaved: *Sagittaria cuneata*
Arrowhead, Broad Leaved: *Sagittaria latifolia*
Arrowhead, Bulltongue: *Sagittaria lancifolia*
Arrowhead Rattlebox: *Crotalaria sagittalis*
Arrowleaf: *Sagittaria latifolia*
Arrowroot: *Balsamorhiza sagittata*, *Thalia geniculata*
Arrow Seed: *Pluchea sericea*
Arrow Weed: *Pluchea sericea*
Arrow Wood: *Cornus canadensis*, *Philadelphus lewisii*, *P.* sp., *Physocarpus capitatus*, *Viburnum acerifolium*, *V. dentatum*, *V. opulus*
Artichoke, Jerusalem: *Helianthus tuberosus*
Artichoke, Wild: *Helianthus maximiliani*
Arum, Arrow: *Peltandra virginica*
Arum, Water: *Calla palustris*
Ash: *Fraxinus* spp., *Ptelea trifoliata*
Ash, Arizona: *Fraxinus velutina*
Ash, Black: *Fraxinus nigra*
Ash, Brown: *Fraxinus nigra*
Ash, Flowering: *Fraxinus cuspidata*
Ash, Fragrant: *Fraxinus cuspidata*
Ash, Green: *Fraxinus pennsylvanica*
Ash, Mountain: see Mountainash
Ash, Oregon: *Fraxinus latifolia*
Ash, Red: *Fraxinus pennsylvanica*
Ash, Singleleaf: *Fraxinus anomala*
Ash, Velvet: *Fraxinus pennsylvanica*
Ash, White: *Fraxinus americana*
Ash, Yellow: *Fraxinus americana*
Aspen, American: *Populus tremuloides*
Aspen, Bigtooth: *Populus grandidentata*
Aspen, Golden: *Populus tremuloides*
Aspen, Large Toothed: *Populus grandidentata*
Aspen, Quaking: *Populus tremuloides*
Aspen, Rocky Mountain: *Populus tremuloides*
Aspen, Trembling: *Populus tremuloides*
Aster: *Aster* spp., *Chaetopappa ericoides*, *Townsendia strigosa*
Aster, Blue: *Aster foliaceus*, *Machaeranthera canescens*, *M. tanacetifolia*
Aster, Common Sand: *Corethrogyne filaginifolia*
Aster, Desert: *Xylorhiza tortifolia*
Aster, Dwarf: *Machaeranthera canescens*
Aster, Golden: *Chrysopsis mariana*, *Heterotheca villosa*, *Pityopsis graminifolia*
Aster, Hairy: *Machaeranthera canescens*
Aster, Hairy Golden: *Heterotheca villosa*
Aster, Hoary: *Machaeranthera canescens*
Aster, Purple: *Aster conspicuus*, *Machaeranthera alta*
Aster, Rayless: *Machaeranthera grindelioides*
Aster, Spiny: *Chloracantha spinosa*
Aster, Tansy: see Tansyaster
Aster, Tanseyleaf: *Machaeranthera tanacetifolia*
Aster, White: *Chaetopappa ericoides*
Aster, Yellow: *Erigeron concinnus*
Au-hu-hu: *Tephrosia purpurea*
Au-ko-i: *Senna occidentalis*
Avens, Calthaleaf: *Geum calthifolium*
Avens, Large Leaved: *Geum macrophyllum*
Avens, Long Plumed: *Geum triflorum*
Avens, Purple: *Geum rivale*
Avens, Three Flowered: *Geum triflorum*
Avens, Water: *Geum rivale*
Avens, White: *Geum canadense*
Avens, Yellow: *Geum aleppicum*, *G. macrophyllum*
Avocado, American: *Persea planifolia*
Awa: *Piper methysticum*
Awapuhi Kuahiwi: *Zingiber zerumbet*
Aweoweo: *Chenopodium oahuense*
'Awikiwiki: *Canavalia galeata*
Azalea, Cascade: *Rhododendron albiflorum*
Azalea, False: *Menziesia ferruginea*
Azalea, Flame: *Rhododendron calendulaceum*
Azalea, Western: *Rhododendron occidentale*
Baby Blue Eyes: *Nemophila menziesii*
Babybonnets: *Coursetia glandulosa*
Baby Sunrose: *Carpobrotus aequilateralus*
Bachelor Buttons: *Centaurea* sp.
Bachelor's Button: *Polygala rugelii*
Balloonbush: *Epixiphium wislizeni*
Ball Root: *Orbexilum pedunculatum*
Balm: *Melissa officinalis*
Balm, Western: *Monardella odoratissima*
Balm of Gilead: *Populus balsamifera*, *P.* ×*jackii*
Balmony: *Chelone glabra*
Balsam: *Abies balsamea*, *Balsamorhiza sagittata*, *Impatiens capensis*
Balsam, Indian: *Lomatium dissectum*
Balsam Apple: *Echinocystis lobata*
Balsam Buds: *Impatiens capensis*
Balsamea, Mexican: *Epilobium canum*
Balsam Herb: *Balsamita major*
Balsam Root: *Balsamorhiza deltoidea*, *B. hookeri*, *B. incana*, *B. sagittata*, *B.* ×*terebinthacea*, *Osmorhiza* sp.
Balsam Root, Arrow Leaved: *Balsamorhiza sagittata*
Balsamroot, Deltoid: *Balsamorhiza deltoidea*
Balsamroot, Hairy: *Balsamorhiza hookeri*
Balsamroot, Hoary: *Balsamorhiza incana*
Balsam Root, Hooker's: *Balsamorhiza hookeri*
Bamboo: *Arundinaria* sp.
Bamboo, Cultivated: *Bambusa* sp.
Bamboo Brier: *Smilax laurifolia*
Bamboovine: *Smilax pseudochina*
Banana: *Musa* sp.
Banana, Paradise: *Musa*×*paradisiaca*
Bandanna of the Everglades: *Canna flaccida*
Baneberry, Red: *Actaea rubra*
Baneberry, Red and White: *Actaea rubra*
Baneberry, White: *Actaea pachypoda*
Bank Flower: *Penstemon ambiguus*
Banner, Golden: *Thermopsis rhombifolia*
Baraprieta: *Caesalpinia gracilis*
Barberry: *Berberis* sp.
Barberry, American: *Berberis canadensis*
Barberry, California: *Mahonia pinnata*
Barberry, Colorado: *Berberis fendleri*
Barberry, Common: *Berberis vulgaris*
Barberry, Desert: *Mahonia fremontii*
Barberry, Dwarf: *Mahonia pumila*
Barberry, Hollyleaved: *Mahonia aquifolium*
Barberry, Red: *Mahonia haematocarpa*
Barberry, Shining Netvein: *Mahonia dictyota*
Barberry, Showy: *Mahonia repens*
Barley, Coast: *Hordeum vulgare*
Barley, Common: *Hordeum vulgare*
Barley, Foxtail: *Hordeum jubatum*

Index of Common Names

Barley, Mediterranean: *Hordeum marinum*
Barley, Mouse: *Hordeum murinum*
Barley, Smooth: *Hordeum murinum*
Barley, Wild: *Hordeum jubatum*
Barratt Willow: *Salix scouleriana*
Barrelcactus, California: *Ferocactus cylindraceus*
Barrelcactus, Candy: *Ferocactus wislizeni*
Barrelcactus, Leconte's: *Ferocactus cylindraceus*
Barrel Cactus, Water: *Echinocactus* sp.
Basketweed: *Rhus trilobata*
Bass-wood: *Tilia americana*
Batflower: *Tacca leontopetaloides*
Bay, California Rose: *Rhododendron macrophyllum*
Bay, Red: *Persea borbonia*
Bay, Swamp: *Persea palustris*
Bay, Sweet: *Magnolia virginiana*, *Myrica* sp.
Bayberry: *Myrica cerifera*, *M. heterophylla*, *M.* sp.
Bay Flower: *Commelina dianthifolia*
Bayhops: *Ipomoea pes-caprae*
Bayonet, Spanish: *Yucca harrimaniae*, *Y. whipplei*
Beach Clover: *Trifolium wormskioldii*
Beach Greens: *Honckenya peploides*
Beachheather: *Hudsonia tomentosa*
Beadruby: *Maianthemum canadense*
Bean, Black Eyed: *Vigna unguiculata*
Bean, Broad: *Vicia faba*
Bean, Buffalo: *Thermopsis rhombifolia*
Bean, Butter: *Phaseolus lunatus*
Bean, Castor: *Ricinus communis*
Bean, Cherry: *Phaseolus vulgaris*
Bean, Coral: *Erythrina herbacea*
Bean, Cranberry: *Phaseolus vulgaris*
Bean, Fuzzy: *Strophostyles helvula*
Bean, Ground: *Amphicarpaea bracteata*
Bean, Horse: *Vicia faba*
Bean, Jumping: *Sapium biloculare*
Bean, Kidney: *Phaseolus vulgaris*
Bean, Lima: *Phaseolus lunatus*
Bean, Little Leaf Horse: *Parkinsonia microphylla*
Bean, Mescal: *Sophora secundiflora*
Bean, Mexican Jumping: *Sapium biloculare*
Bean, Navy: *Phaseolus vulgaris*
Bean, Pole: *Phaseolus vulgaris*
Bean, Screw: *Prosopis glandulosa*, *P. pubescens*
Bean, Sea: *Mucuna gigantea*
Bean, Sieva: *Phaseolus lunatus*
Bean, Slimleaf: *Phaseolus angustissimus*
Bean, Tepary: *Phaseolus acutifolius*
Bean, Wild: *Amphicarpaea bracteata*, *Apios americana*, *Phaseolus acutifolius*, *P. angustissimus*, *Strophostyles helvula*
Bean Salad: *Streptopus roseus*
Bearberry: *Arctostaphylos alpina*, *A. nevadensis*, *A. uva-ursi*, *Lonicera involucrata*
Beard Tongue: *Penstemon barbatus*, *P. fendleri*, *P. fruticosus*, *P. jamesii*, *P. linarioides*, *P.* sp., *P. virgatus*, *Phlox caespitosa*
Beard Tongue, Blue: *Penstemon pachyphyllus*
Beardtongue, Broadbeard: *Penstemon angustifolius*
Beardtongue, Bush: *Keckiella breviflora*

Beardtongue, Chilean: *Penstemon pruinosus*
Beardtongue, Cutleaf: *Penstemon richardsonii*
Beardtongue, Eastern Smooth: *Penstemon laevigatus*
Beardtongue, Gilia: *Penstemon ambiguus*
Beard Tongue, Hairy: *Penstemon fruticosus*, *P. laevigatus*
Beardtongue, James's: *Penstemon jamesii*
Beardtongue, Large: *Penstemon grandiflorus*
Beard Tongue, Narrow: *Penstemon angustifolius*
Beard Tongue, Sharp Leaved: *Penstemon acuminatus*
Beard Tongue, Slender: *Penstemon gracilis*
Beard Tongue, Small Flowered: *Keckiella breviflora*
Beardtongue, Thickleaf: *Penstemon pachyphyllus*
Beard Tongue, Torrey: *Penstemon barbatus*
Beardtongue, Upright Blue: *Penstemon virgatus*
Beardtongue, Woodland: *Nothochelone nemorosa*
Bear Root: *Lomatium macrocarpum*
Bear's Food: *Lomatium orientale*
Bear's Foot: *Smallanthus uvedalia*
Beautyberry, American: *Callicarpa americana*
Beaver Tail: *Opuntia basilaris*
Bedstraw, Fendler's: *Galium fendleri*
Bedstraw, Fragrant: *Galium triflorum*
Bedstraw, Licorice: *Galium circaezans*
Bedstraw, Marsh: *Galium tinctorium*
Bedstraw, Northern: *Galium boreale*
Bedstraw, Oneflower: *Galium uniflorum*
Bedstraw, Rough: *Galium asprellum*
Bedstraw, Scented: *Galium triflorum*
Bedstraw, Shining: *Galium concinnum*
Bedstraw, Small: *Galium tinctorium*, *G. trifidum*, *G. triflorum*
Bedstraw, Stiff Marsh: *Galium tinctorium*
Bedstraw, Sweet: *Galium triflorum*
Bedstraw, Threepetal: *Galium trifidum*
Beebalm, Lemon: *Monarda citriodora*
Beebalm, Mintleaf: *Monarda fistulosa*
Beebalm, Pony: *Monarda pectinata*
Beebalm, Scarlet: *Monarda didyma*
Beebalm, Spotted: *Monarda punctata*
Beebalm, Wildbergamot: *Monarda fistulosa*
Beeblossom, Scarlet: *Gaura coccinea*
Beebrush, Wright's: *Aloysia wrightii*
Beech: *Fagus grandifolia*
Beech Drops: *Epifagus virginiana*
Beefsteakplant: *Perilla frutescens*
Bee Plant: *Cleome serrulata*, *Scrophularia californica*
Bee Plant, Rocky Mountain: *Cleome serrulata*
Beet: *Beta vulgaris*
Beetleweed: *Galax urceolata*
Bee Weed, Rocky Mountain: *Cleome serrulata*
Bee Weed, Sonora: *Cleome multicaulis*
Beggar Lice: *Hackelia virginiana*
Beggar Louse: *Desmodium paniculatum*
Beggartick, Crowned: *Bidens coronata*
Beggartick, Smooth: *Bidens laevis*
Bellflower: *Campanula parryi*, *Uvularia sessilifolia*

Bellflower, American: *Campanulastrum americanum*
Bellflower, Arctic: *Campanula uniflora*
Bellflower, Bluebell: *Campanula rotundifolia*
Bellflower, Blue Marsh: *Campanula aparinoides*
Bellflower, One Flowered: *Campanula uniflora*
Bellflower, Parry's: *Campanula parryi*
Bellflower, Small Bonny: *Campanula divaricata*
Bellflower, Tall: *Campanulastrum americanum*
Bellwort: *Disporum smithii*, *Uvularia grandiflora*
Bellwort, Large Flowered: *Uvularia grandiflora*
Bellwort, Perfoliate: *Uvularia perfoliata*
Bellwort, Sessile Leaved: *Uvularia sessilifolia*
Beloperone: *Justicia californica*
Benne Plant: *Sesamum orientale*
Bentgrass, Upland: *Agrostis perennans*
Bergamot, Wild: *Monarda fistulosa*
Betony: *Stachys rothrockii*
Betony, Wood: *Pedicularis bracteosa*, *P. canadensis*, *P. lanceolata*, *P. racemosa*
Bicknell's Thorn: *Crataegus chrysocarpa*
Big Root: *Balsamorhiza sagittata*, *Ipomoea leptophylla*, *Marah oreganus*
Bilberry: *Vaccinium cespitosum*, *V. vitis-idaea*
Bilberry, Bog: *Vaccinium uliginosum*
Bilberry, Mountain: *Vaccinium membranaceum*
Bilberry, Red: *Vaccinium parvifolium*
Bilberry, Tall: *Vaccinium cespitosum*
Bilsted: *Liquidambar styraciflua*
Bindweed, Chaparral False: *Calystegia occidentalis*
Bindweed, Field: *Convolvulus arvensis*
Birch, Black: *Betula lenta*, *B. nigra*
Birch, Bog: *Betula nana*
Birch, Canoe: *Betula papyrifera*
Birch, Cherry: *Betula lenta*
Birch, Downy: *Betula pubescens*
Birch, Glandulose: *Betula pumila*
Birch, Gray: *Betula populifolia*
Birch, Low: *Betula pumila*
Birch, Northern Dwarf: *Betula nana*
Birch, Paper: *Betula papyrifera*, *B. pubescens*
Birch, Red: *Betula nigra*
Birch, River: *Betula nigra*
Birch, Rocky Mountain: *Betula occidentalis*
Birch, Shrub: *Betula nana*
Birch, Sweet: *Betula lenta*
Birch, Water: *Betula occidentalis*
Birch, Western: *Betula occidentalis*
Birch, Western Paper: *Betula papyrifera*
Birch, White: *Betula papyrifera*, *B. populifolia*, *B.* sp.
Birch, Yellow: *Betula alleghaniensis*, *B.* sp.
Birdrape: *Brassica rapa*
Bird's Beak, Bushy: *Cordylanthus ramosus*
Bird's Beak, Wright's: *Cordylanthus wrightii*
Bird's Foot, Small Leaved: *Amorpha nana*
Biscuitroot, Barestem: *Lomatium nudicaule*
Biscuitroot, Bigseed: *Lomatium macrocarpum*
Biscuitroot, Canby's: *Lomatium canbyi*
Biscuitroot, Carrotleaf: *Lomatium dissectum*
Biscuitroot, Cous: *Lomatium cous*

Biscuitroot, Desert: *Lomatium foeniculaceum*
Biscuitroot, Fernleaf: *Lomatium dissectum*
Biscuitroot, Geyer's: *Lomatium geyeri*
Biscuitroot, Gray's: *Lomatium grayi*
Biscuitroot, Indian: *Lomatium piperi*
Biscuitroot, Nineleaf: *Lomatium triternatum*
Biscuitroot, Northern: *Lomatium farinosum*
Biscuitroot, Northern Idaho: *Lomatium orientale*
Biscuitroot, Nuttall's: *Lomatium nuttallii*
Biscuitroot, Parish's: *Lomatium nevadense*
Biscuitroot, Wyeth: *Lomatium ambiguum*
Bishop's Cap: *Mitella diphylla, M. nuda*
Bistort, American: *Polygonum bistortoides*
Bistort, Mountain Meadow: *Polygonum bistorta*
Bitter Ball: *Tagetes micrantha*
Bittercress, Limestone: *Cardamine douglassii*
Bitternut Hickory: *Carya cordiformis*
Bitter Perfume: *Monarda fistulosa*
Bitter Root: *Lewisia* spp., *Lomatium dissectum*
Bitterroot, Columbian: *Lewisia columbiana*
Bitterroot, Oregon: *Lewisia rediviva*
Bitterroot, Pigmy: *Lewisia pygmaea*
Bittersweet: *Celastrus scandens, Solanum dulcamara*
Bittersweet, Climbing: *Celastrus scandens*
Blackberry: *Empetrum nigrum, Rubus* spp.
Blackberry, Allegheny: *Rubus allegheniensis*
Blackberry, Arctic: *Rubus arcticus*
Blackberry, California: *Rubus ursinus, R. vitifolius*
Blackberry, Canada: *Rubus canadensis*
Blackberry, Common: *Rubus vitifolius*
Blackberry, Cut Leaf: *Rubus laciniatus*
Blackberry, Dwarf Red: *Rubus pubescens*
Blackberry, Evergreen: *Rubus laciniatus*
Blackberry, Hawaii: *Rubus hawaiensis*
Blackberry, High Bush: *Rubus allegheniensis, R. argutus*
Blackberry, Himalayan: *Rubus discolor*
Blackberry, Long: *Rubus allegheniensis*
Blackberry, Pacific: *Rubus ursinus*
Blackberry, Pacific Trailing: *Rubus ursinus*
Blackberry, Sand: *Rubus cuneifolius*
Blackberry, Sawtooth: *Rubus argutus*
Blackberry, Shrubby: *Rubus fruticosus*
Blackberry, Smooth: *Rubus canadensis*
Blackberry, Tall: *Rubus* sp.
Blackberry, Thornless: *Rubus canadensis*
Blackberry, Trailing: *Rubus ursinus*
Blackberry, Wild: *Rubus allegheniensis, R. procumbens, R.* sp., *R. ursinus, R. vitifolius*
Blackberry, Yankee: *Rubus frondosus*
Blackberry Root: *Rubus hispidus*
Blackberry Tree: *Rubus fruticosus*
Blackbrush (or Blackbush): *Coleogyne ramosissima*
Black Cap: *Rubus leucodermis, R. occidentalis*
Black Eyed Susan: *Rudbeckia hirta*
Black Head: *Rubus leucodermis*
Black Jack: *Berchemia scandens*
Black Root: *Baptisia tinctoria*
Blackroot, Wand: *Pterocaulon virgatum*

Black Sampson: *Echinacea pallida*
Blacksnakeroot, Canadian: *Sanicula canadensis*
Blacksnakeroot, Clustered: *Sanicula odorata*
Blacksnakeroot, Pacific: *Sanicula crassicaulis*
Blacksnakeroot, Small's: *Sanicula smallii*
Black Weed: *Baptisia* sp.
Bladder Campion: *Silene latifolia*
Bladderfern, Brittle: *Cystopteris fragilis*
Bladderfern, Lowland: *Cystopteris protrusa*
Bladdernut: *Staphylea trifolia*
Bladder Pod: *Cleome isomeris, Lesquerella* spp., *Physaria didymocarpa*
Bladder Pod, Double: *Physaria didymocarpa, P. newberryi*
Bladderpod, Douglas's: *Lesquerella douglasii*
Bladderpod, Fendler's: *Lesquerella fendleri*
Bladderpod, Mid: *Lesquerella intermedia*
Bladderpod, Straight: *Lesquerella rectipes*
Bladderpod Spiderflower: *Cleome isomeris*
Bladder Seaweed: *Halosaccion glandiforme*
Bladderwrack: *Fucus gardneri*
Blanket Flower: *Gaillardia pinnatifida*
Blazingstar, Adonis: *Mentzelia multiflora*
Blazingstar, Bractless: *Mentzelia nuda*
Blazingstar, Bushy: *Mentzelia dispersa*
Blazingstar, Cutleaf: *Mentzelia laciniata*
Blazing Star, Desert: *Mentzelia multiflora*
Blazing Star, Dotted: *Liatris punctata*
Blazingstar, Grass: *Mentzelia gracilenta*
Blazingstar, Roughstem: *Mentzelia puberula*
Blazing Star, Small Flowered: *Mentzelia albicaulis*
Blazingstar, Smoothstem: *Mentzelia laevicaulis*
Blazingstar, United: *Mentzelia congesta*
Blazingstar, Whitebract: *Mentzelia involucrata*
Blazing Star, Whitestem: *Mentzelia albicaulis*
Bleeding Heart: *Dicentra formosa*
Blisterwort: *Ranunculus recurvatus*
Blood Root: *Sanguinaria canadensis*
Bloodroot Bush: *Jatropha cardiophylla*
Bluebeadlily, Western: *Clintonia andrewsiana*
Bluebeadlily, White: *Clintonia umbellulata*
Bluebeadlily, Yellow: *Clintonia borealis*
Bluebell: *Campanula parryi, Campanulastrum americanum, Mertensia ciliata, M. virginica*
Bluebell, Various Leaved: *Phacelia heterophylla*
Bluebells, Mountain: *Mertensia ciliata*
Blue Bells, Papago: *Dichelostemma pulchellum*
Bluebells, Virginia: *Mertensia virginica*
Blueberry, Alaskan: *Vaccinium ovalifolium*
Blueberry, Alpine: *Vaccinium uliginosum*
Blueberry, Arctic: *Vaccinium uliginosum*
Blueberry, Black: *Vaccinium membranaceum*
Blueberry, Bog: *Vaccinium uliginosum*
Blueberry, Canada: *Vaccinium myrtilloides*
Blueberry, Cascade: *Vaccinium deliciosum*
Blueberry, Dwarf: *Vaccinium angustifolium, V. cespitosum*
Blueberry, Early: *Vaccinium ovalifolium*
Blueberry, Grey: *Vaccinium ovalifolium*
Blueberry, High Bush: *Vaccinium corymbosum*

Blueberry, Late: *Vaccinium corymbosum*
Blueberry, Low Sweet: *Vaccinium angustifolium*
Blueberry, Lowbush: *Vaccinium angustifolium*
Blueberry, Mountain: *Vaccinium cespitosum, V. membranaceum*
Blueberry, Oval Leaved: *Vaccinium ovalifolium*
Blueberry, Shiny: *Vaccinium myrsinites*
Blueberry, Small: *Vaccinium myrtillus*
Blueberry, Sour Top: *Vaccinium myrtilloides*
Blueberry, Swamp: *Vaccinium cespitosum*
Blueberry, Thin Leaved: *Vaccinium membranaceum*
Blueberry, Velvet Leaf: *Vaccinium myrtilloides*
Bluebowls: *Gilia rigidula*
Blue Brush: *Ceanothus griseus, C. thyrsiflorus*
Bluebunch Wheatgrass: *Pseudoroegneria spicata*
Blue Buttons: *Cynoglossum grande*
Blue Cohosh: *Caulophyllum thalictroides*
Blue Curls: *Trichostema lanceolatum*
Blue Curls, Wooly: *Trichostema lanatum*
Blue Dicks: *Dichelostemma pulchellum*
Blue Eyed Grass, Eastern: *Sisyrinchium atlanticum*
Blueeyed Grass, Mountain: *Sisyrinchium montanum*
Blueeyed Grass, Narrowleaf: *Sisyrinchium angustifolium*
Blueeyed Grass, Nash's: *Sisyrinchium nashii*
Blueeyed Grass, Needletip: *Sisyrinchium mucronatum*
Blueeyed Grass, Prairie: *Sisyrinchium campestre*
Blue Eyed Grass, Stout: *Sisyrinchium angustifolium*
Blueeyed Grass, Western: *Sisyrinchium bellum*
Blueeyed Grass, White: *Sisyrinchium albidum*
Blue Eyed Mary, Smallflower: *Collinsia parviflora*
Blue Eyed Mary, Violet: *Collinsia violacea*
Blue Fieldmadder: *Sherardia arvensis*
Blue Flag: *Iris missouriensis, I. versicolor*
Blue Flag Iris: *Iris virginica*
Bluegrass, Plains: *Poa arida*
Bluegrass, Sandberg: *Poa secunda*
Bluegrass, Skyline: *Poa fendleriana*
Bluejoint: *Calamagrostis canadensis, Elytrigia repens*
Bluestar, Woolly: *Amsonia tomentosa*
Blue Stem, Big: *Andropogon gerardii*
Bluestem, Broomsedge: *Andropogon virginicus*
Bluestem, Bushy: *Andropogon glomeratus*
Bluestem, Colorado: *Pascopyrum smithii*
Bluestem, Florida: *Andropogon floridanus*
Bluestem, Little: *Schizachyrium scoparium*
Bluestem, Silver: *Bothriochloa saccharoides*
Bluets: *Houstonia caerulea, H. rubra, H. wrightii*
Blue Weed: *Echium vulgare*
Bluewitch Nightshade: *Solanum umbelliferum*
Bluewood, Hooker's: *Condalia hookeri*
Boa Kelp: *Egregia menziesii*
Boneset: *Eupatorium perfoliatum*
Boneset, False: *Brickellia eupatorioides*
Boneset, Rough: *Eupatorium pilosum*
Borage: *Cryptantha flava*

Index of Common Names

Bottle Brush Grass: *Elymus hystrix*
Bottlebrush Squirreltail: *Elymus elymoides*
Bottle Gourd: *Lagenaria siceraria*
Bottle Weed, American: *Eriogonum elatum*
Bouncing Bet: *Saponaria officinalis*
Bowman's Root: *Porteranthus trifoliatus*
Box Elder: *Acer negundo*
Box Thorn: *Lycium andersonii, L. fremontii, L. torreyi*
Boxwood, Oregon: *Paxistima myrsinites*
Bracken Fern, American (or Western): *Pteridium aquilinum*
Brackenfern, Southern: *Pteridium caudatum*
Bracket Fungus: *Fomes* sp., *Ganoderma applanatum, Polyporus harlowii*
Bractscale: *Atriplex serenana*
Brake Fern: *Pteridium aquilinum, Woodwardia radicans*
Brake Fern, Purple Cliff: *Pellaea atropurpurea*
Brake Fern, Western: *Pteridium aquilinum*
Bramble, Five Leaved: *Rubus pedatus*
Branched Noseburn: *Tragia ramosa*
Brass Buttons: *Erigeron aphanactis*
Breadfruit: *Artocarpus altilis*
Breadroot: *Lomatium ambiguum, Pediomelum esculentum*
Breadroot, Indian: *Pediomelum esculentum, P. hypogaeum*
Breadroot, Largebract Indian: *Pediomelum cuspidatum*
Breeches, Woolly: *Amsinckia spectabilis*
Briar: see also Brier
Briar, Cat: *Smilax bona-nox, S.* sp., *S. tamnoides*
Brickellbush, California: *Brickellia californica*
Brickellbush, Narrowleaf: *Brickellia oblongifolia*
Brickellbush, Plumed: *Brickellia brachyphylla*
Brickellbush, Tasselflower: *Brickellia grandiflora*
Bride's Bonnet: *Clintonia uniflora*
Bride's Bouquet: *Chaenactis douglasii, C. stevioides*
Bride's Feathers: *Aruncus dioicus*
Brier, Bamboo: *Smilax laurifolia*
Brier, Cat: *Smilax bona-nox, S. tamnoides*
Brier, Green: see Greenbrier
Brier, Sasul: *Smilax* sp.
Brier, Sawr: *Smilax glauca*
Brier, Sun: *Aralia spinosa*
Bristlemallow: *Modiola caroliniana*
Brittlebrush, White: *Encelia farinosa*
Brittlebush, Acton's: *Encelia virginensis*
Brittlebush, Button: *Encelia frutescens*
Brittlebush, White: *Encelia farinosa*
Broad Flag: *Iris* sp.
Brodiaea: *Brodiaea* spp., *Dichelostemma pulchellum, Triteleia grandiflora, T. laxa*
Brodiaea, Blue Flowered: *Triteleia laxa*
Brodiaea, Common: *Dichelostemma pulchellum*
Brodiaea, Golden: *Triteleia ixioides*
Brodiaea, Hyacinth: *Triteleia hyacinthina*
Brodiaea, White: *Triteleia hyacinthina*
Brodiaea, White Flowered: *Triteleia peduncularis*

Brome, California: *Bromus carinatus*
Brome, Mountain: *Bromus marginatus*
Brome, Nodding: *Bromus anomalus*
Brome, Soft: *Bromus hordeaceus*
Brome Grass, Downy: *Bromus tectorum*
Brookfoam, Coastal: *Boykinia occidentalis*
Broom, Desert: *Baccharis sarothroides*
Broom, Indian: *Baccharis sarothroides*
Broom, Turpentine: *Thamnosma montana*
Broomcorn: *Sorghum bicolor*
Broomrape, Clustered: *Orobanche fasciculata*
Broomrape, Cooper's: *Orobanche cooperi*
Broomrape, Desert: *Orobanche cooperi*
Broomrape, Louisiana: *Orobanche ludoviciana*
Broomrape, Manyflowered: *Orobanche ludoviciana*
Broomrape, Sand: *Orobanche ludoviciana*
Broom Straw: *Andropogon glomeratus*
Broomweed: *Amphiachyris dracunculoides, Baccharis sarothroides, Gutierrezia sarothrae*
Brown Eyed Susan: *Gaillardia aristata*
Browneyes: *Camissonia claviformis*
Brownfoot: *Acourtia wrightii*
Bubble Seaweed: *Leathesia difformis*
Buckbean: *Menyanthes trifoliata*
Buck Berry: *Cornus canadensis, Shepherdia argentea*
Buck Brush: *Ceanothus cuneatus, C. fendleri, C. oliganthus, C. sanguineus, C. velutinus, Purshia tridentata*
Buckeye, California: *Aesculus californica*
Buckeye, Fetid: *Aesculus glabra*
Buckeye, Ohio: *Aesculus glabra*
Buckeye, Red: *Aesculus pavia*
Buckeye, Sweet: *Aesculus flava*
Buckeye, Yellow: *Aesculus flava*
Bucknut: *Simmondsia chinensis*
Buckroot: *Pediomelum canescens*
Buckthorn: *Ceanothus fendleri, Frangula caroliniana, Rhamnus alnifolia, R. cathartica, R. crocea, R.* sp., *Ziziphus obtusifolia*
Buckthorn, Alder Leaved: *Rhamnus alnifolia*
Buckthorn, California: *Frangula californica*
Buckthorn, Carolina: *Frangula caroliniana*
Buckthorn, Cascara: *Frangula purshiana*
Buckthorn, Common: *Rhamnus cathartica*
Buckthorn, Holly: *Rhamnus crocea*
Buckthorn, Pursh's: *Frangula purshiana*
Buckthorn, Red: *Frangula rubra*
Buckthorn, Redberry: *Rhamnus crocea*
Buckthorn, Woolly: *Sideroxylon lanuginosum*
Buckwheat: *Eriogonum* spp, *Fagopyrum esculentum*
Buckwheat, Abert's: *Eriogonum abertianum*
Buckwheat, Angle Stemmed: *Eriogonum angulosum*
Buckwheat, Annual: *Eriogonum annuum*
Buckwheat, Arrowleaf: *Eriogonum compositum*
Buckwheat, Bailey: *Eriogonum baileyi*
Buckwheat, California: *Eriogonum fasciculatum*
Buckwheat, Crispleaf: *Eriogonum corymbosum*

Buckwheat, Cushion: *Eriogonum ovalifolium*
Buckwheat, Davidson's: *Eriogonum davidsonii*
Buckwheat, Divergent: *Eriogonum divaricatum*
Buckwheat, Eastern Mojave: *Eriogonum fasciculatum*
Buckwheat, Gray California: *Eriogonum fasciculatum*
Buckwheat, Hooker's: *Eriogonum hookeri*
Buckwheat, James's: *Eriogonum jamesii*
Buckwheat, Ladder: *Eriogonum insigne*
Buckwheat, Longstem: *Eriogonum elongatum*
Buckwheat, Mojave: *Eriogonum fasciculatum*
Buckwheat, Naked: *Eriogonum nudum*
Buckwheat, Parsnipflower: *Eriogonum heracleoides*
Buckwheat, Rock: *Eriogonum sphaerocephalum*
Buckwheat, Rockjasmine: *Eriogonum androsaceum*
Buckwheat, Rose and White: *Eriogonum gracillimum*
Buckwheat, Roundleaf: *Eriogonum rotundifolium*
Buckwheat, Seaside: *Eriogonum latifolium*
Buckwheat, Slender: *Eriogonum microthecum*
Buckwheat, Slenderleaf: *Eriogonum leptophyllum*
Buckwheat, Snow: *Eriogonum niveum*
Buckwheat, Sulphurflower: *Eriogonum umbellatum*
Buckwheat, Tall: *Eriogonum tenellum*
Buckwheat, Tall Woolly: *Eriogonum elatum*
Buckwheat, Wand: *Eriogonum roseum*
Buckwheat, Wild: *Eriogonum alatum, E. compositum, E. corymbosum, E. fasciculatum, E. heracleoides, E. microthecum, E. niveum, E. rotundifolium, E.* sp.
Buckwheat, Wright: *Eriogonum wrightii*
Buckwheat, Yucca: *Eriogonum plumatella*
Buena Mujer: *Mentzelia* sp.
Buffalo Berry: *Elaeagnus commutata, Shepherdia argentea, S. canadensis, S. rotundifolia*
Buffalo Berry, Canada: *Shepherdia canadensis*
Buffalo Berry, Russet: *Shepherdia canadensis*
Buffaloberry, Roundleaf: *Shepherdia rotundifolia*
Buffalo Berry, Silver: *Shepherdia argentea*
Buffalo Berry, Thorny: *Shepherdia argentea*
Buffalo Bur: *Solanum rostratum*
Buffalograss, False: *Monroa squarrosa*
Buffalo Nut: *Pyrularia pubera*
Bugbane, Black: *Cimicifuga racemosa*
Bugbane, Carolina: *Trautvetteria caroliniensis*
Bugbane, False: *Trautvetteria caroliniensis*
Bugler, Scarlet: *Penstemon centranthifolius*
Bugleweed: *Lycopus asper, Lycopus uniflorus*
Bull Kelp: *Nereocystis luetkeana*
Bully, White: *Sideroxylon salicifolium*
Bulrush, Alkali: *Scirpus maritimus, S. robustus*
Bulrush, American: *Scirpus americanus*
Bulrush, Cloaked: *Scirpus pallidus*
Bulrush, Giant: *Scirpus tabernaemontani*
Bulrush, Great: *Scirpus acutus, S. tabernaemontani*
Bulrush, Hardstem: *Scirpus acutus*
Bulrush, Nevada: *Scirpus nevadensis*
Bulrush, Panicled: *Scirpus microcarpus*

Bulrush, Prairie: *Scirpus robustus*
Bulrush, River: *Scirpus fluviatilis*
Bulrush, Saltmarsh: *Scirpus maritimus*
Bulrush, Small Flowered: *Scirpus microcarpus*
Bulrush, Small Fruit: *Scirpus microcarpus*
Bulrush, Softstem: *Scirpus tabernaemontani*
Bumblebee Plant: *Pedicularis* sp.
Bum Branch: *Pinus edulis*
Bunchberry: *Cornus canadensis, C. suecica, C. unalaschkensis*
Bundleflower, Prairie: *Desmanthus illinoensis*
Burdock: *Arctium lappa, A. minus, A.* sp., *Xanthium strumarium*; see also Burrdock
Burning Bush: *Euonymus atropurpurea*
Burr, Little: *Cardamine diphylla*
Burrdock: see also Burdock
Burrdock, Greater: *Arctium lappa*
Burrdock, Lesser: *Arctium minus*
Burrobush, Singlewhorl: *Hymenoclea monogyra*
Bursage, Silver: *Ambrosia chamissonis*
Burweed: *Ambrosia chamissonis*
Bushmint, Comb: *Hyptis pectinata*
Bushy Bird's Beak: *Cordylanthus ramosus*
Bustic, Willow: *Sideroxylon salicifolium*
Butter and Eggs: *Linaria vulgaris*
Butterballs: *Eriogonum ovalifolium*
Butterbean: *Phaseolus lunatus*
Butterbur: *Petasites frigidus*
Buttercup, Alkali: *Ranunculus cymbalaria*
Buttercup, Blister: *Ranunculus sceleratus*
Buttercup, Bristly: *Ranunculus hispidus*
Buttercup, Bulbous: *Ranunculus bulbosus*
Buttercup, California: *Ranunculus californicus*
Buttercup, Celeryleaf: *Ranunculus sceleratus*
Buttercup, Creeping: *Ranunculus repens*
Buttercup, Five Petaled: *Ranunculus occidentalis*
Buttercup, Graceful: *Ranunculus inamoenus*
Buttercup, Hooked: *Ranunculus uncinatus*
Buttercup, Lapland: *Ranunculus lapponicus*
Buttercup, Little Flowered: *Ranunculus uncinatus*
Buttercup, Littleleaf: *Ranunculus abortivus*
Buttercup, Pallas: *Ranunculus pallasii*
Buttercup, Pennsylvania: *Ranunculus pensylvanicus*
Buttercup, Rocky Mountain: *Ranunculus cymbalaria*
Buttercup, Sagebrush: *Ranunculus glaberrimus*
Buttercup, Swamp: *Ranunculus hispidus*
Buttercup, Tall: *Ranunculus acris*
Buttercup, Western: *Ranunculus occidentalis*
Butterfly Weed: *Asclepias tuberosa, Gaura parviflora*
Butternut: *Juglans cinerea*
Butterweed: *Conyza canadensis*
Butterwort, Common: *Pinguicula vulgaris*
Butterwort, Small: *Pinguicula pumila*
Butterwort, Yellow: *Pinguicula lutea*
Buttonbush: *Cephalanthus occidentalis*
Button Eryngo: *Eryngium yuccifolium*
Buttons, Spoonshape Barbara's: *Marshallia obovata*
Cabbage: *Brassica oleracea, Sabal palmetto*

Cabbage, Skunk: *Lysichiton americanus, L.* sp., *Symplocarpus foetidus*
Cabbage, Squaw: *Caulanthus inflatus*
Cabbage, Thickstem Wild: *Caulanthus crassicaulis*
Cabbage, Wild: *Brassica oleracea, Caulanthus coulteri*
Cabbage Palmetto: *Sabal palmetto*
Cacao: *Theobroma cacao*
Cactus, Ball: *Mammillaria* sp.
Cactus, Barrel: *Echinocactus* sp., *Ferocactus coulteri, F. cylindraceus, F.* sp., *F. wislizeni*
Cactus, Beavertail: *Opuntia basilaris*
Cactus, Brittle Prickly Pear: *Opuntia fragilis*
Cactus, Bunch: *Echinocereus engelmannii*
Cactus, Candy Barrel: *Ferocactus wislizeni*
Cactus, Cane: *Opuntia imbricata, O.* sp., *O. whipplei*
Cactus, Chandelier: *Opuntia imbricata*
Cactus, Cholla: *Opuntia whipplei*; see also Cholla
Cactus, Christmas: *Opuntia leptocaulis*
Cactus, Cottontop: *Echinocactus polycephalus*
Cactus, Counterclockwise Nipple: *Mammillaria mainiae*
Cactus, Crimson Hedgehog: *Echinocereus coccineus*
Cactus, Cushion: *Escobaria missouriensis, E. vivipara*
Cactus, Fish Hook: *Ferocactus wislizeni, Mammillaria dioica, M. grahamii, Neomammillaria* sp.
Cactus, Gearstem: *Peniocereus striatus*
Cactus, Giant: *Carnegia gigantea, Pachycereus pringlei, Stenocereus thurberi*
Cactus, Graham's Nipple: *Mammillaria grahamii*
Cactus, Hedgehog: *Echinocereus engelmannii, E. fendleri, E. rigidissimus, E.* sp.
Cactus, Hedgewood: *Echinocereus coccineus*
Cactus, Horned Toad: *Mammillaria mainiae*
Cactus, Kingcup: *Echinocereus triglochidiatus*
Cactus, Many Spined: *Opuntia polyacantha*
Cactus, Missouri Foxtail: *Escobaria missouriensis*
Cactus, Mojave Mound: *Echinocereus polyacanthus*
Cactus, Nipple: *Mammillaria grahamii, M. wrightii*
Cactus, Organ Pipe: *Stenocereus thurberi*
Cactus, Pencil: *Opuntia ramosissima*
Cactus, Pinkflower Hedgehog: *Echinocereus fendleri*
Cactus, Prickly Pear: *Opuntia engelmannii, O. fragilis, O. phaeacantha, O. polyacantha, O.* sp.; see also Pricklypear
Cactus, Rainbow: *Echinocereus rigidissimus*
Cactus, Saints: *Echinocereus engelmannii*
Cactus, Scarlet Hedgehog: *Echinocereus coccineus*
Cactus, Slender: *Opuntia leptocaulis*
Cactus, Strawberry: *Mammillaria dioica*
Cactus, Sunset: *Mammillaria grahamii*
Cactus, Sweetpotato: *Peniocereus greggii*
Cactus, Turkey: *Opuntia leptocaulis*
Cactus, Walkingstick: *Opuntia spinosior*
Cactus, Wright's Nipple: *Mammillaria wrightii*
Cactus Apple: *Opuntia engelmannii*
Cactus Pear, American: *Opuntia* sp.
Calabash, Long Handled: *Lagenaria siceraria*

Calabazilla: *Cucurbita foetidissima*
Calabrash: *Lagenaria siceraria*
Calamus: *Acorus calamus*
Calla, Wild: *Calla palustris*
Caltrop: *Tribulus terrestris*
Caltrop, California: *Kallstroemia californica*
Camas, Atlantic: *Camassia scilloides*
Camas, Black: *Camassia quamash*
Camas, Blue: *Camassia leichtlinii, C. quamash*
Camas, Common: *Camassia quamash*
Camas, Death: *Zigadenus elegans, Z. paniculatus, Z. venenosus*
Camas, Great: *Camassia leichtlinii*
Camas, Large: *Camassia leichtlinii*
Camas, Little White: *Lomatium farinosum*
Camas, Meadow: *Camassia* sp.
Camas, Poison: *Zigadenus venenosus*
Camas, Small: *Camassia quamash*
Camas, White: *Camassia* sp., *Lomatium canbyi*
Camass: see Camas
Camomile: *Anthemis* sp.; see also Chamomile
Camomile, Rayless: *Matricaria discoidea*
Camphorweed, Stinking: *Pluchea foetida*
Campion, Bladder: *Silene latifolia*
Campion, Douglas's: *Silene douglasii*
Campion, Drummond's: *Silene drummondii*
Campion, Gregg's: *Silene laciniata*
Campion, Menzies's: *Silene menziesii*
Campion, Mexican: *Silene laciniata*
Campion, Moss: *Silene acaulis*
Campion, Pringle's: *Silene douglasii*
Campion, Scouler's: *Silene scouleri*
Campion, Starry: *Silene stellata*
Campion, White: *Silene latifolia*
Canaigre: *Rumex hymenosepalus*
Canarygrass, Carolina: *Phalaris caroliniana*
Canarygrass, Littleseed: *Phalaris minor*
Cancer Root: *Boschniakia hookeri, Orobanche fasciculata*
Cancer Root, Yellow: *Orobanche fasciculata*
Canchalagua: *Centaurium exaltatum, C. venustum*
Candlewood, American Desert: *Fouquieria splendens*
Candyroot: *Polygala lutea, P. rugelii*
Candytuft, Wild: *Thlaspi montanum*
Cane: *Arundinaria gigantea, Leymus condensatus, Phragmites australis*
Cane, Common: *Phragmites australis*
Cane, Giant: *Arundinaria gigantea*
Cane, Hauve: *Phragmites australis*
Cane, River: *Arundinaria gigantea*
Cane, Sugar: *Saccharum officinarum*
Cane, Switch: *Arundinaria gigantea*
Cane Head: *Sanicula canadensis*
Cane Leaves: *Aletris farinosa*
Cankerweed: *Prenanthes serpentaria*
Cantaloup (or Cantaloupe): *Cucumis melo*
Caraway: *Carum carvi*
Caraway, Wild: *Perideridia gairdneri*
Cardinal, Red: *Erythrina herbacea*

Index of Common Names

Cardinal Flower: *Lobelia cardinalis, L. siphilitica*
Cardon: *Pachycereus pringlei*
Careless Weed: *Amaranthus caudatus, A. palmeri*
Carpenter's Square: *Scrophularia marilandica*
Carrion Flower: *Smilax herbacea*
Carrizo: *Phragmites australis*
Carrot: *Lomatium dissectum*
Carrot, Domesticated: *Daucus carota*
Carrot, Indian: *Perideridia pringlei*
Carrot, Wild: *Daucus carota, D. pusillus, Lomatium macrocarpum, Perideridia gairdneri*
Carve (or Carved) Seed: *Glyptopleura marginata*
Cascara: *Frangula californica, F. purshiana, F. rubra*
Cassena (or Cassine): *Ilex vomitoria*
Cassine, Dahoon: *Ilex cassine, I. vomitoria*
Catberry: *Nemopanthus mucronatus*
Catbriar: *Smilax bona-nox, S. sp., S. tamnoides*
Cat Brier: *Smilax bona-nox, S. tamnoides*
Catchfly: *Silene acaulis, S. douglasii, S. laciniata, S. menziesii, S. scouleri*
Catchfly, Night Flowering: *Silene noctiflora*
Catchfly, Red Mountain: *Silene campanulata*
Catclaw: *Acacia greggii*
Catgut: *Tephrosia virginiana*
Catnip: *Nepeta cataria*
Catnip Noseburn: *Tragia nepetifolia*
Cat's Claw: *Acacia greggii, A. sp.*
Cat's Ear: *Calochortus amabilis, C. tolmiei*
Catseye, Brenda's Yellow: *Cryptantha flava*
Catseye, James's: *Cryptantha cinerea*
Catseye, Sanddune: *Cryptantha fendleri*
Catseye, Silky: *Cryptantha sericea*
Catseye, Tawny: *Cryptantha fulvocanescens*
Catseye, Thicksepal: *Cryptantha crassisepala*
Cat's Foot: *Gamochaeta purpurea*
Cat's Foot, Lesser: *Antennaria howellii*
Cat Tail: *Typha angustifolia, T. domingensis, T. latifolia, T. sp.*
Cat Tail Flag: *Typha angustifolia, T. latifolia*
Cat Tail, American: *Typha latifolia*
Cat Tail, Broad Leaved: *Typha latifolia*
Cat Tail, Common: *Typha latifolia*
Cat Tail, Narow Leaved: *Typha angustifolia*
Cayenne, Bird: *Capsicum annuum*
Cayenne Pepper: *Capsicum annuum*
Cedar: *Calocedrus decurrens, Datura wrightii, Juniperus californica, J. communis, J. monosperma, J. occidentalis, J. osteosperma, J. sp., J. virginiana, Thuja excelsa, T. occidentalis, T. plicata*
Cedar, Alaska: *Chamaecyparis nootkatensis*
Cedar, Atlantic White: *Chamaecyparis thyoides*
Cedar, Eastern Red: *Juniperus virginiana*
Cedar, Giant: *Thuja plicata*
Cedar, Ground: *Lycopodium complanatum*
Cedar, Northern White: *Thuja occidentalis*
Cedar, Pacific Red: *Thuja plicata*
Cedar, Port Orford: *Chamaecyparis lawsoniana*
Cedar, Red: *Juniperus occidentalis, J. scopulorum, J. sp., J. virginiana, Thuja plicata*
Cedar, Red Berry: *Juniperus pinchotii*
Cedar, Rocky Mountain Red: *Juniperus scopulorum*
Cedar, Salt: *Tamarix gallica*
Cedar, Western Red: *Thuja plicata*
Cedar, White: *Chamaecyparis lawsoniana, C. thyoides, Thuja occidentalis, T. plicata*
Cedar, Yellow: *Chamaecyparis nootkatensis*
Celandine, Greater: *Chelidonium majus*
Celery: *Apium graveolens*
Celery, Indian: *Angelica arguta, Lomatium ambiguum, L. nudicaule, L. triternatum*
Celery, Wild: *Angelica lucida, Apiastrum angustifolium, Apium graveolens, A. sp., Cymopterus acaulis, C. bulbosus, C. montanus, Lomatium nudicaule, Peucedanum sp.*
Centaury, Charming: *Centaurium venustum*
Centaury, Desert: *Centaurium exaltatum*
Centaury, Muhlenberg's: *Centaurium muehlenbergii*
Century Plant, American: *Agave americana*
Century Plant, Palmer's: *Agave palmeri*
Century Plant, Schott's: *Agave schottii*
Cereus, Hedgehog Crimson: *Echinocereus coccineus*
Cereus, Night Blooming: *Peniocereus greggii*
Cerro Hawthorn: *Crataegus erythropoda*
Chaetopappa, Golden: *Pentachaeta aurea*
Chamise: *Adenostoma fasciculatum*
Chamiso: *Atriplex canescens*
Chamiza: *Chrysothamnus nauseosus*
Chamomile: see also Camomile
Chamomile, American Field: *Chamaemelum nobile*
Chamomile, False: *Matricaria discoidea*
Chamomile, Stinking: *Anthemis cotula*
Chantarelle: *Cantharellus cibarius*
Charlock: *Brassica rapa, Moricandia arvensis, Sinapis arvensis*
Chastetree, Simpleleaf: *Vitex trifolia*
Cheat Grass: *Bromus tectorum*
Checkermallow, Dwarf: *Sidalcea malviflora*
Checkermallow, New Mexico: *Sidalcea neomexicana*
Cheese Weed: *Malva parviflora*
Cherry, American Wild: *Prunus serotina*
Cherry, Bird: *Prunus pensylvanica*
Cherry, Bitter: *Prunus emarginata*
Cherry, Black: *Prunus serotina, P. sp.*
Cherry, Domesticated: *Prunus sp.*
Cherry, Gound: see Ground Cherry
Cherry, Holly Leaved: *Prunus ilicifolia*
Cherry, Indian: *Prunus ilicifolia*
Cherry, Pin: *Prunus pensylvanica*
Cherry, Quinine: *Prunus emarginata*
Cherry, Red: *Prunus pensylvanica*
Cherry, Red Wild: *Prunus sp.*
Cherry, Sand: *Prunus pumila*
Cherry, Wild: *Oemleria cerasiformis, Prunus emarginata, P. pensylvanica, P. serotina, P. sp., P. virginiana*
Cherry, Wild Black: *Prunus serotina*
Cherry, Wild Red: *Prunus pensylvanica*
Chervil, Spreading: *Chaerophyllum procumbens*
Chestnut: *Aesculus glabra, Castanea dentata, Castanopsis chrysophylla*
Chestnut, American: *Castanea dentata*
Chestnut, Horse: *Aesculus californica, A. hippocastanum*
Chia: *Salvia columbariae*
Chickadee Plant: *Heterotheca villosa*
Chickweed, Bering: *Cerastium beeringianum*
Chickweed, Big: *Cerastium fontanum*
Chickweed, Common: *Cerastium fontanum, Stellaria media*
Chickweed, Field: *Cerastium arvense*
Chickweed, Mouse Ear: *Cerastium beeringianum*
Chicory: *Cichorium intybus, Stephanomeria tenuifolia*
Chicory, Desert: *Pyrrhopappus carolinianus, P. pauciflorus*
Child Developer: *Zeuxine strateumatica*
Chile (or Chili) Pepper: *Capsicum annuum*
Chilicothe: *Marah horridus*
Chilicothe, Spanish: *Marah macrocarpus*
Chiltepine: *Capsicum annuum*
China Berry: *Melia azedarach, Sapindus saponaria*
Chinabrier: *Smilax bona-nox*
Chinchweed: *Pectis papposa*
Chinkapin, Allegheny: *Castanea pumila*
Chinkapin, Sierran: *Castanopsis sempervirens*
Chinkapin Oak: *Quercus muehlenbergii*
Chinquapin: *Castanea pumila, Castanopsis chrysophylla*
Chinquapin, Bush: *Castanopsis sempervirens*
Chinquapin, Giant: *Castanopsis chrysophylla*
Chinquapin, Golden Leaf: *Castanopsis chrysophylla*
Chittam: *Frangula purshiana*
Chives, Wild: *Allium schoenoprasum*
Chocolate Root: *Geum rivale*
Chocolate Tips: *Lomatium dissectum*
Chokeberry: *Prunus virginiana*
Chokeberry, Black: *Aronia melanocarpa*
Chokecherry: *Prunus americana, P. sp., P. virginiana*
Chokecherry, Black: *Prunus virginiana*
Chokecherry, Western: *Prunus virginiana*
Cholla: *Opuntia acanthocarpa, O. arbuscula, O. echinocarpa, O. engelmannii, O. imbricata, O. ×kelvinensis, O. leptocaulis, O. sp.*
Cholla, Arizona Pencil: *Opuntia arbuscula*
Cholla, Branched Pencil: *Opuntia ramosissima*
Cholla, Buckhorn: *Opuntia acanthocarpa*
Cholla, Club: *Opuntia clavata*
Cholla, Jumping: *Opuntia bigelovii, O. fulgida*
Cholla, Many Colored Tree: *Opuntia versicolor*
Cholla, Pencil: *Opuntia arbuscula, O. versicolor*
Cholla, Staghorn: *Opuntia echinocarpa, O. versicolor*
Cholla, Teddybear: *Opuntia bigelovii*
Cholla, Thistle: *Opuntia tunicata*
Cholla, Tree: *Opuntia imbricata*
Cholla, Vari-colored: *Opuntia versicolor*

Cholla, Whipple: *Opuntia whipplei*
Christmas Berry: *Heteromeles arbutifolia*
Chufa: *Cyperus esculentus*
Chuparosa: *Justicia californica*
Cicely: see Sweet Cicely
Cinchweed Fetidmarigold: *Pectis papposa*
Cinquefoil, Arctic: *Potentilla nana*
Cinquefoil, Bearded: *Potentilla crinita*
Cinquefoil, Common: *Potentilla simplex*
Cinquefoil, Dwarf: *Potentilla canadensis*
Cinquefoil, Five Finger: *Potentilla simplex*
Cinquefoil, Gland: *Potentilla glandulosa*
Cinquefoil, Graceful: *Potentilla gracilis*
Cinquefoil, Northwest: *Potentilla gracilis*
Cinquefoil, Norwegian: *Potentilla norvegica*
Cinquefoil, Pacific: *Argentina egedii*
Cinquefoil, Pennsylvania: *Potentilla pensylvanica*
Cinquefoil, Rough: *Potentilla norvegica*
Cinquefoil, Shrubby: *Pentaphylloides floribunda*
Cinquefoil, Silverweed: *Argentina anserina*
Cinquefoil, Sulphur: *Potentilla recta*
Cinquefoil, Tall: *Potentilla arguta*
Cinquefoil, Woolly: *Potentilla hippiana*
Citron: *Citrus medica*
Clammy Weed: *Polanisia dodecandra*
Clapweed: *Ephedra antisyphilitica*
Clasping Leaved Twisted Stalk: *Streptopus amplexifolius*
Clasping Venus's Lookingglass: *Triodanis perfoliata*
Clearweed: *Pilea pumila*
Cleavers: *Galium aparine*
Cleavers, Small: *Galium tinctorium*
Cliffbrake, Birdfoot: *Pellaea mucronata*
Cliffbrake, California: *Pellaea mucronata*
Cliffbrake, Purple: *Pellaea atropurpurea*
Cliffbush: *Jamesia americana*
Cliffrose, Stansbury: *Purshia stansburiana*
Cloudberry: *Rubus chamaemorus*
Clover: *Sedum laxum*, *Trifolium* spp.
Clover, Alsike: *Trifolium hybridum*
Clover, Beach: *Trifolium wormskioldii*
Clover, Branched Indian: *Trifolium dichotomum*
Clover, Bull: *Trifolium fucatum*
Clover, Bur: *Medicago polymorpha*
Clover, Bush: *Lespedeza capitata*
Clover, Clammy: *Trifolium obtusiflorum*
Clover, Cow: *Trifolium wormskioldii*
Clover, Creeping White: *Trifolium repens*
Clover, Cup: *Trifolium cyathiferum*
Clover, Elk: *Aralia californica*
Clover, Foothill: *Trifolium ciliolatum*
Clover, Notchleaf: *Trifolium bifidum*
Clover, Owl: *Orthocarpus densiflorus*, *O. faucibarbatus*, *O. luteus*, *O. purpurascens*, *O. purpureoalbus*
Clover, Pepper: *Trifolium fucatum*
Clover, Pinpoint: *Trifolium gracilentum*
Clover, Prairie: see Prairieclover
Clover, Rancheria: *Trifolium albopurpureum*
Clover, Red: *Trifolium pratense*
Clover, Smallhead: *Trifolium microcephalum*

Clover, Sour: *Melilotus indicus*
Clover, Spanish: *Lotus unifoliolatus*
Clover, Springbank: *Trifolium wormskioldii*
Clover, Suckling: *Trifolium dubium*
Clover, Swamp: *Trifolium variegatum*
Clover, Sweet: *Melilotus indicus*, *M. officinalis*, *M.* sp.
Clover, Tick: *Desmodium incanum*
Clover, Tomcat: *Trifolium willdenowii*
Clover, Tree: *Trifolium ciliolatum*
Clover, White: *Trifolium repens*, *T. wormskioldii*
Clover, White Sweet: *Melilotus officinalis*
Clover, Whitetip: *Trifolium variegatum*
Clover, Wild: *Marshallia obovata*, *Trifolium wormskioldii*
Clover, Yellow Sweet: *Melilotus officinalis*
Clubflower: *Cordylanthus wrightii*
Club Moss: *Huperzia selago*, Lycopodiaceae, *Lycopodium annotinum*, *L. clavatum*, *L.* sp.
Club Moss, Fir: *Huperzia selago*
Clubmoss, Rare: *Lycopodium obscurum*
Club Moss, Running: *Lycopodium clavatum*
Club Moss, Shining: *Huperzia lucidula*
Club Moss, Stiff: *Lycopodium annotinum*
Club Moss, Tree: *Lycopodium obscurum*
Cluster Lily: *Triteleia grandiflora*
Clusterlily, Elegant: *Brodiaea elegans*
Clusterlily, Harvest: *Brodiaea coronaria*
Clusterlily, Vernalpool: *Brodiaea minor*
Clustervine, Ovalleaf: *Jacquemontia ovalifolia*
Coccinia: *Echinocereus* sp.
Cochal: *Myrtillocactus cochal*
Cockerell: *Mahonia repens*
Cocklebur, American: *Xanthium strumarium*
Cocklebur, American Spiny: *Xanthium spinosum*
Cocklebur, False: *Ambrosia ambrosioides*
Cockleburr, Canada: *Xanthium strumarium*
Cockleburr, Rough: *Xanthium strumarium*
Cockscomb: *Amaranthus cruentus*
Cocoa: *Theobroma cacao*
Coconut: *Cocos nucifera*
Cocoplum: *Chrysobalanus icaco*
Coco Yam: *Colocasia esculenta*
Coffee, Indian: *Senna tora*
Coffee Berry: *Frangula californica*
Coffee Senna: *Senna occidentalis*
Coffee Tree, Kentucky: *Gymnocladus dioicus*
Cohosh, Black: *Cimicifuga racemosa*
Cohosh, Blue: *Caulophyllum thalictroides*
Coldthread: *Coptis trifolia*
Colic Root: *Aletris farinosa*
Coltsfoot: *Tussilago farfara*
Coltsfoot, Arctic Sweet: *Petasites frigidus*
Coltsfoot, Arrow Leaved: *Petasites sagittatus*
Coltsfoot, Northern Sweet: *Petasites frigidus*
Coltsfoot, Palmate: *Petasites frigidus*
Coltsfoot, Sweet: *Petasites frigidus*
Columbine, Chiricahua Mountain: *Aquilegia triternata*
Columbine, Colorado Blue: *Aquilegia coerulea*

Columbine, Crimson: *Aquilegia formosa*
Columbine, Garden: *Aquilegia* sp.
Columbine, Mancos: *Aquilegia micrantha*
Columbine, Red: *Aquilegia elegantula*, *A. formosa*
Columbine, Sitka: *Aquilegia formosa*
Columbine, Van Houtte's: *Aquilegia eximia*
Columbine, Western: *Aquilegia formosa*
Columbine, Western Red: *Aquilegia elegantula*
Columbine, Wild: *Aquilegia canadensis*
Columbo Root: *Frasera caroliniensis*
Comb Bushmint: *Hyptis pectinata*
Comfrey: *Symphytum officinale*
Comfrey, Wild: *Cynoglossum virginianum*
Commandra: *Geocaulon lividum*
Compass Plant: *Silphium laciniatum*
Compassplant, California: *Wyethia angustifolia*
Coneflower, Cutleaf: *Rudbeckia laciniata*
Coneflower, Eastern Purple: *Echinacea purpurea*
Coneflower, Green Headed: *Rudbeckia laciniata*
Coneflower, Green Prairie: *Ratibida tagetes*
Cone Flower, Narleaf Purple: *Echinacea angustifolia*
Coneflower, Orange: *Rudbeckia fulgida*
Coneflower, Pale Purple: *Echinacea pallida*
Cone Flower, Prairie: *Ratibida columnifera*
Cone Flower, Purple: *Echinacea angustifolia*, *E. pallida*, *E. purpurea*, *E.* sp.
Coneflower, Upright Prairie: *Ratibida columnifera*
Cone Flower, Yellow: *Ratibida columnifera*
Cone Headed Liverwort: *Conocephalum conicum*
Congo Root: *Orbexilum pedunculatum*
Consumption Plant, Indian: *Lomatium nudicaule*
Coontie: *Zamia pumila*
Copperbush: *Elliottia pyroliflorus*
Coralbead: *Cocculus carolinus*
Coralberry: *Symphoricarpos orbiculatus*
Coraldrops, White River: *Besseya plantaginea*
Coral Root: *Corallorrhiza maculata*
Cordgrass, Prairie: *Spartina pectinata*
Cordgrass, Smooth: *Spartina alterniflora*
Coriander: *Coriandrum sativum*
Corn: *Zea mays*
Corn Salad: *Valerianella locusta*
Corn Smut: *Ustilago zeae*
Cornel, Alternate Leaved: *Cornus alternifolia*
Cornel, Dwarf: *Cornus canadensis*
Cornel, Lapland: *Cornus suecica*
Cornel, Silky: *Cornus amomum*, *C. sericea*
Cornel, Stiff: *Cornus foemina*
Cornfield Creasy: *Barbarea verna*, *B. vulgaris*
Coryphantha: *Coryphantha* spp., *Escobaria vivipara*
Cosmos: *Cosmos parviflorus*
Cosmos, Wild: *Hymenopappus newberryi*
Costmary: *Balsamita major*
Cota: *Anthemis* sp., *Thelesperma filifolium*, *T. longipes*, *T. megapotamicum*
Cotton, Arizona: *Acourtia wrightii*
Cotton, Levant: *Gossypium herbaceum*
Cotton, Thurber's: *Gossypium thurberi*
Cotton, Upland: *Gossypium hirsutum*

Cotton, Wild: *Gossypium* sp., *G. thurberi*
Cotton Batting Plant: *Gnaphalium stramineum*
Cottongrass, Arctic: *Eriophorum callitrix*
Cottongrass, Red: *Eriophorum russeolum*
Cotton Grass, Tall: *Eriophorum angustifolium*
Cottongrass, White: *Eriophorum scheuchzeri*
Cotton Thorn: *Tetradymia axillaris*
Cottonthorn, Mojave: *Tetradymia stenolepis*
Cotton Weed, American: *Anaphalis margaritacea*
Cottonwood, Black: *Populus angustifolia, P. balsamifera*
Cottonwood, Common: *Populus fremontii*
Cottonwood, Eastern: *Populus deltoides*
Cottonwood, Fremont: *Populus fremontii*
Cottonwood, Lanceleaf: *Populus ×acuminata*
Cottonwood, Narrowleaf: *Populus angustifolia*
Cottonwood, Plains: *Populus deltoides*
Cottonwood, Rio Grande: *Populus deltoides*
Cottonwood, Valley: *Populus deltoides, P. fremontii*
Cough Plant: *Antennaria* sp.
Cough Root: *Lomatium dissectum*
Cous: *Lomatium cous, Lomatium simplex*
Covenas: *Dichelostemma pulchellum*
Cowbane, Spotted: *Cicuta maculata*
Cowbane, Stiff: *Oxypolis rigidior*
Cowpea: *Vigna unguiculata*
Cowpea, Hairypod: *Vigna luteola*
Cowslip: *Caltha palustris*
Cowslip, Virginia: *Mertensia virginica*
Coxcomb: *Amaranthus cruentus*
Coyotebrush: *Baccharis pilularis*
Coyote's Rope: *Clematis lasiantha*
Coyote's Tail: *Cirsium pastoris*
Crab Apple: *Malus* spp., *Sorbus americana*
Crab Apple, American: *Malus coronaria*
Crabapple, Oregon: *Malus fusca*
Crab Apple, Pacific: *Malus fusca*
Crab Apple, Prairie: *Malus ioensis*
Crabapple, Southern: *Malus angustifolia*
Crabapple, Sweet: *Malus coronaria*
Crab Apple, Western: *Malus fusca*
Crab Apple, Wild: *Malus coronaria, M. fusca, M. ioensis, M.* sp.
Crabgrass, Carolina: *Digitaria cognata*
Crabgrass, East Indian: *Digitaria setigera*
Crab Grass, Texan: *Schedonnardus paniculatus*
Cranberry, Bog: *Vaccinium oxycoccos*
Cranberry, Highbush: *Viburnum edule, V. opulus*
Cranberry, Large: *Vaccinium macrocarpon*
Cranberry, Low Bush: *Vaccinium vitis-idaea*
Cranberry, Mountain: *Vaccinium vitis-idaea*
Cranberry, Northern Mountain: *Vaccinium vitis-idaea*
Cranberry, Small: *Vaccinium oxycoccos*
Cranberry, Tree: *Viburnum opulus*
Cranesbill: *Geranium atropurpureum, G. lentum, G. richardsonii*
Cranesbill, Incised: *Geranium oreganum*
Cranesbill, Spotted: *Geranium maculatum*
Cranesbill, Western Purple: *Geranium atropurpureum*
Cranesbill, Wild: *Geranium maculatum*
Crazyweed, Cold Mountain: *Oxytropis campestris*
Crazyweed, Lambert's: *Oxytropis lambertii*
Cream Bush: *Holodiscus discolor*
Cream Cup: *Platystemon californicus*
Creases: *Barbarea vulgaris*
Creaseys: *Lepidium virginicum*
Creasy, Cornfield: *Barbarea verna, B. vulgaris*
Creeper, Virginia: *Parthenocissus quinquefolia, P. vitacea*
Creosote (or Creosoto) Bush: *Larrea tridentata*
Cress: *Rorippa palustris, Rorippa sinuata, R. teres*
Cress, Cow: *Lepidium campestre*
Cress, Early: *Barbarea verna*
Cress, Marsh: *Rorippa islandica*
Cress, Sand: *Cistanthe monandra*
Cress, Western Yellow: *Rorippa curvisiliqua*
Cress, Winter: *Barbarea orthoceras*
Cress, Yellow: see Yellowcress
Crinklemat, Matted: *Tiquilia latior*
Crinkle Root: *Cardamine concatenata, C. diphylla*
Crocus, Prairie: *Pulsatilla patens*
Cross Vine: *Bignonia capreolata*
Crowberry: *Empetrum nigrum*
Crowfoot: see also Crow's Foot
Crowfoot, Abortive: *Ranunculus abortivus*
Crowfoot, Bristly: *Ranunculus pensylvanicus*
Crowfoot, Desert: *Ranunculus cymbalaria*
Crowfoot, Hooked: *Ranunculus acris, R. recurvatus*
Crowfoot, Small Flowered: *Ranunculus abortivus*
Crowfoot, Water Yellow: *Ranunculus flabellaris*
Crowfoot, Whitewater: *Ranunculus aquatilis*
Crownbeard, Golden: *Verbesina encelioides*
Crownbeard, Virginia: *Verbesina virginica*
Crownbeard, White: *Verbesina virginica*
Crownscale: *Atriplex coronata*
Crownvetch, Purple: *Coronilla varia*
Crow's Foot: *Cardamine diphylla, Ranunculus* spp.; see also Crowfoot
Crucifixion Thorn: *Canotia holacantha*
Crucillo: *Ziziphus parryi*
Cucumber: *Cucumis sativus, Cucumis* sp.
Cucumber, Bur: *Sicyos angulatus*
Cucumber, Guadeloupe: *Melothria pendula*
Cucumber, Indian: *Medeola virginiana*
Cucumber, Oneseed Burr: *Sicyos angulatus*
Cucumber, Wild: *Echinocystis lobata, E.* sp., *Marah fabaceus*
Cucumber Tree: *Magnolia acuminata*
Cudweed: *Gnaphalium stramineum*
Cudweed, Glandular: *Corethrogyne filaginifolia*
Cudweed, Lobed: *Artemisia ludoviciana*
Cudweed, Marsh: *Gnaphalium uliginosum*
Cudweed, Smallhead: *Gnaphalium microcephalum*
Cudweed, Winged: *Gnaphalium viscosum*
Cudweed, Wright's: *Gnaphalium canescens*
Culver's Physic: *Veronicastrum virginicum*
Culver's Root: *Veronicastrum virginicum*
Cumin: *Cuminum cyminum*
Cupgrass: *Eriochloa aristata*
Cup Plant: *Silphium perfoliatum*
Currant, American Black: *Ribes americanum*
Currant, Black: *Ribes americanum, R. hudsonianum, R. laxiflorum, R. mescalerium, R. wolfii*
Currant, Black Wild: *Ribes americanum*
Currant, Blood: *Ribes sanguineum*
Currant, Blue: *Ribes bracteosum*
Currant, Buffalo: *Ribes aureum*
Currant, Chaparral: *Ribes malvaceum*
Currant, Cultivated: *Ribes rubrum*
Currant, Golden: *Ribes aureum*
Currant, Gooseberry: *Ribes montigenum*
Currant, Hudson Bay: *Ribes hudsonianum*
Currant, Mescalero: *Ribes mescalerium*
Currant, Missouri: *Ribes aureum*
Currant, Northern Black: *Ribes hudsonianum*
Currant, Northern Red: *Ribes triste*
Currant, Prickly: *Ribes lacustre*
Currant, Red: *Ribes cereum, R. rubrum, R. triste*
Currant, Red Flowered: *Ribes sanguineum*
Currant, Rock: *Ribes cereum*
Currant, Shinyleaf: *Ribes cruentum*
Currant, Skunk: *Ribes bracteosum, R. glandulosum*
Currant, Squaw: *Ribes cereum*
Currant, Sticky: *Ribes viscosissimum*
Currant, Stink: *Ribes bracteosum*
Currant, Swamp: *Ribes lacustre*
Currant, Trailing: *Ribes laxiflorum*
Currant, Wax: *Ribes cereum*
Currant, Western Black: *Ribes hudsonianum*
Currant, Whisky: *Ribes cereum*
Currant, White Flowered: *Ribes laxiflorum*
Currant, Wild: *Ribes aureum, R. cereum, R. glandulosum, R. sanguineum, R.* sp.
Currant, Wild Black: *Ribes americanum, R. hudsonianum*
Currant, Wild Blue: *Ribes laxiflorum*
Currant, Wild Red: *Ribes rubrum, R. triste*
Currant, Wild Sierra: *Ribes nevadense*
Currant, Wolf's: *Ribes wolfii*
Currant, Yellow: *Ribes aureum*
Cushaw: *Cucurbita moschata*
Cynthia: *Krigia biflora*
Cypress: *Cupressus* sp., *Sequoia sempervirens, Taxodium* spp.
Cypress, Bald: *Taxodium distichum*
Cypress, Heath: *Lycopodium sabinifolium*
Cypress, Piute: *Cupressus nevadensis*
Cypress, Pond: *Taxodium ascendens*
Cypress, Yellow: *Chamaecyparis nootkatensis*
Daggerpod: *Phoenicaulis cheiranthoides*
Dahoon: *Ilex cassine*
Dahoon Cassine: *Ilex cassine, I. vomitoria*
Daisy, Cutleaf: *Erigeron compositus*
Daisy, Easter: *Townsendia strigosa*
Daisy, English (or Lawn): *Bellis perennis*
Daisy, Mesa: *Machaeranthera pinnatifida*
Daisy, Ox Eye: *Chrysothamnus* sp., *Leucanthemum vulgare*

Daisy, Paper: *Psilostrophe sparsiflora*
Daisy, White: *Leucanthemum vulgare*
Daisy, Wild: *Erigeron foliosus, Leucanthemum vulgare*
Dandelion: *Malacothrix glabrata, Taraxacum* spp.
Dandelion, California: *Taraxacum californicum*
Dandelion, Common: *Taraxacum officinale, T.* sp.
Dandelion, Desert: see Desertdandelion
Dandelion, Dwarf: *Krigia biflora*
Dandelion, False: *Pyrrhopappus pauciflorus*
Dandelion, Goat: *Pyrrhopappus carolinianus*
Dandelion, Mountain: *Agoseris aurantiaca, A. glauca, A. retrorsa*
Dandelion, Twoflower Dwarf: *Krigia biflora*
Dandelion, Wild: *Microseris nutans*
Darnel: *Lolium temulentum*
Darning Needles, Devil's: *Clematis virginiana*
Datil: *Yucca baccata*
Day Flower: *Commelina erecta, C. dianthifolia*
Death Camas (or Camass): *Zigadenus elegans*
Death Camas, Foothill: *Zigadenus paniculatus*
Death Camas, Meadow: *Zigadenus venenosus*
Deathcamas, Nuttall's: *Zigadenus nuttallii*
Deer Brush: *Ceanothus integerrimus*
Deer's Ears: *Frasera speciosa, F.* sp.
Deer's Tongue: *Frasera speciosa;*
Deer Tongue: *Chelone* sp.
Deer Weed: *Lotus scoparius, Polygonum aviculare*
Deerweed, Common: *Lotus scoparius*
Deerweed, Silky: *Lotus procumbens*
Desert Calico: *Loeseliastrum matthewsii*
Desert Candle: *Caulanthus inflatus*
Desertchicory: *Pyrrhopappus pauciflorus*
Desertchicory, Carolina: *Pyrrhopappus carolinianus*
Desert Christmas Tree: *Pholisma arenarium*
Desertdandelion, California: *Malacothrix californica*
Desertdandelion, Fendler's: *Malacothrix fendleri*
Desertdandelion, Smooth: *Malacothrix glabrata*
Desertdandelion, Sowthistle: *Malacothrix sonchoides*
Desert Gourd: *Cucurbita foetidissima*
Desert Gum: *Lygodesmia* sp.
Desert Parsley, Canby's: *Lomatium canbyi*
Desert Parsley, Gray: *Lomatium macrocarpum*
Desert Parsley, Gray's: *Lomatium grayi*
Desertparsley, Great Basin: *Lomatium simplex*
Desertparsley, King: *Lomatium graveolens*
Desert Parsley, Naked: *Lomatium nudicaule*
Desert Parsley, Nevada: *Lomatium nevadense*
Desert Parsley, Piper's: *Lomatium piperi*
Desertparsley, Wasatch: *Lomatium bicolor*
Desertparsley, Watson's: *Lomatium watsonii*
Desert Plume: *Stanleya pinnata*
Desert Ramona: *Salvia dorrii*
Desert Thorn: *Lycium pallidum, L. torreyi*
Desertthorn, Arizona: *Lycium exsertum*
Desertthorn, Fremont's: *Lycium fremontii*
Desert Thornapple: *Datura discolor*

Desert Trumpet: *Eriogonum inflatum*
Desert Trumpets: *Linanthus nuttallii*
Devil's Bite: *Liatris scariosa*
Devil's Claw: *Martynia* sp., *Proboscidea althaeifolia, P. louisianica, P. parviflora, P.* sp.
Devil's Club: *Oplopanax horridus*
Devil's Darning Needles: *Clematis virginiana*
Devil's Gut: *Cassytha filiformis*
Devilshorn: *Proboscidea althaeifolia*
Devil's Shoestring: *Tephrosia virginiana, Yucca filamentosa*
Devil's Walking Stick: *Aralia spinosa, Oplopanax horridus*
Dewberry: *Rubus canadensis, R. flagellaris, R. hispidus, R. pubescens, R.* sp., *R. trivialis, R. ursinus*
Dewberry, Arizona: *Rubus arizonensis*
Dewberry, Bristly: *Rubus hispidus*
Dewberry, Northern: *Rubus flagellaris*
Dewberry, Pacific: *Rubus vitifolius*
Dewberry, Southern: *Rubus trivialis*
Dewflower, Nakedstem: *Murdannia nudiflora*
Dill: *Sanguinaria canadensis*
Dipper Gourd: *Lagenaria siceraria*
Dittany: *Cunila marina*
Dock, Arctic: *Rumex arcticus*
Dock, Bitter: *Rumex obtusifolius*
Dock, Clustered: *Rumex conglomeratus*
Dock, Curly: *Rumex crispus*
Dock, Fewleaved: *Rumex paucifolius*
Dock, Golden: *Rumex maritimus*
Dock, Great Water: *Rumex orbiculatus*
Dock, Green: *Rumex conglomeratus*
Dock, Mexican: *Rumex salicifolius*
Dock, Mountain: *Rumex paucifolius*
Dock, Narrow Leaved: *Rumex salicifolius*
Dock, Pale: *Rumex altissimus*
Dock, Patience: *Rumex patientia*
Dock, Sand: *Rumex hymenosepalus, R. venosus*
Dock, Sheep Sorrel: *Rumex acetosella*
Dock, Sour: *Rumex acetosa, R. aquaticus, R. arcticus, R. crispus, R.* sp.
Dock, Swamp: *Rumex verticillatus*
Dock, Veiny: *Rumex venosus*
Dock, Violet: *Rumex violascens*
Dock, Water: *Rumex altissimus, R. orbiculatus*
Dock, Western: *Rumex aquaticus*
Dock, Willow: *Rumex salicifolius*
Dock, Winged: *Rumex venosus*
Dock, Yellow: *Rumex crispus*
Dock Root: *Arctium minus*
Dodder, Bigfruit: *Cuscuta megalocarpa*
Dodder, California: *Cuscuta californica*
Dodder, Chaparral: *Cuscuta californica*
Dodder, Common: *Cuscuta gronovii*
Dodder, Compact: *Cuscuta compacta*
Dogbane, Intermediate: *Apocynum ×floribundum*
Dogbane, Spreading: *Apocynum androsaemifolium, A. cannabinum*
Dogbane, Velvet: *Apocynum cannabinum*
Dog Feet: *Disporum trachycarpum*

Dog Fennel: *Anthemis cotula*
Dogfennel, Garden: *Chamaemelum nobile*
Dogfennel, Green: *Matricaria discoidea*
Dog Hobble: *Leucothoe axillaris*
Dogweed: *Thymophylla acerosa*
Dogwood, Alternate Leaved: *Cornus alternifolia*
Dogwood, Brown: *Cornus glabrata*
Dogwood, Bunchberry: *Cornus canadensis*
Dogwood, Creek: *Cornus glabrata, C. sericea*
Dogwood, Dwarf: *Cornus canadensis*
Dogwood, Flowering: *Cornus florida, C. nuttallii*
Dogwood, Gray: *Cornus racemosa*
Dogwood, Mountain: *Cornus nuttallii*
Dogwood, Pacific: *Cornus nuttallii*
Dogwood, Panicled: *Cornus racemosa*
Dogwood, Red: *Cornus amomum, Cornus florida*
Dogwood, Red Osier: *Cornus sericea*
Dogwood, Rough: *Cornus asperifolia*
Dogwood, Rough Leaved: *Cornus drummondii*
Dogwood, Round Leaved: *Cornus rugosa*
Dogwood, Silky: *Cornus amomum*
Dogwood, Stiff: *Cornus foemina*
Dogwood, Toughleaf: *Cornus asperifolia*
Dogwood, Western: *Cornus sericea*
Doll's Eyes: *Actaea pachypoda*
Doubleclaw: *Proboscidea parviflora*
Dove Weed: *Croton texensis*
Dove Weed, American: *Croton setigerus*
Dragon, Green: *Arisaema dracontium*
Dragon Head: *Dracocephalum parviflorum*
Dragonhead, False: *Physostegia parviflora*
Dragon Root: *Arisaema dracontium*
Dropseed: *Sporobolus cryptandrus, S. flexuosus, S. wrightii*
Dropseed, Giant: *Sporobolus giganteus*
Dropseed, Mesa: *Sporobolus flexuosus*
Dropseed, Prairie: *Sporobolus heterolepis*
Dropseed, Sand: *Sporobolus cryptandrus*
Dropseed, Spike: *Sporobolus contractus*
Drops of Gold: *Disporum hookeri*
Dropwort, Twiggy Water: *Oenanthe* sp.
Drymary, Fendler's: *Drymaria glandulosa*
Duckbill: *Pedicularis procera*
Duckweed, Star: *Lemna trisulca*
Dudley: *Salix gooddingii*
Dulse: *Rhodymenia palmata*
Dunebroom: *Parryella filifolia*
Durango Root: *Datisca glomerata*
Durum Wheat: *Triticum durum*
Dustymaiden, Douglas's: *Chaenactis douglasii*
Dustymaiden, Steve's: *Chaenactis stevioides*
Dutchman's Breeches: *Dicentra cucullaria*
Dutchman's Pipe: *Aristolochia californica*
Eardrops, Golden: *Dicentra chrysantha*
Earth Star: *Geastrum* sp.
Eel Grass: *Zostera marina*
Eggs, Scrambled: *Corydalis aurea*
Ekaha Kuahiwi: *Asplenium nidus*
Elder, American: *Sambucus canadensis, S. cerulea*
Elder, Blue: *Sambucus cerulea*

Index of Common Names

Elder, Coastal American Red: *Sambucus racemosa*
Elder, European Black: *Sambucus nigra*
Elder, European Red: *Sambucus racemosa*
Elder, Florida: *Sambucus canadensis*
Elder, Marsh: *Iva xanthifolia*
Elder, Pacific Red: *Sambucus racemosa*
Elder, Pale: *Sambucus cerulea*
Elder, Red: *Sambucus racemosa*
Elderberry: *Sambucus canadensis, S. cerulea, S. racemosa, S.* sp.
Elderberry, American: *Sambucus racemosa*
Elderberry, Black: *Sambucus racemosa*
Elderberry, Blue: *Sambucus cerulea*
Elderberry, European Red: *Sambucus racemosa*
Elderberry, New Mexican: *Sambucus cerulea*
Elderberry, Red: *Sambucus racemosa*
Elderberry, Scarlet: *Sambucus racemosa*
Elecampane: *Inula helenium*
Elephant Ear: *Xanthosoma atrovirens*
Elephant Head: *Pedicularis attollens, P. centranthera, P. groenlandica*
Elephant Tree: *Bursera microphylla*
Elk Berry: *Ribes lacustre*
Elk Food: *Lithospermum incisum*
Elkweed, Desert: *Frasera albomarginata*
Elkweed, White: *Frasera montana*
Elm, American: *Ulmus americana, U. rubra*
Elm, Red: *Ulmus rubra*
Elm, Rock: *Ulmus thomasii*
Elm, Slippery: *Ulmus rubra*
Elm, White: *Ulmus americana*
Eltrot: *Heracleum sphondylium*
Erythronium Stalk: *Platanthera leucostachys*
Estafiata: *Artemisia frigida*
Eugenia, Simpson: *Myrcianthes fragrans*
Eulophus: *Perideridia bolanderi*
Euphorbia: *Euphorbia* spp., *Chamaesyce fendleri*
Evening Primrose: see Primrose
Evening Star: *Mentzelia laevicaulis*
Everlasting, California: *Gnaphalium viscosum*
Everlasting, Many Headed: *Gnaphalium obtusifolium*
Everlasting, Mouse Ear: *Gnaphalium obtusifolium*
Everlasting, Pearly: *Anaphalis margaritacea*
Everlasting, Plantain: *Antennaria plantaginifolia*
Everlasting, Spoonleaf Purple: *Gamochaeta purpurea*
Everlasting, Tall: *Antennaria anaphaloides*
Everlasting, White: *Gnaphalium microcephalum*
Everlasting, White Stemmed: *Gnaphalium* sp.
Explorer's Bush: *Ceanothus oliganthus*
Eyebane: *Chamaesyce maculata, C. nutans*
Eyeberry: *Rubus pubescens*
Eyelike Hardweed: *Aster linariifolius*
Fairy Bells, Hooker's: *Disporum hookeri*
Fairybells, Largeflower: *Disporum smithii*
Fairybells, Roughfruit: *Disporum trachycarpum*
Fairyduster: *Calliandra eriophylla*
Fairyfan, Diamond: *Clarkia rhomboidea*
Fairyfan, Elegant: *Clarkia unguiculata*

Fairyfan, Twolobe: *Clarkia biloba*
Fairyfan, Winecup: *Clarkia purpurea*
Fairy Lantern: *Disporum smithii*
False Aloe: *Manfreda virginica*
False Azalea: *Menziesia ferruginea*
False Boneset: *Brickellia eupatorioides*
False Box: *Paxistima myrsinites*
False Buffalograss: *Monroa squarrosa*
False Bugbane: *Trautvetteria caroliniensis*
False Caraway, Western: *Perideridia gairdneri*
False Chamomile: *Matricaria discoidea*
False Cocklebur: *Ambrosia ambrosioides*
False Cymopterus: *Pseudocymopterus montanus*
False Dandelion: *Pyrrhopappus pauciflorus*
False Dandelion, Leafy Stemmed: *Pyrrhopappus carolinianus*
False Dragonhead: *Physostegia parviflora*
False Flax: *Camelina microcarpa*
False Foxglove, Fern Leaved: *Aureolaria pedicularia*
False Foxglove, Smooth: *Aureolaria virginica*
False Gromwell: *Onosmodium molle*
False Hellebore: *Veratrum californicum, V.* sp., *V. viride*
False Indigo: *Amorpha fruticosa, Baptisia alba, B.* sp., *B. tinctoria*
False Indigo, Blue: *Baptisia australis*
False Lily: *Maianthemum canadense*
False Lupine: *Thermopsis macrophylla*
False Mallow: *Sphaeralcea coccinea*
False Mastic: *Sideroxylon foetidissimum*
False Mesquite: *Calliandra humilis*
False Mitrewort: *Tiarella cordifolia*
False Nettle: *Laportea canadensis*
False Pennyroyal: *Hedeoma drummondii, H. nana, H. pulegioides*
False Pennyroyal, Rough: *Hedeoma hispida*
False Rhubarb: *Rheum rhaponticum*
False Sisal: *Agave decipiens*
False Solomon's Seal: *Maianthemum racemosum, M. stellatum, Smilacina* sp.
False Spikenard: *Maianthemum racemosum*
False Tarragon: *Artemisia dracunculus*
False Toadflax: *Geocaulon lividum*
False Yarrow: *Chaenactis douglasii*
Farewell to Spring: *Clarkia amoena, C. biloba, C. purpurea*
Fat Solomon: *Maianthemum racemosum*
Fawnlily, Giant White: *Erythronium oregonum*
Fawnlily, Mahogany: *Erythronium revolutum*
Fawnlily, Midland: *Erythronium mesochoreum*
Fawnlily, Oregon: *Erythronium oregonum*
Featherplume: *Dalea formosa*
Fennel, Dog: *Anthemis cotula, Chamaemelum nobile, Matricaria discoidea*
Fennel, Sweet: *Foeniculum vulgare*
Fern, Amaumau: *Sadleria cyatheoides*
Fern, American Wild: *Dennstaedtia punctilobula*
Fern, Appalachian Shoestring: *Vittaria lineata*
Fern, Beard: *Vittaria lineata*
Fern, Bird's Foot: *Pellaea mucronata*

Fern, Bird Claw: *Pellaea mucronata*
Fern, Birdnest: *Asplenium nidus*
Fern, Bladder: *Cystopteris protrusa*
Fern, Bracken: *Pteridium aquilinum, P. caudatum*
Fern, Brake: *Pteridium aquilinum, Woodwardia radicans*; see also Cliffbrake
Fern, Brittle: *Cystopteris fragilis*
Fern, Chain: *Woodwardia fimbriata, W. radicans*
Fern, Chamisso's Man: *Cibotium chamissoi*
Fern, Christmas: *Polystichum acrostichoides*
Fern, Cinnamon: *Osmunda cinnamomea*
Fern, Cliff: *Woodsia neomexicana*
Fern, Cliffbrake: see Cliffbrake
Fern, Cloak: *Argyrochosma fendleri*
Fern, Deer: *Blechnum spicant*
Fern, Female: *Athyrium filix-femina*
Fern, Fendler's False Cloak: *Argyrochosma fendleri*
Fern, Five Finger: *Adiantum aleuticum, A. pedatum*
Fern, Flowering: *Osmunda regalis*
Fern, Fragrant Shield: *Dryopteris fragrans*
Fern, Giant: *Acrostichum danaeifolium*
Fern, Gold: *Pentagrama triangularis*
Fern, Grape: *Botrychium virginianum*
Fern, Gray Royal: *Osmunda regalis*
Fern, Hay Scented: *Dennstaedtia punctilobula*
Fern, Interrupted: *Osmunda claytoniana*
Fern, Kunth's Maiden: *Thelypteris kunthii*
Fern, Lady: *Athyrium filix-femina*
Fern, Leather: *Polypodium scouleri*
Fern, Licorice: *Polypodium glycyrrhiza, P. virginianum*
Fern, Lip: *Cheilanthes covillei, C. fendleri, C. wootonii*
Fern, Maiden: *Asplenium trichomanes*
Fern, Maidenhair: *Adiantum aleuticum, A. capillus-veneris, A. jordanii, A. pedatum, A.* sp.
Fern, Male: *Dryopteris filix-mas*
Fern, Marsh Shield: *Thelypteris palustris*
Fern, Meyen's Twinsorus: *Diplazium meyenianum*
Fern, New Mexico Cliff: *Woodsia neomexicana*
Fern, Oak: *Gymnocarpium dryopteris*; see also Oakfern
Fern, Ostrich: *Matteuccia struthiopteris*
Fern, Pala: *Marattia* sp.
Fern, Parsley: *Cryptogramma sitchensis*
Fern, Rattlesnake: *Botrychium virginianum*
Fern, Resurrection: *Pleopeltis polypodioides, Polypodium incanum*
Fern, Royal: *Osmunda regalis*
Fern, Sensitive: *Onoclea sensibilis*
Fern, Serpent: *Phlebodium aureum*
Fern, Shield: *Dryopteris arguta, D. carthusiana, D. cristata, D. expansa, D.* sp., *Thelypteris kunthii*
Fern, Spiny Shield: *Dryopteris campyloptera*
Fern, Spiny Wood: *Dryopteris campyloptera*
Fern, Spreading Wood: *Dryopteris carthusiana*
Fern, Sweet: *Comptonia peregrina*
Fern, Sword: *Polystichum munitum*
Fern, Tea: *Pellaea mucronata*
Fern, Virginia Grape: *Botrychium virginianum*

Fern, Walking: *Asplenium rhizophyllum*
Fern, Wood: *Dryopteris campyloptera, D. expansa, D. marginalis, D.* sp., *Woodsia neomexicana*; see also Woodfern
Fernbush: *Chamaebatiaria millefolium*
Fernweed, Woolly: *Pedicularis lanata*
Fescue, Alpine: *Festuca brachyphylla*
Fescue, Idaho: *Festuca idahoensis*
Fescue, Sixweeks: *Vulpia octoflora*
Fetid Marigold: *Dyssodia papposa, Pectis angustifolia*
Fetter Bush: *Leucothoe axillaris*
Fever Bush: *Garrya elliptica, Garrya* sp.
Feverfew, American: *Tanacetum parthenium*
Feverfew, Santa Maria: *Parthenium hysterophorus*
Feverwort: *Triosteum perfoliatum*
Fiddle Neck: *Amsinckia* sp., *A. tessellata, Lotus scoparius*
Fiddleneck, Bristly: *Amsinckia tessellata*
Fiddleneck, Douglas's: *Amsinckia douglasiana*
Fiddleneck, Purple: *Phacelia californica*
Fiddleneck, Tarweed: *Amsinckia lycopsoides*
Fiestaflower, White: *Pholistoma membranaceum*
Fig: *Ficus carica*
Fig, Sea: *Carpobrotus aequilateralus*
Fig, Strangler: *Ficus aurea*
Figwort: *Scrophularia californica, S. marilandica*
Figwort, Hare: *Scrophularia lanceolata*
Figwort, Lanceleaf: *Scrophularia lanceolata*
Filaree: *Erodium cicutarium, Erodium* sp.
Fir, Alpine: *Abies grandis, A. lasiocarpa*
Fir, Amabilis: *Abies amabilis*
Fir, Balsam: *Abies balsamea, A. fraseri, A. lasiocarpa, A.* sp.
Fir, Common: *Abies* sp.
Fir, Douglas: *Pseudotsuga menziesii*
Fir, Fraser's: *Abies fraseri*
Fir, Grand: *Abies grandis*
Fir, Great Silver: *Abies grandis*
Fir, Joint: see Jointfir
Fir, Lowland: *Abies grandis*
Fir, Noble: *Abies procera*
Fir, Pacific Silver: *Abies amabilis*
Fir, Red: *Pseudotsuga menziesii*
Fir, Silver: *Abies amabilis, A. concolor*
Fir, Sub Alpine: *Abies lasiocarpa*
Fir, White: *Abies concolor, A. grandis*
Fire Bush: *Euonymus atropurpurea*
Firecracker Plant: *Dichelostemma ida-maia*
Fire Weed: *Epilobium angustifolium, E. latifolium*
Firewheel: *Gaillardia pulchella*
Fishpoison: *Tephrosia purpurea*
Five Eyes, Greenleaf: *Chamaesaracha coronopus*
Five Finger: *Potentilla arguta, P. glandulosa, P.* sp.
Five Finger, Marsh: *Comarum palustre*
Five Leaf Vine: *Parthenocissus quinquefolia*
Flag Root: *Acorus calamus*
Flag, Blue: *Iris missouriensis, I. versicolor*
Flag, Broad: *Iris* sp.
Flag, Larger Blue: *Iris versicolor*
Flag, Soft: *Typha latifolia*
Flag, Sweet: *Acorus calamus*
Flag, Wild: *Iris missouriensis, I. setosa*
Flannelbush: *Fremontodendron californicum*
Flatsedge, Bearded: *Cyperus squarrosus*
Flatsedge, Chufa: *Cyperus esculentus*
Flatsedge, Fendler's: *Cyperus fendlerianus*
Flatsedge, Fragrant: *Cyperus odoratus*
Flatsedge, Redroot: *Cyperus erythrorhizos*
Flatsedge, Schweinitz's: *Cyperus schweinitzii*
Flatsedge, Smooth: *Cyperus laevigatus*
Flat Top: *Eriogonum plumatella*
Flax, Blue: *Linum lewisii*
Flax, Common: *Linum usitatissimum*
Flax, False: *Camelina microcarpa*
Flax, Orange Flowered: *Linum puberulum*
Flax, Perennial Blue: *Linum lewisii*
Flax, Plains: *Linum puberulum*
Flax, Prairie: *Linum lewisii*
Flax, Rocky Mountain: *Linum lewisii*
Flax, Southern: *Linum australe*
Flax, Stiffstem: *Linum rigidum*
Flax, Western Blue: *Linum lewisii*
Flax, Wild: *Linum lewisii*
Flax, Yellow: *Linum australe, L. puberulum, L. rigidum*
Fleabane: *Conyza canadensis, Erigeron* spp.
Fleabane, Aspen: *Erigeron speciosus*
Fleabane, Beautiful: *Erigeron formosissimus*
Fleabane, Cut Leaved: *Erigeron compositus*
Fleabane, Daisy: *Erigeron strigosus*
Fleabane, Desert Yellow: *Erigeron linearis*
Fleabane, Hoary: *Erigeron canus*
Fleabane, Largeflower: *Erigeron grandiflorus*
Fleabane, Leafy: *Erigeron foliosus*
Fleabane, Marsh: *Pluchea foetida, P. sericea, Senecio congestus*
Fleabane, Navajo: *Erigeron concinnus*
Fleabane, New Mexico: *Erigeron neomexicanus*
Fleabane, Philadelphia: *Erigeron philadelphicus*
Fleabane, Prairie: *Erigeron strigosus*
Fleabane, Rayless Shaggy: *Erigeron aphanactis*
Fleabane, Rough: *Erigeron asper*
Fleabane, Shaggy: *Erigeron pumilus*
Fleabane, Spreading: *Erigeron divergens*
Fleabane, Sprucefir: *Erigeron eximius*
Fleabane, Subalpine: *Erigeron peregrinus*
Fleabane, Thread Leaved: *Erigeron filifolius*
Fleabane, Trailing: *Erigeron flagellaris*
Fleabane, Tufted: *Erigeron caespitosus*
Fleabane, Western Daisy: *Erigeron bellidiastrum*
Flixweed: *Descurainia sophia*
Floating Heart: *Nymphoides cordata*
Flytrap, Venus: *Dionaea muscipula*
Foamflower, Heartleaf: *Tiarella cordifolia*
Foamflower, Threeleaf: *Tiarella trifoliata*
Fogfruit: *Phyla cuneifolia*
Fogfruit, Lanceleaf: *Phyla lanceolata*
Fogfruit, Turkey Tangle: *Phyla nodiflora*
Forget Me Not: *Myosotis* sp.
Forget Me Not, Bay (or Small Flowered): *Myosotis laxa*
Four o'Clock, American: *Mirabilis californica*
Four o'Clock, Bigelow's: *Mirabilis bigelovii*
Four o'Clock, California: *Mirabilis californica*
Four o'Clock, Colorado: *Mirabilis multiflora*
Four o'Clock, Dwarf: *Mirabilis pumila*
Four o'Clock, Greene's: *Mirabilis greenei*
Four o'Clock, Heartleaf: *Mirabilis nyctaginea*
Four o'Clock, Many Flowered: *Mirabilis multiflora*
Four o'Clock, Mountain: *Mirabilis oblongifolia*
Four o'Clock, Narrowleaf: *Mirabilis linearis*
Four o'Clock, Scarlet: *Mirabilis coccineus*
Four o'Clock, Smooth Spreading: *Mirabilis oxybaphoides*
Four o'Clock, Trailing: *Allionia incarnata, Mirabilis oxybaphoides*
Four o'Clock, Wild: *Mirabilis multiflora, M. nyctaginea*
Four o'Clock, Winged: *Mirabilis alipes*
Foxglove: *Aureolaria flava, A. laevigata, A. pedicularia, Digitalis purpurea*
Foxglove, Downy Yellow False: *Aureolaria virginica*
Foxglove, Entireleaf Yellow False: *Aureolaria laevigata*
Foxglove, Purple: *Digitalis purpurea*
Foxglove, Slenderleaf False: *Agalinis tenuifolia*
Foxglove, Smooth Yellow False: *Aureolaria flava*
Foxglove, Wild: *Penstemon grandiflorus*
Foxtail: *Hordeum jubatum, H. murinum*
Foxtail, Marsh (or Shortawn): *Alopecurus aequalis*
Fremontia: *Fremontodendron*
Frijole: *Phaseolus vulgaris*
Fringecup: *Tellima grandiflora*
Fringedpod, Sand: *Thelesperma subnudum*
Fringepod, Sand: *Thysanocarpus curvipes*
Fritillary: *Fritillaria lanceolata, F. pudica, F.* sp.
Frost Root: *Erigeron philadelphicus, E. pulchellus*
Frost Weed: *Helianthemum canadense, Verbesina virginica*
Fuchsia, American Wild: *Keckiella cordifolia*
Fuller's Teasel: *Dipsacus fullonum*
Fuzzybean: *Strophostyles helvula*
Fuzzy Weed: *Artemisia dracunculus*
Gale, Sweet: *Myrica gale*
Galleta: *Hilaria jamesii*
Gall of the Earth: *Prenanthes serpentaria, P. trifoliolata*
Gamble Weed: *Sanicula crassicaulis*
Gamot, Mountain: *Cymopterus montanus*
Garlic: *Allium falcifolium, A. sativum*
Garlic, Field: *Allium canadense, A. vineale*
Garlic, Meadow: *Allium canadense*
Garlic, Wild: *Allium canadense, A. lacunosum, A. vineale*
Garlic, Wild Field: *Allium canadense*
Gayfeather, Dense: *Liatris spicata*
Gayfeather, Dotted: *Liatris punctata*
Gayfeather, Sharp: *Liatris acidota*
Gayfeather, Spike: *Liatris spicata*

Index of Common Names

Gaywings: *Polygala paucifolia*
Gentian, Closed: *Gentiana affinis, G. andrewsii*
Gentian, Fourpart Dwarf: *Gentianella propinqua*
Gentian, Fringed: *Gentianopsis crinita*
Gentian, Green: *Frasera speciosa*
Gentian, Horse: *Triosteum perfoliatum*
Gentian, Plain: *Gentiana alba*
Gentian, Pleated: *Gentiana affinis*
Gentian, Prairie: *Gentiana affinis*
Gentian, Slender Rose: *Sabatia campanulata*
Gentian, Stiff: *Gentianella quinquefolia*
Gentian, Swamp: *Gentiana douglasiana*
Gentian, Yellowish: *Gentiana alba*
Geranium, Rock: *Heuchera americana*
Gerardia, Slender: *Agalinis tenuifolia*
Ghost Ear Fungus: *Exobasidium* sp.
Gianttrumpets: *Macromeria viridiflora*
Gilead: *Populus balsamifera*
Gilia: *Gilia sinuata, Ipomopsis aggregata, I. multiflora*
Gilia, Blue: *Gilia* sp., *Ipomopsis longiflora*
Gilia, Ciliate: *Linanthus ciliatus*
Gilia, Long Flowered: *Ipomopsis longiflora*
Gilia, Many Flowered: *Ipomopsis multiflora*
Gilia, Scarlet: *Ipomopsis aggregata*
Gilia, White: *Ipomopsis congesta, I. longiflora*
Ginger, Bitter: *Zingiber zerumbet*
Ginger, Indian: *Asarum canadense*
Ginger, Wild: *Asarum canadense, A. caudatum, A.* sp., *Hexastylis arifolia*
Ginseng: *Panax quinquefolius, Panax* sp.
Ginseng, Dwarf: *Panax trifolius*
Glasswort: *Arthrocnemum subterminale, Salicornia virginica, Sarcocornia pacifica*
Glasswort, American: *Salicornia virginica*
Glasswort, Parish's: *Arthrocnemum subterminale*
Glasswort, Slender: *Salicornia maritima*
Glasswort, Virginia: *Salicornia virginica*
Globe Flower, Spreading: *Trollius laxus*
Globemallow, Copper: *Sphaeralcea angustifolia*
Globemallow, Desert: *Sphaeralcea ambigua*
Globemallow, Emory's: *Sphaeralcea emoryi*
Globemallow, Fendler's: *Sphaeralcea fendleri*
Globemallow, Gooseberryleaf: *Sphaeralcea grossulariifolia*
Globemallow, Gray: *Sphaeralcea incana*
Globemallow, Munro's: *Sphaeralcea munroana*
Globemallow, Red: *Sphaeralcea coccinea*
Globemallow, Slippery: *Sphaeralcea digitata*
Globemallow, Smallflower: *Sphaeralcea parvifolia*
Goat Nut: *Simmondsia chinensis*
Goat's Beard: *Aruncus dioicus, Tragopogon porrifolius, T. pratensis*
Goat's Beard, Sylvan: *Aruncus dioicus*
Golden Alexander: *Zizia aurea*
Goldenbanner: *Thermopsis macrophylla*
Goldenbush: *Ericameria linearifolia, E. palmeri*
Goldenbush, Alkali: *Isocoma acradenia*
Goldeneggs: *Camissonia ovata*
Goldeneye, Longleaf False: *Heliomeris longifolia*

Goldeneye, Showy: *Heliomeris multiflora*
Goldenfleece: *Ericameria arborescens*
Golden Hala Pepe: *Pleomele aurea*
Goldenhead: *Acamptopappus sphaerocephalus*
Goldenhills: *Encelia farinosa*
Golden Lantern: *Calochortus pulchellus*
Golden Mariana: *Chrysopsis mariana*
Goldenrod, American Western: *Solidago nemoralis*
Goldenrod, Anisescented: *Solidago odora*
Goldenrod, Bog: *Solidago uliginosa*
Goldenrod, Broad Leaved: *Solidago flexicaulis*
Goldenrod, California: *Solidago californica*
Goldenrod, Canada: *Solidago canadensis*
Goldenrod, Coast: *Solidago spathulata*
Goldenrod, Downy Ragged: *Solidago squarrosa*
Goldenrod, Dwarf: *Solidago simplex*
Goldenrod, Dyersweed: *Solidago nemoralis*
Goldenrod, Early: *Solidago juncea*
Goldenrod, Elm Leaved: *Solidago ulmifolia*
Goldenrod, Fragrant: *Euthamia graminifolia*
Goldenrod, Giant: *Solidago gigantea*
Goldenrod, Heath: see Heathgoldenrod
Goldenrod, Late: *Solidago gigantea*
Goldenrod, Low: *Solidago nemoralis*
Goldenrod, Mountain: *Solidago multiradiata*
Goldenrod, Nevada: *Solidago spectabilis*
Goldenrod, Rigid: *Solidago rigida*
Goldenrod, Rock: *Petradoria pumila*
Goldenrod, Rough: *Solidago rugosa*
Goldenrod, Showy: *Solidago speciosa*
Goldenrod, Southern: *Solidago confinis*
Goldenrod, Spike Like: *Solidago spathulata*
Goldenrod, Stiff: *Solidago rigida*
Goldenrod, Stout Ragged: *Solidago squarrosa*
Goldenrod, Threenerve: *Solidago velutina*
Goldenrod, Wrinkleleaf: *Solidago rugosa*
Goldenrod, Zig Zag: *Solidago flexicaulis*
Golden Seal: *Hydrastis canadensis*
Golden Star: *Bloomeria crocea*
Golden Thread: *Coptis trifolia*
Goldentop: *Lamarckia aurea*
Golden Trumpet: *Angadenia berteroi*
Goldenweed: *Ericameria nana*
Goldenweed, Cespitose: *Stenotus lanuginosus*
Goldenweed, Cutleaf: *Machaeranthera canescens*
Goldenweed, Slender: *Machaeranthera gracilis*
Goldfern, Western: *Pentagrama triangularis*
Goldfields: *Lasthenia californica, L. glabrata*
Gold Root: *Hydrastis canadensis*
Goldstar: *Hypoxis hirsuta*
Gold Thread: *Coptis macrosepala, Coptis trifolia*
Gold Wire: *Hypericum concinnum*
Gooseberry, Appalachian: *Ribes rotundifolium*
Gooseberry, Black: *Ribes divaricatum, R. lacustre*
Gooseberry, Black Swamp: *Ribes lacustre*
Gooseberry, California: *Ribes californicum*
Gooseberry, Canadian: *Ribes oxyacanthoides*
Gooseberry, Chaparral: *Ribes quercetorum*
Gooseberry, Coastal Black: *Ribes divaricatum*
Gooseberry, Desert: *Ribes velutinum*

Gooseberry, Eastern Wild: *Ribes rotundifolium*
Gooseberry, European: *Ribes uva-crispa*
Gooseberry, Gummy: *Ribes lobbii*
Gooseberry, Hairystem: *Ribes hirtellum*
Gooseberry, Hillside: *Ribes californicum*
Gooseberry, Idaho Black: *Ribes oxyacanthoides*
Gooseberry, Inland: *Ribes oxyacanthoides*
Gooseberry, Missouri: *Ribes missouriense*
Gooseberry, Orange: *Ribes pinetorum*
Gooseberry, Plateau: *Ribes velutinum*
Gooseberry, Prickly: *Ribes cynosbati*
Gooseberry, Red: *Ribes* sp.
Gooseberry, Rock: *Ribes quercetorum*
Gooseberry, Sierra: *Ribes roezlii*
Gooseberry, Smooth: *Ribes missouriense, R. oxyacanthoides*
Gooseberry, Spreading: *Ribes divaricatum*
Gooseberry, Sticky: *Ribes lobbii*
Gooseberry, Straggly: *Ribes divaricatum*
Gooseberry, Swamp: *Ribes lacustre*
Gooseberry, Trumpet: *Ribes leptanthum*
Gooseberry, White Stemmed: *Ribes inerme*
Gooseberry, Wild: *Ribes americanum, R. aureum, R. cereum, R. divaricatum, R. hudsonianum, R. lacustre, R. missouriense, R. oxyacanthoides, R. viscosissimum*
Gooseberry Vine: *Ribes uva-crispa*
Goosefoot, Blite: *Chenopodium capitatum*
Goosefoot, California: *Chenopodium californicum*
Goosefoot, Desert: *Chenopodium pratericola*
Goosefoot, Fetid: *Chenopodium graveolens*
Goosefoot, Fremont: *Chenopodium fremontii*
Goosefoot, Jerusalem Oak: *Chenopodium botrys*
Goosefoot, Marshland: *Chenopodium humile*
Goosefoot, Mealy: *Chenopodium incanum*
Goosefoot, Narrowleaf: *Chenopodium leptophyllum*
Goosefoot, Nettleleaf: *Chenopodium murale*
Goosefoot, Nevada: *Chenopodium nevadense*
Goosefoot, Red: *Chenopodium rubrum*
Goosefoot, Tasmanian: *Chenopodium carinatum*
Goosefoot, Watson's: *Chenopodium watsonii*
Goosefoot, White: *Chenopodium album*
Goose Grass: *Galium aparine, Polygonum aviculare*
Goosegrass, Feathery: *Equisetum* sp.
Goose Tongue: *Plantago maritima*
Gourd: *Cucumis* sp., *Cucurbita foetidissima, Lagenaria siceraria*
Gourd, American Wild: *Cucurbita foetidissima*
Gourd, Bottle: *Lagenaria siceraria*
Gourd, Buffalo: *Cucurbita foetidissima*
Gourd, Desert: *Cucurbita foetidissima*
Gourd, Dipper: *Lagenaria siceraria*
Gourd, Fingerleaf: *Cucurbita digitata*
Gourd, Missouri: *Cucurbita foetidissima*
Gourd, Wild: *Cucurbita foetidissima*
Gourd Vine: *Lagenaria siceraria*
Grama, Blue: *Bouteloua gracilis*
Grama, Hairy: *Bouteloua hirsuta*
Grama, Matted: *Bouteloua simplex*

Grama, Side Oats: *Bouteloua curtipendula*
Grama Grass, Blue: *Bouteloua gracilis*
Grama Grass, Hairy: *Bouteloua hirsuta*
Grama Grass, Mat: *Bouteloua simplex*
Grama Grass, Tall: *Bouteloua curtipendula*
Grape, Arizona Wild: *Vitis arizonica*
Grape, California Wild: *Vitis californica*
Grape, Calloosa (or Calloose): *Vitis shuttleworthii*
Grape, Canyon: *Vitis arizonica*
Grape, Domesticated: *Vitis vinifera*
Grape, Downy: *Vitis cinerea*
Grape, Fox: *Vitis labrusca, V. vulpina*
Grape, Frost: *Vitis vulpina*
Grape, Graybark: *Vitis cinerea*
Grape, Holly: *Mahonia fremontii, M. repens*
Grape, Little Muscadine: *Vitis munsoniana*
Grape, Mountain: *Mahonia aquifolium*
Grape, Munson's: *Vitis munsoniana*
Grape, Muscadine: *Vitis munsoniana, V. rotundifolia*
Grape, Northern Fox: *Vitis labrusca*
Grape, Oregon: *Mahonia aquifolium, M. nervosa, M. repens, M. sp.*
Grape, Possum: *Cissus verticillata, Vitis cinerea*
Grape, Red Shank: *Vitis aestivalis*
Grape, River Bank: *Vitis riparia, V. vulpina*
Grape, Sand: *Vitis rupestris*
Grape, Summer: *Vitis aestivalis*
Grape, Valley: *Vitis girdiana*
Grape, Wild: *Vitis arizonica, V. californica, V. cinerea, V. girdiana, V. labrusca, V. sp., V. vulpina*
Grapefruit: *Citrus ×paradisi*
Grass, Alkali: *Puccinellia distans*
Grass, American Dune: *Leymus mollis*
Grass, Arrow: *Triglochin maritimum*
Grass, Barnyard: *Echinochloa crus-galli, E. cruspavonis*
Grass, Bead: *Paspalum setaceum*
Grass, Bear: *Bothriochloa saccharoides, Nolina erumpens, N. microcarpa, N. sp., Tradescantia virginiana, Xerophyllum sp., X. tenax, Yucca filamentosa, Y. glauca*
Grass, Bear: *Bothriochloa saccharoides, Nolina erumpens, N. microcarpa, N. sp., Tradescantia virginiana, Xerophyllum sp., X. tenax, Yucca filamentosa, Y. glauca*
Grass, Beard: *Andropogon floridanus*
Grass, Bermuda: *Cynodon dactylon*
Grass, Blue Eyed: *Sisyrinchium albidum, S. angustifolium, S. bellum, S. campestre, S. mucronatum, S. nashii*
Grass, Blue Grama: *Bouteloua gracilis*
Grass, Blue Joint: *Elytrigia repens*
Grass, Blue Star Flower: *Sisyrinchium angustifolium*
Grass, Blue Stem: *Andropogon gerardii, Pascopyrum smithii*
Grass, Bottle Brush: *Elymus hystrix*
Grass, Bristle: *Pennisetum glaucum*
Grass, Brome: *Bromus tectorum*
Grass, Broom: *Andropogon* sp.

Grass, Buffalo: *Bouteloua gracilis, Buchloe dactyloides*
Grass, Bunch: *Echinochloa crus-galli, Muhlenbergia rigens, Oryzopsis hymenoides, Poa secunda, Pseudoroegneria spicata, Sporobolus airoides*
Grass, California Vanilla: *Hierochloe occidentalis*
Grass, Cane: *Phragmites australis*
Grass, Carizzo: *Phragmites australis*
Grass, Cheat: *Bromus tectorum*
Grass, Cockspur: *Echinochloa crus-pavonis*
Grass, Cotton: see Cottongrass
Grass, Crab: see Crabgrass
Grass, Crouch: *Elytrigia repens*
Grass, Cut: *Scirpus microcarpus*
Grass, Deer: *Lotus scoparius, Muhlenbergia rigens*
Grass, Dropseed: *Sporobolus contractus, S. cryptandrus*
Grass, Eastern Bottlebrush: *Elymus hystrix*
Grass, Eel: *Zostera marina*
Grass, Elk: *Xerophyllum tenax*
Grass, Fescue: *Festuca brachyphylla, F. idahoensis, Vulpia octoflora*
Grass, Fetch: *Vicia villosa*
Grass, Fluff: *Erioneuron pulchellum*
Grass, Foxtail: *Hordeum* sp.
Grass, Fringed Brome: *Bromus ciliatus*
Grass, Galleta: *Hilaria jamesii*
Grass, Giant Rye: *Leymus condensatus*
Grass, Giant Wild Rye: *Leymus cinereus*
Grass, Goose: *Galium aparine, Polygonum aviculare*
Grass, Grama: see Grama; and Grama Grass
Grass, Grapevine Mesquite: *Panicum obtusum*
Grass, Green: *Leymus mollis*
Grass, Gulf Cockspur: *Echinochloa crus-pavonis*
Grass, Hair: *Deschampsia cespitosa, D. danthonioides, Muhlenbergia pungens*
Grass, Heller's Rosette: *Dichanthelium oligosanthes*
Grass, Hog: *Sonchus oleraceus*
Grass, Horse Tail: *Equisetum arvense, E. hyemale*
Grass, Indian Rice: *Oryzopsis hymenoides*
Grass, Johnson: *Sorghum halepense*
Grass, Joint: *Elytrigia repens*
Grass, June: *Koeleria macrantha*
Grass, Long Tongue Mutton: *Poa fendleriana*
Grass, Love: *Eragrostis mexicana, E. secundiflora*
Grass, Lyme: *Elymus canadensis, Leymus mollis*
Grass, Manna: *Glyceria obtusa, Puccinellia distans*
Grass, Marsh Reed: *Calamagrostis canadensis*
Grass, Meadow: *Poa arida*
Grass, Merlin: *Polymnia canadensis*
Grass, Mesquite: *Bouteloua curtipendula, Muhlenbergia wrightii, Panicum obtusum*
Grass, Moccasin: *Melothria pendula*
Grass, Muhly: see Muhly
Grass, Mutton: *Poa fendleriana*
Grass, Narrow Leaf: *Sporobolus contractus*
Grass, Narrow Leaved: *Sisyrinchium montanum*
Grass, Nodding Fescue: *Festuca subverticillata*
Grass, Nut: *Cyperus esculentus, C. rotundus, C.* sp.

Grass, Oat: see Oat
Grass, Onion: *Melica bulbosa*
Grass, Openflower Rosette: *Dichanthelium laxiflorum*
Grass, Orange: *Hypericum gentianoides*
Grass, Panic: *Dichanthelium laxiflorum, D. oligosanthes, D. strigosum, Panicum capillare, P. hirticaule, P. sp., P. urvilleanum, Paspalidium geminatum*
Grass, Pepper: *Capsella bursa-pastoris, Descurainia pinnata, Lepidium densiflorum, L. lasiocarpum, L. montanum, L. nitidum, L. sp., L. thurberi, L. virginicum, Perideridia kelloggii*
Grass, Pine: *Calamagrostis rubescens*
Grass, Porcupine: *Stipa spartea, S. speciosa*
Grass, Poverty: *Sporobolus cryptandrus*
Grass, Quack: *Elytrigia repens*
Grass, Rabbitsfoot: *Polypogon monspeliensis*
Grass, Rat: *Scirpus* sp.
Grass, Rattlesnake: *Glyceria canadensis*
Grass, Reed: *Cinna arundinacea, Phragmites australis*
Grass, Reed Canary: *Phalaris arundinacea*
Grass, Reed Meadow: *Catabrosa aquatica*
Grass, Rescue: *Bromus catharticus*
Grass, Rib: *Plantago lanceolata*
Grass, Rice: *Oryzopsis hymenoides, Piptatherum miliaceum*
Grass, Ripgut: *Bromus diandrus*
Grass, Rood Reed: *Cinna arundinacea*
Grass, Rosette: *Dichanthelium laxiflorum, D. oligosanthes, D. strigosum*
Grass, Roughhair Rosette: *Dichanthelium strigosum*
Grass, Rush: *Sporobolus cryptandrus*
Grass, Rye: *Elymus glaucus, E. sp., Leymus cinereus, L. condensatus, L. mollis, L. temulentum, L. triticoides*
Grass, Salt: *Distichlis spicata*
Grass, Sand: *Calamovilfa gigantea, Oryzopsis hymenoides*
Grass, Sand Bunch: *Oryzopsis hymenoides*
Grass, Saw: *Cladium mariscus*
Grass, Scribner's Rosette: *Dichanthelium oligosanthes*
Grass, Scurvy: *Cochlearia officinalis*
Grass, Sea: *Phyllospadix torreyi, Ruppia* sp.
Grass, Silk: *Pityopsis graminifolia*
Grass, Sleepy: *Stipa robusta*
Grass, Slender Hair: *Deschampsia danthonioides*
Grass, Slew: *Beckmannia syzigachne*
Grass, Slough: *Beckmannia syzigachne, Carex barbarae, Spartina pectinata*
Grass, Small Panic: *Dichanthelium oligosanthes*
Grass, Smilo: *Piptatherum miliaceum*
Grass, Snake: *Ipomoea sagittata*
Grass, Spear: *Stipa comata, Stipa speciosa*
Grass, Squaw: *Xerophyllum tenax*
Grass, Squirrel Tail: *Elymus elymoides, Hordeum jubatum*

Index of Common Names

Grass, St. Augustine: *Stenotaphrum secundatum*
Grass, Star: *Aletris farinosa*
Grass, Sugar: *Glyceria fluitans*
Grass, Surf: *Phyllospadix scouleri, P. serrulatus, P. torreyi*
Grass, Sweet: see Sweet Grass
Grass, Sweet Vernal: *Anthoxanthum odoratum*
Grass, Three Awn: *Aristida purpurea*
Grass, Three Cornered: *Scirpus americanus*
Grass, Tule: *Juncus* sp.
Grass, Turkey: *Verbena officinalis*
Grass, Turtle: *Phyla nodiflora*
Grass, Vanilla: *Hierochloe odorata*
Grass, Vernal: *Anthoxanthum odoratum*
Grass, Vine Mesquite: *Panicum obtusum*
Grass, Water: *Echinochloa crus-galli*
Grass, Weeping Alkali: *Puccinellia distans*
Grass, Western Rye: *Elymus glaucus*
Grass, Wheat: *Agropyron* sp., *Elymus trachycaulus, Pascopyrum smithii, Pseudoroegneria spicata, Triticum aestivum*
Grass, White: *Xerophyllum tenax*
Grass, Whitlow: see Whitlowgrass
Grass, Whooping Cough: *Iresine diffusa*
Grass, Wire: *Aristida purpurea, Juncus balticus, J. mexicanus, J. tenuis*
Grass, Witch: *Panicum capillare*
Grass, Wool: *Scirpus cyperinus*
Grass, Worm: *Chenopodium ambrosioides*
Grass, Yelloweyed: *Xyris ambigua, X. caroliniana*
Grass Nut: *Brodiaea* sp., *Dichelostemma pulchellum, Triteleia laxa*
Grass Nut, Climbing: *Dichelostemma volubile*
Grass Nut, Purple Flowered: *Dichelostemma pulchellum*
Grass Nut, White Flowered: *Triteleia hyacinthina*
Grass Nut, Yellow Blossom: *Triteleia ixioides*
Graythorn: *Ziziphus obtusifolia*
Greasebush: *Glossopetalon spinescens*
Greasewood: *Adenostoma fasciculatum, A. sparsifolium, Purshia tridentata, Sarcobatus vermiculatus*
Greenbrier: *Smilax auriculata, S. bona-nox, S. pseudochina, S. rotundifolia, S.* sp.
Greenbrier, Bristly: *Smilax tamnoides*
Greenbrier, California: *Smilax californica*
Greenbrier, Cat: *Smilax glauca*
Greenbrier, Common: *Smilax rotundifolia*
Greenbrier, Earleaf: *Smilax auriculata*
Greenbrier, Hispid: *Smilax tamnoides*
Greenbrier, Laurel: *Smilax laurifolia*
Greenbrier, Roundleaf: *Smilax rotundifolia*
Greenbrier, Saw: *Smilax bona-nox*
Greendragon: *Arisaema dracontium*
Greenleaf Five Eyes: *Chamaesaracha coronopus*
Greenmolly: *Kochia americana*
Greenstripe: *Amaranthus acanthochiton*
Greenthread, Hopi Tea: *Thelesperma megapotamicum*
Greenthread, Longstalk: *Thelesperma longipes*
Greenthread, Stiff: *Thelesperma filifolium*
Gromwell, European: *Lithospermum officinale*
Gromwell, False: *Onosmodium molle*
Gromwell, Manyflowered: *Lithospermum multiflorum*
Gromwell, Narrow Leaf: *Lithospermum incisum*
Gromwell, Western: *Lithospermum ruderale*
Ground Cedar: *Lycopodium complanatum*
Ground Cherry, Clammy: *Physalis heterophylla*
Groundcherry, Fendler's: *Physalis hederifolia*
Ground Cherry, Inedible: *Physalis lanceolata*
Groundcherry, Lanceleaf: *Physalis lanceolata*
Ground Cherry, Large Bladder: *Physalis viscosa*
Groundcherry, Longleaf: *Physalis longifolia*
Groundcherry, Mexican: *Physalis philadelphica*
Groundcherry, New Mexican: *Physalis subulata*
Ground Cherry, Prairie: *Physalis lanceolata*
Ground Cherry, Purple Flowered: *Quincula lobata*
Groundcherry, Sharpleaf: *Physalis acutifolia*
Groundcherry, Starhair: *Physalis viscosa*
Groundcherry, Virginia: *Physalis virginiana*
Ground Cherry, Wright's: *Physalis acutifolia*
Groundcone, California: *Boschniakia strobilacea*
Groundcone, Northern: *Boschniakia rossica*
Groundcone, Vancouver: *Boschniakia hookeri*
Groundcover Milkvetch: *Astragalus humistratus*
Ground Hemlock: *Lycopodium clavatum, Taxus baccata, T. brevifolia, T. canadensis*
Ground Nut: *Apios americana, Claytonia lanceolata, C. multicaulis*
Ground Pine: *Lycopodium complanatum, L. dendroideum, L. obscurum*
Groundpine, Savinleaf: *Lycopodium sabinifolium*
Groundpine, Tree: *Lycopodium dendroideum*
Groundsel, Arrowleaf: *Senecio triangularis*
Groundsel, Broom: *Senecio spartioides*
Groundsel, Bush: *Senecio flaccidus*
Groundsel, Douglas: *Senecio flaccidus*
Groundsel, Lobeleaf: *Senecio multilobatus*
Groundsel, New Mexico: *Senecio neomexicanus*
Groundsel, Ragwort: *Senecio multicapitatus*
Groundsel, Threadleaf: *Senecio flaccidus*
Groundsel Tree: *Baccharis emoryi, B. pilularis*
Grouseberry: *Vaccinium scoparium*
Guava: *Psidium guajava*
Gum, Black: *Nyssa aquatica, N.* sp., *N. sylvatica*
Gum, Desert: *Lygodesmia* sp.
Gum, Spruce: *Pinus rigida*
Gum, Sweet: *Liquidambar styraciflua*
Gumbo Limbo: *Bursera simaruba*
Gum Bully (or Bumelia): *Sideroxylon lanuginosum*
Gum Plant: *Grindelia hallii, G. integrifolia, G. nana, G. robusta, G.* sp., *G. squarrosa*
Gum Weed: *Grindelia* spp., *Madia gracilis*
Gumweed, Curlycup: *Grindelia squarrosa*
Gumweed, Curlytop: *Grindelia nuda*
Gumweed, Great Valley: *Grindelia camporum, G. robusta*
Gumweed, Hairy: *Grindelia humilis*
Gumweed, Hall's: *Grindelia hallii*
Gumweed, Idaho: *Grindelia nana*
Gumweed, Pillar False: *Vanclevea stylosa*
Gumweed, Pointed: *Grindelia fastigiata*
Gumweed, Puget Sound: *Grindelia integrifolia*
Gumweed, Reclined: *Grindelia decumbens*
Gypsyflower: *Cynoglossum officinale*
Gypsyweed: *Veronica officinalis*
Haa: *Antidesma pulvinatum*
Hackberry, Common: *Celtis occidentalis*
Hackberry, Netleaf: *Celtis laevigata*
Hackberry, Rough Leaved: *Celtis occidentalis*
Hackberry, Western: *Celtis laevigata*
Hackelia, Spreading: *Hackelia diffusa*
Hairgrass, Annual: *Deschampsia danthonioides*
Hairgrass, Purple: *Muhlenbergia pungens*
Hairgrass, Tufted: *Deschampsia cespitosa*
Hairybeast Turtleback: *Psathyrotes pilifera*
Hala: *Pandanus tectorius*
Hala-pepe: *Pleomele aurea*
Hame: *Antidesma pulvinatum*
Handsome Harry: *Rhexia virginica*
Hapuu: *Cibotium chamissoi*
Harbinger of Spring: *Erigenia bulbosa*
Hardhack: *Spiraea douglasii, Spiraea tomentosa*
Hard Tack: *Cercocarpus montanus*
Hardweed, Eyelike: *Aster linariifolius*
Harebell: *Campanula rotundifolia*
Harebell, California: *Asyneuma prenanthoides*
Harebell, Southern: *Campanula divaricata*
Haresfoot Pointloco: *Oxytropis lagopus*
Hare's Tail: *Eriophorum callitrix*
Harlequinbush: *Gaura hexandra*
Harvestbells: *Gentiana saponaria*
Harvestlice: *Agrimonia parviflora*
Hau-hele: *Hibiscus* sp.
Hau-kae-kae: *Hibiscus tiliaceus*
Hauve Cane: *Phragmites australis*
Haw: *Asclepias tuberosa, Crataegus douglasii, C. pruinosa, C.* sp., *C. submollis*
Haw, Black: *Crataegus douglasii, Viburnum lentago, V. prunifolium*
Haw, Dotted: *Crataegus punctata*
Haw, Possum: *Viburnum nudum*
Haw, Red: *Crataegus chrysocarpa, C. columbiana*
Hawberry, Black: *Crataegus* sp.
Hawksbeard, Fiddleleaf: *Crepis runcinata*
Hawksbeard, Largeflower: *Crepis occidentalis*
Hawksbeard, Longleaf: *Crepis acuminata*
Hawksbeard, Naked Stemmed: *Crepis runcinata*
Hawksbeard, Siskiyou: *Crepis modocensis*
Hawksbeard, Slender: *Crepis atribarba*
Hawkweed: *Hieracium cynoglossoides, H. fendleri, H. pilosella, H. scabrum, H. scouleri, H.* sp., *H. venosum*
Hawkweed, Canada: *Hieracium canadense*
Hawkweed, Houndstongue: *Hieracium cynoglossoides*
Hawkweed, Mouseear: *Hieracium pilosella*
Hawkweed, Rough: *Hieracium scabrum*
Hawkweed, Yellow: *Hieracium fendleri*

Hawthorn, Arnold: *Crataegus mollis*
Hawthorn, Bigfruit: *Crataegus macrosperma*
Hawthorn, Black: *Crataegus columbiana, C. douglasii*
Hawthorn, Cerro: *Crataegus erythropoda*
Hawthorn, Columbia: *Crataegus columbiana*
Hawthorn, Dotted: *Crataegus punctata*
Hawthorn, Douglas's: *Crataegus douglasii*
Hawthorn, Fireberry: *Crataegus chrysocarpa*
Hawthorn, Hawaii: *Osteomeles anthyllidifolia*
Hawthorn, Littlehip: *Crataegus spathulata*
Hawthorn, Pear: *Crataegus calpodendron*
Hawthorn, Quebec: *Crataegus submollis*
Hawthorn, Red: *Crataegus columbiana*
Hawthorn, River: *Crataegus rivularis*
Hawthorn, Waxyfruit: *Crataegus pruinosa*
Hay, Red: *Andropogon gerardii*
Hay, Swamp: *Carex atherodes, Carex obnupta*
Hay, Wild: *Bouteloua gracilis, Muhlenbergia rigens, Oryzopsis hymenoides, Sporobolus contractus*
Hazel: *Corylus americana, C. cornuta, C.* sp.
Headache Weed: *Clematis hirsutissima*
Heal All: *Prunella vulgaris*
Heart Leaf: *Hexastylis arifolia, H. virginica*
Hearts-a-bustin-with-love: *Euonymus americana*
Heart Seed: *Polygonum pensylvanicum*
Heath, Rose: *Chaetopappa ericoides*
Heather, Arctic White: *Cassiope tetragona*
Heather, Red Mountain: *Phyllodoce empetriformis*
Heather, Western Moss: *Cassiope mertensiana*
Heather, White Arctic Mountain: *Cassiope tetragona*
Heather, White Mountain: *Cassiope mertensiana*
Heathgoldenrod, Chaparral: *Ericameria brachylepis*
Heathgoldenrod, Cliff: *Ericameria cuneata*
Heathgoldenrod, Cooper's: *Ericameria cooperi*
Heathgoldenrod, Dwarf: *Ericameria nana*
Heathgoldenrod, Narrowleaf: *Ericameria linearifolia*
Heathgoldenrod, Palmer's: *Ericameria palmeri*
Heathgoldenrod, Parish's: *Ericameria parishii*
Heathgoldenrod, Rabbitbush: *Ericameria bloomeri*
Hedgehog: *Echinocereus fendleri*
Hedgehog Mushroom: *Dentinum repandum*
Hedgehog, Salmon Flowered: *Echinocereus polyacanthus*
Hedgenettle, California: *Stachys bullata*
Hedgenettle, Great: *Stachys ciliata*
Hedgenettle, Marsh: *Stachys palustris*
Hedgenettle, Rothrock's: *Stachys rothrockii*
Hedgenettle, Smooth: *Stachys tenuifolia*
Hedgenettle, Whitestem: *Stachys albens*
Hei: *Carica papaya*
Heliotrope, Wild: *Heliotropium curassavicum, Phacelia crenulata, P.* sp.
Hellebore, American False: *Veratrum viride*
Hellebore, American White: *Veratrum californicum*
Hellebore, Indian: *Veratrum viride*
Helleborine, Giant: *Epipactis gigantea*
Hemlock: *Picea glauca, Thuja occidentalis, T. canadensis, T. caroliniana, T. heterophylla*
Hemlock, Black: *Tsuga mertensiana*
Hemlock, Canada: *Tsuga canadensis*
Hemlock, Carolina: *Tsuga caroliniana*
Hemlock, Coast: *Tsuga heterophylla*
Hemlock, Crag: *Tsuga caroliniana*
Hemlock, Eastern: *Tsuga canadensis*
Hemlock, Ground: *Lycopodium clavatum, Taxus baccata, T. brevifolia, T. canadensis*
Hemlock, Mountain: *Tsuga mertensiana*
Hemlock, Poison: *Cicuta maculata, Conium maculatum*
Hemlock, Spotted Water: *Cicuta maculata*
Hemlock, Water: see Water Hemlock
Hemlock, Western: *Tsuga heterophylla*
Hemlockparsley, Chinese: *Conioselinum chinense*
Hemlockparsley, Pacific: *Conioselinum gmelinii*
Hemlockparsley, Rocky Mountain: *Conioselinum scopulorum*
Hemp: *Cannabis sativa*
Hemp, Indian: *Apocynum androsaemifolium, A. cannabinum, A.* sp.
Hemp, Mountain: *Apocynum cannabinum*
Hempvine (or Hempweed): *Mikania batatifolia*
Herb of the Cross: *Verbena officinalis*
Herb Sophia: *Descurainia sophia*
Hercules's Club: *Aralia spinosa, Zanthoxylum clava-herculis*
Heronbill: *Erodium cicutarium*
Hickory: *Carya alba, C. laciniosa, C. ovata, C. pallida, C.* sp.
Hickory, Bitternut: *Carya cordiformis*
Hickory, Mockernut: *Carya alba*
Hickory, Pignut: *Carya glabra*
Hickory, Sand: *Carya pallida*
Hickory, Shag Bark: *Carya ovata*
Hickory, Shellbark: *Carya ovata*
Hoarhound: *Marrubium vulgare*
Hoarhound, Water: *Lycopus americanus*; see also Water Horehound
Hoarhound, Wild: *Eupatorium pilosum*
Hoary Pea: *Tephrosia purpurea*
Hoarypea, Florida: *Tephrosia florida*
Hoarypea, Sprawling: *Tephrosia hispidula*
Hoary Puccoon: *Lithospermum canescens*
Ho-a-wa: *Pittosporum* sp.
Hobble Bush: *Viburnum lantanoides*
Hognut: *Amphicarpaea bracteata*
Hog Peanut: *Amphicarpaea bracteata*
Hogweed: *Amaranthus retroflexus*
Hogweed, Little: *Portulaca oleracea*
Ho-i: *Dioscorea bulbifera*
Ho-i-o: *Diplazium meyenianum*
Holdback, James's: *Caesalpinia jamesii*
Ho-le-i: *Ochrosia compta*
Hollowstomach: *Cryptantha* sp.
Holly, American: *Ilex opaca*
Holly, California: *Heteromeles arbutifolia*
Holly, English: *Ilex aquifolium*
Holly, Mountain: *Ilex aquifolium, Nemopanthus mucronatus*
Holly, Sea: *Eryngium alismifolium*
Holly, Western: *Mahonia aquifolium*
Hollycock, Wild: *Sidalcea malviflora*
Hollyhock: *Alcea rosea*
Hollyhock, Desert: *Sphaeralcea emoryi*
Honeydew: *Horkelia californica*
Honey Locust: *Gleditsia triacanthos*
Honey Mesquite: *Prosopis glandulosa*
Honeysuckle, American Fly: *Lonicera canadensis*
Honeysuckle, Arizona: *Lonicera arizonica*
Honeysuckle, Bearberry: *Lonicera involucrata*
Honeysuckle, Bush: *Diervilla lonicera*
Honeysuckle, Butters: *Lonicera dioica*
Honeysuckle, California: *Lonicera hispidula*
Honeysuckle, Chaparral: *Lonicera interrupta*
Honeysuckle, Coast: *Lonicera involucrata*
Honeysuckle, Dwarf Bush: *Diervilla lonicera*
Honeysuckle, Fly: *Lonicera canadensis, L. oblongifolia*
Honeysuckle, Glaucous: *Lonicera dioica*
Honeysuckle, Japanese: *Lonicera japonica*
Honeysuckle, Johnston's: *Lonicera subspicata*
Honeysuckle, Limber: *Lonicera dioica*
Honeysuckle, Orange: *Lonicera ciliosa*
Honeysuckle, Pink: *Lonicera hispidula*
Honeysuckle, Purpleflower: *Lonicera conjugialis*
Honeysuckle, Swamp: *Lonicera involucrata*
Honeysuckle, Swamp Fly: *Lonicera oblongifolia*
Honeysuckle, Twinberry: *Lonicera involucrata*
Honeysuckle, Twining: *Lonicera dioica*
Honeysuckle, Utah: *Lonicera utahensis*
Honohono: *Murdannia nudiflora*
Hop: *Humulus lupulus*
Hopbush: *Dodonaea viscosa*
Hop Hornbeam: *Ostrya virginiana*
Hops: *Humulus lupulus*
Hop Tree: *Ptelea trifoliata*
Hop Vine: *Humulus lupulus*
Horehound, American: *Marrubium vulgare*
Horehound, Water: *Lycopus americanus, L. asper, L.* sp.*, L. virginicus, Marrubium vulgare*
Horehound, White: *Marrubium vulgare*
Horkelia: *Ivesia gordonii*
Hornbeam: *Carpinus caroliniana*
Hornbean: *Carpinus caroliniana*
Horsebean: *Vicia faba*
Horsebrush: *Tetradymia canescens*
Horsebrush, Hairy: *Tetradymia comosa*
Horsebrush, Longspine: *Tetradymia axillaris*
Horsebrush, Mojave: *Tetradymia stenolepis*
Horseflyweed: *Baptisia tinctoria*
Horsehound: *Marrubium vulgare*
Horsemint: *Monarda didyma, M. pectinata, M. punctata, M.* sp.*, Monardella villosa*
Horsenettle, Carolina: *Solanum carolinense*
Horsenettle, Fendler's: *Solanum fendleri*
Horsepurslane: *Trianthema portulacastrum*
Horseradish: *Armoracia rusticana, Rorippa nasturtium-aquaticum*

Index of Common Names 907

Horse Tail: Equisetaceae, *Equisetum arvense, E. hyemale, E. laevigatum, E. palustre, E. pratense, E.* sp., *E. sylvaticum, E. telmateia*
Horse Tail, Branchless: *Equisetum hyemale, E. laevigatum, E.* sp.
Horse Tail, Field: *Equisetum arvense, E. hyemale*
Horse Tail, Giant: *Equisetum telmateia*
Horse Tail, Large: *Equisetum hyemale*
Horsetail, Marsh: *Equisetum palustre*
Horsetail, Meadow: *Equisetum pratense*
Horse Tail, River: *Equisetum laevigatum*
Horse Tail, Small: *Equisetum laevigatum*
Horsetail, Smooth: *Equisetum laevigatum*
Horse Tail, Wood: *Equisetum sylvaticum*
Horse Weed: *Conyza canadensis*
Hound's Tongue: *Cynoglossum grande, C. officinale, C. virginianum*
Huckleberry, Alaska: *Vaccinium ovalifolium*
Huckleberry, Bear: *Gaylussacia ursina*
Huckleberry, Black: *Gaylussacia baccata, Vaccinium ovatum*
Huckleberry, Black Mountain: *Vaccinium membranaceum*
Huckleberry, Blue: *Vaccinium membranaceum, V. ovalifolium V.* sp.
Huckleberry, Blueleaved: *Vaccinium deliciosum*
Huckleberry, California: *Vaccinium ovatum*
Huckleberry, Coast: *Vaccinium ovatum*
Huckleberry, Dwarf: *Vaccinium cespitosum*
Huckleberry, Early: *Gaylussacia baccata*
Huckleberry, Evergreen: *Vaccinium ovatum*
Huckleberry, Fool's: *Menziesia ferruginea*
Huckleberry, Mountain: *Vaccinium membranaceum*
Huckleberry, Oval Leaved: *Vaccinium ovalifolium*
Huckleberry, Red: *Vaccinium ovalifolium, V. parvifolium*
Huckleberry, Thin Leaved: *Vaccinium membranaceum*
Hummingbird: *Castilleja applegatei, C. parviflora*
Hummingbird Trumpet: *Epilobium canum*
Hyacinth, Pine: *Clematis baldwinii*
Hyacinth, Wild: *Dichelostemma pulchellum, Triteleia grandiflora*
Hyssop: *Hyssopus officinalis*
Hyssop, Blue Giant: *Agastache foeniculum*
Hyssop, Blue Water: *Bacopa caroliniana*
Hyssop, Bog: *Bacopa caroliniana*
Hyssop, Fragrant Giant: *Agastache foeniculum*
Hyssop, Giant: *Agastache foeniculum, A. nepetoides, A. pallidiflora, A. scrophulariifolia, A. urticifolia*
Hyssop, Lavender: *Agastache foeniculum*
Hyssop, Nettleleaf Giant: *Agastache urticifolia*
Hyssop, New Mexico Giant: *Agastache pallidiflora*
Hyssop, Purple Giant: *Agastache scrophulariifolia*
Hyssop, Water: *Bacopa caroliniana*
Hyssop, Yellow Giant: *Agastache nepetoides*
Icaco Coco Plum: *Chrysobalanus icaco*
Ice Cream, Indian: *Shepherdia canadensis*
'Ie'ie: *Freycinetia arborea*

Ihi-ai: *Portulaca oleracea*
Iliee: *Plumbago zeylanica*
Ilima: *Sida* sp.
Ili-oha Laau: *Conyza canadensis*
Incienso: *Encelia farinosa*
Indian Balm: *Pluchea* sp.
Indian Balsam: *Lomatium dissectum*
Indian Biscuitroot: *Lomatium piperi*
Indian Bread Root: *Pediomelum esculentum*
Indian Breadroot: *Pediomelum hypogaeum*
Indian Broom: *Baccharis sarothroides*
Indian Carrot: *Perideridia pringlei*
Indian Celery: *Angelica arguta, Lomatium ambiguum, L. nudicaule, L. triternatum*
Indian Cherry: *Prunus ilicifolia*
Indian Coffee: *Senna tora*
Indian Consumption Plant: *Lomatium nudicaule*
Indian Corn: *Zea mays*
Indian Cucumber: *Medeola virginiana*
Indian Cucumberroot: *Medeola virginiana*
Indian Cup: *Sarracenia purpurea*
Indian Cup Plant: *Silphium perfoliatum*
Indian Ginger: *Asarum canadense*
Indian Gum Plant: *Stephanomeria spinosa*
Indian Hellebore: *Veratrum viride*
Indian Hemp: *Apocynum androsaemifolium, A. cannabinum, A.* sp.
Indian Ice Cream: *Shepherdia canadensis*
Indian Lettuce: *Claytonia perfoliata, C. sibirica*
Indian Maize: *Zea mays*
Indian Milkvetch: *Astragalus australis*
Indian Millet: *Oryzopsis hymenoides*
Indian Mulberry: *Morinda citrifolia*
Indian Mustard: *Brassica juncea*
Indian Paint: *Castilleja parviflora, Lithospermum incisum*
Indian Paintbrush: see Paintbrush
Indian Parsley: see Parsley, Indian
Indian Parsnip: *Pteryxia terebinthina*
Indian Pepper: *Saururus cernuus*
Indian Physic: *Porteranthus stipulatus*
Indian Pink: *Spigelia marilandica*
Indian Pipe: *Monotropa uniflora*
Indian Plum: *Oemleria cerasiformis*
Indian Poke: *Veratrum viride*
Indian Posey: *Barbarea vulgaris*
Indian Posy: *Anaphalis margaritacea*
Indian Potato: *Apios americana, Hoffmannseggia glauca, Solanum tuberosum*
Indian Rhubarb: *Darmera peltata, Heracleum maximum*
Indian Rice: *Zizania aquatica, Zizania palustris*
Indian Rice Grass: *Oryzopsis hymenoides*
Indian Rushpea: *Hoffmannseggia glauca*
Indian Spikenard: *Aralia racemosa, Maianthemum racemosum*
Indian Spinach: *Monolepis nuttalliana*
Indian Tea: *Aster umbellatus, Ephedra* sp., *Ephedra viridis*
Indian Tea Tree: *Ledum groenlandicum*

Indian Tobacco: *Lobelia inflata, Nicotiana attenuata, N. quadrivalvis, N. rustica, N.* sp.
Indian Turnip: *Arisaema triphyllum, Lithospermum incisum, Pediomelum esculentum, P. hypogaeum, Psoralidium lanceolatum, P. tenuiflorum*
Indian Walnut: *Aleurites moluccana*
Indian Warrior: *Pedicularis densiflora*
Indian Wheat: *Plantago ovata, P. patagonica, P.* sp.
Indian Woodoats: *Chasmanthium latifolium*
Indigo, Blue Wild: *Baptisia australis*
Indigo, False: *Amorpha fruticosa, Baptisia alba, B.* sp., *B. tinctoria*
Indigo, Largeleaf Wild: *Baptisia alba*
Indigo, Longbract Wild: *Baptisia bracteata*
Indigo, Prairie: *Baptisia alba*
Indigo, Wild: *Baptisia* sp., *Baptisia tinctoria*
Indigobush, Desert: *Amorpha fruticosa*
Indigobush, Dwarf: *Amorpha nana*
Ink Berry, American: *Phytolacca americana*
Ink Weed: *Suaeda moquinii*
Insideout Flower, Golden: *Vancouveria chrysantha*
Insideout Flower, White: *Vancouveria hexandra*
Iodine Bush: *Allenrolfea occidentalis*
Ionactis: *Aster linariifolius*
Ipecac, American: *Euphorbia ipecacuanhae, Porteranthus stipulatus, P. trifoliatus*
Ipecacuanha: *Euphorbia ipecacuanhae*
Ipu Awaawa: *Cucurbita maxima*
Iron Skyrocket: *Ipomopsis laxiflora*
Ironweed: *Vernonia glauca, V. noveboracensis, V.* sp.
Ironweed, Drummond's: *Vernonia missurica*
Ironweed, Missouri: *Vernonia missurica*
Ironweed, New York: *Vernonia noveboracensis*
Ironwood: *Carpinus caroliniana, Forestiera pubescens, Holodiscus discolor, Olneya tesota, Ostrya virginiana*
Islay: *Prunus ilicifolia*
Ivy, Ground: *Glechoma hederacea*
Ivy, Poison: *Toxicodendron pubescens, T. radicans, T. rydbergii*
Iwaiwa: *Asplenium pseudofalcatum*
Jack in the Pulpit: *Arisaema triphyllum*
Jacob's Ladder: *Smilax herbacea*
Jacobsladder, Elegant: *Polemonium elegans*
Jamestown Weed: *Datura wrightii*
Jasmine, Rock: see Rockjasmine
Jasmine, Yellow: *Gelsemium sempervirens*
Jerusalem Thorn: *Parkinsonia aculeata*
Jewelweed: *Impatiens capensis, I. pallida*
Jewelweed, Spotted: *Impatiens capensis*
Jilla Flower: *Cosmos* sp.
Jilly Flower: *Ipomopsis aggregata*
Jimmyweed: *Isocoma pluriflora*
Jimpson Weed: *Datura wrightii*
Jimson Weed: *Datura discolor, D. ferox, D.* sp., *D. stramonium, D. wrightii*
Job's Tears: *Coix lacryma-jobi*
Joe Pye Weed: *Eupatorium maculatum, E. purpureum*
Jointfir, Arizona: *Ephedra fasciculata*

Jointfir, California: *Ephedra californica*
Jointfir, Longleaf: *Ephedra trifurca*
Jointfir, Nevada: *Ephedra nevadensis*
Jointfir, Torrey's: *Ephedra torreyana*
Joint Grass, Blue: *Elytrigia repens*
Jojoba: *Simmondsia chinensis*
Joshua Tree: *Yucca brevifolia*
Juba's Bush: *Iresine diffusa*
Jug, Little Brown: *Hexastylis arifolia*
Jujube, Parry's: *Ziziphus parryi*
Jumpingbean, Mexican: *Sapium biloculare*
Jumpseed: *Polygonum virginianum*
June Berry: *Amelanchier alnifolia, A. arborea, A. canadensis, A. laevis, A. sanguinea, A.* sp., *A. stolonifera, Symphoricarpos occidentalis*
Juniper: *Draba helleriana, D. reptans, Juniperus* spp.
Juniper, Alligator: *Juniperus deppeana*
Juniper, Alpine: *Juniperus communis*
Juniper, American: *Juniperus californica*
Juniper, California: *Juniperus californica*
Juniper, Colorado: *Juniperus scopulorum*
Juniper, Creeping: *Juniperus communis, J. horizontalis*
Juniper, Dwarf: *Juniperus communis*
Juniper, Ground: *Juniperus communis*
Juniper, Low: *Juniperus communis*
Juniper, Mountain: *Juniperus communis*
Juniper, One Seed: *Cupressus* sp., *Juniperus monosperma*
Juniper, Pinchot's: *Juniperus pinchotii*
Juniper, Prickly: *Juniperus communis*
Juniper, Rocky Mountain: *Juniperus scopulorum*
Juniper, Shrub: *Juniperus communis*
Juniper, Sierra: *Juniperus occidentalis*
Juniper, Trailing: *Juniperus chinense*
Juniper, Utah: *Juniperus osteosperma*
Juniper, Virginia: *Juniperus virginiana*
Juniper, Western: *Juniperus occidentalis*
Jusum: *Chenopodium ambrosioides*
Kaa: *Cyperus* sp.
Ka-e-e: *Mucuna gigantea*
Kalakalaioa: *Caesalpinia bonduc*
Kalo: *Colocasia esculenta*
Ka-mano-mano: *Cenchrus calyculatus*
Kamolelaulii: *Polygonum densiflorum*
Kamolelaunui: *Ludwigia bonariensis*
Kanawao: *Broussaisia arguta*
Kanawao Ulaula: *Broussaisia arguta*
Kanawao-keokeo: *Cyrtandra* sp.
Kau-no-a: *Cassytha filiformis*
Kava: *Piper methysticum*
Kawa'u: *Styphelia tameiameiae*
Kelp: *Nereocystis luetkeana*
Kelp, Bladderwrack: *Fucus gardneri*
Kelp, Boa: *Egregia menziesii*
Kelp, Bull: *Nereocystis luetkeana*
Kelp, Giant: *Macrocystis luetkeana, M. integrifolia, Nereocystis luetkeana*
Kelp, Short: *Alaria marginata, Costaria costata, Lessoniopsis littoralis, Postelsia palmaeformis*

Key Lime: *Citrus aurantifolia*
Ki: *Cordyline fruticosa*
Kihapai: *Catharanthus roseus*
Kikania: *Desmodium sandwicense*
King of the Meadow: *Thalictrum pubescens*
King Solomon's Seal: *Polygonatum biflorum*
Kinnickinnick: *Arctostaphylos uva-ursi*
Kinnikinik: *Arctostaphylos uva-ursi*
Kinnikinnic: *Arctostaphylos uva-ursi*
Kinnikinnick: *Arctostaphylos uva-ursi, Cornus amomum, C. sericea*
Kinnikinnick, Small: *Linnaea borealis*
Kinnikinnik: *Arctostaphylos uva-ursi, Cornus amomum*
Knotweed, Bushy: *Polygonum ramosissimum*
Knotweed, Curlytop: *Polygonum lapathifolium*
Knotweed, Denseflower: *Polygonum densiflorum*
Knotweed, Douglas's: *Polygonum douglasii*
Knotweed, Japanese: *Polygonum cuspidatum*
Knotweed, Johnston's: *Polygonum douglasii*
Knotweed, Marshpepper: *Polygonum hydropiper*
Knotweed, Ovalleaf: *Polygonum arenastrum*
Knotweed, Prostrate: *Polygonum aviculare*
Knotweed, Silversheath: *Polygonum argyrocoleon*
Knotweed, Water: *Polygonum amphibium*
Koa: *Acacia koa*
Koali: *Ipomoea cairica*
Kokio: *Hibiscus* sp.
Kokolau: *Bidens amplectens*
Koli Ulaula: *Ricinus communis*
Kookoolau: *Bidens* sp.
Kouse: *Lomatium cous*
Kowali Awa: *Ipomoea indica*
Kowali Pehu: *Merremia dissecta*
Kuawa: *Psidium guajava*
Kukaepuaa: *Digitaria setigera*
Kukui: *Aleurites moluccana*
Kumquat: *Fortunella* sp.
Kunti: *Zamia pumila*
Kupalii: *Peperomia* sp.
Labrador Tea: *Ledum ×columbianum, L. groenlandicum, L. palustre*
Labradortea, Bog: *Ledum groenlandicum*
Labradortea, Marsh: *Ledum palustre*
Labradortea, Western: *Ledum glandulosum*
Ladiestresses, Hooded: *Spiranthes romanzoffiana*
Ladiestresses, Shining: *Spiranthes lucida*
Ladies' Tresses, Slender: *Spiranthes lacera*
Lady Fern: *Athyrium filix-femina*
Lady's Slipper: *Cypripedium, C.* spp., *Platanthera orbiculata*
Lady's Slipper, Mountain: *Cypripedium montanum*
Lady's Slipper, Pink: *Cypripedium acaule*
Lady's Slipper, Ram's Head: *Cypripedium arietinum*
Lady's Slipper, Small Yellow: *Cypripedium parviflorum*
Lady's Slipper, Stemless: *Cypripedium acaule*
Lady's Slipper, Yellow: *Cypripedium parviflorum, C. pubescens*
Lady's Thumb: *Polygonum persicaria*

Lambkill: *Kalmia angustifolia*
Lamb's Quarter: *Amaranthus retroflexus, Chenopodium album, C. incanum, C. leptophyllum, C. murale, C.* sp.
Lamb's Quarter, Desert: *Chenopodium incanum*
Lamb's Quarter, Narrow Leaved: *Chenopodium leptophyllum*
Lance Wood: *Ocotea coriacea*
Larb: *Arctostaphylos uva-ursi*
Larch: *Larix laricina, Larix occidentalis*
Larkspur: *Consolida ajacis, Delphinium* spp.
Larkspur, Blue: *Delphinium hesperium*
Larkspur, Carolina: *Delphinium carolinianum*
Larkspur, Carrotleaf: *Delphinium tenuisectum*
Larkspur, Clark Valley: *Delphinium geraniifolium*
Larkspur, Coastal: *Delphinium decorum*
Larkspur, Dwarf: *Delphinium tricorne*
Larkspur, Foothill: *Delphinium hesperium*
Larkspur, Little: *Delphinium bicolor*
Larkspur, Menzies's: *Delphinium menziesii*
Larkspur, Mountain: *Delphinium scaposum*
Larkspur, Nelson: *Delphinium menziesii*
Larkspur, Nuttall's: *Delphineum nuttallianum*
Larkspur, Prairie: *Delphinium carolinianum*
Larkspur, Red: *Delphinium nudicaule*
Larkspur, Rocky Mountain: *Delphinium scopulorum*
Larkspur, San Bernardino: *Delphinium parryi*
Laurel: *Magnolia virginiana, Umbellularia* sp.
Laurel, Alpine: *Kalmia microphylla*
Laurel, American: *Kalmia polifolia*
Laurel, Bog: *Kalmia polifolia*
Laurel, California: *Umbellularia californica, U.* sp.
Laurel, Great: *Rhododendron maximum*
Laurel, Mountain: *Broussonetia papyrifera, Kalmia latifolia, Umbellularia californica*
Laurel, Sheep: *Kalmia angustifolia*
Laurel, Swamp: *Kalmia polifolia*
Lavender, Desert: *Hyptis emoryi*
Laver, Red: *Porphyra* sp.
Lavender, Sea: see Sealavender
Lawndaisy: *Bellis perennis*
Lead Plant: *Amorpha canescens*
Leadwort: *Plumbago zeylanica*
Leafcup, Hairy: *Smallanthus uvedalia*
Leaf Cup, Small Flowered: *Polymnia canadensis*
Leafcup, Whiteflower: *Polymnia canadensis*
Leaf Cup, Yellow: *Smallanthus uvedalia*
Leaftail: *Pericome caudata*
Leatherfern: *Acrostichum danaeifolium*
Leather Flower: *Clematis bigelovii, Clematis viorna*
Leather Leaf: *Chamaedaphne calyculata*
Leather Leaf Polypody: *Polypodium scouleri*
Leather Root: *Hoita macrostachya*
Leatherweed: *Croton pottsii*
Leatherwood: *Dirca palustris*
Lechuguilla: *Agave lechuguilla*
Leek, House: *Sempervivum* sp., *S. tectorum*
Leek, Wild: *Allium tricoccum*
Lemita: *Rhus trilobata*

Index of Common Names 909

Lemon: *Citrus limon*
Lemonade Berry: *Rhus integrifolia*
Lenscale: *Atriplex lentiformis*
Lentil: *Lens culinaris*
Leopardbane: *Arnica acaulis*
Lettuce, Biannual: *Lactuca ludoviciana*
Lettuce, Bitter: *Lactuca virosa*
Lettuce, Blue: *Lactuca biennis, L. tatarica*
Lettuce, Branch: *Saxifraga micranthidifolia, S. pensylvanica*
Lettuce, Canada: *Lactuca canadensis*
Lettuce, Canker: *Pyrola elliptica*
Lettuce, Chalk: *Dudleya pulverulenta*
Lettuce, Garden: *Lactuca sativa*
Lettuce, Indian: *Claytonia perfoliata, C. sibirica*
Lettuce, Miner's: *Claytonia perfoliata, C. sibirica, C. spathulata*
Lettuce, Mountain: *Ligusticum canadense, Saxifraga micranthidifolia*
Lettuce, Prickly: *Lactuca canadensis, L. sativa, L. serriola*
Lettuce, Purple: *Lactuca tatarica*
Lettuce, Rock: *Dudleya lanceolata*
Lettuce, Rough White: *Prenanthes aspera*
Lettuce, Sea: *Ulva lactuca*
Lettuce, Squaw: *Claytonia cordifolia, C. parviflora, C. perfoliata*
Lettuce, Tall: *Lactuca biennis, Prenanthes altissima*
Lettuce, Tall Blue: *Lactuca biennis*
Lettuce, White: *Prenanthes alba*
Lettuce, Wild: *Claytonia perfoliata, Lactuca canadensis, L. tatarica, L. virosa*
Lettuce, Wire: see Wirelettuce
Lichen: Alectoriaceae, Cladoniaceae, Lecanoraceae, Parmeliaceae, *Parmelia physodes*, Peltigeraceae, Stereocaulaceae, Stictaceae, Teloschistaceae, Usneaceae
Lichen, Curled Shield: *Cetraria cucullata*
Lichen, Dogtooth: *Peltigera canina*
Lichen, Flat: *Parmelia saxatilis, Peltigera aphthosa, Stereocaulon paschale*
Lichen, Old Man's Beard: *Usnea longissima*
Lichen, Shield: *Cetraria crispa*
Lichen, Tree: *Alectoria* sp., *Sticta glomulifera, Usnea* sp.
Lichen, Veined: *Peltigera* sp.
Lichen, Wolf: *Letharia vulpina*
Lichen, Yellow: *Evernia* sp.
Licorice: *Glycyrrhiza glabra, G. lepidota, Hedysarum alpinum*; see also Liquorice
Licorice Root: *Glycyrrhiza lepidota, Hedysarum alpinum*; see also Liquorice Root
Licoriceroot, California: *Ligusticum californicum*
Licoriceroot, Canadian: *Ligusticum canadense*
Licoriceroot, Canby's: *Ligusticum canbyi*
Licoriceroot, Celeryleaf: *Ligusticum apiifolium*
Licoriceroot, Fernleaf: *Ligusticum filicinum*
Licoriceroot, Gray's: *Ligusticum grayi*
Licoriceroot, Hultén's: *Ligusticum scothicum*
Licoriceroot, Porter's: *Ligusticum porteri*

Licoriceroot, Scottish: *Ligusticum scothicum*
Life Everlasting, Fragrant: *Gnaphalium obtusifolium*
Lilac: *Ceanothus* sp., *Syringa vulgaris*
Lilac, California: *Ceanothus oliganthus, C.* sp.
Lilac, Coast: *Ceanothus thyrsiflorus*
Lilac, Mountain: *Ceanothus cuneatus*
Lilac, Wild: *Ceanothus leucodermis*
Lily: *Camassia leichtlinii, Lilium canadense, Sagittaria latifolia*
Lily, Arrowhead: *Sagittaria latifolia*
Lily, Avalanche: *Erythronium grandiflorum*
Lily, Canada: *Lilium canadense*
Lily, Chocolate: *Fritillaria lanceolata*
Lily, Clinton's: *Clintonia borealis*
Lily, Cluster: see Cluster Lily
Lily, Columbian: *Lilium columbianum*
Lily, Corn: *Clintonia borealis, Veratrum californicum*
Lily, Cow: *Nuphar lutea*
Lily, Crag: *Echeandia flavescens*
Lily, Desert: *Hesperocallis undulata*
Lily, Dogtooth: *Erythronium grandiflorum*
Lily, Easter: *Erythronium oregonum, E. revolutum*
Lily, Elk: *Frasera speciosa*
Lily, Eureka: *Lilium occidentale*
Lily, False: *Maianthemum canadense*
Lily, Fawn: see Fawnlily
Lily, Glacier: *Erythronium grandiflorum*
Lily, Green Banded Mariposa: *Calochortus macrocarpus*
Lily, Habenaria: *Piperia elegans, P. unalascensis*
Lily, Kamtchatka: *Fritillaria camschatcensis*
Lily, Leopard: *Fritillaria atropurpurea*
Lily, Lonely: *Eremocrinum albomarginatum*
Lily, Longray Triplet: *Triteleia peduncularis*
Lily, Mariposa: see Mariposa; and Mariposa Lily
Lily, Mignonette Scented: *Triteleia grandiflora*
Lily, Mount Diablo Globe: *Calochortus pulchellus*
Lily, Orange Red: *Lilium philadelphicum*
Lily, Orangecup: *Lilium philadelphicum*
Lily, Panther: *Lilium columbianum*
Lily, Philadelphia: *Lilium philadelphicum*
Lily, Pine: *Xerophyllum tenax*
Lily, Redwood: *Lilium rubescens*
Lily, Rush: *Hastingsia alba*
Lily, Sage: *Leucocrinum montanum*
Lily, Sand: *Mentzelia albicaulis*
Lily, Sego: *Calochortus gunnisonii, C. leichtlinii, C. macrocarpus, C. nuttallii*
Lily, Short: *Calochortus amabilis*
Lily, Sierran Tiger: *Lilium parvum*
Lily, Small Tiger: *Lilium parvum*
Lily, Snake: see Snakelily
Lily, Spring: *Erythronium mesochoreum*
Lily, Tiger: *Fritillaria atropurpurea, Lilium columbianum, L. pardalinum, L. parvum*
Lily, Torrey's Crag: *Echeandia flavescens*
Lily, Triplet: *Triteleia peduncularis*
Lily, Water: see Water Lily

Lily, Western Wood: *Lilium philadelphicum*
Lily, White Rush: *Hastingsia alba*
Lily, Wild: *Lilium philadelphicum*
Lily, Wild Yellow: *Lilium canadense*
Lily, Wood: *Lilium columbianum, Lilium philadelphicum*
Lily, Yellow Avalanche: *Erythronium grandiflorum*
Lily, Zephyr: *Zephyranthes* sp.
Lily of the Valley, Wild: *Maianthemum canadense, M. dilatatum*
Lime: *Citrus aurantifolia*
Lime, Key: *Citrus aurantifolia*
Lime, Wild: *Zanthoxylum fagara*
Linden: *Tilia americana*
Lingon: *Vaccinium vitis-idaea*
Lions Foot: *Prenanthes serpentaria*
Lipfern, Beaded: *Cheilanthes wootonii*
Lipfern, Coville's: *Cheilanthes covillei*
Lipfern, Fendler's: *Cheilanthes fendleri*
Lippa, Wright: *Aloysia wrightii*
Liquoria: *Glycyrrhiza lepidota*
Liquorice: *Glycyrrhiza glabra*; see also Licorice
Liquorice Root: *Hedysarum alpinum, H. boreale*; see also Licorice Root
Liquorice Root, Wild: *Glycyrrhiza lepidota*
Littlebrownjug: *Hexastylis arifolia*
Live Forever: *Sedum telephioides, S. telephium*
Liveforever, Chalk: *Dudleya pulverulenta*
Liveforever, Lanceleaf: *Dudleya lanceolata*
Liverleaf: *Hepatica nobilis*
Liverwort: *Bazzania trilobata*, Conocephalaceae, Jungermanniaceae
Liverwort, Cone Headed: *Conocephalum conicum*
Lizard's Tail: *Anemopsis californica, Saururus cernuus*
Loco, Haresfoot Point: *Oxytropis lagopus*
Locoweed: *Astragalus allochrous, A. calycosus, A. ceramicus, A. cyaneus, A. kentrophyta, A. miser, A. pachypus, A. purshii, A.* sp., *Oxytropis campestris, O. lambertii, O. monticola, O.* sp.
Locoweed, American: *Astragalus mollissimus*
Locoweed, Lambert: *Oxytropis lambertii*
Locoweed, Matthew: *Astragalus mollissimus*
Locoweed, Purple: *Oxytropis lagopus, O. lambertii*
Locoweed, Yellowflower: *Oxytropis monticola*
Locust, Black: *Robinia pseudoacacia, R.* sp.
Locust, Bristly: *Robinia hispida*
Locust, Eastern: *Robinia pseudoacacia*
Locust, Honey: *Gleditsia triacanthos*
Locust, New Mexican: *Robinia neomexicana*
London Rocket: *Sisymbrium irio*
Long Trumpet: *Eriogonum nudum*
Loosestrife: *Lysimachia quadrifolia, Lythrum alatum*
Loosestrife, California: *Lythrum californicum*
Loosestrife, Purple: *Lythrum salicaria*
Loosestrife, Spiked: *Lythrum salicaria*
Loosestrife, Tufted: *Lysimachia thyrsiflora*
Loosestrife, Whorled: *Lysimachia quadrifolia*
Lopseed: *Phryma leptostachya*

Lote Bush: *Ziziphus obtusifolia*
Lotus, Yellow: *Nelumbo lutea*
Lousewort, Attol: *Pedicularis attollens*
Lousewort, Bracted: *Pedicularis bracteosa*
Lousewort, Canadian: *Pedicularis canadensis*
Lousewort, Dwarf: *Pedicularis centranthera*
Lousewort, Elephanthead: *Pedicularis groenlandica*
Lousewort, Giant: *Pedicularis procera*
Lousewort, Sickletop: *Pedicularis racemosa*
Lousewort, Swamp: *Pedicularis lanceolata*
Lousewort, Wooly: *Pedicularis lanata*
Lovage: *Ligusticum apiifolium, L. canbyi, L. filicinum, L. scothicum*
Love Grass: *Eragrostis mexicana, E. secundiflora*
Love-lies-bleeding: *Amaranthus caudatus*
Love Seed: *Lomatium foeniculaceum*
Lungwort: *Mertensia maritima*
Lupine, Arctic: *Lupinus arcticus*
Lupine, Beach: *Lupinus affinis, L. albifrons, L. arboreus, L. densiflorus, L. luteolus, L. nanus, L. polyphyllus, L. succulentus, L. versicolor, Trifolium wormskioldii*
Lupine, Bigleaf: *Lupinus polyphyllus*
Lupine, Blue: *Lupinus littoralis, L. nootkatensis, L. polyphyllus, L. sericeus, L.* sp.
Lupine, Broad Leaved: *Lupinus latifolius, L. polyphyllus*
Lupine, Bush: *Lupinus arboreus*
Lupine, Chinook: *Lupinus littoralis*
Lupine, Dwarf: *Lupinus kingii*
Lupine, False: *Thermopsis macrophylla*
Lupine, Fleshy: *Lupinus affinis*
Lupine, Hollowleaf Annual: *Lupinus succulentus*
Lupine, Intermountain: *Lupinus pusillus*
Lupine, Kellogg's Spurred: *Lupinus caudatus*
Lupine, King's: *Lupinus kingii*
Lupine, Lyall: *Lupinus lyallii*
Lupine, Manycolored: *Lupinus versicolor*
Lupine, Mountain: *Lupinus lyallii*
Lupine, Nootka: *Lupinus nootkatensis*
Lupine, Pale Yellow: *Lupinus luteolus*
Lupine, Riverbank: *Lupinus rivularis*
Lupine, Rose: *Lupinus densiflorus*
Lupine, Rusty: *Lupinus pusillus*
Lupine, Seashore: *Lupinus littoralis*
Lupine, Short Stem: *Lupinus brevicaulis*
Lupine, Silky: *Lupinus sericeus*
Lupine, Silver: *Lupinus albifrons*
Lupine, Silvery: *Lupinus argenteus*
Lupine, Sky: *Lupinus nanus*
Lupine, Sulphur: *Lupinus sulphureus*
Lupine, Sundial: *Lupinus perennis*
Lupine, Whitewhorl: *Lupinus densiflorus*
Lupine, Wyeth's: *Lupinus wyethii*
Lupine, Yellow: *Lupinus luteolus*
Lupine, Yellow Flowered: *Lupinus arboreus*
Lyreleaf Greeneyes: *Berlandiera lyrata*
Madder, Blue Field: *Sherardia arvensis*
Madrona (or Madrone, or Madroño): *Arbutus menziesii*

Madwoman's Milk: *Euphorbia helioscopia*
Mahogany, American: *Cercocarpus montanus*
Mahogany, Birch Leaf: *Cercocarpus montanus*
Mahogany, Mountain: *Cercocarpus ledifolius, C. montanus, C.* sp.
Mahogany, True Mountain: *Cercocarpus montanus*
Maia: *Musa×paradisiaca*
Maiden Fern: *Asplenium trichomanes*
Maidenhair, Aleutian: *Adiantum aleuticum*
Maidenhair, California: *Adiantum jordanii*
Maidenhair, Northern: *Adiantum pedatum*
Maidenhair Fern: *Adiantum aleuticum, A. capillus-veneris, A. jordanii, A. pedatum, A.* sp.
Maile: *Alyxia oliviformis*
Maile-kaluhea: *Coprosma* sp.
Maile-kuahiwi: *Alyxia oliviformis*
Maize: *Zea mays*
Ma-ko-u: *Peucedanum sandwicense*
Mallow: *Malva neglecta, M. parviflora, M.* sp., *Modiola caroliniana, Sphaeralcea ambigua*
Mallow, Alkali: *Malvella leprosa*
Mallow, American: *Malva neglecta*
Mallow, Bull: *Malva nicaeensis*
Mallow, Cheeses: *Malva neglecta*
Mallow, Common: *Malva neglecta*
Mallow, Copper: *Sphaeralcea angustifolia*
Mallow, Creeping Rock: *Malva* sp.
Mallow, Desert: *Sphaeralcea ambigua*
Mallow, False: *Sphaeralcea coccinea*
Mallow, Glade: *Napaea dioica*
Mallow, Globe: see Globemallow
Mallow, Marsh: *Malva moschata*
Mallow, Musk: *Malva moschata*
Mallow, Purple: *Callirhoe involucrata*
Mallow, Purple Poppy: *Callirhoe involucrata*
Mallow, Red Globe: *Sphaeralcea coccinea*
Mallow, Rose: *Hibiscus moscheutos*
Mallow, Scarlet: *Sphaeralcea coccinea*
Mallow, Star: *Eremalche exilis*
Mallow, White: *Eremalche exilis*
Mallow Ninebark: *Physocarpus malvaceus*
Mamaki: *Pipturus* sp.
Manawanawa: *Vitex trifolia*
Mandrake: *Podophyllum peltatum*
Manena: *Melicope cinerea*
Mango: *Mangifera indica*
Mangrove, American: *Rhizophora mangle*
Mangrove, Button: *Conocarpus erectus*
Mangrove, Red: *Rhizophora mangle*
Manioc: *Manihot esculenta*
Manna Grass: *Glyceria obtusa, Puccinellia distans*
Mannagrass, Atlantic: *Glyceria obtusa*
Mannagrass, Water: *Glyceria fluitans*
Man of the Earth: *Ipomoea pandurata*
Man Root: *Ipomoea pandurata, Marah fabaceus*
Manroot, California: *Marah fabaceus*
Manroot, Coastal: *Marah oreganus*
Manroot, Cucamonga: *Marah macrocarpus*
Man Root, Hill: *Marah oreganus*
Man Root, Old: *Marah oreganus*

Manroot, Sierran: *Marah horridus*
Manzanita, Arizona: *Arctostaphylos* sp.
Manzanita, Bigberry: *Arctostaphylos glauca*
Manzanita, Blue: *Arctostaphylos viscida*
Manzanita, Del Norte: *Arctostaphylos×cinerea*
Manzanita, Eastwood: *Arctostaphylos glandulosa*
Manzanita, Great: *Arctostaphylos glauca*
Manzanita, Green: *Arctostaphylos patula*
Manzanita, Greenleaf: *Arctostaphylos patula*
Manzanita, Hairy: *Arctostaphylos columbiana*
Manzanita, Hoary: *Arctostaphylos canescens*
Manzanita, Large: *Arctostaphylos tomentosa*
Manzanita, Mariposa: *Arctostaphylos viscida*
Manzanita, Parry: *Arctostaphylos manzanita*
Manzanita, Pine Mat: *Arctostaphylos nevadensis*
Manzanita, Pointleaf: *Arctostaphylos pungens*
Manzanita, Pringle: *Arctostaphylos pringlei*
Manzanita, Sandmat: *Arctostaphylos pumila*
Manzanita, Small: *Arctostaphylos pumila*
Manzanita, Sticky Whiteleaf: *Arctostaphylos viscida*
Manzanita, Whiteleaf: *Arctostaphylos manzanita*
Manzanita, Woollyleaf: *Arctostaphylos tomentosa*
Ma-o: *Abutilon incanum*
Maple, Big Leaf: *Acer macrophyllum*
Maple, Black Sugar: *Acer nigrum*
Maple, Box Elder: *Acer negundo*
Maple, Broad Leaf: *Acer macrophyllum*
Maple, Douglas: *Acer glabrum*
Maple, Drummond's: *Acer rubrum*
Maple, Dwarf: *Acer glabrum*
Maple, Green: *Acer pensylvanicum*
Maple, Hard: *Acer saccharinum, A. saccharum*
Maple, Manitoba: *Acer negundo*
Maple, Mountain: *Acer spicatum*
Maple, New Mexico: *Acer glabrum*
Maple, Red: *Acer rubrum*
Maple, Rock: *Acer saccharum*
Maple, Rocky Mountain: *Acer glabrum*
Maple, Shrub: *Acer glabrum*
Maple, Silver: *Acer saccharinum*
Maple, Soft: *Acer rubrum, A. saccharinum, A. saccharum*
Maple, Striped: *Acer pensylvanicum*
Maple, Sugar: *Acer saccharinum, A. saccharum*
Maple, Vine: *Acer circinatum*
Maple, White: *Acer alba*
Marbleseed: *Onosmodium molle*
Mare's Tail: *Equisetum* sp., *Hippuris tetraphylla, H. vulgaris*
Mariana, Golden: *Chrysopsis mariana*
Marigold, Aztec: *Tagetes erecta*
Marigold, Bur: *Bidens laevis*
Marigold, Desert: *Baileya multiradiata*
Marigold, Fetid: *Dyssodia papposa, Pectis angustifolia, P. papposa*
Marigold, Fig: *Carpobrotus aequilateralus*
Marigold, Licorice: *Tagetes micrantha*
Marigold, Marsh: *Caltha leptosepala, C. palustris, Ranunculus lapponicus*
Marijuana: *Cannabis sativa*

Index of Common Names

Mariola: *Parthenium incanum*
Mariposa: *Calochortus gunnisonii*
Mariposa, Golden: *Calochortus aureus*
Mariposa, Nuttall: *Calochortus nuttallii*
Mariposa, Sagebrush: *Calochortus macrocarpus*
Mariposa, Smoky: *Calochortus leichtlinii*
Mariposa, Yellow: *Calochortus aureus, C. luteus*
Mariposa Lily: *Calochortus aureus, C. catalinae, C. concolor, C. flexuosus, C. gunnisonii, C. macrocarpus, C. nuttallii, C. palmeri, C.* sp., *C. venustus, C. vestae*
Mariposa Lily, Butterfly: *Calochortus venustus*
Mariposa Lily, Coast Range: *Calochortus vestae*
Mariposa Lily, Goldenbowl: *Calochortus concolor*
Mariposa Lily, Gunnison's: *Calochortus gunnisonii*
Mariposa Lily, Palmer's: *Calochortus palmeri*
Mariposa Lily, Santa Catalina: *Calochortus catalinae*
Mariposa Lily, White: *Calochortus venustus*
Mariposa Lily, Winding: *Calochortus flexuosus*
Mariposa Lily, Yellow: *Calochortus luteus*
Marlberry: *Ardisia escallonoides*
Marshlocks: *Comarum palustre*
Marshmallow, Wild: *Hibiscus moscheutos*
Marshmarigold: see Marigold, Marsh
Marshpennywort: *Hydrocotyle umbellata*
Mask Flower: *Mimulus tilingii*
Mastic, False: *Sideroxylon foetidissimum*
Matchbrush: *Gutierrezia sarothrae*
Matchweed: *Gutierrezia microcephala, G. sarothrae*
Matchweed, San Joaquin: *Gutierrezia californica*
Matchwood: *Gutierrezia sarothrae*
Matrimony Vine: *Lycium pallidum*
Matted Crinklemat: *Tiquilia latior*
Mayflower: *Epigaea repens*
Mayflower, Canada: *Maianthemum canadense*
Maypop: *Passiflora incarnata*
Maystar: *Trientalis borealis*
May Weed: *Anthemis cotula*
Mayweed, Disc: *Matricaria discoidea*
Meadow Beauty: *Rhexia virginica*
Meadowparsnip: *Thaspium barbinode*
Meadow Rue, Early: *Thalictrum dioicum, T. thalictroides*
Meadow Rue, Fall: *Thalictrum pubescens*
Meadow Rue, Fendler: *Thalictrum fendleri*
Meadowrue, Fewflower: *Thalictrum sparsiflorum*
Meadow Rue, Flat Fruited: *Thalictrum sparsiflorum*
Meadow Rue, Purple: *Thalictrum dasycarpum*
Meadow Rue, Sierra: *Thalictrum polycarpum*
Meadow Rue, Western: *Thalictrum occidentale*
Meadow Salsify: *Tragopogon pratensis*
Meadow Sweet: *Petrophyton caespitosum, Spiraea salicifolia*
Meadow Sweet, Broad Leaved: *Spiraea alba*
Meadow Sweet, Narrow Leaved: *Spiraea alba*
Meadow Sweet, White: *Spiraea alba*
Meadow Sweet, Willow Leaved: *Spiraea salicifolia*
Melic: *Melica imperfecta*
Melon, Water: *Citrullus lanatus*
Melon, Wild: *Cucurbita foetidissima*

Mercury: *Acalypha virginica*
Mesamint: *Pogogyne douglasii*
Mescal: *Agastache pallidiflora, Agave americana, A. decipiens, A. deserti, A. palmeri, A. parryi, A. schottii, A.* sp., *A. utahensis, Echinocactus williamsii, Nolina bigelovii, Sophora secundiflora, Yucca whipplei*
Mescal Agave: *Agave* sp.
Mescal Bean: *Sophora secundiflora*
Mescalero Currant: *Ribes mescalerium*
Meskwaki Potato: *Solanum tuberosum*
Mesquite: *Prosopis chilensis, P. glandulosa, P.* sp., *P. velutina*
Mesquite, Cat's Claw: *Acacia greggii*
Mesquite, False: *Calliandra humilis*
Mesquite, Honey: *Prosopis glandulosa*
Mesquite, Screw (or Screwbean): *Prosopis pubescens*
Mesquite, Velvet: *Prosopis velutina*
Mesquite, Vine: *Panicum obtusum*
Mesquite Grass: *Bouteloua curtipendula, Muhlenbergia wrightii*
Mexican Frijole: *Phaseolus vulgaris*
Mexican Jumpingbean: *Sapium biloculare*
Mexican Squawroot: *Conopholis alpina*
Mexican Yellowshow: *Amoreuxia palmatifida*
Mezeron: *Daphne mezereum*
Miami Mist: *Phacelia purshii*
Mile a Minute Vine: *Ipomoea cairica*
Milfoil: *Achillea millefolium*
Milfoil, Water: see Watermilfoil
Milkmaids: *Cardamine californica*
Milkpea: *Galactia volubilis*
Milkvetch, American: *Astragalus americanus*
Milkvetch, Canadian: *Astragalus canadensis*
Milkvetch, Crescent: *Astragalus amphioxys*
Milkvetch, Cyanic: *Astragalus cyaneus*
Milkvetch, Giant: *Astragalus giganteus*
Milkvetch, Groundcover: *Astragalus humistratus*
Milkvetch, Groundplum: *Astragalus crassicarpus*
Milkvetch, Halfmoon: *Astragalus allochrous*
Milkvetch, Hayden's: *Astragalus bisulcatus*
Milkvetch, Indian: *Astragalus australis*
Milkvetch, Painted: *Astragalus ceramicus*
Milkvetch, Patterson's: *Astragalus pattersonii*
Milkvetch, Polar: *Astragalus polaris*
Milkvetch, Prairie: *Astragalus adsurgens*
Milkvetch, Prostrate Loco: *Astragalus miser*
Milkvetch, Rushy: *Astragalus lonchocarpus*
Milkvetch, Sandstone: *Astragalus sesquiflorus*
Milkvetch, Slender: *Astragalus gracilis*
Milkvetch, Speckledpod: *Astragalus lentiginosus*
Milkvetch, Spiny: *Astragalus kentrophyta*
Milkvetch, Stinking: *Astragalus praelongus*
Milkvetch, Thickpod: *Astragalus pachypus*
Milkvetch, Timber: *Astragalus convallarius*
Milkvetch, Torrey's: *Astragalus calycosus*
Milkvetch, Weedy: *Astragalus miser*
Milkvetch, Woody: *Astragalus kentrophyta*
Milkvetch, Woolly: *Astragalus mollissimus*

Milkvetch, Woollypod: *Astragalus purshii*
Milk Vine: *Gonolobus* sp.
Milkvine, Prairie: *Matelea cynanchoides*
Milkvine, Star: *Matelea biflora*
Milkvine, Texas: *Matelea producta*
Milkweed: *Apocynum cannabinum, Asclepias cryptoceras, A. eriocarpa, A. erosa, A. fascicularis, A. involucrata, A. latifolia, A. perennis, A.* sp., *A. speciosa, A. subverticillata, A. syriaca, A. verticillata, Chamaesyce nutans, Euphorbia corollata*
Milkweed, American: *Asclepias syriaca*
Milkweed, Aquatic: *Asclepias perennis*
Milkweed, Big: *Asclepias syriaca*
Milkweed, Broadleaf: *Asclepias latifolia*
Milkweed, Butterfly: *Asclepias tuberosa*
Milkweed, California: *Asclepias californica*
Milkweed, Desert: *Asclepias erosa, A. subulata*
Milkweed, Dwarf: *Asclepias pumila*
Milkweed, Eared: *Asclepias auriculata*
Milkweed, Four Leaved: *Asclepias quadrifolia*
Milkweed, Green: *Asclepias viridiflora*
Milkweed, Hall: *Asclepias hallii*
Milkweed, Heartleaf: *Asclepias cordifolia*
Milkweed, Horsetail: *Asclepias subverticillata*
Milkweed, Large: *Asclepias speciosa*
Milkweed, Lowland: *Asclepias* sp.
Milkweed, Mexican: *Asclepias fascicularis*
Milkweed, Mojave: *Asclepias nyctaginifolia*
Milkweed, Narrow Leaf: *Asclepias fascicularis*
Milkweed, Narrow Leaved: *Asclepias stenophylla*
Milkweed, Pallid: *Asclepias cryptoceras*
Milkweed, Plains: *Asclepias pumila*
Milkweed, Poke: *Asclepias exaltata*
Milkweed, Purple: *Asclepias cordifolia*
Milkweed, Rush: *Asclepias subulata*
Milkweed, Showy: *Asclepias speciosa*
Milkweed, Swamp: *Asclepias incarnata*
Milkweed, Tall: *Asclepias exaltata*
Milkweed, Western: *Asclepias speciosa*
Milkweed, Whorled: *Asclepias fascicularis, A. subverticillata, A. verticillata*
Milkweed, Wiry: *Asclepias fascicularis*
Milkweed, Woollypod: *Asclepias eriocarpa*
Milkweed Vine: *Sarcostemma cynanchoides*
Milkwort, Horned: *Polygala cornuta*
Milkwort, Orange: *Polygala lutea*
Milkwort, Racemed: *Polygala polygama*
Milkwort, Sea: *Glaux maritima*
Milkwort, Sierran: *Polygala cornuta*
Milkwort, White: *Polygala alba*
Milkwort, Whorled: *Polygala verticillata*
Milkwort, Yellow: *Polygala rugelii*
Millet, Indian: *Oryzopsis hymenoides*
Millet, Pearl: *Pennisetum glaucum*
Miner's Lettuce: *Claytonia perfoliata, C. sibirica, C. spathulata*
Mint: *Agastache urticifolia, Mentha* spp., *Monarda* sp., *Phacelia linearis, Pycnanthemum* spp., *Satureja douglasii*

Mint, American Wild: *Mentha arvensis, M. canadensis*
Mint, Aromatic: *Monardella* sp.
Mint, Bush: *Hyptis pectinata, Poliomintha incana*
Mint, Canada: *Mentha arvensis, M. canadensis*
Mint, Coyote: *Monardella odoratissima, M. villosa*
Mint, Downy Wood: *Blephilia ciliata*
Mint, Elk: *Agastache foeniculum*
Mint, Field: *Mentha arvensis, M. canadensis*
Mint, Horse: *Monarda fistulosa*
Mint, Marsh: *Mentha arvensis*
Mint, Mealy: *Pycnanthemum flexuosum*
Mint, Mountain: *Pycnanthemum californicum, P. incanum, P. virginianum*
Mint, Mustang: *Monardella lanceolata*
Mint, Rosemary: *Poliomintha incana*
Mint, Sierra: *Pycnanthemum californicum*
Mint, Tule: *Mentha arvensis*
Mint, Water: *Mentha ×piperita*
Mint, Wild: *Mentha arvensis, M. canadensis*
Mint, Wooly Field: *Mentha canadensis*
Mintleaf Beebalm: *Monarda fistulosa*
Mission Bell: *Fritillaria pudica*
Missionbells, Kamchatka: *Fritillaria camschatcensis*
Missionbells, Scarlet: *Fritillaria recurva*
Missionbells, Spotted: *Fritillaria atropurpurea*
Missionbells, Yellow: *Fritillaria pudica*
Mistletoe, American: *Phoradendron leucarpum*
Mistletoe, American Dwarf: *Arceuthobium americanum*
Mistletoe, Colorado Desert: *Phoradendron macrophyllum*
Mistletoe, Creosote: *Phoradendron* sp.
Mistletoe, Desert: *Phoradendron californicum*
Mistletoe, Digger Pine Dwarf: *Arceuthobium occidentale*
Mistletoe, Douglas Oak: *Phoradendron leucarpum*
Mistletoe, Juniper: *Phoradendron juniperinum*
Mistletoe, Mesquite: *Phoradendron californicum*
Mistletoe, Oak: *Phoradendron leucarpum*
Mistletoe, Pacific: *Phoradendron villosum*
Mistletoe, Pineland Dwarf: *Arceuthobium vaginatum*
Mistletoe, Small: *Arceuthobium americanum, A. campylopodum, A. vaginatum*
Mistletoe, Western Dwarf: *Arceuthobium campylopodum*
Miterwort, Naked: *Mitella nuda*
Miterwort, Threeparted: *Mitella trifida*
Mitrewort: *Mitella diphylla, M. trifida*
Mitrewort, False: *Tiarella cordifolia*
Mnium, Badge: *Plagiomnium insigne*
Moccasin Flower: *Cypripedium acaule*
Mocker Nut: *Carya alba*
Mock Orange: *Cucurbita foetidissima, Philadelphus lewisii, P. microphyllus*
Mock Parsley: *Apiastrum angustifolium*
Mock Pennyroyal: *Hedeoma drummondii, H. nana*
Mock Vervain: *Glandularia bipinnatifida, G. wrightii*

Monkeyflower, Crimson: *Mimulus cardinalis*
Monkeyflower, Eastwood's: *Mimulus eastwoodiae*
Monkeyflower, James's: *Mimulus glabratus*
Monkeyflower, Musk: *Mimulus moschatus*
Monkeyflower, Orange Bush: *Diplacus aurantiacus*
Monkeyflower, Ringen: *Mimulus ringens*
Monkey Flower, Scarlet: *Mimulus cardinalis*
Monkeyflower, Seep: *Mimulus guttatus*
Monkeyflower, Southern Bush: *Diplacus longiflorus*
Monkeyflower, Subalpine: *Mimulus tilingii*
Monkshood, Columbian: *Aconitum columbianum*
Monkshood, Fischer: *Aconitum fischeri*
Monkshood, Larkspurleaf: *Aconitum delphiniifolium*
Moonseed: *Menispermum canadense*
Moonwort: *Botrychium virginianum*
Moorwort: *Andromeda polifolia*
Moosewood: *Acer pensylvanicum, Dirca palustris*
Morelle: *Morchella* sp.
Mormon Tea: *Ephedra californica, E. fasciculata, E. nevadensis, E.* sp., *E. torreyana, E. trifurca, E. viridis*
Morningglory, Darkeye: *Ipomoea tiliacea*
Morningglory, Oceanblue: *Ipomoea indica*
Morning Glory, Piute: *Calystegia longipes*
Morningglory, Saltmarsh: *Ipomoea sagittata*
Morningglory, Shaggy Dwarf: *Evolvulus nuttallianus*
Morning Glory Bush: *Ipomoea leptophylla*
Mosquito Bills: *Dodecatheon hendersonii*
Moss: Aulacomniaceae, *Aulacomnium* sp., *Barbula unguiculata*, Bryaceae, *Bryum capillare*, Dicranaceae, *Dicranum* sp., Hylocomiaceae, Leucobryaceae, Leucodontaceae, Mniaceae, *Octoblephorum albidum*, Polytrichaceae, Pottiaceae, Rhytidiaceae, Sphagnaceae, *Sphagnum* sp.
Moss, Black: *Alectoria fremontii, A. jubata*
Moss, Caribou: *Alectoria nigricans, A. nitidula, A. ochroleuca, Cornicularia divergens*
Moss, Club: *Huperzia selago*, Lycopodiaceae, *Lycopodium annotinum, L. clavatum, L.* sp.
Moss, Hair: *Polytrichum commune*
Moss, Hair Cap: *Plagiomnium juniperinum*
Moss, Hanging: *Antitrichia curtipendula*
Moss, Maidenhair: *Alectoria sarmentosa, Usnea longissima*
Moss, Peat: *Sphagnum* sp.
Moss, Spanish: *Ramalina menziesii, Tillandsia usneoides*
Moss, Tree: *Evernia vulpina*
Moss, Wood: *Dicranum bonjeanii*
Moss, Wolf: *Evernia vulpina, Letharia vulpina*
Mossy Cup: *Quercus macrocarpa*
Motherwort: *Leonurus cardiaca*
Mountain Ash, American: *Sorbus americana*
Mountain Ash, European: *Sorbus aucuparia*
Mountainash, Gray's: *Sorbus sitchensis*
Mountainash, Greene: *Sorbus scopulina*
Mountainash, Northern: *Sorbus decora*
Mountain Ash, Sitka: *Sorbus sitchensis*

Mountainash, Western: *Sorbus sitchensis*
Mountainbalm, Green: *Monardella viridis*
Mountainbalm, Mustang: *Monardella lanceolata*
Mountainbalm, Shelton's: *Monardella villosa*
Mountainbalm, Sierran: *Monardella candicans*
Mountainbells: *Stenanthium occidentale*
Mountainheath: *Phyllodoce empetriformis*
Mountain Mahogany: see Mahogany
Mountainmint, Appalachian: *Pycnanthemum flexuosum*
Mountainmint, Hoary: *Pycnanthemum incanum*
Mountainmint, Virginia: *Pycnanthemum virginianum*
Mountainmint, Whiteleaf: *Pycnanthemum albescens*
Mountain Misery: *Chamaebatia foliolosa*
Mountaintrumpet, Largeflower: *Collomia grandiflora*
Mountaintrumpet, Narrowleaf: *Collomia linearis*
Mouse Ear, Yellow: *Lesquerella douglasii*
Mouse Ear Chickweed: *Cerastium beeringianum*
Mouse Ear Everlasting: *Gnaphalium obtusifolium*
Mouseear Hawkweed: *Hieracium pilosella*
Mousetail, Arizona: *Myosurus cupulatus*
Mousetail, Bristle: *Myosurus aristatus*
Mousetail, Tiny: *Myosurus minimus*
Mudwort: *Limosella aquatica*
Mugwort, California: *Artemisia douglasiana, A. vulgaris*
Mugwort, Mexican: *Artemisia ludoviciana*
Mugwort, Suksdorf's: *Artemisia suksdorfii*
Mugwort, Western: *Artemisia ludoviciana*
Mugwort, White: *Artemisia ludoviciana*
Muhly, Foxtail: *Muhlenbergia andina*
Muhly, Mat: *Muhlenbergia richardsonis*
Muhly, Mexican: *Muhlenbergia mexicana*
Muhly, New Mexico: *Muhlenbergia pauciflora*
Muhly, Pine: *Muhlenbergia dubia*
Muhly, Plains: *Muhlenbergia cuspidata*
Muhly, Pull Up: *Muhlenbergia filiformis*
Muhly, Sandhill: *Muhlenbergia pungens*
Muhly, Spike: *Muhlenbergia wrightii*
Muhly, Spiny: *Muhlenbergia pungens*
Mulberry, Black: *Morus nigra*
Mulberry, French: *Callicarpa americana*
Mulberry, Indian: *Morinda citrifolia*
Mulberry, Paper: *Broussonetia papyrifera*
Mulberry, Red: *Morus rubra*
Mulberry, Texas: *Morus microphylla*
Mulberry, White: *Morus alba*
Mulberry, Wild: *Morus* sp.
Mule Ears: *Wyethia amplexicaulis, W. angustifolia, W. glabra, W. mollis*
Mule Ears, Wooly: *Wyethia mollis*
Mulesears, Coast Range: *Wyethia glabra*
Mulesears, Humboldt: *Wyethia longicaulis*
Mulesears, Southern: *Wyethia ovata*
Mule's Fat: *Baccharis salicifolia*
Mullein, Great: *Verbascum thapsus*
Mullein, Turkey: *Croton setigerus*

Index of Common Names

Mullein Nightshade: *Solanum donianum*
Mullen: *Verbascum thapsus*
Muscadine: *Vitis munsoniana, V. rotundifolia*
Mushroom, Deer: *Agaricus silvicola*
Mushroom, Field: *Agaricus campestris*
Mushroom, Hedgehog: *Dentinum repandum*
Mushroom, Orange Peel: *Peziza aurantia*
Mushroom, Oyster: *Pleurotus ostreatus*
Mushroom, Timber: *Boletus edulis*
Muskat Root: *Acorus calamus*
Muskmelon: *Cucumis melo, C. sp.*
Musk Plant: *Mimulus moschatus*
Musquash Root: *Cicuta maculata*
Mustard: *Brassica* spp., *Descurainia pinnata, Hirschfeldia incana*
Mustard, Black: *Brassica nigra*
Mustard, Charlock: *Sinapis arvensis*
Mustard, Field: *Brassica napus*
Mustard, Hedge: *Descurainia incana, D. pinnata, Sisymbrium officinale*
Mustard, Indian: *Brassica juncea*
Mustard, London Rockets: *Sisymbrium irio*
Mustard, Rape: *Brassica rapa*
Mustard, Shortpod: *Hirschfeldia incana*
Mustard, Tansy: *Descurainia incana, D. obtusa, D. pinnata, D. sophia, D. sp.*
Mustard, Tumble: *Sisymbrium altissimum*
Mustard, White: *Sinapis alba*
Mustard, Wild: *Brassica nigra, B. sp., Sisymbrium irio, S. officinale*
Mustard, Wormseed: *Erysimum cheiranthoides*
Mustard, Yellow: *Brassica rapa, Moricandia arvensis*
Myrtle, Sand: *Leiophyllum buxifolium*
Myrtle, Wax: *Myrica cerifera*
Nagoonberry: *Rubus arcticus*
Nailwort: *Paronychia jamesii*
Nannyberry: *Viburnum lentago*
Nau-paka: *Scaevola sericea*
Needleandthread: *Stipa comata*
Needlegrass, Desert: *Stipa speciosa*
Needlegrass, New Mexico: *Stipa neomexicana*
Nettle: *Tragia ramosa, Urtica* spp.
Nettle, American Stinging: *Urtica dioica*
Nettle, Bear: *Urtica dioica*
Nettle, Bush: *Laportea canadensis*
Nettle, California: *Urtica dioica*
Nettle, Dog: *Urtica urens*
Nettle, Dwarf: *Urtica urens*
Nettle, False: *Laportea canadensis*
Nettle, Hedge: see Hedgenettle
Nettle, Horse: *Solanum carolinense, S. fendleri*
Nettle, Lyall's: *Urtica dioica*
Nettle, Slender: *Urtica dioica*
Nettle, Slim Stinging: *Urtica dioica*
Nettle, Stinging: *Urtica dioica*
Nettle, Western: *Urtica dioica*
Nettle, Wood: *Laportea canadensis*
Niger Weed: *Sphaeralcea angustifolia*
Nigger-toe: *Brodiaea* sp.
Nightshade, Bittersweet: *Solanum dulcamara*
Nightshade, Black: *Solanum douglasii, S. nigrum*
Nightshade, Blue: *Solanum elaeagnifolium*
Nightshade, Bluewitch: *Solanum umbelliferum*
Nightshade, Broadleaf Enchanter's: *Circaea lutetiana*
Nightshade, Buffalobur: *Solanum rostratum*
Nightshade, Chaparral: *Solanum xantii*
Nightshade, Climbing: *Solanum dulcamara*
Nightshade, Cut Leaved: *Solanum triflorum*
Nightshade, Deadly: *Solanum nigrum*
Nightshade, Greenspot: *Solanum douglasii*
Nightshade, Hoe: *Solanum physalifolium*
Nightshade, Larger Enchanter: *Circaea lutetiana*
Nightshade, Mullein: *Solanum donianum*
Nightshade, Purple: *Solanum xantii*
Nightshade, Silver: *Solanum elaeagnifolium*
Nine Bark: *Physocarpus capitatus, P. malvaceus, P. opulifolius*
Nioi: *Coprosma* sp.
Niu: *Cocos nucifera*
Noni: *Morinda citrifolia*
Noseburn, Branched: *Tragia ramosa*
Noseburn, Catnip: *Tragia nepetifolia*
Noyau Vine: *Merremia dissecta*
Nuholani: *Eucalyptus* sp.
Nut Grass: *Cyperus esculentus, C. rotundus, C. sp.*
Nutmeg, California: *Torreya californica*
Nut Plant: *Comandra umbellata*
Oak, American Live: *Quercus virginiana*
Oak, American Red: *Quercus rubra*
Oak, Bear: *Quercus ilicifolia*
Oak, Black: *Quercus emoryi, Q. kelloggii, Q. velutina*
Oak, Blackjack: *Quercus marilandica*
Oak, Blue: *Quercus douglasii, Q. oblongifolia*
Oak, Bur: *Quercus macrocarpa*
Oak, California Black: *Quercus kelloggii*
Oak, California Live: *Quercus agrifolia*
Oak, California Scrub: *Quercus dumosa*
Oak, California White: *Quercus lobata*
Oak, Canyon: *Quercus chrysolepis*
Oak, Canyon Live: *Quercus chrysolepis*
Oak, Cherrybark: *Quercus pagoda*
Oak, Chestnut: *Quercus prinus*
Oak, Chinkapin: *Quercus muehlenbergii*
Oak, Coast Live: *Quercus agrifolia*
Oak, Deer: *Quercus sadleriana*
Oak, Douglas: *Quercus douglasii*
Oak, Emory: *Quercus emoryi*
Oak, Engelmann: *Quercus engelmannii*
Oak, Gambel: *Quercus gambelii*
Oak, Garry: *Quercus garryana*
Oak, Gray: *Quercus garryana, Q. grisea, Q. rubra*
Oak, Hill's: *Quercus ellipsoidalis*
Oak, Interior Live: *Quercus wislizeni*
Oak, Interior Scrub: *Quercus wislizeni*
Oak, Jerusalem: *Chenopodium botrys*
Oak, Large Scrub: *Quercus wislizeni*
Oak, Laurel: *Quercus laurifolia*
Oak, Live: *Quercus agrifolia, Q. chrysolepis, Q. turbinella, Q. virginiana, Q. wislizeni*
Oak, Maul: *Quercus chrysolepis*
Oak, Mexican Blue: *Quercus oblongifolia*
Oak, Mountain White: *Quercus garryana*
Oak, Northern Pin: *Quercus ellipsoidalis*
Oak, Northern Red: *Quercus rubra*
Oak, Nuttall: *Quercus texana*
Oak, Oregon: *Quercus garryana*
Oak, Palmer: *Quercus dunnii*
Oak, Peninsula: *Quercus peninsularis*
Oak, Pin: *Quercus ellipsoidalis, Q. palustris*
Oak, Post: *Quercus laurifolia, Quercus stellata*
Oak, Pungent: *Quercus pungens*
Oak, Red: *Quercus falcata, Q. pagoda, Q. rubra, Q. sp., Q. texana*
Oak, Rocky Mountain: *Quercus gambelii*
Oak, Rocky Mountain Shin: *Quercus ×pauciloba*
Oak, Rocky Mountain White: *Quercus gambelii*
Oak, Running: *Quercus pumila*
Oak, Scrub: *Quercus dumosa, Q. dunnii, Q. ilicifolia, Q. marilandica, Q. ×pauciloba, Q. pungens, Q. sp., Q. turbinella, Q. wizlizeni*
Oak, Scrub Tan: *Lithocarpus densiflorus*
Oak, Shin: *Quercus ×pauciloba, Q. garryana*
Oak, Shingle: *Quercus imbricaria*
Oak, Shrub Live: *Quercus turbinella*
Oak, Small Bushy Scrub: *Quercus* sp.
Oak, Southern Red: *Quercus falcata*
Oak, Spanish: *Quercus falcata*
Oak, Swamp: *Quercus bicolor*
Oak, Swamp White: *Quercus bicolor, Q. rubra*
Oak, Tan: *Lithocarpus densiflorus*
Oak, Tan Bark: *Lithocarpus densiflorus*
Oak, Utah: *Quercus gambelii*
Oak, Valley: *Quercus lobata*
Oak, Valley White: *Quercus lobata*
Oak, Valparaiso: *Quercus chrysolepis*
Oak, Water: *Quercus douglasii, Q. nigra, Q. phellos*
Oak, Wavyleaf: *Quercus ×pauciloba*
Oak, White: *Quercus Q. bicolor, Q. alba, Q. engelmannii, Q. gambelii, Q. garryana, Q. lobata, Q. rubra*
Oak, Willow: *Quercus phellos*
Oak, Yellow: *Quercus muehlenbergii*
Oakfern, Pacific: *Gymnocarpium disjunctum*
Oakfern, Western: *Gymnocarpium dryopteris*
Oat, Cultivated: *Avena sativa*
Oat, Indian Wood: *Chasmanthium latifolium*
Oat, Slender: *Avena barbata*
Oat, Wild: *Avena barbata, A. fatua, A. sp.*
Ocean Spray: *Holodiscus discolor, H. dumosus*
Ocotillo: *Fouquieria* sp., *F. splendens*
Odora: *Porophyllum gracile*
'Oha Wai Nui: *Clermontia arborescens*
Ohia: *Syzygium malaccense*
'Ohi'a: *Metrosideros polymorpha*
Ohia-ohia-a-pane: *Metrosideros polymorpha*
Ohio, Pride of: *Dodecatheon pulchellum*
Okoleomakili: *Vigna luteola*
Olapa: *Cheirodendron gaudicchaudii*
Old Man Root: *Marah oreganus*

Index of Common Names

Old Man's Whiskers: *Geum triflorum*
Olena: *Curcuma longa*
Olive, Russian: *Forestiera pubescens*
Olona: *Touchardia latifolia*
Onion, Aspen: *Allium bisceptrum*
Onion, Autumn: *Allium stellatum*
Onion, Bolander's: *Allium bolanderi*
Onion, Broadstemmed: *Allium platycaule*
Onion, Bulbil: *Allium geyeri*
Onion, Coastal: *Allium dichlamydeum*
Onion, Cultivated: *Allium cepa*
Onion, Desert: *Allium cernuum, A. macropetalum*
Onion, Domesticated: *Allium cepa*
Onion, Douglas's: *Allium douglasii*
Onion, Drummond's: *Allium drummondii*
Onion, Field: *Allium cernuum*
Onion, Fool's: *Triteleia hyacinthina*
Onion, Garden: *Allium cepa*
Onion, Geyer's: *Allium geyeri*
Onion, Glassy: *Allium hyalinum*
Onion, Hooker's: *Allium acuminatum*
Onion, Largeflower Wild: *Allium macropetalum*
Onion, Manyflower: *Allium pleianthum*
Onion, Mexicali: *Allium peninsulare*
Onion, Nevada: *Allium nevadense*
Onion, Nodding: *Allium cernuum*
Onion, Oneleaf: *Allium unifolium*
Onion, Pacific: *Allium validum*
Onion, Pitted: *Allium lacunosum*
Onion, Scytheleaf: *Allium falcifolium*
Onion, Shortstyle: *Allium brevistylum*
Onion, Small: *Allium parvum*
Onion, Tapertip: *Allium acuminatum*
Onion, Textile: *Allium textile*
Onion, Twincrest: *Allium bisceptrum*
Onion, Twinleaf: *Allium anceps*
Onion, White: *Allium geyeri*
Onion, Wild: *Allium acuminatum, A. bisceptrum, A. bolanderi, A. canadense, A. cepa, A. cernuum, A. dichlamydeum, A. drummondii, A. geyeri, A. hyalinum, A. macropetalum, A. peninsulare, A. schoenoprasum, A. sp., A. stellatum, A. textile, A. unifolium, A. validum, Cymopterus sp.*
Onion, Wild Desert: *Allium macropetalum*
Oniongrass: *Melica bulbosa*
Ookow: *Dichelostemma pulchellum*
Opium Poppy: *Papaver somniferum*
Orache: *Atriplex powellii*
Orange: *Citrus sinensis*
Orange, Mock: *Cucurbita foetidissima, Philadelphus lewisii, P. microphyllus*
Orange, Osage: *Maclura pomifera*
Orange, Sour: *Citrus aurantium*
Orange, Sweet: *Citrus sinensis*
Orangegrass: *Hypericum gentianoides*
Orchid, Bog: *Platanthera leucostachys, P. ×media, P. sparsiflora, P. stricta*
Orchid, Bracted: *Coeloglossum viride*
Orchid, Fairyslipper: *Calypso bulbosa*
Orchid, Greater Purple Fringed: *Platanthera grandiflora*
Orchid, Green Addersmouth: *Malaxis unifolia*
Orchid, Hillside Bog: *Piperia elegans*
Orchid, Intermediate Bog: *Platanthera ×media*
Orchid, Large Bracted: *Coeloglossum viride*
Orchid, Large Purpled Fringe: *Platanthera grandiflora*
Orchid, Lesser Purple Fringed: *Platanthera psycodes*
Orchid, Longspur: *Habenaria odontopetala*
Orchid, Modoc Bog: *Platanthera stricta*
Orchid, Purple Fringed: *Platanthera grandiflora, P. psycodes*
Orchid, Rein: *Coeloglossum viride, Habenaria odontopetala, Piperia unalascensis, Platanthera dilatata*
Orchid, Round Leaved: *Platanthera orbiculata*
Orchid, Slender Bog: *Platanthera stricta*
Orchid, Slender Bog: *Platanthera stricta*
Orchid, Soldier's: *Zeuxine strateumatica*
Orchid, Yellow Fringed: *Platanthera ciliaris*
Orchid, Yellow Widelip: *Liparis loeselii*
Orchis, Rein: *Coeloglossum viride, Platanthera dilatata*
Orchis, Stream: *Epipactis gigantea*
Orpine: *Sedum divergens, S. sp., S. spathulifolium*
Orpine Stonecrop: *Sedum debile*
Osage Orange: *Maclura pomifera*
Osha: *Ligusticum filicinum, Ligusticum porteri*
Oso Berry: *Oemleria cerasiformis*
Owl Clover: *Orthocarpus densiflorus, O. faucibarbatus, O. luteus, O. purpurascens, O. purpureoalbus*
Owlsclaws: *Dugaldia hoopesii*
Ox Eye: *Heliopsis helianthoides*
Ox Eye Daisy: *Chrysothamnus sp., Leucanthemum vulgare*
Oxytrope: *Oxytropis nigrescens*
Oysterleaf: *Mertensia maritima*
Oyster Plant: *Tragopogon porrifolius*
Pachistima, Myrtle: *Paxistima myrsinites*
Pagodaplant: *Blephilia ciliata*
Paintbrush, Alaska Indian: *Castilleja unalaschcensis*
Paintbrush, Attenuate Indian: *Orthocarpus attenuatus*
Paintbrush, Common Red Indian: *Castilleja miniata*
Paintbrush, Denseflower Indian: *Orthocarpus densiflorus*
Paintbrush, Exserted Indian: *Orthocarpus purpurascens*
Paintbrush, Green Indian: *Castilleja miniata*
Paintbrush, Harsh Indian: *Castilleja hispida*
Paintbrush, Harsh Indian: *Castilleja hispida*
Paintbrush, Largeflower Indian: *Castilleja stenantha*
Paintbrush, Lesser Indian: *Castilleja minor*
Paintbrush, Marshmeadow Indian: *Castilleja lineata*
Paintbrush, Mexican: *Castilleja integra*
Paintbrush, Mountain Indian: *Castilleja parviflora*
Paintbrush, Northwestern Indian: *Castilleja angustifolia*
Paintbrush, Red: *Castilleja sp.*
Paintbrush, Scarlet Indian: *Castilleja coccinea, C. miniata*
Paintbrush, Sessile: *Castilleja sessiliflora*
Paintbrush, Texas Indian: *Castilleja foliolosa*
Paintbrush, Thompson's Indian: *Castilleja thompsonii*
Paintbrush, Wavyleaf Indian: *Castilleja applegatei*
Paintbrush, Wholeleaf Indian: *Castilleja integra*
Paintbrush, Wyoming Indian: *Castilleja linariifolia*
Paintbrush, Yellow: *Orthocarpus luteus*
Paintbrush, Yellow Indian: *Castilleja thompsonii*
Painted Cup: *Castilleja coccinea, C. linariifolia, C. parviflora, C. sp.*
Paintedcup, Downy: *Castilleja sessiliflora*
Paka: *Nicotiana glauca, Nicotiana tabacum*
Pa-la-a: *Odontoglossum chinensis*
Pa-la-pa-la-i: *Microlepia setosa*
Palm, Blue: *Glaucothea armata*
Palm, California (or Fan): *Washingtonia filifera*
Palm, Sea: *Postelsia palmaeformis*
Palma: *Yucca torreyi*
Palmetto, Cabbage: *Sabal palmetto*
Palmetto, Dwarf: *Sabal minor*
Palmetto, Saw: *Serenoa repens*
Palmilla: *Yucca elata*
Palo Blanco: *Acacia willardiana*
Paloverde: *Parkinsonia florida, P. microphylla, Prosopis glandulosa*
Paloverde, Blue: *Parkinsonia florida*
Paloverde, Yellow: *Parkinsonia microphylla*
Panic Grass: *Dichanthelium laxiflorum, D. oligosanthes, D. strigosum, Panicum bulbosum, P. capillare, P. hirticaule, P. obtusum, P. sp., P. urvilleanum, Paspalidium geminatum*
Panicum, Scribner's: *Dichanthelium oligosanthes*
Pa-ni-ni: *Opuntia tunicata*
Pansy: *Viola bicolor, Viola pedunculata*
Papago Blue Bells: *Dichelostemma pulchellum*
Papaya: *Carica papaya*
Paperflower, Greenstem: *Psilostrophe sparsiflora*
Paperflower, Woolly: *Psilostrophe tagetina*
Paradise Banana: *Musa ×paradisiaca*
Paradise Plant: *Daphne mezereum*
Parilla, Yellow: *Menispermum canadense*
Parosela: *Dalea sp.*
Parosela, Slender: *Dalea enneandra*
Parsley: *Aletes anisatus, Conioselinum chinense, Petroselinum crispum*
Parsley, American: *Apium sp.*
Parsley, Chinese: *Coriandrum sativum*
Parsley, Desert: see Desertparsley
Parsley, Hemlock: see Hemlockparsley
Parsley, Indian: *Aletes acaulis, A. anisatus*
Parsley, Mock: *Apiastrum angustifolium*
Parsley, Prairie: *Lomatium dissectum, L. simplex, L. triternatum*

Index of Common Names

Parsley, Rocky Mountain Indian: *Aletes anisatus*
Parsley, Spring: see Springparsley
Parsley, Stemless Indian: *Aletes acaulis*
Parsley, Water: *Oenanthe sarmentosa*
Parsley, White Flowered: *Lomatium orientale*
Parsley, Wild: *Ligusticum grayi, Lomatium utriculatum, Musineon divaricatum*
Parsnip: *Lomatium dissectum, Pastinaca sativa*
Parsnip, Cow: *Heracleum maximum*
Parsnip, Indian: *Pteryxia terebinthina*
Parsnip, Meadow: *Thaspium barbinode*
Parsnip, Poison: *Cicuta maculata*
Parsnip, Water: see Waterparsnip
Parsnip, Wild: *Angelica lucida, Ligusticum scothicum, Lomatium dissectum, Pastinaca sativa*
Parsnip, Woollyhead: *Sphenosciadium capitellatum*
Partridge Berry: *Mitchella repens*
Partridgefoot: *Luetkea pectinata*
Partridge Vine: *Mitchella repens*
Pasqueflower, American: *Pulsatilla patens*
Pasqueflower, White: *Pulsatilla occidentalis*
Passion Flower: *Passiflora incarnata*
Patata (or Patota): *Monolepis nuttalliana*
Pa-u-o-hiiaka: *Jacquemontia ovalifolia*
Paw, Terrapin: *Prunella vulgaris*
Pa-wale: *Rumex giganteus*
Pawpaw: *Asimina triloba*
Pea: *Lathyrus eucosmus, L. graminifolius, L. nevadensis, Parryella filifolia, Pisum sativum*
Pea, Beach: *Lathyrus japonicus*
Pea, Blackeyed: *Vigna unguiculata*
Pea, Buffalo: *Astragalus crassicarpus*
Pea, Butterfly: *Clitoria mariana*
Pea, Chick: *Cicer arietinum*
Pea, Deer: *Erythrina herbacea*
Pea, Garden: *Pisum sativum*
Pea, Golden: *Thermopsis rhombifolia*
Pea, Hoary: *Tephrosia florida, T. hispidula, T. purpurea*
Pea, Indian Rush: *Hoffmannseggia glauca*
Pea, Partridge: *Chamaecrista fasciculata, C. nictitans, Senna marilandica*
Pea, Scurf: see Scurf Pea
Pea, Turkey: *Sanicula tuberosa*
Pea, Wild: *Lathyrus japonicus, L. lanszwertii, L. ochroleucus, L. palustris, L. polymorphus, L. sp., L. venosus, L. vestitus, Vicia americana*
Peach, Cultivated: *Prunus persica*
Peach, Desert: *Prunus andersonii*
Peach, Wild: *Prunus andersonii*
Peanut: *Arachis hypogaea, Astragalus ceramicus*
Peanut, Hog: *Amphicarpaea bracteata*
Peanut Butter Plant: *Glyptopleura marginata*
Pear: *Pyrus communis*
Pearl Millet: *Pennisetum glaucum*
Pearly Everlasting: *Anaphalis margaritacea*
Pear, Prickly: see Pricklypear
Pearly Pussytoes: *Antennaria anaphaloides*
Pear Thorn: *Crataegus calpodendron*
Peat Moss: *Sphagnum* sp.

Peavine: *Lathyrus eucosmus*
Peavine, Aspen: *Lathyrus lanszwertii*
Peavine, Bonneville: *Lathyrus brachycalyx*
Peavine, California: *Lathyrus jepsonii*
Peavine, Cream: *Lathyrus ochroleucus*
Peavine, Grassleaf: *Lathyrus graminifolius*
Peavine, Manystem: *Lathyrus polymorphus*
Peavine, Nuttall's: *Lathyrus nevadensis*
Peavine, Pacific: *Lathyrus vestitus*
Peavine, Sea: *Lathyrus japonicus*
Peavine, Slenderstem: *Lathyrus palustris*
Peavine, Veiny: *Lathyrus venosus*
Pecan: *Carya illinoinensis, C.* sp.
Pelotazo: *Abutilon incanum*
Pembina: *Viburnum opulus*
Pencilflower: *Stylosanthes biflora*
Penny Cress: *Thlaspi arvense*
Pennycress, Alpine: *Thlaspi montanum*
Pennyroyal: *Hedeoma nana, H. pulegioides, H.* sp., *Mentha arvensis, Piloblephis rigida*
Pennyroyal, American: *Hedeoma pulegioides*
Pennyroyal, Drummond: *Hedeoma drummondii*
Pennyroyal, Mock: *Hedeoma drummondii, H. nana*
Pennyroyal, Mountain: *Monardella odoratissima*
Pennyroyal, Rough: *Hedeoma hispida*
Pennyroyal, Wild: *Piloblephis rigida*
Pennywort: *Obolaria virginica*
Pennywort, Marsh: *Hydrocotyle* sp.
Pennywort, Water: *Hydrocotyle umbellata*
Pennyworth: *Obolaria virginica*
Penstemon, Heartleaf: *Keckiella cordifolia*
Penstemon, Woodland: *Nothochelone nemorosa*
Pentstemon: *Penstemon* spp.; see also Penstemon
Peony, American Wild: *Paeonia brownii*
Peony, Western: *Paeonia brownii*
Peony, Wild: *Paeonia brownii, P. californica*
Pepper, Black: *Piper nigrum*
Pepper, Bush: *Berberis canadensis*
Pepper, Cayenne: *Capsicum annuum*
Pepper, Chili: *Capsicum annuum*
Pepper, Indian: *Saururus cernuus*
Pepper, Poor Man's: *Lepidium virginicum*
Pepper, Spur: *Capsicum annuum*
Pepper, Water: *Polygonum hydropiper*
Pepper, Wild: *Capsicum annuum*
Pepper and Salt: *Erigenia bulbosa*
Pepperidge: *Nyssa sylvatica*
Peppermint: *Artemisia dracunculus, Mentha canadensis, M. ×piperita*
Peppernut: *Umbellularia californica*
Pepper Root: *Cardamine concatenata, C. diphylla*
Peppervine: *Ampelopsis cordata*
Pepperweed, Common: *Lepidium densiflorum*
Pepperweed, Desert: *Lepidium fremontii*
Pepperweed, Field: *Lepidium campestre*
Pepperweed, Menzies's: *Lepidium virginicum*
Pepperweed, Mountain: *Lepidium montanum*
Pepperweed, Shaggyfruit: *Lepidium lasiocarpum*
Pepperweed, Shining: *Lepidium nitidum*
Pepperweed, Thurber's: *Lepidium thurberi*

Pepperwood: *Umbellularia californica, U.* sp.
Periwinkle, Madagascar: *Catharanthus roseus*
Perkysue: *Tetraneuris argentea*
Persicaria, Carey's: *Polygonum careyi*
Persicaria, Dock Leaved: *Polygonum lapathifolium*
Persicaria, Hartwright's: *Polygonum amphibium*
Persicaria, Pennsylvania: *Polygonum pensylvanicum*
Persicaria, Swamp: *Polygonum amphibium*
Persimmon: *Diospyros virginiana*
Persimmon, Mexican (or Texas): *Diospyros texana*
Peruvian Bark: *Cinchona calisaya*
Peyote: *Lophophora williamsii*
Pia: *Tacca leontopetaloides*
Pickerel Weed: *Pontederia cordata*
Pickleweed: *Allenrolfea occidentalis*
Piedmont Staggerbush: *Lyonia mariana*
Pigeon Berry Poke: *Phytolacca americana*
Pigeon Plum: *Coccoloba diversifolia*
Pigeonwings: *Clitoria mariana*
Pigweed: *Amaranthus blitoides, A. cruentus, A. fimbriatus, A. hybridus, A. palmeri, A. powellii, A. retroflexus, Chenopodium album, C. graveolens, C. leptophyllum, C. nevadense, C. rubrum, C.* sp.
Pigweed, Amaranth: *Amaranthus retroflexus*
Pigweed, Prostrate: *Amaranthus albus*
Pigweed, White: *Chenopodium album*
Pigweed, Winged: *Cycloloma atriplicifolium*
Pi-li-kai: *Ipomoea tiliacea*
Pilot Weed: *Silphium laciniatum*
Pimpernel: *Anagallis* sp.
Pimpernel, Yellow: *Taenidia integerrima*
Pin Cushion: *Chaenactis glabriuscula*
Pincushion, Perennial (or Santolina): *Chaenactis santolinoides*
Pincushion, Snowball: *Mammillaria grahamii*
Pincushionplant, Cotulaleaf: *Navarretia cotulifolia*
Pincushionplant, Hollyleaf: *Navarretia atractyloides*
Pine, American Scrub: *Pinus monticola*
Pine, Bishop: *Pinus muricata*
Pine, Black: *Pinus contorta*
Pine, Bristlecone: *Pinus aristata*
Pine, Bull: *Pinus ponderosa*
Pine, California Foothill: *Pinus sabiniana*
Pine, Coulter's: *Pinus coulteri*
Pine, Digger: *Pinus sabiniana*
Pine, Eastern White: *Pinus strobus*
Pine, Ground: *Lycopodium complanatum, L. dendroideum, L. obscurum, L. sabinifolium*
Pine, Jack: *Pinus banksiana*
Pine, Jeffrey: *Pinus jeffreyi*
Pine, Knob Cone: *Pinus attenuata*
Pine, Limber: *Pinus flexilis*
Pine, Loblolly: *Pinus taeda*
Pine, Lodgepole: *Abies grandis, Pinus contorta, P. ponderosa*
Pine, Lodgepole Black: *Pinus contorta*
Pine, Long Tag: *Pinus echinata*
Pine, Murray Lodgepole: *Pinus contorta*
Pine, Northern Scrub: *Pinus banksiana*

Pine, Norway: *Pinus resinosa*
Pine, Nut: *Pinus albicaulis, P. edulis, P. monophylla*
Pine, Piñon: *Pinus edulis, P. monophylla,*
  P. quadrifolia, P. sabiniana
Pine, Pinyon: *Pinus monophylla, P. quadrifolia*
Pine, Pitch: *Pinus rigida*
Pine, Ponderosa: *Pinus ponderosa*
Pine, Prince's: *Chimaphila maculata, C. menziesii,*
  C. umbellata
Pine, Red: *Pinus resinosa*
Pine, Rock: *Pinus ponderosa*
Pine, Screw: *Pandanus tectorius*
Pine, Scrub: *Ostrya virginiana, Pinus banksiana,*
  P. contorta, P. monticola, P. virginiana
Pine, Shore: *Pinus contorta*
Pine, Shortleaf: *Pinus echinata*
Pine, Silver Barked: *Pinus albicaulis*
Pine, Slash: *Pinus elliottii*
Pine, Spruce: *Pinus glabra, P. virginiana*
Pine, Sugar: *Pinus lambertiana*
Pine, Sweet: *Abies* sp.
Pine, Table Mountain: *Pinus pungens*
Pine, Tall Lodgepole: *Pinus contorta*
Pine, Virginia: *Pinus virginiana*
Pine, Western: *Pinus monticola*
Pine, Western White: *Pinus flexilis, P. monticola*
Pine, Western Yellow: *Pinus ponderosa*
Pine, White: *Pinus monticola, P. strobus*
Pine, White Bark: *Pinus albicaulis*
Pine, Yellow: *Pinus echinata, P. ponderosa*
Pineapple: *Ananas comosus*
Pineapple Weed: *Matricaria discoidea*
Pine Drops: *Pterospora andromedea*
Pinesap: *Monotropa hypopithys*
Pinetorum: *Androsace* sp.
Piney Weed: *Hypericum gentianoides*
Pingue (or Pingwing): *Hymenoxys richardsonii*
Pink, American Moss: *Phlox subulata*
Pink, Marsh: *Sabatia campanulata*
Pink, Rose: *Sabatia angularis*
Pink Plume: *Geum triflorum, Polygonum bistorta*
Pink Root: *Ivesia gordonii*
Pinkroot, West Indian: *Spigelia anthelmia*
Pinkroot, Woodland: *Spigelia marilandica*
Piñon: see Pine, Piñon; and Pine, Pinyon; see also Pinyon
Pinweed: *Lechea minor, Lechea* sp.
Pinyon: see Pine, Pinyon; and Pine, Piñon
Pinyon Groundsmoke: *Gayophytum ramosissimum*
Pipe Stem: *Clematis lasiantha*
Pipe Vine: *Aristolochia macrophylla, A.* sp.
Pipeweed: *Eriogonum inflatum*
Pipissewa (or Pipsissewa, or Pipsewa): *Chimaphila umbellata*
Pitahaya: *Stenocereus thurberi*
Pitahaya Agria: *Machaerocereus gummosus*
Pitahayita: *Peniocereus striatus*
Pitcher Plant: *Sarracenia purpurea*
Pitcherplant, California: *Darlingtonia californica*
Plantain, American: *Plantago major*

Plantain, Blackseed: *Plantago rugelii*
Plantain, Bracted: *Plantago aristata*
Plantain, Broad Leaved: *Plantago major*
Plantain, English: *Plantago lanceolata*
Plantain, Green: *Plantago major*
Plantain, Ground: *Plantago major*
Plantain, Heartleaf: *Plantago cordata*
Plantain, Largebracted: *Plantago aristata*
Plantain, Mexican: *Plantago australis*
Plantain, Narrowleaf: *Plantago lanceolata*
Plantain, Rattlesnake: *Goodyera oblongifolia,*
  G. pubescens, G. repens
Plantain, Robin's: *Erigeron pulchellus*
Plantain, Rugel's: *Plantago rugelii*
Plantain, Sagebrush: *Plantago patagonica*
Plantain, Seashore: *Plantago macrocarpa*
Plantain, Swarf: *Plantago virginica*
Plantain, Virginia: *Plantago virginica*
Plantain, White: *Antennaria plantaginifolia*
Plantain, Woolly: *Plantago patagonica*
Plantain Lily: *Hosta lancifolia*
Pleurisy Root: *Asclepias tuberosa*
Plum, American: *Prunus americana*
Plum, Canada: *Prunus nigra*
Plum, Chickasaw: *Prunus angustifolia*
Plum, Ground: *Astragalus crassicarpus, A.* sp.
Plum, Hog: *Ximenia americana*
Plum, Indian: *Oemleria cerasiformis*
Plum, Klamath: *Prunus subcordata*
Plum, Low: *Prunus gracilis*
Plum, Oklahoma: *Prunus gracilis*
Plum, Pigeon: *Coccoloba diversifolia*
Plum, Red: *Prunus subcordata*
Plum, Sierra: *Prunus subcordata*
Plum, Wild: *Cerasus crenulata, Prunus americana,*
  P. nigra, P. sp., *P. subcordata*
Plum, Wild Red: *Prunus subcordata*
Plum, Wild Yellow: *Prunus americana*
Po-he-po-he: *Hydrocotyle poltata*
Po-hue-hue: *Ipomoea pes-caprae*
Poison, Fly: *Amianthium muscitoxicum*
Poison Camas: *Zigadenus venenosus*
Poison Hemlock: *Cicuta maculata, Conium maculatum*
Poison Ivy: *Toxicodendron pubescens, T. radicans, T. rydbergii*
Poison Oak: *Toxicodendron diversilobum, T. pubescens*
Poison Parsnip: *Cicuta maculata*
Poison Sanicle: *Sanicula bipinnata*
Poison Sego: *Zigadenus nuttallii*
Poison Snake Plant: *Asclepias tuberosa*
Poison Sumac: *Toxicodendron vernix*
Poisonvetch: *Astragalus racemosus*
Poison Water Hemlock: *Cicuta douglasii, C. maculata, C. virosa*
Poke: *Phytolacca americana*
Poke, Indian: *Veratrum viride*
Poke, Pigeon Berry: *Phytolacca americana*
Pokeberry: *Phytolacca americana*

Pokeweed: *Phytolacca americana*
Polypody: *Polypodium glycyrrhiza, P. virginianum*
Polypody, California: *Polypodium californicum*
Polypody, Golden: *Phlebodium aureum*
Polypody, Leather Leaf: *Polypodium scouleri*
Polypody, Leathery: *Polypodium scouleri*
Polypody, Rock: *Polypodium virginianum*
Polypody, Western: *Polypodium hesperium*
Pomegranate: *Punica granatum*
Pomme Blanche: *Pediomelum esculentum*
Pomme de Prairie: *Pediomelum esculentum*
Pond Lily: *Nuphar lutea, Nymphaea* sp.
Pondlily, Rocky Mountain: *Nuphar lutea*
Pond Lily, Yellow: *Nuphar lutea, N.* sp.
Pond Scum: *Spirogyra* sp.
Pondweed: *Menyanthes trifoliata, Potamogeton diversifolius, P. natans, P.* sp.
Poodle Dog Bush: *Turricula parryi*
Popcornflower, Arizona: *Plagiobothrys arizonicus*
Popcornflower, Fulvous: *Plagiobothrys fulvus*
Popcornflower, Mojave: *Plagiobothrys jonesii*
Popcornflower, Rusty: *Plagiobothrys nothofulvus*
Poplar: *Liriodendron tulipifera, Populus grandidentata, P.* sp., *P. tremuloides*
Poplar, Aspen: *Populus tremuloides*
Poplar, Balm (or Balsam): *Populus balsamifera*
Poplar, Carolina: *Populus deltoides*
Poplar, English: *Populus nigra*
Poplar, Jack's: *Populus ×jackii*
Poplar, Lombardy: *Populus nigra*
Poplar, White: *Populus alba*
Poppy, California: *Eschscholzia californica, E.* sp.
Poppy, Golden: *Eschscholzia californica*
Poppy, Leafy White Prickly: *Argemone polyanthemos*
Poppy, Matilija: *Romneya coulteri*
Poppy, Opium: *Papaver somniferum*
Poppy, Parish: *Eschscholzia parishii*
Poppy, Prickly: *Argemone munita, A. polyanthemos;* see also Pricklypoppy
Poppy, Thistle: *Argemone hispida*
Poppy, Tree: *Dendromecon rigida*
Poppy, White: *Papaver somniferum*
Poppy, Yellow: *Eschscholzia californica*
Poque: *Boschniakia rossica*
Poreleaf: *Porophyllum gracile*
Port Orford Cedar: *Chamaecyparis lawsoniana*
Possumhaw: *Viburnum nudum*
Potato: *Solanum tuberosum*
Potato, Deer's: *Liatris acidota, Liatris* sp.
Potato, Eskimo: *Claytonia tuberosa, Hedysarum alpinum*
Potato, Hog: *Hoffmannseggia glauca*
Potato, Indian: *Apios americana, Hoffmannseggia glauca, Solanum tuberosum*
Potato, James Wild: *Solanum jamesii*
Potato, Meskwaki: *Solanum tuberosum*
Potato, Native: *Solanum fendleri*
Potato, Ojibwe: *Solanum tuberosum*
Potato, Squaw: *Perideridia oregana*

Index of Common Names 917

Potato, Sweet: *Ipomoea batatas*
Potato, White: *Solanum tuberosum*
Potato, Wild: *Claytonia lanceolata, Hedysarum alpinum, Helianthus tuberosus, Oxypolis rigidior, Solanum fendleri, S. jamesii, S. triflorum, Triteleia grandiflora*
Poverty Threeawn: *Aristida divaricata*
Poverty Weed, American: *Iva axillaris*
Povertyweed, Nuttall's: *Monolepis nuttalliana*
Prairieclover, Canyonlands: *Dalea flavescens*
Prairieclover, Compact: *Dalea compacta*
Prairieclover, Dwarf: *Dalea nana*
Prairieclover, Golden: *Dalea aurea*
Prairieclover, Nineanther: *Dalea enneandra*
Prairie Clover, Purple: *Dalea purpurea*
Prairieclover, Silky: *Dalea villosa*
Prairie Clover, Slender White: *Dalea candida*
Prairieclover, Violet: *Dalea purpurea*
Prairie Clover, White: *Dalea candida*
Prairieclover, Woolly: *Dalea lanata*
Prester John: *Arisaema triphyllum*
Prettyface: *Triteleia ixioides*
Pricklyleaf: *Thymophylla pentachaeta*
Pricklypear: see also Cactus, Prickly Pear
Pricklypear, Beavertail: *Opuntia basilaris*
Pricklypear, Brownspined: *Opuntia parryi*
Pricklypear, Dollarjoint: *Opuntia chlorotica*
Pricklypear, Golden: *Opuntia basilaris*
Pricklypear, Grizzlybear: *Opuntia erinacea*
Pricklypear, Hairspine: *Opuntia polyacantha*
Pricklypear, Kelvin's: *Opuntia ×kelvinensis*
Prickly Pear, Plains: *Opuntia polyacantha*
Prickly Pear, Tall: *Opuntia whipplei*
Pricklypear, Texas: *Opuntia engelmannii*
Pricklypear, Tulip: *Opuntia phaeacantha*
Pricklypear, Twistspine: *Opuntia macrorhiza*
Pricklypoppy, Crested: *Argemone polyanthemos*
Pricklypoppy, Flatbud: *Argemone munita*
Pricklypoppy, Rough: *Argemone hispida*
Pride of Ohio: *Dodecatheon pulchellum*
Primrose, Brook Spike: *Boisduvalia stricta*
Primrose, Butte: *Oenothera cespitosa*
Primrose, Common Evening: *Oenothera biennis*
Primrose, Crownleaf Evening: *Oenothera coronopifolia*
Primrose, Denseflower Spike: *Boisduvalia densiflora*
Primrose, Desert Evening: *Oenothera primiveris*
Primrose, Evening: *Boisduvalia densiflora, Calylophus lavandulifolius, Camissonia brevipes, C. claviformis, C. tanacetifolia, Musineon divaricatum, Oenothera* spp.
Primrose, Fourpoint Evening: *Oenothera rhombipetala*
Primrose, Fragrant Evening: *Oenothera cespitosa*
Primrose, Hairy Evening: *Oenothera villosa*
Primrose, Hooker's Evening: *Oenothera elata*
Primrose, Narrowleaf Evening: *Oenothera fruticosa*
Primrose, Pale Evening: *Oenothera pallida*
Primrose, Rhombic Evening: *Oenothera rhombipetala*
Primrose, Shortfruit Evening: *Oenothera brachycarpa*
Primrose, Small Evening: *Oenothera perennis*
Primrose, Stemless Evening: *Oenothera triloba*
Primrose, Tansyleaf Evening: *Camissonia tanacetifolia*
Primrose, Tufted Evening: *Oenothera cespitosa*
Primrose, Upright Evening: *Boisduvalia stricta*
Primrose, White Evening: *Oenothera albicaulis*
Primrose, Yellow Evening: *Oenothera elata*
Primrosewillow, Carolina: *Ludwigia bonariensis*
Primrosewillow, Savannah: *Ludwigia virgata*
Prince's Pine: *Chimaphila umbellata*
Prince's Pine, Little: *Chimaphila menziesii*
Prince's Pine, Striped: *Chimaphila maculata*
Prince's Plume: *Stanleya albescens, S. pinnata*
Privet, Eastern Swamp: *Forestiera acuminata*
Privet, Florida: *Forestiera segregata*
Privet, Wild: *Forestiera pubescens*
Psoralea, Few Flowered: *Psoralidium tenuiflorum*
Psoralea, Silver Leaf: *Pediomelum argophyllum*
Psoralea, Slender Flowered: *Psoralidium tenuiflorum*
Puccoon: *Lithospermum canescens, L. caroliniense, L. incisum, L. ruderale*
Puccoon, Hairy: *Lithospermum caroliniense*
Puccoon, Hoary: *Lithospermum canescens*
Puccoon, Narrow Leaved: *Lithospermum incisum*
Puffball: *Bovista pila, Bovista plumbea, Bovistella* sp., *Lycoperdon* sp.
Pukiawe: *Styphelia tameiameiae*
Pumpkin: *Cucurbita maxima, C. moschata, C. pepo*
Pumpkin, Cheese: *Cucurbita moschata*
Pumpkin, Cushaw: *Cucurbita moschata*
Pumpkin, Field: *Cucurbita pepo*
Pumpkin, Large Pie: *Cucurbita pepo*
Pumpkin, Pumpion: *Cucurbita pepo*
Pumpkin, Wild: *Cucurbita foetidissima*
Puncturevine: *Tribulus terrestris*
Purplemat: *Nama demissum*
Purselane: *Portulaca oleracea*; see also Purslane
Pursely: *Amaranthus blitoides*
Purslane: *Portulaca oleracea, P.* sp.
Purslane, Erect: *Suaeda calceoliformis*
Purslane, Milk: *Chamaesyce glyptosperma*
Purslane, Sea: *Honckenya peploides*
Pusley: *Portulaca oleracea*
Pusley, Chinese: *Heliotropium curassavicum*
Pussypaws: *Cistanthe monandra*
Pussy Toes: *Antennaria parvifolia, A. rosea, A. rosulata, A.* sp.
Pussytoes, Field (or Howell's): *Antennaria howellii*
Pussytoes, Kaibab: *Antennaria rosulata*
Pussytoes, Pearly: *Antennaria anaphaloides*
Pussy Toes, Pink (or Rosy): *Antennaria rosea*
Pussytoes, Small-leaf: *Antennaria parvifolia*
Pussytoes, Stoloniferous: *Antennaria dioica*
Putty Root: *Aplectrum hyemale*
Pygmy Weed: *Dudleya* sp.
Pyrola: *Pyrola elliptica*
Pyrola, One Flowered: *Moneses uniflora*
Quack Grass: *Elytrigia repens*
Quail Brush: *Atriplex lentiformis*
Quail Bush: *Atriplex lentiformis*
Quamash: *Camassia scilloides*
Queen Anne's Lace: *Daucus carota, D. pusillus*
Queen of the Meadow: *Eupatorium purpureum*
Queen of the Meadows: *Spiraea salicifolia*
Queen of the Prairie: *Filipendula rubra*
Queen's Cup: *Clintonia uniflora*
Queen's Delight: *Stillingia sylvatica*
Quinine: *Cinchona calisaya*
Quinine, Wild: *Parthenium integrifolium*
Quinine Cherry: *Prunus emarginata*
Rabbitbells: *Crotalaria rotundifolia*
Rabbit Brush: *Chrysothamnus depressus, C. greenei, C. nauseosus, C. parryi, C.* sp., *C. viscidiflorus*
Rabbit Brush, Broadscale: *Chrysothamnus nauseosus*
Rabbit Brush, Douglas: *Chrysothamnus viscidiflorus*
Rabbitbrush, Green: *Chrysothamnus viscidiflorus*
Rabbitbrush, Greene's: *Chrysothamnus greenei*
Rabbit Brush, Heavy Scented: *Chrysothamnus nauseosus*
Rabbitbrush, Howard's: *Chrysothamnus parryi*
Rabbit Brush, Little: *Chrysothamnus viscidiflorus*
Rabbitbrush, Parry's: *Chrysothamnus parryi*
Rabbit Brush, Rubber: *Chrysothamnus nauseosus*
Rabbit Brush, Sticky Flowered: *Chrysothamnus viscidiflorus*
Rabbit Bush: *Chrysothamnus nauseosus, C. parryi, C.* sp., *C. viscidiflorus*
Rabbitbush Heathgoldenrod: *Ericameria bloomeri*
Rabbit Foot: *Lespedeza capitata*
Rabbit's Foot: *Polypogon monspeliensis*
Rabbit Thorn: *Lycium pallidum*
Raccoon Liver: *Stenandrium dulce*
Racine Blanche: *Lomatium cous*
Radish: *Raphanus sativus*
Raffia, Wild: *Juncus textilis*
Ragweed, Ambrosia Leaf Burr: *Ambrosia ambrosioides*
Ragweed, American: *Ambrosia artemisiifolia*
Ragweed, Annual: *Ambrosia artemisiifolia*
Ragweed, Cuman: *Ambrosia psilostachya*
Ragweed, Flatspine Burr: *Ambrosia acanthicarpa*
Ragweed, Giant: *Ambrosia trifida*
Ragweed, Silver Burr: *Ambrosia chamissonis*
Ragweed, Slimleaf Burr: *Ambrosia tenuifolia*
Ragweed, Small: *Ambrosia artemisiifolia*
Ragweed, Western: *Ambrosia psilostachya*
Ragwort, Fendler's: *Senecio fendleri*
Ragwort, Golden: *Senecio aureus*
Ragwort, Seaside: *Senecio pseudoarnica*
Ragwort, Small's: *Senecio anonymus*
Ragwort, Tansy: *Senecio jacobaea*
Ragwort Groundsel: *Senecio multicapitatus*
Ramona: *Salvia dorrii*

Ramp (or Ramps): *Allium tricoccum*
Ram's Horn: *Proboscidea louisianica*
Rape: *Brassica napus*
Rape Mustard: *Brassica rapa*
Raspberry, American Red: *Rubus idaeus*
Raspberry, Arizona Red: *Rubus idaeus*
Raspberry, Black: *Rubus leucodermis, R. occidentalis*
Raspberry, Dwarf: *Rubus arcticus, R. pubescens*
Raspberry, Flowering: *Rubus odoratus*
Raspberry, Grayleaf Red: *Rubus idaeus*
Raspberry, Mountain: *Rubus chamaemorus*
Raspberry, Purple Flowered: *Rubus odoratus*
Raspberry, Red: *Rubus aculeatissimus, R. idaeus*
Raspberry, Snow: *Rubus nivalis*
Raspberry, Strawberryleaf: *Rubus pedatus*
Raspberry, Trailing Wild: *Rubus pedatus*
Raspberry, Western: *Rubus leucodermis*
Raspberry, Whitebark: *Rubus leucodermis*
Raspberry, Wild: *Rubus idaeus, R. occidentalis, R. odoratus, R. procumbens*
Raspberry, Wild Black: *Rubus occidentalis*
Raspberry, Wild Red: *Rubus idaeus*
Ratany: *Krameria grayi*
Ratany, Littleleaf: *Krameria erecta*
Ratbane: *Chimaphila umbellata*
Rattan Vine: *Berchemia scandens*
Rattle Box: *Crotalaria rotundifolia, C. sagittalis*
Rattle Pod, American: *Astragalus americanus*
Rattle Pod, Black: *Baptisia bracteata*
Rattle Pod, Little: *Astragalus canadensis*
Rattlesnake Master: *Eryngium aquaticum, E. yuccifolium, Liatris laxa*
Rattlesnake Plantain: *Goodyera oblongifolia, G. pubescens, G. repens*
Rattlesnake Root: *Prenanthes alata, P. altissima, P. alba*
Rattlesnake Weed: *Chamaesyce albomarginata, C. ocellata, Daucus pusillus, Echinacea angustifolia, Hieracium venosum*
Rattle Weed: *Astragalus convallarius, Oxytropis lagopus*
Rattleweed, Many Colored: *Astragalus allochrous*
Red Bell: *Fritillaria recurva*
Redberry: *Rhamnus crocea*
Red Bud: *Cercis canadensis, C.* sp.
Redbug: *Cercis canadensis*
Redcardinal: *Erythrina herbacea*
Red Maids: *Calandrinia ciliata*
Red Root: *Amaranthus albus, Ceanothus americanus, Lachnanthes caroliana*
Red Root, Small: *Ceanothus herbaceus*
Redscale: *Atriplex rosea*
Redshank: *Adenostoma sparsifolium*
Redstem: *Ammannia coccinea*
Redwood: *Sequoia sempervirens*
Reed, Bur: *Sparganium eurycarpum*
Reed, Carrizal (or Carrizo): *Phragmites australis*
Reed, Common: *Juncus effusus, Phragmites australis*
Reed, Giant: *Arundo donax*

Reed, Giant Bur: *Sparganium eurycarpum*
Reed, Giant Sand: *Calamovilfa gigantea*
Reed, Rood: *Cinna arundinacea*
Reed, Sand: *Calamovilfa gigantea, C. longifolia*
Reed, Stout Wood: *Cinna arundinacea*
Rein Orchid: see Orchid, Rein
Resurrection Flower: *Lewisia rediviva*
Rhubarb: *Rheum rhabarbarum, Rheum rhaponticum*
Rhubarb, Alaska Wild: *Polygonum alpinum*
Rhubarb, American Wild: *Rumex hymenosepalus*
Rhubarb, Eskimo: *Polygonum alpinum*
Rhubarb, False: *Rheum rhaponticum*
Rhubarb, Indian: *Darmera peltata, Heracleum maximum*
Rhubarb, Wild: *Heracleum maximum, Polygonum alpinum, Rumex crispus, R. hymenosepalus*
Ribbonwood: *Adenostoma sparsifolium*
Ribgrass: *Plantago lanceolata*
Rice: *Oryza sativa*
Rice, Indian: *Zizania aquatica, Z. palustris*
Rice, Jungle: *Echinochloa colona*
Rice, Mountain: *Oryzopsis hymenoides*
Rice, Wild: *Chasmanthium latifolium, Zizania aquatica, Z. palustris*
Rice Grass: *Piptatherum miliaceum*
Ricegrass, Indian: *Oryzopsis hymenoides*
Rice Root: *Fritillaria camschatcensis, F. lanceolata*
Rich Weed: *Collinsonia canadensis, Pilea pumila*
Rockbrake, Sitka: *Cryptogramma sitchensis*
Rock Cress: *Arabis drummondii, A. fendleri, A. puberula*
Rockcress, Drummond's: *Arabis drummondii*
Rock Cress, Early Flowering: *Arabis perennans*
Rockcress, Fendler's: *Arabis fendleri*
Rock Cress, Few Flowered: *Arabis sparsiflora*
Rock Cress, Hoelboell's: *Arabis holboellii*
Rock Cress, Kamchatka: *Arabis lyrata*
Rockcress, Lyrate: *Arabis lyrata*
Rockcress, Perennial: *Arabis perennans*
Rockcress, Sicklepod: *Arabis sparsiflora*
Rockcress, Silver: *Arabis puberula*
Rockcress, Tower: *Arabis glabra*
Rocket, London: *Sisymbrium irio*
Rockgoldenrod, Grassy: *Petradoria pumila*
Rock Harlequin: *Corydalis sempervirens*
Rockjasmine, Pygmyflower: *Androsace septentrionalis*
Rockjasmine, Western: *Androsace occidentalis*
Rockjasmine Buckwheat: *Eriogonum androsaceum*
Rock Spiraea: *Holodiscus discolor, H. dumosus*
Rock Spiraea, Rocky Mountain: *Petrophyton caespitosum*
Rood Reed Grass: *Cinna arundinacea*
Rosary Babybonnets: *Coursetia glandulosa*
Rosary Root: *Apios americana*
Rose, Bald Hip: *Rosa gymnocarpa*
Rose, Bristly Nootka: *Rosa nutkana*
Rose, California Wild: *Rosa californica*
Rose, Carolina: *Rosa carolina*

Rose, Cluster: *Rosa pisocarpa*
Rose, Dwarf: *Rosa gymnocarpa*
Rose, Dwarf Wild: *Rosa virginiana*
Rose, Early Wild: *Rosa blanda*
Rose, French: *Rosa gallica*
Rose, Interior Wild: *Rosa woodsii*
Rose, Nootka: *Rosa nutkana*
Rose, Pasture: *Rosa carolina*
Rose, Prairie: *Rosa arkansana*
Rose, Prickly: *Rosa acicularis*
Rose, Rocky Mountain: *Rosa woodsii*
Rose, Smooth: *Rosa blanda*
Rose, Swamp: *Rosa palustris*
Rose, Tundra: *Pentaphylloides floribunda*
Rose, Virginia: *Rosa virginiana*
Rose, Wild: *Rosa acicularis, R. arkansana, R. californica, R. gymnocarpa, R. nutkana, R. palustris, R. pisocarpa, R.* sp., *R. woodsii*
Rose, Wood: *Rosa gymnocarpa*
Rose, Wood's Wild: *Rosa woodsii*
Roseberry: *Rosa* sp.
Roseflower Stonecrop: *Sedum laxum*
Rosegentian, Slender: *Sabatia campanulata*
Rosemallow, Crimsoneyed: *Hibiscus moscheutos*
Rosemary, American Wild: *Trichostema lanatum*
Rosemary, Bog: *Andromeda polifolia*
Rosemarymint, Hoary: *Poliomintha incana*
Rose Pink: *Sabatia angularis*
Roseroot: *Sedum rosea*
Rosilla: *Helenium puberulum*
Rosin Weed: *Grindelia* sp., *G. squarrosa, Silphium compositum*
Rosin Weed, Entire Leaved: *Silphium integrifolium*
Rosinweed, Fremont's Western: *Calycadenia fremontii*
Rosinweed, Kidneyleaf: *Silphium compositum*
Rosinweed, Wholeleaf: *Silphium integrifolium*
Rosy Twistedstalk: *Streptopus roseus*
Roughfruit Berry: *Rubus lasiococcus*
Roughfruit Fairybells: *Disporum trachycarpum*
Rubber Plant, Colorado: *Hymenoxys richardsonii*
Rubber Plant, Metcalfe: *Hymenoxys richardsonii*
Rubber Plant, South American: *Dugaldia hoopesii*
Rubber Rabbit Brush: *Chrysothamnus nauseosus*
Rubberweed: *Hymenoxys bigelovii, H. richardsonii*
Rue, American Garden: *Tephrosia virginiana*
Rue, Bush: *Cneoridium dumosum*
Rue, Desert: *Thamnosma montana*
Rue, Fringed: *Ruta chalepensis*
Rue, Garden: *Ruta graveolens*
Rue, Goat's: *Tephrosia virginiana*
Rue, Meadow: see Meadow Rue
Rush, Baltic: *Juncus balticus*
Rush, Basket: *Juncus textilis*
Rush, Bog: *Juncus* sp.
Rush, Cattail: *Typha latifolia*
Rush, Dagger Leaved: *Juncus ensifolius*
Rush, Dudley's: *Juncus dudleyi*
Rush, Mertens's: *Juncus mertensianus*
Rush, Mexican: *Juncus mexicanus*

Index of Common Names 919

Rush, Moor: *Juncus stygius*
Rush, Pacific: *Juncus effusus*
Rush, Path: *Juncus tenuis*
Rush, Poverty: *Juncus tenuis*
Rush, Salt: *Juncus lesueurii*
Rush, Scouring: see Scouring Rush
Rush, Soft: *Juncus effusus*
Rush, Spike: see Spikerush
Rush, Swordleaf: *Juncus ensifolius*
Rush, Toad: *Juncus bufonius*
Rush, Torrey's: *Juncus torreyi*
Rush, Wire: *Juncus balticus*
Rye, Giant: *Leymus condensatus*
Rye, Giant Wild: *Leymus cinereus*
Rye, Stout Nodding Wild: *Elymus canadensis*
Rye, Wild: *Elymus canadensis, E. glaucus, E. sibiricus, E.* sp.*, Leymus cinereus, L. triticoides, Sitanion* sp.
Rye Grass: *Elymus glaucus, E.* sp.*, Leymus cinereus, L. condensatus, L. mollis, L. temulentum, L. triticoides*
Sacahuista: *Nolina microcarpa*
Sacapellote: *Acourtia microcephala*
Sacaton: *Sporobolus wrightii*
Sacaton, Alkali: *Sporobolus airoides*
Saddlebush: *Mortonia sempervirens*
Safflower: *Carthamus tinctorius*
Sage, American Black: *Salvia mellifera*
Sage, Antelope: *Eriogonum jamesii*
Sage, Bastard: *Eriogonum wrightii*
Sage, Big: *Artemisia tridentata*
Sage, Black: *Ambrosia psilostachya, Salvia mellifera*
Sage, Blue: *Artemisia cana, Salvia dorrii, S. lyrata*
Sage, Bud: *Artemisia spinescens*
Sage, Bur: *Ambrosia acanthicarpa, A. tenuifolia*
Sage, Chicken: *Artemisia nova*
Sage, Fringed: *Artemisia frigida*
Sage, Garden: *Salvia officinalis, S.* sp.
Sage, Grayball: *Salvia dorrii*
Sage, Green: *Artemisia campestris, A. dracunculus*
Sage, Kitchen: *Salvia officinalis*
Sage, Little Wild: *Artemisia frigida*
Sage, Lyre Leaved: *Salvia lyrata*
Sage, Man: *Artemisia ludoviciana*
Sage, Pasture: *Artemisia frigida*
Sage, Pitcher: *Lepechinia calycina*
Sage, Prairie: *Artemisia ludoviciana*
Sage, Purple: *Salvia dorrii*
Sage, Rocky Mountain: *Artemisia tridentata*
Sage, Silver: *Artemisia cana, A. filifolia*
Sage, Sweet: *Artemisia dracunculus, A. frigida, Poliomintha incana*
Sage, Thistle: *Salvia carduacea*
Sage, Western: *Artemisia ludoviciana*
Sage, White: *Artemisia ludoviciana, Krascheninnikovia lanata, Salvia apiana*
Sage, Wild: *Artemisia frigida, A. ludoviciana, A.* sp.
Sage, Winter: *Krascheninnikovia lanata*
Sagebrush, Alaska: *Artemisia alaskana*
Sagebrush, Basin: *Artemisia tridentata*

Sagebrush, Big: *Artemisia* sp.*, A. tridentata*
Sagebrush, Black: *Artemisia nova, A. tridentata*
Sagebrush, Boreal: *Artemisia arctica*
Sagebrush, Bud: *Artemisia spinescens*
Sagebrush, California: *Artemisia californica*
Sagebrush, Idaho Threetip: *Artemisia tripartita*
Sagebrush, Mexican White: *Artemisia ludoviciana*
Sagebrush, Mountain: *Artemisia frigida*
Sagebrush, Sand: *Artemisia filifolia*
Sagebrush, Short: *Artemisia tripartita*
Sagebrush, Silver: *Artemisia cana, A. filifolia*
Sagebrush, Three Lobed (or Three Toothed): *Artemisia tripartita*
Sagebrush, Western: *Artemisia campestris*
Sagebrush, Wright: *Artemisia carruthii*
Sage Bush: *Artemisia tilesii*
Sagewort, Absinth: *Artemisia absinthium*
Sagewort, Carruth's: *Artemisia carruthii*
Sagewort, Douglas's: *Artemisia douglasiana*
Sagewort, Dragon: *Artemisia dracunculus*
Sagewort, Field: *Artemisia campestris*
Sagewort, Foothill: *Artemisia ludoviciana*
Sagewort, Forked: *Artemisia furcata*
Sagewort, Louisiana: *Artemisia ludoviciana*
Sagewort, Mountain: *Artemisia ludoviciana*
Sagewort, Pasture: *Artemisia frigida*
Sagewort, Prairie: *Artemisia frigida*
Saguaro (or Sahuaro): *Carnegia gigantea*
St. Andrew's Cross: *Hypericum hypericoides, H.* sp.
St. Anthony's Turnip: *Ranunculus bulbosus*
St. Augustine Grass: *Stenotaphrum secundatum*
St. John's Wort, Coastalplain: *Hypericum brachyphyllum*
St. John's Wort, Dotted: *Hypericum punctatum*
St. John's Wort, Great: *Hypericum ascyron*
St. John's Wort, Marsh: *Triadenum virginicum*
St. John's Wort, Pale: *Hypericum ellipticum*
St. John's Wort, Peelbark: *Hypericum fasciculatum*
St. John's Wort, Scouler's: *Hypericum scouleri*
St. John's Wort, Spotted: *Hypericum punctatum*
St. Johnswort, Virginia Marsh: *Triadenum virginicum*
St. John's Wort, Western: *Hypericum scouleri*
St. Peter's Wort: *Hypericum crux-andreae, H. multicaule*
Salal: *Gaultheria ovatifolia, G. shallon*
Salmonberry: *Rubus chamaemorus, R. parviflorus, R.* sp.*, R. spectabilis*
Salmon's Eye, Summer: *Ranunculus occidentalis*
Salsify: *Tragopogon porrifolius, T. pratensis, T.* sp.
Salt Brush: *Atriplex confertifolia, A. obovata*
Salt Brush, Fourwing: *Atriplex canescens*
Salt Bush: *Atriplex* spp.*, Chenopodium incanum, Suaeda arborescens*
Saltbush, Australian: *Atriplex semibaccata*
Saltbush, Big: *Atriplex lentiformis*
Salt Bush, Bracted: *Atriplex serenana*
Saltbush, Cattle: *Atriplex polycarpa*
Salt Bush, Desert: *Atriplex polycarpa*
Salt Bush, Four Winged: *Atriplex* sp.

Salt Bush, Fourwing: *Atriplex canescens*
Saltbush, Garrett's: *Atriplex garrettii*
Salt Bush, Hoary: *Atriplex canescens*
Saltbush, Mound: *Atriplex obovata*
Saltbush, Nuttall's: *Atriplex nuttallii*
Saltbush, Sack: *Atriplex saccaria*
Saltbush, Shadscale: *Atriplex confertifolia*
Saltbush, Silverscale: *Atriplex argentea*
Salt Bush, Spiny: *Atriplex confertifolia*
Saltbush, Torrey's: *Atriplex torreyi*
Saltbush, Wedgescale: *Atriplex truncata*
Saltbush, Wheelscale: *Atriplex elegans*
Saltbush, Wright's: *Atriplex wrightii*
Saltgrass, Inland: *Distichlis spicata*
Salt Tule: *Scirpus acutus*
Salt Weed: *Atriplex confertifolia*
Saltweed, Powell's: *Atriplex powellii*
Saltweed, Tumbling: *Atriplex rosea*
Saltwort: *Salsola australis*
Samphire: *Salicornia maritima*
Sandcherry: *Prunus pumila*
Sanddune Catseye: *Cryptantha fendleri*
Sanddune Skyrocket: *Ipomopsis gunnisonii*
Sandfood: *Pholisma sonorae*
Sandmat, Contura Creek: *Chamaesyce ocellata*
Sandmat, Fendler's: *Chamaesyce fendleri*
Sandmat, Geyer's: *Chamaesyce geyeri*
Sandmat, Graceful: *Chamaesyce hypericifolia*
Sandmat, Hoary: *Chamaesyce lata*
Sandmat, Ribseed: *Chamaesyce glyptosperma*
Sandmat, Smallseed: *Chamaesyce polycarpa*
Sandmat, Spotted: *Chamaesyce maculata*
Sandmat, Squaw: *Chamaesyce melanadenia*
Sandmat, Threadstem: *Chamaesyce revoluta*
Sandmat, Thymeleaf: *Chamaesyce serpyllifolia*
Sandmat, Variable: *Chamaesyce multiformis*
Sandmat, Whitemargin: *Chamaesyce albomarginata*
Sandplant: *Honckenya peploides*
Sandpuffs: *Tripterocalyx carnea*
Sandreed, Giant: *Calamovilfa gigantea*
Sandreed, Prairie: *Calamovilfa longifolia*
Sand Verbena, Coastal: *Abronia latifolia*
Sand Verbena, Desert: *Abronia villosa*
Sand Verbena, Fragrant: *Abronia fragrans*
Sand Verbena, Fragrant White: *Abronia elliptica*
Sand Verbena, Pink: *Abronia villosa*
Sand Verbena, Snowball: *Abronia fragrans*
Sand Verbena, Transmontane: *Abronia turbinata*
Sand Verbena, White: *Abronia elliptica, A. turbinata*
Sandweed: *Hypericum fasciculatum*
Sandwort, Ballhead: *Arenaria congesta*
Sandwort, Eastwood's: *Arenaria eastwoodiae*
Sandwort, Fendler's: *Arenaria fendleri*
Sandwort, Mojave: *Arenaria macradenia*
Sandwort, Prickly: *Arenaria aculeata*
Sandwort, Seabeach: *Honckenya peploides*
Sandwort, Spreading: *Arenaria lanuginosa*
Sandyseed Clammyweed: *Polanisia dodecandra*
Sangre de Cristo: *Jatropha cardiophylla*

Sanicle: *Sanicula marilandica, S.* sp.
Sanicle, Poison: *Sanicula bipinnata*
Sanicle, Purple: *Sanicula bipinnatifida*
Santolina Pincushion: *Chaenactis santolinoides*
Sarsaparilla (or Sarsaprilla): *Aralia nudicaulis*
Sarsaparilla, Black: *Aristolochia macrophylla, Cocculus carolinus*
Sarsaparilla, Bristly: *Aralia hispida*
Sarsaparilla, White: *Passiflora incarnata*
Sarsaparilla, Wild: *Aralia nudicaulis*
Sarsaparilla, Yellow: *Menispermum canadense*
Sarvice (or Sarvis) Berry: see Serviceberry
Saskatoon: *Amelanchier alnifolia*
Sassafras: *Sassafras albidum*
Sasul Brier: *Smilax* sp.
Satinleaf: *Chrysophyllum oliviforme*
Sauwi: *Panicum sonorum*
Savory, Summer: *Satureja hortensis*
Sawbrier: *Smilax glauca*
Sawgrass: *Cladium mariscus*
Saxifrage: *Fendlera rupicola, Saxifraga ferruginea*
Saxifrage, Anise Burnet: *Pimpinella anisum*
Saya: *Amoreuxia palmatifida*
Scaldweed: *Cuscuta gronovii*
Scentbottle: *Platanthera dilatata*
Scorpionweed, California: *Phacelia californica*
Scorpionweed, New Mexico: *Phacelia neomexicana*
Scorpionweed, Smallflower: *Phacelia dubia*
Scouring Rush: *Equisetum arvense, E. hyemale, E. laevigatum, E.* sp., *E. variegatum*
Scouring Rush, Common: *Equisetum hyemale*
Scouring Rush, Dwarf: *Equisetum scirpoides*
Scouring Rush, Smooth: *Equisetum laevigatum*
Scouring Rush, Stone: *Equisetum hyemale*
Scouringrush, Variegated: *Equisetum variegatum*
Scouring Rush, Western: *Equisetum hyemale*
Scrambledeggs: *Corydalis aurea*
Screw Bean: *Prosopis glandulosa, P. pubescens*
Screwpine, Tahitian: *Pandanus tectorius*
Scullcap: see Skullcap
Scurf Pea: *Pediomelum cuspidatum, P. hypogaeum, Psoralidium lanceolatum, P. tenuiflorum*
Scurf Pea, Hoary: *Pediomelum canescens*
Scurfpea, Breadroot: *Pediomelum esculentum*
Scurfpea, Lemon: *Psoralidium lanceolatum*
Scurfpea, Silverleaf: *Pediomelum argophyllum*
Scurfpea, Slimflower: *Psoralidium tenuiflorum*
Scurvygrass: *Cochlearia officinalis*
Seabean: *Mucuna gigantea*
Sea Blite: *Suaeda calceoliformis, S. moquinii, S.* sp.
Sealavender, California: *Limonium californicum*
Sealavender, Carolina: *Limonium carolinianum*
Sealavender, Mediterranean: *Limonium vulgare*
Seasonvine: *Cissus verticillata*
Seaweed: *Fucus* sp., *Phyllospadix* sp., *Porphyra laciniata*
Seaweed, Bladder: *Halosaccion glandiforme*
Seaweed, Bubble: *Leathesia difformis*
Seaweed, Edible: *Porphyra abbottae, P. perforata*
Sea Wrack: *Fucus gardneri, Fucus* sp., *Zostera marina*
Sedge, Basket: *Carex aquatilis, C. obnupta*
Sedge, Beaked: *Carex rostrata*
Sedge, Blackroot: *Scirpus maritimus*
Sedge, Broad Leaved: *Carex platyphylla*
Sedge, Broom: *Andropogon virginicus*
Sedge, Douglas's: *Carex douglasii*
Sedge, Drooping: *Carex prasina*
Sedge, Fescue: *Carex brevior*
Sedge, Fewseed: *Carex oligosperma*
Sedge, Flat: see Flatsedge
Sedge, Fox: *Carex vulpinoidea*
Sedge, Knotted Spike: *Eleocharis equisetoides*
Sedge, Mendocino: *Carex mendocinensis*
Sedge, Mountain: *Carex concinnoides*
Sedge, Nebraska: *Carex nebrascensis*
Sedge, Northwest Territory: *Carex utriculata*
Sedge, Northwestern: *Carex concinnoides*
Sedge, Plantain Leaved: *Carex plantaginea*
Sedge, Santa Barbara: *Carex barbarae*
Sedge, Sitka: *Carex aquatilis*
Sedge, Slough: *Carex atherodes, C. obnupta*; see also Sledge
Sedge, Smallwing: *Carex microptera*
Sedge, Spike: *Eleocharis equisetoides, E. geniculata*
Sedge, Sun: *Carex inops*
Sedge, Swamp: *Carex exsiccata*
Sedge, Tall: *Carex obnupta*
Sedge, Water: *Carex aquatilis*
Sedge, Western Fox: *Carex vicaria*
Sedge, Western Inflated: *Carex exsiccata*
Seedbox: *Ludwigia virgata*
Seepweed: *Suaeda calceoliformis, S. moquinii*
Seepweed, Desert: *Suaeda suffrutescens*
Seepweed, Pursh: *Suaeda calceoliformis*
Sego: *Calochortus nuttallii*
Sego, Poison: *Zigadenus nuttallii*
Self Heal: *Prunella vulgaris*
Seneca Root: *Polygala senega*
Senna: *Senna tora*
Senna, American: *Senna marilandica*
Senna, Coffee: *Senna occidentalis*
Senna, Wild: *Senna hebecarpa*
Sensitive Plant, American Wild: *Senna hebecarpa*
Sensitive Plant, Maryland Wild: *Senna marilandica*
Sensitive Plant, Wild: *Chamaecrista nictitans*
Septicweed: *Senna occidentalis*
Serviceberry, Allegheny: *Amelanchier laevis*
Serviceberry, Canadian: *Amelanchier canadensis*
Serviceberry, Common: *Amelanchier arborea*
Service Berry, Cusick's: *Amelanchier alnifolia*
Service Berry, Flowering: *Amelanchier alnifolia*
Service Berry, Northwestern: *Amelanchier alnifolia*
Service Berry, Pacific: *Amelanchier alnifolia*
Serviceberry, Pale: *Amelanchier pallida*
Service Berry, Prune Leaved: *Amelanchier utahensis*
Serviceberry, Roundleaf: *Amelanchier sanguinea*
Serviceberry, Running: *Amelanchier stolonifera*
Serviceberry, Saskatoon: *Amelanchier alnifolia*
Serviceberry, Utah: *Amelanchier utahensis*
Service Berry, Western: *Amelanchier alnifolia, A. pallida*
Service Berry Tree: *Amelanchier arborea*
Service Tree: *Pyrus* sp.
Sesame: *Sesamum orientale*
Seven Bark: *Hydrangea arborescens*
Shadblow: *Amelanchier pallida*
Shad Brush: *Amelanchier canadensis*
Shad Bush: *Amelanchier canadensis, Corylus americana*
Shadscale: *Atriplex canescens, A. confertifolia*
Shadscale Saltbush: *Atriplex confertifolia*
Shaggyfruit Pepperweed: *Lepidium lasiocarpum*
Shaggytuft, Sweet: *Stenandrium dulce*
Shawnee Salad: *Hydrophyllum virginianum*
Sheepberry: *Viburnum lentago*
Sheep Laurel: *Kalmia angustifolia*
Sheep Sorrel: *Oxalis stricta, O. violacea, Rumex acetosa, R. acetosella*
Sheep Sorrel Dock: *Rumex acetosella*
Shelf Fungus: *Fomes igniarius, F.* sp., *Fomitopsis pinicola, F.* sp., *Ganoderma* sp., *Polyporus* sp.
Shepherd's Purse: *Capsella bursa-pastoris*
Shield Fern: *Dryopteris arguta, D. carthusiana, D. cristata, D. expansa, D.* sp., *Thelypteris kunthii*
Shield Lichen: *Cetraria crispa*
Shin Leaf: *Pyrola americana, P. asarifolia, P. elliptica*
Shin Leaf, Waxflower (or White Veined): *Pyrola picta*
Shooting Star: *Dodecatheon hendersonii, D. jeffreyi, D. pulchellum*
Sida, Round Leaved: *Malvella leprosa*
Silkgrass, Narrowleaf: *Pityopsis graminifolia*
Silktassel, Pallid: *Garrya flavescens*
Silktassel, Wavyleaf: *Garrya elliptica*
Silk Tassel Bush: *Garrya elliptica, G. flavescens*
Silkweed: *Asclepias syriaca*
Silverbells: *Halesia carolina*
Silverberry: *Elaeagnus commutata*
Silverbush, Charleston Mountain: *Argythamnia cyanophylla*
Silvergreen: *Adenocaulon bicolor*
Silver Plant: *Eriogonum ovalifolium*
Silverpuffs, Cutleaf: *Microseris laciniata*
Silver Weed: *Argentina anserina, A. egedii*
Simmon: *Diospyros virginiana*
Single Delight: *Moneses uniflora*
Sisal, False: *Agave decipiens*
Skeleton Plant: *Lygodesmia juncea*
Skeletonplant, Largeflower: *Lygodesmia grandiflora*
Skeletonweed, Beaked: *Shinnersoseris rostrata*
Skeleton Weed, Rush: *Lygodesmia juncea*
Skeleton Weed, Spiny (or Thorn): *Stephanomeria spinosa*
Skullcap, Blue: *Scutellaria lateriflora*
Skullcap, California: *Scutellaria californica*
Skullcap, Hairy: *Scutellaria elliptica*
Skullcap, Hoary: *Scutellaria incana*
Skullcap, Mad Dog: *Scutellaria lateriflora*
Skullcap, Marsh: *Scutellaria galericulata*

Index of Common Names 921

Skullcap, Small: *Scutellaria parvula*
Skunk Berry: *Rhus trilobata*
Skunk Bush: *Rhus aromatica, R. trilobata*
Skunk Cabbage: *Lysichiton americanus, L.* sp., *Symplocarpus foetidus*
Skunk Weed: *Croton texensis, Navarretia* sp.
Sky Rocket: *Ipomopsis aggregata*
Skyrocket, Iron: *Ipomopsis laxiflora*
Skyrocket, Sanddune: *Ipomopsis gunnisonii*
Sledge, Slough Grass: *Carex nebrascensis*; see also Sedge
Sleepingplant: *Chamaecrista fasciculata*
Sleepydick: *Ornithogalum umbellatum*
Sleepygrass: *Stipa robusta*
Slim Solomon: *Maianthemum stellatum*
Slip Root: *Cypripedium acaule*
Sloe, Stag Bush: *Viburnum prunifolium*
Slough Grass: *Beckmannia syzigachne, Carex barbarae, Spartina pectinata*
Slough Grass Sledge: *Carex nebrascensis*
Slough Sedge: *Carex atherodes, C. obnupta*
Smart Weed: *Polygonum argyrocoleon, P. bistortoides, P. hydropiper, P. lapathifolium, P. persicaria, P. punctatum, P.* sp.
Smartweed, Carey's: *Polygonum careyi*
Smartweed, Dotted: *Polygonum punctatum*
Smartweed, Longroot: *Polygonum amphibium*
Smart Weed, Pennsylvania: *Polygonum pensylvanicum*
Smart Weed, Water: *Polygonum amphibium, P. punctatum*
Smartweek: *Polygonum hydropiper*
Smilograss: *Piptatherum miliaceum*
Smoke Brush: *Ceanothus cuneatus*
Smoke Bush: *Ceanothus cuneatus, Psorothamnus polydenius*
Snakeberry: *Maianthemum dilatatum*
Snake Chief Root: *Manfreda virginica*
Snakelily, Congested: *Dichelostemma pulchellum*
Snakelily, Ida May's: *Dichelostemma ida-maia*
Snakelily, Roundtooth: *Dichelostemma multiflorum*
Snakelily, Twining: *Dichelostemma volubile*
Snake Plant, Poison: *Asclepias tuberosa*
Snakeroot: *Aristolochia serpentaria, Eryngium yuccifolium, Osmorhiza* sp.
Snakeroot, Bitter: *Eryngium aquaticum*
Snakeroot, Black: *Cimicifuga racemosa, Sanicula marilandica, S. odorata, S. smallii*
Snakeroot, Bur: *Sanicula canadensis*
Snakeroot, Button: *Eryngium aquaticum, E. yuccifolium*
Snakeroot, Clustered: *Sanicula odorata*
Snakeroot, Fragrant: *Ageratina herbacea*
Snakeroot, Sampson: *Orbexilum pedunculatum*
Snakeroot, Seneca: *Polygala senega*
Snakeroot, Virginia: *Aristolochia serpentaria*
Snakeroot, Western: *Ageratina occidentalis*
Snakeroot, White: *Ageratina altissima*
Snakeshade: *Solanum nigrum*
Snakeweed: *Chrysothamnus depressus, C. greenei, C. viscidiflorus, Gutierrezia microcephala, G. californica, G. sarothrae, G.* sp., *Petradoria pumila*
Snakewood, Bitter: *Condalia globosa*
Snakewood, Button: *Eryngium alismifolium, E. aquaticum, Liatris punctata, L. spicata*
Sneezeweed: *Dugaldia hoopesii, Helenium autumnale, H. microcephalum*
Snowball Pincushion: *Mammillaria grahamii*
Snowberry: *Symphoricarpos albus, S. longiflorus, S. mollis, S. occidentalis, S. oreophilus, S.* sp.
Snowberry, Creeping: *Gaultheria hispidula*
Snowberry, Desert: *Symphoricarpos longiflorus*
Snowberry, Garden: *Symphoricarpos albus*
Snowberry, Parish's: *Symphoricarpos oreophilus*
Snowberry, Western: *Symphoricarpos occidentalis*
Snowberry, Whortleleaf: *Symphoricarpos oreophilus*
Snow Brush (or Snowbush): *Ceanothus velutinus*
Snowdrop: *Plagiobothrys fulvus*
Snow on the Mountain: *Euphorbia marginata*
Soap Berry: *Sapindus saponaria, Shepherdia canadensis, S.* sp.
Soapolallie: *Shepherdia canadensis*
Soap Plant: *Chenopodium californicum, Chlorogalum angustifolium C. parviflorum, C. pomeridianum*
Soap Root: *Chlorogalum pomeridianum, C.* sp., *Yucca baccata, Y.* sp.
Soapweed: *Nolina bigelovii, Yucca baccata, Y. elata, Y. glauca*
Soapweed Yucca: *Yucca glauca*
Soapwort, American: *Saponaria officinalis*
Sochan: *Rudbeckia laciniata*
Softhair Marbleseed: *Onosmodium molle*
Softstem Bulrush: *Scirpus tabernaemontani*
Soft Wood: *Canotia holacantha*
Solomon, Slim: *Maianthemum stellatum*
Solomon's Seal: *Maianthemum racemosum, Polygonatum biflorum, Streptopus amplexifolius*
Solomon's Seal, Downy: *Polygonatum pubescens*
Solomon's Seal, False: *Maianthemum racemosum, M. stellatum, Smilacina* sp.
Solomon's Seal, Feather: *Maianthemum racemosum*
Solomon's Seal, Hairy: *Polygonatum pubescens*
Solomon's Seal, Large: *Polygonatum biflorum*
Solomon's Seal, Small: *Polygonatum biflorum*
Solomon's Seal, Star Flowered: *Maianthemum stellatum*
Solomon's Seal, Twoleaf False: *Maianthemum dilatatum*
Solomon's Seal, Western: *Maianthemum racemosum*
Soopolallie: *Shepherdia canadensis*
Sorrel, Garden: *Rumex acetosa*
Sorrel, Ladies': *Oxalis corniculata, O. montana, O. stricta*
Sorrel, Purple: *Oxalis violacea*
Sorrel, Redwood: *Oxalis oregana*
Sorrel, Sheep: *Oxalis stricta, O. violacea, Rumex acetosa, R. acetosella*
Sorrel, Sour: *Oxalis stricta*
Sorrel, Wild: *Rumex* sp.
Sorrel, Wood: *Oxalis montana*
Sotol: *Dasylirion wheeleri*
Sourberry: *Rhus trilobata*
Sour Bush: *Rhus trilobata*
Sourwood: *Oxydendrum arboreum*
Sowthistle, Common: *Sonchus oleraceus*
Sow Thistle, Field: *Sonchus arvensis*
Sow Thistle, Spiny Leaved: *Sonchus asper*
Sowthistle Desertdandelion: *Malacothrix sonchoides*
Spanish Bayonet: *Yucca harrimaniae, Y. whipplei*
Spanish Chilicothe: *Marah macrocarpus*
Spanish Dagger: *Yucca aloifolia*
Spanish Needles: *Bidens bipinnata, Palafoxia arida*
Spearmint: *Mentha canadensis, Mentha spicata*
Spectacle Fruit: *Wislizenia refracta*
Spectacle Pod: *Dimorphocarpa wislizeni*
Speedwell: *Veronica americana, V. officinalis, V. peregrina*
Speedwell, Thyme Leaf: *Veronica serpyllifolia*
Spice Bush: *Calycanthus occidentalis, Lindera benzoin*
Spice Wood: *Lindera benzoin, L.* sp.
Spider Bush: *Oemleria cerasiformis*
Spiderflower, Bladderpod: *Cleome isomeris*
Spiderflower, Slender: *Cleome multicaulis*
Spiderflower, Yellow: *Cleome lutea*
Spiderling, Erect: *Boerhavia erecta*
Spiderwort, Longbract: *Tradescantia bracteata*
Spiderwort, Pinewoods: *Tradescantia pinetorum*
Spiderwort, Virginia: *Tradescantia virginiana*
Spiderwort, Western: *Tradescantia occidentalis*
Spikenard, American: *Aralia racemosa*
Spikenard, California: *Aralia californica*
Spikenard, False: *Maianthemum racemosum*
Spikenard, Indian: *Aralia racemosa, Maianthemum racemosum*
Spikerush, Beaked: *Eleocharis rostellata*
Spikerush, Common: *Eleocharis palustris*
Spikerush, Sand: *Eleocharis montevidensis*
Spikesedge, Canada: *Eleocharis geniculata*
Spikesedge, Jointed: *Eleocharis equisetoides*
Spikeweed, Fitch's: *Hemizonia fitchii*
Spinach, Cattle: *Atriplex polycarpa*
Spinach, Indian: *Monolepis nuttalliana*
Spinach, Strawberry: *Chenopodium capitatum*
Spindle Tree: *Euonymus europaea*
Spinystar: *Escobaria vivipara*
Spiraea: *Aruncus dioicus, Luetkea pectinata, Spiraea* spp.; see also Spirea
Spiraea, Rock: *Holodiscus discolor, H. dumosus*; see also Rock Spiraea
Spirea, Coast: *Holodiscus discolor*
Spleenwort, Lacy: *Asplenium horridum*
Spleenwort, Maidenhair: *Asplenium trichomanes*
Spoonwort: *Cochlearia officinalis*
Spring Beauty, Bering Sea, *Claytonia acutifolia*

Springbeauty, Carolina: *Claytonia caroliniana*
Springbeauty, Great Basin: *Claytonia umbellata*
Springbeauty, Heartleaf: *Claytonia cordifolia*
Springbeauty, Lanceleaf: *Claytonia lanceolata*
Spring Beauty, Narrow Leaved, *Claytonia virginica*
Springbeauty, Pale: *Claytonia spathulata*
Spring Beauty, Siberian, *Claytonia sibirica*
Springbeauty, Streambank: *Claytonia parviflora*
Spring Beauty, Tuberous, *Claytonia tuberosa*
Springbeauty, Virginia: *Claytonia virginica*
Spring Beauty, Western, *Claytonia lanceolata*
Spring Cress, Bulbous, *Cardamine rhomboidea*
Spring Cress, Purple, *Cardamine douglassii*
Spring Gold, *Lomatium utriculatum*
Springparsley, Alpine False: *Pseudocymopterus montanus*
Springparsley, Bulbous: *Cymopterus bulbosus*
Springparsley, Fendler's: *Cymopterus acaulis*
Springparsley, Globe: *Cymopterus globosus*
Springparsley, Longstalk: *Cymopterus longipes*
Springparsley, Mountain: *Cymopterus montanus*
Springparsley, Plains: *Cymopterus acaulis*
Springparsley, Purple: *Cymopterus purpureus*
Springparsley, Purplenerve: *Cymopterus multinervatus*
Springparsley, Sweetroot: *Cymopterus newberryi*
Springparsley, Widewing: *Cymopterus purpurascens*
Spruce, Black: *Picea glauca, P. mariana*
Spruce, Colorado Blue: *Picea pungens*
Spruce, Douglas: *Pseudotsuga menziesii*
Spruce, Engelmann: *Picea engelmannii*
Spruce, Norway: *Picea abies*
Spruce, Red: *Picea rubens*
Spruce, Sitka: *Picea sitchensis*
Spruce, Tideland: *Picea sitchensis*
Spruce, White: *Picea glauca*
Spruce Gum: *Pinus rigida*
Spruce Pine: *Pinus glabra, P. virginiana*
Spurge: *Chamaesyce albomarginata, C. glyptosperma, C. melanadenia, C. ocellata, C. polycarpa, C. serpyllifolia, Croton pottsii, Euphorbia* sp.
Spurge, Flowering: *Euphorbia corollata*
Spurge, Geyer's: *Chamaesyce geyeri*
Spurge, Mojave: *Euphorbia incisa*
Spurge, Rocky Mountain: *Euphorbia robusta*
Spurge, San Francisco Mountain: *Euphorbia lurida*
Spurge, Thyme Leaf: *Chamaesyce serpyllifolia*
Spurge, Toothed: *Euphorbia dentata*
Spur Pepper: *Capsicum annuum*
Squash, Crookneck: *Cucurbita moschata*
Squash, Hubbard: *Cucurbita maxima*
Squash, Ojibwe: *Cucurbita maxima*
Squash, Roman: *Cucurbita pepo*
Squash, Turban: *Cucurbita maxima*
Squash, White Scallop: *Cucurbita pepo*
Squash, Wild: *Cucurbita foetidissima*
Squash, Winter: *Cucurbita maxima*
Squashberry: *Viburnum edule*
Squaw Berry: *Rhus trilobata*
Squaw Bush: *Lycium fremontii, Rhus trilobata, Salix lucida*
Squawroot, Mexican: *Conopholis alpina*
Squawthorn: *Lycium torreyi*
Squaw Vine: *Stylosanthes biflora*
Squaw Weed, American: *Rhus trilobata*
Squirrel Tail: *Achillea millefolium, Elymus multisetus, Hordeum jubatum*
Squirreltail, Big: *Elymus multisetus*
Squirreltail, Bottlebrush: *Elymus elymoides*
Squirrel Tail Grass: *Elymus elymoides, Hordeum jubatum*
Stag Bush Sloe: *Viburnum prunifolium*
Stagger Bush: *Lyonia fruticosa, L. mariana*
Stansbury: *Phlox longifolia*
Stansbury Cliffrose: *Purshia stansburiana*
Starflower: *Eriastrum densifolium, Maianthemum stellatum, Trientalis borealis*
Stargrass: *Aletris farinosa*
Stargrass, Yellow: *Hypoxis hirsuta*
Star of Bethlehem: *Ornithogalum umbellatum*
Starwort, Tuber: *Pseudostellaria jamesiana*
Steeple Bush: *Spiraea tomentosa*
Stick Leaf: *Mentzelia involucrata, M. multiflora, M. pumila, Mentzelia* sp.
Stickseed: *Cryptantha crassisepala, Hackelia floribunda, H. hispida, H. virginiana, Lappula occidentalis, Mentzelia laevicaulis*
Stickseed, Desert: *Lappula occidentalis*
Stickseed, European: *Lappula squarrosa*
Stickseed, Flatspine: *Lappula occidentalis*
Stickseed, Manyflower: *Hackelia floribunda*
Stickseed, Showy: *Hackelia hispida*
Stickseed, Spreading: *Hackelia diffusa*
Stickseed, Valley: *Mentzelia affinis*
Stickseed, Virginia: *Hackelia virginiana*
Stickweed: *Lappula squarrosa*
Stickweed, Valley: *Mentzelia affinis*
Stickyseed: *Blennosperma nanum*
Stickywilly: *Galium aparine*
Stone Crop: *Sedum debile, Sedum divergens, S. lanceolatum, S. rosea, S.* sp., *S. spathulifolium, S. stenopetalum*
Stonecrop, Allegheny: *Sedum telephioides*
Stone Crop, Broad Leaved: *Sedum spathulifolium*
Stonecrop, Ditch: *Penthorum sedoides*
Stonecrop, Entireleaf: *Sedum integrifolium*
Stonecrop, Roseflower: *Sedum laxum*
Stonecrop, Roseroot: *Sedum rosea*
Stonecrop, Spearleaf: *Sedum lanceolatum*
Stonecrop, Wormleaf: *Sedum stenopetalum*
Stoneroot: *Collinsonia canadensis*
Stone Seed: *Lithospermum incisum, L. ruderale*
Storksbill: *Erodium* sp.
Strangle Weed: *Cuscuta* sp.
Straw, Broom: *Andropogon glomeratus*
Strawberry, American: *Fragaria vesca*
Strawberry, Barren: *Waldsteinia fragarioides*
Strawberry, Beach: *Fragaria chiloensis*
Strawberry, Blue Leaved Wild: *Fragaria virginiana*
Strawberry, Broad Petaled Wild: *Fragaria virginiana*
Strawberry, California: *Fragaria vesca*
Strawberry, Coastal: *Fragaria chiloensis*
Strawberry, Desert: *Echinocereus fendleri*
Strawberry, European Wood: *Fragaria vesca*
Strawberry, Golden Spined: *Echinocereus engelmannii*
Strawberry, Pacific Beach (or Pacific Coast): *Fragaria chiloensis*
Strawberry, Scarlet: *Fragaria virginiana*
Strawberry, Virginia: *Fragaria virginiana*
Strawberry, Wild: *Fragaria ×ananassa, F. chiloensis, F.* sp., *F. vesca, F. virginiana*
Strawberry, Wood: *Fragaria vesca*
Stretchberry: *Forestiera pubescens*
String Plant: *Psoralidium lanceolatum*
Sue, Perky: *Tetraneuris argentea*
Sugarberry: *Celtis laevigata*
Sugar Bush: *Rhus ovata*
Sugarcane: *Saccharum officinarum*
Sumac, Aromatic: *Rhus aromatica, R. trilobata*
Sumac, Black: *Rhus copallinum*
Sumac, Dwarf: *Rhus copallinum, R. glabra*
Sumac, Flameleaf: *Rhus copallinum*
Sumac, Fragrant: *Rhus aromatica*
Sumac, Ill Scented: *Rhus trilobata*
Sumac, Lemonade: *Rhus integrifolia, R. trilobata*
Sumac, Littleleaf: *Rhus microphylla*
Sumac, Poison: *Toxicodendron vernix*
Sumac, Skunkbush: *Rhus trilobata*
Sumac, Small Leaved: *Rhus microphylla*
Sumac, Smooth: *Rhus glabra*
Sumac, Stag Horn: *Rhus hirta, R.* sp.
Sumac, Sugar: *Rhus ovata*
Sumac, Sweet: *Rhus trilobata*
Sumac, Three Leaved: *Rhus trilobata*
Sumac, Upland: *Rhus glabra*
Sumac, White: *Rhus aromatica, Toxicodendron vernix*
Sumac, Wild: *Rhus trilobata*
Sumac, Winged: *Rhus copallinum*
Summer's Darling: *Clarkia amoena*
Sumpweed: *Iva xanthifolia*
Sun Bonnet: *Chaptalia tomentosa*
Sun Brier: *Aralia spinosa*
Sunbright: *Talinum parviflorum*
Suncup, Froststem: *Camissonia multijuga*
Suncup, Golden: *Camissonia brevipes*
Sundew, Pink: *Drosera capillaris*
Sundew, Roundleaf: *Drosera rotundifolia*
Sundrops, Hartweg's: *Calylophus hartwegii*
Sundrops, Lavenderleaf: *Calylophus lavandulifolius*
Sundrops, Smaller: *Oenothera perennis*
Sunflower: *Balsamorhiza* spp., *Cordylanthus* sp., *Helianthus* spp., *Wyethia angustifolia, Xanthium* sp.
Sunflower, Balsam Root: *Balsamorhiza sagittata*
Sunflower, Cusick's: *Helianthus cusickii*
Sunflower, Fewleaf: *Helianthus occidentalis*
Sunflower, Giant: *Helianthus giganteus*

Sunflower, Hopi: *Helianthus* sp.
Sunflower, Little: *Helianthella uniflora*
Sunflower, Maximilian: *Helianthus maximiliani*
Sunflower, Mountain Bush: *Encelia virginensis*
Sunflower, Nuttall: *Helianthus nuttallii*
Sunflower, Pale Leaved: *Helianthus strumosus*
Sunflower, Parry's Dwarf: *Helianthella parryi*
Sunflower, Prairie: *Helianthus petiolaris*
Sunflower, Saw Tooth: *Helianthus grosseserratus*
Sunflower, Serpentine: *Helianthus bolanderi*
Sunflower, Showy: *Helianthus niveus*
Sunflower, Spring: *Balsamorhiza sagittata*
Sunflower, Tall Swamp: *Helianthus giganteus*
Sunflower, Thin Leaved: *Helianthus decapetalus*
Sunflower, Western: *Helianthus anomalus*
Sunflower, White: *Wyethia helianthoides*
Sunflower, Wild: *Helianthus annuus, H. petiolaris,*
  H. sp., *Wyethia angustifolia*
Sunflower, Woolly: *Eriophyllum ambiguum,*
  E. lanatum
Sunrose, Baby: *Carpobrotus aequilateralus*
Supple Jack: *Berchemia scandens*
Surfgrass, Scouler's: *Phyllospadix scouleri*
Surfgrass, Toothed: *Phyllospadix serrulatus*
Surfgrass, Torrey's: *Phyllospadix torreyi*
Swampfire, Pacific: *Sarcocornia pacifica*
Swamp Root, American: *Anemopsis californica*
Swarf Plantain: *Plantago virginica*
Sweet after Death: *Achlys triphylla*
Sweet Cicely: *Osmorhiza berteroi, O. brachypoda,*
  O. claytonii, O. depauperata, O. longistylis,
  O. purpurea, O. sp.
Sweetcicely, California: *Osmorhiza brachypoda*
Sweet Cicely, Mountain: *Osmorhiza occidentalis*
Sweet Cicely, Smooth: *Osmorhiza longistylis*
Sweet Cicely, Smoother: *Osmorhiza longistylis*
Sweet Cicely, Western: *Osmorhiza berteroi,*
  O. occidentalis
Sweet Cicely, Woolly: *Osmorhiza claytonii*
Sweet Clover: *Melilotus indicus, M. officinalis, M.* sp.
Sweet Coltsfoot: *Petasites frigidus*
Sweet Corn: *Zea mays*
Sweet Crabapple: *Malus coronaria*
Sweet Flag: *Acorus calamus*
Sweet Gale: *Myrica gale*
Sweet Grass: *Anthoxanthum odoratum, Catabrosa*
  aquatica, Glyceria fluitans, Hierochloe hirta,
  H. odorata
Sweet Grass, Alpine: *Hierochloe alpina*
Sweet Grass, California: *Hierochloe occidentalis*
Sweet Grass, Common: *Hierochloe hirta, H. odorata*
Sweet Grass, Northern: *Hierochloe hirta*
Sweet Gum: *Liquidambar styraciflua*
Sweetleaf: *Symplocos tinctoria*
Sweetpepperbush: *Clethra acuminata*
Sweet Potato: *Ipomoea batatas*
Sweet Potato, Wild: *Potentilla nana*
Sweet Shaggytuft: *Stenandrium dulce*
Sweet Shrub: *Calycanthus floridus*
Sweetroot: *Osmorhiza berteroi, O. occidentalis*

Sweetroot, Bluntseed: *Osmorhiza depauperata*
Sweetroot, Clayton's: *Osmorhiza claytonii*
Sweetroot, Longstyle: *Osmorhiza longistylis*
Sweetroot, Purple: *Osmorhiza purpurea*
Sweetroot, Western: *Osmorhiza occidentalis*
Sweetroot Springparsley: *Cymopterus newberryi*
Sweetvetch: *Hedysarum alpinum, H. boreale*
Sycamore, American: *Platanus occidentalis*
Sycamore, California: *Platanus racemosa*
Sylvan Goat's Beard: *Aruncus dioicus*
Syringa: *Philadelphus lewisii, Syringa* spp.
Tallow Wood: *Ximenia americana*
Tamarack: *Larix laricina*
Tamarisk, Athel: *Tamarix aphylla*
Tan (or Tan-bark) Oak: *Lithocarpus densiflorus*
Tangerine: *Citrus reticulata*
Tansy: *Tanacetum vulgare*
Tansyaster, Hoary: *Machaeranthera canescens*
Tansyaster, Lacy: *Machaeranthera pinnatifida*
Tansyaster, Smallflower: *Machaeranthera parviflora*
Tanseyleaf Aster: *Machaeranthera tanacetifolia*
Tansy Mustard: *Descurainia incana, D. obtusa,*
  D. pinnata, D. sophia, D. sp.
Tansy Ragwort: *Senecio jacobaea*
Tapertip Onion: *Allium acuminatum*
Tapioca: *Manihot esculenta*
Tararack: *Larix laricina*
Taro: *Colocasia esculenta*
Taro, Giant: *Alocasia macrorrhizos*
Tarragon, Wild: *Artemisia dracunculus*
Tarry Smelling Weed: *Madia elegans*
Tar Weed: *Grindelia squarrosa, Hemizonia fascicu-*
  lata, H. luzulifolia, H. sp., *Holocarpha virgata,*
  Madia elegans, M. glomerata, M. gracilis,
  M. sativa, M. sp., *Trichostema lanceolatum*
Tarweed, Chile: *Madia sativa*
Tarweed, Cleveland's: *Hemizonia clevelandii*
Tarweed, Clustered: *Hemizonia fasciculata*
Tarweed, Coastal: *Hemizonia corymbosa*
Tarweed, Fitch's: *Hemizonia fitchii*
Tarweed, Grassy: *Madia gracilis*
Tarweed, Hayfield: *Hemizonia luzulifolia*
Tarweed, Mountain: *Madia glomerata*
Tarweed, Showy: *Madia elegans*
Tarweed, Yellowflower: *Holocarpha virgata*
Tarweed Fiddleneck: *Amsinckia lycopsoides*
Tassel Flower: *Arnoglossum atriplicifolium*
Tassel Flower, Red: *Dalea purpurea*
Tasselflower Brickellbush: *Brickellia grandiflora*
Tea: *Camellia sinensis*
Tea, Bitter: *Kalmia angustifolia*
Tea, Brigham: *Ephedra fasciculata, E. torreyana*
Tea, Country: *Ledum groenlandicum*
Tea, Desert: *Ephedra californica, E. nevadensis,*
  E. viridis
Tea, Eskimo: *Ledum palustre*
Tea, Hudson Bay: *Ledum palustre*
Tea, Indian: *Aster umbellatus, Ephedra* sp., *Ephedra*
  viridis, Ledum groenlandicum
Tea, Jersey: *Ceanothus americanus, C. herbaceus*

Tea, Labrador: see Labrador Tea
Tea, Mexican: *Chenopodium ambrosioides,*
  Ephedra viridis
Tea, Mormon: *Ephedra californica, E. fasciculata,*
  E. nevadensis, E. sp., *E. torreyana, E. trifurca,*
  E. viridis
Tea, Mountain: *Ephedra viridis*
Tea, Navajo: *Anthemis* sp., *Thelesperma*
  megapotamicum
Tea, New Jersey: *Ceanothus americanus, C. fendleri,*
  C. herbaceus
Tea, Oswego: *Monarda fistulosa*
Tea, Prairie: *Croton monanthogynus*
Tea, Teamster's: *Ephedra antisyphilitica,*
  E. nevadensis
Teaberry: *Gaultheria humifusa, G. procumbens*
Teaberry, Eastern: *Gaultheria procumbens*
Teaberry, Western: *Gaultheria ovatifolia*
Tea Plant: *Thelesperma megapotamicum*
Teasel: *Dipsacus fullonum*
Telegraphweed: *Heterotheca grandiflora*
Telesonix, James's: *Boykinia jamesii*
Tepary: *Phaseolus acutifolius*
Terrapin Paw: *Prunella vulgaris*
Thelypody, Entireleaved: *Thelypodium integri-*
  folium
Thelypody, Mountain Mock: *Pennellia micrantha*
Thelypody, Wright's: *Thelypodium wrightii*
Thimble: *Rubus odoratus*
Thimbleberry: *Rubus canadensis, R. occidentalis,*
  R. parviflorus
Thimbleweed: *Anemone cylindrica, A. virginiana*
Thistle: *Carduus* sp., *Cirsium* spp.
Thistle, American Star: *Centaurea americana,*
  C. melitensis
Thistle, Bull: *Cirsium brevistylum, C. vulgare*
Thistle, Cainville: *Cirsium calcareum*
Thistle, California: *Cirsium californicum*
Thistle, Canada: *Cirsium arvense*
Thistle, Clustered: *Cirsium brevistylum*
Thistle, Cobweb: *Cirsium occidentale*
Thistle, Cotton: *Onopordum acanthium*
Thistle, Dwarf: *Cirsium drummondii*
Thistle, Eaton's: *Cirsium eatonii*
Thistle, Edible: *Cirsium edule*
Thistle, Elk: *Cirsium scariosum*
Thistle, Fewleaf: *Cirsium remotifolium*
Thistle, Field: *Cirsium discolor*
Thistle, Hooker's: *Cirsium hookerianum*
Thistle, Lavender: *Cirsium neomexicanum*
Thistle, Leafy: *Cirsium congdonii*
Thistle, Little: *Cirsium altissimum*
Thistle, Maltese Star: *Centaurea melitensis*
Thistle, Meadow: *Cirsium scariosum*
Thistle, Mountain: *Cirsium undulatum*
Thistle, New Mexico: *Cirsium neomexicanum*
Thistle, Pale: *Cirsium pallidum*
Thistle, Parry's: *Cirsium parryi*
Thistle, Plumed: *Cirsium drummondii,*
  C. undulatum

Index of Common Names

Thistle, Red: *Cirsium occidentale*
Thistle, Rosette: *Cirsium congdonii*
Thistle, Rothrock's: *Cirsium rothrockii*
Thistle, Russian: *Salsola australis*
Thistle, Scotch Cotton: *Onopordum acanthium*
Thistle, Scottish: *Cirsium vulgare*
Thistle, Short Styled: *Cirsium brevistylum*
Thistle, Snowy: *Cirsium pastoris*
Thistle, Star: *Centaurea americana, C. melitensis*
Thistle, Tall: *Cirsium altissimum*
Thistle, Tioga: *Cirsium tioganum*
Thistle, Wavy Leaved: *Cirsium undulatum*
Thistle, White: *Cirsium hookerianum*
Thistle, Yellow: *Cirsium horridulum*
Thistle, Yellow Spined: *Cirsium ochrocentrum*
Thorn, Bicknell's: *Crataegus chrysocarpa*
Thorn, Pear: *Crataegus calpodendron*
Thorn Apple: *Crataegus* sp., *Datura wrightii*
Thornapple, Chinese: *Datura ferox*
Thornapple, Desert: *Datura discolor*
Thornapple, Sacred: *Datura wrightii*
Thorn Bush: *Crataegus rivularis*
Thorn of Christ: *Castela emoryi*
Thorn Skeletonweed: *Stephanomeria spinosa*
Thoroughwort: *Ageratina herbacea*
Thoroughwort, Lateflowering: *Eupatorium serotinum*
Thread Leaf: *Senecio flaccidus*
Threadleaf, Manyflower False: *Schkuhria multiflora*
Thread, Needle and: *Stipa comata*
Threeawn, Fendler: *Aristida purpurea*
Threeawn, Poverty: *Aristida divaricata*
Threeawn, Red: *Aristida purpurea*
Three Square: *Scirpus pungens*
Thyme, Wild: *Thymus praecox*
Ti: *Cordyline fruticosa*
Tibinagua: *Eriogonum nudum*
Tickseed, Bigelow's: *Coreopsis bigelovii*
Tickseed, Golden: *Coreopsis tinctoria*
Tickseed, Leavenworth's: *Coreopsis leavenworthii*
Tickseed, Stiff: *Coreopsis palmata*
Tickseed, Tall: *Coreopsis tripteris*
Tickseed, Trefoil: *Desmodium nudiflorum, D. perplexum*
Tick Trefoil, Canada: *Desmodium canadense*
Ticktrefoil, Hawaii: *Desmodium sandwicense*
Tick Trefoil, Illinois: *Desmodium illinoense*
Ticktrefoil, Nakedflower: *Desmodium nudiflorum*
Ticktrefoil, Panicledleaf: *Desmodium paniculatum*
Ticktrefoil, Perplexed: *Desmodium perplexum*
Ticktrefoil, Showy: *Desmodium canadense*
Tickweed: *Coreopsis leavenworthii*
Tidy Tips: *Layia glandulosa*
Tietongue: *Coccoloba diversifolia*
Timbleberry: *Rubus parviflorus*
Timothy: *Phleum pratense*
Tinker's Weed: *Triosteum perfoliatum*
Tiny Mousetail: *Myosurus minimus*
Tiplant: *Cordyline fruticosa*

Toadflax, Bastard: *Comandra umbellata*
Toadflax, False: *Geocaulon lividum*
Toadflax, Yellow: *Linaria vulgaris*
Toadshade: *Trillium sessile*
Tobacco, Aztec: *Nicotiana rustica*
Tobacco, Bigelow's: *Nicotiana quadrivalvis*
Tobacco, Cleveland's: *Nicotiana clevelandii*
Tobacco, Commercial: *Nicotiana tabacum*
Tobacco, Coyote's: *Nicotiana attenuata, N. rustica, N. trigonophylla*
Tobacco, Cultivated: *Nicotiana tabacum*
Tobacco, Desert: *Nicotiana trigonophylla*
Tobacco, Hopi: *Nicotiana attenuata*
Tobacco, Indian: *Lobelia inflata, Nicotiana attenuata, N. quadrivalvis, N. rustica, N.* sp.
Tobacco, Ladies': *Gnaphalium californicum*
Tobacco, Native: *Nicotiana attenuata, N. rustica*
Tobacco, Palmer Wild: *Nicotiana trigonophylla*
Tobacco, Rabbit: *Gnaphalium obtusifolium, Pterocaulon virgatum*
Tobacco, Rabbit's: *Pterocaulon virgatum*
Tobacco, Tex-mex: *Nicotiana plumbaginifolia*
Tobacco, Tree: *Nicotiana glauca*
Tobacco, Wild: *Nicotiana attenuata, N. quadrivalvis, N. rustica, N.* sp., *N. trigonophylla*
Tobacco, Woman's: *Antennaria plantaginifolia*
Tobacco, Yaqui: *Nicotiana tabacum*
Tobacco Brush: *Ceanothus velutinus*
Tobacco Root: *Valeriana edulis*
Tolguacha: *Datura wrightii*
Toloache: *Datura wrightii*
Tomatilla: *Lycium pallidum*
Tomatillo: *Lycium pallidum, Physalis philadelphica*
Tomato: *Lycopersicon esculentum*
Tomato, Ground: *Physalis subulata*
Tomato, Husk: *Physalis pubescens*
Tomato, Wild: *Physalis* sp., *Solanum triflorum*
Tomato Flower: *Rosa acicularis*
Tongue, Beard: *Penstemon barbatus, P. fendleri, P. fruticosus, P. jamesii, P. linarioides, P.* sp., *P. virgatus, Phlox caespitosa*
Toothache Tree: *Zanthoxylum americanum*
Tooth Cress, Large: *Cardamine maxima*
Toothwort: *Cardamine diphylla, Dentaria* sp.
Toothwort, Crinkled: *Cardamine diphylla*
Toothwort, Cutleaf: *Cardamine concatenata*
Toothwort, Large: *Cardamine maxima*
Torote: *Jatropha cardiophylla*
Torote Amarillo: *Jatropha spatulata*
Torote Blanco: *Bursera microphylla*
Torote Prieto: *Bursera laxiflora*
Touch Me Not, Pale: *Impatiens pallida*
Touch Me Not, Spotted: *Impatiens capensis*
Touch Me Not, Wild: *Impatiens capensis, I. pallida*
Touristplant: *Dimorphocarpa wislizeni*
Tovinal: *Lotus strigosus*
Toyon: *Heteromeles arbutifolia*
Trailplant, American: *Adenocaulon bicolor*
Trefoil, Bird's Foot: *Lotus wrightii*
Trefoil, Mearns's Birdsfoot: *Lotus mearnsii*

Trefoil, Prairie: *Lotus unifoliolatus*
Trefoil, Tick: see Tick Trefoil
Trefoil Tickseed: *Desmodium nudiflorum, D. perplexum*
Triodia: *Tridens muticus*
Triphysaria: *Orthocarpus faucibarbatus*
Tripletlily, Longray: *Triteleia peduncularis*
Trompillo: *Solanum elaeagnifolium*
Trumpet, Hummingbird: *Epilobium canum*
Trumpet, Long: *Eriogonum nudum*
Trumpetflower, Evening: *Gelsemium sempervirens*
Trumpet Vine: *Bignonia capreolata*
Tuberose, Wild: *Leucocrinum montanum*
Tubular Green Alga: *Enteromorpha intestinalis*
Tulare: *Cladium mariscus, Cladium* sp.
Tule: *Scirpus acutus, S. americanus, S. nevadensis, S. robustus, S.* sp., *S. tabernaemontani*
Tule Grass: *Juncus* sp.
Tule Mint: *Mentha arvensis*
Tulip, Globe: *Calochortus pulchellus*
Tuliptree: *Liriodendron tulipifera*
Tumblegrass: *Schedonnardus paniculatus*
Tumblemustard, Tall: *Sisymbrium altissimum*
Tumblemustard, Westwater: *Thelypodiopsis elegans*
Tumbleweed: *Amaranthus blitoides, Cycloloma atriplicifolium, Salsola australis*
Tumbleweed, Silver: *Atriplex argentea*
Tumbleweed, White: *Amaranthus albus*
Tumbling Saltweed: *Atriplex rosea*
Tuna: *Opuntia ficus-indica*
Tupelo, Water: *Nyssa aquatica*
Turban, Yellow: *Eriogonum pusillum*
Turkish Rugging: *Chorizanthe staticoides*
Turmeric: *Curcuma longa*
Turnip: *Brassica napus, B. rapa, B.* sp., *Frasera speciosa*
Turnip, Indian: *Arisaema triphyllum, Lithospermum incisum, Pediomelum esculentum, P. hypogaeum, Psoralidium lanceolatum, P. tenuiflorum*
Turnip, Prairie: *Pediomelum esculentum*
Turnip, St. Anthony's: *Ranunculus bulbosus*
Turnip, Wild: *Brassica napus, Penthorum sedoides*
Turpentine Broom: *Thamnosma montana*
Turpentine Wavewing: *Pteryxia terebinthina*
Turpentine Weed: *Trichostema lanceolatum*
Turtle Back: *Psathyrotes annua, P. ramosissima*
Turtle Head: *Chelone glabra*
Twayblade: *Liparis loeselii*
Twiberry, Black: *Lonicera involucrata*
Twiggy Water Dropwort: *Oenanthe* sp.
Twinberry: *Myrcianthes fragrans*
Twinberry, Black: *Lonicera involucrata*
Twinberry, Red: *Lonicera utahensis*
Twineberry, Black: *Lonicera involucrata*
Twinevine, Hartweg's: *Sarcostemma cynanchoides*
Twin Flower: *Linnaea borealis*
Twinleaf: *Jeffersonia diphylla*
Twinpod: *Physaria chambersii*
Twinpod, Common: *Physaria didymocarpa*
Twinpod, Newberry's: *Physaria newberryi*

Index of Common Names

Twisted Stalk: *Streptopus amplexifolius, S. roseus*
Twisted Stalk, Clasping: *Streptopus amplexifolius*
Twisted Stalk, Purple (or Rosy): *Streptopus roseus*
Twistflower, Heartleaf: *Streptanthus cordatus*
Uhaloa: *Waltheria indica*
Uhi-keokeo: *Dioscorea* sp.
Uhiuhi: *Caesalpinia kavaiensis*
Ulei: *Osteomeles anthyllidifolia*
Ulu: *Artocarpus altilis*
Umbrella Leaf: *Diphylleia cymosa*
Umbrella Plant: *Eriogonum annuum, Mirabilis nyctaginea*
Umbrella Plant, Parsnip Flower: *Eriogonum heracleoides*
Umbrella Plant, Subalpine: *Eriogonum umbellatum*
Umbrella Plant, Yellow: *Eriogonum flavum*
Umbrella Wort: *Mirabilis coccineus, M. nyctaginea*
Umbrella Wort, Heart Leaved: *Mirabilis nyctaginea*
Umbrella Wort, Narrow Leaved: *Mirabilis linearis*
Umbrellawort, Parish's: *Tauschia parishii*
Umbrellawort, Southern: *Tauschia arguta*
Unicorn Plant: *Proboscidea althaeifolia, P. louisianica, P. parviflora*
Unicorn Root: *Aletris farinosa*
Uwala: *Ipomoea* sp.
Valerian, Greek: *Polemonium reptans*
Valley Tassels: *Orthocarpus attenuatus*
Valley Vinegarweed: *Lessingia glandulifera*
Vanilla Leaf: *Achlys triphylla*
Vasevine: *Clematis viorna*
Vegetable Oyster: *Tragopogon porrifolius, T.* sp.
Velvetweed: *Gaura parviflora*
Venus Flytrap: *Dionaea muscipula*
Venus's Looking Glass: *Triodanis perfoliata*
Verbena, Lemon: *Phyla lanceolata*
Verbena, Sand: see Sand Verbena
Vervain: *Glandularia bipinnatifida, Verbena hastata*
Vervain, American Wild: *Verbena hastata*
Vervain, Blue: *Verbena hastata*
Vervain, Hoary: *Verbena stricta*
Vervain, White: *Verbena urticifolia*
Vetch: *Astragalus* sp., *Vicia americana, V. caroliniana, V. melilotoides, V.* sp., *Wislizenia refracta*
Vetch, American: *Vicia americana*
Vetch, Blue: *Vicia americana*
Vetch, Carolina: *Vicia caroliniana*
Vetch, Crown: *Coronilla varia*
Vetch, Deer: *Lotus humistratus, L. mearnsii, L. scoparius, L. wrightii*
Vetch, Giant: *Vicia nigricans*
Vetch, Milk: see Milkvetch
Vetch, Narrow Leaved: *Vicia sativa*
Vetch, Purple: *Vicia americana*
Vetch, Racemed Poison: *Astragalus racemosus*
Vetch, Sweet: *Hedysarum alpinum, H. boreale*
Vetch, Winter: *Vicia villosa*
Vetchling, Bush: *Lathyrus eucosmus*
Vetchling, Creamy: *Lathyrus ochroleucus*
Vetchling, Marsh: *Lathyrus palustris*

Vinegar Weed: *Trichostema lanceolatum*
Vinegarweed, Valley: *Lessingia glandulifera*
Violet, American Dog: *Viola conspersa*
Violet, Arrow Leaved: *Viola sagittata*
Violet, Birdfoot: *Viola pedata*
Violet, Blue: *Viola cucullata, V.* sp.
Violet, Bog: *Viola nephrophylla*
Violet, California Golden: *Viola pedunculata*
Violet, Canada: *Viola canadensis*
Violet, Creamy: *Viola striata*
Violet, Dog: *Viola conspersa*
Violet, Dogtooth: *Erythronium grandiflorum, E. oregonum*
Violet, Downy Yellow: *Viola pubescens*
Violet, Evergreen: *Viola sempervirens*
Violet, Golden: *Viola pedunculata*
Violet, Green: *Hybanthus concolor*
Violet, Hookedspur: *Viola adunca*
Violet, Marsh: *Viola epipsila*
Violet, Marsh Blue: *Viola cucullata*
Violet, Northern Bog: *Viola nephrophylla*
Violet, Roundleaf Yellow: *Viola rotundifolia*
Violet, Smoothish Yellow: *Viola pubescens*
Violet, Star: *Dalibarda repens*
Violet, Striped Cream: *Viola striata*
Violet, Sweet White: *Viola blanda*
Violet, Twoflower: *Viola biflora*
Violet, White: *Viola blanda*
Violet, Wild: *Viola* sp.
Violet, Wood: *Viola sempervirens*
Violet, Yellow: *Viola adunca, V. pubescens, V. rotundifolia*
Viper's Bugloss: *Echium vulgare*
Virgin's Bower: *Clematis columbiana, C. ligusticifolia, C. occidentalis, C. virginiana*
Visnaga: *Ferocactus wislizeni*
Wahoo: *Euonymus atropurpurea*
Wake Robin: *Trillium grandiflorum*
Wakerobin, Giant: *Trillium chloropetalum*
Wakinakim: *Juniperus communis*
Walink: *Polypodium virginianum*
Walking Stick, Devil's: *Aralia spinosa, Oplopanax horridus*
Wallflower, Nakedstem: *Parrya nudicaulis*
Wallflower, Parry's: *Parrya nudicaulis*
Wallflower, Plains: *Erysimum asperum*
Wallflower, Rough: *Erysimum asperum*
Wallflower, Sanddune: *Erysimum capitatum*
Wallflower, Shy: *Erysimum inconspicuum*
Wallflower, Western: *Erysimum asperum, E. capitatum*
Wallflower, Wormseed: *Erysimum cheiranthoides*
Walnut, Arizona: *Juglans major*
Walnut, Black: *Juglans hindsii, J. major, J. nigra*
Walnut, California: *Juglans californica*
Walnut, English: *Juglans regia*
Walnut, Hinds's Black: *Juglans hindsii*
Walnut, Indian: *Aleurites moluccana*
Walnut, Native: *Juglans major*
Walnut, White: *Juglans cinerea*

Wand Blackroot: *Pterocaulon virgatum*
Wapato: *Sagittaria cuneata, S. latifolia*
Wappatoo: *Sagittaria cuneata*
Water Hemlock, Douglas's: *Cicuta douglasii*
Water Hemlock, Mackenzie's: *Cicuta virosa*
Water Hemlock, Poison: *Cicuta douglasii, C. maculata, C. virosa*
Water Hemlock, Western: *Cicuta douglasii*
Water Hoarhound: *Lycopus americanus*
Water Horehound: *Lycopus asper, L.* sp., *L. virginicus, Marrubium vulgare*
Waterhorehound, American: *Lycopus americanus*
Water Horehound, Northern: *Lycopus uniflorus*
Waterhorehound, Virginia: *Lycopus virginicus*
Waterhyssop, Blue: *Bacopa caroliniana*
Water Knotweed: *Polygonum amphibium*
Waterleaf, Bluntleaf (or Broad Leaved): *Hydrophyllum canadense*
Waterleaf, Pacific: *Hydrophyllum tenuipes*
Waterleaf, Virginia: *Hydrophyllum virginianum*
Waterleaf, Western: *Hydrophyllum occidentale*
Waterleaf, White: *Hydrophyllum fendleri*
Water Lily: *Nuphar lutea, Nymphaea odorata, N.* sp.
Water Lily, Great Yellow: *Nelumbo lutea, Nuphar lutea*
Water Lily, Sweet Scented (or Sweet White): *Nymphaea odorata*
Water Lily, White: *Nymphaea odorata, N.* sp.
Water Lily, Yellow: *Nuphar lutea, Nymphaea* sp.
Water Mannagrass: *Glyceria fluitans*
Watermelon: *Citrullus lanatus*
Watermilfoil, Shortspike: *Myriophyllum sibiricum*
Watermilfoil, Spike: *Myriophyllum spicatum*
Watermilfoil, Whorlleaf: *Myriophyllum verticillatum*
Waterparsnip, Cutleaf: *Berula erecta*
Waterparsnip, Hemlock: *Sium suave*
Water Root: *Arnica acaulis*
Waterthread Pondweed: *Potamogeton diversifolius*
Water Tupelo: *Nyssa aquatica*
Waterweed, Canadian: *Elodea canadensis*
Waterwillow, Thickleaf: *Justicia crassifolia*
Wauke: *Touchardia latifolia*
Wavewing, Turpentine: *Pteryxia terebinthina*
Waxberry: *Symphoricarpos albus*
Wax Work: *Celastrus scandens*
Westwater Tumblemustard: *Thelypodiopsis elegans*
Wheat: *Triticum aestivum, Triticum durum*
Wheat, Durum: *Triticum durum*
Wheat, Narrowleaf Cow: *Melampyrum lineare*
Wheat, Polish: *Triticum polonicum*
Wheat, Sonora: *Triticum aestivum*
Wheat Grass: *Agropyron* sp., *Triticum aestivum*
Wheatgrass, Bluebunch: *Pseudoroegneria spicata*
Wheat Grass, Slender: *Elymus trachycaulus*
Wheatgrass, Western: *Pascopyrum smithii*
Whiskerbrush: *Linanthus ciliatus*
Whitethorn, Chaparral: *Ceanothus leucodermis*
Whiteweed: *Lithospermum ruderale*
Whitlowgrass, Carolina: *Draba reptans*

Whitlowgrass, Heller's: *Draba helleriana*
Whitlowgrass, Mountain: *Draba rectifructa*
Whitlowgrass, Yellowstone: *Draba incerta*
Whitlowwort, James's: *Paronychia jamesii*
Whooping Cough Grass: *Iresine diffusa*
Whorlgrass, Water: *Catabrosa aquatica*
Whortleberry: *Vaccinium myrtillus, V. ovalifolium, V. parvifolium, V. scoparium, V.* sp.
Whortleberry, Big: *Vaccinium membranaceum*
Whortleberry, Grouse: *Vaccinium scoparium*
Whortleberry, Red: *Vaccinium parvifolium, V. scoparium*
Whortleaf Snowberry: *Symphoricarpos oreophilus*
Wicopy: *Dirca palustris*
Widowsfrill: *Silene stellata*
Wild Flag: *Iris missouriensis, I. setosa*
Wild Ginger: *Asarum canadense, A. caudatum, A.* sp., *Hexastylis arifolia*
Wildginger, British Columbia: *Asarum caudatum*
Wild Ginger, Canada: *Asarum canadense*
Wild Ginger, Western: *Asarum caudatum*
Wild Heliotrope: *Heliotropium curassavicum, Phacelia crenulata, P.* sp.
Wild Parsley: *Ligusticum grayi, Lomatium utriculatum, Musineon divaricatum*
Wild Rice: *Chasmanthium latifolium, Zizania aquatica, Z. palustris*
Wildrice, Annual: *Zizania aquatica*
Wildrice, Northern: *Zizania palustris*
Wildrye, Basin: *Leymus cinereus*
Wildrye, Beardless: *Leymus triticoides*
Wildrye, Blue: *Elymus glaucus*
Wildrye, Canada: *Elymus canadensis*
Wildrye, Siberian: *Elymus sibiricus*
Wi-li-wi-li: *Erythrina sandwicensis*
Willow: *Cornus* sp., *Salix* spp.
Willow, Alaska Bog: *Salix fuscescens*
Willow, American: *Salix washingtonia*
Willow, Arroyo: *Salix lasiolepis*
Willow, Balsam: *Salix pyrifolia*
Willow, Barratt: *Salix scouleriana*
Willow, Bayberry: *Salix myricoides*
Willow, Beaked: *Salix bebbiana*
Willow, Bebb: *Salix bebbiana*
Willow, Black: *Salix nigra, Salix* sp.
Willow, Bog: *Salix fuscescens, S. pedicellaris*
Willow, Broad Leaved: *Epilobium latifolium*
Willow, Coastal Plain: *Salix caroliniana*
Willow, Coyote: *Salix exigua*
Willow, Crack: *Salix fragilis*
Willow, Desert: *Chilopsis linearis*
Willow, Diamond: *Salix eriocephala, S. prolixa*
Willow, Dune: *Salix hookeriana*
Willow, Dusky: *Salix melanopsis*
Willow, Dwarf: *Salix humilis*
Willow, Dwarf Gray: *Salix humilis*
Willow, Felty Leaved: *Salix alaxensis*
Willow, Gooding's: *Salix goodingii*
Willow, Gray: *Salix bebbiana, S. humilis, S. scouleriana, S.* sp.

Willow, Heartleaf: *Salix cordata*
Willow, Hinds's: *Salix hindsiana*
Willow, Hoary: *Salix candida*
Willow, Hooker's: *Salix hookeriana, S. lucida*
Willow, Least: *Salix rotundifolia*
Willow, Littletree: *Salix arbusculoides*
Willow, Longleaf: *Salix interior*
Willow, Mackenzie's: *Salix prolixa*
Willow, Missouri River: *Salix eriocephala*
Willow, Mountain: *Salix humilis*
Willow, Narrow Leaved: *Salix exigua, S. melanopsis*
Willow, Northwest Sandbar: *Salix sessilifolia*
Willow, Nuttall: *Salix scouleriana*
Willow, Pacific: *Salix lucida*
Willow, Peach Leaved: *Salix amygdaloides, S. irrorata, S. lucida*
Willow, Piper's: *Salix hookeriana*
Willow, Prairie: *Salix humilis*
Willow, Primrose: see Primrosewillow
Willow, Pussy: *Salix discolor*
Willow, Red: *Cornus sericea, Salix bonplandiana, S. lucida*
Willow, River: *Salix alaxensis*
Willow, Sage: *Salix candida, S. humilis*
Willow, Sand Bar: *Salix exigua, S. hindsiana, S. interior, S. sessilifolia*
Willow, Scouler: *Salix scouleriana*
Willow, Seep: *Baccharis salicifolia, B.* sp.
Willow, Shining: *Salix lucida*
Willow, Silky: *Salix sericea*
Willow, Silver: *Salix exigua*
Willow, Silvery: *Salix candida*
Willow, Sitka: *Salix sitchensis*
Willow, Southern: *Salix caroliniana*
Willow, Sura: *Salix planifolia*
Willow, Tealeaf: *Salix planifolia*
Willow, Tracy: *Salix lasiolepis*
Willow, Tundra: *Salix planifolia*
Willow, Velvet: *Salix sitchensis*
Willow, Water Strings: *Salix caroliniana*
Willow, Weeping: *Salix babylonica*
Willow, White: *Salix alba, Salix hindsiana*
Willow, Wild Weeping: *Salix amygdaloides*
Willow Bustic: *Sideroxylon salicifolium*
Willow Herb: *Epilobium ciliatum*
Willow Herb, Great: *Epilobium angustifolium*
Willow Herb, Northern: *Epilobium ciliatum*
Willow Herb, Small: *Epilobium minutum*
Willow Herb, Tall: *Epilobium brachycarpum*
Willowweed, Autumn: *Epilobium brachycarpum*
Willowweed, Coast: *Epilobium ciliatum*
Willowweed, Small: *Epilobium minutum*
Wind Flower: *Anemone multifida*
Windmills, Trailing: *Allionia incarnata*
Winecup Fairyfan: *Clarkia purpurea*
Winterberry: *Ilex verticillata*
Wintergreen: *Chimaphila maculata, Gaultheria procumbens, Orthilia secunda, Pyrola asarifolia, P. elliptica, P.* sp.
Wintergreen, American: *Pyrola americana*

Wintergreen, Creeping: *Gaultheria procumbens*
Wintergreen, Flowering: *Polygala paucifolia*
Wintergreen, Greenflowered: *Pyrola chlorantha*
Wintergreen, Liverleaf: *Pyrola asarifolia*
Wintergreen, One Flowered: *Moneses uniflora*
Wintergreen, Pink: *Pyrola asarifolia*
Wintergreen, Sidebells: *Orthilia secunda*
Wintergreen, Spotted: *Chimaphila maculata*
Wintergreen, Whiteveined: *Pyrola picta*
Wire Grass: *Aristida purpurea, Juncus balticus, J. mexicanus, J. tenuis*
Wirelettuce, Brownplume: *Stephanomeria pauciflora*
Wirelettuce, Desert: *Stephanomeria runcinata*
Wirelettuce, Narrowleaf: *Stephanomeria tenuifolia*
Wirelettuce, Rod: *Stephanomeria virgata*
Wirelettuce, Small: *Stephanomeria exigua*
Wire Rush: *Juncus balticus*
Witch Grass: *Panicum capillare*
Witchgrass, Fall: *Digitaria cognata*
Witch Hazel: *Hamamelis virginiana*
Witch's Hair: *Alectoria sarmentosa*
Witch's Moneybags: *Sedum telephium*
Witherod: *Viburnum nudum*
Wolfberry: *Lycium andersonii, L. exsertum, L. fremontii, L. pallidum, L.* sp., *Symphoricarpos occidentalis*
Woman's Head: *Eupatorium serotinum*
Woodbalm: *Lepechinia calycina*
Wood Betony: *Pedicularis bracteosa, P. canadensis, P. lanceolata, P. racemosa*
Woodbine: *Parthenocissus quinquefolia, P.* sp., *P. vitacea*
Wood Fern: *Dryopteris campyloptera, D. expansa, D. marginalis, D.* sp., *Woodsia neomexicana*
Woodfern, Coastal: *Dryopteris arguta*
Woodfern, Crested: *Dryopteris cristata*
Woodfern, Fragrant: *Dryopteris fragrans*
Woodfern, Marginal: *Dryopteris marginalis*
Woodfern, Mountain: *Dryopteris campyloptera*
Woodfern, Spinulose: *Dryopteris carthusiana*
Woodlandstar, Common: *Lithophragma affine*
Woodmoss: *Dicranum bonjeanii*
Woodnettle, Canadian: *Laportea canadensis*
Wood Reed, Stout (or Sweet): *Cinna arundinacea*
Wood Rose: *Rosa gymnocarpa*
Woodrush, Common: *Luzula multiflora*
Wood Sorrel: *Oxalis corniculata, O. drummondii, O. montana, O. oregana, O. stricta, O. violacea*
Woollygrass, Low: *Erioneuron pulchellum*
Woollysunflower: *Eriophyllum ambiguum, E. lanatum*
Woollyweed: *Hieracium scouleri*
Woolstar, Desert: *Eriastrum eremicum*
Woolstar, Giant: *Eriastrum densifolium*
Woolstar, Great Basin: *Eriastrum sparsiflorum*
Woolstar, Lavender: *Eriastrum filifolium*
Woolstar, Wand: *Eriastrum virgatum*
Worm Grass: *Chenopodium ambrosioides*
Worm Root: *Apocynum cannabinum*

Wormseed: *Chenopodium ambrosioides*
Wormweed: *Artemisia ludoviciana*
Wormwood, Biennial: *Artemisia biennis*
Wormwood, Canada: *Artemisia campestris*
Wormwood, Coastal: *Artemisia suksdorfii*
Wormwood, Dragon: *Artemisia dracunculus*
Wormwood, European: *Artemisia absinthium*
Wormwood, Field: *Artemisia campestris*
Wormwood, Fringed: *Artemisia frigida*
Wormwood, Kamtschat's: *Artemisia vulgaris*
Wormwood, Oahu: *Artemisia australis*
Wormwood, Pacific: *Artemisia campestris*
Wormwood, Pasture: *Artemisia frigida*
Wormwood, Silvery: *Artemisia filifolia*
Wormwood, Tilesius's: *Artemisia tilesii*
Woundwort: *Stachys mexicana, S. palustris*
Wrack, Sea: *Fucus gardneri, Fucus* sp.
Wrinklefruit: *Tetraclea coulteri*
Yam, Air: *Dioscorea bulbifera*
Yam, Coco: *Colocasia esculenta*
Yam, Fiveleaf: *Dioscorea pentaphylla*
Yam, Wild: *Dioscorea villosa*
Yamp: *Perideridia gairdneri*
Yampa (or Yampah): *Perideridia gairdneri*
Yampah, Adobe: *Perideridia pringlei*
Yampah, Bolander's: *Perideridia bolanderi*
Yampah, Gairdner's: *Perideridia gairdneri*
Yampah, Kellogg's: *Perideridia kelloggii*

Yarrow, American: *Achillea millefolium*
Yarrow, Boreal: *Achillea millefolium*
Yarrow, California: *Achillea millefolium*
Yarrow, False: *Chaenactis douglasii*
Yarrow, Golden: *Eriophyllum lanatum*
Yarrow, Siberian: *Achillea sibirica*
Yarrow, Western: *Achillea millefolium*
Yarrow, Woolly: *Achillea millefolium*
Yarrow, Yellow: *Eriophyllum confertiflorum*
Yaupon: *Ilex vomitoria*
Yautia Amarilla: *Xanthosoma atrovirens*
Yellowcomet: *Mentzelia affinis*
Yellowcress, Alpine: *Rorippa alpina*
Yellowcress, Creeping: *Rorippa sylvestris*
Yellowcress, Curvepod: *Rorippa curvisiliqua*
Yellowcress, Hispid: *Rorippa palustris*
Yellowcress, Northern Marsh: *Rorippa islandica*
Yellowcress, Southern Marsh: *Rorippa teres*
Yellowcress, Spreading: *Rorippa sinuata*
Yellowdicks: *Helenium amarum*
Yelloweyed Grass: *Xyris ambigua, X. caroliniana*
Yellow Herb: *Gutierrezia sarothrae*
Yellow Nicker: *Caesalpinia bonduc*
Yellow Parilla: *Menispermum canadense*
Yellowrocket, American: *Barbarea orthoceras*
Yellowrocket, Early: *Barbarea verna*
Yellowrocket, Garden: *Barbarea vulgaris*

Yellow Root: *Rumex crispus, Xanthorhiza simplicissima*
Yellowroot, Shrub: *Xanthorhiza simplicissima*
Yellowshow, Mexican: *Amoreuxia palmatifida*
Yellowturbans: *Eriogonum pusillum*
Yellow Wood: *Cladrastis kentukea*
Yerba Buena: *Satureja douglasii*
Yerba de Pasmo: *Baccharis pteronioides*
Yerba Golondrina: *Chamaesyce polycarpa*
Yerba Mansa: *Anemopsis californica*
Yerba Santa: *Anemopsis californica, Erigeron foliosus, Eriodictyon angustifolium, E. californicum, E. crassifolium, E. lanatum, E. tomentosum, E. trichocalyx*
Yerbasanta, California: *Eriodictyon californicum*
Yerbasanta, Hairy: *Eriodictyon trichocalyx*
Yerbasanta, Narrowleaf: *Eriodictyon angustifolium*
Yerbasanta, San Diego: *Eriodictyon lanatum*
Yerbasanta, Thickleaf: *Eriodictyon crassifolium*
Yerbasanta, Woolly: *Eriodictyon tomentosum*
Yew, American: *Taxus canadensis*
Yew, Canada: *Taxus canadensis*
Yew, Japanese: *Taxus cuspidata*
Yew, Pacific: *Taxus brevifolia*
Yew, Western: *Taxus brevifolia*
Youth on Age: *Tolmiea menziesii*
Zarzabacoa Comun: *Desmodium incanum*
Zinnia, Abert's Creeping: *Sanvitalia abertii*